AA002357

International Symposium on Power Electronics and Control Engineering (ISPECE 2018)

Journal of Physics: Conference Series Volume 1187

Xi'an, China
28-30 December 2018

Part 1 of 3

ISBN: 978-1-5108-8673-5
ISSN: 1742-6588

Printed from e-media with permission by:

Curran Associates, Inc.
57 Morehouse Lane
Red Hook, NY 12571

Some format issues inherent in the e-media version may also appear in this print version.

This work is licensed under a Creative Commons Attribution 3.0 International Licence.
Licence details: http://creativecommons.org/licenses/by/3.0/.

No changes have been made to the content of these proceedings. There may be changes to pagination and minor adjustments for aesthetics.

Printed with permission by Curran Associates, Inc. (2026)

For permission requests, please contact the Institute of Physics
at the address below.

Institute of Physics
Dirac House, Temple Back
Bristol BS1 6BE UK

Phone: 44 1 17 929 7481
Fax: 44 1 17 920 0979

techtracking@iop.org

Additional copies of this publication are available from:

Curran Associates, Inc.
57 Morehouse Lane
Red Hook, NY 12571 USA
Phone: 845-758-0400
Fax: 845-758-2633
Email: curran@proceedings.com
Web: www.proceedings.com

TABLE OF CONTENTS

VOLUME 1

Preface

Peer Review Statement

POWER ELECTRONIC EQUIPMENT AND SYSTEM

Study of the Mechanism of Tangent Bifurcation in Voltage Mode Controlled DCM Buck Converter 1
Lingling Xie, Fangzheng He, Shuwei Situ

Research on Transformer Fast OLTC System 10
Jinlong Tu, Yongsheng Mi

Research on the Design of the Fuzzy Control System of Full Bridge DC Converter.. 18
Wenda Liu, Shutian Liang, Chuxue Hao

Measurement of Inrush Current in Transformer Based on Optical Current Transducer.................................... 27
Chu Lei, Guo Zhizhong, Chen Yue, Wang Guizhong

Study of the Standard Sine Wave Frequency Conversion Power Supply Based on Analog and
Digital Integrated Control.. 35
Wenbo Jia, Hongda Zhang, Dewen Zhang, Kepeng Tao, Sicong Li, Wanlin Guan

Optimal Installation of Distributed Generators Based on an Enhanced Harmony Search Algorithm 46
Ke Ji, Wentao Wang, Xingong Wang, Zheyu Wei, Jian Yang, Hua Qin

Self-Adaptive Control of Rotor Inertia for Virtual Synchronous Generator in an Isolated Microgrid 51
Hanghang Zeng, Hongsheng Su

Software Consistency Checking Method for Distribution Terminal Based on Chaotic Map 57
Yaokun Wang, Ying Liang, Changkai Shi, Shilei Guan

Three-Level Generalized Discontinuous Pulse-Width Modulation Strategy Considering Neutral
Point Potential Balance... 63
Wang Jikang, Liu Jianzheng, Wang Yi, Guo Zhichao

Research on Intelligent Charging System Technology of Automobile Group.................................... 73
Kedi Yan

Coordination Between Converter-Based Wind Turbines and Synchronous Generators During Inertia
Control.. 78
Fei Jin, Xiaoliang Liu, Guoqiang Li, Jingli Liu, Weichao Li, Jinxin Zheng, Renzhang Liu

Research of a High Voltage and High Value Resistors Standard Device... 87
Xiang Zhou, Xinli Cao, Qianghu Yan, Min Lei

Modeling and Control of DC Microgrid System Based on Hydrogen Production Load.................................... 93
Yingjun Guo, Zhe Shi, Yajie Guo, Fanyi Deng, Hexu Sun

A Novel State of Health Estimation Method for Lithium-Ion Battery in Electric Vehicles............................ 102
Jie Fan, Yuan Zou, Xudong Zhang, Hongwei Guo

The Key Design Technology of Successive Approximation Analog-to-Digital Converter to Improve Efficient and Precision 108

Cheng Xie, Yueyue Chen, Jianjun Chen

Research on Reliability Assessment of Thyristor in HVDC Converter Valve 114

Ning Liang, Jiachen Tian, Cuicui Liu, Yating Gou, Fang Zhuo, Feng Wang

Smart Grid and Electric Power Informatization 123

Jiyuan Ren, Zhe Wang, Zhen Luo, Fuyang Liu

Construction of Power Industry Corpus Based on Data Mining and Machine Learning Intelligent Algorithm 130

Liujun Zhao, Weizheng Kong, Qiuling Wang, Lihua Song

Online Identification Method of Induction Motor Parameters Based on Rotor Flux Linkage 135

Xuefei Li, Yang Kang, Hui Wang, Jie Chen, Fei Gao

Design of Power Consumption Tester for HPLC Power Line Carrier Communication Module 144

Liu Mouhai, Zhang Chenglin, Liu Xiangbin, Chen Hao, Yang Maotao, Chen Fusheng, Shen Liman

Research on Power Enterprise Network Security Solution 152

Shang Yanwei, Lin Qiang, Fengzhi Peng

Morphological Analysis of Optocoupler Accelerated Degradation Test Data 156

Xuangong Zhang, Xihui Mu, Huizhi Li

Research on Overvoltage Distribution of HVDC Converter Valve in Special Environment............. 162

Ning Liang, Cuicui Liu, Yating Gou, Jiachen Tian, Fang Zhuo, Feng Wang

Calculation of Electrical Stress Distribution and Influencing Factors Analysis of HVDC Converter Valve in Special EMP Environment 168

Gao Chong, Zhou Jianhui, Zhang Jing, Gou Yating, Liu Cuicui, Tian Jiachen, Zhuo Fang, Wang Feng

Numerical Simulation of Internal Flow in Direct Burning Coal-Fired Hot Flue Gas Furnace.......... 174

Zhang Zhenwei, Wang Peng

Virtual Synchronous Generator Grid Connected Control Method Based on Virtual Impedance...................... 181

Bao-Zhu Shao, Guan-Feng Zhang, Jun-You Yang, Fei-Fei Gao, Feng Sun, Qing-Song Zhao

Design of Memory Test System for Measuring Transmission Lines Galloping 188

Yulong Li, Jiaxin Liu, Yunfei Jia, Yu Liu

Application of Hybrid Conjugate Gradient Algorithms in Inverse Problems of Electromagnetic Tomography... 195

Li Liu, Wang Lei, Wang Zhanjun

Probabilistic Modeling of Output Characteristics Based on ECM Algorithm for Wind Farms........ 200

Tingxiang Liu, Xiaoying Zhang, Kun Wang, Wei Chen, Xiaolan Wang

Study on the Influence of Insulator on the Coupling Effect of Transmission Tower-Line System.................. 208

Qingshui Gao, Shi Liu, Yi Yang, Chu Zhang

The Impact Research of Delay Time in Steam Turbine DEH on Power Grid 214

Ge Jin, Sun Cai, Shaoxiang Deng

Investigation on the Relationship Between Winding Wire Size and Total Loss of BLDC 221
Jinshun Hao, Shuangfu Suo, Yiyong Yang, Yang Wang, Wenjie Wang

Chaos Control of Bi-Directional DC-DC Converter by Resonant Parametric Perturbation Method in
a DC Microgrid 228
Lihua Wang, Xue Ni, Yue Hu, Rui Zhang

Application Research of Multi-Source Information Fusion Technology in Power Network Fault
Diagnosis 239
Shuxin Liu, Enmin Zhao, Yanjun Zhang, Jing Li, Liang Zhang, Yundong Cao

Research on Power Quality Acquisition and Reconstruction Method Based on Compressed Sensing 246
Bo Yuan

A Study of Simulation on Relationship Between Young's Modulus of Cable Joints and Interface
Pressure Based on Finite Element Method 252
Zhengbing Tian, Liying Sun, Jing Wang, Xiaoye Bai

Effect of Heat Shield on the Heating Efficiency in MOCVD Chamber by Resistive Heating 258
Lili Zhao, Zhiming Li, Jincheng Zhang, Runqiu Guo, Ligen Lu, Lansheng Feng

Study on Basic Experiment and Optimization Prediction Model of Orthogonal Electrolytic
Machining of Film Cooling Hole in High Temperature Nickel-Based Alloy Blades 263
Yulan Hu, Weiguang Hao, Guoqiang Liu, Yajie Liu, Zilin Liu

Multi-Objective Reactive Power Optimization of Hybrid AC/DC Power System Considering Power
System Uncertainty 268
Liu Wenxue, Xing Luhua, Ma Changhui, Li Wenbo

Optimization of Charging Method for Scaled EVs 276
Xincheng Zhang, Zhizhen Liu, Yan Cao, Lijin Duan, Guoshen Tang, Wenchang Liu

Operation Quality Evaluation of Power Communication Network Based on Business QOS
Indicators 282
Geng Zhang, Ke Wang, Yang Wang, Yanan Wang, Chen Xiangzhou

A Stator Flux Calculation Method for Permanent Magnet Synchronous Motor in 60° Coordinate
System 290
Xiao Tang, Zhi Zhang, Chang Liu, Caishen Wang, Zhiping Wang

Analysis of Marketing Strategy of Electricity Selling Companies in the New Situation 296
Taorong Gong, Dezhi Li, Yuting Liu, Guanhong Wang, Haoming Zhu

Study on Zero Drift of Charge Amplifier Based on MOSFET 3N165 and OPA LF356N 305
Zongjin Ren, Cong Sun, Jun Zhang, Lulu Ma, Kai Zhao

Study on Radial Vibration of Circular Piezoelectric Ceramic 315
Yazhou Mao, Jianxi Yang, Haojie Liu, Yonggang Liu, Wenjing Xu

A Method of Power Flow Calculation Considering New FACTS and HVDC 323
Quan Chen, Xiaoming Dong, Haifeng Li, Tao Jin, Jun Yi, Jingzhe Tu

Combined Heat and Power Optimal Dispatch Considering Wind Power Uncertainty 331
Chunlong Chen, Xueshan Han, Wenbo Li

Analysis of Power System Vulnerability Considering Multiple Disturbances Corresponding to Information and Physics 337
Chaochao Wang, Hua Sun, Xiaoming Dong

Influence of Ni-Cu-La-B-Coated Glass Fiber on Conductivity and Electromagnetic Shielding Performance of Coatings 346
Denggao Guan, Yang Liu, Dehao Hu, Rundong Zhou, Xueqi Guo, Yan Wang, Yao Sun, Guanli Xu, Jinghui Lin, Chuanmin Sun

A Renewable Energy Assessment Model Considering the Effect of Frequency Regulation 352
Huang Xuxiang, Han Xueshan, Li Wenbo

The Comparison of Thermal Characteristics of AC Cable and DC Cable 359
Ruiqi Zhang, Hua Sun, Xiaoming Dong

Research on the Protection Range of Bird Droppings of 110kV Transmission Line Based on ANSYS Maxwell 366
Hao Zhang, Renfei Che, Wen Du, Xiaocheng Meng, Juntao He

A Novel Aggregation Method for Doubly Fed Wind Farm 372
Lulu Li, Xueshan Han, Wenbo Li, Zhe Jiang

Research on Substation Perimeter Isolation Based on Phased Array Radar and Multi-Video Fusion Technology 379
Yuancheng Zhu, Zhipeng Lei, Wei Zheng, Hongfeng Ma, Rongzhen Xia, Deyu Song

Fault Causes Identification for Transmission Lines Based on HHT and PNN 386
Wen Du, Renfei Che, Hao Zhang, Xiaocheng Meng, Juntao He

Transmission Line Insulator Fault Detection Based on Ultrasonic Technology 394
Zheng Yao, Xin Yu, Jianchun Yao, Wei Sui, Xiaochen Yu

Insulation Defect Detection of Solid Insulating Material Based on Nanosecond Pulse Voltage 401
Yang Jianzhong, An Shengdongt, Fan Yongqiang, Yin Jianbo, Yue Yonggang

Dynamic Variance Equalization Planning Optimization Method for Power Grid System Protection Communication Network 409
Dongliang Gao, Taorui Luo, Peizhe Xin, Yudong Wang, Jun Lu, Peng Cai

System for Real-Time Transmission Control of AC Motor Temperature Data Based on Linux System 416
Li Hengjie, Wu Long, Chen Wei, Qiao Zhen

Comparison Between UofC Model and Ionosphere-Free Combination Model in PPP 422
Y. C. Zhan

Dynamic Reactive Power Compensation and Harmonic Suppression of Optical Storage Microgrid Control in Natural Coordinates 427
Guangyao Jia, Shudong Wang, Huiying Song, Yanrong Mao, Almadhehagi Luai

Comparison and Analysis of X86 Server and Minicomputer Application in Power Enterprises 434
Yang Bo, Wei Jun, Wang Hua, Wang Gang

Risk Assessment of Power Communication Network Based on LM-BP Neural Network 447
Yanan Wang, Ke Wang, Ran Zhang, Qiao Xue, Xiangzhou Chen, Geng Zhang

Research on Remote Meter Reading Scheme and IoT Smart Energy Meter Based on NB-IoT Technology 453
Xingyuan Fan, Chun Zhou, Ying Sun, Jinyang Du, Ying Zhao

Optimal Configuration of Optical Storage Microgrid Under Demand-Side Response Based on Cooperative Game 461
Wang Shudong, Mao Yanrong, Jia Guangyao, Song Huiying, Qiu Jinliang, Ding Ting

Permittivity Model for GNSS-R Telemetry Wetlands 471
Cao Xinliang, Ren Xincheng

Design of Continuous Automatic Wire-Feeding Device Based on Electric Explosive Wire 479
Ying An-Wen, Yang Jia-Zhi, Yin Xin-Zhe

Calculation and Verification of Voltage Drop When Starting Tunnel Axial-Flow Fan 485
Enshi Wang, Bihui Huang

A Novel Searching Method of Fault Chains for Power System Cascading Outages Based on Quantitative Analysis of Dynamic Interaction Between System and Components 491
Jintao She

INTELLIGENT CONTROL SYSTEM AND MECHANICAL DESIGN

Warehouse Design Model for Shuttle Based Storage and Retrieve System 502
Bin Tian, Yingying Wu

Research on Control Strategy of a Buck-Type Harmonic Injection Three-Phase Rectifier 510
Qiang Jia, Panwei Ma, Yingchuan Qi, Dong Wang, Yang Cheng

Research and Analysis of MRC and IRC Algorithm Based on L TE System 518
Jian Jin, Shan Jinjie, Zhou Jian

Rotor-Mechanical Coupled Fault Feature Extraction Based on Second-Order Blind Identification 524
Feng Miao, Ruzhi Feng, Xianli Wang

Separating for Nonlinear Mixed Rotor Fault Signals Based on Adaptive Particle Swarm Optimization 530
Feng Miao, Xianli Wang, Wei Guo

A Survey of Knowledge-Based Intelligent Fault Diagnosis Techniques 536
Sanchuan Xu

Structure and Simulation of Roadway Disaster Simulation Control System for High Temperature Smoke Drill 542
Guo Jikun, Zhang Rui

Combination of CNN with GRU for Plate Recognition 548
Fucheng You, Yangze Zhao, Xuewei Wang

Methods to Solve Salt &Pepper Noise, and Frame Dropping of Timed Address Event Representation Vision Sensor 553
Lu Yu, Zhonghe Chen, Yun Hao

A Double-Channel Iterative NFXLMS Algorithm Used in Horizontal Vibration Isolation 562
Zhang Chi, Yin Wensheng

Projectile Velocity Measurement System Based on PVDF and Data Processing Method 568
Xiaoxiao Chen, Ping Song, Yayu Zhai

Some New Results on the Finite-Time Control and Its Application to a Chemical Reactor System 574
Ziteng Guo, Caisheng Wei

Dynamic Weighing System Based on Internet of Things Technologies 581
Jing Gang Cui, Jiong Mu, Ke Cheng Liu, Yu Zhu

Application of Weighted Fusion Algorithm in Air Tightness Detection Device 589
Xiaoxiao Chen, Ping Song, Yayu Zhai

Design of Intelligent Commutation Switch System Based on HPLC Carrier Scheme 596
Liu Mouhai, Tan Haibo, Yang Maotao, Chen Hao, Wu Zhiyong, Peng Haijun

Research on Four Axis Manipulator Trajectory Tracking Based on RBF Neural Network Algorithm 602
Luo Long

Design of Multifunctional Intelligent Security Robot Based on Single Chip Microcomputer 607
Liu Penghou, Chen Haichao, Du Yanzhe

Multi-Lane Detection using CNNs and a Novel Region-Grow Algorithm 613
Yi Sun, Jian Li, Zhen P. Sun

A Survey of Cloud Computing Access Control Technology 620
Minghao Wang

The Torsion Bars System Reliability Analysis with Failure Mode in Crawler Vehicle 625
Yi Liu, Pengbo Mou, Bin Zhang

Simulation and Life Prediction of Gear Meshing Process of Gearbox of a Crawler Vehicle 629
Pengbo Mou, Yi Liu, Jian Zhou

Study on Cargo-Swing Reduction of General Gantry Crane using Hybrid Optimal Input Shaper 634
Tuo Jianzhi, Du Peng, Zhang Ke, Zhang Wei, Xie Zonghua

Design and Analysis of the Leveling Hydraulic System of the Combine Harvester 640
Heng Wang, Shukun Cao, Xiangqian Xu, Tao Han, Hejia Guo

Research on Fast Self-Learning Improvement of ADRC Control Algorithm for Film Thickness
Control System 646
Liao Xue-Chao, Zhou You, Chen Zhen-Huan

Research and Analysis of Intelligent RGV Based on Dynamic Scheduling Optimization Model 653
Zheng Wang, Zeyu Zhou, Jiawei Liu

Research on Intelligent Near-Power Early Warning System for Mechanical Vehicles 662
Yingjing Wang, Jiaxin Liu, Yunfei Jia

Design and Research of a Aero Engine Operating Status Monitoring System 668
Wei Lin, He Li-Qing, Wan Yang, Chen Hua-Jie

Motion Recognition Based on Kinect for Human-Computer Intelligent Interaction 673
Xun Pang, Bin Liang

Parameter Optimal Design and Simulation of Power System of Electric Vehicle Based on AVL-CRUISE.. 680
Jianwei Ma

Parameter Design and Simulation Analysis of Power System in Plug-In Hybrid Vehicle............................... 687
Jianwei Ma

Effect of Injection Compression Process Parameters on Residual Stress of Products Based on Numerical Simulation.. 693
Junjie Zhu, Yanfang Chen, Wenhan Huang, Qiurong Zhang, Xiaoming Liao, Yizhi Huang, Zhiwen Qiu

Multiphysics Modelling of Warm Shot Peening of AISI 4140 Steel 701
Wang Cheng, Wang Long, Wang Chuanli

Study on the Control System of Agaricus Bisporus Picking Robot.. 710
Hu Xiaomei, Pan Zhaoren, Yang Shuzhen, Yu Tao

Design and Application of Visual System in the Agaricus Bisporus Picking Robot 716
Hu Xiaomei, Wang Chuan, Yu Tao

Application of Levitation Frame with Mid-Set Air Spring on Maglev Vehicles 723
Min Zhang, Ma Weihua, Shihui Luo

Study on Preparation Methods of Copper-Based Composites.. 732
Nianlian Li, Vanessa Bouchart, Pierre Chevrier, Hongyan Ding

Study on the Performance of the Wind Turbine Airfoil with Icing.. 738
Wang Long, Wang Cheng, Cheng Jie, Wang Chuanli, Li Liang

Research on Straightness Error Detection and Quality Control of Multi-Crankshaft Bores for Large Medium Speed Engine Block.. 745
Jiebin Yang, Xingxing Li, Chang'An Hu, Wanze Li, Jinwei Zhang

Study on Pressure Pulsation Suppression of Reciprocating Pump 751
Bin Li, Song Guo, Biao-Hua Cai, Zhao-Cun Shi

Design of Automated Guided Vehicle for Conveying Objects 759
Denggui Wang, Xinling Ma

A Hydraulic Fault Diagnosis Method Based on IMF Entropy Feature Fusion... 768
Liu Min, Huang Jie, Xianhai Sun

Natural Gesture Control of a Delta Robot using Leap Motion... 772
Xuchong Zhang, Ruiqiu Zhang, Liang Chen, Xianmin Zhang

Research on Manufacturing Technology of Thin-Walled Parts of Fe105 Metal Based on Laser Cladding 780
Tianbiao Yu, Yiting Bao

Research on Electrolyte Jet Assisted Laser Micromachining Technology 787
Yulan Hu, Weiguang Hao, Guoqiang Liu, Yajie Liu, Zilin Liu

Thermal Barrier Coating Processing Based on Improved Ant Colony Algorithm Process Optimization and Verification 795
Yulan Hu, Guoqiang Liu, Weiguang Hao, Yajie Liu, Zilin Liu

Experimental Study on Regression Model of Ultraviolet Laser Processing Thermal Barrier Coating Based on Response Surface Method 802
Yulan Hu, Guoqiang Liu, Weiguang Hao, Yajie Liu, Zilin Liu

Shape Optimization of Hook for Marine Crane 808
Yang Ji, Hu Wang, Hai-Quan Chen, Ming-Xuan Guo, Jun-Jie Wu

Design and Experimental Study of Vibration Reducing Experimental Device for Magneto-Rheological Elastomer 814
Tieshan Zhang, Zhong Ren

A Simple Safety Control Method for PSS Critical Gain Test 822
Siyuan Guo, Shoushou Zhang, Weijun Zhu, Li Li, Jinbo Wu

Design of Seat Clamping Device for Automobile DOF Shaker 826
Jian Zhang, Jianyu Yao, Chao Wen, Denggui Wang, Lingxia Wang

Contour Error Control of X-Y Platform Based on Nominal Model in Polar Coordinate System 833
Guirong Wang, Xinman Gong, Yingqi Li

Analysis of the Pressure Expansion of Bridge Plug Tools and Packers by Equivalent Material Method 841
Lanwen Wang, Xuanyu Sheng, Jiayue Sheng

A New Numerical Force Analysis Method of CBR Reducer with Tooth Modification 848
Xiaoxiao Sun, Liang Han

Influence of Electropulsing Treatment on Residual Stresses and Tensile Strength of As-Quenched Medium Carbon Steel 856
Pan Long

High Accuracy Numerical Simulation on 3D Weld-Pool Shape of Large Parts 861
Zhu Yonggang, Zuo Yanhong

Experimental Study on Cutting Force Comparison Between Inner Cooling and Outer Cooling in Zig-Zag Milling 870
Umair Riaz, Can Liu, Guangyu Tan, Guanghui Li, Ningxia Yin, Muhammad J. Saeed

Study on Vibration Reduction of Crane Monitoring System 875
Wang Jian, Zhang Yong Kui, Yan Xin

Research on Intelligent Communication System for Circuit Breaker Condition Monitoring 882
Qinghong Deng, Hao Zhang, Minfu Liao, Haoxue Zhang, Yifan Fu, Lujie Gai

Research on Energy Saving and Consumption Reduction Technology of Underground Gas Storage Compressor 888
Guan Tong, Liu Guiqiang, Wang Jinxiu, Wang Ping, Sun Dandan, Gao Shan, Liu Pai, Wu Qiang

Load and Stress Distribution of Thread Pair and Analysis of Influence Factors 894
Shikun Lu, Dengxin Hua, Yan Li, Fang Y. Cui, Pengyang Li

Study on the Effect of Temperature on Dynamic Characteristics of Rotor System with Straight Crack 905
Tengfei Kuaia, Changfang Zhao, Jie Ren, Guigao Le

Mechanical Productivity Design and Mechanical Process Analysis Framework Construction.........................911
Cui Li, Chen Hong' Bo

VOLUME 2

Simulation and Experimental Research on the Influence of Tool Geometries on the Cutting Force of High Temperature Alloy 917
Fengyun Yu, Chunyu Yan, Shaobing Guo, Shiwei Ma, He Wang

Research on Extraction and Analysis of Characteristic Conditions of Hub Motor for Electric Vehicles 924
Guohui Yang, Shuo Zhang, Chengning Zhang, Yongxi Yang

Design and Experimental Research of Expansion-Anchorage Device in Deepwater Pipeline......................... 934
Lan Zhang, Chengqiang Zhao, Liquan Wang, Cong Ru, Zhengbin Zhu

Study on Wear Mechanism of Diamond Particles in the Cutting of Pipeline Steel 941
Lan Zhang, Zhengbin Zhu, Liquan Wang, Chengqiang Zhao, Cong Ru

Analysis of Impact Characteristics of Diamond-Beaded Rope and Its Influence on Cutting Efficiency and Life 948
Lan Zhang, Cong Ru, Liquan Wang, Zhengbin Zhu, Chengqiang Zhao

Analysis of Mesh Stiffness of Herringbone Gear Considering Modification... 955
Wei Liu, Lunqin Duan, Runjiao Wang, Yanyan Wang

Modeling of the Stiffness of Corrugated Cardboard Considering Material Non-Linear Effect........................ 967
Jilin Ran, Changli Liu

Numerical Analysis on Fluid-Solid Coupling Cooling of Minimal Surface Lattice Structure 973
Zheng Yinzheng

Preparation and Characterization of Composite Resin Containing Anion Powder.. 984
Lili Ding, Jingli Zhu, Qian Zhao, Bin Liu

Research on Fuzzy PID Control of Forearm of Tunnel Steel Arch Mounting Machine 990
Zhi-Yuan Wang, Shi-Cheng Hu

Early Fault Diagnosis of Rolling Bearing Based on Lyapunov Exponent.. 995
Guo Qingjun, Li Yang

The Effect of Longitudinal Shock Absorber on the Vibration Response of Train-Bridge Coupling System in the Articulated Train 1000
Jie Ding

Simulation and Experiment of Passive Orbit Disconnected Support 1006
Zihan Gao, Yidu Zhang, Qiong Wu

Research on Trajectory Tracking and Vibration Suppression of a Smart Flexible-Joint-and-Link Space Manipulator... 1013
Zefeng Liu, Ruifeng Sun, Minwei Li

Effect of Ultrasonic Treatment on Morphology and Microwave Absorption Performance of ZnO Spheres 1029
Wei Huang, Shicheng Wei, Yi Liang, Bo Wang, Yuwei Huang, Yujiang Wang, Binshi Xu

Effect of Different Volume Fraction Magnetorheological Fluids on Its Shear Properties 1036
Sun Huimin, Zhu Xuli, Liu Nannan, Mou Jiefeng, Li Liang, Li Shixu

Research on Butt Joint of Ultrafine Grained Steel of Manual Arc Welding 1043
Yan Wang

Fiber Diameter Measuring Method of Textile Materials Based on Phase Information 1049
Wen Wang, Fang Zhang, Zhitao Xiao, Lei Geng, Jun Wu

Mechanism Analysis of Ferromagnetic Resonance of Electromagnetic Voltage Transformer in
Neutral Ungrounded System 1057
Tingran Sheng, Tongxin Han

Fast Aerial UAV Detection using Improved Inter-Frame Difference and SVM 1063
Li Xiaoping, Lei Songze, Zhang Boxing, Wang Yanhong, Xiao Feng

Discussion About Artificial Intelligence's Advantages and Disadvantages Compete with Natural
Intelligence 1071
Xiaofei Teng

Development of Artificial Intelligence and Effects on Financial System.. 1078
Minzhen Xie

A Variable Step-Size Adaptive Notch Filter for Frequency Estimation using Combined Gradient
Algorithm 1084
Huiyue Yang, Yaqing Tu, Ming Li

Smart Home System Based on Deep Learning Algorithm 1092
Yanfei Peng, Jianjun Peng, Jiping Li, Ling Yu

Bounded Noises Estimation Based on Cognitive Radio in Distributed Fusion System 1100
Shuyuan Wang, Ting Wang, Yijing Wang

Datacentre TCP Protocol of Centralized Window Control... 1108
Chunlin Wang, Jianyong Sun, Wanjin Xu, Xiaolin Chen

Gas Packaging Container Based on ANSYS Finite Element Analysis and Structural Optimization
Design... 1115
Zhiqin Yin, Tongdan Su, Ming He

Influence of Double Stealth Aircraft Approach Forward Support Cooperative Jamming on Radar
Detection Performance 1121
Lei Bao, Chun-Yang Wang, Hong-Bing Li, Juan Bai, Ming Tan

MD-UÇON: A Multi-Domain Access Control Model for SDN Northbound Interfaces................................ 1130
Rui Chang, Zhaowen Lin, Yi Sun, Jie Xu

Design Research on Information Coding System Under the Concept of Agile Manufacturing 1138
Zitian Liao

Research on Milling Force Prediction Model Based on Improved Particle Swarm Optimization
Algorithm 1142
Liu Ling, Qi Weiwei, Liu Tingting

Optimization of Adaptive Handover Algorithm Based on Distributed Antenna in LTE-R............................ 1147
Ziwei Yang, Yufu Zheng, Jing Jing

A Game-Theory Approach Based on Genetic Algorithm for Flexible Job Shop Scheduling Problem............1153
Li Nie, Xiaogang Wang, Fangyu Pan

An Improved Hybrid Structure Multi-Classification Support Vector Machine 1159
Zhang Xiaoyan, Wang Qiuqiu

Hand-Eye Calibration for Flexible Manipulator...1166
Jiahao Li, Xiuzhi Li, Ao Dun, Songmin Jia, Yanjun Sun

Bounds on the Total Signed Domination Number of Generalized Petersen Graphs $P(n,3)$............1171
Hong Gao, Yanan Yin, Yuansheng Yang

Change Impact Analysis of Complex Mechanical Product Based on Complex Network Theory 1177
Na Zhang, Yu Yang

Influence of Variable Slip Frequency Control Strategy on Tractiv E Performance 1183
Min Zhang, Ma Weihua, Shihui Luo

Study on Detection System of Grooved Rail Based on Inertial Measurement - Laser Triangulation
Comprehensive Algorithm...1190
Hong-Yuan Zhan, Kai-Feng Guo, Yong-Jun Xie

Mimic Defense Structured Information System Threat Identification and Centralized Control 1199
Bo Zhang, Weichao Li, Xin Sun, Yufeng Zhao

SMC Chaos Control of a Novel Hyperchaotic Finance System using a New Chatter Free Sliding
Mode Control 1209
Guoliang Cai, Yanfeng Ding, Qiaoling Chen

Project Evaluation and Analysis of Metrological Verification Regulation Based on Fuzzy
Comprehensive Analysis Method..1215
Zhu Jing, Liu Jiaxin

Safety Adaptability of Engine Retarder (Jacobs) on Long Downhill of Expressways 1222
Liu Jiaxin, Zhu Jing

Design of Information System Vulnerability Governance Platform Based on Distributed Asset
Acquisition and Vulnerability Verification Radar..1228
Guiquan Shen, Jiangang Lu, Wuqiang Shen, Jingzhi Huang, Weiyan Ji

Standard Architecture of China Intelligent Bus Systems...1234
Xianglong Liu

Distributed Scalable Abstract Reasoning Based on Dl-Lite 1240
Bin Xia

Cloud Resource Adaptive Scheduling Framework and Optimization Strategy Based on Swarm
Intelligence 1246
H. W. Zhao, S. Zhang, Y. Ruan, X. H. Jing

Research on Energy-Saving Lighting Control System of Tram Station Based on Traffic and
Passenger Flow Information..1251
Binjie Xiao

APPLICATION OF COMPUTER NETWORK AND INFORMATION TECHNOLOGY

The Optimization of Networking Method for the System Protection Communication Networks
Based on the Delay Analysis 1256
Yanan Wang, Ke Wang, Ran Zhang, Geng Zhang, Yang Wang

The Research of Ship Yaw Detection Method Based on Virtual Navigation Channel 1267
Juanjuan Shao, Shu Zhou, Xin He, Xinzheng Zhang, Xindong Liu

Configuration Generation of Aircraft Swarm Based on Communication Distance Constraint 1273
Qixi Fu, Xiaolong Liang, Jiaqiang Zhang, Lyulong He, Chuangchuang Zhu

Parameter Estimation Algorithm and Application in Industry Design 1282
Mei Yun, Jiang Haiyang, Lin Ying, Wang Fang, Nie Qimeng

A Multi-Model Estimation of Distribution Algorithm 1286
Li Hao

Pruning the Deep Neural Network by Similar Function 1292
Hanqing Liu, Bo Xin, Senlin Mu, Zhangqing Zhu

Application of Internet Segmentation Research Based on Natural Language Processing Technology
in Enterprise Public Opinion Risk Monitoring 1300
Di Liu, Jiangwen Su, Lihua Song, Zhen Qiu

A Gesture Recognition Algorithm Based on Threedimensional Projection and Direction Chain
Code 1308
Jie Li

Model and Design of High Temperature and Thermal-Proof Garment using Genetic Algorithm 1315
Zhang Maoyi, Sun Lele, Jia Haolin

The Design of Analog Signal Communication System Based on Visible Light 1330
Z. N. Zhang, H. Y. Y. Hua

Simulation Study of Dispersion Compensation in Optical Communication Systems Based on
Optisystem 1336
Peng Xia, Li-Hua Zhang, Yao Lin

The Comparison of Crowd Counting Algorithms Based on Computer Vision 1342
Zhaoqing Wang, Qishu Deng, Yusheng Zhao

Detector Design Based on MIMO OTA Test 1348
Zhaoqing Wang

Research on a Fusion Gait Real-Time Recognition Algorithm 1353
Zhi-Qiang Zhao, Meng Li, Ming-Ji Deng, Zheng Zhu, Xin Shi

Application Research of Denoising and Super Pixel Algorithm in Image Processing 1359
Qian Sun, Li Xin, Hanxu Gao, Faliang Chang, Zengshun Zhao

Evaluation Method and Experimental Study on Stationarity of High-Precision Linear Motion 1365
Yifan Zhou, Xingbao Liu, Yangqiu Xia, Shaopeng Cai

The Design of Image Depth Information Extraction Algorithm Based on Joint Bilateral Filtering 1373
Lin Yong

Research on Light-Small Lens Structure Design and Weight Reduction Optimization Based on Neural Network 1378
Shubing An, Minlong Lian, Yiliang Liu, Shaofan Tang

The Theoretical Development and Prospect of Two-Dimensional Topological Insulators 1386
Yichen Zhang

Calculation Formula of Positioning Error Based on Three Dimensions and Four Datum 1395
Quanli Han, Hongqiang Wan, Donchen Han

Application and Realization of Ray Tracing in Network Planning of Wireless Private Network 1401
Yonghua Zhang, Juanjuan Sun

Sparse Manifold Learning Based on Laplacian Matrix 1406
Wenjing Li

Web Advertisement Detection using Naive Bayes 1412
Xin Deng, Lunqing Hou, Fei Wang

Path Planning in Mobile Wireless Sensor Networks 1418
Kezhuang Wu, Junbin Liang

Defect Detection and Recognition Based on ADABOOT-SVM Integrated Model 1425
Zhikai Liang, Shaozhong Cao, Yukun Tan

Unmanned Visual Localization Based on Satellite and Image Fusion 1433
Xiaodan Yang

Cooperative Warp of Two Discriminative Features for Skeleton Based Action Recognition 1441
Zheng Sun, Xing Guo, Wei Li, Zhengyi Liu

Anomaly Detection Algorithm Based on FCM with Improved Krill Herd 1450
Chen Rui, Zhang Fengbin, Xi Liang

An Improved Parallelization of K-Means Algorithm Based on HADOOP 1461
Yizhuo Guo

Remote Sensing Image Building Extraction Based on Deep Convolutional Neural Network 1470
Yang Q. Yi, Wang W. An, Ma X. Dong

Speaker Identification Based on Deep Learning in FX iDeal System 1477
Yan Ke, Li Na, Chen Yutinge

The Improvement of K-NN Classifier with GA-Based Weight-Tunning Method.. 1483
Wei Jin

A Route Optimization Model Based on Link State Awareness in SDN 1488
Ran Li, Zhaowen Lin, Jie Xu, Yi Sun

Numerical Simulation of Deep Learning Algorithm for Gas Explosion in Confined Space 1494
Li Qizhong, Wang Ye, Yangjia, Wang Zhongqi

Study on Temporal and Spatial Patterns of Brain in Emotional State Based on Steady State Visual Evoked Potentials 1503
Kai Yang, Ying Zeng, Li Tong, Bin Liu, Xiyu Song, Bin Yan

Parameter Analysis of Stepped Frequency Pulses Frequency Diverse Array Radar......................................1510
Ming Tan, Chun-Yang Wang, Hong-Bing Li, Juan Bai, Lei Bao

WebVOS-A WebGIS Application for Volunteer Observation Ships..1519
Honghai Zhu, Yu Yu, Shibo Chu

Improvement of LDA Topic Mining Algorithm and Its Application in Short Text......................................1524
Kai Li, Chunmei Li

A Security Model Based on Intelligent Decision...1531
Shiping Xu, Ying Zhou, Ronghua Guo, Jiawei Du, Zhe Liu

A New Clustering Validity Index Based on K-Means Algorithm...1536
Xiangru Hou

Object Detection Based on the Improved Single Shot MultiBox Detector 1543
Songmin Jia, Chentao Diao, Guoliang Zhang, Ao Dun, Yanjun Sun, Xiuzhi Li, Xiangyin Zhang

Medical Image Fusion Based on Statistical Modeling 1549
Huaijing Qu, Hengbin Wang, Hongkui Xu

Automatic Integration Testing Through Collaboration Diagram and Logic Contracts.............................1554
Yi Sun, Xiaohua Yang, Jie Liu, Tonglan Yu, Zhuoran Xu, Zhiqiang Wu, Zhi Chen

Multipurpose IP-Based Space Air-Ground Information Network 1562
Abid Murtaza, Liu Jianwei

ChanDet: Detection Model for Potential Channel of iOS Applications 1575
Guomiao Zhou, Ming Duan, Qi Xi, Hao Wu

Characterizing of Strong Normalization for $\Lambda\mu$-Calculus...1583
Xinxin Shen, Kougen Zheng

Signature Handwriting Identification Based on Generative Adversarial Networks 1594
Siyue Wang, Shijie Jia

Object Detection on Underground Low-Quality Images...1599
Qi Mu, Zhiqiang He, Yankui Liu, Yu Sun

ArchiMate Customization and Architecture Repository Management Practices: For a Technology-Intensive Enterprise...1604
Baobao Ding, Tong Wu, Yingming Yang, Liang Dou, Tiancheng Jin

Sea Clutter Suppression Based on Sea Spikes Identification and Matrix Completion 1612
Zhiyu Shao, Jiangheng He, Shunshan Feng

Exponential Stability for a Class of Nonlinear Singular Markovian Jump Systems with Time-Delay 1618
Daixi Liao, Shouming Zhong, Shaohua Long

A New Molecular Encryption Model Based on Microfluidic Techniques..1623
Pengcheng Ma, Heng Sun

Multi-Peak and Power Cooperative Detection Algorithm to Detect Forwarded Spoofing Interference Signals of BOC Modulation Receivers 1629
Zhiying Wang, Jiawen Wang, Yidong He

Construction of on-Campus 3D Model Based on GIS Technology and OpenGL 1636
  Limei Chen

HTTP Tunnel Trojan Detection Model Based on Deep Learning.. 1645
  Yubo He, Yuefei Zhu, Wei Lin

Design and Implementation of Connect6 Intelligent Game System... 1656
  Zengyu Cai, Chunfeng Du, Xiaoshuang Guo, Jianwei Zhang

Automated Recognition of Retinopathy of Prematurity with Deep Neural Networks 1662
  Yifan Wang, Yuanyuan Chen

Research on Gait Recognition Algorithm Based on Multiinformation Perception....................................... 1667
  Huabin Liu, Zongmiao Dai, Yongkang Zheng

The Combination of Neural Network and "question Matching" Improves the Correct Rate of
Grassland Degradation Decision 1675
  Li Chunmei, Pi Wei, Dong Suo, Li Zhao

The Design and Development of Simulation System for Broad Band Wireless Communication 1682
  Lijuan Gao, Yu He

A Deep Learning Approach for Vehicle and Driver Detection on Highway... 1688
  Peihua Lv, Yang Zhang, Xiaobo Lu, Di Zhou

Robust Modeling for Fleet Assignment Problem Based on GASVR Forecast.. 1696
  A. J. Su, W. D. Yang, C. Zhang, M. X. Kong

An Improved SVM Web Page Classification Algorithm 1703
  Xun-Yi Ren, Chen Shi, Dan Zhang, Wen-Si Wang

The Application of Convolutional Neural Network in Security Code Recognition 1710
  Jingtian Gu

Depth Enhancement with Improved Inpainting Order and Smoothing Method 1720
  Kang Yi, Yuting Zhao, Yiyan Lei, Jian Pan

Recognition of Speed Signs in Uncertain and Dynamic Environments 1733
  Zhilong Zhu, Gang Xu, Hongmei He, Juanjuan Jiang, Tao Wang

Structure Analysis and Generation of X.509 Digital Certificate Based on National Secret 1739
  Hua Jiang, Gang Zhang, Jinpo Fan

Research on Parking Detecting Analysis Based on Projection Transformation and Hough Transform 1745
  Xuemei Yu, Yaojie Sun

Research on the Efficiency of Beijing-Tianjin-Hebei Airport Group Based on System Dynamics............... 1750
  C. Y. Wang, W. W. Wu, J. Zhang

The Application of Alternating Direction Method of Multipliers on l_1-Norms Problems 1758
  Yanchen He

Pre-Flight Rerouting Combining A* Algorithm and AHP Under Severe Weather 1775
  Ding Wencan, Sui Dong

Attacking Intel UEFI by using Cache Poisoning... 1782
  Dong Wang, Wei Y. Dong

A Systematic Review: Road Infrastructure Requirement for Connected and Autonomous Vehicles (CAVs).. 1788
Yuyan Liu, Miles Tight, Quanxin Sun, Ruiyu Kang

Ship Detection and Tracking in Nighttime Video Images Based on the Method of LSDT 1801
L. Liu, G. Liu, X. M. Chu, Z. L. Jiang, M. Y. Zhang, J. Ye

A Novel Approach to Multi-Resolution Technique for Fast Pattern Recognition 1812
Xiaochun Liu, Xiaohua Ding, Xiang Zhou, Wen Cui

Research on PLC Information Model Based on UML Class Diagram 1820
Xu R. Yu, Yun X. Zu, Wei H. Li

A New Improved Simplified Particle Swarm Optimization Algorithm 1827
Haikuan Liu, Dachao Yue, Lei Zhang, Zhiyuan Li, Dawei Jiang

VOLUME 3

Research on Precise Maintenance Method for Green Belt of Municipal Road Based on UAV Image Sequence.. 1834
T. Duan, P. C. Hu, L. Z. Sang, L. Wang

TPO-MAC: Traffic-Priority-Based Opportunistic MAC Protocol for Multi-Channel Cognitive Radio Networks 1839
Gao Q. Shen, Lei Lei, Zhi L. Li

Opportunistic Routing with Available Bandwidth Assurance for High Dynamic UAV Swarms.................... 1846
Zhi L. Li, Lei Lei, Gao Q. Shen

Image Flame Recognition Algorithm Based on M-DTCWT.. 1853
Wenxia Bao, Zhongzhi Zhong, Ming Zhu, Dong Liang

Defect Prediction Model for Object Oriented Software Based on Particle Swarm Optimized SVM............. 1859
Yanan Wang, Ran Zhang, Xiangzhou Chen, Shanjie Jia, Huixia Ding, Qiao Xue, Ke Wang

Research on Knowledge Graph Application Technology 1869
Yong Chen, Xueli Chen

Predicting Failures in Hard Drivers Based on Isolation Forest Algorithm using Sliding Window................. 1877
Tinglei Zhang, Endong Wang, Dong Zhang

A New Crossover Algebra of GA for Solving the Degree Constrained Minimum Spanning Tree Problems... 1883
Hui Li, Kai Shi, Huanran Li, Sheng Lin, Guangping Xu, Shuangxi Li

Research on Hybrid Recommendation Model Based on PersonRank Algorithm and TensorFlow Platform 1889
Guangqi Wen, Chunmei Li

Improvement of LDA Algorithm Based on Microblog Short Text Hotspot Analysis..................................... 1898
Kai Li, Chunmei Li

Extraction of Cerebral Hemorrhage and Calculation of Its Volume on CT Image using Automatic Segmentation Algorithm... 1908
Nian Wang, Fei Tong, Yongcheng Tu, Hemu Chen, Yong Zhou, Jun Tang

Thinking and Exploration on the Teaching of the Course of "College Computer Foundation" Under the Mode of "Internet Plus" 1914
Tang Xianfang, Jia Yachao, Zhang Ru

Prediction of the Anti-Inflammatory Mechanism of Clematis Chinensis Based on Network Pharmacology 1919
Xulong Huang, Junjie Hao, Yuqing Liang, Yuanmin Wang, Juan Kong, Xiangpei Wang, Feng Xu, Hongmei Wu

Streaming Information Transmission Based on OPC UA 1927
Liu Wei, Zu Yunxiao, Li Weihai

A New Point-Weighting Finite-Difference Modelling for the Frequency-Domain Wave Equation 1932
D. S. Cheng, B. W. Chen, J. J. Chen, X. S. Wang

Research on Layout Optimization of Express Parcel Transportation Network Distribution Center Based on Node Operation Process 1939
Jun Han, Shiwei He, Mingkai Bi, Zilong Song

Using Markov Constraint and Constrained Least Square Filter to Develop a Novel Method of Passive Terahertz Image Restoration 1949
Yuanmeng Zhao, Xiao Sun, Cunlin Zhang, Yuejin Zhao

A Novel Dual-Band Video Fusion Algorithm using Fast Lookup-Tables: Toward Naturalistic Color 1955
Yuanmeng Zhao, Weiqi Jin, Lingxue Wang

Terahertz / Visible Dual-Band Image Fusion Based on Hybrid Principal Component Analysis 1962
Yuanmeng Zhao, Yulong Qiao, Cunlin Zhang, Yuejin Zhao, Hong Wu

A Camouflage Effect Detection Model for Fixed Targets 1967
Yang Xin, Xu Weidong, Xiang Lei, Zhu Wannian, Tian Jiyao

Welding Defect Signal Extraction Technology Based on OMP Algorithm 1974
Qi Ailing, Lei Haijun

Calculation Method of Cross Section Area of Collapsing Dangerous Rock Based on Parallel Binocular Vision 1979
Ping Gan, Zhenzhen Ruan, Yang Wang, Lin Huang, Yanyun Li, Yangfan Huang

Automatic Measurement Algorithm of Scoliosis Cobb Angle Based on Deep Learning 1987
Yongcheng Tu, Nian Wang, Fei Tong, Hemu Chen

A Robust $\mathcal{H}_\infty$ Approach of In-Flight Calibration for UAVs with Low-Cost IMU 1994
He Gao, Chao Gao, Guorong Zhao

Invisible Information Transmission System of Visible Light Based on Interleaved Code 2001
Yuting Meng, Yunpeng Hu, Mingchao Li, Yanqun Tang

Application of Deep Convolution Neural Network in Automatic Classification of Land Use 2009
Xiaodong Ma, Guang Yang, Qunyi Yang

Research on IP Address Allocation of Tactical Communication Network 2014
Wang Huitao, Yang Ruopeng, Wufan, Zou Xiaofei

The Priority Assignment of Messages Effects on Delay Performance in VANET 2021
Jin Tian, Qianru Han

An Improved Adaptive Weighted Median Filter Algorithm 2030
  Yuqin Song, Jun Liu

Research on Long-Distance Hand Recognition Based on Depth Information 2036
  Yuyang Fu, Lanfang Miao, Zhifei Li

A Fault Repair Method for Workstation Cluster Based on Probabilistic Model Checking............................ 2042
  Xi Wang, Ting Chen, Chengtian Ouyang

Anomaly Detection for Time Series using Temporal Convolutional Networks and Gaussian Mixture
Model 2048
  Jianwei Liu, Hongwei Zhu, Yongxia Liu, Haobo Wu, Yunsheng Lan, Xinyu Zhang

Automatic Detection of Follicle Ultrasound Images Based on Improved Faster R-CNN 2058
  Tianlong Zeng, Jun Liu

Research on Network Security of Campus Network 2065
  Min Huang, Wanbo Luo, Xing Wan

DATA MINING AND ANALYSIS

Online Route Planning for Cooperative Area Coverage Search of Aircraft Swarm 2071
  Yueqi Hou, Xiaolong Liang, Jiaqiang Zhang, Ye Li, Jiong Wang

AdaptiveSLA: A Two-Stage Scheduling Framework for SLA Profit Maximization in Multi-Tenant
Database 2079
  Yang Ji, Zhang Lin, Tang Rong

Research on Reporting Scheme of Grading Stop-and-Recharge Event of Low-Voltage Acquisition
Terminal... 2089
  R. Huang, Chenglin Zhang, L. Ouyang, M. Liu, X. Liu, Y. Su, H. Chen

Progress and Trends in Mobile Cloud Computing Research 2098
  Wang Xiao-Shu

Review of Research on Blockchain Application Development Method.. 2103
  Yue Zeng, Yue Zhang

Research on the Random Corresponding of Privacy Data Mining in the Association Rules of Cloud
Computing 2109
  Baofeng Hui, Guoqing Jia, Shanji Chen

Discussion on Supplier Selection in the Selection of Large Civil Passenger Aircraft................................. 2113
  Aiping Yin, Zhangkang, Zhenxing Gao, Duoduo Zhuanga

Analysis on Error Compensation for Integrated Navigation Based on Forgotten Kalman Filter 2117
  Ling Yao, Hua Zhang, Xiaoping Huang

Construction and Analysis of Campus Knowledge Payment Platform Under the Wave of Big Data 2123
  Ran Wu, Qiong Wu

The Mobile Payment Based on Public-Key Security Technology.. 2127
  Jiabin Sun, Nan Zhang

The Advisable Technology of Key-Point Detection and Expression Recognition for an Intelligent Class System 2134
Yusheng Zhao, Haotian Yan, Zhaoqing Wang

Research on the Interaction Between Language and Economy 2140
Yusi Wu, Jie Liu, Ping Wang, Yisheng Wang, Peng Zhang, Hongyan Xi

An Expertise-Enhanced Collaborative Filtering Method for Keywords Recommendation in Searching Engine Marketing 2147
Yuan Zhu

Analyse the Data Tendency in the Public Opinion Monitoring System 2152
Jian Li

Analysis and Design of Course Website for Software Testing Based on SPOC 2160
Jiujiu Yu

Product Information Modeling Based on Polychromatic Sets and Scheme Optimum Selection for Conceptual Design 2166
Li Qiang, Xu Bin, Li X. Qiang

Research of a Method for Synchronized Phasor Data Transmission Based on IEC61850 2176
Shanqiang Feng, Aijun Hou, Bochuan Gu

The Design and Creation of an Interactive E-Book: "Book of Answer" 2183
Xinyue Gui

Encrypted Tag Design for RFID Systems 2189
Shida Yu

Automated Sentiment Analysis of Text Data with NLTK 2195
Jiawei Yao

A Study of Speech Feature Extraction Based on Manifold Learning 2203
Cheng Hao, Ma Xin, Yugong Xu

Research and Improvement of CHI Feature Selection in Sentiment Analysis 2211
Li Danyang, Fan Huimin

A Research on the Innovation and Development of Micro and Small Sci-Tech Enterprises 2216
Wang Guanhui

Mode of Freight Rolling and Collecting and Distributing Based on Cloud Logistics Platform 2221
Wang Xiaokun, Wang Lei, Wang Hanyan

Research on the Integration Demonstration System of Space Information Network Based on TDRSS 2232
Dandan Fan, Mengyue Qiu, Xiaoshen Xu, Ming Chen

The Resource Aggregation and Integration Platform for Shared Development of the Direct Bank 2238
Qi Wang, Zhongshi Zhang, Xiaotong Zhang, Qing Zhao

Implementation of Interactive Classroom Design Based on WI-FI Service 2245
Chen Shuai, Zhang Shuifeng, Wu Tianfang

Research on the Challenges and Innovation Models of Public Management in the Age of Big Data 2252
Wang Cong, Su Xiangguo, Lou Ziqiang

Applications Research of Cluster Analysis in Chinese Acupuncture Therapy 2257
Xiaoche Feng

Prediction of Alzheimer's Disease Based on Bidirectional LSTM 2262
Qiao Pan, Shiyu Wang, Junhao Zhang

A Survey of the Key Technology of Software Vulnerability Mining 2267
Dansheng Lin, Xiaoquan Wu, Qinqin Wu, Ye Liu, Jijun Zeng, Jinjun Tan

Research on Security Evaluation of Government Cloud Platform Based on Fuzzy Analytic
Hierarchy Process 2272
Jing Chen, Feng Zhang

Big Data Mining for Investor Sentiment 2278
Tao Cen, Qianqian Chu, Renke He

Research on Frequency Hopping Synchronization Strategies Based on TOPSIS Method 2283
Xiaoyu Cai, Zhanjun Jiang, Qianru Liu, Haoqiang Shi

An Event Detection Method Based on Association Link Network 2291
Lin Sun, Weijun Yang, Xinhuai Tang

A Comparative Study of Customer Complaint Prediction Model of Time Series, Multiple Linear
Regression and BP Neural Network 2297
Xin Xu, Zhijie Sun, Li Wang, Jun Fu, Chao Wang

Anomaly Detection Based on PMF Encoding and Adversarially Learned Inference 2304
Lin Zhang, Wentai Yang, Hua Gan, Meng Li, Xiaoming Wang, Gang Liang

Mimic Defense System Security Analysis Model 2315
Li Qinyuan, Han Jiajia, Sun Xin, Qin Junning, Zhang Bo

The Present of Education Big Data Research in China: Base on the Bibliometric Analysis and
Knowledge Mapping 2323
Cheng Jiang, Qi Wang, Wu Qing, Leiye Zhu, Si Cheng, Haibin Wang

Method for Creating, Updating and Maintaining a Case Library of Service Business Guidance 2329
Xin Xu, Zhijie Sun, Jun Fu, Li Wang, Feng Xie

Design and Implementation of Body Quality Index App Based on Android 2335
Zhao Limei, Ma Xiaotie

Emotional Analysis of Public Opinions in Colleges and Universities:Based on Naive Bayesian
Classification Method 2342
Qingjia Wang, Kun Liu, Kun Ma

Dual-Channel Supply Chain Sale Strategies with Return Guarantee 2347
He Jingshi

Summary of Recommendation System Development 2355
Liu Liling

An Effective Method for Forest Fire Smoke Detection 2360
Luxing Qin, Xuehui Wu, Yichao Cao, Xiaobo Lu

Demand Analysis of Material Reserve Optimization 2368
Jian Gao, Shuming He, Xing Song, Liang Zhang

Mining Spatiotemporal Characteristics of Car-Sharing Demand 2371
Jing-Jing Tian, Dong-Fan Xie, Fu-Jun Ding

An Empirical Study on Regional Logistics Competitiveness in Guangdong 2377
Chunshang Wu

Evaluation Effect of Internet Word of Mouth and Application of Big Data 2384
Te Ma

Purchase Decision Under Network Environment and Application of Big Data 2388
Te Ma

Research on the Evaluation Index System of Intelligent Railway Passenger Station 2393
Shaofu Lin, Qianwen Wei, Sibin Xia

Research and Analysis on the Top Design of Smart Railway 2401
Shaofu Lin, Yafang Jia, Sibin Xia

Study on Vulnerability Rating of the Intelligent and Connected Vehicle's Cybersecurity 2408
Yangyang Liu, Zhao Wang, Yanan Zhang, Peiji Shi, Xuebin Shao

Analysis of Spatio-Temporal Distribution Characteristics of Passenger Travel Behaviour Based on
Online Ride-Sharing Trajectory Data 2421
Xianlei Dong, Lingyu Wang, Beibei Hu

Research on Road Traffic Flow Status Based on Survival Analysis 2429
Beibei Hu, Doudou Lin, Qibo Sun, Xianlei Dong

Research on the Management and Maintenance of Infrastructures in Fog Section of Motorway
Based on the MOT Model 2436
X. L. Li, Q. H. Song, B. M. Tang, C. Li

Research on Semantic Prediction Analysis of Tibetan Text Based on Word2Vec 2446
Ding Hai-Lan, Yu Hong-Zhi, Qi Kun-Yu

Research on the Characteristics of Bitcoin Price Fluctuations Based on ARCH Effect 2453
Juan Wang, Feng Tian, Jie Fu

Design and Realization of Scenes of 3D Virtual Digital Library 2460
Wang Li

End-to-End Speech Synthesis for Tibetan Lhasa Dialect 2466
Lisai Luo, Guanyu Li, Chunwei Gong, Hailan Ding

Research on China's Transport Connectivity Along Corridors of the Belt and Road Initiative 2472
Jiao Wenwen, Li Ran, Li Sicong

A Study on Location of Logistics Hubs of Hub-and-Spoke Network in Beijing-Tianjin-Hebei
Region 2478
Xiaoyu Xin, Nan Yu, Xiang Chao

Blockchain-Based Intelligent Hospital Security and Data Privacy Construction 2488
Qiuzi Huang, Shuyu Chen, Hui Zhao, Junhao Wen

Research and Analysis on the Transformation of Road Passenger Transport Industry 2494
Lili Chen, Zhijun Yi, Xiaofeng Di

Study on the Operation Mode of Suburban Railway at Home and Abroad and the Inspiration to Beijing 2499
Luxi Peng, Chao Wang, Jianhong Wu

Application Scenarios Based on SDN: An Overview 2506
Tong Li, Jinqiang Chen, Hongyong Fu

Overview of End-to-End Speech Recognition 2515
Song Wang, Guanyu Li

Air Traffic Management Process Quality Assessment Model Based on Improved Fuzzy Matter Element Analysis 2519
Pan Wei-Jun, Ren Jie, Wang Run-Dong

Real-Time Estimation of Urban Rail Transit Passenger Flow Status Based on Multi-Source Data 2528
Zhengping Tao, Jinjin Tang

Comprehensive Assessment of Green Development Level for Urban Rail Transit Enterprises Based on ANP and Entropy Weight Method 2537
Ying Tang

Simulation Study on Emergency Evacuation of Metro Stations in Fire Degradation Mode 2544
Xi Jiaojiao, Li Jin

The Evolutionary Game Analysis of Incentive Mechanism for Crowd Sensing of Public Environment 2551
Qiang Zhang, Qingqing Zhang, Xueyan Liu, Jian Dai, Xujuan Zhang

Study on Tibetan Word Vector Based on Word2vec 2559
Ning Yang, Guanyu Li, Hailan Ding, Chunwei Gong

Analysis of Seismic Anomalies of the Jiuzhaigou Earthquake 2565
Xiaoying Jiang, Xiangzeng Kong, Gongde Guo

Research on the Development Orientation and Thinking of New Media in State-Owned Enterprises 2572
Wang Han, Wang Youzi, Xia Liyu

Analysis of the Situation Faced by New Media Propaganda in State-Owned Enterprises 2578
Wang Han, Xia Liyu, Wang Youzi

A New Linguistic Decision Making Method—FLM-VIKOR 2584
G. R. Zhang, W. D. Zeng

Design and Application Research of VR/AR Teaching Experience System 2592
Baiqiang Gan

Application of Blockchain in Document Certification, Asset Trading and Payment Reconciliation 2597
Xingxiong Zhu, Dong Wang

Study on the Relationship Between the Sharing Rate of Vehicle Exhaust Pollution and the Quantity of Possession 2602
Baiyu Chen, Da Fu, Yuanyuan Yang, Xiao Li

Research on Route Planning of Aerial Photography of UAV in Highway Greening Monitoring 2609
T. Duan, P. C. Hu, L. Z. Sang

Research on Specific Eye Movement Mode of Qualified Railway Driver 2615
    Chenzhe Sun, Guanglei Zhang, Xiujun Zhai

Research on Big Data Decision-Making Support Platform of New Energy Bus 2624
    Hao B. W. Feng

Linked Data Crowdsourcing Quality Assessment Based on Domain Professionalism 2629
    Lu Yang, Li Huang, Zhenzhen Liu

Research on Quantitative Risk Assessment Method of Packaged Cargoes Carried by Ship Based on
Online Dynamic Big Data Fusion Technology 2638
    Ru Lan, Wen Chang, Wei Shen, Zhigang Jia

An Analysis Method for Error Propagation Reachability of Component-Based Software 2649
    Mingqi Fan, Min Song, Zhengxian Wei, Wanfeng Mao

Thoughts on the Orientation of Mathematics Education in Colleges and Universities 2656
    Liang Shushuang

Study on Chinese Technical Economy and Global Social Responsibility 2662
    Huanping Zhang, Maohua Li, Zoltán Zéman

Research on Semantic Information Retrieval Model of Bamboo & Rattan Domain Based on Query
Extension 2669
    Lin Peng, Ming-Ming Lai, Xin Zhang

Research on Semantic Information Retrieval Model of Bamboo Rattan Domain Based on Semantic
Relevance 2674
    Lin Peng, Xu-Peng Kou, Lin-Nan Yang

Emergency Event Matching using Hierarchical Blocking Method 2679
    Chang Wen, Yu Liu

Research and Construction of Yunnan Plant Vertical Retrieval System 2686
    Lin Peng, Zhi-Run Ma, Xin Zhang

A Social Network Water Army Detection Model Based on Artificial Immunity 2692
    Hanhua Zhang, Tao Li, Yuqiao Wang

Design and Implementation of Ontology Knowledge Base of Endemic Genera of Seed Plants in
Yunnan Province 2701
    Lin Peng, Xu Li, Lin-Nan Yang

The Applications of the Edge Detection on Medical Diagnosis of Lungs 2706
    Wei Xu, Jinping Li, Hongwei Jia

A Novel Rating Style Mining Method to Improve Collaborative Filtering Algorithm 2712
    Wei Yang, Sheng H. Guo, Chun J. Zhang

Study on Traffic Organization and Work-Zone Optimization of Four-Lane Freeway Reconstruction
and Expansion 2720
    Xuanyu Ye, Yazhen Chen, Chen Chen

Review on Studies of Machine Learning Algorithms 2727
    Peiyuan Xu

Spatial Spillover Effects of the Real Estate Industry on Economic Development -From Destocking Perspective.. 2734
Xiaoqian Liu, Qinqin Zhou, Chang'An Wang

Application of Big Data in Forecasting Traffic Flow 2741
Luo Wanbo, Wan Xing, Huang Min

Financial Risk Analysis and Early Warning Research Based on Data Mining Technology 2746
Yan Hou, Ziyan Yuan

Author Index

ISPECE 2018 Preface

During December 28-30, 2018, the 2018 International Symposium on Power Electronics and Control Engineering (ISPECE 2018) was successfully held in the well-known historical and cultural ancient city of China—Xi'an. The goal of the conference was to provide an excellent international academic forum for all the researchers and practitioners. Any papers and topics that concentrates on the fields of power electronics and control engineering were broadly welcomed.

Besides, the high-quality keynote talks were presented by professors from different areas including Prof. Yonghong Peng of the University of Sunderland, Prof. Jiang Yan of North China University of Technology and Prof. Yinglei Song of Jiangsu University of Science and Technology. It was a great chance for those that are of interest to the topics to interact with the experts to get their advice and consult as well as collaboration opportunities.

ISPECE 2018 proceeding tends to collect the most up-to-date, comprehensive and worldwide state-of-art knowledge and research on the related fields of power electronics and control engineering. All the accepted papers have been submitted to strict peer review by expert referees, and selected based on originality, significance and clarity for the purpose of the conference. The proceeding consisting of 398 papers would be published in Journal of Physics: Conference Series (ISSN:1742-6588), and submitted to EI Compendex, Scopus, Inspec and CPCI for indexing

We would like to acknowledge all of those who supported ISPECE 2018. Each individual and institutional help was very important for the success of this conference. Especially we would like to thank the organizing committee for their valuable advices in the organization and helpful peer review of the papers.

Prof. Guobin Xu
The Chair of ISPECE 2018

Content from this work may be used under the terms of the Creative Commons Attribution 3.0 licence. Any further distribution of this work must maintain attribution to the author(s) and the title of the work, journal citation and DOI.

Published under licence by IOP Publishing Ltd

Organizing Committees

Conference Chairs

Prof. Guobin Xu, Guangdong Academy of International Academic Exchange, China
Prof. FOO Check-Teck, Nanyang Technology University, Singapore
Prof. Zhenxiong Chen, The Australian National University, Australia

Publication Chair

Dr. Dong Xie, University of Auckland, New Zealand

Local Committee

Prof. Zhenxiong Chen, The Australian National University, Australia
Prof. Jiangling Li, Vice-President of Lingnan Normal University, China
Prof. Yongjie Li, Guangdong Academy of International Academic Exchange, China
Prof. Quanwen Liao, Xiamen University, China
Prof. Danming Lin, Shantou University
Prof. Wenquan Ling, Jinan University, China

International Technology Committee

Dr. Caisheng Wei, Northwestern Polytechnical University, China
Dr. Yingjie Yin, Chinese Academy of Sciences, China
Prof. Hamid Reza Karimi, Politecnico di Milano, Italy
Dr. Algo Carè, Universita' degli Studi di Brescia, Italy
Dr. Jan Kubíček, VŠB – Technical University of Ostrava, Czech Republic
Prof. Waldo Hasperué, National University of La Plata, Argentina
Prof. Roberto Montemanni, University of Applied Sciences of Southern Switzerland
Prof. Jihad Hasan Jabali Asad, Palestine Technical University, Palestine
Dr. Rui Melicio, University of Evora, Portugal
Dr. Vishnukumar S, Chairman, Computer Society of India, Trivandrum Chapter/Assistant Professor, Mar Baselios College of Engineering and Technology, Thiruvananthapuram, Kerala, India
Prof. Jianhua Wu, Nanchang University, China
Prof. Zhou Yue, Hunan University of Humanities, Science and Technology, China
Prof. álvaro Rocha, University of Coimbra, Portugal
Dr. Qingbin Song, Macao University of Science and Technology, China
A.Prof. Wenxu Yan, Jiangnan University, China
Prof. Dumitru TOADER, Politehnica University Timisoara, Romania
Prof. SERGIO, University of Pisa, Italy
Dr. Xu Jun, Wuhan University of Technology, China
Dr. Wei Yang, Beihang University, China
Dr. Yang Xue, Shanghai University of Electric Power, China
A.Prof. Jun Zhu, Guangxi Normal University, China
Dr. Chuanyang Li, Tsinghua University, China
Prof. Yigang He, Hefei University of Technology, China
Prof. Xingyuan Tong, Xi'an University of Posts and Telecommunications, China
Prof. Xuan WANG, Shanxi Normal University, China
Prof. Dong Yin, University of Science and Technology of China, China

A.Prof. Zhengyong Yu, huaian vocational college of information technology, China
A.Prof. Hong Ma, Nanjing Vocational Institute of Industry Technology, China
Dr. Jian Huang, South China University of Technology, China
Dr. Ruipeng Ning, East China Normal University, China
Dr. Pengpeng Sun, Chang'an University, China
Prof. Hongwei Li, Southwest Petroleum University, China

ISPECE

IOP Publishing

Peer review statement

All papers published in this volume of *Journal of Physics: Conference Series* have been peer reviewed through processes administered by the proceedings Editors. Reviews were conducted by expert referees to the professional and scientific standards expected of a proceedings journal published by IOP Publishing.

Content from this work may be used under the terms of the Creative Commons Attribution 3.0 licence. Any further distribution of this work must maintain attribution to the author(s) and the title of the work, journal citation and DOI.
Published under licence by IOP Publishing Ltd

ISPECE

IOP Publishing

Study of the Mechanism of Tangent Bifurcation in Voltage Mode Controlled DCM Buck Converter

Lingling Xie[1,2*], Fangzheng He[2] and Shuwei Situ[3]

[1] School of Electric Power, South China University of Technology, Guangzhou, Guangdong, 510641, China

[2] School of Electrical Engineering, Guangxi University, Nanning, Guangxi, 530004, China

[3]Haihong electric CO., LTD, Kaiping, Guangdong, 529339, China

*xielingling1318@163.com

Abstract. Tangent bifurcation is a special bifurcation in nonlinear dynamic systems. The investigation of the mechanism of the tangent bifurcation in voltage mode controlled buck converters operating in discontinuous conduction mode (DCM) is performed. The one-dimensional discrete iterative map of the buck converter is derived. Based on the tangent bifurcation theorem, the conditions of producing the tangent bifurcation in DCM buck converters are deduced mathematically. The mechanism of the tangent bifurcation in DCM buck is exposed from the viewpoint of nonlinear dynamic systems. The tangent bifurcation in the DCM buck converter is verified by numerical simulations such as discrete iterative maps, bifurcation map and Lyapunov exponent. The simulation results are in agreement with the theoretical analysis, thus validating the correctness of the theory.

1. Introduction

In recent years, the research in chaos exhibited in the field of power electronics has been the hot spots. DC-DC converters are a kind of piece-wise and nonlinear system. They exhibit various bifurcation and chaos behavior under some operating conditions, such as period-doubling bifurcation [1]-[5], Hopf bifurcation [6]-[8], border collision bifurcation [9]-[11], tangent bifurcation [12]-[13] and chaos behavior [14]-[18]. Bifurcation is a complex structure in nonlinear system. The chaos is characteristic of non-repeat, uncertainty and is extreme sensitive to initial conditions. These nonlinear phenomena make the nonlinear dynamic characteristics of DC-DC converter more complex. Deep investigation of these nonlinear phenomena is of great benefit to understanding the nonlinear behavior and practical design.

The period-doubling bifurcation in DC-DC converters is reported in many published papers. On the other hand, the tangent bifurcation, which is a special bifurcation, has been less investigated. The most studies of tangent bifurcation mainly focus on the numerical simulation modeling. The main approaches used for simulation include bifurcation diagram, Lyapunov exponent. The essential mechanism causing tangent bifurcation was not analyzed in these simulation methods. However, no rigorous attempts have been made to analyze formally the essential mechanism leading to the tangent bifurcation in DC-DC converters.

Buck converters in voltage mode controlled mode are a kind of important converters with wide applications. Although the work in [13] gives no theoretical insights into the underlying cause of tangent

Content from this work may be used under the terms of the Creative Commons Attribution 3.0 licence. Any further distribution of this work must maintain attribution to the author(s) and the title of the work, journal citation and DOI.

Published under licence by IOP Publishing Ltd

bifurcation in such system, it does prompt the important question of what mechanism may give rise to tangent bifurcation behavior. This paper attempts to answer to this question in the light of the theories of nonlinear dynamic systems. The investigation of the mechanism of the tangent bifurcation in voltage mode controlled buck converters operating in continuous conduction mode (DCM) is deeply studied. In fact, there are strict stability criteria and the conditions leading to the tangent bifurcation in mathematics based on the theories of nonlinear dynamic systems [14]-[15]. Based on the tangent bifurcation theorem, the conditions leading to the tangent bifurcation in the discrete iterative model of the buck converter are demonstrated mathematically. Discrete iterative maps, bifurcation diagram, Lyapunov exponent are done to analyze the mechanism and evolution of leading to the tangent bifurcation. The simulation results are in agreement with the theoretical analysis, thus validating the correctness of the theory. The methods proposed in the paper can also be suitable to analysis of the tangent bifurcation and chaos of other kinds of converter circuits.

2.Discrete Iterative Map of a Buck Converter

The schematic diagram of voltage mode buck converter is shown in Fig.1. The main circuit consists of a switch S, a diode D, a capacitor C, an inductor L and the load resistor R. The controlled circuit consists two comparators, a feedback proportional gain k. X is the expected output voltage, D is the duty cycle in steady state. All the components in the buck converter circuit are ideal, no parasitic effects are considered.

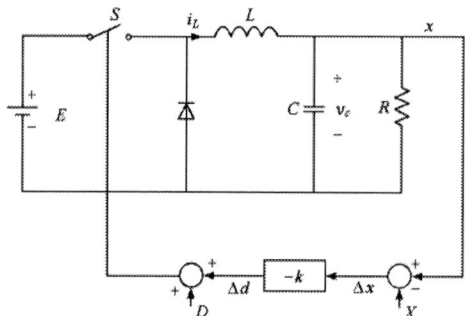

Fig.1. Circuit configuration of Voltage mode controlled buck converter

The buck converter operates in discontinuous conduction mode. Hence, there are three circuit states depending on whether S is closed or open. Assume that the circuit is at the switch state 1 when the switch S is off and diode D is on, at the switch state 2 when S is on and D is off. The two switch states toggle periodically, and at the switch state 3 when S and D are off. The three switch states toggle periodically.

The buck converter is controlled under the voltage mode. The discrete iterative model of the buck converter can be derived as follows[7]:

$$x_{n+1} = f(k, x_n) = \alpha x_n + \frac{\beta h(d_n)^2 E(E - x_n)}{x_n} \tag{1}$$

Where x_n is the voltage across the capacitor at t=nT(n=0,1,2,3,…).

$$\alpha = 1 - \frac{T}{RC} + \frac{T^2}{2C^2 R^2} \tag{2}$$

$$\beta = \frac{T^2}{2LC} \tag{3}$$

$$d_n = D - k(x_n - X) \tag{4}$$

$$h(d_n) = \begin{cases} 0, & d_n < 0 \\ 1, & d_n > 1 \\ d_n, & 0 \le d_n \le 1 \end{cases} \tag{5}$$

Based on the discrete iterative equation, the quadratic and three times discrete iterative model of the buck converter can be derived.

The quadratic discrete iterative can be written in the form of

$$x_{n+2} = f(k, x_{n+1}) = f^{(2)}(k, x_n) = \alpha x_{n+1} + \frac{\beta h(d_{n2})^2 E(E - x_{n+1})}{x_{n+1}} \tag{6}$$

Where

$$d_{n2} = D - k(x_{n+1} - X) \tag{7}$$

$$h(d_{n2}) = \begin{cases} 0, & d_{n2} < 0 \\ 1, & d_{n2} > 1 \\ d_{n2}, & 0 \le d_{n2} \le 1 \end{cases} \tag{8}$$

The other parameters are the same as (2)-(3).

The three times discrete iterative can be calculated by the following equation

$$x_{n+3} = f(k, x_{n+2}) = f^{(3)}(k, x_n) = \alpha x_{n+2} + \frac{\beta h(d_{n3})^2 E(E - x_{n+2})}{x_{n+2}} \tag{9}$$

Where,

$$d_{n3} = D - k(x_{n+2} - X) \tag{10}$$

$$h(d_{n3}) = \begin{cases} 0, & d_{n3} < 0 \\ 1, & d_{n3} > 1 \\ d_{n3}, & 0 \le d_{n3} \le 1 \end{cases} \tag{11}$$

The other parameters are the same as (2)-(3).

3. The Conditions Leading to Tangent Bifurcation

The circuit parameters of the buck converter are listed in Table 1.

Table 1. Circuit parameters

Circuit Components	Values
Switching period T	333.33μs
Input Voltage E	33V
Load Resistor R	12.5Ω
Inductor L	208μH
Capacitor C	222μF
Output volgabe X	25V
Duty cycle D	0.2874

Based on the table 1, the α and β in (2)-(3) can be calculated, the results are as follows: $\alpha=0.8872$, $\beta=1.2$.

3.1 A Theorem of Tangent Bifurcation

Theorem 1 [14]-[15] (Tangent Bifurcation). Assume that f is a C^2 function from R^2 to R. We write $f(x, \mu) = f_\mu(x)$. Assume that there is a bifurcation value μ^* that has a fixed point x^* with derivative equal to one.

(a). $f(x^*, \mu^*) = x^*$.

(b). $f_{\mu^*}{}'(x^*) = 1$.

(c). The second derivative $f_{\mu^*}{}''(x^*) \neq 0$, so the graph of f_{μ^*} lies on one side of the diagonal for x near x^*.

(d). The graph of f_μ is moving up or down as the parameter μ varies, or more specifically, $\frac{\partial f}{\partial \mu}(x^*, \mu^*) \neq 0$.

The tangent bifurcation takes place in the nonlinear system at the fixed point (x^*, μ^*).

3.2 A fixed point

The fixed point of the three times discrete iterative is calculated by the following equation

$$f^{(3)}(k, x_n) - x_n = 0 \tag{12}$$

3.3 Instability boundary

The Instability boundary of the converter is calculated by the following equation

$$\frac{\partial}{\partial x} f^{(3)}(k, x_n)\Big|_{x=x^*} = 1 \tag{13}$$

by substituting of circuit parameters into (13)-(14), we have

$$k^* = 0.2166, \quad x_1^* = 23.76, x_2^* = 26.23 \ x_3^* = 29.57$$

The results show that there are 3 fixed point in the three times discrete iterative of DCM voltage mode controlled buck converter. They are $x_1^* = 23.76, x_2^* = 26.23 \ x_3^* = 29.57$ 。 In addition, when $k = k^* = 0.2166$, these three fixed points are exactly tangent to the diagonal line, and the slope at the tangent point is exactly $+ 1$.

From (9), the partial derivative can be worked out

$$\frac{\partial}{\partial k} f^{(3)}(k, x_n) = \alpha \frac{\partial x_{n+2}}{\partial k} + \frac{\partial}{\partial k}\left(\frac{\beta h (d_{n3})^2 E(E - x_{n+2})}{x_{n+2}} \right) \tag{14}$$

Substituting of (1)-(11) and circuit parameters into (15), gives

$$\frac{\partial}{\partial k} f^{(3)}(k, x_n)\Big|_{k=0.2116, x_1^* = 23.76} = 17.398 \neq 0$$

$$\frac{\partial}{\partial k} f^{(3)}(k, x_n)\Big|_{k=0.2116, x_2^* = 26.23} = -3.3150 \neq 0$$

$$\frac{\partial}{\partial k} f^{(3)}(k, x_n)\Big|_{k=0.2116, x_3^* = 29.57} = 19.694 \neq 0$$

From (9), the second partial derivative can also be worked out

$$\frac{\partial^2}{\partial x_n{}^2} f^{(3)}(k, x_n) = \alpha \frac{\partial^2 x_{n+2}}{\partial x_n{}^2} + \frac{\partial^2}{\partial x_n{}^2}\left(\frac{\beta h (d_{n3})^2 E(E - x_{n+2})}{x_{n+2}} \right) \tag{15}$$

Substituting of (1)-(11) and circuit parameters into (16), gives

$$\frac{\partial^2}{\partial x_n{}^2} f^{(3)}(k, x)\Big|_{k=0.2116, x_1^* = 23.76} = 7.738 \neq 0$$

$$\left.\frac{\partial^2}{\partial x_n^{\,2}} f^{(3)}(k,x)\right|_{k=0.2116,\,x_2^*=26.23} = 75.177 \neq 0$$

$$\left.\frac{\partial^2}{\partial x_n^{\,2}} f^{(3)}(k,x)\right|_{k=0.2116,\,x_3^*=29.57} = -4.612 \neq 0$$

In summary, the voltage mode controlled buck converter operating in DCM satisfies the hypothesis of theorem 1. (a). There are fixed points, $x_1^* = 23.76$, $x_2^* = 26.23$ $x_3^* = 29.57$. (b). When k=0.2166, the system is at the instability boundary, which leading to bifurcation. (c). In the fixed points, the partial derivatives of parameter k are not equal to 0. (d). In the fixed points, the second derivative is not equal to 0. Therefore, the discrete iterative map of $f^3(x_n)$ undergoes the tangent bifurcation at the fixed point, and the tangent bifurcation behavior occurs in this system.

4.Simulations

The evolution process of tangent bifurcation in voltage mode controlled DCM buck converter with voltage feedback gain as parameter is verified by bifurcation diagrams, Lyapunov exponent and discrete iterative maps. The The mechanism of tangent bifurcation in the voltage mode controlled buck converter operating in DCM is analyzed.

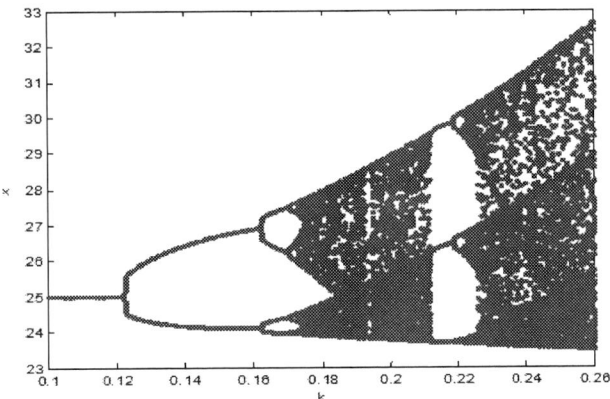

Fig.2 Bifurcation diagram with k as parameter

The horizontal direction is k which is between 0.1 and 0.26, the vertical direction is the output voltage X which ranges from 23V and 33V. The bifurcations, subharmonics and chaotic behavior are indicated in the diagram. As shown in Fig.2, the buck converter goes through period 1, period 2 and eventually exhibits chaos. The period-1 solution is stable until k=0.12 whereupon a period doubling bifurcation takes place. The converter eventually goes to chaos when k=0.173. It can be interestingly observed that a small periodic window, which also exhibits period doubling cascade, is embedded in the chaos region. In the periodic widow, the converter experiences period-3 to period-6 and so on just above k=0.2116. The phenomenon that system transits from chaos to period 3 is known as tangent bifurcation.

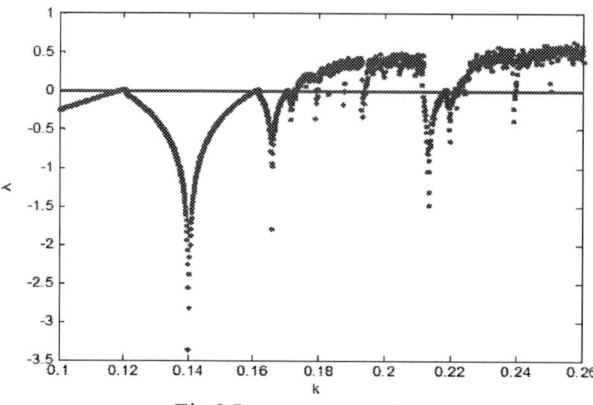

Fig.3 Lyapunov exponent

In Fig.3, the larger of the Lyapunov exponents is plotted as a function of the parameter k over the same range as in Fig.3. It is well known that the presence of chaos is signaled by positive Lyapunov exponent. A negative Lyapunov exponent is characteristic of dissipative (non-conservative) systems, which exhibit point stability. A Lyapunov exponent of zero is characteristic of a cycle-stable system. In this case, the orbits maintain their separation. The tangent bifurcation will be happened when the Lyapunov exponent is changed from the started positive value to zero then to negative value. At k=0.12, where the fixed point changes from attracting to repelling and an attracting periodic orbit is born, the Lyapunov exponent is 0. Just above k=0.173, the Lyapunov exponent is positive, which means that the system is chaotic. This is the same range in which the bifurcation diagram given in Fig.2 showed a whole interval. For larger values of k, above 0.2116, there is another short parameter interval in which there is an attracting period-3 orbit and the Lyapunov exponent is negative. Therefore, the tangent bifurcation will be happened.

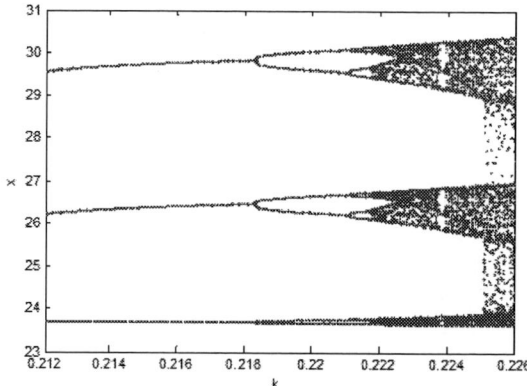

Fig.4 Bifurcation diagram with 0.2116<k<0.226

The Bifurcation diagram with 0.2116<k<0.226 is shown in Fig.4. When 0.2116<k<0.2182, there is an attractive periodic 3 orbit. The bifurcation takes place at k=0.2182 where the period 3 bifurcates to period 6. The system enters chaos for k>0.226.

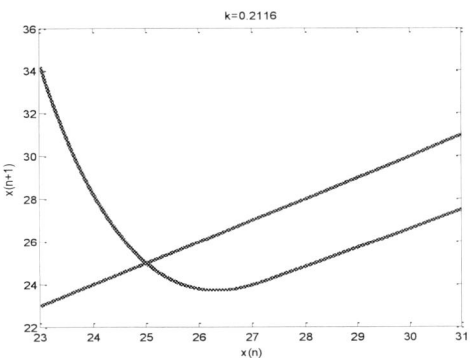

Fig.5 Discrete iterative map of $f(k, x_n)$ with k=0.2116

Fig.5 is the discrete iterative map of $f(k, x_n)$ with k=0.2116. It can be seen that there is an intersection point between the iterative curve and the diagonal line. The slope at this intersection is less than -1. Therefore, this fixed point is stable.

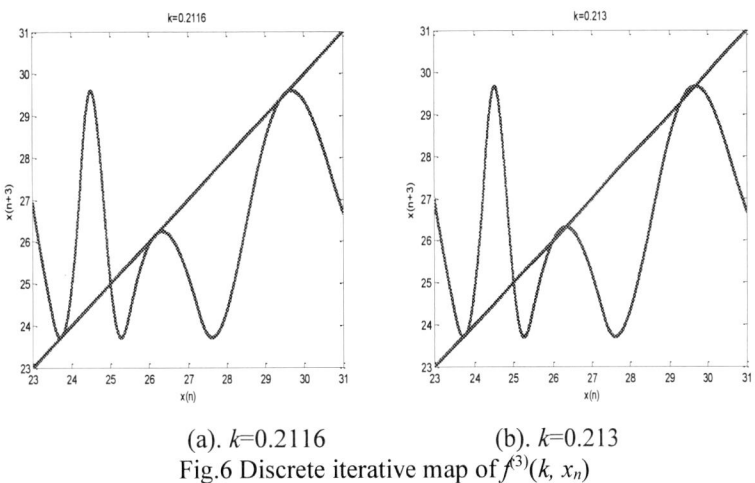

(a). k=0.2116   (b). k=0.213
Fig.6 Discrete iterative map of $f^{(3)}(k, x_n)$

The Discrete iterative map of $f^{(3)}(k, x_n)$ is shown in Fig.6. when k=0.2116, there are four intersection points between the three times iterative curve and the diagonal line. One fixed point is the one in iterative map. The other three fixed points are tangent to the diagonal line and the slope at the tangent point is + 1. The buck converter is at the instability boundary, the bifurcation will occur. When k>0.2116, the three points of tangent will cross the diagonal. Thus, there will be six fixed points. The slope of three fixed points and diagonals is greater than 1, they are not stable fixed points. The slope of another three fixed points and diagonals is less than 1, they are stable fixed points.

5. Conclusion
The mechanism of tangent bifurcation in the voltage mode controlled buck converter operating in DCM is explored in this paper. Based on the discrete iterative map of the buck converter, the one-dimensional discrete iterative maps of $f(x_n)$ and $f^{(3)}(x_n)$ have been derived. By the tangent bifurcation theorem, it is demonstrated in mechanism that the tangent bifurcation will happen inevitably in the buck converter. The discrete iterative maps, bifurcation diagram with voltage feedback gain k as parameter, Lyapunov exponent are used to verify the phenomenon. It has been shown that tangent bifurcation does exist for this system. The method presented in the paper provides the theoretical basics for analyzing the tangent

bifurcation and chaos. It has generality and can be also used to analyze the tangent bifurcation of other kinds of DC-DC converters.

Acknowledgments
The project sponsored by Natural Science Foundation of Guangxi Province (2014GXNSFBA118277), Promotion Project of Guangxi Young Teachers(2017KY0030).

References

[1] Xie Lingling, Gong Renxi, Li jiyong. "Analysis of the dynamical characteristics of the interleaved boost converter in maximum power point tracking for photovoltaic power,". Proceediof the Chinese Society of Electrical Engineering. vol.33,no.6, pp.38-45, Feb. 2013

[2] Ghosh A, Banerjee S. "Study of complex dynamics of DC-DC buck converter,".International Journal of Power Electronics, vol.8,no.8, pp.323-330, 2017.

[3] Al-Hindawi M M, Abusorrah A, Al-Turki Y, et al."Nonlinear dynamics and bifurcation analysis of a boost converter for battery charging in photovoltaic applications,"International Journal of Bifurcation and Chaos, vol.24,no.11, pp.371-391, 2014.

[4] Bocheng Bao, Zhang Xi, Jianping Xu, Jinping Wang. "Critical ESR of output capacitor for stability of fixed off-time controlled buck converter," Electronics Letters, Vol. 49, No. 4, pp. 287-288, Feb. 2013.

[5] Tse,C.K, "Flip bifurcation and chaos in the three-state boost switching regulars," IEEE Trans. Circuits Systems-I, vol.41, no.1, pp.16-23, Jan.1994

[6] Zhioua M, El Aroudi A, Belghith S, et al. "Modeling, dynamics, bifurcation behavior and stability analysis of a DC-DCboost converter in photovoltaic system,"International Journal of Bifurcation and Chaos, vol.26,no.11, pp.1650-1655, 2016.

[7] Xuemei Wang, Bo Zhang, Dongyuan Qiu. Mechanism of period-doubling bifurcation in DCM DC-DC converter [J]. Acta Physica Sinica, 2008, 57(5):2728-2736

[8] Deivasundari P S,Uma G, Sathi R. "Experimental verification of Hopf bifurcation in pulse-width modulated inverter fed cage induction motor drive system,"IET Power Electromics, vol.7,no.7, pp.340-349,2013.

[9] XIE. L.L, GONG. R.X.,ZHUO. H.Z., et al. "Investigation of the mechanism of period-doubling bifurcation in voltage mode controlled buck-boost converter ,". Journal of Electrical Engineering & Technology, vol.6,no.4, pp.519-526.2011.

[10]Lei B, Xiao G, Wu X, et al. "Bifrucation analysis in a digitally controlled H-bridge grid-connected inverter,"International Journal of Bifurcation and Chaos, vol.24,no.1, pp.1450002,2014.

[11]Tong Y N, Li C L, Zhou F."Synchronization control of single-phase full bridge photovoltaic grid-connected inverter,"vol.127,no.4, pp.1724-1728,2016.

[12]Singha A,Kapat S,Banerjee S, et al." Nonlinear analysis of discretization effects in a digital current mode controlled boost converter," IEEE Journal on Emerging and Selected Topics in Circuits and Systems, vol.5,no.3, pp.336-344,2015.

[13]Zhou Yufei, Chen Junning, "Tangent bifurcation and intermittent chaos in current-mode controlled boost converter," Proceedings of the CSEE, vol.25, no.1, pp.23-26, Jan.2005

[14]Hao Bolin, Starting with parabolas, an introduction to chaotic dynamics: Shanghai Scientific and Technological Education Publishing House, 1993

[15]Robinson R C, An introduction to dynamical systems: continuous and discrete: Pearson Prentice Hall.2004

[16]Xie Ling-Ling, Gong, Ren-Xi,Zhuo Hao-ze, et al. "Investigation of tangent bifurcation in voltage mode controlled DCM boost converters,". ACTA PHYSICA SINICA. Vol.61, no.5, pp.1-7, Mar. 2012.

[17]Aroudi EI A. "Prediction of subharmonic oscillation in switching converters under different control strategies,". IEEE Transactions on Circuits and Systems II: Express Briefs, Vol.61, no.11, pp.910-914,2014 .

[18]Rodriguez E,Aroudi El A,Guinjoan F, et al."A ripple-based design-oriented approach for predicting fast-scale instability in DC–DC switching power supplies,"IEEE Transactions on Circuits and Systems I: Regular Papers, Vol.5, no.1, pp. 215-227,2012.

Research on Transformer Fast OLTC System

Jinlong Tu[1], Yongsheng Mi[1]

[1] Institute of mechanical and electrical technology, Nanjing Vocational Institute of Transport Technology, Nanjing, Jiangsu, 218888, China

e-mail: jddq2011@163.com, miyongsh@126.com

Abstract. On-Load Tap Changer(OLTC) transformer with adaptive voltage regulation is the common used method of voltage control in power grid. The study on the switching process of OLTC shows that the transition process is dependent on the structure of the load voltage regulating switch. In this paper, the related equivalent circuit and the calculation formula of the impact current are given. Aiming at the problem of transition state switching in OLTC, a tap-changer control method without transition state is proposed and simulated with inductive load. We have given the electrical main circuit and the measured waveforms of no transition state voltage regulation. We use MCT801D intelligent module to develop a non-contact voltage regulator power supply, which can switch taps rapidly under inductive load, and the output current is continuous without impact.

1. Introduction

In power system, in view of the characteristics of large voltage landing, high line loss and poor power quality in power supply and distribution networks, considering factors such as economy, stability and reliability, on the basis of traditional distribution transformer, On-Load Tap-Changer (OLTC) equipment is added to the optimal scheme of voltage regulation[1]. As an important transmission and transformation equipment, the power transformer is very important for the safe and stable operation. However, the current research shows that the static result of adjusting the OLTC variable is one of the main factors affecting the voltage stability[2]. In addition, because OLTC often needs to change tap, the failure rate is high, accounting for more than 27% of the total transformer fault[3], which affects the safe and stable operation of the power system.

Since people began to study the stability of the power system, the OLTC transformer has been paid special attention. With the further research, it is generally believed that OLTC plays an important role in the voltage stability problem and has published the related paper[4-7].

At present, the research on the influence of load voltage on the voltage stability of power system is mostly based on the traditional OLTC of mechanical switch taps. Due to the long transition time of the tap changer and the generation of the circular current, the OLTC has a great influence on the voltage stability of the power system. Some contactless OLTC that use additional resistors or reactors will also generate circular current, affecting voltage stability. The fast OLTC technology for non contactless switches without additional resistance or current limiting reactor has a fast transition from one stable state to another, and the shortest can be converted to a single tap at a time of a voltage circumferential wave, and this non transition fast OLTC technique can be used for different load properties. Implementation of different control strategies has positive significance for power system voltage stability.

Content from this work may be used under the terms of the Creative Commons Attribution 3.0 licence. Any further distribution of this work must maintain attribution to the author(s) and the title of the work, journal citation and DOI.
Published under licence by IOP Publishing Ltd

2. Transformer on-load regulating principle

2.1. OLTC principle
The principle of transformer on-load voltage regulation is to draw several taps from the windings on one side of the transformer, and switch from one tap to another to change the number of effective turns of windings to adjust the output voltage.

2.2. OLTC switch
At present, most of the OLTC installed in the transformer are mechanical, and the switching unit, as shown in Figure 1, has the characteristics of large voltage regulating range and constant adjusting process. Because of the central task of the OLTC on-load current conversion, it is considered as the heart of the load regulating transformer.

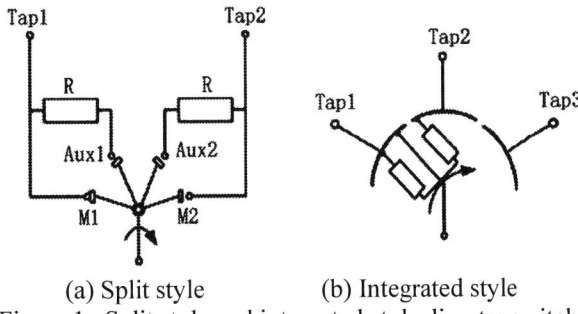

(a) Split style   (b) Integrated style

Figure 1. Split style and integrated style diverter switch.

As the only regular component of transformer, the reliability of OLTC directly determines whether the transformer can operate safely and reliably. The OLTC transform must meet 2 basic conditions: during the process of changing the tap, it is necessary to ensure the continuity of the load current, that is, it is impossible to open; during the process of changing the tap, we must ensure that there is no short circuit between the branches. Therefore, in the process of changing taps, it is necessary to bridge two taps simultaneously at a certain instant to ensure the continuity of load current. In the 2 connection taps, the resistance must be inserted to limit the circulating current and ensure that there is no short circuit between the branches. This allows the transition from one tap to the next tap. Changing the test waveform of the tap switch can describe the transition process of the tap changer during the switching process. The DC test and AC test waveforms are shown in Figure 2[8], respectively. It can be seen that the output voltage of the tap-changer has an obvious transition process during the switching process.

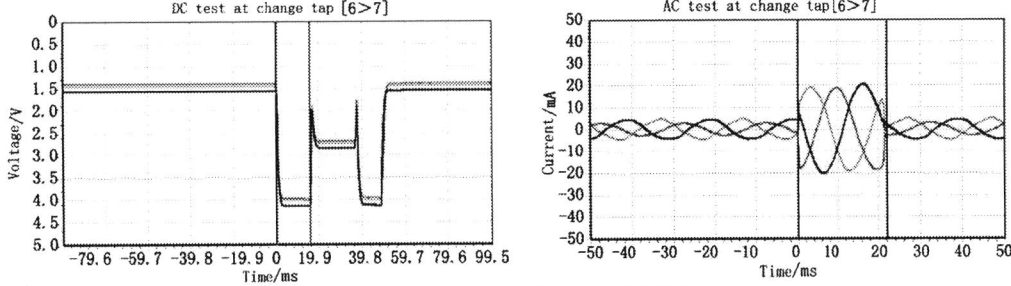

Figure 2. Test waveform of OLTC.

Hybrid on-load tap-changer is composed of mechanical contact and thyristor[9], in which thyristor is used as an electronic switch to change the tap-changer transition. There is no arc in the switching process, and the current passing through the current switching process between the reverse shunt thyristor and the circuit breaker during the voltage regulation is shown in Figure 3. It can be seen that there is transient process in the process of tap changer.

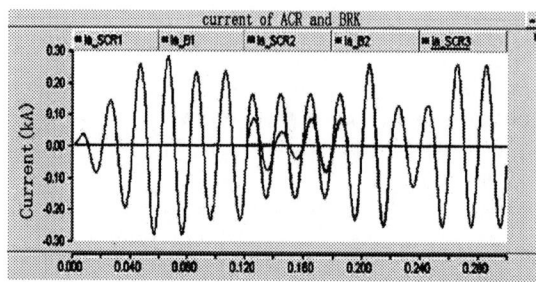

Figure 3. Hybrid OLTC switch current waveform.

The traditional transformer on load tap changer adopts mechanical on load tap changer to realize on load voltage regulation through a transition switch. In this way, although the on load voltage regulation can be realized, the instantaneous change of voltage and current will cause impact and transient process. The arc will erode the contact surface during the process of changing the tap, and affect the life of the contact and the existence of arc. The switching speed is slow, the response time is long, the failure rate is high, and the maintenance amount is large, so that the voltage regulation time can not be accurately controlled.

The existence of these problems also requires the development of a new OLTC to improve the safety and reliability of its use. The new type of on-load tap changer has mechanical improved type, auxiliary coil type and power electronic switch type. Although no arc is realized, the operation of voltage regulation mode is simple, but it will produce harmonic, so that the reliability is poor and the response speed of voltage regulation is not improved.

Since 1990s, people have studied the use of power electronic technology and MCU control technology to control the power electronic devices to switch off and change the number of transformer turns, to realize load voltage regulation. The triac is a semi controlled power electronic device, which triggers the conduction and turns off at the zero current of AC current. It can be used as an actuator of an on load tap changer to solve the problems existing in mechanical tap changers. Some of the main components of SSR are triacs. However, an automatic on load tap changer with solid state relay as a contactless tap changer is used. During the process of changing the tap, the regulating winding and the solid state relay will form a closed loop, which will lead to a circulating current in this loop. If the circulating current exceeds the allowable value, the solid state relay will be damaged.

3. OLTC control strategy and equivalent circuit analysis

3.1. OLTC control strategy

At present, most of the OLTC installed in the transformer are mechanical, and the failure rate accounts for more than 27% of the total transformer fault, mainly because OLTC often needs to change tap, so the failure rate is high. In order to reduce the number of actions of on-load tap changer, the document proposed the OLTC control strategy[10], its control model, as shown in Figure 4.

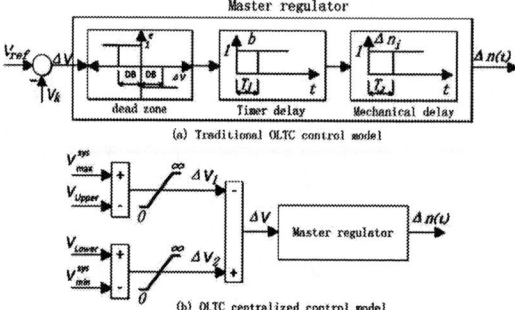

Figure 4. OLTC control model.

3.2. OLTC equivalent circuit

In order to realize the transformer fast on-load voltage regulation, it is necessary to use non-contact switch devices (such as thyristor) to switch the taps, and the transition resistance or transition reactor can not be used. The equivalent circuit of the transformer is as shown in Figure 5 when the load voltage is regulated[11]. The equivalent resistance and equivalent inductance of the load are R'_L and L'_L in the diagram.

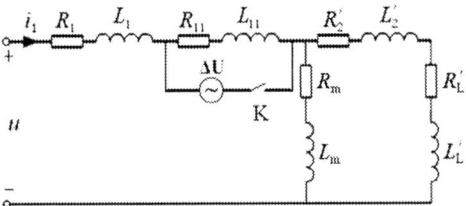

Figure 5. OLTC transformer equivalent circuit schematic.

3.3. The formation of the impact current of OLTC

Because transformer usually works in the near saturation section of magnetization curve, its voltage regulation dynamic characteristic is nonlinear, so it is difficult to quantitatively analyze the actual working process. In the ideal state of neglecting the exciting current, according to figure 5, the expression of the impact current can be obtained.

$$i_1 = I_m \sin(\omega t + \theta - \alpha) + A e^{-\frac{R}{L}t} \tag{1}$$

Where:

$$R = R_1 + R_{11} + R'_2 + R'_L$$
$$L = L_1 + L_{11} + L'_2 + L'_L$$
$$\alpha = \arctan(\omega L / R)$$
$$I_m = \frac{U_m}{\sqrt{(\omega L)^2 + R^2}} \tag{2}$$

When changing the tap of the transformer, the steady state impulse current value is normal when switching accurately. The impact current is very easy to damage the thyristor switch device.

4. Transformer Fast OLTC System

4.1. Characteristics of fast OLTC System

The main characteristic of high performance fast on load tap changer is that there is no transition process when switching tap changer. The load regulating device of a transformer containing a mechanical load tap switch has a certain time transition process, because the inertia of the mechanical parts is larger and the speed is slower. Some contactless load tap changers have load voltage regulator. Although researchers have done a lot of research without impact and transition state, the reliability of the transformer is low in practical application, and a group of transition devices have to be added. In a word, in the existing load regulating technology, the loop short circuit current between the two ends of the winding in the shift voltage regulation process is an unavoidable problem. It not only brings the impact and running loss to the transformer itself, but also affects the stable running state of the power load[12]. Especially under inductive load, thyristor components are prone to malfunction and lead to voltage regulation failure, which seriously affects the reliability of transformer on-load voltage regulation[13].

For a transformer with a thyristor as a contactless switch, a voltage regulation product is used as a contactless switch. When the voltage zero point is used to switch the control method of the tap, the infinite current element in the main circuit can only be applied to the pure resistance load or the rectifier load. In this kind of load, the current is zero when the voltage is zero, and the thyristor devices turn off naturally at the zero point of the current, so it is convenient to switch the taps at the zero point of the voltage to realize the voltage shift switching without the transition process.

However, the power system is complex and changeable, and it will be subject to various kinds of interference at any moment, which will affect the accuracy of detection. For the inductive or capacitive load, the current is not zero when the voltage is zero. If the zero point switching method without transition state is used to control the current, the impact circulating current will inevitably be produced. When the circulating current is too large, the equipment will be damaged and the system can not achieve high performance and high reliability. At this time, not only the anti-interference measures are taken on the circuit and software, but also the integrated voltage and current zero point detection and analysis technology is needed to achieve high performance and high reliability switching without transition process.

4.2. High performance OLTC simulation of change tap at inductive load

Under the R-L load, the switch is switched between the two taps of the transformer with a contactless switch, and the switching time is at the zero point of the current. At this time, the voltage and current waveform produced by the computer simulation is shown in Figure 6. The simulation waveform is shown in the figure, which changes the tap every 40 ms. The red curve 1 is the voltage waveform and the black curve 2 is the current waveform.

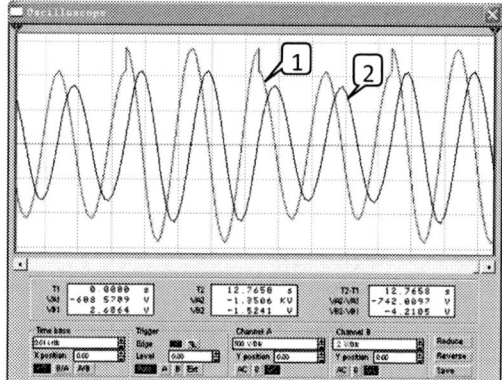

Figure 6. Voltage and current simulation waveforms of OLTC.

4.3. OLTC main circuit and control method without transition state

A main circuit and control diagram of a transformer on load tap changing power supply without transient process are shown in Figure 7. The core control unit uses the MCT801D module with voltage and current zero point accurate detection and analysis technology. The output signal is isolated and amplified to drive the S1 to S4 triac, control the switching of transformer taps, and complete the load voltage regulation.

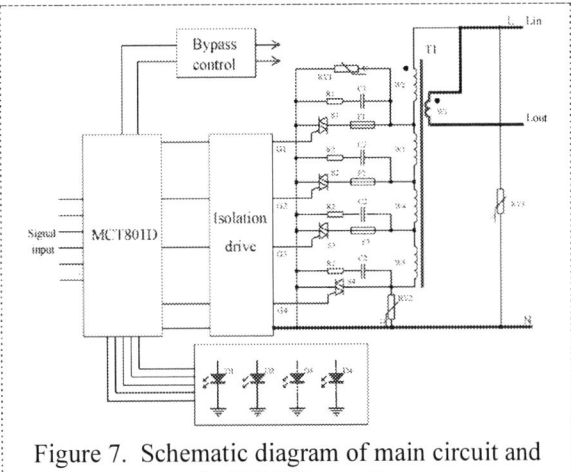

Figure 7. Schematic diagram of main circuit and control of OLTC transformer.

The switching control method of transformer tap changer without transition state is to detect the zero point of voltage and current accurately according to the measured load property and load voltage value. Then, interference filtering algorithm is adopted to avoid interference signals, and voltage switching points and handover timing are calculated. In order to switch the taps, the software automatically selects the switching of the zero point or zero point of the current according to the control algorithm, first closes the current thyristor, then opens the prepared thyristor, switches to the required taps, and completes the load voltage regulation.

This kind of transformer without transition state can automatically adapt to the load of capacitance, inductance and HID lamp. It can keep the current continuously and no impact current. It will not cause the HID lamp to be extinguished or re - light, effectively prolonging the service life of the lamps.

In order to ensure the reliable switching of thyristor and the safe operation of the thyristor, the rising rate of voltage at both ends of the thyristor must be reduced by du/dt and the current rising rate of di/dt of the thyristor passing through the thyristor at the moment of opening. The purpose of limiting du/dt and di/dt is to use parallel resistance capacitance circuit at both ends of thyristor.

4.4. Voltage and current waveform under inductive load
Products made according to the principle of circuit 7, the waveform of the actual output voltage current taken by the Tektronix TDS2002 digital oscilloscope at the 10ms/Div scanning rate under the inductance load is changed at each 40ms, and the Yellow curve 1 is the voltage wave and the blue curve 2 is the current wave form.

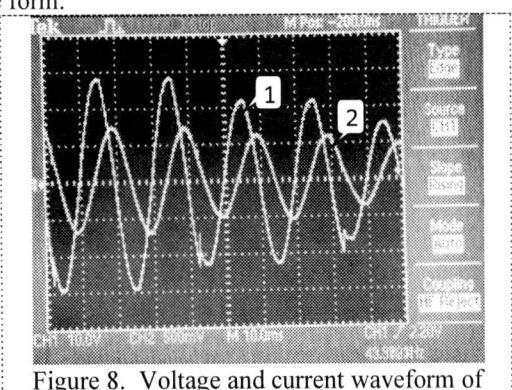

Figure 8. Voltage and current waveform of inductive load.

From figure 8, it can be seen that under the inductance load, there is no transition process when the tap is changed, the output current is continuous and no impact, and the output has no overvoltage or voltage drop.

4.5. Test and Application

According to figure 7, the MJN3 type lighting energy saving power source produced by the circuit is designed, and a set of intelligent monitoring software is designed. In order to verify the correctness of switching control method for transformer taps without transition state, a fast automatic test program is embedded in the software. When the normal inspection test is done before the product is released, the test will be carried out automatically at the speed of 10 thousand times per hour. In laboratory tests, a rapid change of tap test was performed 60 thousand times, 90 thousand times, and 180 thousand times per hour. In a long period of time in a variety of loads (no load, high pressure sodium lamp load, high pressure mercury lamp load, hybrid lamp load, resistance load, inductance load, capacitance load) and the actual use of mass products on the road lighting site, the cumulative shift times more than 300 million times. It is proved that the product can work reliably under various conditions and has stable performance.

5. Conclusion

In view of the current problem of transition state switching in OLTC, a high performance non transition OLTC switch control method is proposed, which is simulated as the most severe inductive load of taps. In the MJN3 contactless tap-changer power supply developed by using MCT801D intelligent module, the correctness of high-performance transition-free OLTC tap-changer control method is verified.

The high performance non transition OLTC switch control method is to detect the voltage and current zero accurately according to the measured load property and the load voltage value. Interference filtering algorithm is adopted to avoid interference signals, and voltage switching points and handover timing are calculated. In order to switch the taps, the software automatically selects the switching of the zero point or zero point of the current according to the control algorithm, first closes the current thyristor, then opens the prepared thyristor, switches to the required taps, and completes the load voltage regulation.

Acknowledgments

This article has been funded by the high level talent research fund project of Nanjing Vocational Institute of Transport Technology (440105001), and the author is here to express his sincere thanks.

References

[1] Song Dongdong, Cheng Lin, Lin Zhifa, et al. (2017) Review of application on power electronic technology in on-load tap changer. Advanced Technology of Electrical Engineering and Energy, 36: 45-55.

[2] Zhu Shanshan. (2017) Influence of OLTC on static voltage stability limit of power system connected DG. Foreign Electronic Measurement Technology, 36: 63-66.

[3] Wang Guan, Liu Jinxin, Zhao Tong, et al. (2017) Mechanical Condition Monitoring of On-load Tap Changers Based on Improved Variational Mode Decomposition. Journal of Hunan University(Natural Sciences), 44: 75-83.

[4] Wang Kaiming, Shu Hongchun, Cao Liping, et al. (2015) Study of OLTC running state evaluation method based on correlation analysis. Power System Protection and Control, 43: 54-59.

[5] Jinlong Tu, Kai Hu. (2015) The Simulation and Practice of High-performance Non-contact Stabilized Voltage Power Supply. In: 2nd International Conference on Information Science and Control Engineering. Shanghai. pp: 996-1000.

[6] Duan Ruochen, Wang Fenghua, Zhou Lidan. (2016) Mechanical Features Extraction of On-Load Tap-Changer in Converter Transformer Based on Optimized HHT Algorithm and Lorentz Information Measure. Proceedings of the CSEE. 36: 3101-3109.

[7] Jinlong Tu，Yanxia Li. (2016) Automatic On Load Voltage Regulating Technology and Energy-saving HID Lamps. In: Advances in Engineering Research. Guangzhou. pp: 0329-0334.

[8] Zhao Jun, Ma Wenhui, Xing Chao, et al. (2017) AC/DC Test Study on Switching Characteristics of Transformer on Load Tap Changer . Hebei Electric Power, 36: 55-57.

[9] Zhang Kai, Yang Shaohui. (2015) Scheme of Areless On-load Voltage Regulation for Power Transformer . Electrical Engineering.10: 55-58.

[10] Zhang Jiaqi, Wang Linchuan. (2017) Coordinated Voltage Control Strategy for Active Distribution Networks. Journal of Northeast Electric Power University, 37: 14-19.

[11] J.L.Tu，C.X.Pan. (2013) Contactless OLTC Technology Based on Inductive Load . Mechanical and Electrical Technology. 392: 484-488.

[12] Han Shanqi. (2016) Distribution Transformer On-Load Regulator . Rural Electrification. 1: 28.

[13] Huang Hui, Du Bo. （2016）Research and Application of Contactless Load Automatic Voltage Regulating Distribution Transformer. China Electric Power(Technology Edition) .1: 53-55.

Research on the design of the fuzzy control system of full bridge DC converter

Wenda Liu [1*], Shutian Liang[2] and Chuxue Hao[3]

[1] Wuhan Institute of Marine Electric Propulsion，Wuhan 430064，China

[2] Wuhan Institute of Marine Electric Propulsion，Wuhan 430064，China

[3] Wuhan Institute of Marine Electric Propulsion，Wuhan 430064，China

[*]Corresponding author's e-mail: liu_wenda@163.com

Abstract. It is difficult to build the mathematical model of a full bridge DC converter in a DC zonal electric distribution system .This paper advanced a design method and the specific design steps of the fuzzy control system of full bridge DC converter . A simulation model was built and the simulation results shown that in the aspect of output voltage control of a full bridge DC converter, the fuzzy control system was obviously superior to PID control system in indexes such as overshoot, response time and the times of oscillation .

1. Introduction

The integrated marine power system integrates the propulsion system and the electric power system , which are traditionally mutually independent, into one system. In the form of electric power, it provides power supply to propulsion load, pulsing load, communication systems, navigation systems and hotel load, so the comprehensive use of power source of the whole ship is realized. The application of integrated power system not only provides power supply to marine load, simplifies the marine power system, boosts the efficiency of marine systems, reduces the noise level of ships and the cost of the ship life cycle, it also conforms the trend of informationization and intellectualization and represents the future trend of marine power system. [1]

Currently the AC power system, which is composed of radial power distribution system and AC generators, remains the mainstream marine power system. The energy of the electric power loads are fed back through cables by a few centralized power distribution center, therefore thousands of cables go through cabins of the ship ,which causes a problem for the laying of the cables. With the increase of the electric load, the power distribution system has become increasingly huge and complicated, causing trouble to the installation, maintenance and protection of the system and setting a limit for the upgrade of the ships. The centralized power distribution system soon reached an extreme and the power distribution system had become a focus in the design and construction of a ship. In real applications, with the increase of the capacity of the power distribution system, the partial electrical isolation becomes increasingly difficult for the ships adopting traditional radial power distribution. In case of accident and damage, the mechanical breaker is unable to isolate the fault swiftly and effectively. The partial fault may influence the whole system or even cause a power loss of the whole ship [2] .

The zonal distribution system differ considerably from the currently adopted radial distribution. The basic component of a zonal distribution system are power transmission bus and zonal distribution

---

Content from this work may be used under the terms of the Creative Commons Attribution 3.0 licence. Any further distribution of this work must maintain attribution to the author(s) and the title of the work, journal citation and DOI.

Published under licence by IOP Publishing Ltd

center. The zonal distribution system provides power supply to the load center of the power supply zones, which obtain power from the main power transmission lines that run through the whole shipboard. From the power electronics conversion devices and power distribution cables ,the load of the zones obtain power and the number of the cables that run through the cabins is greatly reduced. By reducing the amount of power cables and power distribution equipments, the cable installation is simplified and the difficulty of the ship construction is reduced, it is also benefit for the modular construction and the reduction of the whole life cost. The modular building enable the electric consuming equipment in a zone to obtain power without having to connect to the other zones during the construction .The installation, commission and test of the marine equipment is much more convenient[3] .

The DC zonal distribution transmit and distribute power with direct current. Comparing to AC system, a DC zonal distribution system can make better use of zonal distribution for the following two reasons[4], one is that a DC zonal distribution adopts power electronics devices for the system protection, which is more capable of isolating the fault than an electric-magnetic breaker and is better for system reestablishment .Another reason is that a DC zonal distribution system can get rid of the speed coupling between a generator and a load motor, which is benefit for the optimization of the prime movers and the load motors.

2. The control model of a DC converter

The DC converter is the core equipment of a DC zonal distribution system, it converts the voltage of a DC bus to another kind of DC voltage(normally with a voltage drop) to supply power to the DC distribution zones. The DC outputted from the DC converter in the DC distribution zones can be converted into DC or AC of an appropriate voltage class that is suitable to the power consumption equipments. As the main power source in the DC distribution zones, the stability of the voltage output of the DC converter is an important technical index[5-6] ,this paper mainly researched on the control system of the output voltage of a DC converter.

The DC converter can be categorized into direct DC converter and indirect DC converter. A direct DC converter is also named a chopper, which converts a direct current into another direct current without an isolation between the input and output ,and an in direct DC converter has an AC link added in the direct DC converter with an isolation by a transformer between the input and output.

One kind of indirect DC converter, the full bridge DC converter, is currently widely used. The basic structure of a full bridge DC converter is shown in figure 1.The inverter circuit ,consisting of four switches, converts the DC to AC that connects to the primary winding of the transformer. By changing the duty cycle of the switches, the mean value of the commutating voltage is changed and the output voltage is changed. By changing the frequency of the switches, the frequency of AC at the primary side of the transformer is changed.

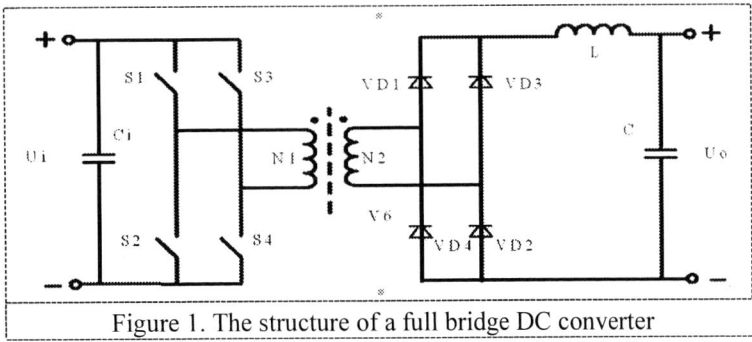

Figure 1. The structure of a full bridge DC converter

When the filter inductance value is great enough ,the filter current is continuous, and the output voltage when the circuit is stable is :

$$U_o = 2n\alpha U_i \qquad (1)$$

Where U_i is the input DC voltage , U_0 is the output DC voltage，n is the ratio of the isolation transformer and α is the duty cycle of the inverter switches.

The structure of the control system of a full bridge DC converter is shown in figure2.

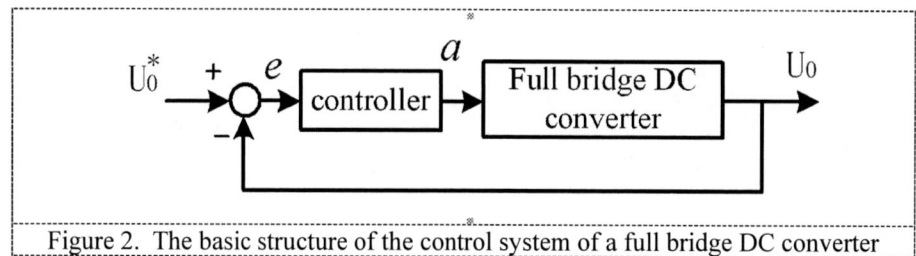

Figure 2. The basic structure of the control system of a full bridge DC converter

Where U_0^* is the ideal output voltage，U_0 is the real output voltage，and e is the error of the voltage.

Equation(1)is the DC voltage output when the circuit is stable，and the transient process of the circuit is related to the parameters of both the full bridge converter and the load. Because the loads in the DC distribution zones include inductive loads, capacitive loads and pure resistive loads, the parameters of the loads varies considerably, meanwhile because the complexity of the running characteristics of the components (i.e.switches and isolation transformers etc)of the full bridge DC converter, it is really difficult to build a precise mathematical model of a full bridge DC converter. Therefore, when engineers adopt the traditional design method based on the precise mathematical model, they use the simplified DC transformer mathematical model ，estimate the load values and design the controller by trial-and-error and the control effect is apparently not ideal.

3. The design of fuzzy control system of a DC converter

The fuzzy control system adopts the fuzzy mathematical approach to simulate the way that the fuzzy logic thinking of human works on the control the object whose precise mathematical model is unknown. Comparing to the traditional control system, it is independent to precise mathematical model and it is an intelligent control system.

The structure of the control system of a full bridge DC converter is shown in figure3[7-8].

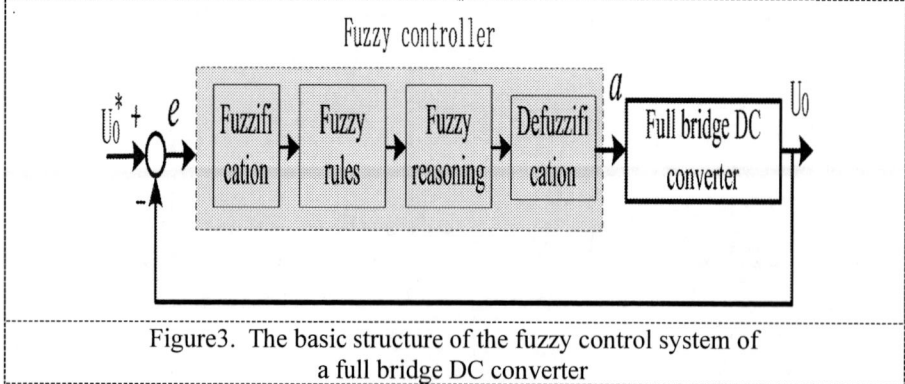

Figure3. The basic structure of the fuzzy control system of
a full bridge DC converter

The fuzzy controller modules includes fuzzification, fuzzy rules,fuzzy reasoning and defuzzification [9].The information processing in a fuzzy controller is shown in figure 4, the error of the system e and its changing rate $\dot{e}$ is multiplied by their corresponding scaling factor k_e and k_{ec} separately ,then we get the domain of error and the change rate of error,which is n_e and n_{ec} separately. Then we obtain the fuzzy set E and EC from the corresponding membership function μ_E and μ_{EC} .Then we get output fuzzy set U according to the fuzzy rules, by fuzzy reasoning we get output domain n_u which is multiplied the scaling factor k_u and we get real output α .

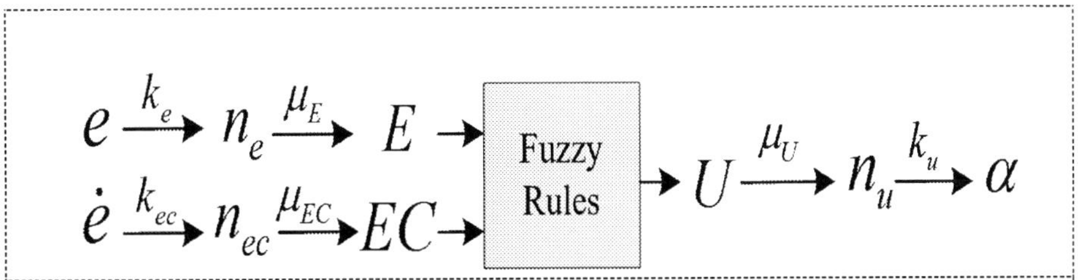

$$e \xrightarrow{k_e} n_e \xrightarrow{\mu_E} E \rightarrow \boxed{\begin{array}{c}\text{Fuzzy}\\\text{Rules}\end{array}} \rightarrow U \xrightarrow{\mu_U} n_u \xrightarrow{k_u} \alpha$$
$$\dot{e} \xrightarrow{k_{ec}} n_{ec} \xrightarrow{\mu_{EC}} EC \rightarrow$$

Figure4. The information processing of a fuzzy controller

In figure4, the fuzzy control rules simulated the control function that human has for controlled object,it is the core of the fuzzy controller.

The following example demonstrate the design steps of the fuzzy control system of a full bridge DC converter. The basic parameters of a full bridge DC converter are as following: inputvoltage=5000V,ideal output voltage=1000V,transformer ratio=0.25,according to equation(1),we obtain the duty cycle of the inverter switches = 40% when the circuit is stable.

3.1 Structural Design

We set the basic domain for e and $\dot{e}$ as [-10, 10], and the basic domain for α as [0, 0.5].

The fuzzy domain of E 、EC is {-5, -4, -3, -2, -1, 0, 1, 2, 3, 4, 5},and The fuzzy domain of U is {-5, -4, -3, -2, -1, 0, 1, 2, 3, 4, 5}.

The fuzzy set of E 、EC is determined as {NB，NM，NS，O，PS，PM，PB} and the fuzzy set of U is determined as {O, L1, L2, L3, L4, L5}.

Hence we determine the scaling factors as $k_e = k_{ec} = 2$, $k_u = 1/10$.

3.2 Fuzzification

The fuzzification is to determine the curves of the each membership function. According to the rules of the membership function and the practical experience we determine the curves as fowllowing.

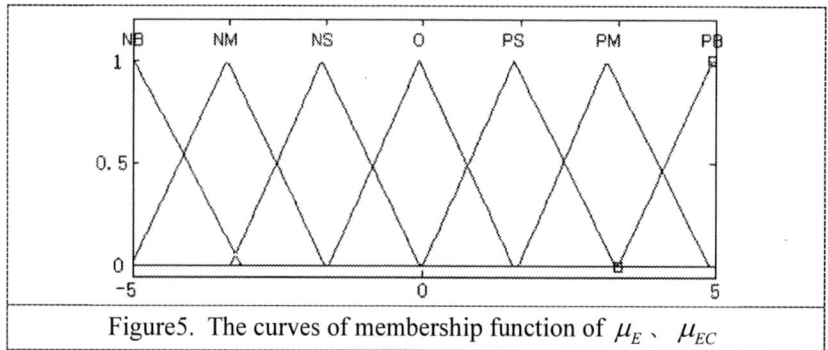

Figure5. The curves of membership function of μ_E、μ_{EC}

3.3 Design of fuzzy rules

By learning from the human-object control experience, we get the table of fuzzy control rules as shown in the following table 1 and the curved surface of rules as shown in figure 6.

Table 1. The table of fuzzy control rules.

U EC E	NB	NM	NS	O	PS	PM	PB
NB	O	O	L1	L2	L3	L3	L4
NM	L1	L1	L2	L3	L3	L4	L5
NS	L2	L2	L3	L4	L4	L4	L5
O	L3	L3	L4	L4	L4	L5	L5
PS	L3	L4	L4	L4	L5	L5	L5
PM	L4	L4	L5	L5	L5	L5	L5
PB	L5	L5	L5	L5	L5	L5	L5

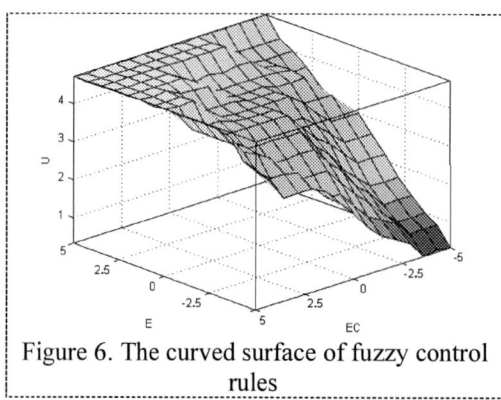

Figure 6. The curved surface of fuzzy control rules

3.4 Defuzzification

We get a fuzzy control variable by the reasoning with fuzzy control rules,which can not be directly used for the control of object. We need to convert the fuzzy variables into precise variables and this conversion process is called defuzzification. Similar to fuzzification, firstly the curves of membership

function μ_U need to be determined, as shown in figure 7. After the membership functions are

determined, we obtain the values of the output domain, which are multiplied by the scaling factor k_u and the real control values are obtained.

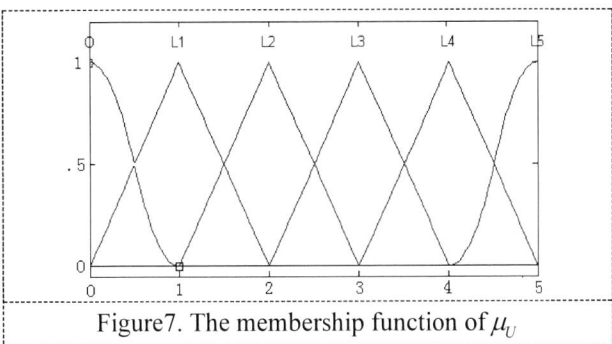

Figure7. The membership function of μ_U

4. Simulation Validation
We built a simulation model and simulated the full bridge DC converter and its control system.

Firstly we built a simulation model of a full bridge DC converter as shown in figure 8, the main parameters of it are listed in table 2.

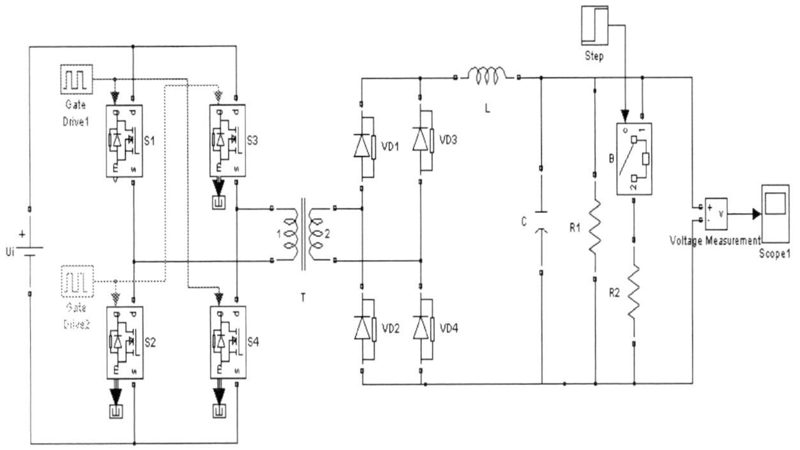

Figure8. The simulation model of a full bridge DC converter

Table2. The parameters of the simulation model of a full bridge DC converter.

Synbols	Names	parameters
Ui	DC voltage source	5000V
S1-S4	MOSFET	
T	Isolation transformer	Ratio=1:0.25
VD1-VD4	Diode	
L	Filter Inductance	L=0.001H
C	Capacitor	C=0.0002F
R1	Load	R=10Ω
R2	Load	R=20Ω
Gate Drive1-2	Trigger Signal	4KHz

The Open-Loop control of a full bridge DC converter, the duty cycle of the inverter switches is 40%, the output voltage curve is shown in figure 9(load R1 first, then load R2 0.01s after when the system is stable) .

From the simulation result we see that the overshoot of the output voltage is obvious ($\sigma \approx 70\%$), the voltage fluctuation is also obvious when the load changes.

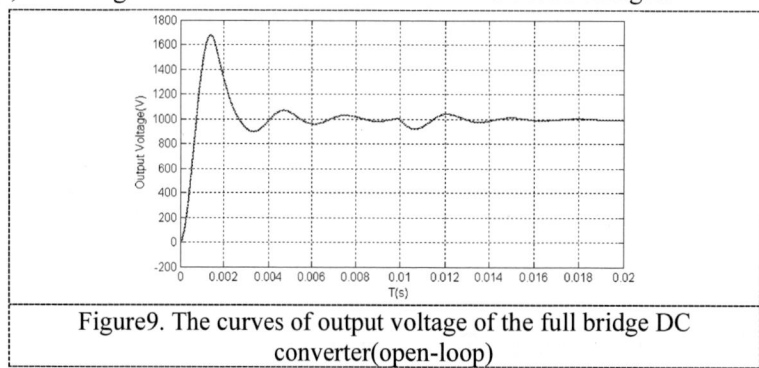

Figure9. The curves of output voltage of the full bridge DC converter(open-loop)

The simulation model after we add fuzzy control system in the full bridge DC converter is shown as in figure 10:

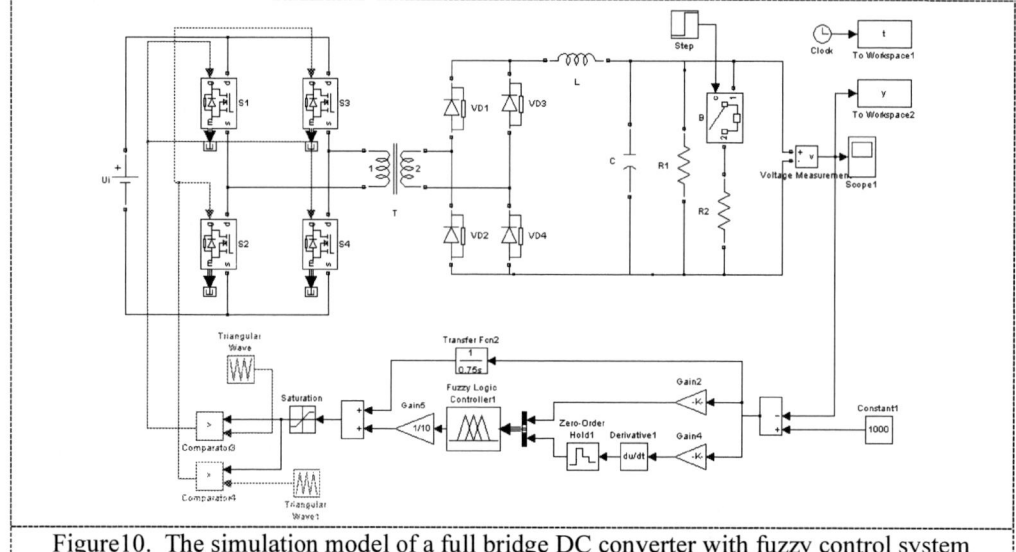

Figure10. The simulation model of a full bridge DC converter with fuzzy control system

We adopt the close-loop fuzzy control on a full bridge DC converter, the output voltage curve of it is shown as in figure11 and figure 12 (load R1 first, then load R2 0.01s after when the system is stable) . The output wave of classic PID control is also drawn in figure 11 for comparison .From the simulation result we see that after the fuzzy control is added in the the full bridge DC converter, the overshoot of the output wave has reduced significantly ($\sigma < 7\%$) , the output wave has less no oscillation, and is more stable during the load changes.

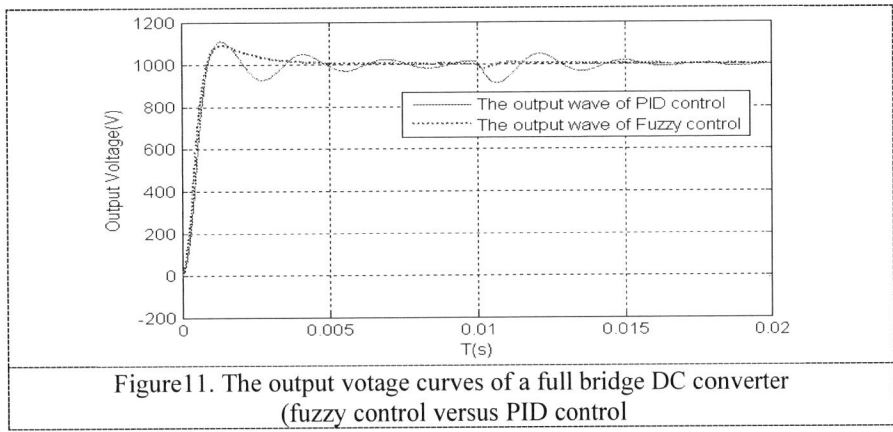

Figure11. The output votage curves of a full bridge DC converter
(fuzzy control versus PID control

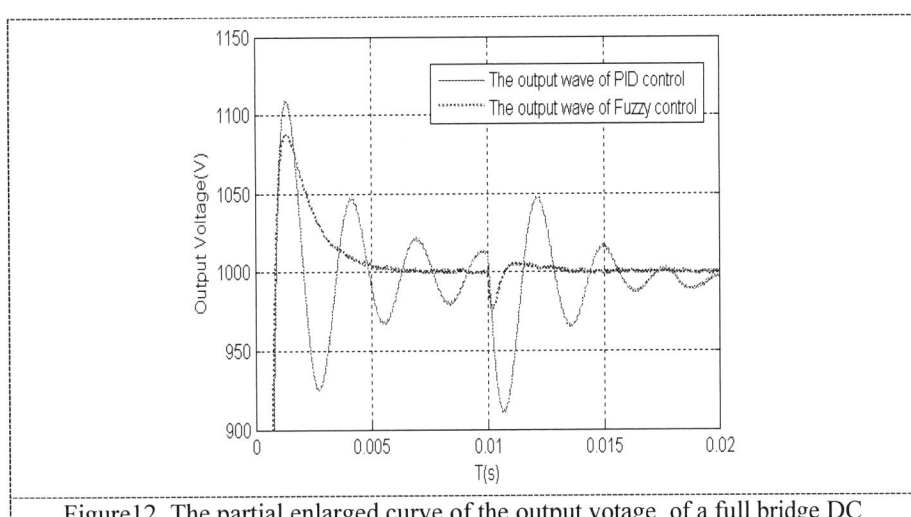

Figure12. The partial enlarged curve of the output votage of a full bridge DC
converter
(fuzzy control versus PID control)

5. Conclusion

The design of the classic control system is based on the mathematical model of the controlled object, the more precise is the model, the better the control system is designed. Two problems emerge out of it, one is that the precise mathematical model is increasingly difficult to build as the complexity of the controlled object increases. Another problem is that even if the precise model is obtained, the designed control system is not applicable in the real control system because of the enormous calculation quantity. In the real application of control system, a fact is found that an experienced operator only inspect and analyze the controlled object by simple meters without knowing the working principle or the mathematical model of the controlled object ,by adjusting the control process through the actuator, the control effect of the complex controlled object is always optimal .The fuzzy control is expressing the operation process of the operator with fuzzy mathematics and using it as a controller. The fuzzy controller do not focus on the mathematical model of the controlled object, and the optimal control of it rely on the "experience of the operator", i.e. the "fuzzy rules" in this paper. Therefore the focus and difficulty of the designing of a fuzzy control system is the designing of the fuzzy rule, which require further study in the future.

References

[1] FU Lijun, LIU Lufeng, WANG,Gang, MA,Fan, YE Zhihao, JI Feng, LIU Luhui ,(2016)The research progress of the medium voltage DC integrated power system in China[J].Chinese Journal of Ship Research，1(1):72-79

[2] CHEN Hong, (2005)DC-ZEDS based shipboard integrated power system. Wuhan Second Ship Design and Research Institute[J]. Ship Science and Technology,27:31-37

[3] CHEN Bo,FU Lijun,YE Zhihao,WANG,Qi,(2004)DC-ZEDS-based Meshy Network[J],Marine Electric and Electronics Engineering，4:193-196

[4] WU Shuang,XIA Li,ZHANG.Chao,WANG,Xinzhi,(2012)Effects of electric propulsion on the stability of DC Zonal Electrical Distribution System[J].Navigation of China，35(4):45-49

[5] ZHAO Yu,CHAI Jianyun,(2007) Current Chopping controlled switched reluctance generator for wind energy applications[J], 47(7):1118-1121

[6] SU Xiaodong,LIANG,Hui,(2008)Research on control method of multiple boost converter Based on Wind Energy Conversion System. Electric Drive，38(8):36-39

[7] LIU Lang, SUN,Peide,WANG Chenglong, ZHAI Fangyu,(2018) The simulation of Cuk chopper circuit based on fuzzy PI control. Information and communication,1:20-23

[8] Visioli A,(2001) Optimal tuning of PID controller for integral and unstable processes[J]. Proceedings of IEE, Part D,148（2）:180-184

[9] LI Shiyong,LI Yan,(2016) Intelligent Control[M] Tsinghua University Press,Beijing

[10] XUE Dingyu,(2016) Computer Aided Control Systems Design Using MAQTLAB Language (Third Edition) Tsinghua University Press ,Beijing

ISPECE   IOP Publishing

IOP Conf. Series: Journal of Physics: Conf. Series **1187** (2019) 022004   doi:10.1088/1742-6596/1187/2/022004

Measurement of inrush current in transformer based on optical current transducer

Chu Lei[1]*, Guo Zhizhong[2] , Chen Yue[1] and Wang Guizhong[1]

[1] Electrical Engineering, Harbin Institute of Technology, Harbin, Heilongjiang, 150001, China

[2] Electrical Engineering, Harbin Institute of Technology at Zhangjiakou, Zhangjiakou, Hebei, 075400, China

*Corresponding author's e-mail: 1403968190@qq.com

Abstract. The accurate measurement of inrush current is a key problem in transformer protection. Traditional electromagnetic current transformer (CT) has nonlinear characteristics of its core. When the transformer generates inrush current, it will be saturated and lead to current distortion, which will lead to transformer differential protection misoperation. Optical current transducer (OCT) is not affected by saturation and can be used to measure and identify inrush current well. This paper analyses the mechanism of inrush current, the reason of CT saturation and the principle of OCT measurement. The simulation and experimental results show that OCT has better measurement quality than traditional CT and superiority to differential protection.

1. Introduction

Transformer is the main equipment in power system and plays an important role in the safe operation of power system. Inrush current has always been a key factor affecting transformer protection. Because the transformer will produce large inrush current when it closes without load. If it is not identified, it will lead to misoperation of differential protection. Therefore, accurate measurement of inrush current is the primary task of transformer protection.

At present, there are two main factors restricting the improvement of transformer protection operation rate: one is how to distinguish the inrush current from the internal fault current. The other is how to avoid the transient unbalanced current caused by external fault leading to differential protection maloperation. Through the current transformer, the transformer protection can sense the current information of the primary side, and then make the differential protection operate correctly [1]. With the increase of power system voltage level and transmission capacity, the transient output current of traditional electromagnetic current transformer is seriously distorted due to the influence of core magnetic saturation. To overcome this distortion, the complexity of relay protection devices will be greatly increased [2]. In addition, the traditional CT also has a small dynamic range, narrow frequency band, susceptible to electromagnetic interference, complex insulation structure, high cost, secondary open circuit will produce high voltage, ferroresonance, flammable explosive, large area and other defects. When transformer fails or no-load switching on, the non-periodic component of transient current on the primary side will cause magnetic saturation of conventional electromagnetic CT and distort the waveform of current on the secondary side of CT. Therefore, it cannot ensure the rapidity, sensitivity, selectivity and reliability of relay protection [3]. Optical current transducer has many advantages, such

Content from this work may be used under the terms of the Creative Commons Attribution 3.0 licence. Any further distribution of this work must maintain attribution to the author(s) and the title of the work, journal citation and DOI.

Published under licence by IOP Publishing Ltd

as no saturation, strong anti-interference ability, longer transmission distance and so on. It has a great inhibition on the above problems.

This paper analyses the saturation characteristics of electromagnetic current transformer, the factors affecting CT saturation and the influence on inrush current identification. Then the basic principle and measurement method of OCT are mathematically modelled. The conclusion that OCT can measure the full waveform of current is obtained. The waveform of inrush current and standard current of two kinds of transformers are compared by simulation. Finally, the waveforms of the two transformers are further analysed by the transient experiment of the inrush current, and the correctness of the theory is verified.

2. Generation mechanism of inrush current

Under the condition of steady operation of transformer, the winding terminal voltage is:

$$u(t) = U_m \sin(\omega t + \alpha) \tag{1}$$

Among them, U_m is voltage amplitude, ω is angular frequency, and α is the initial phase. In order to simplify the reasoning, the inductance calculation formula is set as: $L = \dfrac{\phi}{i}$. Without considering transformer leakage reactance and winding resistance R, the relationship between voltage u and magnetic flux density Φ is:

$$\frac{d\phi}{dt} + R\frac{\phi}{L} = U_m \sin(\omega t + \alpha) \tag{2}$$

The size of the inductor is nonlinear, varying with the saturation of the core. But since $\dfrac{d\phi}{dt}$ is much larger than the size of the inductor L after the voltage is applied, the inductor L can be set as a very small constant here. This simplification has no great effect on the solution of the equation. The formula for calculating the flux linkage can be obtained as follow:

$$\phi(t) = \phi_m \sin(\omega t + \alpha - \varphi) + [\phi_r + \phi_m \sin(\alpha - \varphi)]e^{-\frac{R}{L}t} \tag{3}$$

Among them, $\phi_m = \dfrac{U_m L}{\sqrt{R^2 + (\omega L)^2}}$ is the amplitude of the steady-state component of the flux linkage.

Φ_r is the remanence of the core. $\varphi = arctg\dfrac{\omega L}{R}$. And because of $\omega L \gg R$, so the attenuation of inrush current in this process is not considered. The formula (3) can be simplified to:

$$\phi_k = -\phi_m \cos(\omega t + \alpha) + \phi_m \cos\alpha + \phi_r \tag{4}$$

The first one is steady-state flux and the other two is transient flux. When considering transformer loss, transient flux will decay with time. Assuming that $\phi_m \cos\alpha$ and ϕ_r are in phase, when the no-load switching half cycle, the core flux reaches the maximum. As shown in Figure 1, the maximum flux will be generated when the voltage zero-crossing point is closed without load, which is much larger than the saturated flux of the transformer.

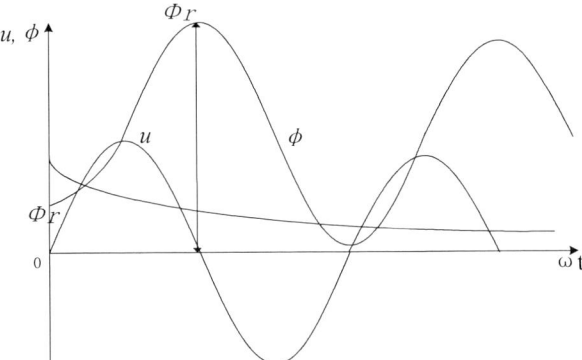

When the transformer is closed at the moment of $u = 0$

Figure 1. The curve of the relation between voltage and magnetic flux.

The flux waveform of transformer is shown in Figure 2 (a) when no load is switched on. The simplified magnetization curve is shown in Figure 2 (b). The magnetic flux φ_k is obtained and the inrush current i_e can be obtained by drawing a simplified magnetization curve. As can be seen from Figure 2, before the core is unsaturated ($\varphi < \varphi_s$), the current i_e is less than i_s and its value can be ignored. But when the core is saturated ($\varphi > \varphi_s$), the current will increase rapidly and the maximum amplitude will reach i_p. As shown in Figure 2 (c), this current is called the inrush current of the transformer and it can reach 6-8 times the rated current [5].

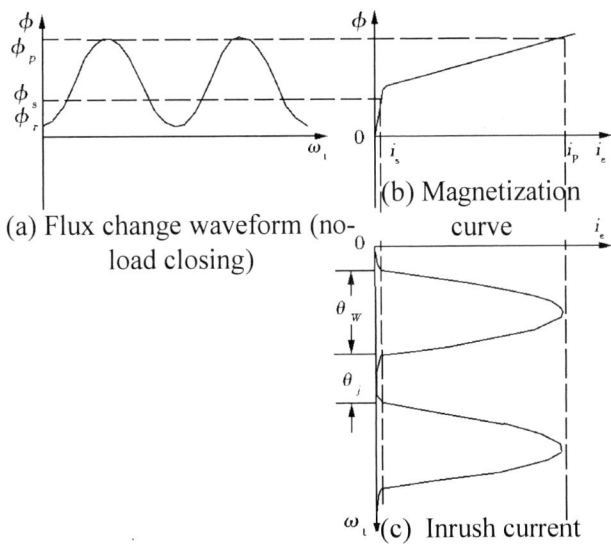

(a) Flux change waveform (no-load closing)

(b) Magnetization curve

(c) Inrush current

Figure 2. Inrush current in unload switching of transformer.

The inrush current is related to the closing angle. When the closing angle α is 0 or π, the inrush current is the largest. As shown in Figure 3, at the same time, the waveform deviates completely from the side of the time axis and is discontinuous. It contains a large number of non-periodic components and a large number of high-order harmonic components. In the high-order harmonic, the second-order harmonic is dominant. The inrush current and fault current can be identified by the harmonic content or the closing angle.

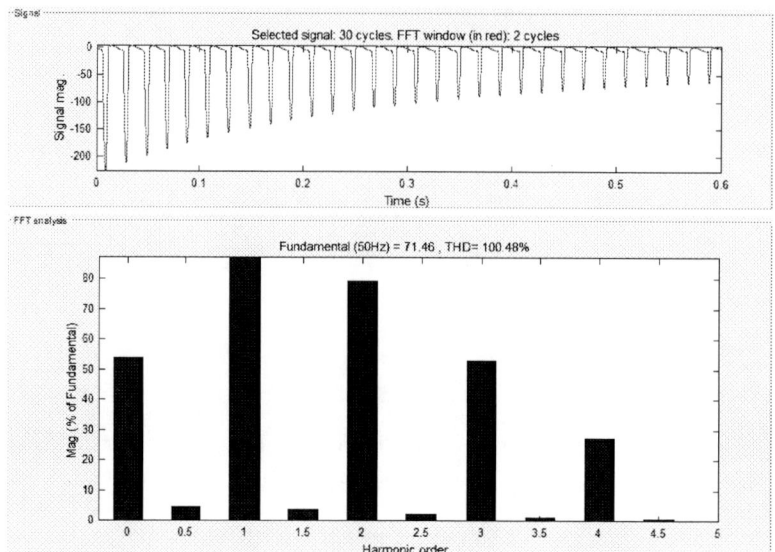

Figure 3. Harmonic analysis of inrush current waveform by using FFT

3. Factors affecting the saturation of traditional CT

The core of the current transformer is a nonlinear component and its magnetization curve is shown in Figure 4. There are many reasons for the magnetic saturation of the core, among which the followings are the main points.

1) The aperiodic component of short-circuit current transient components. Generally speaking, the larger the aperiodic component is, the more severe the saturation is.

2) Saturated time constant of current transformer. The bigger the saturated time constant is, the easier it is to saturate.

3) Secondary side load and its load characteristics. The greater the load is, the easier it is to saturate.

4) Remanence of iron core. If the residual magnetic direction of the core coincides with the flux direction of the fault current, the saturation is easier.

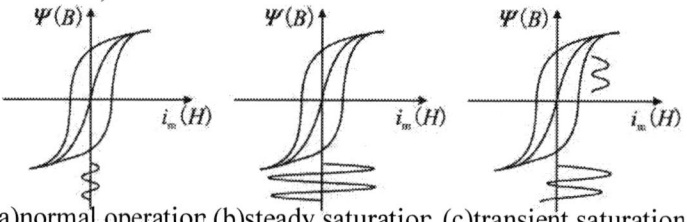

a)normal operation (b)steady saturation (c)transient saturation

Figure 4. Magnetic characteristic curves.

In the factors that cause the saturation of the current transformer, the non-periodic component is the most important. When fault occurs, the fundamental cause of secondary current distortion caused by current transformer saturation is the nonlinear characteristics of the transformer core, most of the aperiodic components cannot be transmitted to the secondary side, thus becoming the excitation current. The larger the aperiodic component and the slower the attenuation, the more serious the saturation of the current transformer. When the flux density of the core reaches saturation, the magnetic conductivity of the core decreases rapidly and the excitation current rises sharply. Most of the primary current becomes the excitation current. The secondary current is seriously distorted, the waveform is defective and the amplitude decreases. In serious cases, the amplitude of the secondary current becomes very small, sometimes even close to zero. For relay protection, the correct action depends firstly on the correct response of the measuring element to the primary system. Because of the iron core in traditional CT, when the fault current exceeds its allowable value, it will produce saturation, which cannot guarantee the accuracy of measurement. Especially in the transient process, due to the influence of non-periodic component of primary current, the current transformer will be in serious transient saturation.

4. Transient characteristics of optical current transducer

Optical current transformers can be based on a variety of optical effects, such as Faraday magneto-optic effect, magnetostrictive effect, piezoelectric effect and so on. Among them, the optical current transformer（OCT）based on Faraday magneto-optic effect is the most fully researched and practical one.

OCT based on Faraday magneto-optic effect indirectly measures the current by measuring the magnetic field caused by the measured current. The principle of Faraday magneto-optic effect is shown in Figure 5. When linearly polarized light passes through a magneto-optic medium under the action of an external magnetic field parallel to its propagation direction, the polarization plane will deflect and the deflection angle can be expressed as:

$$\theta = \mu \cdot V \cdot \int_L \overline{H} \cdot \overline{dl} \tag{5}$$

Among them, μ is the permeability of Faraday magneto-optic materials. V is the Verdet constant of magneto-optic materials, which is related to the characteristics of the medium, the wavelength of the light source, the external temperature. H is the magnetic field intensity acting on magneto-optic materials. L is the optical path length of polarized light passing through magneto-optic materials.

Since the deflection angle of polarized light cannot be measured directly, the unmeasurable deflection angle signal is converted into the measurable polarized light intensity signal in OCT's implementation. According to Marius's law, the output light intensity is expressed as:

$$J_o = J_i \cos^2 \beta \tag{6}$$

Where β is the polarization angle between the polarizer emitted polarized light and the polarizer emitted polarized light. J_i is the input light intensity.

In order to maximize the intensity of polarized light emitted from the polarizer, the angle β between the start and the detector is usually set to $\pi/4$. Because the Faraday rotation angle is very small, the output intensity can be further expressed as:

$$J_o = J_i(1 - \sin 2\theta) \approx J_i(1 - 2\theta) \tag{7}$$

Because J_i and θ are unknown quantities, in order to solve two quantities from an equation, we can use the single optical path method.

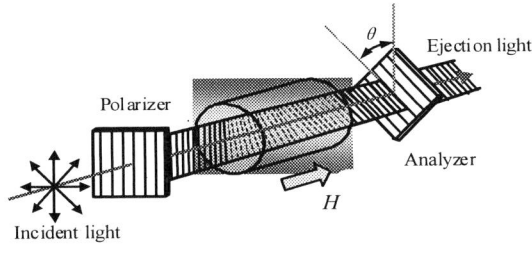

Figure 5. Schematic diagram of single optical path method.

Because the deflection angle of polarized light cannot be measured directly, polarization detector is used to transform it into light intensity signal to detect. The reference coordinate system is used to make the polarizer light transmitting shaft parallel to the X axis, and the angle between the polarizer and the polarizer shaft is $\pi/4$. The photoelectric vector generated by the light source is E_i, and the Jones vector can be expressed as $\begin{bmatrix} E_{ix} \\ 0 \end{bmatrix}$ after the polarizer. And $E_{ix} = \frac{1}{\sqrt{2}}\sqrt{E_{ix}^2 + E_{iy}^2}$

The Faraday rotation characteristics of magneto optical medium can be expressed by Jones matrix as:

$$F = \begin{pmatrix} \cos\theta & -\sin\theta \\ \sin\theta & \cos\theta \end{pmatrix} \tag{8}$$

According to the Jones algorithm, the output vectors after the Faraday medium and the analyzer are:

$$E_o = \begin{pmatrix} \cos\pi/4 & -\sin\pi/4 \\ \sin\pi/4 & \cos\pi/4 \end{pmatrix}\begin{bmatrix} E_{ix} \\ 0 \end{bmatrix}\begin{pmatrix} 1 & 0 \\ 0 & 0 \end{pmatrix} = \begin{pmatrix} \cos\pi/4 & \sin\pi/4 \\ -\sin\pi/4 & \cos\pi/4 \end{pmatrix}\begin{pmatrix} \cos\theta & -\sin\theta \\ \sin\theta & \cos\theta \end{pmatrix}\begin{bmatrix} E_{ix} \\ 0 \end{bmatrix} = \frac{E_{ix}}{\sqrt{2}}\begin{bmatrix} \sin(\theta+\pi/4) \\ \sin(\theta-\pi/4) \end{bmatrix} \tag{9}$$

The relationship between the output light intensity J_o and the input intensity J_i is:

$$J_o = \frac{1}{2}J_i(1+\sin 2\theta) \tag{10}$$

Among them, the AC component is $J_o = \frac{1}{2}J_i\sin 2\theta$. The DC component is $J_o = \frac{1}{2}J_i$.

Single optical path detection method uses filter circuit to detect AC component and DC component respectively. In order to eliminate the influence of input light intensity fluctuation, modulation quantity is

$$m = J_{Ac}/J_{Dc} = \sin 2\theta \tag{11}$$

When the modulation system is relatively small, $m \approx 2\theta$.

Suppose the primary input current is $i_1 = I_1 e^{-\frac{t}{T_1}} - I_1\cos\omega t$. The secondary current can be expressed as:

$$i_2 = K(I_1 e^{-\frac{t}{T_1}} - I_1\cos\omega t) \tag{12}$$

According to (12), the response time of OCT can be neglected, the DC component can be transferred correctly, the steady state of each harmonic can be transmitted correctly and the secondary current can be attenuated by one time during fault removal. However, in the actual implementation process, the measurement accuracy will be affected by temperature, magnetic field interference, vibration and other factors. But these factors are within the error range for OCT steady state and transient measurements. Therefore, OCT has a great advantage over traditional CT, especially when there are a lot of aperiodic components in transient faults.

5. Simulation analysis

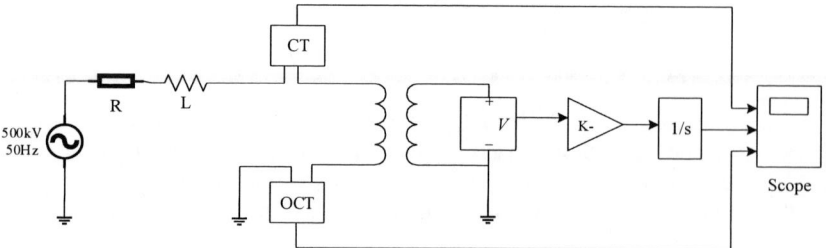

Figure 6. System simulation model.

Using Simulink and power system component library in MATLAB software, transformer simulation model can be easily established. Considering the same physical characteristics of three-phase system, a single-phase system model is adopted here. The simulation model of MATLAB system is shown in Figure 6.

The transformer is connected to 500kV system with rated capacity of 450MVA (single phase is 150MVA) and rated ratio is 500kV/230kV. The primary and secondary windings consider the effects of resistance, leakage inductance and remanence. The transformer is designed by using the model with saturation characteristic and the hysteresis characteristic design tool of MATLAB. The saturated CT module is represented by piecewise linear representation of the magnetization curve and remanence is taken into account. The voltage level is 500kV, and the transformer ratio is 2000/5. The primary load (R, L) is used to control the time constant. The circuit breaker closing time is used to control the aperiodic component. When the circuit breaker closes at the zero-crossing point, the aperiodic component is the largest and CT's saturation is the most serious. The secondary load is adjustable resistance. Output the waveform of OCT with proportional link and delay link. Figure 7 is the simulation waveform of the system. From top to bottom, there are the original waveform of inrush current, the waveform measured by OCT (secondary side) and the waveform measured by traditional CT (secondary side).

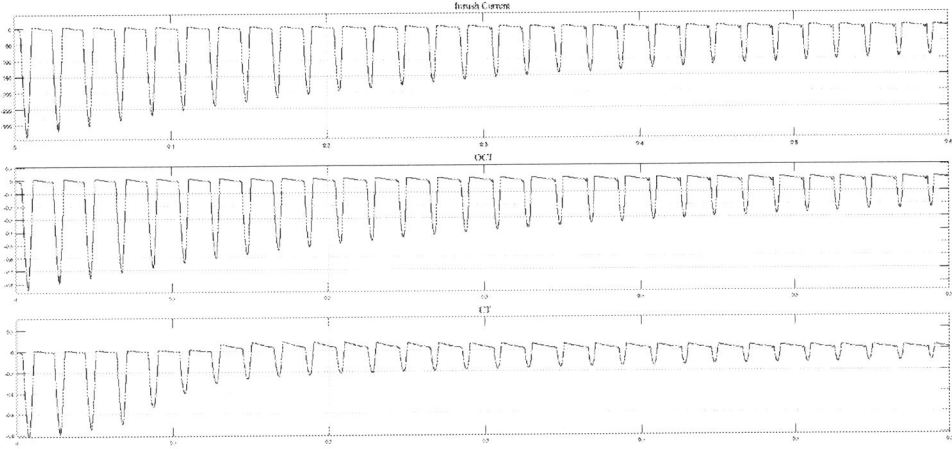

Figure 7. System simulation waveform.

The transmission performance of CT plays an important role in the reliability, sensitivity and speed of protection. It can be seen that when the traditional CT saturates, the secondary current will be distorted and then cannot accurately reflect the inrush current waveform of the primary. When this distorted current flows into the protection device, it may lead to maloperation or rejection of the differential protection. The saturation of CT will result in larger transmission error for primary value and incorrect operation of transformer differential protection. In the transient process, the short-circuit current or inrush current has a high non-periodic component, it is very likely to cause serious saturation and current distortion.

OCT is not affected by transient faults and can completely retain the inrush current information, in which the harmonic components and the closing angle can be completely retained and then lay a foundation for identifying the inrush current. Therefore, OCT can accurately measure the full waveform of inrush current in principle, especially in transient state and its excellent accuracy is far superior to traditional CT.

6. Experimental verification

In order to verify the ability of optical current transducer (OCT) to reflect the inrush current, an on-site operation experiment of OCT was carried out at Hushitai test site of Northeast Electric Power Academy. Two optical current transducers are connected in series in 220kV side circuit of Hushi test-bed transformer (40MVA/220kV). By means of transformer impulse closing test, the operating condition and transient characteristics of OCT under switching vibration and inrush current of transformer are detected. The waveforms of inrush current output from two OCTs are checked to be consistent with those of traditional CT, compared with recorded inrush current waveforms.

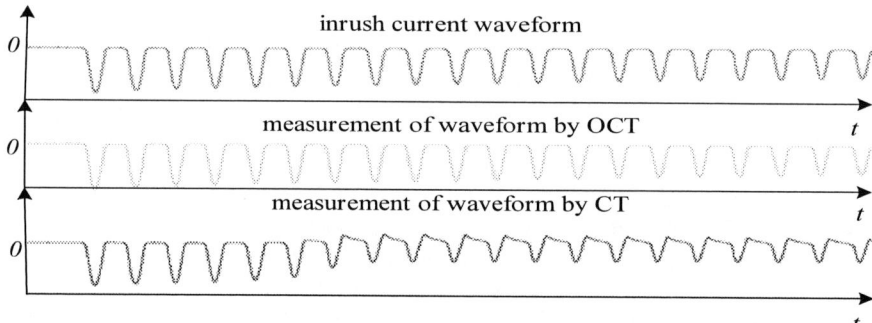

Figure 8. Fifth closing current waveform (expanded waveform).

A total of 5 operations were performed. The interval was 10 minutes and the experimental waveforms were shown in Figure 8. The first and second waveforms are 2 OCT test waveforms and third waveforms are traditional CT test waveforms.

The fifth impact found that the seventh cycle began, optical current transformer and traditional transformer waveforms are greatly different. The OCT's waveforms are correct, the traditional CT waveforms are error due to aperiodic component transmission and saturation. It is proved by experiments that the OCT has the advantage of transient measurement compared with the traditional CT.

7. Conclusion

This paper firstly analyses the inrush current containing a large number of non-periodic components, which will saturate the traditional CT transient and result in the failure to accurately obtain the inrush current. The transformer protection will maloperate or reject. Because of its own basic principles and measurement methods, OCT completely free from the impact of saturation can be thoroughly solve a series of problems caused by CT saturation. The saturation of the two is simulated by simulation software and tested by experiment. It is proved that OCT has excellent transient measurement quality and can realize the measurement information of current "full waveform", which is of great significance to the progress of power system measurement and protection.

References

[1]   Hu Xiaoguang, Wang Zhe, Yu Wenbin. Simulation and analysis of transient process of current transformer [J]. Journal of Power System and Automation, 2001,13(4): 12-15.
[2]   Chen San Yun. [J] analysis of relay protection caused by CT saturation. Power grid technology, 2002, 26 (3): 85-87.
[3]   Qu Yanhua. Application of optical current transformer in relay protection [D]. North China Electric Power University, 2005.
[4]   Teng Lin, Liu Wanshun, Li Guicun, et al. Optical Current Transformer and Its Application in Relay Protection [J] Power Grid Technology, 2002, 26 (1): 31-33.
[5]   He Jia Li. Principles of relay protection for power system: Fourth Edition [M]. Beijing: China Electric Power Press, 2010.
[6]   MAO P L, BO Z Q. Protection of teed transmission circuit using a new directional comparison technique[C] Beijing: International Conference on Power System Technology Proceedings, 1998: 1111-1115.
[7]   YU Wen-bin. Research on temperature characteristic of light intensity of optical current transducer[D]. HARBIN: Harbin Institute of TECHNOLOGY, 2005.

ISPECE

Study of the standard sine wave frequency conversion power supply based on analog and digital integrated control

Wenbo Jia[1], Hongda Zhang[2], Dewen Zhang[3], Kepeng Tao[4], Sicong Li[5] and Wanlin Guan[6]

[1]State Grid Heilongjiang Electric Power Co. Ltd. Electric Power Research Institute, Harbin 150030, China;

[2]State Grid Heilongjiang Electric Power Co. Ltd. Electric Power Research Institute, Harbin 150030, China;

[3]State Grid Heilongjiang Electric Power Co. Ltd. Electric Power Research Institute, Harbin 150030, China;

[4]Guangdong Power Grid Co. Ltd. Jiangmen Power Supply Bureau, Jiangmen 529000, China;

[5]State Grid Shandong Electric Power Co. Ltd. Jining Power Supply company Electric Power Research Institute, Jining 272023, China;

[6]State Grid Heilongjiang Electric Power Co. Ltd. Electric Power Research Institute, Harbin 150030, China

Communication author: Wenbo Jia E-mail address: 112024889@qq.com

Telephone number:18182801466

Abstract: In this article, a standard sine wave frequency conversion power supply based on analog and digital integrated control is proposed. First of all, the article introduces the topology structure and principle involved in this frequency conversion power supply. The system of frequency conversion power supply was designed based on the DSP2812[1] made by TI company. Next, it is stated that the method of analog-digital combination control simplifies the inexact algorithm, under the digital control, of calculating the carrier wave and the PWM driving signal of power device generation at the point of modulation wave intersection. In addition, it is also stated that the analog-digital integration control is equipped with a good performance in dynamic control response and a harmonic content that can effectively drop down the output voltage of the controller; the analog-digital integration control can realize the required motor control performance at a low cost, improve the operation efficiency of the motor, and suppress various secondary harmonic currents of the motor. At the end, the voltage and the input waveform of the electric currents of the system are under analyzed through a test simulation experiment on the power source of the variable frequency power supply system; at the same time, the input voltage waveforms corresponding to different frequencies are also involved in discussion, while the corresponding harmonic waves are analyzed.

Content from this work may be used under the terms of the Creative Commons Attribution 3.0 licence. Any further distribution of this work must maintain attribution to the author(s) and the title of the work, journal citation and DOI.
Published under licence by IOP Publishing Ltd

1. Introduction

Between countries do the state grid standards vary in a certain degree, due to which a large-scale promotion is conducted for the application and development of power supply with variable frequency. And the power grid state in each country is simulated by means of power supply. Thus, it is required for the equipment and power supply to provide such frequency integrated with purity, reliability, low harmonic distortion, and high stability, and such a sine wave power output with a regulated voltage.

In this article, the focus is attached to the introduction of the general development of the power supply technology and the common control methods for power supplying, which comes up with the significance of this subject. Besides, the foreign and domestic research status, and the working principles, the current development state and application of the variable frequency power supply, as well as the systemic structure of this article are also summarized by this paper. In combination with the actual needs of this research, the scheme to achieve a power supply with variable frequency and variable voltage is demonstrated on the basis of the technology of power supply with variable frequency and variable voltage and its development trend; additionally, the control principle--the basic principle of PWM [2] control--is also elaborated in detail. The power supply with variable voltage and variable frequency is designed by both software and hardware, and put into stimulation analysis, system debugging, and experiments by Simulink. For the waveform of each constituting circuit of the system, a comparison study is performed to control the voltage amplitude and output frequency of the test power supply, and to deeply analyze the harmonic waves of the output voltage of the system at the same time.

2. System structure and control method of variable frequency power supply

2.1 AC/DC circuit

In the conversion of alternating current to direct current, the relevant functional requirements can also be accomplished by dint of the APFC. The APFC circuit can output a stable voltage under the control of the grid voltage waveform. Generally, the output voltage is set to 400V. Combining with the loss of the line and the device, it is possible to complete an output of 220V AC and also to increase the power factor and to reduce the effects of harmonic currents. For this end, it is determined by the system to adopt an APFC circuit. For example, Figure 2-1.

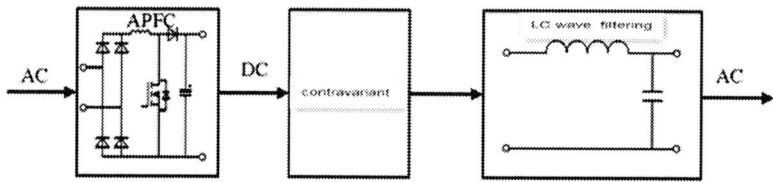

Figure 2-1. Structure diagram of the system used in this paper

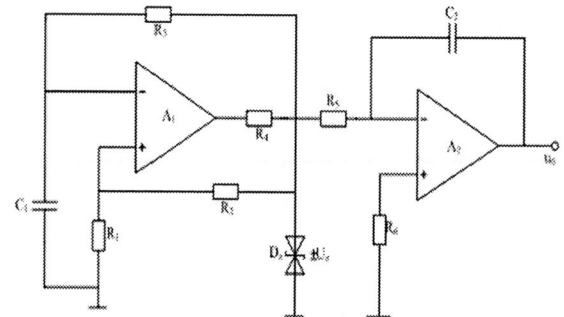

Figure 2-2. Triangle wave circuit principle

2.2 Principles of analog and digital integrated control

In the Figure 2-2 of the principle of triangular wave circuit, on the left side of the circuit is a square wave generator that serves to generate a square wave, and concurrently, on the right side is an integrator system via which the square wave forms a triangular wave. At time t, the positive input of the op amp is higher than the negative input, and thus the op amp outputs a high-level of current (the positive peak of U_Z). The positive input voltage of the operational amplifier is $U_Z \bullet R_2 / (R_1 + R_2)$. The high electric level charges the C1 through R3, so as to slowly increase the capacitor voltage; not until the negative input of the op amp is higher than the voltage of the positive input terminal, is the output of the op amp inverted, that is, the current output is low (negative peak - U_Z). The electric level on the positive input end is $-U_Z \bullet R_2 / (R_1 + R_2)$ which is still lower than that of the negative input voltage. Capacitor C1 discharges the output of the op amp through R3, and its voltage is gradually reduced. Till the capacitor voltage drops below the positive input terminal, the output of the op amp is flipped again; through multiple times of recycling and repeating, the square wave is formed by oscillations. The frequency of the square wave depends on the product of R3 and C1.

Principle of the integrator yielding a triangular wave. The circuit on the right is an integrating circuit. According to the virtual ground principle, the positive and negative input voltages of the op amp should be the same, that is, zero at the same time. If the square wave voltage is added to resistor R5 as a load and it is an electricity at high level, the current on the R5 is a constant current (U_Z/R5) that also charges for the capacitor C2. In this state, the constant current of charging causes the voltage across the capacitor C2 to increase at a constant rate. After a half square-wave period, the input flips into a low-level electricity. The C2 discharges through R5, and the current stays constant (U_Z/R5). In such a condition, the constant current of the discharge causes the voltage across C2 to decrease at a constant speed, and the voltage input and output values across the capacitor C2 are equal. By repeating the above steps in a recycling manner, a triangular wave output will be produced accordingly.

In Figure 2-3, the DSP controller generates a modulated wave, and the triangular wave circuit forms a carrier. With these two parts, a PWM output waveform is formed via a comparator to supply for the driving circuit, and then to ensure the reliability of the inverter bridge and guarantee its stable operation through driving the inverter circuit.

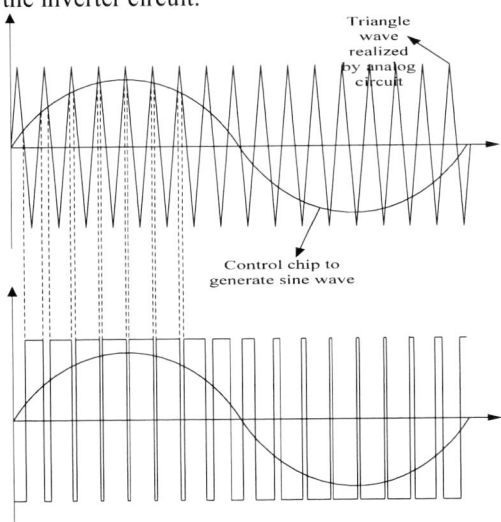

Figure 2-3. SPWM waveform

3. Circuit design of the hardware of frequency conversion power.

3.1 Principle schematic of main circuit of variable frequency power supply

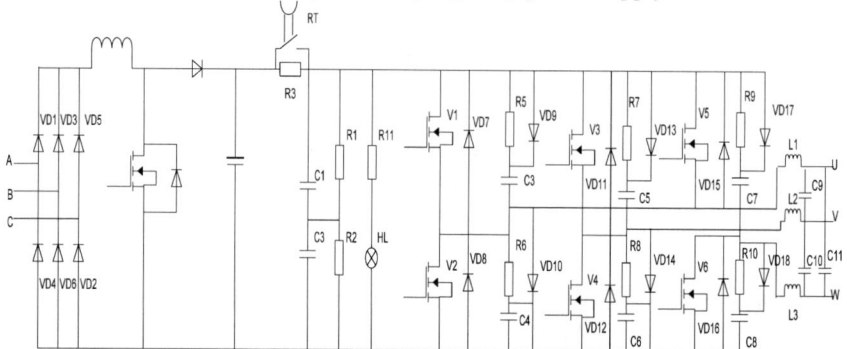

Figure 3-1.Circuit diagram of variable voltage variable frequency power supply system

In the whole circuit, for example, shown in Figure 3-1, the functions of each part of the main circuit are as follows: (1) APFC circuit: The grid input power is converted into pulsating DC power through the APFC circuit and input to the next intermediate filter segment. The circuit is started and the capacitor in the initial state needs to be charged. Thanks to the high capacity of the capacitor, the charging current will be increased within a short time. Thus, if no treatment is performed, the rectifier bridge and the input side fuse will show errors, and a trip will be brought up even on the input side. For this end, R3 should be linked in parallel to control the charging current in the case where the main circuit is connected. And the touch point of RT should be short-circuited so as to enable the R3 to short-circuit and save the electricity. (2) Rectifier circuit: The input electric energy is converted into direct current by means of three-phase bridge type uncontrollable rectification, and is input to the next intermediate filtering section. (3). Intermediate filter circuit: On the strength of capacitive filtering, the AC component in the output current of the rectifier circuit is filtered out, while the DC component is retained only. Because the grid impact and the output will affect the input current, causing the input side voltage and current to fluctuate, the capacitors C1 and C2 should be connected in series to increase the rectifier bridge capacity and reduce the occurrence of circuit problems resulted from this state. By the aid of R1, R2 series grading resistors, the capacitor is subjected to voltage equalization. R11 and HL are connected in parallel to detect the performance of the circuit. (4). Inverter circuit: By feat of the three-phase inverter bridge formed by the MOSFETs of V1-V6, the DC power output from the filter circuit is converted into an alternating current at a predetermined frequency and voltage. Six sets of RCDs constitute a snubber circuit MOSFET [3] which can be effective in small power inverters; however, an over-high switching frequency of the switch tubes will give rise to the error of excessive voltage or excessive current in the process of transition, thereby affecting the safety of the switch tube. Corresponding to the possible problems, it is supposed to adopt the DRC adsorption circuit [4]to protect the tubes. (5) Output filter circuit: The distortion of the output waveform is reduced as much as possible to resemble it a sine wave with the help of LC output filter circuit,

3.2 Design of control circuit

The core of the DC/AC inverter control circuit falls down to the DSP2812. The circuit contains three branch circuits, which are a sampling signal conditioning circuit, an output voltage and current feedback circuit, and an optocoupler isolation driving circuit, respectively, as shown in Figure 3-1.

The most important function of the DSP2812 is ascribed to its capability of generating a good SPWM [5] signal. And the four MOSFETs of the two bridge arms can be switched under the action of the driver circuit.

With the action of the transformer and signal conditioning circuit, the output voltage and output current can pass through the DSP2812 and enter the AD sampling module. The applicable SPWM signal is outputted by a voltage feedback modulation. According to the current [6] that is fed back, is the need of initiating a current protection determined.

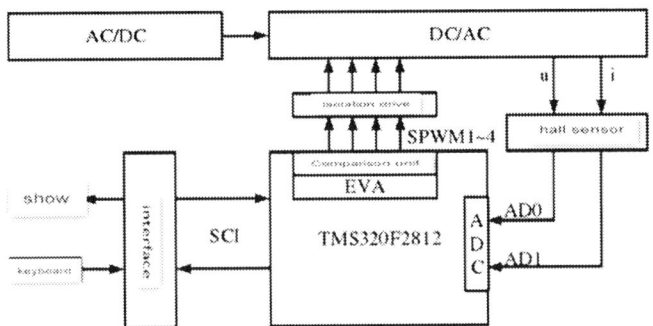

Figure 3-2. Diagram of the control circuit principle

4. System experiment and results analysis

4.1 Simulation results

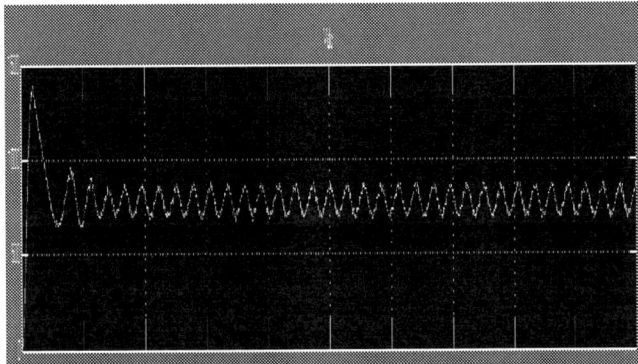

Figure 4-1. Output voltage waveform after rectification and filtering

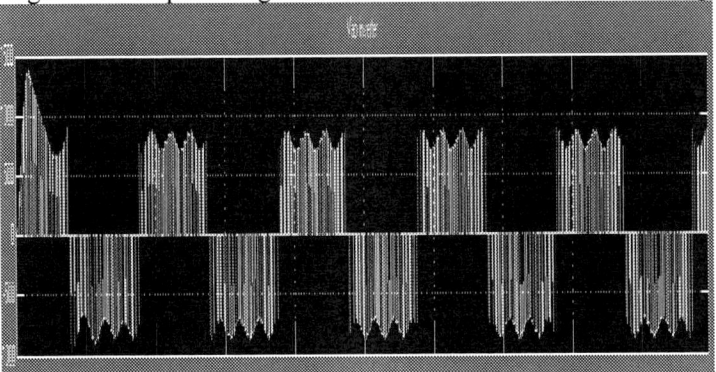

Figure 4-2. Inverter output voltage waveform

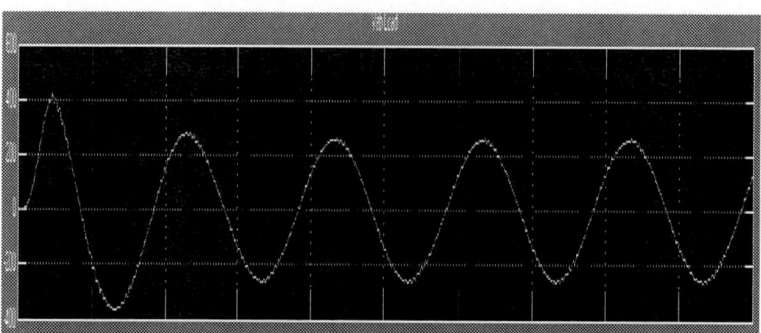

Figure 4-3.Output voltage waveform after inverter filtering

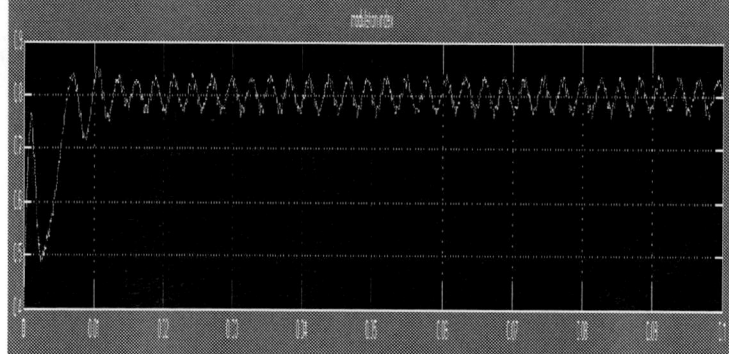

Figure 4-4. Modulation factor

To sum up, in this subject will a variable voltage frequency[7] conversion system based on the digital and analog [8] integrated control for variable voltage frequency conversion system be designed, and an effective simulation model will be established to match the software and hardware corresponding to the DSP design, by which the intelligent control method will be more practical and conducive to the design and development of a control system with high performance, as shown in Figures 4-1 to 4-4 .

4.2 SPWM drive waveform and spectrum analysis

The equipment used for the system test includes: Tektronix oscilloscope TPS2024B, VICTOR portable multiple-purpose meter VC9807A.As shown in Figure. 4-5, the SPWM driving waveform is displayed under the oscilloscope. Figure 4-6 and 4-7 illustrate the analog and digital integrated control SPWM drive signal boards and their PCB diagrams.

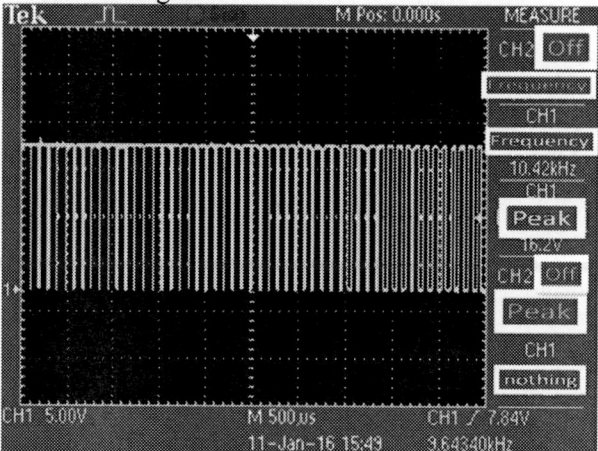

Figure 4-5. SPWM drive waveform

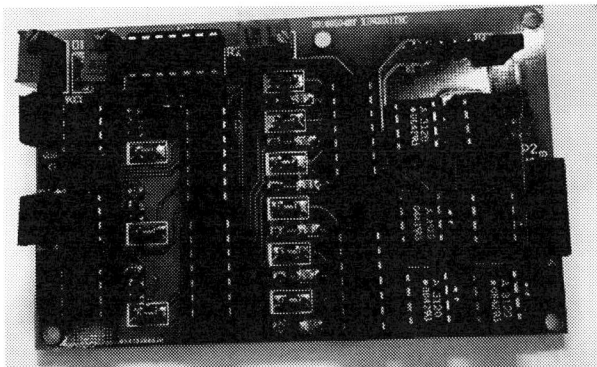

Figure 4-6. Analog and digital integrated control SPWM driver board

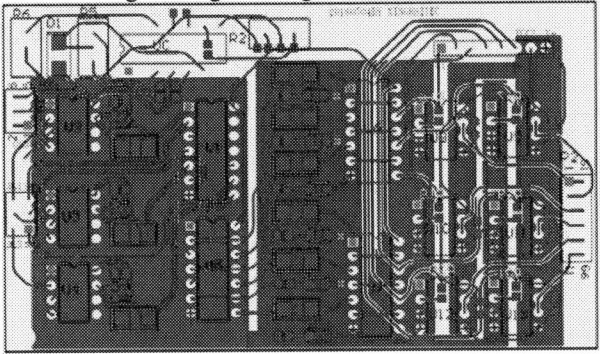

Figure 4-7. PCB diagram of analog and digital integrated control SPWM driver board

As shown in the following Figures of 4-8 and 4-9, they are respectively the waveform three-phase sine wave modulation signal outputted by the D/A module of DSP, and the waveform of sine wave signal emitted by analog triangle wave circuit.

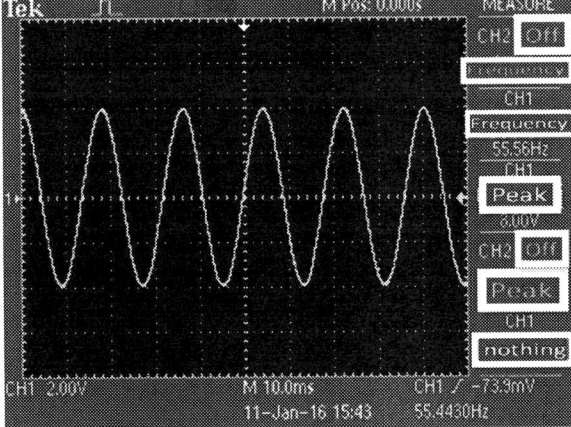

Figure 4-8. Waveform of modulation signal

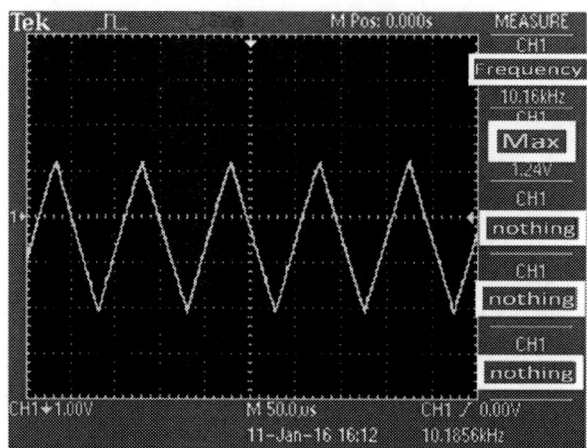

Figure 4-9. Signal waveform diagram of analog triangle wave circuit

Next, with an entity construction of the SPWM generation circuit, the analog SPWM, digital SPWM, and digital-to-analog SPWM waveforms are put into a Fourier decomposition analysis, and the designed details of the circuit structure will be described in the Chapter 4. To grant the SPWM parameters with relativity, these parameters are uniformly set as followings: the modulation depth is set to 0.8 and the carrier frequency is rated as 10KHz. Since that the sine analog circuit is fixed by the parameters of RC crystal oscillator circuit, the modulation wave frequency has been set to 54.6 Hz, and the corresponding SPWM wave is analyzed by Fourier analysis as shown below:

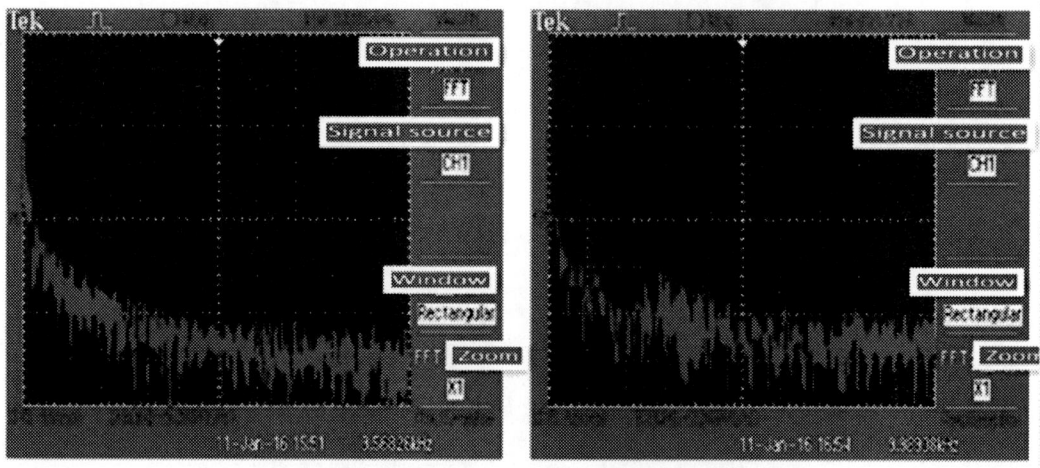

a) Pure analog mode   b) Pure digital mode

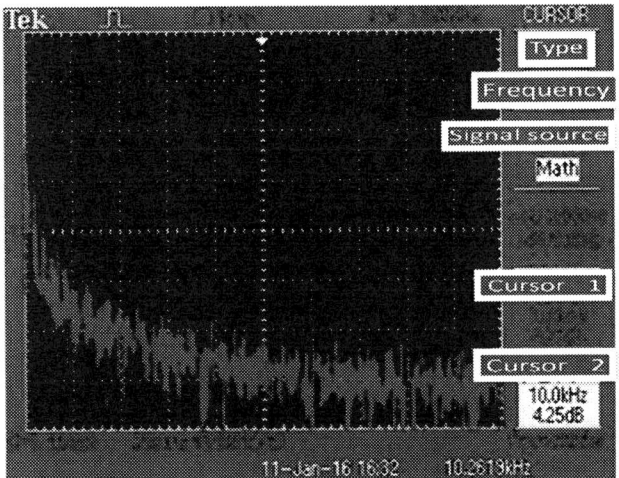

Figure 4-10. Fourier analysis of SPWM waveform

As shown in the combo Figure 4-10, the Figure a represents the Fourier analysis of the SPWM waveform generated by the pure analog circuit, the Figure b shows the Fourier analysis of the SPWM waveform sent by the event manager of the DSP [9], and the Figure c refers to the Fourier analysis of the SPWM wave obtained by the sine wave from the D/A of the DSP and the triangular wave generated by the analog circuit through the voltage comparator. In this figure, the horizontal axis coordinate stands for the frequency, the vertical cursor refers to the switching frequency of 10 kHz, and the vertical axis coordinate is on behalf of the decibel that is proportional to the amplitude. The harmonic frequency amplitude distribution of the analog-digital integrated system is similar to the harmonic frequency amplitude distribution of the pure analog system, while the harmonic frequency amplitude of the pure digital method distributed near the switching frequency is larger. Identical with the pure analog approach, the modulation method of digital-analog integration verifies the superiority of the digit-analog integration. In Figure 4-11 is a combination of analog and digital spectrum analysis.

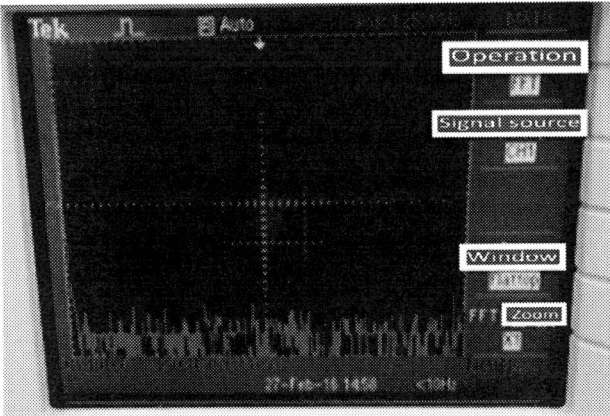

Figure 4-11. Spectrum analysis of the filtered SPWM wave

4.3 Analysis of experiment results and summary of experience
Through the above experiment, it has been found in our obtained results that the designed program in this article is feasible and able to reach the designed technical requirement and an output voltage waveform of a better quality. Besides, the dynamic performance and stability of the system has also been verified through tests.

In the process of system debugging, a number of experienced lessons have been also harvested, from which some parts in need of attention have been summed up.

First of all, it should be paid attention to that the oscilloscope needs to be equipped with an isolation probe when the test wave is under debugging. In particular, it is also required to be cautious when one is engaged in a multipath measurement. In case the operation fails to be proper, it may incur burnout to the circuit board, the oscilloscope, and other equipment and materials.

Secondly, the first step of debugging is to debug the control circuit, so as to ensure that the control circuit can work normally, and then to connect the control circuit into the main circuit.

Thirdly, the chip should be located as close as possible to the output feedback voltage resistor. Moreover, the driver chip should be also placed approximately to the switch tube, in order to minimize the driven line between these two.

5. Conclusion

A design of a standard sine wave frequency conversion power supply based on analog and digital integrated control has been completed in this article according to the technical requirements. At the same time, the design of main circuit and the selection of components were introduced in detail, and additionally, the control circuit has also been designed. In the last part of this article, the system is put into test and analysis, through which the designed program of this article is verified. Of this article, the main work listed below has been accomplished, and the following conclusions are also obtained through the experimental demonstration:

(1)The rectification has reached an effect of reducing input harmonics and improving the power factor of the input terminal by adopting APFC technology.

(2)Compared with a pure analog, the digital control and analog control avoided several shortcomings involved in the pure analog, which includes a large demand of discrete components, low circuit reliability, an increase of system cost, the high difficulty in system debugging, the hard troubleshooting for faults, the vulnerability to the ambient temperature, as well as the aging problems.

(3)The combination of digital control and analog control [10] is featured with a good dynamic control response and an effective performance to reduce the harmonic wave content of the controller output voltage. While retaining the advantages of high precision and fast dynamic response of the original analog control, it is also introduced in the flexibility and controllability of the digital control. The modulation pattern of the digital-analog combination possesses an effect which is similar to the one of pure digital system. Furthermore, this combination replaced the triangular wave modulation programme in the control chip, due to the achievement triangular wave modulation by analog circuit, which to a certain extent saved the on-chip resources and improved the running speed of the control chip.

(4)With a good working performance, the power rectification can be put into a stable operation for a long time, which provides a stable voltage for the inverter part and satisfies the power output requirement. The drive circuit can realize the signal amplification well, and can drive the MOSFET to be powered on and off. The wave filtering effect was good. Both of the waveform and amplitude of the output voltage meet the requirement. In addition, a good effect on the frequency modulation and voltage regulation has been realized under the control of DSP [11].

Acknowledgment

This research was funded by the Science and Technology Project of State Grid which named Research on interoperability of communication and magnetic coupling structure of electric vehicle wireless power transfer systems.

References

[1] Texas Instruments. (2005)TMS320F2812 Digital Signal Processors Data Manual.
[2] Jong-Lick Lin. (2002)A new approach of dead-time compensation for PWM voltage inverters. IEEE Transactions on circuits and System I: Fundamental Theory and Applications,

49(2):476-483.

[3] Liu Fengjunn,(2000) Serveral Compensation Methods for the Influence of SPWM Inverter Dead Zone [C]. In: The 7th Academic Conference of the Institute of Power Electronics, China Electro-Technical Society. 213-222.

[4] Zhang Quanzhu, Huang Chengyu, and Deng Yonghong. (2009)Matlab Simulation Study of Inverter IGBT Absorption Circuit[J]. Electric Drive Automation, 31(6):27-31.

[5] Liu Liang and Deng Minggao. (2005) A New Method to Compensate the Dead Zone Effect of PWM Inverter[J]. Power Electronics Technology, 39(6):123-125.

[6] Zhou Jiemin. (2012) Theory and Design of Switching Power Supply [M]. Beijing Aerospace University Press,Beijing, 70-73.

[7] Huang Lipei and Sunkai. (2016) Simulation Analysis of High Voltage Motor Frequency Conversion Test Power Supply [J]. Group Technology and Production Modernization, 2(2): 12-15.

[8] Xue Dingyu and Chen Yangquan, (2002) System Simulation Technology and Application Based on MATLAB/SIMULINK [M]. Tsinghua University Press, Beijing.

[9] Hyun Rok Cha, Kyoo Jae, Shin Young ju Seo. (2011) Design of Outer Rotor IPM type PMSM for 3 Wheel Electric Vehicle[C]. In: IEEE International Conference on Electrical Machines and Systems, London, The United Kingdom, 1-3.

[10] Huang Junnian and Wang Li. (2010) Analysis of PID Control Principle and Application [J]. Silicon Valley, 12: 109-111.

[11] Lu Qi. (2005)Research and Design of Digital Programmable AC Power Supply Based on DSP [D]. Shanghai Jiao Tong University. Shanghai.

Optimal installation of distributed generators based on an enhanced harmony search algorithm

Ke Ji[1,3], Wentao Wang[2], Xingong Wang[1], Zheyu Wei[1], Jian Yang[1] and Hua Qin[1]

[1] Hohhot Power Supply Bureau of Inner Mongolia Power (Group) Co., Ltd., No.74 North Tongdao Road, Hohhot, China;

[2] School of Electrical Engineering, Beijing Jiaotong University,No.3 Shangyuancun, Haidian District, Beijing, China;

[3] rockjike@hotmail.com

Abstract. The demand of reducing global greenhouse gas emissions and restructuring electricity market has led to an increase in the use of distributed generation. Distributed generators (DGs), which are being connected to utility distribution networks, can improve power supply reliability. This paper describes how an enhanced harmony search algorithm can be used to find the optimal number of distributed generators in a distribution network. Different system parameters and scenarios are chosen to run the algorithm. The results indicate the correctness and availability of the proposed algorithm.

1. Introduction

Distributed generation (DG) will become more and more important in the future electricity distribution system. This tendency is increased by the demand of protecting the environment, the liberalization of the energy market and restructuring electricity market. Recently, many distributed power generation systems have been installed in demand area, and have been directly connected to the distribution system, since there are many advantages such as substitution of large scale generators, extension of facility expansion schedule and loss reduction, etc. On the other hand, some complicated problems may occur in such a power distribution system: voltage increase at the end of a feeder, demand supply unbalance in a fault condition, power quality decline or voltage wave distort in demand side. Over the years, some papers which discuss about the optimal allocation and operation of the distributed generators have been published [1-4]. Das [5] has presented an algorithm based on the heuristic rules and fuzzy multi-objective approach for optimizing network configuration. The drawback with this algorithm is criteria for selecting membership functions for objectives are not provided. Borges and Falcao [6] have presented a technique to evaluate the impact of DG size and placement on losses, reliability and voltage profile of distribution networks. Wang and Nehrir [7] have proposed an analytical technique for optimally allocating distributed generation units in a radial distribution system that minimized power losses. The proposed technique considered different types of load profiles with varying time loads and distributed generation output while also took into account technical constraints, such as feeder capacity limits and voltage profile.

In this paper, a new distribution network planning method based on an enhanced harmony search algorithm to dispatch DGs has been presented. The planner counts on the distributed generators to compensate for the conventional power stations in preference to reducing polluting gases emissions. In

---

Content from this work may be used under the terms of the Creative Commons Attribution 3.0 licence. Any further distribution of this work must maintain attribution to the author(s) and the title of the work, journal citation and DOI.

Published under licence by IOP Publishing Ltd

addition, the optimal numbers of DGs under the electricity demand of different scale cities are obtained.

2. Problem formulation

The problem is to determine allocation and size of DGs which minimize the pollution emissions under the condition that ensure regular power supply. We intend to acquire optimized amount of distributed power and traditional coal-fired power plant via an optimized model which focus on environmental factor.

The traditional harmony search algorithm includes procedures is listed below:

a. Initialization;

b. New vector generation;

c. Harmony memory update.

We improve harmony search algorithm by combining it with swarm pattern search algorithm. The improved harmony search algorithm overcomes local convergence while harmony updating process.

Objective function:

$$\text{Min } \rho = \rho(\mathbf{C}) = \frac{\sum_i e_i C_i}{\sum_{nc} e_0 C_0} \qquad (1)$$

$$nc = \frac{\sum_i C_i p_i}{p_0} \qquad (2)$$

$$s.t. \begin{cases} N_i > 0 \\ \sum_i N_i p_i > P_{LD} \end{cases} \qquad (3)$$

$C = [C_1, C_2, ..., C_n]$ is solution vector, C_i (i=0, 1, 2, ..., n) is number of each kind of power;

e_i (i =0, 1, 2...) is emission of each power source (traditional coal-fired power plant, wind power generator, fuel cells, photovoltaic cells, etc)

e_0: coal-fired power plant emission;

e_1: wind power generator emission;

e_2: fuel cells emission;

e_3: photovoltaic cells emission;

p_i (i =0, 1, 2...) is power contribution of each power source(traditional coal-fired power plant, wind power generator, fuel cell, photovoltaic cell, etc)

p_0: coal-fired power plant output power contribution;

p_1: wind power generator output power contribution;

p_2: fuel cells output power contribution;

p_3: photovoltaic cells output power contribution;

P_{LD} is the average electricity load of the city.

The main steps are as follows:

Step1, parameters initialization: supposing harmony memory is made up of solution vector $\left\{ \mathbf{N}_i(N_{i1}, N_{i2}..., N_{in}) \mid \mathbf{N}_i \in R^n, i = 1,2,...S \right\}$, $\mathbf{N}_i^{best}(p_{i1}^{best}, p_{i2}^{best}..., p_{in}^{best})$ is the best iterative solution of $\mathbf{N}_i$.

$\mathbf{S}_{best}(s_1, s_2..., s_n)$ is the best solution in harmony memory. H_{size} is harmony memory size. The maximum iterative time is I_m, meanwhile, $PAR \in (0,1)$、W imply tuning parameter.

Step2, update all vectors in harmony memory utilizing pattern search formulations below:

$$\mathbf{N}_i^{k+1} = \mathbf{N}_i^k + h\mathbf{e}_i$$

$\mathbf{e}_i$ is an randomly generated unit vector, h ranges (0.03,0.2).

If $\rho(\mathbf{N}_i^{k+1}) < \rho(\mathbf{N}_i^k)$

$$\mathbf{N}_i^{best} = \mathbf{N}_i^{k+1}$$

$$\mathbf{N}_i^{k+1} = \mathbf{N}_i^k + \alpha(\mathbf{N}_i^{best} - \mathbf{N}_i^k)$$

If $\rho(\mathbf{S}_{best}) < \rho(\mathbf{N}_i^{best})$

$$\mathbf{S}_{best} = \mathbf{N}_i^{best}$$
$$k = 1, 2, ... I_m$$
$$i = 1, 2, ... S$$
$$0.25 < \alpha < 0.5$$

Step3, tuning harmony memory vectors: producing two random variables R and $R_{ti} \in (0,1)$. If $R_{ti} < PAR$,

$$\mathbf{N}_i^{k+1} = \mathbf{N}_i^k + R \cdot W$$

Otherwise,

$$\mathbf{N}_i^{k+1} = \mathbf{N}_i^k + R(\mathbf{N}_i^{best} - \mathbf{N}_i^k)$$

Step4，terminal condition：If I_m reaches certain amount the previously set，end the iterative process; or else return to step 2.

3. Numerical examples

In this section, the algorithm is tested by using an example distribution system model in which traditional coal-fired power plant (CF), photovoltaic cells (PV), wind power generation systems (WP) and fuel cells (FC) coexist to meet the electricity demand of different scale cities. The DGs' optimal placement is determined in order to reduce polluting gases emissions. Finally, we get the optimal number of the DGs in distribution network under the electricity demand of different scale cities.

The parameters for DGs and coal-fired power plant are shown in Table 1, and electricity demand of different scale cities is shown in Table 2.

Table 1. The parameters for DGs and coal-fired power plant (the numbers of average output power and polluting gases emissions are the numbers of units of the DGs and coal-fired power plant).

The types of DGs	Average output power	Polluting gases emissions
PV	1.2 kW	335 m³/year
WT	1.5 MW	123.5 m³/year
FC	200 kW	1350 m³/year
CF	60 MW	1490000 m³/year

Table 2. The electricity demand of different scale cities (based on a resident population of 2013).

City scale	Average power load	City scale	Average power load
100,000	95,000 kW	3,000,000	4,690 MW
500,000	650,000 kW	10,000,000	9,820 MW
1,000,000	1,570 MW	20,000,000	17,600 MW

Fig.1 shows polluting gases emissions of distribution network with DGs are less than traditional coal-fired power plant distribution network. From the figure, it can be found that introduce of DGs is beneficial for the environment. The calculation results of the DGs' optimal numbers under the electricity demand of different scale cities are shown in table 3.

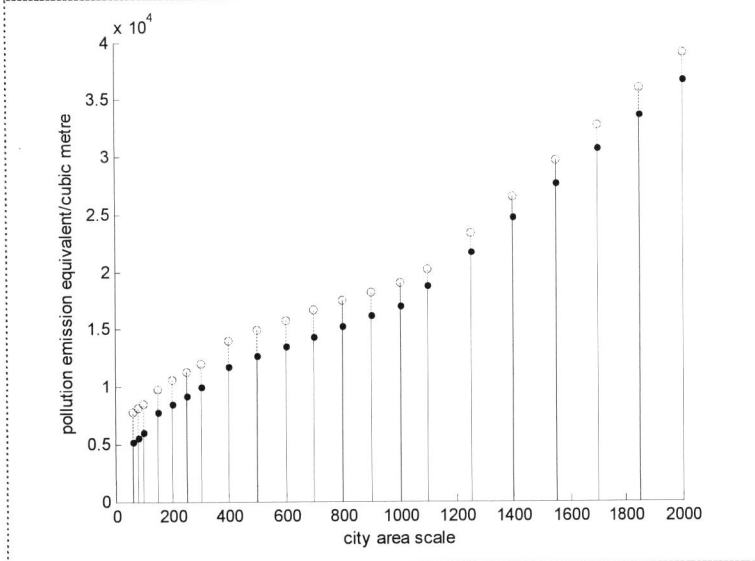

Figure 1. Comparison of polluting gases emissions of distribution network with and without DGs. (Red circle represents original pollution emission when power grid just contains coal-fired power plant, and black dot represents optimized pollution emission when power grid contains both coal-fired power plant and distributed power sources

Table 3. The calculation results of the DGs' optimal number under the electricity demand of different scale cities

City scale	PV	WT	FC	CF
20,000,000	25,005	1,590	430	246
10,000,000	16,050	803	201	120
3,000,000	6,015	397	69	65
1,000,000	3,105	165	35	31
500,000	1,700	103	19	22

4. Conclusion

In this paper, a new distribution network planning method based on an enhanced harmony search algorithm to dispatch DGs has been illustrated. It is found that introduce of DGs is beneficial for the environment. The planner counts on the distributed generators to compensate for the conventional power stations in preference to reducing polluting gases emissions. The proposed scheme is quite useful for the network planning, which can obtain the optimal numbers of DGs to achieve the minimum polluting gases emissions under the electricity demand of different scale cities.

References

[1] Rosehart W, Nowicki E 2002 Optimal placement of distributed generation. *In Proceedings of the 14th Power Systems Computation Conference, Sevilla, 24-28 June 2002*, session 11, paper 2, 5p

[2] Celli G, Ghiani E, Mocci S, Pilo F 2005 A multi-objective evolutionary algorithm for the sizing and the sitting of distributed generation. *IEEE Trans. Power Syst.* **20(2)** 750

[3] Kim KH, Song KB, Joo SK, et al 2008 Multi-objective distributed generation placement using fuzzy goal programming with genetic algorithm. *Eur. Trans. Elect. Power* **18(3)** 217

[4] Singh RK, Goswami SK 2009 Optimum siting and sizing of distributed generations in radial and networked systems. *Elect. Power Compon. Syst.* **37(2)** 127

[5] Das D 2006 A fuzzy multi-objective approach for network reconfiguration of distribution systems. *IEEE Trans. Power Del.* **21(1)** 202

[6] Borges CLT, Falcao, DM 2003 Impact of distributed generation allocation and sizing on reliability, losses, and voltage profile. *IEEE Bologna Power Tech Conference Proceedings* **2** 23

[7] Wang C, Nehrir MH 2004 Analytical approaches for optimal placement of distributed generation sources in power systems. *IEEE Trans. Power Syst.* **19(4)** 2068

Self-adaptive control of rotor inertia for virtual synchronous generator in an isolated microgrid

Hanghang Zeng[1*], Hongsheng Su[1]

[1]School of Automation & Electrical Engineering, Lanzhou Jiaotong University, Lanzhou, Gansu, 730070, China

*Corresponding author's e-mail: 1252524667@qq.com

Abstract. As an effective way to solve distributed generation, microgrid is getting increasingly extensive development. However, most of distributed generators are connected to the microgrid through power electronic converters, which makes the whole system present a small inertial network and the stability of the system face severe challenges. Virtual synchronous generator (VSG) technology has been widely applied in enhancing system stability because it enables inverters to mimic the outer characteristics of synchronous generator. Subsequently, aiming at the deficiency of traditional VSG control based on fixed rotor inertia, a self-adaptive control of rotor inertia for VSG is proposed in this paper. This method can change the inertia of the system and improve the stability of the system by adjusting the virtual rotor inertia adaptively. Finally, the effectiveness of the proposed control method is validated in MATLAB/Simulink environment.

1. Introduction

With the increasing energy crisis and environmental pressure, distributed energy, such as wind energy and solar energy, has attracted increasingly attention[1]. As an effective carrier of integrated distributed generation, microgrid connects distributed generation into the system through power electronic converters. These power electronic devices respond quickly, almost no inertia, and do not participate in frequency modulation and voltage regulation of the system[2]. Therefore, how to improve the stability of the microgrid has become an urgent problem.

In traditional power grids, the rotor of synchronous generator can contain a lot of kinetic energy owing to its mechanical rotational inertia. When disturbance occurs in the power grid, the rotor kinetic energy can be used to exchange energy with the power grid to provide greater inertial support for the system and maintain the stability of the system[3]. It is of great significance to enhance the stability of microgrid if we can learn from the operation experience of conventional grid and make grid-connected inverters mimic the operating characteristics of synchronous generator. For this reason, some scholars put forward the concept of VSG[4-5]. The basic idea of VSG is to make grid-connected inverters mimic the inertia, primary frequency modulation and primary voltage regulation characteristics of synchronous generator through control strategy[6].

Over the years, research on VSG control has attracted wide attention and development. Paper [7-8] refer to mechanical and electromagnetic equation of synchronous generator to control grid-connected inverter, which makes the inverter can match the synchronous generator in mechanism and external characteristics. This kind of control is called virtual synchronous generator technology, which is expected to play an important role in the future active distribution network and microgrid. In [9], the local linearization model of synchronous generator is introduced into traditional active power-

Content from this work may be used under the terms of the Creative Commons Attribution 3.0 licence. Any further distribution of this work must maintain attribution to the author(s) and the title of the work, journal citation and DOI.

Published under licence by IOP Publishing Ltd

frequency droop control, which mimics the inertia, damping property and primary frequency modulation of synchronous generator on the basis of droop control. A small signal model of microgrid based on VSG and droop control is established in [10], by comparing and analyzing the transient response of two models, it is concluded that VSG control not only has the steady-state effect of droop control, but also can provide additional virtual inertia to improve the dynamic stability. An adaptive inertial control strategy is proposed in [11-12], different rotor inertia is selected according to the acceleration and slip of VSG, which can not be realized by traditional synchronous generator.

Compared with the fixed rotor inertia of traditional synchronous generator, the virtual rotor inertia of VSG can be selected adaptively according to the actual situation. In this paper, a self-adaptive control of rotor inertia for VSG is proposed to ensure good dynamic performance and stability of the system. Finally, the effectiveness of the proposed control method is verified by simulation.

2. Basic principle of the VSG

The essence of VSG is to make the inverters mimic outer characteristics of synchronous generator by control strategy, which mainly includes the main circuit and control system. Fig 1 shows the basic topology of VSG.

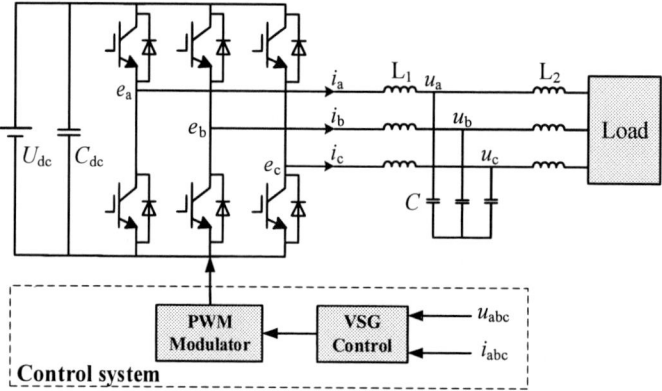

Fig 1. Basic topology of VSG

The current research principally focuses on the classical second-order model, including the electromagnetic part and the mechanical part. The electromagnetic part is modelled by the stator electrical equation as follows:

$$L\frac{\mathrm{d}i_{abc}}{\mathrm{d}t}=e_{abc}-u_{abc}-Ri_{abc} \tag{1}$$

Equation (1) mainly considers the voltage-current relationship of the stator circuit, but does not reflect its flux linkage and inherent electromagnetic features. The electromagnetic model of synchronous inverter proposed by Professor Qingchang Zhong in [9] fully considers the electromechanical and transient characteristics of synchronous generator, enhances the coupling between virtual stator and rotor, and reflects the characteristics of synchronous generator better. The electric and flux equations between the stator and rotor as follows:

$$e_{abc}=M_{f}i_{f}\dot{\varphi}A-M_{f}\frac{\mathrm{d}i_{f}}{\mathrm{d}t}B \tag{2}$$

$$A=\begin{bmatrix} \sin\varphi & \sin(\varphi-2\pi/3) & \sin(\varphi-4\pi/3) \end{bmatrix}^{T} \tag{3}$$

$$B=\begin{bmatrix} \cos\varphi & \cos(\varphi-2\pi/3) & \cos(\varphi-4\pi/3) \end{bmatrix}^{T} \tag{4}$$

where M_f is mutual inductance, i_f is excitation current, and φ is rotor angle.

The mechanical part is modelled by the rotor motion equation as follows:

$$\begin{cases} J\dfrac{d\omega}{dt}=T_{\mathrm{m}}-T_{\mathrm{e}}-D\left(\omega-\omega_0\right) \\ d\theta/dt=\omega \end{cases} \tag{5}$$

where J is the rotor inertia of VSG, T_{m} and T_{e} are mechanical torque and electromagnetic torque respectively, ω and ω_0 are the actual angular velocity and rated angular velocity respectively and D is the damping coefficient. Equation (5) shows that J provides inertial support for power and frequency dynamic processes of the system, while D provides the ability to damp power oscillations.

3. VSG control

3.1. Traditional VSG control

VSG control is divided into two parts: active power-frequency control and reactive power-voltage control. Active power-frequency control is actually to mimic the speed governor of synchronous generator, which is used to characterize the droop feature of active power and frequency. Fig 2 is an active power-frequency control block diagram.

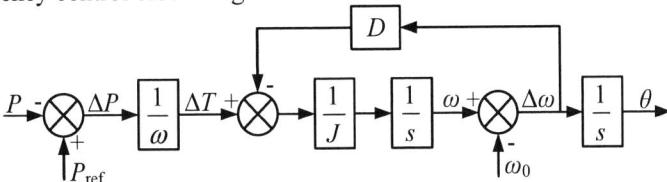

Fig 2. Active power and frequency control of VSG

Where P and P_{ref} are measured and rated values of active power respectively and θ is the phase angle of output voltage. As can be seen from Fig 2, the difference of active power determines the difference of torque, thus controlling the change of frequency. The relationship between $\Delta\omega$ and ΔT is expressed by parameters J and D. These two parameters determine the performance of VSG controller.

Reactive power-voltage control is actually to mimic the excitation regulation function of synchronous generator, which is used to characterize the droop feature of reactive power and voltage amplitude. Fig 3 is a reactive power-voltage control block diagram.

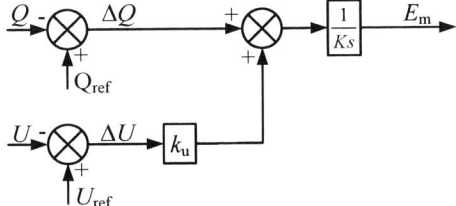

Fig 3. Reactive power and voltage control of VSG

Where Q and Q_{ref} are measured and rated values of reactive power respectively, and E_{m} is the amplitude of output voltage. As can be seen from Fig 3, reactive power-voltage control adjusts the amplitude of the output voltage by the reactive power deviation and the voltage deviation, and k_{u} determines the voltage regulation capability of VSG.

3.2. Self-adaptive control of rotor inertia for VSG

According to equation (5), when the rotor inertia J is a larger value, the change rate of angular velocity in transient process can be slowed down. Moreover, when the angular velocity ω deviates from the rated value after disturbance, the saltation of system frequency can be avoided and the dynamic stability of the system can be improved. When the rotor inertia J is a smaller value, the response of the system is fast. When ω is restored to the rated value, the transition time can be reduced, which is beneficial to the system stability. In order to synthesize the both advantages, we want rotor inertia J to be larger when the system angular velocity ω deviates from the rated value, and smaller when ω restores to the rated value.

Based on the above analysis, a self-adaptive control of rotor inertia for VSG is proposed in this paper, in which the value of rotor inertia J change adaptive by the following equation:

$$J = J_0 + k\left(\omega - \omega_0\right)\frac{d\omega}{dt} \tag{6}$$

Where J_0 is the initial value of virtual rotor inertia in steady state, and k is a constant greater than 0 indicating the accommodation coefficient. From equation (6), when $\Delta\omega = \omega - \omega_0$ and $d\omega/dt$ have same sign, it means that the system angular velocity ω is gradually deviating from the rated value, so it is necessary to increase the value of J. On the contrary, when $\Delta\omega$ and $d\omega/dt$ have opposite sign, it means that the ω is gradually restoring to the rated value and the value of J needs to be reduced.

Combining equation (6) and (5), the rotor motion equation of the proposed control method can be derived:

$$\left(J_0 + k*\Delta\omega\frac{d\Delta\omega}{dt}\right)\frac{d\Delta\omega}{dt} = T_m - T_e - D\Delta\omega \tag{7}$$

Equation (7) is a quadratic equation of one variable with respect to $d\Delta\omega/dt$. Two solutions of the equation can be obtained by formula of root. According to the above analysis, it is known that the product of $\Delta\omega$ and $d\omega/dt$ could be positive or negative. Therefore, after discarding the root which does not satisfy the condition, the only solution of the equation as follow:

$$\frac{d\Delta\omega}{dt} = \frac{2\left(\Delta T - D*\Delta\omega\right)}{J_0 + \sqrt{J_0^2 + 4k*\Delta\omega\left(\Delta T - D*\Delta\omega\right)}} \tag{8}$$

Where $\Delta T = T_m - T_e$. Since the value of $d\omega/dt$ is equal to $d\Delta\omega/dt$, the solution of $d\Delta\omega/dt$ can be brought into equation (6) to eliminate the differential term in the proposed control method, which avoids the influence of system noise on the control and is conducive to the system stability. Fig 4 is the control block diagram of the control method proposed in this paper.

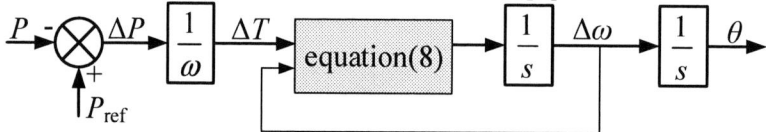

Fig 4. Self-adaptive control of rotor inertia for VSG

4. Simulation Studies

To verify the effectiveness of the self-adaptive control of rotor inertia for VSG proposed in this paper, a simulation experiment was carried out based on MATLAB/Simulink software platform. The main parameters of the system are shown in table 1.

Table 1. System Parameters

Parameter	Symbol	Value
Rated frequency	f_0	50Hz
Voltage reference	U_{ref}	311V
Input voltage	U_{dc}	800V
Active power reference	P_{ref}	10kW
Switching frequency	f_{sw}	10kHz
Initial rotor inertia	J_0	15kg·m^2

4.1. Simulation case 1

In order to analyze the response of the control method proposed in this paper when load changes, the load increases from 10kW to 12kW at 0.5s and restores to the initial value at 2.5s.

Fig 5 (a) shows the variation curve of the system frequency under different rotor inertia. In the figure, J_1=2 indicates that the rotor inertia J takes a small value, J_3=15 indicates that J take a large value, and J_2 represents J with self-adaptive control proposed in this paper. In these three cases, the value of D is fixed. Fig 5 (b) is the curve of the rotor inertia J corresponding to the control method proposed in this paper during load variation.

As can be seen from Fig 5, when the load varies, the system frequency decreases first and then returns to the initial stable state. Compared with taking a fixed value for J, the rotor inertia J with self-adaptive control is larger than the initial value J_0 in the process of the system frequency f reducing and deviating from the rated value, and smaller than the initial value J_0 in the process of f rising to the rated value.

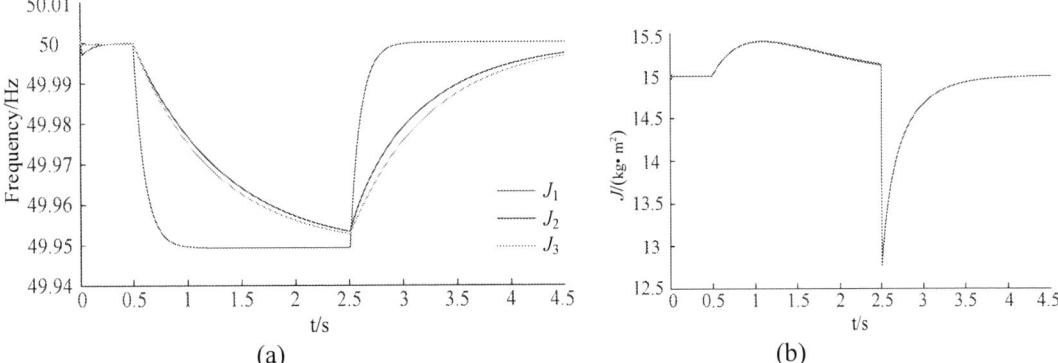

(a)   (b)

Fig 5. Simulation results of frequency and rotor inertia under load variation

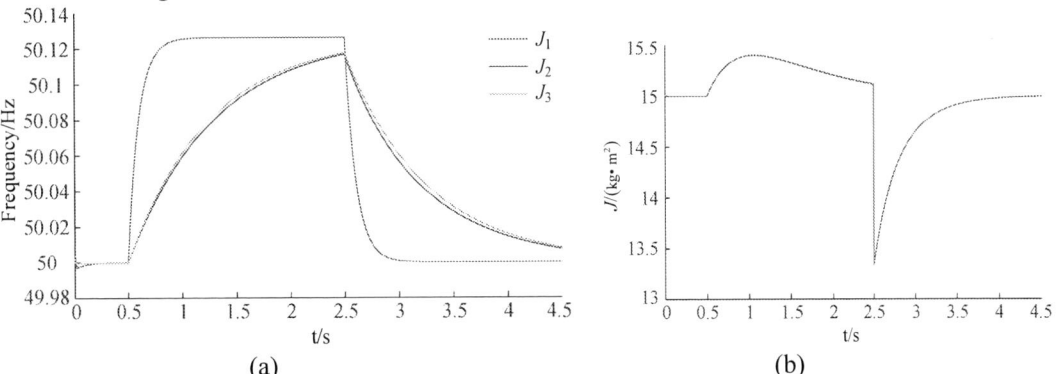

(a)   (b)

Fig 6. Simulation results of frequency and rotor inertia when active power reference changes

4.2. Simulation case 2

Similarly, in order to analyze the response of the proposed control method under step change of the active power reference, the P_{ref} varies from 10kW to 15kW at 0.5s, and then restores to the initial value at 2.5s. Fig 6 are variation curves of the corresponding frequency and rotor inertia J.

As can be seen from Fig 6, when the active power step varies, the system frequency rises first and then returns to the initial stable state. With the control method proposed in this paper, the rotor inertia J with self-adaptive control is larger than the initial value J_0 in the process of the system frequency f rising and deviating from rated value, and smaller than the initial value J_0 in the process of f reducing to the rated value.

5. Conclusion

This paper has proposed a self-adaptive control of rotor inertia for VSG on the basis of the conventional VSG control strategy, which can improve the frequency stability in an islanded microgrid. This control method regulates the value of rotor inertia J adaptively and simultaneously according to the relationship between the angular velocity ω and its deviation and variation rate. At last, Simulation studies in MATLAB/Simulink indicates that the trend of change of angular velocity is more reasonable and practical under disturbance, thus the system stability has a further enhancement.

Acknowledgments

The author would like to thank my mentor, Professor Hongsheng Su, for his careful guidance and kind care. And we thank Lanzhou Jiaotong University for its strong support. The simulation platform is provided by School of Automation & Electrical Engineering.

References

[1] Yang, X., Song, Y., Wang, G., Wang, W. (2010) A comprehensive review on the development of sustainable energy strategy and implementation in china. IEEE Transactions on Sustainable Energy, 1(2): 57-65.

[2] LÜ, Z.P., Sheng, W.X., Zhong, Q.C., Liu, H.T., Zeng, Z., Yang, L., Liu, L. (2014) Virtual Synchronous Generator and Its Applications in Micro-grid. Proceedings of the CSEE, 34(16): 2591-2603.

[3] Cheng, C., Yang, H., Zeng, Z., Tang, S.Q., Zhao, R.X. (2015) Rotor Inertia Adaptive Control Method of VSG. Automation of Electric Power Systems, 39(19): 82-89.

[4] Driesen, J., Visscher, K. (2008) Virtual synchronous generators. In: IEEE Power and Energy Society General Meeting - Conversion and Delivery of Electrical Energy in the 21st Century. Pittsburgh. pp. 1-3.

[5] Zheng, T.W., Chen, L.J., Chen, T.Y., Mei, S.W. (2015) Review and Prospect of Virtual Synchronous Generator Technologies. Automation of Electric Power Systems, 39(21): 165-175.

[6] Wu, H., Ruan, X.B., Yang, D.S., Chen, X.R., Zhong, Q.C., LU, Z.P. (2015) Modeling of the Power Loop and Parameter Design of Virtual Synchronous Generators. Proceedings of the CSEE, 35(24): 6508-6518.

[7] Chen, Y., Hesse, R., Turschner, D., Beck, H.P. (2011) Improving the grid power quality using virtual synchronous machines. In: International Conference on Power Engineering, Energy and Electrical Drives. Malaga. pp. 1-6.

[8] Zhong, Q.C., Weiss, G. (2011) Synchronverters: inverters that mimic synchronous generators. IEEE Transactions on Industrial Electronics, 58(4): 1259-1267.

[9] Du, W., Jiang, Q.R., Chen, J.R. (2011) Frequency Control Strategy of Distributed Generations Based on Virtual Inertia in a Microgrid. Automation of Electric Power Systems, 35(23): 26–31+36.

[10] Liu, J., Miura, Y., Ise, T. (2015) Comparison of dynamic characteristics between virtual synchronous generator and droop control in inverter-based distributed generators. IEEE Transactions on Power Electronics, 31(5): 3600-3611.

[11] Alipoor, J., Miura, Y., Ise, T. (2013) Distributed generation grid integration using virtual synchronous generator with adoptive virtual inertia. In: IEEE Energy Conversion Congress and Exposition. Denver. pp. 4546-4552.

[12] Alipoor, J., Miura, Y., Ise, T. (2015) Power system stabilization using virtual synchronous generator with alternating moment of inertia. IEEE Journal of Emerging & Selected Topics in Power Electronics, 3(2): 451-458.

Software Consistency Checking Method for Distribution Terminal based on Chaotic Map

Yaokun WANG, Ying LIANG, Changkai SHI and Shilei GUAN

(China Electric Power Research Institute, Haidian District, Beijing, 100192 China)

Corresponding author's e-mail: wangyaokunvip@sina.com

Abstract: Distribution terminal system software is the core carrier of its function realization and performance stability. In the process of development and operation and maintenance of distribution terminals, software versions are confusing and inconsistent with the archived versions due to technical upgrades, version management, and regulatory deficiencies. It makes some functions unable to operate normally, performance indicators can not be met, equipment on-line rate decreased, and seriously affecting the operation and maintenance and practicality of distribution automation system. In order to strengthen the software version control of distribution automation terminal equipment, the feature information of software version was extracted by analysing the executable files of terminal software, and then the priority was divided according to the feature information. The MD5 algorithm and CRC32 checking algorithm were used to calculate the comparison factor, which then was encrypted by using pseudo random code generated by Logistic chaotic map. The model and method of software version consistency detection for distribution terminal were established.

1. Introduction

Distribution terminal is the key equipment to improve the level of distribution automation, and plays an important role in the operation and maintenance management of distribution network [1-2]. At present, there are many manufacturers of distribution terminal equipment, but the manufacturing level is uneven, and the quality of products is quite different. In the process of development and operation and maintenance of distribution terminals, the inconsistency between the actual software version and the archived version and the arbitrary modification of the software due to technical upgrade, version management, lack of supervision and other reasons make some functions unable to operate normally and performance indicators can not be met, adversely affecting the practicability of distribution automation. According to statistics, more than 10% of the distribution terminals are offline for more than 7 days per month due to software failure. Therefore, in order to effectively ensure the practicality of distribution automation system, it is particularly important to strengthen the consistency detection and control of software version of distribution automation terminal equipment and prevent software from arbitrary modification.

At present, a more complete consistency detection method has been developed in the field of software version control abroad. Comparatively speaking, domestic is not mature enough in this field, but has also made some achievements. Digital signature, CRC verification, MD5 verification and other technologies have been used to a certain extent. However, most of the existing technologies are based on the consistency comparison model of the terminal software source code, and depend too much on the software development platform, programming language and operating system [3]. In practice, the terminal manufacturer is reluctant to disclose its source code for the sake of technical secrecy, which

---

Content from this work may be used under the terms of the Creative Commons Attribution 3.0 licence. Any further distribution of this work must maintain attribution to the author(s) and the title of the work, journal citation and DOI.

Published under licence by IOP Publishing Ltd

makes the consistency comparison model based on source code less practical. In addition, due to the high security and reliability requirements of the power industry, there is no special version consistency testing method for distribution terminal software.

Therefore, on the basis of meeting the above requirements, this paper designed a version consistency detection method of distribution terminal software based on chaotic encryption algorithm..

2. HASH algorithm and chaotic encryption

2.1 MD5 algorithm and CRC check

HASH algorithm, also known as hash function, is widely used in software consistency comparison. The algorithm can map any file to a fixed length hash value, that is, if the two hash values are different, their corresponding original files are different, so we can judge whether the file has been modified by comparing the hash values. MD5 message digest algorithm is the most widely used hash function, which maps any file to a 128-bit (16-byte) hash value, and the process is irreversible [4].

However, theoretical analysis shows that the hash value and the original file are not one-to-one correspondence, different original files may generate the same hash value, this situation is called collision. The probability of collision is related to the specific HASH algorithm. For example, MD5 algorithm calculates the original file and generates 128bit hash value, so there are only $2 \wedge 128$ different hash values, and the original file is infinite in theory, so using hash value to judge the file, inconsistent conclusions are certain, and consistent conclusions are equivocal.

In order to reduce the collision rate of MD5 algorithm, MD5+CRC32 check is used to check the key information. In this paper, CRC checking is converted into binary system by serializing the data splicing which will be checked. CRC32 (n=32) checking code is calculated and saved. The data which will be checked is converted into binary system after serializing the same splicing. The 32-bit CRC checking code is then spliced at the end of the splicing, and the cyclic division is performed. Check whether the check is correct according to whether the remainder is 0.

2.2 Chaotic encryption

Chaos is a deterministic, stochastic-like process with divergence and non-periodicity, which is very sensitive to system parameters and initial values [5].

Chaotic mapping can obtain pseudo-random sequence or pseudo-random code similar to random sequence, which is called chaotic sequence. The correlation of chaotic sequences is very weak and has good white noise characteristics. The generation process of chaotic sequence is very simple. In theory, a long chaotic sequence can be obtained by iteration of chaotic mapping, and the long-term evolution result of chaotic sequence is unpredictable [6]. This makes chaotic system of great cryptographic value, and is very suitable for sequence information encryption. The application of chaotic system to encryption can simplify the encryption process [7].

Chaotic encryption is to encrypt plaintext by using chaotic sequence generated by chaotic system. The encryption process is reversible, and the plaintext signal is extracted by the same chaotic system[8]. The general process of chaotic encryption is shown in the Figure 1.

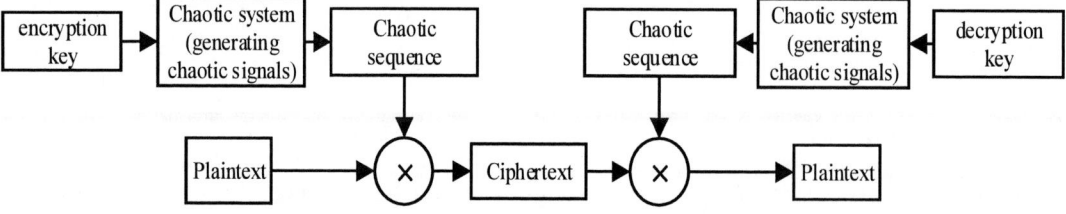

Figure 1 Chaotic Encryption and Decryption Process

Logistic mapping is a kind of classical mapping in chaotic system, and it is a dynamical system originated from population statistics[9]. The equation is:

$$X_{n+1} = \mu X_n \left(1 - X_n\right)$$

The two parameters that affect the equation are the initial value X0 and the system parameter μ. The results show that when the following two conditions are satisfied, the Logistic function works in a chaotic state, and the resulting sequence is nonperiodic, divergent and pseudo-random.

（1）$3.5699456 < \mu \leqslant 4$；

（2）$0 < X_0 < 1$。

3. Distribution terminal software version consistency model

3.1 Software feature information

Usually, the source code of the terminal software can not be obtained when testing the version of the terminal software, so it is necessary to obtain the characteristic information about the software version from the executable file of the terminal running. Select the file name, content, size, file modification time, release date and release / change description as the feature information of the software version.

Among them, the size and content of the software can directly reflect the state of the source code of the software, if the source code changed, the size and content will certainly change. The name, modification time and release date of the software reflect the archive information of the software, which is an important external feature of the software. Release / change description does not directly reflect the source code status of the software, and this feature information is only used to prompt the field acceptor of software changes and modifications, and to remind some installation and debugging considerations.

Because the consistency of terminal software is represented differently by different feature information, it can be prioritized according to the importance of feature information, which can be divided into three levels: Key information, Important information and General information. The degree of importance ranges from high to low in order of K, I and G. K-level information can directly characterize the consistency of software versions, that is, if K-level information does not match, the software has been modified; I-level information can not directly characterize the consistency of software versions, but as an important information, when I-level information does not match, alarms should be generated and confirmed by the detector; when G-level information does not match, there is no alarm, but will prompt the scene installer that G level information has changed.

3.2 Software version consistency model

3.2.1 Consistency comparison factor

Because of the large number of software version feature information and the diversity of form and content, it is not conducive to direct comparison. Therefore, HASH algorithm is considered to calculate the hash value of software feature information, and the calculated hash value is used as the comparison factor of software version detection.

In order to simplify the processing, multiple feature information of the same priority is spliced in a specific order after obtaining the corresponding hash values by corresponding algorithms. The importance level corresponding to each feature information and the encryption algorithm used are shown in the Table.1.

Table.1 Priority and Encryption of Feature Information

Serial number	Name	priority	encryption algorithm
1	name	I	MD5
2	content	K	MD5+CRC32
3	Size	K	MD5+CRC32
4	Modification time	I	MD5
5	Release date	I	MD5
6	Release / change description	G	CRC32

According to the length of the output string of the encryption algorithm, MD5 outputs 16 bytes (128 bits) and CRC32 outputs 4 bytes (32 bits). Therefore, three different hash values of different lengths are obtained by calculation: K-level hash value 40 (16+4+16+4) bytes, I-level hash value 48 (16+16+16) bytes, and G-level hash value 4 bytes. Hash values are used as comparison factors in software version consistency checking. By comparing these hash values, we can judge whether the software has changed.

3.2.2 Consistency ratio factor encryption

After the distribution terminal software passes the test, the version consistency comparison model will be sent to the terminal manufacturer for comparison in site acceptance. However, it is not advisable to send the consistency comparison factor directly to the manufacturer. In order to prevent the comparison factor from being tampered illegally and the errors occurring in the transmission process, it is necessary to encrypt the generated consistency comparison factor.

Using the Logistic chaotic map, the initial value X0 and the system parameter μ are determined, and the chaotic sequence $\left(X_1 \text{、} X_2 \text{、} X_3 \text{L} \right)$ is generated. The encrypted sequence with a length of 40 is selected from the Mth data and is marked as $M = \left(X_M \text{、} X_{M+1} \ X_{M+2} \text{L} \ X_{M+39} \right)$. The encrypted sequence with length 48 is selected as $N = \left(X_N \text{、} X_{N+1} \ X_{N+2} \text{L} \ X_{N+47} \right)$ from the Nth data, and the encrypted sequence with length 4 is selected as $L = \left(X_L \text{、} X_{L+1} \text{、} X_{L+2} \text{、} X_{L+3} \right)$ from the Lth data. The five parameters of X0, M, N and L are saved as Logistic encryption keys.

Each element in the three encryption sequences of M, N and L is represented as finite precision binary numbers. The formula is as follows:

$$ a = \sum_{n=0}^{m-1} 2^{-(n+1)} b_n $$

In the formula, a is a decimal encryption sequence; b_n is binary data; m is binary data's length.

In this paper, the first 8-bit binary number of b_n is chosen to represent the three encrypted sequences, which are transformed into 320 bit, 384 bit and 32 bit respectively.

The binary encryption sequence will be spliced and compared with the original comparison factor, and new data is obtained. The three new data sets are used as a consistency comparison model and sent to the manufacturer in two-dimensional code or some other form, which is the digital proof of the terminal software and used for comparison in the field acceptance. The specific process is shown in Figure 2.

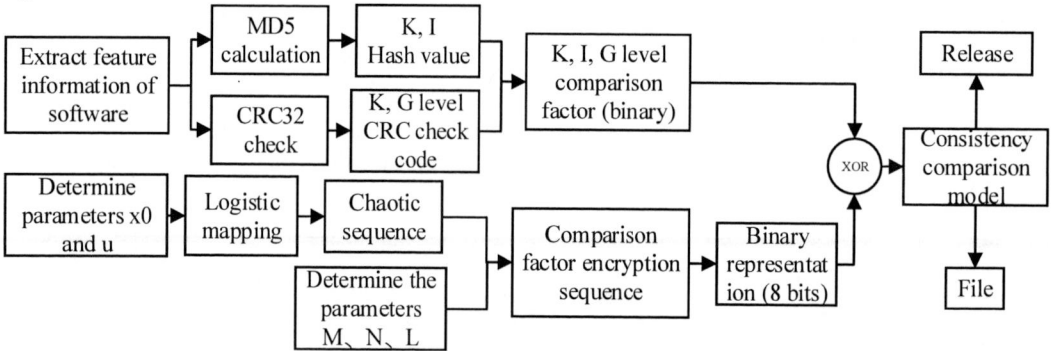

Figure 2 Comparison Factor Encryption Process

3.3 Software version consistency comparison method

According to the process of distribution terminal detection and field acceptance, a terminal field inspection and acceptance method based on the version consistency comparison model of terminal software is designed. The flow chart is shown in Figure 3.

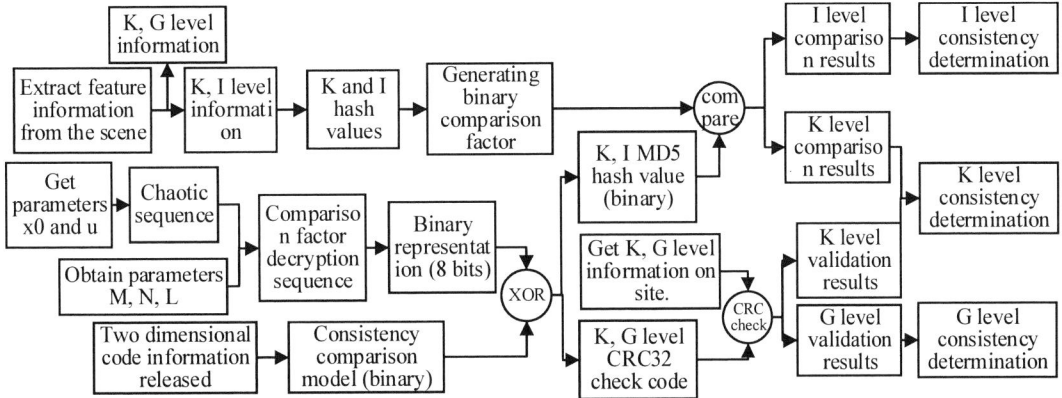

Figure 3 Field Acceptance Process

The on-site acceptance personnel respectively make corresponding operation according to the results of the conformance judgment. If K-level information is inconsistent, the terminal software is considered tampered with; if G-level information is inconsistent, it should be considered that the software has undergone general modification, and the field personnel should pay attention to its prompt information; if I-level information is inconsistent, it is considered that the software has undergone important modification, it should be submitted to the inspector to confirm whether the modification is legal. The I level information processing flow is shown in Figure 4.

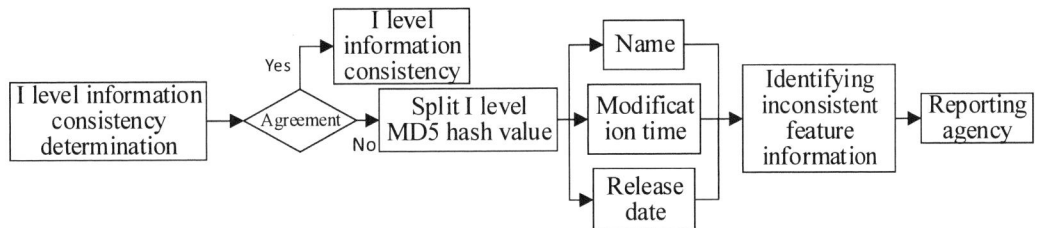

Figure 4 I Level Information Processing Flow

4. Summary

Based on the analysis of software status of distribution   automation terminal, the requirement of software consistency detection and control was put forward. The software feature information was extracted without knowing the source code, and the comparison factor based on MD5 algorithm and CRC32 was established according to the priority. In order to prevent illegal tampering with the comparison factor, a consistency comparison model of distribution terminal software was established by using chaotic encryption algorithm. Through analysis, the practicability of the model was verified. It is proved that the consistency detection method of terminal software designed in this paper plays an important role in improving the quality control capability of distribution terminal equipment, maintaining the consistency between the actual software version and the filed version, and preventing the illegal tampering of the software. The terminal equipment runs stably and promotes the practicability of distribution automation system.

References

[1]   Cong W, Sheng Y.R, Xian G.F. （2018） Distributed power recovery method based on intelligent distribution terminal.Automation of Electric Power Systems.,42(15):77-85.

[2] Han G.Z, Xu B.Y,SuoNanJiaLe.(2012) Realization of automatic discovery technology in distribution termina.Automation of Electric Power Systems.36(18):82-85.

[3] Lv Y.J, Xu W.J, Liu Y.(2016) Design of software recording and comparison system for smart meter based on cryptography technology. Power System Technology.40(11):3604-3608.

[4] Zhang Y.Z, Zhao Yi,Tang X.B. (2008)Research on MD5 algorithm,Computer Science.07:295-297.

[5] Zhang B. (2005)Nonlinear chaos in power electronic converters and its application.Transactions of China Electrotechnical Society.12:1-6+12.

[6] Thompson J M T,Stewart H B.(1986)Nonlinear dynamic and chaos.John Wiley& Sons Ltd.

[7] Jia H.J,Yu Y.X,Wang C.S. (2001)Chaotic phenomena in power system and related research Proceedings of the CSEE.07:27-31.

[8] Qiu S.S,Chen Y.F,Wu M. (2002)Some problems of chaotic secure communication and a new scheme of chaotic encryption. Journal of South China University of Technology(Natural Science Edition). 11:75-80.

[9] Wang D.S, Cao L. (1995)Chaos, Fractal and Its Application. China University of Science and Technology Press, Hefei.

Three-level Generalized Discontinuous Pulse-width Modulation Strategy Considering Neutral Point Potential Balance

WANG Jikang[1], LIU Jianzheng[1], WANG Yi[2], GUO Zhichao[3]

[1] Department of Electrical Engineering, Tsinghua University, Beijing 100084, China;

[2] Beijing Sifang Automation CO. LTD., Beijing 100085, China;

[3] Beijing Electric Power Automatic Equipment CO. LTD., Beijing 100044, China

Wangjikang95@163.com

Abstract. Traditional discontinuous pulse width modulation (DPWM) cannot be efficiently used in some inverters like active power filters (APFs) whose currents are difficult to predict. Therefore, this paper proposes a new generalized discontinuous pulse width modulation (GDPWM) strategy considering neutral point potential balance for neutral-point clamped (NPC) three-level inverters. The core of the proposed strategy is the optimal efficiency modulation strategy. The optimal efficiency modulation strategy determines the clamp mode in real time based on the current flowing through the converter. This strategy minimizes switching losses by clamping the converter to the phase with the largest absolute value of current. At the same time, according to the neutral point potential shift, the neutral point balance strategy and the optimal efficiency modulation strategy are switched to achieve the balance between the switching loss and the neutral point potential. The proposed strategy can be efficiently applied in NPC three-level APFs to minimize the switching losses without the need of predicting the currents and to make neutral point potential balance. In this paper, the proposed modulation strategy is theoretically analyzed, and simulation and experimental verification are carried out.

1. Introduction

The three-level topology has many advantages such as less output harmonics, easy to use for higher voltage conditions, and less filter inductance. It is widely used in APFs, photovoltaic power generation [1]~[2] and so on. As the voltage and current level of the grid-connected device increases, the switching loss also increases. To ensure the safety of the converter, the switching frequency of the converter is limited to a certain range. Thereby it is especially important to reduce switching losses and increase the equivalent switching frequency. DPWM is a modulation strategy that can be applied to three-level converters and effectively reduce switching losses [4]. Due to the many advantages of this modulation strategy, it is also widely used in APFs [3], inverter [4] and many other fields. This modulation strategy can reduce switching losses by reducing the frequency of actions per switching device while the equivalent switching frequency is constant.

Many scholars around the world have done a lot of research on traditional DPWM strategies. Many papers introduce a variety of different traditional three-level DPWM strategies. Literature [5] introduced its principle and implementation method. And Literature [6] compared the switching losses and harmonic characteristics of these different DPWM strategies. It can be seen that the common feature of the traditional DPWMs is that the clamp interval is fixed, and only related to the position of the voltage

Content from this work may be used under the terms of the Creative Commons Attribution 3.0 licence. Any further distribution of this work must maintain attribution to the author(s) and the title of the work, journal citation and DOI.

Published under licence by IOP Publishing Ltd

vector. The neutral point potential fluctuation problem is an inherent problem of the NPC three-level topology. In order to make the converter of this structure safe and stable, it is necessary to ensure that the neutral point potential balance [8]. Literature [9], [10] introduced a method to solve the three-level neutral point balance problem by two different DPWMs switching methods. There are few papers on how to achieve an optimal efficiency modulation strategy for the case where the current cannot be predicted. Only the literature [11] proposed a GDPWM method for the two-level APF to clamp in the maximum phase of the current, and similar modulation methods for the three-level converter have little research. In this paper, a GDPWM strategy considering the neutral point potential balance is proposed for the characteristics of three-level converters with difficult current prediction. The modulation strategy can well balance the control of the neutral point potential balance and the optimal efficiency modulation strategy. so that the converter achieves optimal efficiency within the allowable range of the neutral point potential fluctuation.

The three-level modulation method generally adopts space vector pulse width modulation (SVPWM) and carrier-based pulse width modulation (CBPWM). According to the literature [12], the two modulation methods are essentially the same. Since SVPWM is more intuitive in describing the modulation methods described herein, SVPWM is used herein to describe the modulation strategy.

2. The basic principle of three-level DPWM

2.1. NPC three-level converter topology and switch vector

The NPC three-level converter topology is shown in Figure 1. As can be seen from the figure, the converter topology consists of four switching tubes with anti-parallel freewheeling diodes and two clamping diodes per phase. Each phase can output three levels, and the phase A is used as an example to define the following. When S_{a1} and S_{a2} are turned on and the output level is $+U_{dc1}$, the switching function S_A is 2; When S_{a2} and S_{a3} are turned on and the output level is 0 , the switching function S_A is 1; When S_{a3} and S_{a4} are turned on and the output level is $-U_{dc2}$, the switching function S_A is 0; Similarly, the definitions of B and C phase switching functions S_B and S_C can be obtained.

For a three-level converter topology, there are three switching states for each phase, and a total of 27 switching states for three phases. Analysis of Figure 2 shows that 27 switch states correspond to 19 switch vectors.

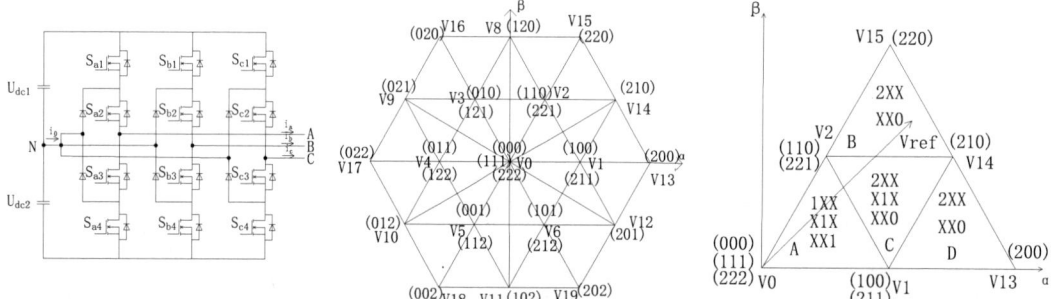

Figure 1. Topology of NPC three-level inverter

Figure 2. Space vector of three-level inverter

Figure 3. Clamp figure of three-level DPWM in sector 1

2.2. The basic principle of three-level DPWM

According to the nearest three vector principle of the three-level space vector modulation method [12]. The DPWM strategy is to select one of the two redundant switching states of the small vector so that one phase is clamped to a certain level during each switching cycle. Compared with continuous PWM, the purpose of reducing switching losses at the same equivalent switching frequency is achieved. However, since there is no complementary small vector complement, the neutral point potential fluctuation is caused. According to this principle, different clampable ways in each small area can be obtained as shown in Figure 3.

3. Three-level neutral point potential balance GDPWM strategy

3.1. neutral point potential balance principle

The neutral point potential balance problem is an inherent problem of the NPC three-level structure. Inconsistent switching characteristics, inconsistent circuit parameters such as capacitors will cause fluctuations in the neutral point potential. The traditional three-level DPWM method does not consider the problem of neutral point potential balance. The neutral point potential imbalance is further exacerbated by the fact that one phase level is clamped and a small vector that compensates for each other cannot be used.

For any phase bridge wall, when the output is 0 level, the current flows into or out of the neutral point of the DC voltage, that is, the neutral point of the two capacitors. This will cause the neutral point potential to shift. The DC neutral point current i_0 is shown as the formula (1):

$$i_0 = i_a \left(1 - |S_a - 1|\right) + i_b \left(1 - |S_b - 1|\right) + i_c \left(1 - |S_c - 1|\right)$$

(1)

Among them, S_a, S_b, and S_c are respectively ABC three-phase switch states; i_a, i_b, and i_c are respectively the three-phase outflow current of the converter ABC.The offset of the neutral potential (ΔU_{dc}) after a switching period T_s from time t is as shown in formula (2):

$$\Delta U_{dc} = \frac{1}{C} \int_t^{t+T_s} i_0(t)dt$$

(2)

$$\Delta U_{dcn} = \frac{1}{C} \int_{T_n} i_a(t)\left(1 - |S_{an} - 1|\right) + i_b(t)\left(1 - |S_{bn} - 1|\right) + i_c(t)\left(1 - |S_{cn} - 1|\right)dt$$

(3)

In summary, (1) and (2) can obtain the change of the neutral point potential in the action time of a certain switch vector as shown in formula (3), where T_n (n can take 1, 2, 3) is the action time of the nth switch vector. And the neutral point potential change in one switching cycle is as shown in formula (4).

$$\Delta U_{dcT_s} = \Delta U_{dc1} + \Delta U_{dc2} + \Delta U_{dc3}$$

(4)

For large vectors, for example, (2,0,0), with (1), the inflow neutral point current is 0. Other large vector inflow neutral point currents are equally available, and large vectors have no effect on the neutral point potential. For the medium vector, for example, (2,1,0) brings in (1), the current flowing into the neutral point is i_b. Other medium vector inflow neutral point currents are equally available. Since the medium vector has no redundant switching state, its influence on the neutral point potential is related to the current. For small vector two redundant switch states, such as (1,0,0), (2,1,1) brought in (1), there are inflow neutral point currents $-i_a$ and i_a respectively. Similarly, the inflow neutral point current of other small vectors can be obtained. The switching states of the small vectors that are redundant with each other have the opposite effect on the neutral point potential.

The following analysis analyzes the influence of the DPWM strategy on the neutral point potential of different switching states in the same region, taking the first sector A region as an example. As shown in Figure 3, There are three clamp modes in Zone A, namely 1XX, X1X and XX1. The switching states of the three clamp modes are respectively brought into (1), (3), and (4), and the influence of the neutral point potential in each clamp mode can be derived. The switching vector used in each clamp mode and its effect on the neutral point potential are shown in Table 1, where T_x, T_y and T_z are the action times of the x, y, and z vectors, respectively. It can be seen from Table 1 that different clamp modes have different effects on the neutral point potential, and this characteristic can be used to formulate a modulation strategy to ensure the neutral point potential balance.

3.2. neutral point potential balanced DPWM modulation strategy

It can be seen from Table 1 that the effect of different clamping modes on the neutral point potential is only related to the integral value of the associated current. Since the time of each switching cycle is very short, we use the current and time product to approximate the integral. The definition of the neutral-point impact factor (NPIF) is shown in formula (5). ΔU_{dcT} can be converted to NPIF as shown in Table 1.

$$NPIF = \left(U_{dc1} - U_{dc2}\right)\left(C \bullet \Delta U_{dcT_s}\right) \tag{5}$$

Table 1 Neutral potential offset of different clamp strategy in A region

Clamp mode	ΔU_{dcT_s}	NPIF
1XX	$\frac{1}{C}\left(-\int_{t_x} i_a dt + \int_{t_y} i_c dt\right)$	$\left(U_{dc1} - U_{dc2}\right)\left(-\bar{i}_a \bullet t_x + \bar{i}_c \bullet t_y\right)$
X1X	$\frac{1}{C}\left(\int_{t_x} i_a dt + \int_{t_y} i_c dt\right)$	$\left(U_{dc1} - U_{dc2}\right)\left(\bar{i}_a \bullet t_x + \bar{i}_c \bullet t_y\right)$
XX1	$\frac{1}{C}\left(\int_{t_x} i_a dt - \int_{t_y} i_c dt\right)$	$\left(U_{dc1} - U_{dc2}\right)\left(\bar{i}_a \bullet t_x - \bar{i}_c \bullet t_y\right)$

Where $\bar{i}_a$, $\bar{i}_b$, $\bar{i}_c$ is the algebraic average of the sampling current and the reference current. Compare the three NPIFs to get the maximum of them. The clamp mode with maximum NPIF is optimal, and the switching state of the mode is selected as the switching vector.

4. Three-level optimal efficiency GDPWM modulation strategy

The core of the modulation strategy proposed in this paper is the optimal efficiency GDPWM strategy. The difference between this modulation strategy and the traditional DPWM strategy is that this modulation strategy can determine the clamp mode in real time according to the current sampling under the premise that the current is unpredictable, so that the switch clamps as much as possible in the phase with the largest absolute value of the current. This reduces switching losses.

4.1. Three-level optimal efficiency GDPWM strategy

As shown in Figure 3, the reference voltage vector can have at least two clamping modes in any one area. There are three clamping modes in the A and C areas. In these two regions, when the absolute value of any phase current is the maximum value of the three phases, the level clamp control of the corresponding phase can be realized in the modulation. In the other two areas, the B and D areas can achieve two-phase level clamping. The above redundant characteristics provide theoretical possibilities for the modulation strategy proposed in this paper.

The basic modulation process of the optimal efficiency GDPWM is as follows. Under the condition that the neutral point potential satisfies the optimal modulation strategy (the method of judging is introduced later). Firstly, according to the sampling current, the phase with the largest absolute value of the current is determined from the three-phase current. Secondly, in the region where the reference voltage is located, the clamp mode is selected based on the principle that the absolute value of the clamp phase current is the largest. That is, the switch vector used is determined. If the reference voltage vector is in a region similar to the B and D regions, which has only two redundant states, and the phase with the largest absolute value of the current is just not clamped, then select a phase with the second largest absolute value of the current as the clamp phase to ensure that the switching loss is minimized.

For example, the absolute value of phase A current is the largest, and the reference voltage vector is as shown in Figure 3. Phase A potential can be clamped to a high level during this switching cycle. Therefore, the selected switching vector is (2,1,0), (2,2,0), and (2,1,1). The above completes a selection process of the switch vector.

4.2. Comparison of switching loss between optimal efficiency GDPWM strategy and traditional DPWM strategies

For inverters operating at a constant DC bus voltage, it is assumed that the switching frequency of the device is infinite and the current varies linearly during turn-on and turn-off of the device. The switching loss at this time is actually proportional to the load current. The switching loss P in one fundamental period can be determined by formula (7) and (8) [6]. Where U_e is the equivalent DC voltage, I_m is the load current amplitude:

$$P = U_e I_m \frac{1}{2\pi} \int_0^{2\pi} f(\theta) d\theta \tag{6}$$

$$f(\theta) = \begin{cases} 0, \text{Clamped} \\ |\cos(\theta + \varphi)|, \text{Not clamped} \end{cases} \tag{7}$$

Define the switching loss function (SLF) as formula (9). Where P_{ref} is the continuous PWM switching loss.

$$SLF = \frac{P}{P_{ref}} = \frac{1}{4} \int_0^{2\pi} f(\theta) d\theta \tag{8}$$

Define the modulation index (MI) as shown in formula (10). Where $U_{dc} = U_{dc1} + U_{dc2}$ is the total voltage on the DC side. MI is 1 on the inscribed circle of the hexagon in Figure 2.

$$MI = \frac{\sqrt{2} |V_{ref}|}{U_{dc}} \tag{9}$$

According to the definition of several traditional DPWM strategies such as DPWM0, DPWM1, DPWM2, DPWM3, DPWMMIN, DPWMMAX, and the principle of efficiency GDPWM strategy in Section 3.1, the SLF diagram of three-phase symmetrical current under different power factors can be made as shown in Figure 4.

It can be seen from the figure that when MI<0.577, the SLF of optimal efficiency GDPWM strategy is stable to 0.5 regardless of the power factor angle. This is because the reference voltage vector is now located in the internal subspace similar to the A region in Figure 3. Therefore, according to the principle of the optimal efficiency GDPWM strategy, the voltage is always clamped to the phase with the largest absolute value of the current. With the increase of MI, the reference voltage vector becomes longer, and there are more and more cases in the area where only two clamping methods are used. In the case where |φ| is large, as the MI increases the SLF will become larger. The SLF reaches the maximum when MI=1 as shown in Figure 4. When -30 ° <| φ | < 30 °, the SLF is kept at 0.5 and does not change with the change of MI.

It can be seen from Figure 4 that the SLF of the optimal efficiency GDPWM is smaller than any other traditional DPWM strategy regardless of the MI and the power factor angle, which is the advantage of the modulation strategy.

4.3. Comparison between the optimal efficiency GDPWM strategy and the traditional DPWM strategies for the fundamental and harmonic modulation waveforms

The traditional DPWM takes DPWM1 as an example, and the clamp mode of DPWM1 is shown in Figure5 [3]. The clamp mode determines the switching vector based only on the position of the voltage vector, independent of the current flowing through the converter.

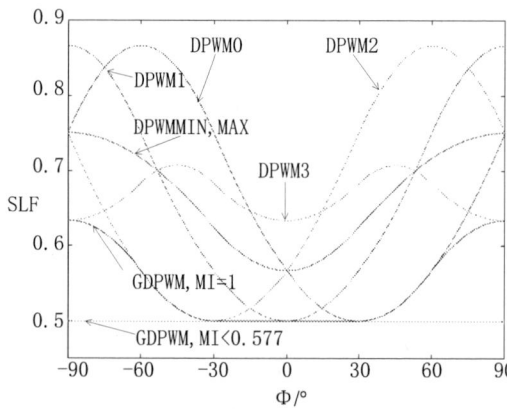

Figure 4. SLF of DPWM and GDPWM

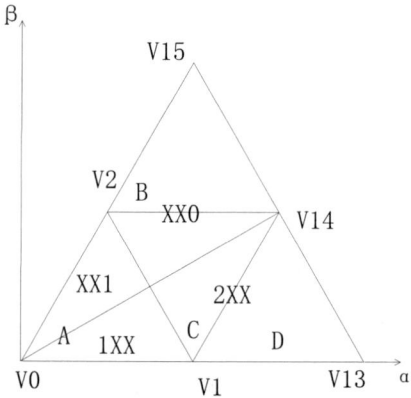

Figure 5. Clamp figure of three level DPWM1 in sector 1

The reactive current is emitted when MI<0.57, and the modulation waveforms of the two modulation strategies are compared as shown in Figure 6. In Figure 6, there are the three-phase current diagram, the A-phase voltage diagram under the DPWM1 strategy, and the A-phase voltage diagram under the GDPWM strategy. In the case of MI<0.57, the voltage vector is always in the internal subspace, and the optimal efficiency GDPWM is always clamped to the phase with the largest absolute value of the current, so the comparison of the two modulation methods is more intuitive. In the waveform diagram of the DPWM1 strategy, it can be seen that the voltage is usually clamped in the region where the absolute value of the current is small. In the waveform diagram of the optimal efficiency GDPWM strategy, it can be seen that the voltage is clamped in the region where the absolute value of the a-phase current is the largest among the three phases. This figure intuitively reflects the advantages of optimal efficiency GDPWM in reducing switching losses.

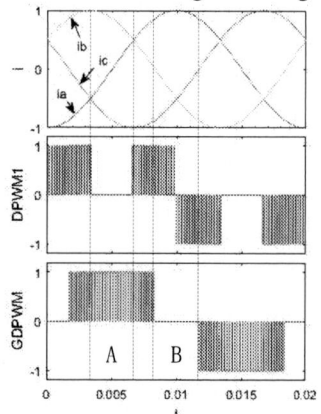

Figure 6. Comparison diagram between DPWM1 and GDPWM of fundamental current

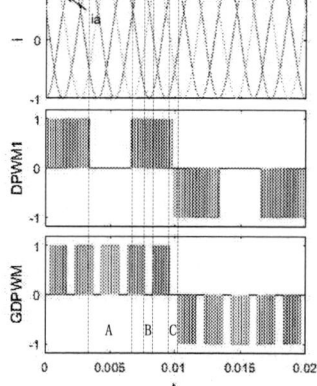

Figure 7. Comparison diagram between DPWM1 and GDPWM of harmonic current

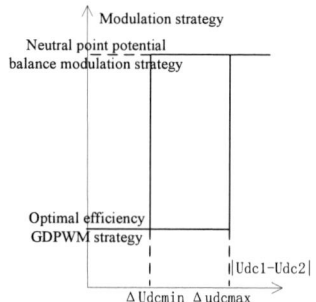

Figure 8. Principle diagram of hysteresis control

Figure 7 shows a case where the modulated wave is a fundamental voltage and the converter emits a harmonic current at a low MI. Here, the fifth harmonic is taken as an example. In Figure 7, there are the three-phase current diagram, the phase A voltage diagram under DPWM1 strategy and the phase A voltage diagram under GDPWM strategy. It can be seen that in one fundamental period, the voltage waveform of the GDPWM strategy has multiple clamp intervals, and each clamp interval corresponds to the interval in which the absolute value of the A-phase current is the largest. The clamp interval of the DPWM1 strategy is fixed in the region near the peak of the reference voltage. The comparison of the two strategies shows that the GDPWM strategy has a lower switching loss.

5. GDPWM modulation strategy considering neutral point potential balance

5.1. Switching method between two modulation strategies
The purpose of the modulation strategy proposed in this paper is to reduce the switching loss as much as possible to improve the efficiency of the device, and to ensure the neutral point potential balance, so the switching between the neutral point balanced modulation strategy and the optimal efficiency GDPWM strategy is required. The neutral point potential offset is used as the determination condition for switching, and the hysteresis comparison method is used for switching. As shown in Figure8, when $|U_{dc1} - U_{dc2}|$ is larger than ΔU_{dcmax}, it switches to the neutral point balance modulation method, and when $|U_{dc1} - U_{dc2}|$ is smaller than ΔU_{dcmin}, it switches to the optimal efficiency GDPWM strategy.

5.2. Vector action time and switch transmission sequence
The action time of the three switch vectors can be calculated from the principle of amplitude-second balance. The transmission of the switch adopts a five-segment type. Due to the continuous switching of the switch vector, the vector transmission sequence of the modulation strategy is performed by dynamic adjustment. After the switch vector is determined, the three switch states in the current switch cycle and the last state of the last switch cycle are compared together to determine the switch transmission sequence with the least change in the switch.

5.3. General Flow of GDPWM Modulation Strategy Considering neutral point Potential Balance
General Flow of GDPWM Modulation Strategy Considering neutral point Potential Balance is as follows. Firstly, according to the method described in Section 4.1, the DC neutral potential is used to determine whether to switch the modulation strategy, and then determine the modulation strategy. Secondly, the vector space region is determined according to the reference voltage.

If the modulation strategy used is an optimal efficiency GDPWM modulation strategy, the switching vector is determined according to the method of Section 3.1; If the modulation strategy used is neutral point potential balance DPWM, the switch vector is determined according to the method of Section 2.3.

Finally, after determining the switching vector, the vector action time and vector transmission sequence are calculated as shown in Section 4.2. This completes a modulation process.

6. Simulation and experimental analysis

6.1. Simulation analysis
In order to verify the effectiveness of the proposed modulation strategy, a three-level converter model using APF control mode was built using PSCAD. The model simulation parameters are shown in Table 2. R1 and R2 are used to simulate the leakage currents of the analog capacitor. The reason for setting the imbalance of the two resistors is to increase the imbalance of the neutral point potential, and then verify the effectiveness of the modulation strategy.

Table 2 Main parameters of grid connected inverter

parameters	value	parameters	value
power voltage, Un/V	380	Capacitor C_1 shunt resistorR_1/Ω	150
DC side voltage, Udc/V	700/1400	Capacitor C_2 shunt resistorR_2/Ω	50/150
Grid side filter inductor, L/uH	325	DC side support capacitor, C_1,C_2/mF	15

Set the DC side voltage to 1400V, that is, MI<0.57. At this time, the reference voltage vector is completely in the internal subspace, and is always clamped to the phase where the absolute value of the current is the largest. The simulation waveform under the condition of no neutral point potential balance control strategy is shown in Figure 9. Figure 9(a) shows the mean value of the fundamental period of the neutral point voltage offset. The neutral point voltage offset gradually increases to a maximum of 108.3V. Figure 9(b) shows the waveform of phase A current and voltage during [1.9s, 2s]. During the period of [1.9s, 2s], the THD was 8.25%, and the second harmonic content was 7.52%, which accounted for the highest proportion.

The simulation waveform of adding the neutral point potential balance control strategy switching under low MI is shown in Figure 10. The hysteresis parameter ΔU_{dcmax} is set to 10V and ΔU_{dcmin} is set to 0.1. Figure 10(a) shows the mean value of the fundamental period of the neutral point voltage offset. The neutral point potential fluctuates within 10V, indicating that the neutral point balance control strategy plays a role. Since the simulation amplifies the neutral point offset factor, the modulation strategy is frequently switched, and the neutral point potential fluctuates greatly. During the period of [1.9s, 2s], the THD was 3.12%, and the second harmonic content was 0.5%. It can be seen that the second harmonic due to the excessive shift of the neutral point potential is eliminated. The above two simulation results verify the effectiveness of the neutral point potential balance control strategy.

The simulated waveform with a DC voltage of 700V and an R2 of 150Ω is shown in Figure 11. The upper and lower diagrams of Figure 11 are the three-phase current waveform and the output voltage waveform of the a-phase converter. At this time, MI>0.57, a part of the reference voltage vector will fall in the outer sector with only two clamping modes, so a part of the clamp region falls on the phase where the absolute value of the current is the second largest. It can be seen that the clamp mode in the figure is consistent with the theoretical analysis.

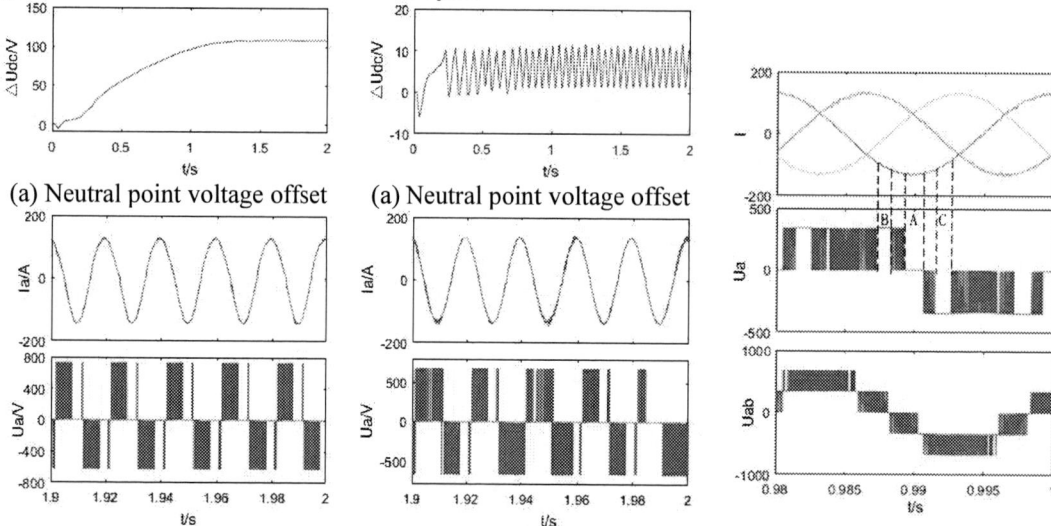

(a) Neutral point voltage offset (a) Neutral point voltage offset

(b) phase A current and voltage (b) phase A current and voltage

Figure 9. Simulation waveforms of low MI without neutral-balancing method

Figure 10. Simulation waveforms of low MI with neutral-balancing method

Figure 11. Simulation waveforms of high MI

The simulation waveform diagram of the case where the DC voltage is 1400V, the resistors R1 and R2 are adjusted to the same 150Ω, and the converter emitting 5th harmonic current is shown in Figure13. The upper and lower diagrams of Figure 12 show the three-phase current value and the output voltage of the a-phase converter, respectively. It can be seen that the a-phase voltage clamp is at the 0 level when the absolute value of the A-phase current is the maximum of the three phases. In a fundamental period, since the emitted current is a 5th harmonic current, the clamp interval is divided into 10 segments.

The above simulation results verify the effectiveness of the proposed modulation strategy in reducing switching losses and neutral point potential balance. The simulation results are consistent with the theoretical analysis.

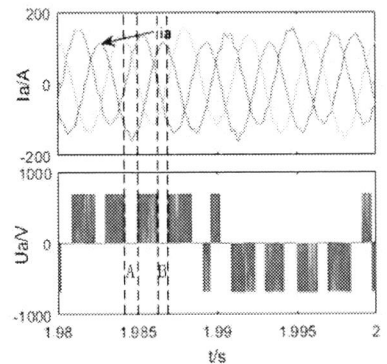

Figure 12. Simulation waveforms of low MI and harmonic current

Figure 13. Experimental platform

6.2. experimental analysis

In order to further verify the correctness of the theoretical analysis and simulation results, the experimental analysis was carried out on the 100A capacity NPC three-level APF prototype experimental platform. The prototype uses Altera's Cyclone IV E: EP4CE115F23C8FPGA as the control system. The experimental platform is shown in Figure 13, and the experimental system parameters are shown in Table 2. There is no capacitor parallel resistance, the switching frequency is 10k, the DC side voltage is 800V, and the neutral point voltage offset is limited to 20V.

The experimental waveform of the fundamental reactive current generated by the prototype using the GDPWM control strategy proposed in this paper is shown in Figure 14. Since MI>0.57, most of the clamped area of U_a is the area with the largest absolute value of I_a, and a part of the clamped area is the area with the second largest absolute value of I_a. The above results are consistent with the simulation results.

The experimental waveform of the fifth harmonic current generated by the prototype using the GDPWM control strategy proposed in this paper is shown in Figure 15. The current in the figure is synthesized by the fifth harmonic and the compensated active fundamental current. From the waveform of U_a, it can be seen that the clamp mode is related to the absolute value of the current. Since MI>0.57, most of the clamped area of U_a is the area with the largest absolute value of I_a, and the part of the clamped area is the area with the second largest absolute value of I_a. The above results are consistent with the simulation results.

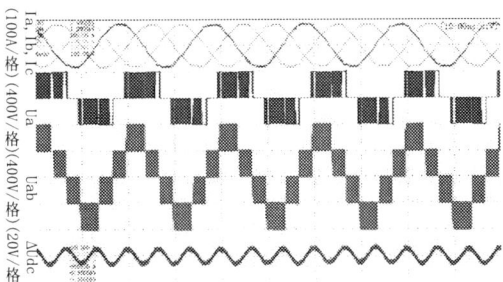

Figure 14. Experimental waveforms of fundamental current

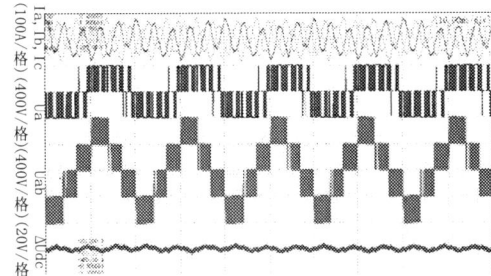

Figure 15. Experimental waveforms of harmonic current

7. Conclusion

In this paper, a GDPWM strategy considering the neutral point potential balance is proposed to reduce the switching loss of the three-level converter under various operating conditions. The modulation strategy selects the optimal level clamp method by comparing the magnitude of the current flowing through the converter while ensuring the balance of the neutral point potential. In turn, the converter achieves minimal switching losses under various operating conditions. This control strategy is particularly effective in reducing the switching losses in the case where the current flowing through the converter changes greatly or is difficult to predict. Simulation and experimental results verify the correctness of the theoretical analysis and the effectiveness of the modulation strategy.

Acknowledgments
This work was supported by 13th Five-Year National Key R&D Program of China(Grant No. 2016YFB0900205).

References
[1] Oleg Vodyakho，Chris C. Mi. Three-Level Inverter-Based Shunt Active Power Filter in Three-Phase Three-Wire and Four-Wire Systems[J]. IEEE Transactions on Power Electronics，2009，24(5)：1350~1363.

[2] H. R. Teymour，D. Sutano，K. M. Muttaqi，et al. Solar PV and Battery Storage Integration Using a New Configuration of a Three-level NPC Inverter with Advanced Control Strategy[J]. IEEE Transactions on Energy Conversion，2014，29(2)：354–365

[3] Lucian Asiminoaei，Pedro Rodriguez，Frede Blaabjerg . Application of Discontinuous PWM Modulation in Active Power Filters[J]．IEEE Transactions on Power Electronics，2008，23(4)：1692-1706.

[4] Wang Hongyan，Deng Yan，Zhao Rongxiang，et al．Switching Loss Minimizing PWM Method for Flying Capacitor Multilevel Inverter[J]．Proceedings of the CSEE，2004，24(8)：51-55.

[5] Zhe Zhang, Ole C. Thomsen, Michael A. E. Andersen. Discontinuous PWM Modulation Strategy With Circuit-Level Decoupling Concept of Three-Level Neutral-Point-Clamped (NPC) Inverter[J]. IEEE Transactions on industrial electronics, 2013, 60(5): 1897~1906.

[6] AN Shaoliang，SUN Xiangdong,，CHEN Yingjuan，et al. A New Generalized Implementation Method of Discontinuous PWM[J]. Proceedings of the CSEE，2012，32(24)：59~66.

[7] REN Kangle，ZHANG Xing，WANG Fusheng，et al. Optimized Design of Discontinuous Pulse-width Modulation and Output Filter for Medium-voltage Three-level Grid-connected Inverter[J]. Proceedings of the CSEE，2015，35(17)：4494~4504.

[8] J. D. Barros,，J. F. A. Silva,，E. G. A. Jesus. Fast-predictive Optimal Control of NPC Multilevel Converters[J]. IEEE Transactions on industrial electronics，2013，60(2)：619~627.

[9] June-Seok Lee，Swungjong Yoo，Kyo-Beum Lee. Novel Discontinuous PWM Method of a Three-Level Inverter for Neutral-Point Voltage Ripple Reduction[J]. IEEE Transactions on industrial electronics，2016，63(6)：3344~3354.

[10] SUN Qingsong，WU Xuezhi，TANG Fen. Three-level Inverter Discontinuous Pulse-width Modulation Strategy Considering Neutral Point Potential Balance[J]. Proceedings of the CSEE，2017，37(0)：1~9.

[11] CHEN Jun，HE Yingjie， WANG Xinyu，et al. Research of the Unity Theory Between Three-level Space Vector and Carrier-based PWM Modulation Strategy[J]. Proceedings of the CSEE，2013，33(9)：71-78.

Research on Intelligent Charging System Technology of Automobile Group

Kedi Yan

Oregon State University
97331

zzyankedi@gmail.com

Abstract: This paper analyzes the smart charging system for dealing with issues related to large parking garages, and analyzes the relevant technical standards of intelligent charging piles application and comprehensive transportation hubs. It mainly includes the number, area and charging method of the charging piles installed in the garages. New forms of construction management is conceived, and the investment and construction are reinforced, apportioning all charging system property rights to investment parties. Simultaneously, the investors take full responsibility for the future management and operation, mainly including the collection of service fees and charging fees.

1. Intelligent charging pile technology

1.1 Alternating current (AC) charging

As for this technology, it generally relies on three-phase AC power to provide the required power for electric vehicles. Since the vehicle-mounted charger and heat dissipation may affect it, the power of charging is small, and the time taken for charging is relatively long. The circuit configuration of a car charger usually has two types. If the correction circuit is added with the power factor in the previous stage, the influence on the harmonics of the power grid will be reduced, and the power factor state is still high.

1.2 Direct current (DC) charging

For this technology, it usually depends on the circuit composition such as control protection and the rectification filter. Among them, only three-phase uncontrollable and pwm-type rectification will utilize rectification technology. The three-phase uncontrollable rectification is relatively low in terms of capital, but the harmonic current is relatively large, requiring the equipment for harmonic control to be installed. The PWM type of rectified power has the characteristics of high efficiency and low harmonic current in conversion, but the capital requirement is relatively high.

1.3 Wireless charging technology

1.3.1 Electromagnetic induction type

This technology relies on two mutual inductance coils to complete wireless charging. When the input current of the coil changes, the magnetic field of the coils on the output side also changes, thereby

Content from this work may be used under the terms of the Creative Commons Attribution 3.0 licence. Any further distribution of this work must maintain attribution to the author(s) and the title of the work, journal citation and DOI.

Published under licence by IOP Publishing Ltd

generating an induced current. In this way, energy is passed to the output end through the input end. This technique of electromagnetic induction has no a long transport distance, but its energy conversion rate is relatively high.

1.3.2 Magnetic field resonance type

Similar to the principle of acoustic resonance, if the resonance frequencies of the two media are the same, energy transfer can be achieved. This type of wireless charging can also achieve simultaneous charging for a large number of devices, but the loss will be relatively high. Two schemes can be considered together to solve the problem of high charging loss (low efficiency) of this technology.

First, a feedback circuit is added at the transmitting end, which is: "parity-time-symmetric circuit incorporating a nonlinear gain saturation element". This component can automatically select the operating frequency according to the transmission distance to maximize the output power [6]. Second, metamaterials are utilized. Metamaterials benefit from its ability to focus magnetic flux, which can increase transmission efficiency and transmission distance. The combination of the two schemes may enable wireless charging to bring a new solution like spring wing to the pure electric vehicle charging problem.

2. Key component nodes analysis

2.1 Charging control box of automobile group

It is a type of outdoor equipment which is designed to integrates charging and power distribution devices and then installed into a sealed, moisture-proof, and rust-proof control box. This type of equipment is characterized by less occupied area and portfolio diversification. The integration of AC and DC can not only achieve AC charging and DC charging at the same time, but also make AC and DC arbitrarily combined to meet various needs. Meanwhile, real-time background monitoring on the charging process can be performed, so once an abnormality is found, the charging process can be automatically adjusted to achieve real-time protection. Real-time communication is maintained with the cloud platform, effectively protecting the charging vehicle and the power grid, so that the electricity utilization is guaranteed.

2.2 Charging terminals

2.2.1 Single-phase AC terminal with a gun

It is equipped with the control box and the charging box to provide the necessary charging interfaces, which is widely used in single-phase AC charging. At the same time, there must be a place where the charging gun is used, and according to its application, the area can be divided into the types of car bumper and wall-mounted. Among them, the wall-mounted type can be installed on the wall surface, with the requirement to avoid the rear end of the charging vehicle during installation. The single-phase AC wall-mounted charging terminal can be installed together with the bracket. For the type of car bumper, it can be used in underground garages and the place where the terrain is high and water is not easy to accumulate [1].

2.2.2 Single-phase AC terminal with seats

It cooperates with a charging port and a control box to provide a corresponding charging interface, mainly used in single-phase AC charging work, and the application place must be equipped with a charging socket.

Its main application is in single-phase AC charging, and it must be equipped with a charging socket. It is divided into two types: wall-mounted and floor-standing according to the application area.

2.2.3 DC terminal with a gun

It cooperates with a charging box to provide a corresponding charging interface. It is mainly used in DC charging and must be equipped with a charging gun. According to the application area, it can be divided into two types of floor and wall-mounted. Among them, the floor type can be classified as single and double-sided in accordance with the site, while based on the different currents it can also be divided into two types of 125A and 250A.

3. Intelligent charging system design scheme

3.1 System configuration scheme

Due to the large difference in charging speed between DC charging and AC charging, a battery electric vehicle (with ordinary battery capacity) needs 7-9 h to be fully charged through the AC charging pile after being fully discharged, while it only requires 2-3 hours to be fully charged through the DC quick charging pile. Therefore, the AC and DC terminals are scientifically matched depending on the big data survey to achieve effective energy utilization and battery protection. For example, in the underground parkings of shopping malls, considering that most people will not park their cars for a long time, the combination ratio of DC and AC terminals can be decided to 8:2. As for the underground parking of the residential areas, takeing into consideration that most people will park here overnight, the ratio of the DC terminals to AC terminals can chose 2:8. Taking a real project as an example, it is equipped with 80 channels of AC and 160 channels of DC terminals, which can allow more than 40 cars to carry out AC slow charging, simultaneously allowing over 160 cars to conduct DC fast charging. The charging function can be turned on by using the mobile phone software to scan the software QR code. In addition, the termination of charging can also be done by means of mobile phone softwares. The overall charging process can be queried by a mobile phone, with a high security. In the next 5-10 years, when the wireless charging is mature, due to its safe, stable and space-saving characteristics, it will gradually replace the AC charging devices, and even supersede the DC charging device if the efficiency problem is improved. The specific configuration scheme can be considered by referring to the following chart.

3.2 Connection scheme

The access power supply of the project is a 0.4kV low-voltage power supply, which needs to be introduced and connected in the substation. For low-voltage power access points, cables need to be installed between the charging devices of one car, which is done by the supplier [2]. The supplier shall formulate a scientific and reasonable engineering installation scheme according to the specific situation of the installation site. This can effectively ensure that the construction work meets the requirements and standards of electric power technology.

3.3 Construction of the cloud platform

It needs to monitor the operational power station information, the charging records and the status of terminal operations in real time. Meanwhile, all power station and terminal information data should be tracked to effectively analyze the data [3]. The cloud platform can not only effectively monitor the voltage, current and charge of the vehicle, but also efficaciously monitor the battery status of the vehicle with BMS.

3.4 Security consideration

For intelligent AC chargers, it needs to provide low-voltage power to the car, and it must have the protection to overvoltage, overload and over-temperature. While for the off-board vehicle DC charger, it is required to select the form of the component design. Among them, the electrical and business parts are required to be independent.

3.5 Intelligent and secure billing system construction

In terms of the charging control unit, the hardware and software interface need to be connected with the input and output components, so that the functions of switching control, data decryption and encryption, billing, human-computer display, etc. of the charging device can be finally realized.

4. Advantages of intelligent charging system

4.1 Pile-free charging

The difference from the conventional charging devices is that the pile-free charging does not need to rely on the charging piles. It transfers charging control and human-computer interaction to the charging box and the cloud respectively, with only one charging terminal in the local charging connection. However, the terminal is resistant to crushing, water repellency, and takes up less space. The pile-free charging transfers the human-computer interaction to the cloud platform, relying on the mobile APP and the cloud platform. As a result, the contact between people and public devices is reduced, making them more personalized and humanized. The pile-free charging can concentrate the control processes to share the communication and power modules, thus reducing the capital investment in the devices. At the same time, it can also make the control module to avoid some damage from the outside world and improve its security.

4.2 Charging system structure module

According to the structure of the product module and the prefabricated production equipment, the site only needs to build up the facility foundation [4]. The construction cycle of a small site is 7 days, and that of a large site is one month. For the intelligent charging technology, its update is extremely fast, and the modular structure is completed very quickly in terms of technology expansion and upgrade, making it consistent with the update of charging technology.

4.3 Intelligent charging

By means of centralized charging, wind energy, solar energy and battery energy storage technology can be effectively combined. In virtue of the function of the vehicle-mounted battery and the charging network, an intelligent microgrid system is established, and the system can operate independently in the grid through the grid-connected operation of the grid. Relying on the operation of the micro-scheduling of the power grid, it will not compete with people for electricity. It will use low-peak electricity to improve the efficiency of the grid. Flexible charging is usually realized on the basis of battery mechanism and characteristics. The charging device is used to determine the current and voltage according to the battery related requirements and the specific conditions, thereby restoring the function of the battery and achieving the purpose of life protection [5]. With the support of flexible charging technology, the traditional form of relying on BMS to determine the charging power is replaced, so that the charging safety is improved and the occurrence of safety accidents is avoided. Based on the collected battery related information, the data on the flexible charging is further determined, and the functions of life protection and battery recovery are realized in charging.

5. Conclusions:

In short, the intelligent charging system of the automobile group belongs to the important project of implementing sustainable development, optimizing energy structure and promoting new energy development. The construction of intelligent charging system can meet the goal of energy structure upgrading, environmental protection, greenhouse gas control and energy conservation in China.

References

[1] Weijie Song, Dongyu Liu. Intelligent Navigation System for Electric Vehicle Charging [J]. Technology Economic Market, 2016(04):17+16.

[2] Yingping Zhu, Xuhong Zhang, Xinying Han. Electric Vehicle Intelligent Charging Dispatching System for Distribution Network Loss Optimization [J]. Guangdong Electric Power, 2016,29(09):94-97+103.

[3] Shenzhen New Energy Vehicle and Charging Equipment Exhibition was Held Successfully [J]. Power World, 2017(06):7.

[4] Shu Su, Jinwen Sun, Xiangning Lin, Jianshan Li. Intelligent Navigation System for Electric Vehicle

Charging [J]. Proceedings of the Chinese Society for Electrical Engineering, 2013, 33(S1):59-67.

[5] Hu Wang, Guifang Gao. Research and Design of Charger of Electric Vehicle [J]. Journal of Changchun University, 2018(04):356-361.

[6] Sid Assawaworrarit, Xiaofang Yu, Shanhui Fan. Robust Wireless Power Transfer Using a Nonlinear Parity-time-symmetric Circuit [J]. Nature, 2017(06):1.

ISPECE   IOP Publishing

IOP Conf. Series: Journal of Physics: Conf. Series **1187** (2019) 022011   doi:10.1088/1742-6596/1187/2/022011

Coordination Between Converter-Based Wind Turbines and Synchronous Generators During Inertia Control

Fei Jin[1], Xiaoliang Liu[1], Guoqiang Li[1], Jingli Liu[1], Weichao Li[2], Jinxin Zheng[2], Renzhang Liu[3]

[1]Weifang Power Supply Company State Grid, Shandong Electrical Power Company, Weifang, China
[2]School of Electrical Engineering, Shandong University, Jinan, China
[3]Penglai Power Supply Company State Grid, Shandong Electrical Power Company, Penglai, China

Weichao Li (corresponding author, E-mail:941917865@qq.com) is with School of Electrical Engineering, Shandong University, Jinan 250061, China.

Abstract: Since the rotor speed of converter controlled wind turbines (WTs) is decoupled with system frequency, large-scale wind power integration into the power system reduces system inertia, and worsens system frequency response after disturbances. Although inertia control of the WT can improve the system frequency response by releasing kinetic energy stored in the rotor of the WTs, it masks the magnitude of the load disturbance and thus delays the response of the synchronous generators (SGs) during primary frequency control. Furthermore, due to the limited kinetic energy, WTs must terminate inertia control at some time, and this termination causes another system frequency drop. This paper proposes a coordination strategy between WTs and conventional SGs. By revising the input signal of the governing system of the SGs, and designing output power for the WTs during inertia control, the system frequency response can be improved. Case study results validate the efficacy of the proposed strategy.

1. INTRODUCTION

With the high penetration of the wind power, modern power systems are faced with more challenges. The system inertia declines because conventional SGs are phased out, and the converter-based WTs cannot naturally contribute to system inertia since their rotor speed is decoupled from the system frequency. The low inertia power system may experience a severe frequency drop after active power imbalance disturbances, and thus have higher risk of under frequency load shedding, or even cascading outages. Therefore, it is necessary to enhance the WTs with the ability to contribute to system inertia in order to guarantee system frequency stability.

Inertia control of WTs has been proposed in recent years [1], Which can provide an emulated inertia after disturbances, and can improve frequency stability. Over the past years, the effect of inertial control on the power system and the way of improving the performance of inertial control have been widely investigated [2].

However, inertia control has some adverse effects to the power system. One of the most significant shortcomings of the existing inertial control is that WTs have fast active power response during system frequency regulation, and this could conceal the actual power imbalance of the system, leading SGs to delay their response when rejecting more mechanical power [3]. Another concern about the inertial control is that WTs have to suddenly switch back to the normal operating mode by decreasing the

Content from this work may be used under the terms of the Creative Commons Attribution 3.0 licence. Any further distribution of this work must maintain attribution to the author(s) and the title of the work, journal citation and DOI.
Published under licence by IOP Publishing Ltd

output power during the system frequency regulation to recover the rotor speed. This causes a significant drop of the system frequency, which is called the *Secondary Frequency Drop* [4].

In [3], the incremental injected active power of WTs is communicated to the conventional SGs so that the governing system can increase the mechanical power more quickly based on the full load imbalance. However, in practice, the proposed method requires high investment on communication system and it is not economical for implementation.

To mitigate or eliminate the secondary frequency drop, some methods have been proposed aiming at smoothing the termination operation of WTs such as dynamic droop-based inertial control [5], and extended state observer (ESO)-based inertia emulation controller (InEC) [6]. Furthermore, it is also possible to mitigate secondary frequency drop by implementing energy storage devices [7]. It can be noticed that during inertia control, the mechanical power of the SGs increases and has an overshoot due to the dynamic of the governing system. This overshoot can be utilized as a compensation of the power drop caused by the termination of the inertia control, and mitigate secondary frequency drop efficiently.

Motivated by the above two aspects, this paper firstly analyses the characteristics of the output power of WTs and SGs, and the system frequency response during inertia control, and proposes a coordination strategy between converter-based WTs and conventional SGs. Applied with the strategy, during inertia control, the SGs can increase the mechanical power based on the full power imbalance of the system, and it is achieved without communication with WTs. Moreover, by designing the output power of WTs, the overshoot of the mechanical power of the SGs can be utilized to mitigate the secondary frequency drop.

2. MODELS AND PROBLEM STATEMENT

A. The simplified model of the SG-WT system

For the convenience of analysis, the SG-WT system model can be simplified as a model which is shown as Fig.1. The input mechanical power of the WT P_{mw} is expressed by,

$$P_{mw} = \frac{1}{2} C_p(\lambda, \beta) \rho A v_w^3 \quad (1)$$

which is a function of air density ρ, the swept area A, the wind speed v_w and the power coefficient C_p. The value of C_p is determined by the tip-speed ratio λ and pitch angle β.

In normal operation mode, the output electrical power of the WT P_{ew} is determined by maximum power point tracking (MPPT) control, and is calculated by,

$$P_{MPPT} = k \omega_r^3 \quad (2)$$

where ω_r is the rotor speed of the WT, and k is the coefficient of the MPPT curve.

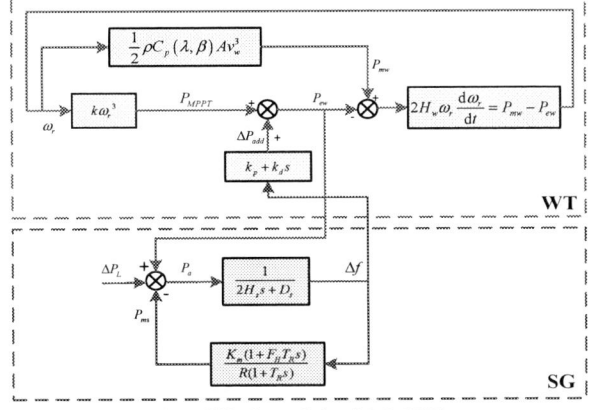

Fig.1 Simplified model of SG-WT system

The model of the SG can be simplified as a low-order SFR model which is proposed in [8], where P_{ms} is the mechanical power of the SG, H_s and D_s are the inertia constant and damping factor of the SG, respectively. K_m, R are the mechanical power gain factor and the regulation factor of the governor of the SG, respectively. F_H is the fraction of total power generated by the High-Pressure turbine of the SG, and T_R is the reheat time constant of the turbine of the SG.

After a disturbance, the system will experience a frequency drop, and the frequency deviation is expressed as Δf. After detects the frequency drop, the wind turbine activates the inertia control by adding an incremental power ΔP_{add} to P_{MPPT}. Synthetic inertia control is a common form of inertia control. It utilizes the system frequency deviation as the input of the control loop, and can replicate the behavior of synchronous generators after the disturbance. The output electrical power of the WT during inertia control is,

$$P_{ew} = P_{MPPT} + \Delta P_{add} = P_{MPPT} - \left(k_p \Delta f + k_d \frac{d\Delta f}{dt} \right) \quad (3)$$

The simplified model of SG-WT system can be used for analyzing characteristics of output power of SGs and WTs during inertia control.

B. The output power of the SG and WF during inertia control

The output power of the SG and WT and the system frequency response after a disturbance with and without inertia control is compared and shown in Fig.2, respectively.

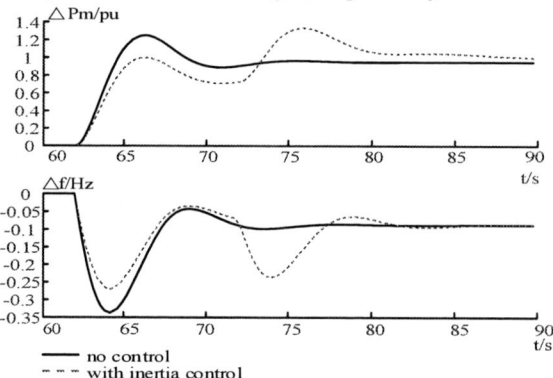

Fig.2 Comparison of system frequency, output power of the SG and WT after a disturbance with and without inertia control

C. Problem Statement

It can be seen from Fig.2 that with inertia control, the system frequency nadir is improved. Nevertheless, the output power of the SG increases slower compared to the scenario without inertia control. The govorning system of the SG increases the output power based on deviation of the rotor speed of the SG, which is coupled with the system frequency. With inertia control, the system frequency deviation is improved, but for the SG, the actual power imbalance is also concealed because its rotor speed deviation is also improved.

It is also shown from Fig.2 that the system has a significant secondary frequency drop, which is caused by the termination of the inertia control. It is observed from simulations that the magnitude of the secondary frequency drop is relavant with penetration of wind power, parameters of inertia control, and the magnitude of disturbance, etc. In some cases, the magnitude of secondary frequency drop may be even severer than the frequency drop caused by the disturbance.

To address the aforementioned two issues, this paper aims at designing a coordination strategy between WTs and SGs, which can avoid the delay of the governing system of the SG, and also mitigate the secondary frequency drop.

3. COORDINATION STRATEGY

The coordination strategy consists of two parts. Firstly, for the SG, the input signal of the governing system of the SG is replaced with "virtual" system frequency deviation which is calculated during inertia control; Secondly, for the WT, the output power of the WT is designed considering the characteristics of the output power of the SG.

A. SG: The "virtual" system frequency deviation

With inertia control, system frequency deviation is improved, thus it also conceals the actual power imbalance. To solve this problem, one approach is to communicate the WT's incremental power to the SG, so that the governing system of the SG can "realize" the actual power imbalance and increase its output power based on that. This approach can effectively accelerate the response of the SG, but with high investment on communication system. Besides, the time delay of communication is also not negligible in large-scale power systems. Therefore, this paper proposes another approach which is free from communication.

The relation between initial rate of change of frequency (RoCoF) of the power system and the magnitude of the disturbance can be expressed as,

$$\Delta P_L = 2H_S \left. \frac{\mathrm{d}\Delta f}{\mathrm{d}t} \right|_{t=0^+} \quad (4)$$

The initial RoCoF can be measured by the SG and its value is not affected by the inertia control of the WT since the activation of the inertia control has a time delay, thus the initial RoCoF of the system is only determined by the inherent inertia provided by the SG. After obtaining the initial RoCoF, the governing system will calculate the actual power imbalance using (4), and then calculate "virtual" system frequency deviation Δf_v using the low-order SFR system, that is,

$$\Delta f_v(t) = \frac{\Delta P_L R}{D_s R + K_m} \left[1 + \alpha e^{-\zeta \Omega_n t} \sin(\Omega_r t + \phi) \right] \quad (5)$$

where,

$$\Omega_n^2 = \frac{D_s R + K_m}{2H_s R T_R} \ , \quad \zeta = \left[\frac{2H_s R + (D_s R + K_m F_H)T_R}{2(D_s R + K_m)} \right] \Omega_n \ , \quad \alpha = \sqrt{\frac{1 - 2T_R \zeta \Omega_n + T_R^2 \Omega_n^2}{1 - \zeta^2}}$$

$$\Omega_r = \Omega_n \sqrt{1 - \zeta^2} \ , \quad \phi = \arctan\left(\frac{\Omega_r T_R}{1 - \zeta \Omega_n T_R} \right) - \arctan\left(\frac{\sqrt{1 - \zeta^2}}{-\zeta} \right) \quad (6)$$

Essentially, this so-called "virtual" system frequency deviation is the system frequency deviation in a scenario where the system has the same disturbance, but the WT is not implemented with inertia control. By replacing the rotor speed with this "virtual" system frequency deviation as the input signal of the governing system of the SG, the SG will avoid the influence of the inertia control and its output power will increase likewise in the scenarios without inertia control.

B. WT: The Output Power During Inertia Control

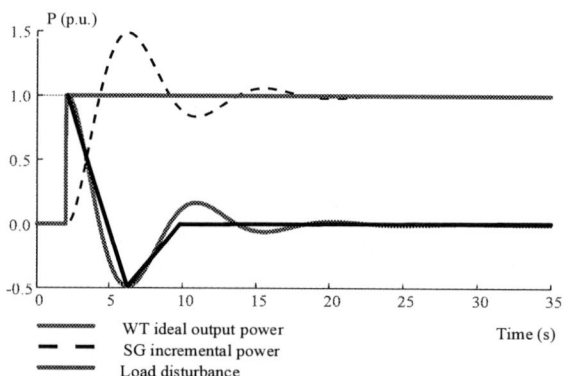

Fig.3 The ideal output power of the WT in load disturbance

The output power of the WT can be designed based on the objective of improving the system frequency response.

From the perspective of the system, the ideal output power of the WT is achieved if it can fully compensate the system power imbalance after the disturbance. In this way, the system will not experience any frequency drop. This ideal output power of the WT ΔP_{wi} can be expressed as,

$$\Delta P_{wi} = \Delta P_L - \Delta P_{ms} \quad (7)$$

where ΔP_{ms} is the incremental mechanical power of the SG, and can be calculated by,

$$\Delta P_{ms} = \frac{K_m \Delta P_L}{D_s R + K_m} \left[1 + \alpha' e^{-\zeta \Omega_n t} \sin(\Omega_r + \phi') \right], \quad \alpha' = \sqrt{\frac{1 - 2T_R \zeta \Omega_n + F_H^2 T_R^2 \Omega_n^2}{1 - \zeta^2}}$$

$$\phi' = \tan^{-1}\left(\frac{\Omega_r F_H T_R}{1 - \zeta \Omega_n F_H T_R} \right) - \tan^{-1}\left(\frac{\sqrt{1 - \zeta^2}}{-\zeta} \right) \quad (8)$$

For the convenience of calculation, the ΔP_{wi} can be approximated as a piecewise linear function (shown as Fig.3), that is,

$$\Delta P_{wi} = \begin{cases} \Delta P_L - \dfrac{\Delta P_{\max}}{t_1} t, & t_0 \leq t < t_1 \\[2mm] \Delta P_L - (\Delta P_{\max} + k_{wi}(t - t_1)), & t_1 \leq t < t_2 \quad (9) \\[2mm] 0, & t \geq t_2 \end{cases}$$

where t_0 is the time when the inertia control is activated, t_1 is the time when the maximum mechanical power of the SG occurs, and t_2 is the time when the mechanical power of the SG reaches its steady state value, $\Delta P_{\max}$ is the value of the maximum mechanical power of the SG. t_1, t_2 and $\Delta P_{\max}$ is given by,

$$t_1 = \frac{n\pi - \phi'}{\Omega_r} = \frac{1}{\Omega_r} \tan^{-1}\left(\frac{\Omega_r F_H T_R}{\zeta \Omega_n F_H T_R - 1} \right), \quad t_2 = t_1 + \frac{\pi}{\Omega_r}, \quad \Delta P_{\max} = \Delta P_m(t_1) \quad (10)$$

The functionality of ΔP_{wi} is to provide a reference of the output power of the WT. Obviously, the output power of the WT cannot be set as ΔP_{wi} since the WT has its constraints on the electric torque and the available kinetic energy. Considering the constraints, the output power of the WT can be determined as,

$$\Delta P_{add} = \begin{cases} \Delta P_{wc} - \Delta P_{MPPT} & \Delta P_{wi} \geq \Delta P_{wc} \,\&\, \omega_r > \omega_{r\min} \\ \Delta P_{wi} - \Delta P_{MPPT} & \Delta P_{wi} \leq \Delta P_{wc} \,\&\, \omega_r > \omega_{r\min} \quad (11) \\ 0 & \omega_r < \omega_{r\min} \end{cases}$$

C. Overview of the Coordination Strategy

The schematic diagram of the proposed coordination strategy is shown as Fig.4.

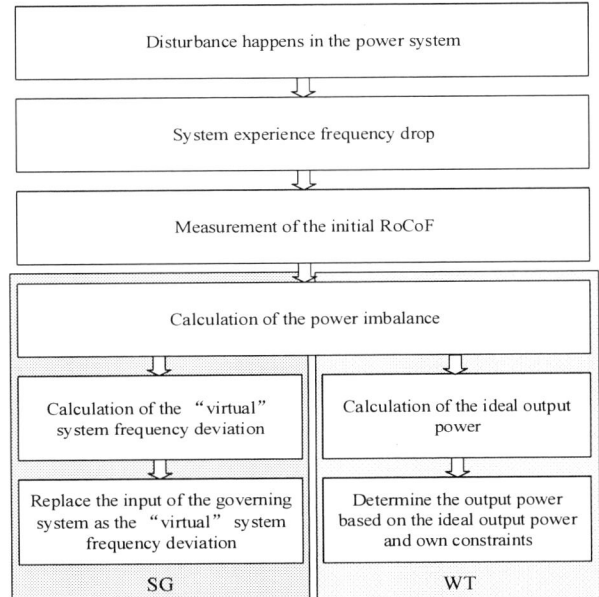

Fig.4 Schematic diagram of the coordination strategy

4. CASE STUDY

To validate the performance of the proposed coordination strategy, case study is performed in a SG-WT system.

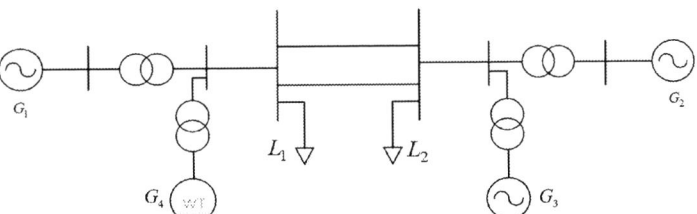

Fig. 5 Simulation system structure diagram

A SG-WT system model is built in the software DIGSILENT/PowerFactory, as shown in Fig. 5. $G_1 \sim G_3$ are thermal power plants, where the active power of G_1 and G_2 are 400MW and the active power of G_3 is 1000MW. G_4 is a wind farm with 200×1.5MW DFIG units. L_1 is a load of 500MW and L_2 is 1000MW. At t=80s, the load L_2 increases by 100MW. The simulations compare system frequency, output power of the WT and SG with inertia control and coordination control during frequency regulation under three different wind speed conditions.

A. Low wind speed

  a) System frequency response

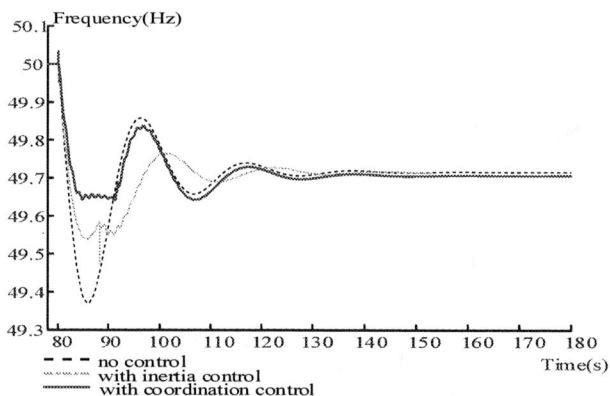

Fig. 6 Comparison of system frequency of the WT during frequency regulation in low wind speed conditions (6.5m/s)

b) Output power of the SG and WT

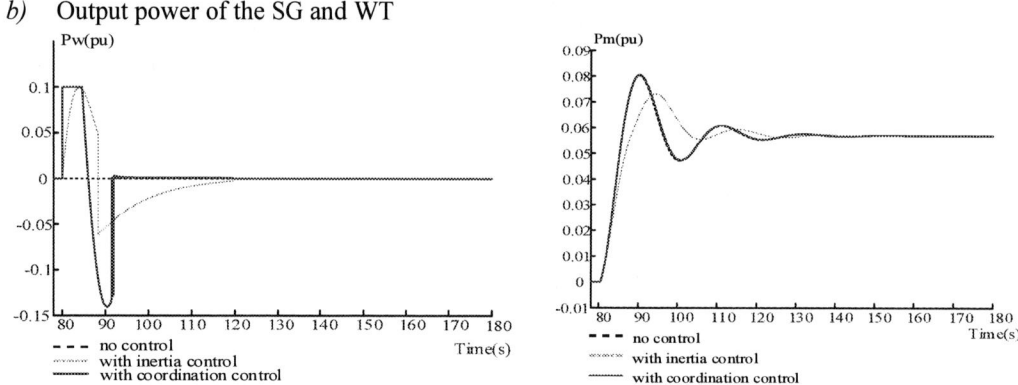

Fig. 7 Comparison of output power of the WT and SG during frequency regulation in low wind speed conditions (6.5m/s)

As shown in Fig. 6 and Fig. 7, in low wind speed zone, the system frequency nadir is improved with inertia control, but the response of the SG is delayed. With the proposed coordination strategy, the response of the SG is not delayed.

B. Medium wind speed
  a) System frequency response

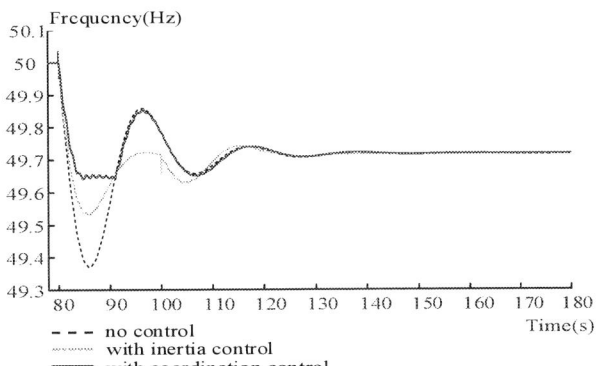

Fig. 8 Comparison of system frequency of the WT during frequency regulation in medium wind speed conditions (7.5m/s)

b) Output power of the SG and WT

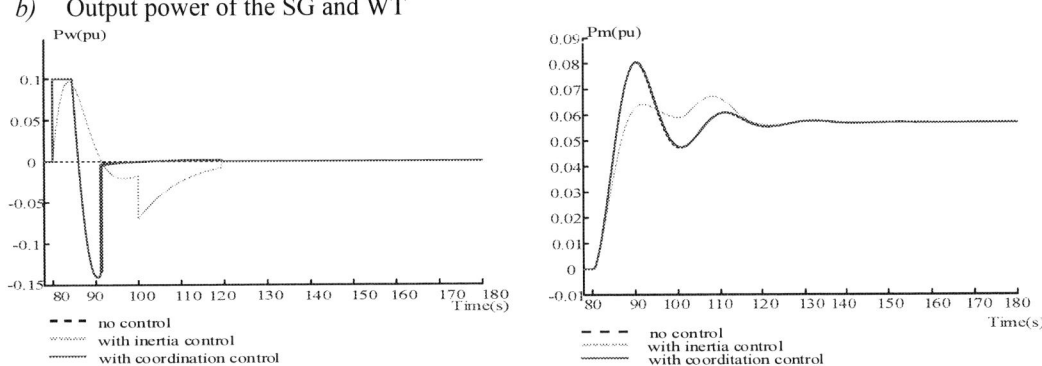

Fig. 9 Comparison of output power of the WT and SG during frequency regulation in medium wind speed conditions (7.5m/s)

It can be seen from Fig. 8 and Fig. 9 that, in medium wind speed zone, the frequency response is also enhanced with proposed coordination strategy, compared with the scenarios without inertia control and with traditional inertia control. Moreover, the delay of the SG's response is also avoided effectively.

C. High wind speed

a) System frequency response

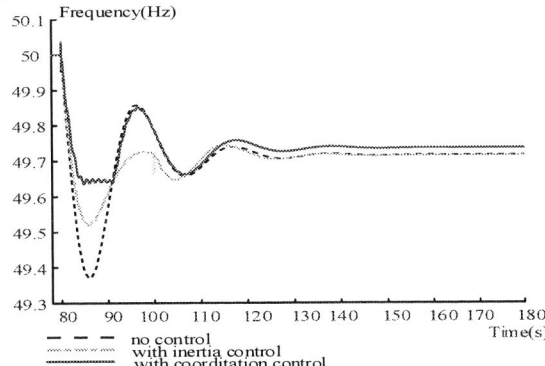

Fig. 10 Comparison of system frequency of the WT during frequency regulation in high wind speed conditions (9.5m/s)

b) Output power of the SG and WT

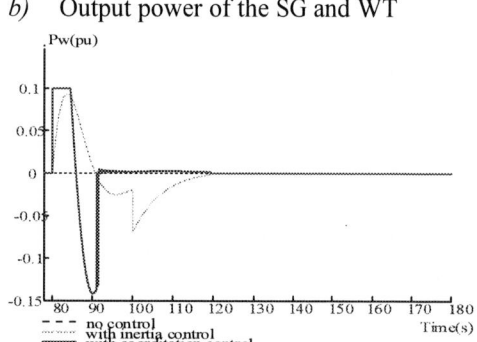 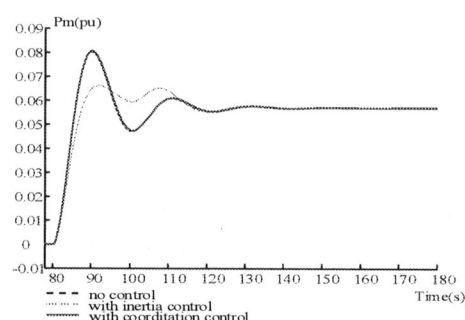

Fig. 11 Comparison of output power of the WT and SG during frequency regulation in high wind speed conditions (9.5m/s)

The performance of the proposed coordination strategy is also validated in high wind speed zone, as is shown in Fig. 10 and Fig. 11. Furthermore, the secondary frequency drop was also improved with the proposed coordination strategy.

5. CONCLUSION

This paper analyzes the characteristics of the WT's and SG's output power as well as system frequency response during inertia control, and proposes a coordination strategy between converter-based WTs and conventional SGs based on the analysis. The proposed coordination strategy can let the SGs increase its mechanical power based on the full power imbalance of the system, and it is achieved without setting up communication between SGs and WTs. Moreover, the proposed coordination strategy can also mitigate secondary frequency drop caused by the termination of the inertia control. Simulation results show that the proposed coordination strategy has satisfactory performance in different wind speed conditions.

REFERENCES

[1] J. Morren, S. W. H. de Haan, W. L. Kling, and J. A. Ferreira, "Wind Turbines Emulating Inertia and Supporting Primary Frequency Control," *IEEE Trans. Power Syst.*, vol. 21, no. 1, pp. 433–434, 2006.

[2] M. Kayikçi and J. V. Milanović, "Dynamic contribution of DFIG-based wind plants to system frequency disturbances," *IEEE Trans. Power Syst.*, vol. 24, no. 2, pp. 859–867, 2009.

[3] J. M. Mauricio, A. Marano, A. Gomez-Exposito, and J. L. Martinez Ramos, "Frequency Regulation Contribution Through Variable-Speed Wind Energy Conversion Systems," *IEEE Trans. Power Syst.*, vol. 24, no. 1, pp. 173–180, 2009.

[4] A. Bonfiglio, M. Invernizzi, A. Labella, S. Member, R. Procopio, and S. Member, "Design and Implementation of a Variable Synthetic Inertia Controller for Wind Turbine Generators," *IEEE Trans. Power Syst.*, vol. 8950, no. c, pp. 1–1, 2018.

[5] M. Hwang *et al.*, "Dynamic Droop – Based Inertial Control of a Doubly-Fed Induction Generator," *Ieee Trans. Sustain. Energy*, vol. PP, no. 99, pp. 1–10, 2015.

[6] F. Liu, Z. Liu, S. Mei, W. Wei, and Y. Yao, "ESO-Based Inertia Emulation and Rotor Speed Recovery Control for DFIGs," *IEEE Trans. Energy Convers.*, vol. 32, no. 3, pp. 1209–1219, Sep. 2017.

[7] L. Miao and C. District, "Coordinated Control Strategy of Wind Turbine Generator and Energy Storage Equipment for Frequency Support Jinyu Wen Wei-jen Lee," *Ieee Trans. Ind. Appl.*, vol. 51, no. 50937002, pp. 1–7, 2014.

[8] P. M. Anderson and M. Mirheydar, "A low-order system frequency response model," *IEEE Trans. Power Syst.*, vol. 5, no. 3, pp. 720–729, 1990.

Research of a high voltage and high value resistors standard device

Xiang Zhou[1], Xinli Cao[2], Qianghu Yan[2] and Min Lei[1*]

[1] China Electric Power Research Institute, Wuhan, Hubei, 430074, China

[2] School of Electrical and Information Engineering, Wuhan Institute of Technology, Wuhan, Hubei, 430074, China

[*]Min Lei's email: leimin@sgepri.sgcc.com.cn

Abstract. With the continuous increase of voltage class and the rapid development of electrical equipment, we need to design a new kind of intelligent high voltage and high value resistors standard metering device which can be used to a much larger range to complete daily verification work. Based on MSP430 single-chip microcomputer, this paper designed the temperature and humidity control module and the high voltage measurement module that the measuring range can be automatically switched. By using this standard device, the measurement of voltage and resistance can reach separately to 10 kV and 200GΩ. At last by measuring the voltages and resistances and testing the temperature and humidity control system, it has indicated that the high voltage and high value resistors standard metering device in this paper meets the requirements.

1. Introduction

Electronic insulation resistance meter, megohmmeter and other high-resistance test equipment are frequently used in the power field and the safety performance test of factories and mines, these equipment are important measurement instruments forcibly verified by the country[1]. The verification and calibration of such electrical test equipment are mainly done by high voltage and high value resistors standard metering device[2-3]. However, the common standard metering device can not realize the automatic control of the calibration, the measurement range is small, the measurement result error is large, the operation is complicated, the calibration efficiency is low and it is difficult to meet the demand[4].

According to JJG1005-2005 "Electronic Insulation Resistance Meter Verification Regulations"[5], with the single-chip controller as the controller and based on the automatic control technology, the modules of the temperature and humidity acquisition and control and the high voltage measurement with automatic range switching have been designed which can effectively complete the 10kV DC voltage measurement and the 100Ω~200GΩ resistance calibration work.

2. Resistance calibration module design

In order to meet the calibration requirements of the metrology department for the insulation resistance meter, the standard resistance is divided into: $(0\sim10)\times10\Omega$, $(0\sim10)\times10^2\Omega$, $(0\sim10)\times10^3\Omega$, $(0\sim10)\times10^4\Omega$, $(0\sim10)\times10^5\Omega$, $(0\sim10)\times10^6\Omega$, $(0\sim10)\times10^7\Omega$, $(0\sim10)\times10^8\Omega$, $(0\sim10)\times10^9\Omega$, $(0\sim10)\times10^{10}\Omega$ and 200GΩ. The maximum allowable error of the standard resistance is ±0.2% for 100Ω~ 0.1MΩ, ±0.5% for 0.1MΩ~10MΩ, and ±1% for 10MΩ~200GΩ.

Content from this work may be used under the terms of the Creative Commons Attribution 3.0 licence. Any further distribution of this work must maintain attribution to the author(s) and the title of the work, journal citation and DOI.

Published under licence by IOP Publishing Ltd

2.1. Standard resistor selection

The most important part of high voltage and high value resistors standard metering device design is to improve the accuracy of standard resistance, which is a key component to optimize the performance of the standard device and ensure the safe production of the power industry. The accuracy of the resistance will be affected by factors such as structure, materials, package and working environment. The actual resistance value will also be related to the temperature coefficient and voltage coefficient[6]. At present, most manufacturers producing high-voltage and large-resistance standard devices use the same material resistance in the design of standard resistance circuits. Although the characteristic of withstanding high voltage of all resistors is guaranteed, different resistance ranges have different demand of accuracy.

Therefore, in the process of designing the resistance circuit of the high voltage and high value resistors standard metering device, the resistance material is selected according to the specific resistance range. When the standard resistance value is small, the accuracy of the resistance is high. The precision resistors of RX70 series with high resistance precision, small temperature coefficient, firm structure and stable performance should be selected. When calibrating high resistance values, the resistance accuracy is relatively low but the resistance should withstand high voltage. The chip glass glaze-film low-sensitivity high-voltage resistors of RI82 series are widely used in high-voltage equipment because of its small size, light weight, high power and ability to withstand large overload capacity.

2.2. Standard resistance multi-gear design

After selecting the standard resistance that meets the requirements, the standard resistors and step switches are used to form a standard internal resistance measuring circuit for calibrating the resistors. The internal resistance network design of the high voltage and high value resistors standard metering device is shown in Figure 1.

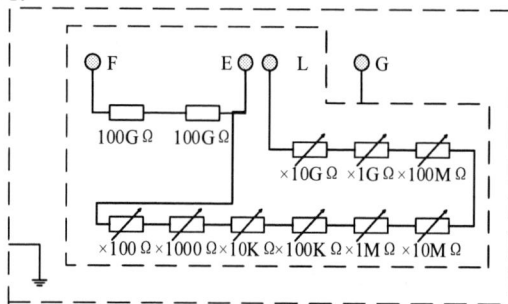

Fig.1 The internal resistance of reference device network diagram

The resistance output is divided into two parts: adjustable resistance and fixed resistance. The 200GΩ resistance is a fixed resistance value and it is calibrated through two ports E and F. The adjustable resistance is calibrated through the E and L ports, and the resistance value of each magnitude is selected by the knob type gear selection switch. The resistance values of each magnitude are connected in series by the equivalent 10 standard resistors. In Figure 1, 9 sliding rheostats are used to replace the function of the step switch in the network, the terminal G is the shield port.

3. High voltage measurement module

The standard device designed in this paper has a nominal voltage range of 250V~10kV, and the six range gears are 250V, 500V, 1kV, 2.5kV, 5kV and 10kV. In order to ensure the measurement range of high voltage and the measurement accuracy of voltage, this paper proposes a high-voltage measurement module with automatic range conversion. The design of the voltage measurement circuit is shown in Figure 2. It consists of the following three parts: input voltage dividing circuit, range automatic conversion circuit, signal processing and amplification circuit, the dot in the figure indicates the remaining three relay control circuits.

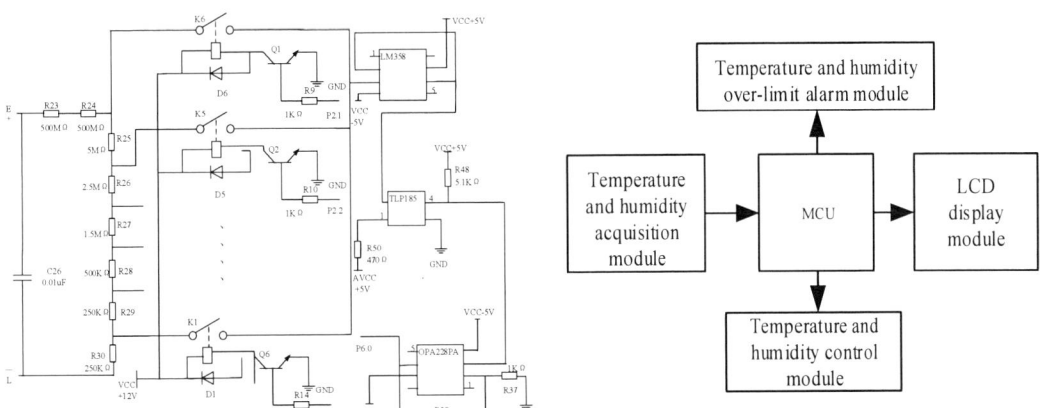

Fig.2 High voltage measuring circuit design diagram

Fig.3 Temperature and humidity control module system diagram

As shown in Figure 2, the measured voltage is input from the E and L ports, then select the range after passing through the voltage dividing circuit. The voltage dividing circuit is mainly composed of two 500MΩ resistors as high voltage side, and then six precision resistors of 5 MΩ, 2.5 MΩ, 1.5 MΩ, 500 KΩ, 250 KΩ, and 250 KΩ are connected in series as low voltage side. The voltage of the full-scale range is reduced to about 2.5V. When the over-range or under-range occurs, the MCU will automatically match the appropriate measurement range through the measurement data. In order to eliminate high frequency interference in high voltage environment, a high voltage ceramic capacitor is connected in series between the high voltage measurement input terminals L and E. At the same time, the voltage output from the relay switch is isolated and buffered by the voltage follower and the linear optocoupler to enhance the stability of the measurement system[7].

The range adaptive function is realized by controlling the six-way relay through the P2.6~P2.1 port level signal of the MCU, and the control port should be low level after the initialization of the MCU, so that the initial state of the relay is normally open, and the circuit is not connected. When the level signal of the MCU port is high, the base of the transistor is input to a high level, so that the transistor is in a saturated conduction state, and the collector becomes a low level. Therefore, the relay coil is energized and the switch is closed. At this time, the measuring circuit works in the corresponding range. In order to prevent the relay from changing from the closed state to the open state and the transient back electromotive force generated by the relay coil will break through the switching device, a diode is connected in parallel on the coil to provide a discharge circuit which increases the safety performance of the system. The output of the voltage dividing circuit is connected with the respective relay control circuit, and the MCU sequentially scans from the high-grade position to the low-level position through the software program until the gear position matching the measured voltage value is found, and then the measurement data is processed and displayed.

4. Temperature and humidity acquisition and control module

The temperature and humidity of the standard device working environment will directly affect the accuracy of the standard resistance and may reduce the insulation performance of the box. Therefore, it is necessary to design the temperature and humidity acquisition and control module to improve the overall performance of the system. This paper adopts a temperature and humidity integrated chip AM2302, which has the characteristics of fast response, low power consumption and strong anti-interference ability. It can collect temperature and humidity detection data at the same time, which simplifies the system hardware design. The system block diagram of the temperature and humidity acquisition and control module is shown in Figure 3.

4.1. The design of AM2302 temperature and humidity detection circuit

The interface circuit between the AM2302 sensor and the MSP430F149 microcontroller is shown in Figure 4.

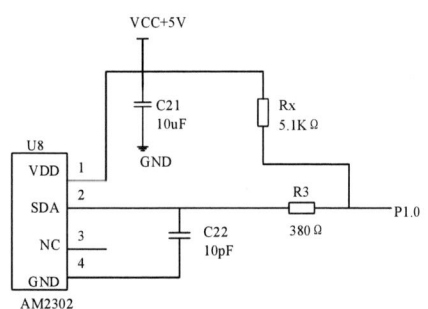

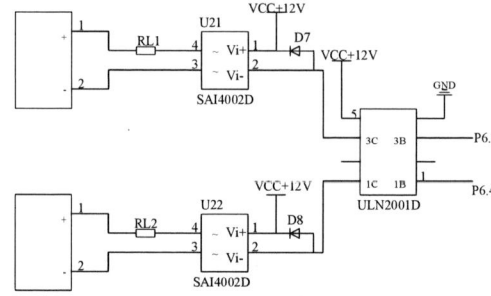

Fig.4 The interface circuit of AM2302 and the single chip

Fig.5 The interface circuit of relay drive circuit and the single chip

The external 5.1kΩ pull-up resistor is designed to keep the high level when the microcontroller is not transmitting data and the bus is idle. The device works in a high voltage environment for a long time. Therefore, a 380Ω resistor and a 10pF capacitor are added between the SDA port and GND to enhance the anti-interference ability. The single bus communication mode is adopted between the sensor and the MCU: after the system is initialized, the AM2302 starts to wait for work and automatically goes to the sleep state; after the user sends a start signal through the master controller, the sensor immediately responds from the sleep mode to the high speed mode. At this time the data line port is kept high, it is ready to detect the external signal when it is in the input mode; after the start signal sent by the MCU is finished, the sensor starts to send the response signal and outputs 40-bit data from the single-bus SDA serial port, in the order: the high digits of humidity data, the low digits of humidity data, the high digits of temperature data, the low digits of temperature data and check digit; complete message collection, measure and record the temperature and humidity data of the environment. After the acquisition, the chip will enter the sleep mode and wait for the next signal to wake up.

4.2. Driving relay circuit design

The peripheral circuit design of the solid-state relay SAI4002D and its interface with the microcontroller are shown in Figure 5.The temperature and humidity control system needs to drive the cooling fan and the heating chip to work separately, so the two solid-state relays are selected to be divided into two circuits for circuit design. The drive function of the SAI4002D is completed by the new three-channel relay driver chip ULN2001D. In this design, channel 1 and channel 3 are used to drive the solid state relays. The control signal of the ULN2001D is the active level signals transmitted from the P6.3 and P6.4 ports of the master.

4.3. The analysis of temperature and humidity control module

The working process of the temperature and humidity control system is as follows: After the system is powered on, the chip port initialization is first completed, and the predetermined temperature and humidity value is set. Then the sensor AM2302 detects the temperature and humidity in the box and transmits the data to the microcontroller. Then, the MCU compares the collected data with the initially set value. When the measured temperature and humidity are not exceeded, the MCU stores the data in the data storage unit and drives the LCD to display the data; when the measured data exceeds the set range, the microcontroller drives the relay control circuit to work. If the temperature is lower than the set value, the controller will drive the heating chip to work and if the detected temperature is higher than the set value, the controller will drive the cooling fan to work. The humidity of the measuring environment is mainly controlled by the humidifying device of the laboratory. When the humidity in the box is higher than the set value, the controller will drive the heating chip and the fan to work at the same time, and place a desiccant in the box to absorb the water vapor.

5. System performance test results

5.1. The voltage measurement module test
After the system debugging work is completed, the test is performed by a standard DC voltage source. The indication error of the standard device high voltage measurement is within the range of ±1% of the maximum allowable error. The test results are shown in Table 1.

Table 1. Voltage test results

Range (V)	Standard values (V)	Measuremen t (V)	Error of indication
×250	250.00	248.75	-0.5%
×500	500.00	498.05	-0.4%
×1000	1000.00	996.28	-0.4%
×2500	2500.00	2514.98	0.4%
×5000	5000.00	4975.36	-0.5%
×10000	10000.00	9972.17	-0.3%

5.2. The temperature and humidity control module test
The goal of the simulation test in this paper is to change the internal environment of the standard box to a temperature of 25°C and a humidity of 70% RH. In a relatively closed laboratory environment, the ambient temperature and humidity are adjusted to 15°C and 90%RH with air conditioning and humidifiers to perform the performance test of the temperature and humidity module, and the standard device is placed more fully in the environment. We set the parameters through the software program and start to record the temperature and humidity measurement data within half an hour. Collect valid data every minute for the first ten minutes, and every two minutes after twenty minutes. The test results are shown in Fig. 6.

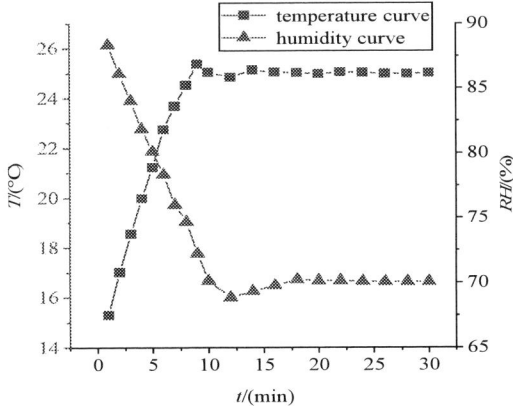

Fig.6 Temperature and humidity module testing

As can be seen from the curve in the figure 6, the designed system takes about 10 minutes to reduce the temperature and humidity of the test environment in the box to a preset value to complete the state transition. After 20 minutes temperature and humidity control, the temperature and humidity indications began to stabilize and the change was small. At the end of the simulation experiment, the state is stable, the error between temperature value and the predetermined value is not more than ±0.3 °C, and the humidity error is not more than 3% RH, indicating that the design of the system meets the requirements.

5.3. Resistance calibration module test

In the standard resistors test, the measured values of all the resistors we get are in the error-allowed range after comparing the measured results with the standard values. The partial verification results are shown in Table 2.

Table 2. Resistance test results

Indicating value (GΩ)	Permissible minimum (GΩ)	Standard value (GΩ)	Permissible maximum (GΩ)	Indicating value (KΩ)	Permissible minimum (KΩ)	Standard value (KΩ)	Permissible maximum (KΩ)	Conclusion P/F
5.000	4.950	5.013	5.050	5.0000	4.9900	4.9941	5.0100	P
8.000	7.920	8.024	8.080	8.0000	7.9840	7.9947	8.0160	P
50.00	49.50	49.72	50.50	50.000	49.900	50.025	50.100	P
80.00	79.20	79.80	80.80	80.000	79.840	80.014	80.160	P

6. Conclusion

Compared to most insulation resistance meter calibration devices, the design of high voltage and high value resistors standard device in this paper has following advantages:

1) Selecting the corresponding resistance model with consideration for the standard resistance structure, material, working environment and specific resistance range can improve the accuracy of the calibration resistor; 2) Using MSP430 single-chip microcomputer as the controller and adding a set of temperature and humidity automatic control module in the standard device can improve the measurement accuracy of the system; 3) Design of high voltage measurement module which can automatically convert the range gear makes the measurement system more intelligent. The test results show that the maximum allowable error of the device designed in this paper is ±0.2% for 100Ω~0.1MΩ, ±0.5% for 0.1MΩ~10MΩ, and ±1% for 10MΩ~200GΩ.

References

[1]   Guan Z, Guan J, Zhao Y, Wang M, Wang X. Research on verification of high voltage and value DC resistors [J].Electrical Measurement & Instrumentation, 2015, 52(S1): 100-103.

[2]   Li J.R, Wang Y.M, Bu C. Research on Precision Calibration Technology of Standard High Voltage and Large Resistance Box [J]. Measurement Technique, 2001(12): 35-36.

[3]   Dong Y. Research on Measurement Uncertainty of DC High-voltage and High-resistance Box Error of Indication by RVD Circuit [J].Control and Instruments in Chemical Industry, 2009, 36(05): 67-69.

[4]   Zhou W, Liu Q.H. Design of High Precision High Voltage High Resistance Calibrater. [J]. Measurement & Control Technology, 2015, 34(12): 5-8.

[5]   JJG1005-2005 Electronic Insulation Resistance Meter Verification Regulation [S]. Beijing: China Metrology Press, 2005.

[6]   Li D.Y, Liu G.J, Li Q, Li H, Hu H.L, Xiong Q.Z, Yang C.Y. Measurement with High Voltage for Temperature Coefficient and Voltage Coefficient of High Precision Resistors [J]. Electrical Measurement & Instrumentation, 2012, 49(10): 20-24.

[7]   Zhen G.Y, Chu J. Design of the measurement circuit about weak-signal amplifying [J]. Electrical Measurement & Instrumentation, 2015, 52(4): 96-100.

Modeling and control of DC microgrid system based on hydrogen production load

Yingjun Guo[1], Zhe Shi[1], Yajie Guo[2], Fanyi Deng[1], Hexu Sun[1]

[1] Hebei University Of Science & Technology, Shijiazhuang050018, China

[2] State Grid Hebei Power Supply Branch, Shijiazhuang050000, China

guoyj_hebust@163.com

ABSTRACT：For the intermittent output of wind turbine and photovoltaic power, aming at the problem of abandoning wind and abandoning light, this paper proposes a DC microgrid system model and control strategy based on hydrogen production load. The DC microgrid system consists of a permanent magnet direct drive wind power system, a photovoltaic power generation system, a buffer energy storage system and an alkaline water electrolysis cell hydrogen production load. The hydrogen production load can adapt to the intermittent fluctuation power supply, make better use of the abandoned wind and light power and be environmentally friendly. Both wind power system and photovoltaic power generation system adopt MPPT control. The battery buffer energy storage device adopts constant voltage charge and discharge control. They are all controlled by their own converters to the DC bus, by producting hydrogen to store the unstable electrical energy generated by distributed generation in the DC microgrid. Based on Matlab/Simulink, results show that the system operates flexibly, the control strategy adopted by the system can maintain the stable operation and load of the system under various power fluctuations and different power supply combinations.

1. Introduction

The DC microgrid has a simple control structure, which can improve the quality of power consumption and reduce power loss. With the increase of DC load, the development advantage is obvious, and the output of power depends on climate and geographical location, so energy storage system is needed [1]. The use of hydrogen to store energy is a suitable option, as a fuel can replace almost all fossil fuel applications [2]. The process of preparing hydrogen by using renewable energy sources such as wind and light is pollution-free and can reduce environmental problems. Electrolyzed water hydrogen can be used as a DC load to adapt to unstable power sources such as wind turbine and photovoltaic.At the same time, it's conducive to solving the problem of abandoned wind and light[3-6].

In recent years, microgrid research combining hydrogen storage systems has been paid attention to in university research. The microgrid system in the literature [7] is an AC microgrid，ach unit in the system has to undergo inverter, which will produce a large loss, and the cell is only activated when the system has a power surplus. The PV/SOFC hybrid power generation system proposed in [8] combines photovoltaic power generation with solid oxide fuel cells and electrolyzers. It is an off-grid system, which also uses PV/SOFC to generate electricity, and the remaining electric energy produces hydrogen storage. The literature [9-10] mainly considers the economics of the system, and studies and optimizes the algorithm of the energy storage unit in the system with the goal of cost optimization. In

Content from this work may be used under the terms of the Creative Commons Attribution 3.0 licence. Any further distribution of this work must maintain attribution to the author(s) and the title of the work, journal citation and DOI.

Published under licence by IOP Publishing Ltd

addition, the literature [11] has studied the modeling and control of the landscape hydrogen integrated energy system.

This paper constructs a DC microgrid system consisting of a permanent magnet direct-drive wind turbine generator, photovoltaic, battery and electrolyzer unit. A control strategy suitable for the system is proposed for the system to operate in two models to achieve stable operation of the system and make full use of distributed power generation.

2. System structure and operation mode
The structure of the DC microgrid system established in this paper is shown in figure 1.

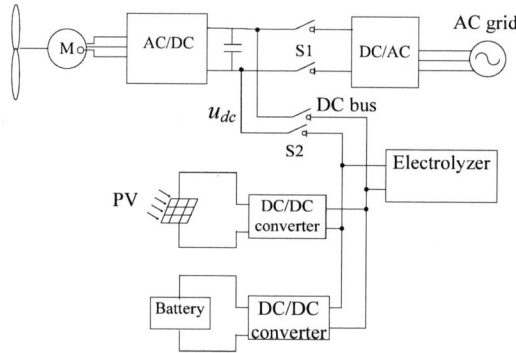

Figure 1. The structure block diagram of dc microgrid based on hydrogen production load.

When the wind level can meet the requirements of grid connection, close S1, disconnect S2, and control the wind power system converter to connect to the grid. Among them, the PV system and the battery system can also prepare hydrogen when the wind power system is connected to the grid to ensure that the electrolytic tank does not stop at the maximum extent. When the wind level does not meet the grid connection requirements or the power system does not allow wind power to be connected to the grid, disconnect S1, close S2, control the wind power system to access the local DC microgrid, and cooperate with the PV system to supply DC hydrogen load. Therefore, the DC microgrid system in this paper can be divided into two modes of operation.

3. System modeling

3.1. Modeling of photovoltaic systems
The equivalent circuit model of the photovoltaic cell selected in this paper is a single diode type, as shown in figure 2.

Obtain the I-U equation for photovoltaic cells [5]:

$$I_{pv} = I_{ph} - I_0 \left\{ \exp\left[\frac{q(U_{pv} + IR_s)}{nkT} \right] - 1 \right\} \tag{1}$$

Where: I_{pv} is the output current of the photovoltaic cell; I_0 is the saturation current; q is the electronic constant ($q = 1.6 * 10^{-19} C$); U_{pv} is the output voltage of the photovoltaic cell; n is the diode characteristic fitting coefficient; k is the Boltzmann constant($k = 1.38 * 10^{-23}$); T is the absolute ambient temperature.

Take I_{ph} equal to short-circuit current I_{sc}, open circuit voltage is expressed as U_{oc}. And let $C_1 I_{sc} = I_0, C_2 U_{oc} = nkT / q$, then equation (1) can be simplified as:

$$I = I_{sc} - C_1 I_{sc} \left\{ \exp\left[\frac{U}{C_2 U_{oc}} \right] - 1 \right\} \tag{2}$$

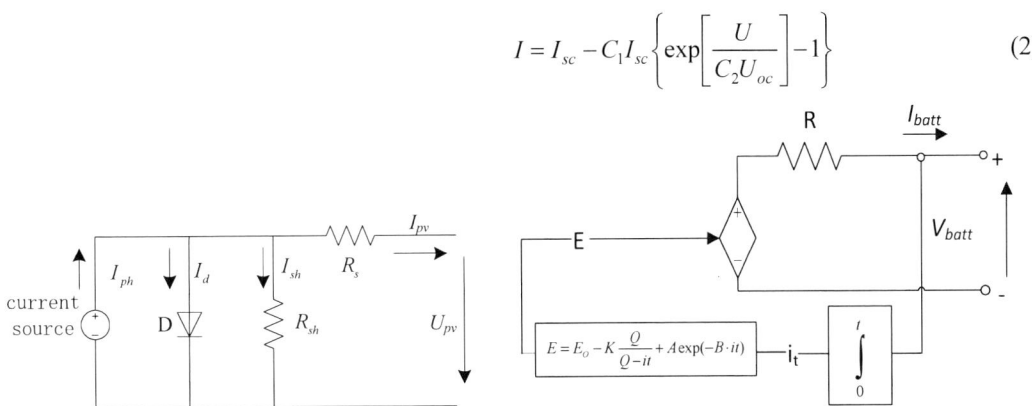

Figure 2. A single diode equivalent circuit of a photovoltaic cell.

Figure 3. Universal model equivalent circuit.

Thus, a mathematical model of photovoltaics for engineering can be obtained, C_1 and C_2 are constants, and since at the maximum power point:

$$\exp\left[\frac{U_m}{C_2 U_{oc}} \right] - 1 \approx \exp\left[\frac{U_m}{C_2 U_{oc}} \right] \tag{3}$$

Can be solved:

$$C_1 = \left(1 - \frac{I_m}{I_{sc}} \right) \exp\left[-\frac{U_m}{C_2 U_{oc}} \right] \quad C_2 = \left(\frac{U_m}{U_{oc}} - 1 \right) \left[I_n \left(1 - \frac{I_m}{I_{sc}} \right) \right]^{-1} \tag{4}$$

3.2. Modeling of wind turbine systems

In the wind power system, the permanent magnet direct-drive wind turbine includes two parts: a wind turbine as a prime mover and a synchronous generator as a generator. Modeling wind turbines [12]:

$$P_m = \frac{1}{2} C_p(\lambda, \beta) \rho A V_w^3 \tag{5}$$

Where: P_m is the mechanical power of the output; ρ is the air density; A is the sweeping area of the blade; V_m is the wind speed; $C_p(\lambda, \beta)$ is the fan efficiency; λ is the tip speed ratio; β is the pitch angle.

The simplified mathematical equations for the voltage, flux linkage and electromagnetic torque of permanent magnet synchronous motor are as follows [13]:

$$\begin{cases} u_{sd} = \dfrac{d\psi_{sd}}{dt} + R_s i_{sd} - \omega_e \psi_{sq} \\ u_{sq} = \dfrac{d\psi_{sq}}{dt} + R_s i_{sq} + \omega_e \psi_{sd} \end{cases} \tag{6}$$

$$\begin{cases} \psi_{sd} = L_{sd} i_{sd} + \psi_f \\ \psi_{sq} = L_{sq} i_{sq} \end{cases} \tag{7}$$

$$T_e = \frac{3}{2} p \left(\psi_{sd} i_{sq} - \psi_{sq} i_{sd} \right) \tag{8}$$

Where: u_{sd}, u_{sq} are generator stator output voltage d, q axis components; ψ_{sd}, ψ_{sq} are the d and q axis components of the generator stator flux linkage; i_{sd}, i_{sq} are generator stator output current d, q axis components; L_{sd}, L_{sq} are d, q axis electronic coil inductance; ω_e is the electrical angular velocity; ψ_f is a rotor flux linkage; R_s is the stator resistance; p is the pole number of the motor.

3.3. Battery modeling

The equivalent model of the lead-acid battery used in this paper is the general model of the battery. The equivalent circuit is shown in figure 3. It consists of a controllable voltage source connected to the battery internal resistance R. Specifically, its mathematical model can be expressed by the following formula [14].

$$V_{batt} = E - R \cdot I_{batt} \tag{9}$$

Where E is a controlled voltage source, which can be expressed as:

$$E = E_0 - K \frac{Q}{Q - \int idt} + A \exp\left(-B \int idt\right) \tag{10}$$

Where: V_{batt} is the battery voltage; E is the no-load voltage; R is an internal resistance; I_{batt} is the battery current; E_0 is the internal potential; Q is the maximum capacity of the battery; $\int idt$ is the amount of discharge; K is the polarization voltage; A is the exponential region voltage amplitude; B is the reciprocal of the time constant of the exponential region.

3.4. Cell modeling

The mathematical model of the electrolyzer used in this paper is based on the experimental experience of the alkaline cell model, which is more suitable for the analysis of the electrical field, including the model of U-I characteristics and the model of hydrogen production rate. The specific description is as follows [8]:

$$U_{elec,cell} = U_{rev} + \frac{r_1 + r_2 T_{elec}}{A} I_{elec} + k_{elec} \ln\left(\frac{k_{T1} + k_{T2}/T_{elec} + k_{T3}/T_{elec}^2}{A} I_{elec} + 1\right) \tag{11}$$

Where: $U_{elec,cell}$ is unit voltage; r_1 and r_2 are ohmic resistance related parameters; T_{elec} is cell temperature; k_{elec}, k_{T1}, k_{T2}, k_{T3} are overvoltage related parameter constants; A is unit electrode area; I_{elec} is output unit current; U_{rev} is reversible open circuit voltage.

Considering the loss of parasitic current in practical applications, the hydrogen production rate can be expressed as follows:

$$q_{H2} = \frac{(I_{elec}/A)^2}{k_{f1} + (I_{elec}/A)^2} \frac{n_c I_{elec}}{2F} \tag{12}$$

Where: k_{f1}, k_{f2} for current efficiency calculation of related parameters.

4. Control of each unit of the system

The control principle is shown in Figure 4. In this system, the wind power system and the photovoltaic system are added to MPPT (Maximum Power Point Tracking) control for making fuller use of energy. The wind power system converter adopts the versatile full-power back-to-back dual PWM topology structure, and the MPPT control is the optimal tip speed ratio method [15]. The machine side converter control method is dynamic indirect current speed control combined with zero d-axis current control and feedforward decoupling control. The grid side inverter adopts grid-based voltage vector control with feedforward decoupling [16,17]. The photovoltaic system interface converter adopts Boost chopper, and the MPPT control is variable step disturbance observation method. As the energy storage unit of the system, the battery can maintain the DC bus voltage within the specified range and maintain the normal operation of the system when the power fluctuates continuously. The interface converter is a bidirectional DC/DC converter, which adopts a constant voltage charging (discharging) electric control strategy [18]. The electrolysis cell can be regarded as a voltage-sensitive nonlinear DC load as an electric device, and the hydrogen gas obtained through the electrolysis cell can be applied in various fields. In addition, fuel cells can convert hydrogen into secondary energy to provide stable power output, which is an important development direction for hydrogen utilization in the future.

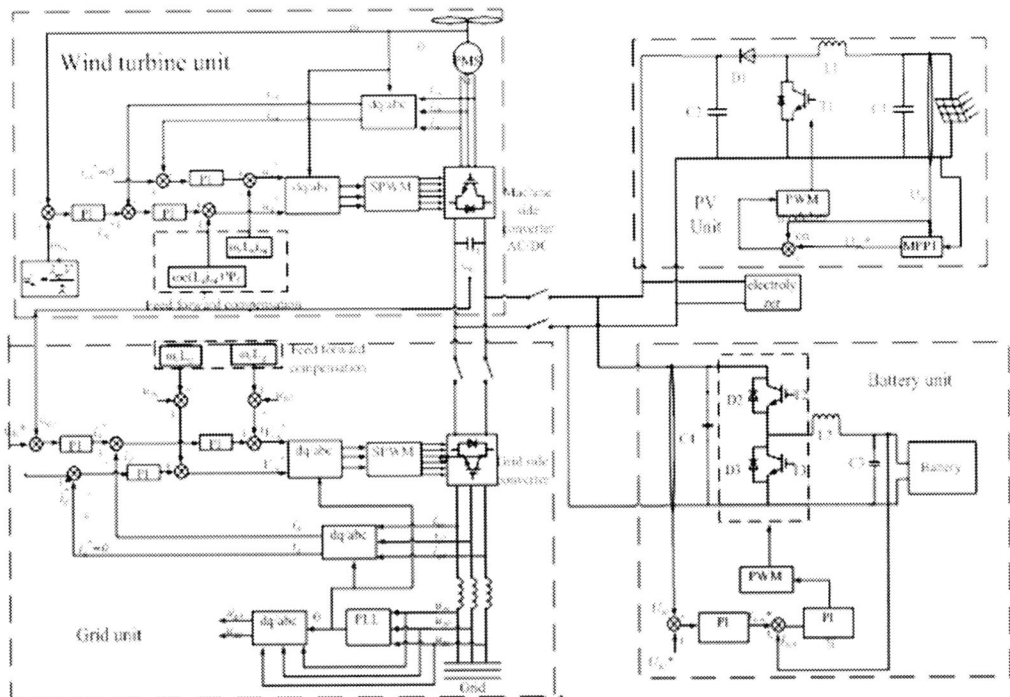

Figure 4. System control block diagram.

5. System simulation verificationr
The main parameters of the module are shown in Tables 1~4.

5.1. Grid-connected operation mode
In the network mode, when the electrical parameters including voltage, frequency, phase, etc. of the direct drive fan meet the requirements of grid connection, the grid-connected switch in the converter is closed by the converter control system. The direct drive fan is connected to the grid for power generation. At this time, the photovoltaic system in the system is more adaptable due to the output form of its direct current and the hydrogen production load. In the case of daytime and no cloud layer is completely blocked, the photovoltaic system can cooperate with the energy storage battery for the hydrogen production load operation.Set the initial wind speed to 8m/s, change the wind speed to 12m/s and 10m/s at 0.8s and 1.5s, and change the light intensity from 1000W/m^2 to 600W/m^2 at 2s. The simulation time is 2.5s. The simulation results are shown in figure 5.

Table 1. Wind turbine parameters.

Wind turbine	Value	Wind turbine	Value
P_N/kW	30	V_N/(m/s)	12
R/m	4	ρ/ (kg/m^3)	1.225
J/(kg·m^2)	12	R_s/(Ω)	0.005
n_p	8	L/(mH)	2

Table 2. Wind power grid connection system parameters.

Inverter section	Value	Grid section	Value
$u_{dc}^{*}/(V)$	800	$U_{abcN}/(V)$	380
$C/(\mu F)$	5000	$f_N/$ (Hz)	50
$f_c/(kHz)$	10	L/(mH)	6

Table 3. Photovoltaic system parameters.

PV	Value	PV	Value	PV	Value
$T_{ref}/°C$	25	$S_{ref}/W/m^2$	1000	$c/°C^{-1}$	0.00288
I_m/A	59.1	U_m/V	437.5	N_s	25
I_{sc}/A	63.3	U_{oc}/V	537.5	P_N/kW	30

Table 4. Battery parameter.

Battery	Value	DC/DC	Value
$U_{nom}/(V)$	120	L/(H)	4.5e-3
$C_{nom}/(Ah)$	250	$C1/$ (F)	1200e-6
$initialSOC(\%)$	80	C2/(F)	1200e-6

The results show that under this control strategy, the wind power output can track the change of wind speed well, the output voltage is consistent with the grid frequency and phase, and the voltage obtained on the DC side is stable and stable when the wind speed changes. For the FFT spectrum analysis of the output grid-connected current after inverter control, the total harmonic distortion rate of the grid-connected current is 0.91%, which is less than 5% of the grid-connected national standard. When the power consumption of the photovoltaic system is reduced at 2s, the battery is changed from the charging mode to the discharging mode to complement the power shortage of the photovoltaic system.

5.2. Off-grid hydrogen production mode

In this mode, the wind power system is integrated into the DC microgrid through the intermediate DC link of the converter, so that the wind power and PV power work together as the power source of the microgrid, and the battery unit as a buffer for energy storage to maintain the stability of the DC bus voltage. Set the simulation time to 2.5s, the initial wind speed is 8m/s, change the wind speed to 12m/s and 10m/s at 0.8s and 1.5s, and change the light from $1000W/m^2$ to $600W/m^2$ at 2s. The result is shown in figure 6.

The simulation results show that the bus voltage of the battery can be stabilized when the wind speed and light intensity conditions change, figure 6 (e) and (f) are the hydrogen production rate waveforms of the load in the system including the battery energy storage device and the energy storage device. It can be seen that the hydrogen production load can stabilize the hydrogen production when the energy storage device is buffered, and the intermittent energy source of the DC microgrid can be better utilized.

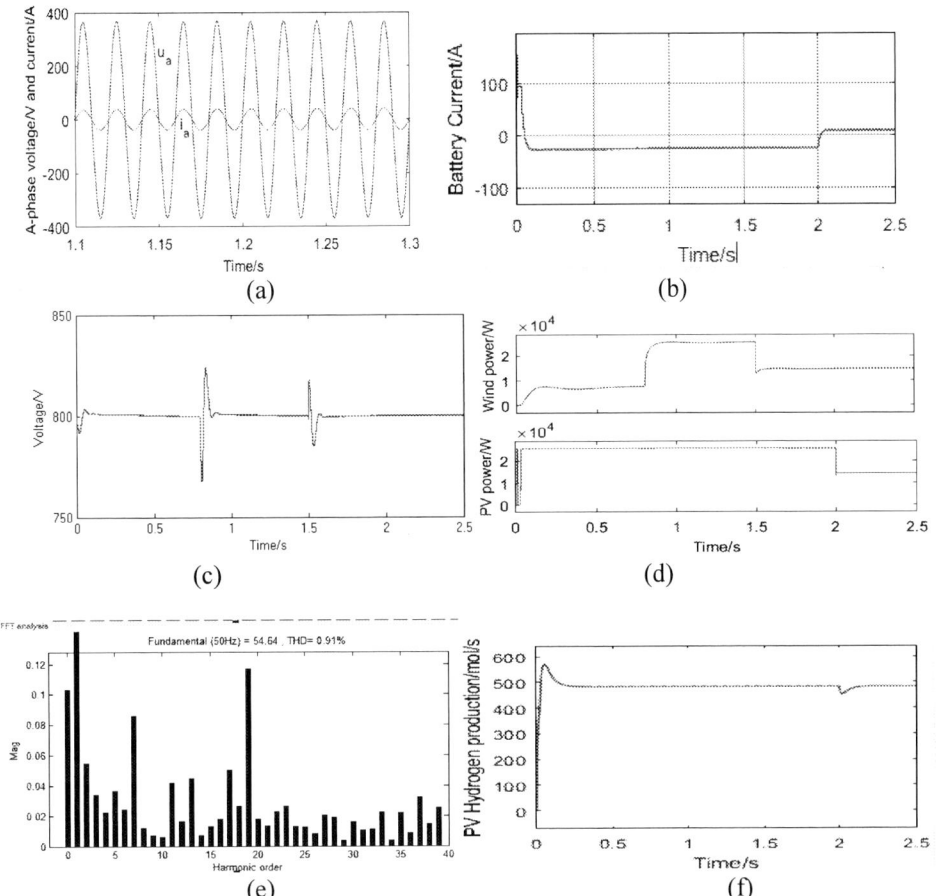

Figure 5. Photovoltaic hydrogen production simulation results in grid-connected mode.

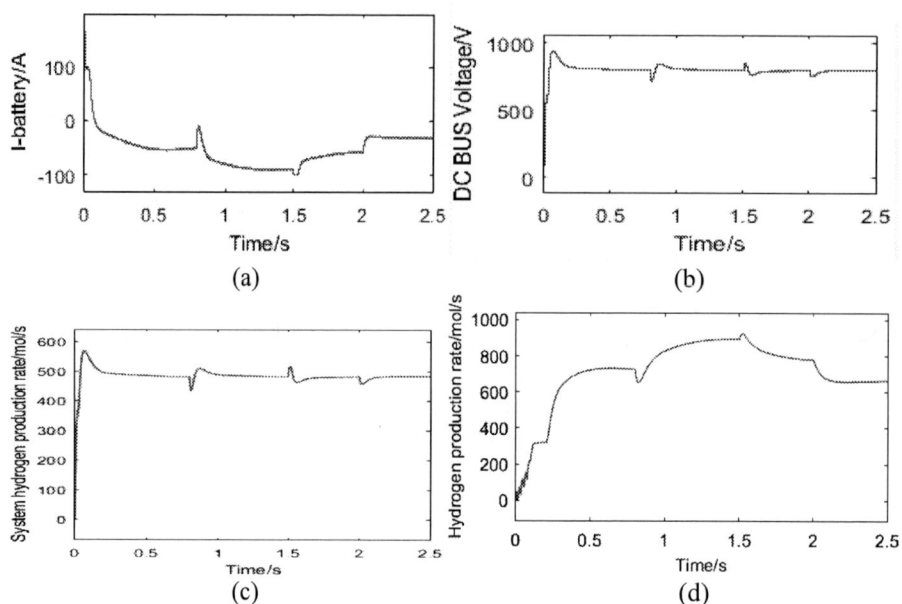

Figure 6. Off-grid hydrogen production simulation results.

6. Conclusion

In this paper, a DC microgrid structure including wind turbine unit, PV unit, battery unit and electrolyzer unit is constructed. The unit is modeled and a suitable control strategy is proposed. And through the simulation results of Matlab/Simulink, it is concluded that the stable three-phase AC power is output in the grid-connected mode and the complete decoupling can be operated under high power factor, and the photovoltaic and energy storage systems can simultaneously prepare hydrogen. In the off-grid hydrogen production mode, the micro-source and the energy storage system can maintain the stability of the bus voltage and the stable operation of the hydrogen production load. However, although the system output power and bus voltage cannot be stabilized when the energy storage system is out of operation, it can also be used to prepare hydrogen if it meets the hydrogen production working conditions. This system operates in a flexible manner and makes better use of the intermittent energy of the DC microgrid.

Acknowledgment

This work is supported by Hebei 'Project Titan' Innovation and Entrepreneurship project fund.Hebei Science and Technology Department project which code 16214510D and Hebei Education Department project which code QN2017313 and QN2016109.

References

[1] Yun yang Research on control strategy of photovoltaic DC microgrid based on hybrid energy storage 2016 Xiamen University

[2] He Du,Hong Lv,Daijun Yang Research progress in simulation of wind solar hybrid power generation system 2017 *Chinese Journal of Power Sources* **43** 173-5

[3] Fengxian Luo Current situation of hydrogen production from renewable energy in the world 2017 *Sino-Global Energy* **22** 25-32

[4] Torreglosa J P, García P, Fernández L M,Jurado F Energy dispatching based on predictive controller of an off-grid wind turbine/photovoltaic/hydrogen/battery hybrid system 2015 *Renewable Energy* **74** 326-36

[5]   Trifkovic M, Sheikhzadeh M, Nigim K,Daoutidis P Modeling and Control of a Renewable Hybrid Energy System With Hydrogen Storage 2014 *IEEE Transactions on Control Systems Technology* **22** 169-79

[6]   Trifkovic M, Sheikhzadeh M, Nigim K, Daoutidis P Hierarchical control of a renewable hybrid energy system 2012 *51st IEEE Conf.on Decision and Control（Hawaii）* pp 6376-81

[7]   Mengzhu Qin,Guoyue Zhang,Donglian Qi Modeling and Simulation of wind energy hydrogen coupling system 2016 *Electronic Technology* **8**

[8]   Wei Guo Modeling and performance simulation of hybrid power system based on PV/SOFC 2014 Shandong University

[9]   Pengfei Liu Research on optimal allocation and energy management of integrated power supply system for wind and hydrogen storage 2017 Zhejiang University

[10] Weiqiang Dong Configuration and battery management of wind solar hybrid power system 2017 Zhejiang University

[11] Guowei Cai,Long Peng,Lingguo Kong Power coordinated control of hybrid photovoltaic power generation system 2017 *Automation of Electric Power Systems* **41** 109-16

[12] Shuhua Bai Application of wind solar combined independent power generation system 2007 Chongqing University

[13] Lei Xiao Research on low voltage ride through technology of direct drive permanent magnet wind power generation system 2009 Hunan University

[14] Tremblay O, Dessaint L A, Dekkiche A I. A Generic Battery Model for the Dynamic Simulation of Hybrid Electric Vehicles 2007 *Vehicle Power and Propulsion Conference(Beijing)* pp 284-9

[15] Bin Wu,Sanmin Wei 2012 *Power conversion and control of wind power generation system*(China:China Machine Press)

[16] Danyang Zhao Coordinated control of wind and solar storage DC microgrid 2015 Southwest Jiaotong University

[17] Wei Cheng Dynamic modeling of photovoltaic and wind power generation systems 2012 Zhejiang University

[18] Fernandez L M, Garcia P, Garcia CA,Torreglosa JP,Jurado F Comparison of control schemes for a fuel cell hybrid tramway integrating two dc/dc converters 2010 *International Journal of Hydrogen Energy* **35** 5731-44

A novel State of Health estimation method for Lithium-ion battery in electric vehicles

Jie Fan[1,2], Yuan Zou[1,2*], Xudong Zhang[1,2*] and Hongwei Guo[2]

[1] National Engineering Laboratory for Electric Vehicles, Beijing Institute of Technology, Beijing 100081, China

[2] School of Mechanical Engineering, Beijing Institute of Technology, Beijing 100081, China

Corresponding author: zouyuan@bit.edu.cn; xudong.zhang@bit.edu.cn.

Abstract. To ensure safe operation of lithium-ion battery, precise estimation of its states like state of health (SOH) is indispensable. This paper proposes a novel SOH estimation method for lithium-ion battery in electric vehicle applications. When it's time to update battery capacity, the estimator extracts historical current and voltage data of last driving event and estimates open circuit voltage (OCV) based on affine projection algorithm. By means of the estimated OCV and the three-dimensional response surface OCV model, the SOH estimator gives the estimation result of the battery capacity. Experiments verify that the proposed estimation method is accurate and robust under different working conditions and different aging states.

1. Introduction

The breakthroughs of lithium-ion battery technology in recent years have pushed its wide applications in automobile industry. Nowadays, lithium-ion batteries, specifically lithium nickel-manganese-cobalt oxide (LiNMC) and lithium iron phosphate (LiFePO$_4$) batteries, have dominated the automotive power battery industry due to higher specific energy, longer time span and lower self-discharging rate.

To ensure safety of electric vehicles and optimize the energy management strategy, it is necessary to realize effective monitoring of power battery packs during vehicle operation. State of health (SOH), as one of the most important parameters for lithium-ion battery, needs to be estimated accurately by the battery management system (BMS).

The SOH describes the degree of battery aging and is usually evaluated by battery capacity. Therefore, the problem boils down to capacity estimation. Existing approaches include incremental capacity analysis (ICA), differential voltage analysis (DVA), empirical model based method, etc. Unfortunately, for ICV/DVA method, the battery needs to be charged or discharged in constant-current condition [1-2]. To construct the empirical model, time-consuming and laborious accelerating aging experiments under various environment conditions are indispensable. Thus, above limitations restrict applications of pertinent methods. A simple while useful capacity estimation method is dividing the accumulated charge by the state of charge (SOC) variation in a certain period of time [3], where the SOC is obtained from an OCV-SOC look-up table. Hence the critical problems are how to estimate OCV accurately and build a robust OCV-SOC look-up table against different levels of battery aging.

This paper proposes a novel capacity estimation method for lithium-ion battery in electric vehicle operation. The three-dimensional response surface based OCV-SOC model [4] is adopted to capture

Content from this work may be used under the terms of the Creative Commons Attribution 3.0 licence. Any further distribution of this work must maintain attribution to the author(s) and the title of the work, journal citation and DOI.
Published under licence by IOP Publishing Ltd

the varying relationship between OCV and SOC at different aging levels of the battery and is combined with particle swarm optimization (PSO) algorithm to give the capacity estimation result. As a state-of-the-art optimal algorithm, the PSO method is able to obtain the global optimum very efficiently. Experiments show that the proposed SOH estimation method is accurate and robust.

2. Battery modelling and OCV estimation

The first-order RC equivalent circuit model (ECM), as shown in figure.1, is widely adopted to depict lithium-ion battery dynamic characteristics due to its good balance between simplicity and accuracy [5]. It is comprised of a voltage source, an ohmic resistance R_o and a RC (polarization resistance R_p and polarization capacitance C_p) network. The correspondence between the HPPC voltage response and components of the battery model is annotated in figure.1 as well. Here the voltage source represents the SOC and capacity-dependent OCV and can be expressed by the three-dimensional response surface model as follows:

$$OCV(SOC, C_n) = \sum_{i=0}^{n_p} c_i \times (SOC)^i \qquad (1)$$

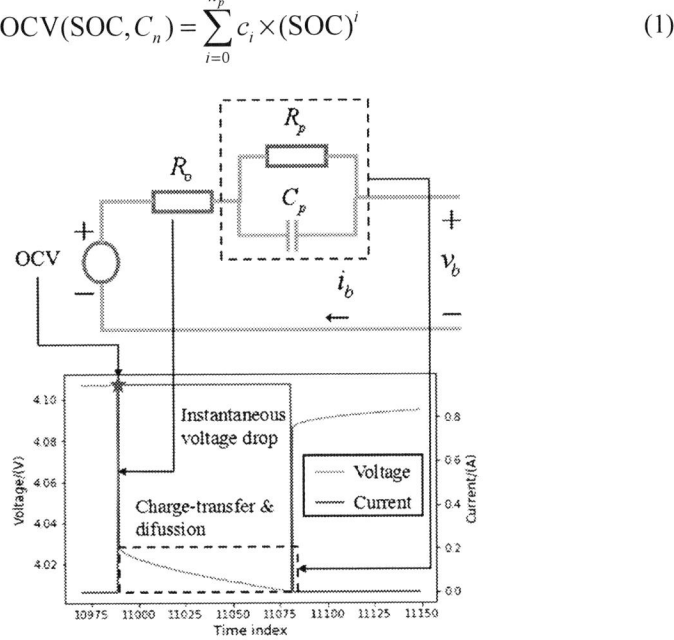

Figure.1. First-order RC equivalent circuit model of the battery

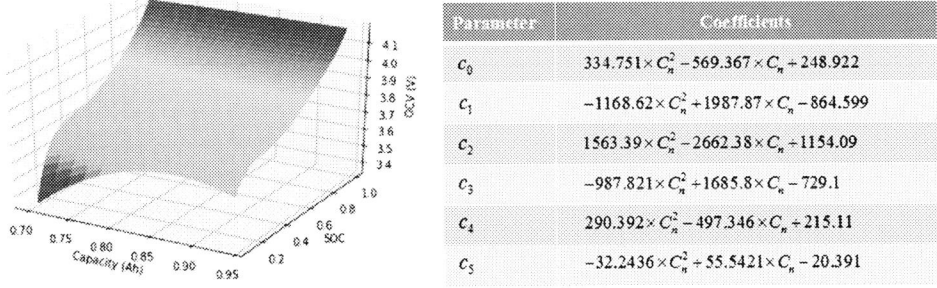

Parameter	Coefficients
c_0	$334.751 \times C_n^2 - 569.367 \times C_n + 248.922$
c_1	$-1168.62 \times C_n^2 + 1987.87 \times C_n - 864.599$
c_2	$1563.39 \times C_n^2 - 2662.38 \times C_n + 1154.09$
c_3	$-987.821 \times C_n^2 + 1685.8 \times C_n - 729.1$
c_4	$290.392 \times C_n^2 - 497.346 \times C_n + 215.11$
c_5	$-32.2436 \times C_n^2 + 55.5421 \times C_n - 20.391$

Figure.2. The three-dimensional response surface model of OCV

In this paper, the polynomial order of OCV with respect to SOC, which is denoted as n_p in equation (1), is set to 5. The polynomial coefficient c_i is two-order polynomial function of capacity C_n. The fitting result of the investigated LiNMC based on HPPC test is shown in figure.2.

According to Kirchhoff's law, the electrical behavior of the adopted ECM can be expressed as:

$$C_p \dot{v}_p + \frac{v_p}{R_p} = i_b \qquad (2)$$

$$v_b = OCV - v_p - i_b R_o \qquad (3)$$

where v_p is defined as the voltage across the RC network, i_b is the outflow current (positive for discharging and negative for charging) and v_b represents the terminal voltage.

By means of z-transformation, the non-linear ECM battery model can be transformed into the following auto regressive exogenous (ARX) model, which is linear and demonstrated below [6]:

$$y_k = \theta_k^T x_k \qquad (4)$$

$$x_k = [-i_b(k), -i_b(k-1), OCV(k-1) - v_b(k-1), 1]^T \qquad (5)$$

$$\theta_k = [\theta_1, \theta_2, \theta_3, \theta_4]^T = [R_o, -R_o + \frac{T_s}{C_t} + \frac{T_s R_o}{C_t R_t}, \frac{T_s}{C_t R_t} - 1, OCV]^T \qquad (6)$$

where k is the step index. T_s is the data sampling period.

In this paper, affine projection (AP) algorithm is used to estimate the OCV. Generally, AP algorithm encompasses a family of configurable algorithms designed to improve the performance of other adaptive algorithms, mainly least mean squares based ones, especially when input data is highly correlated [7], which is just the case for the battery model due to the "memory property" introduced by the RC network. The AP algorithm is summarized as follows.

Let $w_L[n]$ represents the L adaptive filter coefficients and $w_L[n]x_L[n]=d[n]$ denotes the linear model, then the update equation is given by:

$$w_L[n] = w_L[n-1-\alpha(N-1)] + \mu A_\tau^T[n](A_\tau[n]A_\tau^T[n] + \delta I)^{-1} e_{N\tau}[n] \qquad (7)$$

$$e_{N\tau}[n] = d_{N\tau}[n] - A_\tau[n]w_L[n-1-\alpha(N-1)] \qquad (8)$$

$$A_\tau[n] = (x_L[n], x_L[n-\tau], \ldots, x_L[n-(N-1)\tau])^T \qquad (9)$$

$$d_{N\tau}^T[n] = (d[n], d[n-\tau], \ldots, d[n-(N-1)\tau]) \qquad (10)$$

where N is the length of the input signal, which is usually called projection order. μ is the update step-size. δ represents the initial offset covariance. α and τ control the interval between the update term and its based term of the coefficients and input signals respectively. Here, we adopt the most frequently used standard AP algorithm where $\delta=0$, $\alpha=0$ and $\tau=1$.

After estimating the parameter matrix θ_k by AP algorithm based on battery current and terminal voltage, the OCV can be obtained by:

$$OCV = \theta_4 \qquad (11)$$

3. PSO based SOH estimator

In this paper, particle swarm optimization (PSO) algorithm is used to find the optimal battery capacity that best fits the constraints imposed by ampere-hour counting and SOC variation during a period. In general, PSO is an efficient and powerful optimization method, inspired by researches about social behavior of bird herds. In PSO, the set of candidate solutions is defined as numerous particles flying around the given predefined space. Each particle will fly with the combined speed aiming at both "local" and "global" best location during update. When the number of iteration reaches preset value or all particles closely converge to one location, the algorithm stops and the location of the best particle, which gives the lowest fitness function value, is the optimization result.

Assuming $X_i=[x_{i1},x_{i2},\ldots,x_{in}]$ and $V_i=[v_{i1},v_{i2},\ldots,v_{in}]$ are the current location and speed of particle i respectively, pbest$_i$=[pbest$_{i1}$,pbest$_{i2}$,...,pbest$_{in}$] is the best location that particle i has once arrived. If $f(X)$

is the fitness function that needs to be minimized, the current best location of particle i is determined by:

$$\text{pbest}_i(t+1) = \begin{cases} \text{pbest}_i(t) & \text{if } f(X_i(t+1)) \geq f(\text{pbest}_i(t)) \\ X_i(t+1) & \text{if } f(X_i(t+1)) < f(\text{pbest}_i(t)) \end{cases} \quad (12)$$

Assuming N is the number of particles, the best location of all particles gbest(t) is called global best location and defined as:

$$\text{gbest}(t) = \min\{f(\text{pbest}_1(t)), f(\text{pbest}_2(t)), \ldots, f(\text{pbest}_N(t))\} \quad (13)$$

The speed and location update equations are as follows:

$$v_{ij}(t+1) = mv_{ij}(t) + c_1 r_1(\text{pbest}_{ij}(t) - x_{ij}(t)) + c_2 r_2(\text{gbest}_{ij}(t) - x_{ij}(t)) \quad (14)$$

$$x_{ij}(t+1) = x_{ij}(t) + v_{ij}(t+1) \quad (15)$$

where $v_{ij}(t)$ and $x_{ij}(t)$ represent the velocity and location component on jth dimension of particle i in tth generation respectively. pbest$_{ij}(t)$ is the best location component on jth dimension of particle i in tth generation. gbest$_j(t)$ represents the global best location component on jth dimension of all particles in tth generation. m is the inertia coefficient. c_1 and c_2 are the cognitive and social parameters respectively. r_1, r_2 are both random numbers in range [0,1]. The main calculation steps of PSO are listed in Table 1.

Table 1. Main steps of PSO

Step 1: Initialize the velocity and location of all particles. Set local best location pbest of each particle as its initial location. Set global best location gbest as the best location of all particles. Set the maximum number of iterations.
Step 2: Adjust velocity and location of each particle according to equation (14) and equation (15).
Step 3: Update pbest of each particle according to equation (12).
Step 4: Update gbest of particle swarms according to equation (13).
Step 5: Check if terminal condition is satisfied. If it is, terminate iteration and return gbest. Otherwise return to step 2.

The proposed SOH algorithm is shown in figure.3. Firstly, part of the measured current and voltage data are extracted from the last driving event. It needs to be noticed that the data length should be long enough so that distinct SOC variation can be observed. Otherwise, the estimated capacity may not be accurate as expected. Then, estimate the OCV based on AP algorithm and record OCV values of some selected sampling points, which are evenly distributed between the start and end points. Considering the initial guess of OCV may be largely deviated from its true value, the first few estimation results of OCV are not reliable. So the start point defined here doesn't refer to the "first point" but the point where the OCV estimation gets close to its true value with high confidence. As shown in figure.3, part of measured data before start point are left out for convergence. At last, estimate the battery capacity based on the three-dimensional response surface OCV model and PSO algorithm, aiming at minimizing the following fitness function:

$$\min_{C_n}\{\sum_{m=1}^{L}[\int_{t_s+(m-1)T}^{t_s+mT} i_b \mathrm{d}t - C_n \times (\text{SOC}_{s+m-1} - \text{SOC}_{s+m})]\}^2 \quad (16)$$

where L is the number of the sampling points. T represents the time interval between adjacent sampling points. In this case, battery capacity C_n is the only optimization variable in the PSO algorithm, where $X=C_n$. Notice that in equation (19), $\text{SOC}_{s+m}(m=1,2,\ldots,L)$ are also functions of battery capacity as they are looked up from OCV-SOC curve, which is picked up from the three-dimensional response surface OCV model according to battery capacity. The concept behind the method is to find the OCV-SOC curve that best fits the equation relationship between $(\text{SOC}_{s+m-1}-\text{SOC}_{s+m})\times C_n$ and

$\displaystyle\int_{t_s+(m-1)T}^{t_s+mT} i_b \mathrm{d}t$, both referring to how much electricity is consumed during a given period. The capacity corresponding to the best suitable OCV-SOC curve is the estimation result.

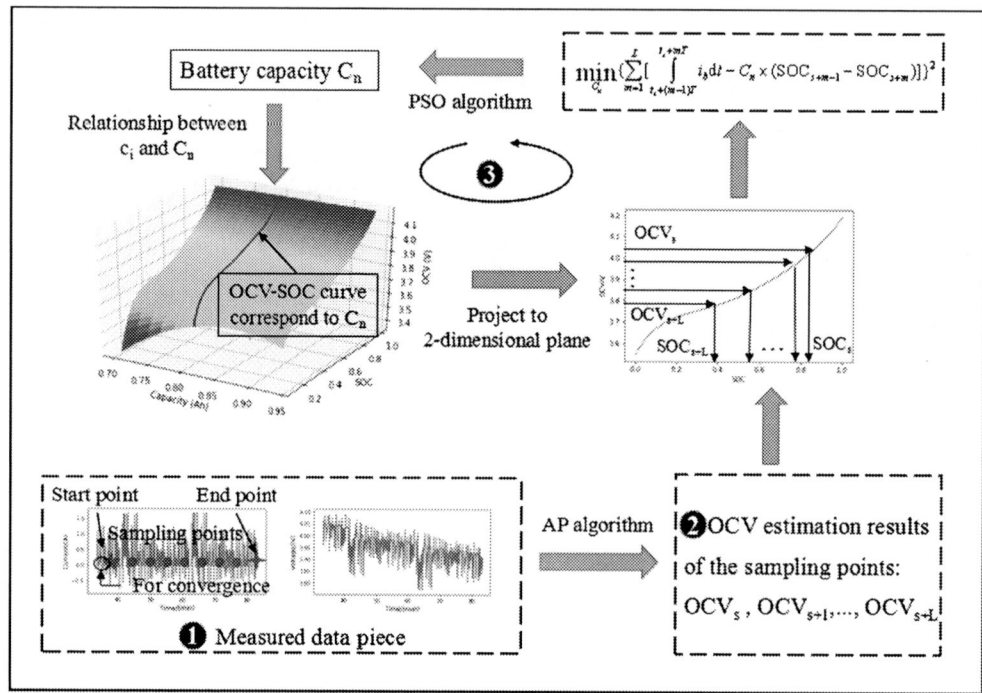

Figure.3. The proposed SOH estimation algorithm

4. Experiment and verification of the proposed SOH estimation method

The battery test data are generated in the test bench consisting of an Arbin BT2000 tester, a thermal chamber, a computer for user-machine interface and a switchboard for cable connection. The voltage, current, temperature of each cell are recorded at the sampling time of 10Hz. The tested batteries were 8 LiNMC battery cells with 0.94Ah nominal capacity, 3.7V nominal voltage. Each cell experienced impedance test and characterization tests (including static capacity test, hybrid pulse test, resistance test, dynamic stress test (DST) and federal urban dynamic schedule (FUDS) test) under 22℃ in different levels of degradation. More details about the battery experiment are available in [5].

Figure.4 demonstrates the validation results of the proposed SOH estimation algorithm. From figure.4 (a), it can be seen that as the number of aging cycle increases, the battery capacity decreases gradually. In addition, whatever the working condition is, the estimated capacity is close to the real value calibrated by coulombic counting method, verifying the robustness of the SOH estimator. To clearly show how accurate the estimation method is, figure.4 (b) plots the relative error between the estimated and real capacity. It is obvious that the relative error is below 1% for all working conditions under different aging states. The mean relative capacity estimation error for HPPC, DST and FUDS at different degradation levels are 0.51%, 0.46% and 0.39% respectively, more accurate than the method reported in [8], which didn't incorporate the influence of capacity on OCV-SOC look-up table and the estimation error can reach 1.66%. The proposed method adopts the three-dimensional response surface model to accurately capture the changing characteristics of OCV-SOC relationship as battery ages and exploits advanced PSO algorithm to find the optimal capacity value that mostly make the SOC change from OCV-SOC look-up table and current integration match. Thus, both the accurate model and optimization algorithm ensure the accuracy of the SOH estimator. At last, it is worthy to mention that the SOH estimator doesn't need to be implemented online considering the slow varying feature of battery capacity as battery degrades.

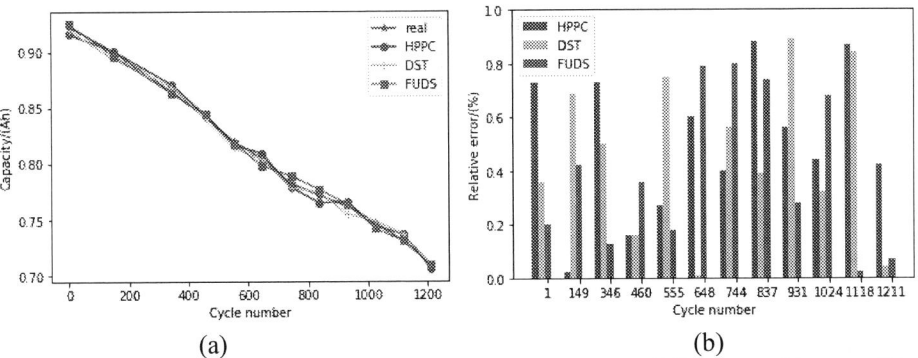

(a)   (b)

Figure.4. Capacity estimation results at different aging states under different working conditions (a) Capacity (b) Relative error

5. Conclusion

This paper presents a novel SOH estimation method for lithium-ion battery in electric vehicle applications. Based on the accurate three-dimensional response surface model of OCV, the SOH estimator is triggered periodically offline to re-calibrate the battery capacity as battery ages. Experiments show the capacity estimation error is below 1% whatever the aging state and working conditions are.

References

[1]   Goh, T., Park, M., Seo, M., Kim, J. G., & Sang, W. K. (2017). Capacity estimation algorithm with a second-order differential voltage curve for Li-ion batteries with NMC cathodes. Energy, 135.

[2]   Riviere, E., Venet, P., Sari, A., Meniere, F., & Bultel, Y. (2015). LiFePO$_4$ Battery State of Health Online Estimation Using Electric Vehicle Embedded Incremental Capacity Analysis. Vehicle Power and Propulsion Conference. IEEE.

[3]   Einhorn, M., Conte, F. V., Kral, C., & Fleig, J. (2012). A method for online capacity estimation of lithium ion battery cells using the state of charge and the transferred charge. IEEE Transactions on Industry Applications, 48(2), 736-741.

[4]   Sun, F., & Xiong, R. (2015). A novel dual-scale cell state-of-charge estimation approach for series-connected battery pack used in electric vehicles. Journal of Power Sources, 274, 582-594.

[5]   Hu, X., Li, S., & Peng, H. (2012). A comparative study of equivalent circuit models for Li-ion batteries. Journal of Power Sources, 198(198), 359-367.

[6]   Bastawrous, H. (2015). Online state of charge and model parameters estimation of the LiFepO4 battery in electric vehicles using multiple adaptive forgetting factors recursive least-squares. Journal of Power Sources, 296(November), 215-224.

[7]   Gonzalez, A., Ferrer, M., Albu, F., & Diego, M. D. (2012). Affine projection algorithms: evolution to smart and fast algorithms and applications. 1965-1969.

[8]   Shen, P., Ouyang, M., Lu, L., Li, J., & Feng, X. (2018). The co-estimation of state of charge, state of health and state of function for lithium-ion batteries in electric vehicles. IEEE Transactions on Vehicular Technology, PP(99), 1-1.

ISPECE   IOP Publishing

IOP Conf. Series: Journal of Physics: Conf. Series **1187** (2019) 022015   doi:10.1088/1742-6596/1187/2/022015

The key design technology of Successive approximation analog-to-digital converter to improve efficient and precision

Cheng Xie[1*], Yueyue Chen[1], jianjun Chen[1]

[1]National University of Defense Technology

Email: 592865823@qq.com,chenyueyue@nudt.edu.cn,1106637684@qq.com

TEL:18252093758

Abstract. Transistor with the advent of the Internet of Things era, high efficiency and high precision SAR ADC (Successive approximation analog-to-digital converter) has become a new research hotspot, and the design has certain challenges. By studying the key techniques for improving the efficiency and accuracy of SAR ADC, it is found that using non-binary DACs can greatly improve efficiency, and resize the size of the key parts of the comparator's transistors, which can greatly improve the precision of SAR ADC. Finally, a 14-bit non-binary SAR ADC was designed by IO transistor (this transistor is connected to 2.5V) and core transistor (this transistor is connected to 1.2V). The ENOB has reached 13.6, SNDR has reached 81.16, SFDR has reached 83.23 through simulation test.

1. Introduction

In the Internet of Things era, every object can be addressed, and every object can communicate, and every object can be controlled[1,2]. This requires constant exchange of information between the natural world and the computer. All signals in the natural world are analog signals, and computers can only process binary digital signals. Therefore, analog-to-digital converters have become a hot research topic[3-7]. The analog-to-digital converter (SAR ADC) of the successive approximation structure has high repetitive utilization of circuit modules, requires less analog circuits, has low circuit complexity[8,9], is simple in design, and has low power consumption and cost. Focus. High-precision, high-speed SAR ADCs are widely used in military, aerospace, medical, and control fields and have high research value.

This paper studies the implementation of a 14-bit 3M/s non-binary ADC and passes the simulation test. The whole circuit of SAR ADC is composed of analog circuit and digital circuit. In order to improve the resolution of the circuit, the sampling switch, non-binary DAC, comparator, etc. are built with IO transistor, and the IO transistor has a higher threshold voltage. And connected to 2.5V, this will make the design a higher margin, more than twice the resolution of all using the core transistor to build the circuit. If all the core transistor are used to build the circuit, then $1LSB=1.2/214=73.24mV$, when using the IO transistor to build the comparator and other circuits, $1LSB=2.5/214=152.59mV$, which means that the comparator only needs to resolve the voltage of 152.59mV. The design difficulty is greatly reduced, and the resolution of the ADC of this architecture is further improved. The SAR logic and clock control module is built with core transistor. The SAR logic is mainly composed of registers, which reduces the power consumption and area of the circuit and greatly increases the sampling speed. The circuit of the core tube and the circuit built by the IO transistor are bridged by the high and low level conversion circuit. The main reason for this part of the circuit design is to make the delay as small as

Content from this work may be used under the terms of the Creative Commons Attribution 3.0 licence. Any further distribution of this work must maintain attribution to the author(s) and the title of the work, journal citation and DOI.
Published under licence by IOP Publishing Ltd

possible, so that the circuit can be completed in a specified period. Generate a digital code. The full circuit design is shown in Figure 1.

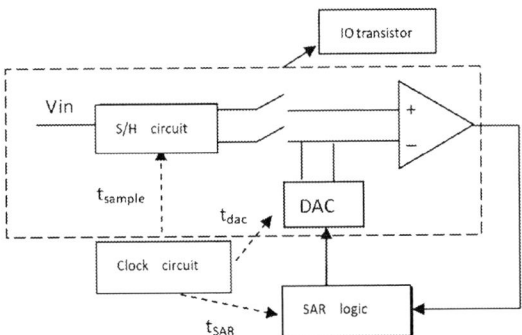

Figure 1 The full circuit design

2. The theory of Non-binary DAC

The DAC for SAR ADCs is based on a charge-extended architecture. The traditional DAC capacitors follow the binary weight, that is, the capacitance from small to large follows the distribution C, 2C, ... 2N-1C, this DAC lineup energy efficiency ratio is not high, the total capacitance will be relatively large, many designs are currently Following the non-binary capacitance ratio, the value of the proportional coefficient a is determined by studying the non-binary characteristic transfer curve. The transfer curves of the three ratios are as follows.

Figure 2(a) is an ideal binary-scale DAC structure, that is, Cn:Cn-1=2, its output is linear, one voltage corresponds to one digital code, and the voltage input is VFS/2, the whole analog conversion The result is 10000.... or 01111...., the final codeword corresponding to each input voltage is separated by one LSB.

Figure 2(b) shows the DAC structure of a capacitor Super-radix-2, ie Cn:Cn-1>2. Its output is non-linear, most of the voltage corresponds to a digital code, and the final codeword corresponding to each input voltage is separated by one LSB. However, when the voltage input is close to VFS/2, the whole analog conversion result will be missed, that is, multiple voltages correspond to one output codeword, so that many analog voltage quantities show the same result, and the information will be lost, even if it is It is impossible to restore the previous information through calibration. It is absolutely not advisable to construct a DAC in this way.

Figure 2(c) shows the DAC structure of a capacitor Sub-radix-2, ie Cn:Cn-1<2. Its output is also non-linear, most of the voltage corresponds to a digital code, and the final codeword corresponding to each input voltage is separated by one LSB. When the voltage input is close to VFS/2, the whole analog conversion result will be out of code, that is, one voltage corresponds to multiple output code words, which causes one voltage to correspond to multiple code words, but other voltages are corresponding. a code word. Some information will be redundant. These redundant input information can be eliminated by calibration. Restore to the information you want before, so that the original analog information is not lost. It is theoretically desirable to build a DAC in this way, and the total capacitance will be much smaller than conventional structures.

In this paper, the DAC structure of Sub-radix-2 is determined. The ratio of the upper and lower capacitors is less than 2. Similarly, the formula can be converted into the following formula:

$$\frac{1-\alpha^N}{1-\alpha} + 1 - 2^{ENOB} - 2.7\theta\sqrt{\frac{1-\alpha^{2N}}{1-\alpha^2} + 1} \geq 0 \tag{1}$$

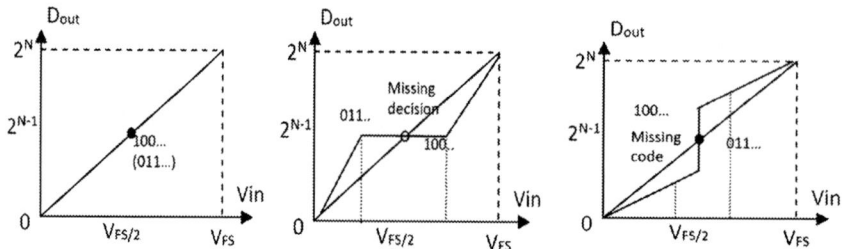

Figure 2 Transfer characteristic curve (a) binary (b) Super-radix-2 (c) Sub-radix-2

θ is very small and can be regarded as 0. Finally, the scale factor required for the target effective number of bits can be calculated according to the calculation. As shown in Table 2.1, it can be seen that the larger the effective number of bits, the larger the required proportional coefficient. This non-binary capacitive DAC array will have a significant bit of 16 bits, and a 16-bit scale factor will be greater than 2, and the non-binary theory will fail.

Table 1 Relationship between effective number of bits and scale factor

ENOB	α
16	2
15	1.91
14	1.85
13	1.78

Finally, according to the calculation, the proportional coefficient required for the target effective number of bits is obtained. As shown in Table 1, it can be seen that the larger the effective number of bits, the larger the required proportional coefficient. This non-binary capacitive DAC array will have a significant bit of 16 bits, and a 16-bit scale factor will be greater than 2, and the non-binary theory will fail.

Applying the above DAC logic timing and improving to a non-binary type, the scale factor is 1.85 and the unit capacitance is 520aF, so the subsequent capacitance is 970aF and 1.8F, and the ratio of the upper and lower capacitors is 1.85. The result of this DAC quantization is non-binary. The output codeword of the last circuit is converted to a binary output by an external circuit. Since the external circuit is used, the overall circuit and various performances of the SAR ADC are not affected. This type of DAC will bring redundant code words to the conversion result, but by studying many existing algorithms, it can be calibrated using a digital perturbation-based algorithm, completely eliminating the adverse effects of redundancy.

At the same time, non-binary DACs bring many benefits, which can eliminate the adverse effects of some process deviation circuit performance. Moreover, using this non-binary structure can save 10 times more power than a DAC with a binary structure, thereby achieving high energy efficiency conversion. The results of the test power consumption simulation are shown in Figure 3 .

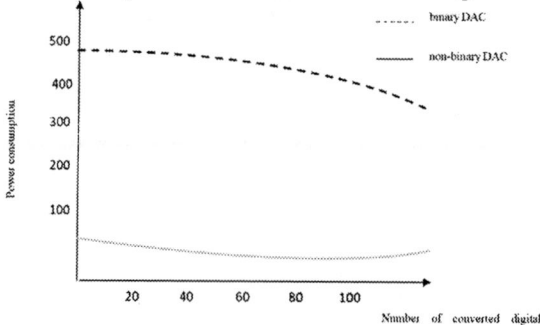

Figure 3 Power consumption comparison between binary DAC and non-binary DAC

3. High precision comparator

The 14-bit 2.5V ADC theoretically needs to recognize a minimum voltage of 160uV. This comparator is built on the simulation platform. The comparator as shown in Figure 4 is amplified in the first stage and latched in the second stage. A number of simulations have been used, mainly to adjust the size of the first stage of the two drive transistor M1 and M2 and the tail current transistor M3 and M4 to increase the gain of the entire first stage, and to increase the size of the second stage input transistor M9 and M10. Through dozens of debugging, the resulting comparator has a minimum comparison voltage of 10uV. Finally, the output of the better swing is obtained. The Monte-carlo simulation of the entire offset voltage and the resolution of the circuit operation. The simulation results are shown in Figure 5.

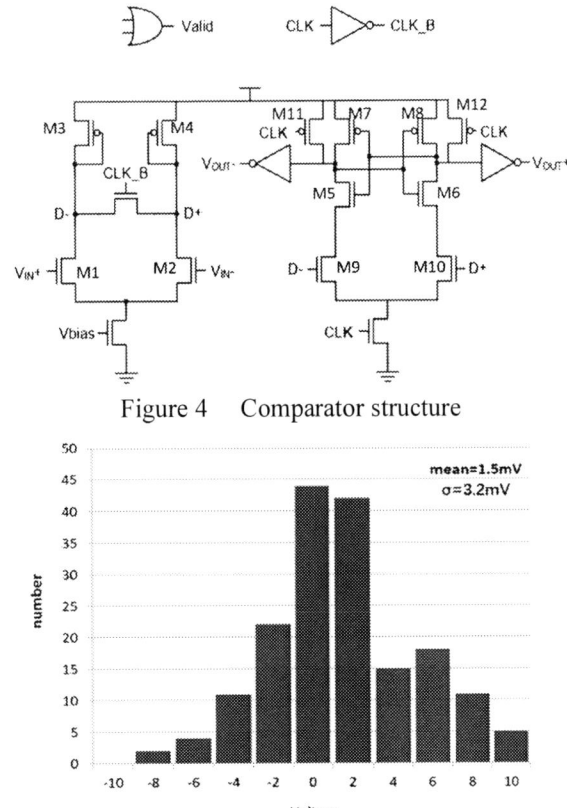

Figure 4   Comparator structure

Figure 5   Monte-carlo simulation of the entire offset voltage

It can be concluded that the offset voltage of the comparator is only 1.5mV, and the deviation of the offset voltage is only 3.2mV, and the obtained performance is quite good.

As can be seen from Figure 6, the comparator can pass the test simulation, the obtained waveform can recognize the voltage of 10uv, and the overall accuracy obtained is very high.

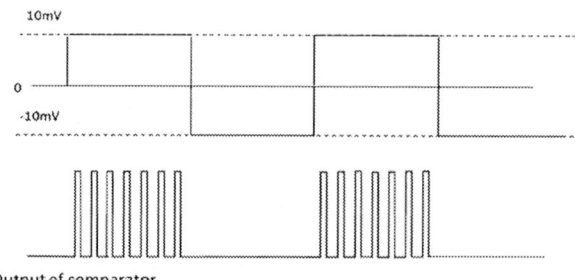
Figure 6 output waveform of Comparator

4.Simulation results

Figure 7 is a full-chip overall circuit. The number 1 represents the sampler, the number 2 represents the DAC, the number 3 represents the disturbance input calibration circuit, the number 4 is the comparator, the number 5 is the high level to low level circuit, the number 6 is the SAR logic circuit, the number 7 is the clock circuit, and the number 8 is a low level to high level circuit, in addition to some other necessary logic operation modules.

By simulating the ADC, the main performance parameters of the SAR ADC were tested by Cadence software based on the CMOS 65nm process. The output signal was converted into a spectrum signal by the fast Fourier transform method to calculate the parameter value. The following parameters are the performance output spectrum of the SAR ADC tested at a supply voltage of 2.5V/1.2V, a target accuracy of 14 bits, and a 3M/s sampling frequency. The result is shown in Figure 8.

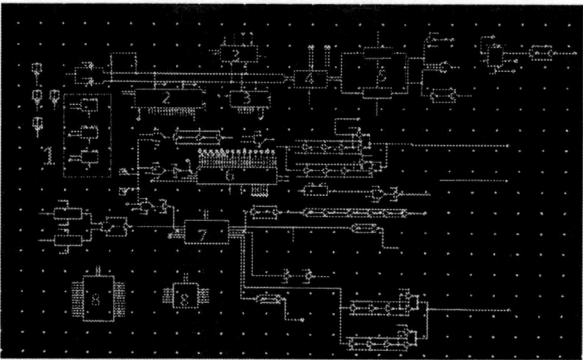
Figure 8 Example of a full circuit implementation diagram

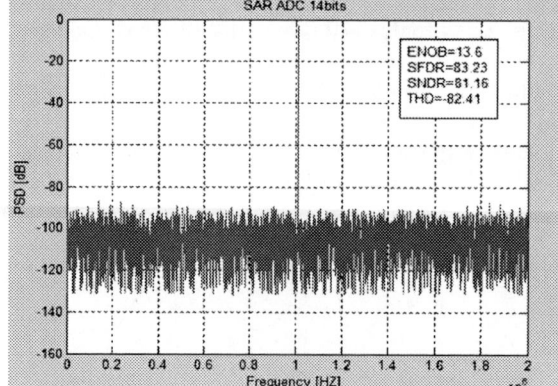
Figure 9 performance test result of SAR ADC

5. conclusion

Under the SMIC65nm process, through the two key design technologies of SAR ADC (non-binary theory and high precision comparator), greatly improving the efficiency and accuracy of the SAR ADC. The whole circuit of 14-bit non-binary SAR ADC is designed by IO transistor and core transistor. The circuit passed the simulation and performance test. ENOB has reached 13.6, SNDR has reached 81.16, SFDR has reached 83.23.

References

[1] Dirk Henrici. 2008. RFID Security and Privacy: Concept, Protocols, and Architectures[M]. Berlin: Springer.

[2] James F.Kurose, Keith W. Ross. 2010. Computer Networking:A Top to Down Approach[M].5th Edition. Boston Mass.:Pearson.

[3] D. Schinkel，E. Mensink,E. Van Tuijl，and B. Nauta,A Double-Tail Latch-Type Voltage Sense Amplifier with 1 8ps Setup+Hold Time[C]．IEEE Int．Solid-State Circuits Conf. (ISSCC)，Dig. Tech. Papers，Feb. 2007，PP. 314—315.

[4] Tasnim B. Nazzal, Soliman A. Mahmoud,n, Mohamed O. Shaker,A 200-nW 7.6-ENOB 10-KS/s SAR ADC in 90-nm CMOS for Portable Biomedical Applications. Microelectronics Journal 56 (2016) 81–96.

[5] S.Brenna,n, A. Bonetti, A. Bonfanti, A.L. Lacaita, An efficient tool for the assisted design of SAR ADCs capacitive DACs. INTEGRATION, the VLSI journal 53 (2016) 88–99.

[6] F. Schembari, G. Bellottia, C. Fiorinia, A 12-bit SAR ADC integrated on a multichannel silicon drift detector readout IC. Nuclear Instruments and Methods in Physics Research A 824 (2016) 353–355.

[7] Shreeniwas Daulatabad, Vaibhav Neema, Ambika Prasad Shah, Praveen Singh, 8-Bit 250-MS/s ADC Based on SAR Architecture with Novel Comparator at 70 nm Technology Node. Procedia Computer Science 79 (2016) 589–596.

[8] Wei Liu, Tingcun Wei, Bo Li, Lifeng Yang, Feifei Xue, Yongcai Hu.A SAR-ADC using unit bridge capacitor and with calibration for the front-end electronics of PET imaging, Nuclear Instruments and Methods in Physics Research A 818 (2016) 9–13

ISPECE   IOP Publishing

Research on Reliability Assessment of Thyristor in HVDC Converter Valve

Ning Liang[1], Jiachen Tian[2*], Cuicui Liu[2], Yating Gou[2], Fang Zhuo[2], Feng Wang[2].

[1]Maintenance&Test(M&T)Center of EHV Power Transmission Company, Guangzhou 510000, China

[2] School of Electrical Engineering, *Xi'an Jiaotong University, Xi'an 710049, China*

Email: liangning@ehv.csg.cn

tianjiachen1995@stu.xjtu.edu.cn

liuc16@stu.xjtu.edu.cn

gouyating@stu.xjtu.edu.cn

zffz@mail.xjtu.edu.cn

fengwangee@mail.xjtu.edu.cn

TEL:15202401505

Abstract: With the development of HVDC transmission technology, the efficiency and reliability of the operation of the converter valve have been widely concerned. Because the core device of the converter valve is thyristor, safe and effective operation of the thyristor is the prerequisite for efficient operation of the converter valve. In this paper, the thermal impedance model is established according to the radial heat dissipation direction of the thyristor. The basic electric heating correspondence is used to describe the rate of heat rise per unit time by the rate of change of the heat capacity temperature at each node. Through simulations and calculations, the heat change of each part of the thyristor during converter valve is in operation is analyzed, and the parts where the thyristor of the converter valve is easily damaged are analyzed. This method takes the influence of temperature on thyristor into consideration and provides a new angle to describe the reliability of the operation of thyristor.

1. INTRODUTION

Long distance power transmission should be realized at low current and high voltage. During the long distance power transmission of alternating current, voltage class should be improved by transformer to ensure the transmission.[1] However, there are around 10 percent losses in the pressurization process of transformer. In order to decrease the losses produced by alternating current in long distance power transmission, the applications of direct current transmission are wider than ever. People pay more attention to ensure the safety in operations of converter valve. The system of HVDC transmission consists of rectifier station, inverter station, transmission line and AC power system. The rectifier station transforms alternating current into direct current. The current flows through transmission line and inverter station transforms direct current into alternating current. Two power systems with different

Content from this work may be used under the terms of the Creative Commons Attribution 3.0 licence. Any further distribution of this work must maintain attribution to the author(s) and the title of the work, journal citation and DOI.
Published under licence by IOP Publishing Ltd

frequency and phase can be combined by DC transmission line to solve the closing problems of power systems with different frequency.

Thyristor is the core device in HVDC converter valve and the persistence of converter valve depends on the reliability of thyristor. Internal structure of thyristor consists of 3 different P-N junctions, which is P_1-N_1-P_2-N_2 structure. When P-N junction gets a balance between multi-sub-diffusion and minority drift, the space barrier zone will form and P-N junction can be regarded as capacity.

Whether it is stationary or overvoltage, the operation of converter valve will be influenced with damage of P-N junctions. Thyristor consists of 3 different P-N junctions, J_1, J_2 and J_3. When the thyristor is conducting, J_1 and J_3 are forward biased and J_2 is reverse biased; when the thyristor is reversed, J_1 and J_3 are reverse biased and J_2 is forward biased. A voltage surge or an increase in temperature does not completely destroy the P-N junction in some cases, but at this time the thyristor has been damaged, which will affect the life and reliability of the converter valve. Through the thermal model construction, the P-N junction inside the thyristor is replaced by multiple parallel equivalent capacitors. If the change of the capacitance voltage changing rate inside the thyristor can be realized, the life and reliability of the converter valve can be improved, even to provide some ideas for online monitoring.

2. Research Background and Failure Mechanism Analysis of HVDC Converter Valve Reliability

From the point of view of mechanical deformation, a method for evaluating the life of thyristors is proposed. According to the formula (9) in text [1], the integral expression of inelastic deformation is one-way and can only be used to analyze the one-way deformations when external force is applied. For the crimp thyristor, in the extreme over-voltage environment, due to the different thermal expansion coefficients of different materials at the crimping gap, when the internal temperature of the thyristor rises, the degree of expansion of different materials is different, and cracks may occur. The deformation at this time is not one-way. The main direction is along the crimping line, and there may be subtle cracks. At this time, the method in the literature cannot handle such a complicated situation [1].

Taking the six-pulse converter valve as an example, the failure mechanism of the converter valve under overcurrent conditions is introduced. Because of the commutation reactance and parasitic inductance of the thyristor, the thyristor has commutating teeth in the commutation process. For example, V_1 and V_6 are the thyristors currently working. V_1 and V_6 commutates to V_1 and V_2 for phase change, at this moment, V_2 and V_6 will be turned on at the same time,. And then B and C are short-circuited in two phases, which generates a huge short-circuit current and is harmful to the converter valve, bringing different kinds of losses. However, the drawback of this document is that no effective solutions have been proposed to limit overcurrent hazards [2].

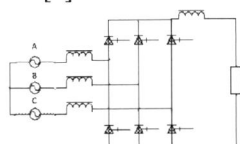

Figure 1 Six-pulse bridge type converter valve

The commutating teeth can raise the voltage significantly, so that the commutation zero of the thyristor comes earlier. For the inverter station in the HVDC transmission line, the thyristor commutation failure may be caused. This problem will increase the reactive power consumed and the harmonic current in the circuit. In this situation, operations of the converter valve are adversely affected [2].

Taking the six-pulse converter as the research object, when V1 is faulty and short-circuited, V_3 is subjected to reverse voltage so that it is turned off. This text regards short-circuit condition as instantaneous zero-state response. The second-order differential circuit model is used to set the damping parameter and to simulate the effects of different parameters on the shutdown voltage, analyzing a number of angles[3].

For the difference of junction temperature of thyristor components in series waterway, a calculation equivalent model of junction temperature of thyristor of valve assembly is established. Seven thyristors are connected in series, and the waterway is constructed according to the method of half-face heat dissipation. This structure can measure the highest temperature Tmax accurately. The temperature rise is calculated through using the heat taken away by the water per minute. This method subtly combines the water cooling system with the thyristor valves and uses PLECS to calculate the final junction temperature. However, this method also has some problems. Firstly, the applicable range is very narrow. Secondly, this method is only for a specific circuit and lacks universality. Moreover, thyristor half-face heat dissipation model cannot guarantee that the water cooling system covers completely in practice. A series of problems such as water impurities and the calculation formula may cause errors in experiments.

Overcurrent and overvoltage have a great influence on the reliability of the converter valve's thyristor. Firstly, the Si layer of the thyristor acts as the main heat-generating layer where the thermal conductivity is low. The heat generated by Si layer is diffused to the Mo layer and the Cu layer partly. The heat build-up causes the junction temperature to rise, which easily causes damage to the P-N junction. Secondly, when the thyristor is turned off and UKA is too high, the thyristor will avalanche breakdown, so that the holes inside P_2 will increase and the electronics inside N_2 will increase. However, they will not recombine after moving so that a huge voltage and an overcurrent damage will be brought to thyristor. Finally, the current has a self-heating effect. When the thyristor is turned on, current flowing through the thyristor will increase the thyristor resistance and the conduction loss. With the losses increasing, the junction temperature of the thyristor rises, which may damage the thyristor.

In order to reduce the damage probability of thyristor, it is necessary to establish a life evaluation model. In this paper, the thermal impedance model is established along the radial heat dissipation direction of the thyristor, which uses the electric heating correspondence. Voltage of the capacity corresponds to temperature rise of the heat capacity. The heat capacity represents the temperature rising relationship with the time. The larger the heat capacity, the faster the temperature rises per unit time. From the angle of the physical processes, the voltage rise of capacity is similar to the temperature rise of heat capacity. The changing rate of the capacity's voltage can correspond to the changing rate of heat capacity's temperature rise. The changing rate of heat capacity's temperature rise mainly reflects the rise of the temperature on the heat capacity per unit time. Under different excitation sources, the larger the parameter, the higher the temperature rises per unit time. The rising rate is large and the effect on the efficiency and stability of the converter valve is higher. In this paper, we use the known thermal resistance and heat capacity parameters to establish the "T" type circuit and Cauer circuit model according to the heat dissipation direction. Through establishing a recursive relationship, this text uses PLECS to simulate these circuits and calculates the temperature rising rate at different nodes. The temperature rise of thyristor is dynamically monitored under the source to determine the safe operation of thyristor.

3. Thermal impedance modeling of thyristor heating layer Si sheet and calculation of heat capacity temperature rising rate under stable working conditions

Under the condition that thyristor is in normal operation, the Cauer model can be used to establish thermal impedance model of the thyristor. However, internal structure of the thyristor is divided into a heat generating layer which can be considered as Si layer and a non-heat generating layer including Mo layer, Cu layer and heat-sink. There are ceramic layers attached to both sides. Therefore, for the heat-generating Si layer, the "T" type circuit can be used to model the thermal resistance circuit. The heat generation of entire thyristor is in the Si layer, so the power current source P(t) can be equally divided into n parts so that each module can uniformly heat up.

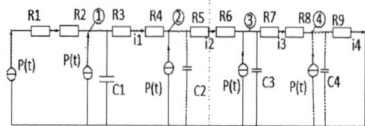

Figure 2 Thermal Si film thermal impedance model

This text takes B as the zero potential reference point. From the left, the positive value of the first capacity is 1 and the second to the right is 2, and so on. It is possible to set a total of n nodes. Write KCL equations for each capacitor node,

$$\frac{P(t)}{n} = i_1 + i_{C_1}; \tag{1}$$

$$\frac{P(t)}{n} + i_1 = i_2 + i_{C_2}; \tag{2}$$

$$\frac{P(t)}{n} + i_2 = i_3 + i_{C_3}; \tag{3}$$

....

$$\frac{P(t)}{n} + i_{n-1} = i_n + i_{C_n}; \tag{4}$$

$$i_n = C\frac{dU_{C_n}}{dt} = CK_n; \tag{5}$$

Add the above n expressions to get

$$P(t) = i_n + C(K_1 + K_2 + \cdots + K_n); \tag{6}$$

In the above formula, P(t) is the power source, in is the heat flow at the n-node, and K_n is the heat capacity temperature rising rate at the n-node.

Through the electro-thermal conversion diagram shown in the above figure, the estimation method of junction temperature is proposed. After the node voltage of each part is obtained, the corresponding junction temperature data can be found by checking the thyristor technical manual and the simulation is performed by using PLECS. The heat flow images of different nodes are made to calculate the heat capacity voltage changing rate of different nodes [5].

Table 1 Si thermal impedance circuit parameters when n=4

Thermal resistance (K/kW)	Thermal resistance in each part (K/kW)	Thermal resistance in each part of "T" type circuit (K/kW)	Thermal capacity (J/K)	Thermal capacity in each part (J/K)
0.19	0.0475	0.02375	35.6	8.9

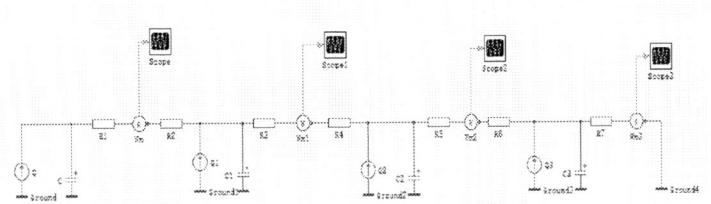

Figure 3 "T" type thermal resistance circuit test chart when n=4

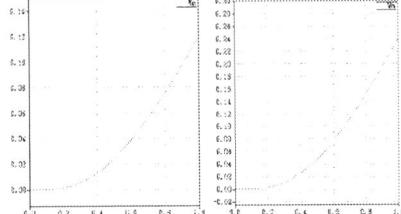

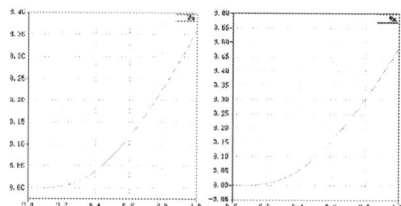

Figure 4 The waveform of i_1 when n=4

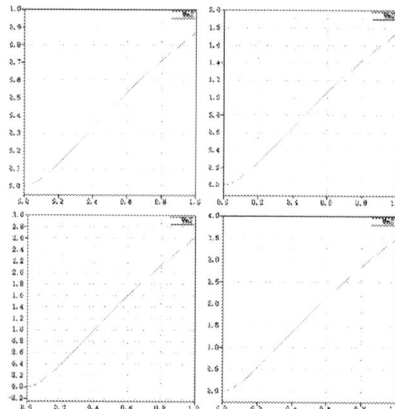

Figure 5 The waveform of i_2 when n=4

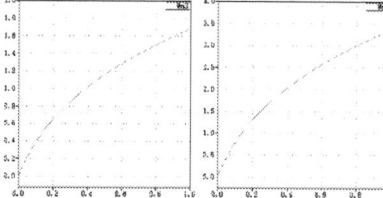

Figure 6 The waveform of i_3 when n=4

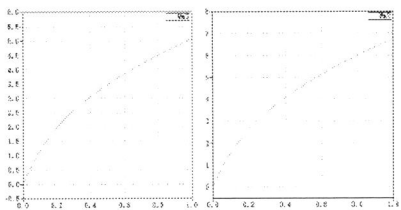

Figure 7 The waveform of i_4 when n=4

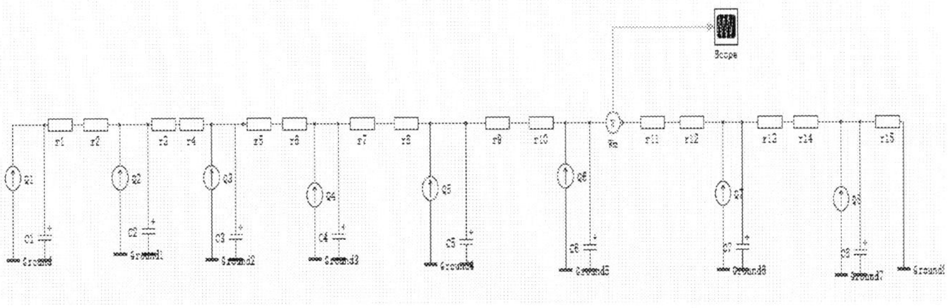

Figure 8 Si plate thermal impedance model when n=8

Table 2 Measurement of i and calculation of K when n=8

P(t)	4kW	8kW	12kW	16kW
i_4	1.7	2.03	3.5	6.7
k_4	0.0191	0.142	0.234	0.2693
i_3	0.87	1.3	2.6	5.1
k_3	0.0573	0.17	0.236	0.2586
i_2	0.38	0.81	1.7	3.4
k_2	0.0832	0.177	0.243	0.259
i_1	0.12	0.39	0.87	1.7
k_1	0.0988	0.181	0.24	0.258

Through the above simulation and formula calculation, the whole thyristor is heated under action of excitation source. Except for the first heat capacity, the temperature rise of heat capacities decreases one by one. However, with increase of the value of power sources, the extents of temperature decrease on following capacities are not obvious. The heat generation of the excitation source and the subsequent heat capacity increase the junction temperature of thyristor. From an electrical point of view, it is equivalent to a capacitor discharge which releases the capacitor current, even if the i_c is very small. The first heat capacity's temperature rise is much larger than the latter heat capacities, so voltage of first capacity in the circuit rises the fastest. The first capacity withstands the highest voltage and is most likely to be broken down, requiring special attentions and protections.

According to the calculation of the basic circuit when n=4, this text measures the heat flow ik and calculates the heat capacity temperature rising rate K of each node when n=8. Under the condition of different segment number n, the reliability of whole thyristor can be based on this assessment.

Table 3 Si thermal resistance circuit parameters when n=8

Thermal resistance (K/kW)	Thermal resistance in each part (K/kW)	Thermal resistance in each part of "T" type circuit (K/kW)	Thermal capacity (J/K)	Thermal capacity in each part (J/K)
0.19	0.02375	0.011875	36.5	4.5625

Table 4 Measurement of i and calculation of K when n=8

P(t)	8KW	16KW	24KW	32KW

i_6	1.75	3.55	5.4	7.1
K_6	0.0056	0.0056	0.1204	0.0113
i_4	0.8	1.6	2.4	3.2
K_4	0.054	0.1082	0.1648	0.2177
i_2	0.28	0.56	0.84	1.12
K_2	0.0944	0.1885	0.2802	0.3755
i_1	0.12	0.24	0.36	0.48
K_1	0.0988	0.1977	0.2966	0.3955

The bigger the n becomes, the higher the accuracy of the description of the circuit describes.

4. Thermal resistance modeling of non-heating layers including Mo, Cu and heat-sink and calculation of capacitance voltage changing rate

This section mainly describes the modeling of the Mo layer and the Cu layer in the non-heating layer, and uses the recursive formula to calculate the thermal parameters.

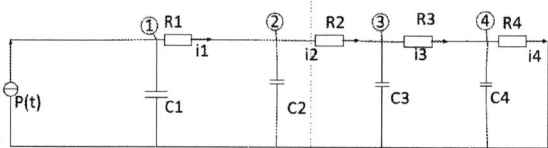

Figure 9 Thermal resistance model structure diagram of Mo layer and Cu layer

When different materials are brought together, contact resistance is generated and contact resistance is generated at the Si-Mo junction and the Mo-Cu junction. The thermal model of the Mo and Cu layers was established through using the Cauer model along the direction of heat dissipation from the thyristors.

Add the excitation source to the port. If the node number closest to the excitation source is 1, you may wish to set a total of n nodes.

$$P(t) = C\frac{dU_{C_1}}{dt} + i_1; \tag{7}$$

$$i_1 = C\frac{dU_{C_2}}{dt} + i_2; \tag{8}$$

$$i_2 = C\frac{dU_{C_3}}{dt} + i_3; \tag{9}$$

....

$$i_{n-1} = C\frac{dU_{C_n}}{dt} + i_n; \tag{10}$$

Add the above n expressions to get

$$P(t) = i_n + C(K_1 + K_2 + \cdots + K_n); \tag{11}$$

Table 5 Thermal resistance circuit parameters of Mo and Cu when n=4

Thermal resistance(K/kW)	Thermal capacity (J/K)	Thermal resistance in each section(K/kW)	Thermal capacity in each section(J/K)
1.89	135	0.4725	33.75

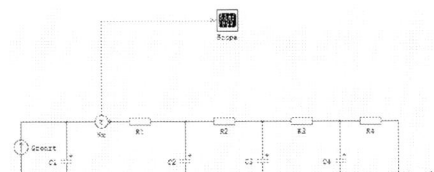

Figure 10 Plate thermal impedance model of Mo and Cu when n=4

Table 6 The value of i and K of each node under different excitation sources

P(t)	5KW	10KW	20KW	30KW
i_1	0.295	0.59	1.2	1.78
K_1	0.1394	0.2788	0.557	0.836
i_2	0.009	0.018	0.036	0.054
K_2	0.0085	0.017	0.034	0.0513
i_3	\	\	\	\
K_3	0.0085	0.017	0.034	0.0513
i_4	\	\	\	\
K_4	0.0085	0.017	0.034	0.0513

It can be obtained from the calculations and simulation results. From the electrical analysis, the first capacity of whole model has the highest withstand voltage, also representing the highest temperature rise on first heat capacity. The position closest to the excitation source suffers from the highest impact, which is the most vulnerable. Starting from the third node, the heat flow of the flow superheat resistor is basically 1% of i_2, which can be negligible. Except for the first heat capacity, the temperature rise of other heat capacities is basically the same, which means that the corresponding capacity's voltage rises is similar.

5. CONCLUSION

By comparing and calculating the temperature changing rate of the above-mentioned Si layer, Mo layer and Cu layer, according to the equivalent model, the heating temperature rise of the Si layer is mainly derived from the exothermic heat of some heat capacities and the power source of each section. The temperature rise of the Mo layer and the Cu layer is not as obvious as that of the Si layer and the temperature is mainly raised by the heat conducted from the Si layer. Since the common point of the two models is that the temperature rise is the highest near the heat source and the corresponding voltage breakdown rate is large, which is easy to damage the thyristor. Corresponding to the thyristor's structure, it can be known from the calculation that the thyristor center of Si piece, Si-Mo junction and the Mo-Cu junction is the most susceptible to damage. Corresponding to the nature of material, due to the different thermal expansion coefficients of different materials at the junction, when there is a large temperature rise, the degree of expansion of different materials is different, which is easy to cause damage to the thyristor. Regardless of the Si layer, the Mo layer or the Cu layer, the temperature rise of the heat capacity during the heat transfer process is basically gradually decreasing, but the temperature rise of the position closest to the heat source is the highest.

For the thyristor structure inside the HVDC converter valve, it is necessary to pay attention to the electric shock and heat change of the thyristor at the time of operation. The analysis method proposed in this paper is only to analyze the change of heat in different levels of the thyristor under transient conditions and junction temperature of thyristor. Real-time monitoring requires further researches and organizations.

Acknowledgment

This work was supported by the EHV Power Transmission Company of the Southern Power Grid project: Research on Lifetime Evaluation of HVDC Converter Valve (CGYKJXM00000027).

Reference

[1] Liang N, Zhang Z.G, Liu C. 2018. Failure Mechanism Analysis and Physics-of-Failure Lifetime Prediction Method for Press-pack Thyristor of Converter Valve in EMP Environment . In: The 2018 International Power Electronics Conference. Niigata. pp. 1157-1161.

[2] Xie T, Zha K.P. (2010) Study on Failure Mechanism of HVDC Valve Caused by Overcurrent in UHVDC Power Transmission Devices. Power System Technology, 10: 71-75.

[3] Xie T, Tang G.F, Zheng J.C. (2012) Analysis on Reverse Voltage Characteristics of HVDC Thyrister in the Fault State. Proceedings of the CSEE. 1:140-146.

[4] Zhang J.B, Huang H, Zhang X. (2017) Calculation Research of Thyrister's Junction Temperature with High Voltage Direct Current Converter Valve. Electrotechnics Electric, 8:24-26

[5] Yang J, Tang G.F. (2013) Study on Equivalent Circuit Model for HVDC Valve Thyrister Junction Temperature Calculation. Proceedings of the CSEE. 15:156-163

Smart Grid and Electric Power Informatization

JiYuan Ren[1], Zhe Wang[1], Zhen Luo[1], Fuyang Liu[1]

[1]Northeast Branch Of State Grid Corporation Of China, No. 1 Yingpan North Street Hunnan New District,Shenyang, Liaoning Province, China, 110000

Abstract. This paper introduces various concepts that relate to information technology and the development of energy transmission and distribution. Key challenges need to be addressed in relation to energy consumption, such as the need to be responsive to current demand, which have been addressed through information technology systems. With the increased connectedness of energy systems, there has also been an increased need to ensure the information security of these systems. The Internet of Things (IoT) concept will be reviewed in relation to the connection of objects in energy systems as well as the concepts of Big Data and Cloud Computing. The former has developed in response to the need to predict energy usage more accurately and the latter offers the advantages of increased failover potential as well as much faster provisioning of enhanced capacity in IT systems to meet consumer demand.

1. Introduction

Electronic power systems need to be able to address demand for electricity and have become automated in order more efficiently to manage transmission and distribution. The introduction of Smart Grid systems has introduced the capacity for bi-directional communication between electricity suppliers and consumers so that real-time demand can be better anticipated. The increasing interconnectedness of energy systems has, however, exposed them to the potential for cyber-attacks, which create a risk to the energy supply with potentially life-threatening consequences. The Smart Grid has also incorporated the IoT concept into energy systems as well as Big Data concepts, in which vast amounts of data is now used, partially through the use of Smart Meters, in order to better estimate customer demand for electricity. These concepts will be addressed in three main parts: first, the automation of electric power systems; second, the issues of information security in power systems; the third part evaluates the use of IoT, Big Data and Cloud Computing in electric power systems.

2. Automotive Operation of Electric Power System

Electronic power systems are comprised of several generating stations, where electricity is generated in response to demand from consumers. The objective of this system is to generate sufficient electrical energy at the best-suited locations and to transmit it to the various load centers and then distribute it to consumers while maintaining quality and reliability (Sivanagaraju, 2009). An important characteristic of electric energy is that it should be used as it is generated; due to the varied demands of consumers (e.g. domestic, industrial, agricultural), the load on the system varies over time. The generating station must, however, be in a state or readiness to meet demand, thus establishing a 'variable load problem' (Sivanagaraju, 2009). Since electric energy cannot be cheaply stored on a large scale, it must be generated to meet demand; it is imperative therefore for electric power utilities to be able estimate load in advance. Load forecasting in electric power systems is critical for the correct planning of power systems and transmission and distribution facilities for the operation, financing, grid formation, personnel requirements and sales of electricity (Sivanagaraju, 2009).

Content from this work may be used under the terms of the Creative Commons Attribution 3.0 licence. Any further distribution of this work must maintain attribution to the author(s) and the title of the work, journal citation and DOI.

Published under licence by IOP Publishing Ltd

Various technologies have been created over the past decades in order to better enable the automation of the ability to forecast load in electric power systems and predict the amount of electrical energy that is required over a given time period. As the global economy becomes more reliant on the sustainable development of energy, a series of problems, such as energy shortages, disconnection and climate change need to be addressed, as well as the need to maximize economies to the consumer (Naveen, Ing, Danquah, Sidhu, & Abu-Siada, 2016). In particular, a recent development which seeks to address some of these current imperatives is the Smart Electrical Power Grid or Smart Grid. A Smart Grid is an electrical network that utilizes both digital and other information technologies to monitor and manage the transport of electricity from the various sources to meet the varying and competing demands of consumers (Naveen et al., 2016). It has been developed to improve mechanisms to increase the information available to predict the need for energy and in turn improve efficiency and sustainability in relation to the use of electrical energy.

3. Information Security of Electric Power System

The introduction of 'Smart Grid' solutions to electricity supply has led to various challenges in relation to cyber security and power system communication systems. While physical threats to electronic power systems are commonly understood, the increasing threats of cyber-attacks to electricity systems need to be addressed (Ericsson, 2010). Power system communication (PSC) systems, supervisory control and data acquisition (SCADA) systems and substations are now interconnected with other systems, leaving them vulnerable to attack through computer networks (Ericsson, 2010). The use of standard products for SCADA and energy management systems (EMS) opens up new possibilities for threats to security. Furthermore increased security threats come from the increasing integration of IT infrastructures at power utilities, including public infrastructures (Dán, Sandberg, Björkman, & Ekstedt, 2012). Increased threats require increased security measures such as Firewalls, and encryption systems that protect information as it is sent across the network. Physical security measures should also be in place to prevent attacks (Kizza, 2017).

Acts of cyber-war or cyber-terrorism can include attacks on critical infrastructures such as electricity power stations. A cyber-attack was carried out in December 2015 which led to service outages to customers in the Ukraine. The cause of the attack was a third party's illegal entry into the Ukrainian Kyivoblenergo, a regional electricity distribution company's computer and SCADA systems. As a result, seven 110 kV and twenty-three 35 kV substations were disconnected for three hours (E-ISAC, 2016). The Ukrainian government claimed that the Russian security services were responsible for the cyber-attack, making it an act of cyber warfare (BBC News, 2017). The attacks were directed at the regional distribution system, as illustrated in figure 1 (Oblenergos are distribution companies) (E-ISAC, 2016).

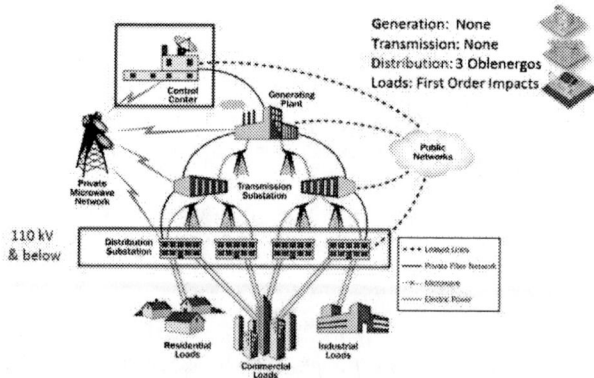

Figure 1: Electrical System Overview Diagram (E-ISAC, 2016)

Cyber-attacks on electricity control systems therefore carry significant risks to a country's power infrastructure, hence the information security of power stations is of increasing importance. However, as the sophistication and interconnectedness of technology increases, attackers gain greater opportunities to do increasing amounts of damage as more and more components of the system are interconnected and controlled through automation.

4. Smart Grid Technology and Information Security

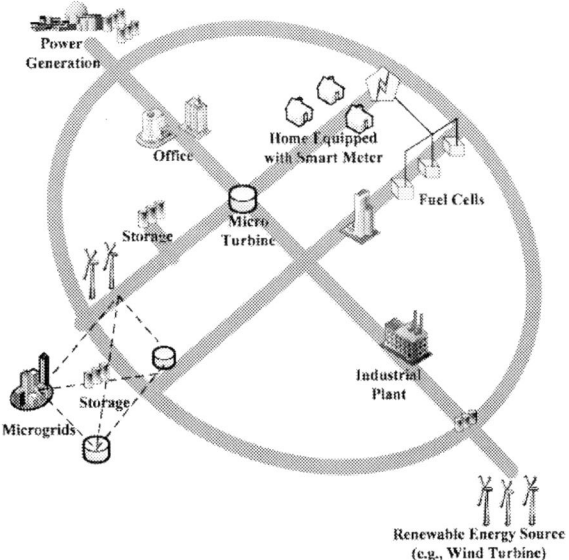

Figure 2: Smart grid landscape and its characteristics (Kayastha, Niyato, Hossain, & Han, 2014)

The smart grid is an innovative energy network (see figure 2) which aims to improve the reliability, cooperation, responsiveness and economy of the conventional grid network. According to NIST, a Smart Grid is "a modernised grid that enables bidirectional flows of energy and used two-way communication and control capabilities that will lead to an array of new functionalities and applications" (Locke & Gallagher, 2010, p. 33). The smart grid includes two-way electrical and data networks through the introduction of advanced metering infrastructure (AMI). The key advantage of the AMI is to provide near real-time metering data, including fault and outage information, which is communicated directly to the utility center (Kayastha et al., 2014).

Smart meters are electric meters that record real time energy consumption and voltage quality, and are installed in the domestic residences (see figure 2) (Naveen et al., 2016). They communicate via the AMI directly to the utility company, thus providing much more detailed and accurate power consumption information than was previously provided through conventional monitoring and billing mechanisms (Kayastha et al., 2014). The advantage of the use of Smart Meters is that they help address the issues of demand response and variable demand discussed above, which are critical to manage predictions and transmission of energy.

5. Cloud computing and IoT with Big Data of Electric Power System

Cloud computing constitutes a model which enables convenient, ubiquitous, on-demand services to a pool of computer resources (Naveen et al., 2016). Cloud computing is a disruptive computing paradigm, and as such it has required some significant changes in many areas of computer systems engineering including data storage, computer architecture, networking, computer security and resource management. Due to its reliance on high speed computer networks, cloud computing has been enabled through the creation of the Internet (Marinescu, 2018).

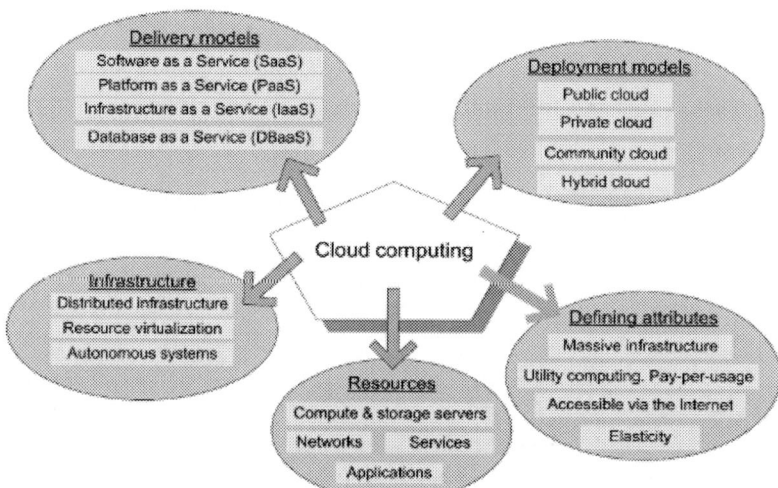

Figure 3: Cloud computing, delivery models, deployment models, infrastructure, attributes, resources (Marinescu, 2018)

Figure 3 illustrates the 4 delivery models of Cloud infrastructure, the 4 deployment models, the three infrastructure options, the resources required to implement cloud and the defining attributes of Cloud computing which distinguishes it from traditional infrastructures where each company or organization hosts their software, platform, infrastructure and database services in-house (Marinescu, 2018).

One of Smart Grid's major concerns is energy management, or the process of monitoring, controlling and conserving energy (Naveen et al., 2016). Cloud Applications have introduced benefits over the conventional Server-Client architectures and thus providing advantages for the Smart Grid. The conventional Master-Slave architecture can be vulnerable to cyber-attacks such as Distributed Denial of Service attacks; it can also suffer from a single point of failure. Cloud allows for load balancing to prevent these problems (Naveen et al., 2016). Traditional infrastructures also are limited to the number of users that can be supported without need to introduce additional hardware: through the use of virtualization, additional services can be added within the Cloud paradigm more seamlessly (Naveen et al., 2016). These advantages thus allow the expansion of the Smart Grid systems to address changes in electrical use by end-use customers in response to price fluctuations (Demand Response) (Naveen et al., 2016)

5.1. Smart Grid Energy System and the Internet of Things

The RFID group has defined the Internet IoT as "the worldwide network of interconnected objects uniquely addressable based on standard communications protocols" (Riggins & Wamba, 2015). The IoT is therefore connected with technologies such as RFID, sensors, actuators, GPS and mobile devices; the integration of these components therefore establishes the basis of the IoT. 'Things' can be sensors, databases, and other devices or software. Sensors include location identifiers such as global positioning system (GPS) and individual identification devices such as radio frequency identification (RFID) tags (Wang & Alexander, 2016). In the Smart Grid environment, multiple devices are implemented such as smart meters, sub-stations, home appliances, sensor notes and communication network devices which are integrated to encompass elements of the IoT (Naveen et al., 2016).

5.2. Big Data and Electronic Power Systems

The application and methodologies applied to very large data sets were developed many years ago in order to forecast electricity load consumption. Recently, developments in monitoring, sensor networks and advance metering infrastructure (AMI) have led to a significant increase in the variety, volume and velocity of available data in relation to electricity distribution and transmission networks (Arghandeh &

Zhou, 2017). To become data driven, it requires a culture shift in organizational and business processes and a holistic approach, and views the transmission and distribution of electricity as a single integrated entity (Arghandeh & Zhou, 2017). Furthermore, while sensors, processing and the visibility they produce have been applied to the transmission system for some time, the real growth they can produce is in the distribution system, where intelligent electronic devices (IED) are proliferating (Arghandeh & Zhou, 2017).

In comparison to conventional grids, the advantages of Smart Grid are manifold. These include its self-healing and recovery functionality, its ability to better incorporate renewable energies, situational awareness and transient stability. These features have been implemented in reliance upon the deployment of smart meter devices and Big Data analytics (Tu, He, Shuai, & Jiang, 2017). The Big Data in Smart Grid is generated from a number of sources. The proliferation of Phasor Measurement Units (PMUs), the advanced meter read (AMR) and other advanced measurement devices such as the Digital Fault Recorder (DFR), the Sequence of Event Recorder (SER) and Intelligent Electronic Devices (IEDs) have introduced vast amounts of data relating to power systems for storage, curation, mining, sharing and visualization (Depuru, Wang, Devabhaktuni, & Gudi, 2011).

Big Data produces both opportunities and challenges for Smart Grid and electronic power systems (Miao, 2014). It has already brought many tangible benefits to electricity users, including increased system stability and reliability, increased asset utilization and efficiency, improved customer experience and satisfaction. This latter effect is considered to have been a result of the implementation of smart meters that enable easier billing, fraud detection, demand response and efficient utilization of energy (Tu et al., 2017). Table 1 (below) illustrates a number of practical applications that have been developed using Big Data analytics to facilitate features of Smart Grid that have been applied by private research groups and corporations.

Table 1. Big Data in Smart Grid applications

Applications	Software	Developer	Description	Reference
Situational Awareness System	FNET/GridEye	Yilu Liu (Lead)	Real-time event detection, location estimation, oscillation detection	(Chai et al., 2016; Liu, 2016)
Wide Area Situational Awareness	SMDA (ver5.0)	Hydro-Quebec	Collects wide area Phasor data in real time and monitors inter-area oscillations	(Basu et al., 2014)
Event Detection and Alarm Management	e-terra 3.0	Alstom	Presents and visualizes disturbances and navigates to the relevant diagnostic data	(ALSTROM, 2012)
Oscillation, Detection and Mitigation	Grid 3P platform	Electric Power Group	Oscillations are observable by fine granular PMU data	(US Department of Energy, 2016)
Power Plant Models Validation	CERTS	BPA & CERTS	BPA engineers calibrate the Colombia Generating Station model without off-line generators	(Overholt, Kosterev, Eto, Yang, & Lesieutre, 2014)
Renewable Resource Integration	DEMS	Siemens	A data-driven system which monitors, manages and intergrades distributed generation and renewable energy in a bulk power system	(Siemens, 2013)

Transient stability and Intruder Detection	WARMAP5000	NARI Technology	Combines real time monitoring data to guard against attack and stability control	(NARI, 2015)

At the same time as these advantages, there are various challenges that have been introduced through the use of Big Data for energy systems. Social power consumption is closely related to Big Data analytics, and the development of Big Data has in itself led to significant growth in systems that require vast energy consumption (Miao, 2014). Smart Grids themselves significantly increase the need for energy consumption; thus, although energy efficiencies may be achieved through the implementation of technology, in another sense electric power consumption is positively correlated with growth in the volume of Big Data (Miao, 2014). Furthermore, the collection and analysis of vast amounts of consumer data lead to Data Protection, security and privacy challenges (Miao, 2014).

6. Conclusion

This paper has reviewed the introduction of a number of information technology systems to electrical energy systems. Due to the increased interconnectedness of systems, there is now an increased need for information security to prevent attacks against vulnerable power supply services. Furthermore, the IoT signifies the way in which the energy grid system is now an interconnected set of components which interoperate to create the Smart Grid system. This Smart Grid system now relies upon concept of Cloud Computing to produce failover and faster provisioning capacity for expansion. Furthermore, Big Data analytics have been used to address the need to better predict demand for energy as well as to improve energy sustainability. Paradoxically, Big Data systems themselves increase the demand for energy.

References

[1] ALSTROM. (2012). *e-terraplatform 3.0 The Power to Adapt*. Retrieved from https://www.gegridsolutions.com/alstomenergy/grid/Global/Grid/Resources/Documents/Automation/NMS/e-terraplatform%203.0%20trans_gene-epslanguage=en-GB.pdf

[2] Arghandeh, R., & Zhou, Y. (2017). Big Data Application in Power Systems. Amsterdam, Netherlands: Elsevier.

[3] Basu, C., Agrawal, A., Hazra, J., Kumar, A., Seetharam, D. P., Béland, J., ... Lafond, C. (2014). Understanding events for wide-area situational awareness. In ISGT 2014 (pp. 1–5). https://doi.org/10.1109/ISGT.2014.6816408

[4] BBC News. (2017, January 11). Ukraine power cut "was cyber-attack." BBC News. Retrieved from https://www.bbc.com/news/technology-38573074

[5] Chai, J., Liu, Y., Guo, J., Wu, L., Zhou, D., Yao, W., ... Patel, M. (2016). Wide-area measurement data analytics using FNET/GridEye: A review. In 2016 Power Systems Computation Conference (PSCC) (pp. 1–6). https://doi.org/10.1109/PSCC.2016.7540946

[6] Dán, G., Sandberg, H., Björkman, G., & Ekstedt, M. (2012). Challenges in Power System Information Security. IEEE Security Privacy, 10(4), 62–70.

[7] Depuru, S. S. S. R., Wang, L., Devabhaktuni, V., & Gudi, N. (2011). Smart meters for power grid — Challenges, issues, advantages and status. In 2011 IEEE/PES Power Systems Conference and Exposition (pp. 1–7). https://doi.org/10.1109/PSCE.2011.5772451

[8] E-ISAC. (2016). TLP: White Analysis of the Cyber Attack on the Ukrainian Power Grid Defense Use Case. Washington DC: SANS ICS. Retrieved from https://ics.sans.org/media/E-ISAC_SANS_Ukraine_DUC_5.pdf

[9] Ericsson, G. N. (2010). Cyber Security and Power System Communication—Essential Parts of a Smart Grid Infrastructure. IEEE Transactions on Power Delivery, 25(3), 1501–1507. https://doi.org/10.1109/TPWRD.2010.2046654

[10] Kayastha, N., Niyato, D., Hossain, E., & Han, Z. (2014). Smart grid sensor data collection, communication, and networking: a tutorial. Wireless Communications and Mobile Computing, 14(11), 1055–1087. https://doi.org/10.1002/wcm.2258

[11] Kizza, J. M. (2017). Guide to Computer Network Security. Springer.

[12] Liu, Y. (2016). Wide-area measurement system development at the distribution level: An FNET/GridEye example. In 2016 IEEE Power and Energy Society General Meeting (PESGM) (pp. 1–1). https://doi.org/10.1109/PESGM.2016.7741185

[13] Locke, G., & Gallagher, P. D. (2010). NIST framework and roadmap for smart grid interoperability standards release 1.0. United States of America: National Institute of Standards and Technology.

[14] Marinescu, D. C. (2018). Cloud Computing: Theory and Practice. Cambridge, MA: Elsevier.

[15] Miao, X. (2014). Big Data and Smart Grid. In Proceedings of the 2014 International Conference on Big Data Science and Computing (pp. 26:1–26:2). New York, NY, USA: ACM. https://doi.org/10.1145/2640087.2644175

[16] NARI. (2015). WARMAP5000. Retrieved October 2, 2018, from http://www.naritech.cn/html/jie166.shtm

[17] Naveen, P., Ing, W. K., Danquah, M. K., Sidhu, A. S., & Abu-Siada, A. (2016). Cloud computing for energy management in smart grid - an application survey. IOP Conference Series: Materials Science and Engineering, 121(1), 012010. https://doi.org/10.1088/1757-899X/121/1/012010

[18] Overholt, P., Kosterev, D., Eto, J., Yang, S., & Lesieutre, B. (2014). Improving Reliability Through Better Models: Using Synchrophasor Data to Validate Power Plant Models. IEEE Power and Energy Magazine, 12(3), 44–51. https://doi.org/10.1109/MPE.2014.2301533

[19] Riggins, F. J., & Wamba, S. F. (2015). Research directions on the adoption, usage, and impact of the internet of things through the use of big data analytics. In System Sciences (HICSS),2015 48th Hawaii International Conference (pp. 1531–1540). IEEE.

[20] Siemens. (2013). DEMS - The Decentralized Energy Management System - Digital Grid - Siemens [WCMS3PortletPage]. Retrieved October 3, 2018, from http://w3.siemens.com/smartgrid/global/en/experts-talk/pages/dems.aspx

[21] Sivanagaraju, S. (2009). Power System Operation and Control. Delhi: Pearson Education India.

[22] Tu, C., He, X., Shuai, Z., & Jiang, F. (2017). Big data issues in smart grid – A review. Renewable and Sustainable Energy Reviews, 79, 1099–1107. https://doi.org/10.1016/j.rser.2017.05.134

[23] US Department of Energy. (2016). Advancement of Synchrophasor Technology. ARRA Projects. Retrieved from https://www.smartgrid.gov/files/20160320_Synchrophasor_Report.pdf

[24] Wang, L., & Alexander, C. A. (2016). Big Data Analytics and Cloud Computing in Internet of Things. American Journal of Information Science and Computer Engineering, 2(6), 70–78.

Construction of power industry corpus based on data mining and machine learning intelligent algorithm

Liujun Zhao[1,a], Weizheng Kong[1,b], Qiuling Wang[2] and Lihua Song[2]

[1] State Grid Energy Research Institute CO., Ltd, Beijing 100000, China;

[2] Fujian Yirong Information Technology CO., Ltd, Fuzhou 350000, China.

[a]zhaoliujun@sgeri.sgcc.com.cn, [b]kongweizheng@sgeri.sgcc.com.cn

ABSTRACT: With the advent of the mobile Internet era, the dissemination and diffusion of information has become faster and faster, and the dissemination and generation of information has also increased exponentially. More and more information is generated and diffused in the Internet, and the subsequent problem is that the collection and determination of information increases with its complexity. Therefore, it is necessary to propose and apply a new method to complete the analysis and processing of the information on the Internet. This paper uses data mining technology and machine learning intelligent algorithm to obtain and classify the information data of the power industry on the Internet, so as to construct the power industry corpus.

1. summarize the application of power industry corpus in industry research and industry development.

Corpus is not only the basic resource of corpus linguistics, but also the main resource of empirical linguistic research methods. It can be applied to dictionary compilation, language teaching, traditional language research, statistical or case-based research in natural language processing, etc. The corpus of the power industry is based on documents in the power industry, social news about the power industry, government announcements, research reports of scientific research institutions and other textual information to study policy and technological trends.

1) role in industry research

Constructing corpus can greatly improve the preprocessing ability of complex industry information, screen out time-sensitive and effective data by machine, and then analyze and study the corresponding trends. Through word frequency analysis, we can get the technical heat. Through geographical graphic analysis, we can get the regional technical preference and application degree. More intuitive expression. It will play a guiding role in future business expansion and research prospects arrangement, which is beyond the traditional methods and individual subjective analysis methods. Moreover, corpus method can extract elements quickly and objectively, so long as the algorithm is properly screened, it can achieve high accuracy. Moreover, the accuracy of prediction and analysis can be well quantified and measured. This is difficult to achieve before the popularity of AI technology. It can be said that the establishment and maintenance of corpus today has become a technology that can not be bypassed in the industry.

2) application in industry development

The main function of corpus construction is to study and analyze the situation of the industry. Based on these data, we can judge and analyze which technology and industry applications are being

Content from this work may be used under the terms of the Creative Commons Attribution 3.0 licence. Any further distribution of this work must maintain attribution to the author(s) and the title of the work, journal citation and DOI.

Published under licence by IOP Publishing Ltd

sought after by the market according to the trend. At the same time, based on the relationship between the correlation and the network hierarchical distribution, the correlation can be representatively expressed.

2. data mining technology based on web crawler technology

1) introduction to web crawler technology

Web crawler technology is an application program or script program that automatically captures Internet information based on certain rules. It is widely used in Internet search engines or other news portals. It can automatically collect all the page content that it can access in order to obtain or update the content of these websites. Search mode. Functionally speaking, crawlers are generally divided into three parts: data acquisition, processing and storage. The traditional crawler starts with one or several URLs of the initial web page and obtains the URLs of the initial web page. In the process of crawling the web page, it constantly extracts new URLs from the current page and puts them into the queue until it meets certain stopping conditions of the system. The workflow of focused crawler is complex. It is necessary to filter topic-independent links according to a certain web page analysis algorithm, retain useful links and put them in the waiting URL queue. Then, it will select the next page URL from the queue according to a certain search strategy, and repeat the process until it reaches a certain condition of the system. In addition, all web pages captured by crawlers will be stored by the system, analyzed, filtered, and indexed for subsequent query and retrieval; for focused crawlers, the analysis results obtained in this process may also provide feedback and guidance for future crawling process.

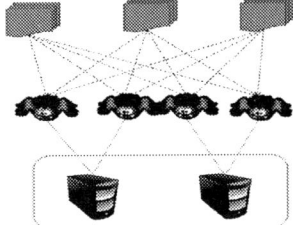

Fig. 1. graphical representation of web crawler

2) construction of web crawler

The basic workflow of web crawler is as follows:

(1) Firstly, according to the information we pay attention to, we adopt appropriate strategies to select some carefully selected seed URLs.

(2) put these URL into URL queue to be grabbed.

(3) Take out the URL queue to be crawled, parse the DNS, get the IP of the host, download the corresponding pages of the URL, and store them in the downloaded webpage library. In addition, put these URL into the URL queue that has been grabbed.

(4) Analyse the URL in the grabbed URL queue, analyze the other URLs, and put the URLs in the queue to be grabbed, so as to enter the next cycle.

(5) add the information concerned to the database.

3) preservation and updating of information data.

Web crawlers crawl down pages that are large text, and can design a storage method to store large-scale data. It should not be appropriate to store it in relational databases such as MySQL or sqlserver. First of all, the pages are relatively independent, basically no relationship, only the simple relationship of URL or describing text corresponding pages, and relational database system in order to support relations and efficient query will increase a lot of additional costs, which is not worth the cost. Moreover, crawlers should be highly efficient in crawling pages. If a relational database is used to store pages, a large number of data will be inserted into I/O in a short time. Insertion is bound to be a bottleneck problem, which is also a big pressure for database maintenance network and physical disk. Therefore, I think it is appropriate to store data text directly. Open-source larbin crawler also uses text-based storage, but it defaults to store a physical file for each page. In my personal opinion, frequent file creation, writing, flush, shutdown, and system overhead are also relatively large. Considering comprehensively, I designed a scheme, that is, a physical file stores multiple pages. In order to support proper search, segmentation and merge operations, the data file will correspond to an index file. In this way, in the operation project, it can be re-indexed in the file, the index file is much smaller than the data file, traversing or querying will be very fast. Not only if, when data is merged, only index files need to be merged, which will be much more convenient. The specific format is shown in Figure 2.

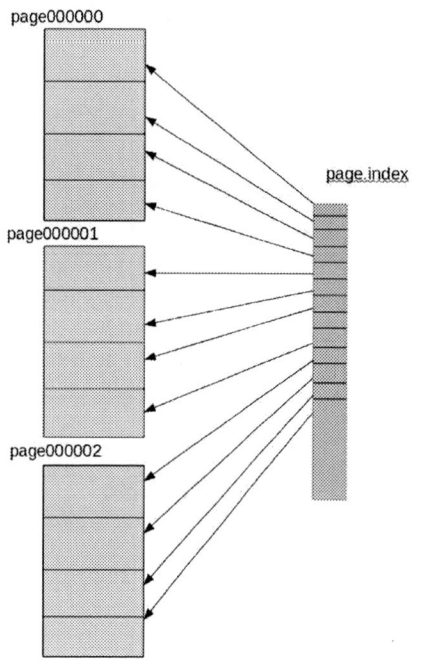

Fig. 2. storage format and structure design

3. Natural Language Processing and multi-level classification of databases

1) Natural Language Processing technology brief introduction

Natural Language Processing (NLP) is a field of computer science, artificial intelligence, and linguistics that focuses on the interaction between computers and human (natural) languages. Therefore, Natural Language Processing is related to the field of human-computer interaction. Natural language processing faces many challenges, including natural language understanding. Therefore, natural language processing involves the area of human-computer interaction. Many challenges in NLP involve natural language understanding, i.e., computers derive from the meaning of human or natural language input, and others involve natural language generation.

Modern NLP algorithm is based on machine learning, especially statistical machine learning. Machine learning paradigm is different from the previous attempt to deal with language. The implementation of language processing tasks usually involves the large set of rules that are directly encoded by hands.

2) participle technology for data classification

In text analysis, word segmentation technology occupies a very critical position. Its main purpose is to divide continuous text into specific lexical elements with meanings. In terms of participle comprehension, Chinese is much more complicated than English. English sentences consist of words, and spaces are used as natural delimiters between words, while Chinese sentences and paragraphs do not have obvious delimiters.

Only words, sentences and paragraphs can be simply delimited by clear demarcation marks. Only words do not have a formal demarcation mark. Although English also has the problem of phrase demarcation, at the word level, Chinese is much more complicated and difficult than English.

These segmented words will be sent to the backstage dictionary for matching, and then the matching results will be brought to the computer, so that the meaning contained in the text can be properly understood.

3) multi level division of word segmentation technology

Using word segmentation technology to establish preliminary partitioning data, at the same time, according to the correlation and frequency of word segmentation, the second or even multiple partitioning is carried out. Specific methods include frequency allocation, weight allocation, correlation degree allocation and logical allocation.

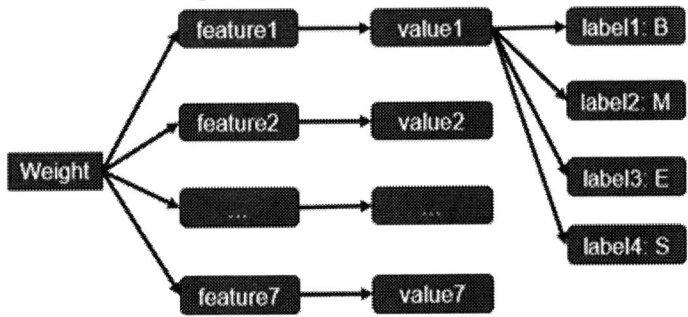

Fig. 3. multilevel division of participles

4. construction and display of knowledge map based on R language

1) introduction to R language text processing technology

Although the ability of processing text in R language is not strong, it can greatly improve the efficiency of work if used properly. At the same time, R language package is a processing ability with high statistical characteristics, so some text operations have to be processed with it. Regular expressions are indispensable for efficient text processing. Although R is inherently inefficient in this respect, it uses regular expressions for most functions dealing with strings.

Regular expressions are expressions used to describe / match a text set. The specific ways of implementation are as follows:

(1) All English letters, numbers and many displayable characters themselves are regular expressions to match themselves. For example,'a'is a regular expression matching the letter'a'.

(2) Some special characters are not used to describe themselves in regular expressions. They have been "escaped" in regular expressions. These characters are called "metacharacters".

Square brackets denote the selection of any one of the square brackets (e.g. [a-z] denotes any lowercase character); ^ Put at the beginning of an expression to denote the beginning of the matching text, and at the beginning of a square bracket to denote any character in non-square brackets; Brackets denote the number of repetitions of previous characters or expressions; | denotes optional items. That is to say, the expression before and after is selected.

(3) using a reference sign (or code change symbol), usually a backslash "/". It should be noted that in R, two backslashes, i.e.'\\', are used. If parentheses are to be matched, they should be written as'\\(\\\\)'

(4) Different languages or applications (in fact, many rules are common) define special metacharacters to represent certain types of characters.

2) knowledge network construction

Using web crawler technology to acquire and update data, at the same time, according to the built machine learning algorithm, natural language processing and word segmentation are carried out. The processed text is segmented twice or more by using R language to construct the database. According to the systematic classification of power industry websites and technical journals, the statistical and relevance-based word segmentation libraries are obtained. The two libraries are analyzed separately, and the first 200 keywords are screened out in R language for display on the web.

3) dynamic network updating and learning

Finally, the automation technology is used to crawl the Internet network data regularly, and a new database is constructed. Compared with the previous version, the words appearing many times and ranking first are added to the corpus. So as to achieve the purpose of dynamic learning.

References:

[1] Zhao Xiaoliang, Liu Yu Zhang. The method of establishing the thesaurus of fiscal classification [J]. library and information guide, 2002, 12 (4): 31-32.

[2] Zou Qimeng, Liu Qing, Yin Xianjun. Establishment method of classification model, keyword selection method and device of SEO thesaurus: CN106294416A [P]. 2017.

[3] Liao Liang. System design and Implementation Based on Bilingual thesaurus retrieval and classification [D]. Kunming University of Science and Technology, 2017.

[4] Liu Kaiying, Guo Bingyan. Natural Language Processing [M]. Science Press, 1991.

[5] Tang Yincai. R language and statistical analysis [M]. higher education press, 2008.

[6] Wu Yingliang, Wei Gang, Li Haizhou. A Chinese word segmentation algorithm based on N-gram model and machine learning [J]. Journal of Electronics and Information, 2001, 23 (11): 1148-1153.

[7] Liu Jun. 1. Summary of research on machine learning algorithms in the field of artificial intelligence [J]. digital communications world, 2018 (1).

ISPECE   IOP Publishing

IOP Conf. Series: Journal of Physics: Conf. Series **1187** (2019) 022019   doi:10.1088/1742-6596/1187/2/022019

Online Identification Method of Induction Motor Parameters Based on Rotor Flux Linkage

Xuefei Li[1], Yang Kang[1], Hui Wang[2], Jie Chen[2] and Fei Gao[3]

[1] CRRC Changchun Railway Vehicles CO., LTD.

[2] Beijing Engineering Research Center of Electric Rail Transportation,

School of Electrical Engineering, Beijing Jiaotong University, 100044 Beijing, China.

[3] Beijing Dahua Radio Instrument CO., LTD.

E-mail: 16121528@bjtu.edu.cn

Abstract. High-performance control of induction motors is inseparable from accurate motor parameters. Online identification of parameters can provide accurate parameters for motor control in real time, so it is particularly important. In this paper, the traditional voltage flux observer is improved. The high-pass filter is connected in series with low-pass filter, and then the amplitude and phase compensation methods are used to obtain the improved voltage-type flux linkage observation method. The current-type flux linkage observation method is also introduced. Then the principle of the model reference adaptive system is studied. Combined with the super-stability theory, an algorithm for identifying the rotor time constant on-line is designed based on the rotor flux linkage. Through the combination of simulation and experiment, the observation effect of two flux linkage observation methods is verified on a 160kW motor. At the same time, the online identification algorithm of rotor time constant is studied under different speeds and different parameters. The results verify its accuracy and effectiveness.

1. Introduction

In recent years, China's high-speed railways have developed rapidly, and emu trains have reached 46% in the proportion of railway-passenger transport[1]. High-performance control of traction motor is vital for efficient and comfortable train operation, and accurate motor parameters could have an important role in it. Off-line identification can only provide initial parameters for motor control. As the train speed and environment change, factors such as temperature, frequency and excitation will all affect the parameters of induction motor[2].Particularly, the rotor-time constant will change greatly. If it is different from the true value of motor in the control of the induction motor, problems like the inaccurate of magnetic field orientation and the incomplete decoupling of stator current can be caused, and thus affect the motor torque output, speed control and the efficient operation of the train[3].

For online identification of rotor time constant of induction motor, scholars at home and abroad have proposed methods such as extended Kalman filter, synovial observer, neural network intelligent algorithm to achieve online parameters identification[4]-[6]. However, these methods are complex and difficult to achieve. In practical applications, the model reference adaptive system (MRAS) is preferred because of simple structure and high efficiency. In this paper, the traditional voltage flux observer is studied and improved, and the current flux observer is introduced. Then a MRAS on-line identification algorithm for parameters based on rotor flux linkage was designed, by combining with

---

Content from this work may be used under the terms of the Creative Commons Attribution 3.0 licence. Any further distribution of this work must maintain attribution to the author(s) and the title of the work, journal citation and DOI.

Published under licence by IOP Publishing Ltd

the principle of MRAS and the characteristics of two kinds of flux observer. Finally, the effect of flux observation and identification algorithm is verified by simulation and experiment on 160kW induction motor.

2. Rotor flux observation of induction motor

The observation methods of rotor flux of induction motor usually include voltage model, current model and hybrid model[7]. In this paper, the improved voltage model and current model will be selected to obtain the accurate rotor flux linkage.

2.1. Traditional voltage rotor flux observer

The expression of the voltage rotor flux observer in the static coordinate system is as follows[8]:

$$
\begin{cases}
\psi_{r\alpha} = \dfrac{L_r}{L_m}\left[\int\left(u_{s\alpha} - R_s i_{s\alpha}\right)dt - \sigma L_s i_{s\alpha}\right] \\[2mm]
\psi_{r\beta} = \dfrac{L_r}{L_m}\left[\int\left(u_{s\beta} - R_s i_{s\beta}\right)dt - \sigma L_s i_{s\beta}\right]
\end{cases}
\tag{1}
$$

In the formula, $\psi_{r\alpha}\psi_{r\beta}$, $u_{s\alpha}u_{s\beta}$ and $i_{s\alpha}i_{s\beta}$ are respectively the flux linkage, stator voltage and stator current component of induction motor in static coordinates, and $R_s, L_r, L_m, L_s, \sigma$ are the stator resistance, rotor inductance, mutual inductance, stator inductance and leakage inductance of the induction motor.

Due to the existence of pure integrator, there is a problem with the integral drift saturation. The usual solution is to replace the integral part with the first order low-pass filter[9]:

$$
\begin{cases}
\psi_{r\alpha} = \dfrac{L_r}{L_m}\left[\dfrac{1}{s+\omega_c}\left(u_{s\alpha} - R_s i_{s\alpha}\right) - \sigma L_s i_{s\alpha}\right] \\[2mm]
\psi_{r\beta} = \dfrac{L_r}{L_m}\left[\dfrac{1}{s+\omega_c}\left(u_{s\beta} - R_s i_{s\beta}\right) - \sigma L_s i_{s\beta}\right]
\end{cases}
\tag{2}
$$

Although the method can eliminate the phenomena of integral saturation to some extent, it can't obtain exact rotor flux due to the low frequency of DC bias and the amplitude and phase errors caused by the low-pass filter method.

2.2. Improved voltage rotor flux observer

Based on the above problems, we connect high and low-pass filters in series, and then use the method of amplitude and phase compensation to obtain exact stator flux. The improved stator flux linkage expression is as follows:

$$
\psi_s = \frac{s}{s+\omega_{cH}}\frac{1}{s+\omega_{cL}}u_s - \frac{s}{s+\omega_{cH}}\frac{1}{s+\omega_{cL}}R_s i_s
\tag{3}
$$

In the formula: ψ_s, u_s, i_s are respectively the stator flux linkage, voltage and current vector, $\omega_{cH}\ \omega_{cL}$ are respectively the cut-off frequencies of high and low-pass filters.

Then the stator flux linkage is compensated for amplitude and phase:

$$
M = \frac{\sqrt{(\omega_e^2 + \omega_{cH}^2)}\sqrt{(\omega_e^2 + \omega_{cL}^2)}}{\left|\omega_e\right|^2}
\tag{4}
$$

$$
\varphi = \arctan(\frac{\omega_e}{\omega_{cL}}) - \arctan(\frac{\omega_{cH}}{\omega_e}) - \frac{\pi}{2}
\tag{5}
$$

Where M and φ are respectively the amplitude and phase compensation values, while ω_e is the electrical frequency.

After the stator flux linkage is known, the rotor flux linkage is obtained by the following formula:

$$\boldsymbol{\psi}_r = (\boldsymbol{\psi}_s - \sigma L_s \boldsymbol{i}_s)\frac{L_r}{L_m} \tag{6}$$

2.3. Current rotor flux observer
The current rotor flux observer is based on the relationship between the stator current and the flux linkage. The flux linkage equation is formulated as follows:

$$\begin{cases} p\psi_{r\alpha} = -\dfrac{1}{T_r}\psi_{r\alpha} - \omega_r\psi_{r\beta} + \dfrac{L_m}{T_r}i_{s\alpha} \\ p\psi_{r\beta} = -\dfrac{1}{T_r}\psi_{r\beta} + \omega_r\psi_{r\alpha} + \dfrac{L_m}{T_r}i_{s\beta} \end{cases} \tag{7}$$

In order to engineering applications, the labrador transform and discretization method can be used to obtain:

$$\begin{cases} \psi_{r\alpha(n+1)} = \dfrac{\Delta t}{T_r + \Delta t}(L_m i_{s\alpha} - \omega_r T_r \psi_{r\beta}) + \dfrac{T_r}{T_r + \Delta t}\psi_{r\alpha(n)} \\ \psi_{r\beta(n+1)} = \dfrac{\Delta t}{T_r + \Delta t}(L_m i_{s\beta} + \omega_r T_r \psi_{r\alpha}) + \dfrac{T_r}{T_r + \Delta t}\psi_{r\beta(n)} \end{cases} \tag{8}$$

Where $\psi_{r\alpha(n+1)}$ and $\psi_{r\alpha(n)}$ are respectively the rotor flux components calculate result of this and last time, Δt, T_r and ω_r are respectively the sampling time, rotor time constant and rotor frequency.

3. Principle and design of model reference adaptive system
The model reference adaptive system is relatively easy to implement and has been widely used for its characteristics like high adaptive speed and high performance.

3.1. Principle of model reference adaptive system
The model reference adaptive system is mainly composed of three parts: reference model, adjustable model and adaptive mechanism[10].

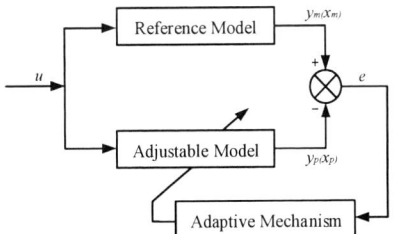

Figure 1. Block diagram of model reference adaptive system (MRAS)

For a system that can establish a mathematical model and unmeasurable parameters, an adjustable model with unknown parameters and a reference model without unknown parameters are designed and made to have output with the same physical meaning. The output error of the two models is adjusted by the adaptive mechanism to generate control signals, then the parameters to be identified in the adjustable model are updated to realize dynamic tracing. Finally, the output error of the two models approaches close to zero, thus the unknown parameters are identified.

3.2. MRAS parameter identification method based on rotor flux linkage

Based on the analyses above, an adaptive system for on-line identification based on rotor time constants of rotor flux linkage can be established.

3.2.1. Adjustable model and reference model
As shown in the figure above, the voltage model without rotor time constant can be used as a reference model, and the current model with rotor time constant can be used as an adjustable model.

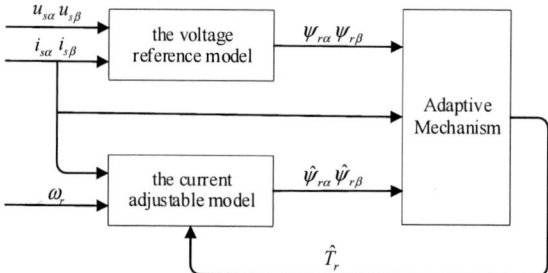

Figure 2. Block diagram of rotor time constant MRAS
identification based on rotor flux linkage

3.2.2. Adjustable model and reference model
The generalized error is defined as: $e_r = \psi_r - \hat{\psi}_r$, plug into the adjustable model and expand by derivation on both sides. The left side of the equation is:

$$\begin{pmatrix} -\dfrac{1}{T_r} & -\omega_r \\ \omega_r & -\dfrac{1}{T_r} \end{pmatrix}\begin{pmatrix} e_{r\alpha} \\ e_{r\beta} \end{pmatrix} - \left(\dfrac{1}{T_r} - \dfrac{1}{\hat{T}_r} \right)\left(\begin{pmatrix} \hat{\psi}_{r\alpha} \\ \hat{\psi}_{r\beta} \end{pmatrix} - L_m \begin{pmatrix} i_{s\alpha} \\ i_{s\beta} \end{pmatrix} \right) \tag{9}$$

Further simplification is obtained:

$$\frac{de_r}{dt} = A e_r - W \tag{10}$$

In the formula: $A = \begin{pmatrix} -\dfrac{1}{T_r} & -\omega_r \\ \omega_r & -\dfrac{1}{T_r} \end{pmatrix}$, $\quad W = \left(\dfrac{1}{T_r} - \dfrac{1}{\hat{T}_r} \right)(\hat{\psi}_r - L_m i_s)$

According to the theory of super stability, the nonlinear link in equation (10) should satisfy Popov integral inequality, and the linear link should be confirmed rigorously.

Make $A=A$, $B=I$, $C=I$, and I is the unit matrix, so that $H(s) = C(sI - A)^{-1}B = (sI - A)^{-1}$, the exist of matrix $P = P^T > 0$, $Q = Q^T > 0$ is the necessary and sufficient conditions of $H(s)$ to be the positive real function of s, and it needs to satisfy:

$$\begin{cases} PA + A^T P = -Q \\ PB = C^T \end{cases} \tag{11}$$

In the formula, let $P = I$ to satisfy $Q = \begin{pmatrix} \dfrac{2}{T_r} & 0 \\ 0 & \dfrac{2}{T_r} \end{pmatrix}$.

The feedback loop must meet the Popov integral inequality, that is to say:

$$\eta(0, t_1) = \int_0^{t_1} v^T(t)w(t)dt \geq -r_0^2 \quad \forall t_1 \geq 0, r_0^2 \geq 0 \tag{12}$$

In the formula: $v(t) = e(t)$.

According to the common mechanism of model reference adaptation, the reciprocal of rotor time constant is turn into the following form of proportional integral:

$$\frac{1}{\hat{T}_r} = \int_0^{t_1} F_1\left(v,t,\tau\right)d\tau + F_2\left(v,t\right)$$ (13)

Plug W into equations (12) and (13), the left side of the equation becomes:

$$\int_0^{t_1} e^T(t)\left(\int_0^{t_1} F_1\left(v,t,\tau\right)d\tau + \frac{1}{T_r}\right)\left(L_m i_s - \hat{\boldsymbol{\psi}}_r\right)dt$$
$$+\int_0^{t_1} e^T(t)F_2\left(v,t\right)\left(L_m i_s - \hat{\boldsymbol{\psi}}_r\right)dt$$ (14)

In order to satisfy $\eta(0,t_1) \geq -r_0^2$, we can respectively take:

$$F_1\left(v,t,\tau\right) = K_i e^T(t)\left(L_m i_s - \hat{\boldsymbol{\psi}}_r\right) \quad K_i > 0$$ (15)

$$F_2\left(v,t\right) = K_p \boldsymbol{e}^T(t)\left(L_m i_s - \hat{\boldsymbol{\psi}}_r\right) \quad K_p > 0$$ (16)

Plug equations (15) (16) into equation (14) can surely meet Popov integral inequality, therefore the feedback system in equation (10) must be asymptotically stable.

By combining equations (13), (15) and (16), the adaptive mechanism of rotor time constant can be obtained as follows:

$$\frac{1}{\hat{T}_r} = K_i \int_0^{t_1} [(\psi_{r\alpha} - \hat{\psi}_{r\alpha})(L_m i_{s\alpha} - \hat{\psi}_{r\alpha}) + (\psi_{r\beta} - \hat{\psi}_{r\beta})(L_m i_{s\beta} - \hat{\psi}_{r\beta})]dt$$
$$+K_p[(\psi_{r\alpha} - \hat{\psi}_{r\alpha})(L_m i_{s\alpha} - \hat{\psi}_{r\alpha}) + (\psi_{r\beta} - \hat{\psi}_{r\beta})(L_m i_{s\beta} - \hat{\psi}_{r\beta})]$$ (17)

In the formula, Ki and Kp are respectively the integral and proportional constants.

4. Analysis and verification by simulation and experiment

Based on the above theoretical analysis, simulation and experiment analysis were carried out by using Matlab/Simulink and induction motor experiment platform respectively. The simulated motor and support motor have the same parameters, and the details are shown in the following table:

Table 1. Parameters of induction motor

Parameter	Value	Parameter	Value	Parameter	Value	Parameter	Value
$R_s(\Omega)$	0.223	$R_r(\Omega)$	0.103	L_{ls}(mH)	0.00158	L_{lr}(mH)	0.002076
L_m(mH)	0.0438	n_p	2	P_e(kW)	160	V_e(V)	1287

4.1. Results of rotor flux observer

The effects of traditional voltage flux observer, improved voltage flux observer and current flux observer were simulated and tested by simulation and experiment respectively. The results are as follows:

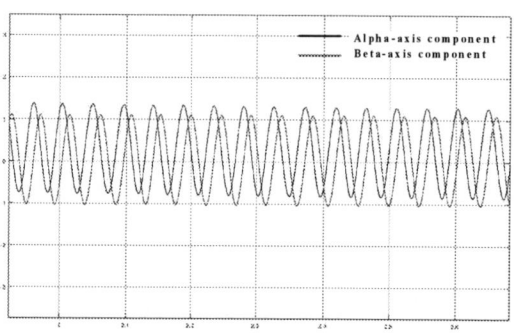

Figure 3. Simulation results of traditional voltage flux observation

As shown in figure 3, the rotor flux component obtained by the traditional voltage flux observer has an obvious problem of DC bias.

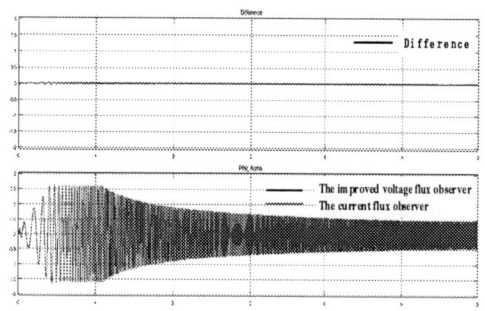

Figure 4. Simulation results of improved voltage and current flux linkage components

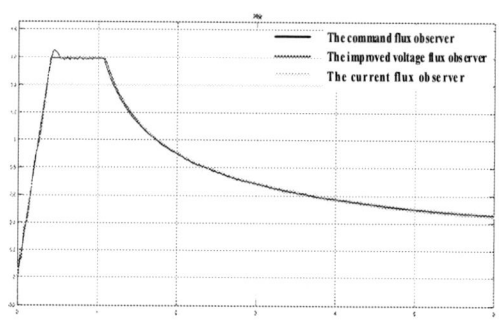

Figure 5. Simulation results of command, current and improved voltage rotor flux linkage

The simulation waveforms of figure 4 and figure 5 show that the improved voltage and current flux observer can both accurately observe the rotor flux linkage, and the two methods have almost no amplitude and phase deviation.

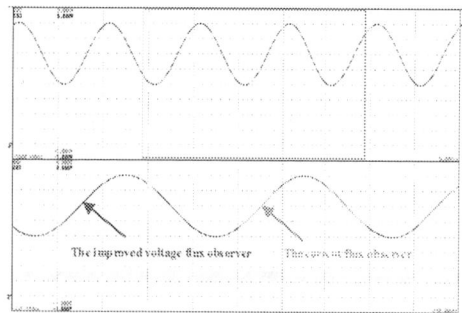

Figure 6. Experimental results of two kinds of flux observations at low speed

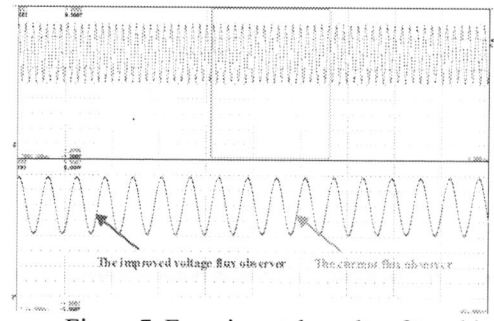

Figure 7. Experimental results of two kinds of flux observations at high speed

Based on simulation, the control algorithm is realized by adopting the TMS320F28335 chip, and the experimental verification was completed on the experimental supporting platform of induction motor. It can be concluded from figure 6 and figure 7, that the two flux observation methods can both obtain the accurate flux linkage with little error at different speeds, laying a foundation for the next step of parameter identification.

4.2. MRAS parameter identification method based on rotor flux linkage

After verifying the two flux observation methods can both obtain the accurate flux linkage though the simulation and experiment, the identification effect of the MRAS parameter identification method based on rotor flux linkage under different conditions is analyzed and studied.

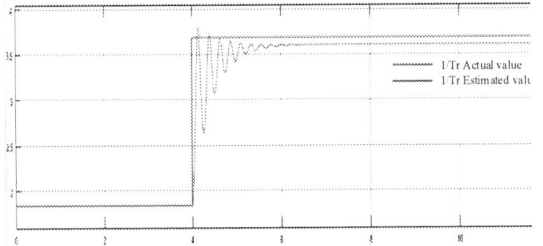

Figure 8. The identification effect of Tr step jump to 0.5Tr at a constant speed of 60Hz

Figure 9. The identification effect of Tr step jump to 1.2Tr at a constant speed of 60Hz

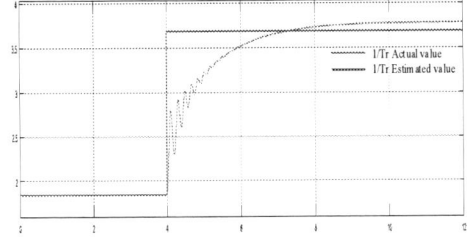 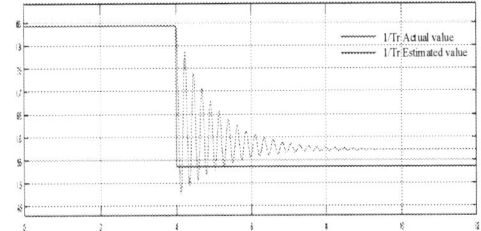

Figure 10. The identification effect of Tr step jump to 0.5Tr at a constant speed of 120Hz

Figure 11. The identification effect of Tr step jump to 1.2Tr at a constant speed of 120Hz

Table 2. Identification results

Speed (s^{-1})	Parameter (s^{-1})	Actual value (s^{-1})	Estimated value (s^{-1})	Errors (s^{-1})
60	$0.5T_r$	0.2272	0.2324	0.023
60	$1.2T_r$	0.5453	0.5584	0.024
120	$0.5T_r$	0.2272	0.2329	0.025
120	$1.2T_r$	0.5453	0.5578	0.023

Firstly, the simulation analysis was carried out with different speed operation conditions and step changes of different parameters. From figure 8-11 and table 2, it can be concluded that the proposed online parameters identification algorithm can quickly identify the actual values of parameters under different speed operation conditions with an error less than 2.5% and has high identification speed and accuracy.

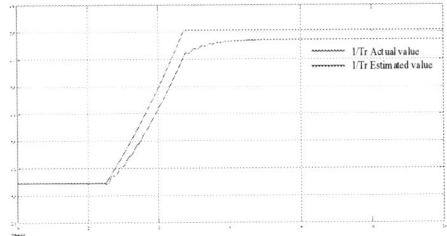 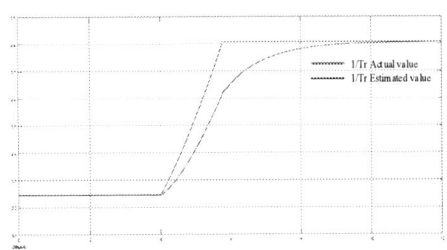

Figure 12. Identification effect of Tr grades to 1.2Tr at a constant speed of 60Hz

Figure 13. Identification effect of Tr grades to 1.2Tr at a constant speed of 120Hz

After that, it can be seen from figure 12 and figure 13, with the gradual change of parameters at different speeds, the algorithm can follow the change well, and when the parameter becomes stable, the maximum error of parameter is only 0.002, and the identification accuracy is higher.

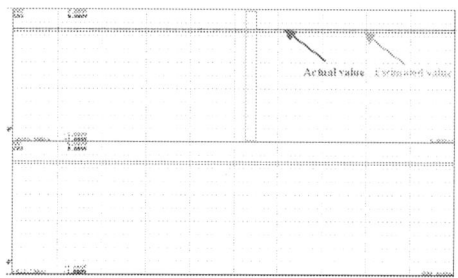 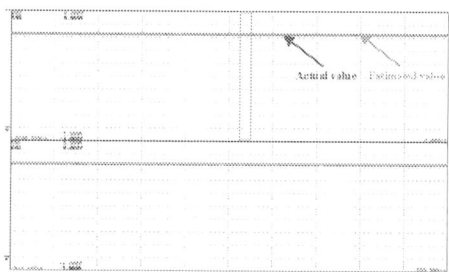

Figure 14. Identification effect of Tr at a constant speed of 60Hz

Figure 15. Identification effect of Tr at a constant speed of 120Hz

Based on the online identification simulation of parameter step and gradient change, the DSP program was programmed and verified by experiments. During the experiment, the motor was operating at a constant speed. After the parameter identification results were stable, the motor temperature was measured and compared with the inherent parameters of the motor. The results are shown in figure 14 and figure 15. By calculation, the Tr identification error at a constant speed of 60 Hz is 0.025, and the Tr identification error at a constant speed of 120 Hz is 0.022. Through the simulation and experimental verification above, it can be concluded that the designed MRAS parameter identification method based on rotor flux linkage is easy to accomplish, and has high identification speed and accuracy.

5. Conclusion

This paper analyzes the shortcomings of the traditional voltage flux observer, and puts forward an improved voltage flux observer which connects high and low-pass filter in series and then compensates in phase and amplitude. After that, the current flux observer is introduced, and made easy to engineering application by transformation. Then combined with the MRAS principle and the characteristics of two kinds of flux observers, the MRAS parameter identification method based on rotor flux linkage is designed, and the adaptive mechanism is designed according to the theory of super stability. On this basis, through simulation and experiment, the effect of flux observation and identification algorithm is verified under different speed conditions and different parameters variations. The results show the magnetic flux observation is accurate, and the algorithm has the characteristics of fast identification, high accuracy and easy implementation, which can provide accurate parameters in real time for improving motor control performance.

Acknowledgments

This work was supported by National Key R&D Program of China 2016YFB1200502.

References

[1] Notice of the State Council on Printing and Distributing the Development Plan for the "13th Five-Year Plan" Modern Comprehensive Transportation System [EB/OL]. (2017-02-03)[2017-02-28]. http://www.gov.cn/zhengce/content/2017-02/28/content_5171345.htm

[2] Rehman H U, Derdiyok A, Guven M K, et al. An MRAS scheme for on-line rotor resistance adaptation of an induction machine[C]// Power Electronics Specialists Conference, 2001. Pesc. 2001 IEEE. IEEE, 2001:817-822 vol.2.

[3] Lü Hao, Ma Weiming, Nie Ziling, Jie Guisheng. Analysis of Induction Machine System Performance Influence About Field-Oriented Inaccuracy [J]. Transactions of China Electrotechnical Society, 2005, 20(8):84-88.

[4] Toliyat H A, Levi E, Raina M. A Review of RFO Induction Motor Parameter Estimation

Techniques[J]. IEEE Power Engineering Review, 2007, 22(7):52-52.

[5] Liu Yan, Qi Xiaoyan. An Luenberger-sliding mode observer parameter identification for induction motors [J]. Electric Machines and Control, 2011, 15(8):93-100.

[6] Wlas M, Krzeminski Z, Toliyat H A. Neural-Network Based Parameter Estimations of Induction Motors[J]. IEEE Transactions on Industrial Electronics, 2008, 55(4):1783-1794.

[7] Wen Xiaoyan. Research on Optimization Control of Train Traction Motor at Low Speed ［D］. Beijing: Beijing Jiaotong University，2012.

[8] Song Wenxiang, Ruan Zhiyong, Yin Yun.Modified rotor flux observer based on statically compensated voltage model [J]. Advanced Technology of Electrical Engineering and Energy, 2012, 31(4):19-23.

[9] Lin J L. A new approach of dead-time compensation for PWM voltage inverters[J]. IEEE Transactions on Circuits & Systems I Fundamental Theory & Applications, 2002, 49(4):476-483.

[10] Chen Fuyang. Adaptive Control [M]. Science Press, 2015.

Design of Power Consumption Tester for HPLC Power Line Carrier Communication Module

Liu Mouhai[1,2], Zhang Chenglin[3], Liu Xiangbin[1,2], Chen Hao[1,2], Yang Maotao[1,2], Chen Fusheng[1,2] and Shen Liman[1,2]

[1] State Grid Hunan Electric Power Company Limited Power Supply Service Center(Metrology Center), Changsha 410004, China.

[2] Hunan Province Key Laboratory of Intelligent Electrical Measurement and Application Technology, Changsha 410004, China.

[3] State Grid Hunan Technical Training Center, Changsha 410004, China.

535246512@qq.com

Abstract. Due to the current situation that the power line carrier communication module power consumption test equipment has slow dynamic power consumption response and low AC power consumption test accuracy, the article deeply analyzes the key technical requirements of the communication module dynamic power consumption test, and focuses on a design of power consumption tester. The paper introduces the working principle of the tester, gives the circuit design scheme of DC power consumption and AC power consumption, and analyzes the calculation methods of static and dynamic power consumption in detail. Finally, the power consumption test results of various test systems are compared, and the test data is analyzed to conclude that the power consumption test data is stable, reliable, and high precision.

1. Introduction

With the development of the national energy strategy, power line communication (PLC) plays an increasingly important role in the construction of the smart grid. The smart grid requires that the electricity information collection system should meet automatic meter reading (AMR), load control, transformer monitoring, power quality remote measurement, security monitoring, time-of-use (TOU) rate, dynamic billing, and other various value-added services, e.g., power line telephones and Internet information services. At the same time, the State Grid put forward the goal of the construction of power consumption information collection system, i.e., the full coverage of power users, the full collection of electricity consumption information, and support for comprehensive electricity tariff control. Traditional narrowband power line communication technology can hardly meet the above requirements. The emergence of highspeed broadband power line communication technology (BPLC) improves the data transmission rate and provides assistance for the smart grid construction project and the four-meter-in-one project promoted by the State Grid[1].

So far, the State Grid and China Southern Power Grid have been bidding for more than 500 million smart meters. Due to the number of carrier modules, their power consumptions are considerable. Hence, it is very important to limit the power consumption of the carrier modules. The power consumption test of BPLC module provides one of the important indicators in the technical specification of power user information collection system. The static and dynamic power consumption of BPLC modules are both specified in the communication unit technical specification[2].

Content from this work may be used under the terms of the Creative Commons Attribution 3.0 licence. Any further distribution of this work must maintain attribution to the author(s) and the title of the work, journal citation and DOI.
Published under licence by IOP Publishing Ltd

Power consumption testing of BPLC modules has two characteristics[3]. On the one hand, dynamic power consumption of BPLC modules changes rapidly. Taking modules from a certain manufacturer as an example，the data transmission rate is low for a narrowband PLC module, which means it will spend longer time sending the same data. For example, the time required by a module with a modulation mode of BPSK for sending an application-layer packet with a length of 16 bytes is 138 ms. However, for the BPLC module, the data transmission rate is high. The time required for sending the same data transmission time is greatly shortened to 2.97 ms. Therefore, the dynamic response of the instrument testing the BPLC module should be very high. On the other hand, high precision is required for testing the ac power consumption. The ac power loss of the PLC carrier module is only about tens of milliwatts, which is very low. So the AC test function of the power consumption meters should work in high precision.

At present, it is difficult for ordinary instruments to meet the above two characteristics. This paper designs a novel device for testing the power consumption of BPLC modules with the above two characteristics taken into account.

2. The working principle of the power consumption tester

The BPLC module power consumption tester adopts modular design, which mainly includes MCU module, power load module, DC power consumption test unit, AC power consumption module, display module and network communication module[4].The DC power consumption test unit and the AC power consumption test unit are designed as two groups, respectively testing the energy meter module and the concentrator module, and the two devices to be tested are simultaneously connected to one power load module. The dynamic power consumption of the device under test can be tested by network meter reading. As shown in Figure 1, the MCU module calculates the data collected by the DC power consumption unit and the AC power consumption, and transmits the test result to the display module. The data can also be transmitted to the PC master computer software through the network communication module. The power consumption of the BPLC module consists of two parts: 12V DC power consumption and 220V AC power consumption. DC power consumption plays a major role, and AC power consumption is generally small, mainly due to the zero-crossing circuit power consumption of the module.

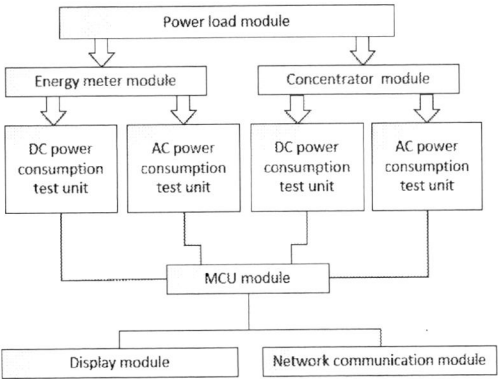

Fig 1. System wiring diagram

2.1 Power load module design

The power line load can affect the power consumption of the BPLC module. After the power supply is connected to the power consumption tester, the power supply must be processed first. Before the AC input to the module, the power supply should be standardized. The line impedance stabilization network can be added to eliminate the influence of the external carrier on the system and provides a stable power line load for the system test, as shown in Figure 2.

After the power supply is transformed, rectified, and filtered, it provides 12V, 5V, and 3.3V for the

ISPECE

IOP Conf. Series: Journal of Physics: Conf. Series **1187** (2019) 022020   doi:10.1088/1742-6596/1187/2/022020

entire tester. The power supply design parameters are as follows: The DC power supply design parameters: the output accuracy is better than 1%, and the voltage stability is less than 1% under different loads (output current 0~0.5A).

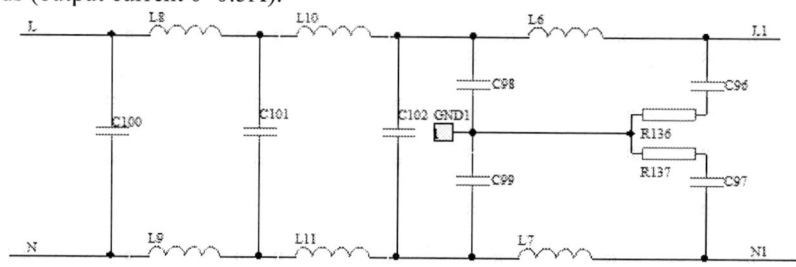

Fig 2. Line impedance stabilization network

2.2 DC power consumption test unit design

A general method for measuring the average DC power consumption is to sample the values of the N sets of instantaneous voltage u_i and instantaneous current i_i at equal intervals with a certain sampling rate in a certain period of time[4], and then calculate the average power P according to the following formula.

$$P = \frac{1}{N} \sum_{i=1}^{N} u_i i_i \qquad (1)$$

Instantaneous power consumption is the multiplication of the voltage acquisition value and the current acquisition value. Voltage sampling: The power supply voltage of the BPLC module is 12V, the measuring range of the ADC sampling chip is 3.3V, which exceeds the AD sampling voltage range. The voltage sampling uses the resistor voltage division method, respectively using 47K and 10K high-precision resistors to make the AD sampling voltage approximately 2.1V and meets the sampling requirements of the ADC chip. Current sampling: According to the relevant technical standards, the static current of the BPLC module is less than 100mA, the dynamic current is less than 500mA, and the current sampling design should reach 500mA with an accuracy of 1mA. As shown in Figure 3, the current sampling is to string the high-precision and low-resistance sampling resistor to the 12V power supply. Using a precision differential op amp, the voltage across the sampling resistor is compared and amplified to output a suitable voltage signal to Single-chip AD. The op amp uses TI's current sampling chip INA282. This chip can amplify the voltage difference across the sampling resistor by a factor of 50. It has the following features: wide common mode range -14V to 80V, offset voltage ±20uV, common mode rejection ratio 140dB, gain error(maximum) ±1.4%, and gain drift(maximum) 0.0005%/°C.

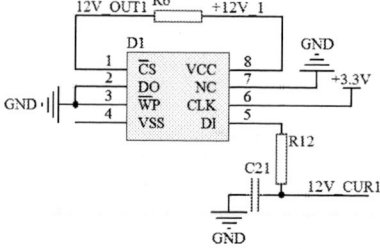

Fig 3. Schematic diagram of current sampling section

2.3 AC power consumption test unit design

The AC power consumption of the general BPLC module is mainly the power consumption of the zero-crossing circuit. The calculation method is the same as that of the formula (1). The AC power consumption is calculated by voltage and current sampling. The AC voltage and current sampling unit

146

use a non-isolated scheme. Using the transmission data with the MCU module requires optocoupler isolation.

The AC power consumption acquisition uses the energy metering chip ESEM16 of Shanghai Eastsoft Microelectronics Co., Ltd.. ESEM16 integrates a high-precision energy metering module microcontroller, including two 24-bit ADCs for current and voltage sampling, a reference voltage source and a dedicated DSP core with active energy metering. It can meter active energy, measure voltage and current RMS, and calculate average active power. In the dynamic range of 1000:1 at 25 ° C, the active energy measurement error is less than 0.1%; in the dynamic range of 500:1, the voltage and current RMS measurement error is less than 1%. The ESEM16 simulation front end consists of two programmable gain amplifiers with current amplification on one side, gains of 1, 2, 4, 8, 16, 32 times. Gains of voltage sampling are 1, 2, 4, and 8 times. After the PGA, two 24-bit AD samples are taken. The chip contains a precision reference voltage source of 1.3V to achieve high-precision voltage sampling. The AC sampling circuit is shown in Figure 4 below.

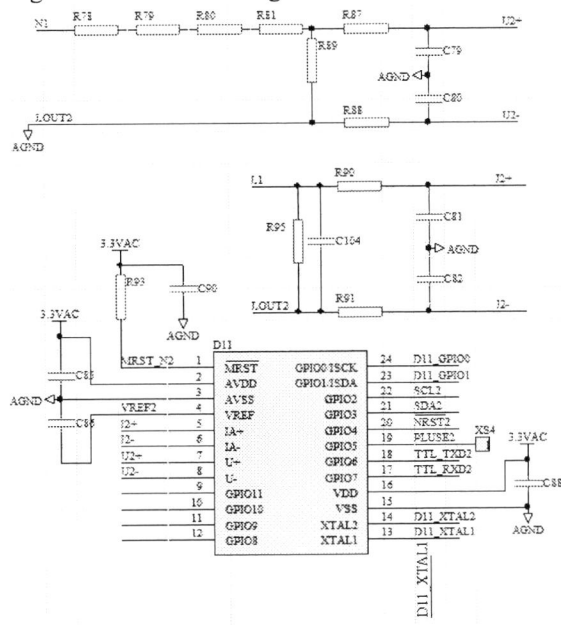

Fig 4. Schematic diagram of AC power consumption test

The voltage sampling adopts the high-precision resistance voltage division method. The ESEM16 voltage sampling requires the input signal to be 100uV to 500mV. According to the effective value of 220V, the design input signal is 100mV. According to the resistance voltage division, It can be calculated that R89=1K, R78=R79=R80=R81=510K. The current sampling uses a high-precision, low-resistance resistor on the power line. The ESEM16 requires a current sampling input signal of 5uV to 25mV. Then Pmin/220×R95>5uV, Pmax/220×R95<25mV. The AC power consumption of the module is mainly at the zero-crossing circuit, and the power consumption is very small. Considering the AC power consumption fluctuation of different manufacturers' modules, the AC power consumption range is set from 1mW to 300mW. According to the above calculation, the sampling resistance is 1.1R<R95<18.3R, and the sampling resistance is selected as 10R.

2.4 MCU module design

The MCU module is the core module of the power consumption tester. The MCU is selected as ST Company's STM32F207ZE. The chip has a high-speed running frequency and application interface. The ADC module has 12-bit measurement accuracy and 30MHz sampling rate, which can meet the BPLC dynamic power consumption test requirements.

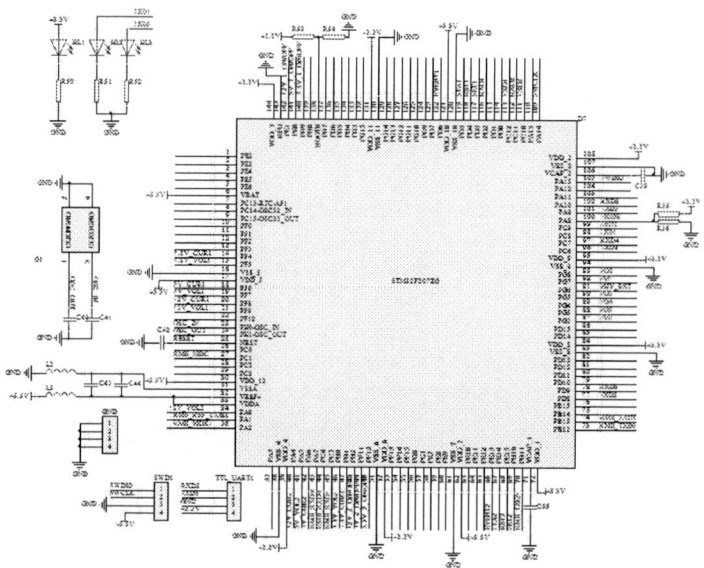

Fig 5. Schematic diagram of MCU module

2.5 Display module and network communication module design

The display module uses a touch display that allows static and dynamic power consumption testing through interface and system interaction. The network communication module can interact with the master computer software to support remote control, and can upload the test data to the PC master computer, so that the power consumption tester can be integrated into other test systems.

3. The implementation of the power consumption tester software

The power consumption tester software consists of two parts: the MCU program and the AC acquisition chip ESEM16 program. The data is interacted through the serial interface, and the serial interface line uses the optocoupler isolation. Figure 6 shows a design block diagram of the test software.

The tester software first initializes the system before sampling begins, and then the system will obtain test instructions from the display or master computer software to determine whether it is static power consumption. If it is static power consumption, it will go directly to the data acquisition and analysis processing flow. If it is dynamic power consumption test, it will enter the networking process. The networking process is similar to the actual courts application. First, the MCU module simulates the virtual energy meter and the virtual concentrator, initializes the concentrator module parameter area, and then downloads the table file and restarts the concentrator module. After that, the system will periodically query the networking status and wait for the networking to complete and enter the dynamic power consumption test process.

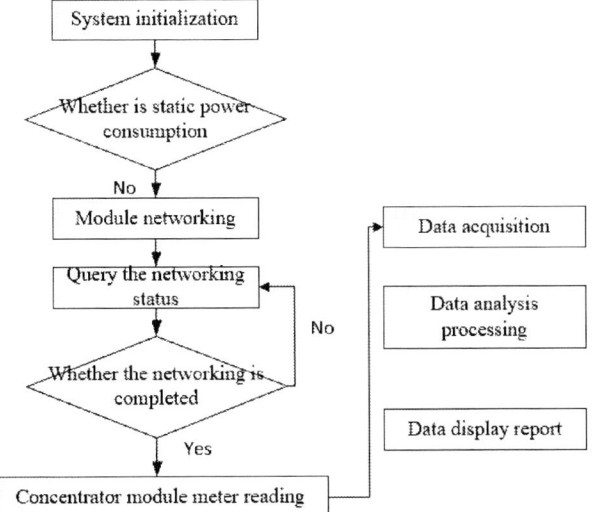

Fig 6. System main program flow chart

3.1 Data acquisition

Data acquisition mainly includes DC voltage and current acquisition and AC voltage and current acquisition. The MUC module can directly acquire DC voltage and current, and the acquired binary data can be calculated as current and voltage through the circuit voltage division formula, so that DC instantaneous power consumption can be calculated. Starting a round of sampling, each channel is acquired 100 times according to the sampling sequence, and each sequence acquires 480 Cycles, so 2 $\times 10 \times 480/30000000=320us$ is needed to complete one round of sampling. To complete sampling, there is a 1000us delay in the program. After completing a round of sampling, the calculation takes about 3us.

The AC acquisition is done by ESEM16 and can be obtained by reading the active power register (EM_PA) value. EM_PA is a 32-bit signed number, expressed in binary complement code form, and the negative number means that the actual power direction is negative. Calculated as follows:

$$PA = \frac{DATA \times k \times Vref^2}{R \times Gi \times Gu \times 2^{31}} \qquad (2)$$

Where DATA is the decimal value of the corresponding active power register value, R is the resistance of the manganese copper shunt (in Ω), k is the voltage divider ratio of the voltage channel (k >1), and G_i and G_u are the PGA gain of current and voltage channel, V_{ref} is the ADC reference voltage (V_{ref} = 1.3V). After the AC acquisition is completed, the test data is sent to the MCU module by monitoring the serial interface DC.

3.2 Data analysis and processing

In the static power consumption test, the system program filters each data in each sample sequence to eliminate the errors, so that the authenticity and usability of the sampled data can be ensured[6]. The software program uses a limiting average filtering algorithm, which can effectively overcome the pulse interference caused by accidental factors[7-8], and significantly eliminate a large number of random noise in the power grid.

The dynamic sending time of the BPLC module is very short. The test data will include dynamic power consumption data and static power consumption data. The filtering processing of the acquired data is different from the static power consumption. The dynamic power consumption test data as shown in Figure 7, the power consumption of the module is much larger than the static power consumption. When processing the dynamic power consumption test data, set a threshold $P_{threshold}$. When it is larger than this threshold, it is considered as dynamic power consumption. When it is less than this threshold, it is considered as static power consumption. After acquiring multiple sets of dynamic power consumption data, the data is filtered to reduce the influence of a large amount of random noise in the power grid on the dynamic power consumption test results.

$$p = \begin{cases} p_{dynamic} & \left(|p| > p_{threshold}\right) \\ p_{static} & \left(|p| < p_{threshold}\right) \end{cases} \qquad (3)$$

Because the power consumption tester has the problem of component accuracy, the test data will be offset. The power tester needs to add a power consumption compensation calibration mechanism. The power consumption compensation calibration of this solution is to add a fixed loss value ΔP to the test result.

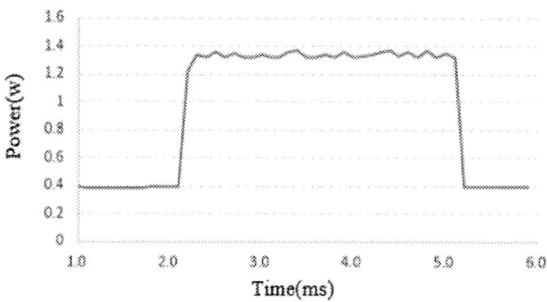

Fig 7. Dynamic power consumption test data

3.3 Data display and remote control

The display module communicates with the MCU module through the serial interface, and can convert the display operation into serial interface command to control the power consumption tester test. Similarly, MCU module sends the test results to the display module, and the display module displays the test results.

The network communication module realizes the remote control of the power consumption tester, communicates with the PC master computer through the TCP/IP protocol. The remote communication application layer protocol is a customized protocol. The protocol has a CRC check on the data to increase the security and stability during the data transmission process.

4. Comparison and analysis of the test results

In order to confirm the accuracy and practicability of the power consumption tester, different test methods are used to compare the static and dynamic power consumption of the energy meter BPLC module. In order to exclude individual differences, this test selects ten energy meter modules for testing. The test results are shown in Table 1.

Table 1. Comparison of measurement results

Test methods	Static power consumption(W)	Dynamic power consumption(W)
A province metrology center	0.391	1.351
A power meter test result	0.395	1.034
Power consumption tester result	0.389	1.343

It can be seen from the test result data that the static power consumption test results are basically the same under different measurement modes. But there is a certain difference when testing dynamic power consumption. The dynamic power consumption tester is consistent with the test results of a province metrology center. The test results of a power meter differ greatly from the other two methods. This is mainly because the dynamic response of a power meter is slow, and the measurement rate cannot keep up with the power consumption change when sending BPLC module.

5. Conclusion

This paper proposes a design method of BPLC power consumption tester, which is mainly composed of MCU module, power load module, DC power consumption test unit, AC power consumption module, display module and network communication module. The tester utilizes a high sampling rate ADC chip and a high-precision AC power metering chip to make DC power consumption measurement response fast and AC power consumption measurement accurate. The measurement results prove that the power consumption tester designed by this scheme is more accurate in dynamic power consumption. In addition, the design scheme is lower in cost, better in portability and integration.

Due to the technical advantages of the tester in portability, low cost and integration, especially the high dynamic response of the tester, and the high-accuracy measurement of AC power consumption, it provides a new kind of accurate measurement of BPLC module power consumption. The new method can be widely used in the power consumption measurement of various BPLC modules.

Author introduction:

Liu Mouhai(1990-), Senior engineer, State Grid Hunan Electric Power Company Limited Power Supply Service Center(Metrology center), Engaged in electric energy metering and acquisition technology research.

References

[1] Song Fang, li Xiaowei, Dang Sanlei, et al. Four Table One Set Copy System Research Based on Power Line Carrier Technology[J]. Times Agricultural Machinery, 2016, 43(05): 52-53

[2] Q/GDW 1374.3-2013, Power user electric energy data acquire system technical specification Part 3: Communication unit[s]

[3] Xu Bin, Wang Bin, Jiang Yuanjian. Power Line Carrier Communication Technology and its Applications in Electric Energy Data Acquisition System[J]. Electrical Measurement & Instrumentation, 2010, 47(7A): 44-47

[4] Fan Xiaoke, Zhang Xiaoqing, Yang Jianming. Development of Electrical Signal DAQ System Based on VI[J]. Electrical Measurement & Instrumentation, 2009, 46(5): 20-23

[5] Zhang Qiuyan, Xu Hongwei, Zhou Ke, et al.. Applications of LabVIEW in power high-precision dynamic measurement of PLC[J]. Electrical Measurement & Instrumentation, 2017, 54(13) :115-119

[6] Li Xingyi. Digital signal processing[M]. Chongqing University Press, 2002

[7] Zhang Weixin, Zhang Jian. Analysis on line noise Characteristic of Low Voltage Grid[J]. North China Electric Power, 2012, (4): 27-30

[8] Rick Bitter, Taqi Mohiuddin. LabVIEW—Advanced Programming Techniques SECOND EDITION[M]. CRCPress, 2010

[9] Zhou Hang, He Wei, Chen Yu, et al. Design of Point to Point Communication Performance Testing Platform for Broadband Power Line Carrier[J]. Electrical Measurement & Instrumentation, 2016, 53(21):100-105

Research on Power Enterprise Network Security Solution

Shang YanWei[1], Lin Qiang[1], FengZhi Peng[1]*

[1]Guangdong Power Grid Co., Ltd IT Information Center，Guangdong,510030, China

Abstract: With the development of scientific and technological productivity, the operation and maintenance of power companies have gradually realized network informationization. However, there are still some risks in the computer network security of power companies. The article first expounds the importance of computer network security and then analyzes the risks existing in the power system computing network and finally puts forward targeted computer network security protection measures.

1. INTRODUCTION

With the advent of the information age, computer networks are increasingly used in power enterprises and they can achieve integrated management in power system automation, monitoring, protection and collection of electricity charges, playing an increasingly important role. At the same time, the national power data communication network also connects various levels of power companies to achieve information sharing. However, due to various factors, the security risks of power enterprise information networks are increasing, network data is lost, operating systems are destroyed and network failures occur from time to time. The safe and reliable operation of power companies is the basic guarantee for ensuring the normal operation of social production and life. How to ensure the computer network security of power companies is an important issue to be solved urgently in the power industry. Based on this, the author discusses this issue[1].

2. THE IMPORTANCE OF COMPUTER NETWORK SECURITY

Computer network security mainly includes two aspects: information security and physical security. Information security refers to network security to ensure the integrity and confidentiality of various network data information. The threat sources of computer networks include viruses, hackers, etc., in which the harm caused by hacking attacks has exceeded the impact of viruses and some attacks are even fatal. Therefore, it is necessary to strengthen the security of computer networks. Only by taking targeted protection measures can we ensure the security and integrity of network information. According to the US Federal Survey, the annual economic losses caused by computer network security in the United States amount to 27 billion yuan per year. The security threat of the domestic Internet is also very serious. According to the 2013 National Emergency Assessment Center's assessment data, 58% of the global "zombie" computers are in China. It can be seen that ensuring the security of computer networks is imperative[2].

3. RISKS OF COMPUTER NETWORKS IN POWER COMPANIES

3.1. TECHNICAL SECURITY RISKS

With the advent of the information age, the dependence of power companies on computer networks has reached an unprecedented scale. Due to the huge security risks of computer networks, the security and stability of power system operations that are over-reliant on computer networks become very fragile. Once the computer network is damaged due to various reasons, the entire power system will be paralyzed. Based on this, power companies must develop comprehensive computer network security protection measures to ensure the safe and stable operation of the power system. However, the current security protection technology has certain pertinence. It is usually designed separately for one or several problems. It is difficult to effectively solve all the problems that arise during operation and it is difficult to provide comprehensive security protection for computer network security. There are certain deficiencies in technology[3]. The frequency of occurrence of various risks is shown in Table 1.

Table 1: Frequency of occurrence of various risks

Years	Technical	Information	Managing
2011	511479	42847	634840
2012	532098	44560	654562
2013	552717	46273	674284
2014	573336	47986	694006
2015	593955	49699	713728
2016	614574	51412	733450
2017	635193	53125	753172
2018	655812	54838	772894
2019	676431	56551	792616
2020	697050	58264	812338
2021	717669	59977	832060

3.2. INFORMATION SECURITY RISK

At present, the reasons for threatening the security of computer networks in power systems include: natural disasters, software vulnerabilities, viruses and hackers. Computer systems are easily affected by natural disasters such as fires, lightning strikes, earthquakes, etc. Once such natural disasters occur, if the protection capacity is insufficient, it is easy to cause loss of power system information. Due to the variety of application network software and different types of vulnerabilities, hackers exploit software vulnerabilities to attack software and ultimately destroy the entire system. Viruses also have an impact on computer network security. Once the network interferes with the virus, it may cause the entire system to run slowly, which may lead to system loss and data loss. In addition, some lawless elements have deliberately destroyed power system data and implemented financial crimes by stealing passwords, etc., which have greatly threatened computer network security[4].

3.3. MANAGING SECURITY RISKS

Management security risks mainly exist in two aspects: first, network management personnel; second, network users. Managers and users are not aware of security. They can use their own accounts or share them with others. They use pirated applications and externally insecure mobile storage media to inadvertently reveal key information such as network passwords and configurations. Problems can easily affect the safe and stable operation of the network.

4. POWER ENTERPRISE COMPUTER NETWORK SECURITY PRECAUTIONS

4.1. USING SWITCHES AND VIRTUAL LAN TECHNOLOGY TO ACHIEVE ACCESS CONTROL

In order to improve the security of the computer network of the power enterprise, the switch and the virtual local area network technology can be adopted to divide the power enterprise into multiple subnets according to different service types and security requirements and the firewall is used to effectively isolate the internal and external networks. For example, the power enterprise's database, email, etc. can be divided into one VLAN1 and the power enterprise external network is divided into another VLAN2. At the same time, the one-way information flow between the two can be controlled, the former can access the latter and the latter cannot Visit the former. In order to ensure the security of the management, the nodes of the unified department of the power enterprise can be divided into the same virtual network and different management rights are assigned to different management personnel. The three-layer switch is used to support the configuration access rights of the unified network. It can effectively isolate broadcast storms and can also complete route routing and reduce delay. Specifically, several measures can be taken: first, control the MAC address port; second, use IP packets for filtering at the network layer; third, register the MAC addresses of all user machines of the power system to ensure that only authorized users can enter[5].

4.2. INFORMATION ENCRYPTION TRANSMISSION

Power enterprise information is vulnerable to hackers in the transmission process. Routers and switches of computer networks are easily attacked by hackers. Based on this, in order to prevent data information from being eavesdropped, modified and leaked during transmission, etc. The transmitted data can be encrypted so that it can be transmitted and stored in cipher text. For the security networking requirements of different data information of the power system, the data transmission can be realized through different network layers. If the data information is transmitted at the physical layer, link encryption can be used, that is, a corresponding encryption device is installed at each communication link port. For example, in the network and transport layer, when the packet enters the packet switch, the related information must be decrypted so that the route of the link can be correctly determined. In this decryption process, the data is also vulnerable to attacks, based on which the network can be transmitted. The layer implements end-to-end encryption to ensure security during the decryption of information. Information encryption algorithm such as formula 1.

$$E(W,B) = \frac{1}{2}\sum_{k=1}^{s2}(t_k - y2_k)^2$$

(1)

4.3. DIGITAL SIGNATURE AND IDENTITY AUTHENTICATION

In the network operating environment, the system and the system, the system and the user, the user, etc. continuously and frequently carry out the information transmission process, it is easy to exist several security risks: First, there is a fake source point identity will be the message Insert into the network; second, confirm or not accept the fake; third, tamper with the content and serial number of the message. In this case, the most effective solution is to authenticate the data modification personnel. The digital signature and certificate technology can be used to ensure the authenticity and validity of the user and the identity of the distributor can be recognized. The receiving aspect can also pass the number. Signatures and other methods to determine whether the data is derived from the integrity of the specified user and data information. In addition, digital signatures can also effectively prevent the negative problem of transmission in this article.

4.4. POWER ENTERPRISE EXTERNAL NETWORK SECURITY ACCESS CONTROL

Power companies use the Internet to achieve effective connections and access to the outside world. In order to ensure the security of information access and transmission in this process, firewall technology is usually adopted. Between the internal and external networks of the power enterprise, the firewall is used to achieve isolation and access control; the access control of different network security domains within the power enterprise also uses a firewall, so as to effectively isolate access control between different networks and avoid A network problem affects another network segment through the local area network. Through this setting, real-time detection and audit of network activities, events and status can be realized and real-time analysis and control of various malicious access control and misoperations can be performed; relevant analysis reports can be obtained according to firewall monitoring and vulnerability detection. According to the content of the report, the power system users are provided with rapid detection of illegal intrusion and provide counterattack means. Therefore, setting the firewall correctly can not only effectively realize the security protection between the internal and external networks, but also provide security protection between different internal network segments and effectively reduce the network security risk of the power system according to the needs of users.

4.5.   LOG MANAGEMENT AND BACKUP DATA

For the power system, in addition to the above-mentioned security protection, the log management and backup data subsystem are the key to ensuring the security of information data. If these two systems are lacking, it is difficult to put into actual operation, so the power system computer network appears. Failure, must first analyze the cause of the problem and can restore the normal operation of the system in the shortest time, that is, the key link of the network system design, log management is mainly to record the operation into the power system, such as database operations, FTP File server, etc. The operation events mainly include: clicking the menu item, clicking the window control, button, drop-down box, etc.; the log table record mainly includes: user name, login time, location, exit time, operation content, etc., so as to be used in future auditing and verification. The power enterprise data is the life of the enterprise. Once the data is lost, the losses caused to the enterprise are immeasurable. To prevent data loss or damage, the database must be archived and backed up in time. Archiving refers to the permanent or restricted saving practice of saving data that is not needed by the database. Data backup is the most basic daily work of the power system. When building a computer network system, you must first set the backup time, backup data selection, etc. to ensure the security of data information.

5.   CONCLUSION

The computer network security of power companies belongs to a complex system engineering, involving advanced security technologies, perfect system management and many other aspects. In order to ensure the safe operation of computer network systems, a multi-level and comprehensive security system must be established. This paper carefully analyzes the security requirements of power companies for computer networks and carefully studies the security, efficiency and convenience of power companies and comprehensively utilizes data encryption, firewalls and virus detection to ensure the security of computer networks.

References

[1]   Sebastian Feyrock. Łukasz Kulesa: Envisioning a Russia-NATO Conflict. Implications for Deterrence Stability. London: European Leadership Network (Euro-Atlantic Security Report), Februar 2018[J]. SIRIUS - Zeitschrift für strategische Analysen,2018,2(3).

[2]   University of Michigan; Patent Issued for Rating Network Security Posture And Comparing Network Maliciousness (USPTO 10,038,703)[J]. Computers, Networks & Communications,2018.

[3]   José De Sousa,Daniel Mirza,Thierry Verdier. Terror networks and trade: Does the neighbor hurt?[J]. European Economic Review,2018,107.

[4]   Tahir Hasan,Tahir Ruhma,McDonald-Maier Klaus. On the security of consumer wearable devices in the Internet of Things.[J]. PloS one,2018,13(4).

[5]   Sergei Shubin,Maria Nesterenko. The Arctic: network diplomacy from the view of global security 2.0[J]. The Polar Journal,2018,8(1).

ISPECE

IOP Publishing

Morphological analysis of optocoupler accelerated degradation test data

Xuangong Zhang[1*], Xihui Mu[2] and Huizhi Li[1]

[1] Department of Ammunition Engineering, Shijiazhuang Campus, Army Engineering University, Shijiazhuang, Hebei 050003, China

[2] Institute of Special Service, Army Research Institute, Shijiazhuang, Hebei 050003, China

*Corresponding author's e-mail: 15132497756@163.com

Abstract. As one of the key components inside the seeker of a certain type of guided missile, optocoupler has attracted much attention for its long storage reliability. In order to accurately assess its storage life, an accelerated degradation test was performed on the optocoupler. Before the processing of the test data, it is necessary to carry out data shape analysis to verify whether the failure mechanism caused by long storage is consistent with the failure mechanism caused by the test. There are many literatures that use mathematical methods to conduct a lot of research on the consistency mechanism of failure mechanism. The method of this paper avoids the lengthy and complicated mathematical derivation, starting from the data morphology analysis, and relies on the failure mechanism obtained by the failure mechanism verification test. The key conclusion that the failure mechanism has not changed is obtained. The method proposed in this paper is portable, and it can provide some reference for the analysis of experimental data of similar products.

1. Description of the test and explanation of the previous conclusions

According to the previously scheduled scheme, the accelerated degradation test uses 10 samples, and the optocoupler is heated in a four-stress step-acceleration degradation test with unequal measurement times. The four stress levels were 70°C, 90°C, 110°C and 120°C, respectively, with 23 measurements, 11 times, 10 times and 9 times. Test parameters were determined as forward voltage drop, reverse breakdown voltage, leakage current, and flip voltage. The overall test time was 424 hours and was tested every 8 hours. The basic test result is that the No.3 optocoupler fails after heating for 408 hours in the fourth stress stage, and the flip voltage cannot be measured; the No.1 optocoupler fails after heating for 424 hours in the fourth stress stage, and the flip voltage cannot be measured. The remaining optocouplers did not fail. From the overall data situation, the degree of change in forward voltage drop, leakage current, and breakdown voltage is relatively intense, and the change in the flip voltage as an indicator parameter is not large. In the next step, data morphological analysis was performed on each parameter of each sample to further obtain its changing trend, failure reason and failure mechanism. Figure 1 is a scatter plot of leakage current over time, where the horizontal axis represents time in hours and the vertical axis represents leakage current in nanoamps. Three vertical lines indicate the division of different temperature stress stages. Due to space limitations, the remaining scatter plots are no longer given.

Content from this work may be used under the terms of the Creative Commons Attribution 3.0 licence. Any further distribution of this work must maintain attribution to the author(s) and the title of the work, journal citation and DOI.
Published under licence by IOP Publishing Ltd

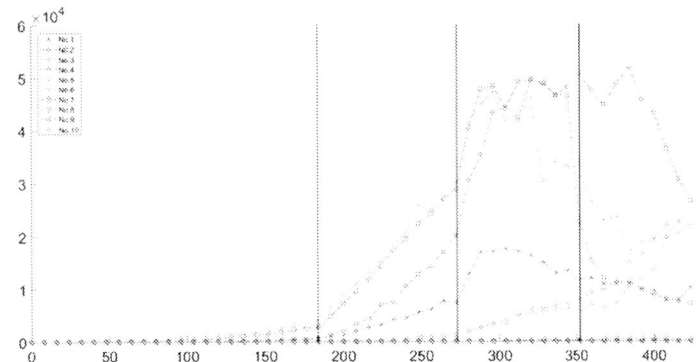

Figure 1 Scatter line diagram of leakage current with time

Through the previous failure mechanism verification test, the following conclusions were obtained: (1) There are two failure mechanisms for optocoupler, namely open circuit failure and leakage failure, respectively. Open circuit failure includes gradual open circuit failure and sudden open circuit failure. (2) The reason for the gradual open circuit failure and the sudden open circuit failure is that the bonding point lead is separated from the bonding point, which is caused by the defect of the bonding process or the poor quality of the transparent adhesive; the leakage failure is caused by the movable ion on the chip to form a surface leakage channel. (3) The key performance parameters are two, namely forward voltage drop and reverse leakage current. (4) The forward voltage drop is affected by two opposite mechanisms, one is an open circuit to increase it, and the other is an increase in leakage current to reduce it. The following data morphological analysis requires these conclusions, so they are given here in advance.

2. Statistical analysis of initial values of samples

The degree of consistency of the sample is reflected in the degree of dispersion of the initial value. In general, the smaller the dispersion of the initial values, the better the consistency of the samples[1]. A frequency histogram was established for the four parameters of the 10 samples, and the mean and standard deviation of the initial values were calculated.

Table 1 The initial value of the parameters and the standard deviation

	μ	σ	$|\sigma/\mu|$
forward voltage drop	2.227	0.0485	0.0218
leakage current	25.18	58.4086	2.3196
breakdown voltage	78.22	11.2392	0.1437
flip voltage	-0.3697	0.0509	0.1376

Through Table 1, the following conclusions can be drawn: (1) The initial value of forward voltage drop is concentrated, although the initial value of leakage current is concentrated, but its polarization is more serious than the forward voltage drop, and the initial value of breakdown voltage and flip voltage is more uniform. (2) $|\sigma/\mu|_{\text{flip voltage}} < |\sigma/\mu|_{\text{breakdown voltage}}$, indicating that the initial value of the flip voltage dispersion is lower than the initial value of the breakdown voltage, but due to they are greater than 0.1, indicating that the initial value of the inversion voltage and the initial value of the breakdown voltage are also relatively discrete with respect to the initial value of the forward voltage drop. (3) The initial value of the leakage current is the one with the largest change among the four parameters, and the $|\sigma/\mu|_{\text{leakage current}}$ is >1, and the dispersion is also the largest.

The above three conclusions indicate that the consistency of the samples is not very satisfactory. The poor degree of consistency reflects the need to improve the production process level of the product. On the other hand, it may induce multiple failure mechanisms at the same time, which increases the difficulty of life assessment.

3. Construction of parameter data morphology analysis table and parametric morphological analysis

3.1. Construction of parameter data morphology analysis table

Before the data shape analysis of the forward voltage drop, leakage current, breakdown voltage and flip voltage, respectively, the *regression* function and the *rcoplot* function in matlab are used to perform the residual analysis to eliminate the abnormal data points, and then through linear fitting[2]:

$$y = k_1 x + k_2$$

the general trend of the curve is determined. When the trend of the data is upward, there is $k_1 > 0$, otherwise $k_1 < 0$. Obviously, when the absolute value of k_1 is larger, the trend of data changes is larger, and vice versa. In addition, the maximum point and the minimum point of the entire curve are obtained, and the change trend of the curve and the degree of oscillation are assisted by judging the difference between the maximum point and the minimum point and the respective positions. Due to the content of the table, due to space, this paper only gives the data morphology analysis table of forward pressure drop.

Table 2 Forward pressure drop data morphology analysis table

	Morphological characteristics	k_1	k_2	*Minimum point*	*Maximum point*	*Overall trend*
1	↗↘↗↘	4.747×10^{-5}	2.221	（0,2.2）	（360,2.33）	↗
2	↗↘↗↘	1.691×10^{-5}	2.205	（16,2.19）	（320,2.23）	↗
3	↘	-1.775×10^{-4}	2.349	（288,2.21）	（40,2.41）	↘
4	↗↘↗↘	-1.016×10^{-5}	2.263	（0,2.21）	（152,2.3）	↘
5	↗↘↗↘	1.96×10^{-5}	2.198	（96,2.19）	（312,2.22）	↗
6	↘↗↘	1.703×10^{-5}	2.22	（168,2.21）	（336,2.24）	↗
7	↗↘↗	1.626×10^{-5}	2.205	（0,2.2）	（216,2.22）	↗
8	↘↗↘↗	5.241×10^{-6}	2.221	（112,2.21）	（320,2.24）	↗
9	↘↗↘↗	5.947×10^{-6}	2.221	（208,2.21）	（312,2.24）	↗
10	↗↘↗↘	9.216×10^{-6}	2.247	（0,2.2）	（64,2.31）	↗

3.2. Parameter morphology analysis and failure mechanism verification

Since the sample has two forms of degradation failure and sudden failure in the failure mechanism verification test, the samples of the formal test are roughly divided into two groups according to the degradation failure and the sudden failure[3]. A significant feature of degradation failure is a significant increase in leakage current, from nA level to μA level. Therefore, according to the data shape analysis table in the previous section, the 1st, 2nd, 3rd, 4th, 5th, and 10th are grouped into one group, and the 6th, 7th, 8th, and 9th are divided into another group.

First, the parameter of the inversion voltage is analyzed. Among the 10 samples, 9 of the inversion voltages are in a downward trend, and one has an upward trend (sample No. 9). The flip voltage is the only parameter with a failure threshold of -1.8V. However, these two trends are very weak, and the magnitude of k_1 is between 10^{-5} and 10^{-7}, and its upward trend is weaker than the downward trend. The maximum difference between the maximum and minimum values is 0.039V (sample No. 4), and the minimum difference is 0.005V (sample No. 9). According to parameter data morphology analysis table, the average value of the inversion voltage is -0.361V. Therefore, flipping the voltage does not characterize the change in optocoupler performance. It is only used as an indicator to reflect the failure of the optocoupler.

Next, the breakdown voltage is analyzed as a parameter. Among the 10 samples, the trend of breakdown voltage showed a polarization. The breakdown voltages of samples 1, 2, 3, 4, 5, and 10 all showed a downward trend, that is, $k_1 < 0$, while samples Nos. 6, 7, 8, and 9 were opposite, and all had an upward trend, that is, $k_1 > 0$. Comparing the morphological analysis table, it can be seen that the data pattern of the breakdown voltage of samples 1, 2, 3, 4, 5, and 10 is roughly opposite to the data pattern of the leakage current, and Table 3 is the correlation coefficient table.

Table 3 Correlation coefficient between leakage current and breakdown voltage

NO.	1	2	3	4	5	6
Correlation coefficient	-0.9637	-0.9315	-0.9978	-0.9578	-0.9795	-0.9479

As shown in the above table, the breakdown voltage of samples 1, 2, 3, 4, 5, and 10 has a strong negative correlation with the leakage current[4], which is consistent with the results observed in the failure mechanism verification test, that is, leakage current's dramatic increase is the key cause of the breakdown voltage.

The breakdown voltages of samples Nos. 6, 7, 8, and 9 all have an upward trend. Compared with k_1 of samples 1, 2, 3, 4, 5, and 10, the absolute value is smaller than the k_1 of the latter. There are approximately 1 to 2 orders of magnitude difference. Although from this point of view, the rising trend of the breakdown voltage is weaker than its downward trend, the k_1 of its rising trend is still 1~2 orders of magnitude higher than the trend of the inverted voltage. Moreover, all four samples have an upward trend, so this is not caused by chance. The reduction in breakdown voltage was previously determined and verified by analytical calculations due to the sharp increase in leakage current. According to the data shape analysis table, all the leakage currents k1 are greater than 0. Therefore, for the above two reasons, the increase of the breakdown voltage is not caused by the leakage current. At the same time, the breakdown voltage of samples Nos. 6, 7, 8, and 9 increases while the forward voltage drop increases. Xiao Shiman and Liu Xin pointed out that when the conductive adhesive is poorly bonded, the forward voltage drop and the reverse breakdown voltage will rise synchronously[5][6]. Therefore, the breakdown voltage of the samples Nos. 6, 7, 8, and 9 was caused by poor adhesion of the conductive paste.

Then analyze the data pattern of leakage current. Since the leakage current of samples Nos. 6, 7, 8, and 9 does not exceed 1 μA, the leakage currents of samples 1, 2, 3, 4, 5, and 10 are mainly analyzed here. As mentioned above, the breakdown voltages of these samples have a strong negative correlation with the respective leakage currents, and the inflection points are generated by 1, 2, 4, and 5, which is consistent with the phenomenon of the failure mechanism verification test. The reason why No. 3 and No. 10 did not produce an inflection point was that the moment of inflection point had not been reached. Therefore, the change of leakage current of these samples is caused by the movable ion pollution, which can correspond to the mechanism explanation of the failure mechanism verification test.

Finally, the parameter of forward pressure drop is analyzed. The forward pressure drop is affected by two opposite mechanisms. According to the analysis results of the bottom test, the gradual open circuit failure will increase the forward voltage drop, and the leakage current will reduce the forward voltage drop. When the two mechanisms act simultaneously, the mechanisms will compete with each other, and the final curve trend will show a trend of strong mechanism. Among the 10 samples, the forward pressure drop of samples 1, 2, 5, 6, 7, 8, 9, and 10 produced an upward trend. Among them, the breakdown voltage of samples Nos. 6, 7, 8, and 9 also rises synchronously. The specific reason has been given in the foregoing, that is, the conductive adhesive is poorly bonded, which is a kind of gradual open circuit failure. The forward pressure drop of samples 1, 2, and 5 also showed an upward trend, but the leakage current of these three samples showed a large change trend, and the increase of leakage current would decrease the forward voltage drop. Therefore, these three samples must have a mechanism of progressive open-circuit failure to increase the forward voltage drop, and this mechanism is stronger than the forward voltage drop reduction mechanism caused by leakage current.

The forward pressure drop of samples No. 3 and No. 4 is special because the forward pressure drop of both is a downward trend. But the downward trend of the two is not caused by the same mechanism. First, analyze sample No. 3. The initial value of forward pressure drop of sample No. 3 was the highest among all samples, and the mean value of forward pressure drop was around 2.22V, but the initial value of forward pressure drop of No. 3 was 2.36V. The forward pressure drop curve of sample No. 3 was always oscillating, and the absolute value of k_1 was the largest in the forward pressure drop curve of 10 samples. The initial value of the forward voltage drop of sample No. 3 differs from the minimum value by 0.15 V, which is obviously not only due to leakage current. Sample No. 3 was the first of the 10 samples to fail. It was tested for forward current and found that it can only withstand a current of about 500 μA. Once this value is exceeded, the current representation quickly returns to zero. Apparently there is an open circuit inside sample No. 3. However, this open circuit is not a gradual open circuit. The gradual open circuit will gradually increase the forward voltage drop, and the failure caused by the gradual open circuit will be able to withstand the current to the mA level[7]. Obviously this is a sudden open circuit. The reasons for sudden open circuit have been pointed out before, including poor bonding process or broken connection lines. There is a case in which the bonding process is poor, that is, the conductive adhesive is not positioned properly, which may cause poor contact between the lead gold ball and the electrode, thereby causing poor current expansion[8]. When the temperature rises, the glue gradually overflows, increasing the contact between the lead gold ball and the electrode, resulting in a decrease in the forward voltage drop[9]. However, after the rubber overflows, the bond strength is lowered, and the bond strength is lowered, causing the lead to come off the electrode, thereby causing a sudden open circuit[10]. Therefore, sample No. 3 was actually caused by a sudden open circuit failure. The forward pressure drop of sample No. 4 also showed a downward trend, but the downward trend was weak. As mentioned above, a sharp increase in leakage current causes a decrease in forward voltage drop. It is clear that the forward pressure drop of sample No. 4 is affected by this mechanism.

4. Conclusion

The samples that failed in this test belong to the sudden open circuit failure. There are two specific manifestations of sudden failure: gradual open circuit failure and sudden open circuit failure. Among them, the No. 3 sample belongs to the sudden open circuit failure, and the No. 1 sample belongs to the gradual open circuit failure. Since the two can be attributed to the same failure mechanism, that is, open circuit, the two forms of expression are uniformly summarized as sudden open circuit failure. The consistency level of the sample is not satisfactory, and the production process needs to be improved. Half of the products in this test produced degraded characteristics. The degradation characteristics are consistent with those observed during the bottom test, and the leakage current changes drastically. The inflection points were observed in samples 1, 2, 4 and 5, which were consistent with the mechanism explanation of the movable ion pollution given by the bottom test. There are two failure modes in the sample, so the time of failure is determined by different mechanisms and ultimately determined by the strong mechanism. Different failure mechanisms do not necessarily occur simultaneously on a single sample. For example, the leakage currents of samples Nos. 6, 7, 8, and 9 do not change drastically. When different failure mechanisms occur in the same sample, the mechanisms compete with each other, but the strength is random and varies from sample to sample.

References

[1]   Mou S S, Wang L L. Accelerated Life Experiment [M]. Beijing: Science Press, 1995.
[2]   Lu C J，Meeker W Q. Using degradation measures to estimate a time-to-failure distribution[J]. Technometrics，1993:161-174.
[3]   Tang S J, Guo X S, Zhou Z F. Modeling and Residual Life Estimation of Step Stress Accelerated Degradation Test[J]. Journal of Mechanical Engineering, 2014, 50(16): 33-39.
[4]   Liu J. Research on step-stress accelerated degradation test of electrical connector[D]. Hangzhou: Ph.D. Thesis of Zhejiang University, 2013:20-28.
[5]   Xiao S M. Optocoupler Package and Related Failure Mechanisms[J]. Semiconductor Technology, 2011, 36(4): 328-330.
[6]   Liu X. Research on Common Failure Modes and Failure Mechanism of LEDs[C]// The 15th

Annual Conference on Reliability, 2010, 122-126.

[7] Si F H. Analysis of Several Common Causes of LED Lamp Failure [J]. Electronic Quality, 2007, 9: 13-14.

[8] Yang S H, LI K L. Analysis of long-term storage degradation characteristics of optocouplers[J]. Electronics Reliability & Environmental Testing, 2013, 31(2): 27-30.

[9] Li H Y, Jin L, Chen C X et al. Long-term storage life of transistor output optocouplers[J]. Semiconductor Optoelectronics, 2014, 35(2): 230-232.

[10] Gao C, X Y Y, Yang D M. Research on storage life evaluation method of plastic encapsulated optocoupler based on POF[J]. Electronic Test, 2018, 2:45-46.

Research on Overvoltage Distribution of HVDC Converter Valve in Special Environment

Ning Liang[1], Cuicui Liu[2*], Yating Gou[2], Jiachen Tian[2], Fang Zhuo[2], Feng Wang[2].

[1]Maintenance&Test(M&T)Center of EHV Power Transmission Company, Guangzhou 510000, China

[2] School of Electrical Engineering, *Xi'an Jiaotong University, Xi'an 710049, China*

Email: liangning@ehv.csg.cn

liuc16@stu.xjtu.edu.cn

gouyating@stu.xjtu.edu.cn

tianjiachen1995@stu.xjtu.edu.cn

zffz@mail.xjtu.edu.cn

fengwangee@mail.xjtu.edu.cn

TEL:18710809616

Abstract. HVDC transmission technology has great advantages in long-distance, large-capacity cross-region power transmission. Thyristor converter valve is the core equipment of HVDC transmission technology, and its reliability directly affects the reliability of the whole system. During the operation of HVDC transmission system, it will face the special operating environment, including high altitude nuclear electromagnetic pulse(HEMP), operating overvoltage, lightning, and steep wave overvoltage. In order to study the reliability of the converter valve under special circumstances, the wideband equivalent circuit model of the key components were built in this paper. On this basis, the overvoltage distribution inside the converter valve under special environments is simulated and analyzed. The results show that the overvoltage distribution inside the converter valve system is related to the characteristic parameters of the special environment, and the distribution trend of the overvoltage of the saturated reactor and the thyristors is also different, so the weak components of the converter valve system under special circumstances can be obtained. It has certain reference value for studying the reliability of the converter valve system under special circumstances.

1. INTRODUTION

As the core equipment of HVDC transmission system, the converter valve has qualified reliability in normal operating conditions.[1] However, in actual engineering, the converter valve will face various abnormal conditions. For example, the converter valve may be subjected to AC and DC systems and overvoltage surges during thunderstorms, and will also experience the HEMP environment. These special operation conditions puts higher demands on the reliability of the converter valve. Therefore, based on the broadband circuit model of each key device and considering some parasitic parameters existing in the valve tower, the Gao-Zhao HVDC transmission project is taken as an example to build a broadband circuit model of HVDC converter valve. The overvoltages of the saturated reactor and the thyristors under special environment were simulated and compared. The purpose of this paper is to study the vulnerable parts in the converter valve under special circumstances through overvoltage simulation analysis, and provide reference for improving the life and the reliability of the converter valve.

2. Structure of the Converter Valve

There are two valve halls in the ±500kV GaoZhao HVDC converter station, with 3 valve towers in each valve hall. And the valve tower is using the hanging layout in the valve hall, as shown in Fig.1. In this project, the single valve tower consists of two left and right sides, six layers above and below and a total of 12 valve modules. Each valve module contains two valve sections, and one single valve contains three valve modules, so one valve tower consists of four single valves in series. In addition, each valve section of the GaoZhao project consists of 13 thyristor levels connected in series with two saturable reactors, and then in parallel with a DC grading capacitor.

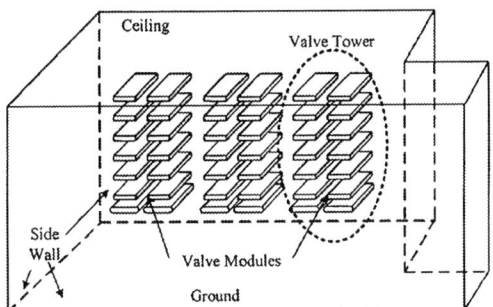

Figure 1 ±500kV GaoZhao converter valve hall layout

3. Wideband Model of Converter Valve

3.1 Wideband model of thyristor level

For a single thyristor device, a capacitor is usually used as the thyristor wideband model. However, EMP is a wide frequency range excitation source, and therefore a more accurate model which could simulate high frequency characteristics is required[1], as shown in Fig.2.

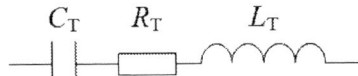

Figure 2. Wideband model of thyristor

Accordingly, the wideband model of thyristor level is shown in Fig.3, in which RSC and CSC form the snubber circuit, RG is the grading resistor.

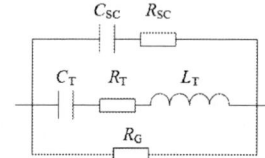

Figure 3.   Wideband model of thyristor level

3.2 Wideband model of saturable reactor

The saturable reactor is a non-linear device and its characteristics is related to its current. The wideband model of saturable reactor is shown in Fig.4[5]. As the structure can be seen in the figure apparently, the parameters considered are as follows, where L is the main inductance, RCu is the ohmic losses, RD is the damping resistance, LLeak is the leakage inductance, REd is the eddy current losses, and CSR is the stray capacitance inside the reactor.

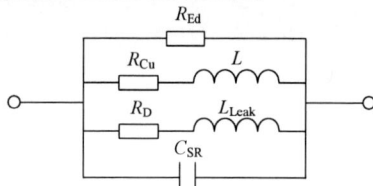

Figure 4.   Wideband model of saturable reactor

3.3 Wideband model of Valve section

The structure of the valve section is shown in Figure 5. It is generally composed of dozens of thyristor levels connected in series, and in the same time in series with one or two saturable reactors, and then connected in parallel with a DC grading capacitor. According to the topology of Fig.5, the wideband circuit model of the converter valve component is combined according to the actual electrical connection mode, and the wideband equivalent circuit model of the single converter valve section is obtained as shown in Fig.6.

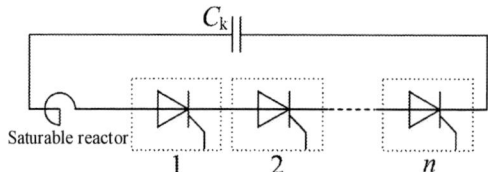

Figure 5. structure of the valve section

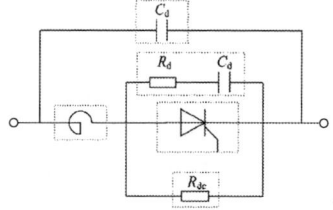

Figure 6. Wideband Equivalent circuit model of valve section

3.4 Wideband Equivalent circuit model of valve tower

There are many stray capacitances in the valve tower of the converter valve. It must be considered when constructing the model of the valve tower. This paper takes the converter valve of the ±500kV GaoZhao HVDC transmission project as an example to carry out the wideband model of the valve tower. The wideband model of the valve section of Fig.6 is connected in electrical order. The wideband model of the converter valve tower with the valve section as the basic unit and including the stray capacitance parameters is shown in Fig.7. The parameters considered are as follows: CMM represents the capacitance between two valve modules on the same layer; CLL represents the capacitance between the upper and lower valve modules; CLG represents the capacitance between each valve module and the ground.

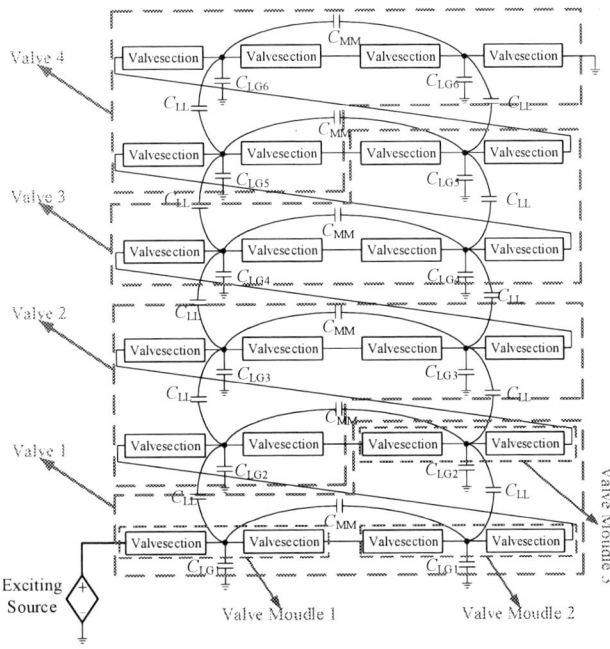

Figure 7. Wideband Equivalent circuit model of valve tower

4. Overvoltage Distribution

According to the basic theory of electromagnetic pulse, the common electromagnetic pulse waveform can be expressed in the form of double exponential function, as shown in equation (1).

$$v(t) = V_0 k(e^{-\alpha t} - e^{-\beta t}) \qquad (1)$$

A relatively reasonable electric field can be given in this function when we calculate. The function also has good application value. The parameters considered are as follows, where V_0 represents the peak voltage of the pulse, k is the correction factor, α and β are the time constants. The following calculations involve HEMP, lightning overvoltage, operating overvoltage and steep wave. The peak voltage of the electromagnetic pulse in this paper is 150kV. The characteristic parameters are shown in Table 1.

Table 1 characteristic parameters of the electromagnetic pulse

type	Pulse rise time	Half pulse width	k	α	β
HEMP	10ns	184ns	1.0502	4×10^6	4.76×10^8
Lightning wave	1.2μs	50μs	1.0372	0.0147×10^6	2.4689×10^6
Operation wave	0.25ms	2.5ms	1.1434	0.033×10^8	0.0115×10^6
Steep wave	0.4μs	1.2μs	85.470	2.1459×10^6	2.2173×10^6

According to the converter valve connection mode of the ±500kV GaoZhao HVDC converter station and the wideband equivalent circuit model of the converter valve tower, the electric stress distribution of the converter valve under special circumstances is simulated. The simulation results as shown in fig.8 (a)- (d).

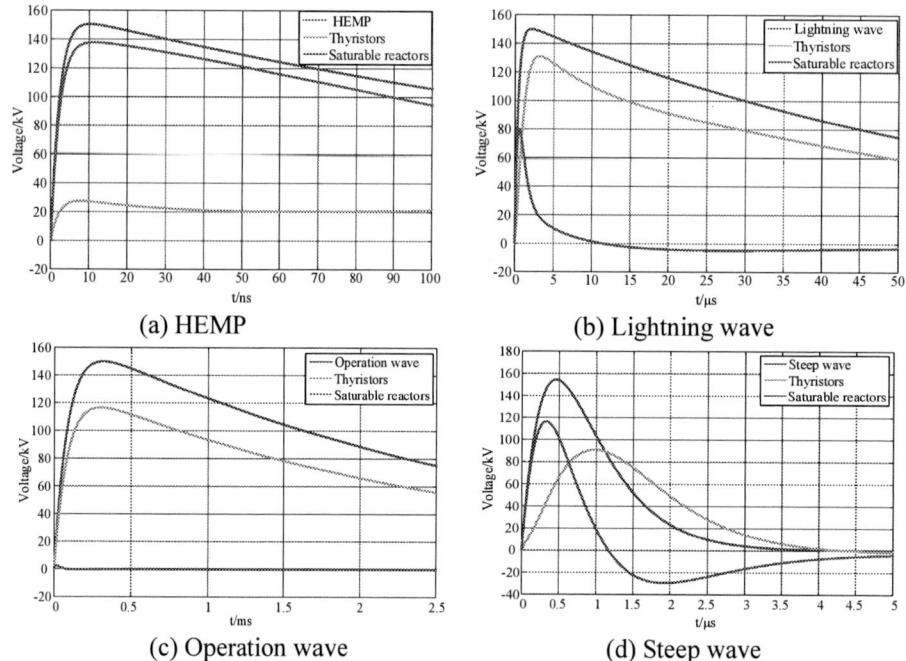

(a) HEMP

(b) Lightning wave

(c) Operation wave

(d) Steep wave

Figure 8. overvoltage distribution of the converter valve in different special environment

It is noted that in the HEMP environment, the saturable reactor is subjected to most of the overvoltage, and the thyristor is subjected to a little overvoltage. In the case of lightning overvoltage, the thyristor is subjected to most of the overvoltage, and the saturable reactor is subjected to a little overvoltage, but the wave front is steep; when the overvoltage is operation wave, the thyristor is almost full of the surge voltage, and the saturated reactor is subjected to a very little voltage. Under the action of steep wave, in the initial stage, the saturable reactor is subjected to most of the overvoltage, and the wave front is steep. However, in the later period, the thyristor is subjected to more overvoltage.

The results show that the HEMP environment has almost no effect on the saturable reactor，the thyristor is the weak component in the converter valve under the HEMP environment, when encountering this environment in the actual project, it is necessary to pay more attention to the reliability of the thyristor.

The influence of the thyristors is extremely small in operation wave, and the excessive overvoltage is endured by the saturable reactor. Therefore, when encountering this environment in actual engineering, it is necessary to pay more attention to the life of the saturated reactor; In the lightning environment, although the saturated reactor is subjected to excessive overvoltage, however, when the electromagnetic pulse acts, the thyristor has a large voltage overshoot. Therefore, this environment requires the thyristor have the ability to absorb the overvoltage, to prevent the pulse from breaking through the thyristor.

5. CONCLUSION

In this paper, based on the wideband circuit model of the key components in the existing converter valve system, the broadband equivalent circuit model of the thyristor converter valve of sorghum DC transmission engineering is built. The model is the research object, and the converter valve in special environment. The overvoltage under the system was simulated and analyzed. The overvoltages of thyristors and saturable reactors under HEMP, lightning wave, operating overvoltage and steep wave electromagnetic pulse were analyzed in detail.

The simulation results show that the weak link inside the converter valve is different under different electromagnetic pulse environment, but the saturated reactor and thyristor are the main equipments to withstand the overvoltage. In order to improve the reliability of the converter valve system under special circumstances, it is necessary to improve the thyristor. And the life of the saturable reactor and the ability to withstand overvoltage surges. The results of this paper have certain guiding significance for analyzing the reliability of actual engineering.

Acknowledgment
This work was supported by the EHV Power Transmission Company of the Southern Power Grid project: Research on Lifetime Evaluation of HVDC Converter Valve (CGYKJXM00000027)

References
[1]   Liu C L, Shuai Q, Qi L, et al. Quantitative analysis of voltage distribution within ±1100kv HVDC converter valve tower under various triasient over-voltage conditions[C]// Lightning Protection. IEEE, 2014:1558-1564.
[2]   Wik M W. International standardization of immunity to high altitude nuclear electromagnetic pulse (HEMP)[C]// International Symposium on Electromagnetic Compatibility. IEEE, 1992:1-2-4/1-2.
[3]   Xie Y Z, Wang Z J, Wang Q S, et al. High altitude nuclear electromagnetic pulse waveform standards: a review[J]. High Power Laser And Particl E Beams, 2003, 15(8):781-787.
[4]   Yang J, Tang G, Cao J, et al. Study on Equivalent Circuit Model for HVDC Valve Thyristor Junction Temperature Calculation[J]. Proceedings of the Csee, 2013, 33(15):156-163.
[5]   Sun H, Cui X, Qi L, et al. Calculation of overvoltage distribution in HVDC thyristor valves[C]// Electromagnetic Compatibility. IEEE, 2010:540-543.
[6]   Qi L, Shuai Q, Cui X, et al. Parameters Extraction and Wideband Modeling of ±1100kV Converter Valve[J]. IEEE Transactions on Power Delivery, 2016, PP(99):1-1.

Calculation of Electrical Stress Distribution and Influencing Factors Analysis of HVDC Converter Valve in Special EMP Environment

GAO Chong[1]   ZHOU Jianhui[1]   ZHANG Jing[1]   GOU Yating[2]   LIU Cuicui[2]
TIAN Jiachen[2]   ZHUO Fang[2]   WANG Feng[2]

[1] Global Energy Interconnection Research Institute co.ltd, Beijing, China

[2] School of Electrical Engineering, Xi'an Jiaotong University, Xi'an, China

gouyating@stu.xjtu.edu.cn

Abstract. The thyristor converter valve is the core equipment in HVDC (high voltage direct current) transmission system, and to evaluate its reliability in special electromagnetic environment is quite important. In this paper, based on the existing wideband model of the key components of the converter valve, the modified wideband model of the converter valve system is built to study the electrical stress of the converter valve under the special electromagnetic environment, and the weak parts of the valve are further studied. The influence of stray parameters in the valve tower on the electrical stress of thyristors is mainly studied. The results show that it is practical and feasible to provide basic conditions and numerical parameters for the reliability evaluation of thyristor components.

1. Introduction

From the 1970s, the research on HVDC converter valve in EMP (electromagnetic pulse) environment mainly focuses on voltage characteristics in foreign countries, including the calculation of overvoltage distribution, analysis of influencing factors, and mathematical approximation of model parameters, etc. [1-2] . A number of domestic universities and research institutes have carried out a series of studies on the electrical stress distribution of the converter valve, especially the voltage stress characteristics, including wideband measurement and modeling of the converter valve components, overvoltage distribution calculation and electromagnetic disturbance characteristics analysis, etc. [3].

The aforementioned studies is mainly on the wideband modeling and voltage stress calculation of the converter valve. The influence of the EMP environment on the reliability of the converter valve can be qualitatively determined, but there is no systematic and quantitative analysis. When the converter valve is in abnormal operating conditions, especially in special environment such as HEMP (high electromagnetic pulse), the stress intensity and environment of the converter valve components are completely different from the normal conditions, and the electrical stress of valves is affected by various stray parameters existing in the valve tower.

Therefore, considering the special working conditions and stray parameters, the electrical stress of each thyristor in the valve is analyzed in detail, which paves the way for the study of the lifetime of thyristor in the special EMP environment, and provides the basic conditions and numerical parameters for the reliability evaluation of converter valve components .

Content from this work may be used under the terms of the Creative Commons Attribution 3.0 licence. Any further distribution of this work must maintain attribution to the author(s) and the title of the work, journal citation and DOI.

Published under licence by IOP Publishing Ltd

2. Modified wideband model of converter valve

To calculate the electrical stress distribution of converter valve, a wide-band circuit model is established necessarily which accords with the spectrum width of the electromagnetic pulse. However, the applicable frequency band of the traditional wideband models can only reach 500kHz-1MHz, which cannot meet the energy band [10kHz, 30MHz] of the HEMP under the Bell laboratory standard[4]. Therefore, the traditional wideband needs to be revised to build a wider modified wideband model.

The model established in this paper is mainly based on the engineering test data of the research group, and is obtained with reference to relevant researches of universities such as North China Electric Power University and Tsinghua University[5].

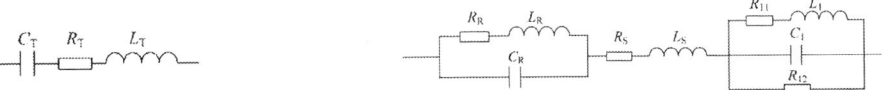

Fig. 1. High-frequency modified wideband model of thyristor

Fig. 2. High-frequency modified wideband model of resistor

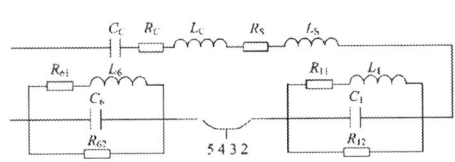

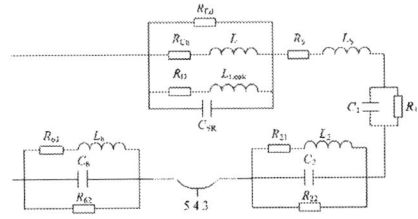

Fig. 3. High-frequency modified wideband model of capacitor

Fig. 4. High-frequency modified wideband model of saturable reactor

2.1. Modified wideband model of thyristor

Thyristor can be approximately equivalent to a junction capacitance in the low frequency band. The wideband model of thyristor studied in this paper is shown in Fig.1. Where C_T is the parasitic capacitance, R_T is the ohmic losses, and L_T is the distributed inductance between thyristor bipolar[6]. The model parameters are obtained based on data provided by the manufacturer[7]. The comparison of impedance frequency characteristics between thyristor measurement and simulation is shown in reference[8]. It is highly coincident below 30MHz, which can meet the spectral requirements of electromagnetic and the special properties at high frequencies.

2.2. Modified wideband model of resistance

The resistor of the converter valve includes the damping resistance of the RC circuit and the grading resistor. The wideband model after high-frequency correction based on the impendance frequency characteristics of the measured resistance is shown in Fig.2, and the comparison curve with the measured characteristics is shown in the reference [9].

2.3. Modified wideband model of capacitor

The capacitor in the converter valve includes the damper capacitor of RC circuit and grading capacitor of valve section. The modified wideband model of capacitor is shown in Fig.3. The comparison between the model and measured impendence frequency characteristic curve is shown in reference [7].

2.4. Modified wideband model of saturable reactor

The saturable reactor is a non-linear device composed of a coil and iron core. It can be approximated as a linear component at high frequency[10]. The wideband model of saturable reactor is shown in Fig.4, where L is the iron core inductance, L_{Leak} is the coil hollow inductance, R_{Cu} is the copper loss, R_{Ed} is the eddy current loss, R_D is the damping resistance and C_{SR} is equivalent to the parasitic capacitance across the reactor.

2.5. Stray capacitances of valve tower

As a multiple-layer structure, the stray capacitance distribution in the valve tower is shown in Fig.5.

The parasitic capacitances considered are as follows: $C_{G1} - C_{G4}$ represent the capacitances between valve module and the ground; C_{G0} and C_{G5} represent the capacitances between the shielding plates and the ground; $C_{L2} - C_{L4}$ represent the capacitances between one valve section and the section above or below; C_{L1} and C_{L5} are the shielding plates capacitance to ground; C_{WB} is the capacitance of the wall bushing which connected the converter transformer.

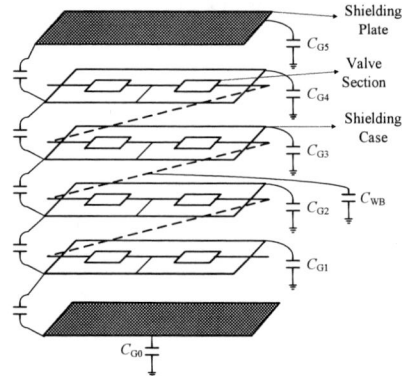

Fig. 5. Stray capacitances of valve tower

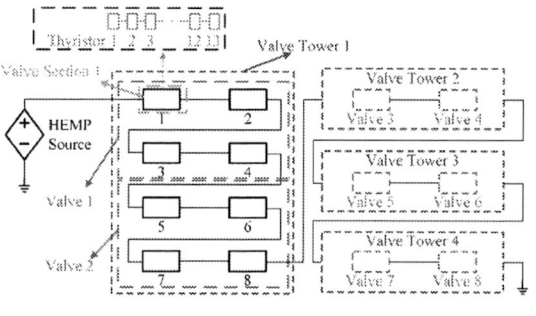

Fig. 6. Simulation diagram of the converter valve system

3. Calculation of electrical stress distribution of converter valve

To accurately obtain the electrical stress of the whole converter valve system in HEMP environment, four single phase valve towers in the bipolar - dual 12 pulsating converter system are taken into account in this paper.

3.1. HEMP waveform and the equation

The functional process of HEMP could be divided into three sections[11], in which the early stage contains broad frequency signals and could perform destructive effects to converter valve. The waveform of HEMP could be simplified and simulated by the following equation[12]:

$$v(t) = V_0 A(e^{-\alpha t} - e^{-\beta t})$$

(1)

where V_0 represents the peak value of pulse, A is the correction factor, α and β are the time constants. The standard of Bell Laboratory is widely used and therefore the following analysis is based on this standard, whose parameters is: $\alpha = 4.0 \times 10^6 s^{-1}$, $\beta = 4.76 \times 10^8 s^{-1}$, $A = 1.052$, $V_0 = 50 kV/m$.

3.2. Electrical stress calculation of converter valves

The simulation diagram of the converter valve system wideband model is shown in Fig. 6. Each valve tower contains two single valves, two valve modules form a single valve, and a valve module contains two valve sections. The structure of the valve tower 1 is shown in detail. Simplified diagrams of the valve towers 2, 3 and 4 are given. In the simulation model, the impulse source is applied at the port where the valve section 1 of the valve 1 is located.

The voltage distribution of valves is shown in Fig. 7. The result indicates that the voltage distribution is extremely non-uniform, while most of the voltage is assumed by the first valve. Valve 2 withstand nearly 10% HEMP, while the rest could be neglected.

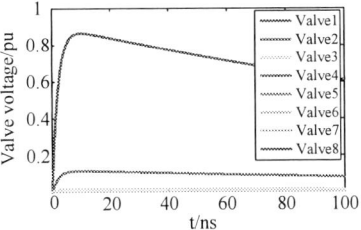
Fig. 7. Valve voltage distribution

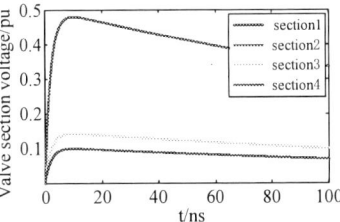
Fig. 8. Valve section voltage distribution

The voltage distribution of different valve sections in a valve is shown in Fig. 8. It can be seen that the voltage distribution of valve sections in the same valve is not uniform, and the valve section closest to the impulse source bear greatest voltage stress. The voltage stress of valve section 2 and 3 coincides highly because it has no stray parameters between the two sections in the established wideband model, therefore, stray parameter is the key factor which cause the uneven voltage stress distribution in converter valves could be proved.

4. Analysis of influencing factors of electrical stress distribution of converter valve

The electrical stress distribution is not identical of different valve sections is caused by the stray capacitances in the valve tower. The stray capacitances are set respectively to 20%, 100%, 500% of the original value, and the three cases are compared as follows.

4.1. Stray parameters of the ground

Stray parameters $C_{G0} - C_{G5}$ are the capacitances to the ground, which affect the voltage distribution of valve sections in valve 1 is shown in Fig. 9. It can be seen that the lager the stray parameters to the ground, the greater the voltage of the thyristor in valve section1 which closet to the electromagnetic pulse source, and the smaller the thyristor voltage in other valve sections.

The result indicates that, with the increase of stray parameters to the ground, the non-uniform degree of electrical stress distribution increases. The electrical stress environment of the thyristor closest to the impulse source of valve 1 becomes the severest, and the electrical stress intensity of the thyristor in other valve sections decreases.

4.2. Stray parameters between valve sections

The influence of stray parameters between valve sections on the thyristor voltage distribution of each valve section is shown in Fig. 10.

Based on the analysis results of voltage stress, it can be concluded that, with the increase of stray parameters between sections, the non-uniform distribution of electrical stress of the converter valve decreases, the electrical stress of the thyristor which far away from the impulse sorce increases, and the electrical stress of other thyristors is more gentle.

ISPECE   IOP Publishing

IOP Conf. Series: Journal of Physics: Conf. Series **1187** (2019) 022024   doi:10.1088/1742-6596/1187/2/022024

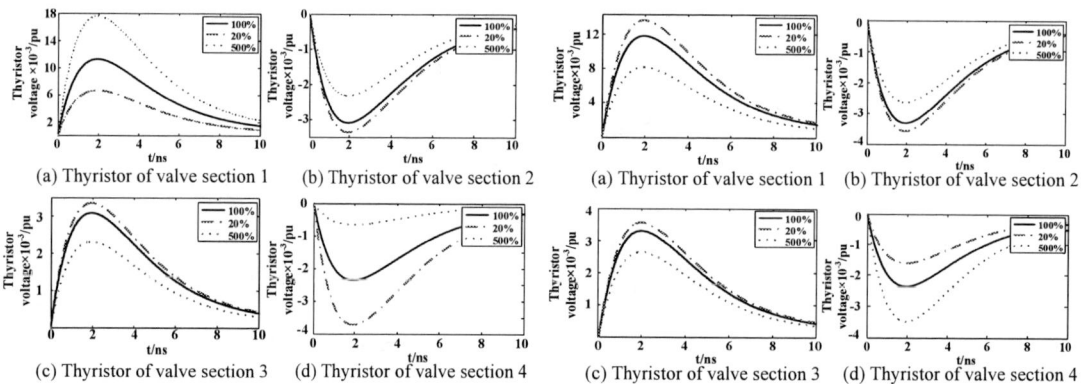

Fig. 9. Thyristor voltage distribution of valve   Fig. 10. Thyristor voltage distribution of valve
sections considering stray   sections considering stray
Parameters to ground   Parameters between valve sections

4.3. Stray parameters of wall bushing

In addition to the stray parameters of valve tower itself, the wall-bushing capacitance of the connecting cable to the converter transformer to the ground should also be considered. The effect of wall bushing capacitance on the thyristor voltage distribution of each valve section in valve 1 is shown in Fig. 11. Compared with the stray parameters to ground and the stray parameters between valve sections, the influence of wall-bushing capacitance is relatively small. When the capacitance increases, the voltage stress of each valve section thyristor slightly increases.

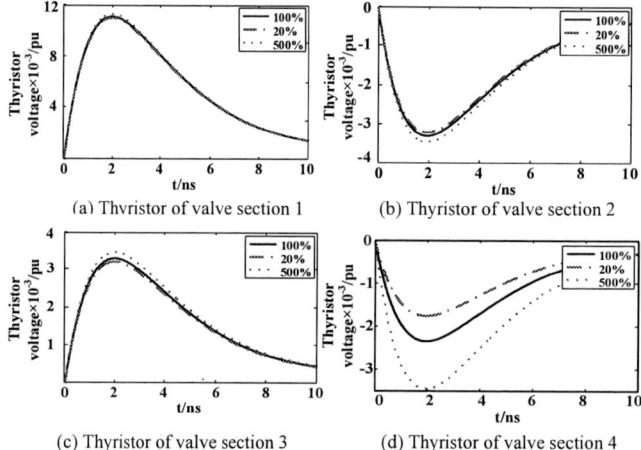

Fig. 11. Thyristor voltage distribution of valve sections
considering stray parameters of wall-bushing

5. Conclusion

Based on the established wideband model of converter valve, the electric stress distribution of valve sections and thyristors in each converter valve is studied with the electromagnetic pulse of special environment as the excitation source. The results show that the stray parameter of valve tower is the main reason for the uneven distribution of electrical stress in converter valve under special environment. The influence of different stray parameters on the voltage stress distribution is analyzed in detail, and the conclusion is drawn that the closer to the impulse source, the greater the electrical stress of the thyristor is. All thyristors of the same valve section are subject to the same electrical stress, and the analysis of the electrical stress of thyristors should be based on the valve section.

Acknowledgment
This work was supported by the State Grid Corporation headquarter project: Research on Mechanism and Evaluation Method of Reliability for Converter Valve in HVDC System (5455ZS160015)

References
[1]   Karady G, Gilsig T. The Calculation of Turn-off Overvoltages in a High Voltage dc Thyristor Valve[J]. IEEE Transactions on Power Apparatus & Systems, 2007, PAS-91 (2): 565-574.
[2]   Karady G, Gilsig T. The Calculation of Transient Voltage Distribution in a High Voltage DC Thyristor Valve[J]. IEEE Transactions on Power Apparatus & Systems, 2007, PAS-92 (3): 893-899.ce
[3]   Liu Jie, Tang Guangfu, Cha Kuipeng, et al. Study on Impulse Transient Voltage Balancing Measures for UHVDC Converter Valves [J]. Chinese Journal of Electrical Engineering, 2016, 36 (7): 1828-1835
[4]   Xie Yanzhao, Wang Zanji, Wang Qunshu, et al. Waveform Standard and Characteristic Analysis of High Altitude Nuclear Explosion Electromagnetic Pulse [J]. Intensive Laser and Particle Beam, 2003, 15 (8): 781-787.
[5]   Yu Zhanqing, He Jinliang, Zhang Bo, et al. Time-domain simulation analysis of conduction disturbance of converter valves in HVDC converter stations [J].China Journal of Electrical Engineering, 2009, (10): 17-23.
[6]   Liu Lei. Broadband Modeling and Electromagnetic Disturbance Characteristics of Valve System in HVDC Converter Station [D]: North China Electric Power University (Hebei), 2008.
[7]   Sun Haifeng. Research on Broadband Circuit Modeling Method and Application of Converter System in Converter Station [D]: North China Electric Power University (Beijing), 2010.
[8]   Dong Fu. +500kV Converter Valve Electromagnetic Disturbance Research [D]: South China University of Technology, 2012
[9]   Sun Haifeng, Cui Xiang, Qi Lei, et al. High Frequency Modeling of HVDC Converter Valve Devices [J]. Journal of Electrical Technology, 2009, 24 (11): 142-148.
[10]   Chen Peng, Cao Junzheng, Wei Xiaoguang, et al. Transient Circuit Simulation Model of Saturated Reactor for HVDC Converter Valve [J]. High Voltage Technology, 2014, 40 (1): 288-293.
[11]   Ianoz M, Nicoara B I C, Radasky W A. Modeling of an EMP conducted environment[J]. IEEE Transactions on Electromagnetic Compatibility, 2002, 38(3):400-413.
[12]   Wik M W. International standardization of immunity to high altitude nuclear electromagnetic pulse (HEMP)[C]// International Symposium on Electromagnetic Compatibility. IEEE, 1992:1-2-4/1-2.

Numerical Simulation of Internal Flow in Direct Burning Coal-fired Hot Flue Gas Furnace

Zhang Zhenwei, Wang Peng[*].

Institute of Mechanical Engineering and Automation, Northeastern University, Shenyang, liaoning110819, China

[*]Corresponding author's e-mail: 812750705@qq.com

Abstract. In this paper, a direct burning coal-fired hot flue gas furnace is taken as the research object. By using the SIMPLEC algorithm and the pressure based implicit solver in the analysis software FLUENT, the temperature, velocity and pressure distribution cloud chart and the CH_4, CO, O_2 and CO_2 content distribution clouds are obtained by the numerical simulation of the gas solid two phases in the direct burning hot flue gas furnace. Through analysis, the high temperature zone in the hot flue gas furnace is the combustion area in the combustion chamber. There is a large amount of volatile matter in this area. The volatiles burn up a lot of oxygen rapidly and release carbon dioxide and heat. At the primary settlement, the radiation temperature of the reflected arch is higher, and more burnt particles are settled. The hot smoke flow in the whole hot flue gas is under negative pressure, and the flow is smoother. The flow in the furnace is relatively gentle, and the flow velocity and turbulence energy are large at the two turns. After the secondary sedimentation, the temperature gradually decreases and the output hot flue gas is substantially clean.

1. Introduction

Hot flue gas furnace is an efficient and energy-saving heating device. It mixes the hot flue gas produced by combustion with air, and then outputs hot flue gas with different temperatures and different cleanliness levels in conjunction with dryer. It dries materials directly in contact with materials, and is widely used in building materials, chemical industry and other fields. In recent years, numerical simulation theory and methods have become a powerful tool for developing combustion technology and guiding the design and performance optimization of combustion devices [1,2]. Xu Minghou et al. [3,4] carried out a large number of numerical simulation and experimental tests on the combustion process of front wall combustion boilers under different working conditions, and the results were not very different. Fan Jianren et al. [5-7] used flow, combustion and heat transfer comprehensive model to simulate the combustion process in boilers with different capacity and combustion modes in detail. The research results provide a theoretical basis for the structural design and improvement of this type of combustion vessel. Through numerical simulation of gas-solid two-phase flow in direct-fired coal-fired hot flue gas furnace, the problems encountered in the process of direct-fired coal-fired hot flue gas operation and the improvement of fuel utilization efficiency are provided. At the same time, it also provides a reference theoretical basis for the structural design and improvement of this type of hot flue gas furnace.

Content from this work may be used under the terms of the Creative Commons Attribution 3.0 licence. Any further distribution of this work must maintain attribution to the author(s) and the title of the work, journal citation and DOI.

Published under licence by IOP Publishing Ltd

2. Structure of Direct-fired Coal-fired Hot Flue Gas Furnace

According to the structure parameters and two-dimensional CAD engineering drawings of the direct-fired coal-fired hot flue gas furnace, the three-dimensional solid model is appropriately simplified and drawn by SolidWorks software, as shown in Figure 1. The analysis model consists of a primary settlement reflection arch, an inlet, a combustion chamber, a secondary settlement windshield, a settlement chamber and an outlet.

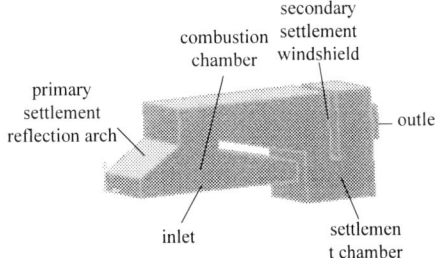

Figure 1. Three-dimensional Model of Direct Burning Hot Flue Gas Furnace

Considering the complexity of the model, in order to get high quality meshes and more accurate analysis results, the model is divided into two parts: segmentation and recombination. According to the structure of hot flue gas stove, the whole model is divided into five regular blocks, while the regular model is easy to be divided into structured grids. The structured grids not only have better quality than unstructured grids, but also have easier convergence of simulation results and reduce calculation errors and distortion of results.

3. Setting of Flow Field Analysis and Solution

3.1. Flow field analysis

The flow process of fuel combustion and gas generation in hot flue gas furnace conforms to the laws of mass conservation, momentum conservation, energy conservation and component conservation.

(1) The mass conservation equation can be expressed as follows: the increase of mass per unit time in a fluid microelement is equal to the net mass flowing into the microelement at the same time interval. The general form of mass conservation equation is:

$$\frac{\partial \rho}{\partial t} + \nabla \cdot (\rho V) = S_m \tag{1}$$

Formula: ρ is density, unit kg/m3, t is time; V is velocity vector, the source term S_m is the mass added to the continuous phase.

(2) The law of conservation of momentum can be expressed as follows: the rate of change of fluid momentum to time in a microelement is equal to the sum of forces acting on the microelement by external forces. According to the expression of this law, the momentum conservation equation can be obtained:

$$\frac{\partial (\rho V)}{\partial t} + \nabla \cdot (\rho V) = -\nabla p + \nabla \cdot (\tau) + \rho g + F \tag{2}$$

Formula: p represents the pressure on the fluid microelement, g and F represents the gravitational and other external volumetric forces acting on the microelement respectively, F also contains other model-related source terms, τ are the viscous stress tensors acting on the surface of the microelement due to the molecular viscous effect.

(3) The essence of the law of conservation of energy is the first law of thermodynamics. The law can be expressed as follows: the increase rate of energy in the microelement is equal to the net heat flux into the microelement plus the work done by the volume force and surface force on the microelement. The energy conservation equation is:

$$\frac{\partial(\rho E)}{\partial t}+\nabla\cdot\left(V\left(\rho E+p\right)\right)=\nabla\cdot\left(k_{eff}\nabla T-\sum_j h_i J_j+\left(\tau_{eff}\cdot V\right)\right)+S_h \qquad (3)$$

Formula: $E=h-p/\rho+V^2/2$ represents the total energy of the bulk microaggregates, $k_{eff}=k_i+k$ represents the effective thermal conductivity, J_j is the diffusion flux of component j, S_h is the heat source term.

(4) For a definite system, the law of conservation of component mass can be expressed as: the change rate of the mass of a chemical component in the system to time is equal to the sum of the net diffusion flux through the system interface and the net productivity of the component formed or disappeared by chemical reaction. According to the expression of this law, the mass conservation equation of components can be obtained:

$$\frac{\partial(\rho V_i)}{\partial t}+\nabla\cdot\left(\rho VY_i\right)=-\nabla\cdot J_i+R_i+S_i \qquad (4)$$

Formula: Y_i is the mass fraction of component i, J_i is the diffusion flux of component i, R_i is the net formation rate of the component consumed or produced by chemical reaction per unit volume of the system in unit time, S_i is the mass source term.

(5) The residual curve of the standard $k-\varepsilon$ turbulence model is easier to converge, and the simulation results are more in line with the actual situation, so the standard $k-\varepsilon$ model is chosen in this paper. Transport equation of standard $k-\varepsilon$ model:

$$\frac{\partial(\rho k)}{\partial t}+\frac{\partial(\rho ku_i)}{\partial x_i}=\frac{\partial}{\partial x_j}\left[\left(\mu+\frac{\mu_t}{\sigma_k}\right)\frac{\partial k}{\partial x_j}\right]+G_k+G_b-\rho\varepsilon-Y_M+S_k \qquad (5)$$

$$\frac{\partial(\rho\varepsilon)}{\partial t}+\frac{\partial(\rho\varepsilon u_i)}{\partial x_i}=\frac{\partial}{\partial x_j}\left[\left(\mu+\frac{\mu_t}{\sigma_\varepsilon}\right)\frac{\partial\varepsilon}{\partial x_j}\right]+C_{1\varepsilon}\frac{\varepsilon}{k}(G_k+G_{3\varepsilon}G_b)-G_{2\varepsilon}\rho\frac{\varepsilon^2}{k}+S_\varepsilon \qquad (6)$$

Formula: G_k is affected by turbulent kinetic energy caused by mean velocity gradient, G_b is affected by turbulent kinetic energy caused by buoyancy, Y_M is the effect of compressible turbulent fluctuation expansion on total dissipation rate, $C_{1\varepsilon}$, $C_{2\varepsilon}$, $C_{3\varepsilon}$ is an empirical constant. In Fluent, $C_{1\varepsilon}$ =1.44, $C_{2\varepsilon}$=1.92, $C_{3\varepsilon}$=0.09, σ_k, σ_ε represent the Prandt number corresponding to turbulent kinetic energy and turbulent energy dissipation rate respectively. In this paper, σ_k is taken as 1.0.

3.2. Setting of Solutions

Four solutions are provided in FLUENT, PISO algorithm is for unsteady compressible fluid, COUPLE algorithm is for pressure and speed coupling. Although it can accelerate the convergence speed, it consumes too much CPU for computer. SIMPLE algorithm is a semi-implicit method for solving pressure coupled equations of incompressible flow field. SIMPLEC algorithm improves and changes the flux correction algorithm, which can make the convergence of the solution process faster. The fluid studied in this paper can be regarded as incompressible fluid, and the flow rate is not high, so SIMPLEC algorithm is chosen. Pressure-based implicit solver is chosen as the solution method. Coal and its volatiles react in a hot flue gas furnace and produce high temperature flue gas. After mixing with air and settling, they are dried. This process conforms to the energy conservation, so the energy model is used in the simulation analysis process.

4. Simulation results and analysis

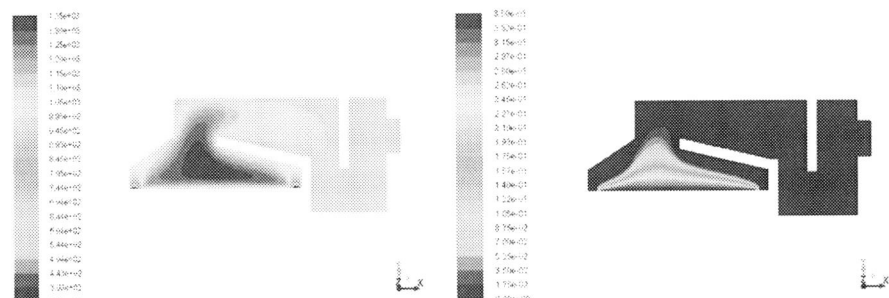

Figure 2. Temperature Distribution Cloud Chart Figure 3. CH$_4$ Content Distribution Cloud Chart

By analyzing the temperature distribution nephogram, it can be concluded that the whole combustion chamber is in a high temperature state, and the high temperature hot smoke flows to the outlet in the form of wave. Because there are a large number of coal volatiles and sufficient oxygen at the entrance, the volatilization separation starts quickly and then the coal is gradually heated, and then the coal begins to burn violently, so the combustion chemical reaction at the entrance is relatively strong, the highest temperature is about 1350K. The temperature of the combustor is about 845K, which is due to the fact that the combustor has no seam coal barrier and some cold air enters the combustor directly. In the coal seam at the entrance, some uneven fine high-temperature particles are blown up by the air. They will continue to burn and release heat with the volatile matter from the combustion chamber, which is transferred to the hot flue gas together with the heat radiated by the coal particles. A large number of high-temperature hot flue gas flows in the furnace. Because of the blockage of the secondary settling windshield, the hot air flow tends to be gentle, and the average temperature in the settling chamber is about 1000K. Finally, the high temperature hot flue gas after sedimentation flows out from the outlet, and the temperature is about 995K.

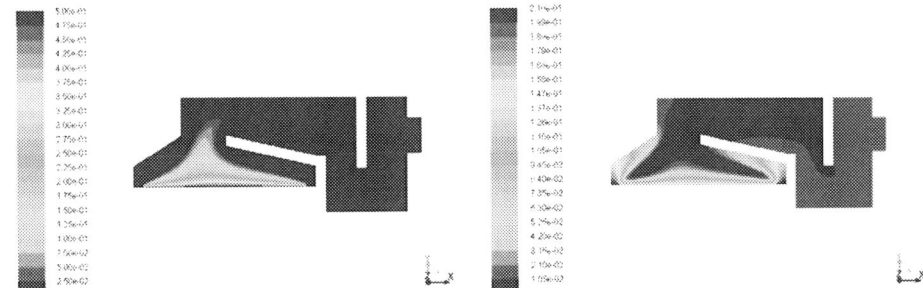

Figure 4. CO Content Distribution Cloud Chart Figure 5. O$_2$ Content Distribution Cloud Chart

The volatiles are mainly composed of hydrocarbons and hydroxides. CH$_4$ and CO with high volatile content were taken as the research objects in this paper. It can be seen from Figure. 3 and Figure. 4 that CH$_4$ and CO in volatile gases react with oxygen in combustion chamber. With the continuous combustion of chemical reactions, CH$_4$ and CO fuels are constantly consumed, and the content of CH$_4$ and CO fuels is higher at the entrance, until the end of the reaction, the content of CH$_4$ and CO fuels gradually tends to zero. The flow trend of combustion chemical reaction is along the outlet direction of hot flue gas furnace, and the mass fractions of CH$_4$ and CO are zero in settling chamber and outlet. In addition, the cloud of CH$_4$ and CO content distribution corresponds to the cloud of temperature distribution in Figure. 2. As the chemical reaction proceeds, the temperature increases and the fuel gas content decreases.

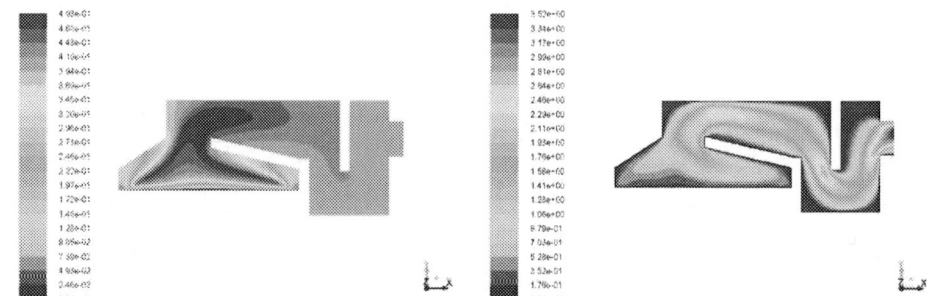

Figure 6. CO_2 Content Distribution Cloud Chart Figure 7. Velocity Distribution Cloud Chart

As can be seen from Figure. 5 and Figure. 6, the oxygen content is the highest at the entrance of the hot flue gas furnace and the CO_2 content is the lowest, because combustion chemical reaction is a process of gradually consuming oxygen to produce CO_2. Therefore, with the continuous reaction, the content of O_2 gradually decreases to the lowest value, the content of CO_2 gradually reaches the highest value and moves towards the outlet direction with the gas, and finally CO_2 diffuses throughout the hot flue gas furnace. The distribution patterns of CO_2 and CO_2 content in Figure. 5 and Figure. 6 also correspond to the distribution in high temperature zone.

Analyzing the velocity distribution nephogram, it can be concluded that the velocity at the entrance is small because the coal fuel layer has some hindrance to the gas. After passing through the fuel layer, a large amount of heat released by fuel combustion makes the temperature of combustion chamber rise sharply, while the space decreases due to the primary settlement reflection arch. In this case, the velocity of hot smoke at high temperature increases gradually. In the furnace, it flows smoothly to the outlet direction. When the high temperature hot smoke gas hits the secondary settling windshield, the velocity decreases, and increases again when passing through the settling chamber with reduced space, and then flows quickly to the outlet. At the exit position, the air velocity is higher, which is caused by the negative pressure at the exit and the smaller cross-section.

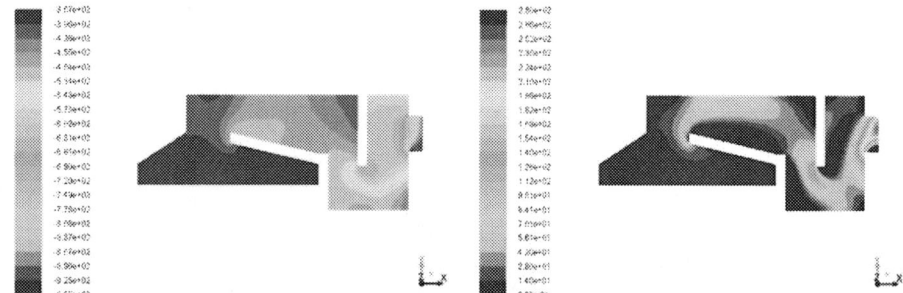

Figure 8. Static Pressure Distribution Cloud Chart   Figure 9. Dynamic Pressure Distribution Cloud Chart

By analyzing the static pressure distribution nephogram of hot flue gas, it can be concluded that the static pressure of the gas in the combustion area of the combustor is larger and more uniform, which fully shows that this area is the combustion chemical reaction area, producing a large number of high temperature gases and filling the whole combustor, corresponding to the temperature distribution nephogram and the CO_2 content distribution nephogram. The non-uniform distribution of static pressure in the furnace is mainly due to the tendency of the hot smoke flow being blocked by the secondary settling windshield wall to move back, which can not flow smoothly to the outlet, resulting in the lower static pressure in the settling chamber.

The static pressure of the whole hot flue gas furnace is between - 1000Pa and - 100Pa, and the whole space is always in a negative pressure state, which helps the hot flue gas generated by combustion to be discharged from the outlet quickly. At the same time, it also prevents the high temperature gas generated by the internal combustion of the hot flue gas furnace from diffusing to the exit of the fire gate, ash clearing door and air regulating valve, which not only ensures the combustion reaction can proceed smoothly in the combustion chamber, but also provides great convenience for the introduction of air from the inlet and the supplement of cold air from the air regulating port.

Dynamic pressure refers to the pressure caused by airflow, which can reflect the level of fluid flow and working force. It is closely related to the flow rate and gas density. It is the pressure that reflects the phenomenon of gas flow, and it is always positive. Analyzing the dynamic pressure distribution nephogram, we can get that the dynamic pressure value is not very large because the gas velocity in the combustion chamber is relatively small. At the exit of the combustion chamber, the dynamic pressure increases gradually until the air flow meets the secondary settling windshield, and then decreases. The dynamic pressure at the exit and the settling chamber is larger. The overall distribution form of dynamic pressure value corresponds to the velocity distribution nephogram in Figure 7, which reflects the flow state of high temperature hot flue gas in the hot flue gas furnace from the side.

5. Conclusion

(1) The combustion reaction of coal mainly occurs in the combustion chamber, and the high temperature zone is concentrated in the combustion area of volatile matter, which causes coal combustion. The fuel layer at the entrance is also in the high temperature zone and transfers heat radiation to the surrounding area. The most affected area by radiation temperature is the first-stage settlement reflection arch, which has a higher radiation temperature. The analysis of temperature field is very helpful to understand the temperature of combustion chamber, the location of high temperature area and the choice of refractories. It can also effectively control the phenomenon of coking and slagging in hot flue gas furnace.

(2) In the whole hot flue gas furnace, the velocity of hot flue gas in combustion chamber is smaller, and it increases at two settling turns, and reaches the maximum at the exit. Through the analysis of velocity distribution nephogram, the size, distribution and flow direction of fluid velocity can be obtained, which can effectively control the cleanliness of hot flue gas when it flows out from the outlet.

(3) The whole hot flue gas furnace is in a negative pressure state. The static pressure of the combustion chamber is the largest, and gradually decreases along the outlet direction. The distribution of static pressure in the furnace is uneven, and the static pressure at the outlet is the smallest. The distribution of dynamic pressure of hot flue gas is basically the same as that of velocity distribution nephogram, whereas the dynamic pressure value is high where the velocity is high. The analysis of pressure nephogram can get the degree of thermal damage and corrosion caused by high temperature hot flue gas on the furnace wall, which is convenient for timely replacement and maintenance.

References

[1] Zhao S X. (2002) Numerical simulation of combustion. Beijing: Science Press, Beijing.
[2] Stopford P J. (2002) Recent applications of CFD modelling in the power generation and combustion industries. Applied Mathematical Modelling, 26: 351-374.
[3] Xu M H, Azevedo J L T, Carvalho M G. (2001) Modeling of a front wall fired utility boiler for different operating conditions. Computer Methods in Applied Mechanics and Engineering, 190: 3581-3590.
[4] Xu M H, Azevedo J L T, Carvalho. (2000) Modelling of the combustion process and NOX emission in a utility boiler. Fuel, 79: 1611-1619.
[5] Fan J R, Sun P, Zheng Y Q. (1999) Numerical and experimental investigation on the reduction of NOX emission in a 600 MW utility furnace by using OFA. Fuel, 78: 1387-1394.
[6] Fan J R, Zha X D, Cen K F. (2001) Study on coal combustion characteristics in a W-shaped boiler furnace J. Fuel, 80: 373-381.

[7]   Fan J R, Sun P, Zha X D et al. (1999) Modeling of combustion process in 600 MW utility boiler using comprehensive models and its experimental validation. Energy & Fuels, 13: 1051-1057.

Virtual Synchronous Generator Grid Connected Control Method Based on Virtual Impedance

Bao-zhu SHAO[1], Guan-feng ZHANG[2], Jun-you YANG[2], Fei-fei GAO[2], Feng SUN[1], Qing-song ZHAO[3]

[1] Electrical Power Research Institute of Liaoning Electric Power Co.,Ltd., Shenyang 110006，China

[2] Shenyang University of Technology，Shenyang 110870，China

[3] Liaoning Dongke Electric Power Company Limited，Shenyang 110003，China

Abstract. The virtual synchronous generator technology is used to simulate the working principle of the synchronous generator, which provides the inertia and damping support for the power grid. At present, the measurement of the phase, amplitude and frequency of the virtual syn-chronous generator is used to measure the output voltage of the distributed generation system. To solve this problem, this paper proposes a virtual synchronous machine grid virtual impedance control method based on the virtual impedance calculation of virtual current in the virtual synchronous machine from the network state and the grid between the assumption of power exchange, puts forward the pre synchronization strategy based on virtual impedance, and gives the simulation results. The results show that compared with the existing methods, the proposed method is more practical and more smooth.

1. Introduction

With the rapid development of the national economy, the problem of lack of energy has become increasingly prominent. The use of clean energy to supplement or replace traditional fossil energy is an important means of ensuring sustainable development in the energy sector. In the field of power systems, distributed power generation, as an important form of clean energy generation, has received strong support from the government and industry and is expected to become an important part of the future power supply.

However, the rapid development of distributed power generation has brought many challenges to the operation of power systems. The existing distributed power generation system uses power electronic devices to be integrated into the power grid, which is more flexible than the traditional power generation system. Meanwhile, it has such disadvantages as low inertia and weak damping, which affects the friendly compatibility of the distributed power generation system with the existing power grid. In response to the above-minded problems, some researchers proposed virtual synchronous machine technology that ensures the stable operation of the system by changing the external characteristics of the grid-connected inverter and absorbing the advantages of synchronous generators. In 2007, the European VCYNC project first proposed the concept of virtual synchronization[1]. Zhong Changqing proposed Synchronverter algorithm to realize the modeling of virtual synchronous machine[2]. Research institutions such as Tsinghua University, the Institute of Electrical Engineering and Hefei University of Technology have contributed to the improvement of system stability in this field[3].

Content from this work may be used under the terms of the Creative Commons Attribution 3.0 licence. Any further distribution of this work must maintain attribution to the author(s) and the title of the work, journal citation and DOI.

Published under licence by IOP Publishing Ltd

On the basis of previous theoretical research, the virtual synchronous machine technology has been put into practice. In 2016, State Grid Corporation transformed the inverter of wind turbine and photovoltaic power generation in Zhangbei Wind Storage and Demonstration Project, which becomes the world's largest virtual synchronous machine demonstration project, and China Electric Power Research Institute distribution station put into use the photovoltaic virtual synchronous machine researched and developed in the new Tianjin Eco-city smart business micro-grid. At present, the technical function of virtual synchronous machine is basically realized. The relevant research has initially shifted from the research and development stage to the optimization stage, and there is room for further research in the aspect of system-level grid-connected application.

The frequency regulation method of micro grid is studied based on virtual synchronous machine to realize different operation modes [4-5]. The charge and discharge strategy of the virtual synchronous machine energy storage unit is studied, and the capacity configuration method is given[6-7]. A wind turbine grid-connected system based on virtual synchronous machine is proposed to achieve a friendly and grid-connected target with stable voltage[8]. The above literatures all focus on the application technology of virtual synchronous machines in existing distributed generation systems. In addition, Based on the future active distribution network operation, the application of virtual synchronous machine technology is discussed[9],and an electric vehicle charging method based on virtual synchronous machine technology is proposed[10]. These research results are established on the same basic principles in terms of the level of inverter control strategy, but the difference lies in the application occasions, which does not involve the improvement of the "ontology" in the application process of virtual synchronous machine, and fails to dig into the flexible characteristics of virtual synchronous machine.

In summary, an improved virtual synchronous machine grid-connecting method is proposed to solve the problem that Additional measuring devices in traditional grid-connecting restrict flexible output of distributed generation systems, avoiding the problems such as that the phase-locked loop has nonlinearity, slow response, and difficult parameter design, and a smoother and more efficient virtual synchronous machine is realized.

2. Virtual synchronous machine synchronous grid-connected grid equivalent model
If the virtual synchronous machine is equivalent to a voltage source, a typical grid-connected system equivalent circuit is shown in Figure 1.

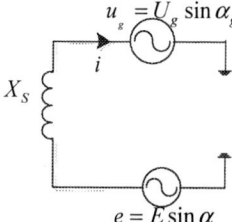

Figure 1.Equivalent circuit diagram

In the above figure, e is the virtual synchronous machine output port voltage; X_s is the synchronous reactance connected to the grid; u_g is the grid voltage.

Under the assumption of a single-machine infinite system, the virtual synchronous machine transmits the active power as shown in equation (1).

$$P = \frac{3U_g E}{2X_s} \sin\left(\alpha - \alpha_g\right) \qquad (1)$$

In the equation, α is the virtual power angle, controlled by the virtual input mechanical torque.
The virtual synchronous machine transmits reactive power as shown in equation (2).

$$Q = \frac{3U_g \left[E \cos(\alpha - \alpha_g) - U_g \right]}{2X} \tag{2}$$

The connection of virtual synchronous machine to the grid needs to ensure the frequency, phase and amplitude of the output voltage and the grid voltage is similar, which is represented by equation (3).

$$\begin{cases} \left| f - f_g \right| \le f_{min} \\ \left| E - U_g \right| \le U_{min} \\ \left| \alpha - \alpha_g \right| \le \alpha_{min} \end{cases} \tag{3}$$

As is shown in the equation, f, E, α is respectively the voltage frequency, amplitude and phase of the virtual synchronous machine; the subscript g is the grid side variable; the subscript min is the set safety threshold, and the safety threshold of the grid connection optimum condition is zero.

It can be seen from equations (1) and (2) that the grid-connected power transfer equation of the virtual synchronous machine mainly depends on the voltage amplitude and phase angle, that is, in order to satisfy the smooth grid connection, the safety threshold needs to satisfy the formula (4).

$$\begin{cases} U_{min} = 0 \\ \alpha_{min} = 0 \end{cases} \tag{4}$$

Substituting the formulas (3) and (4) into the formula (1) and (2) respectively, we obtains the formula (5) as follows.

$$\begin{cases} P = 0 \\ Q = 0 \end{cases} \tag{5}$$

It can be seen that the synchronous synchronization of the virtual synchronous machine can be further understood as a conclusion: the virtual power and the reactive power of the virtual synchronous machine transmitted to the grid are zero.

The output power of the virtual synchronous machine is based on the assumption of the virtual synchronous machine grid-connected condition, but the actual situation is that the virtual synchronous machine has no output power in the off-network state. In order to give a grid-connected control strategy of virtual synchronous machine that satisfies the zero output power, it is necessary to virtualize the power exchange between the two. Therefore, this paper proposes pre-synchronization control strategy based on virtual impedance, as shown in the following section.

3. The Synchronous Grid-connected Control Strategy of Virtual Synchronous Machine

Currently, the virtual synchronous machine system is composed of a DC power supply and a DC/AC inverter. The DC power supply provides the power required by the virtual synchronous machine, and the operation control method adopted by the inverter control unit is the key to the virtual synchronous machine technology.

The topology of the virtual synchronous machine is the same as that of the general three-phase DC/AC inverter, thus not necessary to be described here. This paper mainly considers the electromagnetic relationship and mechanical characteristics of the virtual synchronous machine simulation, as well as its external characteristics such as active frequency modulation and reactive voltage regulation.

The mathematical model of the virtual synchronous machine is shown in equation (6).

$$\begin{cases} v = -Ri - L_m\left(di/dt\right) + e \\ e = M_f i_f \theta \sin\theta \\ J\theta = T_m - T_e - D\theta \\ P = \langle i, M_f i_f \theta \sin\theta \rangle \\ Q = -M_f i_f \langle i, \cos\theta \rangle \theta \end{cases} \qquad (6)$$

Where, T_m Is the virtual mechanical torque; D Is a virtual damping coefficient,; M_f is mutual inductance; v, i is to measure voltage and current; T_e is the virtual electromagnetic torque; i_f is the excitation current; θ is the virtual power angle; e is the output electromotive force.

The basic control of the virtual synchronous machine includes two aspects: virtual speed control and virtual excitation control. The virtual speed control is based on the mathematical model of the virtual synchronous machine and the typical active frequency drooping characteristic, and the double loop structure of the inner loop of the power outer loop frequency is used to simulate the active-frequency characteristics of the synchronous machine and the inertia and damping links. The virtual excitation control does not involve the aforementioned mathematical model, and mainly takes into account the external characteristics of the virtual synchronous machine, so the conventional reactive-voltage droop control structure is adopted.

It can be seen from the foregoing that the purpose of designing the virtual impedance is to virtualize the power exchange between the virtual synchronous machine and the power grid, and propose a control strategy that satisfies the virtual synchronous machine output power of zero.

Therefore, considering the characteristics of droop control, the pre-synchronization control method proposed in this paper is shown in Figure 2.

The grid-connected process of virtual synchronous machine is as follows:

In the off-grid state, the Sa switch is closed, the Sc switch is off, and the B1 switch is closed. Considering that the virtual synchronous machine's off-grid actual output current is zero, we assume that there is a virtual impedance between the output port voltage and the grid voltage, and the virtual current can be calculated as shown in equation (7).

$$i_s = \frac{e - u_g}{L_1 s + R_1} \qquad (7)$$

The active power and reactive power of the virtual output can be calculated from the above formula to satisfy the power exchange assumption between the virtual synchronous machine and the grid.

Equation (8) can be satisfied by controlling the output voltage amplitude and phase angle of the virtual synchronous machine.

$$\begin{cases} T_m = T_e \\ Q_{set} = Q \end{cases} \qquad (8)$$

If P_{set} and Q_{set} are set to zero, the grid-connected condition that the zero power exchange showed by Equation (5) is satisfied. At this time, the grid-connected operation can achieve smooth grid connection.

In the grid-connected state, the B_2 switch is closed and the virtual synchronous machine enters the grid-connected operation mode.

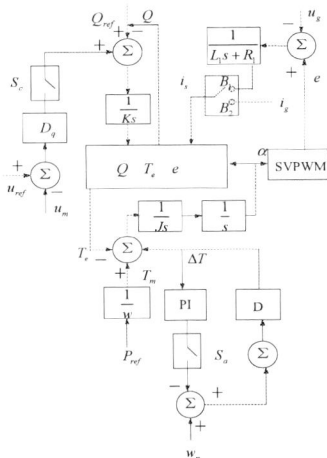

Figure.2 The grid-connected control block diagram of virtual synchronous machine

Where, i_g is the virtual synchronous machine output current; i_s is the virtual current, and L_1 and R_1 form the virtual impedance.

4. Engineering case analysis

Based on the above research, this paper is to verify the simulation by using Matlab/Simulink software. Some parameters are shown in Table 1.

Table 1.Simulation parameters

Power voltage	220V
Supply voltage	400V
Filter inductance	2mH0.1Ω
Filter capacitor	40μF
Line impedance	2mH0.1Ω
Switching frequency	6.4kHz

The simulation system is shown in Figure 3.

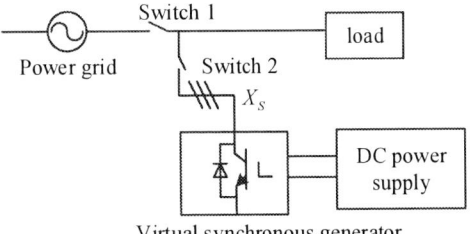

Figure 3. Simulation system

The comparison between the grid-connected control method proposed in this paper and the direct grid-connected method is as follows:

There is a phase difference between the output voltage and the grid voltage when the virtual synchronous machine is running with load, as shown in Figure 4.

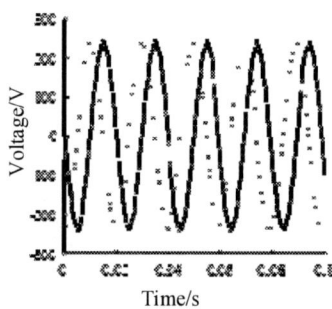 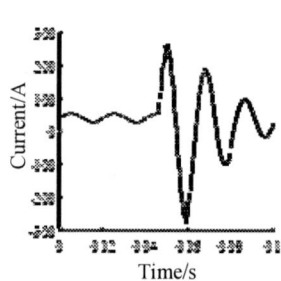

Figure 4. Virtual synchronous machine
output voltage vs grid voltage

Figure 5. Virtual synchronous machine
output current

The direct grid connection will generate an inrush current with a large amplitude. As shown in Figure 5, it has an adverse effect on the stable operation of the virtual synchronous machine and cannot meet the requirements of smooth grid connection.

Using the virtual impedance-based grid-connected control method, the virtual synchronous machine tracks the grid voltage phase by continuous adjustments, as shown in Figure 6.

After adopting the pre-synchronization control strategy proposed in this paper, we can see the output voltage of the virtual synchronous machine is consistent with the grid voltage, and there is no inrush current after the grid-connected operation, which satisfies the smooth grid-connected demand, as shown in Fig. 7.

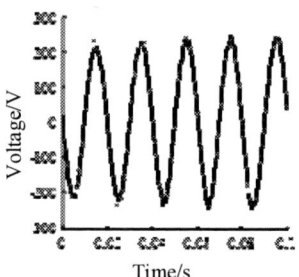 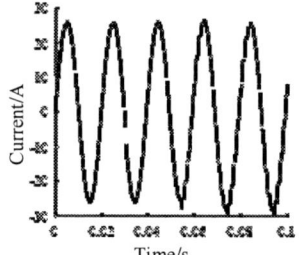

Figure 6. Virtual synchronous machine
output voltage vs grid voltage

Figure 7. Virtual synchronous machine
output current

5. Conclusion

In order to solve the problem that the additional measurement device in traditional grid-connecting restricts the flexible output of distributed generation system, this paper proposes a grid-connected control strategy of the virtual synchronous machine based on virtual impedance:

By designing the virtual impedance link to make the virtual synchronous machine output power zero, the controller itself can be used to synchronize with the power grid before the closing, without the help of a special synchronization unit, and the smooth synchronization of the virtual synchronous machine is realized.

Compared with the traditional virtual synchronous machine grid-connected method, the proposed pre-synchronization method is simpler and more effective, avoiding the problems, such as nonlinear phase-locked loop, slow response and difficult parameter design.

References

[1] Espi J M, Castello J, GarcíA-Gil R, et al. 2011 An Adaptive Robust Predictive Current Control for Three-Phase Grid-Connected Inverters. *IEEE Transactions on Industrial Electronics*, vol 58 pp 3537-3546.

[2] Zhong Q, Weiss G. 2011 Synchronverters: Inverters That Mimic Synchronous Generators *IEEE Transactions on Industrial Electronics* vol 58 pp 1259-1367.

[3] Xie Yongliu, Cheng Zhijiang, Li Yongdong, et al. 2014 Dual closed-loop control strategy of inverter in micro-grid island mode *Renewable Energy* vol 32 pp1632-1638.

[4] Du Y, Guerrero J M, Chang L. Modeling. 2013 analysis, and design of a frequency-droop virtual synchronous generator for microgrid applications *2013 IEEE ECCE Asia Downunder* pp. 643-649.

[5] Du Yan, Su Jianhui, Zhang Liuchen, et al. 2013 A mode adaptive microgrid FM control method *Proceedings of the CSEE* vol 33 pp 67-75.

[6] Vassilakis A, Kotsampopoulos P, Hatziargyriou N. 2013 A battery energy storage based virtual synchronous generator *2013 IREP Symposium of Bulk Power System Dynamics and Control* pp. 1-6.

[7] Haixin Wang, Junyou Yang, Zhe Chen, etal. 2018 Gain scheduled torque compensation of PMSG-based wind turbine for frequency regulation in an isolated grid *Energies* vol 11 pp 1623-:d153709.

[8] Zhong Q, Ma Z, Ming W. 2015 Grid-friendly wind power systems based on the technology *Energy Conversion and Management* vol 89 pp719-726.

[9] Zhao Bo, Wang Caisheng, Zhou Jinhui. 2014 Current Status and Future Development of Active Distribution Network *Automation of Electric Power Systems* vol 38 pp125-135.

[10] Lü Zhipeng, Liang Ying, Zeng Zheng. 2014 Fast Charge Control Method for Electric Vehicles Using Virtual Synchronous Motor Technology *Proceedings of the CSEE* vol 34 pp 4287-4294.

Design of Memory Test System for Measuring Transmission Lines Galloping

Yulong Li[1], Jiaxin Liu[2], Yunfei Jia[3] and Yu Liu[4]

[1] Department of Flight Vehicle Design, School of Mechanical Engineering, NanJing University of Science and Technology,Nanjing,China

[2] State Grid Electric Power Research Institute of Liaoning Electric Power Co., Ltd., Shenyang, China

[3] Department of Instrument Science and Technology, School of Mechanical Engineering, NanJing University of Science and Technology,Nanjing,China

[4] Department of Instrument Science and Technology, School of Mechanical Engineering, NanJing University of Science and Technology,Nanjing,China

Abstract. A mechine in order to warn of transmission lines galloping, reduce the loss caused by transmission lines galloping. Aiming at a transmission system, a set of lines galloping storage and testing device is developed. The device takes STM32 as the control core to realize signal conditioning, data storage, serial communication and other functions. In the design of the software, the method of single external trigger and multiple internal trigger is adopted to effectively avoid the false trigger phenomenon in weak oscillation, so as to test and store the acceleration of the wire when it gallopings violently. The acceleration sensor and the storage test device directly install on electric wire, and it can detect and record the acceleration data in the process of wire galloping.After the test, the recovery device uses the upper computer to read the detected data and display the waveform, which can provide important experimental basis for the theoretical analysis and parameter setting of the lines galloping.

1. Preference

Project Supported by Science and Technology Project of State Grid Liaoning Electric power Supply Co.,Ltd. (2018-YF32). In recent years, with the social economy develop fast, the growing demand for electricity, power grid scale expands unceasingly in our country. There is a problem to be solved: in recent years, our country affected by extreme weather, transmission lines covered by ice and snow, lines galloping accident is more and more frequent, and the trend of increase year by year, seriously affected the safety and stable operation of power grid. Especially along with our country in recent years, a number of such as newly built the three gorges project, 1000kV uhV ac, uhv dc + 800kV power transmission project. These lines are general in terrain complex by region, climate changeable, under the certain condition of climate,galloping accidents extremely easily, conductor wave, large amplitude makes wire tension increased significantly, could damage to tower, even make the tower down.Therefore, reliable device is needed to measure the acceleration of power line galloping, which can provide important experimental basis for theoretical analysis and parameter setting of power line galloping later. The research on the occurrence mechanism, propagation model and online monitoring of wire galloping is of great significance for strengthening the bearing capacity of tower and pole structure and ensuring the safe operation of power grid in galloping areas.

Content from this work may be used under the terms of the Creative Commons Attribution 3.0 licence. Any further distribution of this work must maintain attribution to the author(s) and the title of the work, journal citation and DOI.
Published under licence by IOP Publishing Ltd

Storage test technology can solve the above problems. In order to solve the above problems and study wire galloping better, a set of wire galloping storage and test system is designed. This device has the advantages of high reliability, small size, long acquisition time, accurate and repeated acquisition and so on. Based on the storage test technology, the method of using one external trigger and multiple internal triggers is added, which effectively solves the phenomenon of false triggering and accurately captures the acceleration of the wire when it is galloping. In other words, the device can collect data in the state of wire galloping, while the wire is not in the state of non-galloping. The storage test system can detect whether the wire is in the process of galloping, collect and store the acceleration data, and the power failure data is not lost.

2. Overall scheme design

The system acceleration sensor has a range of ± 500g. The storage test device is designed for the sensor and installed on the power line. The test system consists of a storage test device (lower computer) and a data analysis computer (upper computer LABVIEW terminal). Among them, the storage and test circuit is mainly composed of sensor signal conditioning module, A/D conversion module, data storage module, power supply module, data communication module and STM32 main control module. After the data collection is completed, it is transmitted back to the upper computer through the serial port, and the acceleration waveform is displayed on the LABVIEW terminal.

2.1. System hardware design

2.1.1. Acceleration sensor type selection and signal conditioning circuit design

Piezoelectric acceleration sensor adopts piezoelectric effect material as pressure sensor, which has the characteristics of wide frequency response range, high sensitivity, good dynamic characteristics, strong anti-interference ability, etc.and can meet the requirements of wire dancing detection. The acceleration sensor of this system is CA-YD-180 piezoelectric sensor of Jiangsu Lianneng electronic technology company, which belongs to two-wire IEPE acceleration sensor. Its main technical indicators are shown in table 1.

Table 1.Main technical indicators of ca-yd-180.

Serial number	Technical parameters	Index
1	Reference sensitivity	10mv/g
2	Frequency range	1~10000Hz
3	Acceleration range	± 500g
4	Working current	2~10mA

The actual figure of acceleration sensor is shown in figure 1.

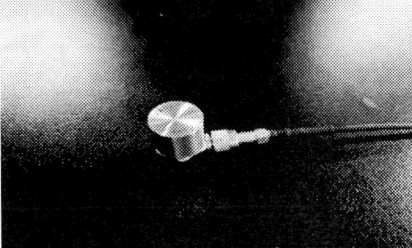

Figure 1. Acceleration sensor picture

The CA-YD-180 sensor requires 2-10mA constant current source power supply. In this paper, a three-terminal adjustable integrated constant current source chip LM134 is used to construct a constant current source circuit.The acceleration sensor is powered by the way shown in figure 2. At this time, the constant current value generated by LM134 is 8.33ma.

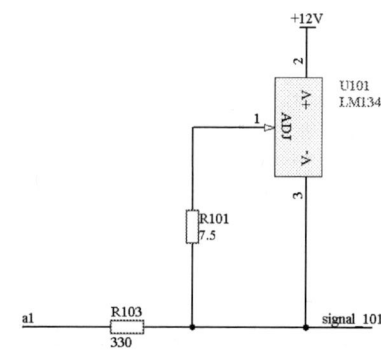

Figure 2. LM134 supplies power to the acceleration sensor

After constant current source power supply, the sensor output is a dynamic signal with dc bias, and RC high-pass filter circuit is adopted to achieve the isolation and direct coupling, the specific circuit is shown in figure 3. The latter stage builds a voltage follower based on operational amplifier LF353, which plays the role of impedance matching.By setting the value of resistance and capacitance, the lower limit cut-off frequency of high-pass filter circuit is determined to be 0.2Hz:

$$f_p = \frac{1}{2\pi(R_{104} + R_{105})C_{101}} \tag{1}$$

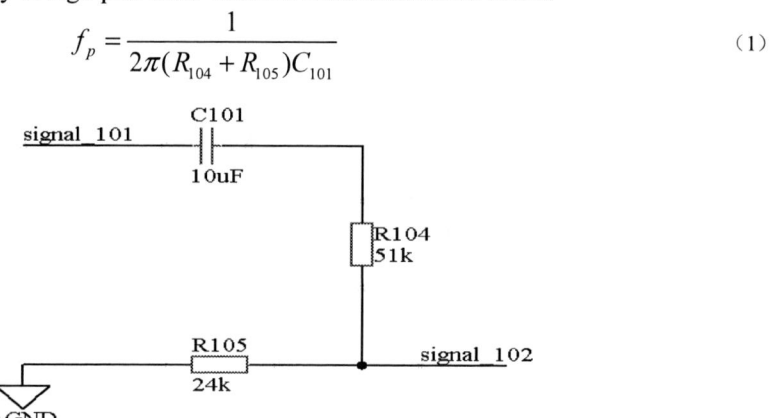

Figure 3. High pass filter circuit

In order to avoid aliasing distortion in the test site, a low-pass filter is set at the back stage of RC high-pass filter circuit, which can effectively eliminate high-frequency interference.

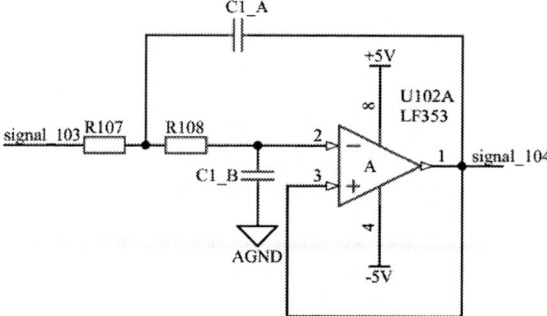

Figure 4. Low pass filter circuit

Based on operational amplifier LF353, this paper designs a classical two-order active butterworth low-pass filter, as shown in figure 4. The cut-off frequency of 12kHz is determined by setting the value of resistance and capacitance:

$$f_c = \frac{1}{2\pi\sqrt{R_{107}R_{108}C_{1\_A}C_{1\_B}}}$$ (2)

2.1.2. Data acquisition and storage circuit design

STM32 is selected as MCU, whose built-in ADC range is 0~3.3v. Since the range of the sensor is -500g~500g, the reference sensitivity of the sensor is about 10mv/g. Therefore, the range of output voltage of the sensor is -5V~5V, which is obviously inconsistent with the range of ADC. Therefore, the corresponding step-down and dc bias circuit needs to be designed to adjust the output signal of the sensor into the range of ADC. Voltage-reduced circuit is shown in figure 3, section 3.1. which reduces the sensor signal into + 1.6 V ~1.6 V (marked signal_104) in the way of these resistances in series. The DC bias circuit is shown in figure 5 below. According to the analysis of the superposition theorem of the circuit: (1) if the input signal of the operational amplifier is grounded, then the circuit output 1.6V dc voltage is connected to the ADC acquisition port; (2)+5V input grounding, the circuit is converted to a voltage follower and a RC high-pass filter circuit in series, and the circuit output is an alternating signal signal_104; (3) if the above two signals are superimposed, the signal signal_104 can be raised by 1.6V. In this way, the sensor signal is adjusted to the range of 0V~+ 3.2v, within the range of ADC.

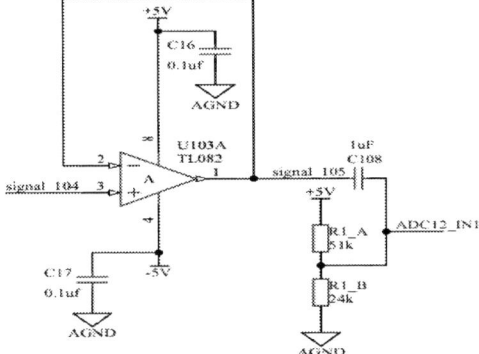

Figure 5. DC bias circuit

Since the frequency of acceleration signal generally does not exceed 10KHz, the sampling rate of ADC and the storage rate of memory chip are not too high. Winbond serial NOR Flash chip W25Q64 (8M bytes) was selected as the memory chip. Its page (256 bytes) write time is approximately, so the Flash chip's storage rate is about 10 times higher than the acceleration signal frequency, enough to store acceleration data.

2.1.3. Power supply design

Battery selection is the core of power supply design, mainly considering the following four factors :

(1) battery volume and weight; (2) the overload resistance of the battery; (3) battery voltage; (4) battery capacity.

Because the storage test device is installed on the power line, the battery is required to be small in size, light in weight and large in capacity.Secondly, the battery bears acceleration load in the process of moving with the wire, so the battery needs to have certain overload resistance. Polymer lithium battery has the advantages of strong impact resistance, light weight, small thickness, customizable shape and so on. This paper selects polymer lithium battery as the power supply battery of the system. The voltage of the battery needs to be combined with the power supply voltage required by each module of the system. In order to reduce the volume of the storage test device, lithium battery with a voltage of 12V was selected in this paper, and the voltages of +5V, + 3.3V and -5V were obtained by stabilizing the voltage chip HT7350, HT7333 and reversing the chip ICL7660, respectively. Finally, the digital circuit is powered by + 3.3v, the constant current source circuit is powered by +12V, and the operational amplifier is powered by. The capacity of the lithium battery is 600mah, and the power consumption current of the circuit board is 60mA after the actual measurement and storage test. It can be calculated that the circuit can work continuously for 10 hours and meet the test time requirements.

2.1.4. Use and interface design

Use method: press 1 to power the system and the green light starts flashing. Press 2 to erase the external flash data. The system starts to pre-collect and wait for the trigger. At this point, the red light flashes once and then goes out. After triggering, acceleration data will be collected. When the system detects that the dancing process is terminated, the collection will be completed. Green light flashing, red light long light, signal collection success. Other status of indicator light indicates collection failure.

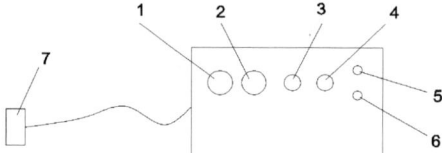

1 power switch 2 trigger button 3 charging port 4 serial port communication interface 5 signal indicator light (red) 6 signal indicator light (green) 7 acceleration sensor

Figure 6. Interface diagram of acceleration test system

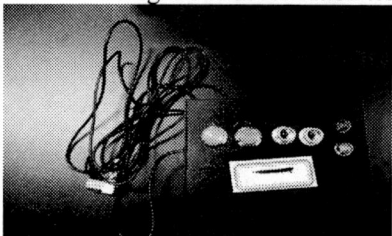

Figure 7. Picture of real product acceleration test system

2.2. System software design

ADC data acquisition is the main part and key function of the test system software, and its program flow is shown in figure 8 below.

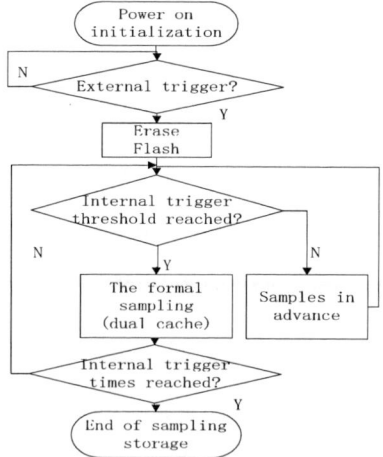

Figure 8. data collection process

After the test system is energized and initialized, wait for external trigger signal. If the external trigger signal arrives, the Flash memory chip is erased, which means that the data collected and stored in Flash last time is cleared.After that, the CPU continuously detects whether the acceleration signal reaches the internal trigger threshold. If not, the collected data will be recorded in the pre-sampling cache in a loop [7]. If the internal trigger threshold is reached, formal sampling is performed, stored in RAM and transferred to Flash in a "double-cached" fashion.

When a formal sampling is completed, the CPU determines whether the total number of internal triggering times has reached. If not, the CPU continues to judge the internal triggering threshold to determine whether the formal sampling is carried out.If the total number of internal triggers is reached, the sampling storage procedure ends. The main innovation of this part of software is one external trigger and multiple internal triggers. Multiple internal triggers can avoid the phenomenon that the installation of storage test device or other uncertain factors such as field interference can lead to false triggering and the acceleration during dancing cannot be recorded. Multiple internal triggers can ensure that the effective acceleration data can be recorded by the storage test device.

This system software uses "double cache" method to realize data storage. So-called double cache storage methods: is in the internal microprocessor, RAM opened two same size of the cache, collecting data to populate the cache first 1, full of later transferred to the external Flash, in the process of transferring data, collect data, but data fill in the cache 2, such as the cache after 2 full can then transfer the data to the external Flash, collect data, in turn, to fill the cache 1 again.Adjust the size of the cache so that data collection and stored procedures are not interrupted.

Since the frequency of acceleration signal generally does not exceed 10KHZ, according to Nyquist sampling theorem, the sampling rate of 100KHZ can meet the requirements of data collection of this system. On the software, the conversion time of STM32 built-in ADC is configured as 1.17us (855KHz). The sampling rate is controlled to be 100KHz by STM32 internal timer, that is, the result of A/D conversion is stored in RAM cache in timer interrupt service program every 10us.

Specifically, we set up two cache BUF1 and BUF2 in STM32 memory RAM, each of which is 1000 bytes. When the sampling rate is controlled by a timer to be 100KHz, it takes 5ms to store 1000 bytes of data into the cache.It took 3.76ms to write 1000 bytes of cached data into Flash by actual measurement.Therefore, it can be known that "double caching" can be implemented with a cache (1000 bytes) as the unit of storage, and the time to write Flash is less than the time to save RAM.

3. Conclusion

In view of the swing test of a transmission line, a special small-volume and anti-overload storage and testing device is designed by using the storage and testing technology and combining the actual object under test with the environment. The test system adopts the method of one external trigger and multiple internal triggers in the software design, which effectively avoids the phenomenon of misfiring during the test and failing to remember the acceleration of wire dancing process.It can provide important experimental basis for theoretical analysis of transmission line movement, transmission line construction and transformation.However, this design requires that the test device can only be read after the end of the test, and there are some limitations in data acquisition efficiency. Then, research on wireless reading is required.

References

[1] Zeng qingpei. Research on conductor dance of high voltage transmission line [D]. Tianjin university, 2012

[2] AnotherQu bingbing. Development of fpga-based solid propellant overload parameter storage and test system [D]. Nanjing university of science and technology, 2013.

[3] Tang xuhui. Finite element model of piezoelectric accelerometer and its application research [D]. Yanjing university, 2012.

[4] Qu bingbing, kong deren, wang fanget al. Design of a solid propellant overload large strain and acceleration storage test system [J]. Journal of missile and guidance, 2013,33 (5) : 126-128.

[5] Qian lihua,Chen yaqi, shen xiaomin et al.Design and application of overload test system for rocket skid [J]. Sichuan journal of military science and technology, 2013,3 (2) : 32-34.

[6] Tu peng, pei dongxing, zhang hongyan, et al. Sampling strategy in storage test technology Research factory j-1 automation instrument,2012,33(11):19-22.

[7] Zhou ruiqing,jia yunfei,pan xiaobin, et al. High overload signal storage for gun launch Design of storage and test system [J]. Measurement and control technology, 2017, 36 (4) : 43-47

[8] Li xiaohong. Sensitivity calibration and transverse effect research of three-axis high g value acceleration sensor [D].North China university, 2011.

ISPECE   IOP Publishing

Application of Hybrid Conjugate Gradient Algorithms in Inverse Problems of Electromagnetic Tomography

Li Liu[1] , Wang Lei[1]，Wang Zhanjun[2]

[1]College of Physics Science and Technology,
Shenyang Normal University,Shenyang 110034,China

[2]Department of computer and mathematics,
Shenyang Normal University,Shenyang 110034,China

Corresponding author: Li Liu，College of Physics Science and Technology,
Shenyang Normal University, Shenyang 110034, e-mail :9981912@qq.com

Abstract. Based on the existing conjugate gradient algorithm, a new MN conjugate gradient method for inverse problems of electromagnetic tomography is proposed. The composition of the electromagnetic tomography system and the characteristics of the positive and inverse problems are described, and the solutions of the inverse problems are given. The existing typical conjugate gradient algorithms and the hybrid conjugate gradient methods using convex combinations are summarized. Based on four typical models of EMT system, PRP, HCG and new MN methods are compared in terms of image reconstruction accuracy, iteration times and iteration time. It is concluded that all three conjugate gradient methods can satisfy the termination conditions. The MN method can reduce the number of iterations and the calculation time, and has excellent numerical performance in practical calculation.

1.Introduction

Electromagnetic tomography (EMT) is a kind of electrical tomography technology based on electromagnetic induction principle. EMT technology has the advantages of non-intrusion and non-contact[1]. It is more suitable for the detection of the internal structure of space in the unavailable or inaccessible environment. It is widely used in medical field and industrial non-destructive testing[2,3]. In view of physical structure, EMT system is generally composed of excitation system, detection system and PC mechanism[2]. The excitation system consists of an excitation signal source circuit and excitation coils. The excitation system mainly forms the excitation magnetic field in the object field space. When the exciting magnetic field encounters a conductive or magnetic medium in the field, it changes and forms a changed object field space. The detection system consists of a detection coil and a detection circuit to obtain the detection values in all directions of the object space for image reconstruction. The function of PC is to form spatial sensitivity coefficient matrix of object field, preprocess the detected value, and design reasonable reconstruction algorithm to realize image reconstruction. In the technical point of view, the research of electromagnetic tomography technology includes two parts[4]: forward and inverse problems. The positive problem is mainly to establish the physical model and sensitivity model of the system, and to study the influence of the change of system parameters, etc. The inverse problem is to reconstruct and display images from measured data.

Content from this work may be used under the terms of the Creative Commons Attribution 3.0 licence. Any further distribution of this work must maintain attribution to the author(s) and the title of the work, journal citation and DOI.

Published under licence by IOP Publishing Ltd

2. Forward and inverse problems

In the sine wave excitation, electromagnetic field of EMT system with the Maxwell equation group of harmonic forms are as follows:

$$\begin{cases} \nabla \times \vec{H} = \vec{J} + j\omega\vec{D} \\ \nabla \times \vec{E} = -j\omega\vec{B} \\ \nabla \cdot \vec{B} = 0 \\ \nabla \cdot \vec{D} = 0 \end{cases} \tag{1}$$

Define the magnetic vector position, which satisfies

$$\nabla \times \vec{A} = \vec{B} \tag{2}$$

According to the distribution of excitation coils in EMT system, the corresponding Neumann boundary conditions are determined.:

$$\begin{cases} \mu^{-1} \cdot \left(\dfrac{\partial^2 \vec{A}}{\partial x^2} + \dfrac{\partial^2 \vec{A}}{\partial y^2} \right) = j\omega\delta\vec{A} \\ \left. \dfrac{\partial \vec{A}}{\partial n} \right|_{x^2 + y^2 = R^2} = \mu_0 \dot{I} \end{cases} \tag{3}$$

The complete mathematical description of EMT system is obtained, where R is the radius of the circumference of the excitation coil, μ_0 is the permeability of air, and $\dot{I}$ is the current density of the excitation coil. In the forward problem, the solution of sensitivity matrix is the key, and it is also an important index to solve the inverse problem. Sensitivity is defined as,

$$s_{ij}(k) = \frac{A_{ij}(k) - A_{ij}(\mu_1)}{A_{ij}(\mu_2) - A_{ij}(\mu_1)} \cdot \frac{1}{\mu_2 - \mu_1} \cdot \omega(k) \tag{4}$$

$$k = 1,2,\ldots\ldots e_0, \qquad i = 1,2,\ldots\ldots\Gamma, \qquad j = 1,2,\ldots\ldots\Gamma - 1$$

where e_0, Γ is The number of subdividing units and coils in the field object. The inverse problem of EMT system can be simplified as follows:

$$sx = b \tag{5}$$

where $s \in R^{m \times n}$, $x \in R^n$, $b \in R^m$, and $m < n$. Formula (5) is an indefinite equation. For the indefinite equations, the direct method can not meet the system requirements. Generally, it is solved by iterative method[4], that is, many times to calculate the objective function corresponding to different x values, to gradually approach the optimal solution.

Conjugate gradient algorithm is recognized as the most effective method to solve large-scale unconstrained linear equations. It has the characteristics of fast convergence and simple algorithm. Formula (5) is optimized into unconstrained continuous differentiable equations.

$$\min f(x) = \min(sx - b) \tag{6}$$

Assuming that $f(x)$ is continuously differentiable[5], d_k is search direction, the iterative formula of conjugate gradient algorithm is

$$x_{k+1} = x_k + \alpha_k d_k \tag{7}$$

$$d_k = \begin{cases} -g_k, & k = 0 \\ -g_k + \beta_k d_{k-1} & k \geq 1 \end{cases} \tag{8}$$

where $g(x) = \nabla f(x)$.

For different conjugate coefficients β_k , various conjugate gradient methods with different numerical experimental results and convergence properties have been developed. The more famous conjugate coefficients β_k [6] are:

$$\beta_k^{FR} = \frac{\|g_k\|^2}{\|g_{k-1}\|^2}, \beta_k^{PRP} = \frac{g_k^T(g_k - g_{k-1})}{\|g_{k-1}\|^2}, \beta_k^{CD} = \frac{\|g_k\|^2}{d_{k-1}^T g_{k-1}}$$

$$\beta_k^{HS} = \frac{g_k^T(g_k - g_{k-1})}{d_{k-1}^T y_{k-1}}, \beta_k^{LS} = \frac{g_k^T(g_k - g_{k-1})}{d_{k-1}^T g_{k-1}}, \beta_k^{DY} = \frac{\|g_k\|^2}{d_{k-1}^T y_{k-1}}$$
(9)

FR method is easy to produce small step size. Under the precise line search, the convergence is different according to the different objective functions. CD method has good convergence effect, but the numerical results are not satisfactory. PRP method and HS method can not achieve both global convergence and experimental data. In order to construct a new algorithm with good convergence property and excellent numerical performance, the conjugate parameter β_k above is directly improved.

3. Hybrid Conjugate Gradient Algorithms

In order to give full play to the advantages of each conjugate method, the conjugate gradient algorithm with different properties is fused.

In the reference[7], PRP algorithm and FR algorithm are combined to form conjugate coefficients by convex combination:

$$\beta_k^{HCG} = \theta_k \beta_k^{FR} + (1 - \theta_k) \beta_k^{PRP}$$
(10)

where $\theta_k \in [0,1]$.The iteration direction generated by this algorithm under standard Wolfe condition is descending.

In the reference[8]，LS algorithm and DY algorithm are combined to form conjugate coefficients by convex combination:

$$\beta_k = \theta_k \beta_k^{DY} + (1 - \theta_k) \beta_k^{LS}$$
(11)

where $\theta_k \in [0,1]$. The advantage of this algorithm is that the iteration direction not only satisfies the famous D-L condition, but also conforms to Newton direction, and this property does not depend on any line search method.

For EMT system, different conjugate gradient algorithms are mixed to extract the maximum parameters, and a new conjugate coefficient is defined as:

$$\beta_k^{MN} = \frac{\|g_k\|^2 - \max\{0, g_k^T g_{k-1}\}}{\max\{\|g_{k-1}\|^2, d_{k-1}^T y_{k-1}\}}$$
(12)

The formula (12) integrates the characteristics of the conjugate gradient coefficient mentioned above. It can not only improve the convergence of the algorithm, increase the step size, but also reduce the number of iterations and shorten the calculation time. The concrete steps are as follows:

Set accuracy $\varepsilon \leq 10^{-7}$.

Step 1: Selection of initial points $x_1 \in R^n$,set $k=1$,and calculate $g_1 = g(x_1)$;

Step 2 : If $\|g_k\| \leq \varepsilon$, then iteration stop. Otherwise, step size $\alpha_k > 0$ is determined by line search conditions;

Step 3: Set $x_{k+1} = x_k + \alpha_k d_k$ and $g_{k+1} = g(x_{k+1})$.

Step 4: Set $d_k = -g_k + \beta_k^{MN} d_{k-1}$, β_k^{MN} calculated by formula (12).

Step5: Set $k = k+1$, jump to Step 2.

The experiment adopts Wofle line search condition and is realized by MATLAB programming. The termination condition of the algorithm is $\| g_k \| \leq 10^{-7}$, or the iteration time of the algorithm exceeds 3600s. The MN algorithm is compared with the basic PRP algorithm and the fusion HCG method. Four typical imaging examples of the object space of EMT system are shown in Table 1. In Table 1, M is the name of the conjugate gradient algorithm; I is the number of iterations of the algorithm to solve a test problem; T is the calculation time of the algorithm to solve a test problem. S is the termination condition of the algorithm, so that $S = 1$, otherwise, S = 0.

As can be seen from Table 1, (1) Either conjugate gradient algorithm can satisfy the termination condition within 600 seconds, and can successfully solve the image reconstruction problem of EMT system; (2) For four typical models, MN method is superior to PRP and HCG methods in terms of iteration times and computation time; （3）When the object is at the edge of the object field, the three conjugate gradient algorithms have the fastest imaging speed, which is related to the high sensitivity around the object field.

Table 1 Numerical results with different original images

original image	M	I	T/s	S
	PRP	78	64.2134	1
	HCG	45	51.4252	1
	MN	34	40.1254	1
	PRP	64	65.7658	1
	HCG	47	56.1543	1
	MN	36	39.7853	1
	PRP	88	71.0346	1
	HCG	78	58.5655	1
	MN	67	49.9987	1
	PRP	80	65.7789	1
	HCG	79	70.3475	1
	MN	45	37.9801	1

4. Conclusion

Based on conjugate gradient method, a new hybrid conjugate gradient algorithm MN method for EMT system is constructed, which has sufficient descent under any line search condition. The algorithm has excellent numerical performance in practical calculation.

Acknowledgments

The authors gratefully acknowledge the helpful comments and suggestions of the anonymous reviewers. This work is supported by the Natural Science Foundation of Liaoning Province, China(NO.20170540817).

References

[1] Griffiths H. Magnetic induction tomography[J]. Measurement Science and Technology, 2001, 12(8): 1126-1131.

[2] Scharfetter H, Lackner HK, Rosell J. Magnetic induction tomography: hardware for multi-frequency measurements in biological tissues[J]. Physiological Measurement, 2001, 22(1): 131-146.

[3] Gursoy D, Scharfetter H. Imaging artifacts in magnetic induction tomography caused by the structural incorrectness of the sensor mode[J]. Measurement Science Technology, 2010, 22(22): 15502-15512.

[4] Vauhkonen M, Hamsch M and CH Igney. A measurement system and image reconstruction in magnetic induction tomography[J]. Physiol. Meas., 2008,29:445-454.

[5] Li Liu, Shao Fuqun. Modified conjugate gradient algorithm for image reconstruction of electromagnetic tomography[J]. Chinese Journal of Scientific Instrument, 2010, 31(3): 655-658.

[6] Qi Changxia. The research of several fused nonlinear conjugate gradient methods[D]. Qinhuangdao: Master's thesis of Yanshan University, 2017.

[7] Saman B K, Reza G. A hybridization of the Polak-Ribiere-Polyak and Fletcher-eeves conjugate gradient methods[J]. Numerical Algorithms, 2015, 68(3): 481-495.

[8] Liu J K, Li S J. New hybrid conjugate gradient method for unconstrained optimization[J]. Applied Mathematics and Computation, 2014, 245: 36-43.

Probabilistic modeling of output characteristics based on ECM algorithm for wind farms

Tingxiang Liu[1], Xiaoying Zhang[1], Kun Wang[2], Wei Chen[1] and Xiaolan Wang[1]

[1]College of Electrical and Information Engineering, Lanzhou University of Technology, Lanzhou 730050, China

[2]State Grid Gansu Electric Power Company Electric Power Research Institute, Lanzhou 730050, Gansu, China

Abstract: With the large-scale development of wind power, grasping the fluctuation characteristics of wind farm output has become a key link in wind power grid-connected operation. Because of the inherent defect in EM(expectation maximization) algorithm that is adopted to model wind farms, the fitting precision of WGMD(weighted Gaussian mixture distribution) is reduced. According to this problem, an improved probabilistic modeling method based on ECM(Expectation Constraint Maximization) algorithm is proposed in this paper. And several common distribution models are compared with simulate the output characteristics of wind farms. Simulation results show that the estimation of model parameters by the ECM algorithm can improve the simulation accuracy of the weighted Gaussian mixture model, that verifying the feasibility and effectiveness of the weighted Gaussian mixture probability model.

1. Introduction

Wind power has been greatly developed as a non- polluting renewable energy source in recent years. However, wind power generation has the characteristics of volatility, intermittence and randomness. Its large-scale interconnection will inevitably increase the uncertainty of the traditional power system operation, scheduling and risk assessment[1]. Moreover, with the continuous increase in the scale of wind farms and the capacity of wind turbines, the proportion of wind power generation in the power grid is increasing, which inevitably brings many difficulties to the planning, design, security and stability analysis of power grids[2]. Therefore, it is great practical significance to study the power output characteristics of the wind farm and establish an accurate probability model of wind power output, which can provide valuable basis for the planning of power system, risk analysis, reliability evaluation and economic dispatch[3].

There are two methods for probabilistic modeling of output characteristics of wind farms. One is based on a random sequence method, which uses the auto- regressive and moving average (ARMA) model in the time series method to simulate the change of the wind speed, and using matrix technology to generate multiple specified wind speed series to simulate the output-related characteristics of the wind farm[4-5]. The other is based on the probability density function. Ding Ming uses Weibull distribution to model the wind turbine power model. According to the relationship between wind speed and wind power, wind power distribution characteristics can be obtained[6]. References [7,8] believes that using Beta distribution to describe wind power output prediction error is more reasonable and the fitting effect is better. However, the above single distribution function model cannot take into account the asymmetry distribution and multi-peak characteristics of the probability

Content from this work may be used under the terms of the Creative Commons Attribution 3.0 licence. Any further distribution of this work must maintain attribution to the author(s) and the title of the work, journal citation and DOI.

Published under licence by IOP Publishing Ltd

distribution of prediction error.

Based on the above analysis, this paper starts with a large number of actual measured historical data from wind farms and establishes a weighted Gaussian mixture distribution (WGMD) for changes in wind power output. The method of estimating WGMD parameters is usually moment estimation method and maximum likelihood estimation method. Yan Hong adopts the traditional EM (Expectation Maximization) algorithm for solving unknown parameters of WGMD[9]. However, the EM algorithm is mainly used to solve unknown parameters under incomplete data, and it is very easy to be affected by the initial conditions, resulting in the calculation result may not be able to converge to the global optimal value, which greatly affects the accuracy of the model[10]. In order to overcome the deficiencies of EM algorithm, this paper uses ECM (Expectation Constraint Maximization) algorithm to estimate the parameters of WGMD.

2. Weighted gaussian mixture distribution

In recent years, Gaussian distribution has been used as a common tool in the statistical, computer and engineering fields for data analysis[11]. WGMD is a model that is weighted by multiple complex data analyses in the case where a single Gaussian distribution cannot effectively represent complex data. It is not limited to a specific assumption of the probability density function form, and any probability density distribution can be approximated by the linear combination of several Gaussian density functions, and the probability density distribution of an arbitrary random variable can be accurately characterized[12].

The weighted Gauss mixture model is used to fit the probability density function of wind farm output as:

$$f(x,\Theta) = \sum_{m=1}^{M} \alpha_m G_m(x/\theta_m) = \sum_{m=1}^{m} \alpha_m \frac{1}{\sqrt{2\pi}\sigma_m} e^{-\frac{(x-\mu_m)^2}{2\sigma_m^2}} \quad (1)$$

Where, $\Theta = \{\theta_m = (\alpha_m, \mu_m, \sigma_m), i = 1, 2, ..., M\}$ is model parameters, M is the number of model parameters, α_m 、 μ_m 、 σ_m^2 are the weights, mean values and variances of the m components of the weighted Gaussian mixture probability model respectively, and x is measured data of wind farm output, furthermore, $\sum_{m=1}^{M} \alpha_m = 1$, $\alpha_m \geq 0$.

For the weighted Gaussian mixture model of wind farm output, the measured data of wind farm output is the observed data, which is called the incomplete data; Whereas each output power of a wind farm is unobservable, it is called lost data. Let $\omega \in (1, 2, ..., M)$ denote the category of wind farm output, and the complete data is $X = \{x_i = (y_i, \omega_i), i = 1, 2, ..., N\}$.

3. Calculate unknown parameters by ECM algorithm

3.1 EM algorithm

EM algorithm is a method of estimating parameters that has developed rapidly and widely in recent years. It is an iterative method proposed by Dempster[10], which simplifies the calculation of maximum likelihood estimation. When maximization is not explicitly expressed, the generalized EM algorithm is given, which is GEM (General Expectation Maximization) algorithm. Meng et al. proposed a special GEM algorithm called ECM algorithm[13], which preserves the simplicity and stability of the EM algorithm.

The EM algorithm is mainly used to solve the problem of maximum likelihood estimation of unknown parameters under incomplete data. Assume that Y_{obs} is the observation data, Y_{mis} is the missing data, θ is the unknown parameter, and the complete data is $Y = (Y_{obs}, Y_{mis})$. Let $f(\theta | Y_{obs})$ denotes the posterior distribution density function of θ based on observation data Y, $f(\theta | Y_{obs}, Y_{mis})$ denotes the posterior distribution density function for θ after adding data Y_{mis}, $f(Y_{mis} | \theta, Y_{obs})$ denotes the conditional distribution density function of potential data Y_{mis} under given θ and observed data Y_{obs}. The purpose of parameter estimation is to observe the mode of the posterior distribution $f(\theta|Y_{obs})$. The EM algorithm remembers $\theta^{(m)}$ as the estimated value of the posterior mode at the beginning of the mth iteration. Then the two steps of the $(m+1)$th iteration are the following:

E step: Impute the expectation of the $\log f(\theta|Y_{obs},Y_{mis})$ conditional distribution of Y_{mis}, then remove Y_{mis} through calculus:

$$Q(\theta|\theta^{(m)},Y_{obs}) = \int \log(f(\theta|Y_{obs},Y_{mis}))f(Y_{mis}|\theta^{(m)},Y_{obs})dY_{mis} \quad (2)$$

M step: Determine $\theta^{(m+1)}$ by maximizing the imputed log-likelihood $Q(\theta|\theta^{(m)},Y_{obs})$:

$$Q(\theta^{(m+1)}|\theta^{(m)},Y_{obs}) = \max_{\theta} Q(\theta|\theta^{(m)},Y_{obs}), \text{ for all } \theta \in \Theta \quad (3)$$

After getting $\theta^{(m+1)}$, an iteration of $\theta^{(m)}$ to $\theta^{(m+1)}$ is formed. The above steps are iterated to $\left\|\theta^{(m+1)}-\theta^{(m)}\right\|$ or $\left\|Q(\theta^{(m+1)}|\theta^{(m)},Y_{obs})-Q(\theta^{(m)}|\theta^{(m)},Y_{obs})\right\|$ is small enough to stop.

3.2 ECM algorithm

The basic idea of the ECM algorithm is to use several simple conditions to maximize the steps instead of the complex M steps in the EM algorithm. Each step of the Conditional Maximization(CM) step is under the constraint of the corresponding parameter, so that the Q function defined by the EM algorithm reaches a maximum[14]. Let $G = \{g_s(\theta); s = 1,2,...,S\}$ be set of $S(\geq 1)$ preselected constraint functions for the parameter θ, The parameter θ is divided into sub-vectors $\theta = (\theta_1,\theta_2,...,\theta_S)$. With ECM, the M-step is replaced by S CM steps in the mth iteration, marked as $\theta^{(m)} = (\theta_1^{(m)},\theta_2^{(m)},...,\theta_S^m)$. In the $(m+1)$th iteration, impute θ_1 make $Q(\theta|\theta^{(m)})$ reach maximum under the condition of a given $\theta_2 = \theta_2^{(m)},...,\theta_s = \theta_s^{(m)}$, and marked as $\theta_1^{(m+1)}$; Again impute θ_2 make $Q(\theta|\theta^{(m)})$ reach maximum under the condition of a given $\theta_1 = \theta_1^{(m+1)}, \theta_j = \theta_j^{(m)}$ $j = 3,...,S$, and marked as $\theta_2^{(m+1)}$. By analogy, after maximizing Sth, $\theta^{(m+1)} = (\theta_1^{(m+1)},\theta_2^{(m+1)},...,\theta_S^{(m+1)})$ is obtained and an iteration is completed.

E step: Impute the expectation of the $\log f(\theta|Y_{obs},Y_{mis})$ conditional distribution of Y_{mis}, then remove Y_{mis} through calculus, as shown in formula (2);

sth CM-step: Find $\theta^{(m+s/S)}$ such that

$$Q(\theta^{(m+s/S)}|\theta^{(m)},Y_{obs}) = \max_{\theta} Q(\theta|\theta^{(m)},Y_{obs}) \quad (4)$$

Where, $\theta \in \Theta_s^{(m)} \equiv \{\theta \in \Theta : g_s(\theta) = g(\theta^{(m+(s-1)/S)})\}$, the next iterate $\theta^{(t+1)} \equiv \theta^{(m+S/S)}$. The rational behind the CM steps is that in problems where maximizing $Q(\theta/\theta^{(m)})$ over $\theta \in \Theta$ is difficult, it may be impossible to choose G so that it is simple to maximize over $\theta \in \Theta_s^{(m)}$ for $s = 1,...,S$.

3.3 WGMD parameter estimation

When the observed value $y = (y_1,y_2,...,y_N)$ of the weighted Gaussian mixture model describing the wind farm output is given, the likelihood function of the probability distribution is:

$$L(\theta) = \sum_{n=1}^{N} \log\{\sum_{m=1}^{M} \alpha_m G_m(y|\theta_m)\} \quad (5)$$

E step: Imputed the conditional expectation of a complete data likelihood function, $\theta^{(t)}$ is the parameter estimate of the tth iteration, $F(\omega|y,\theta^{(t)})$ is the posterior probability of ω:

$$Q(\theta|\theta^{(t)}) = E\{\log f(y,\omega|\theta)|y,\theta^{(t)}\}$$
$$= \sum_{n=1}^{N} F(\omega|y,\theta^{(t)}) \log f(y,\omega|\theta) \quad (6)$$

CM step: Imputed $\theta^{(t+1)} = (\theta_1^{(t+1)},\theta_2^{(t+1)},...,\theta_s^{(t+1)})$ such that

$$Q(\theta^{(t+1)} \mid \theta^{(t)}) = \max_{\theta} Q(\theta \mid \theta^{(t)}) \qquad (7)$$

1thCM: Calculate $\theta_1^{(t+1)} = \arg\max Q(\theta \mid \theta_1^{(t)})$ through a given $\theta_2 = \theta_2^{(t)},...,\theta_s = \theta_s^{(t)}$;

2thCM: Calculate $\theta_2^{(t+1)} = \arg\max Q(\theta \mid \theta_2^{(t)})$ through a given $\theta_1 = \theta_1^{(t+1)}, \theta_j = \theta_j^{(t)}, j=3,...,s$;

After (S-1)th calculations;

SthCM: Calculate $\theta_s^{(t+1)} = \arg\max Q(\theta \mid \theta_s^{(t)})$ through a given $\theta_1 = \theta_1^{(t+1)},...,\theta_{s-1} = \theta_{s-1}^{(t+1)}, \theta_s = \theta_s^{(t)}$.

According to Bayes rule[11],such that:

$$\omega_{mn} = \frac{\alpha_m G_m(y_n \mid \theta_m)}{f(y_n \mid \theta_m)} \qquad (8)$$

According to the method in [15], α_m is solved as:

$$\alpha_m = \frac{1}{N}\sum_{n=1}^{N}\omega_{mn} \qquad (9)$$

4. Evaluation index

Comparing the error parameter of a model with the actual fitting effect, there are root mean squared error (RMSE), the sum of squares due to error (SSE) and the coefficient of determination (R-square). The closer RMSE and SSE are to 0, the better the model fits the original data. And the R-square value range is [0,1], its the closer to 1, the better the interpretation of the original data, the better the model fitting effect.

RMSE: Also called the standard error, which reflects the error between the fitted data and the original data. Calculated as follows:

$$RMSE = \sqrt{\frac{1}{n}\sum_{i=1}^{n}(\hat{y}_i - y_i)^2} \qquad (10)$$

SSE: The sum of the squared errors of the corresponding points of the fitting data and the original data. Calculated as follows:

$$SSE = \sum_{i=1}^{n}(\hat{y}_i - y_i)^2 \qquad (11)$$

R-square: Characterizes the fit of the data by changes in the data. Calculated as follows:

$$R - square = \frac{\sum_{i=1}^{n}(\hat{y}_i - \overline{y}_i)^2}{\sum_{i=1}^{n}(y_i - \overline{y}_i)^2} \qquad (12)$$

Where, $\hat{y}_i$ is the sample data; y_i is the original data; $\overline{y}_i$ is the average value of the original data.

5. Example illustrating ECM

In this paper, the probability density distribution characteristic of wind power fluctuation is obtained by measuring the output power data of wind farm in a certain area abroad. A probability model of a weighted Gaussian mixture model is established to compare the fitting effect of other single distribution function models and verification of its effectiveness in the study of wind power characteristics. The EM algorithm and ECM algorithm were used to estimate the parameters of the Gaussian mixture model, and the ECM algorithm was verified the accuracy of the model.

According to the wind farm power distribution and probability density distribution curve, Weibull distribution, t location-scale distribution and normal distribution are used to fit wind power fluctuation characteristics. The fitting effect diagram is shown in Figure 1.

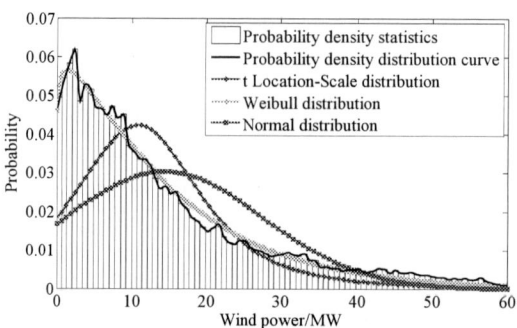

Figure 1. Single distribution function model fitting effect

As can be seen from Figure 1, the single distribution function cannot accurately fit the "tailing" property of the probability density distribution characteristics of wind power fluctuations, and the fitting effect is poor. Therefore, when a single wind farm has fluctuating random characteristics, the output fluctuation characteristics of the wind farm group will no longer satisfy a single specific distribution function curve.

For WGMD modeling, target distributions are usually modeled using no more than 5th-order models. Estimate the probability distribution of wind power output based on measured data, and optimize the fitting effect by increasing the number of Gaussian distributions. This paper uses the 2nd and 5th order WGMD to get the wind power probability density distribution curve, as shown in Figure 2.

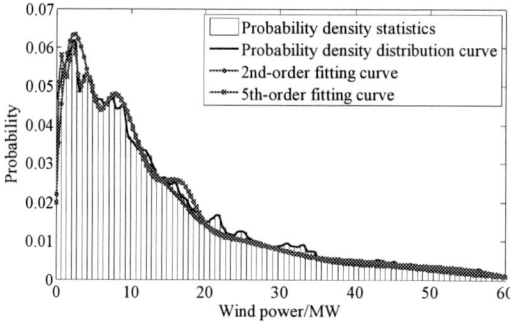

Figure 2. WGMD fits wind power probability density distribution curve

As can be seen from Figure 2, the 2nd-order and the 5th-order WGMD can well fit the probability density distribution curve of wind power fluctuations, and with the increase of the weighted order, the fitting accuracy is also improving. Table 1 gives the index values of the evaluation of the fitting effect to quantify the fitting effect of the two models and the commonly used distribution functions.

Table 1. Fitting effect index value

Fitting model	I_{RMSE}	I_{SSE}	$R-square$
t distribution	0.4	25.05	0.57
Normal distribution	0.8	21.67	0.63
Weibull distribution	0.25	12.95	0.88
2nd-order WGMD	1.8×10^{-3}	5.1×10^{-4}	0.92
5th-order WGMD	5.2×10^{-4}	3.6×10^{-5}	0.98

As shown in Table 1, the I_{SSE} values of the three commonly used distribution functions are all large, I_{RMSE} is much larger than 0, and the fitting error is relatively large. In comparison, the $R-square$ of

Weibull distribution reaches 0.88, and the fitting effect is better, and I_{SSE} and I_{RMSE} are the minimum among the three distributions. Comparing the I_{SSE} and I_{RMSE} index values of WGMD are close to 0, $R-square$ index values are all above 0.9, in which the $R-square$ value of fifth-order WGMD reaches 0.98. The above analysis shows that when characterizing wind power based on probability density function, WGMD is more suitable for fitting wind power fluctuation characteristics, and the fifth-order WGMD fitting effect is better.

In the parameter estimation, because the EM algorithm calculates the fifth-order model parameters, the parameter estimation error is large, the local optimization results are obvious, and it is easy to appear in the iteration loop. Therefore, the EM algorithm and ECM algorithm are used to compare the parameters of the third-order WGMD. The probability density function of wind power based on the third-order Gaussian mixture model is shown in formula (13). The parameters are estimated as shown in Table 2.

$$f(x) = \alpha_1 \frac{1}{\sqrt{2\pi}\sigma_1} e^{-\frac{(x-\mu_1)^2}{2\sigma_1^2}} + \alpha_2 \frac{1}{\sqrt{2\pi}\sigma_2} e^{-\frac{(x-\mu_2)^2}{2\sigma_2^2}} + \alpha_3 \frac{1}{\sqrt{2\pi}\sigma_3} e^{-\frac{(x-\mu_3)^2}{2\sigma_3^2}} \quad (13)$$

Figure 3 uses the probability density distribution of the model determined by the EM algorithm and the ECM algorithm respectively.

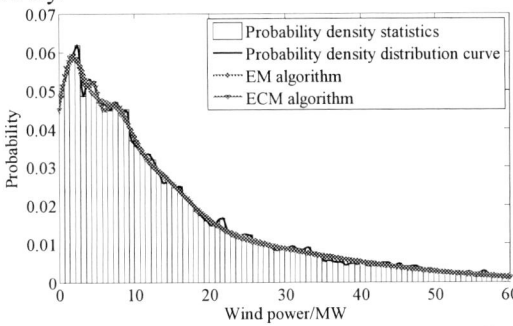

Figure 3. EM algorithm and ECM algorithm fitting distribution curve

Table 2. Wind power probability model parameter estimation

Model parameters	α_1	α_2	α_3	μ_1	μ_2	μ_3	σ_1	σ_2	σ_3
EM algorithm	0.27	0.51	0.19	33.67	25.06	18.79	5.06	10.93	3.22
ECM algorithm	0.26	0.55	0.22	29.77	28.07	15.66	5.37	12.61	3.18

As shown in Figure 3, the probability distribution of WGMD based on ECM algorithm and actual wind power is more consistent and the residual is smaller. The residuals at 0-20 MW based on the EM algorithm are larger, and the wind farm power is mostly concentrated here. Therefore, the EM algorithm has a much poorer overall ECM fitting effect. The evaluation index values of the two algorithms are shown in Table 3.

Table 3. Accuracy evaluation of wind power model

Fitting model	I_{RMSE}	I_{SSE}	$R-square$
EM algorithm	4.2×10^{-4}	5.5×10^{-5}	0.95
ECM algorithm	3.6×10^{-4}	1.3×10^{-5}	0.99

The EM algorithm is used to estimate the unknown parameters, which can easily be affected by the initial conditions, so that the calculation results can not converge to the global optimal value. Instead, the ECM algorithm uses conditional maximization steps to replace the maximum step of EM, so that every step of the maximization is under the corresponding parameter constraints, so that the Q function defined by the EM algorithm reaches the maximum, the optimization parameters of the model can be accurately evaluated.

6. Conclusion

In this paper, the probability density function of the WGMD fitted wind power output fluctuation characteristics is established. The parameter estimation of WGMD is carried out by ECM algorithm, and a wind farm is used as an example to verify. The concrete results are as follows:

(1) With the large-scale development of wind power, the volatility of wind farm output has become a key link in wind power grid operation. This paper analyzes the measured data of a wind farm in a foreign country, establishes WGMD and compares traditional modeling methods. It is verified that WGMD is very suitable for describing wind farms with randomness and volatility, and significantly improves the simulation accuracy of wind farm output characteristics.

(2) For the parameter estimation of the weighted Gaussian mixture distribution model, the commonly used EM algorithm is vulnerable to the initial value and falls into the local optimum. This paper used the ECM algorithm to replace the one maximization of the EM algorithm by multiple condition maximization. The disadvantages of the EM algorithm are avoided, so that the final parameter estimation converges to the global optimum and the model accuracy is guaranteed.

Acknowledgements

The authors also want to give their thanks to the Natural Science Foundation of China. (No1. 51867015, No2.51767017)

References:

[1] Y S Xue, X Lei, F Xue, et al. A review on impacts of wind power uncertainties on power systems[J]. PCSEE, **34(29)**, 5029-5040(2014).

[2] Y Qiao, Z X Lu. Wind farms active power control considering constraints of power grids[J]. AEPS, **33(22)**, 88-93(2009).

[3] H F Liang, D W Cao, B Liu, et al. Modeling and error analysis of wind farm probability distribution model[J]. Journal of North China Electric Power University, **44(3)**, 8-14(2017).

[4] S Y Xie, X L Wang, Y Qu, et al. Influence of different wind speed simulation methods on power system reliability assessment under the framework of segmentation multi-objective risk analysis[J]. PCSEE, **33(31)**, 81-90(2013) .

[5] R. Billinton. Time-series models for reliability evaluation of power systems including wind energy[J]. MICROELECTRON RELIAB, **36(9)**, 1253-1261(1996).

[6] M Ding, Y C Wu, L J Zhang. Study on calculation method of wind speed probability distribution parameters in wind farms[J]. PCSEE, **25(10)**, 107-110(2005).

[7] H Bludszuweit, J A Dominguez-Navarro, A Llombart. Statistical analysis of wind power forecast error[J]. IEEE T POWER SYST, **23(3)**, 983-991 (2008).

[8] A Fabbri, T Román, J Abbad, et al. Assessment of the cost associated with wind generation prediction errors in a liberalized electricity market[J]. IEEE T POWER SYST, **20(3)**, 1440-1446(2005).

[9] H Yan, L C Sun, X Q Chang. The probabilistic modeling of the fluctuation characteristics of wind power output in Xinjiang [J]. POWER SYSTEM TECHNOLOGY, **38(6)** , 1616-1620(2014).

[10] A P Dempster, N M Laird, D B Rubin. Maximum likelihood from incomplete data via the EM algorithm[J]. J R STAT SOC, **39(1)**, 1-38(1977).

[11] H X Gao. *Applied Multivariate Statistical Analysis*[M]. Beijing: Peking University Press(2005).

[12] Y Zhao, J Wang Y J Xiong, et al. Gaussian mixture model of random variables in power system reliability assessment[J]. AEPS, **40(1)**, 66-71,80 (2016).

[13] X L Meng, D B Rudin. Maximum likelihood estimation via the ECM algorithm: A general framework[J]. BIOMETRIKA, **80(2)**, 267-278 (1993).

[14] X L Meng, D B Rudin. Maximum likelihood estimation via the ECM algorithm: Computing the asymptotic variance[J]. STAT SINICA, 55-75(1995).

[15] D M Titterington, A F M Smith. *Statistical analysis of finite mixture distribution*[M]. New York: John Wiley & Sons, 179-191(1985).

Study on the influence of insulator on the coupling effect of transmission tower-line system

Qingshui Gao[1,2], Shi liu[1,2], Yi Yang[1,2] and Chu Zhang[1,2]

[1]Electric Power Research Institute of Guangdong Power Grid Co. Ltd, 510080 Guangzhou, P.R. China

[2]Guangdong Diankeyuan Energy Technology Co., Ltd, 510080, Guangzhou, P.R. China

[a]Corresponding author: 254410806@qq.com

Abstract. There is a coupling between transmission tower and lines. Traditionally, it is assumed that an insulator is in a tight connection with tower. In fact, both tight and loose connections exist. This paper studies the influence of connection state on system dynamics. The tower-line system rigged with the adjustable connection state of insulator was built. We used modal test method to study the coupling effect. The influence of the line is mainly in the longitudinal and torsional mode. It has less influence on the lateral mode. The damping increases for tower-line system and is the largest for loose connection state. The loose connected system's first three modal frequencies are almost the same as that of the tower. However, the tight connected system's longitudinal and torsional frequency decreases. The difference caused by connection state of insulator should be considered. Modelling method under different connection state should be different. The traditional finite element method can be used for tight connection. Under loose connection state, the insulator-line system can be regarded as a tuned mass damper attached to the tower.

1. Introduction

The span of large transmission tower is longer. The mass of the line is not negligible compared with the weight of transmission tower. The internal tension of line is also large, which has a great influence on the dynamic characteristics of transmission tower-line system. In [1], a strong coupling effect between the tower and the line for long span transmission tower was found. In [2], results show that the longitudinal natural frequency of transmission tower decreases with the increase of mass ratio of line / tower. In [3], the simplified vibration model considering the line in the form of an additional mass was adopted and the influence of line on tower-line coupling effect was researched. In [4-11], the multiple particle model was used to consider the influence of line on vibration mode and frequencies of high voltage large span transmission towers. In [12-13], the influence of insulators on the dynamic characteristics of transmission towers under different wind directions and velocities was discussed. The finite element method was used to set up the analysis model. All of the above models assumed that the insulator was tightly connected to the transmission tower. In fact, there are tight connection and loose connection between insulators and towers. The influence of the connection state between the insulator and the transmission tower was not considered.

A test rig was built to study the influence of insulator connection state on the dynamic characteristics of transmission tower-line system. The connection state of the insulator with the tower can be changed. The coupling effect of transmission tower-line system was studied by theory and

Content from this work may be used under the terms of the Creative Commons Attribution 3.0 licence. Any further distribution of this work must maintain attribution to the author(s) and the title of the work, journal citation and DOI.
Published under licence by IOP Publishing Ltd

experiment. The results can provide guidance for vibration control design of transmission tower-line system.

2. Test method for tower-line system

Figure. 1 shows the transmission tower-line system test rig. The tower is a cup type. Its height is 1.80 m, the head width is 0.80 m, and the root opening width is 0.40 m. The tower is composed of angle steel and flat steel. The cross section of the tower main body is square. The total weight of the transmission tower is 18.2 kg. The insulator is 0.13 m long. The connection state between insulators and towers can be changed.

Figure 1. Transmission tower-line system test rig

The transmission line adopts ZR-BVR type multi core copper wire cable with single layer plastic insulation. The cross-sectional area is 16 mm², and the unit length is 0.175 kg/m. It is 1.55 m away from the ground, the length of the single span is 5 m, and the sag is 0.50 m. One end of the line is hung on the insulator, and the other end is fixed on the wall.

We used the modal test method to research the dynamic characteristics of tower-line coupling system in case of two lines suspended in the tower. By applying impact force on the tower with a hammer, the frequency response function was obtained by the method of single point excitation and multi-point vibration measurement. The modal parameters of tower and tower-line system, such as frequency, damping and mode shape, were identified by the modal analysis software ME'scope.

3. Test results analysis

Figure 2 shows the frequency response function of tower without lines. Three peaks in the curve correspond to the longitudinal, the torsional and the transversal modes, respectively.

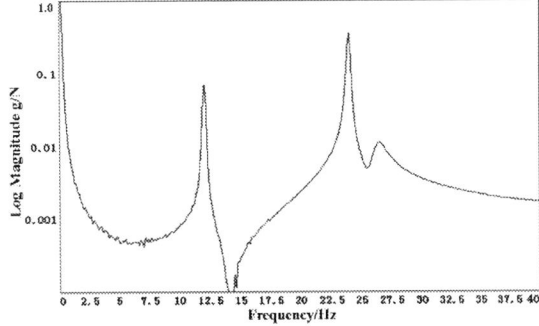

Figure 2. Frequency response function of tower

Table 1 shows the first three modal frequencies and damping of the tower and the tower-line coupling system. We found that

(1) The influence of the suspended lines is mainly in the longitudinal mode and the torsional mode. It has less influence on the lateral mode.

(2) The modal damping increases for the tower-line system, especially in the longitudinal direction. The longitudinal modal damping under loose connection state is larger than that of tight connection state.

(3) The system first three modal frequencies are almost the same as that of the tower under the loose connection state. However, under the tight connection state, it leads to the decrease of the longitudinal and the torsional frequency.

Table 1. Modal parameters of tower and the tower-line coupling system

Parameter	State	Direction		
		longitudinal	torsional	lateral
Frequency/ Hz	Tower	12.25	25.25	26.9
	Tight	10.63	13.63	26.9
	Loose	12.38	25.25	26.9
Damping	Tower	0.302	0.174	2.02
	Tight	1.91	2.77	2.44
	Loose	3.57	2.57	2.19

The tower-line coupling effect is different in different connection state between the insulator and the tower. Considering insulators only in tight connection state may lead to large modelling error.

4. Modelling of Insulator at different state

Modelling method should be different if the connection state between the insulator and the tower is different.

4.1 Tight connection state

In the finite element modelling of tower-line system, the beam-rod hybrid element and the cable element are used to model the tower and line, respectively.

In the case of tight connection, system stiffness matrix and mass matrix are established by combining the stiffness matrix $\mathbf{K}_{t-l}$ and the mass matrix $\mathbf{M}_{t-l}$ of the tower and line in the global coordinate system

$$\mathbf{K}_{t-l} = \sum_{i=1}^{n_t} \mathbf{K}_t^{(i)} + \sum_{j=1}^{n_l} \mathbf{K}_l^{(j)}$$

$$\mathbf{M}_{t-l} = \sum_{i=1}^{n_t} \mathbf{M}_t^{(i)} + \sum_{j=1}^{n_l} \mathbf{M}_l^{(j)}$$

(1)

where n_t, n_l are the element number of tower and line. The insulator is simulated by a bar element with prestress. Its stiffness matrix is

$$\mathbf{K} = \mathbf{K}_e + \mathbf{K}_g + \mathbf{K}_\sigma \qquad (2)$$

where $\mathbf{K}_e$ is the elastic stiffness matrix, $\mathbf{K}_g$ is the displacement nonlinear stiffness matrix, $\mathbf{K}_\sigma$ is the stress nonlinear stiffness matrix. The elastic stiffness matrix and the displacement nonlinear stiffness matrix can be derived from the shape and material properties of the rod. The stress nonlinear stiffness matrix can be obtained from the prestress calculation model shown in Figure. 3. Assuming that the tensile force exerted by the line on the insulator is the gravity of the line between the two insulators, the corresponding stress can be calculated.

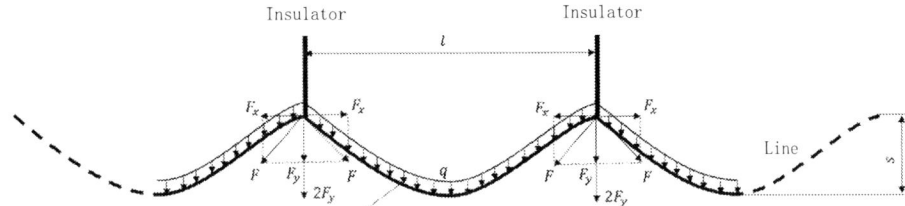

Figure 3. The prestress model of insulator

Table 2 shows the calculated frequencies of the tower and the tower-line coupling system under tight connection. By comparing the results of tables 1 and 2, we can find that

(1) The calculated frequencies are the same as those obtained experimentally.

(2) The influence of tower-line coupling effect on system modal frequencies in three directions is consistent with the experimental results. That is, the influence is mainly in the longitudinal and the torsional modes. The modal frequencies of these two modals are reduced.

Table 2. The Calculated modal frequency of single tower and the tower-line coupling system

System	Longitudinal	Torsional	Lateral
Tower/Hz	12.31	22.05	24.33
Tower-line system/Hz	10.02	12.08	23.42

Suspension line has a comprehensive influence on the additional stiffness and additional mass of the tower. For longitudinal bending and torsional modes, the line increases system mass in the corresponding direction, while the additional stiffness caused by the tension of the line has a relatively small effect. It results the decrease of frequencies in the corresponding modes. In the transverse direction, the effect of additional stiffness compensates for the effect of additional mass, so that the line has little effect on the frequency.

4.2 Loose connection state

Under the loose connection state, insulators can be regarded as a pendulum suspended on transmission tower. The insulator-line system works like a kind of tuned mass damper (TMD) as shown in Figure 4.

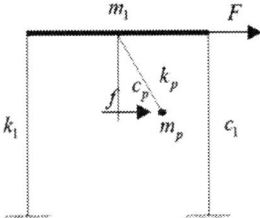

Figure 4. Model of a structure with additional pendulum

Under the linear small angle assumption, the control force F_{spd} generated by pendulum motion can be simplified to

$$F_{spd} = \frac{m_p g}{l}(x_p - x_1) \qquad (3)$$

where m_p is the weight of line, g the acceleration of gravity, x_p, x_1 the displacement of pendulum and structure relative to ground, l the length of the pendulum.

The equation of motion of pendulum is

$$m_p \ddot{x}_p + \frac{m_p g}{l}(x_p - x_1) = 0 \qquad (4)$$

When the damping is not considered, the motion equation of the structure is

$$m_1\ddot{x}_1 + k_1 x_1 = F_{spd} + f \qquad (5)$$

where f is external force, m_1, k_1 are the mass and stiffness of the structure, respectively.

The equation of the structure with additional pendulum is

$$\begin{bmatrix} m_1 & 0 \\ 0 & m_p \end{bmatrix} \begin{Bmatrix} \ddot{x}_1 \\ \ddot{x}_p \end{Bmatrix} + \begin{bmatrix} k_1 + k_2 & -k_2 \\ -k_2 & k_2 \end{bmatrix} \begin{Bmatrix} x_1 \\ x_p \end{Bmatrix} = \begin{Bmatrix} f \\ 0 \end{Bmatrix} \qquad (6)$$

where $k_2 = m_p g / l$.

The two natural frequencies corresponding to the above systems are

$$\omega_n^2 = \frac{\omega_1^2}{2}[1 + a \pm \sqrt{(1+a)^2 - 4(\frac{\omega_2}{\omega_1})^2}] \qquad (7)$$

where ω_1, ω_2 are the natural frequencies of the structure and the pendulum, respectively.

$$\omega_1^2 = \frac{k_1}{m_1}, \omega_2^2 = \frac{k_2}{m_p}, \quad a = (1 + \frac{m_p}{m_1})(\frac{\omega_2}{\omega_1})^2 \qquad (8)$$

Taking the longitudinal vibration of the rig as the example, the corresponding frequencies are

$$\omega_1 = 76.93 \, \text{rad/s}, \ \omega_2 = 6.91 \, \text{rad/s} \qquad (9)$$

From (8), we can find that

$$\omega_2 \ll \omega_1, a \approx 0 \qquad (10)$$

Thus, the tower-line system natural frequency

$$\omega_n \approx \omega_1 \qquad (11)$$

In case of loose connection of insulator, the natural frequencies of tower is less influenced by line.

When the tower vibrates under external excitation, the inertia mass of the TMD device (insulator-line system) absorbs the vibration energy of the controlled mode of the tower. The damping increases. Therefore, the system damping is larger if the insulator is loosely connected with tower.

5. Conclusions

The influence of connection state of insulator with tower on the dynamic characteristics of transmission tower-line system was studied by means of modal test and analysis.

The conclusions are as follows

(1) The connection state between insulator and tower has a certain effect on the dynamic characteristics of the tower-insulator-line system. Considering insulators only in tight connection state with tower is not appropriate. It may lead to large modelling error.

(2) Modelling method under tight connection state and loose connection state should be different. Under the tight connection state, the tower, insulator and line can be modelled by finite element method respectively. The system stiffness and mass matrices can be obtained by combining the stiffness and mass matrices of the tower, line and insulator in the global coordinate system. However, under loose connection state, the system composed of insulator and line can be regarded as a tuned mass damper attached to the tower.

(3) The influence of the lines is mainly in the longitudinal mode and the torsional mode. It has less influence on the lateral mode.

(4) The modal damping increases for the tower-line system, especially in the longitudinal direction. The modal damping under loose connection state is larger than that of tight connection state.

(5) Under the loose connection state, system first three modal frequencies are almost the same as that of the tower. However, under the tight connection state, it leads to the decrease of the longitudinal and the torsional frequency.

Acknowledgment

The paper was supported by China Southern Power Grid Company Science and Technology Project(GDKJQQ20153008)

References

[1] Zhai Changhai, Wu Gang, Li Shuang, et al. Tower-Lines In-Plane Coupling and TMD Seismic Control of Large Crossing Transmission Tower Lines System. Journal of Vibration Engineering,25(4):431-438(2012).

[2] Liang Shuguo, Zhu Jihua, Wang Lizheng. Analysis of Dynamic Characters of Electrical Transmission Tower-Line System with A Big Span. Earthquake Engineering and Engineering Vibration,23(6):64-69 (2003).

[3] Li Hongnan, Shi Wenlong, Jia Lianguang, Limitation of Effects of Lines on In-plane Vibration of Transmission Towers and Simplified Seismic Calculation Method. Journal of Vibration and Schock,23(2):1-8(2004).

[4] Shu Aiqiang, Wu Haiyang, Zou Lianghao, et al. Analysis of 3-D Dynamic Characteristics of High-Voltage Long-Span Transmission Tower-Line System. China Civil Engineering Journal,43 (Suppl.): 224-229(2010).

[5] Ozono S, Maeda J, Makino M. Characteristics of In Plane Free Vibration of Transmission Line Systems. Engineering Structures,10(3):272-280 (1998).

[6] Liang Feng, Li Li, Yin Peng. Investigation on Numerical Model of Electrical Transmission Tower-Line System with A Big Span. Journal of Vibration and Shock,26(2):61-67(2007).

[7] Li Hongnan, Wang Suyan, Ban Wenzhuo. Seismic calculation of large span transmission tower system. Special structure,15(3):47-50(1998).

[8] Cao Meigen, Zhou Fulin, Xu Zhonggen, et al. Simplified Model Research of Combo Steel Tube Tower for Large Crossing Transmission Lines. Steel Construction,20(6):27-32(2005).

[9] Wang Qianxin, Lu Ming, et al.Reasonable Computation Schema for Earthquake-Resistant Analysis of the System Consisting of Transmission Line and Its Supporting Towers.Earthquake Engineering and Engineering Vibration,9(3):81-90(1989).

[10] Shi Wenlong, Jia Lianguang, Li Hongnan.The Effects of Power Lines on Transmission Tower Under Vertical Vibration. Journal of Shenyang Arch. And Civ. Eng. Univ. (Natural Science),17(4): 262-264(2001).

[11] H. Yasui, H. Marukawa, Y. Momomura, et al. Analytical study on wind-induced vibration of power transmission towers. Journal of Wind Engineering and Industrial Aerodynamics,83:431-441(1999).

[12] Y. Momomura, H. Marukawa, T. Okamura, et al. Full-scale measurements of wind-induced vibration of a transmission line system in a mountainous area. Journal of Wind Engineering and Industrial Aerodynamics,72:241-252(1997).

[13] Yang Jingbo, Li Zheng, Jiang Jun.Dynamic Characteristics of Wind-Induced Coupled Vibration Between Conductor and Tower of Transmission Line.Journal of Vibration, Measurement & Diagnosis, 28(2): 47-152(2008).

[14] Zuo Hesheng, Peng Yuying, *Modal Analysis of Vibration Experiment.*Beijing: China Railway Press, 148-153(2003).

The Impact Research of Delay Time in Steam Turbine DEH on Power Grid

Ge JIN, Sun CAI, Shaoxiang DENG

Electric Power Research Institute of Guangdong Power Grid Co., Ltd.,Guangzhou 510080, China

Author: Ge JIN (1982 -), Male, Hubei, Master, Senior engineer, Mainly engaged in research on coordination and simulation of machine network.Tel:13925176745, Email:happyjinge@163.com.

Email: happyjinge@163.com, pingguonothing@163.com, 13600005471@139.com,

TEL:13925176745

Abstract: The effects of different link's delay time in steam turbine DEH (Digital Electric Hydraulic Control System) on electric power unit are not exactly the same. Through the analysis of the main controller of the steam turbine governor, electro-hydraulic mechanism,steam turbine and the single-machine infinity model, we establish the single-machine infinite power system with simulink. The delay time of the electric power feedback, the speed feedback, the oil actuator and the displacement sensor link in the model is changed to explore the influence of different delay times on the power output of the power system. The results show that the electric power feedback, the speed feedback, the Oil actuator and the delay time of the displacement sensor will increase the overshoot of the electric power response curve and increase the time to reach the stability, but their influence degree.is different, the level of influence from low to high is: speed feedback delay, electric power feedback delay, displacement sensor delay ,Oil actuator delay.

1. Introduction

In recent years, with the increasing demand of electric power in China, the scale of power system is becoming larger and larger, and the power transmitted on the main transmission lines is also increasing. However, the operating conditions of the whole power system are getting worse, and low-frequency oscillation occurs frequently, which has become one of the important factors that affect the safe operation of the power grid.

The inadequate ability to deal with sudden load changes is the main factor affecting the stability of the power grid, and the primary frequency can effectively solve this problem, because it can quickly suppress the frequency and power fluctuations caused by load changes under disturbance. In this way, the control level of power quality and grid frequency can be improved [1-5].

With the development of power system, more and more attention has been paid to the stability of power grid. Many factors can affect the stability of power grid, some of them not only cause power grid oscillation, but also may lead to major blackouts. In order to improve it's stability, different factors need to be further studied.

Zhang Jiawei [6] et al. Studied the effect of primary frequency modulation on isolated network control mode in high frequency condition. Dynamic simulation model of thermal power units is established by RTDS. The simulation results show that, the lower frequency limit of primary frequency modulation can effectively reduce the maximum frequency of the isolated network. Sheng Kai [7] proposed a power control strategy based on internal model control under the existing control system conditions and the operation characteristics and characteristics of the actual unit. The simulation results show that this strategy can improve the control quality of the existing system.

Li Yanghai [8] studied the influence of turbine control parameters on power grid low-frequency oscillation from the angle of mechanical damping torque. A single machine infinite bus system model and a multi machine system model are established. This article proposed an effective parameter tuning method, which take mechanical damping coefficient as one of the evaluation criteria of power grid stability.

In addition, in various indicators of the performance of turbine control system, speed inequality, delay rate and load variation amplitude also Have important influence on performance of primary frequency modulation and the transient stability of power grid.

In summary, researchers have done lot of research on the influence of power stability, whether from the grid side or the turbine side. However, there is still a lack of research results on the effect of each link's delay time of DEH system on power stability. This article established a single machine infinite integral power system model, and discussed the influence of delay time of different links on the stability of power system By simulating the delay time of each link in power frequency control mode of steam turbine DEH system.

2. Typical mathematical model of steam turbine DEH system

Digital electro-hydraulic control system (DEH) is usually used to control the power and speed (frequency) of the turbine when it's connected to the grid. The speed difference is obtained by subtracting the actual speed value and the given speed value. When the speed difference exceeds the set dead zone, the power deviation is calculated according to the unequal rate of speed regulation. The power deviation is added to the set value of the power, and the actual value of the power is obtained. After the given value is compared with the actual power, a power difference is output to the PID controller for subsequent calculation. When DEH is in remote control mode, the controller exists in CCS system. Because the control principle is similar, its mathematical model can also refer to DEH power control mode.

DEH operating instructions under different control modes are shown in Figure 1:

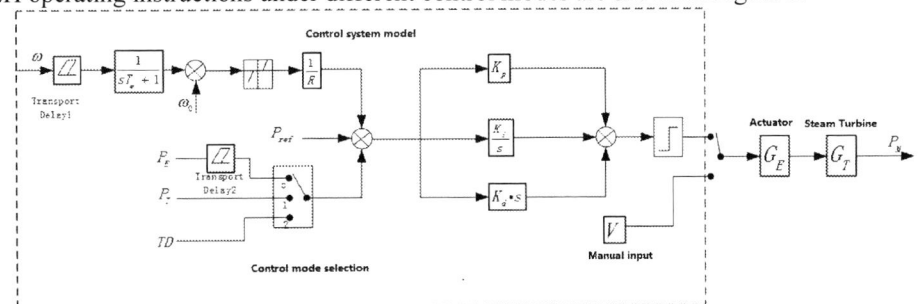

Ω - Actual speed, ω_0 - Speed setting value, s - Laplasse operator, T_ω - Time constant of speed transmitter, R - Unequal rate of speed regulation, K_p - PID ratio coefficient, K_i - PID integral coefficient, K_d - PID differential coefficient, *Transport Delay1* - Delay time of speed feedback, *Transport Delay2* - Delay time of electric power feedback.

Figure.1 Schematic diagram of different control modes of DEH.

In the power feedback mode, the active power PE is compared with the given power, and the difference value of power is sent to the power regulator for the next calculation. The command signal of the valve is output to the electro-hydraulic servo. The opening of the valve is controlled by the actuator to change the steam intake and output the required power.The control valve command signal is obtained, and this command is output to the electro-hydraulic actuator, the opening of the control valve is changed, thus the steam intake of the turbine is changed, and the required power is sent out.

Under the power feedback mode, the following transfer functions can be derived:

$$-P_M = \left(\omega \cdot \frac{1}{sT_\omega + 1} \cdot \frac{1}{R} + P_E \right) G_{pid} G_E G_T = G_{GOV1} \cdot \omega + G_{GOV2} \cdot P_E$$

(1)

Where P_E represents active power, P_M represents mechanical power, G_{pid} represents transfer function of PID in power control, G_E represents transfer function of electro-hydraulic servo mechanism, G_T represents transfer function of steam turbine, G_{GOV1} represents transfer function of Speed, G_{GOV2} represents transfer function of power.

The command signal output from DEH system is converted into the opening of the regulating door by the actuator, which determines the output power of the unit. The actuator is composed by electro-hydraulic servo and oil actuator[10-14], Its model is shown in Figure 2.

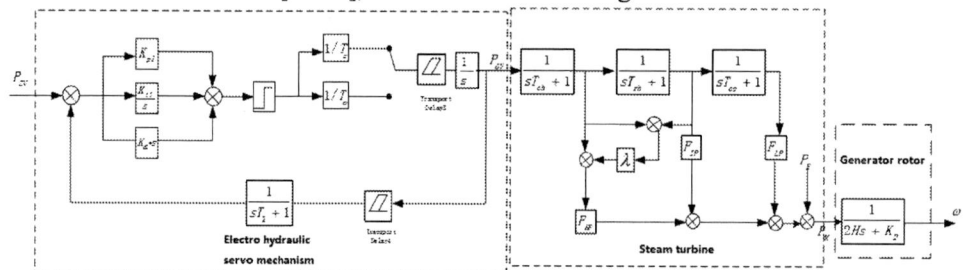

K_{dl} - Differential coefficient of PID in electro-hydraulic servo system, T_C - Close time constant of hydraulic actuator, T_O - Open time constant of hydraulic actuator, T_{ch} - Steam chamber volume time constant, T_{rh} - Reheater volume time constant, T_{co} - Volumetric time constant for intake pipe of low pressure cylinder, F_{HP} - Share of the high pressure cylinder power in total output power, F_{IP} - Share of the medium pressure cylinder power in total output power, F_{LP} - Share of the low pressure cylinder power in total output power, λ - Natural overshoot coefficient of high pressure cylinder power, H - Rotor inertia time constant, K_D - Damping torque coefficient, *Transport Delay3*-Action delay time of hydraulic actuator, *Transport Delay4*-Delay time of displacement sensor.

Figure.2 Mathematical model of the actuator and the steam turbine.

Ignoring the nonlinear links, the following transfer functions can be derived.

$$G_E = \frac{P_{GV}}{P_{CV}} = \frac{\left(sK_{p1} + K_{i1} \right)}{s^2 T_O (sT_2 + 1) + (sT_{p1} + K_{i1})}$$

(2)

$$G_T = \frac{P_M}{P_{GV}} = \frac{((1+\lambda)(1+T_{rh}s)(1+T_{co}s)F_{hp} + (F_{ip} - \lambda F_{hp})(1+T_{co}s) + F_{lp})}{((1+T_{ch}s)(1+T_{rh}s)(1+T_{co}s))}$$

(3)

$$(P_M - P_E)\frac{1}{2H + K_D} = \omega$$

(4)

The power grid model adopts the Single machine infinite bus network model provided in reference [15]. As shown in Figure 3.

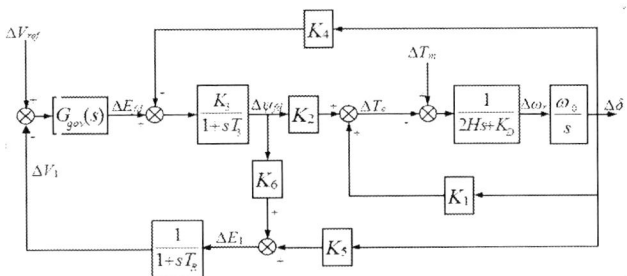

K1,K2,K3,K4,K5,K6 - Ratio coefficient, ΔTm - PID integral coefficient of Electro-hydraulic servo, ΔTe - Variation of electromagnetic power, ΔE1 - Variation of voltage sensor input, Δψfd - Flux variation of excitation loop, ΔEfd - Variation of exciter output voltage, TR - Time constant of voltage sensor, T3 - Time constant of excitation circuit.

Figure.3 Mathematical model of a single machine infinity power grid.

In the simulation, the typical parameter in Table.1 and Table.2 is used [8]:

Table.1 Typical parameters of steam turbine regulating system .

Symbol	λ	T_{ch}	T_{rh}	T_{co}	F_{HP}	F_{IP}	F_{LP}	K_p	K_i	K_d	R	T_ω	K_{p1}	K_{i1}	K_{d1}	T_0	T_2
Typical parameters	0.8	12	0.1	1	0.32	0.68	0.44	1	0.05	0	20	0.02	9	0	0	1.33	0.02

Table.2 Typical parameters of a single machine infinity model.

Symbol	K_1	K_2	K_3	K_4	K_5	K_6	K_A	H	K_D	T_R	T_3
Typical parameters	1.591	1.5	0.333	1.8	0.12	0.3	200	3	0	0.02	1.91

3. Simulation Research

The turbine model, single-machine infinite-bus model, electro-hydraulic converter model and governor model are connected to establish a single-machine infinite-bus model. According to the current technical standards, the dead zone of primary frequency modulation is set at 0.0333 Hz, i.e. the standard unitary value (+0.000666). The step perturbation of the speed signal with a unit value of 0.001666 is given in 4-6 seconds, The control mode is power-frequency feedback mode. The model parameters are set according to table 4 and table 5. The speed feedback, electric power feedback, oil drive action and delay time of displacement sensor are respectively changed. The simulation time is set as 60 seconds to study the influence of different delay time on power stability under the condition of stable load within 60 seconds stipulated by industry standard. For convenience of observation and comparison, the abscissa and ordinate coordinates in all the following figures are the simulation time, and the ordinate coordinates are the scalar unitary values of the increment of the electric power response.

When the power feedback delay time is 0.01 seconds, 0.05 seconds and 0.1 seconds respectively, the influence on power stability is shown in Figure 4.

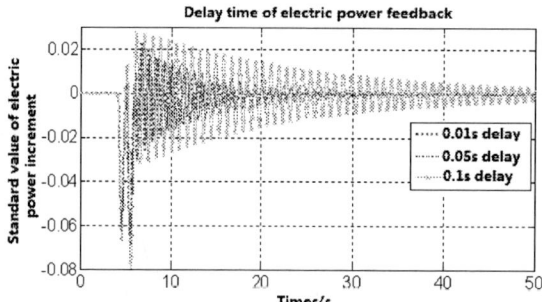

Figure.4 The influence of electric power feedback delay on the power stability.

As Fig.4 shown, when there is a delay in the power feedback, with the increase of the delay time, the disturbance caused by the curve delay signal increases, the electric power oscillation intensifies, the curve attenuates slowly, and the time to reach stability prolongs.

The influence on power stability is shown in Fig. 5 when the speed feedback delay time is 0.01 seconds, 0.1 seconds, 0.3 seconds and 0.5 seconds respectively.

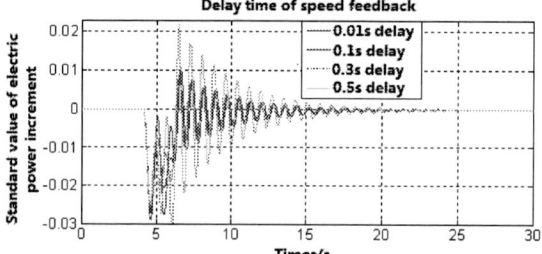

Figure.5 Effect of speed feedback delay on power stability.

As Fig.5 shown, the curves coincide basically when the speed feedback delay time less then 0.3S, and the influence of the speed feedback delay is small. As the delay time increase, the curve oscillation intensifies and the stable time increases.

Fig. 6 shows the effect on power stability when the action delay time of the oil motor is 0.01 seconds, 0.03 seconds and 0.05 seconds respectively.

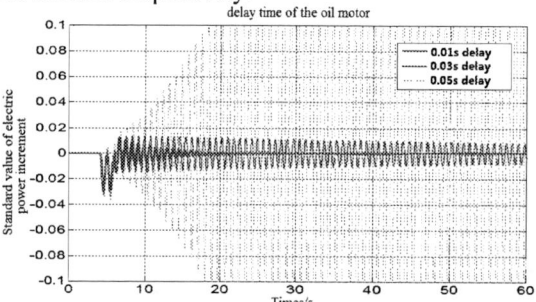

Figure.6 Influence of oil drive action on power stability.

As Fig.6 shown, with the increase of the delay time, the fluctuation amplitude of the curve increases, the electric power oscillation intensifies, and the time to reach stability also increases with the increase of delay time. When the delay time of the oil drive action is 0.03 seconds, the electric power curve oscillates at the same amplitude and can not reach a stable state.

Fig.7 shows the effect of the delay time of the displacement sensor on the power stability at 0.01 seconds, 0.03 seconds and 0.05 seconds, respectively.

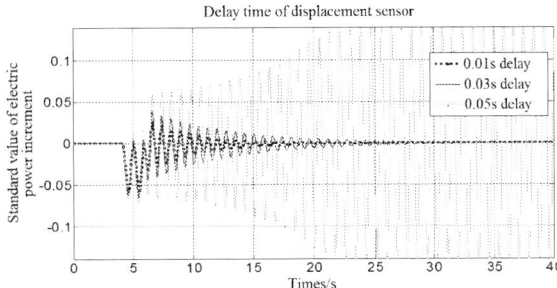

Figure.7 Influence of delay of displacement sensor on power stability.

As shown in Fig.7, the overshoot of the curve increases with the increase of the delay time, and the time to reach stability increases with the increase of the delay time. When the delay time of the displacement sensor is greater than or equal to 0.05 seconds, the power oscillation is obviously forced, and the amplitude increases with the delay time, resulting in serious instability of the system.

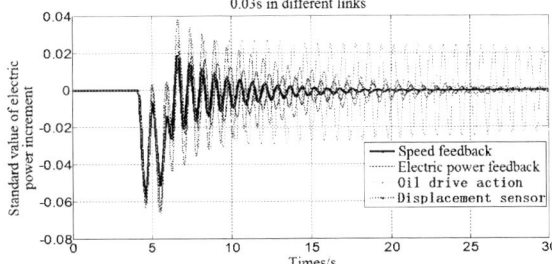

Figure.8 Influence of delay of displacement sensor on power stability.

Table.3 Comparison of grid stability effects of various links under 0.03s delay

0.03s in different links	Overshoot	Frequency of fluctuation	Stabilization time
Speed	0.019	11	17s
Electric power	0.021	19	24s
Oil drive action	0.028	Infinity	>60s
displacement sensor	0.039	23	26s

As Fig. 8 and Table.3 shown, the power curves of the same delay in different links are different. When the delay is 0.03s, the influence on power stability arrives from small to small in order: speed feedback delay, power feedback delay, displacement sensor delay and oil motor action delay.

4. Conclusions

The results show that, the effects of electric power feedback, speed feedback, oil motor and displacement sensor delay on the power stability are similar. The increase of delay time will aggravate the oscillation of electric power response curve, increase the time to reach stability and deteriorate the stability, but the influence degree of different link delays is different. The sensitivities from small to large are: speed feedback delay, electric power feedback delay, displacement sensor delay and oil motor delay.

Through study the influence of each link's delay time on the power stability of the whole system, it is found that the robustness of different links to the delay is obviously different, serious delay may even lead to low frequency oscillation. This research result can not only lay a foundation for the follow-up theoretical research, but also can focus on restraining the delay which has a great impact on the grid when there are delays in different links. The next step is to carry out the research on the method of restraining the delay in the power grid.

Acknowledgment

Fund Project: China Southern Power Grid Corporation Science and Technology (GDKJXM0000007)

References

[1] SONG, X.L., LIU, Z.X., LI, Y.Z., et al. Analysis on Speed Governing System Model for Fossil-Fuel Generating Plant and Its Application in Power System Stability Simulation[J]. Power System Technology, 2008, 32(23): 44-49.

[2] CAI, S., DENG, X.W., FENG, Y.X., et al. Influence on Low Frequency Oscillation by Parameters of Speed Governing System of Prime Mover[J]. Guangdong Electric Power, 2016, 29(3): 0033-05.

[3] JIN, G., DENG, S.X.. Impact on Frequency Modulation Stability of Sequence Valve Management Logic[J]. Guangdong Electric Power, 2013, 26(6):0060-03.

[4] GE, R., DONG, Y., LV, Y.C.. Analysis of Large-Scale Blackout in UCTE Power Grid and Lessons to be Drawn to Power Grid Operation in China[J]. Power System Technology, 2007, 31(2): 1-6.

[5] GAO, X., ZHUANG, K.Q., SUN, Y.. Lessons and Enlightenment from Blackout Occurred in UCTE Grid on November 4,2006[J]. Power System Technology, 2007, 31(1): 25-31.

[6] ZHANG, J.W., TANG, S.L.. Analytical Study on Primary Frequency Limiter in Control of High Frequency Isolated Network[J]. Computer Simulation, 2014, 31(1): 136-140.

[7] SHENG, K.. Power Control System for Thermal Power Generator Units Based on the Inner Model Control[J]. Journal of Engineering for Thermal Energy and Power, 2013, 28(6): 616-621.

[8] LI, H.Y.. Technology research on suppressing low-frequency oscillation of power system on steam turbine side[D], Huazhong University of Science & Technology, 2016.

[9] DL/T824-2002 Performance acceotance guide of steam turbine electro-hydraulic control system.

[10] Q/CSG114003-2012 Guide for Modeling and Testing of Generator Prime Mover and Governor.

[11] WANG, Y., WANG, Y.S., LI Zhonghua. Model and Nonlinear Analysis of 300MW Steam Turbine Speed Control System[J]. Turbine Technology, 2007, 49(1): 0017-04.

[12] SHENG, K., ZHU, X.X.. Model of reheat condensing turbine and its governing system:simulation and verification[J]. Thermal Power Generation, 2014, 43(3): 0087-05.

[13] XU, Y.H., MA, C., CAI Sun. Suppression Measures for Low Frequency Oscillation of Governing System Side Based on Filter[J]. Guangdong Electric Power, 2014, 27(12): 0047-04.

[14] ZHANG, Y.C., XU, Y.H.. Steam Turbine Control System Model for Analyzing Low Frequency Oscillation of Power System[J]. Guangdong Electric Power, 2013, 26(12): 0009-04.

[15] Wang, H.F.. Phillips-Heffron model of power systems installed with STATCOM and applications[J]. IET Proceedings - Generation Transmission and Distribution, 1999, 146(5): 521-527.

ISPECE

IOP Publishing

Investigation on the relationship between winding wire size and total loss of BLDC

Jinshun Hao[1], Shuangfu Suo[2,a], Yiyong Yang[1], Yang Wang[1] and Wenjie Wang[2]

[1]School of Engineering and Technology, China University of Geosciences, Beijing 100083, China

[2]State Key Laboratory of Tribology, Tsinghua University, Beijing, 100084, China

[a] Corresponding author: sfsuo@mail.tsinghua.edu.cn

Abstract. The wire size has a significant influence on the total loss of BLDC. In order to determine the relationship between the wire size and the total loss, and to find the wire size which minimizes the total loss of BLDC, this paper proposes a method based on wire diameter to calculate the total loss, and the finite element method is used to verify it. Using this method to analyse a motor model, it is found that with the increase of wire diameter, the total loss of BLDC shows a trend of decreasing and then increasing, and there is a wire size which minimizes the total loss of BLDC.

1. Introduction

Because of high power density and excellent speed performance, BLDC has been used in many areas[1]. But at this stage, in some special applications, such as aerospace, robotic drive and other conditions, higher power density is required. However, increasing power density will inevitably lead to the rise of motor temperature, and the temperature rise of the motor is one of the main factors which restrict the increase of motor power density [2].

Researchers have done extensive researches on the temperature rise and found that measures for solving the problem are mainly divided into two categories: 1. Improving the cooling capacity of the motor; 2. Reducing the losses of the motor.

In terms of improving the cooling capacity of the motor: a thermal-transfer structure based on the heat pipe has been proposed, which can increase the heat dissipation area of the motor and then promote motor cooling [3]; some nozzles are used in a permanent motor to cool the stator end windings [4]. And in the slots of motors which use flat wires, some space is available and used as water channel to cool the coils [5].

In terms of reducing motor losses: some attempts have been made to optimize the slot/pole ratio to reduce losses of motors [6-7]; and some new control methods have been proposed to improve the motor efficiency [8-9].

However, the research of motor loss reduction at this stage mainly focuses on the optimization of the slot/pole ratio and control methods, and no attention has been paid to the influence of the winding diameter on BLDC losses. It can be known from basic formulas of the BLDC that while keeping the slot fill factor and the rated power constant, the change of wire size will influence winding resistance and then affect the copper loss; meanwhile the rated speed will change and then influence the core loss.

In this paper, a method based on wire diameter to calculate the total loss of BLDC was proposed. By comparing the total losses of BLDC with different wire sizes, the wire size which minimizes total

Content from this work may be used under the terms of the Creative Commons Attribution 3.0 licence. Any further distribution of this work must maintain attribution to the author(s) and the title of the work, journal citation and DOI.

Published under licence by IOP Publishing Ltd

loss can be found. And the proposed method is verified by the finite element method with a constant rated power BLDC.

2. Mathematical model

2.1 Model solution method

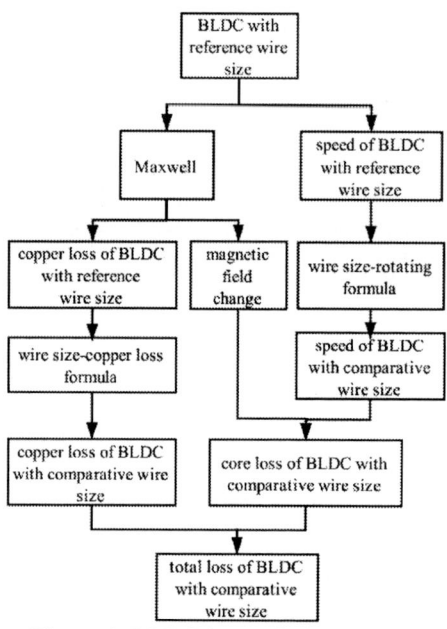

Figure 1. BLDC loss solution process

For a BLDC with reference wire, Maxwell is used to model and analyse it to obtain the data of copper loss and magnetic field change, and then, according to the data of the magnetic field, the core loss can be obtained by taking Bertotti's core loss model, thereby obtaining the total loss of the motor (copper and core). As shown in Figure 1, for the motor with a comparative wire diameter, the copper loss and rotating speed can be obtained by "wire size-copper loss formula" and "wire size-speed formula", and combined with the magnetic field change of the motor with reference wire, the core loss can be calculated; then, the total loss is obtained. Comparing the calculations, the best wire diameter that can minimize the total loss of the BLDC can be obtained.

2.2 Wire size-speed formula

First, it is assumed that the slot full factor T is constant when the BLDC uses different wire sizes.

$$T = \frac{nd^2}{S} \qquad (1)$$

n—conductors per slot
d——wire size
S——slot area

Set d_0 as reference wire size, and d_i as the size of the wire numbered i, then:

$$x = \frac{d_i}{d_0} \qquad (2)$$

Set L_i be the cross-sectional area of the wire numbered i, then:

$$\frac{A_i}{A_0} = x^2 \qquad (3)$$

Since the slot fill factor T is constant, the total volume of armature windings in slots is constant. Therefore:

$$\frac{L_i}{L_0} = \frac{1}{x^2} \qquad (4)$$

L_i——The total length of the effective conductor when the wire is numbered i

However, due to the limitation of the wire size, different specifications of wire can only achieve similar slot fill factor; taking this factor into consideration, let:

$$K = \frac{T_i}{T_0} \qquad (5)$$

Solve Equations (2), (3), (4) and (5) simultaneously:

$$\frac{L_i}{L_0} = \frac{K}{x^2} \qquad (6)$$

At the same time, ignoring the armature reaction, let the rated power P of the BLDC is constant when the wire size changes, there are:

$$P = BI_0 L_0 v_0 = BI_i L_i v_i \qquad (7)$$

B—— Air gap flux density(T)
v—— Rotor line speed(m/s)
I_i—— Armature current with winding numbered I (A)

Solve Equations (6) and (7) Simultaneously :

$$(I_0 v_0) x^2 / K = I_i v_i \qquad (8)$$

At the same time, the armature current calculation formula is :

$$I = \frac{U - BLv}{R} \qquad (9)$$

U——voltage(V)
R——Armature resistance(ohm)

Solve Equations (8) and (9) Simultaneously :

$$v_i = \frac{x^2 U + \sqrt{(x^2 U)^2 - 4BL_0 v_0 K(U - L_0 v_0)}}{2BL_0 K} \qquad (10)$$

From (10), when the rated speed of the motor with the wire size d_0 is known as v_0, and the slot full factor remains basically unchanged, v_i can be obtained, and both motors have the same rated power.

2.3 Wire size-copper loss formula
The formula for the wire resistance R is

$$R = \rho \frac{L}{A} \qquad (11)$$

ρ——Specific resistance

L——Conductor length

S——Conductor cross-sectional area

Solve Equations (3), (6) and (11) Simultaneously :

$$P_{cui} = P_{cu0}(\frac{v_1}{v_i})^2 =$$

$$P_{cu0}(\frac{2BL_0v_0}{x^2U + \sqrt{(x^2U)^2 + 4BL_0v_0K(U - BL_0v_0)}})^2 \qquad (12)$$

From (12), if the rated power and slot full factor keep constantly, the copper loss P_{cui} can be obtained when wire size d_i varies.

2.4 Core loss calculation

Regarding the calculation of motor core loss, researchers at this stage generally adopted the core loss solution model proposed by Bertotti[10] in 1988 :

$$\begin{cases} P_{Fe} = P_h + P_e + P_{ex} \\ P_h = K_h f^2 B_m^h \\ P_e = K_e f^2 \int_0^{2\pi} \left(\frac{dB(\theta)}{d\theta}\right)^2 d\theta \\ P_{ex} = K_{ex} f^{1.5} \int_0^{2\pi} \left|\frac{dB(\theta)}{d\theta}\right|^{1.5} d\theta \end{cases} \qquad (13)$$

Since the above core loss model is based on the permanent magnet synchronous motor, the condition is that the magnetic density changes sinusoidally. However, the magnetic density change of the BLDC is close to the square wave. Therefore, the researchers now mainly use the orthogonal decomposition-Fourier decomposition to process the change data of the magnetic density of the BLDC, and then use the above core loss calculation model to solve the core loss[11-12]: firstly, the BLDC magnetic density is decomposed into Bt and Br in the tangential and normal direction; secondly, Bt and Br are subjected to Fourier decomposition; and then the decomposed results are substituted into the above core loss solving model to calculate the core loss :

$$P_{Fe} = \sum_{i=1}^{n} (K_h f B_{xmi}^h + K_e f^2 B_{xmi}^2 + K_{ex}(f B_{xmi})^{1.5}) +$$

$$\sum_{i=1}^{n} (K_h f B_{ymi}^h + K_e f^2 B_{ymi}^2 + K_{ex}(f B_{ymi})^{1.5}) + \qquad (14)$$

When using the above model to calculate the core loss of the BLDC, a complete stator tooth is first selected and divided into three parts : the yoke, the tooth body and the tooth top. Some points are evenly selected in each part and the core loss value of each point is calculated. Then the average core loss of the points in each area replace the average core loss in each area to calculate the core loss. The locations of three parts and the points are shown in Figure 2.

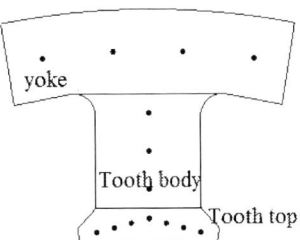

Figure 2. Stator tooth division and point positions

In this paper, for the BLDC with the reference wire size, the magnetic density change of each point which is obtained by Maxwell's analysis at rated speed is substituted into the formula (14) to calculate the motor core loss. Meanwhile, motor speeds calculated by formula (10) and the magnetic density change data of the BLDC with reference wire are used to calculate the core loss of BLDC with comparative windings.

3. Comparison of motor losses

3.1 Motor model

The quantitative analysis between wire size and losses is based on an existing BLDC, and Table 1 shows the main parameters of the motor.

Table 1. main parameters of the motor

Motor parameter	Value
Rated power	4000W
Rated voltage	270V
Stator outer diameter	160mm
Pole/slot	16/18
Slot full factor	70.4%
Stator material	3WW250
comparative wire sizes (mm)	1.938, 1.829, 1.628, 1.537, 1.45
Reference wire size	1.725mm

3.2 Comparison of results

The copper losses of the BLDC are calculated by finite element method and formula (12). As shown in Figure 3: the results obtained by the two methods are well fitted, and the copper loss decreases gradually with the increase of wire diameter.

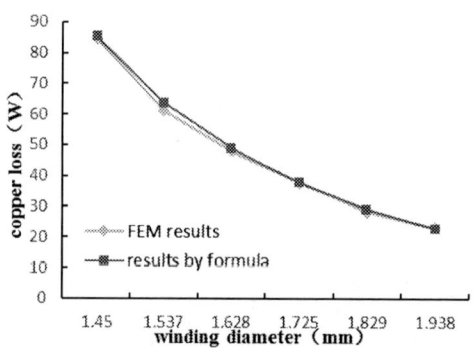

Figure 3. Copper loss calculated by different methods

Two different methods are used to calculate the core losses of the motors: (1) Maxwell software is used to model and analyze the BLDC with the reference wire and comparative wires, so as to obtain the rated speeds as well as the the changes of magnetic density of the selected points on a complete tooth, and then the core losses are calculated based on formula (14); for the motor with reference wire, core loss is calculated based on the method(1); meanwhile, for the BLDC with comparative wires, rated speeds are calculated based on formula(10), then the speeds as well as the magnetic field data of the BLDC with reference wire are used to calculated core losses based on formula(14).As shown in Figure 3: the results calculated by the two methods are in good agreement with each other; and with the increase of wire size, the core loss increases gradually.

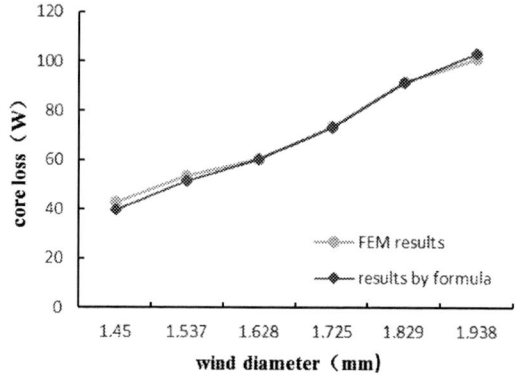

Figure 4. Core loss calculated by different methods

Total loss $P_{loss} = P_{Fe} + P_{cu}$, and as shown in Figure 4: the difference between the results obtained by the formula method and the finite element method is not more than 1.5%; in addition, with the increase of the wire diameter, the total loss of the BLDC first decreases and then increases; and for a BLDC, there is a wire size which minimizes the total loss of it; for the BLDC in this paper, the wire size is 1.628 mm.

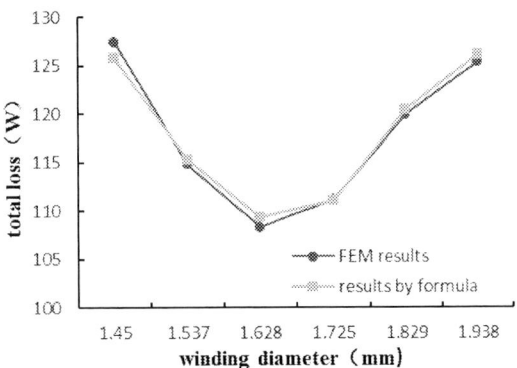

Figure 5. Tore loss calculated by different methods

4. Conclusions

In this paper, a method for quickly finding the wire size which can minimize the total loss of the BLDC is proposed : with this method, total losses of BLDC can be calculated quickly; by comparing these data, it is possible to quickly find the best wire size, thereby reducing the heat dissipation pressure of the BLDC, and helping to further increase the BLDC power density.

In addition, this paper analyzes the motor model and finds:

(1) With the increase of wire size, the copper loss of BLDC gradually decreases, and the core loss gradually increases;

(2) With the increase of wire size, the total loss of BLDC shows a trend of decreasing and then increasing.

(3) For a BLDC, there is a wire size which minimizes the total loss of it.

References

[1] G. Suresh Babu, M. Murali Krishna, B. Vikram Reddy, EPES, **4**, 9 (2015)
[2] M. Zhang, Doctoral dissertation, NWPU (2016)
[3] L. Li, J. Zhang, C. Zhang, J. Yu. TAS, **26** (2016)
[4] A.M. EL-Refaie, J. Alexander, S. Galioto, P.B. Reddy, K.K. Huh, P. Bock, X. Shen, Industry Applications IEEE Transactions on, **50**, 3235 (2014)
[5] M. Schiefer, M. Doppelbauer. *IEMDC*, 1820 (2016)
[6] Y. Fu, M. Takemoto, S. Ogasawara, K. Orikawa, *IEMDC*. 1 (2017)
[7] L. Yu, Y. Rui, J. Sun. MET, **41**, 166 (2018)
[8] G. Jie, W. Ma, J. Geng, Z. Nei, H. Wu, Proceeding of the CSEE, **26**, 131 (2006)
[9] J. Zhao, B. Tan, L. Ding, P. Qu, MET, **41**, 147 (2018)
[10] G. Bertotti. IEEE Transactions on Magnetics, **24**, 621 (1988)
[11] Y. Huang, Q. Hu, J. Zhu, EMCA, **34**, 6 (2007)
[12] W. Zhang, Y. Wan, Q. Wang, K. Cao, J. Hu, MICROMOTORS, **49**, 17 (2016)

ISPECE   IOP Publishing

Chaos control of bi-directional DC-DC converter by resonant parametric perturbation method in a DC microgrid

Lihua Wang*, Xue Ni, Yue Hu and Rui Zhang

School of Electronic Communication & Physics, Shandong University of Science & Technology, Qingdao, Shandong, 266590, China

* Corresponding author: wanglihua7141@163.com

Abstract. Nonlinear phenomena and the control method of the bi-directional DC-DC converter in different operating modes are studied to solve the problem of unstable operation of the bi-directional DC-DC converter in a DC microgrid. The piecewise switch model is built by analyzing the topology and the working process of the bi-directional DC-DC converter as well as employing Matlab/Simulink software. The evolution of the bi-directional DC-DC converter from steady to bifurcation and chaos is got by simulating the piecewise switch model with reference current and output voltage as bifurcation parameters. Finally, the resonant parametric perturbation is applied to control the chaos in bi-directional DC-DC converter, and thereby the stability of the bi-directional DC-DC converter in a DC microgrid has been improved.

1. Introduction

With the development of distributed DC power supply and electrical energy storage equipment in the microgrid, DC microgrids have been highly concerned. As an important component of DC microgrid, bi-directional DC-DC converter also has developed and been widely used[1]. Bi-directional DC-DC converter, which is a strong nonlinear system, may appear irregular behaviors such as bifurcation and chaos when the circuit parameters are varied[2]. Therefore, knowing when and why bifurcation or chaos occurs in a bi-directional DC-DC converter and giving effective control in time will certainly help to improvement design and enhancement performance[3].

At present, most studies about bi-directional DC-DC converter had mainly focused on circuit topology and control strategy[4], but the study on chaotic characteristic is relatively few[5]. This paper, which takes bi-directional buck-boost converter as an object, focuses on nonlinear phenomena by using bifurcation diagrams[6], time-domain waveforms, phase portraits, resonant parametric perturbation method. Bi-directional DC-DC converter is developed on the basis of unidirectional DC-DC converter, and therefore there is a strong connection between the two kinds of converters on chaos analysis and control. But unlike unidirectional DC-DC converter, the inductance value of the circuit is unique because the bi-directional DC-DC converter belongs to common circuit. Therefore, engineers need to adjust the parameters reasonably in the analysis and design process to get the ideal effect.

2. Control strategies and modeling strategies

The main circuit topology of bi-directional DC-DC converter is shown in Figure 1.

Content from this work may be used under the terms of the Creative Commons Attribution 3.0 licence. Any further distribution of this work must maintain attribution to the author(s) and the title of the work, journal citation and DOI.
Published under licence by IOP Publishing Ltd

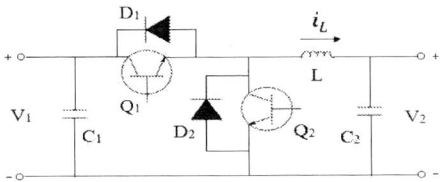

Figure 1. The main circuit topology of bi-directional DC-DC converter

Bi-directional DC-DC converter has three operation modes[7]: Buck mode, Boost mode and alternate mode. The alternate mode is equivalent to the alternation between Buck mode and Boost mode in a short time, and therefore this paper concentrates on the first two operation modes. Besides, Q_1 and Q_2 work in a complementary PWM manner. The inductor current i_L is therefore a triangle wave and the discontinuous inductor current, which occur in unidirectional DC-DC converter, will not occur over a period of time.

The converter mainly adopts voltage feedback control in Buck mode, whereas it adopts current feedback control in Boost mode, as shown in Figure 2. A judgment circuit is added to the circuit. If $i_L > 0$, the bi-directional DC-DC converter is in Buck mode. If $i_L < 0$, the bi-directional DC-DC converter is in Boost mode.

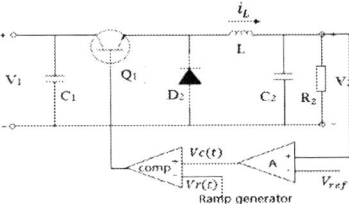

(a)Control circuit in Buck mode

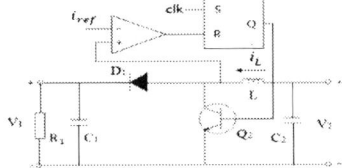

(b)Control circuit in Boost mode

Figure 2. Control circuit of bi-directional DC-DC converter

Assuming that it is in state 1 when the switch Q_1 is closed and it is in state 2 when the switch Q_1 is opened, the state equations in Buck mode can be established.

$$\text{State 1：} \quad \dot{x}=A_1x+B_1V_1 \qquad (1)$$

$$\text{State 2：} \quad \dot{x}=A_2x+B_2V_1 \qquad (2)$$

where x donates the state variable, i.e., $x = [V_0, i_L]^T$, $V_0 = V_2$, the A_1's, A_2's, B_1's and B_2's are the system matrices.

The unified expression can be obtained by introducing the integrating factor $1/s$ and the switching function δ.

$$\begin{cases} i_L = \dfrac{-V_o + V_1 \times \delta}{sL} \\ V_o = \dfrac{-V_o/R_2 + i_L}{sC_2} \end{cases} \qquad (3)$$

Similarly, the state equations in Boost mode can be described by:

$$\begin{cases} i_L = \dfrac{V_2 - V_o \times \delta}{sL} \\ V_o = \dfrac{-V_o/R_1 + i_L \times \delta}{sC_1} \end{cases} \qquad (4)$$

where $V_0 = V_1$.

The piecewise switch model is built based on (3) and (4) by employing Matlab/Simulink software[8]. The structure of the Simulink model is given in Figure 3.

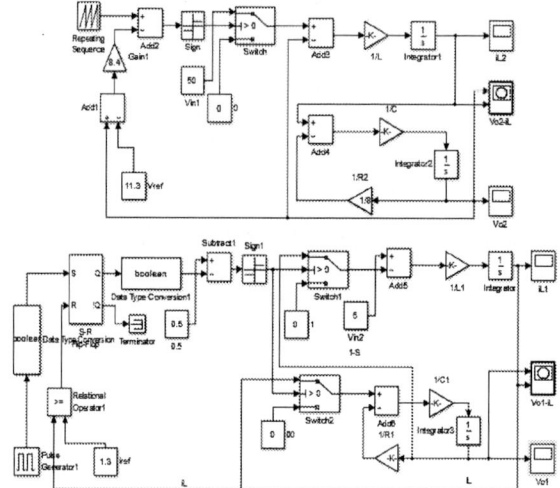

Figure 3. Piecewise switch model of bi-directional DC-DC converter

Assuming that the circuit components are ideal devices, the bi-directional DC-DC converter is analyzed based on the parameters of Table 1 and Table 2.

Table 1. Circuit parameters in Buck mode

Circuit components and parameters	Values
Inductance L	20mH
Load resistance R_2	8Ω
Capacitance C_2	22uF
Magnification times	8.4
Input voltage V_1	10V
Reference voltage V_{ref}	11.3V
Lower threshold of the ramp signal	3V
Upper threshold of the ramp signal	8V
Switching period T	400us

Table 2 Circuit parameters in Boost mode

Circuit components and parameters	Values
Load resistance R_1	20Ω
Capacitance C_1	30uF
Input voltage V_2	5V
Reference current i_{ref}	0.2A-2.5A

Switching period T	100us

3. Nonlinear analysis

3.1 Theoretical analyses on nonlinear behaviors

The nonlinear behavior of bidirectional DC-DC is strongly affected by input voltage V_1 or reference current i_{ref}, according to the Buck or Boost mode of the converter, respectively[9]. Therefore, the bifurcation diagrams of inductor current i_L in Buck mode and Boost mode are got with V_1 and i_{ref} as bifurcation parameters, respectively. The bifurcation diagrams are shown in Figure 4.

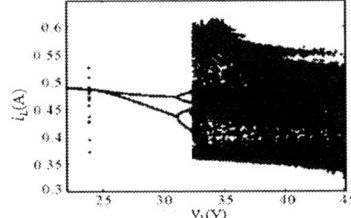

(a)Bifurcation diagram in Buck mode

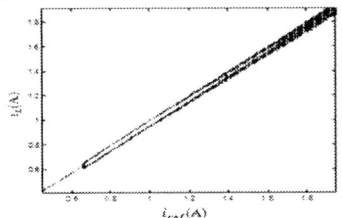

(b) Bifurcation diagram in Boost mode

Figure 4. Bifurcation diagram of bi-directional DC-DC converter

With the change of V_1, the inductor current i_L presents different performance in Buck mode as shown in Figure 4(a). The first bifurcation occurs at $V_1 = 26V$ and the converter enters into period-2 operation. As V_1 is continuously increased to 31V, the converter bifurcates to period 4. When the value of V_1 is further increased, the converter enters into chaotic regime.

With the change of i_{ref}, the inductor current i_L presents different performance in Boost mode as shown in Figure 4(b). The first bifurcation occurs at $i_{ref} = 0.7A$ and the converter enters into period-2 operation. As i_{ref} is continuously increased to 1.3A, the converter enters into chaotic regime and there is no apparent period-4 operation.

3.2 Simulation results on nonlinear behaviors

It is assumed that the value of V_1 or i_{ref} is varied while other parameters are fixed. The piecewise switch model is then simulated and its inductor current i_L and output voltage V_o are sampled. Simulation results in Buck mode are shown in Figures 5, 6, 7 and 8.

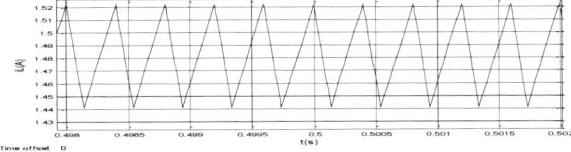

(a) Inductor current waveform

231

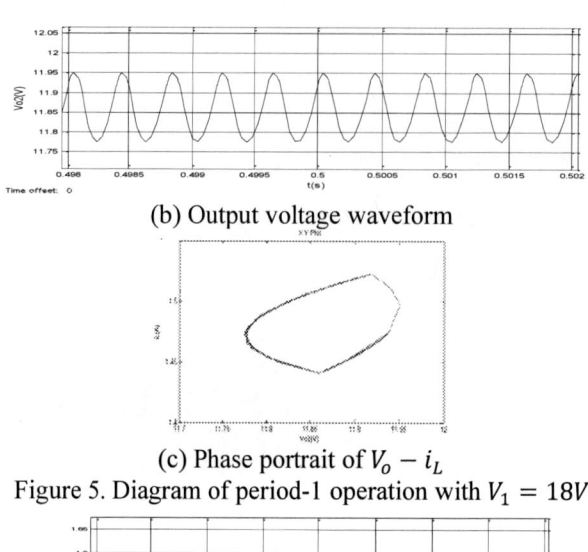

(b) Output voltage waveform

(c) Phase portrait of $V_o - i_L$

Figure 5. Diagram of period-1 operation with $V_1 = 18V$

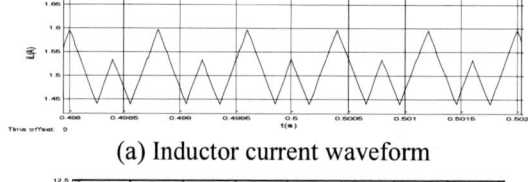

(a) Inductor current waveform

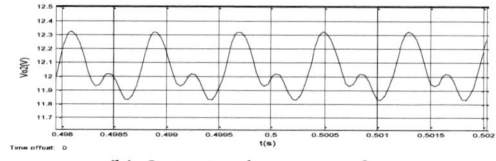

(b) Output voltage waveform

(c) Phase portrait of $V_o - i_L$

Figure 6. Diagram of period-2 operation with $V_1 = 25V$

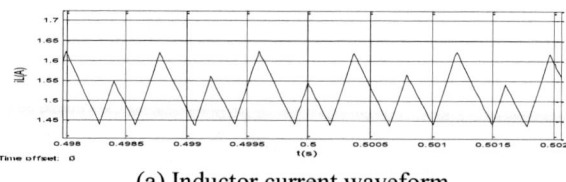

(a) Inductor current waveform

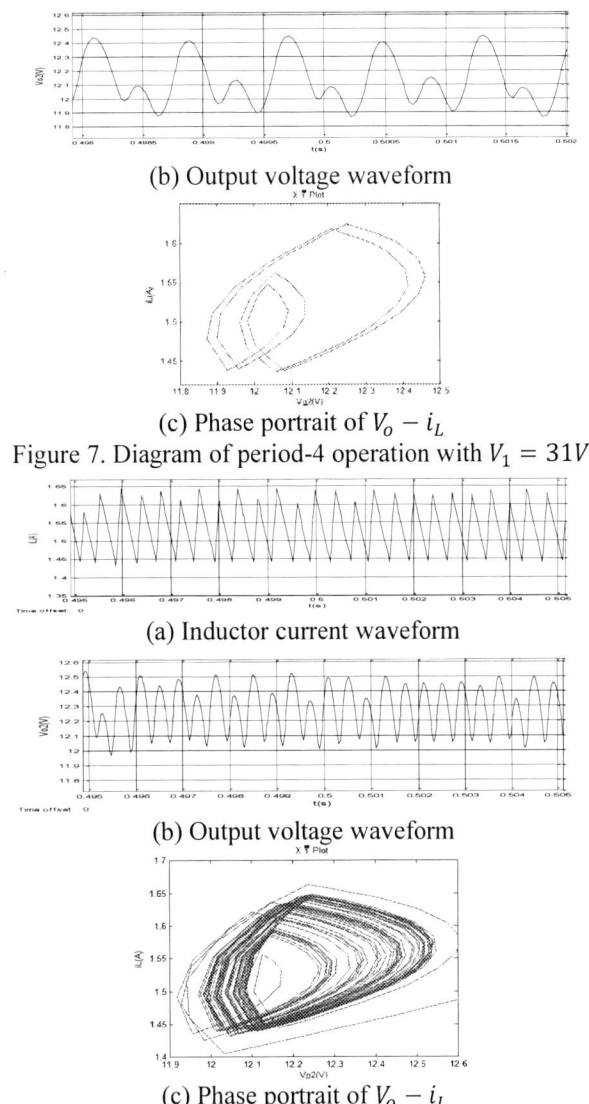

(b) Output voltage waveform

(c) Phase portrait of $V_o - i_L$

Figure 7. Diagram of period-4 operation with $V_1 = 31V$

(a) Inductor current waveform

(b) Output voltage waveform

(c) Phase portrait of $V_o - i_L$

Figure 8. Diagram of chaotic operation with $V_1 = 50V$

In Figure 5, it can be observed that when $V_1 = 18V$, the inductor current i_L waveform is a periodic steady zigzag wave and the output voltage V_o is a periodic steady sine wave. The corresponding phase portrait exhibits only one loop. Thus it can be seen that, the converter is in period-1 orbit.

In Figure 6, it can be observed that when $V_1 = 25V$, the inductor current i_L and the output voltage V_o waveforms present bifurcation but they are still periodic The corresponding phase portrait also presents bifurcation. Thus it can be seen that, the converter is in period-2 orbit.

In Figure 7, it can be observed that when $V_1 = 31V$, four peaks appear in both the inductor current i_L waveform and the output voltage V_o waveform. The corresponding phase portrait exhibits two limit cycles. Thus it can be seen that, the converter is in period-4 orbit.

In Figure 8, it can be observed that when $V_1 = 50V$, the inductor current i_L and the output voltage V_o waveforms behave randomly. The corresponding phase portrait is distributed in a certain range randomly. Thus it can be seen that, the converter is working in the chaotic state.

The simulation results in Boost mode are shown in Figures 9, 10 and 11.

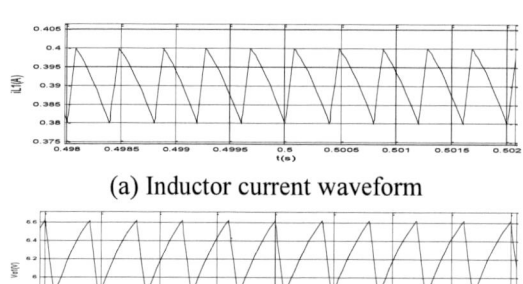

(a) Inductor current waveform

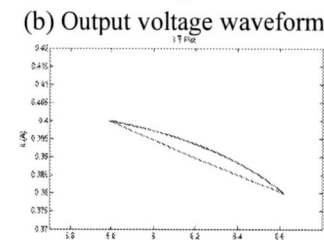

(b) Output voltage waveform

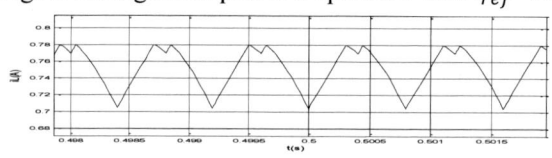

(c) Phase portrait of $V_o - i_L$

Figure 9. Diagram of period-1 operation with i_{ref}=0.4A

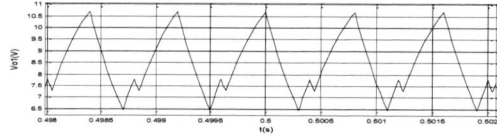

(a) Inductor current waveform

(b) Output voltage waveform

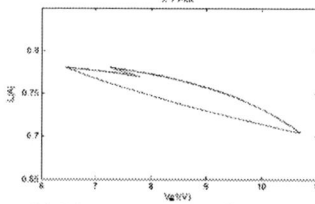

(c) Phase portrait of $V_o - i_L$

Figure 10. Diagram of period-2 operation with $i_{ref} = 0.78A$

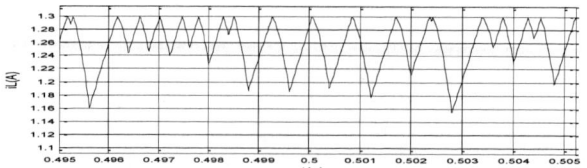

(a) Inductor current waveform

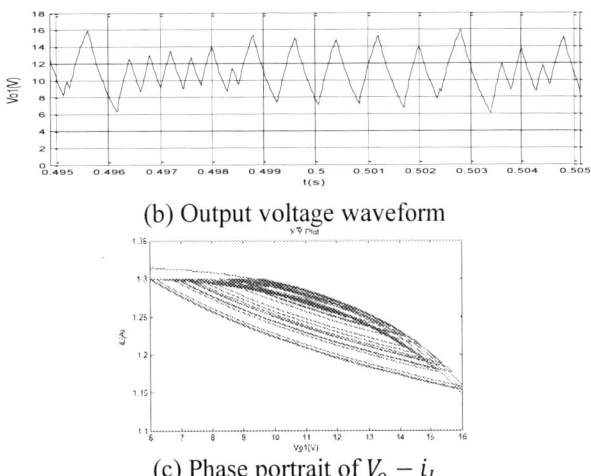

(b) Output voltage waveform

(c) Phase portrait of $V_o - i_L$

Figure 11. Diagram of chaotic operation with $i_{ref} = 1.3A$

In Figure 9, it can be observed that when $i_{ref} = 0.4A$, the inductor current i_L and the output voltage V_o waveforms are periodic with their periods equal to the switching period. The corresponding phase portrait is a finite closed curve with only one loop. Thus it can be seen that, the converter is in period-1 orbit.

In Figure 10, it can be observed that when $i_{ref} = 0.78A$, the inductor current i_L and the output voltage V_o waveforms present bifurcation but they are still periodic with their periods equal to 2 times the switching period. The corresponding phase portrait also presents bifurcation. Thus it can be seen that, the converter is in period-2 orbit.

In Figure 11, it can be observed that when $i_{ref} = 1.3A$, the inductor current i_L and the output voltage V_o waveforms behave randomly. The corresponding phase portrait is distributed in a certain range randomly. Thus it can be seen that, the converter is working in the chaotic state.

4. Control of chaos
Resonant parametric perturbation can suppress chaos by giving perturbations to a parameter at appropriate frequencies. The principle of parameter selection is that the parameter can strongly affect the system and can be easily varied.

4.1 Simulation of chaos control in Buck mode
In Buck mode, the reference voltage V_{ref} is chosen as the perturbation parameter to achieve the chaos control. Essentially we replace V_{ref} by the perturbed reference voltage V_{ref}', i.e.,

$$V_{ref}' = V_{ref} + A sin(2\pi f + \varphi) \qquad (5)$$

The converter is in a chaotic state when $V_1 = 50V$ and chaos control is therefore carried out at this time. The harmonic signal superimposed in (5) should be set to the same frequency and phase as the periodic ramp signal $V_r(t)$. The parameters of the effective perturbation required can be obtained according to the stability criteria of Buck converter[10]. When the perturbation amplitude A is about 20A and the optimal phase φ is about 0, the chaotic converter can be controlled to work on the period-1 orbit. The Simulink model with chaos control is shown in Figure 12. The simulated inductor current i_L time-domain waveform after chaos control is shown in Figure 13.

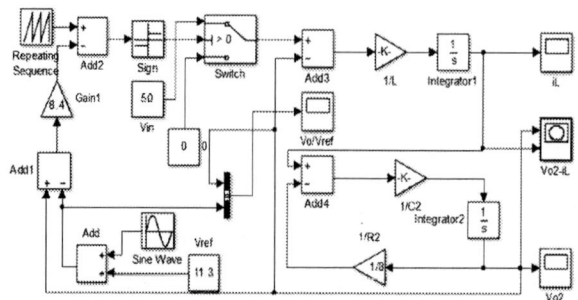

Figure 12. Chaos control switch model in Buck mode

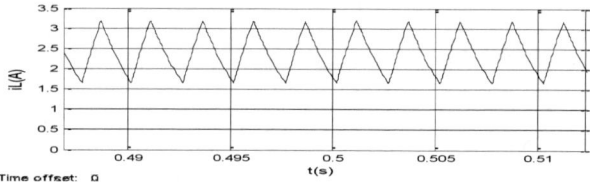

Figure 13. Inductor current waveform after chaos control

From Figure 13, inductor current waveform becomes single periodic state after chaos control, which indicates that the system operates in period-1 steady state after adding chaos control.

4.2 Simulation of chaos control in Boost mode

In Boost mode, i_{ref} is chosen as the perturbation parameter to achieve the chaos control. Essentially we replace i_{ref} by the perturbed reference current i_{ref}', i.e.,

$$i_{ref}'=i_{ref}+A\sin(2\pi f+\varphi) \qquad (6)$$

The converter is in a chaotic state when $i_{ref} = 1.3A$ and chaos control is therefore carried out at this time. The parameters of the effective perturbation required can be obtained according to the stability criteria of Boost converter[10]. When the perturbation amplitude A is about 0.15A and the optimal phase φ is about 6, the chaotic converter can be controlled to work on the period-1 orbit. The Simulink model with chaos control is shown in Figure 14. The simulated inductor current i_L time-domain waveform after chaos control is shown in Figure 15.

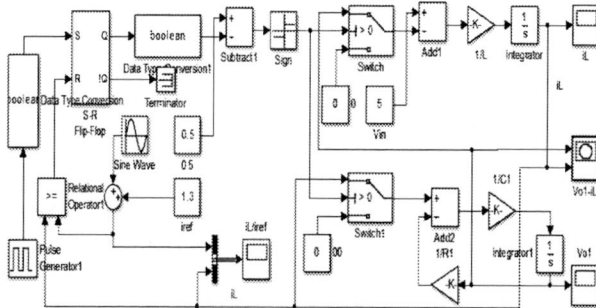

Figure 14. Chaos control switch model in Boost mode

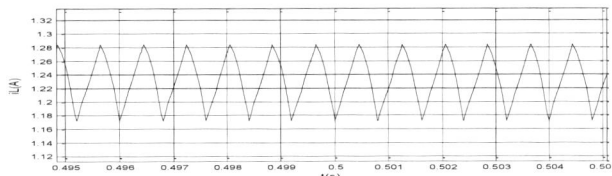

Figure 15. Inductor current waveform after chaos control

From Figure 15, inductor current waveform becomes single periodic state after chaos control, which indicates that the system operates in period-1 steady state after adding chaos control.

5. Conclusions

Nonlinear phenomena and the control method of bi-directional DC-DC converters are studied in this paper. Several conclusions can be drawn as follows,

1) In Buck mode, the system changes with the change of input voltage V_1. The first bifurcation occurs at $V_1 = 26V$ and the converter enters into period-2 operation. As V_1 is continuously increased to 31V, the converter bifurcates to period-4. When the value of V_1 is further increased, the converter enters into a chaotic regime.

2) In Boost mode, the system changes with the change of reference current i_{ref}. The first bifurcation occurs at $i_{ref} = 0.7A$ and the converter enters into period-2 operation. As i_{ref} is continuously increased to 1.3A, the converter enters into the chaotic regime and there is no apparent period-4 operation.

3) The stability of the bi-directional DC-DC converter has been improved through using resonant parametric perturbation method. In Buck mode, when the perturbation amplitude A is about 20A and the optimal phase φ is about 0, the chaotic converter can be controlled to work on the period-1 orbit. In Boost mode, when the perturbation amplitude A is about 0.15A and the optimal phase φ is about 6, the chaotic converter can be controlled to work on the period-1 orbit.

Acknowledgment

This Project (ZR2018MF005) was supported by the Natural Science Foundation Shandong Province, China.

References

[1] Y.G.Yan. *Bi-directional DC-DC Converters*[M]. Phoenix Science Press, 3-9, (2004).

[2] E.Lenz, D.J.Pagano. Nonlinear Control for Bidirectional Power Converter in a dc Microgrid [J]. Ifac Proceedings Volumes, **46**, 359-364, (2013).

[3] C.K.Tse, M.D.Bernardo. Complex Behavior in Switching Power Converters [J]. P IEEE, **90**, 768-781, (2002).

[4] X.H.Xu, C.B.Zheng, C.G.Hu, et al. Design of Bi-directional DC-DC Converter[C]. C IND ELECT APPL, 2283-2287, (2016).

[5] Y.Mei, X.Q.Li. A New Control Method of Balancing Inductor Current for Interleaved Parallel Bi-directional DC-DC Converter[C]. IPEMC, 2988-2992, (2016).

[6] X.R.Li, Y.L.Zhu, R.Shen. Control of chaos of the linear-mode switched-capacitor converter[J]. Journal of Xidian University, **42**, 110-114, (2015).

[7] K.Suresh, R.Arulmozhiyal. Design and Implementation of Bi-directional DC-DC Converter for Wind Energy System[J]. CS, **07**, 3705-3722, (2016).

[8] Z.H.Li, G.H.Zhou, X.T.Liu, et al.Dynamical modeling and analysis of buck converter operating in pseudo-continuous conduction mode[J]. ACTA PHYS SIN-CH ED, **64**, 209-218, (2015).

[9] A.M.Harb, S.M.Harb, I.E.Batarseh. Chaos and Bifurcation of Voltage-Mode-Controlled Buck Dc-Dc Converter with Multi Control Parameters[J]. INT J SIMUL MODEL, **30**, 472-478, (2010).

[10] L.H.Wang. Study on Chaos Theory Applied in Photovoltaic DC Microgrid System[D]. Beijing Jiaotong University, 69-84, (2017).

ISPECE   IOP Publishing

IOP Conf. Series: Journal of Physics: Conf. Series **1187** (2019) 022034   doi:10.1088/1742-6596/1187/2/022034

Application Research of Multi-source Information Fusion Technology in Power Network Fault Diagnosis

Shuxin Liu[1], Enmin Zhao[1], Yanjun Zhang[2], Jing Li[1], Liang Zhang[3], Yundong Cao[1]

[1] Institute of Electrical Apparatus New Technology and Application, Shenyang University of Technology, 110870, Shenyang, China

[2] Liaoning Provincial Electric Power Co., Ltd. Electric Power Research Institute, 110006, Shenyang, China

[3] State Grid Liaoning Electric Power Co., Ltd. Anshan Power Supply Company, 114001, Anshan, China

[a]Corresponding author: jhonminmin@163.com

Abstract. In this paper, a power grid fault diagnosis method based on multi-source information fusion technology is proposed for grid faults. Firstly, the wavelet transform is used to analyze the electrical quantity information of the fault recorder and PMU, and the three fault reliability characterizations of the components are obtained. Secondly, the information fusion based on the improved D-S evidence theory is carried out for each fault characterization, and the comprehensive fault credibility of the components is obtained, and the fault criterion is given. Finally, the fuzzy C-means method is used to make the diagnosis decision of the faulty component, and the final diagnosis result is obtained. The simulation results show that the method combines the electrical quantities of the fault recorder and the PMU to obtain accurate fault diagnosis results, which can effectively reduce the impact of protection and switch refusal misoperation on power grid fault diagnosis. This method has a good application prospect in power system fault diagnosis.

1. Introduction

In the event of a grid failure, the dispatch center will receive a large amount of fault information, including switching quantities and electrical quantities [1]. At present, the power grid fault diagnosis method is mainly based on the switch quantity information, including expert system, neural network, Bayesian network, Petri net, analytical model, etc [2]. For example, Zaihua Li combines expert system and data mining technology, and designs the mining models for simple faults and complex cascading failures in literature [3]. Mingwei Sun uses the temporal Bayesian knowledge base theory to diagnose the power grid fault, which effectively solves the problem of high fault probability value of non-faulty components in literature [4]. The above papers make fault diagnosis based on the switch quantity information, and achieve certain effects, but in the event of actual faults, the protection and circuit breakers may be mis-moved and rejected, and sometimes it is difficult to obtain accurate results based on the fault diagnosis of the switch information [5]. At present, power grid fault diagnosis for electrical quantity and multi-source information fusion technology has become a research hotspot at home and abroad.

Content from this work may be used under the terms of the Creative Commons Attribution 3.0 licence. Any further distribution of this work must maintain attribution to the author(s) and the title of the work, journal citation and DOI.

Published under licence by IOP Publishing Ltd

In summary, this paper obtains the current information and voltage information of the component from the fault recorder, obtains the voltage information of the component from the PMU measurement point, and uses the wavelet transform to analyze the three faultworthiness of the component. Then the improved D-S evidence theory is used to integrate the information, and the comprehensive fault credibility of each component is obtained. Finally, the fuzzy C-means method is used to make the diagnosis decision of the faulty component, and the final diagnosis result is obtained.

2. Electrical quantity feature extraction

This paper analyze and process voltage and current information from the recorder and analyze and process voltage information from the PMU.

2.1 Extracting the moment of failure

The recorded data obtained by the power grid dispatching center is with the fault moment. For the simulated data, in order to accurately obtain the fault time, the fault current of the fault phase is processed by db3 wavelet[6]. this paper uses the high frequency coefficient on the first scale of db3 wavelet to find the corresponding modulus maxima, and the corresponding sampling point as the fault moment. Figure 1 shows the fault phase current data obtained from the power grid dispatching center, and the fault time is extracted by the above method. The time at which the fault occurred (signal abrupt moment) is the 173th sampling point, which is very close to the 175th sampling point marked by the recorder, indicating that the method is valid.

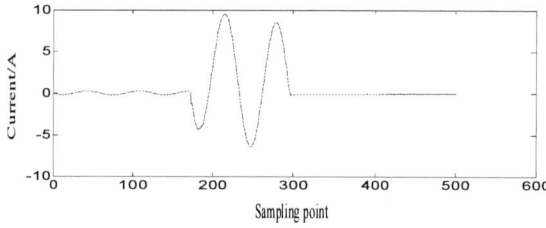

Figure 1. The Fault current Diagram collected by Power Network Recorder

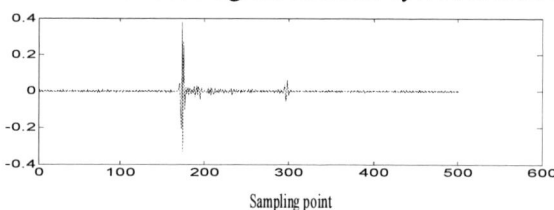

Figure 2. The waveform diagram of wavelet decomposition for fault current

2.2 Energy credibility

Wavelet transform and coefficient reconstruction of the fault signal, let E_1, E_2...E_z, E_{z+1} be the wavelet energy distribution of the signal on z scales.

Assume that the fault signal of the $i(i = 1,...,m)$ th line is $x_i(n)$ when a fault occurs. Find the total wavelet energy of a waveform period after the recorder current signal fails. Take the maximum value of its three phases as the result E_i. Define energy credibility $E^i = (e_1, e_2,..., e_m)$, Its elements represent the relative level of support for each line fault.

$$e_i = \frac{E_i}{\sum_i^m E_i} \tag{1}$$

2.3 Singular credibility

Singular Value Decomposition:The singular value decomposition of any $m \times m$ order matrix A can be expressed as $A = U\Lambda V^T$, Where, U and V are $m \times m$ order and $n \times n$ order orthogonal matrices; $\Lambda = diag(\lambda_1, \lambda_2, ..., \lambda_t)$ is diagonal matrix, $t = \min(m,n)$, Its non-negative diagonal elements are arranged in descending order, which is the singular eigenvalue of A.

When the grid fails, The obtained voltage signal is subjected to 3-layer wavelet decomposition using db4 wavelet, and three reconstruction coefficients are respectively obtained, and the three reconstruction coefficients are formed into a matrix to reflect the detailed information of the voltage signal. Then singular value decomposition on this matrix is performed, Let $\Lambda_i = diag(\lambda_1, \lambda_2, ..., \lambda_t)$ be the singular feature matrix of the i-th component of the system.Let $S_i = \sum_{i=1}^{t} \lambda_i / t$. Define singular credibility $S^i = (s_1, s_2, ..., s_n)$. Where, s_i is obtained in a similar way to equation (1). The singular credibility of the entire voltage signal of the recorder is calculated, and participate in information fusion as a body of evidence.

2.4 Energy Distortion Credibility

After the grid fails, the magnitude of the fault line voltage changes more obviously than the non-fault line, reflecting this characteristic with the energy distortion credibility. The total wavelet energy of one waveform period before and after the fault on the voltage signal obtained by the PMU is calculated. Let the total energy of the wavelet before the line fault be E_b, and the total energy of the wavelet after the fault be E_a, define the energy distortion:

$$D_i = \frac{E_b}{E_a} \qquad (2)$$

The maximum value obtained by the voltage of each phase of the line is taken as the energy distortion of the line. Define the energy distortion confidence $D^i = (d_1, d_2, ..., d_n)$. Where, d_i is obtained in a similar manner to equation (1). Indicates the degree of support for each line failure.

3. FAULT FEATURE INFORMATION FUSION

In practical applications, D-S evidence theory has many shortcomings, such as Zadeh paradox, one-vote veto, robustness and fairness. In this paper, energy credibility, singular credibility, and energy distortion credibility are used as evidence bodies. According to the degree of conflict between evidences, the evidence with weak conflicts is fused with Dempster synthesis rules. Evidence of strong conflicts is fused by using the mean weighting method below to improve the D-S evidence theory. At the same time, according to the reliability of the three faultworthiness defined, the credibility is taken as 0.9, 0.85, and 0.8.

Mean weighting method to improve D-S evidence theory belongs to the improvement of synthesis rules between evidences. For evidence of conflict, Temporarily do not consider the conflict between the two pieces of evidence, and wait until the next piece of evidence comes, then continue to merge [7]. Proceed as follows:

Step1: Read two pieces of evidence m1 and m2;

Step2: Calculate the conflict factor K to determine whether there is a conflict. If it is conflict evidence, go to Step 4: Otherwise, proceed to the next step;

Step3: According to the Dempster synthesis rule, the evidence is merged, and the process proceeds to Step 5;

Step4: Process the conflict evidence according to the mean weighting method, and proceed to the next step;

Step 5: Make a decision;

4. Diagnostic decisions and diagnostic procedures

4.1 Diagnostic decision based on fuzzy C-means clustering

The improved D-S evidence theory was used to obtain the comprehensive fault reliability of the line. On this basis, in order to determine the faulty line, the obtained comprehensive fault credibility is processed by a classification algorithm-fuzzy C-means clustering method (FCM).

This article divides the line into fault and non-fault 2 categories and Set the FCM class membership threshold to 0.5. Since the reliability of the comprehensive fault is classified, the center value of V is large as the fault class. Whether the line is determined to be faulty in the fault class depends on the comparison of the u_{ij} in the U with the class membership threshold, and the greater than the class membership threshold is determined as the final fault line.

4.2 Grid fault diagnosis process

As mentioned above, the specific steps of the multi-source information fusion technology used in power grid fault diagnosis are as follows.

Step1: The voltage and current recording data collected by the fault recorder after the fault occurs from the protection fault information management system (RPMS) is obtained, and the voltage recording data from the PMU measuring point is obtained;

Step2: Based on wavelet transform and energy spectrum analysis, the reliability characterization of electrical faults is obtained, including energy credibility, singular credibility, and energy distortion credibility;

Step3: The above three credibility levels are constructed as evidence bodies, and information fusion is performed on them by using improved D-S evidence theory;

Step4: Perform fuzzy C-means clustering analysis on the fusion results;

Step5: Obtain the final grid fault diagnosis result;

5. Example calculation

The IEEE39 node system is simulated with PSCAD (as shown in Figure 3), The example analysis is based on the following four assumptions.

1) In the PSCAD simulation model of the IEEE39 node system, the line length is taken as 100 km.

2) PMU configuration: The PMU is configured at sections 3, 8, 10, 16, 20, 23, 25, 29 using the method proposed in literature[8] to achieve complete observability of the power system state.

3) The class membership value of FCM is taken as 0.5.

4) Multi-resolution wavelet transform for processing electrical quantities is performed using MATLAB "db4" wavelet.

The simulation duration is 2s. The single-phase ground fault occurs on the line at 1s, and the fault duration is 0.2s. In order to see the clarity, Figure 4 shows a section of the fault signal intercepted.

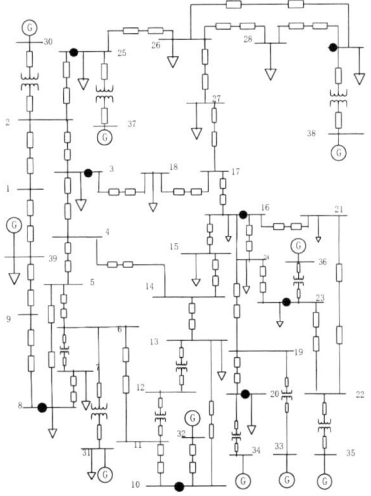

Figure 3. IEEE 39-bus system

Assume that line L_{9-39} (L_{9-39} is the line between node 9 and node 39) has a single-phase short-circuit fault. Using the method of this paper for fault diagnosis, all lines in the system should be processed. Due to the large number of lines, select 10 representative lines (L_{9-39}, L_{1-39}, L_{5-8}, L_{3-18}, L_{6-11}, L_{16-17}, L_{16-19}, L_{19-33}, L_{28-29}, L_{22-23} these 10 lines correspond to the 1-10 numbers in Table 1）, Due to the limited space, the three-phase voltage and current waveforms of L_{9-39} and L_{1-39} are selected as shown in Fig. 4. From top to bottom respectively L_{9-39} current, L_{9-39} voltage, L_{1-39} current, L_{1-39} voltage.

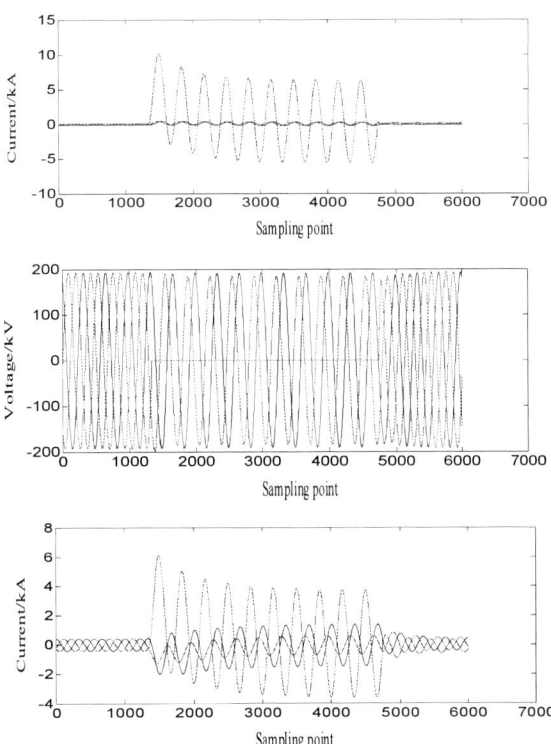

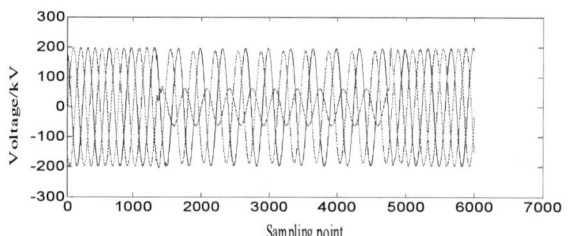

Figure 4. The Circuit voltage and current waveform diagram obtained by simulation

Energy credibility, singular credibility, and energy distortion credibility for 10 lines is evaluated (Corresponding to T1, T2, T3 in Table 1, respectively), The maximum value of the three phases is taken as the desired result. As shown in Table 1.

Table 1. The Line information fusion results

line	Fault feature extraction			Fusion result	diagnostic result
	T_1	T_2	T_3	T^1	T^2
1	0.4351	0.1919	0.8499	0.8645	1
2	0.1532	0.1712	3.47×10^{-5}	0.0464	0.0085
3	0.0515	0.0235	5.24×10^{-6}	0.0087	2.72×10^{-4}
4	0.0066	0.0780	4.39×10^{-6}	0.0061	6.19×10^{-4}
5	0.0131	0.0476	5.18×10^{-6}	0.0050	7.77×10^{-4}
6	0.1410	0.0544	6.59×10^{-6}	0.0259	0.0021
7	0.0176	0.0416	4.02×10^{-6}	0.0053	7.35×10^{-4}
8	0.0272	0.0627	3.74×10^{-6}	0.0084	3.03×10^{-4}
9	0.0265	0.0618	4.87×10^{-6}	0.0082	3.26×10^{-4}
10	0.0280	0.0673	6.94×10^{-6}	0.0089	2.47×10^{-4}
μ	0.1	0.15	0.2	0.0125	

For the fusion result T1, the fuzzy C-means method is used for the diagnosis decision, the iteration stop error is set to 10^{-5}, and the number of iterations is 100 times. The first type of center value is 0.0123, the second type of center value is 0.8636, and the second type is selected as the fault set. The diagnosis result T^2 in Table 1 is the membership degree of the line for the second category. After the membership degree is compared with the threshold value, it can be determined that L_{9-39} has a fault.

In this paper, the electrical quantity information before and after the fault is used to determine the faulty line. If the switch quantity information is considered, the other line may be diagnosed as a faulty component if the circuit breaker refuses to move. Therefore, the direct use of electrical quantity information, the use of its directness and accuracy, in the grid fault diagnosis has a certain advantage over the switch.

6. Conclusion

(1) This paper uses the directness and timeliness of electrical quantity to analyze the grid fault, and analyzes the electrical quantity data collected by the fault recorder and PMU. Starting from the change of voltage and current before and after the fault, three kinds of fault support credibility for the line are proposed, which are energy credibility, singular credibility and energy distortion credibility.

(2) This paper combine the three fault support credibility on the basis of improving the D-S evidence theory, and obtain the comprehensive fault support credibility of the line fault. Finally, the fuzzy C-means method is used for diagnosis decision. Based on the PSCAD simulation example, the method effectively diagnoses the faulty line.

References

[1] Cheng Xuezhen, Lin Xiaoxiao, Zhu Chunhua, et al. Power system fault analysis based on hierarchical fuzzy Petri net considering time association character[J]. Transactions of China Electrotechnical Society, 32(14):229-237(2007)

[2] Guo Chuangxin, Gao Zhenxing, Liu Yi, et al. Hierarchical fault diagnosis approach for power grid with information fusion using multi-data resources[J]. High Voltage Engineering, 36(12): 2976-2983(2010)

[3] Li Zaihua, Bai Xiaomin, Zhou Ziguan, et al. Method of power grid fault diagnosis based on feature mining[J]. Proceedings of the CSEE, 30(10): 16-22(2010)

[4] Sun Mingwei, Tong Xiaoyang, Liu Xinyu, et al. A power system fault diagnosis method using temporal Bayesian knowledge bases[J]. Power System Technology, 38(3): 715-722(2014)

[5] Gu Xueping, Liu Daobing, Sun Haixin, et al. Acquisition of power system fault diagnosis information from SCADA System[J]. Power System Technology, 36(6): 64-70(2012)

[6] Zheng Huazhen, Yue Quanming, Yu Weiyong, et al. Location of fault moment by wavelet in ultra-high-voltage nehvork based on fault record data[J]. Power System Technology, 29(19): 33-38(in Chinese)(2005)

[7] Lv Yuejing, Song Xiangbo, Zhang Lei, et al. A weighted improved D-S evidence reasoning algorithm[J]. Computer Application and Software, 28(10): 30-33(2011)

[8] Peng Chunhua. Optimal PMU placement based on immune BPSO algorithm and topology observability[J]. Transactions of China Electrotechnical Society, 23(6): 119-12(2008)

ISPECE   IOP Publishing

Research on Power Quality Acquisition and Reconstruction Method Based on Compressed Sensing

Bo YUAN[1]

[1]Economic and Technology Research Institute of State Grid Hebei Electric Power Company, Shijiazhuang, China,050021

[a]Corresponding author: yuanbo7396@163.com

Abstract. The power quality acquisition method based on compressed sensing (CS) has become an inevitable trend to solve the problem of explosive growth in power quality data. In order to solve the problems that the signal type is not comprehensively considered and the method of power quality acquisition is not universal, which are difficult to adapt to the actual system construction, this paper studies the method of power quality information acquisition and reconstruction method based on CS from a new perspective. Based on power quality acquisition and reconstruction framework using CS, this paper deeply studies the sparse basis, measurement matrix and reconstruction algorithm, classifying and analyzing all kinds of power quality signals as the result. Finally, through the experimental analysis, the construction scheme for three-element of CS is given, which is suitable for power quality acquisition. All the work done in this paper is intended to provide a reference for follow-up research and actual system construction.

1. Introduction

With the access of new "source-load" resources such as electric vehicles and renewable energy, power quality acquisition faces two problems which are increasing collection points and dense collection intervals. Under the dual pressure of power quality data's explosive growth and equipments' efficient use, compressed sensing is widely used in power quality acquisition and analysis[1-5]. CS completes the two processes of data acquisition and data compression at the same time. Compared with the traditional compression method, sampling-end device has very low complexity in compression and saves storage space, which is very suitable for power quality acquisition systems. Reference [1] uses CS in power quality acquisition and conducted preliminary exploration. Reference [2] studies the identification of various power quality signals. Reference [3] summarizes the research of CS in power quality, and points out that the adaptive three-element of CS is key issue in system building of power quality acquisition. Reference [4] improves the two compression sensing elements which are measurement matrix and reconstruction algorithm according to the characteristics of harmonic signals. Reference [5] optimizes the element of sparse representation method for the power quality signal.

With the deepening of research, the optimization methods for three-element of CS are continuously proposed[1-5], but the following drawbacks are existing. Firstly, the improved methods are developed for a specific and classic method, which is not universal. Secondly, in the actual system construction in the future, mature and classic methods are easier to adopt as core algorithms, but there are less comprehensive and comparative researches on classical algorithms. Thirdly, in the research of power quality acquisition, CS is mostly directed to specific signals and lacks comprehensiveness.

Content from this work may be used under the terms of the Creative Commons Attribution 3.0 licence. Any further distribution of this work must maintain attribution to the author(s) and the title of the work, journal citation and DOI.

Published under licence by IOP Publishing Ltd

In order to overcome the shortcomings of the above research, this paper uses a comprehensive perspective to study the power quality acquisition and reconstruction method based on CS. Based on the classification and modeling of various power quality signals, the three-element of CS (sparse basis, measurement matrix and reconstruction algorithm) are deeply studied. According to different signal characteristics, we analyze the effects of CS using different three-element, and then formulate the three-element construction scheme of CS in order to be suitable for future power quality acquisition systems.

2. Related Models

2.1 Mathematical model of power quality signal

The IEEE Standards Coordination Committee classifies power quality issues into two categories: steady-state power quality (including harmonics/inter-harmonics, voltage fluctuation/flicker, notch and noise, 4 categories in total) and transient power quality (including voltage dip, voltage swell, voltage interruption, pulse transient, oscillating transient, voltage cut, voltage spike, 7 categories in total). According to the power quality IEC series standards IEC-61000-1~IEC-61000-6, we can get of all 11 kinds of power quality data, which can be found in IEC-61000-1~IEC-61000-6 and we are no longer giving formulas of mathematical models here.

Table 1. Mathematical model of CS

Model type	Model formula	Explanation
Model of Compressed Measurement	$y = \Phi x = \Phi \Psi s = \Theta s$	Vector x is the $N \times 1$ dimensional original signal and y is the $M \times 1$ dimensional compressed sample vector ($M \ll N$); x needs to satisfy that coefficient s (sparse vector) is K-sparse under the sparse basis.
Model of Signal Reconstruction	$\begin{cases} \min \quad \|s\|_{l_p} \\ s.t. \quad y = \Phi \Psi s \end{cases}$	When $0 \leq p < 1$, the Model of Signal Reconstruction becomes an NP-hard problem. When $p \geq 1$, it translates into a convex optimization problem, and bringing infinite solutions. The reconstruction model is mainly divided into 2 types: the minimized l_0 norm and the minimized l_1 norm.

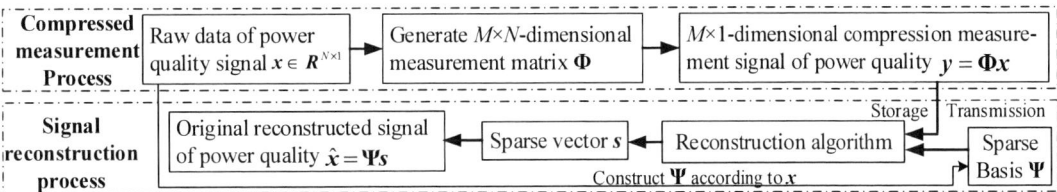

Figure 1. Power quality information collection process based on CS

2.2 Mathematical model of compressed sensing

Sparse basis, measurement matrix and reconstruction algorithm constitute the three-element of CS. The mathematical models of CS[6-7] are shown as Table 1.

3. Power Quality Acquisition and Reconstruction Method Based on CS

3.1 Power quality acquisition based on CS

Power quality acquisition system constructed by CS can divide into 2 processes: compressed measurement and signal reconstruction. Power quality acquisition process based on CS is shown in Figure 1.

As shown in the figure, there are three key problems in the whole process—sparse basis, measurement matrix and reconstruction algorithm, which are the three-element of CS. In addition to considering the requirements of CS theory, the design of these three key issues should also consider the characteristics of power quality data. y is the front-end sampled value, the process for obtaining y is implemented in the analog signal, that is, compression is integrated into the sample.

3.2 Research on sparse basis

Commonly-used sparse basis include fixed orthogonal transform (FOT) basis, multi-scale geometric analysis, and redundant dictionary, among which FOT basis is more suitable for one-dimensional signals. Compared with the redundant dictionary, FOT has simple structure and its implementation is more flexible. Therefore, this paper firstly determines the FOT as the applicable sparse basis, including Discrete Fourier Transform (DFT) basis, Discrete Cosine Transform (DCT) basis and sparse basis based on Wavelet Transform (WT). The signal in power system is a sinusoidal signal based on power frequency 50Hz. Various power quality data models can be considered as superimposing certain interfering signals on the basis of the sinusoidal signal, whether the interfering signal has short-term characteristics (such as pulse, oscillating) or low energy characteristics (such as harmonics). DFT basis and DCT basis are best suited for sinusoidal signals while DCT is better than DFT in implementation effect and degree of difficulty. Therefore, we considered that DCT basis should be selected as the sparse basis.

3.3 Research on measurement matrix

In order to accurately reconstruct signal, the measurement matrix must satisfy Restricted Isometry Property (RIP), and the random matrix can satisfy the RIP with great probability[6-7]. At present, typical random matrices are mostly dense matrices, including Gaussian matrix, partial Hadamard matrix, Bernoulli matrix, partial orthogonal matrix (such as partial fast Fourier transform, PFFT), and Toeplitz matrix. Gaussian measurement is the most widely used in power quality because its incoherence with orthogonal sparse matrices. Compared with dense measurement matrix, one kind of matrices called sparse matrix can greatly save sampling cost in sampling-end equipments through reducing complexity of compression, and can solve implementation difficulty of hardware[8]. This paper proposes to select the Binary Sparse matrix as the measurement matrix in power quality acquisition [8].

Definition 2.1 (Binary Sparse matrix) An $M \times N$-dimensional matrix Φ. if each element of its column contains only μM positions with an element value of 1, and remaining elements are all zero, and μM non-zero element-positions are random, the matrix Φ is called Binary Sparse matrix. μ is the sparsity rate, $\mu \ll 1$.

Compared to dense measurement matrices, Binary Sparse matrix have the following advantages: (1) Great sparsity reduce complexity of measurement in sampling-end, which see Table 2 for details. Due to $\mu \ll 1$, the complexity is greatly reduced. (2) The hardware of sampling-end is easy to implement, and its method can adopt as Analog Information Converter (AIC).

Table 2. Comparison of Binary Sparse measurement and dense measurement

	Binary Sparse	**Dense**
Compression complexity	$O~(\mu MN)$	$O~(\mu MN)$
Storage data of matrix	$2\mu MN$	MN

3.4 Research on reconstruction algorithm

The reconstruction algorithms of CS can be divided into 3 categories, which are greedy algorithm based on l_0 norm, convex optimization algorithm based on l_1 norm and some other combined reconstruction algorithms such as iterative threshold algorithm. The greedy algorithm gradually approximates the signal through multiple iterative searches, including Orthogonal Matching Pursuit (OMP), Compressive Sampling Matching Pursuit (CoSaMP), Subspace Pursuit (SP), Gradient Pursuit (GP), and so on. The convex optimization algorithm is based on the Basis Pursuit problem and has high complexity, including Gradient Projection for Sparse Reconstruction (GPSR), Spectral Projected Gradient (SPG), and so on. The Iteration Hard Threshold (IHT) and Fast Iteration Shrink Threshold Algorithm (FISTA) are representative of the iterative threshold algorithm.

This paper considers various factors to determine the applicable reconstruction algorithm. In all typical algorithms, convex optimization algorithm has higher reconstruction precision while has slow convergence and poor anti-noise performance. Among them, SPG has higher reconstruction accuracy than other convex optimization algorithms. With the development of greedy algorithms, GP algorithm takes the convergence speed and reconstruction effect into account. Its computation time is better than the OMP which is typical fast algorithm, and the reconstruction accuracy is similar to SPG. Therefore, this paper believes that the GP is used as the main algorithm, and SPG is adopted as the reconstruction algorithm for power quality signal with poor sparsity.

4. Experimental Analysis Based on CS for all Power Quality Signals

4.1 Performance of sparse basis
We analyzed the reconstruction effects of all 11 power quality signals with different sparse basis, and give the reconstruction effects of partial steady-state and transient power quality data with DCT basis, DFT basis and DWT basis, shown as Figure 2. The reconstruction errors with the DCT basis and DFT basis are almost identical and they are significantly better than the reconstruction errors with DWT basis, which is consistent with the conclusions previously mentioned herein.

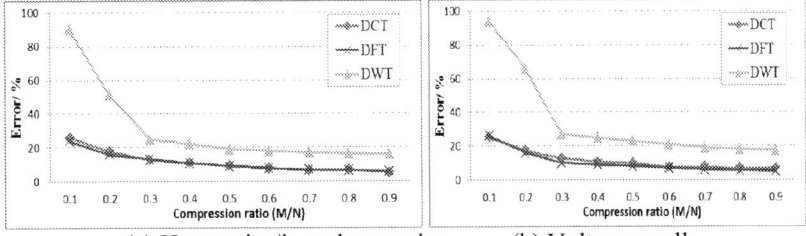

      (a) Harmonics/inter-harmonics   (b) Voltage swell

Figure 2. Reconstruction error with different sparse basis

4.2 Performance of measurement matrix
The power quality data is measured by different measurement matrices, and reconstruction effects with different measurement matrices are compared. The signal noise of reconstruction (SNR) of various signals is shown in Figure 3 and Figure 4. Uniformly, the sparse basis is DCT and reconstruction algorithm is OMP.

It can be seen from the figure that the SNRs with Binary Sparse measurement and with Gaussian measurement are better than other measurement methods by about 3~20 dB. In the Gaussian matrix and Binary Sparse matrix which have better effects than others, the implementation and sampling complexity of the Binary Sparse are obviously better than Gaussian matrix, which is consistent with the conclusions mentioned earlier.

4.3 Performance of reconstruction algorithm
The power quality data is reconstructed by different algorithms, and different effects are compared. The SNR of various signals is shown in Figure 5 and Figure 6. Uniformly, the sparse basis is DCT and the measurement is Binary Sparse matrix. It can be seen from the figure that the convex optimization algorithm is better than others. In all kinds of power quality signals, SPG and GP are obviously superior to other algorithms while IHT and FISTA are the worst. In addition to 4 type transient signals (pulse transient, oscillating transient, voltage cut and voltage spike), reconstruction effect of GP in other power quality signals has similar effects to the SPG, SNR of which is within 1 dB compared with SPG (SPG is slightly better than GP), and GP is superior to some convex optimization algorithms such as GPSR. Therefore, this paper considers that SPG algorithm is used to 4 transient power quality signals including oscillating transient, pulse transient, voltage cut and voltage spike. The other 7 power quality signals adopt GP algorithm.

4.4 Analysis of compression ratio
The smaller compression ratio (*M/N*) will bring the worse reconstruction effect. We analyzed various signal reconstruction effects using Binary Sparse measurement, GP algorithm and DCT basis. When

M/N is less than 30%, the reconstruction error is greatly increased. And when the *M/N* is 40%, the SNR is greater than 60 dB, which exceeds the requirement. Therefore, the proposed compression ratio is 30%~40%.

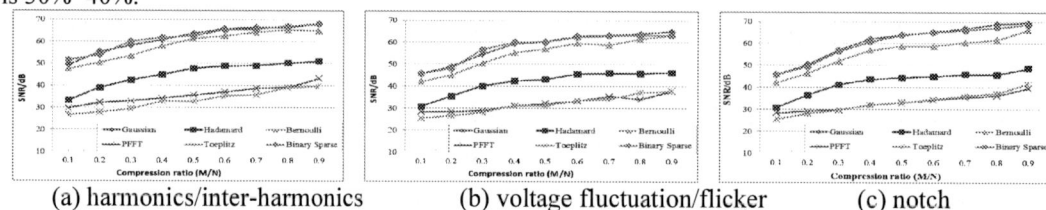

(a) harmonics/inter-harmonics　　(b) voltage fluctuation/flicker　　(c) notch

Figure 3. Reconstruction SNR with different measurement matrix in steady-state power quality

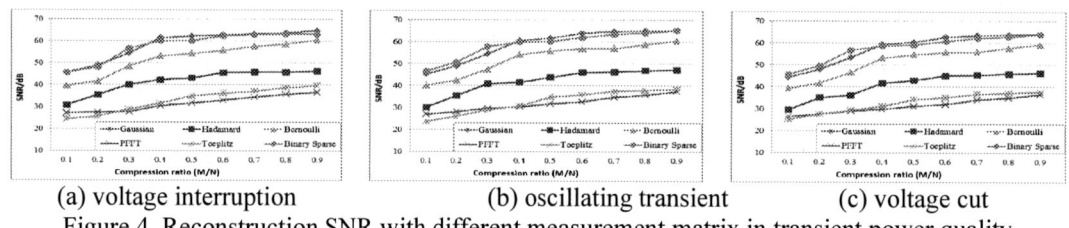

(a) voltage interruption　　(b) oscillating transient　　(c) voltage cut

Figure 4. Reconstruction SNR with different measurement matrix in transient power quality

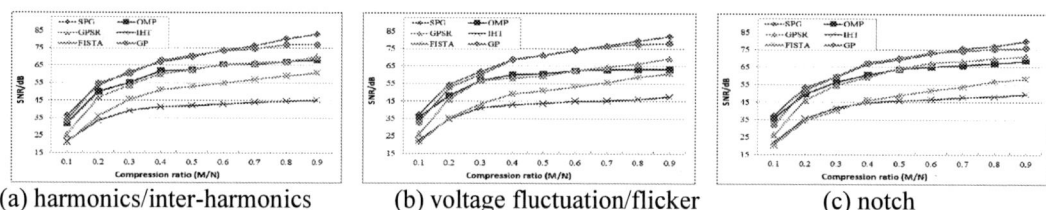

(a) harmonics/inter-harmonics　　(b) voltage fluctuation/flicker　　(c) notch

Figure 5. Reconstruction SNR with different reconstruction algorithm in steady-state power quality

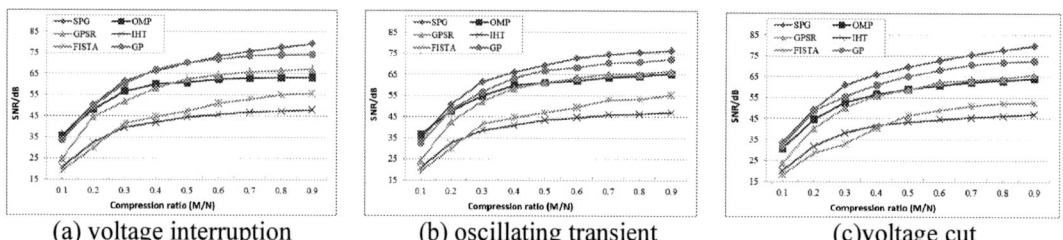

(a) voltage interruption　　(b) oscillating transient　　(c)voltage cut

Figure 6. Reconstruction SNR with different reconstruction algorithm in transient power quality

5. Summary

For the future trend of building power quality acquisition system based on CS, we systematically studied the construction of key elements in CS for power quality acquisition, and make up for the shortfall that current researches are lack of comprehensiveness and applicability. The recommended scheme of power quality acquisition and reconstruction based on CS in this paper is as follows.

Table 3. The recommended scheme of power quality acquisition and reconstruction based on CS

Sparse basis	DCT
Measurement matrix	Binary Sparse measurement matrix
Reconstruction algorithm	SPG: oscillating transient, pulse transient, voltage cut and voltage spike
	GP: other kinds of power quality
Compression ratio	30%~40%

References

[1] X. W. Wang, L. Wang, G. J. Miao, Y. B. Lu, D. Han, H. J. Wan, Y. Zhao, Power System Technology **36**, 3 (2012)

[2] S. Y. Cao, C. H. Dai, Y. F. Zhu, W.R. Chen, Power System Protection and Control, **45**, 3 (2017)

[3] Y. F. Zhu, C. H. Dai, W. R. Chen, Z. Y. He, Proceedings of the CSU-EPSA, **27**, 1 (2015)

[4] B. Yuan, Y. Wang, H. Shao, Y. Wang, C. H. He, T. Yang, Automation of Electric Power Systems, **167**, 20(2016)

[5] Y. Liu, W. Tang, B. Q. Liu, Transactions of China Electrotechnical Society, **33**, 15 (2018)

[6] D. L. Donoho, IEEE Transactions on information theory, **52**, 4 (2006)

[7] E. J. Candès, M. B. Wakin, IEEE signal processing magazine, **25**, 2 (2008)

[8] A. Gilbert, P. Indyk, Proceedings of the IEEE, **98**, 6 (2010).

A Study of Simulation on Relationship Between Young's Modulus of Cable joints and Interface Pressure Based on Finite Element Method

Zhengbing Tian[1], Liying Sun[2a], Jing Wang[2] and Xiaoye Bai[1]

[1]ZTT Cable Accessories Co., Ltd.,226000 Jiangsu, China

[2] Beijing Technology and Business University, School of Materials Science and Mechanical Engineering Sciences, 100048 Beijing, China

[a]Corresponding author: sunliying1102@126.com

Abstract. The interface pressure between cable and cable joint is of essential importance to ensure the electrical transmission safety. As there is no accurate formula to derive the exact interface pressure and the measurement of the interface pressure is quite complicated or may exist errors that cannot be ignored, it is difficult to design a cable joint that can meet all kinds of requirements. In this paper, series of finite element models are established by Ansys. They are based on a 110kV crosslinked polyethylene (XLPE) cable with cross-section area of 1600 mm² which is connected by its cable joint made of silicon rubber. The distribution of the interface pressure value along axis is studied. The relationship between the Young's Modulus of the material of the cable joint is also analysed according to the models. Facts prove that the value the of interface pressure is changed along axis, and it is bigger in the middle of the cable joint. The interface pressure is directly proportional to the value of the cable joint's Young's Modulus. Using finite element method is of instructive significance and can save money and energy in the design of a cable joint.

1. Introduction

High voltage cable joint (Figure 1) plays a key role in connecting cables together to lengthen the cable lines for long distance power transmission. It is usually made by silicon rubber which are variable in Young's Modulus. According to different materials and installation methods, cable joints can be divided into thermal shrinkage type, wrapped type, pouring type, prefabricated type and cold shrinkage type. However, due to the composite interface and electric field stress concentration within the cable joint, it is generally known that cable joints have become the typical part of operational malfunction and weak points of high voltage distribution lines [1-3].

Content from this work may be used under the terms of the Creative Commons Attribution 3.0 licence. Any further distribution of this work must maintain attribution to the author(s) and the title of the work, journal citation and DOI.
Published under licence by IOP Publishing Ltd

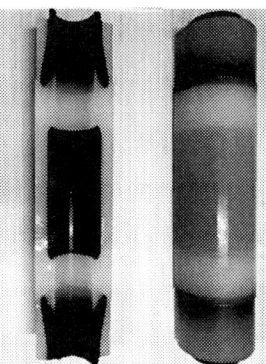

Figure 1. Cable joint

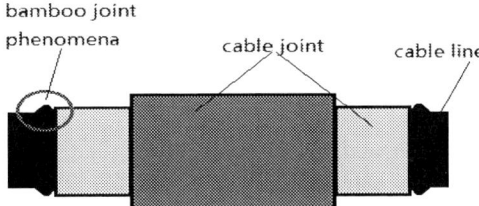

Figure 2. Bamboo joint phenomena caused by too much interface pressure between cable and cable joint.

Among factors which influence the quality of the assembly of cables and cable joints, controlling the interfacial pressure within an appropriate range is of essential importance [4-6]. Because it may not meet the electrical strength requirements and cable insulation strength if the interface pressure is too small, and it may cause difficulties in installation or damage called bamboo joint phenomenon to the cable shown in figure 1 if the interface pressure is too large.

The interface pressure between cable and cable joints is affected by many factors such as the structure of the cable joints, the interference between the cable and the cable joints, the dimensions and thickness of the cable joints, and the Young's Modulus of the materials. Because of the variable cross-section of the cable joints, there is no accurate formula to calculate this interface pressure. At present, the main methods of measuring the interface pressure between cable and cable joints are built-in sensor method, photo elasticity method, on-site electrical measurement etc. But all these methods above are complex or exist measurement errors such as damaging the continuity of the structure because of the insert sensor, different from the actual structure of cables, and all needing to produce the cable joints and assemble them with the cables before the measurement process [7]. Therefore, they are not practical for the design stage of the cable joints.

In this paper, we derived the interface pressure and its distribution along axis positions between the cable and cable joint using finite element method, and studied the relationship between the interface pressure and the Young's Modulus of different materials of the cable joints-silicon rubber. It is helpful for estimating the interface pressure without producing the cable joints before with sufficient measurement accuracy. Therefore, it is useful for the designation of the cable joint in the aspects of structure, size, interference value and so on to create an appropriate interface pressure.

2. Establishment of finite element model

In this study, axisymmetric simulation models are established by using the finite element analysis software Ansys. The models are based on the size of a 110kV crosslinked polyethylene (XLPE) cable with cross-section area of 1600mm^2 connected by the cable joint matching with the cable which is made of silicon rubber. The model is showed in figure 3.

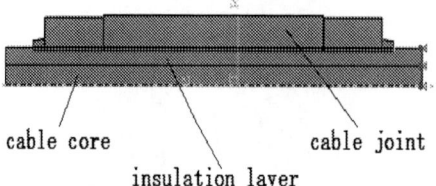

cable core   cable joint

insulation layer

Figure 3. Model of cable and cable joint.

In this model, we simplified the cable into two parts, the part reflecting the cable-core and the part representing the insulation layer. The cable-core is made of copper. So, its Young's Modulus of the material is 119GPa and the Poisson's Ratio is 0.33. The material behaviour of insulation layer is set up referring to XLPE with the Young's Modulus at 150MPa and the Poisson's Ratio at 0.5. And through reading papers about silicon rubber cable joints, we found that the Young's Modulus of materials widely used in cable joints were from 0.58MPa to 1.35MPa [2.8.9]. Therefore, to study the effect on interface pressure produced by different Young's Modulus of the materials and to give a guidance in practical design process for material selection, the material parameters of the cable joint are set up as silicon rubber with a range of Young's Modulus from 0.5MPa to2.0MPa wider than consulted values in data above and the Poisson's Ratio at 0.5. The Young's Modulus increase 0.1MPa in each model.

In the model, to ease the calculation difficulty and accelerate the convergence, the insulation layer is tied to the cable core. The contact relationship between the insulation layer and the cable joint is interference fit. The amount of the interference is constant at 4.5mm in radial direction. The models of these three parts are built according to the actual dimensions before assembling. To reflect the interference fit, we tick the button for automatic shrink fit. According to Saint-Venant's Principle [9], it will not influence the stress if we fix a bottom surface. Thus, one of the bottoms of the cable is in displacement in these models as the boundary condition to eliminate the rigid body displacement.

After all these settings, the model is mashed. The size of each element should not be too large or too small, or the interface pressure value may be inconsistent with actual ones. After many times of attempts, the whole model is shown in figure 4.

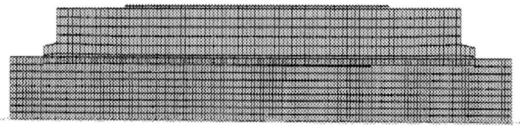

Figure 4. Meshed model.

3. Results of simulation

The distribution of the interface pressure in cable joint with Young's Modulus of 2.0MPa is shown in figure 5 as an example.

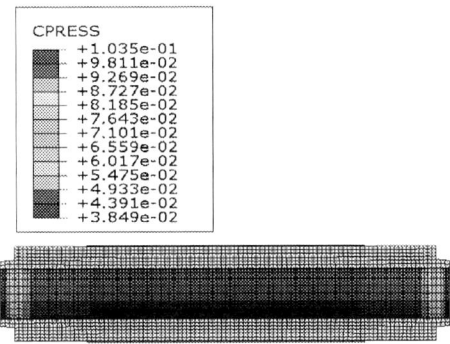

Figure 5. Distribution of interface pressure in the whole cable joint with Young's Modulus at 2.0MPa.

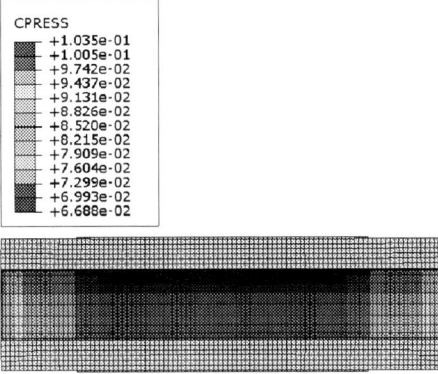

Figure 6. Distribution of interface pressure in the main part of the cable joint with Young's Modulus at 2.0MPa.

As the pressure around the edges is quite small contrasted with the main part of the model, to display a clearer distribution of the interface pressure, after cutting up both ends of the cable joint, the main part's interface pressure is shown in figure 6.

The values of relationship between interface pressure and axial positions of some different Young's Modulus cable joints are extracted from some model and shown in figure 7

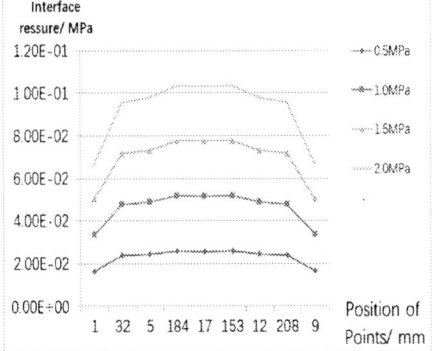

Figure 7. Relationship between interface pressure and axial position

From the results shown above, because the structure of the cable joint is symmetric, the interface pressure is also symmetric along the middle. It can be known that interface pressure is gradually changed along the axis, and it is bigger in the middle part of the cable joint and smaller at the two ends, but the maximum value is not in the exact middle. It can be inferred that it is because the cable joint is thicker in the middle but the connection part is right in the middle and it is weaker than the main part of the joint. As the minimum value appears at the two ends of the cable joints, and the interface pressure here is one of the important parts in the cable joint, it is needed to pay extra attention to the design of the cable joint at the two ends of it.

What we can also find in this figure is that the variety of the interface pressure along axis is lighter in the cable joints with smaller Young's Modules than cables joints with larger Young's Modulus. This law is instructive in the structure design in the aspect of difference in thickness.

In order to study the influence caused by the Young's Modulus and the interface pressure, there is a line chart that drawn to figure change of the maximum value of the interface pressure and the minimum one of each model with different Young's Modulus. The results are shown in figure 8.

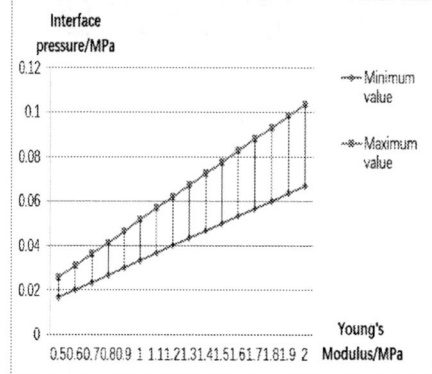

Figure 8. Interface pressure of cable joints with materials with different Young's Modulus.

From this chart, we can know that the maximum value and the minimum value of interface pressure are both directly proportional to the Young's Modulus of the materials. This change indicates that if the size of a cable joint is limited to a certain range, it is possible to change the interface pressure by selecting a different material with different Young's Modulus. It also defined an optional range of material selection for this certain kind of cable joints.

4. Conclusions

In this paper, axisymmetric simulation finite element model is established. Through finite element method, the interface pressure caused by interference fit of the cable and cable joint is calculated. The conclusion can be drawn as follows:

The interface pressure between 110kV cable and this certain kind of cable joint is gradually changed along axis. And the middle part of the pressure is bigger than that of the ends. But the maximum value is not right in the middle. The variation of the interface pressure along axis is smoother in the models with smaller Young's Modulus of cable joints. These conclusions can be referred in the structure design of cable joints.

The calculation results show that both the maximum one and the minimum one of the interface pressure value between the cable and its cable joint is directly proportional to Young's Modulus of the materials of cable joints. This means if the size of cable joints is limited to a certain range, it is a feasible method to control the pressure by selecting different kinds of materials of different Young's Modulus.

Numerous disciplines such as Elastic Mechanics, Electrical Science, and Material Science are involved in the design of cable joint. There is no accurate formula that can reflect the whole design in all factors accurately. So finite element method is a more economical method that can query the needed parameters of an exact product and can also help to optimize product design without producing it before.

References

[1] Z. D. JIA, Y. J. ZHANG, W. N. FAN, et al, Analysis of the Interface Pressure of Cold Shrinkable Joint of 10 kV XLPE Cable, *High Voltage Engineering,* **43**, 661-665 (2017)

[2] Z. JIAO and Q. XIA, Statistics and Analysis on the High-Voltage Power Cable Faults, *Distribution & Utilization,* **28**, 64-66 (2011)

[3] S. LIU, J. K. PENG, S. z. CHEN, et al, Simulation on the Relationship between Shrink Range and Interfacial Pressure in the HV Cable Joint, *Electric Wire & Cable*, **1**, 38-40 (2013)

[4] P. L. WANG, Electrical Field and Interface Pressure Control in HV Cable Accessories Design, *Electric Wire & Cable*, **5**, 1-5 (2011)

[5] S. LIU, J. K. PENG, X. WANG, et al, Relationship Between Permanent Deformation and Shrink Range in Cable Accessory, *Electric Wire & Cable*, **2**, 49-55 (2013)

[6] J. J. CUI, D. YU, X. WANG, et al, Research Progress of Measuring Method for Interface Pressure Between Cable and Accessory, *Insulating Materials*, **51**, 1-6, (2013)

[7] S. LIU, J. k. PENG, S. Z. CHEN, et al, Simulation of Interfacial Pressure between XLPE Cable and Silicon Rubber Prefabricated Joint Coupled with High Temperature, *Electric Wire & Cable*, **1**, 10-14 (2014)

[8] H. Y. WANG, J. M. ZHANG, Y. C. Lü, et al, Design and Verification of the Stress on Terminal Interface of 220 kV XLPE Cable Dry Type GIS, *High Voltage Engineering*, **47**, 105-110 (2011)

[9] Z. L. XU, *Concise Course of Elasticity* (Fourth Edition), (2013)

Effect of heat shield on the heating efficiency in MOCVD chamber by resistive heating

Lili Zhao[1], Zhiming Li[1,a], Jincheng Zhang[2], Runqiu Guo[3], Ligen Lu[1] and Lansheng Feng[3]

[1]Shandong Provincial Key Laboratory of Network based Intelligent Computing, School of Information Science and Engineering, University of Jinan, 250022 Jinan, China

[2]State Key Discipline Laboratory of Wide Band-Gap Semiconductor Technology, School of Microelectronics, Xidian University,710071 Xi'an, China

[3]School of Mechano- Electronic Engineering, Xidian University, 710071 Xi'an, China

[a] Corresponding author: ise_lizm@ujn.edu.cn

Abstract. In the conventional resistive heating systems, the heating efficiency of the resistance heating system is too low. In this work, by adding heat shield in the traditional heating structure. The effects of factors such as the number of layers, spacing and thickness of the heat shields on the heating efficiency of the heating system with a diameter of 12 inches are investigated. The results show that the heat shields significantly improve the heating efficiency compared to the heating systems without heat shields. When the number of layers of the heat shields is 3, the efficiency of the heating system is increased by about 26%.

1. Introduction

With the technological maturity of semiconductor devices and their wide application in various aspects, people pay more and more attention to the quality of semiconductor crystal materials [1-5]. Metal organic chemical gas phase epitaxy (MOCVD) is one of the important technologies for the preparation of nitride semiconductor materials [6-8]. In MOCVD devices, the reaction chamber is required to provide suitable temperature for the film growth and the temperature variation range is small throughout the growth process. Therefore, the design of the reaction chamber, especially the design of the heating system, will directly affect the quality of the crystal film [9]. The domestic and foreign researchers expect to design a superior reaction chamber to meet the growth requirements of crystal films.

References[10-12] shows that the position of the electromagnetic coil and the related electrical parameters is change in the simulation experiment. The article analyses the rule of radial temperature change on the surface of graphite disc, optimizes the structure of reaction chamber and improves the heating efficiency. In reference [13], the MOCVD reaction chamber model was heated by resistance wire, and it was found that the average temperature of graphite base changed linearly with the vertical distance from heater to base, and the heating structure was optimized. Literature [14] proposed two design schemes, array control heater and non-uniform resistance module, which were used to regulate the temperature distribution in the reactor, and both achieved fine adjustment of the temperature distribution in the reactor.

Content from this work may be used under the terms of the Creative Commons Attribution 3.0 licence. Any further distribution of this work must maintain attribution to the author(s) and the title of the work, journal citation and DOI.

Published under licence by IOP Publishing Ltd

In conventional resistive heating systems, the heating efficiency of the resistance heating system is low. In this paper, heat shields under the susceptor are placed. And the influence of factors such as the number of layers, spacing and thickness of the heat shields on the heating efficiency of the heating system with a substrate whose diameter is 12 inches are investigated. The purpose is to improve the heating efficiency of the heating system and to provide theoretical and technical references for the development of large-sized MOCVD reactor.

2. Model of MOCVD reactor

MOCVD epitaxial growth mechanism is complex, and the heating efficiency is influenced by many parameters, such as the heating parameters, the structure and the gas flow. The finite element method is usually used to establish the mathematical model, and the heat conduction model is coupled with the fluid model. In this work, the fluid and heat conduction equations are referred to manuals of the COMSOL multiphysics.

Heat conduction equation:

$$\rho C_p \frac{\partial T}{\partial t} + \nabla(-k\nabla T) = Q - \rho C_p \vec{u}\nabla T \qquad (1)$$

Fluid equation:

$$\rho \frac{\partial \vec{u}}{\partial t} - \nabla \eta \left(\nabla \vec{u} + (\nabla \vec{u})^T\right) + \rho \vec{u}\nabla \vec{u} + \nabla p = 0 \qquad (2)$$

Where ρ is density, C_p is specific heat, T is temperature, k is material thermal conductivity, Q is resistance heating, $\vec{u}$ is gas velocity, η is gas viscosity coefficient, and p is pressure. In order to improve the calculation speed in the simulation experiment, the following assumptions are made for the model without affecting the main simulation results [15].

(1) regardless of the rotation of the susceptor;
(2) chemical reactions between gases are ignored;
(3) assume that the gas is in an ideal state;
(4) the wall temperature is constant;
(5) the thermal expansion of the materials of the various materials is not considered.

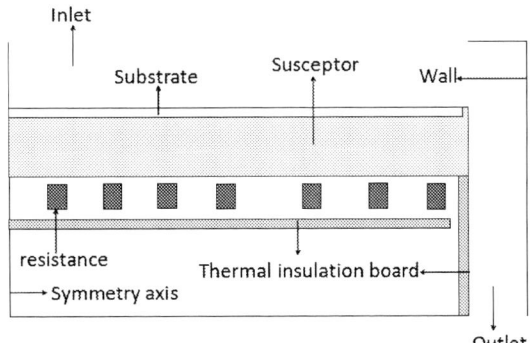

Figure.1 Geometrical model of MOCVD reactor

The MOCVD reactor to be modeled is shown in Figure.1. The reaction chamber is axisymmetric, which including the chamber walls, the inlet and outlet, heating resistances, the substrate and the heat shields. The material of the susceptor is silicon graphite, and the material of the heating resistances and the heat shields are tungsten and molybdenum, respectively. The susceptor height is 25.4mm, the radius of the substrate is 152.6mm. The power densities of the heating resistors from left to right are 7.5E+007, 7.5E+007, 7.5E+007, 8E+007, 9E+007, 1E+008, 1E+008, 1.2E+008, respectively. The length of the heat shields are all the same value-- 148.7 mm and the thickness of each shield is 2 mm. The spacing between the shields is 2 mm.

The gases in the reactor include, H2 (carrier gas) NH3 (N source) and TMGa (Ga source), which are mixed into the reaction chamber from the inlet. The flow rates of H2, NH3 and TMGa are 40slm, 10slm and 10sccm, respectively, which remain unchanged in the simulation. In addition, the boundary condition of the wall temperature is set to be 27℃.

3. Results and analysis

3.1. Effect of the thickness of the shield on heating efficiency

In order to reduce the heat loss caused by the thermal radiation, a layer of heat shield is added under the resistances in the heating system.Under the condition that the length of the shield is kept constant, and the thickness of the heat shield is gradually increased, the effect of heat shield on the heating efficiency is analyzed.

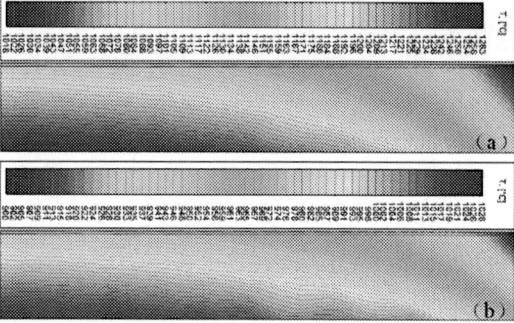

Figure.2 Temperature distributions of two susceptor: new susceptor (a) and conventional susceptor (b)

Figure.2 shows the temperature distribution contours of the susceptor with one heat shield (a) and the traditional susceptor (b). It can be seen from the figure that the temperature distribution in the susceptor is basically the same, and the high temperature region is mainly Focusing on the lower part of the base, the temperature at the edge of the base is relatively low. Due to the radiation effect of the heat shield, the temperature of the base substrate including the edge portion is greatly increased as a whole. Which means that the increase of the heat shield can improve the heating efficiency of the heating system and improve the uniformity of the temperature field in the reaction chamber.

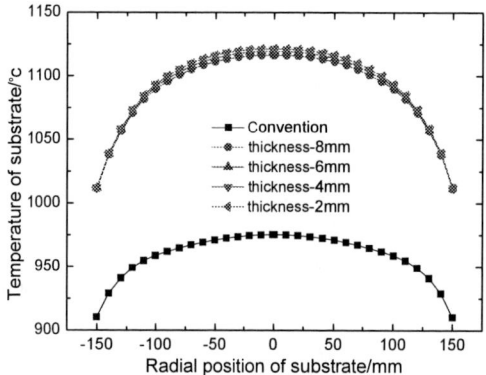

Figure.3 Temperature distributions of substrate in the heating structure with different thicknesses of heat shield and without shield

Figure.3 shows the temperature distributions of substrate in the heating structure with different thicknesses of the heat shield and without the heat shield. .It is found that the average temperature of the substrate in the structure without the shield is 960℃, while the average temperatures of the substrate are 1134℃,1095℃, 1093℃and 1091℃, under the conditions of the thicknesses of the shield are 2 mm, 4mm, 6mm and 8mm, respectively, which shows that the heating efficiency of the structure with the shield is higher than that without the shield. In addition, as the thickness of the heat shield increases, the heating efficiency is slightly reduced. The combination is the actual situation, the thickness of the heat shield to take 2mm is more appropriate.

3.2 Effect of the number of heat shield layers on heating efficiency
On the basis of the above calculation, the thickness of the heat shield is kept at 2 mm and the spacing is 2 mm. The number of layers of the heat shields is gradually increased, and the influence of the number of shield layers on the substrate temperature is investigated.

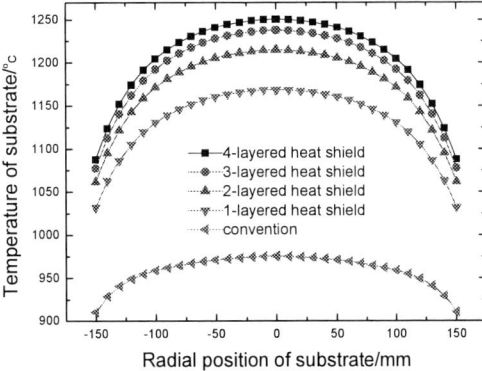

Figure.4 Temperature distributions of substrate in heating structure with different layered shields

Figure.4 shows the temperature distributions of the substrate under the conditions of the heating structure with different layered shields. It can be seen that the substrate temperature is directly proportional to the number of shield layers. Meanwhile, with increasing of the number of shield layers, the substrate temperature decreases. When the number of heat shield layers is changed from 1 to 4, the average temperatures of the substrate are 1134℃, 1175℃, 1196℃ and 1208℃, respectively, which means that the heating efficiency is increased by 18%, 22%, 25% and 26% , respectively. Taking the changes of the substrate temperatures with the shield layers and the growth temperature into account, three-layered shields can meet the actual need of film growth.

3.3 Effect of heat shield spacing on heating efficiency
The thickness of the each shield is kept at 2 mm, and the number of the shield layers is 3. The influence of the spacing of the shields on the temperature is studied.

Figure.5 shows the temperature distributions of substrate in the heating structure with different spacings of heat shields. It is found that the changes of the substrate temperature are not obvious with the increase of the heat shield spacings, which indicates that the heat shield spacings have less influence on the substrate temperature and heating efficiency.

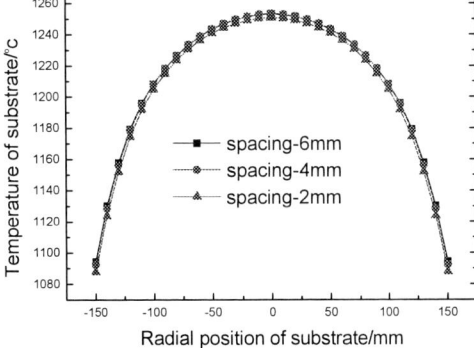

Figure.5 Temperature distributions of substrate in heating structure with different spacings of heat shields

4. Conclusions

In the MOCVD chamber, the resistance heating system without the heat shields has low heating efficiency, because that part of the heat generated by the heating resistance is taken away by the gas flow and the radiation in the reactor. In order to improve the heating efficiency of the heating system, the heat shields are placed under the resistances in the MOCVD reactor to improve the heating efficiency. By studying the influence of the thicknesses, the number of shield layers and the spacings of each shield layer on the substrate temperature, it is found that the shields significantly improves the heating efficiency compared with the system without shields. And the number of layers is directly proportional to the heating efficiency, but the thicknesses and spacings of the shields have little effect on the heating efficiency. The calculation results show that the heat shield can meet the heating requirement when the layer number is 3.

Acknowledgments

This work was financially supported by National Key R&D Program of China(NO.2017YFB0404202).

References

[1] Surender S, Prabakaran K, Loganathan R, et al, NED. J CRYST GROWTH. *Effect of growth temperature on InGaN/GaN heterostructures grown by MOCVD.* J. 468 (2016)

[2] Loganathan R, Balaji M, Prabakaran K, et al, NED. J MATER SCI-MATER EL. *The effect of growth temperature on structural quality of AlInGaN/AlN/GaN heterostructures grown by MOCVD.* J. **26**, 5373 (2015)

[3] Lumbantoruan F, Shrestra N M, Chang E Y, IEEE. ICSE. J. *Investigation of TMAl preflow to the properties of AlN and GaN film grown on Si(111) by MOCVD.* 20 (2014).

[4] Wengjin Cheng, Xiaoliang Gong, Fengwu Chen, et al, CHN. Equip Electron Product Mfr. *Epitaxial growth of AlN by high temperature MOCVD.* J. **4**, 43 (2017).

[5] Boting Liu, Ping Ma, Xilin Li, et al. CHN. CHINESE PHYS LETT. *Influence of Al Preflow Time on Surface Morphology and Quality of AlN and GaN on Si (111) Grown by MOCVD.* J. **5**, 34 (2017)

[6] Ran Zuo, Hong Zhang, CHN. CHIN LIGHT & LIGHTING. *Progress in the Research of Chemical Reaction during the III-Nitride MOCVD Growth.* J. 10, 4 (2017).

[7] Dobrzanski L, Pakula D, Staszuk M, LON. Springer London. *Chemical Vapor Deposition in Manufacturing.* M. 2755 (2014)

[8] Capper P, Irvine S, Joyce T, US. Springer US. *Epitaxial Crystal Growth: Methods and Materials.* M. 231 (2017).

[9] Ran Zuo, CHN. National MOCVD Conference. *Optimal design and optimal growth parameters of MOCVD reactor.* C. National MOCVD Conference. (2010)

[10] Li Z, Li J, Jiang H, et al, NED. J CRYST GROWTH. *Effect of thermocouple position on temperature field in nitride MOCVD reactor.* J. **368**, 29 (2013)

[11] Longquan Xu, Song Fang, Zihan Tang , et al, CHN. CHIN J LUMIN. *Research on Heating Uniformity of MOCVD Heating Device.* J. **38**, 220 (2017)

[12] Longquan Xu, Xinwei Liu, Zihan Tang, et al, CHN. MECH SCI TECHNOL AERO ENG. *Optimization Design of GaN-MOCVD Induction Heating Device.* J. 12, 1 (2017)

[13] Yuxuan Qu, Bin Wang, Shigang Hu, et al, CHN. J CENT SOUTH U. *Analysis and design of resistance-wire heater in MOCVD reactor.* J. **21**, 9, 3518 (2014)

[14] H Xia, D Xiang, P Mou, CHN. J Mechanical S T. *The array-control heater and non-uniform resistance module design for regulating the temperature profile in a reactor chamber.* J. **29**, 2, 593 (2015)

[15] Geiser J. Polymer. *Multiscale modeling of chemical vapor deposit Polymersion (CVD) apparatus: simulations and approximations.* J. **5**, 1, 142(2013)

ISPECE   IOP Publishing

Study on Basic Experiment and Optimization Prediction Model of Orthogonal Electrolytic Machining of Film Cooling Hole in High Temperature Nickel-Based Alloy Blades

Yulan Hu[1], Weiguang Hao[1], Guoqiang Liu[2], Yajie LIu[2], Zilin Liu[1]

[1]School of Information Science and Engineering, Shenyang Ligong University, Shenyang, China

[2]School of Mechanical Engineering, Shenyang Ligong University, Shenyang, China

[a] Corresponding author: author@e-mail.org

Abstract. The orthogonal test was designed based on high temperature nickel-based alloy Inconel718. The variance analysis was carried out on the test results. The significant factor was determined by the significance test. The model structure was trained by the electrolytic processing test data, and finally the prediction model of the momentum-adaptive learning BP Neural Network is established. The model is used to predict the pore size of stainless steel micropores processed under different processing parameters. The results show that the model has a prediction error of less than 5% and has a strong predictive power.

1. EXPERIMENTAL EQUIPMENT AND MATERIALS

Based on the electrolytic processing method and mechanism, the experimental system for electrolytic processing was constructed as shown in Fig. 1. The experimental system includes an electrolysis motion control portion, an electrolyte constant temperature control portion, and an electrolyte circulation control portion. With servo drive, the feed is stable. The electrolysis power supply uses a high frequency pulse power supply.

The material was made of Inconel 718, a high-temperature nickel-based alloy commonly used for turbine blades, and 50 specimens of 10 mm × 20 mm were cut by a fiber laser cutter. Before processing, the test piece is cleaned and dried by an ultrasonic cleaner.

Figure 1. Electrolytic processing experimental system

Content from this work may be used under the terms of the Creative Commons Attribution 3.0 licence. Any further distribution of this work must maintain attribution to the author(s) and the title of the work, journal citation and DOI.
Published under licence by IOP Publishing Ltd

2. MAIN FACTORS AFFECTING ELECTROCHEMICAL MACHINING

Electrochemical machining [6] is a method based on the principle of electrochemical dissolution of metal anode to machine and shape a workpiece with a shaped cathode tool. Electrical parameters, electrolyte parameters, workpiece characteristics, cathode and other parameters (as shown in Figure 1) are the main factors affecting the processing quality.

The prediction model is studied with the easily measured and easily controlled machining parameters such as current value I, electrolytic voltage U, pulse power frequency F, pulse power duty ratio C, initial machining gap D and electrode feed speed V, etc.

3. ORTHOGONAL EXPERIMENTAL DESIGN AND RESULTS ANALYSIS

Based on orthogonal experiments, this paper finds how to quickly find significant factors from many influencing factors, and analyzes the influence of single factors on the response of each processing parameter. The experiment used 6 factors and 5 levels of experiments, a total of 50 groups of experiments. Three wells were processed in each set of experiments and the mean values were taken. Process parameter level table 1 is as shown:

Analysis of experimental results.

Regression variance analysis of inlet roundness. The F-value of the inlet roundness model is 2.44, the corresponding p-value is $0.0404 < 0.05$, and the model is established. It is indicated that the regression equation of the inlet roundness is generally significant, the power supply pulse frequency plays a general role, and the remaining items have less influence on the inlet roundness.

Table 1 Process parameter level table

factor	Level				
	1	2	3	4	5
A.electrolysis voltage （0-24V）	4	7	10	13	16
B.Current value （0-5A）	3	3.5	4	4.5	5
C.Pulse power frequency (1-10kHz)	1	3	5	7	9
D.Pulse power supply duty cycle （%）	30	40	50	60	70
E.Initial machining gap （μm）	40	45	50	55	60
F.Electrode feed rate （μm）	6	8	10	12	14

Regression variance analysis of positive side clearance. The F-value of the positive-side gap model is 16.59, and the corresponding P-value is 0.0001, and the model is established. It is indicated that the regression equation of positive side clearance is the most significant, and the power supply voltage and power supply duty ratio play a major role. Power supply pulse frequency is generally significant, the positive side clearance plays a general role. The remaining items have less effect on positive side clearance.

Analysis of the regression variance of the cone degree. The F-value of the positive-side gap model is 9.99, and the corresponding P-value is 0.0001, and the model is established. It is indicated that the regression equation of taper has the most significant effect on the power supply voltage and the duty ratio of power supply. The pulse frequency of the power supply is generally significant, and it has a general effect on taper.

4. BP NEURAL NETWORK MODEL DESIGN

4.1 BP network structure design

A BP neural network with a single hidden layer is used to establish a predictive model of the relationship between processing parameters and the size of the processed micropore. The number of input layer nodes of the BP network is m = 6, corresponding to the six processing parameters mentioned above; the number of nodes of the output layer is n = 3, corresponding to the upper surface diameter D1 (mm) of the micropore and the surface aperture D2 of the micropore (Mm) and electrolytic processing time t(s); ωij is the weight of the jth input layer node to the i-th hidden layer node, and ωki is the weight of the i-th hidden layer node to the kth output layer node ,as shown in Figure 2.

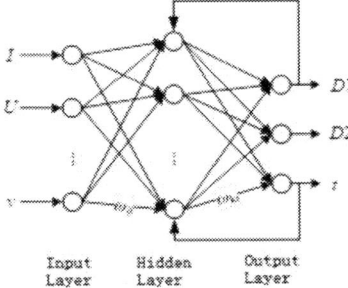

Figure 2. Typical three-layer BP neural network structure

The number of nodes in the middle hidden layer of the BP network has an important influence on the learning and calculation of the network. The number of hidden layer nodes is too small, the network is too simple to meet the accuracy requirements; the number of hidden layer nodes is too large, the training time is too long, and it is easy to over-fitting so that the generalization ability is poor. Determine the number of hidden layer nodes based on the following empirical formula:

$$h = \sqrt{m+n} + b \qquad (1)$$

In the formula, h is the number of hidden layer nodes, and b is an arbitrary number from 1 to 10.

4.2 Design Test

The electrolytic part of the micro-machining system is used for electrolytic machining test. The system mainly consists of gas-assisted laser machining module, high-frequency microsecond pulse electrolytic power supply, thermostatic electrolyte system, sidewall insulated tubular electrode and low-concentration acidic passivating electrolyte.

Considering the cost during the test, a bare electrode with a diameter of 0.6 mm, a NaNO 3 solution with a concentration of 8 %, an electrolyte of 21 ℃, a 1Cr18Ni9 Ti stainless steel sheet with a thickness of 0.59 mm for the anode workpiece, etc. were used. For the six input layer parameters that need to be tested, the orthogonal test method is used in the range of effective parameters, which can greatly reduce the number of tests and make the tests more representative and persuasive. The parameter design is shown in Table 2.

Table 2 Parameter Design in Effective Range

NO	U/V	I/A	f/kHz	c/%	d/μm	v/μm
1	8	3	1	30	40	10
2	9	3.5	3	40	45	11
3	10	4	5	50	50	12

| 4 | 11 | 4.5 | 7 | 60 | 55 | 13 |
| 5 | 12 | 5 | 9 | 70 | 60 | 14 |

4.3 Model training and prediction results analysis

The parameters in Table 1 were designed by orthogonal test method. 50 sets of test data were used to train the improved BP neural network, and the remaining 10 sets of data were used to test the trained BP neural network prediction model. During the training process, it is found that when the number of nodes in the hidden layer is 10, the prediction effect is the best, and the value of each parameter in the improved BP neural network is determined. Among them, the adaptive learning rate initial value $\eta 0 = 0.035$ and $0.015 \leq \eta 1 \leq 0.2$, the momentum factor initial value $mc0 = 0.65$ and $0.65 \leq mc \leq 0.9$, $\alpha = 0.001$. The target error $E0 = 4$, when the target error is too small, there will be over-fitting; when the target error is too large, the higher training fit will not be achieved, both of which will greatly reduce the prediction accuracy. A comparison of the predicted and experimental values of the processing time is shown in Figure 3.

At this point, the value of the prediction accuracy is completed. The above process is repeated ten times, and the grouping conditions are different each time,the calculated inlet diameter simulation error is 5.04%, the exit diameter simulation error is 4.15%, and the electrolytic machining time prediction error is 3.29%. The prediction error is within ±15%, compared with the average prediction error of 5.97% obtained by Zheng Xu et al. [8] in optimizing the process parameters in this field, and Wang Lei et al. [9] has an obvious prediction error of 12% in this field. Improve the effect. The prediction results are basically consistent with the actual processing

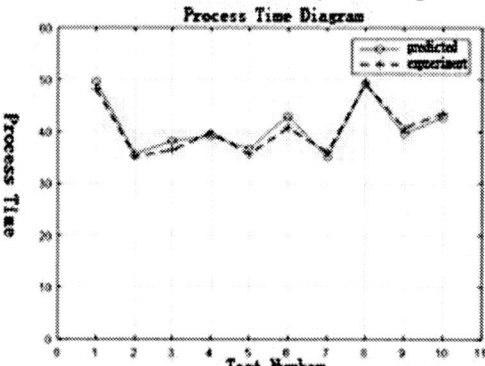

Figure 3. Processing time prediction comparison

results, and the validity of the momentum-adaptive learning BP neural network electrolytic machining prediction model is verified.

5. Conclusion

The orthogonal test of the blade film cooling hole was carried out by the experimental processing system built by the independent experiment. The results show that the power pulse frequency has a general effect on the inlet roundness. The power supply voltage and the power supply duty cycle have a major influence on the positive side clearance. The voltage and power supply duty cycle have a major influence on the taper, and the power pulse frequency is generally significant, which has a general effect on the taper. A model for predicting the parameters of electrolysis machining by momentum-adaptive learning BP neural network is proposed. The main processing parameters are selected to determine the model structure of the BP neural network. The orthogonal test was designed by using the existing test system device, and the momentum-adaptive learning BP network was trained and tested with the test results. It has been verified that the prediction model has an accuracy of 95% for the diameter and processing time of the electroporation micropores. The use of intelligent algorithms such as machine learning and artificial neural networks to solve complex nonlinear problems in the field of practical engineering has obvious advantages. The model of this paper has certain guiding significance for future production.

Acknowledgments

The work was supported by the National Natural Science Foundation of China(NO.61672360).

References

[1] Wei Shaobin, Lu Feng, He Limin, et al. Research progress in thermal barrier coating preparation technology and ceramic layer materials [J]. Thermal Spray Technology, 2013, 5(1): 31-37.

[2] Liu Chao, Peng Wei, Zhang Wei, et al. Progress in the application of thermal barrier coating technology[J]. Aerospace Manufacturing Technology, 2012, (4): 10-13,26.

[3] Li Guangchao, Bai Shusheng, Wu Dong. Research Progress of the Effect of Film Hole Shape on Turbine Blade Film Cooling[J]. Thermal Power Engineering, 2010, 25(6): 581-585.

[4] Gong Wei, Zhu Wei, Qu Ningsong, et al. Study on the stability of tube electrode electrolysis process engineering engineering[J]. Journal of Mechanical Engineering, 2010, 46(11): 179-184.

[5] Sun Jianjun, Li Zhiyong, Xiao Chuanwu. Electrochemical processing of aero-engine air film cooling holes [J]. Electroplating and Finishing, 2015.34 (11): 626-631.

[6] Yan Hongjuan, Li Yongzhi, Cui Qingwei, Yan Fengjie, Zheng Guangming, Li Li.Study on the basic experimental study on the electrolytic film processing of high temperature nickel-base alloy turbine blade film cooling hole[J].Process and Testing,2018,(3):134-142 .

[7] Schmidhuber J. Deep learning in neural networks: An overview[J].Neural Netw, 2014, 61:85-117.

[8] Hecht-Nielsen. Theory of the backpropagation neur al network[C]//International Joint Conference on Ne ural Networks. IEEE, 2002:445-445.

[9] Hadley A J, Krival K R, Ridgel A L, et al. Neura l Network Pattern Recognition of Lingual–Palatal Pressure for Automated Detection of Swallow[J]. D ysphagia, 2015, 30(2):176-187.

[10] LIU Li, HUO Liqing, LU Hongru, et al. The Use of Fuzzy BackPropagation Neural Networks for the Early Diagnosis of Hypoxic Ischemic Encephalopat hy in Newborns[J]. Biomed Research International, 2013, 2011(3):349490.

[11] Khurodze R. Development of the Learning Process for the Neuron and Neural Network for Pattern Re cognition[J]. American Journal of Intelligent System s, 2015, 5(1):34-41.

Multi-objective reactive power optimization of hybrid AC/DC power system considering power system uncertainty

Liu Wenxue , Xing Luhua, Ma Changhui and Li Wenbo

State Grid Shandong Electric Power Company Electric Power Research Institute .
250001 Jinan, China

[a] Corresponding author: Liu Wenxue liu_wenxue@126.com

Abstract. With the High Voltage Direct Current (HVDC) transmission is connected to the AC system, the uncertainty of converter reactive power consumption has brought certain influence to the reactive power optimization of the power grid. For the uncertainty of converter reactive power consumption and load, the DC reactive power and load are expressed by trapezoidal fuzzy parameters, and the multi-objective fuzzy reactive power optimization model of AC / DC interconnected system is established which optimizes the total network loss and node voltage offset of the AC / DC transmission system. The multi-objective optimization is converted into single-objective optimization by using the objective membership function and the fuzzy algorithm of multi-objective nonlinear objective programming, and the biogeography-based optimization algorithm is used to solve the model. The results of extended 39-node system example show that the proposed model and algorithm are suitable for voltage and reactive power control optimization under the condition that the converter reactive power consumption and load are fuzzy parameters.

1. Introduction

Reactive power optimization of power system is to make the active power network loss minimum, the voltage level best or the reactive power reserve capacity optimal by adjusting the generator terminal voltage, adjustable transformer ratio and the switching of compensation equipment under the premise of meeting the system constraints such as power flow equation, node voltage and generator reactive power output[1]. At present, reactive power optimization is mainly focused on the AC system[2-3]. Tabu search method is applied to reactive power optimization of AC power grid and compared with genetic algorithm[2]. Reference [3] proposes the gradient optimization algorithm to solve the reactive power optimization of AC power system.

HVDC transmission is widely connected to the power system because of its advantages such as long distance transmission and large capacity. Due to the power electronic characteristics of converter stations, it is necessary to absorb a large amount of reactive power from the AC system, which brings adverse effects to the reactive power flow of power system. Therefore, the system network loss and voltage quality can be effectively improved by optimizing and coordinating reactive power and voltage control measures in AC/DC transmission system, and the adverse effects of HVDC transmission can be reduced to a certain extent. Therefore, reactive power optimization of AC / DC transmission systems has been attracted widespread attention[4-7]. Reference [4] proposes the right singular vector index of singular value decomposition method, and the weak node of voltage stability in AC/DC system is identified and used as reactive power compensation point. Reference [5] presents a hybrid algorithm, which uses genetic algorithm to deal with discrete variables and interior point

---

Content from this work may be used under the terms of the Creative Commons Attribution 3.0 licence. Any further distribution of this work must maintain attribution to the author(s) and the title of the work, journal citation and DOI.
Published under licence by IOP Publishing Ltd

method to deal with continuous variables, effectively simplifying the traditional AC-DC hybrid power flow calculation method. An improved normalized planar constraint method is proposed to transform multi-objective optimization into a single objective optimization problem for reactive power optimization of AC/DC power systems[6] . Reference [7] presents a multi-objective reactive power optimization control method for AC/DC systems considering the detailed loss characteristics of converter stations.

The above studies mainly focus on the AC/DC reactive power optimization model under deterministic conditions, but there are a lot of uncertainties in power system [8]. Therefore, reactive power optimization of AC / DC system considering system uncertainties should be the direction of future research. Based on the normal distribution characteristics of load errors, a multi-objective reactive power optimization model for AC/DC systems considering load uncertainty is proposed by using the method of bilevel programming theory[9]. However, the load is often fuzzy, and the expression of fuzzy parameters is more suitable [10-12]. Moreover, the uncertainty in the operation of the converter station is not considered in the above paper[13].

Based on the above research, for the uncertainties of load and reactive power consumption of converter stations in AC/DC system, a multi-objective fuzzy optimization model of AC/DC reactive power is constructed with the total network loss and voltage offset as the multi-objective.The biogeography-based optimization algorithm is used to optimize the model. The results of the example show the feasibility of the proposed model and algorithm.

2. Fuzzy characteristics of AC/DC power system

2.1 Fuzzy characteristics of load
Real-time data are needed in reactive power optimization of power system. Because of the large number of nodes or investment, the real-time data of each node load can not be obtained in power grid operation. Some data can only be obtained from historical data or forecast data, so load data can only be regarded as fuzzy information of current operation data. Trapezoidal membership function is widely used in power system because of its rich theoretical basis and accord with people subjective judgment of thing in reality [10-12]. The trapezoid membership function of load is expressed by the predicted value (or historical value) and the four proportional parameters.

$$P = (P_1, P_2, P_3, P_4) = P_{fc}(w_1, w_2, w_3, w_4) \qquad (1)$$

where, P_{fc} represents the reference value (predicted or historical value) of load, and w_k is proportional parameter , $k = 1, 2, 3, 4$.

2.2 Fuzzy characteristics of DC
The AC system and DC system are mainly connected by converter. According to the principle of converter, the formula of reactive power consumption by converter as follows:

$$Q_{dc} = P_{dc} \tan \phi \qquad (2)$$

$$\tan \phi = \frac{(\pi / 180)\mu - \sin \mu \cos(2\alpha + \mu)}{\sin \mu \sin(2\alpha + \mu)} \qquad (3)$$

$$\mu = \cos^{-1}[U_d / U_{di0} - (X_c / \sqrt{2})(I_d / E_{11})] - \alpha \qquad (4)$$

$$U_d / U_{di0} = \cos \alpha - (X_c / \sqrt{2})I_d / E_{11} \qquad (5)$$

where, U_{di0}、P_{dc}、Q_{dc}、ϕ、μ、X_c、I_d、α、E_{11}、U_d are ideal no-load voltage, DC side power, reactive power consumptionr, power factor angle, commutation angle, per-phase commutation reactance, DC operating current, rectifier trigger angle, valves side winding no-load voltage , DC voltage.

It can be seen from the above formula that it needs to obtain reactive power from the AC system no matter whether the converter station is in rectifier or inverter operation state, that is, the converter station is a reactive load for the AC system. Moreover, the reactive power consumed by the converter station is not only affected by the active power, but also related to many operating parameters. The reactive power consumption calculated by the above formulas is all calculated under certain equipment, operation and control parameters. In actual operation, the converter voltage is uncertain because of some parameters, such as bus voltage uncertainty and converter voltage, in addition to the given operational control and other deterministic parameters, when calculating the reactive power consumption of the converter station. Impedance tolerance of converter, inaccuracy of control parameters caused by some measurement data and control errors will lead to uncertainty of reactive power consumption of converter. Ref. [13] analyzes the reactive power consumption of converter under different working conditions and considerations in detail, as shown in Table 1.

Table 1. Reactive power consumption by rectifier side under normal operation.

Operation mode	Consideration of factors	Reactive power(Mvar)
Rated operating mode bipolar, $\pm 500\,kV$ $3000\,MW$	without consideration	1492
	considering control range	1635
	and considering measurement error	1651
	considering all factors	1705

It can be seen that rectifier have different reactive power consumption under different operating conditions . Therefore, the trapezoidal membership function can be used to represent the uncertainty of reactive power dissipated by four different parameters, and the reference value can be expressed by the mean of the middle two values.

3. Multi-objective reactive power fuzzy optimization model

The multi-objective fuzzy chance-constrained
reactive power optimization model makes use of the reliability theory measure parallel to the probability measure in probability theory, which can make up for the shortage of the possibility measure to express the risk and is widely used[12] . Based on the above model, a simpler expectation model is proposed to solve the complex problem of objective function [14] .

Under the condition of credible constraints, the mathematical model of minimizing the uncertain objective function is minimized. The expected model is as follows

$$\begin{cases} \min E\left[f\left(\chi,\xi\right)\right] \\ s.t \\ \mathrm{Cr}\left[g_i\left(\chi,\xi\right)\right] \le 0\, i=1,2...,\,p \end{cases} \quad (6)$$

where, χ is decision vector, ξ is fuzzy variable, $f\left(\chi,\xi\right)$ is uncertain objective function, $g_i\left(\chi,\xi\right)$ are constraint function.

The bjective function $F = \min E[f_{loss}, f_{VD}]$

$$\begin{cases} f_{loss} = \sum_{i,j \in N_L} g_{ij}(U_i^2 + U_j^2 - 2U_i U_j \cos\theta_{ij}) \\ f_{VD} = \sum_{i \in N_{PQ}} (\dfrac{U_i^{spe} - U_i}{U_i^{max} - U_i^{min}})^2 \end{cases} \quad (7)$$

where, f_{loss} is loss of total network, f_{VD} is voltage offset. U_i、 U_j、 U^{spec}、 U^{max}、 $U^{min}$are the voltage amplitude, voltage reference value, upper limit and lower limit of the voltage of node i . N_L、 N_{PQ} are the set of transmission lines and the set of PQ nodes.

Equality constraint, the AC power flow equation:

$$\begin{cases} P_i - U_i \sum_{j=1}^{N} U_j (G_{ij}\cos\theta_j + B_{ij}\sin\theta_j) = 0 \ i \in N_{PQ} \\ Q_i - U_i \sum_{j=1}^{N} U_j (G_{ij}\sin\theta_j - B_{ij}\cos\theta_j) = 0 \ i \in N_{PV}, N_{PQ} \end{cases} \qquad (8)$$

where, P_i、Q_i are the active power and reactive power, G_{ij}、B_{ij} are the conductance and the admittance between the nodes i、j, N_{PV} are the PV node set.

The power balance equation of the AC node connected to the converter station is

$$\begin{cases} \Delta P_z = P_z^s - P_{z(dc)} - P_z(U,\theta) = 0 \\ \Delta Q_z = Q_z^s - Q_{z(dc)} - Q_z(U,\theta) = 0 \end{cases} \qquad (9)$$

where, the formula z represents the AC node connected with the converter station, P_z、Q_z are the active and reactive power of the AC side respectively, and P_z^s、Q_z^s are the active and reactive power given by the node respectively. $P_{z(dc)}$、$Q_{z(dc)}$ are the active and reactive power of DC system is injected into the AC system.

Control variable constraint:

$$\begin{cases} U_{Gi}^{\min} \le U_{Gi} \le U_{Gi}^{\max} \ i = 1,2,\cdots,N_G \\ Q_{Ci}^{\min} \le Q_{ci} \le Q_{Ci}^{\max} \ i = 1,2,\cdots,N_C \\ T_i^{\min} \le T_i \le T_i^{\max} \ i = 1,2,\cdots,N_T \end{cases} \qquad (10)$$

where, N_G、N_C、N_T are the number of generator nodes, the number of compensation capacitors and adjustable transformers, $U_{Gi}^{\min}$、$U_{Gi}^{\max}$、$Q_{Ci}^{\min}$、$Q_{Ci}^{\max}$、$T_i^{\min}$、$T_i^{\max}$ are respectively the upper and lower bounds of corresponding control variables. State variable constraint:

$$\begin{cases} \mathrm{Cr}(U_i^{\min} \le U_i \le U_i^{\max}) \ge \alpha, i = 1,2,...,N_{PQ} \\ Cr(Q_{Gi}^{\min} \le Q_{Gi} \le Q_{Gi}^{\max}) \ge \alpha, i = 1,2,.....N_G \end{cases} \qquad (11)$$

where, $Q_{Gi}^{\min}$、$Q_{Gi}^{\max}$ the upper and lower limits of the reactive power output of the corresponding generators respectively.

4. Solution of multi-objective fuzzy optimization model

4.1. Fuzzy power flow
There are fuzzy variables in the objective function and constraints of the fuzzy value model. How to get the expected value and credibility of the fuzzy power flow is the key to solving the model. The fuzzy membership function is used to solve the credible distribution of state variables in this paper. The calculation process is as follows:

(1)   Formulas (2), (3), (4), (5), (8) and (9) are used to form the power flow equations of AC/DC systems.

(2)   Without considering the fuzzy variables of the system, the power flow of AC/DC system with reference value is solved by Newton method[15]. and V_d、θ_d、P_{lossd}、V_{VDd} are the reference value of node voltage, voltage phase angle, power loss, voltage offset. The subscript d is the reference value of the corresponding variable.

(3)   Solution of fuzzy increment of active and reactive power injected into nodes ΔP、ΔQ.

(4)   Using the Jacobi matrix of Newton power flow algorithm to solve the fuzzy increment of node voltage and phase angle ΔP、ΔQ.

$$\begin{cases} \Delta \theta \approx \mathrm{H}\Delta P \\ \Delta V \approx \mathrm{L}\Delta Q \end{cases} \qquad (12)$$

(5) Solving fuzzy variables V、θ.

(6) Solving fuzzy variables Q、V_{VD} and P_{loss}.

$$\Delta y \approx \sum \frac{\partial y}{\partial x_j} \Delta x_j \qquad (13)$$

where, y are reactive power output, voltage deviation and network loss of power system, and x_j and y fuzzy variables for corresponding systems. After solving the fuzzy increment corresponding to the state variable, the membership function of the objective function is solved by expression 14.

$$y = y_d + \Delta y \qquad (14)$$

(7) The reliability distribution of state fuzzy variable and the expected value of objective function fuzzy variable are solved by the relationship between membership function, reliability measure and expected value[14].

4.2 Multi-objective processing strategy

Pareto optimal frontier is the optimal solution set in the feasible domain, because of its large amount of calculation, and in practical problems, decision makers often make decisions according to demand which increases the amount of calculation. Therefore, this paper uses fuzzy programming to transform multi-objective optimization into single objective optimization [16].

Fuzzy programming needs to solve two key problems: 1) Fuzzy processing is achieved by establishing membership function of objective function; 2) Fuzzy operator is used to synthesize different objectives, and then the overall satisfaction is formed to obtain Pareto optimal solution. In this paper, linear functions are used to construct membership functions. If the decision maker is not satisfied with the local effective solution, the effective solution can be updated by changing the objective membership value until the satisfactory solution is obtained.

4.3 Biogeography-based optimization algorithm

The biogeography-based optimization algorithmis

(BBO)is an optimization algorithm based on swarm intelligence which is just put forward in 2008. It has the advantages of fewer parameters, simple calculation and fast convergence speed[17].

BBO algorithm simulates the mechanism of species migration between habitats.For solving the optimization problem, BBO constructs multiple habitats randomly as the initial solution of the optimization problem, and interacts information through species migration between habitats to improve the habitat. The species diversity of the land is improved, and the HSI of the habitat is improved, thus obtaining the optimal solution of the problem.

5. Example analysis

5.1 Description of the test systems

To verify the effectiveness of the proposed method, an improved IEEE 39-bus system is used to verify the proposed method. The system wiring diagram is shown in Figure 3. HVDC transmission system are introduced. The rectifier side is node 18, the inverter side is node 6. The IEEE 39 extended example is shown in Figure 1. The DC transmission system is controlled by constant current and constant power. The initial DC current is 1.0 p.u, the arc extinguishing angle is 8.0 degrees, and the upper and lower limits of node voltage are 1.10 and 0.92 respectively.

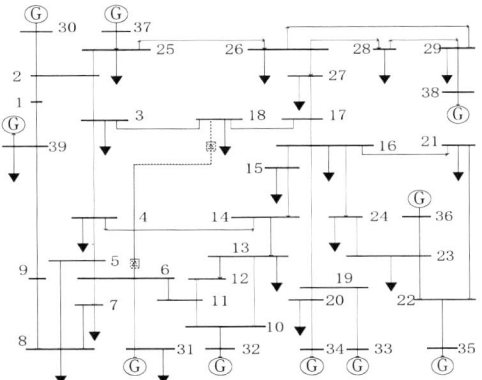

Figure 1. IEEE 39 extended system

5.2 Analysis of optimization results

As shown in Table 2, the proposed uncertainty model is superior to the deterministic model in terms of network loss and voltage offset expectations. Fig. 2 compares the membership functions of the objective function of the uncertain model and the deterministic model. It can be seen that this model is more applicable in uncertain environment.

Table 2. Comparison of the results of the two models

Model	The network loss(pu.)	The voltage offset(pu.)
The proposed uncertain model	0.9231	1.2273
The deterministic model	0.9285	1.2652

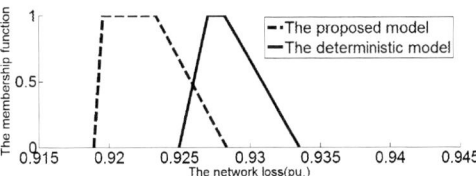

Figure 2. Comparison of membership functions of network loss of different models

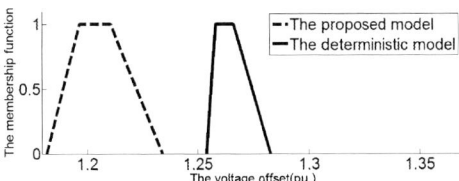

Figure 3. Comparison of membership functions of voltage offset of different models

Fig. 4 indicates the comparison of node voltages before and after optimization.The voltage of nodes before optimization is lower, especially the converter nodes(8 node and 16 node) connected with DC which near the lower limit of voltage. After optimization, the voltage quality of nodes is obviously improved .

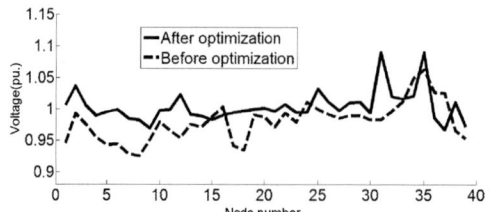

Figure 4. Comparison of node voltage before and after   optimization

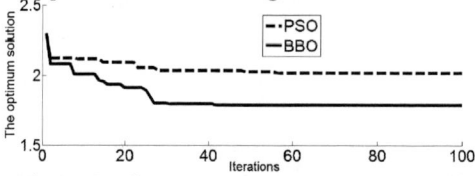

Figure 5. The convergence curve of algorithms

Figure 5 shows the convergence curve of algorithms that particle swarm optimization(PSO) algorithm is easy to fall into local optimum and get local minimum. Although BBO converges slowly, it improves the global search ability and is superior to the PSO in search accuracy. Therefore, the algorithm used in this paper has better convergence and robustness.

6. Conclusions

Considering the influence of uncertainties on power system, this paper establishes the expected value model based on the credibility theory, and transforms the multi-objective function into a single objective function by using the fuzzy goal method of fuzzy programming theory, and solves the problem by using the BBO algorithm. A feasible reactive power optimization control scheme is obtained. The example analysis shows that the expected value model and algorithm proposed in this paper can effectively reduce the active power loss of the system and improve the voltage quality of the nodes.

References

[1] Grudinn N. Reactive power optimization using successivequadratic programming method, *IEEE Trans. on Power Systems* **4**, 13(1998)

[2] Yutian L. Li M. Reactive power optimization basedon Tabu search approach, *Automation of Electric Power Systems* **2**, 24(2000)

[3] Hong Y.Sun D .Lin S.et al. Multi-year multi-caseoptimal VAR planning, *IEEE Trans on PS* **4**, 5(1990)

[4] Tao D.Huaqiang Li.Pei F. Reactive Power Optimization of AC /DC System Based on Singular Value Decomposition and Interior Point Method, *Transactions of China Electrotechnical Society* **2**，24(2007)

[5] Ping Jiang .Le L. Combining Internal Point Method and Genetic Algorithm for AC / DC System Reactive Power Optimization, *High Voltage Technology* **3**, 41(2015)

[6] Qing L .Mingbo L. Liuqing Y. INNC Method for Multi-Objective Reactive Power Optimization of AC / DC Interconnected Power Grid, Proceedings of the CSEE 7,34(2014)

[7] Xin L . Zhibin Y. et al. Multi-objective reactive power optimal control of AC-DC systems including power losscharacteristics of converter stations，*Power System Protection and Control* 9,45(2017)

[8] Hongxin L. Study on voltage stability and reactive power optimization of power system considering uncertainties， Wu Han:Huazhong University of Science and Technology,2013.

[9] Hong F. Xinibin J. et al. Multi-objective reactive power optimization of hybrid AC/DC power system considering load uncertainty，*Electrical Measurement & Instrumentation 10*,55(2018)

[10] Junying S.Dichen L. Runping C. A fuzzy model and approach for power system reactive power optimization，*Power System Technology* 3,25(2001)

[11] Kun Y. Yijia C. Xingying C. et al. Reactive power and voltage optimization of the district grid with distributed generators, *Automation of Electric Power Systems* 8, 35(2011)

[12] Wenxue L. Jun L.Zhihao Y. et al. Multiobjective fuzzy chance constrainer optimal reactive power flow based on credibility theory, *Transactions of China Electrotechnical Society* 21, 30(2015)

[13] Wanjun Z. HVDC transmission engineering technology, Beijing: China power press, 2004

[14] Baoding L.Uncertain planning and application, Beijing:Tsinghua university press, 2003

[15] Zheng X. Dynamic behavior analysis of AC / DC power system,Beijing:Machinery industry press, 2004

[16] Jiuping X. Theory and method of multiobjective decision making, Beijing: Tsinghua university press press, 2005

[17] SIMON D．Biogeography-based optimization, *IEEETransaction on Evolutionary Computation* 6，1(2008)

Optimization of Charging Method for Scaled EVs

Xincheng Zhang[1], Zhizhen Liu[1,a], Yan Cao[2], Lijin Duan[1], Guoshen Tang[1] and Wenchang Liu[1]

[1]Departments of Electrical Engineering, Shandong University, Jinan, China

[2] Departments of Intelligent Manufacturing, Shandong Technician Institute, Jinan, China

[a]Corresponding author: liuzhizhen@sdu.edu.cn

Abstract. At present, EVs(electrical vehicles) charging strategies are mainly focusing on reasonable optimization of EVs' charging. However, previous researches have rarely studied optimizing the charging power of each EV, and the algorithms are always too complicated to implement. Based on this, this paper examines the lowest total load charging method and proposes the automatic breaking charging method which do not rely on complicated algorithms. Through Monte Carlo simulation, the simulation results of the two charging methods are compared respectively, and it is verified that the automatic breaking charging method has a better effect on peak load shifting and it can contribute to the reduction of energy loss.

1. Introduction

Recently, under the centralized control method, the orderly charging behavior of EVs is environmental, efficiency and clean[1]. EV charging behavior which adopts an orderly optimization control, takes advantage of EV as a distributed energy storage element. And then it can achieve the purpose of peak load shifting and stabilizing the output fluctuation of renewable energy, playing an auxiliary role in peaking and frequency modulation in power grid[2].

When the scale of EVs increase, the users' charging behaviors will reflect greater randomness, which will not only increase the complexity of optimization calculation, but also lead to the problem of distribution transformer capacity over the limited. In [3], the electric vehicles with different permeability is randomly connected to the IEEE33 node system through simulation to obtain the node voltage variation curve. The research results show that when the electric vehicle penetration rate is gradually increased from 0 to 70%, there is a risk that the end of the line with a heavier load will have a voltage limit. The access of electric vehicles has an impact on the safe and economic operation of the distribution network and the planning of the power grid. The most important of these is the impact on network loss[4-5] and transformer operation[6-7]. Literature [4] calculates the change of network loss after large-scale electric vehicles access. The calculation results show that as the electric vehicles access volume increases from 0% to 35%, the network loss gradually increases from 4.07% to 4.25%. The literature [6] shows through simulation that when the electric vehicles with a permeability of 50% is disorderly charged in the summer for one day, the distribution transformer will lose 313 minutes of life.

V2G (Vehicle-to-Grid) is a technology that electric vehicles access to the grid. The "access" here is controlled and orderly. It means that there is a purposeful and planned two-way information and energy exchange between the electric vehicles and the grid. The core of V2G is to require electric vehicles to recharge electricity when the electricity is low. While at the peak of power consumption, as a flexible and controllable power supply, EVs supply power to the grid through inverter, providing auxiliary services for power system to "cut peaks and fill valleys"[8-9].

The above-mentioned ordered charging strategies mainly optimize the charging sequence of electric vehicles, but there are few studies on optimizing the charging power of each electric vehicle. In order to meet the users' charging demands, without upgrading the original residential distribution transformer ,this paper examines the reverse recursion charging method and proposes the automatic breaking charging method. By observing the simulation results of the two charging methods, we can see that both of them can decrease the peak-valley ratio of total load, and the automatic breaking charging method is more effective.

2. Factors affecting the charging of EVs

2.1 Charging characteristics of lithium battery

At present, lithium battery is the mainly power battery used in EV. Considering the efficient use of the battery, the charging process commonly adopts 3-stage charging method in order to balance the safety and rapidness of charging process. The method divided the process into three parts based on fitting the characteristic curve of the battery charge and discharge, balanced current charging, balanced voltage charging and float charging.

Take battery which is 4.1V as example. Firstly, in the early stage of charging, if the battery voltage is less than 2.9V, then charging circuit enter the current limiting stage, using a constant rate of 0.1C to charge, with 1C represents the battery is full of 1 hours in the ideal state. The current limiting stage is mainly to avoid the impact of current when the battery voltage is too low. And then the charging mode is converted to balanced charging, adopting constant current to speed up the charging, when the monomer battery voltage rises to saturation voltage(4.1V/4.2V), the capacity of the charge is close to 40% -70% of total capacity at this point. Next, adopting the constant voltage to charge with charge current gradually decreasing. When the current down to 0.1C, it is therefore rational to consider that the battery is full. After that, charging process will enter the floating stage, charging with 0.01C until the end of the charge. Fig 1 shows the three-stage curve diagram of the battery.

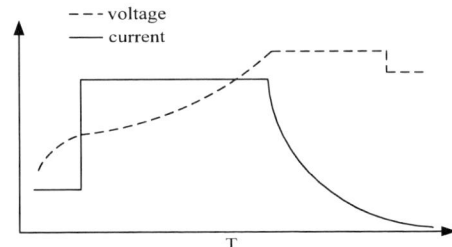

Fig 1. The three-stage curve diagram of the battery diagram

2.2 The PDF(probability density function) of private Evs' return time and daily mileage

According to a research report of National Household Travel Survey (NHTS) proposed by the U.S. Department of Transportation (DOT) in 2009[10], taken all private vehicles into account, during a certain day, 14% of EVs are out of use, 43.5% of EVs drive less than 20 miles (about 32km), 83.7% of EVs drive less than 60 miles (about 97km).Usually most of EVs have returned to the district at 10 pm. It can be seen from the above statistics that generally private EVs have enough time to charge up the batteries at home.

Equation (1) is the PDF of returning time of private EVs return time. Equation (2) is the PDF of the daily mileage of private EVs.

$$f_s(x) = \begin{cases} \dfrac{1}{\sqrt{2\pi}\sigma_s}\exp[-\dfrac{(x+24-\mu_s)^2}{2\sigma_s^2}], \\ 0 < x \le \mu_s - 12 \\ \dfrac{1}{\sqrt{2\pi}\sigma_s}\exp[-\dfrac{(x-\mu_s)^2}{2\sigma_s^2}], \\ \mu_s - 12 < x \le 24 \end{cases} \qquad (1)$$

In equation (1), the independent variable x represents a certain time between [0,24], the expectation value μ_s=17.47, standard deviation σ_s=3.41.

$$f_D(x_L) = \frac{1}{x_L \sigma_D \sqrt{2\pi}}\exp[-\frac{(\ln x_L - \mu_D)^2}{2\sigma_D^2}] \qquad (2)$$

In equation (2), x_L is the daily mileage(unit: mile), the expectation value μ_D=3.2, standard deviation σ_D=0.88.

Based on the lithium battery characteristics, the approximate charging time can be expressed in equation (3).

$$T_C = \frac{LW_{100}}{100 P_C \eta} \qquad (3)$$

In equation (3), L is the daily mileage, W_{100} is the electricity consumption per 100 kilometers, and P_C is the charging power.

3. Optimization model of EV

3.1. Charging demand model of a single EV
Assuming that all the battery types of EVs are lithium batteries. To simplify the study, the charging power is approximated to c/'onstant power throughout the charging process. Figure 2 shows a general graph of battery power variation during charging.

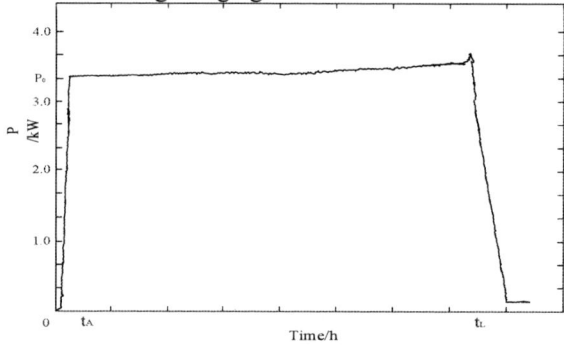

Fig 2. Battery charging power graph

Due to adopting the constant power model, the charging load of the EV-- $P_{ev(i)}(t)$ is

$$P_{ev(i)}(t) = P_0 (t_A < t < t_L) \qquad (4)$$

As shown in (4), t_A and t_L represent the actual start and end time of battery charging, and P_0 is the power of a single EV.

The relationship between required charging time T_{need} and charging power Q_{need} is shown in (5)

$$T_{need} = \frac{Q_{need}}{P_0} \qquad (5)$$

When EV arrives at charging station, it is normally expected that battery can achieve the desired power status at the time of picking up the car. Thus, the charging process should satisfy two following constraints.

1） Charging time constraint

$$T_{need} \leq T_L - T_A \qquad (6)$$

In the formula, T_{need} is the actual duration of charge; T_L is the leaving time; T_A is the arriving time.

2）Battery SOC(State Of Charge) constraint

$$SOC_{min} \leq SOC_A \qquad (7)$$

$$SOC_E \leq SOC_L \leq SOC_{max} \qquad (8)$$

SOC_{min} is the lower limit of battery SOC, by contrast SOC_{max} is the upper limit of SOC. SOC_A and SOC_L represent the initial and finial SOC of charging process and SOC_E is the expected battery SOC.

3.2 Charging method analysis

Based on the typical daily load curve of resident community, this article divides a day into 96 control periods. The intervals between each period were 15 minutes. Assuming that the basic load curve of the kth(i=1,2,…，96) period was P_k. This article separately adopts the lowest total load charging method and the automatic breaking charging method to control the EV charging behavior.

3.2.1 Lowest total load charging method

The total load of power grid of the kth period is consisted of M EVs charging load P_k and superposition of basic load P_0.

$$P_{sumk} = P_k + \sum_{i=1}^{m} p_0 \qquad (9)$$

After each EV participates in charging, re-calculating the total load of power grid in the 96 period, to find out the lowest point of the total load $T_{min\,j}$ and arrange the charging time for next car. Specifically, when the mth EV's charging time was arranged, the minimum value of the total load could be shown as (12):

$$\min P_{sumk} = \min \sum_{k=1}^{96} (P_k + \sum_{i=1}^{m} p_o) \qquad (10)$$

The lowest total load charging method adopts the principle of preferred arrangement to choose charging start time of EV (it refers to the beginning of this time period), after finding the period T_{minj} which includes the lowest load point in charging process. That is to say, EV will adopt off-peak charging as much as possible.

The advantage of this method is that it can make full use of the time-of-use price mechanism and arrange the electric vehicles to charge as much as possible in the valley of the load curve. There is no doubt that it will bring more benefits to users. However, if there are too many electric vehicles gathered and charged at the valley, it may lead to a new peak in the load curve, which will have a bad influence. Therefore, when applying this method, it is necessary to consider the influence of the number of electric vehicles being charged.

3.2.2 Automatic breaking charging method

Automatic breaking charging method starts or ends the EVs charging behavior automatically according to the corresponding algorithm within each time period. When the total load of distribution network reaches the set value at some period of time, charging management system will stop one or a few EVs charging behavior automatically. On the contrary, when it is below the set value, charging management system will start one or a few EV charging behavior.

This method fully considers the impact of the charging load of the electric vehicles on the power grid at any time, and achieves the goal of controlling the load curve by setting the threshold value, and at the same time taking into account the interests of the user. By setting the size of the threshold value, the degree of emphasis on the grid and the users can be reflected. In terms of control effect, this method can be optimal, but its degree of scheduling capability for the power grid and related supporting equipment are correspondingly higher.

3.2.3 Comparison of two charging methods

Relative to ordered charging, disordered charging means that when the users need to recharge EVs, they would immediately connect to the electricity grid. Considering the travel habits, users typically concentrate charging during 16:00 to 21:00. In contrast, the lowest total load charging method and the automatic breaking method arrange charging during night based on different strategies. Specifically, the charging time distribution is shown in Fig 3.

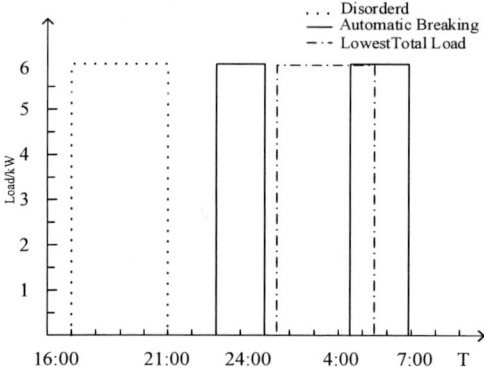

Fig 3. Comparison of two charging methods

4. Example analysis

4.1. Simulation parameters setting

This paper uses a certain district as an example to prove the validity of both methods. The district (including 200 units, average size is 100 square meters) total load includes daily base load and EV charging load. This article assumes that the permeability of EV is 50%. That is to say, this district has 100 EVs. According to the current development situation of EVs, this article makes the following assumptions about the simulation scenario.

- The batteries of EVs are all lithium batteries, and their capacities are 32kW·h, their charging power is $P_c = 7kW$

- The capacity of distribution transformer is 960kVA, the average power factor of load is 0.85, and the upper limit of its active power is $P_{TM} = 960*0.85*0.95 = 775.2kW$

- The initial SOC of EV adopts research results from a brand, as shown in the figure 4.

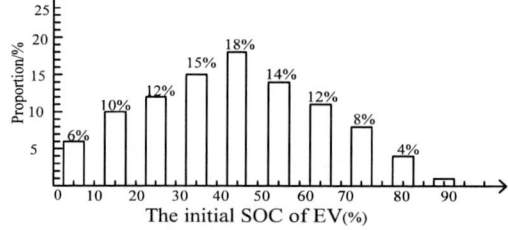

Fig 4. The initial SOC of each EV

- The default expectation, which is set by the users each time, is 1 all the time.

4.2. Simulation results and analysis

Based on the above-mentioned assumptions, in this paper, the Monte Carlo method is used to simulate the charging of 100 electric vehicles in three different situations to observe power grid load curve of different charging method. The simulation results are shown in Fig 5.

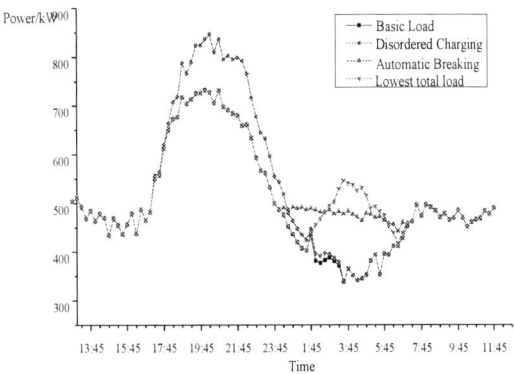

Fig 5. Power grid load curve of different charging mode

From Fig 5, we can see that in the lowest total load charging method, charging behavior concentrated in the load valley, which effectively avoids the peak load during the day. But the concentration of charging led to a new peak at night. In the automatic breaking method, charging load uniformly distributed throughout on-peak period. In consideration of that the more flat the load curve is, the more the network loss reduces, the automatic breaking method has an advantage over the lowest total load charging method, and it can cut peak and fill valley more effectively.

5. Conclusion

Based on the development status and prospect of EVs, as well as some rational assumptions, this paper proposes the automatic breaking method and examines the lowest total load charging method. Through the simulation and analysis, we find that two kinds of coordinated charging methods can both decrease the peak-valley ratio of total load effectively. In particular, the valley period will be smoother under the mode of automatic breaking method.

References

[1] Energy Conservation and New Energy Vehicle Industry Development Plan(2012-2020).
[2] Y.M. Wang, Research on the Soft Environment of Arctic Wind Energy Development[J]. Electric Power, **49**(2016)
[3] L.D Zhang, et al. Study on the influence of electric vehicles random charging on distribution network[J] Journal of Electric Power Science and Technology, **31**(2016)
[4] L.Z Xu, et al. The impact of electric vehicle charging load on Danish distribution system[J]. AEPS, **35**(2011)
[5] J.S Chen, et al. Strategies for Electric Vehicle Charing with Aiming at Reducing Network Losses[J]. Proceedings of the CSU-EPSA. **24**(2012)
[6] J.L Sun, et al. Impact of Plug-in Electric Vehicles on the Operating Life of Distribution Transformer[J]. High Voltage Engineering. **41**(2015)
[7] Y.N Cai, Analysis of the Influence of Large-scale Electric Vehicle Charging on the Life of Distribution Transformer[D]. Chongqing University(2016)
[8] L.L Ma, et al. Summary of Research on Influence of Electric Vehicle Charging and Discharging on Power Grid[J]. Power System Protection and Control. **3**(2013)
[9] W Pan, et al. Review of coordinated control strategy between electric vehicle and power grid[J]. Power Demand Side Management. **4**(2013)
[10] L.T Tian, et al. A Statistical Model for Charging Power Demand of Electric Vehicles[J]. Power System Technology, **34**(2010)

Operation Quality Evaluation of Power Communication Network Based on Business QOS Indicators

Geng Zhang[1], Ke Wang[2,1], Yang Wang[1], Yanan Wang[1], Chen Xiangzhou[1]

[1] China Electric Power Research Institute, Beijing,100089, China

[2] North China Electric Power University, Beijing,100089, China

[a]Corresponding author: happyandluck_wk@163.com

Abstract. The development of the power communication network has brought new challenges to the comprehensive evaluation technology of the service quality and the business risk assessment technology. The existing service quality evaluation system does not take the service subnet indicator of the power communication network into consideration systematically, and it is difficult to meet the comprehensive evaluation requirements of the service quality of the whole network. Based on the characteristics of the power communication network, this paper constructs and stratifies its business QOS indicators. Then the indicators of each layer are introduced in detail, and the weight of each indicator in each layer is found by using the method of the analytic hierarchy process. Furthermore, the normalized value of each indicator is found by using the conversion function based on the analytic hierarchy process to make the defined indicator connotation close to the actual operational data and strengthen the comparability between the indicator in this paper. Finally, the operating quality of the power communication network is evaluated by the normalized value and the weight of each indicator combined with the example. The aim is to provide a theoretical reference for the development of the grid business risk management to ensure that the power communication network can operate in a low-risk state .

1. Introduction

With the development of the power grid, especially the accelerated construction of the smart grid, the demand for communication services in the power system is getting higher and higher. On the one hand, the service quality requirements of some special services have increased; on the other hand, with the addition of new energy sources, the scale of distributed energy and micro-grid has been expanding, which has led to many new business types in the power communication network[1].This makes it difficult for traditional business evaluation index systems to content the needs of differentiated communication services. The main performance is that the business evaluation indicators are difficult to express in a unified way, which brings great difficulties to the establishment of the comprehensive evaluation index system. Moreover, with the rapid development of communication technology and the increasing complexity of network scale, the existing evaluation indicators are not updated synchronously according to network requirements and technologies. The evaluation technology and evaluation effect lag behind the development status of power communication networks[2]. This will inevitably lead to a decline in the indicators of the power communication network, resulting in a decline in the performance of the power system business, and the serious deterioration of the service quality will not meet the requirements for safe, stable and reliable operation of the power system[3] Therefore, this paper will comprehensively analyze and construct the business QOS indicators, and

Content from this work may be used under the terms of the Creative Commons Attribution 3.0 licence. Any further distribution of this work must maintain attribution to the author(s) and the title of the work, journal citation and DOI.

Published under licence by IOP Publishing Ltd

use the analytic hierarchy process and numerical transformation to convert the value of the business QOS indicator into a normalized value with high contrast. Through the analysis of weights and normalized values, each indicator is evaluated to provide guidance for the risk assessment of the grid to improve the ability of the grid to operate stably and reliably.

2. Business QOS indicator

Grid operation requires multiple types of communication service support, and different service types have different QoS requirements. This makes different services have different effects on the operation of the grid, showing different importance. Common business types are as follows:1) Control business: Such as relay protection, safety and stability control, dispatching data network, power transmission and transformation status detection, substation comprehensive monitoring, distribution network operation monitoring, distribution network automation, etc.2) Switched network service: Such as administrative calls and dispatching phones,etc[4].3) Management business: Such as marketing business management system, communication intelligent management system, customer contact system, 95598 and fault repair management, power quality management system, customer electricity information collection, power market trading operations.4) Information business: Such as data (disaster disaster) centers, SG-ERP and conference television systems, etc.

Because too many types of services are bad for data collection and monitoring, which can cause incomplete infirmation easily. And the operational quality evaluation mainly adapts to the normal state of the network, and the degree of satisfaction of the performance indicators in this state, rather than the fault and failure state [5] Therefore, this paper does not include all the above-mentioned services into the secondary indicators when constructing the business QOS indicator system, Instead, it summarizes the above-mentioned services and abstracts the services common of the above services into data services, audio services and flows. Media business and use it as a secondary indicator. At the same time, with the continuous advancement of the IP process, packet switching will become the main form of bearer network implementation. To adapt to this feature, the indicator system of this topic focuses on the quality and risk of packet switched networks. Therefore, when selecting the three-level indicator, the three indicators, such as transmission delay, packet loss rate and congestion rate, are used as the third-level indicator. The indicator architecture diagram is shown in Figure 1.

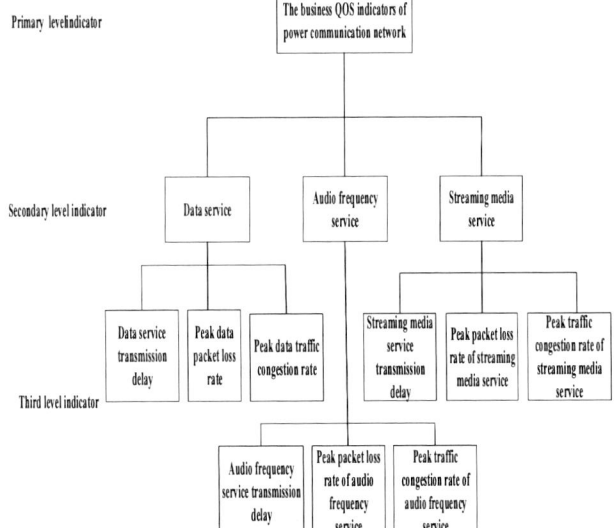

Figure 1 Business QOS indicator architecture

2.1 Business QOS single indicator

2.1.1 Transmission delay of data service, audio frequency service and streaming media service

The overall expression of the transmission delay of the data service, the audio frequency service, and the streaming media service is:

$$D_d = T_d + t'_d = \left(0.49L + T_{sd} + nT_p + T_{dd} + T_{tr}\right) + \qquad \sum_{i=1}^{n} t'_d \qquad (1)$$

Where $0.49L$ indicates the propagation delay of the optical cable length L in the path. T_{sd} indicates the fixed processing delay of the source node. T_{dd} indicates the fixed processing delay of the destination node, including the jitter buffer time. T_p indicates the fixed processing delay of the path intermediate node, indicating the number of intermediate nodes; T_{tr} indicates the transmission delay of the data packet in the path, which is related to the data rate and the packet length. The portion of parentheses in the formula (1) represents a fixed delay component. t'_d indicates the random delay of data queuing, scheduling, and waiting in the intermediate node.

2.1.2 Peak packet loss rate of data service, audio frequency service, and streaming media service

The peak packet loss rate of data service, audio frequency service, and streaming media service is an indicator to measure the reliability of quasi-real-time and non-real-time transmission of these three services. This indicator indicates the maximum packet loss rate for data service, audio frequency service, and streaming media service during a given observation period. The expression of the peak packet loss rate is:

$$\beta_{p\_max} = MAX(\bigcup_{i=1}^{NUM} \frac{1-n_i}{m_i}). \qquad (2)$$

Where n_i is the number of packets correctly received for the i-th data transmission, m_i is the total number of packets transmitted for the i-th data transmission, and $(1-n_i)/m_i$ is the packet loss rate for the i-th data transmission. NUM is the total number of data transfers in the observation time window. The transmission loss rate is a dynamically changing random variable. In order to facilitate the analysis and calculation in engineering, the maximum value method is used to indicate the transmission loss rate of three services.

2.1.3 Peak congestion rate for data service, audio frequency service, and streaming media service

The peak congestion rate of data services, audio services, and streaming services is an indicator to measure the congestion of these three types of services. During a given observation period, the indicator represents the maximum probability of transmission congestion, and its expression is:

$$P_{c\_max} = MAX\{P_r(S_i = C)\}. \qquad (3)$$

Where $(S_i=C)$ indicates a congestion state. The occurrence of congestion state is related to many factors, and the congestion probability $Pr(S_i=C)$ has strong dynamics and changes with time. This indicator refers to the maximum probability of congestion in the three services within a given observation time window.

3. Analytic Hierarchy Process and conversion function

3.1 Analytic Hierarchy Process

Analytic Hierarchy Process (AHP) refers to the decomposing of elements related to the overall goal of decision-making into goals, criteria, and programs. On this basis, qualitative and quantitative analysis methods are used. This method is a hierarchical weighted decision analysis method proposed by American operations researcher Pittsburgh University professor Saty in the early 1970s [6].

The AHP calculation's steps are as follows :

3.1.1 Establish a hierarchical model

The goal of decision-making, the factors considered (decision-making criteria) and the decision-making objects are divided into the highest layer, the middle layer and the lowest layer according to the mutual relationship, and establishing a hierarchical structure diagram . The highest level is the problem to be solved. The lowest level refers to the alternative when making decisions. The middle layer refers to the factors considered and the criteria for decision making. For the adjacent two layers, the upper layer is called the target layer, and the lower layer is the factor layer [7]. This article corresponds to the three-level indicator of business QOS.

3.1.2 Construction judgment matrix A

When determining the weights between the various factors at each level, all factors are compared with each other, and relative scales are used for comparison to minimize the difficulty of comparing factors with different natures and improve the accuracy. For a certain criterion, the schemes under it are compared in pairs and rated according to their importance level [8]. Where a_{ij} is the comparison result of the importance of the element i and the element j, and the reference standard of the value is shown in Table 1.

Table 1. Scale table.

The importance of factor i compared with factor j	Quantitative value (scale)
Equally important	1
Slightly important	3
Stronger important	5
Strongly important	7
Extremely important	9
Intermediate value of two adjacent judgments	2, 4, 6, 8

Table 1 lists the nine importance levels and their valuation, also known as scales. A matrix formed by the results of the pairwise comparison is called a judgment matrix. The nature of the judgment matrix is:

$$a_{ij} = \frac{1}{a_{ji}} \qquad (4)$$

3.1.3 Hierarchical ordering and calculation of weight

In this paper, we use the sum and product method to calculate the weight. First, normalize each column element of A to get the general term of the column normalization element is:.

$$\overline{a_{ij}} = \frac{a_{ij}}{\sum_{i=1}^{n} a_{ij}} \qquad (5)$$

Then, the normalized matrix of the columns is summed to obtain a column vector whose general term is:

$$\overline{w_i} = \sum_{j=1}^{n} \overline{a_{ij}} \qquad (6)$$

Finally, the column vector is normalized to obtain the weight vector W, and the element general term is:

$$w_i = \frac{\overline{w_i}}{\sum_{i=1}^{n} \overline{w_i}} \qquad (7)$$

3.2 Conversion function

In order to further enhance the comparability between indicators, this paper introduces a conversion function. According to the physical meaning and assignment principle of the index, combined with the nature of the index, we use the generalized Logistic function as the transfer function of this paper to realize the conversion of the index evaluation value to the normalized index value [9]..

The expression of the generalized Logistic function is:

$$f(t) = \frac{1}{1+\exp(-B(t-M))} \qquad (8)$$

Where t is the evaluation value of the indicator, and the two parameters B and M are determined by the physical meaning of the specific indicator and the comparability principle of the comprehensive evaluation. And according to the physical meaning of the indicator, the generalized Logistic function is divided into two categories, which are positive (P) and negative (N). Each type of conversion function corresponds to an indicator property.

Positive-type conversion function is abbreviated as P function, and its expression is:

$$f_p(t) = \frac{1}{1+\exp(-B(t-M))} \qquad (9)$$

Key point value: $f_p(-\infty) = 0$, $f_p(M) = 0.5$, $f_p(+\infty) = 1$. $f_p(t)$ is used to achieve numerical conversion of positive indicators.

Negative-type conversion function is abbreviated as N function, and its expression is:

$$f_N(t) = 1 - \frac{1}{1+\exp(-B(t-M))} \qquad (10)$$

Key point value: $f_N(-\infty) = 1$, $f_N(M) = 0.5$, $f_N(+\infty) = 1$, $f_N(t)$ is used to implement the numerical conversion of the negative indicator.

4. Instance verification

According to the above discussion, the second-level indicators of the service QOS indicator are data service, audio frequency service and streaming media service. The third-level indicator is data service transmission delay, peak data packet loss rate, peak data traffic congestion rate, audio frequency service transmission delay, peak packet loss rate of audio frequency service, peak traffic congestion rate of audio frequency service, streaming media service transmission delay, peak packet loss rate of streaming media service and peak traffic congestion rate of streaming media service. According to the actual survey results of the power communication network, the judgment matrices of the second-level indicators and the third-level indicators are shown in Table 2,3.

Table 2. AHP judgment matrix of business QOS second-level indicators.

Judgment matrix	Data service	Audio frequency service	Streaming media service
Data service	1	3	5
Audio frequency service	1/3	1	3
Streaming media service	1/5	1/3	1

According to formula (5), (6), (7), the weight vector of the QOS secondary indicators of the three types of services can be obtained as $[0.61 \quad 0.29 \quad 0.10]^T$

Table 3. AHP judgment matrix of business QOS third-level indicator

Judgment matrix	Transmission delay	Peak packet loss rate	Peak congestion rate
Transmission delay	1	1/7	1/5
Peak packet loss rate	7	1	3
Peak congestion rate	5	1/3	1

Using the AHP method, the weight vector of the business QOS third-level indicator is [0.07 0.65 0.28]^T. According to the actual project status, the weight vector of the data service, audio frequency service and streaming media service indicators is also [0.07 0.65 0.28]^T. In this way, the weight vector of the nine third-level indicators of the service network is. [0.043 0.397 0.171 0.020 0.189 0.081 0.007 0.064 0.028]^T.

According to the relevant data in the operation data of the power communication network and the maintenance records, the evaluation values of all the third-level indicators can be calculated by the equations (1), (2), and (3). Through calculation the average transmission delay of the data service is 60ms, the peak packet loss rate is 10^{-5}, and the peak congestion probability is 10^{-3}. The QoS index values of audio frequency service and streaming media service can be obtained in the same way.

According to the large difference between the values of these third-level indicators, it is not conducive to analysis and comparison. This paper compares the indicators from the perspective of indicator weight and normalized value. The weight of the third-level indicator has been obtained above. The index will be normalized by the conversion function of equation (8). Since these nine third-level indicators have the same characteristics, that is, the increase of the index value will lower the evaluation result of the overall target, so these indicators are negative indicators and need to be transformed by the negative-type conversion function [10]. Therefore, the calculation is performed using the equation (10). According to the performance indicators of various types of business and the basic principles of operational quality evaluation.

The conversion function of the delay class indicator is determined as:

$$f_{N-D}(t) = 1 - \frac{1}{1+\exp(-0.055(t-50))} \quad (11)$$

The conversion function of the peak packet loss rate and the peak congestion rate is:

$$f_N(t) = 1 - \frac{1}{1+\exp(-1.1(\lg t+5))} \quad (12)$$

Therefore, the data collection calculation results and the transformed results of the third-level indicator are shown in Table 4.

Table 4. The value and conversion of the third-level indicators

the third-level indicators	The weight of third-level indicator	Data collection calculation result	Converted value
Data service transmission delay,	0.043	60ms	0.366
Peak data packet loss rate	0.397	1.00E-04	0.250
Peak data traffic congestion rate	0.171	1.00E-05	0.500
Audio frequency service transmission delay	0.020	30ms	0.750
Peak packet loss rate of audio frequency service	0.189	1.00E-04	0.250
Peak traffic congestion rate of audio frequency service	0.081	1.00E-04	0.250
Streaming media service transmission delay	0.007	20ms	0.839
Peak packet loss rate of streaming media service	0.064	1.00E-05	0.500
Peak traffic congestion rate of streaming media service	0.028	1.00E-06	0.750

After the transformation, the index value varies between [0,1], and various indicators have strong comparability. The radar chart can be used to visually represent the values of the converted indicators, as shown in Figure 2.

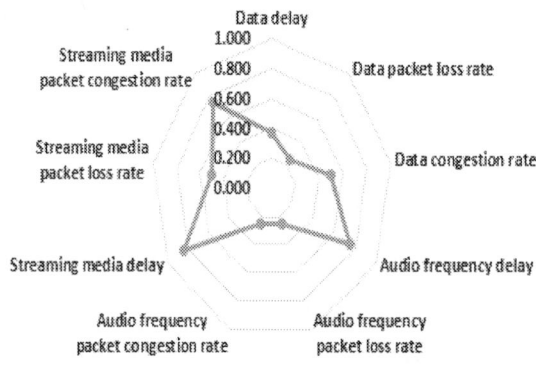

Figure 2 The third-level indicators of business QOS

As can be seen from Figure 4-1 and Table 4-3, the weighted results calculated by theory are consistent with the normalized values calculated from the actual data. That is, the transmission delay of the streaming media service has the smallest weight value, but the normalized value is the largest, indicating that the transmission delay of the streaming media service has the least impact on the performance of the service network. The peak value of the packet loss rate of the data service has the largest weight value, and the normalized value is the same as the peak value of the audio frequency service packet loss rate and the peak value of the audio frequency service congestion rate, indicating that the three QOS indicators have the greatest impact on the performance of the service network. Based on the above-mentioned index values, when formulate the risk management and control measures for the grid business, we should consider more factors that have a greater impact on the performance of the service network, such as the packet loss rate and congestion rate of the audio frequency service, and the packet loss rate of the data service. For the factors that have a small impact, you can take a little care when making measures. This will save network resources while ensuring that the power communication network can operate in a low-risk state.

5. Conclusion

In view of the phenomenon that the existing power communication network's index system can't keep up with the development of the actual power industry, this paper proposes a set of index system for the service network in the power communication network. In order to make the proposed system hierarchical, the paper divides the business network index system into three levels, and introduces in detail the characteristics and calculation methods of each indicator in each layer. Further, in order to compare the degree of influence of each indicator on the service network, an analytic hierarchy process and a transfer function are introduced. By using these two algorithms, the weights and normalized values of the indicators are respectively obtained. Then the paper analyzes and evaluates the impact degree of each indicator in the service network from the weight and normalized value of each indicator. The level of impact of each indicator on the performance of the service network is found out. Through the evaluation of each level, theoretical guidance can be provided for the formulation of risk management measures for the grid business.

Acknowledgments

This paper was supported by the science and technology project from State Grid Corporation of China:" The Research on Distributed Simulation and Optimization Technology of the Power Communication Optical Transmission Network(XX71-18-006)"

References

[1] X Xiong,L M Meng,TS, **4**:20-23,(2006)
[2] Y Zhou,CSTI, **9(19)**:61-62,(2015)

[3] X Zhou,ST, **8**:1-2,(2015)
[4] H D Du,AA, **7**:103-108.(2013)
[5] Y Song,J G Xu,F Ye,AI, **184**:162-164.(2015)
[6] J S Yuan,H S Gao,Y Q Song,HVE, **35(4)**:960-964.7(2009)
[7] Y J Xie,Y T Wang,W Li,EP, **50(10)**:22-27,(2017)
[8] J Wu,S Y Tang,Y J Tan,CSCS, **11(1)**:77-86,(2014)
[9] H S Gao, J X Ran, H Y Xie,ICA,**2008**:5801-5806(2008)
[10] K Jiang,Y Zeng,B Deng,ICNC, **2013**:1595-1599(2013)

A Stator Flux Calculation Method for Permanent Magnet Synchronous Motor in 60° Coordinate System

Xiao TANG*, Zhi ZHANG, Chang LIU, Caishen WANG, Zhiping WANG

School of Electrical Engineering & Intelligentization, Dongguan University of Technology, Dongguan, China

*Corresponding author: tangx@dgut.edu.cn

Abstract. The direct torque control of permanent magnet synchronous motor is with a greater concern. The space vector modulation mode algorithm can greatly improve the torque ripple of traditional direct torque control. The 60° coordinate system used in the space vector pulse width modulation algorithm can make it simpler. The accurate observation of stator flux is the premise of direct torque control. A simple and effective method is to adopt the voltage model based on low-pass filter. In order to ensure the consistency of the algorithm, the form of calculating the stator flux based on low pass filter is deduced in 60° coordinate system.

1. INTRODUCTION

Permanent magnet synchronous motor direct torque control (PMSM-DTC) has received more and more attention [1-4]. The traditional direct torque control has simple structure, but with the drawbacks of large torque ripple and changing switching frequency. To solve the problems, the SVPWM mode DTC (SVM-DTC) structure is a better choice [5-7]. But the SVM-DTC system becomes more complicated. The authors previously studied SVPWM algorithm and its application in PMSM-DTC in 60° coordinate system. The 60° coordinate system is very suitable for SVPWM algorithm due to its characteristics. And then the system is simplified to some extent [8].

In order to ensure the consistency of the algorithm and in-depth study of 60° coordinate system application in DTC, this paper will analyze the representation of a PMSM stator flux observation method in 60° coordinate system.

2. 60° COORDINATE SYSTEM AND COOR-DINATE TRANSFORMATION

The 60° coordinate system is called *gh* coordinate system and its relationship with the two-phase stationary vertical *αβ* coordinate system is shown as below.

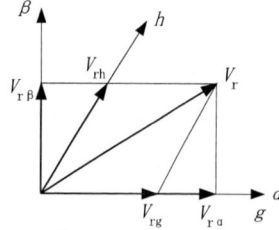

Figure 1. *αβ* coordinate system and 60° coordinate system

The following expressions can be obtained from figure 1.

$$\begin{cases} V_{r\alpha} = V_{rg} + V_{rh}\cos 60^{\circ} \\ V_{r\beta} = V_{rh}\sin 60^{\circ} \end{cases} \tag{1}$$

That is

$$\begin{cases} V_{r\alpha} = V_{rg} + \dfrac{1}{2}V_{rh} \\ V_{r\beta} = \dfrac{\sqrt{3}}{2}V_{rh} \end{cases} \tag{2}$$

Equation (2) can be changed as

$$\begin{cases} V_{rg} = V_{r\alpha} - \dfrac{1}{\sqrt{3}}V_{r\beta} \\ V_{rh} = \dfrac{2}{\sqrt{3}}V_{r\beta} \end{cases} \tag{3}$$

Equation (3) is written in matrix form as

$$\begin{bmatrix} V_{rg} \\ V_{rh} \end{bmatrix} = \begin{bmatrix} 1 & -\dfrac{1}{\sqrt{3}} \\ 0 & \dfrac{2}{\sqrt{3}} \end{bmatrix} \cdot \begin{bmatrix} V_{r\alpha} \\ V_{r\beta} \end{bmatrix} = C \cdot \begin{bmatrix} V_{r\alpha} \\ V_{r\beta} \end{bmatrix} \tag{4}$$

where C is the transformation matrix from the $\alpha\beta$ coordinate system to the gh coordinate system. If the three-phase coordinate system is transformed directly to the gh coordinate system and named the transformation matrix as D, the following equation could be gotten:

$$\begin{bmatrix} V_{rg} \\ V_{rh} \end{bmatrix} = D \cdot \begin{bmatrix} V_a \\ V_b \\ V_c \end{bmatrix} = C \cdot C_{32} \begin{bmatrix} V_a \\ V_b \\ V_c \end{bmatrix}$$

$$= \frac{2}{3}\begin{bmatrix} 1 & -1 & 0 \\ 0 & 1 & -1 \end{bmatrix}\begin{bmatrix} V_a \\ V_b \\ V_c \end{bmatrix} \tag{5}$$

where C_{32} is the transformation matrix from the three-phase coordinate system to the $\alpha\beta$ coordinate system. That is

$$C_{32} = \frac{2}{3}\begin{bmatrix} 1 & -\dfrac{1}{2} & -\dfrac{1}{2} \\ 0 & \dfrac{\sqrt{3}}{2} & -\dfrac{\sqrt{3}}{2} \end{bmatrix} \tag{6}$$

3. THE OBSERVATION METHOD OF STA-TOR FLUX BASED ON LOW-PASS FILTER IN 60° COORDINATE SYSTEM

3.1 Voltage model stator flux observation

There are two commonly used stator flux observation methods for PMSM which named "voltage model" and "current model" respectively. The parameters of the motor are more in the current model, and the rotor position information is needed in order to make the coordinate transformation. Compared with the current model stator flux observation method, the voltage model involves less motor parameters, which is done in stationary coordinate system and does not need to do coordinate transformation. The voltage model is as

$$
\begin{cases}
\psi_\alpha = \int (U_\alpha - I_\alpha R) dt \\
\psi_\beta = \int (U_\beta - I_\beta R) dt
\end{cases}
\tag{7}
$$

where ψ_α and ψ_β are the stator flux components, U_α and U_β are the stator voltage components, I_α and I_β are the stator current components, R is the stator resistance.

3.2 Voltage model stator flux observation based on low-pass filter

Because of the initial value problem and the error accumulation problem, the pure integral voltage model could not be used in practical applications. A simple and effective solution is to replace the integrator with a low-pass filter (LPF) in the voltage model [9]. The transfer function of low-pass filter G_{LPF} is as

$$
G_{LPF} = \frac{1}{s + \omega_c}
\tag{8}
$$

where the ω_c is the cut-off frequency of the LPF. The stator flux observer based on LPF is shown as the following figure.

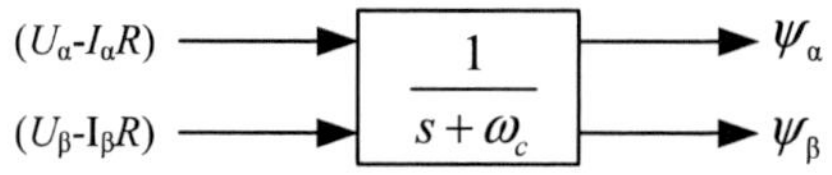

$$
(U_\alpha - I_\alpha R) \longrightarrow \boxed{\frac{1}{s + \omega_c}} \longrightarrow \psi_\alpha
$$
$$
(U_\beta - I_\beta R) \longrightarrow \qquad\qquad \longrightarrow \psi_\beta
$$

Figure 2. Stator flux observer based on low-pass filter

3.3 The compensation for low-pass filter

The LPF can solve the problem of pure integrator in the practical application, but it also leads to amplitude and phase errors of stator flux observation with the actual flux. The errors will affect the performance of the motor. It is needed to compensate for the LPF in order to make the observed flux is consistent with the actual value. There are many kinds of compensation methods, and the idea is to make the stator flux observation results after compensation are consistent with the results of pure integration. A compensation method is proposed to process the input signal of the LPF in literature [10]. The results are as follows:

$$
\begin{cases}
E_\alpha = E_\alpha' + \dfrac{\omega_c}{\omega} E_\beta' \\
E_\beta = E_\beta' - \dfrac{\omega_c}{\omega} E_\alpha'
\end{cases}
\tag{9}
$$

where the E_α' and E_β' are the input components of the LPF without compensation processing which equal the stator voltage component $U_{s\alpha}$ and $U_{s\beta}$ minus the value of the stator resistance voltage drop respectively. The E_α and E_β are the modified input components of the LPF with compensation. The stator flux is calculated as

$$
\begin{cases}
\psi_\alpha = \left[(U_\alpha - I_\alpha R) + \dfrac{\omega_c}{\omega} (U_\beta - I_\beta R) \right] \cdot \dfrac{1}{s + \omega_c} \\
\psi_\beta = \left[(U_\beta - I_\beta R) - \dfrac{\omega_c}{\omega} (U_\alpha - I_\alpha R) \right] \cdot \dfrac{1}{s + \omega_c}
\end{cases}
\tag{10}
$$

It will cause the stator flux with compensation to be too large at the beginning of motor starting due to the small angular velocity. It is needed to make amplitude limiting. It is often to make $|\omega_c/\omega| < 3$ in the above equations.

3.4 Representation of the stator flux observation in 60° coordinate system

First, the voltage and current components in the three-phase coordinate system are transformed to the expressions in 60° coordinate system according to the transformation rules. Then the flux expressions will be changed to the representation in 60° coordinate system. According to (3), (10) will be changed as (11).

$$
\begin{cases}
\psi_g + \dfrac{1}{2}\psi_h = \left\{\left[(U_g + \dfrac{1}{2}U_h) - (I_g + \dfrac{1}{2}I_h)R\right] + \dfrac{\omega_c}{\omega}(\dfrac{\sqrt{3}}{2}U_h - \dfrac{\sqrt{3}}{2}I_hR)\right\} \cdot \dfrac{1}{s+\omega_c} \\[4mm]
\dfrac{\sqrt{3}}{2}\psi_h = \left\{(\dfrac{\sqrt{3}}{2}U_h - \dfrac{\sqrt{3}}{2}I_hR) - \dfrac{\omega_c}{\omega}\left[(U_g + \dfrac{1}{2}U_h) - (I_g + \dfrac{1}{2}I_h)R\right]\right\} \cdot \dfrac{1}{s+\omega_c}
\end{cases}
\tag{11}
$$

Then the stator flux components in 60° coordinate system are as (12).

$$
\begin{cases}
\psi_g = \left\{(U_g - I_gR) + \dfrac{\omega_c}{\omega}\left[\dfrac{1}{\sqrt{3}}(U_g - I_gR) + \dfrac{2}{\sqrt{3}}(U_h - I_hR)\right]\right\} \cdot \dfrac{1}{s+\omega_c} \\[4mm]
\psi_h = \left\{(U_h - I_hR) - \dfrac{\omega_c}{\omega}\left[\dfrac{2}{\sqrt{3}}(U_g - I_gR) + \dfrac{1}{\sqrt{3}}(U_h - I_hR)\right]\right\} \cdot \dfrac{1}{s+\omega_c}
\end{cases}
\tag{12}
$$

4. SIMULATION RESULTS

The stator flux observation method in 60° coordinate system is verified by simulation based on MATLAB. The parameters of the PMSM and control in the simulation model are shown in Table 1.

Table 1. Parameters of the PMSM and Control.

DC-bus voltage	U_{dc}	200V
Reference stator flux	ψ_{sref}	0.12Wb
Permanent magnet flux	ψ_f	0.1 Wb
Number of pole pairs	p	4
Stator resistance	R	0.7 Ω
d-axis inductance	L_d	1.5mH
q-axis inductance	L_q	1.5mH
Sampling period	T_s	$100\,\mu s$

The simulation results are as follows.

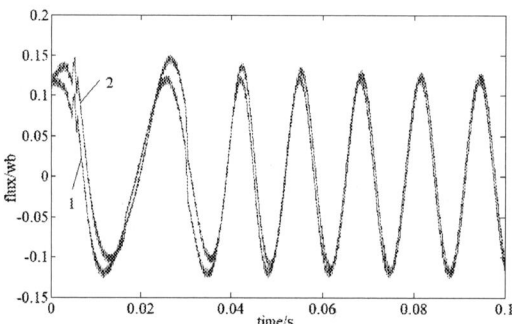

Figure 3. Stator flux based on LPF without compensation and pure integration

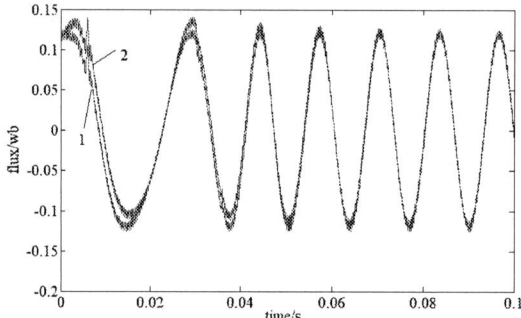

Figure 4. Stator flux based on LPF with compensation and pure integration

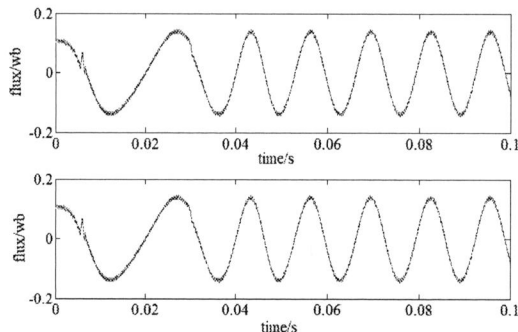

Figure 5. Stator flux in 60° coordinate system

The curves are the α-components of stator flux in Fig.3 in which the Curve 1 is based on LPF without compensation and the Curve 2 is based on pure integration. It can be seen there are amplitude and phase errors between the outputs of the two kinds of stator flux observations. The errors disappear in Fig.4 in which the LPF is with compensation. Because of the incomplete compensation, the two curves are not consistent in the beginning cycles due to the limit value of $|\omega_c/\omega|$ in (9). The β-components of stator flux are with the same conclusions as in Fig.3 and Fig.4.

The waveforms in Fig.5 are the g-components of stator flux, the above one is the changed result from the α-component and β-component of stator flux, the nether one is calculated according to (12). The two curves are completely consistent.

5. CONCLUSION

In order to apply the 60° coordinate system in the PMSM-DTC, this paper analyzes the representation of voltage model flux observation method based on low-pass filter in 60° coordinate system. The correctness of stator flux observation method is illustrated through theoretical analysis and simulation.

ACKNOWLEDGMENT

This work was financially supported by Distinguished Young Teacher Project of Education Department of Guangdong Province (YQ2015156), the Industry-University-Research Collaboration of Dongguan Dongcheng District, China and Dongguan Greenway Technology Co., Ltd in China.

REFERENCES

[1]   L. Zhong, M. F. Rahman, Y. Hu, et al. "Analysis of direct torque control in permanent magnet synchronous motor drives," *IEEE Transaction on Power Electronics*, vol. 12, pp. 528-535, 1997.

[2]   H. Jia, Y. He, "Variable structure sliding mode control for PMSM DTC," *Transactions of China Electrotechnical Society*, vol. 21, pp. 1-6, 2006.

[3]   Y. Zhou, J. Zhong, J. Zhou, "A novel sliding mode variable structure direct torque control for permanent magnet synchronous motor drive system," *Journal of Power Supply*, vol. 1, pp. 79-84, 2011.

[4]   Y. Zhou, G. Chen, "Predictive DTC strategy with fault-tolerant function for six-phase and three-phase PMSM series-connected drive system," *IEEE Transactions on Industry Electronics*, vol. 65, pp. 9101-9112, 2018.

[5]   D. Sun, Y. He, "Space vector modulated based constant switching frequency direct torque control for permanent magnet synchronous motor," *Proceedings of the Chinese Society for Electrical Engineering*, vol. 25, pp. 112-116, 2005.

[6]   Y. Zhang, J. Zhu, W. Xu, et al. "A simple method to reduce torque ripple in direct torque-controlled permanent-magnet synchronous motor by using vectors with variable amplitude and angle," *IEEE Transactions on Industry Electronics*, vol. 58, pp. 2848-2859, 2011.

[7]   D. WANG, T. Yuan, et al. "Reduction of torque and flux ripples for robot motion control system based on SVM-DTC," *37th Chinese Control Conference (CCC)*, pp. 5572-5576, 2018.

[8]   X. Tang, X. Yang, S. Zhao, et al. "SVPWM over-modulation algorithm based on 60° coordinate system," *Journal of South China University of Technology (Natural Science Edition)*, vol. 42, pp. 27-33, 2014.

[9]   J. Xu, Y. Xu, J. Feng, et al. "Direct torque control of permanent magnet synchronous machines based on modified integrator," *Transactions of China Electrotechnical Society*, vol. 21, pp. 1-6, 2006.

[10]   Z. He, Y. Liao, D. Xiang, "Improvement of low-pass filter algorithm for stator flux estimator," *Proceedings of the Chinese Society for Electrical Engineering*, vol. 28, pp. 61-65, 2008.

Analysis of Marketing Strategy of Electricity Selling Companies in the New Situation

Taorong Gong[1,2], Dezhi Li[1,2], Yuting Liu[3,a], Guanhong Wang[1], Haoming Zhu[4]

[1]China Electric Science Research Institute Co., Ltd., China

[2]Beijing Key Laboratory of Demand Side Multi-Energy Carriers Optimization and Interaction Technique, China

[3]North China Electric Power University, China

[4]Northeast Electric Power University, China

[a] Corresponding author: alice.liu@ncepu.edu.cn

Abstract. With the liberalization of the electricity selling side, the electricity selling market has changed from a single seller market to a buyer's market, that is, the buyer can buy electricity through multiple sellers, and the competition in the electricity selling market has gradually increased. Successful marketing strategy can enlarge the popularity of power selling companies, attract more users, and make them invincible in the competition. This paper first analyses the marketing strategies that can be adopted in the competition. Subsequently, through the system dynamics model, the benefits brought by the strategy are studied and simulated to verify the effectiveness of the strategy. Finally, this paper concludes that the strategies that electricity sales companies can adopt are: diversified products, differential pricing, broadening channels and deepening cooperation, and using big data analysis to promote users' electricity consumption behavior.

1. Introduction

With the transformation of China's economic system from a planned economy to a socialist market economy, enterprises focus more on marketing. On March 15, 2015, "Some Opinions of the Central Committee of the CPC and the State Council on Further Deepening the Reform of the Electric Power System" marked the beginning of a new round of reform of the electric power system, which meant that the companies directly in contact with users were transformed from power grid companies to newly established power selling companies. Due to the short time of selling electricity companies, there is not enough research foundation for them. Therefore, this paper decides to use Vensim software to analyse the impact of different marketing strategies on the electricity sales of power companies.

The research on power marketing in China has been carried out before the opening of power selling side. The main research focuses on the business optimization of power supply enterprises, State Grid [1]. In terms of marketing of electricity sales companies, the marketing strategies of foreign electricity sales companies can provide some references for newly established electricity sales companies in China. Through the analysis of the marketing strategy of American power companies, it is concluded that Chinese power companies should occupy the market through abundant products, develop users' use habits and form product dependence [2]. Germany's marketing package includes price package

Content from this work may be used under the terms of the Creative Commons Attribution 3.0 licence. Any further distribution of this work must maintain attribution to the author(s) and the title of the work, journal citation and DOI.
Published under licence by IOP Publishing Ltd

marketing, comprehensive energy marketing, green marketing, client marketing and internet preferential marketing [3]. China's electricity sales companies can learn from the current actual situation. EDF's marketing strategy in France brings enlightenment to China's electricity sales companies: it should attract customers through differentiated marketing strategies [4]. The service model of British power supply enterprises is designed according to customer orientation, and the market share of enterprises is maintained by customer satisfaction [5]. China's power sales companies should establish their customer satisfaction survey system in line with international standards, and improve the service system according to the survey results.

China has not yet carried out the research on the marketing strategy analysis of power selling companies, while system dynamics is suitable for studying the dynamic trend of complex information feedback systems, and can be used to analyse the relationship between various factors. In this paper, the system dynamics model is used to analyse the marketing of the power selling company, and the impact of different marketing strategies on the revenue of the power selling company is analysed.

2. Marketing strategy of electricity selling company

2.1 Product and service strategy

The main product of the power company is electricity. In addition to selling electricity, sales companies can also provide products and services for users.

2.1.1 Smart meter

China's smart grid has entered a comprehensive construction stage. The demand for smart meters at the user end will increase dramatically. Smart meters no longer need manual meter reading. The historical purchase data can be saved. It is convenient for customers to inquire and for power companies to carry out power analysis.

2.1.2 Electricity consumption analysis

For commercial or industrial users with large power consumption, the monthly power consumption analysis provided by power selling companies can help users to strengthen load management. By analyzing the characteristics of users' electricity consumption, power selling companies can provide users with power packages and save costs for users.

2.1.3 Demand response service

For power users, demand response service can increase energy efficiency, reduce the cost of enterprise operation and family life; for power selling companies, it can reduce the peak period of electricity purchase, improve power supply reliability and service level; for society, it can rationally allocate power resources, promote the coordinated development of the economy, promote the use of electricity equipment, and increase the demand for energy-efficient equipment.

2.1.4 The "Internet +" service mode

Power mobile terminal transmits power consumption, data analysis, power load and real-time electricity price to users. At the same time, power mobile terminal and intelligent household appliances are interconnected to realize one-button switch, which can not only reduce the peak load of power grid, but also save users' expenses.

2.1.5 Green power

Overseas electricity companies have already sold clean energy. More than 800 of Germany's more than 1100 electricity companies have offered green power packages. The proportion of end users who buy green power has also exceeded 15%. More than one third of the residents have chosen the green power tariff scheme. It shows that users are willing to pay for cleaner energy with higher price. Green Mountain Energy in the United States shows the amount of carbon emissions that its users have reduced on its official website and converts it into the amount of trees planted. This intuitive display is conducive to users' understanding of the benefits of clean energy to environmental protection.

2.2 Price differentiation strategy

2.2.1 TOU price
Prices vary at different periods, but prices remain stable over the same period. In a week, there are three kinds of electricity prices: peak, flat and low. This pricing model comes from the electricity demand predicted by the market, and it will be set up one month ahead of schedule. Users reduce electricity consumption during peak hours, and the sale company gives them a discount or compensation.

2.2.2 Combined package pricing
Users can choose a set meal according to the amount of electricity they use each month, or they can choose a suitable set meal according to different usage habits, so as to get a price discount. Electricity companies can offer various combination packages on their own APP, and give discounts to users who pay for the packages in advance on the APP.

2.3 Channel diversification strategy

2.3.1 Wholesale channels
Electricity selling companies can not only sell electricity to resident users, but also sell electricity to other power selling companies and large users in large quantities.

2.3.2 Retail channels
Electricity sales companies use the opening of business halls to provide electricity sales services for users, while competing with other power sales companies to compete for users. Electricity sales companies sell electricity through websites to achieve one-click online power purchase with provincial users.

2.3.3 Chain operation
The sale of electricity companies through the establishment of chain agencies to achieve a wider range of electricity sales.

2.3.4 Cross-border cooperation
In the future, the electricity sales industry will become a new cross-border industry, which includes the theories of energy, finance, science and technology, applied mathematics, psychology and so on, and forms a multi-dimensional integration of profit models.

2.4 Promotion strategy

2.4.1 Staff promotion strategy
Personnel of power selling companies enter the community to carry out training on electricity use and energy conservation, publicize the concept of cleanliness and environmental protection, and carry out popular science of corresponding services on demand side.

2.4.2 Big data marketing strategy
Electricity sales companies directly provide energy companies with user information and grid investment information in order to obtain revenue. Based on the analysis of big data, this paper puts forward energy-saving suggestions and personalized packages for users.

2.4.3 Advertising strategy

Electricity companies use existing network sales platforms to publish contract information and service information, expand online and offline electricity sales business, publish advertisements in search engines, portals, social media, publicize and promote through local newspapers, television, radio and other traditional media, and use sponsors or titles to put advertisements.

3. Strategy analysis

3.1 System causality analysis

In this section, we analyse the causality of variables between strategies. The arrows in the figure below represent the causality, while the positive and negative symbols represent the positive and negative effects respectively.

3.1.1 Product and service strategy

As can be seen from Figure 1, the installation of smart meters is the premise of power consumption analysis. The analysis of power consumption is the premise of demand response service development. At the same time, "Internet +" service is the bridge between users and Load aggregator, and is the foundation of demand response. Peak load and valley filling serve as demand response, and its role can increase the total revenue of the sales company. Before demand response, peak demand is too high, peak time electricity price is relatively high, and voltage is unstable. Demand response services provided by power selling companies or load aggregators will reduce peak-time demand, thereby reducing peak-time electricity sales and increasing low-trough electricity sales.

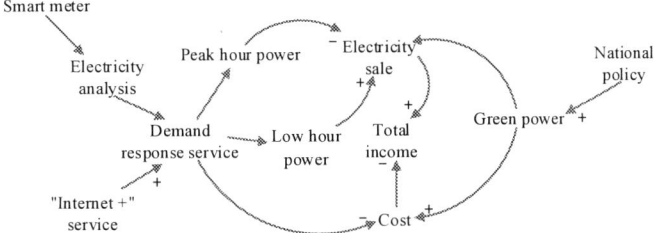

Figure 1. Relationship between product and service strategy, electricity consumption and cost

The development of green power requires the pre-investment of the power selling company, including the cost of green power and the cost of development and maintenance of mobile terminals. In the short run, green power can attract users with environmental awareness to buy electricity from the power sales company. In the long run, green power is beneficial to the future, laying the foundation for ecological harmony.

3.1.2 Price differentiation strategy

It can be seen from Fig. 2 that both the TOU price and the combined package can increase the electricity consumption. Time-of-use tariff includes peak and trough tariffs. Higher tariff in peak period will reduce electricity sales, lower tariff in trough period and increase electricity sales. Price regulation is also a way to regulate the load of power grid. Combo packages attract users to increase electricity sales through their customized customization and relatively preferential prices. Affordable electricity prices can effectively improve customer satisfaction, increase the reputation and popularity of sales companies, retain old users, attract new users, and increase sales.

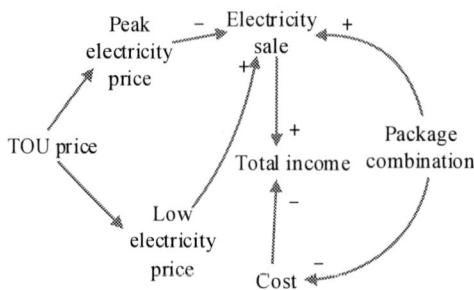

Figure 2. Relationship between price strategy, cost and electricity consumption

3.1.3 Channel diversification strategy

The same sales company can sell electricity in multiple distribution areas in the province, which can increase sales channels. Cross-border cooperation is to cooperate with other brands, expand visibility, and provide power for cooperative brands. It can not only broaden the channels of electricity sales, but also increase the volume of electricity sales. Wholesale and retail channels can't deepen their depth, but can broaden their scope, which requires power companies to carry out customer relationship management. Old users' satisfaction can bring new users and thus increase electricity sales.

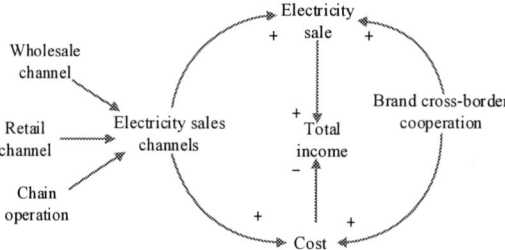

Figure 3. Relationship diagram of channel strategy with cost and electricity sales

3.1.4 Promotion strategy

As can be seen in Figure 4, staff promotion and advertising can effectively enhance the popularity of electricity companies, but also accompanied by cost consumption. The selling company should pay attention to the cost in the process of popularization, and prevent the risk of shortage of funds. In cooperation with big data platforms, power companies can provide users with power analysis, professional power-saving solutions, and help users to conduct power-saving management, which will increase the cost of power companies. Although it may not necessarily enhance its popularity, the power sales company will provide users with analysis and power consumption programs, which will enable users to fully understand their own situation, develop good electricity use habits, and improve the quality of power sales company users.

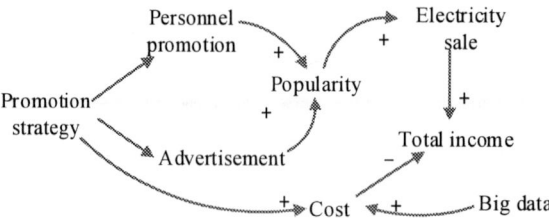

Figure 4. Relationship between promotion strategy and cost, electricity sales

3.2 Establishment of system model

3.2.1 Determination of system boundary

Table 1. Determination of system boundary.

System variables	Specific variables
Product and service strategy	Smart meter input quantity
	Peak time compensation amount
	Low preferential amount
	Green power subsidy
Price differentiation strategy	Difference between package price and general price difference
	TOU price peak low price difference
Channel diversification strategy	Wholesale channel discount price
	Chain operation input cost
	Brand cooperation investment amount
Promotion strategy	Advertising investment amount
	Number of sale staff
	Big data analysis cost
	Big data ad delivery accuracy
	Advertising ratings
Other variables	Cost
	Total income
	Electricity sale

3.2.2 System causality diagram

After determining the boundary of the system, the relationship between the elements in the system is analysed, and the causality diagram is obtained. The power marketing system of the power selling company is affected by many factors. This paper only considers the impact of different marketing strategies on the total revenue of the power selling company, and draws the causality diagram of the marketing system as shown in Figure 5.

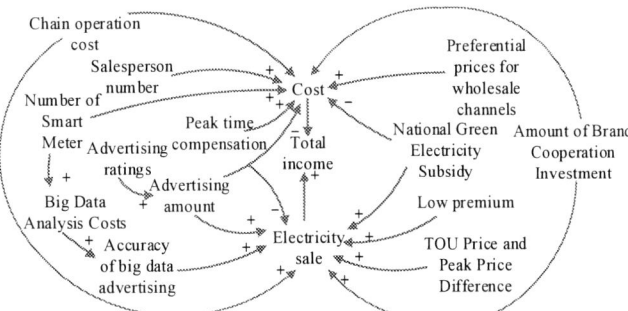

Figure 5. Causality diagram of marketing system

3.2.3 Establishment of system model

Based on the analysis of the causality among the above strategies, the storage flow chart of the marketing system of the power selling company is established by using the system dynamics software

Vensim, as shown in Fig. 6. Some of the parameters of the model are shown in Table 3. Part of the equation of the model is shown in Table 4.

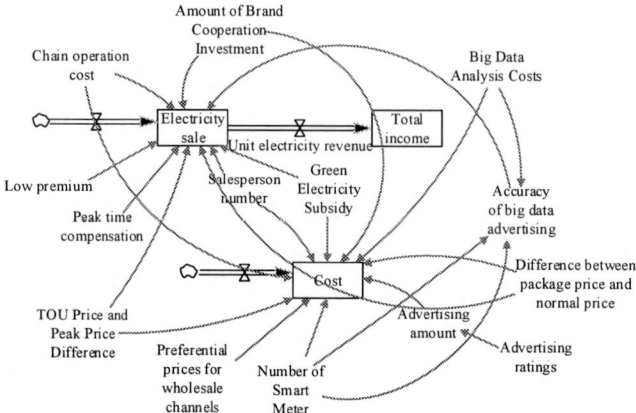

Figure 6. Flow chart of marketing system of power marketing company

Table 2. Parameter description table of marketing system dynamics model of power marketing company.

Model parameter	Representative meaning
P_z	Smart meter input quantity
P_M	Peak time compensation amount
V_M	Low valley discount amount
G_V	Green power subsidy
D_P	Difference between package price and general price difference
D_T	TOU price peak low price difference
M_P	Wholesale channel discount price
C_C	Chain operation input cost
C_{BC}	Brand cooperation investment amount
C_A	Advertising investment amount
M_N	Number of sale staff
C_{BD}	Big data analysis cost
A_{BD}	Big data ad delivery accuracy
A_R	Advertising ratings
V	Unit price
C	Cost
V_t	Total income
S	Electricity sale

Table 3. Partial equations for the marketing system dynamics model of the electricity sales company.

Serial number	Mathematical expression
1	$V_t = S*V - C$
2	$S = S_0 - \alpha*P_M + \beta*Y_M + \gamma*G_V + \delta*D_P + \varepsilon*D_T + \theta*C_C + \vartheta*C_{BC} + \mu*C_A + \rho*M_N + \sigma*A_{BD}$
3	$C = C_0 + a*P_z + C_{BD} + C_C + C_{BC} + C_A - G_V + b*D_{P+} D_{T+} M_P + c* M_N$
4	$A_{BD} = e* P_z + f*C_{BD}$
5	$C_A = g* A_R$
6	INITIAL TIME=1
7	FINAL TIME=12

8	TIME STEP=1
9	UNIT OF TIME: MONTH

4. Model simulation analysis

In this paper, a power company is taken as a sample for simulation analysis. The starting time is January, the ending time is December, and the step size is 1. The simulation considers the effects of compensation amount, peak-valley price difference, brand cooperation investment amount and advertising investment amount on the total income change, and draws the following figure. Figure 7 (a) is the calculation model, and Figure 7 (b) is the output result. From Figure 7 (b), we can see the impact of several marketing methods on the total revenue. With the increase of input, the total revenue increases.

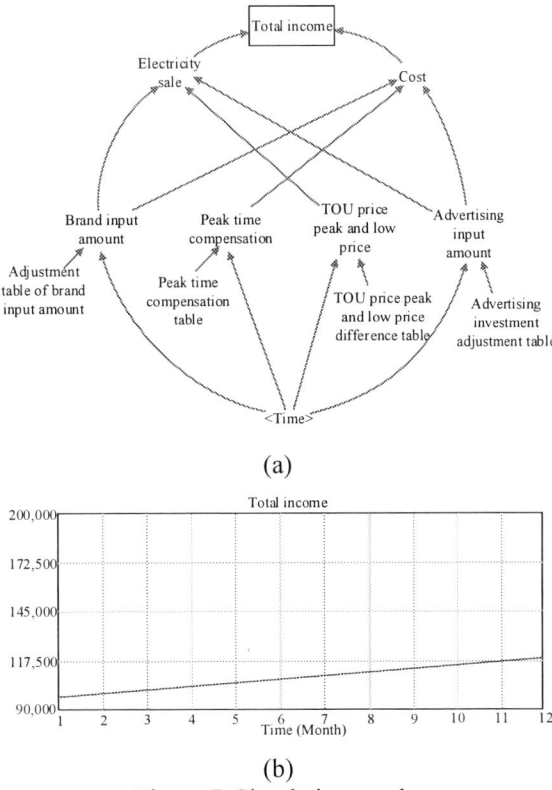

(a)

(b)

Figure 7. Simulation results

5. Conclusion

With the start of a new round of power system reform in China, the market of power selling side has gradually opened up, and power selling companies have entered into people's lives. Electricity companies should not stick to the middle price difference by buying and selling electricity. They should fight price war instead of always lowering the price. Electricity companies should increase marketing efforts, cultivate user habits, establish brand image, and make long-term plans for the follow-up development. It can be sold from the model simulation diagram. The company should strengthen the following aspects.

5.1 Rich products

Electric power is the main product of power selling companies. The potential of China's power selling companies in personalized product customization and series of value-added services still needs to be tapped. According to the big data, the power selling company should deeply analyse the users' electricity consumption habits and design different products and value-added services for different types of users. Let users fully understand their power consumption and electricity usage, and help users develop the habit of saving electricity. It can provide users with comprehensive services such as demand side response, electricity consultation and one key payment.

5.2 Different prices

Before adjusting the electricity price, the power selling company should subdivide the user groups and provide different prices and services for different users. Environmental users are willing to pay more for green electricity, and economical users are willing to reduce the cost of electricity. For environmentally friendly users, power companies can provide relatively high-priced green energy for them, and recommend smart home services to them, so that users can experience the convenience of green and intelligent life. For the energy-saving users, the power sales company can customize the energy-saving scheme for them, and provide them with energy-saving renovation services and electricity consulting services.

5.3 Broaden channels

"Cross-border" is no longer a strange term for Chinese enterprises nowadays. Electricity companies should not only strengthen cooperation with water supply, heating, gas supply and other service enterprises, but also cooperate with scientific research institutions and small and medium-sized enterprises to enhance their popularity and increase electricity sales at the same time. Off-line stores can increase the interaction between power companies and users, solve the problem of power consumption for users, but also is a kind of "advertisement" without paying advertising fees. Potential users can deepen their understanding of power companies through stores.

Acknowledgement

This research was financially project supported by national Power Grid Corp headquarters: Research on operation simulation and effectiveness evaluation of electricity market (YDB17201600102)

This work was supported by the State Grid Corporation of China under the project title: "The Improved Core Analysis Algorithms and Utilities for Smart Grid Big Data" (520940180016)

References

[1] Xiaoyan Zhang, SCT, **15** (2014)

[2] Li Liu, Min Cao, Yongxiu Bai, Fujian Forum **8**,7 (2017)

[3] Wei Ding, Xiaobing Zhou, Dong Xie, Yu Liao. Hubei Electric Power, **40**, 3 (2016)

[4] Li Ma, Xiaoxuan Zhang, Zhe Wei, Song Xue, Su Yang, Junming Tu,SPST, **9**, 4 (2015)

[5] Linghao Zhang, Yan Zhang, Mei Hua, Hui Qian, Tianyi Liu, PDM, **14** (2012)

ISPECE   IOP Publishing

IOP Conf. Series: Journal of Physics: Conf. Series **1187** (2019) 022044   doi:10.1088/1742-6596/1187/2/022044

Study on Zero Drift of Charge Amplifier Based on MOSFET 3N165 and OPA LF356N

Zongjin Ren[1], Cong Sun[1], Jun Zhang[1,a] , Lulu Ma[1], Kai Zhao[1]

[1]Mechanical engineering school, DaLian University of Technology, China

[a] Corresponding author: Zhangj@dlut.edu.cn

Abstract. The compensation of zero drift of charge amplifier is of great significance to its application in testing field. In this paper, the zero drift theory of charge amplifier used in an integrated piezoelectric torque dynamometer is studied, which lays a theoretical foundation for the design of high cost-effective zero drift compensation circuit. The charge amplifier is designed based on MOSFET 3N165 and LF356N operational amplifier. The analysis of the charge conversion circuit of this charge amplifier and the establishment of a reasonable equivalent circuit model are the key to design the compensation circuit. In this paper, the equivalent circuit model of charge conversion is established for this charge amplifier, and the theoretical formula of zero drift at 25°C~40°C is derived. By comparing theoretical value of zero drift with measured value, the compensation ratio coefficient ε of junction temperature is introduced to modify the formula. Finally, the correctness of the revised formula is verified by experiments. The result shows that the modified formula is applicable to the actual zero drift prediction in the experimental process, which lays a theoretical foundation for the design of zero drift compensation circuit for the integrated piezoelectric torque dynamometer with this charge amplifier.

1. Introduction

With the rapid development of modern science and technology, piezoelectric sensors for measuring non-electrical physical quantities have been more and more widely used in various technical fields. The output of piezoelectric sensor needs to be converted by a charge amplifier (charge-voltage conversion) before it is used for subsequent amplification and data processing. Therefore, charge amplifier is an indispensable secondary instrument for measuring with piezoelectric sensor and its main function is to convert charge signal into voltage signal. So, the design of charge amplifier with good performance is of great significance to the measurement with piezoelectric sensor. However, when measuring static or low-frequency charge signals with charge amplifier, the instrument itself will have a stability problem, namely zero drift, which causes instrument's output to be unstable. This problem limits the application of charge amplifier and piezoelectric sensor in static or low-frequency long-term measurement[1], and it seriously affects the processing results of data processing results. The presence of zero drift will cause the drift signal and the effective signal voltage to be indistinguishable. In severe cases, the drift voltage will even drown the effective signal voltage, making the charge amplifier unable to work properly. Therefore, the zero drift of the charge amplifier must be considered while designing good performance charge amplifier. A smaller zero drift is an important index of high performance charge amplifier. At present, the problem of eliminating zero drift of charge amplifier is studied both at home and abroad.

Content from this work may be used under the terms of the Creative Commons Attribution 3.0 licence. Any further distribution of this work must maintain attribution to the author(s) and the title of the work, journal citation and DOI.

Published under licence by IOP Publishing Ltd

Zhende Hou and Ruiting Gao of Tianjin University have proposed a method to eliminate the zero drift of charge amplifier through the feedback compensation network in circuit[2]. However, they do not consider the specific ambient temperature and have their limitations in use. Chunfei Zhang and Jiarong Luo, have proposed a method to remove zero drift with software[3] which is written in object-oriented C++ language. The reusability of code, extensibility of function and portability of system are better[4-5], but it is not applicable for long-term measurement. ZhenxingLi , ZhiwuXuan, and Hongzhou Xu, have proposed a method of sliding the segmental average to remove zero drift when dealing with vibration signals[6].The sliding average method can effectively remove the step-type zero drift, which can improve the accuracy of data processing results and the reliability of spectrum analysis. The problem is that the selection of the width of the window function and the setting of the threshold value require the experimental data processing personnel to make human judgments based on the actual data, but there is no general method and theoretical basis. Miran Milkovic of General Electric Company of the United States proposed a zero drift compensation method for long-term operation of an operational amplifier or integrator. Compared with the chopper-stabilized compensation method[7], this method requires no external components and recovers faster under overload conditions[8], but the system has lower cost performance and higher cost. The series of charge amplifiers 5015A produced by Kistler from Switzerland, has a large measuring range, wide operating frequency and zero-point automatic correction function[9]. It has a good effect of removing zero drift and meets the needs of long-term measurement, but the cost is usually 8~10 times the price of similar products in China. And the introduction of the zero drift compensation part in the manual is not specific enough to be used for reference.

A high cost-effective zero drift compensation circuit can be designed for the charge amplifier of the integrated piezoelectric torque dynamometer. The expression of zero drift on time and temperature is obtained by collecting the zero drift signal, which lays a theoretical foundation for achieving the goal of eliminating zero drift. The charge amplifier is based on the 3N165 MOSFET and the LF356N integrated operational amplifier. In this paper, the charge conversion circuit is analyzed for this type of charge amplifier to establish a reasonable equivalent circuit model. Then the corresponding expressions are listed by the Kirchhoff's law[10], and the zero drift formula is deduced, and the formula is further modified. This is the key to design the zero drift compensation circuit of the charge amplifier.

2. Modeling the core of a charge-transfer circuit
An equivalent circuit model is established for the charge amplifier studied in this paper. In the process of model building, the schematic diagram of the charge-transfer circuit is given first, and its principle is analyzed. Then, the equivalent circuit analysis of the two core components of the dual P-channel enhancement MOSFET model 3N165 and the integrated operational amplifier model LF356N are carried out respectively. Finally, the whole circuit model of the core of the charge-transfer circuit is established. For LF356N, the rate of increase in internal input bias current varies with temperature. Generally speaking, the bias current raises with increasing temperature, but at low temperatures, other sources of leakage in the operational amplifier may change this trend. Such stray leakage may have different temperature characteristics. Little is known about leakage below room temperature. However, more attention is paid to leakage at room temperature and above it in the research at present. Combining with the operating conditions of charge amplifier and data of LF356N, the temperature range studied in this paper is between 25°C and 40°C.

2.1 The schematic diagram of charge-transfer circuit
The charge-transfer circuit is the core circuit of the charge amplifier. The schematic diagram of the charge-transfer circuit is shown in Fig.1, which is based on the principle of integrator. The core components are the dual P-channel enhancement MOSFET 3N165 and integrated operational amplifier LF356N. Where Q is the input charge signal, R_d is the insulation resistance of piezoelectric sensor and C is the feedback capacitor. With LF356N as the core of the integration circuit, R_d and C constitute the integral conversion circuit, which converts the charge signal Q into voltage signal. The

first channel D1, G1 and S1 of the MOSFET 3N165 mainly serves to improve the input resistance of the operational amplifier and can make it up to 10^{14} Ω. The second channel D2, G2 and S2 of the MOSFET 3N165 and sliding rheostat $R4$ and resistance $R5$ constitute a control circuit to regulate the negative input offset voltage of LF356N, so as to ensure that the output Vo of the whole circuit is zero when the charge input is zero.

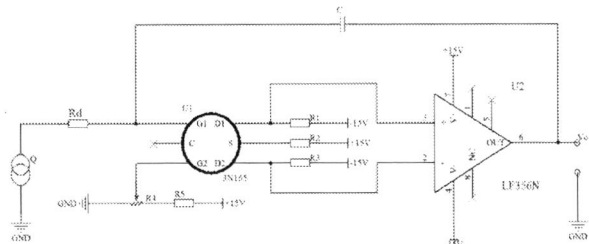

Figure 1. The schematic diagram of the charge-transfer circuit.

2.2 Model of charge-transfer equivalent circuit

In order to simplify the analysis and keep its generality, assuming that the input charge signal Q is zero, then the output Vo of the charge conversion circuit is equated to zero drift. In order to meet the actual working conditions of components, the offset voltage Vos and input bias current Ib of integrated operational amplifier LF356N should be taken into account here (Due to the accuracy problem of the components controlling the input offset voltage of LF356N with the second channel of 3N165 as the core, the offset voltage Vos still exists after adjustment. In order to simplify the analysis, the unbalanced voltage Vos before adjustment is still used for analysis. Therefore, the negative input of LF356N can be used as direct grounding). The equivalent circuit of charge-transfer is shown in Fig.2, in which the direction of the diagram is the direction of current and voltage direction. Among them, where R is the first channel equivalent resistance of MOSFET 3N165, whose value is about $10^{14}\Omega$, Rd is the insulation resistance of piezoelectric sensor, C is the feedback capacitor, Vos and Ib are the input offset voltage and input bias current of integrated operational amplifier LF356N, Ri is the input impedance of LF356N and K is the open-loop gain of LF356N.

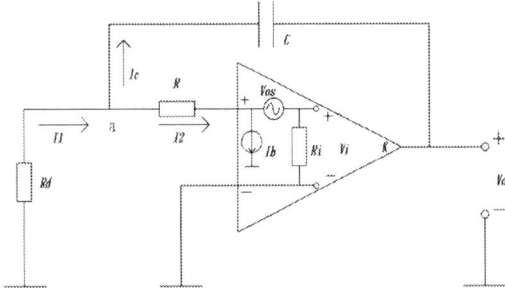

Figure 2. The equivalent charge-transfer circuit.

2.3 Derivation of drift formula under single sensitivity

Based on equivalent charge-transfer circuit, the Kirchhoff Voltage Law(KVL) is applied. And it is applied at point a in Fig.2. The Kirchhoff voltage law at a point is expressed as:

$$I_C = I_1 - I_2 \qquad (1)$$

$$I_1 = -\frac{V_a}{R_d} \qquad (2)$$

$$I_2 = I_b + \frac{V_i}{R_i} \qquad (3)$$

$$V_o = -KV_i \qquad (4)$$

$$V_a = V_{os} + V_i + I_2 R \qquad (5)$$

$$V_o = -\frac{\int_0^t I_c dt}{C} + V_a \qquad (6)$$

The six formula of simultaneous (1), (2), (3), (4), (5), and (6) are:

$$V_o = -\frac{KR_i}{(R_i+R+KR_i)C}\int_0^t \left(\frac{R_i+R_d+R}{KR_dR_i}V_0 - I_b - \frac{V_{os}+RI_b}{R_d}\right)dt + \frac{(V_{os}+I_bR)KR_i}{R_i+R+KR_i}$$

$$(7)$$

At a certain temperature, the last term of formula (7) is constant and can be eliminated as a systematic error. After removing the constants in formula (7), t is derived from both sides of the formula (7):

$$V_o^{'} = -\frac{KR_i}{(R_i+R+KR_i)C}\left(\frac{R_i+R_d+R}{KR_dR_i}V_o - I_b - \frac{V_{os}+RI_b}{R_d}\right) \qquad (8)$$

When $t=0$, $Vo=0$ is the initial condition of equation (8), and the solution is:

$$V_o = -\frac{KR_i}{R_i+R+KR_i)C}\left(e^{\frac{R_i+R_d+R}{CR_d(R_i+R+KR_i)}t} - 1\right) \qquad (9)$$

For a charge amplifier at a certain temperature under a single sensitivity, Vo in Formula (9) is an exponential function of t. Because the power supply voltage of the charge conversion stage circuit is +15V, the zero drift extremum will not exceed +15V. For the MOSFET 3N165 and integrated operational amplifier LF356N, the other variables are known except Vos and Ib which need to be selected according to temperature. Among them, where K is $10^5 \Omega$, Ri is $10^{12}\Omega$, R is $10^{14}\Omega$, Rd is $10^{13}\Omega$ and C is determined by a single sensitivity, and it is 10000pF. Vos and Ib need to be selected by combining LF356N datasheet with charge-transfer circuit.

3. Experiment of zero drift

The zero-drift measurement experiment is carried out for the target charge amplifier in this paper to explore the relationship between the calculated theoretical value and the actual value. Vos and Ib in the formula (9) need to be selected on the basis of the temperature, consequently, it be controlled in the experiment. In this experiment, the temperature is controlled by the thermostat, and the temperature of the incubator is monitored by the thermocouple temperature detector, so that the temperature error value is within plus or minus 0.5°C. To simulate the actual connection of charge amplifiers, the input terminals need to be connected to a piezoelectric sensor. Finally, the zero drift signal is transmitted to the host computer on the PC terminal through the data acquisition card, and then the data is exported. In this experiment, when collecting the zero drift value of charge amplifier, the continuous acquisition time of one experiment is 20min.

Figure 3. Experimental diagram of zero drift acquisition.

The physical map of the zero drift collection of this experiment is shown in Fig.3. The input terminal of the charge amplifier is connected with a piezoelectric sensor to simulate the high input impedance in actual usage and the output terminal is connected with the data acquisition card, connected with the PC host computer for data acquisition. And the grounding end of the charge amplifier is reliably grounded to ensure its normal operation. Charge amplifier is placed in the thermostat and heated in the experiment. The temperature change in the thermocouple is observed by thermocouple thermometer, so as to change the ambient temperature.

Because the temperature can directly affect the *Vos* and *Ib* values, zero drift experiments were carried out at 25°C, 30°C, 35°C and 40°C respectively for the reasons mentioned above. Fig. 4, Fig. 5, Fig. 6 and Fig. 7 are data comparison diagrams of experimental data and theoretical formula (9) calculation at different temperatures respectively.

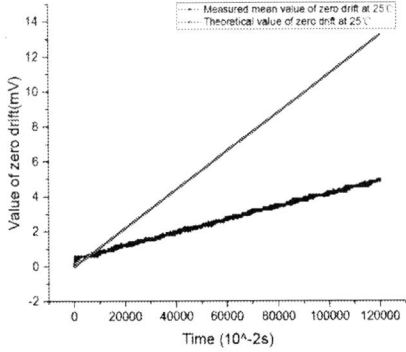

Figure 4. Comparison of measured zero drift and theoretical zero drift at 25°C.

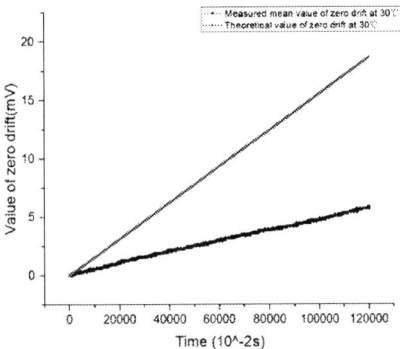

Figure 5. Comparison of measured zero drift and theoretical zero drift at 30°C.

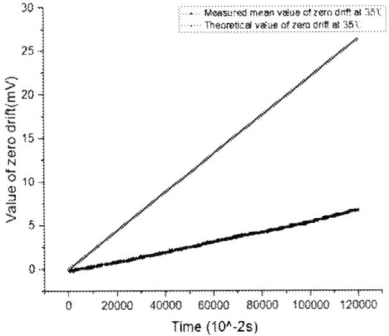

Figure 6. Comparison of measured zero drift and theoretical zero drift at 35°C.

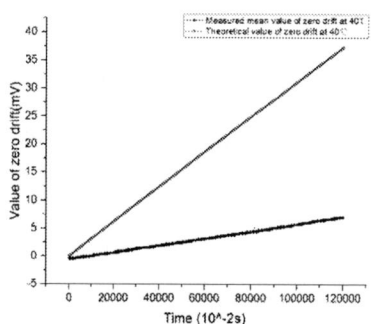

Figure 7. Comparison of measured zero drift and theoretical zero drift at 40°C.

The difference between the measured zero drift value and the theoretical zero drift value at four temperatures is compared and analyzed. It is found that there are different multiplier relations among them. This reason is that there are different junction temperatures of transistors in LF356N at different temperatures, which leads to the change of bias current of operational amplifier. In order to improve the accuracy of the theoretical formula, it is necessary to further modify the theoretical formula (9) at different temperatures.

4. Correction of theoretical formula

According to the problem that the theoretical value of zero drift has different ratio relationship with the measured value at different temperatures, the relevant data are analyzed. A coefficient ε is introduced on the basis of formula (9) and expressed in formula (10), which refers to the junction temperature compensation ratio required by integrated operational amplifier LF356N at different operating temperatures. The data analysis at four temperatures is shown in Table 1, and the compensating ratio coefficients of junction temperature at each temperature are obtained.

$$V_o = -\varepsilon \frac{KR_i(V_{os} + I_bR + I_bR_d)}{R_i + R_d + R}[e^{-\frac{R_i + R_d + R}{CR_d(R_i + R + KR_i)}t} - 1]] \quad (10)$$

Table 1. Junction temperature compensation rate coefficient at different temperatures.

Temperature (°C)	Coefficient of junction temperature compensation ratio ε
25	0.389
30	0.313
35	0.245
40	0.182

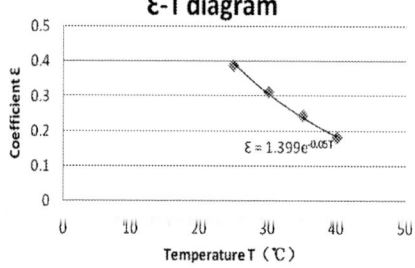

Figure 8. Junction temperature compensation rate coefficient and temperature relationship diagram.

After obtaining the compensation ratio coefficient of junction temperature at different temperatures, the data are fitted as shown in Fig.8, and the functional relationship of ε and temperature T is obtained. Comparing the coefficients ε' calculated from the formula with that ε obtained from the previous data analysis, the following table 2 is obtained. It can be seen from Table 2 that the maximum difference.

The analysis coefficient ε of the compensation coefficient and the calculated value of the compensation coefficient ε' is only 0.012. And the

Table 2. Comparison of calculated value and analytical value of junction temperature compensation rate coefficient.

Temperature (°C)	Analytical value of compensation coefficient ε	Calculated value of compensation coefficient ε'	Deviation
25	0.389	0.401	0.012
30	0.313	0.312	-0.001
35	0.245	0.243	-0.002
40	0.182	0.189	0.007

temperature is 25°C. At 25°C, the maximum drift of zero drift is less than 0.4mV, which is within the allowable range of actual measurement. Therefore, the relationship is consistent with the actual situation and can be used for correction in equation (9). The final form of the correction formula is formula (11).

$$V_o(T,t) = -1.399e^{-0.05T} \frac{KR_i(V_{os}+I_bR+I_bR_d)}{R_i+R_d+R}[e^{-\frac{R_i+R_d+R}{CR_d(R_i+R+KR_i)}t} - 1] \quad (11)$$

Where T is the ambient temperature, the unit is °C, t is time, the unit is s, and the remaining values are selected as described above.

According to the correction formula(11), the theoretical formulas at 25°C, 30°C, 35°C and 40°C are modified to obtain a comparison chart of the theoretical zero drift value and the actual zero drift value, which are shown in Fig.9, Fig.10, Fig.11, and Fig.12. By analyzing the difference between the corrected zero drift value and the actual value, it can be seen that the maximum error value at 25°C, 30°C, 35°C and 40 °C is 0.5mV, 0.2mV, 0.5mV and 0.5mV respectively, and these errors are allowed in the actual measurement under this sensitivity. Therefore, formula(11) can be used as a zero-drift expression formula for charge amplifiers based on MOSFET 3N165 and LF356N operational amplifiers in the range of 25°C to 40°C.

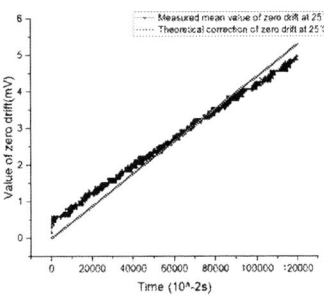

Figure 9. Comparison of zero drift of measured zero drift and correction formula at 25°C.

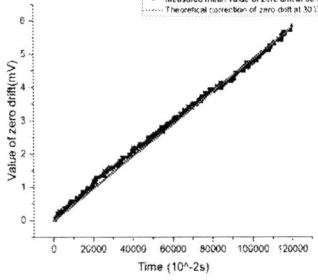

Figure 10. Comparison of zero drift of measured zero drift and correction formula at 30°C.

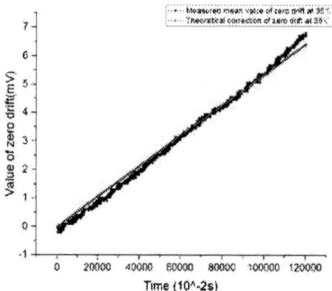

Figure 11. Comparison of zero drift of measured zero drift and correction formula at 35°C.

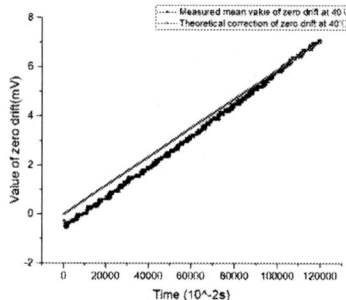

Figure 12. Comparison of zero drift of measured zero drift and correction formula at 40°C.

5. Verification experiment of correction theoretical formula

Experiments under the same conditions are carried out to verify the correctness of the revised theoretical formula after correcting the theoretical formula. The ambient temperature is set at 28°C, and the modified formula is verified experimentally. The other conditions are the same except that the ambient temperature need to be changed by the incubator. The temperature is controlled in the range of 28±0.5°C by the incubator and thermocouple temperature detector. One experiment is still 20minutes. The comparison between measured experimental data and theoretical formula (9) are shown in Fig.13.

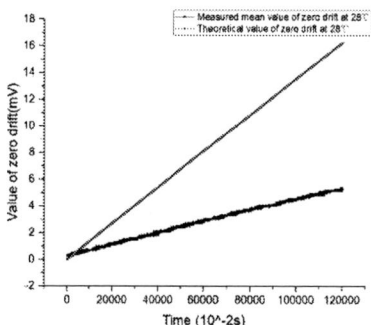

Figure 13. Comparison of measured zero drift and theoretical zero drift at 28°C.

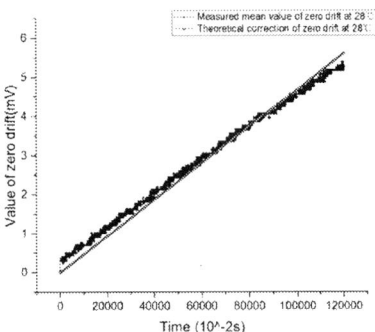

Figure 14. Comparison of zero drift of measured zero drift and correction formula at 28°C.

It can be seen from Fig.13 that the measured zero drift value has a large difference from the theoretical zero drift, so it needs to be corrected by using the correction formula. According to the coefficient $\mathcal{E}$ of compensation ratio of junction temperature introduced in the previous paper, the theoretical zero drift value at 28 °C was corrected, and the comparison chart between the measured zero drift and the theoretical correction formula at 28°C is shown in Fig.14. After correction, the maximum difference between the measured zero drift and the corrected theoretical formula is only 0.5mV, which meets the actual measurement requirements. It is further shown that the introduced coefficient $\mathcal{E}$ of junction temperature compensation ratio conforms to the zero drift compensation law of this type of charge amplifier.

6. Conclusion

In this paper, the zero-drift of the charge amplifier of the charge-transfer circuit is analyzed based on a charge-transfer circuit of 3N165 MOSFET and LF356N integrated operational amplifier. The equivalent circuit model of charge conversion stage is established and the theoretical formula of zero drift of this type of charge amplifier is derived. After the comparison between the theoretical and measured values of zero drift, the coefficient $\mathcal{E}$ of compensation ratio of junction temperature is introduced and the theoretical formula is corrected. Finally, the verification experiment was carried out by the zero drift value at 28°C. The results show that the correction theoretical formula is applicable for the actual zero drift prediction. The results of this study provide a reliable theoretical basis for the application of this charge amplifier to remove zero drift on a rotary torque dynamometer. Based on this theory, the compensation circuit can be used to compensate for the zero drift of this charge amplifier.

Acknowledgment

This project is supported by National Natural Science Foundation of China(No.51675084 and No.5147 5078), the Aeronautical Science Foundation Funded Projects (No. 20160163001) and the Basic Scienti fic Research Project of Central University (No. DUT17GF211).

References

[1] V. Radeka, S. Recia. PF. Manfre. IEEE. *Trans on Nuclear Science*. **40**,4,pp.744-749(1993).

[2] H. Zhende, G. Ruiting. Piez.Acous.**4**.pp.30-32+24(1991).

[3] Z. Chunfei, L. Jiarong. Com.Meas.Cont. 7.pp. 684-686(2004).

[4] B. Eckel. USA:*Prentice Flall*, (1995).

[5] Z. Zhiying. *Modern software engineering*. (2000).

[6] L. Zhenxing, X. Zhiwu, X. Hongzhou. Str.Env. **4**.pp.61-64(2008)

[7] M. Milkovic. United States Patent. US4808942-A(1989).

[8] E. A. Goldberg. RCA Review. pp.296-300(1950).

[9] X. Jing. *Design of a novel charge amplifier*.(2008).

[10] L. Li, Z. Jialiang, Z. Xiaoqiong, K. Jun, X. Pengsheng, L. Qiuyang. Appl.Mecs.Mats.**487**.pp. 435-439(2014).

Study on Radial Vibration of Circular Piezoelectric Ceramic

Yazhou Mao[1,a], Jianxi Yang[1], Haojie Liu[2], Yonggang Liu[1] and Wenjing Xu[3]

[1]School of Mechatronics Engineering, Henan University of Science and Technology, Luoyang 471003, China

[2]Luoyang LYC bearing CO., Ltd, Luoyang Henan 471000, China

[3]Luoyang Railway Information Engineering School , Luoyang Henan 471000, China

[a]Corresponding author: myzlcc@163.com

Abstract. The core component of the piezoelectric transducer is piezoelectric ceramic oscillator. The vibration of the piezoelectric ceramic oscillator is affected by piezoelectric ceramics. In this paper, the radial vibration is analyzed by using ANSYS. The variation of voltage and strain displacement, poisson's ratio and resonance frequency are similar, and they are approximately linear. Moreover, the normalized resonant frequency gradually decreases with the thickness diameter ratio of $t>=1$ and $t<=1$. The accuracy of ANSYS to solve the resonance frequency of the radial vibration of piezoelectric ceramics is proved by experiment, theoretical calculation and ANSYS simulation results.

1. Introduction

Transducer is an important device for energy conversion. It can realize mutual conversion between electrical energy and mechanical energy[1]. Thus, it is able to applied to engineering practice[2]. The core component of piezoelectric transducer is piezoelectric ceramic, which has been widely used in transducer for its advantages of high accuracy and low energy consumption.

Chen[3] et al., used ANSYS software to study the resonance frequency of circular piezoelectric ceramic vibration by Admittance-Frequency screening method. Li[4] et al., studied the changed law of radial vibration of piezoelectric ceramic with different radii by using ANSYS software. Fan[5] et al., pointed that the piezoelectric ceramic model was developted for simulated investigation, and results provided theoretical guidance for improving transducer performance. Xiang[6] et al., had been actively conducting research on resonance frequency equation of piezoelectric ceramic ring. Jia[7] et al., deduced the resonance and anti-resonance frequency formula of the equivalent circuit of ultrasonic transducer. Qin[8] et al., had deduced the resonant frequecy of piezoelectric ring crystals, and the correctness was verified by experiments. Liu and Mao[9-10] et al., studied the vibration performance of piezoelectric ceramic plates.

In this work, the vibration of radial piezoelectric ceramic is analyzed. By comparing theoretical and experimental results with ANSYS simulation, the correctness of ANSYS analysis results of piezoelectric ceramic resonance frequency is verified.

2. Finite element analysis theory of piezoelectric ceramic coupling field

Piezoelectric ceramic has piezoelectricity. The main description of piezoelectricity is the interaction between mechanical quantity (stress tensor T and strain tensor S) and electric quantity (electric field strength E and electric displacement vector D). In ANSYS, the second kinds of piezoelectric equations usually is chosen, and the equation is shown as follows:

$$\begin{cases} T = c^E S - e^S E \\ D = e^E S + \varepsilon^S E \end{cases} \qquad (1)$$

Where, c^E is piezoelectric ceramic elastic matrix. e^E is piezoelectric ceramic stress matrix. e^S is the transposed matrix of e^E. ε^S is the dielectric constant.

Modal analysis is used to determine the vibration characteristics of structure. The equation for ANSYS to dealt with dynamic problems as follows:

$$M\ddot{u} + C\dot{u} + Ku = F \qquad (2)$$

Where, M、C and K are the mass matrix, composite damping and stiffness matrix, respectively. F is external load.

For the undamped linear system, the expression can be changed to next equation:

$$M\ddot{u} + Ku = 0 \qquad (3)$$

The solution of expression (3) is shown as following:

$$u = \phi_i \cos \omega_i t \qquad (4)$$

Where, φ_i is the formation eigenvector corresponding to the i order mode. ω_i is the natural frequency of the i order mode (unit: rad/s). t is time (unit: s).

Substituting expression (4) into (3), then:

$$\left(K - M\omega_i^2 \right)\phi_i = 0 \qquad (5)$$

The characteristic equation of vibration equation is:

$$\det \left| K - M\omega_i^2 \right| = 0 \qquad (6)$$

For calculating natural frequency is shown as follows:

$$f_i = \frac{\omega_i}{2\pi} \qquad (7)$$

Where, the unit of f_i is Hz, i.e. rpm.

3. Theoretical analysis of resonance frequency

The piezoelectric ceramic is a device for realizing electromechanical conversion, which can realize the mutual conversion between machine and electricity (positive piezoelectric effect) or electric machine (reverse piezoelectric effect). An axially polarized piezoelectric ceramic is shown in Figure 1.

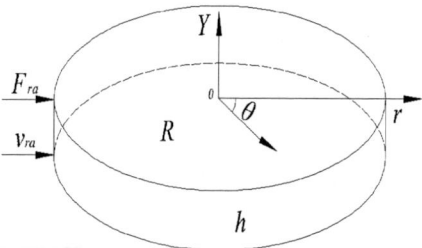

Figure 1. Geometric schematic of the piezoelectric ceramic radial vibration

In Figure 1, R and h are radius and thickness, respectively. F_{ra} and v_{ra} represent the vibration velocity and radial force, respectively. The motion equation of the piezoelectric ceramic radial vibration is shown as follows:

$$\rho \frac{\partial^2 \xi}{\partial t^2} = \frac{\partial T_r}{\partial r} + \frac{T_r - T_\theta}{r} \qquad (8)$$

Where, ρ is the density. r is the radial coordinate. $\xi=\xi_{ra}exp(jwt)$ is the radial vibration displacement. T_r is the radial stress. T_θ is the tangential stress. The piezoelectric equation is shown as follows

$$\begin{cases} S_r = s_{11}^E T_r + s_{12}^E T_\theta + d_{31}E_z \\ S_\theta = s_{12}^E T_r + s_{11}^E T_\theta + d_{31}E_z \\ D_z = d_{31}T_r + d_{31}T_\theta + \varepsilon_{33}^T E_z \end{cases} \quad (9)$$

Where, d_{31} is piezoelectric strain constant. s^E_{11} and s^E_{12} are piezoelectric flexible constant, respectively. ε^T_{33} is permittivity. E_z is polarization direction in axial direction.

According to mechanical vibration characteristics of piezoelectric ceramic, the radial force and tangential force of piezoelectric ceramic are shown as follows

$$T_r = \frac{1}{2}\left(\frac{S_r - S_\theta}{s_{11}^E - s_{12}^E} + \frac{S_r + S_\theta - 2d_{31}E_3}{s_{11}^E + s_{12}^E} \right) = \frac{1}{s_{11}^E}\left(\frac{S_r + v_{12}S_\theta}{1 - v_{12}^2} - \frac{d_{31}E_3}{1 - v_{12}} \right) \quad (10)$$

$$T_\theta = \frac{1}{2}\left(\frac{S_r + S_\theta - 2d_{31}E_3}{s_{11}^E + s_{12}^E} - \frac{S_r - S_\theta}{s_{11}^E - s_{12}^E} \right) = \frac{1}{s_{11}^E}\left(\frac{S_\theta + v_{12}S_r}{1 - v_{12}^2} - \frac{d_{31}E_3}{1 - v_{12}} \right) \quad (11)$$

Substituting (10), (11) into (8) can be obtained:

$$\rho \frac{1}{E_r} \frac{\partial^2 \xi}{\partial r^2} = \frac{\partial^2 \xi}{\partial r^2} + \frac{1}{r}\frac{\partial \xi}{\partial r} - \frac{\xi}{r^2} \quad (12)$$

Where, E_r is elastic constant, $E_r = s^E_{11}/(s^E_{11} - s^E_{12})(s^E_{11} + s^E_{12}) = 1/s^E_{11}(1 - v^2_{12})$. $v_{12} = -s^E_{12}/s^E_{11}$ is poisson's ratio.

After finishing formula (12) can be obtained:

$$\frac{\partial^2 \xi}{\partial r^2} + \frac{1}{r}\frac{\partial \xi}{\partial r} - \frac{\xi}{r^2} + k^2\xi = 0 \quad (13)$$

Where, $k=\omega/v$ is wave number. $\omega=2\pi f$ is angular frequency. v is radial vibration velocity, $v_2=Y^E_0/\rho(1-\sigma^2)$.

The expression (13) represents Bessel equation, and the solution of this equation is shown as follows:

$$\xi = AJ_1(kr) + BY_1(kr) \quad (14)$$

Where, $J_1(kr)$ and $Y_1(kr)$ represent the first-order and second-oder Bessel functions. While, A and B represent constants. The radial vibration velocity and displacement can be obtained as follows:

$$\xi = AJ_1(kr) \quad (15)$$

$$v = j\omega AJ_1(kr) \quad (16)$$

According to the geometric schematic of the piezoelectric ceramic radial vibration in Figure 1.

$$v\big|_{r=R} = -v_{ra} \quad (17)$$

After finishing, the above formulas (16) and (17) can be got the constant A:

$$A = \frac{j}{\omega} \frac{v_{ra}}{J_1(kR)} \quad (18)$$

Based on the above expressions, the radial force T_r is obtained as follows:

$$T_r = v_{ra} \frac{(1-v_{12})\dfrac{J_1(kr)}{r} - kJ_0(kr)}{j\omega J_1(kR)s_{11}^E(1-v_{12}^2)} - \frac{d_{31}E_3}{s_{11}^E(1-v_{12})} \quad (19)$$

According to the geometric schematic of the piezoelectric ceramic radial vibration in Figure 1.

$$F_{ra} = -T_r\big|_{r=R} S_a \quad (20)$$

Where S_a represents piezoceramic superficial area, $S_a=2\pi ah$. After simultaneous sorting of equations (18)、(19) and (20) can be obtained:

$$-F_{ra} = -j\rho U_r S_a \left(\frac{1-v_{12}}{kR} - \frac{J_0(kR)}{J_1(kR)} \right) v_{ra} - \frac{2\pi a d_{31} E_3 h}{s_{11}^E (1-v_{12})} \quad (21)$$

The expression (21), after simplification and sorting:

$$F_{ra} = Z_r v_{ra} + A U_3 \quad (22)$$

Where, U_3 represents voltage, $Z_{ra}=\rho U_r S_a$, $U_3 = E_3 h$, $Z_r = jZ_{ra}\{(1-v_{12})/kR - J_0(kR)/J_1(kR)\}$, $A = 2\pi a d_{31}/s_{11}^E(1-v_{12})$, n represents mechatronic conversion coefficient.

According to electric characteristics of piezoelectric ceramics, it can be obtained by combining expressions (10)、 (11) and (12), then:

$$D_3 = \frac{d_{31}}{s_{11}^E + s_{12}^E} kAJ_0(kR) + \varepsilon_{33}^T E_3 - \frac{2d_{31}E_3 d_{31}}{s_{11}^E + s_{12}^E} \quad (23)$$

The charge Q and current I flowing through the piezoelectric ceramic are shown as follows：

$$Q = 2\pi \int D_3 R dR$$
$$= \frac{2\pi d_{31}}{s_{11}^E + s_{12}^E}(AaJ_1(kR)) + \pi R^2 \left(\varepsilon_{33}^T E_3 - \frac{2\pi d_{31}E_3 d_{31}}{s_{11}^E + s_{12}^E} \right) \quad (24)$$

Where, $C_{0r} = \varepsilon_{33}^T S\{1-2d_{31}^2/\varepsilon_{33}^T(s_{11}^E+s_{12}^E)\}/h$, $S=\pi R^2$ represents piezoelectric ceramic area.

The admittance of piezoelectric ceramic is:

$$Y = \frac{I}{U_3} \quad (25)$$

The resonance frequency equation is expressed as follows:

$$kRJ_0(kR) = (1-v_{12})J_1(kR) \quad (26)$$

The resonant frequency equation is a transcendental equation. If, the solution of the transcendental equation is $R(n)$, the expression for piezoelectric ceramic radial vibration resonance frequency f_n is shown as follows:

$$f_n = \frac{R(n)}{2\pi R} \sqrt{\frac{1}{\rho s_{11}^E (1-v_{12}^2)}} \quad (27)$$

Where, n is vibration order of piezoelectric ceramic.

4. Results and discussion

4.1 The relationship between displacement and voltage

In Figure 2, the experimental result and the simulation result are basically identical. The experimental result is slightly higher than the simulation result, but the error between them is less than 8%. Thus, it is proved that there is an approximate liner relationship between piezoelectric ceramic thickness and voltage. While, the approximate linear relationship meets the following requirements: $y=au+b$.

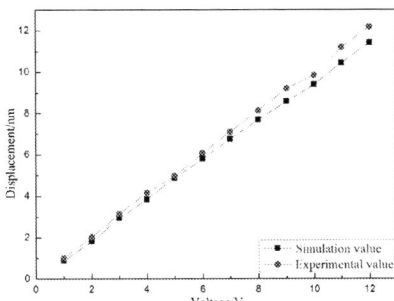

Figure 2. The displacement varying with voltage

In Figure 3, with the voltage loaded on the surface of piezoelectric ceramic, the maximum displacement of piezoelectric ceramic varies linearly with the voltage, and the displacement is gradually increase with the increase of voltage. Piezoelectric ceramic at the same voltage will vibrate more violently and larger amplitude as the thickness is thinner. The amplitude of piezoelectric ceramic is symmetrical distribution about 0V voltage. Because of displacement is a vector, the difference of voltage polarity only changes the direction of piezoelectric ceramic amplitude, but it can not change the magnitude of its amplitude. The larger the voltage, the more intense vibration of piezoelectric ceramic plate. Therefore, the strain of piezoelectric ceramic can be realized by controlling the voltage loaded on the piezoelectric ceramic plate surface.

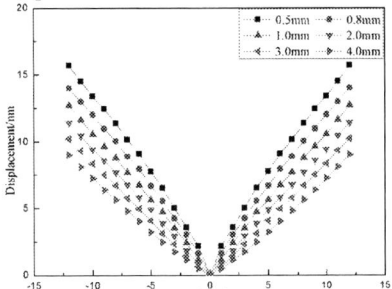

Figure 3. The relationship between voltage and displacement

4.2 The relationship between poisson's ratio and resonant frequency
In order to investigate the relationship between poisson's ratio and resonant frequency. The modeling process is repeated in ANSYS and piezoelectric ceramic with different poisson's ratios are analyzed, as shown in Figure 4. In Figure 4, there is an approximate linear relationship between the poisson's ratio and the resonant frequency in the range of poisson's ratio v=0.27~0.45 selected for study, i.e. the resonant frequency is gradually increase. The resonance frequency at poisson's ratio v=0.45 is 1.14 times of that at v=0.27. Therefore, the sensitivity of resonance frequency is obviously influenced by poisson's ratio.

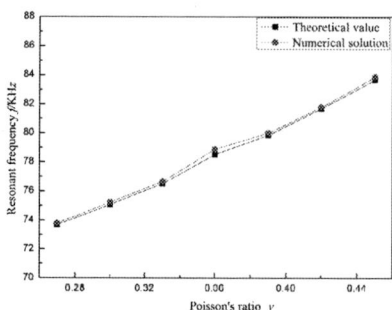

Figure 4. The resonant frequency varying with poisson's ratio

4.3 The relationship between thickness diameter ratio and resonant frequency

Piezoelectric ceramics are created in ANSYS. The thickness-diameter ratio t is defined as $t=h/R$. Normalized resonant frequency is shown in Figure 5 and 6. Wherein, the normalized resonance frequency f of the piezoelectric ceramic is the ratio of the currently resonance frequency to maximum resonance frequency.

In Figure 5, when thickness-diameter ratio ($t<=1$), normalized resonance frequency of numerical solution, theoretical value and experimental value decrease gradually with the increase of thickness-diameter ratio. The resonance frequency of thickness-diameter ratio $t=0.1$ is about seven times of $t=1$. It shows that the smaller t, the higher the resonance frequency value. It provides theoretical guidance for the selection of piezoelectric ceramic plate.

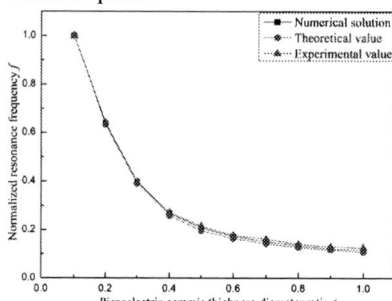

Figure 5. The relationship between thickness diameter ratio t ($t<=1$) and normalized resonance frequency

In Figure 6, the relationship between thickness diameter ratio ($t>=1$) and resonance frequency is the normalized resonance frequency of numerical solution, theoretical value and experimental value decrease gradually with the increase of thickness-diameter ratio. The change regulation of thickness-diameter ratio $t>=1$ and $t<=1$ is similar, but the slope of the curves is different. Moreover, the normalized resonance frequency of piezoelectric ceramic plate with thickness-diameter ratio $t=1$ is about 1.13 times of $t=10$. It shows that the increase of thickness-diameter ratio of the piezoelectric ceramic has little effect on the normalized resonant frequency.

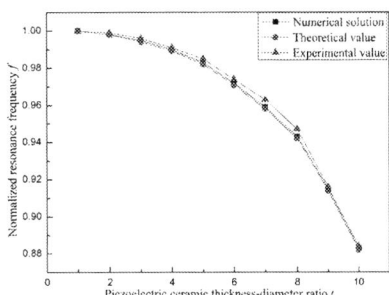

Figure 6. The relationship between thickness diameter ratio t ($t\geq=1$) and normalized resonance frequency

In Table 1, the numerical solution, theoretical value and experimental value are represented by f_1, f_2 and f_3 respectively. The error between numerical solution, theoretical value and experimental value are represented by Δ_1 and Δ_2, i.e. $\Delta_1=|f_1\text{-}f_2|/f_2$ and $\Delta_2=|f_1\text{-}f_0|/f_0$. In Table 1, the Δ_1 and Δ_2 value are both very small and the Δ_1 value is less than the Δ_2. It indicates that the numerical solution is closer to the theoretical value. Although the Δ_2 value is greater than the Δ_1, but the Δ_2 value is less than 2%. It not only indicates that ANSYS simulation result can be fully meet the need of engineering applications, and but also ANSYS is used to solve the resonant frequency of the accuracy and effectiveness of piezoelectric ceramic plate.

Table 1. The error analysis of numerical solution, theoretical value and experimental value

h	R	f_1/KHz	f_2/KHz	f_0/KHz	Δ_1/%	Δ_2/%
2	15	76.615	76.507	77.528	0.14	1.18
3	30	38.727	38.253	39.329	0.98	1.53

5. Conclusions

(1) There is an appropriate linear relationship between voltage and displacement, poisson's ratio and resonant frequency. When thickness-diameter ratio is $t\geq=1$ and $t\leq=1$, the normalized resonance frequency of piezo-electric ceramic plate is similar to the change curve of thickness-diameter ratio, and the normalized resonance frequency decreases with the increase of thickness ratio.

(2) The results of numerical solution, theoretical value and experimental value analysis indicate that the numerical solution is closer to the theoretical value. Although the value of Δ_1 is lower than that of Δ_2, the error value is less than 2%. Therefore, the results obtained by ANSYS software are real, effective and conform to the actual reality.

References

[1] Y.F Tang, S.Y. Lin. Rectangular Bending Vibration of the Piezoelectric Ultrasonic Transducer. Journal of Shanxi Normal University (Natural Science Edition), 44(4):44-48 (2016)

[2] K.T Chang, H.C. Chiang, C.W. Lee. Design and Implementation of a Piezoelectric Clutch Mechanism Using Piezoelectric Buzzers. Sensors and actuators A: Physical, 141(2):515-522 (2008)

[3] S. Chen, J. Yang. Analysis of Radial Vibration Modes of Round Piezoelectric Ceramics Based on ANSYS. Computer Simulation, 26(3):318-321 (2009)

[4] F.X. Li, G.J. Li. Simulation of Radial Vibration of Piezoelectric Ceramic with Axial Polarization. Coal Technology, 35 (02):285-286 (2016)

[5] X.M. Fan, S.W. Ma, X. Zhang, et al. Simulation Analysis of Piezoelectric Ceramic Chip PZT Based on ANSYS. Piezoelectric & Acoustooptics, 36(3),416-420 (2014)

[6] X. Yang, L.K. Wang. Analysis of Radial Thickness Vibration of Piezoelectric Ceramic Ring. Piezoelectric & Acoustooptics, 34(4):561-564 (2012)

[7] L.Y. Jia, G.B. Zhang, T.T. Shi, et al. The Cylindrical Composite Piezoelectric Ceramic Transducer Polarized in Tangential Direction. Journal of Nanjing University (Natural Sciences), 51(6):1153-1159 (2015)

[8] L. Qin, L.K, Wang, T.X, Dong, et al. Analyses on Radial Vibration of Piezoelectric Discs Stack. Journal of Vibration and Shock, 29(12):162-165 (2010)

[9] Y.G. Liu, Y.Z. Mao, D.Y. Li, et al. Study on Radial Vibration of Circular Piezoelectric Ceramics. Coal Technology, 36(5):310-312 (2017)

[10] Y.Z. Mao. Numerical Analysis on Crack and Fault Diagnosis of Wind Turbine Blade. Henan University of Science and Technology (2017)

ISPECE   IOP Publishing

IOP Conf. Series: Journal of Physics: Conf. Series **1187** (2019) 022046   doi:10.1088/1742-6596/1187/2/022046

A Method of Power Flow Calculation Considering New FACTS and HVDC

Quan Chen[1], Xiaoming Dong[1,a], Haifeng Li[2], Tao Jin[2], Jun Yi[3] and Jingzhe Tu[3]

[1]School of Electrical Engineering, Shandong University, Jinan, China

[2]State Grid Jiangsu Electric Power Company, Nanjing, China

[3]China Electric Power Research Institute, Beijing, China

[a]Corresponding author: 1654116735@qq.com

Abstract. Focusing on the influence of new FACTS or HVDC, this paper proposes a method to calculate their power flow of modern power system. The proposed method is carried out by transforming new FACTS or HVDC into equivalent buses like PQ buses in AC. Compared with alternating iterative method and unified iterative method, the proposed method has advantages in convergence property, preparatory work and reusability. This paper used the proposed method, through theoretical deduction of the equations concerning DC system, to complete the power flow calculation of system with HVDC. Through case studies, the results proved that the proposed method has good convergence property and faster computing speed than unified iterative method.

1. Introduction

In modern power system and energy internets, more and more power electronic devices and new ways of transmission or structures are applied. These new components will increase the buses which have different characters. The global energy internet must be enough stable, and have large transmission capacity, which needs electrical network automation providing the base.

The application of FACTS has been considered as an essential trend to improve the stability of system and enlarge the capacity of power transmission [1]. The development and use of FACTS controllers in power transmission system have led to many applications, not only to improve the voltage and transient angle stability of the existing power networks but also to provide operation flexibility of power systems, as presented in [2].

In recent years, China has built ten High Voltage Direct Current Transmission (HVDC) and four Ultra High Voltage Direct Current Transmission (UHVDC), which are under operation, and there will be more HVDC to be put into operation, according to the plan of State Grid Corporation of China.

These new components will increase the difficulty of network modeling and pose increased challenge to the analysis of stability of power system. Power flow calculation (PFC), is a fundamental calculation of other important research. Thus, how to create a unified structure for PFC needs considering carefully.

Faced with new components (FACTS or HVDC) in AC power system, the usual ways to calculate the power flow are with unified iterative method and alternating iterative method. Sheykholeslami has used the unified iterative method to complete the PFC, focusing on the presence of Dynamic Flow Controller in [3]. Paper from [4] has presented the process of a unified iterative solution of PFC. It considers the wind farms and FACTS devices. For HVDC, researchers have completed the PFC

Content from this work may be used under the terms of the Creative Commons Attribution 3.0 licence. Any further distribution of this work must maintain attribution to the author(s) and the title of the work, journal citation and DOI.
Published under licence by IOP Publishing Ltd

through unified iterative method in [5, 6] and paper from [7] improved the unified power flow algorithm by focusing on the changes of buses type.

It should be mentioned that the software OpenDSS has completed much work on the PFC about distribution networks including converter-interfaced DGs. In this paper, the method EPQ is not same with the solution of OpenDSS. The differences rely on below: the EPQ is a method for the entire system, including the FACTS themselves, and the method needs to calculate the operation paraments of FACTS at the same time. The EPQ does not transform the FACTS into pure PQ buses and it will consider the relations between FACTS and AC system which may make the paraments in AC system influence the power injected into AC system. Its purpose is to cope with many different FACTS at an entire system, not only for one type of devices.

This paper mainly consisted of two parts. First part, presented in section 2, proposed a method, named EPQ, to calculate the power flow, and compared it with the alterative iterative method and unified iterative method. Second part, presented in section 3, used EPQ method, after theoretical deduction and analysis of the LCC station, to calculate the power flow of system with HVDC, which is practiced in a case.

2. Methods of power flow calculation
In this part, a method for PFC considering new FACTS is proposed, and the comparison of the traditional two method of PFC with the method is completed.

2.1 Alternating iterative method
For the alternating iterative method, its main advantage against unified iterative method is that, in the iteration the elements and its construct methods of Jacobian matrix are not changed, so this method only needs modifying the power equations. If the numerical oscillation or divergence occurs, the method does not convergence. This method no longer has the quadratic convergence, which is the characteristic of the traditional Newton-Raphson method for computing power flow.

2.2 Unified iterative method
For the unified iterative method, its main advantages rely on its characteristic of quadratic convergence which is preserved from traditional Newton-Raphson. This method combines power equations from the original system and the new equations from new components including control equations and basic equations, to create the Jacobian matrix to complete the PFC. These equations are nonlinear and the unified iterative method uses the principle of the traditional Newton-Raphson method. Thus, this method has quadratic convergence. However, compared with the original PFC, unified iterative method needs expanding the Jacobian matrix for the new variables and the equations of these variables. New variables from new components need considering setting the initial values, while the iterative convergence relays on the initial values. The new equations differ from the original power equations and there may be some situations where modified matrix is pathological and unsolved, which need further processing [8]. Thus, this method requires much work to be done to complete the calculation.

2.3 Equivalent PQ Buses method
Through transforming the new FACTS or HVDC into equivalent PQ buses (EPQ) in AC system, this method can complete the PFC conveniently. The EPQ could be controlled by variables from AC system but not from new FACTS or HVDC on expression, as shown in Fig. 1.

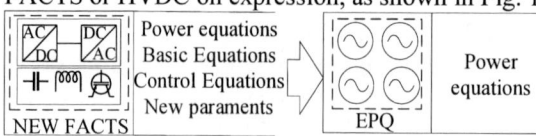

Figure 1. The schematic of EPQ method.

New FACTS is controlled by its own control strategies and basic characteristics from its structure. They also have some restricts on the operation range. EPQ method should analyze the new FACTS, get the basic equations and control equations, and furtherly analyze the reactive and active power injected into devices. Purpose of EPQ is to get an expression of reactive and active power which only concerns variables from AC system, or variables from the control strategies and basic characteristics that can be determined before PFC, as shown in Fig. 2.

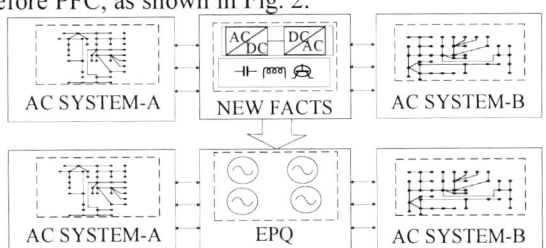

Figure 2. Influence of EPQ method on the variables in PFC.

The detailed process will be presented through using the HVDC with LCC stations as example in this paper. In [9], VSC converter has been equivalent to a controllable voltage source in series with transformer impedance to complete PFC. Thus, the proposed method has built some foundation on the past research, but it is not concluded and proposed systematically.

2.4 Comparison of PFC Method

As shown in Table 1, unified iterative method and EPQ method have quadratic convergence, while the alternating iterative method's convergence has not mathematical proof and in practice its convergence do not better than unified iterative method. To apply certain method to complete the PFC, the alternating iterative method nearly do not need doing external work after modeling the new FACTS, while unified iterative method needs designing the structure of Jacobian, setting the initial values and considering the influence of control strategy on the Jacobian. The EPQ method need to do the theoretical deduction to get the expression of reactive and active power, which may be very difficult, and according to the results of the theoretical deduction, the relation between injected power of new FACTS and the variables from AC system should be analyzed to modify the Jacobian. But after this work is done, even if new FACTS was added into network, the unified iteration just need do the same work for new FACTS based on previous work. Alternating iterative method also just analyzes the new FACTS. But unified iteration needs reconsider the structure of Jacobian and considering the influence of control strategy on the Jacobian. For example, literature [6] focused on the PFC of system incorporating HVDC with LCC and VSC station, while the PFC of system with one type of HVDC, LCC or VSC station, has been completed in the other paper before. Thus, faced with more new FACTS, there will be more work to be done again.

Table 1. Comparison of three methods for PFC.

Comparison	Methods		
	Alternating iteration	Unified iteration	EPQ
Convergence	No mathematical proof	Quadratic	Quadratic
Preparation	Little	Much	Much
Reusability	High	Low	High

Here the difference of alternating iterative method and EPQ method should be explained further, because that, under certain situation the two method has the same process. In the whole iteration of PFC, the EPQ do not change the variables from new FACTS while the alternating iterative method do. Usually the active and reactive power injected into new FACTS in the alternating iterative method will change with the variables changing. The difference is shown in Fig. 3.

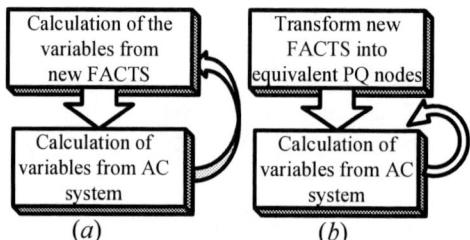

Figure 3. Difference of alternating iteration and EPQ. (a) is the process of alternating iteration and (b) is the process of the iteration of EPQ.

3. EPQ METHOD FOR HVDC
In this part, the EPQ method is applied in a system with HVDC using LCC stations.

3.1 Captions/numbering
The relations concerning LCC station are given in (1).

$$\begin{cases} V = KU\cos\theta - XI \\ V = \gamma KU\cos\varphi \\ I_a = \gamma KI \end{cases} \qquad (1)$$

where all parameters in this paper are per unit quantities. V and I respectively denote the voltage and current in DC system; U denotes the voltage of AC bus connected with transformer. K is the value of transformer ratio; θ represents the converter commutation control angle; X indicates the equivalent converter reactance; γ is a parameter corresponding to the commutation effect, usually set as 0.995; φ is the power factor angle. Equivalent structure of LCC station is shown in Fig. 4, where δ denotes the phase angle of AC bus.

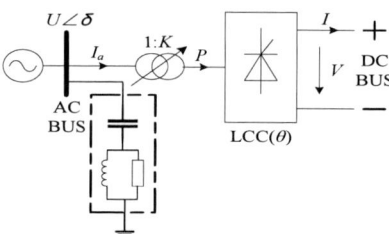

Figure 4. Equivalent Structure of LCC station.

For LCC station, the operating condition of each converter depends on two individual control variables, D-axis control variable and E-axis control variable [6]. Under practical operation condition, most transformers' ratio (K) is usually set with specified integers which correspond to alternative tap positions. In calculation, when constant θ is used and calculation is done, θ should be recalculated by rounding the ratio of transformer to nearest available integer and turning constant θ into constant K. Then the constant quantities for control strategy are elaborated and shown in Table 2.

Table 2. Control strategies of LCC station.

D-axis Control	E-axis Control
P, V, I	K, θ

where P is the active power. Moreover, P, V and I are defined as D-axis control variables, while K and θ belong to E-axis control variables.

3.2 Transform HVDC to Equivalent Nodes
The discussion of different control strategies' effect on Equivalent PQ Buses will be divided into two parts.

3.2.1 D-axis control strategies of HVDC

For an entire DC system, there must be at least one inverter choosing constant voltage as its *D*-axis control, and under this condition, not matter whether other converters' *D*-axis control strategy are constant *P* or *I*, the voltage and injected active power can be calculated by (2).

$$I_k = \sum_{j=1}^{n} g_{kj} V_j; \quad P_k = I_k V_k \qquad (2)$$

where, *n* is the number of converters and *g* is the equivalent conductance.

Therefore, the voltage and injected active power of HVDC system in static status can be obtained based on the admittance matrix of HVDC.

3.2.2 E-axis control strategies of HVDC

(1) When *E*-axis control is constant θ, the power injected by DC into AC could be expressed as (3). Q_k could be obtained through combining (1) and (3), and deviation of power equations can be expressed as (5).

$$P_k = V_k I_k; \quad Q_k = V_k I_k tg\varphi_k \qquad (3)$$

$$Q_k = P_k \sqrt{\gamma^2 \left(V_k^2 + P_k X_k \right)^2 \Big/ \left[V_k^4 \cos^2(\theta_k) \right] - 1} \quad (4)$$

$$\begin{cases} \Delta P_{tk} = P_{tk} - U_k \sum_{j=1}^{n} U_j \left(G_{kj} \cos\delta_{kj} + B_{kj} \sin\delta_{kj} \right) - (\pm P_k) = 0 \\ \\ \Delta Q_{tk} = Q_{tk} - U_k \sum_{j=1}^{n} \left(U_j G_{kj} \sin\delta_{kj} - B_{kj} \cos\delta_{kj} \right) - (Q_k) = 0 \end{cases} \qquad (5)$$

where, P_k, Q_k are scalar values, positive, the sign $\pm$ is + for rectifier and $-$ for inverter, and $\delta_{ij} = \delta_{ij} - \delta_{ij}$. Compared with Jacobian used in AC PFC, new Jacobian for PFC using Equivalent PQ Buses needn't modifying.

(2) When *E*-axis control is constant *K*, *Q* could be obtained through combining (1) and (3), as shown in (6). Deviation of power equations are expressed by (5), while Jacobian need modifying by using (8).

$$Q = P \sqrt{\gamma^2 K^2 U^2 / V^2 - 1} \qquad (6)$$

$$dQ/dV = \gamma^2 K^2 PU \Big/ \sqrt{V^2 \gamma^2 K^2 U^2 - V^4} \qquad (7)$$

$$\begin{aligned} L_{ii} &= 2U_i^2 B_{ii} - \gamma^2 K^2 PU_i \Big/ \sqrt{V^2 \gamma^2 K^2 U_i^2 - V^4} \\ &\quad - U_i \sum_{j \in i, j \neq i} U_j (G_{ij} \sin\delta_{ij} - B_{ij} \cos\delta_{ij}) \end{aligned} \qquad (8)$$

where, subscript *i* indicates the AC bus number, V denotes the voltage of DC connected with AC bus *i*.

3.3 Process of the PFC

The process of the PFC in system with HVDC is shown in Fig. 5, in which initializing variables usually choose the flat start. After analysing the control strategies of LCC station, the DC system can be transformed into PQ buses and Jacobian will be modified to use common AC power flow calculation.

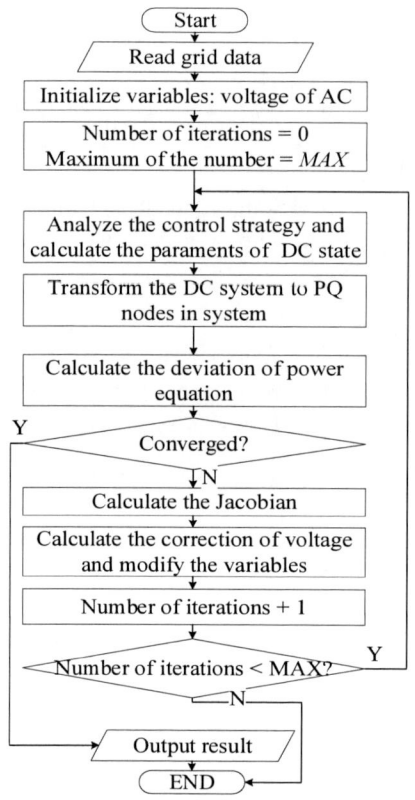

Figure 5. The flow chart of PFC using EPQ.

3.4 Study Case of PFC with HVDC

3.4.1 Convergence property of two methods

The calculation data source is the 22 buses system of China Electric Power Research Institute, and data could be got from [10]. By adding the DC part to original system, this paper got the grids for the case, as shown below in Fig. 6.

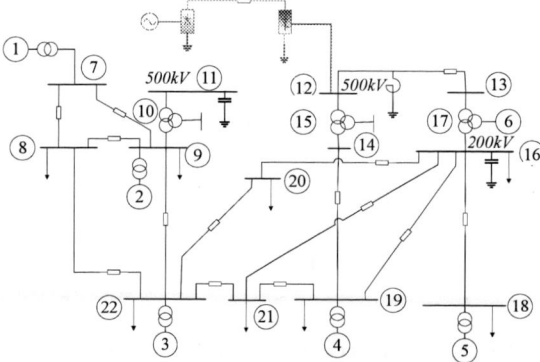

Figure 6. System with HVDC for study case

Equivalent converter reactance: two LCC' equivalent reactance both are 0.013. Control parameter of rectify: active power is 1.5 and commutation control angle is 24.07 degree. Control parameter of inverter: DC voltage is 0.9384 and commutation control angle is 23.92 degree.

Table 3. Results of case using unified iterative method.

Variables	Times				
	1	2	3	4	5
$U(12)$	1.1251	1.0533	1.04	1.0454	1.0454
$\delta(12)(rad)$	-0.2594	-0.2406	-0.24	-0.2435	-0.2435
Maxdelt	10.3402	-2.769	-0.23	0.0023	3.4e-07

Table 4. Results of case using unified iterative method.

Variables	Times				
	1	2	3	4	5
$U(12)$	1.1236	1.0531	1.04	1.0454	1.0454
$\delta(12)$ (rad)	-0.2322	-0.2388	-0.24	-0.2435	-0.2435
Maxdelt	6	-2.7555	-0.23	0.0023	3.3e-07

In Table 3 and 4, the Maxdelt is the max deviation of power equations.

After 5 iterations, the PFC converged separately under unified iterative method and EPQ method, with the condition that convergence precision within 10^{-5} The processes of iteration of two method are shown in Table 3 and 4, in which Maxdelt is the max deviation of power equations. Through Table 3 and 4, it can be found that the EPQ and unified iterative method both have good convergence property.

3.4.2 Calculation speed of two methods

To compare the two method's efficiency, this paper transformed the IEEE39-10Gen, IEEE162-17Gen, IEEE300 through adding HVDC apart. Using C# to write programs to compute the PFC, results were shown in Fig. 7. The calculation time is recorded and shown in Fig. 7, where the results of C1 are calculated with EPQ method and results of C2 are for the case using integrated iteration method. The test computer configuration specified as follows. The CPU, memory and storage are Intel Core i5-4200H @ 2.60GHz Dual-Core Quad-threading, 8 GB 1867 MHz and Sandisc SSD 500G. From the results, it can be found that the EPQ consumed less time to complete the PFC than unified iterative method.

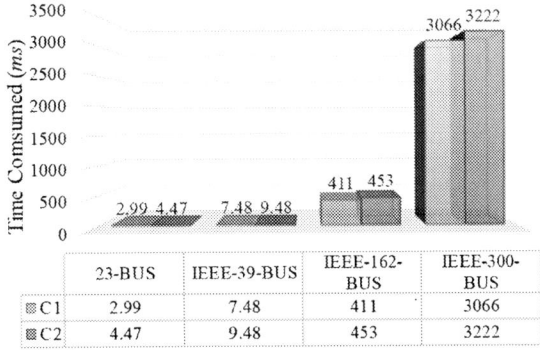

Figure 7. Time consumed for PFC.

4. Conclusion

The development of modern power system and energy internets requires the help of new FACTS and HVDC, which as a new component will increase in types and quantity. In recent years, more researchers devoted time into modelling for new FACTS and calculating the power flow of system with them. When new FACTS is invented and considered to be incorporated into original system, the work of power flow calculation may need redoing. This slows the development of modern power system, the application and research of new FACTS. Therefore, this paper proposed a method (EPQ) to calculate the power flow calculation. The purpose of the method is to remove the new components from AC system. Compared with the alternating iterative method and unified iterative method, EPQ method has more advantages. In a case, this paper uses EPQ to complete the PFC with HVDC, and results imply that compared with unified iterative method, EPQ has same convergence property and faster speed of computation.

Acknowledgment

This work was supported by the Science and Technology Project of State Grid Corporation of China (Research on Dynamic Behavior Mechanism and Coordination Control Measures of Sending and Receiving AC system of 10000 MW level Hierarchical UHVDC).

References

[1] B. Stott, "Review of load-flow calculation methods," Proceedings of the IEEE, vol. **62**, no. 7, pp. 916-929 (1982).

[2] Q. Le-Cao, T. Tran-Quoc, and A. Nguyen-Hong, "Study of FACTS device applications for the 500kV Vietnam's power system," in Transmission and Distribution Conference and Exposition, 2010 IEEE PES, pp. 1-6 (2010).

[3] A. Sheykholeslami, A. R. Ahmadi, S. A. N. Niaki, and H. Ghaffari, "Power flow modeling/calculation for power systems with Dynamic Flow Controller," pp. 1-5 (2008).

[4] I. B. Jaoued, T. Guesmi, and H. H. Abdallah, "Power flow solution for power systems including FACTS devices and wind farms," in International Conference on Sciences and Techniques of Automatic Control & Computer Engineering - Sta, pp. 136-139 (2014).

[5] M. Baradar, M. Ghandhari, D. V. Hertem, and A. Kargarian, "Power flow calculation of hybrid AC/DC power systems," in Power and Energy Society General Meeting, pp. 1-6 (2012).

[6] J. Dou, B. Zhang, R. Chai, Z. Guan, and C. Liu, "Unified iterative power flow algorithm of AC/DC networks incorporating hybrid HVDC," in IEEE International Conference on Environment and Electrical Engineering, pp. 1-6 (2016).

[7] J. Wang, C. Li, and Q. Wang, "An improved power flow algorithm using equation changing method for AC/DC power system with VSC-HVDC," in Power and Energy Engineering Conference, pp. 1913-1917 (2016).

[8] Z. Hongyue, "Research on Power Flow with FACTS and New Energy Resources," Master, South China University of Technology (2013).

[9] Z. Yu, Z. Wu, Z. Huang, and S. Wang, "A general power flow calculation method for power systems with VSC FACTS based on InterPSS and automatic differentiation," in International Conference on Electric Utility Deregulation and Restructuring and Power Technologies, pp. 262-268 (2016).

[10] D. X. Qing, "A new algorithm and its application for the probabilistic load flow of the power system," Master, North China Electric Power University (2006).

Combined Heat and Power Optimal Dispatch Considering Wind Power Uncertainty

Chunlong Chen[1], Xueshan Han[1,a] and Wenbo Li[2]

[1]Key Laboratory of Power System Intelligent Dispatch and Control of Ministry of Education (Shandong University), Jinan, China

[2]State Grid Shandong Electric Power Research Institute, Jinan 250003, Shandong Province, China

[a]Corresponding author: sduchen1994@163.com

Abstract. Large-scale integration of wind power is a clean alternative to power generation, but the uncertainty of wind power has become a huge dispatch challenge, which is much more serious in the northeast of China. For this reason, heat storage is configured to decouple thermoelectric constraint of CHP unit, releasing its flexibility and improving its peak-shifting ability. A random variable is introduced to represent wind power deviations firstly, then a chance constraint economic dispatch model of an aggregation unit containing wind power and CHP unit is established to explain the standby effect of heat storage on wind power uncertainty. Finally, a case study is conducted to verify the performance of the heat storage. The example results show that when configured with heat storage, the CHP unit have broadened its adjustable electrical output range, so more reserve capacity is released to cope with wind power uncertainty.

1. Introduction

Large-scale integration of wind power is a clean alternative to power generation, how to cope its uncertainty and ensure the economic and safe operation of power system has been a huge dispatch challenge. During heating period of northeast district, a time coincident with high wind power generation, the ability of the CHP unit to further reduce output is limited by its thermoelectric coupling constraint, causing insufficient peak-shifting capacity and much wind power curtailment. For this reason, a heat storage is configured to decouple thermoelectric constraint of CHP unit. The CHP unit external characteristics with and without heat storage are analysed in [1]. It is pointed out that the peak-shifting ability of CHP unit is determined by its heat load level. An economic dispatch containing heat storage was conducted in [2]-[5], showing that more wind power integration can be achieved owing to the existence of heat storage or/and electrical boiler, but they took no consideration to wind power uncertainty. Based on multiple wind power forecast scenarios, an optimal dispatch model for an aggregation unit with wind power uncertainty taken into consideration is proposed in [6], which can make full use of the flexibility provided by heat storage, but it made no explanation for the standby effect of heat storage on wind power uncertainty.

In this paper, based on random variable to represent wind power deviations, a chance constraint economic dispatch model of an aggregation unit containing heat storage is proposed to study the standby effect of heat storage responding to the wind power uncertainty. The model can conduct

Content from this work may be used under the terms of the Creative Commons Attribution 3.0 licence. Any further distribution of this work must maintain attribution to the author(s) and the title of the work, journal citation and DOI.
Published under licence by IOP Publishing Ltd

scheduling operation strategy for the aggregation unit. Moreover, case is studied and analysed to verify the performance of the heat storage.

2. The Wind Power Output Model

In this section, based on the predicted value of wind power output, a random variable ΔP_w^t is used to indicate the deviation between the actual output of wind power and the predicted value considering wind power uncertainty. The deviation obeys the normal distribution with variance of σ^2 [7]

$$\Delta P_w^t \sim N[0, \sigma^2(P_{wf}^t)] \qquad (1)$$

where P_{wf}^t is the predicted value of wind power (MW), and σ^2 is a function of P_{wf}^t.

The actual output of wind power at time t can be expressed as Equation (2)

$$P_w^t = P_{wf}^t + \Delta P_w^t \qquad (2)$$

where P_w^t is the actual output of wind power (MW),

3. Aggregation Unit Economic Dispatch Model

The general structure of the aggregation unit is shown in Figure 1.

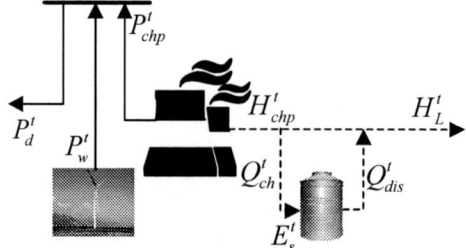

Figure 1 General structure of the aggregation unit.

The aggregation unit has only one bus with wind generator, CHP unit and electrical load connected. As responding to wind power uncertainty, heat storage device is configured in thermal power plant.

3.1. Objective function

The system operator aims at minimizing the aggregation unit cost, so the objective function can be expressed as Equation (3).

$$\min \ C = \sum_{t=1}^{T} [a(P_{chp}^t)^2 + bP_{chp}^t + cP_{chp}^t H_{chp}^t + d(H_{chp}^t)^2 + eH_{chp}^t + f] \qquad (3)$$

where C is the total cost of the aggregation unit, representing the operation cost of the CHP where a, b, c, d, e and f are the cost coefficients of CHP; P_{chp}^t and H_{chp}^t are the electric power and the heat power of the CHP respectively (MW); t represents the discrete time index and T is the amount of periods.

3.2 Constraints

3.2.1 Power Equality constraints

An aggregation unit aiming at realizing distributed autonomy must satisfy the load demand firstly, in which the power equation constraints can be written as Equation (4)

$$\begin{cases} P_{chp}^t + P_{wf}^t = P_d^t \\ H_{chp}^t - Q_{ch}^t + Q_{dis}^t = H_L^t \end{cases} \qquad (4)$$

where P_d^t and H_L^t are local electrical load power and heat load power respectively (MW); Q_{ch}^t and Q_{dis}^t are charge power and discharge power of the heat storage device respectively (MW).

3.2.2 Thermoelectric coupling constraints for the CHP units

As the only coupling point of the two types of energy, the thermoelectric coupling relationship of the CHP unit makes multi-energy complementary be possible, which can be expressed as Equation (5)

$$\begin{cases} 0 \le P_{chp}^{min} - c_v H_{chp}^t \le P_{chp}^t \le P_{chp}^{max} - c_v H_{chp}^t \\ 0 \le H_{chp}^t \le [P_{chp}^t - P_{chp}^{max} + (c_v + c_m)H_{chp}^{max}]/c_m \end{cases} \quad (5)$$

where c_v and c_m are the characteristic parameters of CHP unit.

3.2.3 Reserve capacity constraint

For the purpose of coping with wind power uncertainties, sufficient reserve capacity must be provided in the aggregation unit, which is determined by unit ramp rate or the flexibility of unit release by heat storage. The reserve capacity constraints can be presented as Equation (6)

$$\begin{cases} R_{up}(H_{chp}^t) = \min\{P_{chp}^{max} - c_v H_{chp}^t - P_{chp}^t, r_{up}\Delta t\} \\ R_{dw}(H_{chp}^t) = \min\{[P_{chp}^t - \max(P_{chp}^{min} - c_v H_{chp}^t, \\ \qquad c_m H_{chp}^t + P_{chp}^{max} - c_v H_{chp}^{max} - c_m H_{chp}^{max})], r_{dw}\Delta t\} \end{cases} \quad (6)$$

where R_{up} and R_{dw} represent reserve capacity expressions with H_{chp}^t as variable; r_{up} and r_{dw} are the ramp rate of CHP unit (MW/h).

3.2.4 Wind power uncertainty probability constraint

The wind power uncertainties can be featured by wind power fluctuation. Due to the existence of prediction errors, the actual wind power may be lower or higher than the forecast value. In order to balance the reserve capacity economics and the safety of power system, reserve capacity must be able to cope with wind power fluctuations at high probability. A probability constraints is established as Equation (7) by introducing chance constrained programming to deal with randomness of the wind power, which can be transformed into an equivalent deterministic problem[8].

$$\begin{cases} f_1 = P_r\{P_{wf}^t - p_w^t > R_{up}(H_{chp}^t)\} \le \mu_1 \\ f_2 = P_r\{p_w^t - P_{wf}^t > R_{dw}(H_{chp}^t)\} \le \mu_2 \end{cases} \quad (7)$$

where $P_r(\cdot)$ represents the occurring probability of event $(\cdot)$. μ_1 and μ_2 are the probability value.

4. Case Study

In order to illustrate the standby effect of heat storage responding to wind power uncertainty, a case study is conducted in this section and the parameters used in the model are listed in Table 1.

The model is solved by Visual Studio 2010 and the optimization results are shown as Figure (3)-Figure (5). The following is the analysis of optimization results.

Table 1. Values of the Parameters

Parameters	Value	Parameters	Value
a	0.000171\$/MW^2	P_{chp}^{min}	150MW
b	0.2705\$/MW	H_{chp}^{max}	350MW
c	11.537	E_s^{max}	200MWh
C_v	0.15	Q_{ch}^{max}	80MW/h
C_m	0.75	Q_{dis}^{max}	80MW/h
P_{chp}^{max}	300MW	r_{up}/r_{dw}	70MW/h

As Figure 2 shows, when configured with heat storage, the CHP unit can broaden its adjustable electrical output range, releasing its flexibility and improving its peak-shifting ability. During the periods in which wind generator has high power output, for example period 0-5 and period 20-24, the

CHP unit can further reduce its electrical power to accommodate more wind power under the premise of satisfying the load demand.

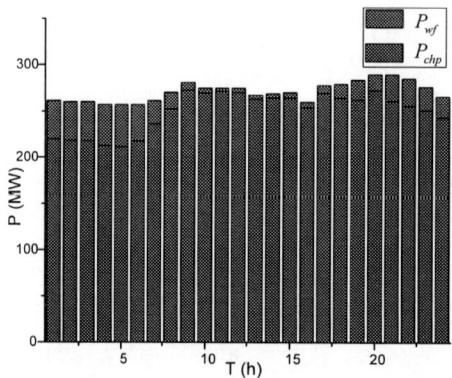

Figure 2 the output of CHP unit and wind power

The comparison of CHP unit's thermal output and thermal load power in the aggregation unit is shown in Figure 3.

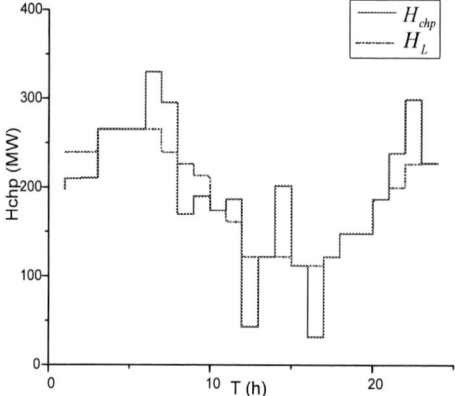

Figure 3 the comparison of CHP unit's thermal output and thermal load power in the aggregation unit

The thermal output of CHP unit is no longer limited by thermal load power owing to heat storage, as shown in Figure 3. The equality constraint of real-time power balance is broken, causing flexible adjustment of thermal power with corresponding operation mode of heat storage shown in Figure 4.

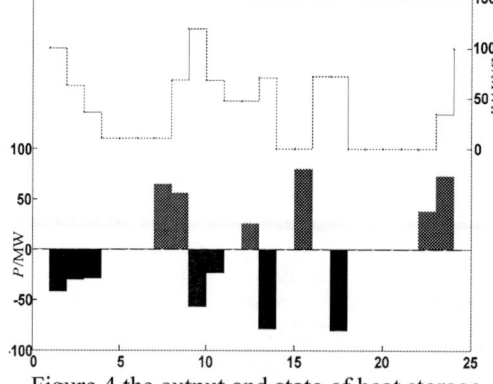

Figure 4 the output and state of heat storage

Figure 4 can better reflect the standby effect of heat storage responding to wind power uncertainty. The green line represents the adjustable electrical output range with heat storage configured and the black line represents the range without heat storage.

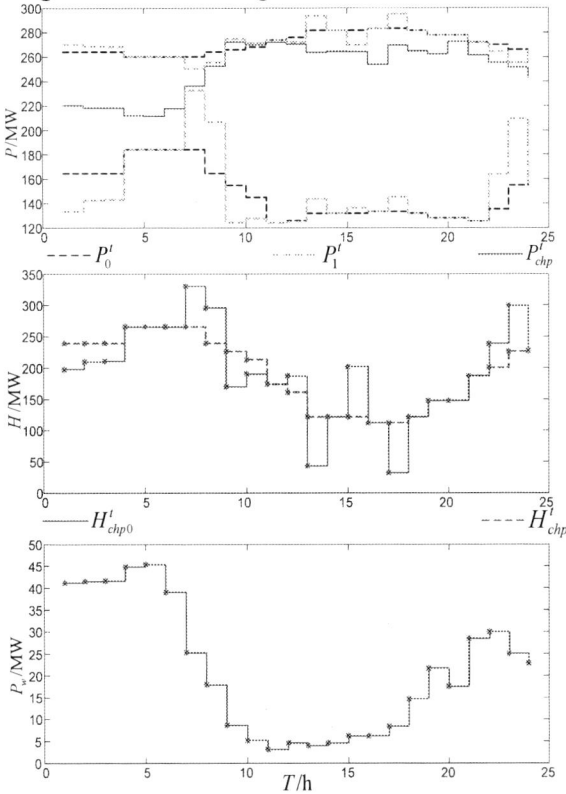

Figure 5 the comprehensive comparison chart

As can be seen from the figure above, in the period of high wind power output, in order to cope with the strong uncertainty of wind power, the heat in the heat storage is consumed to meet the demand of heat load which can be seen in Figure 4, decoupling thermoelectric coupling constraint, releasing more flexibility and thereby providing more wind power reserve capacity. When the wind power is low, the required spare capacity is small, it is necessary to supplement the heat in the heat storage device, preparing for the next heat release.

5. Conclusion

An aggregation unit chance constraint economic dispatch model is presented in this paper, in which the standby effect of heat storage on wind power uncertainty is considered. A random variable is introduced to represent wind power uncertainty. The optimization results show that when configured with heat storage, decoupling thermoelectric coupling constraint to some extent, the CHP unit can broaden its adjustable electrical output range, so more reserve capacity is released to cope with wind power uncertainty effectively, ensuring active power coordination with electrical network.

Acknowledgment

This work is supported by the Science and Technology Foundation of
SGCC(SGSDDK00KJJS1600061).

References

[1] Q. Li, T. Y. Chen, H. X. Wang, et al. Analysis on Peak-load Regulation Ability of Cogeneration unit with Heat Accumulator, Automation of Electric Power Systems. **38(11)**, 34 (2014)

[2] Y. Cui, Z. Chen, G. G. Yan, et al. Coordinated Wind Power Accommodating Dispatch Model Based on Electric Boiler and CHP With Thermal Energy Storage. Proceedings of the CSEE. **36(15)**. 4072(2016)

[3] L. Chen, F. Xu, X. Wang, et al. Implementation and Effect of Thermal Storage in Improving Wind Power Accommodation. Proceedings of the CSEE. **35(17)**. 4283(2015)

[4] Q. Lv, T. Y. Chen, H. X. Wang, et al. Combined Heat and Power Dispatch Model for Power System With Heat Accumulator. Electric Power Automation Equipment. **34(5)**. 79(2014)

[5] J. Yu, H. B. Sun, X. Y. Shen, Optimal Operating Strategy of Integrated Power System With Wind Farm, CHP Unit and Heat Storage Device, Electric Power Automation Equipment. **37(6)**. 139(2017)

[6] Y. H. Dai, L. Chen, Y. Min, Optimal Dispatch for Joint Operation of Wind Farm and Combined Heat and Power Plant With Thermal Energy Storage, Proceedings of the CSEE, **37(12)**. 3470(2017)

[7] Y.P. Li, N. Feng, Y. Cui, et al. Security Constrained Unit Commitment Problem Considering Wind Power Uncertainty and Flexible Load, Electric Power Construction. **38(2)**, 129, (2017).

[8] P. Yan, Combined Optimization Scheduling of Wind and Fire Power System Considering the Feasibility of the Results [D]. (2016)

Analysis of Power System Vulnerability Considering Multiple Disturbances Corresponding to Information and Physics

Chaochao Wang[1,a], Hua Sun[2] and Xiaoming Dong[1]

[1]Key Laboratory of Power System Intelligent Dispatch and Control of the Ministry of Education, Shandong University, Jinan, China

[2]Department of Electrical Automation, Shandong Labor Vocational and Technology College, Jinan, China

[a]Chaochao Wang: smartwcc@163.com

Abstract. With the continuous development of smart grid, the automation degree of power system continues to improve. The number of measurement-calculating and decision-controlling units of the power grid has greatly increased, and the scale of the power information network has become larger and larger. The stable operation of the power system is inseparable from the real-time dispatch of its information system. As the key infrastructure, power system has a huge impact on national security, economic development and social stability, and has become one of the key targets of terrorist destruction. In view of the fact that information network and power network integrate with each other currently, this paper focus on the functions of the power control center, modeling information systems together with physical systems. Considering that the power system has suffered multiple disturbances corresponding to information and physics, the quantitative analysis method is used to simulate the dynamic evolution process after the accident, and the loss of load is used as the evaluation index to analyze the vulnerability of the power system, to find the vulnerable components of the power grid. So effective protection measures can be taken to reduce the loss of the power system after this type of destruction. The constructed model is solved by mixed integer nonlinear programming method. IEEE Reliability Test System-24 verifies the feasibility and effectiveness of this model.

1. Introduction

The smart grid is a composite system composed of a virtual information network and a physical entity network, which is characterized by close coordination between the information system and the physical system [1]. The information system collects the operational data of the power system and sends it to the power control center for analysis. The power control center makes real-time decisions and controls the power system to ensure the safe and stable operation. At present, international terrorism threats, military conflicts and other unstable factors appear frequently. Power system as a hub for mutual transformation of various energy [2], is a key infrastructure of the country and has a tremendous impact on national security, economic development, and social stability. It has become the focus of destruction [3].

The current destruction methods toward the power grid are mainly divided into two types: the first one is to damage the primary equipment of the power system directly [4], mainly for the deliberate destruction of generators, substations, transmission lines, nodes and even some important loads. This type of destruction will cause one or more power equipment to exit running, thus changing the

Content from this work may be used under the terms of the Creative Commons Attribution 3.0 licence. Any further distribution of this work must maintain attribution to the author(s) and the title of the work, journal citation and DOI.

Published under licence by IOP Publishing Ltd

topology of the power network, affecting the transmission and distribution of power seriously, and may even cause grid disassociation and chain fault, causing wide rage of power cut. Another type of destruction is that terrorists use advanced network technologies to invade the power information network and destroy the functions of the information system [5]. Since the control and coordination of the physical equipment largely depend on the information system, the destruction on the information system may lead to a complicated physical interaction process and threaten the security of the power system ultimately. Reference [6] puts forward a new standard for the power systems vulnerability analysis in response to the physical damage caused by natural disasters and malicious destruction, and proposes how to formulate emergency treatment plans so that the grid can restore stable operation as soon as possible. Reference [7] defines information destruction in power systems, classifies information destruction from different perspectives, and analyses some typical scenarios. Reference [8] shows that injecting bad data into information systems deliberately can lead to control center makes wrong decisions, which can not only interfere with the safe operation of the power grid, but also enable the terrorists to obtain illegitimate benefits. Reference [9] simulates destruction against information systems, causing the power system to lose security and stability directly, and proposes corresponding security defense measures. At present, most of the references consider the impact of a single destruction way on the power system, and do not model and analyze the physical system together with the information system. Compared with physical destruction, information destruction has the characteristics of low cost and concealment [10], and the damage to the power system may be more serious. Due to the information system is tightly coupled to the physical system, problems in any one system may lead to serious accidents.

This paper considers the interaction between information systems and physical systems, and considers that the power system has suffered physical destruction, thus some primary equipment of the power system was destroyed and exited operation. At the same time, if the power system suffered information destruction, the function of the power control center is invalid, so the generator output and load of each node of the power grid cannot be adjusted optimally. If the power system did not suffer information destruction, the control center can adjust the generator output and load optimally, with the goal of minimizing the load loss of the power system. Based on the topology analysis algorithm, this paper proposes a DC power flow method, which analyses the dynamic evolution process of power system accident quantitatively, and uses the load loss as the evaluation index of the accident after the disturbances related to information and physics. Finding the vulnerable components of the grid by establishing an analytical model of power grid vulnerability, effective protection measures can be taken to reduce the loss of the power system. The constructed model is solved by mixed integer nonlinear programming method. IEEE Reliability Test System-24 verifies the feasibility and effectiveness of this model.

2. Principles of power system vulnerability analysis

2.1 Dynamic evolution and simulation of power system suffering from disturbances related to information and physics

After the power system is destroyed, if the short-term transient process can be neglected, only concern about power system's steady state after the accident, the power system steady state method can be used to analyse problems [11]. The dynamic evolution process can be described as: when some primary equipment (generators, transformers, power lines, etc.)are physically destroyed and then exit operation, causing changes in the topology of the power grid, further causing problems such as power flow transfer, branch power limit and power imbalance. If the information system is destroyed at the same time, it is considered that the power control center loses the optimal adjustment function. The relay protection device will cut off the overrun branch. At this point, the topology of the power grid will change again, causing a series of cascading faults. If the information system is not destroyed, the power control center optimizes with the goal of minimizing the load loss, and adjusts from the power supply and the load synergistically to ensure the power balance of the power grid while eliminating the power over limit, thus avoiding the occurrence of cascading fault. The above dynamic evolution process is simulated using the steps of Figure 1:

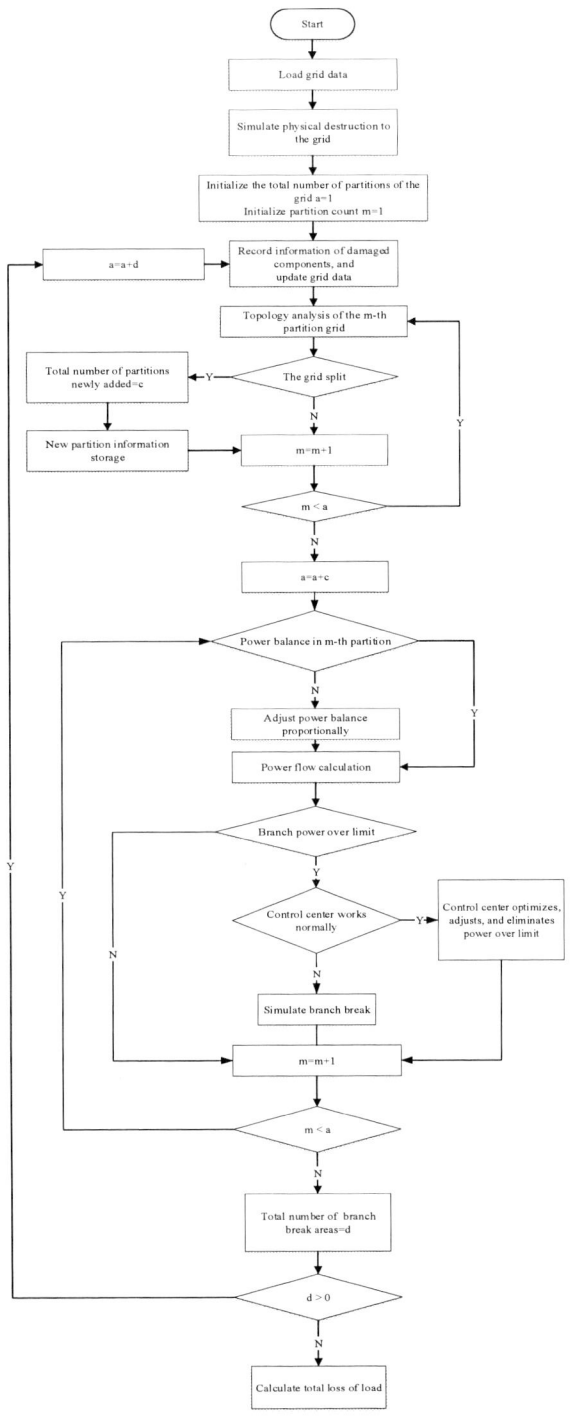

Figure 1. The simulation steps of power system disturbances.

The description of Figure 1 is as follows:

(1) Set the components operating status. Set grid components exit operation which being destroyed.

(2) Power system topology analysis. Through topology analysis, obtain the number of disconnected power islands (subsystems) after the grid is destroyed and obtain the power, network and load of each power island.

(3) Perform the following analysis for each power island (subsystem) which includes both the generator and the electrical load:

(3.1) Simulate power flow distribution. Firstly, proportional adjustment of power to ensure power balance. Then calculate power flow to determine whether there is a branch overload, if yes, enter (3.2), and otherwise enter (4).

(3.2) Judge the cascading fault. If the control center is invalid, it loses operational capability. All overloaded branches are disconnected, then enter (3.3); if the control center is valid, it has operational and control capabilities. The control center eliminates the overload of the branch according to the strategy of minimizing the loss of the island load, and then enters (4).

(3.3) Subsystem topology analysis. It is concluded that the power island is further decomposed into several unconnected new power islands (new subsystems) due to cascading fault, and obtain the power, network and load of each new power island.

(4) Judge whether each power island (subsystem) containing the generator and the power load is calculated, if yes, enter (5), otherwise return (3).

(5) For each new power island (new subsystem) containing the generator and load, re-simulate the power flow distribution, cascading fault judgment and subsystem topology analysis until there is no cascading fault (no new power islands appear).

(6) Judge whether each new power island (new subsystem) containing the generator and the power load is calculated, if yes, enter (7), otherwise return (5).

(7) Statistics of total load loss.

The above vulnerability analysis problem considers that terrorists launch a destruction related to information and physics on the power system, to achieve the purpose of destroying the power supply. According to this method, the load loss of each power island can be obtained, and the total load loss is used as the vulnerability assessment index of the power system. The damaged components, which cause the most load loss, are the vulnerable components of the power grid and need to be protected.

2.2 Planning model for power system vulnerability analysis

First, define the variable of availability of the power control center:

$$\eta = \begin{cases} 1, & \textit{Control center is valid} \\ 0, & \textit{Control center is not valid} \end{cases} \quad (1)$$

In (1), η represents the availability of the control center. Element 1 means the control center works normally and element 0 means the control center loses its function.

●Objective function

This planning problem is based on the generators, power lines, substations and bus nodes as the target of destruction in the power system. The decision variable is the generator output and load power of each node under the control of the control center (if the corresponding node is controllable); the objective is to minimize the total load loss:

$$\text{Object} = \min\{P_G, P_L\} \sum_{i \in L} \Delta P_{Li} \quad (2)$$

In equation (2): P_G, P_L represent the decision variable controlled by the system operator, which is the vector expression of the generator's active output and load power of each node, and its dimension is equal to the total number of nodes of the grid. ΔP_{Li} is the amount of loss of load i.

● Constraints

After the components of power grid are physically damaged, taking into account the action of the relay protection device to cause the damaged components to exit the operation, the constraints of the power system topology changes are as follows:

$$\begin{cases} \delta_j^{Gen} \in \{0,1\} \\ \delta_l^{Line} \in \{0,1\} \\ \delta_n^{Bus} \in \{0,1\} \\ \delta_s^{Sub} \in \{0,1\} \end{cases} \qquad (3)$$

$$\begin{cases} Y_l = \left(1 - \delta_l^{Line}\right) * \left(1 - \delta_{o(l)}^{Bus}\right) * \left(1 - \delta_{d(l)}^{Bus}\right) * \\ \prod_{s|l \in L_s^{Sub}} \left(1 - \delta_s^{Sub}\right) * \prod_{l'|l' \in L_l^{Par}} \left(1 - \delta_{l'}^{Line}\right) \qquad (4) \\ Y_l \in \{0,1\} \end{cases}$$

$$\begin{cases} H_j = \left(1 - \delta_j^{Gen}\right) * \left(1 - \delta_{n(j)}^{Bus}\right) \\ H_j \in \{0,1\} \end{cases} \qquad (5)$$

In equation (3), $\delta^{Gen}, \delta^{Line}, \delta^{Bus}, \delta^{Sub}$ are vectors representing the functional state of the generators, power lines, nodes, and substations respectively in the power system. The vector dimension is respectively equal to the number of generators, power lines, nodes, and substations. Element 1 represents the fault of the power component and element 0 represents the normal operation of the power component without being destroyed. In equations (4) and (5), Y, H are vectors respectively representing the availability of power lines and generators under the influence of the topology of the power system. The vector dimension is equal to the number of power lines and generators. Element 0 represents the component is not available and Element 1 represents the component is available. Equation (4) means that the power line l is not available when the following conditions occur: the power line l is destroyed, or the first node or last node of the power line l is destroyed, or the substations connected to the power line l is destroyed, and if the power line l is one of the multiple lines of the same pole, the multiple lines of the same pole will generally exit operation together after destruction. Y_l is the availability of power line l. $\delta_{o(l)}^{Bus}$ and $\delta_{d(l)}^{Bus}$ respectively are the functional states of the first node and last nodes connected to the power line l, all of which are 0-1 variables; L_s^{Sub} is a set of power lines connected to the substation s, L_l^{Par} is a set of other power lines running on the same pole with the power line l. Equation (5) means that the generator j is not available when the following conditions occur: the generator j is destroyed or the node connected to the generator j is destroyed. H_j means the availability of the generator j, and $\delta_{n(j)}^{Bus}$ means the functional state of the node connected to generator j, all of which are 0-1 variables.

The equality constraints:

$$P_G - P_L = B\theta \qquad (6)$$

Equation (6) is the DC power flow equation [12]. P_G is a vector representing the generator active power of each node, P_L is a vector representing the load active power of each node, B is the DC power flow susceptance matrix, and θ is a vector representing the node phase angle.

$$P_l = Y_l * \frac{1}{x_l} * \sum_{n \in N} A_{ln} \theta_n \qquad \forall l \in D \qquad (7)$$

Equation (7) is the power flow equation of the power lines. P_l is the active power of the power line l, x_l is the reactance of the power line l, A_{ln} is the element of the line-node correlation matrix, θ_n is the phase angle of the node n, N is a set of system nodes, D is a set of power lines

$$\sum_{n \in N} \sum_{j \in G_n} \left(P_{Gj} + \eta * \Delta P_{Gj} \right) = \sum_{n \in N} \sum_{i \in L_n} \left(P_{Li} + \eta * \Delta P_{Li} \right) (8)$$

Equation (8) is the power balance equation. P_{Gj} is the active output of generator j, ΔP_{Gj} is the active power adjustment amount of generator j under the control of control center, G_n is the set of all generators which connect to node n; P_{Li} is the active power of load i, ΔP_{Li} is the power loss of load i, L_n is the set of all loads which connect to node n in the system.

$$-Y_l * P_l^{\max} \le P_l \le Y_l * P_l^{\max} \qquad \forall l \in D \qquad (9)$$

$$H_j * P_{Gj}^{\min} \le P_{Gj} + \eta * \Delta P_{Gj} \le H_j * P_{Gj}^{\max} \quad \forall n \in N, \forall j \in G_n \quad (10)$$

$$0 \le \Delta P_{Li} \le P_{Li} \qquad \forall i \in L \qquad (11)$$

Formula (9) represents the power limits of power lines, and $P_l^{\max}$ is the upper limit of the active power transmission value of the power line l. Formula (10) represents the active power output limits of the generator j. $P_{Gj}^{\max}$ and $P_{Gj}^{\min}$ are the upper and lower limits of the active power output of the generator j. Formula (11) is the active power loss constraint of load i, and L is the set of all loads in the system.

3. Case analysis

3.1 IEEE reliability test system-24(RTS-24) analysis

To verify the effectiveness of the proposed model, a simulated destruction was performed on the IEEE RTS-24. Its network structure shown as Figure 2:

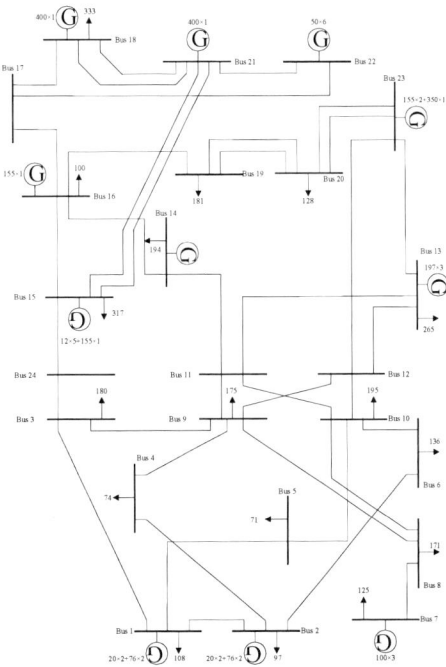

Figure 2. The structure of IEEE RTS-24.

3.1.1 Vulnerability analysis of power system under multiple disturbances of information and physics

First, simulating information destruction on power system, so the control center function is invalid. Then sample the primary equipment and simulate physical damage. The load loss and its proportion of the total load under each destruction scheme (the total load is 2850 MW) shown in Table 1:

Table 1. The load loss under different destruction schemes.

Scheme number	Damaged components	Loss of load(MW)	Load loss ratio (%)
1	Lines:16-19,20-23A,20-23B	309	10.84
2	Lines:11-13,12-13,12-23,14-16,15-24	1086	38.11
3	Node:13; Lines:7-8,12-23,16-17,15-21A,15-21B,20-23A,20-23B	1638	57.47
4	Nodes:13,15,18; Lines:12-23,16-17,20-23A,20-23B	2257	79.19
5	Nodes:2,13,15,18; Lines:12-23,16-17,20-23A,20-23B	2378	83.44
6	Nodes:2,13,15,16,18; Lines:7-8,12-23,20-23A,20-23B	2533	88.88
7	Nodes:2,7,13,15,16,18; Lines:12-23,20-23A,20-23B	2658	93.26

It can be seen from Table 1 that when the ratio of load loss reaches 70% or more, nodes 13, 15 and 18 appear as destruction targets, which are vulnerable nodes of the grid. Lines 12-23 and double-circuit lines 20-23 almost in all destruction schemes, lines 7-8 and 16-17 also appear more often, which are vulnerable branches of the grid. Regardless of the destruction scheme, the vulnerable nodes and branches of the power grid are essentially unchanged. Through the above analysis, the security protection of these nodes and branches should be strengthened to reduce the loss of power system when it is destroyed.

3.1.2 Comparison of the load loss under two types of destruction
Select destruction schemes 2 and 4 in Table 1. Under the premise that the control center is valid and the control center is invalid, simulate physical destruction on the power grid and compare the load loss, as shown in Table 2:

Table 2. Comparison of load loss.

Scheme number	Damaged components	Loss of load(MW)	
		Control center invalid	Control center valid
2	Lines:11-13,12-13,12-23,14-16,15-24	1086	853
4	Nodes:13,15,18; Lines:12-23,16-17,20-23A,20-23B	2257	2022

It can be seen from Table 2 that for the same destruction scheme, the load loss is significantly reduced under the optimal adjustment of the control center, so the control center is also the important target that needs to be protected. In the process of power restoration, the control center should first be restored to normal operation in order to minimize loss.

4. Conclusion
Based on the background of high integration of information system and physical system, this paper proposes a vulnerability analysis model of power system subjected to disturbances related to information and physics, and solves this kind of problem by mixed integer nonlinear programming method. The loss of load is used as the evaluation index to analyse the vulnerability of the power system. Applying the model to analyse different destruction schemes, it can effectively find the vulnerable nodes and branches, and the corresponding protective measures can be taken to reduce the loss and improve the operational safety of power system.

Acknowledgment
This paper was supported by the technology project of China Guangxi Power Grid Company Limited (Development of Auxiliary Decision System for Fault State Model and Operational Risk Assessment of Nuclear Power Grid, Project Number: 0002200000029827).

References
[1] C.Y. Dong, J.H. Zhao, F.Z. Wen, Y.S. Xue. From Smart Grid to Energy Internet: Basic Concepts and Research Framework. J. AEPS, **38**, (2014)
[2] Q.L. Guo, S.J. Xin, H.B. Sun, et al. Modelling and comprehensive safety assessment of cyber-physical power system: driving force and research concept. J. PCSEE, **36(6)**, (2016)
[3] Z. Wang, M. Liu. Questitative analysis on the risk terrorist destruction in electric power system. J. CPS, **2(3)**, (2006)
[4] H.Z. Huang. Model and method for vulnerability analysis of power systems under threat of malicious destruction. D. NCEPU, (2012)
[5] M. Ding, X.J. Li, J.J. Zhang. Influence of Network destruction on SCADA on Power System Reliability. J. PSPC, (2018)
[6] D.Q. Ding. Reducing the physical vulnerability of power system to deal with natural disasters and malicious destruction. J. CEP, **42(6)**, (2009)

[7] Y. Tang, Q. Chen, M. Li, et al. Review of Network destruction Research in Power Information Physics Fusion System Environment. J. AEPS, **40(17)**, (2016)

[8] S. Liu, S. Mashayekh, D. Kundur, T. Zourntos and K. Butler-Purry. A Framework for Modeling Cyber-Physical Switching destruction in Smart Grid. J. IEEE Transactions on ETC, **1**, (Dec. 2013)

[9] M. Ni, W. Yan, R. Bai, et al. Thoughts on Anti-malicious Information destruction in Power System. J. AEPS, (2016)

[10] N. Liu, X.H. Yu, J.H. Zhang. Network Synergy destruction: Deduction and Enlightenment of Ukrainian Blackout Event. J. AEPS, **40(06)**, (2016)

[11] T.Q. Liu. *Modern Power System Analysis-Theories and Methods,* (China Electric Power Press, 2007)

[12] X.F. Wang. *Analysis of Modern Power System,* (Science Press, 2003)

ISPECE IOP Publishing

Influence of Ni-Cu-La-B-coated Glass Fiber on Conductivity and Electromagnetic Shielding Performance of Coatings

Denggao Guan[1,a], Yang Liu[1], Dehao Hu[1], Rundong Zhou[1], Xueqi Guo[1], Yan Wang[1], Yao Sun[2],Guanli Xu[2], Jinghui Lin[1],and Chuanmin Sun[2]

[1]College of Materials and Chemistry & Chemical Engineering, Chengdu University of Technology, Chengdu, Sichuan, 610059, China

[2]College of Earth Sciences, Chengdu University of Technology, Chengdu, China,610059

[a] Corresponding author: gdg@cdut.edu.cn

Abstract. A novel electromagnetic environmental pollution protective coating with high performance was prepared with Ni-Cu-La-B-coated glass fiber and nickel powder as composite fillers. Its conductivity,shielding effectiveness was discussed. The results showed that the coatings with a thickness of 300 μm containing 6 wt% of Ni-Cu-La-B-coated glass fibers had the lowest resistivity of 0.58 Ω·cm and best shielding effectiveness ranging from 51.45 dB to 62.18 dB in 0.3-1 000 MHz frequency band.

1. Introduction

With the rapid development of science and technology, a colorless, tasteless, invisible electromagnetic radiation pollution which is known as the "invisible killer" has been increasingly concerned by all walks of life. The harm of electromagnetic radiation pollution on economic construction, national defense security and social production and living is more and more serious[1-2]. To research and develop all kinds of high cost performance of electromagnetic shielding materials has become one of the important research topic in the field of materials science and engineering. In recent years, electromagnetic environmental pollution protective coatings have become a kind of important electromagnetic shielding materials because of their low cost, simple technology, flexible application and strong adaptability [3-4]. However, the cost and conductivity of fillers in the coatings have become two key factors restricting the coatings cost performance. So, the stduy for cost performance of conductive filler on preparation, characterization, the effect on electrical conductivity and electromagnetic wave shielding performance of shielding coating has important practical significance.

As an energy-saving and environment-friendly surface metallization treatment technology, electroless plating plays an increasingly important role in the preparation of cost-effective conductive composite filler due to its low cost, simple process and good coating quality [5-6]. A low density, low cost, conductive and Ni-Cu-La-B-coated glass fiber was developed by electroless plating [7]. In this paper, a novel electromagnetic environmental pollution protective coating with high performance was prepared using the fibers and nickel powders as composite fillers and Epoxy resin as binders. Furthermore, its conductivity, shielding effectiveness was discussed. This study is helpful to improve the added value and utilization of the glass fiber. This also provides a new idea and method for the development of low-cost electromagnetic environmental pollution protective coatings.

Content from this work may be used under the terms of the Creative Commons Attribution 3.0 licence. Any further distribution of this work must maintain attribution to the author(s) and the title of the work, journal citation and DOI.
Published under licence by IOP Publishing Ltd

2. Experimental

2.1 Materials

Nickel powder (Chengdu Nuclear 857 New Materials co., LTD) with grain diameter of 0.2 to 0.4 microns contained 99.50 wt % nickel. Ni-Cu-La-B-coated glass fibers contained 41.36 wt % nickel, 2.32 wt % copper, 0.5 wt % lanthanum and 0.5 wt % boron. The length of fiber glass was 10-200 microns, the diameter was 2-30 microns, and the length diameter ratio was 5-100. Moreover, glass fibers, epoxy resin, dedicated thinner, additive etc. were all homemade. The basic composition of electromagnetic wave shielding composite coating was shown in table 1.

Table 1. Basic formula of electromagnetic shielding coatings.

Coating ingredient	Composite filler content	Epoxy resin content	Coating thinner	Coating additive
Mass fraction	70-90	10-30	Add according to coating viscosity	Micro-addition

2.2 Coating preparation

The coating with high performance was prepared with Ni-Cu-La-B-coated glass fiber and nickel powder as composite fillers, and epoxy resin as binder. Secondly, the fibers and nickel powders were coupled by 2.5 wt % of titanate coupling agent in order to improve the filler dispersion in the resin. Thirdly, the fiber content in the filler was respectively 0 wt% , 1 wt%, 2 wt%, 3 wt%, 4 wt%, 5 wt%, 6 wt%, 7 wt% and 8 wt%. Then, the filler was respectively mixed evenly and added to the epoxy resin, and a small amount of solvent was added to dilute it to a viscous state.The mixture was ground 2-4 times with a three-roller grinder. And then, the thinner and other additives were put into the mixture and mixed evenly. The amount of thinner was added according to the coating viscosity. It was measured with the coated-4 cup viscosity tester. The viscosity of the coating was 10-30 seconds. The coating was encapsulated for later use.

2.3 Spraying coating

The preparation method of the coating was by spray method. The outer diameter of the plate was 115 mm, the inner diameter was 10 mm and the thickness was 2mm. The above prepared coating was sprayed over the surface of PVC plastic substrate.

When spraying, firstly, the base board was fixed. Secondly, air compressor pressure was adjusted to to 3 ~ 6 Mpa, and the distance between the nozzle and the workpiece was adjusted to 20 ~ 40 cm. And then, the electromagnetic environmental pollution protective coating was evenly sprayed on the substrate. The film thickness was 300 microns.

2.4 Coating curing

After the coating surface drying, coating curing was divided into two stages. The first stage was the artificial drying curing processing. The substrate whose spraying coating has been completed was put into the oven drying 30 minutes at 80-100 °C, and then took out. The second stage was curing at room temperature for 24-48 hours.

2.5 Performance testing and analysis

Resistivity test of the coating was carried out with a digital multimeter after the film was completely cured.

The shielding effectiveness of the coatings was tested by coaxial flange method in accordance with Chinese standard of SJ20524-95, i.e., "Measuring Methods for Shielding Effectiveness of Materials". Schematic diagram of the coaxial flange test device of shielding effectiveness of the shielding coating was shown in Figure 1. The operating frequency of coaxial testing device is in the range of 0.3~1500 MHz, and its impedance is 50 Ω. The value of tested shielding effectiveness was higher than 100 dB dynamic range.

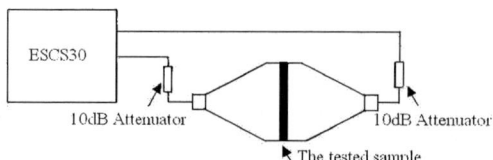

Figure 1. Schematic diagram of shielding effectiveness test device of the shielding coating.

Adhesion test of the coating was performed according to Chinese standard of GB/T 9286-1988, i.e., "Paints and Varnishes-cross Cut Test for Films". Pencil hardness of the coating was tested according to Chinese standard of GB/T 6739-1996, i.e., "Determination of Film Hardness by Pencil Test". Impact resistance determination of the coating was conducted according to Chinese standard of GB/T 1732-1993, i.e., "Determination of Impact Resistance of Film". Wear resistance of the coating was tested following Chinese standard of GB/ GB1768-1979, i.e., "Method of Test for Abrasion Resistance of Paint Films". Drying time test of the coating was conducted according to Chinese standard of GB/T 1728-1979, i.e., "Methods of Test for Drying Time of Coatings of Paints and Putties".

3 Result analysis and discussion

3.1 Coating resistivity
The coating resistivity test results are shown in Figure 2.

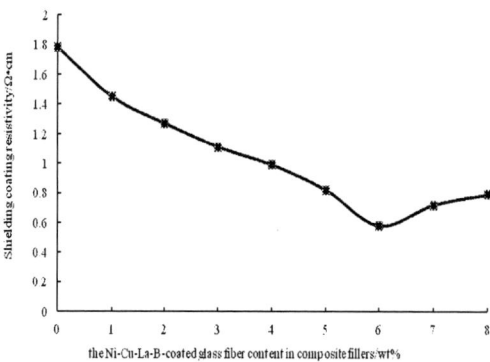

Figure 2. Effect of Ni-Cu-La-B-coated glass-fiber content on the coating resistivity.

In Figure 2, for the flake nickel powder/acrylic resin electromagnetic shielding composite coating, the coating resistivity achieves the minimum of 1.77 Ω•cm with a thickness of 300 μm. However, for the Ni-Cu-La-B-coated glass fibers/flake nickel powder/acrylic resin electromagnetic shielding composite coating, the coating resistivity decreases first and then rises with the increase of Ni-Cu-La-B-coated glass fibers content. The resistivity achieves the minimum when the fibers content is 6 wt% in the composite fillers. At this time, the minimum resistivity of the composite coating is 0.58 Ω•cm with a thickness of 300 μm.

According to the conductive theory of this kind of filled conductive composite coating, the coating conductivity mainly depends on the quality of conductive network formed in the coating by composite fillers and polymer[8-10]. The composite particle's physical contact with each other and electron tunneling effects are the two major factors for the conductive coating. The effects between conductive filler particles in the coating is promoted when the content of Ni-Cu-La-B-coated glass fibers is 6 wt% in the composite fillers. This is beneficial to form relatively dense conductive network structure because of the bridge filling effect of different length of the conductive glass fibre, so as to improve the electrical conductivity of the coating. Thus, the right adding amount of Ni-Cu-La-B-coated glass fibers has a significant effect on the conductivity of the coating.

3.2 Shielding effectiveness

The electrical conductivity of composite coatings have important effects on electromagnetic shielding performance. For this kind of filled conductive electromagnetic shielding coating, the better the conductivity, the greater the electromagnetic shielding effectiveness. So, the shielding effectiveness of two kinds of coatings with or without 6 wt% Ni-Cu-La-B-coated glass fibers are tested by coaxial flange method in accordance with Chinese standard of SJ20524-95. The test results are shown in table 2.

Table 2. Shielding effectiveness of two kinds of coatings [dB].

Frequency /MHz	Coating without Ni-Cu-La-B-coated glass fibers	Coating with 6 wt% Ni-Cu-La-B-coated glass fibers
0.3	48.92	60.63
5	50.85	62.18
30	49.74	61.07
50	51.24	56.61
100	49.87	56.57
200	49.72	52.24
300	45.89	54.36
500	40.55	52.73
800	40.76	51.45
1000	40.13	52.40

In table 2, electromagnetic shielding effectiveness of flake nickel powder/acrylic resin composite coating is only 40.55 ~ 51.24 dB, the whole shielding effectiveness is low in the scope of 0.3 to 1 000 MHz band. When the content of Ni-Cu-La-B-coated glass fibers is 6 wt% in the composite fillers, the composite coating has the highest electromagnetic shielding effectiveness, its value reach 51.45 ~ 62.18 dB in the same frequency rang. Thus it can be seen its shielding efficiency increases significantly and are generally better than that of no adding Ni-Cu-La-B-coated glass fibers.

From above banalysis, to add 6 wt% of Ni-Cu-La-B-coated glass fibers helps to form high quality conductive network in the coating, and this makes the electric performance of coating to achieve the best. This is mainly because the coating containing 6 wt% of Ni-Cu-La-B-coated glass fibers has a more dense conductive networ，better conductivity and stronger electromagnetic wave loss capacity. This is mainly because the electromagnetic energy is loss by electromagnetic wave absorption, reflection and diffuse reflection effect after the incident electromagnetic wave acts on the coating. So, the right adding amount of Ni-Cu-La-B-coated glass fibers also has a significant effect on electromagnetic shielding effectiveness of the coating.

3.3. Environmental properties

The test results of main environmental properties of the coatings containing 6 wt% of Ni-Cu-La-B-coated glass fibers are shown in Table 3.

Table3. Main environmental properties of the coatings.

Test items	Test standard	Test results
Adhesion / level	GB/T9286-1988	1
Pencil hardness /H	GB6739-1986	5
Impact strength /(kg·cm)	GB/T1732-1993	50

Wear resistance(Gravity) /(g/cm2)	GB1768-1979	0.001
Surface dry time, min(25°C)	GB/T1728-1979	30
Actual drying time,h(25°C)	GB/T1728-1979	12

In table 3, the main environmental properties of the electromagnetic shielding coatings containing 6 wt% of Ni-Cu-La-B-coated glass fibers are as follows: the adhesion force is level 1, the pencil hardness is 5 h, the impact strength is 50 kg·cm, the wear resistance (gravity) is 0.001 g/cm^2, the surface dry time (25 °C) is 30 min, and the actual drying time (25 °C) is 12 h. The main environmental properties of the coatings indicators have reached Chinese army standard GJB 2604-1996, i.e., "military electromagnetic shielding coatings general specification of the relevant provisions".

4. Conclusion

A novel electromagnetic environmental pollution protective coating with high performance is developed with Ni-Cu-La-B-coated glass fiber and Nickel powder as composite fillers. Its conductivity, shielding effectiveness and main environmental properties are discussed. The stduy results showed that the right adding amount of Ni-Cu-La-B-coated glass fibers has a significant effect on conductivity and electromagnetic shielding effectiveness of the coating.

For flake nickel powder/acrylic resin electromagnetic shielding composite coating, the coating resistivity achieves the minimum of 1.77 Ω·cm with a thickness of 300 μm. For Ni-Cu-La-B-coated glass fibers/flake nickel powder/acrylic resin electromagnetic shielding composite coating with a thickness of 300 μm, the best addition amount of Ni-Cu-La-B-coated glass fibers is 6 wt%, and the coating resistivity chieves the minimum of 0.58 Ω·cm.

The electromagnetic shielding effectiveness of flake nickel powder/acrylic resin composite coating is low and only reaches 40.55~ 51.24 dB in the scope of 0.3 to 1000 MHz band. When the content of Ni-Cu-La-B-coated glass fibers is 6 wt% in the composite fillers, the electromagnetic shielding effectiveness of Ni-Cu-La-B-coated glass fibers/flake nickel powder/acrylic resin electromagnetic shielding composite coating has significantly increased to 51.45 ~ 62.18 dB in the same frequency rang.

The main environmental properties of the coatings have reached Chinese army standard of military electromagnetic shielding coatings general specification of the relevant provisions (GJB 2604-1996).

This offers a new idea for the development of low-cost electromagnetic environmental pollution protective coatings. This study is helpful to enhance the added value of the glass fiber and utilization, and raise the level of electromagnetic environmental pollution protection.

Acknowledgement

This work was financially supported by the Key Research Projects of Science and Technology Department of Sichuan Province of China (2017GZ0392), and the Science and Technology Huimin Projects of Science and Technology Bureau of Chengdu of Sichuan Province of China (2015-HM01-00387-SF).

References

[1] Y.F. Zhao, *Electromagnetic Pollution in Modern Environment*. (China electro. Indus. press, 2003)
[2] S,Q. Zhang, P. Zhang, Indust. Safet. Envir. Protect.,2008,34(3):30-32
[3] J.H. Lee, D.K. Kang, S.H. Kim, M.J. Son, J.W. Choi, D.K. Choi, J.P. Choi;C. Aranas, J. Materio.4,4(2018)
[4] D. G. Guan, G.L. Xu, Y. Sun, C. M. Sun, J.H. Lin, W.J. Xu, T. Chen, Y.H. Cui, Y. Xu, and H. Yan, Russ. J. Appl. Chem. 87,8 (2014)
[5] D. G. Guan, C.M. Sun, G,L. Xu, J.H. Lin, S.H. Chen, Mater. Rev. 7, (2011)

[6] C. Wu, X.X. liu, J. Huang, J. electropla. & environ. Protec. 30,2 (2010)

[7] D. G. Guan, China invention patent, ZL 201410010339.X, (2016)

[8] X.L.Zhang, B. Lu. *EMC Design under the Complex Electromagnetic Environment*. (China Gansu People's press, 2006)

[9] L.P. Wu, Y.Z. Li, B.J. Wang, Z.P. Mao, H. Xu, Y. Zhong, L.P. Zhang, X.F. Sui, Mater. & Desi.. **159**,(2018)

[10]J. Lee, B. M. Jung, S.B. Lee, S.K. Lee, K.H. Kim: Appl. Sur. Sci., 415,1 (2017)

ISPECE   IOP Publishing

IOP Conf. Series: Journal of Physics: Conf. Series **1187** (2019) 022050   doi:10.1088/1742-6596/1187/2/022050

A Renewable Energy Assessment Model Considering the Effect of Frequency Regulation

Huang Xuxiang[1], Han Xueshan[1,a] and Li wenbo[2]

[1]Key Laboratory of Power System Intelligent Dispatch and Control of Ministry of Education (Shandong University), Jinan, China

[2]State Grid Shandong Electric Power Research Institute, Jinan 250003, Shandong Province, China

[a]Corresponding author: hxxsdu@163.com

Abstract. Under the background of energy Internet, a high proportion of renewable energy is connected to the power system, and its volatility and randomness make the flexibility of the power system itself increasingly important in dealing with the fluctuation of net load. In order to evaluate the limit capacity of renewable energy consumption under the given power structure and the future absorption prospect of regional power grid, a renewable energy evaluation model considering the effect of frequency regulation is proposed in this paper. This method fully considers the role of frequency regulation in dealing with the uncertainty of renewable energy, establishes the absorption index from the perspective of absorption capacity and absorption efficiency, and comprehensively evaluates the absorption capacity of renewable energy from the perspective of frequency modulation, slope climbing and other time scales. Finally, the model is verified by a ten-machine system.

1. Introduction

In the context of energy Internet, renewable energy generation has become the main power generation form of the future power system. China proposes to achieve the goal of 85% renewable energy generation by 2050 [1], and Europe even proposes the idea of 100% renewable energy [2]. A high proportion of renewable energy is connected to the power system, and its volatility and randomness break the power balance mode of conventional unit tracking load, making it difficult for conventional unit to track the fluctuation of net load, and generating the limit of renewable power supply and load cutting capacity, which run counter to the purpose of developing renewable energy. Therefore, it becomes an important subject to evaluate the absorption capacity of renewable energy under the current power structure and provide references for the planning of renewable power.

Most existing studies evaluate the absorption capacity of the system from the perspective of operation simulation, which is mainly divided into two categories: (1) time-series simulation method [3-5] : it mainly conducts scene simulation through unit combination and economic scheduling, and evaluates the absorption capacity according to whether there is load loss and the limit of renewable power supply; (2) flexibility evaluation [6-9] : evaluate the receptivity of renewable energy from the perspective of flexibility. Such studies tend to consider the abundance of system flexibility in a given time profile considering the existing renewable energy capacity.

All the above studies are analyzed from one point or one section, and it is difficult to find the internal law of the proportion of installed capacity and power generation of renewable energy, so the

---

Content from this work may be used under the terms of the Creative Commons Attribution 3.0 licence. Any further distribution of this work must maintain attribution to the author(s) and the title of the work, journal citation and DOI.

Published under licence by IOP Publishing Ltd

absorption potential of the system for renewable energy cannot be obtained intuitively. In this regard, this paper first analyzes the absorption mechanism of renewable energy with different capacities after being connected to the system, and sets up three absorption scenarios according to the measures to deal with the uncertainty of renewable energy: natural absorption, standby absorption and auxiliary absorption. The absorption capacity of renewable energy was evaluated comprehensively from different time scales, such as frequency modulation and slope climbing.

2. UC model considering frequency regulation effect
Existing wind farms and photovoltaic farms are equipped with a prediction system, which reports the prediction information in real time. Based on this, the dispatching department arranges the starting and stopping plan of the next day's conventional units. Variables to be optimized include: 1) starting and stopping variables of conventional units (controllable units of thermal power, hydropower, etc.) at all times $I = \{I_i^t \mid i \in 1 \cdots N, t \in 1 \cdots T\}$ (0 represents shutdown, 1 represents opening); 2) the output base point of conventional units i at t, P_i^t; 3) the output base point of the renewable power source at t, that is P_r^t, the maximum output value of renewable energy that the system can absorb.

2.1 Evaluation indicators
This paper studies the relationship between the proportion of renewable energy capacity and the actual operation proportion. Therefore, this section establishes an evaluation index system from the two aspects of absorption capacity and absorption efficiency to evaluate the saturated capacity and economic saturated capacity of renewable energy that can be accessed in the future years under the current power supply structure of the system.

2.1.1 Absorption ability
The principle of analyzing this index is that, regardless of economy and operating efficiency, conventional units give way to renewable energy and give full play to the absorption potential of the system.

System access to a certain amount of renewable energy, due to its output volatility and randomness, cause system can't fully accepting renewable energy output value, abandon the abandoned light phenomenon is inevitable. The electricity will be renewable power and cutting load phenomenon is referred to as rejection, therefore absorption ability can use renewable energy to reject rate meet certain signs of maximum capacity threshold.

$$\alpha = \alpha_1 \cdot \frac{\sum_{t=1}^{T} \Delta P_r^t}{\sum_{t=1}^{T} P_{r,f}^t} + \alpha_2 \cdot \frac{\sum_{t=1}^{T} \Delta P_L^t}{\sum_{t=1}^{T} P_{L,f}^t} \qquad (1)$$

ΔP_r^t and ΔP_L^t are respectively the limit of renewable power supply and the cutting load in the period t; , respectively, $P_{r,f}^t$ and $P_{L,f}^t$ are the predicted values of renewable power supply and load; α_1, α_2 respectively represent the weight coefficient of the influence of wind and light abandoning and load cutting on the rejection rate.

2.1.2 Absorption efficiency
This paper mainly measures the absorption efficiency from the economic perspective. The main purpose of renewable energy access to the power grid is to replace thermal power, reduce the cost of power generation, in order to achieve the effect of energy conservation and emission reduction; But at the same time, the addition of fans, photovoltaic panels and other devices will increase the fixed cost, and the operation process will also cause the decline of the operation efficiency of conventional units, and increase the operation cost. Therefore, reasonable evaluation of the value of renewable energy is a core problem of high proportion of renewable energy grid connection.

This paper studies the relationship between renewable power capacity and absorption efficiency, so the contribution rate of unit capacity η is defined to characterize the emission reduction efficiency of renewable energy to the system under different access capacities.

$$\eta = \frac{C_o - C_1}{W_L \cdot C_r} \qquad (2)$$

$$C_1 = \sum_{t=1}^{T} \sum_{i=1}^{N} \left[f(P_i^t) \cdot I_i^t + f(I_i^t) \right] + f(C_r) \qquad (3)$$

$$C_0 = \sum_{t=1}^{T} \sum_{i=1}^{N} \left[f(P_i^t) \cdot I_i^t + f(I_i^t) \right] \qquad (4)$$

$$f(P_i^t) = a_i (P_i^t)^2 + b_i P_i^t + c_i \qquad (5)$$

$$f(C_r) = \frac{C_r \cdot d \cdot (1 - \gamma\%)}{D_Y} \qquad (6)$$

C_o, C_1 is the total power generation coal consumption of the system before and after the access of renewable energy, W_L is the total power generation, and C_r is the installed capacity of renewable energy; $f(P_i^t)$ is the power generation cost of conventional units, $f(I_i^t)$ is the start-stop cost, $f(C_r)$ is the fixed cost of renewable power converted to the cost of each day, the straight-line depreciation method for simple calculation; a_i, b_i, c_i and are the coefficients of each part of the power generation cost function respectively; d is the unit capacity cost, $\gamma\%$ is the expected net residual value rate, and D_Y is the expected number of days of use.

2.2 Evaluation UC model

The optimization is carried out on the principle of "maximum acceptance of the output value obtained from the prediction of renewable energy $P_{r,f}^t$", to ensure the minimum power limit of renewable energy, and the optimization objective is

$$\min \sum_{t=1}^{T} (P_{r,f}^t - P_r^t) \qquad (7)$$

Including constraints:

1)power balance:

$$\sum_{i=1}^{N} P_i^t = (P_L^t + \Delta P_L^t) - P_r^t \qquad (8)$$

In order to ensure that the model has a solution, the relaxation quantity is added, that is, the load loss ΔP_L^t in the period t is satisfied

$$0 \le \Delta P_L^t \le P_L^t \qquad (9)$$

2)climbing constraints of conventional units:

$$P_i^t - P_i^{t-1} \le [1 - I_i^t \cdot (1 - I_i^{t-1})] \cdot r_i \cdot \Delta t + I_i^t \cdot (1 - I_i^{t-1}) \cdot R_i \cdot \Delta t \qquad (10)$$

$$P_i^{t-1} - P_i^t \le [1 - I_i^{t-1} \cdot (1 - I_i^t)] \cdot r_i \cdot \Delta t + I_i^{t-1} \cdot (1 - I_i^t) \cdot R_i \cdot \Delta t \qquad (11)$$

r_i is the regulating speed of the conventional unit, R_i is the starting and stopping speed of the conventional unit, and Δt is the interval of time.

3) reserve range:

$$\Delta P_{UP} \ge \Delta P_{L,f}^t + \Delta P_{r,f}^t - \Delta P_L^t \qquad (12)$$

$$\Delta P_{DOWN} \le \Delta P_{r,f}^t + \Delta P_{L,f}^t - \Delta P_r^t \qquad (13)$$

ΔP_{UP}, ΔP_{DOWN} are the up-regulation and down-regulation. The left side of the above equation is the power regulation effect of the system, and the right side is the net load fluctuation.

Scenario 1:

Natural absorption: For the regions with high load level, the unbalance that can be adjusted by primary frequency modulation is large, and the system shows great flexibility. In the early stage of the development of renewable power supply, namely, the access capacity is small, the intermittent characteristics are not obvious, the short-term fluctuation is small, and only one frequency modulation can meet the rejection rate requirements.

$$\Delta P_{UP} = -(\sum_{i=1}^{N} KG_i + KL^t) \cdot \Delta f_{down}^t \qquad (14)$$

$$\Delta P_{DOWN} = -(\sum_{i=1}^{N} KG_i + KL^t) \cdot \Delta f_{up}^t \qquad (15)$$

$$0 \le \Delta f_{up}^t \le \Delta f_{max} \qquad (16)$$

$$\Delta f_{min} \le \Delta f_{down}^t \le 0 \qquad (17)$$

KG_i is the frequency regulation effect coefficient of the unit i; KL^t is the frequency regulation effect coefficient of load; Δf_{up}^t, Δf_{down}^t are the upward and downward variation of the frequency; Δf_{max}, Δf_{min} respectively are the maximum upward and downward changes.

Scenario 2:

Reserve absorption:With the increase of the access capacity of the renewable power supply, the standby demand cannot be met only by primary frequency modulation.

$$\Delta P_{UP} = -(\sum_{i=1}^{N} KG_i + KL^t) \cdot \Delta f_{down}^t + \sum_{i=1}^{N} \Delta P_{up,i}^t \qquad (18)$$

$$\Delta P_{DOWN} = -(\sum_{i=1}^{N} KG_i + KL^t) \cdot \Delta f_{up}^t - \sum_{i=1}^{N} \Delta P_{down,i}^t \qquad (19)$$

4) reserve response speed:

up-regulation $\Delta P_{up,i}^t$, down-regulation $\Delta P_{down,i}^t$

$$0 \le \Delta P_{up,i}^t \le r_i \cdot \Delta \tau \qquad (20)$$

$$0 \le \Delta P_{down,i}^t \le r_i \cdot \Delta \tau \qquad (21)$$

5) conventional unit output constraint:

$$P_i^t + \Delta P_{up,i}^t - KG_i \cdot \Delta f_{down}^t \le P_{max,i} \cdot I_i^t \qquad (22)$$

$$P_i^t - \Delta P_{down,i}^t - KG_i \cdot \Delta f_{up}^t \ge P_{min,i} \cdot I_i^t \qquad (23)$$

$\Delta \tau$ is the standby response time, which depends on the fluctuation speed of renewable energy; $P_{max,i}$, $P_{min,i}$ are the maximum , minimum output of the first unit.

6)other constraints:

In addition to the above constraints, there are conventional unit output constraints, renewable energy output value constraints, minimum start-stop time constraints, and conventional unit start-stop status constraints [10].

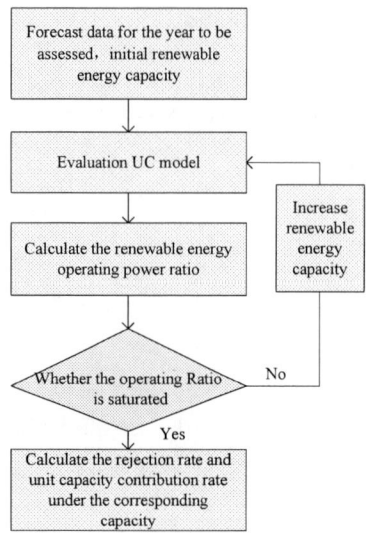

Figure 1. Evaluation flow chart

3. Example

In this paper, 10 conventional units are taken as examples for simulation, and the parameters of the units are shown in literature [10].

Three typical days were taken and the exclusion threshold was set as $\alpha = 10\%$, $\alpha_1 = \alpha_2 = 0.5$, that is, the power limit ratio and load cut ratio of permissible renewable energy were 10%.

(1) Absorption ability

In the case of different renewable power access capacity, the expected value of the total power generation proportion and the expected value of the rejection rate in three typical days can be obtained through simulation operation, as shown in figure 2 and figure 3.

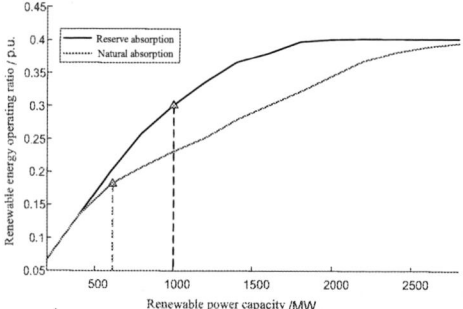

Figure 2. Renewable energy capacity and operating ratio

As can be seen from figure 2, when the installed capacity reaches 1820MW, the operation proportion reaches saturation, but at this time, the exclusion rate reaches 40.24%, exceeding the set threshold $\alpha = 10\%$. As shown in figure 3, the corresponding saturation capacity is 1150MW, which has been marked out in figure 2, and the corresponding operation proportion is about 30.14%.

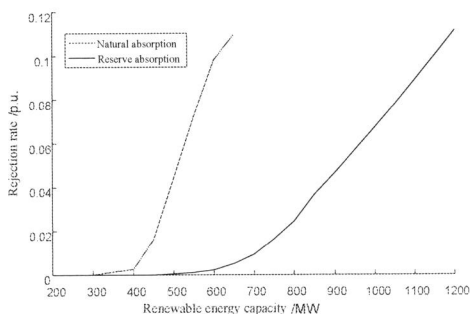

Figure 3. Renewable energy capacity and rejection rate

Natural given scenario: the installed capacity of more than 400 mw, due to its limited ability to cope with uncertainty, resulting in a large number of abandoned discard wind, when the installed capacity of 2800 mw, the running of saturated, but the rejection rate is 62.05%, more than half of the renewable energy output has been abandoned. Figure 3 shows, the saturated capacity corresponding to $\alpha = 10\%$ is 650MW, and the operation proportion in figure 3 is 18.63%.

(2) absorption efficiency

Calculating the contribution rate of unit capacity η, the absorption efficiency under the two absorption scenarios can be obtained, as shown in figure 4.

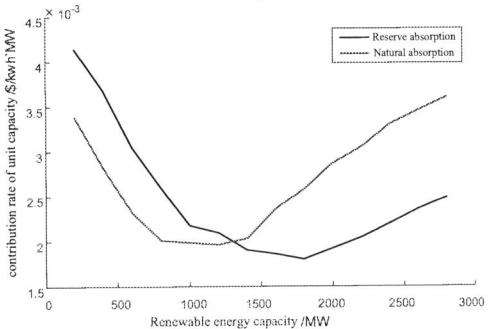

Figure 4. The contribution rate of unit capacity under different capacity

In the case of natural absorption, the benefit of emission reduction is the best when the capacity reaches 790MW. However, based on the analysis in figure 3, the exclusion rate exceeds 10%. Therefore, based on good absorption, that is, before 650MW, the benefit of energy saving and emission reduction of renewable power increases with the increase of installed capacity. In the case of reserve absorption, when the capacity reaches 1850MW, the benefit is the highest, which is much larger than the economic saturation capacity in the case of natural absorption, that is, the system can more economically accept renewable energy.

4. Summary

This paper establishes a renewable energy assessment model that takes into account the frequency regulation capability. The internal law between renewable energy capacity and operation ratio is obtained through operation simulation.Based on the indexes of absorption capacity and absorption efficiency, this paper evaluates the current situation and potential of absorption of renewable energy at the present stage, and provides a reference for the planning of renewable energy.

Acknowledgment

This work is supported by the Science and Technology Foundation of SGCC(SGSDDK00KJJS1600061).

References

[1] Energy Research Institute of the National Development and Reform Commission. China 2050 High Proportion of Renewable Energy Development Scenarios and Paths [R]. Beijing: Energy Research Institute of the National Development and Reform Commission, (2015).

[2] Schellekens G，Battaglini A，Lilliestam J，et al. 100% Renewable electricity：a roadmap to 2050 for Europe and North Africa[R]. London，UK：Price water house Coopers，(2010).

[3] Kang Chongqing, Jia Wenzhao, Xu Yaoyao, et al. Capability Evaluation of Wind Power Accommodation Considering Security Constraints of Power Grid in Real-time Dispatch [J]. *Proceedings of the CSEE*, (33(16): 23-29,2013).

[4] Jia Wenzhao, Kang Chongqing, Li Dan, et al. Evaluation method of wind power capacity based on wind power forecasting[J]. *Power System Technology*,(36(8):69-75,2012).

[5] Li Hongzhong,Lv Zhenbang,Zhu Jiaming, et al. Dynamic evaluation method of wind power consumption considering economics[J]. *Power System Technology*,((4):1261-1268,2017).

[6] Lu Zongxiang,Li Haibo,Qiao Ying.Evaluation and balance mechanism of power system flexibility for high-ratio renewable energy grid-connected[J].*Proceedings of the CSEE*,(37(1):9-19,2017).

[7] Lannoye E,Flynn D，O'Malley M．Transmission，variable generation，and power system flexibility[J].*IEEE Transactions on Power Systems*,(30(1)：57-66,2015).

[8] Nosair H，Bouffard F. Flexibility envelopes for power system operational planning[J]. *IEEE Transactions on Sustainable Energy*，(6(3)：800-809,2015).

[9] Zhao Jinye，Zheng Tongxin，Litvinov E. A unified framework for defining and measuring flexibility in power system[J]. *IEEE Transactions on Power Systems*，(31(1)：339-347,2016).

[10] Li Benxin, Han Xueshan, Jiang Zhe, et al. Combined model and analysis of thermal power units with intermittent wind power[J]. *Power System Technology*, (41(5): 1569-1575,2017).

ISPECE   IOP Publishing

The Comparison of Thermal Characteristics of AC Cable and DC Cable

Ruiqi Zhang[1,a], Hua Sun[2], Xiaoming Dong[1]

[1]Key Laboratory of Power System Intelligent Dispatch and Control of the Ministry of Education, Shandong University, Jinan, China

[2]Department of Electrical Automation, Shandong Labor Vocational and Technology College, Jinan, China

[a]Ruiqi Zhang: 1214748608@qq.com

Abstract. Electro-thermal coupling theory can be used as an effective tool to analyse the thermal characteristics of power cables. As the carrier for transmitting electrical power, cables can convey alternating current(AC) as well as direct current(DC) and have different thermal characteristics under different current forms. This paper analyses the ampacity of electric cables based on electro-thermal coupling, and compares the influence of the thermal resistance of different materials in the cable. Case study reveals the thermal characteristics under different current forms by simulating temperature variations of the two cables.

1. Introduction

Electro-thermal coupling was introduced for unground cables based transmission grids and thermal behaviour is simulated dynamically[[1]]. An algorithm estimating the temperature evolution of power cables was proposed in [[2]]. Its thermal resistivity and specific heat of the cable surroundings are varied as functions of the moisture content. Then, a combination of the heat balance equations with a power flow model is presented to reveal and exploit the potential transfer capability of cable[[3]]. In recent years, some single-circuit XLPE cables were converted into DC operation and they were still in experimental state[[4]]. In terms of the conversion, a number of challenges need to be overcome. The medium-voltage direct current technology is deployed to address these issues[[5]]. A modified 2D model is set up to simulate the thermal field distribution of a 320 kV DC cable, and based on that, the electric field distribution in the XLPE insulation is studied by the finite element method without consideration of space charge effect[[6]]. Space charge characteristics of the three kinds of XLPE after aging for different months were measured by pulsed electro-acoustic (PEA) method. Space charge distributes evenly in the specimens with a constant temperature, while hetero-charge accumulates in the specimens under temperature gradient[[7]]. A three core high-voltage underground cable used for urban power networks is analysed by means of Finite Element Method. The electro-thermal analysis permits the full description of the cable behavior not only in terms of electric and magnetic performances, but also regarding the effect of the Joule heating on the surrounding ambient[[8]]. Regarding the thermal modelling, the volume-element method is used in order heat transfer equations on a two-dimensional axisymmetric cable model to be solved and temperature distribution over space and time to be determined. For the modelling of the electric part, a constant-parameter finite-difference time-domain method is utilized to calculate voltages and currents over space and time as well[[9]]. A procedure is proposed for life estimation of high voltage AC cables in real operating

---

Content from this work may be used under the terms of the Creative Commons Attribution 3.0 licence. Any further distribution of this work must maintain attribution to the author(s) and the title of the work, journal citation and DOI.
Published under licence by IOP Publishing Ltd

conditions, applied to XLPE-insulated high voltage AC cables and subjected to two typical stepwise-constant daily load cycles differing as to the load severity. The application shows that cable life is very sensitive to load cycles, thermal transients and electro-thermal synergism, aspects that all deserve attention for estimating accurately the life expectancy of high voltage AC cables in service[[10]]. This paper makes a comparison of dynamic thermal modelling for AC cable and DC cable. Moreover, the ampacity of power cable can be obtained when dynamic behavior is ignored. Balanced temperatures are simulated corresponding to different line current. Eventually, the main results are summarized and key conclusions are recapped.

2. Thermal Characteristics of Cable

2.1 Thermal Model

The power cables have a complex hierarchical structure as depicted in Fig. 1. Taking the single-core XLPE insulated cable as an example, the structure can be roughly divided into conductor, insulating dielectric, metallic screen and jacket. In order to adapt to different laying requirements, it exists in various models corresponding to different conductor (copper or aluminum), metallic screen (aluminum or steel) and jacket (PVC or PE) for designers to choose the most appropriate version.

The single-core YJV model cable (XLPE insulation, PVC jacket and copper conductor) is selected in the paper. Since this cable for laying in the cable trench or buried directly in loose soil, it is widely used in urban power grid. The thermal model can be obtained according to its equivalent hot circuit shown in Fig. 2.

Conductor DC resistance as shown in (1) can be approximately expressed as a linear function of θ_{cc} in accordance with the algorithm of electro-thermal coupling of transmission cables.

$$r = \frac{\rho_{20}}{A}\left[1 + \alpha(\theta_{cc} - \theta_l)\right]k_1 k_2 k_3 \qquad (1)$$

Where the proportion coefficient, ρ_{20} and α, is dependent on the conductor material properties.

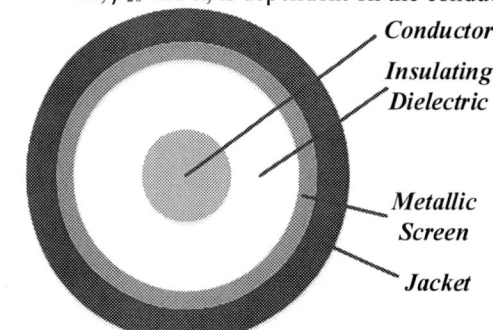

Fig. 1. Hierarchical structure of electric cable.

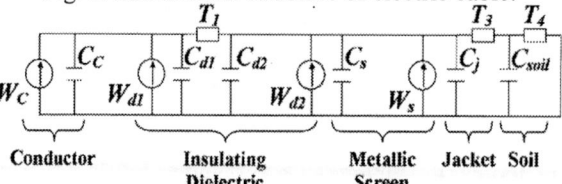

Fig. 2. Equivalent hot circuit of transmission cable.

However, as observed by (2), allowing for skin effort and proximity effect, conductor resistance will be increased when transmitting alternating current.

$$R = r(1 + Y_S + Y_P) \qquad (2)$$

Then, active power loss with respect to AC and DC can be obtained by (3).

$$W_C = \begin{cases} I^2 R \,(AC) \\ I^2 R \,(DC) \end{cases} \qquad (3)$$

Dielectric loss is the active power loss consumed in insulation when transmitting alternating current. It can heat conductor and decrease the ampacity of the line, moreover, significantly affect thermal insulation breakdown. The relation is given in (4).

$$W_d = 2\pi f C_e U^2 \tan\delta \qquad (4)$$

Due to the circulation existing in the metallic screen, a little energy consumed in the screen. Accordingly, the detailed expression of W_s is given in (5), which is dependent on W_c linearly.

$$W_s = \lambda W_c \qquad (5)$$

According to the hot circuit in Fig. 2, AC cable thermal balance differential equations are expressed in (6). These equations reveal the thermal coupling between cable structure layers[[1]]. In addition, W_c and W_d are related with line current and phase voltage respectively, electro-thermal coupling law is reflected evidently.

$$\begin{cases} (C_c + 0.5 C_d)\dfrac{d\theta_{cc}}{dt} = \dfrac{1}{T_1}(\theta_{cs} - \theta_{cc}) + (W_c + 0.5 W_d) \\[2mm] (C_s + 0.5 C_d + C_j)\dfrac{d\theta_{cs}}{dt} = \dfrac{1}{T_1}\theta_{cc} - (\dfrac{1}{T_1} + \dfrac{1}{T_3})\theta_{cs} + \dfrac{1}{T_3}\theta_{ce} + (W_s + 0.5 W_d) \\[2mm] C_{soil}\dfrac{d\theta_{ce}}{dt} = \dfrac{1}{T_3}\theta_{cs} - (\dfrac{1}{T_3} + \dfrac{1}{T_4})\theta_{ce} + \dfrac{1}{T_4}\theta_{soil} \end{cases} \quad (6)$$

In comparison with AC cables, dielectric loss and screen loss don't occur in DC cables. When conveying direct current, there is no circulation in insulation dielectric and metallic screen. Then, DC cable thermal balance differential equations[[2]] is elaborated in (7).

$$\begin{cases} (C_c + 0.5 C_d)\dfrac{d\theta_{cc}}{dt} = \dfrac{1}{T_1}(\theta_{cs} - \theta_{cc}) + W_c \\[2mm] (C_s + 0.5 C_d + C_j)\dfrac{d\theta_{cs}}{dt} = \dfrac{1}{T_1}\theta_{cc} - (\dfrac{1}{T_1} + \dfrac{1}{T_3})\theta_{cs} + \dfrac{1}{T_3}\theta_{ce} \\[2mm] C_{soil}\dfrac{d\theta_{ce}}{dt} = \dfrac{1}{T_3}\theta_{cs} - (\dfrac{1}{T_3} + \dfrac{1}{T_4})\theta_{ce} + \dfrac{1}{T_4}\theta_{soil} \end{cases} \quad (7)$$

2.2 Ampacity of Cable

Especially for steady-state electro-thermal coupling algorithm which focus on slow varying demands in analysing long-term processes of power systems, the deviation of line temperature to time variable, $d\theta_{cc}/dt$, $d\theta_{cs}/dt$ and $d\theta_{ce}/dt$, are assumed to equal to 0, ignoring the heat balance dynamics. Thus, detailed expression of ampacity for AC and DC is given in (8). Specially, for fear of thermal breakdown, the maximum allowable temperature for DC cables is smaller than that for AC cables.

$$I_m = \begin{cases} \sqrt{\dfrac{\theta_{max} - \theta_{soil} - W_d\left(0.5 T_1 + T_3 + T_4\right)}{R\left[T_1 + (1+\lambda)(T_3 + T_4)\right]}} \\[6mm] \sqrt{\dfrac{\theta_{max} - \theta_{soil}}{r\left(T_1 + T_3 + T_4\right)}} \end{cases} \qquad (8)$$

3. Case Study

In this section, concerned parameters are set on the basis of actual YJV model cable, which can be found in the Table I. Especially, the maximum allowable temperatures for AC cable and DC cable are defined as 90℃ and 70℃. Therefore, according to these data and the algorithm, the ampacities for them are equal to 748A and 663A separately. As observed in (8), with the decrease in thermal resistance including T_1, T_3 and T_4, the ampacities for them both increase gradually. Then, their relationships are depicted in Fig. 3.

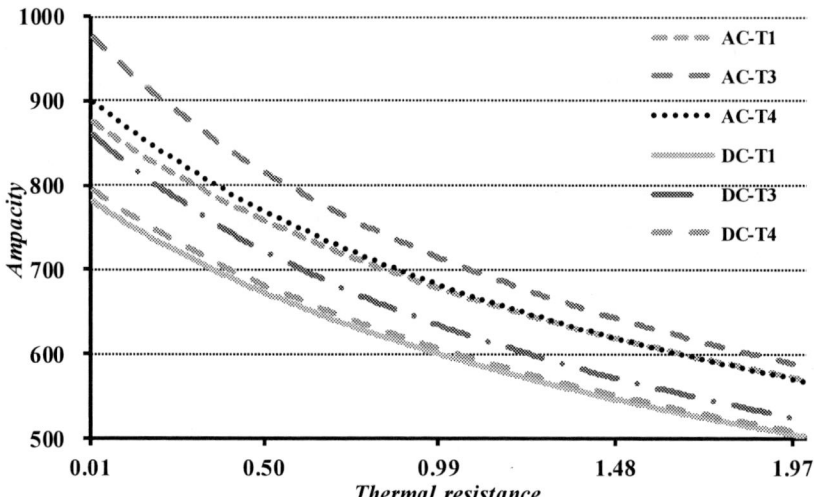

Fig. 3. Relationship between ampacity and thermal resistance.

As can be seen from Fig. 3, ampacity of AC cable is larger than that of DC cable as a whole. Obviously, the lower jacket thermal resistance is, the more contribution made to improve ampacity.

Table I. COEFFICIENTS FOR THE CABLE

Parameter	Value
ρ_{20}	$1.7241 \cdot 10^{-8}$
A	$4 \cdot 10^{-4}$
α	$3.93 \cdot 10^{-3}$
k_1	1.0
k_2	1.007
k_3	1.01
Y_s	0.023
Y_p	0.0002
C_e	$4.09 \cdot 10^{-10}$
$tan\delta$	0.002
U	35000
λ	0.046
C_c	3.5
C_d	1.2
C_s	3.5
C_j	1.5
C_{soil}	2.0
T_1	0.550
T_3	0.800
T_4	0.600
θ_{max}	90(AC) or 70(DC)
θ_{soil}	25
θ_l	20

Assuming that the operating current keeps constant, finally a thermal balance will be achieved. Corresponding to different line current, balanced temperatures of AC cable and DC cable perform distinctions observably. When transmitting the same current, AC cable temperatures are usually higher than DC cable. With the gradual increase in current, every material temperature obtains a nonlinear rise. Evidently, these are all illustrated in Fig. 4.

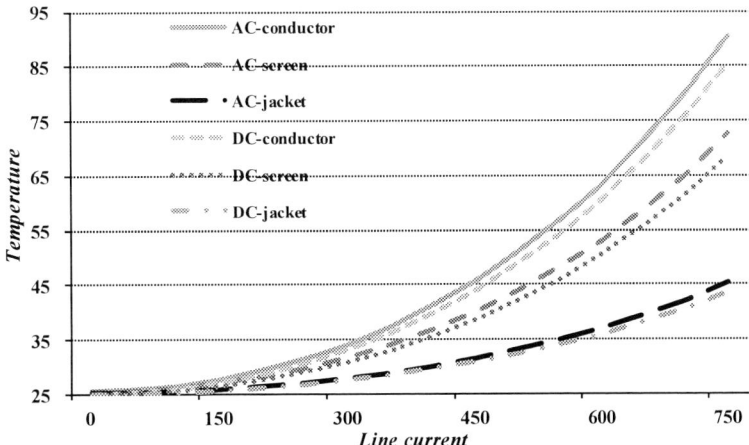

Fig. 4. Relationship between temperature and line current.

4. Conclusions

Firstly, electro-thermal coupling is used as an effective tool to calculate the temperatures of every cable material. It reveals the thermal coupling between cable structure layers.

Secondly, if dynamic balance is ignored, ampacities of AC cable and DC cable can be obtained. According to the equations, they increase differently when different thermal resistances are reduced.

Finally, due to the dielectric loss and screen loss, AC cable temperatures are higher than DC cable temperatures. Compared with the jacket temperature, conductor temperature and screen temperature are much larger.

Nomenclature

I Conductor current (A).

I_m Ampacity of cable (A).

ρ_{20} Copper resistivity at 20℃ (Ω·m).

A Conductor cross-sectional area (m²).

α Temperature coefficient of resistance (1/℃).

k_1 Twist factor.

k_2 Cabling factor.

k_3 Compaction effect coefficient.

Y_s Skin effect coefficient.

Y_p Proximity effect coefficient.

C_e Cable capacitance (F).

δ Dielectric loss angle (degree).

U Voltage between conductor and ground (V).

λ Circulation loss coefficient.

r Conductor DC resistance (Ω/m).

R Conductor AC resistance (Ω/m).

W_c Active power loss at the cable (W/m).

W_d Dielectric loss (W/m).

W_s	Metallic screen loss (W/m).
C_c	Conductor volume heat capacity (J/(m^3·℃)).
C_d	Insulating dielectric volume heat capacity (J/(m^3·℃)).
C_s	Metallic screen volume heat capacity (J/(m^3·℃)).
C_j	Jacket volume heat capacity (J/(m^3·℃)).
C_{soil}	Soil volume heat capacity (J/(m^3·℃)).
T_1	Insulating dielectric thermal resistance (℃/W).
T_3	Jacket thermal resistance (℃/W)
T_4	Surroundings thermal resistance (℃/W)
θ_{max}	Maximum allowable temperature (℃).
θ_{soil}	Soil temperature (℃).
θ_1	Reference temperature (℃).
θ_{cc}	Conductor temperature (℃).
θ_{cs}	Metallic screen temperature (℃).
θ_{ce}	Jacket temperature (℃).

References

[1] R. Olsen, J. Holboell and U. Stella Gudmundsdóttir, "Electrothermal Coordination in Cable Based Transmission Grids." IEEE Trans. Power Syst., Vol. 28, No.4, Nov. 2013.

[2] R. Olsen, George J. Anders and J. Holboell, "Modelling of Dynamic Transmission Cable Temperature Considering Soil-Specific Heat, Thermal Resistivity, and Precipitation." IEEE Trans. Power Deliv., Vol. 28, No. 3, Jul. 2013.

[3] Mengxia Wang, Xueshan Han, "Electro-thermal Power Flow Calculation Considering Thermal Properties of Cable." Automation of Electric Power Systems, Vol. 40, No. 11, Jun. 2016.

[4] Ying Liu, Xiaolong Cao and Mingli Fu, "The Upgrading Renovation of an Existing XlPE Cable Circuit by Conversion of AC Line to DC Operation." IEEE Trans. Power Deliv. Vol. 32, No. 3, Jun. 2017.

[5] J. Yu, K. Smith and M. Urizarbarrena, "Initial designs for ANGLE-DC project: challenges converting existing AC cable and overhead line to DC operation." 24th International Conference & Exhibition on Electricity Distribution(CIRED), ISSN:2515-0855, Jun. 2017.

[6] Ying Liu, Shendong Zhang and Xiaolong Cao, "Simulation of Electric Field Distribution in the XLPE Insulation of a 320 kV DC Cable under Steady and Time-Varying States." IEEE Trans. Diele. and Elec. Insul. Vol. 25, Nov. 3, Jun. 2018.

[7] Xia Wang, Quanyu Liu and Xiaoyang Zhang, "Study on Space Charge Behavior of XLPE after Long-term Aging under Temperature Gradient and DC Stress." 2016 International Conference on Condition Monitoring and Diagnosis(CMD), Xi'an, China, Sept. 2016.

[8] S. Conti, E. Dilettoso and Santi A. Rizzo, "Electromagnetic and thermal analysis of high voltage three-phase underground cables using finite element method." 2018 IEEE International Conference on Environment and Electrical Engineering and 2018 IEEE Industrial and Commercial Power Systems Europe (EEEIC/I&CPS Europe), Palermo, Italy, Jun. 2018.

[9] Dimitrios I. Doukas, Andreas I. Chrysochos, and Theofilos A. Papadopoulos, "Coupled Electro-Thermal Transient Analysis of Superconducting DC Transmission Systems Using FDTD and VEM Modelling." IEEE Trans. Appl. Super., Vol.27, No.8, Dec. 2017.

[10] G. Mazzanti, "The combination of electro-thermal stress, load cycling and thermal transients and its effects on the life of high voltage ac cables", IEEE Trans. Diele. and Elec. Insul. Vol. 16, No. 4, Aug. 2009.

ISPECE

IOP Publishing

Research on the Protection Range of Bird Droppings of 110kV Transmission Line Based on ANSYS Maxwell

Hao Zhang[1], Renfei Che[1,a], Wen Du[1], Xiaocheng Meng[1,2] and Juntao He[1,3]

[1]Key Laboratory of Power System Intelligent Dispatch and Control of the Ministry of Education, Shandong University, Jinan; 250061, China

[2]State Grid Jiangsu Electric Power Company Technician Training Center, Suzhou; 215004, China

[3]Zhengzhou Power Supply Company, Zhengzhou; 450006, China

[a]Corresponding author: cherenfei@sdu.edu.cn

Abstract. The falling of bird droppings near the insulator of overhead transmission lines will cause distortion of the electric field distribution near the insulator, resulting in air gap breakdown and flashover. Studying the area where there may be a risk of breakdown near the insulators in such a situation is of great significance for bird damage prevention of overhead transmission lines. Based on the finite element method, this paper established the 110kV transmission line model using ANSYS Maxwell finite element software and calculated the variation of the electric field distribution around the insulator when the bird droppings fall. Using the air gap average breakdown electric field strength as the criterion to judge whether breakdown occurs or not, the risk area of the air gap breakdown is determined, thereby determining the protection range of the transmission line for the flashover caused by the falling of bird droppings.

1. Introduction

With the continuous extension of the power grid and the improvement of the ecological environment in recent years, the activities of birds near the overhead transmission lines are increasing, and the number of transmission line faults caused by bird activities has increased significantly. The causes of bird-related line faults include bird droppings flashover, bird nesting fault, bird predation fault, and bird flight fault. Among them, bird droppings flashover is the main cause of bird-related line faults [1-3]. The bird droppings flashover is divided into the flashover caused by bird droppings contaminating the insulator and the flashover caused by the falling of bird droppings. Among them, the process of flashover caused by the falling of droppings is the formation and elongation of the bird droppings channel, the severe distortion of electric field around the insulator, and the breakdown of air gap [4]. Therefore, this paper used ANSYS Maxwell software to establish the model of bird droppings near the transmission line to analyse the distortion of the spatial electric field when the bird droppings fall, to determine the risk area of the air gap breakdown, and to determine the protection range against bird droppings. It provides reference for the installation of the bird-related fault prevention devices on transmission lines, which is conducive to the safe and stable operation of the power system.

Content from this work may be used under the terms of the Creative Commons Attribution 3.0 licence. Any further distribution of this work must maintain attribution to the author(s) and the title of the work, journal citation and DOI.
Published under licence by IOP Publishing Ltd

2. The criterion of air gap breakdown

The air gap between the bird droppings and the insulator fittings can be approximately considered as a rod-rod gap or a rod-plate gap, and the relationship between the gap power-frequency breakdown voltage and the gap distance is as shown in Figure 1.

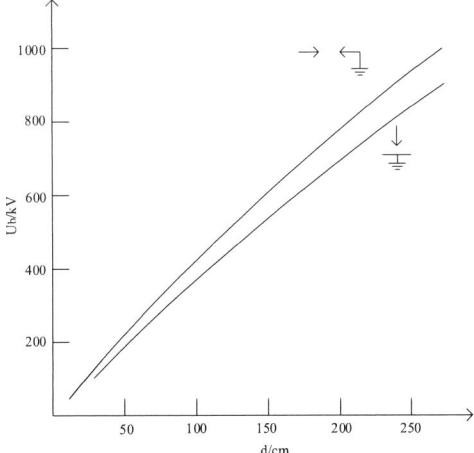

Figure 1. Relationship between power-frequency breakdown voltage and distance of rod-rod and rod-plate air gap.

It can be seen from the figure that when the gap length is less than 1 m, the power frequency breakdown voltage of the two gaps is proportional to the gap distance, and when the distance is the same, the power frequency breakdown voltage is approximately equal. Therefore, according to the above figure, it can be determined that the average breakdown electric field strength near the insulator in the presence of bird droppings is 4 kV/cm, and it can be determined whether the air gap breaks down by comparing with the simulated electric field strength of air gap.

3. Principle of finite element method

The main methods for measuring the electric field distribution of insulators are on-site measurement, analytical method and finite element method. The on-site measurement method is costly and vulnerable to external factors; the analytical method is complex and time-consuming, and it is difficult to analyse geometric models with complex structures using this method; the finite element method can adapt to the complexities of field boundary geometry and media physical property variation. It has great advantages in solving complex insulator electric fields, and many simulation calculation software based on finite element method has appeared recently, which improves the calculation efficiency and saves a lot of time.

The finite element method is a numerical calculation method based on the variation principle. First, the differential equation mathematical model is transformed into the corresponding variation problem, that is, the functional extremum problem. Second, the disparity difference is used to discretize the variation problem to the extremum problem of ordinary multivariate functions, which ultimately comes down to a set of multivariate algebraic equations. Last, the equations are solved to get the numerical solution of the original problem.

4. Simulation process

In this paper, ANSYS Maxwell finite element simulation software was used to calculate the electric field distribution near the insulator when bird droppings fall. The simulation steps of the software are creating and importing models, setting material properties and boundary conditions, meshing, solving and post-processing.

4.1. Model building and property setting

Bird-related line faults occur mostly in transmission lines with voltage levels of 110kV and above. Therefore, this paper selected 110kV overhead transmission lines for modeling and simulation. Bird droppings fall near the middle phase insulator.

The overall model includes towers, insulators, wires, connecting fittings, bird droppings, etc.

The tower is the 67-Z3-type linear tower. The tower size and simplified model are as shown in Figure 2.

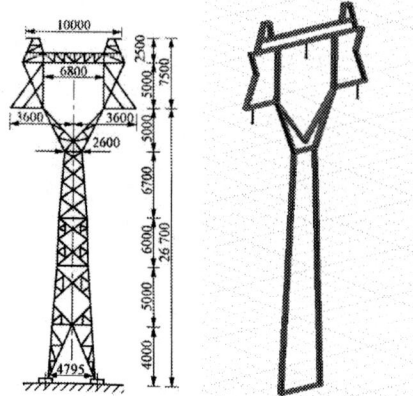

Figure 2. Tower size and model.

The insulator is the FXBW4-110/160 composite insulator, and there are grading rings at both ends. The dielectric constant of the shed is 3.5. The size and model of the insulator are as shown in Figure 3.

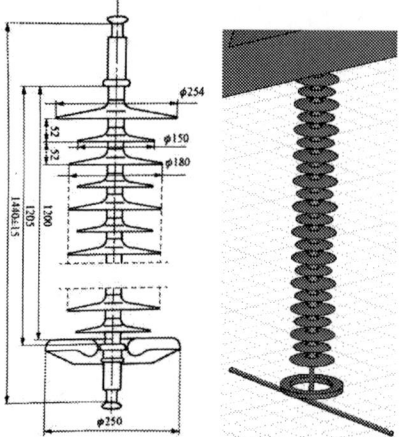

Figure 3. Insulator size and model.

The model of the connecting fittings is simplified and replaced with cuboids. The wires are single-split with a diameter of 21.66 mm. Since bird droppings have a certain viscosity and electrical conductivity, the bird droppings are considered as a slender cylindrical conductor for the convenience of analysis. The radius is 3 mm and the ends are rounded.

The low-voltage side fittings of insulators and the tower are loaded with zero potential. For middle phase, the high-voltage side fittings of insulator are loaded with a high potential to 69.86 kV. For the other two phases, the high-voltage side fittings of insulators are loaded with potential to -34.98 kV. The bird droppings are regarded as conductors and the potential degree of freedom is coupled.

4.2. Meshing

The meshing results of air, tower and insulator are as shown in Figure 4.

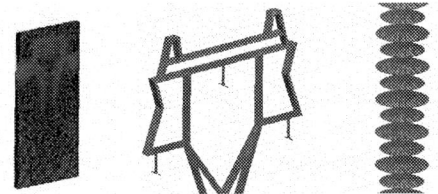

Figure 4. Meshing results.

After meshing, the model can be solved, and the spatial electric field and potential distribution can be obtained through post-processing.

5. Protection range

The specific process of flashover caused by bird droppings can be divided into two steps [5]. First, the bird droppings form a floating potential when they fall, in a certain region, the electric field strength between the lower side of bird droppings and the high-voltage side fittings of insulator exceeds the breakdown field strength that the air gap can withstand, causing the lower side of air gap to break down. Second, after the air gap breakdown, the bird droppings turn into high potential. If the electric field strength between the upper side of bird droppings and the low-voltage side fittings of insulator exceeds the breakdown field strength that the air gap can withstand, the upper side of air gap will also break down, and the insulator will have a strong breakdown and flashover.

Therefore, when the bird droppings fall, there is a risk of breakdown in the air gap both in upper and lower side. Through simulation and calculation, the position and length of bird droppings can be determined when two sides of air gap critically break down, and the risk area of breakdown near the insulator when the bird droppings fall can be determined. Thereby determining the protection range of the transmission line for the flashover caused by the falling of bird droppings.

5.1. Critical breakdown position

If the electric field strength between the lower side of bird droppings and the high-voltage side fittings of insulator exceeds the breakdown field strength, the air gap will break down, and the potential of bird droppings will be raised to 69.86 kV. At this time, if the electric field strength between the upper side of bird droppings and the low-voltage side fittings of insulator exceeds the breakdown field strength, both sides of air gap will break down. According to the breakdown field strength of 4 kV/cm, it can be calculated that the critical breakdown distance of the upper air gap is 175 mm after the bird droppings are raised to a high potential.

Keep the air gap distance between the upper side of the bird droppings and the low-voltage side fittings of the insulator at 175 mm, and the falling position is 125 mm away from the centre of the insulator. Change the length of the bird droppings so that the lower side of the bird droppings is closer to the high-voltage fittings of insulator. The software simulation results showed that when the distance between the lower side of bird droppings and the high-voltage side fittings of insulator was 125 mm, the potential of bird droppings was 20.53 kV and the electric field strength of the lower side of air gap was 3.95 kV, which means the lower side of air gap critically break down. The simulation result of potential distribution near the insulator is as shown in Figure 5. It can be seen from the figure that the bird droppings increased the potential difference of the air gap, making breakdown possible.

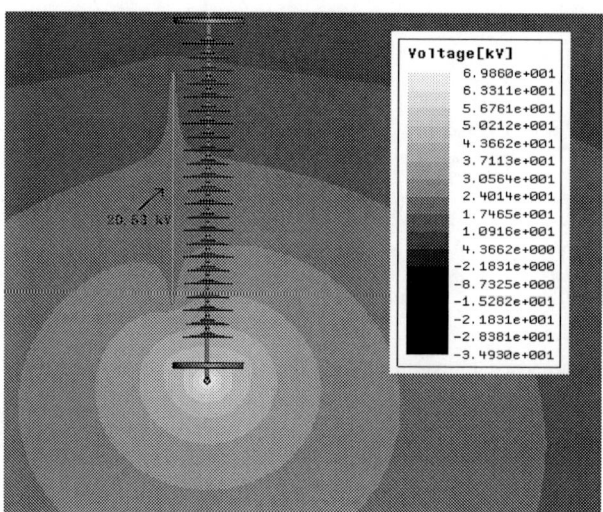

Figure 5. Potential distribution near the insulator.

By changing the horizontal distance from the falling position of the bird droppings to the centre of the insulator, the relationship between the falling position and the length of bird droppings when the air gap critically break down is simulated as shown in Figure 6. Since the minimum shed radius of the insulator is 90 mm, the shortest horizontal distance is 90 mm. As can be seen from the figure, the shortest length of bird droppings when the air gap critically break down is 880 mm.

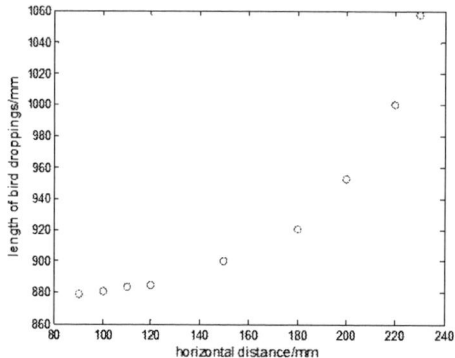

Figure 6. The relationship between the falling position and the length of bird droppings.

5.2. Protection range

It is known from the above that the critical breakdown distance between the upper side of bird droppings and the high-voltage side fittings of insulator is 175 mm, and the critical breakdown distance between the lower side of bird droppings and the low-voltage side fittings of insulator is 125 mm. When the distance between the two sides of bird droppings and two sides of fittings is less than the critical breakdown distance, the breakdown may occur. According to the critical breakdown distance, the front view of the three-dimensional risk area of the air gap breakdown near the insulator when the bird droppings fall can be drawn as shown in Figure 7. Area 1 and area 2 are fan-shaped with radii of 175 mm and 125 mm. When the upper side of bird droppings is in the area 1 and the lower side is in the area 2, there is a risk of breakdown in the air near the insulator. Therefore, for insulator flashover caused by bird droppings, when installing protective devices such as bird guards, it should be ensured that the protection range is larger than or equal to the range of risk areas shown in Figure 7.

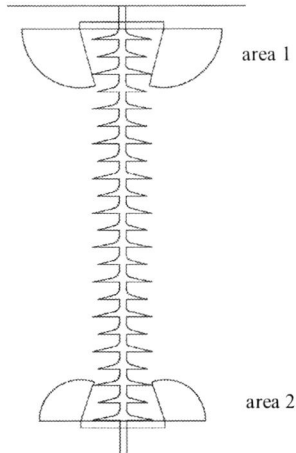

area 1

area 2

Figure 7. Risk area of breakdown.

6. Conclusion

In this paper, the model of bird droppings near the 110kV transmission line was established by using ANSYS Maxwell, and the spatial electric field and potential distribution were obtained by simulation. According to the simulation results, the critical breakdown distance of air gap between insulator and bird droppings was determined, thereby the protection range was determined, which has certain reference significance for the installation of the bird fault prevention and control devices on transmission lines.

References

[1] Q. Zhou, Q. Guo, The Analysis of Bird Harm and Discussion on Its Counter Measure, HEP, **32**, 11-13 (2008)

[2] H. Liu, Y. Liao, Anti-bird Damages Measures Research on High Voltage Transmission Lines of Shandong Power Grid, SEP, **41**, 17-20 (2014)

[3] S. Wang, Z. Ye, Analysis of Bird Damage Accidents on Overhead Transmission Lines and Prevention Techniques, HVA, **47**, 61-67 (2011)

[4] X. Liang, S. Wang, Z. Chen, L. Wang, Composite Insulator Flashover Caused by Bird Dropping and Unknown Reasons, PST, **25**, 13-16 (2001)

[5] D. Huang, J. Yu, Y. Zhang, H. Zhang, J. Xia, Y. Kuang, J. Ruan, Z. Chen, Study on Discharge Mechanism of Suspension Insulator String of 220kV AC Transmission Line with External Object Approaching, POTC, **34**, 4161-4170 (2014)

A Novel Aggregation Method for Doubly Fed Wind Farm

Lulu Li[1], Xueshan Han [1,a], Wenbo Li[2] and Zhe Jiang[2]

[1]Key Laboratory of Power System Intelligent Dispatch and Control of Ministry of Education (Shandong University), Jinan, China

[2]State Grid Shandong Electric Power Research Institute, Jinan 250003, Shandong Province, China

[a]Corresponding author: xshan@sdu.edu.cn

Abstract. With the increasing penetration of wind power, many problems like frequency and voltage stability appear frequently, so the power grid gradually requires that the wind farm can be dispatched and controlled. In this paper, a new aggregation method is proposed to calculate the aggregated active and reactive power regulation range of a doubly-fed wind farm, which provides a dispatching reference for the power system control center. In this method, voltage security and different operation states of wind turbines and internal power losses of the wind farm are considered. Finally, the actual aggregated P-Q regulation range of the wind farm that can ensure operation safety is obtained. A wind farm with 25 doubly fed induction generator (DFIG) wind turbines is used to verify the proposed aggregation method.

1. Introduction

With the increasing installed capacity of wind power generation, the connection between wind farm and the power grid becomes more and more compact. The external characteristics of the wind farm will have an important impact on the safety and control strategy of the power grid. The uncertainty of wind power causes a series of problems, such as voltage fluctuation, frequency stability, power flow inversion. Therefore, wind farms should have certain self-regulation and support capabilities, and cannot rely on the support of the power grid. Therefore, wind farms should have the ability to adjust their active and reactive power, and to respond the dispatching of power system control center.

At present, the common wind farm aggregation modelling methods can be divided into full aggregation and semi aggregated techniques. In [1] the semi aggregated method is adopted, which consists of all wind turbines and one equivalent generator. While in [2] the full aggregated method is adopted, which consists of one equivalent wind turbine and one equivalent generator, and an equivalent wind speed of all wind turbines is calculated. A rotor side converters aggregated method is proposed in [3], which focuses on the problem that operating states of wind turbines in wind farm is totally different. All the above studies aggregate the wind farm to one equivalent generator, so their aggregated method is too ideal in some cases. For the power grid, the power system control center is concerned about the external characteristics of the wind farm, rather than the specific characteristics of all wind turbines. If we get the aggregated active and reactive power regulation range of the wind farm, we will meet the dispatching requirements of power system control center.

DFIG has a strong ability to adjust reactive power [4,5], and many studies pay attention to this reactive power adjustment ability [6,7]. To calculate the aggregated P-Q regulation range, many previous studies ignore the topology and voltage constraints of the wind farm, and algebraically add

Content from this work may be used under the terms of the Creative Commons Attribution 3.0 licence. Any further distribution of this work must maintain attribution to the author(s) and the title of the work, journal citation and DOI.
Published under licence by IOP Publishing Ltd

the reactive power regulation range of each wind turbine, as shown in [8]. This kind of method is unreasonable for large wind farms.

Aiming at above problems, this paper proposes a novel method to calculate the aggregated P-Q regulation range of a doubly fed wind farm, which provides a reference for power system control center. Section 2 contains the overall description of the active and reactive control range of DFIG. Section 3 sets up the mathematical model that calculates the aggregated P-Q regulation range. In section 4, the model is simplified, and the specific solution flow to calculate the aggregated P-Q regulation range of a wind farm is given. Case studies are presented and discussed in section 5.

2. P-Q regulation range of DFIG

The P-Q regulation range of DFIG can be deduced from the steady-state model of DFIG. The reactive power regulation capability of DFIG is composed of two parts, stator reactive power and grid side converter reactive power. The reactive power regulation capability of stator side is limited by the maximum current of stator and rotor, while the reactive power regulation capability of grid side converter is limited by its capacity. The reactive power regulation range of DFIG (i.e., P-Q regulation range), can be presented as $Q_{max,Wi}$ and $Q_{min,Wi}$, as shown in (1) and (2).

$$Q_{max,Wi} = -1.5U_s^2/X_s +$$
$$\sqrt{\left(1.5X_mU_sI_{r\max}/X_s\right)^2 - \left(P_{Wi}/\omega_{r,Wi}\right)^2} + \quad (1)$$
$$\sqrt{S_{c\max}^2 - \left[\left(1-\omega_{r,Wi}\right)P_{Wi}/\omega_{r,Wi}\right]^2}$$

$$Q_{min,Wi} = -\sqrt{\left(1.5U_sI_{s\max}\right)^2 - \left(P_{Wi}/\omega_{r,Wi}\right)^2} -$$
$$\sqrt{S_{c\max}^2 - \left[\left(1-\omega_{r,Wi}\right)P_{Wi}/\omega_{r,Wi}\right]^2} \quad (2)$$

where U_s denotes stator single-phase voltage, $I_{s\max}$ and $I_{r\max}$ are stator and rotor maximum single-phase current. $\omega_{r,Wi}$ is the rotor speed of DFIG Wi. $S_{c\max}$ is maximum capacity of grid side converter. X_s and X_m are stator reactance and excitation reactance of DFIG.

In addition to equations (1) and (2), the P-Q regulation range of DFIG should also be limited by the maximum active power $P_{W\max}$.

Given $\omega_{r,Wi}$ and U_s, the P-Q regulation range of DFIG is shown in the gray area in Figure 1. $-1.5U_s^2/X_s$ represents inductive reactive power absorbed by DFIG. Stator and rotor current limitation, reactive power regulation capability of grid side converter, and the limitation of $P_{W\max}$, constitute the P-Q regulation range of DFIG.

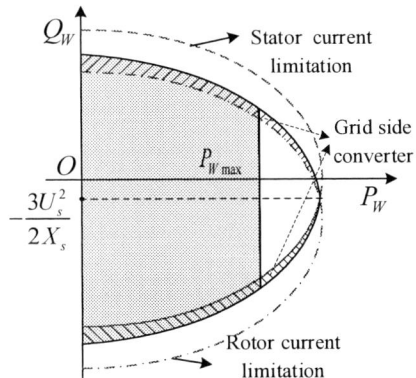

Figure 1. The P-Q regulation range of DFIG

Equations (1) and (2) show that the P-Q regulation range of DFIG is not only depended on P and Q, but also influenced by $\omega_{r,Wi}$ and U_s. The P-Q regulation range is approximate to an ellipse, so $\omega_{r,Wi}$ will impact semimajor axis of this ellipse. When $\omega_{r,Wi}$ is increased or decreased, the ellipse will be elongated or shortened. U_s will impact the dot and radius of this ellipse. Therefore, when calculating the aggregated P-Q regulation range of the wind farm, it is necessary to take into account the state of the rotor speed and terminal voltage of each DFIG.

3. Mathematical modelling

To obtain the aggregated P-Q regulation characteristic of the wind farm, it is necessary to consider the topology, voltage limitation and power flow loss of the wind farm. If all wind turbines of the wind farm output their maximum reactive power, voltages of some wind turbines may exceed the normal level, and may lead to generators outage. In this section, the mathematical model to calculate the aggregated P-Q regulation range is established. The objective is to maximize and minimize the reactive power output of the wind farm. The objective function is expressed as follows:

$$\max \quad F_1 = Q_{PCC} \qquad (3)$$

$$\min \quad F_2 = Q_{PCC} \qquad (4)$$

where Q_{PCC} is the reactive power output at point of common coupling (PCC) of the wind farm.

Constraints are as follows :

(1) Power flow equations :

$$U_j^2 = U_i^2 - 2\left(r_{ij}P_{ij} + x_{ij}Q_{ij}\right) + \left(r_{ij}^2 + x_{ij}^2\right)I_{ij}^2 \qquad (5)$$

$$P_{ij} = \sum_{k:j \to k} P_{jk} + r_{ij}I_{ij}^2 - P_{Wj}, \quad j \in N \qquad (6)$$

$$Q_{ij} = \sum_{k:j \to k} Q_{jk} + x_{ij}I_{ij}^2 - Q_{Wj}, \quad j \in N \qquad (7)$$

$$I_{ij}^2 = \frac{P_{ij}^2 + Q_{ij}^2}{U_i^2} \qquad (8)$$

where N is the set of wind farm nodes. P_{ij} and Q_{ij} are the line active and reactive power from bus i. P_{Wj} is the active power of DFIG at bus j. Q_{Wj} is the reactive power of DFIG at bus j. U_j is the voltage of bus j. r_{ij} and x_{ij} are the resistance and reactance of line ij.

(2) Node voltage magnitude limits :

$$\underline{U}_i \leq U_i \leq \bar{U}_i, \quad i \in N \qquad (9)$$

where $\underline{U}_i$ and $\bar{U}_i$ are the minimum and maximum voltage limits of bus i.

(3) The P-Q regulation range of each DFIG :

$$0 \leq P_{Wi} \leq P_{W\max,i}, \quad i \in N_W \qquad (10)$$

$$Q_{\min,Wi} \leq Q_{Wi} \leq Q_{\max,Wi}, \quad i \in N_W \qquad (11)$$

where $P_{W\max,i}$ is the maximum wind power of DFIG at bus i, and it can be get though wind power prediction. N_W is the set of DFIG wind turbines.

(4) Total active power output limit of the wind farm :

$$P_{PCC} = P_{set} \qquad (12)$$

where, $P_{set} = 0, \cdots, P_{PCC,\max}$ is the total active power output limit which is set point-by-point in advance. Every P_{set} corresponds with a maximum and a minimum reactive power output.

This model fully considers the voltage safety of the wind farm internal nodes, as well as the different operating states of wind turbines. Through ultra-short-term wind power prediction, the maximum active power of each wind turbine at one point in the future can be obtained, and then the aggregated P-Q regulation range characteristic of the wind farm can be calculated. It should be noted that the total active power output of the wind farm is given point by point, and this optimization model just calculate one point of the aggregated P-Q regulation range curve. Therefore, the aggregated P-Q regulation range of double-fed wind farm at a certain moment can be obtained by calculating multiple points.

4. Model simplification and solution

4.1. Model simplification

The mathematical model in section 3 is nonlinear and difficult to solve, and in this section, simplifications are proposed to make the model solved conveniently and quickly. The simplifications include the linearization of power flow equations and the P-Q regulation range of each DFIG. After these simplifications, a linearized model is obtained.

Noting the power flow constraint represented by equations (5) - (8), the nonlinear term is I_{ij}^2, so if we linearize I_{ij}^2, the linearized power flow equations will be obtained. The Taylor expansion of the power flow equations is carried out around the operating point $\left(P_{ij0},Q_{ij0},U_{i0}\right)$ with the second-order and higher-order terms ignored

$$I_{ij}^2 \approx \frac{1}{U_{i0}^2}\left(2P_{ij0}P_{ij} + 2Q_{ij0}Q_{ij} - \frac{P_{ij0}^2 + Q_{ij0}^2}{U_{i0}^2}U_i^2 \right) \quad (13)$$

It is noted that the power flow constraint only contains the square term of voltage, so we use V_i to replace U_i^2. And then, the power flow equations are linearized.

When the rotor speed and terminal voltage are given, the P-Q regulation range of DFIG is approximate to an ellipse, which can be piecewise linearized. To consider the impact of the terminal voltage and rotor speed, measurement values of the last time are used to replace the rotor speed and terminal voltage in equation (11), and then the piecewise linearization of the P-Q regulation range is carried out. By selecting appropriate points, it can be divided into 5 line segments, as shown in Figure 2. Results show that the linearization carried out by 5 segments has high precision.

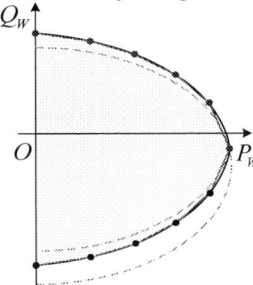

Figure 2. The piecewise linearization of the P-Q regulation range of DFIG

4.2. Solving process

When solving the optimization model, P_{set} is set several times, which leads to cycle calculation of the model. By setting suitable number of P_{set} and the simplification in section 4.1, the speed of calculation can be increased. The solving process is summarized as Figure 3.

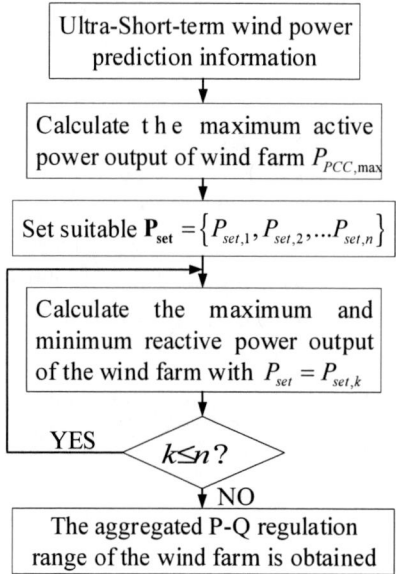

Figure 3. The solving process of the aggregated P-Q regulation range

5. Case study

In this section, a wind farm, including 25*1.5 MW doubly fed wind turbine generators, was used for the case study. Its topology structure is shown in Figure 4. Each DFIG is connected to 35kV lines through 690V/35kV transformer (not shown in Figure 4). Parameters of DFIG used in this paper are given in Table 1. Voltage limits are set to be within the range 0.95-1.05 p.u. In this case study, the voltage of high voltage network is set to be 1.0 p.u.

In this section, the historical wind power data is used, and each DFIG of the wind farm have different wind speed. The maximum wind power of each DFIG is given in Figure 5. At this moment, the average wind speed of the wind farm is high, and the maximum active power of each DFIG is different. To contrast the calculation result, method A is used to calculate the aggregated P-Q regulation range of the wind farm, which is algebraic addition of all wind turbines. The P-Q regulation range result of method A is shown in Figure 6. Using method of this paper, the aggregated P-Q regulation range of the doubly fed wind farm at this point is calculated, and is also shown in Figure 6.

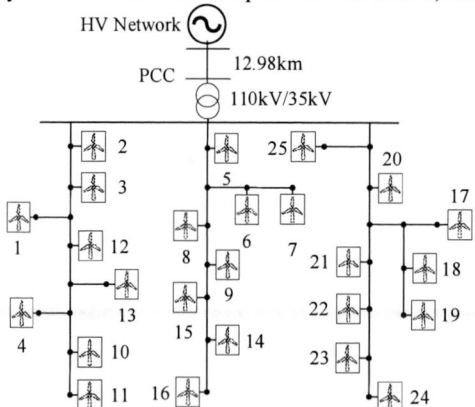

Figure 4. Topology of the wind farm

Table 1. Parameters of DFIG

Parameter	Value	Parameter	Value
Rated voltage	690 V	Excitation reactance	1.720 Ω
Rated capacity	1.5 MW	I_{smax}	1565.8 A
Stator reactance	1.773 Ω	I_{rmax}	1649.5 A

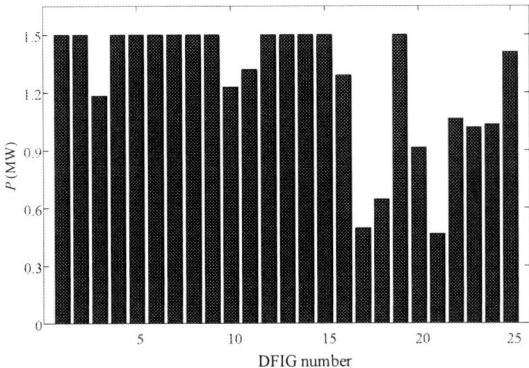

Figure 5. The maximum wind power of each DFIG

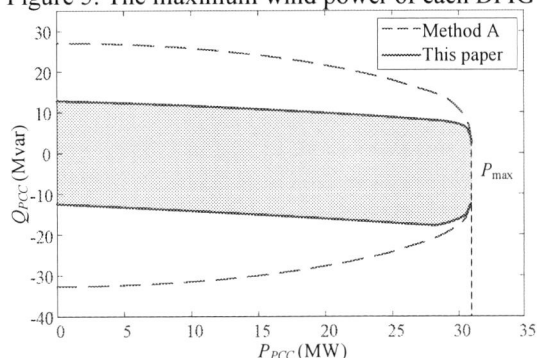

Figure 6. The aggregated P-Q regulation range

From Figure 6, it can be observed that the aggregated P-Q regulation range of the wind farm calculated by this paper's method, is obviously smaller than the range calculated by method A. Furthermore, the variation characteristic of the ability to absorb reactive power is different. The aggregated P-Q regulation range obtained by method A is similar to a single DFIG's P-Q regulation range, and this method doesn't consider voltage limits of wind farm internal nodes. The method of this paper fully considers the voltage level of every DFIG and nodes inside the wind farm. Figure 6 shows that using this paper's method, when P_{PCC} is small, the lagging reactive power output ability of the wind farm is weaken. The reason is that when all wind turbines output the lagging reactive power, the voltage will be raised, and to ensure voltage security, the lagging reactive power output should be limited. But as P_{PCC} increasing, the P-Q regulation range of each DFIG will be the main limit to Q_{PCC}. Using this paper's method, the capacitive reactive power output ability of the wind farm increases first and then decreases with the increasing of P_{PCC}. If wind turbines output the capacitive reactive power, the voltage will be reduced. As P_{PCC} increasing, the voltage is rised, and the capacitive reactive power output ability is heightened. But when P_{PCC} is closed to rated capacity, capacitive Q_{PCC} will decrease because of the P-Q regulation range limit of each DFIG.

Results show that the actual aggregated P-Q regulation range of the wind farm is obviously smaller when considering the voltage security and topology inside the wind farm. This aggregated P-Q characteristic has more practical reference significance for power system control center.

6. Conclusion

In this paper, a novel method to calculate the aggregated P-Q regulation range of doubly fed wind farm is proposed. It is important that voltage security inside the wind farm and different operation states of wind turbines should be considered. If all wind turbines output their maximum reactive power, low voltage or overvoltage will appear inside the wind farm, which is very unsafe to generators. This method will provide a credible dispatching range of the wind farm for power system control center.

Acknowledgment

This work is supported by the Science and Technology Foundation of SGCC(SGSDDK00KJJS1600061).

References

[1] L. M. Fernández, J. F, J. R. Saenz. Aggregated dynamic model for wind farms with doubly fed induction generator wind turbines. Renewable Energy, **33(1)**, 129 (2008).

[2] A. J. Xia, Z. X. Lu, Y. Min, et al. An aggregated model of wind farm composed of doubly fed induction generators. Power System Technology, **39(7)**, 1879 (2015).

[3] Y. Q. Jin, P. Ju, X. P. Pan. Analysis on controller aggregation method for equivalent modelling of DFIG-based wind farm. Automation of Electric Power Systems, **38(3)**, 19 (2014).

[4] Q. Liu, Z. M. Wang. Reactive power generation mechanism & characteristic of doubly fed variable speed constant frequency wind power generator. Proceedings of the CSEE, **31(3)**, 82 (2011).

[5] H. Zhang, Y. C. Zhang, D. W. Yang. Two-vectors-based model predictive direct power control of doubly fed induction generator for grid connection and power regulation. Transactions of China Electrotechnical Society, **31(5)**, 69 (2016).

[6] A. H, A. M. Coordinated reactive power management in power networks with wind turbines and FACTS devices. Energy Conversion & Management, **52(7)**, 2575 (2011).

[7] H. Y. Liu, P. Guan. Optimal control of doubly fed induction generator systems. Electrical Engineering, **17(3)**, 13 (2016).

[8] J. Cao, R. L. Zhang, G. Q. Lin, et al. A voltage control strategy for wind farms using doubly fed induction generator wind turbines. Automation of Electric power Systems, **33(4)**, 87 (2009).

Research on substation perimeter isolation based on phased array radar and multi-video fusion technology

Yuancheng Zhu[1], Zhipeng Lei [1], Wei Zheng[1], Hongfeng Ma[1], Rongzhen Xia[2] and Deyu Song[2,a]

[1] State Grid Yingkou Electric Power Supply Co., Ltd., 115000,Yingkou, China

[2] State Grid Shenyang Electric Power Supply Co., Ltd.,110004, Shenyang, China

[a]Corresponding author: songdeyu_cool@126.com

Abstract: This paper presents a perimeter isolation technique based on phased array radar and multiple video fusion perimeter isolation technique, aiming at the disadvantages of high false alarm rate and easy to be affected by the environment of traditional perimeter security system of substation. It uses the advantages of phased array radar, such as long detection distance, high anti-jamming ability, fast detection speed and high accuracy, and combines the multi-camera collaborative recognition technology with the multi-video data fusion technology as the core to realize the effective fusion of mobile target recognition information.

1. Instructions

Along with unceasing progress of automation level of the substation and unceasing application of unattended substation, the security and the information request of the substations also unceased enhance. Perimeter isolation is the first line to prevent intruders from climbing and crossing and to ensure security. It is the most important link to realize closed-loop management of security prevention and it's very important and indispensable to prevent invasion.

The existing perimeter target detection technology has experienced the first generation of target detection technology that is based on visual information and the second generation of target detection technology that is characterized by non-video signal analysis and detection[1]. The first generation of anti-intrusion target detection technology is mainly based on video monitoring technology. As a visual information perception and acquisition method of a scene, video monitoring technology requires manual full-time monitoring, so that monitoring personnel are easy to get tired and cannot give an alarm in advance or in real time. More often than not, it acts as an afterthought. The second generation of target detection technology includes high voltage electric pulse fence, active infrared correlation technology, tension fence, microwave sensor, static electric field sensor, leak cable[2]. They all belong to the target detection technology that based on the analysis of characteristics of signal. While they solved the problem of artificial monitoring and real-time alarm, there are still some disadvantages, such as high false alarm rate, high missing report rate and significantly affected by environments.

This paper presents a perimeter isolation technique based on phased array radar and multiple video fusion perimeter isolation technique. The phased array radar is regarded as the core sensor of the perimeter system, and the electronic system based on visual information is regarded as the detection sensor of the perimeter system. Multi-sensor fusion is divided into three levels: data fusion between detection sensor and core sensor, feature level fusion based on detection sensor and core sensor, and

Content from this work may be used under the terms of the Creative Commons Attribution 3.0 licence. Any further distribution of this work must maintain attribution to the author(s) and the title of the work, journal citation and DOI.

Published under licence by IOP Publishing Ltd

decision level fusion between detection sensor and core sensor. In this way, the distance, angle, speed and other parameters of the target can be checked comprehensively and quickly, and the linkage can be carried out with video.

2. Research on phased array radar technology

2.1 Fundamentals of phased array

Phased array antenna[3] is composed of many radiation elements, the feed phase of each element can be flexibly controlled and change the wavefront. The array elements of phased array antenna are generally 100-10000, each element is followed by a controllable phase shifter. Changing the phase shift of each phase shifter will change the relative feed phase between the array elements and the wave front direction of antenna radiated electromagnetic waves. The schematic diagram of the array antenna is shown in Figure 2.1 below:

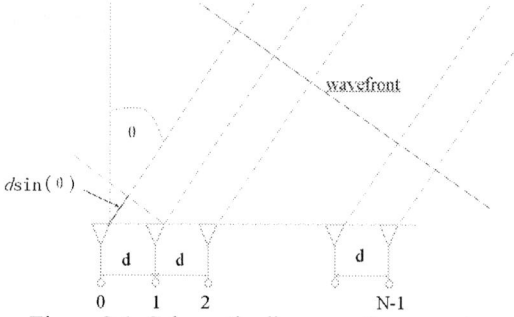

Figure 2.1. Schematic diagram of array antenna

The spacing between the antenna array elements is d, and the angle between the target azimuth and the normal vector of the antenna array surface is θ. The echo phase difference of adjacent array elements is ψ, the wave path-difference is $d\sin\theta$, and the phase difference caused by the wave path-difference is:

$$\psi = \frac{2\pi}{\lambda} d \sin \theta \qquad (2.1)$$

Consider the distance, suppose n-antenna elements are evenly spaced, constant amplitude supply, the sum of the radiation field vectors at some point in the direction θ is:

$$E(\theta) = E \sum_{k=0}^{N-1} e^{jk\psi} \qquad (2.2)$$

If the supply phase difference of each array element is 0, the above formula can be used to study the direction diagram of array antenna. Use the geometric series summation formula, euler formula and formula 2.2,we get normalized antenna pattern:

$$F_a(\theta) = \frac{\sin\left[\frac{\pi N d}{\lambda} \sin \theta\right]}{N \sin\left[\frac{\pi d}{\lambda} \sin \theta\right]} \qquad (2.3)$$

$F_a(\theta)$ calls array factor or array factor. If the antenna elements are not uniformly radiating into space at all angles, the direction diagram is $F_e(\theta)$, the array direction diagram becomes:

$$F(\theta) = F_a(\theta) F_e(\theta) \qquad (2.4)$$

Where $F_e(\theta)$ is called the matrix factor.

2.2 Phased array antenna scanning

The relationship between supply phase difference and equivalent wave path difference is shown in figure 2.2 below:

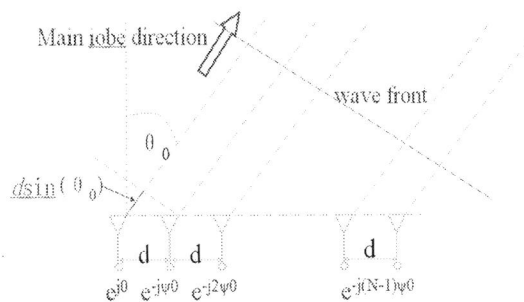

Figure 2.2. Schematic diagram of array antenna scanning

When the supplying phase of the array antenna in the figure decreases according to ψ_0, the beam direction is θ_0. By changing ψ_0, phased array scanning can be realized. The antenna direction is shown as follows:

$$F_a(\theta) = \frac{\sin\left[\frac{\pi N d}{\lambda}\left(\sin\theta - \sin\theta_0\right)\right]}{N\sin\left[\frac{\pi d}{\lambda}\left(\sin\theta - \sin\theta_0\right)\right]} \qquad (2.5)$$

The condition that no grating flap is satisfied is:

$$\frac{d}{\lambda} < \frac{1}{1 + |\sin\theta_0|} \qquad (2.6)$$

3. Research on multi-video fusion technology

Intelligent video surveillance system[4] not only needs to accurately extract moving objects from videos, but also needs to know some characteristics of objects, such as position, size, shape, histogram, speed and direction of motion, so as to analyze the behavior of moving objects and judge whether abnormal or suspicious behavior occurs. The data level fusion algorithm in this paper mainly fuses the position, color histogram[5] and speed of moving objects in multiple cameras. The overall structure is shown in Figure 3.1 below:

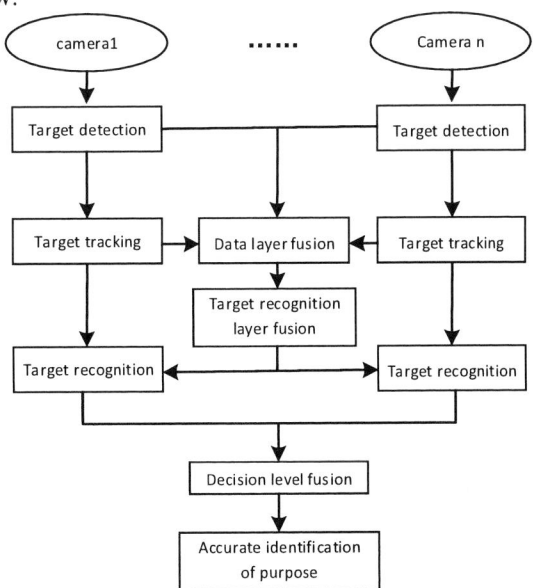

Figure 3.1. Multi-video collaborative recognition system structure

Each single camera starts from acquiring scene information, and first carries out target detection. In order to locate the target continuously, it is necessary to track the target accurately. This process needs to utilize the feature information of the target. The trajectory and duration of moving target can be accurately acquired by target tracking. Finally, the moving target is identified and analyzed. The process is divided into three levels: the moving target detection in low-level, the target tracking in middle-level and the target recognition and analysis in high-level. At the same time, multiple single cameras do not work in isolation, but continue to fuse local information to achieve the goal of collaborative analysis.

Suppose there are N cameras in the monitoring area, and p represents a specific one, $1 \le p \le N$. Suppose there are several targets in camera P, and the target set is denoted as $O = \{O_1, O_2, \cdots, O_M\}$, i represent a specific target, $1 \le i \le M$. Each new target is extracted and saved by the camera tracking it for the first time. In this paper, target position, size and speed are selected as the main features. It is assumed that target I of P in the camera is marked as q_i^p、 v_i^p、 h_i^p. The target is first captured by a single camera and marked (position, shape, speed, color histogram, size). Suppose the unique feature vector $T_i^p\left(h_i^p, s_i^p, w_i^p\right)$ of target I in camera P is obtained through data fusion of data layer, where s_i^p is the shape parameter and w_i^p is the target size parameter. Meanwhile, the unique feature vector $T_d^c\left(h_d^c, s_d^c, w_d^c\right)$ of target d in camera c can also be obtained. Pattern recognition clustering analysis[6] is adopted to achieve the handover of goals. By calculating the Bhattacharyya distance to measure their similarity, the formula is shown in 3.1:

$$sim(h_i^p, h_d^c) = \sqrt{1 - \sum_j \frac{\sqrt{h_i^p(j) \cdot h_d^c(j)}}{\sum_j h_i^p(j) \cdot \sum_j h_d^c(j)}} \qquad (3.1)$$

If the target matches exactly, the value of this parameter is 0, or it is 1.

For the direction parameters, formula 3.2 is used to measure the similarity between them:

$$sim(s_i^p, s_d^c) = \sqrt{\left[1 - \left(s_i^p\right)^2\right] \cdot \left[1 - s_d^c\right]^2} + s_i^p \cdot s_d^c \qquad (3.2)$$

If the targets are the same, the value of the similarity function is 1.

For the size parameters, formula 3.3 is used to measure the similarity between them:

$$sim(w_i^p, w_d^c) = 1 - \left(1 - w_d^c\middle/w_i^p\right)^2 \qquad (3.3)$$

Combined with the above formula, the comprehensive similarity degree function of the two moving targets is defined as:

$$sim(T_i^p, T_d^c) = \left(1 - sim(h_i^p, h_d^c)\right) \cdot sim(s_i^p, s_d^c) \cdot \qquad (3.4)$$
$$sim(w_i^p, w_d^c)$$

Set the threshold T(0<T<1),if $T < sim(T_i^p, T_d^c) < 1$, it is considered that target i in camera P and target d in camera C are the same target, the target handover is completed.

On the basis of successful target handover, parameters of the same target in multiple cameras need to be fused to obtain a more accurate description of the target.

For one frame image, take the average position of the same moving target in multiple cameras as the accurate position of the moving target, then the accurate position $q_i(x_i, y_i)$ of the *ith* target is:

$$q_i = \frac{1}{N} \sum_{p=1}^{N} q_i^p \qquad (3.5)$$

Where:

$$x_i = \frac{1}{N} \sum_{p=1}^{N} x_i^p \qquad (3.6)$$

$$y_i = \frac{1}{N} \sum_{p=1}^{N} y_i^p \qquad (3.7)$$

In multiple monocular cameras, gaussian distribution modeling and motion template algorithm[7] are used to obtain the motion contour. The direction parameter s_i^p of the target is obtained from the aspect ratio of the contour of the moving target, so the direction parameter of the fusion *ith* target is the gaussian average of the target parameters in multiple cameras. That is:

$$s_i = \frac{1}{N}\sum_{p=1}^{N} s_i^p \qquad (3.8)$$

The moving speed of a moving target can be measured by the ratio of the distance that the target moves against the time it takes to pass the distance. The distance the target moves is the length of the final position of the target relative to the initial position, so the velocity of eye I in camera p can be expressed as $v_i^p = s_i^p / t_i^p$, where s_i^p and t_i^p respectively represent the motion distance of the *ith* target in the p-th camera and the time taken to pass through this distance.

For the *ith* target in the p camera, each m frame is used as a computing unit. Assuming that the first frame in which the moving target appears in the camera is frame h, in the first frame m, the moving distance of the moving target is $s_{i\,h,h+m}^p$ and the moving time is $t_{i\,h,h+m}^p$, then the moving speed of the target in this period is:

$$v_{i\,h,h+m}^p = s_{i\,h,h+m}^p / t_{i\,h,h+m}^p \qquad (3.9)$$

By analogy, the remaining velocity can be obtained:

$$v_{i\,h+m,h+2m}^p = s_{i\,h+m,h+2m}^p / t_{i\,h+m,h+2m}^p$$

$$v_{i\,h+2m,h+3m}^p = s_{i\,h+2m,h+3m}^p / t_{i\,h+2m,h+3m}^p$$

$$\cdots\cdots\cdots\cdots$$

$$v_{i\,h+(j-1)m,h+jm}^p = s_{i\,h+(j-1)m,h+jm}^p / t_{i\,h+(j-1)m,h+jm}^p$$

$$v_{i\,h+jm,h+(j+1)m}^p = s_{i\,h+jm,h+(j+1)m}^p / t_{i\,h+jm,h+(j+1)m}^p$$

$$\cdots\cdots\cdots\cdots$$

It can be concluded that the modified speed of the *ith* target in the p-th camera is:

$$v_i^p = \begin{bmatrix} v_{i\,h,h+m}^p + v_{i\,h+m,h+2m}^p + \ldots \\ + v_{i\,h+(j-1)m,h+jm}^p + v_{i\,h+jm,h+(j+1)m}^p \end{bmatrix} / (j+1) \quad (3.10)$$

Where $(j+1)m \le n$, n is the total frame number of video collected by the camera.

Using the method to obtain the moving velocity of the *ith* target in the other n-1 cameras, so the accurate speed of fusion of the *ith* target is:

$$v_i = \frac{1}{N}\sum_{p=1}^{N} v_i^p \qquad (3.11)$$

Due to the different angles of multiple cameras and the different light intensity of different angles, the color histogram obtained locally varies to a certain extent. In this paper, the color histogram of a single camera is modified by integrating the color histogram obtained by different cameras. And the modified histogram can also be used as an auxiliary means of target tracking handover. Firstly, the color histogram equalization of the same moving target in multiple cameras is processed. The gray scale transformation function used in equalization selects the gray scale cumulative distribution function of image[8], which has the function of uniform gray scale stretching. Secondly, each equalized histogram is matched and fused to obtain the final accurate color histogram of the moving target.

4. Application in perimeter isolation of substation

The phased-array radar and multi-video fusion security warning system of substation mainly include the front-end phased-array radar and camera, computer network and wiring, back-end server and alarm management system. When the abnormality is determined, the acoustooptic alarm device in the management alarm area is triggered to feed back to the background monitoring general platform in the first time. The flow chart is shown in figure 4.1 below:

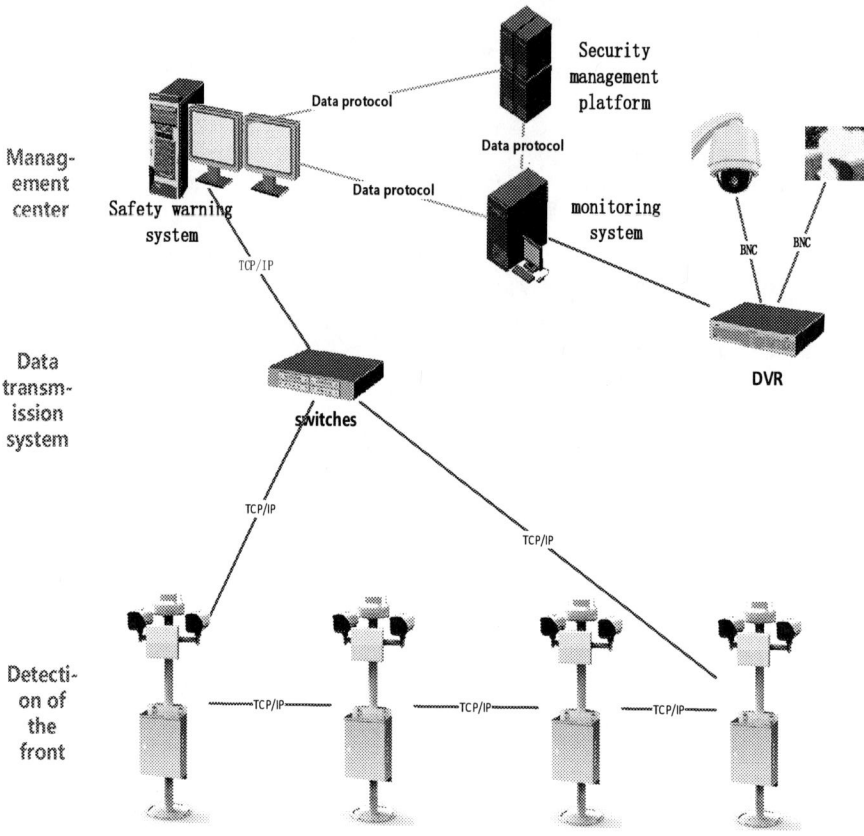

Figure 4.1. Intelligent security system

After receiving the alarm, a variety of alarm information will be displayed on the plane, prompting the operation and maintenance personnel through voice, icon, etc., and the abnormal situation will be synchronously transmitted to the control center, to provide unified command and treatment of the on-duty personnel.

5. Conclusion

This paper presents a phased array radar and visual fusion technology to detect targets and achieve alarm. It can accurately identify all kinds of objects in the alarm area. It also has the great economic and environmental significance in solving the safety problem of substation.

Acknowledgement

Fund project: Science and technology project of State Grid: Research on perimeter defense technology of phased array radar(2018YF-39)

References

[1] A.L.Goetsch,G.D. Detweiler, R. Puchala, Conditions to test electric fence additions to cattle barb wire fence for goat containment, Journal of Applied Animal Research,J.1,(2012)

[2] H. Bowen, Application comparison and analysis of perimeter alarm system of joint station, Chemical management,J.4,14(2017)

[3] Z. Guangyi, Z. Yujie, Phased array radar technology,M.15(2007)

[4] L. Pengyi, Intelligent video analysis technology and application, Network security technology and application,J.**12**,127(2018)

[5] F.S. Mohamad, A.A. Manaf, S. Chuprat, Histogram matching for color detection: A preliminary study, ITSim2010,C.(2010)

[6] T. Dongming, Cluster analysis and its application,D.16-17(2010)

[7] W.Qilong, Research on image classification method based on gaussian distribution modeling,D.18-20(2018)

[8] X.Hongkui, H.Xiao, Image grayscale transformation combined with false edge extraction and histogram analysis, Optics and Precision Engineering,J.**4**,539(2017)

Fault causes identification for transmission lines based on HHT and PNN

Wen Du[1], Renfei Che[1,a], Hao Zhang[1], Xiaocheng Meng[1,2] and Juntao He[1,3]

[1]Key Laboratory of Power System Intelligent Dispatch and Control of the Ministry of Education, Shandong University, Jinan; 250061, China

[2]State Grid Jiangsu Electric Power Company Technician Training Center, Suzhou; 215004, China

[3]Zhengzhou Power Supply Company, Zhengzhou; 450006, China

[a]Corresponding author: cherenfei@sdu.edu.cn

Abstract. In this paper, Hilbert-Huang transform(HHT) is introduced into the application of fault cause identification for transmission lines. This method is composed of empirical mode decomposition (EMD) and Hilbert transform. It can analyze the characteristics of signal frequency quantitatively and accurately[1]. By analyzing the information of four common faults from fault recorder, the Hilbert marginal spectrum and the Hilbert time-frequency spectrum can be considered as the feature of different fault causes which can be classified by a probability neural network (PNN). The results show that the method can effectively identify the cause of the fault and has a high accuracy.

1. Introduction

Transmission line faults is one of the frequent grid accidents, which is harmful to the operation of the power system and the reliability of the power supply. Affected by complicated geographical and climatic environment, transmission lines are vulnerable to various forms of damage such as lightning strikes, birds and beasts, pollution, and external forces [2,3]. Therefore, it's necessary to study the characteristics of various faults causes and distinguish them, which can provide support for professionals to find and eliminate faults timely.

At present, fault recorders have been widely used in power grids of different voltage levels, which can effectively record fault information. In this paper, the HHT method is used to analyze the recorded information and the PNN is used as a classifier to construct a fault cause identification system, which achieves the expected goal.

2. The Hilbert-Huang transform

The Hilbert-Huang transform is a new method for analyzing nonlinear, non-stationary signals and has been widely used in engineering. HHT is capable of accurate time-frequency and spectral analysis with high resolution. The Hilbert-Huang transform consists of two steps. First, empirical mode decomposition(EMD) is performed to obtain a finite number of intrinsic mode functions(IMFs). Then, the Hilbert transform is applied to these intrinsic mode functions to obtain the Hilbert time-frequency spectrum and Hilbert marginal spectrum.

Content from this work may be used under the terms of the Creative Commons Attribution 3.0 licence. Any further distribution of this work must maintain attribution to the author(s) and the title of the work, journal citation and DOI.
Published under licence by IOP Publishing Ltd

2.1 The empirical mode decomposition

The intrinsic mode function calculated by empirical mode decomposition must meet the following two conditions:

①The number of extreme points and zero-crossings must be equal or differ at most by 1. ②At any instant, the mean value of the envelope defined by the local maxima and the local maxima is zero[1].

Considering a given signal S(t), the symmetric envelopes can be obtained respectively by the local maxima and minima of the signal data series. Then by calculating the margin $M_1(t)$ of the upper and lower envelopes a new signal $H_1(t)$ is obtained by

$$H_1(t)=S(t)-M_1(t) \qquad (1)$$

If $H_1(t)$ does not satisfy the above two conditions, then $H_1(t)$ is regarded as a new signal, and the margin $M_{11}(t)$ is again calculated to obtain the $H_{11}(t)$

$$H_{11}(t)=H_1(t)-M_{11}(t) \qquad (2)$$

This cycle will stop when a $H_{1K}(t)$ can satisfy the IMF conditions. Then the first intrinsic mode function is

$$I_1(t)=H_{1K}(t) \qquad (3)$$

The residual signal $S_1(t)$ is obtained by

$$S_1(t)=S(t)-I_1(t) \qquad (4)$$

By repeating the above process, a series of intrinsic mode functions can be obtained, and the original signal can be described as

$$S(t) = \sum_{j=1}^{N} I_j(t) + S_N(t) \qquad (5)$$

N is the number of intrinsic mode functions, $I_j(t)$ is the jth intrinsic mode component, and $S_N(t)$ is the residue.

2.2 The Hilbert transform

The Hilbert transform is performed on the intrinsic mode functions obtained by EMD, so that the instantaneous frequency of the original signal has physical meaning, and then the Hilbert time-frequency spectrum and the Hilbert marginal spectrum can be obtained. For any intrinsic mode function $I_j(t)$, its Hilbert transform is as follows

$$v_j(t) = \frac{1}{\pi} \int_{-\infty}^{+\infty} \frac{I_j(\tau)}{t-\tau} d\tau \qquad (6)$$

Take $I_j(t)$ as the real part and $v_j(t)$ as the imaginary part to construct a signal

$$z_j(t) = I_j(t) + iv_j(t) \qquad (7)$$

Instantaneous amplitude $A_j(t)$ and instantaneous frequency $f_j(t)$ are obtained by

$$\theta_j(t) = \tan^{-1} \frac{v_j(t)}{I_j(t)} \qquad (8)$$

$$f_j(t) = \frac{1}{2\pi} \frac{d\theta_j(t)}{dt} \qquad (9)$$

3. Fault current analysis and feature extraction by HHT

In this paper, the zero-sequence current is taken as the research object, and the faults are all single-phase-to-earth ones. The fault causes include lightning, wildfire, crane collision and tree contact.

3.1 Hilbert marginal spectral feature

The Hilbert marginal spectrum $H(f)$ is obtained by integrating the Hilbert time-frequency spectrum $H(f, t)$ over the time domain. It reflects the statistical amplitude distribution of the data in the frequency domain, compared with the Fourier spectrum [4,5]. Typical marginal spectrums of the above four faults are shown in Figure 1.

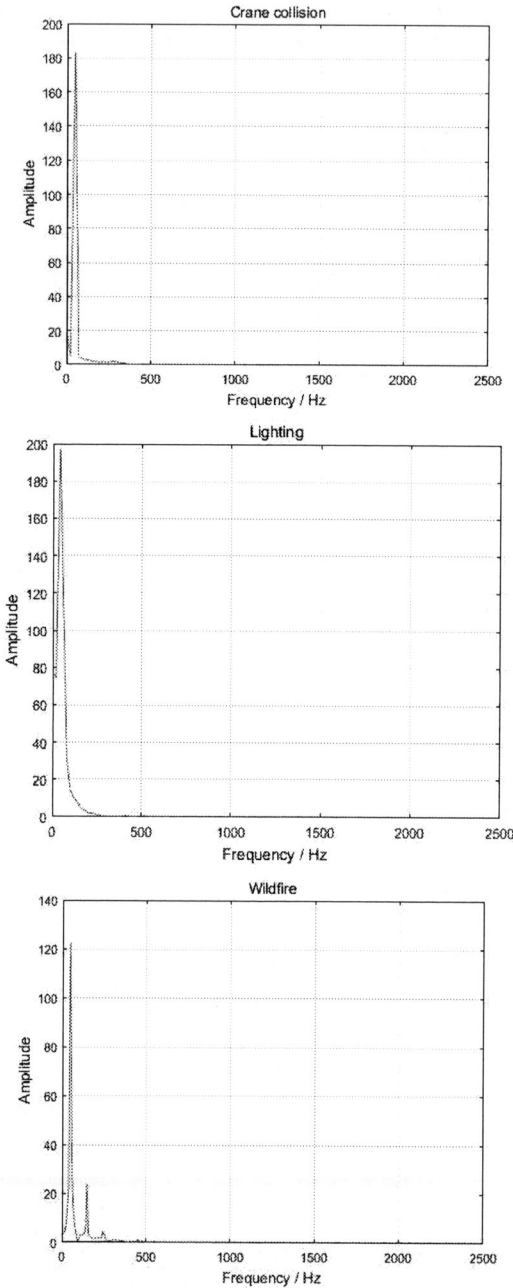

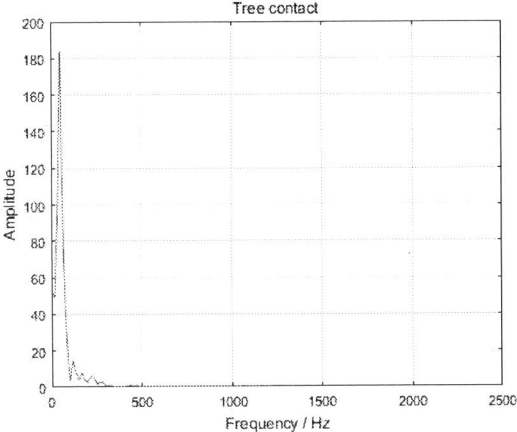

Figure 1. Hilbert marginal spectrums of four faults.

The fault current frequency of crane collision is mainly concentrated at 50Hz, which means it is a metallic ground fault. The lightning fault is also approximate a metallic ground fault, but its DC content is higher, almost no high frequency harmonics. The fault resistance of wildfire and tree-caused fault are nonlinear. Therefore, their fault currents contain high frequency harmonics. Besides, the DC content is less.

3.2 Hilbert time-frequency spectral feature

The Hilbert time-frequency spectrum clearly indicates the relationship between signal frequency and time. The abscissa reflects the distribution of the signal in the time domain and the ordinate reflects its distribution in the frequency domain[7,8]. Typical time-frequency spectrums of four kinds of faults are shown in figure 2.

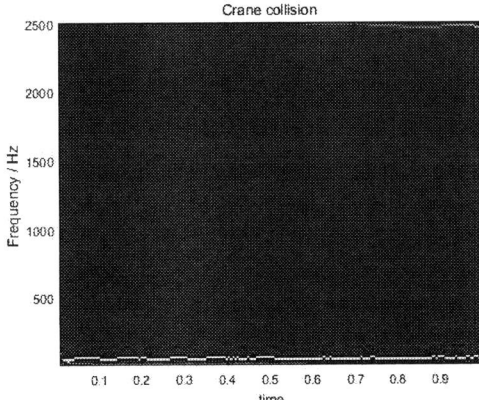

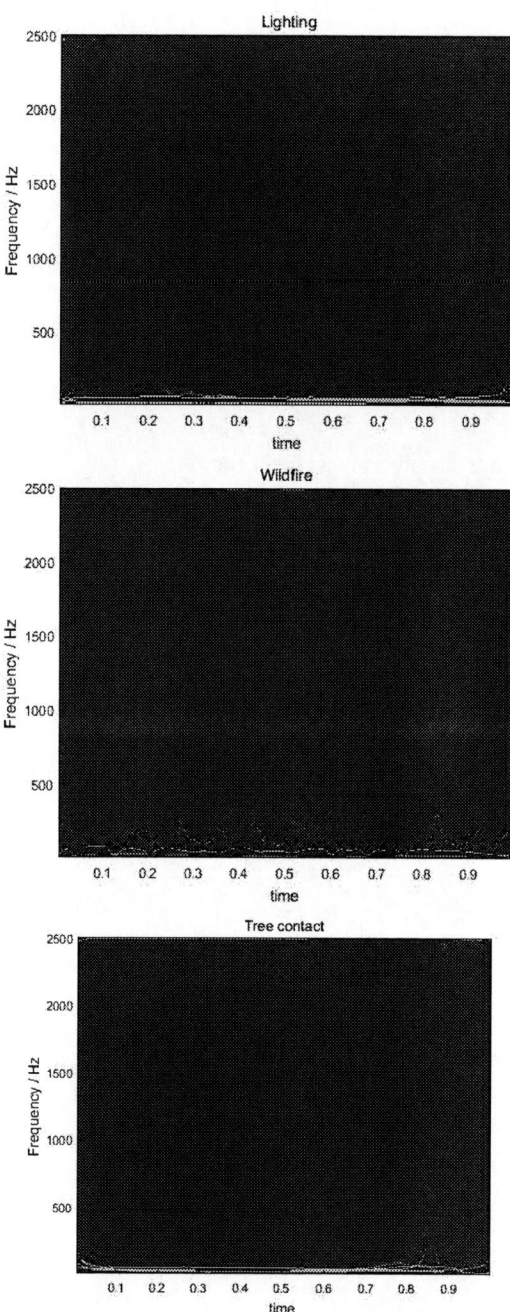

Figure 2. Hilbert time-frequency spectrum of four faults.

The time-frequency spectrums of crane collision and lighting fault are pretty similar. The signal energy during the fault is mainly concentrated at 50Hz. The wildfire fault has high frequency components during the whole fault, which may be related to the continuous combustion of the flame. The high-frequency components of the tree-caused fault are mainly distributed at the beginning and the end of the spectrum.

3.3 Spectral feature quantization

Take the ratio of DC component to the fundamental component as DC content, and the third harmonic to the fundamental component as harmonic content. The more dispersed the time-frequency spectrum, the larger the information entropy is. Therefore, we can use information entropy of image to describe the feature of time-frequency spectrums. Quantitative representation of features is shown in Table 1.

Table 1 Quantitative representation of features

Fault cause	DC content	Harmonic content	Entropy
Crane collision	<10%	<5%	1.75-1.85
Lighting	20%-40%	<10%	1.8-1.9
Wildfire	<10%	10%-25%	> 2
Tree contact	5%-15%	5%-20%	1.8-1.9

4. PNN and classification results

4.1 PNN network structure

PNN is a branch of radial basis network. In essence, it is a supervised neural network classifier based on Bayesian minimum risk criterion. It has the advantages of fast learning process, classification and good fault tolerance.

PNN is generally composed of a four-layer structure: an input layer, a pattern layer, a summation layer, and an output layer. The input layer is responsible for passing the input feature vector to the network. The pattern layer calculates the degree of matching between the input feature vector and each mode in the training sample. Then the Euclidean distance is sent to the Gaussian function to obtain the pattern layer output. The summation layer is responsible for connecting the pattern layer units of each class. The output layer outputs the category with the highest score in the summation layer [13-15]. In this paper, three kinds of features are used to form feature vectors, and there are four fault categories. The corresponding PNN network structure is shown in Figure 3.

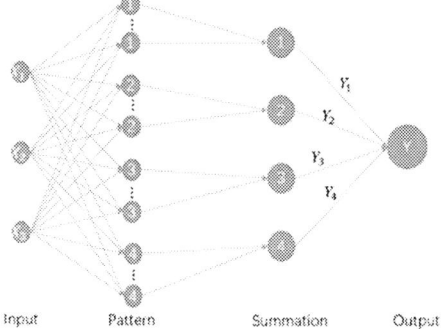

Figure 3. PNN network structure.

4.2 Algorithm process of PNN

①Normalize the training samples. Suppose there are m training samples, the training sample matrix is

$$X=\begin{bmatrix} x_{11} & x_{12} & \cdots & x_{1n} \\ x_{21} & x_{22} & \cdots & x_{2n} \\ \vdots & \vdots & \cdots & \vdots \\ x_{m1} & x_{m2} & \cdots & x_{mn} \end{bmatrix}=\begin{bmatrix} \mathbf{x}_1 \\ \mathbf{x}_2 \\ \vdots \\ \mathbf{x}_m \end{bmatrix} \quad (10)$$

The normalized coefficient matrix is

$$B^{T}=\left[1\bigg/ \sqrt{\sum_{k=1}^{n}x_{1k}^{2}} \quad 1\bigg/ \sqrt{\sum_{k=1}^{n}x_{2k}^{2}} \quad \cdots \quad 1\bigg/ \sqrt{\sum_{k=1}^{n}x_{mk}^{2}} \right] \quad (11)$$

The normalized training sample is

$$C_{m \times n} = B_{m \times 1}[1 \; 1 \cdots 1]_{1 \times n} \cdot X_{m \times n} \qquad (12)$$

② The Euclidean distance between the training sample matrix and the corresponding elements of the test sample matrix is

E=

$$\begin{bmatrix}
\sqrt{\sum_{k=1}^{n}|d_{1k}-c_{1k}|^2} & \sqrt{\sum_{k=1}^{n}|d_{1k}-c_{2k}|^2} & \cdots & \sqrt{\sum_{k=1}^{n}|d_{1k}-c_{mk}|^2} \\
\sqrt{\sum_{k=1}^{n}|d_{2k}-c_{1k}|^2} & \sqrt{\sum_{k=1}^{n}|d_{2k}-c_{2k}|^2} & \cdots & \sqrt{\sum_{k=1}^{n}|d_{2k}-c_{mk}|^2} \\
\vdots & \vdots & \cdots & \vdots \\
\sqrt{\sum_{k=1}^{n}|d_{pk}-c_{1k}|^2} & \sqrt{\sum_{k=1}^{n}|d_{pk}-c_{2k}|^2} & \cdots & \sqrt{\sum_{k=1}^{n}|d_{pk}-c_{mk}|^2}
\end{bmatrix} \qquad (13)$$

c_{ij} is the normalized training sample matrix element, and d_{ij} is the normalized element of the sample to be tested.

③Activate the Gaussian function neuron to get the initial probability matrix (σ=0.1) as

$$P = \begin{bmatrix}
e^{-\frac{E_{11}}{2\sigma^2}} & e^{-\frac{E_{12}}{2\sigma^2}} & \cdots & e^{-\frac{E_{1m}}{2\sigma^2}} \\
e^{-\frac{E_{21}}{2\sigma^2}} & e^{-\frac{E_{22}}{2\sigma^2}} & \cdots & e^{-\frac{E_{2m}}{2\sigma^2}} \\
\vdots & \vdots & \cdots & \vdots \\
e^{-\frac{E_{p1}}{2\sigma^2}} & e^{-\frac{E_{p2}}{2\sigma^2}} & \cdots & e^{-\frac{E_{pm}}{2\sigma^2}}
\end{bmatrix} \qquad (14)$$

④Suppose there are m samples, a total of c classes, the number of samples is the same as k, and the initial probability that each sample of the summation layer belongs to each class is

$$S = \begin{bmatrix}
\sum_{l=1}^{k}P_{1l} & \sum_{l=k+1}^{2k}P_{1l} & \cdots & \sum_{l=m-k+1}^{m}P_{1l} \\
\sum_{l=1}^{k}P_{2l} & \sum_{l=k+1}^{2k}P_{2l} & \cdots & \sum_{l=m-k+1}^{m}P_{2l} \\
\vdots & \vdots & \cdots & \vdots \\
\sum_{l=1}^{k}P_{pl} & \sum_{l=k+1}^{2k}P_{pl} & \cdots & \sum_{l=m-k+1}^{m}P_{pl}
\end{bmatrix} \qquad (15)$$

⑤Calculate the probability of fault type

$$\text{prob}_{ij} = S_{ij} \bigg/ \sum_{l=1}^{c} S_{il} \qquad (16)$$

4.3 Identification results of fault cause

Trained by 76 samples and tested by 43 samples, the classification results by PNN are shown in Table 2. The recognition rates of the four faults are all above 80%, and the overall recognition rate is 88.4%.

Table 2 Identification results of fault cause

Fault cause	Samples	Identified	Rate%
Crane collision	11	9	81.8
Lighting	14	12	85.7

Wildfire	9	9	100
Tree contact	9	8	88.9
Sum	43	38	88.4

5. Conclusion

This paper describes the principle of HHT and analyzes the zero-sequence current, which will result in the Hilbert marginal spectrum and Hilbert time-frequency spectrum used as features. On this basis, PNN for classification and identification has achieved the expected goals. In the follow-up work, it is necessary to dig deeper into other fault features, optimize the representation of features, and identify faults types not covered in this paper effectively.

References

[1] J. Wang, Q. Yang, L. Chen, W. Sima, High Voltage Engineering, **38,** 2068 (2012)
[2] Y. Hu, K. Liu, T. Wu, Y. Liu, Z. Su, High Voltage Engineering, **40**, 3491 (2014)
[3] T.Wu, Y. Hu, J. Ruan, High Voltage Engineering, **37**, 1115 (2011)
[4] X. Peng, X. Li, S. Yao, G. Qian, Southern Power System Technology, **6**, 43 (2012)
[5] H. Li, Y. Zhang, J. Yu, Hydropower Automation And Dam Monitoring, **38**, 67 (2014)
[6] L. Lin, H. Yan, X. Yue, F. Hu, Southern Power System Technology, **8**, 47 (2008)
[7] L. Yang, Y. Zhu, Power System Protection and Control, **43**, 62 (2015)
[8] W. Zhang, C. Wang, Automation of Electric Power Systems, **31**, 34 (2007)

Transmission line insulator fault detection based on ultrasonic technology

Zheng Yao[1], Xin Yu[1,a], Jianchun Yao[2], Wei Sui[1] and Xiaochen Yu[1]

[1]State Grid Liaoning Power Co., Ltd. Dandong Power Supply Company, China

[2]Beijing Boneng Electric Co., Ltd., China

[a]Corresponding author:Xin Yu@zhnganxiejingli@163.com.org

Abstract. With the continuous development of modern society, the construction of transmission network has been rapidly advanced, and the reliability of power supply has been fundamentally improved, but the workload of detecting insulators has also increased. In order to further reduce the tripping rate of the transmission line, the insulator fault detection work cannot be delayed. However, the traditional detection method has low efficiency, poor safety and high cost, which cannot meet the requirements. This paper proposes a method of applying ultrasonic technology to line insulator detection by referring to the application of ultrasonic technology in the United States and South Korea. The method realizes fault detection of the insulator by collecting, processing and amplifying the ultrasonic signal.

1. Introduction

At present, there are many methods for measuring the insulation of the transmission line, including two types of contact measurement and non-contact measurement. Contact measurement mainly uses a contact instrument to measure physical quantities such as voltage, current, and insulation resistance. The most widely used non-contact measurement is temperature measurement, such as infrared sub-imager. Due to cracking of the insulator, damage to the tree, damage to the insulation, and other faults, the method of judging the fault point by temperature measurement is hardly found [1].

Insulator detection is mainly used by artificial climbing towers, which is dangerous and time consuming. Through the understanding of insulator faults, finding the most effective detection method for insulator faults will be of great significance to the safe operation of the entire grid. If there is an instrument, it can patrol the high-voltage transmission line insulators and power supply equipment for long distances, predict potential faults, and detect the operation status of the insulators through long-distance and continuous power, avoiding one-on-one inspection, blindly overhauling and wiping the insulators. It can save a lot of manpower and material resources, and can also improve the health of equipment.

The ultrasonic detecting method uses an ultrasonic sensor to detect a signal generated when a partial discharge of a deteriorated insulator occurs, and the sound wave and the ultrasonic wave emitted from the insulator are used to determine the deterioration degree. The instrument can transmit the laser beam at a long distance to lock the target to be measured. Due to the unique transmission characteristics of the ultrasonic wave, it is not easily affected by the environment and the weather. It is the most effective method for detecting various faults of the insulator at a long distance. Because the method can be carried out at a long distance and charged, it is safe, reliable and efficient, and the application of this method for detecting insulators can greatly save the operation and maintenance cost of the line [2].

2. Current status of ultrasonic technology application at home and abroad

Ultrasonic testing is a comprehensive technology that includes many disciplines based on physics, electronics, mechanics and materials science. It is widely used in many fields such as industry, medical equipment and marine detection. Ultrasonic testing is a type of non-destructive testing, that is, the use of ultrasonic waves to detect the current state of a workpiece or device. From the end of the 19th century to the beginning of the 20th century, after the piezoelectric effect and the anti-piezoelectric effect were discovered in physics, the method of generating ultrasonic waves by using electronic technology was solved. Ultrasound has been widely used in flaw detection, medical treatment, ultrasonic inspection, and measurement. In terms of distance, in recent years, it has also been applied in power detection.

The United States and South Korea first applied ultrasonic testing technology to the detection of defects in power equipment, and achieved significant gains in the ultrasonic field. It was determined that ultrasonic equipment would generate ultrasonic waves when discharge, flashover, and breakdown occurred [3]. The frequency range of the ultrasonic wave is 35 kHz - 40 kHz. Due to the different equipment, the frequency of the transmission is slightly different, but it is basically within this range. Therefore, after research and experiment, the corresponding ultrasonic sensor has been developed to solve the problem of signal acquisition. After several years of continuous improvement and development, it has achieved good results in the United States and South Korea, especially in Korea, which has been applied extensively and played a major role [4].

3. Common methods and disadvantages of transmission line insulator detection

3.1 Direct observation

The most common method for physical structural defects outside the insulator is the direct observation method, which uses direct observation of the insulator with optical instruments such as binoculars or by means of a telescope to find various common surface defects such as insulating sheaths, umbrella groups, fittings, etc. There is no damage to the parts. However, the direct observation method is not reliable enough and is not accurate enough. Sometimes it is necessary to check the tower and it is impossible to know the internal fault of the insulator [5].

3.2 Distributed voltage measurement

The main feature of the deteriorated insulator is that the insulation resistance is lowered, resulting in a low or even zero sharing voltage. Using this feature, the insulator resistance state can be known by comparison with the standard voltage distribution of the normal insulator string. However, this method requires live contact measurement and needs to be tested under good weather conditions.

3.3 AC withstand voltage method

This method judges the resistance state of the insulator by using the deteriorated insulation performance of the insulator and the reduction of the withstand voltage level. This method is the most intuitive and authoritative reference method for testing the effectiveness of other methods. However, this method cannot be measured on site in the field, and the tested insulator needs to be taken off and taken to a special test site.

3.4 Ultraviolet imaging

A small but relatively stable surface partial discharge can result in carbonized channels or electrical erosion of the composite insulator shed and jacket. When the surface of the composite insulator forms a carbonization channel, its service life is greatly reduced, even in the short-term insulation breakdown. The electronic ultraviolet optical detector can be used to electrically detect the carbonization channel formed on the surface of the composite insulator due to partial discharge and the ultraviolet light emitted by the charged particles in the electrical erosion. Partial discharge occurs when a conductive carbonization channel is formed on the surface of the composite insulator. This method is used to detect the partial discharge of the composite insulator, and it is required to be tested at night and in a normal temperature environment, and the effect is better when the detection is performed in an environment with high humidity or even rainfall. However, the test results are easily misjudged by the observation angle, and the detection equipment is also relatively expensive, and cannot be equipped with a large number of transmission line operation and maintenance departments. Therefore, the inspection work cannot be carried out according to the regulations [6].

4. Transmission line insulator fault detection based on ultrasonic technology

4.1 Analysis of application principle of ultrasonic technology

It can be known from the insulator discharge mechanism that no acoustic signal is emitted when the insulator is not discharged. With the development of the discharge program, the discharge is gradually enhanced, and the sound wave emission signal is from nothing, from weak to strong. The sound wave signal can be regarded as a point sound source and propagates to the surroundings in the form of a spherical wave in the air medium [7]. Since the energy of the acoustic signal is part of the energy released by the discharge, there must be a quantitative relationship between the energy of the acoustic signal and the energy of the discharge. A large number of ultrasonic waves will occur when the power line insulators are exploding. Such ultrasonic waves are inaudible to the human ear, the human ear audible frequency is below 20 kHz, and the ultrasonic frequency is between 20 kHz and 200 kHz [8].

4.2 Technical method for detecting insulators of transmission lines by ultrasonic technology

4.2.1 Ultrasonic signal acquisition

The ultrasonic signal can be realized by the wave concentrator, so that more ultrasonic waves are reflected by the interface and concentrated, and received by the ultrasonic sensor, so that more ultrasonic high-frequency signals can be obtained.

Insulator failure Partial discharge phenomena are accompanied by the generation and emission of ultrasonic waves, and the frequency of the ultrasonic signals emitted is concentrated between 20-40 kHz and 80-140 kHz. In technical applications, two different types of ultrasonic sensors can be used to receive ultrasonic signals in two frequency bands.

4.2.2 Ultrasonic signal band selection and filtering

Experiments have shown that insulators produce a very broad continuous spectrum at the time of discharge, with frequencies ranging from a few hertz to hundreds of thousands of hertz. When the sound wave propagates in the medium, as the propagation distance increases, the energy gradually decays, and the attenuation law of sound pressure and sound intensity is:

$$P_x = P_0 e^{-\alpha x} \qquad (1)$$

$$I_x = I_0 e^{-2\alpha x} \qquad (2)$$

Where: P_0、I_0 respectively refers to the sound pressure and sound intensity of the sound source; P_x、I_x respectively refers to the sound pressure and sound intensity from the sound source x; x refers to the distance between the sound wave and the sound source; α is the attenuation coefficient, the unit is Np/m.

The fault insulator test in the laboratory shows that the sound wave in the 25-60 kHz band is the strongest when the insulator has the same intensity current leakage; the same is true when the insulator current leakage changes.

Through many experiments at the engineering site, it was finally confirmed that it is best to select the ultrasonic signal of 40 kHz for analysis, which can be used as the center frequency of ultrasonic signal acquisition.

4.2.3 Amplification of ultrasonic signals

The acquired signal needs to be amplified, but it is found that the amplified signal is easily distorted. Finally, through the multi-stage precision amplification of the front, middle and rear, and AD conversion is placed in the second-stage amplification, the audio output is placed in the three-stage amplification, which not only ensures the signal amplification without distortion, but also ensures the reliability of the output.

The acquired ultrasonic signal is first amplified by the amplifier circuit built in the host computer, and then the signal of the useful frequency segment is selected by the band pass filter circuit to be analyzed and judged by the host circuit, and the ultrasonic signal is filtered before and after as shown in Figure 1.

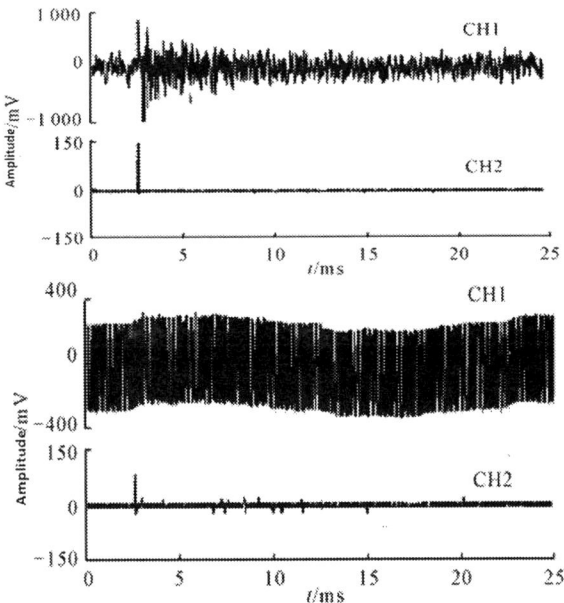

Figure 1. Ultrasonic signal diagram before and after filtering

The host circuit is a signal analysis judgment composed of a single trigger and an overall control circuit. The workflow is shown in Figure 2.

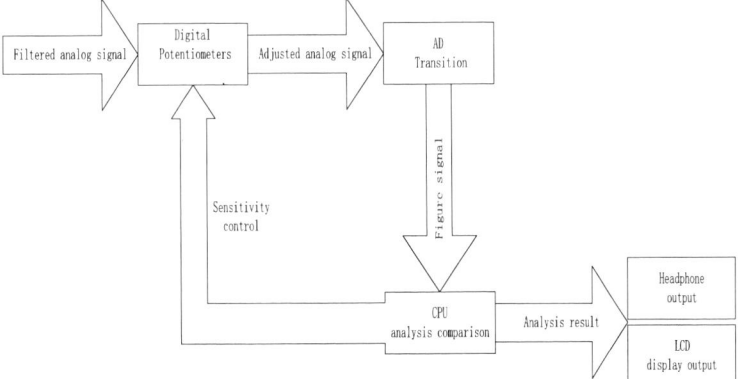

Figure 2. Working flow chart of the host circuit

4.2.4 Host hardware and software design

The ultrasonic online diagnostic system software realizes real-time analysis and intelligent diagnosis of data in the field data collection. After the main device completes the detection work, the front-end data may be imported into the ultrasonic online diagnostic analysis system. The system will read and transfer the data, and automatically and intelligently analyze the data; then the system reads the data into the analysis software. Reprocessing to facilitate analysis of the map.

Through the hardware and analysis software, the ultrasonic signal is finally audibly converted, as shown in Figure 3.

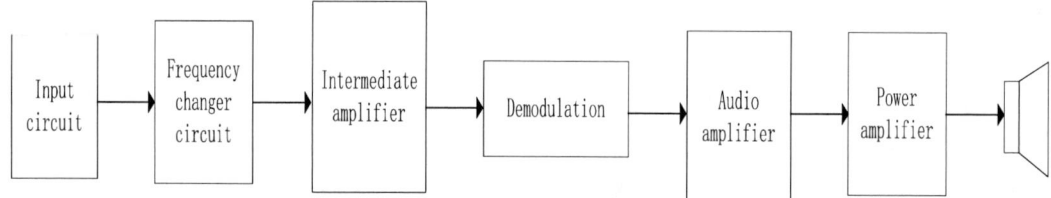

Figure 3. listening conversion process

4.2.5 Ultrasonic probe design

Ultrasonic probes can be divided into piezoelectric type, magnetostrictive type, electromagnetic type according to their working principle, and piezoelectric type is most commonly used. Piezoelectric ultrasonic waves are piezoelectric oscillators that generate a free-running signal after voltage is applied.

Using the design of the dual-probe collector, the acquisition of ultrasonic signals in different frequency bands can enhance the intensity and sensitivity of the detection. The ultrasonic sensor is integrated with the preamplifier. It is necessary to pay attention to the factors of signal interference: one is the interference of the electric field on the piezoelectric ceramic; the other is the influence on the amplifying element. The frequency of the power frequency electric field is low, the effect of the magnetic field can be neglected, and the method of electrostatic shielding is used to prevent interference.

5. Application case

Successfully researched and developed the software for the ultrasonic portable detection device, which has a good effect in the practical application of the Dandong site. The following is a case application of Dandong Power Supply Company:

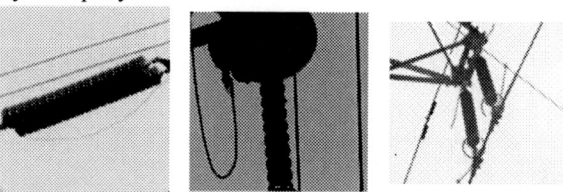

Figure 4. Dandong power supply line insulator real map

6. Conclusion

The ultrasonic detection technology that has emerged in recent years can analyze and test substation equipment and transmission equipment. This technology is enabled in the power grid maintenance process to quickly detect defects existing in the power equipment without power failure, and then determine the defect level and specific location based on the spectrum analysis. New technologies make defect detection more flexible. Inspection efficiency and personnel safety have been greatly improved.

Through on-site and laboratory research and application, we can basically derive three types of typical insulator ultrasonic maps. These maps are important for implementing on-line detection of insulators, quickly locating, and determining the type of fault. Fig. 5 is an ultrasonic waveform emitted by a good insulator, which is characterized by uniform waveform, small amplitude, and almost uniform wave peak. Fig. 6 is an ultrasonic waveform emitted by an insulator at a critical value, characterized in which the waveform is dense, the local amplitude is large, the peak value is uneven, and the local waveform is obviously oscillated. Fig. 7 is an ultrasonic waveform emitted by an insulator having a value of zero, which is characterized by dense waveforms, large amplitudes, and large and uniform peaks.

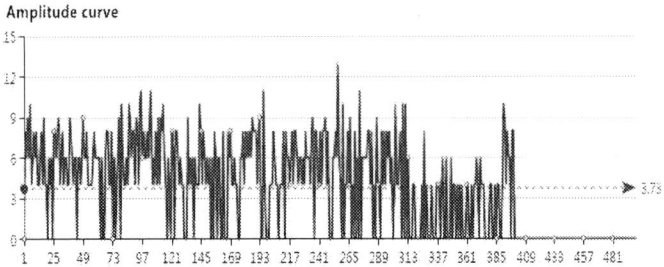

Figure 5. Ultrasonic waveform from a good insulator

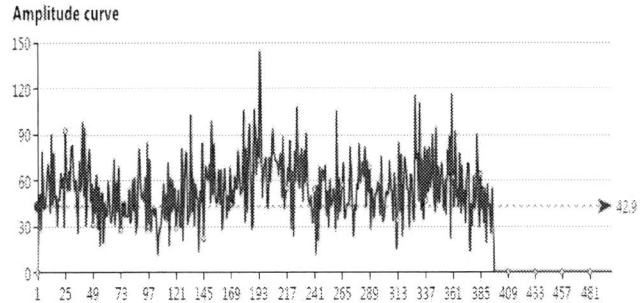

Figure 6. Ultrasonic waveform from the insulator at the critical value

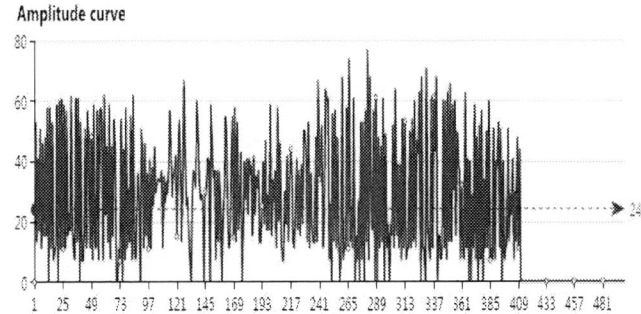

Figure 7. Ultrasonic waveform from an insulator with zero value

References
[1] L.M. Wang,*On-line detection and analysis of leakage signal of faulty high voltage insulator.* Southeast University, (2005)
[2] P. Nie, X.Z. Zhao, Y. Zhang, J.F. Bai ,.*Development of On-line Monitoring System for Insulation of High Voltage Electrical Equipment.* Journal of Northeast Electric Power University,**01**(2000)
[3] Z.D. Jia, W. Zhang,. *Foreign body caused flashover on composite insulator in wet and contaminated condition.* High Voltage Engineering, **8**,36 (2010)

[4] Z.Y. Wang, Q. Wang, *Research on the AC electric-field distribution along contaminated polymer insulators*. High Voltage Apparatus, **4**,6 (2010)

[5] Y.C. Cheng, C.W. Li, X.J. Shen, R.H. Chen, *Comparison of Several Methods for Detection of Composite Insulators*.High Voltage Engineering,**06** (2004)

[6] Ampol Tungkanawanich, Zen-Ichiro Kawasaki, et al. *Ground Fault Discrimination based on Wavelet Transform using Artificial Neural Networks*. T.IEE Japan, **10** (2000)

[7] C.Z. Xie, Z.W. Du. *Analysis of Operational Characteristics of Composite Insulators under Serious Contamination Conditions*.High Voltage Engineering, **07** (2004)

[8] J.F. Li, B.S. Li, S.T. Zhao. *A measuring method fortransmission line sag based on computer vision*.Sustainable Power Generation and Supply Conference Provisional Proceedings. Nanjing, China: UK-China Network of Clean Energy Research, (2009)

ISPECE   IOP Publishing

Insulation Defect Detection of Solid Insulating Material Based on Nanosecond Pulse Voltage

Yang Jianzhong[1], An Shengdongt[1], Fan Yongqiang[2], Yin Jianbo[2] and Yue Yonggang[2,a]

[1]Inner Mongolia electric power (group) co., LTD, 010020, Xilin south road, saihan district, Hohhot, Inner Mongolia, China

[2]Inner Mongolia ultra-high voltage power supply bureau, jinchuan development zone, Hohhot, Inner Mongolia, China

[a]Corresponding author: Yue Yonggang (1979-), male, senior engineer, master candidate. Main research direction: new technology and method for fault diagnosis of power transmission and transformation equipment; insulation aging and life evaluation of electrical equipment. hvyue@163.com

Abstract. In this paper, a partial discharge detection platform with repeated frequency nanosecond pulse voltage is built. Its output voltage is 30kV, the pulse front is 100ns, the half-width of the voltage is 120ns, and the maximum repetition frequency is 10 kHz. Under the excitation of nanosecond pulse voltage, the defects in the test sample produce partial discharge, the number of partial discharge is consistent with the number of applied pulses, the repeatability of discharge is good, the discharge quantity and the initial discharge voltage are easy to be counted, the partial discharge parameters are easy to be extracted. Under continuous pulses excitation, the development of defect of the sample can be judged by the change process of the discharge amount. Therefore, the nanosecond pulse voltage can be used to detect the insulation defect of the sample, and it is helpful to study the destruction process and mechanism of the solid insulating material under partial discharge.

1. Introduction
In recent years, solid insulation materials have made great progress, and a variety of polymer and ceramic materials with excellent performance have been widely used in power system equipment. However, the processing and assembly process inevitably leads to defects on surface and inside in the material. Under the action of electric field, partial discharge occurs in the defects. Partial discharge exists for a long time, which will lead to the deterioration of material insulation and ultimately lead to the breakdown of the body breakdown insulation. Solid dielectric partial discharge detection is a kind of defect detection method widely used in power system. In order to prevent and control accidents better, partial discharge detection of insulating parts is very important. The defect detection of solid media can be carried out under a variety of power supply operating environments. At present, the research on defect detection characteristics under normal voltage such as DC and AC（power frequency voltage） is in-depth, and various detection and evaluation systems are mature [1-4]. Under the condition of the same voltage level, partial discharge under pulse voltage causes much less harm to insulation than power frequency voltage and dc voltage [5]. There are few researches on the application of pulse voltage to defect detection, and the application of nanosecond pulse voltage to defect detection has not been seen yet.

The nanosecond pulse source based on semiconductor opening switch (SOS) has the advantages of high repetition rate, high reliability, long life and so on [6], in the process of exploring the field of civil

Content from this work may be used under the terms of the Creative Commons Attribution 3.0 licence. Any further distribution of this work must maintain attribution to the author(s) and the title of the work, journal citation and DOI.

Published under licence by IOP Publishing Ltd

applications, it has been successfully applied to the production of industrial ozone [7], the research on the biological effects of electromagnetic pulse [8], SOS nanosecond pulse source has a huge development potential in the field of civil. Based on SOS nanosecond pulse source, this paper explores its application in partial discharge detection. The nanosecond pulse voltage partial discharge detection platform was built, and the output voltage of the pulse source was applied to the solid insulation samples of different defect types. By changing the output voltage amplitude and frequency of the pulse source, the defect detection of the insulation samples was carried out, and the characteristics of the nanosecond pulse voltage in detecting the defects of solid insulation materials were studied.

2. Principle and structure of nanosecond pulse source

Nanosecond pulse source is composed of primary charging unit, magnetic pulse compression unit and SOS. The principle is shown in figure 1. The primary charging power supply Udc charges the primary energy storage capacitor C1 through IGBT-1, diode VD1, inductance L1 and the primary winding of the pulse booster transformer PT. The charging current simultaneously excites the magnetic core of the pulse transformer. Charging is completed, the main switch IGBT-2 is closed, C1 is discharge, the high voltage capacitor C2 is charged by the PT, the charging current is pumped semiconductor opening switch (SOS), after the completion of the charging PT core saturation, C2 discharges through the secondary side of the PT and reversely pumping SOS, when reverse current maximum, SOS truncation quickly, because the PT vice winding inductance current can't mutations, the reverse current is transferred to the load resistance, produces high voltage pulse output load resistance R [9-10].

Pulse booster transformer PT undertakes the functions of pulse booster and pulse compression in the working process. The time-sharing trigger circuit controls the on-off timing sequence of IGBT-1 and IGBT-2, and completes the pulse output on the circuit charge and discharge and load resistance. Udc control pulse source output pulse amplitude, The trigger signal controls the output pulse frequency.

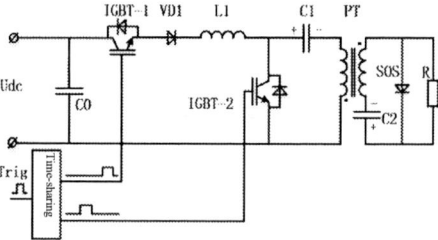

Figure 1. Schematic Diagram of Pulse Source Circuit

3. Test platform and test method

The platform for detecting partial discharge of repetitive frequency nanosecond pulse voltage is shown in figure 2, including Trigger Supply, Nanosecond Pulse power Source, HV probe, Rogowski Coil, Oscilloscope, Column-column electrode, Current-limiting R and Test sample. The test sample was organic film, which was closely connected with the Column-column electrode through the Ptfe-support, and the pulse source output was respectively connected to the upper and lower electrodes of the Column-column electrode. The diameter of the upper electrode of the column-column electrode was 6mm, and the diameter of the lower electrode was 25mm.

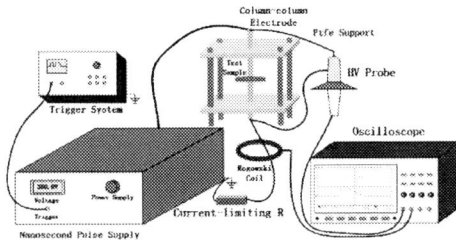

Figure 2. Partial Discharge Detection Platform

A trigger is used to trigger a repeated frequency nanosecond pulse source and to output a specific frequency, continuous or a specific number of triggering signals. Tektronix P6015A, 75MHz bandwidth, partial pressure ratio 1000:1, can be used to measure the peak pulse of 40kV. The coil is Pearson 6595, with a sensitivity of 0.5V/A and a bandwidth of 150MHz. SOS heavy frequency nanosecond pulse source is developed by ourselves, and its output waveform has excellent stability, high reliability and no local discharge. When Udc is determined, the output voltage amplitude of the pulse source varies with the change of the load resistance. The resistance load can choose different resistance values to adjust the output pulse amplitude. The maximum output voltage of the developed pulse source is 30kV. Figure 3 shows the output voltage amplitude of the pulse source, with the amplitude ranging from 10kV to 30kV. Figure 4 shows the waveform record under continuous operation of 1 kHz, and figure 5 shows the envelope of 30,000 pulse waveforms after 5min operation at 100Hz heavy frequency.

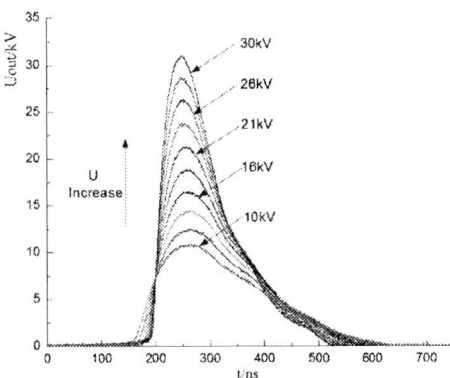

Figure 3. Pulse Voltage Amplitude

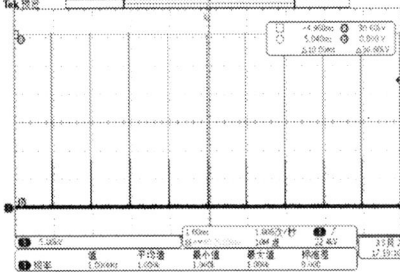

Figure 4. 1kHz Waveform Recording

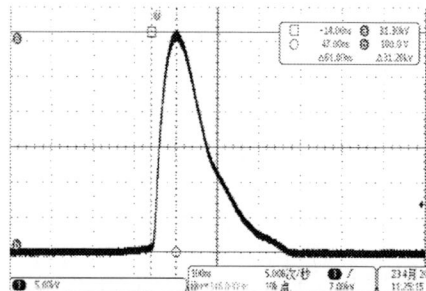

Figure 5. Waveform Envelope Diagram of 30000 Pulses

The sample is capacitive load, and the equivalent circuit is shown in figure 6. The output voltage of the pulse source is applied to both ends of the sample through the Column-column electrode, and a charging current will flow through the sample circuit. When the output voltage of the pulse source is increased, flashover or breakdown of the sample will occur, and discharge will occur. Charge changes will occur at both ends of the sample, and a large current mutation will occur in the test circuit connected with the sample. When the pre-flashover occurs, partial discharge will be generated, the discharge quantity is small, and the current mutation in the loop is small. The defect can be judged by the change of current in the circuit during discharge. Experiment, using the partial discharge experiment dedicated Rogowski Coil current signal measuring loop, Rogowski Coil is ideal means of nanosecond pulse current is measured, the measured current in the process of the sample(s) can well reflect the insulation breakdown current changes [11], try loop low voltage side of series resistance R, limit flashover or breakdown occurs large current in the circuit. The voltage at both ends of the sample is measured by a HV probe.

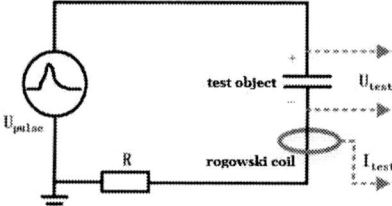

Figure 6. Equivalent Circuit of Test Sample

4. Defect partial discharge experiment

As shown in figure 7, the organic film of zero defect and defective sample(s) have been made: (a) two pieces of sound insulation film, (b) two pieces of insulation film, the upper insulating film in good condition, the lower groove, (c) two pieces of insulation film, the upper insulating film in good condition, the lower disconnect, (d) three pieces of thin film, insulation on the lower layer insulation film in good condition, the middle tier disconnect. The thickness of a single film is 0.5mm.

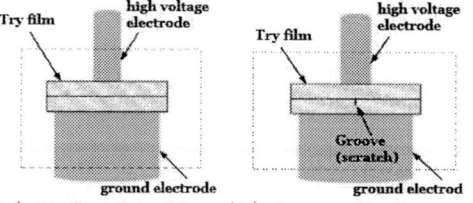

（a）Defect-free film （b）Lower film grooves

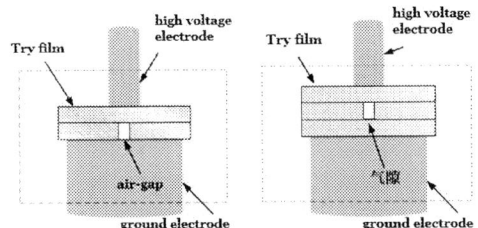

（c）Under film disconnection （d）Intermediate film break

Figure 7. Defective Sample Models

Pulse voltage was applied to the sample, and the current waveform on the sample (a) and (b) was the same, without current mutation. When the voltage amplitude increased to 30kV, there was still no obvious discharge phenomenon, as shown in Figure 8. When the pulse voltage is 25kV, the discharge pulse superimposed on the charging current can be seen on sample (c), as shown in Figure 9. When the pulse voltage is 23kV, obvious discharge pulses can also be seen on the sample (d), as shown in Figure 10. Therefore, the pulse voltage can clearly detect the fracture defects in the sample.

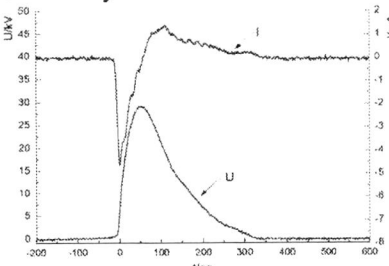

Figure 8. Two Intact Insulating Films

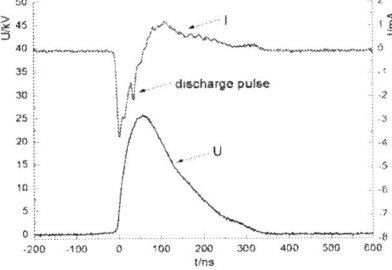

Figure 9. Two Insulating Films, the Lower Layer Disconnected from the Middle

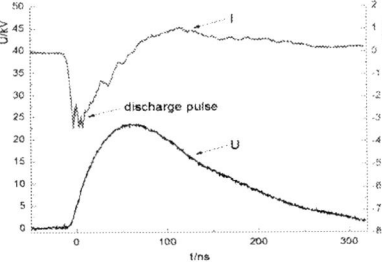

Figure 10. Three-layer Insulation Films with Middle Layer Disconnected from the Middle

At the same pulse voltage, 50Hz continuous repetition frequency experiment was carried out on the sample (c). The change of sample discharge current in the process of pressurization is shown in Figure 11. At the beginning, the defects of the sample were small, and there were small spike pulses in the current waveform of the sample. After the pulse voltage is continuously applied for a period of time, the spike on the current waveform will increase and become severe at the same time, indicating that the defect of the sample becomes larger. After aging for a long time, the circuit current surges, and the sample is broken down, and the insulation cannot be restored. It can be seen from the above experiments that the defects in the insulation wafer can be detected by using pulse voltage excitation, and the partial discharge will become more severe with the increase of defects.

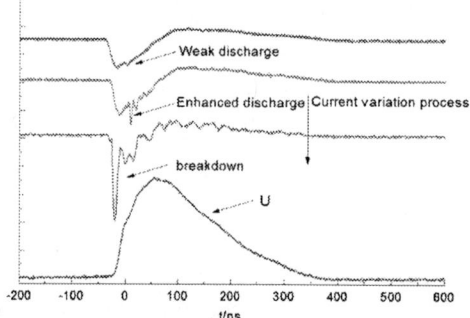

Figure 11. Partial Discharge Waveforms of Defective

Different amplitude voltages are applied to the sample (d). With the gradual increase of pulse amplitude, the sample generates discharge current, reduces voltage amplitude, and the sample discharge disappears. As shown in Figure 12, during the voltage rise, the position of discharge pulse gradually moves forward. When the voltage amplitude reaches about 19kV, the discharge pulse position is at the peak voltage of 50%. With the decrease of voltage, the position of discharge pulse gradually moves backward, and the discharge pulse appears at the drop edge of voltage when it is lower than a certain value. The discharge disappears when the voltage continues to drop to about 13.5kV.

The discharge quantity produced by partial discharge of the same sample under different voltages is stable, and the discharge location is different, but the corresponding initial voltage of partial discharge is the same.

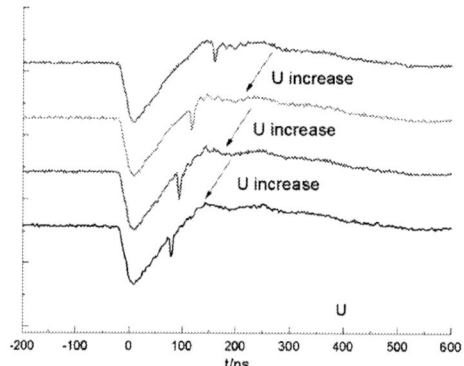

Figure 12. Pulse Positions of Partial Discharge

In order to mark partial discharge signals more accurately, coupling capacitance was added to the original measurement circuit. Two Rogowski Coil were used to measure the pulse current signals of the sample circuit and coupling capacitance circuit respectively. The circuit principle is shown in Figure 13. When the sample (c) is applied with voltage, the polarity partial discharge pulse signal of the sample loop is opposite to the current signal of the coupling loop, so the determined partial discharge signal can be distinguished by the direction of the pulse current, as shown in Figure 14, the discharge pulse in the opposite direction can be clearly seen.

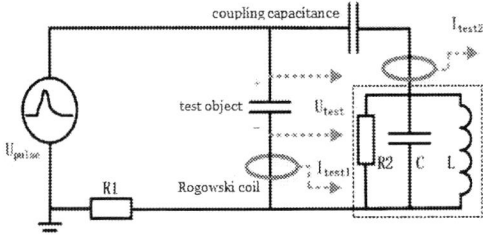

Figure 13. Schematic Diagram of High-frequency PD Test System

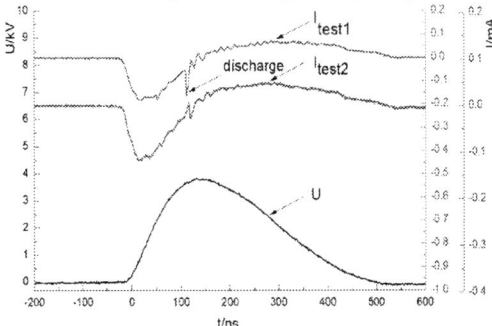

Figure 14. Pulse Current Signals of Main Circuit and Coupling Circuit

5. Experimental results

(1) Nanosecond pulse voltage experiments were carried out on organic films with different defects. By measuring the changes of current waveforms in the circuit, the defects of organic film materials could be judged.

(2) When the defective sample is under continuous high frequency voltage, its insulation will gradually age, leading to the gradual increase of defects and the gradual increase of local discharge. The change of local discharge can reflect the deterioration process of insulation material defects. The severity of defects can be judged by the local discharge.

(3) Under different voltage amplitudes, the same defect has good repeatability of partial discharge and stable discharge, so the initial voltage of partial discharge of the sample can be determined.

(4) A pulse voltage corresponds to a discharge, the charging current and partial discharge current signals are easy to distinguish, the experimental results can be observed in real time, and the discharge is easy to be counted.

6.Conclusion

Nanosecond pulse voltage testing platform is simple and can be used to detect defects in insulating materials. The defects of insulating materials and the severity of defects can be judged by the current waveform of the test circuit. The pulse width and width of nanosecond pulse voltage enable it to run at high frequency and repeat frequency. The discharge times and discharge quantity of samples under continuous pulse can be accurately captured and the initial discharge voltage can be highly accurate. Under continuous pulse, the changing process of sample discharge can be observed and recorded in real time, which is helpful to study the failure process and mechanism of solid insulation under partial discharge.

Acknowledgment

Project supported by science and technology project of Inner Mongolia Power(Group) Co., Ltd.(20170106, 201801024).

Author:

Yang jianzhong (1980-), male. Main research direction: electrical equipment operation management, yangjianzhong@impc.com.cn

Shengdong an (1980-), male. Main research direction: operation management of electrical equipment, anshengdong@impc.com.cn

Fan Yongqiang (1987-), male. Main research direction: engineer, master candidate, main research direction: maintenance and management of primary equipment state of transformer substation, 980579@qq.com

Yin Jianbo (1973-), male. Main research direction: maintenance and management of primary equipment state of transformer substation, yinjinabo@impc.com.cn

Yue Yonggang (1979-), male, senior engineer, master candidate. Main research direction: new technology and method for fault diagnosis of power transmission and transformation equipment; insulation aging and life evaluation of electrical equipment. hvyue@163.com

References

[1] Guo Jun, Wu Guangning, Zhang Xueqin, et al. The Actuality and Perspective of Partial Discharge Detection Techniques[J]. Transactions of China Electrotechnical Society. 2005(02): 29-35.

[2] Li Junhao, Han Xutao, Liu Zehui, et al. Review onPartial Discharge Measurement Technology of Electrical Equipment [J]. High Voltage Engineering, 2015, 41(08): 2583-2601.

[3] Kang Qiang, Gu Yu, Xu Yang, et al. Review of DC Partial Discharge Detection Technology[J]. Southern Power System Technology, 2015, 9(10): 69-77.

[4] Peng Fadong, Liu Bin, Pang Xiaofeng, et al. Analysis of theequivalence of partial discharge, withstand voltageand power frequency of XLPE cable under damped oscillatory wave[J]. High Voltage Apparatus, 2013, 49(07): 116-121.

[5] Ge Jingquan. Partial discharge measurement [M]. MacHin-ery Industry Press, 1984.62-68.

[6] Wang Gang, Su Jiancang, Ding Yijie, et al. High repetitionrate pulse generator based on SOS and LTD technology[J]. High Power Laser and Particle Beams, 2014, 26(04): 109-113.

[7] Chen Hongbin, Meng Fanbao, Li Aiping, et al. Development of SOS-Based Pulsed Power Source [J]. High Voltage Engineering, 2005(09): 56-58.

[8] Zhao Meilan, Cao Xiaozhe, Wang Dewen, et al. High Field Strength EMP with Its Applied Study in Biology [J]. Journal of National University of Defense Technology, 2000(S1): 17-20.

[9] Ding Yijie, Hao Qingsong, Su Jiancang, et al. 8MW, 10kHz Pulse Generator Based on Semiconductor Disconnect Switch [J]. High Power Laser and Particle Beams, 2009, 21(10): 1575-1578.

[10] Zhang Hao, Li Shengli, Huang Heyan, et al. Development and improvement of practical high-voltage pulse power supply [J]. Power Supply Technology, 2015, 39(02): 400-402.

[11] Wang Hao, Zhang Shichang, Yan Ping, et al. Measurement of nanosecond pulse current using self-integrating Rogowski coil[J]. High Power Laser and ParticleBeams, 2004(03): 399-403.

ISPECE

IOP Publishing

Dynamic Variance Equalization Planning Optimization Method for Power Grid System Protection Communication Network

Dongliang GAO[1], Taorui LUO[1], Peizhe XIN[2], Yudong WANG[2], Jun LU[3], Peng CAI[3]

[1]State Grid Sichuan Economic Research Institute, Sichuan, China

[2]State Grid Economic and Technological Research Institute CO., LTD, Beijing, China

[3]School of Electric and Electronic Engineering, North China Electric Power University, Beijing, China

[3] 877593046@qq.com

Abstract. To improve the universality of the backbone communication network architecture model in Smart Grid, this paper proposed a dynamic variance equilibrium planning optimization method for the system protection service in the power grid. Firstly, based on the two-layer architecture model of the power backbone communication network, a network equilibrium optimization model is constructed. Secondly, based on the network equilibrium optimization model, a simulation experiment method with the node usage ratio as the optimization target and the delay in the transmission process as the constraint condition is designed. Finally, the model was quantified through simulation experiments. The simulation results show that the proposed optimization model has excellent applicability in network balance planning of grid protection services.

1. Introduction

The power communication network is a communication network dedicated to the operation and management of power systems, with obvious industrial characteristics and special requirements for safety and reliability[1]. With the continuous improvement of the automation level of power systems, the power communication network plays an increasingly important role in power production and power dispatching. Power systems are high-tech and intensive industries that require high reliability and power information security. This puts higher demands on the power communication network. How to provide high-quality and reliable information services for power system operation and production based on existing communication facilities is a new challenge for power system communication services[2]. Another solution may be Optical Transport Network (OTN) technology. OTN technology is based on wavelength division multiplexing and WDM technology. It guarantees similar SDH protection and management and maintenance functions through frame structure and overhead processing similar to SDH. The OTN architecture consists of an optical layer and an electrical layer. Both the optical layer and the electrical layer network have their own monitoring and management capabilities, and the optical layer and the electrical layer network have better survivability.

Content from this work may be used under the terms of the Creative Commons Attribution 3.0 licence. Any further distribution of this work must maintain attribution to the author(s) and the title of the work, journal citation and DOI.
Published under licence by IOP Publishing Ltd

Many works have been done on the network architecture and smart grid and its related modeling optimization[3-6]. In [3], a latency constrained dynamic routing algorithm is proposed in optical transport network for smart grid systems, in which the influence of spectrum slot size is revealed on traffic latency and blocking probability in grid communication networks. In [4], a label setting algorithm has been presented to solve the problem that the k-shortest path problem, which is given that departure and arrival are constrained within specified time windows. In [5], an optimal delay-based virtual topology is designed using integer linear programming for the smart-grid power backbone communication network, which can achieve superior smart-grid network performance. In [6], an energy-efficient multicast tree construction protocol is presented for real-time data streaming, which considers using real time estimated routing delay from source node to other nodes.

Given the fact that the power related services requires considering Quality-of-Service (QoS) in the power grid system protection communication network, this paper proposes a dynamic variance equalization planning optimization modelling. The rest of the paper is organized as follows. In section 2, the dynamic variance equalization planning modeling and the Dynamic Variance Equalization (DVE) algorithm is described. Section 3 discusses the simulation experiment. Finally, the conclusion is drawn in Section 4.

2. Dynamic variance equalization modelling

2.1 OTN two-layered architecture
The network technology in the power grid selects OTN , the basic architecture to optimize the channel equalization planning of the system protection service. The typical layered architecture of OTN is shown in Figure 1.

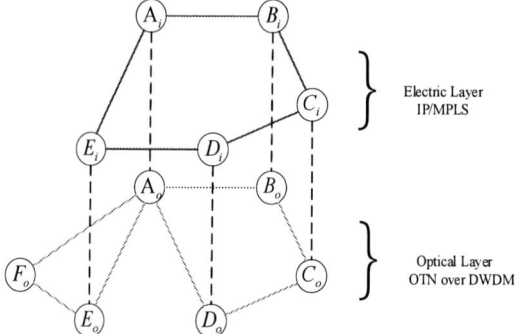

Figure 1. General model of two-layer architecture.

Shown in Figure 1, the electrical layer implements system protection services by mapping parameters of communication capacity and links of electrical layers to the optical layer. In addition, the upper interface in the electrical layer is used to construct the logical path, while the lower interface in the optical layer is used to allocate the associated physical wave channel for data transmission through QoS control.

2.2 Dynamic variance equalization model
The dynamic variance equalization model for optimizing the OTN network planning is constructed. The proposed model considers the time-delay as constraint. Moreover, the optimal goal considers the average node access usage ratio, which supports the system protection services. Suppose there are N_s^e grid-system-protection services (short for grid-service) in the related OTN communication network with nodes' total number as P, a grid-service ID is symbolized as d, which is from 1 to N_s^e.

Assume M_i is the number of the access capacity for each Node i in the ONT Network, and ζ_i^o, which is set to 1, represents one access capacity for each Node i in the optical layer. L_o represents the transmission link in the optical layer. If δ_{dl}^{o} takes 0 or 1 representing whether the given service d transports from or to the Node, Node i related access usage ratio R_i in the optical layer may be formulated as (1).

$$R_i = \frac{1}{M_i} \sum_{l=1}^{L_o} \delta_{dl}^{o} \zeta_i^{o} \qquad (1)$$

Therefore, the optimization model for the dynamic variance equalization model may be formulated as (2).

$$\min \quad f = \left(\frac{\sum_{i=1}^{P} (R_i - \overline{R})^2}{P} \right)^{1/2} \qquad (2)$$

$$s.t. \quad T_d \leq T_{upper} \qquad (2a)$$

$$R_i \leq R_{upper} \qquad (2b)$$

In (2), $\overline{R}$ is the average node usage ratio for the OTN network, which may be calculated by the mathematic-mean operation of R_i. The restraint (2a) represents the time-delay restraint for the grid-service d (generally less than 60ms). In (2a), T_d is the transmission total time-delay for the service d including the electric layer time-delay and the optical layer time-delay, while T_{upper} is the upper limitation of the transmission time-delay for the grid-service d. The restraint (2b) represents the node access usage ratio restraint for the grid-service d (generally less than 50%). In (2b), R_{upper} is the upper limitation of the node access usage ratio.

2.3 DVE flow chart
The flow chart of the proposed DVE algorithm to implement the proposed model is shown in Figure 2.

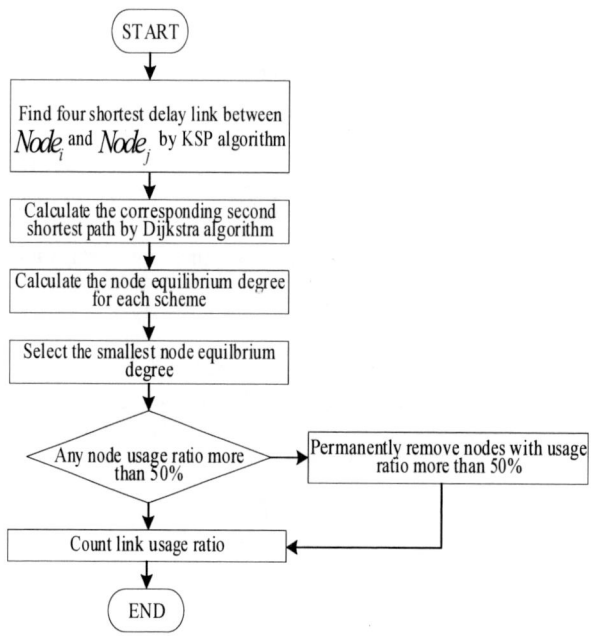

Figure 2. DVE flow chart.

Firstly, for a service between $Node_i$ and $Node_j$, four shortest paths are obtained by K shortest paths(KSP) algorithm. For each candidate scheme, to implement the physical 1+1 backup, the nodes passing through the candidate path are deleted, and the second shortest path that satisfies the physical 1+1 backup is obtained by the Dijkstra algorithm. The node equilibrium degree of each scheme is calculated, and the scheme with the smallest node equilibrium degree is selected to update the node usage ratio. At the same time, if the service is added and the node usage ratio exceeds 50%, the corresponding node is temporarily deleted, and then the next set of nodes is solved.

3. Simulation and analysis

3.1. Experimental parameters settings

The simulation experiment is carried out according to the DVE algorithm mentioned above. In the simulation experiment, the total delay includes the transmission delay and the node forwarding delay. The simulation network is the cost 239 network of Figure 3. 70 random services are generated in the simulation, and services are added sequentially by DVE algorithm and KSP algorithm respectively. KSP algorithm is the comparison algorithm, which directly generates two shortest link path between $Node_i$ and $Node_j$. The node usage ratio and node equilibrium degree are counted for each 10 pieces of service added. Finally, the usage ratio of each node in the added network is analyzed.

ISPECE   IOP Publishing

IOP Conf. Series: Journal of Physics: Conf. Series **1187** (2019) 022058   doi:10.1088/1742-6596/1187/2/022058

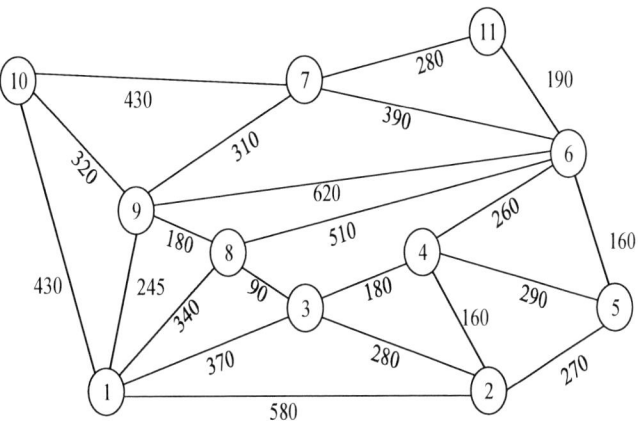

Figure 3. COST239 network topology.

The simulation experimental parameters of OTN technology are set as follows: T_G represents the optical cable delay with the value as 5 us per 100 kilometers, while T_z represents the direct connection delay with the value as 200 us. Moreover, T_y represents the mapping delay with the value as 40us and T_q represents the canceling mapping delay with the value as 40 us. In the following part, we compare the KSP algorithm with our proposed DVE algorithm in terms of node usage ratio and node equilibrium degree.

3.2. Node equilibrium degree experiment
Firstly, the experimental comparison of the node balance of the proposed DVE algorithm and KSP algorithm when adding services is carried out. The simulation results are shown in Fig. 4. The X axis represents the number of added services, ranging from 0 to 70, and the node equilibrium degree is recorded every 10 times in the simulation. The Y axis represents the node equilibrium degree.

It can be seen from Figure 4. that: (a) As the number of services in the network increases, the node equilibrium degree of the proposed DVE algorithm and KSP algorithm increases. Because with the number of services increasing, the network complexity increases, and the network balance decreases. (b) As the number of services increases, the proposed DVE algorithm is more gradual than the KSP algorithm. This shows that the proposed algorithm has more advantages than the KSP algorithm in adding new services, which can reduce the fluctuation of the network when adding services. (c) When adding the same amount of services, the balance of the DVE algorithm is significantly better than the KSP algorithm, and the more services, the more obvious the advantage.

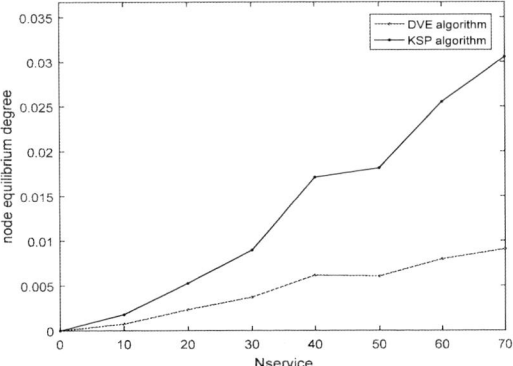

Figure 4. Node equilibrium degree.

413

3.3. Node equilibrium degree experiment

Secondly, when 70 services are added to the network, the two algorithms are simulated. The experimental results are shown in Figure 5. and Table 1. In Figure 5, the X axis represents the node number, ranging from 1 to 11. The Y axis represents the node usage ratio, between 0 and 1. In Table 1, $\overline{R}_{node}$ represents the average node occupancy, $\overline{\sigma}_{node}$ represents the node balance, and Maximum and Minimum represent the maximum and minimum values of the two algorithms.

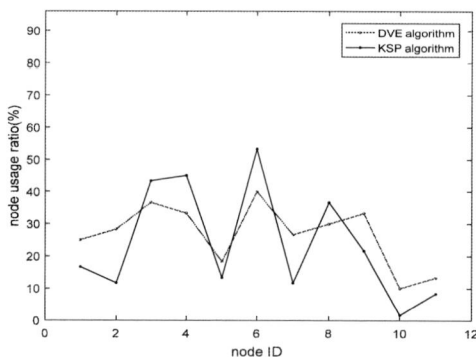

Figure 5. Node usage ratio (Nservice =70).

Table 1. Comparison of statistical results of node usage of DVE and KSP algorithms.

Algorithm	$\overline{R}_{node}$	$\overline{\sigma}_{node}$	Maximum	Minimum
DVE	0.27	0.009	0.40	0.10
KSP	0.24	0.030	0.53	0.02

It can be seen from Fig. 5 and Table I: (a) When the number of added services is 70, the fluctuation of the node usage ratio of the proposed DVE algorithm is smaller than that of the KSP algorithm. (b) The maximum node usage ratio of the KSP algorithm is 53%, and the minimum value is 2%. (c) The DVE algorithm has a maximum node usage ratio of 40% and a minimum of 10%. (d) The node equilibrium degree of the proposed DVE algorithm is lower than that of the KSP algorithm. At the cost, the average node usage ratio is slightly higher than the KSP algorithm.

3.4. Node equilibrium degree experiment

Finally, when adding services, the average node usage ratio of each stage is simulated. The experimental results are shown in Figure 6. In the figure, the X axis represents the number of added services, from 0 to 70. The Y axis represents the average node usage ratio of 11 nodes for each ten services added.

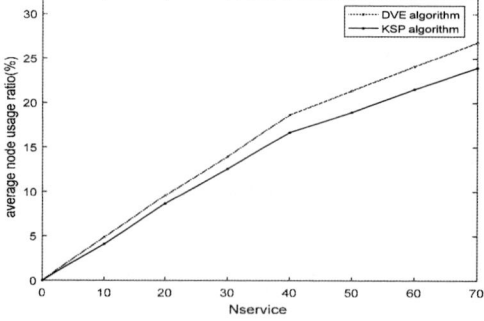

Figure 6. Average node usage ratio.

Table 2. Comparison of statistical results of average node usage and node equilibrium degree.

$N_{service}$	10	20	30	40	50	60	70	Average
Increment	18%	10%	10%	11%	12%	12%	12%	13%
Reduction	58%	55%	58%	64%	66%	69%	70%	63%

It can be seen from Figure 6 and Table 2: (a) When the number of services increases, the average node usage ratio of the proposed DVE algorithm in each stage is slightly higher than the KSP algorithm. (b) In Table II, Increment represents the increment of the average node usage ratio of the DVE algorithm relative to the KSP algorithm, and Reduction represents the decrement of the DVE algorithm node equilibrium degree with respect to the KSP algorithm. It can be seen that although the average node usage ratio of the DVE algorithm increases, the advantage of node equilibrium degree is obvious. It can be seen that the DVE algorithm is relatively better than the KSP algorithm.

4. Conclusion

This paper proposes dynamic variance equalization planning optimization modeling for power grid system protection communication network, which is suitable for adding new services to an existing network Through simulation and analysis, it can be seen that the dynamic variance equalization planning optimization model proposed in this paper can meet the power system protection delay specified in the power system protocol to meet the requirement of less than 50ms. And our DVE algorithm has better equalization performance than KSP algorithm.

Acknowledgements

This work is funded by China State Grid Science and Technology Project (No. SGSCJY00GHJS1800018 and No. SGXT0000ASJS1700054).

References

[1] M. N. Mohr Warp, I. Antonovich, etc. Energy Efficient Partition-light path Scheme for IP over WDM Core Network. Procedia Computer Science, 2015, 52: 324- 348.

[2] K. D. Dambulla, F. M. About, H. T. Chua. Impact of SRS and XPM on the Performance of IP Traffic over a WDM Ring Network. Journal of Optical Communications, 2007, 28(3): 231- 246.

[3] Fan Bo Meng, Tai Yi Fu, Jun, etc. Performance Evaluation of Ethernet Passive Optical Network for Smart Grid. Applied Mechanics and Materials, 2014, 356- 360.

[4] Amit Kumar Garg. An efficient fault localization or detection mechanism for high speed optical networks[J]. Optic - International Journal for Light and Electron Optics, 2013, 124(21).

[5] Pagadian Shashikant, Yilmaz Melittin, Alluri Prayut. Smart-Grid Backbone Network Real-Time Delay Reduction via Integer Programming. IEEE transactions on neural networks and learning systems,2016, 27(8): 231- 219.

[6] Konstantinos N. Androutsopoulos, Konstantinos, etc. Solving the k -shortest path problem with time windows in a time varying network. Operations Research Letters, 2008, 36: 1393- 1404.

System for real-time transmission control of AC motor temperature data based on Linux system

Li Hengjie，Wu Long，Chen Wei，Qiao Zhen

(School of electrical engineering and information engineering, Lanzhou university of technology， Lanzhou city, Gansu province，730050)

About the Author:

Li Hengjie (1981-), male, Xi'an, Shaanxi, Ph.D., associate professor, master tutor. Research direction: embedded systems and new energy

Wu Long (1993-), male, ethnic (Han), native (Hebei, Dingzhou), master, research direction: motor intelligent control manager

Name of the manuscript contact: Wu Long

Work unit: Lanzhou University of Technology

Contact address: No. 287, Langongping Road, Qilihe District, Lanzhou City, Gansu Province

Phone: 18203249145

Zip code: 730050

Abstract: In order to study the temperature change of the motor, in the Linux system environment, the fuzzy control query table is established and the membership function corresponding to the element is displayed;the motor temperature is determined to transmit data in real time, and the data is aggregated to form a database and a rule base. Through the real-time summary analysis of the motor temperature data, the fuzzy decision is made, and then the CUPS sets the Linux printer, and the motor temperature parameter is analyzed at any time for more intuitive paper surface comparison analysis. Relying on historical data to establish a big data platform, through the control algorithm to achieve data comparison analysis, you can determine the deviation of the motor operating temperature and accurately diagnose possible faults.

1. Introduction

In modern industrial automation systems, the control of the motor is related to the operational safety and stability of the entire system. With the rapid development of industrial and information technology and the deepening of motor research, real-time monitoring and control of the motor is required to ensure the stable operation of the whole system and reduce the loss of the motor itself. Normally, there are two ways to monitor the motor temperature. The first is to place a PT100 platinum resistor in the stator winding of the motor, and the lead wire is connected to the temperature transmitter. Output industry standard current 4-20mA, after which various types of microcontrollers will collect and analyze it [10]. But in reality, the microcontroller collects data that is significantly and significantly beyond the motor winding temperature. The second is to estimate the temperature at which the motor operates by varying the resistance of the stator windings with temperature. In the frequency conversion control, special processing is carried out by combining hardware and software, and a relatively accurate motor temperature change can be obtained. However, the calculation of the resistance must be aware of the voltage and current. For the entire industrial control system, interference and fluctuations are ubiquitous and ubiquitous. Although the accuracy is improved relative to the first method, it still cannot meet the requirements of the industry standard.

In this paper, based on the uncertainty of the motor winding temperature change, temperature acquisition is carried out, and the real-time data transmission is combined with the Linux system. The data platform is constructed based on historical experience data, and the data is compared and analyzed, and the motor operation data can be reported independently. The fuzzy control is combined with the PID control to form the system controller, so as to obtain accurate motor operating temperature data and anti-interference ability.

2. temperature collection

2.1. Combination of PID control and fuzzy control
PID control is one of the most commonly used algorithms in industrial control. The PID control law is effective for most industrial controlled objects, especially for linear constant systems. Most of the stator winding coils on the market today are wound with copper wire. For the notification material, its resistance is linear with temperature. The fuzzy control system has the characteristics of simple and fast implementation. In order to change the staticity of the fuzzy control, the PID control is combined with the fuzzy control. Because the control mode of the motor is different under different environmental conditions, the multi-mode control is tested with different control methods within the allowable range of the input variables. When the deviation is relatively large, pure proportional control is adopted; when the deviation is smaller than a certain; When the value is controlled, the fuzzy control is used; when the deviation belongs to the language value "zero", the PI control is used [6]. This produces the following formula 1:

$$u = \begin{cases} K_p e & |e| > e_0 \\ u_F & |e| \leq e_0 \cap e \notin 0 \\ K_p e + K_i \sum e & e \in 0 \end{cases}$$

2.2. Design of motor temperature fuzzy control algorithm
The degree of heat resistance of the motor depends on its insulation rating. In this paper, the insulation grade A is taken as an example. The ambient temperature is 40 °C, and the motor casing temperature should be within 60 °C. Static 60 ° C as a set value, the motor temperature measured for the first time is, then the motor temperature deviation is

It is passed as an input variable to the fuzzy controller.

The trigger voltage is the output variable of the fuzzy controller. Since the voltage directly controls the supply voltage of the motor, it is called the control amount.

It is known from the investigation of the fuzzy rule that value can be described by the positive and negative values of the deviation; the value of the machine temperature lower than the set value (high) can be distinguished by the magnitude of the deviation; The machine temperature is equal to the set value and the deviation can be recognized as zero.

In view of the above expression distinction, the input-level output variables use the following language values:

Negative big, negative small, zero, small, big

Or can be written as

NB, NS, O, PS, PB

among them:

NB: Negative Big

NS:Negative Small

O: Zero

PS:Positive Small

PB:Positive Big

For ease of understanding, only the bias is considered in the fuzzy domain, which can be understood as a set of 7 elements. They are -8, -4, -2, 0, 2, 4, 8. The collection can be expressed as

=\{-8,-4,-2,0,2,4,8\}

The analogy is that the fuzzy domain of the control quantity is

=\{-8,-4,-2,0,2,4,8\}

By defining the above language values in the fuzzy domain, the corresponding membership function can be obtained. As shown below:

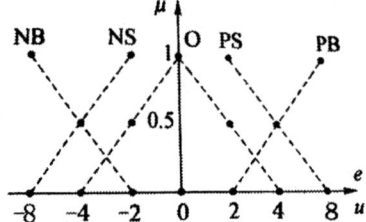

Figure 1 Membership function of the language value

The language value membership results corresponding to the elements -8, -4, -2, 0, 2, 4, 8 are as follows:

PB： 0,0,0,0,0,0.5,1

PS： 0,0,0,0,1,0.5,0

O: 0,0,0.5,1,0.5,0,0

NS： 0, 0.5, 1, 0, 0, 0, 0

NB： 1, 0.5, 0, 0, 0, 0, 0

It can be known from the linguistic value membership degree data that the control of the motor temperature can meet the control requirements of the industrial automation system, and the accuracy is further improved. The design of the fuzzy control algorithm is explained by the description of the temperature control system, which lays a foundation for further research on the fuzzy controller design method.

3. Real-time data transmission under Linux system

3.1. Fuzzy Control Data Query Table

In the actual industrial production automation system, the correspondence between the deviation and deviation variation and the control output is stored in the computer in the form of a fuzzy control table. When the fuzzy controller is in normal operation, the industrial control opportunity first determines the control output according to the quantized value of the deviation and the variation amount of the deviation obtained by sampling, and then multiplies the control output by the optimized scale factor. This will get the control output of the controlled object. The system structure is shown below:

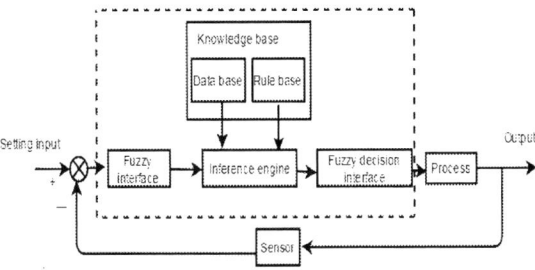

Figure 2 Basic structure of fuzzy control

The fuzzy control data lookup table is established under the offline situation, so it has no influence on the real-time operation of the fuzzy controller, and also meets the control requirements of real-time monitoring.

3.2. Linux system database

One of the best places in Linux is its multitasking environment, multi-user [14]. The control system can separately set up the file database in the motor of the automation system, so that not only the motor temperature data but also the comprehensive data of the motor can be recorded, so as to grasp the real-time running state of the motor. The data is synchronously written to disk according to the command statement sync of the Linux system. In reality, the general account can also use sync, but the general account user can only update their own data when updating the data, root can update the data of the entire system. This makes it possible to build a Linux central database on top of FCS. In order to facilitate the review of the data, the file is recorded as a log and the log file is to be security set.

In order to facilitate Linux user management, the server of the log file is related. In the syslog.conf file, the recorded log file data can be transferred to a remote host or printer. In doing so, even if the hacker deletes /var/log/, it doesn't matter. The important data is directly recorded on the printer, which increases the security and reliability of the system.

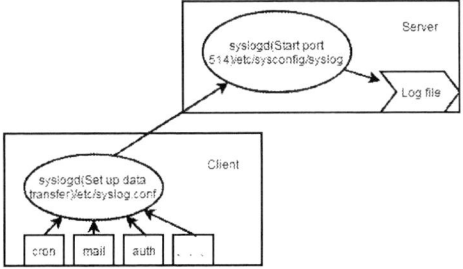

Figure 3 Log file server architecture

Printers are an important tool for us to record feedback data at any time. We can use CUPS to set up a Linux printer [15]. The Linux host also serves as a dialogue for the Printing server. CUPS supports online, so that we can establish a protocol channel and connect different computers through the protocol channel. In addition to the inherent body of the host, the printer must also be able to support Linux, as well as install the print component under the Linux system. As shown in the figure below:

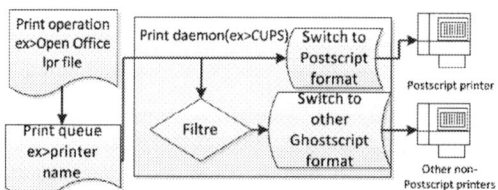

Figure 4 Schematic diagram of each component of the print behavior

4. conclusions

Using the Linux system to monitor the database in real time, build a knowledge base for the temperature parameters of the AC motor and the rest of the industrial detection parameters, establish a rule base through the combination of fuzzy control and PID, and then achieve good control effect through the inference engine. The Linux system database establishes different folders to establish an online system with the distributed control of the industrial site, records the running state of the motor in the form of logs, and realizes the secure transmission of data through the components of the printing behavior. Relying on historical data to establish a big data platform, through the control algorithm to achieve data comparison analysis, you can determine the deviation of the motor operating temperature and accurately diagnose possible faults.

Acknowledgment

Fund Project: Gansu Basic Research Innovation Group Project (18JR3RA133); National Natural Science Foundation Project (51767017)

References

[1] Z.R.Radakovic, V.M.Milosevic, S.B.Radakovic.Application of Temperature Fuzzy Controller in an Indirect Resistance Furnace. Applied Energy, 2002:167-182

[2] St. Boschert, P. Dold, K. W. Benz. Modelling of the Temperature Distribution in a Three-zone Resistance Furnace: Influence of Furnace Configuration and Ampoule Position. Journal of Crystal Growth, 1998: 140-149

[3] Edited by Wang Jiaxuan. Regulators and actuators. Beijing: Tsinghua University Press, 2001: 68-72, 47-57

[4] Xu Xuefeng editor. Sensor Transmitter Measurement and Control Instrumentation. Beijing: Mechanical Industry Press, 1998: 223

[5] Xie Xinmin, Ding Feng, ed. Adaptive Control System. Beijing: Tsinghua University Press, 2002: 1-5, 158-161

[6] Li Shiyong. Theory and Application of Fuzzy Control and Intelligent Control. Harbin: Harbin Institute of Technology Press, 1990:78-80

[7] Zhang Shouyuan, Li Lihong. Exploration of fuzzy tuning of PID control parameters in dynamic process. Mining and Metallurgy Engineering, 2000, (2): 22-24

[8] Zhang Jianmin, Wang Tao, Wang Zhongli. The principle and application of intelligent control. Beijing: Metallurgical Industry Press, 2003: 60-64

[9] Xie Yuandan, Xia Shuyan. Relay self-tuning method for PID regulator parameters. Control and decision making, 1993, 8(1): 77-79

[10] Xie Yuandan, Yu Sheng. Comparison of several algorithms for PID self-tuning. Automation and Instrumentation, 1993, (4): 20-24

[11] Wang Yang, Yan Jian. A Self-tuning Method for PID Parameters Based on Recursive Parameter Estimation. Information and Control, 1996, (4): 182-185

[12] Zhang Huaguang, He Xiqin. Fuzzy Adaptive Control Theory and Its Application. Beijing: Beijing University of Aeronautics and Astronautics Press, 2002: 167-169, 26-28

[13] Jen Yang. Chen. Rule Regulation of Sliding Mode Controller Design: Direct Adaptive Approach. Fuzzy Sets and Systems, 2001, 120: 159-168

[14] Lin Huizhen, "Red Hat Linux Server Configuration and Application" People's Posts and Telecommunications Press 2006.1

[15] Yue Hao "Linux Operating System Tutorial" Mechanical Industry Press 2005.4

ISPECE   IOP Publishing

IOP Conf. Series: Journal of Physics: Conf. Series **1187** (2019) 022060   doi:10.1088/1742-6596/1187/2/022060

Comparison between UofC Model and Ionosphere-free Combination Model in PPP

Y C Zhan[1]

[1]School of Information and Electronics, Beijing Institute of Technology, Beijing, 100081, China

zyc_1544@163.com

Abstract. GPS Navigation Satellite System is a satellite-based positioning system, whose accuracy and reliability depend on the number of observable satellites. The precise point positioning (PPP) in GPS navigation satellite system is a technique which can provide the centimeter-level positioning by a dual-frequency receiver and thus the research on precise point positioning is of great significance. In this paper, the ionosphere correction models of precise point positioning (PPP) are studied, including ionosphere-free combination model and UofC model. The static GPS positioning data obtained from the experiment is used and we compare the positioning accuracy of the two models. The result shows that the positioning precision using UofC model is decimeter grade, while the positioning precision using ionosphere-free combination model is centimeter level.

1. Introduction

GPS system is used widely in our daily life, especially in the field of positioning and measurement. With the fast developing of satellite positioning technique, receiver users expect not only more on accuracy, real-time and integrity of positioning, but also convenience and cost. However, in traditional GPS static positioning, the accuracy of the absolute positioning can be up to 20m, which cannot satisfy the demand of precision navigation and measurement[1]. The PPP technique provides a precision positioning method which can be used in the GPS navigation satellite system. Since the PPP approach was firstly proposed to realize the single station positioning with fixed precise orbit solutions and Doppler satellite observations by R.R. Anderle in early 1970's, it has experienced the development from dual-frequency to single-frequency and to multi-frequency, from CPS single system to GNSS multi systems, from fuzziness floating-point solution to ambiguity fixed solution and from post processing to real-time processing. For GPS users, carrier phase and pseudo-range observations from one GNSS receiver are utilized, combined with the product of high accuracy satellite orbit and clock error correction, as well as model correction and parameter estimation, correcting errors from satellite, signal path and receiver[2]. And then, the high accuracy positioning can be realized. This process is called PPP technique.

Based on different processing methods towards ionosphere and clock errors of receivers, different PPP positioning models were proposed in these years. In early years, an ionosphere-free combination model was utilized, which used dual frequency pseudo-range and phase observation values, constituting the pseudo range observation equation and carrier phase observation equation of the ionosphere respectively[3][4][5]. In 2001, Gao Yang Professor proposed a UofC model to solve the problem of large pseudo-range ionospheric combination noise. UofC model used the phase observation equation of the dual frequency ionospheric combination and the two-half frequency observation equation with pseudo range carrier frequency[6].

Content from this work may be used under the terms of the Creative Commons Attribution 3.0 licence. Any further distribution of this work must maintain attribution to the author(s) and the title of the work, journal citation and DOI.
Published under licence by IOP Publishing Ltd

In this paper, two methods of ionospheric correction are compared, UofC model and ionosphere-free combination model. By the means of comparing the two models, it is helpful when choosing ionosphere correction model.

2. PPP observation equations

In the technique of PPP, observations are pseudo-range and carrier-phase. Pseudo-range and carrier-phase observation equations in meters are:

$$P_{i,F}^{k} = \rho_i^k + c \cdot (dt_i - \delta t^k) + T_i^k + I_{i,F}^k + dm_{i,F}^k + e_{i,F}^k \quad (1)$$

$$L_{i,F}^{k} = \rho_i^k + c \cdot (dt_i - \delta t^k) + T_i^k - I_{i,F}^k + \lambda_F N_{i,F}^k + \delta m_{i,F}^k + \varepsilon_{i,F}^k \quad (2)$$

Where:

P, L: represent pseudo-range and carrier-phase observation in meters. Superscript k represents satellite,subscript i represents receiver,subscript F represents frequency;

$\rho = |r_i - r^k|$: represents the three-dimensional distance of a signal from a satellite to a satellite navigation receiver antenna

$r_i = [x_i, y_i, z_i]^T$ and $r^k = [x^k, y^k, z^k]^T$ represent the position vectors of the receiver antenna and the satellite respectively.

dt and δt represent clock correction of receiver and satellite respectively in seconds.

T tropospheric delay in meters

$I_{i,F}^k = \alpha TEC / f_F^2$: Ionospheric delay in meters,in which α is a constant,TEC is total electron density of signal propagation path,f_F is signal frequency.

dm and δm multipath error in pseudo-range and carrier-phase propagation channels

N: phase ambiguity in cycles

λ: carrier wavelength in m/cycle

c: propagation speed of light in vacuum in m/s

e and ϵ: pseudo-range noise and carrier phase noise respectively in meters.

3. The parameters and error correction of PPP

Concentrating on analyze and compare the ionospheric correction models, two of the ionospheric correction methods are UofC model and ionosphere-free combination model.

3.1. UofC

Ionospheric delay in pseudo-range and carrier-phase observation equations is equal in quantity and opposite in direction. The model of UofC establishes thought pseudo-range and carrier-phase observations summing up in half-rations.
Mathematical expressions are:

$$P_{(i,1)}^{k} = 1/2(P_{(i,1)}^k + L_{(i,1)}^k) = \rho_i^k + c(dt_i - \delta t^k) + T_i^k + I_{(i,1)}^k + dm_{(i,1)}^k + e_{(i,(P1,L1))}^k \quad (3)$$

$$P_{(i,2)}^{k} = 1/2(P_{(i,2)}^k + L_{(i,2)}^k) = \rho_i^k + c(dt_i - \delta t^k) + T_i^k + I_{(i,2)}^k + dm_{(i,2)}^k + e_{(i,(P2,L2))}^k \quad (4)$$

$$L_{(i,F)}^{k} = \frac{1}{(f_1^2 - f_2^2)}(f_1^2 L_1 - f_2^2 L_2) = \rho_i^k + c(dt_i - \delta t^k) + T_i^k + \frac{c(f_1 N_{(i,1)}^k - f_2 N_{(i,2)}^k)}{f_1^2 - f_2^2} + \frac{f_1^2 \varepsilon_{(i,1)}^k - f_2^2 \varepsilon_{(i,2)}^k}{f_1^2 - f_2^2} + \frac{f_1^2 \varepsilon_{(i,1)}^k - f_2^2 \varepsilon_{(i,2)}^k}{f_1^2 - f_2^2} \quad (5)$$

UofC model takes the advantage that ionospheric delay is equal in size and opposite in direction in pseudo-range and carrier-phase observation equations, so that it can eliminate the error. As for parameter estimation, it introduces an extra parameter of ambiguity.

3.2. Ionosphere-free combination model

Let $\alpha = \dfrac{f_1^2}{f_1^2 - f_2^2}$ and $\beta = \dfrac{f_2^2}{f_1^2 - f_2^2}$, then multiply α, β to observation equations of two frequency respectively, subtract each other, we can acquire equations as follows:

$$L_c = \frac{1}{(f_1^2 - f_2^2)}(f_1^2 L_1 - f_2^2 L_2) = \rho_i^k + c(dt_i - \delta t^k) + T_i^k + \frac{c(f_1 N_{(i,1)}^k - f_2 N_{(i,2)}^k)}{f_1^2 - f_2^2} + \frac{f_1^2 \delta m_{(i,1)}^k - f_2^2 \delta m_{(i,2)}^k}{f_1^2 - f_2^2} + \frac{f_1^2 \varepsilon_{(i,1)}^k - f_2^2 \varepsilon_{(i,F)}^k}{f_1^2 - f_2^2} \tag{6}$$

$$P_c = \frac{1}{f_1^2 - f_2^2}(f_1^2 P_1 - f_2^2 P_2) = \rho_i^k + c(dt_i - \delta t^k) + T_i^k + \frac{f_1^2 dm_{(i,1)}^k - f_2^2 dm_{(i,2)}^k}{f_1^2 - f_2^2} + \frac{f_1^2 e_{(i,1)}^k - f_2^2 e_{(i,2)}^k}{f_1^2 - f_2^2} \tag{7}$$

4. Static PPP experiment

In this section, the experiment of two models of PPP was done. Experiment was carried out on a top of building in Beijing, located in 39°57′N, 116°19′E on March 3th, 2018. It observed 12400 epochs, dual-frequency receiver used GPS L1 and L2 frequency. Satellite cut-off elevation angle was 10°. Troposphere delay model used tropospheric zenith delay model, receiver clock error correction looked as white noise approximately. In order to analyze the static positioning accuracy, we used observation data in RINEX format, meanwhile adopted precise ephemeris from IGS to correct the errors.

PPP result saved in RINEX format files in the version of 3.02, accompanied by precise ephemeris. Results were transformed into local Cartesian coordinate system, with three axes in E, N, U terms, which represent east, north and up respectively.

Dealing with ionosphere-free combination model, having converged, result is shown as figure 1:

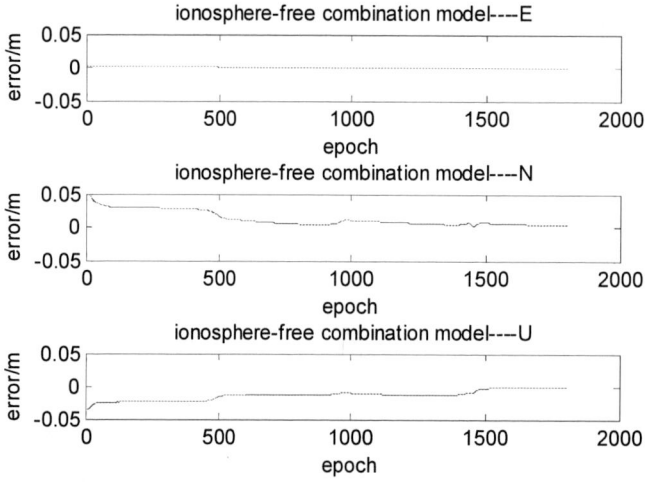

Figure 1. positioning errors dealt with ionosphere-free combination model

While, dealing with UofC model, having converged, result is shown as figure 2

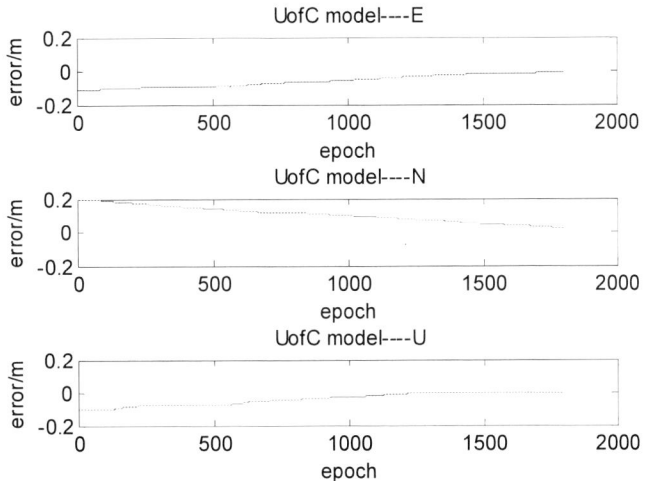

Figure 2. positioning errors dealt with UofC model

In order to compare positioning errors from two models, results after convergence are dealt with 2σ standard deviation. Result of two methods are list as follows:

Table 1. positioning result dealt with 2σ standard deviation.

	UofC	ionosphere-free combination
E(2σ)	0.1057	0.0018
N(2σ)	0.1914	0.0302
U(2σ)	0.0988	0.0245

From list 1, when positioning with ionosphere-free combination model, errors after convergence are within 4cm, which is centimeter-level. As for UofC model, errors after convergence are within 20cm, which is decimeter-level.

5. Conclusion
GPS is the first and most widely-used global satellite navigation system. PPP technique can provide centimeter-level positioning result. Based on the theoretical research on PPP, this paper compared the models of ionosphere-free combination and UofC. We did an experiment to compare the two models with dual-frequency GPS receiver, also using precise ephemeris from IGS. This experiment shows that, in static PPP experiment, PPP with ionosphere-free combination model is centimeter-level, while UofC model is decimeter-level. It also helps us with PPP when choosing models to eliminate ionosphere errors.

References
[1] Li Benyu. Research on GPS/GLONASS presice point positioning technique model and algorithm[M]. Shandong Agricultural University. 2010
[2] Zhang Xiaohong, Li Xingxing, Li Pan. Review of PPP and GNSS applications[J]. Acta Geodaetica et Cartographica Sinica. 2017,46 10):1399-1407
[3] ZUMBERGE J F, HEFLIN M B, JEFFERSON D C, etal. Precise Point Positioning for the efficient and Robust Analysis of GPS Data from Large Networks[J]. Journal of Geophysical

Research,1997,102(B3): 5005-5017

[4] KOUBA J, HEROUX P. Precise Point Positioning Using IGS Orbit and Clock Products[J]. GPS Solutions, 2001, 5(2): 12-28. DOI: 10.1007/PL0012883.

[5] YE Shirong．Theory and Its Realization of GPS Precise Point Positioning Using Un-differenced Phase Observation [D]．Wuhan: Wuhan University, 2002.

[6] GAO Y, SHEN X．Improving Ambiguity Convergence in Carrier Phase G based Precise Point Positioning [C]∥Proceedings of the 14th International Technical Meeting of the Satellite Division of The Institute of Navigation (ION GPS 2001). Salt Lake City, UT: Salt Palace Convention Center, 2001:1532-1539.

Dynamic Reactive Power Compensation and Harmonic Suppression of Optical Storage Microgrid Control in Natural Coordinates

Guangyao Jia[1] ,Shudong Wang[1], Huiying Song[1] ,Yanrong Mao[1] and AL-MADHEHAGI LUAI

[1]College of Electrical and Information Engineering, Lanzhou University of Technology, Lanzhou，Gansu, 730050, China

*Corresponding author's e-mail: jiagy3631@foxmail.com

Abstract. In order to solve the influence of load fluctuation on the power quality of the grid when the distributed photovoltaic power generation system is connected to the grid side, and to reduce the complexity of the control system, this paper proposes the grid-connected control of the microgrid with reactive power compensation and harmonic suppression in natural coordinates. This strategy maintains the DC bus voltage stability in a battery capacity state switching converter control method. Two kinds of grid-connected inverter power control strategies with reactive power compensation are designed for different working conditions, which realizes independent control of active and reactive power of grid-connected inverters, eliminating coordinate transformation and grid phase detection. The experimental results show that the proposed control strategy can achieve the smooth and efficient flow of the optical storage microgrid system, achieve the purpose of reactive power compensation, and has a good dynamic response.

1. Introduction

With the development of distributed power sources and microgrids, distributed generation technology is becoming more and more competitive in the power grid. As a kind of clean energy, solar power generation network has a higher proportion in the power system with its unique advantages. Photovoltaic power generation converts the DC power of the PV array into AC power that is in phase with the same frequency of the grid and feeds it to the grid, and ensures a high grid-connected power factor and improves resource utilization. However, the intermittent and random nature of solar energy itself has led to a series of problems such as large fluctuations in photovoltaic power generation voltage and frequency, power imbalance, poor power quality, and difficulty in grid connection. At present, the application scale of the optical storage power generation system is getting larger and larger. How to improve the application efficiency of the photovoltaic unit and reduce its influence on the stable operation of the system and the poor power quality are issues that must be considered.

In order to realize the unified control of grid-connected power generation, reactive power compensation and harmonic suppression, the system must be able to emit reverse-phase reactive power and harmonic current while detecting the reactive power and harmonic current of the grid load to offset the reactive power of the load. The effect of wave current on the grid. In [3], in order to reduce the influence of load reactive power and harmonics on the power grid, it is proposed to integrate photovoltaic grid-connected power generation and reactive power compensation to form a

Content from this work may be used under the terms of the Creative Commons Attribution 3.0 licence. Any further distribution of this work must maintain attribution to the author(s) and the title of the work, journal citation and DOI.

Published under licence by IOP Publishing Ltd

photovoltaic grid-connected power regulation system, but the system harmonic current is not suppressed. Literature [5] proposed a grid-connected control scheme for optical storage power generation system with reactive dynamic compensation capability. This scheme achieves the purpose of independent control of active and reactive power through Parker transformation, but its coordinate transformation and grid phase detection are relatively complicated. Literature [8] proposes a new definition of generalized reactive current and reactive power in the abc coordinate system still applicable to grid voltage distortion, and gives a method for detecting and compensating generalized reactive current.

Based on the above analysis, this paper proposes the grid-connected control of the microgrid with reactive power compensation and harmonic suppression in natural coordinates. The strategy is based on the instantaneous reactive power theory to detect the reactive and harmonic currents of the system, and achieves an effective and fast response of reactive power compensation and active filtering. At the same time, in order to reduce the complexity of the system, a grid-connected control strategy of optical storage micro-grid with reactive dynamic compensation in natural coordinate system is proposed. According to the state of charge of the energy storage battery, two grid-connected inverter power control is adopted, namely PQ. Power control and VQ power control. The independent control of the active and reactive power of the grid-connected inverter is realized, and the coordinate transformation and the grid phase detection are omitted. In this paper, the combination of photovoltaic power generation control with reactive power compensation and active filter control can not only effectively perform photovoltaic power generation, improve power supply quality and reduce power loss, but also save investment in corresponding equipment and broaden photovoltaic grid-connected power generation. The scope of application.

2. Instantaneous reactive power and harmonic current detection

There are many methods for detecting instantaneous reactive current. In [4], the principle of instantaneous reactive power and harmonic current detection in transient reactive power theory is used to detect the fundamental reactive power and harmonic current of the system. In [6], based on FFT 's high-precision harmonic detection algorithm, accurate detection of non-integer harmonics is proposed. Because the response requirements of the inverter are relatively fast and both are instantaneous values, this paper detects the reactive current of the system based on the instantaneous reactive current instantaneous current detection method, which not only meets the requirements of the fast response of the grid-connected inverter, but also reduce the complexity of the control system. For the transient current detection method of instantaneous reactive power theory, see the literature [4].

Let the three-phase grid voltage be symmetrical and positive. The three-phase currents i_a, i_b, and i_c are transformed by C_{32} and C to obtain the current active component i_p and the reactive component i_q in the $p - q$ rotating coordinate system, respectively.

$$\begin{bmatrix} i_\alpha \\ i_\beta \end{bmatrix} = C_{32} \begin{bmatrix} i_a \\ i_b \\ i_c \end{bmatrix} \quad (1)$$

In the middle $C_{32} = \sqrt{\dfrac{2}{3}} \begin{bmatrix} 1 & -\dfrac{1}{2} & -\dfrac{1}{2} \\ 0 & \dfrac{\sqrt{3}}{2} & -\dfrac{\sqrt{3}}{2} \end{bmatrix}$

$$\begin{bmatrix} i_p \\ i_q \end{bmatrix} = C \begin{bmatrix} i_\alpha \\ i_\beta \end{bmatrix} \quad (2)$$

In the middle $C = \begin{bmatrix} \sin \omega t & \cos \omega t \\ -\cos \omega t & -\sin \omega t \end{bmatrix}$

Where ω is the power angle frequency. Calculated by the above formula, the current components i_p and i_q in the $p-q$ rotating coordinate system are obtained.

When detecting reactive current and harmonics, Bip can calculate the fundamental active current components i_{apf}, i_{bpf} and i_{cpf} of the detected current.

$$\begin{bmatrix} i_{apf} \\ i_{bpf} \\ i_{cpf} \end{bmatrix} = C_{23} C^{-1} \begin{bmatrix} Bip \\ 0 \end{bmatrix} \quad (3)$$

In the middle $C^{-1} = \begin{bmatrix} \sin \omega t & -\cos \omega t \\ -\cos \omega t & -\sin \omega t \end{bmatrix} \quad C_{23} = \sqrt{\dfrac{2}{3}} \begin{bmatrix} 1 & 0 \\ -\dfrac{1}{2} & -\dfrac{\sqrt{3}}{2} \\ -\dfrac{1}{2} & -\dfrac{\sqrt{3}}{2} \end{bmatrix}$

The sum of the harmonic component of the detected current and the fundamental reactive component is:

$$i_{aq} = i_a - i_{apf} \quad i_{bq} = i_b - i_{bpf} \quad i_{cq} = i_c - i_{cpf} \quad (4)$$

On the basis of the completion of reactive power and harmonic current detection, the detected value of each phase is used as the command value of the reactive current component. The current feedback control of the system satisfies the reverse phase requirement of the phase of the control current, and the system generates reversed reactive power and Harmonic current, which in turn achieves reactive power compensation and harmonic suppression.

3. Grid-connected control of optical storage microgrid

3.1. System structure of optical storage microgrid

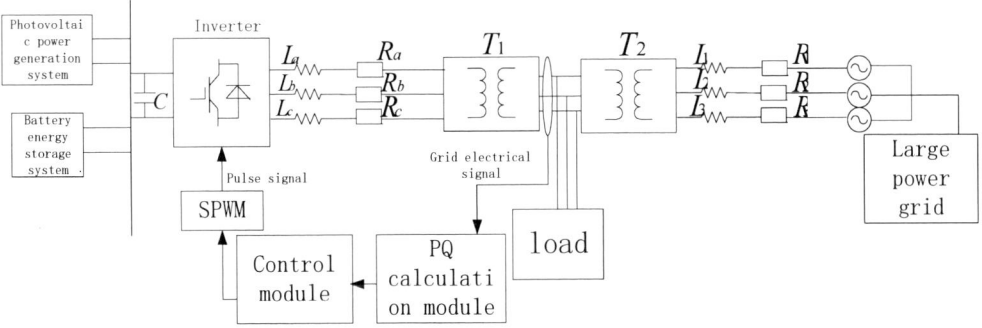

Figure 1 System structure of optical storage microgrid

The system structure of the optical storage microgrid system connected to the grid is shown in Figure 1. The system consists of photovoltaic system, energy storage system, converter, transformer, load and distribution network.

3.2. Energy storage system

The energy storage system is connected in series through a plurality of battery modules to obtain a higher voltage level, and is boosted and regulated by a bidirectional DC/DC converter. The energy storage system has two control modes, namely voltage regulation control and constant current control. The voltage regulation control adjusts the power balance of the energy storage battery charging and discharging power control system to maintain the constant DC voltage; the constant current control passes a given small As the reference value of the energy storage discharge current and the reference value of the charging current, the value is charged and discharged by the constant current of the energy storage battery to ensure the normal use of the battery. In this paper, a general-purpose battery is used as the energy storage device. The output voltage of the energy storage battery is:

$$V_b = V_0 + R_b i_b - K \frac{Q}{Q + \int i_b dt} + C \exp\left(B \int i_b dt\right) \quad (5)$$

$$SOC = 100\left(1 + \frac{\int i_b dt}{Q}\right) \quad (6)$$

Where: R_b is the resistance of the battery; V_b and V_0 are the battery output voltage and open circuit voltage; i_b is the charging current, K is the battery polarization voltage; Q is the battery capacity; K, B and C are constant.

3.3. Control of reactive power compensation system in natural coordinates

1) Assume that in a three-phase symmetrical AC system, the grid direction is defined as the active power direction, and its instantaneous amplitude is selected as the reference value. The instantaneous amplitude of the grid voltage is:

$$U_s = \sqrt{\frac{2}{3}\left(u_{sa}^2 + u_{sb}^2 + u_{sc}^2\right)} \quad (7)$$

Then the three-phase active unit components are:

$$v_a = \frac{u_{sa}}{U_s}, \quad v_b = \frac{u_{sb}}{U_s}, \quad v_c = \frac{u_{sc}}{U_s} \quad (8)$$

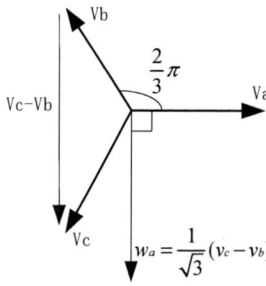

Figure 2 Phasor relationship between unit active and reactive components

The unit reactive component orthogonal to the active component is:

$$w_a = \frac{1}{\sqrt{3}}\left(v_c - v_b\right), \quad w_b = \frac{1}{\sqrt{3}}\left(v_a - v_c\right), \quad w_c = \frac{1}{\sqrt{3}}\left(v_b - v_a\right) \quad (9)$$

The three-phase active current components are:

$$i_{pa} = i_p v_a, \quad i_{pb} = i_p v_b, \quad i_{pc} = i_p v_c \quad (10)$$

The three-phase reactive current components are:

$$i_{qa} = i_q w_a, \quad i_{qb} = i_q w_b, \quad i_{qc} = i_q w_c \quad (11)$$

Then the three-phase reference value of the microgrid grid-connected inverter output is:

$$i_a^* = i_{pa} + i_{qa}, \quad i_b^* = i_{pb} + i_{qb}, \quad i_c^* = i_{pc} + i_{qc} \quad (12)$$

2) Operation control of grid-connected inverter

In order to convert the direct current output from the photovoltaic power generation system and the energy storage system into an alternating current through an inverter to realize grid-connected power generation, the inverter should adopt a corresponding control method.

(1) PQ power control. The collected voltage and current signals are obtained by the PQ calculation module to obtain the real-time measured value of the power, and are respectively made to be different from the given values of the active power and the reactive power, and the inner loop active current command value and the reactive current command value are obtained through the PI link. Then, it is respectively multiplied by the active component and the reactive component of the three-phase current, and then the corresponding vectors are added to obtain the three-phase reference current value of the inner loop. The current inner loop reference current i_{abc}^* is compared with the measured current i_{abc}, and the pulse signal is sent to the grid-connected inverter through the $SPWM$ modulation link, so that the inverter active and reactive output can be realized. The active and reactive control signals of this control strategy are taken from the inverter output active and the grid side input reactive power, so that the inverter can output the active power and control the fixed input reactive power of the grid respectively. The inverter outputs the corresponding reactive power according to the fluctuation of the load reactive power. The measured value in the whole process is the instantaneous value of the power grid connected to the load side, and the dynamic real-time tracking compensation effect is achieved.

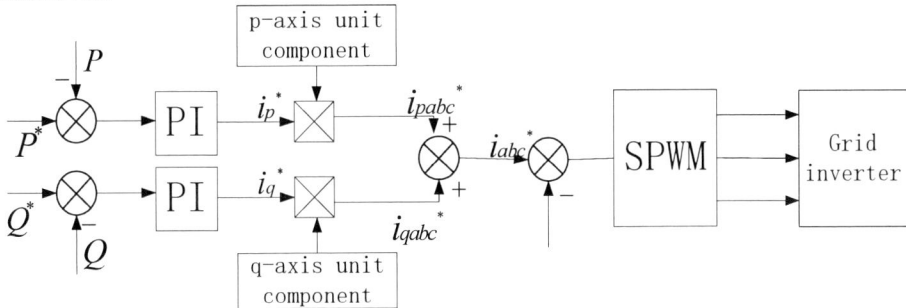

Figure 3 PQ power control flow chart

(2) VQ power control. When the optical storage microgrid system is started, if the energy storage system is in the constant current control mode and the grid-connected inverter adopts PQ control, the DC bus voltage is in an uncontrolled state, and it is difficult to ensure the input power and output of the inverter. The power is consistent, causing the system to crash. In order to ensure stable operation of the system, the DC bus voltage must be stabilized.

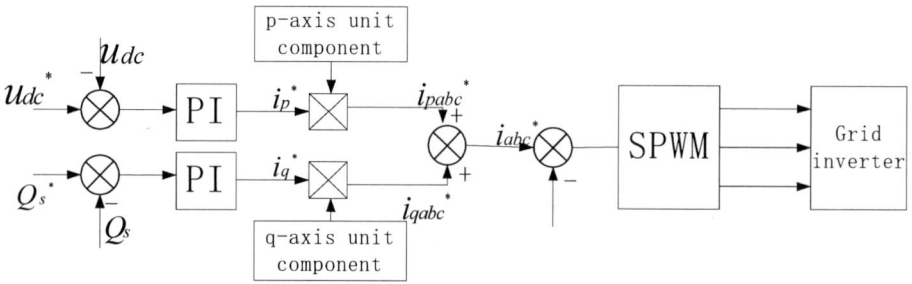

Figure 4 VQ power control flow chart

The power is consistent, causing the system to crash. In order to ensure stable operation of the system, the DC bus voltage must be stabilized. Therefore, when the capacity $SOC < 5\%$ or $SOC > 95\%$ of the energy storage battery is used, the control method as shown in Fig. 3 is employed. By controlling the DC bus voltage, the stability of the DC bus voltage is maintained, and the balanced flow of the grid-connected inverter power is ensured, so that the system operates stably.

4. Simulation results

In order to verify the effectiveness of the control strategy, this paper builds the optical storage microgrid system shown in Figure 1. In this model, the light intensity is $1000W / m^2$, and the maximum power output of photovoltaic power generation is $15kW$. The rated voltage of the battery is $200V$ and the capacity is $300Ah$. The initial state of the battery capacity SOC is 30%. There is no additional reactive power compensation device, and the simulation duration is $5S$. The rated voltage of the load is $500V$. In the initial state, there is a $10kW$ work load, and the reactive load is .At $2.5S$, the active load is unchanged and the reactive load is increased to $9kVar$. Compared with conventional unit power factor grid-connected control under the same conditions. The simulation results are shown in the following figure: Figure 5 shows the RMS effective value of the load line voltage under normal unit power factor, the reactive output on the inverter side and the large grid side. Figure 6 shows the RMS value of the load line voltage, the reactive side of the inverter side and the large grid side under the control strategy of the grid-connected inverter.

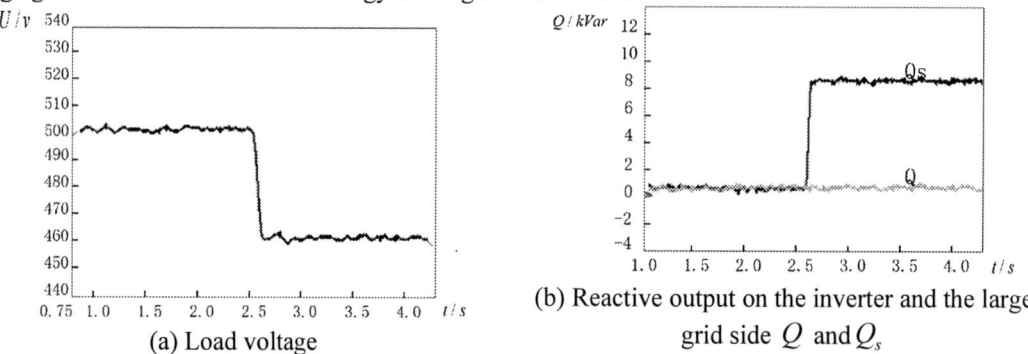

(a) Load voltage

(b) Reactive output on the inverter and the large grid side Q and Q_s

Figure 5 results under the general control strategy

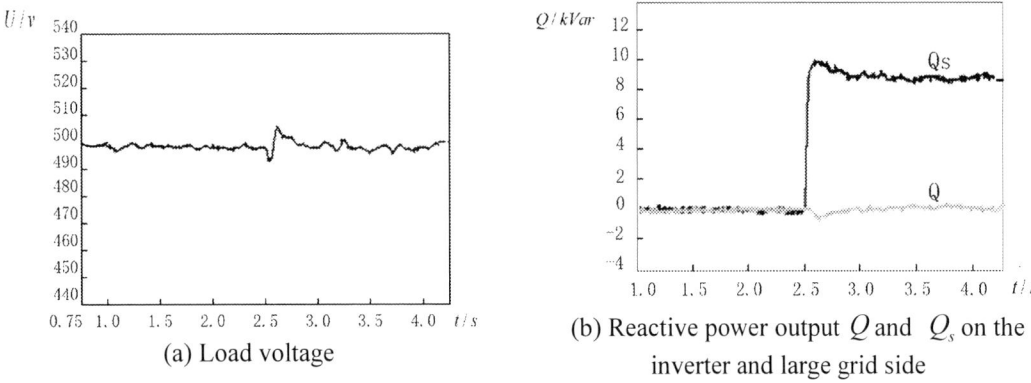

(a) Load voltage

(b) Reactive power output Q and Q_s on the inverter and large grid side

Figure 6 results under the new control strategy

5. Conclusion

The simulation results show that for the conventional control strategy, when the reactive load changes, the voltage of the common coupling point changes greatly due to the transmission of reactive power, which is not conducive to providing reliable power quality. Comparing the simulation waveforms obtained by using two different control strategies, it can be known that for the new control strategy, and it has a certain reactive power compensation capability, which can effectively ensure the stability of the common coupling point voltage.

References

[1] Wang Zhaoan, Yang Jun, Liu Jinjun. Harmonic suppression and reactive power compensation [M]. Beijing: Mechanical Industry Press, 1998.

[2] Zhang Xing, Cao Renxian. Solar photovoltaic grid-connected power generation and its inverter control. Beijing: Mechanical Industry Press, 2013.1-26.

[3] Wang Haining, Su Jianhui, Ding Ming, et al. Photovoltaic grid-connected power regulation system[J]. Proceedings of the CSEE, 2007, 27(2): 75-79.

[4] Wang Haining, Su Jianhui, Zhang Guorong, Ding Ming. Research on Control of Photovoltaic Grid-Connected Power Regulator with Reactive Power Compensation and Harmonic Suppression[J]. Journal of Solar Energy, 2006, 27(6): 540-544.

[5] Li Bin, Fan Shoulu, Tian Xiaohe, et al. Grid-connected control scheme of reactive power compensation for optical storage power generation system [J]. Journal of Tianjin University: Natural Science and Engineering Technology Edition, 2013, 46(11): 977-983.

[6] HILLC, SUCHMC, CHEND, et al. Battery energy storage for enabling integration of distributed solar power generation [J]. Smart Grid, IEEE Transactions on, 2012, 3(2) : 850-857.

[7] Xue Wei, Yang Rengang. High-precision harmonic detection algorithm based on FFT[J]. Proceedings of the CSEE, 2002, 22(12): 106-110.

[8] Yin Bo, Chen Yunping. Definition and compensation of generalized reactive current and power in abc coordinate system[J]. Power grid technology,2003,27(7): 43-51.

[9] Cao Jun, Hu Likun, Lu Zhilin, et al. Grid-connected control of optical storage microgrid with reactive dynamic compensation[J]. Journal of Guangxi University, 2015, 40(6): 1424-1430.

[10] Cao Jun. Research on grid-connected control strategy of optical storage micro-grid with reactive dynamic compensation [D]. Guangxi: Guangxi University, 2016.

Comparison and Analysis of X86 Server and Minicomputer Application in Power Enterprises

Yang Bo[1], Wei Jun[1], Wang Hua[1] and Wang Gang[1].

State Grid Gansu Provincial Electric Power Company Information and Communication Company. No. 629, Xijin East Road, Qilihe District, Lanzhou City, Gansu Province

E- mail: 24207812@qq.com

Abstract. With the transition from "centralized" to "distributed" infrastructure and the continuous development of cloud platform construction, the disadvantages of the traditional small-sized rack have emerged. By comparing and analyzing the performance of X86 server and minimal machine, we evaluated the performance of X86 and minicomputer in different applications combined with the actual situation of Gansu Electric Power Company of the State Grid, and then carried out the application migration. The results show that the X86 architecture platform has the ability to replace the minicomputer architecture in a reasonable design in the data center.

1. Introduction

In recent years, the State Grid Corporation of China has proposed the construction plan of "big data, cloud computing, Internet of Things and mobile Internet", and the informatization construction aims to achieve "one platform, one system, multiple channels, and micro-applications." The transformation of the infrastructure of the basic hardware and software environment from "centralized" to "distributed" is an important part of supporting the company's construction. Based on its inherent architectural advantages, minicomputer has excellent performance in processing performance and RAS features, and is widely used in various application systems in the power industry. Although the company has vigorously promoted the construction of x86 servers instead of minicomputers, there are still a large number of existing systems running on minicomputers, which have brought many hidden dangers for the future cloud computing and distributed architecture construction. Currently, based on the advantages of the performance, reliability and cost performance of the X86 cluster architecture, the traditional centralized architecture is becoming more and more clear to the X86 cluster architecture and the distributed architecture [1]. By comparing and analyzing the performance and application of X86 server and minicomputer in X86 system, this paper illustrates the possibility of X86 server replacing minicomputer.

2. Performance Analysis of X86 Servers and Minicomputer

The industry has a variety of indicators to evaluate the performance of the host computer. Representative examples include TPC-C/TPC-E, SPECjbb/SPECjEnterprise, and SAPS, which correspond to database, JAVA and ERP service processing capabilities.

Content from this work may be used under the terms of the Creative Commons Attribution 3.0 licence. Any further distribution of this work must maintain attribution to the author(s) and the title of the work, journal citation and DOI.
Published under licence by IOP Publishing Ltd

2.1. TPC-C/TPC-E Performance Test

TPC-C is an industry-standard benchmarking project designed to measure the performance and scalability of online transaction processing (OLTP) systems. This benchmark project will test database functions including queries, updates, and queued small batch transactions. TPC-C testing is widely used to measure the performance of the overall system built by the server and the client under the C/S environment, which is made by the TPC (Transaction Processing Corp). It simulates simple database and business logic [2].

TPC-C tests the number of tasks per minute processed by the system, and the unit is tpmC(transactions per minute), which C refers to the C benchmark program in TPC. It is defined as the number of new orders processed by the system per minute. It should be noted that while processing new orders, the system also handles other four types of transaction requests. A new order request is unlikely to exceed 45% of all transaction requests, so when a system's performance is 1000tpmC, the number of requests that it actually handles is more than 2000 per minute.

In recent years, TPC has launched TPC-E, which is also a database transaction processing test that simulates asecurities trading system. As with TPC-C, it actually measures the performance of server and database software dealing with online query transaction processing (OLTP). TPC-E test results must provide tpsE value, which is how many TPC-E database transactions are completed per second (transaction per second). Compared with TPC-C test, TPC-E adds application server layer to test model building, and increases the complexity of database structure.

The TPC-C tests are mainly before 2011 and lacks newer CPU test data. The following table shows the TPC-C performance data of some hosts:

Table 1. TPC-C performance data.

HostModel	Configuration	TPC-C Value (tpmC)
HP DL580G5	Xeon X7350 2.93GHz, 4CPU(16Core, 16Threads)	407,079
IBM x3950 M2	Xeon X7350 2.93GHz, 8CPU(32Core, 32Threads)	841,809
IBM x3950 M2	Xeon X74602.67GHz, 8CPU(48Core, 48Threads)	1,200,632
IBM x3850 X5	Xeon E7-88702.40GHz, 4CPU(40Core, 80Threads)	3,014,684
Sun Server X2-8	Xeon E7-8870 2.40GHz, 8CPU(80Core, 160Threads)	5,055,888
HP RX5670	Itanium2 1.5GHz, 4CPU(4Core, 4Threads)	121,065
HP Superdome	Itanium2 1.5GHz, 64CPU(64Core, 64Threads)	658,277
HP RX6600	Itanium2 1.6GHz, 4CPU(8Core, 16Threads)	372,140
HP Superdome	Itanium2 1.6GHz, 64CPU(128Core, 256Threads)	4,092,799
IBM p5 570	POWER5 1.9GHz, 2CPU(4Core, 8Threads)	203,439
IBM p5 570	POWER5 1.9GHz, 32CPU(64Core, 128Threads)	3,210,540
IBM p570	POWER6 4.7GHz, 2CPU(4Core, 8Threads)	404,462
IBM p595	POWER6 5.0GHz, 32CPU(64Core, 128Threads)	6,085,166
IBM Power 780	POWER74.14GHz, 2CPU(8Core, 32Threads	1,200,011

From the above data, we can see:

• Due to the low clock frequency and the small number of single-CPU cores, Itanium2's performance is comparable to that of the contemporary Xeon (RX6600 vs DL580 G5). In recent years, the performance of X86 architecture has been improved rapidly, and now it has developed to 28 cores in a single CPU. But Itanium architecture is growing slowly, from Itanium2 to Itanium3, the number of single-CPU cores has only doubled. Therefore, we can predict that the current performance of the minicomputer based on Itanium CPUs has lagged behind the similarly configured X86 server.

- There is a nonlinear relationship between host performance and CPU quantity, such as the 8-way E7 8870provides about 80% more performance than the 4-way (x3850 vs X2-8).
- If there is a high requirement for the performance of a single host, the selection of a high-end model of a minicomputer is a better choice at that time. Because more than 8-way X86 servers require third-party controllers, only a few vendors, such as SGI, were able to offer such products before 2010.

Selecting 8 X86 servers, 4 Itanium and 2 Power minicomputers, comparing the TPC-C performance curve of these architecture hosts from 2004 to 2010, it can be seen that the performance of X86 servers improved fastest and the performance of minicomputers improved slower.

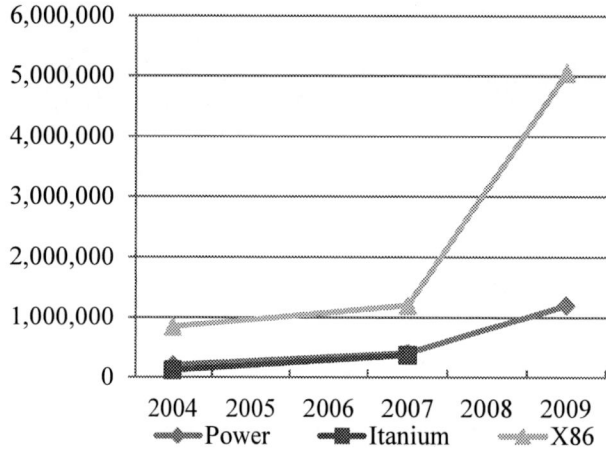

Figure 1. The growth trend of TPC-C performance of different architecture hosts.

The TPC-E test data is mainly generated after 2008 and based on the X86 architecture, with only a small amount of minicomputer performance data:

Table 2. TPC-E performance data.

Host Model	Configuration	tpsE
Lenovo SR650	Xeon Platinum 8180 2.50GHz, 2CPU(56Core, 112Threads)	6,598.36
Lenovo x3850 X6	Xeon E7-8890 v4 2.20GHz, 4CPU(96Core, 192Threads)	9,068.00
Lenovo x3650 M5	Xeon E5-2699 v4 2.20GHz, 2CPU(44Core, 88Threads)	4,938.14
Lenovo x3950 X6	Xeon E7-8890 v3 2.50GHz, 8CPU(144Core, 288Threads)	11,058.99
Lenovox3850 X6	Xeon E7-8890 v3 2.50GHz, 4CPU(72Core, 144Threads)	6,964.75
Lenovo x3950 X6	Xeon E7-8890 v2 2.80GHz, 8CPU(120Core, 240Threads)	9,145.01
IBM x3850 X6	Xeon E7-4890 v2 2.80GHz, 4CPU(60Core, 120Threads)	5,576.27
IBM x3650 M4	Xeon E5-2697 v2 2.70GHz, 2CPU(24Core, 48Threads)	2,590.93
IBM x3850 X5	Xeon E7-8870 2.40GHz, 8CPU(80Core, 160Threads)	5,457.20
IBM x3850 X5	Xeon E7-4870 2.40GHz, 4CPU(40Core, 80Threads)	3,218.46
IBM x3650 M4	Xeon E5-2690 2.90GHz, 2CPU(16Core, 32Threads)	1,863.23
Unisys ES7000	Xeon X7460 2.67 GHz, 8CPU(48Core, 48Threads)	1,165.56
NEC Express5800/1320Xf	Itanium 9150N 1.6 GHz, 32CPU(64Core, 64Threads)	1,126.49

Through the test results, the current X86 architecture CPU continues to maintain a faster performance improvement. The same as the 2-way server, the performance of the Platinum 8180 host is 3.54 times that of the E5-2690 host. Although there is a lack of test results for the latest generation of minicomputers, combined with the data of TPC-C and TPC-E, it can be inferred that the current high-end X86 servers already have the large-scale database application processing capabilities that the previous generation of high-end minicomputers can afford.

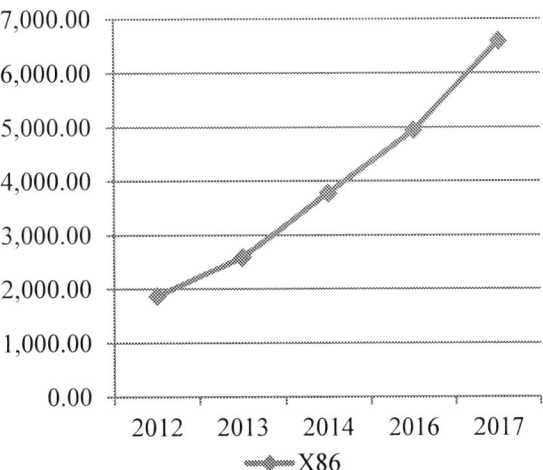

Figure 2. The growth trend of TPC-E performance of X86 servers.

2.2. SPECjbb/SPECjEnterprise Performance Test

SPEC (the Standard Performance Evaluation Corporation) is a global, authoritative third-party application performance testing organization, which aims to establish, modify, and identify a series of standards for performance evaluation of server applications. This test is currently one of the industry's standard and authoritative benchmarks and has been supported and participated by many international software and hardware vendors such as Intel, IBM, HP, DELL, SUN, HUA-WEI and so on. Because it embodies the performance and cost index of the software and hardware platform, it has been selected by key industries such as finance, telecommunications, and securities as an authoritative selection indicator for IT systems.

SPECjbb (Java server benchmark) is a SPEC testing tool to evaluate the performance of server-side JAVA. SPECjbb evaluates the performance of server-side JAVA by simulating a three tier C/S system (primarily middletier). This test software runs JVM (Java virtual machine), JIT (Just-In-Ti-me) compiler, fragment collection, threads and other tasks of the operation system. It also measures the performance of CPU, Cache, memory and SMP. SPECjbb reflects the situation of real-world application systems by providing new, enhanced workloads that run in an object-oriented manner.

SPECjbb contains multiple versions, each with different indicators [3]. The latest version is SPECjbb2015, and MultiJVM will test multiple JVMs in one operating system. The max-jOPS performance index score reflects sustainable system-wide throughput, and the critical-jOPS performance index score measures throughput in the case of limited response time.

Table 3. Comparison of SPECjbb2015 performance between minicomputers and X86 servers.

Host Model	CPU Model	max-jOPS	critical-jOPS
HP DL120 G9	Xeon E5-2699 v3 2.3GHz, 1CPU(18Core. 36Threads)	47,334	9,876
IBM S812LC	Power8 3.49GHz,1CPU(10Core, 80 Threads)	44,883	13,032
Inspur NF5280M4	Xeon E5-2699A v4 2.4GHz, 2CPU(44Core, 88 Threads)	122,829	39,635
Oracle SPARC T7-1	SPARC M7 4.13GHz, 1CPU(32Core, 256 Threads)	120,603	60,280

The previous table shows that the current X86 servers and the Power minicomputers have the same performance in the JAVA application with the same number of CPUs, but there is still a gap compared to the SPARC minicomputer. Since most of these applications can improve performance with simple scale-out, the performance requirements for a single host arenot high, so minicomputers are rarely used.

SPECjEnterprise (J2EE Application Performance Benchmark) tests the performance of J2EE application server by simulating the automotive supply chain system. The SPECjEnterprise test results must provide the EjOPS value, that is, how many enterprise JAVA operations per second (Enterprise Java Operation Per Second), and the current version is SPECjEnterprise2010.

Table 4. Comparison of SPECjEnterprise 2010 performance between minicomputers and X86 servers.

ost Model	CPU Model	EjOPS
IBM Power 780	Power7 3.86GHz, 8CPU(64Core, 256Threads)	16,646.34
Lenovo x3650 M5	Xeon E5-2697 v3 2.6GHz, 2CPU(28Core, 56Threads)	19,282.14
Oracle Server X6-2	Xeon E5-2699 v4 2.2GHz, 2CPU(44Core, 88Threads)	27,803.39

As can be seen from the above table, the X86 server has fully surpassed the Power minicomputer in terms of J2EE performance.

2.3. For SAPS Performance Test

The SAPS benchmark is a benchmarking tool designed specifically for SAP ERP enterprise resource management applications, and the related vendors must pass SAPS test performance as a standard indicator for SAP server configuration. The SAP benchmarking organization publishes various kinds of benchmarks, among which the SAP SD 2-Tier benchmark is used to measure the performance of different hardware vendors plus databases to execute the sales and distribution (SD) module of the SAP enterprise resource management application. The SAP SD 2-tier structure benchmark applies the application server and database server to the same physical server. The test results will be standardized into the SAPS (SAP Application Performance Standard) value of SAP SD application module. SAPS is a hardware-independent performance indicator. The 100 SAPS value is equivalent to 2000 fully business processed order line items per hour in the SAP SD application definition. Each commercial processing order item includes new order creation, invoice generation, order display, change of delivery content, goods entry, listing of orders, and receipt generation. From a technical point of view, it is equivalent to 2400 SAP transactions per hour or 6000 conversations per hour (console change) plus 2,000entry operations per hour.

Table 5. Comparison of SAPS performance between minicomputers and X86 servers.

Host Model	CPU Model	SAPS
IBM Power S824	Power8 3.52GHz, 4CPU(24Core, 192Threads)	115,870
IBM Power E870	Power8 4.19GHz, 8CPU(80Core, 640Threads)	436,100
IBM Power 750	Power7 3.55GHz, 4CPU(32Core, 128Threads)	94,730
IBM Power 760	Power7+ 3.41GHz, 8CPU(48Core, 192Threads)	139,220
IBM Power 780	Power7+ 3.72GHz, 12CPU(96Core, 384Threads)	311,720
IBM Power 795	Power7 4.0GHz, 16CPU(128Core, 512Threads)	384,330
HP RX6600	Itanium 2 90501.6Ghz, 4CPU(8Core, 16Threads)	10,780
HP Superdome	Itanium 2 90501.6Ghz, 16CPU(32Core, 64Threads)	28,200
HP Superdome	Itanium 2 90501.6Ghz, 32CPU(64Core, 128Threads)	46,380
Oracle SPARC T7-2	SPARC M74.133Ghz, 2CPU(64Core, 512Threads)	168,600
Oracle SPARC M7-8	SPARC M74.133Ghz, 8CPU(256Core, 2048Threads)	713,480
Oracle SPARC M6-32	SPARC M63.6Ghz, 32CPU(384Core, 3072Threads)	793,930

Oracle SPARC M5-32	SPARC M53.6Ghz, 32CPU(192Core, 1536Threads)	472,600
HPE DL380 G10	Xeon Platinum 8180M 2.5Ghz, 2CPU(56Core, 112Threads)	154,020
Dell R940	Xeon Platinum 8180 2.5Ghz, 4CPU(112Core, 224Threads)	341,100
Fujitsu 3800B	Xeon Platinum 8180 2.5Ghz, 8CPU(224Core, 448Threads)	562,100
Lenovox3650 M5	Xeon E5-2699 v42.2Ghz, 2CPU(44Core, 88Threads)	110,670
Dell R930	Xeon E7-8894 v42.4Ghz, 4CPU(96Core, 192Threads)	213,900
Lenovo x3950X6	Xeon E7-8894 v42.4Ghz, 8CPU(192Core, 384Threads)	421,230
Dell R730	Xeon E5-2699 v32.3Ghz,2CPU(36Core, 72Threads)	87,600
Cisco UCS C460 M4	Xeon E7-8890 v32.5Ghz, 4CPU(72Core, 144Threads)	159,130
Lenovo x3950 X6	Xeon E7-8890 v32.5Ghz, 8CPU(144Core, 288Threads)	330,930
Dell R720	Xeon E5-2697 v22.7Ghz, 2CPU(24Core, 48Threads)	54,120
Cisco UCS C460 M4	Xeon E7-4890 v22.8Ghz, 4CPU(60Core, 120Threads)	133,820
IBM x3950 X6	Xeon E7-8890 v22.8Ghz, 8CPU(120Core, 240Threads)	271,080
Cisco UCS B200 M3	Xeon E5-26902.9Ghz, 2CPU(16Core, 32Threads)	35,680
HP DL560 G8	Xeon E5-46502.7Ghz, 4CPU(32Core, 64Threads)	69,550
IBM x3850 X5	Xeon E7-88702.4Ghz, 8CPU(80Core, 160Threads)	140,720

As can be seen from the above table, in terms of SD application, the performance of Itanium processor is poor, and the performance of 2-way E5 V2 or 4-way E7 has exceeded 32-way Superdome. The performance of the 8-way Xeon E7 V4 is equivalent to that of the same number of Power 8, while the latest Xeon Platinum performance has surpassed Power 8.

2.4. State Grid Actual Business TPS Test

In order to get better rankings, vendors usually provide products with high configuration in all aspects when they participate in the above tests such as TPC, SPEC, and SAPS. They are often not exactly consistent with the configuration used in the real environment [4], so these data cannot fully respond to the actual effects of the two types of hosts. Therefore, the State Grid Corporation of China organized comparison tests between X86 servers and minicomputers. By deploying the PMS 2.0 system database in the hosts under these two architectures, the performance difference of the maximum transaction processing capability (TPS) of the hardware platform in the actual business system of the State Grid is verified.

Test methods:

- The 8-wayX86 servers and the 4-way P570 minicomputers are used tode ploy two-node RAC, and database is installed to PMS2.0 system. Using100, 200, 300, 400, 500 and other different concurrent users to perform test model standard scripts, it can test the device respectively to achieve the maximum transaction processing capability (TPS) value, and record the response time of each service module and the performance index of the device under test such as CPU, memory, and IO.

- Scale the two-node RAC horizontally to three nodes and repeat the above test.

When an 8-way X86 server (Xeon E7-8870 2.40GHz, 80Core, 160 threads) is used as a database, with the increase of concurrent users, the load change and transaction processing capabilities are shown as the following able.

Table 6. X86 server load and transaction processing data.

Concurrent Users	2 Nodes				3 Nodes				
	CPU Utilization			TPS Value	CPU Utilization				TPS Value
	Node1	Node2	Average		Node1	Node2	Node3	Average	
100	1.54	6.74	4.14	25.63	4.73	2.81	2.64	3.39	25.87
200	8.4	8.48	8.44	51.11	7.07	5.09	5.13	5.76	51.57
300	16.49	12.93	14.71	74.46	10.33	8.17	7.42	8.64	76.5
400	27.97	26.72	27.345	91.79	14.47	11.71	10.97	12.38	100.67
500	45.5	42.16	43.83	106.31	21.26	17.65	16.7	18.54	121.43
Max(700)	62.97	67.28	65.125	119.51	37.91	36.28	34.22	36.14	140.38

When the 4-way P750 minicomputer (Power7, 32Core, 128 thread) is used as a database, with the increase of concurrent users, the load changes and transaction processing capabilities are shown as the following table.

Table 7. Minicomputer load and transaction processing data.

Concurrent Users	2 Nodes				3 Nodes				
	CPU Utilization			TPS Value	CPU Utilization				TPS Value
	Node1	Node2	Average		Node1	Node2	Node3	Average	
100	15.01	14.41	14.71	25.32	10.72	8.77	12.31	10.6	25.01
200	28.28	30.39	29.34	49.87	20.86	17.29	22.65	20.27	49.93
300	44.79	42.28	43.54	74.95	31.55	25.98	32.5	30.01	75.02
400	58.9	57.69	58.3	98.13	38.98	37.3	43.12	39.8	99.01
500	67.3	65.23	66.27	119.23	50.28	47.61	50.59	49.49	122.98
Max(700)	79.67	77.69	78.68	153.62	56.49	50.11	60.74	55.78	152.19

Combining the data of the above two tables, when the number of concurrent users is relatively small, the overall performance of the 8-way X86 server is superior to about 30% of the P750 minicomputer with the advantages of the number of its CPU kernel. With the increase of the number of concurrent users, the P750 minicomputer has the performance advantage of its single-core CPU, and the performance difference with the 8-way X86 server gradually becomes smaller.

3. Performance Evaluation of X86 Server Substitutes for Minicomputer

3.1. Performance Estimation
Since the performance of minicomputers and X86 servers in different types of business applications is not the same, the following are respective evaluations of the existing business application types of Gansu Electric Power Company.

3.1.1. Database applications
Since the HP minicomputers currently used by Gansu Electric Power Company are all equipped with Itanium2 processors, and comprehensively compare TPC-C performance data, Itanium processors are expanded from 4 ways to 64 ways, with performance data of 372, 140 and 4,092,799 respectively. Considering the expansion efficiency and nonlinear growth of the NUMA architecture, assuming no more than 64 CPUs, the processor every doubling, the database application performance average growth index of N1:

- From $372,140 * N1^4 = 4,092,799$, the following can be derived: $N1 \approx 1.82$.
- It is estimated that the TPC-C performance of 8-way CPU (equivalent toRX8640) is as follows:

- 372,140*1.82≈677,295.

The TPC-C performance of 16-way CPU (equivalent to a single Superdome partition) is as follows:

- 372,140*1.82^2≈1,232,677.

According to the TPC-C and TPC-E performance data of the 8 Xeon E7-8870 2.40GHz host in section I (Sun Server X2-8, IBM x3850 X5, TPC-C and TPC-E are 5,055,888 and 5,457.20 respectively), assuming that the actual performance of the two hosts is the same, then the ratio performance scores of TPC-C and TPC-E is M:

- M=5,055,888/5,457.20≈926.46.

Assuming that the performance of X86 architecture hosts is similar under TPC-C and TPC-E, and the ratio of the two scores is m. Then the TPC-C performance value of E series servers can be estimated.

Table 8. E series server TPC-C performance estimates.

Configuration	TPC-E Value	TPC-C Value
Xeon E7-8890 v4 2.20GHz, 4CPU(96Core, 192Threads)	9,068.00	8,401,139
Xeon E5-2699 v42.20GHz, 2CPU(44Core, 88Threads)	4,938.14	4,574,989
Xeon E7-8890 v3 2.50GHz, 8CPU(144Core, 288Threads)	11,058.99	10,245,712
Xeon E7-8890 v3 2.50GHz, 4CPU(72Core, 144Threads)	6,964.75	6,452,562
Xeon E5-2699 v3 2.30GHz, 2CPU(36Core, 72Threads)	3,772.08	3,494,681
Xeon E7-8890 v2 2.80GHz, 8CPU(120Core, 240Threads)	9,145.01	8,472,486
Xeon E7-4890 v22.80GHz, 4CPU(60Core, 120Threads)	5,576.27	5,166,191
Xeon E5-2697 v22.70GHz, 2CPU(24Core, 48Threads)	2,590.93	2,400,393
Xeon E7-8870 2.40GHz, 8CPU(80Core, 160Threads)	5,457.20	5,055,888
Xeon E7-48702.40GHz, 4CPU(40Core, 80Threads)	3,218.46	2,981,774
Xeon E5-2690 2.90GHz, 2CPU(16Core, 32Threads)	1,863.23	1,726,208

As can be seen from the above table, the performance of single partition (16 CPU) of the Superdome is comparable to that of the 2-way E5 server.

For IBM minicomputer, TPC-C performance data can be estimated by rPerf (Relative performance) [5]. RPerf is an estimate of business processing performance derived from the IBM analytic model, which simulates some operations of the system, such as CPU, cache, and memory, but does not simulate the input / output operations of the disk and network.

Reference to the rPerf value of a Power minicomputer is IBM Power Systems Performance Report(www-03.ibm.com/systems/power/hardware/reports/system_perf.html).

Table 9. Power7 minicomputer rPerf value.

Kerne1 Number	Power7 3.50GHz	Power7+ 4.14GHz	Power7+ 3.80GHz	Power7+ 4.42GHz	Kerne1 Number	Power 8 4.0GHz

8	95.29	115.86	112.5	126.1	12	256.8
16	180.90	226.97	219.3	245.7	24	500.7
24	261.19	326.24	315.0	353.0	48	976.4
32	338.58	425.50	410.8	460.3		
64	-	-	729.3	817.1		

According to the TPC-C value of the known device (POWER7 4.14GHz, 2CPU, 8Core, TPC-C value is 1,200,011), the TPC-C values of other devices are estimated in accordance with the same principle of rPerf, and the formula is:

$$TPCC1 = (rPerf1 * TPCC2)/rPerf2 \qquad (1)$$

Table 10. Power7 minicomputer TPC-C performance estimates.

Kerne l Number	Power7 3.50GHz	Power7 4.14GHz	Power7+ 3.72GHz	Power7+ 4.42GHz	Kerne l Number	Power 8 4.0GHz
8	986,959	1,200,011	1,116,530	1,306,071	12	2,659,786
16	1,873,658	2,350,824	2,143,987	2,544,819	24	5,185,962
24	2,705,255	3,379,006	3,059,583	3,656,170	48	10,112,988
32	3,506,816	4,407,083	3,976,215	4,767,522		
64	-	-	7,553,668	8,463,050		

As can be seen from the above table, the performance of the 4-way Power 7 (32 core) minicomputer is slightly lower than that of the 8-way E7 X86server performance (80 core and TPC-C value is 5055888).

The performance of the Power 7 minicomputer is equivalent to the E5/7 V2 series X86 server with the same number of CPU [6]. The performance of the Power 8 minicomputer is equivalent to the E7 V3 series X86 server with twice the number of CPU.

3.1.2. SAP and ERP applications

Comprehensive comparison of SAPS performance data, the SAPS value of 16-way Itanium processor is 28,200, lower than 2-way Xeon E5-2690 2.9Ghz server (SAPS value is 35,680). Therefore, the performance of E5Series 2-way X86 servers is totally beyond that of the RX8640 (8-way Itanium processor).

The performance of 4-way, 8-way Power 7 minicomputers has been lower than the corresponding CPU number of E7 V2 X86 servers. And the performance of 16-way Power 7 minicomputers is lower than 8-way E7 v4 X86 servers.

The performance of the 4-way Power8 minicomputer is comparable to the 8-way E7 X86 server. And he performance of the 8-way Power 8 minicomputer is comparable to that of the 8-way Platinum 8180 X86 server.

3.1.3. Other applications

Non-database and SAP ERP applications lack targeted performance indicators and data. Referring to TPC, SPE-C and SAPS performance data, we can estimate that the performance of E5 Series 2-way X86 servers is equal to or surpass that of the 16-way Superdome minicomputers. And the performance of Power 7 Series minicomputers is comparable to that of the same number of E5/7 v2.

3.2. Analysis of the Measured Data of Performance

When the PMS2.0 system respectively uses a 2-node X86 server and a minicomputer as a database, with the increase of concurrent users, the load changes and transaction processing capabilities are shown as the following table.

Table 11. Comparison of load and transaction processing capability between 2-node X86 server and minicomputer.

Node	Concurrent users	P750		8-way X86 server		8-way X86/P750	
		avg_CPU%	TPS	avg_CPU%	TPS	CPU Comparison	TPS Comparison
2 Nodes	100	14.71	25.32	4.14	25.63	28.14%	101.22%
	200	29.34	49.87	8.44	51.11	28.77%	102.49%
	300	43.54	74.95	14.71	74.46	33.79%	99.35%
	400	58.30	98.13	27.35	91.79	46.91%	93.54%
	500	66.27	119.23	43.83	106.31	66.14%	89.16%
	700	78.68	153.62	65.13	119.51	82.77%	77.80%

As can be seen from the above table, the 8-way X86 server with 2 nodes, compared with the P750 minicomputer, has the same transaction processing ability for 100 to 300 concurrency, but the CPU utilization of the 8-wayX86 server is only about 30% of theP750 minicomputer. Between 400 and700 concurrent users, the performance of the transaction processing of the 8-way X86 server is less than 20% of the P750 minicomputer, but the CPU utilization of the minicomputer CPU is higher than about 20% of the 8-way X86 server.

When the 3-node X86 server and the minicomputer are used as the database respectively, with the increase of concurrent users, the load change and transaction processing ability are shown as the following table.

Table 12. Comparison of load and transaction processing capability between 3-node X86 server and minicomputer.

Node	Concurrent users	P750		8-way X86 server		8-way X86/P750	
		avg_CPU%	TPS	avg_CPU%	TPS	CPU Comparison	TPS Comparison
3 Nodes	100	10.60	25.01	3.39	25.87	32.01%	103.43%
	200	20.27	49.93	5.76	51.57	28.44%	103.29%
	300	30.01	75.02	8.64	76.50	28.79%	101.98%
	400	39.80	99.01	12.38	100.67	31.11%	101.67%
	500	49.49	122.98	18.54	121.43	37.45%	98.74%
	700	55.78	152.19	36.14	140.38	64.78%	92.24%

As can be seen from the above table, comparing the 3-node 8-way X86 server with the P750 minicomputer, the transaction processing ability is equal when within 100 to 700 concurrent users, but the CPU utilization of the 8-way X86 server is only 30% of the P750 minicomputer.

3.3. Recommended Configuration

Based on the above estimation and actual test data, the following table lists the corresponding configurations that can be used to replace minicomputers.

Table 13. Performance evaluation of minicomputer migration.

c	Database	SAP	others
8-way Itanium	2-way E5	2-way E5	2-way E5
16-way Itanium	2-way E5	2-way E5	2-way E5

2-way Power 7	2-way E5 v2	2-way E5 v2	2-way E5 v2
4-way Power 7	4-way E7 v2	4-way E7 v2	4-way E7 v2
8-way Power 7	8-way E7 v2	8-way E7 v2	8-way E7 v2
4-way Power 8	8-way E7	8-way E7	-
8-way Power 8	-	8-way Platinum	-

Although the performance has been approached to or even beyond the minicomputer, but the stability of the X86 server still can not compete with the minicomputer [7], especially the 2-way, 4-way X86 server, is not designed to run the key core business. For various business systems currently running on HP minicomputers of Gansu Electric Power Company, if they are migrated to 2-way X86 servers, the system stability is difficult to meet. However migration to more stable 8-way servers will cause serious waste of host resources. Therefore, it is recommended to migrate the applications which performance requirements can be meet by two-way X86 servers to the computing integrated machine. While ensuring stability, it is also possible to allocate resources on demand and reduce the cost of use.

For an Oracle database that can be migrated to a 4- or 8-way X86 server, it is recommended to use a 3-node RAC mode to ensure system reliability.

4. Application Server Selection and Migration Case

4.1. Selection Principles of X86 Server and Minicomputer

Informatization construction is inseparable from the excellent basic platform system [8]. Whether it is to build a strong regulatory integration system to ensure security, economy, high-quality, and efficient operation of the power grid, or to build a unified, efficient and intensive information-based production management platform, as well as unified, intelligent, interactive, and efficient information-based marketing management platform, all need strong infrastructure support [9].

In the selection of the application system server, we should start from some key factors, and make a reasonable selection by analyzing these factors on the minicomputer and the X86 server [10].

Table 14. Comparison of type selection index between minicomputer and X86 server.

Index	Minicomputer	X86 Server
Reliability	IBM Power and HP Integrity minicomputers are highly reliable, which can minimize planned and out of schedule downtime and ensure continuous and uninterrupted operation of the system.	The reliability of different brands is uneven, and there is still a certain gap on the whole from IBM or HP minicomputer.
Extensibility	The ability to scale vertically is strong. For large databases that value this capability, minicomputers are a better choice.	The economy of horizontal expansion is better. Middleware and other applications can be constantly stacked to ensure performance and reliability. X86 server is a good choice.

| Overall Performance | 1、 Minicomputers based on the latest Power8 or SPARC M7 chip has high CPU performance and vertical expansion capability, and the overall performance is outstanding. 2、 Minicomputer based on Itanium chip has poor performance | 1、 The performance of the ordinary 8-way X86 server is higher, but there is still a certain gap from the high-end minicomputer. 2、 The integrated machine based on X86 architecture can provide higher overall performance. |
| Economical Efficiency | High price, low cost performance | Low price, high cost performance |

As can be seen from the above table, the main difference between minicomputers and X86 architecture servers lies in their reliability and extensibility.

Table 15. Vertical expansibility and reliability.

	High Reliability	Next Highest Reliability	Medium Vertical Reliability
High vertical expansibility	Minicomputer	Prioritize X86 Server	Prioritize X86 Server
Next highest vertical expansibility	Give priority to X86 Server	X86 Server	X86 Server
Medium vertical expansibility	X86 Server	X86 Server	X86 Server

From the point of view of comprehensive importance, we should consider the selection basis of the hosts by the deployment mode of the application system.

Table 16. Application patterns and comprehensive importance.

Comprehensive Importance	Distributed Application Mode	Independent Application Mode	OLTP Feature Database Schema	OLAP Feature Database Schema	Real-time Database Mode	Application and Data Integration Mode
High	X86 Server	Minicomputer or X86 Server	Minicomputer or X86 Server	Minicomputer or X86 Server	X86 Server	X86 Server
Next highest	X86 Server	X86 Server	X86 Server	X86 Server	X86 Server	X86 Server
Medium	X86 Server	X86 Server	X86 Server	X86 Server	X86 Server	X86 Server
NOrmal	X86 Server	X86 Server	X86 Server	X86 Server	X86 Server	X86 Server

From the perspective of comprehensive importance and application deployment model [11], it is possible to initially determine the roughly corresponding target of the servers. When carrying out X86 servers replacement of minicomputers, in order to achieve practical results, it is necessary to combine the business characteristics of the migrated application, and analyze it from the main application system [12]. The following takes an application migration example to illustrate the operability of X86 server instead of minicomputer.

4.2. Migration Case
The power quality monitoring platform database system of the State Grid Corporation, its operating environment is two P750 minicomputer plus Oracle 10g. Due to the limitation of the old hardware and

software resources for a long time, the performance of the system is relatively low, which can not meet the rapid development of business applications in recent years. To this end, the State Grid Corporation of China carried out database migration, which migrated from the original minicomputer environment to the X86 server platform, and up-graded the version of the Oracle database to the Oracle 11g and generated the ADG.

Because of the cross version and cross-platform of database migration, Oracle 11g online transfer table space (Xtts) is used for data migration technology to ensure the security and reliability of the migration process [13]. By replacing two P750 minicomputers with three 4-way X86 servers, the problem of insufficient performance of the power quality database is solved while guaranteeing the stability of the system.

5. Conclusion

In this paper, the performance comparison and evaluation of X86 server and minicomputer are carried out in combination with the various types of business types and application scenarios of the traditional data center minicomputer architecture, and the selection principle of different applications for the minicomputer and the X86 server is discussed. According to the actual situation of Gansu Electric Power Company, the application migration of X86 servers to replace minicomputers is carried out. The results show that X86-based architecture has the capability to replace the traditional minicomputer architecture in a variety of business types and application scenarios in the data center.

References

[1] Wei Huang and Jie Zhang 2013 Application analysis of minicomputers and x86 servers in CRM Telecom Engineering Technics and Standardization. 26(6): 83-88.
[2] TPC BEN CHMARK C Standard Specification Revision 5.11[OL]. http://www.tpc.org/TPC_Documents_Current_Versions/pdf/tpc-c_v5.11.0.pdf. February, 2010.
[3] TPC BENCHMARK E Standard Specification Version 1.14.0[OL]. http://www.tpc.org/TPC_Documents_Current_Versions/pdf/TPC-E_v1.14.0.pdf. April, 2015.
[4] TPC-C Advanced Sort Results List (V2.2) [OL]. http://www.tpc.org/tpcc/results/tpcc_advanced_sort.asp. 2018.
[5] TPC-E Advanced Sort Results List (V2.2) [OL]. http://www.tpc.org/tpce/results/tpce_advanced_sort.asp. 2018.
[6] SPECjbb2015 Benchmark User Guide[OL]. https://spec.org/jbb2015/docs/userguide.pdf. 2015.
[7] SPECjbb2015 Results[OL]. https://spec.org/cgi-bin/osgresults?conf=jbb2015;op=dump;format=csvdump. 2015.
[8] SPECjEnterprise2010 User's Guide[OL]. https://spec.org/jEnterprise2010/docs/UsersGuide.html. 2010.
[9] SPECjEnterprise2010 Results[OL]. https://spec.org/cgi-bin/osgresults?conf=jent2010;op=dump;format=csvdump. 2010.
[10] Measuring in SAPS[OL]. https://www.sap.com/about/benchmark/measuring.html.
[11] SAP SD Standard Application Benchmark Results, Two-Tier Internet Configuration[OL]. global.sap.com/solutions/benchmark/sd2tier.epx.
[12] The relative performance metric for Power Systems servers[OL]. https://www-03.ibm.com/systems/power/hardware/notices/rperf.html.
[13] IBM Power Systems Performance Report[OL]. https://www-03.ibm.com/systems/power/hardware/reports/system_perf.html. February, 2017.

Risk Assessment of Power Communication Network Based on LM-BP Neural Network

Yanan Wang[1], Ke Wang[2, 1], Ran Zhang[2, 1], Qiao Xue[1], Xiangzhou Chen[1], Geng Zhang[1]

[1]China Electric Power Research Institute, Beijing 100192, China

[2]North China Electric Power University, Beijing 102206, China

* Corresponding Author: Ran Zhang; email: happyandluck_wk@163.com

ABSTRACT: In order to predict the operational risk of the power communication network more comprehensively and accurately, this paper presents a BP neural network model based on Levenberg-Marquardt (LM) algorithm. Firstly, establishing a set of indicators that reflect the operational characteristics of the power communication network. These indicator data are used as input to the model. Then, the operating efficiency of the traditional BP neural network is improved by the LM algorithm. The research results show that the model is simple , the forecast performance is stable and the accuracy is high, which provides an effective theoretical basis and modeling method for the prediction of power communication network risk.

1.Introduction

The power communication network is connected with all links of the power system. It is an important infrastructure for power grid dispatching automation, network operation marketization and management informationization. And it is an important guarantee for safe, reliable, economical and efficient operation of the power grid. Therefore, the risk assessment of the power communication network is conducive to improving the operation level of the power communication network and ensuring safe, reliable and stable operation of the power communication network[1].

At present, the more mature risk assessment methods mainly include analytic hierarchy process, fuzzy evaluation method and neural network, etc. However, due to the complexity of the power communication network and the uncertainty of the influencing factors, these methods are difficult to achieve satisfactory results in the accuracy of risk assessment. Moreover, the phenomenon of indicator design redundancy is common in the risk assessment of power communication networks, which increases the complexity of the algorithm to some extent. This paper proposes an evaluation method based on LM-BP neural network for these problems. Firstly, a set of risk assessment indicators for power communication networks was established. Then, using principal component analysis method, the redundant condition attribute in the index system is deleted, the input dimension is reduced, and then the BP neural network improved by LM algorithm is used for training and testing, and the weights of various factors of the risk network are obtained. Finally, the method is simulated, and the simulation results are compared with the evaluation values of the traditional BP neural network algorithm and the actual evaluation values[2]. The final results show that the new method has simple input, short training time and high precision.

Content from this work may be used under the terms of the Creative Commons Attribution 3.0 licence. Any further distribution of this work must maintain attribution to the author(s) and the title of the work, journal citation and DOI.

Published under licence by IOP Publishing Ltd

2.Method principle

2.1.Principle of BP neural network algorithm

Neural networks can learn from a large amount of discrete data and extract domain knowledge. And express this knowledge as network connection weights and thresholds to reflect the structure of complex systems. The error back propagation (BP) neural network's model is simple and algorithm is mature, so it is one of the most widely used neural network models[3]. BP neural network model is a multilayer feed-forward neural network. Its network topological structure includes: input layer, hidden layer and output layer. BP neural network has a high degree of nonlinear mapping capability that can be used to solve complex problems in internal mechanisms. The topological structure of the BP neural network is shown in figure 1.

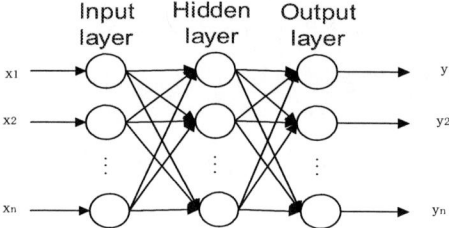

Figure 1. BP neural network diagram

The BP neural network can learn the input samples, and the data flows into the neurons through the input layer. After the calculation of the hidden layer and the output layer, data will flow out from the neural network[4]. The output value of the output layer corresponds to the predicted value of the network. If the error between the predicted value and the expected value does not meet the accuracy requirement, then enter the reverse transfer error phase. Each layer of neural network adjusts the weights and thresholds of each layer according to the gradient descent method.[5] The process of forward and backward information transmission and the reverse of error is actually the process of continuously adjusting the weight and threshold of each layer of neural network. This process last until the output value of the output layer meets the accuracy requirement, or the number of iterations reaches the preset number of learning times. The output signal for each activated hidden layer neuron is:

$$I_j = \sum_{i=1}^{n} u_{ij} x_i - \alpha_j \quad j = 1,2 \dots n \tag{1}$$

$$y_j = f(I_j) \quad j = 1,2,\dots,n \tag{2}$$

Where u_{ij} is the connection weight between the input layer neuron i and the hidden layer neuron j, α_j is the threshold of the hidden layer neuron j, f(x) is the activation function of the hidden layer neuron. This paper uses the Sigmoid function which is $f(x) = \frac{1}{1+e^x}$.[6]

Similarly, the output signal of the neural network output layer can be obtained as:

$$I_j = \sum_{j=1}^{m} v_{ij} t_j - \beta_j, j = 1,2,\dots m \tag{3}$$

$$y_{kj} = f(I_j) \quad j = 1,2,\dots m \tag{4}$$

Where v_{ij} represents the connection weight of the i-th hidden layer neuron and the j-th output layer neuron.β_j is the threshold of the output layer neuron;f(x) is the activation function of the output layer neuron.

2.2.LM algorithm

Traditional BP neural network algorithm uses the steepest descent algorithm for learning. This is an iterative algorithm for the least squares estimation of nonlinear regression model parameters. This method has the disadvantages of slow training speed and weak global search ability[7]. Therefore, this paper uses the Levenberg-Marquardt (LM) algorithm to improve it. LM algorithm is an algorithm combining gradient descent algorithm with Gauss-Newton algorithm. The LM algorithm is a fast algorithm that utilizes standard numerical optimization techniques. This method has the local convergence characteristics of the Gauss-Newton method and the global characteristics of the gradient descent algorithm[8]. Since the LM algorithm utilizes approximate second-order derivative information, the efficiency of the algorithm is better than that of the gradient algorithm. The following is a brief description of the LM algorithm optimization neural network weights. Generally, the BP neural network uses the mean square error as a performance evaluation method, and compares the neural network output value y_i with the target value y_{real} for adjusting the weight and the threshold.

$$E(x) = \frac{1}{2}e_i^2(x) = \frac{1}{2}\sum_{i=1}^{n}(y_{real} - y_i)^2 \tag{5}$$

Among them, y_{real} is the expected output value of the neural network, y_i is the actual output value of the BP neural network, and E(x) is the error.

Suppose x^k represents the vector consisting of the connection weight between neurons and the threshold of neurons at the k-th iteration. The vector $x^{(k+1)}$ consisting of the new weight and threshold can be obtained by the following method:

$$x^{(k+1)} = x^k + \Delta x \tag{6}$$

Where Δx is the amount of change in the weight and threshold. According to the Gauss-Newton method, Δx can be obtained:

$$\Delta x = -[J^T(x)J(x)]^{-1}J^T(x)e(x) \tag{7}$$

The LM algorithm is an improvement on the Gauss-Newton algorithm. The weight and threshold adjustment methods are as follows:

$$\Delta x = -[J^T(x)J(x) + \alpha I]^{-1}J^T(x)e(x) \tag{8}$$

Where J is the Jacobian matrix, $J = (\frac{\partial e_i(x)}{\partial x_j})_{i \times j}$. It can be seen from equation (8) that when $\alpha=0$, it is the Gauss-Newton algorithm. When the value of α is large, it is close to the gradient descent algorithm. At each iteration of the algorithm, the value of α will also become smaller, so that it is similar to the Gauss-Newton method when approaching the error target[9]. Because the LM algorithm utilizes approximate second-order derivative information, it is much faster than the gradient descent method. And $[J^T(x)J(x) + \alpha I]$ is positive definite, so the solution of equation (8) always exists. α is a tentative parameter in actual calculation. For a given α, if the Δx obtained can reduce the error function, the value of α is decreased; if the error function is increased, the value of α is increased.

3.Instance analysis

3.1 The risk indicator system of power communication network

The power communication network is a dynamic and complex network, and its operating state is affected by many factors. The operational risk assessment of power communication networks is a vital technology in routine maintenance. In this paper, considering the difficulty of obtaining various indicators, combining with the actual situation of the power communication network, fully satisfies the principles of comprehensive, scientific and practicality of the indicator design, from the four aspects of equipment failure rate, business operation, operation and maintenance, management and environmental to establish a risk assessment indicator system for power communication networks.

Table 1. The risk indicator of power communication network

First level indicator	Second level indicator(A1)
	Cable failure rate(A2)
	SDH Equipment failure rate(A3)
	PCM Equipment failure rate(A4)
Equipment failure factor(A)	Switch device failure rate(A5)
	Carrier failure rate(A6)
	Power failure rate(A7)
	Average service life of equipment(A8)

	Serious equipment failures(A9)
Business operation factor(B)	Business affected quantity(B1)
	Business dual channel rate(B2)
	Average business interruption time(B3)
Operation and maintenance management factor(C)	Number of faults(C1)
	Failure mean time to repair(C2)
	Maintenance not completed(C3)
Environmental risk factor(D)	Thunder(D1)
	Wind power(D2)
	Human Factors(D3)
	Other factors(D4)

The two indicators D1 and D2 are determined by the meteorological data of the year, and the rest of the data are obtained based on the actual operational data. The risk value is quantified by experts based on empirical evaluation. Due to the difference in the dimension of the data, normalization is required before the neural network training. The normalization formula is:

$$I_{ij} = \frac{I_{ij} - minI_i}{maxI_i - minI_i} \qquad (9)$$

Where I_{ij} is the normalized processing result of the j-th statistical data of the i-th index; $minI_i$ and $maxI_i$ respectively represent the minimum and maximum values of the i-th index.

3.2. Network structure
In the case where the hidden layer node is sufficient, the three-layer neural network is sufficient to approximate a nonlinear function with arbitrary precision. Therefore, this paper adopts a three-layer network structure, and the number of input nodes and output nodes are 18 and 1 according to actual conditions. Relatively speaking, the number of hidden layer nodes is more difficult to determine. There are fewer hidden nodes, and the learning process may not converge; if there are too many hidden nodes, over-fitting may occur[10]. This paper determines the number of hidden layer nodes based on empirical formulas($k = 2n + 1$), where n is the number of input nodes.

3.3. Simulation verification
According to the electricity operation statistics of a province for one year, the probability of occurrence of each indicator in the secondary indicators of table 1 can be calculated. The second level indicators corresponding to the four primary indicators of equipment failure rate, business operation, operation and maintenance management and environmental risk are added together and the mean value is obtained, and the first-level indicator values of the four items are obtained. We input four primary indicators as network data sets into the neural network for training. We can Compare the difference between the traditional BP neural network and the BP neural network improved by the LM algorithm in the training steps, analysing the superiority of using the improved BP algorithm of LM algorithm in the risk assessment of communication network. Set the mean square error of this method to ε =0.001. This paper uses the data collected in the first half of a province as the training set, and the data collected in the second half as the verification set. The simulation results are shown in table 2 and figure 2. According to the simulation statistics, the system error using the LM algorithm reaches 10^{-6} for 14 times, while the traditional gradient descent algorithm has used more than 12,000 times when the error is 10^{-4}. It can be seen that the training speed of the LM-BP neural network is faster.

Table 2. Algorithm to estimate risk value and actual risk value

Month / Risk value	7	8	9	10	11	12
BP neural network algorithm simulation risk value	0.4301	0.3441	0.4486	0.1599	0.3399	0.2056

LM-BP neural network algorithm simulation risk value	0.4400	0.3433	0.4509	0.1607	0.3461	0.2144
Actual risk value in the second half of the year	0.4432	0.3430	0.4613	0.1687	0.3521	0.2148

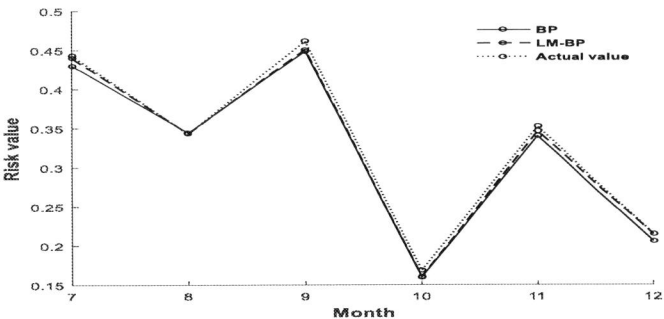

Figure 2.The estimated of BP,LM-BP and actual values

Moreover, according to the simulation results, using the simulation data and the actual data to find the variance of the two algorithms, the simulation results shown in figure 3 can be obtained. It can be seen from figure 3 that compared with the traditional BP neural network, the LM-BP neural network algorithm has less error in the risk assessment of the power communication network, and the evaluation is more accurate.

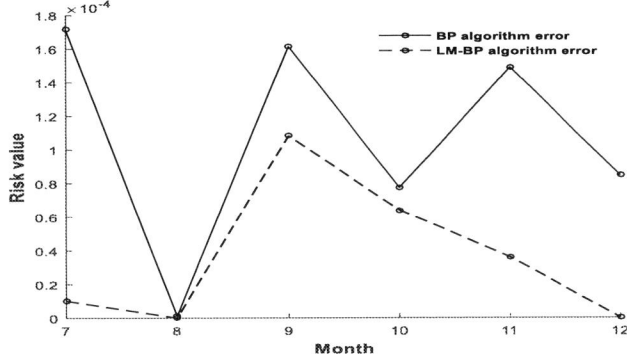

Figure 3.Comparison of error values between BP and LM-BP algorithm

4.Conclusion

This paper proposes an improved BP neural network based on LM algorithm to evaluate the risk of power communication network. The new algorithm inherits the strong fitting and fault tolerance of BP neural network. At the same time, since the LM algorithm is an improved form of the quasi-Newton algorithm, the new algorithm has both the local characteristics of the Gauss-Newton algorithm and the global advantages of the gradient algorithm. Finally, experiments show that the improved BP neural network based on LM algorithm is better than the traditional BP neural network,which effectively improves the computational performance and generalization ability of the model.

Acknowledgments

This paper was supported by the science and technology project from State Grid Corporation of China:"Research on Distributed Simulation and optimization technology of optical transmission network for Electric power communication(5442XX180003-XX71-18-006)"

References

[1] Jiao Aihong,Yuan Lizhe 2012 *Industrial Control and Electronics Engineering (ICICEE)*, 2012 International Conference on,2012

[2] Qiang Wang,Kin Keung Lai,Dongxiao Niu,Qian Zhang 2012 *Business Intelligence and Financial Engineering (BIFE)*, 2012 Fifth International Conference on,2012

[3] Yuanchun Ding, Falu Weng, Jinping Yu 2011 *Advanced Computer Control (ICACC),*3rd International Conference on,2011.

[4] Yun Lin, Yuan Zhang 2012 *Natural Computation (ICNC)*, 2012 Eighth International Conference on,2012.

[5] Xingyu Chen,Huanwei Chen,Can Wang,Haifeng Wu 2018 *J.Information & Communications,* **20**(04):165-66

[6] Xu Niu 2018 *J Electronic World*, 30(13): 96-98.

[7] Xuefeng Feng,Jun Gong,Xiaoyi Lu 2018 *J.Modern Computer*, **46** (24): 50-54.

[8] Wenxiong Chen,Xi Zhu,Xuelin Chen 2018 *J. Groundwater* **40** (05): 115-117+130.

[9] Qu Bo,Xiaofei Sun, Xinhe Zhang 2018 *J Journal of Agricultural Engineering* **34**(18): 44-50.

[10] Xiangyu Xing 2018 *J.China Automation Society Control Theory Professional Committee.* **10**(2):159-163.

Research on Remote Meter Reading Scheme and IoT Smart Energy Meter Based on NB-IoT Technology

Xingyuan Fan[1], Chun Zhou[1], Ying Sun[1], Jinyang Du[1] and Ying Zhao[1]

[1]Guangzhou Power Supply Bureau Co., Ltd, Guangzhou, 510620, China

523812935@qq.com; zhouc@guangzhou.csg.cn;

410050938@qq.com; dujy@guangzhou.csg.cn; zhaoy@guangzhou.csg.cn

Abstract. The energy meter supporting the NB-IoT standard protocol is designed to be connected to the NB-IoT network through the protocol conversion of the remote communication module itself, and directly connected to the existing centralized meter reading automation system. The acquisition terminal device is omitted, and direct communication between the main station and the energy meter is realized. The end-to-end standardized protocol reduces the complexity of the centralized meter reading platform architecture and improves the system's openness and scalability. It realizes end-to-end visual management and real-time monitoring, improving the safety, reliability and fault resistance of power collection. The overall scheme can reduce the large amount of human and material costs of system operation and maintenance.

1. Introduction

The low-voltage centralized meter reading system automatically collects, transmits and manages electricity measurement data, fundamentally changes the traditional mannered reading model, and realizes automatic meter reading and settlement[1]. However, some problems are found along with the on-site operation, such as the system function scalability is not good, meter reading time is long and the real-time performance is insufficient[2], abnormal environment interference causes terminal offline failure, which easily leads to acquisition failure[3]. Narrow Band Internet of Things technology (NB-IoT) is a revolutionary technology that has the advantages of supporting massive connections, deep coverage, and low power consumption. It is suitable for new applications of IoT such as metering and monitoring[4, 5]. The paper studies the design of a new generation of meter-based IoT meter reading scheme through NB-IoT technology. The direct communication between the energy meter and the master station system is realized, the intermediate equipment metering terminal is omitted, the complexity of the existing concentrated copy platform architecture is reduced, the centralized management and real-time monitoring capability of the energy meter is improved, and the collection coverage rate is effectively improved.

2. The Current Low-voltage Centralized Meter Reading System Architecture and Existing Problems

The low-voltage centralized meter reading system is a centralized reading and reading of electric energy data and related electricity information of a plurality of energy meters by a metrology automation system through a remote communication channel, and can remotely implement a network meter reading system for disconnecting and power transmission control. It consists of the user energy

---

Content from this work may be used under the terms of the Creative Commons Attribution 3.0 licence. Any further distribution of this work must maintain attribution to the author(s) and the title of the work, journal citation and DOI.

Published under licence by IOP Publishing Ltd

meter, the acquisition terminal, the concentrator, the communication channel and the master station of the metrology automation system [6]. The system architecture is shown in Figure 1. The data is transmitted to the metrology automation master station through the channel and connected to the marketing system to realize automatic meter reading. At present, the communication network is mainly composed of two parts: the downlink acquisition network includes PLC, RS485 bus, wireless ad hoc network, etc., to realize smart energy meter and concentrator communication; the uplink backhaul network includes GPRS/CDMA, 4GLTE, etc., to realize concentrator and master station system communication; two parts of the network can adopt different networking technologies [7].

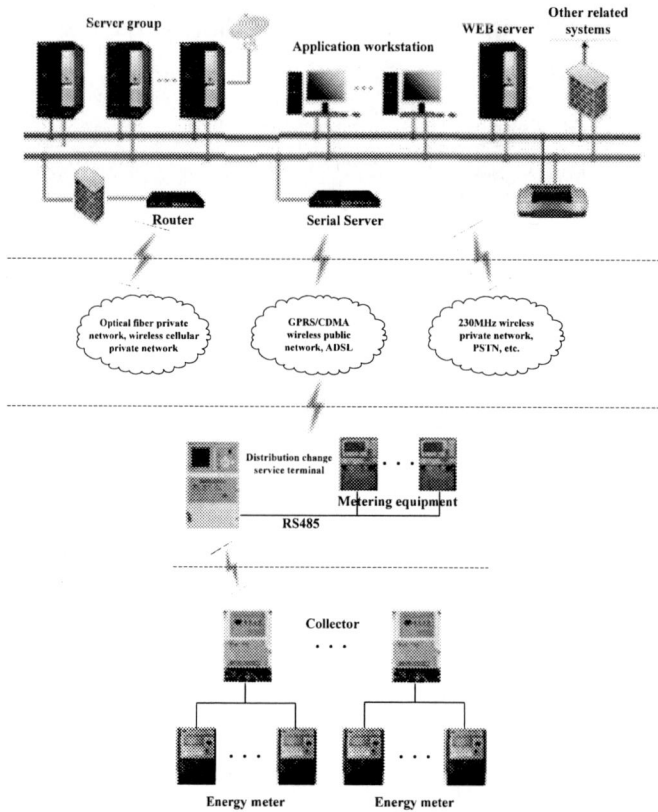

Figure 1. Current low-voltage centralized meter reading system architecture.

With the use of metrology automation systems in recent years, some problems have been encountered: (1) Downstream acquisition networks, most of them are private communication protocols of manufacturers, which are difficult to interconnect and achieve unified management; (2) Currently using non-standardized non-open protocols, the network link conversion of equipment such as concentrators and collectors is required, resulting in poor scalability of the system architecture and complex operation and maintenance; and the non-tunnelled carrier channel and the communication front-end machine are added in the middle, making it difficult to realize the end-to-end visual network connection from the end energy meter to the main station server; (3) The acquisition communication platform is formed by a segment of the network link, and the application layer communication is reachable, but the network layer cannot communicate end-to-end, and the communication status and operation status of each energy meter cannot be monitored in real time. When the fault occurs, the feedback cannot be made in time, and a fault occurs. It is necessary to send people to conduct on-site investigation, which is inefficient and wastes manpower.

3. Design of Remote Meter Reading Scheme Based on NB-IoT Technology

Narrow Band Internet of Things (NB-IoT) has the characteristics of low cost, low power consumption, low speed and wide coverage, and the number of concurrent connections of a single base station can be increased, and spectrum resources can be utilized efficiently [8]. Secondly, NB-IoT coverage sensitivity is 20 decibels stronger than traditional technology, with over 100 times coverage enhancement; with ultra-low power consumption, NB-IoT consumes only 1/10 of 2G. The application of NB-IoT technology in the field of smart grid will help to improve the reliable transmission and comprehensive coverage of the power collection service. The smart energy meter, which is the power terminal of the smart grid, will become an important part of the new application of the Internet of Things for the smart grid.

The meter reading architecture based on NB-IoT technology is shown in Figure 2. It mainly includes the master station layer (management system, database, system firewall, etc.), network layer (NB-IoT base station), IoT smart energy meter. In this architecture, the smart energy meter directly accesses the NB-IoT network by loading a remote communication module that supports the NB-IoT standard protocol communication, eliminating the link between the collector and the concentrator in the original system architecture, and transforming the indirect acquisition into direct acquisition. Data uploading enables end-to-end transmission, and the technical advantages are also reflected in:

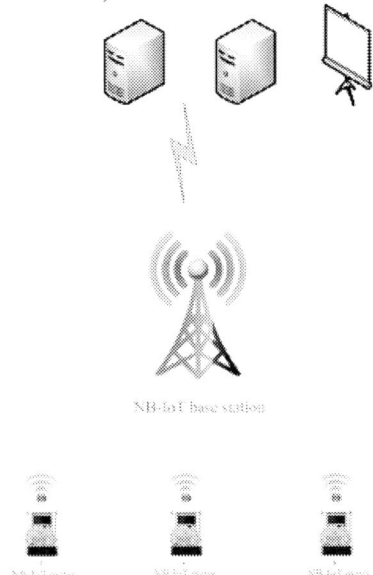

Figure 2. Remote meter reading
application architecture based on NB-IoT.

(1) Direct collection

The metering terminal device is omitted, and the master station directly collects the energy meter, reads the data, and issues parameters or instructions. It meets the frequency, success rate and multi-service application of power collection; at the same time, the terminal equipment is omitted, which saves a lot of equipment cost and operation and maintenance cost for power collection;

(2) High capacity

NB-IoT has 50~100 times higher uplink capacity than 2G/3G/4G, supports massive connection, and solves the problem of base station saturation in current power meter reading system;

(3) Improve collection coverage

NB-IoT boosts 20dB gain compared to LTE, which is equivalent to 100 times higher transmit power (100 times better coverage), and can overcome different levels of signal attenuation such as 144dB, 154dB and 164dB according to coverage level; the deep coverage capability solves the problem that the current meter reading system has a partial signal difference or no signal, and improves the acquisition coverage and success rate;

(4) Low power consumption

The low power consumption solves the problem that the output power of the single-phase meter in the current power meter reading system is small, and the GPRS module cannot be directly loaded. The lower the power consumption, the longer the life and the longer the meter replacement cycle, which can save a huge amount of manpower and material resources;

(5) Concurrency and two-way interaction

The NB-IoT service rate is low, the acquisition delay can be tolerated, the concurrent services are supported, and the coverage distance is long. The high-frequency real-time power data collection can be further supported. In addition to the power enterprise collecting base station information, the NB-IoT technology is adopted. The base station can also receive the enterprise side information and synchronously send the information to each household energy meter to realize two-way interaction between the user and the power company.

4. Key Technologies of IoT Smart Energy Meter Based on NB-IoT Technology

4.1. IoT Smart Energy Meter System Design and Function

The main function of the IoT smart energy meter is measurement and data transmission. The meter design adopts the "double MCU" scheme, including the base meter part and the NB-IoT communication module part. The base meter is the management MCU, which realizes the basic functions of measurement and storage. It can be configured with different uplink communication modules, and supports NB-IoT standard protocol remote communication module. Replacing the external module in the original charge control energy meter, it can quickly connect to the NB-IoT network, and the module supports power failure detection and reporting function. The design is shown in Figure 3 below.

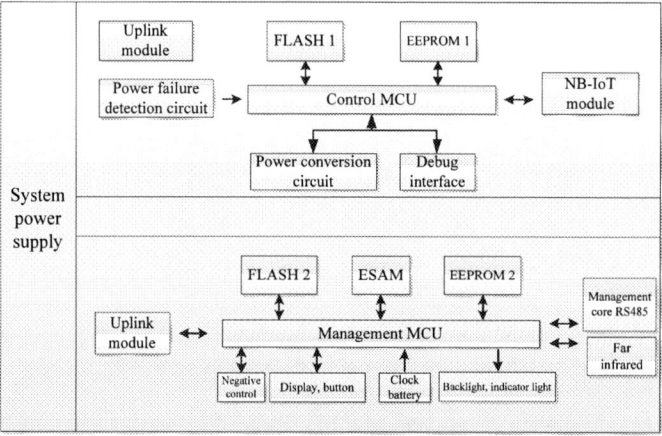

Figure 3. IoT smart energy meter system design block diagram.

4.2. NB-IoT Communication Module and Interface Design

4.2.1. Module Design

NB-IoT remote communication module can adapt to different manufacturers' energy meters, and the energy meter interface circuit adopts fixed interface definition and interface protocol, respectively +5V, +3.3V, GND, RXD, TXD, RST, after one adaptation, the subsequent use of any module does not require re-adaptation development. The energy meter remote communication module includes the main MCU, the power conversion circuit, the memory chips (DATAFLASH, EEPROM), the NB-IoT module, the hardware watchdog circuit, the energy meter interface circuit, the debug interface circuit, etc. The system block diagram is shown in Figure 4. The main module design description is as follows:

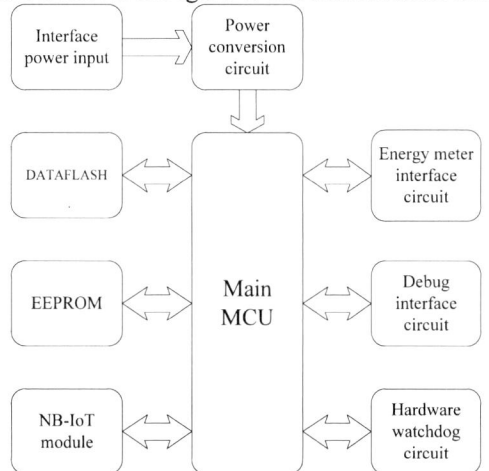

Figure 4. NB-IoT communication module design.

(1)The main MCU functions as the core of the module to achieve the following functions:

1) Initialize the NB-IoT module by AT command, establish link, link maintenance and data response;

2) Reading the configuration parameters according to the communication protocol and the system, and responding to the data interaction between the energy meter and the master station;

3) Store the relevant configuration parameters into the EEPROM to avoid loss of parameter power failure, support remote upgrade, and store the contents in DATAFLASH;

4) Respond to the debugging commands of the debug interface and the printout of key information.

(2) The power conversion circuit is responsible for converting the voltage of the input power source into a voltage suitable for the system, and ensuring the stability of the output power source;

(3) The memory chips (DATAFLASH, EEPROM) respond to the storage parameters such as the configuration parameters of the main MCU and the writing and reading of the upgrade data;

(4) The module is configured with NB-IoT module, and data interaction can be performed through the NB-IoT network;

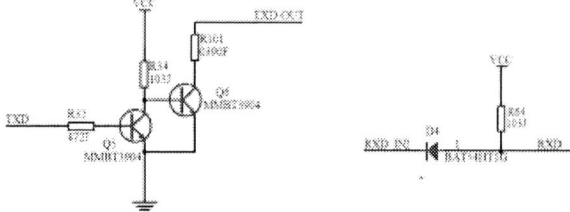

Figure 5. Energy meter interface circuit.

(5) The energy meter interface circuit is responsible for matching and data transmission with other interface circuits. All output interfaces adopt triode open-drain mode, and the triode output terminal is connected in series with a 100 ohm resistor to effectively prevent damage to the device caused by sharp pulses generated during frequent insertions. The input terminal uses a diode protection circuit to effectively adapt the interface circuits of different levels of different energy meters, as shown in Figure 5 below.

4.2.2. Principles of Key Tasks

Key tasks include meter reading task, uplink communication task, freezing task, active reporting task, and power failure reporting task.

(1)Principle of meter reading task

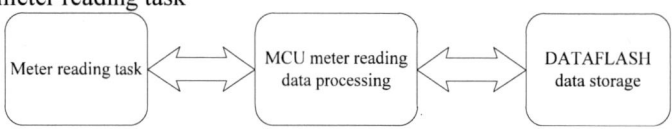

Figure 6. Meter reading task.

The MCU reads the base meter data according to the module reading time interval through the 645 protocol. After the data is successfully read, the data is transmitted to the protocol parsing task, and the related data is extracted according to the protocol, and then stored in the DATAFALSH, as shown in Figure 6.

(2)Principle of uplink communication task

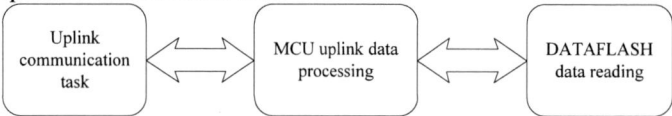

Figure 7. Uplink communication task.

The MCU reads the information according to the master station, performs protocol analysis, reads the relevant information from the DATAFLASH to respond, and sends the data through the remote communication module, as shown in Figure 7.

(3)Principle of freezing task

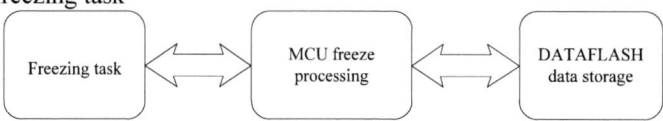

Figure 8. Freezing task.

The MCU will perform daily freezing, monthly freezing and other important data freezing according to the requirements of the meter reading in the specified time. The relevant data will be stored in DATAFLASH according to a certain format, as shown in Figure 8.

(4)Principle of active reporting task

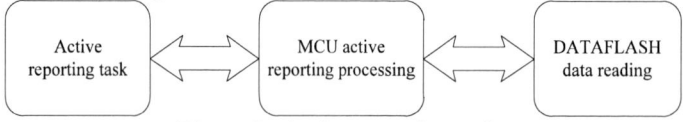

Figure 9. Active reporting task.

The task report can be set according to the protocol. After the active reporting task is triggered, the MCU reads the relevant data in the DATAFLASH, and then reports it to the master station according to the protocol, as shown in Figure 9.

(5)Principle of power failure reporting task

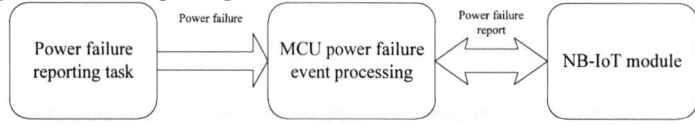

Figure 10. Power failure reporting task.

The meter provides a special power failure judging circuit for the module power supply 12V port. When the voltage drops to a certain threshold, the power failure signal is output. After receiving the power failure signal, the main MCU is judged to be a real power failure after entering the power failure interference, enters the power failure event processing task, performs data framing according to the China Southern Power Grid protocol, and uses the super capacitor backup power supply to report in real time through the NB-IoT module. The process is shown in Figure 10.

4.2.3. Interface Definition with Base Meter

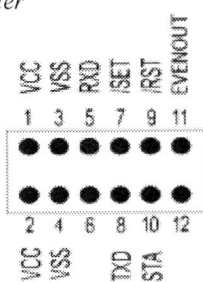

Figure 11. Schematic diagram of the weak electric interface of the communication module.

The weak electric interface of the NB-IoT remote communication module adopts 2×6 double-row pins as the connecting piece, and the weak electric interface of the energy meter adopts 2×6 double-row sockets as the connecting piece, as shown in Figure 11 below, the definition of the weak electric interface pins on the meter side. The open-drain endurance of the energy meter and communication module is 5.5V, the low-level current drive capability of all output interfaces is ≥2mA, the ground voltage should be ≤0.4V when driving the load current of 2mA; in the case of VCC power supply carrying (single-phase meter 125mA), the ripple Vp-p of the VCC power supply should be less than 1‰; the communication interface is isolated from strong power.

4.3. Uplink Communication Protocol

The uplink communication protocol follows the China Southern Power Grid "Q/CSG 11109004-2013 Metrology Automation Terminal Uplink Communication Protocol", which defines the frame format, data coding and transmission rules for data transmission between the master station and the terminal of the metrology automation system. It is suitable for point-to-point, multipoint-partyline and point-to-multipoint communication. It is suitable for the master station to perform the master-slave question and answer mode for the terminal and the terminal active upload mode communication.

NB-IoT remote centralized meter reading communication module, the module completely follows the smart energy meter specification in terms of physical characteristics and energy meter data interface, which can limit the traditional energy meter RS485 communication transmission between a single meter and the module. Therefore, the inherent unreliable limitations of RS485 serial communication are effectively avoided, and the reliability of the entire network communication and the success rate of meter reading are greatly improved. Uplink through the protocol conversion of the communication module itself, after connecting to the NB-IoT network, it can directly access the existing centralized meter reading and operation network system, thereby achieving the expected direct communication between the master station and the meter. Reduce the complexity of the existing centralized meter reading platform architecture through end-to-end standardized protocols, and improve the system's openness and scalability.

5. Test and Field Application Results Analysis

Based on the above design, the IoT smart energy meter based on NB-IoT technology has the tested performance in line with power consumption, power supply voltage, overvoltage, climate impact test, electrostatic discharge immunity test, electrical fast transient disturbance test required by the China Southern Power Grid enterprise standard "Q/CSG 1209003-2015 Single Phase Electronic Charge Control Energy Meter Technical Specification" experimental requirements.

According to the remote meter reading scheme, the sample meter is installed in the field for trial operation, and the 15-minute timed reporting task is set. The real-time data such as power, voltage, current, etc. are reported through the NB-IoT network, the daily freezing data is reported at 0:30 daily, and the uploaded data is complete and accurate; when the main station has a demand, the NB-IoT network can also actively call the relevant information of the meter, and the power failure information can be reported in time after the power failure. During the 6 months of operation, communication is normal and safe.

6. Conclusions

Based on the NB-IoT technology, the paper designs a remote meter reading scheme. This scheme eliminates the link between the collector and the concentrator in the original meter reading system architecture, and transforms the indirect acquisition into direct acquisition. The design and development of the IoT smart energy meter based on NB-IoT technology is mainly studied. The system includes the uplink module and the base meter. The base meter is compatible with different uplink modules. Based on the NB-IoT standard protocol, the remote communication module can be developed to adapt to different manufacturers' energy meters. The hardware design, interface design and main function design of the module are analyzed. The physical characteristics and data interface of the module completely follow the energy meter specification. The function can realize meter reading, uplink communication, freezing, active reporting, power failure, etc.; the uplink communication protocol is in accordance with the existing uplink communication protocol of the metrology automation terminal, and is transmitted through the protocol conversion of the communication module itself, and directly connected to the centralized meter reading system after accessing the NB-IoT network. After being certified by a third-party testing organization, the meter is accurate in measurement, normal in communication, and the performance meets the standard requirements. After half a year of on-site trial operation, the operation is stable, the function is reliable, and the coverage signal is good. The research and pilots laid the foundation for the comprehensive application and promotion of NB-IoT smart energy meters, and opened a new era of intelligent meter reading for Internet of Things applications.

Acknowledgements

This research was financially supported by China Southern Power Grid Corporation Science and Technology Project (Grant No. 080006KK52160001).

References

[1] Shen Shangfeng 2016 Research and application of low-voltage centralized meter reading system for distribution network users. *South China University of Technology*.

[2] Li Xiangwen and Zhu Hua 2013 Analysis of the practical status and countermeasures of low-pressure centralized meter reading system. *Electronic Production*, vol 21, pp 191-191.

[3] Li Yuantao 2017 Discussion on the status and management of low-pressure centralized meter reading system. *Communication World*, vol 5, pp 188-189.

[4] Liu Yi, Kong Jiankun, Niu Haitao, et al. 2016 Discussion on Narrow Band Internet of Things Technology. *Communications Technology*, vol 49(12), pp 1671-1675.

[5] Dai G H and Jun-Hua Y U 2016 Research on NB-IoT background, standard development, characteristics and the service. *Mobile Communications*, vol 24(09), pp 112-116.

[6] Wang Fan 2016 Discussion on several networking methods and maintenance of low-voltage centralized meter reading system. *Science and Technology Innovation*, vol 36, pp 54-55.

[7] Li Juan, Hu Xiaoling and Li Zigang 2016 Analysis of energy consumption test for Narrow Band Internet of Things. *Telecommunications Network Technology*, vol 8, pp 65-67.

[8] Osama ElGarhy 2018 Increasing efficiency of resource allocation for D2D communication in NB-IoT context. *Procedia Computer Science*, vol 130, pp 1084-1089.

ISPECE   IOP Publishing

IOP Conf. Series: Journal of Physics: Conf. Series **1187** (2019) 022065   doi:10.1088/1742-6596/1187/2/022065

Optimal configuration of optical storage microgrid under demand-side response based on cooperative game

Wang Shudong, Mao Yanrong, Jia guangyao, Song huiying, Qiu jinliang, Ding ting

School of Electrical Engineering and Information Engineering, Lanzhou University of Technology, Gansu Lanzhou 730050, China

Abstract. In the power market environment, considering the influence of the demand-side response and energy storage system on the microgrid, the joint optimization and configuration of the system through a cooperative game approach is proposed, and a time-of-use price is proposed when the demand-side user transfers the load appropriately. The microgrid operation mode is used to maximize the revenue and optimal reliability of the microgrid. Firstly, the users of the load transfer, the users and the objective functions and models of the energy storage system under the time-shared price were established; secondly, the three parties were jointly optimized by cooperative game, and the iterative algorithm was used to find the three parties to optimize the Nash equilibrium (optimal configuration scheme). The system achieves optimal returns and optimal reliability. The model and algorithm were applied to a practical photovoltaic microgrid system, and the effectiveness of the model and algorithm was verified.

1. Preface

At this stage, large-scale photovoltaic power generation appears to be difficult to integrate and absorb, and the light is severely discarded[1]. The main reason is that the PV penetration rate is low[2]. Therefore, how to increase the penetration rate of PV and ensure the reliability of the power supply of the system has become a research hotspot. With the improvement of the electricity market, more and more users are participating in the demand-side response. The impact of user response behavior on optical storage microgrids has also become a research hotspot. The energy storage device will reduce the abandonment rate, but the energy storage device is expensive and the configuration capacity will affect the economic efficiency of the microgrid. Therefore, it is very important to find a suitable energy storage capacity. How to combine demand-side with energy storage and optimize configuration is the focus of this article. At present, domestic and foreign scholars have done some researches on the optimal configuration of microgrids. The literature[3] discusses the optimal allocation of different investment entities under the competitive game model, but the user side does not consider it to be comprehensive. Literature[4] proposed a demand-side response model of transferable load, taking into account the closeness of photovoltaic power generation and load, without considering the peak-to-valley difference, and the reliability of microgrid. Literature[5] proposed a demand-side response model for time-of-use tariffs, taking into account the influence of peak-to-valley differences in load and improving the reliability of power supply, but it will reduce the PV penetration rate and require more energy storage devices to be added to the system. The gains brought about some losses. The demand-side response method based on game theory is discussed in literature[6]. This method is only for the study of scheduling, and the solution set obtained may not be the optimal solution set.

Content from this work may be used under the terms of the Creative Commons Attribution 3.0 licence. Any further distribution of this work must maintain attribution to the author(s) and the title of the work, journal citation and DOI.

Published under licence by IOP Publishing Ltd

The demand-side response is to point out the incentive mechanism and price information for the electricity market. Users change their original power consumption patterns and load usage patterns so as to achieve the coordination of supply and demand interests[7]. Therefore, the way in which the user on the demand-side can transfer the load and the time-of-use price response of the user are proposed. The goal of shifting the load is to increase the PV penetration rate and reduce the use of energy storage devices, but the difference in load peaks and valleys will increase and the reliability of electricity consumption will decrease. The goal of the time-shared tariff response is to reduce the load peak-to-valley difference and improve the reliability of electricity use. However, the penetration rate of PV will decrease, energy storage devices will need to be added, and the economy will be affected. The cooperative game plays an important role in optimizing the configuration as a mathematical theory and method for studying cooperation and competition. Based on this, in the demand-side, a cooperative game method is used to optimize the configuration of the two response modes and energy storage, and an iterative algorithm is used to find the optimal configuration of the tripartite, so as to maximize the revenue and optimal reliability of the microgrid.

2. Micro network system joint configuration optimization model establishment

2.1. Transfer demand-side response

The demand-side response mode in the case of a transfer load refers to the purpose of reaching closer to the photovoltaic power generation power by changing the useable time of the transferable load. The effect of transferring the load is shown in figure 1. The microgrid collects photovoltaic power generation data and user load data, and re-arranges the operating time for the load that can be transferred according to the set method for the transferable load. A part of the load that can be transferred will be transferred to photovoltaic power generation for a sufficient period of time, so that the photovoltaic power generation and the load are closer in time sequence, thereby reducing the use of energy storage devices[8]. This method changes the operating characteristics of the load from the demand-side.

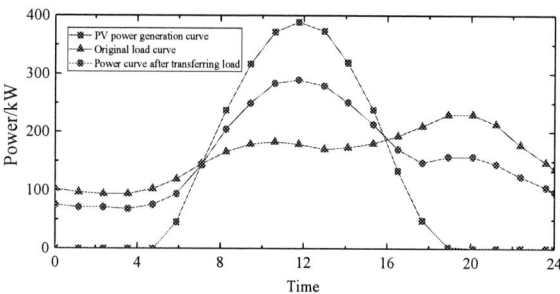

Figure 1. Transfer of load effect picture

1) Transfer load objective function

The goal of the transfer load is to make the photovoltaic power generation and load curve closer in time sequence, that is, the load is closer to the photovoltaic level after the transfer. The expression is:

$$\begin{cases} Q_A = \sum_{t=1}^{T} \left| L(t) - P_{pv}(t) \right| \\ L(t) = L_{bef}(t) + L_{in}(t) - L_{out}(t) \end{cases} \tag{1}$$

Where: Q_A is the load transfer close to the photovoltaic power; T is the transfer load cycle, usually 24 hours; $P_{PV}(t)$ is the t period of photovoltaic power generation; $L(t)$, $L_{bef}(t)$, $L_{in}(t)$, $L_{out}(t)$ are the time to transfer the load after the capacity, transfer the load before the capacity, load transfer capacity and transfer capacity.

2）Transfer load model

$$
\begin{cases}
L_{in}(t) = \sum_{k=1}^{N_L} x_k(t) P_{lk} + \sum_{h=1}^{h_{max-1}} \sum_{k=1}^{N_{la}} x_k(t-h) P_{(h+1).k} \\
L_{out}(t) = \sum_{k=1}^{N_L} y_k(t) P_{lk} + \sum_{h=1}^{h_{max-1}} \sum_{k=1}^{N_{la}} y_k(t-h) P_{(h+1).k}
\end{cases}
\tag{2}
$$

Where: N_L is the number of transferable loads; N_{la} is the number of transfer loads whose running time is greater than one cycle; h_{max} is the maximum value of the duration of the supply of transferable loads; $x_k(t)$ is the number of the k types of loads that are started to run at time t. $y_k(t)$ is the number of the k type of load transfer starting at the t time; P_{lk} is the capacity of the l time of the k type of transfer load.

3）Transfer load constraints

$$
\begin{cases}
L_{in.min} \le L_{in}(t) \le L_{in.max} \\
L_{out.min} \le L_{out}(t) \le L_{out.max}
\end{cases}
\tag{3}
$$

In the formula: $L_{in.min}$ and $L_{in.max}$ are the minimum power and maximum power transferred into the load; $L_{out.min}$ and $L_{out.max}$ are the minimum power and maximum power respectively. When the load is transferred, the demand-side response needs to be performed according to the specified transfer load limit.

2.2. Time-of-use price demand-side response
Time-of-use price refers to the level of load response on the demand-side of the system. Each day is divided into peaks, flats, and valleys. Different periods of time are used to implement different rates. The implementation of peak-to-trough time-of-use price is an effective demand-side response method, which improves the reliability of the microgrid by cutting peaks and filling valleys. Because the electricity price level not only affects the load at this moment, but also affects the load at other moments, the user response can be divided into single-period response and multi-period response[9]. Because the balance of electricity price and electricity price elasticity coefficient can fully characterize the user's response behavior, the electricity fee is higher at the peak period, the user can properly reduce the load or transfer the load to the valley period, at the moment the user buys electricity lower cost Increased user satisfaction[10].

1）Time-of-use price response objective function

$$
W_i = (\max R_i - \min R_i)
\tag{4}
$$

In the formula: W_i is the peak-to-valley difference of the user's electricity in the i period after the demand response. In order to achieve the reliability requirement, the difference between the peak and the valley is as small as possible; $\max R_i$ and $\min R_i$ are the maximum values of the demand electricity in the i period after the time-of-use electricity price is implemented. And minimum value.

2) Time-of-use price response load model
The comprehensive response of the user at i time after the implementation of time-of-use electricity price is the comprehensive load response model based on different price elasticity.

$$
\begin{cases}
R_i = R_{i0}(1 + \rho_{ii}\dfrac{I_i - I_{i0}}{I_{i0}} + \displaystyle\sum_{\substack{j=1 \\ j \neq i}}^{24} \rho_{ij}\dfrac{I_j - I_{j0}}{I_{j0}}) \\[4mm]
\rho_{ii} = \dfrac{\Delta R_i / R_{i0}}{\Delta I_i / I_{i0}} \\[4mm]
\rho_{ij} = \dfrac{\Delta R_i / R_{i0}}{\Delta I_j / I_{j0}}
\end{cases}
\tag{5}
$$

In the formula: R_i is the amount of electricity responded to by the user at time i after the time-of-use electricity price is applied; R_{i0}, I_{i0}, and I_{j0} are the i time original electricity amount, original electricity price, and the original electricity price of the user's response during the time of j; ρ_{ii} and ρ_{ij} are electricity prices from Elasticity coefficient and cross-elasticity coefficient; ΔR_i, ΔI_i, and ΔI_j are the magnitudes of changes in the user's response power at the time of i, the size of the power price changes, and the size of the user's response power price changes in the j period.

3）Time-of-use price response load constraints

$$
R_{i.\min} \leq R_i \leq R_{i.\max}
\tag{6}
$$

Where: $R_{i.\min}$ and $R_{i.\max}$ are the maximum and minimum values of the user's response to electricity after the time-of-use price is applied. This formula indicates that the time-of-use price demand response should be performed within the specified limits, which are determined by the nature of the load.

2.3. Comprehensive User Response Model

The user comprehensive response model refers to the demand-side integrated response method that considers the time-of-use price in the case of a transfer load. As the user's PV penetration increases after the load is transferred, the load peak-to-valley difference increases. On the other hand, considering the reduction in load peak-to-valley difference in time-of-use tariff, user satisfaction increased. Therefore, some users will adopt a comprehensive response mode, that is, to perform a time-of-use tariff response after the load is transferred.

1）User comprehensive response objective function

$$
H = Q_A + W_i
\tag{7}
$$

Where: H is the sum of the near-load power of the PV and the peak-to-valley difference of the user load, that is, the smaller the H is, the better the overall response of the user.

2）User comprehensive response load model

$$
D = L_{in} + R_i - L_{out}
\tag{8}
$$

Where: D is the user's overall response load power. The user side performs comprehensive demand response constraint conditions such as the above-mentioned transfer load response model and time-of-use price response model, and will not be described here.

2.4. Battery Energy Storage Model

Since the battery still occupies a large proportion in the energy storage application at this stage, this article only discusses the impact of the energy storage battery on the demand-side. In operation, the stored energy of the battery energy storage system is measured by the state of charge $S_{\mathrm{SOC}}(t)$. The expression is as follows:

$$
S_{soc}(t+1) = S_{soc}(t) - \frac{E_{BESS}(t)}{V_{BESS}}
\tag{9}
$$

In the formula: $E_{BESS}(t)$ is the charging and discharging size of the battery energy storage system at the momen t, $E_{BESS}(t)$ is negative for charging, $E_{BESS}(t)$ is positive for discharging, and $V_{BESS}(t)$ is the total capacity of the battery energy storage device.

2.5. Optimization Indicators

This article discusses the demand-side response system model that considers time-of-use tariffs under transfer load. The total cost of the micro-grid is the annual cost of the photovoltaic system, the annual cost of the energy storage system, the annual cost of the energy-storage bi-directional converter module, the user's transfer load compensation, photovoltaic compensation, and the maintenance and operation cost. The cost of operation and maintenance includes the cost of maintenance, management, labor, and related upgrades for reasonable expenses in daily operations.

1）Annual Net Profit of Micronet System

$$
\begin{cases}
C_{net} = C_1 R - C_0 \\
C_1 = \sum_{t=1}^{T} [e_d(t)P_d(t) + e_e(t)P_e(t) + e_{PV}(t)P_{PV}(t) + e_l(t)P_l(t) - e_i(t)P_i(t)]\Delta t \\
C_0 = C_{PV} + C_B + C_C \\
C_{PV} = Q_{PV}[I_{PV}\frac{r_0(1+r_0)^m}{(1+r_0)^m - 1}] + u(A) \\
C_B = Q_B[I_B R_E \frac{r_0(1+r_0)^m}{(1+r_0)^m - 1}] + u(B) \\
C_C = P_C[I_C \frac{r_0(1+r_0)^m}{(1+r_0)^m - 1}] + u(C)
\end{cases}
\tag{10}
$$

In the formula: C_{net} is the annual net profit of the micro-net system; C_1 and C_0 are the annual returns of the micro-grid system and the annual cost of the micro-net system investment; R is the similar day, taking 270 similar days, equivalent to get the annual system revenue; $e_d(t)$、$e_e(t)$、$e_{PV}(t)$、$e_l(t)$、$e_i(t)$ respectively the user electricity price, the photovoltaic electricity price, the photovoltaic subsidy electricity price, the user transfer load subsidies and the microgrid purchase price from the grid, where $e_d(t) = e_i(t)$; $P_d(t)$、$P_e(t)$、$P_{PV}(t)$、$P_l(t)$、$P_i(t)$ respectively for the user The load power, the power of the micro-grid network, the amount of the photovoltaic power generation, the total amount of user transfer load, and the power consumption of the micro-grid to the grid. C_{PV}、C_B、C_C are the annual cost of the photovoltaic system, the annual cost of the energy storage system, and the annual cost of the energy storage converter module; Q_{PV} and Q_B are the system photovoltaic capacity and the energy storage system capacity; I_{PV}、I_B、I_C are photovoltaic modules, respectively. Unit price, unit price of energy storage battery, and unit price of energy storage converter; R_E is the number of energy storage battery replacement; $u(A)$、$u(B)$、$u(C)$ are the annual operation and maintenance costs of photovoltaic, energy storage and energy storage converters; r_0 is the discount rate.

2) Photovoltaic permeability

The PV penetration rate refers to the proportion of the users' use of PV power in the whole year's load, expressed as

$$
S_{new} = \frac{Q_{PV,one} + Q_{PV,BESS}}{Q_{load,all}}
\tag{11}
$$

Where: S_{new} is the PV penetration rate; $Q_{PV,one}$ is the PV direct supply load capacity; $Q_{PV,BESS}$ is the size of the PV generation capacity after storage and recharge of the battery; $Q_{load,all}$ is the total user load.

3. Model Solving

3.1. Game Theory

In the optical storage microgrid, the users considering the transfer of load, the users considering the time-of-use tariff response and the interests of the energy-storage investors are related to each other, and there is a certain competition and restriction relationship. In addition, consider the user's comprehensive demand response, that is, the user's interest relationship between the demand response of the interest price and the energy storage under the transfer load. How to enable the three parties to cooperate with each other or the user's comprehensive response and energy storage to cooperate with each other in order to achieve the maximum net income of the photovoltaic microgrid, and to meet the microgrid reliability and maximum penetration of photovoltaic energy. This article adopts a cooperative game approach that enables three-way or two-way configurations to reach Nash equilibrium. Under the cooperation game, the three-party strategy is the percentage of the transferable load capacity, the percentage of the load response capacity under the time-shared price, and the capacity of the energy storage system. The two strategies are the user's comprehensive response capacity percentage and the capacity of the energy storage system. Under the given constraint conditions, the optimal values of the respective optimization goals are pursued, so that the system goals are optimized and the Nash equilibrium under the parties is finally achieved. There are four possible alliance modes for the three-party cooperation game, namely, the cooperation mode between any two alliances and the other party and the three parties forming the total alliance. The following uses [{Q,W},{E}],[{ Q, E}, {W}], [{Q}, {W, E}] and [{Q, W, E}] represent these four cooperative game modes. The two-party cooperation game has only one mode, and [D, E] represents this cooperative game mode. In the following, [{Q,W},{E}] is taken as an example to give a tripartite game strategy model. The two-party model will be given later. Among them, Q represents the user who transfers the load, W represents the user who responds to the time-of-use tariff response, D represents the user who considers the time-shared price response under the transfer load, and E represents the accumulator.

1)[{Q,W},{E}]game strategy model

Participants: {Q,W},{E}

Policy Set: $S_{QW}=[P_{Q,min},P_{Q,max};P_{W,min},P_{W,max}]$,$S_E=[P_{E,min},P_{E,max}]$

Information set: load, electrical parameters, economic parameters, power, etc.

Objective function: $I_{QW}(P_Q, P_W, P_E)$, $I_E(P_Q, P_W, P_E)$

Among them, P_Q, P_W, and P_E are the percentages of the transferable load capacity, the percentage of the load response capacity under the time-shared price, and the number of energy storage batteries; $P_{Q, min}$, $P_{W, min}$, $P_{E, min}$ are the minimum percentages of the two parties respectively. The minimum number of batteries; $P_{Q, max}$, $P_{W, max}$, $P_{E, max}$ are the maximum capacity percentage of the two parties and the maximum number of energy storage batteries; I_{QW} is the objective function of the alliance of users under the transfer load and time-shared price; I_E is the energy storage objective function. The target function, ie, the size of the return, is related to its own strategy, its opponent strategy, and its set parameters. The objective function of this article uses the system's total objective function, namely S_{new} and C_{net}. If there is a Nash equilibrium point (P_Q^*, P_W^*, P_E^*) in the above-mentioned cooperative game model, according to the definition of Nash equilibrium, it is expressed that (P_Q^*, P_W^*) and P_E^* are each other under the opponent's choice of optimal strategy. The optimal countermeasure is that the transfer load user, the time-of-use electricity price user, and the energy storage party can achieve the maximum benefit in the sense of Nash equilibrium. The other three cooperative game modes are similar to this mode and will not be described here.

2) [D, E] Game Strategy Model

Participants: D, E

Policy set: $S_D = [P_{D, min}, P_{D, max}]$, $S_E = [P_{E, min}, P_{E, max}]$

Information set: load, electrical parameters, economic parameters, power, etc.

Objective function: $I_D (P_D, P_E)$, $I_E (P_D, P_E)$

Among them, P_D is the percentage of the user's comprehensive response load capacity; $P_{D, min}$ and $P_{D, max}$ are the minimum and maximum percentages of the user's comprehensive response load capacity; I_D is the user's comprehensive response load objective function. The target function, ie, the size of the return, is related to its own strategy, its opponent strategy, and its set parameters. The objective function of this article uses the system's total objective function, namely S_{new} and C_{net}. If there is a Nash equilibrium point $(P_D{}^*, P_E{}^*)$ in the above cooperation game model, and according to the definition of Nash equilibrium, it is expressed that both $P_D{}^*$ and $P_E{}^*$ are optimal strategies under the optimal strategy selected by the other party, that is, the strategy combination. Under the user's comprehensive response and energy storage can achieve the maximum benefit in the sense of Nash equilibrium.

3.2. Solving steps

For the above game model optimization problem, this paper uses an iterative search algorithm to solve.

Step 1: Input raw data and parameters. The data for initializing the game model mainly includes the size of the load, the size of the light, and the price of electricity.

Step 2: Establish a game model. According to the model design method described in the previous section, an optimization model based on cooperative game is established.

Step 3: Set the initial value of the equilibrium point. The equilibrium point initial value $(P_{Q, 0}, P_{W, 0}, P_{E, 0})$ and $(P_{D, 0}, P_{E, 0})$ are randomly selected in the strategy space of each decision variable.

Step 4: Each game alliance independently optimizes the decision. The results of the j-th round of optimization for each league in the game are $(P_{Q, j}, P_{W, j}, P_{E, j})$ and $(P_{D, j}, P_{E, j})$. In the j-th round of optimization, each coalition passes the optimization results $(P_{Q, j-1}, P_{W, j-1}, P_{E, j-1})$ and $(P_{D, j-1}, P_{E, j-1})$ of the previous round. The optimization algorithm obtains the optimal combination of strategies $(P_{Q, j}, P_{W, j}, P_{E, j})$ and $(P_{D, j}, P_{E, j})$. which is

$P_{Q,j}=\mathrm{argmax}I_Q(P_Q,P_{W,j-1},P_{E,j-1})$;

$P_{W,j}=\mathrm{argmax}I_W(P_{Q,j-1},P_W,P_{E,j-1})$;

$P_{E,j}=\mathrm{argmax}I_E(P_{Q,j-1},P_{W,j-1},P_E)$;

$P_{D,j}=\mathrm{argmax}I_D(P_D,P_{E,j-1})$;

$P_{E,j}=\mathrm{argmax}I_E(P_{D,j-1},P_E)$.

Step 5: Information sharing. Share each player's strategy.

Step 6: Determine if the system finds a Nash equilibrium. If each game participant has the same optimal solution obtained in the adjacent two times, ie $(P_{Q,j},P_{W,j},P_{E,j})=(P_{Q,j-1},P_{W,j-1},P_{E,j-1}) = (P_Q{}^*, P_W{}^*, P_E{}^*)$ and $(P_{D,j}, P_{E,j}) = (P_{D, j-1}, P_{E, j-1}) = (P_D{}^*, P_E{}^*)$. According to the definition of Nash equilibrium, it can be considered that the game under this strategy combination reaches the Nash equilibrium point. If a Nash equilibrium point is found, step 7 is entered and the result is output; if Nash equilibrium is not reached, step 4 is returned.

Step 7: Output system Nash equilibrium points $(P_Q{}^*, P_W{}^*, P_E{}^*)$ and $(P_D{}^*, P_E{}^*)$. Considering the influence of the initial value on the solution of the equilibrium point, if the algorithm does not converge, you can reselect the initial value in step 3.

4. Analysis of examples

Select the load data of a typical PV microgrid typical day. The maximum photovoltaic power in a typical day is 368kW, and the maximum load power is 313kW. Here, a typical daily-seasonal proportionality factor K is introduced. The typical day-seasonal proportionality coefficient refers to the typical daily load data multiplied by a proportional coefficient as the seasonal load data. In the summer, the typical daily ratio coefficient $K_x=1$; in the winter, the typical daily ratio coefficient

K_d=1.2; in the autumn, the typical daily ratio coefficient K_q=0.8; in the spring, the typical daily ratio coefficient K_c=0.8. The coefficient of proportionality in different seasons may slightly change with the fluctuation of load, but it cannot exceed the constraint range, ie

$$K_{min} \leq K \leq K_{max} \tag{12}$$

Where: K_{min} and K_{max} are the maximum and minimum values of the seasonal scale factor.

Based on the above model and algorithm, the Nash equilibrium calculation result of the tripartite cooperative game is shown in Table 1:

Table 1. Cooperative game Nash equilibrium calculation results

Mode number	Game mode	P_Q*/%	P_W*/%	P_E*/N	Q_A/kW	W_i/kW	S_{new}/%	C_{net}/Ten thousand yuan
1	{Q,W,E}	6.4	3.6	230	40	102	89	-21
2	{Q,W},{E}	6.2	3.3	202	35	89	93	-17
3	{Q,E},{W}	5.5	4.1	261	42	85	88	-23
4	{Q},{W,E}	6.8	2.5	272	33	110	95	-24

From the above data, we can see that when [{Q,W},{E}] is used in the game mode, that is, the cooperative game model of time-shared electricity price load response and energy storage system at the Nash equilibrium point ($P_Q{}^*$, $P_W{}^*$, $P_E{}^*$) can make the system maximize profit. Compared to the Nash equilibrium points in the other three cases: the [{Q},{W,E}] model has the closest load to the photovoltaic power, that is, the photovoltaic power generation is closest to the load power, because of the transferable load percentage $P_Q{}^*$. Compared with other modes, the mode has the best PV penetration rate, but the energy storage battery configuration $P_E{}^*$ is also more than other modes. Because the energy storage battery is expensive, the system revenue is the lowest in this mode; the difference in load peak and valley in [{Q,E},{W}] mode is the smallest, because the percentage of $P_W{}^*$ considered in this mode is more than other modes, so the peak-to-valley difference is the smallest. However, in this mode, the energy storage capacity is configured more, so the revenue of the micro-grid is poor compared to other modes; the load in the [{Q,W,E}] mode is close to the photovoltaic power, peak-to-valley load difference, and system revenue are not optimal; [{Q,W},{E}] model system under the best return, the load close to the photovoltaic power is only less than [{Q},{W,E}] mode, the load peak difference is only less than [{Q,E} , {W}] model, because load transfer and load response under time-of-use tariffs can form a cooperative and complementary relationship to a certain extent, making PV permeability is improved, reducing the capacity of the energy storage device configured to make the system economical optimal. Through the above analysis, the [{Q, W}, {E}] model can basically meet the requirements of the economic reliability of the system. Compared with the other three methods, the use of [{Q, W}, {E}] cooperative game method can enable the micro-grid system to achieve better expected optimization goals. Then, we can compare the difference between the user's comprehensive response mode and the transferable load response mode alone and the time-of-use price response mode alone. As shown in Table 2:

Table 2. Comparison of three modes

Model name	P_Q*/%	P_W*/%	P_D*/%	P_E*/个	Q_A/kW	W_i/kW	S_{new}/%	C_{net}/Ten thousand yuan
Q	9.5	—	—	173	9	185	97	-28
W	—	5.8	—	354	118	55	82	-34
{D},{E}	—	—	9.5	202	35	89	93	-17

From Table 2, it can be seen that adopting the [{D}, {E}]] mode is more profitable than using the transferable load mode Q alone and the time-of-use price-demand response mode alone. In this mode, although the PV penetration rate is slightly less than the Q mode, the system load peak-to-valley difference is much smaller than that of the Q mode. This is because Q has more transferable load and there is no load response under time-shared electricity price, although the energy storage capacity ratio in Q This model is less, but considering the factors such as user income under time-of-use price, the [{D},{E}] model has better economics and reliability; the system load of the [{D},{E}] model Although the peak-to-valley difference is slightly larger than the W mode, the system PV is closer to the load and the PV penetration rate is better. This is because there are more load-sharing times in W, and there is no transferable load, so [{D},{E}] model can reduce the configuration of energy storage capacity for better system benefits. Through the above analysis, the [{D}, {E}] model has better economic reliability.

5. Conclusions

In order to achieve a more stable and economical operation of the photovoltaic microgrid, this paper establishes a microgrid system model considering the time-shared tariff response under load transfer, and uses the cooperative game method to solve the optimized configuration of the model. The results show:

1) The grid-connected optical storage microgrid can achieve system stability or economy when considering the demand-side response. This paper lists two demand-side response methods. Under the condition of transferable load, the PV penetration rate can be improved, but this will increase the load peak-to-valley difference; in the case of time-shared electricity prices, the load peak-to-valley difference can be reduced, but the PV penetration rate decreases.

2) Due to the above reasons, this paper uses cooperative game to solve two kinds of demand-side response modes and optimal configuration of energy storage system, uses comprehensive response model and optimal configuration of energy storage system, and uses iterative algorithm to find the Nash equilibrium point (system Excellent configuration). The validity of the proposed model and algorithm is verified by practical examples.

References

[1] Zachar, M., & Daoutidis, P. (2017). Microgrid/macrogrid energy exchange: a novel market structure and stochastic scheduling. IEEE Transactions on Smart Grid, PP(99), 1-1.

[2] Tushar, M. H. K., Assi, C., & Maier, M. (2017). Distributed real-time electricity allocation mechanism for large residential microgrid. IEEE Transactions on Smart Grid, 6(3), 1353-1363.

[3] Zhang, D. D., Tong, Y. B., Jin, X. M., & Liang, J. G. (2016). Optimal allocation of energy storage in micro-grid considering demand response. Power Electronics.

[4] Yang, M., & Han, X. (2015). Reseach on real-time scheduling strategy for micro-grid operation in island mode based on the demand-side response. Modern Electric Power.

[5] Philippou, N., Hadjipanayi, M., Makrides, G., Efthymiou, V., & Georghiou, G. E. (2015). Effective dynamic tariffs for price-based demand-side Management with grid-connected PV systems. PowerTech, 2015 IEEE Eindhoven (pp.1-5). IEEE.

[6] Li, M. A., Liu, N., Zhang, J., Lei, J., Zeng, Z., & Liu, W. (2016). Optimal operation model of user group with photovoltaic in the mode of automatic demand response. Proceedings of the Csee.

[7] Qadrdan, M., Cheng, M., Wu, J., & Jenkins, N. (2016). Benefits of demand-side response in combined gas and electricity networks. Applied Energy, 192.

[8] Aghajani, G. R., Shayanfar, H. A., & Shayeghi, H. (2017). demand-side management in a smart micro-grid in the presence of renewable generation and demand response. Energy, 126, 622-637.

[9] Huang, T., Xiyuan, M. A., Lei, J., Aidong, X. U., Guo, X., & Peng, L. I., et al. (2015). Optimal operation of household user-side microgrid considering time-of-use price and demand response. Southern Power System Technology.

[10] Hai, L. U., Peng, X., Zhang, B., Zhou, J., & Chen, X. (2017). Operation strategy of time-of-use electricity price for demand-side considering output uncertainty of grid-connected distributed energy resource. Electric Power Construction.

Permittivity model for GNSS-R telemetry wetlands

CAO Xinliang[1,a] , REN Xincheng[1]

[1]School of Physics and Electronics Information, Yan'an University, Yan'an 716000, China

[a] Electronic mail: caoxinliang874@163.com

Abstract: The GNSS-R (Global Navigation Satellite System-reflectometry) wetland remote sensing is achieved by receiving a reflected signal. Based on the relationship between reflected energy and permittivity of the reflective medium, the inversion model is built for GNSS-R terrestrial humidity remote sensing, which is the terrestrial humidity dielectric model of the L-band satellite navigation signal left-polarized wave component model on the basis of the analytical expression of vertical polarization. This paper thereby provides theoretical and methodological support for the detection and inversion of GNSS-R soil moisture.

1. Introduction

With the development of spatial information technology, China has independently built, and operated Beidou satellite navigation system (COMPASS), which has become an important part of the global navigation satellite system. However, its further application in the field of detection still needs to be explored. COMPASS can transmit signals in the L-band and S-band (where the L-band B1 and B2 frequencies are allowed to transmit the navigation service signal), providing an open, highly stable source of microwave radiation for microwave detection. Global Navigation Satellite System Microwave Remote Sensing (GNSS-Reflections, GNSS-R) [1] and its detection serves are an alternative solution.

At present, GNSS-R remote sensing is mainly focused on the detection of terrestrial humidity [2-4], snow cover, vegetation water content [5], forest cover, land surface imaging [6] and so on. Meanwhile, the SNR multipath delay phase inversion model of soil moisture and the "non-parametric statistical estimation model" of the snow surface parameters [7] aiming to obtain the surface parameters of the accurate snowfall weather have been put forward. In these models, the study between the permittivity and the reflected energy is the key to GNSS-R remote sensing.

The remote sensing of GNSS-R wetland is realized by capturing the reflected signals. The polarization of the reflected signals, the permittivity of the reflection medium and the height of the wave source are the main factors influencing the energy of the reflected signals. COMPASS launches the signals characterized by left-axis polarization in a right-handed way [8]. In this paper, the permittivity model of the reflected wave of L-band satellite navigation signal is obtained by studying the correlations among the polarization of the signal, the permittivity of the reflection medium and the wave height.

2. The Reflectivity of GNSS-R Wave for Detection

Different echoes of different polarizations are the result of the interaction between electromagnetic waves and ground objects.

Content from this work may be used under the terms of the Creative Commons Attribution 3.0 licence. Any further distribution of this work must maintain attribution to the author(s) and the title of the work, journal citation and DOI.

Published under licence by IOP Publishing Ltd

After the GNSS-R wave is reflected by the land surface, circularly polarized wave spins to reverse. Its reflection coefficient of the vertical and horizontal components is not the same, and the amplitude ratio and the phase difference of the reflected wave are changed before the elliptically polarized wave is synthesized. When the radiation source frequency is increased, the amplitude ratio and the phase difference are changed more significantly, as smooth surface reflection coefficient depends on the frequency, polarization, incident angle of wave and electrical properties of the surface [9].

GNSS-R wetland detection is to obtain the most important parameter, the reflection coefficient, by using the surface reflection signal data directly, and the reflection coefficient can be further associated with the complex permittivity, thus the correlation between the land reflection signal and complex permittivity can be achieved.

The surface reflectance of the GNSS reflected signal is the ratio of the total reflected power to the direct signal power, i.e., the total reflectivity is

$$\Gamma_{total} = \frac{P_{rt}}{P_i} = \frac{1}{2}(|\Re_{HH}|^2 + |\Re_{VV}|^2) \tag{1}$$

Where P_{rt} is the total reflected power. P_i is the incident power, namely, the direct power. Γ_{total} is the total reflectivity.

The reflectivity is a physical quantity directly related to the reflection coefficient, which is a normalized parameter for the transmitter cross-sectional area of the irradiated surface area per unit area. In order to convert the reflectivity and the incident angle (the satellite grazing angle $\gamma = 90^o - \theta$) to the relationship with the satellite elevation angle (elevation angle θ), the reflectivity in formula (1) can be expressed as:

$$\Gamma = |\Re(\theta)|^2 e^{-h \sin^2 \theta} \tag{2}$$

Here, $h[= (4\pi / \lambda)\sigma]$ is the roughness parameter, σ is the standard deviation for the height of the table for flat surface of the land, according to Rayleigh criteria ($\sigma < \lambda / 8\cos\theta$), h is 0, namely:

$$\Gamma = |\Re(\theta)|^2 = |\Re(\gamma)|^2 \tag{3}$$

For purely right circularly polarized GNSS signals, the vertical reflectance Γ_{VV} and horizontal polarization reflectance Γ_{HH} can be expressed as:

$$\Gamma_{VV} = \Re_{VV}^2 \tag{4}$$
$$\Gamma_{HH} = \Re_{HH}^2 \tag{5}$$

L-band frequency dispersion to water is very strong. For a purely circularly polarized GNSS wave, the left-axis component reflectivity can be expressed as:

$$\Gamma_{RL} = \Re_{RL}{}^2 = \frac{1}{4}(\Re_{VV} - \Re_{HH})^2 = \frac{(\varepsilon-1)^2 \sin^2 \theta (\varepsilon - \cos^2 \theta)}{(\varepsilon \sin\theta + \sqrt{\varepsilon - \cos^2 \theta})^2 (\sin\theta + \sqrt{\varepsilon - \cos^2 \theta})^2} \tag{6}$$

And the left-handed circular component reflectivity can be expressed as:

$$\Gamma_{RR} = \Re_{RR}{}^2 = \frac{1}{4}(\Re_{VV} + \Re_{HH})^2 = \frac{(\varepsilon-1)^2 \cos^4 \theta}{(\varepsilon \sin\theta + \sqrt{\varepsilon - \cos^2 \theta})^2 (\sin\theta + \sqrt{\varepsilon - \cos^2 \theta})^2} \tag{7}$$

Generally, the relative permittivity of wetlands varies with the volume of humidity, usually in a larger range of greater than 4. Fig.1 shows the relationship curve between the relative permittivity $\varepsilon \in [2,40]$ and the altitude angle at the altitude angle $\theta = \dfrac{5\pi}{12}$.

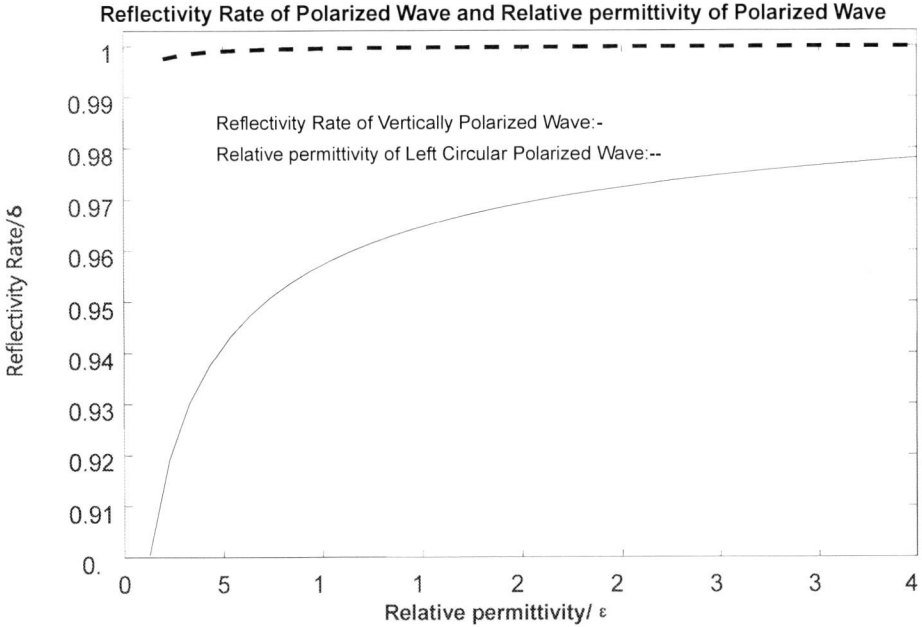

Fig. 1. $\varepsilon - \delta$ relationship curve of relative permittivity change

It can be seen that the reflectivity component tends to increase further when the relative permittivity (ε) increases at the larger altitude angle. Therefore, the reflection signal is obtained in a way to receive the main polarization component as its means.

3. Sensing Ground Surface Object Theory Based on GNSS-R Wave

3.1 Theoretical Derivation of Inversion Model

If the radiation is a linearly polarized wave, the reflected signal is dominated by the vertical polarization component under the condition of a large altitude angle, and the permittivity of the reflector can be detected by receiving the vertically polarized wave. By combining (4), the results are as follows:

$$\varepsilon^2 \sin^2 \theta (\frac{1-\Re_{VV}}{1+\Re_{VV}})^2 - \varepsilon + \cos^2 \theta = 0 \qquad (8)$$

The equation has two solutions. As formula (9), one solution can be taken to apply in the corresponding practical circumstances.

$$\varepsilon = \frac{1 \pm \sqrt{1 - 4\sin^2\theta \times \cos^2\theta \times (\frac{1 - \Re_{VV}}{1 + \Re_{VV}})^2}}{2\sin^2\theta \times (\frac{1 - \Re_{VV}}{1 + \Re_{RVV}})^2} \qquad (9)$$

The estimation of soil moisture is mainly based on the ratio of the reflected wave to the direct wave-related power (the normalized power of the reflected signal). For the value of $\Re_{VV}, \theta$ in (14), $\Re_{VV}$ can be obtained from the direct wave and the reflected wave data output from the receiver, as well as the angle θ from the positioning data.

As with the right-handed polarization signal transmitted by Beidou satellite, under the condition of high altitude of wave source, the reflected signal of the right-handed polarization signal is dominated by the left-handed polarization component, making it difficult to solve the relationship of $\Re_{RL} \sim \varepsilon$ directly from equation (6). Therefore, the relation between $\Re_{VV}$ and $\Re_{RL}$ needs to be discovered.

The following expressions are given by (7)

$$\frac{\Re_{VV}}{\Re_{RL}} = \frac{(\varepsilon\sin\theta - \sqrt{\varepsilon - \cos^2\theta})(\sin\theta + \sqrt{\varepsilon - \cos^2\theta})}{(\varepsilon - 1)(\sin\theta\sqrt{\varepsilon - \cos^2\theta})} = 1 - \frac{\cos^2\theta}{\sin\theta\sqrt{\varepsilon - \cos^2\theta}} = 1 - \Delta(\theta) \qquad (10)$$

Thus, it can be simplified as

$$\Re_{VV} = (1 - \Delta(\theta))\Re_{RL} \qquad (11)$$

Here, $\Delta(\theta) = \dfrac{\cos^2\theta}{\sin\theta\sqrt{\varepsilon - \cos^2\theta}}$

By combining with (9), we can get the permittivity expression, as follows

$$\varepsilon = \frac{1 \pm \sqrt{1 - 4\sin^2\theta \times \cos^2\theta \times (\frac{1 - (1 - \Delta(\theta))\Re_{RL}}{1 + (1 - \Delta(\theta))\Re_{RL}})^2}}{2\sin^2\theta \times (\frac{1 - (1 - \Delta(\theta))\Re_{RL}}{1 + (1 - \Delta(\theta))\Re_{RL}})^2} \qquad (12)$$

This is the ground permittivity detection model of the GNSS-R circular polarization wave.

The models expressed by (9) and (12) are theoretical models related to the Fresnel reflection coefficient. And the actual detection is achieved by the relative power of the reflected signal (i.e., the reflectivity).

3.2 The Inversion Method of Soil Moisture Based on Permittivity
The complex permittivity of soil surface sand can be expressed as follow:

$$\varepsilon' = \varepsilon - j\omega'' = \frac{k}{\varepsilon_0} - j\frac{\sigma}{\omega\varepsilon_0} \qquad (13)$$

Here, ε' is the complex permittivity for the surface of the land, ε is the real part of the relative complex permittivity, k is the absolute permittivity, σ is the conductivity, ε_0 is the permittivity of the free space, ω is the frequency of the electromagnetic, and ε reflects refraction and reflection of the radio wave to the land surface.

The Beidou's signal wavelength $\lambda = 0.19217$m, and the earth's conductivity σ changes within the range of 10^{-4} - 10^{-3}. The relative complex permittivity of the imaginary part varies between 0.0011-0.011, which can be ignored compared to the real part. Thus:

$$\varepsilon' \approx \varepsilon = \frac{k}{\varepsilon_0} \qquad (14)$$

There is a certain relationship between the permittivity of the terrestrial humidity and the reflected energy [9]. For 100% pure loess, the permittivity can be simplified as

$$\varepsilon = 2.863 + 3.462 m_V + 119..639 m_V^2 \qquad (15)$$

Here, m_V is the sand water volume (i.e., humidity).

The greater the permittivity of the scattering medium is, the stronger the ability to reflect the microwave and the smaller the penetration effect could be. During the surface humidity microwave detection, the main factor affecting the electrical constant is the surface water content of the shallow surface. The electrical constant increases linearly with the target surface water content. So when the electromagnetic wave reflectivity is higher, the echo becomes stronger and the microwave penetration is weaker. Another factor that affects the electrical constant is the conductivity of the surface soil surface. The loss and attenuation of microwave energy is a function of the ground target conductivity and wavelength. When the water content of the ground target is constant, there would be higher frequency (the shorter the wavelength), greater energy attenuation in the soil and the lower transmittance. In conclusion, the difference in soil water content directly leads to changes in the permittivity, which in turn affects the reflectivity and transmittance of electromagnetic waves.

The permittivity of the reflection area is the ratio of the wave power obtained by the direct wave and by reflection received from the output of the receiver. The test model is established by obtaining the angle from the positioning data and the permittivity of the wetland when combined with the theoretical model[9].

$$\varepsilon = \Gamma^{-1}(\Gamma_0) = \Gamma^{-1}(\frac{P_r}{P_d}) \qquad (16)$$

Here Γ_0 is the reflectance corresponding to the soil the permittivity ε, P_r is the reflected signal power, and P_d is the direct signal power.

Based on this test model, and combined with the inversion of soil-to-water volume, humidity, the reflectivity is the target parameter of the GNSS-R land-based receiver, and its value depends not only on the permittivity of the reflective medium, but also in relation to the geometry of the test equipment and the satellite position. These geometries can be converted into "Altitude angle". In order to analyze the relationship between reflectivity and altitude angle under different permittivity, when the altitude angle $\theta = \frac{5\pi}{12}$ and the relative permittivity $\varepsilon \in [2,40]$, the $\Gamma - \varepsilon$ simulation curve is shown as Fig.2.

ISPECE

IOP Publishing

IOP Conf. Series: Journal of Physics: Conf. Series **1187** (2019) 022066   doi:10.1088/1742-6596/1187/2/022066

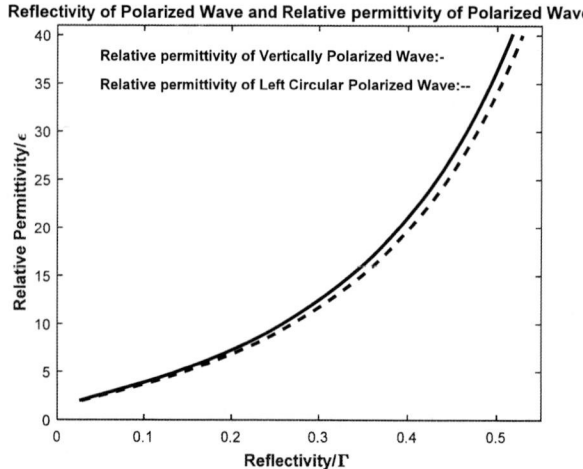

Fig.2. $\Gamma - \varepsilon$ Relation curve under certain altitude angle

Fig.2 reflects that under a certain altitude angle, the greater the permittivity is, the more sensitive to the reflectivity it would be. When the reflectivity is large, a small error will easily lead to a large measurement error of the permittivity.

The Beidou/GPS usually emits a right-handed circularly polarized signal whose reflected signal is left-handed circularly polarized, which aims to detect the humidity of the earth. The relative permittivity varies with a wide range of humidity, and is related to the source location and the geometric distribution of the test system. In order to describe the corresponding relationship between the reflectivity of the left circularly polarized wave and the relative permittivity under the geometric distribution of different constellations and test systems, the relationship is shown in Fig.3. under the conditions of the altitude angle $\theta = \dfrac{\pi}{12}$ 、 $\theta = \dfrac{\pi}{3}$ 、 $\theta = \dfrac{5\pi}{12}$.

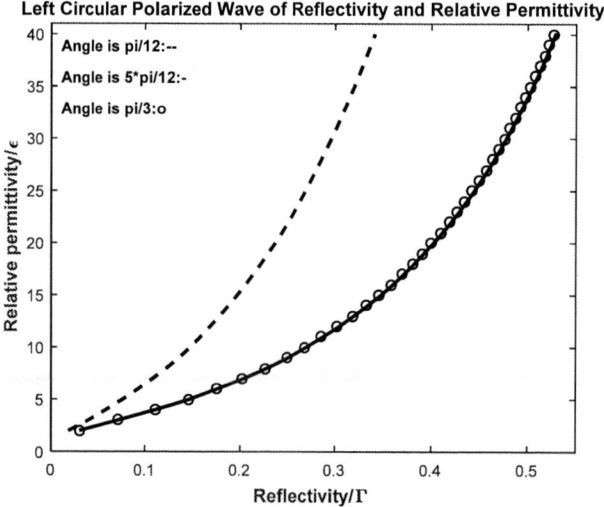

Fig. 3. Relations between right-handed rotating circular polarized wave
reflection and permittivity under different heights

It can be seen from Fig.3 that, when the soil moisture is same (i.e., the permittivity is constant), the greater the altitude angle is, the larger left circularly polarized wave reflectivity is; when the reflectivity is the same, larger the incident angle (the altitude angle is smaller) represents larger permittivity. Moreover, when the altitude angle is over $\frac{\pi}{3}$, the effects of altitude angle on the relationship between the reflectivity and the permittivity are not significant. In order to take the number of visible satellites into account and to receive the reflected signal energy as much as possible, the height of the angle should be suitable for the value of the altitude angle.

4. Discussion

GNSS earth humidity detection is the inverse problem of navigation and positioning system. Based on the analysis of the influence of soil composition on the reflection Fresnel reflection coefficient, this paper proves that the distinguishing degree of signal reflectance between the surface of the loess and the sand is not obvious, and the humidity is the main factor affecting the strength of the reflected signal, which serves as the solid ground for simplifying the complex components of the soil and the rationality of the earth's surface humidity detection with "pure loess" instead of earth. Based on the polarization characteristics of the reflected signal in GNSS-R land-based remote sensing, this paper concludes that the polarization of the large-angle corners is mainly a vertical polarization, while polarization is mainly about left-handed circular components. Moreover, the reflectivity increases with increasing humidity at the same altitude angle.

When the simulated permittivity $\varepsilon = 4$ and altitude angle-$\theta \geq \frac{\pi}{3}$, the main reflection energy and the main polarization wave ratio are more than 60%; when the altitude angle-$\theta = \frac{5\pi}{12}$, δ is more than 90%. When the altitude angle is over $\frac{\pi}{3}$, the altitude angle has no significant effect on the relationship between the reflectivity and the permittivity.

5. Conclusion

In this paper, the theoretical model and test model of the permittivity of GNSS are established by the theoretical analysis of the relationship between permittivity, reflectance and the influence of wave source angle on permittivity. The model shows that the larger the permittivity is, the more sensitive the reflectivity is. And it is easy to cause a large permittivity measurement error. When the altitude angle is over a certain value, the altitude angle on the relationship between the reflectivity and permittivity is not significant. There is a suitable angle of incidence and altitude angle for different permittivity GNSS detection. Therefore, by preliminary estimation of the permittivity, the geometric layout of the detection equipment and satellite constellation selection can be made based on the model, which proves practical significance to improve the measurement accuracy.

Acknowledgements

The author acknowledges the funding support from the National Natural Science Foundation of China (No.61661049), Science and Technology Program of Shaanxi Province (No. 2016GY-138) and Science and Technology Program of Education Department of Shaanxi（No.16JK1851）.

References

[1] WANG Yan , YANG Dongkai, HU Guoying, et al 2009 *GNSS World of China (in Chinese),* Vol. 34, p. 7.
[2] Egido A, Paloscia S, Motte E, et al 2014 *IEEE Journal of Selected Topics in Applied Earth Observation sand Remote Sensing*, Vol. 7, p. 1522.

[3] Valencia E,Zavorotny V U,Akos D M, et al 2014   *IEEE Transactions on Geoscience and Remote Sensing,* Vol.52, p. 3924.

[4] Wan W，Li H，Hong Y，Chen X W ,et al   2015   *Journal of Remote Sensing (in Chinese),* Vol.19, p. 882.

[5] WEI Wan, KRISTIN M L, ERIC E C, et al   2015   *GPS Solutions,* 2015, Vol.19, p.237.

[6] Rodriguez-Alvarez N, Camps A, Vall-llossera M, et al   2011 *IEEE Transactions on Geoscience and Remote Sensing,* Vol.49 , p.71.

[7] JIN Shuanggen, NASSER N   2014   *Advances in Space Research,* Vol.53,p.1623.

[8] WuXuerui，Li Ying   2012   *Advances in Earth Science,* Vol.27, p.895.

[9]ZOU Wenbo, ZHANG Bo, HONG Xeubao, et al 2016   *Acta Ġeo daeticaet Cartographica Sinica (Chinese),*Vol.45, p.199.

Design of continuous automatic wire-feeding device based on electric explosive wire

Ying An-wen[1,2,a], Yang Jia-zhi[1,2,b], and Yin Xin-zhe[1,2,c]

[1]Department of Information Science and Engineering, GuiLin University of Technology, GuiLin, GuangXi, China 541004

[2]Guangxi Key Laboratory of Embedded Technology and Intelligent System, Guilin University of Technology, Guilin, China 541004

[a]yinganwen921004@sina.com, [b]jiazhi_yang@126.com, [c]yxz1993@qq.com

Abstract. Continuous automatic wire-feeding device is a device that delivers metal wires required for electric explosion spraying to the designated position in vacuum environment, thus replacing manual wire-feeding on the equipment. The control system of the continuous automatic wire-feeding experimental device designed in this paper mainly includes three parts: the main control system, the man-machine interface and the upper computer. The main control system is composed of a single-chip microcomputer c8051f020, which realizes the control of components such as the motor drivers, relays, sensors and etc. outputs the collected signals and communicates with the human-machine interface. The human-machine interface consists of an LCD display and an IO expansion board, which can respond to the operator's instructions and display the experimental data. The host computer communicates with the main control system, sends commands and receives the feedback information of the main control for analysis and processing. The experimental results show that the device realizes automatic wire-feeding control with high degree of automation, which improves the efficiency and reliability of the electric explosion spraying experimental device.

1. Introduction

Automatic wire-feeding is an important part of electric explosive wire spraying. At present, there are two problems in general spraying equipment. One is that the spraying material of the metal wire explosion is not high in utilization; the other is that the spraying material cannot be automatically replaced, and it needs to be manually completed before spraying, which has high labor intensity and low efficiency [1-2]; As the center of the automatic control system, single chip computer plays a key role in the automatic control system. The connection between the single-chip microcomputer and the external equipment is usually realized through a serial communication interface[3]. The upper computer of the experimental device communicates with the single-chip microcomputer through RS485. The interface has a simple structure, long communication distance and strong anti-interference ability[4]. On the basis of existing experiments, an automatic wire-feeding spraying device was designed and manufactured in this paper. Based on the spraying equipment, the material utilization rate and reliability of the wire and the length control of the wire at both ends of the electrode tip are improved and optimized.

Content from this work may be used under the terms of the Creative Commons Attribution 3.0 licence. Any further distribution of this work must maintain attribution to the author(s) and the title of the work, journal citation and DOI.

Published under licence by IOP Publishing Ltd

2. Design of Continuous Automatic Wire-feeding Device

The automatic wire-feeding experimental device is mainly composed of three parts: an XY moving platform, a continuous wire-clamping integrated device and a main control circuit. As is shown in Figure 1, the two motors on the XY platform control the workpiece mounting plate under the XY platform to move on a horizontal surface to ensure that each spray can cover the specified area, and then achieve repeated spraying, which can make the thickness of the metal coating more uniform.

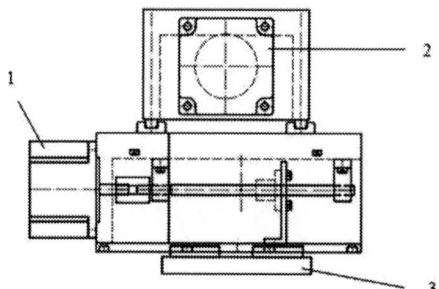

1-X axis motor; 2-Y axis motor;3-Workpiece mounting plate;

Figure 1. XY mobile platform

As is shown in Figure 2, the main components of the continuous wire clamp device consists of a moving component mounting plate, a moving chuck insulating mounting plate, an electrode chuck, an electrode wire guide pipe socket, an upper roller mounting plate, a lower roller mounting plate, a mounting bottom plate, the linear guide rail mounting plate, the left and right chuck insulating mounting brackets and the fixed chuck insulating mounting plate, etc. Installation of the bottom plate is mainly used to build the whole wire-feeding fixture in order to provide support for the device. Linear guide rail mounting plate fixes moving parts mounting plate through slide rail, and can drive it back and forth by stepping motor. The moving parts are used to fix the insulation mounting plate of the movable chuck; the movable chuck mounting plate is used to fix a left fixed chuck mounting bracket and a left movable chuck mounting bracket which can move longitudinally; the circular groove under the left fixed chuck mounting bracket will place permanent magnet, and the left movable chuck mounting bracket will place an adjustable voltage electromagnet. By changing the voltage and electrode direction of the electromagnet, The magnet's attraction and push-off can be realized. The electrode chuck is fixed on the left and right chuck insulating mounting brackets, and the electrode chucks at both ends of the wire are controlled to be pulled or pushed open by the back and forth movement of the left and right movable chuck insulating mounting brackets. Insulation mounting plate with fixed chuck is equipped with right movable chuck and right fixed chuck mounting bracket, on which the principle of right chuck mounting bracket is the same as that of left chuck mounting bracket. The upper roller mounting plate and the lower roller mounting plate fix the upper and lower rollers respectively. The movement of the wire between the upper and lower rollers is realized by the stepping motor driving the lower roller to rotate while driving the upper roller.

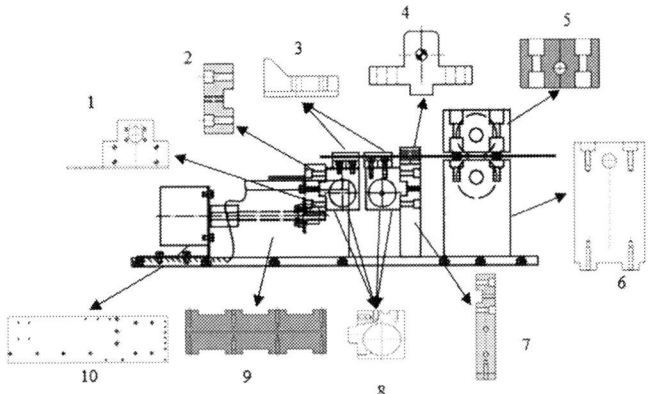

1-moving component mounting plate;2-moving chuck insulating mounting plate;3-electrode chuck;4-electrode wire guide pipe socket;5-upper roller mounting plate;6- lower roller mounting plate;7-fixed chuck insulating mounting plate;8-the left and right chuck insulating mounting brackets;9-the linear guide rail mounting plate;10-mounting bottom plate;

Figure 2. Continuous wire-feeding clamp structure diagram

The main control system of the automatic wire-feeding device uses the c8051f020 single-chip microcomputer as the control board to collect sensor information and control motor drive, relay switch and high-voltage reed relay, and perform data transmission with human-machine interface.

The schematic diagram of the main control system is shown in Figure 3.

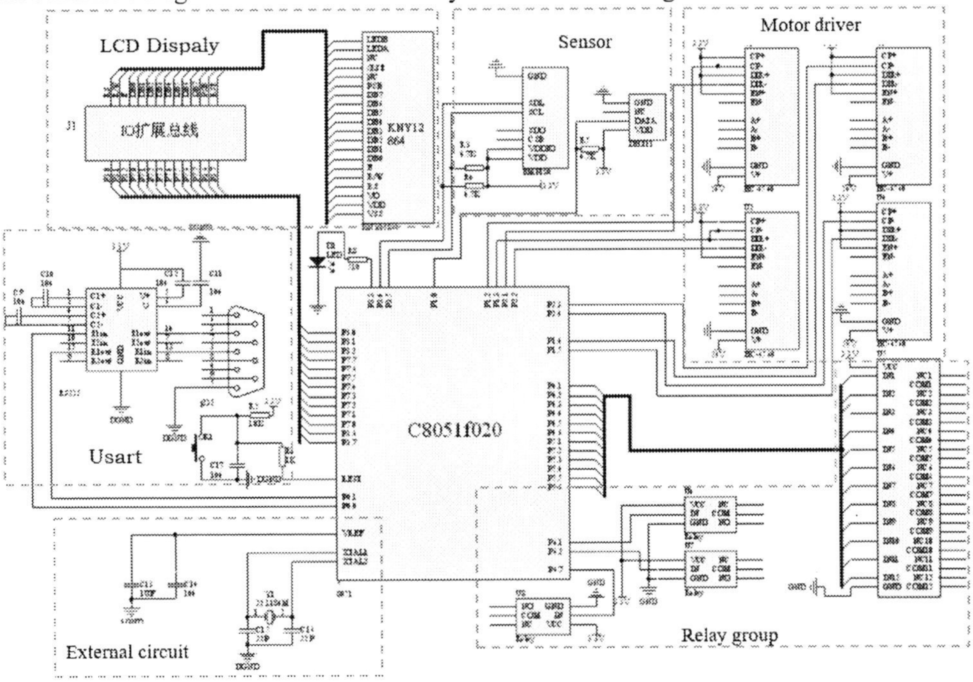

Figure 3. Schematic diagram of master control system

3. The Realization Process of Automatic Wire-feeding Device

The automatic wire-feeding device transports the wire through the motor-controlled roller to the left and right chucks for fixing, in order to realize the electric explosion spraying. The automatic wire-feeding includes the current automatic wire-feeding mechanism, control circuit, high-voltage terminal, etc. and the control circuit adopts the 8051f020 core board. The work flow is shown in Figure 4.

As is shown in Figure 4, after the experimental device is built and detecting whether each module is powered on, power on the PC and the main control chip, and turn on the upper computer, and send the command with it to the main control chip to test whether the main control module can be used. After testing, instructions are sent to the main control chip through the function module of the upper computer. The main control module sends the execution commands to the specified module end, observes the operation status of the module and the data sent by the sensor module displayed on LCD in real time. After all the modules have been tested correctly, adjust the wire to the appropriate position through the step-by-step debugging section on the host computer. Start the continuous wire-feeding button on the upper computer, and then the roller drive motor starts and begins to feed. When the length of the wire exceeds the length of the left fixed chuck to reach a predetermined length, the roller drive motor stops, the electromagnet installed in the left movable chuck mounting bracket (slidable) is forwardly guided, which is attracted by the opposite sex of the permanent magnets fixed in the left fixed collet mounting bracket. And then the two chucks clamp the wire. The reciprocating motor starts and pushes the right chuck closer to the left chuck. When the distance between the left and right chucks reaches the preset value, the reciprocating motor stops. The electromagnet installed in the right-moving chuck mounting bracket is directed. The right-moving chuck and the right-setting chuck clamp the wire beyond a part of the left chuck. The electromagnet installed in the left-moving chuck mounting bracket is directed in the opposite direction, and the two chucks bounce off. The reciprocating motor starts and drives the right chuck away from the left chuck. When the wire reaches

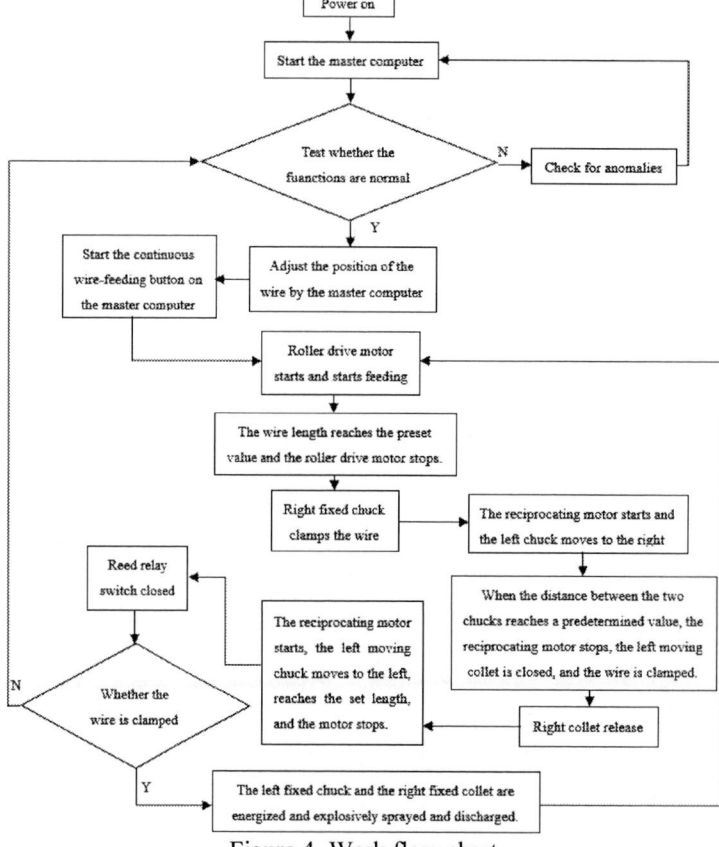

Figure 4. Work flow chart

the set length, the reciprocating motor stops. The electromagnet in the installation bracket of the left-moving chuck is in forward direction. The left-fixed chuck and the left-moving chuck are closed and clamped with metal wire. The dry-reed relay switch is opened to test whether the metal wire is clamped by the two chucks. If it is not clamped, read back check and repeat the above steps. On the contrary, the left and right chucks are electrically sprayed and discharged , and one spray is completed. Repeat the above steps to complete multiple sprays.

4. Requirements for Control System of Automatic Wire-feeding Device
The control system consists of main controller, upper computer, conventional relay, high-voltage dry-reed relay, stepper motor driver, sensor and other modules. The functions of the system are as follows:
- The working mode of the control system: Three switchable working modes are set up, including manual mode, automatic mode and single-step mode. The single-step mode and manual mode are mainly used to detect faults, debug the whole working process, whether the wire is clamped or not, and the execution of a single action. The automatic mode is mainly used in the actual research and testing process to realize the cooperative work of the moving parts of the automatic wire-feeding control system and continuous wire-feeding.
- Fault detection function: firstly, the automatic wire traveling system can automatically detect the failure of each module; secondly, it can detect whether the wire is clamped or not, and when the wire breaks or does not clamp, the signal received by the single-chip microcomputer is fed back to the upper computer to display the alarm information; thirdly, it can detect whether the main executing device is working normally. When abnormal happens, the system automatically alarms and stops, and displays abnormal information.
- Emergency stop function: In case of abnormal situation, you can click the emergency stop button on the host computer to stop all movement of the device.
- Debugging function: In the case of failure or before the device is ready to run, the debugging plate can be activated by the debugging button on the host computer to test whether the moving parts and the control chip can work normally.

5. Safety Design of Automatic Wire-feeding Device
Since the automatic wire-feeding device involves high voltage and strong current, and there is a complicated electromagnetic environment inside, which will generate strong electromagnetic interference to the low voltage control circuit, the safe operation of the whole system should be guaranteed when designing the control system of the experimental device. The measures taken include good grounding system, shielding of low voltage circuits, and isolation of high and low voltage[5].
- Grounding: The whole automatic wire-feeding device uses metal cabinet to surround it, isolate the inside and outside, except that, the metal cabinet is grounded; Other devices and circuitry that need to be grounded are centrally installed with grounded copper bars to connect the ground to ensure that high voltage capacitors can discharge outside the isolation network using buttons completely isolated from the cabinet.
- Shield: In the cabinet where the automatic wire-feeding device is located, there is strong electromagnetic interference, in order to ensure the reliability of the main control circuit. The main control circuit is placed outside the metal cabinet, and the data communication between the main control system and the automatic wire-feeding device is realized through the air plug.
- Isolation: In order to ensure the reliable operation of high and low voltage equipment, the circuit control signals and trigger signals are isolated by optocouplers. The power supply of the trigger circuit and the human-machine interaction terminal circuit is isolated and powered by an isolation transformer.

6. Conclusion
In this paper, an automatic wire-feeding device is designed for the research on electric explosion of mental wire. The device has the following characteristics:

- The execution efficiency is high, and the entire automatic wire-feeding is in the closed vacuum metal container, which is higher than the speed of manual wire-feeding, and meanwhile, the spraying efficiency is improved.
- The use of MCU collaborative control saves cost and manpower, and improves work efficiency.

It has high degree of automation, strong execution and simple operation.

Acknowledgment

This work was financially supported by the Natural Science Foundation of China (51167004), GuangXi Natural Science Foundation (2013GXNSFBA019250,2011GXNSFA08022) , GuangXi key Laboratory Fund of Embedded Technology and Intelligent System (Grant No. 2018B-04), Project of Education Department of Guangxi Zhuang Autonomous Region (No. KY2016YB195).

References

[1] Zhan-yong Gao. Research and development of automatic winding mechanism [J]. Textile Accessories,2012,(39):242-258.

[2] You-wei Wang,Zi-xue Qiu,Guo-wei Wang. The Design of Thermal Fuse Automatic Feeding Machine Based on PLC[J]. Modular Machine Tool & Automatic Manufacturing Technique,2015,4: 115-118.

[3] Xiao-tao Xu.Implementation on Serial Communication Based on MCS-51 Microcontroller[J]. Computer & Network, 2010, (19): 51-53.

[4] WANG Min. Applications of Multi-serial Port Communication for MCS-51 Single Chip Computer [J]. Electronic Technology, 2014, 56-57.

[5] Jia-zhi Yang,Yi-chen Deng,Peng Xiao. Design of experimental equipment control system for electrical exploding wire opening switches [J]. Laboratory Science, 2017, 20 (4): 44-46.

Calculation and verification of voltage drop when starting tunnel axial-flow fan

Enshi Wang[1], Bihui Huang[1]

[1]CCCC Second Highway Consultants Co., Ltd., Wuhan 430052, Hubei, China

Abstract. In the electromechanical design of highway tunnels, the calculation of voltage drop during fan start-up is an important part of electrical design, which is related to the safe operation of the whole system, green and energy-saving. The voltage drop involves the setting of substations, and ultimately affects the tunnel engineering scheme. In this study, a theoretical analysis, by calculating the voltage drop when the axial-flow fan starts was implemented. The method was used to calculate and verify the Milashan tunnel section of the 318 National Highway from Linzhi to Lhasa. The method can provide reference for peer designer in similar electromechanical design.

1.Introduction

The electrical equipment of the highway tunnel is powered by AC 380/220V power supply. The normal operation of electrical equipment determines the safety, reliability and comfort of highway tunnel operations.

Tunnels are divided into extra-long tunnels, long tunnels, medium tunnels and short tunnels according to their lengths. Tunnels are classified as follows in table 1.

Table 1. Tunnel classification.

Tunnel classification	Extra-long tunnel	Long tunnel	Medium tunnel	Short tunnel
Tunnel length L(m)	L>3000	3000≥L>1000	1000≥L>500	L≤500

According to the requirements of *Guidelines for Design of Ventilation of Highway Tunnels* and the engineering experience, the single-tunnel two-way traffic tunnel, which is generally larger than 5000m, adopts sectional smoke exhaust or ventilation, and an axial-flow fan is generally installed. Other tunnels generally adopt a longitudinal ventilation scheme to set a smaller-flow jet fan. The axial-flow fan is generally installed in or outside the tunnel. Whether the high-power equipment can start normally will affect the tunnel substation and the layout of the air shaft. This paper makes a theoretical analysis of the calculation of the voltage drop during the starting of the tunnel axial-flow fan and calculates and verifies the Milashan tunnel in the Milashan tunnel section of the 318 National Highway from Linzhi to Lhasa.

2. Specification and requirements

According to *Code for Design of Electric Distribution of General-purpose Utilization Equipment GB50055-2011*, when the motor starts, its terminal voltage should be able to guarantee the starting torque required by the machine, and the voltage fluctuation caused in the power distribution system should not hinder the normal operation of other electrical equipment. When the AC motor starts, the voltage on the distribution busbar shall comply with the following regulations.

Content from this work may be used under the terms of the Creative Commons Attribution 3.0 licence. Any further distribution of this work must maintain attribution to the author(s) and the title of the work, journal citation and DOI.

Published under licence by IOP Publishing Ltd

1) The distribution busbar is connected with lighting or other loads sensitive to voltage fluctuations. When the motor is frequently started, it should not be lower than 90% of the rated voltage; When the motor is not frequently started, it should not be lower than 85% of the rated voltage.

2) Unconnected lighting or other loads sensitive to voltage fluctuations on the distribution bus shall not be less than 80% of the rated voltage.

3) When no other electrical equipment is connected to the distribution bus, it can be determined according to the conditions for ensuring the starting torque of the motor; for low-voltage motors, the voltage of the contactor coil should be kept below the release voltage.

3. The definition, calculation and significance of voltage drop

The voltage deviation of the power supply and distribution system refers to the difference between the actual operating voltage U of each point of the system and the nominal voltage U_N of the system under normal operating conditions, expressed as a percentage, namely:

$$\delta U = \frac{U-U_N}{U_N} \times 100\% \tag{1}$$

When the motor starts, the voltage drop is caused in the power distribution system. The difference between the effective value U of the voltage before starting and the effective value U_{st} of the voltage at the start is called voltage drop, and is expressed by the relative value (ratio to the nominal voltage U_n of the system) or percentage:

$$\Delta u_{st} = \frac{U-U_{st}}{U_n} \ or \ \Delta u_{st} = \frac{U-U_{st}}{U_n} \times 100\% \tag{2}$$

The calculation and verification of voltage drop is an important part of the design of power supply and distribution system. In many projects, due to the calculation error or uncalculating of voltage drop leading to the failure of power supply system, making the equipment cannot start normally, or choosing a larger wire cross section leads to waste of resources. In the design process of the project, the calculation of the voltage drop at the start of the motor is a key part of the electromechanical engineering design.

4. Theoretical calculation

According to *Industrial and Civil Power Distribution Design Manual*, the calculation formula of the bus and motor voltage at the start of the motor is as follows. The calculation is shown in Figure 1.

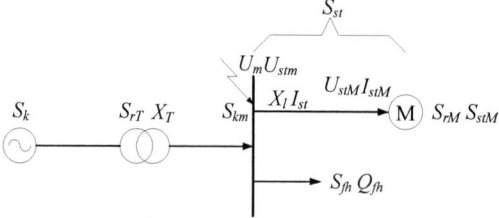

Figure 1. Calculation circuit.

Rated input capacity of the starting circuit:

$$S_{st} = \frac{1}{\frac{1}{S_{stM}} + \frac{X_l}{U_m^2}} \tag{3}$$

Busbar voltage relative value:

$$u_{stm} = \frac{S_{km}+Q_{fh}}{S_{km}+Q_{fh}+S_{st}} \tag{4}$$

Motor terminal voltage relative value:

$$u_{stM} = u_{stm}\frac{S_{st}}{S_{stM}} \tag{5}$$

The rated input current of the starting circuit:

$$I_{st} = u_{stm}\frac{S_{st}}{\sqrt{3}U_m} \tag{6}$$

The symbols are as follows:

S_k -----Transformer primary side short circuit capacity, MVA;

S_{rT}----Rated capacity of the transformer, MVA;

x_T----The relative value of the reactance of the transformer is taken as the relative value of the impedance voltage u_T;

S_{km}---Bus short circuit capacity, MVA;

$$S_{km} = \frac{S_{rT}}{x_T + \frac{S_{rT}}{S_k}} \tag{7}$$

Q_{fh}----Pre-loaded reactive power, M var; On the secondary side busbar of the transformer,$0.6(S_T - 0.75S_{rM})$ is available. If the preload isS_{fh}, its power factor is$cos\,\varphi_{fh}$, then$Q_{fh} = S_{fh}\sqrt{1 - cos^2\,\theta_{fh}}$.

X_l----Wire penetration or line reactance of $\leqslant 10$ kV cable, Ω is 0.08L,When the long line is included in the resistance factor, the copper core wire has a value of $(0.08 + 6.1/S)L$;When no more than 150mm^2, it has a value of $(18.3/S)L$, the copper core wire has a value of $(0.08 + 10/S)L$; When no more than 240mm^2, it has a value of $(30/S)L$, 0.08 is changed to 0.09 when used for cross-linked polyethylene cable, and 0.08 is changed to 0.07 when 6 kV insulation;

S-----Cross section of wire or cable core, mm^2;

L-----Line length, km;

S_{rM}----Motor rated capacity, MVA, its value is$\sqrt{3}U_{rM}I_{rM}$;

S_{stM}---Motor rated starting capacity, MVA, its value is$k_{st}S_{rM}$;

k_{st}----Motor rated starting current multiple, $k_{st} = I_{st}/I_{rM}$;

S_{st}----Rated input capacity of the starting circuit when the motor is started, MVA;

I_{st}----Rated input current of the starting circuit when the motor is started,kA;

I_{stM}---Motor starting current,kA;

U_m----Bus nominal voltage, kV, $U_m = U_n$.

5. Example calculation

The method in this study was used in the Milashan Tunnel Section of the 318 National Highway from Linzhi to Lhasa. The Mila Mountain Tunnel is about 5,720 meters long. It is equipped with an underground axial-flow fan unit and two axial-flow fans. The power is 315kW, the substation is about 50 meters away from the fan, and two 800kVA transformers are installed in the substation. According to the method, the voltage drop of the substation was calculated when the tunnel axial-flow fan is started, and the design specifications requirement was verified.

The known conditions are as follows.

- Transformer primary side short circuit capacity: $S_k = 200MVA$.
- Transformer capacity: $S_{rT} = 800kVA = 0.8MVA$.
- Transformer reactance relative value: $x_T = 6\% = 0.06$.
- Copper busbars from transformer to low voltage distribution cabinet and low voltage distribution cabinet $L_1 = 12m$.
- Low-voltage cabinet to tunnel axial-flow fan cable $L_3 = 50m$.

Tunnel fan motor technical parameters are following.

- Rated power: $P_{rM} = 315kW = 0.315MW$.
- Rated current: $I_{rM} = 580A = 0.58kA$.
- The tunnel fan adopts the soft start mode, and the starting current is controlled at 2~3 times of the rated current. This calculation is considered according to the maximum multiple.
- Starting current: $I_{st} = 3I_{rM} = 1.74kA$.
- Other load on the low voltage cabinet bus (power factor is 0.8): $P_{fh} = 36kW$.

The power distribution system is shown in Figure 2:

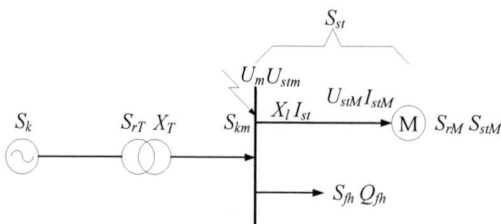

Figure 2. Tunnel Fan Power Distribution System.

5.1 Basic data calculation

According to the known conditions, the wire cross section of the line, the line impedance and the secondary side short circuit capacity are selected.

1) Calculation of rated current on the secondary side of the transformer.

$$I_{rT} = \frac{S_{rT}}{\sqrt{3}U_n} = \frac{800kVA}{\sqrt{3} \times 0.38kV} = 1216A$$

2) Section selection and impedance calculation of copper busbar on secondary side of transformer

According to the rated current calculation of the secondary side of the transformer, the current carrying capacity of the copper busbar on the secondary side of the transformer should be greater than the rated current. The copper busbar of the low voltage cabinet of this project has a specification of $4 \times (80 \times 8) + 1 \times (50 \times 5)$. According to the *Industrial and Civil Power Distribution Design Manual* (third edition), the resistance value of the copper busbar unit length of this specification can be found as $0.031m\Omega/m$, Reactance value is $0.17m\Omega/m$, then.

$$R_{L_1} = 0.031m\Omega \times 12m = 0.00372\Omega$$
$$X_{L_1} = 0.017m\Omega \times 12m = 0.00204\Omega$$
$$Z_{L_1} = \sqrt{R_{L_1}{}^2 + X_{L_1}{}^2} = 0.00424\Omega$$

3) Calculation of impedance of low-voltage cabinet to tunnel fan cable L_2

According to the load capacity of the tunnel fan, the cross section of the low-voltage cabinet to the tunnel fan cable L_2 is $2 \times (3 \times 185 + 1 \times 95)$ when the current carrying capacity and normal operating voltage loss are satisfied. According to the manual, the cross-section cable has a resistance value of $0.1004m\Omega/m$per unit length and a reactance value of $0.07203m\Omega/m$, then.

$$R_{L_2} = 0.1004m\Omega/m \times 50m \times \frac{1}{2} = 0.00251\Omega$$
$$X_{L_2} = 0.07203m\Omega/m \times 50m \times \frac{1}{2} = 0.0018\Omega$$
$$Z_{L_3} = \sqrt{R_{L_3}{}^2 + X_{L_3}{}^2} = 0.0031\ \Omega$$

4) Calculation of active, reactive, and apparent power of other loads connected to the bus of the same environmentally controlled electrical control cabinet.

$$P_{fh} = 36kW$$
$$S_{fh} = \frac{P_{fh}}{cos\,\varphi} = 45kVA = 0.045MVA$$
$$Q_{fh} = S_{fh} \times sin\,\varphi = 0.027Mvar$$

5) Start-up power calculation of tunnel fan

$$S_{rM} = \sqrt{3}U_{rM}I_{rM} = 0.3817MVA$$

According to the starting current multiple of the tunnel fan, the rated starting capacity is

$$S_{stM} = k_{st} \times S_{rM} = 1.1451MVA$$

6) Calculation of short-circuit capacity at the secondary side busbar of the transformer.

$$S_{km} = \frac{S_{rT}}{x_T + \frac{S_{rT}}{S_k}} = 12.5 MVA$$

5.2 Calculation and verification of voltage drop when tunnel fan starts

The voltage at the start of the tunnel fan drops. The input capacity of the circuit at the substation when the tunnel fan starts can be calculated.

$$S_{st} = \frac{1}{\frac{1}{S_{stm}} + \frac{Z_{L2}}{U_m^2}} = 1.124 MVA$$

The relative voltage of the busbar on the substation when the tunnel fan starts is:

$$u_{stm1} = \frac{S_{km} + Q_{fh}}{S_{km} + Q_{fh} + S_{st}} = 91.77\%$$

When the tunnel fan starts, the voltage value on the busbar of the substation is:

$$U_{stm1} = u_{stm1} \times U_m = 0.35 kV$$

The relative voltage at the motor terminals when the tunnel fan starts is:

$$u_{stM} = u_{stm} \times \frac{S_{st}}{S_{stM}} = 90\%$$

The voltage at the motor terminals when the tunnel fan starts is:

$$U_{stM} = u_{stM} \times U_m = 0.342 kV$$

5.3 Results

According to the second section of Section 2.2.2 of the *Code for Design of Electric Distribution of General-purpose Utilization Equipment* GB50055-2011, there are lighting or other loads sensitive to voltage fluctuations. When the motor starts frequently, it should not be lower than the rated voltage. 90%; when the motor is not frequently started, it should not be lower than 85% of the rated voltage. The busbar is connected to the lighting load, according to the provisions of the *Code for Design of Electric Distribution of General-purpose Utilization Equipment* GB50055-2011, the tunnel fan is a type of equipment that is not frequently started, so the voltage on the distribution bus should not be lower than 85% of the rated voltage. According to the calculation, the calculated voltage at the busbar of the low-voltage distribution cabinet at the start of the tunnel fan is 91.77% of the rated voltage, so the requirements are met.

In summary, the voltage drop value of the tunnel axial-flow fan at the start of the tunnel of the Mira Mountain Tunnel in the Milashan Tunnel section of the 318 National Highway from the Linzhi to Lhasa section of the 318 National Highway meets the requirements of the specification and passes the verification.

If the voltage drop value of the tunnel fan does not meet the requirements during the calculation and verification process, some measures need to be taken, such as increasing the cable cross section to reduce the line impedance and setting a short-distance substation to reduce the voltage drop. The tunnel axial flow fan is started to meet the specification requirements.

6. Conclusion

In highway tunnel electrical and mechanical meters, the calculation of voltage drop at the start of high-power equipment is an important part of electrical design, which is related to whether the entire system can operate safely, and involves the layout of substations, ultimately affecting civil engineering. It is expected that the calculation of this project case can provide reference for designers in the same project.

References

[1] GB50055-2011 Code for Design of Electric Distribution of General-purpose Utilization Equipment [S].

[2] Yang Yue. Power Supply and Distribution System [M]. Beijing: Science Press, 2007.

[3] Ren Yuanhui. Industrial and Civil Power Distribution Design Manual (Third Edition) [S]. Beijing: China Electric Power Press, 2005.

[4] Ren Yuanhui. Industrial and Civil Power Distribution Design Manual (Fourth Edition) [S]. Beijing: China Electric Power Press, 2016.

[5] Li Xing, Wang Zhidong, Liu Xiaodong, et al. Voltage fluctuation of power supply and distribution system and selection of starting mode of large motor[J]. China Mine Engineering, 2008, 37(4): 43.

[6] Wang Yanli, Xie Wei, Lu Nan. Research on the setting scheme of follow-up step-down substation in metro station[J]. Electric Railway, 2010, (3): 44.

ISPECE IOP Publishing

A novel searching method of fault chains for power system cascading outages based on quantitative analysis of dynamic interaction between system and components

Jintao She

State Grid Fujian Information&Telecommunication company, Fuzhou, 350001

841831635@qq.com

Abstract: Because the traditional searching method of fault chains for cascading outages usually only consider theover-load protection or simulate the action of protection and control deterministically, a novel searching method of fault chains for power system cascading outages which is based on quantitative analysis of dynamic interaction between system and components is proposed. Based on fault chain model, an evaluating index is established. This index is used to quantize the action situation of protection and control of key elements. Furthermore, according to the index, the action probability of protection and control of key elements under critical condition is figured out to confirm subsequent events of fault chains for cascading outages. Finally, high risk fault chains are screened out based on the risk index. Taken some actual power system under the summer peak load condition in 2017 as an example, the simulation result proves reliability and effectiveness of the proposed method.

1. Introduction

In recent years, there have been many power system blackouts at home and abroad, such as blackouts in the United States and Canada in 2003, blackouts in Western Europe and blackouts in Italy; blackouts in Brazil in 2009 and 2011; two blackouts in India in the summer of 2012 [1]; and a blackout in Turkey on March 31, 2015[2]. Although China has not experienced the above-mentioned serious blackouts, there have also been several local blackouts caused by cascading failures, such as the Central China Power Grid blackouts on July 1, 2006[3]. These blackouts have a very serious impact on people's daily life and economy, and even threaten social stability.

The development of blackouts is a process of deterioration and intensification of power system operation, which starts with a small number of simple component failures and develops through a series of complex interruptions. Previous analysis shows that cascading failures often accompany the development of blackouts, and cascading failures play a role in boosting the occurrence of blackouts [4-5]. According to Murphy's law, if an event has the possibility of happening, although the possibility may be very low, it will happen in a certain situation. Similarly, although the occurrence of cascading failure blackout is a minimal probability event, once it occurs, the direct loss is huge, and the indirect loss is even more difficult to estimate [6]. Therefore, it is necessary to establish an appropriate cascading failure model, study the evolution mechanism of cascading failure and master the development law of cascading failure [7].

At present, there are many kinds of chain fault analysis models with different functions[8]. According to the different starting points of modeling, cascading failure analysis models of power system can be roughly divided into two categories: power flow/stability calculation based research

Content from this work may be used under the terms of the Creative Commons Attribution 3.0 licence. Any further distribution of this work must maintain attribution to the author(s) and the title of the work, journal citation and DOI.

Published under licence by IOP Publishing Ltd

methods and network topology based research methods[9]. Among them, the research methods based on power flow/stability calculation mainly include the related research on self-organizing criticality of complex systems and the search strategy of cascading failure modes[10]. The research on self-organizing criticality of complex systems mainly includes OPA model and its improved model[11-12], hidden fault model [13], Manchester model [14], cascade fault model [15] and branching process model [16]. The research on pattern search strategy mainly includes fault tree analysis method and fault chain model [17]. The research method based on network topology mainly uses complex network theory and combines the characteristics of power grid itself to study cascading failures of power grid from the overall point of view, and to seek the structural root of cascading failures [18]. The model based on self-organizing criticality of complex systems can reveal the mechanism of cascading failures and quantify the risk of cascading failures, but it can not simulate the reaction mechanism of cascading failures in detail, and it is difficult to provide decision-making information for operators and planners. The development path and final results of cascading failures can be visually expressed by mode search strategy. Among them, the model of accident chain can not only simulate the reaction mechanism of cascading fail The study of macro-phenomena of large blackouts can also simulate the cascading failure response mechanism in detail, which provides an engineering application method for cascading failure control research, but its determination and calculation in subsequent links

The key of cascading failure analysis based on accident chain is how to search for high-risk accident chain. Because of the large scale of power grid and the variety of accident inducements, the research focuses on how to improve the search speed of accident chain and search for the key links in the follow-up [20-21]. Literature [22] From the perspective of preventing major blackouts, a fuzzy comprehensive assessment method for cascading failure risk of power system based on accident chain is proposed. The method calculates the occurrence probability according to the quantitative evaluation value of the factors affecting the accident chain, and uses the severity index to evaluate the consequences of the accident chain from the perspective of system transient safety. Fuzzy comprehensive evaluation method is used to evaluate and classify the risk of accident chain, and the definition of risk level and the corresponding prevention and control guidance scheme are given. Literature [23] proposes a prediction index of the intermediate link of the accident chain, which considers both the probability importance of the accident and the structural importance of the equipment in the network. The grey correlation degree algorithm based on maximum deviation is used to synthetically consider all kinds of indicators and construct a fault chain search model for cascading failures in power grids. The model based on complex network theory can analyze the overall system behavior [24], but it generally ignores the unique characteristics of power grid, such as electrical characteristics, operation characteristics and power flow characteristics, and is not fully applicable to the security and stability analysis and control of power grid.

In order to solve the problem that the traditional cascading failure chain search method can not take into account the uncertainty of protection control, this paper presents a method for identifying cascading failure chain in power system based on dynamic interactive quantitative analysis of system and components. This method is based on the accident chain model and considers the existing two or three protection configurations in the process of time domain simulation. The uncertainties of protection control are quantified by the action evaluation index of key component protection control, and the action probability of critical component protection control is calculated to determine the follow-up events of cascading failure chain. In addition, according to the action probability of protection control and its parameters, the following events of cascading failure chain are determined. The control cost after operation, the risk assessment of accident chain, and the screening of high-risk chain of cascading failures provide a reference for the prevention and control of cascading failures.

2.Accident Chain Theory

Accident chain theory originated from safety science and has been developed and applied at present.In many engineering fields. Accident Chain Theory in Electric Power System Large blackouts are caused by many intermediate links, and these are the main causes of blackouts.The intermediate link is not inevitable, but with the operation of the power grid.A series of factors, such as status, management and external factors, are related. Intermediate links are triggered, resulting in cascading failures. Defining the Set of Power Network Accident Chains and the Mathematical Table of Section k Accident Chains Darwinian

$$L = \left\{ \vec{L_1}, \vec{L_2}, ..., \vec{L_n} \right\} \quad (1)$$

$$L_k = \left\{ T_{k1}, T_{k2}, ..., \overrightarrow{L_{km_k}} \right\} \quad (2)$$

In the formula: n is the number of power network accident chains; m_k is numbers of interlinks of accident chain $\vec{L_k}$; T_{kl} is the link l in the accident chain k , $l=1,2,...,m$. The logic represented by the above accident chain and the set of accident chains. The compilation relationship can be illustrated by the accident tree shown in Figure 1.

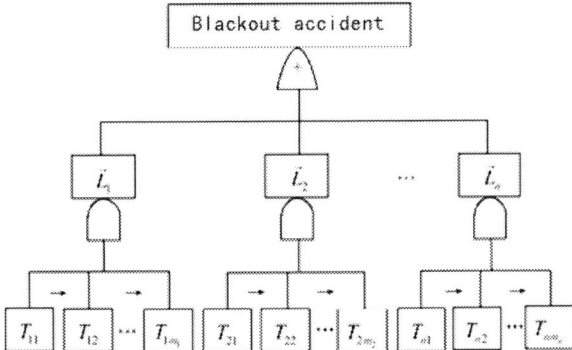

Fig. 1 Logical relationship of fault chains

The chain fault model based on accident chain provides a new control idea for the prevention and control of cascading failures. It only needs to cut off one or more links of the chain, instead of controlling all links, to prevent the occurrence of blackouts. The method of cutting off one or more links of the accident chain is to monitor the factors affecting the accident involved in this link, so as to avoid the failure of this link. According to this control idea, it is necessary to know the accident chain of power system beforehand. Therefore, how to quickly generate a reasonable accident chain has become the difficulty and focus of the theoretical model based on the accident chain.

3. Chain Search Model for Cascading Fault Based on Quantitative Analysis of Dynamic Interaction between System and Components

3.1 Protective Control Action Evaluation Index
There are two main criteria for the operation of protection or control devices in power systems: the first one can be described as (E_c, T_c) , E_c is the threshold value of the electrical quantity of protection and control devices, and can generally be voltage, current, frequency and active power, etc. T_c is the allowable duration time, such as that of generators.

The second can be described as (Q,N), Q is an event, N is the threshold number of protection control actions, such as UHV continuous commutation failure blocking DC protection.

Time domain simulation can accurately simulate the action of protection and control devices and search the accident chain. However, it is difficult to consider some uncertain factors, especially the operation of protection and control in critical conditions, if the protection and automatic device can only be simulated with certainty.

In order to quantify the action of protection control, the two-dimensional criterion of protection control device action is expressed by one-dimensional index, that is, the evaluation index of key component protection control action. The calculation formula is as follows. For the first protection control, its evaluation index is calculated as equation (3).

$$\lambda = \left[E_e - \left(E_c - kT_c \right) \right] \times 100\% \quad (3)$$

Formula:

E_e is the maximum/minimum electrical quantity monitored in the dynamic process (such as voltage, current, active power and frequency); E_c and T_c are the threshold value and allowable duration of the protection control device respectively; k is the duration of electrical offset is converted into the conversion factor of electrical quantity.

When the values are different, the physical meanings are as follows:

$\lambda \geq 0$, indicating that the protective control device does not meet the operating conditions and is far from the operating boundary.

$\lambda \approx 0$ is close to 0, indicating that the protective control device is in a critical state of operation.

$\lambda < 0$, indicating that the protective control device meets the operating conditions.

For the second protection control, the evaluation index is calculated as follows.

$$\lambda' = \frac{N-n}{N} \quad (4)$$

In the formula（4）,n is the number of actual actions; N is the number of threshold set for protective actions.

According to the above analysis and the evaluation index of protection control action, the calculation formula of protection control action probability can be defined as follows.

$$P(\lambda) = \begin{cases} P & \lambda \leq 0 \\ P - \dfrac{\lambda}{\lambda_{ref}}(P-p) & 0 < \lambda \leq \lambda_{ref} \\ P & \lambda > \lambda_{ref} \end{cases} \quad (5)$$

In the formula（5）, λ is the index value of protection and control action evaluation and λ_{ref} is the threshold value of ref are set for protection and control action evaluation, P is the probability of hidden fault (when no hidden fault is considered, P=0), P is the probability of correct action (when no hidden fault is considered P=1), P and P can be obtained by statistical data. Therefore, the relationship between the evaluation index of protective control action and its action probability is shown in Fig. 2.

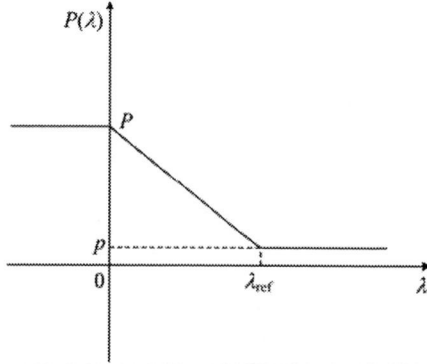

Fig. 2 Relationship between λ and $P(\lambda)$

3.2 Accident Chain Search Process

Combining with the evaluation index of key component protection control action, the proposed accident chain search process. The process shown in Figure 3 consists of the following steps:

1) According to the real-time operation data of power grid, the initial event sets are determined

based on expert experience, such as important main transformer N-1 and N-2 faults, key transmission line N-1 and N-2 faults and three-phase short-circuit single-phase rejection faults.

2) Starting from an event in the initial event set, the time domain simulation analysis is used to determine whether the load loss exceeds the set threshold. If it exceeds, it enters step 5 or step 3.

3) Judging whether there is a protective control action, if the action, the protective control action will be the next event, step 2, or step 4.

4) Calculate the evaluation index and the corresponding action probability of the key component protection control action. If the evaluation index of the key component protection control action is less than the set threshold value, the protection control with the minimum index value is selected.

Braking as the next event, go to step 2, otherwise go to step 5.

5) Calculate and judge whether the risk value of the current accident chain is greater than the set threshold value. If it is greater, save the accident chain and go to step 6, otherwise go directly to step 6.

If the current accident chain develops to m level, the formula for calculating the risk value of the accident chain is as follows

$$R = \sum_{i=1}^{m} p\left(d_i \mid d_1 d_2 ... d_{i-1}\right) C_i \quad （6）$$

in formula （6）: $p\left(d_i \mid d_1 d_2 ... d_{i-1}\right)$ is conditional probability for the occurrence of level i events; C_i is control cost for Level I Events.

6) Judge whether the search of accident chain is over, if it is over, output the set of accident chains and end the method, otherwise step 1) Select the next initial event to continue the search of accident chain.

4. Analysis of three examples

4.1 Research subjects
In this paper, the planned grid structure of a practical power grid in 2017 is taken as the research object, and the high-risk cascading failure chain which may occur when different types of faults occur in the power grid under peak load in summer is analyzed.

4.2 Initial Fault Set
The initial fault set analyzed in this paper mainly includes the N-2 fault of 500 kV line in the actual power grid and the single-phase fault of 500 kV line with three-phase short circuit. The simulation analysis of all the initial faults shows that all the N-2 faults of 500 kV line can maintain the stability of the system, and the probability of cascading faults is very small. Therefore, cascading faults are not further analyzed. Three-phase short circuit single-phase malfunction of partial lines. In order to simplify the calculation, combined with the initial fault characteristics, the following line faults are selected as the initial fault concentration faults, as shown in Table 1.

Table 1 Original fault set to be analyzed

Accident chain sequence	Initial fault	Initial Fault Occurrence Probability
1	Liandu-Ouhai	0.001
2	Lithosphere-phoenix instrument	0.001
3	Wuning-Danxi	0.001
4	Jinhua-Yongkang	0.001

4.3 Cascading Fault Chain Simulation
The threshold value of load loss is 30% of the total network load, the threshold value of protective control action index is 0.5, and the threshold value of accident chain risk is 10,000 yuan. Only the generation and load control costs are calculated, and the control cost of machine cut-off is 0.25 million yuan/MW, and the control cost of load cut-off is 10,000 yuan/MW. The simulation details are as

follows.

4.3.1 Accident Chain 1

The chain of cascading failures starts from the single-phase rejection fault of three-phase short circuit of Liandu-Ouhai line; after the failure, Binjin and Lingshao DC all fail to commutate continuously for three times (as shown in Fig. 3), the power is reduced by 15,000 MW; the bus positive sequence voltage of the two DC lines is shown in Fig. 4; after Lingshao and Binjin DC bipolar blocking, the frequency of the power grid continues to drop to 49.25 Hz (as shown in Fig. 5). As shown above, trigger the low frequency load shedding advance wheel action, the total load shedding is about 6200 MW. After the low frequency load shedding, the power grid frequency restores stability, the system reaches a new equilibrium point, and the cascading failure ends.

As shown in Figure 7, the risk of the chain of cascading failures is $99.5 million.

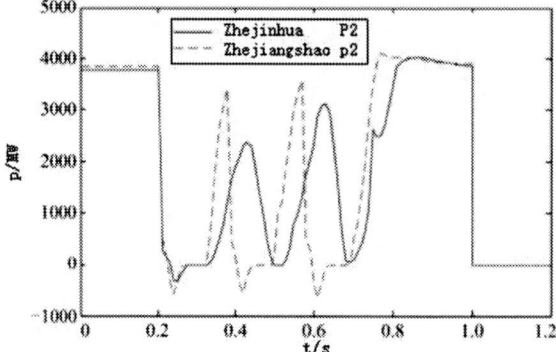

Fig. 3 Active power curve of Ling-Shao and Bin-Jin DC lines

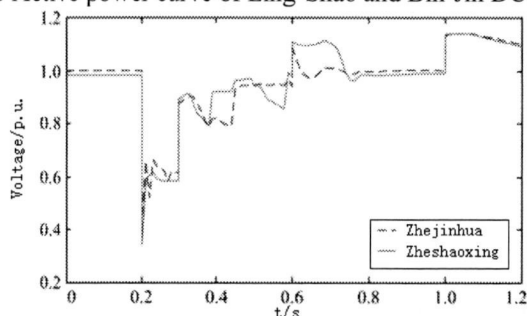

Fig. 4 Bus positive-sequence voltage curve of Ling-Shao and Bin-Jin DC lines

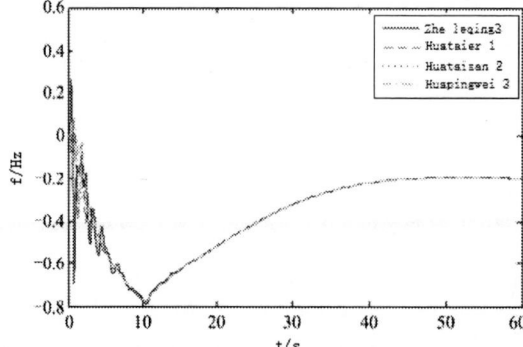

Fig. 5 Frequency curve of some generators

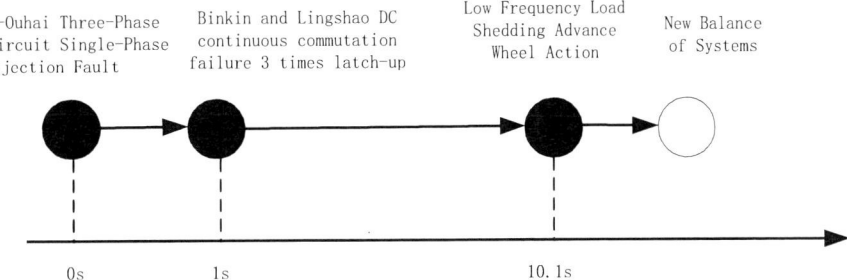

Fig. 6 Schematic of high-risk cascading outage fault chain 1

4.3.2 Accident Chain 2
Chain 2 of cascading faults starts from the single-phase rejection fault of three-phase short circuit of Cangyan-Fengyi line; after the faults, Lingshao DC continuous commutation fails three times, Binkin DC continuous commutation fails two times (as shown in Fig. 7), and the positive sequence voltage of the buses of the two DC lines is shown in Fig. 8.

According to formula (4) and formula (5), the evaluation index of Bingjin DC protection control action is 1/3 and the probability of bipolar blocking is 2/3. After Lingshao and Bingjin DC bipolar blocking, the total loss power is 15,000 MW, and the frequency of power grid continuously drops to 49.25 Hz (as shown in Figure 10), triggering low frequency load shedding ahead of time and reducing load totally about 6200 MW. After low frequency load shedding, the frequency of power grid restores stability. The system reaches a new equilibrium point and cascading failure ends (as shown in Figure 11). The risk of cascading failure chain is 726,000 yuan.

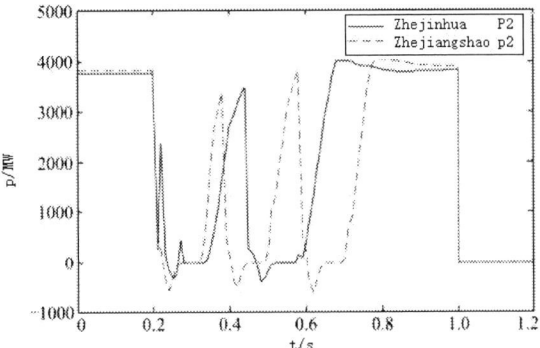

Fig. 7 Active power curve of Ling-Shao and Bin-Jin DC lines

Similarly, accident chain 3 and accident chain 4 can be obtained by simulation. Their schematic diagrams are shown in figs. 12 and 13.

In addition, the appendix gives the simulation results of another high-risk cascading failure chain in the actual power grid, quantifying the key line low-voltage disconnection operation index and its operation probability by formula (3) and formula (5).

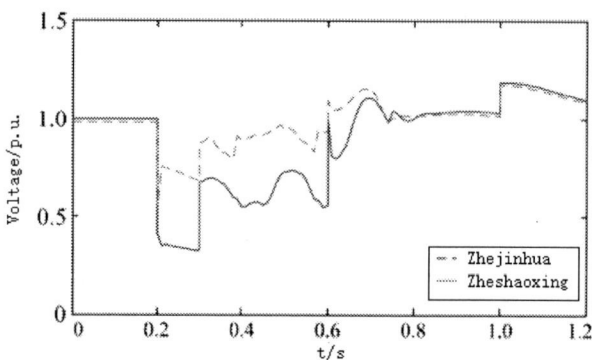

Fig. 8 Bus voltage of Ling-Shao and Bin-Jin DC lines

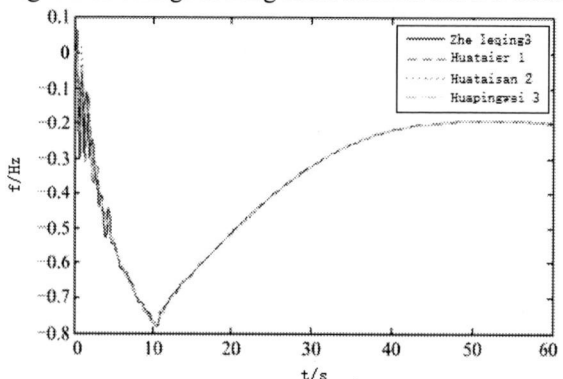

Fig. 9 Frequency curve of some generators

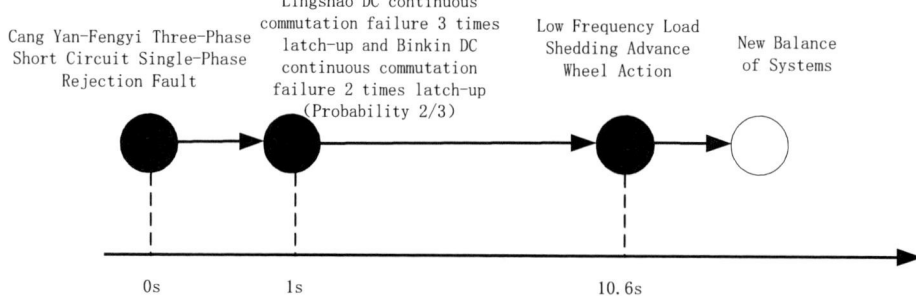

Fig. 10 Schematic of high-risk cascading outage fault chain 2

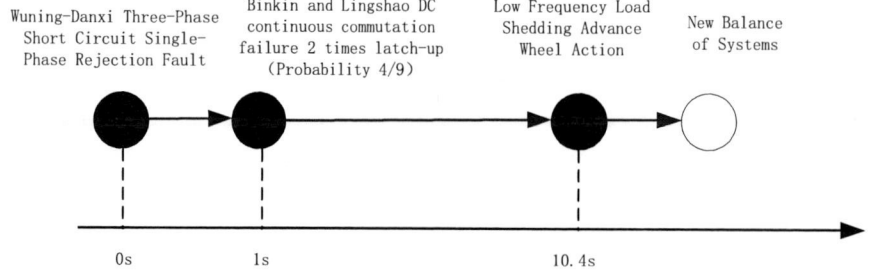

Fig. 11 Schematic of high-risk cascading outage fault chain 3

ISPECE   IOP Publishing

IOP Conf. Series: Journal of Physics: Conf. Series **1187** (2019) 022069   doi:10.1088/1742-6596/1187/2/022069

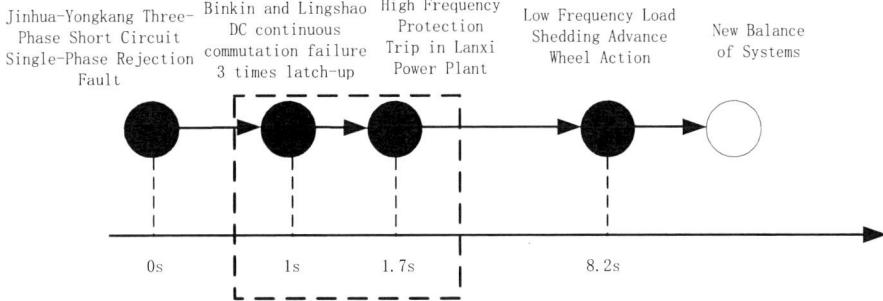

Fig. 12 Schematic of high-risk cascading outage fault chain 4

4.3.3 Simulation results analysis
The simulation results of the chain of cascading failures mentioned above are analyzed and the following conclusions can be drawn:

1) Under typical operation mode, the single-phase fault of three-phase short circuit of 500 kV transmission line may lead to cascading failures with high risk, and the key events of high-risk cascading failure chain are Lingshao and Binkin DC bipolar blocking failures.

2) As shown in Fig. 6, the chain of cascading failures starts from the three-phase short-circuit single-phase rejection fault of Liandu-Ouhai line, experiences the failure of continuous commutation between Binkin and Lingshao DC, DC blocking, low frequency load shedding and finally reaches a new equilibrium point. From the search process of the accident chain, we can see that this method has the ability of traditional accident chain search. It can accurately simulate the deterministic action of protection control device and search the high-risk chain of cascading failures caused by the action of protection control.

3) As shown in Fig. 10, the chain of cascading faults starts from three-phase short-circuit single-phase rejection fault of Cangyan-Fengyi line, experiences three bipolar blockades of Lingshao DC continuous commutation failure, two bipolar blockades of Binkin DC continuous commutation failure, and the probability of bipolar blockade is 2/3. Load shedding at low frequency can finally reach a new equilibrium point. The analysis shows that, unlike the traditional simulation process which can not continue to search for follow-up events, the probability of the protection control device in critical state is calculated, so that the situation after the action is simulated in time domain, and the high-risk chain of accidents is searched. The search way of the chain of high-risk chain of cascading failures is increased, and the chain set of high-risk chain of cascading failures is completed.

5. Conclusions and Prospects
Based on the accident chain model and time domain simulation, combined with the actual operation conditions of the power grid and the configuration of the existing two or three lines of defense, and considering the interaction between the changes of the electrical quantity of the power grid and the protection control action, this paper proposes an evaluation index of the key component protection control action based on the dynamic interactive quantitative analysis of the system and the components. Through this index, the probability function of protection control action of key components is established, which can well simulate the uncertain action of protection control in critical state and increase the search path of chain of cascading failures. Based on risk index, chain of cascading failures with high risk can be searched, which can truly reflect the actual risk of large power grids, so as to guide the formulation of chain of cascading failures. Lock fault prevention and control strategy provides a certain research basis. By simulating a real power grid under peak load in summer of 2017, the effectiveness of the cascading failure chain search method proposed in this paper is verified.

The key component protection control action evaluation index proposed in this paper can be used

ISPECE   IOP Publishing

IOP Conf. Series: Journal of Physics: Conf. Series **1187** (2019) 022069   doi:10.1088/1742-6596/1187/2/022069

to simulate most of the protection control action, but for some complex protection control (such as fan low voltage crossing) needs to be further studied. In addition, it is too simple to use linear function to simulate the action probability of protection control, which needs further study according to the actual situation.

Reference

[1] FANG Yongjie. Application of emergency control toreduce risk of system collapse triggered by power transmission interface tripping: thinking on the India power blackouts[J]. Automation of Electric Power Systems,2013, 37(4): 1-6.

[2] LI Baojie, LI Jinbo, LI Hongjie, et al. Analysis of Turkish blackout on March 31, 2015 and lessons on China power grid[J]. Proceedings of the CSEE, 2016, 36(21):5788-5795.

[3] JIN Bo, XIAO Xianyong, CHEN Jing, et al. A method of risk assessment considering protection failures and dynamic equilibrium of power grid[J]. Power System Protection and Control, 2016, 44(8): 1-7.

[4] Liu Wenying，Yang Nan，Zhang Jianli，et al. Complex grid failure propagating chain model in consideration of adverse weather[J]. Proceedings of the CSEE，2012，32(7)：53-60(in Chinese).

[5] CAI Ye, CAO Yijia, TAN Yudong, et al. Influences of power grid structure on cascading failure based on standard structure entropy[J]. Transactions of China Electrotechnical Society, 2015, 30(3): 36-43.

[6] ZHANG Jingjing, WANG Zhengyu, DING Ming, et al.The security correction strategy in ac and dc hybrid power system[J]. Power System Protection and Control, 2016,44(11): 90-96.

[7] BI Ruyu, LIN Tao, CHEN Rusi, et al. Influences of FACTS element with energy storage on cascading failures[J]. Transactions of China Electrotechnical Society,2016, 31(9): 50-57.

[8] PETERSEN P F, JOHANNSSON H, NIELSEN A H.Investigation of suitability of cascading outage assessment methods for real-time assessment[C] // PowerTech, 2015 IEEE Eindhoven, 2015: 1-5.

[9] LIU Youbo, HU Bin, LIU Junyong, et al. Power system cascading failure analysis theories and application I——related theories and applications[J]. Power SystemProtection and Control, 2013, 41(9): 148-155.

[10] LIU Wenying, DAN Yangqing, ZHU Yanwei, et al.Research on physical indicators to identify power system self-organized critical state[J]. Transactions of China Electrotechnical Society, 2014, 29(8): 274-280, 288.

[11] CARRERAS B A, LYNCH V E, DOBSN I, et al. Critical points and transitions in an electric power transmission model for cascading failure blackouts[J]. Chaos, 2002,12(4): 985-994.

[12] GONG Yuan, MEI Shengwei, ZHANG Xuemin, et al.An improved OPA model considering planning and self-organized criticality analysis[J]. Power System Technology, 2014, 38(8): 2021-2028.

[13] CHEN Jie, THORP J S, DOBSON I. Cascading dynamics and mitigation assessment in power system disturbances via a hidden failure model[J]. International Journal of Electrical Power & Energy Systems, 2005,27(4): 318-326.

[14] NEDIC D P, DOBSON I, KIRSCHEN D S, et al. Criticality in a cascading failure blackout model[J]. International Journal of Electrical Power & Energy Systems, 2006,28(9): 627-633.

[15] DOBSON I, CARRERAS B, NEWMAN D E. A probabilistic loading-dependent model of cascading failure and possible implications for blackouts[C] //Hawaii International Conference on System Sciences,Hawaii, 2003.

[16] DOBSON I, CARRERAS B, NEWMAN D E. Branching process models for the exponentially increasing portions of cascading failure blackouts[C] // Hawaii International Conference on System Sciences, Hawaii, 2005.

[17] DING Ming, XIAO Yao, ZHANG Jingjing, et al. Risk assessment model of power grid cascading

failures based on fault chain and dynamic fault tree[J]. Proceedings of the CSEE, 2015, 35(4): 821-829.

[18] FAN Wenli, LIU Zhigang. An overview on modeling of cascading failures in power grids based on complex system[J]. Automation of Electric Power Systems, 2012,36(16): 124-131.

[19] LUO Yi, WANG Yingying, WAN Wei, et al. Fault chains model for cascading failure of grid[J]. Automation of Electric Power Systems, 2009, 33(24): 1-5.

[20] WANG Ansi, LUO Yi, TU Guangyu, et al. Vulnerability assessment scheme for power system transmission networks based on the fault chain theory[J]. IEEE Transactions on Power Systems, 2011, 26(1): 442-450.

[21] MA Zhiyuan, SHI Libao, YAO Liangzhong, et al. Study on the modeling and search strategy of event chain for cascading failure in power grid[J]. Proceedings of the CSEE, 2015, 35(13): 3292-3302.

[22] WANG Yingying, LUO Yi, YU Guangyu, et al. Risk assessment of cascading failures in power system based on fault chain and fuzzy comprehensive evaluation[J].Proceedings of the CSEE, 2010, 30(S): 25-30.

[23] WANG Tao, WANG Xingwu, GU Xueping, et al. Power system fault chain model and simulation based on probability and structural importance[J]. Electric Power Automation Equipment, 2013, 33(7): 51-56.

[24] CAI Zexiang, WANG Xinghua, REN Xiaona. A reviewof complex network theory and its application in power systems[J]. Power System Technology, 2012, 36(11):114-121.

[25] BO Zhiqian, LIN Xiangning, WANG Qingping, et al.Developments of power system protection and control[J].Protection and Control of Modern Power Systems, 2016,1(1): 1-8. DOI 10.1186/s41601-016-0012-2.

Warehouse Design Model for Shuttle Based Storage and Retrieve System

Bin Tian[1], Yingying Wu*

[1] Department of Logistics Engineering, Shandong University, Jinan, Shandong, 250061, China

*Yingying Wu's e-mail: sophia.wu@sdu.edu.cn

Abstract. Recently, SBS / RS system has been widely used and studied. This paper presents a mathematical model of travel time and cost for shuttle based storage and retrieve systems (SBS/RS), through which we can decide the optimized configuration and picking efficiency of the warehouse. By considering the operating characteristics, such as acceleration and deceleration and assuming that the storage units are uniform distributed and using probability theory, we obtain a time and cost function for a dual command storage and retrieve procedure. Finally, we validate our analysis by numerical examples and sensitivity analysis. The results show that the optimal configuration and efficiency of the warehouse are mainly affected by several key indicators, such as the width of each storage unit, the height of each shelf, etc.

1. Introduction

Automated warehouse technologies are evolving rapidly, from AS / RS systems to AVS / RS to the recently introduced SBS / RS, and many companies are transforming their traditional warehouses with automation equipment. Automation technology undoubtedly improves the storage and retrieve efficiency and shortens picking time, resulting in more benefits.

Based on the relationship of the shuttle and the storage tier, there are usually two main situations: tier-captive and tier-to-tier. In a tier-captive configuration each floor has a shuttle and the shuttle can only move in this tier. However, in a tier-tier configuration, a shuttle may move from one tier to another tier through the lifts. According to the picking process, it can be divided into single command and dual command. In a single command, it only needs to storage or retrieve a tote at once. In a dual command, it needs to storage first and then retrieve the totes. In a typical AVS / RS warehouse, after receiving the picking order, the lift first arrives at the floor where the goods are located, and then the shuttle moves to the position where the goods are stored, moves the goods to the buffer position, and then the conveyer conveys the goods to the lift. Finally, the lift will deliver the goods to the I / O point. At this point, a picking process is completed. Inventory is the opposite process. What we usually care about is the efficiency of storage and retrieve of warehouses and the configuration of warehouses, such as the characteristics of shelves, and the cost.

In this paper, we first propose a time function for a lift and a shuttle to complete a dual command storage and retrieve procedure based on the knowledge of probability and motion. The shuttle is tier-captive. Next, we set up a cash cost function for the warehouse. Finally, by combining the above two, we get the optimal decision of the warehouse configuration and verify it with the data. Different from the past, we do not analyse our model through simulation, but through numerical analysis and theoretical derivation, and finally via programming.

Content from this work may be used under the terms of the Creative Commons Attribution 3.0 licence. Any further distribution of this work must maintain attribution to the author(s) and the title of the work, journal citation and DOI.
Published under licence by IOP Publishing Ltd

The paper is organized as the following manner: Section 2 gives the literature review. In section 3, the mathematical travel time model and cost model of a dual command cycle for a SBS/RS system are presented, which consider a discrete model closer to the actual situation. In the Section 4, we validate our model through numerical analysis and sensitivity analysis and get some useful revelations. Finally, we concluded the paper in Section 5.

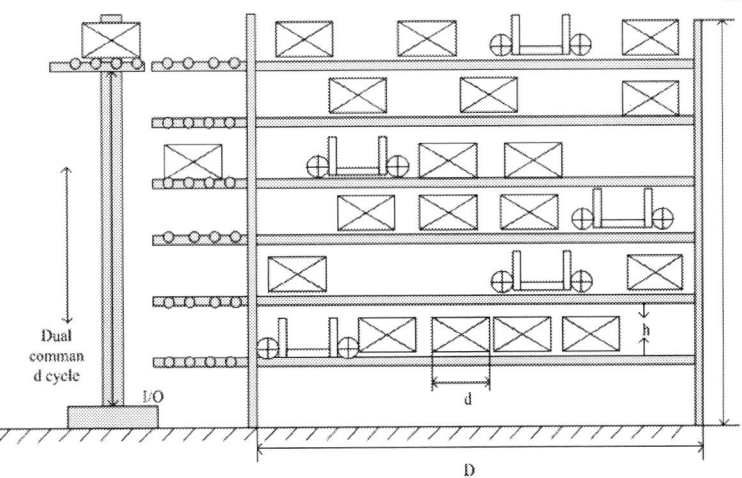

Fig.1 the SBS/RS storage rack

2. Literature review

Malmborg[1] first proposed Analytical conceptualizing tools based on the features of modeling expected performance for an AVS/RS system. Since then, AVS / RS system has been widely used and studied.

There are two main research areas of AVS/RS. Some of them mainly focus on the travel time model of the AVS/RS system. Malmborg[2] considered the travel time of a dual command cycle as well as the system utilization and throughput capacity. Fukunari M, Malmborg CJ[3]analyzed the cycle time based on the iterative computational scheme exploiting random storage assumptions. Some approximation strategies for transaction waiting times are proposed by Zhang et al[4], in which the approximations based on the variance of the transaction inter-arrival times can be adjusted dynamically. Kuo et al[5] provided a practical means of predicting key aspects of system performance based on five design variables. Different from others they think over about the cost. Recently, Tone, Leher et al[6] built the travel time model of both single and dual command cycles for SBS/RS system. They assumed the storage locations are uniformly distributed and proved their model through simulation. However, Tone Lerher[7] established a travel time model where the single-deep racks are replaced by the double-deep racks. So there are more places to storage goods.

Of course, there are many studies which pay more attention to the performance and design of the warehouse, Miki Fukunari, Malmborg[8]studied a queuing approach to estimate the performance of the AVS/RS systems by using opportunistic interleaving. Ekren, B Y[9] et al focused on the rack configuration of AVS/RS. They observe the change of picking time by changing the number of tiers, aisles, bays and setting up a regression model. Then, Ekren, B Y[10] et al compared the performance of the two systems ,AVS/RS and CBAS/RS, five performance measures are considered, such as average flow time, device utilization, waiting time in queue, average number of jobs waiting in queue and cost. Finally, they found the AVS/RS system is more efficiency than the traditional one. After the regression model, Ekren, B Y[11] et al applied the DOE in the analytical model and measure the performance of the system by simulation. Gino Marchet[12] et al established an analytical model to evaluate the performance of the warehouse, however, they just considered the cycle time and waiting time. Both the performance and design decision were analyzed by Debjit Roy[13] et al , the model also contained the allocation of resources to zones, and the vehicle assignment rules.

Different from the past, we considered both the travel time and cost, and established a mathematical model without simulation, programmatically solve the optimal configuration of the warehouse. And when the problem was complex enough, we used heuristics to find the possible best solution. Finally, the effect of each parameter on the optimal decision is obtained.

3. Analytical model for SBS/RS

In this section we mainly propose a travel time model of dual command based on the probability and physical movement knowledge. Later on, the total cost of the SBS/RS will be presented. Considering the above two cost model, both time and cash, we will get an optimal configuration of the SBS/RS warehouse.

3.1. Notations and assumptions
The notations used in this paper are presented below:

k: number of the columns $\qquad$ m: number of the tiers

n: number of the single-deep racks $\qquad$ t_i: time consuming for case i. i=1,2,3,4.

d: length of each storage unit $\qquad$ h: height of each tier

d_{max}: the maximum height of the racks $\qquad$ h_{max}: the maximum length of the racks

w_s: width of each storage unit $\qquad$ w_a: width of the aisle

a_y: accelerated/retarded velocity of the lift $\qquad$ v_y: the maximum velocity of the lift

a_x: accelerated/retarded velocity of the shuttle $\qquad$ v_x: the maximum velocity of the shuttle

P: unit time cost $\qquad$ c_s: cost of each storage unit

N_1: number of columns the shuttle required to accelerate to the maximum speed

N_2: number of tiers the lift required to accelerate to the maximum speed

c_a: cost of each shuttle $\qquad$ c_t: cost of each lift

c_m: cost of unit area

All the hypotheses in this article:

(1). The number of layers and columns is enough to allow the lift and shuttle to accelerate to maximum speed. That is: $m>N_2+2$, $k>N_1+2$.

(2). The distance the lift and shuttle required to accelerate to their maximum speed is greater than the height of the each tier or the length of each storage unit. That is: $v_y^2/2a_y > h, v_x^2/2a_x > d$

(3). The warehouse requires at least 10000 storage units. That is: $2mkn > 10000$

3.2. Dual command travel time model
In the dual command travel, after receiving the storage command the lift reaches the ith floor of the rack with a tote first from the I/O point which usually is in the ground and middle of the single-deep rack, the lift unloads the tote and the shuttle receives it, takes it to the xth column of the ith tier, which of course is an empty storage position. Then a retrieve command is sent to the lift, it moves to the next jth tier. At the same time the shuttle starts to pick the tote at the yth column of the jth tier then unload it on the lift. Finally, the lift carries the tote to the I/O point.

3.2.1. Lift travelling from I/O point to the ith tier
The time that the lift needs from the I/O point to the ith tier lies on the distance of them. In this model we assume the distance required for the lift to accelerate to the. highest speed exceeds the height of the single tier, Namely, $v_y^2/2a_y > h$ So the lift needs at least $[v_y^2/2ha_y]=N_2$ tiers. So we could get the relationship between the vertical velocity and the time (see figure 2). The time-tier function also can be gained as follow:

$$t_1 = \begin{cases} \sqrt{4ih/a_y} & i \le N_2 \\ ih/v_y + v_y/a_y & N_2 < i \le m \end{cases} \quad (1)$$

Then the expected time the lift needs from I/O point to the ith tier is as follow:

$$E(t_1) = [\sum_1^{N_2} \sqrt{4ih/a_y} + \sum_{N_2+1}^{m} (ih/v_y + v_y/a_y)]\frac{1}{m} \quad (2)$$

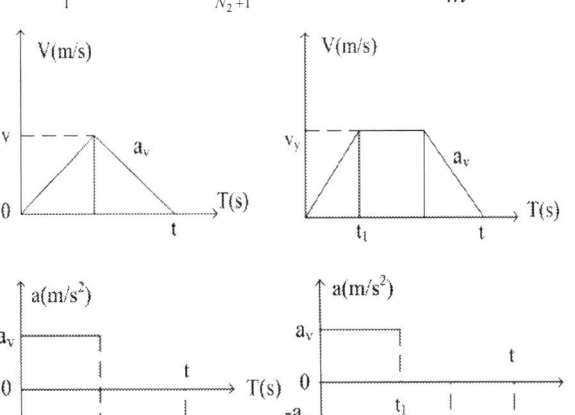

Fig.2 vertical velocity-time relationship

3.2.2. Lift travelling from the ith tier to the jth tier

The time the lift takes from the ith to jth tier is similar to the above, which can also be get:

$$tl_{ij} = \begin{cases} \sqrt{4|i-j|h/a_y} & |i-j| \le N_2 \\ |i-j|h/v_y + v_y/a_y & N_2 < |i-j| \le m \end{cases} \quad (3)$$

There are m^2 cases in total. The expected time of it can be calculated by the matrix method. The matrix of the time from the ith to jth tier is:

$$T_{2ij} = \begin{pmatrix} 0 & tl_{12} & . & . & . & tl_{1m} \\ tl_{21} & 0 & . & . & . & . \\ . & . & . & . & . & . \\ . & . & . & . & . & . \\ tl_{m-1,1} & . & . & . & . & tl_{m-1,m} \\ tl_{m1} & . & . & . & tl_{m,m-1} & 0 \end{pmatrix}$$

Since T_{2ij} is a symmetric matrix, so the mathematical expectation of it can be shown as follow:

$$E(t_2) = \sum_{i=1}^{m-N_2-1} [\sum_{j=i}^{i+N_2} \sqrt{4(j-i)h/a_y} + \sum_{j=i+N_2+1}^{m} ((j-i)h/v_y + v_y/a_y)]\frac{2}{m^2} + \sum_{i=m-N_2}^{m} (\sum_{j=i}^{m} \sqrt{4(j-i)h/a_y})\frac{2}{m^2}$$

3.2.3. Lift travelling from the jth tier to the I/O point

Apparently, this period is exactly the reverse of the first case, namely lift traveling from I/O point to the ith tier. So the expected time it takes can be deduced directly.

$$E(t_3) = [\sum_{1}^{N_2} \sqrt{4ih/a_y} + \sum_{N_2+1}^{m} (ih/v_y + v_y/a_y)]\frac{1}{m} \qquad (4)$$

3.2.4. Shuttle travelling from buffer position to the ith column
Similar to the lift, the time that the shuttle used from the buffer position to the ith column relies on the distance between them. We assume the distance needed for the shuttle to accelerate to the highest speed exceeds the height of the single column, that is : $v_x^2/2a_x > d$. Hence, the lift needs at least $[v_x^2/2da_x] = N_1$ columns. So we could get the horizontal velocity-time relationship The time-column function also can be gained as follow:

$$t_4 = \begin{cases} \sqrt{4id/a_x} & i \le N_1 \\ id/v_x + v_x/a_x & N_1 < i \le k \end{cases} \qquad (5)$$

Then the expected time the lift needs from buffer position to the ith column is as follow:

$$E(t_4) = [\sum_{1}^{N_1} \sqrt{4id/a_x} + \sum_{N_1+1}^{k} (id/v_x + v_x/a_x)]\frac{1}{k} \qquad (6)$$

3.2.5. Shuttle travelling from the ith column to the jth column
First, the time the shuttle takes from the ith to jth column is similar to the lift, which can be seen as follow:

$$ts_{ij} = \begin{cases} \sqrt{4|i-j|d/a_x} & |i-j| \le N_1 \\ |i-j|d/v_x + v_x/a_x & N_1 < |i-j| \le k \end{cases} \qquad (7)$$

There are k^2 cases in total. The time(from the ith column to the jth column) matrix is:

$$T_{5ij} = \begin{pmatrix} 0 & ts_{12} & . & . & . & ts_{1k} \\ ts_{21} & 0 & . & . & . & . \\ . & . & . & . & . & . \\ . & . & . & . & . & . \\ ts_{k-1,1} & . & . & . & . & ts_{k-1,k} \\ ts_{k1} & . & . & . & ts_{k,k-1} & 0 \end{pmatrix}$$

So the mathematical expectation of $ts_{i,j}$ is:

$$E(t_5) = \sum_{i=1}^{k-N_1-1} [\sum_{j=i}^{i+N_1} \sqrt{4(j-i)d/a_x} + \sum_{j=i+N_1+1}^{k} ((j-i)d/v_x + v_x/a_x)]\frac{2}{k^2} + \sum_{i=k-N_1}^{k} (\sum_{j=i}^{k} \sqrt{4(j-i)d/a_x})\frac{2}{k^2}$$

3.2.6. Shuttle travelling from the jth column to the buffer position
It also can be treated as a reverse of the travelling from the buffer position to the ith column. Hence, the expected time it takes is:

$$E(t_4) = [\sum_{1}^{N_1} \sqrt{4id/a_x} + \sum_{N_1+1}^{k} (id/v_x + v_x/a_x)]\frac{1}{k} \qquad (8)$$

At this point, we have completed all the discussion of the travel time in a dual command storage and retrieve process for a SBS/RS system. So the total travel time is:
$E(t) = E(t_1) + E(t_2) + E(t_3) + E(t_4) + E(t_5) + E(t_6)$,namely:

$$\sum_{i=1}^{m-N_2-1}[\sum_{j=i}^{i+N_2}\sqrt{4(j-i)h/a_y}+\sum_{j=i+N_2+1}^{m}((j-i)h/v_y+v_y/a_y)]\frac{2}{m^2}+\sum_{i=m-N_2}^{m}(\sum_{j=i}^{m}\sqrt{4(j-i)h/a_y})\frac{2}{m^2}+[\sum_{1}^{N_1}\sqrt{4id/a_x}+\sum_{N_1+1}^{k}(id/v_x+v_x/a_x)]\frac{2}{k}$$

$$\sum_{i=1}^{k-N_1-1}[\sum_{j=i}^{i+N_1}\sqrt{4(j-i)d/a_x}+\sum_{j=i+N_1+1}^{k}((j-i)d/v_x+v_x/a_x)]\frac{2}{k^2}+\sum_{i=k-N_1}^{k}(\sum_{j=i}^{k}\sqrt{4(j-i)d/a_x})\frac{2}{k^2} \quad (11)$$

3.3. Cash cost model

In this part, we mainly consider the cash cost of racks, shuttles, lifts, area of storage. It is obvious that there are $2*m*k*n$ storage position, $m*n$ shuttles, n lifts and the total area of storage is:

$(w_a+2w_s)*k*d*n$, So the corresponding total cash cost is:

$$TC_1=c_s*2mkn+c_a*mn+c_t*n+c_m(w_a+2w_s)*k*d*n \quad (9)$$

In fact, the cost of the warehouse also includes many hidden parts, which may be difficult to measure, such as operating costs, labor costs, other facilities costs, etc. However, this has little effect on our decision of warehouse configuration or the two are not relevant. Consider the above two situations, the total cost (both time and cash) of a SBS/RS warehouse is: $E(t)*p+TC_1$. We consider the time cost together with the cash cost, which is quite different from the past, and decide the optimal configuration of the warehouse by minimizing the total cost of it.

4. Numerical examples and sensitivity analysis

4.1. Numerical examples

(1). Assuming d=0.8m, h=0.5m, $d_{max}=40m$, $h_{max}=10m$, $a_y=2m/s^2$, $v_y=3m/s$, $a_x=3m/s^2$, $v_x=3m/s$, $w_a=0.5m$, $c_s=30\$$, $c_a=7000\$$, $c_m=50\$$, $c_t=30000\$$. Thus, the best configuration of the warehouse is : 18tiers, 47 columns, 6 single-deep racks and the total cost is 1243630\$.

(2). Assuming all parameters keep constant except d=1.2m. Thus, the best configuration of the warehouse is: 19tiers, 33 columns, 8 single-deep racks and the total cost is 1608190\$.

(2). Assuming all parameters keep constant except $d_{max}=60m$. Thus, the best configuration of the warehouse is: 17tiers, 74 columns, 4 single-deep racks and the total cost is 902681\$.

Based on the above examples, we can see the total cost drops sharply when the maximum length of the warehouse increases, which mainly because it can contain more storage unit in one tier. However, when the length of unit storage increases the total cost rises dramatically. The reason is that each layer can accommodate fewer storage units. So it is vital for decision maker.

4.2. Sensitivity analysis

In this paper, we first propose a travel time model and cash cost model of dual command storage and retrieve process for a SBS/RS system warehouse to analyses the best decision of the warehouse configuration. Next, by minimize $E(t)*p+TC_1$ we calculated and get the numerical examples through C++ program. (See Fig 3). We observe the optimal decision and the total cost by changing the value of each parameter and get some management implications.

From these figures we can observe that the number of tiers augments first then keeps later and the number of single-deep racks rises when the length of unit column increases, but the number of columns decreases. Since we assume that the total number of storage units is not less than 10000, as the length of unit column increases each tier can accommodate fewer storage units. Therefore, the number of columns reduces and the number of tiers increases. However, when the length of a storage unit exceeds a threshold, the increase of the number of tiers will reach a limit, so the storage requirement can be satisfied by adding a single-deep rack.

For the height and length of the warehouse, their changes lead to an increase in the number of tiers and columns. Similar to the total cost, while the number of single-deep racks will decrease. The changes in the accelerated velocity and the maximum speed of the lift and shuttle seem to have no effect on the optimal decision, while the total cost will be reduced. This is because changes in speed and acceleration only affect the total expected travel time. Since the size of the warehouse we are considering is not large enough, the cost of such a short time period to total cost is insignificant. However, we mainly often also care about the picking efficiency of the warehouse, or the time it takes to complete a dual command process.

The impact of shuttle or lift maximum speed and acceleration on storage and retrieve time is easy to understand. Time reduces when the speed increase, the faster it accelerates the little time it needs. But the fact is not always the case, when the speed is high, the car will move to farther place to access the totes, the total time actually is increased.

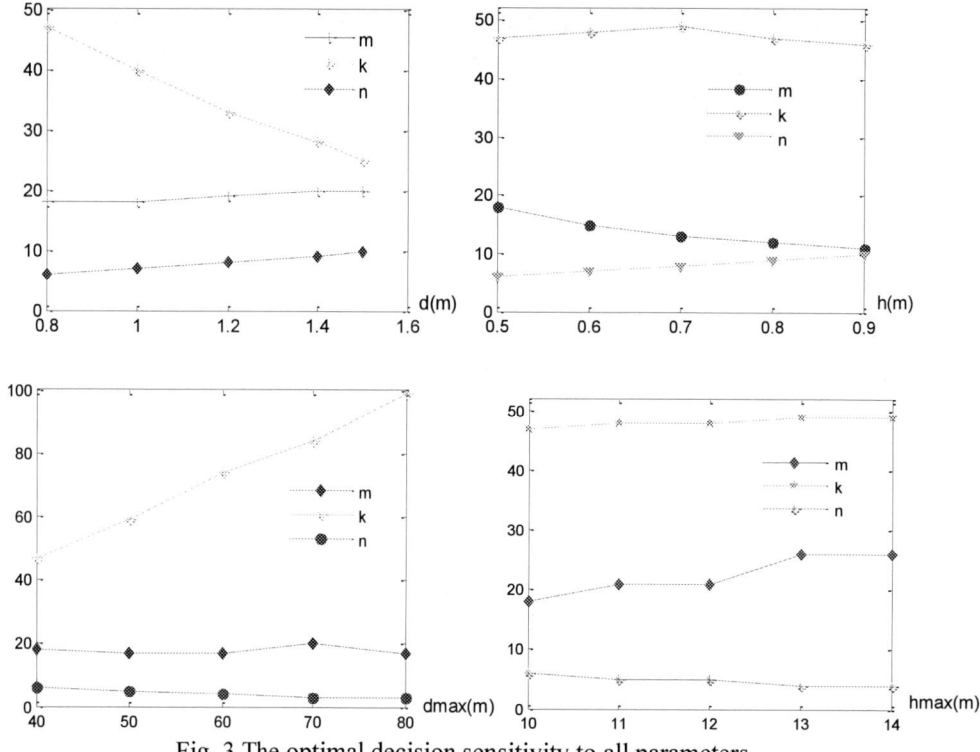

Fig. 3 The optimal decision sensitivity to all parameters

5. Conclusion

This study focused on the design model of a warehouse based on the SBS/RS system, in which the shuttle was tier-captive and executed a dual command at once. By considering the cost and travel time, we obtain the total cost function and get the optimal warehouse configuration by solving the function

First, we assume that each storage location is equally uniform distributed and based on the kinematics of the shuttle and the lift we propose a travel time model. Then, considering the expectations of travel time, use its expectation to represent the average time of storage and retrieve. Next, we also consider the cash cost of the warehouse, including storage costs, area costs, shuttle costs and lift costs. Finally, we programmatically calculate the optimal configuration of the warehouse for a given situation and the efficiency of storage and retrieve.

Of course, our model can be more complex in future. For example, we can change the single-deep rack to double-deep. The shuttles can be tier-to-tier. Each aisle has multiple lifts. The lift executes single and dual commands alternately, or considers the time the shuttle and the goods waits in queue

references

[1] Malmborg CJ. (2002). Conceptualizing Tools for Autonomous Vehicle Storage and Retrieve Systems. International Journal of Production Research, 40 (8):1807–1822.

[2] Malmborg, CJ (2003). Interleaving Dynamics in Autonomous Vehicle Storage and Retrieve Systems. International Journal of Production Research, 41 (5):1057–1069.

[3] Fukunari M, Malmborg CJ. (2008). An efficient cycle time model for autonomous vehicle storage and retrieve systems. Int J Prod Res, 46(12):3167–3184.

[4] Zhang. LA, Krishnamurthy. Malmborg, CJ and SS, Heragu. (2009). Variance-Based Approximations of Transaction Waiting times in Autonomous Vehicle Storage and Retrieve Systems. European Journal of Industrial Engineering, 3 (2):146–169

[5] Kuo, P. Krishnamurthy, A. and Malmborg, CJ. (2007). Design models for unit load storage/retrieve system using autonomous vehicle techniques and resource conserving storage and dwell point policies. Applied Mathematical Modeling, 31: 2332–2346.

[6] Tone Lerher, Banu Y, Ekren, Goran Dukic, Bojan Rosi. (2015). Travel time model for shuttle-based storage and retrieve systems. Int J Adv Manuf Technol, 78:1705-1725.

[7] Tone Leher. (2016). Travel time model for double-deep shuttle-based storage and retrieve systems. International Journal of Production Research, 54:2519-2540.

[8] Miki Fukunari, Malmborg. (2009). A network queuing approach for evaluation of performance measures in autonomous vehicle storage and retrieve systems. European Journal of Operation Research, 193:152-167.

[9] Ekren B Y and S S Heragu. (2010). Simulation-Based Regression Analysis for the Rack Configuration of an Autonomous Vehicle Storage and Retrieve System. International Journal of Production Research, 48 (21): 6257–6274.

[10] Ekren B Y, and S S Heragu. (2011). Performance Comparison of Two Material Handling Systems: AVS/RS and CBAS/RS. International Journal of Production Research, 50 (15): 4061–4074.

[11] Ekren B Y, S S Heragu, A Krishnamurthy, and C J Malmborg. (2010). Simulation Based Experimental Design to Identify Factors Affecting Performance of AVS/RS. Computers & Industrial Engineering, 58 (1): 175–185.

[12] Gino Marchet, Marco Melacini, Sara Perotti and Elena Tappia. (2012). Analytical model to estimate performances of autonomous vehicle storage and retrieve systems for product totes. International Journal of Production Research, 50 (24): 7134-7148.

[13] Debjit Roy, Ananth Krishnamurthy, Sunderesh S. Heragu, Charles J. Malmborg. (2012). Performance analysis and design trade-offs in warehouses with autonomous vehicle technology. IIE Transactions, 44 (12): 1045-1060.

ISPECE   IOP Publishing

IOP Conf. Series: Journal of Physics: Conf. Series **1187** (2019) 032002   doi:10.1088/1742-6596/1187/3/032002

Research on control strategy of a buck-type harmonic injection three-phase rectifier

Qiang Jia*, Panwei Ma, Yingchuan Qi, Dong Wang, Yang Cheng

Air Force Early Warning Academy, Wu Han, Hu Bei, 430019, China

*Corresponding author's e-mail: 512422798@qq.com

Abstract. This paper proposes a buck-type harmonic injection three-phase rectifier to mitigate the high voltage stress of the rear switch of the conventional boost-type power factor correction (PFC) circuit. It develops a control strategy composed of improved double loop control pattern and voltage feedforward pattern. Firstly, the working principle of the buck-type harmonic injection three-phase rectifier is described. Then, it demonstrates the advantage of improved control strategy. Afterwards, the controller's corresponding involved parameters are optimized. Finally the theoretical analysis and simulation results show that the circuit can output the stable low ripple voltage with increased power factor and decrease the harmonic content of input side current.

1. Introduction

At present, with the development of power electronics, various converter equipment has caused serious harmonic pollution to the power grid, and the harmonic pollution will cause serious damage to the electrical equipment [1-2]. Therefore, the PFC technology is generated, which make the input side current tend to be sinusoidal and improve the power factor of the circuit. Common PFC circuits are mainly divided into boost-type and buck-type [3]. Boost-type PFC Circuit inductor on the input side, with boost characteristics, continuous inductor current, simple control, but its high output voltage, the post-stage switch tube stress requirements, power consumption is not conducive to the design of the rear stage circuit, these limit its development [4-6]. Buck-type PFC is able to achieve low voltage output, low stress requirements for post-stage pipes, and inrush current protection, which are becoming more and more interesting with increasing power requirements [7-8].

The article [9] describes in detail the feasibility and effectiveness of the three harmonic midpoint injection principle, which provides the idea for the following research, but the paper does not combine with the specific circuit to further study. Literature [10] in the traditional three-phase six-switch PWM rectifier circuit be based on the addition of two-way switch tube, three harmonic injection to achieve the PFC, but the circuit controllable devices too much, difficult to control. The principle and control strategy of Buck-type harmonic injection rectifier are studied in literature [11], but its control strategy is larger and the response speed is slower when the load changes.

Based on the detailed analysis of the above literatures, this paper mainly analyzes the working principle and current injection circuit of three-phase buck harmonic injection rectifier, and puts forward an improved double closed-loop control mode, which is verified by simulation, the control effect is better, the unit power factor can be realized, the input current harmonic content is improved, achieves low ripple voltage output.

Content from this work may be used under the terms of the Creative Commons Attribution 3.0 licence. Any further distribution of this work must maintain attribution to the author(s) and the title of the work, journal citation and DOI.

Published under licence by IOP Publishing Ltd

2. Main circuit topological structure

The topology of the three-phase Buck-type harmonic injection rectifier is shown in Figure 1. It mainly includes input LC filter part, harmonic current injection part and buck rectifier part. The main function of the input filter is to filter out the high frequency harmonic components in the input current, and to achieve the purpose of harmonic current injection through reasonable control of the conduction timing of the three sets of bidirectional switching tubes, and to realize the sinusoidal input current and improve the power factor of the circuit. The buck rectifier part can achieve buck and stabilize the low ripple voltage by controlling the high frequency switching tube output.

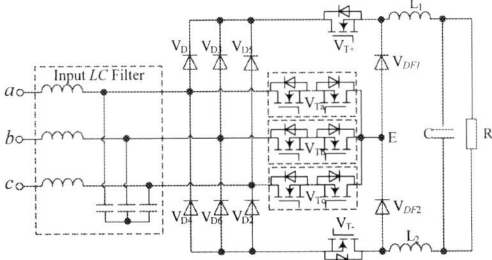

Fig1 Topological structure of three-phase buck-type harmonic injection rectifier

3. principle Analysis

For the rectifier circuit, the injection current acquisition is mainly two channels, namely the external supply and the internal supply of the circuit itself. The former is more intuitive but complex and cost-efficient, and the second method is chosen here [12]. It is found that for three-phase Buck-type harmonic injection rectifier, the current at point E is an ideal three harmonic source, which can be used as the pulsation source of three harmonic circulations injected into the input terminal.

To facilitate analysis, the following assumptions are made:

(1). Both the active switching element and the diode are considered ideal switches, ignoring their conduction pressure drop;

(2). Ignore the low-frequency voltage drop on the input filter inductor, that $v_{Ca,b,c} = v_{a,b,c}$;

(3). The output inductor current is constant.

The current injection path of the rectifier is analyzed using interval $v_a > v_b > v_c$ as an example. Figure 2 is a simplified equivalent circuit diagram.

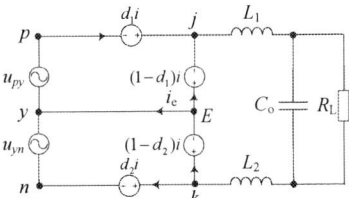

Fig.2 Three-phase buck-type harmonic injection rectifier Equivalent circuit at $v_a > v_b > v_c$

For three-phase Buck-type harmonic injection rectifier, its output voltage v_o is the function of two-phase line voltage, the output voltage of the rectifier bridge is not controlled by the minimum value of the effect, therefore, according to the following formula can be seen, three-phase buck harmonic injection rectifier output voltage can achieve a full range of voltage adjustable.

$$0 < v_o < \sqrt{\frac{3}{2}} v_{l-l,rms} \tag{1}$$

In the formula 1, $v_{l-l,rms}$ is the effective value of the line voltage, therefore, the Formula 1 can be further simplified as the phase voltage peak form, that:

$$0 < v_o < \sqrt{\frac{3}{2}} v_{l-l,rms} = \frac{3}{\sqrt{2}} v_{phase} = \frac{3}{2} \hat{v}_{phase} \qquad (2)$$

In the formula 2, v_{phase} is the phase voltage of the three-phase supply, and $\hat{v}_{phase}$ is the phase voltage peak.

Available from the formula 2:

$$0 < \frac{2}{3} \frac{v_o}{\hat{v}_{phase}} < 1 \qquad (3)$$

As a result, the $2v_o / 3\hat{v}_{phase}$ in the formula 3 can be defined as the modulation ratio M, that:

$$M = \frac{2}{3} \frac{v_o}{\hat{v}_{phase}}, \quad 0 < M < 1 \qquad (4)$$

Assuming that the admittance of the three-phase circuit is G, the three-phase input current is:

$$\begin{cases} i_a = G v_{aN} = I_m \sin \varpi_0 t \\[2mm] i_b = G v_{bN} = I_m \sin(\varpi_0 t - \frac{2\pi}{3}) \\[2mm] i_c = G v_{cN} = I_m \sin(\varpi_0 t + \frac{2\pi}{3}) \end{cases} \qquad (5)$$

For a three-phase buck rectifier, the output current ripple is ignored, and the output filter inductor current is:

$$I_{DC} = \frac{3}{2} G \frac{V_m^2}{v_o} \qquad (6)$$

Where V_m is the amplitude of the phase voltage and v_o is the output voltage.

The duty ratios of the switching Tubes V_{T+} and V_{T-} are d_1 and d_2, respectively:

$$d_1 I_{DC} = i_a, \quad d_2 I_{DC} = -i_c \qquad (7)$$

Integrated the formula 5,6,7 can get:

$$\begin{cases} d_1 I_{DC} = \frac{2}{3} \frac{v_o}{V_m^2} v_{aN} \\[3mm] d_2 I_{DC} = -\frac{2}{3} \frac{v_o}{V_m^2} v_{cN} \end{cases} \qquad (8)$$

For symmetric three-phase Y-junction circuits, $i_a + i_b + i_c = 0$, so the injected current i_e is:

$$i_e = (1 - d_2) I_{DC} - (1 - d_1) I_{DC} = i_a + i_c = -i_b \qquad (9)$$

From the above analysis can be obtained, the circuit can be achieved by the e-point generated three harmonic current injected into the input current, so that the input current continuous, as long as the use of reasonable control strategy and modulation method can be very good to achieve the input current sinusoidal.

4. System control Strategy improvement and design

4.1 Traditional system control strategy

The traditional three-phase Buck-type harmonic injection rectifier control system structure can be divided into two parts, based on the double closed-loop pi adjustment as a negative feedback error signal, the rectifier bridge arm ends of the voltage as a feedforward signal, then the two do poor, and then the triangular wave modulated high-frequency switch control signal, the structure of the diagram shown in Figure 3(a).

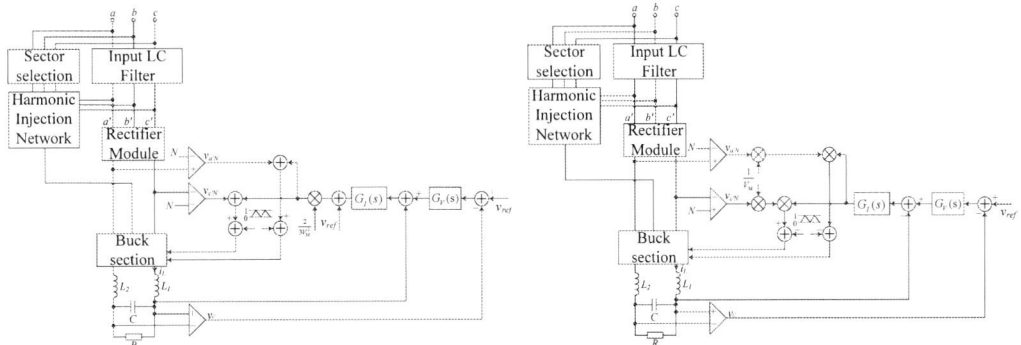

(a) The traditional Control structure (b) The improved control structure
Fig.3 The traditional Control structure The improved control structure of
three-phase buck-type harmonic injection rectifier

Through analysis and simulation, it is found that the control mode can modulate the control signal of the switch tube, and realize the function of three-phase Buck-type harmonic injection rectifier, but the control mode mainly has the following disadvantages:

1. This control mode is the Rectifier bridge arm at both ends of the voltage of the unit ratio $2/3V_M^2$ to the double closed-loop PI control negative feedback signal, this way will greatly reduce the negative feedback error signal amplitude, not very good to play the advantages of double closed-loop PI control;

2. Under this control mode, the output voltage will have a large ripple, and the output voltage is not stable;

3. Through the simulation found that the negative feedback signal and the feedforward signal to do poor, to control the high-frequency switching tube, the modulation signal produced in this way and the bidirectional switch control performance is poor, so that the AC side input current harmonic content is higher.

4.2 Improved system control strategy

Based on the shortcomings of the traditional control strategy, this paper improves the double closed-loop control strategy of three-phase Buck-type harmonic injection rectifier. The structure of the control system presented in this paper is shown in Figure 3(b).

The proposed control methods mainly include the following improvements:

1. The rectifier bridge arm end voltage to the unit ratio $1/V_M$ to the double closed-loop PI control negative feedback signal multiplication, this way is the rectifier bridge arm end voltage unit, which will better use the double closed-loop PI control negative feedback signal, play the advantages of pi control, so that the dual closed-loop PI control participate in the adjustment of the dynamic, The performance of static characteristics is superior;

2. The signal generated by the Feedforward control signal and the double closed loop control is changed from the traditional control to the multiplication, the improvement after simulation shows that the ripple of the DC output inductor current will be greatly reduced, thus the output voltage is more stable and its ripple is almost zero;

3. Improve the modulation of high-frequency switching tube, and according to the different duty ratio led to different circuit working state, respectively, the interleaving modulation and synchronous modulation method, the simulation found that interleaving modulation can better reduce the output voltage ripple, synchronous modulation can better reduce the input current harmonic content.

4.3 Control system Design
1. Feedforward Network Design

After sampling the voltages at both ends of the rectifier bridge arm in this design, the voltage at both ends of the bridge arm of the rectifier bridge must be normalized according to the sawtooth Crest

Peak V_T which is involved in modulation, where it is divided by the amplitude V_M of the three-phase phase voltage.

2. Double closed loop design

Combined with Figure 3(b), and the circuit of the two-way switching tube part of the simplification, you can obtain a double-loop control system schematic diagram, as shown in Figure 4.

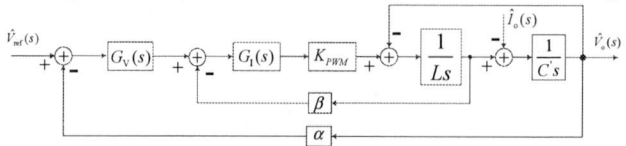

Fig.4 Schematic diagram of a double closed-loop control system

In Figure 4, $C' = C_o / 2$, $K_{PWM} = v_{in\_max} / V_T$, v_{in_max} is the maximum input voltage, V_T is the peak of the sawtooth crest, and $G_V(s)$ and $G_I(s)$ are the correction links of the voltage outer ring and the current inner loop respectively, and α and β respectively are the voltage loop and current loop feedback coefficients.

$$G(s) = \frac{\hat{v}_o(s)}{\hat{v}_{ref}(s)}$$

$$= \frac{G_V G_I K_{PWM}}{LC's^2 + G_I K_{PWM}\beta C's + G_V G_I K_{PWM} + 1} \tag{10}$$

According to the open-loop transfer function, the characteristic root equation of closed loop transfer function can be obtained:

$$D_1(s) = A_4 s^4 + A_3 s^3 + A_2 s^2 + A_1 s + A_0 \tag{11}$$

Which:

$$\begin{cases} A_4 = LC' \\ A_3 = K_{PWM} K_{iP}\beta C's^3 \\ A_2 = K_{PWM} K_{iI}\beta C' + K_{PWM} K_{vP} K_{iI}(1+\alpha) \\ A_1 = K_{PWM} K_{vI} K_{iP}(1+\alpha) + K_{PWM} K_{iP} K_{iI}\alpha + K_{PWM} K_{vP} K_{iI} \\ A_0 = K_{PWM} K_{vI} K_{iI}(1+\alpha) \end{cases} \tag{12}$$

The stability of closed-loop system is mainly determined by the distribution of its closed-loop poles on the S-plane [13], and the dominant poles of the system are:

$$s_{1,2} = -\xi\varpi_n \pm j\varpi_n\sqrt{1-\xi^2} \tag{13}$$

For a 4-order system, the further two poles are farther away from the dominant poles, the more stable the system, and the two non-dominant poles of the system are:

$$s_{3,4} = -n_{1,2}\xi\varpi_n \quad n_{1,2} \in 5{\sim}10 \tag{14}$$

According to the desired ideal pole and non-dominant pole, the ideal characteristic root equation of the system can be obtained:

$$D_1(s) = \left(s^2 + 2\xi\varpi_n s + \varpi_n^2\right)\left(s + n_1\xi\varpi_n\right)\left(s + n_2\xi\varpi_n\right)$$

$$= s^4 + B_3 s^3 + B_2 s^2 + B_1 s + B_0 \tag{15}$$

The overshoot σ and rise time t_p of the system can be expressed separately as:

$$\begin{cases} \sigma = e^{-\pi\xi/\sqrt{1-\xi^2}} \\ t_p = \dfrac{\pi}{\varpi_n\sqrt{1-\xi^2}} \end{cases} \tag{16}$$

According to the actual experience of the project, set the system's overshoot $\sigma = 0.05$, rise time $t_p = 0.001$, the damping ratio of the system according to the upper formula is $\xi = 0.7$, the natural oscillation frequency is $\varpi_n = 4350$. Take two non-dominant pole coefficients $n_1 = 5$, $n_2 = 8$, take the DC output filter inductance $L = 305\mu H$, filter capacitance $C = 470\mu F$, the above parameters into the formula 16, and according to the ideal characteristics of the root equation and the actual characteristics of the relationship between the root equation, you can find the voltage outer ring and the current internal loop PI controller parameters are: the current inner loop proportional control, $K_{iP} = 8.7 \times 10^{-3}$, the voltage outer ring is proportional integral control $K_{vP} = 78.33$, $K_{vI} = 9.33 \times 10^4$.

5. Simulation Results Analysis
Based on the principle analysis of three-phase Buck-type harmonic injection rectifier and the parameters shown in the table below, the three-phase Buck-type harmonic injection rectifier is simulated by Simulink under power frequency sinusoidal three-phase voltage input.

The parameters of the circuit are selected as: Switching frequency f_P=36 kHz, Output voltage v_o =400V, filter inductance $L_a=L_b=L_c$=200 μH, filter capacitance $C_a=C_b=C_c$=4 μF, DC inductance L=305 μH, output capacitance C_o=470 μF.

The output voltage waveform and the input voltage and input current waveform of a phase of the system are shown in Figure 5.

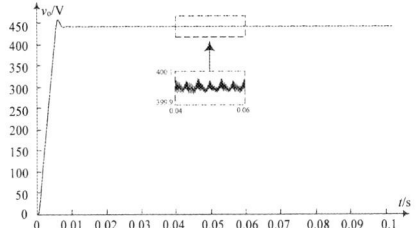 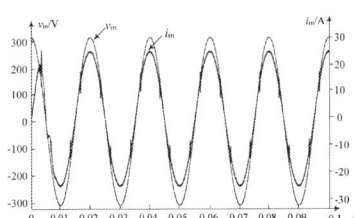

(a) The output voltage waveform   (b) The input voltage and input current waveform of a phase

Fig.5 The output voltage waveform and the input voltage and input current waveform of a phase

As can be seen from Figure 5(a), when the output power is 8kW, the output voltage can be stabilized in 400V, the output voltage ripple is ± 0.06, for the output voltage of 400V, the ripple can be completely ignored, to achieve a low ripple voltage output, and its overshoot is 3.75%, the adjustment time is 0.01s, with good dynamic response. As can be seen from Figure 5(b), harmonic injection realizes the improvement of power factor, reduces the input current harmonic content, the phase-a input current is realized by FFT, its harmonic content is 3.5%, satisfies the industry standard. However, because the two-way switch operation in the low-frequency state, in the two-phase switching of the moment, the injection current will be interrupted, so that in each injection instantaneous current fluctuations.

In order to verify the correctness of the control mode and the rationality of the controller design, the circuit load is changed at 0.05s, the power of the circuit is changed from 8kW to 5.3kW, and the output voltage is shown in Figure 6(a). As can be seen from the diagram, when the load changes, the voltage overshoot is 0.002s, can quickly stabilize in 400V, and the adjustment time is only E, indicating that the circuit has a good fast recovery characteristics and strong stability. As can be seen from Figure 6(b), the input current can also be very small fluctuations in the fast recovery of the sine, the unit power factor, the phase a input current FFT analysis of its harmonics accounted for the fundamental percentage is 3.45%.

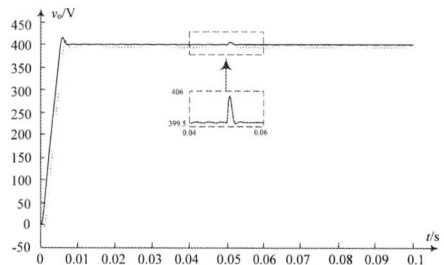 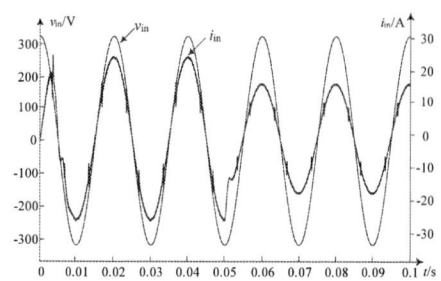

(a) The output voltage waveform   (b) The input voltage and input current waveform of a phase

Fig.6 The output voltage and the input voltage and current waveform of a phase at variable load

6. Conclusion

Aiming at the problem of large voltage stress of the rear switch in the traditional three-phase boost-type PFC Circuit, a three-phase Buck-type harmonic injection rectifier is studied, and an improved control mode of double closed-loop control and voltage Feedforward is proposed, which realizes the voltage regulation control of the rectifier and improves the overall performance of the rectifier. The simulation results show that the power factor of the circuit is higher, the harmonic content of the AC side is low, the stable low ripple voltage can be output, and the dynamic response characteristic is better when the load is changed.

References

[1] Pliesman, Billis, (2010) Murray. Switching power supply design. Electronic industry Press, Beijing.

[2] Wu J, Sun S, Song W, etc. (1988). Power system harmonics. Water Conservancy and Power Press, Beijing.

[3] Johann W. K，Thomas F. (2013) The Essence of Three-Phase PFC Rectifier Systems-Part I. IEEE Transactions on Power Electeonics.28(1).176-198.

[4] Liu C, Li H, Sun Z, etc. (2016) Research on multilevel high voltage cascaded boost DC converter. Electric Drive. 46 (4). 30-36

[5] Li W. (2017) Design and research of boost power factor correction circuit. Zhejiang University. 2-8.

[6] Michael L, Johann W. K, Josef D. (2017) Sinusoidal Input Current Discontinuous Conduction Mode Control of the VIENNA Rectifier. IEEE Transactions on Power Electronics. 32(11). 8800- 8812.

[7] TB Soeiro，GJM De Sousa，MS Ortmann. (2014) Three-phase unidirectional buck-type third harmonic injection rectifier concepts .In: IEEE Applied Power Electronics Conference & Exposition-apec. Fort Worth, Texas.928-934.

[8] Wu H. (2016) Research and design of buck active power factor correction LED driver controller. Zhejiang University. 7-11.

[9] Jun I. (2008) A Novel Three-Phase PFC Rectifier Using harmonic Current Injection Method. IEEE Transactions on Power Electronics .23(2).715-722.

[10] Mancu. F V，T. Soeiro，J. M. (2012) Comparative Evaluation of Bidirectional Buck-Type PFC Converter Systems for Interfacing Residential DC Distribution Systems to the Smart Grid. In: Conference of the IEEE Industrial Electronics Society. Montreal, Quebec.5153-5160.

[11] Chen R，Yao Y. (2015) Inhibiting Mains Current Distortion for SWISS Rectifier –a Three-phase Buck-type Harmonic Current Injection PFC Converter. In: Applied Power Electronics Conference & Exposition. Charlotte, North Carolina.1850- 1854.

[12] Li X. (2007) Research on harmonic suppression technology based on three-harmonic injection. Zhejiang University. 16-18.

[13] Hu S. (2013) Automatic Control Principle (6th edition). Science Press, Beijing.

Research and analysis of MRC and IRC algorithm based on L TE system

Jian Jin，Shan Jinjie，Zhou Jian

(Anhui Institute of Construction Technicians, Hefei, 230000)

Corresponding author: **Shan Jinjie** (13615693385, Email: shan.jingjie@163.com). He received his M.S. degree, and currently serves as a lecturer, conducting intelligent control, image processing and embedded systems researches.

Abstract: This paper mainly studies the theoretical analysis of MRC and IRC algorithm. The users at the edge of the community will suffer much more frequency interference which comes from neighborhood users, especially in the dense urban area. To improve uplink throughput, MRC adaptive algorithm and IRC adaptive algorithm can improve the performance effectively. Through the simulation results of the two algorithms, we can come to the conclusion that the IRC algorithm is better than the MRC algorithm in the case of greater interference. If the interference is larger, choose IRC , otherwise, choose MRC. The two algorithms have different performances under different jamming and different SNR.

1.Introduction

The downlink of LTE system adopts OFDMA technique and the sub-carriers of multiple users that the downlink distributes demonstrate completely perpendicular planes while its uplink employs DFT-S-FDMA and synchronization technology to ensure that there exists the orthogonality in frequency resource distributed among the users. In other words, there is little interference in frequency resource of the inside neighborhood and the interference of LTE system mainly comes from the same frequency resource in the neighboring cells with the same frequency. Additionally, the same frequency interference of neighboring cells refers that the expected cells are disturbed easily by the edge user or the base station, including the interference of uplink and downlink signals (containing four types). When the expected cells and neighboring cells are in the downlink slots, there exists the interference among the base stations; when the expected cells are in the downlink slots and neighboring cells are in the uplink slots, the interference mainly comes from the crossed time slots among the users; when the expected cells are in the uplink slots and neighboring cells are in the downlink slots, the interference is caused by the base stations; when the expected cells and neighboring cells are in the uplink slots, users are easy to produce interference. Therefore, it is necessary to adopt relevant Interference Rejection Combining [1] to guarantee the performance of TD-LTE same frequency network and improve its frequency and efficiency.

2.Algorithm analysis

Employing multi-antenna configuration, the receiving end of base station provides relevant conditions to the merging for basebank via using signal. Due to the different channel fading that the signal comes to different receiving antennas, multi-receive antennas are used to render received diversity, thus increasing the strength of signals, whether it is useful signal or interference signal. Configurating

Content from this work may be used under the terms of the Creative Commons Attribution 3.0 licence. Any further distribution of this work must maintain attribution to the author(s) and the title of the work, journal citation and DOI.
Published under licence by IOP Publishing Ltd

reasonable weighted coefficient for receiving antennas could enhance the ability of receiver to prevent fading and interference so that the wireless network, capacity as well as data rate could be improved. The multi-antenna configuration of receiving end could be adopted to construct received diversity or process the interference rejection [2].

Maximum Ratio Combining (MRC), the most common received diversity technology, could adequately capture the gains of each received branch, whose performance is optimal [4]. However, when there exists strong interference among users, MRC regards the interference as noise during the process of merging, which will lead to the rapid deterioration of detection performance since the interference has played a leading role in receiving signal. Therefore, it is necessary to consider restraining the interference among the users during the checking of uplink [4]. IRC algorithm could estimate and constrain the interference, thus enhancing the checking performance and improving the throughput of uplink. What Interference Rejection technology needs to consider is the time and spatial attribute of interfered signal [4]. The IRC technology determines the weighted coefficient in accordance with the channel, spatial noise as well as the covariance matrix of interference, which not only considers the power value of interference, but also takes the dependency in time and space of interference into account [4] [8]. The interference model shown in Figure 1 is determined to effectively evaluate the performances of IRC scheme in the aspect of datalink layer.

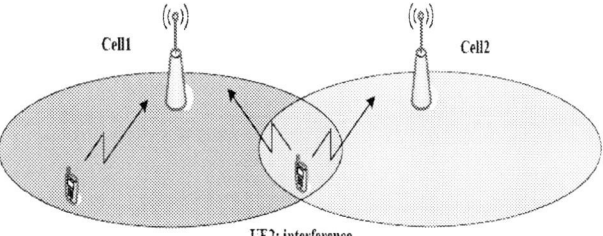

Fig.1 Interference model

Located in the Cell 1, UE1 is the expected users while the UE2 is the interfered users located within the overlapping area between Cell 1 and Cell 2 which is synchronous with Cell 2. The same time-frequency resources of UE1 and UE2 are allocated via schedule and the dependent fading that UE1 and UE2 experience finish the overlapping in the side of base station. The number of the interfered users and power in co-channel interference are approximately simulated by adjusting SIR.

Assume that the semidiameter of cell is smaller, the signal of multi-cell UE that the base station receives is almost the same, the length of CP is much larger than the delay spread of wireless channel and the fading that each subcarrier experiences could be considered as flat fading, the equivalent mathematics model in frequency domain could be expressed as followed:

$$r = H \cdot s + H_0 \cdot s_0 + H_1 \cdot s_1 ... + H_{k-1} \cdot s_{k-1} + N \qquad (1)$$

where H represents the channel response from the expected received users to base station; H_i represents the channel response from the interfered users i of neighboring cells to base station of expected cells; i =0...K-1, K represents the total number of users in the interfered cells (the time-frequency resources of interfered users and expected users are the same) ; s represents sending signal of expected users; s_i represents the sending signal of interfered users.

Assume $I = \sum_{i=0}^{k-1} H_i \cdot s_i$, the equation (1-1)could be expressed as follows:

$$r = H \cdot s + \sum_{i=0}^{k-1} H_i \cdot s_i + N = H \cdot s + I + N \qquad (2)$$

Based on the knowledge of signal processing, it is known that the smallest checking matrix in error function $e^H e = (r - H \cdot s)^H (r - H \cdot s)$ is $\hat{s} = (H^H \cdot H)^{-1} H^H r$ (it also is MRC checking matrix $w = (H^H \cdot H)^{-1} H^H$) when error budget $(r - H_i s_i)$ of each component has the same but irrelevant variances. When there is no interference I, each component has the same but irrelevant variances so that the MRC checking is optimal. However, when there exists the interference, IRC checking matrix, the new checking matrix, is constructed since the MRC checking matrix could not meet the requirements of the assumption due to the dependency of I+N, especially the relatively strong dependency among each antennas because of the smaller space between the receiving antennas.

Making $\sigma^2 R_e$ become the covariance matrix of $e=I+N$ and $R_e = PP^H$, $e=(r-H \cdot s)$, $\varepsilon = P^{-1} e$, $x = P^{-1} r$, the equation (2-2) is expressed as follows:

$$x = P^{-1} r = P^{-1}(H \cdot s + e) = (P^{-1} \cdot H) \cdot s + \varepsilon \qquad (3)$$

Covariance matrix of ε variate is

$$var(\varepsilon \times \varepsilon^H) = var((P^{-1} e) \cdot (P^{-1} e)^H) = var(P^{-1} e \cdot e^H P^{-H})$$
$$= P^{-1} \cdot var(e \cdot e^H) \cdot P^{-H} = \sigma^2 P^{-1} PP^H P^{-H}$$
$$= \sigma^2 I \qquad (4)$$

Due to the same but irrelevant variance of ε each component, the expression of equivalent matrix channel $\tilde{H} = P^{-1} \cdot H$ is as follows:

$$x = \tilde{H} \cdot s + \varepsilon \qquad (5)$$

Because of the same but irrelevant variance of ε each component, the smallest checking matrix expression of error function $(x - \tilde{H} \cdot s)^H (x - \tilde{H} \cdot s)$ is expressed as follows:

$$\hat{s} = ((P^{-1} \cdot H)^H \cdot (P^{-1} \cdot H))^{-1} \cdot (P^{-1} \cdot H)^H \cdot (P^{-1} \cdot r)$$
$$= (H^H \cdot R_e^{-1} \cdot H)^{-1} \cdot H^H \cdot R_e^{-1} \cdot r \qquad (6)$$

And the IRC checking matrix is $w = (H^H \cdot R_e^{-1} \cdot H)^{-1} \cdot H^H \cdot R_e^{-1}$ where R_e represents interference covariance matrix and/or noise covariance matrix.

From the above analysis, it is found that MRC $w = (H^H \cdot H)^{-1} \cdot H^H$ is the best checking matrix when there is no interference. If the MRC could not meet the requirements of testing under the interference condition, IRC $w = (H^H \cdot R_e^{-1} \cdot H)^{-1} \cdot H^H \cdot R_e^{-1}$ is the best checking matrix. The MRC and IRC algorithm could be be compared via two rules of the algorithms.

Successive interference cancellation zero forcing(ZF) rule:

$$\hat{s} = argmin(r - H \cdot s)^H (r - H \cdot s) = argmin \| r - H \cdot s \|_2^2 \qquad (7)$$

Minimum mean square error (MMSE) rule[2]:

$$\hat{s} = argminE(\| r - H \cdot \hat{s} \|_2^2) = \arg \min_w E[(s-wr)(s-wr)^H] \qquad (8)$$

Compared with MMSE, ZF algorithm is simpler and easier to perform but it requires relatively high SNR since the ZF, offering receipt signal the inverse to multiply channel matrix, could eliminate the interference caused by other users. Generally speaking, when the coefficient of channel matrix is smaller than 1, its inverse is greater than 1, namely, it will amplify the noise when multiplying the divisor that is greater than 1.

3. Simulation analysis

There are two kinds of situations when interference exists: one is low SNR, and the other one is high SNR. It is researched which method is better in low SNR situation to inhibit signals from interference; which method comes out with the best interference effects, when the divergence of the receiving angle of arrival becomes greater, that is, the directions of users enlarges; IRC and MRC algorithms will be influenced if the number of users increases and the directions of interference enlarges. Those two algorithms are compared in greater interference situations.

Table 1 Simulation parameters.

parameters	values
system bandwidth	20MHz
channel environment	SCME-B 5Hz
antenna polarization	4+4 dual polarization
antenna separation	0.5λ
simulation environment	PUSCH
occupied RB	6
MCS MCS	MCS5,MCS26
interference	1.none 2. 1 interference
Number of RB	15

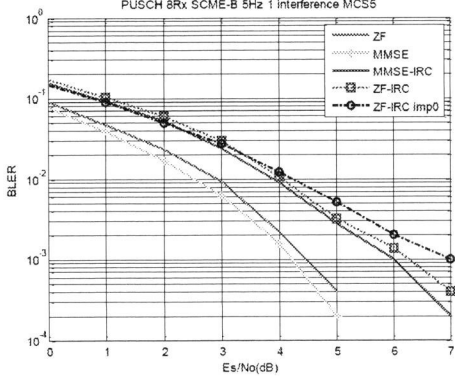

Fig.2 Performance of each receiving algorithm in no SNR

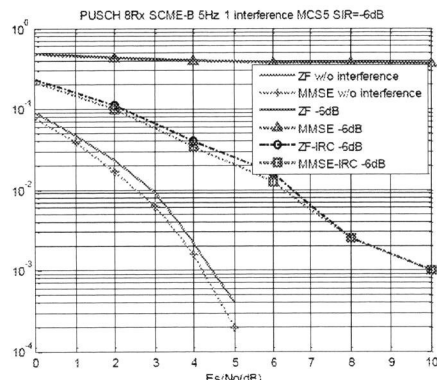

Fig.3 Performance of each receiving algorithm in SIR=-6dB

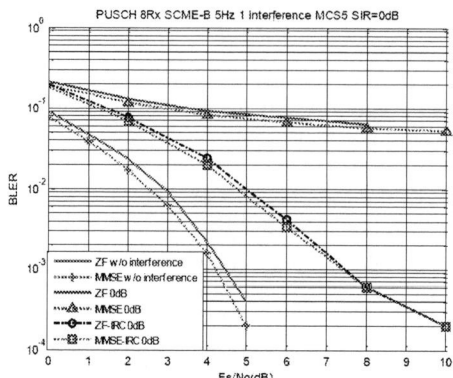

Fig.4 Performance of each receiving algorithm in SIR=-6dB ratio

From the above simulation results, it can be seen that:

If there is no interference, two rules of MRC, MMSE and ZF perform similarly, while the twos of IRC deteriorate by 1.2dB, due to the impossible exact accuracy of IRC interference estimation.

If there is interference:

In the low SNR situation, the detection performance of IRC is superior to that of MRC, because MRC simply takes interference as noise, with no consideration to interference signals, while IRC can effectively inhibit interference signals, consequently improving its detection.

In the high SNR situation, the performance of MRC gradually improves, mainly because of the increasing SIR and closing to the none interference situation. IRC, for its requirement of calculating the interference covariance matrix, which goes with inaccuracy, performs worse in this situation.

4.Conclusions

Based on the above simulation and analysis, the application of IRC algorithm possesses the following four characteristics:

IRC algorithm is applied to the low SNR situation. The detection performance of IRC is far better than that of MRC in that situation, for it with inhibition signals from interference, which is not in part of MRC calculation of MRC.IRC algorithm has better effects on inhibiting interference, with increasing divergence of receiving angle of arrival, i.e. separating directions of users. Taking SCME-B channel and MCS valued 5 as examples, SIR increases by 4dB at the point of 10% BLER, when the angle between the interference and the user is 60 degree, compared to 15 degree.The directions of receiving interference will become separating, if the number of users increases. The IRC inhibition capability will inhibit the increasing number of users and boost the total power of interfering signals, consequently worsening the performance of IRC, for example, the performance difference between two interference users and one is 6-7dB.

IRC algorithm is superior to MRC in the greater interference situation. If the interference is larger, then choose IRC, otherwise, choose MRC. On the basis of the above analysis, the two algorithms function is differently according to different interference and SNR. Thus, MRC and IRC adaptive algorithms gain greater theoretically.

Acknowledgment

Fund Project: Key project of Education Department of Anhui (KJ2016A306)

References

[1]Diao C. Analysis on China Mobile 4G network interference and discussion on the solutions[J]. INFORMATION AND COMMUNICATIONS, 2015(01):241-242.

[2]S J Grant, J K Cavers. Performance Enhancement Through Joint Detection of Cochannel Signals Using Diversity Arrays.IEEE Trans.August 1998, 46 (8):1038-1049.

[3] Han Q, Ge W, Chang Y. MMSE-IRC precoding algorithm based on CoMP-JT scheme[J]. Mobile communications, 2017,41(16):81-85+89.

[4] Zheng Y. Research of uplink interference mitigation in long term evolution[D]. Beijing University of Posts and Telecommunitions, 2010.

[5] Chen Y. Research and application of interference estimation and interference suppression technology[D]. Beijing University of Posts and Telecommunitions, 2017.

[6] Feng Q. Research on key techniques of interference suppression based on network assisted in LTE-A system[D]. University of Electronic Science and Technology of China, 2016.

[7] Tian M. Research on the algorithm of channel estimation in LTE-A system with inter-cell interference[D]. Chongqing University of Posts and Telecommunications, 2017.

[8] Fang M. Research on multi-flow HSDPA system performance and key technology[D]. Beijing University of Posts and Telecommunitions, 2013.

[9] Ao X. Analysis on the performance of MRC techniques in micro-diversity system in mobile communications[J]. Modern science & technology of telecommunications, 2015(01).

[10] Han X. Research on MIMO receiver in the downlink of LTE[D]. Xidian University, 2009.

ISPECE   IOP Publishing

Rotor-Mechanical Coupled Fault Feature Extraction Based on Second-order Blind Identification

Feng Miao[1*], RuZhi Feng[2] and XianLi Wang[1]

[1] School of Physical and Electrical Information, Luoyang Normal University, Luoyang,Henan, 471022, China

[2] Henan Mechanical and Electrical Vocational College, Zhengzhou, Henan, 451191, China

*miaofeng3699@163.com

Abstract. Noise reduction usually is conducted before analysis of mechanical fault feature, which could damage effective signals. This article proposes an algorithm of blind source separation based on the second-order statictics. The method focuses on noise separation rather than noise removal. So there are no harms to effective signals. This idea might provide a new way for noise reduction. The algorithm of blind source separation based on the second-order statistics blind identification is applied to seismic data. The results show that the algorithm is effect, noises are separated and re-moved, and the rotor fault feature is picked up.

1. Introduction

The extraction of mechanical failure features in the noise environment has always been a complex problem[1-3]. In the context of unknown noise, it is often difficult to extract effective fault characteristics if the effects of noise are ignored. Therefore, the influence of noise must be taken into account in the mechanical fault diagnosis based on blind separation theory. In the literature [4-9], the vibration signal is de-noised and separated by wavelet transform, wavelet packet, wavelet filtering, autocorrelation and blind separation method, and then the fault signal is extracted from the separated signal. Literature [10] were carried out noise reduction, and carried out simulation and experimental research. In the separation of mechanical fault sources, although the use of noise reduction pretreatment method has achieved some success, there are some shortcomings, first of all, different working conditions noise is different, using different de-noising method will come to different separation results.

In order to solve this problem, this paper applies the blind separation theory to the noise reduction process. The key is to separate the noise rather than eliminate the noise. Therefore, it does not lose the effective signal when separating the noise, and provides a new method for noise processing. In this paper, a self-extraction method based on second order blind identification is proposed. This method introduces the concept of gradient change rate in the original self-extraction algorithm, which effectively reduces the noise. Through the simulation and the actual rotor vibration data processing shows that this algorithm effectively curb the noise, improve the accuracy of sampling data.

Content from this work may be used under the terms of the Creative Commons Attribution 3.0 licence. Any further distribution of this work must maintain attribution to the author(s) and the title of the work, journal citation and DOI.
Published under licence by IOP Publishing Ltd

2. Signal model

Blind source separation problem refers to the process of recovering individual components only by observing signals without knowing the parameters of the source signal and the transmission channel. The hybrid model is represented as [8, 9]:

$$x(k) = Hs(k) + n(k) \tag{1}$$

Where $x(k) = [x_1(k), x_2(k), ..., x_M(k)]^T$ is a M-dimensional random observation vector In the case of noisy interference, H is a mixed matrix of $M \times N$ with an unknown full rank; $s(k) = [s_1(k), s_2(k), ..., s_N(k)]^T$ is an N-dimensional source signal, the components of the source signal are assumed to be statistically independent and contain at most one Gaussian noise, otherwise they cannot be separated; $n(k) = [n_1(k), n_2(k), ..., n_M(k)]^T$ is an M-dimensional noise signal.

In this article, unless otherwise stated, the following assumptions are made:

(1) The mixed matrix $A \in R^{m \times n}$ full rank;

(2) The source signal is irrelevant to the space-time domain, and the time-domain associated zero mean random signal;

(3) The source signal changes in variance, is the second order non-stationary, or smooth signal;

(4) the source signal and noise are independent, can be airspace color, time domain white, that is

$$E(n(k)n^T(k-p)) = \delta_{po} R_n(p)$$

Where δ_{po} is the Kronecker δ function ; R_n is an arbitrary M * N-order matrix.

3. Algorithm Based on Second Order Blind Identification

In the above hypothesis, where the observed signal vector $x(k)$ is satisfied for the correlation function matrix of nonzero delay p :

$$R_x(0) = E\{x(k)x^T(k)\} = HR_x(0)H^T + R_n(0) \tag{2}$$

$$R_x(p) = E\{x(k)x^T(k-p)\} = HR_x(p)H^T \tag{3}$$

Where $R_x(0) = E\{x(k)x^T(k)\}$ and $R_x(p) = E\{x(k)x^T(k-p)\}$ are nonzero different elements diagonal matrix. When the number of observed signals n is greater than the number m of the source signal, when the noise covariance matrix $R_m = R_n(0) = E\{x(k)x^T(k)\} = \sigma_n^2 I_m$ special form (When the signal-to-noise ratio is high, the noise variance σ_n^2 is estimated from the mean singular value of $R_x(0)$ or the singular value of m-n $R_x(0)$), the unbiased estimation of the covariance matrix of formula (2) is as follows

$$\overline{R}_x(0) = R_x(0) - \sigma_n^2 I_m = HR_s(0)H^T \tag{4}$$

In order to identify the mixed matrix H, we can estimate the diagonalization of $\hat{R}_x(p)$ and $\hat{\overline{R}}_x(0)$ by the covariance matrix of Eq. (3) and (4).

The second-order blind identification algorithm (the number of sensors is greater than the number of source signals) is as follows:

(1) Estimate the correlation matrix of the sensor signal

$$\hat{R}_x(0) = \sum_{k=1}^{N} x(k)x^T(k) \Big/ N \tag{5}$$

(2) Calculate the SVD or EVD of $\hat{R}_x(0)$, ie

$$\hat{R}_x(0) = U_x \sum_x V_x^T = V_x \Lambda_x V_x^T = V_s \Lambda_s V_s^T + V_N \Lambda_N V_N^T \tag{6}$$

Where $V_S = [v_1, v_2, \cdots, v_n] \in R^{m \times n}$ contains the eigenvectors corresponding to the n main eigenvalues of $\Lambda_s = diag\{\lambda_1 \geq \lambda_2 \geq \cdots \geq \lambda_n\}$. Similarly, the matrix $V_N \in R^{m \times (m-n)}$ contains $m-n$ noise eigenvalues $\Lambda_s = diag\{\lambda_{n+1} \cdots \geq \lambda_n\}$ (where $\lambda_n \rangle \lambda_{n+1}$) corresponds to the noise eigenvector. There are usually typical relationships between eigenvalues $\lambda_1 \geq \lambda_2 \geq \cdots \lambda_n \rangle \lambda_{n+1} \approx \cdots \approx \lambda_m$ (m> n).

(3) Estimate the white noise variance σ_n^2 for $m-n$ least significant eigenvalues or singular values. And do pre-whitening processing, such as

$$\overline{x}(k) = \widehat{\Lambda}_s^{-1/2} V_S^T x(k) = Qx(k)$$

Where: $\widehat{\Lambda}_s = diag\{(\lambda_1 - \widehat{\sigma}_n^2), (\lambda_2 - \widehat{\sigma}_n^2), \ldots, (\lambda_n - \widehat{\sigma}_n^2)\}$.

(4) In the case of delay $p \neq 0$, we estimate the covariance matrix of $\overline{x}(k)$, and carry out SVD, that is, $\widehat{R}_{\overline{x}}(p) = \dfrac{1}{N} \sum_{k=1}^{N} \overline{x}(k)\overline{x}^T(k-p) = U_{\overline{x}} \Sigma_{\overline{x}} V_{\overline{x}}^T$

(5) whether there is a difference in the singular value corresponding to the diagonal matrix $\Sigma_{\overline{x}}$ of the delay p. If there is no significant difference, select a different delay p and return to step (3). If there is a significant difference between the singular values and the distance is far from each other, then the estimated mixing matrix is

$$\widehat{H} = Q^+ U_{\overline{x}} = V_S \widehat{\Lambda}_S^{1/2} U_{\overline{x}}$$

It is estimated that the source signal with noise is

$$y(k) = \widehat{s}(k) = U_{\overline{x}}^T \overline{x}(k) = U_{\overline{x}}^T \widehat{\Lambda}_S^{1/2} V_S^T x(k)$$

4. Simulation research and analysis

4.1. Evaluation Criteria
In order to evaluate the effect of signal separation effectively, the similarity coefficient is used as the evaluation index of the difference between the separation signal and the source signal. The similarity coefficient is defined as:

$$\xi_{ij} = \xi(y_i, s_j) = \left| \sum_{t=1}^{M} y_i(t)s_j(t) \right| \Big/ \sqrt{\sum_{t=1}^{M} y_i^2(t) \sum_{t=1}^{M} s_j^2(t)}$$

In the formula, when, is constant, , that is, the separation of the signal amplitude difference; when the separation signal and the source signal is independent of each other . It is also said that if the similarity coefficient matrix is equal to 1 for each row and each column and only one element is close to zero, then the separation effect of the algorithm is considered to be ideal.

4.2. Simulation
In order to validate the method proposed in this paper, we first simulate the simulation. According to the known vibration signal model of the rotating machinery, the simulation source is constructed as follows:

$$\begin{cases} s_1(t) = \sin(0.5t) \\ s_2(t) = \sin(0.3t)\cos(10t) + \sin(2t) \\ s_3(t) = 4rand \end{cases}$$

Sampling frequency 1000Hz, sampling length of 256 points, aliasing matrix H;

$$H = \begin{bmatrix} 0.965 & 0.334 & 0.718 \\ 0.666 & 0.523 & 0.778 \\ 0.726 & 0.273 & 0.081 \end{bmatrix}$$

Generating a mixed signal H shown in Figure 2

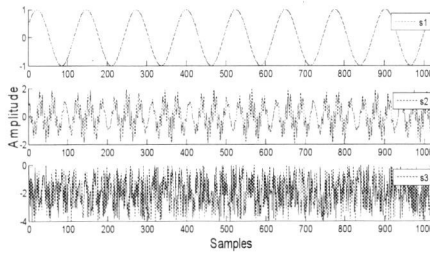

Fig. 1 The simulative signal

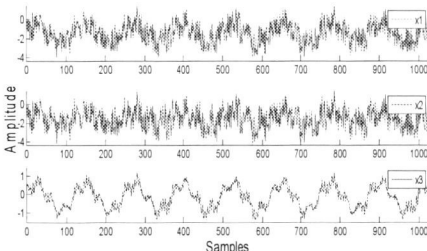

Fig. 2 The mixture of simulative signal

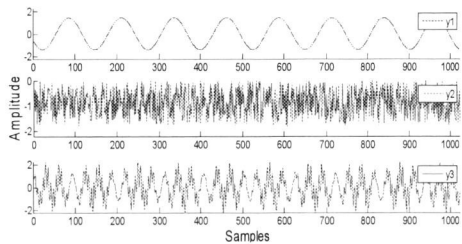

Fig. 3 The separated signal using second-order statistics blind identification

Table 1 3 similarity coefficient of source signals

Signal vector	Similarity coefficient		
	ξ_1	ξ_2	ξ_3
y_1	1.00000	0.00011	0.00023
y_2	0.02132	1.00000	0.00693
y_3	0.01321	0.07685	1.00000

From the mixed signal in Figure 2, it is difficult to distinguish the characteristics of the observed signal. As can be seen from Figure 3, the separated signal and the source signal are basically the same, only the amplitude is inconsistent, it is because the blind signal separation Amplitude and order of the inconsistency; at the same time random noise signal is a good separation. As can be seen from Table 1, the maximum value of the similarity coefficient between the separation signal and the source signal is 1, the similarity is close to 100%, and the separation effect is achieved.

5. Experiment

In order to verify the application of the second - order blind identification in the double - rotor mechanical fault signal separation, this paper uses the algorithm to analyze the real double - rotor rotor fault vibration signal. The experimental sampling frequency is 5000Hz, the sampling point is 5120, the speed is 2800r / min. Three-way sensor installation location shown in Figure 4.When the rotor rubbing fault and imbalance fault exist at the same time, the three sensor signals were shown in Figure 5; through the second-order blind identification after separation of the time domain signal shown in Figure 7. It is difficult to determine the specific fault of the rotor from the time-domain waveform of the observation signal and the separated signal. For the comparison of the complex vibration characteristics of the two-span rotor before and after the separation, the data signal is analyzed before and after the separation, and the spectrum of the observed signal Figure 8 shows the spectrum of the isolated signal.

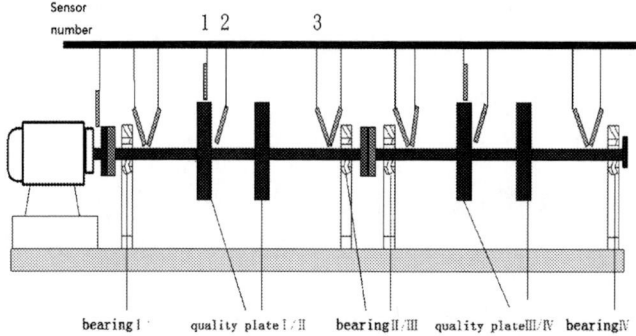

Fig. 4 Three-way sensor installation location

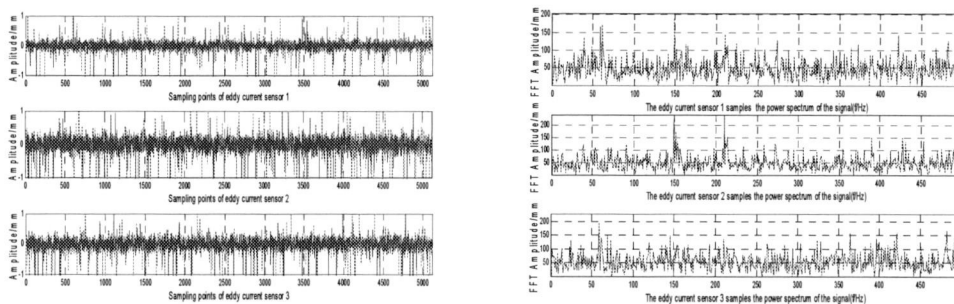

Fig. 5 The vibration signal for rotor with multi-faults   Fig.6 FFT of vibration signal

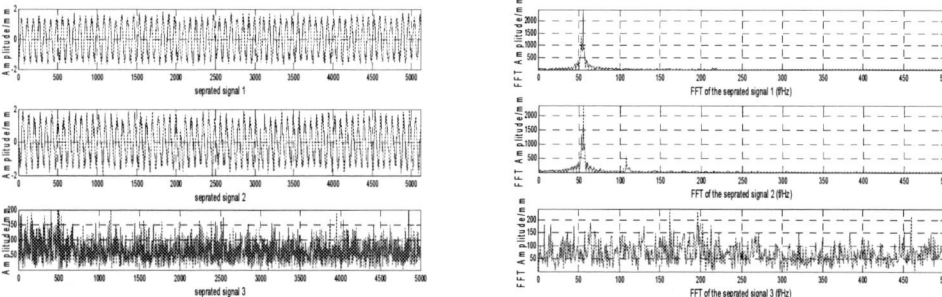

Fig.7 The separated signal   Fig. 8 FFT of the separated signal

It can be seen from Figure 6, The vibration signal is completely covered by the noise, and the fault features of the rotor cannot be reflected. As can be seen from the first picture in fig.8, the tributes of 50HZ are highlighted and the signals of other frequencies are contained; in the two figure in addition to the fundamental frequency spectrum peak obviously, 2 ×, 3 × also have a small peak exists, the performance of the imbalance of the frequency characteristics; the fifth figure shows that the signal in the time domain and frequency domain showed randomness, can be judged as noise signal.

6. Conclusion

In the case of mechanical fault feature extraction based on blind separation theory, the influence of noise has always been an unsolved problem. Due to the complexity of the mechanical conditions, it is difficult to separate the mechanical failure. In this paper, the blind separation theory is applied to denoising, the key is to separate the noise rather than eliminate the noise, so the noise is not lost when the effective signal for noise processing provides a new method. The simulation results show that the algorithm is a good way to separate the random noise and achieve a good separation result. Finally, the proposed method is applied to the fault feature extraction of the double-span rotor. The experimental results also verify the validity of the proposed method. The study of this paper provides an effective method for rotor fault feature extraction in noise environment.

Acknowledgments

The research described in this paper was supported by Foundation of He'nan Educational Committee (16A470021) and key scientific and technological project of Henan Province (172102210097).

References

[1] Z. Peng, Y. He, F. Chu, Vibration signal analysis and feature extraction based on reassigned wavelet scalogram, Journal of Sound and Vibration 253 (5) (2002) 1087-1100.

[2] Z.N. Li, Q.Q. Ding, Z.T. Wu, C.J. Feng, G.B. Yang, Blind system identification and fault diagnosis, Journal of Zhejiang University (Engineering Science) 37 (2) (2003) 215-220.

[3] Miao Feng, Zhao Rongzhen. Separating for nonlinear mixed rotor fault signals with violent pulse interferences. Journal of Vibration, Measurement& Diagnosis,(2014)34(4):625-630.

[4] Lei Yanbin,Li Shunming,Men Xiuhua,et al. Separating mixed rotor vibration signals based on auto-correlation de-nosing[J]. Journal of Vibration and Shock.(2011)30(1):218-222.

[5] Li Zhinong,Liu Weibing,Yi Xiaobing. Underdetermined blind source separation method of machine faults based on local mean decomposition . Journal of Mechanical Engineering, 2011, 47(7) : 97-102.

[6] Wang Jian guo,Li Jian,Wan Xu dong. Fault feature extraction method of rolling bearings based on singular value decomposition and local mean decomposition [J] .Journal of Mechanical Engineering, 2015, 51(3) : 104-110

[7] Antoni J. Blind separation of vibration components: Principles and demonstrations[Jj. Mechanical Systems and Signal Processing, 2005,19(6): 1166-1180.

[8] A.Ghazdali,M.ElRhabi,H.Fenniri. Blind noisy mixture separation for independent/dependent sources through a regularized criterion on copulas, Signal Processing,131 (2017) 502-513.

[9] Theodor D. Popescu. Blind separation of vibration signals and source change detection-Application to machine monitoring. Applied Mathematical Modelling 34 (2016) 3408-3421.

[10] A Ypma, A Leshem. Blind Separation of Machine Vibration with Bilinear Forms.in Proceeding of ICA-2000, Helsinki, June, 2000, 405-410.

ISPECE

IOP Publishing

Separating for Nonlinear Mixed Rotor Fault Signals Based on Adaptive Particle Swarm Optimization

Feng Miao[1*], XianLi Wang[1] and Wei Guo[1]

[1] School of Physical and Electrical Information, Luoyang Normal University, Luoyang, Henan, 471022, China

*miaofeng3699@163.com

Abstract. The performance of existing nonlinear mechanical failure signal separation methods is affected by the non-linear contrast function that is selected according to the distribution of original signals. To solve this problem, a blind source separation algorithm based on adaptive particle swarm optimization is proposed, which takes the negentropy of mixtures as a contrast function. The inertia weight factor depends on the negentropy, which can improve the contradiction between the convergence speed and the performance of separated signals. The simulation results verified the effectiveness of the proposed method. Finally, some mixed rotor vibration signals were separated successfully using the proposed method.

1. Introduction

Blind source separation (BSS) technology is a new method developed in the 1980s. It is a process based on the statistical characteristics of the source signal. Only the observed mixed signal returns the unknown source signal. It is an artificial neural network, statistical signal processing and information theory combined with the method [1]. At present, many different algorithms have been developed. These algorithms involve the selection of nonlinear functions, but the choice of function model depends on the probability density property of the signal source [2-6]. In practical applications, the probability density of the source signal is generally unknown before the signal is separated, especially for the super-Gaussian signal and the sub-Gaussian mixed signal, the separation ability of the blind separation algorithm often depends on the selection of the nonlinear function [7,8]. In the rotating mechanical vibration test, the signal transmission is often affected by the complexity of the internal structure of the system and the transmission process and other factors, in its existence there are more complex nonlinear process, seriously affecting the accuracy and reliability of mechanical fault diagnosis, different separation algorithms will produce different separation effects [9].

Aiming at this problem, a method of mechanical fault feature extraction based on adaptive particle swarm optimization is proposed. In this method, the negative entropy of the observed signal is chosen as the objective function, and the inertia factor can be adjusted adaptively by observing the state of the signal, which can effectively overcome the contradiction between the signal recovery quality and the convergence speed. Through the separation of the simulation signal, the separation of the output signal and the simulation signal consistency. Finally, the fault signal separation is successfully realized by the method, and the effectiveness of the proposed method is verified.

Content from this work may be used under the terms of the Creative Commons Attribution 3.0 licence. Any further distribution of this work must maintain attribution to the author(s) and the title of the work, journal citation and DOI.
Published under licence by IOP Publishing Ltd

2. Blind source signal separation

The blind source separation problem refers to the process of recovering individual components only by observing signals without knowing the parameters of the source signal and the transmission channel. The hybrid model is expressed as [1]:

$$y(t) = As(t) + n(t) \tag{1}$$

Where $y(t) = [y_1(t), y_2(t), ..., y_M(t)]^T$ is a N-dimensional random observation vector in the presence of noise, A is a hybrid matrix of an unknown full rank $M \times N$, $s(t) = [s_1(t), s_2(t), ..., s_N(t)]^T$ is an M-dimensional source signal, and each component $s_i(t)$ in the source signal is assumed to be statistically independent and contains up to A Gaussian noise, otherwise not be separated; $n(t) = [n_1(t), n_2(t), ..., n_M(t)]^T$ is a M-dimensional noise signal.

3. Blind Source Separation Algorithm Based on APSO

The particle swarm optimization algorithm was first proposed by Dr. Eberhart and Dr. Kennedy in Ref. [5], which was a group-based optimization tool with global optimization capabilities. Suppose that in a D-dimensional search space there are N random particles, where the position of each particle represents a potential solution. Particles in the iterative process, by tracking two extreme values to adjust themselves: the first is the particle itself to find the optimal solution, known as the individual extreme P, can also be seen as the flight experience of particles; the other extreme is all Particle group to find the optimal solution, known as the global extreme value G, also known as group experience. The Particle swarm optimization algorithm is manipulated using the following formula:

$$v_{ij}^{t+1} = v_{ij}^{t} + c_1 r_1 \times (pt_{ij}^{t} - x_{ij}^{t}) + c_2 r_2 \times (gt_{ij}^{t} - x_{ij}^{t}) \tag{2}$$

$$x_{ij}^{t+1} = x_{ij}^{t} + v_{ij}^{t+1} \tag{3}$$

Where t is the current number of iterations; c_1、c_2 is the learning factor and is the normal number used to adjust the traction between the optimal position and the global optimal position. r_1、r_2 is a random number between $[0,1]$;

Where the independence of the signal is chosen from the negative entropy to measure, Comon [1] proves that the negative entropy of the multivariate can be expressed as

$$J_i(y_i) \cong \frac{1}{12} k_3^2(y_i) + \frac{1}{48} k_4^2(y_i) + \frac{7}{48} k_3^4(y_i) - \frac{1}{8} k_3^2(y_i) k_4(y_i) \tag{4}$$

Where $k_3(y_i)$ is the third-order cumulant: $k_4(y_i)$ is the fourth-order cumulant. If the sampling signal signal distribution is symmetrical probability distribution, then $k_3 = 0$, then

$$J_i(y_i) \cong \frac{1}{48} k_4^2(y_i) \tag{5}$$

The fourth-order cumulant $k_4(i)$ is normalized to

$$k_4(y_i) = \frac{E(y_i^4)}{E(y_i^2)^2} - 3 \tag{6}$$

$k_4(y_i)$ is the kurtosis of the signal; $k_4(y_i) = 0$ signal is a gaussian signal; $k_4(y_i) < 0$ signal is owed gaussian signal; $k_4(y_i) > 0$ signal is a super gaussian signal; Because $k_4^2(y_i) > 0$, the bigger $k_4^2(y_i)$, the stronger the non-gauss.

In this paper, the negative entropy of the sampled signal is used as the objective function of the particle swarm optimization algorithm.

$$f(y) = \sum_{i=1}^{n} \frac{1}{48} k_4^2(y_i) \tag{7}$$

Under the constraints of $E(yy^T) = I$, for the separation matrix W, the larger $f(y)$, the greater the independence of the separation signal, the better the source separation effect.

The number of particles is n, the fitness of the optimal particles is f_{max}, and the fitness of the particles P_i in the iteration is f_i. The average fitness of the particle group is $\overline{f}_i = \frac{1}{n} \sum_{i=1}^{n} f_i$, and The fitness of the particle that is better than $\overline{f}_i$ is equal to $\overline{f}_i$, the inertia factor can be adjusted by f_i and f_{max}. The formula is [8]

$$w = w_0 \exp(-a \cdot \frac{f_i}{f_{max}})$$

In the formula, the selection of a and w_0 has a great influence on the performance of the algorithm. If a is too small, the adjustment ability of equation (8) is insufficient; if a is too large, the algorithm is easy to fall into local optimum;

In summary, APSO-based blind source separation algorithm can be described as:

1) Whitening and centralizing the sampled signal;

2) initialize, set the learning factor c_1、c_2, and w_0、a, randomly generate a certain number of separation matrix as the initial particles, and randomly generate the particle movement speed;

3) Push Y according to the separation formula $Y = WX$, and center and whiten the Y, and push the adaptive value of the particle by the objective function.

4) Update the individual optimal p_i of each particle and the global optimal g_i of the whole population; then update the position and velocity of the particle according to (3) and (4)

5) To determine whether to meet the termination conditions. If it is satisfied, go to step 6); otherwise, go to step 3) and continue iterating.

6) Output the final global optimal g_i, the algorithm is running.

4. Simulation research and analysis

4.1. Evaluation Criteria
In order to evaluate the effect of signal separation effectively, the similarity coefficient is used as the evaluation index of the difference between the separation signal and the source signal. The similarity coefficient is defined as:

$$\xi_{ij} = \xi(y_i, s_j) = \left| \sum_{t=1}^{M} y_i(t) s_j(t) \right| \Bigg/ \sqrt{\sum_{t=1}^{M} y_i^2(t) \sum_{t=1}^{M} s_j^2(t)}$$

In the formula, when , is constant, , that is, the separation of the signal amplitude difference; when the separation signal and the source signal is independent of each other, . It is also said that if the similarity coefficient matrix is equal to 1 for each row and each column and only one element is close to zero, then the separation effect of the algorithm is considered to be ideal.

4.2. Simulation
I Rotor in the presence of rubbing, cracks, not the middle of the fault, the sensor is often the signal collected aliasing fault signal and AM FM signal. Simulation test in the choice of four kinds of signals, sampling length of 1024, the source signal production function is as follows:

$$s(t) = \begin{pmatrix} s_1(t) \\ s_2(t) \\ s_3(t) \\ s_4(t) \end{pmatrix} = \begin{pmatrix} n(t) \\ 3\sin 0.4t\cos 10t \\ \sin 3t + \sin 6t + \sin 10t \\ \sin 2t \end{pmatrix}$$

Source signal $s_1(t)$ analog noise signal, $s_2(t)$ simulation frictional fault characteristic signal, $s_3(t)$ simulation misalignment fault characteristic signal, $s_4(t)$ analog period vibration signal; $s_1(t)$、$s_2(t)$、$s_3(t)$、$s_4(t)$ time domain waveform shown in Figure 1. The source signal is generated by random mixing of the observed signal $x_1(t)$、$x_2(t)$、$x_3(t)$、$x_4(t)$, the mixing mode is linear superposition, and the time domain waveform of the mixed signal is shown in Fig.

In the simulation experiment, the particle swarm optimization algorithm and the adaptive particle swarm optimization algorithm are used to separate the mixed signal, and the separation signals are shown in Fig. 3 and Fig.

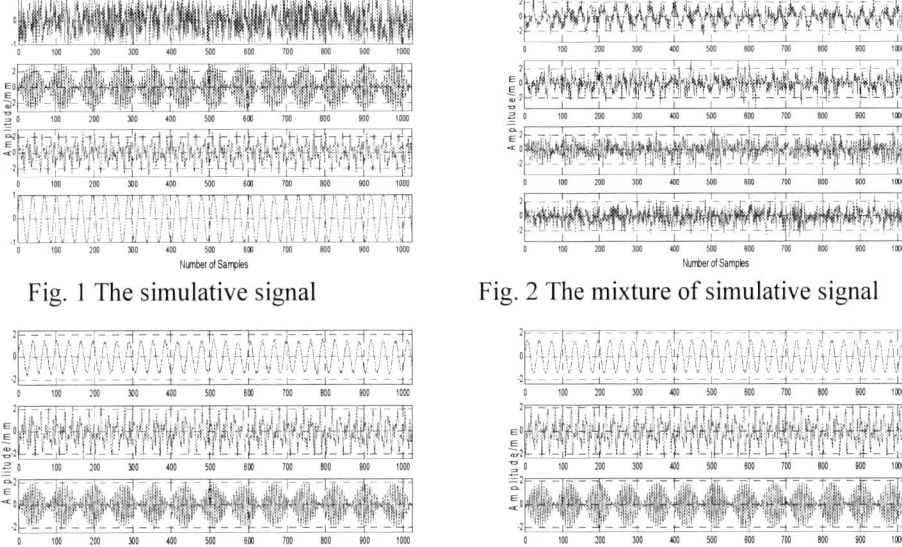

Fig. 1 The simulative signal   Fig. 2 The mixture of simulative signal

Fig. 3 The separated signal using PSO   Fig. 4 The separated signal using APSO

Compared with Fig. 3 and Fig. 4, it can be seen that the source signal separated by the general particle swarm algorithm does not reflect the source signal well. The improved adaptive particle swarm optimization algorithm reflects the waveform information of the source signal well. Although the source signal and the separation signal in the amplitude and order there is inconsistent, but does not affect the identification of signal characteristics. The simulation results show that the adaptive particle swarm optimization algorithm is used to quantify the similarity between the signal and the source signal. The similarity coefficient matrix:

$$\xi = \begin{bmatrix} 0.0001 & 0.0214 & 0.0051 & \underline{0.9997} \\ 0.0031 & 0.0572 & \underline{0.9996} & 0.0327 \\ 0.0861 & \underline{1.0000} & 0.0008 & 0.0141 \\ \underline{1.0000} & 0.0013 & 0.0012 & 0.0069 \end{bmatrix}$$

The underline factor is expressed as the correlation coefficient between the separation signal and

ISPECE   IOP Publishing

IOP Conf. Series: Journal of Physics: Conf. Series **1187** (2019) 032005   doi:10.1088/1742-6596/1187/3/032005

the source signal. It can be seen from ξ that the similarity coefficient of the separation signal and the source signal based on the particle swarm optimization algorithm is very high, which shows that the improved adaptive particle swarm optimization algorithm can accurately extract the source signal characteristics from the mixed signal.

5. Analysis of actual rotor vibration signal

In order to verify the separation performance of the above algorithm to the measured aliasing vibration signal, the vibration signal of the measured aliasing rotor is analyzed in this paper. As the rotor in the rotation process, there may be a number of potential source signals, such as the bearing ball vibration signal, the axis of the axial vibration signal and noise signals, and the sensor is measured at the same time, so the sensor measured signal is aliased Vibration signal. In order to meet the assumption that the number of sensors in the blind source separation is greater than or equal to the number of source signals, four sensors are used for the measurement. Rotor in the rotation process speed is about 2800 r / min, sampling frequency of 5000Hz. The time domain signal after direct separation by the APSO algorithm is shown in Fig7. In order to compare the complex vibrations of the rotor before and after the separation, the data of the data signals before and after the separation are analyzed, and the different characteristics of the signal before and after the separation are observed from the frequency domain. The spectrum before and after separation is shown in Fig. 6 and Fig8.

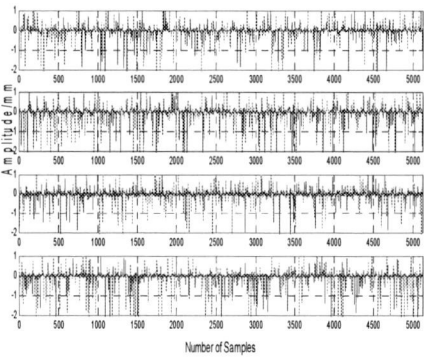

Fig. 5 The vibration signal for rotor with multi-faults

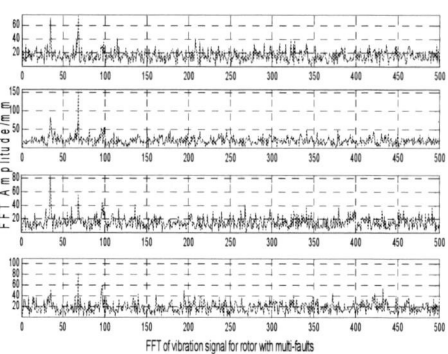

Fig.6 FFT of vibration signal for rotor with multi-faults

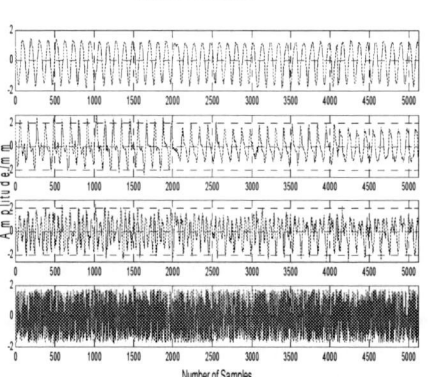

Fig.7 The separated signal using APSO

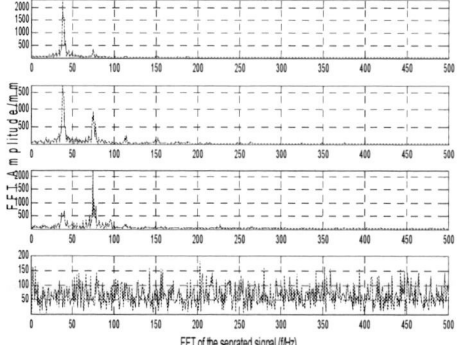

Fig. 8 FFT of the separated signal using APSO

It can be seen from Figure 6, the rotor rotation frequency and its frequency 2 ×,3 × and other frequency characteristics in the order spectrum is more obvious, you can see the fault signal is aliased together. However, it can be seen from the order spectrum of Fig. 8 that only the 1 × spectral peak is very obvious in the first graph, and the unequal frequency division is shown in the second graph. As

534

can be seen in the third graph, there are $1 \times$ and $2 \times$ spectral peak , and the $2 \times$spectral peak is obviously stronger than the $1 \times$ spectral peak, showing the rubbing characteristics; the fourth graph in the time domain shows randomness, you can determine the signal as a noise signal. The above analysis shows that the APSO algorithm can get a better separation effect under the condition of unbalanced - rub - impact - misalignment.

6. Conclusions

The simulation results show that: 1) The established blind source separation algorithm based on adaptive particle swarm optimization is needed to select the nonlinear function by observing the signal state adaptively. Adjust the inertia factor, effectively overcome the signal recovery quality and convergence speed between the spear. 2) The results of numerical simulation show that the source signal can be effectively separated based on the adaptive particle swarm optimization and has high stability. 3) The separation of the fault signal in the case of unbalanced - rub - impact - misalignment is achieved, and a good separation effect is obtained. It is feasible to isolate the rotor fault signal separation method.

Acknowledgments

The research described in this paper was supported by Foundation of He'nan Educational Committee (16A470021) and key scientific and technological project of Henan Province (172102210097).

References

[1] Comon P. and Jutten C., Handbook of Blind Source Separation: Independent Component Analysis and Applications, Elsevier, Oxford: 2010.

[2] Z. Peng, Y. He, F. Chu, Vibration signal analysis and feature extraction based on reassigned wavelet scalogram, Journal of Sound and Vibration 253 (5) (2002) 1087-1100.

[3] ZHANU Xining, LEi Wei, Li Bing. Bearing fault de-tection and diagnosis method based on principal compo-nent analysis and hidden Markov model[J].Journal of Xi'an Jiaotong University, 2017, 51(6):1-7, 109.

[4] GUO X L, YANG T T .Gesture recognition based On HMM-FNN model using a Kinect [J]. Journal On Multimodal User Interfaces, 2016, 11(1):1-7.

[5] LEI Yaguo, HE Ghcngjia, GI Yanyang. Fault diagno-sis based on novel hybrid intelligent model[J].Chi-nese Journal of Mechanical Engineering, 2008, 44(7):112-117.

[6] YANU Lidong, GU Yu, ZHANU Ming. Application of ant colony optimization in speech signal feature se-lcction[J].Computer Simulation, 2016, 33(2):409-412, 417.

[7] Miao Feng, Zhao Rongzhen.Separating for nonlinear mixed rotor fault signals with violent pulse interferences. Journal of Vibration,Measurement& Diagnosis,(2014)34(4):625-630.

[8] Lei Yanbin,Li Shunming,Men Xiuhua etal. Separating mixed rotor vibration signals based on auto-correlation de-nosing[J]. Journal of Vibration and Shock.(2011)30(1):218-222.

[9] Ye Hongxian,Yang Shixi,ang Jianxin. Mechanical vibration source number estimation based on EMD-SVD-BIC[J].Journal of Vibration, Measurement & Diagnosis,2010,30(3):330-334.

A SURVEY OF KNOWLEDGE-BASED INTELLIGENT FAULT DIAGNOSIS TECHNIQUES

Sanchuan Xu[1*]

[1]Department of Mechanical Engineering, The Hong Kong Polytechnic University, Kowloon, Hong Kong

*Corresponding author's e-mail: clark.xu@connect.polyu.hk

Abstract. With the development of information technologies, more and more real-time data can be obtained from production and operation process. Thus, how to extract effective information from these massive data, so as to carry out in-depth statistics and mining of faults, and gradually explore the faults laws and causes are crucial for intelligent factories. In recent years, a variety of statistical learning and data analysis methods have been used in fault diagnosis. Due to the complex structure, multi-source failure and suddenness of the industrial production system, the combination of empirical knowledge and mechanism principles can solve various fault problems. This paper summarizes several commonly used fault diagnosis methods, and focuses on knowledge-based intelligent fault diagnosis, including first-order logic knowledge representation method, production knowledge representation method, framework knowledge representation method, object-oriented knowledge representation method and Semantic-based knowledge representation methods.

1. Introduction

Fault diagnosis is a sub-area of control engineering, which is a multidiscipline intersection product. As far as current research is concerned, Modern Control Theory, Signals and Systems, pattern recognition and other disciplines have been applied into fault diagnosis. The development of fault diagnosis technology will greatly depend on the mentioned disciplines. The basic concept of fault diagnosis is to recognize the failures of the system and determine their types and location in time when the failures occur by monitoring the system.

The whole diagnosis system consists of three parts. They are named fault detection, fault recognition and fault recovery respectfully, which is shown in Figure 1. The concepts of these three parts are described in detail as below.

1) Fault detection: determine whether fault occurs in time. Should there be a fault, the alarm would go off and certain action would be taken to prevent further catastrophic consequence.

2) Fault recognition: after the detection of fault, the module should confirm which fault occurred, which component is dysfunctional. Then confirm the type, location, hierarchy and occurring time of the fault, at the same time, identify the cause of the fault.

3) Fault recovery: establish corresponding countermeasures according to possible adverse effects to ensure the normal operation of the system or to avoid catastrophic errors in system.

In this article, we will focus on the detection and recognition of the fault.

Content from this work may be used under the terms of the Creative Commons Attribution 3.0 licence. Any further distribution of this work must maintain attribution to the author(s) and the title of the work, journal citation and DOI.
Published under licence by IOP Publishing Ltd

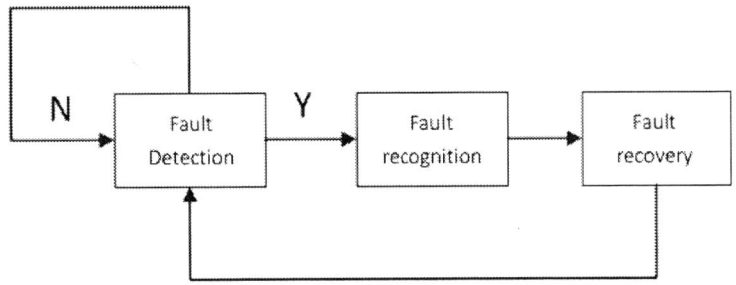

Figure 1 Fault Diagnosis System

2. Classification of the fault diagnosis methods in industrial process

At present, the methods of fault diagnosis can be generally divided into three categories: analytical model driven method, signal processing driven methods, and knowledge driven methods.

Analytical model-driven methods can be traced back to the 1971 doctoral thesis "Failure Accommodation in linear systems through self-reorganization" published by Massachusetts Institute of Technology, Beard. Its basic idea is: before and after the fault occurs, the system is performing normal and error respectively. An analytical model is established while the system is performing normally. By comparing the model output with the measured output of the system, a fault indicator called the "residual" is generated. This indicator, if the fault occurs, will affect the measure output of the system, resulting in the residual exceeding the threshold. At this point, it can be determined that a fault has occurred. By further analysis and evaluation of the residual, it is possible to provide information about fault separation and estimation information. The downside of this approach is the need to obtain accurate mathematical models of the system, which is often difficult to achieve in practice.

For many actual systems, it is difficult to accurately establish an analytical model of the diagnostic object. At this time, the signal model can be used to directly extract the fault feature information. Common signal processing driven fault diagnosis methods include: fault diagnosis method based on wavelet transform, δ operator-based method, principal component analysis method, method for detecting fault using Kullback information criterion and so on [2].

Wavelet transform is a branch of applied mathematics developed in the late 1980s. Ye et al. studied its application in dynamic fault detection and simulated it to obtain satisfactory results. It is a promising fault diagnosis method. Xiao Deyun et al. constructed a fault detection filter based on the δ operator. However, as it is unable to eliminate fault information, the filter may still send alarm after the fault's disappearance, which makes it difficult to generalize in practice. Sang et al. introduced the KPCA (kernel principal component analysis) method into the dynamic fault detection of nonlinear systems and compared it with the traditional PCA method. The simulation results show that the false positive rate is significantly reduced. Kumamaru et al. introduced the Kullback information criterion into fault detection. The basic idea is to compare the KDI (Kullback Discrimination Information) with the set threshold to detect system faults.

In recent years, due to the rapid development of artificial intelligence and computer technology, fault diagnosis methods based on the knowledge have gradually attracted people's attention. This method does not require knowledge of the exact mathematical of the object. It can be mainly divided into: neural network fault diagnosis method, fuzzy fault diagnosis method, expert system fault diagnosis method, support vector machine fault diagnosis method and so on. Yan Mingzhong et al. used the neural network for the fault diagnosis of underwater robots and compensated for the fault conditions to ensure the stable operation of the robot. Hacene Habbi et al. applied the fuzzy mathematics method to the fault diagnosis of aircraft heat exchangers. The results showed that it could effectively detect and isolate the sensor and actuator faults of heat exchanger. Lin Jiliang et al. used the support vector machine for the fault classification of mobile robots. The experimental results show that the method of using wavelet transform to extract the feature vectors, then classifying the faults using the support vector machine classification have a very good result.

It is worth mentioning that various fault diagnosis methods are not independent but can cooperate with each other. Figure 2 shows the various fault diagnosis methods and their categories.

3. Knowledge-based fault diagnosis methods

Knowledge-based fault diagnosis methods can effectively make use of expert knowledge and experience to make judgements. In some fields, when constructing the fault ontology, the researchers would model the relationship between the fault phenomenon and the cause, and then use the ontology reasoning technology to diagnose. However, in the actual fault diagnosis process, there is usually an uncertain relationship between the fault phenomenon of the equipment to be inspected and the cause of the fault. To better meet the needs of knowledge-based fault diagnosis, this paper reviews the research status of knowledge representation, fault knowledge construction and fault knowledge representation and acquisition.

In the field of artificial intelligence, the focus of Knowledge Representation is to use computers to realize the formal expression and information acquisition of knowledge. Therefore, knowledge representation methods and systems can be used to solve complex problems.

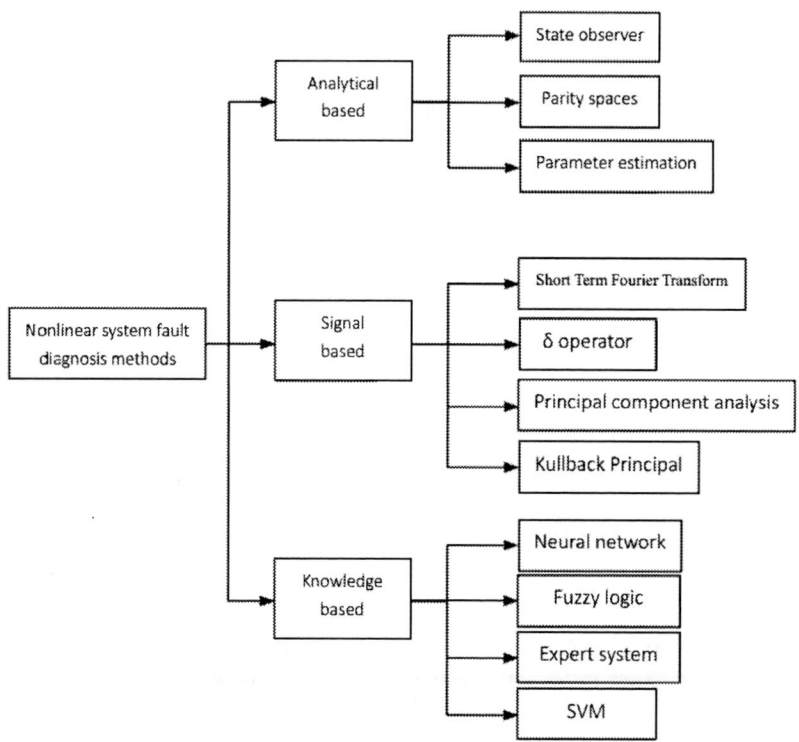

Figure 2 Fault Diagnosis Methods Categories

(1) First-order logic knowledge representation
Predicate logic knowledge representation is the earliest representation method applied to artificial

intelligence. Its purpose is to apply the logic arguments in mathematics to express the law of human thinking activities. So far, the predicate logic knowledge representation method is the most accurate formal language. Liang et al. used the synchrophasor measurement data collected from a power grid to form diagnosis rules for wide-area fault detection [11]. Based on the mined knowledge, three common types of short circuit faults, single-line-to-ground, line-to-line, and three-phase faults, are identified in the power grid.

(2) Production knowledge representation

The expression of the production knowledge representation is: if A then B. The relationship between concept and concept established by production knowledge representation is also the basis of knowledge representation. In the process of development, fuzzy knowledge representation and reasoning can be effectively processed through the improvement of technology. With the fusion with particle swarm optimization algorithm, the function of weighted fuzzy reasoning is improved. Deng et al. conducted fault diagnosis of electronic protective equipment based on CBR (case-based reasoning) by using K-NN Similarity Algorithm and Semantic Feature Vector pattern [12].

(3) Framework knowledge representation

The framework representation can express the internal structural relationship of knowledge and the connection between knowledge. It also can represent the inheritance relationship between knowledge. This is the same with the way of thinking when humans observe things. It has the characteristics of strong adaptability, high generality, good structure and flexible reasoning. Liu et al. developed an object-frame knowledge model which is made up of state-object, test-object and rule-object or repair-object, then used the forward chaining strategy to implement fault reasoning for a meteorological vehicle system [4].

(4) Object-oriented knowledge representation

Object-oriented knowledge representation is a combination of static attributes and dynamic operations. It conforms to people's habitual thinking mode of understanding and analysing problems, and has the characteristics of modularity, encapsulation, inheritance, polymorphism and easy maintenance. Since object-oriented knowledge represents structural and operational characteristics, this method has been applied in various fields such as product design and production, fault diagnosis, biomedicine, and chemistry. Dattatraya et al. used an object-rule structure to represent the procedural knowledge in complex electronics systems, ARM processor boards and large embedded systems and demonstrates the approach's effectiveness in fault diagnosis [3].

(5) Semantic network knowledge representation

The semantic network knowledge representation system matches the network segment of the problem with the knowledge base network segment and finds the solution of the problem according to the structure matchmaking. Semantic network knowledge representation can define relationships between objects artificially at any time as needed. It is structural, natural, associative and non-strict. Martin et al. applied ontologies to model the correlations, constraints, and dependencies among different system parts, and used ontology reasoning to identify the root cause of faults in complex products [10]. Niu et al. used ontology to represent the fault phenomena and properties, and further modelled the fault diagnosis problem as a bipartite graph match problem, which was efficient for fault diagnosis of complex system [1].

In the field of fault diagnosis, different methods are used to describe fault diagnosis knowledge, such as production, oriented object, script, semantic network, Petri net, ontology, causal network, and bond graph. However, the complexity of fault diagnosis makes it difficult to achieve the desired effect by any single method. Therefore, researchers try to solve the problem of multi-knowledge expression through different methods. For example, Son et al. adopted the method of fusing multiple expression methods; the concept of the Expert System Shell was proposed; the cross-hybrid model was proposed, and the knowledge interoperability of large-scale knowledge bases was proposed.

4. Conclusion and future work

Industrial production line is a complex system integrating mechanical, electrical, hydraulic and pneumatic pressures. Its fault occurrence has the characteristics of multi-source, correlation, suddenness and randomness. The acquisition cannot meet the needs of machine tool fault diagnosis with a large number of related characteristics and monitoring blind spots. Therefore, the fault diagnosis of machine tools inevitably requires human participation. For machine tools with a wide distribution area, most of the fault knowledge comes from machine tool design and manufacturing enterprises and users. Therefore, the knowledge of fault diagnosis has the characteristics of dispersion. In the process of fault diagnosis of a certain machine tool, the fault feature analysis must rely on relevant theory and other diagnostic history experience as a reference. Therefore, the integration of human-machine synergy knowledge, theoretical knowledge and maintenance and maintenance experience knowledge is of great significance for the fault diagnosis of machine tools. In the process of using fault diagnosis knowledge, with the improvement of machine tools and the improvement of diagnostic techniques, the content and structure of its knowledge are constantly being improved.

Knowledge of complex equipment fault diagnosis includes design knowledge, manufacturing knowledge, operational knowledge, etc., and there is still a lot of knowledge that is not clear whether it is related. There are differences in the characteristics and expressions of each knowledge. Therefore, constructing a multi-dimensional knowledge representation method that is compatible with basic knowledge, supporting knowledge, supporting algorithm knowledge, and process knowledge is one of the important issues to be considered in the next step.

References

[1] Niu, Qiang, et al. (2009) "A method of fuzzy reasoning based on semantic similarity and bipartite graph matching." *Artificial Intelligence and Computational Intelligence. AICI'09. International Conference on*. Vol. 4. IEEE.

[2] Carlo. C. (2015) "A survey of fault diagnosis and fault-tolerant techniques—Part II: Fault diagnosis with knowledge-based and hybrid/active approaches." *IEEE Transactions on Industrial Electronics*.

[3] Kodavade, D. Vishnu, and Apte. S. D. (2012) "A universal object-oriented expert system frame work for fault diagnosis." *International Journal of Intelligence Science* 2.03: 63.

[4] Liu, B., Duan M., and Zhao G., (2011) "An object frame knowledge representation approach for fault diagnosis expert system." *Future Computer Sciences and Application (ICFCSA), 2011 International Conference on*. IEEE.

[5] Miguelanez, E., et al. (2008) "Fault diagnosis of a train door system based on semantic knowledge representation.": 27-27.

[6] Liu, S.Y., et al. (2018) "Fault Diagnosis of Water Quality Monitoring Devices Based on Multiclass Support Vector Machines and Rule-Based Decision Trees." *IEEE Access* 6: 22184-22195.

[7] Dai, X., and Gao. Z. (2013) "From model, signal to knowledge: A data-driven perspective of fault detection and diagnosis." *IEEE Transactions on Industrial Informatics* 9.4: 2226-2238.

[8] Dong, J., et al. (2017) "Joint Data-Driven Fault Diagnosis Integrating Causality Graph with Statistical Process Monitoring for Complex Industrial Processes." *IEEE Access* 5: 25217-25225.

[9] Wang, L., et al. (2015) "Knowledge representation and general Petri net models for power grid fault diagnosis." *IET Generation, Transmission & Distribution* 9.9: 866-873.

[10] Martin M., Zoitl A., and Moser. T. (2010) "Ontology-based fault diagnosis for industrial control applications." *Emerging Technologies and Factory Automation (ETFA), 2010 IEEE Conference on*. IEEE.

[11] Liang, X.D., Wallace S.A., and Nguyen. D. (2017) "Rule-based data-driven analytics for wide-area fault detection using synchrophasor data." *IEEE Transactions on Industry Applications* 53.3: 1789-1798.

[12] Deng, X.Y., Luo R., and Li J.S., (2015) "Similarity matching algorithm of equipment fault

diagnosis based on CBR." *Software Engineering and Service Science (ICSESS), 2015 6th IEEE International Conference on.* IEEE.

[13] Zhang, Q., and Yao. Q.Y. (2018)"Dynamic Uncertain Causality Graph for Knowledge Representation and Reasoning: Utilization of Statistical Data and Domain Knowledge in Complex Cases." *IEEE transactions on neural networks and learning systems* 29.5: 1637-1651.

ISPECE

IOP Publishing

Structure and Simulation of Roadway Disaster Simulation Control System for High Temperature Smoke Drill

Guo Jikun, Zhang Rui

(School of Electric &Control Engineering, Heilongjiang University of Science & Technology, Harbin, Heilongjiang Province, 150022, China)

Abstract: In view of the outstanding problems of the training environment in the high-temperature smoke environment, such as failing to meet the actual combat requirements, poor effect and low efficiency, an integrated simulation exercise system of high-temperature smoke for coal mine emergency rescue is established, which mainly includes the high-temperature smoke control system and the smoke control system. The high temperature drill control system adopts PLC control temperature algorithm and introduces PID fuzzy control rule. The simulation results show that high accuracy of temperature and flue gas concentration can be achieved through PLC fuzzy PID control. The system can simulate the hot smoke environment and achieve the purpose of the actual exercise.

1. Introduction

At present, the main disasters in China's coal mines are the explosion disaster caused by coal and gas outburst and the fire smoke caused by spontaneous combustion of coal seams [1]. In order to adapt to the harsh environment of high temperature and heavy smoke in coal mines during the catastrophic period, high temperature must be carried out regularly. Dense smoke exercise training, but the current high temperature, thick smoke environment is the original means of construction, the effect is poor, and can not be effectively controlled according to needs. The survey found that the current disaster environment simulation control system generally has backward modeling methods [2], and the problem that smoke is difficult to control and cannot achieve rapid automation compensation. In order to improve the quality of emergency rescue training and improve the ability of ambulance personnel to adapt to high temperature and smoke environment, it is urgent to conduct in-depth research on the automatic control system of temperature and smoke in high temperature exercise smoke.

2. High temperature smoke drill roadway control system

The high-temperature smoke drill roadway control system consists of three parts: the roadway monitoring system, the data processing system and the roadway control system [3]. The data processing system mainly processes the data collected and provides data support for system management. The function of the roadway monitoring acquisition system is to send the detected parameters of the collected high-temperature exercise roadway to the central control system; the main function of the high-temperature exercise roadway control system is to control Equipment and equipment in the roadway guide the entire system.

The high-temperature smoke drill roadway shall have the conditions for implementing high-temperature exercises and smoke exercises, and shall also have a quick smoke exhausting device after the exercise, as follows:

Content from this work may be used under the terms of the Creative Commons Attribution 3.0 licence. Any further distribution of this work must maintain attribution to the author(s) and the title of the work, journal citation and DOI.
Published under licence by IOP Publishing Ltd

(1) High temperature disaster drills. During the high temperature disaster exercise, the temperature should be controlled at 60 °C, the insulation performance in the roadway should be intact, and the fan power should be controlled to control the temperature of the high temperature exercise lane.

(2) Smoke disaster drills. During the smoke disaster drill, the smoke machine is used as the smoke generating device in the roadway. The smoke emitted is smoke. It should be non-toxic, harmless, non-polluting, and slightly smaller than the air. It is controlled by the configuration software of the console. The concentration of smoke is controlled by the length of continuous smoking. Visibility in the training room is generally close to zero within 10 minutes. The smoke emitted by the smoke generating device is non-corrosive, non-toxic and free of residual gas. At the same time, the smoke generating device should have a multi-point timing smoke generating function.

(3) Ventilation and exhaust. After the exercise, the system quickly controlled the smoke by controlling the ventilation fan, and the smoke was discharged from the roadway through the exhaust port.

3. Disaster Simulation Control System

3.1. High temperature exercise control system

3.1.1 Heating method

Assume that the high temperature exercise roadway is 25 m long, 2 m wide and 3 m high. The total volume in the high temperature zone is 150 m^3. According to the design requirements, the indoor temperature should be raised from 20 °C to 60 °C within 30 min. The temperature in the roadway is 20 °C, and the heat required to heat up to 60 °C is

E=pvc=1.29×150×1 000×313=60 565 500 J≈16.82KW/h

Where: v —— the volume of air;

c——the specific heat capacity of the air;

p ——air density;

ΔT——Temperature difference .

Since the roadway is long and underground, the dissipation factor is 50% [4], the heating equipment efficiency is 90%, the heating time requirement is 30 min, and the total power required is 74.8 KW/h. According to the existing heating equipment, in order to meet the requirements of high efficiency and environmental protection, four 20 kw heating devices are selected to be placed in the four corners of the high temperature exercise roadway.

Temperature control adopts PLC-based solid state relay PID discrete control mode, as shown in Figure 1. The temperature in the high temperature zone is controlled at about 60 °C, and the control accuracy is within 1 °C [5]. Due to the PID discrete control method, the heating fan is always started. The setting parameters can be changed according to actual needs. When the temperature is maintained, the electric fan operates at half power, and the temperature is automatically kept within the required range. At the same time, the temperature of the monitoring point in the roadway and the temperature outside the high temperature zone are displayed on the display outside the high temperature zone.

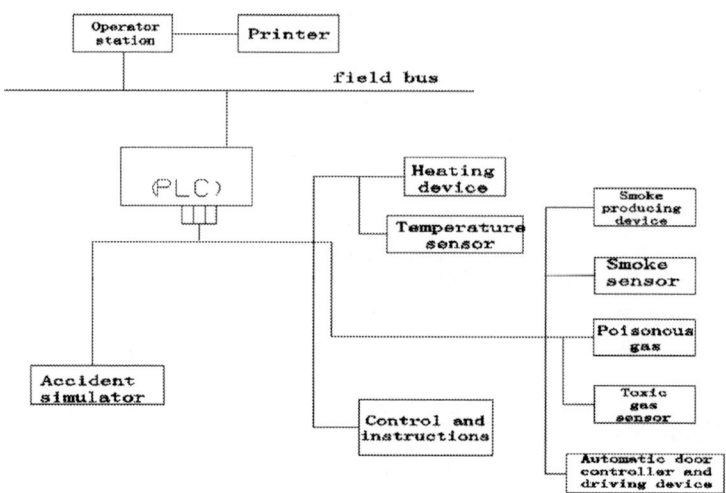

Fig.1 PLC control principle

3.1.2 Temperature adjustment algorithm

When the data measured by the temperature sensor is fed back to the host controller, the PLC controller is used to adjust the temperature of the roadway. The temperature variation rate, integral time and differential time in the roadway are controlled by linear combination. The autocorrelation control state equation of temperature sensing data in high temperature drills is calculated.

$$v(u) = L_q(f(u) + \frac{1}{T_J} \int_0^u f(u)\mathrm{d}u + T_E \frac{\mathrm{d}f(u)}{\mathrm{d}u} \tag{1}$$

Where: L_q——PLC logic programming control temperature change rate;

T_J——Integration time;

T_E——Differential time;

The exercise lane temperature control equation is expressed as

$$\Delta v(l) = \theta_q \Delta f(l) + \theta_j f(l) + \theta_e [\Delta f(l) - \Delta f(l-1)] \tag{2}$$

Where: $\Delta v(l)$——Adjust the output of the output.

$\Delta f(l)$——The error of current system temperature and target temperature.

$\Delta f(l-1)$——The error value of system temperature and target temperature for the last sampling.

θ_q——Tunnel temperature.

θ_j——Integral parameter of temperature acquisition in drill roadway.

θ_e——Differential parameters of temperature control in maneuver roadway.

By using the method of normal correlation state feature extraction and combining the advantages of automatic fuzzy matching of PID control regulation performance, a simple analytical control rule is introduced in fuzzy control theory, which can be expressed as follows:

$$U = aE + (1-a)E_C, a \in [0,1] \tag{3}$$

Where: U —— the amount of control;

E_C ——control volume adjustment factor;

a —— Configure weight coefficient

The fuzzy controller control deviation is compared. When the stability control deviation is large, the coarse adjustment is first performed. When the stability control deviation is small, the system is finely adjusted [6]. The temperature accuracy is maintained by the temperature adaptive adjustment of the high temperature exercise roadway. The control rules are shown in Table 1, and the temperature adaptive adjustment tracking curve is shown in Fig. 2.

Table 1. High temperature drill roadway temperature control rules

NB	NM	NS	ZO	PS	PM	PB	NB
NB	NS	ZO	NM	NM	NM	NS	NS
NM	NS	ZO	NM	NM	NS	ZO	PS
NS	NM	NM	NS	NS	ZO	PS	PM
ZO	NM	NS	NS	ZO	PS	PS	PM

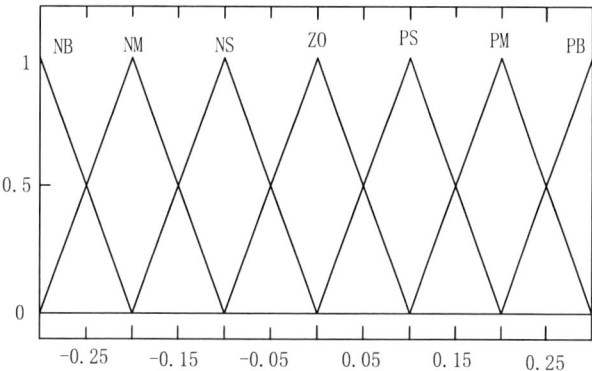

Fig. 2 Temperature control tracking curve

3.2. Smoke Exercise Control System

The smoke generation system consists of a smoke machine and a visibility detection sensor, and the smoke emitted by the smoking device is a non-toxic and harmless gas [7]. The concentration of smoke in the sides of the roadway is automatically controlled by the PLC. Assume that the length of the smoke lane is 25 m, and a smoke machine is placed every 5 m to divide the drill lane into five areas. Through the district to refine the flue gas compensation, each regional hood is responsible for the smoke concentration inside the area to maintain the visibility accuracy. Protection [8]. When the smoke concentration is maintained, the smoke concentration is set indirectly through the time of the smoke generation and the interval of the smoke generation.

3.3. Simulation

In order to verify the feasibility and effectiveness of the temperature fuzzy PID control algorithm, Simulink was used to simulate the temperature control system. First, enter the fuzzy open editor in the working window of Matlab, then add the input and output membership function to edit, select the triangle membership function, then add the control rules of the parameters, then compile the fuzzy controller in Matlab, and finally Establish a system simulation model, as shown in Figure 3.

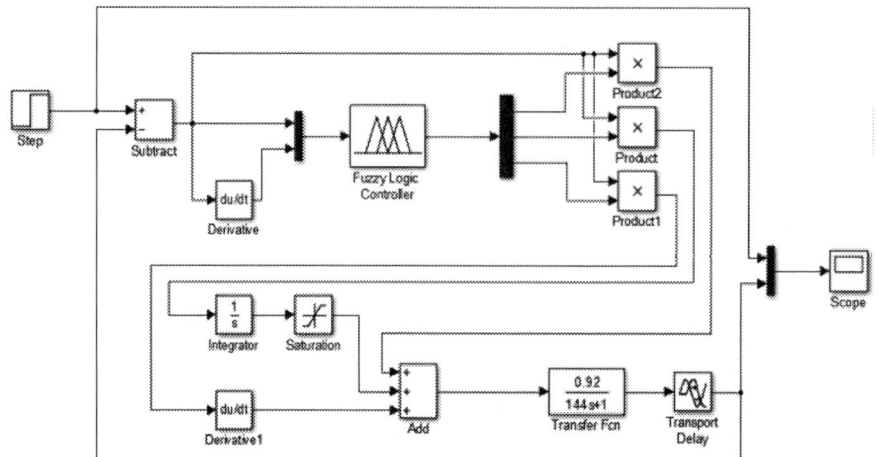

Fig.3 Simulation diagram of temperature control system

Adjusting the PID initial parameters yields a simulation curve as shown in Figure 4. As can be seen from Figure 4, the adjustment time is at most about 180 s, the overshoot is % = 0, and the steady state error is zero.

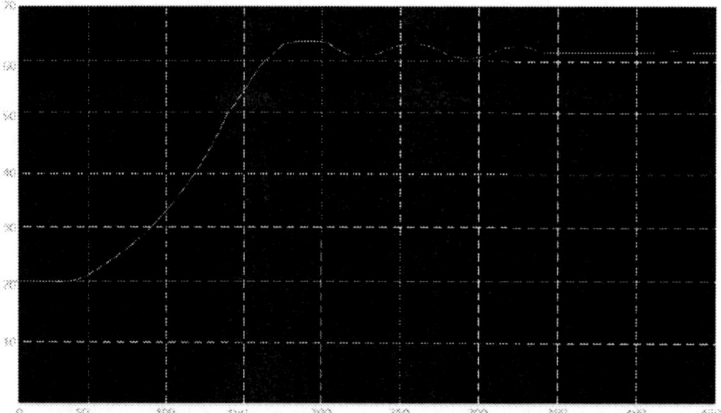

Fig.4 Response curve of temperature control system

The simulation results show that the fuzzy PID control method can achieve the performance indexes required for temperature control such as short adjustment time, overshoot and zero steady state error. It has better temperature control performance and meets the temperature control accuracy requirements of high temperature exercise roadway.

4. Conclusion

(1) The roadway disaster simulation control system of emergency rescue drill under high temperature and smoke environment mainly includes high temperature exercise control system and smoke exercise control system.

(2) Based on PLC fuzzy PID control, an automatic control system for controlling temperature smoke is proposed, and an algorithm based on PLC to adjust temperature and smoke is given.

(3) Temperature fuzzy PID control system can realize temperature control requirements with short mediation time, overshoot and zero steady state error.

(4) The system can realistically simulate the high-temperature smoke environment and achieve the purpose of actual combat exercises. However, the research on the simulation system of other disasters

and the mine rescue team training system, equipment and technology research and development, sensor feedback data accuracy, and standardization construction must be further studied and improved.

Reference
[1]   Li Xuecheng. China's coal mine safety [M]. Beijing: Coal Industry Press, 1998.
[2]   Wang Daoqing, Jia Qiwen, Tian de Yu. Mine rescue [M]. Xuzhou: China University of Mining and Technology press, 2002.
[3]   Liu Mao, Wu Zongzhi. Introduction to Emergency Rescue - Emergency Rescue System and Plan [C]. Beijing: Chemical Industry Publishing House, 2004.
[4]   Hu She-rong, Jiang Dacheng.Research status and Prevention Countermeasures of spontaneous combustion of coal seam[J].Chinese Journal of Geological Disaster and Prevention, 2000.11(4): 69-71.
[5]   Wang Jiefan, Li Wenjun. China Coal Mine Accidents and Expert Comments [M]. Beijing: Coal Industry Publishing House, 2001.
[6]   Yang Daming. Current situation of mine rescue in China [J]. contemporary miners, 2002. (4): 11-11
[7]   Wang Xian Zheng. New technology of coal mine safety [M]. Beijing: Coal Industry Press, 2002.
[8]   Wang Deming. Mine ventilation and safety [M]. Xuzhou: China University of Mining and Technology press, 2005.

Combination of CNN with GRU for Plate Recognition

Fucheng You[1], Yangze Zhao[2], Xuewei Wang[3]

[1]Beijing Institute of Graphic Communication. No. 1, Xinghua Street (two section), Daxing District, Beijing, China

[2]Beijing Institute of Graphic Communication. No. 1, Xinghua Street (two section), Daxing District, Beijing, China

[3]Beijing Institute of Graphic Communication. No. 1, Xinghua Street (two section), Daxing District, Beijing, China

1263108136@qq.com, 1263108136@qq.com, 1263108136@qq.com,

Abstract. License plate recognition has been a hot topic.Most of the existing license plate recognition solutions are mainly implemented through character segmentation and then recognition.However,these methods have been lacking in robustness and character segmentation has been a difficult problem to be solved perfectly.This thesis boils down the problem of character recognition to a problem of sequential learning.The convolution neural network is used for feature extraction to describe the high-level semantics of the image, and the GRU neural network is used as the sequence learning device to effectively model the internal relations of the sequence.Considering that the output sequence cannot be aligned with the input feature frame sequence, we use structured Loss.A background (Blank) category is also introduced to absorb the obfuscation of adjacent characters.The experimental training set of the paper is more than 10,000 plate data sets in the nearly real scene produced by human, and the test results of 99% accuracy can be achieved on hundreds of test sets.

1. Introduction

Automatic License Plate Recognition (ALPR) has been a frequent topic of research ((1), (2),(3)) due to many practical applications, such as automatic toll collection, traffic law enforcement, private spaces access control and road traffic monitoring.Traditional license plate recognition includes license plate character segmentation and character recognition.The main purpose of character segmentation is to segment the image of license plate character into individual characters to recognize it.(4)At present, the methods of character segmentation mainly include image vertical projection feature segmentation, matching algorithm based on template library and connected domain segmentation.The algorithm based on template library matching is too slow to meet the requirement because it needs to traverse every possible location.However, the connected domain algorithm has poor segmentation effect due to fuzzy characters, sticky character borders and close rivets.This paper combines CNN and GRU.We put the license plate recognition problem((5), (6),(7).(8),(9)) as a sequence modeling problem, through the use of sliding window slip license plate area, in order to each image of sliding window to extract the convolution features, so you can get the whole license plate image sequence of convolution characteristics, ((10),(11),(11),(13))then using RNN (Recurrent Neural Network) Network and CTC (Connectionist Temporal Classification) method to get the final no segmentation recognition as a result, the process is simple and high in accuracy under the condition of stability.The following sections will introduce the network model, data set, results and summary of this article in turn.

Content from this work may be used under the terms of the Creative Commons Attribution 3.0 licence. Any further distribution of this work must maintain attribution to the author(s) and the title of the work, journal citation and DOI.

Published under licence by IOP Publishing Ltd

2. Model Architecture

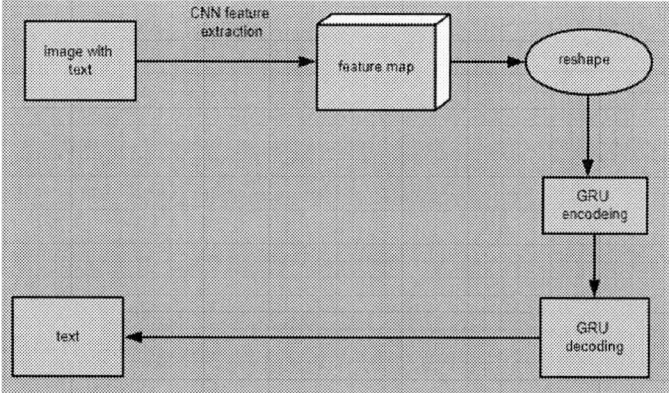

Figure 1 Model Architecture

As is depicted above figure 1,Firstly, image is fed to CNN to extract image features. The next step is to apply Recurrent Neural Network to these features followed by the special decoding algorithm. This decoding algorithm takes LSTM outputs from each time step and produces the final labeling.The detailed architecture will be described as below.

2.1. Feature Extraction

Model receives source image and extract image features by CNN.CNN produces tensor with shape 32*16*16.Due to equipment limitations and lack of data, we only designed smaller ones,including only two convolutional layer and two maxpooling layer.The feature map height equals 32,width equals to 8,and the depth is 16.

2.2. Reshape

Now we do reshape operation.It depends on different situations.Because we have to apply the feature to FC (fully connected layer) so we have to make feature match FC layer. we apply fully connected layer followed by softmax layer and get the vector of 32 elements. This vector contains probability distribution of observing alphabet symbols at each LSTM step.The choice will depend on the specific task.

2.3. Encoding

Encoding process transform feature vector into probability distribution.We apply GRU as encoding algorithm.the input feature vector is 32*32 and each vector of 32 elements is fed in GRU and outputs vector of 512 elements.So final shape of the early stage of GRU is (32,512).And we feed the reshape vector into two identical GRU and add them to 1 dimension.After that we feed the addition result into another two same GRU network.This time we concatenate the two vectors of shape (32, 512).So we get output (32,1024).The final probability distribution of observing alphabet symbols will be got by FC and activation layer.

2.4. Decoding

Decoding process is pretty simple.On the above diagram we have eight vectors of probabilities at each LSTM time step. Let's take most probable symbol at each time step. As a result we obtain the string of eight characters—one most probable letter at each time step. Then we have to glue all consecutive repeating characters into one. In our example two "e" letters are glued to single one. Special blank character allows us to split symbols that are repeated in the original labeling. We added blank symbol to the alphabet to teach our neural network to predict blank between such case symbols. Then we remove all blank symbols.The detailed architecture is displayed as figure 2.And the result is as figure 3.

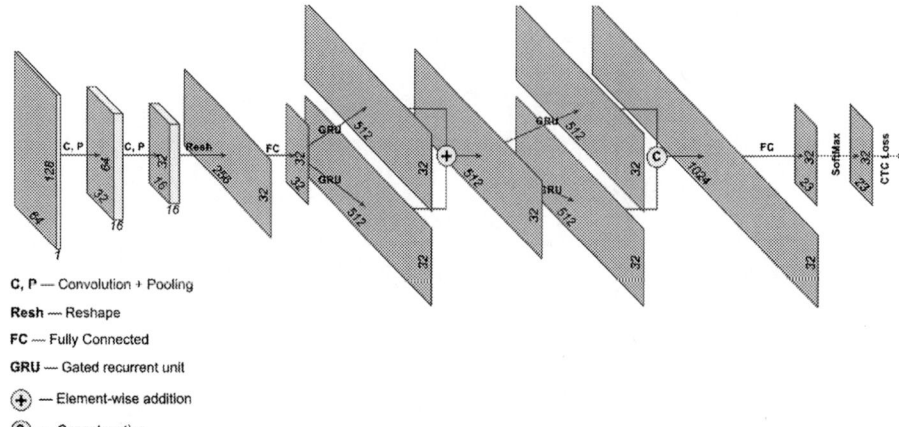

C, P — Convolution + Pooling

Resh — Reshape

FC — Fully Connected

GRU — Gated recurrent unit

⊕ — Element-wise addition

ⓒ — Concatenation

Figure 2 Detailed Architecture

Predicted: T617ME73
True: T617ME73

Figure 3 result sample

3. Dataset

On paper the experimental training sets Supervisely more than ten thousand copies of artificial approximate real scenarios license plate data sets, the style of the data set has been shown above.There are many Chinese characters in China, and it takes time and energy to arrange and obtain license plates. For convenience, the data set is made by taking foreign license plates as an example, namely the combination of English and Numbers.The experimental results of this paper have achieved 99% accuracy on hundreds of test sets.The images look like figure 4.

Figure 4 plate

4. Result

The data set used in this experiment is artificial data set, so it is not compared with other algorithm results.The results obtained in the test set reached 99% of the test accuracy and achieved real-time monitoring effect, about 200ms per image.

5. Conclusion

This paper discusses the combination of CNN and GRU and considers the license plate recognition problem as a series modeling problem.Slip license plate area, through using the sliding window for each image of sliding window to extract the convolution features, so you can get the whole license plate image sequence of convolution characteristics, and then will get maps into LSTM encoder to encode, probability distribution vector collection to get letters, no segmentation to identify the final result by decoder implementation, process simple, and in the case of high accuracy remained stable.Finally, the test set shows superior results, indicating that the network generalization presented in this paper is good.

The future work mainly focuses on the use of Faster Rcnn for vehicle license plate detection in deep reference learning. Under the condition that the recognition accuracy remains unchanged, the location of the license plate in the image is detected.In order to make

After the results is more persuasive, experimental data set will attempt to adopt the License Plate detection in the authoritative SSIG License Plate Character Segmentation Database data set.At the same time, in order to achieve better generalization effect and reduce computing power, we use the migration learning method to directly extract features under the picture data set of real scenes, and only train the GRU encoding and decoding process after feature extraction.

References

[1]   S. Du, M. Ibrahim, M. Shehata, and W. Badawy, "Automatic license plate recognition (ALPR): A state-of-the-art review," IEEE Transactionson Circuits and Systems for Video Technology, vol. 23, no. 2, pp. 311–325, Feb 2013.

[2]   C. Gou, K. Wang, Y. Yao, and Z. Li, "Vehicle license plate recognition based on registrations and restricted Boltzmann machines," IEEE Transactions on Intelligent Transportation Systems, vol. 17, no. 4, pp.1096–1107, April 2016.

[3]   O. Bulan, V. Kozitsky, P. Ramesh, and M. Shreve, "Segmentation and annotation-free license plate recognition with deep localization andfailure identification," IEEE Transactions on Intelligent Transportation Systems, vol. 18, no. 9, pp. 2351–2363, Sept 2017.

[4]   Amin Safaei,Hongying L. Tang,Saeid Sanei. Real-time search-free multiple license plate recognition via likelihood estimation of saliency[J]. Computers and Electrical Engineering,2016,56.

[5]   Tejendra Panchal,Hetal Patel,Ami Panchal. License Plate Detection Using Harris Corner and Character Segmentation by Integrated Approach from an Image[J]. Procedia Computer Science,2016,79.

[6]   C. Gou, K.-F. Wang, Y.-J. Yao et al., "Vehicle license plate recognition based on extremal regions and restricted Boltzmann machines", IEEE Trans. Intell. Transp. Syst., vol. 17, no. 4, pp. 1096-1107, 2016.

[7]   O. Bulan, V. Kozitsky, P. Ramesh, M. Shreve, "Segmentation-and Annotation-Free License Plate

[8]   Recognition With Deep Localization and Failure Identification", IEEE Transactions on Intelligent Transportation Systems, no. 99, pp. 1-13.

[9]   Samuel A. Babatunde. Review of strengthening techniques for masonry using fiber reinforced polymers[J]. Composite Structures,2017,161.

[10]   Jaber Shabanian,Pierre Sauriol,Jamal Chaouki. A simple and robust approach for early detection of deflui dization[J]. Chemical Engineering Journal,2017,313.

[11]   Khalid Aboura,Rami Al-Hmouz. An Overview of Image Analysis Algorithms for License Plate Recognition[J]. Organizacija,2017,50(3).

[12]   Martijn H.J. Hulsmans,Mark van Heijl,R. Marijn Houwert,Tim K. Timmers,Ger van Olden,Egbert Jan M.M. Verleisdonk. Anteroinferior versus superior plating of clavicular fractures[J]. Journal of Shoulder and Elbow Surgery,2016,25(3).

[13]   Yu Wang,Xiaojuan Ban,Jie Chen,Bo Hu,Xing Yang. Corrigendum to "License plate recognition

based on SIFT feature" [Optik 126 (2015) 2895–2901][J]. Optik - International Journal for Light and Electron Optics,2016,127(1).

Methods to Solve Salt &Pepper Noise, and Frame Dropping of Timed Address Event Representation Vision Sensor

Lu YU[1], Zhonghe Chen[1], Yun HAO[1]

[1]Zhonghuan Information College Tianjin University of Technology, Tianjin, 300380, China

E-mail: yuluyouxiang@sina.com

Abstract: Address event representation (AER) vision sensor which only outputs visual information of a pixel with light intensity change, eliminates redundant information radically. Compared with the traditional vision sensor, it has the advantages of high frame rate, low data volume and high dynamic range Firstly, this paper studies the method of eliminating salt & pepper noise in AER vision sensor. By design of arbitrator, the salt and pepper noise can be removed directly from the source through 5×5-nearest neighbor algorithm and cross window-nearest neighbor algorithm. This method not only reduces the output data and preserve the details of the image, but also removes the salt and pepper noise effectively, Secondly, this paper solves the problem of "frame dropping" caused by the time delay and the non-synchronization of ON/OFF events and quantization process. The image is corrected by moving vector which is obtained by Mean Shift target tracking algorithm, and then cross window-nearest neighbor algorithm is used to remove noises. "Filling frame" is completed and a better image is obtained.

1. Introduction

Traditional vision sensor takes "frame" as the basic unit of image, which cannot meet the requirements of high frame rate, low data volume, high dynamic range and high transmission rate[1-2]. Address event representation (AER) vision sensor based on bionic vision, which only outputs visual information of a pixel with light intensity change, eliminates redundant information radically[3-4]. It uses asynchronous output and sparse representation to obtain images with high frame rate, low data volume and high dynamic range. According to advantages above, AER vision sensor is especially suitable for high-speed target shooting and tracking, target recognition, acquisition of high resolution image and other fields[5-6].

AER vision sensor is also defective. In the image follow-up processing, the image information acquired by AER vision image sensor needs to be transformed into "frame" for processing, which causes the redundancy eliminated at the source to be restored in the process of image processing，Moreover, arbitrator is needed due to the centralized output of the data conflict. This results in the shielding delay in quantization process and the "frame dropping" problem[7].

2. The structure and work flow of AER vision sensor

Figure 1shows the structure of AER vision sensor. It is composed of pixel units, row/column control and arbitrator, address/event encoder, external image processing and data transmission control unit[8].

Content from this work may be used under the terms of the Creative Commons Attribution 3.0 licence. Any further distribution of this work must maintain attribution to the author(s) and the title of the work, journal citation and DOI.

Published under licence by IOP Publishing Ltd

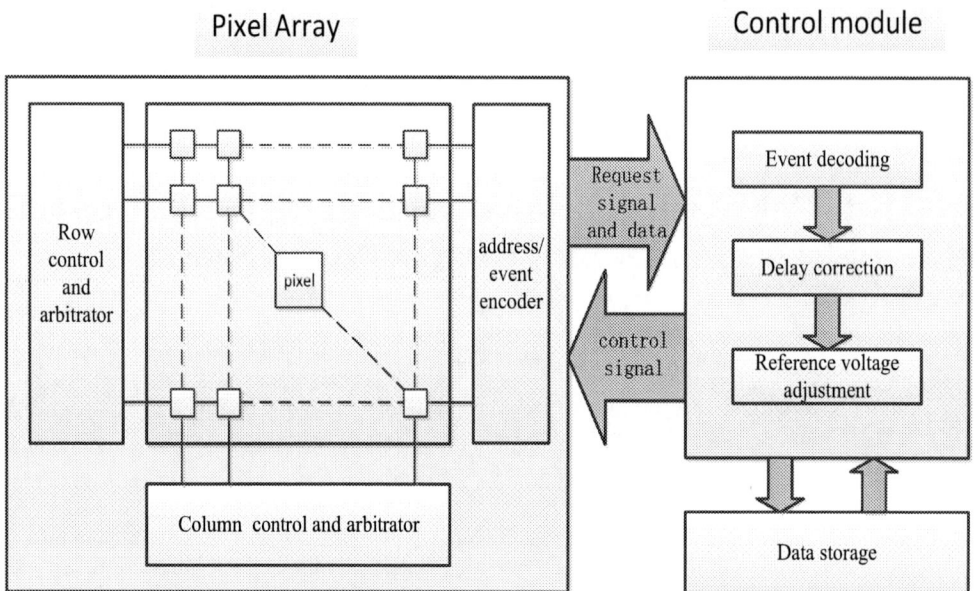

Fig.1 The structure of AER vision sensor

Figure 2shows the structure of a pixel unit, which consists of a CD (change detector) unit and a double sampling PWM circuit unit. CD unit can sense the change of light intensity. Double sampling PWM circuit unit quantizes the light intensity[9].

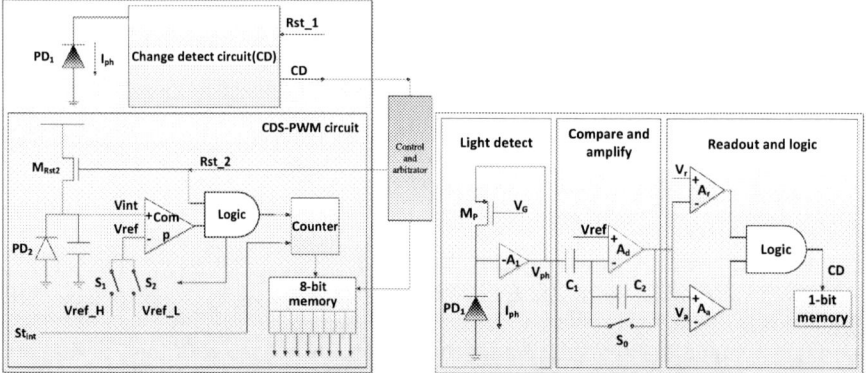

(a)The structure of a pixel   (b)The structure of CD
Fig.2 The structure of a pixel unit

2.1Light detect

When the light intensity changes, the CD unit in a pixel judges whether the light intensity changes beyond the threshold set. If the change of light intensity exceeds threshold, time pulses will be generated. If it does not exceed the threshold, the pixel will not output any data and return to ready state. It is the foundation which AER vision image sensor generate "frameless" output based on.

2.2 Row/column control and arbitrator

The event pulse sends a request signal to the row arbitrator. When the arbitrator confirms that the row is selected, it returns the response signal. If row response signal is received, all pixels on the row emit a column request signal RH (ON event, intensity-enhanced event) or signal RL (OFF event, intensity-weakened event) to the column   arbitrator, according to different event type.

If there is only one column request signal emitted, the pixels in the column are directly selected and output quantization of light by CD unit to address/event encoder. When there are several column request signals emitted at the same time, the column arbitrator will choose only one pixel, and transmits response signal to make it output quantization of light, while the pixels in the other column are in standby mode waiting next arbitration. Then address/event encoder encodes the address of the pixel and quantization of light, and also gives time stamp on it. In this way, the address, light intensity and time stamp of a pixel is gotten and are emitted to control module.

2.3Quantization of light in a pixel
As shown in the figure2, when a pixel gains control of the bus, the double-sampling PWM circuit unit generates two pulses according to different light intensity. The time between the pulses is proportional to the current of light intensity. The counter quantizes the time difference and saves the quantization results in the pixel memory waiting for output.

2.4Output and processing of quantized data
The quantized data, 8 bit, are send to control module, regarded as a whole with row/column addresses and time stamps.

In conclusion，AER vision sensor does not have the concept of "frame" to sample image information. Each pixel works independently and outputs asynchronously and serially, eliminating redundant information from the source.

However, the image processing of AER vision sensor always restores it back to the "frame" image, which does not take full advantage of its sparse representation and low data volume.

3.   Eliminating salt & pepper noise
Salt & pepper noise is a form of noise sometimes seen on images. It is also known as impulse noise. This noise can be caused by sharp and sudden disturbances in the image signal. It presents itself as sparsely occurring white and black pixels. An effective noise reduction method for this type of noise is a median filteror a morphological filter.For reducing either salt noise or pepper noise, but not both, a contra harmonic mean filter can be effective.

Rely on the ON/OFF event is generated by CD unit in AER visual image sensor, the arbitrator can judge whether there is salt & pepper noise in the pixel. If yes, double-sampling PWM circuit unit is not needed to quantize and there is no output. And the quantization of light intensity is still the value before. If no, double-sampling PWM circuit unit quantizes light intensity and outputs it. And the quantization of light intensity is current data.

In this paper, two methods are used to determine whether the pixels of the event output by arbitrator to remove salt and pepper noise from the source.

3.1 5×5-nearest neighbor algorithm
When one pixel (i, j)sends a ON/OFF event request signal to row and column arbitrators, the arbitrators statistics the number of events of 16 pixels in the periphery of the 5×5pixels, which is yellow pixels in Figure 3.

If the number of pixels occurred ON/OFF event is less than or equal to 8, no matter how much ON/OFF events occur in the 3×3 nearest neighbor, which is green pixels in Figure 3, the pixels in the 3×3 nearest neighbor and itself don't output data. The quantization of light is still as same as before in the external register. In other cases, the pixels in the 3×3 nearest neighbor and itself output data, and then the quantization of light is replaced by output data of CD unit at this moment.

i-2,j-2	i-2,j-1	i-2,j	i-2,j+1	i-2,j+2
i-1,j-2	i-1,j-1	i-1,j	i-1,j+1	i-1,j+2
i,j-2	i,j-1	i,j	i,j+1	i,j+2
i+1,j-2	i+1,j-1	i+1,j	i+1,j+1	i+1,j+2
i+2,j-2	i+2,j-1	i+2,j	i+2,j+1	i+2,j+2

Fig.3Pixel position in 5×5-nearest neighbor algorithm

3.2 Cross window-nearest neighbor algorithm

When one pixel (I , j)sends a ON/OFF event request signal to row and column arbitrators, the arbitrators statistics the events' number of up, down, lift, and right of the pixel. If the number of pixels occurred ON/OFF event is less than or equal to 3,the pixels in cross window-nearest neighbor and itself don't output data. The quantization of light is still as same as before in the external register. In other cases, the pixels in the cross window-nearest neighbor and itself output data, and then the quantization of light is replaced by output data of CD unit at this moment.

3.3 Simulation result

Figure 4 shows the simulation result of elimination of salt and pepper noise. Fig.4(a)shows the first frame without noise.Fig.4(b)shows the second frame with salt and pepper noise. In order to simulate the process of AER vision image sensor, the image of second frame is obtained by translating the first frame along - 45 degrees, and adding salt and pepper noise with a noise coefficient of 0.1. Fig.4(c)shows the image after median filter from the second frame.Fig.4(d)shows ON/OFF event output of the second frame in AER mode, which represented as the difference between the first frame and the second frame. When there is no event to output, zero is stored. When there is ON/OFF event to output, one is stored.Fig.4(e)shows the second frame gotten by cross window-nearest neighbor algorithm. Fig.4(f)shows the second frame gotten by 5×5-nearest neighbor algorithm.

When ON/OFF event rate is 46.19%, the output rate in median filter is 100%, in cross window-nearest neighbor algorithm is 0.93%, in 5×5-nearest neighbor algorithm is 12.85%.

Fig.4Simulation results of eliminating salt and pepper noise

Compared with median filter for image acquisition only processing a single image, the two methods proposed in this paper are based on two consecutive frames. As shown in Fig.4, the two methods provide better image quality, especially in maintaining the details of the image. Furthermore, cross window-nearest neighbor algorithm depresses noise better than 5×5-nearest neighbor algorithm, and outputs less data.

4. "Frame dropping" caused by the time delay and its solution

Part 2 of this paper describes the work flow of AER visual image sensor. It can be found that a pixel can't quantize light after it sent ON/OFF event signal until response signal returns by the arbitrator. When the intensity of light changes strongly, there are much more pixels sending request signal. The beginning time of quantizing process is much later than ON/OFF event, caused by waiting response signal returned by the arbitrator. In this way, the quantification of light intensity isn't in the time ON/OFF event occurring, but in the time response signal returning. As a result, there will be "losing frames" problems caused by time delay. This problem is particularly prominent in high-speed moving objects shooting and large pixel array vision image sensors.

AER vision image sensor only outputs visual information of a pixel with light intensity change, so there is no "frame" concept."Frame dropping" refers to the value of the next light intensity quantization cycle replacing the value in present moment.

Figure5 shows "frame dropping". Fig.5(a)-(c) are c images without time delay in which the car is moving down the right side unknown its angle and speed. When "frame dropping" happens in the second frame, some light intensity of its pixels is replaced by the intensity of the third frame. As shown in Fig.5(d), it is evident that there are distinct noise points similar to salt and pepper noise on moving objects, and these points are mainly concentrated in areas with more details.

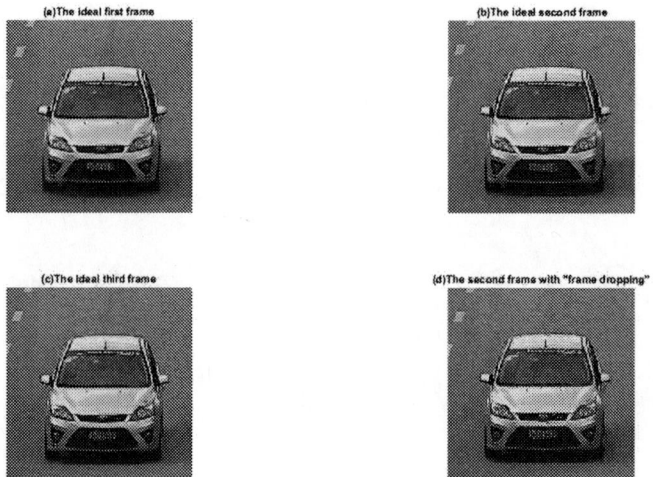

Fig.5"Frame dropping"

Because of the correlation between the frames of AER vision sensor, the method of video repair can be used to solve the problem of frame dropping. The algorithm is shown in Figure 6.

1.Mean Shift target tracking algorithm[10] is used to process the first and second frames to obtain the moving vectors including the direction and distance of the moving object.

2. Moving and clipping the first frame image according to the moving vector obtained in the previous step, the first frame image is corrected.

3.Cross window-nearest neighbor algorithm is used to the new first and second frames, and "frame dropping" reduces just like the way eliminating salt & pepper noise.

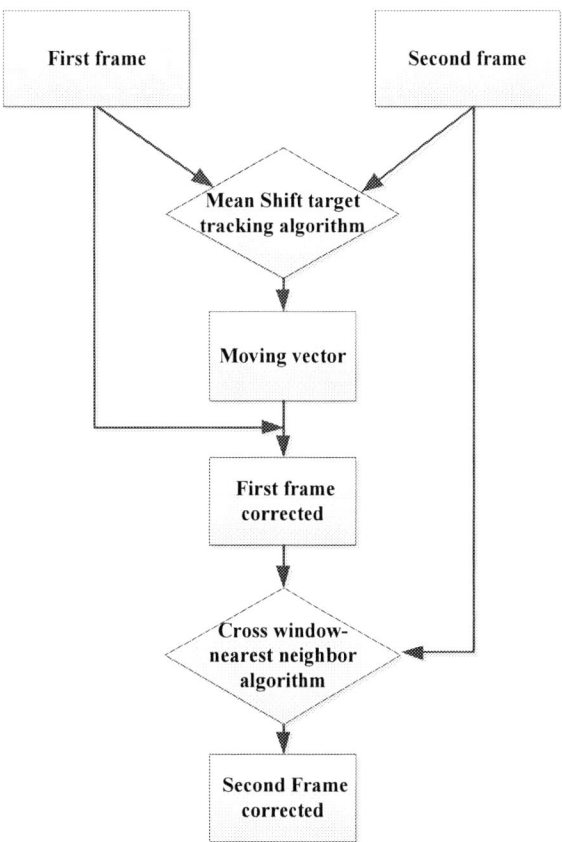

Fig.6 Algorithm of "frame dropping" correction

Figure7 shows the simulation result in reducing frame dropping. Fig.7(a) shows the second frame with frame dropping, where there are several blurred target contour and multiple noise points.Fig.7(b) shows the simulation result of mean filter. The new light intensity in the second frame is average value of the initial second frame, the new first frame and the new third frame corrected by Mean Shift target tracking algorithm. It can be clearly seen that, because the moving vector is not necessarily 100% accurate, there will be a problem of ghosting image (part of dotted line is obvious). At the same time, the noise point has been improved, but the improvement effect is not obvious.Fig.7(c) shows the simulation result of the new second frame by cross window-nearest neighbor algorithm after the new first frame corrected by Mean Shift target tracking algorithm. The problem of ghosting image is eliminated effectively, further more frame dropping is corrected well.

In addition, the algorithm to get moving vectors in this paper is Mean Shift target tracking algorithm. Other algorithm can also be used to obtain moving vectors. The key is acquiring exact moving vectors to correct the first frame.

Fig.7 The simulation result in reducing frame dropping

5. Conclusion

This paper is based on the research of AER vision image sensor. Firstly, this paper studies the method of eliminating salt & pepper noise in AER vision sensor. By design of arbitrator, the salt and pepper noise can be removed directly from the source through 5×5-nearest neighbor algorithm and cross window-nearest neighbor algorithm. This method not only reduces the output data and preserves the details of the image, but also removes the salt and pepper noise effectively. Secondly, this paper solves the problem of "frame dropping" caused by the time delay and the non-synchronization of ON/OFF events and quantization process. The image is corrected by moving vector which is obtained by Mean Shift target tracking algorithm, and then cross window-nearest neighbor algorithm is used to remove noises. "Filling frame" is completed and a better image is obtained.

References

[1] Belbachir A N, Hofstatter M, Litzenberger M, et al. High-Speed Embedded-Object Analysis Using a Dual-Line Timed-Address-Event Temporal-Contrast Vision Sensor[J]. IEEE Transactions on Industrial Electronics, 2011, 58(3):770-783.

[2] Belbachir A N, Hofstätter M, Litzenberger M, et al. High-precision shape representation using a neuromorphic vision sensor with synchronous address-event communication interface[J]. Measurement Science & Technology, 2009, 20(10):104007.

[3] Posch C, Matolin D, Wohlgenannt R. A QVGA 143 dB dynamic range frame-free PWM image sensor with lossless pixel-level video compression and time-domain CDS [J]. IEEE Journal of Solid-State Circuits, 2011, 46(1):259-275.

[4] Posch C, Matolin D, Wohlgenannt R. High-DR frame-free PWM imaging with asynchronous AER intensity encoding and focal-plane temporal redundancy suppression [C]. IEEE International Symposium on Circuits and Systems, Paris, 2010: 2430 - 2433.

[5] Posch C, Matolin D, Wohlgenannt R. High-DR frame-free PWM imaging with asynchronous AER intensity encoding and focal-plane temporal redundancy suppression [C]. IEEE International Symposium on Circuits and Systems, Paris, 2010: 2430 - 2433.

[6] Matolin D, Wohlgenannt R, Litzenberger M, et al. A load-balancing readout method for large event-based PWM imaging arrays [C]. Proceedings of 2010 IEEE International Symposium on Circuits and Systems (ISCAS), Paris, 2010: 361 – 364.

[7] J Xu，D Li，S Yao. A Time Error Correction Method Applied to High-Precision AER Asynchronous CMOS Image Sensor [J]. Journal of Signal Processing Systems, 2014, 75(1):1-13.

[8] Yu Lu, Yao Suying, Xu Jiangtao. An Implementation Method of Real-Time Vision Sensor Based on Address Event Representation [J]. Acta Optica Sinica, 2013(1):251-257.

[9] Yu L, Hao Y, Chen Z, et al. A Time Error Model for Correlated Double Sampling PWM Pixel[C]//

International Conference in Communications, Signal Processing, and Systems. Springer, Singapore, 2016:669-679.

[10] Jiangtao Xu, Mengxing Zhang, Shi Yan, et al. A Method to Solve the Side Effects of Dual-Line Timed Address Event Vision System [J]. Journal of Circuits Systems & Computers, 2015, 24(03):1550028.

ISPECE   IOP Publishing

A Double-channel iterative NFXLMS algorithm used in Horizontal Vibration Isolation

ZHANG Chi[1], YIN WenSheng[2]

[1]The department of Mechanical Engineering, Tsinghua University, Beijing, China

[2]The department of Mechanical Engineering, Tsinghua University, Beijing, China

yinws@tsinghua.edu.cn

Abstract. In order to solve the multi-directional coupling problem of the vibration isolation platform caused by the partial load, a novel adaptive feed-forward control method is proposed. In the paper, we mainly used the Double-channel iterative NFXLMS algorithm to get a better performance of vibration isolation. And the control system is not sensitive to the error from the model identification. Then the algorithm is validated by the simulation and experiments. The results show that the performance of the vibration isolation system is mainly improved in the low frequency.

1. Introduction

1.1. Introduction

The performance of the precision motion control systems, such as lithography machines, are dramatically influenced by the external caused vibrations [1]. In order to preserve the high accuracy of these systems, precision vibration isolation system is widely used. And the active vibration isolation system can use sensors and controllers to achieve a better vibration isolation performance in the wide frequency band [2] [3]. In the field of vibration isolation, the most commonly adopted control method is the feedforward algorithm and feedback algorithm. Various methods such as PID control [4], adaptive control [5], neural network [6], and Data-driven algorithm [7] have been applied in the active vibration isolation system to attenuate vibration. But most of these methods focus on the vertical vibration isolation.

Metal springs are widely used in vibration isolation platforms for its capture of small size, simple structure and low cost. On the other hand, the horizontal stiffness of the metal spring varies with the amount of compression [8], so that the control model change with it and are coupled with each other, which can cause the traditional model-based control methods fail. And the Non-model based algorithm may be too complex with high computational cost.

Based on the results of the horizontal dynamics modeling of the isolation platform, a novel adaptive feedforward control is proposed, using the FXLMS algorithm with dynamic objective function and variable step size. The algorithm not only can better adapt to the difference of system identification between the isolation modules, but also solve the problem of coupling.

1.2. Dynamics model of the platform

The simplified model of the existing metal spring isolator is shown in the Figure1. The platform is connected to the base by four modules. And each module is supported by a metal spring and equipped

---

Content from this work may be used under the terms of the Creative Commons Attribution 3.0 licence. Any further distribution of this work must maintain attribution to the author(s) and the title of the work, journal citation and DOI.

Published under licence by IOP Publishing Ltd

with two motors that controls horizontal and vertical active control. Vibration is transmitted from the spring, forming the Primary Channel, and the motor model is the Secondary Channel.

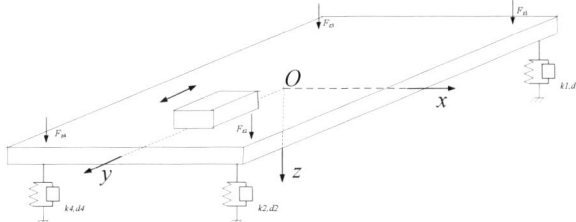

Figure 1. The simplified structure of isolation platform.

Analyzing the x-direction movement of a random point A on the vibration isolation platform, its velocity can be expressed as follows

$$\dot{x}_A = \dot{x}_o + r_{oA} \times \dot{\gamma} \tag{1}$$

Where $\dot{x}_o$ is the x-direction velocity of center of the rotation. $\dot{\gamma}$ is the rotational velocity of the platform around the Z axis. r_{oA} is the distance from point A to the center of rotation. The second term in the equation is the component of the platform's rotational velocity in the x direction. When the partial load exists, the center of rotation and the center of mass do not coincide. And then the difference in the secondary channel of each module is amplified, which introduces the rotation coupling. Therefore, it is necessary to consider both the rotation coupling and the x-direction vibration for the control.

The dynamic model of the vibration isolation system in the horizontal direction can be established according to H. X. Zeng [5]. Moreover, the transfer function of secondary channel of the horizontal x direction and the coupling channel between the x direction and the γ direction are both second-order models similarly, as the following form

$$S(z) = \frac{a_2 s + b_2}{a_1 s^2 + b_1 s + c_1} \tag{2}$$

2. Design of Feedforward Control for the platform

2.1. FXLMS algorithm
The structure of the FXLMS feedforward algorithm applied in the field of vibration isolation [9] is shown in the Figure 2.

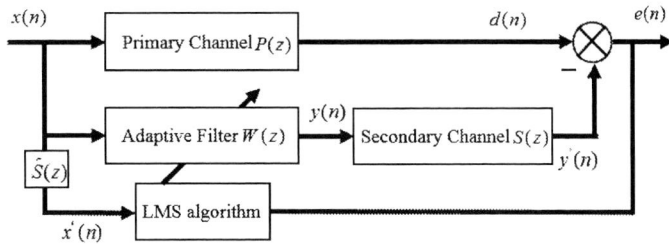

Figure 2. The structure of LMS algorithm in vibration isolation.

The main method is to add an adaptive FIR filter W(z) in front of the secondary control channel $S(z)$, so the control signals $y'(n)$ can be formulated as follows:

$$y'(n) = W(z) \cdot x(n) \cdot S(z) \tag{3}$$

Each parameter of the filter W(z) can be adjusted online by the LMS algorithm to minimize the output vibration of the vibration platform after the feed-forward control. The platform vibration output is the result of control signal combined with the vibration transmitted by the primary channel. The objective function optimized by the LMS algorithm is the 2-norm of the vibration output, as shown in equation (4)

$$J = \frac{1}{2} e^2(n) = \frac{1}{2}\left(d(n) - y'(n)\right)^2 \tag{4}$$

It is very difficult to solve the optimal filter coefficients of minimize the objective function directly. Therefore, the main method is to dynamically update the filter coefficients by the steepest descent method. Also it has been proved in the related literature[10] that as long as the phase error of the secondary channel model is not up to 90°, the identification accuracy of the secondary channel does not affect the convergence of the FXLMS algorithm.

2.2. Double-channel iterative NFXLMS algorithm
We introduce a double iterative FXLMS algorithm to solve the coupling problem, and the control structure block diagram is shown in Figure 3.

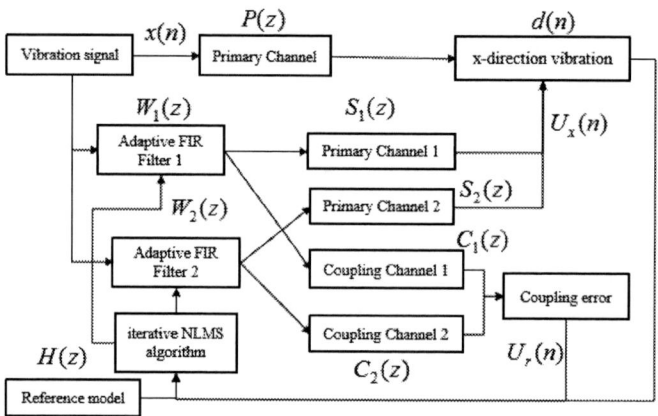

Figure 3. The structure of Double-channel iterative NFXLMS algorithm

For the horizontal x-direction vibration, the active control is mainly implemented by two motors in a diagonal position of the platform. Because of the influence of the partial load, the motor signal control of each module not only controls the x-direction vibration of the centroid, by the Secondary channel $S_1(z)$ $S_2(z)$, but also introduces the rotational coupling, by which we called Coupling channel $C_1(z)$ $C_2(z)$. Then the signal that controls the movement of the centroid and that introduces the rotational coupling are formulated as

$$U_x(n) = x(n) \cdot W_1(z)S_1(z) + x(n) \cdot W_2(z)S_2(z) \tag{5}$$

$$U_x(n) = x(n) \cdot W_1(z)S_1(z) + x(n) \cdot W_2(z)S_2(z) \tag{6}$$

Thus, the x-direction vibration of the platform is $e_x = d(n) - U_x(n)$. And the rotation coupling error of the platform is $e_\gamma = U_\gamma$. The vibration e_x is selected as the optimization objective function of the FIR filter No. 1, thus:

$$J_1 = \frac{1}{2} e_x^2(n) = \frac{1}{2}\left(d(n) - U_x(n)\right)^2 \tag{7}$$

The design of adaptive FIR filter No. 1 is mainly used to suppress the x-direction vibration signal. , By using the variable adaptive convergence factor, the design of FIR filter No. 2 aims to make the objective function as the following form

$$J_2 = \frac{1}{2} e_2^2(n) = \frac{\sigma_x}{2} e_x^2(n) + \frac{\sigma_\gamma}{2} e_\gamma^2(n) \tag{8}$$

In the equation, σ_x, σ_y are the proportional components of the rotation coupling and the x-direction vibration in the objective function. And σ_x, σ_y dynamically change with the vibration.

The FXLMS algorithm uses a reference model for the control channel, and it has been mentioned previously that the coupling channel has an approximate form to the transfer function of the secondary channel. So we set the reference function in the same second-order model $H(z)$. Then the two filter convergence factors were designed separately using the NLMS algorithm with iterative variable step size μ_1 μ_2.

$$W(n+1) = W(n) - \frac{\mu}{2}\nabla J(n) = W(n) + \frac{\mu}{x(n)x(n)^T + \delta}e(n)x(n)H(z) \tag{9}$$

According to the fact that the noise in the vibration system is irrelevant, the autocorrelation estimation can be introduced to eliminate the noise interference. The specific method for updating the convergence factor $\mu_1(n)$ of the No. 1 Filter is as follows

$$p(n) = \alpha p(n-1) + \beta e_x(n-1)e_x(n) \tag{10}$$

$$\mu_1(n) = \gamma \mu_1(n-1) + (1-\gamma)p^2(n) \tag{11}$$

Where $p(n)$ is the autocorrelation estimation of the x-direction vibration of the platform, each parameter should be satisfied $0 < \gamma, \alpha, \beta < 1$. Therefore, a fast convergence of the step size factor can be achieved, and the steady-state error is small as well.

For the No. 2 adaptive channel, the instantaneous ratio of the rotational coupling and the translational vibration $\theta = e_y / e_x$ is calculated using the online method, and the convergence factor $\mu_2(n)$ is formulated as follows

$$\mu_2(n) = \gamma \mu_2(n-1) + (1-\gamma)p^2(n)(1-\theta+\theta^2) \tag{12}$$

Then the objective function J_2 of the No. 2 Filter is:

$$J_2 = \frac{1-\theta}{2}e_x{}^2(n) + \frac{\theta}{2}e_y{}^2(n) \tag{13}$$

When the ratio θ is larger, the coupling angle error is larger. And then the control component of the rotation in the objective function get a greater impact, so that a better fast convergence can be achieved. And when the power ratio θ is small, the controller can concentrate on the x-direction vibration.

Through the design of the above two FIR adaptive filters, the platform gets a better performance with smaller rotation coupling error and smaller x-direction vibration at the same time.

3. Control Simulation and Experiment

3.1. Simulation in Matlab

The algorithm is simulated in Matlab and Simulink. The both adaptive FIR filters are designed in 3 orders. And the Secondary channel $S(z)$, the Primary channel $P(z)$, and the actual identification channel $H(z)$ all adopt a second-order model. The models are set as follows

$$S(z) = \frac{4es}{ms^2 + 4cs + 4k} \quad P(z) = \frac{4cs + 4k}{ms^2 + 4cs + 4k} \quad H(z) = \frac{4e's}{m's^2 + 4c's + 4k'} \tag{14-16}$$

Where m, k, c, e denote the mass, stiffness coefficient, damping coefficient and the coefficient of actuator of the vibration isolation platform. And m', k', c', e' denote the corresponding identification parameter. The models are simplified for the simulation and the values of parameters are shown in the Table. 1.

Table 1. Value of the Parameters

Parameter	Value
m	30kg
k	11500N/m
e	0.5N/ V

The input signal is a 1-100HZ sweep signal and more low frequency signal components are added to ensure adequate excitation. Then the simulating result of the x-direction error combined with the rotation coupling output is obtained. The control results are shown in Figure 4 and Figure 5.

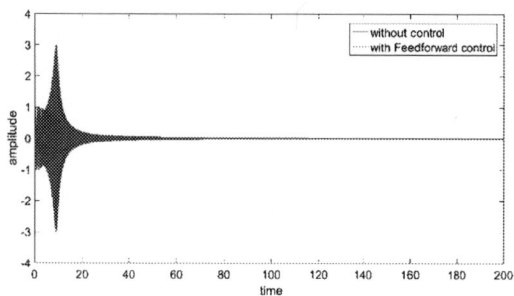

Figure 4. The output signal with control and without control.

Figure 5. Transmissibility curve with feedforward and feedback control.

It can be seen from the figure that the convergence speed of the novel algorithm is fast while the steady-state error is small. Further, initial vibration isolation frequency of the vibration transmissibility curve is down to about 7 Hz.

3.2. The experiment in the platform

The feedforward control method proposed in this paper has been applied to a mental spring active vibration isolation system, which is shown in Figure 6. And we focus on horizontal vibration in experiments.

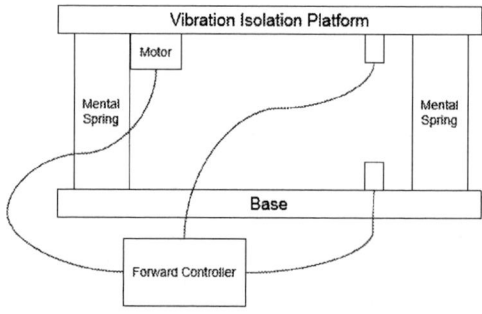

Figure 6. The mental spring active vibration isolation platform

Figure 7. The diagram of control experiments

Simplified diagram of the experiment is shown in the Figure 7. During experiments, feedforward controller parameters are updated online based on ground vibration and platform vibration measured by velocity sensor. Then the controller yields the corresponding control signal driving voice coil motor to achieve vibration isolation. The results of vibration isolation from 2Hz-100Hz are shown in Figure 8. As the figure shown, vibration is attenuated at lower frequencies. The maximum vibration attenuation is near 10Hz, about -25 dB.

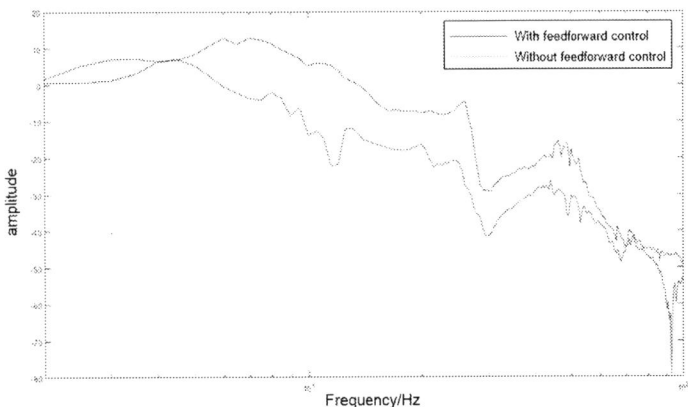

Figure 8. The Vibration transmissibility curves.

4. Conclusion

A double-channel iterative NFXLMS algorithm is proposed in the paper to solve the coupling problem of the vibration isolation platform. It is based on the FXLMS algorithm and it can simultaneously suppress the rotation coupling and x-direction error of the platform. Finally, the method is effectively implemented on the vibration isolation table.

References

[1] D. Karnopp, M. J. Crosby, & R. A. Harwood. (1974). Vibration control using semi-active force generators. Transactions of the Asme Journal of Engineering for Industry, 96(2), 619-626.

[2] Kang, M. S. (2003). Optimal feedforward control of active magnetic bearing system subject to base motion. Control Applications, 2003. CCA 2003. Proceedings of 2003 IEEE Conference on (Vol.1, pp.748-753 vol.1). IEEE.

[3] C. Collette, S. Janssens, & K. Artoosand C. Hauviller. (2010). Active vibration isolation of high precision machines. Diamond Light Source Proceedings, 1(MEDSI-6), 451-456.

[4] Yoshioka H, Murai N. AN ACTIVE MICROVIBRATION ISOLATION SYSTEM. Proceedings of the 7th International Workshop on Accelerator Alignment, 2002: 388-401.

[5] H. X. Zeng, (2014). Research of an active vibration control method with position stabilization function,.M.S. thesis, Dept. Mech. Eng.,Tsinghua Univ., Beijing, China.

[6] Zhou Zhenhua, Chen Xuedong, Zhou Bo. Feedforward compensation in vibration isolation system subject to base disturbance. Journal of Vibration and Control, 2015, 21(6): 1201-1209.

[7] Zhang, Z., & Yin, W. (2017). Data-driven feedforward control on active vibration isolation system. International Conference on Control, Automation and Systems, 2017.

[8] Yang, G., Xiao, S., & Zhang, W. (2010). Analysis on the lateral stiffness of the helical circle spring. China Railway Science.

[9] Pu, H., Luo, X., Jiang, W., Dong, K., & Chen, X. (2010). Modelling and control of hybrid vibration isolation system for high-precision equipment. IEEE International Conference on Control and Automation (pp.2152-2157). IEEE.

[10] Kuo, S. M., Morgan, D. R. Active noise control: a tutorial review. Proceedings of the IEEE 2006, 87(6): 943 – 973.

ISPECE   IOP Publishing

Projectile Velocity Measurement System Based on PVDF and Data Processing Method

Xiaoxiao Chen[1], Ping Song[1*] and Yayu Zhai[1]

[1]Key Laboratory of Biomimetic Robots and Systems (Ministry of Education) Beijing Institute of Technology, Beijing, 100081, China

Email: sping2002@bit.edu.cn

Abstract. Time of flight (TOF) algorithm is mainly used to measure the velocity of a projectile. TOF algorithm measures the time required for the projectile to fly over a fixed distance. We use poiy vinylidene fluoride (PVDF) to measure TOF. Based on the analysis of the signal generated by the projectile penetrating PVDF film, a projectile penetrating unit recognition algorithm and a projectile velocity calculation method are presented. In projectile penetrating unit recognition algorithm, we filter all signals generate by PVDF, and identify the projectile penetrating unit. Then, the time when the projectile reaches two PVDF films is determined by the threshold method, and the TOF is obtained, so the flight speed of the projectile is obtained. It is proved that this method can distinguish the penetrating units of the projectile correctly and calculate the velocity of the projectile accurately.

1. Introduction

Research into measuring the velocity of a projectile has been going on for years. One method is to use TOF to measure the velocity of a projectile. TOF has been used in many other scenarios [1,2,3]. In projectile velocity measurement, Y He used the system based on the laser beam interruption to calculate the velocity of the muzzle [4]. The position and time were obtained, the average velocity between the two barriers was determined. M Singh proposed a system to accurately determine the speed of a projectile (bullet) by measuring the time of flight between two parallel laser screens [5]. They used a system with two parallel laser screens to accurately determine the speed of a projectile by measuring the time of flight between laser screens. G Wang proposed a design of an optical fiber-based velocity measurement system [6]. The measurement principle was based on Doppler effect and heterodyne detection technique. They deduced the relationship between the projectile velocity and the instantaneous frequency (IF) of the optical fiber-based system output signal. By using this relationship, they could get the speed of the projectile. G H Yang used magnetoresistive sensor and coil target combination method to real-time measure projectile muzzle velocity [7]. M Courtney put forward a simple method for using a PC soundcard to accurately measure bullet velocity [8]. The result that recorded time between bullet blast and the bullet hitting the target minus the time that sound return from the target to the microphone was the time of flight for the bullet. J Yu used a structure made by two laser light sources, two optical detectors (OD) and two reflectors to measure the velocity of the projectile [9]. In addition to being able to measure velocity, the device could also measure the position of the projectile. J B Jordan developed an equation for the velocity as a function of the FSP mass and the depth of penetration into Celotex recovery media [10]. Using this function, they could compute the velocity of the projectile. However, some of these methods are influenced by the environment and some are expensive. Therefore, this paper proposes a speed measurement system based on PVDF.

Content from this work may be used under the terms of the Creative Commons Attribution 3.0 licence. Any further distribution of this work must maintain attribution to the author(s) and the title of the work, journal citation and DOI.

Published under licence by IOP Publishing Ltd

ISPECE

IOP Publishing

IOP Conf. Series: Journal of Physics: Conf. Series **1187** (2019) 032011   doi:10.1088/1742-6596/1187/3/032011

2. Method

2.1. Experiment device

The overall diagram of the experiment device is shown in figure 1. The experiment device mainly includes PVDF piezoelectric film, guide rails, support structure, signal conditioning circuit, signal acquisition circuit computer and sky screens. The device consists of two PVDF. Two PVDF are fixed back and forth on the guide rail with the support structure. PVDF piezoelectric film is connected to the signal conditioning circuit, the output of the signal conditioning circuit is connected to the input end of the signal collection circuit, and the output of the signal collection circuit is connected to the computer.

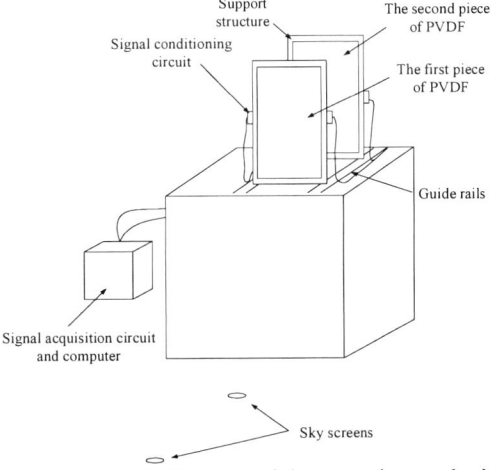

Figure 1. Overall diagram of the experiment device.

The PVDF films used in this paper have a layered structure as shown in figure 2.At the same time, each PVDF consists of several small PVDF units, as shown in figure 3.

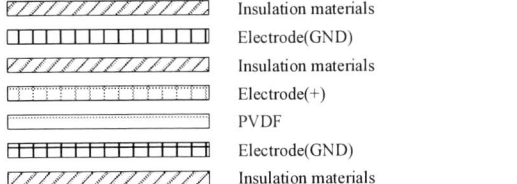

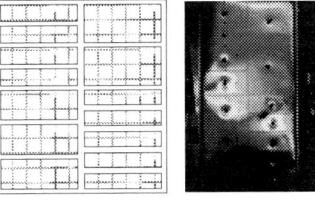

Figure 2. The layered structure of PVDF   Figure 3. The unit structure of PVDF

When PVDF is compressed, the electrode (+) and the electrode (GND) will produce equal positive and negative charges. Since the charge signal is not conducive to direct collection, the signal conditioning circuit converts the charge signal into voltage signal, and the core circuit of the signal conditioning circuit is shown in figure 4.

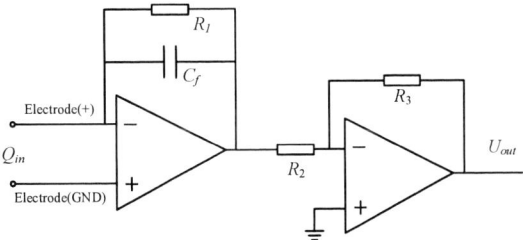

Figure 4. The core circuit of the signal conditioning circuit

According to figure 4, we have following equation.

569

$$U_{out} = \frac{R_3}{R_2} \cdot \frac{Q_{in}}{C_f} \tag{1}$$

The adjusted signal is connected to the signal acquisition circuit for acquisition and transmission to the computer. The acquisition circuit was mainly constructed by NI pxie-6358, with a sampling rate of 1M and a sampling time of 0.2s. The computer recognizes the penetrating units through the data collected, and processes the data corresponding to the projectile to obtain the time difference between the two PVDF films before and after the projectile arrives, so as to calculate the velocity of the projectile. Then, the time when the projectile reaches two PVDF films is determined by the threshold method, and the TOF is obtained, so the flight speed of the projectile is computed.

2.2. Penetrating unit recognition algorithm

The ideal experiment result is that the signal is generated only on the PVDF units where the projectile hits, and the other PVDF units do not output signals. But the real results are different. On the basis of multiple experiments, the experiment results have the following characteristics: 1. The penetrating units produces pulse signals which are mainly of two types (Figure 5a and Figure 5b), one signal is a simple impulse signal and the other signal has an exponential attenuation signal at the back end of the pulse. 2. Unpenetrated PVDF units sometimes produce interference signals, especially those that have been penetrated by the projectile before. Therefore, a penetrating unit recognition algorithm is proposed. In this algorithm, the signal of each unit is filtered by low pass filtering, then set the threshold to determine the number of pulses, if the number of pulses is one, this unit is penetrating unit.

2.3. Particle velocity calculation method

After the above steps, we get the penetrating unit and extract the corresponding data. The velocity of the projectile is calculated respectively by taking 20% and 80% of the rising edge as the time of arrival.

3. Results

3.1. Results of penetrating unit recognition algorithm

Figure 5 shows the output signal waveform of 4 units in an experiment. Figure 5a and 5b are the output signals of the penetrating units, figure 5c is the output signal of the PVDF unit with interference signal, and figure 5d is the output signal of the PVDF unit without interference signal.

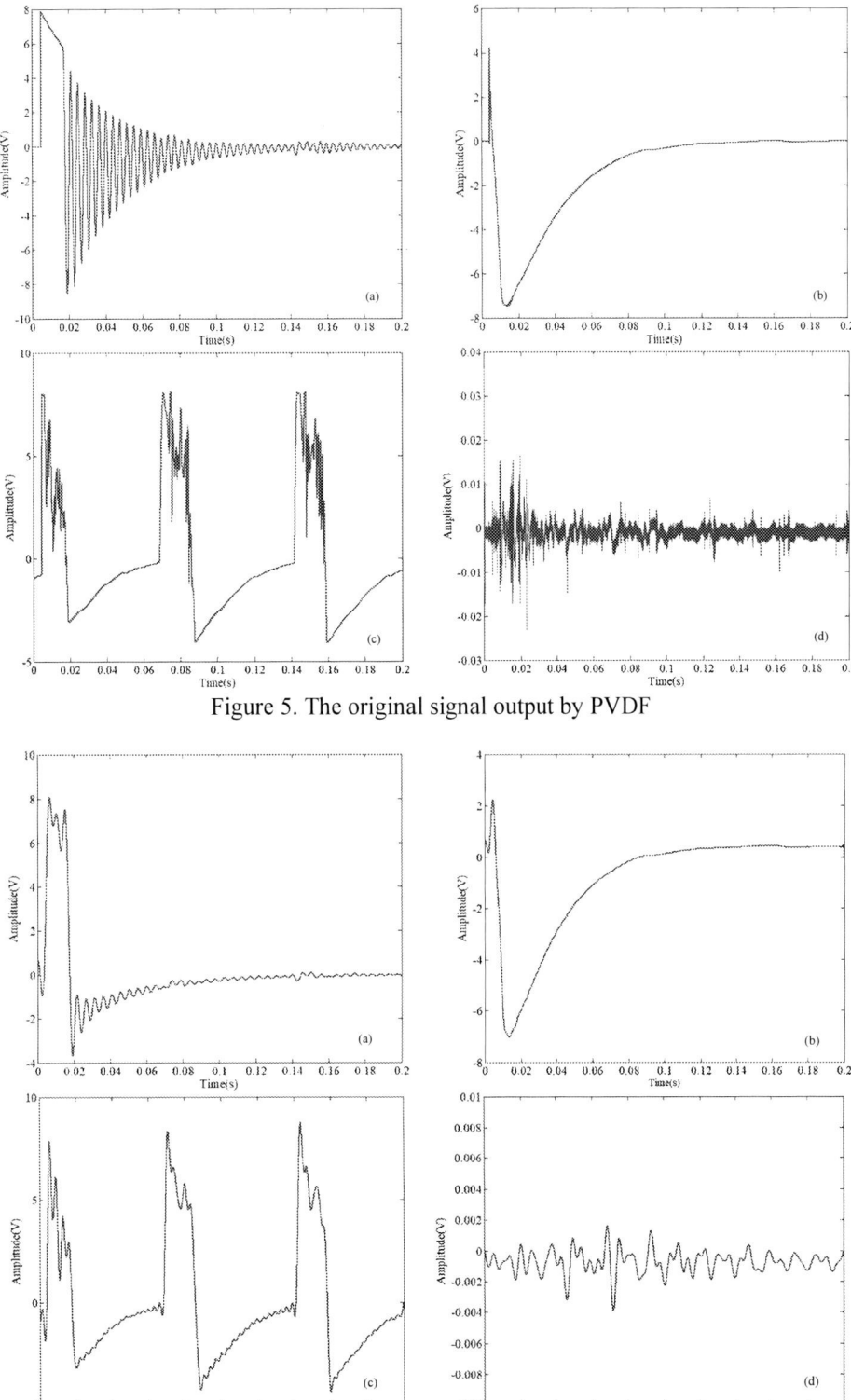

Figure 5. The original signal output by PVDF

Figure 6. The filtered signal

By analysing the spectrum of all signals, we find most of the signals are concentrated at low frequencies with a small amount of high-frequency noise. Filter signals below 250Hz. The signal obtained after low-pass filtering is as follows.

As can be seen from the figure 6, the filtered waveform becomes smooth and white noise is suppressed. Figure 6a is the most obvious. The signal with exponential attenuation at the back end of the pulse is filtered.

The threshold method is adopted for the recognition of the penetrating unit. When the signal crosses the threshold line twice, it is considered that the signal is generated by the penetrating unit. When the signal does not cross the threshold line or cross the threshold line greater than twice, it is considered that the signal is not generated by the penetrating unit. For the experiment in this paper, the threshold line U=1V is adopted, so it can be seen that PVDF units corresponding to figure 6a and 6b are penetrating units.

3.2. Results of projectile velocity

The velocity of the projectile can be calculated after determining the penetrating units. In this paper, two threshold selection methods are used to estimate the velocity of the projectile. The threshold is 20% of the maximum and 80% of the maximum. The velocity of the projectile was estimated in 10 experiments, and the estimated value and true value of the projectile velocity were shown in the table 1. It can be seen from the table 1 that both threshold methods can calculate the velocity. The error between the velocity calculated with the threshold value of 20% of the maximum value and the real value is small, which is more suitable for the calculation method of velocity.

Table 1. The results of ten experiments.

Number of experiments	20% of the maximum (m/s)	80% of the maximum (m/s)	Real value (m/s)
1	722.0333	710.8140	737.3
2	724.1904	729.0638	734.4

Table 1. The results of ten experiments. (continued)

3	720.5491	719.9771	731.9
4	729.1315	721.3024	734.4
5	733.7466	705.2806	735.5
6	732.7274	713.3891	734.6
7	726.7425	712.0294	734.2
8	723.7014	726.3892	737.3
9	725.1310	721.3111	735.1
10	715.5321	701.4436	732.0
Average	725.34853	716.10003	734.67
Error	-1.2688%	-2.5277%	

4. Discussion

The penetrating unit recognition algorithm can extract penetrating units effectively. In the case of determining the penetrating units, the velocity can be calculated by threshold method. It can be seen from the results that the accuracy of the speed calculated with the maximum value of 20% as the threshold is higher than the speed calculated with the maximum value of 80% as the threshold.

5. Conclusions

In this paper, a velocity measurement system based on PVDF is presented. At the same time, a data processing method is proposed. The system and method can calculate the velocity of the projectile. Taking 20% of the maximum value as the threshold value, the error is -1.2688%.

Acknowledge

This work was partially supported by the National Defense Basic Scientific Program of China (JCKY2016208B008). We are also very grateful to the reviewers for their useful opinions and suggestions, which have improved representativeness.

References

[1]   T K Ghosh, S Pal, T Sinha, S Chattopadhyay, K S Golda and P Bhattacharya 2005 *Nucl. Instrum. Methods A* 540 285.

[2]   E M Kozulin, A A Bogachev, M G Itkis, I M Itkis, G N Knyazheva, N A Kondratiev, L Krupa, I V Pokrovsky and E V Prokhorov 2008 *Instrum. Exp. Tech.* 51 (1) 44.

[3]   V Priola and M J Brannan 2003 *Meas. Sci. Technol.* 14(1), 1.

[4]   Y He, S Song, Y Guan, C Cheng, W Dai, X Qiu and Y Li 2015 *IEEE Trans. Plasma Sci.* 43(5) 1647-1651

[5]   M Singh 2007 *Optical Engineering* 46(4) 4303

[6]   G Wang, J Sun and Q Li 2014 *Rev. Sci. Instrum.* 85(8) 351-356.

[7]   G H Yang and X M Zhang 2015 *Transducer and Microsystem Technology* 34(2) 76-78.

[8]   M Courtney and B Edwards 2006 arXiv preprint physics/0601102.

[9]   J Yu, X Wang and Y Li 2009 *Symposium on Photonics and Optoelectronics* 1-4.

[10]   J B Jordan and C J Naito 2010 *Int. J. Impact Eng.* 37(5) 530-536.

ISPECE   IOP Publishing

IOP Conf. Series: Journal of Physics: Conf. Series **1187** (2019) 032012   doi:10.1088/1742-6596/1187/3/032012

Some new results on the finite-time control and its application to a chemical reactor system

Ziteng Guo[1], Caisheng Wei[2*]

[1]School of Textile and Material Engineering, Dalian Polytechnic University

[2]School of Astronautics, Northwestern Polytechinical University

*corresponding author's e-mail: 1097853117@qq.com, yundiqiuyu@163.com

Abstract. In this paper, a novel finite-time control scheme is proposed for a class of non-strict feedback systems with guaranteed preassigned output tracking performance. First, a finite-time convergent performance function is proposed in the prescribed performance control structure. Then, based on the proposed performance function, a finite-time stable controller is devised. Compared with the existing finite-time control schemes, the fractional state or output information and discontinuous phenomenon is avoided totally. Finally, application to a chemical reactor system is organized to validate the effectiveness of the proposed control scheme.

1. Introduction

The past few decades have witnessed the fast development of finite-time control theory due to its widely potential applications in real systems like robotic system, chemical reaction system and etc (e.g., see [1]-[3] and references therein). The main way to achieve finite-time stability is via sliding mode control (SMC) based technique. Namely, fractional power state or output information and symbolic functions are widely used to construct the relevant control schemes. Although effective, the usage of fractional power state or output information makes the relevant controller pretty complex due to the highly computational burden. Moreover, using symbolic function used will make the controller discontinuous, which is not easily achievable in real systems. To conquer the foregoing two inherent limitations, SMC-based technique should be avoided.

In recent years, prescribed performance control (PPC) has gained considerable attention due to its prominent advantage in quantitatively charactering the transient and steady-state performance of the controlled systems (e.g., see [4]-[5] and reference therein). In the existing works, the performance function used in the PPC structure is in an exponential form, which means the controlled system will converge to its equilibrium point exponentially. However, if the performance function is finite-time convergent, the relevant controlled system will be finite-time stable. By following this idea, in this paper, we first propose a novel finite-time convergent performance function. Then, based on the newly proposed performance function, finite-time convergent controller is devised in the PPC structure. Compared with the SMC-based finite-time control schemes, the foregoing inherent limitations will be avoided totally.

The rest of this paper is organized as follows. In Section 2, the problem formulation is stated. Section 3 shows the finite-time controller design with its stability analysis. In Section 4, application to a chemical reactor system is organized to validate the effectiveness of the proposed control scheme. Some conclusions are drawn in Section 5.

Content from this work may be used under the terms of the Creative Commons Attribution 3.0 licence. Any further distribution of this work must maintain attribution to the author(s) and the title of the work, journal citation and DOI.
Published under licence by IOP Publishing Ltd

2. Problem formulation

The nonstrict feedback system considered in this paper is expressed by

$$\begin{cases} \dot{x}_1 = x_2 + f_1(\boldsymbol{x}) \\ \quad \vdots \\ \dot{x}_n = bu + f_n(\boldsymbol{x}) + d \\ y = x_1 \end{cases} \tag{1}$$

where $\boldsymbol{x} = [x_1, x_2, ..., x_n] \in \mathbb{R}^n$, $y \in \mathbb{R}$ are the system state vector and output, respectively. $f_i(\boldsymbol{x}) \in \mathbb{R}$ $(i = 1, 2, ..., n)$ is a nonlinear function, which is unknown. $u, d \in \mathbb{R}$ denote the control input and unknow bounded external disturbance, respectively. $b \neq 0$ is the control gain, which is known.

For system (1), the control objective is twofold: (i). The expected output reference y_r can be tracked under the designed controller with guaranteed preassigned tracking performance; (ii). The tracking error system is finite-time convergent in the presence of unknown nonlinearities and external disturbance.

Remark 1. As presented in Section 1, one can find that the finite-time stability can be achieved via using SMC-based technique in the existing works. However, the relevant controller is pretty tedious owing to the usage of fractional power state or output feedback and symbolic function. Thus, in this paper, a different way to obtain the finite-time stability is proposed, which conquers the inherent drawbacks of SMC-based technique.

Priori to showing the controller design, some preliminary knowledge used in this paper is shown as follows.

2.1. Preliminary knowledge of radial basis function neural network (RBFNN)

Radial basis function neural network (RBFNN) has been widely applied to approximate smooth nonlinear functions with arbitrary approximation accuracy [7]. Thus, for an unknown nonlinear function $f(X) : \mathbb{R}^N \to \mathbb{R}$, it can be approximated by

$$f(X) = \boldsymbol{\mathcal{W}}^{*T} \boldsymbol{\varphi}(X) + \delta(X) \tag{2}$$

where $X = [X_1, X_2, ..., X_N] \in \mathbb{R}^N$ is the input of a RBFNN. $\boldsymbol{\mathcal{W}}^* = \left[\mathcal{W}_1^*, \mathcal{W}_2^*, ..., \mathcal{W}_\ell^* \right]^T \in \mathbb{R}^\ell$ is the optimal weight vector of a RBFNN, which is obtained by

$$\boldsymbol{\mathcal{W}}^* = \arg \min_{\mathcal{W} \in \Omega_{\mathcal{W}}} \left[\sup_{X \in \Omega_X} \left| f(X|\mathcal{W}) - f(X) \right| \right] \tag{3}$$

where $\Omega_{\mathcal{W}}, \Omega_X$ are the relevant compact sets of $\boldsymbol{\mathcal{W}}, X$, respectively. $\delta(X)$ is the approximation error, which can be made sufficiently small via using sufficient nodes of RBFNN. $\boldsymbol{\varphi}(X) = \left[\varphi_1(X), \varphi_2(X), ..., \varphi_\ell(X) \right]^T \in \mathbb{R}^\ell$ is the kernel function (ℓ is the number of nodes in the RBFNN), which is usually chosen in the Gaussian form, i.e.,

$$\varphi_i(X) = \exp\left(-\frac{(X - \zeta_i)^T (X - \zeta_i)}{2\xi_i^2} \right) \tag{4}$$

where $\zeta_i = [\zeta_{i,1}, \zeta_{i,2}, ..., \zeta_{i,N}] \in \mathbb{R}^N$, $\xi_i \in \mathbb{R}$ is the respective center and width of the Gaussian function.

Remark 2. Widespread applications indicate that the optimal weight vector $\boldsymbol{\mathcal{W}}^*$ is bounded.

3. Novel finite-time convergent controller with guaranteed tracking performance

In this section, the finite-time convergent controller design is divided into three parts, namely, finite-time convergent performance function design, finite-time convergent controller design and stability analysis. To achieve finite-time convergence, a novel practically finite-time convergent performance function is first proposed in the following.

3.1. Finite-time convergent performance function design

In the PPC structure, performance function is a vital one to characterize the transient and steady-state performance of the controlled system. In this paper, the performance function $\rho(t)$ is derived by

$$\dot{\rho}(t) = \begin{cases} -\eta_0\left(\rho(t)-\rho_\infty\right)^{\frac{m_1}{n_1}} - \eta_0\left(\rho(t)-\rho_\infty\right)^{\frac{p_1}{q_1}}, \left(\rho(t)-\rho_\infty\right) > 1 \\ 0, \left(\rho(t)-\rho_\infty\right) = 1 \\ -\eta_0\left(\rho(t)-\rho_\infty\right) - \eta_0\left(\rho(t)-\rho_\infty\right)^{\frac{p_1}{q_1}}, \left(\rho(t)-\rho_\infty\right) < 1 \end{cases} \tag{5}$$

where $\rho_0 = \rho(0) > \rho_\infty > 0$ are the initial and final states of the performance function $\rho(t)$. $m_1 > n_1$, $p_1 < q_1$ are positive odd integers. η_0 is a positive constant. From Eq. (5), one can find that $\lim_{|\rho(t)-\rho_\infty|\to 1^+}\dot{\rho}(t) = \lim_{|\rho(t)-\rho_\infty|\to 1^-}\dot{\rho}(t) = \lim_{|\rho(t)-\rho_\infty|=1}\dot{\rho}(t) = 0$. Thus, the performance function $\rho(t)$ is continuous in the time domain. Accordingly, one can obtain the following property for the performance function.

Property 1. Performance function $\rho(t)$ is positive and will converge to ρ_∞ within finite time.

Proof. The proof of **Property 1** is organized as follows. Firstly, define a new variable $\varpi = \left(\rho(t)-\rho_\infty\right)^{1-p_1/q_1}$, then, the first and third parts of Eq. (5) can be rewritten as

$$S_1 : \dot{\varpi} + \frac{q_1-p_1}{q_1}\eta_0\left(\varpi^{\frac{m_1 q_1 - p_1 n_1}{n_1(q_1-p_1)}} + 1\right) = 0; \quad S_2 : \dot{\varpi} + \frac{q_1-p_1}{q_1}\eta_0\left(\varpi+1\right) = 0 \tag{6}$$

For brevity, define $\hbar = \left[(m_1-n_1)q_1\right]/\left[(q_1-p_1)n_1\right]$. Inspired by [8], solving Eq. (6) yields the relevant convergence time T_0, i.e.,

$$\lim_{\varpi_0\to\infty} T_0(\varpi_0) = \lim_{\varpi_0\to\infty} \frac{q_1}{(q_1-p_1)\eta_0}\left(\int_0^1 \frac{1}{\varpi+1}d\varpi + \int_1^{\varpi_0} \frac{1}{\varpi^{1+\hbar}+1}d\varpi\right) < \lim_{\varpi_0\to\infty} \frac{q_1}{(q_1-p_1)\eta_0}\left(\ln 2 + \int_1^{\varpi_0} \frac{1}{\varpi^{1+\hbar}}d\varpi\right) \tag{7}$$

$$< \frac{q_1}{(q_1-p_1)\eta_0}\left(\ln 2 + \frac{1}{\hbar}\right) = \frac{n_1}{(m_1-n_1)\eta_0} + \frac{q_1}{(q_1-p_1)\eta_0}\ln 2$$

Based on Eqs. (6) and (7), one can easily obtain that there exists an upper bound for the convergence of system (5). Namely, $\lim_{t\to T_0}\rho(t) = \rho_\infty > 0$. Consequently, **Property 1** is proved. ∎

Remark 3. From Eqs. (5)-(7), one can find that the proposed performance function is a continuous one only with respect to time. Thus, it can be predefined by the users easily.

3.2. Finite-time convergent controller design with guaranteed output tracking performance

For system (1), the output tracking error is defined as $e = y - y_r$, wherein y_r is the desired output reference command. Before moving on, the following two assumptions are given.

Assumption 1. The desired output reference command y_r is smooth and its first derivative is known.

Assumption 2. The state variables of system (1) are available for measurement.

Remark 4. With consideration of that the reference command is designed by the user, thus **Assumption 1** is easily satisfied. As for **Assumption 2**, there are many effective measurement devices and techniques available. Thus, it is reasonable.

To guarantee the output tracking performance, based on the proposed performance function in subsection 3.1, the following prescribed performance is given

$$-\rho(t) < e < \rho(t) \tag{8}$$

By applying backstepping control technique, three steps are involved in the following controller design.

Step 1. When $i = 1$, we choose the following Lyapunov function

$$V_1 = \frac{1}{2}\frac{z_1^2}{1-z_1^2} + \frac{1}{2}\tilde{\mathcal{W}}_1^T\tilde{\mathcal{W}}_1 \tag{9}$$

where $z_1 = e / \rho(t) \in (-1,1)$ is the standard tracking error. $\tilde{\mathcal{W}}_1 = \mathcal{W}_1^* - \hat{\mathcal{W}}_1$ is the estimated weight vector error to be determined later ($\mathcal{W}_1^*$, $\hat{\mathcal{W}}_1$ denote the optimal and estimated weight vectors of RBFNN, respectively). Based on system (1), the derivative of V_1 equals to

$$\dot{V}_1 = \frac{z_1}{\left(1-z_1^2\right)^2}\dot{z}_1 + \tilde{\mathcal{W}}_1^T\dot{\tilde{\mathcal{W}}}_1 = \frac{z_1}{\rho(t)\left(1-z_1^2\right)^2}\left(\dot{e} - \frac{\dot{\rho}(t)}{\rho(t)}e\right) + \tilde{\mathcal{W}}_1^T\dot{\tilde{\mathcal{W}}}_1 = \frac{z_1}{\rho(t)\left(1-z_1^2\right)^2}\left(x_2 + f_1(x) - \dot{y}_r - \frac{\dot{\rho}(t)}{\rho(t)}e\right) + \tilde{\mathcal{W}}_1^T\dot{\tilde{\mathcal{W}}}_1 \tag{10}$$

Define the coordinate transformation $z_2 = x_2 - s_1$, wherein, s_1 is the approximation for the first virtual controller α_1, which is given later. Based on the preliminary knowledge in subsection 2.1, the unknown nonlinear function $f_1(x)$ can be approximated by a RBFNN. Thus, Eq. (10) equals to

$$\dot{V}_1 = \frac{z_1}{\rho(t)\left(1-z_1^2\right)^2}\left(z_2 + \alpha_1 - \dot{y}_r + \mathcal{W}_1^{*T}\varphi_1(x) + \delta_1 + s_1 - \alpha_1 - \frac{\dot{\rho}(t)}{\rho(t)}e\right) + \tilde{\mathcal{W}}_1^T\dot{\tilde{\mathcal{W}}}_1$$

$$= \frac{z_1}{\rho(t)\left(1-z_1^2\right)^2}\left(z_2 + \alpha_1 - \dot{y}_r + \hat{\mathcal{W}}_1^T\varphi_1(x) + \tilde{\mathcal{W}}_1^T\varphi_1(x) + \delta_1^* - \frac{\dot{\rho}(t)}{\rho(t)}e\right) + \tilde{\mathcal{W}}_1^T\dot{\tilde{\mathcal{W}}}_1 \tag{11}$$

Where $\delta_1^* = \delta_1 + s_1 - \alpha_1$. For Eq. (11), the following inequality holds

$$\frac{z_1}{\rho(t)\left(1-z_1^2\right)^2}\left(z_2 + \delta_1^*\right) \le \frac{2z_1^2}{\rho^2(t)\left(1-z_1^2\right)^4} + \frac{1}{4}z_2^2 + \frac{1}{4}\delta_1^{*2} \tag{12}$$

Thus, the first virtual controller α_1 is devised as

$$\alpha_1 = -k_1 z_1 - \hat{\mathcal{W}}_1^T\varphi_1(x) + \dot{y}_r + \frac{\dot{\rho}(t)}{\rho(t)}e - \frac{2z_1}{\rho(t)\left(1-z_1^2\right)^2} \tag{13}$$

where k_1 is the positive control gain. The corresponding adaptive scheme for the estimated weight vector is

$$\dot{\hat{\mathcal{W}}}_1 = -\mu_1\hat{\mathcal{W}}_1 + \frac{z_1}{\rho(t)\left(1-z_1^2\right)^2}\varphi_1(x) \tag{14}$$

where $\mu_1 > 0$ is a constant. By considering $\dot{\tilde{\mathcal{W}}}_1 = \dot{\mathcal{W}}_1^* - \dot{\hat{\mathcal{W}}}_1 = -\dot{\hat{\mathcal{W}}}_1$, thus, substituting Eqs. (13) and (14) into (11) gets

$$\dot{V}_1 \le -\frac{k_1 z_1^2}{\rho(t)\left(1-z_1^2\right)^2} + \frac{1}{4}z_2^2 + \frac{1}{4}\delta_1^2 + \frac{z_1}{\rho(t)\left(1-z_1^2\right)^2}\tilde{\mathcal{W}}_1^T\varphi_1(x) + \tilde{\mathcal{W}}_1^T\left(\mu_1\hat{\mathcal{W}}_1 - \frac{z_1}{\rho(t)\left(1-z_1^2\right)^2}\varphi_1(x)\right)$$

$$= -\frac{k_1 z_1^2}{\rho(t)\left(1-z_1^2\right)^2} + \frac{1}{4}z_2^2 + \frac{1}{4}\delta_1^{*2} + \left(\mathcal{W}_1^* - \hat{\mathcal{W}}_1\right)^T\mu_1\hat{\mathcal{W}}_1 \le -\frac{k_1 z_1^2}{\rho(t)\left(1-z_1^2\right)^2} + \frac{1}{4}z_2^2 + \frac{1}{4}\delta_1^{*2} - \frac{\mu_1}{2}\hat{\mathcal{W}}_1^T\hat{\mathcal{W}}_1 + \frac{\mu_1}{2}\mathcal{W}_1^{*T}\mathcal{W}_1^* \tag{15}$$

Step 2. When $2 \le i \le n-1$, we define the relevant coordinate transformation as $z_i = x_i - s_{i-1}$, wherein, s_{i-1} is output of the a low-pass filter used to approximate the derivative of the virtual controller.

This is also referred to as dynamic surface control in the existing works like [5]. $s_j\ (j=1,...,n-2)$ is derived from

$$\varepsilon_j \dot{s}_j + s_j = \alpha_j \quad \left(s_j(0) = \alpha_j(0)\right) \tag{16}$$

The filter estimation error is defined as $\chi_j = \alpha_j - s_j$. The relevant Lyapunov function is given by

$$V_i = V_{i-1} + \frac{1}{2}z_i^2 + \frac{1}{2}\tilde{\mathcal{W}}_i^T \tilde{\mathcal{W}}_i \tag{17}$$

where $\tilde{\mathcal{W}}_i = \mathcal{W}_i^* - \hat{\mathcal{W}}_i$ is the estimated weight vector error to be determined later ($\mathcal{W}_i^*$, $\hat{\mathcal{W}}_i$ denote the optimal and estimated weight vectors of RBFNN, respectively). Based on Eq. (1) and *Step 1*, taking the derivative of V_i yields

$$\begin{aligned}
\dot{V}_i &= \dot{V}_{i-1} + z_i \dot{z}_i + \tilde{\mathcal{W}}_i^T \dot{\tilde{\mathcal{W}}}_i = z_i\left(\dot{x}_i - \dot{s}_{i-1}\right) + \dot{V}_{i-1} + \tilde{\mathcal{W}}_i^T \dot{\tilde{\mathcal{W}}}_i \\
&= z_i\left(x_{i+1} + f_i(x) - \dot{s}_{i-1}\right) + \dot{V}_{i-1} + \tilde{\mathcal{W}}_i^T \dot{\tilde{\mathcal{W}}}_i = z_i\left(z_{i+1} + s_i + f_i(x) - \dot{s}_{i-1}\right) + \dot{V}_{i-1} + \tilde{\mathcal{W}}_i^T \dot{\tilde{\mathcal{W}}}_i \\
&= z_i\left(z_{i+1} + \alpha_i + \chi_i + \hat{\mathcal{W}}_i^T \varphi_i(x) + \tilde{\mathcal{W}}_i^T \varphi_i(x) + \delta_i - \dot{s}_{i-1}\right) + \dot{V}_{i-1} - \tilde{\mathcal{W}}_i^T \dot{\hat{\mathcal{W}}}_i
\end{aligned} \tag{18}$$

Thus, the relevant virtual controller α_i and the adaptive scheme for $\hat{\mathcal{W}}_i$ are expressed by

$$\alpha_i = -k_i z_i - \frac{1}{4}z_i - \frac{1}{4}z_i - \hat{\mathcal{W}}_i^T \varphi_i(x) + \dot{s}_{i-1}, \ \dot{\hat{\mathcal{W}}}_i = -\mu_i \hat{\mathcal{W}}_i + z_i \varphi_i(x) \tag{19}$$

Substituting Eq. (19) into (18) yields

$$\begin{aligned}
\dot{V}_i &= \dot{V}_{i-1} + z_i \dot{z}_i + \tilde{\mathcal{W}}_i^T \dot{\tilde{\mathcal{W}}}_i = z_i\left(\dot{x}_i - \dot{s}_{i-1}\right) + \dot{V}_{i-1} + \tilde{\mathcal{W}}_i^T \dot{\tilde{\mathcal{W}}}_i \\
&= z_i\left(x_{i+1} + f_i(x) - \dot{s}_{i-1}\right) + \dot{V}_{i-1} + \tilde{\mathcal{W}}_i^T \dot{\tilde{\mathcal{W}}}_i = z_i\left(z_{i+1} + s_i + f_i(x) - \dot{s}_{i-1}\right) + \dot{V}_{i-1} + \tilde{\mathcal{W}}_i^T \dot{\tilde{\mathcal{W}}}_i \\
&= z_i\left(z_{i+1} + \alpha_i + \chi_i + \hat{\mathcal{W}}_i^T \varphi_i(x) + \tilde{\mathcal{W}}_i^T \varphi_i(x) + \delta_i - \dot{s}_{i-1}\right) + \dot{V}_{i-1} - \tilde{\mathcal{W}}_i^T \dot{\hat{\mathcal{W}}}_i \\
&\leq -\frac{k_1 z_1^2}{\rho(t)\left(1-z_1^2\right)^2} - \sum_{j=2}^{i} k_j z_j^2 - \frac{1}{2}\sum_{j=1}^{i}\mu_j \hat{\mathcal{W}}_j^T \hat{\mathcal{W}}_j + \frac{1}{4}z_{i+1}^2 + \frac{1}{4}\sum_{j=1}^{i}\delta_j^{*2} + \frac{1}{2}\sum_{j=1}^{i}\mu_j \mathcal{W}_j^{*T}\mathcal{W}_j^*
\end{aligned} \tag{20}$$

where $\delta_j^* = \chi_j + \delta_j$ is the lumped error.

Step 3. When $i = n$, we define the relevant coordinate transformation as $z_n = x_n - s_{n-1}$, wherein, s_{n-1} is given in Eq. (16). Then, the relevant Lyapunov function is chosen as

$$V_n = V_{n-1} + \frac{1}{2}z_n^2 + \frac{1}{2}\tilde{\mathcal{W}}_n^T \tilde{\mathcal{W}}_n \tag{21}$$

Similar to *Steps 2* and *3*, the derivative of V_n equals to

$$\begin{aligned}
\dot{V}_n &= \dot{V}_{n-1} + z_n \dot{z}_n + \tilde{\mathcal{W}}_n^T \dot{\tilde{\mathcal{W}}}_n = z_n\left(bu + f_n(x) + d - \dot{s}_{n-1}\right) + \dot{V}_{n-1} - \tilde{\mathcal{W}}_n^T \dot{\hat{\mathcal{W}}}_n \\
&= z_n\left(bu + \hat{\mathcal{W}}_n^T \varphi_n(x) + \tilde{\mathcal{W}}_n^T \varphi_n(x) + d + \delta_n - \dot{s}_{n-1}\right) + \dot{V}_{n-1} - \tilde{\mathcal{W}}_n^T \dot{\hat{\mathcal{W}}}_n
\end{aligned} \tag{22}$$

The relevant actual control input u and the adaptive scheme for $\hat{\mathcal{W}}_n$ are

$$u = \frac{1}{b}\left(-k_n z_n - \frac{1}{4}z_n - \frac{1}{4}z_n - \hat{\mathcal{W}}_n^T \varphi_n(x) + \dot{s}_{n-1}\right), \ \dot{\hat{\mathcal{W}}}_n = -\mu_n \hat{\mathcal{W}}_n + z_n \varphi_n(x) \tag{23}$$

Substituting Eqs. (20) and (23) into (22) gets

$$\dot{V}_n \leq -\frac{k_1 z_1^2}{\rho(t)\left(1-z_1^2\right)^2} - \sum_{j=2}^{n} k_j z_j^2 - \frac{1}{2}\sum_{j=1}^{n} \mu_j \hat{\mathcal{W}}_j^T \hat{\mathcal{W}}_j + \frac{1}{4}\sum_{j=1}^{n} \delta_j^{*2} + \frac{1}{2}\sum_{j=1}^{n} \mu_j \mathcal{W}_j^{*T} \mathcal{W}_j^* \qquad (24)$$

where $\delta_j^* = d + \delta_1$.

3.3. Stability Analysis

Based on the recursive controller design in subsection 3.2, the following theorem is attained.

Theorem 1. The output reference command can be tracked within finite time under the devised controller in Eq. (23). And the prescribed output tracking performance can be achieved. All the closed-loop signals are uniformly ultimately bounded.

Proof. Based on Eq. (24), one can obtain that there exists two positive constants θ_0, ω_0 such that the following inequality holds

$$\dot{V}_n \leq -\theta_0 V_n + \omega_0 \qquad (25)$$

where $\theta_0 = \min\left\{2k_1 / \rho_0, 2k_j, \mu_j \left(j = 2,3,...,n\right)\right\}$, $\omega_0 = \max\left\{1/4\sum_{j=1}^{n} \delta_j^{*2} + 1/2\sum_{j=1}^{n} \mu_j \mathcal{W}_j^{*T} \mathcal{W}_j^*\right\}$. Thus, the output

tracking error system is uniformly ultimately bounded. Accordingly, one can obtain that the standard output tracking error z_1 will converge to the following compact set

$$\frac{1}{2}\frac{z_1^2}{1-z_1^2} \leq \frac{\omega_0}{\theta_0} \Rightarrow |z_1| \leq \sqrt{\frac{2\omega_0}{2\omega_0 + \theta_0}} \Rightarrow |e| \leq \sqrt{\frac{2\omega_0}{2\omega_0 + \theta_0}}\rho(t) < \rho(t) \qquad (26)$$

As shown in Eq. (26), one can find that the output tracking error will be kept into the performance envelope in the whole time domain. Due to the finite-time convergence of the proposed performance function, it is easy to obtain that the output tracking error will converge to the final tracking error bound within finite time. Thereby, **Theorem 1** is proved. ∎

4. Numerical Simulations

To validate the effectiveness of the proposed control scheme, the application to a chemical reactor system- continuously stirred tank reactor (CSTR) is organized, wherein, the CSTR system is given by

$$\begin{cases} \dot{x}_1 = -x_1 + D_a\left(1-x_1\right)\exp\left(\dfrac{x_2}{1+x_2/\kappa_0}\right) \\[2mm] \dot{x}_2 = bu - \left(1+b\right)x_2 + B_0 D_a\left(1-x_2\right)\exp\left(\dfrac{x_2}{1+x_2/\kappa_0}\right) + d \end{cases} \qquad (27)$$

where D_a, κ_0, B_0 represent Damökhler number, activated energy and heat of reaction [9]. In the simulation, $D_a = 0.072, \kappa_0 = 20, B_0 = 8, b = 0.3$. The relevant simulation parameters are chosen as: $k_1 = 10, k_2 = 100, \eta_0 = 0.15, m_1 = q_1 = 5, n_1 = p_1 = 3, \rho(0) = 5, \rho_\infty = 0.2, \varepsilon_1 = 0.01, x_1(0) = 0.15, x_2(0) = -0.3$, $\mu_1 = \mu_2 = 2$. The initial RBFNN weights $\hat{\mathcal{W}}_j (j = 1,2)$ are randomly chosen in the interval [-1.5, 1.5]. The total nodes of the RBFNN are 20, the parameters involved in the Gaussian function $\zeta_{j,i}, \xi_{j,i} (j = 1,2, i = 1,2,...,20)$ are randomly chosen in the interval [-1, 1]. The expected output reference is $y_r = 0.5$. The corresponding simulation results are given in Figs. 1 and 2.

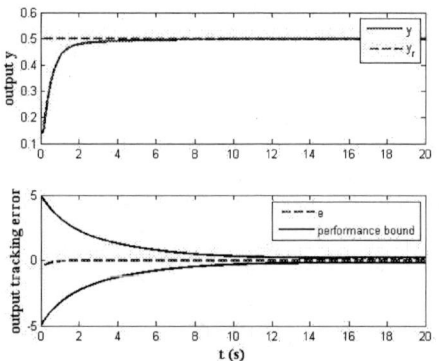

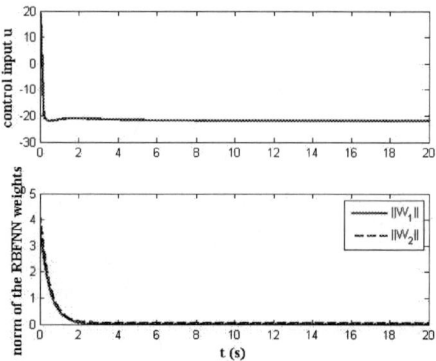

Figure 1. Output trajectory and tracking error   Figure 2. Control input and RBFNN weights

As shown in Figs. 1 and 2, one can find that the desired output reference command can be tracked with guaranteed prescribed performance within 10 seconds. Thus, the proposed control scheme is effective in dealing with the output tracking problem. Figure 2 shows that the RBFNN weights will converge to the relevant optimal ones quickly. So the proposed adaptive scheme is stable. To sum up, the proposed control scheme and the adaptive scheme are effective.

5. Conclusion
In this paper, a novel finite-time control scheme is proposed in the prescribed performance control structure. Compared with the existing works, the fractional power state or output feedback is avoided, which makes the relevant controller easily achievable online. Application to a chemical reactor system validates the effectiveness of the proposed control scheme.

References
[1]   Yu, S., Yu, X., Shirinzadeh, B., & Man, Z. (2005). Continuous finite-time control for robotic manipulators with terminal sliding mode. Automatica, 41(11), 1957-1964.
[2]   Amato, F., Ariola, M., & Cosentino, C. (2010). Finite-time control of discrete-time linear systems: analysis and design conditions. Automatica, 46(5), 919-924.
[3]   Xu, Y. (2017). Robust finite-time control for autonomous operation of an inverter-based microgrid. IEEE Transactions on Industrial Informatics, 13(5), 2717-2725.
[4]   Bechlioulis, C. P., & Rovithakis, G. A. (2008). Robust adaptive control of feedback linearizable MIMO nonlinear systems with prescribed performance. IEEE Transactions on Automatic Control, 53(9), 2090-2099.
[5]   Li, Y., & Tong, S. (2015). Prescribed performance adaptive fuzzy output-feedback dynamic surface control for nonlinear large-scale systems with time delays. Information Sciences, 292, 125-142.
[6]   Wei, C., Luo, J., Dai, H., Yin, Z., & Yuan, J. (2017). Low-complexity differentiator-based decentralized fault-tolerant control of uncertain large-scale nonlinear systems with unknown dead zone. Nonlinear Dynamics, 89(4), 2573-2592.
[7]   Song, Y., & Guo, J. (2017). Neuro-adaptive fault-tolerant tracking control of Lagrange systems pursuing targets with unknown trajectory. IEEE Transactions on Industrial Electronics, 64(5), 3913-3920.
[8]   Ni, J., Liu, L., Liu, C., Hu, X., & Li, S. (2017). Fast fixed-time nonsingular terminal sliding mode control and its application to chaos suppression in power system. IEEE Transactions on Circuits and Systems-II: Express Briefs, CIRCUITS AND SYSTEMS—II: EXPRESS BRIEFS, 64(2), 151-155.
[9]   Chang, W. D. (2013). Nonlinear CSTR control system design using an artificial bee colony algorithm. Simulation Modelling Practice and Theory, 31, 1-9.

| ISPECE | IOP Publishing |

Dynamic weighing system based on Internet of Things technologies

Jing Gang Cui[1,2], Jiong Mu[1,2,*], Ke Cheng Liu[1,2] and Yu Zhu[1,2]

[1]College of Information Engineering,

[2]Key Laboratory of Agricultural Information Engineering of Sichuan Province

Sichuan Agricultural University

Ya'an, Sichuan, China.

[*]Corresponding Author: jmu@sicau.edu.cn

Abstract: This paper studies and implements dynamic weighing system based on the Internet of Things (IOT) technology, which is used to detect the weight changes of animals easily and conveniently, in order to achieve the purpose of accurate feeding. The system is mainly composed of dynamic gravity data acquisition subsystem, RFID identification, network communication, web management, terminal database and BP neural network. The weight of animals is collected in real time, and the purpose of precise feeding is realized through the dynamic accurate measurement and storage of the animal weight. The system is simple, practical and stable.

1. Introduction

As a major industry in China, animal husbandry is one of the bases of national economic development. At present, various intelligent breeding equipment needs to be improved in automation and intelligence, and there is still a certain gap between the precise control of breeding process and that of foreign countries. How to achieve targeted and precise feeding mode through monitoring the weight of each animal is conducive to effective scientific feeding management and effective feed saving [1]. Therefore, how to achieve dynamic and accurate weight monitoring in the process of animal breeding is one of the key points to solve the problem.

In recent years, with the development of monitoring technology, intelligent weighing system is mainly applied in three fields [2]:The first category is mainly applied in the field of material circulation, such as logistics companies, supermarkets, mines and other industries where materials are frequently in and out. The second type is mainly used in the industrial production field. For example, chemical plants, food plants and other plants need to mix different ingredients according to a certain ratio. At this time, intelligent data collection is needed to process and transmit information. The third kind is mainly applied in the management level of enterprises, which is the extension of the second kind of application. For example, in some batching production enterprises, the intelligent weighing system runs through the raw material procurement, production, sales, management and other aspects.

With the development of global e-commerce, the development of the logistics industry is strong, and about 70% of the world's goods need to go through the weighing process. The requirements for the weighing system are different due to different application environments [3]. How to efficiently and accurately weigh is the challenge of various weighing systems. This paper mainly studies how to

Content from this work may be used under the terms of the Creative Commons Attribution 3.0 licence. Any further distribution of this work must maintain attribution to the author(s) and the title of the work, journal citation and DOI.

Published under licence by IOP Publishing Ltd

conveniently carry out real-time and accurate monitoring of animal weight in the animal feeding environment. This study uses the existing dynamic weighing theory to design a convenient and practical dynamic weighing system.

2. General design of system

2.1. System function design

2.1.1. System framework.
The dynamic weighing system consists of the following six subsystems: gravity data acquisition subsystem, network communication subsystem, web management subsystem, database subsystem, data processing subsystem, and RFID identification subsystem.

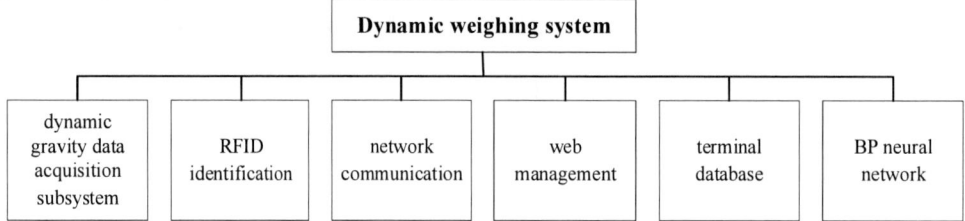

Figure 1: The Constituent is of dynamic weighing system

The gravity data acquisition subsystem collects the weight of the moving object by adopting the gravity sensor of the point array layout. The network communication subsystem stipulates the specific protocol and the related interface standards for communication between the parts of the system to ensure the stable interaction. The database subsystem is responsible for storing the data collected by the gravity data acquisition subsystem to ensure the persistence of the experimental data;The data processing subsystem optimizes the data collected by the gravity data collection subsystem through the corresponding algorithm. The web management subsystem is responsible for the visual management of the data collected by the gravity data acquisition subsystem, which can be used to operate the background data and related visual presentation through the web management subsystem. The RFID identification subsystem can distinguish different individuals in the experiment through the ear tag and accurately control the individual data.

2.1.2. System Modbus interface design.
Modbus protocol is a de-facto standard protocol in industrial automation. As the size and complexity of industry systems increase rapidly, the importance of real-time communication protocols arises as well[4]. And Modbus RTU is a simple and robust master-slave protocol that accepts the integration of a master with up to 247 slaves into a bus topology[5], and its physical interface complies with the specifications of RS-485 and RS-232.

Universal Serial Bus (USB) is a new personal computer interconnection protocol, developed to make the connection of peripheral devices to a computer easier and more efficient. It reduces the cost for the end user, improves communication speed and supports simultaneous attachment of multiple devices (up to127) RS232[6]. The system adopts HIRS-10-C10-FMModbus to carry out A/D conversion of gravity sensor data, convert analog quantity into digital quantity, and transmit it through RS-485 or RS-232. The design is of system hardware.

2.2. Hardware system overview
The hardware platform is mainly composed of object detector, namely weighing platform, Modbus signal conversion circuit, hostcomputer, display, ground inductor, etc.

When the detected object moves on the weighing platform, it first processes the measured data through the hardware system. The initially measured weight signal needs to be converted into a voltage signal by the pressure sensor, and the converted signal is very weak, almost to millivolt level. Therefore, the data amplification circuit is needed to amplify the signal size suitable for the hardware system to process. The collected data is analog signal, and the analog quantity is converted into digital quantity by A/D conversion circuit through Modbus digital junction box and sent to the host for software processing.

The design of the process flow of the system is as follows:

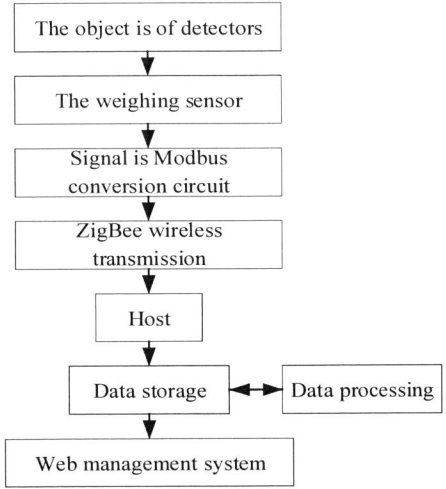

Figure 2: Weighing system is processing flow

2.3. Introduction of each hardware subsystem of the system

2.3.1 Gravity data acquisition subsystem

ZigBee is a specification formalized by the IEEE 802.15.4 standard for low-power low-cost low-data-rate wireless personal area networks[7]. It is very suitable for the requirements of data acquisition by hardware in the project, so ZigBee is chosen as the wireless transmission channel. The weighing sensor is actually a device that converts mass signals into measurable electrical and output it. In the process of weighing, the accuracy of the sensor will decline with the development of time. For the sensor itself, the performance of the sensor used in different environments varies greatly, and the price difference is also large [8]. Factors influencing the long-term stability of the weighing sensor include aging, fatigue and environment. As the operating environment of this system is in the field of animal husbandry and breeding, the environmental cleanliness is relatively poor and the humidity is relatively high. Therefore, the requirements on the operating environment of the sensor are relatively high, and the reliability and stability of the sensor should be taken into consideration.

The gravity sensor is a single-point acquisition device. The collected data cannot be directly obtained by weight, and the value data can only be obtained by special processing. Moreover, the collected object is active. Therefore, this paper proposes a multi-point acquisition array layout, as shown in the following figure:

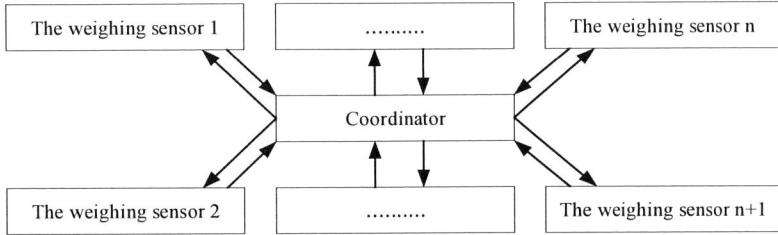

Figure 3: The composition is of gravity data acquisition subsystem

2.3.2. RFID identification subsystem

Existing RFID read-write system was based on cable transmission, there are the positioning ofthe read-write device, poor flexibility short data transmission distance, higher cost of its equipment, and other short-comings in its design[9]. wireless signal transmission is introduced in order to improve the transmission distance. For example, for the RFID chip in the 2.4ghz transmission band, the distance can be up to 10 meters [10]. RFID is combined with ZigBee network. RFID tag relies on wireless signal for identification. The identification of payload information P can be customized according to actual conditions, such as P={ID, gender, weight value}. Applied to this project, the independence of individual information is guaranteed, and the precise control of each individual is realized. As shown in the following figure:

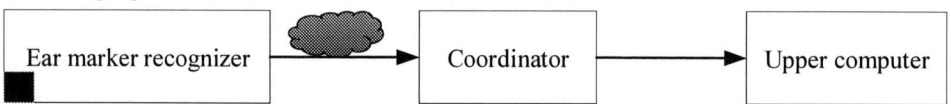

Figure 4: RFID chip data receive process

2.3.3. Network communication subsystem

Zigbee protocol stack is the specific implementation form of the protocol. The protocol stack is an interface between the protocol and the user. Developers use the protocol stack to realize wireless data sending and receiving. After the data collected by the sensor is transmitted to the digital junction box, the digital junction box transmits the corresponding signal to the RS-485 transposition serial interfaceequipment. The RS-485 serial interface is adopted to design the conversion circuit, which realizes the conversion of the collected data from analog signals to digital signals in the standard ModbusRTU format, and transmission through ZigBee wireless network.

The network communication subsystem is shown as follows:

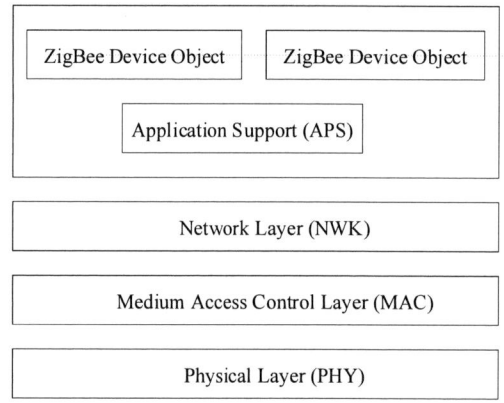

Figure 5: Network communication architecture

3. System software design

3.1. Software system overview

The hardware system transmits the collected data to the upper computer through ZigBee wireless network. The upper computer sends the collected data to the database management system for storage. Then the collected data is preprocessed and divided into train set and test set. The BP neural network is trained by using the train data, and a set of BP neural network weights that meet the accuracy requirements are obtained, and a reasonable network model is established.

3.2. Introduction of software subsystem

3.2.1. Web management subsystem

The Web management subsystem is developed in PHP based on B/S architecture. The function of the Web management subsystem is to manage the data collected by the sensor uniformly, and enable the data to be displayed in the browser for the convenience of data analysis when modeling.

3.2.2. Database subsystem

In order to facilitate data processing, this system adopts MySQL database. The subsystem mainly completes the storage of the raw data collected by the sensor, the training dataset and test dataset for the modeling service.

Meanwhile, in order to improve the storage efficiency of the database, we take the following measures to optimize the database:

(1)   While the database of large product requires high reliability and concurrency, InnoDB as the default MySQL storage engine is a better choice than MyISAM.

(2)   The schema, tables, and fields of the organization database are used to reduce the I/O overhead, keep related items together and plan ahead, so that performance can be maintained at a high level as the volume of data grows. Design the data table need to minimize its space, the table's primary key as short as possible.

(3)   Increasing the buffer pool size at InnoDB allows queries to be accessed from the buffer pool rather than via disk I/O. The buffering indicator is adjusted to the best level by adjusting the system variable innodb_flush_method.

(4)   Periodically review slow query logs and optimize the query mechanism to take full advantage of caching to reduce disk I/O. Use the OPTIMIZETABLE statement to reorganize the table and compress any space that might be wasted.

3.2.3. Data processing subsystem

The data processing subsystem is the core module of the system. In this module, the data collected by the system is processed through the BP neural network. Through the processing of the training dataset, a reasonable network model is established, to achieve better test performance.

In the process of dynamic weighing, dynamic weighing is mainly affected bythe value of five sensors, therefore, five values were selected as input parameters. Dynamic weighing is to estimate the static weight, so the static weight is selected as the output parameter, thus forming a BP neural network with multiple inputs and single outputs. We first learned its scope based on the empirical formula, and then used the train dataset to repeatedly train the BP neural network, then analyzed and considered the training effect of each network, and finally obtained the number of nodes in the hidden layer of BP neural network. The empirical formula is as follows:

$$H = \sqrt{I + O} + a, a \in [1,10] \tag{1}$$

Where, H is the number of nodes in the hidden layer, I is the number of nodes in the input layer, O is the number of nodes in the output layer, and a is a constant.

The BP neural network model is established as follows:

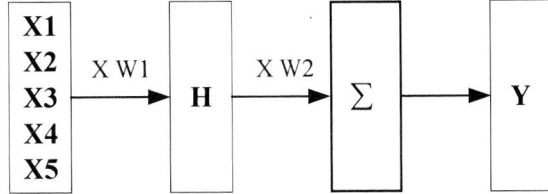

Figure 6: The model is of BP neural network

In the figure, X1 to X5 are the input parameters of BP neural network, them are the values corresponding to the five sensors. Y is the output parameters of BP neural network, namely static weight, and W1 is the weight vector matrix connecting the input layer to the hidden layer. W2 is the weight vector matrix connecting the hidden layer to the output layer. In the BP neural network designed in this system, W1 is the vector matrix of 5x10 and W2 is the vector matrix of 10x1.

Through repeated training, the hidden layer chooses tansig, a hyperbolic tangent function, as the activation function, while the output layer chooses purelin, a linear function, as the activation function. The expressions of the two activation function are as follows:

$$f(x) = \frac{1-e^{-x}}{1+e^{-x}} \quad (-1 < f(x) < 1) \tag{2}$$

$$f(x) = \begin{cases} -1, x \leq -1 \\ x, -1 < x < 1 \\ 1, 0 \leq x \end{cases} \tag{3}$$

In order to ensure the stability of the training process, their selection also affects the training process of the whole network. If it is too large, the adjustment quantity of the weight will become larger, which will produce shock phenomenon and make the network unstable. Otherwise, the network convergence speed will be slow, but the network error value will tend to the minimum. Therefore, learning rate is selected as small as possible, which not only makes the network training stable, but also makes the output error of the network reach the preset requirement. Its selection range is about 0.01 and 0.07. According to the actual situation of the dynamic weighing system, the initial weight range of BP neural network is set as [-1,1], and the learning rate is set as 0.01. In the network, the training steps, training errors and other parameters should be set, and the actual situation of training data should be considered for setting.

In BP neural network training, mean square error is used to measure the network effect. For all the training data, the expression of the mean square error function of BP neural network is as follows:

$$E(m) = \frac{1}{2}\sum_{k=1}^{P} E_k(m) = \frac{1}{2}\sum_{k=1}^{P}\sum_{l=1}^{L}(e_l^k(m))^2 = \frac{1}{2}\sum_{k=1}^{P}\sum_{l=1}^{L}(C_l^k - Y_l^k(m))^2 \tag{4}$$

In the above equation, C is the predicted value of the network, Y is the actual output value of the network, and P and L are the number of input and output samples respectively.

4. Algorithm Design

The system code is written by matlab. The main function is to build the neural network model and train and test it on the data set. We first initialize the network weight and neuron threshold, prepare the training data, and input the training data to the neural network. Then in forward propagation, the training data go through hidden neurons and output the data results in the output neurons. The weights and thresholds of each layer are corrected according to the error function. Until the termination condition is met. Finally, the model is verified with the test data, and the error is obtained. If the requirement is not met, the model is returned to the previous step, and the model is modified again to be trained again.

5. Data test

The system is tested on the integrity of database, interface function, user interface and system integration to ensure that the system has good performance and can withstand certain load strength and stable operation.

5.1. Data collection

During the experiment, the real static weight of the experimental animal or human body was measured in a static state, and then the animal or human body was moved on the dynamic weighing platform. The weighing platform can collect dynamic data 8 times per second (the specific collection times can be set by parameters), forming the training data of the neural network.

Table 1: Structure of data is collected

Actual Weight	sensor1	sensor 2	sensor 3	sensor 4	sensor 5
57850	141383	153153	28820	46294	198929
57850	125064	129928	64884	60162	187338
57850	122323	129483	82580	60963	177302
……	……	……	……	……	……

5.2. Data processing

We take the collected nearly 6,000 bars as the training data set. First, after preprocessing the train data, it is transmitted to BP neural network. After the training of the train data, the weight is updated with the back propagation algorithm, and finally the processing model of BP neural network is established.

After establishing the model, we randomly selected 4 groups of animal data and 6 groups of human data as the test data set, and compared the data results of dynamic weighing data processed by BP neural network with the real data tested under static conditions. The average error rate is 0.016kg.

Table 2: Test data is error rates

Data type	Actual weight	Data group number	Measuring weight	Average error rate	Total average error rate
Animal1	4.5kg	568	4.51kg	0.014kg	0.016kg
Animal2	78.5kg	604	7.74kg		
Animal3	8.2kg	462	8.31kg		
Animal4	12.2kg	478	12.31kg		
human1	58.4kg	600	59.39kg	0.018kg	
Human2	59.1kg	748	60.16kg		
Human3	59.6kg	655	58.53kg		
Human4	60.6kg	489	61.69kg		
Human5	57.85kg	563	58.56kg		
Human6	61.25kg	408	62.41kg		

6. Conclusion

This paper mainly studies how to improve the precision of the dynamic weighing system, and focuses on how to use ZigBee wireless communication, BP neural network and other technologies to process the data, to achieve the accurate acquisition of the weight data of the breeding objects in the animal husbandry farms, so as to improve the intelligent breeding management level. The average error rate obtained in this study will increase slightly with the weight of the actual object. The main reason is the influence of elastic deformation of the sensor. In the later stage, the replacement of the sensor with better stability will be considered. In addition, we will try to deepen and improve the model of neural network in order to obtain higher precision effect.

Acknowledgement

This paper was supported by Student's Platform for Innovation and Entrepreneurship Training Program and Key Laboratory of Agricultural Information Engineering of Sichuan Province, Sichuan Agricultural University.

References

[1] Yu Jie Fei 2016 *Research on intelligent feeding system design and feeding control algorithm* ()Heilongjiang: Harbin engineering university.
[2] Da Si Chen 2016 *Design and implementation of intelligent ingredient weighing system.* Xiamen university, Fujian .
[3] Hong Liang Gao and Na zhao 2013 Dynamic weighing system *Weighing apparatus* **2** 49-50.
[4] BS Kim, D Lee and TH Choi 2016 Performance evaluation for Modbus/TCP using Network Simulator NS3 *Tencon IEEE Region 10 Conference* 1-5.
[5] GBM Guarese, FG Sieben, T Webber, MR Dillenburg and C Marcon 2013 Exploiting Modbus Protocol in Wired and Wireless Multilevel Communication Architecture *Computing System Engineering* 13-18.
[6] V Madhurima and N Suguna 2013 Design and Implementation of RS232 to Universal Serial Bus Protocol Converter Using FPGA *University of Hull* **2**(3).

[7] YY Shih, WH Chung, PC Hsiu and AC Pang 2013 A Mobility-Aware Node Deployment and Tree Construction Framework for ZigBee Wireless Networks *IEEE Transactions on Vehicular Technology* **62** (6) 2763-2779.

[8] B Wang, X Gu, L Ma and S Yan 2017 Temperature error correction based on BP neural network in meteorological wireless sensor network *International Journal of SensorNetworks* **23**(4) 265.

[9] H Wang 2015 The Design of Mobile RFID Read-Writable Terminal for Logistics Warehousing Application *Chinese Journal of Electron Devices.*

[10] Piersol kurt and Gudan ken 2016 Programmatic control of RFID tags *Ricoh Company: US201615214373*, 07-19.

Application of Weighted Fusion Algorithm in Air Tightness Detection Device

Yang Liu, Jie Gao, Linggai Zhang

Chengyi University College, Jimei University, Xiamen, Fujian, 361021, China

*Corresponding author's e-mail: 11872130@qq.com

Abstract. Gastightness testing plays an increasingly important role in ensuring the quality and performance of sealed products and the safety of their use. In this paper, a gas tightness detection scheme based on carbon dioxide gas sensor is proposed. The adaptive weighted fusion algorithm is used to fuse the weighted data collected by multiple carbon dioxide sensors to compensate for the error of single sensor test. The method is applied to leak detection of tracer gas, with high detection accuracy and wide range to meet production demand.

1. Introduction

With the continuous progress of industrial technology, the application field of air tightness detection device is more and more extensive. From aerospace to production and life, any potential minor leakage may lead to unqualified products or even lead to safety accidents. At present, there are many kinds of leak detection devices in the market, their principles and methods of use are different, and the results obtained by optimizing the information obtained are different, thus affecting the detection efficiency. In this paper, a multi-sensor weighted fusion algorithm is proposed. This method is applied to tracer gas leak detection, which has high detection accuracy and wide range, and meets the production requirements.

There are many factors affecting the accuracy of air tightness detection. It is difficult to determine the exact functional relationship between leakage and each factor. With the development of modern control technology, people apply various intelligent algorithms to data processing to improve the accuracy of air tightness detection. In this paper, a multi-sensor adaptive weighted fusion algorithm is proposed. Carbon dioxide is used as a tracer gas, and then the information collected by multiple carbon dioxide sensors is fused with weighted data to compensate for the test error of a single sensor.

2. System scheme

The air tightness testing experimental system is controlled and processed by computer. The controller is composed of a data acquisition card for signal acquisition, transmission and processing. The AI channel of the data acquisition card is used to transmit the signal value collected by the gas pressure sensor and the carbon dioxide sensor. The AO channel outputs the voltage to the electric proportional valve for regulation. Detection of air pressure, DO signal is used to control the opening and closing of the external solenoid valve and control the indicator light for sound and light alarm [57]. The control block diagram of the detection system is shown in Figure 1. Considering the number of test channels and the performance of data acquisition card, Yanhua High Precision Multifunctional PCI-1711 data acquisition card is selected to complete the signal acquisition and control.

Content from this work may be used under the terms of the Creative Commons Attribution 3.0 licence. Any further distribution of this work must maintain attribution to the author(s) and the title of the work, journal citation and DOI.

Published under licence by IOP Publishing Ltd

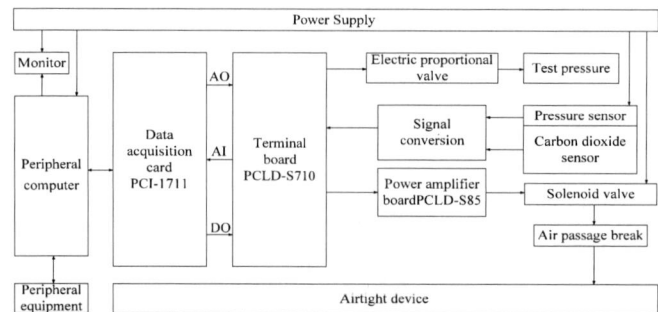

Figure 1 System structure diagram

The principle of gas tightness detection based on carbon dioxide sensor is to calculate the leakage rate by measuring the change of carbon dioxide concentration in the container. According to the experimental purpose and requirement of the gas tightness testing scheme based on carbon dioxide sensor, the software flow of automatic testing is determined, which includes initialization of data acquisition card, zero adjustment of sensor calibration, gas path control, data acquisition and processing, display and storage of test results, etc.

3. Principle of leakage test

3.1. leakage rate formula

The content of each component in the atmosphere is usually expressed as volume concentration and mass volume concentration. When carbon dioxide is used as tracer gas, the leak rate detected cannot be equivalent to the air leak rate, so the results need to be converted equivalently. The equivalent leakage rate of the air tightness detection scheme designed in this paper is shown in formula (1):

$$Q' = \sqrt{\frac{M_1}{M_2}} \frac{(C_2 - C_1) * (V_1 - V_2) * 60}{t * 10^6} \tag{1}$$

In the formula, M_1 is the molar mass of some gas component.

M_2 is the average molar mass of air.

V_1, V_2 indicates the volume of the container and the volume of the detected parts.

C_2, C_1 indicates the initial concentration value of tracer gas in the container and the final concentration value at the end of detection, unit ppm;

t indicates detection time, unit s.

3.2. leakage diffusion model analysis

The air tightness detection scheme designed in this paper is aimed at the micro leakage situation, that is, the uniform continuous leakage model with small change of air pressure. Therefore, the Gauss plume diffusion model is used for theoretical analysis. The Gauss plume diffusion model can be used to describe the concentration distribution of gas diffusion released from a persistent gas leakage point. In a stable detection environment, the leak point is taken as the virtual coordinate origin and the right-hand coordinate system is established. The horizontal (main direction) of gas diffusion is x-axis, the vertical (y-axis) is Y-axis and the vertical (z-axis) is z-axis. Without considering the spatial obstacles, noise and airflow, the expressions of concentration function at each point of the Gauss plume diffusion model can be obtained, as shown in equation (3):

$$C(x, y, z) = \frac{Q}{2\pi\sigma_y\sigma_z} \exp\left[-\frac{1}{2}\left(\frac{y^2}{\sigma_y^2} + \frac{z^2}{\sigma_z^2}\right)\right] \tag{2}$$

$$C(x,y,z) = \frac{Q}{2\pi\sigma_y\sigma_z}\exp\left(-\frac{y_2}{2\sigma_y^2}\right)\left\{\exp\left(-\frac{(z-\mathrm{H})^2}{2\sigma_Z^2}\right) + \exp\left(-\frac{(z+\mathrm{H})^2}{2\sigma_Z^2}\right)\right\} \tag{3}$$

In the formula, H : the height of the leakage source is m .

$C(x,y,z)$: gas concentration at coordinate point (x,y,z), unit $mg\big/m^3$;

Q : gas leakage rate, unit $m^3\big/s$;

σ_y, σ_z : diffusion coefficient of gas in Y and Z directions, unit m .

4. Weighted data fusion algorithm

When the measured part leaks, the gas diffuses through the leak hole into the detection vessel to form an uneven concentration field. The carbon dioxide concentration curve obtained by a single sensor can not reflect the change of carbon dioxide concentration in the detection unit, but reflects the change of carbon dioxide concentration in the local space, so it can not. The detection result from a single sensor represents the leakage rate of the detected system. In this paper, several carbon dioxide gas sensors are distributed in different locations of the detection container to collect data, and then weighted data fusion is carried out on the multi-sensor information. Finally, a difference in concentration representing the change in the concentration of the entire test vessel is obtained, thereby making up for the deficiency of the single gas sensor being susceptible to other environmental gases, and optimizing the experimental scheme.

In this paper, the optimal weighted data fusion algorithm is used, that is, the optimal fusion value $\hat{x}$ is obtained when the total mean square error is the smallest and the corresponding weighted values of carbon dioxide gas sensors in n different locations are different. Assuming that the measurement value of sensor $S_1, S_2, \dots, S_n$ is $X_1, X_2, \dots, X_n$ and the detection results are independent of each other, the corresponding detection results of sensor P at the first time can be recorded as $x_p(i)$. According to the measured data provided by the sensor, the standard deviation corresponding to each single measurement of the sensor is $\sigma_1, \sigma_2, \dots, \sigma_n$, and the corresponding weights are $w_1, w_2, \dots, w_n$, respectively. Weighted data fusion can get the optimal data fusion value $\hat{X}$ with a minimum mean square error of σ^2 . In this paper, the weight of each sensor's measurement value is calculated by the actual measurement results. From the analysis of the actual measurement results, the influence of sensor placement on the experimental results has been solved. The model structure of the optimal weighted data fusion algorithm is shown in Figure 2 .

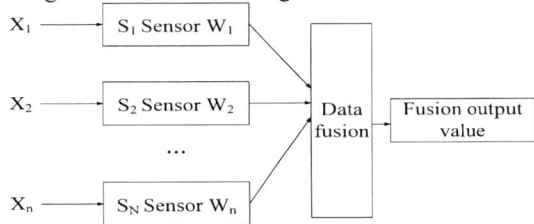

Figure 2. optimal weighted data fusion algorithm

In weighted data fusion, the state estimation and weight of the measured objects meet the requirements of formula (4) respectively.

$$\hat{x} = \sum_{i=1}^{n} w_i x_i, \sum_{i=1}^{n} w_i = 1 \tag{4}$$

The total mean square error is:

$$\sigma^2 = E\left[(x-\hat{x})^2\right] = E\left[\sum_{i=1}^{n} w_i(x-x_i)\right]^2$$

$$= E\left[\sum_{i=1}^{n} w_i^2(x-x_i)^2 + 2\sum_{\substack{i=1,j=1 \\ i}}^{n} w_i(x-x_i)w_j(x-x_j)\right] \qquad (5)$$

Because x_i is independent of each other and is unbiased for X, so:

$$E\left[(x-x_i)(x-x_j)\right] = 0(i,j=1,2,......n, i \neq j) \qquad (6)$$

Therefore, the total mean square error is:

$$\sigma^2 = E\left[\sum_{i=1}^{n} w_i^2(x-x_i)^2\right] = \sum_{i=1}^{n} w_i^2\sigma_i^2 - \lambda(\sum_{i=1}^{n} w_i - 1) \qquad (7)$$

When the total mean square error σ^2 is the smallest, the Lagrangian conditional extremum algorithm is used to construct an auxiliary function to calculate the value of w_i :

$$\sigma^2 = E\left[\sum_{i=1}^{n} w_i^2(x-x_i)^2\right] = \sum_{i=1}^{n} w_i^2\sigma_i^2 - \lambda(\sum_{i=1}^{n} w_i - 1) \qquad (8)$$

Set up equations:

$$\begin{cases} \dfrac{\partial f}{\partial w_i} = 2w_i\sigma_i^2 - \lambda = 0(i=1,2......n) \\ \dfrac{\partial f}{\partial \lambda} = 1 - \displaystyle\sum_{i=1}^{n} w_i = 0 \end{cases} \qquad (9)$$

The required conditions can be obtained:

$$w_i = -\frac{\lambda}{2\sigma_i^2}, \sum_{i=1}^{n} w_i = 1 \qquad (10)$$

From the formula (10), the total mean square error σ^2 value can be obtained at the minimum, and the corresponding weights of each sensor are:

$$w_i = \frac{\sigma_i^{-2}}{\displaystyle\sum_{i=1}^{n} \sigma_i^{-2}} \qquad (11)$$

The resulting fusion output value is:

$$X = \sum_{i=1}^{n} w_i * X_i \qquad (12)$$

5. Analysis of experimental results

In this paper, three carbon dioxide sensors are used to collect data at different locations in the detection container. The standard deviation and weight of each sensor are calculated by calling the standard deviation sub-VI and weight sub-VI which are independently compiled. The weighted data fusion is carried out. Finally, the leakage rate is calculated by using the carbon dioxide concentration value obtained from the fusion. In the sequence of equal precision test results, the formula for calculating the standard deviation of single measurement of the sensor is shown in equation (13):

$$\sigma = \sqrt{\frac{\sum_{i=1}^{n} v_i^2}{n-1}} \tag{13}$$

In this paper, the experimental conditions are designed to meet the experimental ambient temperature of (20 5 C) and the output pressure of the electric proportional valve is set to 100 kPa. The volume of the test vessel is 600 ml, the volume of the tested piece is 100 ml, the time of pumping is 10 seconds, the time of charging is 10 seconds, and the time of stabilization is 50 seconds. Three carbon dioxide sensors were placed in different positions in the detection container, and the sampling period was 0.01s. The results of the three sensors are measured every 25s, and the test data are shown in Table 1. According to the weighted fusion algorithm, the standard deviations of each sensor are 108.2, 113.7 and 112.8, and the corresponding weights are 0.36, 0.33 and 0.34, respectively. The synthetic concentration is shown in Table 1.

Table 1. Formatting sections, subsections and subsubsections.

time/s	S1/ppm	S2/ppm	S3/ppm	Synthesis/ppm
0	483.59	482.91	483.31	483.28
25	542.87	519.95	522.49	528.42
50	665.23	660.40	657.69	661.11
75	730.27	727.29	727.34	728.30
100	762.01	765.01	762.22	763.08
125	774.90	779.84	778.82	777.86
150	785.74	787.40	789.68	787.61
175	796.72	800.58	799.31	798.87
200	796.48	799.60	798.29	798.13
225	797.92	802.83	799.32	800.03
Concentration difference/ppm	314.33	319.92	316.01	316.77
Leakage rate ml/min	0.184	0.185	0.184	0.184

The output value C_i of the carbon dioxide sensor shown in Table 1 shows that after 20 seconds, the concentration value of the carbon dioxide sensor in the detection container rises rapidly, and the growth rate decreases with time. The sensor value tends to be stable around 180 seconds. After the detection, the standard of the carbon dioxide concentration difference ΔC is found in the formula (1). The leakage rate of leakage is 0.184ml/min. The relative error is 2.95% compared with the standard leakage rate of the standard leak rate of 0.19 ml/min under the test conditions.

50 repeated tests were carried out according to the above test conditions. The results of CO_2 concentration difference ΔC are shown in Table 2.

Table 2. reproducibility test results of 100k Pa

Serial number	Difference value/ppm	Serial number	Difference value/ppm	Serial number	Difference value/ppm	Serial number	Difference value/ppm	Serial number	Difference value/ppm

1	361.77	11	320.66	21	315.17	31	326.46	41	325.32
2	319.28	12	318.38	22	330.81	32	319.65	42	317.66
3	325.66	13	317.65	23	316.26	33	325.15	43	320.28
4	318.12	14	315.45	24	315.56	34	324.24	44	324.14
5	315.24	15	319.52	25	318.16	35	314.65	45	329.45
6	324.06	16	315.61	26	320.47	36	324.53	46	315.08
7	315.14	17	320.24	27	320.24	37	321.78	47	320.29
8	319.69	18	318.74	28	327.85	38	320.60	48	320.69
9	317.16	19	316.87	29	317.88	39	320.17	49	318.98
10	318078	20	323.23	30	315.25	40	319.16	50	322.58

According to the test data shown in Table 2, the average $\overline{x} = 318.93\,ppm$, total standard deviation 3.499, sample standard deviation S=3.5354 and relative standard deviation less than 10% can be calculated, which shows that the repeatability of the experiment is guaranteed, and fully verifies the practical feasibility of the detection method proposed in this paper.

6. Conclusion

In this paper, the weighted data fusion algorithm is applied to the gas tightness detection technology, and a gas tightness detection scheme based on carbon dioxide sensor is proposed.The test piece filled with carbon dioxide gas is sealed in the test container. When the test piece leaks, the leakage rate of the test piece is detected by measuring the change of carbon dioxide gas concentration in the test container. In this paper, a multi-sensor weighted data fusion algorithm is proposed. Without any prior knowledge of sensor measurement data, the minimum variance data fusion value can be fused using the measurement data provided by sensors, thus improving the accuracy of measurement data.

Fund projects： Research Projects for Young and Middle-aged Teachers in Fujian Province；
Item number:JA15652；

entry name: Development of QM100 Series Intelligent Air Tightness Testing Machine

References

[1]　Zeng Cheng-zhou.Research and Development of Leak Detecting System Based on Different Pressure Principle[D].Hangzhou: Zhejiang University, 2012.

[2]　Peng Guang-zheng,Ji Chun-hua,Ge Nan. Current Statues and Future Development of Air Tightness Detection Technique[J].Machine tools and hydraulic pressure, 2008,36(11): 172-174.

[3]　Ji Zeng-lian.Research and Design of Air Leak Testing System[D].Dalian: Dalian Jiaotong University,2008.

[4]　Yan Shi-ping. Real Time Monitoring Method of CO2 Concentration and Its Experimental Research[D].Changsha: Hunan University, 2011.

[5]　Zhang Yuan-yuan,Zhang Ju-wei,Shang Si-si ect.Research and Application on Diffusion Model of Leakage gas[J].Contemporary chemical industry,2013,42(4): 507-509.

[6]　Xiao Jian-ming,Chen Guo-hua,Zhang Rui-hua.Research on Algorithm of Diffusion Area for Gauss Plume Model[J].Computer and Applied Chemistry, 2010, 23 (6): 559-564.

[7] Wen Hao,Dong Xiao-rui, Ma Yu-cheng. The Research of the Database Connecting Methods in Lab VIEW based on ADO[A]. In: 2010 International Conference on Computer Application and System Modeling[C]. Taiyuan, China, 2010: 229-233.

Design of Intelligent Commutation Switch System Based on HPLC Carrier Scheme

Liu Mouhai[1,2], Tan Haibo[1,2], Yang Maotao[1,2], Chen Hao[1,2], Wu Zhiyong[1,2], Peng Haijun[1,2]

[1]State Grid Hunan Electric Power Company Limited Power Supply Service Center(Metrology Center), Changsha 410004, China.

[2]Hunan Province Key Laboratory of Intelligent Electrical Measurement and Application Technology, Changsha 410004, China.

1376466188@qq.com

Abstract. China's low-voltage transmission lines are mostly three-phase four-wire power supply networks. Due to the large number of single-phase power users which are dispersedly located, there are different levels of three-phase load imbalance problems in most distribution areas. With the improvement of the national economy, the electricity consumption at the grassroots level has increased, and the problem of unbalanced three-phase loads of the low-voltage power grid has become increasingly prominent. The article introduces a product which can handle unbalanced three-phase problems in the low-voltage distribution area. It can quickly and accurately detect the three-phase unbalance problem of the low-voltage distribution system. The three-phase intelligent commutation switch system can adjust the single-phase load without power cut in real time ensuring that the three-phase load in the station is in a relatively balanced state. This product can effectively reduce transformer loss and line loss caused by three-phase load imbalance, and avoid single-phase over-current, terminal low-voltage, etc. It can also prevent numerous safety hazards caused by three-phase unbalance. This paper introduces the harm of three-phase unbalance, the working principle and system of the commutation switch, and the advantages of the HPLC scheme to highlight the intelligence and program advantages of the commutation switch.

1. Introduction

In the medium and low voltage distribution network system, there are a large number of single-phase, asymmetrical, non-linear, and impact loads. Due to the poor design of the early power grid, a large number of single-phase loads are concentrated in one or two phases. These unbalanced loads can cause three-phase imbalance in the power distribution system, resulting in an imbalance of three-phase voltage and current in the power supply system.

Due to uneven load distribution, the natures of the load are also inconsistent, resulting in insufficient reactive power and unbalanced load in the low-voltage power supply system.

2. Definition of three-phase imbalance

First, Three-phase unbalance refers to the inconsistency of three-phase current (or voltage) amplitude in the power system, and the amplitude difference exceeds the prescribed range. The main cause of the three-phase imbalance is the unbalanced three-phase load, which belongs to the fundamental wave load allocation problem.

Content from this work may be used under the terms of the Creative Commons Attribution 3.0 licence. Any further distribution of this work must maintain attribution to the author(s) and the title of the work, journal citation and DOI.
Published under licence by IOP Publishing Ltd

Article 8.7.4 of The State Grid Corporation Enterprise Standard (Q/GDW519-2010) Rules for Operation of Distribution Networks stipulates that the formula for calculating unbalance degree is as follows: (maximum current-minimum current)/maximum current x 100%.

National Grid Standard: The unbalance of three-phase load should not exceed 15%. For the three-phase transformer with a small amount of single-phase load, and the neutral line current should not exceed 25% of the rated current.

2.1. The harm of three-phase imbalance

The title is set 17 point Times Bold, flush left, unjustified. The first letter of the title should be capitalized with the rest in lower case. It should not be indented. Leave 28 mm of space above the title and 10 mm after the title.

At present, China's low-voltage power grids generally adopt three-phase four-wire system and the distribution transformer is connected by Yyn0. Because of the large number of single-phase loads and the asynchronous of power consumption, three-phase unbalanced operation of distribution transformers is inevitable. Three-phase voltage or current asymmetry will cause a series of hazards to power generation, transmission, distribution equipment and electrical equipment in the power system.

- increased line loss. When the current passes through the conductor, the resistance of conductor will cause power loss. The line loss increases as the imbalance degree increases.
- increased the active power loss of distribution transformer. The existing 10/0.4kV low-voltage distribution transformers are mostly Y/yn0 connected. When the secondary side load is unbalanced and there is zero sequence current, and the primary side can not flow because there is no neutral lead-out line. When the zero-sequence current is too large resulting in the excessive zero-sequence magnetic current. The excessive drift of neutral point will cause some phase voltage to be too high, which will lead to the saturation of the core and greatly increase the iron loss.
- reducing the output of the distribution transformer. Transformer capacity is designed and manufactured according to three-phase load balance conditions. Its three-phase winding structure and performance are consistent. The rated capacity of each phase is equal, and the maximum allowable output is limited by the rated capacity of each phase. When the three-phase load is unbalanced, its maximum output can only be limited to the rated capacity of the largest phase of the three-phase load, and the relatively affluent capacity of the lighter load can reduce the output of the transformer. The output reduction of the transformer is related to the balance degree. The output reduction increases as the unbalance degree increases. It also affects the utilization of transformer equipment.
- the output power of the motor is affected and the winding temperature is raised. The three-phase voltage asymmetry caused by the three-phase load imbalance will generate a reverse-rotating magnetic field in the stator of the induction motor, and the motor operates under the action of the forward and reverse two-order rotating magnetic fields. Since the positive sequence rotating magnetic field is stronger than the reverse order rotating magnetic field, the motor rotation direction is unchanged. However, since the rotor reverse sequence impedance is small and the reverse sequence current is large, the reverse sequence magnetic field and the reverse sequence current will generate a large braking torque, which will reduce the output power of the motor and increase the temperature of the winding, which jeopardizes the safe operation of the motor.

2.2. Traditional Method to Manage Three-phase Imbalance

2.2.1. Principle overview. Adjust the load by manually changing the line. This is the most common method but requires great manpower input. And it needs to cut off the power and can't adapt to the change pattern of the load for a long term.

2.2.2. Phase-to-phase capacitance compensation. By connecting parallel capacitors phase to phase and compensating reactive power to the two phases, some active power will be transmitted while compensating reactive power according to Wang's theorem. The capacitor compensation cost is low, and the system failure does not affect the user's power supply. However, the strength of the active adjustment is small, and it is easy to cause reactive power over-compensation; the single-phase flow can only be used in each case, and the utilization rate of the capacitor is low; the balance on the distribution side is solved, but the line side is still in an unbalanced state, and the line loss cannot be reduced.

Power electronic type three-phase load can automatic adjusting device. The low-voltage static idle compensation SVG and the active power filter APF detect the unbalanced components, reactive components and harmonic components of the load through the detection technique, and use the inverter technology to generate the unbalanced, reactive, and harmonic components that need to be compensated. It can comprehensively solve problems such as reactive power, harmonics, voltage fluctuations and three-phase load imbalance in the power distribution area. However, this method is costly and can only achieve an approximate balance of the three-phase current of the low-voltage side of the transformer, which is not a real load balance.

3. Introduction to intelligent commutation switch system

The intelligent commutation switch system is a set of products used to control the three-phase unbalance of the low-voltage distribution area. It is suitable for three-phase four-wire 380V/220V low-voltage power distribution system, which can quickly and accurately detect the three-phase unbalance problem of low-voltage power distribution system. Through three-phase intelligent commutation switch system, the single-phase load is adjusted in time without power cut. The three-phase load in the station is in a relatively balanced state. This product can effectively reduce transformer loss and line loss caused by three-phase load imbalance, and suppress single-phase over-current, terminal low-voltage, etc. It can also reduce numerous hazards caused by three-phase unbalance.

The intelligent communication switch system uses the power line carrier to communicate which the intelligent master control switch acts as the host and the commutation switch act as the slave. The communication module uses HPLC carrier module, the main control switch uses an HPLC route carrier module, and the commutation switch uses an HPLC single-phase carrier module for communication.

Advantages of HPLC carrier technology:
- The carrier of HPLC is based on TCP/IP network technology which has been widely verified. It has perfect data protection and verification in link layer and network layer, which is far more better than all kinds of lightweight node organization and relay algorithms.
- High speed carrier communication of HPLC can complete data transmission in a very short time, which can greatly reduce the impact of sudden interference. Even if a communication fails, it can be quickly retransmitted to ensure data reliability.
- Most of the carrier chips of HPLC are based on 32-bit core of high performance and DSP technology, which have advantages in technical level and performance.
- In addition to data encryption in the application layer, high-intensity encryption algorithms such as DES, 3DES and AES are supported by the carrier of HPLC in the link layer, and data communication is of high security.
- Even in the communication distance where narrowband carriers have more advantages, the high-performance modulation methods such as OFDM and perfect relay networking mechanism can fully meet the current application needs of most stations.
- High performance, fast speed and strong expansion capability of carrier communication of HPLC can load many network applications, but its cost is not much higher than that of narrowband carrier, so it has the advantage in cost.

3.1. Consist of system

The system consists of the intelligent commutation terminal (responsible for load monitoring and automatic commutation control) and several commutation switch units (responsible for the operation of load commutation) as shown in fig1. Intelligent commutation terminal monitors the three-phase current of distribution transformer low-voltage outgoing line in real time. If the unbalance degree of three-phase load of distribution transformer low-voltage side exceeds the limit within a certain monitoring period, the intelligent commutation terminal reads the real-time data of current and phase sequence of distribution transformer low-voltage outgoing line and all load branches of commutation switching unit, carries out optimization calculation and sends out optimal commutation control instructions, according to which each commutation switch unit completes the specified commutation process.

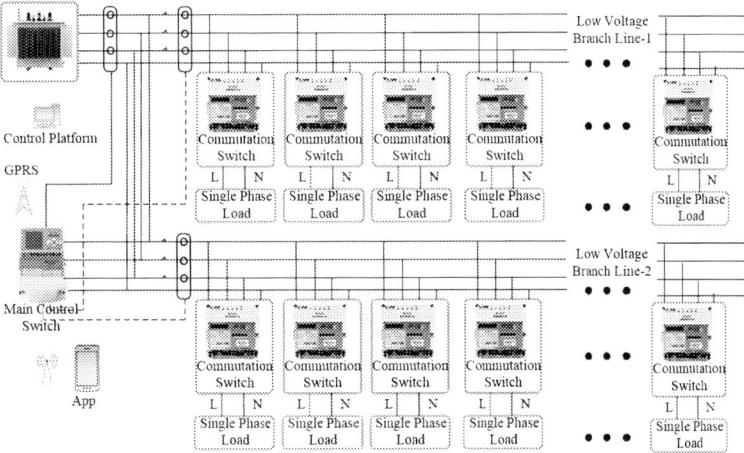

Fig 1. System topology System function

3.2. System function

- Automatic balancing of three-phase load: monitors three-phase unbalance in real-time and automatically adjusts three-phase load according to unbalance degree. Commutation time is less than 20 msec, without power cut. It will not cause the reset and restart of common electrical appliances, nor will it cause damage to electrical appliances.
- Reduce transformer loss: make the transformer in a symmetrical operation state, effectively reducing transformer loss.
- Reduce line loss: effectively reduce the neutral line current, thereby reducing the neutral line loss and phase line loss.
- Solve the problems of low or over-voltage: Solve the problems of low or over-voltage caused by three-phase unbalance, and avoid burning down electric equipment or affecting the normal operation of electric equipment due to over-voltage.

3.3. System advantages

- Maintenance-free and management-free: no special maintenance and management is required after the system has been put into operation, saving manpower and material resources and improving efficiency.
- Automatic commutation and not need to cut power; automatic commutation, and no manual participation is required; commutation time is less than 20 msec, and will not lead to power interruption.
- Reliable phase-to-phase short-circuit prevention technology; reliable hardware blocking technology to prevent multiple phase sequences from being connected at the same time; multiple software algorithms to intelligently prevent phase-to-phase short circuit.

- The switching element does not consume electricity, and the power consumption of the device is small; the permanent magnet relay mechanism is used and operated without voltage; the power consumption of the device is ≤8W.

4. Principle of the technology

4.1. Principle of Balance

4.1.1. Principle overview. A main control switch is installed at the beginning of each branch to monitor the three-phase unbalance condition and issue an adjustment command; a commutation switch is installed at the front end of the branch to monitor the load condition of the self-loaded circuit. According to the commutation command issued by the main control switch, the corresponding commutation operation is performed.

4.1.2. Intelligent networking. The commutation switch uses the power carrier to communicate. The main control switch adopts the HPLC routing module, and the commutation switch adopts the HPLC single-phase carrier module, improving the communication efficiency and reliability.

Each master switch is only responsible for communicating with the commutation switch of its same branch. One branch constitutes a subsystem, the master switch acts as the master, and the commutation switch acts as the slave.

For the mode of multi-meter switch common network, the system develops a unique intelligent networking mechanism-preemptive time-sharing communication mechanism, which avoids the conflict and interference of carrier communication between different branches and realizes the function of intelligent networking [9].

4.2. Principle of no-power commutation

The operating principle of non-power-down commutation is to complete the phase sequence switching in a very short time, and the basic basis is as follows:

4.2.1. Action components. The commutation switch uses a permanent magnet relay as the action element. Based on the characteristics of high load capacity, low power consumption, fast moving speed, low loss, reliable operation and low cost of permanent magnet relay, the commutation switch realizes the function of non-power-off commutation, which will not cut off power and ensures the quality of power supply.

There are lots of theoretical investigations and actual tests showing that the 30ms power-down time will not cause the non-load sensitive power equipment to lose power, and the commutation switch commutation time is less than 20ms, which meets the application requirements.

4.2.2. Zero-crossing commutation. To ensure longevity, the commutation switch uses zero-crossing switching technology to minimize damage to the operating components. The zero-crossing cutting technology is based on the principle of "current zero-crossing, voltage zero-crossing input", which can achieve the extremely small impact and extremely small electric arc.

4.3. Commutation Algorithm Principle.

When the unbalance degree of the branch in the main control switch exceeds the set value, the balancing logic algorithm will be started. The system is based on the balance algorithm principle of mathematical recursive logic: when the unbalance degree of the main control switch surpasses the set value, it will start the balance logic algorithm. Each user's power load is different, and the unbalance degree of each branch is different. Based on the principle of mathematical recursive logic algorithm, the system combines the balance demand with the load size of each commutation switch, carries out logical combination operation, and solves the optimal strategy.

The principle of balancing algorithm based on branch balancing strategy is that after the optimal strategy is calculated by the main control switch in each branch, the commutation switch in the branch will be ordered to perform the corresponding commutation operation, thus realizing the branch balancing. Each branch in the platform reaches three phase equilibrium state, and the three-phase balance of the transformer can be realized.

The main conclusions of this paper are as follows:

- The product basically conforms to the technical principles and requirements to be followed in the design, production and selection of the three-phase load unbalance automatic regulating device (commutation switch type) of the State Grid. Its advantages are: a) realizing three-phase balance based on single-phase load of switching terminal, and solving three-phase unbalance problem from the original cause; b) realizing three-phase balance between distribution transformer and line, effectively reducing line loss; c) not affecting user's power consumption when switching phase.

- Communication based on the carrier of HPLC has characteristics of high communication performance, fast speed and strong expansion capability. The equipment has been tested in many places, and the effect of three-phase unbalance control is obvious.

References

[1] Chang Tie-yuan, Wang Yang, Sun Hui-jia. The Design of Smart Meter with Power Quality Monitoring[J]. Electrical Measurement & Instrumentation, 2012, 49(12): 74-77

[2] Zou Hang, He Wei, Chen Yu, Liu Bing, Zhao Junhong, Zhang Huanghuang. Design of point to point communication performance testing platform for broadband power line carrier[J]. Electrical Measurement & Instrumentation, 2016, 53(21): 100-105

[3] Hsu, S.-M. and Czarnecki, L.S. (2003)Adaptive Harmonic Blocking Compensator. IEEE Transactions on Power Delivery, 18, 895-902

[4] Huang Shengli, Li Hongtao. Research on Three Phase Unbalanced Loads Based on Solid-State Intelligent Switching-Phase Switch[J]. Low Voltage Apparatus, 2016(5):13-17

[5] Feng Yongjun, Li Mingxia. Intelligent exchange switching system based on narrow-band high-speed dual mode communication[J]. Manufacturing Automation, 2017, 39(11):154-156

[6] Lin Jian- yu. The Design for Smart Metering Based on Spread Spectrem Communications[J]. Electrical Measurement & Instrumentation, 2010, 47(S2): 57-60

[7] Yang Rui, Sun Jiandong, Jiang Tiewei. Research and Realization of Low Voltage Simulation Environment for Demarcation Load Switch in Distribution Network[J]. Low Voltage Apparatus, 2014(20):33-37

[8] Xu Renheng, Qu Jingzhi, Li Dixing. Research on the full life cycle management system of smart meters[J]. Electrical Measurement & Instrumentation, 2017,54(01):67-70

[9] Tian Haiting, Zhong Kan, Ju Hanji. Research on electric power information acquisition system data analysis and monitor technology[J]. Electrical Measurement & Instrumentation,2015, 52(S1): 1-3

ISPECE   IOP Publishing

Research on four axis manipulator trajectory tracking Based on RBF Neural Network Algorithm

Luo Long

Mechanical & Electrical Department Guangzhou Institute of Technology

No. 465, HuanShi Road, Yuexiu Dist., Guangzhou 510075, P. R. China

Abstract. 1. According to the characteristics of strong coupling tracking and highly nonlinear of four axis stamping robot, this paper based on neural network control theory proposes a manipulator trajectory tracking method based on RBF neural network. The algorithm is that the neural network as a mechanical arm joint servo controller realizes the fast tracking of mechanical arm movement posture. The simulation results show that this algorithm can improve the effectiveness and accuracy of mechanical arm trajectory tracking.

1. Introduction
Robot as a comprehensive means of extending and expanding people's physical and mental will implement the "automation" in the contemporary highest sense. The application and popularization of the robot is changing the human production mode, life style and way of fighting. Stamping robots can be used for automobile, motor, household electrical appliances industry, and stamping equipment constitute a single automatic punching machine and machine automatic stamping production line. Stamping robot abroad has developed rapidly in recent years. Countries such as Britain and America in the stamping production stamping widening the use of robots further improve the single machine and stamping production line automation. Stamping robot in the development of our country is still in its infancy, most of the production enterprises also are in the use of robots. 5. Many enterprises have not even introduced manipulator, also in manual operation.

2.Stamping robot structure
Four axis series stamping robot is consist of two rotary joints and two mobile joints. The two rotary joint axis parallel, and are used for robot main body and the rotation of the end executor motion.; Move up and down two mobile joints used in the robot main body and telescopic motion robot arm. Basic parameters such as the basic parameters of the robot are shown in table 1

Table1. Basic parameters of stamping robot

number	Technical indicators	stamping robot
1	Arm form	Four axis
2	Vertical range / mm	400mm
3	Level range / mm	1000
4	Grab parts maximum weight /kg	8kg

Content from this work may be used under the terms of the Creative Commons Attribution 3.0 licence. Any further distribution of this work must maintain attribution to the author(s) and the title of the work, journal citation and DOI.
Published under licence by IOP Publishing Ltd

Each robot joint transmission way is as follows: joint 1, the rotation of the motor is transferred by synchronous belt, ball screw mechanism will motor rotational motion into straight lifting movement; 2 joints by synchronous belt transfer the rotation of the motor to the input end of the harmonic reducer, the output of the harmonic reducer connection rotation and robot arm, implementation main body rotation. 3 joints are connected by a synchronous belt transfer the rotation of the motor to in synchronization with the components of the movement, to realize the telescopic linear motion of the robot arm; Joint 4 motor machine of decelerate of planet of meter rotary motion and synchronous belt drive.

Four joint robot driven by ac servo motor, combined with the feature of the structure of the robot, in the periodic motion of the robot, according to the parameters of the drive motor and the structure characteristics of the robot body, calculate each joint movement speed and output force (moment) limit, as shown in table 2.

Table2. Table Basic parameters of joints

parameter	axis	Technical indicators
Range of motion	S	（±140°）
	L	（-40°、+85°）
	U	（-15°、+70°）
	T	（±360°）
Maximum speed	S	148.8°/s
	L	99°/s
	U	148.8°/s
	T	225°/s

3. The robot dynamics model

In 1955, Denavit and Hartenberg in "ASME Journal of Applied Mechanics" published a paper, then use that this paper for the representation and modeling for the robot, and deduced the equation of motion, this has become a robot and the robot motion modeling method of standard. Denavit and Hartenberg (D_H) model of the robot modeling connecting rods and joints of a very simple method, can be used in any robot configuration, regardless of the structure of the robot how to order and complexity. Assuming that the robot is composed of a series of joints and connecting rod. These joints may be sliding (linear) or rotation (rotation), they can be placed in any order and in any plane. The length of the connecting rod can be arbitrary (including zero), it can be bent or distorted, may also at any plane. So any set of joints and connecting rod can constitute a want to model and represent the robot.

Based on d-h notation for each joint to specify a reference coordinate system, then, to determine from a joint to the next joint (a coordinate system to the next) for transformation. If will from the base to the first joint, again from the first joints to the second joint until in the end all the transformation of a joint together, we get the total transformation matrix of the robot.

In the case of robot arm of a connecting rod. On both ends of the connecting rod n relevant section n and n + 1. The connecting rod by two geometric parameters: the connecting rod length and Angle of twist. Because at the ends of the connecting rod joints respectively have their own joint axis, normally the two axis is spatial straight lines in different planes, then the two straight lines in different planes of the common normal of long an is the connecting rod length, the Angle between two straight lines in different planes n is the connecting rod torsion Angle.

On the base of the robot, can start from the first joint transform to the second joint, and then to the third... To the robot's hand, and ultimately to the end of the actuator. If convert each defined, it can be said many transformation matrix..   Total transformation between the base of the robot and the hand is as follows:

$$^{R}T_{H} = {}^{R}T_{1}\,{}^{1}T_{2}\,{}^{2}T_{3} \cdots {}^{n-1}T_{n} = A_{1}A_{2}A_{3} \cdots A_{n} \tag{1}$$

Where n is the number of joints.

Through right by said four movement of four matrix transformation matrix will be given A four, in turn, the movement of the matrix A. Because all the transformation is relative to the current coordinate system (that is, they are relative to the current local coordinate system to measure and implementation), so all of the matrix are right. And the results are as follows:

$$^nT_{n+1} = A_{n+1} = Rot(z,\theta_{n+1}) \times Tran(0,0,d_{n+1}) \times Tran(a_{n+1},0,0) \times Rot(x,a_{n+1})$$

$$= \begin{bmatrix} C\theta_{n+1} & -S\theta_{n+1} & 0 & 0 \\ S\theta_{n+1} & C\theta_{n+1} & 0 & 0 \\ 0 & 0 & 1 & 0 \\ 0 & 0 & 0 & 1 \end{bmatrix} \times \begin{bmatrix} 1 & 0 & 0 & 0 \\ 0 & 1 & 0 & 0 \\ 0 & 0 & 1 & d_{n+1} \\ 0 & 0 & 0 & 1 \end{bmatrix} \times \begin{bmatrix} 1 & 0 & 0 & a_{n+1} \\ 0 & 1 & 0 & 0 \\ 0 & 0 & 1 & 0 \\ 0 & 0 & 0 & 1 \end{bmatrix} \times \begin{bmatrix} 1 & 0 & 0 & 0 \\ 0 & C\alpha_{n+1} & -S\alpha_{n+1} & 0 \\ 0 & S\alpha_{n+1} & C\alpha_{n+1} & 0 \\ 0 & 0 & 0 & 1 \end{bmatrix}$$

$$A_{n+1} = \begin{bmatrix} C\theta_{n+1} & -S\theta_{n+1}C\alpha_{n+1} & S\theta_{n+1}S\alpha_{n+1} & a_{n+1}C\theta_{n+1} \\ S\theta_{n+1} & C\theta_{n+1}C\alpha_{n+1} & -C\theta_{n+1}S\alpha_{n+1} & a_{n+1}S\theta_{n+1} \\ 0 & S\alpha_{n+1} & C\alpha_{n+1} & d_{n+1} \\ 0 & 0 & 0 & 1 \end{bmatrix} \tag{2}$$

4. Stamping robot trajectory tracking based on RBF neural network

The main purpose of the robot trajectory tracking control system is through a given drive moment of each joint, makes the robot's position, speed and so on ideal state variables to track the given trajectory. With general mechanical system, when determine the structure and the mechanical parameters of the robot, the dynamic characteristics of mathematical model to describe the dynamics equation. Therefore, the design method of the automatic control theory can be used provided, using the method based on mathematical model of the robot controller is designed. But in actual engineering, due to the robot system is a nonlinear and uncertainty, it is difficult to get accurate mathematical model of the robot. Using the neural networks, which can realize the accurate approximation of the unknown part of the robot dynamics equation, so as to realize control without model.

In the structure of RBF network, as $\mathbf{X} = [x_1, x_2, \dots x_n]^T$ the input vector of the network. A radial basis vectors $\mathbf{H} = [h_1, \cdots, h_m]^T$ of the RBF network, including hj for the gaussian basis function:

$$h_j = \exp(-\frac{\|\mathbf{X}\text{-}\mathbf{C}_j\|^2}{2b_j^2}), j = 1, 2, \cdots m \tag{3}$$

The first node of network center vector to $\mathbf{C}_j = [c_{j1}, \cdots, c_{jn}]$, $i = 1, 2, \cdots, n$。

Assumes that weights $\mathbf{W}$, approximating function $\mathbf{f}(\mathbf{x})$ ideal RBF network output is:

$$\mathbf{f} = \mathbf{W}\mathbf{h}(\mathbf{x}) + \varepsilon(\mathbf{x}) \tag{4}$$

Among them $\mathbf{W}$ is the weight vectors of the network, $\mathbf{h} = [h_1, h_2 \cdots h_n]$, $\varepsilon(\mathbf{x})$ is approximation error, $\varepsilon(\mathbf{x}) < \varepsilon_N(\mathbf{x})$.

Set n joint manipulator equation is:

$$M(q)\ddot{q} + C(q,\dot{q})\dot{q} + G(q) + F(\dot{q}) + \tau_d = \tau \tag{5}$$

The M（q）order for positive definite inertia matrix. $C(q,\dot{q})$ to order inertia matrix, which G（q）$n \times 1$ order inertia vector, $F(\dot{q})$ friction, τ d for the unknown external disturbance, τ as control input.

The tracking error is:

$$e(t) = q_d(t) - q(t)$$

Define the error function is:

$$r = \dot{e} + \Lambda e$$

Among them, $\Lambda = \Lambda^T > 0$

$$\dot{q} = -r + \dot{q}_d + \Lambda e$$

$$M\dot{r} = M(\ddot{q}_d - \ddot{q} + \Lambda \dot{e}) = M(\ddot{q}_d + \Lambda \dot{e}) - M\ddot{q}$$

$$= M(\ddot{q}_d + \Lambda \dot{e}) + C + G + F + \tau_d - \tau$$

$$= M(\ddot{q}_d + \Lambda \dot{e}) - Cr + C(\dot{q}_d + \Lambda e) + G + F + \tau_d - \tau$$

$$= -Cr - \tau + f + \tau_d$$

Where $f(x) = M(\ddot{q}_d + \Lambda \dot{e}) + C(\dot{q}_d + \Lambda e) + G + F$

In practical engineering, the f model uncertainties are unknown, therefore, need to f is approximated to uncertainties.

5. Simulation discussion

Choose robot manipulator system, its dynamic model is:

$$M(q)\ddot{q} + C(q,\dot{q})\dot{q} + G(q) + F(\dot{q}) + \tau_d = \tau$$

Where

$$M(q) = \begin{bmatrix} p_1 + p_2 + 2p_3 \cos q_2 & p_2 + p_3 \cos q_2 \\ p_2 + p_3 \cos q_2 & p_2 \end{bmatrix}$$

$$V(q,\dot{q}) = \begin{bmatrix} -p_3 \dot{q}_2 \sin q_2 & -p_3(\dot{q}_1 + \dot{q}_2)\sin q_2 \\ p_3 \dot{q}_1 \sin q_2 & 0 \end{bmatrix}$$

$$G(q) = \begin{bmatrix} p_4 g \cos q_1 + p_5 g \cos(q_1 + q_2) \\ p_5 g \cos(p_1 + p_2) \end{bmatrix},$$

$$F(\dot{q}) = 0.02 \operatorname{sgn}(\dot{q}), \quad \tau_d = [0.2\sin(t) \quad 0.2\sin(t)]^T \tag{6}$$

$\text{Taking } P = [p1,\ p2,\ p3,\ p4,\ p5] = [2.9, 0.76, 0.87, 3.04, 0.87], b = 0.20, z = [e\ \dot{e}\ q_d\ \dot{q}_d$

$\ddot{q}d]$。 Using Simulink and S function to the design of the control system, the system simulation results are shown below:

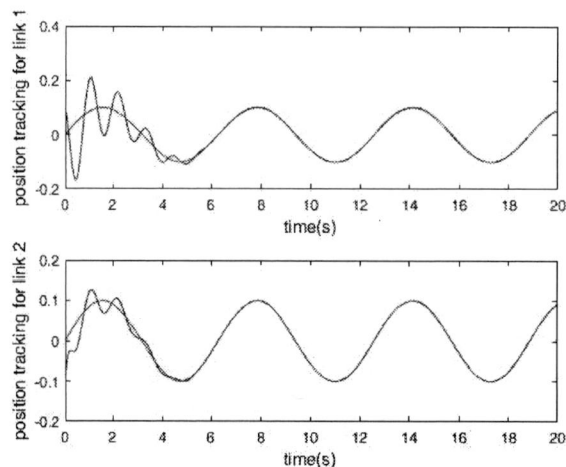

Figure 1 trajectory tracking

6. Conclusions

Due to the mechanical arm joint control system has the characteristics of nonlinear and parameter changes, the traditional control method based on linear time-invariant systems are difficult to obtain ideal control effect. Therefore, only using the advanced control method can improve dynamic characteristic of the controlled object, and improve the quality of the control. 11.   The simulation case shows that the adaptive control method adopted in this paper is effective and feasible, and has a certain reference value in mechanical arm joint control. In the simulation, figure 1 shows the mechanical arm joint Angle on the desired trajectory tracking performance. So the neural network control good intelligent control used in the mechanical arm.

Acknowledgments

This work was financially supported by project of Education department of Guangdong province science and technology (2017GKTSCX048).The authors also gratefully acknowledge the helpful comments and suggestions of the reviewers, which have improved the presentation.

References

[1] J.Park,I.W.Sandberg. Universal Approximation Using Radial-Basis-Function Networks [J] Neural Computation,1991,3(2):246-257

[2] F.L. Lewis ; A. Yesildirek ; Kai Liu .Multilayer neural-net robot controller with guaranteed tracking performance. [J] IEEE Transactions on Neural Networks, 1996,7(2): 388 - 399

[3] Liu jinkun.Robot control system design and MATLAB simulation the basic design method [M]. Tsinghua university press .2016

[4] Zeyan Hu, Xiaoguang Zhou, Shimin Wei. The modeling and controller design of an angular servo robot based on the RBF neural network adaptive control[M], Proceedings of the 2014 International Conference on Advanced Mechatronic Systems,2014

[5] M. M. Fateh1, S. M. Ahmadi and S. Khorashadizadeh. Adaptive RBF network control for robot manipulators[J]. Journal of AI and Data Mining,2014,2(2): 159-166.

ISPECE   IOP Publishing

Design of Multifunctional Intelligent Security Robot Based on Single Chip Microcomputer

Liu Penghou[1] Chen Haichao[2] Du Yanzhe[1]

[1]Qingdao Institute of Technology 266300

[2]Qingdao Hualu Hangsheng Automation Co Ltd, 266000

Abstract: This paper mainly introduces the intelligent robot with security function to analyze the advantages of the single-chip microcomputer and the security, and the combination of the two to the robot. Through the design and analysis of the multi-function intelligent security robot home version of the single-chip microcomputer, this paper presents the advantages of the multi-function intelligent security robot of the single-chip microcomputer.

1. Introduction

With the advancement of society, the computer technology industry, the control technology industry, and the development of information processing technology and sensor technology have made a qualitative leap. Intelligent robots have been used from a few high-end companies to consume and use them, whether they are industrial or electronic, or some commercial areas, and play an extremely important role. This is also the trend of artificial intelligence research and development. Through years of in-depth research, senior experts have made new breakthroughs in the field of robotics, especially the voice control. It has been able to integrate the continuous hidden Markov model perfectly with digital signal processing to help people control the robot through voice. In sensor ranging, intelligent robots with multiple sensors to quickly measure the distance between obstacles not only have convenient voice control and recognition functions, but also have intelligent recognition obstacles.

2. Microcontroller Overview

The single chip microcomputer is a chip with an integrated circuit, which can refine the ultra-large type integrated circuit, and the components of the single chip are composed of an arithmetic unit, a controller, a memory, and an input and output device. The process of executing the program command by the single-chip microcomputer is to complete the process of issuing the command by the staff. This process is formed by the step-by-step transfer of the instructions from the initial coding to the instructions in the system. One command is equivalent to a basic operation, and the instruction executed by the single-chip microcomputer is executed. That is, all the basic operations are completed. Different MCUs will have different characteristics, and there must be some changes in the command. If you want the MCU to complete the special instructions of the staff, you need to arrange the instructions in advance. These well-arranged instructions are all gathered together. It is a program. These programmed programs need to be pre-stored in a hard disk with savings function, which we call memory. The memory is also composed of a large number of storage units. These units are like the households of a household in the building. Each household room has a corresponding command stored in it. The address number is the house number of the room. Find the corresponding storage unit. When the MCU works, it passes these instructions, then finds the address number first, then transfers the

Content from this work may be used under the terms of the Creative Commons Attribution 3.0 licence. Any further distribution of this work must maintain attribution to the author(s) and the title of the work, journal citation and DOI.

Published under licence by IOP Publishing Ltd

address number to the system to start the system operation related content.

3. Intelligent Security Application

Intelligent security mainly refers to the informationization, image transmission and fast storage methods presented after the service upgrade. With the rapid development of society, the gap between intelligent security technology and network information technology has gradually disappeared.

The rapid development of intelligent security technology is obvious to all, especially in various corporate residences and some commercial uses. In addition, in recent years, the state has encouraged the promotion of digital intelligent security. In order to make the promotion of intelligent security more smooth, the Ministry of Construction and the Ministry of Public Security have successively signed a number of relevant documents to strengthen the intelligence of enterprises and residential communities. Security facilities.

Since the beginning of the new century, our country has begun to implement the "Technology and Innovation" project, and the most important one of them is to make the security and tranquility of enterprises and residential communities strengthen the intelligence of enterprises and residential communities with scientific and technological means. Construction of safety and security facilities. Intelligent security showed the world its usefulness in 2003. A SARS in 2003 can be said to be unprepared for China. From the beginning of several cases to hundreds of cases, in order to reduce the spread of SARS, the relevant government called for everyone to go out as little as possible, staying in the air at home. Staying at home all day and night is naturally boring and powerless, but with the help of a digital network, people can "see" at home, get news from all over the world, and even shop and learn online. Community forums like Weibo and Weibo need to operate security facilities inside the community, build defense systems, and handle problems and alarms for problems that may be discovered. Through this SARS incident, people are more aware of the importance of security robots and the development of related intelligent security information. It can be said that this time SARS has played an important role in promoting intelligent security advancement. Later, the 2008 Beijing Olympic Games made people realize the importance and convenience of intelligent security.

4. Single-chip Multi-function Security Robot

According to the researchers, unlike a human security officer, this safety patrol robot is like a moving electronic eye. Through infrared thermal imaging technology, it can observe some video monitoring dead angles, or the abnormal situation of the place can not be seen by the naked eye, and can judge the external temperature, in the anti-theft, anti-fire, the ability is very prominent.

At the 2016 Shenzhen High-Tech Fair, the exhibition of various robots constantly refreshed people's traditional cognition. Coupled with the rapid development of single-chip multi-functional security robots in recent years, slowly single-chip multi-function security robots have also entered the lives of ordinary people, such as the "Security Patrol Robot" organized by Shenzhen Zhongzhi Kechuang Robot Co., Ltd. In order to let security personnel better understand and master the single-chip multi-function security robot, Zhike Chuang Robot Co., Ltd. specially held this event, hoping to show the arrival of the era of "robot and security".

The traditional "human defense + physical defense" security system actually has common loopholes. For example, the fixed camera has a blind spot for monitoring. At night or during low visibility and rainy season, the camera often has problems, and the maintenance work is not convenient. In addition, the labor costs are increasing every year, the cost of security personnel is getting bigger and bigger, many security personnel are reluctant to work in the same unit for too long, the mobility of personnel is increasing, and the recruitment of personnel is becoming more and more difficult. However, the new era of single-chip multi-function security robots has arrived. The single-chip multi-function security robot has solved these problems very well. After the new concept of "robot + security" has landed, it not only affects the security industry, but also makes people realize the robot. In fact, it can be linked to our daily lives. The single-chip multi-functional security robot has transformed the traditional security system into modern intelligent security, which also indicates another major advancement in China's security.

Zhongzhike is a subsidiary of China Security Technology Co., Ltd., relying on the parent company's years of exploration and security in the security field, accumulating the mature technology and deep security system operation service foundation, first proposed "robot + security" in China. The modern science and technology intelligent security concept, and take the lead in spending a lot of manpower and material resources to carry out related research and development, the hard work pays off, finally developed on the basis of cloud technology, has been affirmed by various domestic industries, and also received strong support from the government. Its self-developed security patrol robot realizes the intensive development of mobile robots, multi-sensing technology and cloud security platform, and is at the forefront of the industry.

Zhongzhike has created a new security solution for industry customers by building a "dynamic and static" three-dimensional security system. It can be combined with the traditional security system and manpower security to form a security operation mode that combines dynamic and static, 360-degree audio monitoring and autonomous patrol. It can also complete a series of three-dimensional security functions such as environment awareness, intelligent alarm, face recognition, etc., to assist security personnel to perform patrol tasks.

The security patrol robot was unveiled at several large-scale exhibitions. Double-creation week, high-tech fair, and Anbo.... In September 2016, it was promoted to the GITEX exhibition in Dubai. It was appreciated by the Dubai Chief Mohammed who came to the exhibition. .

At present, the security patrol robot has taken the lead in the "job" work of Huawei Putian Industrial Park, combined with the original security system, to build a dynamic and static intelligent security monitoring system to achieve all-weather, all-round, full-autonomous unattended patrol of the park. The management of personnel, vehicles and services provides three-dimensional guarantee without dead ends, effectively reducing the labor intensity of security personnel, reducing operating costs, improving response speed, and improving the automation and intelligence level of security inspections in the park.

Fig. 1 Security patrol robot

5. Design of Multifunctional Intelligent Security Robot for Single Chip Microcomputer
The following article will take the home-type MCU multi-function security robot as an example to analyze the design of MCU multi-function security robot.

5.1 Design principle
In order to integrate home smart security functions with home service functions, robot manufacturers

have developed a home service robot that can provide security indoors while meeting the simple needs of people in their daily lives. The multi-function intelligent security robot of the home version of the MCU can use the moving wheels at the bottom of the robot to move freely indoors. Because the intelligent security system has been installed inside the robot, the installed security system is combined with the electronic home access control, electronic window grid, infrared sensor and other external home service equipment in the multi-function intelligent security robot of the single-chip microcomputer. Functional intelligent security robots can replace people's better indoor hygiene, while also providing a safe and comfortable environment for the host.

5.2 Serial communication parameter setting

The serial communication device function used by the intelligent security software mainly sets the serial port number, data bit and check digit used by CccmmSet-tingDlg, and automatically saves the settings in the registry. There are two main storage methods, one is joining the global function to complete the relevant registry to improve the reading and writing work within the communication parameters. Another method is to add a message such as IDOK control, transfer the sent message to the function through the control, and fill in the registry to summarize.

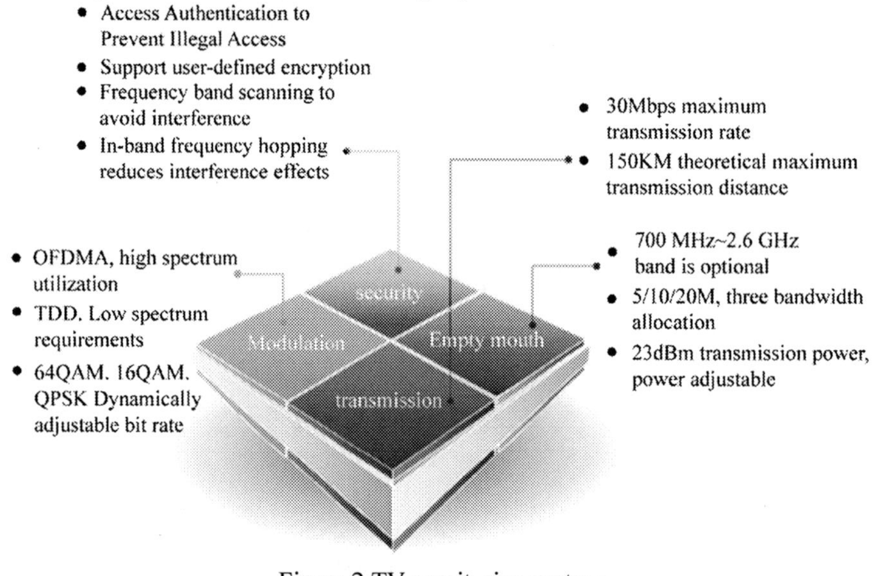

- Access Authentication to Prevent Illegal Access
- Support user-defined encryption
- Frequency band scanning to avoid interference
- In-band frequency hopping reduces interference effects

- 30Mbps maximum transmission rate
- 150KM theoretical maximum transmission distance

- OFDMA, high spectrum utilization
- TDD. Low spectrum requirements
- 64QAM. 16QAM. QPSK Dynamically adjustable bit rate

- 700 MHz~2.6 GHz band is optional
- 5/10/20M, three bandwidth allocation
- 23dBm transmission power, power adjustable

security
Modulation
Empty mouth
transmission

Figure 2 TV monitoring system

5.3 Real-time monitoring

In order to ensure that the alarm signal of the arming trigger is obtained, the timer is set in the program setting. The setting time of the timer cannot be too long, and it can be several tens of milliseconds or several seconds. Try not to be too short or too long, because if If the fixed timing is too short, the computer may not be able to accept and process the response quickly. If the set time is too long, it will directly affect the capture of the specific aging dynamics of the event. In order to ensure that the set timing time can be accurate, it can be set by the SETIMER function, and the robot can be monitored and set by the SETIMER function.

5.4 Real-time alarm

In the event that an indoor alarm is required, the alarm of the home service robot is divided into a multi-function intelligent security robot installed by the single-chip microcomputer. The alarm sounds the alarm sound to scare off the intruder and the multi-function intelligent security robot of the single-chip microcomputer uses the serial port to link the GSM mobile phone to the owner. Send a message to remind the owner of the alarm.

6. Principles to Be Followed Before And After Design

6.1 Rationality principle

It ensures that intelligent machine equipment can have reasonable design from internal system to external assembly, and can meet the special needs of users. It can be fine-tuned for different industries of different users, and has open software and hardware interfaces, which is convenient for users to self-simple. The external information is connected.

6.2 The principle of advancement

The current computer and communication technology is developing at a high speed. The requirements for computers are not simply at this stage. It is also necessary to continuously carry out relevant explorations, and constantly explore the space that can be developed, so that the single-chip microcomputer can go further and further.

6.3 Principle of practicality

Unity, practicality, and sustainability.

6.4 Reliability and stability principles

TV monitoring room is currently a large-scale retail shop, community and school's favorite security system, can run 24 hours a day, while the storage space for a certain time range is to save video recording, so that people can turn Check the previous video. In particular, it is convenient for investigators to obtain surveillance videos to find prisoners.

6.5 The principle of scalability

Scalability is mainly manifested in the ability to develop horizontally and vertically at the same time in one thing. Horizontally, the security monitoring system can continuously expand the output capacity to meet the needs of personnel, but it does not affect people's normal operation. The vertical expansion is mainly manifested in the compatibility of the security monitoring system, and can be used for secondary development of the customer to facilitate the operation of the customer.

6.6 System security and confidentiality principles

The problem of the security of the TV monitoring system is the problem that needs to be considered in the security industry at present. The whole system data needs to be safe and error-free and can be managed according to the level. It is necessary to carry out special hierarchical protection of key data, and it also needs to be done well. Relevant operation records are convenient for later personnel to find. Like the public security department, it is necessary to protect the image transmission and do a good job of confidentiality. At present, the rapid development of computer and communication technology makes the design of the system not only to take full advantage of the current state-of-the-art technology, but also to consider that with the further development of technology, new technologies can be continuously incorporated into the system, so that the system is always full of vitality. Always maintain technological advancement.

7. Conclusion:

In order to use the multi-function intelligent security robot of the single-chip microcomputer, it is necessary to develop the technology of security and single-chip microcomputer to the best, and to pave the way for the combination of the two through the common place between the two. In designing single-chip multi-functional intelligent security robots, we must adhere to the sustainable development route, and take into account the follow-up green energy development of single-chip multi-functional intelligent security robots. Only with long-term vision can the single-chip multifunctional intelligent security robots go further and further.

Acknowledgment

Scientific and Technological Planning Projects of Colleges and Universities in Shandong Province （J18KA389）

References

[1] Intelligent robot control system design based on single chip microcomputer [J]. Jiang Hongfa. Intelligent robot. 2018(02)

[2] Design of Intelligent Robot Based on Single Chip Microcomputer[J]. Liu Tianzhao. Science and Technology Information. 2012(36)

[3] The application of single chip microcomputer in intelligent robots[J]. Li Weiqian, Wu Yaobin. Fujian Computer. 2012(09)

[4] Research on the application of single-chip microcomputer control system in intelligent robots [J]. Lu Guoce. Electronic Production. 2015(04)

[5] Application Research of Single Chip Microcomputer Technology in Robot Control System[J]. Chen Guiyin. Automation Application. 2017(05)

Multi-Lane Detection Using CNNs and A Novel Region-grow Algorithm

Yi Sun[1], Jian Li[1] and Zhen Ping Sun[1]

[1]National University of Defense Technology, Changsha, China

1171272375@qq.com, lijian316@163.com, 13974913933@139.com

Abstract. In this paper, we propose a novel approach to detect lanes robustly in different scenarios. Theoretically, HT (Hough Transform) can extract straight lines efficiently from images, but it cannot distinguish whether the straight lines are belong to lane-markings. To solve this problem, we integrate a simple convolutional neural networks modified from Lenet and geometry constraints to distinguish the types of lines. And in the following part, a region-grow algorithm is adopted in this paper to fit lanes, which is realized by growing local ROI (Region-of-Interest) gradually. We use RANSAC to extract the main direction in local ROI and guide the growth of the ROI. At the same time, in order to ensure right growth of ROI when the lane is dashed, we use a DVF (Direction Vector Field, modified from GVF [19]), which has large capture range and is generated by the edge-direction map. Tests of the proposed approach under kinds of conditions are discussed.

1. Introduction

Lane detection algorithms play an important role in ADAS (Advanced Driver Assistance System), not only it is significant for lane-keeping tasks, but it can tell the traffic rules represented by lane markings on structured roads. Research in this domain mainly focus on how to extract the correct lane features (such as edge feature, line feature) and line fitting, which are the key problem to ensure the robustness of the algorithm.

We use HT to extract line segments in this paper, which is used in many works. However, efficient post-process is necessary for eliminating the noisy line segments in its result, because a few line segments extracted by HT are not belong to traffic lane. Geometry constraints for post-process were used in [12, 17, 18], such as vanishing-point-based constraints, road width constraints and so on. Unfortunately, this kind of constraint usually failed to deal with some line segments on rails, appearance features shall be used to improve the performance in post-process. We have designed a simple CNNs (Convolutional Neural Networks) modified from Lenet [20] to extract feature of the result of HT and purify it. Patches used by CNNs are proposed by line segments. Sometimes, although the global scene changes, the scenarios in proposal-region actually are changeless, so that the generalization risk is limited.

To make our algorithm fitting the lane more accurate at distance, a region-grow algorithm is adopted in our work which is used to choose reasonable edge points for fitting, and is realized by growing local ROI gradually from bottom to top in image. We extract the main direction, which is provide by RANSAC and DVF, in local ROI and guide the growth of the next ROI. We can obtain a series of ROIs growing along the road direction through this method, then we can extract edge points for fitting from these ROIs.

Related Works are introduced in the second part of this paper, part three describes the proposed CNNs,

---

Content from this work may be used under the terms of the Creative Commons Attribution 3.0 licence. Any further distribution of this work must maintain attribution to the author(s) and the title of the work, journal citation and DOI.

Published under licence by IOP Publishing Ltd

and a novel Region-grow algorithm which is constrained by RANSAC and DVF is introduced in the fourth part.

2. Related Works
The extraction of lane features, and lane fitting are important problems in a lane detection system. Several approaches were proposed in the past few years to solve these issues.

2.1 Extracting Lane Feature
Edge operators such as Canny are usually used to extract the edge features [4, 12, 18]. In order to improve the sign-to-noise ratio in edge-map, Yoo H *et al.* [4] strengthen the lane-markings feature by training a LDA model, Gaikwad V *et al.* [12] propose a brightness stretching function PLSF which makes lane-marking more clearly. Pollard E *et al.* [14] combine two different extraction algorithm and make use of different local threshold, which makes the extraction algorithm perform well. Geometry constraints have been used to filter false lines in [12, 17, 18]. Machine learning methods are also used to improve the detection performance, Kim [21] uses classifier to judge whether a patch includes the lane-marking, these patches are provided by small sliding windows. Zhao K *et al.* [7] use the magnitude and direction of gradients to validate lines.

Convolutional neural networks often achieves very good results when extracting features. In [2], a semantic segmentation networks is trained by using the ground truth generated by high-precision map after an on-line calibration. In [9], CNN and RNN are combined for extracting image structures and lane-markings, Lee *et val.* [15] introduce a multi-task learning networks based on predicting vanishing point location, which can extract lane-markings under complex conditions.

2.2 Curve-Fitting Methods
Zhou S *et val.* [1] and Ozgunalp U *et val.* [6] use the road tendency information provided by the vanishing point to estimate an optimal parameters of the curve model they used. Wang Y *et val.* [14] divide the image into several parts, and extract control points from each part, finally fitting these points by B-snake. Kim [21] uses RANSAC algorithm to fitting curve. In [16], this problem is transformed into an optimal connection problem between super-pixels, and they use CRF (Conditional Random Field) to select the optimal connection. Neven D *et val.* [11] turn lane detection task into an instance segmentation task, which solve both feature-extraction and curve-fitting at the same time.

3. Line extraction with HT and classification with CNNs
In this paper, we use the edge-direction of edge points to narrow the voting range of edge points when using HT according to equation (1). Actually, it is proved to be good for reducing noise in most cases. In equation (1), (x, y) is the location of edge point $P(x, y)$ in image, and ϕ is the edge direction of $P(x, y)$, v represents the half width of voting range, we set (ρ, θ) as the Hough space.

$$\rho = x \cdot \cos\theta + y \cdot \sin\theta \quad \theta \in [\phi - v, \phi + v] \tag{1}$$

It is worth mentioning that we calibrate the edge direction of $P(x, y)$ according to equation (2), and get a new edge-direction map $D(x, y)$. In equation (2), it shows that the edge-direction of $P(x, y)$ is same as the slope k of the line segments it belong to. $D(x, y)$ will be used to calculate DVF in part 4.

$$y = k \cdot x + b$$
$$D(x, y) = k \tag{2}$$

Convolutional neural networks is a good feature extractor, which can free us from designing complicated handicraft features and can often extract better features under supervision. To eliminate false lines extracted by HT, we propose a CNNs which is modified from Lenet to classify patches provided by line segments. Table 1 shows the structure of our CNNs.

Table 1. Structure of CNNs.

Index	1	2	3	4	5	6
Layer	data	conv+relu	pooling	conv+relu	interp	conv
Index	7	8	9	10	11	12
Layer	pooling	conv	pooling	Inner-Product	Inner-Product	Prob

We choose two endpoints of straight line as diagonal ends to get the patches described in figure 2. These patches will be used as the input of us networks. We have compared handicraft features composed of HOG and RGB feature with feature extracted by CNNs. We extract two different features from 5000 patches, and use PCA to reduce feature dimensions for visualization, which is displayed in figure 1. We also test both two different ways on our test dataset in different scenarios, the result shows that CNNs do better than handicraft features combined with SVM, and it also have a good generalization performance.

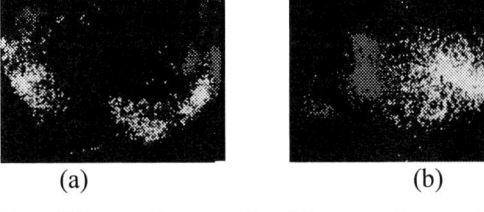

(a)　　　　　　　　(b)

Figure 1. Two different Features (positive samples are drawn in red, negative samples are drawn in cyan): (a) handicraft feature composed of HOG and RGB feature; (b)Feature extracted by CNNs

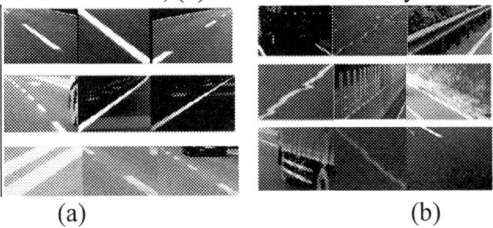

(a)　　　　　　　　(b)

Figure 2. Samples: (a) positive samples, (b) negative samples.

Table 2. Accuracy of two features on classification mission in different scenarios. HOG and RGB features are used by SVM.

	Easy condition	Poor visibility	Overexposure	Night	Background	Occlusion
SVM	0.822	0.634	0.687	0.737	0.651	0.764
CNNs	0.948	0.952	0.953	0.986	0.965	0.981

4. Curve fitting based on RANSAC and Direction Vector Field

We propose a region-grow algorithm based on RANSAC and Direction Vector Field, which are used to guide the growth of ROI. This kind of ROI will grow from the top of each line segment extracted by Hough Transform. RANSAC algorithm is used to extract local line segment $(l : y = k \cdot x + b)$ in ROI, this line segment l provide a guidance for the growth of next ROI. We define $P1(x_1, y_1)$, $P2(x_2, y_2)$, $P3(x_3, y_3)$ and $P4(x_4, y_4)$ are four endpoints of the ROI, we can get new ROI by updating these endpoints according to equation (3), where w is the width of ROI, Δs is the height, and $(\hat{x}_1, \hat{y}_1)$ is the last location of $P1$. Figure 3 shows this process.

$$\{P1, P3\} \in \{(x, y) \| y = k \cdot x + b - w\}$$
$$\{P2, P4\} \in \{(x, y) \| y = k \cdot x + b + w\}$$
$$x_3 = x_4 = \hat{x}_1 \tag{3}$$
$$x_1 = x_2 = x_3 + \Delta s$$

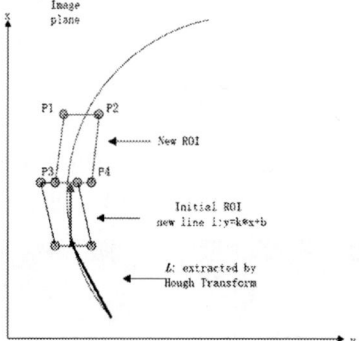

Figure 3. Region-grow constrained by RANSAC.

However, sometimes there might not exist edge points such as some parts in a dashed lane, to solve this problem, we create a *directional vector field* (DVF) V to replace RANSAC algorithm, which have a larger influence range than edge-point. This DVF are modified from GVF [20], it will provide general trend guidance for the growth of next ROI.

$$V = [u, v] \tag{4}$$

The initial value of V is provided by $D(x, y)$. We can get the final DVF by minimize E in equation (5). We can obtain the Euler equation of equation (5), and then obtain the following iterative formulas (6), where t represents the iterative times:

$$E = \iint u_x^2 + u_y^2 + v_x^2 + v_y^2 dxdy \tag{5}$$

$$u_{t+1} = \nabla^2 u + u_t \tag{6}$$
$$v_{t+1} = \nabla^2 v + v_t$$

V will guide the updating of ROI when there is no enough edge points for RANSAC, a new line $l : y = k \cdot x + b$ is extracted from ROI by using V according to equation (7), where N is the number of pixels in ROI. Then $P1(x_1, y_1)$, $P2(x_2, y_2)$, $P3(x_3, y_3)$ and $P4(x_4, y_4)$ can be calculated by equation (3).

$$k = \frac{1}{N} \sum_{i=1}^{N} \frac{u_i}{v_i} \tag{7}$$
$$b = -k \cdot \hat{x}_1 + \hat{y}_1 + w$$

Finally, we can extract sets of edge points $\{(x_i, y_i) \| i = 1, 2, ..., n\}$ from ROIs gotten by region-grow algorithm, equation (8) shows the curve model used in this paper and finally we use least square method to fitting this points described by equation (9).

$$y = a \cdot x + b + \frac{c}{x - vx} \tag{8}$$

$$E = \sum_{i=1}^{n}(y_i - a \cdot x_i - b - \frac{c}{x_i - vx})^2 \quad (9)$$

5. Experiment

We test our algorithm in different scenarios, using a dataset provided by Xi'an Jiaotong University [22]. Classification result is showed by figure 4, where the noisy lines are marked with green, and the true lanes are marked with red color. We have chosen 36 different scenarios for test, among which 14 roads are easy and 22 roads are challenging such as occluded, broken road.

Table 3 shows the TPR/FPR of our algorithm in easy and challenging road, it also compare the different performance when we use classification networks only and when we incorporate the CNNs and geometry constraints mentioned by [12, 17, 18], obviously, accuracy of our classification algorithm is improved. The TPR and FPR are calculated by equation (10).

$$TPR = \frac{TP}{TP + FN}$$
$$FPR = \frac{FP}{TN + FP} \quad (10)$$

Figure 4. Performance of our classification networks under different conditions

Table 3. TPR/FPR distribution of our algorithm tested in different sections.

	No Geometry Constraints(Easy Road)	Geometry Constraints(Easy Road)	No Geometry Constraints(Challenging Road)	Geometry Constraints(Challenging Road)
TPR	0.961	0.984	0.891	0.924
FPR	0.018	0.007	0.049	0.006

Figure 5 shows the DVF generated by method described in Section 4, which use quiver to represent the directional vector flow V. This kind of field provide a larger guidance range for region-grow especially when there is no edge points. And figure 6 shows the results of region-grow only using RANSAC when the lane is dashed and has a large curve, apparently, our region-grow algorithm failed under this kind of condition. Figure 7 shows the results integrating RANSAC and DVF, we can see the performance of region-grow is improved obviously. We show the final result of the whole algorithm proposed in this paper in figure 8.

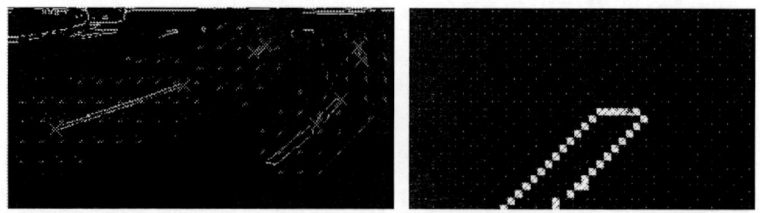

Figure 5. Direction vector field: the direction is expressed by quiver

Figure 6. Result of Region-grow without DVF.

Figure 7. Result of Region-grow with DVF.

Figure 8. Result of approach proposed in this paper

6. Conclusions

In this paper, we propose using CNNs and Geometry Constraints to choose right lines from the result of Hough Transforms. The CNNs proposed in this paper is modified from Lenet, which takes little computing resources. By combining the Geometry Constraints and classification, we could eliminate noisy lines effectively. In order to fitting curve line better at a distance, we use region-growing algorithm to get edge-points used for fitting, this region-growing algorithm is guided by local direction, which is provided by RANSAC and Vector-Field. Then, we can use least square method to fitting these points, finally we get the curve model of the lane. Method introduced in this paper performs well in kinds of scenarios, but it still has some drawbacks, for example, the Vector-field is easily disturbed by noisy edges, then the wrong force-field will misleading the direction of the region-grow, and the FNR of the line classification will be supposed to reduce.

Acknowledgement

This work was supported by NSFC Grants 61473303

References

[1] Zhou S, Jiang Y, Xi J, et al 2010 A novel lane detection based on geometrical model and Gabor filter *Intelligent Vehicles Symposium*, pp 59-64.

[2] Behrendt K and Witt J 2017 Deep learning lane marker segmentation from automatically generated labels *International Conference on Intelligent Robots and Systems. IEEE,* pp 777-782.

[3] Kim J, Kim J, Jang G J, et al 2017 Fast learning method for convolutional neural networks using extreme learning machine and its application to lane detection *Neural Networks*, **87** 109-121.

[4] Yoo H, Yang U and Sohn K 2013 Gradient-Enhancing Conversion for Illumination- Robust Lane Detection. *IEEE Transactions on Intelligent Transportation Systems*, **14(3)** 1083-1094.

[5] Sivaraman S, Trivedi M and M 2013 Integrated Lane and Vehicle Detection, Localization, and Tracking: A Synergistic Approach[J]. *IEEE Transactions on Intelligent Transportation Systems*, **14(2)** 906-917.

[6] Ozgunalp U, Fan R, Ai X, and Dahnoun N 2017. Multiple lane detection algorithm based on novel dense vanishing point estimation. *IEEE Transactions on Intelligent Transportation Systems*, **18(3)** 621-632.

[7] Zhao K, Meuter M, Nunn C, et al 2012 A novel multi-lane detection and tracking system *Intelligent Vehicles Symposium. IEEE*, pp 1084-89.

[8] Ju H Y, Lee S W, Park S K, et al 2017 A Robust Lane Detection Method Based on Vanishing Point Estimation Using the Relevance of Line Segments[J]. *IEEE Transactions on Intelligent Transportation Systems* **PP(99)** 1-13.

[9] Li J, Mei X, Prokhorov D, et al 2017 Deep Neural Network for Structural Prediction and Lane Detection in Traffic Scene[J]. *IEEE Transactions on Neural Networks & Learning Systems*, **28(3)** 690-703.

[10] Long J, Shelhamer E and Darrell T 2014 Fully Convolutional Networks for Semantic Segmentation[J].*Transactions on Pattern Analysis & Machine Intelligence*, **39(4)** 640-651.

[11] Neven D, De Brabandere B, Georgoulis S, et al 2018 Towards End-to-End Lane Detection: an Instance Segmentation Approach[J] *preprint* arXiv/1802.05591.

[12] Gaikwad V and Lokhande S 2015 Lane Departure Identification for Advanced Driver Assistance[J].*Transactions on Intelligent Transportation Systems* **16(2)** 910-918.

[13] Wang Y, Teoh E K and Shen D 2004 Lane detection and tracking using B-Snake[J]. *Image & Vision Computing*, **22(4)** 269-280.

[14] Pollard E, Gruyer D, Tarel J P, et al 2011 Lane marking extraction with combination strategy and comparative evaluation on synthetic and camera images[C] *International IEEE Conference on Intelligent Transportation Systems*. pp 1741-46.

[15] Lee S, Kim J, Yoon J S, et al 2017 Vpgnet: Vanishing point guided network for lane and road marking detection and recognition[C] *International Conference on Computer Vision*, pp 1965-73.

[16] Hur J, Kang S N and Seo S W 2013 Multi-lane detection in urban driving environments using conditional random fields[C] *Intelligent Vehicles Symposium* pp 1297-02.

[17] Lee C and Moon J H 2018 Robust Lane Detection and Tracking for Real-Time Applications[J]. *IEEE Transactions on Intelligent Transportation Systems*, **PP(99)** 1- 6.

[18] Niu J, Lu J, Xu M, et al 2016 Robust Lane Detection using Two-stage Feature Extraction with Curve Fitting[J]. *Pattern Recognition*, **59(C)** 225-233.

[19] Xu C and Prince J L 1997 Gradient vector flow: A new external force for snakes *Computer Vision and Pattern Recognition*, pp 66-71.

[20] Lecun Y L, Bottou L, Bengio Y, et al 1998 Gradient-Based Learning Applied to Document Recognition. *Proceedings of the IEEE*, **86(11)** 2278-2324.

[21] Kim Z W. 2008 Robust lane detection and tracking in challenging scenarios[J]. *IEEE Transactions on Intelligent Transportation Systems*, **9(1)** 16-26.

[22] http://trafficdata.xjtu.edu.cn/index.do

A Survey of Cloud Computing Access Control Technology

Minghao WANG

Dalian vocational & technical college

116035

Abstract. Cloud computing access control technology originated in the 1970s. The initial goal of this technology is to meet the requirements of the primary server for data access rights, identify the identity of the visitor through relevant procedures, and then set access rights based on the authentication result. And it also committed to protect important data and prevent the main server from being illegally invaded. With the development of technology, cloud computing access control technology has been widely used in computer systems, which has played a good role. This article will briefly discuss cloud computing access control technology research and present personal insights.

1. Introduction

With the rapid development of network platforms, cloud computing technology is widely used in computer management systems. At the same time, it also brings various cloud security issues. In order to solve cloud security problems and protect cloud resources, cloud computing access needs to be set up the control technology. This article will briefly introduce the basic concepts of cloud computing, discuss cloud computing security management issues, analyze cloud computing access control technologies in traditional mode, and discuss about the access control technologies in cloud environments.

2. Basic Concepts of Cloud Computing

The term cloud computing originated from English Cloud Computing and belongs to a new network service management model. From a development perspective, as early as the 1960s, American computer scientist McCarthy (John) proposed the assumption that computing power would be provided to all users like hydropower resources. This idea became the beginning of cloud computing [1] . Moreover, McCarthy (John) is a pioneer in the field of artificial intelligence. He designed the table processing language in 1958 and proposed the concept of processing characteristics of tree structure (for calculation). These studies are the development of cloud computing technology. It laid the foundation of theoretical foundation and technical research. Later, scholars in different fields of research said that the scale of cloud computing is very large, there is no clear boundary, the location is extremely vague, and it has dynamic stretching characteristics. Amazon, a cloud computing technology business giant, once defined cloud computing as EC2, the "elastic computing cloud" in grid computing mode. In addition, in the 21st century network era, the business community is mostly accustomed to using cloud patterns to represent the network, which is one of the important reasons why contemporary network computing technology is called "cloud computing technology." It should be noted that the interpretation of cloud computing in different fields is different. The National Institute of Technology (NIST) said that cloud computing technology is a paid service model that meets different needs. This mode mainly provides users with convenient and available professional networks. The access activity can meet the network information needs of the user, and allows the user to enter the computing resource sharing pool through the cloud service interaction (in general, the internal resources of the computing resource sharing pool mainly include the network, the storage, the software, the server, and other services). IBM, the IT business giant, believes that cloud computing technology is a modern consumer delivery model. In the process of consumption and delivery, cloud computing technology integrates computer technology, information technology and various business services. Users can choose the procurement mode according to their specific needs, access network resources, consumer subjects are not only people, but also equipment or programs. Consumer goods mainly include computing resources, servers, storage capabilities, business services and other resources.

From the analysis of service structure, cloud computing has three layers of service subsystems. The first layer is SPI mode. Its English name is Infrastructure as a Service, Iaa S. This mode is mainly used to provide infrastructure services like host, storage, network and Various hardware services; the second layer is Paa S, the English full name is Platform as a Service, that is, service platform, such as identity authentication, service bus, workflow, access control, data mining, etc.; the third layer is Saa S, the full name of English Software as a Service, namely software services, such as communication services, mail delivery and content management [2].

From the perspective of technical support system, cloud computing has a key technology, namely intelligent scheduling technology, which not only can rationally adjust resource dynamics, comprehensively monitor data, migrate business dynamics, but also has the function of scheduling large-scale data. Quickly match cloud resources and reduce or increase CPU computing units, cloud servers, and storage space as needed.

On the other hand, cloud computing technology has seven characteristics: First, the scale is large. Cloud computing technology can scientifically dispatch cloud resources, integrate massive cloud resources, and eventually form a large-scale cloud resource pool, thereby continuously strengthening cloud service capabilities, computing power, and storage capabilities. In addition, cloud services can provide users with service resources, computing power and storage performance that traditional computers cannot provide to meet the needs of different users. Second, it needs to be virtualized. In the information age, users can use the Internet to obtain all valuable cloud resources and use the corresponding terminals to obtain the required services without locating the specific location of the hardware architecture and cloud technology. Moreover, cloud technology is specific, but not tangible. Therefore, it is not limited by geographical location, and it can provide various services to users through virtualization technology. Third, the reliability is good. The services provided by cloud computing technology are more reliable than traditional technology service models. In short, cloud computing technology combines a variety of valuable data and copy fault tolerance technologies with homogeneous compatibility technologies to back up valuable data and ensure the completeness, continuity and reliability of stored data. Fourth, it should be multifunctional. Compared with other features, the versatility of cloud computing technology is not specific to a specific application, but is reflected in the application of a variety of different service structures, and to ensure that a variety of applications can be effectively run with the support of the cloud platform. Fifth, the service is flexible. Cloud computing technology and traditional computer technology services have a significant difference - cloud computing technology can provide a good flexible service, which is scalable and fast, meets the user's growth needs, and automatically expands or reduces the size of cloud resources, it can continuously improve the utilization of cloud resources. Sixth, the convenience is strong.

Seventh, the effectiveness is high. The 21st century cloud computing technology belongs to a modern network business service model with high centralized management functions, which can greatly reduce data management costs, improve cloud resource utilization, improve computer technology operation mode, and assist users to smoothly enter services. The webpage, which saves a lot of time, experience and cost, shows that cloud computing technology has good effectiveness.

3. Cloud Computing Security Management Issues
From a narrow point of view, the cloud computing security management problem mainly comes from

the intrusion of network viruses. There are five main features, namely, concealment, contagious, destructive, stimulating and unpredictable [3]. Among them, concealment means that the presence, infection and destruction of computer data are not easy to be found; infectivity means that most network viruses can self-replicate under the corresponding conditions, and the infection speed is very fast, if it cannot be cleared in a short time. The virus will affect the entire network system; destructive means that once a virus program is attached to the currently running program, the running program will be infected, thereby affecting the entire network system, destroying the contents of the disk file, and illegally deleting the data. Deliberately tampering with files, occupying a large amount of storage space, resulting in disk formatting and data loss; the essential nature of stimulating is conditional control. Under normal circumstances, the types of viruses are different, and the excitation conditions controlled by the outside are also different, but as long as the computer network system environment can meet the conditions of virus transmission, and the virus program will be further activated, resulting in a paralysis of the computer network system; the unpredictability mainly means that the virus spreads much faster than the anti-virus software, in short, There is currently no anti-virus software that can clear the site. Network virus. From the operation of the virus, computer network viruses can be divided into three parts: virus boot program, virus infection program and virus disease program. Among all viruses, Trojan virus is the most common virus file at present. It is different from the general network virus. This virus does not copy and multiply itself, and does not interfere with other files, but it will provide open hackers for Trojans. The host's portal, thus arbitrarily destroying and stealing computer files, passwords, stock accounts, bank accounts, etc., and even remotely controlling the host being hosted, the Trojan virus is seriously jeopardizing the safe operation of modern networks. In this regard, it is necessary to set up a perfect protection system and security password by means of cloud computing technology, to avoid network erosion by viruses, to do network security monitoring work, thereby continuously improving computer performance, reducing power configuration, speeding up computer startup speed and running speed, and comprehensively doing good computer network maintenance work.

4. Cloud Computing Access Control Technology in Traditional Mode

From a macro perspective, there are three kinds of cloud computing access control technologies in the traditional mode. The first one is Discretionary Access Control, which enables the subject to directly manage the object, and assists the owner to select the access control requirements. Secondly, Discretionary Access Control (Discretionary Access Control) has two implementation modes, namely access control matrix and access control list. The former belongs to a mature and complete control scheme. The scheme mainly represents the access control strategy through matrix form. Use rows to represent the subject and columns to represent the object. Table 1 is the list of access control matrices:

Table 1 Access Control Matrix List

	object resource 1	object resource 2	object resource 3
main part 1	R	W	R、W
main part 2	W	R、D、W	D
main part 3	R、W	R、D	R

It can be seen from Table 1 that the object resources in the access control matrix are controlled by the subject and are realized by the subject.

An access control list is a permission table in computer storage that is used to express access rights to individual files and folders. Moreover, the access control list has the right to access the object, and can specify the specific steps and processes of the subject operation object.

The second type is the Mandatory Access Control Technology System (MAC). The full name of the English is Mandatory Access Control. This system is mainly used to access user control requirements and set security fixed policies and management rights for computers. In addition, the core of the Access Control Technology System (MAC) is to protect the computer information system, avoid data leakage, and set the access basis for resources. Moreover, the Mandatory Access Control Technology System (MAC) sets up security control and "up-write" and "read-down" execution to ensure that the subject and the object can be safely executed and operated [4].

The third is Role-Based Access Control, which enables organic separation of access subjects and objects, and builds a Role-Based Access Control model to perform authorized operations through mapping, strengthen access rights management.

5. Access Control Technology in Cloud Environment

5.1 T-ABAC Model Concept
The basis of the T-ABAC model (Trust and Attribute-Based Access Control) is the ABAC model. Under the support of the cloud environment, the T-ABAC model mainly uses the ABAC model method to implement dynamic access control and fine-grained access control. From the microscopic point of view, the T-ABAC model has many attributes, including subject attributes, object attributes, trust attributes, action attributes and environment attributes. These attributes can be combined to meet the fine-grained access requirements in the cloud environment. [5].

5.2 T-ABAC Model Combination
The T-ABAC model combination is divided into three modules, namely user authentication module, access judgment module and trust evaluation module. The user authentication module is mainly used to process cloud user identity authentication management services. In the authentication process, cloud users are used. You must log in to the cloud computing environment to access cloud resources. The cloud platform will play the role of authentication technology. The user identity is legal or not, and illegal users are prohibited from accessing resources. The access judgment module is mainly used to process access requests of cloud users, and provides and restricts access according to the content of the request. The trust evaluation module is an adjustment to the trust attribute and can enhance the management of the attribute access mechanism. In addition, the T-ABAC model belongs to the quintuple, which are S, O, E, A, and T, respectively, where S is the abbreviation of subiect (subject), O is the object (object), and E is the environment (environment), A On behalf of action, T is the abbreviation of trust, these five combinations play their respective roles.

5.3 User Behavior Evidence Collection Technology
User behavior evidence collection technology is mainly for cloud user login, access request and access content records. This technology can regulate user behavior, strengthen access rights management, prohibit illegal users from accessing confidential information, and maintain the security of computer systems and data resources.

6. Conclusion
In summary, the new network service management model of cloud computing originated in the 1960s. With the development of technology, the technology is widely used in various fields to effectively enhance cloud service capabilities, computing power and storage capabilities. To comprehensively improve the access control technology in the cloud environment, it is necessary to construct a good T-ABAC model, refine the combination module, and continuously optimize the user behavior evidence collection technology.

References
[1] Wang Shulan. Research on attribute-based access control technology in cloud computing [D]. Shenzhen University, 2016 (06:1).

[2] Liu Zhengnan. Research on access control model based on user behavior evaluation in cloud environment [D]. Northwest A&F University, 2016(05:2).

[3] Feng Lixiao. Research on attribute-based access control method in cloud computing environment

[D]. Inner Mongolia University of Science and Technology, 2014 (06:3).

[4] Deng Xiaohong. Research on IMS-based cloud computing service access control technology [D]. Northeastern University, 2017 (05:4).

[5] Li Wenxue. Research on access control evaluation technology of cloud computing platform [D]. Harbin Institute of Technology, 2013 (06:5).

ISPECE

IOP Publishing

IOP Conf. Series: Journal of Physics: Conf. Series **1187** (2019) 032020   doi:10.1088/1742-6596/1187/3/032020

The Torsion Bars System Reliability Analysis with Failure Mode in Crawler Vehicle

Yi Liu, Pengbo Mou, Bin Zhang

Mechanics Institute, bengbu, An Hui, China

mpb2011@163.com

Abstract. The traditional reliability analysis of mechanical systems was not considered the correlation between components, which leads to inaccurate analysis. In this paper, Copula function was applied to reliability analysis of the track vehicle torsion bar system, and a system reliability model which failure mode of a track vehicle torsion bar correlated was established. The relationship between trip and system reliability was obtained, and the superiority of Gumbel Copula system reliability model in analyzing the reliability of mechanical system was verified, which provided a new idea for reliability analysis and reliability sensitivity analysis of mechanical system.

1. Introduction

A torsion bar was an important buffer of a crawler vehicle suspension parts, due to installation error, unreasonable design, and the different hit from the road, lead to the failure of torsion bar rarely caused by a single failure mode in use process, was the failure mode mixing, and failure modes correlated. At present, in terms of reliability research and life prediction, the correlation between mechanical systems could not be accurately expressed, and the analysis results were deviated greatly by independent assumptions, so that the time of system failure could not be accurately predicted, which led to the delay of maintenance time and increased the probability of failure. Based on the analysis of the torsion bar reliability was introduced to describe the nonlinear correlation between mechanical parts of copulas function, for the reliability of mechanical system analysis provided a new method.

2. Determination of edge distribution function of torsion bar system

A crawler vehicle had 10 torsion bars, because of the bad working conditions of the crawler vehicle, torsion bar fracture was common failures encountered in engineering applications, the torsion bar failure mode was mainly by the high cycle fatigue damage and low cycle fatigue damage caused, for a crawler vehicle, a torsion bar fails, will made the crawler vehicle couldn't run normally, therefore, a torsion bar system could hypothesis for the series system. The structure diagram of the serial system was shown in figure 1.

Figure 1. Series system structure diagram.

The three-parameter weibull distribution had a good fitting effect on mechanical system parts. It was assumed that the life distribution function of the i torsion bar was shown in equation (1).

Content from this work may be used under the terms of the Creative Commons Attribution 3.0 licence. Any further distribution of this work must maintain attribution to the author(s) and the title of the work, journal citation and DOI.

Published under licence by IOP Publishing Ltd

$$F_i(t) = 1 - R_i(t) = 1 - \exp(-\frac{t_i - \gamma_i}{\eta_i})^{m_i} \tag{1}$$

Where, m_i was the shape parameter of the i torsion bar, $m_i > 0$, η_i was the scale parameter of the i torsion bar, $\eta_i > 0$, γ_i was the position parameter of the i torsion bar, also known as the minimum life parameter, which meant that it will not fail before γ_i, (when $\gamma_i = 0$, it degenerated into a two-parameter distribution). t_i was the working life, $t_i \geq \gamma_i$. $F_i(t)$ was the marginal distribution function of the i torsion bar.

3. Established the reliability model of torsion bar system

Copula connect function contained many distribution families, including Gumbel Copula connect distribution function had high sensitivity in the end, maintenance personnel wanted to get the system reliability near the tail, in order to determined whether you need vehicle maintenance, ensured reliability of the vehicle in the next run, thus the Gumbel Copula connect reliability model could meet the needs of the maintenance personnel.

The Gumbel Copula function took its generating function as $\varphi(u) = (-\ln(u))^{1/\theta}$, The multidimensional distribution function and density function could be obtained from formula (1), as shown in formula (2) and (3) respectively.

$$C(u_1, u_2, \cdots, u_n; \theta) = \exp\left\{-\left[(-\ln u_1)^{1/\theta} + (-\ln u_2)^{1/\theta} + \cdots + (-\ln u_n)^{1/\theta}\right]^{\theta}\right\} \tag{2}$$

$$c = \frac{\partial C(u_1, u_2, \cdots, u_n; \theta)}{\partial u_1 \partial u_2 \cdots \partial u_n} \tag{3}$$

Where, u_i was the failure distribution function $F_i(t)$ of the driving gear at the i gear position; θ was the relevant parameter of each random variable, $\theta \in (0,1]$. when $\theta = 1$, The variable u_i was independent, that was $C(u_1, u_2, \dots u_n; 1) = \prod_{i=1}^{n} u_i$; when $\theta \to 0$, $\lim_{\theta \to 0} C(u_1, u_2, \dots u_n; \theta) = \min(u_i)$, The variable u_i tended to be completely correlated. The established degree of reliability model for torsion bar of a crawler vehicle was shown in equation (4).

$$
\begin{aligned}
R^s(t) &= P(T_1 > t, \cdots, T_{12} > t) = 1 - P(\bigcup_{j=1}^{12} T_j \leq t) = 1 - \sum_{j=1}^{12} P(T_j \leq t) + \sum_{1 \leq j < \cdots \leq 12} P(T_1 \leq t, \cdots, T_{12} \leq t) \\
&= 1 - \sum_{i}^{12} F_i(t) + \sum_{1 \leq k < j \leq 12} (F_k(t) \cdots 0, \cdots 0, \cdots, F_j(t)) - \cdots + (-1)^n \sum_{1 \leq k < j < \cdots < l \leq 12} (F_1(t), \cdots, F_{12}(t)) \\
&= 1 - \Delta_{F_1}^1 \cdots \Delta_{F_{12}}^1 C(F_1(t), F_2(t), \cdots, F_{12}(t))
\end{aligned}
\tag{4}
$$

Where in （4）, $C(F_1(t), F_2(t), \cdots, F_{12}(t))$ was Copula function; $F_i(t)$ was the failure distribution function; $F_i = 1 - R_i$, Δ According to difference.

Due to the malfunction of any torsion bar with a crawler vehicle will be unable to run normally, and in the process of investigation found no any two trouble occurs at the same time, and crawler vehicle working conditions bad, so the low cycle fatigue damage fracture of torsion bar accounted for 80% of torsion bar all trouble, so in this paper only considered the failure mode correlation between the torsion bar by the low cycle fatigue damage, system reliability model could be simplified as binary model, $u_1(t)$ namely failure rate of torsion bar caused by low cycle fatigue damage fracture, $u_2(t)$

namely failure rate of the other torsion bars caused by low cycle fatigue damage fracture, The model could be simplified as shown in equation (5).

$$
\begin{aligned}
R^s(t) &= P(T_1 > t, T_2 > t) = 1 - P(\bigcup_{j=1}^{2} T_j \le t) = 1 - \sum_{j=1}^{2} P(T_j \le t) + \sum_{1 \le j < k \le n}^{n} P(T_1 \le t, T_2 \le t) \\
&= 1 - u_1(t) - u_2(t) + C(u_1(t), u_2(t)) \\
&= 1 - u_1(t) - u_2(t) + \exp-\{[-\ln u_1(t)]^{1/\theta} + [-\ln u_2(t)]^{1/\theta}\}^{\theta}
\end{aligned}
\tag{5}
$$

4. Parameter estimation of torsion bar system reliability model

The maximum likelihood estimation method was adopted for the failure distribution function of a single torsion bar, and the maximum likelihood estimation value of m_i, η_i, γ_i, and the three parameters could be obtained by MATLAB. By solving the parameters of the distribution function model, the maximum likelihood estimation was shown in table 1.

Table 1. m_i, η_i, γ_i Parameter estimation.

Failure Unit	γ_i	η_i	m_i
$u_1(t)$	566.79	13336.15	3.28
$u_2(t)$	378.46	15337.18	2.87

Correlation coefficient θ could be estimated through MATLAB programming by using the estimation method of base-large likelihood function or non-parametric estimation method based on a large number of data collected in the survey. The results of parameter estimation were substituted into Gumbel Copula function respectively, and the relationship between torsion bar system reliability, mileage and parameters was obtained, as shown in figure 2. In this paper, the collected data were programmatically estimated $\theta = 0.38$, and the relationship between its reliability and kilometer were shown in figure 2.

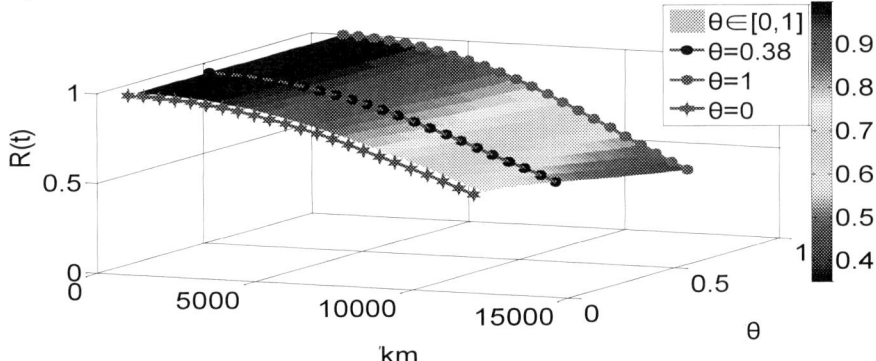

Figure 2. Reliability relation diagram of torsion bar system.

When $\theta \to 1$, the components of a system tend to be independent of each other; when $\theta \to 0$, Components tend to be completely correlation to each other. In the reliability analysis of the system, when the subsystems were completely correlated, the system had the highest consistency. When the subsystems were independent of each other, the system had the maximum randomness. For the series system, the better the consistency, the higher the reliability. Therefore, the reliability was the highest when the subsystems were completely correlated, while the reliability was the lowest when the subsystems were independent of each other, i.e:

$$
R^0(t) \le R^s(t) \le R^1(t)
\tag{6}
$$

It could be seen from the figure that the system reliability model considering correlation could reflect the actual situation more accurately than the traditional reliability analysis.

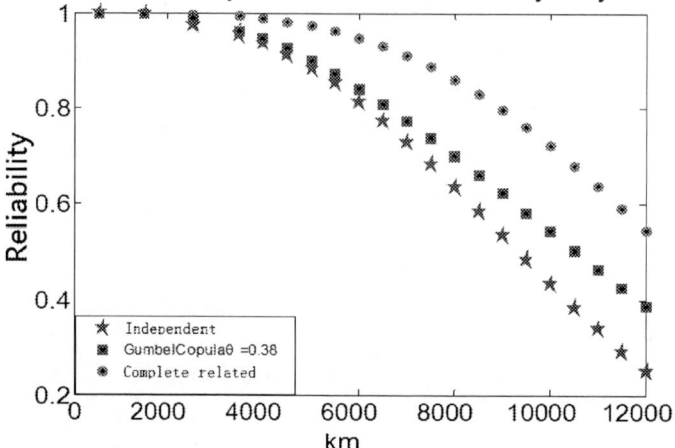

Figure 3. Scatter diagram of characteristic value of reliability of torsion bar system.

With the increase of mileage, assuming the components were independent of each other and assumptions components associated in full, torsion bar system reliability increased gradually, the deviation in the range of 12000 km, the reliability of the maximum deviation was 0.2201, if in accordance with the previous system components were independent of each other or completely correlation to system reliability analysis and forecast, the deviation between the two was more and more big, the resulting reliability maintenance personnel of the torsion bar without an accurate grasp, or the reliability of the same system prediction results were too good, or predict the results were too conservative, resulting in vehicle maintenance or less maintenance.

5. Conclusion

In this paper, the Copula function was introduced into the reliability analysis of the torsion bar system of crawler vehicles. The reliability model of the torsion bar system based on Gumbel Copula function was established. The complex multiple integral operations were replaced by simple Copula function operations, which reflected the practicability of the established system reliability model. The advantage of Gumbel Copula function to solve the problem of system reliability of components with correlation is verified, and the dynamic relationship between trip and system reliability is obtained. By comparing the prediction results with the survey results, it is verified that the Gumbel Copula system reliability model can solve the problem of large deviation between the reliability prediction of mechanical system and the actual situation, and provides an important technical means for the accurate solution of the reliability of mechanical system.

References

[1]   Roger B, Nelsen. An Introduction to Copula[M]. *New York: Springer press*. 2005.

[2]   Nelsen, Roger B, Úbeda Flores, The Lattice-theoretic Structure of Sets of Bivariate Copulas and Quasi-copulas. *C R Acad Sci Paris Ser I. In press*[M], 2005.

[3]   Chengmin-He, Wei Wu. Reliability Model of Mechanical System Based on Copula and Its Application [J], *Acta Armamentarii*. 2012.

[4]   Wende-Yi. Model and Application of Financial Risk Dependency Structure Based on Copula Theory [M]. *China Economic press*. 2011: 15-16.

Simulation and Life Prediction of Gear Meshing Process of Gearbox of A Crawler Vehicle

Pengbo Mou, Yi Liu , Jian Zhou

Mechanics Institute, Bengbu, An Hui, China

mpb2011@163.com

Abstract. A crawler vehicle gearbox was easy to invalidation, stress change of the gear tooth meshing process was difficult to accurately measure, and the life prediction was not accurate. this paper established a finite element model of gear meshing, the analysis of the gear teeth began to come into contact with the complete separation of highly nonlinear dynamic process, and the simulation results of test, and established the life prediction model by nCode Designife software, and estimated the gear fatigue failure position and life. Compared with the actual results verified the feasibility of dynamic analysis method of gear meshing, for gear strength check, the reference for the optimization design and fatigue life prediction.

1. Introduction

Gears in the transmission process, due to the friction and wear, meshing impact and the influence of such factors as easy to invalidation, all invalidation modes in tooth, the tooth root bending fatigue invalidation accounted for the largest share, followed by the tooth surface contact fatigue failure. Second gears of a crawler vehicle were involute spur gears, using the theory of the ball or cylinder contact analysis before the tooth contact state at a certain moment, already could not comprehensively reflect the changing process of tooth surface contact state. In this paper, by using ABAQUS/Explicit dynamic Explicit finite element method, the contact stress and fatigue bending stress of the second gear of a crawler vehicle transmission with time were analyzed, and the life of the gear was predicted.

2. Gearbox load spectrum obtained

In order to simulated the stress changes on the tooth surface and the tooth root during the gear meshing process under actual conditions, dynamic simulation was carried out on the established virtual prototype of a crawler vehicle. The simulation time was set at 1s and the step length was 0.001s, and the load spectra of the torque and angular velocity of the driving gear and passive gear under the second-grade E-class road surface were obtained. See figure 1 and figure 2.

Through wavelet analysis, filter out the vibration produced by interference signal, the second driving gear angular velocity and the passive gear torque as shown in figure 3 and figure 4 shows, from the diagram, in 0.35 ~ 0.45 s gear because of the influence of road roughness mutations made torque, due to the mutation load was the main cause of gear failure, therefore, The angular velocity and torque of the driving gear and passive gear within 0.35~0.45s were taken as the boundary conditions of finite element.

Content from this work may be used under the terms of the Creative Commons Attribution 3.0 licence. Any further distribution of this work must maintain attribution to the author(s) and the title of the work, journal citation and DOI.

Published under licence by IOP Publishing Ltd

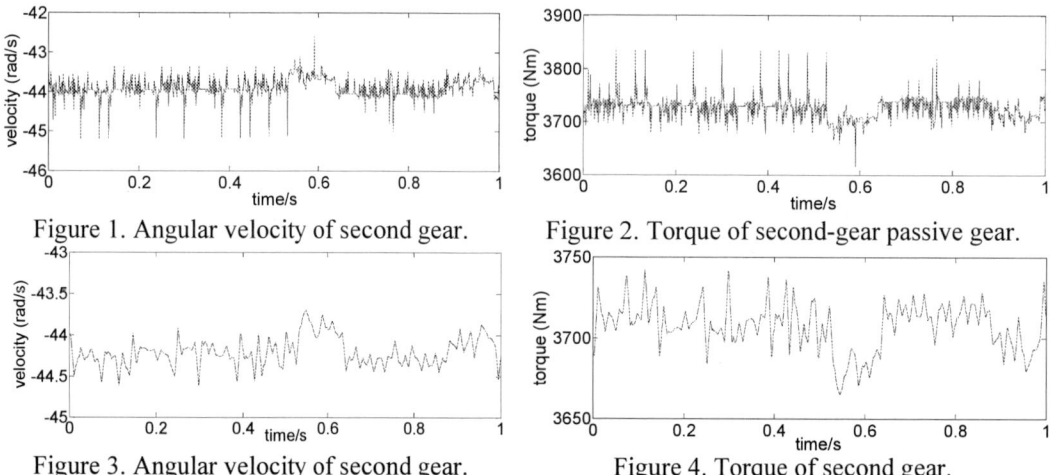

Figure 1. Angular velocity of second gear.

Figure 2. Torque of second-gear passive gear.

Figure 3. Angular velocity of second gear.

Figure 4. Torque of second gear.

3. Finite element model establishment and result analysis

The driving system virtual prototype was simulated from 0.35~0.45s to obtain the angular velocity and torque of the driving gear and passive wheels added to the boundary conditions of ABAQUS. At the same time, the active wheel speed was set as 0.14rad/s, and the simulation duration was 0.1s. Finite element simulation analysis was conducted on gear meshing. The stress cloud diagram as a function of time was shown in Figures 5~8.

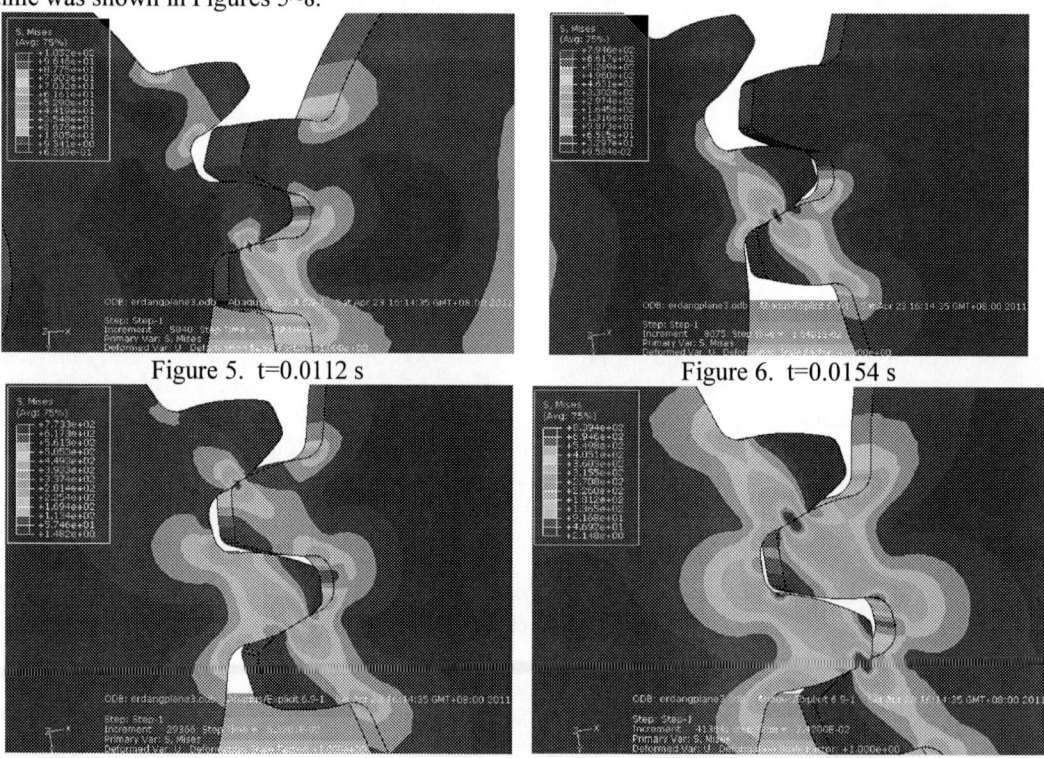

Figure 5. t=0.0112 s

Figure 6. t=0.0154 s

Figure 7. t=0.0532 s

Figure 8. t=0.0742 s

It could be obtained from the analysis in Figure 5~8 that in the transmission process of gears, the contact force generated when the gears first contact was large, which led to the bending and deformation of the tooth root of passive gear. As the driving gear rotates, the meshing position changed from the tooth root to the tooth end of the driving gear and from the tooth end to the tooth root of the passive gear, causing the contact stress position to change along the tooth surface, and the bending stress of the tooth root occurs in the whole process of meshing. Therefore, in the process of gear meshing, tooth root was prone to fatigue failure, tooth surface due to collision, friction and other reasons, easy to produce tooth surface peeling.

It could be seen from the stress cloud diagram that the stress of the gear tooth was very small and basically zero in the non-meshing condition. In the case of meshing, the stress value increases rapidly and reached the peak in a very short time. This showed that the stress in the meshing process of gear teeth was similar to the dynamic stress under the impact load.

The number of teeth of the second gear of a crawler vehicle transmission was 20, the number of teeth of the passive gear was 28, the modulus was 9mm, the manufacturing accuracy was level 6, and the load impact was medium impact. According to the Hertz contacted stress calculation formula(shown in Formula 1):

$$\sigma_{\max} = \sqrt{\frac{\omega_n}{\pi \rho_{red}} \frac{1}{\frac{1-v_1^2}{E_1} + \frac{1-v_2^2}{E_2}}} \tag{1}$$

where, v_1, v_2 was poisson's ratio of two cylinders, E_1, E_2 was the elastic modulus of two cylinders, ρ_{red} was the radius of comprehensive curvature, ω_n ω_n was the normal force on the length of the unit contact line of the cylinder.

It was calculated that when the driving gear and passive gear started to contact, that was, the contacted stress of the passive gear near the tooth tip and the active gear near the tooth root was the maximum, and the value was 989.5 MPa.

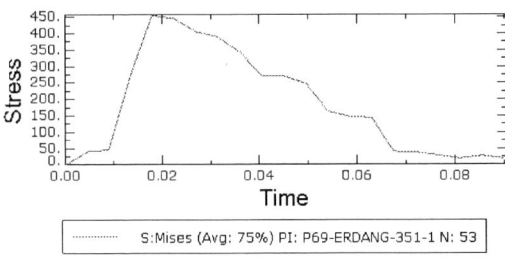

Figure 9. Root bending stress diagram of driving gear under tension

Figure 10. Bending stress diagram of passive gear root under pressure

Fig 9 and 10 from the stress nephogram and bending stress graph could be seen that, in the process of the whole tooth mesh, tooth side by extrusion, the other side by stretching, stress concentration in the fillet part, and the tooth root were subjected to bending stress and tooth started contacted as the tension side of the driving gear and driven gear tooth root bending stress from zero to maximum pressure side, but as time extended, stress decreases, until it was zero, this suggests that the gear meshing process, the tooth root was the part most prone to fatigue failure. Calculation formula of tooth root stress of spur gear(shown in Formula 2):

$$\sigma_F = \frac{F_t}{bm_n} K_A K_V K_{F\beta} Y_{Fa} Y_{Sa} Y_\beta \tag{2}$$

Determine parameter values by referring to the manual, $K_A = 1$, $K_V = 1.514$, $K_{F\beta} = 1.17$, $Y_{Fa} = 1$, $Y_{FS} = 4.28$, was substituted into equation (2), it could be obtained that under the load of 3700 Nm meshing torque, the maximum bending stress of the tooth root of the passive gear was 357.4 MPa, and the maximum bending stress of the tooth root of the active gear was 443.6 MPa.

For the maximum stress simulation results, Hertz contact ED theory calculation results and bending stress calculation results were compared, as shown in table 1.

Table 1. Comparison results of theoretical value and simulation value.

	Passive gear		Driving gear	
	σ_{max}	σ_F	σ_{max}	σ_F
Theoretical value	989.5	357.4	989.5	443.6
Simulation value	1031.6	361.5	996.7	454.6

σ_{max} (MPa) was the contact stress of tooth surface.

σ_F (MPa) was the bending stress of tooth root.

The simulation results were compared with the theoretical values. Considering the existence of friction force and the fact that both torque and speed were load spectra simulating the actual road conditions, both the contact stress and bending stress values were greater than the theoretical values. This research method was closer to the actual value and had certain reference value.

4. Gear life prediction

ABAQUS/Explicit simulation results in higher fatigue analysis software nCode Designlife life prediction and fatigue failure parts in the gear was 20Cr2Ni4A alloy steel material, after carburizing and quenching and low temperature tempering heat treatment, surface hardness was not under HRC57, core hardness was HRC34 ~ 45, reliability was 99%, the tensile strength limit was 1175 MPa, modulus of elasticity for the MPa.

To conservative projections for gear life, the heat treatment of residual compression stress in this should not be considered, from access to the complete separation of gear dynamic process of cyclic loading, according to the linear cumulative damage theory, the simulation given gear fatigue failure of the parts in advance as the root, and the driving gear than passive gear failure in advance, the small gear failure in advance.

According to the conversion formula (3) between the second gear speed and stroke of a crawler vehicle, the driving distance of the second gear when fatigue failure occurs could be converted.

$$S = N \frac{S'}{1000} \tag{3}$$

Where, S' was the driving distance corresponding to the load time history; S was the mileage corresponding to the fatigue life of parts; N was the cycle number of fatigue life.

By a crawler vehicle transmission ratio, the second driving gear turned a circle, the vehicle moved 0.57 m, but the passive gear turned a circle, the vehicle moved 0.8 m, according to software analysis, the survival rate 99%, the driving gear tooth root after 2.523×10^6 cycles fatigue failure occurs, the passive gear tooth root after 4.4×10^6 cycles fatigue failure occurs, calculated by the type (3), the driving distance before invalidation of the second driving gear and passive gear:

$$S_1 = N \frac{S'}{1000} = 2.523e6 \frac{0.57}{1000} = 1438.11 km \qquad S_2 = N \frac{S'}{1000} = 4.4e6 \frac{0.8}{1000} = 3520 km$$

According to the research and statistical analysis, the second usage accounted for 20% of the usage of the gear, the convert of the crawler vehicle mileage, under the second gear to run about 1600 ~ 1900 km dedendum fatigue failure occurs, consistent with actual situation, and as a result, a crawler vehicle can be determined through simulation analysis method of overhaul time, has certain engineering application value.

5. Conclusion

By ABAQUS/Explicit dynamic analysis method, to the height of the gear meshing process by simulating the nonlinear problem, and compared with the theoretical value and the simulation value by tooth surface contact stress and tooth root bending stress, and using the fatigue life of simulation compared with survey data. The results showed that the dynamic analysis was carried out on the gear meshing process to closer to the actual situation, the result more accurate, the fatigue life of the resulting results consistent with the research results, to predict the gearbox life provides an important reference.

References

[1] Guoyun-Li, Dengyun-Zhou.Simulation Analysis of Dynamic Contact Process of Gear Meshing[J]. *Journal of Lanzhou University of Ttechnology*. 2012(4):27~30.

[2] Jiebang-Zhou,Bailin-Zheng. Dynamic Analysis of Fiber Structure Collision Based on Central Difference Method [M]. *Mechanics Quarterly*. 2011, 32(3),466-472.

[3] Jing Cheng. Dynamic Simulation and Life Study of Transmission of Heavy Duty Vehicle Based on Finite Element Method [D]. *Jilin University*. 2009:56-61.

[4] Naishi-Cheng. Reducer and Transmission Design and Selection Manual [M]. *Machinery Industry Press*. 2001:83-123.

[5] Liang Shen. Simulation Analysis of Gear Fatigue Life and Root Crack [D]. *Chongqing University*. 2011:30-35.

Study on cargo-swing reduction of general gantry crane using hybrid optimal input shaper

TUO Jianzhi[1,a], DU Peng[1], ZHANG Ke[1], ZHANG Wei[1] and XIE Zonghua[1]

[1]Weifang Engineering Vocational College, Weifang City, Shandong Province, China

[a]tuojianzhi@qq.com

Abstract. The gantry crane is widely used, but easy to accidents, which mostly due to the invalid control of cargo-swing. Based on the appropriate dynamic model and the characteristics of gantry crane, the Hybrid Optimal (HYO) input shaper is designed by interpolation method. HYO input shaper combines the advantages of Zero Vibration Derivation(ZVD) input shaper on the slightly cargo-swing and the Extra Insensitivity(EI) input shaper on the greatly cargo-swing. The comprehensive property of HYO input shaper is better than ZVD and EI input shaper. HYO input shaper effectively suppresses the cargo-swing of gantry crane within 0.018 rad, largely avoids the accidents caused by unstable control of the gantry crane.

1. Introduction

The 3-dimensional transportation of the cargo makes the gantry crane to be a typical variable-parameter flexible mechanical system with high control difficulty. Establishing a reasonable dynamic model and accurate analyzing its dynamic characteristics are the basis for effectively controlling the operation of gantry cranes, according to the dynamic principle, the nonlinear dynamic model of the gantry crane is established and reasonable linearization is carried out.

The input shaping technology originated from the Posicast technique proposed by Smith et al. in 1957, which aims to suppress the oscillatory component, such as flexible spacecraft [1]. In the 1990s, the input shaping technology was used to control flexible spacecraft by Singhose, Derezinski and Singer, they proposed several extra insensitive input shapers to further improve the control performance of the input shaper [2,3]. In recent years, LI Minzhi and LIANG Chunyan designed the input shaper to realize the cargo-swing control of the crane based on the optimization idea [4]. DONG Mingxiao and MEI Xuesong study on the time-delay filtering theory and its engineering application [5]. At present, researches on this technology for cargo-swing reduction of gantry crane has not been found.

2. Control model and dynamic analysis of gantry crane

2.1. Control model

The gantry crane pulls the cargo through the wire rope to complete the lifting, transportation and placing. The general type A5-75/20t double-girder gantry crane is studied, as shown in Fig. 1. The span of the main cart is 26m, the lifting heights of the main and auxiliary hooks are 11m and 13m. The rated lifting speeds of the main and auxiliary hooks are 4.6m/min and 9.23m/min, the rated running speeds of the trolley and the main cart are 38.4m/min and 40.62m/min. The working level is A5. The main cart moves 1-dimensionally, the trolley moves 2-dimensionally, drags the cargo through the rope, during the start and stop, the cargo-swings due to the inertia.

Content from this work may be used under the terms of the Creative Commons Attribution 3.0 licence. Any further distribution of this work must maintain attribution to the author(s) and the title of the work, journal citation and DOI.
Published under licence by IOP Publishing Ltd

The Cartesian coordinate system $\{x_0, y_0, z_0\}$ and spherical coordinate system $\{\mathbf{e}_l, \mathbf{e}_\theta, \mathbf{e}_\phi\}$ are established[5], as shown in Fig. 2. The origin O_0 of the Cartesian coordinate system is taken at one end of the main cart, the coordinates of O in the Cartesian coordinate system are (x, y, z), the spherical coordinates of the cargo are defined as (l, θ, ϕ). The degrees of freedom of the gantry crane is 5, therefore, the 5 independent generalized coordinates are defined as x and y indicating the displacements of the trolley on the x-axis, the y-axis direction, the lifting length of the wire rope l, ϕ and θ, which are used to determine the particle location of the system [5].

 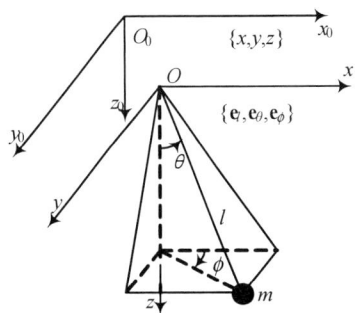

Figure 1. Type A 5-75/20t double-girder gantry crane.

Figure 2. Cargo-swing diagram of Gantry crane.

The generalized forces corresponding to x, y, and l are defined as Q_x、 Q_y and Q_l, indicating the driving force of the trolley, main cart and the lifting force of the wire rope.

According to the Lagrange motion equation

$$\frac{d}{dt}\left(\frac{\partial L}{\partial \dot{q}_k}\right) - \frac{\partial L}{q_k} = Q_k \qquad (k = 1,2,3,4,5) \tag{1}$$

$$L = T - V \tag{2}$$

$$T = \frac{1}{2}m[(\dot{x} + \dot{l}\sin\theta\cos\phi - l\dot{\phi}\sin\theta\sin\phi + l\dot{\theta}\cos\phi\cos\theta)^2 + (\dot{y} + \dot{l}\sin\phi + l\dot{\phi}\cos\phi)^2 +$$
$$(\dot{l}\cos\phi\cos\theta - l\dot{\phi}\cos\theta\sin\phi - l\dot{\theta}\sin\theta\cos\phi)^2] + \frac{1}{2}M_1\dot{x}^2 + \frac{1}{2}M_2\dot{y}^2 \tag{3}$$

$$V = -mgl\cos\theta\cos\phi \tag{4}$$

The where k is the number of degrees of freedom, q_k is the generalized displacement of the system, Q_k is the generalized force of the system, L is the Lagrange function. T is the kinetic energy of particle system, including the kinetic energy of the main cart, trolley and cargo. V is the potential energy of the particle system, including the potential energy of the cargo only.

The nonlinear dynamic model of the gantry crane is established. In order to facilitate the effective control of the gantry crane in engineering, according to the linearization theory, near the cargo balance position $\theta = 0°$, $\phi = 0°$ and swing in a small range, the nonlinear model is reasonably linearized, then the linearized model is

$$M_1\ddot{x} + b_x\dot{x} - mg\theta + ml\ddot{\theta} = Q_x \tag{5}$$

$$M_2\ddot{y} + b_y\dot{y} - mg\phi + ml\ddot{\phi} = Q_y \tag{6}$$

$$m(\ddot{l} - g + \ddot{x}\theta + \ddot{y}\phi) + b_l\dot{l} = Q_l \tag{7}$$

$$l\ddot{\theta} + 2\dot{l}\dot{\theta} + g\theta = -\ddot{x} \tag{8}$$

$$l\ddot{\phi} + 2\dot{l}\dot{\phi} + g\phi = -\ddot{y} \tag{9}$$

where M_1, M_2 and m are the masses of the trolley, main cart and cargo; b_x, b_y and b_l are the equivalent damping coefficients of the trolley, main cart and hoisting motion; g is the gravitational acceleration; $\dot{l}$ and $\ddot{l}$ are the cargo lifting speed and hoisting acceleration; $\dot{\theta}$, $\dot{\phi}$ and $\ddot{\theta}$, $\ddot{\phi}$ are the angular velocity and acceleration; $\dot{x}, \dot{y}$ and $\ddot{x}, \ddot{y}$ are the corresponding velocity and acceleration.

2.2. Dynamic analysis of gantry crane
The model is composed of the motion equations of the trolley and main cart, Eqs. (5), (6), the equations of hoisting and cargo motion, Eqs. (7), (8) and (9). The cargo motion are about 2-order oscillation of the cargo-swing angle θ and ϕ, which describes the kinematic relationship between the trolley, main cart and cargo motion. The accelerations of the trolley and main cart are the inputs, and the cargo-swing angle is the output. The driving forces Qx, Qy and Ql drive the trolley, main cart and lifting mechanism, the wire rope is a flexible body, therefore, the lifting motion makes the system a weak damping flexible system containing a rigid mode. The system is time-varying 2-order nonlinear system, state variables are coupled to each other. The frequency of the cargo-swing $\omega_n = \sqrt{g/l}$ is related to the length of the rope, the damping ratio $\xi = \dot{l}/\sqrt{gl}$ is related to the length of the rope and lifting speed, the amplitude of the swing is related to the acceleration of the trolley.

During the simulation, it is required to lift the cargo from the ground to 1.25m, then maintain this height, and finally release the cargo to the ground. The gantry is required to accelerate for 2s, run at a constant speed for 17s, then decelerate for 2s, then stop.

The motion simulation of the gantry crane model is based on MATLAB. The oscillating angles θ and ϕ of the gantry crane nonlinear model and the linearized model have very similar variations in the two directions, as shown in Fig. 3.

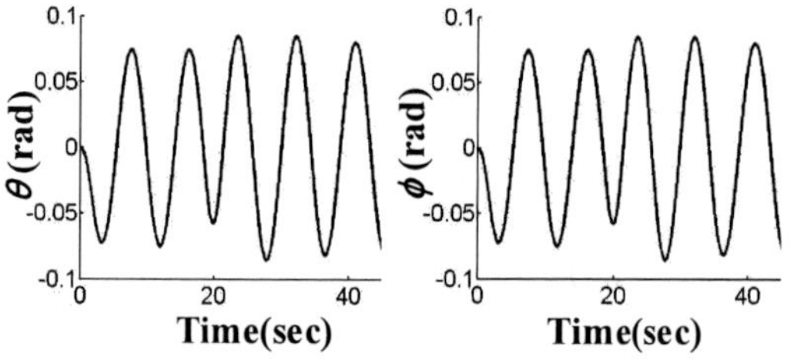

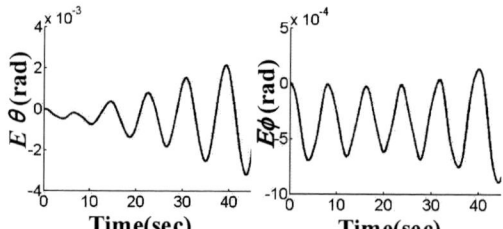

Figure 3. Cargo-swing angle of the simulation model of gantry crane.

Figure 4. The errors of cargo swing angle between the nonlinear and linear model.

Respectively solving the cargo-swing angle errors between the two models, as shown in Fig. 4, the maximum linear error is on the order of 10^{-3}, it is negotiable in engineering, so the linear model can accurately describe the dynamic performance of the gantry crane to a certain extent, has the advantage of being physically convenient to implement.

3. Hybrid optimal input shaping

3.1. Input shaping

Input shaping is a control strategy which convolves the reference command with the shaper pulse sequence, uses the resulting shaping command to drive the system. In the case of reasonable design of the shaper pulse amplitude and time lag, the system vibration can be effectively suppressed. After the end of the action of the input shaper, the ratio of the amplitude of the system unit impulse response between with and without the shaper control is called residual vibration [5].

ZVD input shaper is designed, the amplitudes at the model frequency and damping ratio are 0, their differentials to ω_n are 0. When the model parameters change in a small range, the ZVD input shaper can effectively suppress the residual vibration of the system [3]. when the control parameters vary widely, the control effect will be greatly reduced, as shown in Fig. 5.

If residual vibration of the shaper is required to be less than a value V_{exp} at the undamped natural frequency ω_n and the damping ratio ξ, to be 0 at some two frequency points on two sides of ω_n, and the differential of V to ω at ω_n is 0, such shaper is called Extra Insensitivity(EI) input shaper [3], which can ensure that the residual vibration of the controlled system is less than the V_{exp} when the system parameters vary widely around ω_n, as shown in Fig. 5.

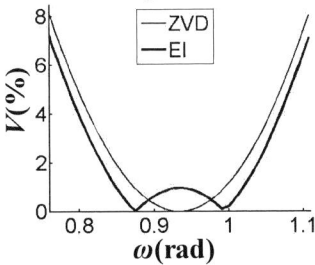

Figure 5. 2-order system residual vibration under 3-pulse ZVD and EI control.

The main parameters affecting residual vibration are frequency ω and damping ratio ξ, which are determined by the lifting height and speed in the gantry crane system. Such parameters change large and small during the different working processes of the gantry crane.

3.2. Hybrid optimal input shaping

Based on ZVD and EI input shaper, the Hybrid Optimal (HYO) input shaper is designed by interpolation method, which combines the advantages of these two input shapers. An interpolation parameter λ ($0<\lambda<1$) is introduced, the time lag and pulse amplitude are

$$t_{iHYO} = \lambda t_{iEI} + (1-\lambda)t_{iZVD} \qquad (10)$$

$$A_{iHYO} = \lambda A_{iEI} + (1-\lambda)A_{iZVD} \qquad (11)$$

Where t_{IZVD}, t_{iEI}, A_{IZVD}, A_{iEI} are action times and amplitudes of the i-th pulse of ZVD and EI shaper.

Within the range of system parameter variation, in order to solve the optimal interpolation parameter value λ, the change of the system parameters follows the average distribution law, the probability density function is

$$f(\omega,\ \xi) = \begin{cases} \dfrac{1}{(\omega_l-\omega_l)(\xi_u-\xi_l)}, & \omega \in [\omega_l,\ \omega_u],\ \xi \in [\xi_l,\ \xi_u] \\ 0, & \omega \notin [\omega_l,\ \omega_u],\ \xi \notin [\xi_l,\ \xi_u] \end{cases} \qquad (12)$$

Where ω_u and ω_l are the upper and lower boundaries of frequency, ξ_u and ξ_l are the upper and lower boundaries of damping ratio.

Combined with parameter changes and residual vibration, the objective function is

$$J = \int_0^1 \int_0^\infty V(\omega, \xi) f(\omega, \xi) d\omega d\xi \qquad (13)$$

Substituting Eqs. (10), (11), (12) into Eq. (13), based on optimization function in Matlab, the special value of λ is obtained which minimize the value of J, at the same time, the residual vibration of the controlled system is also minimized. The parameters of HYO can be obtained by substituting such λ into Eqs. (10), (11). It can be seen that HYO only needs to solve the minimum value of a function with a variable λ, which has a more concise optimization process than the general optimization.

3.3. Simulation results of hybrid optimal input shaping

In the normal working condition, the range of rope length is from 1.52m to 12.52m, the hoisting speed is from 0 to 4.6m/min, planning the operation mode that the trolley accelerates for 2s, moves with a uniform speed of 14s and decelerates for 2s. The parameters are substituted into the cargo-swing model for simulation, designing a 3-pulse ZVD input shaper

$$F(s) = 0.2570 + 0.4960e^{-3.3630s} + 0.2400e^{-6.7310s} \qquad (14)$$

Taking V_{exp} 10%, designing a 3-pulse EI input shaper

$$F(s) = 0.2630 + 0.4810e^{-3.3539s} + 0.2660e^{-6.7260s} \qquad (15)$$

Designing a 3-pulse HYO input shaper according to Eqs. (10), (11), (12), (13), (14), (15)

$$F(s) = 0.2499 + 0.4830e^{-3.3656s} + 0.2566e^{-6.7310s} \qquad (16)$$

When the gantry crane parameters change slightly, the HYO control performance is close to the ZVD input shaper, the residual swing angle of the cargo is controlled within 0.02 rad. The EI input shaper controls the residual swing angle within 0.08 rad, as shown in the Fig. 6.

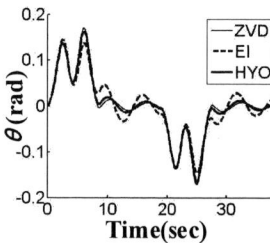 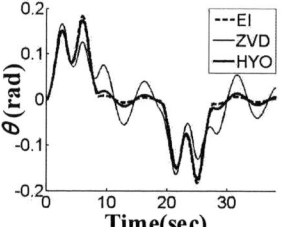

Figure 6. Cargo-swing angle of gantry crane controlled by the input shapers when the parameters change slightly.

Figure 7. Cargo-swing angle of gantry crane controlled by the input shapers when the parameters change greatly.

When the parameters change greatly, the HYO input shaper control performance is closer to the EI input shaper, the residual swing angle of the cargo is controlled within 0.018 rad, the ZVD input shaper controls the residual swing angle within 0.051 rad, as shown in Fig. 7.

4. Conclusions

According to the analysis of the linearized model and the simulation results, the moving directions of the trolley and main cart are perpendicular to each other, the driving motor only controls the respective motions and does not affect each other. Then, the motions in such directions can be decoupled, the effect on the cargo motion in the corresponding direction is similar. To study the movement of the gantry crane, it is only necessary to study the horizontal movement of the trolley or main cart in the x-axis or y-axis direction in combination with the lifting motion, which simplifies the space pendulum motion of the cargo into a plane pendulum motion.

The linearized model of the gantry crane can be used as the control object, the motion characteristics of the main cart and trolley are studied separately, then it is easy to realize the effective and stable control of the gantry crane.

In view of the frequent changes of the gantry crane parameters, taking the advantages of ZVD and EI input shaper, the interpolation method is used to design HYO input shaper to suppress the cargo-swing of the type A5-75/20t double-girder gantry crane. The control results show that the control effect of HYO input shaper is better than ZVD and EI input shaper, which achieves the original design purpose that effectively suppressing the cargo swing of gantry crane, largely avoiding the accidents caused by unstable control of the gantry crane.

References
[1]   Smith O J M. Posicast Control of Damped Oscillatory Systems[A].Proceedings of the IRE[C], 1957: 1249~1255.
[2]   Singer N C, Seering W P. An Extension of Command Shaping Methods for Controlling Residual Vibration Using Frequency Sampling[A]. IEEE International Conference on Robotics and Automation. Nice[C], France, 1992: 800~805.
[3]   Singhose W, Derezinski S, Singer N. Extra-Insensitive Shapers for Controlling Flexible Spacecraft[A]. AIAA Guidance, Navigation, and Control Conf[C], Scottsdale, AZ, 1994.
[4]   LI Minzhi, JIA Qing, LIANG Chunyan. Hybrid Time Filter Design in Flexible Systems[J]. JOURNAL OF SHANGHAI JIAO TONG UNIVERSITY. 2001. 35(8):1117-1120.
[5]   DONG Mingxiao, MEI Xuesong. Time-delay Filtering Theory and Its Engineering Application[M]. Beijing: Science Press, 2008:94-153.
[6]   Jianzhi Tuo, Wei Sun and Peng Du. Research on multi-point temperature detection and time display system[A]. 2018 IOP Conf. Series: Materials Science and Engineering[C], 452: 042116.

Design and analysis of the leveling hydraulic system of the combine harvester

Heng Wang[1], Shukun Cao[1,*] Xiangqian Xu[2], Tao Han[2] and Hejia Guo[2]

[1]School of mechanical Engineering, University of Jinan, Jinan, China

[2]Shandong Gold Dafeng Machinery Co. , Ltd. , Jining, China

*Corresponding author e-mail: caoshukun@126. com

Abstract: The grain combine harvester is a large and complex agricultural machine, and the transmission system is an important part of the combine harvester. The effect of the transmission system has a significant impact on the performance of the combine harvester. Hydraulic systems play an increasingly important role in harvester drive systems. The body leveling hydraulic system generally has overflow loss and heavy throttling loss leading to hydraulic energy loss. There is a power mismatch between the power source and the load, and the power mismatch causes the energy loss problem. At present, there is less research on the leveling of harvesting locomotives in China. In this paper, reasonable power matching is carried out for the components of the grain combine harvester body, and the hydraulic drive scheme of the functional module that meets the requirements is designed. The parameters of the hydraulic system are optimized by AMESim simulation software, and the subsequent combine harvester drive System design provides a reference.

1. Introduction

In the mountainous and hilly areas of the northwest, due to the complex topography, the harvesting machine is affected by the unevenness of the ground during field operations, coupled with the real-time changes in the car body's own feeding volume, granary quality, header attitude, and drum speed[1]. The centroid changes, which leads to the real-time change of the oil and gas suspension cylinder load[2]. The harvester is prone to rollover phenomenon. The rollover reduces the efficiency, reliability and safety of the harvester, causing major economic losses such as casualties[3]. We design a set the hydraulic system for automatic leveling of the combine harvester body is very necessary[4].

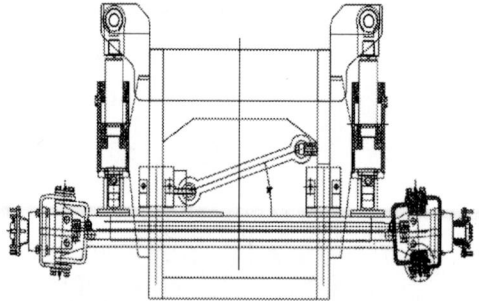

Figure 1 Oil and gas suspension structure layout

Content from this work may be used under the terms of the Creative Commons Attribution 3.0 licence. Any further distribution of this work must maintain attribution to the author(s) and the title of the work, journal citation and DOI.
Published under licence by IOP Publishing Ltd

In this paper, we designed a concrete combiner that can be automatically leveled by the vehicle body. The connected oil and gas suspension system is adopted. The layout of the oil and gas suspension structure is shown in Figure 1. In terms of anti-rollover performance, the combined oil and gas suspension can significantly improve the anti-rollover performance of the vehicle, reduce the roll angle generated when the vehicle turns, make the vehicle more stable, and the hydraulic system is simple. We designed a hydraulic system for the automatic leveling of the combine harvester body to use a connected oil and gas suspension system.

2. Working principle and design of the car body leveling hydraulic system

2.1   Working principle of the car body leveling hydraulic system |
In view of the huge difference in load bearing between the front and rear axles of the harvester, when the left and right leveling is leveled, the one-sided hydraulic cylinder cannot be lifted synchronously. The reason for the analysis is that the flow control is not carried out on the oil inlet path of the suspension cylinder, so that the suspension cylinder is entered. The flow rate is different. We can install the flow control valve on the oil inlet to ensure the same flow at both ends. Second, we can't eliminate the influence of the accumulator on the load and discharge caused by the load change. For this, due to the suspension of the oil and gas During the flat process, the load of each hydraulic cylinder is different, then the accumulator can only be cut off during the leveling, that is, the accumulator is first cut off during the raising process, and the suspension is extended by the control of the flow valve, because there is no energy storage. The influence of the device can be controlled by the flow valve to simultaneously raise and lower the suspension cylinder on one side. However, after the leveling, due to the difference between the pressure in the accumulator and the pressure in the oil circuit, when the accumulator is reconnected, the hydraulic cylinder changes rapidly due to the large difference in pressure, causing the hydraulic cylinder to shake. In this paper, a pressure tracking valve is installed in the hydraulic control system. When the accumulator is cut off, the pressure in the accumulator is always kept the same as the pressure in the oil circuit during the leveling process.

The grain combine harvester designed in this paper uses a hydraulic cylinder base mounted on both the frame and the axle. The innovative non-rigid connection of the traditional harvester is adopted, and the hydraulic suspension cylinder is used to connect the frame and the axle. The hydraulic cylinder of the oil and gas suspension not only has the effect of adjusting the posture of the vehicle body, but also the effect of damping the oil body in the flow process due to the small oil port inside the hydraulic cylinder of the oil and gas suspension. The connection between the frame and the axle is shown in Figure 2.

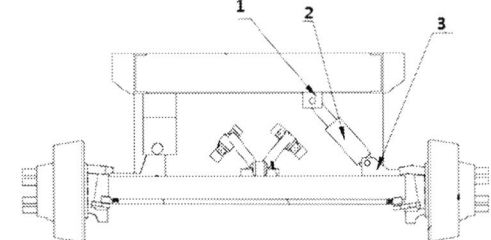

1-Workshop, 2-Hydraulic Cylinder, 3-Hydraulic Suspension
Figure 2. Frame and Axle

2.2   Design of car body leveling hydraulic system

2.2.1 Selection and calculation of components for car body leveling hydraulic system
According to the overall layout of our harvester, the full load of the harvester is around 14 tons! The front axle bears 10 tons when the harvester is fully loaded, and the rear axle has a load of 4 tons when

ISPECE   IOP Publishing

IOP Conf. Series: Journal of Physics: Conf. Series **1187** (2019) 032023   doi:10.1088/1742-6596/1187/3/032023

the harvester is fully loaded. At the same time, the working pressure of the harvester hydraulic cylinder is 7-16Mpa, because the harvester is tilted and leveled, in the extreme position, When all the weights are concentrated on one hydraulic cylinder, it is assumed that all the weights are vertically pressed onto a single hydraulic cylinder. According to the same, the inner diameter of the harvester hydraulic cylinder is 105mm. According to the data provided by Xinhua Hydraulic Company, we select the inner diameter of the hydraulic cylinder to be 110mm. When the hydraulic cylinder is in the neutral position, the hinge center distance is 833 mm. The hydraulic cylinder structure diagram and parameters are shown in figure 3.

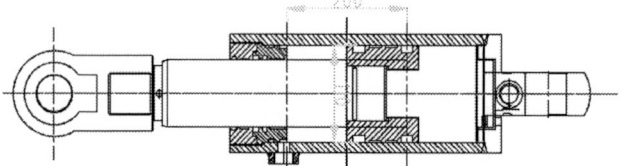

Figure . 3 Hydraulic cylinder structure and parameters

2.2.2 Design of hydraulic control system for car body leveling
According to the functional requirements, this paper draws the hydraulic system control of the header as shown in figure 4:

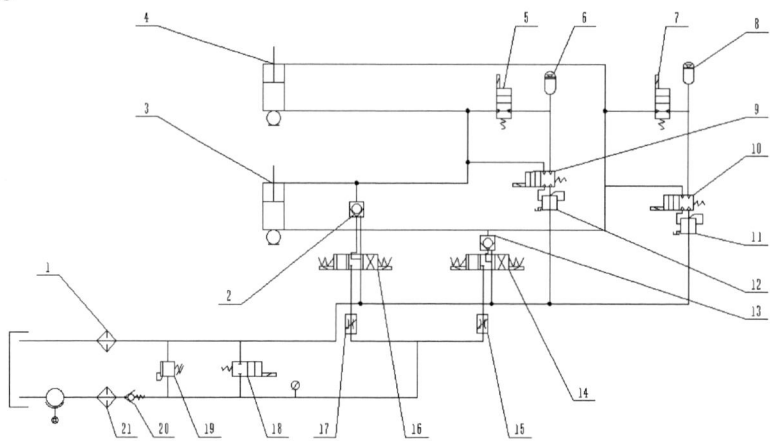

1, 21-filter 2, 13, 2-way valve 3, 4-hydraulic cylinder 5, 7, 18-two-position two-way electromagnetic reversing valve 6, 8 -Energy device 9, 10 - two four-way electromagnetic reversing ball valve 11, 12 - pressure tracking valve 14, 16 - three four-way electromagnetic reversing valve 15, 17-speed control valve
Figure 4. Constant flow system for connected hydro-pneumatic suspension

The hydraulic principle of the whole system is: firstly close the solenoid valves 5, 7, cut off the accumulator and the suspension cylinder, and simultaneously open the solenoid valves 9, 10, 14, 16 and the high-pressure oil pushes the suspension cylinder to rise, passing the speed control valve The control of 15, 17 makes the flow rate into the single-side suspension cylinder the same, then the suspension cylinder on one side is extended synchronously, so that the side of the harvester is synchronously raised, and the solenoid valves 9, 10 are opened during the raising process. By engaging the pressure tracking valves 11, 12, the pressure in the accumulator can always follow the pressure in the suspension cylinder, the pressure in the accumulator is the same as the pressure in the suspension cylinder. When it is raised into position, the solenoid valves 9, 10, 14, 16 lose power and the solenoid valves 5, 7 are energized, the high pressure port is closed, and the accumulator is connected to the system. At this time, the pressure in the accumulator and the suspension cylinder are inside. The pressure is equal, and smooth switching can be achieved to ensure the stability of the whole vehicle.

The working principle of the common hydraulic control check valve in figure. 4 is: when the hydraulic control port has the control oil pressure, the pressure oil pushes the piston, and then pushes the poppet valve spool to open, so that the oil P1 to P2 and P1P2 are connected, when the liquid When the oil pressure of the oil control port K is zero, the function of the normal check valve is the same, the oil P1 to P2 are connected, and the P1P2 is not connected. 2Car body leveling hydraulic system AMESim modeling .

AMESim is a software from France's IMAGINE for simulation analysis of fluid power, mechanical, thermal fluids and control systems. AMESim uses a graphical model based on physical models to provide users with a rich library of component applications. Modeling in AMESim can directly select the model of the component from the AMESim component library, or use the model in the HCD library to build the components you need [5].

In this paper, the hydraulic system piping is assumed to be rigid, and the length of the pipeline is not considered [6]. The simulation model of the vehicle body leveling hydraulic system established in the AMESim. environment according to the above-mentioned car body leveling hydraulic system principle is shown in figure. 5.

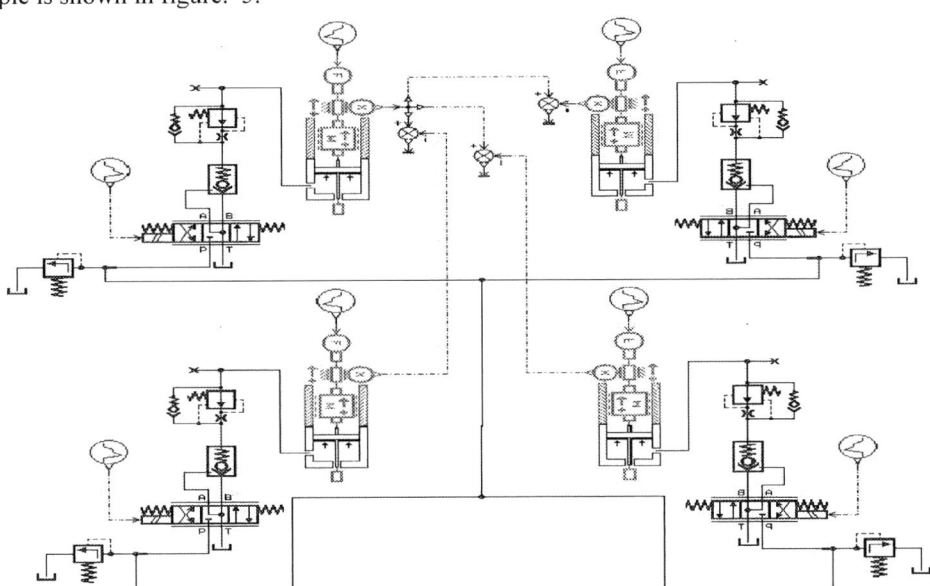

Figure 5. Simulation model of vehicle body leveling hydraulic system

When the grain combine harvester is leveling the car, the working conditions that need to be leveled mainly include slopes, brakes, and sharp turns. The eccentric load of the harvester is mainly caused by the offset of the position of the center of gravity of the harvester. The AMESim simulation verifies the correctness of the leveling hydraulic system by simulating the two working conditions of the model, observing the pressure of the hydraulic cylinder and the eccentric load of the hydraulic cylinder. The pressure and displacement of the hydraulic cylinder of the combine harvester during the load change are shown in figure. 6. The pressure and displacement of the hydraulic cylinder of the combine harvester when the eccentric load changes are shown in figure. 7.

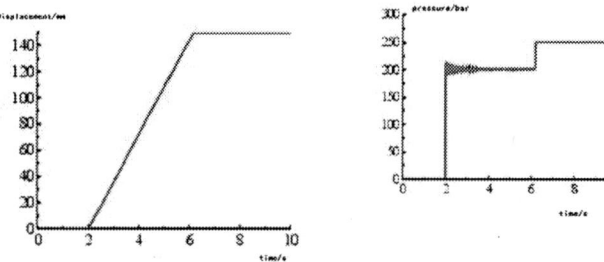

(a) displacement curve   (b) Pressure curve

Figure 6. Pressure and displacement changes of the hydraulic cylinder when the load changes continuously.

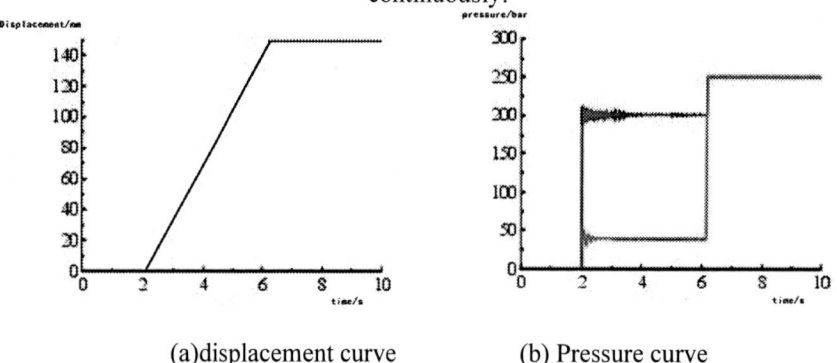

(a)displacement curve   (b) Pressure curve

Figure 7. Pressure and displacement changes of the hydraulic cylinder when the eccentric load changes

As shown in figure6 and figure7, during the change of the load and the eccentric load, it can be seen that the displacement values of the four support points under the two working conditions are basically the same, and there is no "virtual leg" phenomenon; the eccentric load condition The pressure of the lower four support points is also equal, and there is no "virtual leg" phenomenon at four points. The simulation results prove that the hydraulic system of the grain combine harvester designed in this paper can meet the requirements of the car body leveling in the actual working process.

3. Conclusions

1. In this paper, the hydraulic system of the combine harvester body is designed, and the hydraulic system is proved to be reliable through simulation and actual work. It provides a reference for the hydraulic design of the similar combine harvester.

2. This paper simulates and analyzes the combined harvesting vehicle body leveling system through AMESim. Through simulation analysis, it can reflect the motion characteristics of the car body leveling more intuitively. Using AMESim software to study the pump speed change to the car body leveling cylinder the effects of telescopic efficiency and motion stability provide a reference for the hydraulic system design of other related models.

3. This chapter determines that the hydraulic control system of the oil and gas suspension adopts the connected oil and gas suspension hydraulic control system, which controls the single cylinder compared with the traditional single valve, and the anti-rollover effect is better. In the case of analyzing the difference of the bearing load of the traditional harvester before and after the traditional connected oil and gas suspension is insufficient, the hydraulic control system of the connected oil and gas suspension based on the speed regulating valve and the pressure tracking valve is proposed.

4. The simulation results prove that the design of the leveling hydraulic system of the harvester body meets the actual working conditions.

Acknowledgments

This work was financially supported by Shandong Science and Technology Major Project, R&D and Industrialization Demonstration of Intelligent Corn Combine Harvester(2015ZDZX10001);2016Shandong agricultural machinery equipment research and development plan, 4YZP-4self-propelled corn harvester optimization upgrade fund.

References

[1] You Lei. Research on design of platform hydraulic leveling system [D]. Xihua University, 2014.

[2] Yang Yujing. Research on automatic leveling control system for heavy-duty flatbeds [D]. Yanshan University, 2015.

[3] Bao Yuanhai. Working principle and precautions for the use of grain combine harvester [1]. Agricultural machinery use and maintenance, 2018 (08): 54.

[4] Meng Jiale. Research on electro-hydraulic automatic leveling system for high adaptive aerial work vehicle [D]. Zhejiang University of Technology, 2015.

[5] Hang Gang, Quan Long. Simulation Analysis of Excavator Rotary Hydraulic System Based on AMESim[J]. Hydraulics & Pneumatics, 2009, 29(05): 49-51.

[6] Cao Peilei. Optimization and Simulation of Heavy Vehicle Suspension System [D]. Jilin University, 2011.

ISPECE   IOP Publishing

Research on fast self-learning improvement of ADRC control algorithm for film thickness control system

Liao Xue-chao[1,2], Zhou You[1,2,3] and Chen Zhen-huan[1,2]

[1] School of Computer Science and Technology, Wuhan University of Science and Technology, Wuhan 430081, P. R. China

[2] Hubei Province Key Laboratory of Intelligent Information Processing and Real-time Industrial System, Wuhan 430081, P. R. China

[3] Correspondence: helloklaus@163.com; Tel.: +86-15071290248

Abstract. In the biaxially stretched film (BOPP) thickness control system, the traditional PID and Active-Disturbance Rejection Controller (ADRC) can't achieve the ideal control effect. The Smith prediction method is used in the essay to establish a discretization model for the BOPP thickness control system. Combining with BP self-learning algorithm, a fast self-learning improved ADRC control algorithm (FSADRC) is proposed. By means of the additional momentum term and the adaptive learning rate method, the nonlinear combination of the ADRC system is adjusted in real time, the optimal control parameters are found, and the parameters are self-tuned. As a consequence, the improved algorithm is applied to the biaxial tensile film thickness control model. The simulation results show that the method has the advantages of high response speed and strong self-adaptive ability, which can effectively improve the control performance of the BOPP thickness control system.

1. Introduction

The thickness uniformity of the Biaxially-oriented polypropylene (BOPP) is one of the important criteria for its quality. If the film's thickness uniformity is not good, the relative deviation will occur at a certain position of the film. If the deviation position remains constant, the film may have some defects such as grooves, hoops or ribs after thousands of layers accumulating, and then cause permanent deformation. So the measurement and control of the film thickness are very important because it directly affects the mechanical properties and performance quality of the film product [1]. And film thickness control is a complex system with the characters of nonlinear, multivariable coupling, time-varying and large lag [2].

Since the film thickness control model cannot be accurately established, the existing control method mainly aims to eliminate errors based on system errors. The representative method is the PID controller, its structure is simple, mature and reliable, and so it is widely used in industrial control. However, for the film thickness high-precision control system, the PID controller has a contradiction between rapidity and overshoot and its anti-interference is poor. The active disturbance rejection control (ADRC) technology [3] is an improved method for PID, it can treat the internal and external interference of the system as the total disturbance so as to proceed observation compensation although the disturbance rejection controller proposed in [4] and [5] can theoretically handle complex control systems, it not only has many parameters but also is difficult to adjust a set of relatively ideal control

Content from this work may be used under the terms of the Creative Commons Attribution 3.0 licence. Any further distribution of this work must maintain attribution to the author(s) and the title of the work, journal citation and DOI.
Published under licence by IOP Publishing Ltd

parameters. The neural network [6] has the strong nonlinear fitting ability and self-learning ability and has a positive effect on parameter optimization, so it is widely used in the control field.

In this paper, combined with the neural network the self-learning method, by means of improving self-learning method, the control parameter of optimal nonlinear combination in the ADRC model can been adjusted more fastly. The control method has be applied to the BOPP film thickness control system, and the Smith predicts method is used to compensate for the large delay in film thickness control. The traditional PID controller, ADRC, self-learning ADRC (SADRC) and the improving method has been realized in MATLAB simulation, the four methods simulation results show that the improved fast self-learning ADRC (FSADRC) algorithm has many advantages such as fast response, short transition process and strong adaptability.

2. Modeling the controlled object

The BOPP production process is as follows: first the raw material is melted by the extruder, then cooled and shaped into the casting film by the cooling roller, then the casting film is stretched vertically and horizontally into the thin film, finally winded up into a film roll by the winder material. According to the production process, a detection feedback link is needed to control the film thickness in a closed loop, and the film thickness value y after the biaxial stretching is fed back to the thickness v of the original open-loop control model to form a closed-loop control system. The flow chart of BOPP production is shown in figure 1.

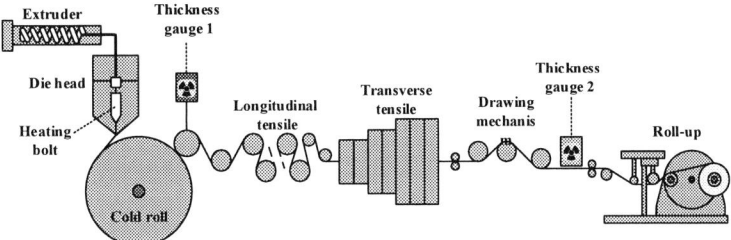

Figure 1. Flow chart of BOPP production.

In this paper, the BOPP thickness control system is simulated by MATLAB. From the BOPP production line, the transfer function of the identified film thickness model is [7]:

$$G(s) = \frac{2.45}{(10.5s+1)(2.5s+1)} e^{-8s} \tag{1}$$

In response to this large delay control problem, Smith proposed a pure lag compensation model, which is based on a compensation link parallel connection with the controller [8]. This compensation link is called the Smith predictor.

Taking the sampling period as 1s, combined with the Smith prediction method, the transfer function formula in equation (1) is transformed into a discrete form, and the form of the BOPP thickness control model is obtained as follows:

$$\begin{cases} y(k) = 1.5795y(k-1) - 0.6094y(k-2) + 0.0397u(k-9) + 0.0337u(k-10) \\ x_m(k) = 1.582x_m(k-1) - 0.6094x_m(k-2) + 0.03972u(k-1) + 0.03368u(k-2) \\ y_m(k) = 1.582y_m(k-1) - 0.6094y_m(k-2) + 0.03972u(k-9) + 0.03368u(k-10) \end{cases} \tag{2}$$

The above discretization equation (2) is the controlled plant model for the subsequent fast self-learning ADRC control model.

3. Improve the ADRC control model

3.1. Self-learning ADRC
Although the NLSEF of ADRC has a fixed structure, it has many parameters and is difficult to analyze and understand. The neural network self-learning technology has strong robustness, memory ability,

nonlinear fitting ability, and powerful self-learning ability [9]. In this paper, the self-learning method is applied to the NLSEF. This control method improves the response speed, tracking accuracy and anti-interference ability. A self-learning nonlinear ADRC model (SADRC) is designed as shown in figure 2, in which the nonlinear error feedback (NLSEF) part adds an e_0 input parameter.

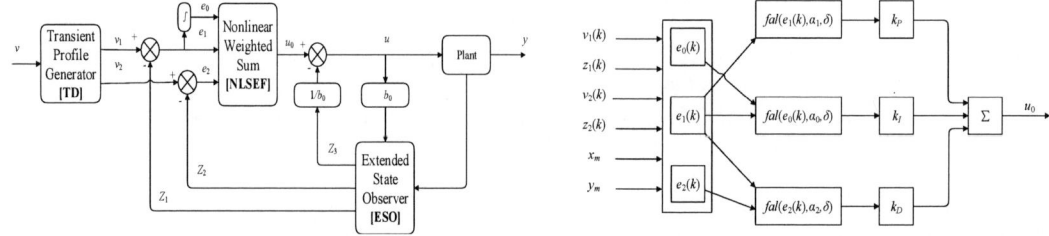

Figure 2. Self-learning nonlinear auto disturbance rejection control system model.

Figure 3. Self-learning nonlinear combination model structure.

For the second-order controlled object of the thickness control model, the discretization of the nonlinear differential tracker is:

$$v_1(k+1) = v_1(k) + hv_2(k), \quad v_2(k+1) = v_2(k) + hfst(v_1(k) - v(k), v_2(k), \delta, h_0) \tag{3}$$

Make:

$$\begin{cases} \lambda_1 = v_1(k) - v(k), \quad d = \delta h, d_0 = dh, \quad \tau = \lambda_1 + hv_2, \quad a_0 = \sqrt{d_2 + 8\delta|\tau|} \\ a = \begin{cases} v_2 + [(a_0 - d)/2]sign(\tau) & if \ |\tau| > d \\ v_2 + \tau/h & if \ |\tau| \le d \end{cases} \end{cases}$$

In the above formula h is the sampling period. The system optimal control function $fst(\cdot)$ has the form:

$$fst(\lambda_1, v_2, \delta, h) = \begin{cases} -\delta sign(a) & if \ |a| > d \\ -\delta(a/d) & if \ |a| \le d \end{cases} \tag{4}$$

Using the system output y and input u, the third-order extended state observer is constructed as follows:

$$e = z_1 - y, \quad \dot{z}_1 = z_2 - \beta_1 e, \quad \dot{z}_2 = z_3 - \beta_2 fal(e, \alpha_1, \delta) + bu, \quad \dot{z}_3 = -\beta_3 fal(e, \alpha_2, \delta) \tag{5}$$

Where z_1, z_2, and z_3 are the states of the observer; β_1, β_2, and β_3 are observer gain coefficients(greater than zero). The nonlinear combined power function $fal(e, \alpha, \delta)$ is designed as:

$$fal(e, \alpha, \delta) = \begin{cases} \dfrac{e}{\delta^{1-\alpha}} & if \ |e| \le \delta \\ |e|^\alpha \mathrm{sgn}(e) & if \ |e| > \delta \end{cases} \tag{6}$$

The nonlinear control law obtained from figure 2 is:

$$\begin{cases} u_0 = k_P fal(e_1(k), \alpha_1, \delta) + k_I fal(e_0(k), \alpha_0, \delta) + k_D fal(e_2(k), \alpha_2, \delta) \\ u = u_0 - z_3/b_0 \end{cases} \tag{7}$$

Where k_P, k_I, k_D are adjustable parameters and let:

$$\begin{cases} e_0(k) = e_1(k-1) + e_1(k), \quad e_1(k) = v_1(k) - z_1(k) - x_m + y_m, \quad e_2(k) = v_2(k) - z_2(k) \\ b_1(k) = fal(e_1(k), \alpha_1, \delta), \quad b_2(k) = fal(e_0(k), \alpha_0, \delta), \quad b_3(k) = fal(e_2(k), \alpha_2, \delta) \end{cases} \tag{8}$$

In neural network self-learning system, e_0, e_1, e_2 are used as input, and u_0 are used as output. $b_1(k), b_2(k), b_3(k)$ are used as the excitation function of the neural network is hidden layer neurons, with k_P, k_I, k_D as the neural network weight, and the structure of the neural network is shown in figure 3. The nonlinear control model has a simple structure, and the three weights k_P, k_I, k_D can be dynamic regulated, so that it has a good nonlinear control effect.

The self-learning process of parameters k_P, k_I, and k_D is as follows:

Let $E(k) = v_1(k) - z_1(k)$, and the neural network output layer error (loss function) are defined as:

$$J = 1/2\, E^2(k+1) \tag{9}$$

In order to minimize the output error, the steepest gradient descent method is used to adjust the neural network weights [10]:

$$k_i(k+1) = k_i(k) + \eta \frac{\partial J}{\partial k_i}, \quad i \in \{P, I, D\} \tag{10}$$

In the above formula, $\partial J / \partial k_i$ $(i = P, I, D)$ is:

$$\frac{\partial J}{\partial k_i(k)} = -E(k+1)b_j(k)z_{u_0}(k), \quad i \in \{P, I, D\} \; and \; j = 1,2,3 \tag{11}$$

Among them:

$$z_{u_0}(k) = \partial z_1(k+1) / \partial u_0(k) \tag{12}$$

From equation (11), both $E(k+1)$ and $z_{u_0}(k)$ are related to the system's future state [11], which makes it difficult to train neural network weights. If the algorithm is convergent, then there must be $|E(k+1)| < |E(k)|$, so you can get:

$$\left| E(k+1) \right| = \lambda E(k), \quad 0 < \lambda < 1 \tag{13}$$

Since λ can be compensated by the learning rate η, $E(k)$ can be used instead of $E(k+1)$. In addition, because $z_{u_0}(k)$ is unknown, it can be approximated by a symbol function, namely:

$$z_{u_0}(k) = sign \frac{z_1(k+1) - z_1(k)}{u_0(k) - u_0(k-1)} \tag{14}$$

Comprehensive (11)-(24) are available:

$$k_P(k+1) = k_P(k) - \eta E(k)b_1(k)z_{u_0}(k), \quad i \in \{P, I, D\} \; and \; j = 1,2,3 \tag{15}$$

In order to avoid excessive power, causing oscillations in the neural network training process, normalizing the weights [12] as follows:

$$k_i(k+1) = \frac{k_i(k+1)}{\sum_{\alpha=1}^{3} \left| k_\alpha(k+1) \right|} \quad (i = 1,2,3) \tag{16}$$

3.2. Fast Self-learning ADRC

In the above, the neural network uses the steepest gradient descent method to adjust the weight. In order to further improve the response speed, tracking accuracy and anti-interference ability, the method of additional momentum term is adopted to design the adaptive mechanism for learning rate, so as to improves the original SADRC algorithm. In this section, the fast self-learning ADRC (FSADRC) algorithm based on dynamic adaptive learning rate is implemented.

3.2.1. Additional momentum term.
The additional momentum term is an optimization method widely used to accelerate the convergence of the gradient descent method. The core idea is that during the gradient descent searching process, if the current gradient descent is the same as the previous' one, the search is accelerated, and vice versa. The parameter update items of the neural network standard BP algorithm are:

$$\Delta w(k) = \eta g(k) \tag{17}$$

Where $\Delta w(k)$ is the parameter adjustment amount of the kth iteration, η is the learning rate, and $g(k)$ is the gradient calculated by the kth iteration.

After adding the momentum term, the parameter update based on the gradient descent is:

$$\Delta w(k) = \eta[(1-\mu)g(k) + \mu g(k-1)] \tag{18}$$

Where μ is a momentum factor (value 0~1). The above formula is also equivalent to

$$\Delta w(k) = \alpha \Delta w(k-1) + \eta g(k) \tag{19}$$

Where α is called the "forgetting factor", and $\alpha \Delta w(k-1)$ represents the adjustment effect of the direction and size information of the previous gradient dropping on the current gradient.

3.2.2. Adaptive learning rate. The additional momentum method faces the difficulty of the learning rate selecting, which leads to a contradiction between convergence speed and convergence. Then the learning rate adaptive adjustment method is introduced, namely:

$$\eta(k) = \sigma(k)\eta(k-1) \tag{20}$$

Where $\sigma(k)$ is the adaptive learning rate factor at the kth iteration, and an expression of $\sigma(k)$ is defined as follows:

$$\sigma(k) = 2^{\lambda} \tag{21}$$

Where λ is the gradient direction, and the expression is:

$$\lambda = sign(g(k)g(k-1)) \tag{22}$$

Combined with the above method of adding momentum terms and adaptive learning rate, equations (19) and (20) can be obtained:

$$\Delta w(k) = \alpha \Delta w(k-1) + \sigma(k)\eta(k-1)g(k) \tag{23}$$

We substitute equation (22) into equation (15), then:

$$k_P(k+1) = k_P(k) - \Delta w_P(k), \quad i \in \{P, I, D\} \tag{24}$$

Among them, P, I, D parameter update items are obtained by:

$$\left. \begin{array}{l} g_i(k) = E(k)b_j(k)z_{u_0}(k) \\ \eta_i(k) = 2^{sign(gi(k)g_i(k-1))}\eta_i(k-1) \\ \Delta wi(k) = \alpha_i \Delta w_i(k-1) + \eta_i(k)g_i(k) \end{array} \right\} i \in \{P, I, D\} \ and \ j = 1,2,3 \tag{25}$$

4. Comparative analysis of experiments

In order to verify the performance of the above control algorithm, the simulation experiment was carried out using the MATLAB simulation platform. The controlled plant model experimentally tested in this paper is the delay model described in equation (2). According to the actual situation of the BOPP production line, the input signal v is taken as

$$v(k) = \begin{cases} 0.01k \ , & k < 500 \\ 5 \ , & k \geq 500 \end{cases} \tag{26}$$

And in order to test the anti-interference ability of the FSADRC controller, the interference signal $d(k) = 0.2$ is added at the 800th sampling time point of the input signal $v(k)$. The four controlled models of PID, ADRC, SADRC, and FSADRC are used to control the plant object.

Figure 4 is a comparative analysis of the experimental simulation results of the four control algorithms. Figure 4(a) is a general comparative diagram of the simulation of the system adjustment process of the four control algorithms, figure 4(b) is the end stage details of the system setpoint changing and figure 4(c) is the interference stage details of the system set value Table 1 is the adjustment process's performance indicators comparison between four control algorithms.

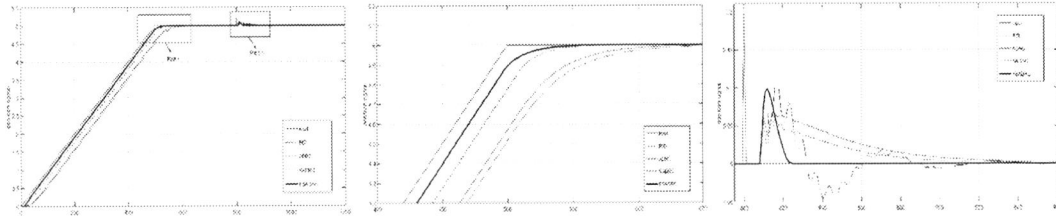

| (a) General comparison of the system adjustment process. | (b) Detail comparison of Part I (end stage). | (c) Detail comparison of Part II (interference stage). |

Figure 4. Comparison of system adjustment processes for four control algorithms.

It can be seen from figure 4(b) that in the end stage of the system setpoint changing, the adjustment of PID controller takes the longest time, and the FSADRC adjustment completion time is the least, it's adjustment speed is the fastest. It can be seen from figure 4(c) that, in the interference stage, the has obvious forward and negative overshoot, The ADRC and SADRC controller has the smaller overshoot, but the adjustment time is also long. While the FSADRC has the shortest adjustment time, only relatively small forward overshoot, and zero steady-state error, and its comprehensive adjustment performance is optimal. The interference is highly anti-interference and robust by FSACRC.

Table 1. Control performance indicators comparison of the four algorithm.

Control algorithm	Start stage	Interference stage		
	Adjustment time (/s)	Overshoot (/%)	Steady-state error (/%)	Adjustment time (/s)
PID	103	56.15	0.909	83
ADRC	85	34.30	0	61
SADRC	58	28.45	0	56
FSADRC	**42**	**49.55**	**0**	**15**

5. Conclusions

In this paper, the improved self-learning algorithm is combined with the ADRC control algorithm. By means of the self-learning ability of the neural network and the nonlinear function approximation ability, the ADRC parameters can be dynamically adjusted in real-time, and the BOPP thickness can be effectively controlled. The simulation results show that comparing with other control methods, the proposed FSADRC method has the advantages of fast response, no overshoot and undershoot, good adaptability and strong anti-interference, which can meet the requirements of many large delay control systems. The performance of the BOPP thickness control system can be effectively improved.

References

[1] Xu Xiaoyong, Sun Yu, Jiang Qinghai. Control system for the casting film thickness based on the virtual instrument[J]. *Manufacturing Automation*, 2012, 34(7):46-49.

[2] Li Yangfan, Jiang Pinquan, Luo Xiaoshu, Li Tinghui, Long Yuanyuan. Application of RBF Neural Network in Film Thickness Control System[J]. *Modern Electronics Technique*. 2010,33(05):147-150.

[3] Liu X Q, Tang L, Zhu L T. Three-motor synchronous control system based on fuzzyactive disturbances rejection control[J]. *Electric Machines & Control*, 2013, 17(4):104-109.

[4] Li Jie, Qi Xiaohui, Xia Yuanqing, Gao Zhiqiang. On Linear/Nonlinear Active Disturbance Rejection Switching Control[J]. *Journal of Automation*, 2016,42(02):202-212.

[5] YE Li-feng, WANG Jing, ZHANG Fei. Research of Smith-AGC based on active disturbance rejection control[J]. *Metallurgical Industry Automation*, 2013, 37(2):40-45.

[6] Liu Jun. The Application of BP Neural Network in Multidimensional Nonlinear Function[J]. *Journal of Shangluo University*, 2014, 28(6):19-22.

[7] Li Yangfan, Jiang Pinquan, Li Tinghui. Simulation of Film Thickness Controlling System Based on MCGS and MATLAB[J]. *Journal of Guangxi Normal University(Natural Science*

Edition), 2010, 28(2):18-21.

[8]   Smith OJM. A controller to overcome dead time[J]. *Isa Journal*, 1959, 6(2):28-33.

[9]   Sheng Zhongbiao, Tong Xiaoyong. The Application of BP Neural Networks in Curve Fitting[J]. *Science Technology and Engineering*, 2011, 11(28):6998-7000.

[10]   Guo Yuedong, Song Xudong. Analysis and improvement of gradient descent method[J]. *Technology Outlook*, 2016, 26(15).

[11]   S. Su and H. Yang, "Single neuron ADRC control based on RBF neural network on-line identification," *Proceedings of the 30th Chinese Control Conference*, Yantai, 2011: 3669-3672.

[12]   Zhu Zhenguo, Tian Songlu. Research on Improvement of Adaptive Learning Rate of BP Neural Network Based on Weight Change[J]. *Computer Systems & Applications*, 2018(7).

Research and Analysis of Intelligent RGV Based on Dynamic Scheduling Optimization Model

Zheng Wang[1], Zeyu Zhou[2], Jiawei Liu[3]

([1]School of Electromechanic Engineering, Hohai University, Changzhou Jiangsu China; [2]School of Business Administration, Hohai University, Changzhou Jiangsu China; [3]School of The Internet of Things Engineering, Hohai University, Changzhou Jiangsu China;)

About the author: Zheng Wang, Email: 1428558627@qq.com, undergraduate; Zeyu Zhou, undergraduate; Jiawei Liu, undergraduate;

Abstract Aiming at the problem of fault-free scheduling in single process, a single process fault-free dynamic scheduling algorithm is proposed by giving several scheduling principles and several lemma. Aiming at the problem of fault-free scheduling of two-process, this paper puts forward the definition and determination principle of "Balance Principle", "Rare CNC" and "m-Center Principle", and gives the algorithm of two-process fault-free dynamic scheduling based on these principles and "single-process fault-free dynamic scheduling algorithm". Aiming at the problem of single process with fault scheduling and the problem of two-process fault scheduling, the number of faulty workpieces is determined according to 1% of the number of workpieces obtained by the corresponding fault-free dynamic scheduling algorithm, and randomly assigned to CNC and the time period of failure, using the finished time of CNC and the working time of RGV, The corresponding fault dynamic scheduling algorithm is given by using the corresponding trouble-free dynamic dispatching algorithm. Finally, this paper validates the operation example, and gives the idea and flow of the optimized all-intelligent RGV dynamic scheduling model algorithm.

1. Research background

Production planning and scheduling is an important means for enterprises to improve production efficiency, reduce production costs and improve market competitiveness. This requires scheduling decision tools to help business decision makers make the right and reasonable decisions in less time. Among them, shop scheduling is a kind of production resource allocation problem which satisfies the constraints of task configuration and process, duration and so on.

The research on the workshop scheduling problem has always been a hot topic in the theoretical circle, but because many workshop scheduling problems belong to NP problem, even some small-scale scheduling problems are difficult to obtain the optimal solution. With the development of computer technology, some intelligent algorithms, such as genetic algorithm and ant colony algorithm, have gradually become the hotspot to solve the problem of shop scheduling, and intelligent algorithm will also be the mainstream of the development of scheduling algorithm in the future.

2. Primarily

A specific intelligent machining system, consisting of 8 computer CNC machine tools (CNC), 1 orbital Automatic guide vehicles (RGV), 1 RGV linear tracks, 1 feeding conveyor belts, 1 feeding conveyor

Content from this work may be used under the terms of the Creative Commons Attribution 3.0 licence. Any further distribution of this work must maintain attribution to the author(s) and the title of the work, journal citation and DOI.
Published under licence by IOP Publishing Ltd

belts and other ancillary equipment. RGV is a driverless, smart car that can run freely on a fixed track. It can automatically control the direction and distance of movement according to the instructions, and bring its own mechanical arm, two manipulator claws and material cleaning tank. It can complete the up and down materials and cleaning material and other work tasks.

The operating flow of RGV is as follows:

(1) After the intelligent machining system is powered on and started, the RGV is in its initial position and all CNC is idle;

(2) When the CNC is idle and the processing operation is completed, it will send the demand signal to RGV;

(3) Once received the demand signal of a CNC, RGV will determine the order of the CNC's loading and unloading work, and sequentially serve for it;

(4) After completing a loading and unloading operation, the cleaning operation will follow up by RGV;

(5) After completing a job task, RGV waits in place if no other job instructions are received.

(6) The system repeats (3) to (5) over and over again until the system stops working and RGV returns to its original position.

This article will analyze the following three specific situations:

(1) Material processing operations of single process, each CNC is installed with the same tool and materials can be processed on any CNC;

(2) Material processing operation of two processes, the first and second processes of each material are processed by two different CNC in turn;

(3) CNC may fail during processing (1%is the given probability of failure), each troubleshooting (manual processing, unfinished material scrapping) needs 10-20 minutes; Once troubleshooting finished, CNC joins the job sequence immediately. It is required to consider the material processing operation of single process and two processes respectively.

3. Model establishment and solution

3.1 semi-intelligent single process fault-free dynamic scheduling Model

3.1.1 Semi-intelligent RGV Dynamic scheduling scheme operating principle

Lemma 1: The position of the CNC-a is $S_a = \left\lceil \dfrac{a}{2} \right\rceil$, (S_a is the position of CNC ordered with number a

(S_a valued 1,2,3,4)), which is the upper integer of the pair.

Definition 1: If the RGV is in the middle of CNC-a and the CNC on the other side of it, the position of the RGV is defined as $S_R = S_a$. (S_R is the position of RGV (valued as 1,2,3,4)).

Lemma 2: When requirement instructions are received from CNC-a, CNC-b, if $(t_{d_{R,a}} + SS_a) >$ $(t_{d_{R,b}} + SS_b)$ (t_n is the time requires RGV to move n length unit), the RGV moves firstly to CNC- b for the job; if $(t_{d_{R,a}} + SS_a) \le (t_{d_{R,b}} + SS_b)$, RGV firstly moves to the CNC-a for the work.

Proof: If RGV moves to CNC-b and then to CNC-a, the total time is $(T_a = t_{d_{R,a}} + SS_a + C + t_{d_{a,b}})$ (C is the time requires to complete the cleaning operation of an item), and if RGV moves to CNC-b and then to CNC-b, the total time is $(T_b = t_{d_{R,b}} + SS_b + C + t_{d_{b,a}})$. If $T_a > T_b$, that is $(t_{d_{R,a}} + SS_a) > (t_{d_{R,b}} + SS_b)$, the RGV firstly moves to the CNC-b for the work, otherwise, RGV firstly moves to the CNC-a.

Definition 2: Distance between RGV and CNC-a is defined as $d_{R,a} = |S_R - S_a|$; Distance between

CNC-a and CNC-b is defined as $d_{a,b} = |S_a - S_b|$.

A specific RGV operating principle is following:

Principle 1: When only 1 CNC requirements instructions are received, RGV moves to the CNC and carries out up and down materials and cleaning operations.

Principle 2: When 2 CNC requirements instructions (CNC-a, CNC-b) are received, let the distance between RGV and these two CNC machines be $d_{R,a}$、$d_{R,b}$.

① If $d_{R,a} = d_{R,b}$, RGV moves to a CNC with an odd number first, and work on it; otherwise, randomly serve any of the CNCs.

② If $d_{R,a} > d_{R,b}$, the job is scheduled according to the following conditions:

Case 1: If b is even, a is odd, and $T_a > T_b$, then RGV first moves to CNC-b and carries on the loading and unloading, cleaning and other operations; If $T_a < T_b$, RGV first moves to CNC-a and carried out operations; If $T_a = T_b$, RGVmoves first to an odd number of CNC and then work.

Case 2: If b is an odd number or both a, b are even, RGV moves to CNC-b first.

Principle 3: When 3 CNC issuance requirements are received (CNC-a, b, c), let the distance between RGV and these 3 CNC machines be $d_{R,a}$、$d_{R,b}$ 与 $d_{R,c}$.

① When the farthest CNC number c is even, remove the CNC, and then the problem transforms into Principle 2.

② When the farthest CNC number c is odd

Case 1: If a or b or both a and b are odd, then eliminate the CNC-c, and then the problem transforms into Principle 2.

Case 2: If both a and b are even, when $d_{R,a} < d_{R,b}$, discard the farther CNC-b, according to principle 2 to determine the order of RGV processing CNC-a and CNC-c; When $d_{R,a} = d_{R,b}$, discard the CNC that is not on the same side with CNC-c, and then according to Principle 2 decided the order for RGV to deal with CNC-c and the CNC of the same side (one of CNC-a, CNC-b).

Principle 4: When 4 or 5 CNC issuance requirements are received, RGV moves first to the closest CNC.

Principle 5: When 6 CNC issuance requirements are received:

① If there is a CNC demand order in the location of RGV, such CNC is served first.

② If there is no demand order from the CNC location of the RGV, select any CNC that is the nearest number with the RGV distance to serve.

Principle 6: When 7 CNC issuance requirements are received, RGV serves the local CNC firstly.

Principle 7: When 8 CNC issuance requirements are received, RGV serves the local CNC with an odd position number firstly.

3.1.2   semi-intelligent RGV algorithm

According to the principles given above, the dynamic scheduling solution algorithm of semi-intelligent RGV is presented:

Semi-intelligent single-process trouble-free algorithm

Step1: After the intelligent machining system is energized, RGV in the initial position in the middle of CNC1 and CNC2, the initial working time of the RGV, that is $time = 0$, RGV continues the first round of job scheduling in the order of

$$CNC1 \to CNC2 \to CNC3 \to CNC4 \to CNC5 \to CNC6 \to CNC7 \to CNC8.$$

Step2: Calculate the working time of the current RGV, $time$; RGV immediately identify the number of CNC that issued the requirements:

① If no other job instructions are received, RGV waits in place until the next job instruction.

② If only 1 CNC issues instruction is received, the dispatch is carried out in accordance with principle 1, then perform Step3;

③ If 2 CNC issues instructions are received, the dispatch is carried out in accordance with principle 2, then perform Step3;

④ If 3 CNC issues instructions are received, the dispatch is carried out in accordance with principle 3, then perform Step3;

⑤ If 4 or 5 CNC issues instructions are received, the dispatch is carried out in accordance with principle 4, then perform Step3;

⑥ If 6, 7 or 8 CNC issues instructions are received, the dispatch is carried out in accordance with principle 5, 6 or 7, then perform Step3;

Step3: When $time \geq 3600 * A$ seconds (A is the maximum time for a continuous job per shift (in hours)), the algorithm stops; otherwise, Step2 is executed.

[Note 1]: The formula for calculating the working time of RGV is ($time = time + t_{d_{R,a}} + SS_a + C$).

[Note 2]: In view of how to distinguish the number of CNCs issued by the RGV, this paper compares the working time of the current RGV, by calculating the expected completion time of the CNC (that is, the sum of the start processing time and the length of the operation). If the expected completion time is less than the working time, such CNC is considered to have sent a request and program needs to count the number of these CNC.

3.2 Semi-intelligent two-process trouble-free dynamic scheduling model

3.2.1 dual-process arrangement of different CNC

The first and second processes of each material are completed by two different CNC, and this article agreed that the second process must be started only if the first process has been completed. Then this article first needs to install 8 CNC with different tools to process different process.

Definition 3: Record x (non-integer) as the number of CNC producing the first process, $8 - x$ (non-integer) is the number of CNC producing the second process.

Definition 4: Record X as the nearest integer from the calculated x, that is $X = [x + 0.5]$. Let $\Delta x = X - x$.

Balance principle: In this paper, the smaller Δx of the two schemes is obtained:

① Let x_1 be the number of CNC in the production of the first process, $8 - x_1$ be that of the second process. If the CNCs in production of the first process are placed in the odd digits, the CNCs in production of the second process are placed in the even digits, then

$$\frac{x_1}{8 - x_1} = \frac{p_1 + SS}{P_2 + SO} \qquad (1)$$

p_1 is the time requires CNC to complete the first process. p_2 is the time requires CNC to complete the second process. SS is the time takes for RGV to load and unload once for CNC1, 3, 5 or 7. SO is the time takes for RGV to load and unload once for CNC 2, 4, 6 or 8.

② Let x_1 be the number of CNC in the production of the first process, $8 - x_1$ be that of the second process. If the CNCs in production of the first process are placed in the even digits, the CNCs in production of the second process are placed in the odd digits, then

$$\frac{x_2}{8 - x_2} = \frac{p_1 + SO}{P_2 + SS} \qquad (2)$$

By comparing $\Delta x_1 = |X_1 - x_1|$ with $\Delta x_2 = |X_2 - x_2|$, select the smaller one as Δx, that is, select the scheme with the smaller difference between x and X. According to this, set the number of CNC producing the first process is X, the number of CNC producing the second process is $8 - X$.

3.2.2 CNC arrangement of two-process Jobs

Definition 5: Let M is the number of CNC in the production of the first process, N is the number of CNC production of the second process. If M<N, note the CNC in production of the first process as "The first type of rare CNC"; If M> N, note the CNC in production of the second process as "The second type of rare CNC". Both "The first type of rare CNC" and "The second type of rare CNC" are called "rare CNC".

Definition 6: Give a graph Q to find its m vertices so that the sum of the distances of these m vertices to the rest of the vertices is the smallest, then these m vertices are called the m-Center vertices of the graph.

m-Center Principle Let number of rare CNC be m, (assumes that rare CNC is placed in odd digits), constructs a graph G, the 8 vertices of the graph correspond to 8 CNC, the distance of any two points CNC-a, CNC-b is defined as $d_{a,b}$. The m vertexes are taken from vertexes numbered odd so that these m vertices are the m-center of graph G. Such m rare CNCs are placed according to the results from the m-center.

In the case of trouble-free dual-process, the arrangement of rare CNC is given according to the m-center principle.

① If 4 rare CNCs exist, arrange these 4 CNC on the same side;

② If 3 rare CNCs exist, according to the m-Center principle, to get the arrangement of 3 rare CNCs is optimal;

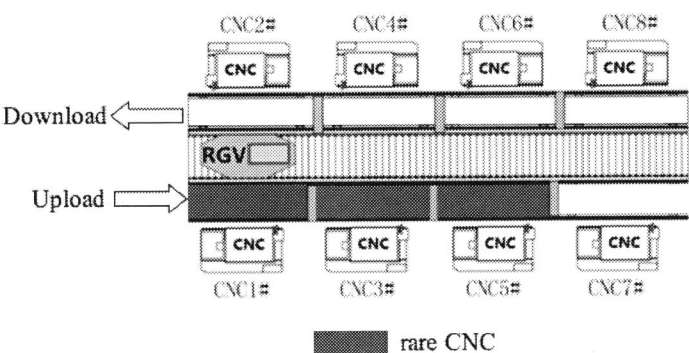

Figure 1. Method to get the arrangement of 3 rare CNCs

③ If 2 rare CNCs exist, according to the m-center principle, get these 2 rare CNC placed in the same side of the middle two CNC;

④If 1 rare CNC exists, this rare CNC is placed within other CNCs according to the m-Center principle.

Semi-intelligent two-process trouble-free algorithm

Step1: According to the balance principle, determine the number of rare CNC (m) and which side of rare CNC is placed (odd side or even side), and determine which process these m rare CNC processing (the first, second way), and then according to the m-Center principle, determine the order of rare CNC;

Step2: After the intelligent machining system is energized, the initial position of the RGV is in the middle of CNC1 and CNC2, the initial working time of the RGV is $time = 0$; RGV first use "semi-intelligent single-process trouble-free algorithm" to serve the CNC in first process;

Step3: Calculate the current time, if $time \geq 3600 * A$, perform Step4; otherwise RGV discriminates the requirements from the CNC:

① If the CNCs in different processes request, use "Semi-intelligent single-process trouble-free algorithm" to serve one of the CNC in the first process; then use "Semi-intelligent single process trouble-free algorithm" to serve one of the CNC in the second process, then perform Step3;

② If no job instructions are received, or if the CNC that are demanding are all in the same process, perform Step3;

Step4: Output RGV scheduling arrangement, and the algorithm stops.

3.3 Semi-intelligent single process with fault dynamic scheduling model

RGV records the time of each CNC starting to work. When the time reaches the expected completion time for an item, if the CNC of that item does not send any demanding signal, RGV assumes that the CNC has failed, and change its expected job completion time to infinity.

Semi-intelligent single process with fault algorithm:

Step1: After the system is powered, the initial position of RGV is set within CNC1 and CNC2 with the initial working time defined as $time = 0$. RGV starts the first round of job scheduling in the order of $CNC1 \rightarrow CNC2 \rightarrow CNC3 \rightarrow CNC4 \rightarrow CNC5 \rightarrow CNC6 \rightarrow CNC7 \rightarrow CNC8$ and the total number of completed materials is $M = 0$;

Step2: RGV immediately identify the number of CNC issue the requirements: if time is more than one of completion time, and RGV did not receive demand signal from that CNC, then the completion time of the CNC is rewritten as infinity and then execute Step2. Otherwise, execute Step3;

Step3: RGV immediately identify the number of CNC issue the requirements:

① If none of job instructions are received, RGV waits until the next job instruction;

② If only 1 CNC issues instructions, the dispatch is carried out in accordance with principle 1, then perform the Step2;

③ If 2 CNC issues are issued, the dispatch is carried out in accordance with principle 2, then perform Step2;

④ If 3 CNC issues are issued, the dispatch is carried out in accordance with principle 3, then perform Step2;

⑤ If 4 or 5 CNC issue instructions, the dispatch is carried out in accordance with principle 4, then perform Step2;

⑥ If 6, 7 or 8 CNC issued instructions, the dispatch is carried out in accordance with principle 5, 6 or 7, then perform Step2;

Step4: If $time \geq 3600*A$, the current total number of completed items is recorded as M and the algorithm stops. Otherwise, execute Step2.

3.4 Semi-intelligent dual process with fault dynamic scheduling model

RGV records the time of each CNC starting to work, too. RGV always compares the current time with the expected time:

① If the item is in the first process, and the current time is equal to the expected completion time, the total production number $M=M+1$;

② If the item is in the first process, and the current time is over the expected completion time, it indicates that the CNC of that item failed before, M=M;

③ If the item is in the second process, and the current time is equal to the expected completion time, M=M;

④ If the item is in the first process, and the current time is over the expected completion time, the total production number $M=M-1$.

Semi-intelligent two-process fault-free algorithm:

Step1: According to the balance principle, determine the number of rare CNC; According to the m-Central principle, determine their arrangement;

Step2: The system is powered on. RGV is in the initial position and all CNC are in idle state. The initial time $time = 0$, and the number of completed material $M = 0$;

Step3: Check whether the working time is over A; if so, the system job stops, the algorithm scheduling ends; if not, the Step4 is executed;

Step4: Check whether CNCs in the first or second channel are sending requirements at the same time; if so, perform Step5; if not, return to Step3

Step5: For CNC in the first process, transform questions into a "semi-intelligent single-process trouble-free algorithm" and schedule according to the principles; Then calculate the current RGV working time;

Step6: Compare the current predicted completion time with the current schedule; if they are equal, then $M = M + 1$; Otherwise, M is not changed;

Step7: For CNC in the second process, transform questions into a "semi-intelligent single-process trouble-free algorithm" and schedule according to the principles; Then calculate the current RGV working time;

Step8: Compare the current predicted completion time with the current schedule; if they are equal, then M is not changed; Otherwise, $M = M - 1$;

Step9: Serve the selected CNC and calculate the working time, then returns the Step3.

4. Simulation results and validity of the algorithm

Table 1: Validation data sheets for the operation parameters of the intelligent machining system (/second)

System operating parameters	Group 1	Group 2	Group 3
Time for RGV to move 1 unit	20	23	18
Time for RGV to move 2 unit	33	41	32
Time for RGV to move 3 unit	46	59	46
Time for CNC to complete a single process	560	580	545
Time for CNC to complete the first process	400	280	455
Time for CNC to complete the second process	378	500	182
Time for RGV to load and unload for CNC1, 3, 5 or 7 once	28	30	27
Time for RGV to load and unload for CNC2, 4, 6 or 8 once	31	35	32
Time for RGV to complete a material cleaning operation	25	30	25

Note: each shift for 8 hours.

CNC sends out the demanding signal after processing, and there is an idle time between waiting for RGV to come forward and feed. In this paper, the principle of designing the dynamic scheduling algorithm is to minimize the average idle time of each CNC. Based on this idea, this paper designs a method to verify the effectiveness of the model: the ideal machining time of CNC ratio the actual machining time.

Calculate the total number of completed materials, total of three groups of given data list is as follows:

Table 2: Simulation results of four algorithms under three sets of data

	Group 1	Group 2	Group 3
Single process without failure	357	337	367
Double process without failure	210	144	187
Single process with failure	352	336	364
Double process with failure	201	140	184

When calculating the effectiveness of the model, the amount of waste caused by the fault is small and can be neglected compared with the total number of materials, so the efficiency value does not change much. Therefore, this paper only calculates the efficiency value of single process without failure and double process without failure, as shown in list below.

Table 3： Two fault-free algorithms are efficient in three sets of data

	Group 1	Group 2	Group 3
Single process without failure	96.63%	95.68%	96.71%
Double process without failure	79.45%	73.77%	78.86%

This paper can be concluded that: in the "single process trouble-free RGV dynamic scheduling algorithm", "double process trouble-free RGV dynamic scheduling algorithm", "single process of RGV containing fault dynamic scheduling algorithm" and "double process of RGV containing fault dynamic scheduling algorithm, the total efficiency of CNC values to achieve a higher standard, namely every CNC can be fully used, and confirms the practicability and validity of the algorithm.

5. Improved Model -- -- -- -- -- - Fully Intelligent RGV Dynamic Scheduling Model
In this paper, a kind of fully intelligent RGV dynamic scheduling model is proposed, namely the RGV knows the whole producing process of CNC and advances their real-time situation (RGV can early moves to the position wait for the next work).

As a result, RGV system requires with a register and internal processor, register is used to continually check the start time for each of the process, and use the processor to calculate the completion time of the material. Consider the minimum combination of 'Time prior material still needs+ Time for RGV to move + Time for working scheduling' for the first job scheduling.

When time reaches expected completion time for a certain material, CNC of that material did not send a signals, the CNC fails, and will be removed from the job queue (change its expected job completion time into infinity).

Fully intelligent RGV dynamic scheduling algorithm
Take single process trouble-free as an example:

Step1: After the system is powered, the initial position of RGV is set within CNC1 and CNC2 with the initial working time defined as . RGV starts the first round of job scheduling in the order of $CNC1 \rightarrow CNC2 \rightarrow CNC3 \rightarrow CNC4 \rightarrow CNC5 \rightarrow CNC6 \rightarrow CNC7 \rightarrow CNC8$.

Step2: Calculate the current working time of RGV, and RGV immediately judge the number of CNC issuing the demand instructions. If none, turn Step3; otherwise, turn Step4;

Step3: Calculate total time of every CNC:

$$T_t = T_r + T_m \tag{3}$$

(T_t is the total time; T_r is the difference between the time and time required by CNC next time; T_m is the time the RGV moved to the CNC.)

RGV choses CNC that needs totally least waiting time to serve, then turn Step2;

Step4: Transform the problem into 'Semi-intelligent single process without failure', namely, according to the number of CNC, schedule in accordance with the principles primarily

Computing new *time'* currently, that is,

$$time' = time + T_m \tag{4}$$

And expecte whether there is a new CNC to send the demand signal before *time'*:

(1) If not, RGV directly to serve primary CNC, then turn Step2;

(2) If yes, renew the CNC machining time p'_{new}

$$p'_{new} = p_{new} + T_r \tag{5}$$

Transform the question into 'Semi-intelligent single process without failure'. According to the number of CNC, schedule in accordance with the principles, then turn Step2

Step5: If $time \geq 3600 * A$, algorithm stops

6. Conclusion

The scheduling of each step of RGV starts from the perspective of making full use of CNC processing time, so that each step of operation is the optimal solution meeting the current situation. In order to achieve the scheduling problem of two processes, this paper considered different situations of the combination of RGV and CNC, and determined the producing proportion of CNC in the first and second process as identical as possible in the same time. In terms of the arrangement of rare CNC, this paper aims at radiating as much as possible to non-rare CNC as each rare CNC can, and calculates the sum of the time it takes for rare CNC to traverse non-rare CNC under different arrangements, and takes the arrangement with the minimum time as the actual arrangement.

However, the scheduling of RGV is only related to the CNC that sent the request signal before, and the CNC that sent the request during the operation cannot affect the scheduling decision, which may cause the decision deviation. This paper also puts forward the improved model in chapter 6, the completion time and decision making in advance can be predicted by the way to overcome this kind of problem. With the automation and intelligence of production, the application of RGV dynamic scheduling in production is more and more extensive, and the more efficient and more stable RGV dynamic scheduling model will also be explored constantly.

Acknowledgements

This paper was financially supported by National College Students Innovation and Entrepreneurship Training Program of China---project number: 201810294084).

References

[1] Hongqu Zhou, Huizhen Zhao. Hybrid discrete particle swarm optimization algorithm in the application of mixed flow assembly line production scheduling [J]. Journal of chongqing university of science and technology (natural science edition), 2015, (3) : 58-64. The DOI: 10.3969 / j.i SSN. 1674-8425 (z). 2015.03.012.

[2] Kang Jiang, Huaitian Zhuang, Kai Xian. The discrete manufacturing system for production scheduling algorithm research and design? [J]. Journal of combination machine tools and automatic processing technology, 2016, (11) : 154-156160. The DOI: 10.13462 / j.carol carroll nki mmtamt. 2016.11.041.

[3] Lijun Fu, Chong Zhou. Elevator group control system of the optimal scheduling simulation [J]. Computer simulation, 2012, (4) : 263-267. DOI: 10.3969 / j.i SSN. 1006-9348.2012.04.064.

[4] Jiyang Ji, Chang'an Zhu. Production scheduling based on Petri net and simulated annealing algorithm [J]. Journal of jiangsu university of science and technology (natural science edition), 2011, 25 (4) : 346-349. The DOI: 10.3969 / j.i SSN. 1673-4807.2011.04.009.

[5] Xiao Xiongwu, Guo Bingxuan, Li Deren, Et al. Multi-view Stereo Matching Based on self-adaptive Patch and Image Grouping for Multiple Unmanned Aerial Vehicle Imagery[J].Remote Sensing,2016,8(2). DOI:10.3390/rs8020089.

[6] Kang Jiang, Kai Xian, Yu Hao. Engineered. Split the MTO enterprises of production scheduling based on the order and the algorithm research [J]. Manufacturing technology and machine tools, 2018, (3) : 143-148. The DOI: 10.19287 / j.carol carroll nki. 1005-2402.2018.03.029.

[7] Shusen Yang, Shunsheng Guo. Production scheduling based on hybrid genetic algorithm research [J]. Journal of mechanical manufacturing, 2017 zhongguo kuangye daxue (10) : 108-111116. DOI: 10.3969 / j.i SSN. 1000-4998.2017.10.031.

Research on intelligent near-power early warning system for mechanical vehicles

Yingjing Wang[1], Jiaxin Liu[2] and Yunfei Jia[3]

[1]Department of Instrument Science and Technology, School of Mechanical Engineering, NanJing University of Science and Technology, Nanjing, China

[2]State Grid Electric Power Research Institute of Liaoning Electric Power Co., Ltd., Shenyang, China

[3]Department of Instrument Science and Technology, School of Mechanical Engineering, NanJing University of Science and Technology,Nanjing,China

Abstract. Base on the electric field strength measurement technology, an early warning system for a mechanical vehicle close to a high voltage electric wire is studied. By analyzing the influence of the distorted electric field, the correction matrix of the probe is obtained by comparing the standard value with the measured value, which greatly improves the correction matrix and the accuracy of the alarm. Through the signal conditioning at the signal collector, the data comparison judgment is realized.

1. Preference

With the development of scientific and technological productivity and the improvement of modern industrialization, the scale of power grid construction has been changing with each passing day. Although power has brought a lot of convenience to people's life production, its own potential harm cannot be ignored. Since the emergence of high-voltage electricity, electric shock accidents have occurred frequently, especially the electric shock trip accident caused by mechanical construction is not uncommon. In order to effectively realize the near-power early warning function of high-voltage transmission line mechanical vehicles, the detection of transmission cables, the monitoring of electric field strength, and the ranging alarm of high-voltage transmission lines are all issues to be considered. At present, the mainstream methods and ideas for dealing with this problem mainly focus on near-inductive cable detection, image recognition, power line laser ranging, radar system high-voltage line ranging, etc. [1].

The near-inductive charge detection electric field uses an induction component to sense the electric field signal in the space. An electric field band composed of charged particles is formed around the AC high-voltage transmission line, and the electric field is stronger as it is closer to the transmission line. Using this principle, the literature [2] achieves the monitoring of the electric field alarm and the distance of the high-voltage transmission line by detecting the electric field change formed on the cable by the wire rope cutting the magnetic field line. Mechanical vehicle electric shock can also be solved by infrared detection image recognition technology.

The literature [3] performs block processing on the collected image, and enhances the contrast between the measured target and the background through technical means, and fits the horizontal power line by computer simulation. In the literature [4], the laser signal emitted by the laser emitter is reflected by the rotation of the polyhedral prism into a fan beam surface laser signal. The emitted fan

---

Content from this work may be used under the terms of the Creative Commons Attribution 3.0 licence. Any further distribution of this work must maintain attribution to the author(s) and the title of the work, journal citation and DOI.

Published under licence by IOP Publishing Ltd

beam laser can cover the high voltage cable within a certain range, so this increases the ability to automatically evade slender obstacles. In the literature [5], the method of radar ranging can identify cables with a diameter of 10 mm, and the detection distance can reach 300-900 m, and the recognition rate can reach 99.5%. Image recognition, laser ranging and radar systems can achieve cable identification, but it is difficult to achieve low cost and easy to use. The method of near-induction is simple in system structure, low in cost but poor in accuracy [6]. In this paper, the measurement principle of spherical sensor is analyzed, the source of measurement error is pointed out, and the influence of distortion electric field is studied. By comparing the standard value with the measured value, the correction matrix of the probe is obtained, which greatly improves the accuracy of measurement. On this basis, a mechanical vehicle near-electrical early warning system with low cost, simple installation and no influence on construction operation has been developed.

2. Spherical sensor measurement principle

The electric field strength measuring sensor uses a spherical sensor, which has the advantage of simple structure, convenient design and accurate measurement. The one-dimensional spherical electric field probe is a two metal hemispherical shell connected by a hollow insulating material, which is equivalent to a capacitance sensor. The housing is two stages of capacitance, and the insulating material is a capacitance medium, as shown in figure 2.1. At this time, as long as a capacitor is connected between the two poles to act as a measuring capacitor, a voltage is formed between the measuring capacitors due to the induced charge. This can be used to reflect the intensity information of the spatial electric field [7].

Under ideal conditions, the spherical probe is placed in a uniform electrostatic field electric field, and the position of the center of the sphere is recorded as O point. Under the electric field of the spherical probe, the surface generates an induced charge. Assuming that the upper shell surface area is S, the induced charge density of the surface is set to $\sigma(t)$.

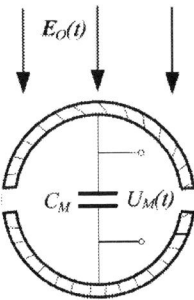

Figure 2.1. Schematic diagram of one-dimensional spherical sensor probe.

The amount of induced charge in the hemispherical shell is:

$$Q(t) = \int \sigma(t)d\text{S} \qquad (2.1)$$

The related literature proves that the amount of induced charge $Q(t)$ generated by the spherical probe in the electric field is proportional to the electric field strength $Eu(t)$ at the O point of the spherical center:

$$Q(t) = kE_u(t) \qquad (2.2)$$

Where k is the proportionality factor. Because the upper and lower housings induce charge, a voltage difference is formed in the measurement capacitor Cm.

$$U_M(t) = KE_u(t) \qquad (2.3)$$

Substitute (2.2) is available:

$$U_M(t) = \frac{Q(t)}{C_M} \qquad (2.4)$$

It can be seen from (2.4) that the electric field strength Eu(t) at which the center of the ball is located can be obtained by measuring the voltage difference Um(t) between the capacitance Cm. This is the theoretical principle of measurement of spherical sensors.

3. Research on near-power early warning sensor

3.1. Spherical sensor measurement principle
In the measurement principle of spherical electric field sensor, the spherical electric field measuring sensor is designed in this test system, as shown in figure 3.1. The probe of the elect ric field sensor utilizes a voltage value converted by a capacitance generated between corresponding electrodes in a space electric field. Then, the electric field strength is measured by calculating the correspondence between the voltage and the field strength.

Figure 3.1. 3D spherical probe making physical map.

3.2. Near-power warning sensor calibration
In order to adjust and calibrate the accuracy and sensitivity of a spherical electric field sensor, a uniform electric field of known size and adjustable intensity is required for testing. The electric field is formed between two aluminum plates with a length and width of 1 m and a thickness of 2.5 mm. Transient uniform electric field can be generated between the electrodes of the two aluminum plates.

During the development of the spherical electric field device, the test capacitors with different capacitance values were tried. After trying different magnifications, the experimental requirements were met when the magnification was 1000 times. The standard electric field strength is measured by the conditioning circuit, and the electric field strength of the standard electric field has been calibrated by the electric field meter. The measured capacitance result is converted into an electric field value through a one-dimensional voltage value, and the final synthesized electric field value is obtained by a three-dimensional electric field synthesis formula to facilitate comparison.

The calibration result is shown in the line graph shown in figure 3.2. It can be seen intuitively that the measured value of the measured capacitance cannot be measured normally, and the measured value cannot be sensed. In 0pF, 120pF, 220pF and 560pF, the measured values of 300pF are the highest, and there is no result exceeds range. Finally, 300pF is selected as the measurement error of the spherical electric field.

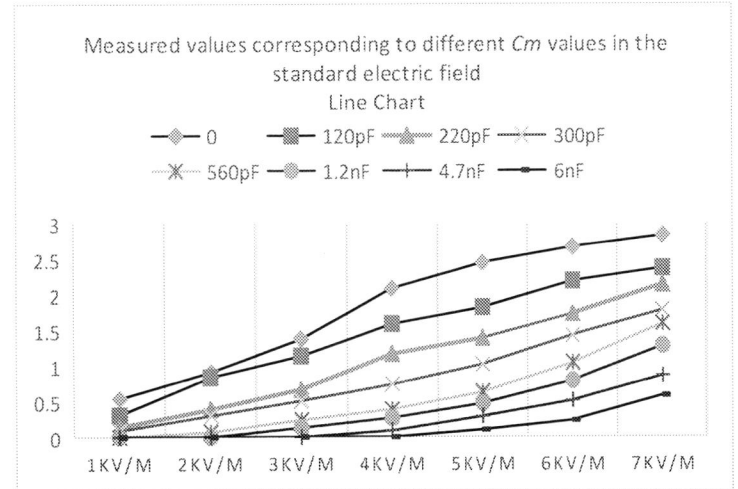

Figure 3.2. Measured value U(V) line graph corresponding to different Cm values.

The calculated residual standard deviation is 0.090, which is the best linearity of the measured capacitance for all different capacitance values.

4. Mechanical vehicle near-power early warning system design

The system is equipped with an infinite communication module, a GPS positioning module and a short message communication module to provide the monitoring station with the location status and data of the dangerous area workers in real time. The signal acquisition module collects the electric field intensity signal through the spherical sensor and performs amplification filtering treatment. Then, the 50Hz power frequency electric field intensity signal is extracted, passed into MCU and compared with the AD conversion. If the electric field strength is greater than the set comparison threshold, an alarm signal is sent back to the communication alarm device. After receiving the signal, the communication alarm device will alarm to indicate that the user equipment is close to strong power. At the same time, the position coordinates of the time are collected by the GPS module, and sent to the supervisor's mobile phone through the short message sending module.

The mechanical vehicle early warning system designed the function of three-speed alarm sensitivity and boundary alarm according to the actual needs of the construction environment. Through the hardware button control and software program to achieve three-speed sensitivity, each file sensitivity has a comparison of the electric field strength, the construction personnel to set according to actual needs. The composition of the near-power early warning system is shown in figure 4.1.

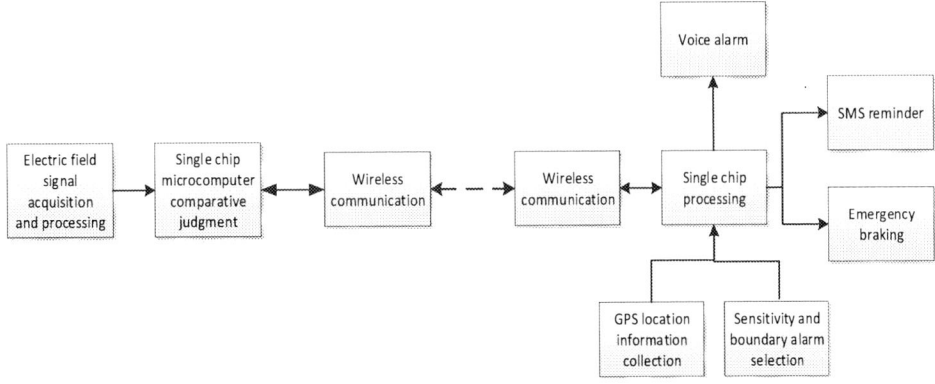

Figure 4.1. Mechanical vehicle near-power system function composition diagram.

5. Software design of near-power early warning system

In order to meet the requirements of the mechanical vehicle near-power system, combined with the design of the hardware circuit, the software is also divided into a signal acquisition part and an alarm communication part. Signal acquisition mainly implements functions such as data acquisition and processing and data transmission. The alarm communication part includes data transmission, GPS information reading processing, short information data transmission, voice module execution, and boundary alarm program design. The overall flow chart is shown in figure 5.1.

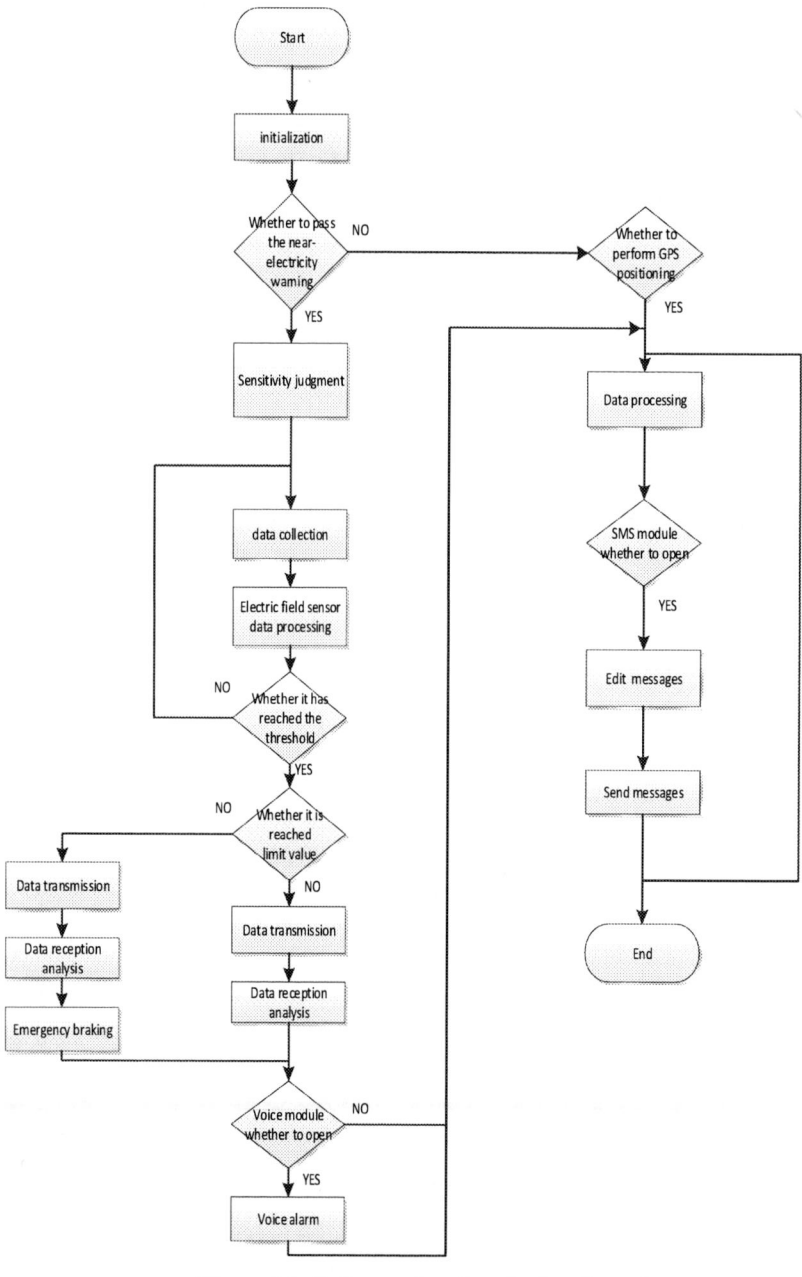

Figure 5.1. Software function design flow chart.

6. Conclusion

(1) By studying the measurement principle of spherical sensor and analyzing the source of its measurement error, it is proved that the error of the measurement result of the spherical sensor is negligible, and the three-dimensional probe can compensate the error when the measurement direction of the probe is inconsistent with the direction of the power line. The electric field distortion caused by the probe can be compensated by the correction matrix, and the correction parameters of the spherical electric field sensor probe are calculated by experimental data.

(2) The mechanical vehicle near-power early warning system has designed three alarm sensitivity and boundary alarm according to the actual needs of the construction environment. The boundary alarm can be analogized to a self-learning type alarm. By pressing the button, the electric field strength of the signal acquisition end can be collected and set to compare the threshold value. If the next time near the same electric field strength, the alarm device will perform a voice alarm.

(3) The system is divided into a signal acquisition terminal and a communication alarm terminal, and data transmission is performed between the two through the NRF905 wireless module. Both modules use the PIC18F87K22 series MCU. The signal collector focuses on signal conditioning, and the MCU realizes data comparison judgment. The communication alarm terminal integrates several modules of wireless communication, voice alarm, GPS positioning, SMS communication and mechanical vehicle emergency braking.

Acknowledgment

Project Supported by Science and Technology Project of State Grid Liaoning Electric power Supply Co.,Ltd.(2018-YF32).

References

[1] Mckinley M.C.,Smith C.V. Jr..Laser interference figure measurement of 60-Hz high voltages using a LiNbO/sub 3/ spherical electro-optic sensor[J].IEEE Photonics Technology Letters,1990,2(6):447-449.

[2] Yunnan Construction Engineering Fifth Construction Co., Ltd. High-voltage line and crane anti-collision device:China, CN201110409844.8[P].2012-6-13.

[3] IEEE Std C95. 6-2002. IEEE Standard for Safety Levels with Respect to Human Exposure to Electromagnetic Fields,0-3 kHz[S]. The Institute of Electrical and Electronics Engineers Inc,2002:

[4] Akimoto, Ikuko (Faculty of Systems Engineering, Wakayama University,Sakaedani 930,Wakayama 640-8510, Japan);Nagao, Kazuhiro;Kan'no, Ken-ichi. Magnetic field dependence of triplet state in iridium complex phosphor. Physica Status Solidi (C) Current Topics in Solid State Physics, v4, n3,p 813-816, 2007,Papers presented at the 10th Euro physical Conference on Defects in Insulating Materials, EURODIM 2006

[5] A Ferrero. Measuring electric power quality: Problems and perspectives [J].Measurement, 2008,41:121-129.

[6] W.R.pFaff. Accuracy of a Spherical Sensor for the Measurement of Threedimensional Electric Fields [J]. Fifth International Symposium on High Voltage Engineering,2010,23:50-57.

[7] G.H.Varttaneourt,S.Carignan,C.Jean.Experience with the Detection of Faulty Composite Insulators on High-voltage Power Lines by the Electric Field Measurement Method. IEEE Trans.1998,13(2)661-666.

ISPECE
IOP Publishing

Design and Research of a Aero Engine Operating Status Monitoring System

Wei Lin, He Li-qing, Wan Yang, Chen Hua-jie

Civil Aviation Flight University of China, Guanghan 618307,China

E-mail address:15883671690@163.com

Abstract. A system for monitoring aero engine speed and discrete operating status signals from the engine is proposed in this paper. The system is mainly composed of MCU technology, sensor signal processing technology and serial communication technology , etc, which is used to monitor engine speed and discrete operating status signals from the engine. The system implements the following functions .On the one hand, the DC voltage signal outputted from the engine speed sensor and the DC voltage signal of the discrete working state of the engine output are collected, and the above information are processed and converted into a digital signal and transmitted to the master computer to obtain the data of engine speed and operating status ; On the other hand, the output current signal is used to output a discrete alarm signal according to the detected signal. Aero engine operating status monitoring system has been tested and all parameters are in accordance with relevant regulations.

1. Introduction

Aero engine rotate speed, temperature, pressure, and oil quantity are important parameters for measuring the performance of an aero engine and the normal operation of the aircraft, which must be monitored. At present, The methods of engine speed monitoring include photoelectric code measuring method [1], magnetic flux velocity measuring method [2], flash speed measuring method [3], and speed measuring method of tachogenerator [4]. The fighter uses a speed measurement based on the speed measuring method of tachogenerator. The measured rotate speed signal is subjected to a signal conditioning circuit to obtain a specific rotate speed. Reasonable design of the signal conditioning circuit is essential for obtaining an accurate rotate speed. In addition, the working state of the engine is obtained by a temperature sensor, a pressure sensor, a fuel amount sensor, etc., subjected to a conditioning acquisition circuit, and then processed by the controlling chip to obtain a discrete alarm signal. In this paper, the engine conditioning acquisition circuit and the working state conditioning acquisition circuit are designed to obtain the rotate speed signal and the working status signal of the 10-way engine, which achieve monitoring of the operational status of this type of aircraft engine.

2. Overall design

The block diagram of the overall design scheme of the system is shown in Figure 1. The PIC18F8722 single-chip microcomputer is selected as the core. The signal output from the engine rotate speed sensor is collected and processed by the rotate speed signal conditioning circuit, and the discrete working state signal output by the engine working state acquisition circuit is used, realizing the output of the discrete alarm signal and the communication of the host computer. The engine working condition monitoring system directly connects the rotate speed sensor and each input and output switch through hard connection.

Content from this work may be used under the terms of the Creative Commons Attribution 3.0 licence. Any further distribution of this work must maintain attribution to the author(s) and the title of the work, journal citation and DOI.

Published under licence by IOP Publishing Ltd

The analog signal outputted by the sensor is sampled, regulated and measured. The converted digital signal is transmitted to the upper computer through the RS-422 communication interface. After the completion of engineering design, high temperature test, low temperature test, vibration test, impact test, acceleration test, etc. are all normal, having practical engineering value.

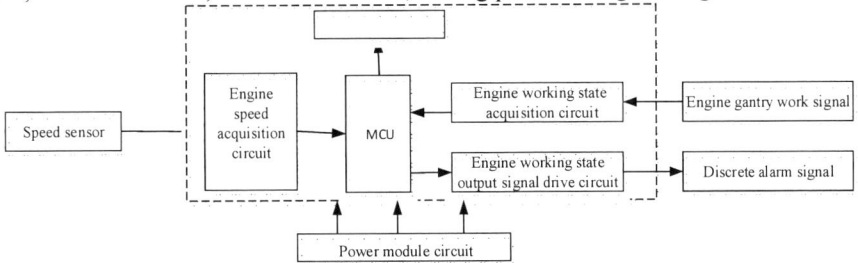

Figure 1. System overall design block diagram.

3. System module design

This article is designed to two parts: system acquisition module and power module. The acquisition modules include a rotate speed signal conditioning circuit, an engine working state acquisition circuit, a single-chip MCU circuit, and a serial communication interface circuit. The following design gives the key circuits.

3.1. Engine rotate speed acquisition circuit design

The rotate speed sensor uses RE0110 DC voltage rotate speed measuring motor. Its output characteristic is DC voltage signal. The output voltage amplitude is proportional to the input rotate speed [U(mv)=2n (rmp)]. Since the rotate speed measurement range is -7000 to 7000r /min (positive and negative), so the corresponding voltage range is -14 to 14V. The AD7890 (which is 12-bit) has an input voltage range of -10 to 10V. Therefore, the input voltage of the rotate speed sensor is divided, considering the matching of the impedance, the divider resistance is 100KΩ/220KΩ and the precision is 1‰. The sampling voltage of the input AD7890 is about -9.625 to +9.625V, which satisfies the input voltage range of AD7890. OP07 has high common mode rejection, high open loop gain, low noise, low output temperature drift, low input offset voltage, and sufficient bandwidth to achieve ideal amplification. Therefore, OP07 is used to form the circuit. In order to prevent the voltage range of the speed measuring motor from fluctuating greatly, and to facilitate the sampling of the rear stage AD, the analog ground and the digital ground are respectively grounded through a single point of 0 Ω. In addition, a second-order low-pass Butterworth filter is applied to the front of the operational amplifier to pass the signal to the AD7890 converter. The AD7890 has fast conversion time, low power consumption, low total harmonic distortion, and built-in tracking/holding amplifier with high rotate speed and flexible serial interface. After the analog-to-digital conversion is completed, the processed speed signal is directly collected by using the PICI8F8722 [5], and the corresponding rotational speed is calculated according to the collected voltage value. The specific engine rotate speed acquisition circuit diagram is shown in Figure 2.

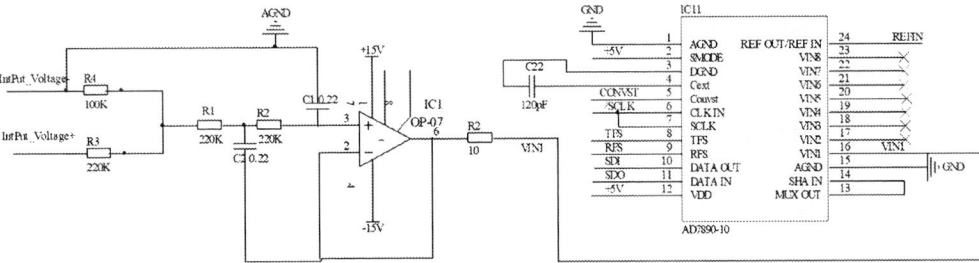

Figure 2. Engine rotate speed acquisition circuit.

3.2. Engine working state acquisition circuit

The discrete operating state of the engine output has 10 signals (collecting 10 engine operating states), all of which are DC voltage signals of 28 VDC. Firstly, the acquisition circuit of the switch input signal uses optocoupler to do electrical isolation and level conversion. After being processed by the inverter, it is sent to the IO port of the acquisition MCU. The discrete acquisition circuit of the specific engine output is shown in Figure 3.

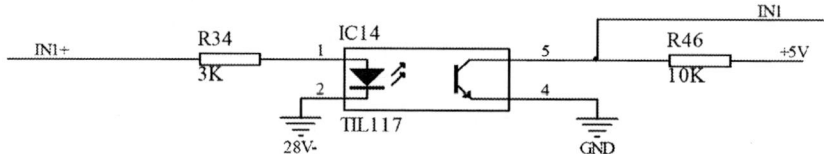

Figure 3. Discrete output circuit of the engine output.

3.3. Engine working state output signal drive circuit design

The engine working state output signal driving circuit is mainly composed of a processing single chip, an isolation circuit and an amplifying circuit. It is processed by the IO port of the processing chip to give the switching output signal, which is processed by the inverter circuit and then sent to the amplifier driving circuit to enhance the load capacity of the circuit. The specific engine operating state output signal driving circuit is shown in Figure 4.

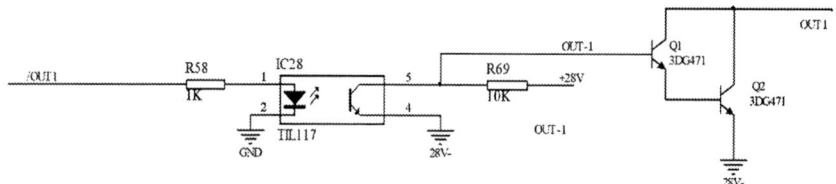

Figure 4. Engine operating state output signal drive circuit.

3.4. Power module circuit design

The DC output voltage of the aviation power is 18-36V with a certain degree of fluctuation. In order to reduce the influence of voltage disturbance on the rear-stage circuit, ensure the reliability, and improve the electromagnetic compatibility of the system, Power module circuit add the fuse at its the input, filtering with the EMI module. The input voltage can be adjusted to a specific voltage by the DC-DC conversion module. Power module circuit has the advantages that are less peripheral circuits, stable output, and reliable operation. The stability of the power and the DC-DC switching generating pulse by itself will have an impact on the system performance. First, it will affect the noise level of the baseline, resulting in inaccurate acquisition of the engine state. On the other hand , it will affect the stability of the amplification parameters of the conditioning circuit. Therefore, on the basis of the pre-reserved filter, the DC-DC rear-stage matching power supply filter unit which is improved, and the matching capacitor is added at the power supply output stage.

The engine operating condition monitoring system requires 15V, -15V, and 5V DC power voltages. The DC/DC power module makes that the externally supplied 28 VDC power converts to the DC power supplying voltage during system operating. The specific power module circuit diagram is shown in Figure 5.

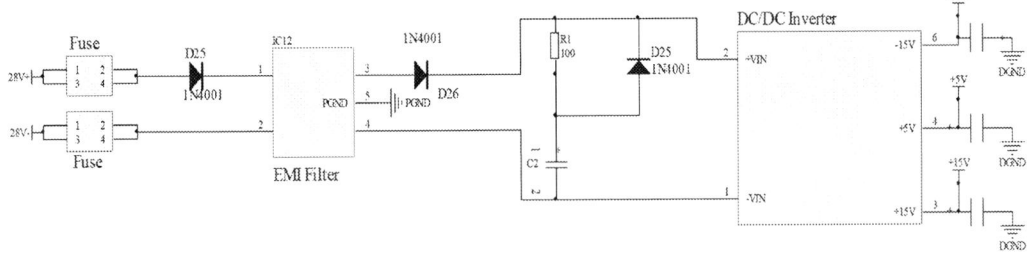

Figure 5. System overall design block diagram.

4. Debugging data and analyzing results

The engine working condition monitoring system adopts a standard air box structure, and its internal structure mainly consists of a partition plate, a board mounting accessory, a circuit board assembly, a power module, a chassis box, and upper and lower covers.

The chassis adopts a frame structure, which is formed by splicing the front and rear covers, the left and right side plates, the front and rear partition plates and the mother board reinforcement frame. The M3 countersunk screws are staggered and connected, and the components constituting the chassis frame are embedded with each other. It ensures structural stability. The chassis case and each circuit board assembly form a unitary structure that is effective against vibration and shock. When the chassis is assembled, the connections between the various parts are in the form of screw staggered connections and the chassis is subjected to vibration in various directions in a vibrating environment, the screws are always axially stressed to ensure the reliability of the connection. Metal frame with motherboard installed on the chassis, which can greatly improve the strength of the chassis and achieve the purpose of anti-vibration and impact. The physical object is shown in Figure 6.

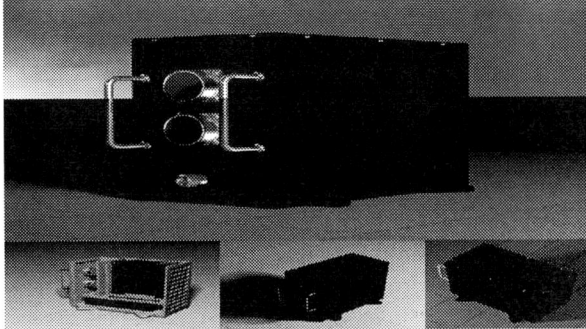

Figure 6. Physical picture.

4.1. Engine rotate speed acquisition circuit debugging

The maximum fluctuation of the sampling of the rotate speed signal conditioning circuit after processing by the Butterworth filter is about 5 bits, and the maximum rotate speed error is about 6.2 r/min. In order to further reduce the fluctuation, the control software is used to process the rotate speed sampling algorithm, and the filtering algorithm is combined with three filtering algorithms: limiting filtering, arithmetic averaging filtering and recursive averaging filtering. Although the filtering algorithm can filter out the interference and fluctuation signals, improper processing will also cause the real-time performance of the signal to deteriorate. Therefore, it is necessary to perform short-time accelerated sampling observation to judge its accelerated performance. The data point line diagram of the 5s acceleration sampling rotate speed bit value is shown in Figure 7. During the acceleration process, the figure shows that sample value rises smoothly and there is no data breakage, which proves that the speed sampling circuit is reliable.

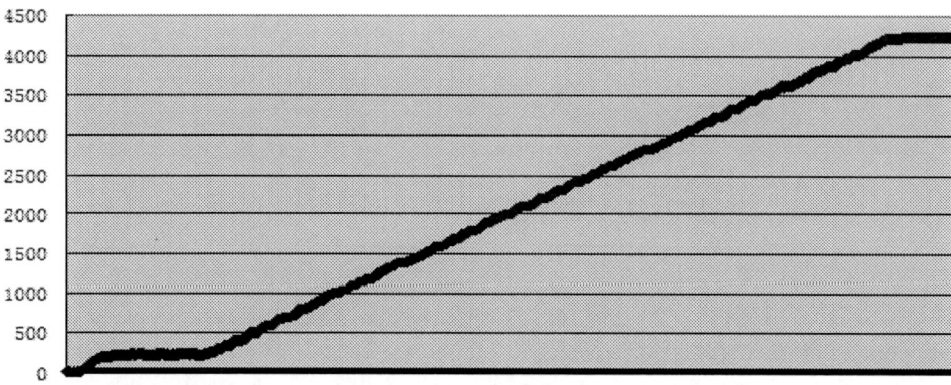

Figure 7. Data point line diagram of 5s acceleration sampling rotate speed bit value.

5. Conclusion

The design and development of the aero-engine working condition monitoring system realizes the monitoring of the engine rotate speed and engine working state, monitors the aircraft's health status, and outputs the engine discrete alarm signal. The rotate speed signal module is tested and the test results are in compliance with relevant regulations. The application of this system to aero engines is of great significance, improving the safety and utility of aviation aircraft.

Acknowledgment

Innovation Project of Civil Aviation Flight University of China :X2018-11

References

[1] ZHU Cairong 2012 *Design and Simulation of Rotate speed Measurement System Based on Optical Encoder* (China: Instrumentation Technology) p 51- 54

[2] LIU Gaojun 2010 *Design of Rotate speed Measurement System Based on PLC Motor* (China: Equipment Manufacturing Technology) p 67-67

[3] ZHANG Wenhai 2005 *Micro-motor low rotate speed test analysis* (China: Micro-motors) p 74-75

[4] ZHOU Xiujun ENG Yulin 2012 *New method of based on PLC motor rotate speed detection and display* (China: Information Technology and Network Security) p 61-63

[5] LIU Lingshun 2009 *Atomatic control component* (China: Beijing University of Aeronautics and Astronautics)

Motion recognition based on Kinect for human-computer intelligent interaction

Xun Pang, Bin Liang*

School of Water Resources and Electric Power, Qinghai University, Xining, China

*Corresponding author e-mail: hdkgigi@163.com

Abstract. At present, human-computer interaction technology has become the focus and hot spot of many scholars. With the deepening of research, facial expression recognition, speech recognition, gesture recognition, face recognition and human motion recognition have become the important content of current human-computer interaction research. In this paper, Microsoft Kinect somatosensory camera is used as the input device of motion, a new method of fast human motion recognition is proposed, and a set of real-time motion recognition and robot control system is designed

1. Introduction

Generally speaking, the human-robot interaction needs to be mediated by computer, so in a sense, this interaction is actually the human-computer interaction. Man-machine interaction is a technical subject that studies the mutual understanding between human and computer, carries out communication and communication, and completes information management, intelligent service and multi-information processing functions for people to the maximum extent [1] [2].

After the man-machine interaction technology experienced the command line interface and the graphical user interface, the natural user interface embodied the user-centered man-machine interaction concept [3].Traditional way of human-computer interaction in the mouse, keyboard, touch screen is given priority to, although these are often necessary equipment of a computer system, but for contact learned too little children and old people learning ability is not strong, operating the equipment to realize the interactions with the robot becomes a difficult thing, so a natural and intuitive human-computer interaction means to cause the extensive concern of the scientific community [4-6].

2. acquisition of depth images and bone data

2.1. Introduction of Kinect depth sensor

2.1.1. Kinect hardware structure

Microsoft got a lot of attention when it released a new Kinect depth sensor in 2010.Originally, the Kinect was designed specifically for the body sense game console Xbox360. People could stand in front of Xbox360 and use voice and body movements to control the game, realizing human-computer interaction. There was no need to operate any controller similar to the handle.

Figure 1 shows the appearance of Kinect. It can be seen from the figure that Kinect has three optical elements. In the middle is the RGB camera, which, like a normal camera, captures space scenes to form color images. Depth of the pair is formed on both sides of the image of the main components, including an infrared emitter and a CMOS infrared receiver and the two devices is the most important

---

Content from this work may be used under the terms of the Creative Commons Attribution 3.0 licence. Any further distribution of this work must maintain attribution to the author(s) and the title of the work, journal citation and DOI.

Published under licence by IOP Publishing Ltd

components, called the formation of depth image, players contour and skeleton tracking function will depend on the camera, the depth device also includes a set of microphone array, can identify the players voice, a motor, can adjust the pitch Angle.

Fig. 1 External structure of Kinect

2.1.2. introduction of Kinect software features

The Kinect for Windows SDK contains a large number of basic libraries and development tools. Developers can use multiple languages to develop Kinect, including C++, C#, VB and other languages. Through the SDK development kit, researchers can obtain important data such as color data flow, depth data flow, and voice information and three-dimensional coordinates of skeletal points from sensor hardware. NUI (Natural User Interface) is a kind of invisible User Interface. People can interact with the machine in the most Natural way, for example, using language and body movements to realize human-computer interaction. The NUI API is the core of the Kinect for Windows, and through the NUI API users can access the color, depth, skeleton and voice data provided by the Kinect hardware to control the Kinect.

The Kinect NUI API mainly includes the following functions :(1) provide driver program to identify the Kinect device connected to PC;

(2) obtain the color data flow and depth data flow from the Kinect camera;

(3) Process the depth data and color data and pass them to PC via USB cable;

(4) Through the machine learning processing of the depth data, the human body is separated and the bone is tracked to transmit the bone number

According to the flow.

(5) obtain audio data from the microphone for speech recognition and sound source localization. The application communicates with Kinect via the NUI API interface. The information transfer diagram is shown in figure 2.

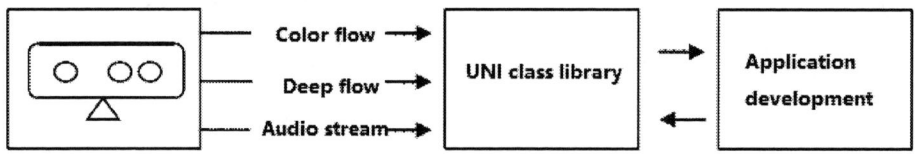

Fig.2 Data stream of Kinect

2.2. Acquisition of depth data

Kinect depth image acquisition is based on a kind of structured Light technology, named as Light Coding by Prime Sense Company. It illuminates the space scene with Light source and encodes the space scene. Is different from common light source device using the light source is called "laser speckle", called light emitter emits the laser emitter surface has a layer of frosted glass, laser penetrates the frosted glass can produce light diffraction, plus an object's surface is rough, the diffraction of random spots formation of the pattern will be different and constantly changing with distance. As long as the space is hit structure light, the object of different distance will make the structure light produce different patterns, record the speckle pattern, through the light source calibration system, can determine the position information of the space object.

2.3. Bone tracking

Kinect can see the three-dimensional world from the depth image, and on this basis, the most critical step is in the depth map

The figure is further separated from the human body contours to form a bone chart from the depth chart. Device on the depth of the image of the "brain" "pixel" assessment of the depth of each pixel of the image scanning, through the image processing technology, computer vision technology for each pixel features identification classification, including edge detection, feature extraction and region segmentation to separate the body contour from the environment, the implementation of active tracking or up to 2 players skeleton of passive tracking for seven players. Skeletal tracking functions are divided into three steps:

(1) The human body depth map was isolated from the complex environmental background;

(2) Divide the human body into 32 parts through machine learning;

(3) Considering the overlapping of human body parts, the machine learning method is further adopted, from the front and side

At the same Angle, 32 parts are determined as 20 nodes according to the pixel features, and the space of each node is given

Three-dimensional coordinates, in meters, are shown in figure 3.

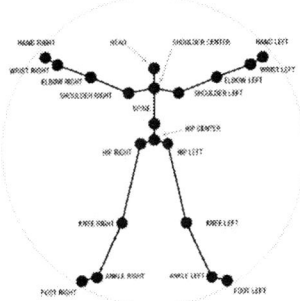

Fig.3 Schematic of human skeleton joints

On the basis of the depth image formed by Kinect, the internal part of the body is segmented by machine learning theory to form bone data, and the three-dimensional coordinates of each node in the Kinect coordinate system are given.Through the study of the feature extraction of specific key points, you can proceed to the description of the action, forming characteristics, and then select the appropriate classification method for action recognition, gesture recognition of this article is based on a device with 20 key points coordinates, as a result, skeletal tracking in action recognition has played a crucial role in [7].

3. based on the action feature extraction of the node

In the pattern recognition problem, the key step is to extract the features of the identified objects. The appropriate feature selection directly determines the success rate of recognition. In the previous chapter, we introduced the skeletal tracking function of Kinect, which provides the three-dimensional coordinates of 20 human nodes.Kinect's depth image frame rate is 30fps, again, providing a stream of 30 frames per second of bone data.This chapter focuses on the extraction of appropriate motion features from a continuous stream of bone data [8].

In this paper, four key points of the right and left hands and the left and right elbows were extracted as the key points to describe upper limb movements. Four key points of the left and right feet and the right and left knee joints were selected as the key points to describe lower limb movements.In a frame of static bone data, the 3d coordinates of 8 points with a total of 24 dimensions can uniquely represent the current state of human body, and the continuous bone frame sequence can uniquely describe an action. For example, 9 frames of bone data can be extracted from a right leg swing motion, and the human skeleton diagram can be drawn as shown in figure 4

Figure 4 Skeleton schematic of right leg confidence right

Some of the movements are sensitive to changes in the Angle of the nodes, such as one hand waving; Some movements are sensitive to the distance changes of the nodes, such as walking, distance between hands and distance between feet. In this chapter, a distance feature extraction method is proposed for the Angle feature of the node and the distance feature of the node pair.In the recognition stage, the Angle feature and distance feature are integrated to enhance the system robustness and improve the recognition rate of actions [9-10].

Simple actions can be represented by the Angle sequence of several nodes, while complex actions are usually described by the mutual relations between some pairs of nodes, such as applause, and the distance between two pairs of nodes is constantly changing. The walk, the distance between your hands and the distance between your feet are constantly changing. Usually, the distance between two points is measured by the Euclidean distance. Compared with the angular features, such as walking and clapping, the distance features between the nodes are more representative.For example, clapping and jogging, the change of distance between the nodes of the two hands is shown in figure 5.The two curves are the waveforms of two movements of different people. It can be seen from the figure that the features of the same type of movements are very similar and the features of different types of movements are different.

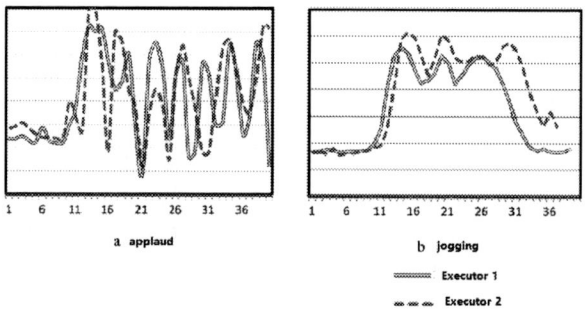

Fig.5 Wave form of handclap and keeper

4. Quick action recognition

Behavior recognition is regarded as the key research object in the field of pattern recognition and artificial intelligence, and has important application value in the field of security monitoring, target tracking and control of intelligent devices. For example, in public places, the use of cameras to collect crowd behavior video can be used to track the target of abnormal behaviors and actions in crowd by means of pattern recognition image processing, so as to maintain normal order and safety in public places. On the control of intelligent equipment, through motion recognition, you can use the body movements to control the robot, intelligent air conditioning, physical sense game machine and other intelligent equipment. Instead of the traditional human-computer interaction, action recognition has a broad prospect of application and development, and some achievements have been made. In this paper, the depth sensor Kinect is used as the motion acquisition device, and the bone model of Kinect is used to realize the recognition of 20 movements, and the recognition results are sent to the robot as instructions, so that the robot can complete corresponding functions according to the instructions and realize human-computer interaction.

Similar to the human nervous system, neural network is composed of many neurons .The neural network has strong learning ability and self-organizing ability through weight connection between neurons, which is suitable for processing large-scale parallel data and has strong robustness, fault tolerance and noise resistance. In the field of action recognition, common neural networks include BP neural network and convolution neural network.BP neural network is a feedforward neural network composed of input layer, intermediate hidden layer and output layer. As shown in figure 6, N is the number of neurons in the input layer of the network, L is the number of neurons in the middle layer of the network, and M is the number of neurons in the output layer of the network.First of all, the training sample data input from the input layer, through the weights of connections between layer and layer, the input samples by different weight through the hidden layer neurons, the final output from the output layer classification training as a result, if the results did not achieve the desired effect, then the actual output and the expected value of difference, namely the error information will feedback to continue training network, expectation until you reach the end of the training.

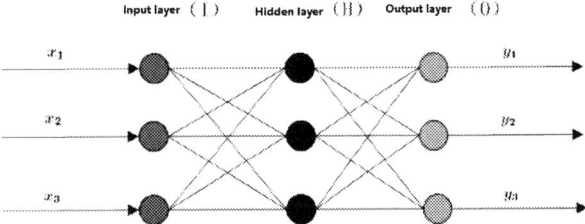

Fig. 6 BP neural network

Like BP network, convolutional neural network is a kind of multilayer feedforward neural network, which includes convolution layer, pooling layer and full connection layer.In the process of action recognition, no matter which kind of network is selected, the feature extraction of the action should be carried out first, the input layer data should be obtained, and the output results can be obtained through network training. It is difficult to achieve the effect of fast action recognition with large training samples and multiple intermediate hidden layers.

5. Interactive system construction and experimental analysis

In this chapter, the skeleton model of Kinect is used to realize the fast recognition of 20 kinds of movements, and the robot is quickly controlled to realize human-computer interaction.This paper introduces the construction of the hardware and software environment of the system of action recognition and the construction of the hardware and software environment of the robot. Finally, the recognition rate, recognition time and robustness of action recognition are verified under different feature extraction methods. A simple experimental demonstration is made for the motion control robot, and the results show that the system has achieved good recognition and control effects.

5.1. System robustness verification

The motion control robot system designed in this paper requires that the system can achieve good recognition effect under different circumstances, including the complexity of the environmental background, the intensity of ambient light, the position change of the tester, the size of the tester, and the successful recognition under the interference of multiple people.

(1) Verification experiment of recognition rate

In order to verify the effectiveness of the motion recognition scheme in this paper, the recognition rate verification experiment was carried out. The experimenter is composed of 10 students of different body types in the research room. Standing within the effective field of vision of Kinect, each movement is performed by 10 students, respectively through two feature extraction methods proposed in chapter 3, and the action recognition experiment is conducted by FDTW algorithm.The recognition rate of each movement and the average recognition rate are shown in table 1.

Tab.1 The recognition rate of six kinds of actions under two feature extraction methods

The Action name	Vector special extraction recognition rate	Angle distance feature fusion recognition rate
Very different raised	97%	97%
applaud	96%	98%
The Left legs Left swing	95%	96%
Kick in front of the right leg	96%	97%
Bend over	96%	98%
Squat	97%	97%
6 kinds of action average recognition rate	96.21%	97.53%

It can be seen from the recognition rate in the table that the action recognition in this paper has a good recognition effect and the average recognition rate is respectively

Reached 96.21% and 97.53%, basically meeting the requirements of controlling the robot.

5.2. Identification time verification experiment

In order to verify the effectiveness of the lower bound function and truncation technique of the proposed FDTW algorithm to improve the speed of action recognition, a recognition time verification experiment was conducted. The general DTW algorithm and the average recognition time of the FDTW algorithm proposed in this paper were respectively calculated under different action types, as shown in figure 7.

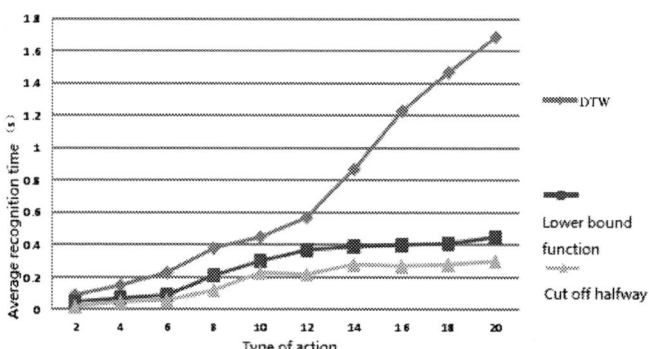

Fig.7 Recognition time verification experiment

It can be seen from the figure that, when the number of action types is no more than 8, the average recognition time of ordinary DTW algorithm is around 0.4s due to the small amount of template data, which can also meet the interaction requirements. However, as the number of action types increases, for example, 20 is designed in this paper

Therefore, the sequence of the action to be tested needs to match and calculate the similarity one by one with the 20 sequences in the action template. In this case, the calculation amount is very large, and the average recognition time rises to 1.7s, which seriously affects the recognition speed, causing delay when controlling the robot. The tester's actions are executed, but they are not immediately recognized, which affects the timeliness of the interaction. The first step of the algorithm is accelerated by the lower bound function, and the average identification time has a significant downward trend. Then, the second step is accelerated by the intermediate truncation technology, and the recognition efficiency of the average identification time drops to about 0.3s, indicating the effectiveness of the FDTW algorithm proposed in this paper.

6. Conclusion

With the introduction of Microsoft Kinect depth sensor, new opportunities have been brought to motion recognition researchers. With its strong bone tracking technology and advantages of being free from interference of natural light, it has made outstanding contributions to the field of motion recognition. In this paper, 20 kinds of custom action recognition were completed based on Kinect bone tracking technology. For the rapid action recognition system in this paper, in addition to the improvement in the matching algorithm, in the aspect of feature extraction, the dimension of feature data is too large, which also causes the problem of slow calculation speed. The principal component analysis (pca) can be used to reduce the dimensionality of feature data, so that the feature data can be simplified without losing the validity of its description, so as to further enrich the action template and meet the needs of society.

References

[1]   Yu Tao Kinect application development in the most natural way to talk to the machine [M]. Beijing: Mechanical Industry Press, 2012

[2]   Dong Shihai. The progress of human-computer interaction and The challenges it faces [J]. Journal of computer-aided Design & computer Graphics 2004,16(1):1-13

[3]   Wang Songlin. Research on gesture recognition and robot control technology based on Kinect [D]. Beijing: Beijing Jiaotong University, 2014

[4]   Meng Ming, Yang Fangbo, Yan Qingshan. Human motion detection based on Kinect depth image information [J]. Journal of Instrumentation, 2015, 36(2): 386-393

[5]   Liu Yang, Shang Zhaowei. Traffic police gesture recognition based on Kinect skeleton information [J]. Computer Engineering and Applications, 2015, 51(3): 157-161

[6]   Wu Zepeng, Guo Lingling, Zhu Mingchao. The Image registration method combining Image information entropy and feature points [J]. Infrared and Laser Engineering, 2013, 42 (10) : 2846-2852.

[7]   Abdulla W H, Chow D, Sin G. Cross-words reference template for dtw-based speech recognition systems[C]// TENCON 2003. Conference on Convergent Technologies for the asia-pacific Region.

[8]   Keogh E, Ratanamahatana C. Exact indexing of dynamic time warping [J]. Knowledge and Information Systems, 2005, 7(3):358-386.

[9]   Wei L, Keogh E, Herle h. Atomic wedgie: Efficient query filtering for streaming times series[C]// IEEE International Conference on Data Mining. IEEE Computer Society, 2005:490-497.

ISPECE   IOP Publishing

Parameter optimal design and Simulation of Power System of Electric Vehicle Based on AVL-CRUISE

Jianwei Ma[1,a]

[1]Department of Automobile Engineering, Xingtai Polytechnic College, Xingtai, Hebei, China

[a]Corresponding author: 94718894qq.com

Abstract. In order to meet demand of dynamic of electric vehicle, based on vehicle design method theorety, power parameters of motor and transmission ratio are designed using method of theoretical calculation, model of vehicle is built and simulated in AVL-CRUISE software, transmission ratio is optimized to improve economic performance. Simulation result shows performance of the whole vehicle is improved. So It lays the foundation for development of E V.

1. INSTRUCTION

The dynamic performance and economy of EV are affected design level of dynamic system directly, the accurate of calculating power of motor and parameter matching between motor and the transmission are key of design[1]. Through theoretical calculation, dynamic demand of motor and range of transmission ratio are determined[2][3]. However, for different motor, the power is same, but the driving characteristic are different, so the demand for transmission ratio is different, and how to ascertain the optimal transmission ratio is a critical on the condition of certain motor [4].

2. DESIGN DEMAND OF DYNAMIC

Table 1. Parameters of vehicle.

Name of Parameter	Values	Units
Kerb mass	980	kg
Gross Mass	1250	kg
Frontal area	2.10	m^2
Drag coefficient	0.35	-
Rolling resistance coefficient	0.015	-
Wheel radius	0.33	mm
Transmission efficiency	96%	-

Content from this work may be used under the terms of the Creative Commons Attribution 3.0 licence. Any further distribution of this work must maintain attribution to the author(s) and the title of the work, journal citation and DOI.
Published under licence by IOP Publishing Ltd

Table2 .Overall performance index

Performance demand	Item	Index
Speed demand	Maximum speed	≥110 km/h
Acceleration performance	0~50km/h	≤7s
	50~70km/h	≤8s
Climbing performance	Maximum value of speed at slop of 6%	≥50 km/h
	Maximum value of speed at slop of 9%	≥20 km/h
	Maximum value of speed at slop of 20%	≥10km/h
Energy of consumption per one hundred kilometers		≤15kWh

3. PARAMETER DESIGN OF POWER SYSTEM

3.1 Parameter design of motor

Power demand can be acquired by the formula (1), formula(2) and formula (3) on 3 kind of states for maximum speed, acceleration and climbing.

$$P_{e1} = \frac{1}{\eta_T}(\frac{mgfu_a}{3600} + \frac{C_D Au_a^3}{76140}) \quad . \tag{1}$$

$$P_{e1} = \frac{1}{\eta_T}(\frac{mgfu_a}{3600} + \frac{C_D Au_a^3}{76140} + \frac{\delta mu_a}{3600}\frac{du}{dt}) \cdot \tag{2}$$

$$P_{e1} = \frac{1}{\eta_T}(\frac{mgfu_a}{3600}\cos\alpha + \frac{mgu_a}{3600}\sin\alpha + \frac{C_D Au_a^3}{76140}) \cdot \tag{3}$$

The drive-line efficiency is η_T, f is rolling resistance coefficient, m is full-load-quality, C_D is the air resistance coefficient, A is the frontal area, u_a is the maximum speed, δ is moment of inertia conversion coefficient, and α is climbing angle.

On the basis of formula (1), formula (2) and formula (3), the maximum dynamic requirement is estimated under kinds of condition, they are shown in Table 3.

Table 3. Estimated value of peak power

Quality requirement	Power of motor（kW）
Maximum speed v_{max} ≥110 km/h	23.75
Acceleration quality 0~50km/h≤7s	40.18
Acceleration quality 50~80km/h≤8s	32.15
Maximum steady speed at slop of 6%≥60 km/h	18.09

Maximum steady speed at slop of 9%≥30 km/h	11.39
Maximum steady speed at slop of 20%≥10 km/h	7.48

To satisfied the dynamic performance, rated power of motor must satisfy the maximum speed, the peak power must satisfy the demand of acceleration quality. From the Table 3, the rated power is more than 23.75kW, peak power of motor is more than 40.18kW, and rated power of motor chosen is 25kW, peak power of power motor chosen is 45kW.

Permanent magnet motor possess characteristic of high power density and high efficiency. Its voltage is between 100V and 400V, maximum speed is between 4000r/min and 10000r/min, constant power factor is 2.25, all parameter of motor designed are shown in Table 4.

Table 4 .Parameters of motor

Parameter	Values
Rated power	25kW
Overload coefficient	1.8
Base speed	3000rpm
Voltage of motor	320V
Peak value of power	45kW
Peak value of torque	14Nm
Peak value of speed	8000rpm

3.2 Design of transmission system

Transmission ratio must satisfy the requirement of largest climbing gradient,and it also must assure the greatly desired maximum speed. Transmission ratio required must satisfy formula (4) and formula(5).

$$i_0 \geq \frac{mg(f \cos\alpha_{max} + \sin\alpha_{max})r}{T_{max}\eta_t}. \quad (4)$$

$$i_0 \leq \frac{0.377 n_{max} r}{v_{max}}. \quad (5)$$

i_0 is gear ratio , r is radius of wheel , T_{max} is the peak torque, n_{max} the maximum speed of revolution of motor. $6.2 \leq i_0 \leq 8.29$ can be calculated .

4. SPEED RATIO OPTIMIZATION

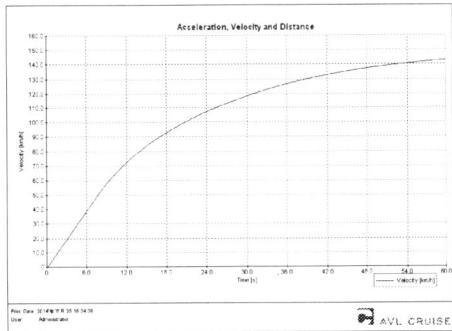

（a）Simulation diagram of acceleration

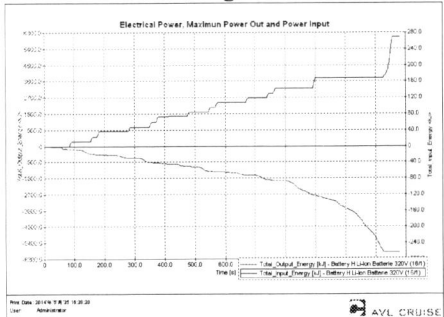

（b）Simulation diagram of energy consumption on NEDC condition

Figure 1.Simulation when transmission ratio is taken as 6.5

4.1 Optimization of speed ratio optimization

The whole vehicle model is built in software of AVL-CRUISE, the transmission ratio is 7.5,7.0,6.5, and vehicle performances are simulated;the simulation diagram of energy consumption and acceleration condition are shown in Figure.1, Figure.2 and Figure.3 under condition of NEDC.

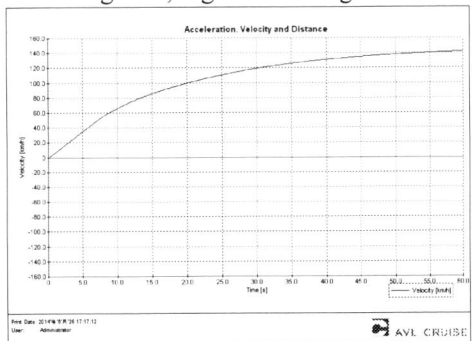

（a）Simulation diagram of acceleration condition

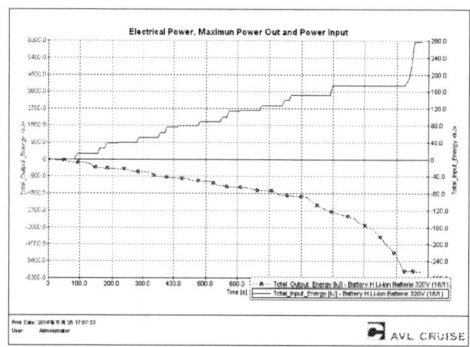

（b）Simulation diagram of energy consumption on NEDC condition

Figure.2 Simulation result when transmission ratio is 7.0

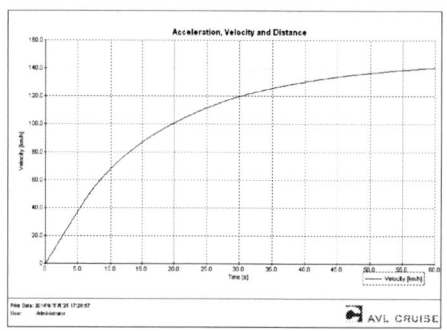

（a）Simulation diagram of acceleration condition

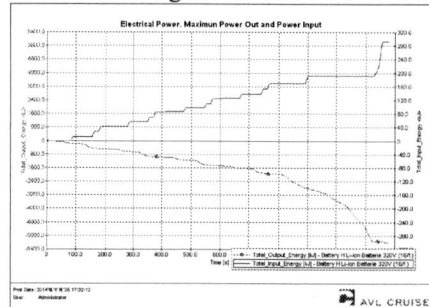

（b）Simulation diagram of energy consumption on NEDC condition

Figure.3 Simulation diagrams when gear ratio is taken as 7.5

From Figure.1, Figure.2, Figure.3, when gear ratio is taken as 6.5,7.0 and7.5, acceleration performances and energy consumption of vehicle are shown in Table 5.

Table 5 The energy consumption and acceleration performance of vehicle

Gear ratio	Acceleration Time		Energy consumption
	0~50km/h	50~70km/h	
6.5	7.76s	6.22s	14.36（kWh/100km）
7.0	7.26s	6.44s	14.42（kWh/100km）
7.5	6.76s	6.31	14.43（kWh/100km）

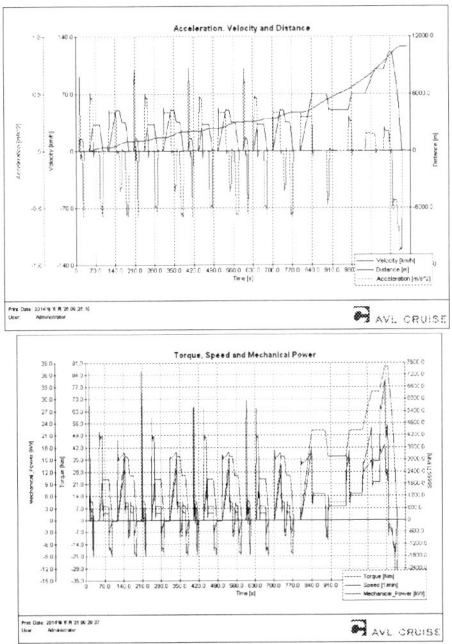

Figure.4 Simulation results of under NEDC condition

From the Table 5, with growing of gear ratio, acceleration performances is improving, and energy consumption also adds. To satisfy the demand of acceleration performance, at same time improving economy, ,the gear ratio is design as 7.25 in the design.

5.THE ANALYSIS OF VEHICLE PERFORMANCE

When the gear ratio is taken as 7.25, on NEDC condition , the whole vehicle performance is shown in Figure.4, dynamic index and economy index are shown in Table 6.

Table 6 Simulation results of power dynamic and economy

Performance demand	Item	Simulation result	Reference Index
Speed demand	Max speed	156.6 km/h	≥120 km/h
Acceleration performance	0~50km/h	6.98s	≤7s
	50~80km/h	6.29s	≤8s
Climbing performance	Maxi-speed at slop of 6%	67 km/h	≥60 km/h
	Max-speed at slop of 9%	58 km/h	≥30 km/h
	Max-speed at slop of 20%	33km/h	≥10 km/h
Energy consumption	14.43 kWh/100km		≤15kWh/100km

From Figure 5, under NEDC condition, the torque range of motor is between 30Nm and 70Nm, speed is from 2000 to 5000r/min, that is the efficient region of motor, the motor designed is rational. In Table 6, it is seen that requirement of dynamic and economy are satisfied.

6. CONCLUSION

The simulation result indicated that in range of gear ratio calculated, gear ratio increases, dynamic also improves, but economy descends,dynamic and economic is mutually contradictory, dynamic improves, at the same time, energy consumption also increase.

Through the power balance equation, parameters of motor is determined and and range of gear ratio is determined too, and gear ratio is optimized according to simulation diagram, in order to ensure power , energy consumption reduces, the whole performance of vehicle is improved greatly, that lay foundation for whole vehicle's development .

REFERENCES

[1] Zh.S.Yu. *Automobile principle* (Machine Press, 2009).
[2] Z.L.Liu, B.L. Zhang. JHUT,**30** (2007).
[3] F.Z.Ji, F.Gao, R. AT, **6** (2005).
[4] 4.B.Zhou. CME,**22** (2005).
[5] Z.F.Wang, C.N. Zhang. *Electricity Drive Theory and Design of Electric vehicle* (Machinery Industry Press, 2012).

Parameter Design and Simulation Analysis of Power System in Plug-in hybrid vehicle

Jianwei Ma[1,a]

[1]Department of Automobile Engineering, Xingtai polytechnic college, Xingtai, Hebei, China

[a]Corresponding author: 94718894r@qq.com

Abstract. According to power system's demands of plug-in hybrid vehicles, though design theory of hybrid vehicles, the parameters of motor, transmission system，engine-generator set and battery other subsystem are designed, and the model of power system is established and simulation analysis is performed using AVL-CRUISE software, simulation results show that all parameters designed can satisfy working demand under the NEDC,so design is very reasonable,which lays foundation for the plug-in hybrid vehicle's development .

1. Introduction

Compared with ordinary hybrid vehicle，the plug-in hybrid vehicle can be charged at working , and also be charged using a outlet, it can running longer only by battery， particularly daily commute distance[1] , plug-in hybrid vehicle can work in pure electrotype, The internal combustion engine can provide additional power at the long distance travel， plug-in hybrid vehicle can use alternative fuel as full as possible , in the new energy vehicle it is with wide prospect[2][3]. contrasting the traditional engine vehicles, plug-in hybrid vehicle has engine-generator set,motor, power battery and many other systems, at early stage of plug-in hybrid vehicle, power， through the accurately designing parameters of power system, coordinating and matching subsystem performance, reliability and economy of plug-in hybrid vehicle can be greatly improved [4].

2. Performance parameters requirements OF Vehicle

Dynamic property of electric vehicles can be comparable with conventional vehicle, because of the constant development of the motor, battery, transmission and control technology[5]. Combining electric vehicle's performance characteristics, according to demand of vehicle performances, basic parameters and vehicle performance index of plug-in hybrid vehicle are shown in Table 1 and Table 2.

Table 1 Parameters of plug-in hybrid vehicle

Name of parameters	Values
Gross Mass	1580(kg)
Kerb mass	1200(kg)

Content from this work may be used under the terms of the Creative Commons Attribution 3.0 licence. Any further distribution of this work must maintain attribution to the author(s) and the title of the work, journal citation and DOI.
Published under licence by IOP Publishing Ltd

Wheel base		2468(mm)
Frontal area		1.97(m$^{2)}$
Centroid height		500(mm)
Drag coefficient		0.284
Rolling resistance coefficient		0.013
Radius of wheel	Static radius	287(mm)
	Dynamic radius	301(mm)

Table 2 Performances index of plug-in hybrid vehicle

Performance demand	Items	Index
Speed demand	Max-speed	≥120 km/h
Acceleration performance	0~50km/h	≤8s
	50~80km/h	≤10s
Climbing slope performance	Maxi-speed at 4% slop	≥70 km/h
	Max-speed at 12% slop	≥40 km/h
	Maximum gradeability	≥20%

3. Parameter design of dynamic system

3.1 Parameter design of motor

the maximum speed, acceleration and climbing of plug-in hybrid vehicle can be calculated though formula (1), formula (2) and formula (3) .

$$P_{e1} = \frac{1}{\eta_T}(\frac{mgfu_a}{3600} + \frac{C_D Au_a^3}{21.15 \times 3600}) \cdot \qquad (1)$$

$$P_{e2} = \frac{1}{\eta_T}(\frac{mgfu_a}{3600} + \frac{C_D Au_a^3}{21.15} + \delta mu_a \frac{du}{dt}) \cdot \qquad (2)$$

$$P_{e3} = \frac{1}{\eta_T}(mgf \cos\alpha u_a + \frac{C_D Au_a^3}{21.15} + mgu_a \sin\alpha) \qquad (3)$$

In the formula, m is full-load quality, η_T is driveline efficiency, f is rolling resistance coefficient, C_D is air resistance coefficient, A is frontal area, u_a is the maximum speed, v is the speed, α is climbing angle, δ is moment of inertia conversion coefficient .

Table 3 Estimated value of peak power of motor

Performance demand	Power of Motor
Max-speed $v_{max} \geq 120$ km/h	20.88 kW

acceleration performance of 0~50km/h≤7s	49.3 kW
Acceleration performance of 50~80km/h≤8s	38.6 kW
Max steady speed at 4% slop ≥70 km/h	19.85 kW
Max steady speed at12% slop o≥40 m/h	24.93 kW
Maxi steady speed at 20% slop ≥5 km/h	4.83 kW

The max power requirement which are estimated are shown in Table 3,though formula (1), formula (2) and formula (3).

To satisfy the power performance of plug-in hybrid vehicle, drive motor's rated power must satisfy the needs of the max-speed, the peak power must satisfy the needs of acceleration performance. In Table 3, the rated power of motor is more than20.88kW, peak power is more than 49.3kW, so the rated power should be taken as 25kW, peak power of motor is taken as 50kW.

Permanent magnet motor possesses the advantage of high power density and high efficiency, its working voltage is between 100V and 400V, maximum rotational speed is between 4000 and 10000r/min, the constant power factor is 2.25, so permanent magnet synchronous motor is adopted, the parameter of permanent magnet synchronous motor designed is shown in Table 4.

Table 4   Motor's Parameters

Parameter	Values
Rated power	25kW
Overloading coefficient	2
Motor voltage	324V
Base speed	3000rpm
Peak torque	160Nm
Peak speed	8000rpm
Peak power	50kW

3.2 Parameters calculation of engine-generator set

If with sufficient electric quantity, plug-in hybrid vehicle will switch off engine, preferentially uses electricity, when power battery's electricity quantity declines to the certain extent , power battery can not continue to supply electrical, power provided by battery can not satisfy need of vehicle, the engine-generator set starts run for providing electricity power to satisfy need of vehicle.

In order to ensure battery's safety, reduce the discharge or charge times of battery, on design, generator begins working, also it can meet the demand of peak power when the charge quantity of battery is less 40%. Engine-generator set designed does not only satisfy need of the peak power, but also satisfy engine-generator running located at efficient district . Based on the analysis and matching work areas, parameters of engine-generator set calculated are shown in Table 5.

Table 5 Parameters calculated of engine-generator set

Engine		Generator	
Max rotational speed	6000rpm	Max rotational speed	6000rpm

Max torque /	115Nm	Voltage	324
Max power	62kW	Max power	50 kW
Displacement	1.5L	Peak torque	160Nm
Max speed	4500rpm		

3.3 Parameters design of power battery

The pure electricity endurance distance has great influence to application performance of plug-in hybrid vehicle, pure electricity endurance distance is distance of driving when battery is charged fully, does not rely on engine recharging. Battery's rated capacity should be counted by formula 4.

$$W_0 = \frac{P_v S}{0.7 v \eta_m} . \tag{4}$$

In formula 4, W_0 is total battery's rated energy , P_v is motor power when velocity is v, S is driving distance when velocity is v. Q is Rated capacity, Q should meets formula 5.

$$Q = \frac{W_0}{U_B} . \tag{5}$$

In formula 5, Q is battery's rated capacity, U_B is voltage.

The plug-in hybrid is demanded driving 90km and the velocity is 40km/h using pure electric in design, according to the formula 4, battery capacity is 10.39kwh. According to formula 5, the battery rated capacity is taken as 32.07Ah.

3.4 Design of Transmission System

According to the requirement, speed ratio needs to meet need of maximum grade-ability, and it ensures the desired max-speed. Transmission ratio must satisfy the formula 6 and formula7.

$$i_0 \geq \frac{mg(f \cos \alpha_{\max} + \sin \alpha_{\max})r}{T_{\max} \eta_t} . \tag{6}$$

$$i_0 \leq \frac{0.377 n_{\max} r}{v_{\max}} . \tag{7}$$

In formula, i_0 is transmission's speed ratio , r is the rolling radius of wheel , $T_{\max}$ is motor's peak torque, $n_{\max}$ is maximum evolution speed of motor. , the i_0 is chosen as 7.05 in design.

4. Simulation Analysis

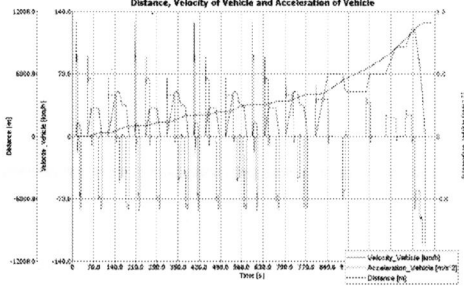

Figure 1. Under condition on NEDC

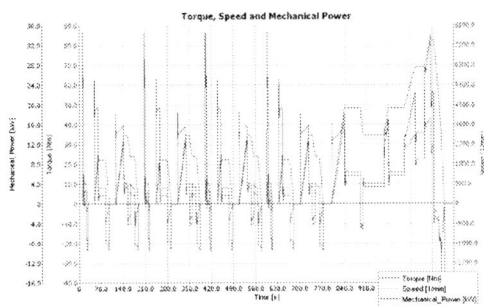

Figure 2.running curve of motor

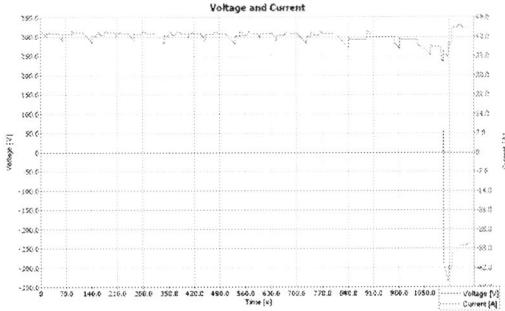

Figure 3.Running curve of engine

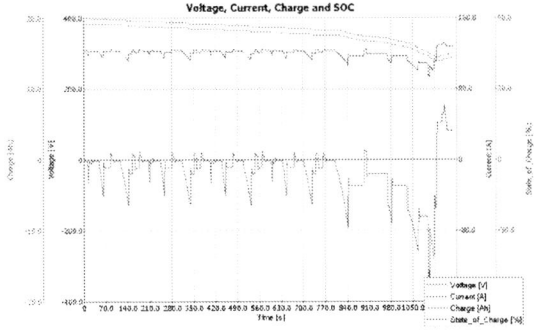

Figure 4. Battery's working curve

Table 6 Simulation result of power performance

Performance demands	Items	Simulation results	Reference scope
Speed demand	Max-speed	128 km/h	≥120 km/h
Acceleration performance	0~50km/h	6.6s	≤7s
	50~80km/h	5.82s	≤8s
Climbing clop performance	Maxi-speed at 4% slop	128 km/h	≥70 km/h

Maxi-speed at 12% slo	83 km/h	≥40 km/h
Max gradeability	23.8%	≥20%

In order to prove the feasibility of the design, power system of plug-in hybrid vehicle is simulated in AVL-CRUISE software, the simulation results are shown in Figure. 1, Figure.2, Figure.3 and Figure.4 under condition of NEDC .

From Figure. 1, Figure.2, Figure.3 to Figure.4, It can be seen , total distance is 10.93km under condition of NEDC, and duration time is up to 1179.8S, max-speed is up to120.5km/h, max acceleration is 1.46m/s^2, max deceleration is -0.953 m/s^2. The motor works between 2000rpm and 7500rpm at the most time, in the beginning the SOC of battery is 60% , from beginning to 1100 seconds, the SOC is approaching to 43%, the engine–generator set starts running automatically, providing power for motor to maintain normal driving of plug-in hybrid vehicle, meanwhile charging for battery. From Table 6, It can be seen that the max-speed of plug-in hybrid vehicle is 128km/h, acceleration time of the plug-in hybrid vehicle is 6.6s from 0 to 50 km/h, acceleration time the plug-in hybrid vehicle is 5.82s from 50 to 80 km/h, maximum grade-ability is up to 23.8%, and power parameters designed satisfy design demand.

5. Conclusion

On the basis of index of plug-in hybrid vehicle, though calculation, the parameters of drive motor, parameters of engine-generator set, parameters of battery,parameters of transmission system and other systems are designed, the model of plug-in hybrid vehicle is built in AVL-CRUISE software, though simulation, power performance under NEDC condition is obtained. Simulation result shows that the plug-in hybrid vehicle designed can satisfy running requirements under NEDC condition, while max-peed, climbing slop performance, acceleration performance satisfy design demands, all parameters designed of motor, engine–generator set, drive battery, transmission and other systems are reasonable, that lay the foundations for development and progress of plug-in hybrid.

References

[1] C. L. Yin. *Typical structures and type* of *New energy vehicles*-(Shanghai Technology and science press,2013)
[2] I. Husain. *Design Fundamental of Electric and Hybrid Vehicles*(Machine Press,2012)
[3] Z.S. Yu. *The Theory of Vehicle* (Machine Press,2009)
[4] N .W. Xue, X.M .Ma, C.F.Pan, JCQJTU,**31**(2012).
[5] S.L. Liu, L.Q. Gu, DST,**22** (2008)

Effect of injection compression process parameters on residual stress of products based on numerical simulation

Junjie Zhu, Yanfang Chen, Wenhan Huang, Qiurong Zhang, Xiaoming Liao, Yizhi Huang and Zhiwen Qiu

School of Electro-mechanical Engineering, Heyuan Ploytechnic, Heyuan, Guangdong517000, China

Author: zhujunjie2006mmxy@163.com

Abstract. With the maximum residual stress of Mises-Hencky as the index, a numerical model for injection compression molding of the disc was established by CAE software. The process parameters of 7 factors including mold temperature, melt temperature, pressure holding time were analyzed by orthogonal experiment. Effect of key process parameters on residual stress was studied by single factor method finally. The results show that effect of mold temperature on residual stress was greatest, followed by melt temperature, compression delay time and compression speed, the residual stress decreases with the increase of mold temperature and melt temperature, increases with the increase of compression delay time, and the effect of compression speed on residual stress is not obvious.

1. Introduction

With the wide application of optical components, higher requirements are put forward for the quality of transparent products. Compared with traditional injection molding technology, injection compression molding has unique advantages in shaping transparent products. Injection compression molding can achieve uniform packing pressure effect, so that the physical properties are more uniform and the residual stress is lower, the warpage and birefringence of product are improved finally. The size and residual stress distribution of products are mainly determined by the history of time, temperature and pressure during the forming process.Therefore, the process conditions such as mold temperature, melt temperature, compression speed and compression time have important effect on the final residual stress and warping deformation[1].Wang Kejian[2] and others investigated the birefringence distribution of optical products by different injection process parameters through a single variable experiment, calculated the residual stress value on the symmetry axis through the stress-photoelastic law, and optimized it. Liu Wenjuan[3]and others used the kriging agent model and the sequence optimization method of EI point criteria to optimize the process parameters of polycarbonate products to reduce the residual stress. Bushko[4-5] had studied the solidification mechanism of large plate forming process and analyzed effect of residual stress on the warpage of plastic parts during solidification process.

This paper takes a disk as an example, injection compression molding process were simulated by CAE numerical simulation software, the quantitative effect of key process parameters on the residual stress of Mises-Hencky after product release was analyzed by orthogonal experiment, in order to study effect of injection compression process parameters on residual stress.

Content from this work may be used under the terms of the Creative Commons Attribution 3.0 licence. Any further distribution of this work must maintain attribution to the author(s) and the title of the work, journal citation and DOI.
Published under licence by IOP Publishing Ltd

2. Numerical calculation theory of residual stress

Residual stress of injection molding parts refers to the sum of all kinds of stresses that are not relaxed in the products after products are unformed and free to contract and deform without external constraints[6].Generally, the residual stress of product is considered as inclusion residual stress and thermal residual stress. In the filling and packing stage, the viscous flow of melt produces flow residual stress, which is related to the molecular orientation of the material, On the other hand, because of the temperature and pressure history of each part of the product in the molding process are different, that lead to uneven cooling contraction and thermal residual stress, which is the main source of product residual stress [1].

Thermal residual stress is mainly determined by temperature distribution, solid-liquid interface location and pressure history, according to the Bolzmann principle, a mathematical model based on the distribution of residual stress induced by thermal stress and compression pressure is used, when residual stress calculated by Moldflow numerical software[8]:

$$\sigma_{ij} = \int_{-\infty}^{t} C_{ijkl} \left(\xi(t) - \xi(t') \right) \partial \varepsilon_{kl} / \partial t' \, dt' \\ - \int_{-\infty}^{t} B_{ij} \left(\xi(t) - \xi(t') \right) dT(t') \tag{1}$$

In this formula, σ_{ij} —Stress tensor; C_{ijkl} —Material mechanical properties tensor; B_{ij} —Material thermal properties tensor; T —Thermodynamic temperature; $\partial \varepsilon_{kl}$ —Coefficient of thermal expansion tensor; $\xi(t)$ —Change function of material temperature over time

$$\xi(t) = \int_{0}^{t} 1 / a_T \, dt' \tag{2}$$

a_T —conversion factor of temperature and time

3. Numerical example simulation

3.1 Finite element analysis model and material selection

This paper takes a disk as the analysis model, the diameter of the disc is 120mm, the thickness is 0.8mm, the number of finite element mesh elements in the analysis model is 6014, and the number of nodes is 3025, the finite element model is shown in figure 1. Molding material is PMMA with the brand of CP61, the main properties of materials were shown in table 1.

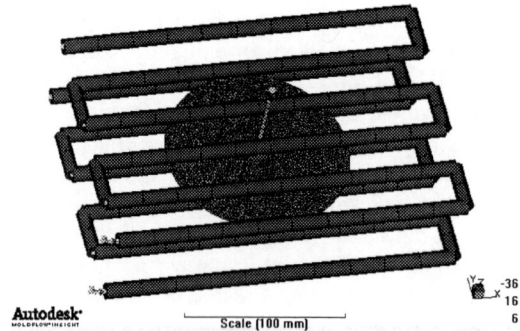

Figure1. Finite element analysis model of disc

Table1. Main properties of PMMA

Melt temperature/℃	200~280
Mold temperature/℃	20~80
Maximum allowable shear stress/MPa	0.26

Shear modulus/MPa	1011
Elasticity modulus/MPa	2740
Poisson's ratio	0.355

3.2 Orthogonal experiment

Combined with actual production experience, the maximum Mises-Hencky residual stress after the mold release was selected as the index, 7 injection compression process parameters including mold temperature, melt temperature, pressure packing time, compression delay time, compression force, compression distance and compression speed were taken as experimental factors, denoted *A, B, C, D, E, F* and *G* respectively. According to the parameters of the injector and recommended parameters range of materials, 3 levels were uniformly taken within the range of values, and the specific experimental factors and levels were shown in table 2.

According to the experimental factors and level in table 2, L18(3^7) orthogonal table is selected for numerical simulation on the CAE software platform, the maximum Mises-Hencky residual stress (denoted *R*)was obtained after simulation, the test results were shown in table 3.

Table2. Factors and levers table

Factors / Levels	*A*/℃	*B*/℃	*C*/s	*D*/s	*E*/t	*F*/mm	*G*/cm/s
1	50	220	8	0.5	150	1	0.1
2	60	230	10	1	170	2	0.2
3	70	240	12	1.5	190	3	0.3

Table3. Orthogonal experimental results

Column number / Experiment number	1 *A*	2 *B*	3 *C*	4 *D*	5 *E*	6 *F*	7 *G*	*R* /MPa
1	1	1	1	1	1	1	1	30.52
2	1	2	2	2	2	2	2	27.84
3	1	3	3	3	3	3	3	26.61
4	2	1	1	2	2	3	3	26.22
5	2	2	2	3	3	1	1	26.17
6	2	3	3	1	1	2	2	25.87
7	3	1	2	1	3	2	3	24.75
8	3	2	3	2	1	3	1	22.72
9	3	3	1	3	2	1	2	23.57
10	1	1	3	3	2	2	1	30.49
11	1	2	1	1	3	3	2	27.83
12	1	3	2	2	1	1	3	26.61
13	2	1	2	3	1	3	2	26.26
14	2	2	3	1	2	1	3	26.08
15	2	3	1	2	3	2	1	25.94
16	3	1	3	2	3	1	2	24.81
17	3	2	1	3	1	2	3	22.65
18	3	3	2	1	2	3	1	23.51

3.3 Data processing and analysis

Orthogonal experiment is try to find the optimal parameter horizontal combination and experimental factors with a few experiments, so the weight of the experimental index should be investigated, However, results have deviation because of the unstable factors, Therefore, S/N ration (Signal /Noise) is adopted to measure the robustness of output characteristics.S/N is the ratio of noise to signal, it has three types: small feature, large feature and visual feature[9-10]. Due to small residual stress characteristics, the smaller the test results, the larger the S /N.The formula for calculating the S/N of small features is shown in (3).

$$\frac{S}{N} = -10Lg\left(\frac{1}{n}\sum_{i=1}^{n} y_i^2\right) \qquad (3)$$

S/N — Signal to noise ratio(dB), y_i—Results of experiment i, n—Number of repetitions per trial, n=1.

In order to observe the effect of various experimental factors on the quality index at different levels, average analysis and range analysis of Mises-Hencky residual stress after S/N ratio transformation of the test results are required, the results are shown in table 4.

Table4. *S/N* average and range of Mises-Hencky residual stress

Levels	Factors						
	A	*B*	*C*	*D*	*E*	*F*	*G*
Average 1	-28.591	-28.650	-28.297	-28.410	-28.178	-28.369	-28.427
Average 2	-28.329	-28.116	-28.239	-28.078	-28.356	-28.347	-28.294
Average 3	-27.476	-28.068	-28.298	-28.246	-28.299	-28.120	-28.112
Range *k*	1.115	0.582	0.059	0.332	0.178	0.249	0.315

According to the calculation formula of S/N ratio of small feature (3), it is a minus function. the higher the S/N, the smaller the corresponding quality index. When mold temperature is at level 3(70℃), melt temperature is at level 3(240℃), pressure packing is at level 2(6s), compression delay time is at level 2(1s), compression force is at level 1(150MPa), compression distance is at level 3(3mm), compression speed is at level 3(0.3cm/s), the Mises-Hencky residual stress of product is minimal,it indicates that $A_3B_3C_2D_2E_1F_3G_3$ is the best process parameter combination to control residual stress. The residual stress distribution of $A_3B_3C_2D_2E_1F_3G_3$ process parameters are shown in Figure 2, the maximum residual stress is 21.24MPa, which is 30.41% lower than the maximum residual stress (30.52MPa) in the experiment scheme.

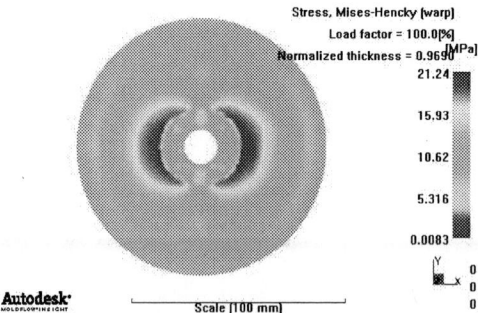

Figure2. Residual stress after optimization

In addition, according to the range *k* of different factors, the order in which the effect of injection compression process parameters on the residual stress of the product can be obtained from large to small is: Mold temperature > Melt temperature > Compression delay time > Compression speed >

Compression distance > Compression force >Pressure packing time.

3.4 Single factor numerical simulation analysis
According to the results of S/N analysis, four key injection-compression process parameters effect on residual stress more obviously were selected(mold temperature, melt temperature, compression delay time, compression speed),the effect of each key technological parameter on the residual stress of products was studied by single factor experiment, according to the range of parameters of the injection machine and the range of molding materials, the level of each process parameter is set as shown in table 5.

Table5. Single factors and levers table

Factor / Level	Mold temperature/℃	Melt temperature/℃	compression delay time/s	compression speed/cm/s
1	30	220	0	0.1
2	40	230	0.5	0.2
Reference	50	240	1	0.3
3	60	250	1.5	0.4
4	70	260	2	0.5

4. Results and discussion
Effect law of the key injection compression process parameters (mold temperature, melt temperature, compression delay time and compression speed) on the residual stress of the product was obtained based on the above numerical simulation and orthogonal experiment.

4.1 Effect of mold temperature on residual stress
Residual stress of product at different mold temperatures is shown in figure 3, residual stress of disk is decreasing gradually. When the temperature from 30℃to 70℃, the residual stress decreased from 29.12 MPa to 22.70 MPa, was reduced by 22.05%. This is mainly because the high temperature melt enters the low temperature mold cavity, the plastic melt will rapidly undergo condensation and hardening. With the increase of mold temperature, the condensation layer will become thinner, cooling efficiency will decrease, and the cooling and solidification time of melt will be longer, which can release the internal stress on the surface of the product more fully, thus the residual stress of the product will decrease.Residual stress distribution with mold temperature of 30℃and 70℃ were shown in figure 4 (a) and (b).

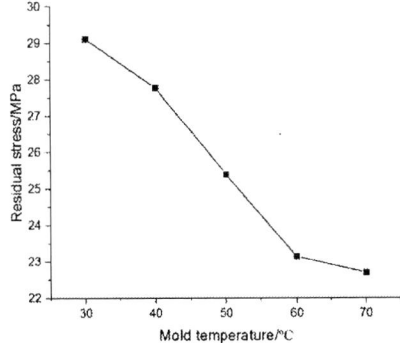

Figure3. Effect of mold temperature on residual stress

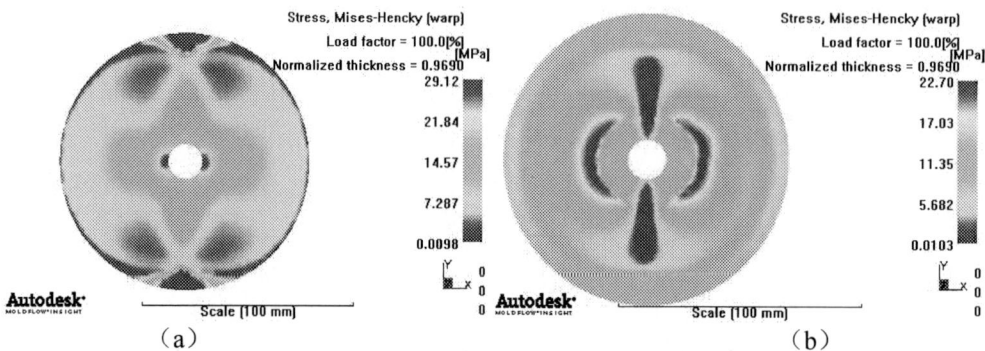

Figure4. Residual stress distribution at different mold temperature

4.2 Effect of melt temperature on residual stress

Residual stress of product at different melt temperatures is shown in figure 5. The residual stress of the disk is decreasing gradually. When the temperature from 220℃ to 260℃, the residual stress decreased from 28.34 MPa to 22.47 MPa, was reduced by 20.71%. According to the PVT property of the material (PMMA), melt temperature rises and thermal contraction is obvious under certain other conditions, On the other hand, the increase of melt temperature improves the fluidity of the melt, so that the solidification time of the melt is prolonged, which results in the increase of feeding amount in the pressure maintaining stage, the decrease of volume contraction and the decrease of residual stress.

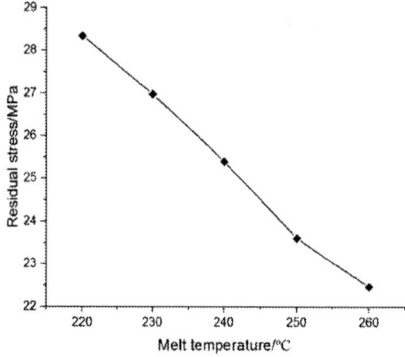

Figure5. Effect of melt temperature on residual stress

4.3 Effect of compression delay time on residual stress

Residual stress of product at different compression delay time was shown in figure 6. The residual stress of the disk shows a gradual upward trend, increasing from 24.17MPa at 0s to 39.28MPa at 2.0s. With the extension of compression delay time, the molecular orientation increases, and the thickness of the condensate layer at the cavity interface increases, In the secondary compression process, the compression mechanism in the cavity affected the equilibrium distribution of stress, the residual stress increased finally[8]. In addition, when the compression delay time is too long, defects such as short shot will be produced due to the gradual condensation of melt, thus affecting the forming quality. Residual stress distribution with compression delay time of 0 s and 3 s are shown in figure 7 (a) and (b), there is obvious short shot defect in the gray area of (b).

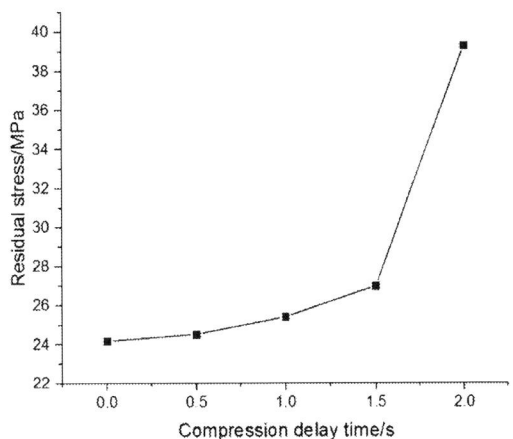

Figure6. Effect of compression delay time on residual stress

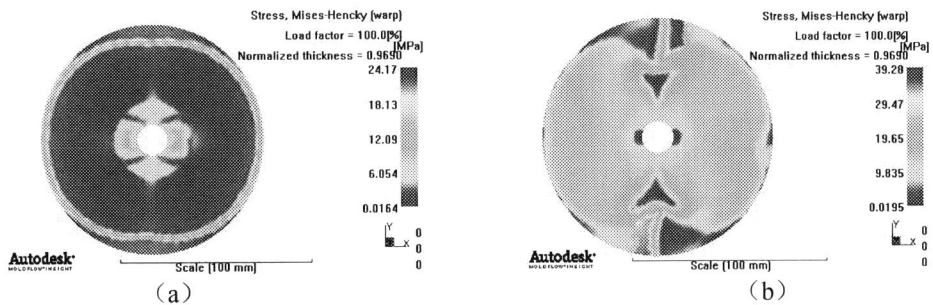

（a）　　　　　　　　　　　　　　　（b）

Figure7. Residual stress distribution at different compression delay time

4.4 Effect of compression speed on residual stress

Residual stress of product at different compression speeds is shown in figure 6. Generally, the residual stress of the disc is increased, but the change is not obvious, 25.38MPa at 0.1cm/s increased by 25.50MPa at 0.5cm/s, the main reason is that compression speed affects the speed and time of secondary flow of plastic melt in mold cavity, with the reduction of compression speed, there are more melts near the gate, the melt is more compact and the internal stress of the product is less,when the compression speed increases, the amount of plastic melt flowing into the end from the gate changes little, so it has little effect on the residual stress of products.

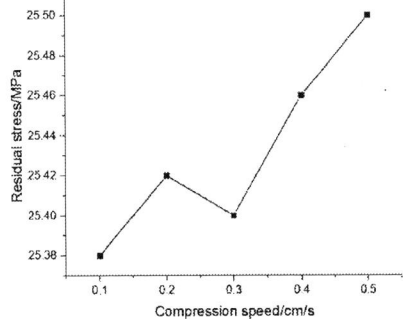

Fig.8 Effect of compression speed on residual stress

5. Conclusions

Effect of different injection compression process parameters on the residual stress of the disc were analyzed in detail through numerical simulation analysis, and the following results were obtained finally:

(1) The degree of effect of injection compression process parameters on the residual stress of the product is from large to small: Mold temperature > Melt temperature > Compression delay time > Compression speed > Compression distance > Compression force >Pressure packing time, mold temperature, melt temperature, compression delay time and compression speed are the main factors affecting the residual stress of products.

(2) The effect of four key process parameters on the residual stress of products was studied by using the single factor method, the residual stress decreases with the increase of mold temperature and melt temperature, increases with the increase of compression delay time, and the effect of compression speed on residual stress is not obvious.

References

[1] G.D. Xi, H.M. Zhou, D.Q. Li, Influence of injection molding parameters on residual stress and shrinkage of products.J. Journal of chemical industry, 58(2007).

[2] K.J. Wang, G.R. Cao, G.W. Yang. Residual stress in injection molding of polycarbonate optical products.J. Polymer materials science and engineering, 34(2018)

[3] W.J. Liu, X.Y. Wang, Z.Li, Optimization analysis of residual stress of injection molded parts based on Kriging agent model.J. Plastics industry, 43(2015)

[4] W.C. Bushko, V.K. Stokes, Solidification of thermoviscoelastic melts(1): Formulation of model Problem.J. Polym.Eng.Sci., 35(1995)

[5] W.C. Bushko, V.K. Stokes, Solidification of thermoviscoelastic melts(2): Effect of processing conditions on shrinkage and residual stresses.J. Polym.Eng.Sci., 35(1995)

[6] Q.Lin, Study on improvement of shear stress and residual stress of injection parts.J. Journal of datong university of shanxi (natural science edition),27(2011)

[7] X.H. Huang, M.Q. Yu, C.X. Zhou, Research progress of polymer injection residual stress.J.Polymer materials science and engineering,34(2018)

[8] L.L. Sun, B.Y. Jiang, L. Chen, Microlens arrays inject the residual stress of compression molding[J]. Journal of central south university (natural science edition), 48 (2017)

[9] L.J. Chen, H.S. Wang, Optimization of the Taguchi method for the process parameters of two-cavity asymmetric thin-walled injection molded parts.J. Plastics industry, 40(2012)

[10] R.L. Liu, Z.H. Hu, Study on multi-objective optimization of process parameters of thin wall injection molded parts based on CAE.J. Plastics science and technology, 42(2014)

Multiphysics Modelling of Warm Shot Peening of AISI 4140 Steel

Wang Cheng, Wang Long[a] and Wang Chuanli

School of Mechanical Engineering, Anhui University of Science and Technology, 168 Taifeng Road, Anhui, Huainan, 232001, China

[a]Corresponding author: nuaawl@126.com

Abstract. Aiming at the effect of the elevated temperature on shot peening, the multiphysics modeling of warm shot peening is carried out by coupling the processes of heat transfer and shot peening. The temperature field and thermal stress field resulted from the analysis of heat transfer are imported into the model of shot peening to simulate the process of warm shot peening. The obtained results show that the maximum temperature is located in the subsurface layer after multiple shot impacts under 100% peening coverage, and an obvious temperature gradient can be found along the material depth; with the increase of heat flux load, the resultant compressive residual stresses in the surface and subsurface decrease, while the depth of the compressive residual stress increases; the peened surface roughness increases only if the heat flux density exceeds a critical value; the predicted residual stresses in the cases of $q = 1 \times 10^7$ W/m^2 and $q = 2 \times 10^7$ W/m^2 are in good agreement with the experimental results.

1. Introduction

Shot peening (SP) is a well-established mechanical surface treatment widely used in the aerospace and automobile industries, which is mainly aimed at improving the fatigue life of metallic components under the service environment [1-2]. In the process of shot peening, a large number of shots with high velocities impact the surface of a metallic component randomly. The indentations surrounded by plastic regions followed by elastic regions are produced in the peened surface layer. The elastic-plastic deformation results in the beneficial compressive residual stresses which can effectively enhance the resistance of the metallic component exposed to fatigue loading [3]. The experimental studies of shot peening have been existed for a long time, and got the rich achievements [4]. With the development of the finite element method and computational power, numerical modeling of shot peening process is accepted by many researchers [5-6]. When compared with the experimental study, numerical simulation costs smaller and studies the mechanism of shot peening more conveniently. In recent years, a lot of models [7-9] of shot peening are proposed to predict the shot peening results, such as residual stress, surface roughness, and work hardening. The numerical study of shot peening effectively promotes the development of shot peening technology.

Warm shot peening (WSP) is thermomechanical treatment technique deriving from the conventional shot peening on which the shot peening process is carried out at elevated temperature [10-11]. Schulze et. al [12-13] built a new device to conduct the experiments of warm shot peening, as shown in Fig.1, the warm shot peening nozzle mixes the hot blast air and shot flow to heat the metallic component in the process of shot peening. Research shows that the larger plastic region and depth of compressive residual stresses can be obtained by warm shot peening [12-14]. However, to

Content from this work may be used under the terms of the Creative Commons Attribution 3.0 licence. Any further distribution of this work must maintain attribution to the author(s) and the title of the work, journal citation and DOI.

Published under licence by IOP Publishing Ltd

our best knowledge, it is a pity that there are little literatures reported on the numerical modeling of warm shot peening. Therefore, the multiphysics modeling of warm shot peening is performed in this paper, and the corresponding products: temperature filed, residual stress field and surface roughness are studied in detail.

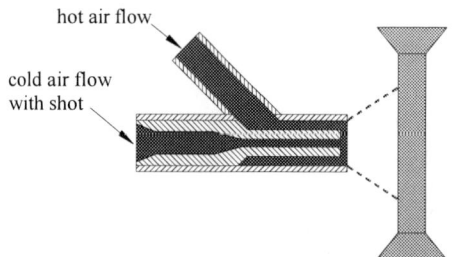

Fig. 1 Warm shot peening of AISI 4140 steel [12-13]

2. Modeling of warm shot peening

2.1 Heat transfer analysis

In order to simulate the heat transfer process of warm shot peening, a three dimensional cylinder-shape finite element model with the diameter of 10mm and height of 4mm is developed, as shown in Fig. 2. Eight-node linear heat transfer brick elements (DC3D8) are used to mesh the finite element model, and the finest element size is 40μm. The initial temperature of the three dimensional model is 300K, and the heat flux load is applied on the center area with the dimension of $2mm \times 2mm$ which is located on the model's top surface. Two constant heat flux densities are $1 \times 10^7 \, W/m^2$ and $2 \times 10^7 \, W/m^2$ respectively. The transient temperature field can be obtained by

$$\frac{\partial T}{\partial t} = \frac{\lambda}{\rho \cdot c} \cdot \nabla^2 T \qquad (1)$$

where T is the temperature field, t is time, λ is the thermal conductivity ($\lambda = 42 \, W/K/m$ for AISI 4140 steel), ρ is density ($7850 \, kg/m^3$), c is specific heat ($580 \, J/(kg \cdot K)$) and ∇ represents the gradient. The convective heat transfer between the outer surface of the cylinder-shape model and the ambient air is conducted by

$$q = h_c \left(T_s - T_a \right) \qquad (2)$$

where q is the heat flux density, h_c is convective heat transfer coefficient, T_s is the temperature of the model's outer surface and T_a is the ambient temperature (300K). The effect of the thermal radiation is ignored.

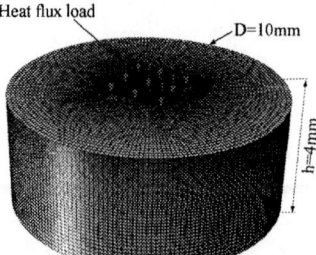

Fig. 2 Heat transfer model

Fig. 3 shows the distributions of temperature and thermal stress induced by the heat flux load before shot peening. The maximums of temperature and thermal stress, which increase with the increasing heat flux density, are both on the top surface of the finite element model, and the obviously gradient can be found. The obtained temperature field and thermal stress field are simultaneously imported into the shot peening model by the means of analytic field, to simulate the process of warm shot peening.

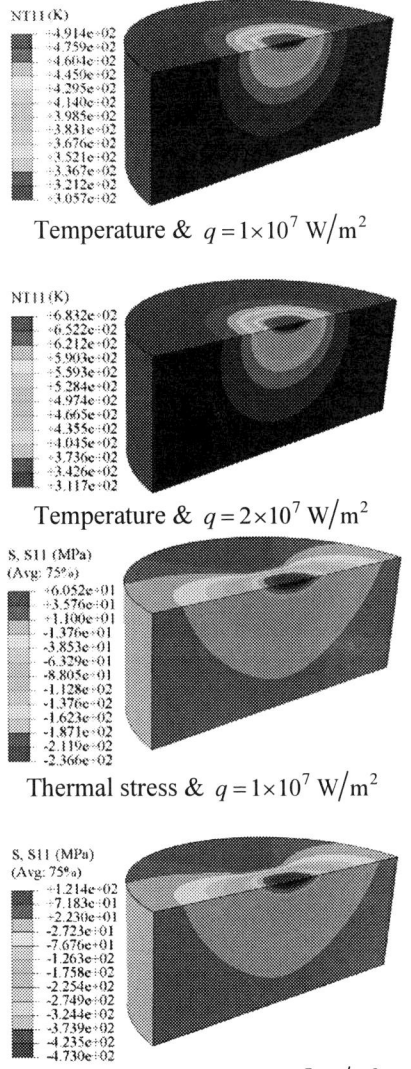

Temperature & $q = 1 \times 10^7 \, \text{W}/\text{m}^2$

Temperature & $q = 2 \times 10^7 \, \text{W}/\text{m}^2$

Thermal stress & $q = 1 \times 10^7 \, \text{W}/\text{m}^2$

Thermal stress & $q = 2 \times 10^7 \, \text{W}/\text{m}^2$

Fig. 3 Gradient distribution of temperature and thermal stress under the heat flux load

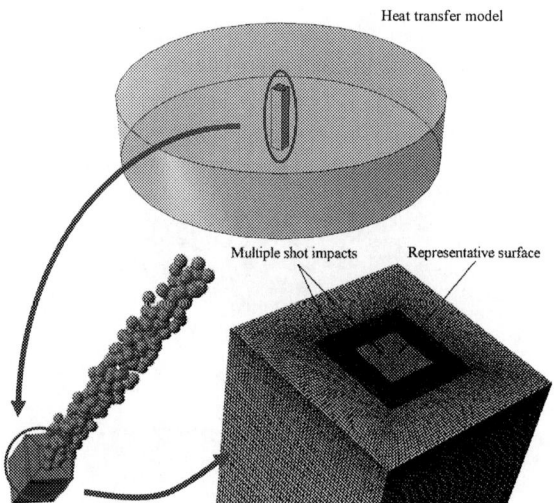

Fig. 4 Warm shot peening model

2.2 Shot peening model

A cuboid-shape finite element model with the dimension of $2mm \times 2mm \times 4mm$, taken from the cylinder-shape heat transfer model, is used to simulate the process of shot peening, as shown in Fig. 4. The shot with the diameter of 0.43mm is treated as the rigid body, considering that little deformation is produced when compared with the cuboid-shape model. The shot initial velocity is applied to the reference point of the rigid body which is at the shot geometric center, and the value of the initial velocity is estimated by [15]

$$v = \frac{16.35 \times p}{1.53 \times k + p} + \frac{29.50 \times p}{1.196 \times R + p} + 4.83 \times p \qquad (3)$$

where p represents the jet pressure (bar), k is the shot mass flow (kg/min) and R is the shot radius (mm). According to the given experimental conditions [12-13], the shot initial velocity is computed as 37.3m/s by Eq. (3). The center area with the dimension of $1mm \times 1mm$ on the top surface of the cuboid-shape model is vertically impinged by the shots, and the bottom surface is completely constrained. Eight-node linear brick elements with reduced integration and hourglass control (C3D8R) are used to mesh the cuboid-shape finite element model, and the finest element size is 10μm to simulate the very high gradient stress and strain.

Two hundred shots are created to simulate the process of shot peening under 100% peening coverage of the treated area, as shown in Fig. 4. The shot peening coverage is defined as the ratio of the area covered by peening indentations to the total treated surface area. [16] The generation of these shots is flexibly constrained by a linear distribution function between two shot centers [17]

$$P_c = \begin{cases} 100\% \times \dfrac{l}{2r} & l \leq 2r \\ 100\% & l > 2r \end{cases} \qquad (4)$$

where P_c represents the probability of shot generation, l is the distance between the randomly generated shot center and the prior shot center, r is the indentation's radii induced by single shot impact. Fig. 5 shows the shot peening coverage with and without the flexible constraint for the experimental conditions [18]. The coverage without the flexible constraint means that the distribution of indentations produced by multiple shot impacts on the peened surface is completely randomly. Obviously, the predicted coverage with the flexible constraint is more close to the experimental results.

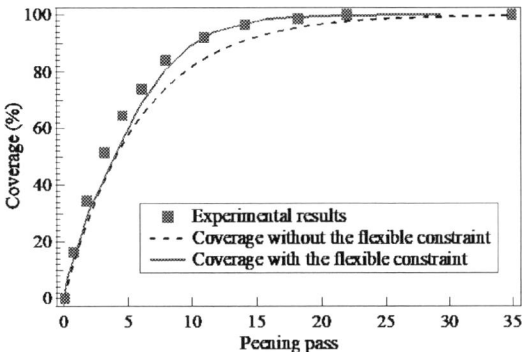

Fig. 5 Shot peening coverages with and without the flexible constraint

Based on the shot peening model, the simulation of warm shot peening process is performed by importing the temperature field and thermal stress field resulted from the heat transfer simulation into the shot peening model as the initial conditions. In order to study the influence of heat flux load on the shot peening, Johnson-Cook model is employed to calculate the dynamic flow stress, which is related to the temperature (T), equivalent plastic strain ($\bar{\varepsilon}^p$) and equivalent strain rate ($\dot{\bar{\varepsilon}}^p$), i.e.

$$\sigma_f = \left[A + B \left(\bar{\varepsilon}^p \right)^n \right] \cdot \left[1 + C \ln \left(\frac{\dot{\bar{\varepsilon}}^p}{\dot{\bar{\varepsilon}}_0^p} \right) \right] \cdot \left[1 - \left(\frac{T - T_r}{T_m - T_r} \right)^m \right] \quad (5)$$

where A, B, C, n and m are material constants, $\dot{\bar{\varepsilon}}_0^p$ is the reference strain rate, T_r is the room temperature and T_m is the melting point. For AISI 4140 steel, A=594Mpa, B=615Mpa, C=0.023, m=0.142, n=1.1611, $T_m = 1800K$, $T_r = 300K$ [19].

3. Result and discussions

3.1. Temperature field
The temperature increment caused by the plastic deformation induced by multiple shot impacts can be calculated by

$$\Delta T = \frac{\eta}{\rho \cdot c} \int_0^{\bar{\varepsilon}^p} \bar{\sigma} d\bar{\varepsilon}^p \quad (6)$$

where η is the converting efficiency from plastic work to heat (ranging generally between 0.9 and 1.0), $\bar{\sigma}$ is the equivalent stress.

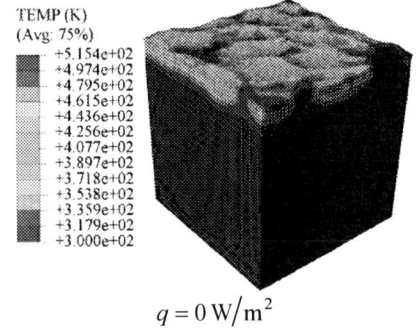

$q = 0 \, \text{W/m}^2$

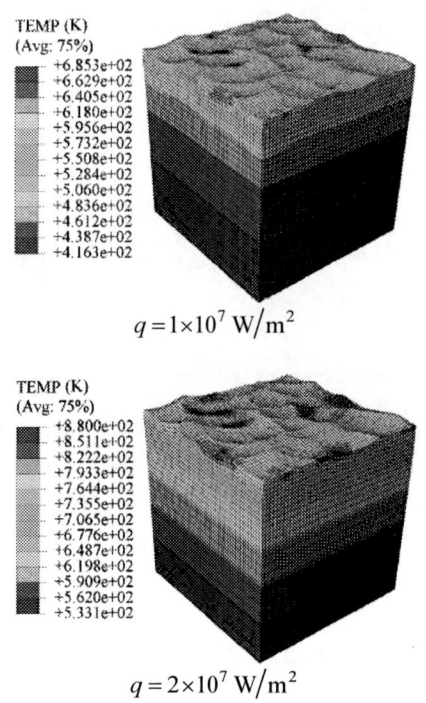

$$q = 1 \times 10^7 \ \mathrm{W/m^2}$$

$$q = 2 \times 10^7 \ \mathrm{W/m^2}$$

Fig. 6 Temperature fields induced by warm shot peening

Fig. 6 shows the temperature field of the representative region with the dimension of $0.5\mathrm{mm} \times 0.5\mathrm{mm} \times 0.5\mathrm{mm}$ induced by multiple shot impacts under different heat flux loads. With the increase of heat flux load, the temperature in the surface and subsurface increase significantly, which is attributed to the severe plastic deformation. Fig. 7 shows the distribution of the temperature along the depth, and the in-depth temperature is the area-averaged value. The maximum temperatures corresponding to the heat flux loads are all located in the subsurface with the distance of 0.1mm from the top surface, and then the temperature decreases along the depth direction, which shows the obvious temperature gradient.

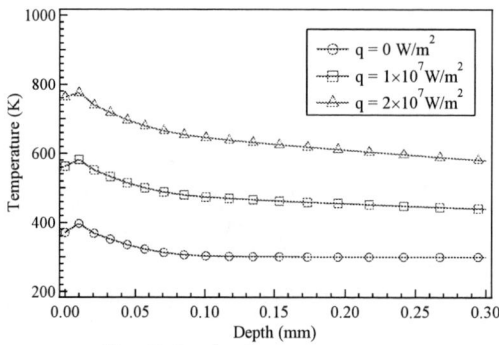

Fig. 7 In-depth temperatures

3.2 Surface roughness

Surface roughness is one of the most common parameters used for evaluation of the shot peening. Peak-to-valley roughness(PV) is defined with Eq. (7) as the distance between the highest peak (R_p) and the lowest valley (R_v) within the sampling length

$$\mathrm{PV} = R_p + R_v \tag{7}$$

The surface topographies after multiple shot impacts under different heat flux loads are shown in Fig. 8, and the z direction displacements of the nodes on the representative area with the dimension of $0.5\text{mm} \times 0.5\text{mm}$ are used to evaluate the peened surface roughness [20]

$$PV = \max(U_z) - \min(U_z) \qquad (8)$$

where $\max(U_z)$ and $\min(U_z)$ are the highest peak and the lowest valley within the reference area. From Fig. 8, the values of PVs are 0.019mm in the case of $q = 0\,\text{w/m}^2$, 0.019mm in the case of $q = 1 \times 10^7\,\text{w/m}^2$, and 0.021mm in the case of $q = 2 \times 10^7\,\text{w/m}^2$, respectively. It is therefore concluded that the PV would increase with the increasing temperature only if the heat flux load exceeds a critical value.

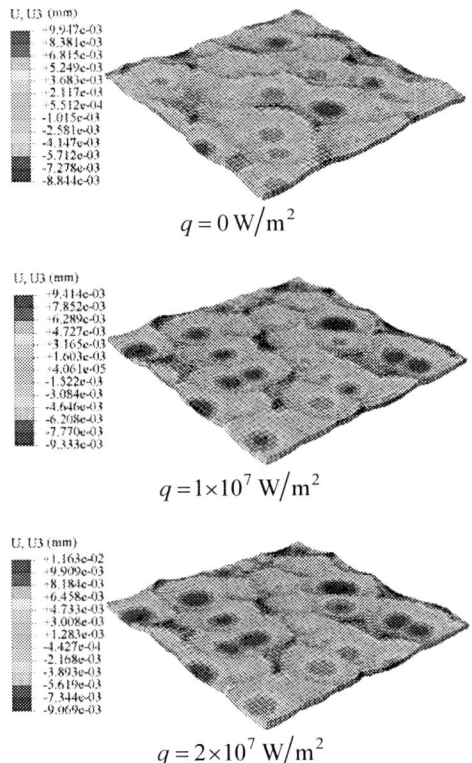

$q = 0\,\text{W/m}^2$

$q = 1 \times 10^7\,\text{W/m}^2$

$q = 2 \times 10^7\,\text{W/m}^2$

Fig. 8 Surface topographies of the reference area after multiple shot impacts

3.3 Residual stresses

As result of the elastic springback of material surrounding the plastic region induced by multiple shot impacts, the residual stress field is produced, as shown in Fig .9. The compressive residual stresses are mainly located in the peened surface and subsurface, and the maximum compressive residual stress decreases with the increase of heat flux load. It should be noted that some tensile residual stresses are produced in the peened surface, which are related to the uneven plastic deformation and surface roughness. In order to study the effect of heat flux load on the resultant residual stresses, the comparisons of the distributions of the area-averaged residual stresses are shown in Fig. 10. It can be clearly seen that, with the increase of heat flux load, the compressive residual stresses in the peened surface and subsurface decrease, while the depth of the compressive residual stresses increases. The predicted residual stresses in the cases of $q = 1 \times 10^7\,\text{w/m}^2$ and $q = 2 \times 10^7\,\text{w/m}^2$ are in good agreement with the experimental results [12-13], which verifies the effectiveness of multiphysics modelling of warm shot peening.

ISPECE   IOP Publishing

IOP Conf. Series: Journal of Physics: Conf. Series **1187** (2019) 032032   doi:10.1088/1742-6596/1187/3/032032

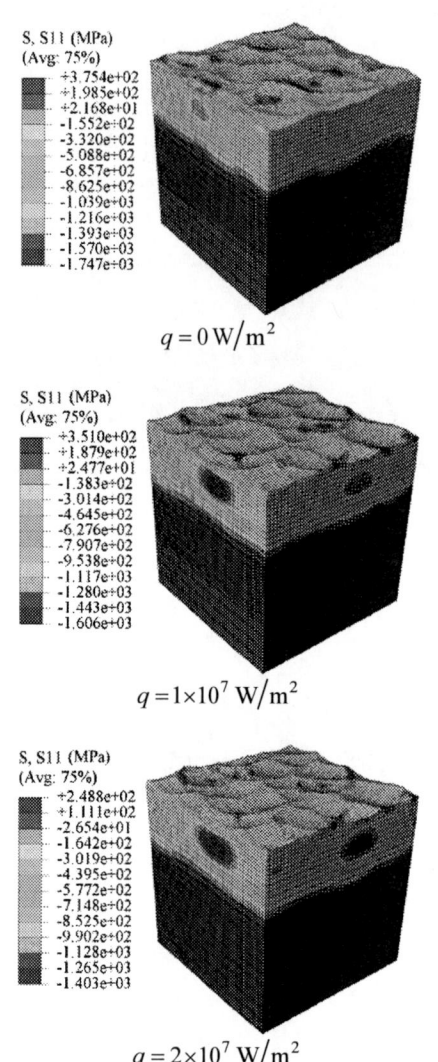

Fig. 9 Residual stress fields induced by warm shot peening

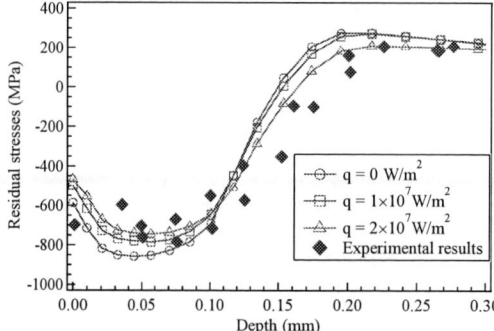

Fig. 10 In-depth residual stresses

4. Conclusions

By coupling the analysis of heat transfer and simulation of shot peening, multiphysics modeling of warm shot peening is carried out, and the obtained conclusions are drawn as following:

(1)   The 100% shot peening coverage with the flexible constraint is more close to the experimental results.

(2)   The maximum temperature induced by warm shot peening is located in the subsurface layer, and the obvious temperature gradient can be found along the depth direction of the peened material.

(3)   The peened surface roughness would increase only if the heat flux load exceeds a critical value.

(4)   With the increase of heat flux load, the compressive residual stresses in the surface and subsurface layers decrease, while the depth of compressive residual stress increases.

(5)   The predicted residual stresses in the cases of $q = 1 \times 10^7 \ \mathrm{W/m^2}$ and $q = 2 \times 10^7 \ \mathrm{W/m^2}$ are in good agreement with the experimental results.

Acknowledgements

This project is supported by National Natural Science Foundation of China (Grant No. 51705002), the Natural Science Foundation of Anhui Province of China (Grant No. 1708085QE123).

References

[1]   E. Maleki, O. Unal, K.R. Kashyzadeh. Surf. Coat. Tech. **344,** 62 (2018)
[2]   T. Klotz, D. Delbergue, P. Bocher, et al. Int. J. Fatigue **110**, 10 (2018)
[3]   M. Chen, H. Liu, L. Wang, et al. Surf. Coat. Tech. **344**, 132 (2018)
[4]   E. Nordin, B. Alfredsson. Exp. Techniques **41**, 433 (2017)
[5]   J. Zhang, S. Lu, T. Wu, et al. Adv. Eng. Softw. **115**, 283 (2018)
[6]   A. Gariépy, H.Y. Miao, M. Lévesque. Adv. Eng. Softw. **114**, 121 (2017)
[7]   F. Tu, D. Delbergue, H. Miao, et al. Surf. Coat. Tech. **319**, 200 (2017)
[8]   D.Y. Hu, Y. Gao, M. Fanchao, et al. Chinese J. Aeronaut. **30**, 1592 (2017)
[9]   C. Wang, J.C. Hu, Z.B. Gu, et al. Chin. J. Mech. Eng. **30**, 344 (2017)
[10] W. Xin, Z. Tao, H.H. Zhao, et al. Rare Metal Mat. Eng. **47**, 1668 (2018)
[11] C. Wang, C. Jiang, M. Chen, et al. Mat. Sci. Eng. A **707**, 629 (2017)
[12] A. Wick, V. Schulze, O. Vöhringer. Mat. Sci. Eng. A **293**, 191 (2000)
[13] R. Menig, V. Schulze, O. Vöhringer. Mat. Sci. Eng. A **335**, 198 (2002)
[14] Y. Harada, K. Mori. J. Mater. Process. Tech. **162**, 498 (2005)
[15] L. Xie, J. Zhang, C. Xiong, et al. Mater. Design **41**, 314 (2012)
[16] T.Q. Pham, N.W. Khun, D.L. Butler. Surf. Eng. **33**, 687 (2017)
[17] C. Wang, L. Wang, X. Wang, et al. Int. J. Mech. Sci. **146**, 280 (2018)
[18] H.Y. Miao, D. Demers, S. Larose, et al. J. Mater. Process. Tech. **210**, 2089 (2010)
[19] M. Agmell, A. Ahadi, J.E. Ståhl. Mech. Mater. 77, 43 (2014)
[20] H.Y. Miao, S. Larose, C. Perron, et al. Adv. Eng. Softw. **40**, 1023 (2009)

Study on the control system of Agaricus bisporus picking robot

Hu Xiaomei[1,a], Pan Zhaoren[1], Yang Shuzhen[1,2] and Yu Tao[1,2]

[1]The Key Laboratory of Intelligent Manufacturing and Robotics, School of Mechatronic Engineering and Automation, shanghai University, shanghai, China

[2]School of Mechatronic Engineering and Automation, Shanghai Second Polytechnic University, shanghai, China

[a] Corresponding author: sufeimasohxm@163.com

Abstract. At present, the Agaricus bisporus industrialized cultivation has formed in the world, and fertilization, feeding, humidity control has been automated. However, no automation has been achieved in the human-like picking process. Therefore, an A. bisporus picking robot is designed and the control system is put forward. The control system of A. bisporus picking robot takes the motion control card as the core and Windows10 as the operating system. The upper computer receives a row of A. bisporus photographs taken by the camera and obtains A. bisporus coordinate points through image processing, and the stepper motor and DC motor are driven by the terminal board of the motion control card to pick the A. bisporus. Moreover, the parallel control system is designed to implement image processing and A. bisporus picking in order to improve the efficiency of A. bisporus picking robot. In addition, semi-closed loop system of stepping motor is adopted to compensate for the error caused by multi-step walking, and the control method of two-step walking of A. bisporus picking robot is proposed to solve some A. bisporus that can not be picked up. The experiment results show that the control system of A. bisporus picking robot runs smoothly and efficiently, and it has very strong application value.

1. Introduction

In recent years, with the rapid development of automation industry, people get rid of the tedious and arduous work, and robots instead of people complete the complex tasks. However, hand picking of A. bisporus is still a main way in the A. bisporus industrialized cultivation process. A. bisporus picking robots can improve work efficiency and reduce labour costs for enterprises, which makes it become a research hotspot.

However, A. bisporus grows overlapping and interconnected, and is easy to be damaged [1]. At present, Research groups in Australia and the United Kingdom have designed the robots to pick A. bisporus, but these robots are still in the development stage.

By our A. bisporus picking tests, picking process is determined and the A. bisporus picking robot is designed. The control system is implemented according to A. bisporus picking process.

2. Work principle

Figure 1 shows platform architecture diagram of the A. bisporus picking robot. It has three areas: (1) vision area is primarily a mobile camera that takes a region of A. bisporus and is driven by a stepper motor with an encoder; (2) picking area is mainly a collecting device that can move the X (long) and

Y (wide) axes, and is also driven by a stepping motor with an encoder. The Z-axis drop is driven by a stepper motor. And through the vacuum pump and the DC motor, the suction cup picks up the A. bisporus by suction and rotation; (3) auxiliary area is used to place various hardware devices, such as electrical cabinet, industrial control computer, vacuum pump and other components.

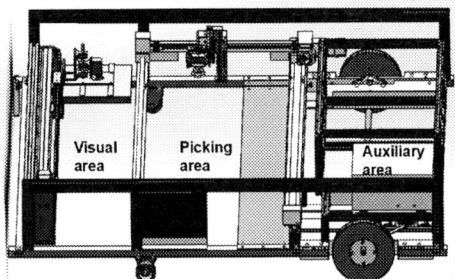

Figure 1. platform architecture diagram of the A. bisporus picking robot

2.1 Robot control process

The control process of A. bisporus picking robot is as follows:

(1) The initialization of parameters is to eliminate the alarm information of emergency stop, clear the variable parameter, and all stepper motors returns to their coordinate origin through the travel switch. After initialization, image processing and A. bisporus picking control process enter their threads [2].

(2) In the thread of image processing, the camera performs shooting and image processing after reaching the specified position. The centre coordinates of the mushroom point are obtained.

(3) A. bisporus picking control process thread gets the array of centre coordinate points of the last shot of the A. bisporus, and the picking mechanism is moved to the corresponding position by extracting a single A. bisporus coordinate point, and it is picked up by rotation and suction. If it is picked up, A. bisporus is moved to the storage location, and then picking mechanism moves to the next point. If the A. bisporus are all picked, the picking mechanism returns to the reference point.

(4) When both threads have finished running, the robot moves forward 140mm until it has reached the end. The A. bisporus coordinates obtained from the last image processing are passed to the picking mechanism.

2.2 Control method of two-step walking

The camera's field of vision is about 240*180mm. The average diameter of the A. bisporus that can be picked is about 40mm. Because the camera edge is distorted greatly and the edge of A. bisporus is usually incomplete, the centre coordinates of these A. bisporus obtained by image processing are inaccurate. According to the distortion experimental data and diameter of the A. bisporus that can be picked, the effective visual area is 200mm * 140mm. So to ensure that all the A. bisporus are captured by the camera, the picking robot can only move forward 140mm each time. And the width of the picking area(140mm) must be the same as the width of the visual effective area, so that the A. bisporus coordinate points in the two areas can correspond to each other.

One-stage walking scheme is to move the robot forward for 140mm after the camera has photographed the A. bisporus, and then the picking mechanism in the picking area will pick the A. bisporus. But this scheme would result in some A. bisporus being unable to be picked. As shown in Fig. 2, the picking area after the move can not coincide with the effective area before the move.

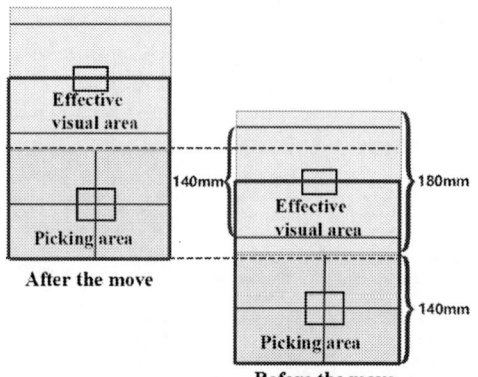

Figure 2. One-stage walking scheme

In order to solve this problem, we adopt a two-stage walking scheme. For this reason, it is necessary to add a blank area between the effective visual area and the picking area, as shown in Figure 3. The effective visual area before the movement can completely coincide with the picking area of the second movement, which ensures that all the A. bisporus are picked up.

The advantage of the two-stage walking scheme is that not only will A. bisporus not be missed, but the blank area can give more mechanical space to install some mechanisms and components.

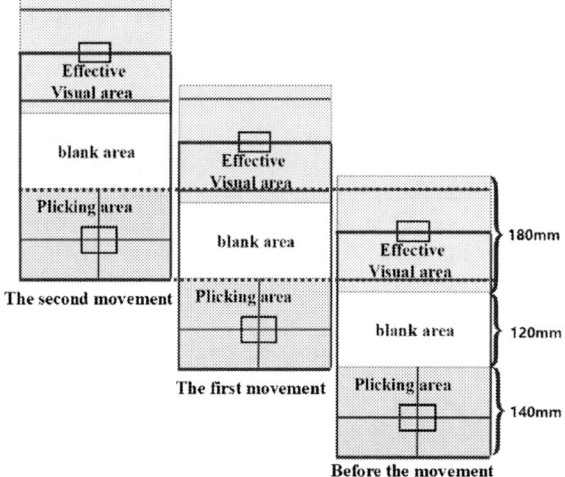

Figure 3. Two-stage walking scheme

3. Control system circuit design

3.1 Integrated system and circuit design

The system is mainly composed of upper computer （PC）, camera, motion control card, touch screen, motor system and so on. The whole system is controlled by the C++ language of the VS platform of the upper computer. The platform is equipped with QT software for man-machine interaction and MySQL software to store data from mushroom coordinates [3]. The triggering mode of industrial camera is software triggering, which is connected with the upper computer through Gigabit Network wire, thus ensuring the stability and rapidity of image transmission. Motion control cards are connected by PCI slots to the upper computer, and data transmission and feedback through port line and terminal plate. By calling the function library instructions provided by the manufacturer, they can control various components and receive the feedback I/ O signals (encoder pulse number, stroke switch and photoelectric switching signal). The touch screen is connected with the upper computer through HDMI, which plays the role of human-machine interaction.

3.2 circuit design

3.2.1 Stepper motor circuit design
Compared with servo motors, stepper motors and its stepping driver are smaller in size than servo motors and its server driver at the same torque, enabling them to be mounted on mobile robot.

To ensure the positioning accuracy of stepping motor, PID algorithm is adopted. The computational amount of incremental PID control algorithm is much smaller than that of position control algorithm [4, 5]. Therefore, incremental PID algorithm and differential encoder are used in semi-closed loop control, and through the upper machine repair error. The expression of incremental PID is as follows:

$$\triangle U_k = K_p(e_k - e_{k-1}) + K_I e_k + K_D(e_k - 2e_{k-1} + e_{k-2}) + U_o \quad (1)$$

in the formula: U_k is the computer output value at the k th sampling; e_k is the input deviation value at the k th sampling; e_{k-1} is the input deviation value at the k th sampling; K_p is the proportionality coefficient; K_I is the Integral coefficient; K_D is the differential coefficient; U_o is the original initial value when starting PID control.

Although the PID algorithm can correct the error of a section, the picking mechanism needs to walk several times in order to pick and store mushrooms, which makes the accumulated error may increase until the picking mechanism returns to the coordinate origin. A comparison system is designed for this, as shown in figure 4.

$$Ep = Pp - Ap \quad (2)$$

$$Es = Ep * Pm \quad (3)$$

By comparing the actual pulse (*Ap*) with the planned pulse (*Pp*), the pulse error (*Ep*) is known. The path of each pulse (*Pm*) multiplied by the pulse error can get the path error *(Es)*. When the path error exceeds the error value, the upper computer sends out instructions to make the X axes and Y axes return to reference and then continue to pick, thus ensuring the picking accuracy.

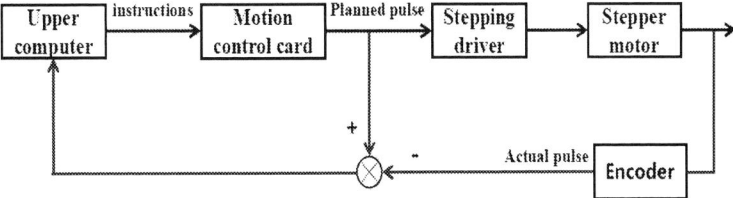

Figure 4. Feedback control model design

3.2.2 DC motor circuit design
The DC motor is used to rotate the sucker so that the mushroom rhizome is disconnected from other rhizomes to increase the success rate of harvesting. Through the terminal board EXO port and two relays to achieve the positive and negative reversal of the DC motor, one end of the motor is connected to port 3 (positive) of No. 2 relay and port 6 (negative) of No. 3 relay, and the other end of DC motor is the same principle. When one of the EXO ports is turned on, the relay is turned on, the positive and negative stages are turned on, and the DC motor is turned on. The wiring mode is shown in Figure 5 of the DC motor positive and negative module.

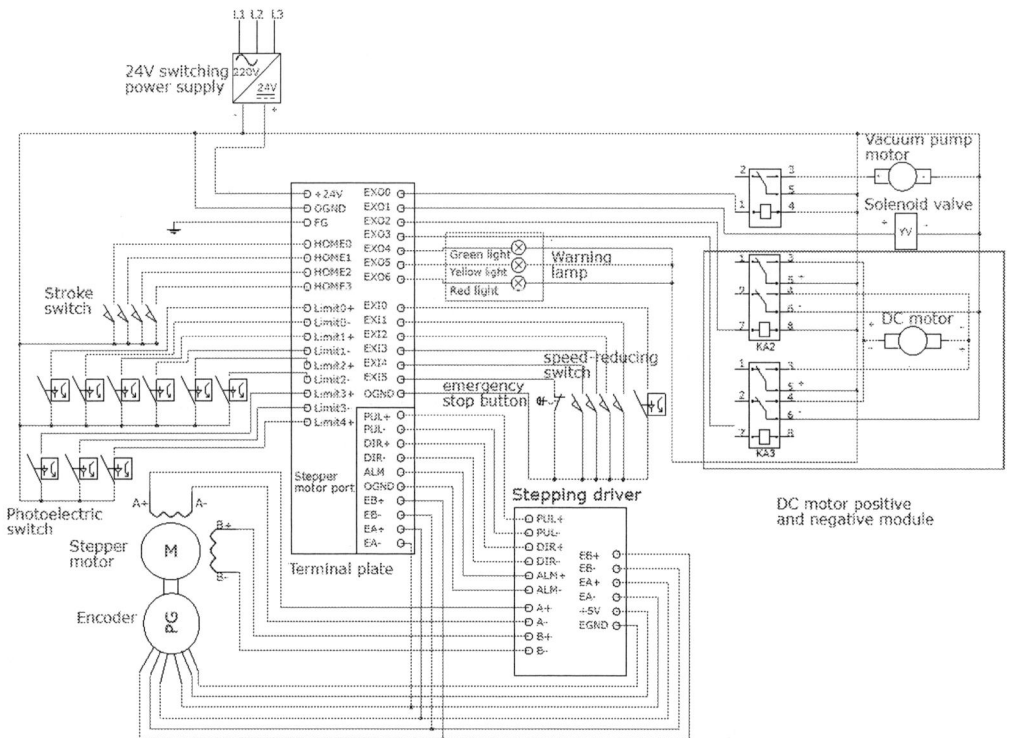

Figure 5. Circuit diagram of A. bisporus picking robot

3.2.3 Component circuit design

In addition to the Z axis, each stepper motor requires two limit switches to ensure that the work slipway driven by the motor does not hit the mechanism. Using photoelectric switch as the limit of work slipway, and the corresponding ports are Limi0 to Limit3. Due to the limited space, the Z axis motor only has a positive limit corresponding to the Limit4 port. The wiring is shown in Figure 5.

Each work slipway needs to return to the original point. Considering the efficiency and accuracy of returning to the origin point, first each work slipway returns to the origin at a faster speed of V1. Then when the slipway hits the speed-reducing switch, the slipway moves slowly at speed V2. Finally, when the slider hits the stroke switch, it moves backwards at V2 speed to the origin point. Therefore, each stepper motor needs a speed-reducing switch and stroke switch to enable the work slipway to return to the origin point accurately and quickly. The corresponding ports of the stroke switch are from HOME0 to HOME3, while the corresponding ports of the speed-reducing switch are from EXI1 to EXI4. The wiring is shown in Figure 5.

An emergency stop button is needed. To be safe, when the emergency stop button is pressed, the power supply is cut off not only by hardware, but also by software. The corresponding port is EXI5. The wiring is shown in Figure 5.

4. Experiments

After the experimental test at the mushroom base, the mushroom picking robot reached the expected requirements, as shown in Figure 6. The success rate of picking a single A. bisporus reached 88%, and the success of picking a group of A. bisporus that overlapped and staggered reached 71%.

Figure 6. Picking experiment of A. bisporus picking robot

The camera can photograph a region of A. bisporus continuously, and the picking mechanism can also pick a region of A. bisporus continuously. Multi-threading of C++11 further improves efficiency.

The two-step walking mode makes the visual coordinate system and the picking coordinate system coincide completely, which ensures that all A. bisporus can be picked up.

The semi-closed loop design of the stepper motor based on PID algorithm ensures the positioning accuracy of the picking mechanism, so that it can pick A. bisporus quickly and accurately.

Based on the VS platform, QT is used as the interface, which makes it possible to interact with the mushroom picking robot and see the current axis and the machine running in real time. When the mushroom picking robot is not working properly, it can give an alarm in time, and the alarm can be removed artificially through the interface.

5. Conclusions

The control system is designed based on the self-developed A. bisporus picking robot. The parallel control system is designed to implement image processing and A. bisporus picking, which improves the efficiency of robots. The semi closed loop system based on PID algorithm has effectively solved the phenomenon of step out and overshoot of stepping motor. Therefore, the A. bisporus picking robot based on this control system has high efficiency and smooth operation, which has great commercial value. Furthermore, the future A. bisporus picking robot can further improve the picking efficiency by adding more than one picking mechanism and reduce the volume of the robot when conditions permit.

References

[1] F. Tarlak, M. Ozdemir, M. Melikoglu, Mathematical modelling of temperature effect on growth kinetics of Pseudomonas spp. on sliced mushroom (A. bisporus),INT J Food Microbiol,266 (2018)

[2] Václav Rek,Ivan Němec. Parallel Computation on Multicore Processors Using Explicit Form of the Finite Element Method and C++ Standard Libraries,P I MECH ENG C-J MEC, 66(2) (2016).

[3] Felipe R. Monteiro,Mário A. P. Garcia,Lucas C. Cordeiro,Eddie B. Lima Filho. Bounded model checking of C++ programs based on the Qt cross‐platform framework, SOFTW TEST VERIF REL, 27(3) (2017).

[4] Nehal M. Elsodany,Sohair F. Rezeka,Noman A. Maharem, Adaptive PID control of a stepper motor driving a flexible rotor, AEJ,50(2)(2010).

[5] Li Jingyi, Wang Huibin, Position control of stepper motor based on PID algorithm, Dual-purpose technology and products, 55-57(2015).

Design and application of visual system in the Agaricus bisporus picking robot

Hu Xiaomei[1,a], Wang Chuan[1] and Yu Tao[1,2]

[1]The Key Laboratory of Intelligent Manufacturing and Robotics, School of Mechatronic Engineering and Automation, Shanghai University, Shanghai, China

[2]School of Mechatronic Engineering and Automation, Shanghai Second Polytechnic University, shanghai, China

[a] Corresponding author: sufeimasohxm@163.com

Abstract. The Agaricus bisporus picking robot has not been commercially applied in the world. Based on the self-developed A. bisporus picking robot, the visual system design is carried out, and aiming at the specificity of the picking object, a measuring method of diameter and center point position based on monocular vision is proposed. Since the unequal heights of the A. bisporus affect the accuracy of visual recognition, the three-dimensional coordinates of the centre point of the A. bisporus are measured by the horizontal movement of the camera and the ellipse-fitting algorithm is proposed to improve the accuracy of position. Moreover, the depth information is used to compensate the error of the diameter measurement of A. bisporus. Through the picking experiment at the A. bisporus planting base, the results prove that the recognition success rate of the visual system is up to 90%.

1. Introduction

Most of A. bisporus planting bases still use labor for picking and sorting in the world. The efficient of manual picking A. bisporus is less, and it is difficult to achieve picking A. bisporus 24/7. Meanwhile, everyone evaluates the quality of A. bisporus by the naked eye, which fatigues people after a long time. In addition, the evaluation criteria of people are different, so that the quality classifications of A. bisporus are not rigorous enough. It is an inevitable trend to develop A. bisporus picking robot to achieve automatic picking and classification. Machine vision technology is a key technology for the A. bisporus picking robot [1].

In recent years, with the rapid development of machine vision technology, it has been widely used in industrial manufacturing, product packaging, logistics, unmanned driving, etc. [2], but the application of machine vision technology in agriculture is full of difficulties and challenges because the growth environment of crops is full of uncontrollable factors[3,4]. A. bisporus have different postures during the growth process. Sometimes, multiple A. bisporus cluster growth and their roots connect together, which will adversely affect the recognition and position of A. bisporus.

According to the growth characteristics of A. bisporus, the A. bisporus picking robot is developed independently and the visual system design is carried out. The visual system measures the three-dimensional coordinates of the centre point of A. bisporus by horizontal movement of monocular camera, and compensates the error of the diameter measurement of A. bisporus by nonlinear model of perspective projection error.

Content from this work may be used under the terms of the Creative Commons Attribution 3.0 licence. Any further distribution of this work must maintain attribution to the author(s) and the title of the work, journal citation and DOI.
Published under licence by IOP Publishing Ltd

2. Design of vision system in A. bisporus picking robot

The overall layout of the picking robot is shown in the Figure 1. There are three areas: visual area, picking area and auxiliary area. The diameter of the A. bisporus and the coordinates of its centre point can be measured in the visual area, and the collecting device can pick up A. bisporus with a diameter larger than 30 mm in the picking area by moving the X (long) and Y (wide) axes. Meanwhile, the positioning error in X and Y coordinate system of the centre point of A. bisporus should be less than or equal to 3 mm to ensure that the device can complete the picking of A. bisporus. Moreover, the diameter measurement error should be less than or equal to 1 mm.

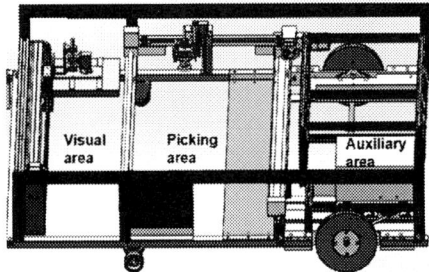

Figure 1. A. bisporus picking robot

2.1. Vision system model of A. bisporus picking robot

According to the principle of the parallel binocular stereo vision [5], the design of the vision system model shown in Figure 2 is made. One camera moves from the position M to the position N along the guideway to obtain two images at different viewpoints, and the distance between the positions M and N is the baseline b for binocular stereo vision.

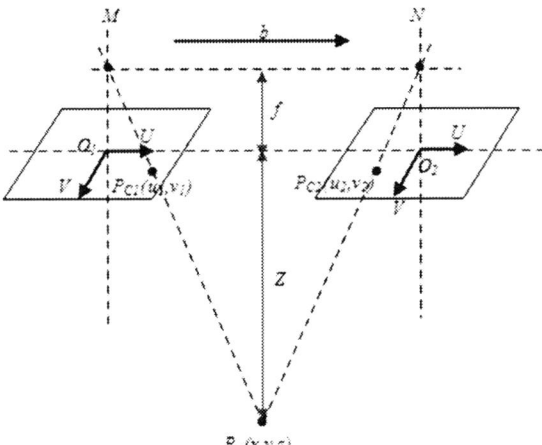

Figure 2. Visual system model of A. bisporus picking robot

The centre point $Pw(x, y, z)$ of an A. bisporus on the culture shelf correspond to $P_{c1}(u_1, v_1)$ and $P_{c2}(u_2, v_2)$ in the image coordinate system of the M and N positions, respectively. If the direction of the guide way is the X-axis direction of the camera coordinate system, then $v_1 = v_2$, the parallax is set as $U = |u_2 - u_1|$, and the focal length is set as f. The coordinates of the centre point of this A. bisporus in the binocular vision system coordinate system are:

$$z = \frac{bf}{U} \tag{1}$$

$$x = \frac{bu_1}{U} = \frac{u_1 z}{f} \tag{2}$$

$$y = \frac{bv_1}{U} = \frac{v_1 z}{f} \qquad (3)$$

2.2 Hardware platform of visual system in A. bisporus picking robot

The visual hardware system mainly includes camera, lens, light source, IPC (Industrial Personal Computer) and their mechanical fixing mechanism. The CPU of IPC is Inter Core i7-3610QE. Its basic frequency is 2.3GHz, the running memory is 4.0GB, and the operating system is Windows 10. The image processing software is Halcon 13, and the images are processed in real-time in Visual Studio 2017. The camera is a 3 million-pixel industrial camera, and the camera's CMOS size is *1/2*″. The focal length of the lens is 4mm. In order to improve the uniformity and quality of illumination, the white light opening backlight is used.

In order to ensure the horizontal movement accuracy of the camera, the sliding table module whose repositioning precision achieving 0.02 mm is adopted, and its effective movement is 1100 mm. As shown in Figure 3, the visual hardware platform is built. The camera and the light source are fixed on the line-glide rail, and the slider moves linearly to perform image acquisition.

Figure 3. The visual hardware platform

The visual system is calibrated by using a 7×7-mark points calibration board whose size is $120mm \times 120mm$ to obtain the parameters of the visual system. In order to improve the calibration accuracy, 16 calibration pictures are used for calibration. The calibration results show that the focal length is 3.9580mm and the mean error of the reconstruction results is 0.0032mm.

2.3 Selection of baseline length

According to the formulas (1), (2) and (3), baseline is a vital parameter for visual accuracy. Δx, Δy and Δz represent the *X, Y, Z*-axis measurement error of the model respectively. Because the positioning error should be less than or equal to 3 mm to ensure that the device can complete the picking of A. bisporus successfully, the positioning error in *X* and *Y* coordinate system of the centre point of A. bisporus is set as Δs, and $\Delta s \leq 3mm$.

$$\Delta s = \sqrt{\Delta x^2 + \Delta y^2}$$
$$= \sqrt{\left(\frac{u_1 \Delta z}{f}\right)^2 + \left(\frac{v_1 \Delta z}{f}\right)^2} \qquad (4)$$
$$= \frac{\Delta z}{f}\sqrt{u_1{}^2 + v_1{}^2}$$

Because the size of the image sensor of the camera is $6.4mm \times 4.8mm$, $|u_1| \leq 3.2mm$, $|v_1| \leq 2.4mm$. It is seen from formula (4) that $\Delta z \leq 2.9685mm$ to meet $\Delta s \leq 3mm$. The camera movement error and parallax error can cause the *Z*-axis measurement error Δz calculated by the formula as follows.

$$\Delta z = \left| z - \frac{(b+\Delta b)f}{U+\Delta U} \right| = \left| \frac{z^2 \Delta U - zf\Delta b}{bf + z\Delta U} \right| \qquad (5)$$

The parallax error and the baseline error are set as ΔU and Δb. $|\Delta U| \leq 0.0032mm$, $|\Delta b| \leq 0.02mm$. According to the actual depth information of A. bisporus in the planting base, the *Z*-axis coordinate can be set as $115mm \leq z \leq 135mm$. It is seen from formula (5) that the baseline length $b \geq 4.2837mm$ to meet $\Delta z \leq 2.9685mm$. The larger the baseline length, the smaller the measurement error is. Meanwhile, the larger the baseline length, the smaller the overlap area of the two pictures from *M, N* positions will be. Thus, the baseline length is set as 5.5mm.

2.4 Visual workflow of A. bisporus picking robot

The visual workflow of A. bisporus picking robot is as follow:

(1)The visual system of A. bisporus picking robot collects images at M and N positions respectively.

(2)The distortion corrections are carried out for the M and N position images.

(3)The two images are transformed from RGB color space to HSV color space, and a simple threshold segmentation algorithm is used to separate the A. bisporus and soil. The segmented regions of the A. bisporus are processed by the corrosion and expansion operator.

(4)XLD (eXtended Line Descriptions) contours are extracted from the Agarius bisporus regions, and the XLD contours are elliptically fitted to calculate the centre point coordinates (u_1, v_1), (u_2, v_2) and diameter of each A. bisporus equal the long axis of the fitted ellipse.

(5)The template matching of each fitted ellipse is carried out, and the parallax U is calculated.

(6)According to the formulas (1), (2) and (3), the spatial coordinates *(x, y, z)* of the centre point of each A. bisporus in the camera coordinate system are calculated.

(7)The error of diameter measurement of A. bisporus is compensated according to the Z-axis coordinate of the centre point of A. bisporus.

3. Visual error correction of the A. bisporus picking robot

3.1. The contour fitting error

The contour fitting error affects the measurement accuracy of the diameter of A. bisporus and the accuracy of centre point positioning. According to the general algebraic formula of ellipse, the ellipse fitting objective function shown as formula (6) can be calculated by the least square method [6].

$$f(A, B, C, D, E, F) = \sum_{i=1}^{N} \left(\begin{array}{c} Ax_i^2 + Bx_iy_i + Cy_i^2 \\ +Dx_i + Ey_i + F \end{array} \right)^2 \quad (6)$$

Among this function: *i=1, 2,.., N* and N is the number of sample points on the contour. However, the contour of A. bisporus fitted is affected easily by the edge discrete points that affect the measurement of the diameter of A. bisporus and the centre point position, as shown in Figure 4.

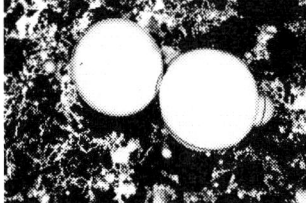

Figure 4. Influence of outliers

In order to make the ellipse fitting algorithm more robust, the method of joining weights can be used to reduce the influence of outliers. The Huber weight function shown as formula (7) is commonly used.

$$\omega(\delta) = \begin{cases} 1 & |\delta| \leq \tau \\ \frac{\tau}{|\delta|} & |\delta| > \tau \end{cases} \quad (7)$$

The parameter δ represents the distance from the outlier to the fitting ellipse, and the greater the distance, the smaller the weight ω will be. τ represents the distance threshold, and the smaller τ, the smaller the weight is given to outlier. The improved ellipse fitting objective function is shown as the following.

$$f(A, B, C, D, E, F) = \sum_{i=1}^{N} \omega(\delta) \left(\begin{array}{c} Ax_i^2 + Bx_iy_i + Cy_i^2 \\ +Dx_i + Ey_i + F \end{array} \right)^2 \quad (8)$$

The results of experiments show that the fitting ellipse of the edge of A. bisporus is ideal when $\tau = 2$. As shown in Figure 5. This method can almost eliminate the contour fitting error of A. bisporus.

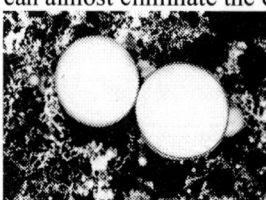

Figure 5. Fitting ellipse with Huber weights

3.2 Diameter error compensation of A. bisporus

According to the nonlinear model of perspective projection error [7], the perspective projection error ΔL of the diameter of A. bisporus is calculated by the formula as follows.

$$\Delta L = \frac{-L}{\frac{H_0 - f}{\Delta H} - 1} \qquad (9)$$

L represents the real diameter of A. bisporus, H_0 represents the distance from the calibration plane to the lens, f represents the focal length of the lens, and ΔH represents the distance from the centre point of the A. bisporus to the calibration plane, and ΔH is calculated according to the Z-axis coordinate of the centre point of A. bisporus by the formula as follows.

$$\Delta H = H_0 - f - z \qquad (10)$$

The error of diameter measurement of A. bisporus can be compensated by the formula as follows.

$$L_C = \tilde{L} + \Delta L = \tilde{L} + \frac{-\tilde{L}}{\frac{H_0 - f}{H_0 - f - z} - 1} \qquad (11)$$

$$= \tilde{L} \frac{2z - H_0 + f}{z}$$

L_C represents the compensated diameter of A. bisporus. $\tilde{L}$ represents the uncompensated diameter of A. bisporus.

4. Experimental results and application analysis

Firstly, the experiment was made in the laboratory by simulating the actual depth of 115mm~135mm of A. bisporus, as shown in the Figure6. In order to measure the error of the vision system, some 3D printed A. bisporus were inserted on a standard board of A0 size. From the reference point of the vision system, the A. bisporus was photographed in five positions to obtain their final coordinates. The accuracy of the vision system is measured by comparing the obtained coordinates by the vision system with the real coordinates on the chessboard.

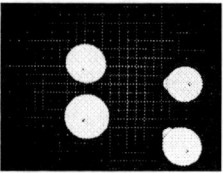

Figure 6. Visual accuracy experiment

In addition, the experiment of picking A. bisporus was made in the edible fungus base of Shanghai Lianzhong, and the success rate of recognition, centre point positioning and picking of the A. bisporus picking robot were statistically analyzed.

4.1. Analysis of experimental results

In the laboratory, a number of A. bisporus were tested at different sizes (ranging from 30 mm to 50 mm in diameter), different positions, different postures (inclined 5, 10, 15, 20, 25 degrees), and different heights. 20 groups of experimental results were counted as shown in Table 1.

Table 1. The experimental results in laboratory

Error	Average	Standard deviation	Success rate
Positioning error	1.82mm	2.04mm	95%
Diameter error	0.2931mm	0.8005mm	90%

4.2 Application analysis

The visual system can recognize $140mm \times 200mm$ effective area in each shot. It can collect images in five positions continuously to recognize the $140mm \times 1000mm$ area of A. bisporus. The recognition and picking experiment was made for $1400mm \times 1000mm$ area of A. bisporus. There were 497 ripe A. bisporus, and the height of ripe A. bisporus was between 40 mm and 60 mm. The diameter of ripe A. bisporus was generally between 30 mm and 50 mm. Total images processing time of the visual system was 150s, and the experimental results were shown in the Table 2.

Table 2. The experimental results in the base

Style	Numble of success	Success rate
Effective recognition	487	97.99%
Effective position	465	93.56%
Effective picking	457	91.95%

5. Conclusion

Aiming at the A. bisporus on the culture shelf, a measuring method of diameter and center point position based on monocular vision is proposed. The spatial coordinates of the centre point of the A. bisporus are measured with the images acquired by monocular camera that can move at two different viewpoints horizontally, and the ellipse-fitting algorithm introduced Huber weights to reduce the influence of outliers is used to improve the accuracy of visual system, and the perspective projection error of the diameter of A. bisporus is compensated according to the Z-axis coordinates. Experiments in the laboratory and the A. bisporus planting base showed that the recognition rate of the visual system of the A. bisporus picking robot was over 90%, which can meet the working requirements of the A. bisporus picking robot.

References

[1] T. Hazisawa, M. Toda, T. Sakoil. Image analysis method for grading raw shiitake Agaricus bisporus. *Incheon*, 46-52(2013)

[2] G. Fantoni, M. Santochi, G. Dini, K. Tracht, B. Scholz-Reiter, J. Fleischer, T.K. Lien, G. Seliger, G. Reinhart, J. Franke, H.N. Hansen, A. Verl, Grasping devices and methods in automated production processes. CIRP Annals- Manufacturing Technology, **63**(2): 679-701 (2014)

[3] R. Xiang, YB. Ying, HY. Jiang. Development of real-time recognition and localization methods for fruits and vegetables in field. Transactions of the Chinese Society for Agricultural Machinery, **44**(11):208-223(2013)

[4] J.R.Cai, X.J.Zhou, Y.L.Li. Recognition of mature oranges in natural scene based on machine vision. Transactions of the CSAE, **24**(1):175-178. (2008)

[5] H.Yu, T.W.Xing, X. Jia. The analysis of measurement accuracy of the parallel binocular stereo vision system. Proceedings of SPIE, 32-35(2016)

[6] W.G.Wang, S.R.Wang, Z.F. Xu, et al. Optimal Ellipse Fitting Algorithm of Least Square Principle Based on Boundary. Computer Technology and Development, **23**(4):67-70(2013)

[7] Z.SUN, Z.P.XU, Y.Q.WANG, et al. Control and compensation of perspective projection error analysis in machine vision measurement. Computer Engineering and Applications, **54** (2) : 266-270(2018)

Application of levitation frame with mid-set air spring on maglev vehicles

ZHANG Min MA Weihua LUO Shihui

Traction Power State Key Laboratory, Southwest Jiaotong University, Chengdu, China

Corresponding author: MA Weihua

Email: mwh@swjtu.edu.cn

Abstract: The vehicle-guideway coupled vibration (VCV) is always an important problem for the medium-low speed maglev vehicle; to alleviate the problem, a newly typed levitation frame with mid-set air spring (LFMAS) is presented, and the lower dynamic interaction of which with the guideway is explained by motion equation. By comparing the running performance of the LFMAS and the levitation frame with end-set air spring (LFEAS), the effectiveness of the LFMAS on reducing the VCV problem is proved. The results show that the resistance required to adjust the same angle of the LFMAS is $2kl^2\theta + 2cl\dot{\theta}$ smaller than that of the LFEAS; the new-generation medium-low speed maglev with LFMAS can reach 121km/h on the Shanghai Lingang test line, and can levitates stably half an hour on the turnout; in all line operating conditions, the vehicle's transverse and vertical stability indicators are less than 2.5, reach the excellent level. The LFMAS exhibits stronger levitation bearing capacity and track adaptability attributing to the release of the ends degree of freedom, which can effectively reduce the influence of the VCV effect on the maglev vehicle.

1. INTRODUCTION

Medium-low speed maglev vehicles have the advantages of small turning radius, strong climbing ability, low noise and comfortable seat, etc., having good prospects and booming trends, and there are 4 operation lines in the domestic and overseas. The VCV problem is always a barrier on the way of the development of the medium-low speed maglev; the running mechanism is too sensitive to the guideway, resulting in levitating failure easily. In order to relieve the VCV effect [1-2] on the vehicle, domestic and foreign scholars have done a lot of studies. Starting from the levitation control algorithm, Cui et al. [3,4] compensated the levitation system, introduced the flux feedback control scheme, analyzed the influence of the levitation control algorithm and the main parameters on the levitation stability, and improved the robustness of the levitation system against load changes. Starting from the track beam structure, Lee et al. [5,6] established a vehicle-bridge coupling numerical model, and discussed the importance of dynamic coupling analysis of maglev vehicle bridges, pointed out that reasonable rail beam structure characteristics and levitation control algorithm have great significance to suppress vehicle-rail coupling vibration.

A lot of significant achievements have been made in the above researches, but the problem of VCV has not been completely solved. High quality and stiffness requirements of the running mechanism to the track makes the track cost higher and the construction process more complicated. Based on the above research, this paper puts forward the research direction of optimizing the running mechanism from the perspective of optimizing the vehicle structure; the team of professor Luo from southwest jiaotong university carried out a decade-long study to explore a new-type running mechanism with better track adaptability and bearing capacity, its low-dynamic-interaction mechanism is explained by the motion equation in this paper; after long-term operation on the test line, the strong levitation bearing capacity and track adaptability of which are been proved.

2. Two kinds of levitation frames

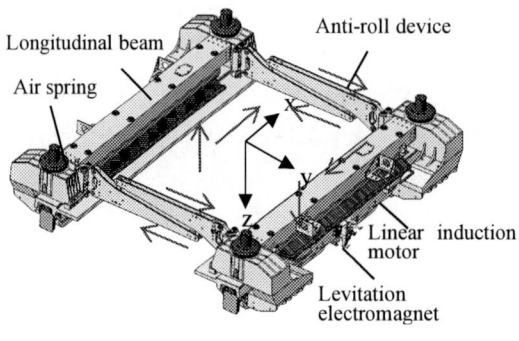

Figure 1. LFEAS

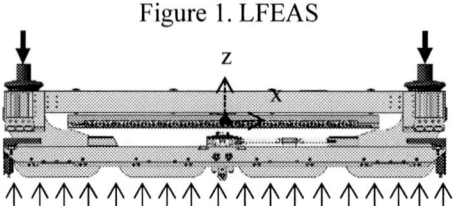

Figure 2. Bearing forces of the LFEAS

The medium-low speed maglev vehicle is mainly composed of two parts: car body and levitation frame. The levitation frames of the current operating vehicles are all LFEAS, in which levitation frame consists of two levitation modules, and the levitation modules are connected by two sets of anti-roll devices, as shown in figure 1; the levitation module mainly comprises longitudinal beam, linear induction motor and levitation electromagnet. Each end of the longitudinal beam has an air spring; one levitation frame has four air springs to support the car body. The LFEAS is sensitive to the mass and stiffness of the track as well as the variation of vehicle load, whose complex structure and large dead weight reduce the carrying capacity of the vehicle, and limit the installation space of the linear induction motor, resulting in insufficient traction capacity.

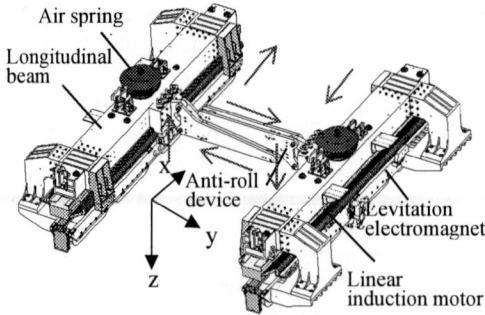

Figure 3. LFMAS

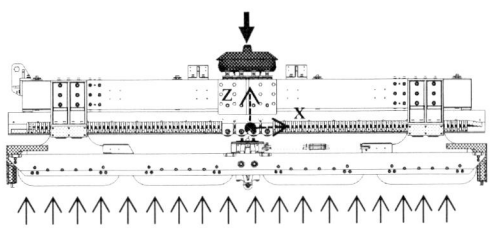

Figure 4. Bearing forces of the LFMAS

The main forces the LFEAS subjected are shown in Fig. 2, from top to bottom, including gravity of the car body whose direction is downward; the normal force generated by linear induction motor may be up or down with the change of the speed and slip frequency; the suction force supplied by levitation electromagnets is upward. The forces generated by the linear induction motor and levitation electromagnets are uniform; the gravity of the car body transferred by air spring can be regarded as concentrated. Considering reducing the number of forces acting on the levitation module to reduce the interaction between the levitation frame and the track, the number of air spring on a levitation module is reduced from two to one, so that the vertical motion of the levitation module ends become freer. Specifically, the structure shows that two air springs at both ends of the levitation module are reduced to one air spring in the middle of the longitudinal beam, as shown in Fig. 3; the forces the levitation module subjected is shown in Fig. 4.

Air springs on the LFMAS is placed at the middle of the longitudinal beam, the linear induction motor installed in the LFMAS can be get longer, resulting in the increasing of the tractive force. The lateral resetting force of the levitation module is mainly provided by the levitation electromagnets, which has the function of self-regulating guided resetting. The anti-roll devices connecting two levitation modules is designed to inhibit the rotation of the levitation module around the x axis and allow two levitation modules to have certain dislocation in the longitudinal, lateral and vertical directions, as shown in Fig. 1 and Fig. 3.In terms of these four degrees of freedom, one set of anti-roll device can achieve the same effect as two sets; when the anti-roll device is two sets, the constraint effect on the rotation degree of freedom of the levitation module around z and y axis is greater than that of one set of anti-roll device. The rotation of the levitation module around z axis mainly occurs when passing curves and be reset by the lateral force supplied by the levitation electromagnets. The rotation around y axis mainly occurs in the levitation adjustment process, and the restraint will hinder the levitation adjustment and increase the burden of the levitation system. Therefore, one set of anti-roll device can achieve the same functionality while reducing the coupling effect of levitation adjustment.

3. Vertical motion equations

3.1 vibration model

The vertical bearing forces of the levitation module from top to bottom are gravity of the car body, normal force of the motor and levitation force of electromagnets. The value of the normal force can be controlled by traction inverter, so its influence can be neglected. The vibration model of the levitation module is shown in Fig 5. The force from the air spring can be considered concentrated. The levitation force is considered uniform, and its local value is closely related to the size of the air gap and the current, as well as the arrangement of the coils.

In the process of starting levitation, the levitation module can be regarded has only translational motion without rotation. At this point, the levitation force is distributed symmetrically along the longitudinal beam; and the module doesn't have eccentric torque due to the uniform distribution of the force; at the same time the equivalent levitation force F_l is applied at the center of the levitation module and e is 0.

ISPECE   IOP Publishing

IOP Conf. Series: Journal of Physics: Conf. Series **1187** (2019) 032035   doi:10.1088/1742-6596/1187/3/032035

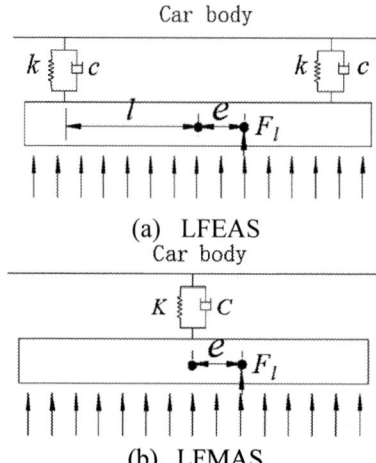

(a) LFEAS

(b) LFMAS

Figure 5. Vibration model of levitation module

In the running process, there are track excitations such as the irregularity of the track and the malposition of the rail joint. When passing through the excitation section, the sensors on the same levitation module feed different air gap size and acceleration to the corresponding controller, resulting in different current values in the four coils of the electromagnet, so as to achieve the purpose of levitation adjustment. The levitation module has both translation and rotation in the adjustment process; generally, F_l is eccentric and e is not 0.When the levitation force on the right side of the levitation module is greater than the left one, e is on the right side of the center point, and vice versa. K is the stiffness of the air spring on LFMAS, and k is that of the LFEAS.

3.2 Motion equation

The biggest difference between the structure of the LFEAS and the LFMAS is the position and number of the air spring on a levitation module; there are two air springs at both ends of the levitation module for LFEAS, while only one air spring in the middle of the levitation module for LFMAS. In order to adapt the rail changes, the levitation module is being adjusted constantly to maintain the air gap size. The levitation module has vertical and lateral movement relative to the track, the lateral malposition is mainly caused by the curve passing; and the vertical movement is the main motion, which is translation motion coupled with rotation. The motion equations of the two kinds of levitation frames are compared in this section, and the coordinate center selected to analyze the equations is the middle of the longitudinal beam.

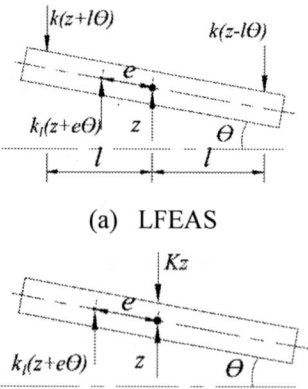

(a) LFEAS

(b) LFMAS

Figure 6. Vertical bearing force of the levitation frame

726

The air spring position of the LFMAS, which is at the center of the longitudinal beam, is taken as the origin of coordinates; the motion equation of the LFEAS is expressed as follows:

$$m\ddot{z} = -k(z-l\theta) - k(z+l\theta) + k_l(z+e\theta) - 2c\dot{z} \quad (1)$$

$$J\ddot{\theta} = k(z-l\theta)l - k(z+l\theta)l + k_l(z+e\theta)e - 2cl\dot{\theta} \quad (2)$$

Simplified as

$$m\ddot{z} = -2kz + k_l(z+e\theta) - 2c\dot{z} \quad (3)$$

$$J\ddot{\theta} = -2kl^2\theta + k_le(z+e\theta) - 2cl\dot{\theta} \quad (4)$$

The matrix is shown as

$$\begin{bmatrix} m & 0 \\ 0 & J \end{bmatrix} \begin{Bmatrix} \ddot{z} \\ \ddot{\theta} \end{Bmatrix} + \begin{bmatrix} 2c & 0 \\ 0 & 2cl \end{bmatrix} \begin{Bmatrix} \dot{z} \\ \dot{\theta} \end{Bmatrix} + \begin{bmatrix} 2k-k_l & -k_le \\ -k_le & 2kl^2 - k_le^2 \end{bmatrix} \begin{Bmatrix} z \\ \theta \end{Bmatrix} = 0 \quad (5)$$

The vertical motion equation of the LFMAS is shown as follows:

$$m\ddot{z} = -Kz + k_l(z+e\theta) - C\dot{z} \quad (6)$$

$$J\ddot{\theta} = k_l(z+e\theta)e \quad (7)$$

The matrix is shown as

$$\begin{bmatrix} m & 0 \\ 0 & J \end{bmatrix} \begin{Bmatrix} \ddot{z} \\ \ddot{\theta} \end{Bmatrix} + \begin{bmatrix} C & 0 \\ 0 & 0 \end{bmatrix} \begin{Bmatrix} \dot{z} \\ \dot{\theta} \end{Bmatrix} + \begin{bmatrix} K-k_l & -k_le \\ -k_le & -k_le^2 \end{bmatrix} \begin{Bmatrix} z \\ \theta \end{Bmatrix} = 0 \quad (8)$$

When the air spring stiffness of the LFMAS is 2 times of that of the LFEAS, that is K = 2k, the equation (8) becomes

$$\begin{bmatrix} m & 0 \\ 0 & J \end{bmatrix} \begin{Bmatrix} \ddot{z} \\ \ddot{\theta} \end{Bmatrix} + \begin{bmatrix} 2c & 0 \\ 0 & 0 \end{bmatrix} \begin{Bmatrix} \dot{z} \\ \dot{\theta} \end{Bmatrix} + \begin{bmatrix} 2k-k_l & -k_le \\ -k_le & -k_le^2 \end{bmatrix} \begin{Bmatrix} z \\ \theta \end{Bmatrix} = 0 \quad (9)$$

As can be seen from the comparison between equation (5) and (9), the vertical translational equations is exactly the same while the rotational equation is different, which means that the resistance needed to adjust the same rotation angle for LFEAS is $2kl^2\theta + 2cl\dot{\theta}$ bigger than that for LFMAS.

When the rotation angle of the levitation module θ is 0, means the levitation module has only translational motion without any rotation; this adjustment state generally exists in the starting levitation process. The motion equation of the LFEAS is:

$$m\ddot{z} = -2k\,z + k_l z - 2c\dot{z} \quad (10)$$

The motion equation of the LFMAS is:

$$m\ddot{z} = -K\,z + k_l z - C\dot{z} \quad (11)$$

When the air spring stiffness of the LFMAS is 2 times of that of the LFEAS, the motion equations of the two kinds of levitation modules are identical in the starting levitation process.

As can be seen from equations (5) and (9), the mass matrix and damping matrix of the vertical motion equations are diagonal, and the stiffness matrix is non-diagonal, which fails to achieve simultaneous decoupling. At the same time, it can be seen that the vertical translational equations of the two kinds of levitation modules are the same, but the rotational equation is different. The resistance needed to adjust the same rotation angle for the LFEAS is $2kl^2\theta + 2cl\dot{\theta}$ bigger than that for the LFMAS. The optimization of the levitation frame structure releases the ends degree of freedom, resulting in the difference of the motion equation of the levitation module; and the force needed to overcome when rotate the same angle is smaller.

4. Application

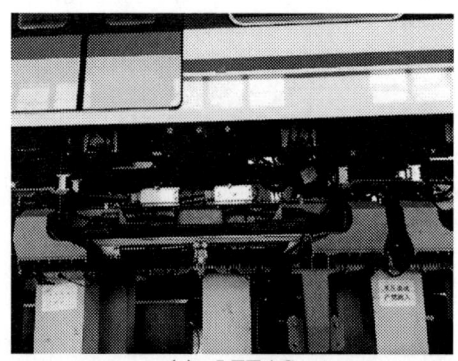

(a) LFEAS

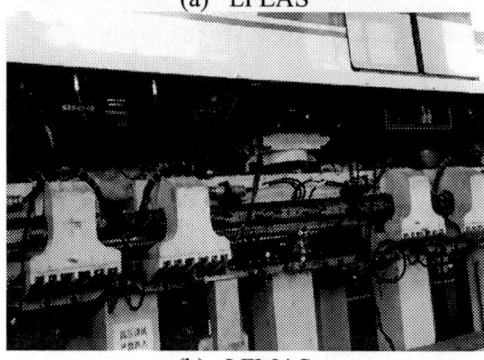

(b) LFMAS

Figure 7. Application of the two kinds of levitation frame

As shown in Fig. 7 (a), the LFEAS has been applied on two maglev operation lines in China (Beijing S1 line and Changsha airport line), but its operating speed is less than 100km/h. The LFMAS independently developed by Southwest Jiaotong university is shown in Fig 7 (b). After more than 2 years of test line operation, it has been proved that the LFMAS has stronger levitation-bearing capacity and track adaptability, and its sensitivity to the track is lower than that of the LFEAS. Limited to the length of the test line, the test speed of the LFMAS on the test line has reached 121km/h, and can levitates stably half an hour on the turnout, which prepare for its marketization and promotion.

In order to test the levitation stability of the LFMAS shown in Fig. 7 (b) with various speeds and line conditions, the line running tests are carried out. On the test line with a total length of 1.7 km, the vehicle passes the R50m curve at the speed of 10 km/h, passes the R75m curve at the speed of 25 km/h, go through the straight line and turnout at the speed of 80 km/h and 100 km/h respectively, and climbs the slope at the speed of 25 km/h. The test results are shown in Fig. 8 and Fig. 9.

Table 1. Evaluation criteria of the vehicle stability indicator

Stability level	Evaluation	Stability indicator
Level 1	Excellent	$W<2.5$
Level 2	Good	$2.5<W<2.75$
Level 3	Pass	$2.75<W<3$

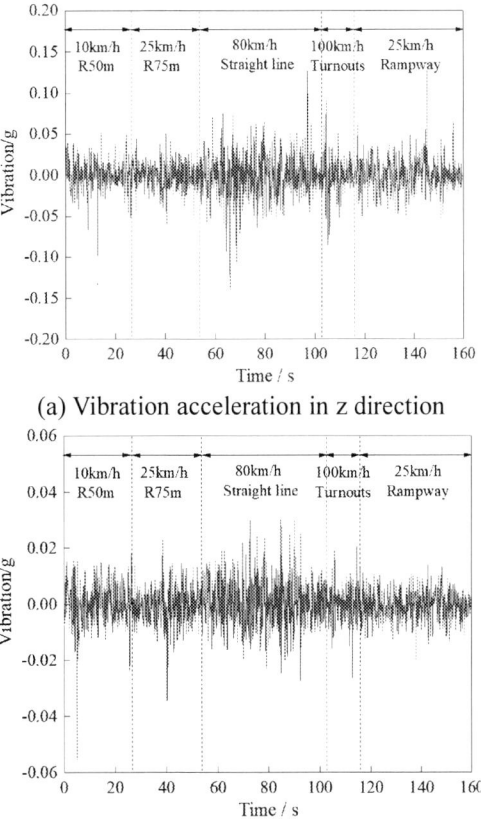

(a) Vibration acceleration in z direction

(b) Vibration acceleration in y direction

Figure 8. Vibration acceleration of the car body

Fig 8 shows the changes of the lateral and the vertical accelerations of the car body. In all operation conditions, including R50 m small radius curve, turnout and the hill with 0.07 slope, the largest vertical acceleration is less than 0.2 m/s^2, and the maximum lateral acceleration is less than 0.1 m/s^2, far less than the comfort requirements [7] of the maglev vehicle which requires that the lateral acceleration is less than 1 m/s^2, vertical additional upward acceleration is less than 0.5 m/s^2, vertical downward additional acceleration is less than 1 m/s^2 .There is no abrupt change of the acceleration when passing the small radius curve line and turnout, which indicates that the LFMAS has better track adaptability, lower sensitivity to the track, and can levitates stably more easier after the decoupling of the ends of the levitation module.

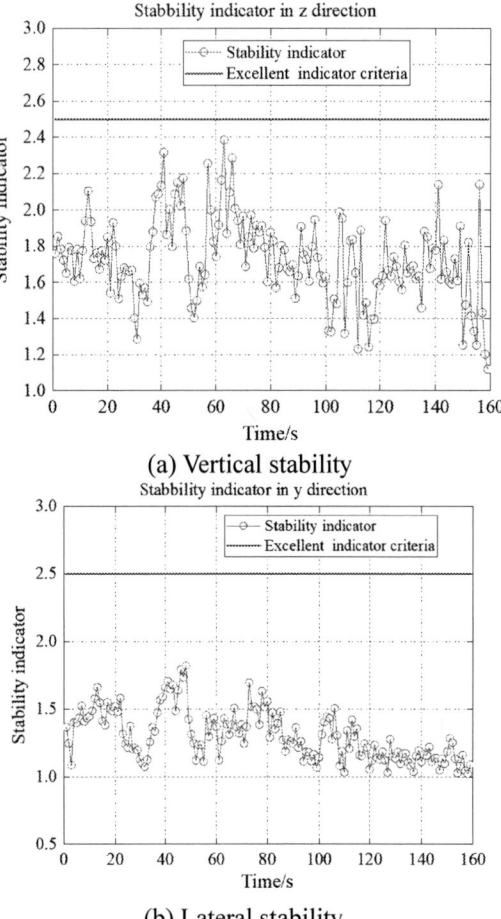

(a) Vertical stability

(b) Lateral stability

Figure 9. Evaluation of vehicle stability

The speed of the vehicle is higher when running in the linear track, so its acceleration amplitude a little larger than other working conditions. Fig. 9 shows the further analysis of the lateral and vertical stability. In the whole line and running conditions, the lateral and vertical stability indicators of the vehicle are less than 2.5, reach the excellent level.

5. Conclusion

1）The optimization of the levitation frame structure releases the ends degree of freedom, resulting in the difference of the motion equation of the levitation module. The resistance required to adjust the same angle of the LFMAS is $2kl^2\theta + 2cl\dot{\theta}$ smaller than that of the LFEAS, which indicates that the LFMAS can adapt the rail more easily in the levitation process.

2）The new-generation medium-low speed maglev adopting the LFMAS can reach 121km/h on the Shanghai Lingang test line, and can levitates stably half an hour on the turnout; The transverse and vertical stability indicators all reach excellent level. The research proves that the LFMAS exhibits stronger levitation-bearing capacity and track adaptability attributing to the release of the ends degree of freedom, can effectively reduce the influence of the VCV effect on the maglev vehicle.

References

[1] LEE H W, KIM K C, JU L. Review of maglev train technologies[J]. IEEE Transactions on Magnetics, 2006, 42(7): 1917-1925.

[2] Zhang M, Luo S H, Gao C and Ma W H. Research on the mechanism of a newly developed levitation frame with mid-set air spring [J]. Vehicle System Dynamics, 2018,1-21

[3] CUI P, LI J, ZHANG K, et al. Design of the suspension controller based on compensating feedck linearization[C]. International Conference on Measuring Technology and Mechatronics Automation, 2010, (1): 1056-1059.

[4] Li J H, Li J. A practical nonlinear controller for levitation system with magnetic flux feedback[J]. Journal of Central South University, 2016, 23(7): 1729-1739.

[5] LEE J S, KWON S D, KIM M Y, et al. A Parametric study on the dynamics of urban transit maglev vehicle running on flexible guideway bridges[J]. Journal of Sound and Vibration, 2009, 328(3): 301-317.

[6] CAI Y, CHEN S S, ROTE D M, et al. Vehicle/guideway dynamic interaction in maglev systems[J]. Journal of Dynamic Systems, Measurement, anc Control, Transactions of the ASME, 1996, 118(3): 526-530.

[7] Lin Z, Zhou D. Brief Introduction of the Track of Shanghai Maglev[J]. China Railway Science, 2003, 24(1): 104-107

Study on preparation methods of copper-based composites

Nianlian Li[1,2], Vanessa Bouchart[1], Pierre Chevrier[1], Hongyan Ding[2]

[1]Ecole Nationale d'Ingénieurs de Metz-Université de Lorraine, CNRS, Arts et Métiers ParisTech, LEM3, F-57000, Metz, France

[2]Faculty of Mechanical and Material engineering, Huaiyin Institute of Technology, 223003, Huaian, China

Abstract: Copper matrix composites are generally used as friction disc materials of the water pump clutch due to their excellent mechanical properties, but they also experience severe wear and relative slip between the friction disc and the impeller ring at a high velocity. To investigate the influence of the different structures of the composite on wear performance, this paper discussed the copper-based composites with three process methods, by contrast, the process method of low-pressure preloading and sintering, then perform a high-pressure could effectively improve the densification, hardness and wear resistance.

1. Introduction

The switchable water pump (SWP) is a component to cool the engine, and significantly affects the service lifetime of the engine. Inside the SWP, there is a device with a similar structure to the wet clutch of the automobile, it is named SWP clutch here. The friction disc and impeller ring are made and used in the application of the SWP clutch. The SWP clutch generally operates with a higher speed and a lighter load in comparison to that of the wet clutch mounted in the automobile [1]. Specifically, the SWP clutch normally operates at a wide range of temperature from -30 °C to 120 °C and pressure from 0 to 0.66 MPa. Generally, the friction disc is connected to the drive shaft, and the impeller ring is connected to the driven shaft [2]. When the temperature of the engine is up to 80 °C, the friction disc and the impeller ring engage and the power is transferred to the driven shaft, leading to the rotation of the impeller and resultant coolant circular motion rapidly to cool the engine. It is found that the achievement of the power transferred course is relied on the engagement and disengagement of the two discs, which is significantly affected by the applied normal pressure, the dynamic coefficient of friction, the velocity, and the temperature of the clutch plates [3]. The friction coefficient of the contacted surface that experiences a severe wear after a long-time cycle and dynamic contact would be considerably reduced, generating a relative slip between the two discs and attendant decline of the torque transmission, especially for the application of a high velocity of the engine. Thus, it can be derived that the wear resistance is considered as a crucial factor in obtaining a high-performance SWP clutch.

A number of research efforts have focused on the friction behavior and wear resistance of the materials for the applications of the wet clutch [4, 5]. Among them, copper and its alloys, generally exhibit a higher thermal conductivity in comparison to other alloys, have performed the combined properties, e.g., high strength, high thermal conductivity and low coefficient of thermal expansion and wear resistance, favoring the sufficient dissipation of the heat generated during friction [6]. Su et al. [7] have produced the copper-based composite with self-lubricated graphite for wet clutch and the excellent lubricating performances have obtained, attributing to the formation of lubricating and

Content from this work may be used under the terms of the Creative Commons Attribution 3.0 licence. Any further distribution of this work must maintain attribution to the author(s) and the title of the work, journal citation and DOI.
Published under licence by IOP Publishing Ltd

transferring films during friction. Ma et al. [8] have disclosed that there is a critical speed where a transition of the friction and wear regimes of the composite formed of the copper-graphite composite with a pin-on-disc configuration, implying an enhancement of the wear properties. Powders were loaded into a graphite mold and sintered in an Ar atmosphere at 900 °C for 1 h under a pressure of 25 MPa to fabricate copper matrix composites reinforced with grapheme, the coefficient of friction of the composite is stable with increasing the applied load [9]. Akbarpour. et al. [10] have studied that the effect of SiC nanoparticles on tribological properties of copper matrix composites were plasma spark sintered in a vacuum atmosphere, indicating that the addition of 4 vol% silicon carbide to copper matrix reduced the wear track depth and with lower plastic deformation during dry sliding wear test. It can be found that the friction properties of the copper-based composites are useful for the SWP wet clutch.

However, the structure is a main factor to influence the properties of copper-based composite, and generally the powder sintered process and the relevant parameters selected could determine the necessary structure. To data, the wet clutch and the brake pads with a considerably low wear loss are highly required in the applications of the SWP wet clutch. In the press study, the copper-based composites with various process methods were prepared through powder sintering to obtain an increased micro-hardness and a reduced wear loss, and give priority satisfying preconditions of SWP wet clutch.

2. Experimental procedures

The copper-based composites consist of copper, iron, silica, graphite and the other elements. Different types of the powder with a same average size of 80 μm were employed in the present study. The QM-ISP04 planetary ball mill (Nanjing Nanda instrument Co.,Ltd, Nanjing) was applied to mix the composite powders in ZrO_2 ceramic jars duration of 4 h with a speed of 200 rpm which to prepare for fabricating the composite samples. The fabrication process included three methods. Method A: Mixed composite powder was pressed into the small-sized pieces with a pressure of 800 MPa at room temperature by the DY-30 tablet press machine, then with the vacuum furnace sintered in an Ar atmosphere at 800 °C for 1 h. Method B: Mixed composite powder was pressed into the small-sized pieces with a pressure of 400 MPa at room temperature by the DY-30 tablet press machine, then loaded into a graphite mold and sintered in an Ar atmosphere at 800 °C for 1 h under a pressure of 135 MPa. Method C: Mixed composite powder was pressed into the small-sized pieces with a pressure of 400 MPa at room temperature by the DY-30 tablet press machine, then with the vacuum furnace sintered in an Ar atmosphere at 800 °C for 1 h. Subsequently, the sintered samples were pressed under a larger pressure of 800 MPa, and then with the vacuum furnace sintered in an Ar atmosphere at 500 °C for 2 h to anneal. Finally, all the samples with a diameter of 10 mm and a thickness of 5 mm were obtained.

The surface. morphologies of the as-produced composites were characterized by the MicroXAM 3D non-contact three-dimensional morphometer (ADE, USA). The Vickers micro-hardness of the samples was measured with an automatic turret micro-hardness tester (HY-1000, China) under a load of 0.3kg with the holding time of 10 sec.

The friction tests were carried out using the Friction wear testing instrument (UMT-2, USA) with a ball-on-plate manner. The GCr15 ball with a diameter of 4mm and a polished end-face was used as the friction pairs of the copper-based composite. The reciprocated amplitude was settled as 1 mm, and the friction tests achieved in the glycol coolant at room temperature with an applied load of 5 N, and an operated speed of 900 rpm for 1 h. The coefficient of friction (COF) of the composite was automatically recorded during friction test. The worn samples were ultrasonically rinsed with alcohol and dried with nitrogen gas, subsequently, the wear extent was measured by the non-contact three-dimensional morphometer.

3. Results and Discussion

3.1 Surface morphologies

Fig. 1 illustrates the three-dimensional surface morphologies and incidental surface roughness of the copper-based composites with various process methods. It was apparent that the different process methods could significantly affected the surface quality of the composites. Method A : the mixed powders with a pressure of 800 MPa at room temperature then were put into the vacuum furnace sintered in an Ar atmosphere at 800 °C for 1 h, the surface with a large amount of shallow zone that was identified as the produced pores was obtained (Fig. 1a), implying a relatively high surface roughness of 0.496 μm (Fig. 1d). Method B: the mixed powders with a pressure of 400 MPa at room temperature, then loaded into a graphite mould and sintered in an Ar atmosphere at 800 °C for 1 h under a pressure of 135 MPa. was used, the surface appeared to be smooth, except for limited fluctuation in a certain area (Fig. 1b), demonstrating a considerably reduced surface roughness of 0.32 μm (Fig. 1d). Method C with a pressure of 400 MPa at room temperature then sintered at 800 °C for 1 h. Subsequently, the sintered samples were pressed under a pressure of 800 MPa, then annealed at 500 °C for 2 h, the surface was observed with the presence of less asperities, due to the porosity resulting from the first sintering was reduced under the second pressure load (Fig. 1c). As a result, the minimal surface roughness of 0.2608 μm was measured (Fig. 1d), indicating a good surface quality of the composite fabricating with method C.

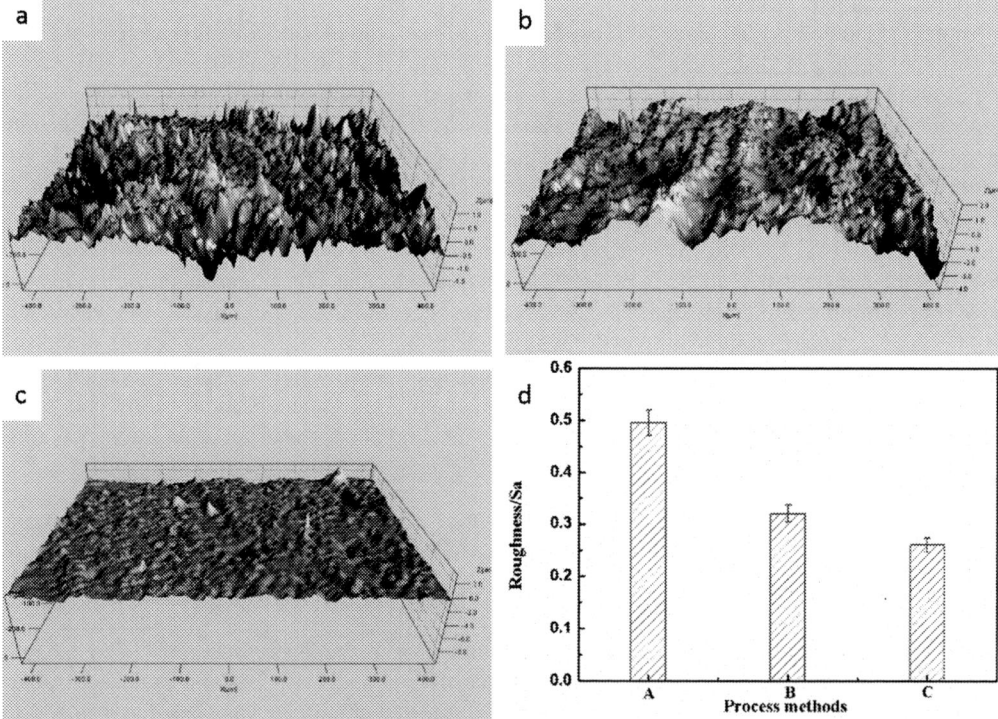

Fig. 1. The characterized 3D morphologies showing the surface features of the copper-based composites with various process methods: (a) method A, (b) method B, (c) method C, and (d) presenting the specific surface roughness of the composite with various process methods.

3.2 Vickers micro-hardness

Fig.2 shows the Vickers micro-hardness indentation morphologies and attendant hardness of the copper-based composites with various process methods. It could be found that the measured Vickers micro-hardness of the composites fluctuated in the range from 84.05 HV to 92.8HV and to 125.76 HV as varying the applied process methods A, B and C. As process method A, for only once press and sintering, presenting more pores leads to poor compactness and lower hardness (Fig. 2d). Process method B chose the second time to press at a high temperature, it could appropriately improve the compactness of the composites. It showed that the value of micro-hardness increased to 92.8HV and the indentation turned to small (Fig. 2b and 2d). However, since graphite mold also produced thermal expansion at a high temperature, the pressure could not be sufficiently transferred to the sample, which would limit the increase in the compactness of the sample. Process method C preloaded with a small load, and then pressed with a larger load after sintering, which effectively eliminated the pores generated after sintering, thereby fully improving the compactness of the sample, and the micro-hardness reached 125.76HV with a small size of the indentation (Fig. 2c and 2d). Therefore, reasonable to conclude that the Vickers micro-hardness of the composite is significantly affected by the combined factors, including the compactness and the homogeneous microstructure.

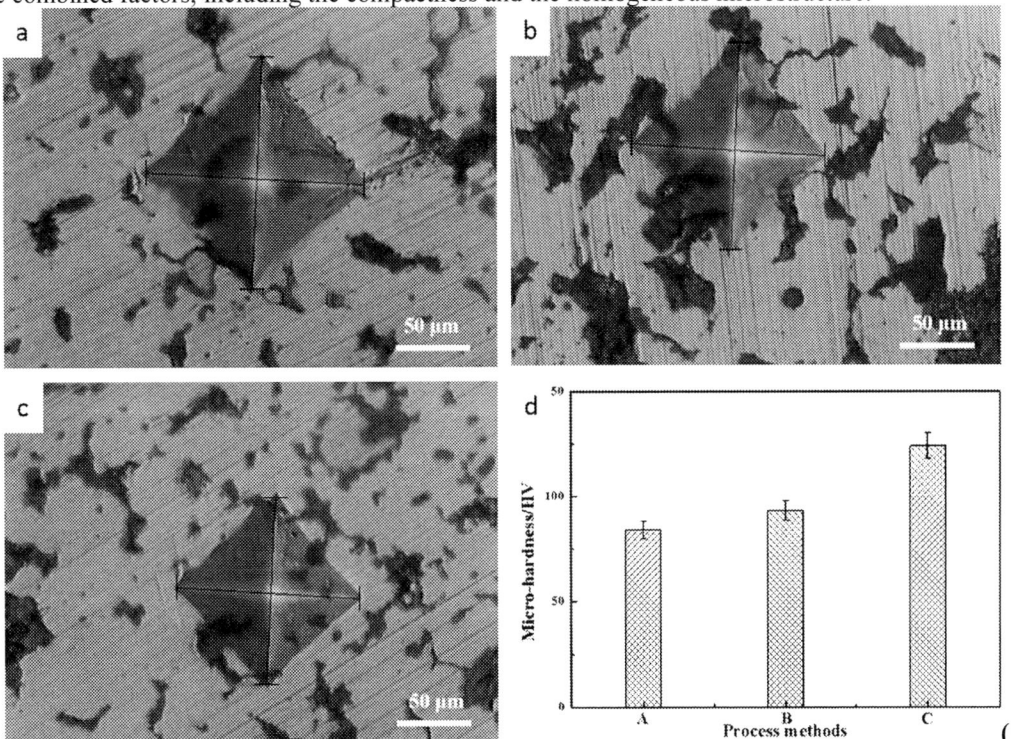

Fig. 2. The optical images representing the surface morphologies of the Vickers micro-hardness of the copper-based composites with various process methods: (a) method A, (b) method B, (c) method C, and (d) illustrating the quantitative relation of the measured Vickers micro-hardness with various process methods.

3.3 Wear morphologies and volume

Fig. 3 gives the characterized 3D morphologies showing the worn surface features of the copper-based composites with various process methods and attendant wear extent and volume. Wear test of the copper-based composites with the various process methods sliding against the counterpart GCr15 in the glycol solution for 1h. It could intuitively found that the wear extent of the copper-based composites involving the width and depth of worn cracks. The values of wear volume exhibited the considerable differences. For the copper-based composite with process method A, the wear volume with a relatively large value of $13.490 \times 10^{-3} mm^3$ was measured. Pressing under a high temperature as process method B, the wear volume of the composite decreased to $10.911 \times 10^{-3} mm^3$. For process method C, the wear volume of the composite rapidly decreased to $7.001 \times 10^{-3} mm^3$, due to preload with less pressure, then increasing the pressure after sintering, which can effectively eliminate the generated pores by sintering. Process method A and method B presented a relatively large amount of wear volume, attributed to the high porosity level and the low Vickers micro-hardness of the composites, indicating a poor wear resistance. Normally, an increase in hardness of the composite is favorable to obtain a lower wear volume [11]. The composite with process method C exhibited a lower wear volume due to its high Vickers micro-hardness, revealing a high wear resistance.

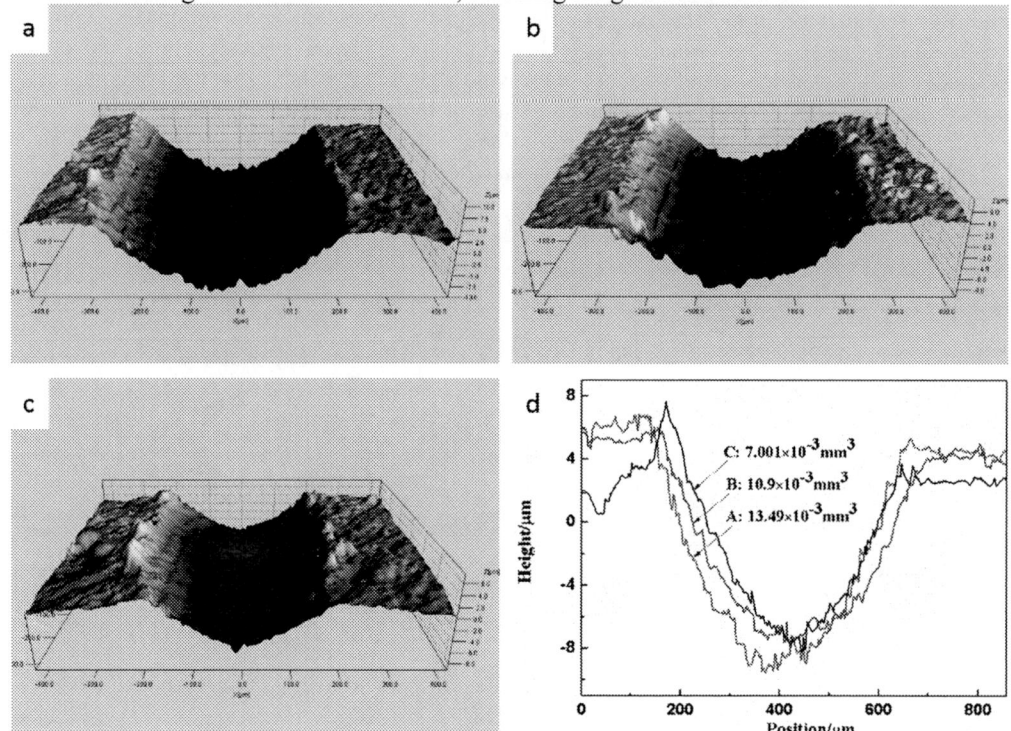

Fig. 3. The characterized 3D morphologies showing the worn surface features of the copper-based composites with various process methods: (a) method A, (b) method B, (c) method C, (d) wear extent and volume.

4. Conclusion
The wear resistance of copper-based composites with the various process methods was investigated with a ball-on-plate friction test under glycol coolant. The influence of materials with various process methods on surface roughness, micro-hardness, and the wear resistance of the composites were discussed. From our work, the major conclusions can be drawn as follows:

1)   The copper-based composites were fabricated by powder sintered process, and the process method A, B, and C all could obtain the composites with uniform composition.

2)   Process method C preloaded with a small load, and then pressed with a larger load after sintering, which effectively eliminated the pores generated in the course of sintering, thereby fully improving the compactness of the composite. The composite also exhibited a lower wear volume due to its high Vickers micro-hardness that revealed a high wear resistance.

References
[1] P. Marklund, R. Maki, R. Larsson, E. Hoglund, M.M. Khonsari, J. Jang, Tribol. Int. **40**, 876 (2007).
[2] S.H. Li, Z.C. Di, L. Cheng, Z.G. Liu, J.C. Piao, D.S. Huang, Z.Y. Zhao, H.T. Cui, Y. Li, Tribol. Int. **102**, 319 (2016).

[3] Q. Zou, C. Rao, G. Barber, B. Zhou, Y. Wang, Wear, **302**, 1378 (2013).

[4] A. Jullien, M.H. Meurisse, Y. Berthier, Wear, **194**, 116 (1996).

[5] W.B. Li, J.F. Huang, J. Fei, L. Cao, C.Y. Yao, Tribol. Int. **94**, 428 (2016).

[6] J. Bernhardt, A. Albers, S. Ott, Tribol. Int. **59**, 267 (2013).

[7] Y.F. Su, Y.S Zhang, J.J. Song, L. T. Hu, Wear 372-373, 130 (2017).

[8] W. Ma, J.J. Lu, Tribol. Lett. **41**, 363 (2011).

[9] X. Gao, H.Y. Yue, E.J. Guo, S.L. Zhang, L.H. Yao, X.Y. Lin, B. Wang, E. H. Guan, J. Mater. Sci Technol. **34**, 1925 (2018).

[10] M.R. Akbarpour, S. Alipour, Cerma. Int. **43**, 13364 (2017).

[11] G. Purcek, H. Yanar, O. Saray, I. Karaman, H.J. Maier, Wear, **311**,149 (2014).

Study on the Performance of the Wind Turbine Airfoil with Icing

Wang Long, Wang Cheng[a], Cheng Jie, Wang Chuanli, and Li liang

School of Mechanical Engineering, Anhui University of Science and Technology, 168 Taifeng Road, Anhui, Huainan, 232001, China

[a]Corresponding author: aust_wangch@163.com

Abstract. If the performance of the blade was properly assessed, which would help improve wind turbine performance, wind turbine blade aerodynamics at low temperatures could severely affect the output power of the wind turbine. The research on the performance of wind turbine blade with icing which had the three different shapes was carried out. The blades flow field and the lift coefficient were got from the free stream wind whose speed was 10m/s at different angles of attack, The Influences on the flow Field and aerodynamic performance were discussed, when the ice shape changed. The results show that:1) Model 1 has the best performance, with the maximum lift coefficient of 1.2853, its stall angle of 14 degrees, and the stalling phenomenon of icing blade is postponed;2) the difference in the icing shapes has less affected on the flow field distribution and the lift coefficient at the small angle of attack, and the distribution of flow field is more sensitive to the ice shape on the upper part of the leading edge at the large angle of attack.3) Ice the irregular icing shape can cause the oscillation phenomenon for the lift coefficient at the angle of attack from 0 to 20 degree.

1. Introduction

As a country of the wind power, wind turbines are used widely in CHINA, which is beneficial to energy supplements. Now, the supply of wind turbines power in northern occupies a large proportion in China's wind energy. But the wind turbines would freeze at low temperature, it is great disadvantage for its power output and safe operation [1-2]. So the aerodynamic problem should be paid more attention to in low temperature environment. Computational fluid dynamics (CFD) [3-5] method can deal with the complex flow problems of the wind turbine. With the rapid development of computer capabilities, CFD plays an increasingly important role on the aerodynamic performance analysis of wind turbines.

Foreign scholars first used numerical methods to make predictions on the formation of ice shapes [6-8]. Domestic scholars have begun to carry out corresponding research in recent years. Zhang Chiyu and others used the turbulence model with transition criterion to study the suction control for the upper surface and leading edge of the supercritical airfoil with icing. The results showed that laminar flow control could delay the transition of the clean airfoil surface, but it was not effective for icing airfoils [9].

Ren Pengfei and others studied the NRELS809 airfoil using business-oriented software called Fluent. The aerodynamic performance of the airfoil with icing in experiment was different from simulation with the software. Further, the aerodynamic performance of airfoil after icing was lower than that of airfoil without icing [10]. Wang Kun and others carried out the numerical simulation for hot air anti-icing system of aircraft and the prediction for the ice ridge. When the hot air anti-icing

Content from this work may be used under the terms of the Creative Commons Attribution 3.0 licence. Any further distribution of this work must maintain attribution to the author(s) and the title of the work, journal citation and DOI.
Published under licence by IOP Publishing Ltd

ISPECE   IOP Publishing

IOP Conf. Series: Journal of Physics: Conf. Series **1187** (2019) 032037   doi:10.1088/1742-6596/1187/3/032037

system opened, the surface temperature could reach up to 308K, finally there were ice ridges on the upper and lower of the wing surface behind the heating area [11].

In this paper, the research on the performance of the three different shapes of wind turbine blade with icing was carried out by using computational fluid dynamics. The flow field and the lift coefficient of blades were got from the wind whose speed was 10m/s from 0 to 20 attack of angle. The Influences on the flow Field and aerodynamic performance were discussed, which could provide a reference for the correct assessment of wind turbine performance in low temperature environment. All calculations were done in ANSYS software, and the number of the total calculation examples was 33.

2. Model and Meshing

2.1 Model

The three different wind turbines models with icing were shown in Fig. 1. According to the different shapes of ice, they were respectively named "model_1", "model_2" and "model_3". The geometric data of these models were derived from the literature [12].

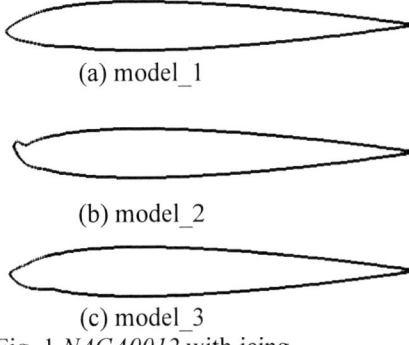

(a) model_1

(b) model_2

(c) model_3

Fig. 1 *NACA0012* with icing

2.2 Meshing

Fig.2 showed the partition of the computational domain. The upper and lower boundary which belonged to the computational domain was 15 times that of the chord length. In order to ensure the orthogonal grids, the front of the computational domain used "C" structure, and the outlet boundary of the fluid was 30 times the length of the chord from the coordinate origin. The boundary condition consisted of the far field and the wall.

Fig. 3 showed the grid of the blade, including the global grid and the local grid for the leading and tail of the wind turbine blade. The first layer grid near the wall was about 1e-5m from the wall, and the total number of grid was about 80, 000.

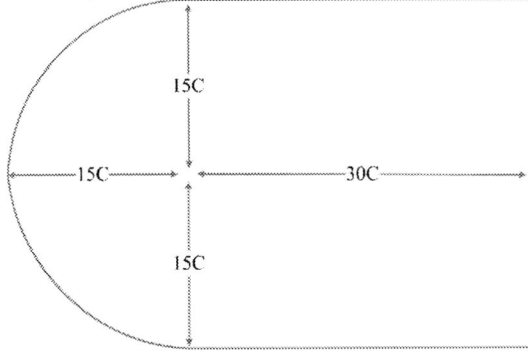

Fig.2 the computational domain

739

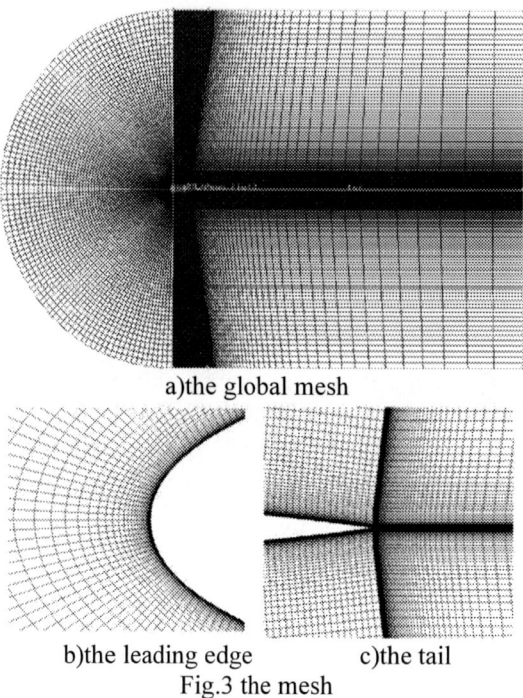

a)the global mesh

b)the leading edge   c)the tail

Fig.3 the mesh

3. Calculation results

Figure 4 showed the pressure coefficient of the upper and lower surfaces of the blade after the convergence of the flow field calculation at 1.49^0 angle of attack. The "Exp" indicated the experimental value [13], and the "CFD" indicated the numerical calculation result.

It could be seen from Fig.5 that the trend of pressure coefficient calculated by the numerical calculation was not only consistent with the experimental, but also the accuracy of the calculated results agreed well with the experimental values.

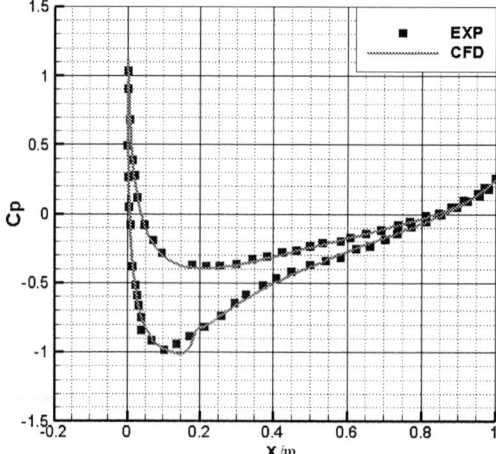

Fig.4 the pressure coefficient comparison of the blade wall

Fig.5 showed the calculation results of the lift coefficient, which was about the three different shapes of wind turbine blade with icing from 0 to 20 degree angle of attack. It could be seen from the figure that the lift coefficient of the blade greatly fluctuated after icing. Model 1 had the highest aerodynamic performance, followed by model 3, and model 2 showed the obvious oscillation characteristics, which is closely related to the shedding vortex of the leading edge. Under the small angle of attack (0~10 degree), by comparing with the lift curve of model 1and model 2, it could be seen that the upper of the leading edge had a great influence on the aerodynamic performance of the blade. The aerodynamic performance of the blade was not sensitive to the icing at the lower of the leading edge.

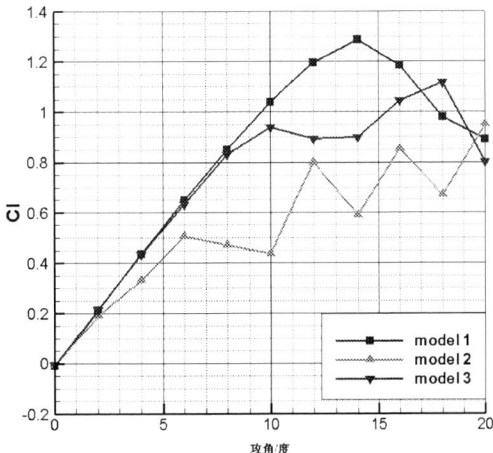

Fig.5 the lift coefficient of three different shapes

The maximum lift coefficient of model 1 was 1.2853 and the stall angle was 14 degree. The maximum lift coefficient of model 2 was 0.9498, which appeared at 18 degree, and it was 26.1% smaller than the maximum lift coefficient of model 1. The maximum lift coefficient of model 3 was 1.1166, which appeared at the 18 degree, and it was 13.13% smaller than the maximum lift coefficient of model 1. The static stall of model 2 was not obvious, and the stability of the lift became worse.

Fig.6 showed the flow field distribution of different blades with icing at 10° angle of attack. It could be seen from Fig. 6 a)、 b) and c)that there was no separation phenomenon between model 1 and model 2 had a large eddy area above the suction surface behind the blade. By observing the state of the flow field, it also could be seen that the regular icing shape had little effect on the distribution of the flow field under the small attack angle. The distribution of the flow field between model 1 and model 3 was similar in general. There was only difference in the flow field at the tail edge.

Because of there was a certain angle between the ice shape and the incoming flow in model 2, the fluid passing through the ice shape would form a recirculation zone, and the vortex phenomenon would occur at the same time.. As the flow went on, the vortex would further expand its influence area, and eventually lead to a larger recirculation zone in the flow field above suction surface of the blade. The aerodynamic performance of the blade decreased, and the lift coefficient reduced to 0.45, which was smaller than the lift coefficient of model 1 and model 3.

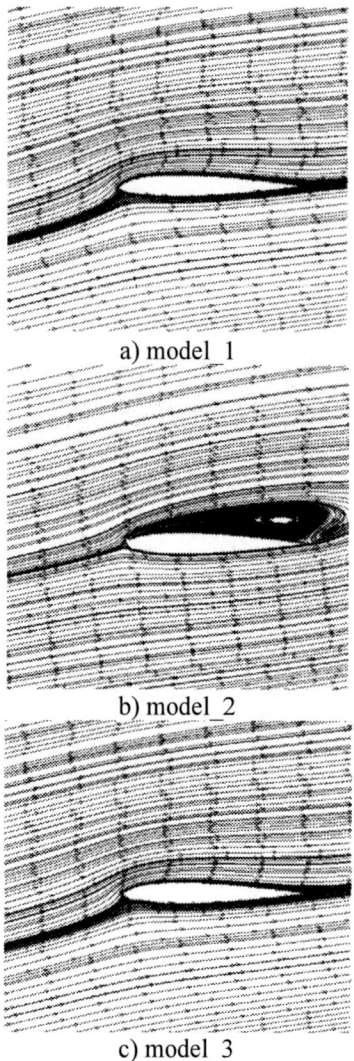

a) model_1

b) model_2

c) model_3

Fig.6 the flow field distribution of different blades with icing at 10° angle of attack

Fig.7 showed the flow field distribution of different blades with icing at 14° angle of attack. It also could be seen from Fig.8 a)、b) and c) that the flow fields of the three models were separated under the larger attack angle. The flow field separation of model 1 was the weakest. Model 2 had a larger vortex area above the suction surface of the blade, with the shedding vortex phenomenon at the blade tail. Model 3 had two relatively weak vortices above the suction surface of the blade. At the large attack angle, both model 1 and model 3 showed a separation phenomenon. Compared to icing in the lower part of the leading edge, the icing in the upper part of the leading edge had a bigger influence on the change of flow field. The flow field disturbance caused by the ice in the leading edge was shown through the downstream flow field. The flow field was more sensitive to the ice shape in the upper part of the leading edge, which could also be seen by comparing the ice shape structure in Fig. 2.

ISPECE   IOP Publishing

IOP Conf. Series: Journal of Physics: Conf. Series **1187** (2019) 032037   doi:10.1088/1742-6596/1187/3/032037

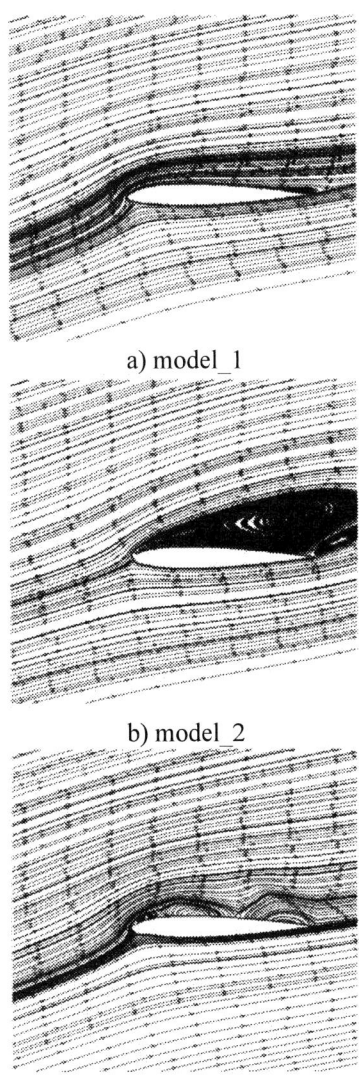

a) model_1

b) model_2

c) model_3

Fig.7 the flow field distribution of different blades with icing at 14° angle of attack

From the observation of the flow field, it could be seen that the existence of vortex region at the back of the blade would reduce the aerodynamic performance under the same angle of attack. The direct performance was the decrease of the lift coefficient. The size of the vortex region at the back of the blade determined the magnitude of the decrease of the lift coefficient.

4. Conclusion

In this paper, by using computational fluid dynamics, the research on the performance of the three different shapes of wind turbine blades with icing was carried out. The flow field and the lift coefficient of the blade were got from the free stream wind whose speed was 10m/s in different angles of attack. The Influences on the flow Field and aerodynamic performance were discussed. The conclusions were as follows.

　　1)　　The numerical simulation of NACA0012 airfoil is carried out by using computational fluid dynamics method. The distribution of pressure coefficient is in good agreement with the experimental results. The numerical results provide a reference for the aerodynamic performance evaluation of the wind blade.

2)   At the attack of angle which was from 0 to 20 degree, the aerodynamic performance of model 1 is the highest, the performance of model 3 is the second, and the performance of model 2 is the worst, whose lift coefficient has an obvious oscillation characteristics. The maximum lift coefficient of model 1 is 1.2853, and the stall angle is 14 degree. The maximum lift coefficient of model 3 is 0.9498, appearing at 18 degree. The static stall of model 2 is not obvious.

3)   The regular ice shape has little effect on the distribution of flow field at small angle of attack. The distribution of the flow field is more sensitive to the ice shape in the upper part of the leading edge under the larger angle of attack. At the same angle of attack, the existence of the vortex region at the suction surface of the blade reduces the aerodynamic performance, and the direct performance is the drop of the lift coefficient when the size of the vortex region at the suction surface determines the magnitude of the decrease of the lift coefficient.

Acknowledgements
This project is supported by National Natural Science Foundation of China (Grant No. 51705002), the Natural Science Foundation of Anhui Province of China(Grant No. 1708085QE123), Young Teachers Fund for Scientific Research of Anhui University of Science and Technology(Grant No. 12664), Introduction of talent fund for Scientific Research of Anhui University of Science and Technology(Grant No.ZY041).

References
[1]   Chinese Academy of Sciences. Beijing: Science Press. (2011)
[2]   HE Dexin. Engineering Sciences.**13**, 6(2011)
[3]   Zhong Wei, Wang Tongguang. Acta Energiae Solaris Sinica. **35,** 9 (2014,)
[4]   Zhou Zhenggui. Acta Aeronautica et Astronautica Sinica. **29,** 2 (2008)
[5]   Wang Long, Zhong Yicheng, Wu Qing, et al. Journal of Aerospace Power. **28,** （2013）
[6]   Shin J, Berkowitz B, Chen H H, et al. Journal of Aircraft. **31,** 2 (1994)
[7]   S. Özgen, and M. Canıbek. Heat & Mass Transfer.**45,** 3(2009)
[8]   Cebeci T, Chen H H, Alemdaroglu N. Journal of Aircraft. **28,** 9 (1991)
[9]   Zhang Chiyu, Xiao Zhixiang,Deng Yiju. Scientia Sinica(Physica, Mechanica & Astronomica). **44,** 9 (2014)
[10] Ren Pengfei, Xu Yu,Song Juanjuan, et al. Journal of Engineering Thermophysics(in Chinese). **35,** 4 (2014)
[11] Wang Kun, Bai Junqiang, Xia Lu, et al. Journal of Aerospace Power(in Chinese).**29,** 11(2014)
[12] Shin J, Bond T H. 30th Aerospace Sciences Meeting and Exhibit, Aerospace Sciences Meetings, Reno, NV,U.S.A. (1992)
[13] Harris C D. Nasa TM-81927, (1981)

ISPECE   IOP Publishing

IOP Conf. Series: Journal of Physics: Conf. Series **1187** (2019) 032038   doi:10.1088/1742-6596/1187/3/032038

Research on straightness error detection and quality control of multi-crankshaft bores for large medium speed engine block

Jiebin Yang [a], Xingxing Li, Chang'an Hu, Wanze Li, Jinwei Zhang

National Institute of Measurement and Testing Technology, Chengdu, 610021, China

[a] Corresponding author: 383710514@qq.com

Abstract. Crankshaft is the key component of internal combustion engine equipment, the machining precision of crankshaft hole has a direct impact on the performance and life of the whole engine. Aiming at the precision machining of crankshaft bore of large medium speed engine block with high precision and multiple crankshaft holes, the inspection tool and straightness measurement method of finishing process on crankshaft holes were studied. Based on the virtual manufacturing technology and the modular design concept of the checking tool and the laser detection and alignment technology, a relationship model describing the matching between the center straightness of the crankshaft hole and the crankshaft deflection is established. A single crankshaft hole shape and position tolerance measurement and multi-crankshaft hole are designed. The special measuring tool for measuring center deflection is verified by experiment to find the compensation law for the deflection of linearity evolution, and should be used in the process of finishing with crankshaft hole. The results of many cutting experiments and laser measurements show that the tool can effectively measure the size deviation and deflection of crankshaft holes, and form a quality control method for crankshaft holes finishing. This study aims to provide a set of measurement methods for crankshaft holes straightness error of large internal combustion engine block..

1. Introduction

Internal combustion engine is a kind of thermodynamic engine which burns the fuel in the engine and converts the heat energy from the fuel into power directly. Generally speaking, it mainly includes reciprocating piston internal combustion engine, rotating piston engine and free piston engine.

Large internal combustion engines are widely used in all kinds of transport vehicles (automobiles, tractors, diesel locomotives, etc.), mining, construction and engineering machinery. The precision grade of crankshaft holes, camshaft holes, cylinder holes and intermediate gear holes in the key parts of the engine block is very difficult to finish machining. Tool and product collisions are easy to occur when machining on the universal large gantry milling machine [1]. As the power output part of diesel engine, crankshaft hole assembly accuracy directly affects the efficiency of the product, and the machining accuracy of multiple crankshaft holes on the same block is the main influence factor of assembly accuracy. Too large or too small crankshaft holes will have a great impact on the service life and performance of the body. Therefore, the necessary prerequisite for accurate machining of multi-crankshaft hole straightness of large block is to detect the straightness in the machining process and select the corresponding compensation according to the measurement results, so as to adjust the follow-up processing [2], so that the measurement and compensation of machining process is particularly important for the precision machining process system of large-sized and high-precision

Content from this work may be used under the terms of the Creative Commons Attribution 3.0 licence. Any further distribution of this work must maintain attribution to the author(s) and the title of the work, journal citation and DOI.

Published under licence by IOP Publishing Ltd

block. The design of measuring tools for machining process around the effectiveness and efficiency has been widely concerned by engineers and technicians.

Zhang Kan developed a special combined boring machine for machining the spindle hole of locomotive engine block, analyzed and determined the technological requirements for repairing the spindle hole of the block and the deflection of the boring bar under two different conditions, designed and manufactured a special combined boring machine for repairing the spindle hole and the corresponding fixture [3]. At present, many manufacturers in China adopt the two-axis plating machine to roughen the camshaft hole of 6-cylinder internal combustion engine block. Due to the improper selection of machining amount, it is easy to produce the camshaft hole which can not completely eliminate the defects caused by rough machining after semi-finishing [4]. Professor Wu Xianming of Wisconsin New University of America is the first one in grinding process. The roundness error of workpiece was compensated successfully by DDS method [5]. The roundness of workpiece taper was reduced from 074 micron to 0 375 micron by modeling and compensating the radial error of spindle of external cylindrical grinder in Wisconsin University [6]. The roundness error of workpiece taper was measured by Kalman filter method in literature [7], and the accuracy of axial workpiece in external cylindrical grinding was effectively improved. But this kind of inspection tool can only be used for final inspection of products, and can not be used as a process inspection tool. In addition, the stability of machine tools, process parameters such as temperature, etc., will also have an impact on the accuracy [8]. Therefore, ensuring that every factor changes in its reasonable range is the key to ensure the quality of crankshaft bore boring.

Based on the above situation, taking the L-type large marine engine block of a certain company as the research object, based on the virtual manufacturing technology, drawing on the modular design concept of the checking tool and the laser detection and alignment technology, the relationship model describing the matching between the center straightness of the crankshaft hole and the crankshaft deflection is established, and the shape and position tolerance of the single crankshaft hole is designed. The special measuring tool for measuring the center deflection of multi-crankshaft hole is verified by experiment to find the compensation law for the deflection of linearity evolution, and should be used in the finishing process with crankshaft hole.

2. Machining accuracy of crankshaft hole for a large engine block
Large internal combustion engine has large structure size, high machining and assembling accuracy requirements, a wide range of processing parts, more process involved, special detection methods, so the requirements of equipment, technology, precision detection and other comprehensive conditions are very harsh. Fig. 1 is the design feature of multiple crank holes on a large internal combustion engine block.

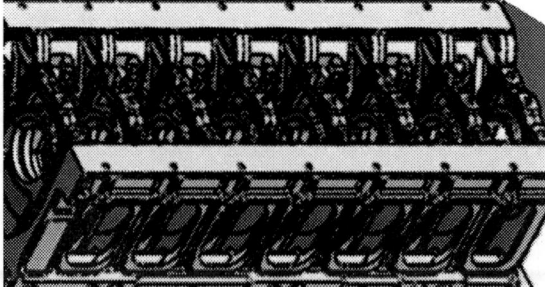

Figure 1. Design model of a large internal combustion engine block.

At present, the processing technology of large internal combustion engines is at a bottleneck stage in China. For one thing, the old-fashioned modular machine tools are used in machining methods, so their universality is poor; for another thing, the precision measuring tools are mostly conventional types, which can not achieve efficient and fast measurement, resulting in high rejection rate in processing foreign high-precision internal combustion engine products. With the development of internal combustion engine units towards super-large, high efficiency and complexity, the detection methods of machining errors and the design of special inspection tools are still the key research fields at home and abroad. Especially for a large internal combustion engine block shown in Fig. 1, the long-distance straightness inspection with multi-crankshaft holes and other characteristics is still the focus of attention, adn the measurement is the key problem that needs to be solved urgently.

The design structure for crankshaft hole of the large internal combustion engine block is shown in Fig. 1. The total length of the crankshaft hole is 5550mm. The diameter of the crankshaft hole is ø400 mm, the thickness is 100 mm, the distance between the two crankshaft holes is 500 mm, and the difference between the two crankshaft holes is 0.03 mm in the height direction. The deflection curve shows that the roughness of the inner hole is Ra1.6, the specific design parameters are shown in Table 1. It can be seen that the straightness tolerance is very strict, coupled with its high roughness requirements, if there is a small amount of deviation in the process, will lead to processing scrap.

Table 1. Design parameters of crankshaft bore.

Margin	mm
Crankshaft hole diameter	Ø400
Crankshaft hole thickness	100
Center distance	500
Height difference	0.03

According to the different models of the whole machine, there are 10 different sizes and specifications of the body structure, but the crankshaft hole center straightness tolerance and crankshaft hole size and tolerance are the same. Therefore, the design of a fixture can be applied to the measurement of the center straightness of the crankshaft holes of the various specifications.

3. Fixture design
In the national standard, the coaxiality tolerance of the axis is defined as "the tolerance zone is the area within a cylindrical surface with a diameter of and the axis of the cylindrical surface is coaxial with the reference axis". It has the following three control elements: a) the establishment of the reference axis; b) the establishment of the axis of the object under test; c) considering the actual work or assembly requirements for flexibility.

The function of the internal combustion engine block is clear and relatively stable. From the processing technology, the large Longmen milling machine can be processed in two states. Although the number of cylinder holes is different in different series, the spacing of all cylinder holes in the middle is the same, and the structural dimensions on both sides have the same distribution characteristics with the cylinder hole surface. Therefore, it is more appropriate to adopt structural modularization to design the processing tooling in two conditions. According to the above modular division of the block structure, the corresponding modular assembly structure is designed on the premise that the large gantry milling machine is selected as the processing equipment, as shown in Figure 2.

ISPECE   IOP Publishing

IOP Conf. Series: Journal of Physics: Conf. Series **1187** (2019) 032038   doi:10.1088/1742-6596/1187/3/032038

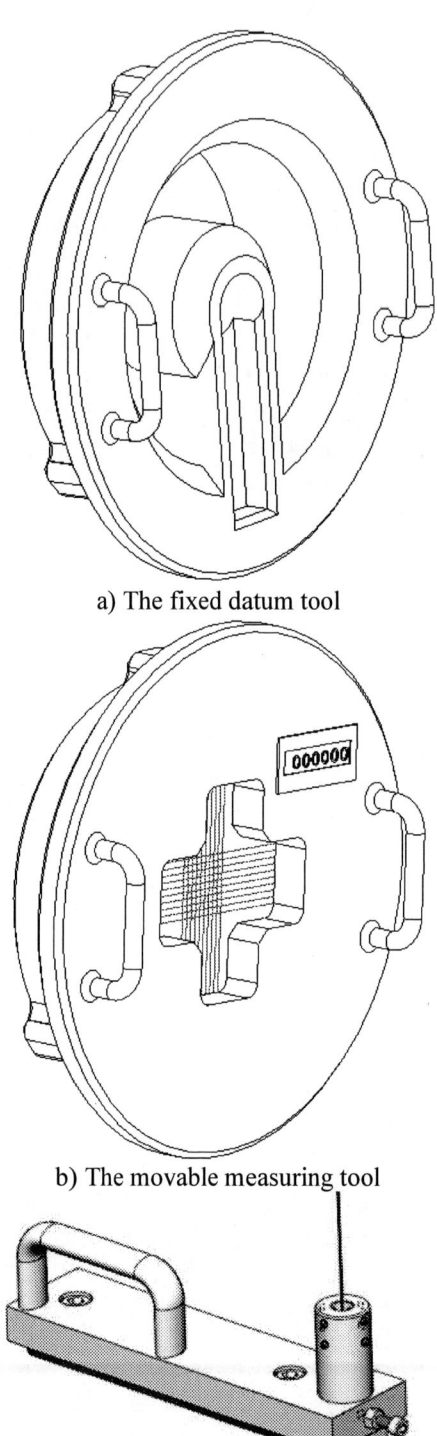

a) The fixed datum tool

b) The movable measuring tool

c) The laser emission and detection device
Figure 2. Special fixture of coaxiality for crankshaft bore .

748

The modular special measuring tool includes three sub-tools, which are fixed datum tool (the first crankshaft hole installed at any end of the measurement is regarded as the measurement datum), movable measuring tool (installed in the crankshaft hole to be detected) and laser measuring tool. The tool can realize the functions of unifying the locating datum of each mounting, satisfying the processing of different specifications of the engine block with the same locating datum, reducing the time of repeated installation and disassembly, and completing the overall processing time of the under-line mounting card. It provides a quick tool for the pipeline processing of the engine block.

4. Testing method for finishing process

In order to ensure the accuracy of measurement, a rectangular groove with a width of 0.3 mm is opened at the same position of the center hole of the benchmark and movable tooling, and a standard thickness measuring block is installed on the same side of the rectangular groove. The standard thickness of the measuring block is 0.2 mm.

The measuring principle adopted in this device is to convert the measurement of coaxiality of two holes into the offset measurement of four points on the upper and lower sides of the center hole of the two tooling, and then convert it into coaxiality. The accuracy of crankshaft hole coaxiality is judged by measuring offset. The precondition of measuring the special measuring tool is to ensure that the crankshaft hole circle runout processing conforms to the design scope. The specific measuring steps are as follows:

4.1 Preparation before measurement

Cleaning the of the tools surface to prevent foreign bodies from entering the inner surface of the tooling and crankshaft hole, otherwise the tooling can not be closely assembled with the parts under test, thus affecting the accuracy of the measurement data.

4.2 Loading card

Firstly, the benchmark fixture is installed in the first crankshaft hole on the left side as shown in Fig. 1, and the gap between the tool and the outer surface of the measured shaft is detected by using a 0.01 mm thickness stopper. If the stopper can be accessed locally and the symmetrical position is found, the fixture and the measured shaft are incorrectly clamped, and the benchmark fixture needs to be tapped straight. If there is only a gap on one side, the plug can enter, which means that the machining circle runout error of the inner surface of the crankshaft hole is large, and the value of the circle runout can be determined by selecting the gradually increasing thickness of the plug in turn. If the 0.01 mm thick plug can not enter any contact position between the tool and the measured shaft, the clamping is completed.

Secondly, the movable tooling is installed in the second position of the crankshaft hole, and the specific installation method and the method of checking the installation are the same as the above method.

4.3 The coaxiality Measurement

The laser measuring unit is placed in the fixed position of the benchmark fixture, and the laser beam is opened. The data displayed on the screen of the movable fixture is read and recorded, that is, the deviation value of the center point of the first crankshaft hole and the second crankshaft hole, which reflects the specific deviation of the center of the two measuring crankshaft holes in the longitudinal and horizontal directions respectively.

Remove the movable tooling from the second crankshaft hole, place it with the next crankshaft hole at one time, read the measurement data displayed on the display screen and record it, and so on until the deviation between all crankshaft holes and the first reference hole is completed.

Because the design requirements only care about the crankshaft hole center in the longitudinal direction of the deviation, so the recorded data of each measurement position is compared and processed, converted the deviation between the adjacent two crankshaft holes, and compared with the design accuracy to determine whether to meet the requirements of processing accuracy, so as to complete the measurement.

5. Conclusion

In this paper, a large internal combustion engine block is taken as the research object, and its finishing quality is measured by designing a special measuring tool for crankshaft hole. It can be seen that the simulation of product machining process based on virtual manufacturing technology can accurately evaluate the machining difficulties, specify effective measures and optimize the machining process, and solve the technical difficulties of machining the block. . Based on the virtual manufacturing technology and the modular design concept of the measuring tool and the laser detection and alignment technology, a special measuring tool is designed for measuring the shape and position tolerance of single crankshaft hole and the center deflection of multi-crankshaft hole. The fixture can effectively measure the size deviation and deflection value of the crankshaft hole.

References

[1] R.S. Zhang , R. Kang, Q.C. Cui. A new method of machining longer taper [J]. Machinery Design & Manufacture, **2006**(03): 91-92 (2006)

[2] Y.P .Research on deep hole machining mechanism and cutting parameter optimization technology for key parts of marine diesel engine[J].Zhenjiang: Jiangsu University of Science and Technology (2014)

[3] K. Zhang. Design and Manufacture of Combination Boring Machine, a Special Equipment for Machining Spindle Hole of Diesel Engine Block [D]. Xi'an University of Technology (2005)

[4] P. Zhou. Processing technology of camshaft bush burning and body camshaft hole of multi-cylinder internal combustion engine[J].Internal combustion engine, **1993** (05): 13-14 (1993)

[5] N.I. Jun. A perspective review of CNC machine accuracy enhancement through real-time error compensation [J]. China Mechanical Engineering, **8** (1): 29-33 (1997)

[6] R.J. Lian. A grey prediction fuzzy controller for constant cutting force in turning[J]. International Journal of Machine Tools & Manufacture, **2005** (45): 1047-1056 (2005)

[7] Q.J. Li. The design and application of universal taper testing tool [J]. Machinery, **2006**(10): 68 (2006)

[8] K.G. Fan, J.G. Yang, X. Yao. Error Compensation for Shaft Parts in Batch Manufacture Based on Newton Interpolation [J]. Journal of Mechanical Engineering, **47**(09): 112-116 (2011)

ISPECE   IOP Publishing

Study on Pressure Pulsation Suppression of Reciprocating Pump

Bin Li [1,a], Song Guo [1], Biao-hua Cai [1] and Zhao-cun Shi [1]

[1]Wuhan Second Ship Design and Research Institute, Wuhan, China

[a]Corresponding author: libinzju@sina.com

Abstract. The reciprocating pump has the problem of excessive pulsating pressure due to the inherent characteristics of the equipment, which causes the system to have large vibration and noise in the process of conveying fluid. Aiming at the problem of reciprocating pump pressure pulsation, the solution of reciprocating pump pressure pulsation suppression is put forward in this paper. The comparison test before and after the system improvement shows that the pulsating pressure inside the system has been significantly suppressed, and the pulsating abatement device adopted has a better application prospect.

1. Introduction

Reciprocating pump is often used to perform bilge discharge and other service tasks due to their excellent self-absorption performance. It has certain advantages for the medium conveying small flow and high head. However, due to the inherent discontinuous flow of the piston during the reciprocating movement, the discharge of the pump outlet is uneven, which causes the pressure pulsation of the system pipeline and makes the system inevitably impact and vibration during the fluid transportation. This may not only cause damage to the pipe and its accessories, but also transfer pipe vibration to the ship through various connectors, resulting in increased vibration and environmental noise of the ship. The piston in the first half of the stroke for acceleration, the latter half of the stroke for deceleration, its displacement is not uniform. Due to the inhomogeneity of instantaneous flow rate, the inertial force is generated by the variable acceleration of fluid in the pipeline. It is found that there are overturning moment, reciprocating inertia force and centrifugal force in the pump group. The unbalance force and fluid pulsation generated by these mechanical movements are the main reasons for the system to produce pressure pulsation.

2. Pressure suppression characteristics

The calculation principle model of the reciprocating pump system is shown in Fig. 1, where P stands for pressure pulsation, Q for flow pulsation, and Z_p for impedance of the pulsation abatement device. The following assumptions are made when calculating the model: the flow pulsation remains unchanged and the pressure pulsation changes after installing the device in the system. Pressure pulsation at the end of the system remains constant and P_0 is zero.

Content from this work may be used under the terms of the Creative Commons Attribution 3.0 licence. Any further distribution of this work must maintain attribution to the author(s) and the title of the work, journal citation and DOI.

Published under licence by IOP Publishing Ltd

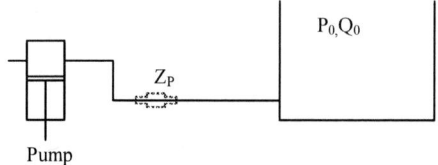

Figure 1. Pressure fluctuation model of the system

It can be obtained by the wave equation of the pipeline:

$$\begin{pmatrix} P_1 \\ Q_1 \end{pmatrix} = \begin{pmatrix} \cos\dfrac{\omega l}{a} & jZ_c \sin\dfrac{\omega l}{a} \\ j\dfrac{1}{Z_c}\sin\dfrac{\omega l}{a} & \cos\dfrac{\omega l}{a} \end{pmatrix} \begin{pmatrix} P_0 \\ Q_0 \end{pmatrix} \qquad (1)$$

$$\begin{pmatrix} P_1' \\ Q_1' \end{pmatrix} = \begin{pmatrix} 1 & 0 \\ \dfrac{1}{Z_p} & 1 \end{pmatrix} \begin{pmatrix} \cos\dfrac{\omega l}{a} & jZ_c \sin\dfrac{\omega l}{a} \\ j\dfrac{1}{Z_c}\sin\dfrac{\omega l}{a} & \cos\dfrac{\omega l}{a} \end{pmatrix} \begin{pmatrix} P_0 \\ Q_0' \end{pmatrix} \qquad (2)$$

- ω is the pulsating angular frequency;
- l is the length of pipe;
- a is the sound velocity in the fluid medium;
- Z_c is the characteristic impedance of the pipeline;
- ρ is the fluid density;
- A is the cross sectional area of the pipeline;
- Z_p is the impedance of the pulsating subduction device;
- K_v is the volume stiffness of the elastomer.

$$P_1 = jZ_c \cdot Q_1 \cdot tg\dfrac{\omega l}{a} \qquad (3)$$

$$P_1' = \dfrac{jZ_c \sin\dfrac{\omega l}{a}\cdot Q_1}{j\dfrac{Z_c}{Z_p}\sin\dfrac{\omega l}{a} + \cos\dfrac{\omega l}{a}} \qquad (4)$$

The value of pulsation suppression:

$$L_p = 20\lg\left(\left|1 - \dfrac{Z_c}{K_v}\omega \cdot tg\dfrac{\omega l}{a}\right|\right) \qquad (5)$$

3. Numerical simulation of system

3.1 Device model

Aiming at the pressure pulsation mechanism of reciprocating pump, the pulsation abatement device can be installed in the system pipeline to suppress the amplitude of pressure pulsation inside the system. Select l is 250mm, L is 500mm, d is 50mm, D is 150mm.

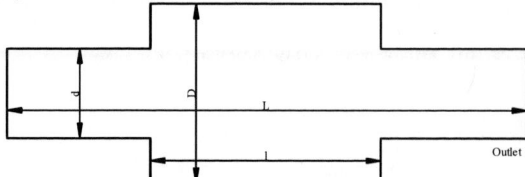

Figure 2. Schematic structure diagram of pulsation abatement device

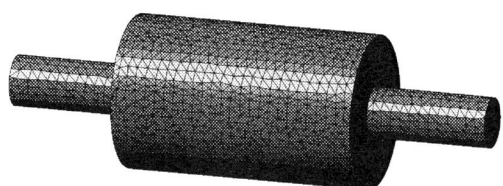

Figure 3. Schematic mesh diagram of pulsation abatement device

The grid size is set to 10mm, the absolute sag is set to 1mm, and the element type is selected as linear. The wave absorber is meshed as Fig 3. The results are as follows:

- 100.0% of the elements valid up to 3788.6Hz;
- 80.0% of the elements valid up to 4262.6Hz;
- 60.0% of the elements valid up to 4548.8Hz;
- 40.0% of the elements valid up to 4903.5Hz;
- 20.0% of the elements valid up to 5306.0Hz.

In order to guarantee the calculation accuracy, the calculation upper limit frequency should not exceed 3788.6Hz.

3.2 Simulation model

3.2.1 Parameter Settings

It is necessary to define a sound absorbing property on the outlet to simulate the non-reflecting boundary. Acoustic impedance of fluid is set as 416.5. The boundary type of outlet is impedance. The simulation frequency is from 10Hz to 3500Hz and the step length is 10Hz.

3.2.2 Sound pressure response

The lower side is the damper inlet and the upper side is the damper outlet, and the sound pressure response at the inlet and outlet of 10Hz to 3500Hz are analysed. The sound pressure response under typical frequency conditions is shown in the following figure.

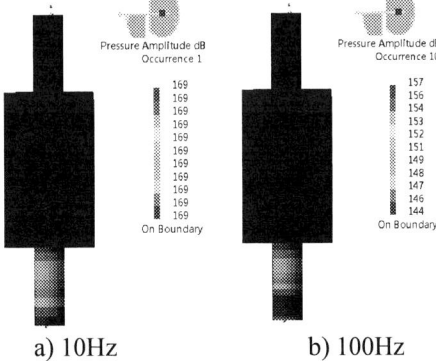

a) 10Hz　　　　　　　b) 100Hz

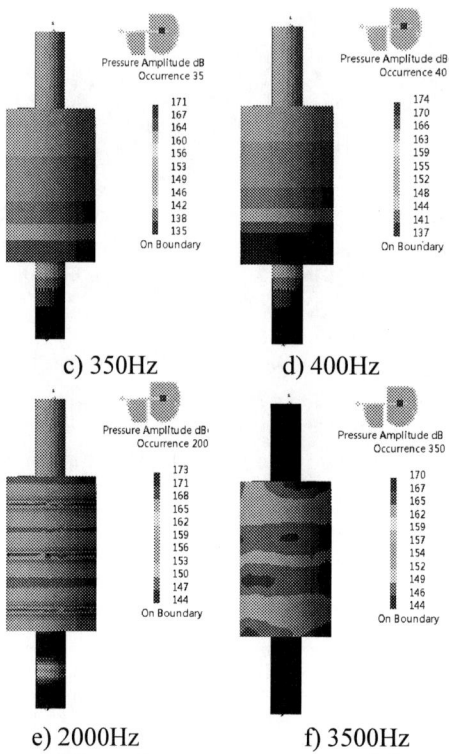

c) 350Hz d) 400Hz

e) 2000Hz f) 3500Hz

Figure 4. Diagram of sound pressure response at inlet and outlet ends

It can be seen from Fig 4 that the pulsation abatement device has a good effect on the intermediate frequency about 350Hz to 1000Hz. The pressure level frequency response curve is shown in Fig 5.

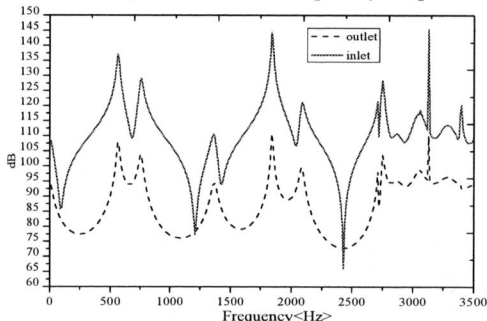

Figure 5. Curve of pressure level frequency response

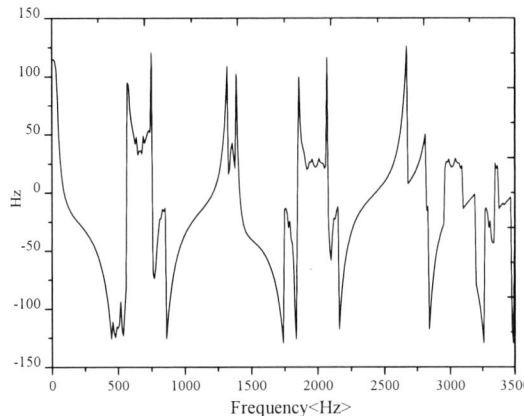

Figure 6 Curve of acoustic pressure level frequency contribution

According to the simulation results, the variation trend of the outlet pressure of the pulsation abatement device is similar to that of the inlet pressure. The result of simulation shows that the pulsation abatement device has pressure pulsation suppression effect on the range of 10Hz to 3500Hz. Low frequency and high frequency contribute less to the acoustic pressure of the pulsation abatement device.

3.2.3 Insertion loss

According to formula 5, the insertion loss of the pulsation abatement device is calculated. The calculation results are shown in the figure below.

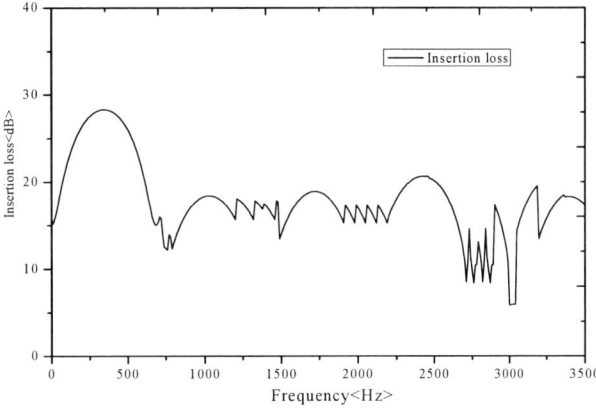

Figure 7. Curve of insertion loss

4. Experiment

4.1 Experiment system

The test system is mainly composed of reciprocating pump, wave damper, ball valve, check valve and pressure sensor, which as shown in Fig 8 below.

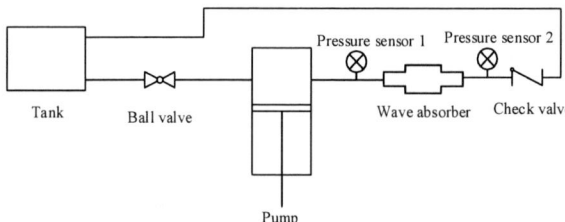

Figure 8. Schematic diagram of test system

Dynamic signal analyzer records the time-domain curve of pressure pulsation by pressure sensor, the frequency domain line spectrum can be obtained by FFT, and the total value of pressure pulsation within a certain frequency range is calculated by the formula below.

$$P_1 = jZ_c \cdot Q_1 \cdot tg\frac{\omega l}{a} \tag{6}$$

● n is the number of frequency;

● P_r is the reference pressure.

4.2 Experiment conditions

During the test, a section of rigid pipe is firstly installed in the test section, and the pressure pulsation value of the system is recorded with the dynamic signal analyser. Then replace the rigid tube of the test section with the pulsation abatement device and repeat the above test. The difference of pressure pulsation value measured. The difference of pressure is the pulsation dampening effect of the pulsation abatement device.

$$\Delta L_{pt} = L_{pt}^G - L_{pt}^T \tag{7}$$

4.3 Experiment results

Fig 9 shows the time-domain comparison curve before and after the installation of the pulsating abatement device measured by the dynamic pressure sensor. It can be seen that the pulsating pressure amplitude of the system before installation can reach to 0.35MPa. After the installation of the pulsating abatement device, the system pulsating pressure is generally lower than 0.05MPa, and the dynamic pressure pulsation inside the installation system is significantly reduced.

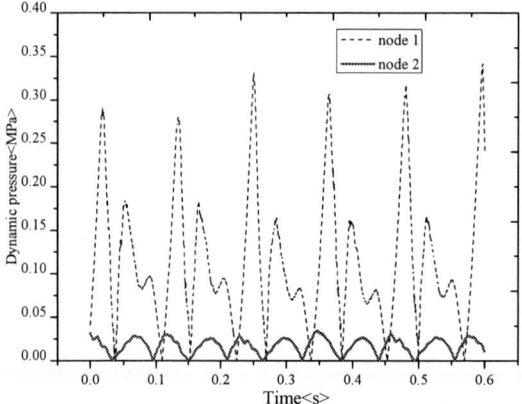

Figure 9. Time domain contrast signal of dynamic pressure

The insertion loss in the frequency domain as shown in Fig 10 is obtained after processing according to formula 6 and formula 7. In the low frequency band below 600Hz, the pulsating abatement device has a relatively good effect on the pressure pulsation suppression of the system, generally above 15dB, which is mainly caused by the low-frequency pressure of the reciprocating pump.

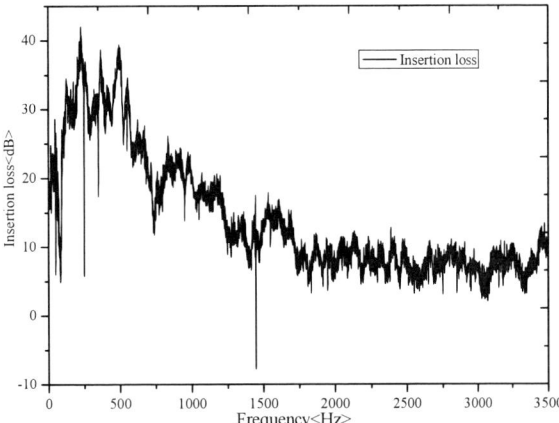

Figure 10. Insertion loss of fluctuation pressure

It is known that the outlet pipe of the reciprocating pump system is 65mm, the crankshaft speed of the bilge pump is 170r/min, water sound speed is 1500m/s, and the elastic volume stiffness is about 1000MPa/m³. Pipe impedance is 4.52e⁸, pulse angular frequency is 70.9. Assuming the pipe length and substitute the above data into formula 7, the pressure pulsation suppression effect can be calculated as:

$$L_{p=0.5} = 20 \lg \left(\left| 1 - \frac{4.52e8}{1000e6} 22.6 \cdot \pi \cdot tg \frac{22.6 \times 5 \cdot \pi}{1500} \right| \right) \quad (8)$$

In the range of 10Hz to 3.5kHz, the pressure pulsation of reciprocating pump system can be suppressed by installing pulsation abatement device. The pressure pulsation value of the system before installation is 98.2dB and after installation is 80.6dB, which can reduce the system pressure pulsation about 17.6dB and is basically consistent with the calculation results. The calculation result of insertion loss is 16.5dB.

5. Conclusions

The pressure pulsation of the system is caused by the inherent discontinuity of the reciprocating pump. In order to suppress the pressure pulsation of the system, this paper mainly works from several aspects. Firstly, the characteristics of internal pressure pulsation and its influencing factors are analysed, and the main technical scheme of the pressure pulsation suppression is determined. Secondly, according to the research plan, the simplified calculation model of the pressure pulsation of the system is established by using the pressure fluctuation equation of the pipeline system, and the suppression effect of pressure pulsation is preliminarily calculated. Thirdly, test verification is carried out for typical working conditions, and the test data are analysed. The research in this paper shows that the wave damper can significantly suppress the fluid pressure pulsation of the reciprocating pump and has a good application prospect.

According to add the pulsation abatement device to the reciprocating pump system can effectively suppress the pulsating pressure in the process of reciprocating pump operation. According to the working frequency of the pump set, the pressure pulsation of the system is effectively reduced, the pressure of the system is stabilized, and the vibration and noise of the system are effectively reduced.

References

[1] Chen Liao-yuan, Zhu Zeng-bao. Studies on the hydraulic characteristic simulation for reciprocating pump-unit. Journal of Anhui University of Science and Technology (Natural Science), 2006, 12(4): 50-54.

[2] J A Ferreira,F Gomes Almeida. Hybrid models for hardware-in-the-loop simulation of hydraulic systems[J]. Proceedings of the Institution of Mechanical Engineers, Journal of Systems and Control Engineering, 2004 218: 465-473

[3] Majid Nabavi, Kamran Siddiqui, Javad Dargahi. Analysis of the flow structure inside the valveless standing wave pump[J]. PHYSICS OF FLUIDS 20, 2008:1-10

[4] Chen Bo, Yang Guoping. Modeling and simulation of axial piston hydraulic Pump[J].Third International Conference on Measuring Technology and Mechatronics Automation, 2011 : 609-612

[5] Yeneha,Miteh.Automatic tool changer gives robots more flexibility. Welding design and fabrieation, 2003(7) : 22

[6] Choong-Hwi LEE, Jae-Eung OH, Yong-Goo JOE, You Yub LEE. The Performance Improvement for an Active Noise Control of an Automotive Intake System under Rapidly Accelerated Conditions[J]. JSME International Journal, No. 03-5038

[7] Yuan Qi-hui. Simulation research on performance of exhaust muffler of automotive based on virtual lab. Chongqing Jiaotong University[D]. 2013.

[8] Rufin Makerewicz. Infuence of ground effect and refraction on road traffic noise[J]. Applied Acoustics,1997.

[9] K.S.Peat. A Numerical Decoupling Analysis of Perforated Pipe Silencer Elements[J]. Journal of Sound and Vibration,1988,123(2): 199 - 212.

[10] Barbieri R, Barbieri N. Finite element acoustic simulation based shape optimization of a muffler[J]. Applied Acoustics, 2006, 67(4) : 346-357.

Design of Automated Guided Vehicle for Conveying Objects

Denggui Wang, Xinling Ma

School of Transportation and Automotive Engineering, PanZhihua University, Pan-Zhihua 617000, China

254601384@qq.com, 1416484791@qq.com

* Corresponding Author: *Denggui Wang*; 254601384@qq.com:.; 18081712617..; 0812-3373338:..

Abstract: STC90C516RD+microcontroller was chosen as the core controller of AGV. The magnetic line was detected by Hall sensor, the tracking function of AGV is realized, the ultrasonic ranging function was used to avoid obstacles, and the differential steering of the car was realized by using the output PWM wave, after installation and debugging, AGV can basically achieve the requirement of conveying objects.

1. INTRODUCTION

Intelligent robots are taking the place of human beings to accomplish these tasks, and can adapt to different environments, not affected by temperature, humidity and other conditions, and fulfill special tasks in dangerous areas and human beings can not intervene, AGV is one of them. AGV is widely used in factory production, national defense, medical industry, accident search and rescue, geological exploration, scientific research and so on. [1] AGV is mainly used in the enterprise, instead of manual or manual mechanical handling, play the role of automatic sourcing and transportation to the destination.

2. GENERAL PLAN DESIGN

The hardware of AGV mainly includes: MCU, trace module, obstacle avoidance and collision avoidance module, acousto-optic alarm module, display module, DC motor drive module, power steering module, power module and car body. Its overall structure is shown in Figure 1. The system used STC90C516RD, a 8-bit microcontroller produced by STC Company, as the microcontroller to realized the motor driving and steering, as well as the signal processing of electromagnetic tracking and ultrasonic ranging obstacle avoidance alarm.

Content from this work may be used under the terms of the Creative Commons Attribution 3.0 licence. Any further distribution of this work must maintain attribution to the author(s) and the title of the work, journal citation and DOI.

Published under licence by IOP Publishing Ltd

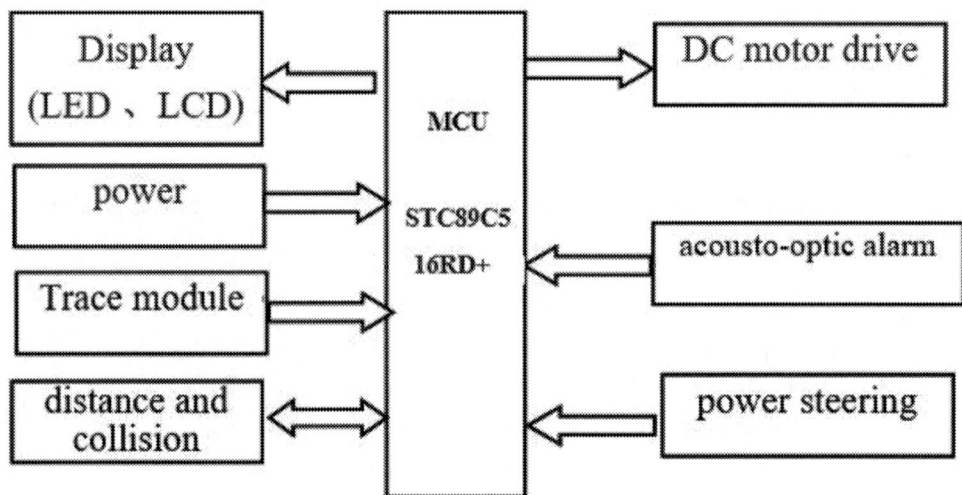

Fig.1System hardware structure diagram

2.1Trace Module

There is a sinusoidal current in the center of the feeding path. according to Biot-Savart law, the changing electric field excites the changing magnetic field, and the changing law of the magnetic field is consistent with the electric field. For a steady current I, a magnetic field is generated around a straight wire of length L[2]. The magnetic induction at point P is as follows:

$$B = \int_{\theta_1}^{\theta_2} \frac{\mu_0 I}{4\pi r} \sin\theta d\theta \tag{1}$$

In equation (1), $\mu_0 = 4\pi \times 10^{-7} N / A^2$, and for wireless long DC conductors, where $\theta_1 = 0, \theta_2 = \pi$, there is:

$$B = \frac{\mu_0 I}{4\pi r} \tag{2}$$

When a coil is wound above an electrified wire, the positive current will cause a change in magnetic flux through the coil loop, and the induced EMF e will be produced in the loop. According to Faraday's law of electromagnetic induction, the magnitude of the induced EMF is proportional to the rate of change of magnetic flux Φ through the wire loop, that is, the magnitude of the induced EMF is proportional to the rate of change of magnetic flux through the wire loop. That is:

$$e = -\frac{d\phi}{dt} \tag{3}$$

If the distance between the center of the coil and the wire is d and the magnetic field distribution is approximately uniform in a small range, the induced EMF in the coil is:

$$e = -\frac{d\phi}{dt} = \frac{k}{d} \frac{di}{dt} \tag{4}$$

k is the area of the coil. Formula (.4) shows that the magnitude of the induced electromotive force in the coil is proportional to the rate of change of the current and inversely proportional to the distance from the center of the coil to the wire. That is, the closer the inductance is to the live wire, the stronger the inductive electromotive force is, whereas the farther the inductance is from the live wire, the weaker the inductive electromotive force is.

The electromagnetic sensor used in this design is 3144 type Hall sensor. A symmetrical electromagnetic sensor is installed in front of the car. When the car deviates from the electrified wire in the course of automatic driving, the electromagnetic sensor in front of the car will produce inductive electromotive force. Ideally, the electrified wire passes right through the robot's central axis, and the electromagnetic sensors on both sides produce the same inductive electromotive force with the same magnitude and direction, the difference being zero. If the car deviates from the direction in the course of driving, the EMF difference will be generated by the electromagnetic sensors on both sides. [3] The EMF difference will be converted into digital signal by AD converter and processed by MCU to adjust its driving direction.

Because the electromotive force directly induced by inductance coil has the characteristics of weak signal and much noise, it is necessary to select frequency, amplify and detect the signal in order to guide the car forward accurately by current magnetic field. The correlation current is shown in Figure 2.

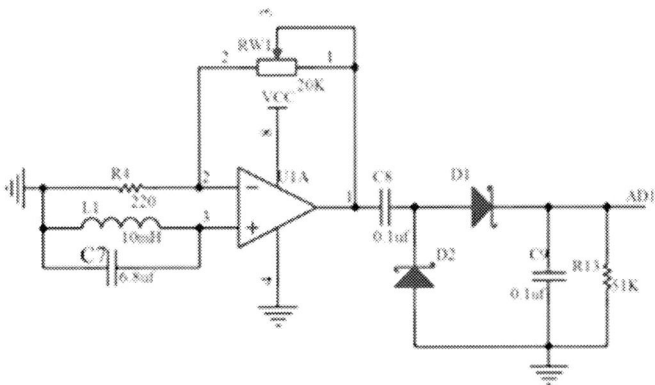

Fig. 2 Circuit of signal processing

2.2Distance and collision avoidance module

The working principle of system ranging is that the transmitting probe continuously emits 40 kHz ultrasonic wave under the control of single chip microcomputer. When it encounters obstacles, it produces reflected wave, and the receiving probe receives the reflected wave signal and converts it into electric signal.

When the first echo signal is obtained, the time difference between transmitting and receiving is Δt, and the distance of obstacle can be calculated according to the $S = C \cdot \Delta t / 2$ (C is the speed of sound).Because the speed of sound in air is related to temperature, if the temperature is constant, the speed of sound will not change. At 0 degrees Celsius, sonic $C_0 = 331.5$m/s. For any temperature T_i, the speed of sound C_i will be $C_i / C_0 = T_i / 273$,that is.

$$C_i = 331.5 / 273 \qquad (5)$$

According to the related literature [5], the sound speed increases about 0.607m/s with the temperature rising one degree, and the relationship between the sound speed C and the field temperature T is obtained as follows:

$$C = 331.5 + 0.607T \qquad (6)$$

so: $\qquad S = (331.5 + 0.607T) \cdot \Delta t / 2 \qquad (7)$

The temperature T in the form is measured by the temperature test module (DSB1820). In a very short period oftime $\Delta t'$, the displacement of the truck is ΔS, in a timely manner:

$$v = \lim_{\Delta t \to 0} \frac{\Delta S}{\Delta t'} \qquad (8)$$

The ultrasonic transmitting circuit of this system is mainly composed of reverser 74AHC04 and ultrasonic transducer The receiving circuit adopts integrated circuit CX20106A[4]. After receiving square wave, CX20106A outputs low level to single chip microcomputer. The circuit is shown in Figure 3.

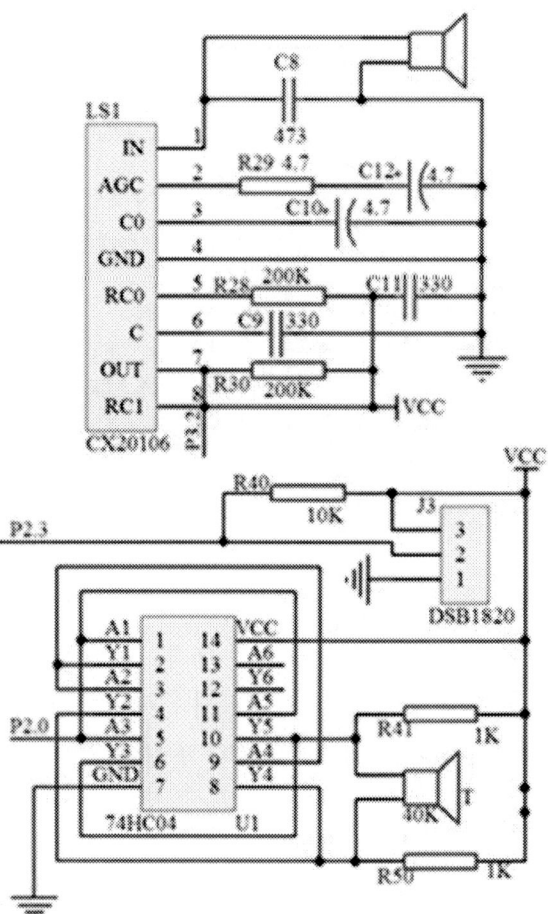

Fig. 3 Circuit of ultrasonic distance measuring

2.3Design of driving steering circuit

2.3.1kinematic analysis

In order to facilitate the kinematics analysis of the car, a coordinate system is first established as shown in Figure 4.

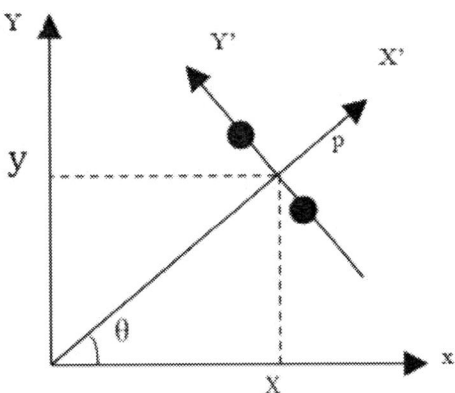

Fig. 4 coordinate system for kinematic analysis

Set up two coordinate systems for the intelligent car. XOY is the world coordinate system, X'PY' is the moving coordinate system of the car, and PX' is the moving direction of the car. At any time, the position and direction of the car in the space, that is, the position of the trolley can be x, y, θ. There are three parameters to represent, where x is the moving component of the car on the X axis, y is the moving component of the car on the Y axis, and θ is the angle between the position of the car at any time and the positive direction of the X axis.[5] The so-called intelligent car differential drive means that the speed of the car's left and right wheels is used to control whether the car is running in a straight line or turning left and right.

Using v_l and v_r to express the speed of the left and right wheels of the car respectively, the calculation formula of the amount of movement of the car in space is as follows:

$$x(t) = \frac{1}{2}\int_0^t [v_r(t) + v_l(t)]\cos[\theta(t)]dt \tag{9}$$

$$y(t) = \frac{1}{2}\int_0^t [v_r(t) + v_l(t)]\sin[\theta(t)]dt \tag{10}$$

$$\theta(t) = \frac{1}{l}\int_0^t [v_r(t) - v_l(t)]dt \tag{11}$$

In equation (11) ,l is the distance between the two driving wheels of the trolley. The following conclusions can be drawn from the above three formulas: when the speed of the two wheels of the car is equal and the direction is the same, the trajectory of the car is a straight line; when the speed of the two wheels of the car is equal and the direction is opposite, the trajectory of the car is rotated around the origin of the moving coordinate system; when the direction and speed of the two wheels of the car are the same. When the size remains constant and the speed difference remains unchanged, the trajectory of the car is circular.

2.3.2 Driving diagram Circuit

The system uses two speed-down DC motors to control the front wheels and the left and right wheels of the car respectively, corresponding to the four-channel output of L298. Optical coupler TLP521-1 is added between L298 and MCU to completely isolate the interaction between the MCU system and the motor circuit, so as to completely eliminate the influence of motor operation on the system. Its circuit is shown in Figure 5.

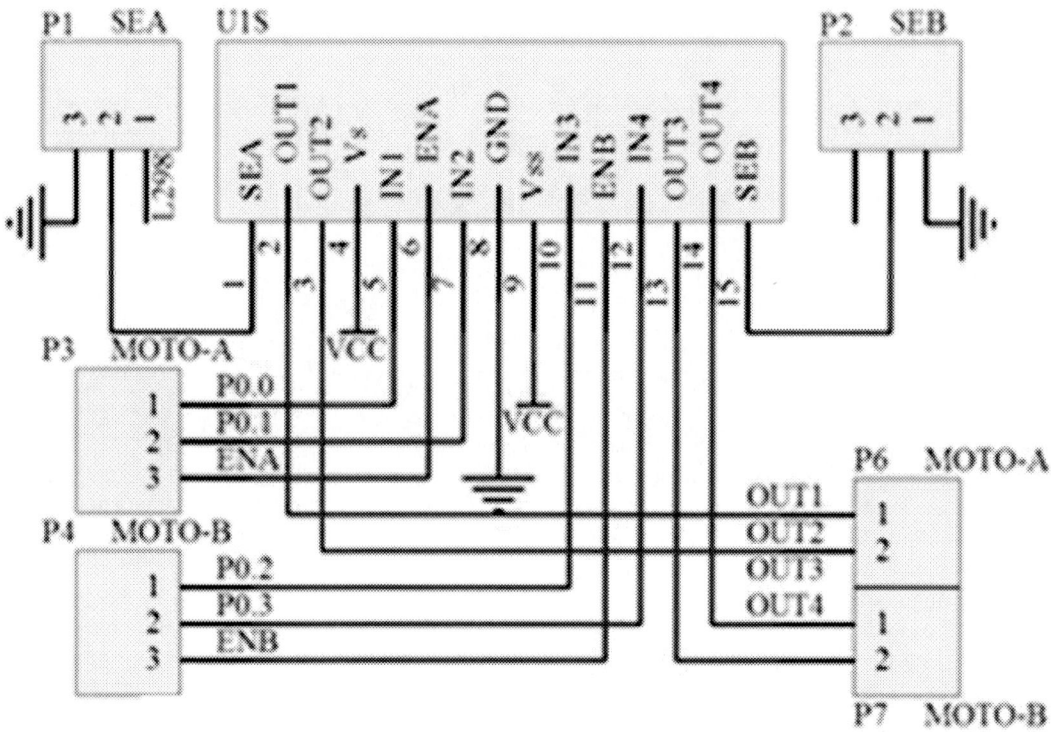

Fig.5 Circuit of driving diagram

2.4Design of display function

The display module mainly includes three parts: LED indicator display, LED digital tube display and LCD display. The indicator is mainly used to indicate the working state of the truck, which is programmed by users according to their needs. The LED digital tube can be used to display four different speed grades and vehicle operation modes of the moving vehicle. LCD SMC 1602A LCM is mainly used to display the distance from the obstacle, the current speed and ambient temperature of the moving vehicle [6]. Software design.

The software of the system is mainly composed of main program, shoe-avoiding subroutine, tracking subroutine, ranging subroutine, display subroutine, infrared decoding subroutine and delay subroutine, etc.

3. SOFTWARE DESIGN IDEAS

When the control system is powered on, relevant signals will be judged. According to the signal conditions, corresponding controls will be made by the intelligent control program. The general idea of the program is shown in Figure 6.

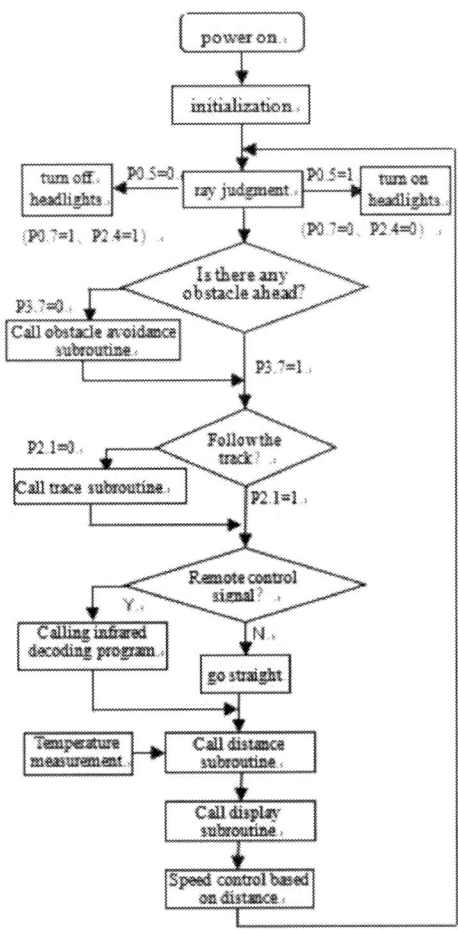

Fig.6 Program flow chart

According to the design idea of the flow chart in Fig. 6, the main program is taken as the main line, and each subroutine is constantly called by the main program according to the level state of the control signal, so as to control the whole running system. The main program main( ) is as follows:

```
Void main (void)        //main program entry
 { bit ExeFlag=0;       //Define executable bit variables
   Init();              //Initialization function
   while(1)             //Program main loop
    { if(RL1==0)        //Ray judgment
       { RightLed=1;
         LeftLed=1;  }
        else
       { RightLed=0;
         LeftLed=0; } }
FontIR();               //Obstacle judgement function
 { if(P2^0==0)          //Judgement tracking function
   {Xunji();}           //Tracking function
    Csbcj();    //Ultrasonic ranging function
    Display(dis);       //Display distance
    Chvast()}}          //Speed regulating function
```

In addition to the subroutine invoked in the main program, the remote signal acquisition program is set in the form of interrupt service subroutine. When P3.3 receives the remote signal, it sends the interrupt request signal to the MCU in time.

4. EXPERIMENTAL ANALYSIS
The experimental prototype is shown in Figure 7.

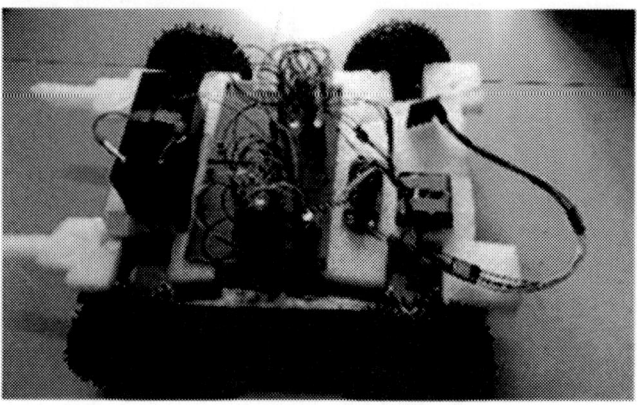

Fig. 7 Prototype vehicle

The experimental prototype is tested. The current temperature, obstacle distance and speed are displayed by LCD. When approaching the obstacle, it can send out the alarm message perfectly and bypass the obstacle or cliff to continue running; If the automatic tracking function is activated, it can drive automatically in the electromagnetic track, with load capacity up to 50Kg, and it also has a good performance in the path with large bending degree.

However, due to the influence of environment temperature and the time taken by operating instructions of Single ChipMicyoco [7], there is a certain deviation between the distance measured and the actual distance, as shown in table 1:

Table 1 detection distance and measured values (unit: cm)

Actual distance	Detection distance	absolute error	Actual distance	Detection distance	absoluterror
20	22.2	2.2	180	181.9	1.9
50	52.2	2.2	230	231.9	1.9
90	92.1	2.1	270	271.8	1.8
130	132.0	2.0	300	201.7	1.7

In order to reduce the distance measurement bias, the linear fitting was adopted to correct the data. In Matlab, the least square method was applied to the linear fitting of first-order polynomial: $y=ax+b$, the coefficient $a=0.9983$ and the coefficient $b=2.2526$ were obtained. The modified result is shown in figure 8:

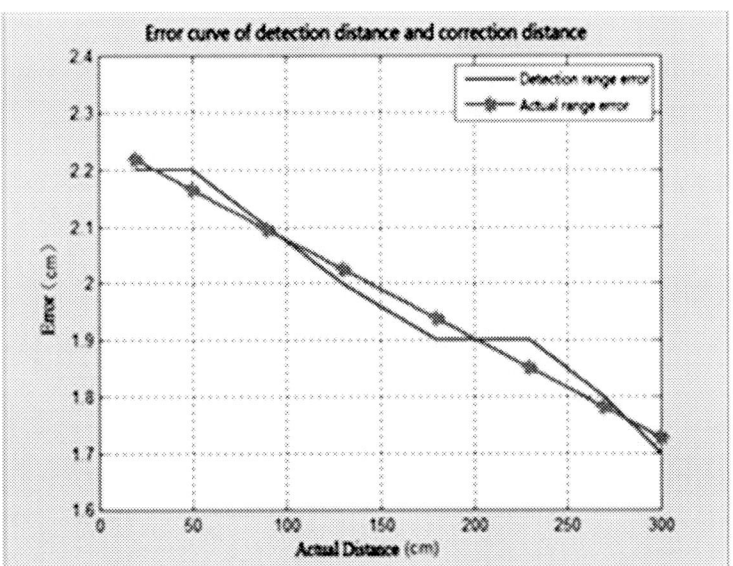

Fig. 8 Error curve of detection distance and correction distance

5. CONCLUSION

To sum up, AGV is a very good hardware platform, and the cost is not expensive, which can be widely used for moving objects in underground mines, automated warehouses, factory workshops and large shopping malls, etc. According to the practical needs of life, some control circuits can be added to complete the task of obstacle avoidance robot, fire fighting robot, transport robot, driverless car and so on. This design has certain popularization and application value.

REFERENCES

[1] J.D. Zhang, Control system of intelligent material handling trolley [P]. Chinese patent: 206178473 2017-05-17.

[2] J. Xing, Differential steering and torque matching of four-wheel drive wheeled robot [D]. Shanghai: tong ji university, 2008.1.

[3] Y.Q. WANG, Design of Anti-collision Warning System of Coal Mine Electric Locomotive Based on AT89C52 Single-chip Microcomputer [J]. Coal Mine Machinery, 2013, 34 (3):248-250

[4] W.CH. Guo, McS-51 microcontroller principle, interface and application [M]. Beijing: electronic industry press, 2013.10-15.

[5] W. Guo, The design and implementation of logistics automatic tracing truck control system [J]. China Foundry Machinery & Technology, 2015, (4):39-41.

[6] L.H. Liu, Simulative and Experimental Study on Stand−alone Solar Photovoltaic Inverter [J]. Journal of Liaoning Institute of Science and Technology,2017,19 (5), 11.

[7] Attaran M. RFID: an enabler of supply chain operations[J]. Supply Chain Management, 2007, 12(4):249-257

ISPECE   IOP Publishing

IOP Conf. Series: Journal of Physics: Conf. Series **1187** (2019) 032041   doi:10.1088/1742-6596/1187/3/032041

A Hydraulic Fault Diagnosis Method Based on IMF Entropy Feature Fusion

Liu Min[1], Huang Jie[2*], Xianhai Sun[2]

[1]Changsha Mystical Bow Information Science And Technology Co.,Ltd, Nanjing 210007,China

[2]School of Field Engineering, Army Engineering University,Nanjing 210007,China

Email of all the authors: 270911114@qq.com;huangjie051501@126.com; 36026435@qq.com

* Corresponding Author: Huang Jie; email: huangjie051501@126.com.

Abstract: A feature extraction and fault diagnosis method based on IMF entropy feature fusion was proposed for external vibration signals in five common fault states of hydraulic equipment: normal, leakage, blockage, air cavity and impact. Firstly, all kinds of signals were decomposed by improved EMD based on frequency cutoff, and effective IMF components were screened, then the fusion features of multiple information entropy were extracted, and then the deep learning method of DBN was adopted for feature learning and status recognition. The experimental results show that this method has high recognition accuracy and can effectively realize multi-fault recognition of hydraulic system.

1. Introduction

The working environment of hydraulic equipment is relatively harsh, and it is often exposed to unstable working state such as high load and strong impact. Therefore, the hydraulic system is most prone to faults in the mechanical equipment. In addition, the hydraulic oil used in the hydraulic system is easy to be polluted by environment and impurities inside the equipment. As a result of the oil pollution, the hydraulic system is prone to blockage, air pockets, leakage and other failures, resulting in damage to the operation of mechanical equipment. The oil leakage will also cause serious pollution and damage to the environment. Therefore, the study of the state monitoring and fault diagnosis technology of the hydraulic system is of great significance to extend the service life of equipment, ensuring product quality, preventing major economic losses, avoiding serious safety accidents and environmental pollution. The core of fault diagnosis of hydraulic system is elimination of disturbance, extraction of effective features and fusion recognition.

2. EMD decomposition and IMF component acquisition

Four typical hydraulic faults, including leakage, blockage, air cavity and impact, are set up through the hydraulic test bench. Empirical Mode Decomposition (EMD)[1,2] is an adaptive processing method based on the characteristic scale of signals, which can decompose a complex non-stationary signal into Intrinsic Mode Function (IMF) [3]and a residual trend Function according to its Intrinsic characteristics. The IMF component contains the intrinsic frequency of the signal, reflecting the inherent volatility within the signal. In theory, IMF components should meet the following conditions: 1) the difference between all extreme points and the number of zeros is not greater than 1; 2) the upper

Content from this work may be used under the terms of the Creative Commons Attribution 3.0 licence. Any further distribution of this work must maintain attribution to the author(s) and the title of the work, journal citation and DOI.
Published under licence by IOP Publishing Ltd

and lower envelope is locally symmetric about the horizontal axis.

Screening flow chart and specific steps are as follows:

(1) Determine all local extremum points of the analysis signal, and use cubic spline interpolation method to fit all maximum points to get upper envelope lines, and similarly fit all minimum points to get lower envelope lines.

(2) Calculate the mean value of upper and lower envelope lines. The average envelope curve $m_1(t)$, Subtract $m_1(t)$ from $x(t)$, Get $h_1(t)$

$$h_1(t) = x(t) - m_1(t) \tag{1}$$

(3) Judge whether $h_1(t)$ meets the two IMF conditions. If yes, then $h_1(t)$ is the first IMF component. $c_1(t) = h_1(t)$; If not, Repeat steps (1)~(2) in place of $x(t)$ with $h_1(t)$, Until the new $h_1(t)$ is an IMF, let's call it $c_1(t)$.

(4) Subtract IMF component $c_1(t)$ from $x(t)$ and get residual $r_1(t)$, Replace the analysis signal $x(t)$ with $r_1(t)$ and repeat steps (1) ~(3) until $r_n(t)$ is a monotonic function.

$$\begin{cases} r_1(t) = x(t) - c_1(t) \\ r_2(t) = r_1(t) - c_2(t) \\ \quad\vdots \\ r_n(t) = r_{n-1}(t) - c_n(t) \end{cases} \tag{2}$$

(5) Finally, signal $x(t)$ is decomposed into n IMF components and a residual value. The IMF component $c_1(t), c_2(t), \cdots, c_i(t)$ represents the components of different frequency bands from high to low, and the residual function $r_n(t)$ reflects the average trend of signals.

$$x(t) = \sum_{i=1}^{n} c_i(t) + r_n(t) \tag{3}$$

It is proved that some problemsi are easy to occur when using EMD directly to process signals. For example, mode mixing[4], endpoint effect problem[5]. The problem of false components and overshoot and undershoot makes IMF components doped with many false components, which is not conducive to feature extraction and accurate identification of faults.

3. Improvement of EMD method based on frequency cutoff

This method takes the minimum characteristic frequency of the signal itself as the decomposed cutoff frequency of the EMD method. When the main frequency component of the decomposed IMF component is less than the cutoff frequency, the decomposition ends. The detailed decomposition steps are as follows:

(1) carry out spectral analysis on decomposed signal $s(t)$ to obtain the effective characteristic frequency of the signal, and select the minimum characteristic frequency as the cutoff frequency of decomposition termination condition, denoted as f_{sd}.

(2) EMD screen $s(t)$. For each decomposition, one IMF component c_i is obtained, and power spectrum analysis is carried out to find the frequency with the maximum amplitude in the frequency component of c_i, which is denoted as f_{max}.

(3) Compare the sizes of f_{max} and f_{sd}. If f_{max} is greater than f_{sd}, go back to step (2) and continue the decomposition; If f_{max} is less than or equal to f_{sd}, the decomposition stops.

(4) Finally, a set of IMF components $c_1, c_2, \cdots, c_n$ and a residual function r_n are obtained, and there are

$$s(t) = \sum_{1}^{n} c_i + r_n \tag{4}$$

4. Feature extraction based on IMF entropy feature fusion

Taking a typical cavitation fault signal in the experimental sample as an example, the extraction process of the fusion IMF entropy feature vector is described. Figure 1(a) is the spectrum diagram of typical normal state signals as a reference, Figure 1(b) is the spectrum diagram of cavitation fault signal, and Figure 2 is the decomposition result.

After EMD decomposition, the cavitation fault signal was decomposed to obtain the IMF component of 11th order and the residual of 1st order, among which, the frequency components of IMF2, IMF3, IMF7 and IMF8 components were the most obvious, and some of the components without obvious frequency components were interference components or false components.

For the effective IMF component group $\{c_i(t), i = 1, 2, \cdots, 6\}$ of the selected air-hole fault signal, energy entropy E_{We}, Singular value entropy E_{Se}, power spectrum entropy E_{Ge}, Hilbert spectrum entropy E_{He}, Hilbert envelope spectrum entropy E_{Be}, fuzzy entropy E_{Fze}.

The fusion entropy feature vector $F = [E_{We}, E_{Se}, E_{Ge}, E_{He}, E_{Be}, E_{Spe}, E_{Ape}, E_{Fze}]$ of the signal is composed of the 8 6-dimensional entropy vectors, and obviously the dimension of F is 48. The fusion entropy feature vectors of all experimental samples were extracted by similar methods.

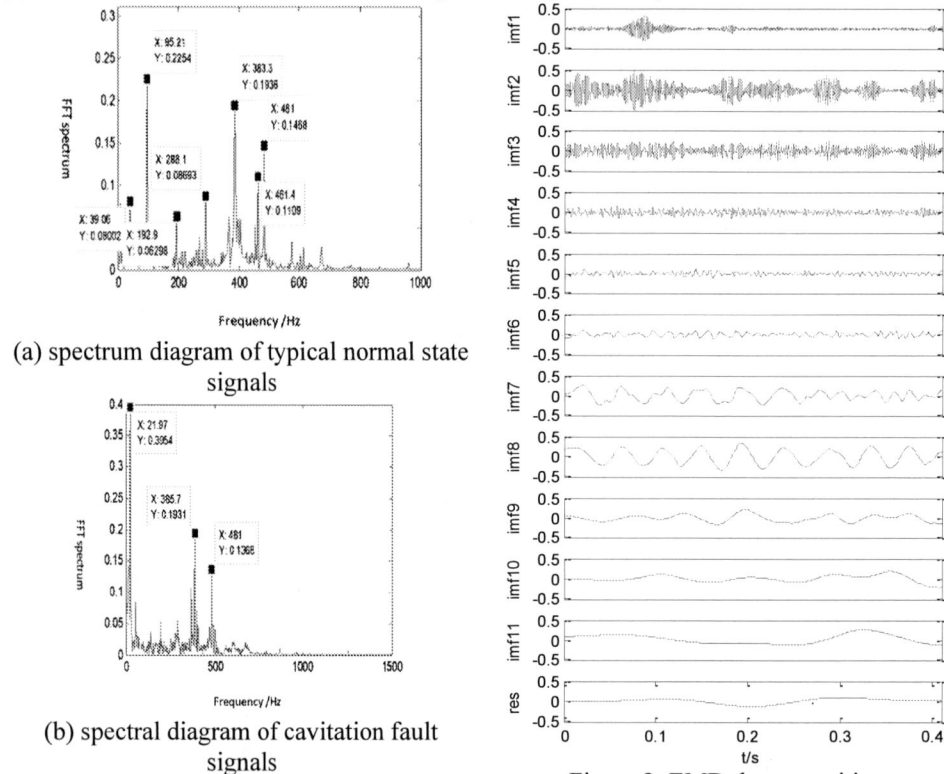

(a) spectrum diagram of typical normal state signals

(b) spectral diagram of cavitation fault signals

Figure 1. Spectral diagram of the signal

Figure 2. EMD decomposition results of cavitation fault signals

5. Classification identification of DBN

The DBN network of 2 hidden layers is selected as the recognition classifier, and the number of hidden layers is 100, so the DBN network structure is 48×100×100×5. According to the feature vectors of the 500 groups of experimental samples extracted from table 1, a total of 350 groups of 70 feature vectors in each state were selected to form the training set, and the remaining 30 groups of feature vectors in each state were 150 groups to form the test set. After normalization processing, as the input of DBN, the classification recognition result of DBN is obtained through training test, as shown in Figure 3.

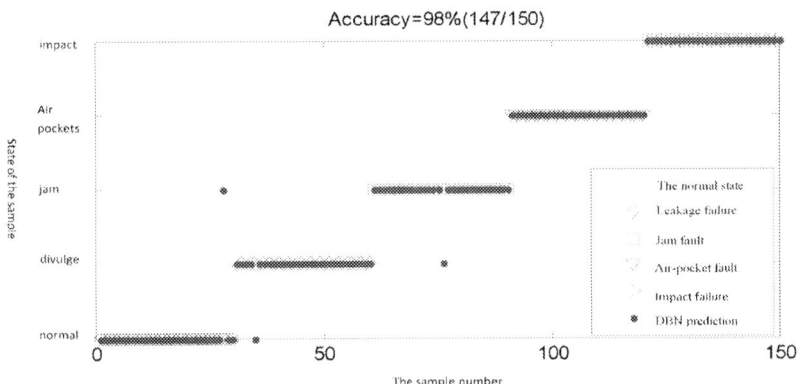

Figure 3. DBN classification recognition result of multiple faults in hydraulic system

From the recognition results, 147 of 150 test samples were accurately identified, and the recognition rate reached 98%, which was quite ideal. The air cavity and impact state with independent features are accurately identified. In general, the IMF entropy feature fusion method combined with DBN can effectively realize the intelligent multi-fault identification of hydraulic system.

6.Conclusion

In this paper, vibration signals of the hydraulic system of five states, including normal, leakage, blockage, air cavity and impact, are taken as the research objects. An intelligent hydraulic multi-fault identification method based on the combination of IMF entropy feature fusion and DBN is proposed. Entropy features of IMF components are extracted by using multi-types of information entropy. Through experimental verification, it can accurately realize intelligent multi-fault identification of the hydraulic system and effectively meet the condition monitoring and fault diagnosis requirements of the hydraulic system.

References

[1] Hu Ai jun , Sun Jingjing , X iang Ling. Mode Mixing in Empirical Mode Decomposition [J]. Journal of Vibration, Measurement & Diagnosis, 2011, 31(4):429-434.

[2] Dina S L, Arturo R M, Bikash C P, et al. A refined Hilbert-Huang transform applications to interarea oscillation monitoring [J]. IEEE Transactions on Power System, 2009, 24(2): 610-620.

[3] Yu Dejie, Cheng Junsheng, Yang Yu. Transformation method of hilbert-huang for mechanical fault diagnosis [M]. Beijing: science press, 2006:24-25.

[4] Hu Aijun, Sun Jingjing, Xiang Ling. Problem of mode aliasing in empirical mode decomposition [J]. Vibration, test and diagnosis,2011, 31(4):429-434.

[5] Cheng Junsheng, Yu Dejie, Yang yu. Treatment of hilbert-huang transformation endpoint effect problem based on support vector regression machine [J]. Journal of mechanical engineering,2006,42(4):23-31.

[6] Rilling G, Flandrin P, Goncalves P. . On empirical mode decomposition and its algorithms[J]. in Proc. IEEE-EURASIP Workshop on Nonlinear Signal and Image Processing NSIP-03, Grado (I, 2003, 3.

ISPECE   IOP Publishing

IOP Conf. Series: Journal of Physics: Conf. Series **1187** (2019) 032042   doi:10.1088/1742-6596/1187/3/032042

Natural Gesture Control of a Delta Robot Using Leap Motion

Xuchong Zhang[1,2,a], Ruiqiu Zhang[1], Liang Chen[1], Xianmin Zhang[2]

[1]School of Design, South China University of Technology, 510006, Guangzhou, Guangdong, China

[2]Guangdong Provincial Key Laboratory of Precision Equipment and Manufacturing Technology, 510640, Guangzhou, Guangdong, China

[a] Corresponding author: sdxczhang@scut.edu.cn

Abstract. Human-Robot Interaction (HRI) is a research hotspot in recent years, most robots used for HRI are serial robots. This paper deals with the natural gesture control of a Delta parallel robot, using the touchless optical sensor Leap motion controller. The kinematics of the Delta robot is firstly described, its inverse kinematic solutions are obtained, and its workspace is plotted. The Leap motion controller is introduced and the key functions used in this research are described, eight gestures are defined. The experimental rigs are established, experiments are successfully carried out and results are discussed. Results show that it is a natural way for users to control the Delta robot with an acceptable accuracy and efficiency, by using the Leap motion controller, even though they have never operated a robot before. This research extends the interaction way between human and robots.

1. Introduction

Human-Robot Interaction is a research hotspot in recent years [1], in which nature interaction is one of the key issues [2]. Leap motion controller is a touchless sensors and is widely used and studied by many researchers since it was firstly developed [3]. Compared with other sensors, such as Kinect or haptic glove, it has the advantage of high accuracy, touchless, small volume, and importantly, very cheap. Guna [4] analyzed the precision and reliability of the leap motion sensor and its suitability for static and dynamic tracking. Xu [5] studied the natural gesture control of a rotor-craft using the Leap motion controller. Ruffaldi [6] and Gunawardane [7] applied the Leap motion controller on the adaptive robotic arm manipulator. In the papers of Artal-Sevil [8] and Vani [9], the Arduino smart car is controlled by the leap motion. In order to enhance the touchless interaction, Nguyen combines a haptic glove with the Leap motion [10], which indicates that this way is feasible when some specific conditions are met, but still with some bottlenecks.

The robots used for HRI are mostly serial robots. Compared with the serial mechanism [11], the parallel mechanism has the advantage of high stiffness, high speed and high accuracy [12, 13]. One representative is the Delta robot, which was firstly developed by Clavel [14] and now is successfully used for factory application and laboratory research [15], its moving platform has 3 translational DOFs. The research area of Delta robots includes the kinematics [16, 17], dynamics, optimal design, control strategy and so on. Codourey [18] proposed a simplified dynamic model of the Delta robot for real time computed-torque control. Liu [19] presented a new design method considering a desired workspace and swing range of spherical joints of a Delta robot. Zhao [20] presented a method for parameter tuning of the fixed gain motion controller of Delta robot based on servo identification. In

---

Content from this work may be used under the terms of the Creative Commons Attribution 3.0 licence. Any further distribution of this work must maintain attribution to the author(s) and the title of the work, journal citation and DOI.
Published under licence by IOP Publishing Ltd

ISPECE
IOP Publishing

IOP Conf. Series: Journal of Physics: Conf. Series **1187** (2019) 032042   doi:10.1088/1742-6596/1187/3/032042

these researches about Delta robot, HRI is not taken into consideration, this concern will be studied in this research.

This research deals with the natural gesture control of a Delta robot using the Leap motion controller, which combines the advantages of both devices. The contributions of this research are that, firstly, a touchless control method is applied to the Delta parallel robot. Secondly, eight natural hand gestures are defined to interact with the robot. Thirdly, experiments are successfully carried out, which verified the effectiveness of the proposed method. This research is a fundamental research for the further study of the HRI of the Delta robot.

This paper is organized as follows, Section 1 gives a brief research background of this topic; Section 2 describes the modelling of the Delta robot, the detailed process of the kinematics is presented, and the workspace of the moving platform is plotted; Section 3 introduces the Leap motion controller, the key functions used in this research are clarified and eight hand gestures are defined; In Section 4, the experimental rigs are introduced, experiments are carried out and the results are discussed; Conclusions are made and future developments are prospected in Section 5.

2. Modelling the Delta robot

The Delta parallel robot is shown in Fig. 1. The robot includes the fixed platform, the moving platform, the active links, and the passive links. The fixed platform and active links are connected by revolute joints $A_1A_2A_3$, the active links and passive links are connected by spherical joints $B_1B_2B_3$, while the passive links and moving platform are connected by spherical joints $C_1C_2C_3$. The global coordinate system O-XYZ is that, the origin point O is the center of triangle $A_1A_2A_3$, the axis OX is parallel with A_2A_3, the axis OY is perpendicular with OX, while axis OZ is perpendicular with plane $A_1A_2A_3$, fitting the right-hand law. The end point P is the center of triangle $C_1C_2C_3$, the moving platform has 3 translational DOFs. In experiment, a gripper actuated by steering gear is attached to the moving platform.

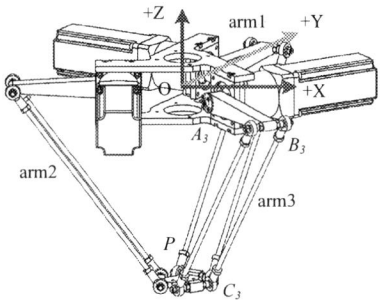

Figure 1. Delta robot.

2.1 Constraint equation

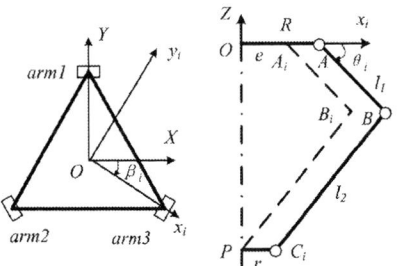

Figure 2. Geometry of a single train.

773

To simplify the modelling process, one of the 3 chains is analyzed, as is shown in Fig. 2. The radius of the circle $A_1A_2A_3$ is referred as R, the radius of the circle $C_1C_2C_3$ is referred as r, the length of active link is referred as l_1, and the length of passive link is referred as l_2. The fixed loop of the bold line in Fig. 2 can be replaced by the dash line, so the vector equation of end point P is

$$\overrightarrow{OP} = \overrightarrow{OA_i} + \overrightarrow{A_iB_i} + \overrightarrow{B_iP} \qquad (1)$$

in which $i=1\sim3$. $\overrightarrow{OP}=(x,y,z)^T$; x, y, z is the global coordinate of end point P; $\overrightarrow{OA_i}=\mathbf{T}\cdot(e,0,0)^T$; $e=R-r$; $\overrightarrow{A_iB_i} = \mathbf{T}\cdot(l_1\cos\theta_i,0,l_1\sin\theta_i)^T$; T is the rotation matrix

$$\mathbf{T} = \begin{bmatrix} \cos\beta_i & -\sin\beta_i & 0 \\ \sin\beta_i & \cos\beta_i & 0 \\ 0 & 0 & 1 \end{bmatrix} \qquad (2)$$

in which θ_i is the input angle of each active link, while β_i is the assemble angle of each chain.

Thus the Equation (1) can be transformed into

$$\overrightarrow{B_iP} = \overrightarrow{OP} - \overrightarrow{OA_i} - \overrightarrow{A_iB_i} = \begin{bmatrix} x-(e+l_1\cos\theta_i)\cos\beta_i \\ y-(e+l_1\cos\theta_i)\sin\beta_i \\ z-l_1\sin\theta_i \end{bmatrix} \qquad (3)$$

As the length of passive link is constant, so the equation above is

$$\begin{aligned} &\left[x-(e+l_1\cos\theta_i)\cos\beta_i\right]^2 + \left[y-(e+l_1\cos\theta_i)\sin\beta_i\right]^2 \\ &+(z-l_1\sin\theta_i)^2 = l_2^2 \end{aligned} \qquad (4)$$

in which $i=1\sim3$. These three equations are the kinematic constraint equations.

2.2 Inverse kinematics

The inverse kinematics is that given the position of end point, to know the inputs. As for the parallel mechanism, the inverse kinematic is simple. The constraint Equation (4) can be transformed into

$$A_i\cos\theta_i + B_i\sin\theta_i = C_i \qquad (5)$$

in which

$$\begin{cases} A_i = 2l_1(e - x\cos\beta_i - y\sin\beta_i) \\ B_i = -2l_1z \\ C_i = l_2^2 - (l_1^2 + x^2 + y^2 + z^2 + e^2) \\ \quad + 2e(x\cos\beta_i + y\sin\beta_i) \end{cases} \qquad (6)$$

Applying the half-angle formulation, Equation (5) is changed into

$$(C_i + A_i)\tan^2(\theta_i/2) - 2B_i\tan(\theta_i/2) + (C_i - A_i) = 0 \quad (7)$$

Thus the solution of the above equation is

$$\theta_i = 2\tan^{-1}\left(\frac{B_i \pm \sqrt{A_i^2 + B_i^2 - C_i^2}}{(C_i + A_i)}\right) \qquad (8)$$

It can be seen that there are 2 solutions for each chain, so there are totally 8 solutions for the robot. In real application, the angle of the active link which is outer side is accepted.

2.3 Workspace

The dimensions of the established Delta robot are that $R=86$, $l_1=140$, $l_2=320$, $r=30$mm. The workspace of the end point can be calculated point by point using the Matlab software, they are plotted in Fig. 3. It can be seen that although the dimensions of the robot are small, the workspace is pretty large, which indicates that the structure of the Delta robot is properly designed. A cuboid space with dimension 400*400*150mm can be found in the workspace. Considering the assembly limitation and rotational range of spherical joints, the real workspace is a little smaller than the ideal one.

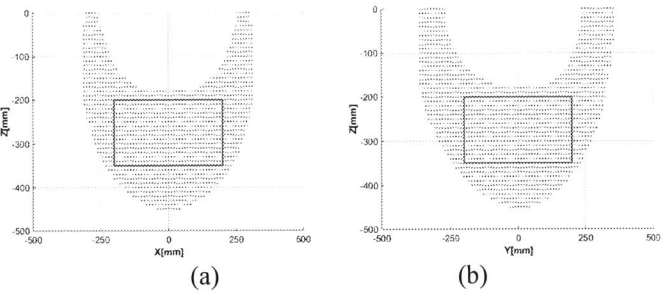

(a)   (b)

Figure 3. Work space of the Delta robot: (a) XOZ plane, (b) YOZ plane.

3. Leap motion controller

The appearance and coordinate system of Leap motion controller is shown in Fig. 4. The Leap motion controller used in this research is the second generation, it uses optical sensors and infrared light, which recognizes and tracks hands and fingers. The device operates in an intimate proximity with high precision and tracking frame rate and reports discrete positions and motion. Its workspace is an inverted pyramid, with a field of view of about 150 degrees, and the working height is 25~600mm, the resolution is up to 1mm [3].

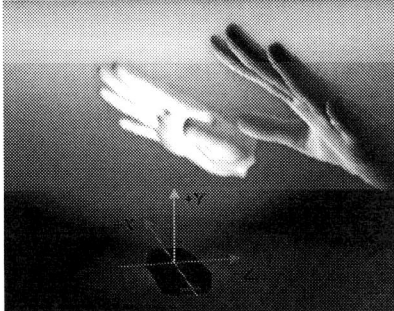

Figure 4. Leap motion controller.

The first key function used in this research is "*hand.palmPosition()*", part of the C++ code is

Frame frame = controller.frame();
HandList hands = frame.hands();
Leap::Vector handCenter = hand.palmPosition();

The code returns a vector *handCenter* representing the palm position in Leap motion coordinate system. In the application of this research, the workspace of the delta robot is limited to be [-80, 80, -80, 80, -380, -260], while the working space of the Leap motion controller is chosen to be [-100, 100, 100, 250, -100, 100], so there should be a mapping matrix between the two coordinates, which can be expressed as

$$\begin{cases} x_D = 0.8x_L \\ y_D = -0.8z_L \\ z_D = 0.8y_L - 460 \end{cases} \qquad (9)$$

in which $(x, y, z)_D$ is the coordinate of moving platform in Delta coordinate system, while $(x, y, z)_L$ is the coordinate of hand palm in the Leap motion coordinate system. There is a scaling ratio 0.8 between the two coordinates.

The second function used is "*hand.sphereRadius()*", which captures the opening status of the palm. It is used to control the steering gear, to decide the opening and closing of gripper, which is fixed at the end of the moving platform. Part of the code is

    float sphereRadius = hand.sphereRadius();

There should be thresholds of the *sphereRadius* to determine the rotate angel of steering gear, when the *sphereRadius* is smaller than the lower threshold, the gripper is closed, while when the *sphereRadius* is larger than the upper threshold, the gripper is opened.

Using the above information, eight gestures are defined to control the robot, as listed in Table 1. The mapping between the hand gesture and platform movement is very straightforward, which means that it is an easy way for users to interact with the robot.

Table 1. Gesture define and robot response.

Hand Gesture	Robot Response
Palm up	Moving platform up
Palm down	Moving platform down
Palm left	Moving platform left
Palm right	Moving platform right
Palm forward	Moving platform forward
Palm backward	Moving platform backward
Palm open	Gripper open
Palm close	Griper close

4. Experiment

4.1 Experimental systems

The experimental rig established in this research is shown in Fig. 1(a), the block diagram is shown in Fig. 5. The system includes the Arthur IPC-610 (CPU-i5, ROM-4G, RAM-128G), the Googol motion controller (GTS-400-PG-PCI), the DELTA servo driver and motor (ASD-B2-0421-B, ECMA-C20604RS), the planetary reducer (PLF060-L1-10-S2-P2), the Delta parallel mechanism, the gripper and the connecting lines. The rated power of servo motor is 400W, the rated torque is 1.27Nm, and the rated speed is 3000r/min. The ratio of the reducer is 10:1, the electronic gear ratio is set carefully to make sure that, 1000 pulses will lead to the active link rotate one degree, which means that the resolution of the input angle is 0.001degree/pulse. The motion controller communicates with IPC through PCI bus line, while the leap motion connects with IPC through USB. The steering gear of gripper is controlled through PWM signal.

ISPECE   IOP Publishing

IOP Conf. Series: Journal of Physics: Conf. Series **1187** (2019) 032042   doi:10.1088/1742-6596/1187/3/032042

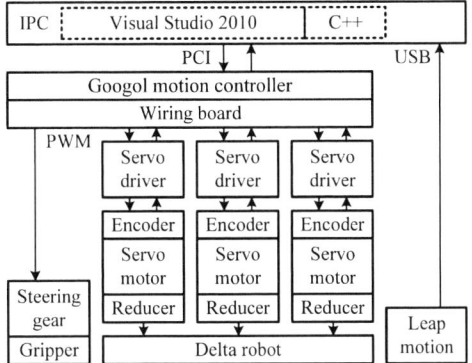

Figure 5. Block diagram of experimental system.

The program is written in visual studio 2010 environment, using the C++ language. The working procedure is as follows, the Leap motion catches the position of the palm, and then it is transformed into the coordinate of the moving platform of Delta robot through Equation (9). The inverse kinematic is carried out, then the input angles are obtained through equation (8). The IPC gives order to Googol motion controller, the motion controller gives order to servo drivers, and the servo drivers give order to servo motors, thus the moving platform moves as expected. The motion mode of the servo is in P2P mode. On the other hand, when the Leap motion detects the opening status of the palm, the PWM signal is produced by the motion controller and sent to the steering gear, thus the opening or closing of the gripper is realized. The whole process keeps going on in real time until the program is stopped.

4.2 Experimental results

Eight gestures are tested in the experiments, the results are shown in Fig. 6, it can be seen that the robot is successfully controlled by the natural gesture using Leap motion.

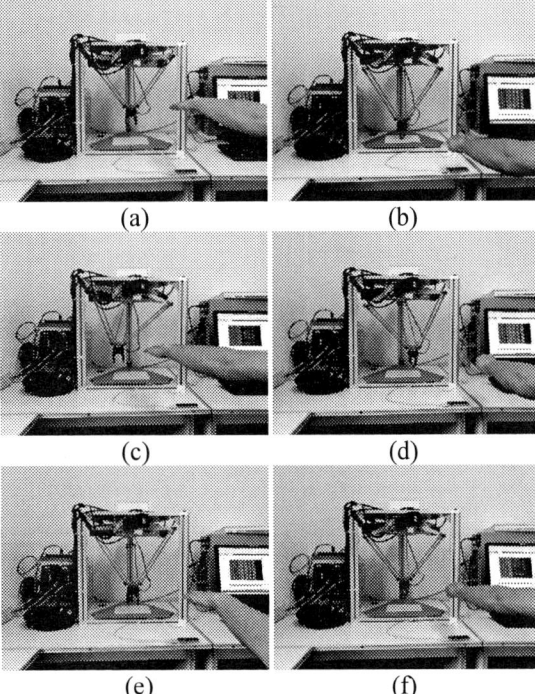

Figure 6. Experimental results: (a) Palm up & opened, (b) Palm down & closed, (c) Palm left & opened, (d) Palm right & closed, (e) Palm forward & opened, (f) Palm backward & closed.

To estimate the accuracy and efficiency of the control method, a picking and placing task is carried out. From point A, an object is picked up, then move through point B, C and D, on each point picking and placing actions are executed, then back to point A and the object is placed. This procedure is carried out by 10 different peoples, the average placing error and time spent are recorded and listed in Table 2.

Table 2. Accuracy and Efficiency of the Experiment.

Point	Action	Success rate	Error [mm]	Time [s]
A(50,50,-300)	Pick	10/10	-	2.2
B(-50,50,-300)	Pick & place	10/10	1.3	4.2
C(-50,-50,-300)	Pick & place	10/10	1.4	4.3
D(50,-50,-300)	Pick & place	10/10	1.3	4.2
A(50,50,-300)	Place	10/10	1.3	2.1

It can be seen from the results that the successful rate is one hundred percent, the average position error is around 1.3mm, and the average time for each picking and placing is about 4.2s. The results show that the users can control the robot successfully with acceptable accuracy and efficiency, even though they have never operated a robot before.

5. Conclusion

This paper deals with the natural gesture control of a Delta robot using the Leap motion controller, which combines the advantages of both devices. The kinematics of the Delta robot is analyzed and experiments are successfully carried out. This research shows the feasibility of touchless natural gesture control of the robot, which allows the user to operate the robot naturally, without knowing the complex principles of the system. This is a fundamental research for HRI, there are still some aspects can be improved, such as the delay of the moving platform with respect to human hands, and the accuracy of the system. Moreover, the robot should have more intelligence to predicate the behavior of the human. These topics are to be considered in the further.

Acknowledgements

This work is supported by the Natural Science Foundation of Guangdong Province (Grant No. 2016A030310420), the Guangdong Provincial Key Laboratory of Precision Equipment and Manufacturing Technology Open Funded Project (PEM201701), the Fundamental Research Funds for the Central Universities of China, South China University of Technology (2017BQ081) and the Reform Project of Undergraduate Teaching and Research of South China University of Technology (Y1180601). These supports are greatly acknowledged.

References

[1] A. Thomaz, G. Hoffman, M. Cakmak. Computational human-robot interaction, Found. Tre. Robot., **4**, 2-3: 105-223 (2016)

[2] G.Q. Liang, X.G. Lan, H.B. Zhang, and X.Y. Chen. Intention-based human robot collaboration, *The 11th International Conference on Intelligent Robotics and Applications, Wuhan, China, LANI10462*, 605-613 (2017)

[3] Leap Motion Inc. API Overview [OL]. *Available: https://developer.leapmotion.com/ [Accessed 9 Oct 2018]* (2018)

[4] J. Guna, G. Jakus, M. Pogačnik. An analysis of the precision and reliability of the leap motion sensor and its suitability for static and dynamic tracking, Sensors, **14**, 2, 3702-3720 (2014)

[5] Y.X. Xu, L. Chen. The attitude control method and realization of micro rotor-craft based on natural interaction, *the 2nd International Conference on Machinery, Materials Engineering, Chemical Engineering and Biotechnology. Part B*: 348-351 (2015)

[6] D. Bassily, C. Georgoulas, J. Güttler, and T. Linner. Intuitive and adaptive robotic arm manipulation using the leap motion controller. *in 41st International Symposium on Robotics; Proceedings of VDE*: 1-7 (2014)

[7] P.D.S.H. Gunawardane, N.T. Medagedara, B.G.D.A. Madhusanka. Control of robot arm based on hand gesture using leap motion sensor technology, Int. J. Robot. Autom., **2**, 1: 1-9 (2015)

[8] J.S. Artal-Sevil, J.L Montañés. Development of a robotic arm and implementation of a control strategy for gesture recognition through leap motion device, *in Proceedings of 2016 Technologies Applied to Electronics Teaching*, Seville, Spain: 1-7 (2016)

[9] L. Vani, R.A. Reddy. Robotic arm manipulation using leap motion controller, Int. J. Adv. Technol. Eng. and Sci., **3**, 1: 95-103 (2015)

[10] V.T. Nguyen. *Enhancing touchless interaction with the leap motion using a haptic glove*, master diss., University of Eastern Finland, Joensuu, Finland (2014)

[11] Q.K. Yu, G.L. Wang, T.Y. Ren, and L. Wu. An efficient algorithm for inverse kinematics of robots with non-spherical wrist, Int. J. Robot. Autom., **33**, 1: 45-52 (2017)

[12] X.C. Zhang, X.M. Zhang, Z. Chen. Dynamic analysis of a 3-RRR parallel mechanism with multiple clearance joints, Mech. Mach. Theory, **78**: 105-115 (2014)

[13] X.C. Zhang, X.M. Zhang. A comparative study of planar 3-RRR and 4-RRR mechanisms with joint clearances, Robot. Cim-Int Manuf., **40**: 24-33 (2016)

[14] R. Clavel. Device for displacing and positioning an element in space, *US Patent No. 4976582*, 1990-12-11 (1990)

[15] I. Bonev. Delta parallel robot—the story of success, *in Newsletter* (2001)

[16] J. Gallardo-Alvarado, A.L. Balmaceda-Santamaría, E. Castillo-Castaneda. An application of screw theory to the kinematic analysis of a Delta-type robot, J. Mech. Sci. Technol., **28**: 3785~3792 (2014)

[17] F.C. Can, M. Hepeyiler, Ö. Başer. A novel inverse kinematic approach for delta parallel robot, Int. J. Mate. Mech. Manuf., **6**, 5: 321-326 (2018)

[18] A. Codourey. Dynamic modeling of parallel robots for computed-torque control implementation, Int. J. Robot. Res., **17**, 12: 1325-1336 (1998)

[19] X.J. Liu, J.S. Wang, K.K. OH, J.W. Kim. A new approach to the design of a DELTA robot with a desired workspace, J. Intell. Robot. Syst., **39**: 209-225 (2004)

[20] Q. Zhao, P.F. Wang, J.P. Mei. Controller parameter tuning of delta robot based on servo identification. Chin. J. Mech. Eng., **28**, 2: 267-275 (2015)

Research on Manufacturing Technology of Thin-walled Parts of Fe105 metal Based on Laser Cladding

Tianbiao Yu[1,a] , Yiting Bao[2]

[1] Mechanical manufacturing, School of mechanical engineering and automation Northeastern University, 110004 Shenyang, China

[2] Mechanical manufacturing, School of mechanical engineering and automation Northeastern University, 110004 Shenyang, China

[a] Tianbiao Yu: tianbiaoyudyx@gmail.com

Abstract. According to some working conditions of thin-walled parts, Fe105 iron-based alloy powder is used to print parts based on laser cladding. Laser cladding, also called 3D Print, is the last manufacturing method in world. The test method of single factor experiment and orthogonal experiment is used to find the best parameters of powder feed rate, laser power, scanning speed and gas flow rate. The research is based on the laser cladding, using advanced equipment, such as, OM, SEM, Micro-hardness, to find out the best parameters of laser machine. Also, by the equipment, the morphology and characteristic of thin-walled parts were studied.

1. Introduction

Laser cladding technology is also known as 3D printing technology, which is the new material processing and manufacturing technology[1], its working principle is that using high energy density of laser beam exposure on the substrate to simultaneously send alloy powder. Powder and substrate surface clad together. Thin-walled parts are widely used in industry and life[2], but in the process of machining, because of the fixed workpiece and forging processing problems, leading to poor shape-accuracy of parts. The property of Fe105 metal powder, such as, large hardness, good wear resistance, low cost, is the reason why it is chosen to be the raw material. The research will adopt the method of laser cladding and use Fe105 alloy power. Single-channel multilayer forming experiment[3] was carried out. The research studied optimization and performance on the process.

2. Experiment and analyse

For this experiment, firstly, single factor experiment should be done to find the scope of the best parameters, after using orthogonal experiment to make sure the parameters.

2.1 Materials and evaluation parameters

Substrate materials is #45 steel and alloy powder is Fe105 metal, the components of it is shown in the table 1. Fe105 metal alloy powder as cladding material, because of its hardness and the ability of cladding layer to resist anti-cracking performance. Antioxidant ability of Fe105 metal is strong.

Laser cladding system adopted in the experiment is constituted by KUKA ZH30/60 robot, powder feeding pump, laser transmitter, optical maser, laser control system and experiment platform.

Content from this work may be used under the terms of the Creative Commons Attribution 3.0 licence. Any further distribution of this work must maintain attribution to the author(s) and the title of the work, journal citation and DOI.
Published under licence by IOP Publishing Ltd

Table 1. Components of Fe105 alloy powder							
Components	C	Cr	Si	Mo	Ni	Mn	B
Mass Fraction	1.5	4.5	1.3	1.5	12	1	2.8

Experiment for the research is single-channel monolayer experiment by laser system. The important influence factors of laser cladding mainly includes scanning speed, laser power, powder feed rate, gas flow rate[4].

For Z axis lifting capacity and laser zoom, because of single-channel monolayer experiment cannot study it and the laser system cannot change zoom, the research is incapable to study it. Single factor experiment using specific data are shown in table 2.

Table 2. The range of single factor experiment

	Value range	Average
Laser power (W)	350/380/410/440/470 /500/530	440
Scanning speed(mm/s)	4/4.5/5/5.5/6/6.5 /7	5.5
Powder feed rate(r/min)	0.3/0.4/0.5/0.6/0.7 /0.8/0.9	0.6
Gas flow rate(L/h)	9/11/13/15/17/19 /21	15

Single-channel monolayer experiment evolution parameters mainly include: the shape coefficient of single-channel cladding layer$\zeta$and cladding layer dilution rate D[5-6].

The shape coefficientζcalculation formula is as follow:
$$\zeta = b/H \qquad (1)$$

H is the height of the cladding layer and b is the width of the cladding layer. The shape coefficient is the ratio of width to height of cladding layer. It is generally believed that when the ratio is 2 or less than 2, the collapsed cladding layer phenomenon may occur. When the ratio is more than 2, the effect of the cladding layer is stable, which can print valuable micro thin-walled parts. But when the ζ is much larger than 2, the cladding remelting area will expand, which is harmful to the cladding layer.

The dilution rate D calculation formula is as follow:
$$D = S2/(S1 + S2) \qquad (2)$$

For the formula, S1 is the meaning of the area on vertical direction along the cross-sectional face of laser cladding layer. S2 is the meaning of the area on vertical direction along the cross-sectional face of substrate material. Because the formula is the calculation of the area, the formula can be converted into the following form:
$$D = \frac{\int h \, db}{\int h \, db + \int H \, db} \qquad (3)$$

Equation (3) can be simplified as the following form:
$$D = h/(h + H) \qquad (4)$$

h is the height of the miscibility area of substrate

Normally, dilution rate D should be stable within a certain range, if the dilution rate D is too small, metal on the surface of the molten pool is too small to be strong and it's harmful to cladding. But if the dilution rate D is too large, it will cause the thickness of the cladding layer. And if the laser energy is too high, it's easy to cause the surface cracking, deformation and collapse. So the dilution rate should be in the safety range[7].

2.2 Results of experiment

2.2.1 Single factor experiment
For the single factor experiment, the research focus on the four parameters points of laser cladding: scanning speed (V_s), laser power (P), powder feed rate (V_f) and gas flow rate (G). With laser cladding system, the results of the single factor experiment are shown in table3, table4, table5 and table6.

Table 3. Single factor experimental results of laser power

	P(w)	V_s(mm/s)	V_f(r/min)	G(L/h)	ζ
1	320				2.836
2	350				2.322
3	380				2.289
4	410	5.5	0.6	15	2.367
5	440				2.352
6	470				3.143
7	500				3.208

The data in the table 3 show that P=470W and P=500W is too large for the experiment. The shape coefficient is much larger than others. And when P=320, P=350,P=380 laser power is too small, the substrate materials cannot completely melt, as well as, irregular feature of vertical direction along the cross-sectional face of laser cladding layer is obvious. Sticky powder phenomenon is also obvious. And it cannot reach the purpose of the laser cladding. So at last, the range of laser power are P=410W, P=440W. And another parameter is the average of them, P=425W.

Table 4. Single factor experimental results of scanning speed

	P(w)	V_s(mm/s)	V_f(r/min)	G(L/h)	ζ
1		4			1.658
2		4.5			1.873
3		5			2.412
4	410	5.5	0.6	15	2.367
5		6			2.262
6		6.5			2.464
7		7			2.541

In the case of other parameters constant, faster scanning speed will reduce the time of the powder stay on the substrate materials, because of that, it can also reduce the balling phenomenon. But when the scanning speed is too fast to make metal melt, the molten pool will be discontinuous.

The data in the table 4 show that shape coefficient is less than 2 when V_s=4, and V_s=4.5.Becaue the scanning speed is slow, leading to metal powder stack on the surface of substrate. The effect of the cladding is deteriorate. But when the V_s=7, V_s=6.5, the scanning speed is too fast, the powder cannot melt completely, sticky powder phenomenon is obvious. Finally, V_s=5, V_s=5.5 compared with V_s=6 that had the better cladding effect. The range of scanning speed are V_s=5, V_s=5.5, and another one is the average of them, V_s=5.25.

Table 5. Single factor experimental results of powder feed rate

	P(w)	V_s(mm/s)	V_f(r/min)	G(L/h)	ζ
1			0.3		2.430
2			0.4		2.585
3			0.5		2.207
4	410	5.5	0.6	15	2.367
5			0.7		2.196
6			0.8		2.603
7			0.9		2.372

Powder feed rate is important for the effect of laser cladding. With the large powder feed rate, powder is too much for the laser, powder cannot absorb the energy of the laser, it cannot melt completely and result of it is cladding layer become worse. However, with the small rate, powder feed system can not determine the output. The powder feed rate of the research isn't certain.

The data in table 5 show that when V_f=0.7, V_f=0.8, V_f=0.9, the rate is so large that some powder cannot melt, the phenomenon of sticky powder is obvious. When V_f= 0.3, V_f=0.4, the shape of the cladding layer cannot be guaranteed. The effect of cladding layer is terrible. Because of that, the range of powder feed rate are V_f=0.5, V_f=0.6, and another one is the average, V_f=0.55.

Table 6. Single factor experimental results of gas flow rate

	P(w)	V_s(mm/s)	V_f(r/min)	G(L/h)	ζ
1				9	2.432
2				11	1.965
3				13	2.128
4	410	5.5	0.6	15	2.367
5				17	2.415
6				19	2.725
7				21	3.157

Shielding gas protect the cladding layer so that it can't be oxidized under the condition of high temperature. But too large gas flow rate may affect the experiment. Because shielding gas may blow away the metal powder and change the landing site of metal powder on the surface of substrate[8].

The data in table 5 show when the G=11, the shape coefficient is less than 2, so it is abandoned. When the G=19 and G=21, the shielding gas flow rate is too large. It may blow away the metal powder and some powder cannot absorb the energy of the laser, because of that, powder can't be stacked on the surface and the shape coefficient much larger than others. So at last, the range of shielding gas flow rate are G=13 and G=15. Another one is average of them, G=14.

2.2.2 Orthogonal experiment

By single factor experiment, research find that cladding layer is effected by these factors, the laser power, scanning speed, powder feeding speed and shielding gas flow rate. Each influence factors had their own way to affect the text. Research selected the three horizontal reference values of each parameters, for that select the optimal solution of laser cladding by Fe105 metal powder.

Research adopt the orthogonal experiment by random grouping, the test is four factors and three levels orthogonal experiment grouping conditions and the results are shown in table 7.

Table 7. Orthogonal experiment table

	P(W)	V_s(mm/s)	V_f(r/min)	G(L/h)	ζ
1	410	5.00	0.50	15	3.394
2	410	5.25	0.55	14	2.914
3	410	5.50	0.60	13	2.870
4	425	5.00	0.50	14	3.251
5	425	5.25	0.55	13	2.896
6	425	5.50	0.60	15	3.147
7	440	5.00	0.55	15	3.178
8	440	5.25	0.60	14	2.589
9	440	5.50	0.50	13	2.695
10	410	5.00	0.60	13	3.015
11	410	5.25	0.50	15	2.685
12	410	5.50	0.55	14	2.526
13	425	5.00	0.55	13	2.963
14	425	5.25	0.60	15	3.028
15	425	5.50	0.50	14	2.528
16	440	5.00	0.60	14	3.792
17	440	5.25	0.50	13	2.267
18	440	5.50	0.55	15	4.203

In theory, when laser cladding print thin-walled part which have radian, the shape coefficient of the cladding layer must more than 3. If the ratio less than 3, the thin-walled parts prone to collapse and bend. But when the shape coefficient is too large, the width of the cladding layer is too thick which will cause the height of cladding layer is insufficient that will affect the printing effect of thin-walled parts[9]. In the table 7, group 1, 4, 6, 7, 10, 14, 16, 18 meet the requirements, but the group 18 is large than others and group 10 and 14 just a little more than 3, so the research rule out them and continue to study group 1, group 4, group 6, group 7, group 16.

Table 8. Cladding layer dilution rate table

	1	4	6	7	16
D	0.418	0.323	0.369	0.381	0.4061

Dilution rate is important for the cladding layer. Due to the effect of high temperature from laser, in the process of cladding, laser can make the cladding layer and substrate a certain extent dilution, and this phenomenon is inevitable. Within the scope of the dilution ratio, the ratio is smaller, the result of the test will be better. Dilution ratio ensure that the performance of cladding layer. But the dilution rate can't be too small. For that dilution rate will lead to the cladding layer and substrate could not form a combination on the surface of substrate.

Cladding layer dilution rate of five groups are shown in the table 8.All of ratio were among from 3 to 5 and dilution rate meet the basic requirements. Group 4 had the smallest dilution rate so the optimal solution is group 4 with P=425W, V_s=5mm/s, V_f=0.50r/min, G=14L/h.

3. Organizational analysis of thin-walled parts

The research use the optimal solution to print a micro thin-walled parts. The parts was processed and it was analysed by electron microscope.

It can be seen in Fig1(a) that at the bottom of the cladding layer, a good metallurgical bonding is formed between cladding layer and the substrate. And there is a clear structural transformation at the bonding interface. There is a thin planar crystal transition structure between the cladding layer and the substrate. Then, the columnar crystal at the bottom of the cladding layer grows epitaxially along the substrate, inversing heat flow direction and extending to the cladding layer.

As the growth distance increases, the crystal size gradually changes from a single thick columnar crystal to a small cross-dendrite.In the middle of the cladding layer, as shown in Fig1(b), the cladding layer is mainly composed of crystallized oriented cross-dendrites and tiny equiaxed crystals. At the top of the cladding layer, the structure is mainly composed of denser and smaller equiaxed crystals, as shown in Fig1(c).It also can be seen the microstructures of cladding surface in Fig1(d)(e)(f).

By observing the microstructure change of the cladding layer from the bottom to the top, the overall microstructure transformation of the cladding layer is homogeneous, the microstructure is dense and small. There is no crystal structure which is abnormal thick or throughout the crystallization zone. And there is no microcracks, blowholes and other defects inside the cladding layer. The cladding layer shows good microstructure characteristics of laser cladding and orientated solidification and excellent forming quality

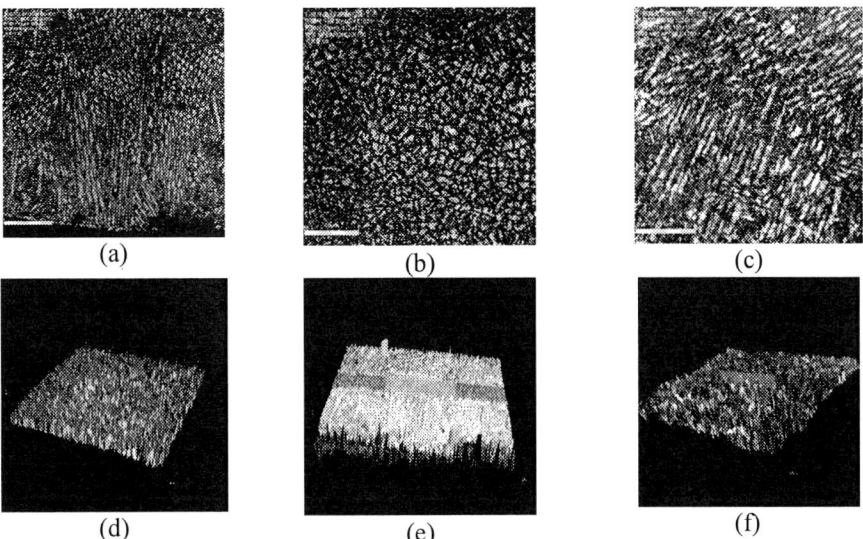

Figure 1. Microstructures of cladding: (a) substrate and bonding interface region (b) intermediate region (c) top region
Microstructures of cladding surface (d) substrate and bonding interface region (e) intermediate region (f) top region

4. Conclusion

The research is based on the laser cladding to print a micro thin-walled parts by Fe105 metal powder. In the process of the experiment, the research adopted single factor and the orthogonal test to find out the optimal solution of the Fe105 metal powder.

The experiment force on four parameters: scanning speed (V_s), laser power (P), powder feed rate (V_f), gas flow rate (G). And by the single factor test, the study found the range of the optimal parameters by the shape coefficient of single-channel cladding layer and cladding layer dilution rate. Within the range, the study does the orthogonal experiment. The study analyse the data by the state of the cladding layer. At the end of experiment, the study find that when P=425W, V_s=5mm/s, V_f=0.50r/min, G=14L/h, the state of the cladding layer is good and also the shape coefficient and dilution rate meet the requirements.

The study also used the electron microscope to analyse the microstructures of cladding. The study found there is no micro-cracks, blowholes and other defects inside the cladding layer.

Acknowledgment

This work was supported by the National Ministry of Industry and Information major special projects research fund (201675514).

References

[1] J.H. Jang, B.D. joo, C.J. Van Tyne, et. al. Characterization of deposited layer fabricated by direct laser melting process. Met. Mater. Int. 19, 3: 497-506 (2013)

[2] J. Lin, Y. Lv, Y. Liu, et. al. Microstructural evolution and mechanical property of Ti-6Al-4V wall deposited by continuous plasma arc additive manufacturing without post heart treatment. J. Mech. Behav. Biomed. Mater. 69: 19-29 (2017)

[3] M.J. Tobar, J.M. Amado, J. Montero, et. al. A Stuudy on the Effects of the Use of Gas or Water Atomized AISI 316L Steel Powder on the Corrosion Resistance of Laster Deposited Material. Phys. Procedia. 83: 606-612 (2016)

[4] S. Bontha, N.W. Klingbeil, P.A. Kobryn, et. al. Thermal process maps for predicting solidification microstructure in laser fabrication of thin-wall structures. Journal of Mechanical Working Technology. 178, 1-3: 135-142 (2006)

[5] T. Durejko, M, Zietala, W. Polkowski, et. al. Thin wall tubes with Fe3Al/ss316L graded structure obtained by using laser engineered net shaping technology. Mater. Des. 63: 766-774 (2014)

[6] N. Kumar, M. Mukherjee, A. Bandyopadhyay. Comparative study of pulsed Nd:YAG laser welding of AISI 304 and AISI 316 stainless steels. Opt. Laser Technol. 88: 24-39 (2017)

[7] G. Miranda, S. Faria, F. Bartolomeu, et. al. Predictive models for physical and mechanical properties of 316L stainless steel produced by selective laser melting. Mater. Sci. Eng., A. 657: 43-56 (2016)

[8] M. Naveed Ahsan, A.J. Pinkerton, R.J. Moat, et. al. A comparative study of laser direct metal deposition characteristics using gas and plasma-atomized Ti-6Al-4V powders. Mater. Sci. Eng., A. 528, 25-26: 7648-7657 (2011)

Research on Electrolyte Jet Assisted Laser Micromachining Technology

Yulan Hu[1], Weiguang Hao[1], Guoqiang Liu[2], Yajie LIu[2], Zilin Liu[1]

[1]School of Information Science and Engineering, Shenyang Ligong University, Shenyang, China

[2]School of Mechanical Engineering, Shenyang Ligong University, Shenyang, China

Abstract. For the traditional laser processing method, processing defects such as recasting layer, microcrack and high concentration of heat affected zone are inevitably generated, which reduces the surface quality and service life of the microstructure. In this paper, the electrolyte jet assisted laser micromachining technology is proposed. This technology applies the laser beam and the corrosive electrolyte jet beam coaxially to the workpiece. At the same time as the laser acts, the corrosive electrolyte jet continuously scour, cool and slightly corrode the processing area. The effect can reduce the thickness of the recast layer, reduce the number of microcracks, and eliminate the high concentration of the heat affected zone. In this paper, the causes of recasting layer in laser processing are analyzed firstly. The material removal mechanism involved in the laser jet assisted laser micromachining and the mechanism of laser coupled electrolyte jet beam are studied. The flow field in the laser jet assisted laser drilling process is established. And the mathematical model of the temperature field; the effects of laser pulse energy, laser repetition frequency, electrolyte concentration and electrolyte jet velocity on the thickness of material recast layer were studied, and the basic processing technology of electrolyte jet assisted laser micromachining was preliminarily mastered, laying the foundation for further research on electrolyte jet assisted laser micromachining technology.

1. Electrolyte jet assisted laser micromachining mechanism

1.1 Mechanism of laser impact force in electrolyte jet

The high heat of the laser causes the surface material of the workpiece to be converted from solid to liquid until a large amount of vapor particles are generated on the surface of the workpiece. The vapor particles collide with each other to propagate away from the surface of the target, and the layer of vapor particles only a few microns away from the surface is called the Knudsen layer. The vapor particles continue to absorb the laser energy and eventually ionize to form a plasma. The generated plasma has a strong absorption effect on the laser. When the absorbed laser energy is sufficiently large, the plasma and the material vapor rapidly expand to propagate at a supersonic speed to generate a high voltage. Shock wave, in the electrolyte jet assisted laser processing, the presence of an electrolyte confinement layer on the surface of the workpiece limits the external expansion of the high-pressure shock wave and generates a strong reaction force to the workpiece. When the laser is strong enough to propagate from the surface of the workpiece at supersonic speeds along the direction of the laser the plasma generation ,the plasma expansion wavefront is called the laser-supported blast wave (LSDW) [1].

In addition to the blast wave reaction force formed by the plasma expansion wave front, the pressure generated by the hollow bubble cavitation in the composite processing process promotes the

Content from this work may be used under the terms of the Creative Commons Attribution 3.0 licence. Any further distribution of this work must maintain attribution to the author(s) and the title of the work, journal citation and DOI.

Published under licence by IOP Publishing Ltd

processing at the same time. The electrolyte jet assists in the laser processing process, the electrolyte constrained layer is heated, and some of the electrolyte vaporizes to form vacuoles, which will be generated up to the surface of the workpiece.140MPa~180MPa, such high impact force will form fine pits on the surface of the workpiece.

In summary, it can be seen that the impact force acting on the surface of the workpiece during the laser jet assisted laser processing is mainly composed of the shock wave reaction force of the Knudsen layer plasma, the hollow bubble cavitation [2] impact force and the jet impact force of the processing region solution, and electrolysis. The mechanism of laser impact force in liquid jet, as shown in Figure 1.

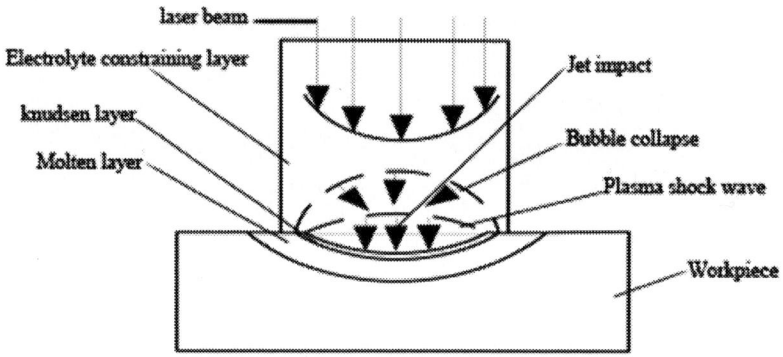

Figure 1 Processing mechanism of laser impact force effect in electrolyte jet

1.2 Laser beam coupled electrolyte jet beam mechanism
The laser beam is emitted by the laser and then polymerized by the convex lens. After the different refraction of the air layer, the glass layer and the electrolyte layer, the laser jet is injected at a certain angle and totally reflected at the interface between the electrolyte and the air, and finally acts on the surface of the workpiece. In order to make the total reflection of the laser beam in the jet of the electrolyte, it is necessary to adjust the distance between the center of the appropriate lens and the upper surface of the window glass and the maximum angle of incidence of the total reflection of the laser in the jet of the electrolyte to meet certain optical transmission conditions. By selecting the appropriate laser focus front radius and the convergence convex lens focal length, it is possible to theoretically achieve total reflection of the laser beam in the electrolyte jet. The laser beam couples the electrolyte jet beam mechanism, as shown in Figure 2.

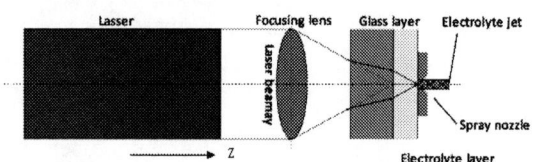

Figure 2. Laser beam coupled electrolyte jet beam mechanism
The mechanism of laser impact force and the mechanism of laser coupled electrolyte jet beam involved in electrolyte jet assisted laser processing are deeply explored, which lays a theoretical foundation for the further study of electrolyte jet assisted laser micromachining technology.

2. Physical model establishment and simulation result analysis of fluid jet assisted laser drilling process flow field
During the laser jet assisted laser processing, the electrolyte jet penetrates into the surface of the workpiece after propagating in the air, which is a two-phase flow problem in the multiphase flow (gas-

liquid two-phase flow). The simulation was performed using the VOF model. The physical model of the jet of electrolyte passing through the air to the surface of the workpiece during the drilling process, as shown in Figure 3.

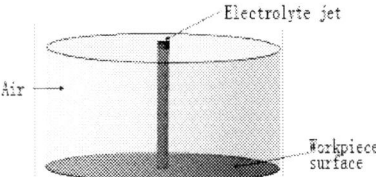

Figure 3 Physical model diagram of gas-liquid two-phase flow in punching process

The flow field simulation is simplified to 1/4 of the physical model, and the calculation cost is reduced under the premise of ensuring accurate calculation. The workbench is used to build the model and mesh, as shown in Figure 4.

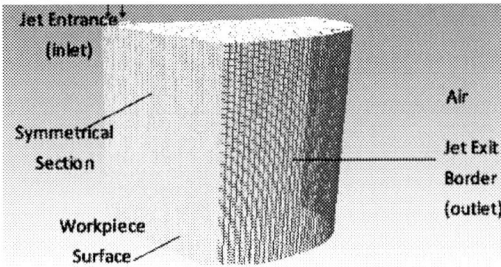

Figure 4 Flow field model and grid

The electrolyte jet inlet is set as a compressible fluid, the inside of the model is an incompressible fluid, and the bottom surface of the model is a wall. The jet velocity direction is the normal direction of the inlet and is perpendicular to the upper surface of the workpiece. According to the established physical model and the set initial conditions, the unsteady implicit separation solution is used to solve the flow field simulation in fluent 14.0, and the flow field simulation results of the electrolyte jet during the machining process are shown in Figure 5.

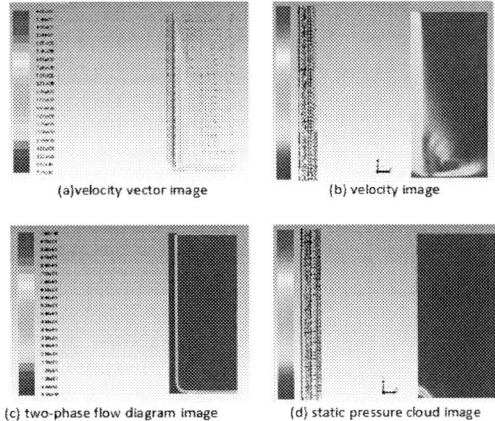

Figure 5 Flow field simulation results

From the graph velocity vector diagram (a) and the velocity cloud diagram (b), the velocity value of the electrolyte jet during processing and the velocity distribution of each region are known. The velocity of the electrolyte jet before contacting the upper surface of the workpiece is uniform in cross section. Maintaining 12m/s, the velocity is gradually reduced as it approaches the surface of the workpiece, and the velocity at the point of contact between the jet axis and the workpiece surface is finally reduced to 0m/s; from the jet pressure cloud diagram (d), the surface pressure of the workpiece is along the intersection with the jet axis gradually decreases outward, and the pressure is maximum at the intersection. This pressure distribution is favorable for the erosion of the melt during processing; the phase diagram of the electrolyte jet and air is as follows (c), the red part is air, blue For the electrolyte jet, the shape changes from a regular cylindrical shape to a diverging shape during the jet transport of the electrolyte, and the thickness of the liquid when contacting the surface of the workpiece is very thin, and the change of the shape of the electrolyte jet is favorable for punching. During the processing, the local area of the processing is cooled, thereby reducing the concentration of thermal stress.

3. Establishment of temperature field physical model of electrolyte jet assisted laser drilling and analysis of simulation results

Electrolyte jet assisted laser microfabrication compared to conventional air laser processing, not only produces various thermal phenomena between the laser and the material (heat transfer, heat convection, heat radiation, etc.), but also the electrolyte jet to the processing area during processing continuous scouring produces impact force effects, forced convection heat transfer effects of electrolyte jets, and the like.When the electrolyte jet assists the laser drilling process, there is only axial relative motion between the jet and the workpiece, and the jet acts perpendicularly on the surface of the workpiece, and the physical model diagram is established, as shown in Figure 6.

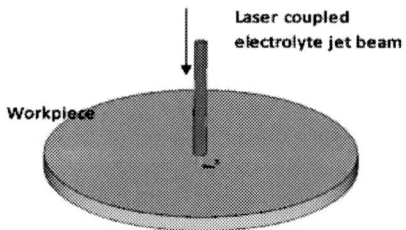

Figure 6 Schematic diagram of the physical model

When using ANSYS to simulate the machining process, the more the number of units is and the smaller the unit size is, the higher the configuration requirements of the computer are. Under the premise of ensuring the correct analysis results, in order to reduce the amount of simulation calculation, this paper builds a model for 1/36 workpieces. In the meshing, the laser spot diameter of the laser reaches the micron level when the electrolyte jet assisted laser drilling is simulated. In order to ensure the accuracy of the solution and the calculation amount in the simulation process, the model is divided into different unit sizes. The overall unit size is 1/3 of the spot diameter, and the electrolyte jet beam influence area is 1/16 of the spot diameter, and a finite element model diagram is obtained, as shown in Figure 7.

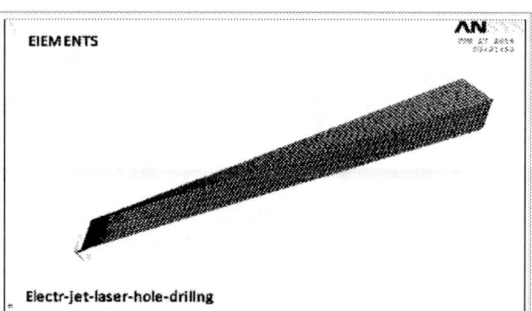

Figure 7 finite element model diagram after meshing

4. Effect of processing parameters on material removal

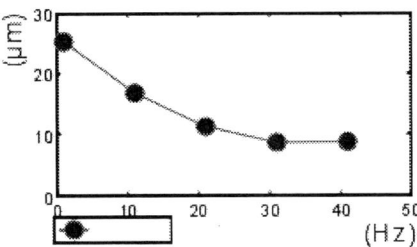

Figure 8 Effect of laser repetition frequency on the thickness of recast layer

It can be seen from the figure 8 that as the repetition frequency of the laser increases, the thickness of the recast layer gradually decreases. This is because as the repetition frequency increases, the number of pulses acting on the workpiece increases, causing the molten material to remain in a molten state for an increased time, and the electrolyte jet is more, the residual melt is sufficiently washed and corroded. But the excessive repetition frequency and the small pulse energy are reduced, which reduces the amount of material removal. Therefore, the appropriate laser repetition frequency should be selected during the composite processing.

It can be seen from the figure that as the laser energy increases, the thickness of the recast layer first decreases. When the laser energy is greater than 300 m J, the thickness of the recast layer increases with the increase of the laser energy. This is because as the laser energy increases, The material in the processing area is enhanced by gas-liquid phase change, and the liquid material remains for a long time. When the electrolyte jet is washed away from the substrate, the residual amount of the melt decreases and the thickness of the recast layer gradually decreases, but when the laser energy is too large At the same time, the gas phase material of the material increases, and the water mist generated when the jet contacts the processing region increases, which

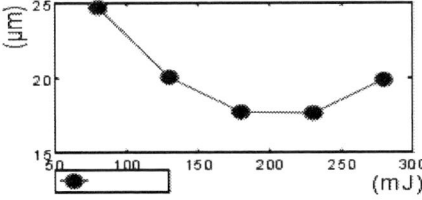

Figure 9 Effect of laser energy on the thickness of recast layer

reduces the hot working effect of the laser, and at the same time, more and more molten material is generated, and the electrolyte jet saturates the melting ability of the molten material. The residual slag will increase continuously, and the thickness of the recast layer of the hole will gradually become larger. From the analysis, it can be seen that in order to reduce the thickness of the recast layer and improve the surface quality during the composite processing, excessive laser energy should not be selected. For composite processing, at the same time, according to different processing purposes, select the electrolyte jet velocity that is compatible with the processing laser energy, when When the processing efficiency purposes, when increasing the laser energy, to improve the proper electrolyte velocity of the jet.

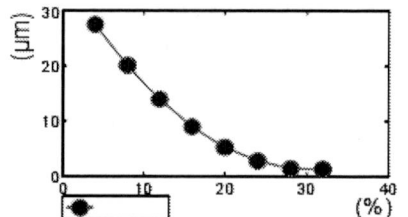

Figure 10 Effect of electrolyte concentration on the thickness of recast layer

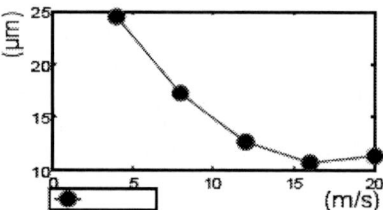

Figure 11 Effect of jet velocity on the thickness of recast layer

It can be seen from the figure that the thickness of the recast layer gradually decreases with the increase of the jet velocity of the electrolyte and the concentration of the electrolyte, because the jet velocity becomes higher, the scouring effect of the melt and the chemical corrosion effect of the melt are enhanced, and the electrolyte concentration is increased. The increase causes the chemical corrosion effect of the melt to increase. As the concentration of the electrolyte increases, the thickness of the cast layer gradually decreases, and finally the chemical corrosion effect is saturated, and the thickness of the recast layer tends to be stable; as the jet velocity reaches 16 m/s, The thickness of the cast layer has increased. This is because when the jet velocity is large, the forced convection effect is too strong, and the melt is cooled too fast. Before it is washed, it begins to condense on the hole wall to form a recast layer, and the thickness of the cast layer is re-cast. Increase; jet velocity and electrolyte concentration too large will make the laser pulse energy attenuation amount larger, reduce the composite processing efficiency, and choose the appropriate electrolyte jet velocity and electrolyte concentration under the premise of ensuring the quality of the machined surface.

5. Conclusion

Based on the laser processing mechanism, the material removal mechanism involved in the laser jet assisted laser micromachining and the laser coupled electrolyte jet beam mechanism were studied. The jet flow field and the perforation temperature during the laser jet assisted laser drilling process were established. The mathematical model of the field has obtained the basic processing technology of the electrolyte jet assisted laser micro-machining technology. The specific work and conclusions are as follows:

(1) This paper analyzes the causes of recasting layer in laser processing and the progress of laser re-casting layer removal at home and abroad, and studies the material removal mechanism and laser-coupled electrolyte jet beam mechanism involved in electrolyte jet assisted laser micromachining;

(2) The mathematical model of the medium jet flow field and the perforation temperature field of the electrolyte fluid assisted laser processing was established, and the iterative calculation was performed on the fluent 14.0 and ANSYS software by the finite element method. The evolution process and the composite of the hole were obtained. The influence of processing parameters on the shape of the hole, the actual drilling process is carried out by using the simulated machining parameters, and the hole shape and size are compared. The consistency of the two is good, and the effectiveness of the simulation is verified. Numerical simulation provides theoretical guidance for the electrolyte jet assisted laser drilling test.

(3) The effects of electrolyte concentration, electrolyte jet velocity, laser pulse energy, laser repetition frequency on material removal and recast layer thickness were investigated. The basic process rules of electrolyte jet assisted laser micromachining were preliminarily obtained. Further research basis of electrolyte jet assisted laser micromachining technology.

Acknowledgments
The work was supported by the National Natural Science Foundation of China(NO.61672360).

References
[1] Lu Jian, He Anzhi. Physics of Interaction between Laser and Materials. Beijing: Mechanical Industry Press, 1996.
[2] Li Yongjian. Study on the mechanism of surface topography during the occurrence of cavitation. [PhD thesis]. Beijing: Tsinghua University, 2009.
[3] Liu Lina, Li Zhiqiang. Application and Development Trend of Special Processing Technology in China[J].Journal of Inner Mongolia University for Nationalities(Natural Science Edition), 2010 (3): 286-288.
[4] Zhu Chengkang. Research on simulation technology of micro-spiral hole electroforming forming [[D]. [Master's thesis]. Zhejiang Industry 2010.
[5] Chen Yuanfang, Xian Yang. Electron beam processing technology and its application [[J], Modern Manufacturing Process, 2009(8), 153-157.
[6] Feng Dongju. Research on the principle and related technology of ultrasonic milling [D]. [PhD thesis]
[7] Xue Baiwen. Research on ultra-high pressure abrasive water jet cutting technology and its application in finishing [[D]. [Master's thesis]. Taiyuan University of Technology, 2005.
[8] Li Guijiang. Study on the performance of 1 Crl 8Ni9Ti stainless steel laser surface alloying coating [D]. [Master's thesis]. Hebei University of Technology, 2009.
[9] Mao Hongwei. Design, Development and Experimental Research of Micro Electrolysis Processing System [D]. [Master's thesis]. Shandong University of Technology, 2012.
[10] D.K.YLow, A.G.Corfe. Effects of assist gas on the physical characteristics of spatter during laser percussion drilling of NIlVIONIC 263 alloy.Applied Surface Science 154-155. 2000. 689-695
[11] Liu Zengwen. Study on erosion mechanism of hard and brittle materials and pre-mixed micro-abrasive water jet polishing technology [D]. [PhD thesis]. Shandong University, 2011.
[12] ZHANG Hua,XU Jiawen,WANG Jiming.Experimental Study on Laser-Current Processing of Ni-based Superalloy Jet Liquid Beam Electrolysis[J].Journal of Materials Engineering,2009, 4: 75-80.
[13] Song Qiang. Fundamental research on laser beam processing assisted by laser beam[D]. Nanjing: Nanjing University of Aeronautics and Astronautics, 2008.
[14] Zhang Hua. Basic research on jet liquid beam electrolysis-laser composite processing [D]. Nanjing: Nanjing University of Aeronautics and Astronautics, 2009.
[15] Qiu Qiangqiang. Study on laser cladding welding performance of nickel-based alloy powder [D]. [Master's thesis]. Changchun University of Science and Technology, 2009.
[16] Zheng Qiguang, editor. Laser advanced manufacturing technology. Hunan: Huazhong University of Science and Technology Press, 2001.
[17] Chen Heming, editor. Laser principle and application. Beijing: Publishing House of Electronics Industry, 2009.
[18] Lu Jian, He Anzhi. Physics of Interaction between Laser and Materials. Beijing: Mechanical Industry Press, 1996.
[19] Xu Rongqing, Chen Xiao, Chen Jianping, et al. Study on the shock wave and cavitation effect of laser ablation of underwater metals. Acta Optica Sinica, 2004, 24(12): 1643-1648.

[20] Chen Xiao. Research on the process and mechanism of interaction between high-power laser and underwater materials. [Ph.D. thesis] Nanjing: Nanjing University of Science and Technology, 2004.

Thermal barrier coating processing based on improved ant colony algorithm Process optimization and verification

Yulan Hu[1], Guoqiang Liu[2], Weiguang Hao[1], Yajie LIu[2], Zilin Liu[1]

[1]School of Information Science and Engineering, Shenyang Ligong University, Shenyang, China

[2]School of Mechanical Engineering, Shenyang Ligong University, Shenyang, China

Abstract. This paper presents a new method for electrolyte-assisted UV laser processing of thermal barrier coating materials. Under a certain thickness of the electrolyte layer, the ultraviolet laser beam is focused on the surface of the workpiece, the laser galvanometer realizes the rapid rotation of the thermal barrier coating material by the ultraviolet laser beam, and the "two-photon absorption" removes the matrix of the thermal barrier coating material, and collapses. The cavitation bubble action avoids the secondary adhesion of the processed material, meets the processing precision requirement (±0.05mm), and realizes the cold processing of the non-recast layer of the thermal barrier coating micropores. The improved ant colony algorithm is used. The minimum diameter difference is the objective function, the upper and lower diameter difference and the upper and lower machining precision are the constraints. The thermal barrier coating processing optimization model is established, and the process parameters are optimized and experimentally verified.

1. Improvement of ant colony algorithm

Ant colony algorithm has been widely used in many fields, but the ant colony algorithm has the disadvantages of stagnation and partial optimality. At present, for the rapid optimization of algorithms and the need to avoid falling into local optimal problems, the corresponding improvement of ant colony algorithm is needed, such as improvement of path selection strategy, improvement of pheromone update strategy, introduction of mutation strategy, addition of random disturbance. Strategies and integration with other algorithms to improve the speed of algorithm optimization.

1.Improvement of pheromone update strategy: Through the discovery of the social behavior of ant population, ants use pheromone to realize information exchange and cooperation between individuals and between individuals and the environment. In the development of ant colony algorithm improvement strategy, the improvement of pheromone has done a lot of research. It is found that the different ways of updating the pheromone have great influence on the performance of the ant colony algorithm, which will affect the convergence speed and optimization ability of the algorithm. However, in the pheromone update, if the global pheromone is updated, the algorithm is not easy to achieve convergence. Conversely, if the pheromone of a single ant loop path is updated, the pheromone on that path will accumulate too much and the algorithm will fall into local optimum. Therefore, the improvement strategy of ant colony algorithm should take these two aspects into consideration, and ensure that the algorithm can quickly find the optimal solution under the condition of convergence. It is usually obtained by integrating global pheromone update and local pheromone update. Satisfying result.

Content from this work may be used under the terms of the Creative Commons Attribution 3.0 licence. Any further distribution of this work must maintain attribution to the author(s) and the title of the work, journal citation and DOI.

Published under licence by IOP Publishing Ltd

2. Improvement of search speed: When searching for ant colony algorithm, positive feedback will gradually increase the concentration of pheromone on the suboptimal solution path, and subsequent ants may choose a suboptimal solution path with high concentration. At the same time, the pheromone is constantly volate, and the concentration of the sub-optimal solution pheromone is increasing, so that the chances of the subsequent ants finding the global optimal solution are gradually reduced, the optimal path cannot be found, the search range is narrowed, and then into the local optimal, the stagnation occurs. Therefore, the search speed is a defect that limits the development of ant colony algorithm. At the same time, the search speed has always been the bottleneck of the application of ant colony algorithm in large-scale optimization problems. Therefore, it is necessary to study the appropriate search speed and search space, and take corresponding improvement measures. To improve the performance of the ant colony algorithm.

3. Improvement of path selection strategy: In order to reduce the probability that the algorithm falls into local optimum and improve the ability of the algorithm to find the optimal, it is necessary to design a suitable path selection strategy. Adjusting the path of the ant colony algorithm is to guide the ant's initial optimization process through a state transition rule, so that it can find a feasible solution faster, and accumulate on this basis to achieve the effect of speeding up the convergence. At the same time, the random search operation is added, which is beneficial to the global optimal solution search of space, and can effectively overcome the shortcomings of the ant colony algorithm which is slow in evolution and easy to fall into local optimum. The ant will select the path of pheromone with a large probability, which makes the search in the initial state may be on a locally short path with a high probability, which makes the ant colony search lose diversity. Therefore, in order to enable the ant to select more paths in the initial optimization process, the pheromone volatilization factor can be set. When the pheromone concentration does not reach a certain threshold, the existence of the optimal solution is ignored, and only when the pheromone reaches the threshold Let the ants tend to accumulate more paths of information under this machine of pheromone. Here, the kth ant can be converted from state i to state j with the following probability:

$$ j = \begin{cases} \arg\limits_{s \in Allowed_k} \max\{\tau_{is}(\eta_{is})^\beta\}, & if\ q \le q_0 \\ J, & otherwise \end{cases} \quad (1\text{-}1) $$

In the middle: q ----- a random number evenly distributed in the interval [0, 1]

qo（0<qo<1）----- a parameter, J is a random variable generated according to the given selection rule.

4. Add random perturbation strategy: The ant colony algorithm has the disadvantages of slow search speed, easy to fall into local optimum, and prone to stagnation. In order to overcome the stagnation in the search process, a random perturbation strategy can be adopted, and the probability of random selection is dynamically adjusted during the evolution process. This increases the chances of path selection and helps to overcome stagnation in the search process. What is usually done in the ant colony algorithm is the combination of deterministic selection and random selection. This deterministic choice causes the ant to always choose the path with the largest transfer coefficient, while the random selection leads to a strong random when calculating the transfer coefficient. Sex. It is the combination of the two that complements each other to make the algorithm have a stronger global search ability and truly improve the algorithm.

5. Fusion with other algorithms: Ant colony algorithm has many advantages. The strong positive feedback ability is a point that can be applied to the improved algorithm. The positive feedback of the formal ant colony algorithm makes the algorithm accumulate a certain amount of pheromone in the later stage. It can speed up the evolution of the algorithm and make the algorithm reach a convergence state quickly. But at the same time, we also found that the ant colony algorithm also has better coupling ability, so we can use the positive feedback of ant colony algorithm and good coupling ability to fuse with other algorithms, which is also a good way to improve ant colony algorithm. The fusion algorithm can show good properties. The fusion of ant colony algorithm and genetic algorithm can not only utilize the fast convergence performance of ant colony algorithm, but also make full use

of the fast global search ability of genetic algorithm to achieve complementarity. The fusion of ant colony algorithm and p-Opt local search algorithm improves the efficiency and accuracy of the best path in each environment. The ant colony algorithm can also be integrated with the neural network, which complements the broad mapping ability of the neural network and the fast global convergence of the ant colony algorithm, presenting characteristics that are not unique to a single algorithm.

In addition, the hybrid algorithm combined with the ant colony algorithm has been greatly developed. The advantages and disadvantages of various algorithms can be used to complement the shortcomings of various algorithms, and the advantages of the respective algorithms can be used to solve practical problems.

2. Thermal barrier coating processing optimization model

The above lower diameter difference is minimized as the optimization target, and the thermal barrier coating processing optimization model is established by taking the process parameters and other experimental results as the constraints. The model can be expressed as:

Min UDDD	% （UDDD）
s.t. $0 \leq ULDD \leq 50$	% （ULDD）
$0 \leq DLDD \leq 50$	% （DLDD）
$0 \leq ULDD \leq 50$	% （UDDD）
$-50 \leq UMA \leq 50$	% （UMA）
$-50 \leq DMA \leq 50$	% （DMA）
$630 \leq x1 \leq 950, x1 = 0.01 * i1$	%laser power factor
$30 \leq x2 \leq 46, x2 = 1*i2$	%Laser pulse frequency
$10 \leq x3 \leq 26, x3 = 1*i3$	%Laser rotary cutting speed
$10 \leq x4 \leq 42, x4 = 1*i4$	%Laser rotary cutting interval
$ij=1,2,3,4, j=1,2,3,4$	

3. Process optimization method

The improved ant colony algorithm is used to optimize the process parameters. The specific steps are as follows:

1.Adaptive pheromone volatilization factor. In the process of searching the path of the ant colony algorithm, if the pheromone strength is always a constant, its size will not have a great impact on the performance of the algorithm; but if the pheromone intensity is dynamically changed during the search process, the performance of the algorithm will be Have an impact. Therefore, in order to avoid the stagnation phenomenon in the search optimal solution process, the pheromone on the optimization path is dynamically adjusted by the adaptive pheromone volatilization factor to improve the global search ability and search speed of the algorithm.

The adaptive variation formula of pheromone volatilization factor is as follows:

$$\rho(t) = \begin{cases} = 0.95 * \rho(t-1) , if \quad \rho(t) \geq \rho_{min} \\ = \rho_{min} \qquad , otherwise \end{cases} \quad (3\text{-}1)$$

In the middle：ρmin The minimum value of the pheromone is expressed, the purpose of which is to prevent the convergence speed of the algorithm from being slow due to the fact that $\rho(t)$ is too small.

2. Adaptive transfer probability. In the process of improving the search path of ant colony algorithm, in order to avoid the ant later choosing the same path to make the algorithm fall into the stagnation state too early, a new adaptive factor is introduced in (3-2) to explore the route of low concentration pheromone through the new incentive, so as to suppress the positive feedback and strengthen the global search capability.

The transfer probability formula is as follows:

$$p(t,i) = \frac{\tau(i) * \exp\left(-\left|\tau(i) - tau\_regu\right|\right)}{\sum_{j=1}^{ant} \tau(j) * \exp\left(-\tau(j) - tau\_regu\right)} \quad (3\text{-}2)$$

Among them, tea_regu is a regulatory factor, usually set to a smaller value than max{tau (I)}. The smaller the value is, the greater the probability of access to smaller pheromone concentration paths are.

3. variable scale chaotic search. Chaos optimization is a new optimization algorithm, which improves the efficiency of stochastic optimization algorithm by using randomness, universality and initial sensitivity of chaotic sequences. The variable scale chaotic search algorithm is integrated into the ant colony algorithm, and a better solution is found near the current global optimal solution. As the ant colony algorithm is carried out, the search range of chaotic operator is gradually reduced, and the aim of improving the search accuracy is to prevent the local optimum in the early stage of ant colony search and improve the search accuracy at the later stage.

The formula of variable scale carrier based on chaotic sequence is as follows:

$$x(t) = \begin{cases} \hat{s} + 2 * r(t) * (ChaosSeq - 0.5) & , p \geq p_0 \\ \hat{s} + 2 * Loc(t) * (ChaosSeq - 0.5) & , p < p_0 \end{cases} \quad (3\text{-}3)$$

In the formula, x (t) is the self variable value after the superposition of chaotic sequence, r(t) is a variable scale chaotic search radius, Loc (t) is a local chaos search radius, and ChaosSeq is a chaotic sequence.

4. Optimization results and verification

The minimum diameter difference is the optimization target, the diameter difference of the length and length, the difference between the length and length, the machining precision and the machining precision are the constraints. The optimization method based on the improved ant colony algorithm is adopted to obtain the optimal process parameters, the optimal value and the experimental results, such as table 4.1.

Table4.1 Optimal process parameters and optimal values

process parameters	Laser power factor	Laser pulse frequency	laser Rotary cutting speed	Laser cutting spacing	ULDD	DLDD	UDDD	UMA	DMA
	(‰)	(kHz)	(mm/s)	(μm)	(μm)	(μm)	(μm)	(μm)	(μm)
Optimization results	950	46	10	28	18	33	48	50	2
experimental result	950	46	10	28	16	35	42	47	2

The optimized process parameters are used to verify the perforation. The optimized process parameters are repeated three times and the average value of the three times is the experimental value. The optimization results are basically the same as the experimental results. The optimization results are compared with the experimental results, such as figure 4.1. The electrolyte assisted UV laser optimization process parameters processing samples, such as figure 4.2, are compared. The error is within the range of allowable value (less than 15%), and the test results meet the requirement of precision (+ 0.05mm). Thus, the optimization method based on improved ant colony algorithm is proved to be efficient to find the optimal experiment scheme.

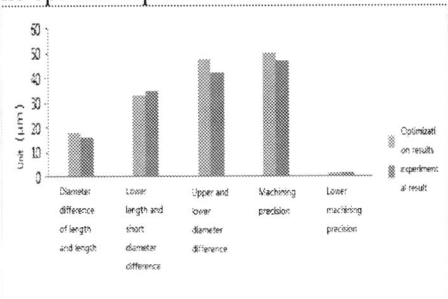

Figure. 4.1 Response optimization, experimental results comparison chart

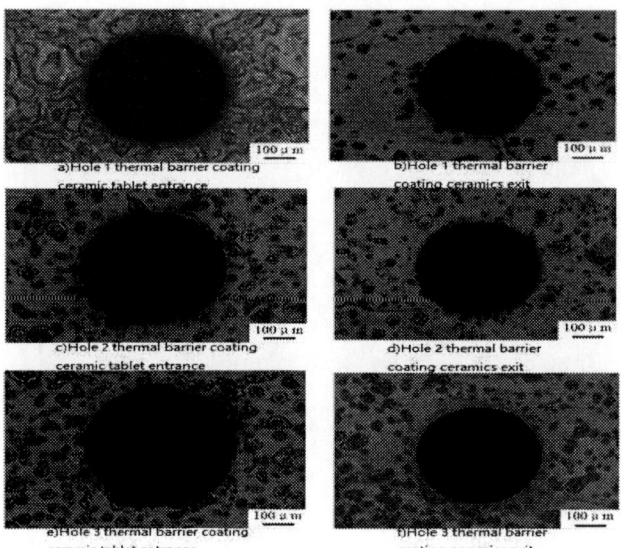

Figure. 4.2 Electrolyte assisted UV laser optimization process parameters processing sample map

5. Conclusion

The ant colony optimization algorithm is introduced and the improved strategy of ant colony algorithm is analyzed. The improved ant colony algorithm was used to optimize the experimental data, and the optimization model of thermal barrier coating process based on improved ant colony algorithm was established. The experimental verification was carried out and good processing results were obtained. The improved ant colony algorithm was used to establish the thermal barrier coating processing optimization model, and the best experimental scheme was found efficiently. The optimization results show that the electrolyte assisted UV laser cutting thermal barrier coating microstructure can obtain better microporous structure. The machining accuracy is ±0.05mm, which meets the process requirements. Through theoretical analysis and process experiment research, it shows that electrolyte-assisted UV laser processing technology has considerable engineering application prospects.

Acknowledgments

The work was supported by the National Natural Science Foundation of China(NO.61672360).

References

[1] Guan Hengrong, Li Meizhen, Sun Xiaofeng, et al. Oxidation and failure of high temperature alloy thermal barrier coatings[J].Acta Metallurgica Sinica,2002,38(11): 1133-1140.

[2] Vassen R, TietzF, StoeverD. New them al barrier coatings based on pyrochlore/YSZ double-layer systems [J]. International of Journal Applied Ceramic Technology, 2004, 1(4): 351-361.

[3] Wei Shaobin, Lu Feng, He Limin, Xu Zhenhua. Research Progress in Preparation Technology of Thermal Barrier Coatings and Ceramic Layer Materials[J]. Thermal Spray Technology, 2013, 5(01): 31-37.

[4] Lai Hongkun. Laser micromachining of metal and thermal barrier coatings using 532nm nanosecond fiber laser [D]. [Master's thesis]. Shanghai: Shanghai Jiaotong University, 2014.

[5] Lai Hongkun, Qi Huan. Laser micromachining of metal and thermal barrier coatings using 532nm nanosecond fiber laser[J]. China Laser, 2013, 40(08): 52-57.

[6] Xu Qingze, Liang Chunhua, Sun Guanghua, Wang Zhihong. Development of Thermal Barrier Coating Technology for Turbine Blades of Foreign Aviation Fan Engines[J]. Aero Engine, 2008, 34(3): 52-56.

[7] Xu Qingze, Liang Chunhua, Sun Guanghua, et al. Development of Thermal Barrier Coating Technology for Turbine Blades of Foreign Aviation Turbofan Engines[J]. Aero Engine, 2008, 34(3): 52-56.

[8] Wei Shaobin, Lu Feng, He Limin, Xu Zhenhua. Research Progress in Preparation Technology of Thermal Barrier Coatings and Ceramic Layer Materials[J]. Thermal Spray Technology, 2013, 5(01): 31-37.

[9] LIU Chao, PENG Wei, ZHANG Wei, et al. Progress in the application of thermal barrier coating technology[J]. Aerospace Manufacturing Technology, 2012(4):10-13.

[10] Gupta U, Nath A K, Bandyopadhyay P P. Laser micro-hole drilling in thermal barrier coated nickel based superalloy[C]// Materials Science and Engineering Conference Series. [11]Materials Science and Engineering Conference Series, 2016.

[11] Perelessantiago V, Washington M, Brugan P, et al. Faster and damage-reduced laser cutting of thick ceramics using a simultaneous prescore approach[J]. Journal of Laser Applications, 2005, 17(4):219-224.

[12] Tangwarodomnukun V, Wang J, Huang C Z, et al. An investigation of hybrid laser–waterjet ablation of silicon substrates[J]. International Journal of Machine Tools & Manufacture, 2012, 56(1):39-49.

Experimental Study on Regression Model of Ultraviolet Laser Processing Thermal Barrier Coating Based on Response Surface Method

Yulan Hu[1], Guoqiang Liu[2], Weiguang Hao[1], Yajie LIu[2], Zilin Liu[1]

[1]School of Information Science and Engineering, Shenyang Ligong University, Shenyang, China

[2]School of Mechanical Engineering, Shenyang Ligong University, Shenyang, China

Abstract. In view of the non-conductivity and brittleness of thermal barrier coating materials, the problems of conventional electromachining and machining can not be used. The processing technology of thermal barrier coating materials is studied. The regression model of processing technology based on response surface method is established. The influence of process parameters on the experimental results is of great significance in the field of processing key components of ceramic coatings. The main research work of this thesis is as follows: The regression equation based on response surface method is established, the parameters of laser processing and rotary cutting are analyzed, and the influence of processing precision of various process parameters is analyzed. The electrolyte assisted UV laser processing is designed and carried out.

1. Process parameter analysis

Electrolyte-assisted UV laser processing is the photothermal effect of laser and matter, and there are many factors influencing it. In order to obtain a better micropore structure and improve the processing accuracy, it is necessary to analyze the laser processing parameters and the laser cutting parameters. Zirconium dioxide ceramics doped with stabilized cerium oxide are widely used in thermal barrier coating materials in aerospace engineering. Therefore, this paper uses a zirconia ceramic piece workpiece doped with partially stabilized yttria for systematic process experiments.

1.1 Laser processing parameter analysis

In the electrolyte-assisted UV laser processing experiment, parameters such as laser wavelength, laser pulse energy, and laser repetition frequency will directly affect the feasibility and processing accuracy of the processing. Therefore, these parameters are analyzed as follows:

Laser wavelength. When the laser is transmitted in a liquid, the laser energy is attenuated due to the absorption and scattering of the solution, and the Langber Beer's law is followed. The spectrum of the absorption length of the laser of different wavelengths in pure water is shown in Figure 1 [40]. The absorption length of pure water to 355 ultraviolet light is about 5m, which is calculated. The attenuation coefficient of laser at 355nm wavelength in pure water is $0.2m^{-1}$, due to the low-concentration pure salt solution prepared by pure water, laser attenuation and pure water. There is no significant difference in attenuation, so the 355 nm wavelength laser has an attenuation coefficient of $0.2 \ m^{-1}$ in the electrolyte.

Content from this work may be used under the terms of the Creative Commons Attribution 3.0 licence. Any further distribution of this work must maintain attribution to the author(s) and the title of the work, journal citation and DOI.

Published under licence by IOP Publishing Ltd

ISPECE   IOP Publishing

IOP Conf. Series: Journal of Physics: Conf. Series **1187** (2019) 032046   doi:10.1088/1742-6596/1187/3/032046

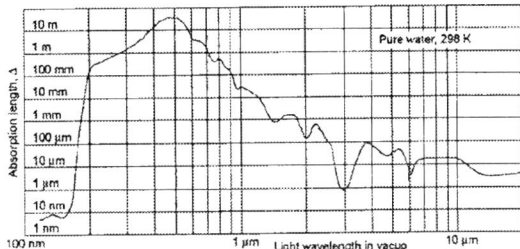

Figure 1 Spectra of absorbance length in pure water without wavelength laser

Laser pulse energy. The laser pulse energy is a main parameter of the pulsed laser. The energy control of the nanosecond ultraviolet laser used in the experiment is adjusted by changing the laser power factor (PWF), that is, the pulse energy is adjusted in steps of "‰". Using a laser power meter to measure the variation of laser pulse power with power factor, the curve of the pulse frequency is 30kHz, 60kHz, 80kHz, as shown in Figure 2, if the laser pulse energy is too low, the unit within the processing area will be If the area power density is too small to melt and vaporize the material, laser processing cannot be performed. However, if the laser single pulse energy is too high, a large amount of bubbles will be generated due to excessive laser power density at the focal point per unit area. Large, affecting the processing effect; therefore, the laser power factor should be reasonably selected according to the actual situation, the laser pulse energy should be adjusted, and better processing results can be obtained under the premise of ensuring the processing efficiency.

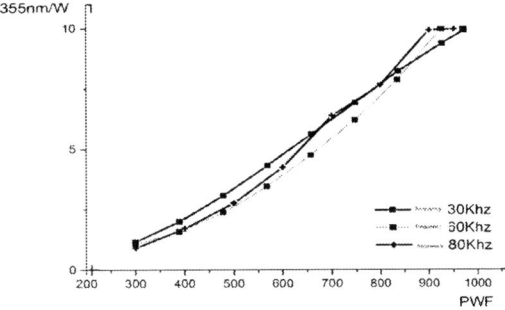

Figure 2 Repetition frequency 30kHz,60kHz,80kHz, pulse energy versus power factor (PWF)

Laser pulse frequency. The 355nm UV laser used in this experimental system has a repetition rate adjustable from 30kHz to 150kHz. When the current is 28A, the chiller is set to $20\,^{\circ}c$, and the laser power factor (PWF) is 950, the measured curve of the laser pulse energy with the repetition frequency is shown in Figure.3. As the repetition frequency increases, the laser pulse energy decreases. Small, due to the attenuation of the electrolyte, the selection of the repetition frequency is too large, the purpose of removing the laser-based material is lost, and the effect of removing the material cannot be achieved; the repetition frequency is too small, the laser pulse energy is large, and the processing precision cannot be guaranteed.

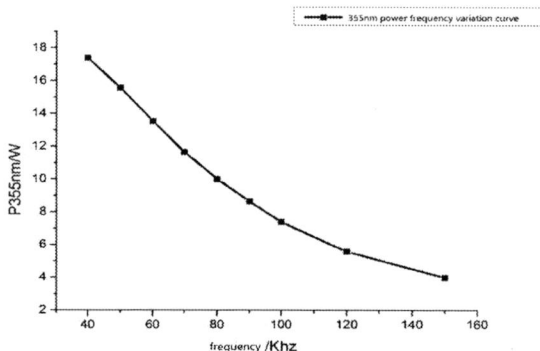

Figure 3 Laser pulse energy with repetition frequency curve

1.2 Analysis of rotary cutting parameters

The laser galvanometer is used to control the focus of the laser beam to perform the rotary cutting micropore test on the processed sample, and the processing area is rotated and processed by the software system. Among them, the repeated test is repeated in the case where the selected micropore diameter is 0.4 mm. Contrast, select the concentric circular cutting method to machine the small holes. Combined with the characteristics of UV laser processing parameters, the laser cutting speed and the turning distance are analyzed.

(1) Rotating speed. The laser cutting speed has a great influence on the circular hole roundness error and the upper and lower diameter difference. The smaller the turning speed, the higher the overlapping rate of the laser spot, the greater the power density of the laser acting on the material, and the smaller the removal of the small hole material. In the case where the turning distance and the number of processing times are selected, the larger the turning speed, the smaller the spot overlap ratio, and the corresponding material removal amount is reduced.

(2) The cutting distance. The laser cutting distance directly changes the laser spot overlap ratio in the direction of the center of the concentric circle, so that the power density per unit area of the laser processing increases, and the removal amount of the laser increases. Therefore, the larger the laser rotation spacing, the smaller the spot overlap rate and the less the material removal.

2. Experimental design

In this experiment, according to the laser electrolysis combined micro-machining test basis, the main process parameters laser power factor, pulse frequency, rotary cutting speed and turning distance are selected as the influence factors, which are respectively coded as x1, x2, x3, x4, and above The range of process parameters is as follows: laser power factor: 630~950, laser pulse frequency: 30kHz~46kHz, laser cutting speed: 10mm/s~26mm/s, laser cutting distance: 10μm~42μm. The main process parameters of electrolyte-assisted UV laser processing were horizontally coded by orthogonal test design method as shown in Table 1.

Table1. Four-factor five-level coding table

factor		Horizontal code / actual value				
		1	2	3	4	5
Laser power factor（‰）	x_1	630	710	790	830	950
Laser pulse frequency（kHz）	x_2	30	34	38	42	46
Laser rotary cutting speed（mm/s）	x_3	10	14	18	22	26
Laser cutting spacing（μm）	x_4	10	18	26	34	42

3. Regression analysis

For the determination of the response value: the upper and lower diameter difference evaluation taper; the upper machining accuracy to evaluate the inlet machining accuracy.

The experimental results were processed by Design Expert 7.1.3 software, and the regression equations of each response value were established, and the influence of process parameters on the experimental results was analyzed.

The regression variance analysis of the upper and lower diameter difference, corresponding to the regression equation, see Formula 1. From the analysis, the upper and lower diameter difference model F value is 3.449409, the corresponding p-value is 0.0015<0.05., the model is established. The regression equation showing the difference of upper and lower diameter is the highest. The first laser cutting speed and the rotary cutting spacing are the highest, the influence on the upper and lower diameter difference is greater, the other items have less effect on the upper and lower diameter difference.

The quadratic regression equation of the difference between the upper and lower diameters:

$$UDDD=315.19337-0.429682*x14.722614*x2-$$
$$4.027824*x3+0.47301*x4+0.003035*x1x2+0.002044*x1x30.001415*x1x40.008375*x2x3+0.004273$$
$$*x2x4+0.041222*x3x4+0.000183*x12+0.038195*x22+0.075584*x32+0.005375*x42 \qquad (1)$$

Analysis of the regression variance of the machining accuracy. The corresponding regression equation is shown in Formula 2. The analysis shows that the F value of the upper machining accuracy model is 3.252048, the corresponding p-value is 0.0023<0.01, and the model is established, which indicates that the regression equation of machining precision is the most significant. The first time laser power factor, the pulse frequency and the rotation tangent spacing interaction term showed the highest remarkable, has the main influence to the upper processing precision. The interaction of pitch interaction, spin-cutting speed and rotation-tangent spacing and power factor two are generally significant, and the accuracy of the machining is generally affected; the remaining items have less impact on the machining accuracy.

The quadratic regression equation of the upper machining accuracy:

$$UMA=235.758597+0.765607*x1+2.653537*x2+0.886572*x36.151811*x40.003582*x1x2+0.0031$$
$$27*x1x3+0.004432*x1x4+0.082021*x2x3+0.088884*x2x4-0.07359*x3x4-0.00047*x12-$$
$$0.043922*x22-0.139001*x32+0.009639*x42 \qquad (2)$$

4. Impact analysis

The effect of process parameters on the difference between the upper and lower diameters is shown in Figure 4.

(1) The difference in upper and lower diameters substantially decreases as the laser power factor increases. At the same time of the increase of the laser power factor, the average laser power increases, and the laser spot size does not change, which will increase the erosion ability of the laser processing, and the ejection capability of the melt is enhanced, thereby reducing the difference in the upper and lower diameters.

(2) The difference between the upper and lower diameters increases slightly with the increase of the laser pulse frequency. The increase of the laser pulse frequency leads to an increase in the laser spot overlap ratio in the laser cutting direction, an increase in the pulse frequency, an increase in the laser erasing ability, an increase in the diameter of the upper surface circle, and a diameter of the lower surface due to the attenuation of the electrolyte. The size does not have a large influence on the diameter of the upper surface circle, resulting in a small increase in the difference between the upper and lower diameters.

(3) The difference in upper and lower diameters increases as the laser cutting speed increases. Under the same conditions of laser pulse frequency and spot size, the larger the laser cutting speed is, the smaller the laser spot overlap ratio are, the smaller the laser power density is, and the corresponding laser erosion amount decreases, so the upper and lower diameter differences increase.

(4) The difference between the upper and lower diameters generally increases as the distance between the laser concentric circles increases. The distance between the concentric circles of the laser increases, the spot overlap ratio on the trajectory is reduced, and the average power density of the laser is reduced, thereby reducing the difference in diameter between the upper and lower sides.

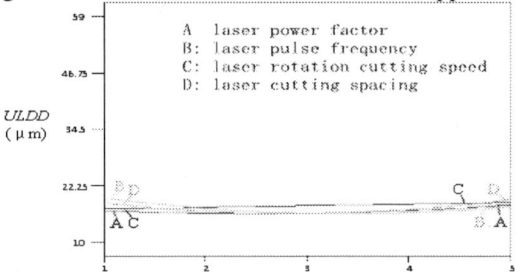

Figure 4 The influence of process parameters on the diameter difference between upper and lower

The effect of process parameters on the upper machining accuracy is shown in Figure 5.

(1) As the laser power factor increases, the upper machining accuracy increases first and then decreases. As the laser power factor increases, the average laser density will increase. As the diameter of the laser spot is almost constant, the amount of laser erosion increases. At the same time as the material is eroded, the effect of the laser cavitation bubble and the attenuation of the laser transmission by the electrolyte. Sex, the upper machining accuracy will increase first and then decrease.

(2) The upper machining accuracy increases as the laser pulse frequency increases. When the laser pulse frequency is increased, the spot overlap ratio in the laser cutting speed direction will increase, the average laser power density in the same processing area will increase, and the laser erasing ability will be enhanced, so that the upper machining accuracy is increased.

(3) As the laser cutting speed increases, the upper machining accuracy increases slightly and then decreases. Under the same laser pulse frequency and spot size, the larger the laser cutting speed, the smaller the spot overlap ratio in the direction of the rotation, at which time the size of the laser erosion plays a leading role, and the upper machining accuracy will increase slightly. When the spot overlap ratio is reduced to a uniform track rotation, a better machined hole shape can be obtained, thereby reducing the upper machining accuracy.

(4) The upper machining accuracy generally decreases as the pitch of the laser concentric circles increases. As the distance between the concentric circles of the laser increases, the laser spot overlap ratio on the concentric circular track decreases, and the laser power density decreases, thereby reducing the upper machining accuracy.

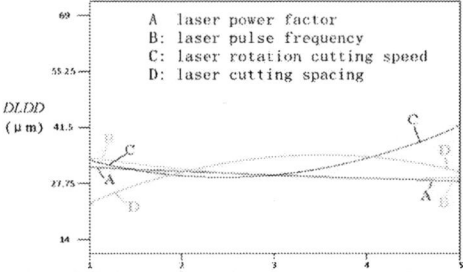

Figure 5 The influence of process parameters on the machining accuracy

5. Conclusion

Based on the mechanism of electrolyte-assisted UV laser processing thermal barrier coating materials, the main experimental parameters of the experiment were analyzed firstly. Secondly, the electrolyte-assisted UV laser processing was designed and carried out, and the regression equation based on experimental results was established. Finally, the influence of various process parameters on the experimental results is analyzed. The processing rules of thermal barrier coatings are summarized, and the optimal experimental scheme is efficiently found to provide an effective method for high-quality thermal barrier coating microstructure processing. It has important significance and application prospects in the processing and manufacturing of key components such as aviation and aerospace with thermal barrier coating.

Acknowledgments
The work was supported by the National Natural Science Foundation of China(NO.61672360).

References
[1] LI Xiaoyu, SUN Huilai, ZHAO Fangfang, NIE Xiaoju. Research on parameter optimization design of femtosecond laser processing SiC model[J]. LASER & IR, 2016, 46(08): 948-952.
[2] Jia Wei, Wang Qingyue, Fu Xing, Hu Xiaotang. Application of Femtosecond Laser in Material Micromachining[J]. Chinese Journal of Quantum Electronics, 2004, (02): 194-201.
[3] Lai Hongkun. Laser micromachining of metal and thermal barrier coatings using 532nm nanosecond fiber laser [D]. [Master's thesis]. Shanghai: Shanghai Jiaotong University, 2014.
[4] Lai Hongkun, Qi Huan. Laser micromachining of metal and thermal barrier coatings using 532nm nanosecond fiber laser [J]. China Laser, 2013, 40(08): 52-57.
[5] ZHAN Caijuan, LI Changzhen, WANG Yuli. Heat transfer analysis and numerical simulation of water jet guided laser drilling[J]. Journal of Applied Lasers, 2009, 29(05): 415-418+422.
[6] Zhang Hua,Xu Jiawen. Modeling and Experimental Investigation of Laser Drilling with Jet Electrochemical Machining[J]. Chinese Journal of Aeronautics,2010,23(4):.
[7] Nguyen M D, Rahman M, Wong Y S. Simultaneous micro-EDM and micro-ECM in low-resistivity deionized water[J]. International Journal of Machine Tools & Manufacture, 2012, 54-55: 55–65.
[8] Hu Xiaotong, Huan Rong. The current situation and development of China's laser industry [J]. Applied Laser, 1990, (03): 97-100.
[9] Avanish Kumar Dubey, Vinod Yadava. Laser beam machining—A review [J]. International Journal of Machine Tools & Manufature 48 (2008): 609-628.
[10] He Fei, Cheng Ya. Femtosecond laser micromachining: a new frontier in the field of laser precision machining [J]. China Laser, 2007, (05): 595-622.
[11] Xie Yijiang, Li Dianjun, Zhang Chuansheng, Guo Weihai. Acousto-optic QC_2 laser[J]. Optics and Precision Engineering, 2009, 17(05): 1008-1013.
[12] CHEN Zhiling, SHI Tielin, LIU Sheng, XIONG Liangcai. Excimer laser microfabrication technology and its application[J]. Progress in Laser and Optoelectronics, 2004, (02): 47-53.

ISPECE
IOP Publishing

Shape Optimization of Hook for Marine Crane

Yang Ji[1], Hu Wang[1], Hai-quan Chen [1,a] , Ming-xuan Guo [1] and Jun-jie Wu [1]

[1]College of Marine Engineering, Dalian Maritime University Dalian, Liaoning, China

[a] Corresponding author: Chen Hai-quan, chenapec@dlmu.edu.cn

Abstract. The hooks are prone to fatigue fracture due to the frequent impact loads on the hooks at sea. In order to further strengthen the structural strength of the hooks, the influence of the three parameters on the hooks is studied, such as the deflection angle of the hooks with large openings, the opening diameter of the hooks and the position of the maximum thickness of the hook walls.The main points are as follows: parameterized modeling of crane hooks, data simulation and analysis with ANSYS, and optimization analysis of hooks' strength, which provide basis for further enhancing reliability of hooks.

1. Introduction

Hook is one of the important components of crane.When the crane is carrying out transportation operations, the hook will be subjected to frequent impact loads. Once there is a failure such as fatigue fracture of the hook, it will cause significant loss of personnel and property. As for the hook of the ship crane, it is mainly used for important tasks such as cargo transportation and transfer, sea supply, launching of underwater operation equipment and recycling. Ships on the sea will be affected by external environment such as wind, waves, currents and self - control, so that the crane hook will be subjected to more severe repeated loading than land operations, which puts forward higher requirements on the strength and fatigue strength of the crane hook and poses great challenges to the safe operation of the crane hook[1].

Under different loads, the crane hook has been optimized and studied. It has been found that the strength of the hook can be enhanced by increasing the thickness of the cross section of the hook and the width of hook [2]. In order to meet the use of crane hooks under more severe working conditions, the influence of changes in the three places on the strength and deformation of hooks was studied under the condition of keeping the maximum thickness of hooks unchanged, including the angle of deflection of hooks, the diameter of hooks and the position of the thickest hooks. Through the optimization numerical analysis of these three parameters, the further optimization of the hook is realized, so that the shape optimization of the hook has more choices and basis.

In this paper, ANSYS software is used to carry out parametric modeling and force analysis on the hook, and structural optimization analysis is carried out on the hook according to the response surface and sensitivity [3], which provides a reasonable basis for further shape optimization of the hook.

2. Hook parametric model and force analysis

The hook is a lifting device on the crane and is also the main bearing component of the crane. The strength of the hook and the rationality of its design are crucial to the safety of the crane's work. Taking the hook bearing 0.5t produced by a factory as an example, the strength analysis of the hook is carried out, the deformation and stress distribution laws of the hook are explored, and its dangerous

Content from this work may be used under the terms of the Creative Commons Attribution 3.0 licence. Any further distribution of this work must maintain attribution to the author(s) and the title of the work, journal citation and DOI.
Published under licence by IOP Publishing Ltd

section is analyzed, which provides a theoretical basis for the study of hook structure optimization and has important engineering significance.

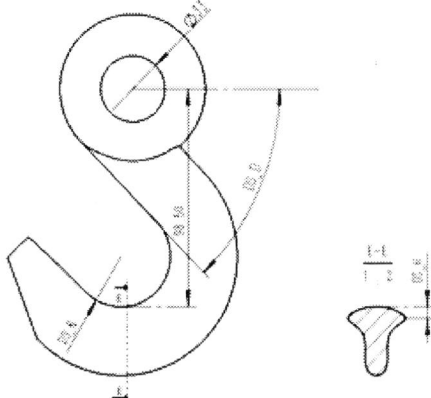

Figure 1. Simplified structure of the hook(Unit: mm)

When modeling the hook, some unimportant details and details that have little influence on the strength of the hook are ignored, thus simplifying the hook model and eliminating interference on the optimization parameters. The structure diagram of crane hook is shown in figure 1. Low alloy steel is selected, with a density of 7800kg/m3, an elastic modulus of 2e11, poisson's ratio of 0.27, and a tensile strength of 450MPa. The hook model is divided into 48911 units and 73168 nodes using tetrahedral grids.

3. Strength analysis of automatic unloading hook

When the hook is working on a crane, the bottom of the hook is subjected to the gravity exerted by the rope. In this paper, the load acting area of the suspension cargo cable to the hook is simulated by projection marks, where in the load applied by the cable to the hook is 5000N.After the hanger lug of the hook is subjected to displacement constraint, the strength of the automatic unloading hook is analyzed. The constraint conditions and loads are shown in figure 2.

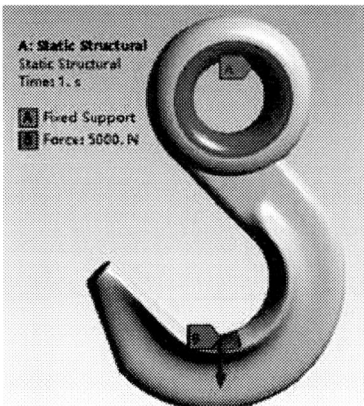

Figure 2. Constrains and loads

Where in the A load is the load caused by the gravity of 500 kg of goods. As for the determination of the action area of the A load, the projection area of the steel cable on the hook in the vertical direction when the crane hook is working is simulated, that is, the contact area between the hook and the rope. At the same time, the projected positioning is associated with the sketch of the hook, so as to ensure that when the hook is optimized later, the load action area will not be affected due to changes in parameters, thus eliminating the interference of the load action area on the result.

The equivalent stress distribution of the hook is shown in figure 3, the maximum equivalent stress value is 115.17 MPa, the distribution of hook shape variables is shown in figure 4, and the maximum shape variable is 0.21289 mm.The maximum equivalent stress occurs at the center of the hook body bend of the hook, which is caused by the tensile force and bending moment of the load at the hook bend, resulting in tensile deformation. The maximum deformation of the hook occurs at the top of the hook opening. Since the maximum deformation is only 0.17362 mm, it can ensure the safety of hanging goods during operation. Under the load, the maximum equivalent stress of the hanger is only 115.17 MPa, and its minimum safety factor is 3.91, which meets the safety requirements of the hanger. Therefore, on the premise of ensuring that the strength and shape variables of the automatic unloading hook meet, the shape of the hook is optimized.

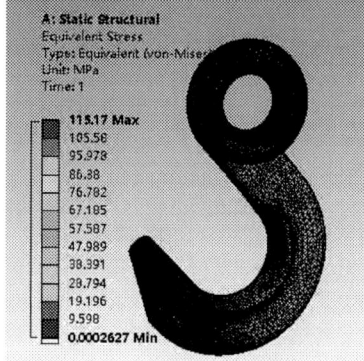

Figure 3. Hook equivalent stress distribution

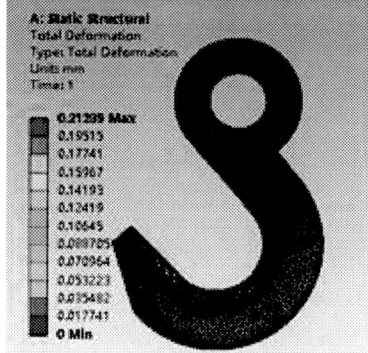

Figure 4. Hook shape variable distribution map

4. Hook structure optimization

4.1 Design variables and output variables

In order to optimize the shape of the hook, shape parameters need to be selected as design variables. The deflection angle DS _ C of the hook, the opening diameter DS_R of the hook, and the position DS _ H at the thickest part of the hook section are selected as design parameters. The specific positions are shown in figure 1 below. Preview the shape change of the hook model within the specified range of the defined parameters. In order to avoid the shape distortion or fracture surface of the hook model within the range of parametechange[4].

At the same time, the maximum value of the equivalent stress of the hook and the maximum form variable are selected as the output parameters. The basis of the optimized design is that the spreader is within the safe range, and at the same time meets the requirements of normal operation, so that the strength of the structure can be further enhanced. When carrying out the test design, the change range of each design variable must be set first. the specific parameter settings are shown in table 1.

Table 1. Parameter design table

Parameter	Symbol	Design parameter range
Hook opening radius/mm	DS_R	25~30
Deflection angle of hook/°	DS_C	41~50
The thickest part of the hook section/mm	DS_H	6~9

4.2 Sensitivity analysis and construction of response surfaces

The midpoint composite design method was used to design the design points, and 15 design points were selected for static analysis and response surface construction of the spread. The parameter sensitivity is shown in Figure 5.

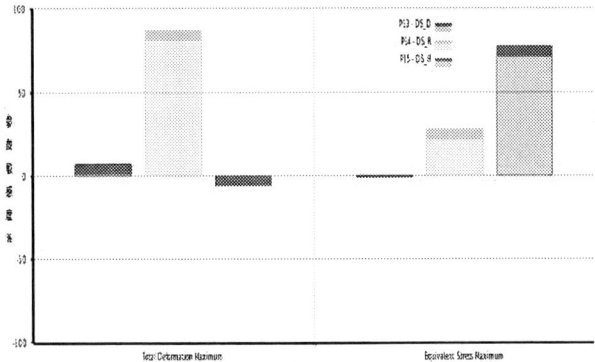

Figure 5. Parameter sensitivity map

4.3 Multi-parameter and multi-objective optimization analysis

Carry out the same structural analysis on the optimized hook, and its analysis structure is shown in figure 6 and figure 7. As can be seen from figure 6, the maximum equivalent stress value of the hook after the sling is optimized is 102.03 MPa, which also occurs in the center of the hook body. Its value is far less than the allowable stress value of the material. The strength of the hook after optimization still meets the safety requirements. As can be seen from figure 7, the maximum shape variable of the hook after optimization is 0.655 mm, which is mainly located at the top of the hook opening. since the shape variable of the hook is very small, the phenomenon of goods separation still does not occur. The comparison results before and after optimization are shown in Table 2.

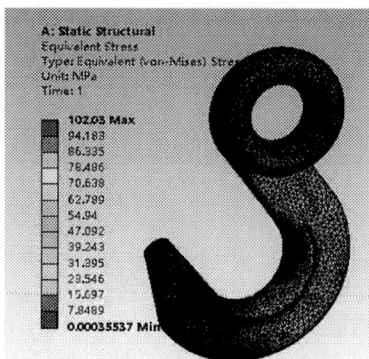

Figure 6. Equivalent stress of spreader

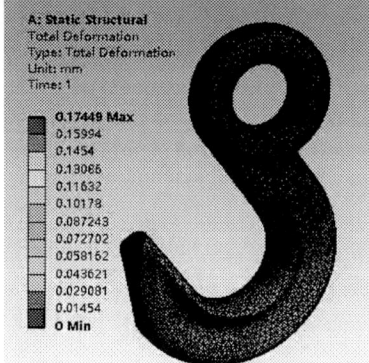

Figure 7. Total deformation of spreader after optimizat

Table 2. Comparison of results before and after optimization

Parameter name	Symbol	Design parameter range	Optimized value
Hook opening radius/mm	DS_R	25~30	25
Deflection angle of hook/°	DS_D	41~50	46.2
The thickest part of the hook section/mm	DS_H	6~9	6.1
Maximum amount of total deformation/mm	Total Deformation Maximum	115.17	102.03
Maximum value of the equivalent stress/Mpa	Total Deformation Maximum	0.21289	0.17419

5. Conclusion

The deflection degree of the hook has a slight influence on the strength and shape variables of the hook, while the opening diameter of the hook has a great influence on the strength and shape variables of the hook. When optimizing the hook, under the condition of maintaining the overall thickness, width and length, it can be considered to appropriately reduce the diameter of the opening of the hook. Among the three parameters, the location of the thickest hook section has the greatest influence on the strength of the hook. In order to enhance the strength of the hook, the strength of the hook can be enhanced by changing the location of the thickest hook section under the condition that other shapes and structures are unchanged. By optimizing the design of the hook through the three parameters of the deflection angle of the hook, the opening diameter of the hook and the position of the thickest section of the hook, the analysis shows that the optimized hook meets the safety requirements in terms of structural strength and shape variables, achieves the expected purpose, and makes the shape of the hook more reasonable.

References
[1] Rui GuangLiu. Research on swing analysis and control of hoisting system of crane ship [D].Tianjin: Tianjin University, （2004）.
[2] Li Shuilang, Gao Wenjie,Mao Pengjun. Research on optimization of hook parameters based on AWE [J]. Mechanical design and manufacturing,2011(12):100-102
[3] Ma BaoSheng. Application of response surface methodology in many practical optimization problems [D].BeiJing: School of mechanical engineering and applied electronics technology, Beijing university of technology, （2007）.
[4] Zhang Yongxing, Zhong Zhipan, Chenke. Finite element analysis and optimization design based on Pro /MECHANICA hook [C]. Anhui association for science and technology annual meeting mechanical engineering annual meeting, （2008）.
[5] Xie Yancai. Sensitivity analysis of mechanical structure reliability based on response surface methodology[D]. Changchun:School of mechanical science and engineering, Jilin university,(2008).

ISPECE   IOP Publishing

Design and Experimental Study of Vibration Reducing Experimental Device for Magneto-rheological Elastomer

Tieshan ZHANG, Zhong REN

China University Of Mining And Technology Yinchuan College, Yinchuant, Ningxia, China

Tieshan ZHANG : zts336699@126.com

Abstract. The current automotive suspension system is fixed due to its rigidity and damping, so the damping effect is not adjustable, which affects the ride comfort.Trying to use magnetorheological elastomer as the core material of the suspension system, the theoretical model of the vehicle vibration damping experimental device excited by vertical vibration is established.According to the theoretical model, the experimental device of automobile vibration reduction system is designed, and the comparison experiment of vibration reduction effect under different conditions is carried out by using the device.The results show that the magnetorheological elastomer is effective in the automobile vibration damping system, and the magnetic field strength is the most important factor affecting the vibration damping effect.

1. Introduction

At present, the damping system in most vehicles is mainly composed of elastic components and rigid components, and its characteristic parameters are fixed, so the vibration damping effect cannot be adaptively adjusted according to the running condition and the road surface condition. On this issue, domestic and foreign scholars have proposed that the vibration reduction effect can be adjusted by the method of active suspension and semi-active suspension,because the active suspension which has complicated structure, high cost and large power consumption requires a high-precision servo mechanism, complex equipment and large external power sources. The semi-active suspension system has defects such as small adjustment range and uncontrollable adjustment ability.For example, Sun Jianming[1] studied the characteristics of existing active suspensions. It is proposed that the application range of active suspension is limited due to high cost, complicated structure and large power consumption, and the LMS control strategy method is adopted to reduce the cost. Zhu Hua[2] analyzed the characteristics of several suspensions in the development trend of semi-active suspensions, and pointed out that their commonality is small adjustment range and low control precision. In order to overcome the complex defects, Kou Farong designed a vehicle suspension system basing on electric hydrostatic actuation, and simplified the suspension structure.In order to solve these contradictions, improve the real-time operational stability and ride comfort of the vehicle. This paper proposes a new type of smart material which is Magneto-rheological elastomer as the main component of automotive vibration damping system. The shear modulus of this material is controlled by the strength of the magnetic field, while the shear modulus affects the stiffness and damping of the material. Therefore, it has an application basis that realizes adjustable vibration damping effect. Based on the above research status, it is proposed to apply magneto-rheological elastomer to automobile vibration reduction[3-4]. For this purpose, an experimental device has been developed and an experimental study on vibration damping effect has been carried out.

Content from this work may be used under the terms of the Creative Commons Attribution 3.0 licence. Any further distribution of this work must maintain attribution to the author(s) and the title of the work, journal citation and DOI.
Published under licence by IOP Publishing Ltd

2. Design of Automobile Vibration Reduction Experimental Device Based on Magneto-rheological Elastomer

2.1 Theoretical model of automobile vibration reduction experimental device
Automotive vibrations include vertical vibration, pitch vibration, roll vibration, and lateral vibration. As the first step in the study of magneto-rheological elastomers for vibration reduction, This paper studies only vertical vibration, and its simplified theoretical model[5] is shown in **Figure 1.**

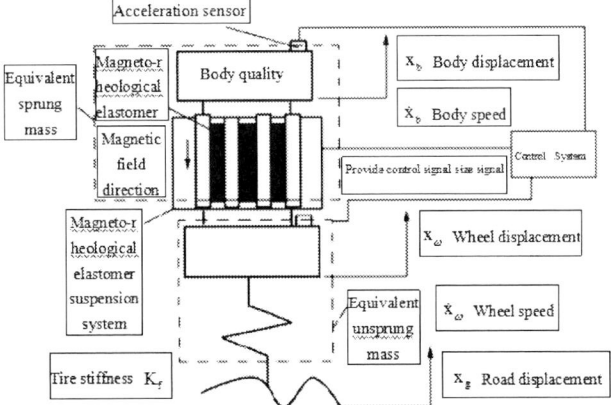

Figure 1. Simplified theoretical model of automobile vibration reduction experimental device
Assumption:The equivalent damping and equivalent stiffness of the part of the magneto-rheological elastomer vehicle suspension system are respectively C, K_S, The quality of the body and wheels is m_b, m_ω, The acceleration of the body and the wheel is $\ddot{x}_b, \ddot{x}_\omega$. According to Newton's second law, the equation of motion of the system can be obtained as:

$$m_b \ddot{x}_b = -C\ (\dot{x}_b - \dot{x}_\omega) - K_s \left[x_b(t) - x_\omega(t) \right] \quad (1)$$

$$m_\omega \ddot{x}_\omega = -C\ (\dot{x}_b - \dot{x}_\omega) + K_s \left[x_b(t) - x_\omega(t) \right]$$
$$- K_t \left[x_\omega(t) - x_g(t) \right] \quad (2)$$

The input stimulus uses a periodic signal whose function is $\omega(t)$.

$$\dot{x}_g(t) = -2\pi f_0 x_g(t) + 2\pi \sqrt{G_0 U_0} \, \omega(t) \quad (3)$$

In the formula: x- Road displacement
g- Road roughness coefficient
U- Vehicle forward speed
F- Natural frequency of the system

Combine formula (1) ~ formula (3),We write the system motion equation and the system input excitation equation in the form of a matrix, which is the state space equation of the system.

$$\dot{X}(t) = AX(t) + FW(t) \quad (4)$$

In the formula: $X(t)$ - System state vector, $X(t) = \left[\dot{x}_g(t), \dot{x}_\omega(t), x_b(t), x_\omega(t), x_g(t) \right]^T$;

$W(t)$ - Periodic signal input matrix, $W(t) = [\omega(t)]$;

$$A = \begin{bmatrix} \dfrac{-C}{m_b} & \dfrac{C}{m_b} & -\dfrac{K_s}{m_b} & \dfrac{K_s}{m_b} & 0 \\ \dfrac{C}{m_\omega} & \dfrac{-C}{m_\omega} & \dfrac{K_S}{m_\omega} & \dfrac{-K_t - K_s}{m_b} & \dfrac{K_t}{m_\omega} \\ 1 & 0 & 0 & 0 & 0 \\ 0 & 1 & 0 & 0 & 0 \\ 0 & 0 & 0 & 0 & -2\pi f_0 \end{bmatrix}$$

$$F = \begin{bmatrix} 0 \\ 0 \\ 0 \\ 0 \\ 2\pi\sqrt{G_0 U_0} \end{bmatrix}$$

According to this model, to know the vibration reduction effect of the suspension system, it is only necessary to compare the displacement and speed of the vibration before and after the vibration reduction，compare x_ω and x_b, OR $\dot{x}_\omega$ and $\dot{x}_b$.

According to the mechanical properties of the magneto-rheological elastomer and the theoretical model structure, the equivalent damping and equivalent stiffness of the theoretical model under different magnetic field strengths are calculated as shown in **Table 1**.

Table 1. Equivalent damping and equivalent stiffness under different magnetic fields

Parameter name	Numerical value		
Magnetic field strength	0 MT	250 MT	500 MT
Equivalent damping $C/N \cdot (m \cdot s^{-1})^{-1}$	543.5	1086.7	1753.5
Equivalent stiffness $K_s/(N \cdot m^{-1})$	9846.2	18067.2	27546.8

The equivalent damping and equivalent stiffness are substituted into the theoretical model, and the rms evaluation method is used to evaluate the damping effect of the theoretical model. The root mean square values under different magnetic field strengths are shown in **Table 2**.

The rms quantification proves that the magneto-rheological elastomer damping theory model is feasible and has good vibration damping effect, but it is necessary to control the appropriate magnetic field strength in order to achieve the best equivalent damping and equivalent stiffness.

Table 2. Root mean square acceleration of acceleration under different magnetic field strengths

Magnetic field size	Body acceleration	Suspension travel	Tire dynamic displacement
0 MT	1.43	29.12	11.98
250 MT	2.68	23.05	11.68
500 MT	1.23	21.23	16.78

2.2 Mechanical structure design of automobile vibration reduction experimental device
Based on the theoretical model of the vehicle vibration reduction experimental device, the mechanical structure of the device is divided into three parts:

(1)Exciting structure

The excitation structure adopts an eccentric cam mechanism, through which the periodic motion is generated as an input excitation signal of the automobile vibration damping experimental device, wherein the eccentricity of the eccentric is 2 mm, so the total stroke of the entire device is 4 mm. The eccentric wheel is connected to the AC speed regulating motor, and the vibration frequency of the device is controlled by the rotation speed of the AC motor.

(2)Magneto-rheological elastomer suspension system damping structure

The general working modes of magnetorheological elastomers are shear and extrusion. The general working modes of magnetorheological elastomers are shear and extrusion. The general working modes of magneto-rheological elastomers are shear and extrusion.When the direction of the force is perpendicular to the direction of the magnetic field passing through the magneto-rheological elastomer, it is called the shear mode.When the direction of the force is parallel to the direction of the magnetic field passing through the magneto-rheological elastomer, it is called the extrusion mode. The working mode of the magneto-rheological elastomer in the damping system designed in this paper adopts the shear mode, and its controllability and adjustment range are better than the extrusion mode.

(3)Data acquisition structure

The author judges the vibration damping effect of the experimental device by comparing the accelerations $\ddot{x}_o$ and $\ddot{x}_b$ before and after the vibration reduction.Therefore, an acceleration sensor is used and connected to the vibration measuring plane.

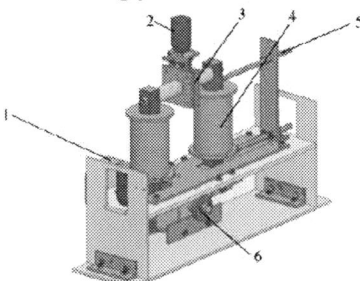

1.AC adjustable speed motor 2.Accelerometer 3.Magneto-rheological elastomer
Figure 2. Structure of automobile vibration reduction experimental device

Designed automobile vibration damping experimental device structure is shown in **Figure 2.** According to the theoretical model, the equivalent sprung masses of 3, 4, and 5 are shown in the figure below, 1, 6Equivalent unsprung mass.

2.3 Design of Data Acquisition System for Automobile Vibration Reduction Experimental Device
The signal to be collected by the experimental system is the voltage signal output by two acceleration sensors. After they are collected, signal conditioning, A/D conversion, signal processing and analysis, vibration signal data storage and image output are required.In order to achieve these functions,Using DSP TMS320F20 815 as the core, design and develop the data acquisition card, and compile the signal processing program, communicate with the computer through RS232, read the vibration signal in the SRAM into the computer, save it as data file, and then draw out before and after vibration reduction. The vibration image is analyzed based on the image to analyze the vibration reduction effect.The functional block diagram of the developed capture card is shown in **Figure 3.**

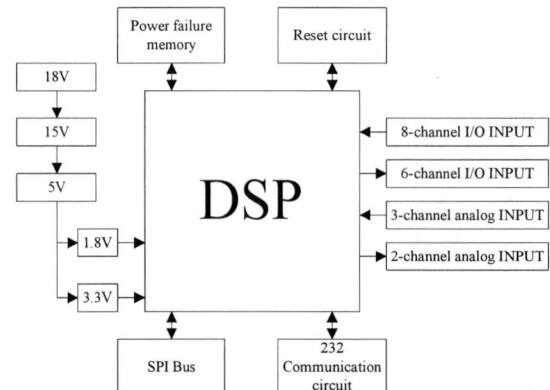

Figure 3. DSP data acquisition card functional block diagram

CCS3.3 as a software development tool, programming to achieve signal acquisition start, A/D conversion, data processing and analysis, save and other operations. According to the system characteristics and functional requirements, the FFT is used to transform the signal. After FIR filtering and spectrum analysis, the vibration signal to be acquired is selected, the data is saved in the SRAM, and the data is read to the computer through the serial communication software[6-7]. In addition, real-time control of system processes and monitoring via image display is required during the acquisition process.

3. Experimental Analysis of Magneto-rheological Elastomer Automobile Vibration Damping Device

Based on the designed experimental device, we carried out experimental research on the system damping effects of different thickness magneto-rheological elastomers, different excitation frequencies and different magnetic field strengths. The root mean square value evaluation method is used to judge the data before and after vibration reduction, analyze the vibration reduction effect, and further explore the key factors affecting the vibration reduction effect, and improve and perfect the experimental system. The experimental content includes the following three aspects.

3.1 Testing of the same magneto-rheological elastomer thickness and excitation frequency, and vibration damping effect under different magnetic fields

The excitation frequency used in the experiment, the speed of the AC motor is 150r/min, The magneto-rheological elastomer has a thickness of 3mm. Comparison curve of vibration acceleration before and after vibration reduction under different magnetic fields is shown in **Figure 4.** The root mean square value is shown in Table 3.

ISPECE   IOP Publishing

IOP Conf. Series: Journal of Physics: Conf. Series **1187** (2019) 032048   doi:10.1088/1742-6596/1187/3/032048

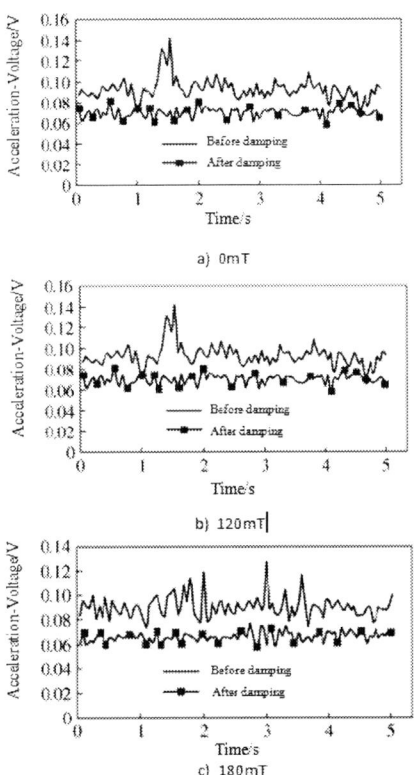

Figure 4. Comparison curve of vibration acceleration before and after vibration reduction under different magnetic field strengths

Table 3. Acceleration (output voltage) rms at different magnetic field strengths

	0 MT	120 MT	180 MT
Before damping	0.091162	0.093216	0.090662
After damping	0.073588	0.069829	0.066751
Damping ratio	19.78%	25.09%	26.57%

As can be seen from **Figure 4** and **Table 3**, The change of the magnetic field has a great influence on the vibration damping effect of the magneto-rheological elastomer. When the magnetic field increases to a certain extent, the influence of the magnetic field on the damping capacity is weakened.

3.2 Vibration damping test of magneto-rheological elastomers with different thicknesses under the same excitation frequency and magnetic field

The AC motor used in the experiment has a rotational speed of 150 r/min and the magnetic field strength is 120 mT.**Table 4** lists the root mean square values of the acceleration at different thicknesses.

Table 4. Acceleration root mean square value of magneto-rheological elastomers with different thicknesses

	3 mm	6 mm
Before damping	0.093216	0.091162
After damping	0.069829	0.066851
Damping ratio	25.09%	26.61%

Visible from Table 4,The effect of the increase in the thickness of the magneto-rheological elastomer on its damping capacity is not affected by the strength of the magnetic field; however, as the thickness increases, the damping capacity increases accordingly.From the structure of the experimental device, due to the increase in thickness, the magnetic field passing through the middle

819

part of the magneto-rheological elastomer is not large on both sides, and the equivalent damping and stiffness in the middle are not as large on both sides, so the shear strength in vibration is not as good. On both sides, these weaken the damping effect of the magneto-rheological elastomer.

3.3 Damping effect of the same elastomer thickness and magnetic field and different excitation frequencies

The magneto-rheological elastomer used in the experiment has a thickness of 6 mm and a magnetic field strength of 180 mT.The root mean square value of the acceleration at different excitation frequencies is shown in **Table 5**. Experiments show that the change of excitation frequency has little effect on the damping capacity of magneto-rheological elastomer, which indicates that the damping capacity of magneto-rheological elastomer is affected by its own structure, damper structure and magnetic field. The influence of other external factors is large.

Table 5. Root mean square value of acceleration at different excitation frequencies

	150 r/min	200 r/min
Before damping	0.093096	0.100574
After damping	0.067429	0.071282
Damping ratio	27.88%	28.91%

Discuss: Three sets of contrast experiments show that the change of the magnetic field has the greatest influence on the damping effect of the magneto-rheological elastomer,but the magnetic field strength needs a suitable range,not the bigger the better. The second most important factor is the thickness of the magneto-rheological elastomer, again the excitation frequency.In addition, from the effect of thickness change, the damping effect is not proportional to the change in thickness, which indicates that the damping effect is highly correlated with the structure inside the magneto-rheological elastomer, but due to the existing experimental conditions. The limit does not reflect this associated impact.One possible explanation is that the internal structure of a 6 mm thick magneto-rheological elastomer is not as strict as 3 mm.The chain arrangement affects the vibration damping effect of the magneto-rheological elastomer.

4. Conclusion

As a new type of smart material, magneto-rheological elastomer has broad prospects in vibration damping applications.However, its application research on vehicle vibration reduction is still at a very preliminary stage.The author establishes a vehicle vibration damping model based on magneto-rheological elastomer, and then establishes an experimental device according to the characteristics of the model, and performs multiple sets of contrast experiments under different conditions to test and verify the vibration damping effect under different conditions. In the comparative analysis, it is concluded that the magnetic field strength is the most critical factor affecting the damping effect, and different magnetic field strengths can obtain different damping effects. Research indicates, the magnetic field strength is the most important factor affecting the application performance of magneto-rheological elastomer. In the future, the application of magneto-rheological elastomer should be further explored and tested in the internal structure of magneto-rheological elastomer and the influence of magnetic field changes. Analysis, combined with the actual application, makes the experimental research gradually move to practical application.

ACKNOWLEDGEMENTS

This work was funded by the Ningxia Natural Science Foundation Project (NZ16239)

References

[1] Sun Jianming. Research on Vehicle Active Suspension System Control Technology [D]. Harbin: Harbin Engineering University, 2003.

[2] Zhu Hua. Research status and development trend of semi-active suspension system [J]. City car Vehicle，2009,(4):38 - 39.

[3] Watson J R, Canton M. Method and Apparatus for Varying the Stiffness of a Suspension Bushing[P]. US Patent 5,609,353,1997.

[4] Stewart W M,et al.Method and Apparatus for Reducing Brake Shudder[P]. US Patent 5,816,587,1998.

[5] Ginder J M, et al.Controllable-Stiffness components based on magneto-rheological elastomers[A]. Proceeding of SPIE, 2000, 3985: 418 - 425.

[6] Yi L, et al. Experimental research on the electrochemical abra-sive belt grinding 0Cr17Ni4Cu4Nb stainless steel[J]. Advances in Materials Manufacturing Science and Technology, 2009, 626 - 627: 617 - 622.

[7] Deng Huaxia, et al. Development of Magneto-rheological Elastomer Frequency Modulated Vibration Absorber[J]. Functional Materials, 2006, (5) :790 - 792.

ISPECE   IOP Publishing

IOP Conf. Series: Journal of Physics: Conf. Series **1187** (2019) 032049   doi:10.1088/1742-6596/1187/3/032049

A Simple Safety Control Method for PSS Critical Gain Test

Siyuan Guo[1,a], Shoushou Zhang[2], Weijun Zhu[1], Li Li[1], and Jinbo Wu[1]

[1] State Grid Hunan Electric Power Company Limited Research Institute, Power System Technology Center, 410007 Changsha, China

[2] Central South University of Forestry and Technology, Bangor College, 410018 Changsha, China

[a] Corresponding author: siyuanguo2001@163.com

Abstract. As an additional control function of the generator excitation system, power system stabilizer (PSS) plays an important role in suppressing the low frequency oscillations of the power system. In this paper, the focus of research lies in PSS field test. For the risk of oscillations that may be triggered in the PSS critical gain test, a simple safety control method is proposed. The critical gain test is performed by temporarily reducing the PSS output limit value to obtain a PSS operating gain more safely. Taking the PSS field test of a grid-connected unit of Hunan Power Grid as an example, the applicability of the method is verified.

1. Introduction

With the interconnection of large-area power grids in China and the large-scale operation of fast and high-gain excitation systems, the increasingly complex system structure and the heavy load of long-distance and large-capacity transmission lines have caused the system damping level decreasing, and increased the risk of dynamic stability of the power system. As an additional control link of the excitation system, the power system stabilizer (PSS) is simple and convenient to implement. It is the most economical and effective technical means to suppress the low-frequency oscillation of the power system [1]-[2].

PSS is a power technology developed in the 1970s. Engineers analyzed the generator phase relation between each quantity in the process of oscillation, and realized that the phase lag characteristic of generator excitation system was the reason why there was an excessively sensitive voltage regulation. Furthermore, the idea of using an additional signal in the excitation system to generate positive damping torque through phase compensation emerged. Later, the Phillips-Heffron model was adopted to analyze the synchronous torque and damping torque of the generator, and the effect of the excitation system was understood from the physical perspective, so as to analyze the mechanism of suppressing low-frequency oscillation [3]. Nowadays, PSS has developed a complete standard system [4] and a complete field test procedure [5].

To determine the critical gain in the field test of PSS, the PSS output gain K_{s1} should be gradually increased until the excitation voltage begins to show an oscillatory trend of increasing amplitude. However, this method has the risk that the increased oscillation will lead to generator disconnection. Aiming at reducing the risk of oscillations that may be triggered in the PSS critical gain test, a simple safety control method is proposed in this paper. By temporarily reducing the PSS output limit value, the critical gain test is performed to obtain a PSS operating gain. Taking the PSS field test of a grid-connected unit of Hunan Power Grid as an example, the applicability of the method is verified.

Content from this work may be used under the terms of the Creative Commons Attribution 3.0 licence. Any further distribution of this work must maintain attribution to the author(s) and the title of the work, journal citation and DOI.
Published under licence by IOP Publishing Ltd

2. PSS Field Test

According to the industry standard, PSS field tests include uncompensated phase-frequency characteristic measurement of the excitation system, the lead-lag time constant setting, critical gain test, step response test on load and inverse-regulation effect test. Here we take the widely used PSS2B as an example.

2.1 PSS2B Mathematical Model

The PSS2B model uses the power and speed signals to synthesize the acceleration power integral signal, and the amplitude limiting output is achieved after the three-stage lead-lag and gain links, as shown in Fig.1.

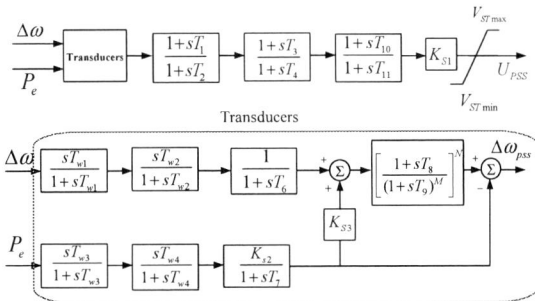

Figure 1. PSS2B Mathematical model

PSS2B has the advantages of easy implementation and low noise. However, its single-frequency structure can only take into account the low and high frequency bands by adjusting the parameters in a balanced way, and the effect of low-frequency oscillation suppression needs to be improved.

2.2 Time Constants Tuning

According to the principle of phase compensation method [3], the additional torque ΔT_{pss} generated by PSS should be in phase with the $\Delta\omega$ axis to generate the maximum positive damping torque. $T_1 \sim T_4$ and $T_{10} \sim T_{11}$ are the three-stage lead-lag phase compensation time constants, which need to be obtained by optimizing algorithm to meet the phase compensation requirements of industrial standard [5].

2.3 Critical Gain Test

The PSS critical gain is influenced by the load level, the PSS configuration in the system and other factors, which is determined by the field test after phase compensation. Increase the gain K_{s1} slowly until oscillation occurs, and take $1/3 \sim 1/5$ of the critical gain as the PSS operating gain.

3. The Safety Control Method for PSS Critical Gain Test

3.1 Introduction to Excitation System

Taking the excitation system of No.3 generator in Kongzhou Hydropower Station of Hunan Power Grid as an example, the PSS field test is carried out. The rated capacity of the generator is 31.11 MVA. The excitation system is self shunt and adopts EXC9200 digital excitation regulator by Guangzhou Qingtian Industrial Co. Ltd.

PID control has a long history and is still one of the most widely used control strategies in industrial process control. The PID control block diagram of EXC9200 is shown in Fig.2, where U_{gd} is terminal voltage reference, U_g is terminal voltage, U_{k_AVR} is excitation control voltage, and PSS_uk is PSS output.

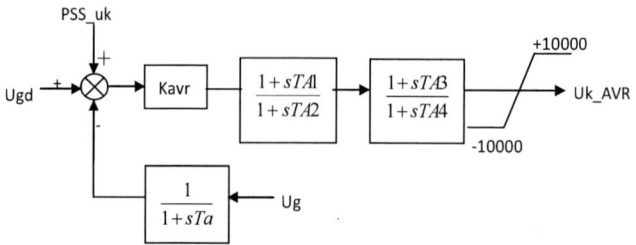

Figure 2. PID control block diagram of EXC9200

The EXC9200 is equipped with PSS2B power system stabilizer, and its mathematical model is shown in Fig.3, where USTmax and USTmin are PSS output limit value. Combined with Fig.2 and Fig.3, the PSS output PSS_uk is superimposed on the voltage reference point, so as to suppress the power oscillation by controlling the terminal voltage. The limit value of PSS is usually around ±10%.

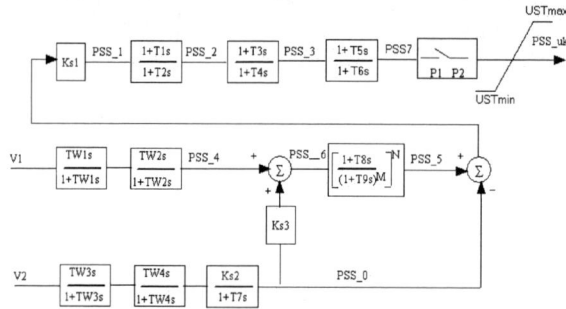

Figure 3. PSS2B model of EXC9200

3.2 The Proposed Safety Control Method

Since the output amplitude limit of PSS determines the effect of PSS suppressing low-frequency oscillation, the oscillation amplitude of the critical gain test can also be limited by controlling the output amplitude limit of PSS. Assume U_s is the absolute value of PSS output limit value. The PSS critical gain is carried out under different U_s, as shown in Fig.4. When PSS is on, increase the gain K_{s1} slowly until oscillation occurs. When PSS is removed, the oscillations disappear. By comparison, when the critical gain is achieved, the larger the limit value of PSS output, the larger the excitation voltage oscillation amplitude be and the faster the divergence be.

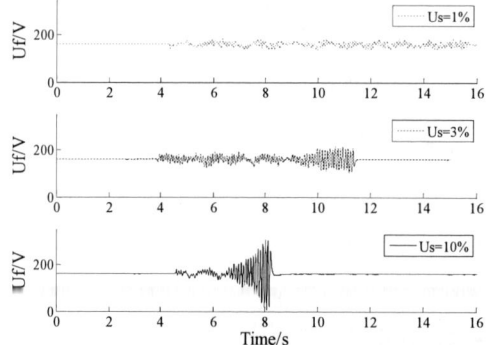

Figure 4. Critical gain test under different PSS output limit value

Fig.5 shows the excitation control voltage observed by EXC9200 Debug software under different PSS output limit value, which verifies the effect of PSS output limit value on oscillation amplitude again.

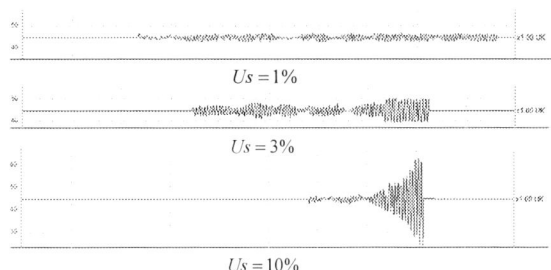

$Us = 1\%$

$Us = 3\%$

$Us = 10\%$

Figure 5. U_{k_AVR} under different PSS output limit value

Set U_s at 3% and resume the critical gain test. It can be seen in Fig.6 that with the increase of K_{s1}, the amplification oscillations gradually appeared but the amplitude is smaller than before the modification of U_s, which reduces the risk of generator disconnection.

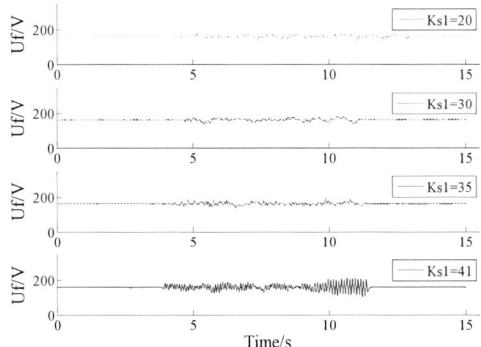

Figure 6. Critical gain test under different $K_{s1}@U_s = 3\%$

4. Conclusion

In this paper, a simple safety control method for PSS critical gain test is investigated. By temporarily reducing the PSS output limit value, the critical gain test is performed to obtain a PSS operating gain. Taking the PSS field test of a grid-connected unit of Hunan Power Grid as an example, the risk of PSS critical gain test is reduced due to PSS output value control.

References

[1] P. Kundur, *Power system stability and control* (McGraw-Hill, New York, 1994)
[2] Q. Liu, *Power system stability and generator excitation control* (China Electric Power Press, Beijing, 2007)
[3] C. Huo, Z. Liu, F. Zhu, Proc. of the CSEE. **35**, 12 (2015)
[4] IEEE-SA Standards Board, *IEEE Recommended Practice forExcitation System Models for Power System Stability Studies* (IEEE Power and Energy Society, New York, 2016)
[5] China Electricity Council, *DL/T 1231-2013 Guide for setting test of power system stabilizer* (National Energy Administration, Beijing, 2013)

Design of Seat Clamping Device for Automobile DOF Shaker

Jian Zhang*, Jianyu Yao, Chao Wen, Denggui Wang, Lingxia Wang

School of Transportation and Automobile Engineering, Panzhihua University, Panzhihua 617000, Sichuan

*Corresponding author: zhangjianpzh@126.com

Abstract. The clamping device is used to connect and fix the vibration table and the test part, as well as transmit vibration excitation in the vibration test. Taking the RX / ZDT-6-200 automobile DOF vibration table as the development platform, a clamping device suitable for a variety of automobile seats is designed based on the analysis of clamping device design key points. The 3D modeling of seat clamping device is completed in the Creo software environment, and its natural frequency is simulated and analyed. It can be seen from the analysis results that the device can effectively avoid resonance. Finally, the clamping device is manufactured and tested. The test results show that the clamping device can meet the requirements of teaching experiment.

1. Introduction

The automobile vibration simulation test is used to simulate the vibration condition of the automobile driving on the actual road in the laboratory and monitor automobile response, as well as test the reliability of automobile components in vibration environment. These tests mainly relies on the experimental equipment of automobile DOF vibration table. In order to ensure the experiment carry smoothly, clamping device are needed to connect and fix vibration table and the specimen. Therefore, clamping device not only has good rigidity, but also has the characteristics of transmitting shock vibration and avoiding resonance. In addition, due to the vibration table mainly tests automobile parts or assemblies, it is better not to destroy the table surface when using the clamping device so that vibration table to undertake more experimental tasks. Considering above factors, a seat clamping device of automobile DOF vibration table is designed for teaching experiments.

2. Key points of clamping device design

Clamping device is the clamp that connects and fixes vibration table with specimen, as well as transmits vibration excitation in vibration test, which makes it different from the ordinary clamp. Compared with ordinary clamping device, the clamping device is in a dynamic vibration environment during the test process, which needs to withstand a large impact load without any additional impact on the seat. So the device is required to have high rigidity and low quality. At the same time, in order to get more accurate test data, the clamping device needs to have the following characteristics:

(1)The clamping device should have a better pre-tightening force to avoid the separation of the seat and the vibration table due to a large load during the experiment.

(2)The rigidity and mass ratio of the clamping device should be large enough to improve the natural frequency.

(3)The first order natural frequency of the clamping device should be as high as the maximum test frequency, to avoid vibration coupling between clamping devices and products.

(4)When the clamping device connects the seat and the vibration table, the connecting points should be evenly distributed to ensure that the impact load on each part of the seat is fairly uniform.

(5)Clamping device damping should be as large as possible, and clamping device motion perpendicular to the excitation direction should be small, so as to avoid disturbing the vibration test.

3. Overall structure design of clamping device

3.1 Clamping device design process

When designing automobile seat clamping device, we should first know the shape of automobile seat，its positioning and installation on the automobile，as well as master the shape and size of platform on the vibration table, so as to ensure space-layout rationality of the clamping device on the platform and restore the installation mode of automobile seat on the automobile. The design process is as folllows:

1)Draw the model diagram of automobile seat clamping device in Creo software.

2)Then establish the finite element model of the clamping device in Creo software.

3)Carry on the modal analysis to get the vibration mode and modal frequency of the clamping device.

4)Judge whether the calculation results meet the test requirements.

If the calculation results meet the test requirements, we can draw 2D drawings and manufacture clamping devices. If it does not meet the design requirements, it is necessary to modify the clamping device partially until it meets the design requirements. The design flow chart of the clamping device is shown in Fig 1.

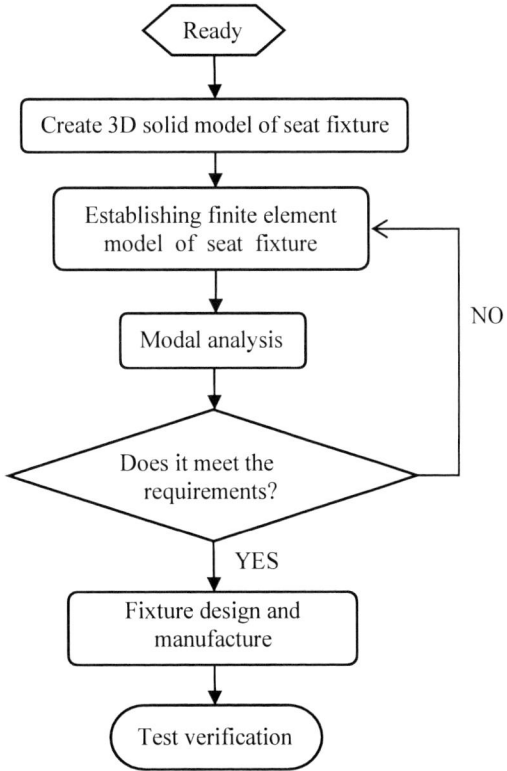

Fig.1. Design Flow Chart of Clamping Device

3.2 Structural design of clamping device

The structure of the clamping device is shown in Fig 2. This clamping device is mainly composed of screw clamping mechanism, T-groove guide rail and the main body of the clamping device, of which 2, 3 and 4 constitute screw clamping mechanism.

In the vibration test of automobile seat, it is necessary to fix the seat completely on the table and limit its six freedom degrees. In order to simulate the road conditions and restore the installation form of the seat on the automobile, the seat is fixed on the T-groove guide rail by four screw holes on the seat bottom plate. On the T-groove guide rail, the seat is positioned by the bolt holes. These bolt holes is processed through the size and position of four screw holes on the seat bottom plate. A sliding groove is arranged at the joint of the T-groove guide rails and the clamping device. Therefore, the clamping of multiple seats can be realized by adjusting the positions of the two T-groove foundation plates.

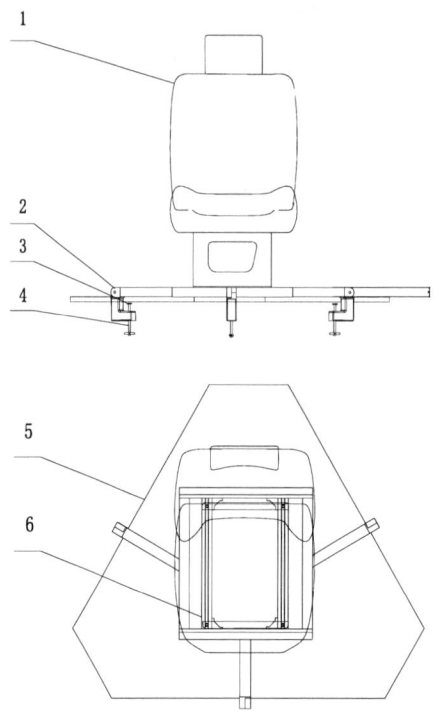

1 .Car seat 2. Lockable movable hinge 3. Swing block
4. M12 screw 5.Upper mesa of platform vibrator
6. T-groove guide rail
Fig.2. Design of Clamping Device

3.3 Materials and processing methods

Table 1. Material Properties

Material	Aluminum	Magnesium	Steel
Modulus of elasticity E/ (G pa)	71.5	44.1	205.8
Density ρ/ (g.cm^{-2})	2.8	1.8	8
$E/\rho \times 10^{6}$(m^2.Pa.kg^{-1})	25.5	24.5	25.7

The clamping device is in a dynamic environment during the test, and there is no or less resonance in the test frequency band. The factor controlling natural frequency is E/ρ, E is elastic modulus, and ρ is material density. Under the same structural conditions, the larger the ratio is, the higher the frequency is, and the material with large damping should be choosen. From Table 1, it can be seen that the value of E/ρ of steel is the largest, but its density is too high, while the value of E/ρ of aluminum is the closest to that of steel, but its density is much smaller than that of steel. Therefore, we should choose aluminum (aluminum alloy) as manufacturing material. In order to improve the universality of the clamping device, the design principle and concept similar to modular fixture are adopted, and the aluminum alloy material with T-groove is used as the manufacturing material. In the preliminary manufacturing model, 4.0T national standard 40 * 40 industrial aluminum profile is used as the production materia.

The main manufacturing methods of clamping device include integral machining, screw connection, casting, welding, bonding, epoxy resin forming, etc. In the above manufacturing methods, the overall machining is preferable. Next is casting, but heat treatment or aging treatment should be carried out to eliminate prestress after casting, screw clamping device has poor high-frequency vibration performance.

The manufacturing method of the clamping device adopts the combination of welding processing and thread connection to give full play to their advantages. The main parts of the clamping device are joined by welding, while some parts are joined by screw connection, so that the clamping device can obtain higher damping. The fixing hole connecting bolt and the vibration table should be a flat-bottomed buried head hole, which aims to shorten thread length and increase stiffness.

4. Simulation analysis and test of three-dimensional modeling

4.1 Overall structure design and modeling of clamping device
In the vertical direction exciting vibration, the composite center of specimen and the clamping device will fall on the center line of the platform as far as possible, so as to avoid the shaking of the platform and lead to the distortion of the vibration waveform. In particular, the clamping device center of gravity or specimen is high, it is more important to design a clamping device with good balance performance. Therefore, the clamping device should be designed as symmetrical and low center of gravity. Such as cube, box, hemispherical and conical.

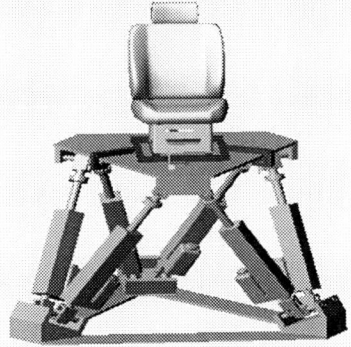

Fig.3. Clamping Device Model Diagram

In the structural design of the clamping device, the shape and dimension of the platform on the vibration table, the locating mode and the fixing mode of the seat are considered comprehensively. Finally, the structure of the left and right symmetry is used, and clamping device center of gravity is ensured to fall on the platform center line, and reduce clamping device center of gravity as far as possible. Complete the three-dimensional modeling of the seat clamping device in Creo software environment. The model diagram of the seat clamping device is shown in Fig.3.

4.2 Modal analysis
In order to test whether the main dynamic performance of the clamping device meets the requirements, it is necessary to carry out modal analysis before the vibration test. If it does not meet the requirements,

relevant measures should be taken to correct it. The first four modes are shown in Fig.4.The final design of the clamping device results are shown in Table 2.

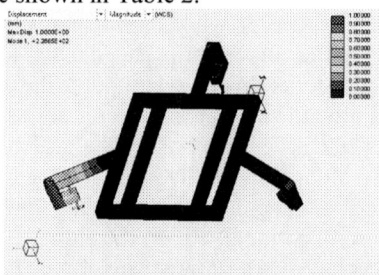

(a) First Mode Vibration of Clamping Device

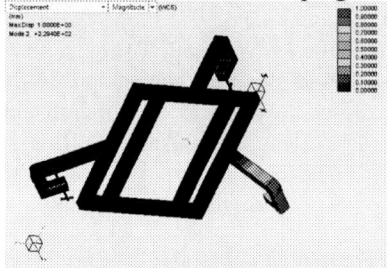

(b) Second Mode Vibration of Clamping Device

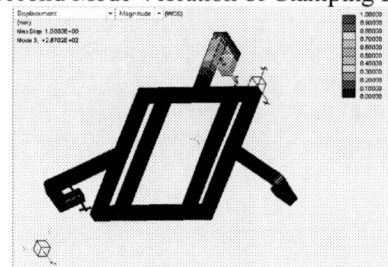

(c) Third Mode Vibration of Clamping Deviced

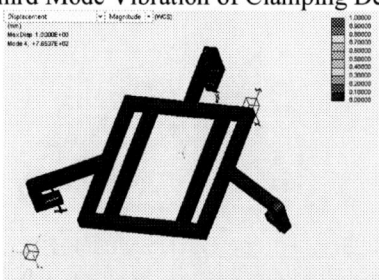

(d) Fourth Mode Vibration of Clamping Device

Fig.4. Fourth Order Vibration Mode Diagram of Clamping Device

Table 2. The First Four Grades Modal Calculation Results of Clamping Device

Grade	Frequency（Hz）
1	228.648
2	229.401
3	267.024
4	765.374

As shown in Table 2, the natural frequencies of the workpiece modes increase with the order of mode shapes. This is because nodes of higher order vibration will increase with the order of mode, the load energy exciting workpiece higher order vibration will gradually weaken, and the excitation of higher order vibration will become more difficult. Therefore, the natural frequency of the workpiece at the first mode is the smallest. But in model research, the influence of low order natural frequency on vibration characteristics is greater than that of high order natural frequency, which means that the accuracy of first order natural frequency is one of the key factors of the reliability in this experiment. The range of test frequency of vibration table for automobile freedom degree is 0-50Hz. According to the analysis results of clamping device, first order natural frequency is 201.591Hz which is about four times the test frequency of vibration table for automobile freedom degree. So, the dynamic performance of the clamping device meets the design requirements.

4.3 Teaching vibration test

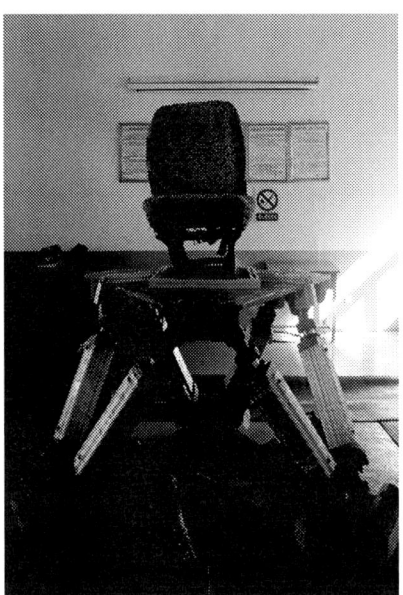

Fig.5. The Real Object of Clamping Device

The clamping device is fixed on the vibration table surface by screw clamping mechanism, and the seat whose translation and rotation are limited in the x y z axis is installed on the clamping device. Fig.5 shows the figure of clamping device. This device has no position excursion or slack shaking in the lasts-forty-minutes vibration test (it's less than the time of experimental teaching) and there is no cracking at the welding point, which meets expectations we design.

5. Conclusions

A car seat clamping device that doesn't destroy table surface is designed and it takes the vibration table for automobile freedom degree as the development platform. The clamping device can be used to clamp a variety of seats by adjusting the position of two T-groove baseplate and it has preferable versatility. The three-dimensional modeling of car seat clamping device was completed in the Creo environment and the clamping device can avoid resonance effectively by natural frequency simulation analysis. According to the design, the real subject made with aluminum was tested. The test results show that the clamping device has expected rigidity and reliability and it can be used in teaching.

Acknowledgment

In this paper, the research was sponsored by 2018
Research Program of Education Office of Sichuan   Province   (Project   No. 18ZB0335) and 2016

Research Project of Panzhihua University (Project No. 2016YB009)

References

[1] Song Wu, Xu Zhang, Xinjun Long ,Qiwei Guo. Dynamic modeling and analysis of hanging modal test system [J/OL]. Acta Aeronautica et Astronautica Sinica:1-8[2018-10-18].http://kns.cnki.net/kcms /detail/ 11.1929.V. 20180502. 1510.014.html.

[2] Youchao Liu, Hui Zhu. Fixture design for vibration impact test of a series of chassis [J]. China Equipment, 2010(01):190.

[3] Shaoming Yu, Guo Wei,Feng Yang,Jing Du. Fixture Design and Practice for Vibration Test [J]. Equipment Environmental Engineering, 2014, 11(02:81-86.

[4] Zheng Ma, Dongqiang Li, Yang Gu, Kai Liu. Design and Experimental Verification of a Special Shaped Vibration Fixture Structure [J]. Equipment Environmental Engineering,2017,14(03):90-94.

[5] Songlin Zheng, Cheng Gao, Zhagen Ma,Wenwei Hu. Design of vibration test fixture for automobile headlight [J].Modern Manufacturing Engineering, 2013(04):62-66.

[6] Ruixuan Wu. Study of Testing Method of Vibration Fixture [J]. Equipment Environmental Engineering, 2010,7(06): 252-255+263.

[7] Bo Zhou. Study on Structural Design and Experiment of the Reliability Text Fixture of the Electrical Equipment of a SRV[D].Hefei University of Technology,2010.

[8] Hua Wen, Ling Wang,Guofu Yin, Haiji Huang ,Wei Bai. Design and Simulation Analysis of Modular Fixture System [J].Tool Engineering,2018,52(04): 136-138.

Contour Error Control of X-Y Platform Based on Nominal Model in Polar Coordinate System

Guirong Wang[1,a], Xinman Gong[2,b], Yingqi Li [3,c]

[1]Departments of Institution of Electrical and Mechanical Engineering, University of China JiLiang, 310018, HangZhou

[2]Departments of Institution of Electrical and Mechanical Engineering, University of China JiLiang, 310018, HangZhou

[3]Departments of Institution of Electrical and Mechanical Engineering, University of China JiLiang, 310018, HangZhou

[a] Guirong Wang: wangguirong314@126.com

[b] Xinman Gong : gxm41672135@163.com

[c] Yingqi Li: 13588012369@163.com

Abstract. In the case of the x-y linear motor platform, the contour error model of the nonlinear dynamic and curve trajectory is relatively complicated, and the traditional control system does not have the problem of systematic adjustment of parameters, and the precision of the contour is affected when machining. In this paper, the dynamic model of the system is modeled under the polar coordinate system, and the polar coordinate contour error nonlinear model of the linear motor x-y platform is obtained. At the same time, the nonlinear model application is based on the theory of sliding mode control, which makes the error of the system close to zero and can have good robustness. The theoretical derivation and simulation platform verification results show that the control system designed in this paper can improve the precision of the contour machining of the linear motor x-y platform.

1. Introduce

NC machine tools are developing to the direction of precision, high speed and compounding. The X-Y platform of linear servo motor has been widely used in high-speed machining equipment, such as fast response, accurate positioning and high reliability.

Many scientists have applied a large number of modern control theories, such as robust control, variable structure control, model reference adaptive control, neural network control and genetic algorithm, to the X-Y platform of linear motor and achieved good results. However, these traditional control methods are based on the Cartesian coordinate system.These methods have some shortcomings.Firstly,the feed-forward, feedback and cross-coupling controllers consider the case of each axis separately. The coupling relationship between the two axes is not clear enough, and there is no systematic method to design a comprehensive controller. Secondly, for complex paths, cross-coupling is used[1]. The linear approximation method is used to reduce the order of the controller, and the traditional method based on the linear model expressed by the transfer function is difficult to be applied to nonlinear systems.

Content from this work may be used under the terms of the Creative Commons Attribution 3.0 licence. Any further distribution of this work must maintain attribution to the author(s) and the title of the work, journal citation and DOI.
Published under licence by IOP Publishing Ltd

In view of the above shortcomings, Shyh-Leh Chen and others put forward the idea of contour control for two-axis system in polar coordinate system[2]. This paper adopts this idea and designs a sliding mode controller based on nominal model to enhance the robustness of the system[2]. The theoretical derivation and simulation results show that the control system designed in this paper can effectively improve the contour machining accuracy of linear motor X-Y platform.

2. Nonlinear Model of Polar Coordinate Contour Error for X-Y Platform

2.1. Mathematical Model of X-Y platform
The dynamic equations of the two-axis linear motor system are defined as follows:

$$M\ddot{q} + V(q,\dot{q}) + \Delta(q,\dot{q}) + dt = T \qquad (1)$$

Where M is the positive definite mass matrix of 2x2;q is the matrix of 2x1, representing the displacement of X-Y axis; $V(q,\dot{q})$ is viscosity friction coefficient; $\Delta(q,\dot{q})$ is the uncertain part of modeling; dt is the external interference to the system; T is the input of the control signal.

2.2. The X-Y platform Model of Linear Motor based on Polar coordinates
For X-Y biaxial systems, the traditional control methods are the contour errors defined in the Cartesian coordinate system, as shown in figure 1, which is the path distance from the actual position to the desired position:

$$\varepsilon(t) = \sqrt{x(t)^2 + y(t)^2} - r_d \qquad (2)$$

The main cause of contour error is the dynamic mismatch of two axes. From formula (2) we can see that the contour error originates from the real-time dynamic position of XY axis, so the controller for contour error should not be designed independently. The controller is designed based on the definition of contour error and the real time XY axis dynamic position information. The method to solve this problem is to model the contour error in polar coordinate system.

In polar coordinate system, the contour error of trajectory can be expressed as radial error $\varepsilon_r(t)$.

$$\varepsilon_r(t) = r(t) - r_d(t) \qquad (3)$$

Where $r(t)$ is the radius of the actual trajectory circle; $r_d(t)$ is the radius of the orbit circle in it.

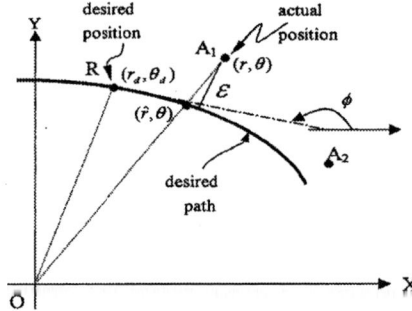

Figure 1. Tracking error and Contour error.

However, for the curve of the non-positive circle trajectory, the angle error $\varepsilon_\theta(t)$ is also needed to represent the contour error completely.

$$\varepsilon_\theta(t) = \theta(t) - \theta_d(t) \qquad (4)$$

Where $\theta(t)$ is the angle of the actual track circle; $r_d(t)$ is the angle of the orbit circle.

In this paper, taking ellipse as an example, we discuss the case of machining nonlinear trajectory of linear motor X-Y platform.

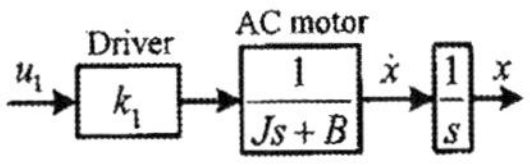

Figure 2. Model of the X positioning stage.

The expression of the trajectory converted to polar coordinate system is: $x(t) = r\cos\theta$, $y(t) = r\sin\theta$.
Owing to:

$$q = \begin{bmatrix} x(t) \\ y(t) \end{bmatrix} = \begin{bmatrix} r\cos\theta \\ r\sin\theta \end{bmatrix}$$

So:

$$\dot{q} = \begin{bmatrix} \cos\theta & -r\sin\theta \\ \sin\theta & r\cos\theta \end{bmatrix} \begin{bmatrix} \dot{r} \\ \dot{\theta} \end{bmatrix} \qquad (5)$$

Order:

$$R = \begin{bmatrix} \cos\theta & -r\sin\theta \\ \sin\theta & r\cos\theta \end{bmatrix}, \dot{P} = \begin{bmatrix} \dot{r} \\ \dot{\theta} \end{bmatrix}, \ddot{P} = \begin{bmatrix} \ddot{r} \\ \ddot{\theta} \end{bmatrix},$$

$$S = \begin{bmatrix} -2\sin\theta & -r\cos\theta \\ 2\cos\theta & r\sin\theta \end{bmatrix} \qquad (6)$$

So there is

$$\ddot{q} = R\ddot{P} + \dot{\theta}S\dot{P} \qquad (7)$$

Order: $A = \dot{\theta}MR^{-1}M^{-1}S + MR^{-1}M^{-1}VR$; $B = MR^{-1}M^{-1}$

From the formula (1), (5), (6) and (7), the dynamical equations of the system in polar coordinate system can be obtained:

$$M\ddot{P} + A\dot{P} + Bdt = BT \qquad (8)$$

3. Controller Design of Contour Error of X-Y Platform

3.1. The structure of control system

As can be seen from figure 3, the control system consists of two controllers, one of which is the sliding mode controller of the actual system, and the other is the controller for the nominal model, which is implemented in polar coordinates:

$$r \to r_d, \theta \to \theta_d \qquad (9)$$

Thus, In the World Coordinate System:

$$x(t) \to x_d(t), y(t) \to y_d(t) \qquad (10)$$

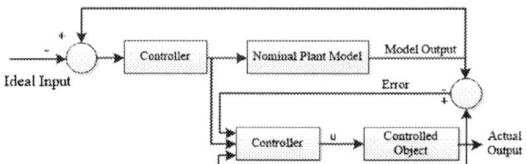

Figure 3.Control system structure

3.2.Design of Contour error Controller

From the kinetic equation (9), the following results can be obtained:

$$M\ddot{P} + A\dot{P} = T + w \qquad (11)$$

Where $T = BT, w = -Bdt$.

P_d is taken as input instruction and $e = P_d - P$ is used as error signal. The sliding surface is designed as follows:

$$S = \dot{e} + Ce, C = diag(c_1, c_2, \cdots c_n), c_i > 0 \qquad (12)$$

Defining Lyapunov function:

$$V = \frac{1}{2} S^T M S \qquad (13)$$

There will be: $\dot{V} = \frac{1}{2} S^T \dot{M} S + S^T M \dot{S} \quad = \frac{1}{2} S^T (\dot{M} - 2A)S + S^T A S + S^T M \dot{S}$

Because M is a constant positive definite matrix, therefore $\dot{M} = 0$ and :

$$\dot{V} = S^T M \dot{S} \qquad (14)$$

From the expression (13):

$$\dot{S} = \ddot{e} + C\dot{e} \qquad (15)$$

At the same time:

$$\ddot{e} = \ddot{P}_d - \ddot{P} \qquad (16)$$

Bring (12) (16) (17) into (15), we can get : $\dot{V} = S^T M \dot{S} = S^T [M(\ddot{P}_d + C\dot{e}) + A\dot{P} - T - w]$.
Accordingly, the control law is designed as:

$$T = M_0(\ddot{P}_d + C\dot{e}) + A_0\dot{P} - w_0 + \Gamma \operatorname{sgn}(S) \qquad (17)$$

Where M_0, A_0 and w_0 are nominal values of M, A and w respectively:
$\Delta M = M - M_0$; $\Delta A = A - A_0$; $\Delta w = w - w_0$
Thereupon: $\dot{V} = S^T [\Delta M(\ddot{P}_d + C\dot{e}) + \Delta A\dot{P} - \Delta w] - \Gamma|S|$
Where $\Gamma = diag(\gamma_1, \gamma_2, \cdots \gamma_n), (\gamma_i > 0)$.
Taking

$$\gamma_i > |\Delta M|_{max} (\ddot{P}_d + C\dot{e}) + |\Delta A|_{max} \dot{P} - |\Delta w|_{max} \qquad (18)$$

So, we can get $\dot{V} \le 0$, When and only as S=0, $\dot{V} = 0$

When $t \to \infty$, $S \to 0$.

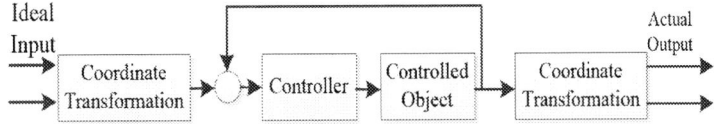

Figure 4. Linear motor platform control block diagram

4. Controller Simulation And Analysis

In this paper, the linear motor platform of Huimusen BJSM-III series is simulated. The parameters are as follows table 1.

Table 1 Platform Parameters

Parameters	Symbol	X-Axis	Y-Axis
Full Thrust(N)	F_p	240.0	625
Sustained Thrust(N)	F_c	76	276
Constant of the machine ($N\sqrt{M}$)	K_m	7.8	32.0
Maximum Speed（m/s）	v	10	6.6
Maximal Acceleration(m/s^2)	a	580	150
Reverse Potential Constant(V$_{peak}$/m/s)	K_e	27.5	51.4
Thrust Constant(N/Arms)	K_f	33.7	59.8
Mover mass(kg)	M	3.7	8.0

The instruction path is an "8" glyph, and the two-axis input instruction is:

$$x_{d1} = \sin\theta \cdot \cos(\cos\theta), x_{d2} = \sin\theta \cdot \sin(\cos\theta)$$

Interference is set to λ sgn(s)，$\lambda = 10$.

The sliding mode coefficient is c1=20; c2=30; $M_0 = 0.8M$, $A_0 = 0.8A$, $w_0 = 0.8w$.

The response curves of each axis are obtained by matlab simulation as follows:

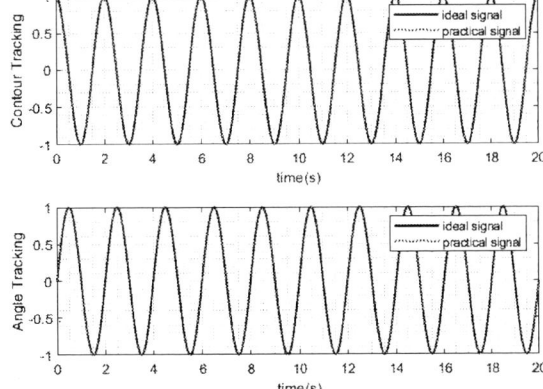

Figure 5. Actual output and Ideal input of X Y

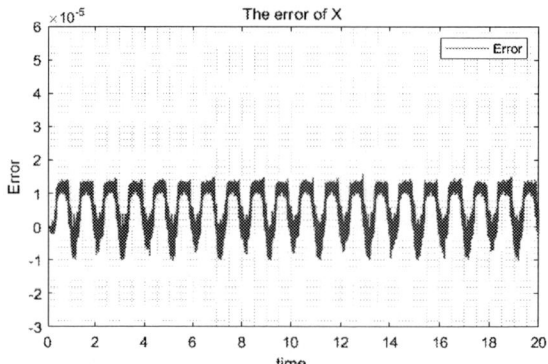

Figure 6.Contour Error of X-axis

As can be seen from figure 5, the actual signal quickly overlaps with the ideal signal.

From figure 6, we can see that the contour error of the X-axis in polar coordinate system is basically maintained in the range of $\pm 1.6 \times 10^{-5}$.

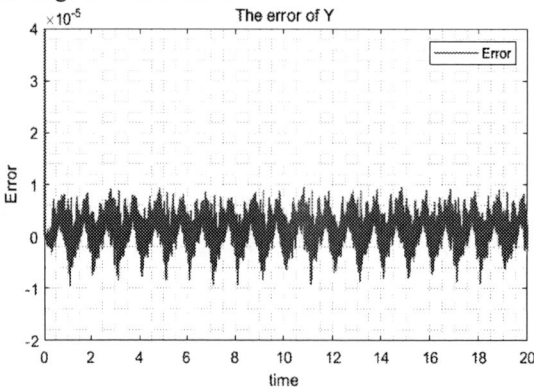

Figure 7.Contour Error of Y-axis

From figure 7, we can see that the system quickly reached a stable range of errors and the contour error of the Y-axis in polar coordinate system is basically maintained in the range of $\pm 1.0 \times 10^{-5}$.

By calculating the mean value of contour error, it is obtained that the average contour error between actual trajectory and instruction trajectory is 3.7×10^{-6} m, which is compared with that in reference [8-9] as follows:

Table 2 Contour error of three Control methods

Control Method	Cross-Coupling Control based on Contour error Vector estimation	Contour control based on coordinate transformation	Contour error Control of X-Y platform based on nominal Model in Polar coordinate system
Input	100mm	100mm	100mm
Average Error	5×10^{-3}	4×10^{-4}	3.7×10^{-6}

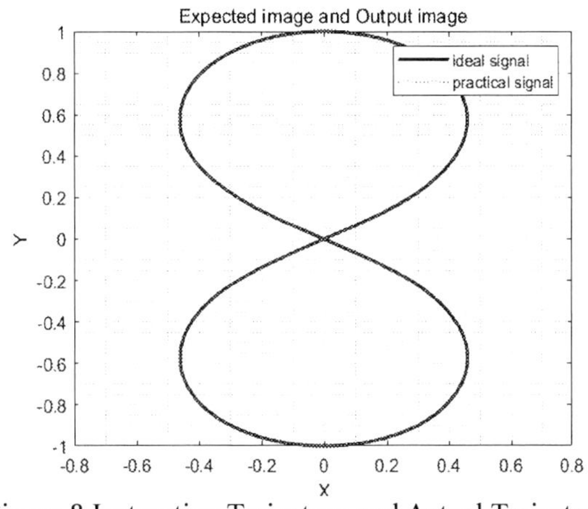

Figure 8.Instruction Trajectory and Actual Trajectory

From figure 9, we can see that actual trajectory and instruction trajectory is mainly coincidence.

The simulation results show that the contour error control of the X-Y platform based on nominal model in the polar coordinate system can not only effectively guarantee the robustness of the servo system, but also can effectively improve the precision of the contour error, and the design of the controller is relatively simple.

5. CONCLUDING

In this paper, the contour error control system of X-Y platform in polar coordinate system is used to realize the synchronous tracking control of the two-axis linear motor platform, and the sliding mode control based on the nominal model is adopted, which not only can effectively guarantee the robustness of the servo system. At the same time, it can improve the precision of contour error effectively, and the controller design is relatively simple. Simulation results show that the designed controller can improve the contour control accuracy of linear motor.

References

[1] Sun Yibiao, long Xi, Jin Fuying, et al. Contour Control of Linear Motor XY platform based on Polar coordinate method[J]. Combined Machine tool and automatic Machining Technology, 2010(6):57-60.

[2] Chen S L, Liu H L, Ting S C. Contouring control of biaxial systems based on polar coordinates[J]. IEEE/ASME Transactions on Mechatronics, 2002, 7(3):329-345.

[3] Lu Jinduo, Liu Jinbo. Coordinated Control of Mechanical Motion Control system of dual Motor Transmission in Polar coordinates[C]. Academic Forum on the Frontier issues of Electrical Technology. 2005.

[4] Zhao Ximei, Guo Qingding. Linear Servo robust tracking Control using DOB and ZPETC to improve Contour Machining accuracy[J]. Transactions of China Electrotechnical Society, 2006, 21(6):111-114.

[5] Liu Jinkun. MATLAB Simulation of sliding Mode variable structure Control [M]. Tsinghua University Press, 2015.

[6] Liu Jinkun, Sun Fuchun. Research and Development of sliding Mode variable structure Control Theory and its algorithm[J]. Control Theory and Application, 2007, 24(3):407-418.

[7] Yeh S S, Hsu P L. Analysis and design of integrated control for multi-axis motion systems[J]. Control Systems Technology IEEE Transactions on, 2003, 11(3):375-382.

[8] Yeh S S, Hsu P L. Estimation of the contouring error vector for the cross-coupled control design[J]. Mechatronics IEEE/ASME Transactions on, 2002, 7(1):44-51

[9] Cheng M Y, Lee C C. Motion Controller Design for Contour-Following Tasks Based on Real-Time Contour Error Estimation[J]. IEEE Transactions on Industrial Electronics, 2007, 54(3):1686-1695.

[10] Lou Y, Meng H, Yang J, et al. Task Polar Coordinate Frame-Based Contouring Control of Biaxial Systems[J]. IEEE Transactions on Industrial Electronics, 2014, 61(7):3490-3501.

[11] Chen S L, Wu K C. Contouring Control of Smooth Paths for Multiaxis Motion Systems Based on Equivalent Errors[J]. IEEE Transactions on Control Systems Technology,2007, 15(6):1151-1158.

[12] Meng H, Lou Y, Chen J. High speed contouring control of biaxial systems based on task polar coordinate frame[M]. 2013.

[13] Shannon, Wang Yu. Study on the method of complete linearization for nonlinear Systems[J]. Control Theory and Application,1997(1):139-143.

[14] Peng Zhenzhou. Design of contour error control system for XY table of NC machine tool driven by linear motor[D]. University of Electronic Science and Technology of China.

Analysis of the Pressure Expansion of Bridge Plug Tools and Packers by Equivalent Material Method

Lanwen Wang[1,a], Xuanyu Sheng[2] and Jiayue Sheng[3]

[1]Shandong University, Weihai, Department of Mechanical, Electrical and Information Engineering, 264209, Shandong, Weihai, P.R.China

[2]Tsinghua University, Department of Mechanical Engineering, 100084, Beijing, P.R.China

[3]University of Wisconsin-Madison, 53711, Wisconsin, Madison, USA

[a]Corresponding author: 13176803577@163.com

Abstract. For the bridge plug tool and packer, the pressure expansion process and differential pressure loading process of the bridge plug and packer have been simulated by finite element method. The numerical structural stress analysis of the bridge plug and the packer are carried out, and the structure of the bell mouth of the cartridge assembly is optimized. The results show that the replacement of the laminated steel structure in the bridge plug/packer with equivalent materials can greatly improve the calculation efficiency, thus completely calculating the pressure expansion and differential pressure loading of the bridge/packer. After the bridge plug/packer expands, the excessive stress is concentrated at the bell mouth. The optimized structure of the elliptical curve at the bell mouth has relatively low stress. This calculation provides theoretical support for the sealing and optimization of bridge plugs and packers.

1. Introduction

Bridge plug tools and packers are sealing devices used in oil and gas wells. Their principle is that the internal pressure is pressurized, the rubber cylinder assembly composed of the inner rubber tube and outer rubber tube made of laminated steel strip and rubber is expanded after being pressed, then the outer rubber tube is in contact with the casing of the well wall. Under the action of internal pressure extrusion and upper and lower pressure difference, the oil and gas wells are sealed [1-2]. In order to understand the load distribution condition of each part of the bridge plug/packer, the pressure expansion process of the bridge plug/packer is simulated by finite element calculation. After the structural stresses calculation, structural optimization is performed on the overstressed parts.

2. Material and Methods

The main working part of the bridge plug and packer is the rubber cylinder assembly (inner rubber tube, outer rubber tube, laminated steel strip, floating head, upper sleeve joint). Therefore, when building the finite element model, the other parts of the bridge plug tool and packer are simplified. The simplified bridge plug tool and packer finite element models are shown in Fig. 1 and Fig. 2, respectively. The bridge plug tool/packer laminated steel strip has an outer diameter ofΦ52 mm, Φ60 mm for the outer rubber tube, and Φ160 mm inner diameter for the outer sleeve that needs to be sealed (hidden in the model).

Content from this work may be used under the terms of the Creative Commons Attribution 3.0 licence. Any further distribution of this work must maintain attribution to the author(s) and the title of the work, journal citation and DOI.
Published under licence by IOP Publishing Ltd

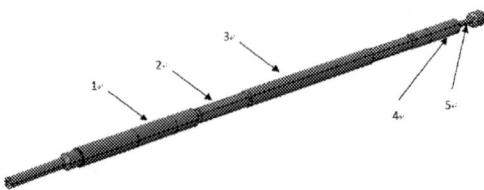

Figure 1. Schematic diagram of finite element model of bridge plug tool
1. Upper sleeve joint 2. Laminated steel strip 3. Outer rubber tube 4. Floating head 5. Mandrel

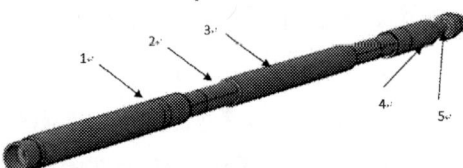

Figure 2. Schematic diagram of finite element model of packer
1. Upper sleeve joint 2. Laminated steel strip 3. Outer rubber tube 4. Floating head 5. Central tube

In the models, the steel sheet structure is replaced by an equivalent material, which can greatly reduce the number of contacts and small-sized local units, thus completely reduce the calculation time of the pressure expansion and the differential pressure loading of the bridge plug/packer [3-6]. Since the steel sheet structure is easily expanded in the radial direction, but the deformation in the longitudinal direction is relatively small, the steel sheet is formed according to the orthotropic material [7]. Orthotropic material means that there are three mutually perpendicular symmetry planes at any point of the material, and the direction perpendicular to the symmetry plane is called the main direction of elasticity. In the main direction of elasticity, the elastic properties of the material are the same, and the axis parallel to the main direction of the elasticity is the elastic main axis or the material main axis, and the three material main axes are represented by 1, 2 and 3. Since the properties of orthotropic materials are different in different directions, it is necessary to set the direction of the material. As shown in Fig. 3, the equivalent structure of the steel sheet is generated in three directions according to the cylindrical coordinates, 1 is the radial direction r, 2 is the circumferential rotation direction φ, and 3 is the height direction z. Therefore, different material parameters can be defined in the three directions.

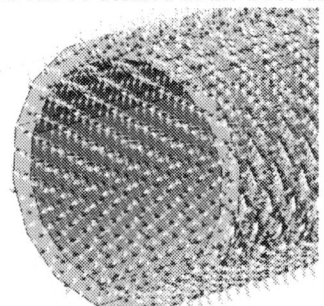

Figure 3. Materials directional distribution of equivalent structural of steel sheets

The main structural material of the bridge plug/packer is selected from No. 304 steel. The performance of the stainless steel material at 150 °C is shown in Table 1.

Table 1. Material properties of No. 304 steel（150℃）

Density	Poisson's ratio	Thermal expansion coefficient	Elastic Modulus
kg/m³	/	K⁻¹	MPa
8030	0.31	16.6E-6	1860000

For rubber materials, the hyperelastic constitutive model parameters are determined by uniaxial tensile, plane tensile and biaxial tensile tests. Poisson's ratio can be determined by experimental data of volume change rate. Commonly used hyperelastic constitutive models include the Neo-Hooke model, the Mooney-Rivlin model, the Yeoh model, the Gent model, the Ogden model, and the Arruda-Boyce eight-chain model. Since the rubber used in the model has exceeded 45% in the initial stage of constrained deformation, it is more suitable to use the Ogden model. The strain energy density function of Ogden model is defined as:

$$W = \sum_{k=1}^{N} \mu_k \left(\frac{\lambda_1^{\alpha_k} + \lambda_2^{\alpha_k} + \lambda_3^{\alpha_k} - 3}{\alpha_k} \right) \qquad (1)$$

where λ_i is the main stretch, k, α_k are material constants (determined by experimental data), and μ_i is the number of terms in the function. The penalty function used in the Ogden formula uses the form of the function used in the Mooney-Rivlin model. The actual strain energy density function is a modified Ogden function, as described by Eq. (2):

$$W = \sum_{k=1}^{N} \mu_k \left(\frac{\lambda_1^{\alpha_k} + \lambda_2^{\alpha_k} + \lambda_3^{\alpha_k} - 3}{\alpha_k} - \ln(J) \right) + \frac{1}{2\alpha} G^2(J) \qquad (2)$$

where J is the ratio of the deformed volume to the undeformed volume, and N is the number of terms in the function, $G(J) = J^2 - 1$, and

$$\frac{4}{\alpha} = \frac{1}{3} \sum_{k=1}^{N} \mu_k \alpha_k \left[\frac{(1+4\nu)}{2(1-2\nu)} \right], \nu = Poisson's\ ratio \qquad (3)$$

A three-item (modified Ogden) model is widely used, and up to four-item models (N=4) can be used.

3. Results

3.1 Bridge plug tool

The complete compression simulation of the bridge plug is carried out based on the model and material parameters defined in the previous section. The requirements for the use of the bridge plug are 16 MPa for loading internal pressure and 16 MPa for differential pressure at the upper and lower ends. The calculated stress distribution results of the bridge plug tool are shown in Fig. 4.

According to the calculation results, the maximum stress of the spindle is 399.2 MPa, which is located at the contact of the upper sleeve joint with the steel strip. The maximum stress of the floating head is 1217 MPa, which is located at the contact of the lower sleeve joint with the steel strip. The outer rubber tube has a maximum stress of 234.6 MPa and is located at the edge of the outer rubber

(a)Overall stress distribution of bridge plug tool

(b)Stress distribution of bridge plug spindle

(c)Stress distribution of floating head

(d)Stress distribution of outer rubber tube

(e)Stress distribution of steel strip

(f)Stress distribution of inner rubber tube

Figure 4. The overall distribution of the bridge plug and the stress distribution of each component

ISPECE   IOP Publishing

IOP Conf. Series: Journal of Physics: Conf. Series **1187** (2019) 032052   doi:10.1088/1742-6596/1187/3/032052

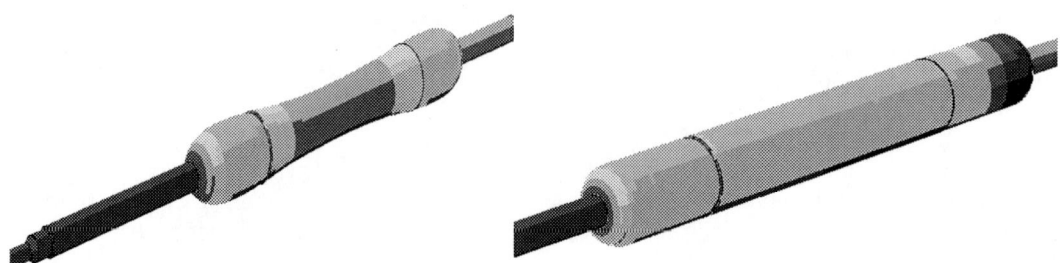

Figure 5. Simulation results of the packer's expansion process

tube. The steel strip has a maximum stress of 1458 MPa and is located at the end of the steel strip welded to the sleeve joint. The inner rubber tube has a maximum stress of 36.9 MPa and is located at the maximum displacement inside the inner rubber tube, that is, at the ends without the outer rubber tube covering limit.

3.2 Packer
The overall calculation process of the packer is similar to the bridge plug[8]. According to the determined model and material parameters, the expanded state of the final simulated packer is shown in Fig.5, which is the result of displacement distribution of the packer as a whole. According to the simulation results, when the load is applied to 1.54 MPa, the structure at the radial maximum displacement of the expansion tube has been in contact with the outer sleeve. When the load is applied to 2.44 MPa, the expansion tube has been fully opened, and the entire radial reaches 160 mm of the outer sleeve.

When the packer is loaded with internal pressure to 16 MPa and the differential pressure between the upper and lower ends is 16 MPa, the stress distribution diagram of the packer is the same as that of the bridge plug. According to the calculation results, the maximum stress of the main structure of the packer is 2019 MPa, which is located at the contact of the upper sleeve joint with the steel strip. The maximum stress of the floating head is 1247 MPa, which is located at the contact of the lower sleeve joint with the steel strip. The outer rubber tube has a maximum stress of 196.2 MPa and is located at the edge of the outer rubber tube. The steel strip has a maximum stress of 954.9 MPa and is located at the end of the steel strip welded to the sleeve joint. The inner rubber tube has a maximum stress of 24.93 MPa and is located in the middle of the inside of the inner rubber tube.

4. Discussion
The analysis results show that the excessive stress in the bridge plug/packer is concentrated at the bell mouth (the position where the sleeve joint is in contact with the steel strip) [9-10]. Here, the steel strip is flared and squeezing strongly with the sleeve joint. According to preliminary calculations, the stress here exceeds the allowable stress of the material, which may cause damage in subsequent use. So optimization here can improve the problem of excessive stress. Taking the packer as an example, the original simplified model is too simple to handle here, and it has a certain deviation from the actual one. The original model is shown in Fig.6.

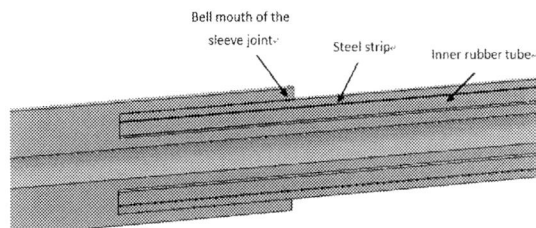

Figure 6. Schematic diagram of unoptimized model

The optimized structure of this location is shown in Fig.7.

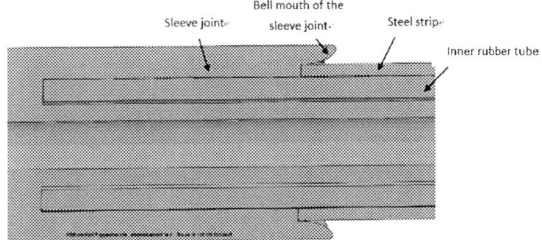

Figure 7. Schematic diagram of the optimization model

The inner arc of the sleeve joint is designed to be elliptical. According to the thickness of the sleeve joint and the contact angle with the steel strip, the semi-major axis a   of the elliptical arc is calculated, and the length of the semi-minor axis b is obtained by the difference between the thickness and the chamfer, as shown in Fig.8.

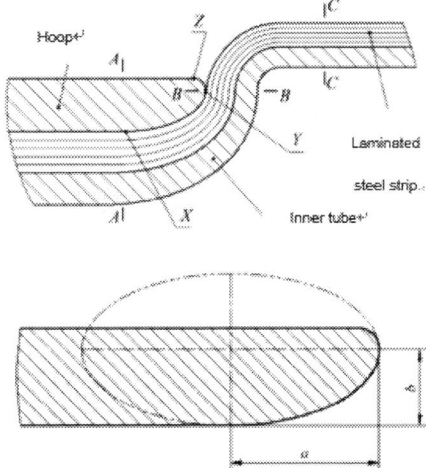

Figure 8. Optimized shape in the sleeve joint

After optimization according to the elliptic curve, since the thickness of the sleeve joint is 6 mm, the chamfering is removed, and the elliptical arc-shaped semi-short axis b is taken as 5 mm. The semi-major axis a is calculated to be 6.41 mm. The stress distribution result of the packer after calculation is shown in Fig. 9.

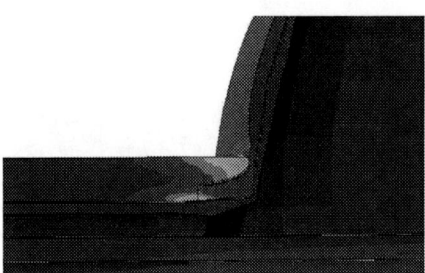

Figure 9. Schematic diagram of optimizing model contact stress distribution

According to ASME standards and GB150 regulations for pressure vessel parts [11-12], stress linearization at the maximum equivalent stress of each component before and after optimization. The film stress P_m and film stress plus bending stress P_m+P_b of each component are obtained, and the calculation results are shown in Table 2.

Table 2. Stress of each optimized component

Optimized component	Upper sleeve joint	Lower sleeve joint	Steel strip at the upper sleeve joint	Steel strip at the lower sleeve joint
Unoptimized P_m （MPa）	520.54	177.08	318.54	279.94
Optimized P_m （MPa）	577.82	159.45	175.37	85.48
Unoptimized P_m+P_b （MPa）	800.33	571.17	717.54	616.15
Optimized P_m+P_b （MPa）	811.79	295.17	280.59	119.8

From the results of Table 2, the elliptical arc-shaped sleeve joint can significantly reduce the stress of the steel strip, because the ellipse is more in line with the deformed shape of the steel strip when in contact. Due to the pressure difference at the lower sleeve joint, the elliptical arc structure also reduces its stress. The bearing at the spindle is large, and the elliptical arc structure is equivalent to reducing the thickness of the upper sleeve joint. When the steel strip is superposed on the internal pressure and the pressure difference, the steel strip is finally expanded more than the lower sleeve joint. So as the contact stress decreases, the overall film stress and bending stress increase. Therefore, the optimization of the elliptical structure at the bell mouth has a good effect on reducing the stress.

5. Conclusions

（1）In this paper, the finite element method is used to establish model of bridge plug and packer in proportion to the real thing. The equivalent material is used to replace the laminated steel strip structure in the bridge plug/packer. This simplified method can greatly improve the computational efficiency, thus fully calculating the pressure expansion and the differential pressure loading of the bridge plug/packer.

（2）The pressure expansion process of the bridge plug and packer is simulated separately, and the stress distribution and maximum stress of each component of the rubber cylinder assembly are calculated.

（3）Optimize the structure according to the elliptic curve at the position of the excessively stressed position in the packer. When selecting the appropriate semi-long axis, except for the upper sleeve joint, which overall film stress and bending stress are increased due to the reduced thickness，the overall film stress and bending stress of the lower sleeve joint, steel strip at the upper sleeve joint and steel strip at the lower sleeve joint are reduced. The optimization of the elliptical structure at the bell mouth has a good effect on reducing the stress.

Acknowledgements

This research was financially supported by the Major National Science and Technology Projects of China (Grant No. 2016ZX05017-002).

References

[1] Jianghan Petroleum Administration Oil Production Technology Research Institute. Packer theory basis and application [M] . First edition, Beijing: Petroleum Industry Press, 1983, 1, 74-83.

[2] Zhang Chengwu. Sealing mechanism of segmented fracturing packer laminated steel sheet expansion cylinder [J]. Petroleum machinery, 2007, 35（3）, 5-7.

[3] Zhang Pu. Elastic performance calculation and optimization design of two-way ply laminate based on homogenization method [C]. The first national seminar on mechanics in the field of aerospace, Volume II, 2004, 337-340.

[4] Chen Xing. Study on Mechanical Properties of Typical Heterogeneous Materials Based on Progressive Homogenization [D]. Huazhong University of Science and Technology, 2015.

[5] Francu, Jan. Homogenization of Linear Elasticity Equations [J]. Aplikace Matematiky, 1982, svazek 27, 96-117.

[6] Bendsøe, Martin Philip & N. Kikuchi. Generating Optimal Topologies in Structural Design Using a Homogenization Method. Computer Methods in Applied Mechanics and Engineering[J]. 71 (1988), 197-224.

[7] Wang Benjin, Homogenization method for crack propagation of orthotropic steel bridge deck [J] . China Journal of Highways, 2017, 30（3）,114-117

[8] Zhu Shaogong, Application of finite element method to analyze the force condition and sealing ability of compression packer [C]. Oil production engineering anthology, 2016, 4, 6-9

[9] Zhang Xiaolin. Comparison of structure optimization and optimization methods for packer rubber [J] Petroleum machinery, 2013, 41（6）, 101-105.

[10] Yang Xiaolong. Optimized design of steel band packer rubber tube hoop bell mouth [J]. Mechanical design and research, 2015, 31（3）, 161-163.

[11] ASME-VIII-1, Pressure vessel construction rules -ASME Boiler and pressure vessel specifications [S].

[12] GB 150.1-2011, Pressure vessel part 1: Common enquiries [S].

ISPECE

IOP Publishing

A New Numerical Force Analysis Method of CBR Reducer with Tooth Modification

XiaoXiao Sun[1], Liang Han[1, a]

[1]School of Mechanical Engineering, Southeast University, Nanjing, P.R. China

[a]Corresponding author: melhan@seu.edu.cn

Abstract. A new one stage cycloidal reducer called China Bearing Reducer (CBR) which has large transmission ratio, high payload, high torsional stiffness, high tilting stiffness and compact size is designed. In this paper, a new force analysis algorithm is proposed to compute contact force of each tooth with modification. The cycloid drive theory and structure of CBR reducer is introduced firstly. Then the steps of force analysis algorithm are described and a numerical example of CBR25 reducer is analysed based on the algorithm. This method of force computing is more accurate than conventional force computing and can help better designing of CBR reducer.

1. Introduction

The cycloid drive has the advantages of large reducer ratio, high torque, high precision, high stiffness and compact structure. It has been widely used in precision reducer, such as RV reducer. In this paper, the CBR reducer is designed and a new numerical force analysis method is proposed.

Many researches had been done by scholars on cycloid-pin gear reducer. Yang and Blanche [1,2] discussed the formulas of cycloid drive and investigated the effect of machining tolerances on backlash and torque ripple. Then they presented an analytical and computer-aided analysis and synthesis of cycloid drives. Teruaki Hidaka et al. [3] proposed a method by using equivalent dynamic model with equivalent error to analyze rotational transmission error. Subsequently, the effects of machining and assembly errors of elements on the rotational transmission error were investigated [4]. Yunhong Meng et al. [5] proposed a mathematical model of 2K-H cycloid-pin reducer with one tooth difference and analyzed the transmission performance of clockwise and counterclockwise. Carlo Gorla et al. [6] proposed an innovative cycloidal speed reducer whose profile is the external offset of an epitrochoid and investigated the structural characteristics and the kinematic principles. Mirko Blagojevic et al. [7] designed a two-stage cycloid gear reducer and analyzed the loading sharing and stress by FEM. Bingkui Chen et al. [8] proposed a new cycloid drive by applying double-enveloping gear theory, establishing the meshing equation, deriving the equation of tooth profile and meshing line and the formula of induced normal curvature. And the superior characteristics of the new conjugated tooth profile is represented by comparison of induced normal curvature with conventional cycloid drives.

The aim of the paper is to investigate the contact force of cycloidal gear reducer. It will be expected to analyze the load distribution among the multiple contact tooth pairs.

2. Cycloid drive and design of CBR reducer

2.1 Cycloid profile and its modification

Content from this work may be used under the terms of the Creative Commons Attribution 3.0 licence. Any further distribution of this work must maintain attribution to the author(s) and the title of the work, journal citation and DOI.
Published under licence by IOP Publishing Ltd

As shown in Fig. 3, an overall fixed coordinate system $S_f(X_f-O_f-Y_f)$ and a moving coordinate system $S_p(X_p-O_p-Y_p)$ fixed to the pin gear are established in the center of the pin gear. And a moving coordinate system $S_c(X_c-O_c-Y_c)$ is established in the center of the cycloid gear. In the initial position, the Y_p axis coincides with the Y_f axis, and the Y_c axis is parallel to the Y_f axis. In the IGTS, the cycloid gear rotates φ_c at the angular speed w_c counterclockwise around the point O_c (that is, Sc rotates φ_c around O_c). According to the relative motion relation, the pin gear will rotate φ_p at the angular speed w_p counterclockwise around the point O_p (that is, Sp rotates φ_p around O_p). So we can get

$$i^H = w_c/w_p = \varphi_c/\varphi_p = z_p/z_c \qquad (1)$$

Where i^H is the transmission ratio of the cycloid gear and the pin gear. z_c is the number of teeth of the cycloidal gear and z_p is the number of teeth of pins.

Based on the theory of gear meshing and differential geometry, the cycloid disc profile in S_c coordinate system can be expressed as

$$
\begin{cases}
x_c = \left(R_p - R_{rp}S_r^{-\frac{1}{2}}\right)\cos\left[(1-i^H)\varphi_p\right] - \left(a - K_1 R_{rp}S_r^{-\frac{1}{2}}\right)\cos\left(i^H\varphi_p\right) \\
y_c = \left(R_p - R_{rp}S_r^{-\frac{1}{2}}\right)\sin\left[(1-i^H)\varphi_p\right] + \left(a - K_1 R_{rp}S_r^{-\frac{1}{2}}\right)\cos\left(i^H\varphi_p\right)
\end{cases}
\qquad (2)
$$

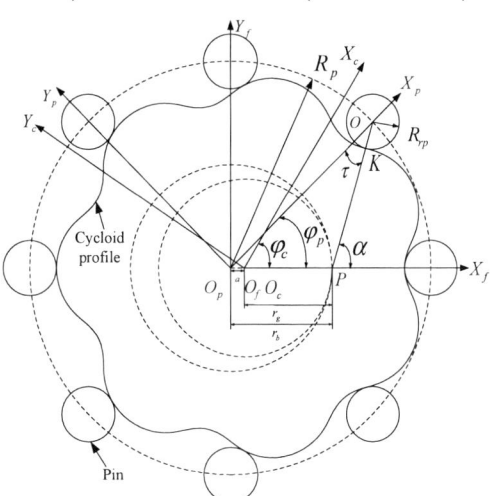

Fig. 1 Profile generation for a cycloidal disc.

To compensate for the errors caused by assembly and manufacturing, it is necessary to modify the cycloidal gear profile. Three ways of tooth modification are commonly used: isometric modification (modifying the pin radius R_{rp}), offset modification (modifying the pin gear radius R_p) and angle rotation modification (modifying the rotation angle φ_c of cycloid gear). The general cycloid gear profile equation with three modification methods can be established as

$$
\begin{cases}
x'_c = \left((R_p + \Delta R_p) - (R_{rp} + \Delta R_{rp})S_r^{-\frac{1}{2}} \right) \cos\left[(1 - i^H)\varphi_p - \delta \right] - \\[4pt]
\quad \dfrac{a}{R_p + \Delta R_p} \left((R_p + \Delta R_p) - z_p \left((R_{rp} + \Delta R_{rp})S_r^{-\frac{1}{2}} \right) \right) \cos\left(i^H \varphi_p + \delta \right) \\[8pt]
y'_c = \left((R_p + \Delta R_p) - (R_{rp} + \Delta R_{rp})S_r^{-\frac{1}{2}} \right) \sin\left[(1 - i^H)\varphi_p - \delta \right] + \\[4pt]
\quad \dfrac{a}{R_p + \Delta R_p} \left((R_p + \Delta R_p) - z_p \left((R_{rp} + \Delta R_{rp})S_r^{-\frac{1}{2}} \right) \right) \sin\left(i^H \varphi_p + \delta \right)
\end{cases}
\tag{3}
$$

2.2 Structure of CBR reducer

CBR reducer is a new type of cycloid gear reducer, which adopts a symmetrical transmission structure. As shown in Fig.2, it is mainly composed of bearing end cover, outputting flange, cross roller bearing, crank bearing, disc connector, turning bearing, crank shaft, pins, cycloid gear, case, outputting bearing and input support. Compared with RV reducer, it has the advantages of simple structure, small number of parts, simple assembly process and low manufacturing cost.

Because of its compact structure, it can achieve the installation dimension of the harmonic drive reducer. So, it can be used in robot joints with space restricted instead of harmonic reducer, and its payload, torsional stiffness and instantaneous impact resistance are several times higher than that of the harmonic drive reducer.

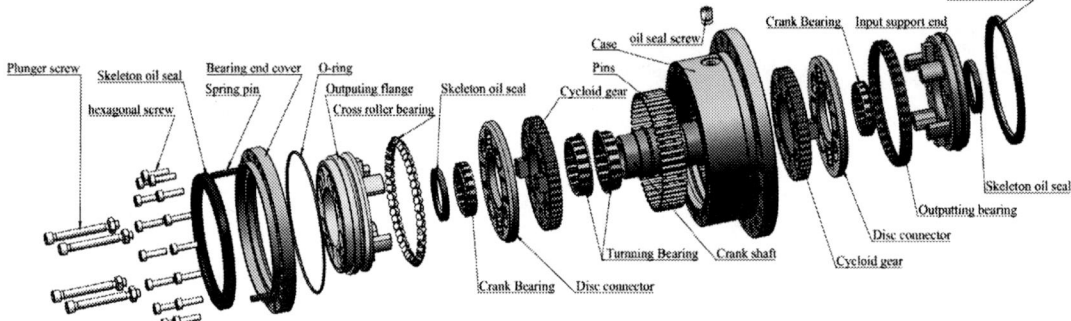

Fig.2 Exploded view of CBR Reducer.

3. Force analysis algorithm with tooth modification

In this section, a new numerical algorithm is proposed to compute contact force. Not only can it analyze initial gap, but also can compute the contact force and simultaneous contact teeth. Below are the detail steps of the algorithm.

(1) First, calculating the initial gap.

The initial gap has been derived by the method of geometric analysis and conventional TCA [9,10]. Here, the initial gap is solved by the new force analysis method, and the main flow is summarized as follows:

Step 1. As shown in Fig.3, the cycloid gear is discretized and divided into n_p equal parts, each corresponding to a pin. Each part is discretized into n points, the coordinates of each point are (x_{ci}, y_{ci}). φ_c is the rotational angle of cycloid gear. According to the modified cycloid equation Eqn. (3), the coordinates (x_{ci}, y_{ci}) of every point on the cycloidal profile are obtained.

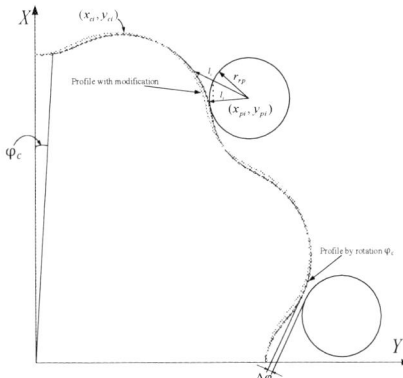

Fig. 3 Discretized points for cycloidal gear profile.

Step 2. Update the value of the rotational angle of the cycloid gear, the $\varphi_c = \varphi_c + d\varphi_c$, and get the coordinates (x'_{ci}, y'_{ci}) of the discrete points.

Step 3. Calculating the coordinates (x_{pi}, y_{pi}) of the center of the pin.

Step 4. Calculating the distance between the center of the pins and the points of the corresponding part of the cycloid gear, $l_i = \sqrt{(x_{pi} - x'_{ci})^2 + (y_{pi} - y'_{ci})^2}$.

Step 5. Judgement condition. If $l_i - R_{rp} < 0$, then stop calculating and extract the value of φ_c. Otherwise, turn to step 2.

Step 6. Outputting the initial gap. The distance between center of other pins and normal direction of cycloid profile minus the radius of the pin is the initial gap $\Delta\varphi_i$.

The new force analysis method uses the numerical method to solve the initial gap. It can avoid the error solved by the approximate geometric method. The result can be obtained accurately by reduce the iteration value $d\varphi_c$ and increase the numbers n of discretized points in the tooth of cycloid gear.

(2) Then, calculating the contact force.

After determining the first contact tooth of the cycloid gear, the cycloid gear continues to rotate, and the pins and the cycloid gear will produce contact force. According to the magnitude of the loading torque, the number of gear teeth that produces contact force will be different. Assuming the loading torque is rated load T_N, the contact force and the number of simultaneous meshing teeth are solved as follows:

Step 1. Calculating the contact stiffness K_m.

The cycloid disc tooth as a non-uniform cantilever beam, according to the beam theory and Hertzian theory [11], the mesh stiffness of one tooth pair of cycloid gear in mesh can be expressed as:

$$K_m = \frac{1}{\frac{1}{K_b} + \frac{1}{K_s} + \frac{1}{K_a} + \frac{1}{K_h}} \tag{4}$$

$$\frac{1}{K_h} = \frac{4(1 - \mu^2)}{\pi EW}, \frac{1}{K_b} = \int_0^d \frac{(x\cos\alpha - h\sin\alpha)^2}{EI_X} dx, \frac{1}{K_s} = \int_0^d \frac{1.2\cos^2\alpha}{GA_x} dx, \frac{1}{K_c} = \int_0^d \frac{\sin^2\alpha}{EA_X} dx$$

The cycloid-pin gear mesh stiffness of CBR25 reducer is calculated according to Eqn. (4) and is shown in Fig.4.

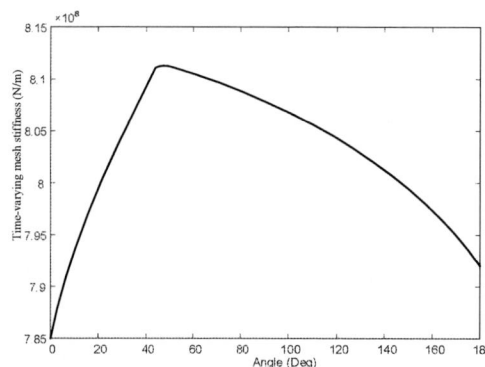

Fig. 4 Time-varying mesh stiffness of singe tooth of cycloid disc.

Step 2. Calculating the maximum nominal contact depth h_{imax}. After obtaining the φ_c value of the initial gap distribution, increasing the φ_c value continually, $\varphi_c = \varphi_c + d\varphi_c$, the pins which meet the condition $l_i - R_{rp} < 0$ engage contact. As shown in Fig. 5, the maximum normal contact depth can be obtained as follow

$$h_{imax} = |l_i - R_{rp}| \tag{5}$$

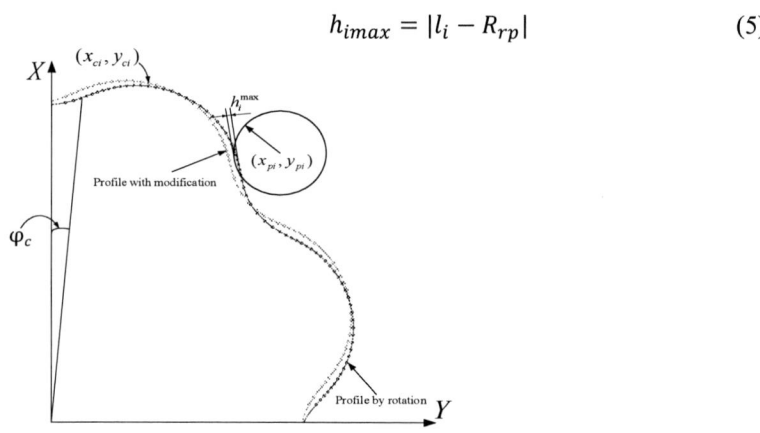

Fig. 5 Maximum nominal contact depth.

Step 3. Calculating the contact force F_i for each tooth,

$$F_i = K_m \cdot h_{imax}. \tag{6}$$

Step 4. Calculating the total torque T_c, as shown in Fig.6, the pins are enumerated, n is the number of simultaneously contact teeth, L_i is the arm length of contact force,

$$T_c = \sum_{i=1}^{n} F_i \cdot L_i = \sum_{i=1}^{n} F_i \cdot (a \cdot z_c \cdot sin\varphi \cdot S_r^{-\frac{1}{2}}) \tag{7}$$

Step 5. Comparing the torque T_c and T_N, if $T_c \leq T_N$, then stop calculating, else go to step 2.

Step 6. Outputting the contact force and simultaneously contact number.

The flowchart of the new force analysis method is shown in Fig. 7.

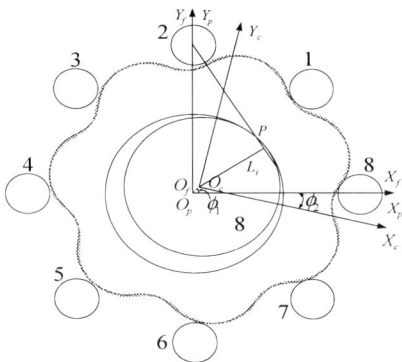

Fig. 6 Coordinate systems for contact force analysis.

4. Numerical examples

Taking the CBR25 reducer as an example, its design parameters are shown in Table 1. The isometric and offset modification are used in cycloidal profile. The output flange of the reducer is applied 100Nm torque. The input crank shaft rotates $2\pi*49$, so the cycloid gear will rotate 2π. We use the new force analysis method described above to calculate the initial gap and contact force of CBR25. The initial gap with respect to input crank angle is shown in Fig. 8 and each tooth contact force in one state is shown in Fig. 9.

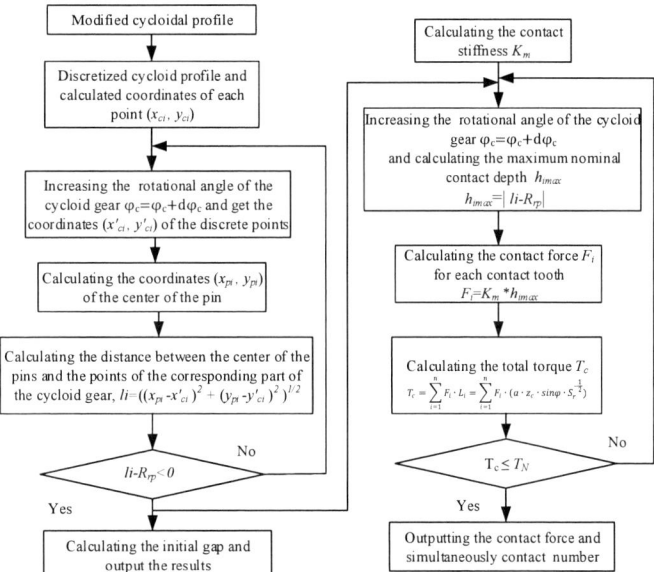

Fig.7 Flowchart of the force analysis method

Table 1. Parameters of CBR25 reducer.

Radius of pin position r_p	29.6	Eccentricity a	0.462
Pin radius r_{rp}	0.975	Short width coefficient K_1	0.7804
Teeth number of Pins z_p	50	Tooth width w_d	6.3
Teeth number cycloid gear z_c	49	Isometric modification ΔR_{rp}	-0.018
Transmission ratio i	49	Offset modification ΔR_p	-0.025

853

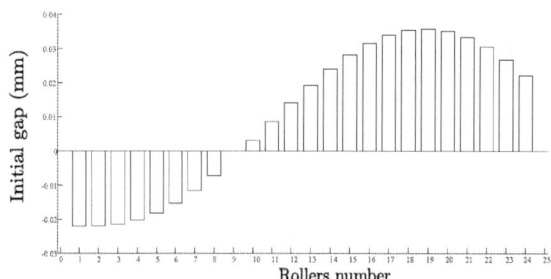

Fig. 8 Initial gap of CBR25 reducer with modifications.

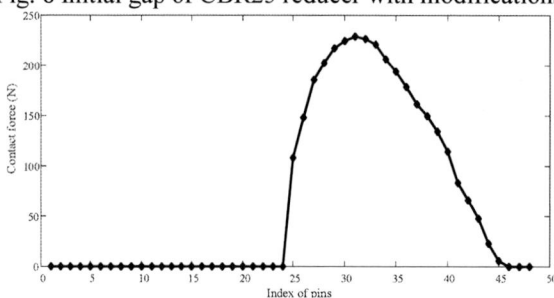

Fig. 9 Each tooth contact force of CBR25 reducer.

5. Conclusion

This article presents the design and a new force analysis method of CBR reducer. CBR is a one-stage reducer with a compact structure and a wide range of installation size. The modifed profle equation of cycloid gear was derived and the algorithmn process of force analysis was described. The numerical example of CBR25 reducer was simulated and the initail gap and contact force of each teeth were obtained. With the help of new force analysis method, the designer can obtain the initial gap and contact force accurately, and hence to improve the design of the CBR reducer in the product development phase.

Acknowledgment

This project is supported by Science and Technology Program of Jiangsu Province (Grant No. SBE2015000030), the Fundamental Research Funds for the Central Universities and the Research Innovation Program for College Graduates of Jiangsu Province (Grant No. KYLX16_0187). These supports are gracefully acknowledged.

References

[1] J.G. Blanche, D.C.H. Yang, Cycloid drives with machining tolerances, J. Mech. Transm. Autom. Des. **111** (1989) 337–344.

[2] D.C.H. Yang, Design and application guidelines for cycloid, Mech. Mach. Theory **25** (5) (1990) 487–501.

[3] T. Hidaka, Hong-You Wang, T. Ishida, K. Matsumoto, M. Hashimoto, Rotational transmission error of K-H-V planetary gears with cycloid gear: 1st report, analytical method of the rotational transmission error, Trans. Jpn. Soc. Mech. Eng. Ser. C **60** (570) (1994) 645–653.

[4] T. Ishida, H.-Y. Wang, T. Hidaka, K. Matsumoto, M. Hashimoto, Rotational transmission error of K-H-V-type planetary gears with cycloid gears: 2nd report, effects of manufacturing and assembly errors on rotational transmission error, Trans. Jpn. Soc. Mech. Eng. Ser. C **60** (578) (1994) 3510–3517.

[5] Y. Meng, C. Wu, L. Ling, Mathematical modeling of the transmission performance of 2K-H pin cycloid planetary mechanism, Mech. Mach. Theory 42 (7) (2007) 776–790.

[6] C. Gorla, P. Davoli, F. Rosa, C. Longoni, F. Chiozzi, A. Samarani, Theoretical and experimental analysis of a cycloidal speed reducer, J. Mech. Des. 130 (11) (2008) 112604.

[7] M. Blagojevic, N. Marjanovic, Z. Djordjevic, B. Stojanovic, A. Disic, A new design of a two-stage cycloidal speed reducer, J. Mech. Des. 133 (8) (2011) 085001.

[8] B. Chen, H. Zhong, J. Liu, C. Li, T. Fang, Generation and investigation of a new cycloid drive with double contact, Mech. Mach. Theory 49 (4) (2012)270–283.

[9] X.X. Sun, L. Han, K. Ma, L. Li, J. Wang, Lost motion analysis of CBR reducer, Mech. Mach. Theory **120** (2018) 89-106.

[10] A. Demenego, D. Vecchiato, F.L. Litvin, N. Nervegna, S. Mancó, Design and simulation of meshing of a cycloidal pump, Mech. Mach. Theory 37 (2002) 311-332.

[11] Z.Y. Ren, S.M. Mao, W.C. Guo, Z. Guo, Tooth modification and dynamic performance of the cycloidal drive, Mech. Syst. Signal Process. 85 (2017) 857-866.

Influence of electropulsing treatment on residual stresses and tensile strength of as-quenched medium carbon steel

Pan Long[1, 2, a]

[1]Research Department of Intelligent Manufacturing Equipment, Nanjing Institute of Technology, Nanjing 211167, China

[2]Jiangsu Provincial Engineering Laboratory of Intelligent Manufacturing Equipment, Nanjing 211167, China

[a]Corresponding author: panlong0229@126.com

Abstract. With the development of the technique of the high energy electric current pulse, the electropulsing treatment (EPT) is used in manufacturing more frequently. The effects of EPT on residual stresses and tensile strength of as-quenched medium carbon steel are investigated by experiments. The results show that EPT can reduce residual stresses, and improve tensile strength of the specimens. With the increase of the current density, the residual stresses decrease, and the tensile strength increases. The increase of the tensile strength is due to the decrease of the residual stresses.

1. Introduction

With the development of the technique of the high energy electric current pulse, EPT is widely used in manufacturing, such as current-pulse-aided rolling [1, 2], current-pulse-aided cold drawing [3], electromagnetic jigsaw [4, 5]. Furthermore, the effects of EPT on the material properties are studied by many researchers. For example, EPT can improve the plasticity of the metals which is called "electro-plasticity" [6-9], affects the ten-sile properties and bending behavior of Ti6Al4V alloy sheet [10], and facilitate the nanocrystallization of the materials [11]. However, there are a little of reports about the effects of EPT on the residual stress. Stepanov and his co-workers did some works about this study [12-15], but they did not study the effects of EPT on the quenching stresses. Hence, the effects of the EPT and tensile strength are studied in this paper, which enriches the results of this research area.

2. Material and Methods

Table 1. Chemical composition of medium carbon steel

Element	C	Si	Mn	P≤	S≤	Fe
W.t %	0.452	0.271	0.442	0.035	0.040	Bal.

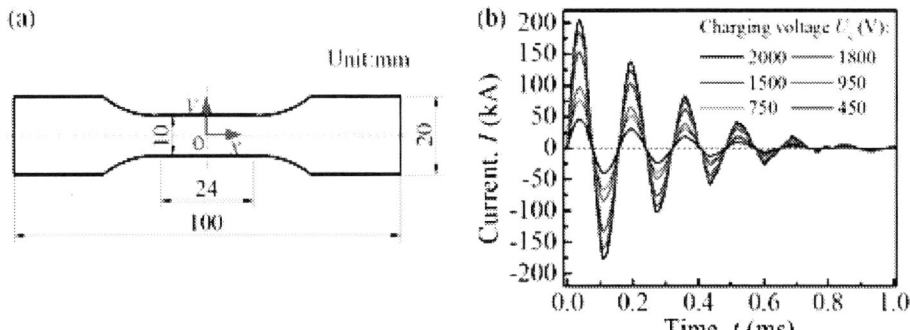

Figure 1. (a) Dimensions of specimen, and (b) waveforms of EPTs at different charging voltage

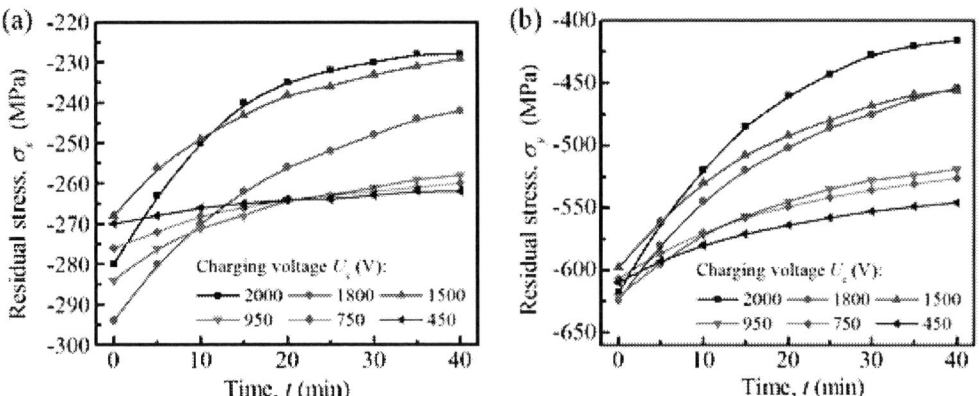

Figure 2 Curves of residual stress (a) σ_x and (b) σ_y verse time at different charging voltage

The material used in this study was commercial medium carbon steel, and its chemical composition was shown in Table 1. The specimens were 1 mm thick plate (see Fig. 1a), and treated by quenching. During the quenching process, temperature of specimens rose up to 850 °C and held at isothermal conditions for 3 minutes until the material was completely austenitized. The specimens were then put into salt brine at 20 °C to form a sufficient quenching quality.

Specimens were treated by high density EPT, which was generated by a device containing a group of high-voltage and large-value capacitors [16]. The capacitance was 400 μF and the charging voltage was from 450 V to 2000 V, with a typical oscillating EPT being generated as shown in Fig. 1b. It was found that the periodic time of oscillations was 1.6×10^{-4} s. Specimens were treated by a current pulse every 4 s for a duration of 40 minutes.

The surface residual stresses at the center point o (Fig. 1a) were evaluated using the laboratory non-destructive residual stress measurement system LXRD of PROTO. The specimens were produced according to ASTM E8M-04 Standard Test Methods for Tension Testing of Metallic Materials [17], the selected target material was Cr, and the diffraction plane was {2 1 1} with diffraction angle 156°. Tensile tests were conducted at a constant strain rate of 1×10^{-3} s^{-1} at room temperature using CMT 5205 electromechanical universal test systems of MTS Systems Corporation.

3. Results and discussions

3.1 Residual stresses

The residual stresses of the as-quenched carbon steel specimens decreased after EPT as shown in Fig. 2. The residual stresses decreased rapidly at the beginning of the treatment, and tended to be stable after some treating time. σ_x was small and its reduction was small; σ_y was large and its reduction was large too. The changes of the residual stresses were different after the specimens treated by different EPT. The decrease rates η (η_x of σ_x, η_y of σ_y) were shown in Fig. 3. The charging voltage was larger, the current density through the specimens was larger, and the decrease rates were larger. It was found that EPT can reduce the residual stresses, and the decrease rates increased with the current density increasing.

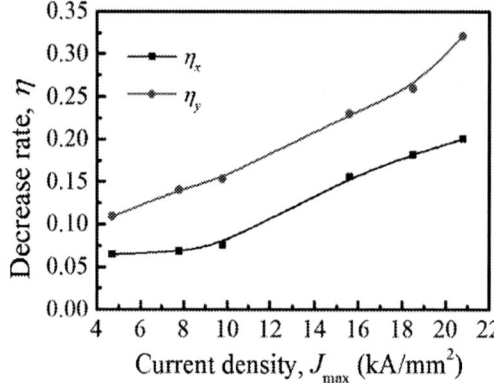

Figure 3. Decrease rate of residual stresses versus maximum current density

3.2 Tensile strength

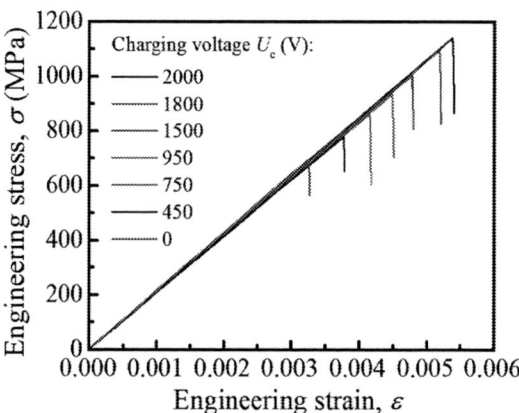

Figure 4. Strain – stress curves of specimens after EPT at different charging voltage

The engineering strain – stress curves of the specimens after EPT as shown in Fig. 4. The EPT of different charging voltage had different effect on the tensile strength. The tensile strength increased with the charging voltage increasing. The result showed that EPT can improve the tensile strength of the as-quenched carbon steel specimens, and the tensile strength increased with the current density increasing.

3.3 Relationship between residual stresses and tensile strength

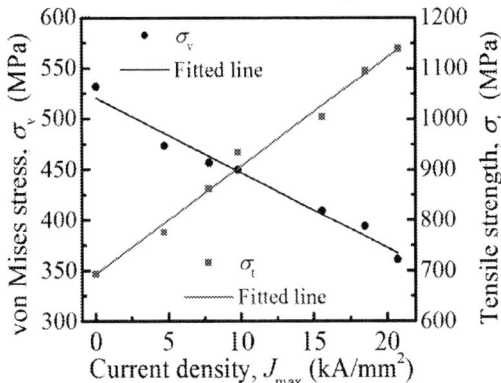

Figure 5. Von Mises stress and tensile strength versus maximum current density

According to the von Mises yield criterion, the von Mises stress σ_v (or equivalent tensile stress) due to the residual stresses is

$$\sigma_v = \sqrt{\sigma_x^2 + \sigma_y^2 - \sigma_x \sigma_y} \ . \qquad (1)$$

The von Mises stress σ_v and the tensile strength σ_t were shown in Fig.5. With the increase of the maximum current density J_{max}, σ_v decreased and σ_t increased, and both had a good linear relationship. During tensile test, when the sum of tensile strength σ_t and residual stress σ exceeds the tensile strength σ_C of samples with no residual stresses, the sample fractures, which directly indicates that the decrease of residual stresses can improve the tensile strength. Hence, the decrease of residual stresses is the primary factor of the increase of the tensile strength.

4. Conclusions

The as-quenched medium carbon steel specimens were treated by EPT of different charging voltage, and the residual stresses and tensile strength were investigated. The relationship between residual stresses and tensile strength was analyzed. The major results and conclusions of this study are summarized as follows:

1)   EPT reduces the residual stresses of specimens, and the decrease rates of residual stresses increase with the current density increasing;

2)   EPT improves the tensile strength of specimens, and the tensile strength increase with the current density increasing;

3)   After EPT, the increase of the tensile strength is primary due to the decrease of the residual stresses.

Acknowledgements

Financial supports from the Scientific Research Staring Founds for the High-level Talents of Nanjing Institute of Technology (YKJ201742), and the Natural Science Foundation of Jiangsu Higher Education Institutions of China (18KJB460015) are gratefully acknowledged.

References

[1]   I.M. Mal Tsev. Russ J. Non-ferr Met, **49**: 175-180 (2008)

[2]   K.M. Klimov, I.I. Novikov. Dokl Phys, **52**: 359-360 (2007)

[3]   G. Tang, J. Zhang, Y. Yan, H. Zhou, W. Fang. J. Mater Proc Tech, **137**: 96-99 (2003)

[4]   P. Kumar, A. Mishra, T. Watt, I. Dutta, D.L. Bourell, U. Sahaym. Procedia CIRP, **6**: 600-604 (2013)

[5] S.A. Baranov, V.I. Staschenko, A.V. Sukhov, O.A. Troitskiy, A.V. Tyapkin. Russ Elec Eng., **82**: 477-479 (2011)

[6] G. Tang, J. Zhang, J. Zhang, M. Zheng, W. Fang, Q. Li. Mater Sci Eng A, **281**: 263-267 (2000)

[7] H. Conrad. Mater Sci Eng A, **287**: 276-287 (2000)

[8] K.M. Klimov, I.I. Novikov. Strength Mater, **16**: 270 (1984)

[9] O.A. Troitskii, V.I. Likhtman. Soviet Phys Dokl, 91 (1963)

[10] Li, Q. Zhou, S. Zhao, J. Chen. Procedia Eng., **81**: 1799-1804 (2014)

[11] V.V. Stolyarov. Mater Sci Eng A, **503**: 18-20 (2009)

[12] G.V. Stepanov, A.I. Babutskii, I.A. Mameev. Strength Mater, **41**: 623-627 (2009)

[13] G.V. Stepanov, A.I. Babutskii, I.A. Mameev, M. Ferraris, V. Casalegno, M. Salvo. Strength Mater, **40**: 452-457 (2008)

[14] G.V. Stepanov, A.I. Babutskii, V.P. Pakhotnykh. Strength Mater, **40**: 629-634 (2008)

[15] G.V. Stepanov, A.I. Babutskii. Strength Mater, **39**: 189-193 (2007)

[16] J.Y. Zheng, W. He, R.J. Shen. Adv Mater Res, **139-141**: 163-166 (2010)

[17] ASTM International. Standard Test Methods for Tension Testing of Metallic Materials. E8M-13a, West Conshohocken, PA (2013)

ISPECE

IOP Publishing

High accuracy Numerical simulation on 3D weld-pool shape of large parts

Zhu Yonggang[1,2] Zuo yanhong[2]

[1]School of mechanical engineering, Anhui Sanlian College, 230061, Hefei, China

[2]School of mechanical and electrical engineering, Anhui jianzhu university, 230601, Hefei, China

[a]Zhu Yonggang: 351921811@qq.com

Abstract. The shape of weld pool and the distribution function of heat source are the premise and foundation of all welding simulations, but many assumptions are given to them which lead to different or even contradictory results. So the model of three-dimensional weld pool should be established without predetermining the shape of weld pool, the shape of free surface and the distribution of heat flux density, moreover it is combined with the thermal model. Through decoupling, the fluid velocity field, the temperature contour and the law of temperature distribution of the weld pool are obtained, the shape of molten pool is also gotten out which is like a revolving body of the pen holder, and the density function of heat flux distribution is fitted out. The internal relationship between simulation results and weld fusion line are presented. Welding experiments of large parts are carried out under the same process conditions. It shows the weld fusion line and the shape of the surface of the weld are in good agreement with the simulation results. This has great practical significance for unifying simulation results, guiding production and improving quality of large welding parts.

1. Preface

Construction machinery mainly includes shovel, transport machinery, lifting, pumping, pile and road machinery, etc. The sales have maintained growth at nearly 100% every year. But the situation of construction machinery manufacturing industry in our country is severe, especially the skeleton of construction machinery such as manipulator arm, bucket and other large structural parts is lagging behind by backwardness of welding technology, and led to second-class product, poor quality and reliability. So the overall performance of construction machinery is extremely unfavorable. The following is the super long arm of the EX230 excavator, shown in Figure 1.

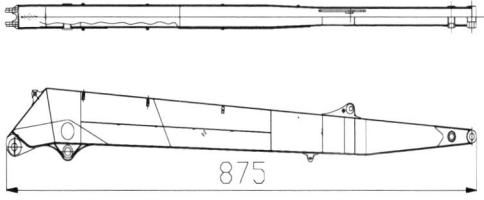

Fig 1. EX230 super long arm of excavator

The first quality control of large-scale structure welding parts is carried out in the stage of product design. Secondly, the numerical simulation method combining physical test is often used to predict

Content from this work may be used under the terms of the Creative Commons Attribution 3.0 licence. Any further distribution of this work must maintain attribution to the author(s) and the title of the work, journal citation and DOI.

Published under licence by IOP Publishing Ltd

and control the deformation of large-scale structure welding parts. The simulation need heat source density distribution, but the premise of heat source density distribution is the molten pool shape.

By now, the widely used heat source model is the Gauss model [1] which can be classified into planar heat source and volume heat source according to their different modes of actions [2], it can also be divided into ellipsoid heat source and double ellipsoid heat source according to the heat density distribution. But the result of simulation is not consistent with the actual weld fusion line well [3]. Particularly in the thick plate welding, the simulation shows that temperature is too high in part [4].

In order to make use of the advantages of the both kinds of heat sources, modern welding simulation is mostly using the combined heat sources model. We combine the two kinds of heat sources by adjusting the ratio artificially to fit the actual welding Cross section shape[5]. And it has significance in the application of middle or thick plate welding [6]. But when the welding conditions change, some adjustments should be made to the combined heat sources, otherwise there will lead to large deviations. So it is very necessary to study the pool shape and heat source distribution function. However, the research in this area is few, and mostly focuses on the two-dimensional flow and temperature fields[7]. In order to study the pool shape and heat source in large parts welding, the super-long arm of EX230 excavator is taken as the engineering background. Model of weld pool is established on hydrodynamics, together with the surface control equation and heat convection transfer equation to study the three-dimensional static weld pool shape and heat flux distribution function. Through large calculation, we get laws for production guidance.

2. Model building

We does not specify the shape of molten pool, the shape of free surface, or the heat flux distribution function in advance to avoid the deviation by presuppositions.

Free surface control equation[8].

$$P_{arc} - \rho g \varphi + C_2 =$$
$$-\gamma \frac{(1+\varphi_y^2)\varphi_{xy} - 2\varphi_x \varphi_y \varphi_{xy} + (1+\varphi_x^2)\varphi_{xy}}{(1+\varphi_x^2+\varphi_y^2)^{3/2}} \qquad (1)$$

φ surface shape function, φ_x, φ_y, is the first-order partial derivative of the shape function to x, y, φ_{xx}, φ_{yy}, is the second-order partial derivative of the shape function to x, y, φ_{xy} is the second-order mixed partial derivative of the shape function to x, y, C2 undetermined constant, γ surface pressure, P_{arc} arc pressure. The method of solution refers to literature 9.

Incompressible continuous equation.

$$\frac{\partial u_x}{\partial x} + \frac{\partial u_y}{\partial y} + \frac{\partial u_z}{\partial z} = 0 \qquad (2)$$

When homogeneous fluid is incompressible, $\rho = const.$

N-S equation is

$$\frac{du_x}{dt} = X - \frac{1}{\rho}\frac{\partial p}{\partial x} + v\left(\frac{\partial^2 u_x}{\partial x^2} + \frac{\partial^2 u_x}{\partial y^2} + \frac{\partial^2 u_x}{\partial z^2}\right)$$

$$\frac{du_y}{dt} = Y - \frac{1}{\rho}\frac{\partial p}{\partial y} + v\left(\frac{\partial^2 u_y}{\partial x^2} + \frac{\partial^2 u_y}{\partial y^2} + \frac{\partial^2 u_y}{\partial z^2}\right) \qquad (3)$$

$$\frac{du_z}{dt} = Z - \frac{1}{\rho}\frac{\partial p}{\partial z} + v\left(\frac{\partial^2 u_z}{\partial x^2} + \frac{\partial^2 u_z}{\partial y^2} + \frac{\partial^2 u_z}{\partial z^2}\right)$$

When fluid is incompressible with constant physical properties and no internal heat, convection-conduction equation is

$$\frac{\partial t}{\partial \tau} + u_x \frac{\partial t}{\partial x} + u_y \frac{\partial t}{\partial y} + u_z \frac{\partial t}{\partial z} = \alpha \nabla^2 t$$

$$= \alpha \left(\frac{\partial^2 t}{\partial x^2} + \frac{\partial^2 t}{\partial y^2} + \frac{\partial^2 t}{\partial z^2} \right)$$

(4)

3. Calculation and results

The calculation model is shown in Figure 2, and length, width and height of cube are all 10mm. The calculation initial conditions boundary conditions are as follows: structure steel material is 16Mn, physical property parameters are shown in table 1, the surface arc acts over is a circle area with 2.5mm radius, the temperature on the circle is 6000^0C , and pressure is 200pa (gauge pressure), the other surface temperature is 300^0C, acting pressure is 0 pa (gauge pressure), just under the pressure of atmosphere. Flow chart of the calculation program is shown in Figure 3.

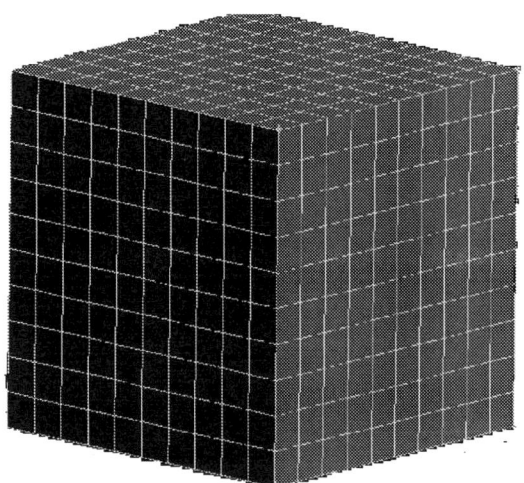

Figure 2. calculation model

The basic parameters are shown in follow table [10]

Table 1. Fluid basic physical parameters

Item value unit	Value	Unit
Density (solid phase)	7600	kg/m^3
Density (liquid phase)	7200	kg/m^3
Latent heat of solidification	300	kj/kg
Heat capacity (solid phase)	680	J/ (kg.^{0}C)
Heat capacity (liquid phase)	840	J/ (kg.^{0}C)
The thermal conductivity	22	W/ (m.^{0}C)
Melting point (solid phase)	1431	^{0}C
Melting point (liquid phase)	1512	^{0}C
Viscosity	0.006	Pa.s

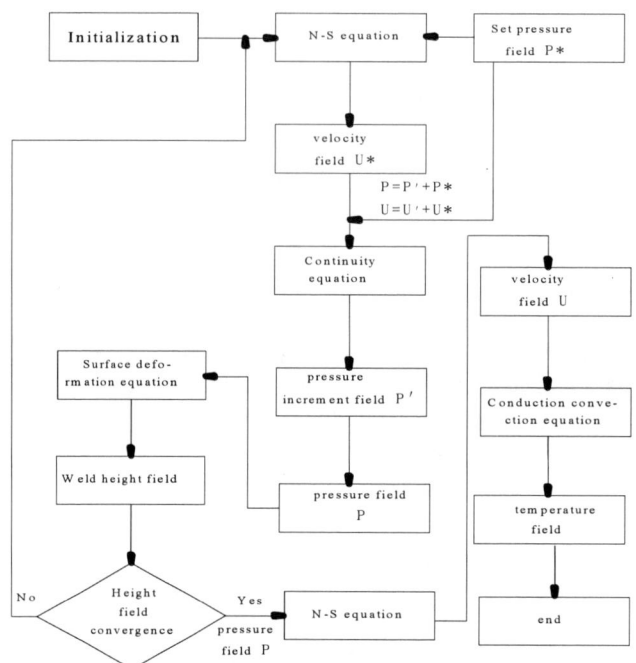

Figure 3. Flow chart of calculation program

Through decoupling, simulation result is obtained as following.

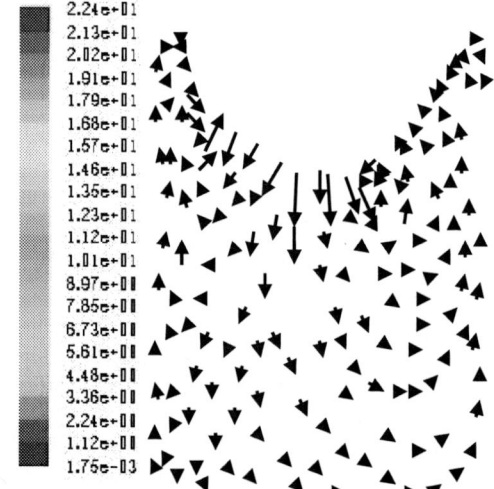

Figure 4. X=5mm section velocity vector diagram

ISPECE

IOP Publishing

IOP Conf. Series: Journal of Physics: Conf. Series **1187** (2019) 032055   doi:10.1088/1742-6596/1187/3/032055

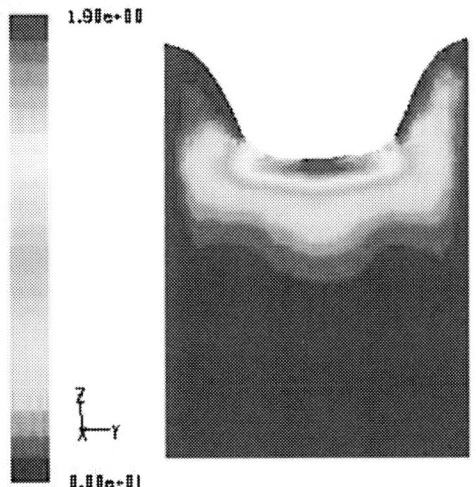

Figure 5. X=5mm section velocity cloud picture

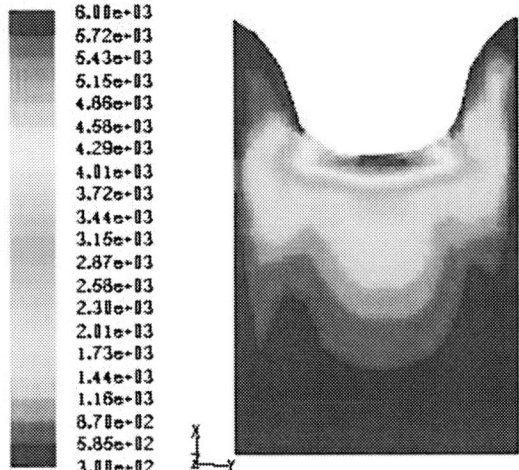

Figure 6. X=5mm section temperature cloud picture

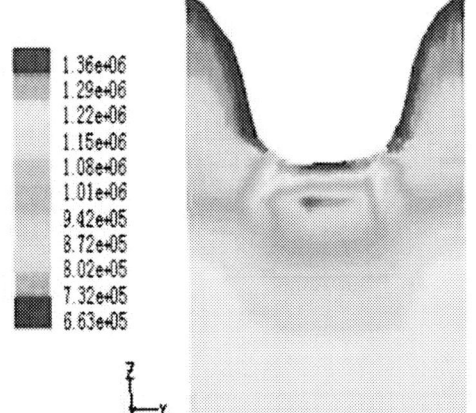

Figure 7. X=5mm section pressure cloud picture

865

4. simulation analysis

The temperature is positively associated to the velocity and flowing direction of fluid. The higher the velocity of flow, the higher the temperature; the slower the velocity of flow, the smaller fluid obtained the energy. Due to the flow of fluid at the bottom is restricted by the high pressure, the downward velocity is little, the upper part of the fluid is open without any restrict, the upward velocity is large. The direction of the edge fluid first flow downwards and then upwards under the high welding pressure,. Therefore, the true shape of the heat source is not an ellipsoid, but a rotating body of a pen holder. While closer to the rotator center, the high temperature layer is thicker and higher; the farther away from the center, the temperature layer is thinner and lower, then thicker the higher again, finally disappears steeply, as shown in Figure 8. With the filling of welding material, the position of weld pool increases taller gradually, and the edge of weld pool will not restrict the pool. When weld pool rises, the top surface of weld pool becomes ellipsoid under gravity operation without supporting solids around the weld pool gradually, as shown in Figure 9. Meanwhile, the simulation shows the shape of weld pool is smaller than the weld section, that is to say, the weld section is not formed by melting-solidification at one time, but melting-solidification repeatedly. So the weld section is formed layer by layer. The shape of the weld section connecting line we usually see is a envelope formed by weld pool overlapping layer by layer, as shown in Figure 10. The welding seam is formed by the superposition of the envelope line of the weld pool in the welding direction, as shown in Figure. 11.

Fig. 8. Heat source section

Figure 9. Heat source section over solid surface.

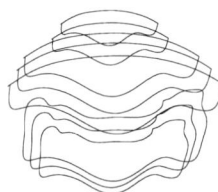

A Section superposition graph

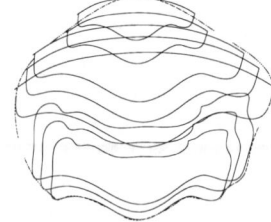

B Superposition graph with envelope

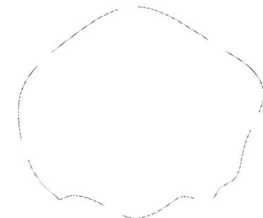

C Superposition envelope of section
Figure 10. weld line

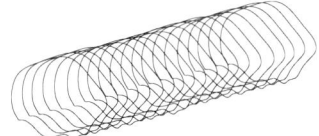

Figure 11. Envelope of section envelope

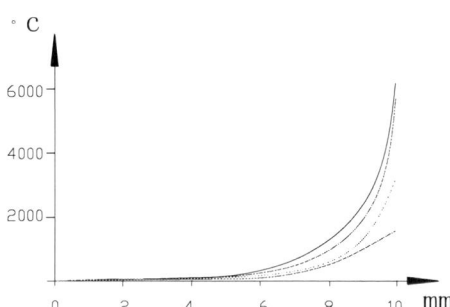

Figure 12. Temperature distribution function in vertical direction through the center point

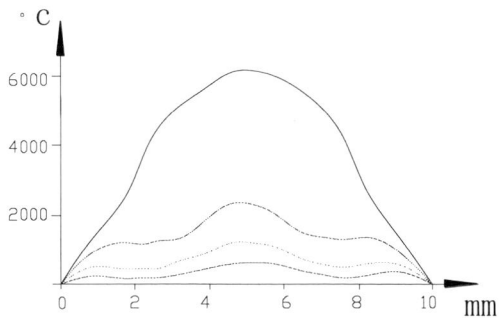

Figure 13. Temperature distribution function in the lateral the vertical direction through the center point

According to the simulation results, as shown in Figures 12 and 13, the heat flux equation of 16Mn steel welding can be fitted out as following equation:

$$q(x,y,z) = \frac{10Q}{\pi r_0^2 d}\left(\frac{d}{d_0}\right)^2 \left(\begin{array}{c} \dfrac{-25(x^2+y^2)}{8r_0^2}(\dfrac{z}{z_0})^{10.23} \\ +3\exp\left(100(\dfrac{z^2}{z_0^2}-1)\right) \end{array} \right)$$

Here, r_0 is arc radius, z_0 the distance between the arc acting surface and the ground, x、y、z is the coordinates of the points in the model, d_0 is the arc diameter, and d is the diameter of the horizontal circle with its point in the model while the center on the central axis.

5. experimental and verification

Referred to the crane girde manufacturing process conditions of large parts, 16Mn steel with thickness 10mm is used as shown in Figure 14. The surface is treated with wire brush for simple rust removal , then welded by manual arc on the seamless board. The welding rod is J422, the current is 110A, and the voltage is 60V. The welding section photographs are compared with the simulation results in Figure 15, showing a high degree of coincidence.

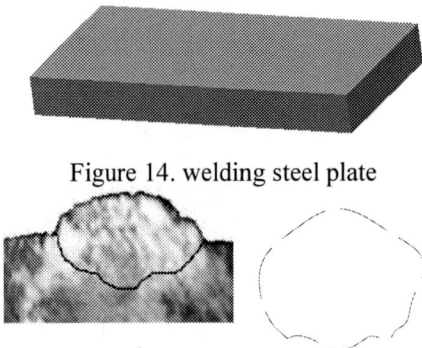

Figure 14. welding steel plate

Figure 15. Comparison of weld photograph and simulation results on seamless board

6. conclusion

(l) Two-dimensional model of molten pool with hypothetical conditions has inherent defects in accuracy. In order to make up for its shortcomings, a three-dimensional static mathematical model of molten pool based on hydrodynamics coupling with heat convection conduction equation is established, which has clear physical meaning and can improve the accuracy of welding analysis.

(2) The weld pool shape obtained by decoupling is a penholder-shaped rotating body, and the size of the weld pool is obviously smaller than that of the actual weld section, which indicates that the actual weld section is formed by multiple melting-solidification of the weld pool.

(3) The heat source density is wavy distribution in the horizontal plane and decreasing distribution in the direction of symmetry line. By analyzing the simulated data, a new heat flux equation of molten pool is fitted out, which provides a reference tool for the analysis of temperature field and stress field of large welding parts.

(4) A welding verification on seamless 16Mn plate under the same welding conditions indicates the experiment is in good agreement with the simulation. The model reflects the basic characteristics of large workpiece welding.

References
[1] Chen Y B. Modern laser welding technology. Beijing: Science Press, 2005, 10.181-207

[2] Zheng Z T. Heat source model and development trend of fusion welding. Welding,2008,4:3-6

[3] Gai D Y, Zhu Y Z, Li Q F, et al. Application of combined heat source model in numerical simulation of welding. Journal of Welding, 2009，5：61-64.

[4] Tso-Liang Teng, Chih-Cheng Lin.Effect of welding conditions on residual stresses due to butt welds [J].International Journal of Pressure vessels and piping.1998,75(12): 857-864.

[5] John Goldak.a new finite element model for welding heat sources [J].metal lurgual transactions, 1984, 15B (2): 299-305.

[6] Wang Min, Dong Zhibe, Yu lan. Numerical simulation of temperature fields for T-joint during TIG welding of titanium alloy [J]. China Weld, 2008, 17 (3): 6.

[7] Wu Chuansong; Meng Xiangmeng; Chen Ji; Qin Guoliang. Advances in Numerical Simulation of Welding Heat Process and Molten Pool Behavior [J]. Journal of Mechanical Engineering, 2018-2 (54) -21

[8] Ohji T, Nishiguchi K. Mathematical modeling of molten pool in arc welding of thin plate [J] International Institute of Welding Document, 1983; 1:212-555

[9] Lin M L, Eagar T. W. Welding Journal, Influence of arc pressure on weld pool geometry [J] 1985; 64:163-s

[10] Wang Jinming; Huo Dan; Chang Yunlong; Bao Changli. Numerical Simulation of Coupled Field in Welding Pool Based on Finite Difference Method [J] Journal of Shenyang Institute of Engineering (Natural Science) 01, 2011

[11] Chen Haizhou. Numerical simulation of incompressible free surface flow by SPH method [D]. Tianjin University, 2008.

[12] Chen Jiaquan, Xiaoshun Lake, Wu Gang, Yang Xinyan. Comparison of heat source models for numerical simulation of welding process [J].Welding technology, 2006, 35 (1): 9-11

Experimental Study on Cutting Force Comparison between Inner Cooling and Outer Cooling in Zig-zag Milling

Umair Riaz[1], Can Liu[1,a], Guangyu Tan[1], Guanghui Li[1] and Ningxia Yin[1], Muhammad Junaid Saeed[1]

[1]Collage of Mechanical and Power Engineering, Guangdong Ocean University, 524088, Zhanjiang, Guangdong, China

[a]Corresponding author: Liucanzj@163.com

Abstract. The cutting force is a key factor that affects the milling performance. The performance of a kind of inner cooling mill with double straight channels developed by us was testified with an index of cutting force. Zig-zag milling experiments were done on CNC milling machine. Cutters with different diameters were carried out at different speeds for inner cooling and outer cooling. The average peak cutting forces per tooth were picked up. The results showed that forces of the outer cooling had been greater than that of inner cooling, and the bigger of the cutter, the more significant of their comparison, which means that inner cooling should be better than outer cooling, and the performance of the cutter was worthy. This work might play important roles in various cutting manufacturing factories and in milling research where inner cooling milling could be used instead of outer cooling.

1. Introduction

The inner cooling method in which we apply the cutting fluid by an internal channel of the tool on the work piece, is an excessive way to save the cutting fluid.

Inner cooling can be used for machining especially when the cutting depth exceeds 3mm. In inner cooling experiment researches, different tool structures (double straight channel, single straight channel and double helical channel) were used to check the cutting forces and tool wear on side milling, the results showed that milling cutter with double straight channel should have had approximately double cooling effect as being compared to double helical channel under cryogenic minimum quantity lubrication [1]. Islam et al. [2] noted that, the effects of inner cooling by cryogenic on the machining of hardened steel. Design, development of rotary liquid nitrogen applicator and investigation of machining performance under cryogenic and its comparison with dry and flood cutting were studied. The effect of the input cutting condition (cutting speed, feed rate and depth of cut) on the output (cutting force, surface roughness and tool flank wear) were analyzed. Result indicated that the following results cryogenic application via rotary applicator demonstrated improved cooling action. Cutting forces was reduced by cryogenic cooling. Liquid nitrogen increased the hardness of tool edges. The cutting speed and feed rate affected surface finish, cutting force and tool performance in various degree. Moreso, inner cooling research, Ferri et al. [3] reported that, the fluids performance, accurate control and influence through the inner geometry to control the heat generation. Results showed that among the three cutting parameters in inner cooling, the chip temperature depended significantly on the depth of cut rather than feed and cutting speed. The development of cutting force model is very important to tools development, optimization, and analysis of milling operations [4], [5].

Content from this work may be used under the terms of the Creative Commons Attribution 3.0 licence. Any further distribution of this work must maintain attribution to the author(s) and the title of the work, journal citation and DOI.
Published under licence by IOP Publishing Ltd

M. Wan., etc. [6] explored the cutting force model of peripheral milling. (Yun & Cho, 2000) [7] Presented the new way to estimate the 3D force coefficients in which the constant cutting force coefficient of the milling cutter and the constant cutting force coefficient of the work piece is determined regardless of the angle of rotation of the tool or the cutting condition. These constants are used to calculate the cutting force. In the article "The impact of cooling methods on the maximum temperature of the processed object during side milling", the highest temperature is measured by thermocouples. The cooling method has been noted to influence the cutting forces coefficient (Nowakowski, Skrzyniarz, & Miko, 2017) [8]. In the paper "state of art of a cooling method for dry machining", it was suggested that the cooling method in dry machining removes heat better. This, in turn, reduces the cutting resistance, focusing on the heat removal at the cutting tool tip, and improves the machining performance compared with wet machining and other methods (Ramachandran et al., 2017) [9]. In the article, "Dry machining: Machining of the future", by P.S. Sreejith, B.K.A. Ngoi presented that dry machining without any fluid is more efficient due to the safety environment. The cost of coolants and lubricant in machining represents 16-20% manufacturing costs [10]. "The possibilities and limitation of dry machining" by Dr. Neil Canter states that for an open-faced operation like milling and boring dry machining were efficient while for closed face machining operation such as drilling, tapping and hole making dry machining is not recommended [11]. The use of minimum quality lubrication (MQL) or no coolant results in lots of subsurface damage and short tool service life (Beer, Özkaya, & Biermann, 2014) [13]. Can et al. [12] noted the performance of 316 stainless steel on High-Speed Milling by experimental study. The analysis concluded that the depth of cut and the feed significantly affected the milling force while speed was insignificant.

Though there are lots of researches on the cutting forces on different milling, additional knowledge is still needed to broaden the data baseline which can assist further researchers in their research work. This study is aimed at performance testing of a kind of self-developed inner cooling tool. The inner cooling tool used in our experiments was developed by our discipline. The content of this paper is to do an experiment, do the analysis of the cutting forces in zig-zag milling through inner and outer cooling method and then conclude. If the cutting forces are less so the milling is better so we can compare the forces of inner cooling and outer cooling. Furthermore, cutting forces for both inner and outer cooling at different diameters and different cutting speeds while the feed, cutting width and cutting depth remains same will be examined.

Table 1. Experimental Parameters

Test No.	1	2	3	4	5	6	7	8	9	10	11	12
Cooling Method	Inner Cooling						Outer Cooling					
Cutter Dia. (mm)	12	16	20	12	16	20	12	16	20	12	16	20
Spindle Speed (rpm)	540	540	540	1080	1080	1080	540	540	540	1080	1080	1080

2. Experiments

All experiments were carried out on a CNC milling machine which had a maximum speed of 15000 rpm. The dynamometer was connected to the strain amplifier. The experimental system is shown in Fig. 2. Information about sampling points and waveform were saved into the personal computer. The cutting material (steel) was fastened on the dynamometer. The milling cutter used for the measurement was flat end mill cutter. The tool specification was D12x35x90, D16x45x100 and D20x50x110. The diameters of the cutter were 12mm, 16mm, and 20 mm with 4 teeth shown in Fig. 1. Tool material was tungsten carbide. Before the experiments, we were needed to design the experiment parameters. We knew that cutting conditions affect cutting forces. It was not possible to observe the reaction of cutting forces to a variety of every parameter, so we were only deal with the spindle speed at different diameters while feed and cutting depth remained same throughout. In three different diameters tool experiments, cutting depth was designed 1mm, the Feed rate of 0.05 mm/tooth. The cutting width was 6mm. We will use zig-zag milling shown in Fig. 2. Firstly, the test # 1-3 and 7-9 experimental reading was done for the inner and outer cooling with a spindle speed 540 rpm. Secondly, the next test # 4-6 and 10-12 was done for the inner and outer cooling with a spindle speed 1080 rpm. Total 4 readings were taken at 12 mm cutter diameter. It was also same for 16mm and 20mm cutter diameters. So we were a total 12 test, 4 were taken at 12mm cutter diameter, 4 were taken at 16mm cutter diameter and 4 were taken at 20mm cutter diameter as shown in Table 1.

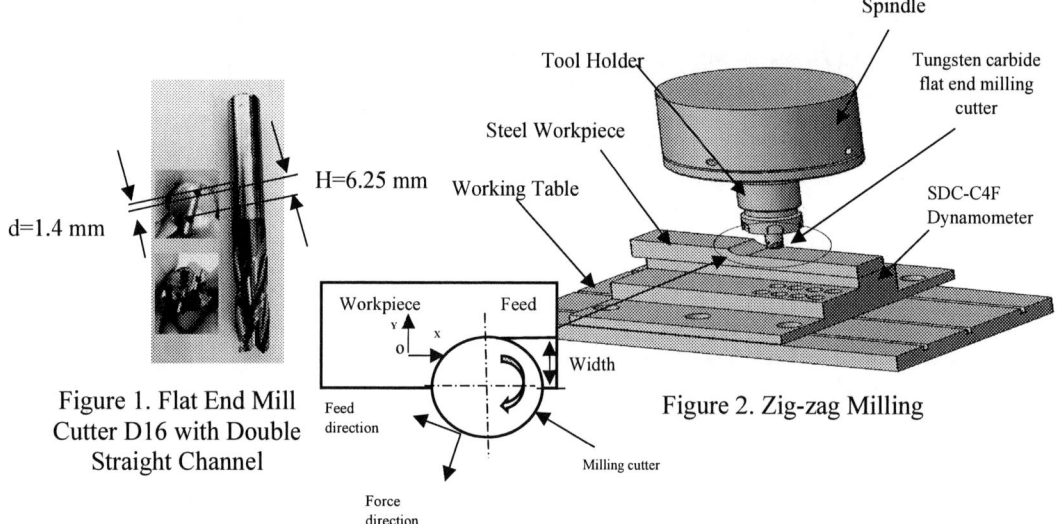

Figure 1. Flat End Mill Cutter D16 with Double Straight Channel

Figure 2. Zig-zag Milling

3. Cutting force analysis

The Subdirectory was analyzed in Matlab software. The noise was filtered by chebyshav's type 2 low filter and the frequency domain spectrum were developed through Fourier transform tool FFT. In this paper for the convenience of the analysis, we will analyze the 1st test (Inner Cooling) and 7th test (Outer Cooling) where the cutter diameter was 12 mm and the rest were similar to this case. In additional arguments, it was not possible to put all spectral analysis in this paper. The average cutting forces were found from the wavelength in the time domain. In this analysis, we will use average forces. We have got from the experiment we don't need to compute the resultant forces because in zig-zag milling the cutter cuts the material in only one direction so for the zig-zag milling cutting force/component force F only in one direction. In certain cases, we were found these average cutting forces in the feed direction(x-direction), in certain cases, the average forces were in the cutting width direction(y-direction). In Fig.3 and Fig.4, Original Noisy Signal shows the waveform in the time domain at cutter diameter 12mm and spindle speed 540 rpm. Firstly wavelength also involve the portion where cutter was not cut the material or was not in contact with the material so we were not needed that section. We were needed to examine when the cutting was stable. So we were needed to cut that section from the graph before observation. We percevied that it's needed +5000 for inner cooling and +20004 for outer cooling to eliminate the unstable portion. Fig.3, Noisy signal indicate the forces in the feed direction after implementing the coding of uncut material at cutter diameter 12mm, spindle speed 540 rpm, Inner Cooling Case. Fig.4, Noisy signal graph shows that forces in the cutting width direction after implementing the coding of uncut material at cutter diameter 12mm, spindle speed 540 rpm, Outer Cooling Case.

Fig.3 and Fig.4, Nosiy signal shows that the wavelength contains vibration/noise. Noise is a very important issue during machining because noise also effects the cutting forces. The frequency of the electromagnetic noise is 50 Hz.

May be this vibration caused by the eccentricity. Eccentricity effects the tool dimension and accuracy. Eccentricity maybe in spindle, cutting tool or tool holder. To resolve this problem we were designed a filter. We were designed a low pass chebyshav's type 2 filter. As for design chebyshev's filter, we needed 4 parameters. 1. Low-Pass Response 2.cut off frequency 3. A number of the filter 4. Band pass ripple. To design the low filter, stopband edge frequency/Cut off frequency should be two times the tool frequency. Tool frequency was observed from the amplitude spectrum.

Fig.3 and Fig.4, Second wavelength shows the filtered waveform in the time domain. We were needed to calculate the cutting forces. So we will enlarge the picture to see the cutter 4 teeth's and wavelength height of 4 teeth's. We were taken any 10 revolutions of the cutting tool and then noted down the height by the aim to find the cutting forces. Similarly, we calculated the tool cutting forces for remaining all reading. Since we have 4 tooth milling cutter, can easily observe in the Fig.3 and Fig.4, 4 tooth wavelength.

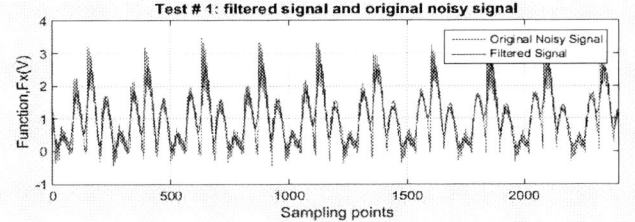

Figure 3. Test # 1 (Inner Cooling Case, d=12mm and Spindle Speed 540 rpm)

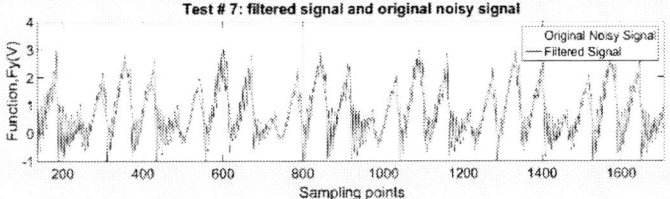

Figure 4. Test # 7 (Outer Cooling Case, d=12mm and Spindle Speed 540 rpm)

Table 2. Results

Speed (rpm)	540			1080		
Cutter Dia. (mm)	12	16	20	12	16	20
Force for Inner Cooling (V)	1.5627	1.0216	0.3063	1.8283	0.9137	0.4260
Force for Outer Cooling (V)	1.5696	1.1017	0.7258	1.8909	1.3488	0.6622

4. Conclusions

For zig-zag milling, the performance of the cutter plays an important role. We can observe from results at 12mm, 16mm, 20mm milling cutter diameter and cutting speeds 540 rpm, 1080 rpm respectively. By increasing cutter diameter inner cooling has been conquering than outer cooling for flat end milling cutter with the double straight channel. So inner cooling is better than outer cooling, especially for the bigger tool.

It was very dominant to find out the performance of an inner cooling tool that newly developed. For this purpose, we conducted different tests and the results concluded that the performance of the newly developed inner cooling tool is very good.

It is also noted that if the spindle speed is not great and other parameters are same then average cutting force are near to each other.

Acknowledgments

This research was supported by the National Natural Science Foundations of China (Grant Number 51375099, 51375100).

References

[1] Zhang, C., Zhang, S., Yan, X., & Zhang, Q, Int. J. Adv. Manuf. Technol., **83**, 975–984, (2016).

[2] Islam, A. K., Mia, M., & Dhar, N. R, Int. J. Adv. Manuf. Technol.,, **90**,11–20, (2017).

[3] Ferri, C., Minton, T., Bin, S., Ghani, C., & Cheng, K, Proceedings of the IMechE., (2013).

[4] Wan, M., Zhang, W. H., Dang, J. W., & Yang, Appl. Math. Model., **34**, 823–836, (2010).

[5] Wan, M., Zhang, W. H., Qin, G. H., Tan, G., Dang, J. W., & Yang, Int. J. of Mach. Tools & Manuf., **47**, 1767–1776, (2007).

[6] Min Wan & Wei-HongZhang, Int. J. of Mach. Tools & Manuf., **49**, 424–432, (2009).

[7] Yun, W. S., & Cho, D. W, Int. J. Adv. Manuf. Technol., **16**, 851–858, (2000).

[8] Nowakowski, L., Skrzyniarz, M., & Miko, E, EPJ Web of Conferences, **143**, 02083, (2017).

[9] Ramachandran, K., Yeesvaran, B., Kadirgama, K., Ramasamy, D., Che Ghani, S. A., & Anamalai, K, MATEC Web of Conferences, **90**, 01015, (2017).

[10] P.S. Sreejith, B.K.A. Ngoi, J. of mat. Pro. technol., **101**, 287-291, (2000).

[11] Canter, N, Tribol. & Lubricat. Technol., 40-44, (2009).

[12] Liu, C., Wu, J. Q., Zheng, W. F., Li, G. H., & Tan, G, Appl. Mech. and Mat., **33**, 408–412, (2010).

[13] Beer, N., Özkaya, E., & Biermann, D. Procedia CIRP, **24**, 49–55, (2014).

ISPECE   IOP Publishing

IOP Conf. Series: Journal of Physics: Conf. Series **1187** (2019) 032057   doi:10.1088/1742-6596/1187/3/032057

Study on vibration reduction of crane monitoring system

Wang Jian[1], Zhang Yong Kui[2] and Yan Xin[3]

[1]Safety and Production Dept, State Power Investment Co. Ltd Tibet Branch, Lhasa 850000, Tibet China

[2]Planning and Development Dept, State Power Investment Co. Ltd Tibet Branch

[3]Electrical Manager, Safety and Production Dept, State Power Investment Co. Ltd Tibet Branch

[a]Wang Jian: 350273761@qq.com

Abstract. Based on the analysis of the vibration factors of the hoist and the vibration control theory, the BS composite anti-vibration suspension structure based on the hoist monitoring system is designed. The vibration damping test of the designed structure is carried out. The experimental results show that the anti-vibration structure designed in this paper is universal and has good effect on vibration isolation.

1. Introduction

The monitoring system of medium and large hoist is the key guarantee for stable and efficient operation of equipment. However, the phenomenon of shaking and vibration is inevitably accompanied by the daily operation of the crane, so that the image collected by the monitor on the hanger is blurred and distorted by the uncertain vibration, which greatly affects the work efficiency and the safety of the work. Therefore, optimizing the link structure between the monitor and the hanger is of great significance for ensuring the stability and efficiency of the monitor and improving the monitoring level of the equipment.

Since 1965, Stewart'sp [1] earliest article has solved the problem of motion coupling using the 6-DOF active vibration isolation device developed by Stewart. And researchers achieved certain results in the field of vibration isolation control. W.S.M. Lau, K.H. Low [2] et al. analyzed the movement of the suspension mass attached to the crane and studied the vibration of the crane hanger. Sun Wei[3] et al. used the finite element software ANSYS to build a model of the damping spring anti-vibration hammer-transmission line, and used the breeze excitation force to load it, and optimized the structural parameters of the damping anti-vibration hammer. Zhao Pengrui [4] applied the TRIZ theory to analyze the vibration generated by the agricultural locomotive seat, innovated the seat suspension structure design, and simulated and analyzed the vibration reduction design of the agricultural locomotive seat the dynamic analysis software ADAMS Innovative results have been achieved in the design of vibration reduction for agricultural locomotive seats. Li Jingchun [5] et al. used the finite element analysis software to analyze the aircraft cabin of the aircraft. Without changing aircraft gun bay main dimensions, the researches innovatively designed the damping vibration damping device applied to the structure. Starting from the vibration source, the vibration damping effect of the aircraft gun cabin structure was obtained. The dynamic response stress value was reduced to about 70% of the original value, and a great achievement was achieved in the structural dynamics of the aircraft.

Content from this work may be used under the terms of the Creative Commons Attribution 3.0 licence. Any further distribution of this work must maintain attribution to the author(s) and the title of the work, journal citation and DOI.

Published under licence by IOP Publishing Ltd

The vibration damping structure adopted in the above research results is mainly based on the supported vibration damping structure. Based on the vibration control principle and the application of crane in actual engineering, this paper develops a suspension damping structure based on the crane monitoring system. A number of experimental data show that the suspension-type anti-vibration structure developed in this paper has good vibration-damping and anti-vibration effect, and it is ideal in the application of the actual crane monitoring system.

2. Vibration characteristics of medium and large hoist

Medium and large hoisting machines are mainly composed of a number of complex mechanical structures. Most of the vibrations are nonlinear vibrations caused by the power system and the transmission system, and are also multi-degree-of-freedom system vibrations. The actions of the large hoist include the change of the pitch angle of the elephant nose frame and the rotation of the fuselage, and the movement of the nose frame, the hanger, and the hook. When the lifting machine is rotating or the boom is telescopic, the hanger mainly swings in the horizontal plane; when it is carrying out the lifting operation, the hanger mainly generates vibration in the vertical direction, and multiple generalized coordinates are required at any instant to fully determine their position. Therefore, the vibration of a large hoist is a combination of various vibration types, and the vibration law is difficult to describe, and the vibration is complicated. The monitoring device is suspended and mounted on the hanger of the hoist through the vibration damping structure. The vibration damping device camera is actually installed in the crane as shown in Fig. 1.

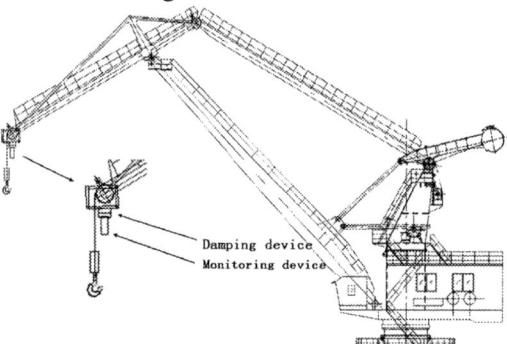

Fig.1 Damping structure installation position

3. Design and analysis of vibration damping device structure

3.1 Vibration isolation design principle

Vibration isolation is used to prevent vibration of the monitoring system of the lifting equipment. Vibration isolation, referred to as vibration isolation, according to the direction of transmission, vibration isolation measures can be divided into two categories [6]:

①Active vibration isolation: to reduce the vibration caused by the disturbance of the object. The purpose is to isolate the source of vibration, the machine itself that is the source of vibration, and to isolate it from the entire foundation in order to reduce its impact on surrounding equipment.

② Passive Vibration Isolation: to reduce the vibration caused by the movement of the pedestal. The goal is to isolate the response, that is, to isolate the entire foundation from the precision instruments and machinery that allow the vibration to be small, in order to influence the surrounding source.

Obviously, passive vibration isolation measures should be used in this project to isolate the vibration of the rack and prevent the vibration of the monitoring equipment installed on the rack.

For the monitoring device installed on the hanger of the medium and large hoist crane, it is subjected to the exciting force in several directions, so the vibration isolation design needs to be based on the multi-degree of freedom system. The vibration isolation device and the base are simplified into a rigid body of mass, and the origin is the coordinate of the center of gravity. The three central principal axes of the rigid body are the x-axis, the y-axis, and the z-axis, as shown in Figure 2. The installation of the vibration isolating device is usually symmetrical in the xoz and yoz planes, and the arrangement of all the vibration isolating devices on the same is the same.

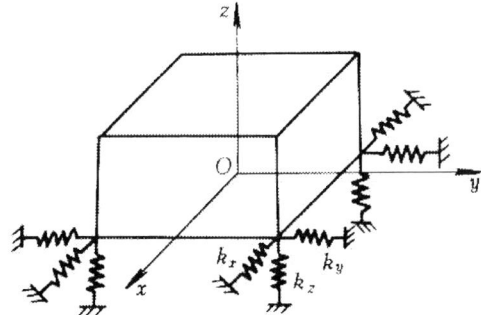

Fig.2 Multi-degree of freedom vibration isolation system

3.2 Damping structure design

Combined with the location and use requirements of the monitoring system installation, the system of Fig.2 can adopt the multi-dimensional form of vibration isolation, damping and vibration reduction to adopt the vibration isolation and vibration isolation design. In this paper, BS (Buffer-Spring) combined anti-vibration suspension structure is proposed. The specific combination of spring, hydraulic buffer and ball hinge is used to isolate the vibration generated by the hanger during the working. The specific structure is shown in Fig.3:

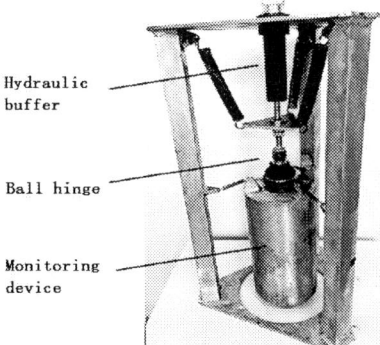

Fig.3 BS combination anti-vibration suspension structure

The main advantages of this structure are as follows:

①The combination of springs, hydraulic shock absorbers and ball hinges is highly feasible and can effectively reduce vibrations in six main directional dimensions.

②The main structure is connected by a ball hinge, which occupies a small installation space.

③Reasonable use of the spring-symmetric structure makes the main structure balanced in three directions and is easy to maintain steady state.

4. Experimental study on vibration of vibration-damping structure

4.1 Experimental equipment and installation plan

In this paper, the QLVC-ZSA1 vibration signal analyzer is selected for the vibration test, as shown in Fig.4:

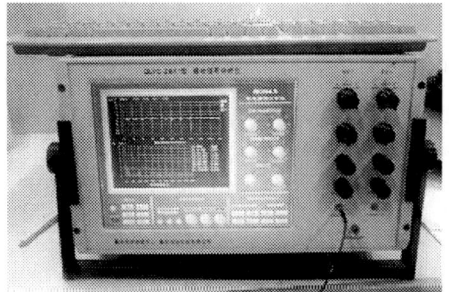

Fig.4 QLVC-ZSA1 vibration signal analyzer

The sensor of the QLVC-ZSA1 vibration signal analyzer is a piezoelectric acceleration sensor whose main principle is the piezoelectric effect: when a force proportional to the vibration acceleration acts on the sensitive core, the surface of the piezoelectric material will generate this pressure. Piezoelectric accelerometers are widely used because of their simple structure, light weight, small size and long service life [7]. The installation and connection of the test equipment are mainly divided into the following steps.

(1) Operators lower the hanger of the hydraulic lift truck to the appropriate height.

(2) Testers install the BS combined anti-vibration suspension structure on the hanger of the hydraulic lifting transport vehicle as required.

(3) Testers fix the sensor on the analyzer channel A to the hanger and connect the signal output to the No. 1 end of the analyzer.

(4) Testers fix the sensor on the analyzer channel B to the hanger, and connect the signal output to the No. 2 end of the analyzer.

The vibration isolation test of the vibration isolation device designed and prototyped in this paper is shown in Fig.5:

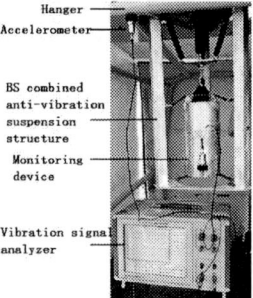

Fig.5 Test equipment and connection

In order to vibrate the entire hanger, it is necessary to excite the cantilever connected to the hanger during the test. At this time, the amplitude of the vibration generated by the monitor fixed to the base is almost the same as the frequency and the vibration generated by the hanger. Therefore, the vibration characteristics of the monitor without the vibration-damping structure on the hanger can be reflected by the vibration characteristics of the hanger itself. Under the premise of ensuring the feasibility and reliability of the test, the percussive method is used to sequentially sample the vibration acceleration signal in several vertical excitation experiments in the vertical direction and the horizontal direction. The sampling frequency is 4000 Hz, and the sampling duration is 2.20 s.

4.2 Test results time-domain analysis

The vibration acceleration curve of the hanger and the monitor under continuous vertical excitation is shown in Fig. 6. It can be seen that the vibration acceleration of the hanger fluctuates greatly under the force condition, the maximum value is 1937.14 mm/s^2, and the effective value is 294.20 mm/s^2; After the combination of the anti-vibration structure of the BS, the vibration acceleration signal of the monitor tends to be gentle, with a maximum value of 104.10 mm/s^2 and an effective value of 59.62

mm/s². The vibration in the vertical direction is efficiently isolated by the combined structure. In the same way, under the continuous horizontal excitation, the vibration acceleration signal of the hanger fluctuates greatly, the maximum value is 463.03 mm/s², and the effective value is 38.02 mm/s²; after the BS combined anti-vibration structure, the vibration acceleration of the monitor. The signal fluctuations also tend to be flat, with a maximum value of 67.98 mm/s² and an effective value of 7.52 mm/s². The signal fluctuations also tend to be flat, with a maximum value of 67.98 mm/s² and an effective value of 7.52 mm/s². It can be seen that the vibration in the horizontal direction can also be effectively isolated.

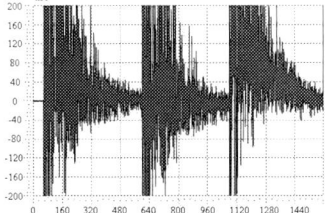

(a)Vibration acceleration of the hanger

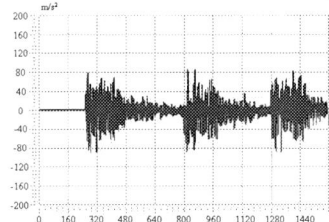

(b)Vibration acceleration of the monitor

Fig.6 Acceleration signal of the device under continuous vertical excitation

4.3 Frequency domain analysis

The amplitude spectrum of the vibration signal of the hanger and monitor under continuous horizontal excitation is shown in Fig.7.

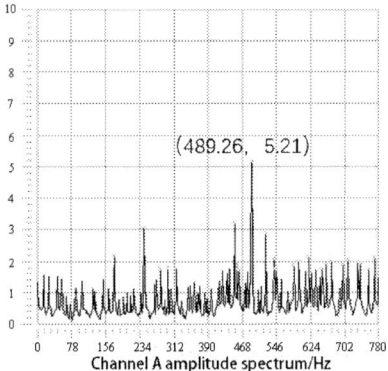

(a) Vibration spectrum of the hanger

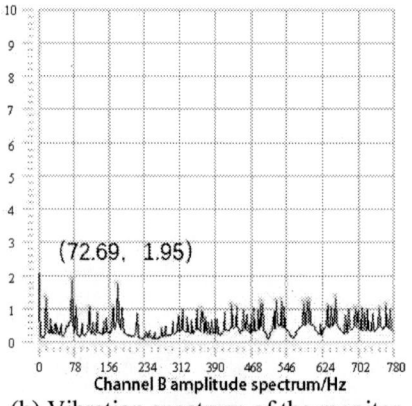

(b) Vibration spectrum of the monitor

Fig.7 Amplitude spectrum of the device under continuous horizontal excitation (a, b)

The maximum spectral value of the pylon is 5.21, and its corresponding vibration frequency is 489.26 Hz. The maximum spectral value of the monitor is 1.95, and its corresponding vibration frequency is 72.69 Hz. The results show that the frequency corresponding to the peak vibration of the two is quite different. Combined with the spectrogram under continuous horizontal excitation, the frequency corresponding to the peak vibration of the hanger under vertical excitation is mainly concentrated at about 520 Hz. The peak corresponding to the peak vibration of the monitor under vertical excitation is about 175 Hz. Therefore, when the amplitude and the vibration frequency are both large, the frequency corresponding to the lifting device hanger is between 400-600 Hz, and the corresponding frequency of the camera mounted on the vibration-damping structure is between 72-175 Hz, both of which occur. The frequency corresponding to the peak vibration and the natural frequency differ greatly, so the two basically do not resonate. Under the experimental conditions, the BS combined anti-vibration suspension structure can effectively reduce the vibration in the frequency range where the hanger generates vibration, which proves the rationality of the structure again.

In addition, due to the limitations of the test conditions, we only use the QLVC-ZSA1 vibration signal analyzer and its own piezoelectric sensor as the test and analysis instrument; taking into account the test safety and other factors, this test uses the height adjustable hydraulic lift. The transport vehicle and its hangers replace the large lifting equipment and its hangers, which is still different from the actual working conditions of the equipment.

5. Conclusion

With the popularization of medium and large lifting equipment, anti-vibration and vibration reduction technology has been paid more and more attention by engineers. The current damping structure is mostly supported vibration damping structure. Based on the actual working environment and vibration control principle of the crane, this paper designs a BS combined anti-vibration suspension damping structure based on the crane monitoring system and conducts corresponding vibration analysis test on the structural sample. Based on the actual working environment and vibration control principle of the crane, this paper designs a BS combined anti-vibration suspension damping structure based on the crane monitoring system and conducts corresponding vibration analysis test on the structural sample.

References

[1] STEWART D. A platform with six degrees of freedom [J]. Proceedings of the Institution of Mechanical Engineers, 1965,180(1): 371-386.

[2] W.S.M. Lau, K.H. Low, "Motion analysis of a suspended mass attached to a crane", Computers & Structures, Vol. 52, pp. 169–178, 1993.

[3] K.W, Hu kunzhi, Study on the Vibration Suppression Performance of Damping Spring Damper [J]. Shaanxi Electric Power,2013,17(6): 40-42

[4] Zhao pengrui. Vibration damping structure design of agricultural vehicle seat based on TRIZ theory [J]. Journal of Chinese Agricultural Mechanization, 2014,35(4): 161-164

[5] Li jingchun, Chen zhongming, He lianzhong. Structure design and study of the plane gun bay's vibration reduction [J]. JOURNAL OF SHENYANG INSTITUTE OF AERONAUTICAL ENGINEERING, 2003,20(2): 5-8

[6] Wu Tianhang, Hua Hong Xing. Mechanical vibration [M]. Beijing: Tsinghua University press, 2014

[7] Liu Pan, Model turbine and aqueduct vibration test and analysis [D]. Baoding: Hebei University of Engineering.2013

ISPECE   IOP Publishing

Research on Intelligent Communication System for Circuit Breaker Condition Monitoring

Qinghong Deng[1], Hao Zhang[2], Minfu Liao[2,a], Haoxue Zhang[2], Yifan Fu[2] and Lujie Gai[3]

[1]State Grid Hunan Electric Power Co., Ltd. Construction Branch, 410000 Changsha, China

[2]School of Electrical Engineering Dalian University of Technology, 116024 Dalian, China

[3]State Grid Dalian Power Electric Supply Company, 116024 Dalian, China.

[a] Corresponding author: LMF@dlut.edu.cn

Abstract. Based on the research of vacuum circuit breaker condition monitoring, a circuit breaker state monitoring intelligent communication system using ZigBee+SWM61850 communication configuration is proposed. Under the IEC61850 modelling specification, the state monitoring intelligent electronic device (IED) model is constructed. Combined with the experimental system, data read and write service and message transmission operation of the insulating gas monitoring logic node are simulated. At the same time, the communication system can avoid signal interference in complex electromagnetic environment, reduce data transmission delay, and share real-time monitoring data to multiple functional nodes. These features are beneficial for the improvement of equipment status warning, protection actions, and comprehensive performance evaluation. The communication system integrates the control command transmission with the on-site measurement mode. Simultaneously, the system uses the monitoring data as a constraint for implementing the adjustment operation mode. And the operational stability of the power equipment can be further improved.

1. Introduction

The normal operation of the circuit breaker is related to the overall stable operation of the power system. [1] With the in-depth promotion of intelligent substations, online monitoring technology for primary equipment status such as transformers and GIS switches in the station has become increasingly mature and stable[2,3].

This paper designs an integrated functional IED model for protection and measurement on integrated circuit breaker. Based on the synchronous sampling technology [4-6], data such as current and voltage are collected by sensors, and the switch position signal is collected by auxiliary contact. After analysing and processing the collected signals, real-time circuit breaker characteristic state monitoring and early warning state evaluation are realized. The focus of this paper is to combine the IEC61850 communication protocol and the Internet of Things (IoT) communication method to construct an intelligent communication system for condition monitoring sampling data sharing and comprehensive evaluation of equipment performance.

Content from this work may be used under the terms of the Creative Commons Attribution 3.0 licence. Any further distribution of this work must maintain attribution to the author(s) and the title of the work, journal citation and DOI.
Published under licence by IOP Publishing Ltd

The goal of this design is to reduce the data transmission interference in complex electromagnetic environment to provide real-time monitoring information for system operation, and to use real-time monitoring data as an important parameter of system operation mode and protection action.

2. Data acquisition system

The vacuum circuit breaker condition monitoring experimental system is shown in Figure 1.The experimental system mainly consists of signal acquisition and processing system, power supply system, chamber and vacuum maintenance system. The voltage on coupling capacitor 6 is handled with the signal processing circuit including a power amplifier, low pass filter and some circuits to improve the loading capacity. The voltage signal is transferred to the processor to calculate effective voltage.

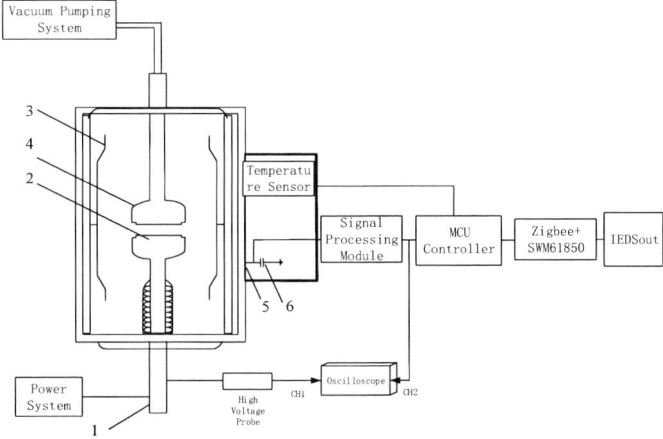

Figure 1. Diagram of condition monitoring experiment

This design uses a coupled capacitive sensor to measure the voltage signal and send the data to the IEDSout by the ZigBee+SWM61850 communication module. The measured data is also converted into a vacuum pressure signal by circuit processing and program analysis, which can implement functions such as circuit breaker insulation gas alarm and protection action.

3. Communication system

The condition monitoring of vacuum circuit breakers is also an important task in the monitoring of the substation equipment. Combined with substation IEC61850 communication system, the construction of an all-round condition monitoring system for vacuum circuit breakers will be indispensable for future state monitoring of substation process layer equipment.

In order to adapt to the complex working environment of the circuit breakers, the current vacuum detection system mainly has two communication modes, the ZigBee wireless communication based on the Internet of Things (IoT), and the scalable intelligent substation system IEC61850 communication.

ZigBee communication has the advantages of low power consumption, various connection forms, relatively long distance, short delay, high capacity, and strong anti-interference ability. The vacuum degree detection in the project is to transmit data at a relatively close distance. The circuit breaker works in a complex electromagnetic environment. ZigBee communication does not require routing and additional power supply, and is not affected by environmental interference. ZigBee communication module consists of CC253X, CC2591 and circuit components.

In order to adapt to the intelligentization of substation equipment, circuit breakers intelligent terminal that fully supports substation digitalization is configured. The entire substation is uniformly modeled under the IEC61850 communication system to improve the intelligent level of power system equipment condition monitoring, data maintenance, fault diagnosis and system integration. In this design, the circuit breaker status monitoring function is classified to construct the intelligent circuit breaker communication model under the IEC61850 standard. The SWM61850 is installed to provide an Ethernet interface for the IEC61850-8-1 communication protocol. The communication module can communicate with the digital substation local monitoring interface and supports remote services such as Telnet and FTP. The vacuum degree message data transmission between the data model and the substation communication server discriminates the vacuum state in real time. The substation communication model based on the IEC61850 is shown in Figure 2.

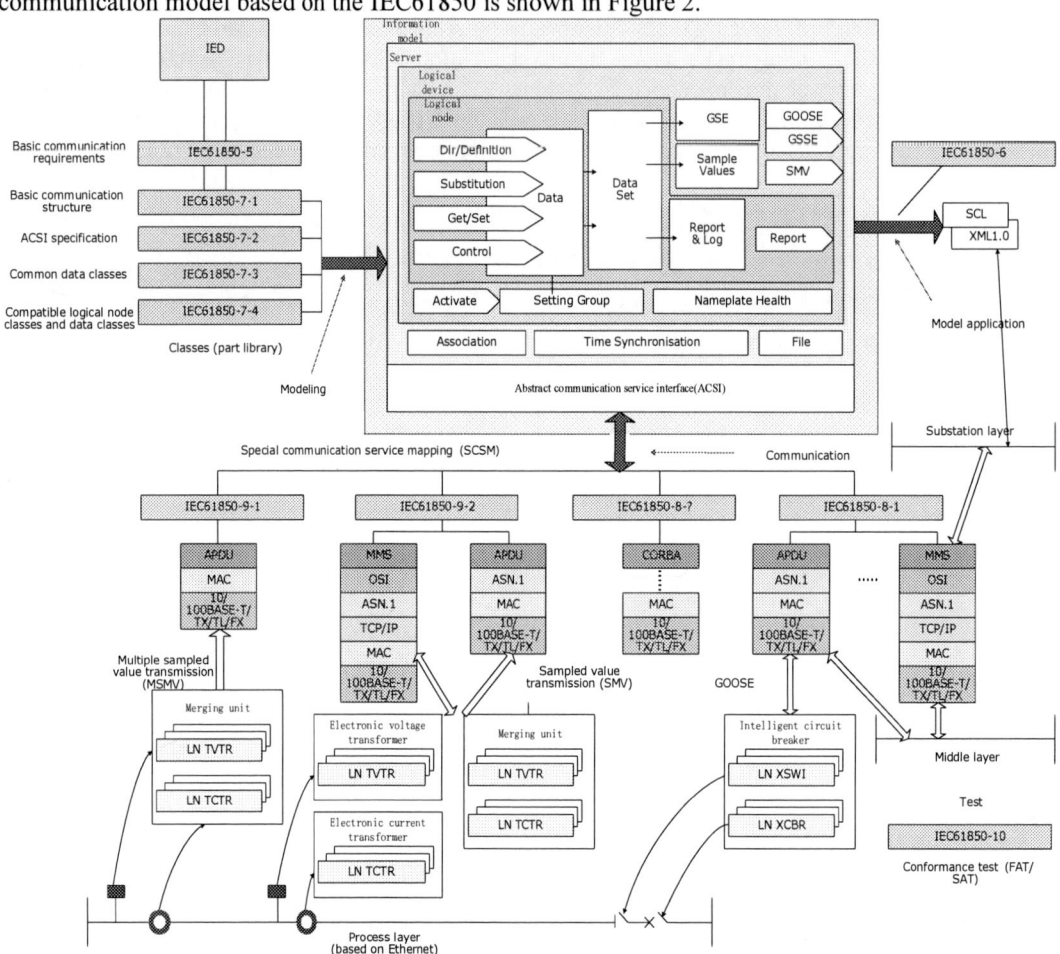

Figure 2. IEC61850 communication system application

In the experimental system, the vacuum pressure is measured in an indirect manner to characterize the state of the insulating gas. We can also monitor other state parameters. In order to meet the substation communication requirements, the intelligent circuit breaker state monitoring IED model is written in Structured Control Language (SCL). The model function structure is shown in Figure 3.The model configures various protection functions of the circuit breaker in logic device PROT. Logic device MEAS performs parameter measurement and state judgment on voltage parameters, current parameters, power factor, arc state, contact wear, gas insulation and other factors in circuit breaker operation. Logic device CTRL configures the interlock function of the circuit breaker, the switch control, the phase selection operation, the fixed point closing and closing functions. Logical device PIGO summarizes the alarm information, the operation command.

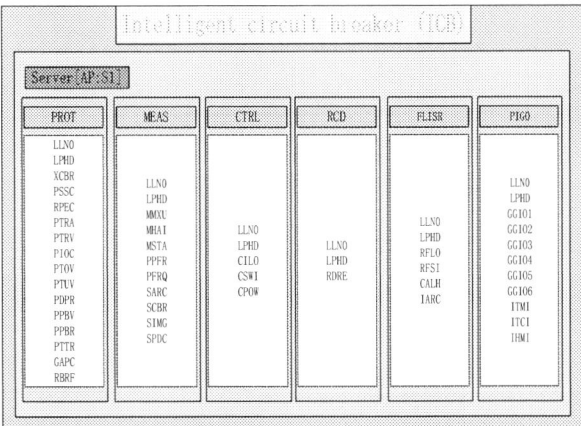

Figure 3. Intelligent circuit breaker function model

The configuration interface of the circuit breaker status monitoring model in the IEDSout software is shown in Figure 4. The IED model includes various functional logic devices-logical node configuration, parameter settings for communication services, and message read-write operations.

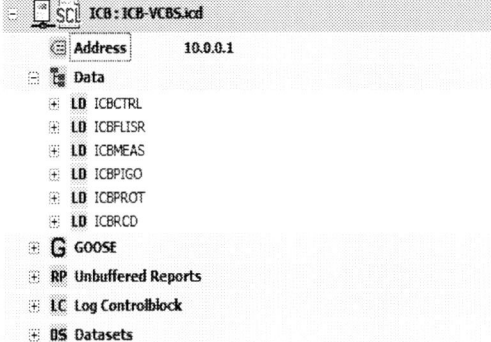

Figure 4. Intelligent circuit breaker IED model

4. Data communication analysis

The built-in protocol conversion SWM61850 can be directly embedded into the terminal intelligent device in the form of a module, and the IEC61850 protocol service can be directly provided externally. In this design, the module will complete the message communication between the IED and the system server. For the vacuum degree monitoring data, the sampled measured value (SMV) message is used to send the data to the IEDSout software circuit breaker state monitoring IED logic node SIMG to determine the current circuit breaker vacuum state. The communication process of data transmission can be analyzed by Ethereal software. For voltage data, SMV is used for sampled data transmission. SMV messages are continuously sent with real-time data collection. According to the range of communication transmission data values, the switch selection control flow will be triggered under the current communication model, and the corresponding mode command is executed to change the switch operation mode. The switch selection mode operation flow and the signal of the detection signal voltage sampling according to IEC61850-9-2 is shown in Figure 5. The signal voltage data is sent in accordance with the protocol IECSMV. According to the contents of the data packet, 0x88ba is the message type of the SMV. According to the IEC 61850-9-2 protocol data format specification, the packet sampling rate is 200 and the number of Application Service Data Unit (ASDU) is 5. The voltage sample value is 2.7V.

```
0000  ff ff ff ff ff ff 01 0c cd 04 00 01 81 00 80 00   ................
0010  88 ba 40 00 02 83 00 00 00 00 60 82 02 77 80 01   ..@.......`..w..
0020  08 a2 82 02 70 30 4c 80 0f 30 30 32 32 30 34 30   ....p0L..0022040
0030  30 38 30 31 34 31 2f 46 82 02 00 20 83 04 00 00   080141/F... ...
0040  00 01 85 01 01 87 2c 00 00 00 00 00 00 00 00 00   ......,.........
0050  00 00 00 00 00 00 00 00 00 00 00 00 00 00 00 00   ................
0060  00 00 00 00 00 00 00 00 00 00 00 00 00 00 00 00   ................
0070  00 00 00 00 30 4c 80 0f 30 30 32 32 30 34 30 30 38   ...0L..002204008
0080  30 31 34 31 2f 46 82 02 00 21 83 04 00 00 00 01   0141/F...!......
0090  85 01 01 87 2c 00 00 00 00 00 00 00 00 00 00 00   ....,...........
00a0  00 00 00 00 00 00 00 00 00 00 00 00 00 00 00 00   ................
00b0  00 00 00 00 00 00 00 00 00 00 00 00 00 00 00 00   ................
00c0  00 30 4c 80 0f 30 30 32 32 30 34 30 30 38 30 31   .0L..00220400801
00d0  34 31 2f 46 82 02 00 22 83 04 00 00 00 01 85 01   41/F..."........
00e0  01 87 2c 00 00 00 00 00 00 00 00 00 00 00 00 00   ..,.............
00f0  00 00 00 00 00 00 00 00 00 00 00 00 00 00 00 00   ................
0100  00 00 00 00 00 00 00 00 00 00 00 00 00 00 30 30   ..............00
0110  4c 80 0f 30 30 32 32 30 34 30 30 38 30 31 34 31   L..0022040080141
0120  2f 46 82 02 00 23 83 04 00 00 00 01 85 01 01 87   /F...#..........
```

Figure 5. Detect voltage sampling message data

In addition to monitoring the pressure of the insulating gas, we can also comprehensively analyze the current insulation properties of the gas based on information such as density and temperature. According to the range of communication transmission data values, the switch selection control flow will be triggered under the current communication model, and the corresponding mode command is executed to change the switch operation mode. The switch selection mode operation flow is shown in Figure 6.

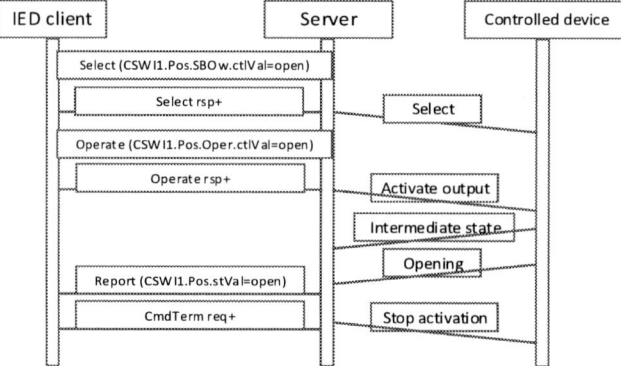

Figure 6. Switch control flow

The SMV packet is a periodic fast packet, which is generally a sampling message sent by the merging unit to the protection and monitoring device. According to the logical node configuration, the vacuum state is divided into four operation modes, mode 1 is safe operation state, mode 2 is the critical pressure state, mode 3 is lock operation and alarm, and mode 4 is device isolation trip. The SIMG vacuum monitoring node obtains the data information in node data transmission process as shown in Table 1.

Table 1. SIMG node data information

LN: SIMG			
		Mod	{{0},{1},{2},{3}}
	CF	CbBeh1	{{0,1.33e-02}, " safe operation"}
		CbBeh2	{{1.33e-02,6.6e-02}, " Critical operation"}
		CbBeh3	{{6.6e-02,9.9e-02}, " Close operation and alarm"}
		CbBeh4	{{9.9e-02,1.01e+05}, " Isolate the device and trip"}
FC	MX	Pres	{{9.16e-04},{2.1426e-03},{3.4268e-02},{6.4785e-2}}
		Tmp	{20.12}
		Den	{1.18e-31 }
	ST	InsAlm	{1.33e-02}
		InsBlk	{6.6e-02}
		InsTr	{9.9e-02}
		PresAlm	{6.6e-02}

According to the function node data information, it can be known from Tab. 1 that when the vacuum pressure is in the interval (0, 1.33e-02), the system executes mode 1; when the vacuum pressure is in the interval (1.33e-02, 6.6e-02), the system executes mode 2; when the vacuum pressure is in the interval (6.6e-02, 9.9e-02), the system executes mode 3; when the vacuum pressure is in the interval (9.9e-02, 1.01e+05), the system executes mode 4 . In addition to monitoring the pressure of the insulating gas, we can also comprehensively analyze the current insulation properties of the gas based on information such as density and temperature. The node data pointed to by DATA-SET will update the function node data according to the reporting period to meet the requirements of real-time monitoring and be associated with the relevant protection function nodes.

5. Conclusion

This paper designs and applies ZigBee+SWM61850 intelligent communication system in the condition monitoring of vacuum circuit breakers. The message transmission between vacuum pressure data and condition monitoring IED monitoring node SIMG is completed. ZigBee+ SWM61850 is a combination of IoT communication and intelligent substation communication system, which can shield complex electromagnetic environment interference and easily realize data sharing among multiple nodes. The intelligent communication system synergizes various functions such as state monitoring, control operation, protection coordination and data sharing, greatly improving the accuracy of data transmission and the stability of equipment operation.

This paper designs the intelligent circuit breaker IED model based on IEC61850, and divides it into six logical devices according to functional requirements. SMV message is used to periodically send sampling data. The device status information is periodically uploaded by using GOOSE message, and the operation control command is sent according to the evaluation. Combined with the vacuum pressure monitoring of vacuum circuit breakers, four protection action intervals are delineated according to the vacuum pressure, which has an important influence on the vacuum state monitoring, protection synergy and intelligent terminal function integration of vacuum circuit breakers.

References

[1] J Li, X Wu Microgrid Monitoring System Based on IEC 61850 High Power Converter Technology J. **21**,26 (2012)

[2] S Hans, G Georges Compact High-voltage Vacuum Circuit Breaker a Feasibility Study IEEE transactions on Dielectrics and Electrical Insulation J. **14**,613 (2007)

[3] X Duan, Z Zhao, J Zou Partial discharge research for the internal pressure of vacuum Interrupters online condition monitoring High Voltage Apparatus J. **8**,30 (2000)

[4] F Frontzek, D Konig Measurement of emission currents immediately after arc polishing of contacts-Method for internal-pressure diagnostics of vacuum interrupters IEEE Transactions on Electrical Insulation J. **28**,700 (1993)

[5] Z Yin, W Liu, Q Yang, Y Qin, D Lin Modeling and Mapping Implementation of A Sampled Value Model Based on IEC 61850 Automation of Electric Power System J. **28**,38 (2004)

[6] G Han, B Xu Modeling of intelligent distribution terminal based on IEC61850 Automation of electric power systems J. **31**,104 (2011)

Research on Energy Saving and Consumption Reduction Technology of Underground Gas Storage Compressor

Guan Tong[1], Liu Guiqiang[1], Wang Jinxiu[1], Wang Ping[2], Sun Dandan[1], Gao Shan[1], Liu Pai[1], Wu Qiang[1]

[1]Engineering Technology Research Institute of Huabei Oilfield Company, Renqiu, China

[2]Underground Gas Store Management Agency of Huabei Oilfield Company, Renqiu, China

cyy_gt@petrochina.com.cn, cyy_liugq@petrochina.com.cn, cyy_wjx1@ petrochina.com.cn ,cy4_wp2@petrochina.com.cn, cyy_sundd@petrochina.com.cn, cyy_gaos@petrochina.com.cn, cyy_liup@petrochina.com.cn, cyy_wq@petrochina.com.cn

* Corresponding Author: Guan Tong; email: cyy_gt@petrochina.com.cn; phone:18713719209

Abstract: Suqiao Gas Storage Group is a seasonal peaking gas storage in North China, where energy consumption cost of the compressor gas injection is more than 30% of the operation cost of the gas storage. In the Suqiao Gas Storage, the main energy consumption equipment is the electrically driven compressor. Aimed at the problem of high energy consumption of compressor, the influencing factors of compressor energy consumption are found out by analyzing the formula of compressor indicated power. According to the gas injection status of the gas storage, the main influencing factors are the inlet pressure, outlet pressure, clearance and compression ratio of the compressor. The relationship between power consumption and various influencing factors is obtained by testing the power consumption of compressor and the main influencing factors. Based on the above analysis results, the optimization model of compressor energy consumption is established, the compressor energy optimization software was compiled by C#, and the calculation results of the software were applied on site, which reduced the energy consumption of the on-site compressor, saved the gas production and operation cost of the gas storage, and provided the basis for scientific storage of gas storage.

1. Introduction

Underground gas storage plays an irreplaceable role in seasonal peak shaving and guaranteeing the safety of gas supply. In recent years, it has been developed vigorously in China. Suqiao gas storage group belongs to the underground gas storage group of seasonal peak shaving type for guaranteeing the use of gas in North China. The gas storage group uses electric drive reciprocating compressor for gas injection, the compressor gas injection energy consumption is the main energy consumption equipment of gas storage, and its electricity consumption cost accounts for more than 50% of the operation cost of gas storage. How to rationally adjust compressor parameters, improve compressor utilization ratio and reduce operation cost is of great significance for reducing cost and increasing efficiency of gas storage[1-3].

Content from this work may be used under the terms of the Creative Commons Attribution 3.0 licence. Any further distribution of this work must maintain attribution to the author(s) and the title of the work, journal citation and DOI.
Published under licence by IOP Publishing Ltd

2. Compressor indicator power calculation

The index of evaluating compressor energy consumption generally refers to the indicator power of compressor, which is the indicator work consumed in unit time. The indicator work of the compressor generally refers to the total work consumed by compressor cylinder in working cycle. The indicator work is expressed in L, the indicator power is expressed in N[4-5].

The calculation formula of indicator work is:

$$L = P_s V_h \lambda_v \frac{k}{k-1} \left[\left(\frac{P_d}{P_s} \right)^{\frac{k-1}{k}} - 1 \right]$$

The calculation formula of indicator power is:

$$N = 1.634 n P_s V_h \lambda_v \frac{k}{k-1} \left[\varepsilon (1 + \sigma)^{\frac{k-1}{k}} - 1 \right]$$

$$\lambda_v = 1 - \alpha (\varepsilon^{\frac{1}{m}} - 1)$$

In the equation: P_s is inlet pressure of cylinder, kgf/cm²;

V_h is stroke volume of cylinder, m³;

λ_v is Volume coefficient, means degree of reduction in utilization of cylinder working volume;
α is relative clearance volume, means ratio of clearance space volume to total cylinder volume;
m is gas expansion index;
k is gas adiabatic exponent;
ε is pressure ratio, means ratio of cylinder exhaust pressure to intake pressure;
n is the speed of compressor;
σ is relative pressure loss,%.

3. Influence factor of compressor power

According to the indicator power formula of compressor, the main influencing factors of compressor power are gas property, compressor inlet pressure, outlet pressure, compression ratio, clearance, cylinder size and speed. according to the gas injection status of the gas storage, the cylinder stroke volume, speed and gas properties of the compressor can not be changed, only the inlet pressure, outlet pressure, clearance and compression ratio are studied.

3.1. Impact of inlet pressure

The rated power of compressor used in Suqiao Gas Storage Station is 4500KW, three-stage compression mode. The adjusting range of inlet pressure is 4.0-4.6MPa. The outlet pressure is 34 MPa fixed and the clearance is closed. The compressor power consumption and gas injection under different inlet pressure are collected through field experiments, and the influence of inlet pressure on power consumption and gas injection volume of compressor is analyzed.

As shown in Fig.1, with the increase of the compressor inlet pressure from 4.0MPa to 4.6MPa, the gas injection volume increases gradually. the gas injection volume is the largest at 4.6MPa. With the increase of inlet pressure 0.1MPa, the gas injection volume increases by 2.7%~8.6%. As shown in Fig. 2, with the increase of the inlet pressure, the power consumption increases, and with the increase of the inlet pressure 0.1MPa, the power consumption increases. 2.5%~11.4%. The increase rate slows down with the increase of the inlet pressure. Because the pressure ratio decreases with the increase of the inlet pressure, the indicating power decreases. At the same time, the indicating power increases with the increase of the inlet pressure. When the pressure ratio is greater than 1.1, the increasing power of the compressor is more than the decreasing power. Therefore, the inlet pressure increases. When the pressure ratio is less than 1.1, the inlet pressure increases and the compressor power decreases[6-7].

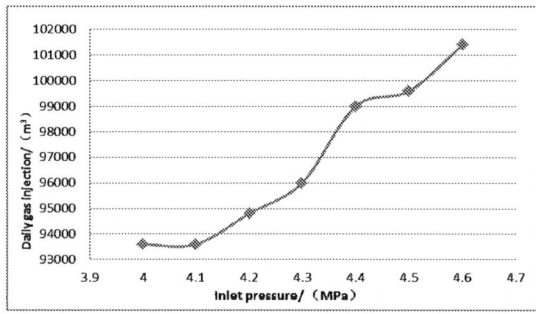

Figure 1. Relation between inlet pressure and daily gas injection of compressor

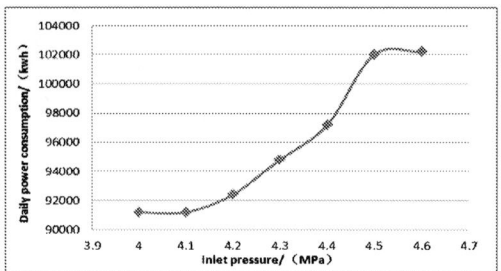

Figure 2. Relation between inlet pressure and daily power consumption of compressor

The gas injection unit consumption of the compressor is calculated by power consumption * electricity price / gas injection. As shown in Figure 3, with the increase of the inlet pressure of the compressor, the gas injection unit consumption of the compressor decreases gradually, and at 4.6MPa, the unit consumption of the compressor is the lowest.

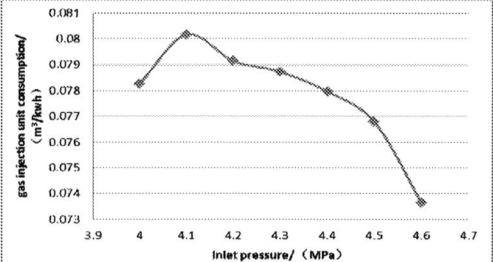

Figure 3. Relation between inlet pressure and the gas injection unit consumption of compressor

3.2. Impact of outlet pressure

According to present work situation of the compressor unit in the gas storage, keep the compressor inlet pressure 4.5MPa unchanged, change the outlet pressure from 24MPa to 35MPa, and collect the corresponding power consumption of a single compressor. As shown in Figure 4, the outlet pressure increases, the indicated power of the compressor increases, and the power consumption increases, and the outlet pressure increases by 0.1MPa each time the outlet pressure increases. Indicative power increases by 1.5%~2.3%. With the increase of outlet pressure, the compressor's air injection decreases. For every 0.1MPa increase of outlet pressure, the gas injection volume decreases by 1%~1.4%, as shown in Fig. 5. According to the relationship between power consumption, air injection and outlet pressure, the unit consumption of gas injection decreases with the increase of outlet pressure, as shown in Fig. 6.

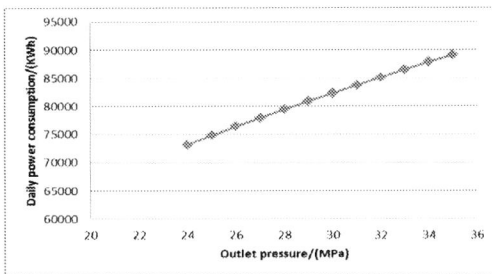

Figure 4. Relation between outlet pressure and daily power consumption of compressor

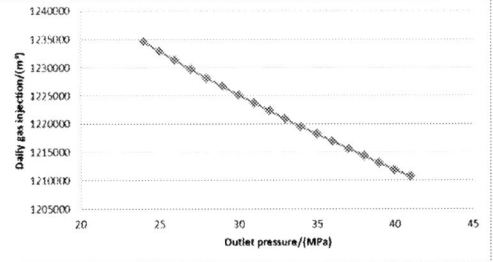

Figure 5. Relation between outlet pressure and daily gas injection of compressor

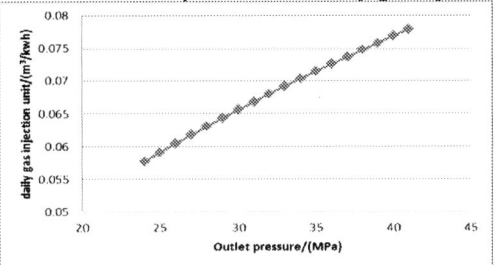

Figure 6. Relation between outlet pressure and daily gas injection unit of compressor

3.3. Impact of clearance

Compressor clearance is fully opened. Comparing with the data of clearance closed, under different inlet pressures, the volume of gas injection affected by clearance is obtained. The inlet pressure rises in the range of 4MPa to 4.6MPa, and the volume of gas injection affected by clearance keeps rising, changing from 170,000 m^3 per day to 210,000 m^3 per day. Keeping the outlet pressure unchanged and the inlet pressure changing from 4MPa to 4.6MPa, the daily power consumption of compressors with full clearance opened and full clearance closed is compared and analyzed. The average daily power consumption of compressors with full clearance opened is about 12 000 kwh/day lower than that with full clearance closed, as shown in Figure 7.

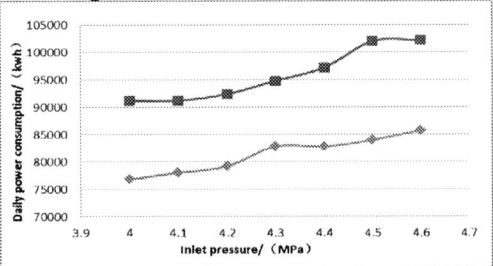

Figure 7. The daily power consumption of compressors with full clearance opened and full clearance closed

3.4. Impact of compression ratio

Through the analysis of the operation data of the compressor unit in 2017, removing the data loss, unit power outages and other accidental factors, the compression ratio and compressor power consumption trend basically, the greater the compression ratio, the greater the power consumption of the compressor power, and the compression ratio is proportional to the actual operation, and the actual situation is consistent with the theoretical formula.

4. Establishment of power consumption optimization model for compressor units and development of power consumption optimization software

By analysis the factors which is affecting the compressor energy consumption, we use compressor inlet pressure as a design variable, taking the minimum gas injection consumption of the compressor as the objective function, Establish an energy optimization model as below:

$$\min d = \frac{f(x) \cdot y}{Q_1}$$

$$st \begin{cases} Q_1 = Q_z \\ x \in [4.0, 4.6] \\ y \in [0,8], \ y \in Z \end{cases}$$

In the equation: "d" is the unit consumption of gas injection, degree/ m^3; "f(x)" is the power consumption of daily gas injection with Su 4 and Su 49 reservoir compressors, KWh; "x" is the inlet pressure, and the value is between 4.0 and 4.6 MPa; "Q1" is the daily gas injection of Su 4 and Su 49, m^3; "y" is the number of compressors which is started in Su 4 and Su 49, taking values from 0 to 8, taking integers; "Qz" is the gas injection of daily production from Su 4 and Su 49, m^3.

Use the collected compressor power consumption, gas injection volume and inlet pressure data to fit a curve. According to the formula of compressor indicating power, the indicated power and the inlet pressure are linear, and by the fitting is obtained a formula about indicating power and inlet pressure while the clearance cleared or the full clearance.

According to the current injection, the equation of the indicated power and the inlet pressure and the compressor injection in full clearance and in fully closed clearance, solving the objective function, finally obtain the compressor working condition with the minimum unit consumption.

Using the energy consumption optimization model of the compressor unit which is already established, compile an energy optimization software by C#, through entering the daily gas injection calculate the best compressor starting scheme, including the number of compressor starting units, inlet pressure, switch of clearance and return, estimated value of the electricity and gas consumption .

5. Field application

In August, the daily gas injection of Su 4 was 2.4 million square meters. In the first half of the month, we injecting gas with experience, running two compressors, the inlet pressure is 4.4 MPa, the clearance is fully closed, the total power consumption is 2931240kW•h; in the second half of the month, using the program which is provided by the energy optimization software, running two compressors, the inlet pressure is 4.2MPa, the clearance is fully closed, after half a month of testing, the total power consumption is 2776980kW • h, the power consumption after optimization in the second half of the month is significantly lower than before. The average optimization is 10284kW•h per day, that means we can save 308520kW•h per month. As the sheet 1 shows.

Table 1.The Power consumption of compressor before and after optimization.

Before optimization		After optimization	
Date	Power consumption（kwh/d）	Date	Power consumption（kwh/d）
20180801	200660	20180816	184800
20180802	193000	20180817	176300
20180803	190200	20180818	186500
20180804	197800	20180819	182400
20180805	193300	20180820	180200
20180806	198900	20180821	188700
20180807	192400	20180822	183200
20180808	192500	20180823	179600
20180809	195400	20180824	187480
20180810	193100	20180825	182600
20180811	197200	20180826	188300
20180812	195300	20180827	189700
20180813	196300	20180828	187600
20180814	197480	20180829	190700
20180815	197700	20180830	188900

6. The conclusion

By analyzing the indicate power formula of compressor, According to the gas injection status of the Gas Storage, we can sure the main influencing factors are the inlet pressure, outlet pressure, clearance and compression ratio of the compressor. So, after analyzing and testing the various influencing factors, the compressor energy optimization model was finally established, and as well the compressor energy optimization software was compiled. The application showed that the daily power consumption was optimized to 10284kW•h when the gas injection was 2.4 million square meters per day. That means, if it is equivalent to 30 days, it can save 308520kW.h a mouth, which just verified the calculation results' reliability of the software. Using energy optimization software can reasonably adjust the compressor parameters, improve the compressor utilization, reduce the operating costs, and it must have great significance to the cost reduction and the efficiency increasing of the Suqiao Gas Storage.

References

[1] Yao Li, Xiao Jun, Wu Qing & Jiang Xuemei. Operation &management and cost analysis of underground gas storage facilities[J]. Natural Gas Technology and Economy, 2016, 10(6):50-54.

[2] Tian Jing, Wei Huan & Wang Ying. Operation and management mode of international underground gas storage[J]. International Petroleum Economics, 2015, 23(12): 39-43.

[3] Yang Ying, Li Shibing, Chen Ziwei, Chen Jiawen, Li Peng. Optimization measures to maintain economical operation of electrically driven compressor during gas injection in underground gas storage[J]. Natural Gas Exploration and Development, 2017, 40(3): 102-106.

[4] Wang Wei, Liu Yang, Sun Wenzhong. Technical analysis of reciprocating compressor energy saving and consumption reduction[J]. Natural Gas Technology and Economy, 2014, 8(1):49-51.

[5] Wang Han, Jiang Yingying. Application of energy saving and consumption reducing technology in reciprocating compressor[J]. Technological innovation and Application, 2015, 30:75.

[6] Su Jianhua, Xu Kefang, Song Deqi. Gathering and handling of natural gas field[M]. Beijing: Petroleum Industry Press,2004.

[7] Wang Wei. Modification of compressor for long distance natural gas pipeline[J]. Plant Maintenance Engineering,2008(6):48-49

Load and Stress Distribution of Thread Pair and Analysis of Influence Factors

Shikun Lu[1,2*], Dengxin Hua[1*], Yan Li[1*], Fang yuan Cui[1], Pengyang Li[1]

[1]Faculty of Mechanical and Precision Instrument Engineering, Xi'an University of Technology, Xi'an 710048

[2]Laiwu Vocational and Technical College, Laiwu 271100

Prof. Dengxin Hua: xauthdx@163.com

Prof. Yan Li: ly-jyxy@xaut.edu.cn

Mrs. Fangyuan Cui: cfyxaut@126.com

Dr. Peng-yang LI: lipengyang@xaut.edu.cn

*Corresponding Author: Shi-kun Lu; email: lushikun@163.com; phone:18263458369.

Corresponding author: Li Yan, Ph.D, Professor and PhD supervisor.

Prof. Dengxin Hua: xauthdx@163.com

Prof. Yan Li: ly-jyxy@xaut.edu.cn

Abstract: Analyzing the axial load(axial force) distribution and the stress distribution of the root of the thread is helpful for accurately predicting the thread failure and optimizing the design of the structure of the nut and the bolt. The existing research mainly analyzes the axial load distribution and stress distribution of the thread by using the two-dimensional(2D) finite element model(FEM), and they have no theoretical support. In this paper, a three-dimensional(3D) finite element model was established, and a theoretical model of the stress calculation of the root of the thread was established. The theoretical calculation results of this paper were compared with the results of the 3D finite element model analysis. The results show that the two are in good agreement with each other, which verifies the consistency between the theoretical model and the 3D finite element model. Then, the stress of the root of the bolt and nut is investigated under the action of axial force. And the factors, such as Young's modulus E, axial total load F, friction coefficient μ, radial thickness H, engaging threads number N, thread defect which affect the load distribution are systematically studied.

1. Introduction

Threaded connections are used to connect parts into a whole [1–2], used to transmit force, torque, or movement, etc. [3] Bolt connections are widely used in various engineering fields, such as aviation, aerospace, machine tools, and precision instruments, and the accuracy of the bolt connection directly affects the quality of equipment assembly. Thread connection stiffness [4-5] also directly affects equipment assembly stiffness, and a high-quality threaded connection can effectively improve equipment grades. For these reasons, it is important to study force distribution and the factors that influence the design of a durable and high-quality thread connection.

Content from this work may be used under the terms of the Creative Commons Attribution 3.0 licence. Any further distribution of this work must maintain attribution to the author(s) and the title of the work, journal citation and DOI.
Published under licence by IOP Publishing Ltd

In order to analyze the thread stress, Yang Qiang et al. [6] established a 2D model, and performed FE analysis of the thread. On this basis, the shape of the nut thread was improved. Zhou Guanghui [7] and others also used the 2D FE analysis method to summarize the influence of the nut structure and the thread design parameters on the axial load distribution of the transmission bolt. Yin YiHui [8] and others used the FE numerical simulation method to consider the elastic-plastic nature of bolts. Kenny and Patterson [9] described a method of measuring thread strain and stress. Sopwith[10] theoretically analyzed the uneven distribution of nut load distribution and discussed a method of improving the uneven distribution of load. Ibrahim Alkan [11] analyzed the effect of preloaded dental implant screw pressure distribution. Kenny [12] reviewed studies of the distribution of load and stress in fastening threads and Miller [13] established a mathematical thread stress analysis spring model, and compared it with the FE and experimented to verify the correctness of the mathematical model. Wang and Marshek [14] proposed an improved thread analysis spring model to predict the load distribution of the thread part of the connector, compared the load distribution of the elastic and yield threaded joints, and discussed the influence of the yield curve on load distribution. In order to study the load distribution, CHEN Haiping [15] discussed the analytical, photo elasticity and 2D FE methods, and the factors, which affect the load distribution are systematically studied, and the main factors were investigated by regression analysis.

2D thread model [15] for FE analysis , is a 3D model of simplification, which ignores a lot of geometric parameters. It is clear that this analysis method can not reflect the actual.

The above study did not establish a theoretical model of the root stress of thread，and they do not also analyze, or fully analyze, the effect of the thread defect, nut diameter, nut force position, joint surface friction coefficient, thread number, and other factors on the distribution of thread load and the stress distribution. In this paper, a theoretical model of the stress calculation of the root of the thread is established, and the 3D FE model is used to study the influence of the thread defect, nut radial dimension, friction coefficient, and the influence of the number of circles on the load distribution of the bolt thread. So the research of this paper is helpful for accurately predicting the thread failure and optimizing the design of the structure of the nut and the bolt.

2. Comparisons of 3D FE Method and Analytic Method

2.1 Analytic method [10] [15-16]

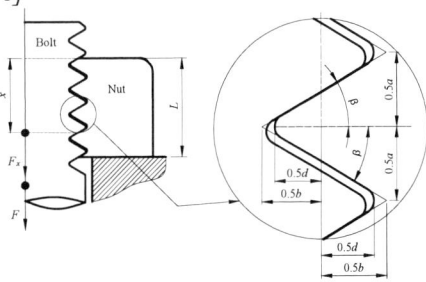

Fig.1 Thread size

As shown in the Fig. 1, the upper surface of the nut as a starting point to establish the coordinate system, the axial load F_x distribution along the x-axis is expressed(SOPWITH method)[10] as

$$F_x = F \frac{\sinh\left[(x/L)\theta_1\right]}{\sinh(\theta_1)} \tag{1}$$

F is the axial total load, and where θ_1 can be calculated from the [10].

YAMATOTO[16] also derives the formula of axial load F_x along the axial direction of any cross section of the nut.(YAMATOTO's method)

$$F_x = F \frac{\sinh(x\lambda)}{\sinh(L\lambda)} \tag{2}$$

where λ can be calculated from the Appendix.

When the value of θ_1 and λ are get, the distribution of the axial force F_x can be calculated by taking them into formulas (1) and (2) respectively. When axial force F_x can be calculated, the axial stress distribution can be calculated. The average stress distribution in the axial direction of the thread is given by the following[17]

$$\overline{\sigma}_{\text{A-stress}}(x) = \frac{F_x}{A_{rcsa}} \tag{3}$$

Where, A_{rcsa} is the radial root cross-sectional area (effective cross-sectional area). So, the average axial stress distribution of bolts and nuts can be expressed as

$$\begin{cases} \overline{\sigma}_{\text{b-a-stress}}(x) = \dfrac{F_x}{A_{bcs-area}} \\[4mm] \overline{\sigma}_{\text{n-a-stress}}(x) = \dfrac{F_x}{A_{ncs-area}} \end{cases} \tag{4}$$

where, $A_{bcs-area}$ is the radial cross-sectional area of the root of the bolt (effective cross-sectional area), and $A_{ncs-area}$ is the root radial cross-sectional area (effective cross-sectional area) of the nut.

Suppose the stress concentration of the maximum stress σ_{max}, and the reference stress σ_n ratio is defined as the stress concentration factor, that is, when the stress is tensile stress, compressive stress, bending stress, the stress concentration factor K_σ can be expressed[17-19]

$$K_\sigma = \frac{\sigma_{max}}{\sigma_n} \tag{5}$$

Therefore, the bolt, nut thread root stress can be expressed as

$$\begin{cases} \sigma_{brs}(x) = K_\sigma \sigma_{\text{b-a-stress}}(x) = \dfrac{K_\sigma F_x}{A_{bcs-area}} \\[4mm] \sigma_{nrs}(x) = K_\sigma \sigma_{\text{n-a-stress}}(x) = \dfrac{K_\sigma F_x}{A_{ncs-area}} \end{cases} \tag{6}$$

Where, K_σ is the thread stress concentration factor.

2.2 FE model

In this paper, ANSYS Workbench is used to establish 3D model for thread connection, and mesh subdivision of thread contact surface, the FE model of screw pair is shown in the following Fig. 3. Here, a triangle thread connection of M4.00×0.80 mm is established. The axial load F is $100N$ and the number N of thread engagement is 8. The load diagram of thread joint surface and total load acting surface is shown as Fig. 2, then mesh the model (see Fig. 3) the FE analysis is carried out (see Fig. 4).

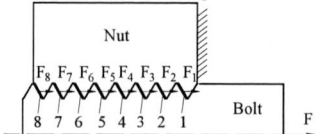

Fig. 2 Thread joint surface and total load acting surface

Table 1 Thread model parameter

Parameter	Nut length: L/mm	Semi-angle of thread: β	Pitch of thread: a/mm	Depth of fundamental triangle of thread: b/mm	Depth of the thread: d/mm
Value	6.38	308	0.8	0.7	0.53
Parameter	Mean diameter of thread: D/mm	Poisson's ratio of nut: σ	Root diameter of nut: D_2/mm	Poisson's ratio of bolt: v_b	Root diameter of nut: E_n/Pa
Value	3.47	0.3	4.08	0.3	2×10^{11}

Parameter	Equivalent outside diameter of nut:D_3/mm	Young's modulus of the bolt: E_b/Pa	Diameter of hole in bolt:D_0/mm	Coefficient of friction:μ
Value	10.09	2×10^{11}	0	0.2

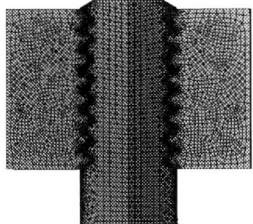

Fig. 3 FE model section

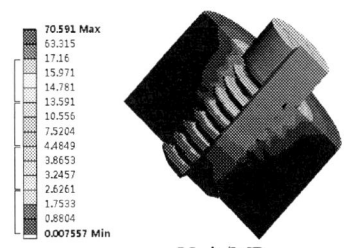

Unit/MPa

Fig. 4 FE analysis results

ANSYS Workbench can be used to extract the axial force of the contact surface.The axial force extraction results of each contact surface are shown in Table 2.

Table 2 The results of the extraction of the axial force of each contact surface.

（M4.0×0.8）

Which circle?	1	2	3	4	5	6	7	8
Force/N	(F1) 40.688	(F2) 22.064	(F3) 13.244	(F4) 8.2098	(F5) 5.56	(F6) 4.0867	(F7) 3.2956	(F8) 2.8532

2.3 Comparison of 3D FEM and Analytic Method

2.3.1 Comparison of Load Distribution

In the SOPWITH's method, substituting the parameter values of Table 1 into equation that from Eq.(1-2) to Eq.(1-7)(See Appendix), θ_1 is calculated and have a value of 4.2349. The following Fig.5 shows the axial load comparison of the SOPWITH's method and the FEM.

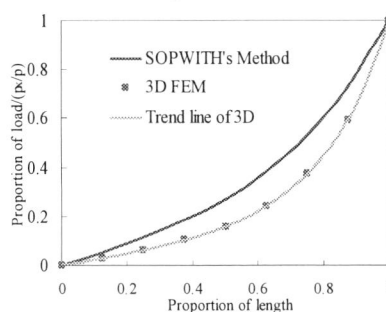

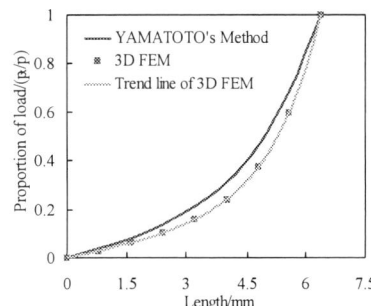

Fig. 5 Comparison of 3D FEM and SOPWITH's method

Fig. 6 Comparison of 3D FEM and YAMATOTO's method

Similar to the above, substituting the parameter values of Table 1 into Eq.(1-9) to Eq.(1-21)(See Appendix), λ can be calculated and have a value of 0.3927. The Figure of the comparison between the YAMATOTO's method of the nut axial load and the 3D FE method is shown (see Fig.5).

From Fig. 5 and Fig. 6, it can be seen that the three-dimensional finite element results of the axial load distribution of threads are basically consistent with the two analytical algorithms (SOPWITH algorithm and YAMATOTO algorithm). Therefore, it is considered that the 3D FE model of thread load distribution proposed in this paper is accurate and reliable.

2.3.2 Comparison of the Stress Distribution

After experiments, the stress concentration factor Kσ of the standard thread root is 3.85[17-19]. The total axial load F is 200N. The results of theoretical calculations and the results of FE analysis for comparison, see Fig.7.

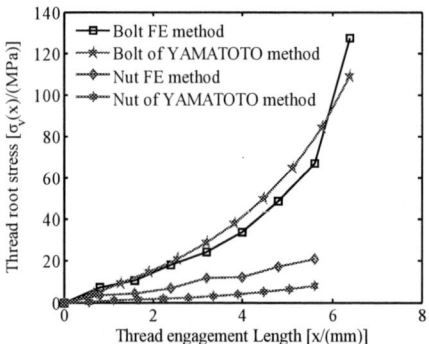

Fig.7 Thread root stress

As can be seen from the Fig.7, the stress deviation of thread root of bolt between FEM and YAMATOTO's method is small. The nut thread root stress deviation between FEM and YAMATOTO's method is a bit big, but the trend is basically the same，Therefore, the FE analysis of the stress results can be considered credible.

3. Stress analyses of nut and bolt

The M4.00×0.80 is established. To improve the efficiency of the analysis, the hexagonal nut is simplified into a round nut, and the outer diameter of the nut is øH=ø4.69, ø5.7, ø6.49, ø10.09 and ø16.09 mm (see Table 3) respectively. The friction coefficient of the thread contact surface is 0.2, and the total axial load F at the end of the bolt is 200N.

The modulus of Young's materials of bolt and nut is 2E+11 Pa, Poisson's ratio is 0.3, the compressive yield strength 2.5E+08 Pa, and the tensile ultimate strength is 4.6E+08 Pa. The meshing of bolt and nut is that the bolt has 272,544 nodes and 65,736 elements, the nut has 77,882 nodes and 16,300 elements. After that, FE analysis is performed. The stress of the root of the bolt thread and the nut are obtained respectively, as shown in Fig. 8 and Fig. 9.

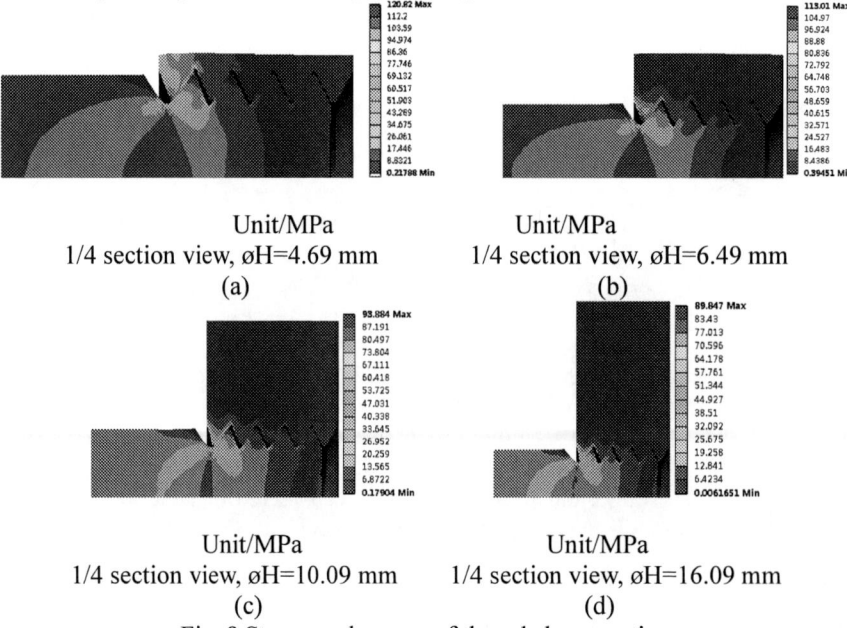

Fig. 8 Stress nephogram of threaded connection

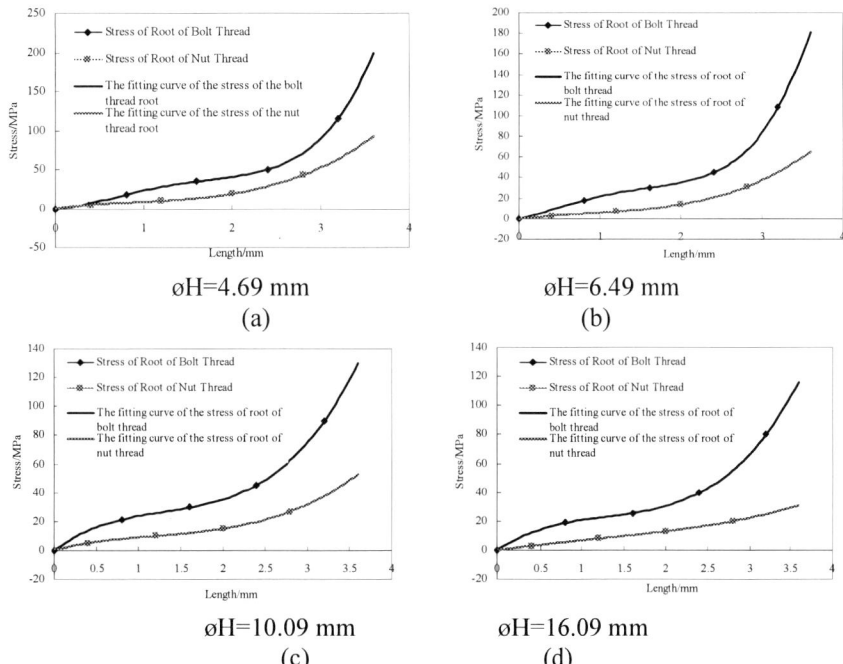

øH=4.69 mm
(a)

øH=6.49 mm
(b)

øH=10.09 mm
(c)

øH=16.09 mm
(d)

Fig. 9 Stress of the root of the bolt thread and the nut thread

According to Fig. 9, when the axial load is the same, the thicker the nut is, the smaller the maximum stress at the root of the bolt and nut thread is. The reason for this phenomenon is that: the nut is thin, thread is easy to slip out (see Fig. 10), the greater the screw deformation, the greater the stress on the first circle is. (see Fig. 9).

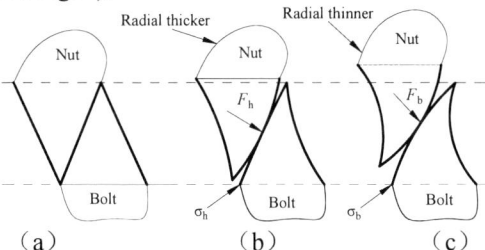

(a) (b) (c)

Fig. 10 Sketch diagram of thread deformation

Fig.9 shows that: The stresses at the root of each pair of threaded bolts are smaller than the root stresses of the corresponding bolt threads.

4. The influence of axial total load F on the force and stress of bolt thread

Take the M4.00×0.80,and øH = 5.7mm, the number of threaded engagement N is 4. Respectively, take the end of the bolt action load F as 100N, 200N, 300N, and 400N. The Young's modulus of bolt and nut of materials is 2e+11Pa, Poisson's ratio is 0.3. The friction coefficient of the thread contact surface is 0.2. The structural parameters and other material parameters of bolt and nut are same as Table 1. Then FE analysis is done, the axial load of circle of bolt threads is obtained respectively, see Fig. 11, and the stress of the root of the bolt thread is obtained (see Figure 12).

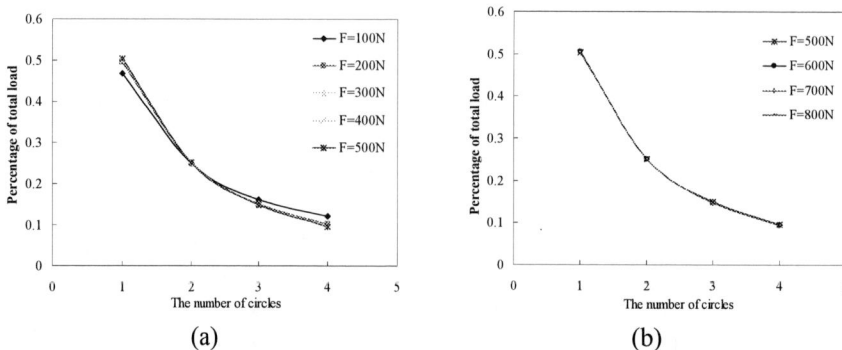

(a)   (b)

Fig. 11 The percentage of the total load of each thread

Figure 11 shows the axial load on the threads of each circle when the bolt thread is subjected to different axial total load. It can see from Fig.11 that, when the material is in the elastic range, the bolt end total loads are 100N, 200N, 300N, and 400N, and the percentage of axial load in each circle in the total axial load is almost unchanged. It is shown that the total load has little effect on the distribution of the force of the thread under the elastic range of the material. With the increase of the axial load, the load of each thread increases at the same rate.

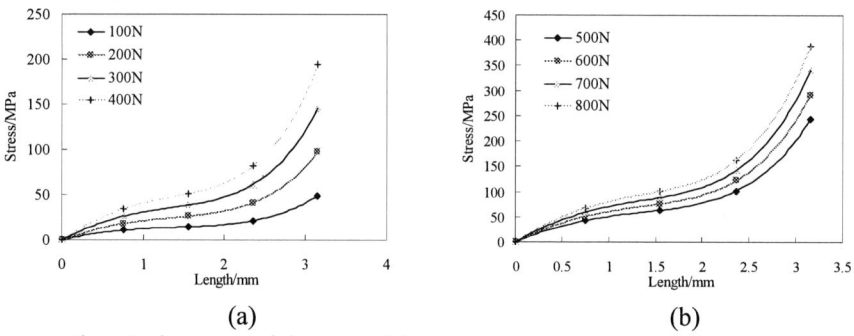

(a)   (b)

Fig. 12 The stress of the root of the bolt thread under different loads F

The following can be seen from Fig. 12: (1)The stress of the root of the bolt thread increases with the increase of the axial load. (2)The stress of the root of each circle is not equal, the stress increment of the root of the fourth circle is the smallest, and the stress of the first circle is the largest.(3) There is an increase in the axial load, the root of the first circle is damaged first.

5. Influence of the friction coefficient on the force and stress distribution of each circle of the thread

In threaded connections, different lubrication conditions result from the use of different lubricants. What is the influence of different lubrication conditions on the load distribution of the bolt teeth? People are still unknown these. In order to investigate the influence of lubrication on the force distribution of bolt thread contacts, in this paper, a FE analysis of thread contact under different lubrication conditions is performed. The bolt load F is 200N, the Young's modulus of the bolt same as the nut, it is $E=2E+11Pa$. The value of bolt and nut structure parameters and other material parameters are the same as those mentioned above 4. The friction coefficients of the threads were taken as from 0 to 1, respectively, and the FE analysis is performed. The axial load of bolt threads is obtained, as shown in figure 13, and the maximum stress of the root of the bolt thread is obtained (see Fig 14).

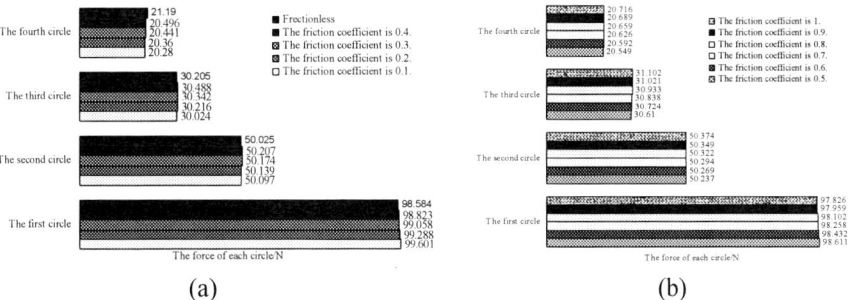

(a)   (b)

Fig. 13 The force of each circle of different friction coefficient

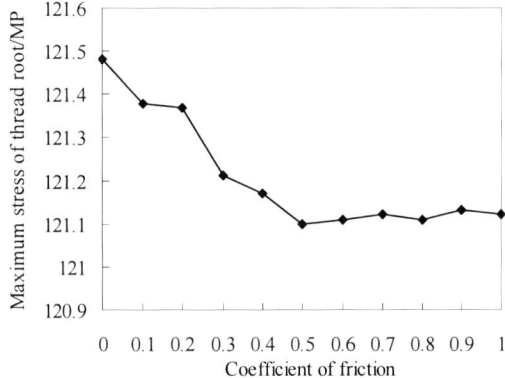

Fig. 14 The maximum stress of the thread of different friction coefficient

The following can be seen from Fig.13: (1) The load on the fourth thread, third and second threads of the bolt is gradually increasing slowly with the increase of friction coefficient. The load on the first thread gradually decreases slowly. The greater the coefficient of friction is, the more the force of each circle tends to be averaged. This shows that the friction coefficient can change the load distribution of the threads. However, because the bolt and nut materials are elastic, no matter the friction coefficient, the load of each thread cannot be equal. (2) It can be seen from the figure that the effect of the friction coefficient on the load distribution of each thread is not obvious. The effect of the friction on the contact surface of thread can be negligible. It can be seen from Figure 14 that the friction coefficient has a little effect on the maximum stress of the thread under the load, as the friction coefficient increases, the maximum stress of the bolt gradually decreases and reaches a fixed value.

6. The influence of the radial thickness of the nut on the force distribution of the bolt threads

Take total load F as $100N$. The Young's modulus of the nut and the bolt is 2E +11 Pa. and the bolt structure parameter selection in accordance with the above 4 and the dimensions of the nut is shown in table 3. The friction coefficient of thread contact surface is 0.2. The FE analysis was performed, and gets the bolt root curve of stress, as shown in Fig. 16, and the axial load of bolt threads is also obtained(see Fig .15).

Table 3

The nuts	a	b	c	d	e
Diameter of nuts øH/mm	4.69	5.7	6.49	10.09	16.09

ISPECE   IOP Publishing

IOP Conf. Series: Journal of Physics: Conf. Series **1187** (2019) 032060   doi:10.1088/1742-6596/1187/3/032060

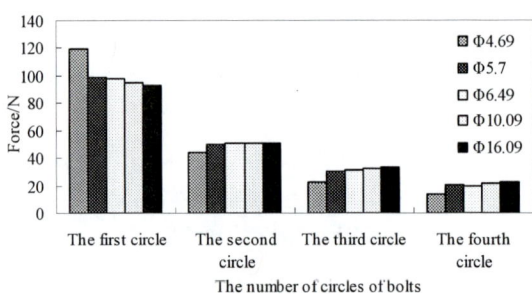
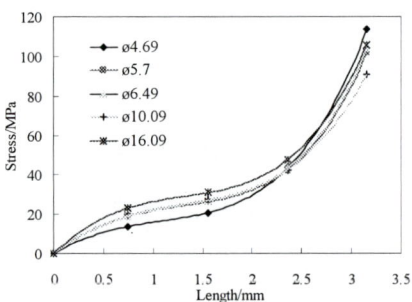

Fig. 15 Distribution of the force of the bolt threads   Fig 16 Stress of the bolt

The following can be seen from Fig. 15: The thickness of the nut is thinner, and the distribution of the load of each coil is more uneven. Conversely, and the thicker the radial thickness of the nut, the more average the load distribution of the bolt thread. The radial thickness of nut has an influence on the load distribution of bolt thread. However, from the general trend, the diameter of the nut radial thickness increases to a certain value, and the impact on the load distribution of the bolt thread is negligible, and the following can be seen from Fig. 16: (1) when the radial thickness of the nut is different, the stress distribution in the diameter of the bolt is uneven, and the stress at the root of the first circle is the maximum, and then gradually decreases. (2) It can also be seen that the smaller the radial thickness of the nut, the less uniform distribution of the stress in the minor diameter of the bolt. Conversely, the greater of the radial thickness of the nut, the more uniform of the stress distribution in the minor diameter of the bolt thread. (3). When the radial thickness of the nut is increased to a certain extent, the trend of the stress distribution in the minor diameter of the bolt is not obvious.

7. The influence of the number of circles on the load distribution

M4.00×0.80 was selected as the study object. The structure parameters of nuts are shown in table 4. The Young modulus of nuts and bolts is 2E+11 Pa. The friction coefficient of the thread contact surface is 0.2, and then the FE analysis is performed.

Table 4

Bolt joints	1	2	3	4
The number of threads in engagement	3	4	6	8

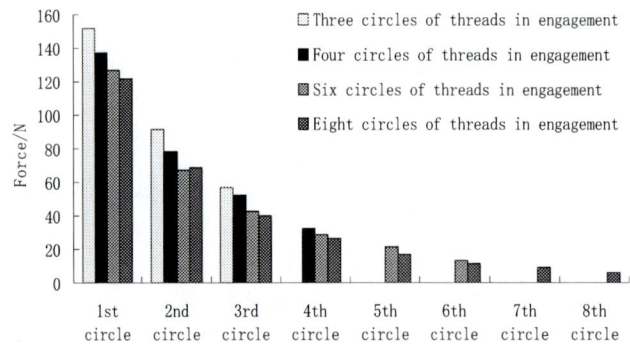

Fig. 17 FE analysis of the five kinds of threaded connections

The following can be seen from Fig. 17: (1) The number of circles of bolt thread affects the load distribution of thread. (2) The distribution of the load in each circle decreases gradually. (3) The load on each circle thread is greater than that of other bolt threads when the bolt is bolted in the three circles. When the bolt is screwed in the four circles, the load on each circle is greater than the load of the bolt circles of the six circles and the eight circles thread. When the bolt is bolted in the six circles thread, the load distributed in each circles thread is greater than the load of the corresponding circles

902

of the bolt thread of the eight circles. (4) It can be seen that with the increase of the number of circles, the load distribution of each circles thread has been reduced. (5) This shows that when the number of thread circles is increased, the thread load at the front circle can be shared, when the number of the circles increases to a certain extent, the number of circles continues to increase, and the change of the load on the front circles will be less obvious.

8. Conclusions

1) 3D FE analysis results of the axial load distribution of the thread are in roughly agreement with the SOPWITH's method and YAMATOTO's method.

2) 3D FE analysis results of the thread root stress distribution of the thread are in roughly agreement with this paper's method.

3) The analysis shows that, usually, in a pair of threads engaged with each other, the bolt root stress is greater than the nut, so the first damage is the bolt, not the nut.

4) In the elastic range, the size of the bolt axial load has little effect on the load percentage of the threads. With the increase of axial load of bolt, the load of each circle is increased by the same proportion, and the percentage of the load of each circle thread is basically unchanged.

5) The Young's modulus has an effect on the load distribution of the bolt threads. If the shape of the nut is not changed, changing the ratio of Young's modulus of the nut to the bolt has a slight effect on improving the distribution of the thread load.

6) The radial thickness of nuts has an effect on the distribution of the load of the threads. The smaller the thickness of the nuts is, the more uneven the load distribution is. From the overall trend, the radial thickness of the nut increases to a certain extent, and the impact of the bolt thread load distribution will be very small.

7) The number of circles in engagement has an effect on the load distribution of each circle. With the increase of the number of circles, the load of each circle is reduced. The number of circles is increased and the load of the original circles can be shared. When the number of circles is increased to a certain degree, the change of the load quantity of the previous circles is not obvious.

8) The coefficient of friction has a slight effect on the force distribution of the threads, but not obvious, and this effect is almost negligible.

Acknowledgment

Projects(51675422,51475366,51475146) supported by the National Natural Science Foundation of China

References

[1] LI Xi-bing, ZHOU Hong-tao,ZHOU Zi-long, et al. Parameter analysis of anchor bolt support for large-span and jointed rock mass[J]. Journal of Central South University of Technology, 2005, 12(4):483-487.

[2] LI Shu-cai, WANG Hong-tao, WANG Qi, et al. Failure mechanism of bolting support and high-strength bolt-grouting technology for deep and soft surrounding rock with high stress [J]. Journal of Central South University of Technology, 2016, 23(2):440-448.

[3] WANG Lian-guo, LI Hai-liang, ZHANG Jian. Numerical simulation of creep characteristics of soft roadway with bolt-grouting support [J]. Journal of Central South University of Technology, 2008, 15(s1):391-396.

[4] Jiang X, Zhu Y, Hong J, et al. Stiffness Analysis of Curvic Coupling in Tightening by Considering the Different Bolt Structures[J]. Journal of Aerospace Engineering, 2016, 29(3):04015076.

[5] Alkatan F, Stephan P, Daidie A, et al. Equivalent axial stiffness of various components in bolted joints subjected to axial loading[J]. Finite Elements in Analysis & Design, 2007, 43(8):589-598.

[6] Yang Qiang，Miao De-hua，Wang Yan-li，Xue Qiang. Influence of nut thread shape and screwing length on bolt root stress[J].Journal of Tianjin University of Technology and

Education.2007(3): 29-32.(In Chinese)

[7]Zhou Xian-hui Sun You-song Zhang Er-wen. Analysis of axial load distribution trend for transmission screw by finite element [J].Machinery Design & Manufacture. 2008(1): 16-18.(In Chinese)

[8] Yin Yi-Hui Yu Shao-Rong, Analysis of Distributions of And Frictional Effect on Axial Force And Stress in a Threaded Bolt[J]. Journal of Mechanical Strength. 2006(4): 524-531. (In Chinese)

[9] Kenny B, Patterson E A. Load and stress distribution in screw threads [J]. Experimental Mechanics, 1985, 25(3): 208-213.

[10] Sopwith D G. The distribution of load in screw threads[J]. Proceedings of the Institution of Mechanical Engineers, 1948, 159(1): 373-383.

[11] Alkan I, Sertgöz A, Ekici B. Influence of occlusal forces on stress distribution in preloaded dental implant screws[J]. The Journal of prosthetic dentistry, 2004, 91(4): 319-325.

[12] Kenny B, Patterson E A. The distribution of load and stress in the threads of fasteners-a review[J]. Journal of the Mechanical Behavior of Materials, 1989, 2(1-2): 87-106.

[13] Miller D L, Marshek K M, Naji M R. Determination of load distribution in a threaded connection[J]. Mechanism and machine Theory, 1983, 18(6): 421-430.

[14] Wang W, Marshek K M. Determination of load distribution in a threaded connector with yielding threads [J]. Mechanism and machine theory, 1996, 31(2): 229-244.

[15] Chen H P, Zeng P, Fang G, et al. Load distribution of bolted joint［J］. Journal of Mechanical Engineering, 2010, 46(9): 171-178 (in Chinese)

[16] Yamatoto A. The theory and computation of threads connection ［M］ Guo K Q, et al, Translator. Shanghai: Shanghai Scientific and Technological Literature Publish House, 1984 (in Chinese)

[17]Peterson,R E. Stress concentration factor [M]. National Defense Industry Press, 1988. (in Chinese)

[18]Aviation Industry Science and Technology Commission. Stress concentration factor manual [M]. Higher Education Press, 1990. (in Chinese)

[19]Nishida Masao, Li Anding. Stress concentration [M]. Machinery Industry Press, 1986. (in Chinese)

| ISPECE | IOP Publishing |

Study on the Effect of Temperature on Dynamic Characteristics of Rotor System with Straight Crack

Tengfei Kuai[a],* and Changfang Zhao*, Jie Ren, Guigao Le

School of Mechanical Engineering, Nanjing University of Science and Technology, Nanjing, 210094, China

* Tengfei Kuai and Changfang Zhao contributed equally to this work and should be considered co-first authors.

[a]Corresponding author: ktf0815@sina.com

Abstract. The effect of temperature on the dynamic characteristics of cracked rotor is studied in this paper. The crack is simulated by the extended finite element method (XFEM). Based on the theory of fracture mechanics, appropriate fracture criteria are selected, and the damage and failure criteria of materials are set up, and the finite element simulation analysis is carried out. The results show that with the increase of temperature, the amplitude of 3X in frequency domain increases most obviously, and the deformation degree of axis trajectory is high. The research results have important reference value and scientific significance for the study of dynamic characteristics of cracked rotor system under high temperature environment.

1. Introduction

Rotor system is the core component of rotating machine, but many rotor systems often operate in harsh environment. It is essential for the effect of temperature factors on the rotor system. In the fields of aviation, aerospace and military industry, there are a large number of harsh environments with strong temperature change. For example, the design speed of turbofan engines is between 10,000 and 20,000 revolutions per minute, and the number of revolutions of turbojet engines is even higher. In the working process of this kind of rotor system, the gas temperature is generally above 2000 ℃, and the highest temperature can be up to 3500 ℃. The rotor system is subjected to a strong temperature variation of several thousand degrees per second. It can be said that it is a kind of typical rotor system which works under strong temperature variation. The working environment of aeroengine is very bad and it runs under high temperature for a long time, which leads to the thermal bending and deformation of the components such as the rotor inside the aero-engine. These problems usually cause larger vibration of the system [1]. It can be seen that the aero-engine under high temperature environment is prone to vibration failure.[2]. Although the effect of temperature on rotor system is considered in the current literature, the effect of temperature on the dynamic characteristics of cracked rotor system is ignored. In reality, many large engines have cracks. But the crack does not affect the normal operation. This kind of engine has always been in high-intensity operation, and there will be unexpected hidden danger of safety. Li Jiukai of Sichuan University studied the influence of temperature and high cycle fatigue behavior on fatigue life of supercritical steam turbine rotor steel[3]. Wang Kun of Huazhong University of Science and Technology studied fatigue damage assessment caused by high temperature and low cycle according to energy theory and process fictitious [4-5]; Zhao Mei of Shanghai Jiaotong University analyzed the diagnostic method of cracked rotor in rotating

Content from this work may be used under the terms of the Creative Commons Attribution 3.0 licence. Any further distribution of this work must maintain attribution to the author(s) and the title of the work, journal citation and DOI.

Published under licence by IOP Publishing Ltd

machinery theoretically and verified it by experiment, but the effect of temperature field on rotor system was not considered in this study [6]. The crack is simulated based on the extended finite element method (XFEM). According to the theory of fracture mechanics, the appropriate fracture criteria are selected, and the damage and failure criteria of materials are set up, and the finite element simulation analysis is carried out. The results have important reference value and scientific significance for the study of dynamic characteristics of cracked rotor system under high temperature environment. It is of great academic value and scientific significance to study the dynamic characteristics of cracked rotor system in high temperature environment.

2. Establishment of finite element model for cracked rotor system

2.1 Simulation of cracks

In this paper, the extended finite element method (XFEM) [7] is used to simulate the crack. The finite element simulation of the cracked rotor system is completed by selecting the appropriate fracture criterion, setting up the damage and failure criterion of the material and reasonable mesh division.

2.1.1 Fracture criteria for cracks

Combined with the actual crack analysis problem and the crack body structure, the maximum energy release rate criterion is chosen as the crack fracture criterion of the simulation model in this paper [8-9]. In practical engineering applications, the loads of mechanical members are usually composite loads under the combined action of multiple loads, and the composite loads can be divided into three kinds of typical crack forms in the case of small deformation [10]. In general, the energy release rate of the composite crack can be expressed by the superposition of the energy release rate of three typical crack forms. The expression for the energy release rate of the crack in the composite form is as follows:

$$G = G_{\text{I}} + G_{\text{II}} + G_{\text{III}} = \frac{(1-v^2)}{2E} K_{\text{I}}^2 + \frac{(1-v^2)}{2E} K_{\text{II}}^2$$
$$+ \frac{(1+v)}{2E} K_{\text{III}}^2 \tag{1}$$

2.1.2 damage and failure criteria for materials

The damage and failure criterion of mechanical structure material determines the failure form and degradation law of material. Therefore, the finite element simulation analysis in this paper needs to pre-set the damage and failure criterion of material [11].

The existing law of damage evolution is mainly divided into two kinds: one is based on displacement and the other is based on fracture ability, which is used to describe the law of material stiffness degradation. The evolution law of fracture ability is mainly controlled by specifying critical fracture ability and correlation coefficient. The law is divided into three kinds: POWER rule, BK rule and Reeder rule. Among them, the POWER rule:

$$(\frac{G_{\text{I}}}{G_{\text{IC}}})^\alpha + (\frac{G_{\text{II}}}{G_{\text{IIC}}})^\alpha + (\frac{G_{\text{III}}}{G_{\text{IIIC}}})^\alpha = 1 \tag{2}$$

2.1.3 mesh generation and setting near cracks

The finite element model of cracked rotor system is divided into 60320 hexahedron elements with 68341 nodes and 11000 dense meshes near the crack. Finite element model meshing is shown in Figure 1.

Figure 1. Finite element model meshing of cracked rotor

XFEM is used to simulate the straight crack, and the maximum principal stress failure criterion is selected as the damage criterion.

2.2 Establishment of simulation model for rotor with straight crack

Figure 2(a) is a straight crack section, Figure 2(b) is the size of a straight crack, and the crack shape is semicircular. The ratio of crack depth to crack depth is h=r/R. In this chapter, the depth of straight crack is r=5 mm (h=1mm).

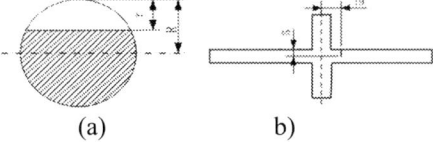

(a)   b)

Figure 2. Rotor system with straight crack

Figure 3 shows the rotor system model after adding a straight crack.

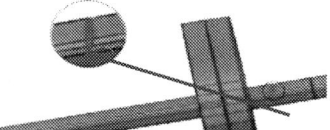

Figure 3. Model of rotor system with straight crack

3. Effect of temperature on vibration response of rotor with straight crack

The vibration response of the rotor system with a straight crack is analyzed at   20 ℃, 300 ℃ and 500 ℃. The rotational speed is 2000rpm and the crack depth ratio is 1.

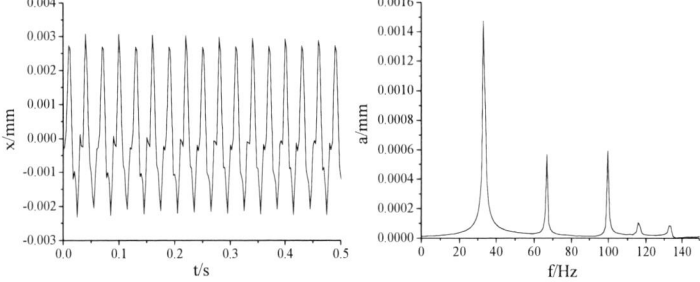

Fgure 4. Time-domain and frequency-domain curves of the *x* direction of a rotor system with a straight crack at 20 ℃

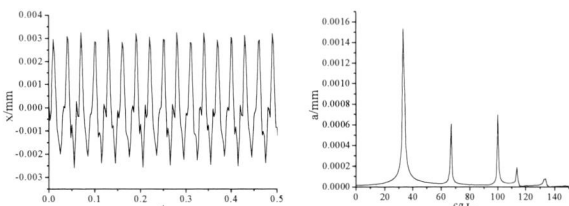

Figure 5. Time-domain and frequency-domain curves of the *x* direction of a rotor system with a straight crack at 300 ℃

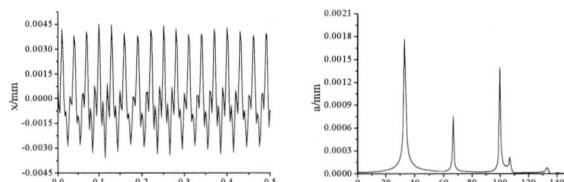

Figure 6. Time-domain and frequency-domain curves of the *x* direction of a rotor system with a straight crack at 500 ℃

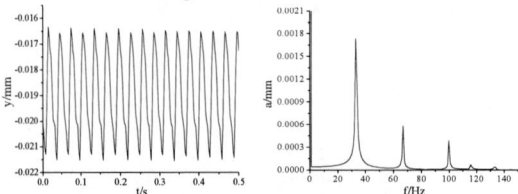

Figure 7. Time-domain and frequency-domain curves of the *y* direction of a rotor system with a straight crack at 20 ℃

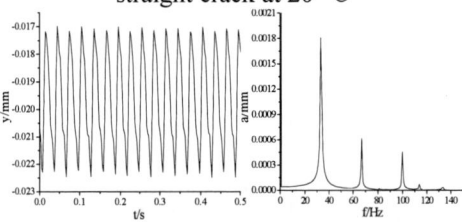

Figure 8. Time-domain and frequency-domain curves of the *y* direction of a rotor system with a straight crack at 300 ℃

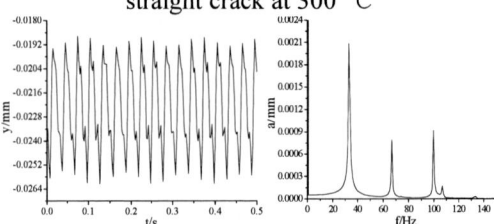

Figure 9. Time-domain and frequency-domain curves of the *y* direction of a rotor system with a straight crack at 500 ℃

Table 1. Frequency doubling amplitudes in *x* and *y* directions at different temperatures

Direction	Temperature /℃	1X amplitude / mm	2X amplitude / mm	3X amplitude / mm
	20	0.00150	0.00058	0.00061
x	300	0.00160	0.00061	0.00070
	500	0.00180	0.00080	0.00140
	20	0.00180	0.00060	0.00040
y	300	0.00183	0.00063	0.00045
	500	0.00210	0.00080	0.00093

Figures 4-9 shows the time-domain and frequency-domain curves of the cracked rotor system in x and y directions at different temperatures, and table 1 shows the frequency-domain frequency-doubling amplitudes of x and y directions at different temperatures.

(1) At different temperatures, the waveforms of the time domain images in both directions are deformed greatly, and both of them appear synthetic vibration. The 1X amplitude is the largest.

(2) With the increase of temperature, the deformation degree of time domain waveform in two directions is larger, and the vibration is more complex. In the direction of x and y, the increase rate of 3X amplitude is larger, 1X is the second, and 2X amplitude is smaller.

(3) When the temperature ranges from 20 ℃ to 300 ℃, the time domain and frequency domain curves change slightly, and the temperature ranges from 300 ℃ to 500 ℃, the time and frequency domain curves vary greatly, and the high-frequency components in the 500 ℃ time-frequency domain diagram increase and the fractional frequency components appear, and the nonlinear characteristics of the rotor system become stronger. If the temperature increases gradually at 500 ℃, the vibration response of the straight crack is more complicated.

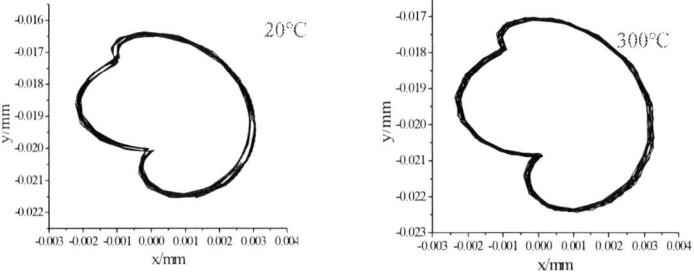

Figure 10. Axis trajectory curve of rotor system with straight crack at 20 ℃ and 300 ℃

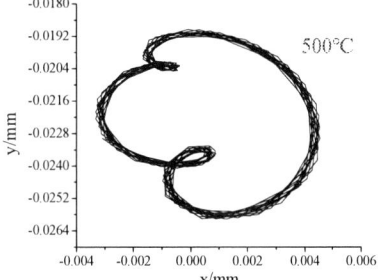

Figure 11. Axis trajectory curve of rotor system with straight crack at 500 ℃

From figures 10-11, we can see that the degree of deformation of the axis trajectory increases with the increase of temperature, and the vibration range becomes larger. At the same time, there are two grooves in the axis track. The degree of depression increases with the increase of temperature, and the figure becomes distorted. Coils appear at grooves.

4. Conclusion

(1)With the increase of temperature, the whole vibration response of the cracked rotor becomes stronger, the vibration range becomes larger, the change of 3X amplitude is the most obvious, the high frequency component increases, the fractional frequency component appears, and the nonlinear phenomenon is enhanced. The vibration response of straight crack is more complicated.

(2)The deformation degree of the axis trajectory increases with the increase of temperature, and the vibration range becomes larger. At the same time, two grooves appear in the axis locus, the degree of depression increases with the increase of temperature, and the distortion is more complicated.

Acknowledgements

This work was supported by National Natural Science Foundation of China (51303081).

References

[1] W. He. *Northeast University,* 9(2012).

[2] X.Z. Zhu, *Power Engineering* **28**, 114-116(2008)

[3] I.I. Kryukov, S.A. Leont'ev, V.S. Platonov and A. I. Rybnikov, *Thermal Stresses* **8**, 351-395(2006)

[4] F.M. Lu, Q.Wei, T and L. Y. Lu, *Journal of Power Engineering* **6**, 443-449,494(2014)

[5] K.Wang, *Huazhong university of science and technology*, 12(2004)

[6] Z.K. Peng, W.T. Peter, F.L. Chu, *Journal of Mechanical Systems,* (2005).

[7] S.Jean-Jacques,*Communications Nonlinear Science Numerical Simulation.* (2009)

[8] Z. Xi *China University of Mining and Technology* (Beijing), (2017)

[9] X. L. Wang, *University of Science and Technology of China,* (2017)

[10] C. Xian. *Chongqing University*, (2014)

[11] L. Zhang, *Chongqing University,* (2015)

Mechanical Productivity Design and Mechanical Process Analysis Framework Construction

Cui Li, Chen Hong'Bo

Qingdao Institute of Technology, 266300

Abstract: The effective development of modern enterprises has fully considered the mechanical productivity, and the dependence on mechanical processes with good applicability has gradually deepened. In this context, in order to achieve sustainable development of the enterprise and meet the development requirements of advancing with the times, it is necessary to strengthen the mechanical productivity design related to it, and pay attention to the efficient use of mechanical processes, so that the relevant production plans can be smoothly implemented. Gradually improve the production level of modern enterprises and increase their production efficiency in practice. Based on this, this paper will systematically expound the design of mechanical productivity and mechanical process, in order to invigorate the better development of the enterprise.

1. Introduction

Focusing on the design of mechanical productivity and analysis of mechanical processes is conducive to maintaining a good production situation of the enterprise, promoting the development of machinery production operations more efficiently, and increasing the technical content in the implementation of relevant production plans. Therefore, it is necessary to combine the changes in the situation in practice and the long-term development requirements of the enterprise, to carry out the design work of mechanical productivity in a targeted manner, and to strictly control the design process, so that the final mechanical productivity design scheme is more perfect and has a good potential value. At the same time, the mechanical process analysis should be strengthened, and the corresponding analysis work should be put in place to promote the efficient development of mechanical production activities to meet the requirements of increasing technical content in the mechanical production process.

2. Pay Attention to The Value of Mechanical Productivity Design and Mechanical Process Application

In order to make the mechanical productivity design work in place and realize the scientific application of mechanical technology, it is necessary to understand the value of mechanical productivity design and mechanical process application. In the meantime, related content includes the following aspects:

2.1 Pay Attention to The Value of Mechanical Productivity Design

In practice, by paying attention to the design of mechanical productivity, its value is as follows: (1) Emphasis on mechanical productivity design can make the implementation of mechanical production plan more significant, increase the production efficiency of enterprises in practice; (2) Pay attention to mechanical productivity The design is beneficial to improve the efficient use of materials and processes with reliable performance, so that the mechanical productivity under the cooperation of materials and mechanical processes can be gradually improved; (3) Emphasis on mechanical productivity design is conducive to better adapt to the situation changes in practice. To achieve sustainable development of modern enterprises and improve the status of machinery production.

2.2 Pay Attention to The Value of Mechanical Process
In practice, by paying attention to the application of mechanical technology, its value is as follows: (1) Emphasis on the application of mechanical technology, can effectively support the implementation of mechanical production planning, effectively deal with mechanical production risks, and avoid the impact on mechanical production efficiency. (2) Emphasis on the use of mechanical processes, which can have a good technical content in the mechanical production process, and gradually improve the mechanical production efficiency and quality of the enterprise in practice; (3) Emphasis on the application of mechanical processes, capable of producing enterprises provide better support, optimize machinery production methods, and gradually improve the comprehensive competitiveness of modern production enterprises.

3. Design Analysis of Mechanical Productivity
In the process of improving mechanical productivity, it is necessary to strengthen its design and clarify the corresponding design points. The main points of mechanical productivity design in practice include the following:

3.1 Pay Attention to The Improvement of Mechanical Parts
Mechanical parts are the main part of the machine and are closely related to the good mechanical productivity. Therefore, in the process of mechanical productivity design, it is necessary to pay attention to the improvement of mechanical parts. The specific performances are as follows: (1) Enhance the improved design consciousness of mechanical parts, and under the cooperation of rich practical experience and professional theoretical knowledge, carry out the targeted design work of mechanical parts, so as to optimize the function of the machine, for the machine Increase productivity to provide the required support; (2) Pay attention to the serialization of mechanical products, standardization of parts and generalization of parts, and control the improved design process of mechanical parts, so that the final mechanical parts are well applied. The functional characteristics meet the requirements of effective design of mechanical productivity; (3) Based on the improved design of mechanical parts, it is also necessary to control the selection process of parts manufactured by cutting method, and improve the utilization efficiency of parts manufactured by non-cutting method, and reduce the error in the application to ensure that it has a good fit with the mechanical equipment, thereby increasing the mechanical productivity and enriching its design content [1].

3.2 Pay Attention to The Improvement of Mechanical Materials
Whether the performance of materials involved in the implementation of mechanical production planning and mechanical parts is reliable, which is related to the level of mechanical productivity. Therefore, in the process of implementing mechanical productivity design work, it is necessary to pay attention to the improvement of mechanical materials as follows: (1) In the process of mechanical material selection, materials with strong plasticity, good toughness, suitable hardness and good thermal conductivity should be selected and used to avoid adverse effects on mechanical productivity due to material quality defects; (2) need to be used When cutting materials, it is necessary to select mechanical materials with good cutting performance, mainly because the surface quality and cutting effect of such materials are good, and the requirements for improvement of mechanical productivity can be met. At the same time, the designer should comprehensively consider the cost performance of the materials involved in the scientific design of mechanical productivity, and optimize the use of mechanical materials under the cooperation of material quality inspection and evaluation mechanism, and promote the mechanical productivity supported by reliable mechanical materials. Higher, perfect related design; (3) Based on the improved design of mechanical materials, it is necessary to control the

design process, analyze the factors affecting the material properties, and carry out targeted design improvement of mechanical materials to maintain the machinery. The application of good materials required in production practice provides the necessary support for the gradual improvement of mechanical productivity. In the process of implementing the mechanical material improvement design work, materials with good toughness and good plasticity should be considered, and with the support of heat treatment methods, the cutting performance of mechanical parts should be optimized to meet the requirements of parts standardization, and finally the mechanical productivity can be improved [2]. The schematic diagram of the machining site is shown in Figure 1.

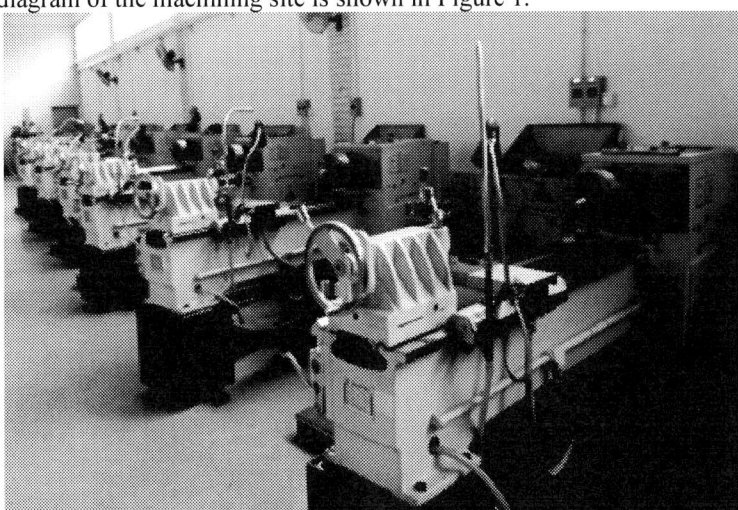

Figure 1 Schematic diagram of the machining site

3.3 Consider the Improvement of Mechanical Process

In the process of implementing the mechanical productivity design work plan, it is also necessary to pay attention to the improvement of the mechanical process, so that the mechanical production work supported by the mechanical process can be carried out more efficiently, and the mechanical productivity is gradually improved. The specific performances are as follows: (1) Based on the mechanical productivity design, it should be based on the consideration of the mechanical process function characteristics, combined with the efficient implementation requirements of the mechanical production plan, implement the mechanical process improvement design work, and scientifically control the design process, which will be effective. The mechanical process is applied in the process of manufacturing mechanical parts, thereby improving the mechanical productivity, and accumulating rich practical experience for its design work; (2) In the process of mechanical process improvement design, the use of automation technology should be strengthened to enhance the mechanical production process. Control effect, eliminate potential safety hazards in production, provide effective technical support for the improvement of mechanical productivity, and achieve scientific response to mechanical production risks; (3) Mechanical process improvement in practice, also need to consider the actual situation that is related to the production of enterprises, and evaluate the application effect of the improved mechanical process, so that it can meet the requirements of mechanical productivity improvement, and broaden the design ideas of mechanical productivity improvement. At the same time, it is necessary to have a correct understanding of the potential application value of mechanical process improvement, so that its practical role in the improvement of mechanical productivity can be fully exerted, laying the foundation for the application level improvement of modern mechanical processes, and maintaining good production efficiency in the process of enterprise development. In practice, the use of different measures in these aspects is conducive to improving the improved design level of mechanical processes and improving the design of mechanical productivity [3].

4. Mechanical Process Analysis

In the process of coping with the changes in the situation in the new era and improving the level of mechanical production, it is necessary to pay attention to the use of mechanical processes and analyze them to understand the relevant content. The mechanical process related content in practice includes the following aspects:

4.1 Consider The Application of Automation Technology

The application of automation technology in mechanical production can improve the level of control work in mechanical production, deepen its automation and promote the formation of mechanical automation production system. At the same time, the automated mechanical production process can automate the production, processing and output of the product, reducing the investment of manpower and reducing the defective rate of the product. In addition, with the support of automation technology, effective control of environmental pollution can be achieved in the mechanical production process, which is an important development trend of mechanical production in the future. Through the scientific application of automation technology in mechanical process analysis, it can also effectively improve the labor conditions of the production workshop, greatly improve the production efficiency of the production workshop, improve the product quality of the enterprise, and increase the economic benefits of the long-term practice of the production enterprise. Therefore, in the process of promoting the development of mechanical technology and comprehensively improving its practical application level, we should pay attention to the application of automation technology, and realize the automatic control of mechanical production process by fully considering and integrating the factors of automatic control theory and computer network. The system promotes the entire mechanical production process to be in a controllable state, reducing its production risk while providing technical support for the sustainable development of mechanical processes [4].

4.2 Pay Attention to The Improvement of Single Piece Production Efficiency

In practice, if the labor time of the mechanical single piece can be shortened, the production efficiency of the mechanical single piece can be improved, the mechanical production efficiency can be greatly improved, and the practical application effect of the mechanical process can be enhanced. Therefore, in the process of implementing the mechanical process analysis work, attention should be paid to the improvement of single piece production efficiency. The specific performance is as follows: (1) Increase the cutting amount and speed up the cutting feed rate. In the process of using mechanical processes, the scientific selection of new tools with reliable performance can speed up the cutting speed in mechanical production practice and increase the feed rate. Therefore, in the process of promoting the development of mechanical processes, it is possible to start with the improvement of mechanical cutting tools and adopt high-speed and powerful cutting technology to shorten the working time of mechanical parts and meet the requirements of improving production efficiency; (2) adopting more Process technology. By reducing the cutting and cutting time of the cutting tool, the time for each piece to be subjected to the cutting process is shortened, thereby achieving the goal of improving the productivity of the single piece; and (3) reducing the machining allowance. Based on the application of mechanical technology, in order to improve the precision of the blank, reduce the machining allowance, and maintain a good single-piece production efficiency, it is necessary to consider the use of advanced mechanical processes such as precision forging and pressure forging. At the same time, advanced fixtures can be used to shorten the loading and unloading time of the workpiece and the auxiliary time under the support of the continuous processing method, so as to shorten the auxiliary time and improve the production efficiency of the single piece, and effectively develop and apply the mechanical process. The level of improvement lays the foundation. The relevant contents of machining precision are shown in Table 1.

Table 1 Related to Machining Accuracy

Accuracy level	Size accuracy range	Ra value range (micron)	processing methods
High precision	IT7-IT6	0.8-0.2	Usually obtained by grinding
Medium precision	IT10-IT9	6.3-3.2	Usually obtained by finishing, milling and planing
	IT8-IT7	1.6-0.8	
Low accuracy	IT13-IT11	25-12.5	Usually obtained by roughing, milling, planing, drilling

(Note: Ra is the average deviation of the contour of one of the evaluation parameters for evaluating the surface roughness of mechanical parts)

4.3 Other Points

(1) In the process of realizing the efficient use of mechanical processes and enhancing its potential application value, it is necessary for production enterprises to pay attention to the combination of advanced processes such as cold extrusion and powder metallurgy, so as to achieve the purpose of improving the utilization efficiency of raw materials and the accuracy of blanks. Improve the mechanical production status while ensuring the effectiveness of mechanical process applications. In the process of mechanical material processing, if the use of special processing technology can be strengthened, it is beneficial to optimize the processing method of mechanical materials and increase the technical content in the processing of mechanical materials such as extra brittle and special hard. In addition, it can improve the scientific use of non-cutting technology, improve the application level of machining technology, and meet its requirements for efficient development of mass production operations [5].

(2) Equipment supervision under the support of mechanical technology can improve the rational allocation of the number of caretakers, the effective setting of automatic alarm function, etc., comprehensively improve the efficiency of the inspection of mechanical production equipment, and achieve effective cost to the production cost of enterprises. Control and provide the required reference information for the development of the mechanical process. At the same time, it is necessary to increase the research work in mechanical process, starting from the aspects of cost economy and application effect, and conducting scientific evaluation in the application of mechanical technology to maintain its good application.

5. Conclusion

In summary, the design of mechanical productivity and the analysis of mechanical processes have important practical reference significance: it can enhance the implementation effect of the mechanical production plan, and provide technical support for the efficient development of related production activities, prompting the production enterprises to practice. It can be in a good state of development. Therefore, in the future, when carrying out research work on mechanical productivity, more attentions should be given to the design and application of mechanical processes, and a comprehensive evaluation of the application effects of mechanical productivity design and mechanical processes should be carried out, so as to stabilize the development of related production enterprises and provide reliable protection. On this basis, it is conducive to increasing China's technological advantages in mechanical production and enriching the practical experience in the development of related production activities.

References

[1] Fu Jun. Reliability Analysis of Mechanical Manufacturing Process [J]. Southern Agricultural Machinery, 2018 (18): 60.

[2] Li Ruipeng, Tang Jianyong. Analysis of advanced machinery manufacturing technology and mechanical manufacturing process [J]. Heilongjiang Science, 2018 (18): 58-59.

[3] Huang Hai. Analysis of modern machinery manufacturing process and precision machining

technology [J]. Science and Technology Innovation, 2018 (25): 179-180.

[4] Zhou Xilai, Sun Tianyu. Research on mechanical productivity design and mechanical process [J]. Private Science and Technology, 2015 (12): 6.

[5] Wang Xin. Research on mechanical productivity design and mechanical process [J]. Silicon Valley, 2014 (01): 141-142.

AUTHOR INDEX

Ailing, Qi 1974
An, Shubing 1378
An, Wang W. 1470
An-Wen, Ying 479
Bai, Juan 1121, 1510
Bai, Xiaoye 252
Bao, Lei 1121, 1510
Bao, Wenxia 1853
Bao, Yiting 780
Bi, Mingkai 1939
Bin, Xu 2166
Bo, Yang 434
Bo, Zhang 2315
Bouchart, Vanessa 732
Boxing, Zhang 1063
Cai, Biao-Hua 751
Cai, Guoliang 1209
Cai, Peng 409
Cai, Shaopeng 1365
Cai, Sun 214
Cai, Xiaoyu 2283
Cai, Zengyu 1656
Cao, Shaozhong 1425
Cao, Shukun 640
Cao, Xinli 87
Cao, Yan 276
Cao, Yichao 2360
Cao, Yundong 239
Cen, Tao 2278
Chang, Faliang 1359
Chang, Rui 1130
Chang, Wen 2638
Changhui, Ma 268
Chao, Xiang 2478
Che, Renfei 366, 386
Chen, B. W. 1932
Chen, Baiyu 2602
Chen, Chen 2720
Chen, Chunlong 331
Chen, H. 2089
Chen, Hai-Quan 808
Chen, Hemu 1908, 1987
Chen, J. J. 1932
Chen, Jianjun 108
Chen, Jie 135
Chen, Jing 2272
Chen, Jinqiang 2506
Chen, Liang 772

Chen, Lili 2494
Chen, Limei 1636
Chen, Ming 2232
Chen, Qiaoling 1209
Chen, Quan 323
Chen, Shanji 2109
Chen, Shuyu 2488
Chen, Ting 2042
Chen, Wei 200
Chen, Xiangzhou 447, 1859
Chen, Xiaolin 1108
Chen, Xiaoxiao 568, 589
Chen, Xueli 1869
Chen, Yanfang 693
Chen, Yazhen 2720
Chen, Yong 1869
Chen, Yuanyuan 1662
Chen, Yueyue 108
Chen, Zhi 1554
Chen, Zhonghe 553
Cheng, D. S. 1932
Cheng, Si 2323
Cheng, Wang 701, 738
Cheng, Yang 510
Chenglin, Zhang 144
Chevrier, Pierre 732
Chi, Zhang 562
Chong, Gao 168
Chu, Qianqian 2278
Chu, Shibo 1519
Chu, X. M. 1801
Chuan, Wang 716
Chuanli, Wang 701, 738
Chunmei, Li 1675
Cong, Wang 2252
Cui, Fang Y. 894
Cui, Jing Gang 581
Cui, Wen 1812
Cuicui, Liu 168
Dai, Jian 2551
Dai, Zongmiao 1667
Dandan, Sun 888
Danyang, Li 2211
Deng, Fanyi 93
Deng, Ming-Ji 1353
Deng, Qinghong 882
Deng, Qishu 1342
Deng, Shaoxiang 214

Deng, Xin	1412	Gan, Baiqiang	2592
Di, Xiaofeng	2494	Gan, Hua	2304
Diao, Chentao	1543	Gan, Ping	1979
Ding, Baobao	1604	Gang, Wang	434
Ding, Fu-Jun	2371	Gao, Chao	1994
Ding, Hailan	2466, 2559	Gao, Dongliang	409
Ding, Hongyan	732	Gao, Fei	135
Ding, Huixia	1859	Gao, Fei-Fei	181
Ding, Jie	1000	Gao, Hanxu	1359
Ding, Lili	984	Gao, He	1994
Ding, Xiaohua	1812	Gao, Hong	1171
Ding, Yanfeng	1209	Gao, Jian	2368
Dong, Ma X.	1470	Gao, Lijuan	1682
Dong, Sui	1775	Gao, Qingshui	208
Dong, Wei Y.	1782	Gao, Zhenxing	2113
Dong, Xianlei	2421, 2429	Gao, Zihan	1006
Dong, Xiaoming	323, 337, 359	Geng, Lei	1049
Dou, Liang	1604	Gong, Chunwei	2466, 2559
Du, Chunfeng	1656	Gong, Taorong	296
Du, Jiawei	1531	Gong, Xinman	833
Du, Jinyang	453	Gou, Yating	114, 162
Du, Wen	366, 386	Gu, Bochuan	2176
Duan, Lijin	276	Gu, Jingtian	1710
Duan, Lunqin	955	Guan, Denggao	346
Duan, Ming	1575	Guan, Shilei	57
Duan, T.	1834, 2609	Guan, Wanlin	35
Dun, Ao	1166, 1543	Guangyao, Jia	461
Fan, Dandan	2232	Guanhui, Wang	2216
Fan, Jie	102	Gui, Xinyue	2183
Fan, Jinpo	1739	Guiqiang, Liu	888
Fan, Mingqi	2649	Guizhong, Wang	27
Fan, Xingyuan	453	Guo, Gongde	2565
Fang, Wang	1282	Guo, Hejia	640
Fang, Zhuo	168	Guo, Hongwei	102
Feng, Hao B. W.	2624	Guo, Kai-Feng	1190
Feng, Lansheng	258	Guo, Ming-Xuan	808
Feng, Ruzhi	524	Guo, Ronghua	1531
Feng, Shanqiang	2176	Guo, Runqiu	258
Feng, Shunshan	1612	Guo, Shaobing	917
Feng, Wang	168	Guo, Sheng H.	2712
Feng, Xiao	1063	Guo, Siyuan	822
Feng, Xiaoche	2257	Guo, Song	751
Fengbin, Zhang	1450	Guo, Wei	530
Fu, Da	2602	Guo, Xiaoshuang	1656
Fu, Hongyong	2506	Guo, Xing	1441
Fu, Jie	2453	Guo, Xueqi	346
Fu, Jun	2297, 2329	Guo, Yajie	93
Fu, Qixi	1273	Guo, Yingjun	93
Fu, Yifan	882	Guo, Yizhuo	1461
Fu, Yuyang	2036	Guo, Ziteng	574
Fusheng, Chen	144	Haibo, Tan	596
Gai, Lujie	882	Haichao, Chen	607

Haijun, Lei 1974
Haijun, Peng 596
Hai-Lan, Ding 2446
Haiyang, Jiang 1282
Han, Donchen 1395
Han, Jun 1939
Han, Liang 848
Han, Qianru 2021
Han, Quanli 1395
Han, Tao 640
Han, Tongxin 1057
Han, Wang 2572, 2578
Han, Xueshan 331, 372
Hanyan, Wang 2221
Hao, Chen 144, 596
Hao, Cheng 2203
Hao, Chuxue 18
Hao, Jinshun 221
Hao, Junjie 1919
Hao, Li 1286
Hao, Weiguang 263, 787, 795, 802
Hao, Yun 553
Haolin, Jia 1315
He, Fangzheng 1
He, Hongmei 1733
He, Jiangheng 1612
He, Juntao 366, 386
He, Lyulong 1273
He, Ming 1115
He, Renke 2278
He, Shiwei 1939
He, Shuming 2368
He, Xin 1267
He, Yanchen 1758
He, Yidong 1629
He, Yu 1682
He, Yubo 1645
He, Zhiqiang 1599
Hengjie, Li 416
Hong'Bo, Chen 911
Hong-Zhi, Yu 2446
Hou, Aijun 2176
Hou, Lunqing 1412
Hou, Xiangru 1536
Hou, Yan 2746
Hou, Yueqi 2071
Hu, Beibei 2421, 2429
Hu, Chang'An 745
Hu, Dehao 346
Hu, P. C. 1834, 2609
Hu, Shi-Cheng 990
Hu, Yue 228

Hu, Yulan 263, 787, 795, 802
Hu, Yunpeng 2001
Hua, Dengxin 894
Hua, H. Y. Y. 1330
Hua, Wang 434
Hua-Jie, Chen 668
Huang, Bihui 485
Huang, Jingzhi 1228
Huang, Li 2629
Huang, Lin 1979
Huang, Min 2065
Huang, Qiuzi 2488
Huang, R. 2089
Huang, Wei 1029
Huang, Wenhan 693
Huang, Xiaoping 2117
Huang, Xulong 1919
Huang, Yangfan 1979
Huang, Yizhi 693
Huang, Yuwei 1029
Hui, Baofeng 2109
Huimin, Fan 2211
Huimin, Sun 1036
Huitao, Wang 2014
Huiying, Song 461
Ji, Ke 46
Ji, Weiyan 1228
Ji, Yang 808, 2079
Jia, Guangyao 427
Jia, Guoqing 2109
Jia, Hongwei 2706
Jia, Qiang 510
Jia, Shanjie 1859
Jia, Shijie 1594
Jia, Songmin 1166, 1543
Jia, Wenbo 35
Jia, Yafang 2401
Jia, Yunfei 188, 662
Jia, Zhigang 2638
Jiachen, Tian 168
Jiajia, Han 2315
Jian, Wang 875
Jian, Zhou 518
Jianbo, Yin 401
Jiang, Cheng 2323
Jiang, Dawei 1827
Jiang, Hua 1739
Jiang, Juanjuan 1733
Jiang, Xiaoying 2565
Jiang, Z. L. 1801
Jiang, Zhanjun 2283
Jiang, Zhe 372

Jianhui, Zhou	168
Jianwei, Liu	1562
Jianzheng, Liu	63
Jianzhi, Tuo	634
Jianzhong, Yang	401
Jiaojiao, Xi	2544
Jiaxin, Liu	1215, 1222
Jia-Zhi, Yang	479
Jie, Cheng	738
Jie, Huang	768
Jie, Ren	2519
Jiefeng, Mou	1036
Jikang, Wang	63
Jikun, Guo	542
Jin, Fei	78
Jin, Ge	214
Jin, Jian	518
Jin, Li	2544
Jin, Tao	323
Jin, Tiancheng	1604
Jin, Wei	1483
Jin, Weiqi	1955
Jing, Jing	1147
Jing, X. H.	1246
Jing, Zhang	168
Jing, Zhu	1215, 1222
Jingshi, He	2347
Jinjie, Shan	518
Jinliang, Qiu	461
Jinxiu, Wang	888
Jiyao, Tian	1967
Jun, Wei	434
Junning, Qin	2315
Kang, Ruiyu	1788
Kang, Yang	135
Ke, Yan	1477
Ke, Zhang	634
Kong, Juan	1919
Kong, M. X.	1696
Kong, Weizheng	130
Kong, Xiangzeng	2565
Kou, Xu-Peng	2674
Kuaia, Tengfei	905
Kui, Zhang Yong	875
Kun-Yu, Qi	2446
Lai, Ming-Ming	2669
Lan, Ru	2638
Lan, Yunsheng	2048
Le, Guigao	905
Lei, Chu	27
Lei, Lei	1839, 1846
Lei, Min	87

Lei, Wang	195, 2221
Lei, Xiang	1967
Lei, Yiyan	1720
Lei, Zhipeng	379
Lele, Sun	1315
Li, Bin	751
Li, C.	2436
Li, Chunmei	1524, 1889, 1898
Li, Cui	911
Li, Dezhi	296
Li, Guanghui	870
Li, Guanyu	2466, 2515, 2559
Li, Guoqiang	78
Li, Haifeng	323
Li, Hong-Bing	1121, 1510
Li, Huanran	1883
Li, Hui	1883
Li, Huizhi	156
Li, Jiahao	1166
Li, Jian	613, 2152
Li, Jie	1308
Li, Jing	239
Li, Jinping	2706
Li, Jiping	1092
Li, Kai	1524, 1898
Li, Li	822
Li, Lulu	372
Li, Maohua	2662
Li, Meng	1353, 2304
Li, Ming	1084
Li, Mingchao	2001
Li, Minwei	1013
Li, Nianlian	732
Li, Pengyang	894
Li, Ran	1488
Li, Shuangxi	1883
Li, Sicong	35
Li, Tao	2692
Li, Tong	2506
Li, Wang	2460
Li, Wanze	745
Li, Wei H.	1820
Li, Wei	1441
Li, Weichao	78, 1199
Li, Wenbo	331, 372
Li, Wenjing	1406
Li, X. L.	2436
Li, Xiao	2602
Li, Xingxing	745
Li, Xiuzhi	1166, 1543
Li, Xu	2701
Li, Xuefei	135

Li, Yan 894
Li, Yanyun 1979
Li, Ye 2071
Li, Yingqi 833
Li, Yulong 188
Li, Zhi L. 1839, 1846
Li, Zhifei 2036
Li, Zhiming 258
Li, Zhiyuan 1827
Lian, Minlong 1378
Liang, Bin 673
Liang, Dong 1853
Liang, Gang 2304
Liang, Junbin 1418
Liang, Li 738, 1036
Liang, Ning 114, 162
Liang, Shutian 18
Liang, Xi 1450
Liang, Xiaolong 1273, 2071
Liang, Yi 1029
Liang, Ying 57
Liang, Yuqing 1919
Liang, Zhikai 1425
Liao, Daixi 1618
Liao, Minfu 882
Liao, Xiaoming 693
Liao, Zitian 1138
Liling, Liu 2355
Liman, Shen 144
Limei, Zhao 2335
Lin, Dansheng 2267
Lin, Doudou 2429
Lin, Jinghui 346
Lin, Shaofu 2393, 2401
Lin, Sheng 1883
Lin, Wei 668, 1645
Lin, Yao 1336
Lin, Zhang 2079
Lin, Zhaowen 1130, 1488
Ling, Liu 1142
Li-Qing, He 668
Liu, Bin 984, 1503
Liu, Can 870
Liu, Chang 290
Liu, Changli 967
Liu, Cuicui 114, 162
Liu, Di 1300
Liu, Fuyang 123
Liu, G. 263, 787, 795, 802, 1801
Liu, Haikuan 1827
Liu, Hanqing 1292
Liu, Haojie 315

Liu, Huabin 1667
Liu, Jianwei 2048
Liu, Jiawei 653
Liu, Jiaxin 188, 662
Liu, Jie 1554, 2140
Liu, Jingli 78
Liu, Jun 2030, 2058
Liu, Ke Cheng 581
Liu, Kun 2342
Liu, L. 195, 1801
Liu, M. 2089
Liu, Qianru 2283
Liu, Renzhang 78
Liu, Shi 208
Liu, Shuxin 239
Liu, Tingxiang 200
Liu, Wei 955
Liu, Wenchang 276
Liu, Wenda 18
Liu, X. 2089
Liu, Xianglong 1234
Liu, Xiaochun 1812
Liu, Xiaoliang 78
Liu, Xiaoqian 2734
Liu, Xindong 1267
Liu, Xingbao 1365
Liu, Xueyan 2551
Liu, Yajie 263, 787, 795, 802
Liu, Yang 346
Liu, Yangyang 2408
Liu, Yankui 1599
Liu, Ye 2267
Liu, Yi 625, 629
Liu, Yiliang 1378
Liu, Yonggang 315
Liu, Yongxia 2048
Liu, Yu 188, 2679
Liu, Yuting 296
Liu, Yuyan 1788
Liu, Zefeng 1013
Liu, Zhe 1531
Liu, Zhengyi 1441
Liu, Zhenzhen 2629
Liu, Zhizhen 276
Liu, Zilin 263, 787, 795, 802
Liyu, Xia 2572, 2578
Long, Luo 602
Long, Pan 856
Long, Shaohua 1618
Long, Wang 701, 738
Long, Wu 416
Lu, Jiangang 1228

Lu, Jun	409
Lu, Ligen	258
Lu, Shikun	894
Lu, Xiaobo	1688, 2360
Luai, Almadhehagi	427
Luhua, Xing	268
Luo, Lisai	2466
Luo, Shihui	723, 1183
Luo, Taorui	409
Luo, Wanbo	2065
Luo, Zhen	123
Lv, Peihua	1688
Ma, Hongfeng	379
Ma, Jianwei	680, 687
Ma, Kun	2342
Ma, Lulu	305
Ma, Panwei	510
Ma, Pengcheng	1623
Ma, Shiwei	917
Ma, Te	2384, 2388
Ma, Xiaodong	2009
Ma, Xinling	759
Ma, Zhi-Run	2686
Mao, Wanfeng	2649
Mao, Yanrong	427
Mao, Yazhou	315
Maotao, Yang	144, 596
Maoyi, Zhang	1315
Meng, Xiaocheng	366, 386
Meng, Yuting	2001
Mi, Yongsheng	10
Miao, Feng	524, 530
Miao, Lanfang	2036
Min, Huang	2741
Min, Liu	768
Mou, Pengbo	625, 629
Mouhai, Liu	144, 596
Mu, Jiong	581
Mu, Qi	1599
Mu, Senlin	1292
Mu, Xihui	156
Murtaza, Abid	1562
Na, Li	1477
Nannan, Liu	1036
Ni, Xue	228
Nie, Li	1153
Ouyang, Chengtian	2042
Ouyang, L.	2089
Pai, Liu	888
Pan, Fangyu	1153
Pan, Jian	1720
Pan, Qiao	2262

Pang, Xun	673
Peng, Du	634
Peng, Fengzhi	152
Peng, Jianjun	1092
Peng, Lin	2669, 2674, 2686, 2701
Peng, Luxi	2499
Peng, Wang	174
Peng, Yanfei	1092
Penghou, Liu	607
Ping, Wang	888
Qi, Yingchuan	510
Qiang, Li X.	2166
Qiang, Li	2166
Qiang, Lin	152
Qiang, Wu	888
Qiao, Yulong	1962
Qimeng, Nie	1282
Qin, Hua	46
Qin, Luxing	2360
Qing, Wu	2323
Qingjun, Guo	995
Qinyuan, Li	2315
Qiu, Mengyue	2232
Qiu, Zhen	1300
Qiu, Zhiwen	693
Qiuqiu, Wang	1159
Qizhong, Li	1494
Qu, Huaijing	1549
Ran, Jilin	967
Ran, Li	2472
Ren, Jie	905
Ren, Jiyuan	123
Ren, Xun-Yi	1703
Ren, Zhong	814
Ren, Zongjin	305
Riaz, Umair	870
Rong, Tang	2079
Ru, Cong	934, 941, 948
Ru, Zhang	1914
Ruan, Y.	1246
Ruan, Zhenzhen	1979
Rui, Chen	1450
Rui, Zhang	542
Run-Dong, Wang	2519
Ruopeng, Yang	2014
Saeed, Muhammad J.	870
Sang, L. Z.	1834, 2609
Shan, Gao	888
Shao, Bao-Zhu	181
Shao, Juanjuan	1267
Shao, Xuebin	2408
Shao, Zhiyu	1612

She, Jintao	491
Shen, Gao Q.	1839, 1846
Shen, Guiquan	1228
Shen, Wei	2638
Shen, Wuqiang	1228
Shen, Xinxin	1583
Sheng, Jiayue	841
Sheng, Tingran	1057
Sheng, Xuanyu	841
Shengdongt, An	401
Shi, Changkai	57
Shi, Chen	1703
Shi, Haoqiang	2283
Shi, Kai	1883
Shi, Peiji	2408
Shi, Xin	1353
Shi, Zhao-Cun	751
Shi, Zhe	93
Shixu, Li	1036
Shuai, Chen	2245
Shudong, Wang	461
Shuifeng, Zhang	2245
Shushuang, Liang	2656
Shuzhen, Yang	710
Sicong, Li	2472
Situ, Shuwei	1
Song, Deyu	379
Song, Huiying	427
Song, Lihua	130, 1300
Song, Min	2649
Song, Ping	568, 589
Song, Q. H.	2436
Song, Xing	2368
Song, Xiyu	1503
Song, Yuqin	2030
Song, Zilong	1939
Songze, Lei	1063
Su, A. J.	1696
Su, Hongsheng	51
Su, Jiangwen	1300
Su, Tongdan	1115
Su, Y.	2089
Sui, Wei	394
Sun, Chenzhe	2615
Sun, Chuanmin	346
Sun, Cong	305
Sun, Feng	181
Sun, Heng	1623
Sun, Hexu	93
Sun, Hua	337, 359
Sun, Jiabin	2127
Sun, Jianyong	1108

Sun, Juanjuan	1401
Sun, Lin	2291
Sun, Liying	252
Sun, Qian	1359
Sun, Qibo	2429
Sun, Quanxin	1788
Sun, Ruifeng	1013
Sun, Xianhai	768
Sun, Xiao	1949
Sun, Xiaoxiao	848
Sun, Xin	1199
Sun, Yanjun	1166, 1543
Sun, Yao	346
Sun, Yaojie	1745
Sun, Yi	613, 1130, 1488, 1554
Sun, Ying	453
Sun, Yu	1599
Sun, Zhen P.	613
Sun, Zheng	1441
Sun, Zhijie	2297, 2329
Suo, Dong	1675
Suo, Shuangfu	221
Tan, Guangyu	870
Tan, Jinjun	2267
Tan, Ming	1121, 1510
Tan, Yukun	1425
Tang, B. M.	2436
Tang, Guoshen	276
Tang, Jinjin	2528
Tang, Jun	1908
Tang, Shaofan	1378
Tang, Xiao	290
Tang, Xinhuai	2291
Tang, Yanqun	2001
Tang, Ying	2537
Tao, Kepeng	35
Tao, Yu	710, 716
Tao, Zhengping	2528
Teng, Xiaofei	1071
Tian, Bin	502
Tian, Feng	2453
Tian, Jiachen	114, 162
Tian, Jin	2021
Tian, Jing-Jing	2371
Tian, Zhengbing	252
Tianfang, Wu	2245
Tight, Miles	1788
Ting, Ding	461
Tingting, Liu	1142
Tong, Fei	1908, 1987
Tong, Guan	888
Tong, Li	1503

Tu, Jingzhe	323	Wang, Shuyuan	1100
Tu, Jinlong	10	Wang, Siyue	1594
Tu, Yaqing	1084	Wang, Song	2515
Tu, Yongcheng	1908, 1987	Wang, Tao	1733
Wan, Hongqiang	1395	Wang, Ting	1100
Wan, Xing	2065	Wang, Wen	1049
Wanbo, Luo	2741	Wang, Wenjie	221
Wang, Bo	1029	Wang, Wen-Si	1703
Wang, C. Y.	1750	Wang, Wentao	46
Wang, Caishen	290	Wang, Xingong	46
Wang, Chang'An	2734	Wang, X. S.	1932
Wang, Chao	2297, 2499	Wang, Xi	2042
Wang, Chaochao	337	Wang, Xiangpei	1919
Wang, Chunlin	1108	Wang, Xianli	524, 530
Wang, Chun-Yang	1121, 1510	Wang, Xiaogang	1153
Wang, Denggui	759, 826	Wang, Xiaolan	200
Wang, Dong	510, 1782, 2597	Wang, Xiaoming	2304
Wang, Endong	1877	Wang, Xingong	46
Wang, Enshi	485	Wang, Xuewei	548
Wang, Fei	1412	Wang, Yan	346, 1043
Wang, Feng	114, 162	Wang, Yanan	282, 447, 1256, 1859
Wang, Guanhong	296	Wang, Yang	221, 282, 1256, 1979
Wang, Guirong	833	Wang, Yanyan	955
Wang, Haibin	2323	Wang, Yaokun	57
Wang, He	917	Wang, Yifan	1662
Wang, Heng	640	Wang, Yijing	1100
Wang, Hengbin	1549	Wang, Yingjing	662
Wang, Hu	808	Wang, Yisheng	2140
Wang, Hui	135	Wang, Yuanmin	1919
Wang, Jiawen	1629	Wang, Yudong	409
Wang, Jing	252	Wang, Yujiang	1029
Wang, Jiong	2071	Wang, Yuqiao	2692
Wang, Juan	2453	Wang, Zhao	2408
Wang, Ke	282, 447, 1256, 1859	Wang, Zhaoqing	1342, 1348, 2134
Wang, Kun	200	Wang, Zhe	123
Wang, L.	1834	Wang, Zheng	653
Wang, Lanwen	841	Wang, Zhiping	290
Wang, Li	2297, 2329	Wang, Zhiying	1629
Wang, Lihua	228	Wang, Zhi-Yuan	990
Wang, Lingxia	826	Wannian, Zhu	1967
Wang, Lingxue	1955	Wei, Caisheng	574
Wang, Lingyu	2421	Wei, Chen	416
Wang, Liquan	934, 941, 948	Wei, Liu	1927
Wang, Minghao	620	Wei, Pi	1675
Wang, Nian	1908, 1987	Wei, Qianwen	2393
Wang, Ping	2140	Wei, Shicheng	1029
Wang, Qi	2238, 2323	Wei, Zhang	634
Wang, Qingjia	2342	Wei, Zhengxian	2649
Wang, Qiuling	130	Wei, Zheyu	46
Wang, Runjiao	955	Weidong, Xu	1967
Wang, Shiyu	2262	Weihai, Li	1927
Wang, Shudong	427	Weihua, Ma	723, 1183
		Wei-Jun, Pan	2519

Weiwei, Qi	1142
Wen, Chang	2679
Wen, Chao	826
Wen, Guangqi	1889
Wen, Junhao	2488
Wenbo, Li	268, 352
Wencan, Ding	1775
Wensheng, Yin	562
Wenwen, Jiao	2472
Wenxue, Liu	268
Wu, Chunshang	2377
Wu, Hao	1575
Wu, Haobo	2048
Wu, Hong	1962
Wu, Hongmei	1919
Wu, Jianhong	2499
Wu, Jinbo	822
Wu, Jun	1049
Wu, Jun-Jie	808
Wu, Kezhuang	1418
Wu, Qinqin	2267
Wu, Qiong	1006, 2123
Wu, Ran	2123
Wu, Tong	1604
Wu, W. W.	1750
Wu, Xiaoquan	2267
Wu, Xuehui	2360
Wu, Yingying	502
Wu, Yusi	2140
Wu, Zhiqiang	1554
Wufan	2014
Xi, Hongyan	2140
Xi, Qi	1575
Xia, Bin	1240
Xia, Peng	1336
Xia, Rongzhen	379
Xia, Sibin	2393, 2401
Xia, Yangqiu	1365
Xianfang, Tang	1914
Xiangbin, Liu	144
Xiangguo, Su	2252
Xiangzhou, Chen	282
Xiao, Binjie	1251
Xiao, Zhitao	1049
Xiaofei, Zou	2014
Xiaokun, Wang	2221
Xiaomei, Hu	710, 716
Xiaoping, Li	1063
Xiao-Shu, Wang	2098
Xiaotie, Ma	2335
Xiaoyan, Zhang	1159
Xie, Cheng	108

Xie, Dong-Fan	2371
Xie, Feng	2329
Xie, Lingling	1
Xie, Minzhen	1078
Xie, Yong-Jun	1190
Xin, Bo	1292
Xin, Li	1359
Xin, Ma	2203
Xin, Peizhe	409
Xin, Sun	2315
Xin, Xiaoyu	2478
Xin, Yan	875
Xin, Yang	1967
Xincheng, Ren	471
Xing, Wan	2741
Xinliang, Cao	471
Xin-Zhe, Yin	479
Xu, Binshi	1029
Xu, Feng	1919
Xu, Gang	1733
Xu, Guangping	1883
Xu, Guanli	346
Xu, Hongkui	1549
Xu, Jie	1130, 1488
Xu, Peiyuan	2727
Xu, Sanchuan	536
Xu, Shiping	1531
Xu, Wanjin	1108
Xu, Wei	2706
Xu, Wenjing	315
Xu, Xiangqian	640
Xu, Xiaoshen	2232
Xu, Xin	2297, 2329
Xu, Yugong	2203
Xu, Zhuoran	1554
Xue, Qiao	447, 1859
Xue-Chao, Liao	646
Xueshan, Han	352
Xuli, Zhu	1036
Xuxiang, Huang	352
Yachao, Jia	1914
Yan, Bin	1503
Yan, Chunyu	917
Yan, Haotian	2134
Yan, Kedi	73
Yan, Qianghu	87
Yang, Guang	2009
Yang, Guohui	924
Yang, Huiyue	1084
Yang, Jian	46
Yang, Jianxi	315
Yang, Jiebin	745

Yang, Jun-You	181	You, Fucheng	548
Yang, Kai	1503	You, Zhou	646
Yang, Li	995	Youzi, Wang	2572, 2578
Yang, Lin-Nan	2674, 2701	Yu, Fengyun	917
Yang, Lu	2629	Yu, Jiujiu	2160
Yang, Ning	2559	Yu, Ling	1092
Yang, Qunyi	2009	Yu, Lu	553
Yang, W. D.	1696	Yu, Nan	2478
Yang, Wan	668	Yu, Shida	2189
Yang, Wei	2712	Yu, Tianbiao	780
Yang, Weijun	2291	Yu, Tonglan	1554
Yang, Wentai	2304	Yu, Xiaochen	394
Yang, Xiaodan	1433	Yu, Xin	394
Yang, Xiaohua	1554	Yu, Xu R.	1820
Yang, Yi	208	Yu, Xuemei	1745
Yang, Yingming	1604	Yu, Yu	1519
Yang, Yiyong	221	Yuan, Bo	246
Yang, Yongxi	924	Yuan, Ziyan	2746
Yang, Yu	1177	Yue, Chen	27
Yang, Yuansheng	1171	Yue, Dachao	1827
Yang, Yuanyuan	2602	Yun, Mei	1282
Yang, Ziwei	1147	Yunxiao, Zu	1927
Yangjia	1494	Yutinge, Chen	1477
Yanhong, Wang	1063	Zéman, Zoltán	2662
Yanhong, Zuo	861	Zeng, Hanghang	51
Yanrong, Mao	461	Zeng, Jijun	2267
Yanwei, Shang	152	Zeng, Tianlong	2058
Yanzhe, Du	607	Zeng, W. D.	2584
Yao, Jianchun	394	Zeng, Ying	1503
Yao, Jianyu	826	Zeng, Yue	2103
Yao, Jiawei	2195	Zhai, Xiujun	2615
Yao, Ling	2117	Zhai, Yayu	568, 589
Yao, Zheng	394	Zhan, Hong-Yuan	1190
Yating, Gou	168	Zhan, Y. C.	422
Ye, J.	1801	Zhang, Bin	625
Ye, Wang	1494	Zhang, Bo	1199
Ye, Xuanyu	2720	Zhang, C.	1696
Yi, Jun	323	Zhang, Chenglin	2089
Yi, Kang	1720	Zhang, Chengning	924
Yi, Wang	63	Zhang, Chu	208
Yi, Yang Q.	1470	Zhang, Chun J.	2712
Yi, Zhijun	2494	Zhang, Cunlin	1949, 1962
Yin, Aiping	2113	Zhang, Dan	1703
Yin, Ningxia	870	Zhang, Dewen	35
Yin, Yanan	1171	Zhang, Dong	1877
Yin, Zhiqin	1115	Zhang, Fang	1049
Ying, Lin	1282	Zhang, Feng	2272
Yinzheng, Zheng	973	Zhang, G. R.	2584
Yong, Lin	1373	Zhang, Gang	1739
Yonggang, Yue	401	Zhang, Geng	282, 447, 1256
Yonggang, Zhu	861	Zhang, Guan-Feng	181
Yongqiang, Fan	401	Zhang, Guanglei	2615

Zhang, Guoliang 1543
Zhang, Hanhua 2692
Zhang, Hao 366, 386, 882
Zhang, Haoxue 882
Zhang, Hongda 35
Zhang, Hua 2117
Zhang, Huanping 2662
Zhang, J. 1750
Zhang, Jian 826
Zhang, Jianwei 1656
Zhang, Jiaqiang 1273, 2071
Zhang, Jincheng 258
Zhang, Jinwei 745
Zhang, Jun 305
Zhang, Junhao 2262
Zhang, Lan 934, 941, 948
Zhang, Lei 1827
Zhang, Liang 239, 2368
Zhang, Li-Hua 1336
Zhang, Lin 2304
Zhang, M. Y. 1801
Zhang, Min 723, 1183
Zhang, Na 1177
Zhang, Nan 2127
Zhang, Peng 2140
Zhang, Qiang 2551
Zhang, Qingqing 2551
Zhang, Qiurong 693
Zhang, Ran 447, 1256, 1859
Zhang, Rui 228
Zhang, Ruiqi 359
Zhang, Ruiqiu 772
Zhang, S. 1246
Zhang, Shoushou 822
Zhang, Shuo 924
Zhang, Tieshan 814
Zhang, Tinglei 1877
Zhang, Xiangyin 1543
Zhang, Xianmin 772
Zhang, Xiaotong 2238
Zhang, Xiaoying 200
Zhang, Xin 2669, 2686
Zhang, Xincheng 276
Zhang, Xinyu 2048
Zhang, Xinzheng 1267
Zhang, Xuangong 156
Zhang, Xuchong 772
Zhang, Xudong 102
Zhang, Xujuan 2551
Zhang, Yanan 2408
Zhang, Yang 1688
Zhang, Yanjun 239

Zhang, Yichen 1386
Zhang, Yidu 1006
Zhang, Yonghua 1401
Zhang, Yue 2103
Zhang, Z. N. 1330
Zhang, Zhi 290
Zhang, Zhongshi 2238
Zhangkang 2113
Zhanjun, Wang 195
Zhao, Changfang 905
Zhao, Chengqiang 934, 941, 948
Zhao, Enmin 239
Zhao, Guorong 1994
Zhao, H. W. 1246
Zhao, Hui 2488
Zhao, Kai 305
Zhao, Li 1675
Zhao, Lili 258
Zhao, Liujun 130
Zhao, Qian 984
Zhao, Qing 2238
Zhao, Qing-Song 181
Zhao, Yangze 548
Zhao, Ying 453
Zhao, Yuanmeng 1949, 1955, 1962
Zhao, Yuejin 1949, 1962
Zhao, Yufeng 1199
Zhao, Yusheng 1342, 2134
Zhao, Yuting 1720
Zhao, Zengshun 1359
Zhao, Zhi-Qiang 1353
Zhaoren, Pan 710
Zhen, Qiao 416
Zheng, Jinxin 78
Zheng, Kougen 1583
Zheng, Wei 379
Zheng, Yongkang 1667
Zheng, Yufu 1147
Zhen-Huan, Chen 646
Zhenwei, Zhang 174
Zhichao, Guo 63
Zhiyong, Wu 596
Zhizhong, Guo 27
Zhong, Shouming 1618
Zhong, Zhongzhi 1853
Zhongqi, Wang 1494
Zhou, Chun 453
Zhou, Di 1688
Zhou, Guomiao 1575
Zhou, Jian 629
Zhou, Qinqin 2734
Zhou, Rundong 346

Zhou, Shu 1267
Zhou, Xiang 87, 1812
Zhou, Yifan 1365
Zhou, Ying 1531
Zhou, Yong 1908
Zhou, Zeyu... 653
Zhu, Chuangchuang 1273
Zhu, Haoming 296
Zhu, Honghai 1519
Zhu, Hongwei 2048
Zhu, Jingli... 984
Zhu, Junjie 693
Zhu, Leiye.. 2323
Zhu, Ming 1853
Zhu, Weijun 822
Zhu, Xingxiong 2597
Zhu, Yu 581
Zhu, Yuan.. 2147
Zhu, Yuancheng 379
Zhu, Yuefei 1645
Zhu, Zhangqing 1292
Zhu, Zheng.. 1353
Zhu, Zhengbin 934, 941, 948
Zhu, Zhilong 1733
Zhuanga, Duoduo 2113
Zhuo, Fang.. 114, 162
Ziqiang, Lou 2252
Zonghua, Xie 634
Zou, Yuan.. 102
Zu, Yun X... 1820

International Symposium on Power Electronics and Control Engineering (ISPECE 2018)

Journal of Physics: Conference Series Volume 1187

Xi'an, China
28-30 December 2018

Part 2 of 3

ISBN: 978-1-5108-8673-5
ISSN: 1742-6588

Printed from e-media with permission by:

Curran Associates, Inc.
57 Morehouse Lane
Red Hook, NY 12571

Some format issues inherent in the e-media version may also appear in this print version.

This work is licensed under a Creative Commons Attribution 3.0 International Licence.
Licence details: http://creativecommons.org/licenses/by/3.0/.

No changes have been made to the content of these proceedings. There may be changes to pagination
and minor adjustments for aesthetics.

Printed with permission by Curran Associates, Inc. (2026)

For permission requests, please contact the Institute of Physics
at the address below.

Institute of Physics
Dirac House, Temple Back
Bristol BS1 6BE UK

Phone: 44 1 17 929 7481
Fax: 44 1 17 920 0979

techtracking@iop.org

Additional copies of this publication are available from:

Curran Associates, Inc.
57 Morehouse Lane
Red Hook, NY 12571 USA
Phone: 845-758-0400
Fax: 845-758-2633
Email: curran@proceedings.com
Web: www.proceedings.com

TABLE OF CONTENTS

VOLUME 1

Preface

Peer Review Statement

POWER ELECTRONIC EQUIPMENT AND SYSTEM

Study of the Mechanism of Tangent Bifurcation in Voltage Mode Controlled DCM Buck Converter 1
Lingling Xie, Fangzheng He, Shuwei Situ

Research on Transformer Fast OLTC System 10
Jinlong Tu, Yongsheng Mi

Research on the Design of the Fuzzy Control System of Full Bridge DC Converter 18
Wenda Liu, Shutian Liang, Chuxue Hao

Measurement of Inrush Current in Transformer Based on Optical Current Transducer 27
Chu Lei, Guo Zhizhong, Chen Yue, Wang Guizhong

Study of the Standard Sine Wave Frequency Conversion Power Supply Based on Analog and
Digital Integrated Control 35
Wenbo Jia, Hongda Zhang, Dewen Zhang, Kepeng Tao, Sicong Li, Wanlin Guan

Optimal Installation of Distributed Generators Based on an Enhanced Harmony Search Algorithm 46
Ke Ji, Wentao Wang, Xingong Wang, Zheyu Wei, Jian Yang, Hua Qin

Self-Adaptive Control of Rotor Inertia for Virtual Synchronous Generator in an Isolated Microgrid 51
Hanghang Zeng, Hongsheng Su

Software Consistency Checking Method for Distribution Terminal Based on Chaotic Map 57
Yaokun Wang, Ying Liang, Changkai Shi, Shilei Guan

Three-Level Generalized Discontinuous Pulse-Width Modulation Strategy Considering Neutral
Point Potential Balance 63
Wang Jikang, Liu Jianzheng, Wang Yi, Guo Zhichao

Research on Intelligent Charging System Technology of Automobile Group 73
Kedi Yan

Coordination Between Converter-Based Wind Turbines and Synchronous Generators During Inertia
Control 78
Fei Jin, Xiaoliang Liu, Guoqiang Li, Jingli Liu, Weichao Li, Jinxin Zheng, Renzhang Liu

Research of a High Voltage and High Value Resistors Standard Device 87
Xiang Zhou, Xinli Cao, Qianghu Yan, Min Lei

Modeling and Control of DC Microgrid System Based on Hydrogen Production Load 93
Yingjun Guo, Zhe Shi, Yajie Guo, Fanyi Deng, Hexu Sun

A Novel State of Health Estimation Method for Lithium-Ion Battery in Electric Vehicles 102
Jie Fan, Yuan Zou, Xudong Zhang, Hongwei Guo

The Key Design Technology of Successive Approximation Analog-to-Digital Converter to Improve Efficient and Precision 108
Cheng Xie, Yueyue Chen, Jianjun Chen

Research on Reliability Assessment of Thyristor in HVDC Converter Valve 114
Ning Liang, Jiachen Tian, Cuicui Liu, Yating Gou, Fang Zhuo, Feng Wang

Smart Grid and Electric Power Informatization 123
Jiyuan Ren, Zhe Wang, Zhen Luo, Fuyang Liu

Construction of Power Industry Corpus Based on Data Mining and Machine Learning Intelligent Algorithm 130
Liujun Zhao, Weizheng Kong, Qiuling Wang, Lihua Song

Online Identification Method of Induction Motor Parameters Based on Rotor Flux Linkage 135
Xuefei Li, Yang Kang, Hui Wang, Jie Chen, Fei Gao

Design of Power Consumption Tester for HPLC Power Line Carrier Communication Module 144
Liu Mouhai, Zhang Chenglin, Liu Xiangbin, Chen Hao, Yang Maotao, Chen Fusheng, Shen Liman

Research on Power Enterprise Network Security Solution 152
Shang Yanwei, Lin Qiang, Fengzhi Peng

Morphological Analysis of Optocoupler Accelerated Degradation Test Data 156
Xuangong Zhang, Xihui Mu, Huizhi Li

Research on Overvoltage Distribution of HVDC Converter Valve in Special Environment 162
Ning Liang, Cuicui Liu, Yating Gou, Jiachen Tian, Fang Zhuo, Feng Wang

Calculation of Electrical Stress Distribution and Influencing Factors Analysis of HVDC Converter Valve in Special EMP Environment 168
Gao Chong, Zhou Jianhui, Zhang Jing, Gou Yating, Liu Cuicui, Tian Jiachen, Zhuo Fang, Wang Feng

Numerical Simulation of Internal Flow in Direct Burning Coal-Fired Hot Flue Gas Furnace 174
Zhang Zhenwei, Wang Peng

Virtual Synchronous Generator Grid Connected Control Method Based on Virtual Impedance 181
Bao-Zhu Shao, Guan-Feng Zhang, Jun-You Yang, Fei-Fei Gao, Feng Sun, Qing-Song Zhao

Design of Memory Test System for Measuring Transmission Lines Galloping 188
Yulong Li, Jiaxin Liu, Yunfei Jia, Yu Liu

Application of Hybrid Conjugate Gradient Algorithms in Inverse Problems of Electromagnetic Tomography 195
Li Liu, Wang Lei, Wang Zhanjun

Probabilistic Modeling of Output Characteristics Based on ECM Algorithm for Wind Farms 200
Tingxiang Liu, Xiaoying Zhang, Kun Wang, Wei Chen, Xiaolan Wang

Study on the Influence of Insulator on the Coupling Effect of Transmission Tower-Line System 208
Qingshui Gao, Shi Liu, Yi Yang, Chu Zhang

The Impact Research of Delay Time in Steam Turbine DEH on Power Grid 214
Ge Jin, Sun Cai, Shaoxiang Deng

Investigation on the Relationship Between Winding Wire Size and Total Loss of BLDC 221
Jinshun Hao, Shuangfu Suo, Yiyong Yang, Yang Wang, Wenjie Wang

Chaos Control of Bi-Directional DC-DC Converter by Resonant Parametric Perturbation Method in
a DC Microgrid 228
Lihua Wang, Xue Ni, Yue Hu, Rui Zhang

Application Research of Multi-Source Information Fusion Technology in Power Network Fault
Diagnosis 239
Shuxin Liu, Enmin Zhao, Yanjun Zhang, Jing Li, Liang Zhang, Yundong Cao

Research on Power Quality Acquisition and Reconstruction Method Based on Compressed Sensing 246
Bo Yuan

A Study of Simulation on Relationship Between Young's Modulus of Cable Joints and Interface
Pressure Based on Finite Element Method 252
Zhengbing Tian, Liying Sun, Jing Wang, Xiaoye Bai

Effect of Heat Shield on the Heating Efficiency in MOCVD Chamber by Resistive Heating 258
Lili Zhao, Zhiming Li, Jincheng Zhang, Runqiu Guo, Ligen Lu, Lansheng Feng

Study on Basic Experiment and Optimization Prediction Model of Orthogonal Electrolytic
Machining of Film Cooling Hole in High Temperature Nickel-Based Alloy Blades 263
Yulan Hu, Weiguang Hao, Guoqiang Liu, Yajie Liu, Zilin Liu

Multi-Objective Reactive Power Optimization of Hybrid AC/DC Power System Considering Power
System Uncertainty 268
Liu Wenxue, Xing Luhua, Ma Changhui, Li Wenbo

Optimization of Charging Method for Scaled EVs 276
Xincheng Zhang, Zhizhen Liu, Yan Cao, Lijin Duan, Guoshen Tang, Wenchang Liu

Operation Quality Evaluation of Power Communication Network Based on Business QOS
Indicators 282
Geng Zhang, Ke Wang, Yang Wang, Yanan Wang, Chen Xiangzhou

A Stator Flux Calculation Method for Permanent Magnet Synchronous Motor in 60° Coordinate
System 290
Xiao Tang, Zhi Zhang, Chang Liu, Caishen Wang, Zhiping Wang

Analysis of Marketing Strategy of Electricity Selling Companies in the New Situation 296
Taorong Gong, Dezhi Li, Yuting Liu, Guanhong Wang, Haoming Zhu

Study on Zero Drift of Charge Amplifier Based on MOSFET 3N165 and OPA LF356N 305
Zongjin Ren, Cong Sun, Jun Zhang, Lulu Ma, Kai Zhao

Study on Radial Vibration of Circular Piezoelectric Ceramic 315
Yazhou Mao, Jianxi Yang, Haojie Liu, Yonggang Liu, Wenjing Xu

A Method of Power Flow Calculation Considering New FACTS and HVDC 323
Quan Chen, Xiaoming Dong, Haifeng Li, Tao Jin, Jun Yi, Jingzhe Tu

Combined Heat and Power Optimal Dispatch Considering Wind Power Uncertainty 331
Chunlong Chen, Xueshan Han, Wenbo Li

Analysis of Power System Vulnerability Considering Multiple Disturbances Corresponding to Information and Physics 337
Chaochao Wang, Hua Sun, Xiaoming Dong

Influence of Ni-Cu-La-B-Coated Glass Fiber on Conductivity and Electromagnetic Shielding Performance of Coatings 346
Denggao Guan, Yang Liu, Dehao Hu, Rundong Zhou, Xueqi Guo, Yan Wang, Yao Sun, Guanli Xu, Jinghui Lin, Chuanmin Sun

A Renewable Energy Assessment Model Considering the Effect of Frequency Regulation 352
Huang Xuxiang, Han Xueshan, Li Wenbo

The Comparison of Thermal Characteristics of AC Cable and DC Cable 359
Ruiqi Zhang, Hua Sun, Xiaoming Dong

Research on the Protection Range of Bird Droppings of 110kV Transmission Line Based on ANSYS Maxwell 366
Hao Zhang, Renfei Che, Wen Du, Xiaocheng Meng, Juntao He

A Novel Aggregation Method for Doubly Fed Wind Farm 372
Lulu Li, Xueshan Han, Wenbo Li, Zhe Jiang

Research on Substation Perimeter Isolation Based on Phased Array Radar and Multi-Video Fusion Technology 379
Yuancheng Zhu, Zhipeng Lei, Wei Zheng, Hongfeng Ma, Rongzhen Xia, Deyu Song

Fault Causes Identification for Transmission Lines Based on HHT and PNN 386
Wen Du, Renfei Che, Hao Zhang, Xiaocheng Meng, Juntao He

Transmission Line Insulator Fault Detection Based on Ultrasonic Technology 394
Zheng Yao, Xin Yu, Jianchun Yao, Wei Sui, Xiaochen Yu

Insulation Defect Detection of Solid Insulating Material Based on Nanosecond Pulse Voltage 401
Yang Jianzhong, An Shengdongt, Fan Yongqiang, Yin Jianbo, Yue Yonggang

Dynamic Variance Equalization Planning Optimization Method for Power Grid System Protection Communication Network 409
Dongliang Gao, Taorui Luo, Peizhe Xin, Yudong Wang, Jun Lu, Peng Cai

System for Real-Time Transmission Control of AC Motor Temperature Data Based on Linux System 416
Li Hengjie, Wu Long, Chen Wei, Qiao Zhen

Comparison Between UofC Model and Ionosphere-Free Combination Model in PPP 422
Y. C. Zhan

Dynamic Reactive Power Compensation and Harmonic Suppression of Optical Storage Microgrid Control in Natural Coordinates 427
Guangyao Jia, Shudong Wang, Huiying Song, Yanrong Mao, Almadhehagi Luai

Comparison and Analysis of X86 Server and Minicomputer Application in Power Enterprises 434
Yang Bo, Wei Jun, Wang Hua, Wang Gang

Risk Assessment of Power Communication Network Based on LM-BP Neural Network 447
Yanan Wang, Ke Wang, Ran Zhang, Qiao Xue, Xiangzhou Chen, Geng Zhang

Research on Remote Meter Reading Scheme and IoT Smart Energy Meter Based on NB-IoT Technology 453

Xingyuan Fan, Chun Zhou, Ying Sun, Jinyang Du, Ying Zhao

Optimal Configuration of Optical Storage Microgrid Under Demand-Side Response Based on Cooperative Game 461

Wang Shudong, Mao Yanrong, Jia Guangyao, Song Huiying, Qiu Jinliang, Ding Ting

Permittivity Model for GNSS-R Telemetry Wetlands 471

Cao Xinliang, Ren Xincheng

Design of Continuous Automatic Wire-Feeding Device Based on Electric Explosive Wire 479

Ying An-Wen, Yang Jia-Zhi, Yin Xin-Zhe

Calculation and Verification of Voltage Drop When Starting Tunnel Axial-Flow Fan 485

Enshi Wang, Bihui Huang

A Novel Searching Method of Fault Chains for Power System Cascading Outages Based on Quantitative Analysis of Dynamic Interaction Between System and Components 491

Jintao She

INTELLIGENT CONTROL SYSTEM AND MECHANICAL DESIGN

Warehouse Design Model for Shuttle Based Storage and Retrieve System 502

Bin Tian, Yingying Wu

Research on Control Strategy of a Buck-Type Harmonic Injection Three-Phase Rectifier 510

Qiang Jia, Panwei Ma, Yingchuan Qi, Dong Wang, Yang Cheng

Research and Analysis of MRC and IRC Algorithm Based on LTE System 518

Jian Jin, Shan Jinjie, Zhou Jian

Rotor-Mechanical Coupled Fault Feature Extraction Based on Second-Order Blind Identification 524

Feng Miao, Ruzhi Feng, Xianli Wang

Separating for Nonlinear Mixed Rotor Fault Signals Based on Adaptive Particle Swarm Optimization 530

Feng Miao, Xianli Wang, Wei Guo

A Survey of Knowledge-Based Intelligent Fault Diagnosis Techniques 536

Sanchuan Xu

Structure and Simulation of Roadway Disaster Simulation Control System for High Temperature Smoke Drill 542

Guo Jikun, Zhang Rui

Combination of CNN with GRU for Plate Recognition 548

Fucheng You, Yangze Zhao, Xuewei Wang

Methods to Solve Salt &Pepper Noise, and Frame Dropping of Timed Address Event Representation Vision Sensor 553

Lu Yu, Zhonghe Chen, Yun Hao

A Double-Channel Iterative NFXLMS Algorithm Used in Horizontal Vibration Isolation 562

Zhang Chi, Yin Wensheng

Projectile Velocity Measurement System Based on PVDF and Data Processing Method 568
  Xiaoxiao Chen, Ping Song, Yayu Zhai

Some New Results on the Finite-Time Control and Its Application to a Chemical Reactor System 574
  Ziteng Guo, Caisheng Wei

Dynamic Weighing System Based on Internet of Things Technologies 581
  Jing Gang Cui, Jiong Mu, Ke Cheng Liu, Yu Zhu

Application of Weighted Fusion Algorithm in Air Tightness Detection Device 589
  Xiaoxiao Chen, Ping Song, Yayu Zhai

Design of Intelligent Commutation Switch System Based on HPLC Carrier Scheme 596
  Liu Mouhai, Tan Haibo, Yang Maotao, Chen Hao, Wu Zhiyong, Peng Haijun

Research on Four Axis Manipulator Trajectory Tracking Based on RBF Neural Network Algorithm 602
  Luo Long

Design of Multifunctional Intelligent Security Robot Based on Single Chip Microcomputer 607
  Liu Penghou, Chen Haichao, Du Yanzhe

Multi-Lane Detection using CNNs and a Novel Region-Grow Algorithm 613
  Yi Sun, Jian Li, Zhen P. Sun

A Survey of Cloud Computing Access Control Technology 620
  Minghao Wang

The Torsion Bars System Reliability Analysis with Failure Mode in Crawler Vehicle 625
  Yi Liu, Pengbo Mou, Bin Zhang

Simulation and Life Prediction of Gear Meshing Process of Gearbox of a Crawler Vehicle 629
  Pengbo Mou, Yi Liu, Jian Zhou

Study on Cargo-Swing Reduction of General Gantry Crane using Hybrid Optimal Input Shaper 634
  Tuo Jianzhi, Du Peng, Zhang Ke, Zhang Wei, Xie Zonghua

Design and Analysis of the Leveling Hydraulic System of the Combine Harvester 640
  Heng Wang, Shukun Cao, Xiangqian Xu, Tao Han, Hejia Guo

Research on Fast Self-Learning Improvement of ADRC Control Algorithm for Film Thickness
Control System 646
  Liao Xue-Chao, Zhou You, Chen Zhen-Huan

Research and Analysis of Intelligent RGV Based on Dynamic Scheduling Optimization Model 653
  Zheng Wang, Zeyu Zhou, Jiawei Liu

Research on Intelligent Near-Power Early Warning System for Mechanical Vehicles 662
  Yingjing Wang, Jiaxin Liu, Yunfei Jia

Design and Research of a Aero Engine Operating Status Monitoring System 668
  Wei Lin, He Li-Qing, Wan Yang, Chen Hua-Jie

Motion Recognition Based on Kinect for Human-Computer Intelligent Interaction 673
  Xun Pang, Bin Liang

Parameter Optimal Design and Simulation of Power System of Electric Vehicle Based on AVL-CRUISE 680
Jianwei Ma

Parameter Design and Simulation Analysis of Power System in Plug-In Hybrid Vehicle 687
Jianwei Ma

Effect of Injection Compression Process Parameters on Residual Stress of Products Based on Numerical Simulation 693
Junjie Zhu, Yanfang Chen, Wenhan Huang, Qiurong Zhang, Xiaoming Liao, Yizhi Huang, Zhiwen Qiu

Multiphysics Modelling of Warm Shot Peening of AISI 4140 Steel 701
Wang Cheng, Wang Long, Wang Chuanli

Study on the Control System of Agaricus Bisporus Picking Robot 710
Hu Xiaomei, Pan Zhaoren, Yang Shuzhen, Yu Tao

Design and Application of Visual System in the Agaricus Bisporus Picking Robot 716
Hu Xiaomei, Wang Chuan, Yu Tao

Application of Levitation Frame with Mid-Set Air Spring on Maglev Vehicles 723
Min Zhang, Ma Weihua, Shihui Luo

Study on Preparation Methods of Copper-Based Composites 732
Nianlian Li, Vanessa Bouchart, Pierre Chevrier, Hongyan Ding

Study on the Performance of the Wind Turbine Airfoil with Icing 738
Wang Long, Wang Cheng, Cheng Jie, Wang Chuanli, Li Liang

Research on Straightness Error Detection and Quality Control of Multi-Crankshaft Bores for Large Medium Speed Engine Block 745
Jiebin Yang, Xingxing Li, Chang'An Hu, Wanze Li, Jinwei Zhang

Study on Pressure Pulsation Suppression of Reciprocating Pump 751
Bin Li, Song Guo, Biao-Hua Cai, Zhao-Cun Shi

Design of Automated Guided Vehicle for Conveying Objects 759
Denggui Wang, Xinling Ma

A Hydraulic Fault Diagnosis Method Based on IMF Entropy Feature Fusion 768
Liu Min, Huang Jie, Xianhai Sun

Natural Gesture Control of a Delta Robot using Leap Motion 772
Xuchong Zhang, Ruiqiu Zhang, Liang Chen, Xianmin Zhang

Research on Manufacturing Technology of Thin-Walled Parts of Fe105 Metal Based on Laser Cladding 780
Tianbiao Yu, Yiting Bao

Research on Electrolyte Jet Assisted Laser Micromachining Technology 787
Yulan Hu, Weiguang Hao, Guoqiang Liu, Yajie Liu, Zilin Liu

Thermal Barrier Coating Processing Based on Improved Ant Colony Algorithm Process Optimization and Verification 795
Yulan Hu, Guoqiang Liu, Weiguang Hao, Yajie Liu, Zilin Liu

Experimental Study on Regression Model of Ultraviolet Laser Processing Thermal Barrier Coating Based on Response Surface Method 802

Yulan Hu, Guoqiang Liu, Weiguang Hao, Yajie Liu, Zilin Liu

Shape Optimization of Hook for Marine Crane... 808

Yang Ji, Hu Wang, Hai-Quan Chen, Ming-Xuan Guo, Jun-Jie Wu

Design and Experimental Study of Vibration Reducing Experimental Device for Magneto-Rheological Elastomer... 814

Tieshan Zhang, Zhong Ren

A Simple Safety Control Method for PSS Critical Gain Test... 822

Siyuan Guo, Shoushou Zhang, Weijun Zhu, Li Li, Jinbo Wu

Design of Seat Clamping Device for Automobile DOF Shaker 826

Jian Zhang, Jianyu Yao, Chao Wen, Denggui Wang, Lingxia Wang

Contour Error Control of X-Y Platform Based on Nominal Model in Polar Coordinate System..................... 833

Guirong Wang, Xinman Gong, Yingqi Li

Analysis of the Pressure Expansion of Bridge Plug Tools and Packers by Equivalent Material Method 841

Lanwen Wang, Xuanyu Sheng, Jiayue Sheng

A New Numerical Force Analysis Method of CBR Reducer with Tooth Modification.................................. 848

Xiaoxiao Sun, Liang Han

Influence of Electropulsing Treatment on Residual Stresses and Tensile Strength of As-Quenched Medium Carbon Steel... 856

Pan Long

High Accuracy Numerical Simulation on 3D Weld-Pool Shape of Large Parts 861

Zhu Yonggang, Zuo Yanhong

Experimental Study on Cutting Force Comparison Between Inner Cooling and Outer Cooling in Zig-Zag Milling.. 870

Umair Riaz, Can Liu, Guangyu Tan, Guanghui Li, Ningxia Yin, Muhammad J. Saeed

Study on Vibration Reduction of Crane Monitoring System 875

Wang Jian, Zhang Yong Kui, Yan Xin

Research on Intelligent Communication System for Circuit Breaker Condition Monitoring........................... 882

Qinghong Deng, Hao Zhang, Minfu Liao, Haoxue Zhang, Yifan Fu, Lujie Gai

Research on Energy Saving and Consumption Reduction Technology of Underground Gas Storage Compressor.. 888

Guan Tong, Liu Guiqiang, Wang Jinxiu, Wang Ping, Sun Dandan, Gao Shan, Liu Pai, Wu Qiang

Load and Stress Distribution of Thread Pair and Analysis of Influence Factors 894

Shikun Lu, Dengxin Hua, Yan Li, Fang Y. Cui, Pengyang Li

Study on the Effect of Temperature on Dynamic Characteristics of Rotor System with Straight Crack 905

Tengfei Kuaia, Changfang Zhao, Jie Ren, Guigao Le

Mechanical Productivity Design and Mechanical Process Analysis Framework Construction 911
  Cui Li, Chen Hong' Bo

VOLUME 2

Simulation and Experimental Research on the Influence of Tool Geometries on the Cutting Force of High Temperature Alloy 917
  Fengyun Yu, Chunyu Yan, Shaobing Guo, Shiwei Ma, He Wang

Research on Extraction and Analysis of Characteristic Conditions of Hub Motor for Electric Vehicles 924
  Guohui Yang, Shuo Zhang, Chengning Zhang, Yongxi Yang

Design and Experimental Research of Expansion-Anchorage Device in Deepwater Pipeline 934
  Lan Zhang, Chengqiang Zhao, Liquan Wang, Cong Ru, Zhengbin Zhu

Study on Wear Mechanism of Diamond Particles in the Cutting of Pipeline Steel 941
  Lan Zhang, Zhengbin Zhu, Liquan Wang, Chengqiang Zhao, Cong Ru

Analysis of Impact Characteristics of Diamond-Beaded Rope and Its Influence on Cutting Efficiency and Life 948
  Lan Zhang, Cong Ru, Liquan Wang, Zhengbin Zhu, Chengqiang Zhao

Analysis of Mesh Stiffness of Herringbone Gear Considering Modification 955
  Wei Liu, Lunqin Duan, Runjiao Wang, Yanyan Wang

Modeling of the Stiffness of Corrugated Cardboard Considering Material Non-Linear Effect 967
  Jilin Ran, Changli Liu

Numerical Analysis on Fluid-Solid Coupling Cooling of Minimal Surface Lattice Structure 973
  Zheng Yinzheng

Preparation and Characterization of Composite Resin Containing Anion Powder 984
  Lili Ding, Jingli Zhu, Qian Zhao, Bin Liu

Research on Fuzzy PID Control of Forearm of Tunnel Steel Arch Mounting Machine 990
  Zhi-Yuan Wang, Shi-Cheng Hu

Early Fault Diagnosis of Rolling Bearing Based on Lyapunov Exponent 995
  Guo Qingjun, Li Yang

The Effect of Longitudinal Shock Absorber on the Vibration Response of Train-Bridge Coupling System in the Articulated Train 1000
  Jie Ding

Simulation and Experiment of Passive Orbit Disconnected Support 1006
  Zihan Gao, Yidu Zhang, Qiong Wu

Research on Trajectory Tracking and Vibration Suppression of a Smart Flexible-Joint-and-Link Space Manipulator 1013
  Zefeng Liu, Ruifeng Sun, Minwei Li

Effect of Ultrasonic Treatment on Morphology and Microwave Absorption Performance of ZnO Spheres 1029
  Wei Huang, Shicheng Wei, Yi Liang, Bo Wang, Yuwei Huang, Yujiang Wang, Binshi Xu

Effect of Different Volume Fraction Magnetorheological Fluids on Its Shear Properties 1036
Sun Huimin, Zhu Xuli, Liu Nannan, Mou Jiefeng, Li Liang, Li Shixu

Research on Butt Joint of Ultrafine Grained Steel of Manual Arc Welding 1043
Yan Wang

Fiber Diameter Measuring Method of Textile Materials Based on Phase Information 1049
Wen Wang, Fang Zhang, Zhitao Xiao, Lei Geng, Jun Wu

Mechanism Analysis of Ferromagnetic Resonance of Electromagnetic Voltage Transformer in
Neutral Ungrounded System 1057
Tingran Sheng, Tongxin Han

Fast Aerial UAV Detection using Improved Inter-Frame Difference and SVM 1063
Li Xiaoping, Lei Songze, Zhang Boxing, Wang Yanhong, Xiao Feng

Discussion About Artificial Intelligence's Advantages and Disadvantages Compete with Natural
Intelligence 1071
Xiaofei Teng

Development of Artificial Intelligence and Effects on Financial System 1078
Minzhen Xie

A Variable Step-Size Adaptive Notch Filter for Frequency Estimation using Combined Gradient
Algorithm 1084
Huiyue Yang, Yaqing Tu, Ming Li

Smart Home System Based on Deep Learning Algorithm 1092
Yanfei Peng, Jianjun Peng, Jiping Li, Ling Yu

Bounded Noises Estimation Based on Cognitive Radio in Distributed Fusion System 1100
Shuyuan Wang, Ting Wang, Yijing Wang

Datacentre TCP Protocol of Centralized Window Control 1108
Chunlin Wang, Jianyong Sun, Wanjin Xu, Xiaolin Chen

Gas Packaging Container Based on ANSYS Finite Element Analysis and Structural Optimization
Design... 1115
Zhiqin Yin, Tongdan Su, Ming He

Influence of Double Stealth Aircraft Approach Forward Support Cooperative Jamming on Radar
Detection Performance 1121
Lei Bao, Chun-Yang Wang, Hong-Bing Li, Juan Bai, Ming Tan

MD-UCON: A Multi-Domain Access Control Model for SDN Northbound Interfaces................................. 1130
Rui Chang, Zhaowen Lin, Yi Sun, Jie Xu

Design Research on Information Coding System Under the Concept of Agile Manufacturing 1138
Zitian Liao

Research on Milling Force Prediction Model Based on Improved Particle Swarm Optimization
Algorithm 1142
Liu Ling, Qi Weiwei, Liu Tingting

Optimization of Adaptive Handover Algorithm Based on Distributed Antenna in LTE-R............................. 1147
Ziwei Yang, Yufu Zheng, Jing Jing

A Game-Theory Approach Based on Genetic Algorithm for Flexible Job Shop Scheduling Problem............1153
Li Nie, Xiaogang Wang, Fangyu Pan

An Improved Hybrid Structure Multi-Classification Support Vector Machine 1159
Zhang Xiaoyan, Wang Qiuqiu

Hand-Eye Calibration for Flexible Manipulator..1166
Jiahao Li, Xiuzhi Li, Ao Dun, Songmin Jia, Yanjun Sun

Bounds on the Total Signed Domination Number of Generalized Petersen Graphs $P(n,3)$.............................1171
Hong Gao, Yanan Yin, Yuansheng Yang

Change Impact Analysis of Complex Mechanical Product Based on Complex Network Theory 1177
Na Zhang, Yu Yang

Influence of Variable Slip Frequency Control Strategy on Tractiv E Performance 1183
Min Zhang, Ma Weihua, Shihui Luo

Study on Detection System of Grooved Rail Based on Inertial Measurement - Laser Triangulation
Comprehensive Algorithm...1190
Hong-Yuan Zhan, Kai-Feng Guo, Yong-Jun Xie

Mimic Defense Structured Information System Threat Identification and Centralized Control 1199
Bo Zhang, Weichao Li, Xin Sun, Yufeng Zhao

SMC Chaos Control of a Novel Hyperchaotic Finance System using a New Chatter Free Sliding
Mode Control 1209
Guoliang Cai, Yanfeng Ding, Qiaoling Chen

Project Evaluation and Analysis of Metrological Verification Regulation Based on Fuzzy
Comprehensive Analysis Method...1215
Zhu Jing, Liu Jiaxin

Safety Adaptability of Engine Retarder (Jacobs) on Long Downhill of Expressways 1222
Liu Jiaxin, Zhu Jing

Design of Information System Vulnerability Governance Platform Based on Distributed Asset
Acquisition and Vulnerability Verification Radar...1228
Guiquan Shen, Jiangang Lu, Wuqiang Shen, Jingzhi Huang, Weiyan Ji

Standard Architecture of China Intelligent Bus Systems...1234
Xianglong Liu

Distributed Scalable Abstract Reasoning Based on Dl-Lite 1240
Bin Xia

Cloud Resource Adaptive Scheduling Framework and Optimization Strategy Based on Swarm
Intelligence 1246
H. W. Zhao, S. Zhang, Y. Ruan, X. H. Jing

Research on Energy-Saving Lighting Control System of Tram Station Based on Traffic and
Passenger Flow Information 1251
Binjie Xiao

APPLICATION OF COMPUTER NETWORK AND INFORMATION TECHNOLOGY

The Optimization of Networking Method for the System Protection Communication Networks
Based on the Delay Analysis 1256
Yanan Wang, Ke Wang, Ran Zhang, Geng Zhang, Yang Wang

The Research of Ship Yaw Detection Method Based on Virtual Navigation Channel 1267
Juanjuan Shao, Shu Zhou, Xin He, Xinzheng Zhang, Xindong Liu

Configuration Generation of Aircraft Swarm Based on Communication Distance Constraint 1273
Qixi Fu, Xiaolong Liang, Jiaqiang Zhang, Lyulong He, Chuangchuang Zhu

Parameter Estimation Algorithm and Application in Industry Design 1282
Mei Yun, Jiang Haiyang, Lin Ying, Wang Fang, Nie Qimeng

A Multi-Model Estimation of Distribution Algorithm 1286
Li Hao

Pruning the Deep Neural Network by Similar Function 1292
Hanqing Liu, Bo Xin, Senlin Mu, Zhangqing Zhu

Application of Internet Segmentation Research Based on Natural Language Processing Technology
in Enterprise Public Opinion Risk Monitoring 1300
Di Liu, Jiangwen Su, Lihua Song, Zhen Qiu

A Gesture Recognition Algorithm Based on Threedimensional Projection and Direction Chain
Code 1308
Jie Li

Model and Design of High Temperature and Thermal-Proof Garment using Genetic Algorithm 1315
Zhang Maoyi, Sun Lele, Jia Haolin

The Design of Analog Signal Communication System Based on Visible Light 1330
Z. N. Zhang, H. Y. Y. Hua

Simulation Study of Dispersion Compensation in Optical Communication Systems Based on
Optisystem 1336
Peng Xia, Li-Hua Zhang, Yao Lin

The Comparison of Crowd Counting Algorithms Based on Computer Vision 1342
Zhaoqing Wang, Qishu Deng, Yusheng Zhao

Detector Design Based on MIMO OTA Test 1348
Zhaoqing Wang

Research on a Fusion Gait Real-Time Recognition Algorithm 1353
Zhi-Qiang Zhao, Meng Li, Ming-Ji Deng, Zheng Zhu, Xin Shi

Application Research of Denoising and Super Pixel Algorithm in Image Processing 1359
Qian Sun, Li Xin, Hanxu Gao, Faliang Chang, Zengshun Zhao

Evaluation Method and Experimental Study on Stationarity of High-Precision Linear Motion 1365
Yifan Zhou, Xingbao Liu, Yangqiu Xia, Shaopeng Cai

The Design of Image Depth Information Extraction Algorithm Based on Joint Bilateral Filtering 1373
Lin Yong

Research on Light-Small Lens Structure Design and Weight Reduction Optimization Based on Neural Network 1378
Shubing An, Minlong Lian, Yiliang Liu, Shaofan Tang

The Theoretical Development and Prospect of Two-Dimensional Topological Insulators 1386
Yichen Zhang

Calculation Formula of Positioning Error Based on Three Dimensions and Four Datum 1395
Quanli Han, Hongqiang Wan, Donchen Han

Application and Realization of Ray Tracing in Network Planning of Wireless Private Network 1401
Yonghua Zhang, Juanjuan Sun

Sparse Manifold Learning Based on Laplacian Matrix 1406
Wenjing Li

Web Advertisement Detection using Naive Bayes 1412
Xin Deng, Lunqing Hou, Fei Wang

Path Planning in Mobile Wireless Sensor Networks 1418
Kezhuang Wu, Junbin Liang

Defect Detection and Recognition Based on ADABOOT-SVM Integrated Model 1425
Zhikai Liang, Shaozhong Cao, Yukun Tan

Unmanned Visual Localization Based on Satellite and Image Fusion 1433
Xiaodan Yang

Cooperative Warp of Two Discriminative Features for Skeleton Based Action Recognition 1441
Zheng Sun, Xing Guo, Wei Li, Zhengyi Liu

Anomaly Detection Algorithm Based on FCM with Improved Krill Herd 1450
Chen Rui, Zhang Fengbin, Xi Liang

An Improved Parallelization of K-Means Algorithm Based on HADOOP 1461
Yizhuo Guo

Remote Sensing Image Building Extraction Based on Deep Convolutional Neural Network 1470
Yang Q. Yi, Wang W. An, Ma X. Dong

Speaker Identification Based on Deep Learning in FX iDeal System 1477
Yan Ke, Li Na, Chen Yutinge

The Improvement of K-NN Classifier with GA-Based Weight-Tunning Method 1483
Wei Jin

A Route Optimization Model Based on Link State Awareness in SDN 1488
Ran Li, Zhaowen Lin, Jie Xu, Yi Sun

Numerical Simulation of Deep Learning Algorithm for Gas Explosion in Confined Space 1494
Li Qizhong, Wang Ye, Yangjia, Wang Zhongqi

Study on Temporal and Spatial Patterns of Brain in Emotional State Based on Steady State Visual Evoked Potentials 1503
Kai Yang, Ying Zeng, Li Tong, Bin Liu, Xiyu Song, Bin Yan

Parameter Analysis of Stepped Frequency Pulses Frequency Diverse Array Radar 1510
Ming Tan, Chun-Yang Wang, Hong-Bing Li, Juan Bai, Lei Bao

WebVOS-A WebGIS Application for Volunteer Observation Ships 1519
Honghai Zhu, Yu Yu, Shibo Chu

Improvement of LDA Topic Mining Algorithm and Its Application in Short Text 1524
Kai Li, Chunmei Li

A Security Model Based on Intelligent Decision 1531
Shiping Xu, Ying Zhou, Ronghua Guo, Jiawei Du, Zhe Liu

A New Clustering Validity Index Based on K-Means Algorithm 1536
Xiangru Hou

Object Detection Based on the Improved Single Shot MultiBox Detector 1543
Songmin Jia, Chentao Diao, Guoliang Zhang, Ao Dun, Yanjun Sun, Xiuzhi Li, Xiangyin Zhang

Medical Image Fusion Based on Statistical Modeling 1549
Huaijing Qu, Hengbin Wang, Hongkui Xu

Automatic Integration Testing Through Collaboration Diagram and Logic Contracts 1554
Yi Sun, Xiaohua Yang, Jie Liu, Tonglan Yu, Zhuoran Xu, Zhiqiang Wu, Zhi Chen

Multipurpose IP-Based Space Air-Ground Information Network 1562
Abid Murtaza, Liu Jianwei

ChanDet: Detection Model for Potential Channel of iOS Applications 1575
Guomiao Zhou, Ming Duan, Qi Xi, Hao Wu

Characterizing of Strong Normalization for Λμ-Calculus 1583
Xinxin Shen, Kougen Zheng

Signature Handwriting Identification Based on Generative Adversarial Networks 1594
Siyue Wang, Shijie Jia

Object Detection on Underground Low-Quality Images 1599
Qi Mu, Zhiqiang He, Yankui Liu, Yu Sun

ArchiMate Customization and Architecture Repository Management Practices: For a Technology-Intensive Enterprise 1604
Baobao Ding, Tong Wu, Yingming Yang, Liang Dou, Tiancheng Jin

Sea Clutter Suppression Based on Sea Spikes Identification and Matrix Completion 1612
Zhiyu Shao, Jiangheng He, Shunshan Feng

Exponential Stability for a Class of Nonlinear Singular Markovian Jump Systems with Time-Delay 1618
Daixi Liao, Shouming Zhong, Shaohua Long

A New Molecular Encryption Model Based on Microfluidic Techniques 1623
Pengcheng Ma, Heng Sun

Multi-Peak and Power Cooperative Detection Algorithm to Detect Forwarded Spoofing Interference Signals of BOC Modulation Receivers 1629
Zhiying Wang, Jiawen Wang, Yidong He

Construction of on-Campus 3D Model Based on GIS Technology and OpenGL 1636
    Limei Chen

HTTP Tunnel Trojan Detection Model Based on Deep Learning.. 1645
    Yubo He, Yuefei Zhu, Wei Lin

Design and Implementation of Connect6 Intelligent Game System... 1656
    Zengyu Cai, Chunfeng Du, Xiaoshuang Guo, Jianwei Zhang

Automated Recognition of Retinopathy of Prematurity with Deep Neural Networks 1662
    Yifan Wang, Yuanyuan Chen

Research on Gait Recognition Algorithm Based on Multiinformation Perception.. 1667
    Huabin Liu, Zongmiao Dai, Yongkang Zheng

The Combination of Neural Network and "question Matching" Improves the Correct Rate of
Grassland Degradation Decision 1675
    Li Chunmei, Pi Wei, Dong Suo, Li Zhao

The Design and Development of Simulation System for Broad Band Wireless Communication 1682
    Lijuan Gao, Yu He

A Deep Learning Approach for Vehicle and Driver Detection on Highway... 1688
    Peihua Lv, Yang Zhang, Xiaobo Lu, Di Zhou

Robust Modeling for Fleet Assignment Problem Based on GASVR Forecast... 1696
    A. J. Su, W. D. Yang, C. Zhang, M. X. Kong

An Improved SVM Web Page Classification Algorithm 1703
    Xun-Yi Ren, Chen Shi, Dan Zhang, Wen-Si Wang

The Application of Convolutional Neural Network in Security Code Recognition 1710
    Jingtian Gu

Depth Enhancement with Improved Inpainting Order and Smoothing Method 1720
    Kang Yi, Yuting Zhao, Yiyan Lei, Jian Pan

Recognition of Speed Signs in Uncertain and Dynamic Environments 1733
    Zhilong Zhu, Gang Xu, Hongmei He, Juanjuan Jiang, Tao Wang

Structure Analysis and Generation of X.509 Digital Certificate Based on National Secret 1739
    Hua Jiang, Gang Zhang, Jinpo Fan

Research on Parking Detecting Analysis Based on Projection Transformation and Hough Transform 1745
    Xuemei Yu, Yaojie Sun

Research on the Efficiency of Beijing-Tianjin-Hebei Airport Group Based on System Dynamics............... 1750
    C. Y. Wang, W. W. Wu, J. Zhang

The Application of Alternating Direction Method of Multipliers on l_1-Norms Problems 1758
    Yanchen He

Pre-Flight Rerouting Combining A* Algorithm and AHP Under Severe Weather.. 1775
    Ding Wencan, Sui Dong

Attacking Intel UEFI by using Cache Poisoning... 1782
    Dong Wang, Wei Y. Dong

A Systematic Review: Road Infrastructure Requirement for Connected and Autonomous Vehicles (CAVs) 1788
Yuyan Liu, Miles Tight, Quanxin Sun, Ruiyu Kang

Ship Detection and Tracking in Nighttime Video Images Based on the Method of LSDT 1801
L. Liu, G. Liu, X. M. Chu, Z. L. Jiang, M. Y. Zhang, J. Ye

A Novel Approach to Multi-Resolution Technique for Fast Pattern Recognition 1812
Xiaochun Liu, Xiaohua Ding, Xiang Zhou, Wen Cui

Research on PLC Information Model Based on UML Class Diagram 1820
Xu R. Yu, Yun X. Zu, Wei H. Li

A New Improved Simplified Particle Swarm Optimization Algorithm 1827
Haikuan Liu, Dachao Yue, Lei Zhang, Zhiyuan Li, Dawei Jiang

VOLUME 3

Research on Precise Maintenance Method for Green Belt of Municipal Road Based on UAV Image Sequence 1834
T. Duan, P. C. Hu, L. Z. Sang, L. Wang

TPO-MAC: Traffic-Priority-Based Opportunistic MAC Protocol for Multi-Channel Cognitive Radio Networks 1839
Gao Q. Shen, Lei Lei, Zhi L. Li

Opportunistic Routing with Available Bandwidth Assurance for High Dynamic UAV Swarms 1846
Zhi L. Li, Lei Lei, Gao Q. Shen

Image Flame Recognition Algorithm Based on M-DTCWT 1853
Wenxia Bao, Zhongzhi Zhong, Ming Zhu, Dong Liang

Defect Prediction Model for Object Oriented Software Based on Particle Swarm Optimized SVM 1859
Yanan Wang, Ran Zhang, Xiangzhou Chen, Shanjie Jia, Huixia Ding, Qiao Xue, Ke Wang

Research on Knowledge Graph Application Technology 1869
Yong Chen, Xueli Chen

Predicting Failures in Hard Drivers Based on Isolation Forest Algorithm using Sliding Window 1877
Tinglei Zhang, Endong Wang, Dong Zhang

A New Crossover Algebra of GA for Solving the Degree Constrained Minimum Spanning Tree Problems 1883
Hui Li, Kai Shi, Huanran Li, Sheng Lin, Guangping Xu, Shuangxi Li

Research on Hybrid Recommendation Model Based on PersonRank Algorithm and TensorFlow Platform 1889
Guangqi Wen, Chunmei Li

Improvement of LDA Algorithm Based on Microblog Short Text Hotspot Analysis 1898
Kai Li, Chunmei Li

Extraction of Cerebral Hemorrhage and Calculation of Its Volume on CT Image using Automatic Segmentation Algorithm 1908
Nian Wang, Fei Tong, Yongcheng Tu, Hemu Chen, Yong Zhou, Jun Tang

Thinking and Exploration on the Teaching of the Course of "College Computer Foundation" Under the Mode of "Internet Plus" 1914
Tang Xianfang, Jia Yachao, Zhang Ru

Prediction of the Anti-Inflammatory Mechanism of Clematis Chinensis Based on Network Pharmacology 1919
Xulong Huang, Junjie Hao, Yuqing Liang, Yuanmin Wang, Juan Kong, Xiangpei Wang, Feng Xu, Hongmei Wu

Streaming Information Transmission Based on OPC UA 1927
Liu Wei, Zu Yunxiao, Li Weihai

A New Point-Weighting Finite-Difference Modelling for the Frequency-Domain Wave Equation 1932
D. S. Cheng, B. W. Chen, J. J. Chen, X. S. Wang

Research on Layout Optimization of Express Parcel Transportation Network Distribution Center Based on Node Operation Process 1939
Jun Han, Shiwei He, Mingkai Bi, Zilong Song

Using Markov Constraint and Constrained Least Square Filter to Develop a Novel Method of Passive Terahertz Image Restoration 1949
Yuanmeng Zhao, Xiao Sun, Cunlin Zhang, Yuejin Zhao

A Novel Dual-Band Video Fusion Algorithm using Fast Lookup-Tables: Toward Naturalistic Color 1955
Yuanmeng Zhao, Weiqi Jin, Lingxue Wang

Terahertz / Visible Dual-Band Image Fusion Based on Hybrid Principal Component Analysis 1962
Yuanmeng Zhao, Yulong Qiao, Cunlin Zhang, Yuejin Zhao, Hong Wu

A Camouflage Effect Detection Model for Fixed Targets 1967
Yang Xin, Xu Weidong, Xiang Lei, Zhu Wannian, Tian Jiyao

Welding Defect Signal Extraction Technology Based on OMP Algorithm 1974
Qi Ailing, Lei Haijun

Calculation Method of Cross Section Area of Collapsing Dangerous Rock Based on Parallel Binocular Vision 1979
Ping Gan, Zhenzhen Ruan, Yang Wang, Lin Huang, Yanyun Li, Yangfan Huang

Automatic Measurement Algorithm of Scoliosis Cobb Angle Based on Deep Learning 1987
Yongcheng Tu, Nian Wang, Fei Tong, Hemu Chen

A Robust $\mathcal{H}_\infty$ Approach of In-Flight Calibration for UAVs with Low-Cost IMU 1994
He Gao, Chao Gao, Guorong Zhao

Invisible Information Transmission System of Visible Light Based on Interleaved Code 2001
Yuting Meng, Yunpeng Hu, Mingchao Li, Yanqun Tang

Application of Deep Convolution Neural Network in Automatic Classification of Land Use 2009
Xiaodong Ma, Guang Yang, Qunyi Yang

Research on IP Address Allocation of Tactical Communication Network 2014
Wang Huitao, Yang Ruopeng, Wufan, Zou Xiaofei

The Priority Assignment of Messages Effects on Delay Performance in VANET 2021
Jin Tian, Qianru Han

An Improved Adaptive Weighted Median Filter Algorithm 2030
  Yuqin Song, Jun Liu

Research on Long-Distance Hand Recognition Based on Depth Information 2036
  Yuyang Fu, Lanfang Miao, Zhifei Li

A Fault Repair Method for Workstation Cluster Based on Probabilistic Model Checking............................ 2042
  Xi Wang, Ting Chen, Chengtian Ouyang

Anomaly Detection for Time Series using Temporal Convolutional Networks and Gaussian Mixture
Model 2048
  Jianwei Liu, Hongwei Zhu, Yongxia Liu, Haobo Wu, Yunsheng Lan, Xinyu Zhang

Automatic Detection of Follicle Ultrasound Images Based on Improved Faster R-CNN 2058
  Tianlong Zeng, Jun Liu

Research on Network Security of Campus Network 2065
  Min Huang, Wanbo Luo, Xing Wan

DATA MINING AND ANALYSIS

Online Route Planning for Cooperative Area Coverage Search of Aircraft Swarm 2071
  Yueqi Hou, Xiaolong Liang, Jiaqiang Zhang, Ye Li, Jiong Wang

AdaptiveSLA: A Two-Stage Scheduling Framework for SLA Profit Maximization in Multi-Tenant
Database 2079
  Yang Ji, Zhang Lin, Tang Rong

Research on Reporting Scheme of Grading Stop-and-Recharge Event of Low-Voltage Acquisition
Terminal... 2089
  R. Huang, Chenglin Zhang, L. Ouyang, M. Liu, X. Liu, Y. Su, H. Chen

Progress and Trends in Mobile Cloud Computing Research 2098
  Wang Xiao-Shu

Review of Research on Blockchain Application Development Method.. 2103
  Yue Zeng, Yue Zhang

Research on the Random Corresponding of Privacy Data Mining in the Association Rules of Cloud
Computing 2109
  Baofeng Hui, Guoqing Jia, Shanji Chen

Discussion on Supplier Selection in the Selection of Large Civil Passenger Aircraft................................ 2113
  Aiping Yin, Zhangkang, Zhenxing Gao, Duoduo Zhuanga

Analysis on Error Compensation for Integrated Navigation Based on Forgotten Kalman Filter 2117
  Ling Yao, Hua Zhang, Xiaoping Huang

Construction and Analysis of Campus Knowledge Payment Platform Under the Wave of Big Data 2123
  Ran Wu, Qiong Wu

The Mobile Payment Based on Public-Key Security Technology... 2127
  Jiabin Sun, Nan Zhang

The Advisable Technology of Key-Point Detection and Expression Recognition for an Intelligent Class System.. 2134
  Yusheng Zhao, Haotian Yan, Zhaoqing Wang

Research on the Interaction Between Language and Economy 2140
  Yusi Wu, Jie Liu, Ping Wang, Yisheng Wang, Peng Zhang, Hongyan Xi

An Expertise-Enhanced Collaborative Filtering Method for Keywords Recommendation in
Searching Engine Marketing 2147
  Yuan Zhu

Analyse the Data Tendency in the Public Opinion Monitoring System 2152
  Jian Li

Analysis and Design of Course Website for Software Testing Based on SPOC........................... 2160
  Jiujiu Yu

Product Information Modeling Based on Polychromatic Sets and Scheme Optimum Selection for
Conceptual Design 2166
  Li Qiang, Xu Bin, Li X. Qiang

Research of a Method for Synchronized Phasor Data Transmission Based on IEC61850 2176
  Shanqiang Feng, Aijun Hou, Bochuan Gu

The Design and Creation of an Interactive E-Book: "Book of Answer" 2183
  Xinyue Gui

Encrypted Tag Design for RFID Systems... 2189
  Shida Yu

Automated Sentiment Analysis of Text Data with NLTK 2195
  Jiawei Yao

A Study of Speech Feature Extraction Based on Manifold Learning 2203
  Cheng Hao, Ma Xin, Yugong Xu

Research and Improvement of CHI Feature Selection in Sentiment Analysis..............................2211
  Li Danyang, Fan Huimin

A Research on the Innovation and Development of Micro and Small Sci-Tech Enterprises.......... 2216
  Wang Guanhui

Mode of Freight Rolling and Collecting and Distributing Based on Cloud Logistics Platform 2221
  Wang Xiaokun, Wang Lei, Wang Hanyan

Research on the Integration Demonstration System of Space Information Network Based on
TDRSS 2232
  Dandan Fan, Mengyue Qiu, Xiaoshen Xu, Ming Chen

The Resource Aggregation and Integration Platform for Shared Development of the Direct Bank.............. 2238
  Qi Wang, Zhongshi Zhang, Xiaotong Zhang, Qing Zhao

Implementation of Interactive Classroom Design Based on WI-FI Service.................................. 2245
  Chen Shuai, Zhang Shuifeng, Wu Tianfang

Research on the Challenges and Innovation Models of Public Management in the Age of Big Data 2252
  Wang Cong, Su Xiangguo, Lou Ziqiang

Applications Research of Cluster Analysis in Chinese Acupuncture Therapy 2257
Xiaoche Feng

Prediction of Alzheimer's Disease Based on Bidirectional LSTM 2262
Qiao Pan, Shiyu Wang, Junhao Zhang

A Survey of the Key Technology of Software Vulnerability Mining 2267
Dansheng Lin, Xiaoquan Wu, Qinqin Wu, Ye Liu, Jijun Zeng, Jinjun Tan

Research on Security Evaluation of Government Cloud Platform Based on Fuzzy Analytic
Hierarchy Process.. 2272
Jing Chen, Feng Zhang

Big Data Mining for Investor Sentiment 2278
Tao Cen, Qianqian Chu, Renke He

Research on Frequency Hopping Synchronization Strategies Based on TOPSIS Method 2283
Xiaoyu Cai, Zhanjun Jiang, Qianru Liu, Haoqiang Shi

An Event Detection Method Based on Association Link Network 2291
Lin Sun, Weijun Yang, Xinhuai Tang

A Comparative Study of Customer Complaint Prediction Model of Time Series, Multiple Linear
Regression and BP Neural Network 2297
Xin Xu, Zhijie Sun, Li Wang, Jun Fu, Chao Wang

Anomaly Detection Based on PMF Encoding and Adversarially Learned Inference.................................. 2304
Lin Zhang, Wentai Yang, Hua Gan, Meng Li, Xiaoming Wang, Gang Liang

Mimic Defense System Security Analysis Model 2315
Li Qinyuan, Han Jiajia, Sun Xin, Qin Junning, Zhang Bo

The Present of Education Big Data Research in China: Base on the Bibliometric Analysis and
Knowledge Mapping 2323
Cheng Jiang, Qi Wang, Wu Qing, Leiye Zhu, Si Cheng, Haibin Wang

Method for Creating, Updating and Maintaining a Case Library of Service Business Guidance 2329
Xin Xu, Zhijie Sun, Jun Fu, Li Wang, Feng Xie

Design and Implementation of Body Quality Index App Based on Android.. 2335
Zhao Limei, Ma Xiaotie

Emotional Analysis of Public Opinions in Colleges and Universities:Based on Naive Bayesian
Classification Method.. 2342
Qingjia Wang, Kun Liu, Kun Ma

Dual-Channel Supply Chain Sale Strategies with Return Guarantee 2347
He Jingshi

Summary of Recommendation System Development 2355
Liu Liling

An Effective Method for Forest Fire Smoke Detection... 2360
Luxing Qin, Xuehui Wu, Yichao Cao, Xiaobo Lu

Demand Analysis of Material Reserve Optimization 2368
Jian Gao, Shuming He, Xing Song, Liang Zhang

Mining Spatiotemporal Characteristics of Car-Sharing Demand 2371
Jing-Jing Tian, Dong-Fan Xie, Fu-Jun Ding

An Empirical Study on Regional Logistics Competitiveness in Guangdong 2377
Chunshang Wu

Evaluation Effect of Internet Word of Mouth and Application of Big Data 2384
Te Ma

Purchase Decision Under Network Environment and Application of Big Data 2388
Te Ma

Research on the Evaluation Index System of Intelligent Railway Passenger Station 2393
Shaofu Lin, Qianwen Wei, Sibin Xia

Research and Analysis on the Top Design of Smart Railway 2401
Shaofu Lin, Yafang Jia, Sibin Xia

Study on Vulnerability Rating of the Intelligent and Connected Vehicle's Cybersecurity 2408
Yangyang Liu, Zhao Wang, Yanan Zhang, Peiji Shi, Xuebin Shao

Analysis of Spatio-Temporal Distribution Characteristics of Passenger Travel Behaviour Based on
Online Ride-Sharing Trajectory Data 2421
Xianlei Dong, Lingyu Wang, Beibei Hu

Research on Road Traffic Flow Status Based on Survival Analysis 2429
Beibei Hu, Doudou Lin, Qibo Sun, Xianlei Dong

Research on the Management and Maintenance of Infrastructures in Fog Section of Motorway
Based on the MOT Model 2436
X. L. Li, Q. H. Song, B. M. Tang, C. Li

Research on Semantic Prediction Analysis of Tibetan Text Based on Word2Vec 2446
Ding Hai-Lan, Yu Hong-Zhi, Qi Kun-Yu

Research on the Characteristics of Bitcoin Price Fluctuations Based on ARCH Effect 2453
Juan Wang, Feng Tian, Jie Fu

Design and Realization of Scenes of 3D Virtual Digital Library 2460
Wang Li

End-to-End Speech Synthesis for Tibetan Lhasa Dialect 2466
Lisai Luo, Guanyu Li, Chunwei Gong, Hailan Ding

Research on China's Transport Connectivity Along Corridors of the Belt and Road Initiative ... 2472
Jiao Wenwen, Li Ran, Li Sicong

A Study on Location of Logistics Hubs of Hub-and-Spoke Network in Beijing-Tianjin-Hebei
Region 2478
Xiaoyu Xin, Nan Yu, Xiang Chao

Blockchain-Based Intelligent Hospital Security and Data Privacy Construction 2488
Qiuzi Huang, Shuyu Chen, Hui Zhao, Junhao Wen

Research and Analysis on the Transformation of Road Passenger Transport Industry 2494
Lili Chen, Zhijun Yi, Xiaofeng Di

Study on the Operation Mode of Suburban Railway at Home and Abroad and the Inspiration to Beijing 2499
Luxi Peng, Chao Wang, Jianhong Wu

Application Scenarios Based on SDN: An Overview 2506
Tong Li, Jinqiang Chen, Hongyong Fu

Overview of End-to-End Speech Recognition 2515
Song Wang, Guanyu Li

Air Traffic Management Process Quality Assessment Model Based on Improved Fuzzy Matter Element Analysis 2519
Pan Wei-Jun, Ren Jie, Wang Run-Dong

Real-Time Estimation of Urban Rail Transit Passenger Flow Status Based on Multi-Source Data 2528
Zhengping Tao, Jinjin Tang

Comprehensive Assessment of Green Development Level for Urban Rail Transit Enterprises Based on ANP and Entropy Weight Method 2537
Ying Tang

Simulation Study on Emergency Evacuation of Metro Stations in Fire Degradation Mode 2544
Xi Jiaojiao, Li Jin

The Evolutionary Game Analysis of Incentive Mechanism for Crowd Sensing of Public Environment 2551
Qiang Zhang, Qingqing Zhang, Xueyan Liu, Jian Dai, Xujuan Zhang

Study on Tibetan Word Vector Based on Word2vec 2559
Ning Yang, Guanyu Li, Hailan Ding, Chunwei Gong

Analysis of Seismic Anomalies of the Jiuzhaigou Earthquake 2565
Xiaoying Jiang, Xiangzeng Kong, Gongde Guo

Research on the Development Orientation and Thinking of New Media in State-Owned Enterprises 2572
Wang Han, Wang Youzi, Xia Liyu

Analysis of the Situation Faced by New Media Propaganda in State-Owned Enterprises 2578
Wang Han, Xia Liyu, Wang Youzi

A New Linguistic Decision Making Method—FLM-VIKOR 2584
G. R. Zhang, W. D. Zeng

Design and Application Research of VR/AR Teaching Experience System 2592
Baiqiang Gan

Application of Blockchain in Document Certification, Asset Trading and Payment Reconciliation 2597
Xingxiong Zhu, Dong Wang

Study on the Relationship Between the Sharing Rate of Vehicle Exhaust Pollution and the Quantity of Possession 2602
Baiyu Chen, Da Fu, Yuanyuan Yang, Xiao Li

Research on Route Planning of Aerial Photography of UAV in Highway Greening Monitoring 2609
T. Duan, P. C. Hu, L. Z. Sang

Research on Specific Eye Movement Mode of Qualified Railway Driver 2615
    Chenzhe Sun, Guanglei Zhang, Xiujun Zhai

Research on Big Data Decision-Making Support Platform of New Energy Bus 2624
    Hao B. W. Feng

Linked Data Crowdsourcing Quality Assessment Based on Domain Professionalism 2629
    Lu Yang, Li Huang, Zhenzhen Liu

Research on Quantitative Risk Assessment Method of Packaged Cargoes Carried by Ship Based on
Online Dynamic Big Data Fusion Technology 2638
    Ru Lan, Wen Chang, Wei Shen, Zhigang Jia

An Analysis Method for Error Propagation Reachability of Component-Based Software 2649
    Mingqi Fan, Min Song, Zhengxian Wei, Wanfeng Mao

Thoughts on the Orientation of Mathematics Education in Colleges and Universities 2656
    Liang Shushuang

Study on Chinese Technical Economy and Global Social Responsibility 2662
    Huanping Zhang, Maohua Li, Zoltán Zéman

Research on Semantic Information Retrieval Model of Bamboo & Rattan Domain Based on Query
Extension 2669
    Lin Peng, Ming-Ming Lai, Xin Zhang

Research on Semantic Information Retrieval Model of Bamboo Rattan Domain Based on Semantic
Relevance 2674
    Lin Peng, Xu-Peng Kou, Lin-Nan Yang

Emergency Event Matching using Hierarchical Blocking Method 2679
    Chang Wen, Yu Liu

Research and Construction of Yunnan Plant Vertical Retrieval System 2686
    Lin Peng, Zhi-Run Ma, Xin Zhang

A Social Network Water Army Detection Model Based on Artificial Immunity 2692
    Hanhua Zhang, Tao Li, Yuqiao Wang

Design and Implementation of Ontology Knowledge Base of Endemic Genera of Seed Plants in
Yunnan Province 2701
    Lin Peng, Xu Li, Lin-Nan Yang

The Applications of the Edge Detection on Medical Diagnosis of Lungs 2706
    Wei Xu, Jinping Li, Hongwei Jia

A Novel Rating Style Mining Method to Improve Collaborative Filtering Algorithm 2712
    Wei Yang, Sheng H. Guo, Chun J. Zhang

Study on Traffic Organization and Work-Zone Optimization of Four-Lane Freeway Reconstruction
and Expansion 2720
    Xuanyu Ye, Yazhen Chen, Chen Chen

Review on Studies of Machine Learning Algorithms 2727
    Peiyuan Xu

Spatial Spillover Effects of the Real Estate Industry on Economic Development -From Destocking Perspective.. 2734
    Xiaoqian Liu, Qinqin Zhou, Chang'An Wang

Application of Big Data in Forecasting Traffic Flow 2741
    Luo Wanbo, Wan Xing, Huang Min

Financial Risk Analysis and Early Warning Research Based on Data Mining Technology 2746
    Yan Hou, Ziyan Yuan

Author Index

| ISPECE | IOP Publishing |

Simulation and Experimental Research on the Influence of Tool Geometries on the Cutting Force of High Temperature Alloy

Fengyun Yu[1, a], Chunyu Yan[1,b], Shaobing Guo[1,c], Shiwei Ma[1,d], He Wang[1,e]

[1]Heilongjiang University of Science and Technology Harbin, 150022，China

E-mail address: [a]1578415766@qq.com, [b]1056757440@qq.com, [c]1551630559@qq.com, [d]652992793@qq.com, [e]346786233@qq.com

Abstract. For the difficult-to-process characteristics of the aero-engine turbine disk plank produced by Nickel-based superalloy GH4169 that appeared in the turning process, the numerical simulation and experimental research were conducted in this paper. The two-dimensional orthogonal cutting model of the Nickel-based superalloy GH4169 was established to conduct the single factor experiment, in which the rake angle and tip radius were as the single variable, and the influence of cutting force caused by the rake angle and tip radius was researched by the finite element analysis software ABAQUS. The correctness of the simulation modal was verified by the comparison of the cutting force data obtained in simulation and the experimental result. The study demonstrates the fact that the back force F_y will decrease while the feed force F_f and the main cutting force F_c increase slowly with the increase of the rake angle. The back force F_y will increase with the increase of the tip radius. In addition, the changing trend of cutting force obtained in the experiment was consistent with that of simulation. Under the experimental conditions, the rake angle of $10°$ and tip radius 0.4mm were suitable processing parameters for machining the Nickel-based superalloy GH4169.

1. Introduction

The high-temperature alloy, for its well high temperature plastic, high temperature oxidation resistance and mechanical properties at room temperature and high temperature, has been widely used in the area of aerospace. It is one of typical difficult machining materials. Domestic and foreign scholars have conducted in-depth research on the problems existing in the high speed cutting process of nickel-based superalloy GH4169[1-2],including two-dimensional and three-dimensional numerical simulation the cutting force influenced by cutting parameters when cutting the high temperature alloy GH4169 with the ceramic tools [3-4]and so on. By the cutting experiment and simulation of cutting iron-based superalloy with coated carbide tools[5], cutting the nickel-based superalloy with the PCBN tools[6], cutting the nickel-based superalloy in high-speed with the PCBN tools[7], the research was conducted for the high temperature alloy cutting performance influenced by such factors as rake angle, tip radius, negative chamfer parameters and PCBN material and so on.

In the paper, based on the result of the single factor experiment for the tool material, the single factor experiment, in which, the rake angle and tip radius were as single factor various, was conducted. The cutting force obtained from the experiment was analyzed and the influence of turning high temperature alloy GH4169 was produced by the tools geometry parameters. In the single factor experiment, in which the tool geometry parameters were as single factors, the correctness of

Content from this work may be used under the terms of the Creative Commons Attribution 3.0 licence. Any further distribution of this work must maintain attribution to the author(s) and the title of the work, journal citation and DOI.

Published under licence by IOP Publishing Ltd

simulation modal established by ABAQUS software was verified, and meanwhile, the influence of cutting force produced by rake angle and tip radius was researched, all of which provide the technical support for the machining of GH4169.

2. The establishment of simulation modal

The essential work, such as the arrangement of seeds grid of high temperature alloy GH4169, element shape set, element type set, meshing techniques and algorithms set, meshing and mesh quality inspection and so on, was finished with the Mesh functional module in ABAQUS/CAE. The ABAQUS grid cell type window is shown in figure 1. The two-dimensional orthogonal cutting model after meshing was shown in figure 2. The grid in the cutting zone was meshed intensively, but the grid far from the cutting zone was thin gradually. The element type of grid after meshing is four node plane strain reduced integral unit CPE4RT.

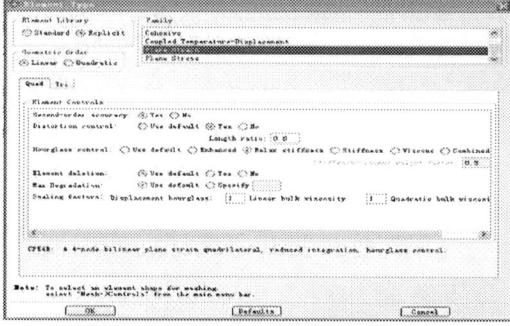

Figure1. Unit type setting window.

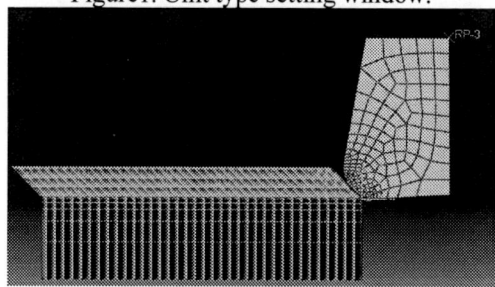

Figure2. Two-dimensional orthogonal cutting grid model.

3. The design for experiment

3.1 The establishment of experimental system

The work-piece size is Bar with Φ155×60;

The material of the tools is Coated Carbide KC5510;

In this paper, the experiment uses external turning superalloy GH4169 by ordinary horizontal lathe CA6140. Due to the outer surface of the purchase is not very smooth, using a dial indicator alignment clamping in the workpiece fixture and detecting workpiece center position. The central axis of the workpiece and machine tool spindle must be maintained a high coaxiality in order to enhance and ensure the machining accuracy of each test. Cutting force is measured by using Kistler 9257B dynamometer system which is shown in Figure 3.

ISPECE   IOP Publishing

IOP Conf. Series: Journal of Physics: Conf. Series **1187** (2019) 032063   doi:10.1088/1742-6596/1187/3/032063

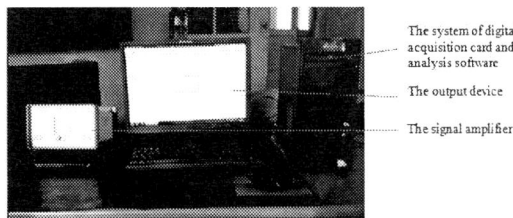

Figure3. The measurement system of cutting force.

3.2 The simulation and experiment condition
The cutting amount: cutting speed:v_c=50m/min.
   The feed rate: f=0.15mm/r;
   The back cutting depth: a_p=0.4mm;
   The tools geometry parameter: The tool relief angle: α=3°.

4. The research for the experiment and simulation of cutting force

4.1 The influence of cutting force caused by rake angle
In the experimental condition mentioned above, the regular pattern of cutting force was studied when the rake angle respectively is 3°, 7°, 10°, 12°. The changing curve of cutting force influenced by the change of rake angle was shown in figure 4, was obtained after processing experimental data by extracting the average value of cutting force in each direction. When the cutting process is in the steady state, the comparison of the back force F_y obtained from experiment and simulation, and was shown in figure 5~figure 8. The compared curve of cutting force was shown in figure 9.

As the figure 4 shown that: with the rake angle increasing, the feed force F_x and the main cutting force F_z performed decreasing slowly, when the rake angle increase to 10°, the feed force F_x and the main cutting force F_z decrease to the minimum value. Being the rake angle 10° as the inflection point, the feed force F_x and the main cutting force F_z would perform the increasing trend slowly.

The back force F_y will increase to 502 N when the rake angle is 3°. When the rake angle increase to 7°, the back force F_y decreased to 486 N.When the rake angle increasing to 10°,the back force F_y decreased to 423 N. When the rake angle increased to 12°, the back force continuously decreased to 254 N. By comprehensively analyzing, the preferred rake angle is 10°.

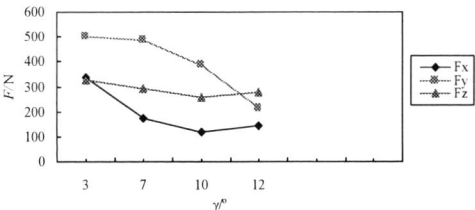

Figure4. Cutting force is affected by cutting tool rake angle.

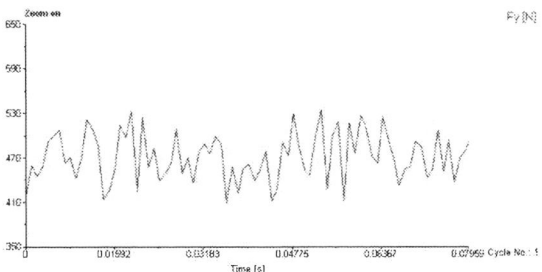

919

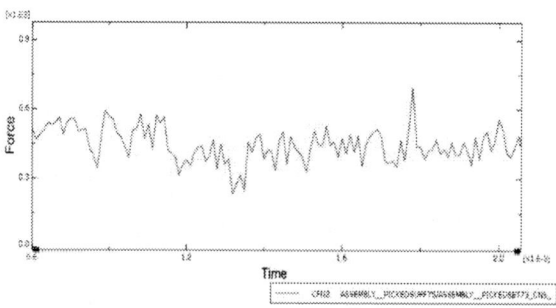

Figure5. The dynamic cutting force changing curve obtained from experiment and simulation ($\gamma=3°$) .

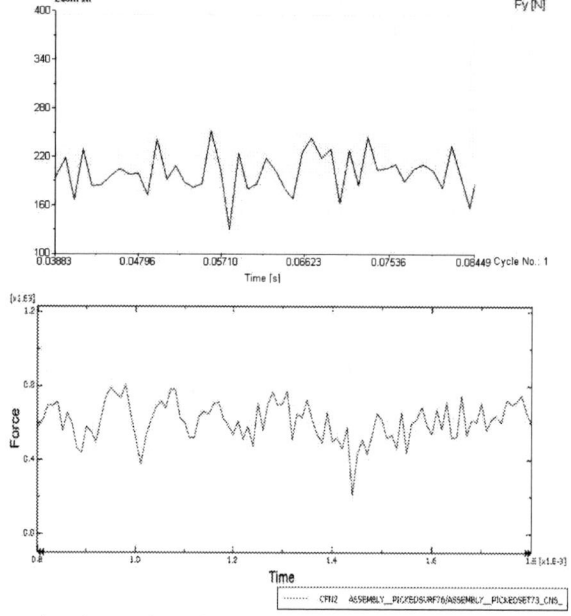

Figure6. The dynamic cutting force changing curve obtained from experiment and simulation ($\gamma=7°$).

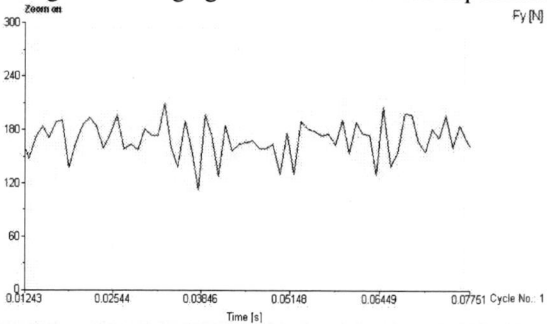

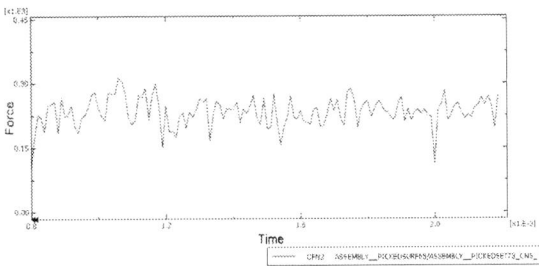

Figure7. The dynamic cutting force changing curve obtained from experiment and simulation ($\gamma=10°$).

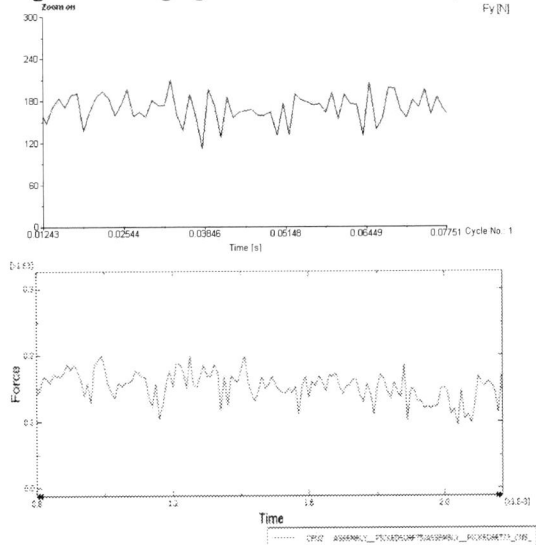

Figure8. The dynamic cutting force changing curve obtained from experiment and simulation ($\gamma=12°$).

From the figure 5, we can see that: the changing trend obtained from simulation and experiment are consistent. Because the blunt edge radius will decrease gradually when the rake angle gradually increased, which will directly lead to the strengthening of cutting action in the edge of tool and the decreasing of produced cutting force. Meanwhile, the shear angle in the cutting process will increase with the rake angle increasing, which also lead to the decrease of cutting force.

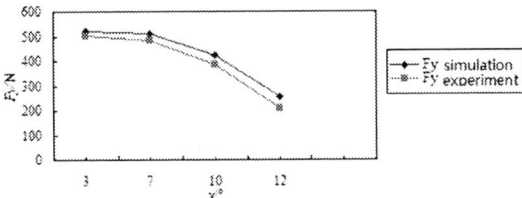

Figure9. Comparison of simulation and experiment of cutting force.

4.2 The influence of cutting force caused by tools tip radius

In the experimental condition mentioned in the third part, the regular pattern of cutting force was studied when the tip radius respectively is 0.2 mm, 0.4 mm, 0.8 mm and 1.2 mm.The changing curve of cutting force influenced by the change of tip radius, shown as figure 10, was obtained after processing experimental data by extracting the average value of cutting force in each direction. When the cutting process is in the steady state, the changing curve of the back force F_y obtained from experiment and simulation, were respectively shown in figure 11($r=0.2$ mm). The compared curve of cutting force was shown in figure 12.

As the figure 8 shown that: in the process of the tip radius increasing from 0.2 mm to 0.4 mm, the feed force F_x and the main cutting force F_z increase slowly, but the back force F_y has little change. When the tip radius increases from 0.4 mm to 1.2 mm, the feed force F_x and the main cutting force F_z almost have little change, but the back force F_y increases gradually. So the optimized tip radius is 0.4 mm.

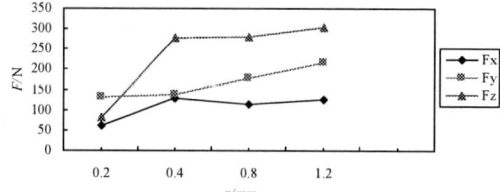

Figure10. Influence of tip radius on cutting force.

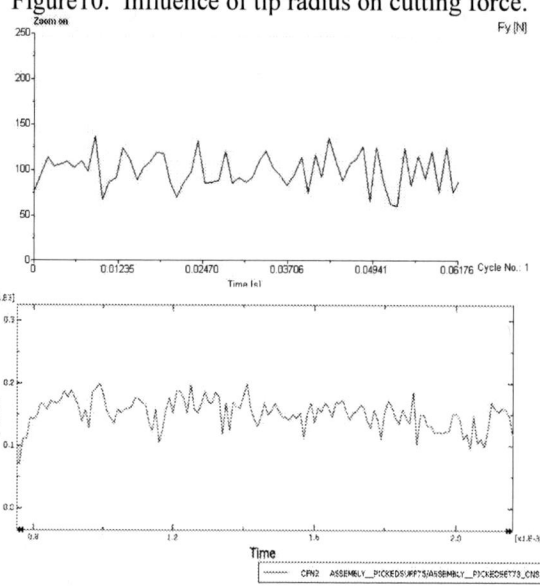

Figure11. The dynamic cutting force changing curve obtained from experiment and simulation(r=0.2 mm).

From the figure 10, we can see that: the changing trend obtained from simulation and experiment are consistent. With the tip radius increasing, the cutting force is increasing. Because the blunt edge radius will increase gradually, in the process of tip radius increasing gradually, which will directly lead to the weakening of cutting action in the edge of tool and the increasing of cutting force. However, in the process of tip radius decreasing gradually, the tip more sharp, the easier to cutting, which lead to the appearance of smaller cutting force. Always, the tip wear is very serious.

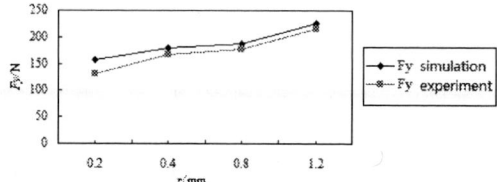

Figure12. Comparison of simulation and experiment of cutting force.

5. Conclusions

In this paper, the influence of cutting force produced by rake angle and the tip radius was researched. From the study, It can be concluded that with the rake angle increasing, the back force F_y decreased gradually but the feed force F_f and the main cutting force F_z increased slowly after the rake angle reflection angle $10°$; With the tip radius increasing gradually, the back force increases gradually.

By the simulation of the finite element analysis software ABAQUS, the law of the cutting force was changed with the changing of rake angle and tip radius was obtained. Meanwhile, by the comparison with the result obtained from experiment and simulation, we could find that the changing trend between them was consistent.

From the research above, the conclusion could be obtained: the rake angle suitable for cutting the high temperature alloy is $10°$ and the tip radius is 0.4 mm.

Acknowledgment

This work was supported by Harbin Science and Technology Bureau Project (2017RAXXJ015) and 2018 Special Fund Project of Basic Scientific Research Bussiness Expenses of Heilong jiang Provincial Undergraduate Universities (Hkdxp201802).

References

[1] X.Teng, T.Wierzbicli and H.Couque, On the transition from adiabatic shear banding to fracture, Mechanics of Materials, Vol.**39**, pp. 107-125 (2007).

[2] Yuan ChongHui, Finite Element Simulation and Analysis for High-speed Machining High temperature alloy GH4169, Qingdao Technological University(2011).

[3] Wu Xia, Yao ChangFeng and Zhao Lei, Study on Surface Roughness in Turning GH4169 Superalloy by Using Ceramic Tool, Aeronautical Manufacturing Technology, Vol.**19**, No.19, pp. 432-439(2012).

[4] Ma TianYu, WANG Yu, Li Yufu, Fu MingMing and Zhang JiaYi, Research on Cutting High Temperature Alloy by Ceramics Tool, Aviation Precision Manufacturing Technology, Vol.**48**, No.1, pp. 38-41, 53(2012).

[5] Feng ZhenXing, Research on Cutting the Iron-based Superalloy with the Coate Carbide Tools，Dalian University of Technolog, Dalian university of Technology(2013).

[6] Song TingKe, Li Man and Zhang Hongtao, Research on the cutting property of polycrystalline carbide boron (PCBN) tools in turning nickel-based superalloy GH4169, Diamond & Abrasives Engineering, Vol.**31**, No.1, pp. 70-73(2011).

[7] Li Chang, Experimental and Simulation Research on High Speed Turning Nickel-based Superalloy with PCBN Cutting Tools, Xiangtan University(2014).

Research on Extraction and Analysis of Characteristic Conditions of Hub Motor for Electric Vehicles

Guohui Yang, Shuo Zhang, Chengning Zhang[a] and Yongxi Yang

National Engineering Laboratory for Electric Vehicles, Beijing Institute of Technology (No. 5, Zhongguancun South Avenue, Haidian District, Beijing, China,100081)

[a]Corresponding author: mrzhchn@bit.edu.cn

Abstract: The traditional electric machine optimization design technology is generally based on continuous power or overload conditions. It is difficult to fully consider the entire operating domain of the machine, and it is difficult to effectively improve the overall efficiency of the motor. In order to achieve an effective improvement of the overall efficiency of the electric motor under the operating conditions of the vehicle, the motor drive system is well matched with the vehicle design. This paper proposes an electromagnetic optimization design technology for multiple working conditions, uses NEDC working conditions to simulate the actual running conditions of the vehicle to the motor, and optimizes the design of the motor with the highest comprehensive efficiency (lowest comprehensive energy consumption) under the working conditions. By using statistical method, more than 1000 working points of the motor in the whole operating range are simplified into 11 characteristic working conditions. By calculating and comparing the overall energy consumption of the motor under NEDC working condition and 11 characteristic points, the effectiveness of the extraction method of characteristic working conditions is verified.

1. Introduction

The electromagnetic loss of the motor under different working conditions is different. The traditional design method is based on the rated working point for the electromagnetic design of the motor. However, the motor for the vehicle does not have a fixed rated working point. The motor has a wide working range and a large torque/speed change. The electromagnetic design method of the motor at the working point is difficult to ensure that the comprehensive efficiency of the motor for the vehicle under the actual operating conditions of the vehicle is the highest, and even the efficiency of the motor at the rated design point is high, and the overall efficiency is low under the actual operating conditions of the vehicle.

In the traditional motor design method, 80% of the area is larger than 80% of the total working area of the motor. This method can be used to evaluate the motor efficiency distribution characteristics. However, it is difficult to ensure a good match between the motor and the whole vehicle. It is difficult to ensure that the motor drive system has a higher overall efficiency in the actual operating conditions of the vehicle. The efficiency of the motor in the low speed and high torque field is relatively low, the motor matching design is unreasonable, and it is easy to cause the motor to work in the low efficiency area for a long time and reduce the efficiency of the motor drive system. In order to improve the overall efficiency of the motor under the actual operating conditions of the vehicle, the actual working condition of the vehicle should be considered in the motor design stage, and the comprehensive efficiency under the working condition is optimal as a performance optimization index in the motor

---

Content from this work may be used under the terms of the Creative Commons Attribution 3.0 licence. Any further distribution of this work must maintain attribution to the author(s) and the title of the work, journal citation and DOI.
Published under licence by IOP Publishing Ltd

design process.

The operating conditions of the automobile are relatively complicated. At present, the general urban and suburban working conditions for electric vehicles have not yet met the standards. Therefore, the typical working conditions of the traditional gasoline vehicles are used, and the NEDC working conditions of both urban and suburban conditions are used as the simulation operation of the vehicle to carry out the hub motor optimization design. Through technical combination, it is found that there are technical problems to be solved in the research of the subject: the motor optimization design is a cycle iterative process, and the NEDC contains 1180 working conditions, even if the motor is configured for more than 1000 finite elements. The time is also very large and unacceptable. How to simplify the working conditions to meet the engineering calculation requirements is a technical problem that needs to be solved.

2. Analysis of vehicle operating conditions
Motor performance indicators in this study:

Rated power ≥10kW, peak power ≥20kW (30s); maximum speed 1400rpm, output rated torque ≥150Nm, peak torque ≥280Nm (30s); torque density ≥9Nm/kg; effective mass power density ≥1.5kW/kg.

2.1 The choice of simulation cycle conditions
The performance of the vehicle has a lot to do with the actual driving conditions. During the actual driving process, the vehicle constantly experiences acceleration and deceleration and uniform speed. In the evaluation and analysis of automobile-related performance, a vehicle speed-time curve that is more suitable for the local actual situation is often selected as the driving condition of the vehicle, so that the analysis result is closer to the actual performance level of the vehicle.

The operating speed of the whole vehicle under NEDC cycle conditions is shown in Figure 1.

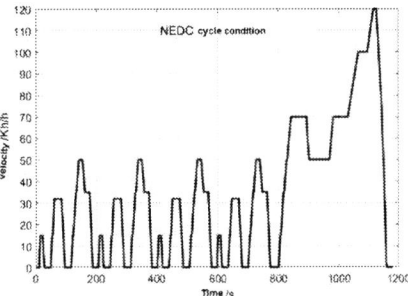

Figure 1 NEDC cycle condition

Under the condition of NEDC, the deceleration condition is simplified to separate the electric brake of the motor, regardless of the mechanical brake. It is assumed that the efficiency of the motor is the same as the efficiency under the driving state, and the generated energy can be completely recovered.

2.2 Calculation of motor operating conditions
According to the basic parameters of the vehicle and the cyclical conditions of the vehicle, the basic equations of the vehicle theory can be used to calculate the speed/torque/power of the motor at each time point t=t_i, respectively, $n(t_i)$, $T(t_i)$, $p(t_i)$, the speed/torque of the motor and the power distribution. Figure 2 shows.

$$\begin{cases} F - mgcos\theta - \dfrac{C_D A v^2}{21.15} - mgf = ma \\ P = \dfrac{Tn}{9549} \\ F = \dfrac{T}{r} \end{cases}$$

（1）

Where, F is the driving force, T is the driving torque, m is the conditioning quality, C_D is the drag coefficient, A is the windward area, n is the wheel speed, and P is the output power;

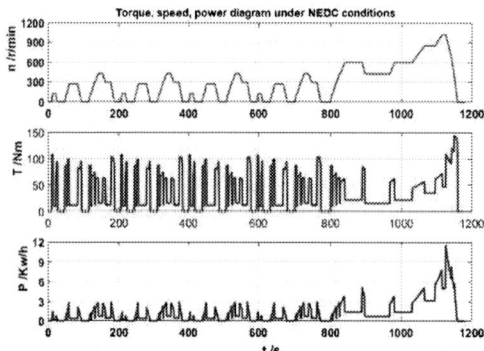

Figure 2 Motor torque and speed distribution

From Fig. 2, it can be seen that in the whole NEDC working condition, the rotational speed, torque and power demand at different time points, obviously the torque demand under the whole working condition is below 150Nm, most of which is concentrated around 100Nm, in the urban cycle Under the circumstance, the power demand of the motor is about 3kW, and the torque and power demand in the suburban high-speed working conditions are relatively large, at around 150Nm and 12KW. Further analysis, in the NEDC working condition, the motor's speed-torque demand is matched, according to the index requirements of the project, the external characteristic requirement curve of the motor is established, and the motor speed-torque demand under the NEDC condition is combined. The demand is shown in Figure 3.

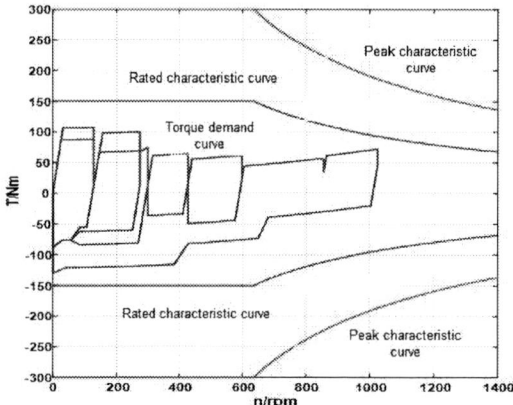

Figure 3 Motor speed-torque demand diagram under NEDC conditions

Since the energy consumption of the electric vehicle drive system is mainly reflected in the loss of the motor, or the energy consumption of the whole vehicle is reflected in the energy consumption of the motor drive system. According to the power $P(t_i)$ and time Δt_i of the motor at each point, the energy E_i in the unit time interval is defined to represent the energy consumption of the motor at that point.

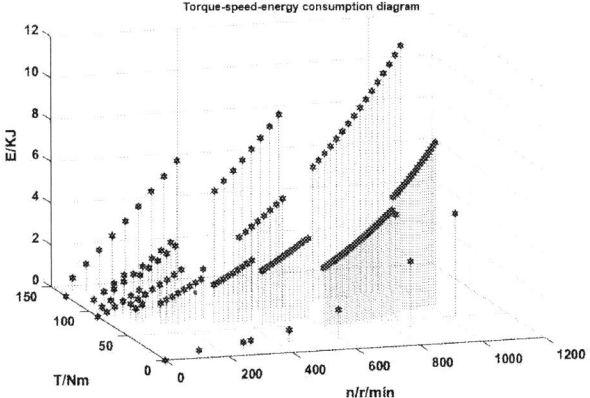

Figure 4 Energy consumption of the motor under the NEDC operating point

$$E_i(t_i) = P(t_i)\Delta t_i \qquad (2)$$

Through the motor torque, speed, power distribution map, the energy consumption of the motor at each working point is established (regardless of the efficiency of each point), as shown in Figure 4.

3. Analysis and study on the extraction method of characteristic working conditions

3.1 Equivalent feature point extraction
Obviously, under the whole cycle condition, the corresponding energy consumption of the motor under different working conditions is different, and the amount of data is huge, which cannot be directly applied to the optimal design of the motor. Analysis of the energy consumption of the motor under various working conditions can be seen: the energy consumption of each working point has a certain regularity, the energy consumption of the motor in the medium and high speed area and the low speed heavy load area is high, and the energy of the low speed light load area The consumption is relatively low, and the distribution of working conditions shows regularity, which is relatively concentrated within a certain range. Project the operating point under the NEDC condition to the external characteristic curve of the motor as shown in Figure 5.

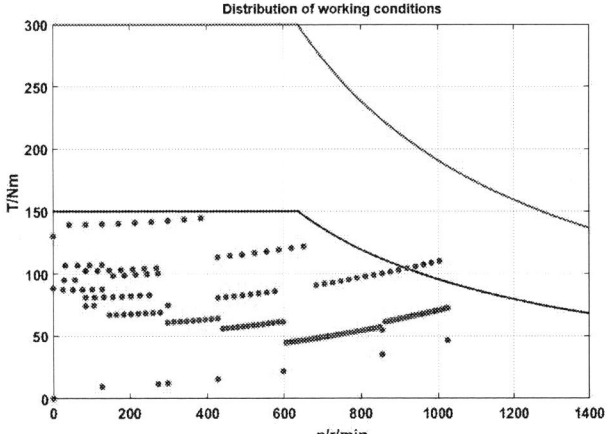

Figure 5 NEDC operating point distribution map

Excluding the several points where the energy consumption is too high, the energy consumption distribution under various working conditions of the motor is shown in Fig. 6. Obviously, in different speeds and torque ranges, the motor energy consumption has a certain regularity, and hundreds of operating points in a certain area have relatively similar energy consumption. To reduce the overall number of operating points, the motor is based on the speed, torque and energy consumption are divided into several areas, and the energy consumption in different areas is processed.

Specific division principle:

(1) The motor is relatively concentrated in the area;

(2) The speed/torque change is not large;

(3) The difference in energy variation of the motor in this area is within a certain range.

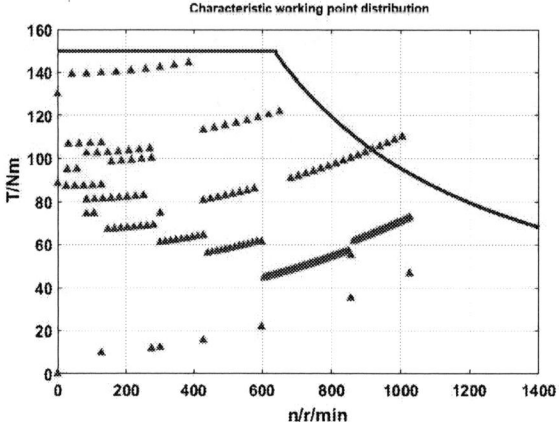

Figure 6 NEDC operating point distribution map

According to the above principle, the working point of the motor under the NEDC condition is divided into 11 regions, and the specific divided regions are shown in Table 2. The number of working points of the motor under low load is small, the energy consumption is low, and two regions are divided according to the speed interval below 40Nm; in the torque range of 40-90Nm and 90-130Nm, the number of motor operating points is large, and different speeds The motor energy consumption in the range is quite different. Therefore, this part of the area needs to be divided into thin sections. It is divided into 7 sections according to energy consumption and speed. In the interval where the torque is greater than 130Nm, the working points are mainly concentrated in two parts, divided into 2 parts. A total of 11 work areas, as shown in Table 1.

Table 1 Speed/torque area division

Region	Tm（Nm）	n（r/min）	Number of working points
1	0-40	0-600	435
2	0-40	600-1400	30
3	40-90	0-200	88
4	40-90	200-400	100
5	40-90	400-600	48
6	40-90	600-1400	77
7	90-130	0-400	88
8	90-130	400-660	8
9	90-130	660-1400	16
10	130-150	0-200	4
11	130-150	200-1400	6

The speed and torque in each area of the motor are analyzed based on statistical methods. The rotational speed and torque center of gravity in each area are calculated. The center of gravity value is used to replace the points in each area. The energy loss is calculated for the total energy under all working conditions. Consumption, used to characterize the overall energy consumption in each area.

The energy consumption in each region is represented by the sum of the energies in the following equation:

$$E_i = \sum_{j=1,2,\ldots}^{N_i} E_{ij} \qquad (3)$$

The centers of angular velocity ω_{mci} and torque T_{mci} are given by:

$$\omega_{mci} = \frac{1}{E_i} \sum_{j=1,2,\ldots}^{N_i} E_{ij} \ \omega_{ij} \qquad (4)$$

$$T_{mci} = \frac{1}{E_i} \sum_{j=1,2,\ldots}^{N_i} E_{ij} \ T_{ij} \qquad (5)$$

Where N_i is the number of operating points in the i-th region.

The energy consumption, rotation speed and torque under various working conditions of the motor are calculated as shown in Table 2.

Table 2 Speed, torque, energy consumption in different areas

Region	n (r/min)	Tm (Nm)	E (kJ)	Percentage
1	472.5046	17.7585	286.2	18.16
2	854.0395	35.1862	94.36	6.32
3	124.3280	80.2412	77.8	5.21
4	314.4887	66.6050	214.3	14.15
5	494.3554	65.7768	157.64	10.55
6	872.6665	57.2749	319.4	22.4
7	182.9250	101.5762	123.2	8.25
8	550.8427	117.7544	53.13	3.56
9	860.6012	100.9883	142.14	9.51
10	128.2133	139.9125	6.25	0.42
11	311.6451	142.8539	22.35	1.5

It can be seen from the analysis in Table 2 that the equivalent point condition represents the energy consumption of the motor under different working conditions and the equivalent speed and torque. The total energy consumed in each working condition accounts for a percentage of the total energy, among which, the torque The energy consumption at the operating point below 40Nm accounts for about 25% of the total energy consumption, of which the low speed (472r./min) light load (about 17.7Nm) accounts for 19% of the total energy consumption, and the motor efficiency at this moment is relatively low. Therefore, it is necessary to pay attention to the motor design process. At medium and low speeds (300-500r/min), the energy consumption of the motor torque below 70Nm accounts for about 25% of the total energy consumption; at medium and high speeds (around 700-900r/min), the medium and high torque (The energy consumption of 60Nm and 100Nm accounts for about 35% of the total energy consumption; for some of the above-mentioned several energy consumption conditions, it is necessary to pay attention to the motor design stage to ensure that the efficiency of the motor is relatively high under this condition. High to achieve a higher overall efficiency of the motor system. The motor consumes about 2% at low speed (100-300r/min) heavy load (about 140Nm), but the efficiency of the motor is low at this time, and it needs to be considered in design.

Through the above 11 feature points, the energy loss of the whole vehicle under the whole NEDC condition is greatly reduced, and the calculation amount of the motor energy utilization rate in the motor design optimization process is greatly reduced, especially when calculating the efficiency by using the finite element method. Ascension is unimaginable.

3.2 Effectiveness Analysis of Equivalent Operating Point

The use of 11 equivalent operating conditions to replace the 1180 operating conditions under the entire NEDC operating conditions, whether the equivalent method is reasonable or not will be highlighted in this section.

(1) Analysis of the compatibility of working conditions

The NEDC operating point and 11 characteristic operating points are projected into the external characteristic curve of the motor, as shown in Fig. 7. The distribution of the speed/torque distribution and the distribution of the characteristic points of the motor under the NEDC operating conditions are compared and analyzed. Obviously, the characteristic point distribution can well represent the working conditions in different speed/torque ranges.

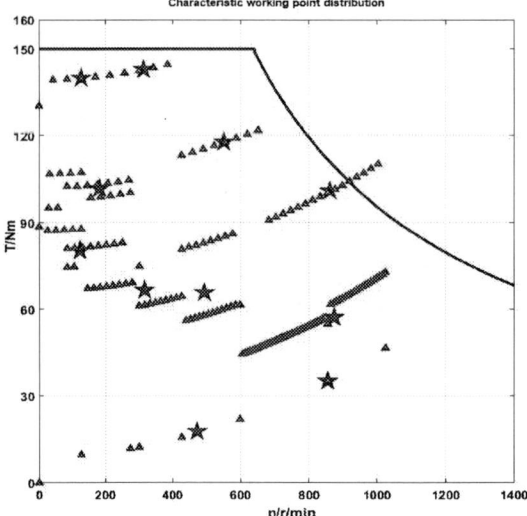

Figure 7 NEDC operating point and characteristic operating point

(2) Energy loss analysis

The energy consumption of 11 feature points is compared with the energy consumption under NEDC conditions to verify the rationality of using the feature point conditions instead of the entire NEDC operating conditions.

The motor efficiency map is constantly changing as the electromagnetic structure parameters of the motor change, and the efficiency changes of different motors at different operating points directly affect the energy consumption. The efficiency map of the motor proposed in this project is a special one, which is assumed to have little change during the motor optimization design process, or is understood to have a fixed efficiency map during the process of changing the electromagnetic parameters of the motor. All use this map as a benchmark.

Obviously, this assumption is not the same as the actual situation, but whether it can reflect certain problems to a certain extent. In the process, you need to make certain compromises and trade-offs. Choosing the right motor efficiency map is the first step in the analysis. The efficiency map of a certain motor selected in the previous design is shown in Figure 8.

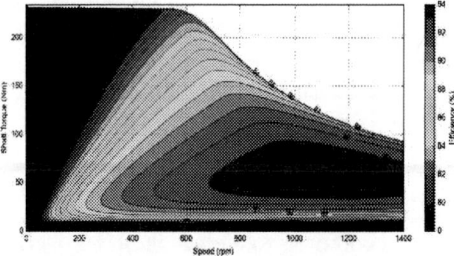

Figure 8 Efficiency map

Calculate the energy consumption of the motor under the entire NEDC characteristic condition using equation (6).

$$EL = \sum_{i=1,2,...}^{N} \omega_m(t_i) T_m(t_i) \Delta t_i (1 - \eta_i)/\eta_i \qquad (6)$$

Where η_j is the efficiency at $\omega_m(t_i)$ and $T_m(t_i)$ in the motor energy efficiency diagram.

The energy consumption of the motor under 11 characteristic point conditions is calculated by formula (7).

$$EL = \sum_{j=1,2,...}^{11} E_{nj}(1-\eta_i)/\eta_j \qquad (7)$$

Where E_{nj} and η_j represent the energy consumption and efficiency at the j-th point, respectively.

The total energy consumption of the motor under NEDC conditions and the energy consumption under 11 characteristic points are calculated by MATLAB as shown in Table 3.

Table 3 Comparison of energy consumption

NEDC energy consumption at all operating points (kJ）	1784.1
Energy consumption of 11 feature points (kJ)	1875.9
Error comparison	4.89%

Obviously, the total loss calculated by the characteristic point working condition can well characterize the loss characteristics of the motor under the whole NEDC condition. In the motor optimization design stage, the characteristic point condition is used for motor performance check, comparative analysis, and iterative Optimize the design of the motor, greatly reducing the amount of calculation in the motor optimization design process.

The alternatives to the working conditions mentioned in this paper are not limited to NEDC operating conditions, and their ideas are still applicable to other typical operating conditions. The difference lies only in the division of specific characteristic working conditions. The division principle of working conditions will affect the representativeness of the characteristic points to the whole to a certain extent. In principle, the more regional, the more representative, the better the better. The energy consumption characteristics of the whole working condition, but the corresponding calculation amount will also increase, so the trade-off between the working conditions should be properly weighed.

In the process of motor optimization design, the variation of electromagnetic parameters must lead to the variation of motor efficiency distribution characteristics, and the distribution of motor efficiency map must be different.

In essence, multi-operating condition electromagnetic optimization design technology is to carry out motor electromagnetic parameter design based on the actual operating conditions of the whole vehicle, or motor efficiency map design movement problem, so as to ensure the motor operating efficiency as high as possible under the comprehensive operating conditions of the whole vehicle.

4. Summary
In this paper, by means of statistical method, more than 1000 operating points of hub motor NEDC for electric vehicles in the whole working condition range are simplified into 11 characteristic working conditions. Moreover, the energy consumption comparison between the whole working condition and characteristic points is calculated to verify the effectiveness of the feature point extraction method. It provides the foundation for improving the comprehensive efficiency of the motor vehicle under the operating condition, and also provides the idea for the motor drive control system to match the good design of the vehicle.

Acknowledgement
This research was financially supported by the National Natural Science Foundation of China (Grant No. 51677005).

References
[1] C. Zhang, S. Zhang, G. Han, H. Liu , Power Management Comparison for a

Dual-Motor-Propulsion System Used in a Battery Electric Bus.[J] IEEE Transactions on Industrial Electronics, **vol.64**, pp3873-3882(2017).

[2] C. Zhang, H. Liu, X. Wu, Y. Yang, X. xin, A Computationally Efficient PM Power Loss Derivation in Thermal Modelling for Surface-mounted Brushless AC PM Machine[C], ICAE2016, **vol.105**, pp2891-2897(2017).

[3] Y. Ji,L. Ren,J. Zhou, Boundary conditions of active steering control of independent rotating wheelset based on hub motor and wheel rotating speed difference feedback.[J] Vehicle System Dynamics, **vol.56**, pp1883-1898(2018).

[4] J.Dai, Z.Zhao,T. Liu,C. Wang,X. Hu, Review of research on AFPM hub motor[J], Journal of Hebei University of Science and Technology, **vol.39**, pp 17-23(2018).

[5] R. Wrobel, P. H. Mellor, M. Popescu, D. A. Staton, Power Loss Analysis in Thermal Design of Permanent Magnet Machines – A Review[J], IEEE Transactions on Industry Applications, **vol.52**, pp 1359-1368(2016).

[6] R. Wrobel, G. Vainel, C. Copeland, T. Duda, D. Staton, and Phil H. Mellor, Investigation of Mechanical Loss Components and Heat Transfer in an Axial-Flux PM Machine, [J], IEEE Transactions on Industry Applications, **vol.14**, pp 3000-3011(2015).

[7] J. Goss, R. Wrobel, P.H. Mellor, D. Staton, The Design of AC Permanent Magnet Motors for Electric Vehicles: A Design Methodology[C], IEEE International Conference on Electric Machines and Drives, pp. 871 – 878(2013).

[8] R. Wrobel, J. Goss, A. Mlot, P.H. Mellor, Design Considerations of a Brushless Open-Slot Radial-Flux PM Hub Motor[J], IEEE Trans. Ind. Appl., **vol. 50**, pp. 1757-1767(2014).

[9] J. Goss, P.H. Mellor, R. Wrobel, D.A. Staton, M. Popescu, The Design of AC Permanent Magnet Motors for Electric Vehicles: A Computationally Efficient Model of the Operational Envelope,[C], 6th IET International Conference on Power Electronics,Machines and Drives, PEMD'12, pp. 1– 6(2012).

[10] X. Wu. Research on Method of Electromagnetic Loss Derivation over the Entire Torque-Speed Envelope and Thermal Field in Permanet Magnet Synchronous Motor[D].Beijing Institute of Technology(2016).

[11] R. Wrobel, D.E. Salt, A. Giffo, P.H. Mellor, Derivation and Scaling of AC Copper Loss in Thermal Modeling of Electrical Machines[J], IEEE Trans. Ind. Elect., **vol. 61**, pp. 4412 – 4420(2014).

[12] P. Zhang, G.Y. Sizov, J. He, D.M. Ionel, N.A.O. Demerdash, "Calculation of Magnet Losses in Concentrated-Winding Permanent-Magnet Synchronous MachinesUsing a Computationally Efficient Finite-Element Method[J], IEEE Trans. Ind. Appl., **vol.49**, pp. 2524 - 2532 (2013).

[13] D. A. Howey, A. S. Holmes, K. R. Pullen, Measurement of Stator Heat Transfer in Air-Cooled Axial Flux Permanent Magnet Machines[C], 35th IEEE Industrial Electronics Annual Conference, IECON'09, pp. 1197 – 1202(2009).

[14] A. C. Malloy, R. F. Martinez-Botas, M. Jaensch, M. Lamperth, Measurement of Heat Generation Rate in Permanent Magnet Rotating Electrical Machines[C], 6th IET International Conference on Power Electronics, Machines and Drives, PEMD'12, pp.1– 6, 2012(2012).

[15] X. F. Ding and C. Mi, Modeling of Eddy Current Loss in the Magnets of Permanent Magnet Machines for Hybrid and Electric Vehicle Traction Application[C] Vehicle Power and Propulsion Conference, pp. 419-424(2009).

[16] K. Yoshida, Y. Hita, K. Kesamaru, Eddy-Current Loss Analysis in PM of Surface-mounted-PMSM for Electric Vehicles[J], IEEE Trans. Magn., **vol. 36**, pp. 1941-1944(2000).

[17] L. J. Wu, Z. Q. Zhu, D. Staton, M. Popescu, and D. Hawkins, Analytical Modelling and Analysis of Open-circuit Magnet Loss in Surface-mounted Permanent Magnet Machines[J], IEEE Trans. Magn., **vol. 48**, pp. 1234-1246(2011).

[18] K. Atallah, D. Howe, P. H. Mellor, and D. A. Stone, "Rotor Loss in Permanent-magnet Brushless AC Machines[J], IEEETrans. Ind. Appl., **vol. 36**, pp. 1612-1618(2000).

[19] P.H. Mellor, R. Wrobel, D. Holiday,A Computationally Efficient Iron Loss model for Brushless AC Machines that Caters for Rated Llux and Field Weakened Operation[C], IEEE International Conference on Electrical Machine and Drives, IEMDC'09, pp. 490-494(2009).

ISPECE

IOP Publishing

Design and Experimental Research of Expansion-Anchorage Device in Deepwater Pipeline

Lan Zhang[1], Chengqiang Zhao[1], Liquan Wang[1] Cong Ru[1] and Zhengbin Zhu[1]

[1]College of Mechanical and Electrical Engineering, Harbin Engineering University, Harbin 150001, China

[a]Corresponding author: 965569816@qq.com

Abstract. Anchoring device is one of the key technologies of deep water pipeline internal packer. In this paper, the structural design of the anchorage device is completed on the basis of the principle of the inclined plane reinforced structure and elastoplastic mechanics, and its three-dimensional model and mechanical model are established. Theoretical research and simulation analysis of the strength of key parts of the anchoring device are carried out, with emphasis on the design of the anchoring block, material selection, anchoring depth between the anchoring block and the inner wall of the pipe, and theoretical and simulation research of the contact stress. The correctness of theory and simulation analysis is verified by anchoring experiment. The research content of this paper can lay a foundation for the design and research of anchoring device.

1. Introduction

In recent decades, with the continuous decrease of onshore and offshore oil and gas resources, the exploration and production of deep-sea oil and gas resources have become an important way to meet the global energy demand [1].Although pipeline oil transfer has always been considered as a safe and efficient way of oil transfer, due to the complex deep-sea environment, pipeline aging and other factors, the submarine pipeline is prone to damage [2-3].The importance of submarine oil pipelines is self-evident, which requires the development of pipeline repairment equipment for regular repairment [4]. Closure in deep water pipeline plays an important role in deepwater pipeline maintenance, which mainly includes the guiding device, sealing device, anchor device, power plant, switching device and hydraulic control system. Anchor device is the core part of the pipeline closure, and anchor block is an important part of anchoring device. Its performance fit or unfit quality directly affects the reliability of the anchor device, which affects the performance of the pipe inner stopper[5-6].This paper mainly introduces the design and research of anchoring device.

However, the existing anchoring device has such problems as unstable force applied in the process of anchoring, greater damage to pipelines, inconvenient replacement of anchoring block, and easy damage of anchoring device parts. The design of the anchor device in this paper is based on the theoretical research of the inclined plane force increasing structure and the principle of elastic-plastic mechanics. The theoretical design model of the main parts of the anchor device is established. The anchor block can be disassembled to facilitate the maintenance and replacement of the anchor block. The design strength is checked by simulation analysis, and the theoretical and simulation results are verified by experiments.

2. Design theory of anchoring device

The anchoring device is mainly composed of 1 octahedral cone, 2 t-shaped guide rail, 3 semicircular block, 4 anchoring block, 5 arc block, 6 external expansion block and 7 cylinder connecting sleeve. The anchor device model is shown in figure 1.

Content from this work may be used under the terms of the Creative Commons Attribution 3.0 licence. Any further distribution of this work must maintain attribution to the author(s) and the title of the work, journal citation and DOI.
Published under licence by IOP Publishing Ltd

ISPECE

IOP Publishing

IOP Conf. Series: Journal of Physics: Conf. Series **1187** (2019) 032065   doi:10.1088/1742-6596/1187/3/032065

Figure 1. Model anchoring device

2.1 Stress analysis of anchoring process

In the process of the sealing device in the deep sea pipeline reaching the final anchoring, the power device is required to provide the power to ensure that the teeth of the anchoring block are embedded in the pipe wall. Figure 2 is to convert the spatial confluence force system into the plane confluence force system. In the figure2, F is the total force provided by the fluid pressure in the pipe and the hydraulic cylinder, and F_N is the positive pressure of the pipe to the anchor block group.

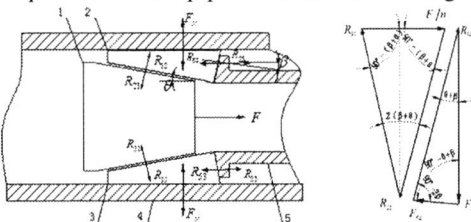

1. Octahedral cone 2. External expansion block 3. Anchor pressure block 4. Pipe 5. Hydraulic cylinder connecting frame

Figure 2. schematic diagram of anchoring stroke force

In figure 2, the triangle of force is obtained by positive mystery theorem:

$$\frac{F/2n}{\sin(\theta+\beta)} = \frac{R_{21}}{\sin[90^\circ-(\theta+\beta)]} \qquad (1)$$

$$\frac{F_N}{\sin(90^\circ-2\beta)} = \frac{R_{12}}{\sin(90^\circ-\theta+\beta)} \qquad (2)$$

Where, F is axial force on the packer; F_N is the positive pressure of the inner wall of the pipe on an anchor block group; n is the number of anchor block groups; β is the angle of friction between contact surfaces; θ is octahedral cone inclination angle; R_{12}, R_{21}, R_{13} R_{31} are the interaction between the external expansion block and the octahedral cone; R_{52}, R_{25}, R_{53} R_{35} are the interaction between the external expansion block and the cylinder connecting sleeve.

Because of $|R_{12}|=|R_{21}|$, formula (3) can be obtained:

$$F = F_N \frac{\sin(90^\circ-\theta+\beta)\sin 2(\theta+\beta)}{\sin[90^\circ-(\theta+\beta)]\sin(90^\circ-2\beta)} \qquad (3)$$

According to the mechanical manual to know that the expansion block and octahedral cone friction coefficient f=0.12, ,The friction Angle is calculated from formula (4), $\beta=6.84$.

$$\beta = \arctan f \qquad (4)$$

According to the graph of positive pressure F_N and inclination Angle θ in figure 3, When the inclination Angle is $\theta=4.5^\circ\text{-}6^\circ$, the positive pressure F_N increases; When the inclination Angle is $\theta=6^\circ\text{-}7.5^\circ$, the positive pressure F_N decreases. But when F_N is too large, result in anchor block embedded conduit wall exceeds the yield limit of pipe, in order to meet the expanding tightly embedded pipe when the pressure is F_N requirement, anchorage axial and radial motion displacement

935

and the choice of the hydraulic cylinder stroke, and to satisfy the requirement of the anchor device self-locking, take structure of slope angle to $\theta = 6.5°$.

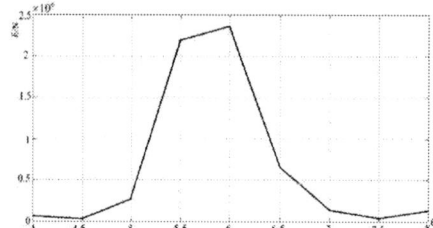

Figure 3. Curve of positive pressure and inclination

The friction Angle and inclination Angle are substituted into formula (1-2) to get $F_N = 2.11F/2n$, from which it is known that the inclined plane structure makes the output force 2.11 times of the input force. Substitute in the data to obtain $F = 2.67 \times 10^6 N$, $F_N = 7.04 \times 10^5 N$, $R_{21} = 1.41 \times 10^6 N$.

3. Design and strength analysis of key parts

3.1 Simulation analysis of external expansion block

Figure 4 shows that the maximum deformation of the external expansion block is 0.06mm, the deformation is very small, and the maximum stress is 459MPa. The external expansion block selects no. 45 steel tempered treatment, and the yield strength meets the requirements.

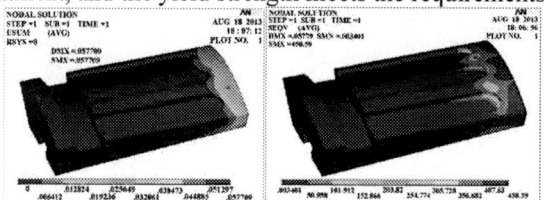

Figure 4. Deformation cloud chart and stress cloud chart of outer bulge block.

3.2 Simulation analysis of octahedral cone

Formula (2) calculates the vertical pressure $R_{21} = 7.04 \times 10^5 N$ on the contact surface of the conical body. ANSYS analysis results of the octahedral cone are shown in figure 5. The maximum deformation of the octahedral cone is 0.185mm and the maximum stress is 340MPa. They are produced at the end of the cone, so the material of the octahedral cone is also 45 steel that has been tempered and chromed.

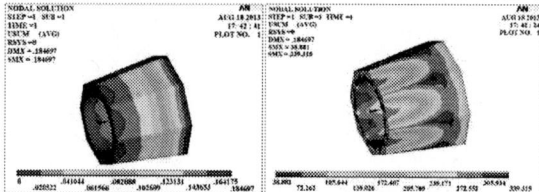

Figure 5. Deformation cloud diagram and stress cloud diagram of octahedral cone

4. Design calculation and simulation analysis of anchor block

After the gate valve is completely closed, the positive pressure of each external expansion block on the pipe wall is $F_N = 7.04 \times 10^5 N$, each external expansion block is arranged with 3 columns of 5 anchor blocks (the Angle between each column is), the positive pressure of each anchor block is as follows:

$$P_1 = \frac{F_N}{5 \times (2\cos 15° + 1)} = 4.8 \times 10^4 N \qquad (5)$$

The tooth shape of anchor block is designed as triangle. The deeper the anchor block is embedded into the pipe wall, the greater the contact area between the pipe wall and the anchor block will be. The anchor block is made of high carbon alloy steel. The tensile strength is $\delta_b = 1320\text{MPa}$, the yield strength is $\delta_s = 600\text{Mpa}$, hardness is HRC60. After calculation, the embedding width of the anchor block can be obtained, as shown in equation (6).

$$B = \frac{P_1}{\pi d \sigma_p} \qquad (6)$$

Where, P_1 is positive pressure on an anchor block, d is the circular tooth tip diameter, B is the width of ring belt, σ_p is stress of anchor block.

The steel used in API standard submarine oil and gas pipelines is X42~X70 (divided by yield limit), and the commonly used material X65 (L450) is taken. When $\sigma_p = \sigma_s$ is taken, the annular bandwidth is $B = 1.46\text{mm}$, and the depth of anchor block embedded in the pipe wall is about h=0.87mm on the triangular section of tooth shape.

4.1 Strength analysis of anchorage block teeth
Under the action of 4.8t positive pressure, the depth of the anchor block embedded in the inner wall of the pipe is 0.87mm.As can be seen from figure 6, the side length L and positive pressure P_1 of the tooth shape can be obtained:

$$L = \frac{h}{\cos 40^\circ} = 1.136\text{mm}$$

$$P_1 = P \sin 40^\circ = 3.10\text{T}$$

Then the compression area S of the anchor block is:

$$S = 2\pi \times L \times d = 167.65\text{mm}$$

From formula (15), it can be known that the stress on the anchor block is:

$$\sigma_p = \frac{P_1}{S} = \frac{3.1 \times 10^4}{167.65} = 184.9\text{MPa}$$

It can be seen from figure 6 that the dangerous section of annular tooth of anchorage block is a-a section, the height of a-a section is b=6mm, and the distance between the tooth tip and a-a section is H=3.5mm. The bending strength of the dangerous section is shown in formula (7), and the shear strength is shown in formula (8) :

$$\sigma_b = \frac{M}{W} = \frac{6FH}{8ab^2} \qquad (7)$$

$$\tau = \frac{F}{8\pi dB} \qquad (8)$$

Where, F is Axial pressure of a single anchor block F=2.23×10⁴N, d is the circular tooth tip diameter, M is bending moment of anchor block, W is a-a section approximates the flexural modulus of the beam.

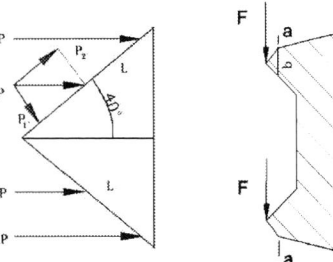

Figure 6. The positive pressure on the anchorage block teeth and the force on the cross section of the teeth

By substituting the data, formula (7) and formula (8) can obtain:

$$\sigma_b = \frac{M}{W} = \frac{6FH}{8ab^2} = 196.4\text{MPa}$$

$$\tau = \frac{F}{8\pi dB} = 161.3\text{MPa}$$

Allowable stress of anchor block:

$$\left[\sigma_{\text{p}}\right] = \frac{\sigma_{\text{s}}}{n_{\text{s}}} = \frac{600}{1.25} = 480\text{MPa}$$

$$\left[\tau\right] = 0.6\left[\sigma\right] = 0.6 \times \frac{\sigma_{\text{s}}}{1.25} = 288\text{MPa}$$

To sum up, the extrusion strength, flexural strength and shear strength of the anchor compression block all meet the design requirements.

4.2 Analysis of contact stress and deformation of anchor block

It is known that the yield limit of pipeline material is 450MPa, and the yield limit of anchor block material is 600Mpa. It is known from figure 7 that the maximum plastic deformation of the pipeline is 0.82mm, and from figure 8 and 9 that the maximum stress of the anchor block is 512Mpa. The maximum strain is 0.004mm, so the strength of the anchor block meets the requirements.

Figure7. Ductal plastic deformation cloud pattern

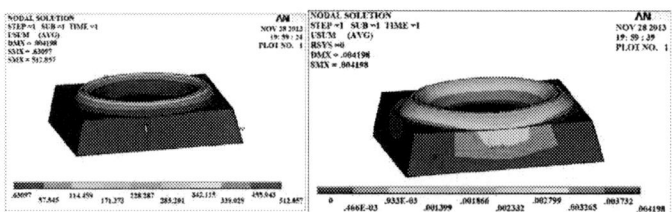

Figure8. Stress and deformation nephogram of anchor block

5. Experimental Study

5.1 Experimental purpose and equipment

The test equipment mainly includes the test plugging device prototype as shown in figure 9. Figure 10 shows the pipeline to be blocked; Syl-40/1.6 manual hydraulic pump is shown in figure 2.6.

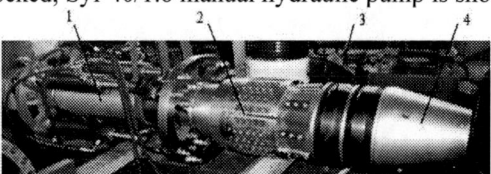

1. Hydraulic drive device 2. Anchoring device 3. Sealing device 4. Guide device

Figure 9. Deep water pipeline internal packer test prototype

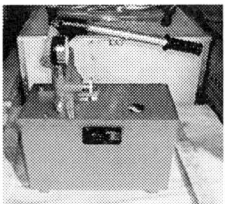

Figure 10. The pipeline to be blocked and manual hydraulic pump

5.2 Experimental steps

To simulate the above process in the land environment, the following experiments are designed:

(1) take a section of oil pipeline as the damaged pipeline to be sealed, one end of which is open for installing the packer, complete the anchoring steps of the packer.

(2) Successively open the pipe and the globe valve of the packer to relieve pressure, press the rod cavity of the hydraulic cylinder of the packer, remove the sealing state of the packer by anchoring, remove the equipment, and observe the anchoring traces on the inner wall of the pipeline.

5.3 Experimental results

During the test, in the process of pressure rising in the oil pipeline, there is no relative displacement between the locating sleeve of the packer and the pipe mouth, and there is no obvious slip in the anchoring trace of the inner wall of the pipeline after removal, indicating that the anchoring of the packer can be reliable. Anchor marks on inner wall of pipe and sealing device test equipment installation as shown in figure 11.

Figure11. Anchor marks on inner wall of pipe and sealing device test equipment installation

6. Conclusion

This paper focuses on the force analysis, design theory and design analysis of anchoring device of deep-sea pipeline internal closure device, as well as the key parts of anchoring device. The depth of anchorage block embedded in pipe and the contact stress between anchorage block and pipe are analyzed theoretically and simulated. The theoretical and simulation results of anchor design are verified to meet the practical engineering requirements by simulating the plugging process of a deep sea packer. The research content of this paper can provide reference basis for the stress, design of anchoring device and the research of anchoring on pipeline damage.

Acknowledgement

This paper is funded by the ocean engineering equipment scientific research project of Ministry of Industry and Information Technology (No.Z16SJENK0033) and the National Natural Science Foundation of China (Grant NO. 5167051260).

References

[1] A. Shukla, H. Karki. Application of robotics in offshore oil and gas industry— A review Part II [M]. North-Holland Publishing Co(2016).

[2] X. Zhu, D. Wang, H. Yeung, et al. Comparison of linear and nonlinear simulations of bidirectional pig contact forces in gas pipelines[J]. Journal of Natural Gas Science & Engineering(2015).

[3] X. Zeng, M. Duan, X. Che. Critical upheaval buckling forces of imperfect pipelines[J]. Applied Ocean Research,(2014).

[4] L. Q. Wang, Z. L. Wei, S. M. Yao, et al. Sealing Performance and Optimization of a Subsea Pipeline Mechanical Connector[J]. Chinese Journal of Mechanical Engineering, 31(1):18(2018).

[5] D. Geng, S. M. Zhang, D. G.Wang, et al. The optimization of the fluke structure of intelligent pipe plug[J]. Oil & Gas Storage and Transportation(2011).

[6] D. Wang, S. P. He, X. Zhang. A Study on Contact Stress of Packer Slip[J]. Journal of Experimental Mechanics(2006).

ISPECE   IOP Publishing

Study on Wear Mechanism of Diamond Particles in the Cutting of Pipeline Steel

Lan Zhang[1], Zhengbin Zhu[1,a], Liquan Wang[1], Chengqiang Zhao[1] and Cong Ru[1]

[1]College of Mechanical and Electrical Engineering, Harbin Engineering University, Harbin 150001, China

[a]Corresponding author: 969699840@qq.com

Abstract. The wear of diamond particles is an important factor that limits the working efficiency of diamond wire saws. In this paper, the physical model of diamond particle is established and the mechanical analysis is carried out. ANSYS finite element analysis software was applied to establishing unworn model and worn model for diamond particles. The cone angle of diamond particle, the protrusion height of diamond particle and the depth of the diamond particles press into the cutting face were analysed to study theirs effect on the shear stress extreme value of diamond particles in the process of cutting underwater pipeline. The results shown that the depth of the diamond particles press into the cutting surface and the cone angle of diamond particle have a great influence on the wear of the diamond particles, and the protrusion height of diamond particle is relatively small for the wear of the diamond particles.

1. Introduction

Wear mechanism of diamond particles is an important research field. In the process of cutting the underwater pipeline by the diamond wire saw machine, the diamond particles are subjected to severe impact, scraping and chip erosion of the material. Therefore, most diamond tools are faced with serious wear, and the wear of diamond particles will directly affect the processing efficiency and processing cost. Tönshoff [1] pointed out that the wear mechanism of diamond particles can be divided into four main forms: adhesive wear, friction wear, diffusion wear and abrasive particle breakage. Ersoy [2] studied the wear of diamond abrasive grains from ten different rock materials, and considered cutting power as a key factor which affects wear characteristics. Aydin [3] studied the wear performance of diamond circular saw blades in granite cutting and determined the effect of various operating variables on specific worn rates. Hui [4] studied the material removal mechanism and abrasive wear characteristics of diamond bead rope cutting rock. It is believed that the material removal mechanism is mainly volumetric fracture. Yilmazkaya [5] studied the effect of line speed and cutting speed on the unit wear and tear of a single wire cutter. Su [6] studied the change of diamond wear shape and blade height during the cutting process. Goel [7] used a molecular dynamics simulation method to study the wear mechanism of diamond tools on single crystal silicon during single-point diamond turning.

Research on the wear characteristics of diamond cutting tools has achieved a lot of results. However, due to the complexity of the problem, most of the work focused on the study of the wear type, worn morphology and worn process of diamond particles. In this paper, the process of sawing the pipeline steel by diamond wire saw is analysed, and the diamond particle is modelled and analysed. Through the finite element software, unworn model and worn model for diamond single-particle were established, and the influence of diamond particle parameters on wear was obtained.

Content from this work may be used under the terms of the Creative Commons Attribution 3.0 licence. Any further distribution of this work must maintain attribution to the author(s) and the title of the work, journal citation and DOI.

Published under licence by IOP Publishing Ltd

2. Force analysis of diamond particle

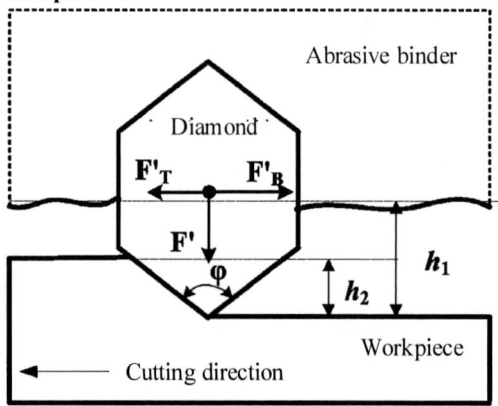

Figure 1. Force state of diamond particle.

Diamond particles on the bead surface directly act on the workpiece material during cutting. The strength of the diamond particles is much greater than the strength of the workpiece material. Under the action of extrusion, the workpiece material yields to achieve the cutting effect. However, after the diamond particles are worn, the contact area is increased, the cutting effect is lowered, and sometimes the abrasive particles are detached from the binder. In order to analyze the cutting effect of diamond abrasive grains, it is assumed that the diamond particle on the bead surface is regular, diamond-shaped structures with the same particle size and diamond particles uniform adhesion to the surface of the binder material. Figure 1 shows the stress state of a single diamond particle on the bead surface. Where F'_T is the axial pulling force along the sawing rope, F'_B is the cutting surface resistance, F' is the normal pressure of the single diamond particle along the cutting surface, h_1 is the average blade protrusion height of the diamond particle, and h_2 is the depth of diamond particle press into the cutting surface, φ is the cone angle of diamond particle.

In the quasi-static state, the depth of the diamond particles press into the cutting surface depends on the normal pressure, so it is necessary to first determine the normal pressure of the single diamond particles. This paper introduces a linear relationship based on the basic theory of nanoindentation testing:

$$F' = HA \qquad (1)$$

Where: H is the indentation hardness of the cutting material, and A is the projected area of the contact surface of the diamond particles and the cutting material.

According to the particle geometry, the specific expression can be obtained as

$$A = \frac{3\sqrt{3}}{4} \tan^2 \left(\frac{\varphi}{2} \right) h_2^2 \qquad (2)$$

Substituting equation (2) into equation (1), the expression of the pressure of a single diamond particle and its indentation depth can be obtained:

$$F' = \frac{3\sqrt{3}}{4} H \tan^2 \left(\frac{\varphi}{2} \right) h_2^2 \qquad (3)$$

Although the diamond particles have higher strength, they are more brittle. And impact and shear are prone to cracking, which causes the diamond particles to break or brittle. During the cutting process of the sawing rope, the end of the diamond particle is subjected to the shear stress caused by the frictional resistance. The resistance of a single diamond particle cutting surface is expressed as:

$$F_B' = \frac{F_3 L}{N_d \pi r_w N L_d L_c} \qquad (4)$$

Where F_3 is the frictional resistance of the cutting surface, N_d is the number of diamond particles per unit area of the bead surface, N is the number of beads that are simultaneously cut, L is the total length of the saw rope, r_w is the outer radius of the bead, and L_d is the axial length of the diamond bead.

The average shear stress expression is:

$$\tau = \frac{F_B'}{A} \qquad (5)$$

According to the equation, h_2 and φ have a great influence on the shear stress.

3. Finite element model

Based on Fig. 1, the finite element model of the problem is established by ANSYS and the finite element analysis is carried out to obtain the influence of different parameters on the force and wear of diamond particles. This paper analyzes and discusses the situation based on unworn and worn.

The parametric APDL language is used to draw the geometric model of the problem. The definition of h_1 is the average protrusion height of the diamond particles, h_2 is the depth of the diamond particles press into the cutting surface, h_3 is the height of worn, and φ is the cone angle of diamond particle. The above four parameters are the parameters to be discussed in the finite element analysis.

The geometric model is meshed with a finer cell size. Based on the Saint-Venant principle, displacement constraints are applied to the side and bottom edges of the steel tube section to statically set the model. At the same time, according to the direction of the tension of the saw rope and the direction of the cutting pressure, the corresponding uniform load is added on the side line and the upper line of the adhesive part, and the mesh and the constraint are as shown in figure 2.

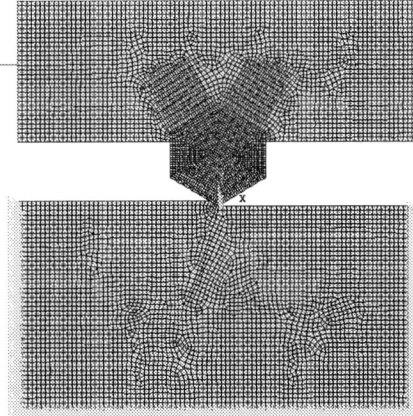

Figure 2. Constraint and load.

4. Results

4.1. Result of unworn diamond

4.1.1 Effect in the particle cone angle φ

Taking h_1=0.4mm, h_2=0.04mm, 0.05mm, 0.06mm, 0.07mm, 0.08mm; φ=60°, 75°, 90°, 105°, 120°. The calculated internal shear stress extreme value of the diamond is shown in Figure 3.

It can be seen from the calculation results that the extreme value of shear stress in diamond decreases with the φ increase. This is mainly due to the increase in the inclination angle of the contact surface between the diamond and the steel pipe. The shear stress extreme point is at the tip position, as is shown in Figure 4. It can be seen that the larger φ plays a significant role in reducing the wear of the diamond.

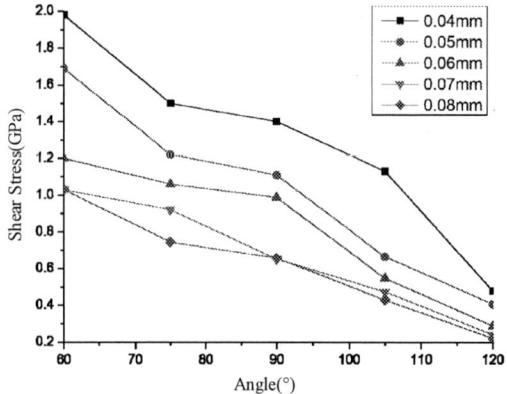

Figure 3. Effect in unworn diamond's shear stress extreme from φ.

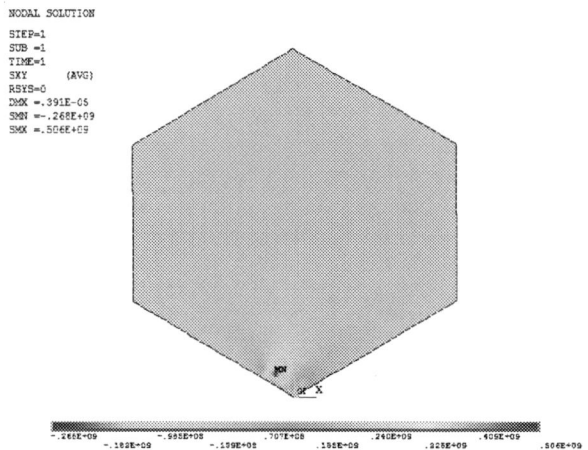

Figure 4. Location of shear stress extreme.

However, if the value of φ is large, the equivalent stress at the cutting point will be reduced, thereby reducing the cutting effect. Therefore, the proper angle φ plays an important role in both the wear resistance and the cutting effect.

In addition, it can be seen from the calculation results that the shear stress extreme value in diamond also decreases with the increase of h_2. This is due to the increased contact area between the diamond and the steel tube.

4.1.2 Effect in the average protrusion height h_1
Taking h_2=0.04mm, h_1=0.30mm, 0.35mm, 0.40mm, 0.45mm, 0.50mm; φ=60°, 75°, 90°, 105°, 120°. The diamond internal shear stress extreme value is shown in the figure 5.

It can be seen from the calculation results that the shear stress extreme value in diamond gradually increases with the increase of h_1, but the increase is relatively small. The shear stress extreme point is at the tip position. Therefore, adjusting the size of h_1 has little effect on the wear of diamond.

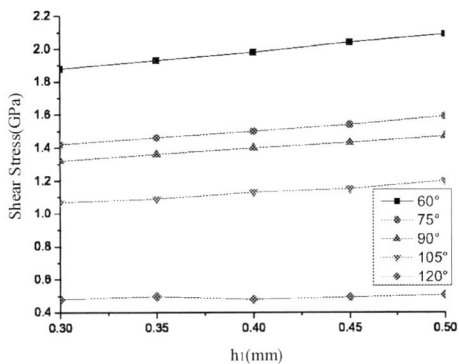

Figure 5. Effect in unworn diamond's shear stress extreme from h_1.

4.2. Result of worn diamond

4.2.1 Effect in the particle cone angle φ

Taking h_1=0.4mm, h_2=0.04mm, φ=60°, 75°, 90°, 105°, 120°; h_3=0.04mm, 0.06mm, 0.08mm, 0.10mm, 0.12mm. The calculated internal shear stress extreme value of the diamond is shown in Figure. 6.

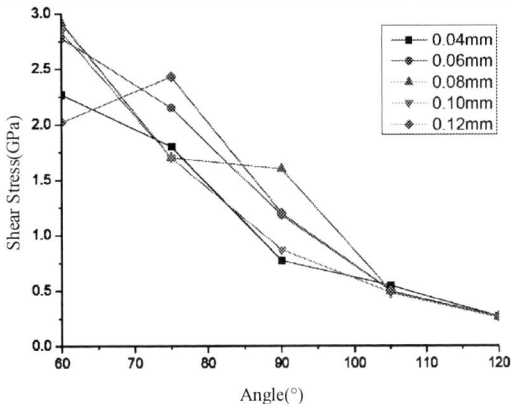

Figure 6. Effect in worn diamond's shear stress extreme from φ.

It can be seen from the calculation results that the shear stress extreme value of diamond decreases with the φ increase. This is mainly caused by the increase in the inclination angle of the contact surface between the diamond and the steel pipe. In addition, the shear stress extremes are greater when worn under the same conditions than when there is unworn. It is shown that the wear resistance of the worn diamond will decrease.

4.2.2 Effect in the depth of the diamond particles press into the cutting surface h_1

Taking h_1=0.4mm, φ=120°, h_2=0.04mm, 0.05mm, 0.06mm, 0.07mm, 0.08mm; h_3=0.04mm, 0.06mm, 0.08mm, 0.10mm, 0.12mm. The calculated internal shear stress extreme value of the diamond is shown in Figure. 7.

It can be seen from the calculation results that the shear stress extreme value in diamond decreases with the increase of h_2. This is mainly caused by the increased contact area between diamond and steel pipe. When h_2=0.04mm, the shear stress extreme point is located on the shallow surface of the diamond near the end of the contact surface. In other cases, the shear stress extreme point is located at the lower end of the contact surface. It can be seen that increasing h_2 can play a certain role in reducing the wear of diamond.

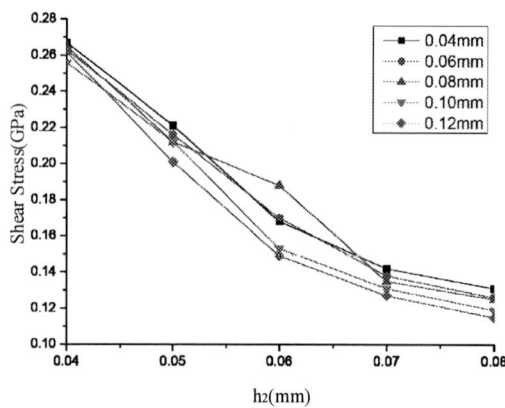

Figure 7. Effect in worn diamond's shear stress extreme from h_1.

4.2.3 Effect in the average protrusion height h_1

Taking h_2=0.4mm, φ=120°, h_1=0.30mm, 0.35mm, 0.40mm, 0.45mm, 0.50mm; h_3=0.04mm, 0.06mm, 0.08mm, 0.10mm, 0.12mm. The calculated internal shear stress extreme value of the diamond is shown in Figure. 8.

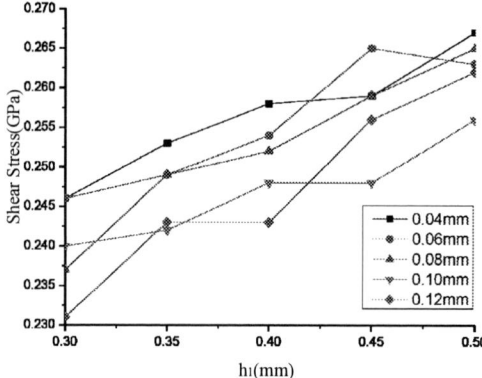

Figure 8. Location of shear stress extreme (φ=120°).

It can be seen from the calculation results that the shear stress extreme value in diamond increases with the increase of h_1, but the increase is small. The shear stress extreme points are located on the shallow surface of the diamond near the end of the contact surface. It can be seen that adjusting h_1 has little effect on the wear of diamond.

5. Conclusion

In this paper, based on the worn and unworn conditions of diamond particles, the effect of the average protrusion height of the diamond particles h_1, the depth of the diamond particles press into the cutting surface h_2, the wear height h_3, and the cone angle of the particles φ were analysed to study the shear stress extreme values of the diamond particles, that is, the impact on the wear of the diamond wire saw.

(1) The extreme value of shear stress in diamond shows an obvious downward trend with the increase of the depth of the diamond particles press into the cutting surface h_2.

(2) The shear stress in diamond increases with the increase of the average protrusion height h_1, but the increase is small.

(3) The extreme value of shear stress in diamond decreases with the increase of the cone angle of the particles φ. The larger φ plays a significant role in reducing the wear of diamond.

(4) The shear stress extreme value is greater when there is worn under the same conditions than when there is unworn. It indicates that the wear resistance of the worn diamond will decrease.

Acknowledgement

This paper is funded by the Fundamental Research Funds for the Central Universities (Grant NO. HEUCFP201848) and the National Natural Science Foundation of China (Grant NO. 5167051260).

References

[1] H.K. Tonshoff, H. Hillman-Apman, J. Asche, Diamond tools in stone and civil engineering industry: cutting principles, wear and applications, Diamond Relat. Mater. **11**, 736–741(2002)

[2] A. Ersoy, I.S. Buyuksagis, U. Atıcı, Wear characteristics of circular diamond saws in the cutting of different hard and abrasive rocks Wear, **258**, pp. 1422-1436(2005)

[3] A. Gokhan, K. Izzet, A. Kerim, Wear Performance of Saw Blades in Processing of Granitic Rocks and Development of Models for Wear Estimation, Rock Mechanics & Rock Engineering, **46**(6), 1559-1575(2013).

[4] H. Huang, G. Huang, X. Xu, An experimental study of machining characteristics and tool wear in the diamond wire sawing of granite, Proc IMechE Part B: J Engineering Manufacture, **227**(7), 943-953(2013).

[5] E. Yilmazkaya, Y. Ozcelik. The Effects of Operational Parameters on a Mono-wire Cutting System: Efficiency in Marble Processing, Rock Mechanics & Rock Engineering, **49**(2), 523-539(2016).

[6] S. Yu , H. Guo, J.B. Wang, Y.J. Zhang, Study on fashion of wearing and height of protrusion of diamond in wire saw, Superhard Material Engineering(2009).

[7] S. Goel, X. Luo, R. L. Reuben, Wear mechanism of diamond tools against single crystal silicon in single point diamond turning process, Tribology International, **57**(57), 272-281(2013).

| ISPECE | IOP Publishing |

Analysis of Impact Characteristics of Diamond-Beaded Rope and Its Influence on Cutting Efficiency and Life

Lan Zhang[1], Cong Ru[1,a], Liquan Wang[1], Zhengbin Zhu[1] and Chengqiang Zhao[1]

[1]College of Mechanical and Electrical Engineering, Harbin Engineering University, Harbin 150001, China

[a]Corresponding author: 172982993@qq.com

Abstract. The impact of wire saw cutting is one of the important factors affecting the cutting of diamond wire saw. It affects not only the efficiency of wire saw cutting, but also the service life of beaded rope. The actual working conditions of the impact of diamond wire saw cutting are analyzed in this paper. The wire saw impact model was built using ANSYS WORKBENCH and the dynamics simulation study was carried out. The stress distribution of the bead rope and maximum stress at different linear velocities are obtained. The results show that as the linear velocity increases, the additional stress on the bead increases. Through experiments, it is concluded that the wear of the bead will increase as the linear velocity increases. The cutting efficiency of the wire saw increases with the increase of the linear velocity in the early stage, but when the speed exceeds a certain speed, the cutting efficiency will decrease rapidly. Studies have shown that the impact of beading will be beneficial to cutting to a certain extent, but when it exceeds a certain limit, the bead wear will increase, which will seriously affect the cutting efficiency and service life.

1. Introduction

Underwater diamond wire saw cutting is a means of underwater mechanical cold processing[1]. Compared with other marine equipment, the diamond wire saw cutting method has strong adaptability to the marine environment, and has the advantages of large diameter of processing pipeline, and can cut metal materials, non-metal materials and composite materials[2]. Due to the non-continuous cutting characteristics of the wire sawing machine, the beaded rope will continuously impact the pipe wall during the working of the wire sawing machine[3]. The slight impact is beneficial to the cutting of the submarine pipeline, but the impact is too large, which will seriously affect the beading rope life and cutting efficiency[4]. There are currently few studies on this phenomenon.

2. Establishment of beaded rope impact model

2.1 Beading condition analysis

When the wire saw machine performs the cutting experiment, taking the round pipe as an example, the impact of the diamond bead and the cutting object is shown in Figure 1.

The figure 1 indicates several probability positions of a single diamond bead at the initial impact with the wall of the tube.

Content from this work may be used under the terms of the Creative Commons Attribution 3.0 licence. Any further distribution of this work must maintain attribution to the author(s) and the title of the work, journal citation and DOI.
Published under licence by IOP Publishing Ltd

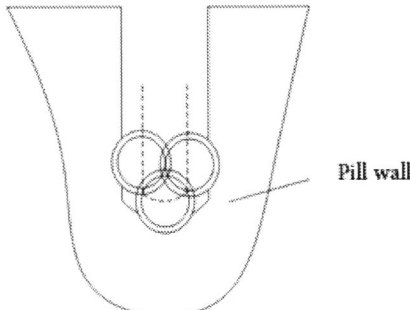

Figure 1. Beaded random position.

Combined with the model established in Figure 1, a single diamond bead is continuously struck against the pipe wall throughout the cutting process, and the impact position and timing are random, mainly acting on the radial motion of the beaded rope. direction. The impact causes the diamond abrasive grains embedded on the substrate to fall off and break, reducing the working efficiency and service life of the wire saw. In the model, it can be considered that the main impact phenomenon is periodic and has the greatest influence on the diamond wire saw[5-6].

2.2 Bead impact model

The individual beads have a small mass, which is negligible here to simplify the model and other forces when hitting the transient state. In the event of an impact, it is assumed that the bead and the bead string are aligned at a sufficiently small length. The length of this beaded rope is assumed to be *l*, and the impact model is established as shown in Figure 2.

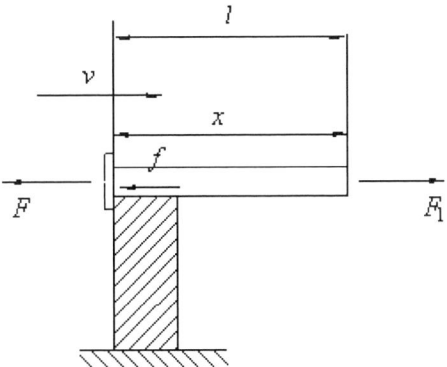

Figure 2. Model of string bead wires impacting.

In the initial state, the bead rope moves at a speed v in the axial direction. When the left side of the bead hits the wall in an instant, the speed of movement cannot be abrupt, and an acceleration is generated. At this time, the force analysis of the bead is performed, and the projection to the bead axis direction can establish the equation (1):

$$F_1 = (F + f) - F_h \qquad (1)$$

Where F_1 is the traction of the beaded rope on the left side of the bead; f is the friction of the tube wdall against the outer surface layer of the beaded rope, and its value is small relative to the other two, which is ignored here; F_h is the mutual impact force between the bead and the tube wall. Then the stress on the bead rope can be written as (2) before the impact occurs.

$$\sigma = \frac{F_1}{A} = \frac{F + f}{A} \approx \frac{F}{A} \qquad (2)$$

Where A is the cross-sectional area along the radial direction of the diamond bead.

At the moment of impact, a single bead can be decomposed into two parts of stress by transient stress σ': one is the stress σ generated by the traction force F_1; the other is the impact stress σ_h caused by the impact force F_h:

$$\sigma' = \sigma + \sigma_h = \sigma + {F_h}/{A} \qquad (3)$$

3. Explicit dynamics finite element simulation of beaded rope impact

3.1 Explicit dynamics simulation model for bead rope impact

Through the analysis of the single diamond bead motion process, combined with the diamond wire sawing machine to cut the 8″ single-layer tube operation process, this chapter uses the explicit dynamics module in Ansys Workbench to simulate the impact effect caused by the beaded rope cutting 8″ single-layer tube. The finite element simulation model is shown in Figure 3.

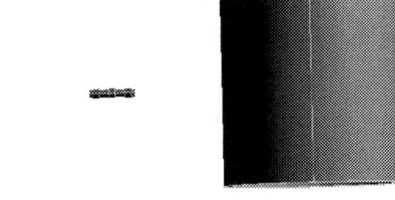

Figure 3. Finite element simulation model.

3.2 Analysis of explicit dynamics finite element simulation results of bead rope impact

It can be seen from figure 4 that when the bead rope is in contact with the pipe wall, the stress at the contact point of the bead is the largest, and the stress on the bead rope near the impact position is the largest. The stress amplitude spread in the opposite direction of the movement direction of the wire saw, the futher away from the impact position, the smaller the stress amplitude is. Due to inertia, the impact mainly affects the next bead, which has little effect on the subsequent bead rope.

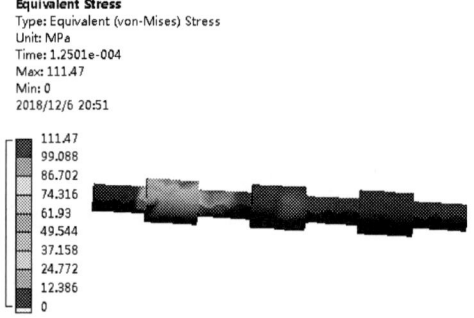

Figure 4. Stress cloud chart.

It can be seen from figure 5 that when the beaded rope moves at a given speed, its initial stress is small; but when the beaded rope collides with the pipe wall, the additional stress will reach a maximum. And the maximum value of the stress increases as the wire speed of the wire saw increases. After the impact, the stress on the bead rope decays rapidly and reaches a stable value.

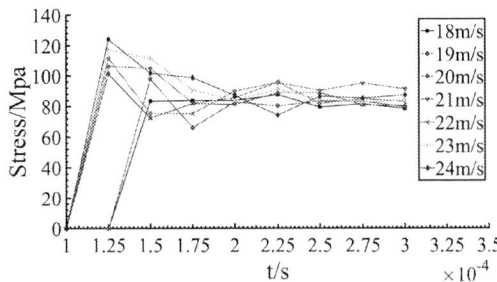

Figure 5. Stress Diagrams at Different Times.

As can be seen from figure 5, since the finite element model initially has a certain distance between the bead and the tube wall, when the bead rope speed is 18m/s and 19m/s, the bead and the tube wall are in contact with the tube wall at 0.000125s. At higher than 19 m/s, the bead is in contact with the tube wall at 0.0001 s. The maximum stress on the beaded rope is the smallest at 18m/s, reaching 83.396MPa, the maximum at 24m/s, and the maximum can reach 124.01MPa. Since the yield strength of the bead matrix is 350 MP, considering its fatigue characteristics, its fatigue strength should be 30% to 35% of its yield strength [7], so the impact stress should be less than 116 MA. As can be seen from the figure 6, the linear speed of the beaded rope should be no more than 23 m/s.

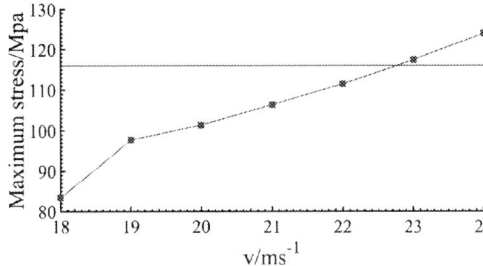

Figure 6. Maximum stress at different linear velocities.

4. Experimental study on the effect of beaded rope impact on cutting

4.1 Experimental equipment and method

The experimental platform is mainly used for underwater pipeline cutting experiments. The outer diameter of the beaded rope used in the experiment is 10.5mm, the diameter of the flywheels on both sides is 0.32m, the center distance is 1.4m, and the tension is 2000N.

To comprehensively study the impact of impact on the wear of different parts of the diamond saw rope, the single bead is divided into three parts: front, middle and back, as shown in figure 7.

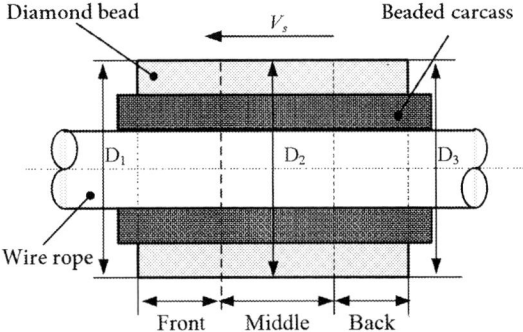

Figure 7. Diamond bead segmentation diagram.

To objectively analyze the wear of the bead, 20 beads were selected as the research object at the same distance. The machine was stopped at intervals of 3 minutes. The diameter of the bead was measured twice in the direction perpendicular to the bead circumference. The average value is recorded as the measured bead diameter value.

4.2 Impact of impact on bead life
Figure 8 shows the SEM photograph of the bead at the end of the cutting. From figure 8, it can be seen that the diameter of the front and rear of the bead is significantly smaller than the middle of the bead. Most of the diamond abrasive grains have fallen off, and the carcass has obvious scratches. The cylindrical shape becomes a waist drum.

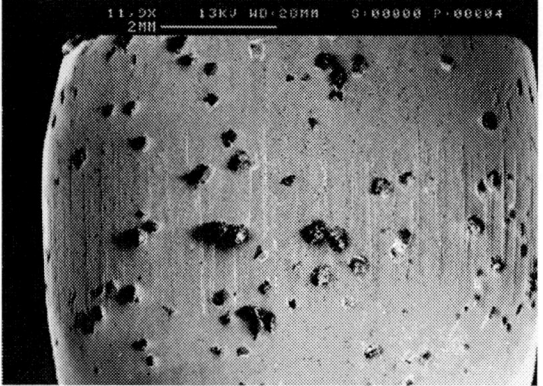

Figure 8. Bead SEM morphology at the end of cutting.

Figure 9, Figure 10, Figure 11shows the variation of the diameter of different parts of the diamond saw rope at different cutting speeds (feed speed 0.25m/h, rope tension 1500N).

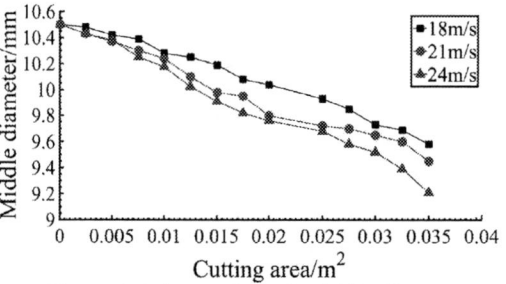

Figure 9. Diamond bead middle diameter.

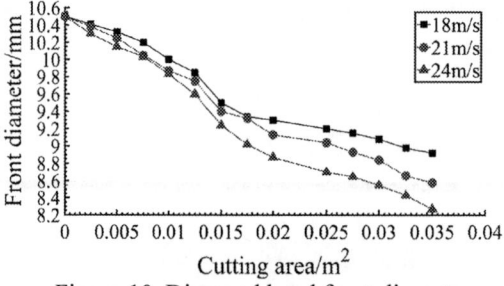

Figure 10. Diamond bead front diameter.

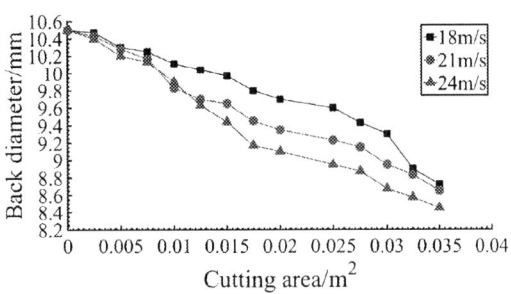

Figure 11. Diamond bead back diameter.

It can be concluded from the figure that the diameters of the front, middle and back parts of the bead are different at different cutting speeds. When the cutting speed is 18m/s, the diameters of the beads in front, middle and back are reduced by 1.59mm, 0.92mm and 1.77mm respectively. When the cutting speed is 21m/s, the diameters of the beads in the front, middle and back are reduced respectively 1.99mm, 1.05mm and 1.89mm; the diameter of the bead in front, middle and back was reduced by 2.10mm, 1.26mm and 1.93mm respectively at the cutting speed of 24m/s. It can be seen that as the cutting speed increases, the wear of the beads increases. The cutting speed has a significant effect on diamond bead wear.

4.3 Impact of impact on cutting efficiency

Figure 12 shows the curve of the cut-away area of the steel pipe with the cutting speed (pressure 50N, feed rate 0.25m/h) in the same time.

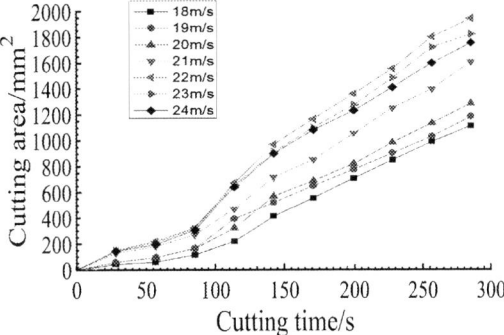

Figure 12. Cutting amount at different cutting speeds.,

It can be seen from Figure 12 that under the condition that the cutting speed is changed, the cutting area of the steel pipe will change. When the cutting speed is less than 22m/s, the cut-out area of the steel pipe increases with the increase of the cutting speed, but when the cutting speed is greater than 22m/s, the increase of the cutting speed reduces the cut-out area of the steel pipe.

5. Conclusion

In this paper, the finite element model of diamond bead rope impact is established. Under the condition that the tension of the bead rope and the constraint are the same, the other parameters are set. Only the influence of the speed of the bead rope on the impact is analyzed. The bead rope cutting life and cutting efficiency experiment were completed, and the experimental results were compared and analyzed. The conclusions are as follows:

(1) The impact of the wire rope on the impact is less than the impact of the diamond bead. The impact of the impact on the operation of the diamond wire saw is mainly due to the influence of diamond beads.

(2) The impact on the beaded rope increases as the speed of the beaded rope increases.

(3) The wear of the bead will increase as the impact increases.

(4) Within a certain impact range, the impact will promote the cutting of the wire saw machine, but as the impact increases, the increase of diamond particle damage will also reduce the cutting efficiency and seriously affect the service life of the bead rope.

Acknowledgment

This paper is funded by the Fundamental Research Funds for the Central Universities (Grant NO. HEUCFP201848) and the National Natural Science Foundation of China (Grant NO. 5167051260).

References

[1] Minghui, Zhang, Z. Yanna. Ocean engineering application progression and prospect, Shanxi Architecture,**42**(01):253-255(2016).

[2] Tonshoff H K, Friemuth T, Hillmann-Apmann H. Diamond Tools for Wire Sawing Metal Component, Diamond and Related Materials.**11**(3):742-748 (2002)

[3] Yuhong, Lian, Q. P. Company. Application of diamond wire saw cutting concrete technology,Petrochemical Industry Technology. **22**(10):113-114 (2015)

[4] Fei W, Bo H, Jinsheng Z, et al. Analysis on vibration characteristics of multi-wire diamond beads sawing machine. Diamond & Abrasives Engineering. **35**(3):5 -9(2015)

[5] Wang H, Zhang L, Meng Q, et al. Study on Characteristic Parameters of Underwater Diamond Wire Saw Cutting Pipe,Machine Tool & Hydraulics. **46**(11):115-118(2018)

[6] R.Garrard,S.R.Pecock and M.Hori. The future role of diamond in the construction industrial, Industral Diamond Review.**2**:121-129P(2001)

[7] Yongkun Z, Chunfeng W, Xing T, et al. Fatigue strength analysis of CSP thin-gauge and high-strength steel, Wuhan Iron & Steel Corporation Technology. **55**(04):24-26.(2017)

ISPECE   IOP Publishing

Analysis of mesh stiffness of herringbone gear considering modification

Wei Liu[1], Lunqin Duan[1*], Runjiao Wang[1] and Yanyan Wang[1]

[1]School of Mechanical and Electronic Control Engineering, Beijing Jiaotong University, Beijing 100044, PR China

*Corresponding author. Tel.: +86-01-51688175

E-mail address: 16121296@bjtu.edu.cn

Abstract. Herringbone gear is a key component in the reducer. Its mesh stiffness is an important parameter in gear dynamics research. In this paper, an improved Velex method is proposed to calculate the mesh stiffness of herringbone gears, and compared with Weber method and Finite Element Method (FEM), the accuracy of the method is proved. Meanwhile, the influence of helix angle and modification coefficient on the stiffness of the gear is analyzed by Weber method. The results show that the two parameters have different degrees of influence on the mesh stiffness of the herringbone gear, which provide a theoretical basis for the design of the herringbone gear transmission. Finally, by the calculation of the stiffness of the herringbone gear, the difference between the herringbone gear and the helical gear is compared and analyzed, and the conclusion that the herringbone gear is not a simple superposition of the helical gear is obtained. Because of the difference between the helical gear and the herringbone gear, it is of great value to the stiffness analysis of the herringbone gear.

1. Introduction

With the wide application of herringbone gear in aerospace, marine and other industrial fields, the dynamic problem has become increasingly prominent. Therefore, the dynamic characteristics of gear transmission has become a hotspot issue for scholars. Among them, the mesh stiffness of the gear is a key issue, and many scholars have carried out a lot of research on it.

Through the literature research, it is found that the calculation methods of gear mesh stiffness can be summarized into three categories: empirical formula, finite element method and analytic method. For the research of spur gear, Wang [1] combines the finite element method with the analytical method to study the mesh stiffness of the spur gear. Yang and Lin [2] proposed that the potential energy integration method is widely applied to the stiffness calculation of spur gear. Liang et al. [3] analyzed the mesh stiffness of the spur gear by means of potential energy method, studied the influence of the crack on the mesh stiffness, and established the mesh stiffness equation of the crack propagation. Ma et al. [4] studied the calculation of mesh stiffness of spur gear with spalling defects. By comparing the gear mesh stiffness under different spalling width, length and position with finite element method, the correctness of the analysis model is verified. Saxena et al. [5] studied the effect of time-varying friction coefficient on the total effective mesh stiffness of spur gears. Based on the static two-dimensional finite element analysis method, Maclennan [6] studied the influence of tooth profile error on the load sharing capacity and mesh stiffness of spur gears.

For the helical gears, domestic and foreign scholars have also done a lot of research. Lin et al. [7] used

---

Content from this work may be used under the terms of the Creative Commons Attribution 3.0 licence. Any further distribution of this work must maintain attribution to the author(s) and the title of the work, journal citation and DOI.

Published under licence by IOP Publishing Ltd

the combination of potential energy method and slicing method to study the influence of friction on the mesh stiffness of helical gears. Wang et al. [8] proposed an improved time-varying mesh stiffness model of helical gear pair to study the effect of axial meshing force component on the time-varying mesh stiffness of helical gear pair. Yu et al. [9] defined two kinds of helical gear pairs based on the relationship between the lateral contact ratio of the helical gear and the overlapping contact ratio, and used the slicing principle to calculate an improved analytical model of the single mesh stiffness of the helical gear. Chang et al. [10] used the finite element method and the local contact analysis of elastic body method to study the analytical method for calculating the mesh stiffness and load distribution of helical gears. Wang et al. [11] studied a model of the influence of tooth deviation and assembly error on the mesh stiffness of helical gears, and approximated the helical gear to a series of relatively small independent straight teeth. The stiffness of the tooth is calculated by potential energy method, and the validity of the model is verified.

The above literature shows that the analysis of gear mesh stiffness is mainly focused on spur gear and helical gear. Through the great efforts of scholars, the theory and method are relatively mature. However, the research on mesh stiffness of herringbone gear is relatively few. Zhu et al. [12] studied the effect of torsional stiffness on load sharing characteristics of herringbone gear transmission system. Wang et al. [13] used the herringbone gear as the helical gear to calculate the stiffness of the herringbone gear transmission system in the vibration characteristic analysis and experimental research. At present, the study of the properties of herringbone gears is to copy the characteristics of the helical gears, and simplifies the herringbone gears as a helical gears, which will cause certain errors to the final result. In summary, there is no systematic study on the mesh stiffness of herringbone gears. Therefore, this paper analyzes the mesh stiffness of the herringbone gear under the condition of modification.

2. Theory

2.1 Improved Velex Method
The calculation method of the meshing stiffness proposed by the French scholar Velex is improved to calculate the meshing stiffness of the herringbone gear [14-15]. The method considers that the mesh stiffness of a pair of gears on the unit contact line of the meshing transmission is constant value, and the length of the total contact line is time-varying, so the product of the stiffness of the unit contact line and the length of the contact line is the mesh stiffness.

The total length of the contact line at t time is $L(t)$. As shown in Fig.1, M is an arbitrary point on the contact line. The mesh stiffness of the unit contact line length at point M is $K(M)$ and the contact line length is dL, then the mesh stiffness of the t time is as follows:

$$k(t) = \int_{L(t)} k(M) \, dL \qquad (1)$$

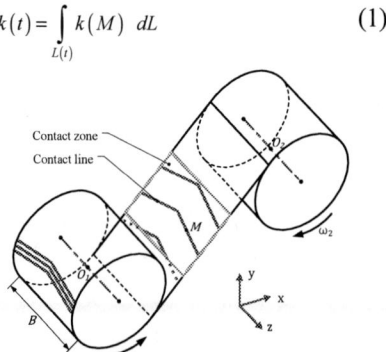

Fig.1. Contact area of herringbone gear

According to the parallel connection of stiffness, the time-varying mesh stiffness of the herringbone gear is calculated as:

$$k(t) = 2k_0 L(\tau) \qquad (2)$$

The dimensionless time $\tau = t/T_m$ was introduced, $T_m = P_{b1}/R_{b1}\omega_1$ is the time when the gear rotates a base circle pitch, then the time-varying contact line length $L(\tau)$ is:

$$L(\tau) = \left(1 + 2\sum_{k=1}^{\infty} Sinc(k\varepsilon_\alpha)Sinc(k\varepsilon_\beta)\cos\left(\pi k(\varepsilon_\alpha + \varepsilon_\beta - 2\tau)\right)\right)\Big|L_m \quad (3)$$

$L_m = B\varepsilon_\alpha/2\cos\beta_b$ is the average contact line length, $Sinc(x) = \sin(\pi x)/\pi x$ is the classic sine cardinal function. Where B is the tooth width of the gear, ε_α is the overlap contact ratio, and β_b is the helix angle of base circle.

The calculation method of mesh stiffness k_0 for the length of unit contact line is as follows:

$$k_0 = \cos\beta\frac{0.8}{q} \quad (4)$$

Where q is the flexibility coefficient and the formula is as follows:

$$q = C_1 + \frac{C_2}{Z_{n1}} + \frac{C_3}{Z_{n2}} + C_4 x_1 + C_5\frac{x_1}{Z_{n1}} + C_6 x_2 + C_7\frac{x_2}{Z_{n2}} + C_8 x_1^2 + C_9 x_2^2 \quad (5)$$

In the formula, the coefficients C_1, C_2, ...C_9 are as shown in Table 1. Z_{n1} and Z_{n2} are the virtual number of teeth of gears respectively. x_1 and x_2 are the modification coefficients of gears respectively.

Table 1. Coefficient values of the flexibility coefficient.

C_1	C_2	C_3	C_4
0.04723	0.15551	0.25791	-0.00635

C_5	C_6	C_7	C_8	C_9
-0.11654	-0.00193	-0.24189	0.00529	0.00182

2.2 Weber Method

According to the elastic mechanics, the gear teeth can be equivalent to a non-uniform cantilever beam. The center of the herringbone gear is considered as original point and the xoy coordinate system is established in Fig.2. It is assumed that the thickness of segment i is H_i, the area is A_i, the inertia distance of the section is I_i, and the distance from the segment i to the load action point along the x-axis direction is L_{ij}. The Poisson's ratio of the tooth material is v, the equivalent elastic modulus is E_e, the standard elastic modulus of the material is E, and the angle between the load direction and the vertical direction is β_j.

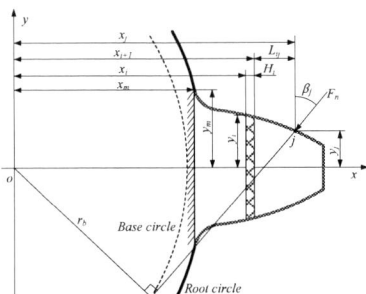

Fig.2. Structure sketch of gear teeth cantilever beam

(1) Tooth bending deformation and shear deformation

The deformation resulting from load F_n at i segment consists of two parts: bending deformation and shear deformation. The deformation δ_{Bij} can be expressed as:

$$\delta_{Bij} = \frac{F_n}{E_e}\left\{\cos^2\beta_j\left[\frac{(H_i^3 + 3H_i^2 L_{ij}^2)}{3I_i} + \frac{12H_i(1+v)}{5A_i}\right] - \frac{(H_i^2 y_j + 2H_i y_j L_{ij})}{4I_i}\sin(2\beta_j) + \frac{H_i}{A_i}\sin^2\beta_j\right\} \quad (6)$$

Among them, E_e is the equivalent elastic modulus, which depends on the width of the teeth. The relationship between E_e and the material's standard modulus of elasticity E is expressed as:

$$E_e = \begin{cases} \dfrac{E}{1-v^2} & (\dfrac{B}{H_p} > 5) \\ E & (\dfrac{B}{H_p} < 5) \end{cases} \qquad (7)$$

Where H_p is the pitch circle width.

The n small segments along the root to the meshing point j are respectively solved and superposed. The deformation amount of the j point caused by the bending deformation and shear deformation of the gear teeth is obtained as follows:

$$\delta_{Bj} = \sum_{i=1}^{n} \delta_{Bij} \qquad (8)$$

(2) Additional deformation at the meshing point caused by elastic deformation of the gear base

As the base of the gear root is also elastic, additional deformation at the meshing point caused by the deformation of the gear base should be considered. According to the analysis of gear by R.W.Cornell [16], the additional deformation of the meshing point caused by the elastic deformation of the gear base is expressed as:

$$\delta_{mj} = \frac{F_n \cos^2 \beta_j}{BE_e} \left\{ \begin{array}{c} \dfrac{16.67}{\pi}(\dfrac{L_f}{H_f})^2 + \dfrac{2L_f}{H_f}\dfrac{1-2v}{1-v} + \\ 1.534[1 + \dfrac{\tan^2 \beta_j}{2.4(1+v)}] \end{array} \right\} \qquad (9)$$

Where, $\qquad L_f = x_j - x_m - y_j \tan \beta_j \qquad (10)$

$$H_f = 2y_m \qquad (11)$$

(3) Contact deformation of gear tooth meshing

The contact deformation δ_{cj} of the meshing point j is determined by the Hertz contact theory. According to the derivation of H.H.Lin, it can be expressed as:

$$\delta_{cj} = \frac{1.275}{E_e^{0.9} b^{0.8} F_n^{0.1}} \qquad (12)$$

Where, b is the half width of the Hertz contact zone.

$$b = \sqrt{\frac{4F_n}{\pi L}\frac{\dfrac{2(1-v^2)}{E}}{\dfrac{1}{\rho_1} + \dfrac{1}{\rho_2}}} \qquad (13)$$

By summing up the above three kinds of deformation, the total normal deformation of gear meshing point j is obtained as follows:

$$\delta_j = \delta_{Bj} + \delta_{mj} + \delta_{cj} \qquad (14)$$

The single tooth normal mesh stiffness for the meshing point j is:

$$K = \frac{F_n}{\delta_j} \qquad (15)$$

The mesh stiffness of the single tooth from the entry to the exit meshing is obtained, and then the overall mesh stiffness of the gear is obtained by superimposing according to the contact ratio.

2.3 Finite Element Method

The mesh stiffness of gears is solved by finite element analysis. The purpose is to solve the deformation of gear teeth. The finite element analysis of the herringbone gear was performed according to herringbone gear parameters of Table 2.

Table 2. The basic parameters of herringbone gear pair.

Parameters	Driving gear	Driven gear
Number of teeth	24	24

Normal module (mm)	3.94
Pressure angle (°)	20
Helix angle (°)	30
Tooth width (mm)	60
Material	42CrMo
Young's modulus (Pa)	2.12×10^{11}
Poisson's ration	0.28
Mass density /(kg/m^3)	7.85×10^3

(1) Hexahedral mesh division

The quality of the mesh and the grid density of the contact area are related to the accuracy of the finite element contact analysis. An accurate herringbone gear model is established in Pro/E, and the model is imported into HyperMesh for hexahedral meshing. For a single tooth model, the quadrilateral 2D mesh is first divided. Since the tooth is the involute tooth profile, it is divided into different regions when dividing the grid. As shown in the Fig.3a, the single tooth is divided into several different regions. Based on the 2D grid, the hexahedral mesh is stretched by the Solidmap command to obtain a higher quality mesh, as shown in the Fig.3b.

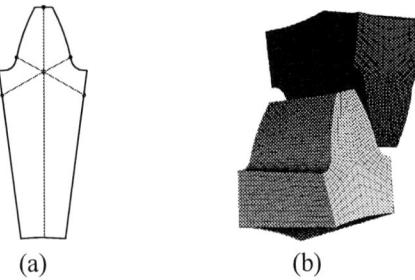

(a)   (b)

Fig.3. Single tooth mesh model

In gear contact analysis, when analyzing the gear mesh stiffness, the gear wheel body stiffness is much larger than the gear tooth stiffness. The non-participating gear teeth have little effect on the gear wheel body stiffness. The number of meshing gear teeth does not affect the gear mesh stiffness analysis. Considering the size of the finite element model of the herringbone gear and the characteristics of the periodic variation of the meshing, the simplified processing is carried out during the research process, and only the partial solid model of the two gears involved in the meshing is established [17], as shown in Fig. 4. It can not only reflect the actual meshing situation, but also save computing resources as much as possible.

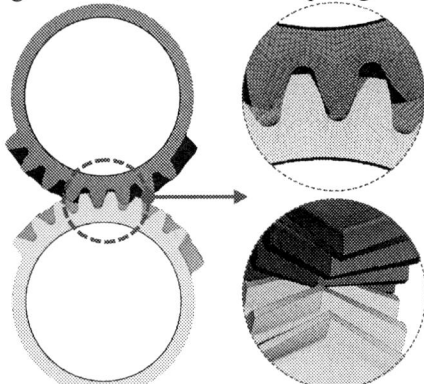

Fig.4. Finite element model and partial enlargement of the gear

(2) Finite element solution

As an excellent representative of CAE software, ABAQUS plays an important role in many fields [18]. Especially for solving the problem of contact nonlinearity, ABAQUS has its own unique advantages. In this paper, ABAQUS is used as a finite element simulation tool for analyzing gear mesh stiffness.

The change values of the rotation angles θ_1 and θ_2 of the reference points of the driving and driven gear rotation centers are respectively measured, and the quasi-static transmission error is obtained:

$$\Delta\theta = \theta_2 - \frac{z_1}{z_2}\theta_1 \tag{16}$$

Where Z_1 is the number of teeth of the driving gear, Z_2 is the number of teeth of the driven gear.

Because the influence of installation error, manufacturing precision and other factors on the system is not considered in the simulation process, the transmission error $\Delta\theta$ can be considered to be caused by the gear load deformation, and the transmission error is converted into the normal deformation amount δ_n:

$$\delta_n = \frac{\Delta\theta \cdot m_n z_2 \cos\alpha_t}{2\cos^2\beta} \tag{17}$$

The tooth surface meshing normal load F_n is:

$$F_n = \frac{2000T}{m_n z_2 \cos\alpha_n \cos\beta} \tag{18}$$

Finally, the normal mesh stiffness K_n of the gear can be obtained as:

$$K_n = \frac{F_n}{\delta_n} \tag{19}$$

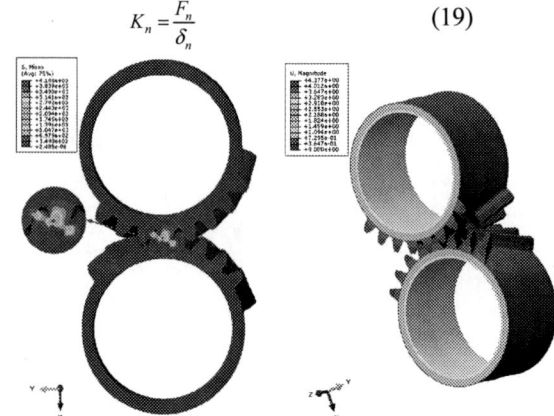

(a) Stress nephogram of the gear   (b) Deformation nephogram of the gear

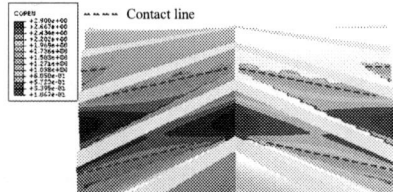

(c) Contact state of the gear

Fig.5. Finite element analysis results of herringbone gear

According to the above three methods, the mesh stiffness curve of the herringbone gear is obtained, as shown in Fig.6. Due to the large contact ratio of the herringbone gears, the gear teeth come into meshing engagement and secede from the meshing gradually, as shown in Fig.5c. Therefore, the mesh stiffness of the herringbone gear changes gently, unlike the mesh stiffness of the spur gear, there is a significant step abrupt change.

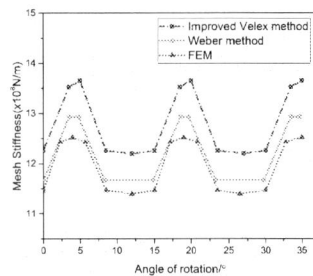

Fig.6. Three kinds of mesh stiffness curves

The three methods are compared and the results are shown in Table 3.

Table 3. Comparison of the results of the three methods.

	Mean mesh stiffness(N/m)	Error
Weber Method	12.10×10^8	——
FEM	11.96×10^8	1.2%
Improved Velex Method	12.93×10^8	6.8%

In the finite element calculation, the quasi-static simulation method is applied. The dynamic impact of the gear teeth will cause the actual deformation to be larger than the theoretical value. When calculating the stiffness, the normal load is a static load, resulting in a smaller final result. For the improved Velex method, the change trend of the mesh stiffness is similar to the other two methods, and the result is 6.8% different from the Weber method, which shows the accuracy of the improved Velex method.

3. Results and discussion

Gear mesh stiffness is a parameter that reflects the meshing state of gears from meshing to disengaging. The research shows that different gear parameters will cause the change of mesh stiffness. According to the three parameters of the helix angle and modification coefficient of the herringbone gear, the Weber method is used to analyze the influence of each parameter on the mesh stiffness. And from the perspective of stiffness, the difference between the herringbone gear and the helical gear is compared.

3.1 Effect of the helix angle

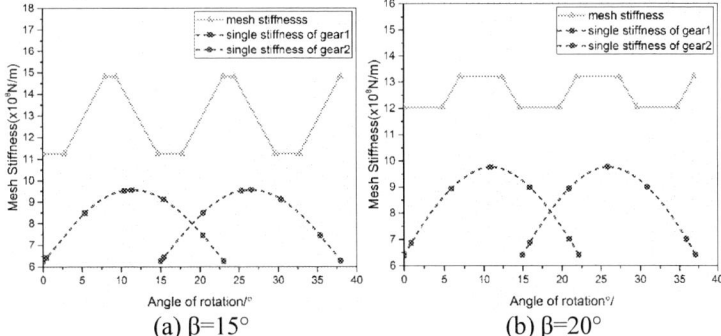

(a) β=15°　　　　　　　　　(b) β=20°

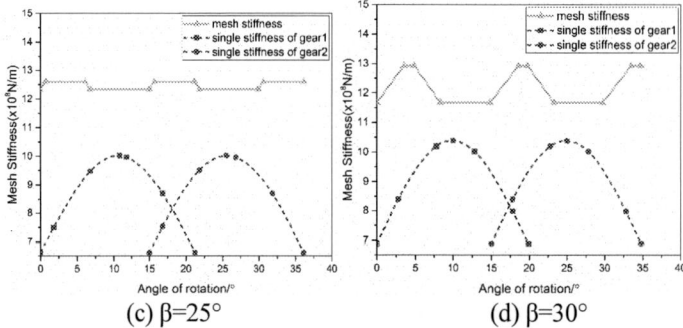

(c) β=25°   (d) β=30°

Fig.7. Relationship between helix angle and stiffness

Through the calculation of Weber method, the curve of gear stiffness with helix angle is obtained, as shown in Fig.7. As a result, the single tooth stiffness and the mesh stiffness of the herringbone gear change with the change of the helix angle. For the single tooth stiffness, as the helix angle of the gear increases, the stiffness of the single tooth increases gradually, and the rate of change of the stiffness curve also increases. This is due to the influence of the overlap contact ratio ε_β of the herringbone gear.

$$\varepsilon_\beta = \frac{B\sin\beta}{\pi m_n} \tag{20}$$

From Eq. (20), it can be obtained that as the helix angle increases, the overlap contact ratio gradually increases in the range of 0 to 45 degrees. The mesh stiffness of herringbone gears varies periodically with time. In order to describe the fluctuation of the mesh stiffness, the fluctuation δ_C of the mesh stiffness of the herringbone gear is defined as [19]:

$$\delta_C = \frac{\Delta C_r}{C_{rm}} \times 100\% \tag{21}$$

Where ΔC_r is the difference between the maximum and minimum values of the mesh stiffness, C_{rm} is the mean of the mesh stiffness.

According to Eq. (21), the calculation result is shown in the Table 4.

Table 4. Fluctuation value of mesh stiffness of herringbone gear.

Helix angle/°	Mesh stiffness /[10^8(N/m)]			Fluctuation value
	Maximum	Minimum	Mean value	
15	14.83	11.25	12.91	27.7%
20	13.21	12.04	12.65	9.2%
25	12.61	12.33	12.45	2.2%
30	12.93	11.67	12.10	10.4%

A series of helix angles is calculated to obtain a scatter plot of helix angle and stiffness fluctuation, as shown in Fig.8.

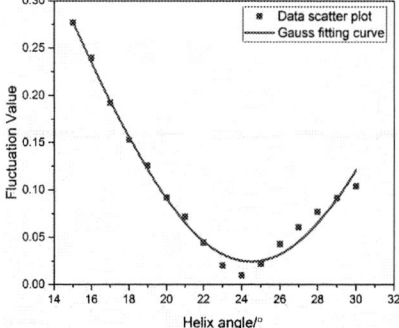

Fig.8. Curve of relationship between helix angle and stiffness fluctuation

The fitting equation is obtained by Gauss curve fitting:

$$y = 0.76 - 0.73e^{-\frac{(x-24.5)^2}{2 \times 10.4^2}} \qquad (22)$$

In the transmission design process of the gear reducer, the fluctuation of the gear mesh stiffness should be controlled. From the fitting equation, the fluctuation of the gear stiffness can be judged, which provides a theoretical basis for the selection of the helix angle of the herringbone gear.

3.2 Effect of the modification coefficient
According to the modification coefficient of gears, there are three forms of gear: positive modification, zero modification and negative modification. According to the classification of transmission, gear can be divided into equal modification transmission, standard transmission, positive transmission and negative transmission.

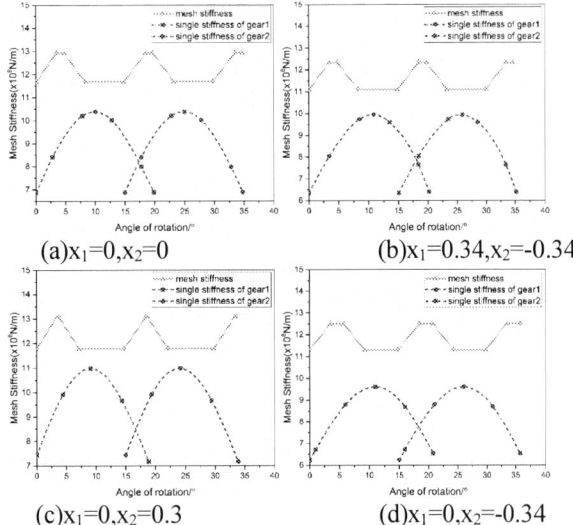

(a)$x_1=0,x_2=0$ (b)$x_1=0.34,x_2=-0.34$

(c)$x_1=0,x_2=0.3$ (d)$x_1=0,x_2=-0.34$

Fig.9. Relationship between modification coefficient and stiffness

Compared with the standard gear transmission, various modifications are analyzed. As shown in Fig.9b, when the gear has equal modification transmission, due to the change of the tooth shape, the stiffness of the gear is decreased, but the change is little, and it can be said that the equal modification transmission has no influence on the stiffness of the gear.

When the gear has positive transmission, as shown in Fig.9c. The positive modification coefficient increases the single tooth stiffness of the gear, and the mesh stiffness also increases, but the fluctuation of the mesh stiffness increases. Because the large mesh stiffness fluctuation is not conducive to achieving the load sharing properties of the reducer [20], the positive modification transmission is not conducive to improving the gear transmission properties.

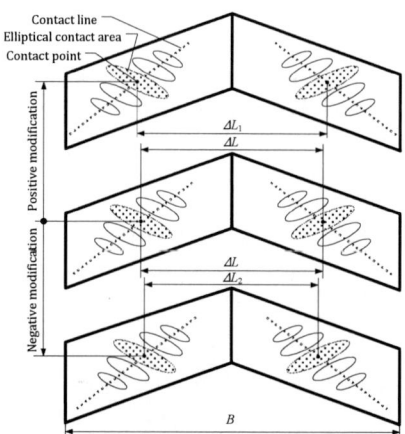

Fig.10. Change of meshing point after modification

When the gear has negative transmission, as shown in Fig.9d. The negative modification coefficient decreases the single tooth stiffness of the gear and the mesh stiffness. But the negative modification not only reduces the center distance, also reduces the fluctuating value of the mesh stiffness. This modification mode is conducive to improving gear transmission properties.

As the gear is modified, its single gear stiffness curve also changes. When a modification occurs, the mesh point of the gear will produce slight axial displacement relative to the standard gear, as shown in Fig.10 [21]. The change of the mesh point before and after the modification, $\Delta L_1 > \Delta L > \Delta L_2$, leads to the difference of the stiffness change rate of the teeth.

3.3 Comparison of herringbone gear and helical gear

For the study of herringbone gears, many scholars have equivalent the herringbone gears to helical gears. They think that the herringbone gears are the superposition of two helical gears. According to the research on herringbone gears, this statement is wrong.

Owing to the helical gear is involved, the helix angle of the gear is changed to 20° and the other parameters remain unchanged. Select the B=120mm herringbone gear and helical gear, B=120mm herringbone gear and B=60mm helical gear as the research model.

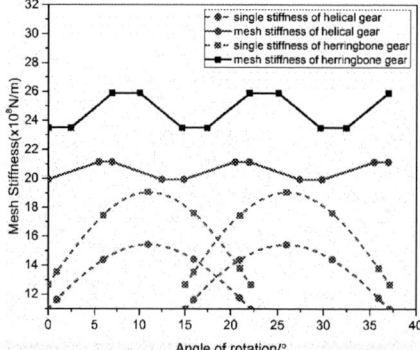

Fig.11. Comparison of the equal width of herringbone and helical gear

As shown in Fig.11, the stiffness of the herringbone gear is significantly larger than that of the helical gear, and the mesh stiffness is naturally larger than that of the helical gear. On the one hand, due to the unique symmetrical structure of the herringbone gear, there is no influence of the axial force, which causes the deformation of the gear teeth to decrease. Under the same load conditions, the stiffness of the herringbone gear is greater than the stiffness of the helical gear. On the other hand, the meshing of the herringbone gear is a process in which both sides enter and gradually withdraw at the same time, while the helical gear is one side meshing. The two meshing modes cause different contact areas. Under the same working conditions, the deformation of the teeth of herringbone gear must be decreased.

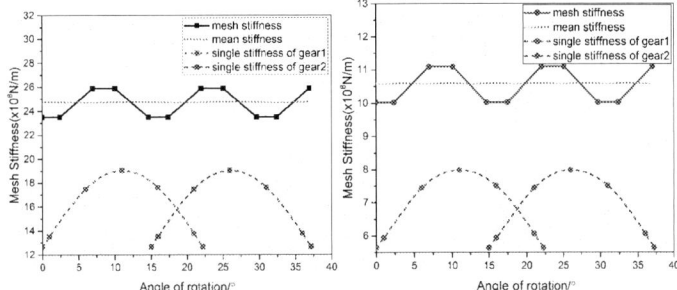

(a) Herringbone gear with B=120mm   (b) Helical gear with B=60mm

Fig.12. Comparison of herringbone and helical gear

Under the same conditions, compared with the pair of B=120mm herringbone gears and the pair of helical gears with B=60mm, as shown in Fig.12, the fluctuation trend of the mesh stiffness of them is the same, but there are huge differences in numerical values. Based on the mean mesh stiffness of the two models, the mean mesh stiffness of the herringbone gear is 24.756×10^8N/m, but the mean mesh stiffness of the helical gear is only 10.59×10^8N/m. The authors have used a large number of examples to illustrate the quantitative relationship between the mean mesh stiffness of the two models, as shown in Table 5. It can be concluded that the mean mesh stiffness of the herringbone gear is 2.3 times that of the helical gear. Therefore, the herringbone gear is not a simple superposition of the helical gear.

Table 5. Comparison of mean mesh stiffness of gears under different factors. (Unit: 10^8N/m)

		Herringbone gear B=120mm	Helical gear B=60mm	Quantitative relationship
Helix angle	16°	25.26	10.88	2.3
	18°	25.19	10.85	2.3
Modification Coefficient	0.235/-0.078	23.52	10.09	2.3
	-0.105/0.032	23.49	10.12	2.3

4. Conclusions

In this paper, three methods are used to calculate the meshing stiffness of the herringbone gear, and the accuracy of the improved method is proved. The influence of the helix angle and the modification coefficient of the herringbone gear on the mesh stiffness is analyzed. Finally, the difference between the herringbone gear and the helical gear is compared.

(1) The mesh stiffness of the herringbone gear is calculated by three methods. Compared with the Weber method, the difference between the two methods is 6.8%, which proves the correctness of the improved Velex method.

(2) The influence of helix angle and modification coefficient of herringbone gear on mesh stiffness is analyzed. The results show that mesh stiffness fluctuation of herringbone gear has a nonlinear relationship with helix angle, and the change trend is similar to Gauss curve. Negative modification is beneficial to improving the load sharing properties of gears. Modification coefficient affects the change of mesh points and causes fluctuation of mesh stiffness.

(3) Comparing the stiffness of the herringbone gear with the helical gear, the difference between herringbone gear and helical gear is proved. The herringbone gear is not the superposition of the helical gear, which provides theoretical guidance for the future research of the herringbone gear.

References

[1] J.D. Wang, Numerical and experimental analysis of spur gears in mesh, Ph.D. Thesis, Curtin University of Technology, Australia, 2003

[2] Yang D C H, Lin J Y. Hertzian Damping, Tooth Friction and Bending Elasticity in Gear Impact Dynamics, Journal of Mechanical Design, 1987, 109(2):189-196

[3] Liang X, Zuo M J, Pandey M. Analytically evaluating the influence of crack on the mesh stiffness of a planetary gear set, Mechanism & Machine Theory, 2014, 76(6):20-38

[4] Ma H, Li Z, Feng M, et al. Time-varying mesh stiffness calculation of spur gears with spalling defect, Engineering Failure Analysis, 2016, 66:166-176

[5] Saxena A, Parey A, Chouksey M. Time varying mesh stiffness calculation of spur gear pair considering sliding friction and spalling defects, Engineering Failure Analysis, 2016, 70:200-211

[6] Maclennan L D. An analytical method to determine the influence of shape deviation on load distribution and mesh stiffness for spur gears, ARCHIVE Proceedings of the Institution of Mechanical Engineers Part C Journal of Mechanical Engineering Science 1989-1996 (vols 203-210), 2002, 216(10):1005-1016.

[7] Han L, Xu L, Qi H. Influences of friction and mesh misalignment on time-varying mesh stiffness of helical gears, Journal of Mechanical Science & Technology, 2017, 31(7):3121-3130

[8] Wang Q, Zhao B, Fu Y, et al. An improved time-varying mesh stiffness model for helical gear pairs considering axial mesh force component, Mechanical Systems & Signal Processing, 2018, 106:413-429

[9] Yu W, Mechefske C K. A New Model for the Single Mesh Stiffness Calculation of Helical Gears Using the Slicing Principle, Iranian Journal of Science & Technology Transactions of Mechanical Engineering, 2018(3):1-13

[10] Chang L, Liu G, Wu L. A robust model for determining the mesh stiffness of cylindrical gears, Mechanism & Machine Theory, 2015, 87:93-114

[11] Wang Q, Zhang Y. A model for analyzing stiffness and stress in a helical gear pair with tooth profile errors, Journal of Vibration & Control, 2015, 23(10):20-3

[12] Gui Y F, Zhu R P, et al. Impact of torsional stiffness on dynamic load sharing coefficient of two-input cylindrical gear split-torque transmission system, Journal of Aerospace Power, 2014, 29(9):2264-2272

[13] Wang F. Dynamic Characteristics Research and Experimental Study on Herringbone Gear Drive System, Northwestern Polytechnical University, 2014

[14] Maatar M, Velex P. An Analytical Expression for the Time-Varying Contact Length in Perfect Cylindrical Gears: Some Possible Applications in Gear Dynamics, Journal of Mechanical Design, 1996, 118(4):586-589

[15] Gu X, Velex P, Sainsot P, et al. Analytical Investigations on the Mesh Stiffness Function of Solid Spur and Helical Gears, Journal of Mechanical Design, 2015,137(6):063301

[16] Cornell R W. Compliance and Stress Sensitivity of Spur Gear Teeth，Journal of Mechanical Design, 1981, 103(2):447-459

[17] Kiekbusch T, Sappok D, Sauer B, et al. Calculation of the Combined Torsional Mesh Stiffness of Spur Gears with Two- and Three-Dimensional Parametrical FE Models，Strojniski Vestnik-Journal of Mechanical Engineering, 2011, 57(11):810-818

[18] Chen Y C, Tsay C B. Stress analysis of a helical gear set with localized bearing contact, Finite Elements in Analysis & Design, 2002, 38(8):707-723

[19] Ding Y F, Lin L, Wu L Y, et al. Research on Variation Rules of Mesh Stiffness Fluctuation for Helical Gears, Mechanical Transmission, 2014(5):24-27

[20] Zhao N, Wang J, Fu C X. Design and Analysis of the Split Torque-Combine Power Transmission System, Machinery Design & Manufacture, 2013(8):28-31

[21] Tran V T, Hsu R H, Tsay C B. Tooth contact analysis of double-crowned involute helical pairs shaved by a crowning mechanism with parallel shaving cutters, Mechanism & Machine Theory, 2014, 79:198-216

ISPECE   IOP Publishing

Modeling of the Stiffness of Corrugated Cardboard Considering Material Non-linear Effect

Jilin Ran*, Changli Liu

School of Mechanical and Power Engineering, East China University of Science and Technology, Shanghai 200237, China

Email: y30160358@mail.ecust.edu.cn, clliu@ecust.edu.cn

TEL: 18221957873

Abstract: This paper studies the application of corrugated cardboard in vibration isolators. Cardboard is known as a highly environment-friendly material. The focus of the study was on modeling of the stiffness of a cardboard or a group of cardboard (cardboard system) along the cardboard's thickness direction which is also the vibration direction. Cardboard in this study is corrugated cardboard. There are two shortcomings in modeling the stiffness of such cardboard in literature: (1) neglect of the material non-linear effect of cardboard, and (2) neglect of the effect of cardboard width even for the situation that cardboard width is greater than cardboard length. A finite element method was applied to overcome these shortcomings - particularly by adopting an orthotropic material constitutive model and using a shell element. In addition, the study also examined the effect of inaccuracy in describing the core shape on prediction of the stiffness of cardboard. The experiment was performed, which shows significant improvements in the prediction accuracy with the new finite element model.

1. Introduction

The corrugated cardboard is widely used in packaging industry because it is cost-effective, easy for recycling, and environment-friendly [1]. A typical corrugated cardboard is illustrated in Figure 1, which is made up of paper and has a sandwich structure, in particular consisting of a corrugated core (flute) and two liners [2].

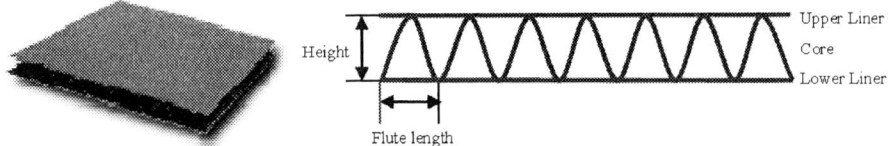

Figure 1. The outlook and structure of a corrugated cardboard

There are several studies on modeling and prediction of the stiffness of corrugated cardboard under compressive loading in its thickness direction and several finite element models were proposed. Lu et al. [3] proposed a finite element model to predict the compressive behavior of the corrugated cardboard under a uniform flat compressive loading. But the error between their analysis result and the experimental result is about 30%. They concluded that the error is caused by the use of inaccurate non-linear constitutive relation for the material used. Krusper et al. [4] refined the Lu's model with a more accurate model. In their model, a linear elastic material property was however assumed. When the displacement over to 0.2mm in their paper, the model predicted result and experimental result can't match..

Content from this work may be used under the terms of the Creative Commons Attribution 3.0 licence. Any further distribution of this work must maintain attribution to the author(s) and the title of the work, journal citation and DOI.

Published under licence by IOP Publishing Ltd

On the other hand, the width of corrugated cardboard and peak load are two important parameters for the application of it for the vibration isolation. Difference in the width parameter could give rise to a significantly different stiffness of the cardboard. They can't overcome the challenge, because the beam element is used in their work. When the width of corrugated cardboard is longer than the length, the beam element is not suitable for modeling of the stiffness. In addition, inaccuracy is also contributed by the constitutive model which did not consider non-linear material property of the cardboard. Thus, it is necessary to present a new finite element model that includes the nonlinear material property and the width effect. That is the main motivation for the present study.

Considering the width effect and the peak load, the study in this paper developed a more accurate finite element model for the stiffness of the corrugated cardboard under compressive loading in its thickness direction. The experiment was conducted to verify the model. The study neglects the humidity and the thermal effect on the cardboard from the environment. The next section gives a detailed description of the proposed finite element model, followed by the experimental verification of the model. There is a conclusion at the end of the present paper.

2. Finite element modeling for stiffness of the cardboard

In this paper, an orthotropic material constitutive model was employed for the two liners and core of the corrugated cardboard. The orthotropic constitutive model consists of two parts: the linear elastic and the nonlinear plastic portions. The linear elastic portion is governed by orthotropic Hooke`s Law, while the plastic portion is governed by a quadratic Hill yield criterion.

2.1 Linear elastic material property

The linear elastic orthotropic constitutive model is represented by [5]:

$$
\begin{bmatrix} \varepsilon_x \\ \varepsilon_y \\ \varepsilon_z \\ \gamma_{xy} \\ \gamma_{xz} \\ \lambda_{yz} \end{bmatrix} = \begin{bmatrix} \dfrac{1}{E_x} & \dfrac{-v_{yx}}{E_y} & \dfrac{-v_{zx}}{E_z} & 0 & 0 & 0 \\ \dfrac{-v_{xy}}{E_x} & \dfrac{1}{E_y} & \dfrac{-v_{zy}}{E_z} & 0 & 0 & 0 \\ \dfrac{-v_{xz}}{E_x} & \dfrac{-v_{yz}}{E_y} & \dfrac{1}{E_z} & 0 & 0 & 0 \\ 0 & 0 & 0 & \dfrac{1}{G_{xy}} & 0 & 0 \\ 0 & 0 & 0 & 0 & \dfrac{1}{G_{xz}} & 0 \\ 0 & 0 & 0 & 0 & 0 & \dfrac{1}{G_{yz}} \end{bmatrix} \times \begin{bmatrix} \sigma_x \\ \sigma_x \\ \sigma_x \\ \tau_{xy} \\ \tau_{xz} \\ \tau_{yz} \end{bmatrix} \tag{1}
$$

Where $\varepsilon_x, \varepsilon_y, \varepsilon_z$ is Strains in the x, y, z direction, $\gamma_{xy}, \gamma_{xz}, \gamma_{yz}$ is Strains in the xy, xz, yz plane, E_x, E_y, E_z is Young`s modulus in the x, y, z direction, v_{xy}, v_{xz}, v_{yz} is Poisson ratio in the xy, xz, yz plane and G_{xy}, G_{xz}, G_{yz} is Shear modulus in the xy, xz, yz plane.

The symmetrical geometry of the cardboard leads to [5]:

$$
\frac{v_{xy}}{E_x} = \frac{v_{yx}}{E_y}, \frac{v_{xz}}{E_x} = \frac{v_{yx}}{E_z}, \frac{v_{yz}}{E_y} = \frac{v_{zy}}{E_z} \tag{2}
$$

There are nine unknown variable, and they are: $E_x, E_y, E_z, v_{xy}, v_{xz}, v_{yz}, G_{xy}, G_{xz}, G_{yz}$. Generally, all these unknown variables are determined by measuring. However, the dimension of the liner and core in the thickness direction is too small to measure some variables. For the cardboard system as shown in Figure 1, the in-plane material parameters (E_x & E_y) was measured by the standard tensile test, while $E_z, G_{xy}, G_{xz}, G_{yz}$ were derived empirically as follows. For the Young's modulus in the thickness direction, it was approximated given by

$$
E_z = E_x / 200 \tag{3}
$$

The shear modulus are approximated by

$$G_{xy} = 0.387\sqrt{E_x E_y}, G_{xz} = E_x / 55, G_{yz} = E_y / 55 \tag{4}$$

For both liner and core, the value of v_{xy}, v_{xz} and v_{xz} were set according to Nordstand [6], which is 0.34, 0.01 and 0.01, respectively. In order to make the result as close as possible to the experimental result, we choose different material parameters corrugated cardboard for testing. Based on this trial and error procedure, we obtain the elastic material parameters for the finite model. Finally, Table 1 lists these elastic parameters.

2.2 Nonlinear plastic material property

In this study, Quadratic Hill yield criterion in ANSYS was used, as the model was simplified by assuming there is no difference in yield strength in tension and compression. The yield criterion was used with the isotropic hardening option, which is given by Equation (5) from [7].

$$f\{\sigma\} = \sqrt{\{\sigma\}^T [M]\{\sigma\}} - \sigma_0\left(\varepsilon^p\right) = 0 \tag{5}$$

Where σ_0 is yield stress in the x direction, ε_p is equivalent plastic strain, $\{\sigma\}$ is yield stress matrix and $[M]$ is plastic compliance matrix.

The plastic compliance matrix $[M]$ can be written as [7]

$$M = \begin{bmatrix} G+H & -H & -G & 0 & 0 & 0 \\ -H & F+H & -F & 0 & 0 & 0 \\ -G & -F & F+G & 0 & 0 & 0 \\ 0 & 0 & 0 & 2N & 0 & 0 \\ 0 & 0 & 0 & 0 & 2L & 0 \\ 0 & 0 & 0 & 0 & 0 & 2M \end{bmatrix} \tag{6}$$

Where F, G, H, L, M and N are material constants that can be determined experimentally. They were defined by [7]:

$$\begin{cases} F = \dfrac{1}{2}\left(\dfrac{1}{R_{yy}^2} + \dfrac{1}{R_{zz}^2} - \dfrac{1}{R_{xx}^2}\right), G = \dfrac{1}{2}\left(\dfrac{1}{R_{zz}^2} + \dfrac{1}{R_{xx}^2} - \dfrac{1}{R_{yy}^2}\right), H = \dfrac{1}{2}\left(\dfrac{1}{R_{xx}^2} + \dfrac{1}{R_{yy}^2} - \dfrac{1}{R_{zz}^2}\right) \\ L = \dfrac{3}{2}\left(\dfrac{1}{R_{yz}^2}\right), M = \dfrac{3}{2}\left(\dfrac{1}{R_{xz}^2}\right), N = \dfrac{3}{2}\left(\dfrac{1}{R_{xy}^2}\right) \end{cases} \tag{7}$$

In the above equations, the yield stress ratios $R_{xx}, R_{yy}, R_{zz}, R_{xy}, R_{yz}$ and R_{xz} can be found by [7]

$$R_{xx} = \frac{\sigma_{xx}^y}{\sigma_0}, R_{yy} = \frac{\sigma_{yy}^y}{\sigma_0}, R_{zz} = \frac{\sigma_{zz}^y}{\sigma_0}, R_{xy} = \sqrt{3}\frac{\sigma_{xy}^y}{\sigma_0}, R_{yz} = \sqrt{3}\frac{\sigma_{yz}^y}{\sigma_0}, R_{xz} = \sqrt{3}\frac{\sigma_{xz}^y}{\sigma_0} \tag{8}$$

Where σ_{ij}^y is the yield stress in the x, y, z, xy, yz and xz direction. Further, the plastic slope of the material after yield point is given by

$$E^{pl} = \frac{E_x E_t}{E_x - E_t} \tag{9}$$

Where E_x is elastic modulus in the x direction and E_t is tangent modulus after the yield point.

In the above equations, we need to determine $\sigma_0, E_t, R_{xx}, R_{yy}, R_{zz}, R_{xy}, R_{yz}$ and R_{xz}. Specifically, the in-plane material parameters $\sigma_0, E_t, R_{xx}, R_{yy}$ were derived from the results of tensile testing, while the values of $R_{zz}, R_{xy}, R_{yz}, R_{xz}$ are related to the values of $\sigma_{zz}^y, \sigma_{xy}^y, \sigma_{yz}^y, \sigma_{xz}^y$, respectively, which are the yield stress in z, xy, yz and xz direction, respectively. According to [8], σ_{zz}^y was about 0.003-0.007 Gpa, while $\sigma_{xy}^y, \sigma_{yz}^y, \sigma_{xz}^y$ are about 0.003 to 0.011 Gpa. In this work, for the liner and core, σ_{zz}^y was set to be 0.007Gpa and 0.004Gpa, and $\sigma_{xy}^y, \sigma_{yz}^y, \sigma_{xz}^y$ were all set to be 0.011 and 0.004 for the first estimate, respectively. Further, $R_{zz}, R_{xy}, R_{yz}, R_{xz}$ were determined according to Equation (8). In order to get the result as close as possible to the experimental result, we also choose different material parameters corrugated cardboard for testing. Based on the trial and error procedure, we obtain the plastic material parameters for the finite model. Finally, these plastic parameters are listed in Table 1.

Table1. Elastic and Plastic material parameters of the liner and core of the corrugated board

Plastic material property	Liner	Core	Elastic material property	Liner	Core
$\sigma_0 (Gpa)$	0.030	0.011	$E_x (Gpa)$	3.200	5.000
$E_t (Gpa)$	2.500	0.010	$E_y (Gpa)$	2.000	1.300
R_{xx}	1.000	1.000	$E_z (Gpa)$	0.016	0.025
R_{yy}	0.350	0.300	v_{xy}	0.340	0.340
R_{zz}	0.230	0.300	v_{yz}	0.010	0.010
R_{xy}	0.635	0.630	v_{xz}	0.010	0.010
R_{yz}	0.635	0.630	$G_{xy} (Gpa)$	1.000	1.000
R_{xz}	0.635	0.630	$G_{yz} (Gpa)$	0.058	0.050
			$G_{xz} (Gpa)$	0.057	0.005

2.3 Finite element model

Generally, the geometry of the core can be approximately as the sinusoidal shape. The comparison of the sinusoidal shape and actual shape is shown in Figure 2 [9]. However, the actual shape was considered in this study, due to the higher accuracy of the model. The dimensions of the card board for a single core were: the width is 7.6mm, the flute length is 7.6mm, the height is 3.6mm and the thickness of the liner and core is 0.23mm. And we use the share the same node to model the connected nodes between the liner and the tip of the core.

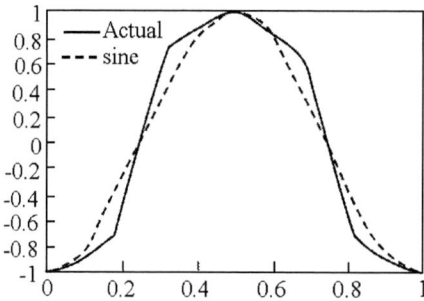

Figure 2. The actual shape and sinusoidal shape

Considering the width effect, we choose the shell element to represent the corrugated cardboard. By definition, a shell is a geometric form where the length and the width are of the same order of magnitude but the thickness of the element is considerably small in comparison with the length and width dimensions. Apparently, the element shell can be equivalent to the liner and core of the corrugated cardboard. We selected the shell 181 element in ANSYS. Because this type element includes reduced integration schemes, and can improve the computationally efficient. In order to simplify the model, we propose the following assumption of the boundary conditions and loads: the friction between the machine and the cardboard is so large that the upper and lower lines can't move in the horizontal direction. This assumption consists with the test result. Loads were applied on the finite element model as shown in Figure 3. The uniform vertical displacement load that applied on the top liner of the corrugated cardboard promises the compressive loads is uniform. As for the boundary condition, the top liner of the corrugated cardboard is constrained in all direction, and the bottom line is constrained in all direction. In addition, the symmetry boundary condition is performed so that a quarter finite element model can represent the full size finite element model. To be specific, the two edges of the corrugated cardboard are constrained only in the horizontal direction because of the symmetry.

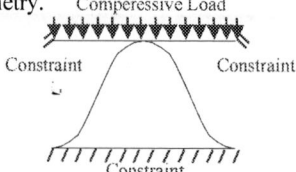

Figure 3. Compressive loads and boundary condition

3. Results and discussions

The length of the corrugated cardboard that this paper studys is 60.8mm(8 flute length), and the width is 38mm. But the length of the finite element model is 30.4mm(4 flute length) and width is19mm due to the symmetry, as shown in Figure 4. The finite element model result and measurement are shown in Figure 5. the finite element model result has excellent coherence to the measurement result until to 0.76mm. Becase the plastic model property is not accurately availabel under the larger displacement.So that not all parameters in the model can be readily determined, which is a challenge and considered for future work.

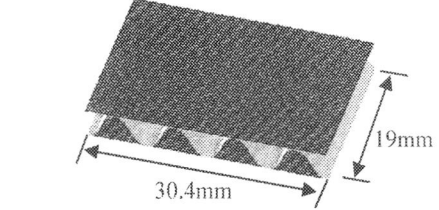

Figure 4. Geometry of FE model for specimen

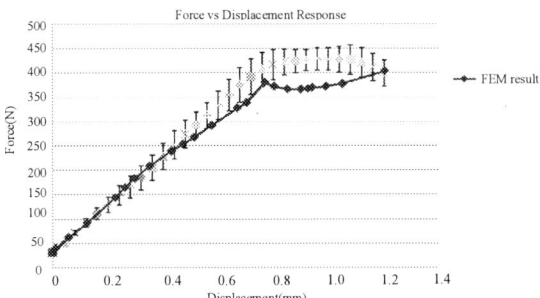

Figure 5. Comparison of the measurement results and FEM results

4. Conclusions

This paper presented a study of the improvement of the finite element model of the corrugated cardboard. An improved finite element model of the corrugated cardboard with a nonlinear orthotropic material constitutive model was presented in ANSYS environment. In particular, the peak load of the corrugated cardboard can be predicted. The experimental validation was conducted, which has shown that the improved finite element model has better accuracy for stiffness prediction than the models in literature.

Acknowledgments

The authors would like to thank the China Natural Science Funds (NSFC, Grant No. 51175179), and the Fundamental Research Funds for the Central Universities for providing financial support for this work.

References

[1] Specter, S. P. (2017) Reuse corrugated cardboard boxes: A new closure device is allowing for repeated and damage-free reuse of corrugated cartons, Modern Materials Handling. 72: 16.

[2] Campell, A. C., Housner, G. W., Bergman, L. A. (2010) Structural control: past, present and future, Journal of Engineering Mechanics 123: 897-971.

[3] Lu, T. J., and Chen, C. (2001) Compressive behavior of corrugated board panels, Journal of Composite Materials. 35: 2098–2126.

[4] Krusper, A., Isaksson, P., and Gradin, P. (2007) Modeling of Out-of-Plane Compression Loading of Corrugated Paper Board Structures, J. Eng. Mech. 133: 1171–1177.

[5] Allansson, A., and Tomas, N. (2003) Stability and collapse of corrugated board, numerical and experimental analysis, Master`s Thesis, Sweden: Division of Structural Mechanics, LTH, Lund University.

[6] Nordstrand, T., Carlsson, L. A. (1997) Evaluation of transverse shear stiffness of structural core sandwich plates, Composite Structure. 37: 145-153.

[7] Newswire PR, (2017) ANSYS 18.1 Expands Pervasive Engineering Simulation, PR Newsire US.

[8] Gooren, L.G, Cherny, R. R. (2016) Mathematical model of a drying process of corrugated cardboard, Polythematic Online Scientific Journal of Kuban State Agrarian University. 115:191-202.

[9] Biancolini, M.E., and Brutti, C. (2010) Numerical and experimental investigation of strength of the corrugated board packages. Packaging Technology and Science 16: 47-60.

Numerical Analysis On Fluid-solid Coupling Cooling Of Minimal Surface Lattice Structure

Zheng Yinzheng

College of Mechanical Engineering,Nanjing University of Science and Technology,Nanjing,Jiangsu Province210094,china

Corresponding author: zhengyin0405@qq.com
Tel: 15951913361

Abstract: This paper established the minimal surface lattice structure cooling model , researched the fluid-solid coupling cooling properties in addition flow by interface function transmit and compared the cooling properties of different minimal surface lattice and bar lattice in the same density. The influence of cooling fluid velocity was considered. The thermal stress and impact stress of the structure was analyzed by sequential coupling method. The result showed that the minimal surface structure had better cooling properties. The bottom temperature decreased with the increase of coolant flow velocity. There was a threshold for different structures, when the flow rate exceeded this value, the bottom surface temperature will not decrease. Stress analysis showed that stress concentration occurred at the junction of lattice and panel. The overall stress decreased first and then increased with the increase of flow rate.

1. Introduction

Aerospace industry has entered a period of rapid development. According to statistics, when hypersonic aircraft fly at an altitude of 30km and the flight Mach number reaches 3, the surface temperature can reach about 600K because of aerodynamic hot , and when the Mach number reaches 5, the temperature is as high as 1200K[1]. The temperature in the scramjet combustion chamber can even reach 3500K[2]. The thermal stress caused by high temperature causes damage to the aircraft structure, and internal instruments and equipment are also prone to potential safety problems due to high temperature. Therefore, thermal protection structure becomes an indispensable link in aircraft design.

Lightweight lattice structure has high specific strength, heat insulation, vibration isolation and other functions[3]. The internally connected pore configuration makes the structure itself an excellent radiator[4]. At present, it has been used in the cooling system of spacecraft shell, aero-engine cooling panel and other cooling systems. Vermaak [5]found that the traditional cooling channel is prone to local overheating, which results in fuel coking and blockage and deteriorates the heat dissipation performance because of its long and narrow and disconnected configuration. A series of numerical analysis and experimental tests were carried out on the heat dissipation performance of pyramid, tetrahedron, Kagome and other lattice structures by Roper[6], joo[7]and lu [8]. and they found that compared with the traditional cooling channel, the lattice structure had better flow characteristics and better heat dissipation effect. Luo[9]compared the heat dissipation performance of common lattice structures and found that kagome lattice has the best heat dissipation effect under the same density.

In recent years, with the continuous development of additive manufacturing technology, the minimal surface lattice structures modeled by parameterized implicit functions have been used.

Content from this work may be used under the terms of the Creative Commons Attribution 3.0 licence. Any further distribution of this work must maintain attribution to the author(s) and the title of the work, journal citation and DOI.

Published under licence by IOP Publishing Ltd

Compared with the Bar type lattice ,it has larger surface-volume ratio, zero mean curvature at all points on the surface, smooth transition and good flow characteristics[11]. Some scholars have researched on its mechanical and thermal properties[12,13].

This paper took the commonly used Gyroid, Dimond and Primitive minimal surface lattice structure and the traditional bar type lattice kagome lattice with the best heat dissipation performance as the research object. The fluid-solid coupled heat dissipation model of lattice sandwich cooling panel with temperature boundary condition was established based on the method of interfacial function transfer.And the heat dissipation performance of the structures with the same density was compared. The stress of the structure is analyzed by sequential coupling method.

2. Lattice heat dissipation structure model

In this paper, with heat dissipation structure of engine combustion chamber as the background, the commonly used Gyroid(G), Dimond(D), Primitive(P) minimal surface lattice and Kagome (K) bar type lattice heat dissipation model were established(figure 1) . The element model is shown in table 1. To facilitate horizontal comparison, the relative densities of the four structures are equal.

Table 1. Lattice element configuration

	Gyroid	Diamond	Primitive	Kagome
Model				

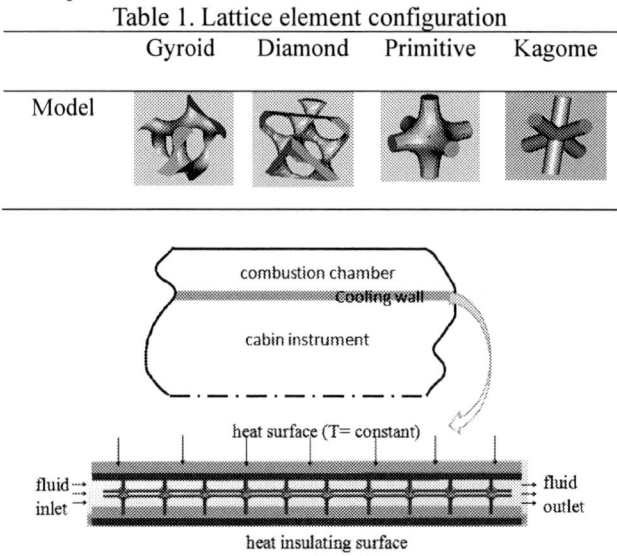

Fig 1. Lattice structure heat dissipation model

The upper panel of the model is a heated layer, connected to the combustion chamber and subjected to continuous temperature boundary conditions. In the middle is the lattice heat dissipation layer, the cooling flow flows in from the left side and flows out from the right side after the lattice structure has fully heat exchange. The lower panel is insulated and connected to the instruments in the cabin. The temperature shall not be too high. The panel size is slightly larger than the lattice to allow the coolant to flow fully. The overall size parameters of the model are shown in table 2. Referring to the actual working condition of the engine combustion chamber, the heating surface temperature is 1000k, and the coolant inlet temperature is 300k[2]. Titanium alloy (TC4) was used for lattice materials, and the thermophysical parameters varied with the temperature, as shown in table 3.

Table 2. Cooling model size

Size of lattice cell	4
volume fraction of lattice	0.18
Thickness of panel	1
Length of model	45
Width of model	9

Table 3 .TC4 material parameters

Temperature（^{0}C）	heat conductivity（w/m.k）	thermal ../../../../ 520-PC/AppData/Local/youdao/dict/Application/7.5.0.0/resultui/dict/?keyword=expansivity（$10^{-6}.^0$C^{-1}）	specific heat（J/(kg.^{0}C）
20	6.8	9.1	611
100	7.4	9.2	624
200	8.7	9.3	653
300	9.8	9.5	674
400	10.3	9.7	691
500	11.8	10.0	703

The coolant adopts aviation kerosene, and its density, viscosity coefficient, thermal conductivity coefficient and specific heat capacity at constant pressure are given by the following fitting function.

$$\rho = 1087.47 - 0.9488 \times T \qquad \mu = 7.2444 \times 10^5 \times T^{-3.5464} \qquad (1)$$

$$\lambda = 0.1993 - 1.9705 \times 10^{-4} \times T \qquad C_p = 705.55 + 4.03 \times T$$

3. Numerical calculation method

Due to the temperature difference between the upper and lower panels, the heat will be transferred from the heating surface along the lattice structure to the bottom surface, while the incoming coolant contact with the lattice structure and take away the heat. The heat transfer mechanisms involved include solid heat transfer, solid-liquid interface conjugate heat transfer and coolant flow heat transfer. The heat transfer of solids follows Fourier law

$$\rho_s C \frac{\partial T_s}{\partial t} = \frac{\partial}{\partial x}(k_s \frac{\partial T_s}{\partial x}) + \frac{\partial}{\partial y}(k_s \frac{\partial T_s}{\partial y}) + \frac{\partial}{\partial z}(k_s \frac{\partial T_s}{\partial z}) \quad (2)$$

Where ρ_s、 C、 k_s are density, specific heat capacity and thermal conductivity of solid respectively.

The coolant is regarded as viscous incompressible fluid, the $k - \varepsilon$ standard turbulence model is adopted, and the continuous equation is

$$\frac{\partial \rho}{\partial t} + \nabla.(\rho U) = 0 \qquad (3)$$

The momentum equation:

$$\frac{\partial \rho U}{\partial t} + \nabla.(\rho U \otimes U) - \nabla.(\mu_{eff} \nabla U) = \nabla. p^{'} +$$
$$\nabla.(\mu_{eff} \nabla U)^T + B \qquad (4)$$

Where, B is the sum of volume forces, μ_{eff} is the effective viscosity, and $p^{'}$ is the modified pressure. The expression is

$$\mu_{eff} = \mu + \mu_t \qquad (5)$$

$$p^{'} = p + \frac{2}{3}\rho k \qquad (6)$$

Where μ_t is turbulence viscosity, and the $k - \varepsilon$ model assumes that turbulence viscosity is related to turbulent kinetic energy and turbulent kinetic energy dissipation

$$\mu_t = C_\mu \rho \frac{k^2}{\varepsilon} \qquad (7)$$

The value k、 ε is directly solved from the turbulent kinetic energy and the turbulent kinetic energy dissipation equation, and the turbulent kinetic energy equation is

$$\frac{\partial(\rho K)}{\partial t}+\nabla.(\rho UK)=\nabla.[(\mu+\frac{\mu_l}{\sigma_k})\nabla K]+P_k-\rho\varepsilon$$

$$\frac{\partial(\rho\varepsilon)}{\partial t}+\nabla.(\rho U\varepsilon)=\nabla.[(\mu+\frac{\mu_l}{\sigma_\varepsilon})\nabla\varepsilon]+\frac{\varepsilon}{K}(C_{\varepsilon 1}P_k-C_{\varepsilon 2}\rho\varepsilon)$$

(8)

Where, σ_k、σ_ε、$C_{\varepsilon 1}$、$C_{\varepsilon 2}$ is a constant.

P_k Is the turbulence product of viscous force and buoyancy, and its equation is

$$P_k=\mu_l\nabla U.(\nabla U+\nabla U^T)-\frac{2}{3}\nabla.U(3\mu_l\nabla.U+\rho k)+P_{kb}$$

(9)

For incompressible fluid, the value of $\nabla.U$ is relatively small and has little influence on the results of the whole equation

The fluid-solid coupling boundary is determined by the transient heat transfer relationship between the fluid and the solid. According to the continuity condition of temperature and heat flow

$$T_s=T_f$$

$$k_s\frac{\partial T_s}{\partial n}=k_f\frac{\partial T_f}{\partial n}$$

(10)

T_s and T_f The temperature of solid and fluid at the boundary respectively.

In order to explore the influence of fluid velocity on heat dissipation effect, inlet velocity boundary condition was set, and the outlet was free flow boundary condition

4. Results analysis

4.1 Heat dissipation performance analysis

For the convenience of description, the coolant flow direction is defined as x direction, the inlet coordinate is 0, and the vertical direction is z direction, as shown in figure 2. Figure 3 shows the temperature cloud in the z-direction of the four lattice structures when the flow rate is 1m/s. When the top surface temperature is 1000K, the bottom surface temperature of G, D and P lattice structures is 750, 728k and 773k respectively, and the cooling effect reaches 25%, 27.2% and 22.7% respectively. Kagome lattice structure has a temperature of 842K, and the cooling effect is only 15.8%.

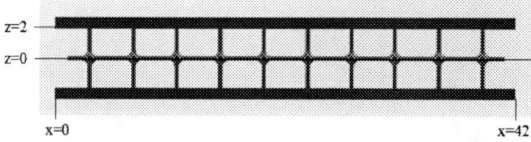

Fig 2. Front view of lattice sandwich heat dissipation model

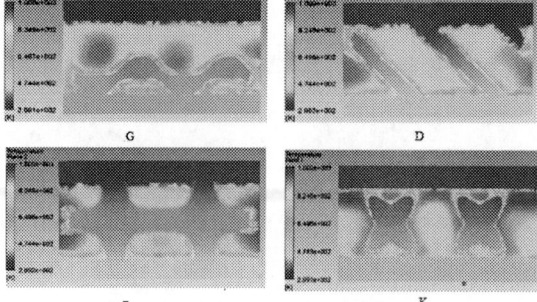

Fig3. Z section temperature cloud

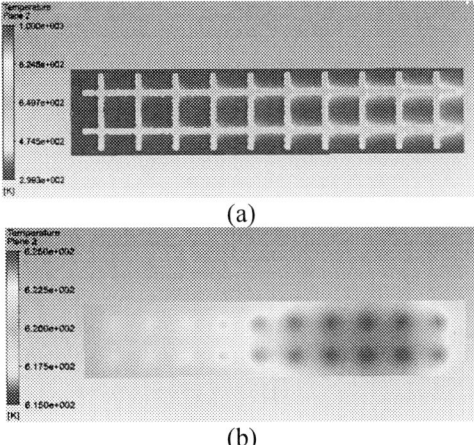

(a)

(b)

Fig 4. Temperature cloud diagram of surface p structure (a) temperature distribution of coolant x-direction section z=0 (b) temperature cloud diagram of floor x-direction section

The temperature of the coolant increases continuously from the inlet to the outlet. Taking the p-curved structure as an example, the temperature increases from 300k to 375k, indicating that the coolant carries away heat through the lattice structure. The bottom temperature is higher in the place which is connected with the lattice and shows a series of hot circles in the temperature cloud. The temperature of the bottom plate increases gradually along the x direction, which is because the temperature of the coolant in this direction increases continuously, resulting in the decrease of heat transfer efficiency near the outlet

Convection heat transfer coefficient and Nusselt number are introduced. The convection heat transfer coefficient is defined as the Heat exchange between solids and liquids per unit area per unit time when temperature change of 1K. Take P structure (figure 5 (a)) as an example, in the area near the inlet, the coolant does not flow sufficiently and the convection heat transfer coefficient is small. After stabilization, the convection heat transfer coefficient presents a regular horseshoe-shaped distribution and the coefficient around the pillar is larger. This is because the coolant flows around the pillar forming a vortex where the velocity of flow is larger than that of other regions, and the local heat transfer performance is improved. It also shows that the lattice structure has the potential to be an excellent heat transfer structure

In order to make the comparison of heat dissipation performance more universal, nusselt number was used .which is defined as

$$Nu = \frac{h.l}{\lambda}$$
(11)

Where, h is the convection heat transfer coefficient, λ is the thermal conductivity of the coolant, and l is the characteristic length of the lattice element. Figure 6 shows the change of Nussel number on the fluid-solid coupling interface of the three structures along the x direction, It can be seen from the figure that the Nusselt number fluctuates within a stable range after reaching a steady state, D surface lattice structure has the largest Nusselt number, the strongest fluid-solid heat transfer performance, followed by G and P surfaces, and Kagome is the smallest.

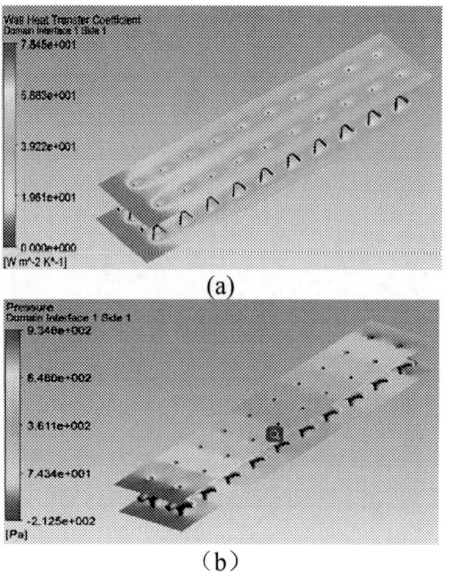

(a)

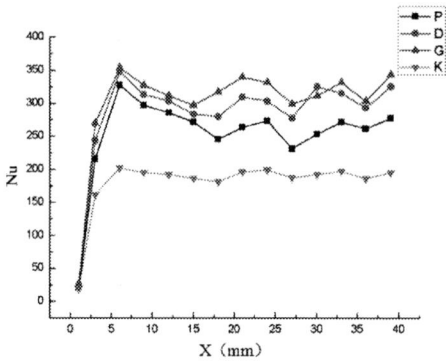

（b）

Fig 5. Convection heat transfer coefficient of p-curved structure and strong cloud diagram of cooling hydraulic pressure (a) convection heat transfer coefficient (b) pressure

Fig 6. Nussel number changes along the x direction

4.2 The influence of flow rate

Figure 7 statistics the average temperature of the bottom surface of the three structures at different flow rates. When the flow rate is close to 0, the bottom temperature is close to the 1000K, which applied on the top surface. The larger the flow rate is, the less obvious the heat dissipation effect is improved by increasing the flow rate. When the flow rate is greater than a certain threshold, the bottom temperature will not decrease. On the one hand, the increase of flow velocity strengthens the flow of local eddy current and enhances the heat transfer effect. On the other hand, the thickness of the boundary layer may increase. Since the velocity in the boundary layer changes exponentially, the proportion of coolant with relatively low velocity in the near-wall area will increase when the velocity is not very high, which has a negative effect on heat transfer. When the speed is large, the two influences can be coordinated to stabilize the temperature, and the increase of the flow rate is at the cost of consuming more power, so the appropriate flow rate should be selected to ensure effective heat dissipation while minimizing the power consumption. As can be seen from the figure7, the optimal heat dissipation velocity of G, D and P structures is about 6m/s, 5m/s and 9m/s, respectively. Kagome structure does not reach the maximum heat dissipation efficiency at 10m/s.

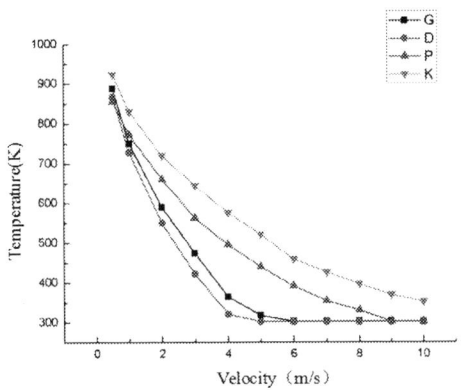

Fig 7. Bottom temperature - coolant velocity diagram

4.3 pressure analysis

In addition to good heat exchange capacity, low pressure loss is another important performance index of coolant. Due to the shape resistance of the bar and the viscous resistance of the panel, the pressure of the coolant will be lost after heat exchange with the lattice structure. The lower the pressure loss is, the less the circulating pumping motion force is demanded. Figure 5 (b) is a cloud diagram of the coolant pressure of the P lattice, and the coolant pressure gradually decreases from the inlet to the outlet. Figure 8 showed the pressure of the three structures along the x direction. The pressure of the coolant from the inlet to the outlet presents a step-down. After entering the stable state, the pressure loss of each structural unit is constant, and the pressure loss of G, D and P structures is 238pa, 271pa and 216pa respectively.

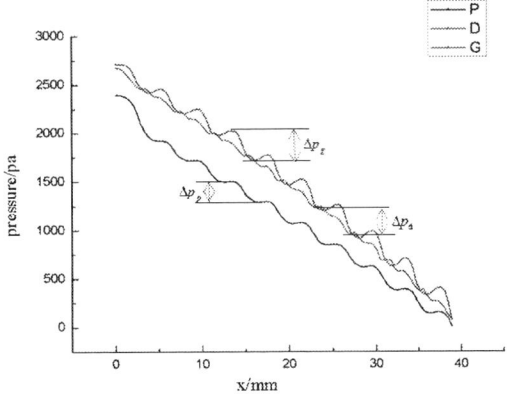

Fig 8. Diagram of variation of cooling fluid strength along the x direction

4.4 stress analysis

Lattice heat dissipation structure stress mainly comes from two aspects: thermal stress caused by temperature rise and the impact stress caused by the impact of the coolant. We calculated them by applying the temperature and pressure of the fluid-solid coupling analysis results as boundary conditions to the structure field

Figure 9 shows the thermal stress cloud diagram of the three structures. The stress distribution at the panel is relatively uniform, and the stress concentration is mainly at the contact between the lattice and the panel. this is because the longitudinal temperature gradient of the lattice structure is large, and the panel limits the longitudinal expansion of the lattice structure, resulting in a large plastic strain at the contact between the lattice and the panel. The thermal stress value decreases exponentially with the increase of flow velocity, and its fitting relationship is shown as follows equation (12).

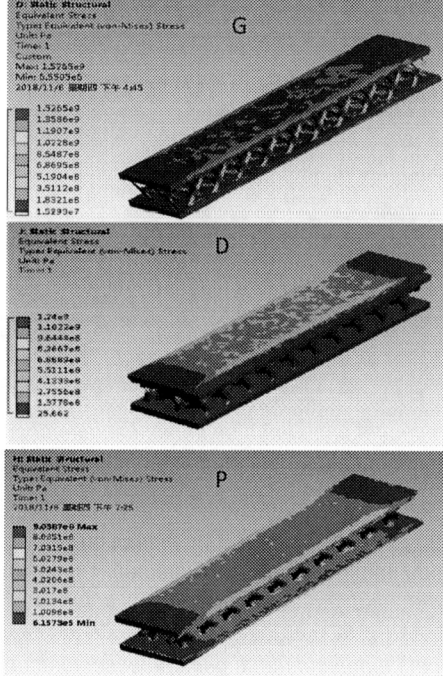

Fig 9. Structure thermal stress cloud diagram

$$p_h = 66.39v^{-0.25} \ldots\ldots\text{G}$$
$$p_h = 105.8v^{-0.34} \ldots\ldots\text{D} \qquad (12)$$
$$p_h = 46.08v^{-0.28} \ldots\ldots\text{P}$$

Figure 10 shows the impact stress cloud diagram of the structure. It can be seen that the stress of the lattice structure gradually decreases from the inlet to the outlet, which is caused by the continuous decrease of fluid pressure. The maximum stress is located at the contact point between the first element and the panel. The relationship between maximum impact stress and flow velocity is fitted as shown in equation (13).

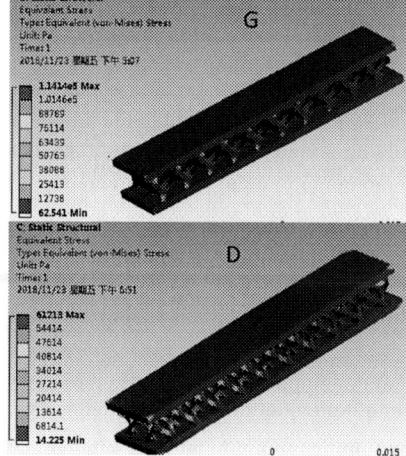

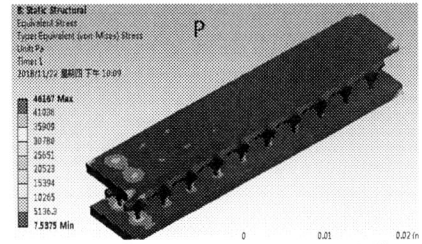

Fig 10. Structure Impact stress cloud diagram

$$p_v = 2.004 \times 10^{-4} \times v^{7.03} \quad\text{.........G}$$
$$p_v = 1.41 \times 10^{-4} \times v^{6.70} \quad\text{...........D} \qquad (13)$$
$$p_v = 2.73 \times 10^{-5} \times v^{6.39} \quad\text{..........P}$$

Finally, the relationship between the structural stress and the coolant flow rate is obtained, as shown in Figure 11. It can be found that the structural stress value decreases first and then increases with the increase of the coolant flow rate. The stress value of G, D and P structures reaches the lowest when the flow rate is 3.4m/s, 4.7m/s and 5.3m/s, respectively. In addition, the maximum stress value of the structure within the flow rate range of the cooling fluid reaching the maximum heat dissipation efficiency is less than the yield strength of titanium alloy (860MPa), which indicates the strength safety of the lattice heat dissipation structure.

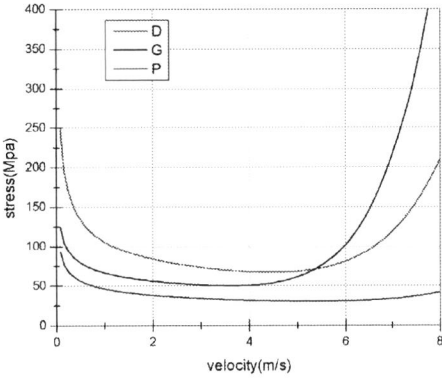

Fig 11. Relationship between structural stress and coolant flow rate

The heat dissipation performance, pressure loss and stress of the three kinds of surface lattice heat dissipation structure are comprehensively compared. Under the same Relative density, the heat dissipation effect of the d structure is the best, and the cooling fluid flow rate is the lowest when the maximum heat dissipation efficiency is reached, the effect of the p surface is the worst. The pressure loss of the p structural element is the least, and the power demand is the least in the cooling liquid circulation process. D surface structure has the maximum stress, up to 268Mpa, and P structure has the minimum stress, up to 104Mpa. In the process of heat dissipation, P structure has the minimum damage to the heat dissipation structure.

Table 4. Comprehensive comparison of performance of minimal surface lattice structure

performance	indicators	Gyroid	Diamond	Primitive
lightweight	Volume fraction	0.18	0.18	0.18
The heat dissipating performance	The average heat transfer intensity Nu	343	325	277
	Average bottom surface temperature	750	728	773

	(velocity 1m/s) (K)			
	Optimum heat dissipation velocity	6m/s	5m/s	9m/s
Power demand	Pressure loss (pa	238	271	216
stress	Maximum stress (Mpa)	137	268	104

5. conclusion

CFD simulation of fluid-solid coupling heat dissipation performance of minimal surface lattice structure and Kagome bar type lattice structure with equal density was carried out in this paper, and the thermal properties of the coolant and the high-temperature thermal properties of titanium alloy were taken into account. The results show that compared with the traditional bar type lattice, the minimal surface lattice structure has better heat dissipation performance due to its larger heat dissipation area and smooth surface configuration. For the commonly used G, D, P three minimal surface lattice structure, D surface heat dissipation performance is the best, but the maximum pressure loss. Before reaching the maximum heat dissipation efficiency, the bottom temperature decreases with the increase of coolant flow rate. The maximum stress is located at the junction of the first lattice element and the panel. The stress value decreases first and then increases with the increase of coolant flow rate .In the future, the influence of relative density and lattice element size on the heat dissipation effect will be further studied, the structure will be optimized and the fluid-solid coupling experiment will be carried out.

Acknowledgements

This research was financially supported by Nanjing University of Science and Technology.

References

[1] Yuan shijian. Lightweight molding technology [M]. Beijing: national defense industry press, 2008.

[2] Ma xifang. Research on coupled heat transfer of scramjet combustor and its cooling structure [D]. Harbin Institute of Technology, 2007.

[3] Gao gao. Active heat transfer and optimal design of multi-functional composite lattice sandwich structure [D]. Harbin Institute of Technology, 2014.

[4] Zhao wangan. Thermal performance analysis and optimization of composite lattice sandwich structure under thermal load [D]. Harbin Institute of Technology, 2010.

[5] Vermaak N, Valdevit L, Evans A G. Materials Property Profiles for Actively Cooled Panels: An Illustration for Scramjet Applications[J]. Metallurgical & Materials Transactions A, 2009, 40(4):877-890.

[6] Roper C S, Fink K D, Lee S T, et al. Anisotropic convective heat transfer in microlattice materials[J]. Aiche Journal, 2013, 59(2):622-629.

[7] Joo J H, Kang K J, Kim T, et al. Forced convective heat transfer in all metallic wire-woven bulk Kagome sandwich panels[J]. International Journal of Heat & Mass Transfer, 2011, 54(25):5658-5662.

[8] Kim T, Hodson H P, Lu T J. Contribution of vortex structures and flow separation to local and overall pressure and heat transfer characteristics in an ultralightweight lattice material[J]. International Journal of Heat & Mass Transfer, 2005, 48(19).4243-4264.

[9] Luo shukun, song hongwei, huang chenguang, et al. Heat fluid-solid coupling response analysis of light-weight lattice active cooling panel [J]. Strength and environment, 2012, 39(2):31-40.

[10] Maskery I, Aboulkhair N T, Aremu A O, et al. Compressive failure modes and energy absorption in additively manufactured double gyroid lattices[J]. Additive Manufacturing, 2017, 16.

[11] D.-J. Yoo, Advanced porous scaffold design using multi-void triply periodic minimal surface models with high surface area to volume ratios, Int. J. Precis.Eng. Mat. 15 (8) (2014) 1657e1666.

[12]Zheng X , Fu Z , Du K , et al. Minimal surface designs for porous materials: from microstructures to mechanical properties[J]. Journal of Materials Science, 2018.

[13]Feng zixin. Model lightweight method based on three-period minimal surface [D]. Dalian university of technology, 2011.

Preparation and Characterization of Composite Resin Containing Anion Powder

Lili Ding, Jingli Zhu, Qian Zhao, Bin Liu [a]

School of Stomatology, Lanzhou University, Lanzhou 730000, China

Lili Ding and Jingli Zhu equally contributed to this work.

[a]Bin Liu: liubin126566@163.com

Abstract. Objective. Anion powder is a kind of cyclic silicate mineral with antibacterial effect. A new type of composite resin containing anion powder was prepared by adding aion powder into the composite resin. Methods. By using mechanical mixing dispersion and ultrasonic dispersion, silane coupling agent was added, and the anion powder was evenly dispersed in Durafill (DF) composite resin at the proportions of 0, 1%, 2%, 4%, respectively. EDS was used to detect the relative content of anion powder in composite resin, and scanning electron microscope (SEM) was used to observe the dispersion uniformity of anion powder. The friction coefficient (COF) of composite resin containing anion powder was measured by UMT-2MT friction and wear tests. The influence of different proportion of anion powder added to the composite resin on its mechanical properties. The next step is to test the antimicrobial property of the composite resin containing anion powder, in order to find the best proportion of anion powder added to the resin. Results Mechanical agitation dispersion and ultrasonic dispersion were applied, silane coupling agent was added, and anion powder of different proportions was successfully mixed into the composite resin. EDS results showed that anion powder was successfully added into the composite resin, and the surface particles were uniformly dispersed by SEM. The friction and wear experiments showed that the addition of anion powder would change the mechanical properties of the composite resin, and the friction coefficient of the composite resin was higher than that of the control group, and the friction coefficient of the composite resin increased with the increase of anion powder content.

1. Introduction

With the development of dental materials, dental filling materials are also developing. From silver amalgam to glass ionomer cement and composite resin, they have been applied in clinic with their respective advantages. At present, the commonly used dental filling material is composite resin, which has good biological safety, good aesthetic effect and clinically acceptable mechanical properties, and is widely used in the repair of various dental defects [1]. In particular, its physical properties such as wear resistance is improved, and clinical dentists operate more conveniently, so the clinical application is more and more extensive.

However, after the repair of photosensitive composite resin, due to the failure of edge sealing, micro-leakage and secondary caries caused by saliva and microbial infiltration around the restoration body are common, which are the main factors leading to the failure of filling [2]. Clinically, the main reason for the loss of the prosthesis after filling is the formation of secondary caries around the material, and about $50 \sim 60\%$ of the prosthesis replacement patients were diagnosed as secondary caries [3]. The long-term use of oral materials has always been the focus of clinicians, which is directly related to the quality of

Content from this work may be used under the terms of the Creative Commons Attribution 3.0 licence. Any further distribution of this work must maintain attribution to the author(s) and the title of the work, journal citation and DOI.
Published under licence by IOP Publishing Ltd

tooth restoration. In recent years, the research and development of antimicrobial oral materials are gradually emerging. Scholars expect to inhibit the adhesion of pathogenic bacteria around oral materials by making the materials have antimicrobial properties, so as to extend the use time of oral materials and improve the repair efficiency [4]. Antimicrobial agents are divided into natural, organic, inorganic and organic polymer antibacterial agents. Inorganic antimicrobial materials have attracted more and more attention due to their excellent properties. Compared with ordinary materials, inorganic antimicrobial materials have the advantages of aging resistance, high temperature resistance, excellent comprehensive performance, stable antimicrobial property and long term [5]. At present, many scholars have paid attention to the addition of inorganic antibacterial agents into oral materials to make them have certain antibacterial properties, so as to reduce the generation of secondary caries after restoration and extend the service life of prosthesis [6]. Inorganic antibacterial agents can be divided into two categories according to the mechanism of their action on microorganisms. One is the combination of metal compounds with antibacterial effects (silver, copper, zinc, etc.) and inorganic carriers (such as zeolite, bentonite, activated carbon, etc.), which are called silver antibacterial agents [7]. The other is the use of photocatalytic substances as antibacterial agents, such as titanium oxide is a widely used photocatalyst, called titanium dioxide photocatalyst system antibacterial agent [8]. Silver, copper, zinc and other metal have antibacterial ability, through the methods of physical adsorption and ion exchange, the silver, copper, zinc and other metal ions (or its) fixed on the different carrier made of different types of silver antibacterial agent, including antibacterial zeolite antimicrobial agent, bentonite, silica gel antibacterial agent, phosphate double salt antibacterial agent, etc. [9].

Anion powder (Negative ions powder) is a kind of annular silicate minerals, composed of SiO_2, FeO, Fe_2O_3, B_2O_3, Al_2O_3, Na_2O, MgO style, Li_2O and MnO, contains trace Cr, Zr, zinc and Ti elements beneficial to human body, known as the vitamin in the air, can adjust the body from the balance between positive ion and negative ion, and can inhibit the aging of human body cell [10]. Zhang and others added the negative ions powder to the denture base resin, which showed that the mechanical properties of the composites increased with the increase of negative ion content [11,12]. Zhang cailing added tourmaline nano-powder and mint extract with hair generating anion function to viscose cellulose spinning solution, and the results showed that the fiber had good antibacterial and anion generating function [13]. Antibacterial composite resin refers to the addition of antibacterial agents in the composite resin, so that its own antibacterial, in a certain period will stick to the above bacteria to kill or inhibit its proliferation. It is generally required that the antimicrobial properties of composite resin products have high efficiency, wide spectrum, good continuity, non-toxic and odorless, and have good biocompatibility and mechanical properties [14]. At the same time, studies have shown that the addition of inorganic fillers into the matrix of composite resin can enhance its performance, reduce the volume change during polymerization, reduce the thermal expansion coefficient, inhibit deformation, improve resin rigidity, surface hardness and wear resistance and other mechanical properties [15]. Based on this, the preparation and characterization of the anion powder added to the photocurable composite resin have been completed in this paper. The next step is to carry out the antibacterial performance test.

2. Methods

2.1 Materials and equipment

2.1.1 Materials (Table 1)

Table 1 The Materials

Name	Manufacturer
Negative ions powder	ultra - fine powder laboratory self -processing
Anhydrous ethanol	Li 'an long bohua pharmaceutical chemical co., LTD

Silane coupling agent	Tianjin chemical reagent co. LTD
N, n-dimethyl formamide (DMF)	Li 'an long bohua pharmaceutical chemical co., LTD
High purity deionized water	Millipore, inc
DF light curing composite resin	Germany heraeus gusha dental co., LTD

2.1.2 Equipment (Table 2)

Table 2 The Equipment

Name	Manufacturer
Scanning electron microscope	SEM, JEOL, 5600, Louis vuitton, Japan
Light curing machine	Dentsply
Ultracentrifuge	Optima L-100XP, the United States
Microhardness tester	Hengyi precision instrument Co. LTD. China
Fourier transform infrared spectrum	ATR-FTIR, IFS66V/S, Bruker
Umt-2mt reciprocating friction meter	CETR corporation, USA

2.2 Prepare composite resin composites containing anionic powder

Anion powder dispersed in the resin matrix material by solution mixing method [16], using anhydrous ethanol as solvent. In the process of mixing, this experiment adopts the method of mechanical stirring and ultrasonic dispersion to prevent particles together.

(1) In the experimental group, negative ion powder was added to the composite resin material in the proportion of 1%, 2% and 4%, respectively.4g of each resin was taken and dissolved with anhydrous ethyl ether. A drop of silane coupling agent was added and mixed evenly with magnetic agitator and ultrasonic oscillation according to the above proportion. Finally, the rotatory evaporator was used to remove diethyl ether and different proportions of anionic powder composite resin were prepared.

(2) Control group: the composite resin sample without anion powder was 4g, and the composite resin without anion powder was prepared according to the same treatment method of the experimental group.

2.3 Scanning electron microscopy

The sample containing 0%, 1%, 2% and 4% anion powder was vertically irradiated by the photocuring machine at $500w \cdot cm^{-2} \cdot 40s$ curing condition to make a plate-shaped solid block, which was sprayed with gold after surface polishing, and then the anion powder dispersion was observed by SEM.

2.4 EDS analysis of the relative content of Al and Si

The relative contents of Al and Si on the surface of the sample containing 0%, 1%, 2% and 4% anion powder were analyzed by electron energy spectrometer.

2.5 Mechanical performance test

The friction properties of composite resin materials containing anion powder were tested in accordance with international standard (ISO 4049-1978) and Chinese dental composite resin material standard (YY 91042-1999).

Preparation of friction test specimens: $20mm \cdot 10mm \cdot 1mm$; Each group has three specimens, and the curing condition is the same as (1). Both ends of the specimen with 600 mesh water sand paper burnish, polishing, make the surface level, the specimens preserved in 37 °C physiological saline for 24 h after the test.

The friction performance: the configured artificial saliva was added into the container as the friction medium, take the steel ball as the grinding head, load 20 N, frequency 2 Hz, friction time 30 min, repeat the test for 3 times for each sample. The friction coefficient when the friction curve of each sample is stable is recorded and the average value is obtained.

2.6 Data statistics
SPSS10.0 data statistical analysis system software was used for one-way anova and t test.

3. Results and discussion

3.1 Scanning electron microscopy (SEM)
The results of scanning electron microscopy (SEM) on the surface of composite resin containing 0%, 1%, 2% and 4% anion powder are shown in Figure 1. Figure 1A shows no anionic particles in the control group. Figure 1B, C and D show that there are many cubic anion powders that have been mixed into the composite resin without agglomeration and dispersed well. The results show that the anion powder was successfully combined with the composite resin and the anion powder could be uniformly dispersed without agglomeration.

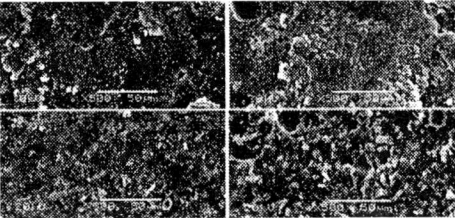

Figure 1. SEM images of anionic powder composite resin with different proportions (0,1%, 2%, 4%)

3.2 EDS
The composition of anion powder is shown in Table 3

Table 3 The main components of anion powder

Composition	Content (wt %)
Al_2O_3	35.10
The SiO_2	34.81
B_2O_3	11.02
MgO style	4.70
Fe_2O_3	10.18
Na_2O	0.91
P_2O_5	0.22
TiO_2	0.26

The relative contents of Al and Si elements on the surface of samples containing 0%, 1%, 2% and 4% anionic powder analyzed by EDS are shown in Table 4.

Table 4 Relative contents of Al and Si elements on the surface of anionic powder composite resin samples

	Al (weight %)	Si (weight %)
0	2.08	4.11
1%	10.25	9.32
2%	12.66	11.25
4%	15.05	14.11

It can be seen from Table 3 that the main composition of anion powder is Al_2O_3 And the SiO_2. EDS detection of composite resin containing anion powder was carried out accordingly. Table 4 showed that with the increase of the proportion of anion powder added, the relative content of Al and Si elements increased, indicating that anion powder had been successfully added into the composite resin.

3.3 Experimental results of friction and wear

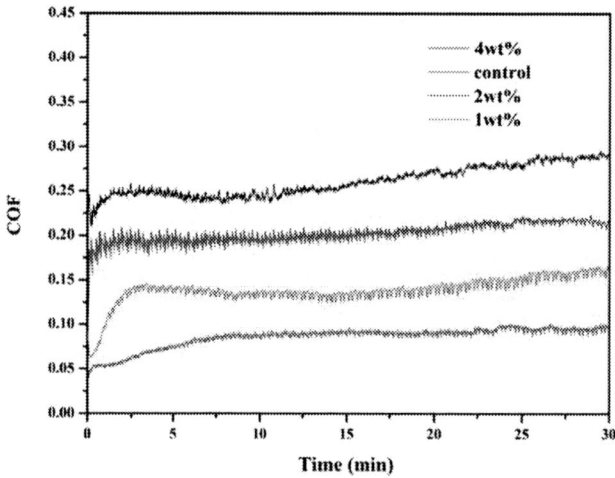

Figure 2 Friction coefficient of composite resin with different proportion of anion powder

Figure 2 shows the original data of friction coefficient of composite resin material containing anion powder and steel ball after grinding and its statistical analysis results. The figure shows the corresponding friction curve, showing the change trend of friction coefficient in the friction process with the friction and wear experiment. The experimental results showed that the friction coefficient of the anion powder changed after adding the composite resin in different proportions, which was higher than that of the control group. After statistical analysis, the friction coefficient of the 1%, 2% and 4% groups were statistically significant compared with that of the control group.

If appropriate concentration of inorganic filler can improve the mechanical properties of the composite resin, but excess can cause decline. The experiment needs to test the appropriate concentration of anion powder. The anion powder added in composite resin in the experiment influenced the mechanical performance. The reason may be that the particle surface combined with resin matrix molecules groups, which may affect the original packing and resin matrix in combination, whether the hypothesis is yet to be further research.

4. Conclusion

The experiment successfully compounds the different proportions of anion powder mixed into the composite resin by mechanical mixing dispersion and ultrasonic dispersion. EDS results showed that anion powder was successfully added into the composite resin, and the surface particles were uniformly dispersed by SEM. The friction test show that the friction coefficient increases with the increase of anion powder content. However, the appropriate concentration of anion powder needs to be further studied. In addition, its long-term antimicrobial properties need to be further studied.

References

[1] R. A. Azeem, N. M. Sureshbabu, Journal of Conservative Dentistry Jcd, **21**(2018)

[2] S. Kubo, A. Kawasaki, Y. Hayashi, Dental Materials Journal, **30**(2011)

[3] F. F. Demarco, K. Collares, F. H. Coelho-de-Souza, M. B. Correa, M. S. Cenci, R. R. Moraes, N. J.M. Opdam, Dental Materials, **31**, 10 (2015)

[4] S. Sakuma, Anti-bacterial composite particles and anti-bacterial resin composition[J]. (2003)

[5] S. T. Khan, A. A. Al-Khedhairy, J. Musarrat, Journal of Nanoparticle Research, **17**, 6(2015)

[6] S. Savas, E. Kucukyılmaz, E. U. Celik, M. Ates, J Oral Sci, **57**, 4(2015)

[7] H. M. Lim, H. J. Park, Current Green Chemistry, **2**(2015)

[8] S. T. Khan, A. A. Al-Khedhairy, J. Musarrat, Journal of Nanoparticle Research, **17**, 6(2015)

[9] Q. H. Tran, V. Q. Nguyen, A. T. Le, Advances in Natural Sciences Nanoscience & Nanotechnology, **4**, 3(2013)

[10] F. Wang, Stabilization and application of ultrafine anion powder[D]. Shenyang university of technology (2016)

[11] Y. Zhang, G. Gao, X. Gou, W. Liu, B. Liu, H. J. Wang, International Conference on Advanced Computer Science and Engineering (ACSE), (2014)

[12] M. Liu, X. Zhang, J. Zhang, Q. Zheng, B. Liu, IOP Conference Series: Materials Science and Engineering, **301** (2018)

[13] C. Zhang, Studies on fiber formation and properties of anion powder/mint extract/viscose cellulose blend [D]. Qingdao university (2009)

[14] M. Uchida, K. Miyake, Y. Kurihara, Antibacterial zeolite particles and antibacterial resin composition, 2013.

[15] Ding J. Abrasion resistance of inorganic filler modified hydroxy-terminated liquid nitrile rubber /epoxy-resin composites[J]. China Synthetic Rubber Industry, 2014.

[16] L. Zhang, Study on Microstructure and Tribological Properties of Plasma Nitriding Forged CoCrMo Alloy[J]. Lubrication Engineering, 2010.

Research on fuzzy PID control of forearm of tunnel steel arch mounting machine

Zhi-yuan WANG[1,2] and Shi-cheng HU[1,2]

[1]College of Mechanical and Electrical Engineering Central South University, 410000 Changsha, China

[2] State Key Laboratory of High Performance Complicated Manufacturing, 410000 Changsha, China

Abstract. Taking the smooth transition of angular velocity as the key point, the forearm vibration reduction control system based on fuzzy PID control is studied to reduce the influence of the forearm vibration on the docking operation of steel arch. According to the fuzzy rules, the control system establishes the correspondence between the real-time rotation angle error of the forearm and the control signal to reduce the self-vibration caused by the sudden change of the speed at the end of the rotation. By establishing the AMESim-Simulink joint simulation model of the forearm rotation system, the application effect of the fuzzy PID control system on the forearm vibration reduction control is studied. The simulation and experimental results show that this control system effectively reduces the sudden change of the speed and the influence of the self-vibration of the forearm, and each small-angle posture adjustment time can be controlled within 3s.

1. Introduction

The tunnel steel arch mounting machine is a construction machine dedicated to the installation of the support structure (steel arch) in the tunnel. The operator only needs to operate the steel arch mounting machine to complete the connection of the steel arch. The operation divided into two actions, one is that the forearm raises the steel arch to a proper position, and the other action is to rotate the steel arch to connect another steel arch in the horizontal plane[1,2]. However, each connection requires multiple operations, but the traditional PID control system has a slower response speed when the error is still large enough, which increases the adjustment time. Moreover, due to the nonadjustable parameters' value, the steady-state error is still quite large.

In view of the problems above, this paper establishes fuzzy control rules by studying fuzzy control theory and combining actual engineering conditions [3], which can realize online self-adjustment of control parameters and improve the response speed in certain error regions and the steady-state accuracy, and shorten each docking time.

2. Fuzzy Control Theory

Fuzzy control is a control strategy based on linguistic control rules and fuzzy logic inference methods. According to the subjective concepts such as experience, professional knowledge and control effect expectation, the designer uses the fuzzy relationship between the three control parameters K_P、K_I、K_D of PID and the error value e and the rate of change of the error ec to write fuzzy control rules [4,5]. Then the system uses these rules to adjust the

Content from this work may be used under the terms of the Creative Commons Attribution 3.0 licence. Any further distribution of this work must maintain attribution to the author(s) and the title of the work, journal citation and DOI.

Published under licence by IOP Publishing Ltd

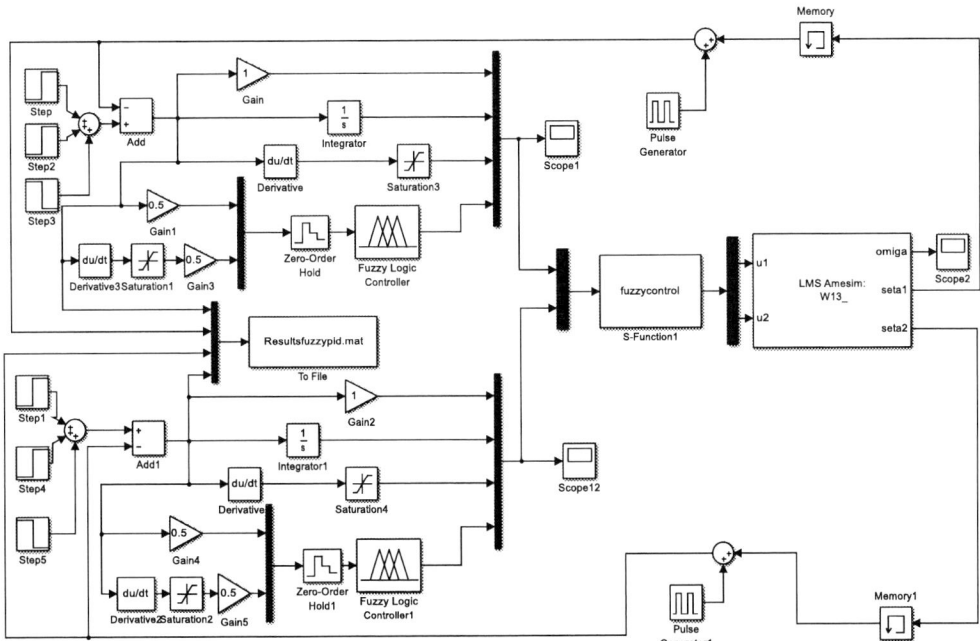

Figure 1.Controller Module

parameters online, that is, construct a suitable vector from e and ec to U to approximate the ideal control effect[6]. The output of the system is:

$$U(t) = K_P^{`} \times e(t) + K_I^{`} \times \int e(t)dt + K_D^{`} \times \frac{de(t)}{dt} \quad (1)$$

Where $K_P^{`}$、$K_I^{`}$、$K_D^{`}$ are parameters' value after adjustment.

Based on the Fuzzy control theory, we used Simulink module to design a controller of the forearm vibration reduction control system which is connecting to AMESim through the interface module[7,8]. In this controller, a Fuzzy Logic Controller is used to perform the fuzzy rules, which can output the adjustment parameters' value to achieve a better control effect. The controller module is shown in Figure 1.

3. Fuzzy PID Control System Modeling

The advantage of AMESim-Simulink co-simulation is that it omits the cumbersome steps of deriving the transfer function of the system. It only needs to establish the simulation model in AMESim software, and then the system will automatically compile the transfer function by using its interface module [9].

In the design process of the tunnel steel arch mounting machine, considering the safety of operators and machine, only one hydraulic cylinder can be operated at the same time, which simplifies the modeling of the servo system. The components related to the forearm rotation system are built in AMESim software, and the interface module is added to intercommunicate the signal between AMEsim and Simulink. The simulation model of the forearm fuzzy control system is shown in Figure 2.

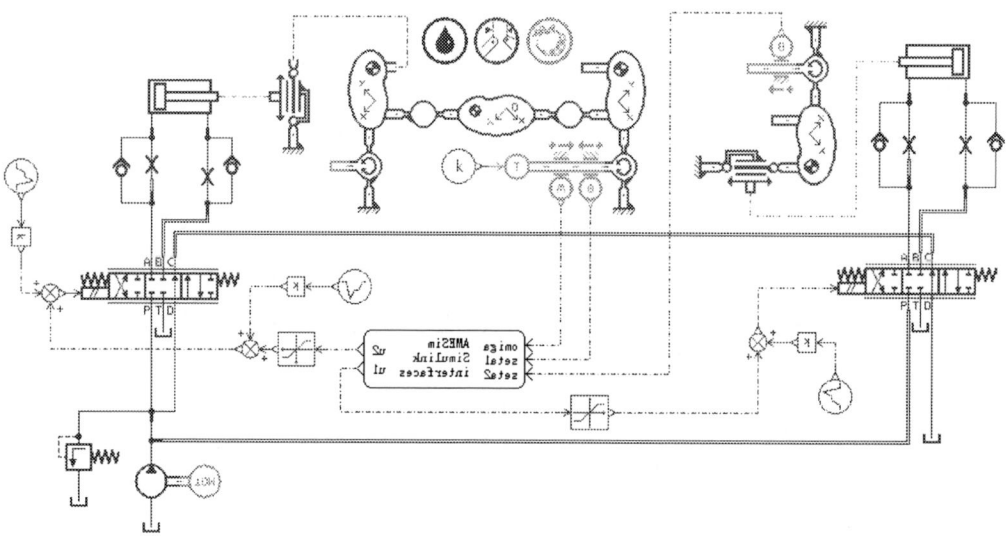

Figure 2. The Forearm Fuzzy Control System Simulation Model

The research idea of this control system is to use a angle sensor to detect the corner signal of the forearm, and then the controller calculates the error value e and the rate of change of the error ec, and last the controller outputs the control current according to the fuzzy control algorithm.

4. Simulation And Experiment

In order to verify the superiority of the fuzzy control system, the fuzzy control system are simulated. According to the actual operation situation, the target value of the forearm angle is simulated by the step signal, and the angle curve and angular velocity curve of the forearm are compared to analyze the effects of the fuzzy control system. The simulation time is set to 20s, and the simulation results are shown in Figure 2 and Figure 3.

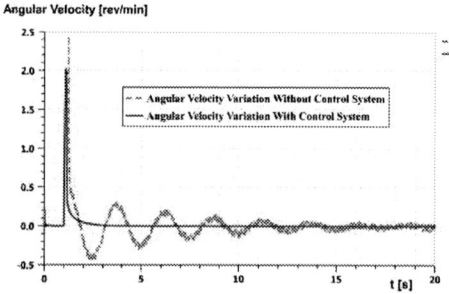

Figure 3.Angular velocity variation curve

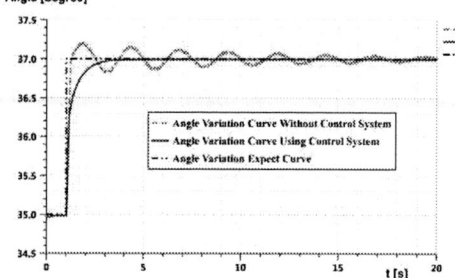

Figure 4. Angle variation curve

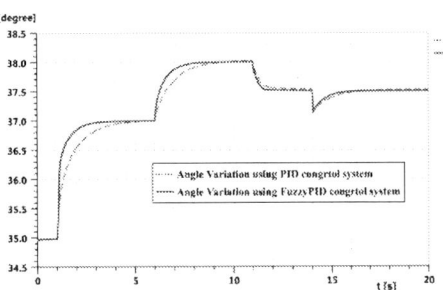

Figure 5. Angle variation curve of different control system

Fig. 3 and Fig. 4 show that although the forearm vibration reduction control system slightly extends the forearm angle response time, the angle is almost no oscillation after the end of the rotation, the adjustment time is almost 0. Instead, when the forearm vibration reduction control system is not used, the forearm vibration is obvious after the end of the rotation and the adjustment time is much longer.

Fig. 5 shows that each small angle adjustment response time can be controlled to about 3s with the help of the fuzzy control system, while the traditional control system requires nearly 5s, what's more, when the forearm rotation system is disturbed, the fuzzy control system can help the forearm return to the steady state faster.

In order to verify the actual control effect of the forearm vibration reduction control system, the tunnel steel arch mounting machine named WXHLC1215 was used as the testing machine. During the experiment, the forearm was rotated at a certain angle to observe the vibration after the end of the rotation, and the response time and adjustment time were recorded, and then we compare and analyze various experimental data to verify the control effect of the control system. The experimental results are shown in Table 1 from where we can see that the use of the control system effectively reduces the vibration of the forearm and shortens the adjustment time of the vibration.

Tab.1　Experimental results of WXHLC1215 tunnel steel arch mounting machine

Target Angle(°)	Control System	Response Time(s)	Vibration Situation	Adjustment Time(s)
2	none	0.22	obvious	65.83
	Fuzzy Control System	3.24	none	0
1	none	0.10	obvious	59.27
	Fuzzy Control System	2.29	none	0

5. Conclusion

1) The forearm vibration reduction control system based on fuzzy control theory effectively reduces the vibration problem caused by excessive residual speed of the forearm, and the posture adjustment process of the forearm is smoother.

2) The forearm vibration reduction control system based on fuzzy control theory can control the response time to about 3s when the forearm is adjusted at a small angle, and there is no vibration adjustment time.

3) The forearm vibration reduction control system based on fuzzy control theory can restore the system to the steady state more quickly to reduce the influence of the external disturbance.

References

[1] ZHANG Mao-yi, WANG Lian-xin. The General Situation and Development Trend of Arch Trolley Domestic and Overseas[J]. Construction Machinery Technology & Management,2018,**31**(03):73-77.

[2] XU Li-ping, ZHANG Yi-fei, REN De-zhi. AMESim-based Motion Simulation and Control of Steel Arch Installation Mechanical Arm[J].Chinese Hydraul-ic&Pneumatics,2014,(02):78-80+84.

[3] SHEN Qi-hao, LIAO Hua, MA Xun, DU Li-wei, LIU Zu-ming, LI Jing-tian. Journal of Yunnan Normal University(Natural Sciences Edition).Journal of Yunnan Normal University(Natural Sciences Edition), 2018,38(05):7-13.

[4] YANG Hong-bin. Study on optimal fuzzy control in the cantilever vibration control[J].Electronic Design Eng-ineering,2016,24(18):104-105+108.

[5] LIU Hui-sen, ZHANG Yu-lian, DONG Quan-lin. Application of Fuzzy PID in Control System of Intelligent Control Valve[J].Machine Tool & Hydr-aulics,2018,46(01):91-96.

[6] FU Tian-tian, ZHU Yu-chun, GU Ya-jun. Research on Fuzzy PID Control of Electro-hydraulic Servo System Based on MATLAB-AMESim Joint Simulation[J].Ma-chine Tool & Hydraulics,2016,44(20):144-146+154.

[7] HE Jian-hai, HU Yi-huai, ZHANG Jian-xia. Hydraulic Control of Sail Based on AMEsim-Simulink Co-Simulation[J].Machine Tool & Hydraulics,2018,46(11):186-189.

[8] LI Qiang-qiang, JIN Bao-quan, GAO Yan, ZHANG Hong-juan. Mill Hydraulic Pressure System Analysis Based on AMESim and Simulink Co-simulation[J].Ch-inese Hydraulics & Pneumatics ,2016(07):18-23.

[9] LIU Shao-long, WEI Cong-mei, DING Ding. Sim-ulation Research of Four Cylinders Synchronous Circuit Based on AMESim/MATLAB[J]. Fluid Power Transmission & Control,2017,(03):34-37.

ISPECE

IOP Publishing

Early Fault Diagnosis of Rolling Bearing Based on Lyapunov Exponent

Guo Qingjun[1], Li Yang[2]

[1] mechanical engineering,Chongqing Jianzhu College,400072Chongqing,China

[2] mechanical engineering,nstitute of Prospecting Technology, 610000Chengdu,China

Corresponding author: Guo qingjun,email: 672139860@qq.com

Absrtact. Lyapunov exponent is an important quantitative index to measure the dynamic characteristics of the system. It represents the average exponential rate of convergence or divergence between adjacent orbits in phase space. Whether there is dynamic chaos in the system can be judged intuitively from whether the maximum Lyapunov exponent is greater than zero. The Lyapunov exponent obtained by the small data method has the characteristics of simple calculation process and accurate calculation results, and it is applied to the fault diagnosis of rolling bearings. By calculating the Lyapunov exponents of bearing rollers, bearing outer rings, bearing inner rings and normal state bearings, and comparing the calculation results under different working conditions, it is concluded that Lyapunov exponents are of great significance in judging the early failure of rolling bearings.

1. CHAOS CHARACTERISTICS

Chaos theory plays an important role in the research of non-linear science. Up to now, non-linear science mainly includes non-linear systems, soliton theory, symbolic dynamics, chaotic dynamics, quantum chaos, fractal geometry and physics. So far, the concept of "chaos" has different understandings and expressions in different fields of science and technology, reflecting the universality of the existence of chaos phenomenon and its unique characteristics in their respective fields. Chaotic phenomena can be regarded as macro-nonlinear systems, which show uncertain or unpredictable characteristics under general conditions [1]. It is the unification of certainty and uncertainty, order and disorder, regularity and irregularity. Chaos has the following characteristics:

(1) Similar randomness

Similar randomness is characterized by random disorder and chaotic trajectory at the macro level, but self-similar infinite nesting at the micro level, so it seems to occur randomly on the surface, but in fact it is determined by certain rules and can be deduced in a short period.[2]

(2) Sensitivity of initial conditions

The relative concept of initial-condition sensitivity and short-period computability is that the system can not be calculated in the long-term scale. Because the small change of initial conditions of the system will lead to large changes in subsequent motion, which reflects another characteristic of chaotic system,   namely, sensitivity to initial conditions.

2. LYAPUNOV EXPONEN

Lyapunov exponent is a characteristic quantity describing attractor characteristics at macro level. It is an important parameter of chaotic system and represents the average index of convergence or

Content from this work may be used under the terms of the Creative Commons Attribution 3.0 licence. Any further distribution of this work must maintain attribution to the author(s) and the title of the work, journal citation and DOI.

Published under licence by IOP Publishing Ltd

divergence of adjacent orbits in phase space dynamic system.

In the one-dimensional dynamic system $F(x_n) = x_{n+1}$, pay attention to $\left|\dfrac{dF}{dx}\right|$ .If $\left|\dfrac{dF}{dx}\right| > 1$, Then the initial two-point iteration will diverge.; If $\left|\dfrac{dF}{dx}\right| < 1$, after the initial two-point iteration, it will be aggregated。 In subsequent iterations，the value of $\left|\dfrac{dF}{dx}\right|$ is dynamic.，In phase space, two orbits are separated and gathered. In order to reflect the degree of separation and gathering of two orbits as a whole, it is necessary to calculate the average number of iterations or time. [3]。

Two points with a distance of ε, after several iterations, the distance becomes:

$$\varepsilon e^{n\lambda(x_0)} = \left|F(x_0 + \varepsilon) - F^n(x_0)\right| \quad (1)$$

If $\varepsilon \to 0$, $n \to \infty$, the upper form becomes

$$\lambda(x_0) = \lim_{n\to\infty}\lim_{\varepsilon\to 0}\frac{1}{n}\ln\left|\frac{F_n(x_0+\varepsilon)-F_n(x_0)}{\varepsilon}\right| = \lim_{n\to\infty}\frac{1}{n}\ln\left|\frac{dF^n(x)}{dx}\right|_{x=x_0} \quad (2)$$

The above formula is independent of the initial value and can be simplified to

$$\lambda = \lim_{n\to\infty}\frac{1}{n}\sum_{j=1}^{n}\ln\left|\frac{dF(x)}{dx}\right|_{x=x_j} \quad (3)$$

λ is the Lyapunov exponent in the prime mover system. It represents the average exponential separation rate caused by each iteration in several iterations. There are three main methods for calculating Lyapunov index, Wolf method, Jacobi method and small data quantity method. This paper mainly uses small data quantity method to calculate Lyapunov index.:

The small data quantity method is a widely used method to obtain Lyapunov exponent. It is an improvement of Wolf's method. It has the advantages of small data amount, high accuracy and simple calculation process. [4]。

Firstly, the phase space is reconstructed, and then the nearest neighbors of the orbits in the phase space are found:

$$dt(0) = \min_{x(\hat{i})}\left\|X(t) - X(\hat{i})\right\|, \quad \left|t - \hat{i}\right| > p \quad (4)$$

In the formula, P is the average period of time series. Its value can be calculated by calculating the reciprocal of the average frequency of energy spectrum. From this, the maximum Lyapunov exponent can be estimated according to the average divergence rate of the nearest neighbor point of each orbit。

According to Sato et al:

$$\lambda_1(i) = \frac{1}{i\Delta t}\frac{1}{(M-i)}\sum_{j=1}^{M-i}\ln\frac{d_j(i)}{d_j(0)} \quad (5)$$

In the formula, Δt is the sample period and $d_j(i)$ is the distance of the j pair of adjacent points in orbit after i discrete time steps.

According to the above formula:

$$d_j(i) = C_j e^{\lambda_1(i\Delta t)}, \quad C_j = d_j(0) \quad (6)$$

Logarithms on both sides of the upper formula

$$\ln d_j(i) = \ln C_j + \lambda_1(i\Delta t) \quad (7)$$

In the formula $j = 1,2,\ldots,M$。The maximum Lyapunov exponent is the slope of the upper curve, which can be obtained by the least square method.

3. Characteristic analysis of test data

Next, the validity of Lyapunov exponent in early fault diagnosis of rolling bearings is verified by experiments. The bearing fault is simulated by manual faults and the whole life test platform of rolling bearing is used in the test.（Figure 1）, Crack faults are manually manufactured on the inner ring, outer ring and rolling body of bearings. The crack size is 0.4 mm wide and 0.2 mm deep.

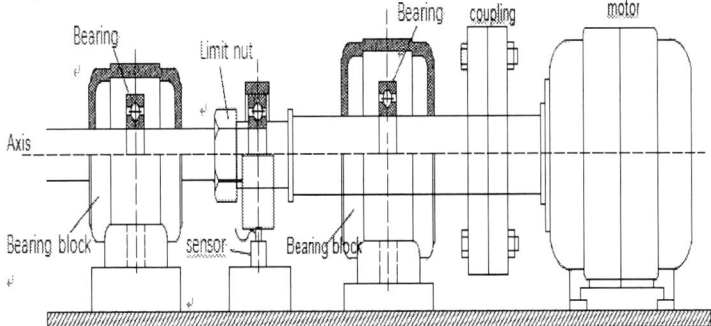

Figure 1.Structural diagram of rolling bearing life test platform

By using the vibration signal data acquisition system, the time-domain vibration signals of rolling bearings under different working conditions are collected, and a number of groups of digital sequences bearing fault information of rolling bearings are obtained. Since the motion of rotating machinery system inevitably shows chaotic state, we can take Lyapunov exponent as a characteristic quantity of the system and calculate different fault types. The maximum Lyapunov exponent of signal sequence under the same working condition is used as the feature of each fault type to identify the fault type.

The Lyapunov exponent is calculated by the small data method. The calculation steps are as follows:

(1) Estimate the average period of the sequence by FFT transform.

(2) Phase space reconstruction of time series. The parameters of phase space reconstruction refer to the previous calculation method.

(3) Find the nearest distance point according to formula (4) to limit the short separation.

(4) Calculate the average distance of adjacent points.

(5) The maximum Lyapunov exponent is obtained by fitting the slope of the straight line according to equation (7), as shown in Fig. 2.

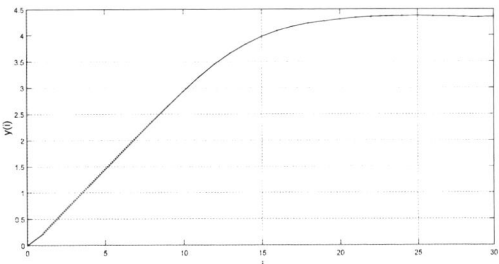

Figure 2.Calculates the Maximum Lyapunov Index

Through the above steps, the Lyapunov exponents of bearing vibration signal sequence under inner ring fault, outer ring fault, rolling element fault and normal state are calculated. Due to space limitation, Table 1 only gives the characteristic parameters of six sets of vibration signal sequences under four working conditions. The first two data names represent the fault type, and the last two represent the fault data groups, as shown in the table.

Table 1 Maximum Lyapunov exponent of sample data (six groups for each failure)

Inner Ring Fault		Outer Ring Fault		Roller Fault		Normal Conditions	
D0101	0.0963	D0201	0.1324	D0301	0.1377	D0401	0.0403
D0102	0.0472	D0202	0.0181	D0302	0.1553	D0402	0.0479
D0103	0.0413	D0203	0.1379	D0303	0.1569	D0403	0.0136
D0104	0.0168	D0204	0.0874	D0304	0.1987	D0404	0.0334
D0105	0.0225	D0205	0.1051	D0305	0.2366	D0405	0.0350
D0106	0.0862	D0206	0.0339	D0306	0.1793	D0406	0.0192

In order to intuitively express the difference and regularity of Lyapunov exponents among different groups of data, the data of each group under various working conditions in Table 1 are made into a three-dimensional histogram (Figure 3). From the graph, we can see the numerical difference and distribution regularity of the largest Lyapunov exponents among different types of faults.

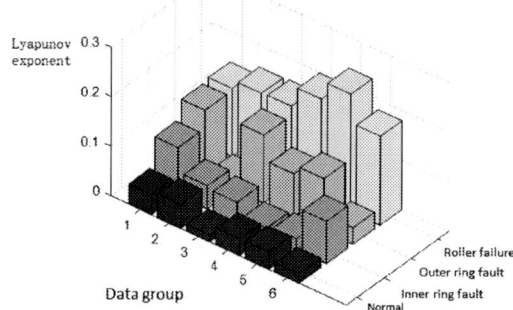

Figure 3. Comparison chart of maximum Lyapunov exponent for each fault type

4. conclusion
The maximum Lyapunov exponent of the vibration signal sequence of rolling element faults is high; the maximum Lyapunov exponent of the vibration signal sequence of outer ring faults is slightly larger than that of inner ring faults; the maximum Lyapunov exponent of the vibration signal sequence under normal working conditions is relatively small. The results show that because of the existence of rolling bearing faults and the different fault location, the chaotic characteristics of each group of data systems are obviously different. So the maximum Lyapunov exponent can be used as a feature for pattern recognition of early fault types of rolling bearings.

Acknowlegements
Found project:Youth Fund Project of Chongqing Jianzhu College, Project Name: Research on Mechanical Fault Diagnosis Based on Wind Power Gearbox Transmission System, contract number: NQ2018006

Reference:
[1] Zhang Shuqing, Wu YueE, Jiao Guanghe. Feasibility analysis of weak signal detection method based on chaos theory [J]. Measurement and control technology, 2002, **21** (7).
[2] Zhang Jinliang, Tan Zhongfu. Mixed prediction method of chaotic time series [J]. System Engineering Theory and Practice, 2013,**33**(3).
[3] Zhang Hailong, Min Fuhong, Wang Enrong. Comparisons on the calculation methods of Lyapunov index [J]. Journal of Nanjing Normal University, 2012, **12** (1).
[4] Yang Yongfeng, Ni Minjuan, Gao Zhe, Wu Yafeng, Ren Xingmin. Selection of parameters for calculating the maximum Lyapunov exponent by small data volume method. Vibration testing and diagnosis, 2012,**32**(3).
[5] Takens, F. (1981). Detecting strange attractors in turbulence. Lecture notes in Math, 898, 361-381.

[6] Wang, C., Kong, F. R., Huang, W. G., Li, C. L., Chen, H. (2014). Application of improved singular value decomposition in bearing fault diagnosis. Journal of Vibration Engineering, (02), 296-303.

[7] Takens, F. (1981). Detecting strange attractors in turbulence. Lecture notes in Math, 898, 361-381.

ISPECE   IOP Publishing

The Effect of Longitudinal Shock Absorber on the Vibration Response of Train-Bridge Coupling System in the Articulated Train

Jie Ding [a]

School of Mechanical Engineering, Lanzhou Jiaotong University, Lanzhou,Gansu,China

[a]Corresponding author: 37194175@qq.com

Abstract: On the basis of analyzing the structural characteristics of articulated trains, directly combined with the method of UM software and importing three-dimensional model into the UM software, the dynamics model of the vehicle-bridge coupling system of three-section articulated train and the simple-supported box girder bridge is established. The dynamics model compares the vehicle-bridge coupled vibration response of the articulated train with or without longitudinal shock absorber and analyses the difference. The results show that the vertical vibration of the end train is stronger than that of the intermediate train; the longitudinal shock absorber can effectively suppress the vibration of the intermediate train body of the articulated train, and the vibration suppression effect on the end body is not obvious; the strength of the coupling effect has little effect on the vertical vibration of the bridge.

1. Introduction

When the train passes the bridge, the vehicle load is transferred to the bridge through the wheel-rail connection and the bridge-rail connection, causing the vibration and deformation of the bridge which will in turn aggravate the vibration response of the vehicle. Therefore, it is of great theoretical value and practical significance to study the dynamic performance of vehicle- track - bridge coupling system. G.Diana, F.Cheli[1] established a vehicle-track-bridge dynamic model based on the elastic track and wheel-rail connection. The simulation results were basically consistent with the measured results. Biondi B et al [2] idealized the train as a series of identical vehicles traveling at constant speed. Rails and bridges are modeled as Bernoulli-Euler beams. They proposed a component modal synthesis method that combines continuous (track and bridge) and discrete (train) substructures. The numerical calculations showed that the method can simultaneously calculate the dynamic response of trains, tracks and bridges with high precision and efficiency. Zhai W M, Cai CB[3] established two dynamic models of train operating on ballasted and ballastless tracks. He considered the effects of orbital structure and wheel-rail interaction on system dynamics in the model and studied the effects of random irregularities of track on the train-orbit-bridge dynamic interaction. The most commonly used vehicle model is the vehicle space vibration model based on the rigid body dynamics hypothesis[4-6]. In this model each train is composed of wheelsets, bogies and car bodies which are all regarded as rigid bodies and are connected with each other by spring and damper.

After the French TGV articulated high-speed trains have achieved great success in terms of technology and commercial use, some countries, such as Germany, Spain, South Korea and China,

---

Content from this work may be used under the terms of the Creative Commons Attribution 3.0 licence. Any further distribution of this work must maintain attribution to the author(s) and the title of the work, journal citation and DOI.
Published under licence by IOP Publishing Ltd

have joined the ranks of the development and operation of articulated high-speed trains. In recent years, the research on the dynamic performance of the articulated high-speed train-bridge coupling system has also attracted the attention of many scholars. Zhang Nan et al[7] established the 115-degree-of-free Thalys articulated high-speed train dynamic model and the Antoing bridge dynamics model, and analyzed the vibration acceleration of the train and the dynamic response of the bridge deflection and acceleration, and it is in good agreement with the field test results. Xia He et al[8] established an articulated vehicle unit model and a coupled dynamic model of finite element bridge model to study the dynamic interaction of articulated high-speed train axles. Zhai Wan Ming[9] established an articulated high-speed train-track vertical coupling dynamics model, and compared the vertical dynamic performance of articulated high-speed trains and non-articulated high-speed trains. The research results show that the articulated high-speed train has good vertical dynamic performance. Wang Fu Tian et al[10] expounded the design principles and design ideas of the articulated high-speed train bogie. Three kinds of software, SUNDYNA, MEDYNA and TPLDYNA, were used to calculate the running stability and curve passing performance of the vehicle. The above studies all use numerical integration to carry out research, but the use of multi-body dynamics software to study the vibration response of articulated high-speed train car-bridge coupling system is still relatively rare.

In this paper, the UM of multi-body dynamics software is used to establish the three-vehicle group train model of the articulated train and the vehicle-bridge coupling system dynamics model of the simple-supported box girder bridge. Based on the train-bridge coupling relationship, considering the high-speed operation of the train on the bridge, the influence of the longitudinal shock absorber on the vibration response of the articulated train-bridge coupling system is studied.

2. Structural features of articulated trains
The articulated high-speed train uses a traditional power bogie in addition to the conventional power bogie. The intermediate vehicle uses an articulated bogie. The bogie supports the end part of the front train and the front part of the rear vehicle. The front and rear ends of the middle vehicle are supported ends and hinged ends respectively. A two-series suspension spring cap is placed on each side of the supporting end wall, and a lower spherical core seat is arranged in the middle. There is no two-series suspension spring cap on the hinge end, and an upper spherical core seat is placed in the middle. The lower spherical core seat of the supporting end and the upper spherical core seat of the hinged end are hinged together, and a part of the vertical load of the hinged end body is transmitted to the supporting end body through the spherical core seat, and then transmitted to the air spring through the spring bearing platform and turn to the frame. It is equivalent that a bogie can support a vehicle, which greatly reduces the number of bogies in the train and reduces the weight of the train. At the same time, the running resistance and vibration noise of the train are greatly reduced as the number of bogies decreases. The vehicles are tightly hinged by a central elastic hinge, so the coupling between the vehicles is more obvious.

3. The establishment of dynamic model of train-bridge coupling system
The articulated train model established in this paper adopts the TGV bolsterless bogie structure. The train consists of three train bodies (one articulated intermediate train body and two end train bodies) supported by four bogies. The vehicle model consists of a multi-rigid system consisting of a body, a bogie and a wheel pair. According to the steps of body, hinge and force element, they are modelled in order from bottom to top. The geometry of the articulated train is directly combined with the method of UM software and importing three-dimensional model into the UM software. The articulated train model of the three train groups is shown in Figure 1.

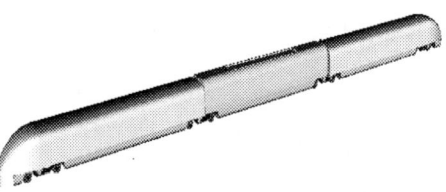

Figure 1. Schematic of articulated train model.

The bridge is a prestressed concrete double-line single-hole simple-supported box (single-box single-chamber) beam bridge with a span of 32m. The standard is Tongqiao (2008) 2322A-II. The track adopts C60 rail, and the track irregularity spectrum adopts the low interference spectrum of German high-speed railway. The ANSYS software was used to build a flexible body model of the bridge. The vehicle rigid body subsystem and the bridge flexible subsystem form an axle-coupled system, and the dynamic model of the articulated train-bridge coupling system is shown in Fig 2.

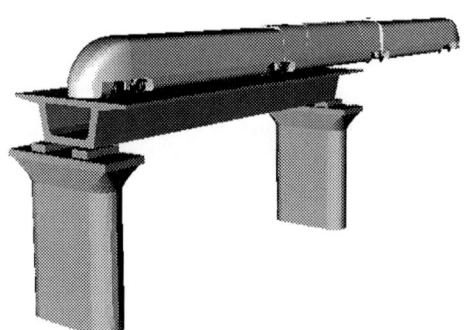

Figure 2 . Schematic of dynamic model of an articulated train-bridge coupling system.

4. The analysis of influence of longitudinal shock absorber on vibration response of train-bridge coupling system

In order to analyze the influence of the longitudinal coupling effect of the articulated train on the vibration response of the train-bridge coupling system, the vibration response of articulated trains and Bridges with and without longitudinal shock absorbers was studied when the train passed through the bridge at a speed of 300 km/h.

It can be seen from Fig.3 that the vibration acceleration of the head vehicle with or without the longitudinal shock absorbers is basically the same, and the longitudinal shock absorbers has little influence on the vibration acceleration of the head vehicle of the articulated train. It can be seen from Fig. 4 that the vertical vibration acceleration of the intermediate vehicle of the articulated train is significantly increased in the absence of the longitudinal shock absorbers, indicating that the longitudinal shock absorbers between the articulated trains has a significant effect on suppressing the vibration of the intermediate train body. According to further analysis in Fig. 3 and Fig. 4, the head train body has only longitudinal damping at the end, and the single side damping has no obvious effect on the vibration of the train body, while the intermediate vehicle body has longitudinal damping at both ends, and both sides are damped. The resulting suppression of vibration has a significant effect.

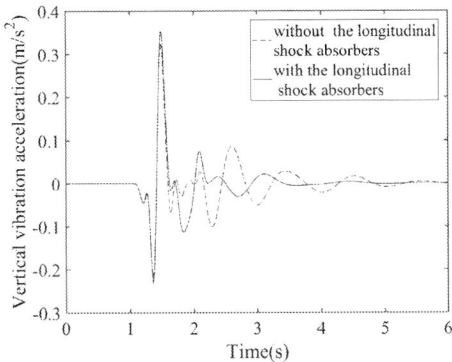

Figure 3. Schematic of the effect of longitudinal shock absorber on the vertical vibration acceleration of the head vehicle of the articulated train.

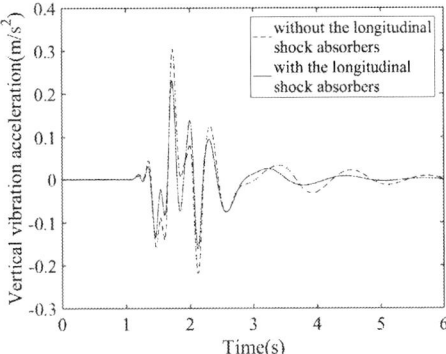

Figure 4. Schematic of the effect of longitudinal shock absorber on the vertical vibration acceleration of the intermediate vehicle of the articulated train.

Fig. 5 shows that the vertical vibration of the head vehicle of the articulated train is larger than that of the middle vehicle, which reflects that the distribution of the acceleration of hinged train body shows a law of large at both ends and small in the middle. Fig. 6 shows the case where the articulated vehicle removes the longitudinal shock absorbers. At this time, the vertical vibration acceleration of the intermediate vehicle is significantly increased, which is relatively close to the vertical vibration acceleration of the head vehicle, but it is still smaller than the vibration acceleration of the head vehicle. Under normal circumstances, the vibration of the intermediate vehicle is weaker than that of the end vehicle. The comfort of the intermediate vehicle is better than that of the end vehicle. The articulated connection can increase the stability of the train and effectively reduce the vibration of the train body.

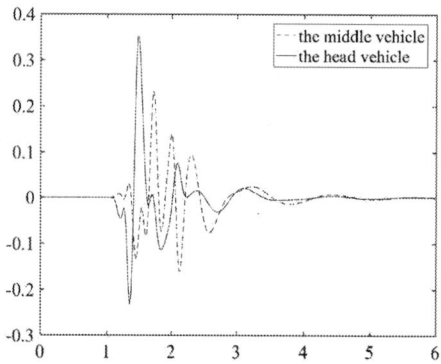

Figure 5. Schematic of comparison of the vertical vibration acceleration between the head vehicle the and the middle vehicle of the articulated train with longitudinal shock absorbers.

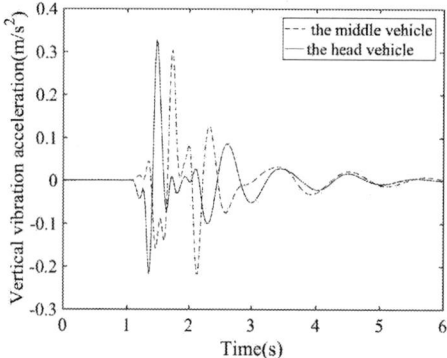

Figure 6. Schematic of comparison of the vertical vibration acceleration between the head vehicle the and the middle vehicle of the articulated train without longitudinal shock absorbers.

The vertical vibration acceleration limit of the bridge span is following: the ballasted track bridge is $[a]_{max} = 0.35g$, the ballastless track and the Mingqiao bridge is $[a]_{max} = 0.50g$. It can be seen from Fig. 7 the longitudinal shock absorber has little effect on the Mid-span vertical vibration acceleration of bridge, indicating that the longitudinal Shock absorber has a limited influence on the coupling between the train and the bridge. The coupling effect between vehicle bodies is less related to the response of bridge vibration. The mid-span vertical vibration of bridge is not caused by longitudinal damping between vehicle bodies, but determined by the performance parameters of the train itself.

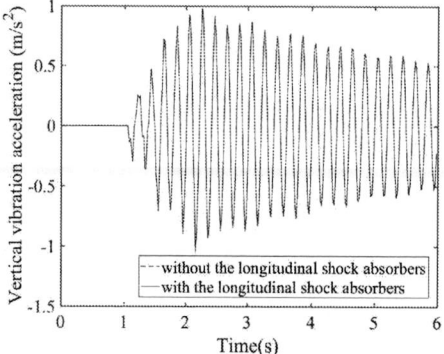

Figure 7. Schematic of the effect of longitudinal shock absorber on the mid-span vertical vibration acceleration of bridge

5. Conclusion

In this paper, the UM of the multi-body dynamics software is used to establish the three-vehicle group train model of the articulated train and the train-bridge coupling system dynamics model of the simply supported box girder bridge. When the train is running on the bridge at a speed of 300 km/h, the influence of the longitudinal shock absorber on the vibration response of the articulated train-bridge coupling system is studied. The following conclusions were obtained:

(1) The vertical vibration of the end of the articulated train is stronger than the vertical vibration of the intermediate vehicle.

(2) The main advantage of the articulated train is that the damping between the vehicle bodies strengthens the coupling between the vehicle bodies, and the vibration of the intermediate vehicle body is effectively suppressed, so that the intermediate vehicle is strongly restrained by coupling and the vibration is weak. However, the damping plays a relatively limited inhibitory role for the head vehicle and the tail vehicle.

(3) The strength of the coupling between the body of the articulated train has limited influence on the vertical vibration of the bridge.

Acknowledgements

This study was supported by the natural fund of GanSu province(No. 148RJZA057)and the young scholars science foundation of Lanzhou Jiaotong university (No. 2013019).

References

[1] G Diana, F Che li. *Vehicle system dynamics*,**18**(1989)
[2] Biondi B, Muscolino G, Sofi A. *Computers & Structures*, **83**(2005)
[3] Zhai W M, Cai C B. *Vehicle System Dynamics*, **37**(2002)
[4] Xia H, Roeck G D, Goicolea J M. *New York: Nova Science Publishers Inc*, (2011)
[5] Zhang N, Xia H, Roeck G D. *Journal of Mechanical Science & Technology*, **24** (2010)
[6] Li H L, Xia H, Soliman M, et al. *Engineering Structures*, **99** (2015)
[7] Zhang N, Xia H, DE ROECK Guido. *Engineering mechanics,* **21**(2004)
[8] Xia H , Zhang N , Roeck G D . *Computers & Structures*, **81** (2003)
[9] Zhai W M. *Journal of the railway.* **4**(1997)
[10] Wang F T, Yang G Z. *Journal of the railway*, **s1** (1997)

Simulation and experiment of passive orbit disconnected support

Zihan Gao[1,2], Yidu Zhang[1,2,a] and Qiong Wu[1,2]

[1]Beihang University, School of Mechanical Engineering & Automation, 100083 Beijing, China

[2]Beijing Engineering Technological Research Centre of High-efficient & Green CNC Machining Process and Equipment, 100083 Beijing, China

[a]Corresponding author: ydzhang@buaa.edu.cn

Abstract. Research has been done on Passive Orbit Disconnect Support (PODS) using the Finite Element Method (FEM) and experiment to study its' property. Through the FEM simulation of static structural analysis and steady-status thermal analysis, the results show that PODS can change its contact status as the force changes, thus change the force and heat transfer path. In the disconnected status, the heat leakage decreases because longer heat transfer path increases the thermal resistance. In the contact status, the PODS can bear larger force load. It is verified that the PODS can adaptively change its contact state as the load change using experiment. The gap between the stem and nut or end cold body can be calculated. This paper can provide some guidance for the PODS applications.

1. Introduction

With the development of the aerospace industry, higher requirements are placed on the operating time of spacecraft in orbit. In order to achieve that goal, one of the methods is to increase the life of the tanks which contain liquid oxygen and hydrogen [1]. Some studies show that the heat loss of the supports between the shell and tank accounts for 67% of the total heat leakage [2]. Thus it is very necessary to explore the property of the support structure.

In 1981, R. T. Parmle researcher designed, analyzed, fabricated and tested the PODS in the NASA-166473 contact in response to the problem of tank support heat leakage [3]. T. Nast and D. Frank found that S-epoxy FRP had the lowest thermal conductivity at the temperature of 50 K or more and it was the optimal low temperature support structure material [4]. Chunliu Yu studied the material selection of the PODS and figured out the order of the influence of the components was launch tube > orbit tube > the other part [5]. Tiangang Wang used Workbench to analyze the thermal-structural performance of the PODS and found that the deformation was the result of combination of the structural stress and thermal stress. Among them, thermal stress was the main factor affecting the deformation [6]. Fangfang Zhuang analyzed two support forms the rod system and shell system and realized that the rod system had better performance of reducing the heat leakage [7].

There are three forms of heat transfer, which are heat conduction, convection and radiation. Some studies had pointed out that heat conduction was the main cause of heat loss. So only heat conduction was considered in this paper. Passive Orbit Disconnected Support is a novel structure which can support the tank while reducing the heat conduction in orbit. It contains two parts which are end cold body as shown in Figure 1 and end warm body. It is worth noting that there are some gaps in PODS

Content from this work may be used under the terms of the Creative Commons Attribution 3.0 licence. Any further distribution of this work must maintain attribution to the author(s) and the title of the work, journal citation and DOI.
Published under licence by IOP Publishing Ltd

which are the key of reducing heat conduction. The gap 1 is between the nut and the stem and the gap 2 is between the end cold body and stem. Sample pieces were made to test the PODS functions. The material of the launch tube and orbit tube are respectively carbon fiber and PEEK, while the other components are 6061 Aluminium Alloy.

2. Theory and Experiment

2.1 Theory of changing force or heat conduction path

Different parts of the structure have different stiffness because of its materials and dimensions. One of the gaps might be reduced and eventually disappears as the force increases. The working status can be divided into three periods according to the status of the gaps. The contact between the stem and the nut is referred to as period 1; the contact between the stem and cold end body is referred to as period 2; it is referred to as period 3 while the gap 1 and gap 2 all exist. Period 1 and period 2 are called contact status and period 3 is called disconnected status.

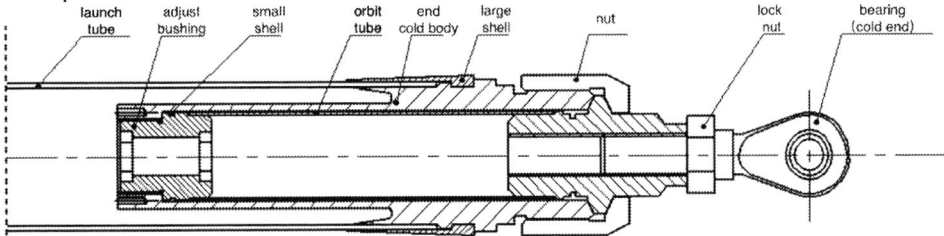

Figure 1. End cold body of the PODS

Taking the axial force load as an example, each part of the structure is deformed due to force. The gap between the stem and cold body end roughly conforms the formula 1. As the force increases, the gap will decrease and eventually disappear when the force hits some certain value. The heat transfer path and the force load path are also changed because of the contact status is changed.

$$\Delta = \sum_{i=1}^{n} \frac{F l_i}{E_i A_i} \qquad (1)$$

Where F stands for force load, l stands for the length of the part, E stands for the modulus of elasticity; A stands for the minimum cross-sectional area of the part and i stands for different part.

The gap size could be calculated using the FEM software Workbench. The dimensions of gap 1 and gap 2 are equal to 0.1mm. The end warm body is fixed and the end cold body is subjected to an axial force ranging from   -2000N to 2000N. The relationship between force and displacement of the end cold body and can be calculated also the stress and strain distributions can be got.

2.2 Theory of reducing heat loss

In orbit, thermal conduction is the main reason of heat leakage. Formula 2 is a function of thermal conductivity. An R can be proposed from formula 2 to represent the thermal impedance as shown in the formula 3. Assuming that the working temperature is constant, increasing the parameter b, decreasing the parameter λ or A can decrease the heat transfer rate Φ. It can be achieved by using lower thermal conductivity material and decreasing the contact area. In the different working period, the PODS is in different contact status. The thermal conductive path will increase while it switches to disconnected status. At the same time the contact area is decreased also.

$$\phi = \frac{T_1 - T_2}{b/(\lambda A)} = \frac{\Delta T}{R} \qquad (2)$$

$$R = \frac{b}{\lambda A} \qquad (3)$$

Where ϕ represents the convective heat transfer rate; T_1 represents the temperature of the higher side; T_2 represents the temperature of the lower side; b represents the length of the PODS; A represents the heat transfer area; λ represents the heat conductivity of the material; R represents the thermal impedance.

In order to explore the characteristic of the PODS in different period, Steady-status thermal analysis is carried out using FEM software Workbench. The thermal property of the PODS in different period is different. The temperature of the end clod body is 3K while the temperature of the end warm body is 300K.

2.3 Principle of test contact

The key of whether the PODS works properly lies in the contact status which can adaptively change along the load. The PODS sample piece is fabricated to verify its function. The working status of the PODS can't be measured directly. Luckily the material of orbit tube is not able to conduct electricity and there are some gaps around stem. It means that the electricity can't be conducted in the disconnected status while it can be conducted in the contact status. The status of the PODS can be judged by measuring the electricity between stem and end cold body. The method principle is shown in Figure 2. Here are some instructions should be followed of the experiment.

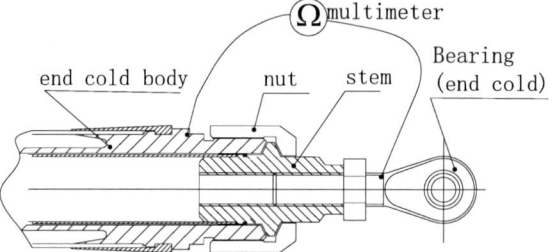

Figure 2. Contact measurement principle

(1) Assemble the PODS and adjust the adjust bushing and nut to make sure the sizes of gap 1 and gap 2 in a certain value.

(2) Connect the probes of the multi electricity meter to the stem and end cold body.

(3) Assemble the PODS to the tensile-compressive force machine while make sure the axis of the PODS along with the direction of force.

(4) Increase the load smoothly and record the force and displacement of the stem.

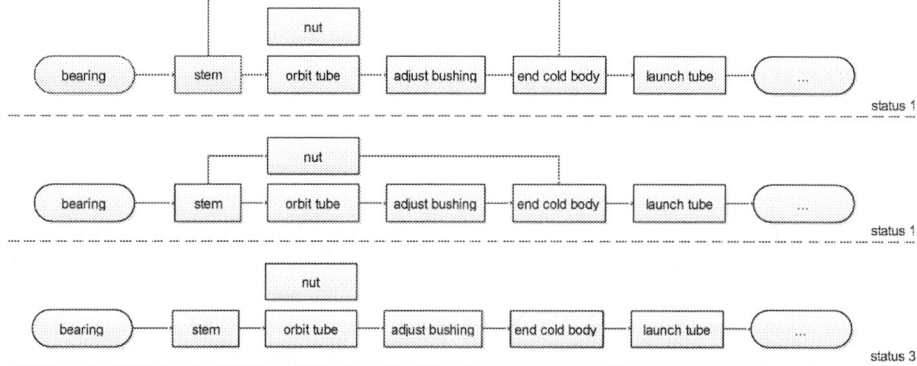

Figure 3. The heat transfer path in different statuses

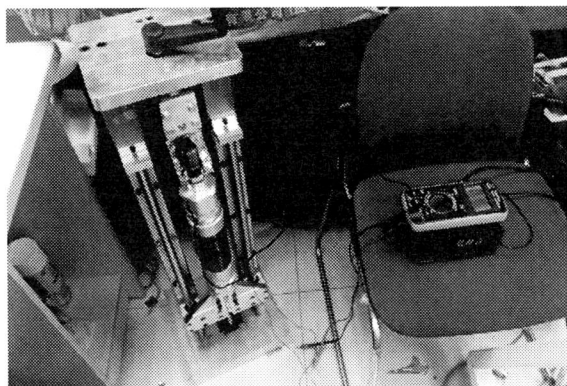
Figure 4. Contact test experiment site

3. Result and discussion

3.1 Variable stiffness of the PODS

Calculate the deformation of the PODS under different force load ranging from -2000N to 2000N and get the curve of force and deformation of stem as shown in Figure 5. It can be seen that the curve is divided into 3 parts corresponding to three different working status of the PODS obviously. Stiffness refers to the ability of the structure to resist elastic deformation when stressed. It can be calculated by the force required for a unit displacement. And that is the slope of the deformation – force curve. The stiffness of the PODS in status 1 and status 2 is larger than its in status 3 because of the contact occurs. The stiffness of the launch tube is smaller than the other parts. When it is working in status 3 the deformation of the PODS is mainly caused by the deformation of the orbit tube. In contrast in the status 1 and status 2, the launch tube can only meet a little part of the load and the contact transfer most of the load.

The deformation of the launch tube and others except orbit tube can be ignored because the stiffness of the PODS in status 3 is much less than the other two statuses. The relationship between the gap and the force is approximately equal to the relationship between the deformation and the force as shown in formula 6. And the gap can be calculated if the force can be measured while the contact occurs.

$$y = 10^8 x - 12297 \qquad (4)$$

$$y = 10^8 x + 13682 \qquad (5)$$

$$y = 6 \times 10^6 x \qquad (6)$$

Where x is the displacement of the stem; y is the force.

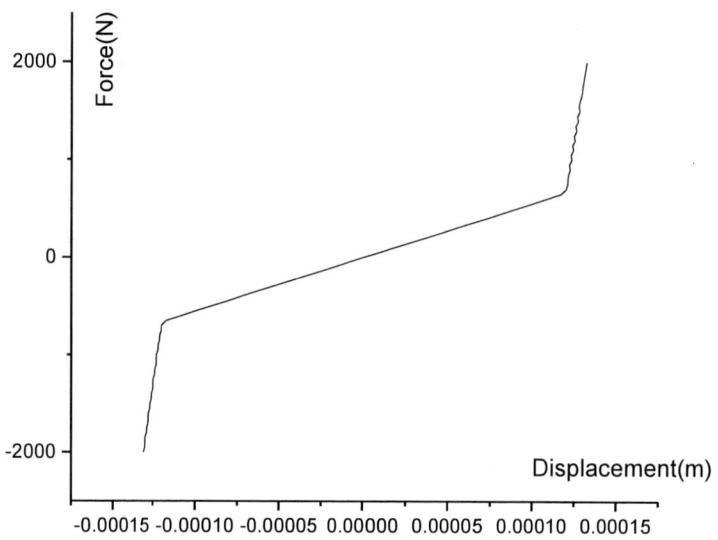

Figure 5. The relationship between displacement and loading force

3.2 Transfer path changes along force changing

The stress distribution and heat flux distribution of different statuses can be obtained by FEM Software Workbench. The stress of conical surfaces of stem and end cold body is larger than the other area of the parts in Figure 6 (a). That means the gap disappears and a contact occurs. The force is divided into two parts. One is through the orbit tube and the other is directly to the end cold body. In this status, the thermal transfer path is also divided into two parts which is the same with force path. The force path and the heat transfer path are similar which are shown in detail in Figure 6. But in status 3 as shown in Figure 6(c) and (f), there is almost the lowest stress and heat flux in the conical surface of the stem.

Comparing the heat transfer path in Figure 3, it can be clearly seen that the path of state 3 is shorter than the other path. Corresponding to the thermal resistance, the thermal resistance of the PODS in status 3 is larger so its thermal conductivity is the worst. The heat flux can be measured in the middle of the launch tube that can be used to characterize the thermal conductivity. The heat loss can be reduced about 71.28% in status 3.

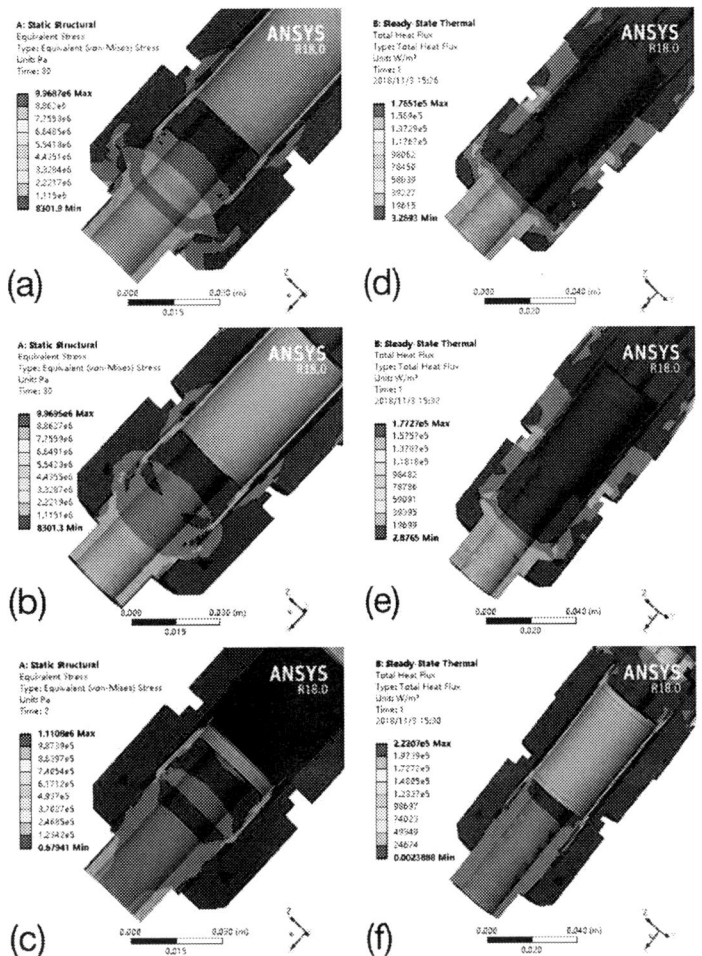

Figure 6. tress and heat flux distribution in different status. (a) represents the stress distribution in status 1; (b) represents the stress distribution in the status 2; (c) represents the stress distribution in status 3; (d) describes the heat flux distribution in status 1; (e) describes the heat flux distribution in status 2; (f) describes the heat flux distribution in status 3.

3.3 Result of experiment

In the tensile loading experiment, during rotating the loading handle, it can be seen that the load increase in the panel of the machine. When the force reaches to 70N, the ohmmeter detects a short circuit. It means a contact occurred. While reversing the handle to apply compressing force the contact occurs when the force reaches to -110N. The contact can occur and disappear automatically while the load changes. The performance of the working status can change adaptively as the load changes. The Figure 7 can be obtained through the experimental data. The contacts occurred at the point A and point B. It can be seen that the forces which caused contact are different because the size of gap 1 and gap 2 were different. It can be calculated by the formula 6. The dimensions of the gap 1 and the gap2 are respectively 0.0112mm and 0.018mm. The gaps can be changed by adjusting the adjusting bushing and nut.

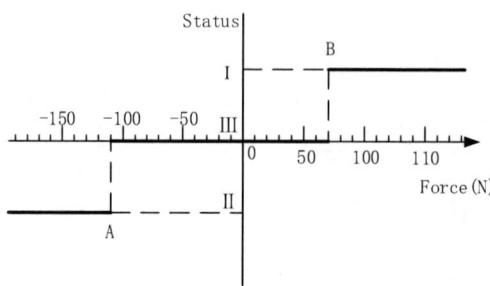

Figure 7. The experiment result of the status changing with the force load changing

4. Conclusion

Through the FEM simulation of static structural analysis and steady-thermal analysis, the results show that the PODS have many characteristics and some conclusion can be got as below.

The stiffness of the PODS in status 1 and status 2 is larger than that in status 3 because of the contact occurs.

PODS can change the contact status as the load changes, thus changing the force and heat transfer path. In the disconnected status, the heat leakage decreases because longer heat conduction path increases thermal resistance. In the contact status, the PODS can bear larger force load.

Verify the auto-change of different working status along various force load. The gap between the stem and nut or end cold body can be calculated by the switching status force. It can provide some guidance for the PODS applications.

References:

[1] Zhang, S.H., et al., Demonstration and Inspiration in Technology of Cryogenic Propellant Long-term Storage and Transfer in Orbit of NASA. Missiles & Space Vehicles, (2017).

[2] Meng, G., et al., Numerical characteristic calculation of passive orbital disconnect strut on side-load effect. Cryogenics & Superconductivity, (2016).

[3] Parmley, R.T., Passive Orbital Disconnect Strut (PODS 3) structural test program. (1985).

[4] Nast, T., D. Frank and K. Burns. Cryogenic Propellant Boil-Off Reduction Approaches. in Aiaa Aerospace Sciences Meeting Including the New Horizons Forum and Aerospace Exposition. (2013).

[5] Yu, C., et al., Optimizing material selection of novel passive orbital disconnect strut. Cryogenics & Superconductivity 41, 25-28(2013).

[6] Wang, T., et al., Thermal-structure coupled analysis of novel passive orbital disconnect strut. Cryogenics & Superconductivity 42, 11-15(2014).

[7] Zhuang, F., et al., Optimization Design of Cryogenic Tank Connection Structure. Manned Spaceflight 22, 160-163(2016).

ISPECE   IOP Publishing

IOP Conf. Series: Journal of Physics: Conf. Series **1187** (2019) 032076   doi:10.1088/1742-6596/1187/3/032076

Research on Trajectory Tracking and Vibration Suppression of a Smart Flexible-joint-and-link Space Manipulator

Zefeng Liu[1,a] **, Ruifeng Sun**[2]**, Minwei Li**[3]

[1,2,3]China Aero-Polytechnology Establishment, 7 Jingshun Road,Chaoyang District,Beijing,China

[a]Corresponding author: 370012109@qq.com

Abstract. The under actuated model of a smart space manipulator with flexible links and joints is established based on the Lagrange equation and assumed model method. Afterwards, using singular perturbation method, the system is decomposed into a slow subsystem and a fast subsystem, which describe the rigid-body motion and flexible vibration, respectively. Then, a modified computed torque controller based on the under actuated model is put forward to realize trajectory tracking of joints. In the meantime, the joint flexibility is compensated real-timely by the motors while the real-time suppression of the link vibration is realized by the embedded piezoelectric actuators. Finally, the simulation results indicate that the proposed composite control strategy can realize the trajectory tracking rapidly and in the same time suppress the vibration caused by flexible joints and links efficiently.

1. INTRODUCTION

In view of the practical need of aerospace industry, there is more and more attention focused on flexible space manipulator, which can be characterized by low consumption, light mass and so on [1-4]. On the other side, compared with the rigid manipulator system, the dynamic modelling and control of flexible manipulator system is much more difficult because of its structure flexibility. Besides, considering that the flexible manipulator works in weightlessness environment and there are no compositions and angle controllers of the base, linear momentum conservation and angular momentum conversation are introduced to the system. As a consequence, the dynamic behaviour of the system becomes more complex.

The structure flexibility of a flexible space manipulator consists of joint and link flexibility. However, even though a considerable amount of researches on the flexible manipulator have been done, a majority of them only take either the joint elasticity or the link flexibility into consideration [5-10]. Moreover, most of the researches, which are based on comprehensive consideration of the link and joint flexibility, are focused on the ground manipulator system while precious few of them are made for the free-floating manipulator [11,-14].

In most of the researches discussed above, it is just by the motors located at the root of links that the vibration of a flexible manipulator is suppressed. Besides, without considering the joint flexibility, some researchers suggest that piezoelectric actuators can be adopted to compensate the link vibration. Therefore, with the comprehensive consideration of joint and link flexibility, it's reasonable to compensate the joint vibration and the link flexibility by motors and embedded piezoelectric actuators, respectively. But unfortunately, this type of research is rare for the present.

We aim to realize the trajectory tracking rapidly and suppress the vibration of the flexible space manipulator effectively by the combination of motors and embedded piezoelectric actuators. To begin

Content from this work may be used under the terms of the Creative Commons Attribution 3.0 licence. Any further distribution of this work must maintain attribution to the author(s) and the title of the work, journal citation and DOI.

Published under licence by IOP Publishing Ltd

with, the under actuated model of a smart space manipulator with n flexible links and n flexible joints is established. Afterwards, with the using of singular perturbation method [5], the full system can be decomposed into a slow subsystem and a fast subsystem, which describe the rigid-body motion and the flexible vibration, respectively. Then, a modified computed torque controller based on the under actuated model is put forward to realize trajectory tracking of joints. In the meantime, the joint flexibility is compensated real-timely by the motors while the real-time suppression of the link vibration is realized by the embedded piezoelectric actuators. The simulations results demonstrate the validity of the composite control strategy proposed in this paper.

2. Dynamic Modelling

Without loss of generality, a smart free-floating manipulator with n flexible links and n elastic joints was taken into consideration. The system consists of a rigid base, flexible links, flexible joints and an inertial payload of mass Mp and inertia Ip. Its structure is shown in Figure 1.

The following coordinate frames are established: the inertial frame which is located at the centroid of the manipulator system (X_d, Y_d), the moving frame associated to the base (X_0, Y_0), the moving frame associated to link i (X_i, Y_i).

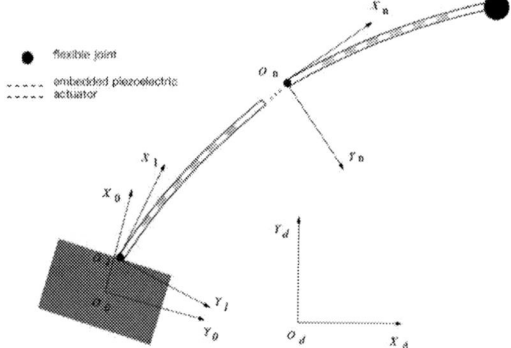

Figure 1. Smart flexible pace manipulator

2.1 Simplified model of an elastic joint

According to the research made by Spon[8], on the assumption of small deformation, the elastic flexible joint can be dynamic simplified as a liner torsion spring which works as a connector between the rotor and flexible link. As is shown in Figure 2, the ith rotor angular position is a_i while the ith link angular position is θ_i. The ith rotor where an input torque τ_i is applied has mass m_i and moment of inertia J_i . k_i is the spring constant of the ith flexible joint.

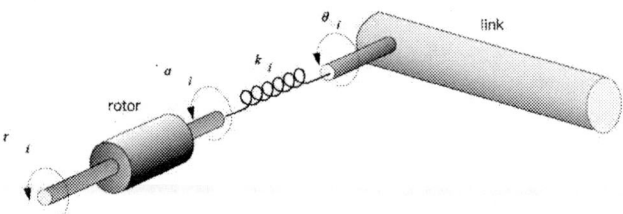

Figure 2. Schematic of flexible joint

2.2 Simplified model of a flexible link

To develop a dynamic model of the flexible manipulator, the following assumptions are made:

1) Each link is long and slender and can be regarded as an Euler-Bernoulli beam.
2) The deflection of each link is much smaller than the rigid displacement.

The assumed mode method is used to describe the deflection $u_i(x,t)$ at a point x on the ith link:

$$u_i(x,t) = \sum_{s=1}^{m} q_s^i(t)\varphi_s^i(x) \qquad (1)$$

where $\varphi_s^i(x)$ and $q_s^i(t)$, respectively, are the assumed mode shape function and mode displacement for the sth natural mode of the ith link and the first m modes are used to described the deflection. In view of the working status of the space manipulator, the mode shape function for the specific beam boundary which takes the mass and the inertia located at the tip into consideration is applied[18, 19].

2.3 Simplified model of a flexible link
It's obvious that the stress distribution will be different when the piezoelectric material imbedded in the matrix material. First of all, as the matrix material and the piezoelectric material are combined closely, the strain of them should be consistent. Besides, the piezoelectric material will deform when the driving voltage is applied, due to its attribute to inverse piezoelectric effect. Therefore, the stress in the piezoelectric material which is a part of the system, on the one hand, is produced by the external load. On the other hand, it is caused by the constraint force of the matrix material. The stress distribution of the embedded piezoelectric material is shown as Figure 3.

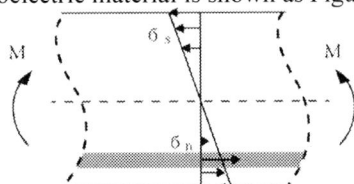

Figure 3. The stress distribution of the embedded piezoelectric material

As the mass and size of the embedded piezoelectric actuator is much smaller than that of the flexible manipulator, the embedded piezoelectric actuator works as concentrated torque ME applied on the flexible link.

Considering the inverse piezoelectric effect together with the contribution from the external load, the piezoelectric equation can be expressed as

$$\varepsilon = d_{31}\frac{U}{2t} + \frac{\sigma_p}{E_p} \qquad (2)$$

where 2t, E_p and d_{31} denote the piezoelectric actuator's height, elastic modulus and piezoelectric constant, respectively. σ_p and ε are the stress and strain of the piezoelectric actuator. U is the infliction voltage on the piezoelectric actuator. As the stress of the matrix material σ_s and σ_p are action and reaction, σ_s can be given as

$$\sigma_s = -\sigma_p = E_p(d_{31}U - \varepsilon) \qquad (3)$$

So, the concentrated torque *ME* can be written as

$$ME = \int_{d-t}^{d+t} \sigma_s z\,dz = \int_{d-t}^{d+t} E_p(d_{31}U - \varepsilon)z\,dz \qquad (4)$$

where z is the distance from the centreline of the link to the stress point. According to the bending theory of beam in mechanics of material, ε can be expressed by *ME* and other parameters. Then, combined with Eq.(4), *ME* is written as

$$ME = \frac{-3bdd_{31}E_s IE_p U}{3E_s I + 6bE_p td^2 + 2bE_p t^3} = K_u U \qquad (5)$$

where b, d respectively, denote the piezoelectric actuator's width and the distance from the neutral surface of the piezoelectric actuator to the centreline of the link. E_S and I are the elastic modulus and inertia moment of the link, respectively. K_U is the constant describing the relationship between the voltage and the moment.

The virtual work applied by these actuators is given as follows:

$$\delta W_i = \sum_{j=1}^{m} \int_{b_j}^{b_j+s_j} ME_j^i \delta \frac{\partial u_i(x,t)}{\partial x} dx = \sum_{j=1}^{m} \int_{b_j}^{b_j+s_j} \sum_{v=1}^{m} ME_j^i \frac{d\varphi_v^i(x)}{dx} \delta q_v^i dx \quad (6)$$

where m is the number of pairs of the piezoelectric actuators, s_j and b_j are the length of the actuator and the minimum distance from the actuator to the fixed point of the link, respectively.

2.4 Dynamic equations of motion

As is shown in Figure 4, the only difference between the frame (X_{i-1}, Y_{i-1}) and the frame (X_i'', Y_i'') is the positions of the original points. Owing to the rigid-body motion of the ith joint and the deflection of the (i-1)th link, (X_i'', Y_i'') is transferred to the (X_i, Y_i).

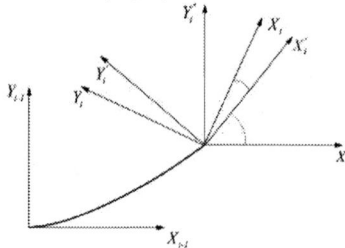

Figure 4. Schematic of coordinate transformation

The rigid transformation matrix, A_i, from (X_i'', Y_i'') to (X_i', Y_i') is expressed as:

$$A_i = \begin{bmatrix} \cos(\theta_i) & -\sin(\theta_i) \\ \sin(\theta_i) & \cos(\theta_i) \end{bmatrix} \quad (7)$$

On the assumption that the deflection of each link is small, the elastic transformation matrix, E_i, from (X_i', Y_i') to (X_i, Y_i) can be given as follows:

$$E_i = \begin{bmatrix} 1 & -\frac{\partial u(x,t)}{\partial x}\big|_{x=l} \\ \frac{\partial u(x,t)}{\partial x}\big|_{x=l} & 1 \end{bmatrix}, E_0 = I_{2\times 2} \quad (8)$$

Consequently, the global transformation matrix, S_i, from the inertial frame (X_d, Y_d) to the flexible coordinates (X_i, Y_i) associated to link i is obtained as:

$$S_i = E_{i-1}A_i \cdots E_0 A_1 A_0 \quad (9)$$

Let $r_i'(x)$ and $r_i(x)$, respectively, be the position vector of the arbitrary point on the ith link with respect to the frame (X_i, Y_i) and the frame (X_d, Y_d). Define the position vector of the ith joint $p_i(x)$. Then we can get the following equations:

$$r_i' = \begin{bmatrix} x \\ u_i(x,t) \end{bmatrix}, p_1 = r_0 + A_0 \times \begin{bmatrix} l_0 \\ 0 \end{bmatrix}, r_i(x) = S_i r_i' + p_i, p_{i+1} = r_i(l) \quad (10)$$

where r_0 is the position vector of the centroid of the base with respect to he frame (X_d, Y_d). Based on the previous discussion, the kinematics of any point along the link is fully characterized.

As the system is subject to the conservation of linear momentum, the centroid is fixed where the origin of the frame (X_d, Y_d) is located. According to the definition of the mass centre, we can get r_0 from the following equation:

$$m_0 r_0 + \sum_{i=1}^{n} \rho \int_0^{l_i} r_i(x)dx + m_p r_p = 0 \quad (11)$$

The total kinetic energy is obtained by the sum of the following contributions:

$$T = T_0 + T_l + T_\alpha + T_p \quad (12)$$

where T_0, T_l, T_α and T_p are the kinetic energy associated to the base, links, joints and payload respectively, which can be written as:

$$T_0 = \frac{1}{2}m_0\dot{r}_0^2 + \frac{1}{2}J_0\theta_0^2, \quad T_l = \sum_{i=1}^{n}\frac{1}{2}\int_0^{l_i}\rho\dot{r}_i'\dot{r}_i\,dx,$$

$$T_\alpha = \sum_{j=1}^{n}\frac{1}{2}(J_j\alpha_i^2 + m_j\dot{r}_j'(0)\dot{r}_j(0)), \qquad (13)$$

$$T_p = \frac{1}{2}m_p\dot{r}_n'(1)\dot{r}_n(1) + \frac{1}{2}J_p(\sum_{i=0}^{n}\dot{\theta}_i^2 + \sum_{i=1}^{n}\dot{u}_i'(l_i,t)\dot{u}_i(l_i,t))$$

In absence of gravity, the total potential energy consists of the following parts:

$$V = V_\alpha + V_l \qquad (14)$$

where V_α and V_l, respectively, are the potential energy of the flexible joints and links, which can be expressed as:

$$V_\alpha = \sum_{j=1}^{n}\frac{1}{2}k_j(\alpha_j - \theta_j)^2, \quad V_l = \sum_{j=1}^{n}\frac{1}{2}\int_0^{l_i}EI_i(\frac{\partial^2 u_i(x,t)}{\partial x^2})^2\,dx \quad (15)$$

where EI_i denotes the flexural rigidity of the ith link.

To avoid the situation that the free-floating manipulator becomes a nonholonomic system because of the introduction of the conservation of angular momentum, the attitude of base is supposed to be controlled by torque. Here, the generalized co-ordinate vector consist of the base angular position θ_0, rotor angular position vector $\alpha = [\alpha_1, \alpha_2, \ldots, \alpha_n]^T$, link angular position vector $\theta = [\theta_1, \theta_2, \ldots, \theta_n]^T$, and modal displacement vector $q = [q_1^1, \ldots, q_m^1, \ldots q_1^n, \ldots, q_m^n]^T$. Using the Lagrange equation of second kind and making the torque applied to the base be zero, the dynamic equations of the holonomic system can be obtained as:

$$J\ddot{\alpha} + K_\sigma(\alpha - \theta) = \tau_{n\times 1} \qquad (16)$$

$$M(\theta,q)\begin{bmatrix}\ddot{\theta}_0 \\ \ddot{\theta} \\ \ddot{q}\end{bmatrix} + F(\theta,\dot{\theta},q,\dot{q}) + \begin{bmatrix}0 \\ -K_\sigma(\alpha - \theta) \\ K_q q\end{bmatrix} = \begin{bmatrix}0_{(n+1)\times 1} \\ \tau'_{mn\times 1}\end{bmatrix} \quad (17)$$

where Eq. (16) and (17) are the dynamic equations of the flexible joins and flexible links manipulator, respectively. The generalised force vector is $\Gamma = [0, \tau_{n\times 1}, \tau'_{mn\times 1}]$, where $\tau'_{mn\times 1}$ is the vector of torque applied to the rotors and $\tau'_{mn\times 1}$ is the vector of generalised force applied to the links by the piezoelectric actuators, which are given as follows:

$$\tau'_{mn\times 1} = [\sum_{j=1}^{m}ME_j^1(\frac{d\varphi_1^1(b_{1j}+s_{1j})}{dx}\varphi_1^1(b_{1j}+s_{1j}) - \frac{d\varphi_1^1(b_{1j})}{dx}), \ldots, \qquad (18)$$

$$\sum_{j=1}^{m}ME_j^1(\frac{d\varphi_m^1(b_{1j}+s_{1j})}{dx} - \frac{d\varphi_m^1(b_{1j})}{dx}), \ldots, \sum_{j=1}^{m}ME_j^n(\frac{d\varphi_1^n(b_{1j}+s_{1j})}{dx} - \frac{d\varphi_1^n(b_{1j})}{dx}),.$$

$$\ldots, \sum_{j=1}^{m}ME_j^n(\frac{d\varphi_m^n(b_{1j}+s_{1j})}{dx} - \frac{d\varphi_m^n(b_{1j})}{dx})]^T = P_{nm\times nm} \times ME_{nm\times 1}$$

Considering Eq. (16) and (17), $M(\theta,q)$ is the inertia matrix, J is a diagonal matrix which consists of inertia moment of joints, $F(\theta,\dot{\theta},q,\dot{q})$ is the vectors reflecting Corioils and centrifugal forces, K_σ is the torsion rigidity matrix of the flexible joins, K_q is the stiffness matrix due to the distributed flexibility links. J, K_σ and K_q are given as:

$$J = \begin{bmatrix} J_1 & & & \\ & J_2 & & \\ & & \ddots & \\ & & & J_n \end{bmatrix}, K_\sigma = \begin{bmatrix} k_1 & & & \\ & k_2 & & \\ & & \ddots & \\ & & & k_n \end{bmatrix},$$

$$K_q = \begin{bmatrix} \int_0^{l_1} EI(\frac{d^2\Phi^1}{dx^2})^T \times (\frac{d^2\Phi^1}{dx^2})dx & & \\ & \ddots & \\ & & \int_0^{l_i} EI(\frac{d^2\Phi^i}{dx^2})^T \times (\frac{d^2\Phi^i}{dx^2})dx \end{bmatrix} \quad (19)$$

where $\Phi^i = \left[\varphi_1^i, \ldots, \varphi_m^i \right]^T$.

3. A SINGULAR PERTURBATION MODEL

Considering the fact that the rigid-body motion and the flexible vibration take place in two time scales, the flexible manipulator system can be decomposed into a slow subsystem and a fast subsystem, which describe the rigid-body motion and the flexible vibration respectively, based on singular perturbation method. Then, on the basis of the singular perturbation model, the control laws of trajectory tracking and vibration suppression can be designed independently so that the robustness of the control system is improved.

Since the inertia matrix $M(\theta,q)$ is positive definite, it can be inverted and denoted by D. M and D can be partitioned and expressed as follows:

$$\begin{aligned} M^{-1} &= \begin{bmatrix} M_{11}(\theta,\dot{\theta},q,\dot{q}) & M_{12}(\theta,\dot{\theta},q,\dot{q}) \\ M_{21}(\theta,\dot{\theta},q,\dot{q}) & M_{22}(\theta,\dot{\theta},q,\dot{q}) \end{bmatrix}^{-1} \\ &= D = \begin{bmatrix} D_{11}(\theta,\dot{\theta},q,\dot{q}) & D_{12}(\theta,\dot{\theta},q,\dot{q}) \\ D_{21}(\theta,\dot{\theta},q,\dot{q}) & D_{22}(\theta,\dot{\theta},q,\dot{q}) \end{bmatrix} \end{aligned} \quad (20)$$

According to the theory of matrix inversion, it may be noted that:

$$M_{11}(\theta,\dot{\theta},q,\dot{q}) = (D_{11}(\theta,\dot{\theta},q,\dot{q}) - D_{12}(\theta,\dot{\theta},q,\dot{q})D_{22}^{-1}(\theta,\dot{\theta},q,\dot{q})D_{21}(\theta,\dot{\theta},q,\dot{q}))^{-1} \quad (21)$$

Substituting $F(\theta,\dot{\theta},q,\dot{q})=[f_1(\theta,\dot{\theta},q,\dot{q}), f_2(\theta,\dot{\theta},q,\dot{q})]^T$ and $\sigma = \alpha - \theta$ into Eq.(16) and (17) gives:

$$J\ddot{\alpha} + K_\sigma\sigma = \tau_{n\times 1}, \quad (22)$$

$$\begin{bmatrix} M_{11}(\theta,\dot{\theta},q,\dot{q}) & M_{12}(\theta,\dot{\theta},q,\dot{q}) \\ M_{21}(\theta,\dot{\theta},q,\dot{q}) & M_{22}(\theta,\dot{\theta},q,\dot{q}) \end{bmatrix} \begin{bmatrix} \ddot{\theta}_0 \\ \ddot{\theta} \\ \ddot{q} \end{bmatrix} + \begin{bmatrix} f_1(\theta,\dot{\theta},q,\dot{q}) \\ f_2(\theta,\dot{\theta},q,\dot{q}) \end{bmatrix}$$

$$+ \begin{bmatrix} 0 \\ -K_\sigma\sigma \\ K_q q \end{bmatrix} = \begin{bmatrix} \mathbf{0}_{(n+1)\times 1} \\ \tau'_{2n\times 1} \end{bmatrix} \quad (23)$$

Hence, $\ddot{\sigma}, \ddot{\theta}, \ddot{q}$ can be determined as follows：

$$\ddot{\sigma} = \ddot{\alpha} - \ddot{\theta} = -J^{-1}K_\sigma\sigma + J^{-1}\tau - \ddot{\theta}, \quad (24)$$

$$\begin{aligned} \ddot{\theta} = &-D_{11}(\theta,\dot{\theta},q,\dot{q})f_1(\theta,\dot{\theta},q,\dot{q}) - D_{12}(\theta,\dot{\theta},q,\dot{q})f_2(\theta,\dot{\theta},q,\dot{q}) \\ &-D_{12}(\theta,\dot{\theta},q,\dot{q})K_q q + D_{11}(\theta,\dot{\theta},q,\dot{q})\begin{bmatrix} 0 \\ K_\sigma\sigma \end{bmatrix} + D_{12}(\theta,\dot{\theta},q,\dot{q})\tau', \end{aligned} \quad (25)$$

$$\begin{aligned} \ddot{q} = &-D_{21}(\theta,\dot{\theta},q,\dot{q})f_1(\theta,\dot{\theta},q,\dot{q}) - D_{22}(\theta,\dot{\theta},q,\dot{q})f_2(\theta,\dot{\theta},q,\dot{q}) \\ &-D_{22}(\theta,\dot{\theta},q,\dot{q})K_q q + D_{21}(\theta,\dot{\theta},q,\dot{q})\begin{bmatrix} 0 \\ K_\sigma\sigma \end{bmatrix} + D_{22}(\theta,\dot{\theta},q,\dot{q})\tau' \end{aligned} \quad (26)$$

To obtain the standard form of the singular perturbation equations, a common scale ε is defined as follows:

$$\varepsilon = 1 / \min(k_1, \cdots, k_i, k_{11}, \cdots k_{1m}, \cdots, k_{m1}, \cdots, k_{mm}) \qquad (27)$$

With this common scale, $\mathbf{K}_\sigma$ and $\mathbf{K}_q$ can be scaled by ε such that $\bar{\mathbf{K}}_\sigma = \varepsilon \mathbf{K}_\sigma, \bar{\mathbf{K}}_q = \varepsilon \mathbf{K}_q$. Then, define $z_\sigma = \boldsymbol{\sigma} / \varepsilon, z_q = q / \varepsilon$. Substituting $\boldsymbol{\sigma} = \varepsilon z_\sigma, q = \varepsilon z_q$ into Eq. (24), (25) and (26) gives:

$$\varepsilon^2 \ddot{z}_\sigma = \ddot{\boldsymbol{\alpha}} - \ddot{\boldsymbol{\theta}} = -\mathbf{J}^{-1} \mathbf{K}_\sigma z_\sigma + \mathbf{J}^{-1} \boldsymbol{\tau} - \ddot{\boldsymbol{\theta}}, \qquad (28)$$

$$\ddot{\boldsymbol{\theta}} = -D_{11}(\theta, \dot{\theta}, \varepsilon^2 z_q, \varepsilon^2 \dot{z}_q) f_1(\theta, \dot{\theta}, \varepsilon^2 z_q, \varepsilon^2 \dot{z}_q) - D_{12}(\theta, \dot{\theta}, \varepsilon^2 z_q, \varepsilon^2 \dot{z}_q) f_2(\theta, \dot{\theta}, \varepsilon^2 z_q, \varepsilon^2 \dot{z}_q) \qquad (29)$$

$$-D_{12}(\theta, \dot{\theta}, \varepsilon^2 z_q, \varepsilon^2 \dot{z}_q) \bar{\mathbf{K}}_q z_q + D_{11}(\theta, \dot{\theta}, \varepsilon^2 z_q, \varepsilon^2 \dot{z}_q) \begin{bmatrix} 0 \\ \bar{\mathbf{K}}_\sigma z_\sigma \end{bmatrix} + D_{12}(\theta, \dot{\theta}, \varepsilon^2 z_q, \varepsilon^2 \dot{z}_q) \boldsymbol{\tau}'$$

$$\varepsilon^2 \ddot{z}_q = -D_{21}(\theta, \dot{\theta}, \varepsilon^2 z_q, \varepsilon^2 \dot{z}_q) f_1(\theta, \dot{\theta}, \varepsilon^2 z_q, \varepsilon^2 \dot{z}_q) - D_{22}(\theta, \dot{\theta}, \varepsilon^2 z_q, \varepsilon^2 \dot{z}_q) f_2(\theta, \dot{\theta}, \varepsilon^2 z_q, \varepsilon^2 \dot{z}_q) \qquad (30)$$

$$-D_{22}(\theta, \dot{\theta}, \varepsilon^2 z_q, \varepsilon^2 \dot{z}_q) \bar{\mathbf{K}}_q z_q + D_{21}(\theta, \dot{\theta}, \varepsilon^2 z_q, \varepsilon^2 \dot{z}_q) \begin{bmatrix} 0 \\ \bar{\mathbf{K}}_\sigma z_\sigma \end{bmatrix} + D_{22}(\theta, \dot{\theta}, \varepsilon^2 z_q, \varepsilon^2 \dot{z}_q) \boldsymbol{\tau}'$$

For the sake of getting the slow subsystem, setting $\varepsilon = 0$ and solving for z_σ, z_q yields:

$$(31)$$

$$\bar{z}_\sigma = \bar{\mathbf{K}}_\sigma^{-1}(\boldsymbol{\tau}_s - \mathbf{J} * \ddot{\boldsymbol{\theta}})$$

$$\bar{z}_q = \bar{\mathbf{K}}_q^{-1} \bar{D}_{22}^{-1}(\theta, \dot{\theta}, 0, 0)(-\bar{D}_{21}(\theta, \dot{\theta}, 0, 0) f_1(\theta, \dot{\theta}, 0, 0)$$

$$-\bar{D}_{22}(\theta, \dot{\theta}, 0, 0) f_2(\theta, \dot{\theta}, 0, 0) + \bar{D}_{21}(\theta, \dot{\theta}, 0, 0) \begin{bmatrix} 0 \\ \boldsymbol{\tau}_s - \mathbf{J} * \ddot{\boldsymbol{\theta}} \end{bmatrix})$$

where over bars are used to indicate that the system with $\varepsilon = 0$ is considered and $\boldsymbol{\tau}_s$ denotes slow torques applied by motors. With substitution of Eq. (31) and $\varepsilon = 0$ into Eq. (29) and using of Eq. (21), the equations of the slow subsystem can be written as:

$$(\bar{\boldsymbol{M}}_{11}(\boldsymbol{\theta}, \dot{\boldsymbol{\theta}}, 0, 0) + \begin{bmatrix} 0 & 0_{1*n} \\ 0_{n*1} & \mathbf{J}_{n*n} \end{bmatrix}) \begin{bmatrix} \ddot{\boldsymbol{\theta}}_0 \\ \ddot{\boldsymbol{\theta}} \end{bmatrix} + \bar{\boldsymbol{f}}_1(\boldsymbol{\theta}, \dot{\boldsymbol{\theta}}, 0, 0) = \begin{bmatrix} 0 \\ \boldsymbol{\tau}_s \end{bmatrix} \qquad (32)$$

It's obvious that the slow subsystem is under actuated.

According to the singular perturbation theory, a fast time scale γ and boundary layer corrections are introduced, which are defined as follows:

$$\gamma = t / \sqrt{\varepsilon}, \eta_1 = z_\sigma - \bar{z}_\sigma, \eta_2 = \varepsilon \dot{z}_\sigma, \beta_1 = z_q - \bar{z}_q, \beta_2 = \varepsilon \dot{z}_q \qquad (33)$$

Then, at the boundary layer ($\varepsilon = 0$), substituting these corrections to the variables z_σ, $\dot{z}_\sigma$, z_q, $\dot{z}_q$ in Eq.(28), (29) and (30), the fast subsystem can be determined as follows:

$$\frac{d\boldsymbol{\eta}_1}{d\gamma} = \boldsymbol{\eta}_2,$$

$$\frac{d\boldsymbol{\eta}_2}{d\gamma} = -\mathbf{J}^{-1} \bar{\mathbf{K}}_\sigma \boldsymbol{\eta}_1 + \mathbf{J}^{-1} \boldsymbol{\tau}_f,$$

$$\frac{d\boldsymbol{\beta}_1}{d\gamma} = \boldsymbol{\beta}_2, \qquad (34)$$

$$\frac{d\boldsymbol{\beta}_2}{d\gamma} = -\bar{D}_{22}(\boldsymbol{\theta}, \dot{\boldsymbol{\theta}}, 0, 0) \bar{\mathbf{K}}_q \boldsymbol{\beta}_1 + \bar{D}_{21}(\boldsymbol{\theta}, \dot{\boldsymbol{\theta}}, 0, 0) \begin{bmatrix} 0 \\ \bar{\mathbf{K}}_\sigma \boldsymbol{\eta}_1 \end{bmatrix} + \bar{D}_{22}(\boldsymbol{\theta}, \dot{\boldsymbol{\theta}}, 0, 0) \boldsymbol{\tau}'$$

where τ_f denotes slow torques applied by motors. To facilitate designing the control law of the fast system, the fast subsystem described by the Eq. (34) is converted into state-space form, which are given as

$$\dot{\boldsymbol{x}}_f = \boldsymbol{A}_f \boldsymbol{x}_f + \boldsymbol{B}_f \boldsymbol{u}_f \qquad (35)$$

where

$$x_f = \begin{bmatrix} \eta_1 \\ \beta_1 \\ \eta_2 \\ \beta_2 \end{bmatrix}, u_f = \begin{bmatrix} \tau_f \\ ME \end{bmatrix}, A_f = \begin{bmatrix} 0 & 0 & \mathbf{I} \\ -\mathbf{J}^{-1}\bar{\mathbf{K}}_\sigma & 0 & 0 \\ \bar{D}_{21_{l_2}}\bar{\mathbf{K}}_\sigma & \bar{D}_{22}(\boldsymbol{\theta},\dot{\boldsymbol{\theta}},0,0)\bar{\mathbf{K}}_q & 0 \end{bmatrix},$$

$$B_f = \begin{bmatrix} 0 & 0 \\ \mathbf{J}^{-1} & 0 \\ 0 & \bar{D}_{22}(\boldsymbol{\theta},\dot{\boldsymbol{\theta}},0,0)\mathbf{P} \end{bmatrix} \qquad (36)$$

herein $\bar{D}_{21_{l_2}}$ is the submatrix of the $\bar{D}_{21}(\boldsymbol{\theta},\dot{\boldsymbol{\theta}},0,0)$.

4. Design of composite control scheme

As is shown in Figure 5, a composite control strategy is adopted. For the slow subsystem, a modified computed torque controller based on the under actuated model is put forward to realize trajectory tracking of joints by the torques τ_s applied by motors; for the fast subsystem, the joint vibration is compensated by the torques τ_f applied by motors while the link vibration is compensated by the torques *ME* applied to the links by the piezoelectric actuators.

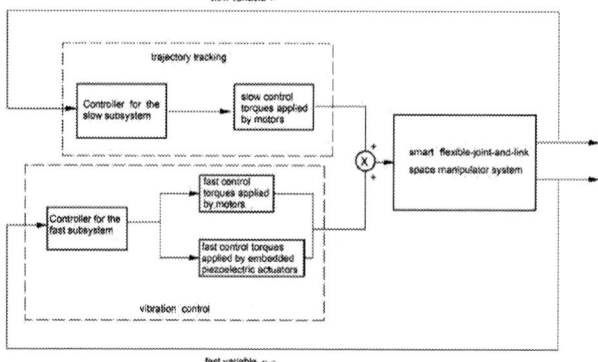

Figure 5. Composite control strategy

4.1 Controller for the slow subsystem

Since there are no items related to the flexibility of the system in Eq.(32), the slow subsystem is used to describe the rigid-body motion of the manipulator. When it comes to the slow controller, all the well-established controller developed for rigid manipulators can be applied. As is a typical controller for rigid manipulators, the computed torque control technique[20] has been widely applied in the ground manipulator systems. However, because of the absence of the torque applied to the base, the space manipulator is an under-actuated system so that the traditional computed torque control technique can't be used any more. Therefore, it's necessary to modify the traditional computed torque control technique and put forward a modified computed torque controller based on the under actuated model. The control system block diagram is shown as Figure 6.

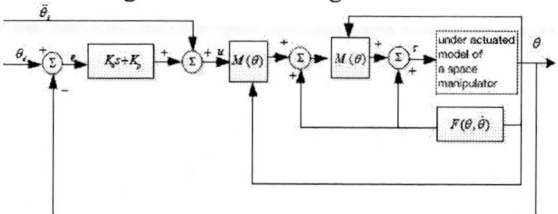

Figure 6. modified computed torque control scheme

With the introduction of the parametric variable δ, based on the Eq. (32), the slow control torque $\boldsymbol{\tau}_s$ is given as:

$$\begin{bmatrix} 0 \\ \boldsymbol{\tau}_s \end{bmatrix} = (\bar{\boldsymbol{M}}_{11} + \begin{bmatrix} 0 & 0_{1*n} \\ 0_{n*1} & \boldsymbol{J}_{n*n} \end{bmatrix}) \begin{bmatrix} \delta \\ \boldsymbol{u}_s \end{bmatrix} + \boldsymbol{f}_1(\theta,\dot{\theta},0,0) \qquad (37)$$

Then, since the inertia matrix $M(\theta,q)$ is invertible, the combined system of Eq. (37) and (32) reduces to:

$$\begin{bmatrix} \ddot{\theta}_0 \\ \ddot{\boldsymbol{\theta}} \end{bmatrix} = \begin{bmatrix} \delta \\ \boldsymbol{u}_s \end{bmatrix} \qquad (38)$$

herein u_s is the reference input which can be written as:

$$\boldsymbol{u}_s = -\mathbf{K}_p e - \mathbf{K}_v \dot{e} + \ddot{\boldsymbol{\theta}}_d \qquad (39)$$

where $\mathbf{K}_P$ and $\mathbf{K}_V$ are the diagonal position and velocity gain matrices of the controller, $\ddot{\boldsymbol{\theta}}_d$ is the desired angular acceleration, e is the tracking error obtained as:

$$e = \boldsymbol{\theta}_d - \boldsymbol{\theta} \qquad (40)$$

Substituting of Eq. (40) into Eq. (38) gives:

$$\ddot{\theta}_0 = \delta$$
$$\ddot{e} + K_p e + K_v \dot{e} = 0 \qquad (41)$$

It can be noticed that e is going to be zero by choosing appropriate $\mathbf{K}_P$ and $\mathbf{K}_V$.

Considering Eq. (37), solving for δ yields:

$$\delta = M_{11_{11}}^{-1}(-M_{11_{12}} \boldsymbol{u}_s - \boldsymbol{f}_1(\boldsymbol{\theta},\dot{\boldsymbol{\theta}},0,0)) \qquad (42)$$

where $M_{11_{11}} = M(1,1), M_{11_{12}} = M(1,2:n+1)$.

4.2 Controller for the fast subsystem

As the fast system (35) is completely controllable, the LQR approach is devised to force its state x_f to zero. By regarding the slow variable θ_0 and $\boldsymbol{\theta}$ as time-varying parameters of the fast subsystem, the fast control torque can be computed real-timely so that the vibration is compensated in real time.

Considering the control effect and input energy simultaneously, the cost function is given as follows:

$$\mathbf{J}_f = \int_0^\infty [x_f^T(t)\mathbf{Q}x_f(t) + \boldsymbol{u}_f^T(t)\mathbf{R}\boldsymbol{u}_f(t)]dt \qquad (43)$$

herein $\mathbf{Q}$ is a semi-positive weighting matrix of the states and $\mathbf{R}$ is a positive weighting matrix of the input. To minimize the JF, the optimal torque uf is given as:

$$\boldsymbol{u}_f = (\boldsymbol{R}^{-1}\boldsymbol{B}^T\boldsymbol{G})x_f(t) \qquad (44)$$

where $\mathbf{G}$ is the solution of the algebraic Riccati equation, which is expressed as:

$$\boldsymbol{A}^T\boldsymbol{G} + \boldsymbol{G}\boldsymbol{A} - \boldsymbol{G}\boldsymbol{B}\boldsymbol{R}^{-1}\boldsymbol{B}^T\boldsymbol{G} + \boldsymbol{Q} = \boldsymbol{0} \qquad (45)$$

Considering Eq. (43), uf consist of the fast control torques $\boldsymbol{\tau}_f$ applied by motors and torques ME applied by the piezoelectric actuators.

5. Simulation results and discussions

The availability of the composite control strategy proposed in this paper is tested by means of simulations for the smart space manipulator with dual flexible links and elastic joints. The parameters of the manipulator system are given in Table 1. Since the low order modes play more important roles in describing the vibration of links than the higher modes do, only the first mode is considered[21, 22] to simplify calculation. Two pairs of embedded piezoelectric actuators which are located at the root of each link respectively are adopted to compensate the vibration of the link[23]. The parameters of the embedded piezoelectric actuator are given as follows: length s=105mm, width b=15mm, height h=0.5mm and the constant describing the relationship between the voltage and the moment K_u=-3.5×10^{-4}N·m/V.

Table 1. Parameters of the manipulator

Parameter	Symbol	Link 1, i=1	Link 2, i=2
Length of links	l_i	2m	1m
Elastic modulus of links	E_i	72GPa	72GPa
Mass density of links	ρ_i	2700kg/m^3	2700kg/m^3
Cross section area of links	S_i	2*4mm^2	2*4mm^2
Stiffness constant	k_i	100Nm/rad	100Nm/rad
Rotor inertia	J_i	0.001 kg·m^2	0.001 kg·m^2
Payload mass	m_p	\	2kg
Payload inertia	J_p	\	0.005 kg·m^2

The simulation results are shown in Figure 7 to Figure 11. As the system abide by the law of momentum conversation, the base is rotated as shown in Figure 7. Figure 8 to Figure 11 shows the tracking effect of the joint 1 and joint 2 as well as the slow control torque applied by the motor 1 and motor 2.Even the link 2 has less moment of inertia, it needs a lager torque because of the influence applied by the link 1.Further, it can be obtained from the simulation results that the modified computed torque controller proposed in this paper can realize the trajectory tracking of the joint without applying a torque to the base rapidly.

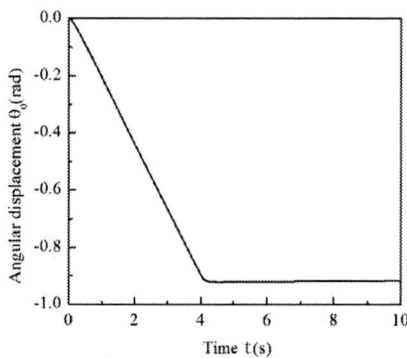

Figure 7. Angular displacement of the base

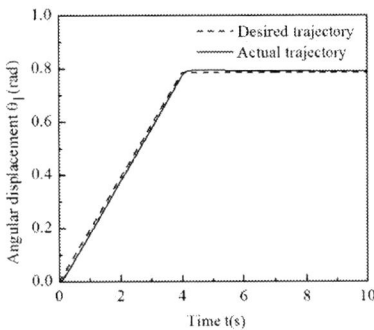

Figure 8. Comparison of joint 1 desired and actual trajectory

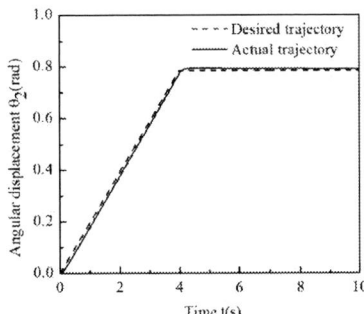

Figure 9. Slow control torque applied by motor 1

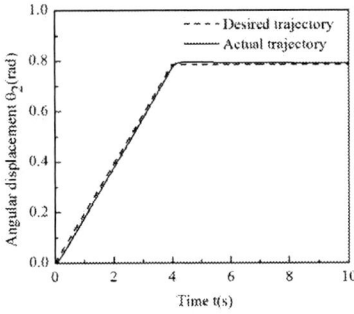

Figure 10. Comparison of joint 2 desired and actual trajectory

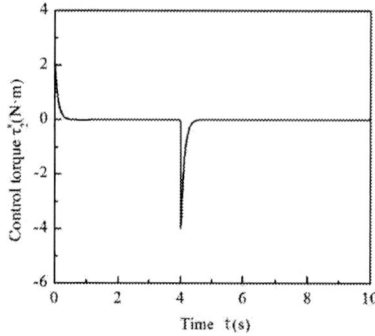

Figure 11. Slow control torque applied by motor 2

It can be seen from Figure 12 to Figure 15 that the joint deflection settle down fast by appropriate fast torques applied by the motors. Figure 16 and Figure 18 are the control effect for the first mode vibration of the links by the embedded piezoelectric actuators, which is controlled by the voltage showed in Figure 17 and Figure 19. The simulation results indicate that there is high frequency vibration of the joints and links during the motion of the manipulator and the residual vibration will last a long time. This vibration will lead to the tip deflection. Therefore, it's obvious that the vibration control must be introduced to the flexible manipulator system. Last but not least, the simulation results show that by the motors and embedded piezoelectric actuators, the real-time suppression of the joint and link are realized, respectively.

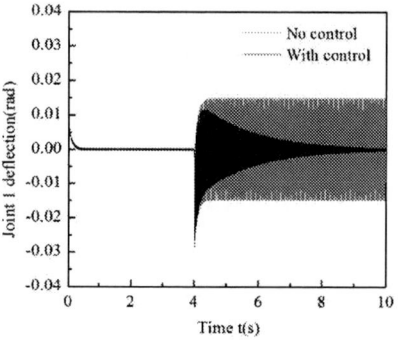

Figure 12. Comparison of joint 1 deflection

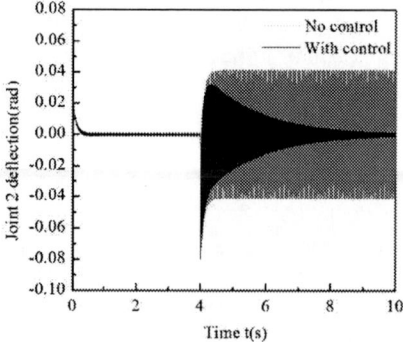

Figure 13. Comparison of joint 2 deflection

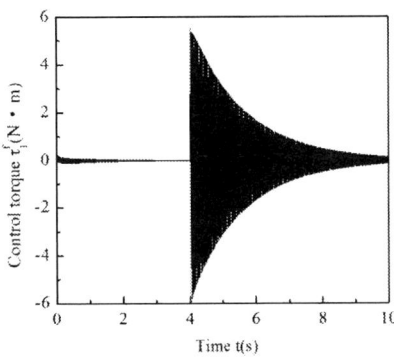

Figure 14. Fast control torque applied by motor 1

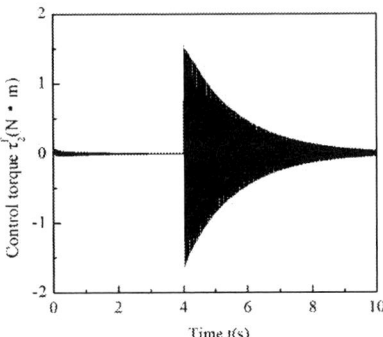

Figure 15. Fast control torque applied by motor 2

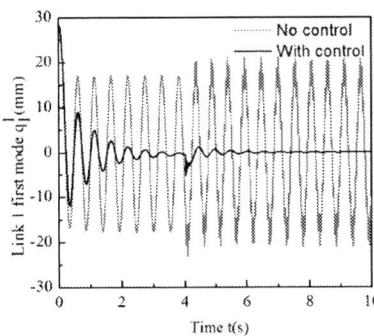

Figure 16. Link 1 first mode trajectorie

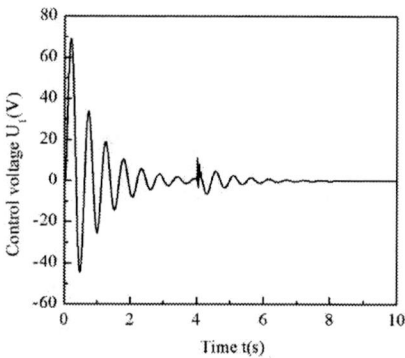

Figure 17. Control voltage applied on the embedded actuators in link 1

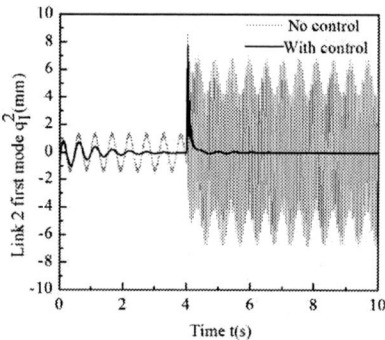

Figure 18. Link 2 first mode trajectories

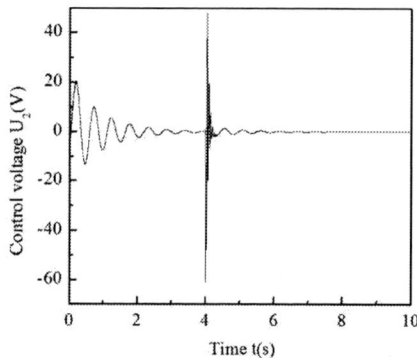

Figure 19. Control voltage applied on the embedded actuators in link 2

In summary, the results from the simulation results indicate that the proposed composite control strategy is able to provide excellent trajectory tracking and in the same time suppress the vibration of flexible joints and links by the motors and embedded actuators efficiently.

6. Conclusions

Using the Lagrange equation and assumed model method, the under actuated model of a smart space manipulator with n flexible links and n flexible joints is established. Based on singular perturbation method, the system can be decomposed into a slow subsystem and a fast subsystem, which describe the rigid-body motion and flexible vibration, respectively. Further, for the slow subsystem, a modified computed torque controller based on the under actuated model is put forward to realize trajectory tracking of joints by motors; for the fast subsystem, the joint flexibility is compensated real-timely by the motors while the real-time suppression of ink vibration is realized by the embedded piezoelectric actuators. The proposed composite control strategy can realize the trajectory tracking rapidly and in the same time suppress the vibration caused by flexible joints and links efficiently.

References

[1] Korayem M H, Rahimi H N, Nikoobin A. Mathematical modeling and trajectory planning of mobile manipulators with flexible links and joints[J]. Applied Mathematical Modelling, **36**(2012):3223-3238.

[2] Feliu V, Pereira E, Díaz I M. Passivity-based control of single-link flexible manipulators using a linear strain feedback[J]. Mechanism and Machine Theory, **71**(2014):191-208.

[3] Hong Z B. Research on Intelligent Control for Rigid and Flexible Space-Based Space Robot System[Dissertation]. Fuzhou University, **169**(2011).

[4] Wu L C. *Flexible-Link Manipulator Modeling, Analysis and Control*[M](Higher Education Press,Beijing, 2012)

[5] Carmelo D C, Arcangelo M. Exact modeling for control of flexible manipulators[J]. Journal of Vibration and Control, **18(10)** (2012),:1526-1551.

[6] Li Y C, Lu Y F. Robust Control for Trajectory Tracking of A Two-Link Flexible Manipulator[J]. Acta Automatica Sinica, **25(03)**(1999),:46-52.

[7] D X, Cai G P. Active Control for Two-Link Flexible Manipulator with Tip Mass[J]. Chinese Journal of Applied Mechanics, **26(4)** (2009),:672-678.

[8] Spong M W. Modeling and Control of Elastic Joint Robots[J]. Journal of Dynamic Systems, Measurement and Control, 1987,**109(4)**:310-318.

[9] Xie L M, Chen L. Motion adaptive sliding mode control and vibration active suppression of free-floating flexible-joint space robot[J]. Chinese Journal of Computational Mechanics, **05**(2013),:647-652.

[10] Salmasi H, Fotouhi R, Nikiforuk P N. A Biologically Inspired Controller for Trajectory Tracking of Flexible-Joint Manipulators[J]. International Journal of Robotics, **27(2)** (2012),:151-162.

[11] Karimzadeha G R V A. Impedance control of a two degree-of-freedom planar flexible link manipulator using singular perturbation theory[J]. Robotica, **24(2)** (2006),:221-228.

[12] Gogate S, Lin Y J. Modeling and control of a deformable-link flexible-joint robot[J]. International Journal of Systems Science, **7(2)** (1994),:237-251.

[13] Bian Y S, Lu Z. Method for Dynamic Modeling of Flexible Manipulators[J]. Journal of Beijing University of Aeronautics and Astronautics, **25(4)** (1999),:486-490.

[14] Liang J., CHen L. Neural network adaptive backstepping control and double flexible vibration active hierarchical suppression of space robot with flexible-joint and flexible-link[J]. Chinese Journal of Computational Mechanics, **04**(2014),:459-466.

[15] Masajedi P, Shirazi K H, Ghanbarzadeh A. Verification of Bee Algorithm Based Path Planning for a 6DOF Manipulator Using ADAMS[J]. Journal of Vibroengineering, **15(2)** (2013),:805-815.

[16] Sangveraphunsiri V, Chooprasird K. Dynamics and control of a 5DOF manipulator based on an H-4 parallel mechanism[J]. International Journal of Advanced Manufacturing Technology, **52(1)** (2011),:343-364.

[17] Liu Y, Li W, Wang Y, et al. Coupling vibration characteristics of a translating flexible robot manipulator with harmonic driving motions[J]. Journal of Vibroengineering, **17(7)** (2015),:3415-3427.

[18] Subudhi B A M S, Morris A S. Dynamic modeling, simulation and control of a manipulator with flexible links and joints[J]. Robotics and Autonomous Systems, **41(4)** (2002),:257-270.

[19] De Luca A, Siciliano B. Closed-form dynamic model of planar multilink lightweight robots[J].

Systems, Man and Cybernetics, IEEE Transactions on, **21(4)** (1991),:826-839.

[20] Huo W. *Robot Dynamics and Control*[M](Beijing: Higher Education Press, 2005)

[21] Lou J Q. Research on Integrated Control of Trajectory Tracking and Vibration Suppression of a Space Flexible Manipulator System Using Piezoelectric Actuators[Dissertation] [J]. Zhejiang University, **129**(2013).

[22] Liao Y H, Li D K, Tang G J. Dynamic Modeling and Simulation of a Free-floating Flexible Manipulator System[J]. Journal of National University of Defense Technology, **32(05)** (2010):29-33.

[23] Wu D. F., Liu A. C., Mai H. C. Study on active vibration control of piezoelectric intelligent flexible beam[J]. Journal of Beijing University of Aeronautics and Astronautics, **30(02)** (2004):160-163.

ISPECE

IOP Publishing

Effect of ultrasonic treatment on morphology and microwave absorption performance of ZnO spheres

Wei Huang, Shicheng Wei, Yi Liang, Bo Wang, Yuwei Huang, Yujiang Wang[a], Binshi Xu

National Key Laboratory for Remanufacturing, Academy of Army Armored Forces, Beijing 100072

[a]Corresponding author: hitwyj@126.com

Abstract. ZnO spheres were prepared by hydrothermal method, following calcination process and ultrasonic treatment. Effect of ultrasonic treatment on morphology and microwave absorption properties of ZnO spheres were studied by X-ray diffraction (XRD), scanning electron microscopy (SEM) and vector network analyzer (VNA). The results show that sea urchin-like ZnO spheres (with ZnO whiskers on the surface) can be obtained by ultrasonic treatment. The microwave absorption properties are improved as the number of ZnO whiskers increases.

1. Introduction

With the development of high technology such as radar detection, satellite communication, aerospace and electronic countermeasure, especially the rise of anti-electromagnetic interference, stealth technology and microwave anechoic chamber in recent years, more and more attention has been paid to the research of electromagnetic wave absorbing materials [1, 2]. With the rapid development of modern electronic industry and information industry, the number of electronic products that produce electromagnetic waves has increased dramatically, which leads to serious electromagnetic radiation and new environmental pollution. Therefore, the study of electromagnetic shielding technology and electromagnetic wave absorbing materials is of great significance in both national defense industry and civil industry [3, 4].

In recent decades, remarkable progress has been made in the synthesis of ZnO. The results show that the piezoelectric, photocatalytic, gas-sensitive and pressure-sensitive properties of ZnO with different morphology are obviously different [5]. In addition, ZnO has the ability to absorb electromagnetic wave, visible light and infrared ray, so it has the potential of multi-band stealth [6-8]. As a special kind of ZnO material, nano-ZnO whiskers can produce quantum size effect and tunneling effect, so they have excellent absorbing properties. There are two main methods for preparing ZnO whiskers [9, 10]:(1) Gasification of zinc powder at high temperature and phase oxidation in an oxygen atmosphere. (2) The zinc powder is mixed with carbon powder and heated in the atmosphere. The first method is the instantaneous oxidation of zinc. The high concentration of zinc oxide and the sharp increase in the number of nuclei results in the poor regularity and low yield of zinc oxide whiskers. The second method is to use the reducibility of carbon to consume oxygen in the surrounding air to meet the growth regulation of whiskers, but in the specific implementation process, a large number of $ZnCO_3$ will be produced to reduce the purity of products. In this paper, spherical ZnO was prepared by hydrothermal method and calcination process from zinc acetate and ammonium bicarbonate. The

Content from this work may be used under the terms of the Creative Commons Attribution 3.0 licence. Any further distribution of this work must maintain attribution to the author(s) and the title of the work, journal citation and DOI.
Published under licence by IOP Publishing Ltd

effects of subsequent physical ultrasound on the morphology and microwave absorption properties of surface whiskers were studied.

2. Experimental section

2.1 Preparation of ZnO sphere

A typical preparation of ZnO sphere was as follows: 0.05 mol $Zn(CH_3COO)_2$ and 0.15 mol NH_4HCO_3 were dissolved in 400 ml distilled water respectively, and stirred for 0.5 h. After all the solvents were totally dissolved, they were transferred to a 500 ml teflon-lined stainless-steel autocalve and heated at 100 °C for 10 h. The resultant products were washed with distilled water for several times, and dried at 60 °C for 4 h (The precursor was $ZnCO_3$). ZnO spheres were obtained by calcination of $ZnCO_3$ at 450 °C for 4 h with a heating rate of 10 °C/min. In order to investigate the effect of ultrasonic process, ZnO spheres were irradiated in an ultrasonic reactor for 20 min and 40 min, respectively. For convenience, without ultrasound treatment, 20 minutes with ultrasound treatment and 40 minutes with ultrasound treatment were named S1, S2 and S3, respectively.

2.2 Characterization and measurement

The structure and morphology of ZnO spheres were studied by employing X-ray diffractometer (XRD, Japan Rigaku D/MAX-cA) using a CuKa radiation (λ=1.5406 Å) and scanning electron microscopy (SEM, Hitachi H-800). The complex permittivity and complex permeability of samples in the frequency range 2-18 GHz were tested by a network analyzer (VNA, N5242A, Agilent) for simulation of reflection loss. The sample was mixed with paraffin at a ratio of 20%. All the tests were carried out at room temperature without special conditions.

To study the microwave absorption properties of ZnO spheres, the reflection loss (R_L) values was calculated using complex permittivity and complex permeability at a range of 2-18 GHz based on the transmit line theory, which was summarized as the following equations [11]:

$$R_L = 20 \lg \left| \frac{Z_{in} - Z_0}{Z_{in} + Z_0} \right| \qquad (1)$$

$$Z_{in} = Z_0 \sqrt{\frac{\mu_r}{\varepsilon_r}} \tanh[\ j \frac{2\pi df}{c} \sqrt{\mu_r \varepsilon_r}\ ] \qquad (2)$$

Among them, ε_r and μ_r are complex permittivity and complex permeability, c is the speed of light in vacuum, f is microwave frequency, d is the thickness of absorber, Z_{in} is the input impedance of absorbing material, and Z_0 is the impedance of free space.

3. Results and discussion

3.1. XRD analysis

The crystallographic structure and phase composition of ZnO were identified by the XRD. As shown in Fig.1, All the diffraction peaks of samples were measured to be 2θ = 31.8 °, 34.6 °, 36.6 °, 47.9 °, 57.1 °, 63.2 °, 68.2 ° and corresponded to (110), (002), (101), (102), (110), (103), (112) planes respectively, which can be readily indexed as the hexagonal wurtzite phase with space group P63mc and lattice parameters of a = 3.25 Å and c = 5.21 Å according to JCPDS card no.36-1451. No impurity peaks were observed, indicating that the purity of the synthetic ZnO was very high. It is noteworthy that the intensity of diffraction peaks of ZnO decreases with the prolongation of ultrasonic treatment time, which indicates that the crystallization degree of the surface of ZnO spheres decreases gradually.

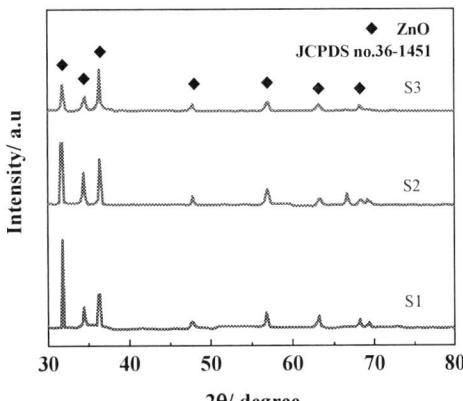

Figure 1. XRD patterns of different samples

3.2 Morphology analysis

In order to observe the surface morphology of ZnO spheres, SEM were carried out and the results were illustrated in Fig.2. It can be seen that the diameter of the prepared ZnO spheres is about 8-12um, without obvious agglomeration, and the surface is smooth. Subsequent ultrasonic treatment has a significant effect on the surface morphology. When the time of ultrasonic treatment was 20 minutes, the surface of ZnO sphere began to dissociate and the debris increased. When the time of ultrasonic treatment is 40 minutes, all the debris on the surface of ZnO sphere is broken into whiskers. This structural evolution from sphere to urchin reduces the crystallization of ZnO surface, which is consistent with the previous XRD analysis.

In addition, a large number of studies have shown that the absorbing properties of materials are closely related to their structures [12, 13]. Compared with S1 and S2, S3 has the largest specific surface area, more defects and vacancies for the sea urchin-like structure, and can provide a large number of scattering sites, which may be conducive to electromagnetic wave absorption.

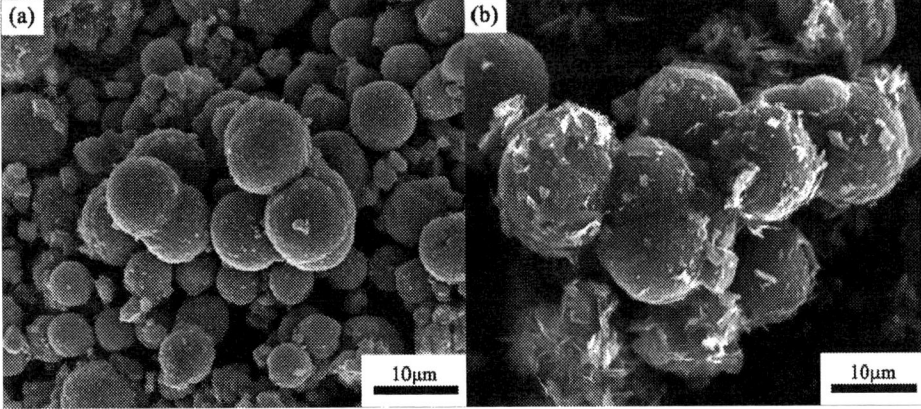

Figure 2. SEM images of different samples:
(a) S1 (b) S2 (c) S3

3.3. Microwave absorption properties
As a microwave absorbing material, ZnO has very low magnetic loss and belongs to dielectric loss material. Therefore, the dielectric properties of ZnO are mainly analyzed in this paper. It can be seen from Fig.3 (a) (b) that the complex permittivity of ZnO increases with the time of ultrasonic treatment, which indicates that its polarization and loss capacity increase gradually. The dielectric loss tangent (tan $\delta_E = \varepsilon''/\varepsilon'$) was calculated based on the data of EM parameters which can be observed in Fig.3(c). The dielectric loss of S1 and S2 varies with frequency in a similar way, with significant differences only at 3.2 GHz, 9.3 GHZ and 15.8 GHz. This is because S1 and S2 have similar morphology. After 40 minutes of ultrasonic treatment, the dielectric loss ability of ZnO has been improved significantly. This is due to the fact that the structure of sea urchin S3 has a large number of whiskers, and the charges can gather at the tip of the needle to form multiple polarization centers, which can cause scattered reflection of electromagnetic waves.

The phenomenon can also be explained by the Debye dipolar relaxation model [14]:

$$\varepsilon - \varepsilon_\infty = \frac{\varepsilon_S - \varepsilon_\infty}{[1 + (j\omega\tau_0)^{1-\alpha}]} \qquad (3)$$

In the formula, τ_0、α、ε_∞、ε_S are respectively the relaxation time, parameter variable, optical frequency dielectric constant, and static dielectric constant. The complex permittivity ε can be expressed by the following formula [15]:

$$\varepsilon = \varepsilon' - j\varepsilon'' = \frac{1}{j\omega C_0 Z} = \varepsilon_\infty + \frac{\varepsilon_S - \varepsilon_\infty}{1 + 2\pi f \tau} \qquad (4)$$

The complex permittivity real part ε' and imaginary part ε'' can be expressed as:

$$\varepsilon' = \varepsilon_\infty + \frac{\varepsilon_S - \varepsilon_\infty}{1 + (2\pi f)^2 \tau^2} \qquad (5)$$

$$\varepsilon'' = \frac{2\pi f \tau (\varepsilon - \varepsilon_\infty)}{1 + (2\pi f)^2 \tau^2} \qquad (6)$$

According to the real part (ε') and imaginary part (ε'') of complex permittivity, the Debye dipole relaxation equation can be expressed as:

$$(\frac{\varepsilon' - \varepsilon_\infty}{2})^2 + (\varepsilon'')^2 = (\varepsilon_S - \varepsilon_\infty)^2 \qquad (7)$$

According to this formula, We can make the Cole-Cole curve of S3 as shown in Fig.3(d). The frequency dependence of ε' and ε'' in S3 can be approximated as a large number of semicircles. Each semicircle corresponds to a Debye relaxation. The Debye semicircle can reflect the spectral characteristics of the dielectric polarization and the extent to which the dielectric deviates from the Debye relaxation. For these reasons, the strong dielectric loss ability of S3 can be determined as multiple relaxation.

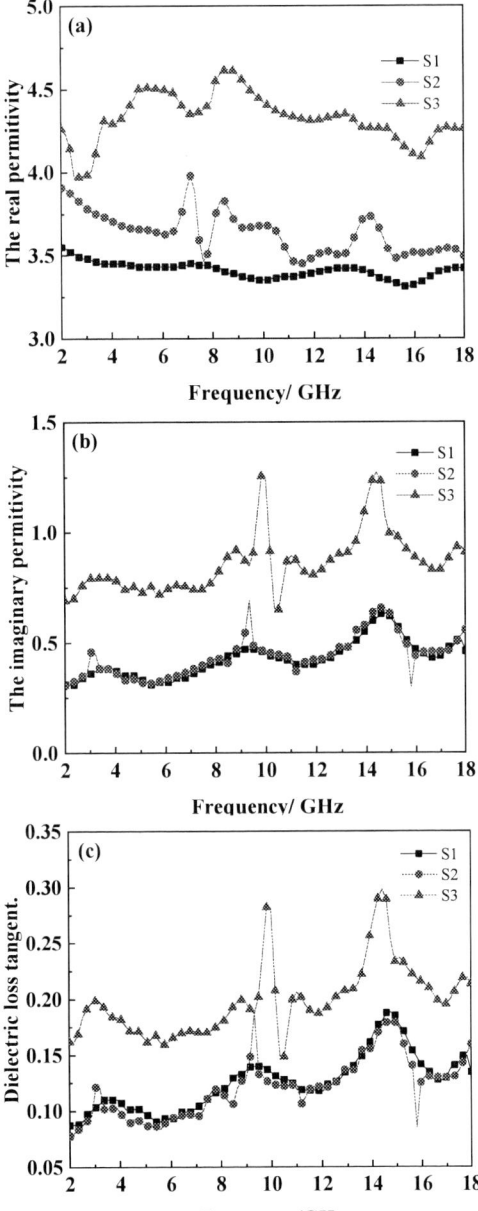

ISPECE   IOP Publishing

IOP Conf. Series: Journal of Physics: Conf. Series **1187** (2019) 032077   doi:10.1088/1742-6596/1187/3/032077

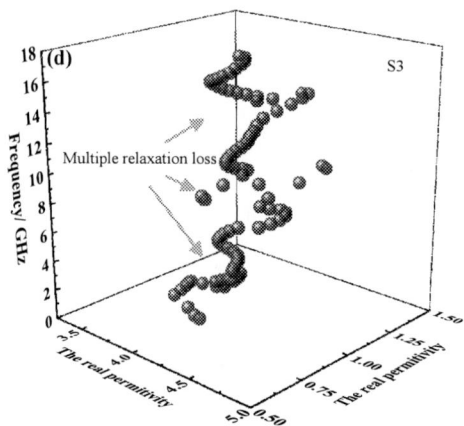

Figure 3. Electrical loss analysis for ZnO :(a) the real permittivity, (b) the imaginary permittivity, (c) dielectric loss tangent and (d) Cole-Cole curve of S3 in the range of 2-18 GHz

Formulas (1) and (2) can be used to calculate the reflective loss of a material at a specific thickness. The simulated reflection losses for the different samples of ZnO spheres in paraffin matrix with the coating thickness of 3.0 mm on a perfect conductor are displayed in Fig.4. As the ultrasonic time increases, the surface whiskers increase, and the absorbing properties of the ZnO spheres increase steadily. S3 has the best absorbing performance, the minimum reflection loss can reach -13.4 dB, and the effective bandwidth ($R_L < -10$dB, equivalent to absorbing 90 % of electromagnetic waves) can reach 1.8 GHz.

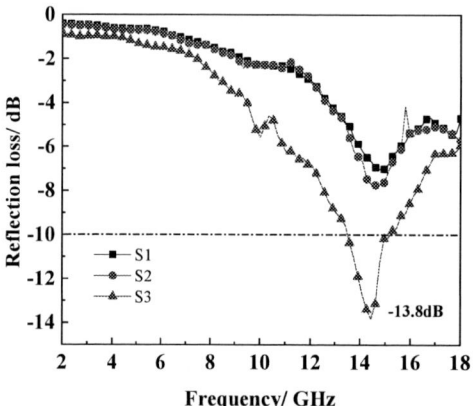

Figure 4. Reflection Loss Diagrams of Different Samples at 3 mm

4. Conclusion

Effect of ultrasonic treatment on morphology and microwave absorption performance of ZnO spheres have been investigated systemically in the present work. From the above results, conclusions are as following. With increases of ultrasonic time, the number of nano-ZnO whiskers on the surface also increases, leading to the microwave absorption properties are improved due to the multiple relaxation losses caused by a large number of whiskers. The work will be helpful for the future development of microwave absorption materials with high absorption properties by ultrasound treatment.

References
[1] L. Kong, X.W. Yin, F. Ye, Q. Li, L.T. Zhang, L.F. Cheng. J. Phys. Chem. C **117**, 2135 (2013)

[2] M.K. Han, X.W. Yin, L. Kong, M. Li, W.Y. Duan, L.T. Zhang, L.F. Cheng. J. Mater. Chem. A **2**, 16403 (2014)

[3] G.L. Wu, Y.H. Cheng, X. Qian, Z.R. Jia, X. Feng, H.J. Wu. Mater. Lett **144**, 157 (2015)

[4] L. Kong, X.W. Yin, Q. Li, F. Ye, Y. Liu, G.Y. Duo, X.W. Yuan, N. Alford. J. AM. CERAM. SOC **96**, 2211 (2013)

[5] E. S. Jang, J.H. Won , S.J. Hwang , J.H. Choy. Adv. Mater **18** , 3309 (2010)

[6] R.M. Mohamed, D. Mckinney, M.W. Kadi, I.A. Mkhalid. Ceram. Int **42**, 2299 (2016)

[7] S.J. Fang, W. Wang, X.L. Yu, H. Hong, Y. Zhong, X.F. Sui. Mater. Lett **143**, 120 (2015)

[8] X.L. Yu, L.P. Zhang, Y. Zhong, H. Xu, Z.P. Mao. A. M.M **692**, 337 (2014)

[9] Z.W. Zhou, L.S. Chu, S.C. Hu. Mat. Sci. Eng. B. Solid **126**, 93 (2006)

[10] Z.W. Zhou, J.J. Liu, S.C. Hu. J Cryst. Growth **276**, 317 (2005)

[11] M. Zhou, X. Zhang, J.M. Wei, S.L. Zhao, L. Wang, B.X. Feng. J. Phy. Chem. C **115,** 1398 (2011)

[12] J. Cao, W.Y. Fu, H.B. Yang, Q.J Yu, Y.Y. Zhang, S.M. Wang. H. Zhao, Y.M. Sui, X.M. Zhou, W.Y. Zhao, Y. Leng, H. Zhao, H. Chen, X.F. Qi. Mater. Sci. Eng. B **175**, 56 (2010)

[13] X. Xuan, H.Y. Zhang, G.X. Zeng, N.G. Bo, C.H. Chan. Mater. Sci. Forum **852**, 1055 (2016)

[14] K.S Cole. J. Chem. Phys **10**, 98 (1942)

[15] W. Feng, Y.M. Wang, J.C. Chen, L. Wang, L.X. Guo, J.H. Ouyang, D.C. Jia, Y. Zhou. Carbon **108**, 52 (2016)

ISPECE   IOP Publishing

IOP Conf. Series: Journal of Physics: Conf. Series **1187** (2019) 032078   doi:10.1088/1742-6596/1187/3/032078

Effect of different volume fraction magnetorheological fluids on its shear properties

Sun Huimin, Zhu Xuli, Liu Nannan, Mou Jiefeng, Li Liang, Li Shixu

School of Mechanical and Electronic Engineering, Shandong University of Science and Technology, Qingdao, 266590, China

Abstract. In order to study the relationship between shear stress and volume fraction and shear rate of MRFs, four kinds of MRFs with different volume ratios were prepared. The shear stress of MRFs at different shear rates was tested by rheometer. The shear stress expression of MRFs is obtained by fitting and analyzing the experimental data. The experimental results show that under the same volume fraction, the shear stress of MRF increases slowly, and the apparent viscosity decreases exponentially with the increase of shear rate.

1. Introduction

Magnetorheological Fluid (MRFs) are a new kind of intelligent material, which is a suspension formed by dispersing micron-sized ferromagnetic particles in liquid polymer[1]. MRFs can flow freely and show the characteristics of Newton fluids without the presence of magnetic field. The magnetorheological effect of MRFs occur in the presence of magnetic field with the ferromagnetic particles gather themselves along the field direction and the fluids turn to solid-like within a few milliseconds. MRFs have not only good abilities of easy control, fast response and reversibility[2,3], but also the advantages of simple preparation, low cost and wide temperature range for use, MRFs can be widely used in vibration absorb mechanism control and other fields[4-6].

The research on MRFs is mainly focus on properties of shearing. Bossis studied the shear stress of MRFs with different volume ratio under different magnetic field, and obtained that the shear stress is proportional to the volume fraction and the square of magnetic field strength[7]. By studying the particle coating process, Liu found that the shear stress and apparent viscosity of the coated MRF increased by half under the same magnetic field strength[8]. Kim'study shows that the shear stress of MRF is directly proportional to the magnetic field strength[9]. Chiranjit Sarkar'study found that the shear stress of MRF was also related to the particle size. The larger the particle size, the better the shear performance[10]. Yi et al obtained similar conclusions through finite element analysis. These studies show that the shear stress of MRF is closely related to volume fraction, magnetic field strength and particle size[11].

In this paper, the rheological behavior of MRFs in presence of magnetic field was studied, and the relationship between shear yield stress and shear rate of MRFs with different volume fraction was analyzing.

2. Paterial preparation and testing

2.1 Sample preparation

The MRFs samples consist of four components: ferromagnetic particles, base carrying fluids, surfactants and additives. The composition, size and volume fraction of ferromagnetic particles have

---

Content from this work may be used under the terms of the Creative Commons Attribution 3.0 licence. Any further distribution of this work must maintain attribution to the author(s) and the title of the work, journal citation and DOI.

Published under licence by IOP Publishing Ltd

great influence on the properties of MRFs. The carbonyl iron powder MRS-MRF-35 that manufactured by Jiangsu Tianyi Ultra-fine Metal Pouder Company was used as the ferromagnetic particles. The shape of the particles is spherical and the average size of the particle is 3.14 micrometer. The matrix carrier is generally selected from a material with small viscosity to facilitate uniform dispersion of the carbonyl iron powder particles Dimethyl silicone oil is selected as the matrix carrier. It can be used in the temperature range of -50 to 300 ° C. Surfactants of sodium dodecyl sulfate was added during the preparation to prevent sedimentation. Polyvinylpyrrolidone used as the dispersant with mass ratio to matrix carrier was 1:10. Fig.1 shows the preparation processes in which a high-speed stirring is carried out using a KQM-X4 planetary ball mill, and the DZF-6020 vacuum drying oven is used for vacuuming. MRFs with ferromagnetic particle volume fractions of 10%, 20%, 30%, and 40% were prepared for subsequent experiments.

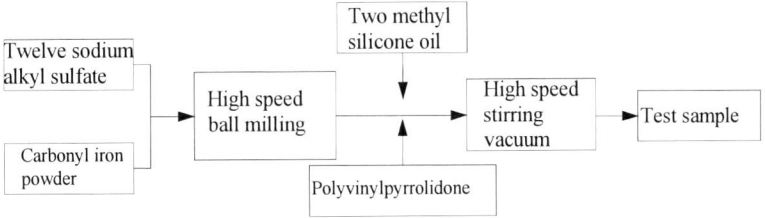

Figure 1. Preparation Processes of MRFs

2.2 Particle structure observation

The particle microstructure of the MRFs were observed at the presence of a uniform magnetic field by Keyence's VHX-600 optical microscope. The self-made magnetic field source was fixed on the microscope's observatory. The magnetic field source is an electromagnetic solenoid with a wide air gap. The magnetic field was modified by DC power. Fig.2 is the observing system and Fig.3 shows the observed pictures. It can be seen that the carbonyl iron powder particles are randomly distributed under the zero field. After the application of the magnetic field, the carbonyl iron powder particles rapidly aggregate into the chain along the direction of the magnetic field.

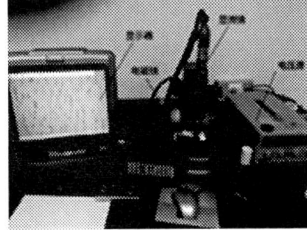

Figure 2. 3-Dimentional microscope

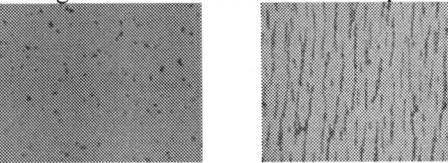

(a) without magnetic field (b) in magnetic field

Figure 3. Micro motions of MRFs

Microscopic view of the internal structure observed by electron microscopy: With the increase of the volume ratio of magnetorheological fluid, based on the minimum energy state, the magnetic particles gradually form a continuous single row of long chains of particles. The arrangement between the particles and the particles is more and more dense, the spacing is reduced, and the magnetic field force between the particles is increased. As the volume ratio continues to increase, a network structure will eventually form, the structure becomes denser, and the macroscopic performance is greater the shear stress. Therefore, as the volume fraction increases, the shear yield stress of the magnetorheological fluid increases at the same shear rate.

2.3 Colour illustrations

The shear properties of MRFs were tested by Aaton Paar Physica MCR 301 with the magnetic component. In the experiments, the shear rate was set in the range of 0-1000s^{-1}, the magnetization current was set to 1A, the temperature was set to 18°C, and the shear time is set to 30 s. The shear stress of MRFs was measured automatically every 0.1s. The effect of different volume fraction on the shear stress was studied by changing the test sample and shear rate in turn.

3. Analysis and discussion

Fig.4 shows the different volume fractions of MRF, and the shear stress varies with shear rate. It can be seen that as the shear rate increases, the shear stress increases gradually and is basically linear. When the volume ratio is 40%, the shear stress increases significantly when the shear rate is between 100s^{-1} and 500s^{-1}. At the same shear rate, the shear stress increases significantly with the increase of volume ratio. The shear stress increases slowly between 20% and 30%. The extent of the increase in the remaining stages is significantly faster. It can be seen that when the shear rate is constant, the magneto-rheological fluid increases with the volume fraction from 10% to 40%, and the yield stress increases from 3.9KPa to 14.2KPa, and the stress increases significantly.

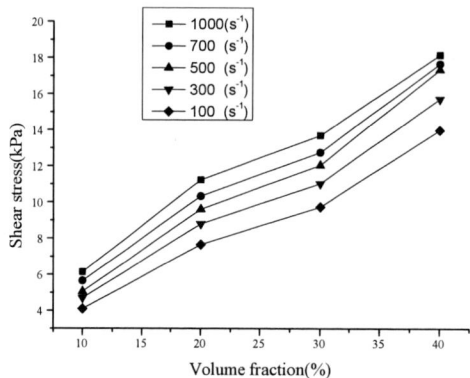

Figure 4. Effect of Shear Rate on Shear Stress of MRFs

According to the research, a generalized Bingham model and a nonlinear model are established, which can accurately reflect the shear thinning phenomenon of MRF[12].

$$\tau = \tau_0(B) + \eta \dot{\gamma} \qquad (1)$$

τ -Shear stress;

$\tau_0(B)$-Shear stress under magnetic field

η -Apparent viscosity

$\dot{\gamma}$ -Shear rate

A double exponential fit was performed on the experimental data.

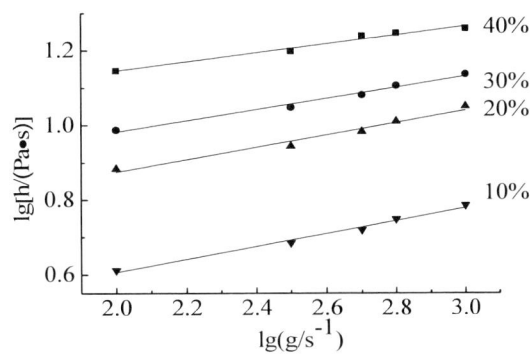

Figure 5. Shear rate fitting curve of shear stress of MRFs

Under the action of an external magnetic field, the magnetorheological fluid has a relatively large shear stress due to the aggregation of its magnetic solid particles to form a certain structure. According to the formula, the shear stress of the magnetorheological fluid is divided into two parts, one is the shear stress generated by the magnetic field force under the action of the magnetic field, and the other part is related to the apparent viscosity and the shear rate.

The shear stress generated by MRFs under the action of magnetic field is further analyzed and studied. By fitting the shear stress with the volume fraction, the relationship between the formula $\tau_0(B)$ and the volume fraction is obtained:

$$\tau_0(B) = 32.913\varphi + 0.5848 \qquad (2)$$

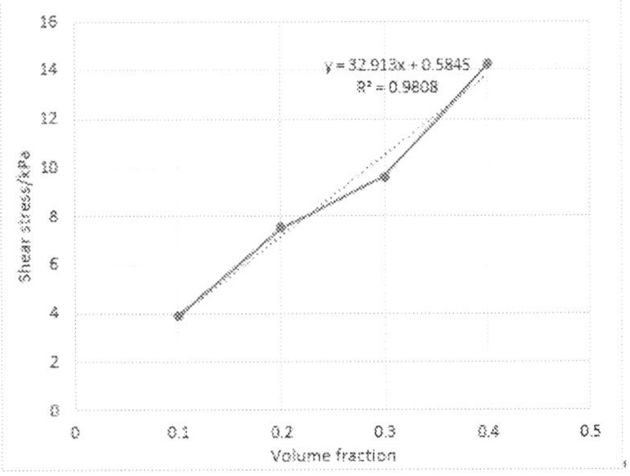

Figure 6. Fitting curve of shear stress with volume fraction

Apparent viscosity is a physical concept that refers to the quotient of the shear stress divided by the shear rate at a given velocity gradient. Therefore, the apparent viscosity can be expressed as:

$$\eta = \eta(\varphi, \dot{\gamma})$$

By fitting the curve of apparent viscosity with volume fraction, the relationship between η and volume fraction is obtained:

$$\eta(\varphi) = 0.3526\ln(\varphi) + 4.1189 \qquad (3)$$

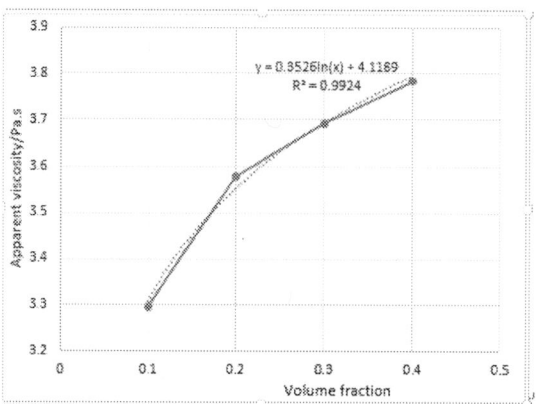

Figure 7. Approximate curve of apparent viscosity with volume fraction

Figure.8 shows the shear rate as a function of MRFs apparent viscosity at different volume fractions. It can be seen from the figure that the shear rate is between 100 s^{-1} and 300 s^{-1}, the apparent viscosity drops rapidly, and the change is slow and tends to be flat after 300 s^{-1}. At the same shear rate, the apparent viscosity increases with increasing volume fraction. When the shear rate is 100s-1 and the volume fraction is increased from 10% to 40%, the apparent viscosity increases from 40.91 to 139.9Pa·s. When the shear rate increases to 1000 s^{-1}, the apparent viscosity is stable at 6.131 and 18.16. The volume fraction of carbonyl iron powder in the magnetorheological fluid increases, and the zero field viscosity of the magnetorheological fluid increases. The mechanism is that the magnetorheological fluid fluid increases its internal friction due to the presence of ferromagnetic particles. Therefore, the viscosity of the magnetorheological fluid becomes large. The higher the density of the ferromagnetic particles, the greater the internal friction per unit volume, resulting in a larger zero field viscosity of the magnetorheological fluid.

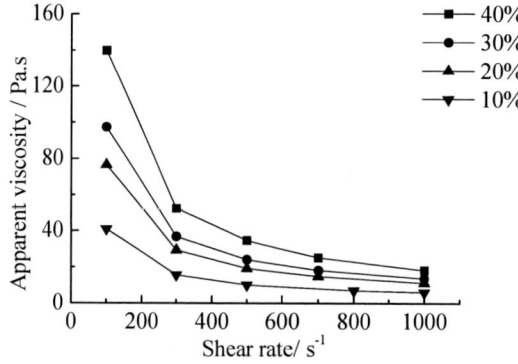

Figure.8 Apparent viscosity changes with shear rate

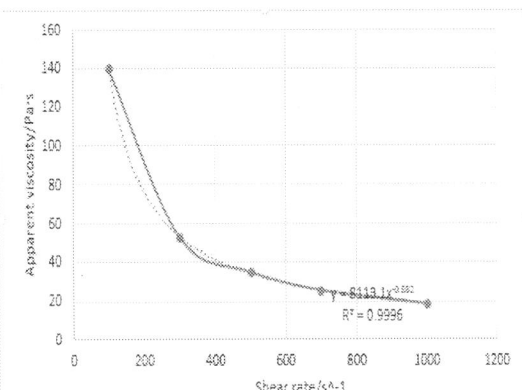

Figure.9 Approximate relationship between apparent viscosity and shear rate

Analysis and calculation of the above data, available

$$\eta(\dot{\gamma}) = 8113.1\dot{\gamma}^{-0.882} \qquad (4)$$

Bring formula (2).(3).(4) into (1) respectively:

$$\tau = \tau_0(B) + \eta\dot{\gamma}$$

$$= 32.913\varphi + 0.5848 + [(0.3526\ln(\varphi) + 4.1189)8113.1\dot{\gamma}^{-0.882}]\dot{\gamma}$$

Therefore, the shear stress is related to the volume fraction and shear rate, which is consistent with the proposed influencing factors, and the test results are in line with the previous theory.

4. Conclusion

Through theoretical and data analysis, the following conclusions can be drawn:

(1) The Bingham model was used to fit the shear stress and shear rate of MRF under different volume fractions. As the volume fraction increases from 10% to 40%, the yield stress of MRF increases from 3.9 kPa to 14.2 kPa. When the volume ratio is constant, the shear stress of MRF increases slowly with the shear rate, but the change is not very obvious.

(2) Through the experimental data, you and the analysis, the calculation formula of the shear stress and volume fraction and shear rate of MRFs is derived, which is the basis for the subsequent stress calculation.

(3) When the volume ratio is constant, the surface viscosity of MRF decreases exponentially with the increase of shear rate.

Acknowledgements

National Natural Science Foundation of China (51575323)

References

[1] Ashtiani M, Hashemabadi S H, Ghaffari A. A review on the magnetorheological fluid preparation and stabilization[J]. Journal of Magnetism & Magnetic Materials, 2015, 374:716-730.

[2] Chen Tong, Ge Jinming, Lin Cheng, Liu Jinsong, Han Yue, Liu Ruitong, Shen Yan. Magnetic hysteresis loss simulation based on magnetostrictive material current transformer [J]. science and technology and engineering, 2018,18 (07): 147-152.

[3] Ashtiani M, Hashemabadi S H, Ghaffari A. A review on the magnetorheological fluid preparation and stabilization[J]. Journal of magnetism and Magnetic Materials, 2015, 374: 716-730.

[4] Wang Fengxiang. Working principle and application of magnetron shape memory alloy actuators. Science, Technology and Engineering, 2003; 3(6): 577-581

[5] Fitrian Imaduddin. A design and modelling review of rotary magnetorheological damper. Materials & Design, 2013 ; 51 (5) : 575 —591

[6] Chen Fei, Tian Zu Wei, Wang Jian. Effects of temperature on properties [J]. functional materials, 2014, 45 (20); 20095-20098.

[7] G Bossis, C Mathis, Z Mimouni, et al. Magnetoviscosity of Micronic Suspensions [J]. Europhysics Letters, 1990, 11(2): 133-137.

[8] Liu Y D, Choi H J. Carbon nanotube-coated silicated soft magnetic carbonyl iron microspheres and their magnetorheology[J]. Journal of Applied Physics, 2012, 111(7):3701.

[9] Pilkee Kim. Analysis of a viscoplastic flow with field- dependent yield stress and wall slip boundary conditions for a magnetorheological (MR) fluid. {journal_en_name}, 2014; 204 (1) : 72 —86

[10] Sarkar C, Hirani H. Effect of Particle Size on Shear Stress of Magnetorheological Fluids[J]. Smart Science, 2015, 3(2): 65-73.

[11] YI Chengjian, PENG Xianghe, SUN Hu. Magneto-rheological fluid microstructure micro-structure magnetization and macro-mechanical analysis based on finite element method[J]. Functional Materials, 2011, 42(8): 1500-1503.

[12] Weng Jiansheng, Hu Haiyan, Zhang Miaokang, et al. Experimental and modeling of rheological mechanical properties of magnetorheological fluids[J]., 2000, 17(03): 1-5.

Research on Butt Joint of Ultrafine Grained Steel of Manual Arc Welding

Yan Wang[1,2,a]

[1]College of Mechanical & Power Engineering, China Three Gorges University, Yichang 443002, China

[2]Hubei Key Laboratory of Hydroelectric Machinery Design & Maintenance, Yichang 443002, China

[a]Corresponding author: wy9867113@163.com

Abstract. 400Mpa grade ultrafine grained steel was welded by manual arc welding with J506 electrode and I type grove butt welding. The microstructure and hardness change of welded joints were also studied. The welded joints were quenching heat treated and the change of microstructure and hardness were studied before and after heat treatment. The result showed that the microstructure grain size in the weld zone and overheated zone increased with the increase of current. Both the weld zone and the overheated zone were coarse ferrite and pearlite, and the weld zone was dendritic morphology. The weld zone had the highest hardness in the welded joint, followed by the overheated zone, the normalized zone and the base metal. The weld joints did not soften. After heat treatment, the microstructure of welded joints were the mixed structure of low carbon martensite, pearlite, ferrite and bainite. The hardness of welded joints were all improved.

1. Introduction

Ultrafine grained steel is new steel. The grain size is micron or submicron which can improve strength and toughness of steel. Compared with other equal strength steel, chemical composition of ultrafine grained steel is low carbon content which is beneficial to the improvement of weldability. Micron ultrafine grained steel is one of structural materials which is widely used in production practice. The study of its weldability is very hot[1-6].

2. Experimental material and parameter

Experimental materials were 4 mm thickness ultrafine grained steel, J506 electrode used in butt welding and J422 electrode used in tack weld. Chemical composition of ultrafine grained steel was shown in Table 1. Manual arc welding has been adopted and welding current of four test blocks was increased gradually. Welding parameters were shown in Table 2. The four samples were quenching heat treated. These samples were heated to 920℃，and heat preservation time was 15 minutes. And these four samples were cooled by water. The change of microstructure and hardness were studied before and after heat treatment.

Table 1. Chemical composition of ultrafine grained steel.

C	Si	Mn	P	S
0.150	0.213	0.800	0.015	0.004

Content from this work may be used under the terms of the Creative Commons Attribution 3.0 licence. Any further distribution of this work must maintain attribution to the author(s) and the title of the work, journal citation and DOI.

Published under licence by IOP Publishing Ltd

Table 2. Welding parameters.

Sample number	1	2	3	4
Welding current (A)	95	110	120	130
Average welding speed (cm/s)	0.6	0.8	0.89	1.2

3. Experimental result and analysis

3.1 Welded joint before heat treatment

The microstructure of base metal and welded joint of sample 1～4 was observed by 500 times metalloscope. Base metal microstructure of sample 1～4 was similar and was fine ferrite and pearlite which was showed in Figure 1. Welded joint was divided into weld zone, overheated zone and normalized zone. The microstructure of weld zone, overheated zone and normalized zone of sample 1～4 were shown in Figure 2～5.

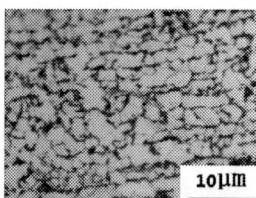

Figure 1. Base metal before heat treatment.

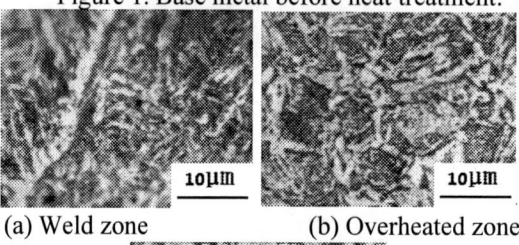

(a) Weld zone   (b) Overheated zone

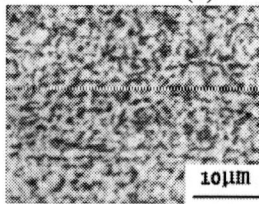

(c) Normalized zone

Figure 2. Welded joint of sample 1 before heat treatment.

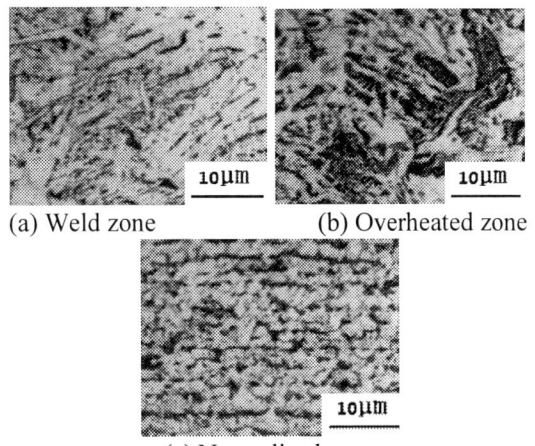

(a) Weld zone   (b) Overheated zone

(c) Normalized zone

Figure 3. Welded joint of sample 2 before heat treatment.

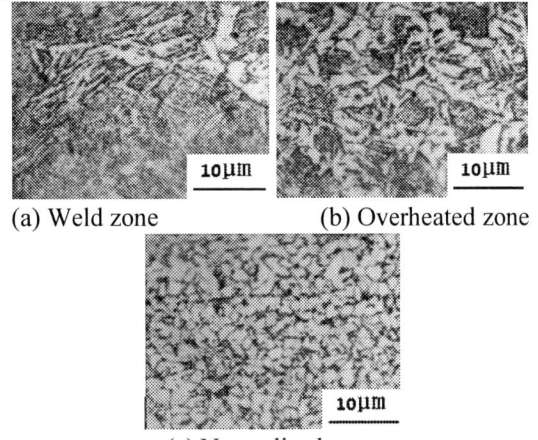

(a) Weld zone   (b) Overheated zone

(c) Normalized zone

Figure 4. Welded joint of sample 3 before heat treatment.

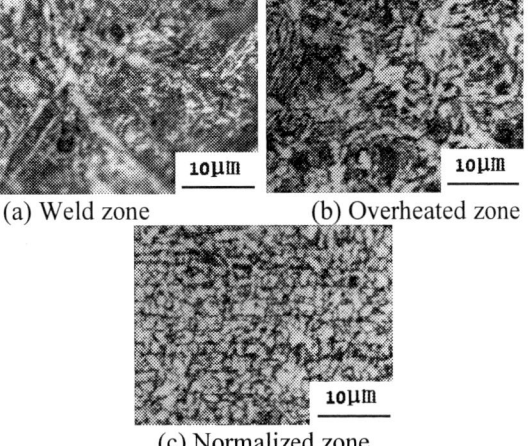

(a) Weld zone   (b) Overheated zone

(c) Normalized zone

Figure 5. Welded joint of sample 4 before heat treatment.

From Figure 2～5, we can know that grain size of weld zone and overheated zone increased with the increase of welding current. From Figure 2(a)～5(a), we can know that there were elongated grains and equiaxed grains in the weld zone. The microstructure of weld zone showed dendritic. The carbon content of weld zone was low, so the microstructure of weld zone were ferrite and pearlite which began precipitating from boundary of austenite and so their grains were coarse. From Figure 2(b)～5(b), we can know that the microstructure of overheated zone were obviously coarse ferrite and pearlite because this zone was overheated by welding thermal cycle. From Figure 2(c)～5(c), we can know that   the normalized zone of sample 1～4 was similar and was finer ferrite and pearlite than base metal.

The hardness of welded joint reflected the microstructure of welded joint. It was important index of harden quenching tendency of ultra-fine grain steel. The microstructure distribution of welded joint were uneven, so the hardness distribution of welded joint were uneven. The hardness of welded joint before heat treatment were shown in Table 3. From Table 3, we can know that the hardness distribution of welded joint were uneven. The hardness of weld zone was the highest in all samples. The hardness of overheated zone was higher than normalized zone. The hardness of normalized zone was higher than base metal. The welded joint did not soften.

Table 3. Hardness of welded joint before heat treatment (HV).

Sample number	Base metal	Normalized zone	Overheated zone	Weld zone
1	170.74	174.58	185.80	196.82
2	156.25	157.03	190.22	191.23
3	162.17	163.42	177.49	217.49
4	163.67	167.12	178.70	203.70

3.2 Welded joint after heat treatment
Metallograph were shown in Figure 6～9.

(a) Base metal   (b) Weld zone

(c) Heat affected zone
Figure 6. Microstructure of sample 1 after heat treatment.

(a) Base metal   (b) Weld zone

(c) Heat affected zone

Figure 7. Microstructure of sample 2 after heat treatment.

(a) Base metal   (b) Weld zone

(c) Heat affected zone

Figure 8. Microstructure of sample 3 after heat treatment.

(a) Base metal   (b) Weld zone

(c) Heat affected zone

Figure 9. Microstructure of sample 4 after heat treatment.

After quenching heat treatment, base metal, heat affected zone and weld zone of sample 1～4 were observed by 500 times metallographic microscope. From Figure 6～9, we can know that the microstructure of base metal, heat affected zone and weld zone were all the mixed structure of low carbon martensite, ferrite, pearlite and bainite after quenching heat treatment. The microstructure of base metal was more uniform.

The hardness of sample 1~4 after heat treatment were shown in Table 4.

Table 4. Hardness of welded joint after heat treatment (HV).

Sample number	Base metal	Weld zone	HAZ
1	380.54	236.14	236.45
2	331.73	268.13	325.69
3	212.89	304.95	335.82
4	356.02	205.04	353.96

From Table 4, we can know that the hardness of base metal, heat affected zone and weld zone were all increased after quenching heat treatment. The increase of base metal hardness was the highest and secondly heat affected zone (HAZ).

4. Conclusion

(1) Welding current and postweld heat treatment were main factors of the quality of welding joint.

(2) Before heat treatment, grain size of weld zone and overheated zone increased with the increase of welding current. The microstructure of weld zone and overheated zone were all coarse ferrite and pearlite and the organization form of weld zone was dendritic. The hardness of weld zone was the highest followed by overheated zone, normalized zone and base metal.

(3) After heat treatment, the microstructure of base metal, heat affected zone and weld zone were all the mixed structure of low carbon martensite, ferrite, pearlite and bainite. The hardness of base metal and welded joint were obviously improved. The increase of base metal was the most.

References

[1] Lei Yi, Yu Shengfu, Xu Xiaofeng, Ordnance Material Science and Engineering **28**, 44 (2005)
[2] Zhang Xifeng, Yuan Shouqian, Wei Yingjuan, Journal of Iron and Steel Research **20**,1(2008)
[3] Lin Zhenou, Hot Working Technology **36**, 3 (2007)
[4] Zhu Jian, Guo kuiwen, Wang Bingxin, Welding Technology **38**, 47(2009)
[5] Xu Chunhua, Zhang Maosen, Materials for Mechanical Engineering **36**, 94 (2012)
[6] Shan Xiaolong, Fang Junfei, He Yizhu, Materials Science and Technology **24**, 87(2016)

Fiber diameter measuring method of textile materials based on phase information

Wen Wang[1,2,a], Fang Zhang[1,2], Zhitao Xiao[1,2], Lei Geng[1,2], Jun Wu[1,2]

[1]School of Electronics and Information Engineering, Tianjin Polytechnic University, Tianjin 300387, China

[2]Tianjin Key Laboratory of Optoelectronic Detection Technology and System, Tianjin 300387, China

[a]wangwen@tjpu.edu.cn

Abstract. For achieving rapid and accurate measurement of fiber diameter, so as to evaluate product performance and guide the improvement of manufacturing process, a method based on phase information for measuring fiber diameter of which the fiber images of textile materials is collected by scanning electron microscope was proposed. Firstly, the improved Kuwahara filter is used to smooth the fiber image and enhance the foreground target. Then the fiber edge was detected by phase consistency method and repaired. Then, the spatial position and direction information of each pixel on the axis of a single fiber were determined, and the boundary pixel points were found by scanning along the vertical direction of the axis. Finally, the fiber diameter is calculated according to the position of border pixel points obtained by scanning. The experimental results show that this method can accurately and effectively measure the fiber diameter. Compared with the standard manual measurement, the fiber with a relative error of less than 2% accounts for 95% of the total fiber.

1. Introduction

Modern textile technology is a new technique for multi-scale processing of the structure of fiber or fiber aggregation. In order to produce textiles that satisfy clothing, decorative and industrial needs, it's important to get information about textile fibers. Fiber diameter equal to the linear density of textile material is an important geometric parameter of the fiber. The size of the fiber diameter has a great influence not only on fiber strength, elongation, stiffness, elasticity, but also on the handfeel of textile, style and the processing of yarn and textile. The smaller fiber diameter can get greater friction of the fiber body and higher yarn forming strength. The contact area between the fibers in the yarn forming section is large and the probability of slipping is low, so the yarn strength is improved.[1][2][3] In addition, in the international trade of textiles, both the seller and the buyer specifies the diameter of the fiber.

At present, there are mainly two methods for measuring fiber diameter. The first method is based on whole fiber image processing, and the second method is collecting a single fiber image for processing. The basic idea of both methods is: firstly, pre-process the image, enhance the image contrast and filter out the noise; then conduct edge detection, refinement and line segment connection; finally measure the diameter according to the repair fiber edge line[4][5][6]. But the current approach doesn't mention the overlap of the fibers or the bonding of the fibers. It's difficult to get a single fiber, because the fibers are dense and adhesions overlap.

Content from this work may be used under the terms of the Creative Commons Attribution 3.0 licence. Any further distribution of this work must maintain attribution to the author(s) and the title of the work, journal citation and DOI.

Published under licence by IOP Publishing Ltd

In the above methods, the poor measurement effect of fiber diameter is due to the low accuracy of identifying a single fiber. How to accurately identify a single fiber, and how to measure as much fiber as possible in a fiber image is the key point to our current research.

In this paper, a method for measuring fiber diameter of textile materials based on phase information is proposed, which is based on the fiber images with magnification of 20000 collected under scanning electron microscope. Phase consistency is used to extract the edge of the fiber image and identify the single fiber. Then, scan along the direction perpendicular to the central axis to find the edge pixel points, so as to obtain the pixel-level width of the fiber.

2. Fiber diameter measurement based on phase information

2.1 Preprocessing

2.1.1 Improved Kuwahara filtering
For the problem that Kuwahara filter is prone to block effect, the shape of sub-blocks can be changed and the local mean can be replaced by local weighted value [7]. Here, an adjacent region around the pixel point is evenly divided into 8 fan-shaped regions, and the improved Kuwahara filter is used to filter the fiber image [8]. Images before and after filtering is shown in Fig. 1:

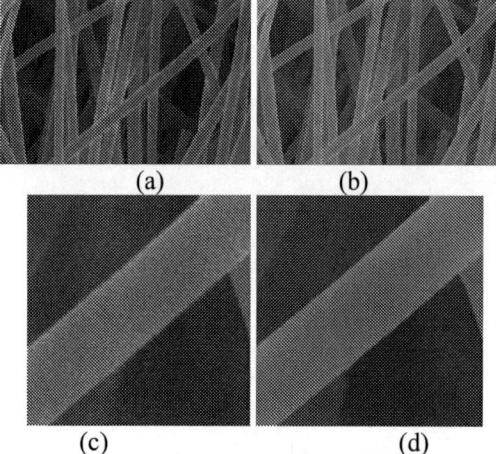

Fig. 1 Comparison before and after filtering. (a) Fiber image before filtering; (b) Fiber diagram after filtering; (c) Local area of fiber image before filtering (d) Local area of fiber image after filtering

2.1.2 Foreground target enhancement
In the process of image acquisition, some fibers in the collected fiber images are blurred due to different focuses. In order to highlight the foreground target with clear edges, this paper adopts mathematical morphology to enhance the foreground target. Add the filtered image to the result after the high hat transformation, and subtract the result after the low hat transformation to obtain the enhanced result:

$$E(f) = f + T(f) - B(f)$$

Here, f is the target image to be operated, $T(f)$ is the output image after high hat transformation, and $B(f)$ is the output image after low hat processing.

After the high-low hat method, the output fiber image is clearer, the foreground target is prominent, and the single fiber is enhanced, as shown in Fig. 2.

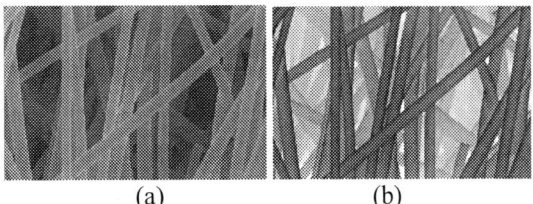
|(a)|(b)|

Fig. 2. Image after high and low hat transformation. (a) Fiber image before transformation; (b) Output image after transformation of high and low hat.

2.2 Edge detection and repair

In order to ensure the accuracy of fiber diameter measurement, the primary task is to accurately locate the fiber edge. On the basis of fully analyzing the characteristics of fiber image, an edge detection algorithm based on phase consistency is introduced in this paper.

2.2.1 Edge detection and refinement based on phase consistency

In edge detection methods, the algorithm based on gray scale and linear filtering is widely used, like the typical Canny and Log operator. But the gradient threshold of these algorithms is difficult to set, especially when gray level difference between the target object and the background is very small, the edge features are difficult to be detected.

Morrone and Owens defined the phases consistency function [9]:

$$P(x) = \max_{\overline{\varphi}(x) \in [0, 2\pi]} \frac{\sum_n A_n(x) \cos(\varphi_n(x) - \overline{\varphi}(x))}{\sum_n A_n(x)}$$

Here, $A_n(x)$ is the amplitude of the nth Fourier component at the point x, and $\varphi_n(x)$ is the phase of the nth Fourier component at the point x, so that the maximum $\overline{\varphi}(x)$ is the average local phase weighted by the amplitude of all Fourier components at the point. Therefore, the point with the maximum phase consistency corresponds to the point with the minimum weighted average transformation of the local phase.

When the 1D signal is extended to 2D image, firstly, the polarizable filter is convoluted with the extension function in the vertical direction, and then convoluted with the filter bank in the horizontal direction. The polarizable filter function is:

$$F(\omega, \theta) = G(\theta) \cdot H(\omega)$$

Here, $G(\theta)$ is the expansion function, constitutes the angular filter; $H(\omega)$ is the filter bank, θ is the direction Angle, ω and is the angular frequency.

In this paper, the following formula is selected to construct angular filter:

$$G(\theta) = \frac{\cos(\Delta\theta) + 1}{2}$$

Here, $\Delta\theta$ is the difference between the directional angle of the angular filter and the pixel angle.

Radial filtering uses Log Gabor wavelet to reflect and process the frequency response of natural image more realistically. It can extract more local frequency information, which is conducive to the calculation of phase consistency. In this paper, the following formula is selected to construct the radial filter:

$$H(\omega) = \exp(\frac{-(\ln(\omega / \omega_0))^2}{2(\ln(\beta / \omega_0))}) \cdot \frac{1}{\sqrt{1 + \varepsilon_2^2 T_N^2(\omega / \omega_0)}}$$

Here, ω_0 is the central frequency of the filter; $T_N(\omega / \omega_0)$ is an order N Chebyshev polynomial; ε_2 is the ripple parameter, and $0 < \varepsilon_2 < 1$.

On the basis of the above analysis, the method steps for edge detection of fiber image with phase information are as follows:

(1)Fourier transform is performed on the acquired fiber image.

(2) Use the above $G(\theta)$ calculation formula to construct the angular filter, and use the filter to filter the fiber image Fourier coefficient in q directions.

(3) Use the above $H(\omega)$ calculation formula to construct a radial filter, and then the fiber image processed by the angular filter on s different scales respectively.

(4) Calculate the sum of the local energy and the amplitude of the Fourier component for each point superimposed on multi-scale multiple directions.

(5) Finally, the phase consistency matrix of fiber image can be obtained by using the results of angular filter and radial filter, and the matrix can be converted into gray image as the output result, as shown in Fig. 3.(a).

The phase consistency gray scale generally reflects the edge profile of the fibers in the fiber image, but it can be found that the fiber edge line width is generally not a single pixel. As there is noise in the image, and the fibers are crossed and overlapped with each other, there will be fiber adhesion in some areas, which will lead to some false edges and broken edges in the edge detection of the fiber image, so this does not achieve the ultimate goal of image edge detection. For this reason, this paper adopts the non-maximum suppression method based on the bilinear interpolation and the adaptive double-threshold method to obtain the two-valued edge graph of a single pixel, as shown in Fig. 3 (b).

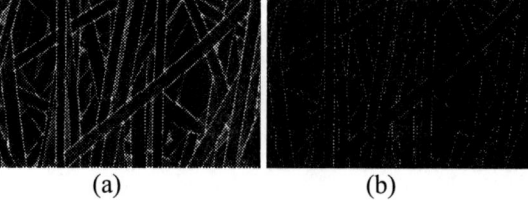

(a)   (b)

Fig. 3 Edge dectection. (a) Edge detected by phase consistency; (b) Fiber edge detection results obtained by this method

2.2.2 Edge repair

Due to the influence of light, some weak edges may be missed in edge detection, and some burrs and short lines may exist after refinement, resulting in fracture and incomplete edge lines. These conditions have a great impact on the identification of a single fiber, and the refined fiber edge image must be repaired and corrected to obtain a relatively complete fiber edge, providing a reliable basis for future measurement [10]. Therefore, in order to obtain a complete fiber edge, it is necessary to take a series of operations to repair and correct the fiber edge, and repair the original incomplete fiber edge segment into a complete fiber edge. Common edge repair includes intersection point removal, burr removal, short line removal and short edge repair.

After edge repair, the fiber edge line is clearer and not confused, as shown in Fig. 4.

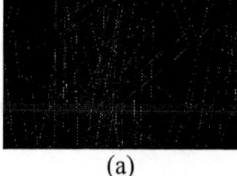

(a)

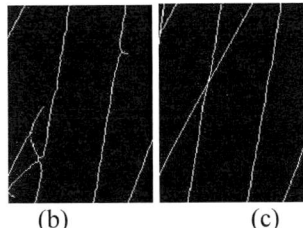

(b)   (c)

Fig. 4 Fiber edge reparation. (a) Fiber edge image after reparation; (b) Local image before reparation; (c) Local image after reparation

2.3 Diameter measurement

2.3.1 Fiber identification

After a series of treatment and repair of the fiber edge, the edge of each fiber is composed of two single-pixel wide lines. The main work of this section is matching all edge lines in the repaired edge image, and the two edges successfully matched are considered as belong to the same fiber, so as to complete the identification of a single fiber. After edge repair, the X and Y coordinates of each point of the fiber edge curve are recorded in an array. The coordinates of each point in the array are fitted by the principle of least square method, and constraints are established to determine whether the two line segments are matched [11][12]. If constraints are satisfied at the same time, the two edges are considered to be matched successfully. Fig. 5 shows the fiber identification and filling result.

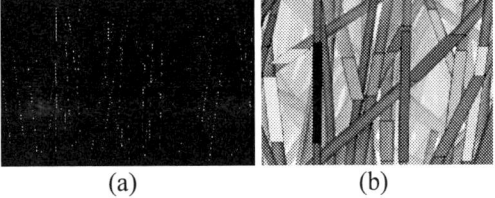

(a)   (b)

Fig. 5 Fiber recognition images. (a) Fiber edge matching results; (b) Fiber recognition results

2.3.2 Fiber diameter measurement

According to the fiber identification results, the fiber central axis is obtained. By using the spatial position and direction information of pixels on the central axis, the edge boundary pixels on both sides of the current point are scanned in the vertical direction of the central axis, so as to obtain the pixel-level width of the fiber.

The specific steps are as follows:

(1) Firstly, the identified single fiber filling area (as shown in Fig. 5(b)) was expanded, so as to ensure that the real fiber edge falls in the filling area;

(2) Calculate the center line of the filled area according to the coordinates of the four vertices of the filled area, and get the central axis of the fiber;

(3) Based on the edge detection results in Fig. 4(a), scan the edge line with the center line obtained in step (2), as shown in Fig. 6;

(4) Calculate the central axis and axis Angle θ according to the Bresenham algorithm. Choose five points in the central axis as a starting point for scanning in the perpendicular direction. (as shown in Fig. 7) P_i is on the central axis and the dotted line is edge of fiber, scan with 1 pixel along a direction perpendicular to the fiber, then find edges on both sides of the boundary coordinates Q_1, Q_2;

(5) The Euclidean distance of Q_1, Q_2 is the fiber pixel diameter.

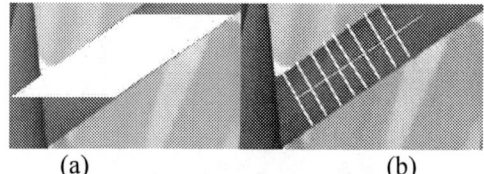

(a)   (b)

Fig. 6 Image of partial fiber diameter filling and measurement. (a) Partial filling image; (b) Partial measurement image

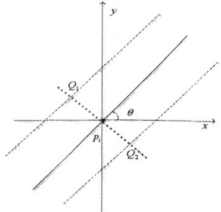

Fig. 7 Fiber diameter measurement diagram

In this paper, 5 points were taken as the starting point of scanning at the interval of each central axis, and 5 fiber diameter measurements were obtained through scanning calculation. Mean values of these 5 measurements were taken as the final diameter measurements of the current fiber. Finally, the actual width of the fiber can be obtained by proportional conversion according to the ruler set in fiber collection.

3. Experimental results and analysis

This paper presents an algorithm for measuring fiber diameter based on phase information. By using this algorithm to process fiber image, a large number of fiber diameter data can be obtained. In order to evaluate the accuracy of fiber diameter data, the measured data were taken as the standard data and compared with the results of this paper.

Traditional ultra-fine fiber diameter measurement is completed by manual measurement of images collected by electron microscopy. In this paper, the manual measurement software image-j is used for manual measurement of fiber images. Then, the accuracy and stability of the measurement data are evaluated. Because the distribution of fibers in textile materials is very complex and special, there is no unified method to determine the diameter of fibers. There is a certain subjectivity in the actual operation process, but it can make up for the deficiency by averaging multiple measurements. The algorithm is evaluated by using the average diameter of multiple measurements as the standard data. In this paper, the mean value of 5 measurements was taken as the standard data of fiber diameter.

3.1 Accuracy analysis

In this experiment, 74 fiber diameters photographed at different magnification ratios under scanning electron microscopy were measured. Part of the measurement results of the method in this paper are shown in Table 1, and the experimental errors were analyzed through two parameters, namely the absolute error between the measured values of the algorithm in this paper and the mean values of the manual measurements σ_1, and the relative error between the measured values of the algorithm in this paper and the mean values of the manual measurements σ_2. Expressed as:

$$\sigma_1 = \left| l_i - \overline{l_m} \right|$$

$$\sigma_2 = \frac{\sigma_1}{\overline{l_m}}$$

Here, l_i is the algorithm measurement result of each diameter, and $\overline{l_m}$ is the average value of manual measurement of each diameter. The experimental results show that 77% of the fibers with a relative error of <1% and 95% of the fibers with a relative error of <2% account for the total fiber. The measurement results are relatively accurate.

Table 1 Partial fiber diameter measurement results

NO.	Manual Measurement/nm						This Method /nm	Absolute Error /nm	Relative Error /%
	1	2	3	4	5	Average Value			
1	829.28	824.77	823.60	823.69	817.57	823.78	824.77	0.99	0.12
2	523.51	524.50	531.62	527.03	516.31	524.59	524.49	0.11	0.02
3	741.62	747.93	739.73	740.09	729.73	739.82	747.28	7.46	1.01
4	519.64	516.40	519.82	519.10	524.59	519.91	520.77	0.86	0.17
5	771.62	769.91	769.73	770.99	765.50	769.55	766.85	2.70	0.25
6	645.59	645.50	646.40	644.95	647.30	645.95	641.50	4.45	0.19
7	1024.23	1024.50	1023.78	1022.34	1020.45	1023.06	1022.91	0.15	0.01
8	754.41	752.79	755.32	756.31	755.50	754.86	749.90	4.96	0.16

3.2 Fiber diameter uniformity analysis

The uniformity of fiber diameter has an important effect on fiber material. After measuring the fiber diameter, this section calculates the variance of the fiber diameter in the figure by taking two fiber pictures as examples. The small variance indicates that the fiber thickness is relatively uniform.

Fig. 8 Fiber diameter identification results (1)

Fig. 9 Fiber diameter identification results (2)

After calculation, the pixel-level variance of fiber diameter in the image shown in Fig. 8 is 123, and the pixel-level variance of fiber diameter in the image shown in Fig. 9 is 98, indicating that the fiber diameter distribution in Fig. 9 is relatively uniform.

4. Summary

A method for measuring fiber diameter of textile materials based on phase information is proposed. Firstly, the improved Kuwahara filter was used to filter the fiber image and enhance the foreground target. Then, the method based on phase consistency was used to detect the fiber edge and repair the edge, including deleting the intersection point, short line and burring, and carrying out edge fracture connection. On this basis, the spatial position and direction information of each pixel on the central axis of the single fiber was identified.

Further research can be carried out in the following aspects in the future.

Firstly, due to the limited conditions for image collection, this paper only measures the fiber image magnified 20,000 times. However, there are still many types of actual fiber images. In the future, more and more complete fiber images need to be collected for measurement.

Secondly, the recognition rate of single fiber in the algorithm of measuring fiber diameter in this paper needs to be improved. In the future work, deep learning technology can be used to segment single fiber to improve the recognition rate of single fiber and further comprehensively measure fiber diameter.

Thirdly, this paper mainly measures the diameter of the fiber with a straight shape. In fact, part of the fiber is bent. In the future, the diameter measurement method of the bent fiber should be studied.

References

[1] Y. Z. Yang, Y. H. Zhang. Improvement of quality of combed yarn and its product development[J]. Journal of Textile Research, 2004, 25(06): 94-97

[2] C. X. Wang. Development of bamboo fiber and cotton blended yarn[J]. Advances in Textile Technology, 2007, 1: 25-26.

[3] K. H. Huang. Development of siluo knitting yarn in tiansi/hemp/wool/silk blended yarn[J]. Shanghai Textile Technology, 2011, 39(11): 42-44.

[4] H.Q. Su, L.L. Wu, T. Chen. Application progress of image processing technology in nonwoven field[J]. Textile Guide, 2013 (12): 65-68.

[5] G.D. Fu, H. Chen, H. Liu, et al. Advances in digital image processing technology in the field of textile and apparel[J]. Silk, 2011, 48(12): 22-25.

[6] S. J. Ren, W. S. Zhang, Y. He, et al. Measurement methods of fiber diameter and curvature based on image analysis[J]. Journal of Image and Graphics, 2008, 13(6): 1153-1158.

[7] G. Papari, P. Campisi, P. L. Callet, et al. Artistic Stereo Imaging by Edge Preserving Smoothing[C]// Digital Signal Processing Workshop and, IEEE Signal Processing Education Workshop, 2009. Dsp/spe 2009. IEEE. IEEE, 2009: 639-642.

[8] W. T. Zhang. Research and implementation of fog sky image restoration technology[D]. Nanjing University of Aeronautics and Astronautics, 2012.

[9] C.T. Wei, Z. X. Zhang, J. Q. Zhang, et al. A method for detecting the characteristics of remote sensing image power lines based on phase consistency[J]. Bulletin of Surveying and Mapping, 2010, (3): 13-16.

[10] Misra D K, Tripathi S P, Singh A. Fingerprint image enhancement, thinning and matching[J]. International Journal of Emerging Trends & Tech in Comp Science (IJETTCS), 2012, 1(2): 17-21.

[11] K. Zhang, L. Yan. Curve fitting around square tube based on least square method[J]. Journal of changshu institute of technology, 2009, 23(2): 70-73.

[12] M.J. Shang, B. T. Zhu, X. Li, et al. An opposites fiber sorting approach based on least square method[J]. Industrial control computer, 2008, 21(5): 64-65.

Mechanism analysis of ferromagnetic resonance of electromagnetic voltage transformer in neutral ungrounded system

Tingran Sheng[1] and Tongxin Han[1*]

[1]China Academy of Railway Sciences, Locomotive & Car Research Institute, 100081, No.2, Daliushu Road, Haidian District, Beijing, China

TEL:13601306271

Email: htx503@126.com@126.com

Abstract. In the neutral point ungrounded power distribution system, due to the nonlinear characteristics of the electromagnetic voltage transformer (PT) excitation inductance, ferromagnetic resonance overvoltage is easily generated under certain conditions, which seriously affects the safe operation of the system. This paper takes the Yecheng substation 6 kV distribution network as the model, and uses ATP-EMPT electromagnetic transient calculation software to theoretically analyze the characteristics and mechanism of ferromagnetic resonance caused by electromagnetic voltage transformer, and points out the main problems of ferromagnetic resonance in PT operation. This article discussed the existing harmonic elimination measures, proposed some effective suppression schemes and demonstrated the effectiveness of some mainstream harmonic elimination methods. The author's work can provide a good reference for the prevention and control of ferromagnetic resonance overvoltage, and it also has high practical value for improving the power supply safety and reliability of the distribution network.

1. Introduction

In a neutral point ungrounded distribution system, there are three types of overvoltage effects that are most severe: lightning overvoltage, arc grounding overvoltage, and ferromagnetic resonance overvoltage. The first two types of overvoltage have relatively clear and effective measures for protection. For example, lightning surges are generally limited by lightning protection devices such as lightning arresters; arc grounding overvoltages are generally connected to arc suppression devices such as arc suppression coils to attenuate its effects and so on.

However, for the ferromagnetic resonance overvoltage caused by the matching of the system capacitance and inductance parameters, although there are many existing harmonic elimination measures, because the mechanism of resonance generation is not fully understood, and the actual operation of the power supply system is complicated and varied, the current suppression, the various measures to eliminate the influence of ferromagnetic resonance are evaluated differently. This paper studies this phenomenon.

2. Ferromagnetic resonance mechanism

In the neutral point ungrounded system, the equivalent circuit diagram of the power supply potential, the phase voltage transformers and the respective ground capacitors is shown in Figure.1. In Figure.1,

$L_A, L_B,$ and L_C are the respective magnetizing inductances, C_0 is the relative capacitance, and E_A, E_B and E_C are three-phase power sources. The relative admittances are:

$$Y_X = j \ (\omega C_0 - \frac{1}{\omega L_X}) \qquad (1)$$

In the normal operation of the electromagnetic voltage transformer, the PT excitation inductance is much larger than the system capacitive reactance, namely:

$$\omega L >> 1/\omega C \qquad (2)$$

The relative ground load is Y_0 and is capacitive, $U_N = 0kV$, that is, the neutral point and zero potential of the power grid coincide, there is no neutral point voltage displacement phenomenon, and the sum of the total charge of the system line to ground capacitance is zero. Therefore, the system does not have a resonance condition during normal operation, and resonance does not occur.

However, when the grid breaker is grounded to the ground bus (especially the light loaded bus line), or a single-phase arcing grounding occurs due to lightning strikes, etc., the system is faulty or disturbed, and the non-fault is added to the electromagnetic voltage. The voltage on the transformer excitation coil is raised from the phase voltage to the line voltage. When the ground fault continues, under the action of the line voltage, the charge flows at the wire and the ground with the ground point as a path to form a capacitor current. When the ground fault is eliminated, the path is cut off, and the relative ground voltage is restored from the line voltage value to the original phase voltage value. The charge of the corresponding line voltage carried by the original non-faulty phase conductor needs to find the path to the ground. Because the original line ground path is cut off, the free charge has to flow to the ground through the grounded winding of the star-connected electromagnetic voltage transformer. If the line-to-ground capacitance is large, there will be a lot of accumulated free charges. In the process of venting to the earth, the excitation coil core will be saturated, and the excitation coil inductance ωL will decrease. When the excitation coil inductance continues to decrease, to $\omega L = 1/\omega C$, a linear resonance is formed between the PT excitation coil and the line-to-ground capacitance, thereby causing the electromagnetic voltage transformer to resonate.

The electromagnetic voltage transformer has an inductive reactance greater than the capacitive reactance ($\omega L > 1/\omega C$) under normal operating conditions, so the system does not have a resonant condition during normal operation and resonance does not occur.

However, in the event of a system failure or disturbance, the PT excitation coil voltage would increase from the phase voltage to the saturation of the excitation coil core, and the inductive reactance wL would reduce. With the inductance of the excitation coil continues to decrease, even drop to $\omega L = 1/\omega C$, a linear resonance is formed between the PT excitation coil and the line-to-ground capacitance, resulting in an overvoltage. This phenomenon is called ferromagnetic resonance.

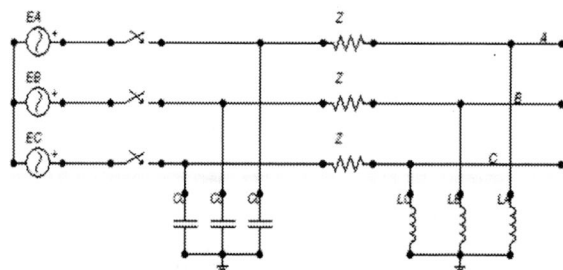

Figure 1. Series resonant circuit diagram

The research set Yecheng substation as the object. Yecheng substation is a neutral point ungrounded system. The grounding parameters of the system mainly include power equipment, cable-to-ground capacitance C_0 and voltage transformer's excitation inductance L, as shown in Figure 2.

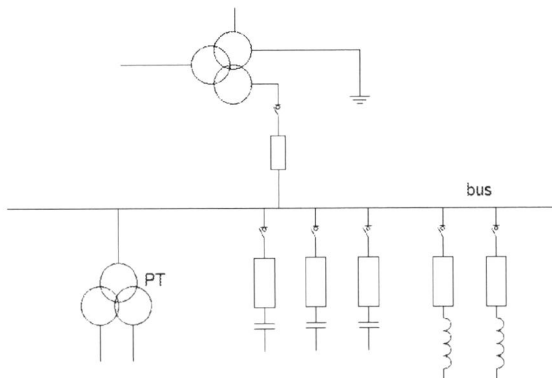

Figure 2. Circuit diagram of Yecheng substation

In normal condition of operation, the magnetizing inductance is very strong, the system impedance is capacitive, and the three phases could basically keep balance. And the displacement voltage U_N of the neutral point of the power grid is close to the value of zero. However, when the system has a phase-to-earth fault due to a wire breakage, a lightning strike or any other reasons, the relative ground voltage U_A would reduce to near zero, and the non-fault relative ground voltage (U_B, U_C) would instantaneously increase, which would be three times the value of the original one. The excitation current of the B Phase and C Phase of the voltage transformer suddenly increases, the excitation coil is saturated, and the equivalent excitation inductance L_A decreases, resulting in unbalanced three-phase impedance, thereby generating a neutral point displacement voltage U_N.

After the B Phase is grounded, the system is disturbed, the B Phase ground fault point flows through the 6kV system ground capacitance current, the sound phase A Phase and C Phase voltage rise to the line voltage, and the transformer A/C Phase primary side field winding The excitation current suddenly increases, causing the excitation winding to become saturated. The saturation inductance of the transformer after saturation becomes smaller. When the system-to-ground inductance matches the parameter of capacitance to ground, that is, when $\omega L = 1/\omega C$, the system would lead to resonance and ferromagnetic resonance would take place.

3. Measures to suppress ferromagnetic resonance

Eliminating resonance generally takes measures from three aspects:

(1) Change the inductance and capacitance parameters of the system by the Peterson curve so that it does not have the resonance matching condition. The measures attributed to this type of harmonic elimination method mainly include: installing a three-phase star capacitor group with neutral point grounding on the busbar, and selecting a voltage transformer with good excitation characteristics to reduce the number of parallel connections in the same network, and the neutral point of the PT high voltage side. String single-phase PT (4PT method), system neutral point would be grounded by arc suppression coil

(2) Increasing the system zero-sequence loop damping resistance, consuming the resonant energy of the zero-sequence loop after resonance. If the neutral point of the PT high voltage side is grounded via a non-linear resistor, the open delta winding of the voltage transformer is connected in series with a damping resistor or a harmonic elimination device.

(3) Change the system wiring method. Change the system to ground via a small resistor, or operate at the moment of resonance, such as cutting off the voltage transformer, disconnecting the neutral point grounding wire of the high voltage side of the voltage transformer or switching the neutral point of the system side to temporary grounding.

A detailed study on several mainstream harmonic elimination methods would be listed as follows:

3.1 System neutral point is grounded by arc suppression coil

ISPECE   IOP Publishing

IOP Conf. Series: Journal of Physics: Conf. Series **1187** (2019) 032081   doi:10.1088/1742-6596/1187/3/032081

The neutral point of the system power supply is grounded via the arc suppression coil. The wiring diagram is shown in Figure 3.

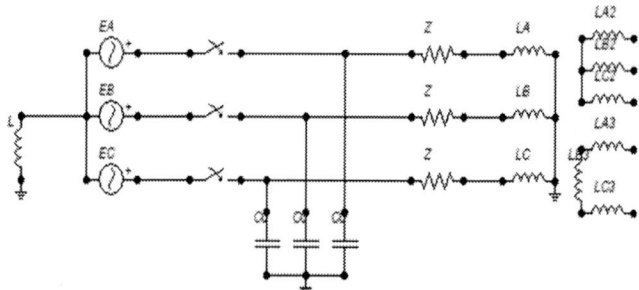

Figure 3. Circuit diagram of system neutral point grounded by arc suppression coil

Grounding the neutral point of the system through the arc suppression coil is equivalent to connecting the arc suppression coil on each phase excitation inductance of the main voltage transformer, because the value of the inductance of the arc suppression coil is much smaller than the inductance of the excitation winding, and the resonance condition would be broken after the parallel connection, the system capacitance. The inductance parameters cannot meet the requirements of resonance, making the resonance difficult to occur. The harmonic elimination of the neutral point of the system through the arc suppression coil preserves all the advantages of the neutral point ungrounded system, and could prevent the occurrence of ferromagnetic resonance as well.

3.2 Voltage transformer high-voltage side neutral point is grounded by single-phase PT
The neutral point of the high voltage side of the voltage transformer is grounded through a single-phase PT. The wiring diagram is shown in Figure 4.

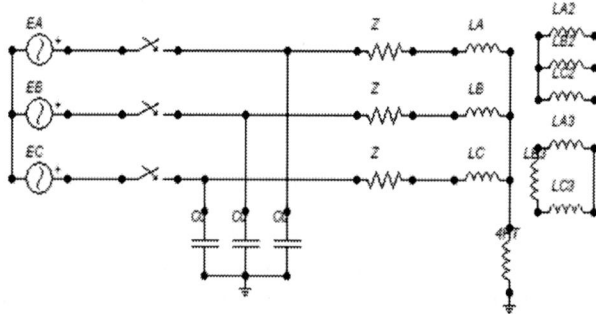

Figure 4. Circuit diagram of voltage transformer high-voltage side neutral point is grounded by single-phase PT

The high-voltage winding side of the main voltage transformer adopts the star connection method; the neutral point of the high-voltage side is grounded to earth by a single-phase voltage transformer(4 PT); the secondary auxiliary winding side of the main voltage transformer adopts the delta connection method, of which the triangular opening is short-circuited; the single-phase voltage transformer's secondarily the winding is on the secondary winding coil of the main voltage transformer.

With this type of wiring, when a single-phase ground fault occurs in the system or other causes cause asymmetrical distribution of the three-phase voltage of the system, the zero-sequence voltage in the system will be proportionally distributed according to the magnitude of the zero-sequence impedance. Since the secondary auxiliary winding of the main voltage transformer is short-circuited by the triangle, the zero-sequence impedance of the main voltage transformer is extremely small, and the zero-sequence voltage is almost entirely borne by the single-phase voltage transformer on the

1060

ground line, which means that three phases' voltages of the main voltage transformer could stay basically around the phase voltage, instead of rising to the line voltage. This phenomenon ensures the system cannot meet the requirements of resonant excitation condition. And, also, the impedance of the single-phased voltage transformer itself has the effect of expanding the energy of resonance.

3.3 Voltage transformer high-voltage side neutral point string nonlinear resistance
The neutral point of the high voltage side of the voltage transformer is grounded via a non-linear resistor. The circuit diagram is shown in Figure 5.

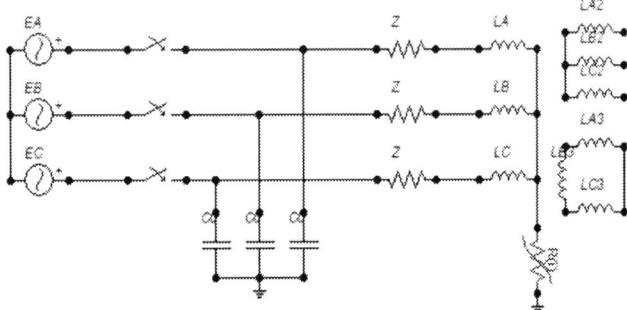

Figure 5. Circuit diagram of system neutral point grounded by nonlinear resistance

The high-voltage winding coil of the main voltage transformer adopts the star connection method; the neutral point of the high-voltage side is grounded to earth through the nonlinear resistor, and the secondary auxiliary winding of the main voltage transformer adopts the delta connection method.

In the high voltage side of the voltage transformer, the neutral point grounding wire is connected to the resistor, which is equivalent to connecting a non-linear resistor in each phase of the three-phased voltage transformer. When the grid is in normal operation, the voltage on the nonlinear resistor is not high. The nonlinear resistor is high in value of resistance, which means resonance is not easy to occur in the initial stage. When a single-phased grounded to earth accident occurs in the system, a high voltage appears on the nonlinear resistor, then the resistance would exhibit a low resistance value. The resistor can share the voltage applied to the voltage transformer, thereby limiting the current in the voltage transformer. The nonlinear resistor could also limit the high-amplitude current, which flows through the voltage transformer when the arc is grounded. Making the current in the high voltage winding coil to a small level, the nonlinear resistor is equivalent to improving the volt-ampere characteristics of the voltage transformer.

4. Conclusion
At present, when relating about the ferromagnetic resonance phenomena in the power system, we tend to pay attention to the governance after the accident while ignoring the process of prevention. In view of the importance of substation power reliability for the safe and stable operation of the system, the ferromagnetic resonance overvoltage threatens the normal operation of the substation, and measures should be done before the accident happens. We should focus on the beginning of the design of the power supply system, and considering about various operating conditions of the substation, and make the substation reliably predicted before it is put into operation, making the equipment of substation as far from the resonance region as possible to avoid resonances; even if it is unavoidable, it can be based on the predicted resonant resonance type as much as possible. Considering about other factors, such as communication systems, the requirements on protection of the personal and electrical equipment, etc., we should take appropriate harmonic elimination measures to ensure reliable operation of the substation.

References
[1] Rudenberg. R, *Transient Performance of Electric Power Systems*., (NY: McGraw-Hill Book

Company, **ch.48,** 1950)

[2] Dugan, R.C., *Electrical power system quality*, (2th edition, McGraw-Hill, 2004)

[3] IEEE Working Group on Modeling and Analysis of Systems Transients Using Digital Programs, Modeling and Analysis Guidelines for Slow Transients—Part III: The Study of ferro-resonance, IEEE TOPD, **VOL. 15,**1 (2000)

[4] Escudero, V., Dudurych, M., and Redfern, M.A., Characterization of ferroresonant modes in HV substation with CB grading capacitors, Electric Power Systems Research, **77, 1506** (2007)

[5] H. A. Peterson, *Transients in Power Systems*.(New York: Wiley, **pp. 206-209,**1951)

[6] A. Rezaei-Zare, M. Sanaye-Pasand, H. Mohseni, S. Farhangi, and R. Iravani, *"Analysis of ferroresonance models in power transformers using Preisach-type hysteretic magnetizing inductance,"* IEEE Trans. Power Del., **vol. 22, no. 2, pp. 919–929**(2007)

[7] D. A. N. Jacobson, *"Examples of ferroresonance in a high voltage power system,"* in Proc. IEEE Power Eng. Society General Meeting, **vol. 2, pp. –1212**(2003)

[8] N. Janssens, V. Vandestockt, H. Denoel, and P. A. Monfils, *"Elimination of temporary overvoltages due to ferroresonance of voltage transformers: Design and testing of a damping system,"* in Proc. CIGRE, **pp. 1–8** (1990)

Fast Aerial UAV Detection Using Improved Inter-frame Difference and SVM

Li Xiaoping[1,a], Lei Songze[2,b], Zhang Boxing[3,c], Wang Yanhong[4,d], Xiao Feng[5,e]

[1]School of Computer Science and Engineering
Xi'an Technological University
86-15289387309, 710021

[2]School of Computer Science and Engineering
Xi'an Technological University
86-18991896239, 710021

[3]School of Computer Science and Engineering
Xi'an Technological University
86-17719755818, 710021

[4]School of Science
Xi'an Technological University
710021

[5]School of Computer Science and Engineering
Xi'an Technological University
 710021

[a]919083845@qq.com [b]lei_sz@163.com [c]1076781906@qq.com [d]29314998@qq.com
[e]544070146@qq.com

ABSTRACT: In order to detect UAV in real time, the paper choose to use a dynamic detection method based on two consecutive inter-frame differences method to extract the region of interest. The position of the target appeared on the image was obtained by the method of two consecutive inter-frame difference, and the UAV was detected by the trained SVM classifier. UAV could be detected quickly and accurately in complex background and in different position and angle circumstances. Compared to the traditional HOG+SVM sliding window detection method, the experimental results show that the detecting speed with the methods is obviously improved when the recognition accuracy is invariable.

1. INTRODUCTION

In recent years, UAV(unmanned aerial vehicle) have been widely used in the military field. They could monitor troops in real time and detect important military targets and find all military regions and carry offensive weapons to attack hostile military regions. In order to avoid the destruction of troops and military regions by hostile UAV, it is necessary to detect UAV in real time.

The main difficulty detecting aerial UAV in real time is that the aerial non-cooperative UAV is easy to be affected by its own motion, which could change the angle, position and structure of the UAV in the video. As a result, the robustness of UAV features is low and it is difficult to detect accurately. The difficulty of target detection is similar to UAV detection, so it is worth studying and

Content from this work may be used under the terms of the Creative Commons Attribution 3.0 licence. Any further distribution of this work must maintain attribution to the author(s) and the title of the work, journal citation and DOI.
Published under licence by IOP Publishing Ltd

practicing in order to obtain accurate and real-time detection method for aerial UAV. In recent years, the target detection method based on machine learning is easy to be used, which mainly includes two steps: region of interest segmentation and target detection. The methods segmenting regions of interest mainly includes moving object detection [2] and static object detection, including optical flow field method [1], background estimation, inter-frame difference method [2], and image segmentation method [17]. Image segmentation is insensitive to illumination and is suitable for scenes with simple background and clear target. The optical flow field method is suitable for all kinds of backgrounds. However, it's caculation need to use too many pixels, large amount of computation and insufficient real-time performance. The method of inter-frame difference is easy to be affected by illumination. To sum up, it is difficult to detect small UAV by image segmentation when the UAV is far away from the camera and the background is complex and the illumination change is weak, and it is easy to misunderstand some complex background.

A real time detection method for aerial UAV using two continuous frame difference method combined with SVM is proposed in the paper. The pixels of the corresponding position of each successive two frame image in the video were subtracted and the absolute value was taken, and then the two adjacent values were performed or calculated, and the absolute value was compared with the predetermined threshold value, if the absolute value was greater than the threshold value, then the location of the absolute value of the frame is recorded and segmented in the gray image. Then it is sent to the support vector machine (SVM) model trained by gradient direction feature vector and Fisher linear discriminant analysis (HOG-FLD) in the methods.

2. SEGMENTING OF REGION OF INTEREST BASED ON INTER-FRAME DIFFERENTIAL METHODS

The improved inter-frame difference method is used in the paper. The inter-frame difference method [6] is to take the absolute value $D_n(x,y)$ for the gray value of the corresponding position of each successive frame in the video, and then the two adjacent values is performed OR operation, and the inter-frame differential image $P_n(x,y)$ is obtained. When the absolute value is greater than the pre-determined threshold value T_3, it is converted to binary image $R_n(x,y)$ to obtain the contour information of UAV. Its specific principles are as follows formula(1) , (2), (3):

$$P_n(x,y) = | f_k(x,y) - f_{k-1}(x,y) | \qquad (1)$$

$$D_n(x,y) = P_n(x,y) | P_{n+1}(x,y) \qquad (2)$$

$$R_n(x,y) = \begin{cases} 0, & D_n(x,y) < T_3 \\ 1, & D_n(x,y) \geq T_3 \end{cases} \qquad (3)$$

Among them, "$|$"is used to OR operation, $k = 2,4,6,8, \cdots 2n$, $n = 1,2,3,4, \cdots n$.

The captured video is captured to a small video containing 90 frames in order to complete the real-time detection in the paper. After the video is processed by inter-frame differential methods, the binary video is saved as a file in avi format and the region of interest is segmented in the later stage. The results processed by the two consecutive frames difference method are shown in figure 1:

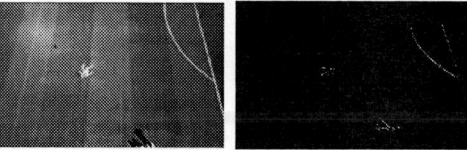

(a)Original Image   (b)Two Successive Inter-Frame Differential Image

Figure 1. Comparison Image.

Because there are many noise points in the binary image and the target is prone to cavitation, the target location to be tagged is too wide and the number of targets to be tagged is not practical. So it is necessary to be processed used the morphological methods on the threshold image. The cavitation is filled by used the closed operation and the adjacent pixels is connected, and the open operation is used to remove the noise and the size of the target is not changed, which makes the target more full and helps to save the location and detect the target in this paper. According to the constructed disk radius of 10 structural elements *se*, the binary image is first closed and then opened in the paper. The principle is respectively shown in expressions (4) and (5) .

$$Xose = (X \oplus se)\Theta se \qquad (4)$$

$$Xose = (X\Theta se)\oplus se \qquad (5)$$

In the formula, X is the binary image to be processed; Θ is the operator for the morphological corrosion operation [6]; and $\oplus$ is the operator for the morphological expansion operation.

Then the connected region marking function (bwlabel) in matlab is selected to save the position of eight connected regions with a threshold of 1, and the region of interest is segmented from the corresponding position of the gray image of the corresponding frame, and the region of interest is normalized to a pixel of 128*64. The segmentation results are shown in figure 2:

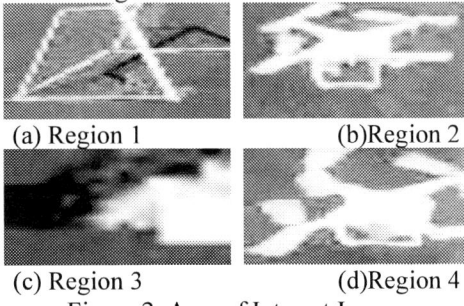

(a) Region 1   (b)Region 2

(c) Region 3   (d)Region 4

Figure 2. Area of Interest Image.

3. TARGET UAV DETECTION BASED ON HOG-FLD FEATURE FUSION AND SVM DETECTION

3.1 Feature Extraction From HOG-FLD Feature Fusion

Because of the strong stability and extensibility of the edge contour feature of the rotor UAV, the HOG [3](Histogram of Oriented Gradient) feature describing the edge contour feature of the object is calculated from the picture. The HOG feature is not easy to be affected by the local minor deformation of the object and it's dimension is large, which is not conducive to the training of the classifier and to meet the real-time requirements of the subject. HOG-FLD [19] feature fusion method is used for feature extraction of image files in the paper. On the basis of the definite contour features, the computation amount is reduced to improve the speed of the algorithm, and the features that are favorable to the classification can be extracted, and the precision rate and processing time of the algorithm can be greatly improved.

The basis of feature extraction is to extract the characteristic of HOG. The idea of the algorithm is to calculate the edged gradient of image to be entered; each image is divided into rectangles with fixed size and equal size as cell containing pixel of m * m; and all cells are divided into 18 directional channels or 9 non-directional channels, and the gradient histogram of all directions is all voted, and the weight used to vote is the gradient value calculated in previous step. Then the fixed blocks of the same size is combined by above-divided units, that contains n * n cells. And the local feature vectors corresponding to each block are normalized so that it make the effect of the experimental from the light from the image reduced, and the HOG feature vectors of the image are combined by the feature vectors of the blocks.

To extract HOG feature, the positive sample image of UAV is firstly grayed and filtered with Gamma correction method to make the image meet the standard requirement that the image preprocessed could make the influence of local shadow and light change reduced. Then the image is divide into a number of cells, they could formed a number of fixed blocks of the same as size. For blocks of the same size in the previous paragraph, the gradient range is classified by the above rules. we can calculate the cell characteristics in these blocks and eventually connect all the blocks and the feature vector of image is obtained.

The formula normalizing the image is as Formula (6); the formula calculating the gradient component of each pixel is as formula (7) and formula (8); the formula calculating the size and direction of the gradient is as formula (9) and formula (10).

$$v_g \leftarrow \sqrt{v_g /(\| v_g \|_1 + \varepsilon)} \qquad (6)$$

$$Gx_{(x,\ y)} = pi_{(x+1,\ y)} - pi_{(x-1,\ y)} \qquad (7)$$

$$Gy_{(x,\ y)} = pi_{(x,\ y+1)} - pi_{(x,\ y-1)} \qquad (8)$$

$$S_{(x,\ y)} = \sqrt{Gx^2{}_{(x,\ y)} + Gy^2{}_{(x,\ y)}} \qquad (9)$$

$$\theta_{(x,\ y)} = \arctan(Gy_{(x,\ y)} / Gx_{(x,\ y)}) \qquad (10)$$

Where, v_g* is the result of histogram normalization, and v_g is the extracted vector histogram; $pi_{(x+1,y)}$, $pi_{(x-1,y)}$, $pi_{(x,y+1)}$ and $pi_{(x,y-1)}$ respectively denote the location of 4 pixel points; $Gx_{(x,y)}$ and $Gy_{(x,y)}$ respectively denote the coordinate position in the horizontal and vertical directions of the two pixels. And $S_{(x,y)}$, $\theta_{(x,y)}$ respectively denote the length of the gradient direction vector and the angle of the gradient direction vector.

In this article, a window including 64 * 128 pixel is used to scan the sample image and the image to be detected using a window including 64 * 128 pixel. The scanning step size is 8 pixels (scanning is in horizontal and vertical direction). The window is divided into cells including 8 * 8 pixel and forms 8 * 16 = 128 units. Then setting up four adjacent units up and down to the left and right as a block of pixels, a window contains 105 blocks of pixels. A 3780 dimensional feature vector named HOG feature description value is generated in a window containing 105 pixel blocks according to the calculation steps of HOG. Its specific HOG algorithm is shown in Figure 3:

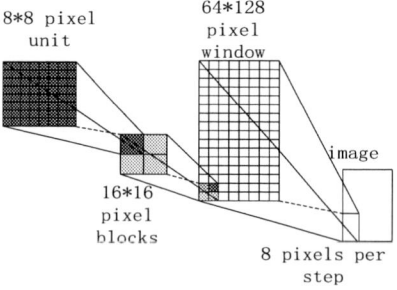

Figure 3. the Principle of HOG Feature.

On the basis of HOG feature, the linear subspace is constructed by using FLD[18]. By calculating the optimal projection matrix, the methods could obtain projection matrix for feature extraction of the training set, and the similarity of its projection vector is taken as the similarity degree S_{cos} of cosine similarity, which purpose is to reduce the intra-class dispersion S_W as much as possible and to increase the inter-class dispersion S_B as much as possible (or that is in the training concentration, to make the sample data of the same UAV as close as possible, and the sample data of different UAV to be far away). In this way, the features with classification ability are extracted. When dealing with a problem that type is c, the mathematical formulas of inter-class dispersion S_B and intra-class dispersion S_W should be defined as follows:

$$S_B = \sum_{i=1}^{c} N_i(\mu_i - \mu)(\mu_i - \mu)^T \qquad (11)$$

$$S_W = \sum_{i=1}^{c} \sum_{x_k \in \Omega_i} (x_k - \mu_i)(x_k - \mu_i)^T \qquad (12)$$

Where μ_i denotes the mean value of class Ω_i; μ means the mean value of total sample; N_i means the number of samples of class Ω_i. The optimal projection matrix W_{opt} could be obtained by solving the optimization problem such as formula (13), where S_W must be a nonsingular matrix (or that is the total number of training samples N is greater than the characteristic dimension of UAV image):

$$W_{opt} = \arg\max_W \frac{|W^T S_B W|}{|W^T S_W W|} \qquad (13)$$

W_{opt} could also be obtained by solving the generalized eigenvalue problem such as the formula (14):

$$S_B W = S_W W \Lambda \qquad (14)$$

In order to solve the problem that the intra-class dispersion matrix S_W is singular, PCA principal component analysis (PCA) is used to reduce the dimension of the feature space (dimensionality reduction to N-cu), and then Fisher linear discriminant analysis (FLD) is used to deal with it. The projection vector y of test sample x is obtained according to formula (15):

$$y = W_{opt}^T x \qquad (15)$$

Cosine similarity S_{cos} is used as the similarity measure of projection vector y. Where the cosine similarity S_{cos} of vector $A = \{a_1, a_2, \cdots, a_n\}$ and $B = \{b_1, b_2, \cdots, b_n\}$ is defined as follows:

$$S_{cos} = \frac{\langle A, B \rangle}{\|A\|_2 \|B\|_2} = \frac{\sum_{i=1}^{n} a_i b_i}{\|A\|_2 \|B\|_2} \qquad (16)$$

3.2 Support Vector Machine

Because the support vector machine (SVM) proposed by Vapnik has the advantages of simple system structure, global optimization, good generalization, and short training and prediction time [9], this paper uses SVM as a machine learning tool to calculate the rule of samples in order to achieve fast and efficient learning sample features and accurate classification purposes. The main idea of SVM is to deal with the linear inseparability of the original space by selecting the kernel function of Polynomial Kernel to correspond the data to the high-dimensional space. When the algorithm is used to realize the two-classification, the sample features such as HOG must be extracted from the original space first, and then the sample features in the original space are represented as a vector in the high-dimensional space. In order to minimize the error rate of the two class classification problems, we need to find a hyperplane that is used to divide the two classes in the high dimensional space.

Let the sample set be x_i, y_i where i=1,2,...,N, $x_i \in R^e$, $y_i \in \{0,1\}$ is the class identifier. Then in e dimensional space, its linear discriminant function is as follows:

$$g(x) = w * x + b \qquad (17)$$

The formula of the classification surface equation is as follows:

$$w * x + b = 0 \qquad (18)$$

After normalization of the discriminant function, the following conditions must be satisfied for the two types of samples:

$$g(x) \geq 1 \qquad (19)$$

The classification interval could be $2/\|w\|$, where the maximum requirement $\|w\|$ for the classification interval is kept to a minimum, all samples must be correctly classified, and the following conditions must be met:

$$y_i [(w * x_i) + b] - 1 \geq 0 \qquad (20)$$

An SVM whose inner product function is $k(x_i, x_j)$ is constructed with the following formulas (which can be understood as the formula for calculating the extreme value of a quadratic function with conditional constraints):

$$Q(a) = \sum_{i=1}^{N} a_i a_j y_i y_j k(x_i, x_j) \qquad (21)$$

Its constraints are expressed as: $0 \le a_i \le C, \sum_{i=1}^{N} a_i y_i = 0$

The formulas of the support vector machine that could be calculated are as follows:

$$f(x) = \text{sgn}(\sum_{i=1}^{N} a_i^* y_i k(x, x_i) + b^*) \qquad (22)$$

Among them, b^* is a constant parameter, indicating the size of the threshold that needs to be classified.

3.3 Obtaining the Sample and Training Classifier Model

Firstly, 500 original images of positive samples are used to statistic the aspect ratio of UAV, and the statistics show that aspect ratio of UAV is 1:2, so the pixels of each sample image are normalized to 64 * 128 to avoid the effect of image size on the recognition effect of the algorithm. Finally 1400 positive sample images and 1400 negative sample images are used to train SVM model in the paper. As described in the HOG-FLD feature part above, the input image all could create 105 pixel blocks which could get a 36 dimensional feature vector respectively, so 36 dimensional feature vector finally could combined into a 3780 dimensional feature vector as HOG feature vector of the input image. Then the extracted HOG vector is used as the input vector of FLD analysis algorithm to cut back the dimension of the feature vector of the whole image. According to the experimental recognition efficiency of of the whole algorithm, the dimension of the vector is adjustable and determined with its parameter changed. Sending the final extracted HOG-FLD feature vector to the SVM model, and the SVM classifier that could detect the input image is if the UAV could be trained .

4. PARAMETER ANALYSIS AND EXPERIMENTAL RESULTS

4.1 Parameter Adjustment of HOG-FLD

In the experiment of extracting feature vector, the parameter k of code of FLD algorithm represent the dimension of the feature vector that needs to be cut back by FLD analysis algorithm, with the change of the parameter k, the dimension of the feature vector is also changing. As shown from Figure 4, the contrast graph of recognition time of the algorithm is obvious with the change of the parameter k of the algorithm. It could be clearly seen from the line chart that when parameter k equals 50, the recognition time of the detected algorithm is the least in the paper.

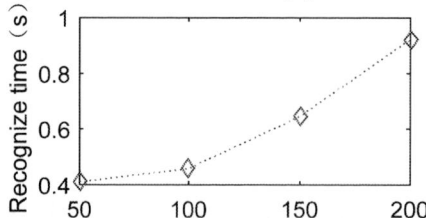

The dimension of the feature vector（k）

Figure 4. the Time Line Chart of with k Changed.

4.2 Experimental Effect of Aerial UAV Detection Algorithm Based on Region of Interest

Experiments show that in the whole algorithm, when the number of pixels in the divided blocks is 8*8, SVM kernel function type is 2, when the threshold of segmentation is k=90 and sigma=10, the algorithm has the best overall recognition effect on accuracy rate and time, and the recognition result is shown in Figure 5:

| (a)Result 1 | (b)Result 2 |
| (c) Result 3 | (d)Result 4 |

Figure 5. Image of Identification Results.

4.3 Comparison of Detection Effect of Traditional HOG-SVM UAV Detection Algorithm

The experimental results of the proposed method are compared with the experimental results of the HOG and SVM method based on image segmentation, and the experimental results are obtained by statistics so that it could be verified its own efficiency. According to the accuracy of one of the evaluation criteria in the machine learning algorithm, N is used to represent the number of regions of interest per test image, TP is used to represent the number of UAVs recognized as UAVs, and TN is used to represent the number of non-UAVs recognized as non-UAVs. The ACC formula for accuracy is:

$$ACC = \frac{TP + TN}{N} \qquad (23)$$

As shown from Table 1, the recognition accuracy rate of the algorithm is close to the latter algorithm and the recognition time of the algorithm is faster than that of the latter algorithm.

Table 1. Result Comparison of the Detection Algorithm

Test Algorithm	ACC	Test Time
HOG_FLD+SVM	92.45%	0.093s
HOG+SVM	92.60%	0.162s

5. CONCLUSION

The method based on inter-frame difference to segment the region of interest is used in the paper. In the testing stage, the acquired regions of interest are input into the trained the model of SVM classifier, which cuts back the recognition time of the whole algorithm in the extent. Extracting the HOG-FLD feature of the input image is easier for training SVM classifiers. In the platform of Matlab, compared traditional HOG+SVM detection algorithm to the algorithm in the paper, the experimental results show that the detection algorithm in the paper is better than the sliding window (HOG+SVM) algorithm to detect UAV in terms of time and accuracy .

ACKNOWLEDGMENTS

Fund projects: Key projects in the Industrial Field of Shaanxi (2016KTZDGY4-09), Scientific Research Program Funded by Shaanxi Provincial Education Department (17JK0364), National Natural Science Foundation of China (61572392)，National Joint Engineering Laboratory of New Network and Detection Foundation (Grant No. GSYSJ2016008).

REFERENCES

[1] J. Barron, D. Fleet, S. Beauchemin. (1944). Performance of optical flow techniques. International Journal of Computer Vision, 12: 42-77.

[2] A. Lipton, H. Fujiyoshi, R. Patil. (1998). Moving target classification and tracking from real-time video. Proc IEEE Workshop on Applications of Computer Vision, 8-14.

[3] DALALN, TRIGGS B. (2005). Histograms of oriented gradients for human detection. Proc IEEE Conference on Computer Vision and Pattern Recognition, 1-8.

[4] Qiang Zhu, Shai Avidan, Mei Chen Yeh, and Kwang Ting Cheng. (2006). Fast human detection using a cascade of histograms of oriented gradients. Proc.IEEE international Conference on Computer Vision and Pattern Recognition.

[5] Suard F, Akotomamonjy A R, Bensrhair A, etal. (2006). Pedestrain Detection Using Infrared Images and Histograms of Oriented Gradients. Proceedings Intelligent Vehicle Symposium, 206-212.

[6] Wang Maosen, Chen Long, Dai Jinsong. (2016). Research on UAV Motion Detection Based on Improved Frame Difference Method. Electricity and Automation, 45: 165-168.

[7] Sun Ting, Qi Yingchun, Geng Guohua. (2016). Moving Target Detection Algorithm Based on Inter-frame Difference and Background Difference. Journal of University, 46: 1325-1329.

[8] Vapnik V N. (1995). The nature of statistical learning theory. Springer-Verlag, New York , 37-69.

[9] Guo Mingwei, ZhaoYuzhou, Xiang Junping, etal. (2014). A Survey of Target Detection Algorithms Based on Support Vector Machine. Control and Decision, 29: 192-200.

[10] Zhang Han, He Dongjian. (2011). Cross-camera Moving Target Detection and Recognition. Automation Technology and Application, 30: 43-46.

[11] Zhai Jiyou, Zhuang Yan. (2017). Significant Detection of Boundary Prior and Adaptive Region Merging. Computer Engineering and Application.

[12] Chen Shanchao, Fu Hongguang, Wang Ying. (2012). Application of An Improved Graph Segmentation method in Tongue Image Segmentation. Computer Engineering and Application, 48: 201-203.

[13] Yan Yu, Song Wei. (2016). Color and Texture Mixed Descriptor Image Retrieval Method. Computer Science and Exploration, 1-8.

[14] Chan Qiwen. (2011). ROI Detection Algorithm for Small Infrared Target in Infared Image Based on Hypothesis Testing. Computer and Modernization, 8: 135-137.

[15] Zhao Zhuxin. (2012). Estimation of Target's Motion Parameters Using Line-scan Camera. Opto-Electronic Engineering, 36-40.

[16] Girshick R, Donahue J, Darrell T, etal. (2014). Rich Feature Hierarchies for Accurate Object Detection and Semantic Segmentation. Computer Science, 580-587.

[17] Wu Dapeng. (2010). Cam-shift Object Tracking Algorithm Based on Inter-frame Difference and Motion Prediction. Opto-Electronic Engineering, 1: 210-213.

[18] Belhumeur P, Kriegman D. (1997). Eigenfaces vs. Fisherfaces: Recognition Using Class Specific Linear Projection. IEEE Transactions on Pattern Analysis and Machine Intelligence, 19: 711-720.

Discussion About Artificial Intelligence's Advantages and Disadvantages Compete with Natural Intelligence

Xiaofei Teng

High School Affiliated to Renmin University, Beijing 100000, P.R. China

Chelsea_bean@yeah.net

ABSTRACT Artificial intelligence and natural intelligence both have their own advantages and disadvantages. This article discusses the situations where artificial intelligence outcompetes natural intelligence, and where natural intelligence outcompetes natural intelligence, by modeling the scene of predation. Via adjusting variables in the model, different circumstances can be simulated and different outcomes can be verified. The results coincident well with the most wildly-accepted theory about the origin of natural intelligence, and can reasonably infer about artificial intelligence's limits under current technology and algorithms.

1. INTRODUCTION

In the past decades, artificial intelligence has grown rapidly. It accomplished things that we can never imagine before, like beating the world champion in a chess game. However, there are still numerous things, which are super easy for human beings, that we consider impossible for artificial intelligence to achieve. So, what makes such big difference between artificial intelligence and natural intelligence? And, what, actually, is the difference? To answer the first question, we have to have a clear view of how artificial intelligence works and how our brains work; and more complicated researches must be conducted in order to answer the second one.

First, a brief introduction of the origin of artificial intelligence will be made, which will serve the purpose to illustrate how today's AI works. The story goes back to the year 1936, when Turing first came up with the idea of an abstract model of a machine, which was later named "The Turing Machine". The Turing Machine can do three things: read the information from a type, write or remove a 0 or 1 on the paper, and move the type whether left or right. Turing considered the model as an "universal machine" because he thought it was able to complete any kind of work that could be written in a mathematical language. He was right, and that made him the father of all computers. The system was all made of 0 and 1, which is called the binary system and is still applied in computers today. Computer science was developed rapidly, but artificial intelligence would never exist if biologists didn't discover that the neuros in our brain fired electronic signals. The discover led to a burst of enthusiasms which was believed that computers could completely simulate the brain's function and human beings were able to create robots that were just as clever as people themselves. The reason was simple: when a neuron receives a signal, whether it fires or not, and that exactly correlates our comprehension to the binary system, which is successfully employed in computers. Of course, those scientists failed. Later researches have proved that our brain works in a way that is far more complex, and the signals fired by neurons are more likely to be mechanic instead of electronic, which creates huge difference between signals. Despite the hopeless facts that were provided by researchers, computer scientists were still optimistic, and for sure, they didn't make any progress in the field, until the 21st century.

Things are different in the new era, when software technic grows in a shocking speed. Because of the number of running speed of the CPU being astronomical, possibility emerges from mission impossible. On the other hand, biology steps forward in a faster pace, and computer scientists are convinced to give up the idea that creates an artificial intelligence which is as clever as human beings. Instead of that, they come up with new ideas, and that's when neuron network and deep learning take the stage. Scientist have long been puzzled by how to "teach" computers, and this time, they make them "learn" on their own. The neuron network is simply a lot of neurons, each has its own variables and functions. Working together, they give an output to particular input. The neuron network is like a black box, which means scientist have no idea what each variable should be, but in fact, that doesn't matter. The running speed of computers are so advanced that they can adjust variables in the neuron network on their own based on huge amount of labeled data. Moreover, if the outcomes aren't as good as expected, just retrain the network. Through the neuron network and deep learning process, the AI can do complex tasks such as NLP, image detection, and video analysis. Some of the tasks require heavy labor, and some of them are even not be capable to do by humans. So, it won't be surprising to find that artificial intelligence plays a very important role in today's society.

While AI has its own advantages, its disadvantages are just as obvious as ever. For instance, most of people can recognize a movie star even if they have only had a glance at his or her new movie on TV. However, thousands of pictures from different perspectives of that star are needed if you want to train an AI to recognize him or her. The function of human brain is known as one-shot learning, whereas the AI's is known as deep learning. We see from the example that our brains work in a more flexible way, and that has something to do with the origin of natural intelligence. The most famous aspect of the origin of human intelligence is that our evolution happened because of the last ice age. In most situations, creatures evolve by natural selection. The cost of intelligence is so high that in most situations natural selection won't give you that. However, natural selections take generations to occur, and it may not work well in extreme climate like the ice age. Professor Steen has an interesting view about natural selection: he considers it just as the function of deep learning, where huge datasets are required to make improvements. "In the ice age," according to Professor Steen, "functions which works in your parents' time may not work in yours, and that's when intelligence outcompete natural selection and the cost pays off.

The question is clear from here. The competition between artificial intelligence and natural intelligence is actually a competition between deep learning method and one-shot learning method. Both advantages and disadvantages of the two methods are obvious; however, it still need to dig deeper into the conditions that influence the performance. In order to do that, a praxiological model is made and different variables are set inside the model.

2. THE PREDATION MODEL
The first model that is made to solve the problem was simple. It simulates a predation so it likes a game. In the model, there is a predator, a prey, and n paths, each with a weight. Each time, the prey chooses the path which weighs the most on its side, and so does the predator. When the prey gets caught, game overs.

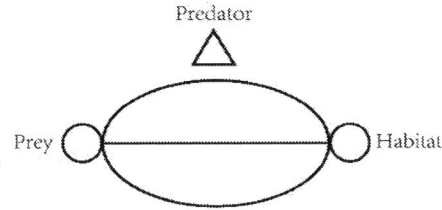

Figure 1. The predation model

But, the influence of intelligence, interactions must be involved, which means the weighs of the paths should change according to the actions the prey and the predator take each time. In order to avoid randomness in the experiment to ensure the outcome is correct, formulas for changing the weight each time are calculated, and as the "degree" of intelligence must be shown clearly, the parameter p_prey and p_pred, which stand for how much the prey's and the predator's next move will be influenced by the former performance, range from 0 to 1.

```cpp
for (int i=1;i<=n;i++)
{
    if (i==x[step])
    {
        a[i][step].w_prey=a[i][step-1].w_prey+(1-2*p)/n;
        a[i][step].w_pred=a[i][step-1].w_pred+1/k;
    }
    else if (i==y[step])
    {
        a[i][step].w_prey=a[i][step-1].w_prey+(2*p-1)/n;
        a[i][step].w_pred=a[i][step-1].w_pred-1/k;
    }
    if (a[i][step].w_prey>prey_max) {prey_max=a[i][step].w_prey;prey_next=i;}
    if (a[i][step].w_pred>pred_max) {pred_max=a[i][step].w_pred;pred_next=i;}
}
```

Figure 2. The formula in the C++ window

To be honest, the first version of the predation model is a complete failure. First, randomness is unignorable in real world. Even natural selection happens because of random mutations in DNA. To remove the extreme situations, the program must be run for multiple times and the relative outcomes must be collected in order to make sure that doesn't seem to go extreme. Also, by adding randomness, whether the algorithm used is proper can be examined, because an outcome which proves to be unstable is more likely to go wrong. Then, the behavior mode should not always be the same for both preys and predators. It was considered to be a side effect to avoid randomness, which makes both the prey and predator extra predictable. And, as the model works as a double-blind game, unchanged behavior mode means that select completely in random will be the best solution and there's nothing to do with intelligence. Finally, as predators are often superior in intelligence and less in number in real situation, a model of zero-sum game shouldn't be used. If the model is designed to put the prey and the predator in equal situations, it must have failed in simulate real predation.

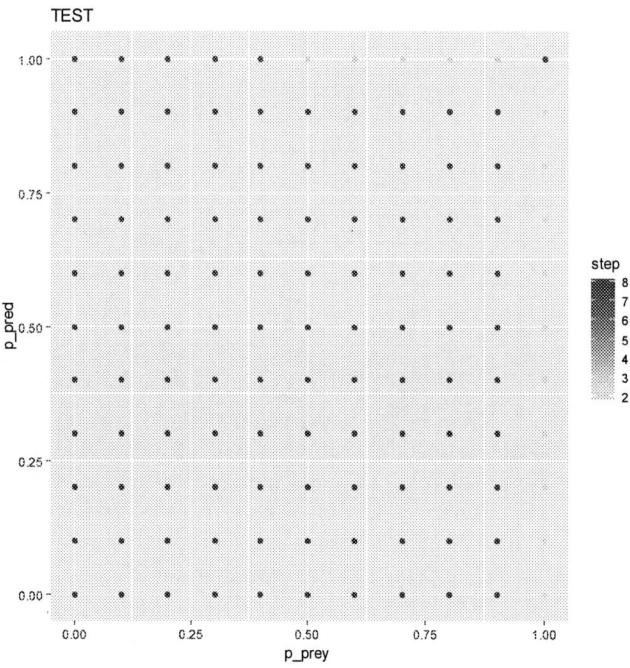

Figure 3. The outcome of the first version of the model

Another problem which is also worth-mentioning is a so called "Traversal Trap". When this occurs under specific set of parameters, the prey always maintains its choice, and the predator goes through all possibilities and the outcome is always n, which is the number of totally routes available.

Based on the failure of the alpha version, changes must be made. To make the simulation seems more like a real situation, variable behavior modes are added to the prey and preference is used instead of exact weight of each route. Also, randomness is added, which means the prey's choice is no longer determined by its preference, but influenced instead. To further improve the changeability of the prey, multiple preys are introduced, which means the preys will have more than one chances before they die out: the preys' preference changes every time a prey is caught or when no prey is caught in t turns (consider t as a period). In conclusion, the preys are given a behavior mode, a total number, and a period at the beginning. There are three designed behavior models in all: none, slight, and strong, each of those has different degree of influence on the random function which decides the final choice of preys each turn. On the other hand, there are also changes made for the predator-a new way to measure its intelligence level. The predator is given a k, which means he will look back in the last k steps taken by the prey to decide its next move. The larger k is, a larger dataset is involved to determine the outcome, thus the predator behaves more like deep learning; in contrast, a smaller k means the predator behaves in a more "one-shot-learning" way. The function of determining the movement is simple majority, which means the predator will always take the route that is visited the most often by the preys in the last k rounds of the game. In the beta version, it shows that under what circumstances a small k will do better and under what circumstances a large k will do better by adjusting the behavior mode of the preys, the period which the preys change their preference, the number of routes available, and the total number of preys; and, correlate these parameters with real meanings to find out the advantages and disadvantages of artificial and natural intelligence. The result is quite the same as predicted:

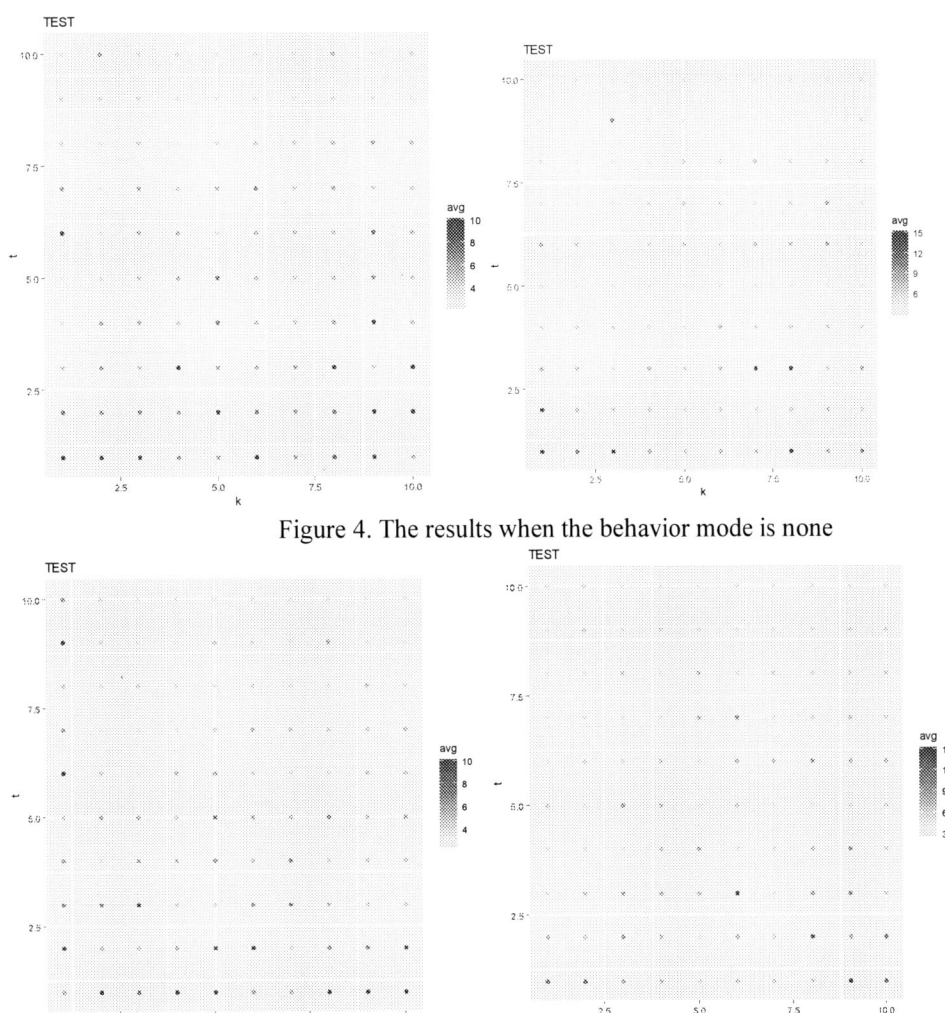

Figure 4. The results when the behavior mode is none

Figure 5. The results when the behavior mode is slight

ISPECE

IOP Publishing

IOP Conf. Series: Journal of Physics: Conf. Series **1187** (2019) 032083   doi:10.1088/1742-6596/1187/3/032083

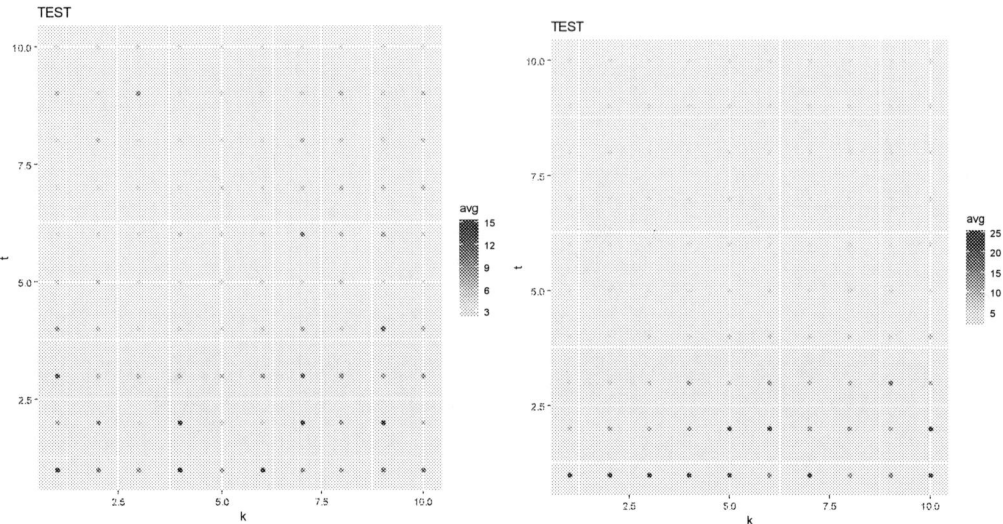

Figure 6. The results when the behavior mode is strong

Take a set of outcomes which the behavior mode is set differently as an example. It is predicted that preys with a small t will be harder to catch and predators with a large k will take less steps to catch all the preys, and those points that do not correlate with this trend is noisy. From the plots above, a fact can be found that the first two pictures contain the most "noisy" points and the last two pictures contains the least. So, it can be concluded that the outcome will be less noisy if the preference of the prey is stronger. And as the behavior mode measures how predictable the prey is and the number of noisy points measures the side effect caused by randomness, it can be known that when the prey is more predictable, randomness has less influence on the creature. Likewise, other conclusions can be drawn from the plots.

3. THE STANDARD MODEL

After all the works above have been done and with all the results that have been collected, more samples were tested; and almost by accident, a plot was made that is almost completely regular, which leaded to a new thought. Through testing datasets, making plots, and analyzing the outcomes, finally the standard model was found, in which not a single point is noisy, all the points are arranged in a perfectly regular way. The dataset is "n=15, m=43, c=strong". The standard model means far more than it looks to. With the standard model, the relationship between any two parameters can be determined by keeping other parameters unchanged, and that should cover all the situations in real hunting scene.

Professor Steen also gave a suggestion about the model. He mentioned that the benefit brought by a higher k seemed to have a linear relationship with k, because higher intelligence always brings more benefits. But in real situations, intelligence is "expensive", when the benefits of intelligence don't match with its cost, creatures stop evolving their brains, just as human did. He suggested that another parameter for the cost of intelligence should be added, which is in proportion to k, and see if it can get a quadratic outcome that indicates an extremely high intelligence is not worthy. The suggestion is interesting and reasonable, but I haven't had time for that.

4. CONCLUSION

From the experiment, it can be known that artificial intelligence specializes at particular fields. When the task is highly repetitive and is not very complex, artificial intelligence outcompetes human being by its efficiency and accuracy. Whereas the true value of intelligence glows when the environment changes rapidly and tasks require highly complexity of critical thinking. A human can do anything an artificial intelligence does, just as we can do all the calculations instead of using a calculator. But the mechanism of computers makes artificial intelligence a lot faster when doing tasks in a certain order and unchanged methods. However, bio scientists still have difficulties understanding the function of our brain. The only thing they can tell is that the brain is a more precise machine which works with more complex physical and chemical functions than a chip in the CPU. So, before mystery of our brain is solved, or new technology and algorithm are employed, "The only thing we have is still stupid AI," according to professor Steen.

REFERENCES

[1] BaiduBaike. "Artificial Intelligence"[EB/OL].[2018-08].
https://baike.baidu.com/item/%E4%BA%BA%E5%B7%A5%E6%99%BA%E8%83%BD/9180?fr=aladdin

[2] BaiduBaike. "Turing Machine" [EB/OL].[2018-08].
https://baike.baidu.com/item/%E5%9B%BE%E7%81%B5%E6%9C%BA/2112989?fr=aladdin

[3] UCLA system "Bash commands". http://vrnewsscape.ucla.edu/cm/Unix_shell_commands

[4] UCLA system "R commands". http://vrnewsscape.ucla.edu/cm/R_commands

Development of Artificial Intelligence and Effects on Financial System

Minzhen Xie

The only suthor Room 701, No52 Hebinbei Road, Conghua District, Guangzhou, Guangdong, China (61)0402557505

Mxie0002@student.monash.edu

ABSTRACT In recent years, due to the rapid development of artificial intelligence (AI) and machine learning, its application has been widely used in many aspects of financial area, as well as significantly impacts financial market, institutions and regulation. The artificial intelligence technology brings enormous change to the entire financial industry, which creates a series of innovative financial services such as intelligent consultant, intelligent lending, monitoring and warning, and intelligent customer service as times required. In this paper, it aims to summarize the development and application of artificial intelligence and machine learning in financial system, as well as its impacts on macroeconomics and microeconomics. In the meantime, it is realised that a series of problems and risks were conducted by artificial intelligence during its use. Lastly, some suggestions and strategies are provided for reasonable usage of artificial intelligence in financial risk management, based on the financial risk management raised by artificial intelligence.

1. INTRODUCTION

As the booming development of Internet and information technology, as well as in the context of Internet-Finance, method of financial data wrangling is not only limited to traditional statistical approach, but also adopt and combine with various information processing technology such as machine learning, which has obtained significant achievements. For example, Support vector regression algorithm and time series model in machine learning are used in the performance measure problem in establishing prediction model, which could improve the accuracy of prediction and financial data analysis. [1]

In the stock investment of financial market, public always wish to grasp the rule behind the transaction, which could be used for analysis and prediction [2]. Investment experts from all around world are also trying to apply different methods of investment analysis and data mining in the amount of stock data, in order to find out potential operating rules and stock trading rule behind the stock market, and to predict the stock market trend, aiming to maximize the profit. Since the stock market is affected by various market and non-market factors, which interact with each other, it is difficult to establish an accurate model to describe the mechanism of internal interaction [3]. Therefore, machine learning, as a "black box" model prediction, is increasingly being used in stock market prediction.

With the increasing of computer computing power, more sophisticated artificial intelligence algorithm can satisfy the need of new power in financial field. More specifically, artificial intelligence is widely used in investment management, algorithm trading, fraud detection, loan and insurance underwriting. Besides, artificial intelligence has profound impacts on financial regulation institution, and will help regulators to determine illegal compliance, evolve from past experience based on supervise

Content from this work may be used under the terms of the Creative Commons Attribution 3.0 licence. Any further distribution of this work must maintain attribution to the author(s) and the title of the work, journal citation and DOI.
Published under licence by IOP Publishing Ltd

transaction with algorithm and analysis of massive amount of data, while new required skills and knowledge for regulators will be presented.

Therefore, this paper aims to summarize the development of artificial intelligence and machine learning in financial field, and the impacts on macroeconomics and microeconomics, as well as providing suggestions for enhancement of financial regulation by using artificial intelligence and machine learning.

2. DEVELOPMENT AND APPLICATION OF AI AND MACHINE LEARNING IN FINANCIAL FIELD

2.1 The Development of Artificial Intelligence in Financial Field

With the rapid development of artificial intelligence technology, AI is widely popularized in financial field. Factors that accelerate the Fintech development, promote the development of artificial intelligence and machine learning in financial field, and drive financial institution to reduce cost, management risk, improve quality of service and increase profit by using AI and machine learning. In early 1960s, one of the algorithms in machine learning-Bayesian Statistics become famous, it has been widely used in financial area until now [4]. Moreover, the reason that Bayesian Theory become popular in financial area, is its application in auditing area. In the auditing field, judgement made by auditor, used to rely on professional knowledge and experience, but different cases have different situations. Various uncertain factors need to be considered in the decision-making of auditing. Therefore, Bayesian model provides objective and rational probability to auditor and help them to make more accurate assessment, as well as reducing the misjudgement caused by auditor's personal emotion [5]. In the initial stage of cooperation between artificial intelligence and financial industry, it focuses on reducing the workload of financial practitioners by computing power of computers. Until 1980s, Expert System (intelligence system based on knowledge) is used in the financial industry to predict the trend of market and provide customize financial plan. A basic Expert System includes six components, which are knowledge base, data base, inference engine, explanation facility, knowledge acquisition and user interface, as figure 1 shown. Hodgkinson and Walker [7] raised a rule-based Expert System to achieve decision-making process, which helps financial intuitions to make decision on credit application of cooperate credit loan. Shue et al. [8] establish Expert System contains domain knowledge base and operational knowledge base, to credit ranking of listed companies in Taiwan stock market. Janulevicius and Goranin [9] built up a risk management Expert System to help middle and small-sized enterprise to solve the problem of lack of access to professional data security analysis due to limited fund.

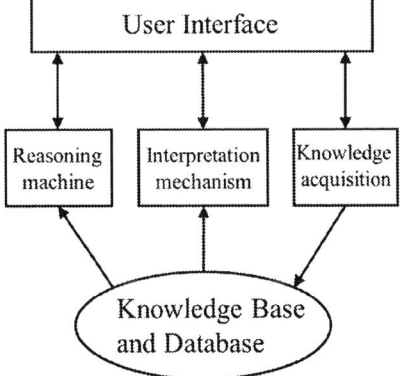

Fig. 1 Basic structure of Expert System

In 1990s, due to improved computing power of computer, a series of artificial technologies are used in financial fraud detection. Artificial intelligence trying to analyze massive amount of data and find out the outlier to determine financial fraud. [10] As the increasing computing power of computer and continuous improvement of artificial intelligence algorithm, the integration of artificial intelligence and financial industry will be higher and higher, and it will also be applied to more aspects in financial field.

2.2 Application of Artificial Intelligence in Financial Field

The application of artificial intelligence and machine learning in financial field can be divide into four aspects. Firstly, it is customer-oriented (front-end) applications, including credit scoring, insurance and customer-oriented service robot; secondly, management level (back-end) applications, including capital optimization, risk management and market impact analysis; thirdly, financial market transactions and portfolio management; lastly, AI and machine learning are used in financial institutions for "RegTech" or financial regulators for "SupTech".

In addition, AI and machine learning are widely used in specific scenarios such as quantitative transactions in the financial field, natural language processing, semantic search and intelligence investment consultants.

3. DATA MINING

The application of data mining is widely used and cover wide range of aspects, especially being used popularized in financial and manufacturing industry. [11] With the advent of the era of big data, data mining is attracting more attention than ever, specific in precision marketing and market analysis. With the benefit of development of data mining technology and computer technology, establishment of data warehouse develop rapidly [12]. Especially in the securities investment of financial markets, investment experts from various countries use different stock analysis methods for data mining from massive stock data, in order to find out the potential operating rules and stock trading rules behind the stock market, and realize forecasting changes in the stock market in the future to achieve the goal of maximizing returns [13-14]. Regular data mining methods include association rule learning, cluster analysis, classification analysis, sequence analysis, deviation detection, prediction analysis, pattern similarity mining and regression analysis. [15-18]

Data mining being used popularity in financial application. Marketing is one of the earliest application of data mining, especially achieving well performance in e-commerce. Data engineer in-deep mining and analyze on customer's browsing data, search data, order records and other behaviour to identify the customer's purchasing behaviour pattern, which achieve significant performance on advertisement and product recommendation. On the aspect of risk analysis and fraud detection, analyzing banks or insurance customer's creditworthiness, asset and loan configuration, to detect fraud behaviour, such as bad debt and fraud insurance. In the manufacture industry, according to the comprehensive analysis of data, to identify the outlier and mining the data to analyze the cause of product failure or expose the unqualified product, which helps the quality inspection engineer to find out the flaws and make corrective measurement. In addition, the financial data generated by most banks and financial institutions is relatively structured and highly reliable, which is advantageous for data visualization and data mining, as well as being applied to approaches like pattern analysis to detect money laundering and other financial crimes.

4. IMPACTS OF ARTIFICIAL INTELLIGENCE ON MICROECONOMIC

The rapid development of artificial intelligence has brought new impacts and vitality to various industries. In addition to the practice field of traditional artificial intelligence- the Internet industry, it also brought new impact and vitality to traditional industries, such as manufacturing and service industries.

1) Artificial intelligence can automate "programmable work". Once the cost of this automation is much lower than the labor cost, industry will not hesitate to use artificial intelligence to replace this part of the labour force. It will result in a large number of "programmable" practitioners unemployed. Moreover, most of the unemployed "programmable" practitioners are low-level skill and educated practitioner, which is difficult to re-educated to satisfy the new position. On the other hand, artificial intelligence brings new vitality to employment, even though AI will replace most of the "programmable

work" practitioners. The demand to "non-programmable work" practitioners will increase. Practitioners with high-level education and skill practitioners will increase in the long term and in large numbers as well.

2) Since the artificial intelligence changes the structure of labor market, it promotes the redistribution of internal income distribution among labor. There will be more significant differences in the income of labor with different skill levels. The change of income distribution also reflects on the component (industry). The industry with less "programmable" work is less affected by artificial intelligence, and the change in income distribution is also less than other industries.

3) Upgrading and transforming with the industry structure of market: every technological innovation brings a new industry replacement and upgrade. From the financial, manufacture in the past to Internet and technology company nowadays. The arrival of artificial intelligence will strengthen the dominant position of technology and Internet companies; therefore, traditional industries must transform and integrate new technology to avoid being eliminated by the wave of development. [19]

4) The impact of artificial intelligence on the innovation ability of enterprises: As the labor force develops toward high technology and high knowledge, enterprises need strong innovation ability and speed to maintain core competitiveness. Enterprise managers must be keenly aware of the cutting-edge technology and be bold to use new technology to innovate products. [19]

5) The impact of artificial intelligence on human resources: As enterprises need innovation to maintain core competitiveness, high-end talent resources will be particularly popular, artificial intelligence expert will become an important asset of enterprises. Companies will also attract talent through more favorable conditions. [19]

5. THE IMPACTS OF ARTIFICIAL INTELLIGENCE ON MACROECONOMIC

1) The impact of artificial intelligence on economic growth: Artificial intelligence is mainly embodying in the automated processing of "programmable" work, thereby by reducing labor costs and improving production efficiency in the production process, it brings the profits. However, it will lead to cost increase in non-automated sector, which will reduce the share of capital return in the economy.

2) The impact of artificial intelligence on industrial organization: The first channel is the direct impact of technology, and second channel is the change of enterprise structure caused by technology. With the development of artificial intelligence, the trend of mergers and acquisitions of downstream enterprises by the large platform enterprises will become more obvious. Large-scale platform enterprises are not competing for direct market profits and shares, but the data resources of entire industry chain, so as to better develop artificial intelligence.

3) The impact of artificial intelligence on trade: Since artificial intelligence has significant impact on factor returns, and changes the relative returns between different elements, so that the dynamic advantages of countries change. High-tech, high-knowledge talents are vital factors in the development of artificial intelligence, therefore, talents will become important targets of trade.

4) The impact of artificial intelligence on GDP: Artificial intelligence can play an important role in e-commerce recommendation system, which could change people's consumption habit and promoting consumption. It can satisfy the requirement of stimulating domestic demand and become a new carriage that drives consumption.

5) The impact of artificial intelligence on public policy: Based on the negative impact of artificial intelligence on the labor market, the government should formulate targeted public policies to ensure that overall social welfare is not impaired and alleviate the pressure caused by income inequality, which could better enjoy the productivity growth brought by artificial intelligence and the economic growth by social stability. For example, after the first industrial revolution, many low-skilled laborers in the UK were unemployed, resulting in uneven income distribution. The British government solved the problem of income inequality by providing free public education for the unemployed and improving the legal

status of the trade union. Future public polices for artificial intelligence can also be: re-education of unemployed people, taxation of robots and raising minimum guaranteed income standards. [20]

6) The impact of artificial intelligence on macroeconomic research methods: Traditional economic research methods focus on the small samples and low-dimensional data, which make the traditional economic model have certain limitations. After applying AI in it, when researching new economic models, it is possible to use large-sample, high-dimensional data to verify. Artificial intelligence will greatly promote the development of economics.

6. CONCLUSION

As a new field full of opportunities and challenges, artificial intelligence is an inevitable outcome of the development of science and technology, but at the same time, there are corresponding challenges in applying artificial intelligence. Therefore, the financial system should completely understand artificial intelligence and make its application system in the financial field more consummate. In order to design a complete artificial intelligence, it is necessary to set guiding principles firstly, which aims to guide the whole process of artificial intelligence development, design, use, management and control, and carefully promote artificial intelligence applications in the field of financial risk management. In the field of financial risk management, the used of artificial intelligence for information collection must follow certain criteria to ensure the legitimacy of information collection and the interests of those who do not harm the information source. The types and strength of artificial intelligence information collection and behavior of information collection are standardized. What's more, we should focus on R&D and application of user information encryption technology and strengthen the application of artificial intelligence in the field of financial risk management.

REFERENCES

[1] F Li, ZH Han, EY Feng. Financial data analysis based on machine learning [J]. Sciences & Wealth, 2016, 6: 624.

[2] WQ Huang, XT Zhuang, S Yao. A network analysis of the Chinese stock market [J]. Physica A Statistical Mechanics & Its Applications. 2009, 388 (14): 2956-2964.

[3] T Preis, DY Kenett, HE Stanley, D Helbing, E Benjacob. Quantifying the behavior of stock correlations under market stress [J]. Scientific Reports, 2012, 2 (7420): 752.

[4] RT Bayes. An essay towards solving a problem in the doctrine of chances [J]. Resonance, 2003, 8: 80-88.

[5] JE Sorensen. Bayesian analysis in auditing [J]. Accounting Review, 1969, 44(3): 555-561.

[6] KC Chen, T Liang. PROTRADER: An expert system for program trading [J]. Managerial Finance, 1989, 15(5): 1-6.

[7] L Hodgkinson, E Walker. An expert system for credit evaluation and explanation [J]. Consortium for Computing Sciences in Colleges, 2003, 19(1): 62-72.

[8] LY Shue, CW Chen, W Shiue. The development of an ontology-based expert system for corporate financial rating [J]. Expert Systems With Applications, 2009, 36(2): 2130-2142.

[9] Justinas Janulevičius, N Goranin. Expert system for data security risk management for SMEs [J]. Science – Future of Lithuania, 2013, 5 (2).

[10] J West, M Bhattacharya. Intelligent financial fraud detection: A comprehensive review [J]. Computers & Security, 2016, 57(C): 47-66.

[11] QH Sun, FX Shen. The application of data mining in the era of big data [J]. Electronic Technology and Software Engineering, 2016 (6): 204.

[12] J Han, M Kamber. Data Mining Concepts and Techniques [M]. Mechanical Industry Press, Beijing, 2001.

[13] R Tsaih, Y Hsu, CC Lai. Forecasting S&P 500 stock index futures with a hybrid AI system [J]. Decision Support Systems, 1998, 23(2): 161-174.

[14] [NK Liu, KK Lee. An intelligent business advisor system for stock investment [J]. Expert Systems, 2010, 14(3): 129-139.

[15] David Hand, Heikki Mannila, Padhraic Smyth. Data Mining Concepts and Techniques [M]. Mechanical Industry Press, 2003.

[16] Rakesh Agrawal, Tomasz Imielinski, Arun Swami. Mining association rules between sets of items in large databases [A]. Proc. ACM SIGMOD Conference, Washington, 1993, [C]: 207-216.

[17] E Cohen, M Datar, S Fujiwara, A Gionis, P Indyk,   R Motwani, JD Ullman, C Yang. Finding interesting associations without support pruning [J]. Knowledge & Data Engineering IEEE Transactions, 2001, 13(1): 64-78.

[18] MJ Zaki. Scalable algorithms for association mining [J]. IEEE Educational Activities Department, 2000, 12(3): 372-390.

[19] S Makridakis. The forthcoming Artificial Intelligence (AI) revolution: Its impact on society and firms. Futures, 2017.

[20] J Cao, YL Zhou. Research progress on the influence of artificial intelligence on economy [J]. Economics Information, 2018, 1: 103-115.

ISPECE

IOP Publishing

A variable step-size adaptive notch filter for frequency estimation using combined gradient algorithm

Huiyue Yang[1,a]*, Yaqing Tu[1,b], Ming Li[1,c]

[1]Army Logistics University of PLA, Chongqing 401311, China

* correspond author: [a]huiyue_yang@163.com

[b]yqtcq@sina.com

[c]limitonly@126.com

Abstract To improve the performance of adaptive notch filter (ANF), a variable step-size ANF using combined gradient algorithm is proposed for frequency estimation. In this method, combined gradient algorithm is designed for improving constringency with both FIR-ANF and IIR-ANF advantages considered. According to the noise influence, a bias correction strategy is established for depressing the bias and MSE of frequency estimation. Additionally, variable step-size is adopted for the balance of convergence and precision. The proposed method process is given. In simulations, we discussed the influence of the ANF parameters and SNR on accuracy. Algorithms of DPG and MPG are carried out as comparisons. Convergences of these methods are also analyzed. Coriolis mass flow meter (CMF) is taken as an application to test the proposed method. Simulation results and CMF application both confirm the availability and superiority of the proposed method.

1. INTRODUCTION

Adaptive estimation of the frequency of a single-tone signal in noise is an important research topic that has varied applications, such as Radar, Sonar, control engineering, communication systems, testing instrument, and so on. Adaptive notch filter (ANF) is one of many choices to serve such applications.

The frequency is estimated in ANF by minimizing the errors of filter, according to the parameters which is adaptive adopted with signal character. It is found in the literature survey that there two types of the ANF, namely, infinite-impulse-response notch filter (IIR-ANF) and finite-impulse-response notch filter (FIR-ANF), which will be both considered in this work. IIR-ANF has some advantages, such as simple structure, little computation and easy realized. But because of its gradient algorithm, IIR-ANF needs a long time for convergence especially when initial frequency of ANF is away from signal frequency. In the last several decades, methods for improving the structure of ANF or adaptive algorithm have been intensively studied. A majority of these methods adopt gradient-based adaptive algorithms, such as the plain gradient(PG)[1], the direct plain gradient(DPG)[2], the modified plain gradient(MPG)[3,4], the modified sign algorithm(MSA)[5], the unbiased plain gradient(UPG)[6] and the unbiased modified plain gradient(UMPG)[7] and so on, are available and suitable for real time frequency estimation. However, extensive studies have shown that the precision of these methods still need to improve as there are inherent biased estimators. [8-11] In addition, there is a trade-off between a small steady state error and a fast convergence [12,13]. A small step-size provides small steady state error

Content from this work may be used under the terms of the Creative Commons Attribution 3.0 licence. Any further distribution of this work must maintain attribution to the author(s) and the title of the work, journal citation and DOI.

Published under licence by IOP Publishing Ltd

but also gives reduction to convergence. On the other hand, large step-size gives the opposite performance.

To serve the problem mentioned above, a variable step-size ANF using combined gradient algorithm (CGA) for frequency estimation is proposed in this paper. In the proposed ANF, CGA is developed to improve the performance, and biases corrected formula is deduced based on the correlation of noise in the processed signal. Synchronously, variable step-size technique is adopted to balance the steady sate error and the convergence.

The rest of the paper is organized as follows. In Sec. II, the principle of the proposed method is introduced in detail. Sec.III displays the process of the proposed method. The simulation and experimental results validating the proposed method are reported in Sec.IV. Sec.V concludes this paper.

2. PRINCIPLE

2.1 Frequency iterative calculation
Assume that the signal waiting for frequency estimation has the form of

$$x(k) = A\cos(\omega_0 k + \theta) + v_0(k), k = 1, 2, ..., N \tag{1}$$

where A, ω_0, θ is the signal amplitude, frequency and phase, respectively. $v_0(k)$ is a additional white noise with zero mean and variance σ_v^2. N is the length of sample data. The system function of adaptive notch filter used in this work is given by

$$H(z, \widehat{\omega}_0) = \frac{N(z, \widehat{\omega}_0)}{D(z, \hat{\omega}_0)} = \frac{1 - 2\cos\widehat{\omega}_0 z^{-1} + z^{-2}}{1 - 2\rho\cos\widehat{\omega}_0 z^{-1} + \rho^2 z^{-2}} \tag{2}$$

where $N(z, \widehat{\omega}_0)$ and $D(z, \hat{\omega}_0)$ are, respectively all zeros and poles systems. ρ is pole radius restricting the bandwidth of ANF, $0 << \rho < 1$. $\widehat{\omega}_0$ is the estimation of frequency ω_0. Accordingly, the iterative formula of frequency calculation is described by the following equation:

$$\widehat{\omega}_0(k+1) = \widehat{\omega}_0(k) - \frac{\mu}{2}\frac{\partial J(\omega(k))}{\partial \widehat{\omega}_0(k)} \tag{3}$$

Based on (2), we can get the relationship that the input $x(k)$ to the out puts $e_1(k)$ and $e_2(k)$ of $N(z, \widehat{\omega}_0)$ and $H(z, \widehat{\omega}_0)$, denoted by

$$e_1(k) = x(k) - 2\cos\widehat{\omega}_0 x(k-1) + x(k-2)$$
$$e_2(k) = e_1(k) + 2\rho\cos\widehat{\omega}_0 e_2(k-1) - \rho^2 e_2(k-2)) \tag{4}$$

The gradient function of IIR-ANF is constructed as $J(\omega(k)) = e_1^2(k)$, while gradient function FIR-ANF is $J(\omega(k)) = e_2^2(k)$.

2.2 Combined gradient
To integrate the advantages of IIR-ANF and FIR-ANF, a combined gradient function is defined by

$$J(\omega(k)) = [e_1(k) + e_2(k)]^2 \tag{5}$$

In computation, $J(\omega(k))$ is obtained according to the equation $\hat{J}(\omega) = \frac{1}{N}[\sum_{k=1}^{N}(e_1(k) + e_2(k))^2]$.

Take the noise into consideration, the average $E[J(\omega)]$ can be derived as follows:

$$E[J(\omega)] = 2A^2(\cos\omega_0 - \cos\omega)^2 + A^2 H^2(\omega)/2$$
$$+ A^2 H(\omega)N(\omega)\cos(\phi_N(\omega) - \phi_H(\omega)) \tag{6}$$
$$+ E[v_N^2(k)] + 2E[v_N(k)v_H(k)] + E[v_H^2(k)]$$

where

$$H(\omega) = \frac{2\left|\cos\omega_0 - \cos\omega\right|}{\sqrt{[(1+\rho^2)\cos\omega_0 - 2\rho\cos\omega]^2 + [(1-\rho^2)\sin\omega_0]^2}}$$

$$N(\omega) = 2(\cos\omega_0 - \cos\omega), \quad \phi_N(\omega) = \begin{cases} \omega_0 & \omega_0 \le \pi/2 \\ \pi + \omega_0 & \omega_0 > \pi/2 \end{cases}$$

$$\phi_H(\omega) = \begin{cases} \arctan\dfrac{(1-\rho^2)\sin\omega_0}{(1+\rho^2)\cos\omega_0 - 2\rho\cos\omega} & \omega_0 \le \pi/2 \\[2ex] \pi + \arctan\dfrac{(1-\rho^2)\sin\omega_0}{(1+\rho^2)\cos\omega_0 - 2\rho\cos\omega} & \omega_0 > \pi/2 \end{cases}$$

$$E[\upsilon_N^2(k)] = 2\sigma^2\left(1 + 2\cos^2\omega\right)$$

$$E[\upsilon_N(k)\upsilon_H(k)] = \sigma^2\left(\frac{1}{\rho^2} + \frac{4\cos^2\omega\rho^2 - (1+\rho)^2}{\rho^2}(1-\rho)^2\right)$$

$$E[\upsilon_H^2(k)] = \sigma^2\left(\frac{1}{\rho^2} - \frac{1-\rho}{1+\rho}\frac{(1+\rho^2)(1+\rho)^2 - 8\rho^2\cos^2\omega}{\rho^2(\rho^4 - 2\rho^2\cos 2\omega + 1)}\right)。$$

In order to evaluate the property of the combined gradient function, $J_1(\omega) = E[\hat{e}_2^2(k)]$ introduced in Ref. [2] and $J_2(\omega) = E[\hat{e}_1(k)\hat{e}_2(k)]$ introduced in [3] are taken as compares. Gradient function curves are shown in Fig.1 when frequency $\omega = \pi$. In simulations, assuming that $A = 1$, $\theta = \pi/6$, $\sigma^2 = 0$, $\rho = 0.95$ and $N = 200$. From the Fig.1, we can see that $J(\omega)$ has a similar form with $J_1(\omega)$ and $J_2(\omega)$, but differs in amplitude and gradient. $J_1(\omega)$ and $J_2(\omega)$ are smoother than $J(\omega)$, while $J(\omega)$ has a faster convergence rate.

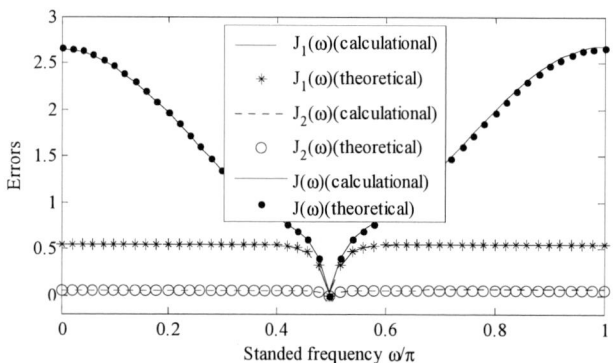

Fig.1 Gradient function curves

2.3 Bias correction

By substituting $J(\omega(k)) = [e_1(k) + e_2(k)]^2$ into Eq. (3), the estimation formula with the combined function can be derived as

$$\omega(k+1) = \omega(k) - \mu[e_1(k) + e_2(k)][g_1(k) + g_2(k)] \tag{7}$$

where $g_1(k) = \dfrac{\partial e_1(k)}{\partial\omega(k)} = 2x(k-1)\sin\omega(k)$, $\quad g_2(k) = \dfrac{\partial e_2(k)}{\partial\omega(k)} \approx 2[x(k-1) - \rho e_2(k-1)]\sin\omega(k)$

Frequency estimation by Eq. (7) is biased as a result of the noise. Therefore, correction for estimation results is needed. We figure out the bias denoted by $R(k) = (3\rho - 5)\sin 2\omega(k)\sigma^2$. As $\sigma^2 \approx E[x(k)e_1(k)]$, σ^2 at the point of k approximates to $\sigma^2(k) \approx x(k)e_1(k)$. Accordingly, Eq. (7) can be rewritten as:

$$\omega(k+1) = \omega(k) - \mu G(k) \tag{8}$$

where $C(k) = (3\rho - 5)\sin 2\omega(k)$, $\qquad G(k) = [e_1(k) + e_2(k)][g_1(k) + g_2(k)] - C(k)x(k)e_1(k)$.

2.4 Variable step-size

Fast convergence and stability variance are two key targets of ANF. In the steady state, the mean of frequency estimation variance is [14]

$$E[\Delta^2(k)] = \frac{Ln(k) + Rn(k)}{L(k)/\mu - Ls(k) - Rs(k)} \tag{9}$$

where $\Delta = \widehat{\omega}_0 - \omega_0$, Ln, Rn, L, Ls, Rs are parameters depended on noise. It's obviously from Eq.(9) that variance deduces with the decrease of step μ, while convergence of the proposed method decelerates at the same time. Simulations shown in Fig.2 validate the conclusion.

Varying step-size according variance is a current strategy for balance the conflict between convergence rate and precision. To reduce the influence of noise correlation, step size updates according Eq.(10).

$$\mu(k+1) = \alpha\mu(k) + \gamma\{e^2(k) + e(k)e(k - D)\} \tag{10}$$

where $e(k) = e_1(k) + e_2(k)$, $\mu_{min} < \mu(n) < \mu_{max}$, $0 < \alpha < 1, \gamma > 0$. D is great than correlation radius of noise and less than correlation radius of input signal time. We regulate the step size according to self-correlation of errors $e(k)e(k - D)$. For this reason, convergence and precision are balanced. Sensitivity index of the method on noise is weakened.

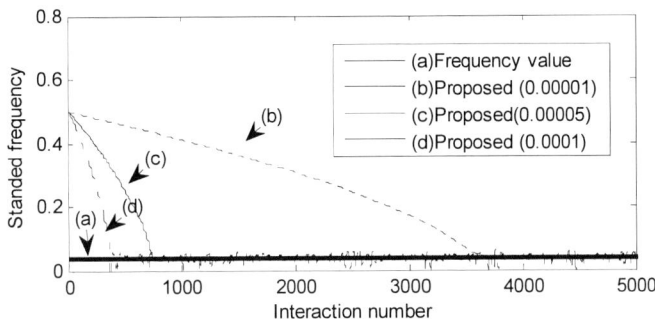

Fig.2 the frequency estimation bias and MSE for ANF at different steps

3. PROCESS

In summary, the process of the proposed method is displayed in Fig.3. In the first place, sample the input signal and initialize the parameters. Second, compute the outputs $e_1(k)$ and $e_2(k)$ according to $N(z, \widehat{\omega}_0)$ and $H(z, \widehat{\omega}_0)$. Then, calculate the differential coefficient $g_1(k)$ and $g_2(k)$ of $e_1(k)$ and $e_2(k)$, respectively. To compensate the error, we then need to calculate $C(k) = (3\rho - 5)\sin 2\omega(k)$ and $G(k)$. And then, we update $\mu(k)$ according to Eq. (10). To guarantee the stability of algorithm, the length of step size $\mu(k)$ should be controlled. The control equation is given by

$$\begin{cases} \mu(k) = \mu_{max}, & \mu(k) > \mu_{max} \\ \mu(k) = \mu_{min}, & \mu(k) \leq \mu_{min} \end{cases} \tag{11}$$

Finally, the frequency $\omega(k+1)$ at the point of $k+1$ can be estimated according to Eq. (8).

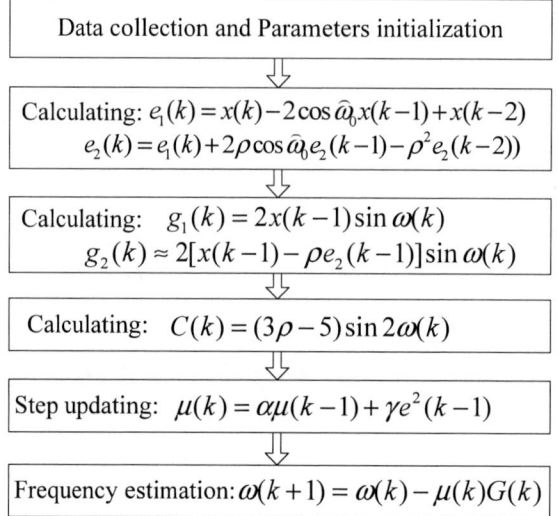

Fig.3 Process of the proposed method

The boxes in the flowchart contain:

Data collection and Parameters initialization

Calculating: $e_1(k) = x(k) - 2\cos\hat{\omega}_0 x(k-1) + x(k-2)$
$e_2(k) = e_1(k) + 2\rho\cos\hat{\omega}_0 e_2(k-1) - \rho^2 e_2(k-2))$

Calculating: $g_1(k) = 2x(k-1)\sin\omega(k)$
$g_2(k) \approx 2[x(k-1) - \rho e_2(k-1)]\sin\omega(k)$

Calculating: $C(k) = (3\rho - 5)\sin 2\omega(k)$

Step updating: $\mu(k) = \alpha\mu(k-1) + \gamma e^2(k-1)$

Frequency estimation: $\omega(k+1) = \omega(k) - \mu(k)G(k)$

4. RESULTS AND ANALYSIS

In this section, the performances of the proposed method are compared with those of the DPG and MPG algorithms in terms of the analytical and simulation results.

4.1 Simulation results

Simulation signals are produced with these parameters: $A = \sqrt{2}$, fixed value θ located in $[0, 2\pi)$, signal to noise ratio SNR=5dB, and frequency ω_0 respectively initialized as 0.01π , 0.05π and 0.1π . Supposing $\rho = 0.9$, we obtain the frequency estimation shown in Fig.4. It can be seen from Fig.4 that the convergence time of the proposed algorithm is obviously shorter than DPG and MPG. Additionally, convergence time of MPG increases rapidly when initialization frequency is far from the signal frequency, while DPG and the proposed algorithm sit by.

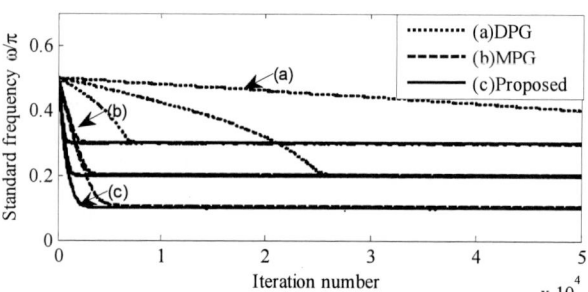

Fig.4 Convergence of frequency estimation

The movement of step size is shown in Fig.5. The ANF owns comparative great step size at the convergence moment, which is good for improving the rate of convergence. Comparatively, the step size localizes in a small range at stability, which guarantees the accuracy of frequency estimation.

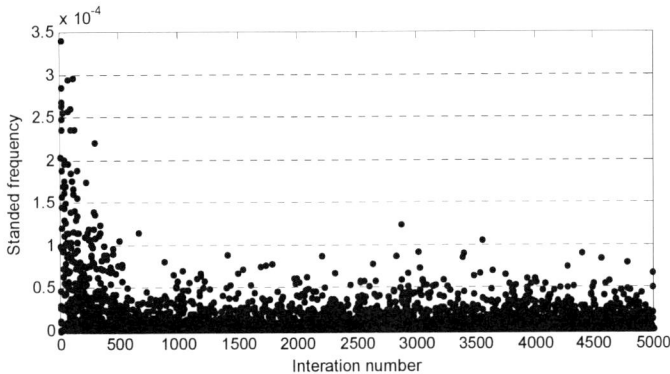

Fig.5 Distribution of step size

To obtain the bias and MSE estimations of these algorithms from the simulations, 20 runs are calculated with the same parameters. Fig.6 shows the bias and MSE estimations. From Fig.6, we can see both the bias and the MSE of the proposed algorithm is lower than MPG.

The bandwidth of ANF depends on ρ. The bias and MSE of frequency estimation at different initializations of ρ are shown in Fig.7. The accuracy of frequency estimation increases as ρ tends to 1. Although computational results have some distortion with theoretical, it still reveals the precision trend of frequency estimation. The precision of MPG decline at the condition of $\rho < 1$, while the proposed algorithm is not sensitive to the initialization of ρ. Therefore, ANF parameters initialization is simplified in the proposed algorithm.

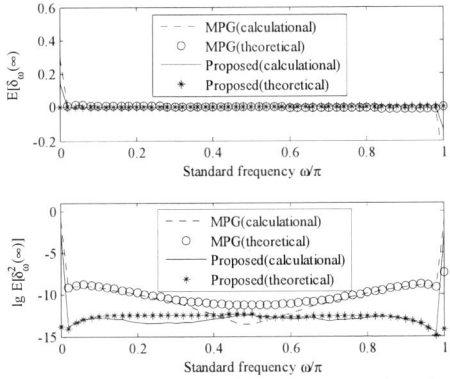

Fig.6 Bias and MSE of frequency estimation

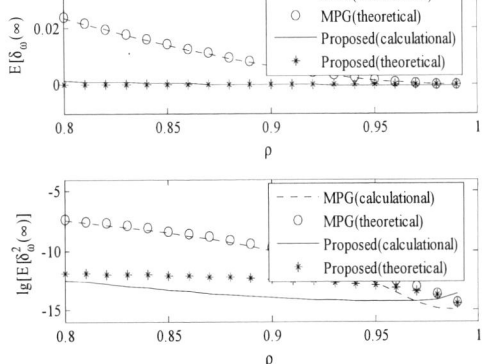

Fig.7 Bias and MSE of frequency estimation with different ρ

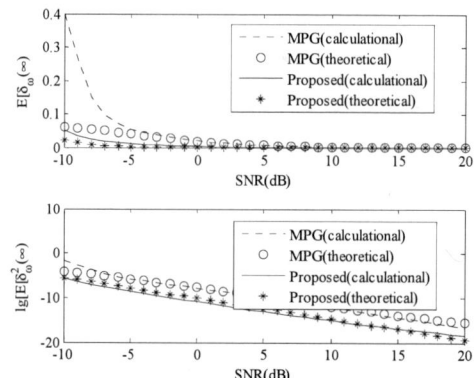

Fig.8 Bias and MSE of frequency estimation with different *SNR*

The bias and MSE of frequency estimation at different SNR are shown in Fig.8. It can be seen from Fig.8 that the precision improves with SNR increases. What's more, the proposed method performances better than MPG at the low SNR condition, which illuminates the anti-jamming character of the proposed method.

4.2 Application in CMF

To experimentally validate the proposed method, we take the Coriolis mass flowmeter (CMF) as an application. CMF calculates the mass flowrate by measuring the frequency and phase difference between two signals detected by electromagnetic sensors. In this work, oscillation signals come from the F200S CMF (with a 1700R transmitter). Sampling frequency is set 20 kHz. As we cannot get the real frequency of CMF, only DPG, MPG and the proposed method are compared in the application, as shown in Fig.9 and Tab.1. As we know, the CMF frequency is fixed at steady flow. Results of the proposed method maintain equable compared with DPG and MPG, which validates the serviceability in application of the proposed method.

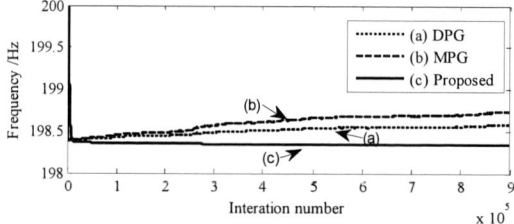

Fig.9 Estimation curves of CMF frequency

Table.1 Frequency estimation under different flow rates

Mass flow rate kg/min	DPG /Hz	MPG /Hz	Proposed method /Hz
2.9	198.5803	198.7230	198.3688
10	198.4521	198.5008	198.3704
82.1	198.5237	198.6224	198.3658
102.2	198.5179	198.6217	198.3540

5. CONCLUSION

In this paper, a variable step-size ANF using combined gradient algorithm is proposed for frequency estimation. Combined gradient algorithm is designed for improving constringency, with both FIR-ANF and IIR-ANF advantages been considered. The bias and MSE of frequency estimation are depressed as bias correction strategy been used in the proposed method. What' more, variable step-size is adopted for the balance of convergence and precision. Simulation results and CMF application validate the availability and superiority of the proposed method compared with DPG and MPG.

ACKNOWLEDGMENT
The study is supported by National Natural Science Foundation of China (NNSFC) (Grant Nos.61871402) and Natural Science Foundation of Chongqing (Grant Nos. CSTC2015jcyjBX0017).

REFERENCES
[1] Xiao Y, Takeshita Y and Shida K.2001.Steady-state analysis of a plain gradient algorithm for a second-order adaptive IIR notch filter with constrained poles and zeros. *IEEE Transactions on Circuits and Systems II: Analog and Digital Signal Processing*, 48,7(Oct.2001),733-740.

[2] Zhou J, and Li G. 2004. Plain gradient-based direct frequency estimation using second-order constrained adaptive IIR notch filter. *Electronics Letters*,40,5(Nov.2004), 351-352.

[3] Punchalard R, Lorsawatsiri A Koseeyaporn J, Wardkein P, and Roeksabut A. 2008. Adaptive IIR notch filters based on new error criteria. *Signal Processing*, 88,3(Mar.2008), 685-703.

[4] Nosan A, and Punchalard R. 2012. A complex adaptive notch filter using modified gradient algorithm. *Signal Processing,*92,6(Jun.2012),1508-1514.

[5] Punchalard R, Wardkein J and Wardkein P.2009. Adaptive IIR notch filter using a modified sign algorithm. *Signal Processing,*89,2(Feb.2009),239-243.

[6] Loetwassana W, Punchalard R, Koseeyaporn J and Wardkein P. 2012. Unbiased plain gradient algorithm for a second-order adaptive IIR notch filter with constrained poles and zeros. *Signal Processing,*90,8(Aug.2012), 2513-2520.

[7] Punchalard R., Nosan A.2011. Bias removal in a modified indirect plain gradient algorithm for adaptive IIR notch filter. *IEEE ECTICON*, 2011,926-929.

[8] Punchalard R.2012. Mean square error analysis of unbiased modified plain gradient algorithm for second-order adaptive IIR notch filter. *Signal Processing*, 92,11(Nov.2012), 2815-2820.

[9] OSman K. 2011. Analysis of the dynamics of a memoryless nonlinear gradient IIR adaptive notch filter. *Signal Processing*, 91,5(Nov.2011),2379-2394.

[10] Yang HY, Tu YQ, Zhang HT and Li M. 2014. Feedback corrected adaptive notch filter for vibration signal frequency tracking. *Chinese Journal of Vibration and Shock*,33,3(Mar.2014),145-149,176.

[11] Li M, Tu YQ, Shen TA and Mao YW. 2014. Extremely frequency direct estimation algorithm based on new adaptive notch filter. *Journal of Vibration and Shock*, 27,5(May.2014),785-793.

[12] Sung J B, Chang W L, Hyeonwoo C, etc.2010. A variable step-size adaptive algorithm for direct frequency estimation. *Signal Processing,*90,4(Apr.2010),2800-2805.

[13] Hong CW. 2012.Variable step size LMS algorithm using squared error and autocorrelation of error. *Procedia Engineering*, 41(Feb.2012),47-52.

[14] Li M, Tu YQ. Shen TA and Yang H Y. 2014. A new frequency estimation method based on adaptive notch filter and its performance analysis. *Acta Electronica Sinica*, 42,1(Jan.2010),49-57.

Smart Home System Based on Deep Learning Algorithm

Yanfei Peng[1,a], Jianjun Peng[2,b]*, Jiping Li[3,c]*, Ling Yu[4,d]

[1]School of Information Science and Engineering, Dalian Polytechnic University, Dalian Liaoning 116034, China +86-18366881989

[2]School of Information Science and Engineering, Dalian Polytechnic University, Dalian Liaoning 116034, China +86-13942061732

[3]College of Mathematics and Informatics, South China Agricultural University, Guangzhou Guangdong 510642, China +86-13138692766

[4]Network Information Center, Dalian Polytechnic University, Dalian Liaoning 116034, China +86-15524599728

[a]343626363@qq.com, Corresponding author: [b]pengjj@dlpu.edu.cn, Corresponding author: [c]li-jiping@163.com , [d]yuling@dlpu.edu.cn

ABSTRACT This paper presents a smart home control system on the strength of human body point cloud data attitude recognition technology. Wherein, the human body point cloud image is obtained by Kinect, and the point cloud image is extracted, and the human body attitude is recognized by Convolutional neural network (CNN) algorithm. The Arduino microprocessor will process the received data to realize the intelligent control of residential electrical appliances in the process. The experimental results show that the system algorithm is able to achieve effective recognition of human attitude and effective control of household appliances and other functions through the analysis and processing of human point cloud images, which features a certain degree of innovation, reliability and practicality.

1. INTRODUCTION

With the development of society and the improvement of living standard, the smart home industry is developing rapidly correspondingly. An attitude recognition technology based on depth sensor has gradually entered the field of vision. In November 2010, the launch of Microsoft Kinect depth sensor provided hardware support for the design and development of smart home based on attitude recognition technology [1]. It is simpler, more natural, and humanized compared with the traditional human-computer interaction system using mouse, keyboard and other modes of operation, and human attitude recognition technology with these new features is playing an important role in human-computer interaction.

Smart home is a kind of living environment for human being, integrating intelligent control and management of various subsystems related to home life through the network using advanced computer technology, network communication technology and other integration of personality needs to make home life more convenient and comfortable. There are lots of literatures recoding the studies of theory and technology of smart home, wherein, literature [2-5] refers to the interactive design pattern theory of somatosensory, that is, obtaining the key points of human skeleton data through Kinect and using to recognizing human attitude by various algorithms. However, the skeleton data is extremely unstable, easy to lose, and the algorithms are mostly threshold judgment, recognition rate need to be improved

Content from this work may be used under the terms of the Creative Commons Attribution 3.0 licence. Any further distribution of this work must maintain attribution to the author(s) and the title of the work, journal citation and DOI.
Published under licence by IOP Publishing Ltd

thereby. Literature [6-7] connects all kinds of smart devices in the home by using wireless Bluetooth technology, through which users are able to directly control the smart devices in real time by using mobile phone client. Nonetheless, the problem of poor user experience still exists. In literature [8], the integration of smart grid and smart home was studied, however, it mainly merely involved the control of household appliances, whereas the research of interactive technology was not mentioned.

In order to solve the problem that the user unable to interact with the machine in a natural way or the low recognition accuracy in the current smart home system, this paper integrates the human body point cloud data attitude recognition technology into the design of the smart home human-computer interaction system, which uses Kinect to obtain the human body depth image and point cloud image. The algorithm is used to extract the features of human point cloud image, and the Convolutional neural network model is imported to recognize and classify human attitude. The processing result was then transmitted to Arduino microprocessor through XBee wireless communication module to realize intelligent control of home appliances.

The contributions of this paper are as follows: Firstly, the human point cloud images obtained by Kinect are not affected by external illumination, and it is capable of effectively protecting the privacy of users. What's more, Kinect point cloud data has higher confidence than skeleton data. Secondly, a more diversified set of human attitude point cloud image data was created, and a Convolutional neural network model was built for training and testing. Thirdly, it improves the rigidity of human-computer interaction and the single control object of the original equipment, contributing to offering more natural human-computer interaction and improving the attitude recognition rate so that the "intelligent" component of smart home will be significantly improved.

2. SYSTEM STRUCTURE AND DESIGN

Smart home control system based on human body point cloud data attitude recognition technology is combined with software Processing and Kinect to obtain depth images in the scene, and three-dimensional point cloud images were generated therein. An algorithm was used to extract image features, which were assigned to trained classifiers and then used to judge human attitude. The determined data is transmitted to the Arduino microprocessor via the wireless communication module XBee to control various household appliances. The overall structure of the system is shown in Figure 1.

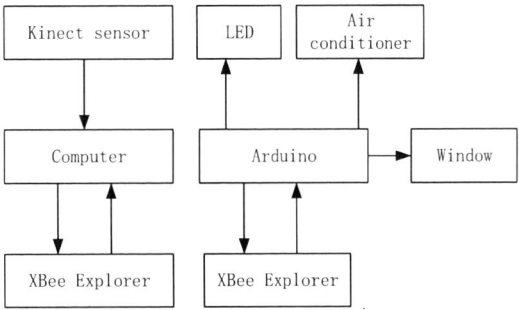

Figure 1. Overall structure diagram of the system.

2.1 Kinect Sensor

The somatosensory device used in this system is a kind of somatosensory peripheral Kinect issued by Microsoft. Kinect is comattituded of color camera, infrared device, microphone array, logic circuit, motor and other parts. Kinect has three cameras, the RGB color camera in the middle, the left and right lenses are infrared transmitters and CMOS infrared cameras comattituded of 3D depth sensors, can obtain color images and depth images at the same time. The microphone array, consisting of four downward-facing built-in microphones. Figure 2 depicts the depth image and the corresponding point cloud image taken by Kinect at night. As we have observed, the system can be used late at night or even at night, because Kinect can extract the point cloud image [10] from a lightless room.

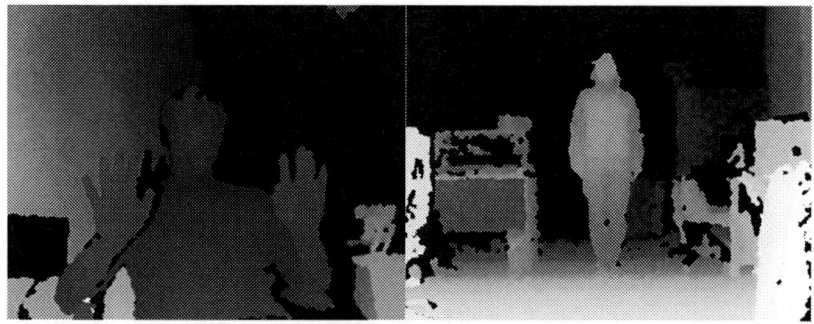

Figure 2. Point cloud image in depth image and real world coordinate system.

2.2 Arduino

Arduino mainly consists of two parts: the hardware part is Arduino circuit board, and the other is Arduino IDE, which is a program development environment stored in the computer. Arduino is an open source hardware development platform with 8 bit ATMEGA328 microprocessor as the core, which provides 14 digital input and output pins and 6 analog input pins. It can support USB data transmission. humans can connect different electronic devices on the I/O port [9], As shown in Figure 3.

Figure 3. Arduino circuit diagram.

3. DESIGN AND IMPLEMENTATION OF SYSTEM ALAORITHM

In this paper, Kinect was used to acquire the point cloud image of human body, and the attitude recognition instructions of various households were defined. The Convolutional neural network model was established, the sample images were captured and labeled, and each kind of attitude training samples was input into CNN to train the model, through which adjusting the model until it turns to be astringency; the output layer was transformed into a soft Max classifier. Input test images to identify the accuracy of validation results. Through this model, human attitude was recognized accurately and efficiently, and finally the desired results were achieved. The algorithm box is shown in Figure 4.

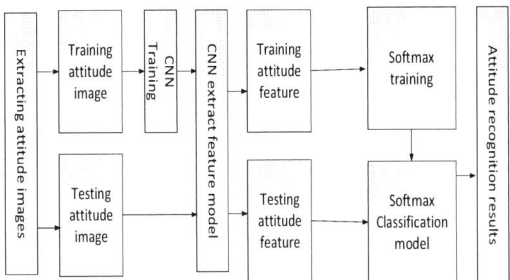

Figure 4. Overall block diagram of algorithm.

3.1 Implementation of Kinect Control Instruction

This novel input method of Kinect provides a new idea for the control of smart home. Firstly, it is necessary to personalize the attitude recognition instructions of all kinds of home, as shown in Table 1. Taking body attitude as input element, the system can understand the command and realize the control of smart home.

Table 1. Control instruction for smart home

Electrical type	Operation instruction	Attitude operation
lamps	Create lamps	raising hands above the head
	Selection lamps	unbending right arm
	Adjust luminance	The distance between the left hand and the head
Air conditioner	Open the air conditioner	body presenting "大" shape
	Turn off the air conditioner	presenting "T" shape
	Refrigeration mode	raising left hand highly, right hands on waist
	Heating mode	arising right hand highly, left hand on waist
Window	Open The window	left hand on waist, laterally raising right hand
	suspend	both hand on the waist
	Close the window	right hand on waist, laterally raising left hand

3.2 Convolutional Neural Network Model Design

Convolutional neural network is a kind of feedforward artificial neural network, which is developed on the basis of traditional neural network. Nowadays, Convolutional neural network has been widely used in image and video recognition, and it has been an appropriate and effective method for many computer vision problems.

In the first few layers of the Convolutional neural network, each node is organized into a three-dimensional matrix. Each node in the first several layers of Convolutional neural network is only connected with some nodes in the upper layer. Plus, a Convolutional neural network is comattituded of input layer, Convolutional layer, pooling layer, full connection layer and Softmax layer as shown in Figure 5. The CNN recognition model designed in this paper has a total of seven layers. Firstly, it comes to the input of the network. CNN network is able to learn the features of two-dimensional image independently. The original image is point cloud image, which can be used as the input of the network directly. The Convolutional operations create 64 features, the first full connection layer will have 384 hidden nodes, the second connection layer will connect 384 hidden nodes to 192 hidden nodes, and finally output 10 classification results.

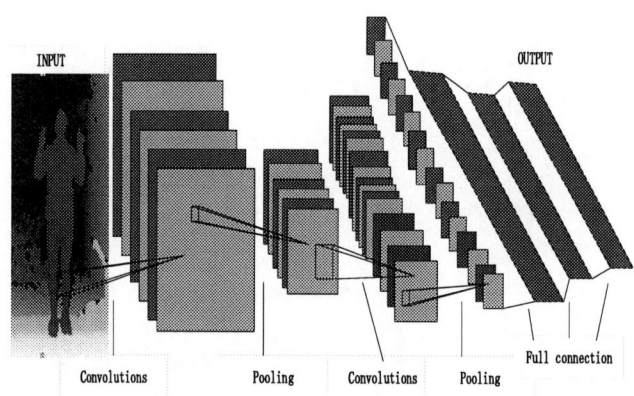

Figure 5. Convolutional neural network architecture diagram.

3.3 Dataset

In this paper, 720 times of image shooting of 6 people and 10 attitudes were collected to build attitude training database. Wherein, the 10 attitudes are raising hands above the head; unbending right arm; body presenting "大" shape; presenting "T" shape; raising left hand highly, right hands on waist; arising right hand highly, left hand on waist; left hand on waist, laterally raising right hand; both hand on the waist; right hand on waist, laterally raising left hand and other attitudes, grouping the images collected to attitude1- attitude10. The image effect of the point cloud image part of collected body attitude by shooting is shown in Figure 6 .

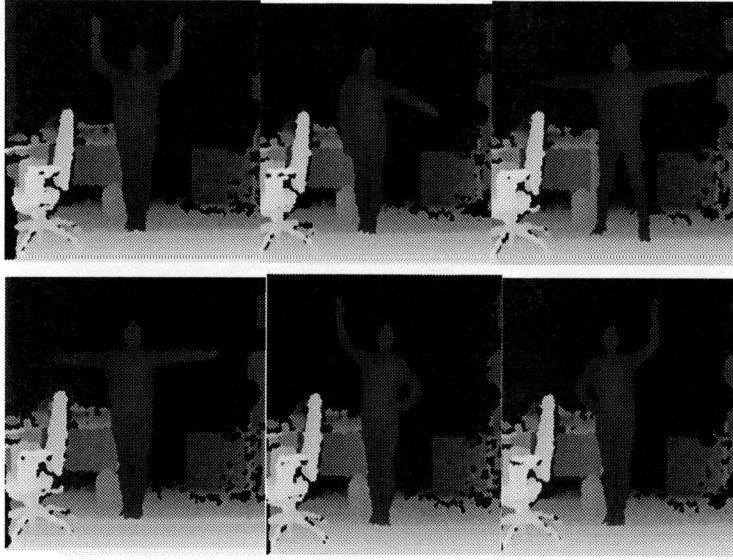

Figure 6. Collection results of point cloud image dataset.

Guide the dataset into a convoluted neural network model for training and testing. Firstly, the network is initialized, and each group of attitude samples in the dataset is input into the CNN network model separately, whereby one attitude is trained in one round. With the increase of the number of iteration rounds, the training and testing accuracy of the model will be improved. Then, the loss function value will be reduced, and it converges and tends to be stable after 800 iterations, as shown in Figure 7, whereby the parameter values of the network is determined.

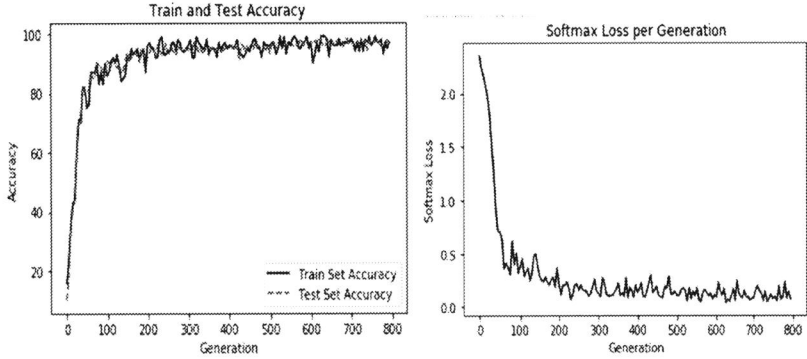

Figure 7. The left figure shows the accuracy of 800 generations of training set and test set, and the right figure shows the softmax loss function of 800 generations of training set and test set.

4. EXPERIMENT AND ANALYSIS

In this paper, the system is tested experimentally. The system designed is deployed in a real home environment, and the Kinect is installed in a suitable location to cover the whole home detection environment. The intelligent home system designed in this paper is capable of processing the collected information and completing the control instructions simultaneously without causing the fluctuation of processing speed. Performance degradation. When someone enters the Kinect sensor's field of vision, the system will automatically start, through a specific operation to open attitude command input, real-time collection of human body information and complete control commands, point cloud image to protect the privacy of users.

4 volunteers participated in the evaluation of the algorithm and system in this paper. Each volunteer performed the above nine attitudes to control different household electrical appliances, each attitude was performed 25 times. Through the training of Convolutional neural network as well as the commissioning of Aduino control system, the experimental results obtained are shown in Table 2. It can be seen that the recognition accuracy reaches up to above 93%, and the expected accuracy is achieved.

Table 2. Accuracy of smart home control

Attitude operation	Operation instruction	Test times	Correct times	accuracy
1. raising hands above the head	Create lamps	100	96	96%
2. unbending right arm	Selection lamps	100	97	97%
3. body presenting "大" shape	Open the air conditioner	100	97	97%
4. presenting "T" shape	Turn off the air conditioner	100	95	95%
5. raising left hand highly, right hands on waist	Refrigeration mode	100	94	94%
6. arising right hand highly, left hand on waist	Heating mode	100	95	95%
7. left hand on waist, laterally raising right hand	Open The window	100	94	94%
8. both hand on the waist	suspend	100	95	95%
9. right hand on waist, laterally raising left hand	Close the window	100	93	93%

This paper designs a smart home system based on Kinect using depth learning algorithm and a more natural human-computer interaction control system, which is able to meet the user's demand for intelligent home environment. The detection accuracy of this method is higher than that of the Kinect skeletal tracking method adopted in [2-5], addressing the problems of unstable data of human skeletal joint points and the situation of easy to get lost. The Convolutional neural network constructed and trained in this paper also witnessed some progresses compared with reference [14]. In addition, compared with the traditional intelligent home control method in literature [6-8], it improved the rigidity of the original equipment and the unitary control object, realizing more natural human-computer interaction, and protecting the privacy of users at the same time.

5. CONCLUSION

This paper introduces a reliable and more natural human-machine interaction smart home system. The point cloud images of human living environment are captured by Kinect, and the features of different human attitude images are extracted. Plus, the Convolutional neural network model was built to recognize human attitude accurately and efficiently. The Arduino microprocessor would conduct real-time processing upon receiving the control instructions, and control the work of household appliances by wireless a well. The experimental results showed that the system has certain practical value. During the operation of the system, there is no need wearing any devices or controllers, instead, it contributes to objects interacting with the machine in a simpler and more natural way; using the Kinect point cloud image for color spectrum detection, which is able to effectively protect privacy; meanwhile, the Kinect infrared camera will not be affected by external light. It can work at any time, improving the detection efficiency.

ACKNOWLEDGMENTS

The authors wish to express their gratitude towards the financial supports from Science and Technology Department of Liaoning province (No. 20170052), Education Department of Liaoning province (No. 2017J047) and China Science Technical Department (No. 2017YFC0821003).

REFERENCES

[1] Qu Chang, Sun Jie, Wang Junze, et al. 2016.Automatic detection of falls in elderly people based on Kinect somatosensory sensor Journal of [J]. sensing technology, 2016,29 (3): 378-383.

[2] Shen Muqi, Qi Shunlan. 2017. Internet of Things smart home system based on somatosensory interaction technology [J]. Modern information technology, 2017, 1 (4): 42-44.

[3] Su Benyue, Wang Guangjun, Zhang Jian. 2013. Smart home system based on somatosensory interaction technology under the Internet of Things [J].Journal of Central South University (Natural Science Edition), 2013, 44 (s1): 181-184.

[4] Yu Zesheng.2017. Research and Design of Smart Home System Based on Kinect attitude Recognition [D]. Liaoning University of Science and Technology.

[5] Guo Zhe, Chen Peitou, Hu Mengkai, et al. 2016. Kinect-based Smart Home System [J]. Modern Electronic Technology, 2016, 39 (18): 149-152.

[6] Hou Yuyi, Yang Dongtao, Liu Yan, et al. 2016. Smart Home Life and Security System Based on Wireless Bluetooth Technology [J]. Journal of Jiaying University, 2016, 34 (5): 36-40.

[7] Li Tao. 2014. Design and implementation of Android based smart home APP [D]. Suzhou: Soochow University.

[8] Naglic M, Souvent A. 2013. Concept of Smart Home and Smart Grids integ ration [C]. Energy ,International Youth Conference on.IEEE, 2013:1-5.

[9] Smisek J, Jancosek M, Pajdla T. 2013. 3D with Kinect[J]. Advances in Computer Vision & Pattern Recognition, 2013, 21(5):1154-1160.

[10] Kepski M, Kwolek B. 2013. Human Fall Detection Using Kinect Sensor[J]. 2013, 226:743-752.

[11] Klemenjak C, Egarter D, Elmenreich W. 2016. YoMo: the Arduino-based smart metering board[J]. Computer Science - Research and Development, 2016, 31(1-2):97-103.

[12] Barbon G, Margolis M, Palumbo F, et al. 2016. Taking Arduino to the Internet of Things: The ASIP programming model[J]. Computer Communications, 2016, s 89 – 90:128-140.

[13] Krauss R. 2016. Combining Raspberry Pi and Arduino to form a low-cost, real-time autonomous vehicle platform[C]// American Control Conference. IEEE, 2016:6628-6633.

[14] Dai XiGuo. 2017. Research on human attitude recognition based on Convolutional neural network [D]. Chengdu University of Technology.

ISPECE

IOP Publishing

Bounded Noises Estimation Based On Cognitive Radio In Distributed Fusion System

Shuyuan Wang[1,a], Ting Wang[2,b], Yijing Wang[3,c]

Student

[1]Chongqing University of Posts and Telecommunications, NO.2,Chongwen Road,Nan'an, District,Chongqing, China 086-17783062047

Teacher

[2] Chongqing University of Posts and Telecommunications, NO.2,Chongwen Road,Nan'an, District,Chongqing, China 086-18516229253

Student

[3] Chongqing University of Posts and Telecommunications, NO.2,Chongwen Road,Nan'an, District,Chongqing, China 086-17723088194

[a]384445765@qq.com, [b]101451686@qq.com, [c]ruizhibei@qq.com

ABSTRACT In distributed fusion system, the close interaction between the network and the physical world is emphasized, and the cognitive radio (CR) of wireless terminal has sufficient intelligence or cognitive ability to effectively improve the communication quality by detecting the history and current conditions of the surrounding wireless environment. Based on the Kalman filter, a channel sensing and conversion mechanism is studied and a new necessary and sufficient condition for MSE stability is deduced. Stable gain is received based on the stable requirements. And the average estimated error covariance is deduced. Finally, the average estimated error covariance with the mechanism is at least 63% less than that without this mechanism by using MATLAB simulation.

1. INTRODUCTION

As a smart radio communication system, cognitive radio technology dynamically detects spatial spectrum through spectrum sensing technology, thereby improving spectrum utilization efficiency [1]. There are many delimiting of cognitive radio. It is generally believed that cognitive radio is an intelligent wireless communication system with the learning ability of artificial intelligence technology, so it can recognize the external environment change. By changing its own working parameters in real time, such as modulation technology, operating frequency, transmission power, etc.

It allows the internal state to adapt the exchange of wireless transmission environment. That is to say, it can make full use of spectrum resources to meet the purpose of reliable communication at any time and any place [2]. One of the significant characteristic of cognitive radio is that it allows unauthorized users use holes opportunistically, by which without interfering the authorized communication.

As we know, the CR devices can be turned into different frequencies and allow the extended links to operate on different bands. This special CR feature will help to reduce co-band interference between

Content from this work may be used under the terms of the Creative Commons Attribution 3.0 licence. Any further distribution of this work must maintain attribution to the author(s) and the title of the work, journal citation and DOI.

Published under licence by IOP Publishing Ltd

the extended links so that the end-to-end throughput may be improved. A number of documents have been published and various aspects of CR network have been studied such as optimal sensing Algorithm, a spectrum selection and scheduling algorithm based on the opportunistic capacity concept, network protocols, and network security.

In this article, we consider the distributed fusion estimation issue for the networked time varying systems with bounded noises based on cognitive radio. Multisensor fusion estiamtion is one of the most important studies in the area of the state estimation fusion. More recently, networked multisensor fusion estimation has found applications in a wide range id areas, such as networked filtering in wireless sensor networks and distributed fusion systems. Under the networked fusion framework, two problems must be taken into account: 1) communication delays and packet dropouts; 2) restrains of sensor energy and sensor communication bandwidth. We can see that information loss is ineluctable because of the above constraints, and for this fusion estimation with incomplete information will degrade the estimation performance. In this artical, we follow the interest of the constraint of sensor energy and the sensor communication bandwidth.

For the lower communication traffic meets the bandwidth constraint, the main idea is to reduce the size of data packets. From this significance we can say that to guarantee a significant fusion estimation performance and various quantization reduction methods have been proposed in [3]. Meanwhile, when designing a estimator in energy-constrained sensor networks, it is not inevitable for sensors to transmit messages at every sampling instants [4]. Although this method cannot handle the bandwidth constrain problem, it can be combined with the key point of packet size reduction to design the fusion estimator in bandwidth and energy constrained sensor network [5]. As we all known, Kalman filtering is the most impactful way to find the optimal estimation of unknown state for the dynamic systems [6]. Figure 1 and Figure 2 show packet losses significantly affecting the estimation performance.

The issue of state estimation stability under a variety of factors such as bounded noises has attracted inrensice sesearch attentions [7]-[14]. [15]Bo Chen et al. study a new local estimator with time-varying gain is designed by solving a class of convex optimization problems such that the square error of the estimator is bounded. The information fusion noise statistics estimators are presented by averaging the local estimators of noise statistics in [16]. More generally, a class packet loss models in the view of semi-Markov chains is studied in [17]. We use CR over multiple channels to enhance the state estimation performance. We propose a CR based channel sensing and switching mechanism(CSSM) for state estimation. We focus on answering whether and how the state estimation can be increased by the new mechanism and the constraint of sensor energy and communication bandwidth. We get the conditions under which the proposed mechanism can improve the estimation performance. A couple of upper and lower bounds for mean square fusion estimation error is authenticated. And the performance melioration is analyzed.

Notations: The $\mathbf{R}^n$ is n-dimensional Euclidean space. $\mathbf{E}[\cdot]$ and $\mathbf{P}\{\}$ manifest the expectation and likelihood of random variable, respectively. For any matrix M, M' and M^H manifest $Tr(\cdot)$ and $\rho(\cdot)$ manifest the trace and spectral radius of a square matrix, respectively. $\lambda_{\min}(Y)$ is the maximum eigenvalues, and $\lambda_{\min}(X)$ is the minimum eigenvalues.

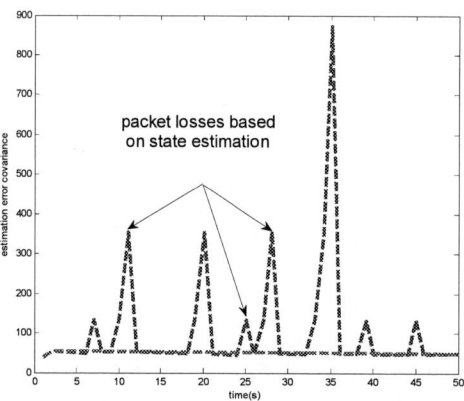

Figure 1. Effect of the packet losses.

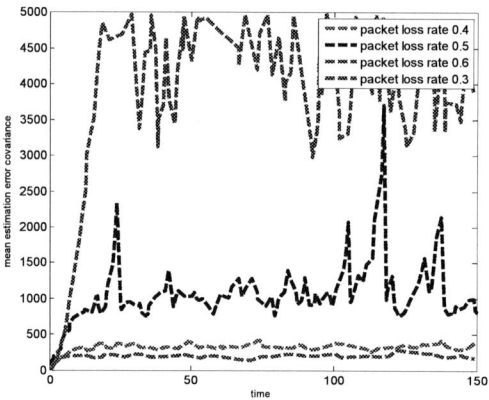

Figure 2. Effect of the packet losses rates

2. PROBLEM STATEMENT

2.1 System Modeling

Consider a linear discrete-time state dynamics described by the following state-space model:

$$x_{k+1} = Ax_k + Ba_k + w_k \quad (1)$$

$$y_k = Cx_k + v_k \quad (2)$$

where $x_k \in \mathbf{R}^n$ manifest the system state, $a_k \in \mathbf{R}^n$ manifest the control action, $w_k \in \mathbf{R}^n$ manifest the system noises., $v_k \in \mathbf{R}^n$ manifest the meterage noise. A, B, C are constant matrices with corresponding dimensions.

Kalman Filtering based State Estimation: we define a random variable $\gamma_k \in \{0,1\}$ to model the wireless communication reliability , $\gamma_k = 1$ indicates the measurement packet is successfully received, $\gamma_k = 0$ indicates the packet is lost.

Based on the measurements, the local estimator is given by following resursive form:

$$\begin{cases} \hat{x}_{k|k-1} = A\hat{x}_{k-1|k-1} + B\partial_k \\ P_{k|k-1} = AP_{k-1|k-1}A' + Q \\ \hat{x}_{k|k} = \hat{x}_{k|k-1} + \gamma_k K_k(y_k - C\hat{x}_{k|k-1}) \\ K_k = P_{k-1|k-1}C'\left(CP_{k|k-1}C' + R\right)^{-1} \\ P_{k|k} = (I - \gamma_k K_k)P_{k|k-1} \end{cases} \quad (3)$$

Based on Equation (3), the predicted error covariance meets the following modified algebraic Riccati equation.

$$P_k = AP_{k-1}A' - \gamma_k AP_{k-1}C'(CP_{k-1}C' + R)^{-1}CP_{k-1}A' + Q \qquad (4)$$

2.2 CSSM

The sensor is equipped with an antenna that can sense the signals in the channels. The sensor scans the authorized channels following the ordered channel index set Ω_m. When sensed idle, it will stop sensing and transmit packet by the first channel. Let $\Omega_m = (1, \cdots, m)$. For each $i \in \Omega_m$, it indicates the state of CH_i at the time when sensor senses CH_i in step k as $s_{i,k} \in \{0,1\}$. CH_i is busy so $s_{i,k} = 1$, $s_{i,k} = 0$, or else. Let $o_{i,k} \in \{0,1\}$ manifests the perception results. If CH_i is sensed busy, so $o_{i,k} = 1$, $o_{i,k} = 0$ otherwise. Once a transmission is completed, the sensor will turn back to CH_0, in order to minimize probable interference to PU.

The following proposition originates from probability transition matrix Φ_i of CH_i where $[\Phi_i]_{jl} =: \mathbf{P}\{s_{i,k+1} = l-1 | s_{i,k} = j-1\}$.

Proposition 1: $\{s_{i,k}\}_{k \geq 0}$ comprises a homogeneous Markov chain with the transition probability matrix

$$\Phi_i = \begin{bmatrix} 1-\alpha_i & \alpha_i \\ \beta_i & 1-\beta_i \end{bmatrix} \qquad (5)$$

where $\alpha_i = \dfrac{\omega_{i,0}}{\omega_{i,1}+\omega_{i,0}}[1 - e^{-(\omega_{i,1}+\omega_{i,0})T}], \beta_i = \dfrac{\omega_{i,1}}{\omega_{i,1}+\omega_{i,0}}[1 - e^{-(\omega_{i,1}+\omega_{i,0})T}]$.

To describe the channel sensing accuracy, we define two probabilities. $p_{d,i} = \mathrm{P}\{o_{i,k} = 0 | s_{i,k} = 0\}$ manifests the correct detection probabilities, $p_{f,i} = \mathrm{P}\{o_{i,k} = 0 | s_{i,k} = 1\}$ manifests the false detection probabilities.

3. CHANNEL CASE

In this portion, the conditions for estimation stability will be discussed. Then we show how the CSSM will expand estimation stability region. Let us define

3.1 The Critical Matric $\Phi\Psi$

Delimiting $B_m(\cdot)$ as follows. $\forall b \in \{1, \cdots, 2^m\}$, there exists the only binary vector $B_m(b) = [b_1, \dots, b_m]$ such that $b = \sum_{j=1}^{m} 2^{j-1}b_j + 1$. Define channel state vector $s_k = [s_{1,k}, \dots, s_{m,k}]'$, $p_k = [p_{1,k}, \dots, p_{2^m,k}]'$. Based on p_k and the hypothesis of the inter-channel independencies.

$$\begin{aligned} [\Phi]_{ij} &= \mathbf{P}\{s_{k+1} = B_m(j) | s_k = B_m(i)\} \\ &= \prod_{l=1}^{m} \mathbf{P}\{s_{l,k+1} = [B_m(j)]_l | s_{l,k} = [B_m(i)]_l\} \end{aligned} \qquad (6)$$

where $i, j \in \{1, \cdots, 2^m\}$. The sensing matrix Ψ is in form $\mathrm{Diag}\{\psi_1, \cdots, \psi_{2^m}\}$ where $\forall b \in \{1, \cdots, 2^m\}$, $\psi_b = \mathbf{P}\{\gamma_k = 0 | s_k = B_m(b)\}$. ℓ_i^s indicates the packet loss rate on CH_i when it is in state $s: \ell_i^s = 1$ if $s = 1$ and $\ell_i^s = \ell_i$ otherwise. Thus

$$\psi_b = \mathbf{P}\{\gamma_k = 0 \,|\, [s_{1,k}, \ldots, s_{m,k}]' = \mathbf{B}_m(b)\}$$
$$= \mathbf{P}\{o_{1,k} = 0 \,|\, s_{1,k} = b_1\}\ell_1^{b_1} + \mathbf{P}\{o_{1,k} = 1 \,|\, s_{1,k} = b_1\}$$
$$\times \mathbf{P}\{o_{2,k} = 0 \,|\, s_{2,k} = b_1\}\ell_2^{b_2} + \cdots + \mathbf{P}\{o_{1,k} = 0 \,|\, s_{1,k} = b_1\}$$
$$\times \cdots \times \mathbf{P}\{o_{m-1,k} = 1 \,|\, s_{m-1,k} = b_{m-1}\}$$
$$\times \mathbf{P}\{o_{m,k} = 0 \,|\, s_{m,k} = b_m\}\ell_m^{b_m} \tag{7}$$
$$+ \mathbf{P}\{o_{1,k} = 0 \,|\, s_{1,k} = b_1\} \cdots \mathbf{P}\{o_{m,k} = 0 \,|\, s_{m,k} = b_m\}\ell_0$$
$$= \sum_{i=1}^{m} \left(\prod_{j=1}^{i-1} \mathbf{P}\{o_{j,k} = 0 \,|\, s_{j,k} = b_1\} \right) \mathbf{P}\{o_{i,k} = 0 \,|\, s_{i,k} = b_1\}\ell_i^{b_i}$$
$$+ \ell_0 \prod_{i=1}^{m} \mathbf{P}\{o_{i,k} = 1 \,|\, s_{i,k} = b_i\}$$

Where the second equality is based upon the definition of the sending sequence Ω_m.

3.2 Stability Analysis

Theorem 1: For system (1) with above channel sending schedule Ω_m, a necessary condition for MSE stability of fusion estimation process with CSSM is

$$\rho(\Phi\Psi)\rho(A)^2 < 1 \tag{8}$$

Moreover, (8) is sufficient if C has full column rank.

3.3 Performance Analysis

Theorem 2: Let $\ell_{\min} = \min\{\ell_0, \ell_1, \ldots, \ell_m\}$. With sensing schedule Ω_m, stability gain η satisfies:

$$\eta = \sqrt{\frac{\ell_0}{\rho(\Phi\Psi)}} \leq \left\{ \frac{1}{\sqrt{\prod_{i=1}^{m}(1-\beta_i)}}, \sqrt{\frac{\ell_0}{\ell_{\min}}} \right\} \tag{9}$$

Let $\bar{\ell}^* = \max\{e_i'\Phi\Psi u, \forall i \in \{1, \cdots, 2^m\}\}$, where e_i is a $2^m \times 1$ vector with $[e_i]_i = 1$ and $[e_i]_j = 0$ for all $j \neq i$, $u = [1, \ldots, 1]'_{1 \times 2^m}$.

Theorem 3: If $\bar{\ell}^* \leq \ell_0$, and initial conditions satisfy $P_0^+ = P_0$, then $Tr(\mathrm{E}[P_k]) \leq Tr(\mathrm{E}[P_k^+])$.

Theorem 4: $\forall k > 0$, $\underline{P}_k \leq P_k \leq \overline{P}_k$, $\mathrm{E}[\underline{P}_k] \leq \mathrm{E}[P_k] \leq \mathrm{E}[\overline{P}_k]$. Moreover

$$Tr(\mathrm{E}[\underline{P}_\infty]) = \sum_{i=0}^{\infty} [\upsilon'(\Phi\Psi)^i u Tr(\underline{W}(o)A^iA^i)] \tag{10}$$

$$Tr(\mathrm{E}[\overline{P}_\infty]) = \sum_{i=0}^{\infty} [\upsilon'(\Phi\Psi)^i u Tr(\overline{W}(o)A^iA^i)] \tag{11}$$

where υ mannifests a changeless state of the vector P_k, $o = 1 - \upsilon'(I - \Phi\Psi)u$.

Remark: Our results can apply to a wide range of scenarios. Considering another channel sensing which differs from Ω_m in which the sensor can directly transmit packets through CH_m if CH_{m-1} is sensed busy. If we hold CH_m as a primordial channel; the scenario the original channel is Markovian and the CSSM is not used can be viewed as special single-authorized-channel case. So the results will be renewed.

4. SIMULATION EXAMPLE

Considering a maneuvering target with the physical process satisfies (1). We use the same parameters as in[18] : $A = \begin{bmatrix} 1.25 & 0 \\ 1 & 1.1 \end{bmatrix}$, $Q = 20I_{2\times2}$. and the sensor meterages are described by Equation(2), where $C = I_{2\times2}$, $R = 2.5I_{2\times2}$, the sample period $T = 1$. $\alpha = 0.537$, $\beta = 0.461$, $l_1 = 0.05$. So the largest eigenvalue of critical matrix $\Phi\Psi$ is $\sigma_2 \approx 0.415 < \ell_0 < \frac{1}{\rho(A)^2}$, the state fusion estima- tion based on the Kalman filter is stable in the mean square sense. However, the case completes different performance according to $Tr(\mathrm{E}[P_k])$. As shown in Figure 3, the CSSM brings about much more fusion estimation performance.

Firstly, we observe the bounds are tight in the simulation case. Without CSSM, it is wide known that fusion estimation error covariance diverges only if ℓ_0 is larger than $\frac{1}{\rho(A)^2} = 0.64$. In contrast, with CH_1 and CSSM, the critical value increases obviously, i.e., the demands for the estimation stability is relaxed. From Figure 4 we demonstrate that CSSM inproves estimation performance when $l_0 \geq 0.34$. When each local estimate $\hat{x}_k(t)$ is sent to the FC due to finit bandwidth and the linited sensor energy. In reality, when $l_0 \geq 0.34$, both $\sigma_2 < l_0$ and $\psi_2(1-\beta)+\psi_1\beta \leq l_0$ are true. Thus, the fusion estimation capability is assured to be enhanced. However, the delineation also manifest that there is a performance descend when $l_0 < 0.34$ such that quanlity of CH_0 is better than that of the CH_1. This can be finished off by strengthening the channel sensing accuracy.

More authorized channels can provide higher opportunities for sensor to triumphantly transmit its measurement packets. The performance bounds and the performance ratio obtained in Theorem 4, where $m = 0$ demonstates the case without CSSM. We can see that: 1) With CSSM, the upper bound and relaxed upper bounds are quite close in all situations. 2) The worst situation property ratio is less than 36.68%. 3) The property is clearly improved by introducing to the CSSM more authorized channels.

5. CONCLUSION

In this paper, the fusion estimation problem with linear process state dynamics has been investigated for networked time-varying fusion system with bounded noises. Based on cognitive radio technology, we propose CSSM mechanism for sensor to opportunistically enter authorized spectrum in data transfer. We develop a new necessary and sufficient condition for fusion estimation stability in mean square sense. At last, simulation results manifest that fusion estimation property is remarkably improved by CSSM.

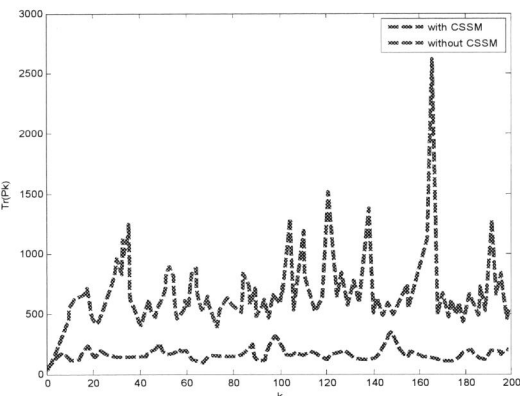

Figure 3. Performance comparison.

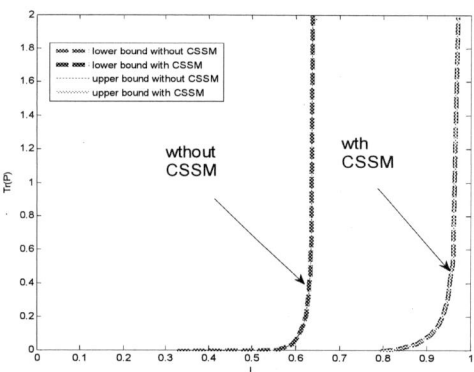

Figure 4. Performance bound comparison.

REFERENCES

[1] Lee W Y, Akyildiz I F. Optimal spectrum sensing framework for cognitive radio networks[J]. IEEE Transactions on Wireless Communications, 2008, 7(10):3845-3857.

[2] Pan M, Li P, Fang Y. Cooperative Communication Aware Link Scheduling for Cognitive Vehicular Networks[J]. IEEE Journal on Selected Areas in Communications, 2012, 30(4):760-768.

[3] Xia Y F, Zhu Y M, Huang K N. Scheme for observation quantization in information estimation fusion[J]. Journal of Sichuan University, 2009, 41(2):211-215.

[4] Pei Y, Liang Y C, Teh K C, et al. Energy-Efficient Design of Sequential Channel Sensing in Cognitive Radio Networks: Optimal Sensing Strategy, Power Allocation, and Sensing Order[J]. IEEE Journal on Selected Areas in Communications, 2011, 29(8):1648-1659.

[5] Wu X, Tian Z. Optimized Data Fusion in Bandwidth and Energy Constrained Sensor Networks[C]// IEEE International Conference on Acoustics, Speech and Signal Processing, 2006. ICASSP 2006 Proceedings. IEEE, 2006:IV-IV.

[6] Sinopoli B, Schenato L, Franceschetti M, et al. Kalman filtering with intermittent observations[J]. IEEE Transactions on Automatic Control, 2004, 1(9):1453-1464.

[7] Guerra P, Puig V, Ingimundarson A. Robust fault detection using a consistency-based state estimation test considering unknown but bounded noise and parametric uncertainty[C]// Control Conference. IEEE, 2015:1595-1601.

[8] Becis-Aubry Y, Boutayeb M, Darouach M. An ellipsoidal state estimation algorithm for nonlinear systems subject to bounded disturbances[C]// European Control Conference. IEEE, 2015.

[9] Foo Y K, Soh Y C, Moayedi M. Linear set-membership state estimation with unknown but bounded disturbances[J]. International Journal of Systems Science, 2012, 43(4):715-730.

[10] Xu X B, Zhang Z, Zheng J, et al. State estimation method based on evidential reasoning rule[C]// Advanced Information Technology, Electronic and Automation Control Conference. IEEE, 2016:610-617.

[11] Chen B, Ho D W C, Zhang W A, et al. Networked Fusion Estimation with Bounded Noises[J]. IEEE Transactions on Automatic Control, 2017, PP(99):1-1.

[12] Xie W, Xia Y. Recursive parameter estimation with bounded noises in the presence of missing outputs and outliers[C]// Control Conference. IEEE, 2014:5167-5172.

[13] Qu X, Zhou J, Tan W. Robust decentralized estimation fusion in energy-constrained wireless sensor networks with correlated noises[J]. Digital Signal Processing, 2018.

[14] Jin X B, Bao J, Zheng H J. Centralized robust fusion estimation in estimation of paper basis weight based on norm-bounded parameter uncertain model[J]. Proceedings of SPIE - The International Society for Optical Engineering, 2010, 7820(1):485-490.

[15] Chen B, Yu L, Zhang W A, et al. Robust Information Fusion Estimator for Multiple Delay-Tolerant Sensors With Different Failure Rates[J]. IEEE Transactions on Circuits & Systems I Regular Papers, 2013, 60(2):401-414.

[16] Zhang M, Xiao-Ling F U, Cui P. Multi-sensor optimal information fusion for time-delay systems with multiplicative noise[J]. Journal of Shandong University, 2010.

[17] Censi A. Kalman Filtering With Intermittent Observations: Convergence for Semi-Markov Chains and an Intrinsic Performance Measure[J]. Automatic Control IEEE Transactions on, 2011, 56(2):376-381.

[18] Sinopoli B, Schenato L, Franceschetti M, et al. Kalman filtering with intermittent observations[J]. IEEE Transactions on Automatic Control, 2004, 1(9):1453-1464.

[19] Wang Z S, Liu W J, Zhen Z Y. Design of Optimal Tracking Controller for Nonlinear Discrete System with Input Delay Using Information Fusion Estimation Method[C]// IEEE International Conference on Control and Automation. IEEE, 2007:2535-2538.

[20] Zhang W A, Ni H, Song H, et al. Distributed information fusion estimation for sensor networks with nonuniform sampling rates[C]// International Conference on Mechatronics and Control. IEEE, 2015:502-505.

ISPECE

IOP Publishing

Datacentre TCP Protocol of Centralized Window Control

Chunlin Wang[1,a] ,Jianyong Sun[1,c], Wanjin Xu[1,d] , Xiaolin Chen[1,b]

[1]School of Information Science & Technology, Chuxiong Normal University, P.R.China

Corresponding author: Xiaolin Chen

[a]wcl@cxtc.edu.cn, [b]chenxl@cxtc.edu.cn ,[c]sunny@cxtc.edu.cn, [d]xwj_cx@cxtc.edu.cn

ABSTRACT There is an important class of applications in data centre called online data-intensive (OLDI) application, such as search engine, electronic shopping, and advertising, which has low latency, high throughput, and high burstiness. For better user experience, the query response delay in the data centre is controlled under 300 milliseconds. DCTCP can better meet the requirements of OLDI, but all the switches in the data centre are required to support RED. For data centre that does not fully support RED switch, TCP protocol can only be used. TCP Incast will appear in the application, which is far from the requirements of OLDI application. This paper proposes the TCP protocol of data centre based on window control. It is in existing switches and network devices,and centralized control makes the send window match with the switch queue cache. In the case of ensuring low latency, it keeps the switch queue in a small range of fluctuation, enabling the data centre to meet low latency, high throughput and high burstiness application requirements. Simulation experiments verify that WCTCP congestion control method can effectively solve the above problems. It is 5 times lower than TCPNewReno packet loss rate, 6 times higher than the network utilization rate, and the network performance is close to the DCTCP protocol.

1. INTRODUCTION

In recent years, the data centre has changed the application mode of computer from the traditional single-machine mode and network mode to the current computing mode the with data centre as the core, and with the emergence of cloud computing service providers such as Alibaba, Microsoft and Google. In data center design, using low-cost switches and servers to build high-performance, manageable and maintainable data center is an ideal goal for all data center operators. On the other hand, with the continuous growth of Internet applications, Internet users expect shorter query response and higher throughput to meet different applications, such as network search, online e-commerce, advertising promotion and online on-demand. This type of appli-

cation is called online data-intensive application (OLDI) [3], which controls the query response delay within the data centre to be less than 300 milliseconds.Online data-intensive application (OLDI) generates different short data stream and long data stream, and expects data centre network to meet the requirements of low latency, sudden fault tolerance, and high throughput of long data stream. After subtracting the typical Internet delay, the total network delay of OLDI within the data center is usually less than 230-300ms, which is also known as soft real-time constraint. If soft real-time constraint exceeds 300ms, it will affect user experience and operator revenue. OLDI soft real-time constraint (300ms) cannot meet the application requirements in traditional 1:1 client and server mode. Therefore, OLDI adopts a tree-based divide-and-conquer algorithm (the calculation mode of N: 1). Each query is decomposed into multiple sub-queries through the control centre, then distributed to different servers

Content from this work may be used under the terms of the Creative Commons Attribution 3.0 licence. Any further distribution of this work must maintain attribution to the author(s) and the title of the work, journal citation and DOI.

Published under licence by IOP Publishing Ltd

to complete the query, and the results of the query are fed back to the control centre. For OLDI application like this, soft real-time constraint is the key metric. To avoid missing soft real-time constraint, the control centre issues incomplete response without waiting for slow query that misses deadline. Soft real-timeconstraint.timeout can negatively impact OLDI application,reducing user experience and affecting the competitiveness of data centre and service providers. Therefore, reducing the missing soft real-time constraint is one of the focuses of data centre research.

2. CURRENT RESEARCH

2.1 Problem Is Introduced

In OLDI application, the client decomposes the query task into multiple sub-tasks assigned to hundreds of servers. When the server finishes the query and returns the query results to the client via the switch, the queue of the switch will be occupied in an instant until the buffer overflow occurs and the packet is lost due to the server sending the packet to the switch at the same time (N: 1). TCP has two ways to handle data retransmission after packet loss: one is to trigger timeout retransmission when the timer goes off. The second is that the sender receives three repeated ACK acknowledgment messages, triggering fast retransmission. Traditional TCP retransmission timer (RTO) is generally no less than 200ms, while data center network environment round-trip delay (RTT) is generally on the order of microseconds. During this period, the server's sending window continuously sends data until the sending window runs out. The network throughput drops dramatically when the switch is idle from the sending window running out to the start of retransmission. This phenomenon is called TCP Incast[6]. The preconditions for the occurrence of TCP Incast include: 1.the network has high bandwidth and low latency,and the cache of the switch is small.2.There is synchronous many-to-one traffic in the network.3.The data traffic on each TCP connection is small. Literature [4-5] all believe that
packet loss leads to timeout retransmission, and that the timeout time of TCP retransmission of RTO does not match the RTT value, which leads to TCP Incast problem. Currently, three methods are commonly used in research to solve TCP Incast: 1. RTO is set according to the environment of the data centre, with reduced retransmission timer, so that it can be retransmitted earlier. 2.Larger switch cache, will increase the operating cost of the data centre. 3. Better congestion control enables better matching between the sending window and the switch cache to reduce packet loss and improve network utilization. In academia and industry, the TCP protocol congestion control of data centre is mainly studied. This paper also focuses on congestion control.

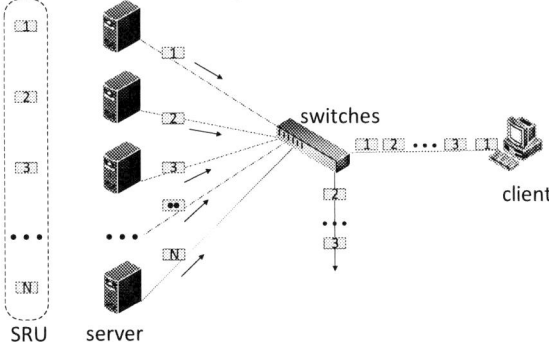

Figure 1. Experimental topology

2.2 The Research Progress

The DCTCP protocol [1] was proposed by Microsoft research institute at the SIGCOMM conference in 2010. The idea of DCTCP is to use explicit congestion notification (ECN). It first monitors the queue length in the switch. If the queue length exceeds the upper limit, the TCP receiver is explicitly notified through the ECN field. The receiving end reduces the sender's sending window through the window field, reduces the sending speed, and reduces the packet entering the switch queue. This equalizes the queue length of the switch to a certain length, reduces packet loss and increases network utilization. In 2011 SIGCOMM, Balajee Vamanan from Purdue university proposed the D2TCP congestion control protocol [2], which has better emergency response and lower latency. D2TCP adopts a new congestion avoidance algorithm, which uses ECN feedback and deadline to modulate the congestion window through gamma correction function, to speed up the delivery speed and priority of services when latency is close to soft real-time constraint, and to reduce soft real-time timeout.

2.3 Problems To Be Solved
The two previous approaches have two drawbacks: 1. All switches in the data centre are required to support ECN. 2. The TCP protocol needs to be modified, and the operating system should support the corresponding TCP protocol. These are hard to do in some smaller data centres. This paper focuses on small data centres that do not support ECN, conducts research around TCP protocol congestion control of data centres, and proposes solutions to the TCP Incast problem. This provides small data

3. The solution

3.1 Algorithm thought
This paper proposes a congestion control algorithm for centralized sending window control. The idea is: in OLDI application, if the sending window is too large, multiple senders will send data to the network continuously. This results in queue overflow, packet loss and idleness of the switch, which reduces network performance. On the other hand, a small sending window reduces network utilization and throughput. In this algorithm, the sum of the sending windows of all service groups is centrally controlled and constrained to match the cache size of the switch and keep the queue length of the switch within a certain range of fluctuation. It not only avoids overflow and packet loss in switch queue, but also ensures high network utilization and low latency. In the algorithm, the total window is initialized first, and the average value of the total window is the upper limit of each server sending window. If no packets are lost, adjust the value of the total window. If retransmission reduces the value of the total window, update the upper limit of the server sending window again. If the data sent by the server is smaller than that of other servers, the upper limit of the sending window of the server is reduced and the upper limit of the sending window of other servers is increased. The size of the total window in this algorithm is one of the researches focuses in this paper. The relationship between the total window, the size of the switch cache and the network utilization in this algorithm is described in figure 2.

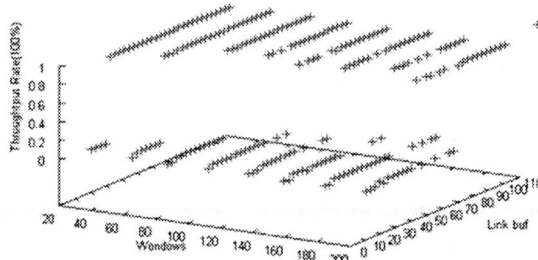

Figure 2. The relationship between throughput and total window size and queue length

3.2 Algorithm description
Start as: upper limit of the total window connected wnd_cap=min[swich_link_buf$_{[1]}$,swich_link_buf$_{[2]}$.....swich_link_buf$_{[n]}$];
swich_link_buf$_{[n]}$ is the number of switch cache in the connection, and the limit value of the connection window is window_ ←0. Connection data N← number of servers

Update the upper limit of window_: check_trans(N){
{
If(connections_number <N){
If（wnd_cap /(connections_number +1)>=2）
{ connections_number ← connections_number +1
window_ ← int(wnd_cap/ connections_number)
new_connection(connections_number)
}
}
For(i←0;i<=N;i++)
{
if(prev_npkts_i>= npkts_i)
{ connections_number ← connections_number -1
window_ ← int(wnd_cap/ connections_number)
}
prev_npkts_i← npkts_i
}
If(N>0) check_trans(N)
}
Update the total size of the wnd_cap sending window：g←0.8, queue_length: current switch queue length

$$\begin{cases} \text{wnd\_cap} = \text{wnd\_cap} - 1 & (\text{queue\_length} > g * \text{swich\_link\_buf}) \\ \text{wnd\_cap} = \text{wnd\_cap} + 1 & (\text{queue\_length} < g * \text{swich\_link\_buf}) \end{cases}$$

4. Evaluation
In the experiment, NS2 is used to conduct simulation experiments on the algorithm, and the experimental structure is shown in figure 1:8 servers, a switch and a client are used in the experiment. The algorithm is evaluated in terms of queue length, query delay, throughput rate and loss rate.

4.1 Queue Length Evaluation
In the queue length evaluation, TCPnewReno and DCTCP, two congestion control algorithms are respectively used for comparison. The switch cache swich_link_buf←16 is set, and the amount of data to be sent is 1.6MB. The change of the switch queue length is observed,as shown in figure 3. After TCPnewReno is transmitted for a period of time, the phenomenon of Incast appears after packet loss, resulting in rapid decline of network utilization. WCTCP algorithm queue always fluctuates between [10,14], which has a smaller queue length than DCTCP algorithm, and the transmission delay is basically the same.

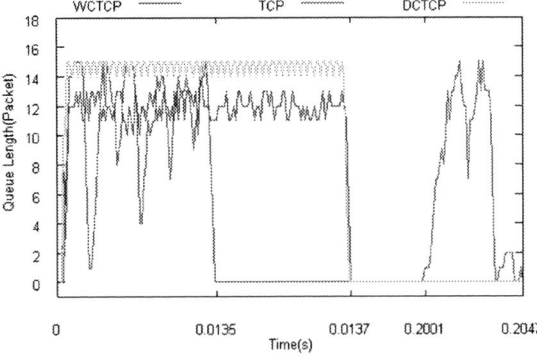

Figure 3. Queue Length

4.2 Delay

In the latency measurement experiment,8KB,24KB,···,157MB data blocks are sent respectively in 1Gb network, as shown in figure 4. In the experiment, since WCTCP and DCTCP algorithms differ very little in transmission delay (10^{-4}), they are highly overlapping in the figure. In data block transmission under 17MB, WCTCP can control the transmission time well below soft real-time constraint (300ms),meeting the requirements of OLDI application for short data stream with low latency.

4.3 Throughput

In the throughput measurement (as shown in figure 5), WCTCP can maintain a high network utilization rate in the network bandwidth of 0.1Gb to 5Gb. Throughput rate is decreasing after 5Gb and consistent with TCP. This is mainly because in the 5Gb and consistent with TCP. This is mainly because in the experimental test, the total switch cache of the algorithm is 100KB, and the service sending window is 18PKT.

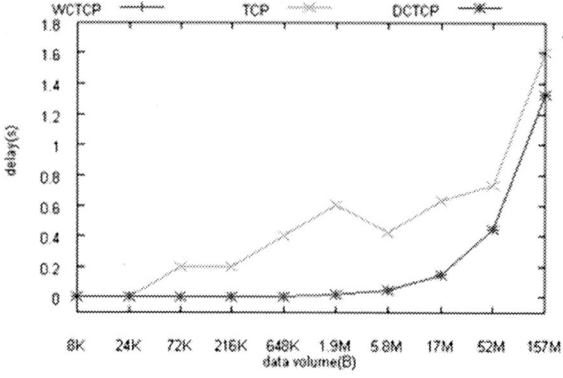

Figure 4. Delay

In a network of more than 10Gb, the mismatch between the switch cache and the sending window is the cause of the decrease in network utilization. Solving this problem can increase the size of the switch cache and the sending window.

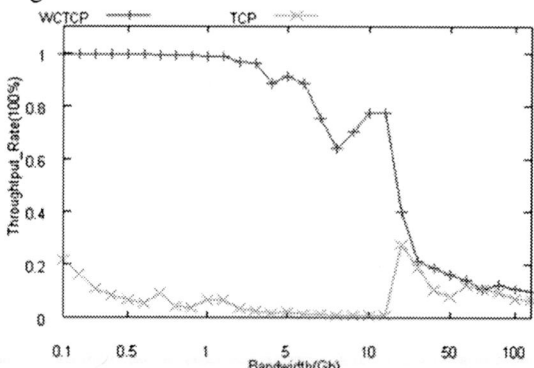

Figure 5. Throughtput

4.4 Loss Rate

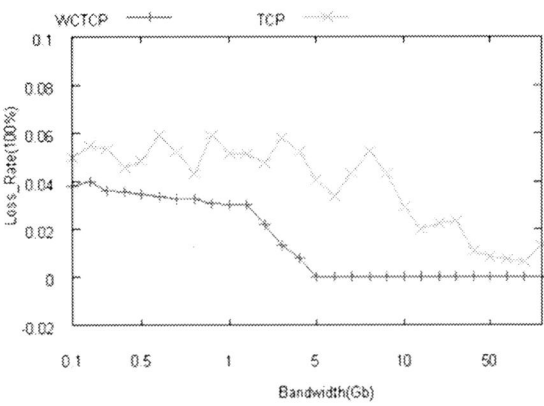

Figure 6. Loss Rate

In the test of packet loss rate, 1.6MB packets are sent, and the packet loss of switch is recorded in the process of the transmission rate increasing from 0.1Gb to 50Gb. The packet loss rate of WCTCP algorithm is gradually decreasing. After 5Gb, the network packet loss rate is close to zero. Experimental results show that it has lower packet loss rate than TCP protocol.

5. CONCLUSION

Simulation experiments verify that WCTCP algorithm is in line with expectations, with low switch queue length, high throughput and low latency. In low-cost data center application, the WCTCP algorithm does not need to support RED switch. The dynamic allocation of the sending window can reduce the network congestion, which is 5 times lower than the packet loss rate of TCPNewReno, 6 times higher than the network utilization rate, and the network performance is close to the DCTCP protocol. In a network of more than 10Gb, more support of sending window and high-performance switch is needed, otherwise the performance drops dramatically.

ACKNOWLEDGMENT

This work is supported in part by Yunnan Provincial Department of Education Project Data Center and the Science and Technology Program of Yunnan Province, China. And the soft timeout transmission control protocol project number and the Grant Number are respectively 2018JS454 and 2017FH001-124

REFERENCES

[1] Alizadeh M, Greenberg A, Maltz D A, et al. Data center tcp (dctcp)[J]. ACM SIGCOMM computer communication review, 2011, 41(4): 63-74.

[2] Vamanan B, Hasan J, Vijaykumar T N. Deadline-aware datacenter tcp (d2tcp)[J]. ACM SIGCOMM Computer Communication Review, 2012, 42(4): 115-126.

[3] Meisner D, Sadler C M, Barroso L A, et al. Power management of online data-intensive services[C]//ACM SIGARCH Computer Architecture News. ACM, 2011, 39(3): 319-330.

[4] Vasudevan V, Phanishayee A, Shah H, et al. Safe and effective fine-grained TCP retransmissions for datacenter

[5] Chen Y, Griffith R, Liu J, et al. Understanding TCP incast throughput collapse in datacenter networks[C]//Proceedings of the 1st ACM workshop on Research on enterprise networking. ACM, 2009: 73-82.

[6] Nagle D, Serenyi D, Matthews A. The panasas activescale storage cluster: Delivering scalable high bandwidth storage[C]//Proceedings of the 2004 ACM/IEEE conference on Supercomputing. IEEE Computer Society, 2004: 53.

[7] Kwan B, Agarwal P, Ashvin L. Flexible buffer allocation entities for traffic aggregate containment: U.S. Patent 8,532,117[P]. 2013-9-10.

[8] Al-Fares M, Loukissas A, Vahdat A. A scalable, commodity data center network architecture[C]//ACM SIGCOMM Computer Communication Review. ACM, 2008, 38(4): 63-74.

[9] Floyd S. RED: Discussions of setting parameters[J]. http://www. aciri. org/floyd/REDparameters. txt, 1997.

[10] Qazi I A, Andrew L L H, Znati T. Congestion control using efficient explicit feedback[C]//INFOCOM 2009, IEEE. IEEE, 2009: 10-18.

[11] Vasudevan V, Phanishayee A, Shah H, et al. Safe and effective fine-grained TCP retransmissions for datacenter communication[C]//ACM SIGCOMM computer communication review. ACM, 2009, 39(4): 303-314.

[12] Xia Y, Subramanian L, Stoica I, et al. One more bit is enough[J]. ACM SIGCOMM Computer Communication Review, 2005, 35(4): 37-48.

[13] Wilson C, Ballani H, Karagiannis T, et al. Better never than late: Meeting deadlines in datacenter networks[J]. ACM SIGCOMM Computer Communication Review, 2011, 41(4): 50-61.

[14] Ellison N B, Steinfield C, Lampe C. The benefits of Facebook "friends:" Social capital and college students' use of online social network sites[J]. Journal of Computer - Mediated Communication, 2007, 12(4): 1143-1168.

ISPECE

IOP Publishing

IOP Conf. Series: Journal of Physics: Conf. Series **1187** (2019) 032089   doi:10.1088/1742-6596/1187/3/032089

Gas Packaging Container Based on ANSYS Finite Element Analysis and Structural Optimization Design

Zhiqin Yin[1], Tongdan Su [2], Ming He [1]

[1]Yunnan Open University,650000;

[2]Central South University,410083

Abstract Gas packaging containers are often used for flammable, explosive, toxic, strong corrosion and compressed gas packaging. The transportation and use environment is more complicated and harsh than other metal packaging containers, and has very high requirements for its structural design. In this paper, taking a certain type of industrial oxygen cylinder as an example, ANSYS Workbench is used to establish the finite element model, and the equivalent linearization treatment method is used to optimize the structural design of the gas cylinder, which provides ideas and methods for the optimization design of the gas packaging container structure.

CLC number: TH49   Document code: A

1. Introduction

Gas packaging containers are an integral part of the packaging industry. They are often used for flammable, explosive, toxic, strong corrosion and compressed gas packaging. The transportation and use environment is more complicated and harsh than other metal packaging containers, so the structure design is very High requirements are attributed to special equipment in China, and relevant laws and regulations are formulated for special management.

As an important industrial material, industrial gas is widely used in the fields of machinery, chemical industry, metallurgy, aerospace, medical and health, and daily life of residents. As a gas container for industrial gas packaging, it is widely used in social production and life. Taking a certain type of industrial oxygen cylinder as an example, this paper uses ANSYS Workbench to carry out finite element analysis of the oxygen cylinder during transportation to obtain the stress value and distribution of the oxygen cylinder during transportation, and apply the equivalent linearity to the existing safety hazards. The structural optimization method is used to optimize the structure design, which provides an idea and method for the optimization design of gas packaging container structure.

2. Structural Analysis of Gas Packaging Containers

2.1 Main Characteristics of Gas Packaging Containers
In addition to the very high quality requirements for the inner surface of the gas packaging container, it should also comply with the national pressure vessel quality standards. The gas cylinder is a kind of gas packaging container. According to the "Cylinder Safety Supervision Regulations", it usually refers to the re-inflatable use under normal environment (-40~60 °C). The nominal working pressure is 1.0~30MPa (gauge pressure). A mobile pressure vessel with a nominal volume of 0.4 to 1 000 L and containing permanent gas, liquefied gas or dissolved gas.

Content from this work may be used under the terms of the Creative Commons Attribution 3.0 licence. Any further distribution of this work must maintain attribution to the author(s) and the title of the work, journal citation and DOI.

Published under licence by IOP Publishing Ltd

2.2 Oxygen Cylinder Structure

The oxygen cylinder is a steel cylindrical high-pressure vessel for storing and transporting industrial oxygen. It is generally made of seamless steel pipe and is a kind of permanent gas cylinder. It is an indispensable material equipment in industrial production. The common oxygen cylinder wall thickness is 5-8mm, and its shape, structure and standard should meet the requirements of GB5099 "Steel Seamless Gas Cylinder" [1]. The research object of this paper is limited to industrial 40L oxygen cylinder, the nominal pressure is 15MPa, and its appearance structure parameters are shown in Table 1:

Table 1 40L Industrial Oxygen Cylinder Construction Parameters

Outer Diameter	Bottle Wall Thickness	Nominal Pressure	Actual Pressure	Material	Length
219mm	5.7mm	15MPa	12.5MPa	32Mn2V	1450mm

Because the oxygen cylinder is subjected to a dangerous accident due to external force, the action point of the external force is usually concentrated on the bottom of the bottle, the shoulder of the bottle and the bottle body, so the bottle mouth and the bottle valve area are simplified. In addition, the bottom position of the bottle is convenient for calculation and is limited to a concave structure. The concave part structure should meet the design requirements of GB5099 [1]:

（1）$S_1 = (2.0 \sim 2.6) \, S$

（2）$S_2 = (1.8 \sim 2.2) \, S$

（3）$S_3 = (2.0 \sim 2.6) \, S$

（4）$r = (0.07 \sim 0.09) \, D_0$

（5）$H = (0.13 \sim 0.16) \, D_0$

Where S is the thickness of the oxygen bottle wall and D_0 is the outer diameter of the oxygen bottle, as shown in Figure 1 [1]:

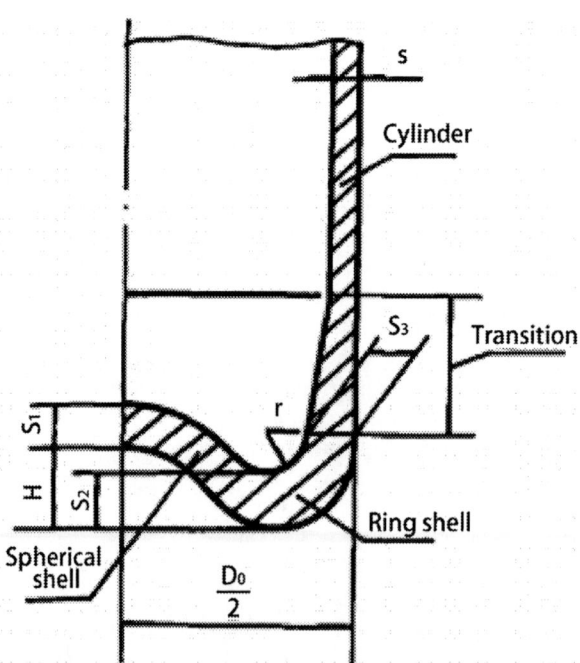

Figure 1 Design Requirements for The Concave Bottom of the Oxygen Cylinder

According to the design requirements, the selected values are shown in Table 2:

Table 2 Selected Model Values							
S (mm)	1/2D₀(mm)	S_1(mm)	S_2(mm)	S_3(mm)	r(mm)	H(mm)	L (mm)
5.7	109.5	14.25	11.4	14.25	17.52	32.85	1450

2.3 Oxygen Cylinder Force Analysis

The oxygen cylinder material is 34Mn2V structural steel, which is a plastic material. We can use the fourth strength theory to judge the static structure analysis result of the gas cylinder. The stress is calculated by the following formula:

$$\sigma_{eq} = \sqrt{\frac{1}{2}[(\sigma_1 - \sigma_2)^2 + (\sigma_2 - \sigma_3)^2 + (\sigma_3 - \sigma_1)^2]} \tag{1}$$

Among them, σ_1、σ_2 and σ_3 are represent the first principal stress, the second principal stress and the third principal stress, respectively.

The main loads that gas cylinders are subjected to during transportation are: gas pressure inside the bottle, external air pressure, acceleration load caused by transportation, and gravity load on the cylinder itself. The normal working pressure of the gas cylinder is 12.5 MPa, assuming it is evenly distributed on the inner surface of the cylinder. The bottle body is affected by the external atmospheric pressure, and the atmospheric pressure is assumed to be 1.01 MPa, which is evenly distributed on the outer surface of the oxygen cylinder. The bottle body is also affected by its own gravity, and the gravitational acceleration is assumed to be 9.8m/s^2 , and the direction is vertically downward. In addition, because the oxygen cylinder is under transport conditions, it is bound to be affected by the acceleration of the car during road travel. Considering the complex and variability of the driving situation of the road during road transportation, it is limited to the special case of the emergency braking of the truck, and its acceleration is 5m/s^2.

3. Finite Element Analysis and Structural Optimization Design of Oxygen Cylinder

3.1 Finite Element Analysis of Oxygen Cylinder

The 40L industrial oxygen cylinder is generally made of 34Mn2V. Due to the trace element alloy vanadium, it has been widely used in the manufacture of high pressure seamless gas cylinders [2]. Its main mechanical properties are as follows [3]:

(1) Density: 7850 Kg/m3
(2) Poisson's ratio: 0.3
(3) Modulus of elasticity (Young's modulus): 185000 MPa

Automatic meshing in the Mechanical module, assuming that the cylinder is upright during transportation, and is constrained by two restraining bands at the cylinder body, and the constraint is firm and reliable, so the cylinder model is set to adopt cylindrical surface restraint gas, the axial and radial directions of the bottle are fixed. After the loading is completed, the finite element model can be solved. The finite element analysis results are shown in Figure 2:

ISPECE   IOP Publishing

IOP Conf. Series: Journal of Physics: Conf. Series **1187** (2019) 032089   doi:10.1088/1742-6596/1187/3/032089

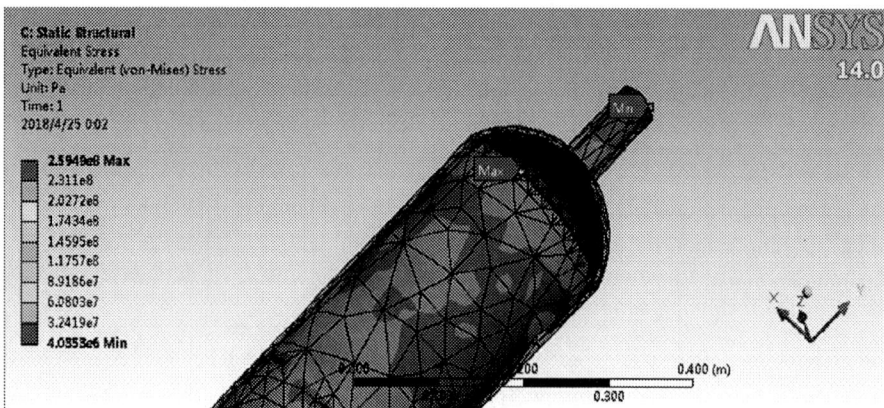

Figure 2 Stress on The Oxygen Cylinder

The maximum stress is 259 MPa and the minimum is 4 MPa. According to the data, the tensile strength of 34Mn2V structural steel σ_b is about 745MPa, and the yield strength σ_y is about 530MPa. According to the "GB150-2011 Pressure Vessel" [4], the allowable stress should take the smaller value between $\frac{\sigma_b}{3.0}$、$\frac{\sigma_y}{1.5}$.

$$\frac{\sigma_b}{2.7} = 276\text{MPa} \qquad (2)$$

$$\frac{\sigma_y}{1.5} = 353\text{MPa} \qquad (3)$$

$$[\sigma] = \frac{\sigma_b}{3.0} = 248\text{MPa} \qquad (4)$$

It can be seen that under the assumption that the stress intensity of the cylinder is very close to the allowable stress, there is a safety hazard, so it is necessary to optimize the structure.

3.2 Oxygen Cylinder Structural Optimization Design
The total stress experienced by the pressure vessel during actual use is the superposition of multiple stresses such as film stress, bending stress, and thermal stress. Different stresses have different discriminant criteria, and a single stress exceeding the allowable value may cause danger. When designing and analyzing the stress condition and structural safety of the pressure vessel, if only the calculated nominal film stress or bending stress is compared with the allowable stress alone to optimize the wall thickness, there is a possibility that exists hidden safety dangers. Therefore, it is necessary to perform equivalent linearization on the pressure vessel.

The linearization process is to calculate the stress distribution curve and linearize it according to the principle of static equivalence, and decompose it into: film stress equivalent to the resultant force and evenly distributed along the thickness of the section; equivalent to the combined force. The bending stress and the peak stress of the section thickness are linearly distributed for checking separately.

The film stress calculation method is as follows:

$$\left(\sigma_{ij}\right)_m = \frac{1}{e}\int_{-\frac{e}{2}}^{\frac{e}{2}} \sigma_{ij}\,dz \qquad (5)$$

The bending stress calculation method is as follows:

$$\left(\sigma_{ij}\right)_b = \frac{12z}{e^3}\int_{-\frac{e}{2}}^{\frac{e}{2}} \sigma_{ij}z\,dz \qquad (6)$$

In the formula, σ_{ij} represents each stress component, $\left(\sigma_{ij}\right)_m$ represents film stress of each stress component, $(\sigma_{ij})_b$ represents the bending stress of each stress component, e indicates the thickness of the cross section, z represents the coordinate system along the path [5].

For the stress equivalent linearization of the cylinder, the point of analysis should be specified first, and then the path of the analysis should be determined to perform the calculation. This paper selects the maximum stress point for analysis.

When using finite element analysis software for analysis, the path should be determined first, and then the stress distribution curve is obtained by fitting the stress at each point on the path to calculate the value of each stress [6]. The selection of the calculation path generally follows the following principles:

(1) Generally selected in the discontinuous part of the geometry, including the maximum point of stress, or may also be selected at the shortest path of the two dangerous surfaces;

(2) In the plate-and-shell type, the mid-surface normal passing through the maximum point of stress is generally selected. For the wall thickness change, a straight line passing through the minimum wall thickness direction of the maximum stress point is generally selected;

(3) Set the path along the direction in which the crack is most likely to expand [5].

The maximum stress calculated in this paper is located at the arc transition between the bottle body and the bottle shoulder, which is the discontinuous part of the geometry mentioned above, so the calculated calculation path is the shortest path along the wall thickness of the maximum stress point. The equivalent linearization treatment results of the oxygen cylinder are shown in Figure 3:

Tabular Data

	Length [m]	☑ Membrane [Pa]	Bending [Pa]	☑ Membrane+Bending [Pa]	Peak [Pa]	☑ Total [Pa]
1	0.	1.366e+008	1.2678e+008	2.5675e+008	1.4021e+005	2.5689e+008
2	1.2163e-004	1.366e+008	1.215e+008	2.5161e+008	1.8582e+005	2.5172e+008
3	2.4326e-004	1.366e+008	1.1622e+008	2.4648e+008	2.3181e+005	2.4655e+008
4	3.6489e-004	1.366e+008	1.1093e+008	2.4136e+008	2.78e+005	2.414e+008
5	4.8652e-004	1.366e+008	1.0565e+008	2.3624e+008	3.2431e+005	2.3626e+008
6	6.0815e-004	1.366e+008	1.0037e+008	2.3113e+008	3.7069e+005	2.3112e+008

Figure 3 Results of Equivalent Linearization of Oxygen Cylinders

Membrane is classified into a total film stress Pm, Bending is summarized as a bending stress P_b, and Peak is a peak stress F. Referring to the standards of other scholars, from a more secure point of view, the total stress value is summarized as (primary film stress) + (primary bending stress), that is, $P_L + P_b$ [7]. With reference to national standards, the safety factors and assessment results of the above several stresses are shown in Table 3:

Table 3 Stress Assessment Results

Type of Stress	Analog Value/MPa	Safety Value/MPa	Evaluation Results
P_m	137	$[\sigma]=279$	Safe
P_L+P_b	127	$1.5[\sigma]=419$	Safe

Under the set conditions, the maximum stress of the oxygen cylinder is at the arc transition between the bottle body and the shoulder of the bottle. Although the stresses are individually checked and meet the safety requirements, the oxygen bottle is close to the allowable value. The maximum stress is optimized. According to the knowledge of engineering mechanics, increasing the radius of the arc is an effective method to reduce stress concentration and improve safety. Therefore, this paper increases the radius of the cylinder arc transition and performs finite element analysis and equivalent linearity in turn. Processing. The calculated results are shown in Figure 4:

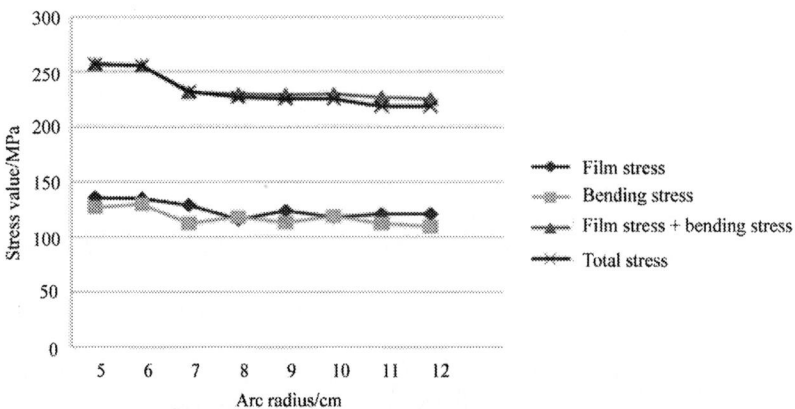

Figure 4 Effect of Arc Radius on Stress

It can be seen from Fig. 4 that increasing the radius of the arc can effectively reduce the stress at the arc transition of the oxygen cylinder, and the film stress, bending stress and total stress decrease with the increase of the radius of the arc.

4. Conclusion

Gas packaging containers are widely used in machinery, chemical, metallurgy, aerospace, medical and health, and daily life of residents. In this paper, ANSYS Workbench is used to analyze the typical gas packaging container-40L industrial oxygen cylinder, and the equivalent linearization method is used to optimize the cylinder structure, which significantly improves the safety of the cylinder transportation process. The packaging container structure optimization problem provides ideas. However, it should be pointed out that the analysis results have certain limitations due to the simplified handling of the cylinder structure during the analysis.

Acknowledgments

Funding Project: Yunnan Province Science and Technology SMEs Technology Innovation Fund Project "High-strength Lightweight Packaging Box Industrialization" (Project No.: 2017EH126) About the author: Yin Zhiqin (1970-), female, associate professor, mainly engaged in packaging printing machinery and process research.

References:

[1] State Council. GB5099-94 steel seamless gas cylinder [S].

[2] Chen Xingyuan. Processing defects of 34Mn2V steel [A]. National Symposium on Material Physical and Chemical Testing and Product Quality Control (Physical Testing Section): 355-359.

[3] Luo Ziqiang,Xing Yingming,Chen Jun.Finite Element Analysis of Physical Explosion Accident of Oxygen Bottle[J].China Special Equipment Safety,2013,29(02):41-44.

[4] State Council. GB150-2011 Pressure Vessel [S]

[5] Chen Ke, Huang Haiyuan, Zheng Hongmei, Ding Shuguang, Yan Qilong, Sun Chenglin. Stress equivalent linearization analysis of drum separator based on Workbench[J]. Fluid Machinery, 2014, 42(09): 31-34 .

[6] Zhuo Gaozhu, Kong Fanjing, Guo Huabo, Han Zhaoqiang. Finite Element Analysis and Equivalent Linearization of Pressure Vessels[J]. Power Generation Equipment, 2008(05): 373-376.

[7] Li Yang, Wang Rongshun, Wang Caili. Stress Analysis and Optimization of Low Temperature Insulated Cylinder under Inertial Force Using ANSYS[J].Cryogenics and Superconductivity,2009,37(11):5-9.

ISPECE   IOP Publishing

Influence of Double Stealth Aircraft Approach Forward Support Cooperative Jamming on Radar Detection Performance

Lei Bao[1,a],Chun-yang Wang[2,b],Hong-bing Li[3,c],Juan Bai[4,d],Ming Tan[5,e]

[1]Graduate College of Air Force Engineering University. Xi'an, Shanxi, 710051, China. 86 17795834063.

[2]Air and Missile Defense College of Air Force Engineering University. Xi'an, Shanxi, 710051, China. 86 13571041693.

[3]Air and Missile Defense College of Air Force Engineering University. Xi'an, Shanxi, 710051, China. 86 13359226696.

[4]Air and Missile Defense College of Air Force Engineering University.
Xi'an, Shanxi, 710051, China.
86 15249042646.

[5]Graduate College of Air Force Engineering University.
Xi'an, Shanxi, 710051, China.
86 18629679068.

[a]2743681405@qq.com, [b]13571041693@189.cn, [c]84574762@qq.com,
[d]b_juan@163.com, [e]tanming_1992@163.com

ABSTRACT Aiming at the single stealth aircraft approaching support jamming, the support jamming suppression ability is insufficient, and if the jamming power is too large, it will be tracked and located. With the increase of hazard coefficient, a model of double stealth aircraft cooperative proximity support jamming is proposed. Based on solving the attitude angle of two aircraft, comparing with the static RCS database in the whole airspace and simulating the time-varying RCS of two aircraft, the influence degree of detection distance of monostatic radar is analyzed and verified by simulation using power co-suppression method. The results show that when the formation of two aircraft is 1 km, the suppression effect of the cooperative suppression and shielding task aircraft is better, which can effectively reduce its RCS, reduce the radar detection distance, improve the cooperative jamming effect and enhance the overall conduct operations level.

Chinese Classification Number: V218; TN972+.1

Document Identification Code: A

1. INTRODUCTION

With the increasing number of modern ground air defense weapons and equipment and the increasing performance and support system, the electromagnetic environment of the battlefield is becoming more and more complex, and the form of cooperative air support jamming and shielding operations is becoming increasingly prominent. It is the inevitable result of the internal development law of system

Content from this work may be used under the terms of the Creative Commons Attribution 3.0 licence. Any further distribution of this work must maintain attribution to the author(s) and the title of the work, journal citation and DOI.
Published under licence by IOP Publishing Ltd

ISPECE   IOP Publishing

IOP Conf. Series: Journal of Physics: Conf. Series **1187** (2019) 032090   doi:10.1088/1742-6596/1187/3/032090

and system confrontation. E.E.Foreman and Rafael A.Acevedo analyzed the advantages of cooperative operations in depth. In the aspects of multi-aircraft cooperative tactics, cooperative target allocation and cooperative detection, Lan Weihua and others [1-2] have studied, deepened the research content and expanded the concept of cooperative air combat.Zhang Yangrui et al[3-4]. Aiming at the key technology of multi-aircraft concomitant cooperative jamming radar network, this paper analyzes the different cooperative suppression/deception modes and jamming effects, and determines the strategy of multi-aircraft concomitant cooperative jamming against radar network. However, these researches mainly focus on the decision-making of cooperative air combat and the design of cooperative jamming modes, the optimization of jamming resource allocation and the evaluation model of various jamming effects. The practical application of tactical scenarios in specific air combat is less involved, and the precise coordination of how the double stealth aircraft is arranged to approach the guidance radar defense and control area. There is no research on the operational style requirement of precise cooperative support for jamming shield carrier to break through enemy air defense zone and the evaluation of detection performance of guidance radar in this method. In view of this, this paper proposes a new method for double stealth aircraft formation detection, which is provided in literature [3-4], identification and successful penetration of aircraft in proximity support cooperative jamming shield, which relieves single aircraft due to suppression power limitation and passive passivity determination. Position tracking can increase the self-threat coefficient, realize joint electronic warfare operations and effectively improve operational effectiveness.

2. DOUBLE AIRCRAFT COOPERATIVE INTERFERENCE SUPPORT MODEL

2.1 Scene Design

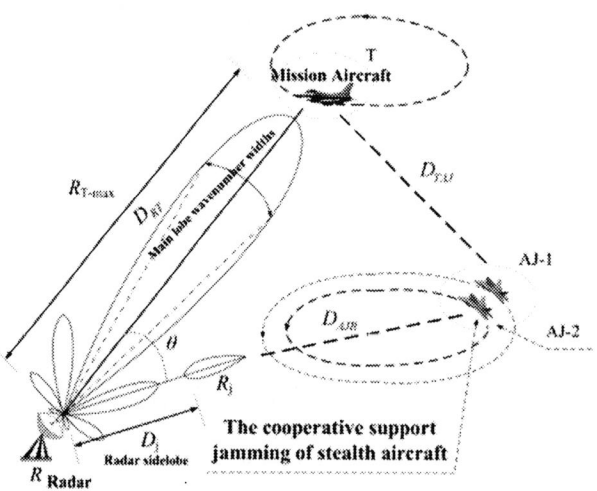

Figure 1. The escort-support jamming scenario of double stealth aircraft.

Figure one is the scene of approaching the anti-jamming zone. The mission plane refers to the group of aircraft carrying out the attack to break through the enemy air defense zone, and the main lobe beam of the ground air defense radar searches the target. At this time, the double stealth aircraft keeps the distance of 1 km to reach the support flight, and the main lobe of jamming equipment is aimed at the side lobe or main lobe of ground air defense radar to suppress jamming to cover the successful penetration of Mission Aircraft formation. $R_{T\text{-max}}$ is the maximum distance of guidanuce radar to the enemy aircraft in the airspace. θ is the angle between the direction of support jamming and the direction of radar main lobe detection when the stealth aircraft approaches the guidance radar. R_j is the stealth aircraft approaching the support jamming distance. D_j is the burning distance to support

1122

interference[5]. AJ-1 and AJ-2 are the immediate spatial locations of stealth aircraft 1 and stealth aircraft 2 respectively. T is the immediate spatial location of the task machine. D_{RT} is the immediate spatial location of the task machine. $(x_T(t), y_T(t), z_T(t))$ is the instant location of task machine. $(x_{AJ_i}(t), y_{AJ_i}(t), z_{AJ_i}(t))$ is the immediate position of stealth aircraft. $(0,0,0)$ indicates that radar is at the origin of the coordinate system. In the triangle composed of the real-time space position of the mission plane, stealth aircraft and guidance radar, the three sides respectively are $|D_{RT}|_t$, $|D_{TAJ}|_t$ and $|D_{AJR}|_t$.

2.2 Analysis of Stealth Aircraft Performance

2.2.1 Calculation of dynamic RCS
(1) Parameter setting of double aircraft setting

Table 1. double aircraft formation setting

Track parameter	Numerical value
Stealth aircraft AJ-1 and AJ-2 endurance speed v/Ma	1.4
Mission aircraft speed v/Ma	1.4
Flight height of stealth aircraft AJ-1 and AJ-2H/km	8
Mission machine penetration height H/km	10
Stealth aircraft AJ-1 flying radius R/km	20
Stealth aircraft AJ-2 flying radius R/km	19
Mission machine turning radius R/ km	40
Flight attitude elevation of double aircraft	$\varphi < 5^0$
Rolling angle of aircraft circling	$\eta = 30^0$

(2) Instant attitude calculation
1) The definition of coordinate system

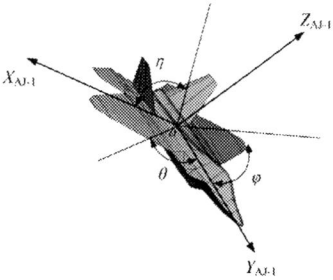

Figure 2. Coordinate modeling of stealth aircraft.

In Figure 4, the stealth aircraft coordinate system ($O - X_{AJ-1}Y_{AJ-1}Z_{AJ-1}$) and the radar coordinate system ($O - X_R Y_R Z_R$). θ, φ and η are the azimuth, pitch and roll angles of the aircraft.

2) Coordinate system conversion

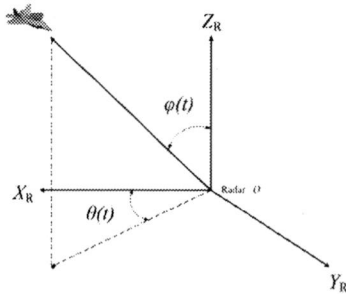

Figure 3. Aircraft body and radar coordinate system conversion.

Figure 3 shows the solution process for understanding the line of sight azimuth and line of sight pitch. The conversion relationship between the radar coordinate system and the body coordinate system [6-9] is

$$\begin{bmatrix} x_{AJ-1}(t) \\ y_{AJ-1}(t) \\ z_{AJ-1}(t) \end{bmatrix} = Q \left(\begin{bmatrix} x(t) \\ y(t) \\ z(t) \end{bmatrix} - \begin{bmatrix} x_R(t) \\ y_R(t) \\ z_R(t) \end{bmatrix} \right) \tag{1}$$

$(x(t), y(t), z(t))$ is the position of any Point in Radar coordinate system. The coordinates in the airframe coordinate system of stealth aircraft is $(x_{AJ-1}(t), y_{AJ-1}(t), z_{AJ-1}(t))$, $(x_R(t), y_R(t), z_R(t))$ is the position of the aircraft corresponding to the coordinates in the radar coordinate system. Q is the transformation matrix from the mission coordinate system to the radar coordinate system. Conversion matrix Q:

$$Q = \begin{bmatrix} \cos\theta(t)\cos\varphi(t) & \sin\varphi(t) & -\sin\theta(t)\cos\varphi(t) \\ \sin\theta(t)\sin\eta(t) - \sin\varphi(t)\cos\eta(t)\cos\theta(t) & \cos\varphi(t)\cos\eta(t) & \sin\theta(t)\sin\varphi(t)\cos\eta(t) + \sin\eta(t)\cos\theta(t) \\ \sin\theta(t)\cos\eta(t) & -\sin\eta(t)\cos\varphi(t) & \cos\theta(t)\sin\eta(t) \end{bmatrix}$$

$$(2)$$

When the radar coordinate origin is substituted into the formula (3), the time-varying attitude angle of sight [12] is expressed as

$$\begin{cases} \theta(t) = \arctan\dfrac{y_{AJ-1}(t)}{x_{AJ-1}(t)} \\ \varphi(t) = \arctan\dfrac{z_{AJ-1}(t)}{\sqrt{x_{AJ-1}^2(t) + y_{AJ-1}^2(t)}} \end{cases} \tag{3}$$

(3) Static database extraction time variant RCS

The FEKO platform is equipped with simulation conditions to calculate the static RCS data of a stealth aircraft (service frequency: 5.8GHz; polarization mode: HH; azimuth range: $0^0 \sim 360^0$; pitch range: $-30^0 \sim 10^0$; step angle: 1^0). In the static RCS two-dimensional database, extract the instantaneous RCS of the line of sight angle. Programming in MATLAB for data processing, drawing dynamic RCS sequence of time-varying graphics.

Figure 4 describes the time-varying dynamic RCS sequence of a single stealth aircraft in the approach forward supports jamming.

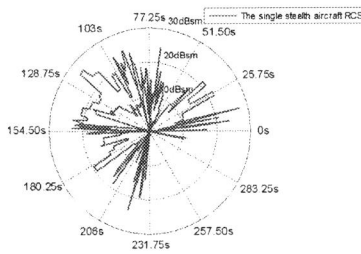

Figure 4. Dynamic RCS sequence of single stealth aircraft

Figure 5 shows the time-varying dynamic RCS sequence of double stealth aircraft during the jamming process.

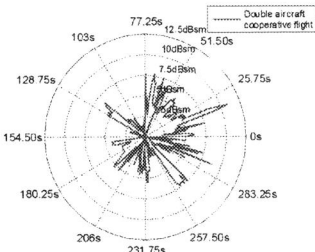

Figure 5. The dynamic RCS sequences of double stealth aircraft coordinated support

The dynamic RCS sequence generated by a single aircraft approaching support interference fluctuates between(-26.813~22.511dB), The dynamic RCS sequence generated by two machine cooperation approaches support interference, and the fluctuation range is between -44.889~12.016dB. It shows that the stealth performance of the two aircraft stealth aircraft is stronger than that of the single aircraft.

3. ANALYSIS OF RADAR DETECTION PERFORMANCE

3.1 Radar detection distance

1) Radar detection distance without interference

Under the premise of setting detection threshold and giving false alarm probability, the detection range of time-varying radar without jamming state is [3]

$$R_{T-\max}(t) = \left[DI \frac{G_t G_r \lambda^2}{(4\pi)^3 k\tau f_r T_e B_w F_w L_z} \frac{P_{av}\sigma}{(\xi)_{\min}} \right]^{1/4} \tag{4}$$

In formula (4), average transmitting power: $P_{av} = P_t \tau f_r$; radar transmit peak power: P_t; pulse width: τ; PRF: f_r; transmit antenna gain and receiving antenna gain: $G_t = G_r$; wavelength: λ; RCS value of aircraft: σ; Boltzmann constant: k; effective noise temperature: T_e; Receiver bandwidth: B_w; Receiver noise figure: F_w; Radar system loss: L_z; coherent accumulation gain: I; Echo target pulse pressure gain: D; minimum detection signal-to-noise ratio: $(\xi)_{\min}$.

2) Radar detection distance of single aircraft jamming

The time variant radar detection range of single stealth aircraft under the condition of jamming support is [3]

$$R_{T-j\max}(t) = \left[\frac{DI}{D_j I_j} \frac{P_t R_j^2(t)}{P_j} \frac{G_t G_r}{G_j(\varphi)G_r(\varphi)} \frac{\gamma_j L_j \sigma}{4\pi L_z} \frac{\xi_j}{(\xi)_{\min}} \right]^{1/4} \tag{5}$$

ISPECE	IOP Publishing

In formula (5), noise signal pulse pressure gain and Coherent accumulation gain: D_j, I_j; transmitter power of jammer: P_j; transmit antenna gain: $G_j(\varphi)$; receiving antenna gain: $G_r(\varphi)$; signal interference loss: L_j; polarization adaptation: γ_j; signal to noise ratio after signal processing: ξ_j.

3) Radar detection distance under the condition of double aircraft cooperative jamming [3]

$$R_{T-cjmax}(t) = \left[T \frac{G_t G_r \sum_{j=1}^{M} \zeta_j}{\xi_0} \middle/ \left[\frac{G_{j1}(\varphi)G_r(\varphi)}{R_{j1}^2(t)} + \frac{G_{j2}(\varphi)G_r(\varphi)}{R_{j2}^2(t)} \right] \right]^{1/4} \tag{6}$$

In formula (6), $T = DIP_t\gamma_j L_j \sigma / 4\pi D_j I_j P_j L_z$, The transmit antenna gain of stealth aircraft AJ-1 is $G_{j1}(\varphi)$; The transmit antenna gain of stealth aircraft AJ-2 is $G_{j2}(\varphi)$ Jamming distance of stealth aircraft AJ-1: $R_{j1}(t)$, Jamming distance of stealth aircraft AJ-2: $R_{j2}(t)$.

Burn-through distance must be taken into account when approaching support interference. For single base radar, it has transceiver antenna. Therefore, the signal power [5] received by the receiver is

$$Rr = P_t + G_t + G_r - 103 - 20\lg(T_f)$$
$$-40\lg(ID_j) + 10\lg(\sigma) \tag{7}$$

In equation (7), Rr is the signal power at the receiver input, the unit is dB; T_f is the transmitted signal frequency, the unit is MHz. Interference distance is ID_j.

The interference power [5] entering the receiver input is

$$I_j = P_j + G_j - 32 - 20\lg(T_f)$$
$$-20\lg(R_j) + G_r' \tag{8}$$

In the formula (8), I_j is the interference power of the receiver at the receiver, and the unit is dB. According to the set scenario, the $J/S = I_j/Rr$ to letter is expressed as

$$I_j/Rr = 71 + G_j + G_r' + 40\lg(ID_j) + P_j - P_t - G_t - G_r$$
$$-20\lg(R_j) - 10\lg(\sigma) \tag{9}$$

After finishing:

$$40\lg(ID_j) = P_t + G_t + 10\lg(\sigma) + 20\lg(R_j)$$
$$+ J/S - 71 - P_j - G_j + G_r - G_r' \tag{10}$$

The firing distance is [7]: $id_j = 10^{\left[40\lg(ID_j)/40 \right]}$

When $id_j < R_j$, stealth aircraft can effectively release noise and suppress interference. The gain of radar antenna can be obtained from the following empirical formula [10]:

$$G_r' = \begin{cases} G_r, & |\theta| \leq \theta_{0.5}/2 \\ K(\theta_{0.5}/\theta)^2 G_r, & \theta_{0.5}/2 < |\theta| \leq 90^0 \\ K(\theta_{0.5}/90^0)^2 G_r, & 90^0 < |\theta| \leq 180^0 \end{cases} \tag{11}$$

In the above formula, $\theta_{0.5}$ is the main lobe width of the radar antenna; $K = 0.04 \square 0.1$ is a constant related to the characteristics of the radar antenna; θ is the angle between the main lobe direction of the radar and the connecting direction of the radar to the jammer.

4. SIMULATION ANALYSIS

The average dynamic RCS of stealth aircraft in jamming is only -5.4dBsm. At present, mission fleet is an aircraft with strong scattering characteristics. Therefore, the radar cross section parameter of the mission aircraft is set to 10dBsm in the simulation analysis.

4.1 Detection distance

(1) Simulation conditions

Radar parameter setting:

Transmitting power /kw: 100; Pohl Seidman constant /J/K:1.38×10-23; Internal noise temperature /K:290; Transmit antenna gain /dB:35; Receiving antenna gain / dB:35; Receiver bandwidth /MHz:5; Noise figure / dB:4; System wastage / dB:5; Minimum detectable SNR / dB:20; Coherent accumulation gain $D=BT$; Signal pulse width/μs :10; Cumulative number of pulses l :64; false-alarm probability: $P_{fa} = 1e - 6$.

Interference parameter setting:

Jamming device AJ-1=AJ-2 power /w($P_{j1} = P_{j2}$):100; Jamming antenna gain /dB:10; Interference bandwidth /MHz:12; Polarization loss:0.5; Total cost of jamming equipment / dB:6; Radar receiving bandwidth /MHz:5; Noise coherent accumulation gain I_j /dB:0; Only considering random noise interference,the pulse pressure gain is $D_j = 0dB$.

Effective release of random noise suppression from single stealth aircraft under proximity support jamming. The time-varying burn through distance is as follows:

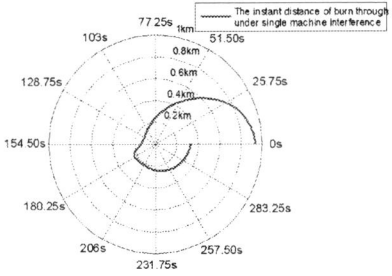

Figure 6. Instant burn distance under single machine interference

Double stealth aircraft can effectively release random noise and suppress interference under close support jamming. The time-varying burn through distance is as follows:

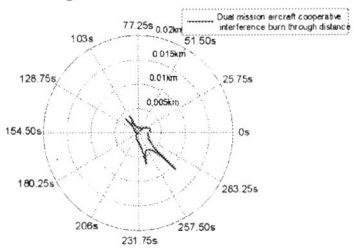

Figure 7. Instant burn distance based on two aircraft cooperative jamming

Comparative analysis is shown in figures 6 to 7. Under the support jamming of single stealth aircraft, the time varying burn through distance is $id_j < 0.95km$, Under the support jamming of double stealth aircraft, the time varying burn through distance is $id_j < 0.011km$, The instant support jamming distance of stealth aircraft is $R_j > 19km$. It can be seen that the suppress distance of stealth aircraft is far greater than the burning distance. Therefore, we can get the real-time distance of radar detection mission aircraft under the support interference.

(2) Instantaneous detection distance between mission aircraft and radar.

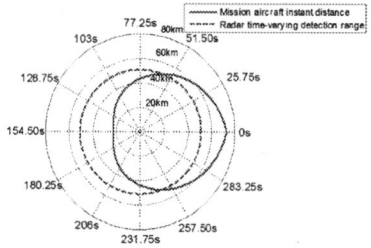

Figure 8. The mission machine is detected by radar Without jamming.

Instantaneous detection distance of radar detection mission under support jamming cover of single stealth aircraft.

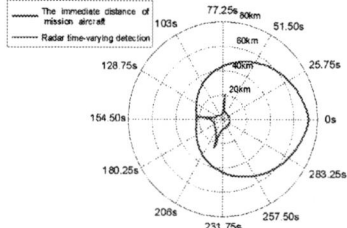

Figure 9. Radar detection distance with standalone support

The radar detect the real time distance of mission aircraft based on cooperative jamming of double stealth aircraft.

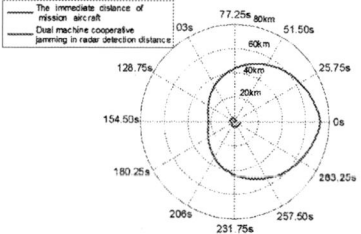

Figure 10. Radar detection distance with two aircraft support jamming

Comparative analysis is shown in figures 8 to 10. If the stealth aircraft does not support jamming, the detection distance of the radar is larger than the real distance of the aircraft. Under support jamming, radar detection performance decreases, and mission aircraft penetration is safer.

5. CONCLUSION

In this paper, based on the background of penetration of double-aircraft approach support jamming shield mission aircraft, the advantages of double-stealth aircraft cooperative approach combat in power cooperative suppression distance are studied, and the corresponding simulation and comparative analysis are made. The simulation results show that the advantages of system-based countermeasure in joint operations are more obvious than that of single-aircraft. It provides a new way for stealth aircraft to cooperate in penetration support operations.

Acknowledgments

Foundation item: National Natural Science Foundation of China under Grant (61601503)

REFERENCES

[1] Weihua, L., Rong, Y. 2005. [J]. Optics. &. control, 12(6): 12-15.

[2] Wencheng, P., Dianjie, H., Wen, Z. 2009. *Force loss lanchester equation based on cooperative engagement*[J]. Operation research and management science, 18(3): 128-131.

[3] Yangrui, Z. 2015. *Research on Key Technologies of Cooperative ECM in Multi-syndrome Jammers for Countering Radar Net. Doctoral dissertation.* Beijing Institute of Technology.

[4] Zhanqiang, L., Lujiang, L., Chunyang, W. 2018. *Stealth Aircraft Escort-support Jamming Influence on Radar Detection Performance*[J]. Journal of Detection & Control, 40(1): 72-79.

[5] ADAMY, D. EW101. 2013. *A first course in electronic warfare* [M]. Translated by WANG Y, ZHU S. Beijing: Publishing House of Electronics Industry, 118-124. (in Chinese).

[6] Chong, D., Zhenghai, X., Shunping, X. 2013. *Analysis for differences between dynamic and static RCS characteristics of radar target* [J]. Journal of Signal Processing, 29(9): 1256-1263. (in Chinese).

[7] Jia, L., Ning, W., Yongjun, X. et, al. 2015. *Dynamic target RCS characteristic analysis under the influence of attitude perturbation*[J]. Systems engineering and electronics, 37(4): 775-781. (in Chinese).

[8] Chong, D., Zhenghai, X., Shunping, X. 2014. *Simulation method of dynamic RCS for non-cooperative targets* [J]. Acta Aeronautica et Astronautica Sinica, 35(5): 1374-1384. (in Chinese).

[9] Zhangsong, S., Zhong, L., Hangyu,W., et, al. 2010. *Method and theory of target tracking and data fusion* [M]. Beijing: National Defense Industry Press. (in Chinese).

[10] Tianquan, N., Jiandong, W., Yian, L. 2010. *Research on target route planning of confrontation of ESJ to radar network* [J]. Acta Armamentarh, 31(12): 1599-1603. (in Chinese).

ISPECE   IOP Publishing

MD-UCON: A Multi-Domain Access Control Model for SDN Northbound Interfaces

Rui CHANG[1,a,*],Zhaowen LIN[2,b],Yi SUN[3,c,*],Jie XU[4,d]

[1]Network and Information Center, Institute of Network Technology, Beijing University of Posts and Telecommunications, Beijing, 100876, China Science and Technology on Information Transmission and Dissemination in Communication Networks Laboratory National Engineering Laboratory for Mobile Network Security (No. [2013] 2685) Network and Information Center, Institute of Network Technology / Institute of Sensing Technology and Business, Beijing University of Posts and Telecommunications. Beijing, 100876, China

[2]Network and Information Center, Institute of Network Technology, Beijing University of Posts and Telecommunications, Beijing, 100876, China Science and Technology on Information Transmission and Dissemination in Communication Networks Laboratory National Engineering Laboratory for Mobile Network Security (No. [2013] 2685) Network and Information Center, Institute of Network Technology / Institute of Sensing Technology and Business, Beijing University of Posts and Telecommunications. Beijing, 100876, China

[3]Network and Information Center, Institute of Network Technology, Beijing University of Posts and Telecommunications, Beijing, 100876, China Science and Technology on Information Transmission and Dissemination in Communication Networks Laboratory National Engineering Laboratory for Mobile Network Security (No. [2013] 2685) Network and Information Center, Institute of Network Technology / Institute of Sensing Technology and Business, Beijing University of Posts and Telecommunications. Beijing, 100876, China

[4]Network and Information Center, Institute of Network Technology, Beijing University of Posts and Telecommunications, Beijing, 100876, China Science and Technology on Information Transmission and Dissemination in Communication Networks Laboratory National Engineering Laboratory for Mobile Network Security (No. [2013] 2685) Network and Information Center, Institute of Network Technology / Institute of Sensing Technology and Business, Beijing University of Posts and Telecommunications. Beijing, 100876, China

[a]changrui@bupt.edu.cn, [b]linzw@bupt.edu.cn, [c]sybupt@bupt.edu.cn, [d]cheer1107@bupt.edu.cn

ABSTRACT In SDN (Software Defined Network) environments, for the security considerations of the upper-layer application behaviors, we need to consider adding an access control mechanism to the northbound interface of the SDN control layer to limit the capabilities of the upper-layer applications, thereby improving the security of the SDN. In addition, considering the performance and management requirements of SDN, access control features including cross-domain support should be considered. In this paper, we proposed an MD-UCON access control

Content from this work may be used under the terms of the Creative Commons Attribution 3.0 licence. Any further distribution of this work must maintain attribution to the author(s) and the title of the work, journal citation and DOI.
Published under licence by IOP Publishing Ltd

model with role mechanism extension based on UCON and role-based access control mechanism. At the same time, we introduced a cross-domain role mapping method to support cross-domain access authorization, thereby enabling the model to be applied to the application of access control for the SDN northbound interface.

1. INTRODUCTION

In recent years, with the development of Software Defined Network (SDN), its advantages in terms of openness, reliability, flexibility, control and scalability have become increasingly prominent, and have been received in industry and academia. SDN decouples the control layer from the data layer of networks, allowing users to dynamically configure the network in a programmed manner, thereby simplifying network management and enabling flexible traffic control, also providing a good platform for core network operation and application innovation. However, SDN technology is not yet mature. The SDN northbound interface directly serves the upper layer application, and various control layer systems are implemented based on different technologies, resulting in a decentralized development direction. There is no unified interface standard, and the mainstream SDN software controllers respectively provide completely different northbound interfaces. Most SDN control layer elements provide northbound interfaces, but lack access control mechanisms for upper layers [1][2].

In the absence of the corresponding security mechanism, the security and stability of the SDN are threatened, including, between the application layer and the control layer, there may be application errors, inconsistent application layer and control layer status; when the control layer passes when a northbound interface is connected to multiple applications, problems such as command conflicts and inconsistent application status may occur between multiple applications. When an application error occurs or application is maliciously controlled, it will also bring unpredictable danger to the network [3][4].

Meanwhile, in recent years, with the continuous improvement in all areas of the network demand, it is necessary to fully consider the network scale when deploying SDN networks. When the network reaches a certain scale, the performance of the SDN controller may become a network performance bottleneck [5], also, the large single-domain network is inconvenient in management. Given that multi-domain networking is an effective solution to scale problems, and cross-domain access requirements are becoming more and more urgent, when researching access control mechanisms, further consideration needs to be given to how to build cross-domain access control mechanisms.

Traditional access control models, including Discretionary Access Control (DAC) [6], Mandatory Access Control (MAC) [7], and Role Based Access Control (RBAC) [8][9][10] can meet the requirements and protect resources from unauthorized access under general traditional scenarios. However, the authorizations of these access control models are based on pre-defined rules, and it is difficult to meet the access control requirements of cross-domain access in dynamic and complex scenarios.

In 2002, Park and Sandhu proposed the Usage Control (UCON) model, which integrates the models of access control, digital rights management and trust management, providing fine-grained, continuous access control ability, and meeting the requirements of access control in complex scenarios [11][12][13].

UCON is an abstract model, and its significance is more to provide a scheme to the modeling, rather than directly describe a specific access control mechanism. In order to meet the requirements of multi-domain SDN northbound interface access control better, we extended the UCON model and proposed the MD-UCON model, achieving the SDN northbound interface access control that supports cross-domain access.

The rest of this article is organized as follows. In Section 2, we showed the current researches related to this article. In Section 3, we briefly introduced the UCON model. In Section 4, we described our main contribution, the MD-UCON model. In Section 5, we summarized the work of this paper.

2. RELATED WORKS

The works related to this paper includes researching on the access control of SDN northbound interface in multi-domain SDN environment, proposing the role extended multi-domain access control model, researching cross-domain role conversion mechanism in multi-domain scenarios.

In previous researching, some research efforts attempted to provide access control for SDN northbound interfaces or SDN upper layer applications to improve SDN security. Canini et al. implemented the security verification of third-party SDN applications by means of attribute checking and symbolic execution. However, the proposed verification is not dynamic monitoring at runtime, and the verification is mainly for a single SDN application scenario [14].

Sherwood et al. sliced and partitioned the control logic of different users on the control layer to provide access control based on certain rules for user access, but this processing method can only isolate the control logic between different users [15].

Wen et al. assigned permissions to the commands that can be used by the application layer, which realized the isolation between the application and the control layer kernels, and improved the security and stability of the control layer. The research focuses more on preventing all upper-layer applications from destroying the underlying network. There is no access control for different policies for multiple applications [16].

In particular, Klaedtke et al. proposed an access control scheme for SDN upper-layer applications [17]. The research proposed to use network users as the subjects, and use information and configurations, including statistics, flow rules, as objects. The rights are defined as the set of reading operations and modifying operations. And, the concept of strategy is introduced. The scheme is largely in line with UCON components, such as subject, object, authority, and authorization. This research provides a reference for the work of this paper.

Since previous researches on SDN access control did not adequately consider the need to support scenarios include multiple domains and multiple SDN applications, especially for requirements in multi-domain scenarios, a further research work is still needed.

3. UCON

The Usage Control (UCON) model was first proposed by Park and Sandhu in 2002. UCON integrates the traditional access control, trust management and digital rights management. In terms of access control, UCON covers traditional DAC, MAC and relatively advanced RBAC. The UCON model extends the traditional access control and defines three decisive factors: Authorization (A), Obligation (B) and Condition (C), and brings continuity and mutability to UCON innovatively. This allows UCON to adapt well to highly dynamic, continuously changing scenes.

The UCON model consists of six elements, including subjects, objects, rights, authorizations, obligations, and conditions. The UCON model can be shown by figure [1]. Subjects are entities that contain attributes, holding the rights to subjects. Objects are entities who also contain attributes, who subjects hold rights on. Rights are privileges on accessing objects, holding by subjects. Authorization are a set of rules that should be satisfied before the subject accessing the object. Obligation is the mandatory requirement that the subject must accomplish before accessing the objects. Conditions is a set of decision factors that are verified before subjects accessing the objects.

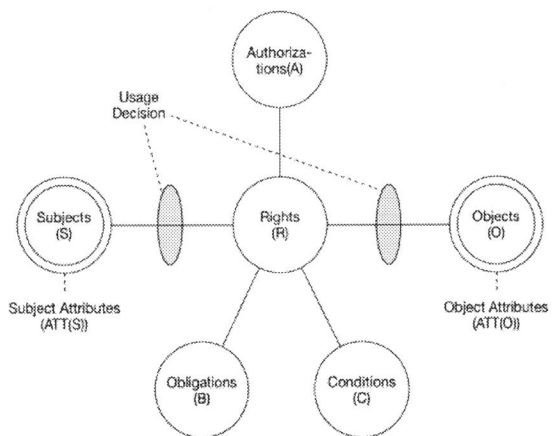

Figure 1. Structure of UCON Model

Compared to traditional access control models, UCON's most outstanding feature is the continuity of decision and mutability of attributes. Continuity of decision means that the control strategy can be executed before and during the access. The control decision component can perform verifications in these two phases, called pre-decision and ongoing-decision. Mutability of attributes means that attributes of subjects and objects can be updated as accessing results. Attribute updates include pre-updates, ingoing-updates, and post-updates. The updating may cause the system to perform actions, including allowing or revoking access rights. Figure [2] shows a completed use process with three phases in time: before-usage, ongoing-usage and after-usage.

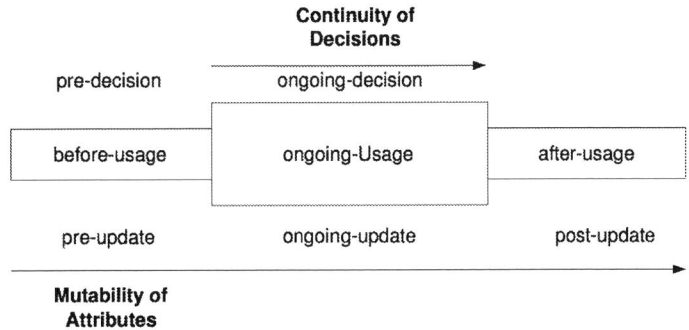

Figure 2. Continuity and Mutability of UCON

In the UCON model, decision factors are composed of authorizations, obligations, and conditions. The UCON model can be classified according to the continuity of decisions and the mutability of attributes. The UCON model can be divided into 24 categories according to the continuity of decision and the variability of attributes. For practical reasons, some models have no practical significance, so there are total of 16 available models. Figure [3] shows all possible models of UCON, where "Y" indicates that the model has practical significance, and "N" indicates that the model has no practical significance. Figure [3] (a) shows the possible combinations of UCON$_{ABC}$ models and their relationships. Figures [3] (b), (c) and (d) show possible scenarios in the UCON$_A$, UCON$_B$, and UCON$_C$ models, respectively.

Table 1. The 16 Basic ABC Models

	0 (immutable)	1 (pre-update)	2 (ongoing-update)	3 (post-update)
preA	Y	Y	N	Y

onA	Y	Y	Y	Y
preB	Y	Y	N	Y
onB	Y	Y	Y	Y
preC	Y	N	N	N
onC	Y	N	N	N

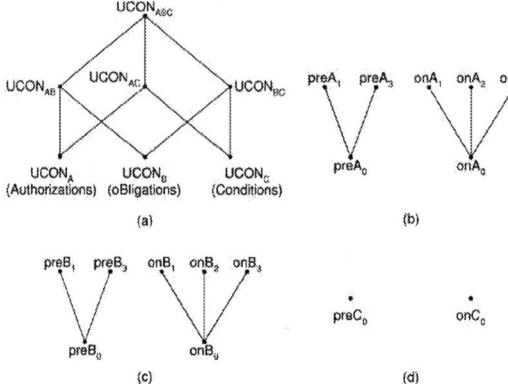

Figure 3. UCON$_{ABC}$ Model Family

4. MD-UCON

4.1 Introduction of MD-UCON

In this section, we proposed a Multi-Domain Usage Control (MD-UCON) model by extending the basic UCON model, thereby supporting cross-domain accessing. First, we introduced the domains entities, which are corresponding the real-world domains, so that the real domains factors can be suitably modeled. Then, we introduced attributes and mechanisms from role-based access control (RBAC), simplifying the complexity of access control mechanism when facing the multi-domain scenarios; Finally, we introduced a cross-domain role mapping mechanism to resolve the problem that the cross-domain access cannot be directly dealt due to the independence of roles from different domains.

4.2 Components of MD-UCON

MD-UCON introduced the concept of domains, which can divide all subjects and objects into finite groups. Therefore, each subject has a property called *subjectDomain* that represents the domain in which the subject resides; each object also has an attribute named *objectDomain* that represents the domain in which the object resides. The domains in the model are entities that can hold some attributes, corresponding to real-world domains. For example, in a general SDN deployment, all network elements managed by the same controller, as well as all upper-layer applications directly accessing this controller, belongs to the same domain. According to whether the domain attribute of the subject and object are the same, the access from subject to object can be divided into two categories, intra-domain access and cross-domain access.

MD-UCON also introduced the concept of role, which is a key attribute of the subject and is used for the authentication process of access control. A role is an entity that can contain some attributes. It has a many-to-many relationship with the subject. Each subject has a property called *subjectRoles* that represents a collection of role attributes held by the subject. When the access occurs, the access control system queries the role attribute set of the access subject. If any of the role attributes meet the permission requirements of the current access, the authentication succeeds. Due to the existing relationship between the role and the subject and the relationship between subject and the domain, the role and the domain also have an indirect many-to-one relationship, that is, a domain may contain a finite number of different role entities. The introduction of roles entities reduces the complexity of subject attributes and authentication rules, while providing support for cross-domain access control.

The extended MD-UCON components and their relationships are shown in Figure [5].

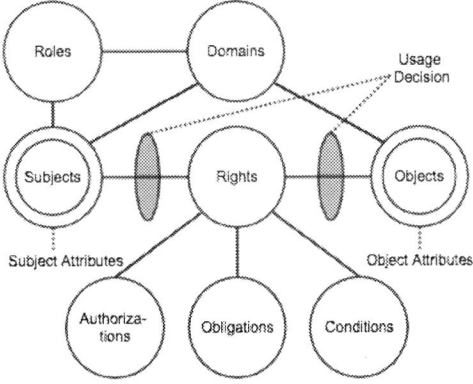

Figure 5. MD-UCON Model

4.3 Cross-Domain Role Mapping

The purpose of the cross-domain role mapping mechanism is to solve the problem that role-based access control systems cannot be directly applied to cross-domain access scenarios. The traditional intra-domain role mapping policy can be divided into three categories. The first is the default policy, that is, any external domain role is mapped to a default role in the local domain. In the common case, the default role is the minimum privilege role of the domain. The second is the explicit policy, that is, separated role association, the administrator clearly defines the mapping rules of the external domain role to the local domain role. The third is the implicit policy, that is, after the transfer association mechanism is introduced, a certain implicit policy can be derived from the explicit policy and the role inheritance relationship.

Considering that in the SDN environment, the default policy obviously does not meet the requirements in terms of flexibility; the explicit policy requires the administrator to clearly know all possible roles in different domains and set mapping rules for all possible roles, so the feasibility is unacceptable in practical applications. The implicit strategy with role inheritance may cause the role conflict problem to occur when the rules are set incorrectly. At the same time, the role inheritance mechanism makes the role adding, editing, deleting and other operations have a high complexity in system implementation, and there is also a large system overhead in terms of performance at runtime, therefore using of the strategy should be carefully considered.

In the access control scenario for the SDN northbound interfaces, the authentication rules and the role attributes of the upper application are generally set by trusted administrators. The upper application only holds an immutable access credential, and cannot forge the role and fraud permissions by itself. Under such premises, we can use a new role mapping strategy. We can pre-set a set of public roles in the global scope and raise requirements for each domain: first, any role in any domain must be able to be mapped to a global role; second, each domain is required to has the ability to deal with global roles - either by setting permissions for all global roles, or providing mapping rules for all global roles. By using a global role which can be understood by any domain as an intermediate role, it is guaranteed that any subject can be understood by the access control system when cross-domain access occurs. Since the number of global roles is limited and usually is relatively small, and role rules are easy to set, resulting in a good feasibility of this strategy. Therefore, we solved the problems that the traditional role mapping strategy is not flexible enough and the complexity is too high to be applied.

4.4 Formal Model Definition

The formal definition of MD-UCON is shown as follows:

1) S represents a finite set of subjects. subjects are entities that initiates access, having role attributes and domain attributes, and may also include general attributes for extension. A user in the SDN network may serve as a subject;

2) O represents a finite set of objects. Objects are the accessed entities, having domain attributes, and may contains general attributes for extension. The statistical information, configuration, flow rules, etc. in the SDN network can be used as objects;

3) D represents a finite set of domains. Domains are entities containing attributes, generally corresponding to the domains in the real world;

4) ROLE represents a finite set of roles. Roles can contain general attributes for extension;

5) R, A, B, and C respectively represent a limited set of permissions, authorization rules, obligations, and conditions, which are the same as the UCON model;

6) subjectDomain: (s: S) $\rightarrow$ D, a function mapping from the subject s to its domain;

7) objectDomain: (o: O) $\rightarrow$ D, a function map from the object o to its domain;

8) subjectRoles: (s: S) $\rightarrow 2^{ROLE}$, a function map from the subject s to the set of role attributes it holds;

9) globalRole: (role: ROLE) $\rightarrow$ ROLE, a function mapping from a domain role to a global role, the specific rules are usually set by the administrator;

10) roleAllowed: (role: ROLE, o: O, r: R), is a Boolean function, indicating whether the role has permission to operate r on the object o, the specific rules are usually set by the administrators;

11) innerDomainAllowed: (s: S, o: O, r: R), is a Boolean function, indicating whether the subject s has the permission to operate r on the object of the same domain in a non-cross-domain scenario, which can be represented as:

$$innerDomainAllowed(s, o, r) = \exists role \in subjectRoles(s) \wedge roleAllowed(role, o, r)$$

12) crossDomainAllowed: (s: S, o: O, r: R), is a Boolean function, indicating whether the subject s has permission to operate r on the object o in a cross-domain scenario, which can be represented as:

$$crossDomainAllowed(s, o, r) = \exists role \in subjectRoles(s) \wedge roleAllowed(globalRole(role), o, r)$$

13) allowed: (s: S, o: O, r: R), is a Boolean function, indicating whether the subject s in general sense has the right to operate r on the object o, combine the two scenarios of non-cross-domain access and cross-domain access, it can be represented as:

$$allowed(s, o, r) = innerDomainAllowed(s, o, r) \vee crossDomainAllowed(s, o, r)$$

5. CONCLUSIONS

In this paper, the MD-UCON model is proposed by extending the UCON model, to support the access control for the SDN northbound interface, and support the cross-domain access control in multi-domain scenarios. The model introduces the domain entity to enable the modeling of the domain in the real world. By introducing the role entities and the corresponding mechanisms, the complexity of the subject attribute is reduced, and the settings of access control rules are enabled. Finally, the cross-domain role mapping mechanism is introduced, achieving access control in a cross-domain access scenario. Compared with the traditional access control model, MD-UCON meets the requirements of SDN northbound interface access control, with higher flexibility and finer granularity of control, while ensuring a lower complexity, therefore it's more suitable for SDN northbound interface access control with multi-domain requirements.

ACKNOWLEDGEMENT

This work is supported by the National Natural Science Foundation of China (Grant no. 61601064), the Fundamental Research Funds for the Central Universities (2018RC55), and the Beijing Talents

Foundation (Grant. no. 2017000020124G062), the Fundamental Research Funds for the Central Universities.

REFERENCES

[1] Pang Tao, Wei Huanyu, Wu Juan, Chen Jian. Research on the Development Status and Trends of SDN Northbound interface. China Internet，2014(9).

[2] SHI Zhikao, ZHU Guosheng. Research on security of software-defined networking. Journal of Computer Applications, 2017, 37(S1):75-79.

[3] Scott-Hayward, S., Natarajan, S., & Sezer, S. (2016). A Survey of Security in Software Defined Networks. IEEE Communications Surveys and Tutorials, 18(1), 623-654. DOI: 10.1109/COMST.2015.2453114

[4] ZUO Q Y, ZHANG H S. Analysis and Research on Network Security for OpenFlow-based SDN[J]. Netinfo Security, 2015, (2):26-32.

[5] Kandula S. Sengupta S. Greenberg A. et al. The nature if data center traffic: measurements & analysis[C]//Proceedings of the 9th ACM SIGCOMM conference on Internet measurement conference. ACM, 2009:202-208

[6] Thuraisingham B. Mandatory Access Control[J]. NFS vulnerabilities, International Symposium on Collaborative Technologies and Systems, 2013, 29(9):114-116.

[7] Moffett J, Sloman M, Twidle K. Specifying Discretionary Access Control Policy For Distributed Systems[J]. Computer Communications, 1990, 13(9):571-580.

[8] Sandhu R S, Coyne E J, Feinstein H L, et al. Role-Based Access Control Models[J]. Computer, 1996, 29(2):38-47.

[9] Sandhu R, Ferraiolo D, Kuhn R. The NIST model for role-based access control: towards a unified standard[C]// ACM Workshop on Role-Based Access Control. ACM, 2000:47-63.

[10] US Department of Commerce, NIST. Role Based Access-Control (RBAC): Features and Motivations[J]. December, 1995.

[11] Park J, Sandhu R. Towards usage control models: beyond traditional access control[C]// ACM Symposium on Access Control MODELS and Technologies. DBLP, 2002:57-64.

[12] Park J, Sandhu R. The UCON ABC usage control model[J]. Acm Transactions on Information & System Security, 2004, 7(1):128-174.

[13] Sandhu R, Park J. Usage Control: A Vision for Next Generation Access Control[J]. Lecture Notes in Computer Science, 2003, 2776:17-31.

[14] Canini M, Venzano D, Peresini P, et al. A NICE way to test OpenFlow applications[C]//Proc. of the 9th USENIX Symp. on Networked Systems Design and Implementation (NSDI), 2012:10.

[15] Sherwood R, Gibb G, Yap K, et al. Can the production network be the testbed?[C]//Proc. of the 9th USENIX Conf. on Operating Systems Design and Implementation (OSDI), 2010:1-6.

[16] Wen X, Chen Y, Hu C. Towards a Secure Controller Platform for OpenFlow Applications[C]//HotSDN, 2013:171-172.

[17] Klaedtke F, Karame GO, Bifulco R, Cui H. Access control for SDN controllers. In: Proc. of the 3rd Workshop on Hot Topics in Software Defined Networking. Chicago: ACM, 2014. 219-220. [doi: 10.1145/2620728.2620773]

Design research on information coding system under the concept of agile manufacturing

Zitian Liao

School of Electronic Engineering
Xidian University
Xi'an 710126
China

Abstract: Agile manufacturing can achieve the characteristics of high quality, low cost, short production cycle to meet the increasingly fierce market competition. Information coding system in agile manufacturing environment is particularly important, which directly determines the quality of information exchange in enterprise dynamic alliance of agile manufacturing. This paper analyzes the requirements of information coding system under agile manufacturing concept, and gives the design points of each module, providing some references for relevant researchers.
Mathematics Subject Classification 2010: 74E99

1. Concept and Features of agile manufacturing

1.1 Concept
Agile manufacturing is a manufacturing paradigm that enterprises adapt to and respond to market changes through efficient response in an unpredictable competitive environment. The key technologies of agile manufacturing are concurrent engineering and virtual enterprise alliance. The advantages of agile manufacturing can gather cross-regional resources, form a certain enterprise alliance, and respond to market demand quickly. Since the concept of agile manufacturing has been put forward, it has been supported and recognized by all sectors of society. The advantage of meeting the market demand makes agile manufacturing become one of the manufacturing modes which have been paid close attention in the century. The core of agile manufacturing is agility, which requires enterprises to respond quickly and enhance their competitive advantages in a complex competitive environment. It is a kind of strategic competitiveness. Agile organizational form is to form an agile enterprise alliance, which takes the core competitiveness of the enterprise as the basis, uses a quick way to quickly seek the same core competitiveness of resources nationwide and even globally, organizes these resources quickly, forms a certain scale of production and manufacturing chain, from the design of products. The origin of operation mode of agile manufacturing is market demand. The market opportunities and challenges are analyzed. The operation model of agile manufacturing is shown as follows.

Content from this work may be used under the terms of the Creative Commons Attribution 3.0 licence. Any further distribution of this work must maintain attribution to the author(s) and the title of the work, journal citation and DOI.
Published under licence by IOP Publishing Ltd

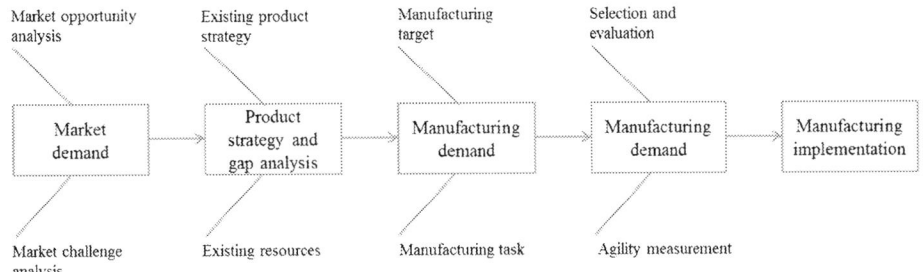

Fig. 1

Operation model of agile manufacturing

1.2 Features

Agile manufacturing is one of the most competitive production modes in this century. Under the background of economic globalization, agile manufacturing can respond quickly and flexibly. It can get the maximum profit with the shortest delivery time and win the market and customers. Agile manufacturing resources can be flexibly organized to form a rapid response to the market. Because the organization of agile manufacturing is flat, the decision-making under this manufacturing mode is rapid and flexible. For external resources, it can be linked together with effective information technology, make full use of network resources for rapid and effective transmission of information, can speed up the dynamic change of resource cooperation. Quality control is only controlled by advanced technology to ensure the high quality of vehicles. Agile manufacturing needs to combine excellent resources from all over the world to cooperate in production and manufacturing. Therefore, the production information exchange and production and operation between cooperative enterprises need to have a certain relationship. Therefore, the enterprise must be a development structure, which can allow more excellent resources to join.

2. Principles and process of information coding system design

2.1 Principles

Scientific information classification coding is becoming more and more important in all walks of life. With the emergence of agile manufacturing, enterprises are no longer confined to internal integration, but must be integrated with external to achieve information sharing. Under the new manufacturing mode, the product is developed in the environment of remote design and manufacturing. Therefore, the design of enterprise information classification and coding system should start from each link of code application in the enterprise, considering both the current situation of the enterprise and the development of the enterprise. Because the classification and coding of information is a difficult, complex and long-term work, it may take years or even longer to establish a scientific and perfect coding system. Therefore, the top-level planning should be done well before the implementation of information classification and coding system, and the formulation of standards should be organized from a global and long-term perspective. The main reason for the distortion of information is that the definition and expression of information are not standardized and inaccurate. Therefore, it is necessary to standardize all kinds of basic data and information in various original documents, reports and other information carriers from the system and process, and formulate enterprise standards, and publish and implement them in enterprises. Standardize the application process of the coding system by supporting the software. We should design a coding system that is compatible with the characteristics of enterprises, and do not copy the coding schemes of other enterprises.

2.2 Process

It is necessary for the relevant personnel of the design, manufacture, procurement and even the cooperative manufacturer to exchange information with each other in a timely manner. At the same time, in the concurrent engineering environment of mold development, it is also necessary to share information and data and resources among the various departments of design and manufacture to overcome various problems caused by geographical and organizational factors. The analysis and design of enterprise information classification and coding system can be carried out according to the following steps by tracking and implementing the information classification and coding system projects of many enterprises.

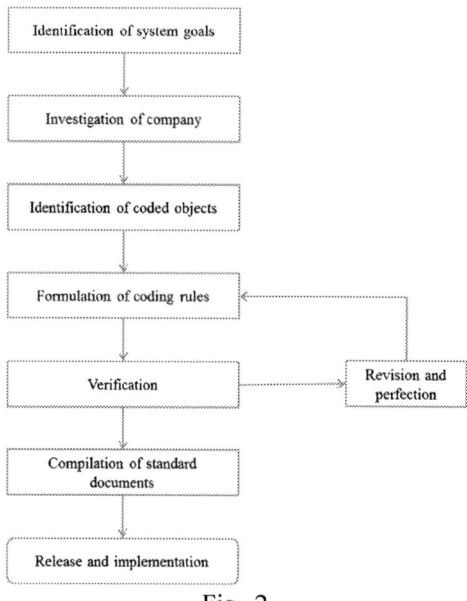

Fig. 2

Analysis and design process of information coding system design

3. Requirement analysis of information coding system design under the concept of agile manufacturing

The information coding system based on agile manufacturing should not only adhere to the basic design principles of information code such as uniqueness, stability, extensibility, applicability and operability, but also have the following characteristics: high flexibility of the system. Coding rules are set entirely by the user-defined, and according to different circumstances, the total number of three-element code segments, each bit, each code value can be modified to form a complete, hierarchical clear rule structure to face the changing products and processes. The structural reconfigurability of the system. In the process of system operation, the system function modules can be expanded and redeveloped according to the changes of environment and the needs of customers, and the system structure can be changed rapidly by constantly updating and adding new information. The system should be designed based on modular principle, and only modular design can meet the requirements of rapid reconfiguration of the system. Agile enterprises are dynamic alliances based on market conditions and their respective advantages. Agile enterprises are the integration of people, processes and technologies in different regions. Agile manufacturing emphasizes a flexible mode of production organization. Information coding system is required to operate across regions to ensure the overall operation of the enterprise.

4. Design of information coding system in agile manufacturing environment

4.1 Code generation module

Code generation includes auxiliary encoding and auditing code. This module encodes the specific information object according to the defined coding rules and initialization contents. During the coding process, the system will judge whether the information object is coded repeatedly, and automatically generate the code according to the coding rules. At the same time, it can strictly guarantee the uniqueness of the code. In addition, the generated code can also be audited to ensure the generation. The correctness of the code can be corrected by error code. According to the change of information object state in agile manufacturing process, information code can be frozen and thawed. Applying for encoding is a process of generating unique encoding according to the rules of the defined encoding object and the user's choice. Application coding function is the most basic function and core function of the whole system.

4.2 Code query module

Code generation includes auxiliary encoding and auditing code. This module encodes the specific information object according to the defined coding rules and initialization contents. During the coding process, the system will judge whether the information object is coded repeatedly, and automatically generate the code according to the coding rules. At the same time, it can strictly guarantee the uniqueness of the code. In addition, the generated code can also be audited to ensure the generation. The correctness of the code can be corrected by error code. According to the change of information object state in agile manufacturing process, information code can be frozen and thawed. Applying for encoding is a process of generating unique encoding according to the rules of the defined encoding object and the user's choice.

4.3 User management module

User management module is mainly used to manage users of the system, including adding users, deleting users, modifying passwords, resetting passwords and mapping roles for users. In this system, the mapping relationship between users and roles is many-to-one relationship. The role management module is mainly used to manage system roles, including adding, deleting and modifying roles, and assigning permissions to roles. The privilege management module is mainly used to grant roles to access and operation privileges to the system.

5. Conclusion

The main conclusions are as follows: (1) The process of information coding system design include identification of system goals, investigation of company, identification of coded objects, formulation of coding rules, verification, compilation of standard documents and release and implementation. (2) The information coding system in agile manufacturing environment must adhere to the basic design principles and should has the highly flexible function, powerful communication function and remote maintenance function. (3) Design of information coding system in agile manufacturing environment the code generation module, the code query module and the user management module.

References

[1] Dubey R, Gunasekaran A. Agile manufacturing: framework and its empirical validation[J]. The International Journal of Advanced Manufacturing Technology, 2015, 76(9-12): 2147-2157.

[2] Leite M, Braz V. Agile manufacturing practices for new product development: industrial case studies[J]. Journal of Manufacturing Technology Management, 2016, 27(4): 560-576.

[3] Balakirsky S. Ontology based action planning and verification for agile manufacturing[J]. Robotics and Computer-Integrated Manufacturing, 2015, 33: 21-28.

[4] Ghobakhloo M, Azar A. Business excellence via advanced manufacturing technology and lean-agile manufacturing[J]. Journal of Manufacturing Technology Management, 2018, 29(1): 2-24.

[5] Uçaktürk A, Uçaktürk T, Yavuz H. Possibilities of Usage of Strategic Business Intelligence systems Based on Databases in Agile Manufacturing[J]. Procedia-Social and Behavioral Sciences, 2015, 207: 234-241.

[6] Sindhwani R, Malhotra V. Modelling the attributes affecting design and implementation of agile manufacturing system[J]. International Journal of Process Management and Benchmarking, 2016, 6(2): 216-234.

[7] Yang H, Baradat C, Krut S, et al. An agile manufacturing system for large workspace applications[J]. The International Journal of Advanced Manufacturing Technology, 2016, 85(1-4): 25-35.

Research on Milling Force Prediction Model Based on Improved Particle Swarm Optimization Algorithm

Liu Ling*, Qi Weiwei and Liu Tingting

School of information technology engineering,Tianjin University of Technology and Education,Tianjin,300222,China

*Corresponding author's e-mail: winterpost@126.com

ABSTRACT: According to the remarkable characteristics of milling force, an innovative method of milling force modeling using improved particle swarm optimization (PSO) fuzzy system based on support vector machine (SVM) is proposed in this paper. The experiment of titanium alloy milling is designed and implemented. The advanced tester is used to measure the milling force. The training data and test data based on the fuzzy system are obtained. The gradient descent algorithm is embedded in the ordinary particle swarm optimization algorithm to obtain the improved particle swarm optimization algorithm. The convergence effect of the improved particle swarm optimization algorithm is obviously better than that of the ordinary particle swarm optimization algorithm. The improved particle swarm optimization (IPSO) based on fuzzy system is applied to the milling force modeling. Finally, the improved particle swarm optimization (PSO), gradient descent algorithm and improved particle swarm optimization (IPSO) are used to train the fuzzy system, and the conclusion that the final training error of the improved particle swarm optimization (IPSO) is the smallest is obtained.

1. Introduction

Cutting force is the key parameter of the object in the process of cutting processing, and its size directly affects the object's cutting heat, cutting deformation and so on. There are three main vertical components of cutting force: main cutting force, back force and feed force. Modeling cutting force can optimize cutting parameters, effectively control machining deformation, and provide reference for calculating the wear degree of cutting tools. Traditional cutting force modeling methods include mechanical modeling, empirical modeling, intelligent modeling and finite element modeling.

Traditional cutting force modeling methods are mostly aimed at steel, iron, aluminum alloy and other easy-to-process materials. This kind of material for titanium alloy processing complex, because it has the tendency of work hardening and high thermal conductivity characteristics in low titanium alloy cutting, outstanding performance for the cutting force, high tool wear degree and the higher cutting temperature. Moreover, between the cutting force and cutting force parameters there is a complicated relationship. It is difficult for the traditional modeling method of cutting force to correctly simulate the mechanical process. Intelligent modeling method of cutting force can handle more complex nonlinear problems, and has become the key research content in modeling method of cutting force.

The cutting force of intelligent modeling, mainly in the application of BP neural network, extracting key data information by using the BP neural network, prediction model of cutting force. In the field of artificial intelligence, the application of fuzzy system modeling method is not common, but because of the fuzzy system can not only obtain accurate data information, also can realize the expert system. In

ISPECE

this paper, the milling process of titanium alloy is based on the improved particle swarm optimization algorithm, and its application in fuzzy system, to realize the prediction of milling force modeling method.

2. Fuzzy system and improved particle swarm optimization algorithm

Mamdani fuzzy system can effectively compensate for the shortcomings of the traditional universal pure fuzzy systems, Mamdani fuzzy system has been widely applied in many fields such as communication, control. This system is the choice of fuzzy support vector machine, at the same time with singleton fuzzifier, average defuzzifier and Gauss function Mamdani fuzzy system, the mathematical expression of Mamdani fuzzy systems as shown in formula (1).

$$f(x) = \frac{\sum\limits_{l=1}^{M} \overset{-1}{y} \left[\prod\limits_{i=1}^{n} \exp\left(-\left(\frac{x_i - \overset{-1}{x_i}}{\sigma_i^l} \right)^2 \right) \right]}{\sum\limits_{l=1}^{M} \left[\prod\limits_{i=1}^{n} \exp\left(-\left(\frac{x_i - \overset{-1}{x_i}}{\sigma_i^l} \right)^2 \right) \right]} \tag{1}$$

Where is the number of fuzzy rules is the number of fuzzy rules; some of the conditions; fuzzy rules is part of the membership function parameters; fuzzy set fuzzy rules is part of the THEN center is the input of fuzzy system value.

A fuzzy system, such as formula (1), is designed to ensure the minimum error value of the following training.

$$e = \sum\limits_{p=1}^{N} \frac{1}{2} \left[f(x_0^p) - y_0^p \right]^2 \tag{2}$$

Which is the system of training error; training data number; belongs to the training data input and output items. To sum up, and said, respectively, in order to make the fuzzy system can fully meet the needs of formula (2) in the training of the minimum error, the need for training of fuzzy system parameters. Usually, people in the process of analyzing problem, the basic conclusion is uncertain, therefore, according to the fuzzy rules and formula (1), can determine the fuzzy system parameters are part of the THEN training center by fuzzy sets, the gradient descent algorithm to train as shown in formula (3).

$$\overset{-1}{y}(q+1) = \overset{-1}{y}(q) - \alpha \frac{\partial e}{\partial \overset{-1}{y}} \tag{3}$$

Suppose:

$$\begin{cases} f = \dfrac{a}{b}, a = \sum\limits_{l=1}^{M} (\overset{-1}{y}\,\overset{-1}{z}), b = \sum\limits_{l=1}^{M} \overset{-1}{z} \\[2mm] \overset{-1}{z} = \prod\limits_{i=1}^{n} \exp\left(-\left(\dfrac{x_i - \overset{-1}{x_i}}{\sigma_i^l} \right)^2 \right) \end{cases}$$

According to the derivation of Gauss compound function, we can get formula (4).

$$\frac{\partial e}{\partial \overset{-1}{y}} = \sum\limits_{i=1}^{N} (f-y) \frac{\partial f}{\partial a} \frac{\partial a}{\partial \overset{-1}{y}} = \sum\limits_{i=1}^{N} (f-y) \frac{1}{b} \overset{-1}{z} \tag{4}$$

The formula (4) into the formula (3), the fuzzy rule learning algorithm THEN fuzzy set center as shown in formula (5).

$$\overset{-1}{y}(q+1) = \overset{-1}{y}(q) - \alpha \sum\limits_{i=1}^{N} (f-y) \frac{1}{b} \overset{-1}{z} \tag{5}$$

Particle swarm algorithm belongs to the group of iterative computation algorithm of artificial intelligence environment, which has high robustness, global ability and fast convergence characteristic search. In order to improve the local search ability of particle swarm algorithm, the gradient descent algorithm into particle swarm algorithm, an improved particle swarm algorithm, the new algorithm for each individual particle produced in accordance with the established practice of gradient descent algorithm to calculate the probability.

3. Milling force prediction model based on fuzzy system

In the theory of metal cutting, metal cutting force is related to many factors, including materials, dosage, workpiece and tool, etc. This paper studies the influence of milling parameters on metal, milling force so selected spindle speed, feed rate, radial cutting depth and the axial depth of cut as input and output of the fuzzy system, the metal milling force said.

The metal milling force experiment is studied in this paper based on the realization of the NC milling machine, using three to the dynamometer of the milling force measurement, the cutter is cemented carbide tool, a total of 4 teeth, diameter of spiral angle of milling cutter.

The workpiece foot length and width and height respectively, and TC18 type workpiece materials, cutting fluid in the process of milling in Figure 1 for the test system of metal cutting force.

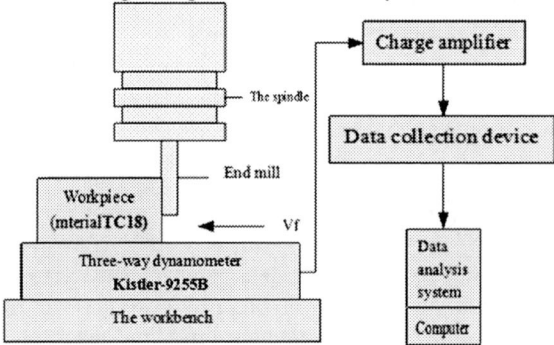

Figure 1. Schematic diagram of cutting force test system

This paper is a study of metal milling force milling force of the finishing process, in accordance with the range of cutting parameters and orthogonal principle, orthogonal experimental design (4 factors and 4 levels), correlation factor and level selection as shown in table 1. During the experiment, the cutting parameters of metal are shown in Table 2. The sample data obtained from experiments can be used as the sample space characteristics, the fuzzy system as training data, at the same time, on the completion of the 4 group and the orthogonal experiment with different experimental principle, the results are shown in Table 3, and the fuzzy system as test data.

Table 1. Factors and levels of experiment

Factor	level			
	1	2	3	4
Spindle speed $n/(r \cdot min^{-1})$	600	800	1000	1200
Feed speed $v_f /(mm \cdot min^{-1})$	100	120	200	250
Radial cutting depth a_e / mm	1.0	1.5	20	2.5
Axial cutting depth a_p / mm	0.2	0.3	0.4	0.5

Table 2. cutting force parameters of experimental data

Experimental number	axial depth a_p / mm	radial depth a_e / mm	spindle speed $n/(r \cdot min^{-1})$	feed speed $v_f /(mm \cdot min^{-1})$
1	0.2	1.0	800	100
2	0.2	1.5	600	150

| 3 | 0.2 | 2.0 | 1200 | 150 |
| 4 | 0.2 | 2.5 | 800 | 200 |

In order to avoid the error produced in the course of the experiment, this paper has carried out 3 experiments in the same way, a total of 48 sets of training data and 12 sets of fuzzy systems, the results are as shown in Table 3. The maximum milling force in each cycle is respectively.

4. Training of fuzzy system based on Improved Particle Swarm Optimization Algorithm

In this paper, by descent algorithm, particle swarm algorithm and improved particle swarm algorithm to train fuzzy gradient system respectively, this performance comparison of three algorithms for numerical algorithm for minimum error. Therefore, the algorithm selects the same parameters, in order to make the training process more clearly, each complete 20 training after the extraction of 1 errors, convergence of gradient descent algorithm and particle swarm algorithm and improved particle swarm algorithm as shown in figure 2.

Table 3. testing data based on fuzzy system

Training data	1	2	...	12
Axial cutting depth a_p/mm	0.2	0.2	...	0.2
Radial cutting depth a_e/mm	1.0	1.5	...	2.5
Spindle speed $n/(r\cdot\min^{-1})$	800	600	...	800
Feed speed $v_f/(mm\cdot\min^{-1})$	100	150	...	200
Milling force F_{xm}/N	17.61	20.10	...	31.26
Milling force F_{ym}/N	48.33	56.33	...	65.31
Milling force F_{zm}/N	18.53	26.16	...	32.56

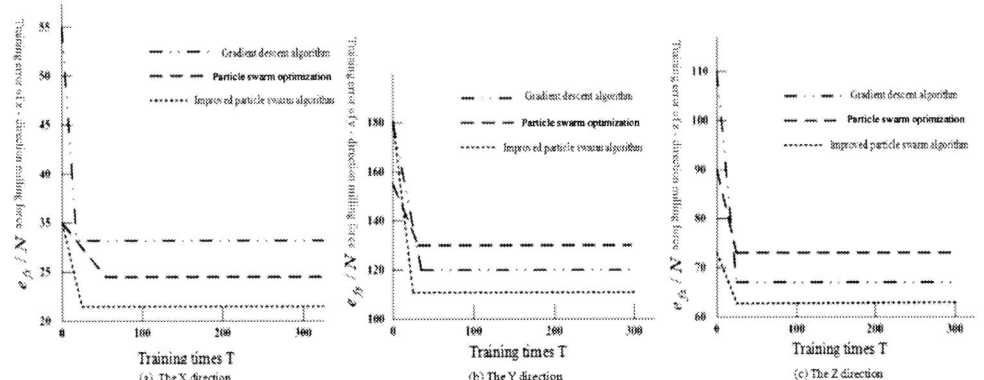

Figure 2. fuzzy system training process of three algorithms

In Figure 2, the milling force convergence of the 3 directions of the integrated display, it can be seen from the figure, the improved particle swarm algorithm to train the fuzzy system's actual convergence is better than gradient descent algorithm and particle swarm algorithm. The milling force direction, the fuzzy system is trained using the improved particle swarm algorithm, the training error is derived, descent algorithm to train the fuzzy system with the gradient, the training error is derived, using particle swarm algorithm to train the fuzzy system, the training error is. The milling force direction, the training error obtained using improved particle swarm optimization algorithm for training, get the error reduction algorithm and particle swarm algorithm based on gradient and. The milling force direction, the training error is obtained by using the improved particle swarm optimization algorithm for training, get the error reduction algorithm and particle swarm algorithm based on gradient and.

Among them, the local search ability of particle swarm algorithm is better than gradient descent algorithm, and the global searching ability of PSO is better than gradient descent algorithm, improved particle swarm algorithm combining the advantages of two algorithms and overcomes the disadvantages of a single algorithm, obtain the optimal searching ability, the minimum training error.

5. Conclusion

In summary, based on the research of milling force prediction model, the gradient descent algorithm into particle swarm algorithm, forming a new improved particle swarm optimization algorithm, using titanium alloy as an example,The training data and test data of the fuzzy system are obtained by measuring instrument. Gradient descent algorithm, particle swarm optimization algorithm and improved particle swarm optimization algorithm are used to train the fuzzy system respectively. The simulation results show that the improved particle swarm optimization algorithm has good convergence performance, which is obviously superior to gradient descent algorithm and particle swarm optimization algorithm.

Acknowledgments

Supported by Tianjin University Students Innovation and Entrepreneurship Training Program Project: Intelligent Vehicle License Plate Recognition System (No. 201710066069)

Reference

[1] Wang Gang, Wan Min, Liu Hu, Zhang Weihong. (2015) Milling force modeling method of particle swarm optimization fuzzy system [J]. Journal of Mechanical Engineering, 13:123-130.

[2] Wang Baosheng, Zuo Jianmin, Wang Mulan. (2017) Modeling of instantaneous milling force and identification of milling force coefficient by particle swarm optimization [J]. Mechanical design and manufacturing,3:63-65.

[3] Niu Xinghua, Yang Zhongbao, Bian Yangqing, Ma Chao, Yin Wenshan, Cui Ying. (2017) Experimental study and modeling of dynamic milling force based on the shape of cutting layer. Manufacturing technology and machine tools,7:112-116.

[4] Li Yingsong, Xia Ping, He Donghuan.(2017) Milling force model of end mill based on Fourier series. Mechanical design and manufacturing,7:256-258.

[5] Zheng Jinxing. (2008) Application of particle swarm optimization artificial neural network in high-speed milling force modeling . Computer integrated manufacturing system,9:1710-1716.

[6] Chung-Liang Tsai, Yunn-Shiuan Liao. (2010) Cutting force prediction in Ball-End Milling with inclined feed by means of geometrical analysis.The International Journal of Advanced Manufacturing Technology,11 :5-8.

[7] Wan M., Zhang W. H. (2006) Calculations of chip thickness and cutting forces in flexible end milling. The International Journal of Advanced Manufacturing Technology,6 :7-8.

[8] Yann Quinsat, Laurent Sabourin. (2016) Optimal selection of machining direction for three-axis milling of sculptured parts. The International Journal of Advanced Manufacturing Technology,9 :11-12.

[9] Lei Zhang, Li Zheng. (2015)Prediction of cutting forces in end milling of pockets.The International Journal of Advanced Manufacturing Technology,8: 38-41

[10] Wang J.J., Chang H.C. (2014) Extracting cutting constants via harmonic force components for a general helical end mill.The International Journal of Advanced Manufacturing Technology,11:58-62.

Optimization of Adaptive Handover Algorithm Based on Distributed Antenna in LTE-R

Ziwei YANG [1,a*], Yufu ZHENG[2] and Jing JING[3]

[1,2,3] School of Electronics and Information Engineering,Lanzhou Jiaotong University,Lanzhou,Gansu,730000,China

[a]*Corresponding author's e-mail: 377866700@qq.com

Abstract. In the high-speed railway, the traditional handover algorithm uses a fixed handover hysteresis threshold, which will result in a lower success rate of the train when the train is traveling at a high speed, and cannot meet the wireless communication needs of the passenger. According to the distribution of antennas in the high-speed railway and the analysis of the handover process and measurement parameters, an adaptive handoff algorithm based on distributed antenna is proposed. For the deployment of antennas in high-speed railways, a distributed approach is adopted, and the RAUs with the best quality are selected to maintain the connection. In the process of switching, the hysteresis threshold value of switching is dynamically adjusted at different positions according to different speeds, and the success rate, interruption probability and average value of switching are simulated and comparedSimulation results show that in the case of distributed antenna deployment, handover threshold is dynamically adjusted according to different speeds to reduce the interruption probability of handover, and the success rate of handover is significantly improved, but the average number of handover will increase correspondingly, which can meet the requirements of passengers for wireless communication system in high-speed railway.

1. Introduction

With the rapid development of high-speed railways in China, passengers are increasingly demanding mobile communication requirements and enjoying good voice and broadband data services. The current GSM-R railway communication can no longer meet the development direction of the future railway two-way, real-time, large-capacity, especially the user's call experience is significantly different.UMTS (Universal Mobile Telecom System) Long Term Evolution (LTE) is a new generation of broadband developed by the 3rd Generation Partnership Project (3GPP) team to meet the needs of users for further improvements in wireless broadband speed. The wireless mobile communication standard, the International Union of Railways (UIC), has clearly stated that railway-specific wireless communications will be directly transitioned from GSM-R to 4G LTE-R [1-6].

Distributed antenna system is applied to railway system to enhance coverage, improve user SNR, and improve wireless communication system performance. Distributed antenna system is a multi-antenna system with a wider range of applications and stronger adaptability, which can better play its role in railway system. [7]. In the scenario of frequent handover of high-speed railways, the speed will lead to a rapid change of path loss. At this time, it is unreasonable to use the same RSRP and RSRQ handover decision delay tolerance for mobile terminals of different speeds. When the train speed is fast, it takes a short time to pass through the overlapping area, so it needs to switch the threshold value of the situation and the time. When the speed is low, it takes a long time to pass through the overlapping area,

Content from this work may be used under the terms of the Creative Commons Attribution 3.0 licence. Any further distribution of this work must maintain attribution to the author(s) and the title of the work, journal citation and DOI.
Published under licence by IOP Publishing Ltd

so the hysteresis tolerance needs to be increased to avoid ping-pong switching caused by premature switching. Therefore, based on the speed characteristics, the adaptive algorithm is used to dynamically adjust the RSRP and RSRQ hysteresis tolerances to optimize the value of the hysteresis tolerance.

2. Handover Planning

In the high-speed scene, effective switching model planning for train switching can make the train switch faster and more efficiently, reducing the probability of failure. The planning of the switching area in this paper is shown in Figure 1:

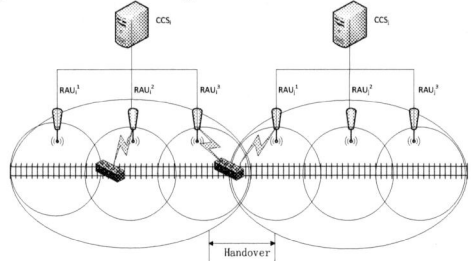 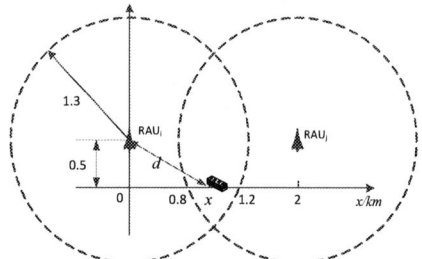

Figure 1. Handover model Figure   Figure 2. Coordinate system model

When the UE traverses the coverage areas of the two CCSs, after receiving the measurement report sent from the radio station, the CCS selects a neighboring RAU with the best received signal quality from the pre-stored neighbor cell list to maintain the connection. The handover request message is sent from the serving CCS to the target CCS. After the target CCS performs access admission control on the user information carried in the handover request message, the handover request confirmation message is sent to the serving CCS. The serving CCS then sends a handover command message to the UE. After receiving the handover command message, it begins to establish an underlying connection with the target RAU on the new wireless channel. The coordinate system model of the switching process is shown in Figure 2:

The distance between the two RAUs is 2km, the base station is 0.5km away from the rails, and the coverage of the two cells is 0.8km - 1.2km, so the sampling starts from 0.5km, the sampling ends at 1.5km, and the switch starts at 0.8km. From the starting point, the horizontal coordinate of the program simulation is [0.8, 1.5].

Table 1 shows the reference variables and reference values in the handover model:

Simulation parameter	symbol	Value
Base station power	P_{eNB}	44dBm
Base station antenna height	H	30m
Train speed	v	120/350km/h
Base station spacing	D	2km
Overlap area length	D_1	0.4km

3. ADAPTIVE HANDVOER ALGORITHM BASED ON DISTRIBUTED ANTENNA

In high-speed rail scenarios, periodic measurement reports cannot be triggered in time due to poor signal quality. In order to avoid this, consider using the position information of the train as a condition for triggering measurement reporting [9]. According to the characteristics of the high-speed railway scene, the following vehicle trajectory is fixed, the direction can be predicted, and the like, when the train travels to the specific position of the base station covering the overlapping area, as the trigger condition for the first reporting. If the base station is deployed in the center of the coverage area of the base station to enhance the signal, the trigger position is selected according to:

$$10\log 10(R_{(\tau,x)}) - 10\log 10(R_{(s,x)}) = H \quad (1)$$

Where $R_{(\tau,x)}$ is the signal quality of the target base station at the point, $R_{(s,x)}$ is the signal quality of the serving base station at the point x, and H is the hysteresis threshold.

3.1 Handover measurements

The handover includes three phases: handover measurement, handover decision, and handover execution.

In the handover measurement phase, the high-speed train needs to perform layer three filtering on the received signal before the handover departure judgment. Layer three filtering can effectively reduce ping-pong switching. The performance of layer three filtering is:

$$F_n = (1-\gamma)F_{n-1} + \gamma M_n, \gamma = Tm/Tu \quad (2)$$

Where F_n is the calculation result of this filtering: F_{n-1} is the filtered measurement result reported by the previous measurement period; M_n is the layer-filtered measurement value (point B); γ is the layer three filter factor, Tm is the layer-one filter period, and Tu is the d-layer three- Filter period[1].

The UE performs related measurement according to the measurement configuration message sent by the CCS, and reports the measurement result to the CCS. This paper chooses the open channel model (Hata model) as:

$$Lbs = 69.5 + 26.16 \times \log 10(f) - 13.82 \times \log 10(hb) - 2[\log(f/28)]^2 - 5.4 + X_d \quad (3)$$

X_d is the same frequency interference: $X_d = \sqrt{(d+10)^2 + (hb)^2/10^6 + (dt)^2/10^6}$, Where d is the location where the handover occurs, $d = v \times t \times n$, a function of speed, t is the switching time, and n is the number of handovers. Normally, $n=2$ is considered, that is, the handover is successful twice.

The handover parameters discussed in this paper are based on RSRP and RSRQ. RSRP refers to a mobile terminal receiving received power from a base station, The larger the RSRP value, the higher the signal strength. RSRQ means that the mobile terminal receives the pilot signal reception quality from the base station, and its expression is as follows: $RSRQ = N\dfrac{RSRP}{RSSI}$, Where a is N coefficient introduced during the measurement process to compensate for the difference in RSRP and RSSI bandwidth[3]. The RSSI is specifically the average of the measured sizes of all received carrier frequencies in the entire network bandwidth. In the conventional handover decision using RSRP, the trigger condition of the handover is that the received target base station signal power is higher than the serving base station by a fixed hysteresis threshold. Since the value of the received signal power is determined by the base station transmit power, path loss, and shadow fading, and the base station transmit power is generally equal, the received signals from the destination base station and the serving base station are respectively: the signal quality of the train receiving the service base station : $P_{(i,x)} = Pt_{eNB} - PL_{(i,x)} - A_{(i,x,\delta)}$ (3), the signal quality of the train receiving the destination base station: $P_{(j,x)} = Pt_{eNB} - PL_{(j,x)} - A_{(j,x,\delta)}$ (4).

3.2 Handvoer trigger probability

During the handover decision phase, the CCS evaluates according to the measurement result reported by the UE, and determines whether to trigger the handover. Switch trigger decision conditions: $P_1 = P_{tr}\{P_{(j,x)} - P_{(i,x)} \geq H\}$ (5) .Where $P_{(j,x)}$ is the signal quality of the target base station at point x, $P_{(i,x)}$ is the signal quality of the serving base station at point x, H is the hysteresis threshold, and is related to the position of the train, A represents the shadow fading obeying the normal distribution, f-means is 0 and the variance is σ.

The condition for triggering the handover is that $RSRP$ received by the in-vehicle relay from the target cell is higher than $RSRP$ received by the current serving cell by a hysteresis H (dB).

$P_h(t)$ is the probability that a handover occurs at time t. $P_{2|1}$ is the probability of switching from the serving cell to the target cell, P_{12} is the probability of switching from the target cell to the serving cell, and $P_1(t)$ and $P_2(t)$ respectively indicate the probability that the current cell is the serving cell and

the target cell. Then the probability of the current cell is: $P_1(t) = P_1(t-1)(1-P_{2|1}(t)) + P_2(t-1)P_{1|2}(t)$,
$P_2(k) = P_1(k-1)P_{2|1}(k) + P_2(k-1)(1-P_{1|2}(k))$ (6)

Then the handover probability can be expressed by: $P_h(t) = P_1(t-1)P_{2|1}(t) + P_2(t)P_{1|2}(t)$ (7)

The link interruption probability is that when the received signal strength of the current serving base station is less than the minimum threshold H of the guaranteed communication, the communication link is interrupted. The lower the probability of the link's outage, the better the switching performance of the system. Here we use P_O to indicate the outage probability of the handover, then the expression of $P_O(t)$ is as follows: (better evaluation of the power allocation scheme in the distributed antenna system.)
$P_O(t) = \Pr\{P_{(i,x)} < T\} + \Pr\{P_{(j,x)} < T\}$ (8)

3.3 Handover success rate

After the train terminal is successfully triggered, it enters the handover execution phase. When the train passes the delay trigger, if the handover condition is still met, the handover is performed. During the handover execution, the average signal strength received by the target base station needs to be met, which is greater than the minimum strength strength threshold of the communication. The CCS controls the UE to handover to the target cell according to the decision result, and finally completes the handover by the UE. Execution rate during train switching: $P_e(t) = \frac{1}{v \cdot t_2} \int_{x_1}^{x_1 + v \cdot t_2} P_{(j,x)}[F_n(j,x) > L] dx$ (9).

Where x_1 is the location at which the base station performs the handover command; t_2 is the time required to switch the execution process [1]. Switching success rate: Since the train may not be successfully switched during the switching process, it is necessary to perform two or even multiple switching to ensure the success of the handover. The probability of successful handover is subject to the binomial distribution. The probability that the train will switch successfully once is $P_{s1}(t) = P_{h1}(t) \cdot P_{e1}(t)$ (10). If one handover is not successful, a second handover is performed, and the success rate of the secondary handover is represented by $P_{s2}(t) = (1 - P_{s1}(t))P_{h2}(t) \cdot P_{e2}(t)$ (11). Then the probability of the train switching from the serving cell to the destination base station is: $P_s(t) = \sum (P_{s1}(t) + P_{s2}(t))$ (12)

4. Simulation Results

4.1 Handover success rate simulation

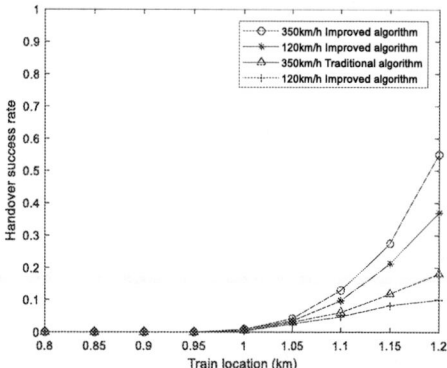
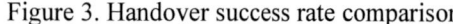

Figure 3. Handover success rate comparison

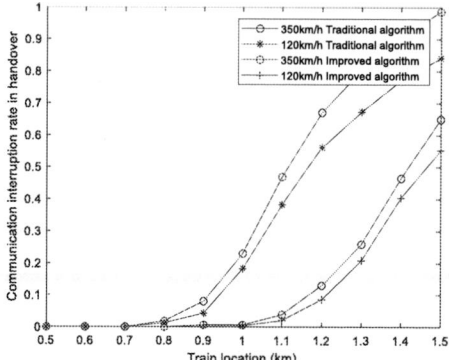

Figure 4. Interrupt probability comparison

Figure 3 shows the switching success rate of the traditional scheme and the improved scheme when the train is in different positions. As shown in the figure, when the speed is 120km/h and 350km/h, the success rate of the improved algorithm is higher than that of the traditional scheme, and it reaches the maximum at 1.2km, which is increased by 2.7 and 3.6 respectively. It is obvious. Because the base is in a different position from the train, the adaptive adjustment of the hysteresis tolerance makes the switching trigger more fast and accurate, so that it can be switched more quickly.

4.2 Interrupt probability simulation
Based on the difference in speed, the delay threshold of the relay is adjusted, and the probability of interruption of the handover between the train and the source base station is as shown in the figure above. Since the position of the train is different at the speed of the no-pass, the probability of interruption of the handover is also different. It can be seen from the above figure that when the speed is 120km/h and 350km/h, the probability of interruption of the improved scheme is reduced compared with the probability of interruption under the traditional scheme. At 1.2km, the probability of interruption is reduced from 0.7 to 0.1. For trains in high-speed environments, the increase in the probability of interruption will greatly reduce the call drop rate of users and increase the user's Qos experience.

5. Conclusion
According to the different speeds of trains in high-speed environment and different locations at the same time, an adaptive handoff algorithm based on distributed antennas is proposed. The deployment of the distributed antenna is used to establish a handover model, and the handover threshold is dynamically adjusted according to the speed. The values of RSRP and RSRQ under the traditional algorithm are compared, and the handover success rate, the interruption probability and the average handover number are compared. Algorithm verification and simulation. The simulation results show that the distributed adaptive handoff algorithm improves the success rate of handover and greatly reduces the interruption probability of handover, which ensures that the current wireless communication system Qos technology of China's railways has a handover rate greater than 99.5% required, but the average number of handovers has increased Therefore, the adaptive algorithm optimization based on distributed antenna is more suitable for the choice of handover in high-speed scenarios, which provides the necessary technical support for the future application of LTE-R system in railway.

References
[1] Chen Yonggang, Li Dewei, Zhang Caizhen.A Speed-based LTE-R Handover Optimization Algorithm[J]. Journal of the China Railway Society,2017,39(7):67-72.
[2] Xu Yan, Zhang Qiang. Research on TD-LTE handover algorithm for high-speed railway mobile communication[J]. Journal of the China Railway Society, 2015, 37(5): 47-51.
[3] Zhang Pu, Wang Junxuan. Research on Handover Algorithm in LTE System[J]. Journal of Xi'an University of Posts and Telecommunications,2010,15(5):1-5.
[4] Wang Xiaoxuan, Yang Tao, Jiang Hailin, Shu Xiaomeng. The Handover Algorithm of Urban Rail Transit Vehicle Communication System Based on TD-LTE[J]. China Railway Science, 2016, 37(3): 109-115.
[5] Li Tai, Li Wei. A Survey of Research on Handover of Mobile Communication Systems in High-Speed Railway Scenes[J]. Communications Technology,2015,48(5):566-572.
[6] Ruan Cheng, Fang Xuming, Yang Chongzhe, Liu Linghui. A Seamless Switching Scheme for GSM-R Redundant Networks Based on Vehicle Dual Antenna[J]. Journal of the China Railway Society, 2012, 34(5): 51-56.
[7] Cao Yuan, Ma Lianchuan, Zhang Yuxi, Mu Jiancheng. Comparison of the success rate of handover between LTE-R and GSM-R[J]. China Railway Science, 2013, 34(6): 117-123.
[8] Mi Gensuo, Ma Shuomei.Application of Speed-triggered Early Switching Algorithm in LTE-R[J].Journal of Electronics & Information Technology,2015,37(12):2 852-2 857.
[9] ZHANG Yi-fan,ZHANG Mu-qing,GE Shun-ming,et al. Optimization of Time-to-trigger Parameter on Handover Performance in LTE High-speed Railway Networks[C].Proceedings of the 15th International Symposium on Wireless personal Multimedia C0mmunications(WPMC).New York: IEEE Press,2012:251-255.

[10] LUAN Lin-lin,WU Mu-qing,CHEN Yu-han,et al.Handover Parameter Optimization of LTE System in Variational Velocity Environment[C].Proceedings of International Conference on Communication Technology and Application(ICCTA 2011). London:IET Prees,2011:395-399.

ISPECE   IOP Publishing

IOP Conf. Series: Journal of Physics: Conf. Series **1187** (2019) 032095   doi:10.1088/1742-6596/1187/3/032095

A game-theory approach based on genetic algorithm for flexible job shop scheduling problem

Li Nie[1]*, Xiaogang Wang[1] and Fangyu Pan [1]

[1] Shanghai Polytechnic University, Shanghai, 201209, P. R. of China

*Corresponding author's e-mail: nieli@sspu.edu.cn

Abstract. In the paper, flexible job shop scheduling problem (FJSP) which joints the objective of maximizing the manufacturer's efficiency and the objective of maximizing the customer's delivery satisfaction is considered. An optimization model based on game theory is put forward for the FJSP. Therefore, the problem of FJSP is transferred into a game, in which all jobs and the manufacturer are regarded as players in the game. The players behave with the objective of maximizing their own profits. The manufacturer wants to minimize the makespan of all the jobs, whereas each job wants to minimize the own tardiness. Eventually they gain the equilibrium. In order to solve the game, Nash equilibrium (NE) searching approach based on genetic algorithm (GA) is designed and developed. The efficiency of the proposed approach is validated on several benchmark instances.

1. Introduction

In modern manufacturing enterprises, many flexible manufacturing systems and numerical control machines are introduced in order to improve the production efficiency [1], which makes the scheduling problem in the shop floor of these manufacturing enterprises more complex because these flexible manufacturing systems and numerical control machines usually can process several types of the operations. It means the scheduling decision maker not only needs to choose the routing for each job according with its processing constraints but also needs to sequence the jobs on each machine. The scheduling problem is usually called flexible job shop scheduling problem (FJSP). Since FJSP is more complicate than job shop scheduling problem (JSP) which it is a NP-hard problem, it is necessary to develop effective optimization technology to eliminate the conflicts between the manufacturing facilities and to take advantages of the flexibility of the manufacturing facilities in order to reduce flow-time and work-in- process and to improve production resources utilization.

The scheduling optimizing approaches for FJSP is increasingly attractive to researchers and practitioners [2]. Many researches focused on the single objective FJSP with different kinds of meta-heuristics such as genetic algorithm [3], tabu search [4], swarm optimization algorithm [5] etc., and makespan was usually used as their scheduling objective in most situations while other objectives such as tardiness may also be taken as the scheduling objective in certain situations [6]. Some researchers focused on the multi-objective flexible job shop problem. The difficulty with the multi-objective problem partially lies in the conflicts between different objectives, and a group of Pareto optimal solutions are often adopted to cope with them. For example, Rahmati et al. (2018) [7] developed Pareto envelope-based selection algorithm (PESA) to solve the multi-objective FJSP which joint maintenance and production planning problem. Another method for the multi-objective problem is to find the equilibrium between the different objectives by game theory based approach. Zhang et al.

Content from this work may be used under the terms of the Creative Commons Attribution 3.0 licence. Any further distribution of this work must maintain attribution to the author(s) and the title of the work, journal citation and DOI.

Published under licence by IOP Publishing Ltd

(2017) [8] considered the FJSP for the manufacturing shop floor to improve energy efficiency and production efficiency. To solve this problem, a dynamic game theory based two-layer scheduling method was developed to reduce makespan, the total workload of machines and energy consumption to achieve multi-objective optimization. Krenczyk and Olender (2015) [9] designed game theory models to solve the FJSP in order to minimize the cycle length and the production cost. Sun et al. (2013) [10] applied non-cooperative game theory with complete information to build new scheduling model for flexible job shop scheduling problem subject to machine breakdown in order to optimize the conflicting objective of robustness and stability simultaneously. Zhan et al. (2012) [11] applied non-cooperative game theory to multi-objective scheduling problem in the automated manufacturing system. The methods mentioned above usually considered the objectives of the efficiency and cost for the manufacturer, such as makespan, machine utilization and so on. There were few researchers to consider the optimization problem for both the manufacturer and customers. In other words, a schedule should consider not only the manufacturer's efficiency and cost objectives, but also the customer's delivery satisfaction.

In this paper, the scheduling problem considering both the manufacturer's efficiency and customer's delivery satisfaction is modeled as a FJSP with the objective of minimizing the makespan and maximum tardiness of all jobs simultaneously. An effective game theory approach which hybridizes the genetic algorithm (GA) has been proposed for the FJSP. This paper is organized as follows: Section 2 gives the game model for the FJSP. NE solution approach based on GA is given in Section 3. Section 4 presents experiments and the results. Lastly, the conclusion is drawn in Section 5.

2. Formulating of FJSP

2.1. Description of FJSP
The FJSP can be stated as follows. A number of production jobs come from different orders of different costumers are to be processed on a number of machines. The sequence of operations for each job is fixed. Each operation can be operated on any of its alterative machines but visit only one of them exactly once. The processing time of each operation on each of its alterative machine is known and deterministic. It is assumed that:
- Each machine can process at most one operation at any time.
- Each operation can be processed only at one machine at a time.
- Operations of all the jobs must be processed in a given order.
- The setup time of any operation is independent of the schedule, fixed, and included in the corresponding processing time.
- Transport time of each job to transfer from one machine to another is neglected.
- All jobs are ready to start at time zero. And there is no new jobs arrive.
- Machines are available all the time.

In order to discuss the problem in detail, the notations listed below are employed:
- There are m machines indexed by M_j ($j = 0, \ldots, m\text{-}1$) and all machines construct the machine set of M.
- There are n jobs to be processed indexed by J_i ($i=0, \ldots, n\text{-}1$) and all jobs construct the job set of J.
- Each job J_i consists of a fixed sequence of operations $O_{i,l}$ ($i=0, 2, \ldots, n\text{-}1$; $l= 0, \ldots, h_i\text{-}1$) and h_i is the number of operations in job J_i.
- A set of alterative machines for operation $O_{i,l}$ is presented by $M_{i,l} \subset M$.
- The processing time for operation l of job i on machine j is denoted as $p_{i,l,j}$ ($i=0, \ldots, n\text{-}1$; $l= 0, \ldots, h_i\text{-}1$; $j=0, \ldots, m\text{-}1$).
- f_i denotes the finish time of job J_i.
- d_i denotes the due date of take J_i.

2.2. Game modeling for FJSP

The FJSP addressed in this paper is modelling into an $N+1$-person non-cooperative game with complete information where each job and the manufacturer act as players in the game. They make the decision of the processing strategies (i.e., selection of suitable machines and the sequence entering the system) to achieve their goal individually. The model is expressed as a tuple: $G = (N+1, S, U)$, where $N+1$ denotes number of the players in the game, S denotes the strategy profile of all players, $S = \{s_0, s_1, \ldots, s_{n-1}, s_n\}$, where s_i denotes the strategy of player i, and U denotes the payoff, $U = \{u_0, u_1, \ldots, u_n\}$, where u_i denotes the payoff of player i.

2.2.1. Players.
In the job scheduling game, n jobs come from different customers' orders is regarded as n players, and the manufacturer is regarded as the $(n+1)^{th}$ player.

2.2.2. Strategies.
Each job has several operations. Each operation should be assigned to a suitable machine in order to optimize the scheduling objectives. s_i represents the strategy set of player i. $(i=0, \ldots, n-1)$, where $(j=0, \ldots, h_i-1)$. Assumed J_i has 3 operations and there are 3 alternative machines for each operation, the number of the possible strategies for the job is 27. $s_n = \{l_1, l_2, \ldots, l_n\}$ means a sequencing of integer of 1 to n.

2.2.3. Payoffs.
In the scheduling game, each job tries to schedule their operations on correspondent machines to be processed to maximize its own profit. On one hand, in the customers' position, each customer's order should be delivered as quickly as possible. On the other hand, in the manufacturer's position, the manufacturing efficiency should be as high as possible. Therefore, in the game, the payoff of player i ($i = 0, \ldots, n-1$) is defined to be the decreasing function of tardiness of job i, and the payoff of play n is defined to be the decreasing function of makespan of all the jobs. It is noticeable that profit of each player is affected not only by its own scheduling strategies but also by the strategies of other players.

$$u_i = u_i(S) = 1/L_i \quad (i = 0, \ldots, n-1) \tag{1}$$

Where $L_i = \max\{0, f_i - d_i\}$

$$u_n = u_n(S) = 1/\max\{f_i \mid i = 0, \ldots, n-1\} \tag{2}$$

2.2.4. Nash equilibrium.
Nash equilibrium (NE) is applied as the solution for $N+1$-person non-cooperative game. A NE point is an $N+1$-tuple of strategies, one for each player, such that anyone who deviates from it unilaterally can impossibly improve its expected payoff. Therefore, a NE point is repressed as $S* = \{s_0^*, s_1^*, \ldots, s_{n-1}^*\}$. It is satisfied the equation $u_i(s_i^*, s_{-i}^*) \le u_i(s_i, s_{-i}^*)$, $i = 0, \ldots, n$, where $s_{-i}^* = \{s_0^*, s_1^*, \ldots, s_{i-1}^*, s_{i+1}^*, \ldots, s_{n-1}^*\}$.

According to the definition above, scheduling jobs are translated into searching the NE of the job scheduling game. In this paper, GA is incorporated into the approach to search the NE point of this game. The GA mechanisms will be described in detail in the next section.

3. Proposed approach based on GA

Since the NP complexity of the game model, GA is used to search for the NE point of this game. In the paper, the traditional process of GA is employed. Firstly, set the parameters for the algorithm. Then, generate an initial population randomly and evaluate every individual's fitness. Thirdly, generate a new generation of the population through applying the genetic operators to the current generation of the population. Finally, judge if the termination criteria are satisfied. If the criteria are satisfied the algorithm is terminated, otherwise, iterations continues by transferring to the third step. Various basic genetic operators, such as roulette selection method, two-point crossover, precedence operation crossover, neighbourhood mutation and swapping mutation are used in the paper. The subsection only focuses on the discussion of encoding and decoding scheme and fitness function because of the limitation of length.

Table 1. Data of a simple FJSP

Job	Ope	Num	Alternative machines			Processing time		
J_0	$O_{0,0}$	2	M_0	M_2	-	5	4	-
	$O_{0,1}$	3	M_0	M_1	M_2	2	4	5
J_1	$O_{1,0}$	2	M_0	M_1	-	4	4	-
	$O_{1,1}$	3	M_0	M_1	M_2	5	2	1
	$O_{1,2}$	2	M_1	M_2	-	2	3	-
J_2	$J_{2,0}$	3	M_0	M_1	M_2	3	2	2
	$J_{2,1}$	2	M_0	M_1	-	1	2	-

3.1. Encoding and Decoding Scheme

Encoding means to describe a feasible solution of the scheduling problem into a "chromosome" of the GA population. Decoding is the inverse process of encoding. Each chromosome is designed to compose of $N+1$ parts if the FJSP problem includes n jobs. As for each job, its correspondent part of the chromosome depicts its strategy to assign its operations to alternative machines. As for the manufacturer, its correspondent part of the chromosome determines the permutation of all the jobs. According to the instance in Table 1, the integer string of {1,0;1,2,0;2,0;0,1,2} is a chromosome example. Since the problem includes 3 jobs, the chromosome is composed of 4 parts. The first part 1,0 is the strategy of the first job J_0. It means that the first operation of the job, i.e. $O_{0,0}$ is assigned to its second alternative machine M_2 and second operation $O_{0,1}$ is assigned to its the first alternative machine M_0. The second part 1,2,0 is the strategy of the second job J_1. It means that the first operation of the job, i.e. $O_{1,0}$ is assigned to its second alternative machine M_1, the second operation $O_{1,1}$ is assigned to its third alternative machine M_2, and the third operation $O_{1,2}$ is assigned to its first alternative machine M_1. The third part 2,0 is the strategy of the third job J_2. It means that the first operation of the job, i.e. $O_{2,0}$ is assigned to its third alternative machine M_2 and second operation $O_{2,1}$ is assigned to its the first alternative machine M_0. The last part 0,1,2 is the strategy of the manufacturer. It means the 3 jobs are released to the system in the sequence of 0,1,2.

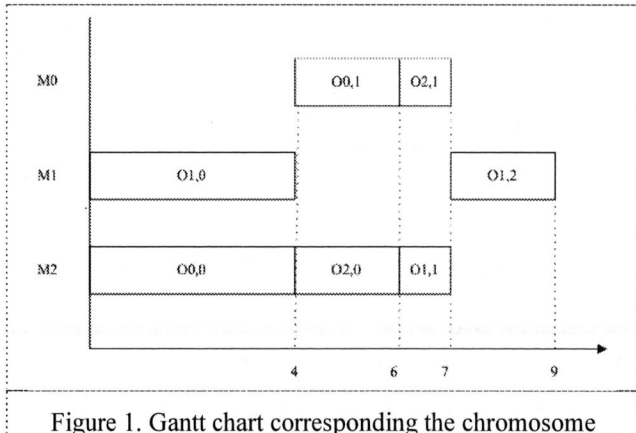

Figure 1. Gantt chart corresponding the chromosome

The Gantt chart corresponding the chromosome is shown in Fig 1. Obviously, according to this encoding and decoding method, it is easy to convert a chromosome into a schedule. And it is also convenient to carry out various genetic operators on the chromosome without generating infeasible offspring.

3.2. Fitness function

In order to evaluate the individual in the population, the fitness function is designed as below:

$$F = 1/\max(f_i \mid i = 0,...,n-1) + 1/\max(L_i \mid i = 0,...,n-1)] \tag{3}$$

Where, F denotes the fitness of an individual. It is found that an individual with a better makespan and better tardiness for each jobs will be assigned a higher fitness and the individual will survive with a higher probability in the iteration.

In the paper, the parameters for GA are listed below: The size of the population is 100, the time of iterations is 100. The probability of crossover and mutation operation is 0.8 and 0.2, respectively.

4. Experiment and results

Some instances in [18] are used as benchmarks in the paper. The due date of each job is generated randomly according to the equation.

$$d_i = r_i + k\underline{p}_i (i = 0,...,n-1) \tag{4}$$

Where k is 0.5 or 1, and $\underline{p}_i$ is the minimum processing time of Job i.

Table 2 summaries the results. The column of "Makespan" and "Tardiness" are optimal result obtained by the approaches. From Table 2, it is found that the pressure of due date increases with the value of k decrease. The proposed approach could make a good balance between makespan and tardiness, which means the interests of customers and the manufacturer gain the equilibrium. For example, the optimal makespans and tardiness of the instance of TFJSP 15*10 are 30 and 23 when $k = 1$. Whereas, the optimal makespan and tardiness are 29 and 27 when $k = 0.5$. Although the latter is worse than the former in the aspect of tardiness, the latter's improve the makespan. The Gantt chart of TFJSP 10*10 and TFJSP 15*10 when $k = 1$ are shown in Fig 2 and 3.

Table 2. Strategy and the payoff under NE.

Instance	$k=1$		$k=0.5$	
	Makespan	Tardiness	Makespan	Tardiness
TFJSP 10*7	19	12	19	15
TFJSP 10*10	14	11	14	12
TFJSP 15*10	30	23	29	27

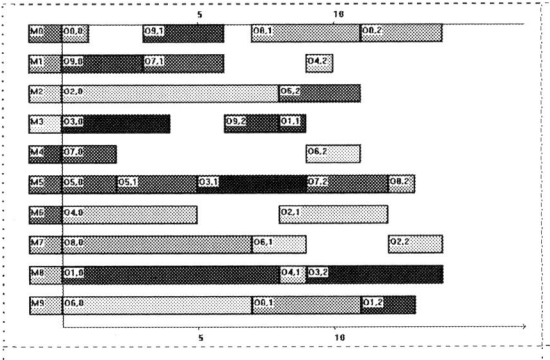
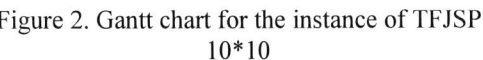

Figure 2. Gantt chart for the instance of TFJSP 10*10

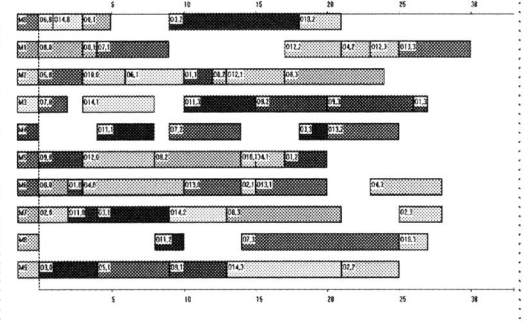

Figure 3. Gantt chart for the instance of TFJSP 15*10

5. Conclusion

It is important to assign jobs to the manufacturing facilities of complex manufacturing systems to optimize the objective of maximizing the manufacturer's efficiency and the objective of maximizing the customer's delivery satisfaction simultaneously. However, the objectives of two aspects are usually conflicting each other. In the paper, the FJSP with the objective of minimizing the makespan and maximum tardiness of all jobs is regarded as a game, and concept and method of game theory is employed to establish a mathematical model and a GA-based algorithm is designed to find the NE solution. Jobs from several costumers' order are players in the game and try to minimize its tardiness. On the other hand, the manufacturer is also regarded as a player in the game and tries to minimize the global makespan. In order to find the NE solution of the scheduling problem, the encoding and decoding scheme is developed and the fitness function is designed base on the principle of GA. The proposed GA-based approach is tested on several instances. It is found that the approach has the ability to schedule jobs with good strategies in both the position of costumers and the manufacturer, which validate its effectiveness. However, in practice there are many other conflict objectives in different position of customers and/or manufacturers. Nash Equilibrium is only one commonly used strategy to solve this kind of problem. Stackelberg strategy is a leader-follower type solution that works well in a situation where one player dominates over the other in the process of decision-making. To use the Stackelberg strategy of GT to develop optimizing approaches for FJSP is one of the future research directions.

Acknowledgments

The authors would like to thank the editor and anonymous referees for their valuable comments. This research is supported by the National Natural Science Foundation of China under Grant No. U1537110 and 51605273 and the National Key R&D Program of China under Grant No. YS2017YFGH000967.

References

[1] Seebacher, G., Winkler, H. (2014) Evaluating flexibility in discrete manufacturing based on performance and efficiency. Int.J.Prod.Econ., 153: 340–351.

[2] Chaudhry, I.A., Khan,A.A. (2016) Aresearch survey: review of flexible job shop scheduling techniques. Int.Trans.Oper.Res., 23: 551–591.

[3] Gu, X., Huang, M., Liang, X. (2017) The improved simulated annealing genetic algorithm for flexible job-shop scheduling problem. In: 6th International Conference on Computer Science and Network Technology, Dalian, pp. 22-27.

[4] Li, X., Gao, L. (2016) An effective hybrid genetic algorithm and tabu search for flexible job shop scheduling problem. Int. J. Prod. Econ., 174: 93-110.

[5] Xu, S., Wu, D., Kong, F., Ji, Z. (2017) Solving Flexible Job-Shop Scheduling Problem by Improved Chicken Swarm Optimization Algorithm. Journal of System Simulation, 29: 1497-1505.

[6] Ning, T., Guo, C., Chen, R., Jin, H. (2016) A novel hybrid method for solving flexible job-shop scheduling problem. Open Cybernetics and Systemics Journal, 10: 13-19.

[7] Rahmati, S.H.A., Ahmadi, A., Karimi, B. (2018) Multi-objective evolutionary simulation based optimization mechanism for a novel stochastic reliability centered maintenance problem. Swarm Evol. Comput., 40: 255-271.

[8] Zhang, Y.F., Wang, J., Liu, Y. (2017) Game theory based real-time multi-objective flexible job shop scheduling considering environmental impact. J. Clean Prod., 167: 665-679.

[9] Krenczyk, D., Olender, M. (2015) Simulation Aided Production Planning and Scheduling Using Game Theory Approach. Applied Mechanics and Materials, 809-810: 1450-1455.

[10] Sun, D.H., He, W., Zheng, L.J., Liao, X.Y. (2013) Scheduling Flexible Job Shop Problem Subject to Machine Breakdown with Game Theory. Int. J. Prod. Res., 52: 3858-3876.

[11] Zheng, X., Zhang, J., Gao, Q. (2012) Application of non-cooperative game theory to multi-objective scheduling problem in the automated manufacturing system. In: International Conference on Automatic Control & Artificial Intelligence IET, Xiamen, pp. 554-557.

[12] Kacem, I., Hammadi, S., Borne, P. (2002) Pareto-optimality approach for flexible job-shop scheduling problems: hybridization of evolutionary algorithms and fuzzy logic. Math. Comput. Simul., 60: 245-276.

ISPECE   IOP Publishing

IOP Conf. Series: Journal of Physics: Conf. Series **1187** (2019) 032096   doi:10.1088/1742-6596/1187/3/032096

An Improved Hybrid Structure Multi-classification Support Vector Machine

Zhang Xiaoyan[1*]**, Wang Qiuqiu**[1]

[1] College of Computer Science, Xi'an University of Science and Technology, Xi'an,71000, China

*Corresponding author's e-mail: zhang_xy@xust.edu.cn

Abstract. In order to improve the speed of multi-class support vector machine, based on One-versus-One SVM, the method of combining hierarchical classification is proposed which can reduce the number of classifiers during training and testing, and use the inter-class separation degree, the intra-class sample distance, and the intra-class sample distance standard deviation as the classification measures to divide the subset of binary classification and then form the binary tree structure. Finally, the 1-v-1 training is performed on the subclasses respectively. Experiments show that compared with the traditional 1-v-1 SVM, this method can effectively shorten the time required for classification and reduce the influence of error accumulation of H-SVMs.

1. Introduction

Support Vector Machine(SVM) is a traditional machine learning algorithm which was originally used in the two-category problem. With the development of market, its multi-category expansion has emerged. At present, there are several methods to extend SVM to the field of multi-class classification, which can be roughly divided into two categories: direct and indirect method. The direct method is the most ideal classification method, and its thought of realization is to find multiple hyperplanes directly, solving the classification demand at one time. This kind of method has higher accuracy. However, the method is limited by the fatal weakness of the complicated implementation process and the large amount of calculation [1]. Now, the implementation of multi-classification mainly adopts the indirect method. The implementation process of the indirect method is simple and the calculation amount is small. The main idea is to transform the original problem into a problem that can be directly solved by multiple SVMs, and finally realize the Multi-classification by the combination of multiple two categories [2][3].

In the development of multi-class support vector machines, the most commonly used classification methods are 1-v-r (short for one-versus-rest), 1-v-1 (short for one-versus-one), DAG-SVM (short for Directed Acyclic Graph-Support Vector Machine) and H-SVMs (short for Hierarchical Support Vector Machines). According to the slowness in training speed of 1-v-1 method and the error accumulation phenomenon of hierarchical support vector machine, this paper proposes a support vector machine model combines H-SVM and 1-v-1, which uses the binary tree structure of H-SVM to divide the original data in the early stage. At the primary classification, considering comprehensively about the inter-class separation degree and the intra-class sample distance, it is proposed to use the intra-class sample distance standard deviation to weaken the outliers, which could separate the categories with higher discrimination and reduce the error rate of the initial classification. After converting the multi-classification problem into the multiple two-category problem, the 1-v-1 training is performed on the sub-category respectively.

Content from this work may be used under the terms of the Creative Commons Attribution 3.0 licence. Any further distribution of this work must maintain attribution to the author(s) and the title of the work, journal citation and DOI.
Published under licence by IOP Publishing Ltd

Therefore, it reduces the number of classifiers and improves computational efficiency while ensuring high classification accuracy.

2. Multi-classification SVM

The SVM was originally designed for the two-class problem and has good robustness，the goal of which is to find a hyperplane in the sample space that can separate the samples by category and maximize the distance from the sample point to the hyperplane. Currently, the main idea of using SVM to solve multi-classification problems is to convert the original problem into a two-category problem that can be directly solved by multiple SVMs, and finally combine multiple two-categories into multiple classifications. Here are two methods [1]:

2.1. 1-v-1 SVMs

One-versus-One. A classifier is trained for each of the two classes in K Classes, with a total of $k(k-1)/2$ classifiers. When classifying, each classifier classifies unknown samples and votes on the corresponding results, and the category that ends up with the largest number of votes is the final classification result. Which was used in the multi-classification of LIBSVM. The method is simple, the precision is high, there is no sample imbalance phenomenon, but the number of classifiers constructed and used is more, and with the increase of the number of categories, the number of classifiers increases exponentially.

2.2. H-SVMs

Hierarchical SVMs, hierarchical support vector machines [7], which divides all categories into two subclasses, and then divides the subclasses into two subclasses, repeating this step until a single category is finally obtained. This method decomposes the original multi-classification problem into a two-category sub-problem. In an ideal case, only $k-1$ classifiers need to be constructed, and only $\log_2 k$ classifiers are used for classification. This is a structure similar to a decision tree. However, because of the exclusion strategy, if a stage classification error exits, the next steps would have no meaning, which easily leads to error accumulation, causing classification performance.

In a hierarchy, the sample set needs to be divided into two subsets at each layer. If the degree of differentiation between the subsets is not enough, it will seriously affect the accuracy of subsequent classification results. Therefore, how to classify subclasses has always been the main research content in H-SVMs. Using the distance between two classes or the distribution of samples within each class to measure the degree of separation between classes [6][9] is a common method when dividing subclasses. However, this method is prone to errors when the number of samples in the sample varies considerably, or when there are outliers in the sample in the class. As shown in Figure 1, where the solid line is located is the optimal division, and S1 and S2 are classified into one subclass. However, due to the existence of two outliers in the S1, resulting in the actual line will be shown in the dotted line of S2, S3 together, which is not the ideal division.

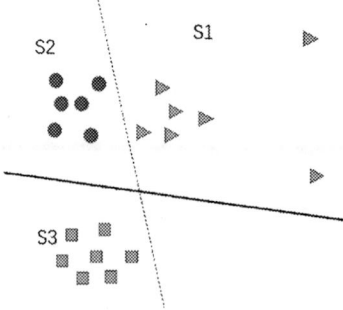

Figure 1. Based on the division of distance between classes when there have outliers

3. Improved Hybrid Structure Multi-class SVM

The 1-v-1 SVM adopts the method of voting, which can solve the problem of overlapping and inseparable classification to a certain extent, and have a higher classification accuracy. However, the number of classifiers used in its training and testing stages is higher and the speed is slower. The number of classifiers of 1-v-1 SVM is related to the number of categories. When there are k categories in the sample, the number of classifiers is k(k-1)/2. As the number k increases, the number of classifiers increases exponentially. The H-SVMs method splits the original multi-classification problem into a series of two-class sub-problems in the structure of a binary tree, and only needs to train k-1 classifiers, that quantity is much smaller than the number of classifiers in 1-v-1 SVM. A support vector machine based on hybrid binary tree structure [4] also shows that this tree structure can improve classification efficiency and reduce training time. If the two classifications are combined, the number of classifiers for subsequent classifications can be significantly reduced.

The use of hierarchical classification will cause problems of error accumulation. If the previous classifier has a classification error, the subsequent ones cannot be corrected, and as the depth of the binary tree increases, the error rate will be higher. Assuming that the error rate of each layer is fixed at e, then the probability that a sample will be correctly classified is:

$$T=\prod_{n=1}^{d}\left(1-e\right) \tag{1}$$

Where d is the number of layers. It can be seen that when the hierarchy is less, the correct rate is higher, so it is necessary to reduce the number of divisions. In fact, as long as a division can effectively reduce the number of subsequent 1-v-1 classifiers. In the best case, only k(k-2)/4+1 classifiers need to be trained, and the test only uses k(k-2)/8+1 classifiers. As shown in Figure 2:

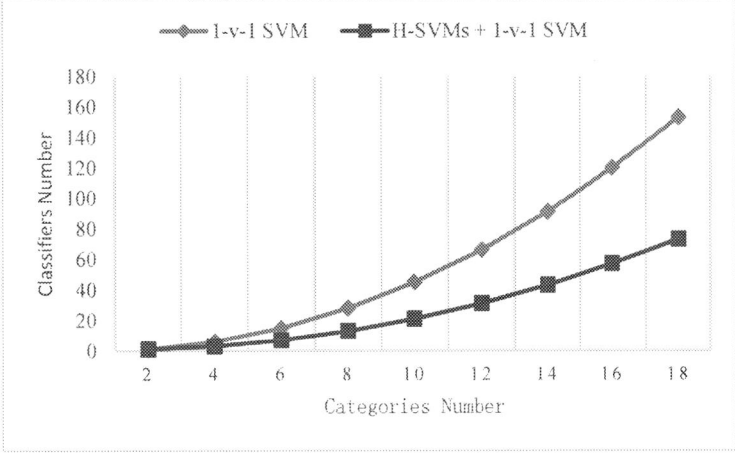

Figure 2. Relationship between the number of categories and the number of classifiers

As can be seen from Figure 2, the number of classifiers of 1-v-1 SVM increases exponentially with the increase of the number of sample categories. However, when adding a hierarchical classification and dividing it only once, the number of initial categories in 1-v-1 training can be reduced, and the number of total classifiers can be effectively reduced.

But how to carry out the first category division is related to the final classification accuracy. If the two types of divisions are unreasonable, classification errors will occur at the hierarchical classification stage, which will directly affect the final accuracy. This requires that when the sample is divided, the more differentiated classes are divided into different classes[3], and the less differentiated classes are divided together. In this paper, a method of considering the sample distance between classes and the sample distribution within the class is proposed, and the original samples are reclassified into two categories.

3.1. Classification Measurement Methods between Classes

For two classes A and B in the original sample set, where $A = \{a_1, \cdots, a_m\}, B = \{b_1, \cdots, b_m\}$. First calculate the average distance within the class, here with A as an example:

1) Calculate the Euclidean distance d_{ij}^A between any two sample points a_i, a_j in the class.

2) Calculate the average distance between a_i and other samples.

$$d_i^A = \frac{1}{m-1} \sum\nolimits_{j=1}^m d_{ij}^A, \, i \neq j \tag{2}$$

3) Calculate the average distance between all samples.

$$d_A = \frac{1}{m} \sum\nolimits_{j=1}^m d_i^A \tag{3}$$

4) Calculate the mean distance standard deviation between all samples.

$$sd^A = \left(\frac{1}{m} \sum\nolimits_{i=1}^m \left(d_i^A - d^A \right)^2 \right)^{\frac{1}{2}} \tag{4}$$

5) Calculate the ratio of the average distance to the standard deviation of the average distance.

$$E^A = \frac{d^A}{sd^A} \tag{5}$$

By calculating the ratio E, the true distribution of samples within a class can be better reflected, and the influence of individual outliers on the distribution of samples within the class is effectively reduced. After calculating the average distance in each category, calculate the distance between classes. Here, take A and B as examples:

1) Calculate the Euclidean distance d_{ij}^{AB} between each sample in A and B.

2) Calculate the average distance between the samples in A and each sample in B.

$$d_i^{AB} = \frac{1}{n} \sum\nolimits_{j=1}^n d_{ij}^{AB}, \, i \in \{1, 2, \cdots m\} \tag{6}$$

3) Calculate the distance between the two classes.

$$d^{AB} = \frac{1}{m} \sum\nolimits_{j=1}^m d_i^{AB} \tag{7}$$

Finally, according to the principle that the average distribution range within the class is small and the distance between classes is small, the classification measures is defined as follows:

$$c^{AB} = d^{AB} + \beta \left(E^A + E^B \right) \tag{8}$$

Where β is the weight, and the weight between the distance between classes and the distance within the class is adjusted. In this way, c^{AB} comprehensively considers the sample distance between classes and the sample distribution within the class. As a measure of dividing subclasses, the smaller the c^{AB}, the more the two classes should be divided together. Otherwise, the two classes with larger c^{AB} should be distinguished. This can improve the classification accuracy of subsequent stages.

3.2. Specific steps for hybrid structure multi-class SVM

The steps of the hybrid structure multi-classification support vector machine combined with the hierarchy structure are as follows:

Define the sample set X, y, where y is the original class label.

Step1: Calculate sd, E, d, and c between classes according to the inter-class classification metric in 3.1.

Step2: Construct a symmetric matrix $C = c^{i,j}$, $i, j \in \{1, \cdots, k\}$ representing the classification metric between the classes.

$$C = \begin{bmatrix} 0 & c^{1,2} & \cdots & c^{1,k-1} & c^{1,k} \\ c^{2,1} & 0 & \cdots & c^{2,k-1} & c^{2,k} \\ \vdots & \vdots & \ddots & \vdots & \vdots \\ c^{k-1,1} & c^{k-1,2} & \cdots & 0 & c^{k-1,k} \\ c^{k,1} & c^{k,2} & \cdots & c^{k,k-1} & 0 \end{bmatrix} \quad (9)$$

Where k is the number of categories, and when $i = j$, $c^{ij} = 0$.

Step3: Sum each row of matrix C and sort by value, classify classes with larger values into one subclass, and the rest are another subclass. The original sample set X is thus divided into two parts X_1, X_2, $X_1 + X_2 = X$.

Step4: Take X_1 as a positive sample and X_2 as a negative sample, and perform the first SVM training to obtain the classifier M_0.

Step5: Perform 1-v-1 SVM training on each class in X_1 to obtain a classifier set M_p.

Step6: For X_2, repeat step 4 and step 5 to obtain the classifier set M_n.

During the test, the test sample T is first classified by the classifier M_0. If the classification result is a positive sample, then M_p is used for classification voting; if it is negative sample, M_n is used for classification voting. Finally, the final classification results are obtained based on the number of votes.

Take the 6 classes label as an example, the entire combination of hierarchical multi-class SVM structure is shown in Figure 3:

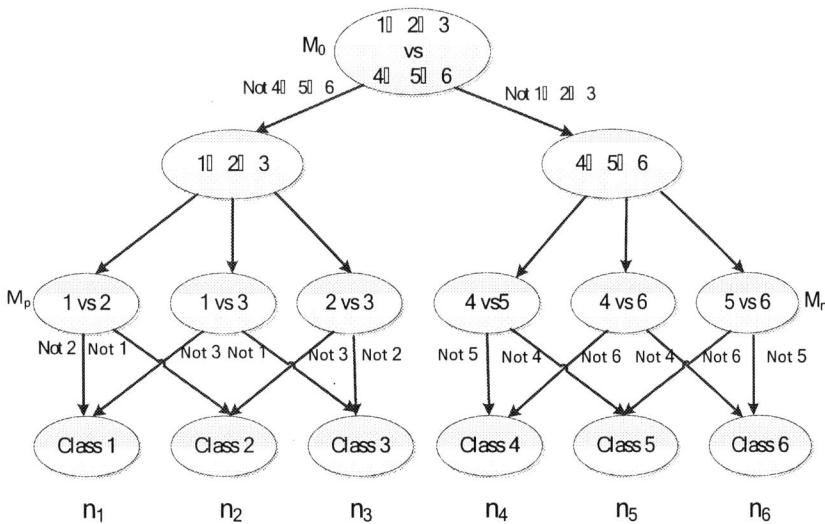

Figure 3. SVM classification combined with hierarchy

4. Experiment and results

In order to verify the performance of the methods studied in this paper that in the case of different categories and data volumes, we performed experiments in 7 datasets which are the wine, glass, iris, pendigits, vowel of the UCI database and vehicle, segment of the Statlog database. The properties of each dataset are shown in Table 1:

Table 1. Dataset Properties

Dataset	Categories Number	Feature Number	Data Volume
Iris	3	4	150
Wine	3	13	178
Vehicle	4	18	846
Glass	6	9	214
Segment	7	19	2310
Vowel	11	10	990
Pendigits	10	16	10992

The environment used in this experiment is: Windows 10 64bit, 8GB RAM, CPU Main frequency 2.60GHz, Python 2.7. The kernel function uses the RBF radial basis kernel function. In H-SVM, subclasses are divided according to the measurement proposed in chapter 3.1. The overall results of the experiment are shown in Table 2 and Table 3:

Table 2. Comparison of classification accuracy between this method, 1-v-1 SVM and H-SVM

dataset	1-v-1 SVM	H-SVM	this method
Iris	100%	100%	100%
Wine	94.8%	93.0%	94.8%
Vehicle	75.8%	75.2%	75.7%
Glass	60.2%	57.8%	59.5%
Segment	88.7%	87.4%	88.3%
Vowel	51.6%	49.7%	51.2%
Pendigits	89.4%	85.1%	87.2%

Table 3. Comparison of the method and the 1-v-1 SVM operation time (ms)

dataset	1-v-1SVM	this method	time variant
Iris	307	248	-19.2%
Wine	429	339	-21.0%
Vehicle	9567	7131	-25.5%
Glass	1562	1142	-26.9%
Segment	25263	18584	-26.4%
Vowel	10479	7899	-24.6%
Pendigits	207889	170244	-18.1%

As can be seen from Table 2, the H-SVM method has the lowest accuracy, especially when the number of categories increases, the performance decreases significantly. This difference is determined by the structure of the hierarchical support vector machine, which can only be reduced by optimizing the method of dividing subclasses. Compared with the 1-v-1 method, the accuracy of this method is slightly lower, but in Table 3, it can be seen that the calculation speed is significantly improved, and the time taken is reduced by about 20%.

Combined with table 1 and table 3, it can be found that with the increase of the number of categories, the operation speed of this method is increasing gradually, but in the Pendigits data set, the increase rate is decreased because the amount of data in the Pendigits data set is relatively large. All the data is used when training the first classifier M_0, so the operation time is longer.

5. Conclusion

Based on the research of 1-v-1 SVM and H-SVMs, combined with the advantages and disadvantages of each, this paper proposes a hybrid structure multi-classification support vector machine combined with hierarchical classification, which reduces the number of classifiers in training while ensuring the classification accuracy, effectively reducing the computation time and improving the speed. The influence of outliers is weakened by using the standard deviation of sample distance in the class, and the classification accuracy is improved. However, this method still has problems that the initial classification error can not be corrected, resulting in a slight loss of classification accuracy. Therefore, what needs to be studied next is how to optimize the hierarchical classification method and structure, further improve the accuracy, or use the sorting method to reduce the number of classifiers and further improve the speed.

References

[1] Liu ZhiGang, Li Deren, et al. An Analytical Overview of Methods for Multi-category Support Vector Machines[J]. Computer Engineering and Applications, 2004, 40(7):10-13.

[2] Sun Shaoyi, Huang Zhibo. A multi-class SVM classification algorithm[J]. MICROCOMPUTER ITS APPLICATIONS，2016, 35(8):12-14.

[3] Duan Xiusheng, Shan Ganlin, Zhang Qilong. Hierarchical Support Vector Machine for Multi-class Classification[J]. JOURNAL OF ORDNANCE ENGINEERING COLLEGE, 2009, 21(1):64-66.

[4] Leng Qiangkui, Liu Fude, Qin Yu-ping. Multi-class Classification Algorithm for SVM Based on Hybrid Binary Tree Structure[J], Computer Science, 2018, 45(5).

[5] Zhou Zhihua. Machine Learning[M].

[6] Tang T, Chen S, Zhao M, et al. Very large-scale data classification based on K-means clustering and multi-kernel SVM[J]. Soft Computing, 2018:1-9.

[7] Wang Xiuhua, Qin Zhenji. A Support Vector Machine Model Based On Hierarchical K-means Clustering[J]. Computer Applications and Software, 2014(5):172-176.

[8] Liang Shengzhuo, Xie Wenxiu，Li Mang. Improved 1-v-1 Multi-class Classification Algorithm[J],Journal of Nanchang University(Natural Science) , 2013, 37(3):287-289.

[9] Tang Faming, Wang Zhongdong, Chen Mianyun. On Multiclass Classification Methods for Support Vector Machine[J]. CONTROL AND DECISION, 2005,20(7):746-749.

Hand-eye calibration for flexible manipulator

Jiahao Li[1,2*], Xiuzhi Li[1,2], Ao Dun[1,2], Songmin Jia[1,2] and Yanjun Sun[1,2]

[1]Faculty of Information Technology, Beijing University of Technology, Beijing 100124, China;

[2]Beijing Key Laboratory of Computational Intelligence and Intelligent System.

Email: 825334787@qq.com

Abstract. This paper studies the determination of the pose relationship between camera which is embedded in the Barrett hand and end-effector of the UR5. It is fundamental for subsequent visual control. Least-square method and Lie group is used in this paper, and we also come up with an algorithm to increase the quantity of data for least-square to improve the accuracy of eye-in-hand calibration result. Firstly, through camera calibration, the intrinsic and external parameters of the camera are obtained. Secondly, we obtain equation of $CX = XD$ by eye-in-hand calibration, and X not only is the matrix we want to compute but also represents the transformation relationship between camera and end-effector. Finally, to evaluate the proposed method, an experiment was designed, and the test result demonstrates the effectiveness of the proposed approach.

1. Introduction

Visual guided localization of industrial robots is an important part of industrial production intellectualization [1]. Accurately establishing the geometric position relationship between the visual sensor and the end-effector of the manipulator is the key to the precisely locating of the target by the manipulator. Hand-eye calibration methods fall into two different categories. One of the hand-eye calibration methods is eye-to-hand [2] and the mounting of camera is independent of the robot in this case. Another is eye-in-hand [3] calibration which means that camera is installed on board and moves with robot. The eye-to-hand configuration faces practical difficulties, such as fixed capture scenes, easy occlusion and so on [2].

This paper deals with the eye-in-hand configuration because it is more flexible for robots to actively grasp objects [4]. In the past decades, many researchers were devoted to studying eye-in-hand calibration. The basic method [5] of robotic eye-in-hand calibration is controlling a known calibration reference for robotic grippers in different observation spaces. Generally speaking, traditional eye-in-hand calibration solves the relative position between two coordinate systems by cumbersome procedures. We come up with a method to resolve this problem, by assembling different group of calibration images to new combinations. As noted in article [6], it is difficult to avoid singularity when computing the equation of matrix. Inspired by [7], if the camera move with a larger offset angle between the optical center during the camera calibration, the problem can be solved.

2. Methodology

2.1 Eye-in-hand calibration system

Content from this work may be used under the terms of the Creative Commons Attribution 3.0 licence. Any further distribution of this work must maintain attribution to the author(s) and the title of the work, journal citation and DOI.

Published under licence by IOP Publishing Ltd

The goal of machine vision calibration is to compute the transformation relationship between image coordinate system and robotic coordinate system which is shown in Figure 1. Camera calibration provides the transformation relationship between the world coordinate system and the camera coordinate system. Hand-eye calibration offers the transformation relationship between the camera coordinate system and the robotic coordinate system, so we should divide the process of hand-eye calibration into two steps, camera calibration and eye-in-hand calibration.

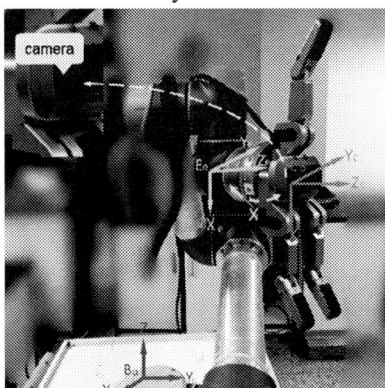

Figure 1. UR5 and Barrett Hand

2.2 Camera calibration

The most popular camera calibration is Zhang's method [8]. This method is highly accurate and easy to implement. Calibration results can be evaluated according the undistorted images via intrinsic parameters.

2.3 Eye-in-hand calibration

Owing to high cost and complex operation, direct measuring [8] approach has not been applied widely. An alternative solution is to compute the transformation parameters between the robot and the camera which is term with called eye-in-hand calibration.

The common method of eye-in-hand calibration is to construct and solve the $CX = XD$ equation by means of three or more pose transformations of the robotic end-effector [7]. The transformation of system coordinate can be demonstrated in Figure 2. Among them, C_{obj} represents the world coordinate system, C_{c1} and C_{e1} represent the camera coordinate system and the end-effector coordinate system before the platform moving, C_{c2} and C_{e2} represent the two coordinate systems after platform moving. Camera can be calibrated with the fixed chessboard at the position of C_{c1} and C_{c2} respectively to obtain its internal and external parameters. The camera external parameter is the relative position between camera and chessboard in C_{c1} and C_{c2} positions, expressed by A and B. If relative transformation between C_{c1} and C_{c2} can be represented by D, the $D = BA^{-1}$.

Due to the matrix E and F of robotic motion parameters can be read by the feedback of controller, the relative position of C_{e1} and C_{e2} are known parameters represented by matrix C. We can acquire matrix E and F, and $C = F^{-1}E$. The matrix of X can be computed by equation of $CX = XD$.

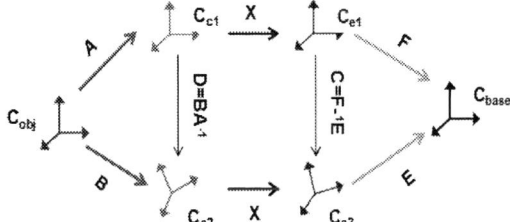

Figure 2. Transformation relationship of system coordinate

The equation of $CX = XD$ can be written as follows:

$$\begin{bmatrix} R_C & t_C \\ 0^T & 1 \end{bmatrix} \begin{bmatrix} R & t \\ 0^T & 1 \end{bmatrix} = \begin{bmatrix} R & t \\ 0^T & 1 \end{bmatrix} \begin{bmatrix} R_D & t_D \\ 0^T & 1 \end{bmatrix} \tag{1}$$

As a result, the final expression can be obtained:

$$R_C R = R R_D \qquad R_C t + t_C = R t_D + t \tag{2}$$

In the upper expression, R_C, R_D, t_C, t_D are known parameters, and R, R_C, R_D are unit orthogonal matrix. R and t are parameters to be computed.

Due to the observation noise in camera calibration process, it is necessary to solve the equation with multiple sets of expression of $CX = XD$ by least squares method in practical measurement. In order to reduce the workload, we come up with an idea of assembling different group of calibration images to new combinations. And in this experiment, we define the left, middle and right orientations of calibration.

In each orientation, the 4 calibration images are collected which is shown in Figure 3, so we can get the 64 equation of $CX = XD$ through 12 calibration results. For example, the No.1 left photo can match with the No.1 to No.4 middle photos and No.1 to No.4 right photos. This step can generate 16 outcomes of X, the four left photos can get 64 outcomes. This method can generate 64 results of equation.

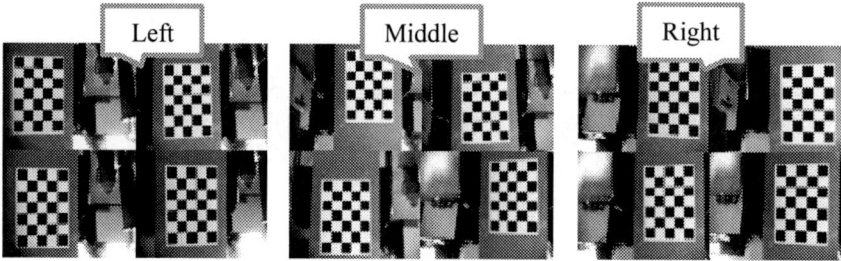

Figure 3. Different group of calibration images

In order to obtain a unique solution to (11), two pairs of (C_i, D_i) whose rotational parts satisfy certain conditions are required.

The previous sections assumed that in determining a unique solution to $CX = XD$ no noise was present in the measured values for C and D. Unfortunately, this assumption is physically unrealistic. A more practical approach is to find some type of "best-fit" solution from a set of noise measurements $\{(C_1, D_1)....,(C_i, D_i)\}$ and to find $X \in SE(3)$ that minimizes an error criterion of the form

$$\eta = \sum_{i=1}^{k} d(C_i \dot{X}, X D_i)$$

d represents the Euclidean distance. Using the knowledge of Lie group theory [9], the above minimization problem can be converted to the least squares fitting problem finally. When there are multiple sets of observations (C_i, D_i), the solution of the R are more accurate.

3. Experiment

The experimental platform consists of the UR5 and Barrett Hand. As shown in Figure 1, the camera is embedded in the palm of Barrett Hand, and Barrett Hand is fixed on the end-effector of UR5. The resolution of captured image is 640 (H) × 480 (V). The employed calibration board has 6 × 4 corner points. Experiment platform is shown in Figure 4.

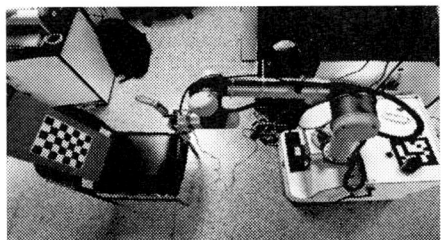

Figure 4. Experiment platform

After 64 sets of equation of $CX = XD$ computing, we can get final result of X matrix as follows:

$$T_X = [3.065, 20.566, 90.568] \quad R_X = \begin{bmatrix} -0.0079 & -0.0091 & -0.0004 \\ 0.1225 & 0.0107 & 0.0059 \\ -0.0017 & -0.0002 & -9.8451 \end{bmatrix} \quad (3)$$

In order to prove the reliability of the experimental results, we choose one corner point of the chess board to locate its position in the UR5-base coordinate system. Firstly, we move the end-effector three times in different position to observe this point by corner matching, at the same time, we can get the distance from camera to this point by laser-range finder embedded in the Barrett Hand. Then we can establish three equations to compute this point that: $(x_1-x_0)^2+(y_1-y_0)^2+(z_1-z_0)^2=d_1^2$; $(x_2-x_0)^2+(y_2-y_0)^2+(z_2-z_0)^2= d_2^2$; $(x_3-x_0)^2 + (y_3-y_0)^2 + (z_3-z_0)^2 = d_3^2$.

Among these equations, A (x_1, y_1, z_1), B (x_2, y_2, z_2), C (x_3, y_3, z_3) is the camera coordinate values in UR5-base coordinate system. P(x_0, y_0, z_0) is selected coordinate frame origin, d_1, d_2, d_3 are values of the distance between camera and the P point. We can obtain P point coordinate value in UR5-base by the above three equation. Then we move camera in another position, comparing d_4 measured by laser-range finder to d_5 which can be computed by distance of camera coordinate point and P coordinate point. We define the $\triangle d$ as the distance error. We also use the point P coordinate value with camera external parameters to compute the P pixel value in camera, comparing it with the P pixel coordinate value of observation. In Table 1, the left column list is number of equations of CX = XD, we also define the $\triangle x$, $\triangle y$, $\triangle z$ as the error of each axis.

Table 1. Error results and analysis

	$\triangle$x(mm)	$\triangle$y(mm)	$\triangle$z(mm)	$\triangle$d(mm)
8	2.65	-10.195	-12.28	16.18
16	13.39	3.797	-7.333	15.73
24	-1.67	8.593	-5.976	10.60
32	-3.41	1.332	-7.68	8.51
40	1.23	1.4	-4.6	4.96
48	-0.1	2.507	-3.876	4.62
56	4.06	1.098	-3.505	5.14
64	1.74	1.9	-3.4	4.26

4. Conclusion

Through eye-in-hand calibration, we can get the position relationship between the camera and the end-effector of UR5, which can better realize the active detection in the grasping process and better make us understand the coordinate transformation relationship between the target object and the UR5. In the process of solving the error problem, we use least squares method and Lie group to eliminate the errors caused by camera calibration. According to the experimental results, we can obtain more accurate values of X by more equation of CX = XD which is shown in Figure 5.

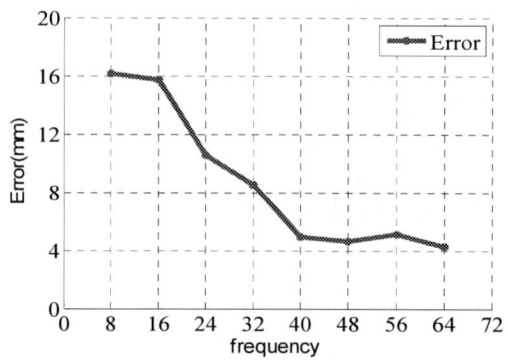

Figure 5. Error of eye-in-hand calibration

Acknowledgments

The research work was supported by 2017 BJUT United Grand Scientific Research Program on Intelligent Manufacturing (040000546317552).

References

[1] Sung, H., Lee, S., & Kim, D. (2013). A robot-camera hand/eye self-calibration system using a planar target. IEEE. International Symposium on Robotics. pp.1-4.

[2] Hu, X., Xie, K., & Peng, T. (2013). Research on direct calibration method of eye-to-hand system of robot. Sixth International Symposium on Precision Mechanical Measurements. International Society for Optics and Photonics.

[3] Delden, S. V., & Hardy, F. (2008). Robotic eye-in-hand calibration in an uncalibrated environment. Journal of Systemics Cybernetics & Informatics. pp.6.

[4] Batista, J., Dias, J., & Almeida, A. T. D. (1993). Monoplanar Camera Calibration - Iterative Multi-Step Approach. British Machine Vision Conference. pp.479--488.

[5] Tsai, R. Y., & Lenz, R. K. (1989). A new technique for fully autonomous and efficient 3d robotics hand/eye calibration. IEEE. Transactions on Robotics & Automation. pp.345-358.

[6] Huang, C., Chen, D., & Tang, X. (2015). Robotic Hand-Eye Calibration Based on Active Vision. IEEE.International Symposium on Computational Intelligence and Design. pp.55-59.

[7] Hongyao Zhang, Li Lun. (2018). Research on robot hand eye calibration and accuracy analysis[J]. Modular Machine Tool and Automatic Machining Technology. pp.69-72.

[8] Fang, X., Qiang, Z., Zou, F., Kai, J., Zhang, Y., & Wang, X. (2017). Error distribution estimation based weighted least square estimation for service robot hand-eye calibration. IEEE. Information Technology and Mechatronics Engineering Conference.

[9] F. C. Park and B. J. Martin. (1994), Robot sensor calibration: solving AX=XB on the Euclidean group. IEEE. Transactions on Robotics and Automation. pp.717-721.

ISPECE

IOP Publishing

IOP Conf. Series: Journal of Physics: Conf. Series **1187** (2019) 032098   doi:10.1088/1742-6596/1187/3/032098

Bounds on the total signed domination number of generalized Petersen graphs $P(n,3)$

Hong Gao[1]*, **Yanan Yin[1]** and **Yuansheng Yang[2]**

[1] Department of Mathematics, Dalian Maritime University, Dalian, Liaoning, 116026, China

[2] School of Computer Science and Technology, Dalian University of Technology, Dalian, Liaoning, 116024, China

*Corresponding author's e-mail: gaohong@dlmu.edu.cn

Abstract. In this paper, the total signed domination number of generalized Petersen graphs $P(n,3)$ is studied. Some total signed dominating functions are constructed. Upon these functions, we get bounds on the total signed domination number for $P(n,3)$ with a small gap.

1. Introduction

Let G be a finite connected simple graph with a vertex set $V(G)$ and an edge set $E(G)$. For an element $x \in V(G) \cup E(G)$, the total closed neighborhood of x, denoted by $N_T[x]$, is $N_T[x] = \{y | y$ is adjacent to x or y is incident with $x, y \in V(G) \cup E(G)\} \cup \{x\}$.

Signed domination and signed edge domination are introduced by Dunbar (1995) [1] and Xu (2001) [2] respectively. Total signed domination is introduced by Lv (2007) [3]. A total signed dominating function of G is a function $f : V(G) \cup E(G) \to \{-1,1\}$, such that $\sum_{y \in N_T[x]} f(y) \geq 1$ for all $x \in V(G) \cup E(G)$. The total signed domination number $\gamma_s^*(G)$ of G is the minimum weight of a total signed dominating function on G. There are many variations of signed domination, so researchers are still studying this subject of graphs [4-6]. Since there are both vertices and edges in the neighborhood, total signed domination is more difficult than signed (vertex) domination and signed edge domination for graphs. Zhou [7] provided the exact value of signed mixed domination number of a complete graph with even order. Shan [8] gave the exact value of total signed mixed domination (signed mixed domination) number of a complete bipartite graph. Generalized Petersen graphs are classical, and many domination parameters of them have been studied [9-10].

In this paper, $\gamma_s^*(P(n,3))$ is studied. Branch and bound method is used to design high-performance algorithms and construct some total signed domination functions. Based on these functions, we get some upper bounds on $\gamma_s^*(P(n,3))$ which are very close to lower bounds.

2. Lower bounds on total signed domination number of $P(n,3)$
According to [3], for any graph G,

$$\gamma_s^*(G) \geq \left\lceil \frac{\delta(G) - \Delta(G) + 1}{\delta(G) + \Delta(G) + 1} (| E(G) | + | V(G) |) \right\rceil_{P(|E(G)|+|V(G)|)} \tag{1}$$

Content from this work may be used under the terms of the Creative Commons Attribution 3.0 licence. Any further distribution of this work must maintain attribution to the author(s) and the title of the work, journal citation and DOI.

Published under licence by IOP Publishing Ltd

ISPECE

and this bound is sharp, where $P(s)$ is defined to be the parity of s, that is, $P(s)$=odd if s is odd and $P(s)$=even if s is even. Based on (1), we can easily obtain lower bounds of $\gamma_s^\bullet(P(n,3))$.

Theorem 1. For any positive integers $n > 6$, we have

$$\gamma_s^\bullet(P(n,3)) \geq \left\lceil \frac{5n}{7} \right\rceil_{P(5n)} . \tag{2}$$

3. Upper bounds on total signed domination number of $P(n,3)$

In this section, we construct some total signed dominating functions. Upon these functions, we get upper bounds on the total signed domination number of $P(n,3)$. The following is the definition for the total signed dominating functions.

Let $G = (V, E)$ be a graph, we define

$$f_n = f(V \cup E) = \begin{pmatrix} f(e_0) & f(e_1) & f(e_2) & \cdots & f(e_{n-1}) \\ f(u_0) & f(u_1) & f(u_2) & \cdots & f(u_{n-1}) \\ f(e_n) & f(e_{n+1}) & f(e_{n+2}) & \cdots & f(e_{2n-1}) \\ f(v_0) & f(v_1) & f(v_2) & \cdots & f(v_{n-1}) \\ f(e_{2n}) & f(e_{2n+1}) & f(e_{2n+2}) & \cdots & f(e_{3n-1}) \end{pmatrix}$$

where u_i, v_i $(0 \leq i \leq n\text{-}1)$ and e_i $(0 \leq i \leq 3n\text{-}1)$ are shown in figure 1 for $P(7,3)$.

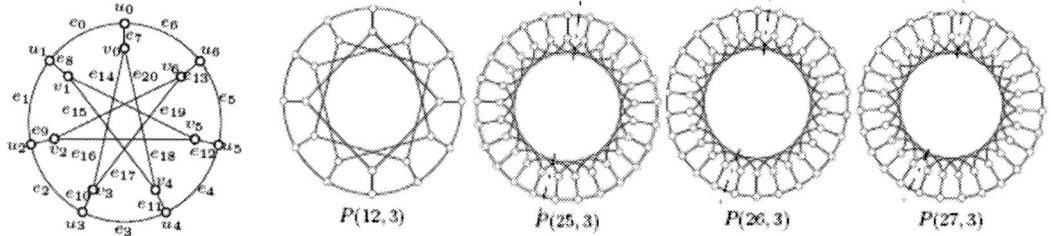

Figure 1. Graph $P(7,3)$.   Figure 2. Graph $P(n,3)$ corresponding to f_n for n=12, 25, 26, 27.

Theorem 2. Let G be a graph $P(n,3)$, and for any integer $n \geq 7$,

$$\gamma_s^\bullet(G) \leq \begin{cases} n - 2\left\lfloor \dfrac{n}{12} \right\rfloor, & n(\text{mod }12) = 0,5,6,7,9,10,11, \\ n - 2\left\lfloor \dfrac{n-12}{12} \right\rfloor, & n(\text{mod }12) = 1,2,3,4,8. \end{cases} \tag{3}$$

Proof.

Case 1. Suppose $n \equiv 0(\text{mod }12)$, $(n \geq 12)$. We construct f_n as follows:

$$f_n = \begin{pmatrix} 1 & 1 & -1 & -1 & -1 & 1 & 1 & 1 & 1 & -1 & 1 & 1 \\ -1 & 1 & -1 & 1 & 1 & -1 & 1 & -1 & 1 & -1 & -1 & 1 \\ -1 & -1 & 1 & 1 & 1 & 1 & -1 & -1 & -1 & 1 & 1 & -1 \\ -1 & 1 & -1 & 1 & 1 & -1 & 1 & -1 & 1 & 1 & 1 & 1 \\ 1 & -1 & 1 & -1 & 1 & 1 & -1 & 1 & -1 & 1 & -1 & 1 \end{pmatrix}^{\frac{n}{12}}$$

Then $w(f_n) = 10 \times \dfrac{n}{12} = n - 2 \times \dfrac{n}{12} = n - 2\left\lfloor \dfrac{n}{12} \right\rfloor$. See figure 2 for $P(12,3)$, where red vertices (edges) stand for $f(x)$=1, and blue vertices (edges) stand $f(x)$=-1. Hence, for $n \geq 12$ and $n \equiv 0(\text{mod }12)$, $\gamma_s^\bullet(G) \leq n - 2\left\lfloor \dfrac{n}{12} \right\rfloor$.

Case 2. Suppose $n \equiv 1(\text{mod }12)$, $(n \geq 13)$. We construct f_n as follows:

$$f_n = \begin{pmatrix} 1 & 1 & -1 & -1 & -1 & 1 & 1 & 1 & 1 & -1 & 1 & 1 \\ -1 & 1 & -1 & 1 & 1 & -1 & 1 & -1 & 1 & -1 & -1 & 1 \\ -1 & -1 & 1 & 1 & 1 & 1 & -1 & -1 & -1 & 1 & 1 & -1 \\ -1 & 1 & -1 & 1 & 1 & -1 & 1 & -1 & 1 & 1 & 1 & 1 \\ 1 & -1 & 1 & -1 & 1 & 1 & -1 & 1 & -1 & 1 & -1 & 1 \end{pmatrix}^{\frac{n-13}{12}} \begin{pmatrix} 1 & 1 & -1 & -1 & -1 & 1 & 1 & 1 & -1 & -1 & -1 & 1 & 1 \\ -1 & 1 & -1 & 1 & 1 & 1 & -1 & -1 & 1 & 1 & 1 & -1 & 1 \\ -1 & -1 & 1 & 1 & 1 & -1 & 1 & 1 & -1 & 1 & 1 & 1 & -1 \\ -1 & 1 & -1 & 1 & 1 & 1 & -1 & -1 & 1 & 1 & 1 & -1 & 1 \\ 1 & -1 & 1 & -1 & 1 & -1 & 1 & -1 & 1 & -1 & 1 & 1 & 1 \end{pmatrix}$$

Then $w(f_n) = 10 \times \dfrac{n-13}{12} + 13 = n - 2 \times \dfrac{n-13}{12} = n - 2\left\lfloor \dfrac{n-12}{12} \right\rfloor$. See figure 2 for $P(25,3)$. Hence, for $n \geq 13$ and $n \equiv 1 \pmod{12}$, $\gamma_s^\star(G) \leq n - 2\left\lfloor \dfrac{n-12}{12} \right\rfloor$.

Case 3. Suppose $n \equiv 2 \pmod{12}$, $(n \geq 14)$. We construct f_n as follows:

$$f_n = \begin{pmatrix} 1 & 1 & -1 & -1 & -1 & 1 & 1 & 1 & 1 & -1 & 1 & 1 \\ -1 & 1 & -1 & 1 & 1 & -1 & 1 & -1 & 1 & -1 & -1 & 1 \\ -1 & -1 & 1 & 1 & 1 & 1 & -1 & -1 & -1 & 1 & 1 & -1 \\ -1 & 1 & -1 & 1 & 1 & -1 & 1 & -1 & 1 & 1 & 1 & 1 \\ 1 & -1 & 1 & -1 & 1 & 1 & -1 & 1 & -1 & 1 & -1 & 1 \end{pmatrix}^{\frac{n-14}{12}} \begin{pmatrix} 1 & 1 & -1 & -1 & -1 & 1 & 1 & 1 & 1 & -1 & 1 & -1 & 1 & 1 \\ -1 & 1 & -1 & 1 & 1 & -1 & 1 & -1 & 1 & -1 & -1 & 1 & -1 & 1 \\ -1 & -1 & 1 & 1 & 1 & 1 & -1 & -1 & -1 & 1 & 1 & 1 & 1 & -1 \\ -1 & 1 & -1 & 1 & 1 & -1 & 1 & -1 & 1 & 1 & 1 & 1 & -1 & 1 \\ 1 & -1 & 1 & -1 & 1 & 1 & -1 & 1 & -1 & 1 & -1 & 1 & 1 & 1 \end{pmatrix}$$

Then $w(f_n) = 10 \times \dfrac{n-14}{12} + 14 = n - 2 \times \dfrac{n-14}{12} = n - 2\left\lfloor \dfrac{n-12}{12} \right\rfloor$. See figure 2 for $P(26,3)$. Hence, for $n \geq 14$ and $n \equiv 2 \pmod{12}$, $\gamma_s^\star(G) \leq n - 2\left\lfloor \dfrac{n-12}{12} \right\rfloor$.

Case 4. Suppose $n \equiv 3 \pmod{12}$, $(n \geq 15)$. We construct f_n as follows:

$$f_n = \begin{pmatrix} 1 & 1 & -1 & -1 & -1 & 1 & 1 & 1 & 1 & -1 & 1 & 1 \\ -1 & 1 & -1 & 1 & 1 & -1 & 1 & -1 & 1 & -1 & -1 & 1 \\ -1 & -1 & 1 & 1 & 1 & 1 & -1 & -1 & -1 & 1 & 1 & -1 \\ -1 & 1 & -1 & 1 & 1 & -1 & 1 & -1 & 1 & 1 & 1 & 1 \\ 1 & -1 & 1 & -1 & 1 & 1 & -1 & 1 & -1 & 1 & -1 & 1 \end{pmatrix}^{\frac{n-15}{12}} \begin{pmatrix} 1 & 1 & -1 & -1 & -1 & 1 & -1 & 1 & -1 & 1 & -1 & 1 & -1 & 1 & 1 \\ -1 & 1 & -1 & 1 & 1 & -1 & 1 & 1 & -1 & -1 & 1 & 1 & -1 & -1 & 1 \\ -1 & -1 & 1 & 1 & 1 & 1 & 1 & 1 & 1 & 1 & 1 & 1 & 1 & 1 & -1 \\ -1 & 1 & -1 & 1 & 1 & -1 & -1 & -1 & 1 & 1 & 1 & -1 & 1 & 1 & 1 \\ 1 & -1 & 1 & -1 & 1 & 1 & -1 & -1 & -1 & 1 & -1 & 1 & -1 & 1 \end{pmatrix}$$

Then $w(f_n) = 10 \times \dfrac{n-15}{12} + 15 = n - 2 \times \dfrac{n-15}{12} = n - 2\left\lfloor \dfrac{n-12}{12} \right\rfloor$. See figure 2 for $P(27,3)$. Hence, for $n \geq 15$ and $n \equiv 3 \pmod{12}$, $\gamma_s^\star(G) \leq n - 2\left\lfloor \dfrac{n-12}{12} \right\rfloor$.

Case 5. Suppose $n \equiv 4 \pmod{12}$, $(n \geq 16)$. We construct f_n as follows:

$$f_n = \begin{pmatrix} 1 & 1 & -1 & -1 & -1 & 1 & 1 & 1 & 1 & -1 & 1 & 1 \\ -1 & 1 & -1 & 1 & 1 & -1 & 1 & -1 & 1 & -1 & -1 & 1 \\ -1 & -1 & 1 & 1 & 1 & 1 & -1 & -1 & -1 & 1 & 1 & -1 \\ -1 & 1 & -1 & 1 & 1 & -1 & 1 & -1 & 1 & 1 & 1 & 1 \\ 1 & -1 & 1 & -1 & 1 & 1 & -1 & 1 & -1 & 1 & -1 & 1 \end{pmatrix}^{\frac{n-16}{12}} \begin{pmatrix} 1 & 1 & -1 & -1 & -1 & 1 & 1 & 1 & 1 & -1 & 1 & 1 & 1 & -1 & 1 & 1 \\ -1 & 1 & -1 & 1 & 1 & -1 & 1 & -1 & 1 & -1 & -1 & 1 & -1 & 1 & -1 & 1 \\ -1 & -1 & 1 & 1 & 1 & 1 & -1 & -1 & -1 & 1 & 1 & -1 & -1 & 1 & 1 & -1 \\ -1 & 1 & -1 & 1 & 1 & -1 & 1 & -1 & 1 & 1 & 1 & 1 & -1 & 1 & -1 & 1 \\ 1 & -1 & 1 & -1 & 1 & 1 & -1 & 1 & -1 & 1 & -1 & 1 & 1 & 1 & 1 & 1 \end{pmatrix}$$

Then $w(f_n) = 10 \times \dfrac{n-16}{12} + 16 = n - 2 \times \dfrac{n-16}{12} = n - 2\left\lfloor \dfrac{n-12}{12} \right\rfloor$. See figure 3 for $P(28,3)$. Hence, for $n \geq 16$ and $n \equiv 4 \pmod{12}$, $\gamma_s^\star(G) \leq n - 2\left\lfloor \dfrac{n-12}{12} \right\rfloor$.

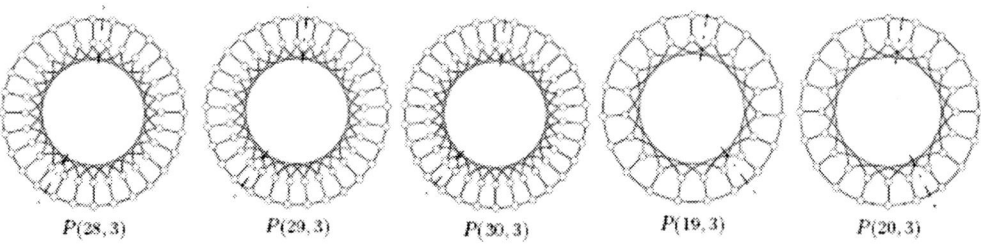

$P(28,3)$ $P(29,3)$ $P(30,3)$ $P(19,3)$ $P(20,3)$

Figure 3. Graph $P(n,3)$ corresponding to f_n for $n=28, 29, 30, 19, 20$.

Case 6. Suppose $n \equiv 5 \pmod{12}$, $(n \geq 17)$. We construct f_n as follows:

$$f_n = \begin{pmatrix} 1 & 1 & -1 & -1 & -1 & 1 & 1 & 1 & 1 & -1 & 1 & 1 \\ -1 & 1 & -1 & 1 & 1 & -1 & 1 & -1 & 1 & -1 & -1 & 1 \\ -1 & -1 & 1 & 1 & 1 & 1 & -1 & -1 & -1 & 1 & 1 & -1 \\ -1 & 1 & -1 & 1 & 1 & -1 & 1 & -1 & 1 & 1 & 1 & 1 \\ 1 & -1 & 1 & -1 & 1 & 1 & -1 & 1 & -1 & 1 & -1 & 1 \end{pmatrix}^{\frac{n-17}{12}} \begin{pmatrix} 1 & 1 & -1 & -1 & -1 & 1 & 1 & -1 & -1 & -1 & 1 & 1 & 1 & 1 & 1 & -1 & 1 & 1 \\ -1 & 1 & -1 & 1 & 1 & -1 & 1 & -1 & 1 & 1 & 1 & -1 & 1 & -1 & 1 & -1 & -1 & 1 \\ -1 & -1 & 1 & 1 & 1 & 1 & -1 & 1 & 1 & 1 & 1 & 1 & -1 & -1 & -1 & 1 & 1 & -1 \\ -1 & 1 & -1 & 1 & 1 & 1 & -1 & 1 & -1 & 1 & 1 & -1 & 1 & -1 & 1 & 1 & 1 & 1 \\ -1 & -1 & 1 & -1 & 1 & 1 & 1 & -1 & 1 & -1 & 1 & 1 & -1 & 1 & -1 & 1 & -1 & 1 \end{pmatrix}$$

Then $w(f_n) = 10 \times \dfrac{n-17}{12} + 15 = n - 2 \times \dfrac{n-5}{12} = n - 2\left\lfloor\dfrac{n}{12}\right\rfloor$. See figure 3 for $P(29,3)$. Hence, for $n \ge 17$ and $n \equiv 5 \pmod{12}$, $\gamma_s^*(G) \le n - 2\left\lfloor\dfrac{n}{12}\right\rfloor$.

Case 7. Suppose $n \equiv 6 \pmod{12}$, $(n \ge 18)$. We construct f_n as follows:

$$f_n = \begin{pmatrix} 1 & 1 & -1 & -1 & -1 & 1 & 1 & 1 & 1 & -1 & 1 & 1 \\ -1 & 1 & -1 & 1 & 1 & -1 & 1 & -1 & 1 & -1 & -1 & 1 \\ -1 & -1 & 1 & 1 & 1 & 1 & -1 & -1 & -1 & 1 & 1 & -1 \\ -1 & 1 & -1 & 1 & 1 & -1 & 1 & -1 & 1 & 1 & 1 & 1 \\ 1 & -1 & 1 & -1 & 1 & 1 & -1 & 1 & -1 & 1 & -1 & 1 \end{pmatrix}^{\frac{n-18}{12}} \begin{pmatrix} 1 & 1 & -1 & -1 & -1 & 1 & 1 & -1 & -1 & -1 & 1 & 1 & 1 & 1 & 1 & -1 & 1 & 1 \\ -1 & 1 & -1 & 1 & 1 & -1 & 1 & -1 & 1 & 1 & 1 & -1 & 1 & -1 & 1 & -1 & -1 & 1 \\ -1 & -1 & 1 & 1 & 1 & 1 & -1 & 1 & 1 & 1 & 1 & 1 & -1 & -1 & -1 & 1 & 1 & -1 \\ -1 & 1 & -1 & 1 & 1 & 1 & -1 & 1 & -1 & 1 & 1 & -1 & 1 & -1 & 1 & 1 & 1 & 1 \\ 1 & -1 & 1 & -1 & 1 & 1 & -1 & 1 & -1 & 1 & -1 & 1 & 1 & -1 & 1 & -1 & 1 & 1 \end{pmatrix}$$

Then $w(f_n) = 10 \times \dfrac{n-18}{12} + 16 = n - 2 \times \dfrac{n-6}{12} = n - 2\left\lfloor\dfrac{n}{12}\right\rfloor$. See figure 3 for $P(30,3)$. Hence, for $n \ge 18$ and $n \equiv 6 \pmod{12}$, $\gamma_s^*(G) \le n - 2\left\lfloor\dfrac{n}{12}\right\rfloor$.

Case 8. Suppose $n \equiv 7 \pmod{12}$, $(n \ge 7)$. We construct f_n as follows:

$$f_n = \begin{pmatrix} 1 & 1 & -1 & -1 & -1 & 1 & 1 & 1 & 1 & -1 & 1 & 1 \\ -1 & 1 & -1 & 1 & 1 & -1 & 1 & -1 & 1 & -1 & -1 & 1 \\ -1 & -1 & 1 & 1 & 1 & 1 & -1 & -1 & -1 & 1 & 1 & -1 \\ -1 & 1 & -1 & 1 & 1 & -1 & 1 & -1 & 1 & 1 & 1 & 1 \\ 1 & -1 & 1 & -1 & 1 & 1 & -1 & 1 & -1 & 1 & -1 & 1 \end{pmatrix}^{\frac{n-7}{12}} \begin{pmatrix} 1 & 1 & -1 & -1 & -1 & 1 & 1 \\ -1 & 1 & -1 & 1 & 1 & -1 & 1 \\ -1 & -1 & 1 & 1 & 1 & 1 & -1 \\ -1 & 1 & -1 & 1 & 1 & -1 & 1 \\ 1 & -1 & 1 & -1 & 1 & 1 & 1 \end{pmatrix}$$

Then $w(f_n) = 10 \times \dfrac{n-7}{12} + 7 = n - 2 \times \dfrac{n-7}{12} = n - 2\left\lfloor\dfrac{n}{12}\right\rfloor$. See figure 3 for $P(19,3)$. Hence, for $n \ge 7$ and $n \equiv 7 \pmod{12}$, $\gamma_s^*(G) \le n - 2\left\lfloor\dfrac{n}{12}\right\rfloor$.

Case 9. Suppose $n \equiv 8 \pmod{12}$, $(n \ge 8)$. We construct f_n as follows:

$$f_n = \begin{pmatrix} 1 & 1 & -1 & -1 & -1 & 1 & 1 & 1 & 1 & -1 & 1 & 1 \\ -1 & 1 & 1 & 1 & 1 & 1 & 1 & 1 & -1 & 1 & -1 & 1 \\ -1 & -1 & 1 & 1 & 1 & 1 & -1 & -1 & -1 & 1 & 1 & -1 \\ -1 & 1 & -1 & 1 & 1 & -1 & 1 & -1 & 1 & 1 & 1 & 1 \\ 1 & -1 & 1 & -1 & 1 & 1 & -1 & 1 & -1 & 1 & -1 & 1 \end{pmatrix}^{\frac{n-8}{12}} \begin{pmatrix} 1 & 1 & -1 & -1 & 1 & -1 & 1 & 1 \\ -1 & -1 & -1 & 1 & 1 & -1 & -1 & 1 \\ -1 & 1 & 1 & 1 & 1 & 1 & 1 & -1 \\ 1 & 1 & 1 & 1 & 1 & 1 & 1 & 1 \\ -1 & -1 & -1 & 1 & -1 & 1 & -1 & 1 \end{pmatrix}$$

Then $w(f_n) = 10 \times \dfrac{n-8}{12} + 10 = n - 2 \times \dfrac{n-20}{12} = n - 2\left\lfloor\dfrac{n-12}{12}\right\rfloor$. See figure 3 for $P(20,3)$. Hence, for $n \ge 8$ and $n \equiv 8 \pmod{12}$, $\gamma_s^*(G) \le n - 2\left\lfloor\dfrac{n-12}{12}\right\rfloor$.

Case 10. Suppose $n \equiv 9 \pmod{12}$, $(n \ge 9)$.

For $n = 9$, we construct f_n as follows:

$$f_n = \begin{pmatrix} 1 & 1 & 1 & 1 & 1 & 1 & 1 & 1 & 1 \\ -1 & -1 & -1 & -1 & -1 & -1 & -1 & -1 & -1 \\ 1 & 1 & 1 & 1 & 1 & 1 & 1 & 1 & 1 \\ 1 & 1 & 1 & 1 & 1 & 1 & 1 & 1 & 1 \\ -1 & -1 & -1 & -1 & -1 & -1 & -1 & -1 & -1 \end{pmatrix}$$

Then $w(f_n) = 9 = n - 2\left\lfloor \dfrac{n}{12} \right\rfloor = 9 - 2\left\lfloor \dfrac{9}{12} \right\rfloor = 9$. See figure 4 for $P(9,3)$. Hence, for $n = 9$,

$\gamma_s^\star(G) \le n - 2\left\lfloor \dfrac{n}{12} \right\rfloor$.

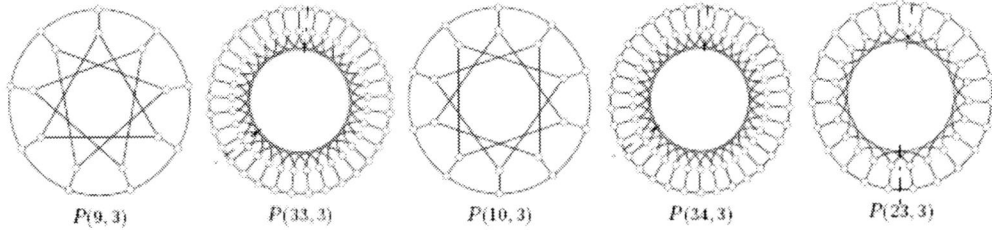

Figure 4. Graph $P(n,3)$ corresponding to f_n for $n = 10, 22, 34$.

For $n \ge 21$, we construct f_n as follows:

$$f_n = \begin{pmatrix} 1 & 1 & -1 & -1 & -1 & 1 & 1 & 1 & 1 & -1 & 1 & 1 \\ -1 & 1 & -1 & 1 & 1 & -1 & 1 & -1 & 1 & -1 & -1 & 1 \\ -1 & -1 & 1 & 1 & 1 & 1 & 1 & -1 & -1 & -1 & 1 & -1 \\ -1 & 1 & -1 & 1 & 1 & -1 & 1 & -1 & 1 & 1 & 1 & 1 \\ 1 & -1 & 1 & -1 & 1 & 1 & -1 & 1 & -1 & 1 & -1 & 1 \end{pmatrix}^{\frac{n-21}{12}} \begin{pmatrix} 1 & 1 & -1 & -1 & -1 & 1 & 1 & 1 & 1 & -1 & 1 & 1 & -1 & -1 & -1 & 1 & 1 & 1 & 1 & -1 & 1 & 1 \\ -1 & 1 & -1 & 1 & 1 & -1 & -1 & -1 & 1 & -1 & -1 & 1 & 1 & 1 & -1 & 1 & -1 & 1 & -1 & 1 & -1 & 1 \\ -1 & -1 & 1 & 1 & 1 & 1 & 1 & -1 & -1 & -1 & 1 & 1 & 1 & 1 & 1 & 1 & -1 & -1 & -1 & 1 & 1 & -1 \\ -1 & 1 & -1 & 1 & 1 & 1 & 1 & 1 & 1 & 1 & 1 & 1 & 1 & 1 & 1 & 1 & -1 & 1 & -1 & 1 & 1 & 1 \\ 1 & -1 & 1 & -1 & 1 & -1 & -1 & -1 & -1 & 1 & -1 & 1 & -1 & 1 & 1 & -1 & 1 & -1 & 1 & -1 & 1 & 1 \end{pmatrix}$$

Then $w(f_n) = 10 \times \dfrac{n-21}{12} + 19 = n - 2 \times \dfrac{n-9}{12} = n - 2\left\lfloor \dfrac{n}{12} \right\rfloor$. See figure 4 for $P(33,3)$. Hence, for $n \ge 21$ and $n \equiv 9 \pmod{12}$, $\gamma_s^\star(G) \le n - 2\left\lfloor \dfrac{n}{12} \right\rfloor$.

Case 11. Suppose $n \equiv 10 \pmod{12}$, $(n \ge 10)$.

For $n = 10$, we construct f_n as follows:

$$f_n = \begin{pmatrix} 1 & 1 & 1 & 1 & 1 & 1 & 1 & 1 & 1 & 1 \\ -1 & -1 & -1 & -1 & -1 & -1 & -1 & -1 & -1 & -1 \\ 1 & 1 & 1 & 1 & 1 & 1 & 1 & 1 & 1 & 1 \\ 1 & 1 & 1 & 1 & 1 & 1 & 1 & 1 & 1 & 1 \\ -1 & -1 & -1 & -1 & -1 & -1 & -1 & -1 & -1 & -1 \end{pmatrix}$$

Then $w(f_n) = 10 = n - 2\left\lfloor \dfrac{n}{12} \right\rfloor = 10 - \left\lfloor \dfrac{10}{12} \right\rfloor = 10$. See figure 4 for $P(10,3)$. Hence, for $n = 10$,

$\gamma_s^\star(G) \le n - 2\left\lfloor \dfrac{n}{12} \right\rfloor$.

For $n \ge 22$, we construct f_n as follows:

$$f_n = \begin{pmatrix} 1 & 1 & -1 & -1 & -1 & 1 & 1 & 1 & 1 & -1 & 1 & 1 \\ -1 & 1 & -1 & 1 & 1 & -1 & 1 & -1 & 1 & -1 & -1 & 1 \\ -1 & -1 & 1 & 1 & 1 & 1 & 1 & -1 & -1 & -1 & 1 & -1 \\ -1 & 1 & -1 & 1 & 1 & -1 & 1 & -1 & 1 & 1 & 1 & 1 \\ 1 & -1 & 1 & -1 & 1 & 1 & -1 & 1 & -1 & 1 & -1 & 1 \end{pmatrix}^{\frac{n-22}{12}} \begin{pmatrix} 1 & 1 & -1 & -1 & -1 & 1 & 1 & 1 & 1 & -1 & 1 & 1 & -1 & -1 & -1 & 1 & 1 & 1 & 1 & -1 & 1 & 1 \\ -1 & 1 & -1 & 1 & 1 & -1 & 1 & -1 & -1 & 1 & 1 & 1 & -1 & 1 & 1 & 1 & -1 & 1 & -1 & 1 & -1 & 1 \\ -1 & -1 & 1 & 1 & 1 & 1 & 1 & -1 & -1 & 1 & 1 & -1 & -1 & 1 & 1 & 1 & 1 & -1 & -1 & -1 & 1 & -1 \\ -1 & 1 & -1 & 1 & 1 & 1 & 1 & -1 & 1 & 1 & 1 & -1 & 1 & 1 & 1 & 1 & -1 & 1 & -1 & 1 & 1 & 1 \\ 1 & -1 & 1 & -1 & 1 & 1 & -1 & -1 & -1 & 1 & -1 & 1 & -1 & 1 & 1 & -1 & 1 & -1 & 1 & -1 & 1 & 1 \end{pmatrix}$$

Then $w(f_n) = 10 \times \dfrac{n-22}{12} + 20 = n - 2 \times \dfrac{n-10}{12} = n - 2\left\lfloor \dfrac{n}{12} \right\rfloor$. See figure 12 for $P(34,3)$. Hence, for $n \ge 22$ and $n \equiv 10 \pmod{12}$, $\gamma_s^\star(G) \le n - 2\left\lfloor \dfrac{n}{12} \right\rfloor$.

Case 12. Suppose $n \equiv 11 \pmod{12}$, $(n \ge 11)$. We construct f_n as follows:

$$f_n = \begin{pmatrix} 1 & 1 & -1 & -1 & -1 & 1 & 1 & 1 & 1 & -1 & 1 & 1 \\ -1 & 1 & -1 & 1 & 1 & -1 & 1 & -1 & 1 & -1 & -1 & 1 \\ -1 & -1 & 1 & 1 & 1 & 1 & -1 & -1 & -1 & 1 & 1 & -1 \\ -1 & 1 & -1 & 1 & 1 & -1 & 1 & -1 & 1 & 1 & 1 & 1 \\ 1 & -1 & 1 & -1 & 1 & 1 & -1 & 1 & -1 & 1 & -1 & 1 \end{pmatrix}^{\frac{n-11}{12}} \begin{pmatrix} 1 & 1 & -1 & -1 & -1 & 1 & 1 & -1 & -1 & 1 & 1 \\ -1 & 1 & -1 & 1 & 1 & -1 & -1 & 1 & 1 & -1 & 1 \\ -1 & -1 & 1 & 1 & 1 & 1 & 1 & -1 & 1 & 1 & -1 \\ -1 & 1 & -1 & 1 & 1 & 1 & -1 & 1 & 1 & 1 & 1 \\ 1 & -1 & 1 & -1 & 1 & 1 & -1 & 1 & -1 & 1 & -1 \end{pmatrix}$$

Then $w(f_n) = 10 \times \dfrac{n-11}{12} + 11 = n - 2 \times \dfrac{n-11}{12} = n - 2\left\lfloor \dfrac{n}{12} \right\rfloor$. See figure 4 for $P(23,3)$. Hence, for $n \geq 11$ and $n \equiv 11 \pmod{12}$, $\gamma_s^*(G) \leq n - 2\left\lfloor \dfrac{n}{12} \right\rfloor$.

By theorem 1 and theorem 2, we have

Theorem 3. Let G be a graph $P(n,3)$, and for any integer $n \geq 7$,

$$\left\lceil \frac{5n}{7} \right\rceil_{P(5n)} \leq \gamma_s^*(G) \leq \begin{cases} n - 2\left\lfloor \dfrac{n}{12} \right\rfloor, & n(\mathrm{mod}\ 12) = 0,5,6,7,9,10,11, \\[2mm] n - 2\left\lfloor \dfrac{n-12}{12} \right\rfloor, & n(\mathrm{mod}\ 12) = 1,2,3,4,8. \end{cases}$$

where $P(s)$ is defined to be the parity of s, that is, $P(s)$=odd if s is odd and $P(s)$=even if s is even.

4. Conclusions

In this paper, we have studied the total signed domination number of $P(n,3)$. According to the definition of total signed domination and the characteristics of $P(n,3)$, we design high-performance algorithms and construct total signed dominating functions. With these functions, we get upper bounds of $\gamma_s^*(P(n,3))$. The gap between lower bounds and upper bounds is very small, about $5n/42$.

Acknowledgments

This work is supported by the Fundamental Research Funds for the Central University, GrandNo: 3132016306.

References

[1] Dunbar, J. E., Hedetniemi, S. T., Henning, M. A., and Slater, P. J. (1995) Signed domination in graphs. Graph Theory. Combinatorcs, and Applications, 1: 311-322.

[2] Xu, B. G. (2001) On signed edge domination numbers of graphs. Discrete Mathematics, 239: 179-189.

[3] Lv, X. Z. (2007) A lower bound on the total signed domination numbers of graphs. Science in China Series A: Mathematics, 50: 1157-1162.

[4] Atapour, M., Bodaghli, A., and Sheikholeslami, S. M. (2018) Twin signed total domination numbers in directed graphs. Ars Combinatoria, 138: 119-131.

[5] Khodkar, A., and Ghameshlou, A. N. (2017) Signed edge domination numbers of complete tripartite graphs: Part One. Utilitas Mathematica, 105: 237-258.

[6] Volkmann, L. (2016) Signed Roman k-domination in digraphs. Graphs and Combinatorics, 32: 1217-1227.

[7] Zhou, Z. W. (2010) Vertex-edge total signed domination number of a complete graph with even order. Acta Math. Appl. Sin., 33: 112-117.

[8] Shan, E.F., and Zhao, Y.C. (2015) Signed mixed dominating functions in complete bipartite graphs. International Journal of Computer Mathematics, 92: 712-721.

[9] Chen, L. L., Ma, Y. B., and Shi, Y. T. (2018) On the [1, 2]-domination number of generalized Petersen graphs. Applied Mathematics and Computation, 327: 1-7.

[10] Wang, H. L., Xu, X. R., Yang, Y.S., and Wang, G. Q. (2015) On the domination number of generalized Petersen graphs $P(ck, k)$. Ars Combinatoria, 118: 33-49.

ISPECE

IOP Publishing

Change Impact Analysis of Complex Mechanical Product Based on Complex Network Theory

Na Zhang, Yu Yang [a]

College of Mechanical Engineering, Chongqing University, NO. 174 Shazhengjie, Shapingba, Chongqing 400044, China

[a] Corresponding author: yuyang@cqu.edu.cn

Abstract. In order to reduce the change impact and prevent avalanche, the paper proposes a method for analyzing the change impact based on complex network theory. Firstly, the complex network theory is used to systematically describe the structure of complex mechanical products. Secondly, the change propagation process is analyzed based on propagation dynamics theory, and then, the change propagation rate and changed nodes ratio are proposed for the analysis of change impact based on network evolution theory and network topology properties. Finally, the proposed method is used to analyze the change propagation and change impact of clutch.

1. Introduction

With the diversification and personalization of customer needs, the changes of customer needs have become an inevitable issue in the development of complex products. Due to a complex mechanical product contains a large number of parts and complicated relationships between parts, A change occur in one part will cause a series of changes in others which connected to it, or even will cause avalanche effect. Changes of product structure will eventually lead to the changes of design tasks, and eventually resulting in design delay and extra costs. In this regard, designers require to respond to changes quickly. The response to changes depends on the analysis of change propagation and change impact. Therefore, research on the change propagation and change impact of complex mechanical product design is of great significance to respond to change, control change costs and improve change efficiency.

At present, the existing analysis methods of change impact include risk matrix [1], axiomatic design matrix [2], design structure matrix [3-8], relation matrix [9,10] and complex network theory [11], etc. Due to the diversity, uncertainty and dynamics of changes propagation, complex network theory provides new ideas for solving the issue of change impact assessment. In this study, a method for analyzing the change impact based on complex network theory. Firstly, a network model of complex mechanical product is constructed. Secondly, the change propagation process is analyzed based on propagation dynamics theory, and then, the change propagation rate and changed nodes ratio are proposed for the analysis of change impact based on network evolution theory and network topology properties. Finally, a case study is presented to illustrate the proposed method.

2. The construction of the network model

The construction of network model is the basis of analyzing change impact. Due to the characteristic of complex mechanical products, the complex network is used to describe the structure of complex mechanical products systematically and quantitatively.

Content from this work may be used under the terms of the Creative Commons Attribution 3.0 licence. Any further distribution of this work must maintain attribution to the author(s) and the title of the work, journal citation and DOI.

Published under licence by IOP Publishing Ltd

In the network model, the parts of complex mechanical product are regarded as nodes, the relationship between parts are regarded edges and the strength of the edges are defined as network weights. Based on this, the network model G of the complex mechanical product can be defined as follows.

$$G = (V, E, W) \qquad (1)$$

where $V = \{v_i, i = 1, 2, ..., n\}$ is the set of nodes, and v_i represents the part i; $E = \{e_{ij}, i = 1, 2, ..., n, i \neq j\}$ is the set of edges, e_{ij} is the edge between node v_i and v_j. $W = \{w_{ij}, i = 1, 2, ..., n, i \neq j\}$ is the set of weights, and w_{ij} is the value of relationship between node v_i and v_j.

The determination of weights is a key step in constructing a network model. In this paper, the weights are determined by considering functional correlations and structural correlations. Therefore, weight is calculated as follows.

$$w_{ij} = \alpha w_{ij}^f + \beta w_{ij}^s \qquad (2)$$

where α and β correspond to w_{ij}^f and w_{ij}^s. w_{ij}^s denotes the structural correlation strength between v_i and v_j. w_{ij}^f represents the functional correlation strength between v_i and v_j.

The weight w_{ij}^f is determined by the functional load, and the value is calculated as follows:

$$w_{ij}^f = \sum_{h=1}^{H} \frac{f_h}{F_h} \qquad (3)$$

where f_h is the functional load that v_j can bear for the realization of function h, F_h is the total functional capacity.

The weight w_{ij}^s is determined by the structural constraints, such as size constraints, shape constraints, etc., and it can be calculated as follows:

$$w_{ij}^s = \sum_{l=1}^{L} w_{ij-l}^s \qquad (4)$$

where w_{ij-l}^s denotes the correlation strength of the l-th kind of structural correlations. L represents the total number of types of structural correlations.

It should be noted that the greater the correlation strength between two nodes, the greater the weight of the node is.

The above analysis shows that the constructed network model is an undirected weighted network. Moreover, it has been proved that the network model belongs to scale-free network. Therefore, the network model which constructed in this paper has the propagation characteristics and evolution characteristics of the scale-free network.

3. The assessment of change impact

3.1. Change propagation process

The changes of customer needs are transformed into the changes of parts. Due to propagation dynamics theory and the evolutionary characteristic of network model, the change propagation can be described as follows: the node that needs to be changed is adding in network as a new nodes and connecting to the other nodes according to rules, and then, the weights are recalculated.

Therefore, combining the evolutionary characteristic and the classical propagation model SI, the change propagation process is analyzed as follows.

(1) The initial network is defined as: $G_0 = (V_0, E_0, W_0)$, where $V_0 = \{v_{0i}, i = 1, 2, ..., n_0\}$, $E_0 = \{e_{0ij}, i, j = 1, 2, ..., n_0\}$, $W_0 = \{w_{0ij}, i, j = 1, 2, ..., n_0\}$, , $i \neq j$.

(2) Suppose that there is a change on node v_{0i} at time t_0, and v_{0i} is transformed to v'_{0i}. As a new node, v'_{0i} is added to the initial network model and connected to other nodes in initial network model according the. Then, a new network model is formed and the weights are recalculated.

(3) At time t_1, change is propagated from node v'_{0i} to other unchanged nodes (such as node v_{0j}) that connected to it. The propagation probability of node v_{0j} is calculated as follows.

$$p_j = \frac{w_{ij}}{\sum_{j \in I(i)} w_{ij}} \qquad (5)$$

where w_{ij} is the weight between v'_{0i} and v_{0j}. $I(i)$ is the set of nodes that connected to v'_{0i}.

(4) At time t_a, the changed nodes at the previous moment are added to the network model as new nodes, and then, repeat step 2 and step 3 until the network evolution is completed.

3.2. The assessment indexes

The change propagation process is essentially the evolution process of network model driven by changes. Therefore, the change impact can be assessed through parameter changes before and after the network evolution. In this paper, the change impact is evaluated from two aspects: change propagation rate and changed nodes ratio.

Change propagation rate refers to the speed that a changed node causes other unchanged nodes to change at a certain time. The greater the degree of the changed node, the bigger the change propagation rate. Change propagation rate $p(t_a)$ can be calculated as follows.

$$p(t_a) = \sum_{j \in I(i)} \frac{w_{ij}}{w_{\max}} \qquad (6)$$

where $w_{\max}$ is the largest weight in the network model.

Changed nodes ratio refers to the proportion of changed nodes in the network at a certain moment. The equation for calculating the changed nodes ratio $r(t_a)$ is as follows.

$$r(t_a) = r(t_{a-1}) + \frac{n_a}{N_a} \qquad (7)$$

where $r(t_{a-1})$ is the changed nodes ratio at time t_{a-1}. N_a is the total number of nodes in network model at time t_a and n_a is the number of nodes that changed at time t_a. The results show that the greater the changed nodes ratio, the greater impact of change is, and the more parts need to be changed.

4. case study

This paper takes the change impact assessment of clutch design as an example. Due to the limitations of the paper length, the ten key parts (as shown in Table 1) of clutch are selected for change impact analysis.

Table 1. The key parts of clutch.

No	Parts	No	Parts
v_1	Flywheel	v_2	Pressure plate
v_3	Clutch cover	v_4	Release bearing
v_5	Spring	v_6	Bearing ring

v_7	Separation hook	v_8	Platen
v_9	Driven shaft	v_{10}	Friction plate

In this section, firstly, the relationship between parts are analyzed and the network model of clutch are constructed. Secondly, the change propagation and the change impact are analyzed and discussed.

According to the design specification and equations (1)-(4), the relationship between parts and the correlation strengths are determined. Based on this, combined with the network construction rules, the network model is drawn by Netdraw (as shown in Figure 1). As shown in Figure 1, the network model contains 10 nodes and 90 edges, moreover, the weights are shown in Figure 1.

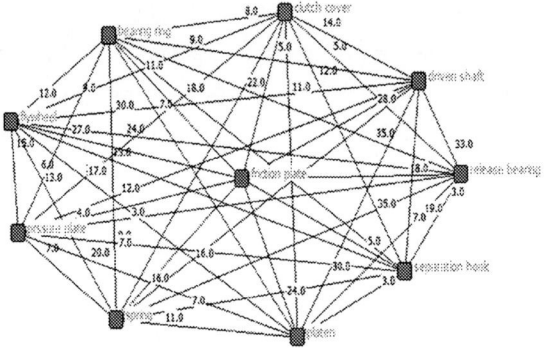

Figure 1. The network model of clutch.

On the basis of the network model, the change propagation and change impact are analyzed. Using the method proposed in section 3, the change propagation rate of all nodes are calculated (as shown in Figure 2).

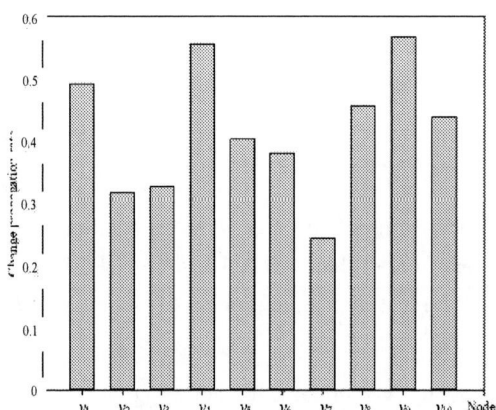

Figure 2. The change propagation rate of all nodes.

Furthermore, the change propagation process of different initially changed node is analyzed and the changed nodes ratio of different initially changed node at different time are calculated (as shown in Figure 3).

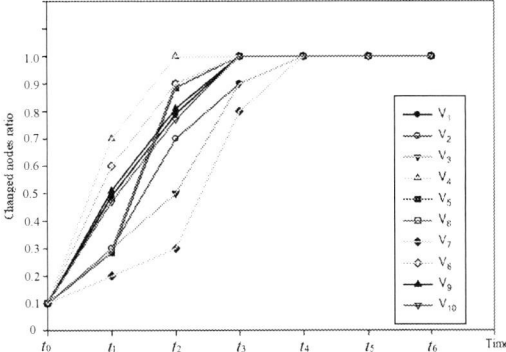

Figure 3. The changed nodes ratio of different initially changed nodes.

Based on the above calculation results, the following conclusions can be drawn:

(1) The change propagation capabilities of different nodes are different. The change propagation capability of a node is determined by the node degree and weight. The greater the node degree and weight, the stronger the change propagation capability of the node is. It can be seen from Figure 2 that driven shaft (v_9) and release bearing (v_4) have the strongest propagation capability, which requires enterprises to fully consider the customer requirements that these parts meet in the early stage of design to avoid changes in the later stages of design.

(2) For different initially nodes, the number of changed nodes are different at different time and the change completion time are different. Therefore, to reduce the change impact scope and control change costs, the companies is need to stop the change propagation at the right time.

(3) When the change propagation period exceeds a certain time, the number of changed parts tends to be stable, that is, all parts are changed. That is to say, as the change propagation cycle increases, any part will change without effective control, and the entire product will change.

5. Conclusions

The analysis of the change impact can provide a basis for companies to make decisions on changes of complex mechanical products. The complex network theory is used to analyze change impact of complex mechanical products in the study. The change propagation process is analyzed based on propagation dynamics theory, and then, the change propagation rate and changed nodes ratio are proposed for the analysis of change impact based on network evolution theory and network topology properties.

The results show that the change propagation capabilities of different nodes are different. The greater the node degree and weight, the stronger the change propagation capability of the node is. For different initially nodes, the number of changed nodes are different at different time and the change completion time are different. In addition, when the change propagates for a certain time, all parts will change, which requires companies to terminate the change at the right time to reduce the impact of the change.

Acknowledgements

This study was supported by National Natural Science Foundation of China under Grant (No. 71571023) and a project supported by graduate research and innovation foundation of Chongqing, China (Grant No. CYB17024)

References

[1] N. Smith, S. Mahadevan, J Spacecraft Rockets **42** 4 (2015)
[2] Y. Ma, G. Chen, G. Thimm, Comput Ind **59** 2 (2008)

[3] H. Seol, J Eng Design **21** 1 (2010)

[4] J. Lee, Y.S. Hong, Res Eng Des **28** 4 (2017)

[5] T. Cohen, S.B. Navathe, R.E. Fulton, Comput Aided Design **32** 5 (2000)

[6] B. Morkos, P. Shankar, J.D. Summers, J Eng Design **23** 12 (2012)

[7] D. Tang, J Mech Eng **46** 1 (2010)

[8] C. Eckert, P.J. Clarkson, W. Zanker, Res Eng Des 15 1 2004

[9] S. Tosserams, A.T. Hofkamp, L.F.P. Etman, J.E. Rooda, Struct Multidiscip O 42 5 (2010)

[10] L. Chen, A. Macwan, S. Li, J Mech Des 129 3 (2007)

[11] N, Zhang, Y. Yang, Y.J. Zheng, J.F. Su, J Intell Manf **6** (2017)

Influence of variable slip frequency control strategy on tractive performance

ZHANG Min, MA Weihua*, LUO Shihui

Traction Power State Key Laboratory, Southwest Jiaotong University, Chengdu,610031, China

Email of all the authors: _zhmlzhm@126.com_; _mwh@swjtu.edu.cn_; _shluo@swjtu.edu.cn_

* Corresponding author: MA Weihua; Email: _mwh@swjtu.edu.cn_; Phone: 13084477943

Abstract: In order to improve the current situation of poor traction ability of medium-low speed maglev train, a variable slip frequency control (VSFC) strategy is proposed based on the constant slip frequency control (CSFC) of linear induction motor (LIM). The control frequency of the starting stage of the vehicle is selected low on the premise of no obvious influence on the levitation performance. After reaching the maximum power point, the traction of the motor is always the maximum that can be exerted under the corresponding speed. Firstly, the mathematical model of linear induction motor is established, and the analytical expressions of traction and normal force are listed; Then the influence of slip frequency on vehicle tractive performance is analyzed by traction calculation, the VSFC strategy is proposed and its advantages are analyzed. The results show that the VSFC strategy can fully suppress the influence of normal force on levitation stability at low speed, and can give full play to the tractive performance of the motor in the whole process.

1. Introduction

The LIM has the advantages of simple structure, good heat dissipation conditions and strong climbing ability, so it is used as the traction equipment of medium-low speed maglev train. The medium-low speed maglev train [1-2] simultaneously adopts electromagnet to achieve suction levitation. The LIM and the levitation electromagnet are respectively installed in the middle and lower part of the levitation module. The motor produces normal force while generating traction, and the existence of the normal force will affect the levitation stability of the vehicle. Therefore, the LIM not only determines the tractive performance of the vehicle but also affects its levitation performance. In order to improve the tractive performance of the motor and effectively control the magnitude of normal force, a lot of researches have been done by domestic and foreign scholars.

On the design and optimization aspect, Bazghaleh et al. [3] developed several different multi-objective functions, which will be used to improve efficiency, power factor, end effect intensity, and motor weight. Employing the derived equations and considering all phenomena involved in the single-sided LIM, Shiri et al. [4] presented a simple design procedure and analyzed the effect of different design variables on the performance of the motor. In the aspect of Analytical computing, Faiz et al. [5] introduced a new idea to account for the longitudinal end effect factor. Meanwhile, other electromagnetic effects, such as transverse edge effects and skin effect, are taken into account. Pai R M et al. [6] developed a per-phase equivalent circuit of the LIM with sheet secondary. In the aspect of

---

Content from this work may be used under the terms of the Creative Commons Attribution 3.0 licence. Any further distribution of this work must maintain attribution to the author(s) and the title of the work, journal citation and DOI.

Published under licence by IOP Publishing Ltd

controlling, accounting for the end effect, a field orientation control scheme is developed by Kang et al. [7]. With the proposed control scheme, the flux-attenuation problem due to the end effect is shown to be resolved in the high-speed range. A decoupled-control method of levitation and propulsion for the single-sided LIM maglev-vehicle was proposed by Yoshida et al. [8].

Various researchers have done a lot of in-depth research on the structural optimization, theoretical calculation and control methods of the LIM, and put forward many meaningful conclusions for the design and control of the LIM, but there is a lack of research on the combination with practical engineering and other systems. The current application status is that the maximum running speed of medium-low speed maglev vehicle is no more than 100km/h. In order to improve the tractive performance of maglev vehicle and improve its market competitiveness, this paper studies the influence of slip frequency on the operation performance of maglev vehicle based on the relevant data of the LIM for the new-generation medium-low speed maglev test vehicle, and proposes a better slip frequency control strategy to provide selection basis for engineering application.

2. LIM for medium-low speed maglev

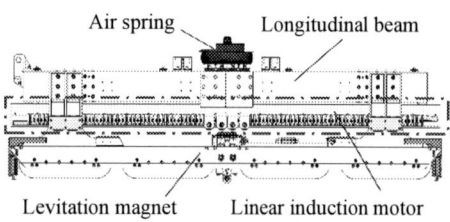

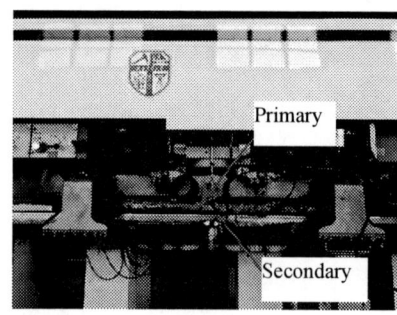

Figure 1. The LIM installed in the levitation module   Figure 2. The LIM installed on the test vehicle

The running mechanism of the medium-low speed maglev vehicle is levitation frame, each levitation frame includes two levitation modules, one of which is equipped with a LIM and a levitation electromagnet, as shown in figure 1.The LIM installed on the vehicle moving with the vehicle is the primary; The secondary of the motor is the track, including the back of the F rail and a layer of about 4mm aluminum plate, as shown in figure 2.The LIM and the levitation electromagnet are installed in the same levitation module, which will inevitably affect each other when they work. In particular, when the motor generates traction, it will generate normal force, which can be represented as suction or repulsion. In the form of suction, the burden of levitation system is increased. As the repulsive force, it can act as a part of levitation force. The magnitude and direction of the force can be controlled by slip frequency.

3. Analytical calculations of traction and normal force

The LIM series equivalent circuit considering the two dynamic side effects is shown in figure 3. Z_1 is the primary resistance plus leakage resistance, Z_{0e} is the secondary equivalent resistance and excitation reactance, $Z_{we} = k_{we} Z_{0e}$ is the impedance caused by the second type transverse end effect, and $Z_{le} = k_{le} Z_{0e}$ is the impedance caused by the second type longitudinal end effect.

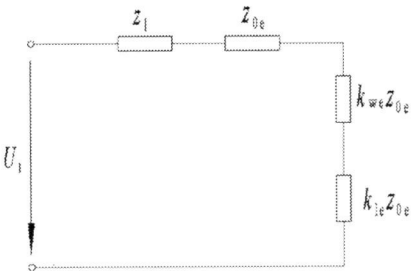

Figure 3. Equivalent circuit of the LIM

The magnetic vector potential can be calculated [9] as follows:

$$\nabla^2 A - \mu_0 \rho \frac{\partial A}{\partial t} + \mu_0 \rho (v \times (\nabla \times A)) = -\mu_0 j_1 \tag{1}$$

The air gap flux density expressions in the x and y directions are as follows:

$$B_x = (k_1 \cosh \alpha y + k_2 \sinh \alpha y)\alpha\, e^{-j\alpha x}; \quad B_y = j(k_1 \sinh \alpha y + k_2 \cosh \alpha y)\alpha\, e^{-j\alpha x} \tag{2}$$

Where $k_1 = \dfrac{\mu_0 J_1}{\alpha} \dfrac{1}{\cosh \alpha \delta_e - j\frac{\alpha\gamma}{s\omega\mu_0}\sinh \alpha \delta_e}$, $k_2 = \dfrac{-j\gamma J_1}{s\omega} \dfrac{1}{\cosh \alpha \delta_e - j\frac{\alpha\gamma}{s\omega\mu_0}\sinh \alpha \delta_e}$, $\alpha = \dfrac{\pi}{\tau}$, $\gamma = k_r \dfrac{\rho}{d}$.

The expression of traction is: $F_x = \dfrac{1}{2} B_y J_1 S$ (3)

The normal force of the motor is: $F_y = F_{ay} + F_{ry} = \dfrac{1}{2\mu_0} B_y^2 S - \dfrac{1}{2} B_y J_1 S$ (4)

At present, the LIM is usually controlled by constant slip frequency, that is, the speed of travelling magnetic field is always faster than the running speed of the vehicle. f is the slip frequency and τ is the motor pole pitch. Table 1 shows the LIM parameters of the new-generation maglev vehicle. Based on the parameters, the variations of the traction and normal force with velocity and slip frequency are analyzed and calculated. Under CSFC, the motor generally includes two stages in the full speed range, the first stage is the constant force region and the second is the constant power region. The primary current in the constant force region remains unchanged, and the supply frequency and voltage increase with the speed, at this stage, the traction of the motor changes little. The constant power region refers to the stage after the motor reaches the maximum power value, in which the motor voltage remains unchanged, the current decreases as the speed increases, the supply frequency increases as the speed increases, and the traction attenuation of the motor is relatively fast, as shown in figure 4a.

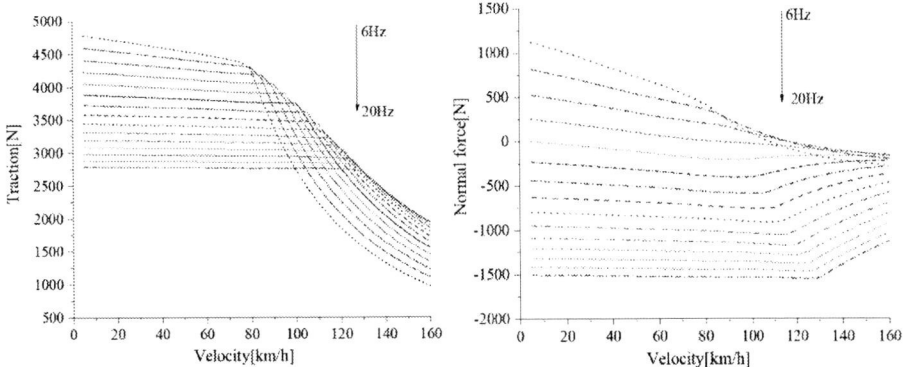

（a）The theoretical calculation value of the traction　（b）The theoretical calculation value of the normal force

Figure 4. The force characteristics of the LIM with different slip frequencies

Table 1. LIM parameters

NAME	SIZE	NAME	SIZE
The length of the motor/mm	2850	The motor capacity/kVA	248
The width of the motor/mm	220	Rated phase voltage/V	212
Pole number	12	Rated phase current/A	390
Pole pitch/mm	220	Air gap/mm	11
Phase number	3	Slip frequency/Hz	8
Number of slots per pole per phase	3	Single phase effective series turns	144
Resistivity of secondary plate / （Ωm）	2.83×10^{-8}	Secondary thickness /mm	4

The frequency setting of the CSFC is generally between 6-20Hz. The LIM normal suction in the starting stage is large when the slip frequency selection is too small, the traction is insufficient when the selection is too big, and the vehicle accelerates slowly. Other parameters remain unchanged in the analytical calculation of the traction and normal force, such as motor structure and air gap size; only the slip frequency is changed, and the change of traction and normal force in the process from 0 to 160km/h at different slip frequency is compared. The calculation results are shown in figure. 4. It can be seen that: 1) the smaller the slip frequency is, the greater the traction is when the motor starts, and the greater the normal force is;2) with the improvement of slip frequency, the speed of the motor to reach the constant power point will increase;3) the lower the slip frequency is, the faster the traction attenuation is at high speed;4) the larger the slip frequency is, the smaller the change ranges of the traction and the normal force are.5) when the slip frequency is 10Hz, the normal force at the start stage of the motor is close to 0.

4. Resistance calculation

The basic resistance of the medium-low speed maglev vehicle only takes into account air resistance, electromagnetic resistance and flow resistance. The calculation formula of resistance [10] refers to the parameters of Japan's HSST train, and the formula is modified by multiplying the correlation coefficient before the constant. The calculation formula is as follows:

$$W_a = \left(k_1 \cdot 1.652 + k_2 \cdot 0.572N\right)v^2 \tag{5}$$

$$W_m = \begin{cases} k_3 \cdot 3.354Mv & v < 5.6\text{m/s}\,(20\text{km/h}) \\ \left(k_4 \cdot 18.22 + k_5 \cdot 0.074v\right)M & v \geq 5.6\text{m/s}\,(20\text{km/h}) \end{cases} \tag{6}$$

$$W_c = k_6 \cdot 41.67 \tag{7}$$

Among the above categories: W_a represents air resistance, W_m represents electromagnetic resistance, and W_c represents flow resistance, unit N; N denotes the number of vehicles, and this paper takes 1 for single vehicle. M stands for train weight, unit t; v denotes the running speed of the train in a still wind state, unit m/s; k_1 to k_6 is the correction coefficient, $k1=1$, $k_2=1.9$, $k_3=1.5$, $k_4=1.5$, $k_5=1.8$, $k_6=2$.

The total resistance calculation formula of the engineering vehicle is: $W = W_a + W_m + W_c$ (7)

5. Tractive performance

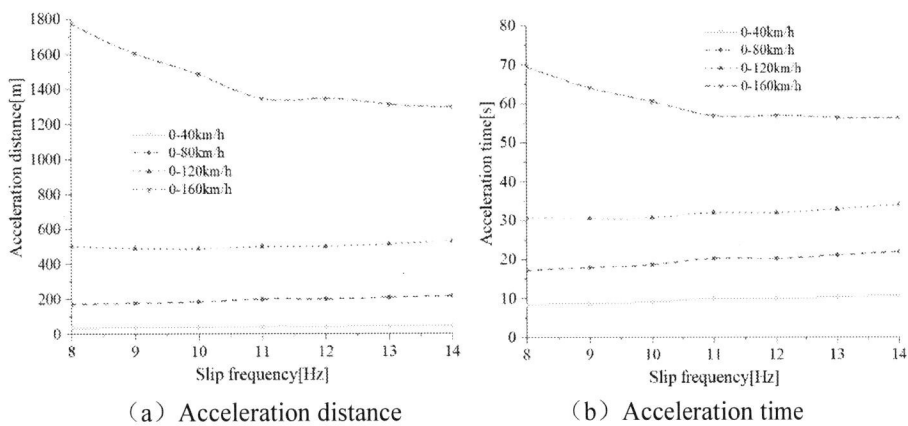

(a) Acceleration distance   (b) Acceleration time

Figure 5. Comparisons of the tractive performance by different CSFC

In the traction calculation in this section, three levitation frame maglev vehicles are adopted, that is, one vehicle includes three levitation frames and six LIMs. The full load vehicle weight is 18t, the velocity range is calculated to be 0-160km/h, and the maximum speed running resistance is calculated to be about 6.4kN.The range of slip frequency used for medium-low speed maglev vehicle is 8-14hz. The acceleration distances and average accelerations of each speed range by different CSFC are calculated. The variation of tractive performance by different CSFC is compared and analyzed. The calculation results are shown in figure 5.

It can be seen from figure 5, when low slip frequency is used, the average acceleration at low speed is larger and the acceleration distance is shorter; The acceleration distance and average acceleration of different slip frequency have little difference between 0 and 120km/h speed intervals. At the speed range of 0-160km/h, the average acceleration controlled by the low slip frequency is significantly reduced, while the acceleration is more stable under the higher slip frequency. Therefore, the slip frequency is relatively small, and the traction capacity of the vehicle is relatively strong in the initial stage, but it decays quickly after reaching the constant power point.

6. VSFC strategy

The design speed of the motor analyzed is 160km/h, so the selected range of traction calculation and control strategy is 0-160km/h. The detailed variable slip control strategy is shown in figure 6. When the speed is less than 90km/h, the constant slip frequency control of 9Hz is selected. Under the air gap of 11mm, the normal force of the LIM is about 250N and the traction is about 4200N, as shown in figure 7 and 8, the influence of normal force caused by the starting slip frequency on levitation stability can be ignored. At the speed range of 90-130km/h, the slip frequency increases linearly to 18Hz. At the range of this region, the maximum output power of the motor remains unchanged and the real constant power area is maintained. At this time, the traction curve of the motor with the speed goes through the constant power transition at each slip frequency. When the speed is greater than 130km/h, the slip frequency remains unchanged at 18Hz, at which time the motor can exert the maximum output force that can be achieved in the speed section.

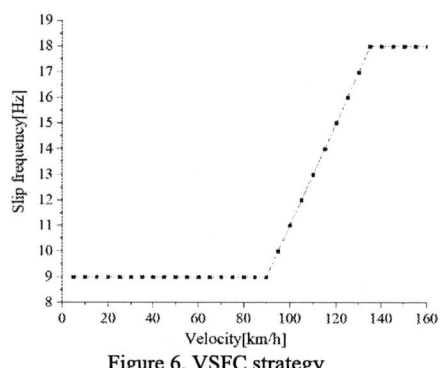

Figure 6. VSFC strategy

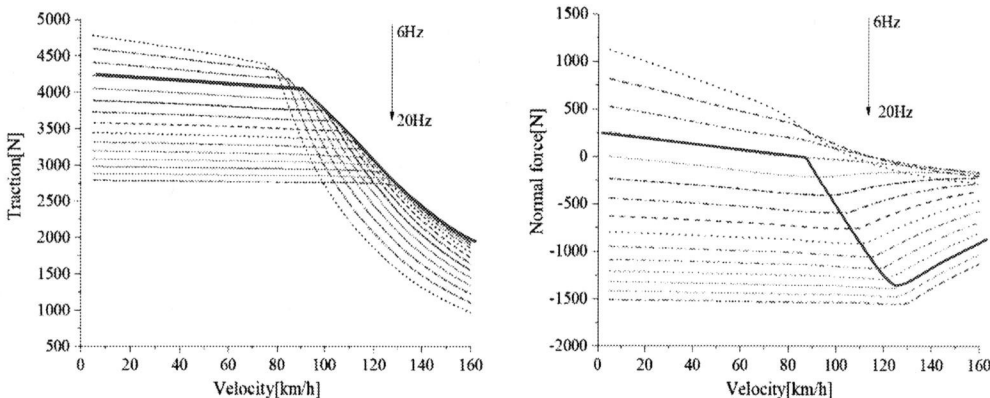

Figure 7. The traction by the VSFC strategy
Figure 8. The normal force by the VSFC strategy

Table 2. Tractive performance under VSFC

	0-40km/h	0-80km/h	0-120km/h	0-160km/h
Accelerating distance/(m)	38.08/36.56	177.61 /171.02	457.58 /486.23	1175.78 /1294.84
average acceleration/(m/s2)	1.27 /1.32	1.24 /1.29	1.15 /1.09	0.90 /0.79

In table 2, the left side data is the tractive performance by the VSFC strategy, and the right side data is the optimal value of the corresponding speed section by the CSFC of 8-14 Hz. It can be seen that when the speed is below 80km/h, the data of constant slip frequency control is better, because the data at this time is the result by 8Hz, but the normal force of the CSFC is 2 times than that of the VSFC strategy. At the high speed ranges of 0-120km/h and 0-160km/h, the average accelerations and acceleration distances by the VSFC strategy are better than the maximum by the CSFC. Therefore, the VSFC strategy has obvious advantages considering comprehensively the tractive performance and the normal force suppression.

7. Conclusion

1）Although the lower slip frequency makes the vehicle starting acceleration larger, the normal force is also larger, which brings a burden to the levitation system. The traction attenuation in the constant power zone is faster, which makes the vehicle difficult to drive at a higher speed. The higher slip frequency makes the starting normal force as the repulsive force, which can reduces the burden of the levitation system, but accelerates slowly at the low speed range and the acceleration time is longer.

2）The constant slip frequency control cannot give full play to the tractive performance of the LIM. In this paper, the VSFC strategy is proposed, the constant force region controls the normal force to a small size; after the maximum power point, the traction of the motor is always the maximum value the LIM can exerts at the corresponding speed. The VSFC strategy can effectively control the starting normal force and improve the tractive performance of the maglev vehicle.

Acknowledgements
This work was supported by the National Natural Science Foundation of China (Grant No.51875483) and the Sichuan Science and Technology Program (Grant No.2018GZ0054).

References
[1] LEE H W, KIM K C, JU L. Review of maglev train technologies[J]. IEEE Transactions on Magnetics, 2006, 42(7): 1917-1925.
[2] Luguang Y. Progress of the maglev transportation in China[J]. IEEE TRANSACTIONS ON APPLIED SUPERCONDUCTIVITY ASC, 2006, 16(2): 1138-1141.
[3] Bazghaleh A Z, Naghashan M R, Meshkatoddini M R. Optimum design of single-sided linear induction motors for improved motor performance[J]. IEEE Transactions on Magnetics, 2010, 46(11): 3939-3947.
[4] Shiri A, Shoulaie A. Design optimization and analysis of single-sided linear induction motor, considering all phenomena[J]. IEEE Transactions on energy conversion, 2012, 27(2): 516-525.
[5] Faiz J, Jagari H. Accurate modeling of single-sided linear induction motor considers end effect and equivalent thickness[J]. IEEE Transactions on Magnetics, 2000, 36(5): 3785-3790.
[6] Pai R M, Boldea I, Nasar S A. A complete equivalent circuit of a linear induction motor with sheet secondary [J]. IEEE Transactions on Magnetics, 1988, 24(1): 639-654.
[7] Kang G, Nam K. Field-oriented control scheme for linear induction motor with the end effect[J]. IEE Proceedings-Electric Power Applications, 2005, 152(6): 1565-1572.
[8] Yoshida K, Shi L, Yoshida T. A proposal of decoupled-control of attractive-normal and thrust forces in a SLIM for maglev vehicle[C]//IMACS. 1999: III. 221- 226.
[9] Long Xialing. The theoretical analysis and electromagnetic design of linear induction motor [M]. Beijing: Science Press, 2006 (in Chinese).
[10] JIAO Yanjun, WEN Yanhui, LIU Shaoke. Uphill traction strategy optimization of middle/low-speed maglev train [J]. Electric Drive for Locomotives,2016(02):37-39+43(in Chinese).

ISPECE
IOP Publishing

Study on Detection System of Grooved Rail Based on Inertial Measurement - Laser Triangulation Comprehensive Algorithm

Hong-Yuan Zhan[1], Kai-Feng Guo[1], Yong-Jun Xie[2]*

[1]School of Electronics and Information Engineering, Jinan University, 519000 Zhuhai, China

[2]Rail Transit Research Institute of Jinan University, 519000 Zhuhai, China

[1]Hong-Yuan Zhan: 115198193@qq.com

[2]Kai-Feng Guo: 627259687@qq.com

[3]Yong-Jun Xie: xieyongjun919@jnu.edu.cn

Abstract. In order to meet the requirement of grooved rail geometric parameter detection, based on the previous research of the project team, this paper combines the principle of inertial measurement and laser triangulation, a comprehensive algorithm to develop a grooved rail detection system for detecting the longitudinal and alignment irregularity. The system is composed of geometric parameter detecting vehicle and upper computer software. Firstly, the outline data of grooved rail are measured by 2D laser sensors, and the motion data of the vehicle are measured by inertial sensors. Then, this paper presents an inertial measurement - laser triangulation comprehensive algorithm and the laser triangulation is used to process the outline data and calculate the gauge of grooved rail. The error of laser triangulation and the attitude angle of vehicle are corrected through inertial measurement method. Finally, the longitudinal irregularity and alignment irregularity are calculated through numerical integration method in the upper computer. The stability and accuracy of the system are verified by designing experiments. The results show that the repeating precision of the irregularity is less than 0.5mm while the detection accuracy is within ±0.7mm.

1. INTRODUCTION

With the development of modern tram, the operating mileage of China's modern trams will continue to increase rapidly and many cities have officially operated trams[1]. Therefore, the safety of trams is becoming more and more important.

However, the detection technology for geometry parameters of grooved rail in Chain is still lagging behind. Compared to grooved rail, i-rail has a more mature detection technology. In recent years, many experts and scholars have begun to deeply research the rail geometry parameter detection. Chen(2014) proposed an i-rail detection technology based on image recognition with two cameras that are parallel and perpendicular to the rail moving forward[2]. With the growing maturity of laser sensor technology, it has been widely used in the field of rail geometry parameter detection. And many new rail detection vehicles in China are using laser sensors as the hardware foundation like GJ-5, GJ-6 and so on. We now take GJ-6 as an example. It uses foreign advanced technology to finish rail detection such as laser sensors technology, RFID mileage positioning technology and other technologies[3]. Although the detection technology of i-rail is becoming more and more mature, the above technologies cannot be directly applied to the field of grooved rail geometric parameters detection through simple

Content from this work may be used under the terms of the Creative Commons Attribution 3.0 licence. Any further distribution of this work must maintain attribution to the author(s) and the title of the work, journal citation and DOI.
Published under licence by IOP Publishing Ltd

modification because of the structural differences between i-rail and grooved rail. In the early stage, the project team had studied the application of laser triangulation method in grooved rail detection, and realized the function of the detection system to detect gauge and abrasion of grooved rail, but the team could not achieve the detection of parameters such as the longitudinal irregularity and alignment irregularity. In this paper, we conduct a study about the longitudinal irregularity and alignment irregularity with the method combined with both inertial measurement and laser triangulation.

The rest of paper is organized as follows: The Section 2 introduces the main measurement methods and the algorithm of the system in this paper. The Section 3 introduces the structure of the system. The Section 4 introduces the designed experiment, and displays the experimental data. And the Section 5 is conclusion.

2. PRINCIPLE AND ALGORITHM

2.1. Detection Principle

2.1.1. Laser Triangulation
In order to meet the requirement of gauge measuring and to improve the precision of calculation results, a direct laser triangulation method with the advantages of one-to-one correspondence of direct spot and position, broad spot and high light intensity concentrated is selected in this paper[4]. The direct laser triangulation method is a kind of laser triangulation methods used in a special situation when the angle of incidence between the laser beam of the sensor and the measured object is 90 degrees, or, when the incident beam is perpendicular to the measured object. The measurement formula is as follow[5]:

$$ y = \frac{x(a-f)\sin\beta}{f\sin\alpha \pm x(1-\dfrac{f}{a})\sin(\alpha+\beta)} $$

In the formula, α is the working angle, that is, the angle between the reflected light OP and normal; β is the angle between the photosensitive element and O'P'; a is object distance; b is image distance; f is focal length; y is the moving distance of the measured object relative to the reference point O; x is the moving distance of the imaging point P' relative to the reference point P.

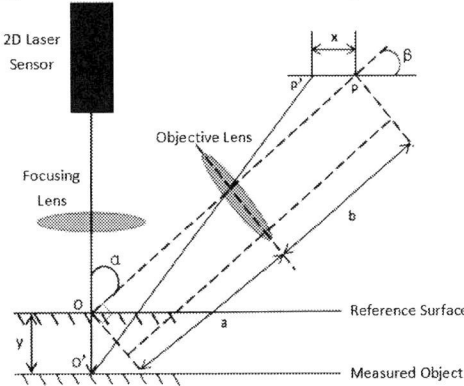

Figure 1. Laser triangulation principle diagram

2.1.2. Algorithm of Quaternion
It is assumed that the coordinate system of grooved rail detecting vehicle is A system, and the geographic coordinate system is N system. The roll, pitch and yaw of vehicle can be obtained by determining the attitude matrix of the A system relative to the N system[6-7]. The attitude differential equation of the grooved rail inspection vehicle expressed by quaternion is as follow:

$$Q = \cos\frac{\theta}{2} + u^N \sin\frac{\theta}{2}$$

In the above formula, u^N is the rotary instantaneous axis and rotary direction; θ is the rotation of grooved rail detecting vehicle. The attitude matrix of the A system relative to the N system C_A^N is obtained through quater-nion. The formula is as follow:

$$C_A^N = \begin{bmatrix} 1-2(q_2^2+q_3^2) & 2(q_1q_2-q_0q_3) & 2(q_1q_3+q_0q_2) \\ 2(q_1q_2+q_0q_3) & 1-2(q_1^2+q_3^2) & 2(q_2q_3-q_0q_1) \\ 2(q_1q_3-q_0q_2) & 2(q_2q_3+q_0q_1) & 1-2(q_1^2+q_2^2) \end{bmatrix}$$

The relation between the attitude matrix and rotation angle can be deduced by the relation between the quater-nion and rotation angle, and then the attitude angle of the grooved rail detecting vehicle can be deduced.

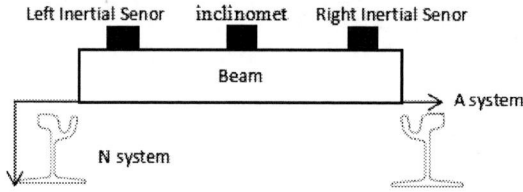

Figure 2. Inertial platform

2.2. Algorithm

2.2.1. Comprehensive Algorithm

Combining inertial measurement method and laser triangulation method, this paper proposes comprehensive algorithm for detecting grooved rail irregularities. First of all, 2D laser sensors collect the outline data of grooved rail. Then, the grooved rail contour feature points are extracted through the project team's mature gauge algorithm[5]. For the convenience of calculation, this paper selects gauge point as the irregularity measurement point. Next, the upper computer find the distance between the irregularity measurement point and the sensor from the data flow returned by laser sensors, which is used to calculate the irregularity. Finally, inertial sensors obtain the motion data of the system, and the correct attitude angle of the grooved rail inspection vehicle is obtained through the quaternion algorithm. The algorithm flow chart is as follow:

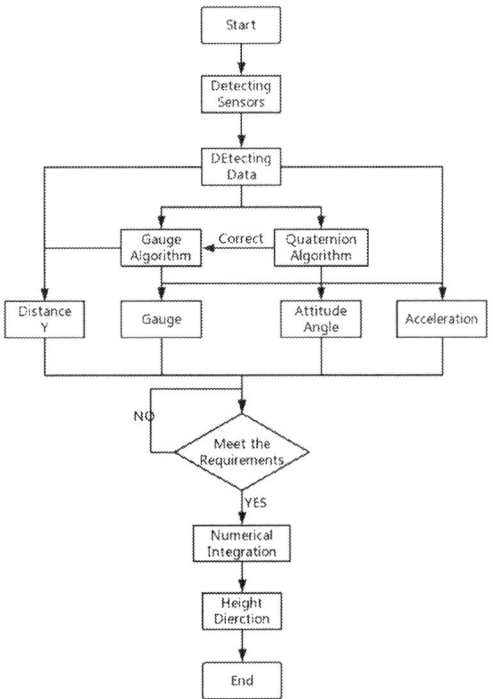

Figure 3. Comprehensive Algorithm flow chart

2.2.2. Longitudinal Irregularity Algorithm

Algorithm of longitudinal irregularity is as follow：

Step.1: Connect the sensors. At first, detect the data returned by inertial sensors and get vertical acceleration Acc of the grooved rail detecting vehicle. Furthermore, the upper computer detects the data returned by 2D laser sensors and gets vertical distance Y between gauge point and sensors.

Step.2: Transmit data. The sensors collect the service data within 10ms of the vehicle's operation and upload it to the upper computer. The upper computer determines whether the data is integrated or not. If so, the data will be segmented and stored.

Step.3: Detect. The upper computer determines whether the data is satisfactory. If not, the computer will reject the data and repeat executing Step.3.

Step.4: Calculate. The distance between the gauge point of the left and right grooved rail and the sensors at time t is $Y1_t$ and $Y2_t$ respectively. Let m and n be two moments and m is less than n. Then, the relative vertical displacement of the left and right grooved rail named W_1 and W_2 can be calculated. Furthermore, the inertial sensors measure the vertical acceleration of the vehicle at equal time intervals in 50ms at 4 points, which was recorded as Acc_1 to Acc_4. Setting up vertical acceleration $Acc_1=0$ of vehicle at first time. The upper computer obtains the vertical displacement Z of the vehicle through numerical integral method. Finally, the upper computer calculates the longitudinal irregularity of grooved rail.

Step.5: Save data.

2.2.3. Alignment Irregularity Algorithm

Based on application of inertial measurement principle to i-rail, this paper explores a method of applying inertial measurement principle and laser triangulation to the detection of grooved rail geometric parameters. The inertial measurement principle is applied to get the formula of alignment irregularity and to realize the detection. The formula of alignment irregularity is as follow:

Left: $Y_L = Z + \dfrac{S}{2} = \iint a_0 dtdt + \dfrac{S}{2}$

Right: $Y_R = Z - \dfrac{S}{2} = \iint a_0 dtdt - \dfrac{S}{2}$

In the formula, Y_L is the alignment irregularity of left grooved rail and Y_R is right; a_0 is modified inertial acceleration; S is gauge. In the set inertial coordinate system, the detection system will draw two orbital curves, namely, the plane strike diagram of the grooved rail. a_0 is the inertial acceleration detected by gyroscope; g is the acceleration of gravity; θ_t is the angle between grooved rail and ground; v is speed of grooved rail inspection vehicle; R is the radius of rail and θ is the angular velocity of θ_t. Then, using g, θ, v, R, θ_t, and S to modify a_0.

In the alignment irregularity algorithm, R cannot be measured simply, but it can be obtained indirectly by calculating the curvature of grooved rail. The formula of the curvature is: K = da/ds . The curvature at a point on the curve is equal to the tangent instant angle da over the instantaneous arc length ds. Based on the definition and formula of curvature, the curvature of grooved rail is equal to the angular velocity of the yaw: K=w. In the formula, K is the curvature and the w is the angular velocity of the yaw. Through the above formula, a_0 can be corrected more accurately.

3. SYSTEM STRUCTURE

The detection system includes software and hardware. In terms of hardware, 2D laser sensors detect the outline data of grooved rail and transmit it to the upper computer through Ethernet communication while the inclinometer, inertial sensors and gyroscope collect the motion data of vehicle. As for software, it has functions of real time data displaying, image rendering, out of limit alarming and so on. The detection system implementation flowchart shows as follow:

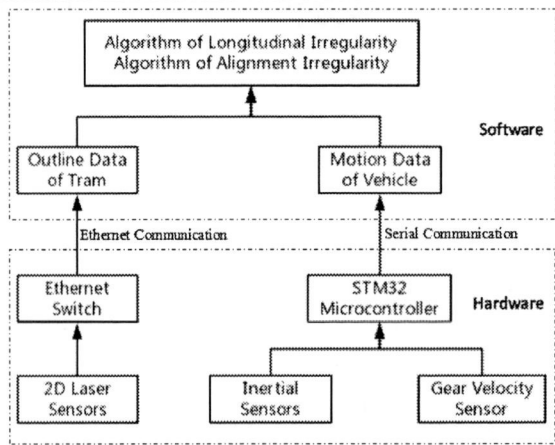

Figure 4. The detection system implementation flowchart

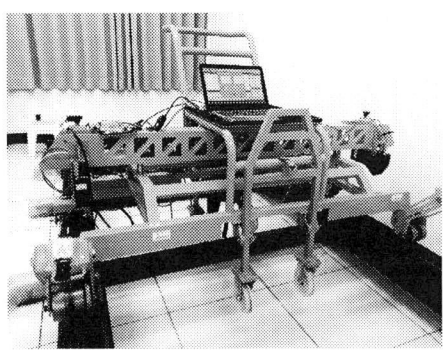

Figure 5. The track inspection vehicle

4. EXPERIMENT

In order to meet the requirement of experiment, we set a 6m long 60R2 grooved rail in the laboratory. According to the standard of i-rail detecting vehicle, it requires that the indication error of the grooved rail irregularity detection system is within ±0.7mm while the grooved rail is shorter than 10m chord, and the repeatable measurement error is less than 0.5mm on the same length of grooved rail[8].

4.1. Verification Experiment of Height

In order to verify the longitudinal irregularity algorithm, this paper takes the grooved rail in the laboratory as the standard grooved rail, which is, the longitudinal irregularity of the grooved rail in the laboratory is nonexistent. So that the precision of the longitudinal irregularity algorithm can be obtained by comparing the ideal longitudinal irregularity(height) and the measure-ment. From the experimental data, it can be seen that the maximum error of system detection is 0.5947 mm, less than 0.7 mm, which meets the requirements of modern grooved rail geometric parameters detection. The data is as follow:

Table 1. The partial experimental data of height

Sequence Number	Experimental Data(unit: mm)		
	Ideal value	Measured value	Measurement error
NO.1	0	-0.0101	0.0101
NO.2	0	-0.0177	0.0177
NO.3	0	-0.0156	0.0156
NO.4	0	0.0901	0.0901
NO.5	0	0.0368	0.0368
NO.6	0	0.0374	0.0374
NO.7	0	-0.0320	0.0320
NO.8	0	-0.0151	0.0151
NO.9	0	0.0206	0.0206
NO.10	0	0.5947	0.5947

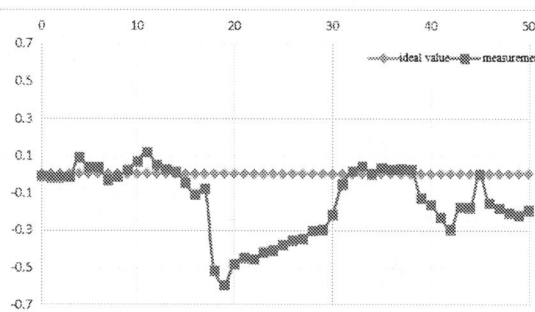

Figure 6.Ideal value and measurement of height

Besides, repeating the experiment to verify whether the algorithm meet the requirement of the repeatable measurement error. From the data, it shows that the maximum error of system detection is 0.137 mm, less than 0.5 mm, which meets the accuracy requirements of repeated experiments. The data is as follow:

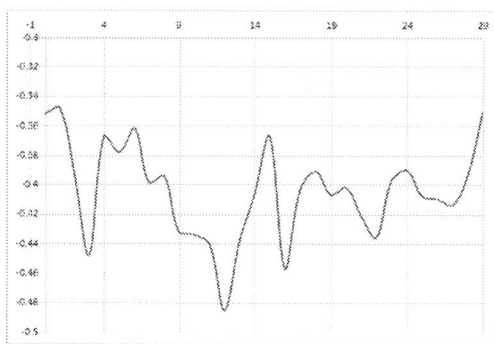

Figure 7. The result of repeating experiment

4.2. Verification Experiment of Rail Direction
In order to verify the alignment irregularity algorithm, this paper takes the grooved rail in the laboratory as the standard grooved rail, which means, the grooved rail in the laboratory is straight and parallel. We now take the left rail as an example, the ideal grooved rail alignment irregularity(direction) and the measurement are compared under the same coordinate system to obtain the precision of detection system. From the experimental data, it shows that the maximum error of system detection is 0.657 mm, less than 0.7 mm, which meets the requirements of modern grooved rail geometric para-meters detection. The data is as follow:

Table 2. The partial experimental data of direction

Sequence Number	Experimental Data(unit: mm)		
	Ideal value	Measured value	Measurement error
NO.1	0	0.1303	0.1303
NO.2	0	0.4230	0.4230
NO.3	0	0.0346	0.0346

NO.4	0	0.0656	0.0656
NO.5	0	0.2320	0.2320
NO.6	0	0.3580	0.3580
NO.7	0	0.0374	0.0374
NO.8	0	0.5218	0.5218
NO.9	0	0.6566	0.6566
NO.10	0	0.2339	0.2339

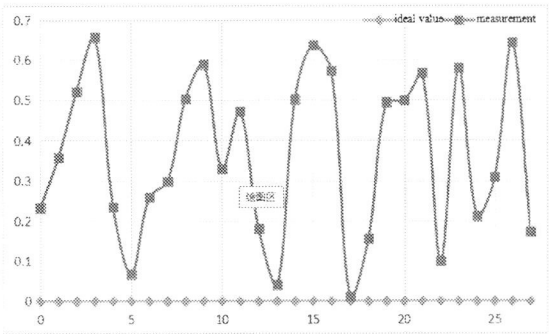

Figure 8. Ideal value and measurement of rail direction

5. CONCLUSIONS

Based on the inertial measurement - laser triangulation comprehensive algorithm, this paper presents a practical and feasible scheme of the longitudinal irregularity and alignment irregularity of grooved rail detection. The original modern tram detection system developed by the project team is perfected and the inertial measurement - laser triangulation algorithm is verified by experiment in this paper. From the experimental data, it shows that the measured value of the alignment irregularity fluctuates greatly. Although the experimental data shows that it has met the requirement of the grooved rail detection, its error is huge. The reason for this may be the vibration of detection vehicle, resulting in a huge error in its attitude data, which makes the algorithm unstable. The next phase of the project team will focus on eliminating the impact of the vibration of detection vehicles.

ACKNOWLEDGMENT

This research was supported by Natural Science Foundation of Guangdong Province, China (NO:2017A-030310184), Zhuhai Collaborative Innovation Center for Efficient Rail Transit Operation, Guangxi Key Laboratory Open Project Foundation (16-380-12-012k), the college students innovation and entrepreneurship training program named Studied on Dynamic Detection and Error Correction Algorithm for Grooved Geometry Parameters Based on Inertial Detection Method(NO: 201810559059).

REFERENCE

[1] Shen, Jingyan. *"Discussion on Development, Criterion and Planning of Tram in China."* Urban Rapid Rail Transit (2015).

[2] Chen, Q., et al. *"Static geometry measurement of high-speed railway tracks by vehicle-borne photogrammetry."* Journal of the China Railway Society 36.3(2014):80-86.

[3] Liu, Lian Ping, et al. *"Development and verification of GJ-6 track inspection system."* Railway Technical Innovation, Vol. 2, (2015):53-56.

[4] Tang, Wen Bin, et al. *"Structure Design and Research of Track Gauge Instrument for Modern Tram."* Railway Standard Design (2018).

[5] Wan, Jin, and Y. Q. Huang. *"Study on Laser Triangulation Method Measurement."* Journal of Sanming University (2006).

[6] Wang, Xuemei, and N. I. Wenbo. *"Measurement Foundation of Railway Track Geometrical Parameters Based on Strapdown Inertial Technique."* Journal of Southwest Jiaotong University (2012).

[7] Qin, Yongyuan. *"Inertial Navigation."* Science Press(2014).

[8] TB/T 3147-2012, *"Inspecting instrument for railway track."*, China Railway Publishing House(2012).

Mimic Defense Structured Information System Threat Identification and Centralized Control

Bo Zhang[1,a], Weichao Li[2,b], Xin Sun[3,c], Yufeng Zhao[4,d]

[1]State Grid Key Laboratory of Information & Network Security
Nanjing, China

[2]State Key Laboratory of Mathematical Engineering and Advanced Computing
Zhengzhou, China

[3]State Grid Zhejiang Electric Power Research Institute
Hangzhou, China

[4]State Key Laboratory of Mathematical Engineering and Advanced Computing
Zhengzhou, China

[a]email: zhangbo@geiri.sgcc.com.cn [b]email: 010liweichao@163.com [c]email:
16526452@qq.com [d]email: 2298869023@qq.com

Abstract: With the speeding up of the informatization process of our national economy and social development, the degree of informatization in our country is getting higher and higher. Information system has become the key infrastructure of our country. The security of these basic information networks and important information systems has been seriously related to national security and social stability, to the collective interests of enterprises, and to the vital interests of the broad masses of the people. Increasing application in improving work efficiency and economic benefits at the same time, the increasingly com-plex information system has also brought greater challenges to operation and maintenance managers. The use of known or unknown vulnerabilities, causing damage to information systems, stealing sensitive information, endangering the normal operation of business systems and other behaviors, has become one of the main causes of various security incidents. How to effectively deal with the vulnerability threat and various vulnerability problems of enterprise information systems, concentrate on the implementation of information security vulnerability warning related risks, and provide solutions to related vulnerability threats, have become an urgent problem to be solved. Therefore, we construct the mimic structured enterprise information system by establishing the threat identification system through unified management to form a comprehensive and secure cyber threat management system.

1. Introduction
With the rapid development of the Internet, the network and web application have brought a fundamental change in people's living and thinking. However, as a key security issue, cyberspace security has become more prominent [1]. The traditional defense, mostly using signature detection, configuration access control label and other static, passive protection strategy. The traditional defense has hysteresis effect on the aspect of coping with unknown vulnerability attacks. To enhance the protection capability of various kinds of vulnerabilities, in recent years, active defense has gradually become the central research issue in cyberspace defense technology [2].

Content from this work may be used under the terms of the Creative Commons Attribution 3.0 licence. Any further distribution of this work must maintain attribution to the author(s) and the title of the work, journal citation and DOI.
Published under licence by IOP Publishing Ltd

Our related work is using dynamic, heterogeneous to achieve structural diversity on active defense model [2]. Zhang et al. introduced a basic construction method and performance index of dynamic and heterogeneous web server [3]. Tong et al. introduced a method of constructing system heterogeneity by using the diversity of software and hardware [4]; Zhang et al. introduced a method of software diversified compilation, which is proposed to resist the threat of attack [5]. Li et al. introduced a heterogeneous tagging method against SQL injection attacks [6]. Ma et al. introduced a similarity solving method, which is proposed to enhance the discriminating and decision-making capabilities of DHR architecture systems [7]. And in another paper, Ma et al. proposed a dynamic scheduling model and method to resist persistent threats [8]. However, the current related work has no effect on Threat Identification and Centralized Control, which may make the identification of attack activities deficient.

2. DHR ARCHITECTURE AND INFORMATION SYSTEM

The DHR architecture is a principle method for implementing mimic defense. In the practical application of the information system based on DHR architecture, in order to ensure the normal security of the enterprise information system, it is usually necessary to configure various functional modules in the mimic defense architecture. Once the configuration is wrong, it will result in Multiple heterogeneous executables generate a collaborative attack vulnerability, which is easily attacked by an attacker to obtain some important data in the information system, causing irreparable damage to the user. Fig. 1 shows the DHR structure with a formalization research [2,3,4].

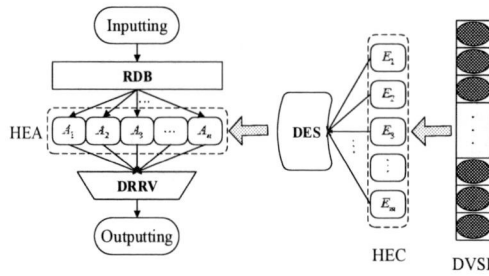

Figure 1DHR architecture.

- The request dispatching and balancing module (RDB) which dispatches and balances user input requests to the appropriate executors.
- The dissimilar virtual server pool (DVSP) that consists of numerous virtual servers to construct heterogeneous.
- The heterogeneous element component (HEC) which is generated in DVSP and waits for scheduling.
- The dynamically executing scheduler (DES) which dynamically chooses heterogeneous elements for the HEA according to the principle of the heterogeneity maximization.
- The heterogeneous executor aggregate (HEA) which receives requests from the RDB, parses them, and passes the executed result to the voter.
- The dissimilar redundant response voter (DRRV) which responses and process requests across the executor aggregate and vote out the inaccurate responses.

In order to ensure the security and validity of system configuration, we provide an information processing method and threat identification and control equipment. After obtaining the data access request, the identity information carried by the system will be verified to be qualified, and then the independent processing results of multiple heterogeneous executives on the same input data can be obtained. The voting result and the processing logs corresponding to the plurality of heterogeneous executors can avoid the illegal operation of illegal users. In addition, by combining the voting results with multiple processing logs, the existence of abnormal heterogeneous executors in multiple heterogeneous executors is analyzed, and abnormal heterogeneous executors are replaced in time, so as to ensure the security, reliability and stability of the system and equipment configuration.

3. CENTRALIZED CONTROL PLATFORM

a) Platform Design

The mimic defense structured enterprise information system and its threat identification and centralized control p-

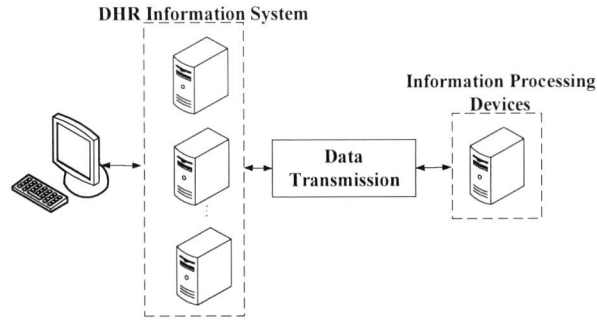

Figure 2 Platform design

latform mainly includes: request dispatcher, response voter, multiple heterogeneous executors, data transmission devices and information processing equipment. Among them, request dispatcher, response voter and multiple heterogeneous executives belong to the application of mimetic defense technology, which is not discussed. The application scope of surgery is not mentioned in this article. The threat identification and centralized control platform includes data transmission devices and information processing devices. Fig. 2 shows the design method of the mimc information system management platform.

According to the functional composition of each module, the centralized management and control platform for threat identification of enterprise information system under the framework of mimic defense can be divided into four functional modules: security audit, security management, operation and maintenance and system upgrade. Through independent information processing interface, data communication with dynamic heterogeneous redundant architecture can be realized. Among them, security audit can

be used to manage the processing logs and authenticate the requesting users, that is, the above authentication module, information acquisition module and exception determination module; security management is mainly used to ensure that the system does not provide redundant network servers, and in the event of abnormal circumstances, to ensure that The module of this function can belong to the part of security management, and the part of operation and maintenance is mainly used to monitor the running state of the system, and automatically generate specific recommendation policies, such as the above authentication loss. The system upgrade part is mainly used to realize the system upgrade, and the upgrade related modules belong to this part.

b) Information Processing

Information processing devices generate corresponding processing logs for each heterogeneous executor to record the processing of input data by the heterogeneous executor. Combining with multiple processing logs and voting results, we can accurately analyze the existence of abnormal heterogeneous executors in multiple heterogeneous executors, that is, to detect the running state of computer equipment system. The abnormal heterogeneous executor can be switched and reconstructed by the authorized administrator

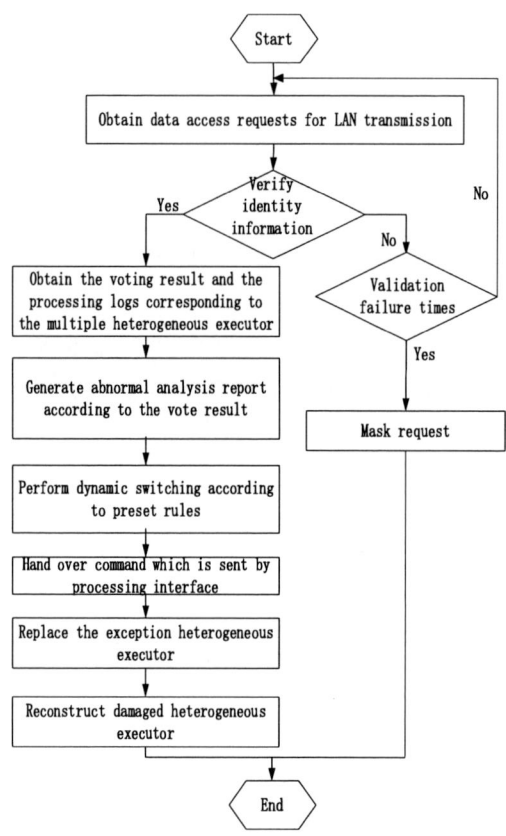

Figure 3 Information processing

promptly and reliably, which ensures the stability of the system. After the heterogeneous executor is attacked, the voter can not accurately judge the abnormal heterogeneous executor, and the system can not switch the abnormal heterogeneous executor in time. The information processing flow is shown in Fig. 3.

In the actual application scenario, it is usually divided into internal network and external network. This implementation can isolate the internal network and external network, and use a relatively secure internal network to achieve data transmission and computer equipment access. The internal network is usually a computer communication network, which is composed of computers, external devices and databases connected to each other in a local geographical area. It is a local area network. It can be connected with other local area networks, databases or processing centers through data communication networks or special data circuits. A wide range of information processing systems, therefore, this embodiment uses a local area network to achieve data transmission, the specific transmission mode is not limited.

c) Data Transmission

The data transmission device mainly includes the information processing interface of the business interface, which can realize the synchronous transmission of the processing logs of multiple heterogeneous executives to improve the work efficiency. For the management of information system by authorized administrator, it can be implemented by independent management interface, which is the information processing interface. It should be noted that the information processing interface is different from the bu-

ISPECE   IOP Publishing

IOP Conf. Series: Journal of Physics: Conf. Series **1187** (2019) 032102   doi:10.1088/1742-6596/1187/3/032102

siness interface of the computer equipment system, that is, the interface used by ordinary users to access the computer equipment. By analyzing the data transmitted by this information processing interface, we set up an independent information processing interface for the management of computer equipment, so as to avoid the attacker posing as authorized administrator to log on computer equipment, steal important data from computer equipment, or destroy the operation of computer equipment system.

It can be seen that the computer equipment which only contain dynamic heterogeneous redundancy architecture have added information processing equipment relative to the transmission computer equipment, which can monitor the running state of the system, alarm and deal with the abnormal situation, set and modify the management strategy, upgrade the system and so on from many aspects, thus ensuring the stability of the operation of computer equipment. In addition, through the security cooperation processing of information processing interface, the vulnerabilities of cooperative attack caused by misconfiguration can be found in time, so that the attacker can not gain access to computer equipment and steal or destroy important data, which can improve the security of computer equipment.

4. THREAT IDENTIFICATION
a)   Association Analysis of Alarm Data

Based on the threat of mimic ruling reasoning methods of abnormal alarm information that created by inconsistency degree of the data flow is too single to trigger some problems, such as the low identification of attack behavior and the difficulty of attack traceability. In order to enhance the accuracy and effectiveness of warning information and facilitate and facilitate the analysis of common malicious attacks, the verdict log and the log record in the actuator container are merged to analyze the attack threat more effectively [9].Fig. 4 is a flow diagram of the generation of abnormal alarm data for the mimic defense system.

Firstly, the system input distributes the request to a set of N actuator containers, which parse and calculate the request, and send their calculation results to a decision module. Then, the ruling module makes a multi-mode ruling on the calculation results of the N actuator containers. According to the degree of inconsistency of the multi-mode ruling, the results with higher degree of inconsistency are marked as abnormal alarm data. When the exception information is recorded in the ruling log, the executor container index and the failure type recorded in the ruling log are used to determine whether the exception occurs in the application log from the log of the specific application, and then the related records of the process running with or without the exception are determined in the system log, and the relevant records of the process running with or without the exception are based on the type of failure recorded in the ruling log. Threat level is divided according to the situation that appears in the application level. Finally, the alarm information is ge-

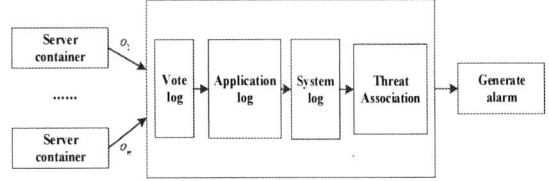

Figure 4 Flow diagram of generating abnormal alarm data

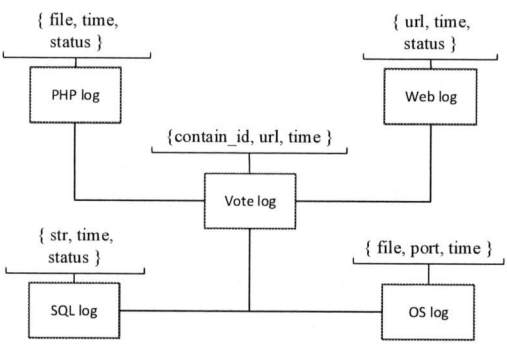

Figure 5 The alarm data correlation diagram

nerated from the multiple determined attack threats. The alarm data correlation diagram is shown in Fig. 5.

Specifically, the decision log recodes decision inform-ation data, such as the ruling inconsistent degree, executive body index, the decision to visit the url, the enforcement of the ruling time; It is recorded in the PHP application log: response time, execution file, PHP parsing result status, and other program code execution data; It is recorded in the Web service log, such as the field of response time, response url, response status, and application service running data. It is recorded in the SQL application log that response time, SQL execution statement, SQL execution result state, and other SQL statements process the result data. Recorded in the operating system log, system files, port operation, running time, such as system call process data. The threat level is established by identifying threat generation at multiple levels. For example, if a Web service uses three executive body of the container mimic defense architecture, each execution body has four layers: application logic layer, Web service layer, data storage layer and system architecture. The abnormal alarm data generated by heterogeneous execution body container at different levels are obtained. According to the threat level, the threat level is from high to low: system architecture layer, data storage layer, Web service layer and application logic layer. The specific threat level is classified according to the inconsistency of response output recorded in the ruling log.

For example, if all three of the actuators are treated the same, no attack threat is considered and the threat level is 0. If the three execution results obtained by the executors are different, it is deemed that the attack is not effective on all three executors, and threat is only found in the application logic layer, then it is considered as level 1 threat. Threats found in the Web services layer are considered as level 2 threats. Threats found in the data storage layer are considered as level 3 threats. Threats identified at the system archite-cture level are considered level 4 threats. If two of the three execution results obtained by the processing of the three executors are the same, and the same two execution results are different from the other execution results, the attack is considered to be effective on one executor, and the attack is considered as a 5-8 threat from the application logic layer of the system to the system architecture layer. The classification of threat level can be adjusted according to the different choice of simulation defense architecture and decision strategy.

b) The classification prediction method of threat

When the ruling log is merged with the multi-level system log, there will be a lot of data information in the log record, including contain_id, url, time, IP, port, file and so on. Therefore, for the purpose of classification prediction of• attack threat, the historical information knowledge base is built according to the feature fields extracted from the fusion log. The deep data information is mined and learned through deep learning, so as to achieve the classification prediction of threats [10].

Sensitive information such as access categories and sources in abnormal access traffic is analyzed according to the data record of the simulated defense threat perception log. In the abnormal access traffic recorded by the mimetic device, it is divided into six categories:

- Attempt of xss attack. Include trying javascript function in the use of abnormal access request, such as "publish/main / 9 / javascript: the history, the back ()", "/ publish/main / 17 / javascript: void (0)", etc.
- Attempt of webshell connection. Include trying .php, .asp, .jsp and other types of abnormal access requests, such as "/plus/mytag_js.php", "/index.asp", etc.
- Attempt of xml utilization. Include exception access requests that try to make use of external xml files, such as "/ rss /bullet _2_0.xml", etc.
- Attempt of directory browsing. Include an attempt directory burst exception access request, such as "/ yishi/","/admin/", etc.
- Attempt of variable utilization. Include exception access requests that try to take advantage of variable operations in the source code, such as "public/column/4664041"? type = 4 & catId = 4694378 & action = list ", etc.
- Attempt of the other exception access. Include abnormal access requests for scanning backup files that may exist on the site, such as "/robot.txt", "/www_cert_org_cn.rar", "/ 4652100.html", etc.

Of the 6 types of abnormal access, there were 29 XSS cross-site attack attempts, 237 webshell connection attempts, 749 XML utilization attempts, 2839 directory browsing atte-

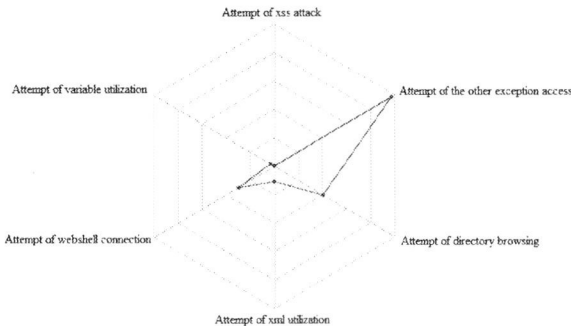

Figure 6 The Classification of threat

mpts, 237 variable utilization attempts and 6,798 other exce-ptions. The threat classification of mimic equipment is shown in Fig. 6.

c)   Visualization of Threat Situation

As early as 1977, John elaborated on the profound impact of visualization on data analysis [11]. For the warning data after the classification of threat level, the information needs to be expressed more clearly and effectively by means of graphics and images. Using data visualization technology, therefore, will each data representation for system situation elements, a large number of data collection system show figure, at the same time, in the form of multidimensional data system situation of each attribute value, from the aspect of different dimensional data analysis, thus further observation and analysis of data [12]. The visualization of simulated threat situation adopts the method of generating situation display diagram, which mainly includes general threat trend diagram, average threat trend diagram, overall request trend diagram, average request trend diagram, threat source distribution diagram and threat rank diagram. By classifying threat data and presenting them according to different display requirements, features are aggregated, and different feature aggregation results are generated according to different contents that need to be counted, and then displayed visually.

Data types can be divided into the following 5 categories in the visualization demonstration of mimic threat situation:

- for data with classification attributes, each classif-ication attribute has a value. Therefore, it is easy to grasp the range of data sets by classifying data. Taking the ranking chart of attacked websites as an example, cluster the domain name information in the threat perception log to obtain the number of threats generated by all different networks. Through the histogram, the numerical value of each classification and the most value in the classification can be clearly observed.

- for data containing overall and partial relationships, it is necessary to be able to clearly express the hierarchical structure of the data and keep all the groups together for analysis. Taking the attack category proportion map as an example, the attack categories recorded in the threat perception log were classified and summarized, the number of attack categories and total attack amount were calculated, and the proportion of each attack category was displayed through the pie chart.

- for data containing nested relationships of subcate-gories, multi-dimensional data is usually needed to be introduced for observation, so as to obtain richer connotation information. Taking the execution body state diagram as an example, if we want to know the attack situation of multiple execution body containers, we need to increase the data dimension of the execution body container category in the attack classification data. The proportion of each attack type in the running state of each execution body container is shown through the multiple ring graphs.

- for data that changes with time, the change of the data at what time, the degree of the change, and the reason for the change of the data trend are all contents that need to be paid attention to by the time series data. Take the attack trend diagram as an example to summarize the number of threats over a period of time, and then connect the number of threats at each time point in a chronological order to form a curve, so as to clearly observe the number of attacks and the trend of attacks.

- for the data affected by spatial changes, the spatial location information expressed in the data is the most important. Taking the display of the attack world map as an example, the longitude and latitude information of the source of the attack is mapped to the two-dimensional world map, and different colors are used to mark the source of the attack in the map. The number of different attack sources can be obviously observed through the differences of colors.

5. EVALUATION

To exclude the risk factors and evaluate the Centralized control platform efficiently, we construct a PHP information system with various vulnerabilities, shut down the front-end filtering function and allow users to attack on the internet. All tests were performed on Intel core i7 CPU clocked at 3.1 GHz on machines with 32 GB RAM running PHP.

Fig. 7 shows that from 11:00 to 12:30, the highest number of visits is 210 times per minute, the lowest number of visits is 0 times per minute, the average number of 26.65 times per minute. The information service system runs normally and smoothly.

Fig. 8 shows that the total number of attacks recorded during the day is 1857 and the attacks started at 9:00 and increased significantly after 11:00. The Centralized control platform can effectively record threats.

6. CONCLUSIONS

This paper provides a method of mimic information system and threat identification and centralized control, whi-

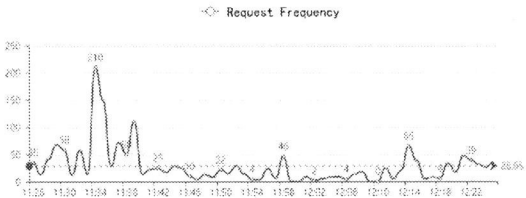

Figure 7 The Request Frequency

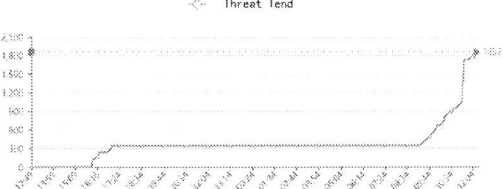

Figure 8 The Threat Trend

ch provides a solution for the implementation of information security vulnerability early warning risk and vulnerability threat analysis. The centralized management and control platform is constructed through the information processing and data transmission interface. The anomaly alarm data association, threat classification prediction and threat situation visualization are used to analyze the anomaly response of the system. In this way, it can effectively help security personnel to analyze and evaluate the effect of mimic defense.

ACKNOWLEDGMENTS

This work is Supported by the science and technology project of State Grid Corporation of China: "Research and application of Mimic Defense Gateway applied to web application for power grid" (Grand No. 52110118001F)

REFERENCES

[1] Scarfone, Karen, Benigni, Dan, and Grance, Tim. 2017. Cyber security standards. *Smart Business Akron/canton.*

[2] Wu JX. 2016. Research on Cyber Mimic Defense. *J. Cyber Security.* 1, 4 (Sep. 2016), 1-10.

[3] ZHANG Z, Ma BL, Wu JX. 2017. The Test and Analysis of Prototype of Mimic Defense in Web Servers. *J. Cyber Security.* 2, 1 (Jan. 2017), 13-28.

[4] Tong Q, Zhang Z, Wu JX. 2017. The Active Defense Technology Based on the Software/Hardware Diversity, *J. Cyber Security.* 1, 1 (Jan. 2017), 1-12.

[5] Zhang YJ, PANG JM, ZHANG Z, WU JX. 2017. Mimic Security Defence Strategy Based on Software Diversity. J. *Computer Science.* 45, 2 (Mar. 2017), 215-221.

[6] Li WC, Zhang Z, Wang LQ. 2018. Improvement in diversify active defense for web application by using language and database heterogeneity. In *Proceedings of the 11th IEEE International Conference on Anti-counterfeiting, Security and Identification* (ASID). IEEE, 30-35.

[7] Ma BL, Zhang Z, Liu JX. 2018. A Similarity calculation method applied to dynamic heterogeneous web server system. *J. Comput Eng Desig.* 1, 282-287.

[8] Ma BL, Zhang Z, Zhu YS. 2017. A formalization research on web server and scheduling strategy for heterogeneity. *Advanced Information Management, Communicates, Electronic and Automation Control Conference* (IMCEC). IEEE, 1447-1451.

[9] Rinzler K, Cino R, Scully B. Situational awareness system security features[J]. 2018.

[10]   Onwubiko C. Security operations centre: Situation awareness, threat intelligence and cybercrime[C]// International Conference on Cyber Security and Protection of Digital Services. IEEE, 2017.

[11] Situational Awareness Using 3D Visualizations[C]// *International Conference on Cyber Warfare and Security* Iccws. 2018.

[12] Lopezcuevas A, Medinaperez M A, Monroy R, et al. FiToViz: A Visualisation Approach for Real-time Risk Situation Awareness[J]. IEEE Transactions on Affective Computing, 2018, PP(99):1-1.

ISPECE
IOP Publishing

SMC Chaos Control of a Novel Hyperchaotic Finance System Using a New Chatter Free Sliding Mode Control

Guoliang Cai[1,2], Yanfeng Ding[1]*, Qiaoling Chen[1]

[1]Institute of Applied Mathematics, Zhengzhou Shengda University of Economics, Business & Management, Zhengzhou, Henan 451191, PR China.

[2] Nonlinear Scientific Research Center, Jiangsu University, Zhenjiang, Jiangsu 212013, PR China.

Corresponding Author: Yanfeng Ding; email: dingdin3696@sina.com

Abstract. In this paper, using a chatter free sliding mode control (SMC) strategy, SMC chaos control of a novel hyperchaotic finance system is discussed. The nonlinear uncertain chaotic systems in the presence have unknown bounded uncertainties and external disturbances. We propose a sliding mode surface with differential operator, which converts the switching term of discontinuous gain function into only one control input. The control input is smooth and time-differentiable. Based on Lyapunov stability theory and sliding mode control method, SMC stability analysis is carried out, and a theorem for designing chattering-free sliding mode control input is obtained. Finally, numerical simulation verifies the effectiveness and correctness of the control strategy.

1. Introduction

Chaos is an interesting physical phenomenon in nonlinear dynamical systems, which has been studied for decades [1-2]. Chaotic control of nonlinear dynamical systems has attracted wide attention in recent years. In the past two decades, many chaotic control methods have been proposed, such as such as the Ott-Grebogi-Yorke (OGY) method [3], LMI-based non-fragile control method [4], passive control method [5], and backstepping design method [6]. Therefore, chaos control and its application have become a research hotspot in the nonlinear fields.

Sliding mode control (SMC) is a variable structure control. A high-speed switch control law is used to derive the system state trajectory onto a specified user-selected surface, i.e., the so-called sliding surface, and to maintain the system state trajectory on the sliding surface for a subsequent time. SMC is a well-known robust trajectory control method, not the state of direct control system [7, 8]. In this paper, In order to solve the above problems, Lyapunov stability theory and SMC technology are combined to prove the reliability of the strategy [9, 10]. Because the control is smooth and differentiable, the chatter is weakened effectively.

In this paper, a new chatter-free sliding control strategy is applied to a novel hyperchaotic finance system [11]. The system that we first introduced is constructed in that background, where the global economic crisis shows the existence of chaos in the finance system in 2007. From the numerical examples, the hyperchaotic finance system can achieve ideal condition quickly through this method.

The structure of this paper is as follows. In Section 2, a novel hyperchaotic finance system is presented. Section 3 presents the design of chattering-free sliding mode controller. The SMC chaos

Content from this work may be used under the terms of the Creative Commons Attribution 3.0 licence. Any further distribution of this work must maintain attribution to the author(s) and the title of the work, journal citation and DOI.
Published under licence by IOP Publishing Ltd

control of the new hyperchaotic financial system with uncertain parameters is studied. A numerical example is given in this section. Finally, the conclusions are given in Section 4.

2. System description

More recently, a new hyperchaotic finance system was constructed by [11]. A fractional dimensional autonomous hyperchaotic system can be described by the following differential equation:

$$\begin{cases} \dot{x} = z + (y - a)x + w \\ \dot{y} = 1 - by - x^2 \\ \dot{z} = -x - cz \\ \dot{w} = -dxy - kw \end{cases} \tag{1}$$

where variable x represents the interest rate in the model, variable y represents the investment demand, variable z is the price exponent and variable w is the average profit margin, a, b, c, d, k are the parameters of the system (1), and they are positive constants. When the parameters are $a = 0.9$, $b = 0.2$, $c = 1.5$, $d = 0.2$ and $k = 0.17$, the four Lyapunov exponents calculated with Wolf algorithm are 0.034432, 0.018041, 0, −1.1499. The Lyapunov fractional dimension of this hyperchaotic system is $D_L = 3.050121$. Figure 1 shows the 3-dimensional phase portraits of hyperchaotic finance system (1). For more detailed analysis of the complex dynamics of the system, please see relative Ref. [11].

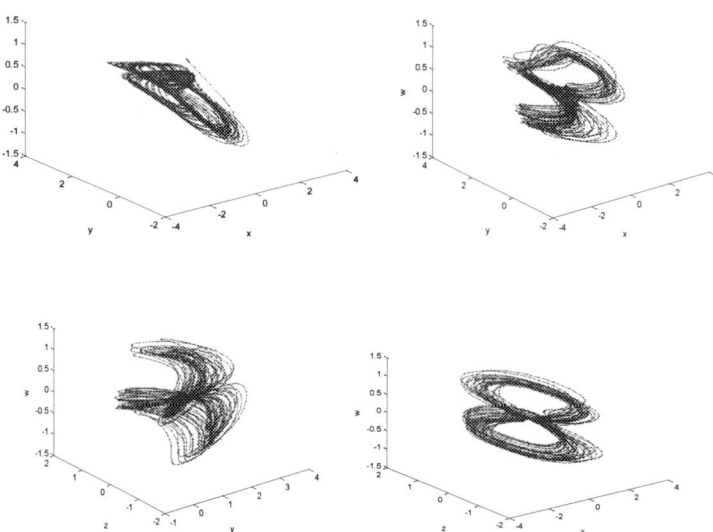

Figure 1. Phase portraits of hyperchaotic finance system (1).

3. SMC chaotic control of a new hyperchaotic finance system

In this part, we mainly discuss the stability of a nonlinear dynamical system. We will give some assumptions, lemma, and conditions to make sure that the system (1) will get stable.

3.1. Chatter free SMC design [9, 10]

Consider a class of controlled hyperchaotic systems described by the following nonlinear differential equations:

$$\dot{x} = Ax + F(t, x) + d(t) + u(t) \tag{2}$$

where $x = [x_1, x_2, \cdots, x_n]^T \in R^n$ is the state vector of the system , $F(t,x): R^+ \times R^n \to R^n$ denotes a nonlinear vector function. $u(t) \in R^n$ denotes the control input vector, $d(t) \in R^n$ denotes the external disturbance. $A \in R^{n \times n}$ is a coefficient matrix. The control goal is to design a chattering-free sliding mode controller under given initial conditions so that the asymptotic stability of the system (2) can be realized, i.e., $\lim_{t \to \infty} \|x(t)\| = 0$, where $\|\cdot\|$ is the Euclidean norm of a vector.

There are two basic steps for SMC design. First, it needs to select a suitable switch surface. Secondly, a control law is needed to ensure the stability of the sliding surface.

Firstly, an integral operator is used to define the sliding surface.

$$S_i(t) = \eta_i \int_0^t x_i(\tau) d\tau + x_i(t) \tag{3}$$

where η_i is a positive constant (i=1,2,…,n). In order to obtain smooth differentiable control input and further weaken the chatter, the switching term of discontinuous symbolic function is transferred to the control input. Using the sliding mode surfaces defined above, we propose the following dynamical sliding mode surface:

$$\sigma_i(t) = \dot{s}_i(t) + \lambda_i s_i(t) \tag{4}$$

where λ_i presents a positive constant (i=1,2,…,n).

From this we can draw:

$$
\begin{aligned}
\dot{\sigma}_i(t) &= \ddot{s}_i(t) + \lambda_i \dot{s}_i(t) = \ddot{x}(t) + (\lambda_i + \eta_i)\dot{x}_i(t) + \lambda_i \eta_i x_i(t) \\
&= A_i \dot{x} + \sum_{j=1}^{n} \frac{\partial F_i(t,x)}{\partial x_j}\dot{x}_j + \frac{\partial F_i(t,x)}{\partial t} + \dot{d}_i(t) + \dot{u}_i + (\lambda_i + \eta_i)(A_i x + F_i(t,x) + d_i(t) + u_i) + \lambda_i \eta_i x_i
\end{aligned}
\tag{5}
$$

where F_i presents the ith row of matrix or vector F.

Secondly, we need to design a sliding mode control strategy to converge the trajectory of the driving system to the sliding mode surface $\sigma_i(t) = 0$ (i=1,2,…,n).

Assumption 1. In equation (2), the uncertain term $d_i(t)$ and the derivate of the uncertain term $\dot{d}_i(t)$ are assumed to be bounded, that is, there exists a positive bounded function $B_i(x)$ and $\overline{B}_i(x)$ making the following inequalities hold: $|d_i(t)| \le B_i(x)$, $|\dot{d}_i(t)| \le \overline{B}_i(x)$ $\forall x \in R^n$ (i=1,2,…,n)

Assumption 2. There exists a positive constant ξ_i satisfying the following inequalities hold:

$$\xi_i > \overline{B}_i(x) + (\lambda_i + \eta_i) B_i(x)$$

where the parameters η_i and λ_i are the same as that in equation (4) (i=1,2,…,n).

Barbalat lemma[12]. If $f(t)$ is nonnegative, integrable (has a finite integral) and uniformly continuous on the interval $[a, +\infty)$, then $f(t)$ tends to 0 as $t \to \infty$.

Theorem 1. For the nonlinear controlled hyperchaotic system (2), the state vector $x(t)$ of the system will converge asymptotically to zero, if the dynamics sliding mode control law is designed as follows:

$$\dot{u}_i = -A_i \dot{x} - \sum_{j=1}^{n} \frac{\partial F_i(t,x)}{\partial x_j}\dot{x}_j - \frac{\partial F_i(t,x)}{\partial t} - (\lambda_i + \eta_i)(A_i x + F_i(t,x) + u_i) - \lambda_i \eta_i x_i - \varepsilon_i sign(\sigma_i) \tag{6}$$

(i=1,2,…,n).

Proof. Construct the Lyapunov function as follows:

$$V = \frac{1}{2}\sum_{i=1}^{n} \sigma_i^2$$

By calculating the derivative of $V(t)$ for time, we get

$$
\begin{aligned}
\dot{V} &= \sum_{i=1}^{n} \sigma_i \dot{\sigma}_i = \sum_{i=1}^{n}\left(\sigma_i\left(\dot{d}_i(t) + (\lambda_i + \eta_i)d_i(t) - \varepsilon_i sign(\sigma_i)\right)\right) = \sum_{i=1}^{n}\left(\sigma_i \dot{d}_i(t) + (\lambda_i + \eta_i)\sigma_i d_i(t) - \varepsilon_i|\sigma_i|\right) \\
&\le \sum_{i=1}^{n}\left(|\sigma_i|\left(\overline{B}_i + (\lambda_i + \eta_i)B_i\right)\right) - \varepsilon_i|\sigma_i| = -\sum_{i=1}^{n}\left(\varepsilon_i - \left(\overline{B}_i + (\lambda_i + \eta_i)B_i\right)\right)|\sigma_i|
\end{aligned}
$$

According to formula (6), we can get $\dot{V} \le 0$.

As a result of $\dot{V} \le 0$, we can obtain that of σ_i on time t is integrable and uniformly continuous on $[0, +\infty)$. According the Barbalat lemma, we get $\sigma_i \to 0$ as $t \to \infty$. So the sliding mode surface is globally asymptotically stable at its equilibrium point $\sigma_i = 0$ (i=1,2,...,n). Therefore, equation (4) is equivalent to $\dot{s}_i = -\lambda_i s_i$, which means that $s_i \to 0$ as $t \to \infty$. So we get

$$x_i(t) = -k_i \int_0^t x_i(\tau)d\tau, \quad (i\text{=}1,2,...,n)$$

When $s_i \to 0$, by differentiability of $x_i(t)$

$$\dot{x}_i(t) = -k_i x_i(t), \quad (i\text{=}1,2,...,n)$$

Therefore, the state vector $x(t)$ of the system (2) will converge asymptotically to zero. The proof is complete.

3.2. SMC control of hyperchaotic finance system

The system is described by the following hyperchaotic finance system [11]:

$$\dot{x} = Ax + F(t,x) \tag{7}$$

where

$$x^{\mathrm{T}} = (x_1, x_2, x_3, x_4), \quad A = \begin{pmatrix} -a & 0 & 1 & 1 \\ 0 & -b & 0 & 0 \\ -1 & 0 & -c & 0 \\ 0 & 0 & 0 & -k \end{pmatrix}, \quad F(t,x) = \begin{pmatrix} x_1 x_2 \\ 1 - x_1^2 \\ 0 \\ -d x_1 x_2 \end{pmatrix}.$$

Considering the characteristics of hyperchaotic finance system, so as state variable x_3 is stabilized to zero, then the others state variables will automatically converge to zero. Therefore, we must design the control input u_3, and then it will be placed on the right side of the third equation in equation (7) so that all states can be stabilized to its origin.

Consider adding an uncertain disturbance $d(t)$ to the right of the system (8). Uncertain hyperchaotic financial systems can be written as follows

$$\dot{x} = Ax + F(t,x) + d(t) + u \tag{8}$$

where $d(t)$ and u be represented as $[0,0,d_3(t),0]^{\mathrm{T}}$ and $[0,0,u_3,0]^{\mathrm{T}}$, representing uncertainties and control inputs, respectively. In addition, it is assumed that $d_3(t)$ satisfies conditions required to ensure that the system defined in has a unique solution in the interval $[t_0, +\infty)$, $t_0 > 0$, for any given initial condition.

In this example, let uncertainty $d_3(t)$ be sin(t), according to the above assumptions, the control parameters could be designed as $k_3 = \lambda_3 = 2$, $B_3(x) = \bar{B}_3(x) = 2$ and $\varepsilon_3 = 10$.

According to (3), the sliding mode surface $s_3(t)$ is designed as follows

$$s_3 = 2\int_0^t x_3(\tau)d\tau + x_3(t)$$

thus

$$\sigma_3 = \dot{s}_3 + 2s_3$$

According to formula (7), the dynamics sliding mode control law is designed as follows:

$$\dot{u}_3 = -\dot{x}_1 - c\dot{x}_3 - 4(-x_1 - cx_3 + u_3) - 4x_3 - 10\mathrm{sign}(\sigma_3)$$

Then system (9) satisfied all conditions in Theorem 1. Thus, the sliding surface of the uncertain hyperchaotic financial system (9) is globally asymptotically stable at its equilibrium point.

To verify the effectiveness of a chatter free SMC method, in the numerical simulations, the initial values of the hyperchaotic system chosen as $[x_1(0), x_2(0), x_3(0), x_4(0)] = [-1, -2, -3, -4]$, the step size

$\tau = 0.001$. Figure 2 shows the sliding mode surface $s(t)$. Figure 3 shows the dynamic sliding mode surface $\sigma(t)$. Figure 4 shows the response of control input $u(t)$ with time t.

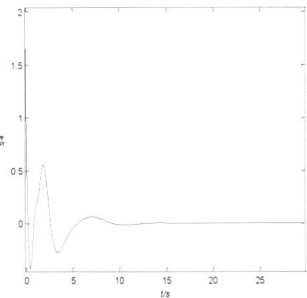

Figure 2. Simulation results of the sliding mode surface $s(t)$.

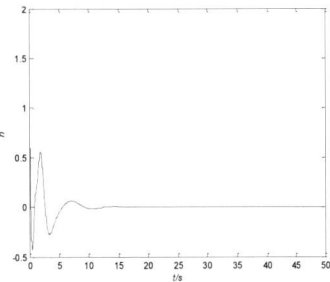

Figure 3. Simulation results of t the dynamic sliding mode surface $\sigma(t)$.

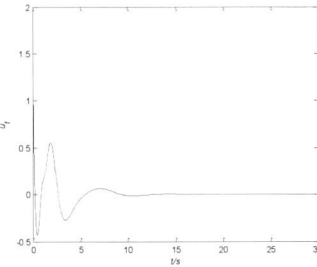

Figure 4. Simulation results of the response of control input $u(t)$ with time t.

4. Conclusions

Aiming at the control problem of hyperchaotic finance system, a new chattering-free sliding mode control strategy is proposed. Based on the Lyapunov stability theory and sliding mode control method, a sliding mode controller is designed from state vector to a desired point in the state space. The simulation results have confirmed the effectiveness of the proposed scheme to control the hyperchaotic finance system. The control is smooth. It shows that the designed controller has lower cost and complexity.

Acknowledgements

This work was supported by the National Social Science Foundation of China (No. 18BJL073), the Key Scientific Research Projects of Higher Education Institutions of Henan Province (Nos. 18A120013, 19A110039), the Social Science Foundation of Educational Department of Henan Province (No. 2019-

ZZJH-202), and the Young Key Teachers Program of Higher Education Institutions of Henan Province (No. 2017GGJS193). Especially, thanks for the support of Zhengzhou Shengda University of Economics, Business & Management.

References

[1] H. Layeghi, M. T. Arjmand, H. Salarieh, A. Alasty, "Stabilizing periodic orbits of chaotic systems using fuzzy adaptive sliding mode control," *Chaos Solitons Fractals*, vol. 37, pp. 1125-1135, 2008.

[2] H. T. Yau, C. L. Chen, "Chatter-free fuzzy sliding mode control strategy for uncertain chaotic systems," *Chaos Solitons Fractals*, vol. 30, pp. 709-718, 2006.

[3] G. L. Cai, S. Zheng, L. X. Tian, " Adaptive control and synchronization of an uncertain new hyperchaotic Lorenz system," *Chin. Phys. B*, vol. 17, pp. 2412-2419, 2008.

[4] A. A. Ahmadi, V. J. Majd, "GCS of a class of chaotic dynamic systems with controller gain variations," *Chaos Solitons Fractals*, vol. 39, pp.1238 – 1245, 2009.

[5] F. Wang, C. Liu, "Synchronization of unified chaotic system based on passive control," *Physica D*, vol. 225, pp. 55 – 60, 2007.

[6] M. T. Yassen, "Controlling, synchronization and tracking chaotic Liu system using active backstepping design," *Phys. Lett. A*, vol. 360, pp.582 – 587, 2007.

[7] S. Dadras, H. R. Momeni, V. J. Majd, "Sliding mode control for uncertain new chaotic dynamical system," *Chaos Solitons Fractals*, vol.41, pp. 1857-1862, 2009.

[8] J. Yan, Y. Yang, T. Chiang, C. Chen, "Robust synchronization of unified chaotic systems via sliding mode control," *Chaos Solitons Fractals*, vol. 34, pp. 947 – 954, 2007.

[9] M. Feki, "Sliding mode control and synchronization of chaotic systems with parametric uncertainties," *Chaos Solitons Fractals*, vol. 41, pp.1390–1400, 2009.

[10] H. Q. Li, X. F. Liao, C. D. Li, C. J. Li, "Chao control and synchronization via a novel chatter free sliding mode control strategy," *Neurocomputing*, vol. 74, pp. 3212-3222, 2011.

[11] H. J. Yu, G .L. Cai, Y. X. Li, "Dynamic analysis and control of a new hyperchaotic finance system," *Nonlinear Dyn.*, vol. 67, pp. 2171–2182, 2012.

[12] K. Gopalsamy, "*Stability andOscillations in Delay Differential Equations of Population*," Kluwer, Academic Publishers, Dordrecht, 1992.

ISPECE   IOP Publishing

IOP Conf. Series: Journal of Physics: Conf. Series **1187** (2019) 032104   doi:10.1088/1742-6596/1187/3/032104

Project Evaluation and Analysis of Metrological Verification Regulation Based on Fuzzy Comprehensive Analysis Method

Zhu Jing[1], LiuJiaxin[2]

[1] NC-MERB of RIOH, No. 8, West Tucheng Road, Haidian District, Beijing,China

[2] RIOH, No. 8, West Tucheng Road, Haidian District, Beijing,China

e-mail: [1] 530170517@QQ.COM [2] 1306279649@QQ.COM

Abstract. The technical specification of highway engineering measurement has been applied as an independent field since 2016. But at present, there is a lack of concrete evaluation methods in the process of project establishment. Within the scope of Highway Engineering Metrology Technical system table, starting from the technical attributes of instruments and equipment such as road engineering, bridge and tunnel engineering, traffic engineering, and combining with the demand of future highway metrology management, it is one of the important means to evaluate the whole process of Highway Engineering Metrology Technical specification scientifically. Based on the fuzzy comprehensive analysis method, this paper tries to simulate the quantitative analysis and evaluation of a metrological verification regulation project, identifies the main factors of the project evaluation, carries out the weight analysis, and finally obtains the quantitative evaluation results. The analysis results show that the evaluation index of metrological verification regulation project is divided into three levels. The first level index is the evaluation of research project. The second level index mainly includes five contents: the necessity of project establishment, the main content and the research scheme, the expected benefit, the research supporting conditions and the risk. The third level index project evaluation content is included in the sub-content. The capacity includes 22 categories, which fully reflects the main factors of project evaluation. The maximum value of the comprehensive quantitative total evaluation element is 0.471, which belongs to the "excellent" evaluation grade, and the project evaluation control effect belongs to the excellent grade.

1. Introduction

The scientific and rational evaluation of scientific research projects is helpful to the scientific selection of projects. Highway Engineering Metrology Technical Specification Project Establishment Evaluation is mostly based on qualitative analysis, as well as other scientific research project evaluation. The comprehensive opinions of appraisal experts are used to make subjective evaluation on each index of project establishment. To a certain extent, this method optimizes the index of project establishment and realizes a more standardized evaluation process, but due to the influence of subjective factors, the evaluation results often have a certain degree of one-sided [1,2]. How to scientifically achieve project evaluation has become one of the research areas of some researchers. At present, some domestic researchers have carried out a number of analysis around the project evaluation. In quantitative analysis, Sun Jing and others have carried out a hierarchical-entropy combination weighting method for the evaluation of agricultural science and technology projects. The innovation value of agricultural development projects is divided into several primary and secondary indicators, and combined with

Content from this work may be used under the terms of the Creative Commons Attribution 3.0 licence. Any further distribution of this work must maintain attribution to the author(s) and the title of the work, journal citation and DOI.

Published under licence by IOP Publishing Ltd

variance maximization. The weight value of weight method is compared and analyzed to verify the feasibility of the method [3]. Zhang Ruifeng and others have constructed the evaluation index of metrological scientific research project, and carried on the index weight analysis under the analytic hierarchy process, and obtained the reasonable result [4]. Based on extension theory, Bao Haijun established an extension evaluation model [5] for scientific research projects.

In this paper, the metrological verification rules for the project as the analysis object, based on fuzzy analytic hierarchy process (AHP) to establish a project evaluation index system for quantitative evaluation, aiming to better guide the practice of such projects through evaluation.

2. The basic principle of Fuzzy Analytic Hierarchy Process(AHP)

2.1. Use AHP to determine index weight.

The overall idea of evaluation is to establish sub-analysis indicators under different research schemes, determine the importance of the indicators with expert scores, and then get the weights of different factors. Finally, the total scores of various schemes are calculated according to the weights of all indicators. The highest total score is the selected target.

The process of AHP is as follows: [4]:

1) Identify the influencing factors of the target.

The main analysis factors are identified according to the selected content. If C is used to represent the set of all the factors, there is $C = \{C_1,\ C_2,\ C_3,...,C_n\}$ a factor.

2) Establish index scoring system.

According to the selected content, the main analysis factors are identified. If C is used to represent the set of all factors, there is $P = \{P_1, P_2, P_3,...P_n\}$, and $P_i(i=1,2,3,...,n)$ is one of the factors.

3) Establish multiple evaluation systems.

Using three-tier index system, namely U_i, U_{ij}, U_{ijk}, respectively, the first, second and third tiers, the specific relationship is as follows:

$$\{U_1, U_2, U_3,...,U_i,...,U_m\}, i=1,2,3,...m \qquad (1)$$

$$\{U_{i1}, U_{i2}, U_{i3},...,U_{ij},...,U_{in}\}, j=1,2,3,...n \qquad (2)$$

$$\{U_{ij1}, U_{ij2}, U_{ij3},...,U_{ijk},...,U_{ijo}\}, k=1,2,3,...o \qquad (3)$$

4) Calculate index weight

The basic content of AHP is used to calculate the index weight. Firstly, all the factors are compared in pairs. Then the judgment matrix is established according to the comparison results between the factors. Then the maximum eigenvalue of the judgment matrix is solved and its consistency is checked. Finally, the weights of the comparative factors are obtained.

Index evaluation system is composed of several levels, and different levels of judgment matrix should be constructed. In a three-level index system, the judgment matrix of each level is calculated according to the level of each level, and then the index weights of each level are obtained. If the index weights of the lower level are multiplied by the index weights of the higher level, the weights of the lower level indexes relative to the higher level indexes can be obtained. Each low level index weight can be expressed quantitatively at the same level.

If A is used to represent the judgment matrix of a certain level, the maximum eigenvalue and eigenvector of A are calculated first, and then the consistency of the judgment matrix is checked. The basic steps of AHP are described in detail.

2.2. Fuzzy comprehensive analysis and evaluation

Fuzzy comprehensive analysis method considers a variety of factors, uses membership function and membership degree to describe the excessive information of intermediary, divides risk based on factor threshold, and obtains quantitative analysis results based on mathematical calculation.The process of fuzzy comprehensive analysis is as follows:

Set up a comprehensive judgement set → The membership degree and membership function of various factors are considered comprehensively →Determining the basic model of evaluation→Determining factors →Weight values between →objects Comprehensive evaluation and analysis. In this paper, the single factor fuzzy evaluation method is adopted, and its concrete calculation process is as follows:

Assuming that r_{ij} is the subordinate degree of i to j, the evaluation results of i relative to factor set can be expressed as follows:

$$R = \{r_{i1}, r_{i2}, ... r_{in}\} \qquad (4)$$

Type (18), R - single factor fuzzy set. The fuzzy sets of factors with n factors are as follows:

$$R = \begin{bmatrix} r_{11} & r_{12} & .... & r_{1m} \\ r_{21} & r_{22} & .... & r_{2m} \\ .... & .... & .... & .... \\ r_{n1} & r_{n2} & .... & r_{nm} \end{bmatrix} \qquad (5)$$

The specific steps of single factor fuzzy evaluation are as follows:

1) establish an evaluation index system U;

2) get index membership degree R;

3) First-order fuzzy comprehensive evaluation: the weight $A = (a_1, a_2, ... a_n)$ obtained by AHP is transformed into B B by R, namely: $B = A \circ R$, "$\circ$" - the synthetic operator in fuzzy method and B - the synthetic evaluation vector.

4) the vector of the comprehensive evaluation is the principle of maximum membership.

The rating level of the project evaluation is: V=[excellent, good, general, qualified, unqualified]

3. metrological verification regulation project evaluation

3.1. Project background

Metrological verification regulations refer to the technical documents with national statutory nature which are used to evaluate the metering performance of measuring instruments and as the basis for verification. It is engaged in metrological verification work of the technical basis, is a national technical regulations to ensure the accuracy and consistency of measuring instruments. There are three kinds of metrological verification regulations: national metrological verification regulations, departmental metrological verification regulations and local metrological verification regulations. Its contents mainly include: the scope of application of the verification regulations; measurement performance; verification items; verification conditions; verification methods; verification cycle and the treatment of verification results. At present, more than 90 items of metrological verification regulations in the field of highway metrology have been published with statistics, and more than 20 items are under study, and they are increasing at the rate of 5-8 items per year. The system table is an important reference material for the declaration of verification regulations, and most of its range professional testing equipment are composite parameters, comprehensive quantity, dynamic on-line measuring equipment. With the development of testing technology, professional testing equipment is replaced quickly, and the operation of equipment is becoming more and more complex. The formulation of corresponding measurement technical specifications should be improved at any time according to the situation. Therefore, it is urgent to evaluate the urgency of setting up a project of instrument and equipment from the instrument and project itself in order to ensure the quality of measurement work continuously.

3.2. Index system of project evaluation for metrological verification regulation

The evaluation index system is divided into three levels: the first level is the evaluation of research projects, the second level is the basic content of the evaluation of research projects, including the necessity of project establishment, main contents and research programs, expected benefits, research support conditions, risk, and the third level is the evaluation content of project establishment. The sub economy of the project is shown in Table 1.

Table 1　Index System of project evaluation

First level evaluation index	Two level index	Three level index
Research project evaluation B	Project necessary B_1	Integration with regional development strategy B_{11}
		Demand of science and technology B_{12}
		Demand of society B_{13}
		Agreement with the formation, upgrading and development of regional industries B_{14}
	Main contents and research plan B_2	Advanced research target B_{21}
		Quality of feasibility study report B_{22}
		Form of research results B_{23}
		The key to solve the problem B_{24}
		Scientific technical route B_{25}
	The expected benefits B_3	Economic benefits B_{31}
		Social benefits B_{32}
		Environmental benefits B_{33}
		Industrialization prospect B_{34}
	Research support conditions B_4	Research conditions of project executing agencies B_{41}
		Management mechanism B_{42}
		The person in charge of the project quality B_{43}
		Comprehensive quality of project team B_{44}
		Financing and budgetary arrangements B_{45}
	Risk B_5	Technical risk B_{51}
		Management risk B_{52}
		Policy risk B_{53}
		Market risk B_{54}

20 experts (experienced senior engineers, project evaluation personnel, government departments, etc.) were invited to evaluate the project evaluation index of the verification regulation. The evaluation results were five grades, namely, excellent, good, general and qualified. Unqualified, the corresponding evaluation criteria are divided into 100 points, 90 points, 70 points, 60 points, 0 points, the final evaluation results will be formed by statistics, as shown in Table 2 below.

3.3. determine the membership matrix and carry out fuzzy comprehensive evaluation.
(1) determining membership degree matrix

According to the results of expert scoring and the relative importance of different evaluation indicators, the membership degree of single factor indicators of different evaluation indicators is obtained by AHP, as shown in Table 3 below.

Table 2　the membership of single factor index of different evaluation indexes

evaluating indicator Grade	excellent	preferably	commonly	pass	Not pass
B_{11}	0.5	0.3	0.2	0	0
B_{12}	0.8	0.2	0	0	0
B_{13}	0.4	0.4	0.2	0	0
B_{14}	0	0	1	0	0
B_{21}	0	0.7	0.2	0.1	0
B_{22}	0.2	0.4	0.3	0.1	0
B_{23}	1	0	0	0	0
B_{24}	0	0.1	0.3	0.2	0.4
B_{25}	0.3	0.4	0.2	0.1	0
B_{31}	0.7	0.3	0	0	0
B_{32}	1	0	0	0	0
B_{33}	0	1	0	0	0
B_{34}	0.8	0.1	0.1	0	0
B_{41}	0.8	0.2	0	0	0
B_{42}	0	0.2	0.4	0.1	0.3
B_{43}	0	0	1	0	0
B_{44}	0	0.2	0.7	0	0
B_{45}	0	0.2	0.4	0.4	0
B_{51}	0.2	0.4	0.4	0	0
B_{52}	0.1	0.2	0.3	0.4	0
B_{53}	0.3	0.3	0.4	0	0
B_{54}	0	0.3	0.2	0.5	0

The matrix of the two level index is obtained from table 3, that is:

$$R_{B1} = \begin{pmatrix} 0.5 & 0.3 & 0.2 & 0 & 0 \\ 0.8 & 0.2 & 0 & 0 & 0 \\ 0.4 & 0.4 & 0.2 & 0 & 0 \\ 0 & 0 & 1 & 0 & 0 \end{pmatrix}, \quad R_{B2} = \begin{pmatrix} 0 & 0.7 & 0.2 & 0.1 & 0 \\ 0.2 & 0.4 & 0.3 & 0.1 & 0 \\ 1 & 0 & 0 & 0 & 0 \\ 0 & 0.1 & 0.3 & 0.2 & 0.4 \\ 0.3 & 0.4 & 0.2 & 0.1 & 0 \end{pmatrix};$$

$$R_{B3} = \begin{pmatrix} 0.7 & 0.3 & 0 & 0 & 0 \\ 1 & 0 & 0 & 0 & 0 \\ 0 & 1 & 0 & 0 & 0 \\ 0.8 & 0.1 & 0.1 & 0 & 0 \end{pmatrix}, \quad R_{B4} = \begin{pmatrix} 0.8 & 0.2 & 0 & 0 & 0 \\ 0 & 0.2 & 0.4 & 0.1 & 0.3 \\ 0 & 0 & 1 & 0 & 0 \\ 0 & 0.2 & 0.7 & 0 & 0 \\ 0 & 0.2 & 0.4 & 0.4 & 0 \end{pmatrix};$$

$$R_{B5} = \begin{pmatrix} 0.2 & 0.4 & 0.4 & 0 & 0 \\ 0.1 & 0.2 & 0.3 & 0.4 & 0 \\ 0.3 & 0.3 & 0.4 & 0 & 0 \\ 0 & 0.3 & 0.2 & 0.5 & 0 \end{pmatrix};$$

(2) fuzzy comprehensive evaluation

The weight of evaluation indicators at all levels is calculated by analytic hierarchy process. Based on the expert group consultation, through many debugging and improving the evaluation data, the second-level index judgment matrix is finally obtained, as shown in Table 3 below.

Table 3 second level index judgement matrix

B	B_1	B_2	B_3	B_4	B_5	W	CR
B_1	1	1/3	1/4	1/2	1/3	0.131	
B_2	3	1	1/2	1/2	1	0.266	
B_3	4	2	1	1/2	3	0.189	0.017
B_4	2	2	2	1	2	0.178	
B_5	3	1	1/3	1/2	1	0.236	

The feasibility judgment matrix of project feasibility can be obtained from table 3. $W = [0.131, 0.266, 0.189, 0.178, 0.236]$;

Similarly, the judgement matrix under the two level index is as follows: $W_{B_1} = [0.301, 0.301.0.277, 0.121]$; $W_{B_2} = [0.102, 0.138.0.302, 0.324, 0.134]$; $W_{B_3} = [0.305, 0.205.0.277, 0.213]$; $W_{B_4} = [0.222, 0.258, 0.192, 0.198, 0.130]$; $W_{B_1} = [0.331, 0.331.0.213, 0.125]$

According to the content of fuzzy comprehensive analysis, the comprehensive evaluation vector is obtained. A , That is,

$$A_{B1} = W_{B1} \cdot R_{B1} = [0.301, 0.301, 0.277, 0121] \cdot \begin{pmatrix} 0.5 & 0.3 & 0.2 & 0 & 0 \\ 0.8 & 0.2 & 0 & 0 & 0 \\ 0.4 & 0.4 & 0.2 & 0 & 0 \\ 0 & 0 & 1 & 0 & 0 \end{pmatrix} = (0.6, 0.33, 0.07, 0, 0);$$

The same reason can be obtained: $A_{B2} = (0.156, 0.768, 0.076, 0, 0)$; $A_{B3} = (0.193, 0.73, 0.077, 0, 0)$; $A_{B4} = (0.123, 0.712, 0.135, 0.03, 0)$; $A_{B5} = (0.137, 0.675, 0.135, 0, 063, 0)$ 。

The matrix of the two level index single factor is synthesized.

$$R_B = \begin{pmatrix} 0.6 & 0.33 & 0.07 & 0 & 0 \\ 0.156 & 0.768 & 0.076 & 0 & 0 \\ 0.193 & 0.73 & 0.077 & 0 & 0 \\ 0.123 & 0.712 & 0.135 & 0.03 & 0 \\ 0.137 & 0.675 & 0.135 & 0.063 & 0 \end{pmatrix}$$

Calculate the overall evaluation index evaluation matrix results: $A = R_B \cdot W = (0.471, 0.322, 0.203, 0.004)$

(3) determine the evaluation results.

Comparing the result of the total evaluation matrix with that of V , the maximum value of the element is 0.471, which corresponds to the grade of "excellent" evaluation. It can be seen that the process of setting up the project achieves the control effect of excellent grade.

4. Conclusions

Based on the fuzzy analytic hierarchy process (FAHP), this paper analyzes the project evaluation of a specific metrological verification regulation, analyzes the weight of the identified evaluation index system, and carries out fuzzy comprehensive calculation, and obtains quantitative evaluation results. The main conclusions of this paper are as follows:

(1) The evaluation index of metrological verification regulations is divided into three levels. The first level is the evaluation of the project. The second level mainly includes five categories, namely, the necessity of the project, the main contents and research programs, the expected benefits, the supporting conditions and the risks. The sub-contents of the evaluation content of the third level include The 22 category. The three level indicators fully reflect the main factors of project evaluation.

(2) the effect of the project control in the metrological verification regulation is excellent grade.

Acknowledgement

Central Public Welfare Scientific Research Institute of basic research special funds(RIOH)

References

[1] MuLianjie.Research on the evaluation index system of investment in health service industry based on AHP and fuzzy comprehensive evaluation [D]. Shanghai Foreign Studies University, 2018.

[2] Zhang Jia Lu. Research on comprehensive evaluation of UAV projects [D]. Beijing University of Posts and Telecommunications, 2011.

[3] Sun Jing, Chen Baofeng. Evaluation of Agricultural Science and Technology Development Project Based on Innovative Value [J].Scientific Research Management, 2017,38 (S1): 453-461.

[4] Zhang Ruifeng, Liang Xingzhong, Wang Lei. Establishment of Evaluation Index System for Metrological Research Projects [J].China Metrology, 2016 (07): 78-80.

[5] Bao Haijun, Xu Bao Gen. Extension comprehensive evaluation of scientific research projects. [J]. Soft science, 2009 (10): 42-45.

[6]metrological verification regulation. Baidu Encyclopedia [DB/OL]. https://baike.baidu.com/item/%E8%AE%A1%E9%87%8F%E6%A3%80%E5% AE%9A%E8%A7%84%E7%A8%8B/8969122?fr=aladdin.

ISPECE

IOP Publishing

Safety Adaptability of Engine Retarder (Jacobs) on Long Downhill of Expressways

Liu Jiaxin, Zhu Jing

Research Institute of Highway No.8 Xitucheng Road, Haidian, Beijing 100088

Abstract. In order to solve the problem that the truck should spray water to cool its brake during run the long downhill on mountain highway in winter, which can lead to icy road and cause traffic accidents. We specially selected the straight truck (6×4) which installs the engine brake (Jacobs), carrying out safety tests on Yaxi highway downhill sections in winter. The tests showed that 30% of vehicle overloading conditions, without the engine brake (Jacobs), relying on the service brake to control downhill speed, when the elevation drop of about 300m, the truck's highest brake drum temperature will rise to a hot recession critical failure temperature (260 °C) above. If you continue driving, it will lead to break failure and loss of control. When the truck going downhill and open the engine brake (Jacobs), due to the brake load is greatly reduced, the brake drum temperature of the truck maintained at 150 °C or less to maintain a good braking performance, and the truck can pass the 51km gap between 1500m long downhill safely. The results of field test show that, when a straight truck goes down a long downhill slope and opens the engine braking (Jacobs), the brake maintains a safe speed without spraying water, and keeps good braking performance. So it can solve the straight truck pass the long downhill sections of mountain highway fundamentally and the brake is not tricking.

1. Introduction

Southwest China is basically a karst landform, referring to the surface of limestone areas common stone buds, stone forests, peak forests, karst ditches, funnels, sinkholes, water caves, erosion depressions, slope valleys, blind valleys and underground karst caves, underground rivers and other cave systems and caves stalactites, stalagmites, pillars, stone waterfalls, etc. For the above unique landforms in the limestone area, the academic circles at home and abroad all named the karst landform plateau in the typical limestone landform area of Yugoslavia, namely "karst landform" (the term "karst landform" was also used in China in the past). The rocks are very hard, the terrain is very complex in many areas, and belongs to hilly areas. Towns are mainly distributed in a few plains and valleys. Urban morphology is greatly influenced by terrain factors. Highway construction in China is advancing to the depth of mountainous areas at present, so restricted by natural conditions, the newly built expressways have steep hills, narrow and sharp roads, natural disasters, low basic conditions for vehicle operation safety, and become accident-prone sections. For example, the Iasi expressway, which has been built and opened to traffic, is located in the Sichuan section of the Beijing Kunming Expressway (G5). The topographic and geological conditions along this section are very complex, the whole process through a total of six ridge crossing, forming three super-long continuous longitudinal slope. Among them, the length of continuous longitudinal slope on the north slope of Niba Mountain, the south slope and the north slope of Tuowu Mountain are 33 km, 26 km and 51 km respectively, ranking the highest in China, with the average longitudinal slope of 2.28%, 2.57% and 2.97% respectively. In addition, there are 22 km above the snow

Content from this work may be used under the terms of the Creative Commons Attribution 3.0 licence. Any further distribution of this work must maintain attribution to the author(s) and the title of the work, journal citation and DOI.

Published under licence by IOP Publishing Ltd

line in the three extra-long longitudinal slope sections, and some sections are also affected by bad weather conditions such as ice, snow, rain and fog at a certain altitude. The operation safety situation is facing severe situation [1].

2. Testing road and vehicle conditions

2.1. test vehicle parameter configuration
Table 1 configuration parameters of test vehicle

Tab.1 The test vehicle configuration parameters

Vehicle Brand	Shanxi Auto Delong F2000	Vehicle Type	SX3255DR434C
Engine Power	254KW/1900rpm	Drive Form	6*4
Gearbox	China Fast Gear 9JS180A	Engine Model	ISME345 30
Preparation Quality / Ton	13	Auxiliary Brake Type	Jacobs
Vehicle Bridge Speed Ratio	5.262	Standard Load Gross Weight / Ton	25

2.2. Test Equipment.
Vehicle performance testing system (Vbox III), infrared temperature measurement system (Optris CT0220), anemometer, stopwatch, etc.

2.3. Test Section

2.3.1.Niba Mountain. 26km on the northern slope of the Niba Mountain is long downhill with 6.7km above the snow line.

2.3.2. Tuowu Mountain. 51km long downhill on the northern slope of Tuowu Mountain, 16.7km above the snow line.

2.3.3.Temperature. The ambient temperature is 10 C.

2.3.4.Wind speed. The maximum wind speed is 8m/s.

3. Engine braking (Jacobs) brake characteristic test
The braking characteristic test is mainly carried out on the flat road. The vehicle is loaded with 19.5 tons (30% overload). It is used to study the deceleration braking ability of the Jacobs engine brake device and determine the engine braking characteristics.

3.1. Test method

3.1.1.Transmission neutral skid test. The vehicle is accelerated to 80km/h, then the transmission is hung in neutral gear, the vehicle is free to glide, and the data of the relationship between speed and time are recorded with the test equipment [2].

The test is conducted in an opposite direction.

3.1.2.Gearshift taxiing test. Accelerate the vehicle to the top speed of 5 gears, turn on the engine brake (Jacobs) switch, and the vehicle glides freely. Record the data of the relationship between the speed and time with the test equipment.

The test is conducted in an opposite direction.
Followed by 6, 7 and 8 stalls, the test is identical to the 5 slip test.

3.2. Engine braking (Jacobs) force test and calculation

According to the above test method, the starting speed (v_0), the terminal speed (v_t), the running time (t), the running distance (s), the wind speed (v_w) and other parameters can be accurately detected. According to the vehicle driving equation, [3]:

$$\delta_i \cdot m \cdot \frac{du}{dt} = F_J + F_f + F_w \dots \dots \quad (1)$$

F_J: Jacobs engine braking force； F_f: Wheel rolling resistance；

F_w: Air resistance；

δ_i: The conversion coefficient of rotation mass of the corresponding gear；

m: Total vehicle mass； $\frac{du}{dt}$: vehicle deceleration.

$$F_J = \delta_i \cdot m \cdot \frac{du}{dt} - (F_f + F_w) \dots \dots \quad (2)$$

$$\frac{du}{dt} = \frac{(v_0 - v_t)}{t} \dots \dots \dots \dots \quad (3)$$

$$F_J = \delta_i \cdot m \cdot \frac{(v_0 - v_t)}{t} - (F_f + F_w) \dots \dots \quad (4)$$

$$F_w = \frac{C_D A u_a^2}{21.15} \dots \dots \dots \dots \dots \quad (5)$$

$$F_f = G \cdot f \dots \dots \dots \dots \dots \quad (6)$$

According to the test results, the relationship between the total braking force of each gear and the vehicle speed can be obtained by calculating the data. See Fig. 1.

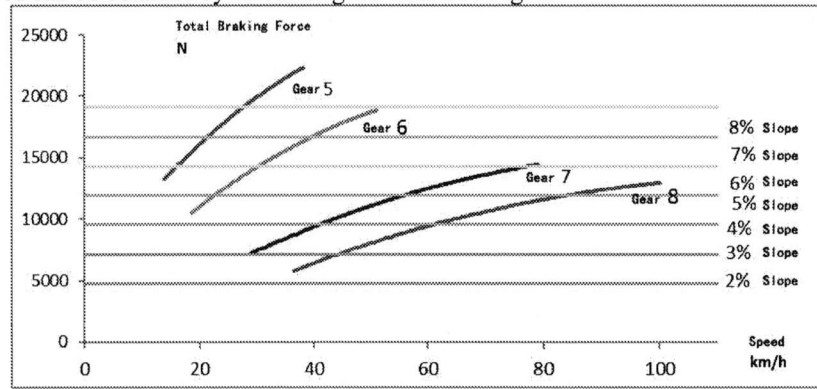

Figure 1 the total braking force curve that each block of engine brake with the speed change

3.3. Engine braking (Jacobs) braking power test and calculation

According to the formula of engine power:

$$P_e = T_{tq}\omega_e = \frac{T_{tq} i_g i_0}{r} u$$

The braking power of the engine's braking (Jacobs) can be calculated, as shown in Table 2 and Figure 2.

Table 2 ISM345 30 engine brake (Jacobs) braking power and driving power

Rotational speed (r/min)	1000	1200	1400	1600	1800	1900
Braking power (kW)	52	73	96	121	147	160
Driving power (kW）	159	214	230	244	251	254
Braking power / driving power	32.7%	34.1%	41.7%	49.6%	58.6%	63.0%

The results show that the driving power of ISM345 30 engine is 254 kW and the braking power is equivalent to 63.0% of the driving power at the rated speed of 1900 r/min. At the maximum speed of 2300 r/min, the braking power is 211kW, equivalent to 83.1% of the rated driving power.

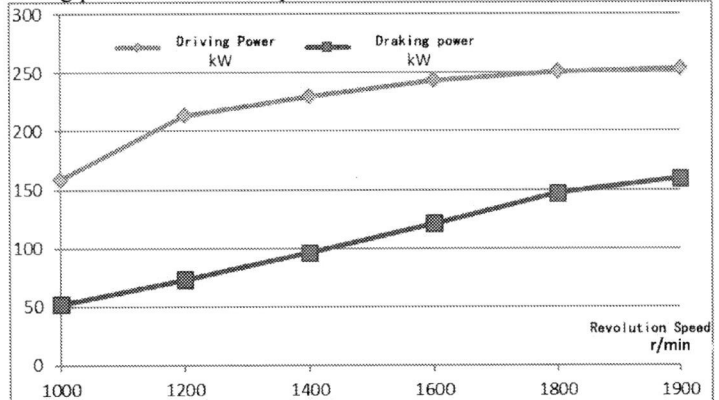

Figure 2 Engine brake (Jacobs) power characteristic curve (ISM345 30)

4. Safety test for long downhill of expressways in Yaxi mountainous area

In order to test the safety degree of long longitudinal slope section of mountain expressway after installing engine brake (Jacobs) on single truck, the research group selected two long downhill sections of Niba Mountain (25.6 km) and Tuowu Mountain (51 km) on Yaxi expressway for field test.

During the test, the total weight of the vehicle is 30% overloaded and the total weight is 32.5 tons. When going downhill, the sprinkler is closed, and the engine brake system is closed and the engine brake system is opened. If necessary, the running brake is used to control the speed of the vehicle.

4.1. Safety test for North Slope of Niba Mountain.

4.1.1.Close the sprinkling device, turn off the engine exhaust brake (Jacobs) and drive downhill. The environmental temperature (NA), the right front hub temperature (RF), the right rear front axle hub temperature (RR1) and the speed of the truck are shown in the figure when the truck is downhill in the 7-8 gear (high grade).

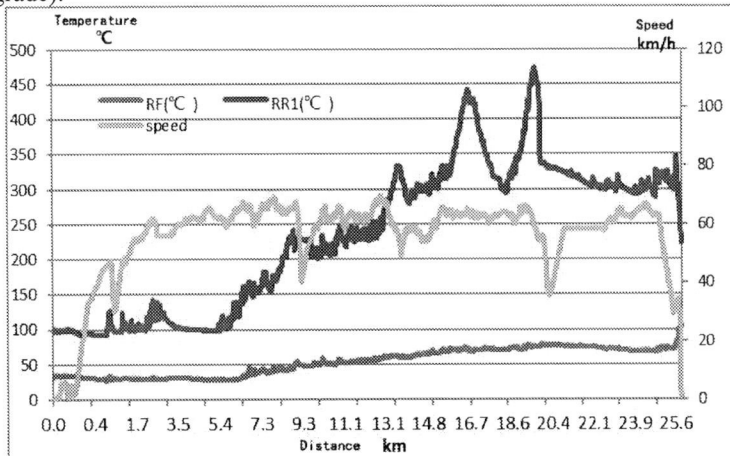

Figure 3 Truck overloaded 30% driving downhill, turn off the water spray and engine brake system, brake drum Temp, speed change curve, north Niba mountain slope

The Figure 3 diagram shows that: the whole downhill process takes 27 minutes and 46 seconds, the highest speed is 69.7 km/h, the average speed is 55.6 km/h, the brake use time accounted for 69.2% of the total travel time; because the front axle load is lighter, so the temperature rise of the right front wheel hub changes less, the range is within 100 degrees Celsius. The latter bridge is relatively large, so the temperature rises faster, and the maximum temperature of the rear axle hub reaches 472.7 degrees.

According to the research conclusion of the western project of the Ministry of Communications, when the brake drum temperature reaches 260 C, the vehicle will lose the ability of emergency braking. Therefore, in the long downhill section of mountainous expressway, the brake efficiency of heavy-duty truck can easily be reduced to the limit value because of the brake overheating under the condition that the brake is not watering and there is no auxiliary brake device, resulting in traffic accidents [4] [5].

4.1.2.Close the sprinkling device and start the engine braking (Jacobs) downhill driving. Under the same loading and downhill gear conditions as in the first test, the engine brake (Jacobs) is turned on when going downhill, and the parameters of the vehicle in the downhill process are shown in the figure.

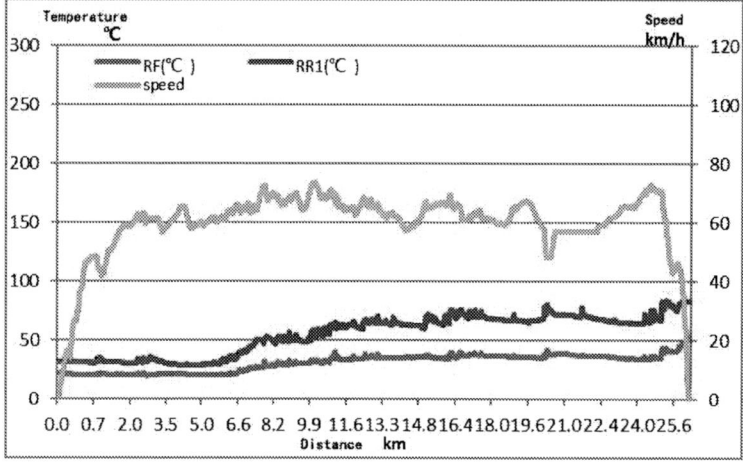

Figure 4 Truck overloaded 30% driving downhill; turn on the water spray and engine brake system, brake drum Temp, speed change curve, north Niba mountain slope

The diagram shows that the whole downhill process takes 26 minutes and 14 seconds, the highest speed is 73.5 km/h, the average speed is 59.4 km/h, and the ratio of brake using time to total driving time is 47.1%. As the use of engine brake, the brake frequency and strength are greatly reduced, so the brake drum temperature also rises very slowly, the right front hub temperature change is small; the range is within 50 degrees Celsius. The maximum temperature of the brake drum of the right rear axle is 149.7 degrees centigrade.

Thus, under the condition of using engine brake (Jacobs), a single truck can ensure the validity of the braking system even if it runs downhill with 30% overload, and make the vehicle safely descend a long slope at a faster speed.

4.2. Safety test on the north slope of Niba Mountain (51km)
In order to further test the reliability of the truck engine brake (Jacobs) on the ultra-long downhill section, the research group conducted a test on the 51 km downhill section of the north slope of Tuowu Mountain on the Yaxi Expressway. During the test, the vehicle is still overloaded by 30%. When going downhill, close the sprinkler, turn on the engine brake (Jacobs) and try not to use the running brake. The test results are as follows:

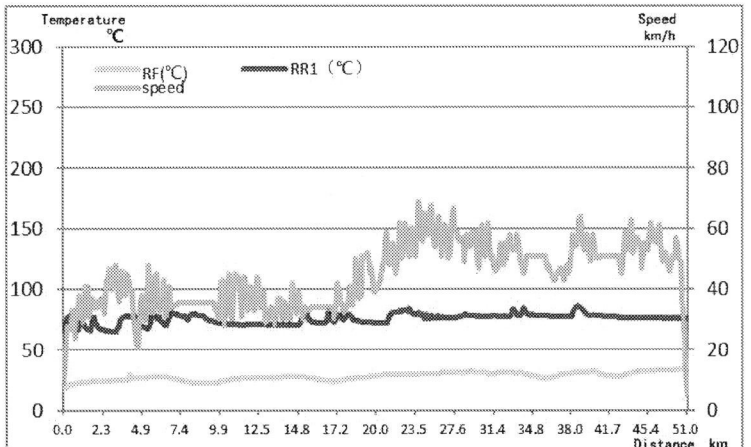

Figure 5 power characteristic curve of engine braking (Jacobs) (ISM345 30)
Truck overloaded 30% driving downhill; turn on the water spray and engine brake system, brake drum
Temp, speed change curve, north Tuowu mountain slope

5. Conclusion

When the truck is loaded standard, the engine brake (Jacobs) can safely pass through the long downhill section of the mountain expressway without watering the brake. It is more suitable to control the speed at about 60 km.

When a single truck is loaded beyond the limit of 30%, the engine brake (Jacobs) can safely pass the long downhill section of the mountain expressway without watering the brake, and the speed of the truck can be controlled at about 50 km.

As there are many manufacturers of engine brake systems in China and the quality of products is different, it is suggested that trucks should walk on mountain expressways by bicycle and store a certain amount of water in the water tank of the sprinkler for occasional use.

Acknowledgement

Central Public Welfare Scientific Research Institute of basic research special funds(RIOH)

References

[1] Wang Jian-bo，Chen Bin. The research on the Special road operating safety technology of Yaxi expressway[R]. Sichuan Vocational and Technical College of Communications,2012:1～3

[2] Di Zhen-hua. The research on Jacobs engine brake tese[D].Xi'an: Chang'an University，2009:25～29

[3] YU Zhi-sheng. Auto theory [M]. Beijing; China Machine Press, 2009:36-45

[4] YU Qiang, Study on downhill continuous braking performance of automobile [D]. Chang'an University, 2000:45～52

[5] Yu Qiang, Chen Yinsan, Ma Jian and so on. Braking downhill ability of bus engine [J]. Journal of Chang'an University (Natural Science Edition), 2003 (3): 95-97 [Function analysis on exhaust brake of bus [J]. Journal of Xi'an Highway University, 2000, 20 (S0): 95-97]

Design of Information System Vulnerability Governance Platform Based on Distributed Asset Acquisition and Vulnerability Verification Radar

Guiquan Shen[1], Jiangang Lu[1], Wuqiang Shen[1], Jingzhi Huang[1], Weiyan Ji[2*]

[1]Information Center, Guangdong Power Grid Company Limited, Guangzhou, China

[2]Guangdong Information Technology Security Evaluation Center, Guangzhou, China

*Corresponding author's e-mail: jiwy@gditsec.org.cn

Abstract. There are a large number of information vulnerabilities in enterprise information systems. Many hackers use vulnerabilities to obtain enterprise or personal information, which greatly jeopardizes enterprise information security. This paper proposes a design method of integrated information system vulnerability governance platform based on distributed asset acquisition and vulnerability verification radar, discusses the business flow and architecture of the vulnerability governance platform, the logical relationship between modules, business function modules, and the technical framework for building platforms.

1. Introduction

In recent years, hackers have been attacking enterprise information systems more and more frequently, and enterprise information system security has become extremely important. The reason that enterprise information systems are attacked is that there are many information security vulnerabilities in information systems. If the vulnerability is exploited or traded on the black market, it is extremely harmful to businesses and users [1]. Therefore, information system vulnerability governance is the most important measure of information security protection, and it's an important guarantee for enterprise information security.

Recent years, some vulnerability scanners were developed for searching information system vulnerabilities. Antunes propose a benchmarking approach to assess and compare the effectiveness of vulnerability detection tools in web services environments [2]. IBM Rational AppScan and HP WebInspect both are Commercial vulnerability scanning tool [3]. They can automatically scan the security vulnerability of Web applications, and help companies enhance the reliability of Web applications. IBM Rational AppScan is a leading suite of automated Web application security and compliance assessment tools that scan for common application vulnerabilities [4]. HP WebInspect performs web application security testing and assessment for today's complex web applications, built on emerging Web 2.0 technologies. HP WebInspect delivers fast scanning capabilities, broad security assessment coverage and accurate web application security scanning results. WSDigger is a free open source tool designed by Foundstone to automate black-box web services security testing (also known as penetration testing). WSDigger is more than a tool, it is a web services testing framework. Version one of this framework contains sample attack plug-ins for SQL injection, cross site scripting and XPATH injection attacks. Doupé proposes a novel way of inferring the web application's internal state machine from the outside—that is, by navigating through the web application, observing differences in output,

Content from this work may be used under the terms of the Creative Commons Attribution 3.0 licence. Any further distribution of this work must maintain attribution to the author(s) and the title of the work, journal citation and DOI.
Published under licence by IOP Publishing Ltd

and incrementally producing a model representing the web application's state [5]. Lukanta developed a vulnerability scanning tool extending an existing open source tool, namely Nikto, to detect session management vulnerabilities [6]. Pelizzi presents a new server-side defense against CSRF attacks, called JCSRF, operates as a server-side proxy, and does not require any server or browser modifications. Thus, it can be deployed by a site administrator without requiring access to web application source code, or the need to understand it [7]. You propose an improved CSRFGuard that the Servlet filter was used to intercept responses, and responses of pages' source codes were stored by a custom response wrapper class to add script tags, so that scripts were automatically inserted [8]. Recently, some enterprise information system vulnerability management platforms have been proposed [9-10].

Although there are many vulnerability scanners, different vulnerability scanners have different strengths and weaknesses. Information system vulnerabilities that a single vulnerability scanner can detect are limited. At the same time, it is not possible to embed an existing vulnerability scanner into an enterprise information system, and it is necessary to manually use an existing vulnerability scanner to test information system vulnerabilities, which is time consuming and laborious, and cannot detect new vulnerabilities in time. Existing Web application technology research can't satisfy the requirement of information system development [11]. Therefore, enterprises urgently need to establish their own vulnerability testing platform, detect enterprise information systems in real time, discover enterprise information system vulnerabilities and fix vulnerabilities in a timely manner.

In view of the above problems, this paper presents the design of information system vulnerability governance platform based on distributed asset acquisition and vulnerability verification radar, discusses the business flow, architecture, business function design and technical framework.

2. Business process design

Enterprise information systems are often composed of multiple information subsystems, and each information subsystem may have security vulnerabilities. The Vulnerability Governance Platform should be able to handle all information subsystem information vulnerabilities. Internet vulnerability publishing platforms often publish new information system vulnerability information. The vulnerability management platform should be able to obtain these vulnerability information and update the platform vulnerability information so that new vulnerabilities in the information system can be discovered in time.

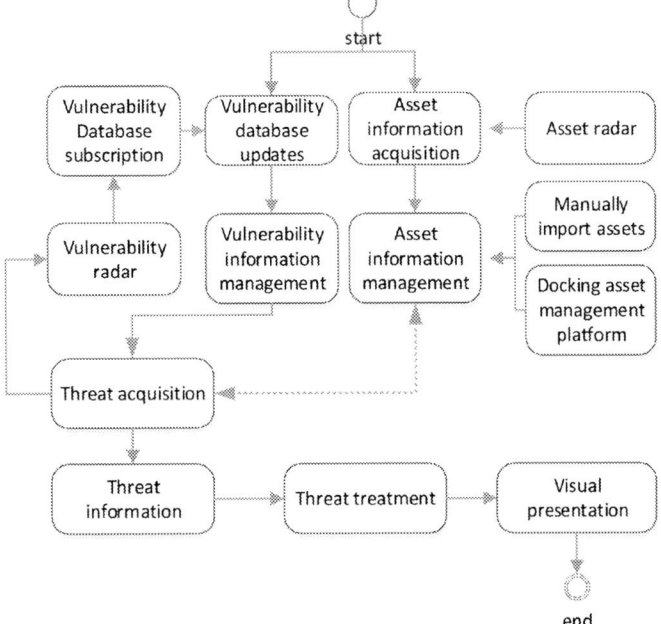

Figure 1. business flow

Business flow of Vulnerability Governance Platform is shown in figure 1. The vulnerability information management module has a common interface to support the connection with third-party vulnerability information platforms to obtain the latest vulnerabilities. The module has an automated inspection module that crawls the vulnerability information published by the Internet vulnerability publishing platform (including but not limited to: CNVD, CNNVD, National Internet Emergency Center website, CVE general vulnerability and disclosure website, etc.) according to the set vulnerability crawling rules. The module also supports batch import or manual entry to add vulnerability information.

Sources of information assets include: 1) Interfacing with existing asset information management related platforms (such as ITSM systems, etc.) to obtain asset-based data. 2) Dispatching the distributed asset collection radar, obtaining asset attributes and fingerprint information, and monitoring asset changes. 3) Support batch import or manual entry to add asset information.

Sources of threats include: 1) Establishing vulnerability scanning tasks on the platform, conducting vulnerability scanning, and exploiting vulnerability. Vulnerability scanning supports different granularity vulnerability scans, including: full vulnerability library scanning, scanning by vulnerability category, specifying scans for specific vulnerabilities, And more.2) When the vulnerability database or asset library changes, obtain the information in the internal network of the information through asset matching and rapid verification of the vulnerabilities in the network, and automatic measures threat levels, and form internal threat intelligence.

Visualization of internal threat intelligence is mainly based on map visualization, visualization based on different latitudes, visualization of vulnerability response based on different time points, recurring visualization based on the same type of vulnerability, visualization based on vulnerability validation, visualization based on the scope of the vulnerability.

3. Architecture of platform

3.1. Overall architecture design

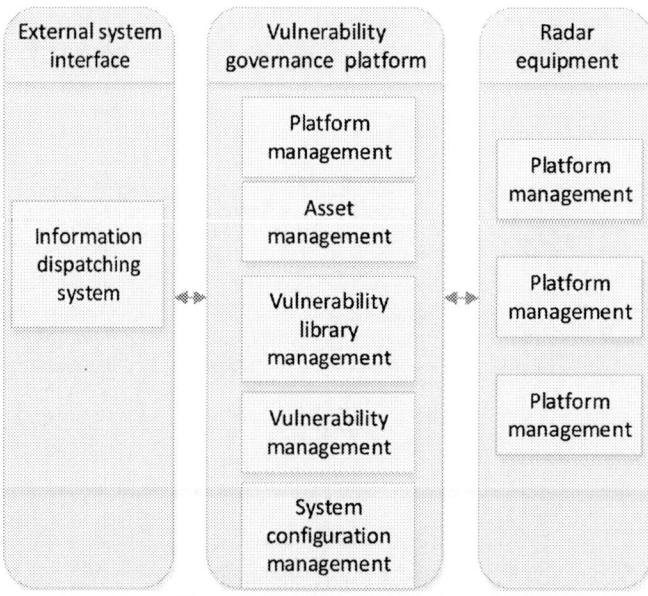

Figure 2. Overall architecture

The architecture of Integrated vulnerability governance platform is shown in Figure 2. The Integrated vulnerability governance platform mainly includes several functional modules such as platform management, asset management, vulnerability database management, vulnerability management, and system configuration management. The Integrated vulnerability governance platform controls the radar equipment information. Vulnerability verification radar detects system security and determines if there is a vulnerability in the system. Asset acquisition radar scan asset attribute information. Radar detection radar verifies vulnerability information by executing poc scripts. The platform docks the external system to obtain external system resources.

3.2. Logical architecture
At the functional level, the system includes process management, data resource management, component management, system management, and unified display management, as shown in Figure 3.

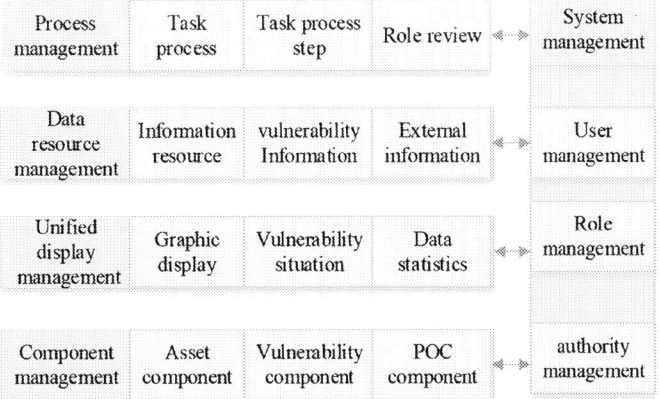

Figure 3. Logical architecture

Process management is invoked through data management, starting, stopping, and suspending various processes to control the definition and execution of processes.

Data management includes vulnerability information data, operational process resource data, and external system submission data, which is provided to the statistical data for statistical data foundation.

Component management provides asset scanning discovery, attribute identification, vulnerability scanning, and POC scanning.

Unified display management provides data statistics for different business types and provides a vulnerability security posture. System management provides users with functional rights and data permission controls, and strictly defines the data resources accessed by each user.

3.3. Functional structure
Business function architecture includes: system management, information asset management, vulnerability database management, vulnerability task management, system threat management which statistics report and internal threat intelligence analysis, as shown in figure 4.

System management includes role management, Authority management, user management, task policy management, and radar configuration management. The platform's asset data comes from three aspects, existing asset management platform, manually imported asset information, and asset information obtained by asset acquisition radar.

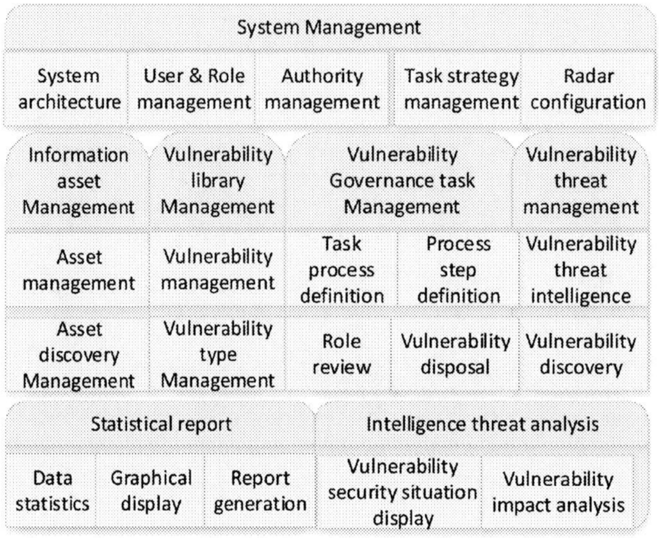

System Management

System architecture	User & Role management	Authority management	Task strategy management	Radar configuration

Information asset Management	Vulnerability library Management	Vulnerability Governance task Management		Vulnerability threat management
Asset management	Vulnerability management	Task process definition	Process step definition	Vulnerability threat intelligence
Asset discovery Management	Vulnerability type Management	Role review	Vulnerability disposal	Vulnerability discovery

Statistical report			Intelligence threat analysis	
Data statistics	Graphical display	Report generation	Vulnerability security situation display	Vulnerability impact analysis

Figure 4. Business function architecture diagram

Vulnerability library information sources include subscribed vulnerability update library, vulnerability information obtained by docking security organization vendors and vulnerability validation radars. After obtaining the latest vulnerability information, the vulnerability validation radar quickly verifies the existing assets, obtains the intranet vulnerability threat, Among them, each area shares vulnerability information and vulnerability disposal suggestions. When a vulnerability threat is discovered, an internal threat information display is formed based on the existing vulnerability information, asset information, and threat information, and is presented by using a visual chart.

3.4. Technical framework

The technical architecture of the system is divided into four levels, namely user interaction layer, functional service layer, technical framework layer and system platform layer.

User interaction layer provides the ultimate human interaction framework. The functional service layer provides customized functional components that support system operation. The technical framework layer provides the general technology needed to implement functional components. The system platform layer provides the operating environment required for a common technical framework.

The technical framework layer provides the underlying component set for the system, ensuring that application system development only needs to focus on the implementation of business logic, without having to care too much about the underlying technology. The set of components provided by the technology framework layer covers all levels of the application system: data framework for data provision, process framework for process definition and control, security components for application security, business framework for business implementation, and UI framework for interface presentation.

The data framework function is divided into two parts: data access and data processing. MyBatis framework is used to access database. It converts database objects into program objects for access. The process framework provides the basic environment for the work approval process for the upper layer, including Workflow Engine, Spring, and Security. The workflow engine is responsible for maintaining the definition of the process and various process instances of the defined process, and is responsible for the flow control and business rule process. The security component provides the system with technical functions such as security filtering, secure transmission, and data security. These functions are completed by Security, SSL, AOP, Model, and Encrypt. The business data of the information system vulnerability governance platform mainly includes: task process management data, system management data, information asset management data, vulnerability database data, vulnerability task management data, POC task data, rule library file data, rule base data, and data dictionary data.

4. Results

To illustrate the applicability and scalability of the cross-regional vulnerability management and control platform, we use six different types of vulnerabilities to detect whether the vulnerability governance platform can detect vulnerabilities and conduct them in a timely manner. These six different types of vulnerabilities are SQL Injection, XPath Injection, Code Execution, Buffer Overflow, Username/Password Disclosure, and Server Path Disclosure.

We artificially created 60 information vulnerabilities in enterprise information systems, including 10 vulnerabilities for each type. The results show that all of these vulnerabilities have been successfully detected by the vulnerability governance platform.

5. Conclusion

The governance of information security vulnerabilities is an important guarantee for corporate information security. The vulnerability governance platform automatically updates vulnerability information from the subscribed vulnerability information website which realizes the early detection and disposal of high-risk vulnerabilities in enterprise information systems, improves the efficiency of vulnerability management and the security of enterprise assets.

Acknowledgments

This work was financially supported by GDKJXM20162130(037800KK52160003).

References

[1] Makino Y, Klyuev V. Evaluation of web vulnerability scanners[C]// IEEE, International Conference on Intelligent Data Acquisition and Advanced Computing Systems: Technology and Applications. IEEE, 2015:399-402.

[2] Antunes N, Vieira M. Benchmarking Vulnerability Detection Tools for Web Services[C]// IEEE International Conference on Web Services. IEEE, 2010:203-210.

[3] Vieira M, Antunes N, Madeira H. Using web security scanners to detect vulnerabilities in web services[C]// Ieee/ifip International Conference on Dependable Systems & Networks. IEEE, 2009:566-571.

[4] Dao T B, Shibayama E, Security Sensitive Data Flow Coverage Criterion for Automatic Security Testing of Web Applications[C]// International Conference on Engineering Secure Software and Systems. Springer-Verlag, 2011:101-113.

[5] Parimala G,Sangeeth a M, Andalpriyadharsini R, Efficient Web Vulnerability Detection Tool for Sleeping Giant-Cross Site Request Forgery[C]// National Conference on Mathematical Techniques and its Applications,2018.

[6] Lukanta R, Asnar Y, Kistijantoro A I, A vulnerability scanning tool for session management vulnerabilities[C]// International Conference on Data and Software Engineering. IEEE, 2014:1-6.

[7] Pelizzi R， Sekar R. A server and browser-transparent CSRF defense for web 2.0 applications[C]// Proceedings of the 27th Annual Computer Security Applications Conference, New York, 2011, pp. 257-266

[8] You J, Guo F, Improved CSRFGuard for CSRF attacks defense on Java EE platform[C]// International Conference on Computer Science & Education. IEEE, 2014:1115-1120.

[9] Gui-Ping L I, Xian Fanyi, Design of security vulnerability detection platform based on Wi-Fi[J]. Information Technology, 2016.

[10] So I G, Setiadi N J, Papak B, et al. Action Design of Information Systems Security Governance for Bank Using COBIT 4.1 and Control Standard of ISO 27001[J]. Advanced Materials Research, 2014, 905:663-668.

[11] Sun Y, Liang D Y, Wang W J, Enhancement of Test Platform for Web Application Security[J]. Applied Mechanics & Materials, 2014, 511-512(5):1205-1210.

| ISPECE | IOP Publishing |

Standard Architecture of China Intelligent Bus Systems

Xianglong LIU

China Academy of Transportation Sciences, Ministry of Transportation of China, No.240, Huixinli Street, Beijing, China.

liuxianglong@live.cn

Abstract. This paper summarizes the ITS standards of China, and designs the standard architecture of intelligent bus system, including the technical specifications on devices, communication protocols, data resource, and application software being implemented. The development efforts will update the intelligent bus system architecture to ensure interoperability among devices, protocols, databases, and application systems; and it will increase reusability of the devices and systems of intelligent bus system; and integrate the operations among bus companies, bus industry, and management agencies.

1. Background

China has significantly increased its urbanization from 20% to 57% in the last 30 years resulted from the economic development in the last several decades, and also the number of vehicles and private cars ownership, which lead to the growing urban traffic congestion and environmental pollution issues.

Starting from the end of 2012, the State Council of China has issued the Transit Priority Policies. To support this national policy, it is very important to establish the needed system standardization to improve the QoS of bus system through the intelligent bus systems. There are several successful examples on how the Advanced Public Transportation System (APTS) standardization can improve the operations around the world. In 1992, APTA established Transit Communications Interface Profile (TCIP). The EU standards organization is also promoting the integrated Fare Management Standards under ISO-TC204-WG8. To further standardize the data communication and data interface, APTA has launched TCIP 3.0 to cover 9 additional public transport business areas, including the Common Public Transportation, Control Center ,Fare Collection, Incident Management, On board, Passenger Information ,Scheduling and Run cutting, Spatial Representation ,Transit Signal Priority.

In China, although the public transit system got rapid development, but there are lack of technical standards on intelligent bus system. This paper summarizes the ITS standards of China, and designs the standard architecture of intelligent bus system, including the technical specifications on devices, communication protocols, data resource, and application software being implemented.

2. Framework of Intelligent Bus System

China intelligent bus system framework is developed to deliver different users' demands. These include the daily bus operation, routine management, and automatic service quality assessment. As shown in Figure 1, the architecture include the five layers, including the sensing layer, communication protocol layer, data layer, application layer, and the user layer.

◇ The sensing layer is mainly focus on using all kinds of sensors to detect or monitor the status of the passengers, buses, bus-stops and bus-lanes and so on, is the foundation of the intelligent system.

◇ The protocol layer is mainly focus on data communication and transferring between sensing

Content from this work may be used under the terms of the Creative Commons Attribution 3.0 licence. Any further distribution of this work must maintain attribution to the author(s) and the title of the work, journal citation and DOI.
Published under licence by IOP Publishing Ltd

layer and data layer, in accordance with the requirements of the protocol accepted by the system.

◇   The data layer is composed of data management, data resource include metadata, data exchange and databases, data management mainly focused on data quality audit and data modelling.

◇   The application layer is about different user's application; mainly include traveller information system for travellers, intelligent dispatching and enterprise resources management for bus companies, quality of service supervising and development evaluation for different level management agencies, and data opened for other user's application such as research agencies or researchers.

◇   The user layer mainly includes interactions among the passengers, bus companies, management agencies and other research agency or personal researchers.

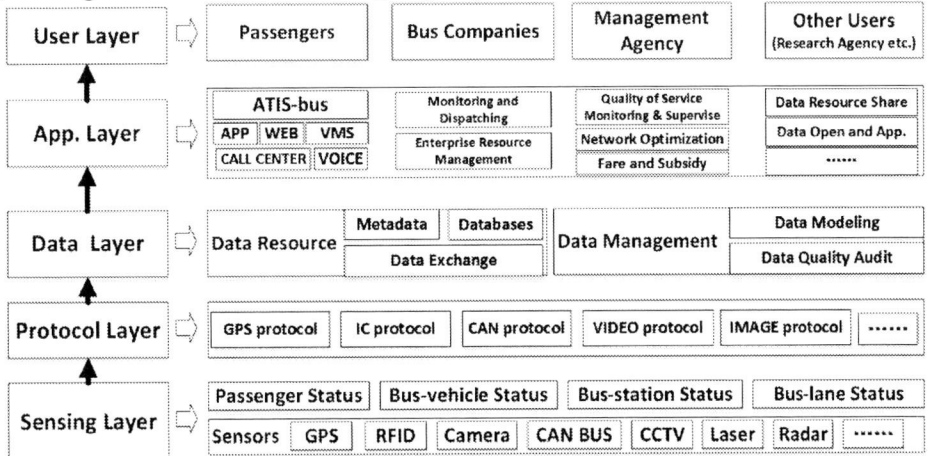

Figure 1. Framework of Intelligent Bus System of China

3. Standard Architecture of Intelligent Bus System

3.1. Standard Architecture Design

The Architecture is developed as based on the function demand analysis and technical framework design of the intelligent bus system in China. Figure 2 illustrates the Standard architecture and related data flow to support the data collection, data management, data analysis, and data distribution of the intelligent bus system. The system architecture is further divided into three sub-architecture layers; including operation and monitoring standard-sub-architecture, data resource standard-sub-architecture, and management and service standard-sub-architecture, respectively.

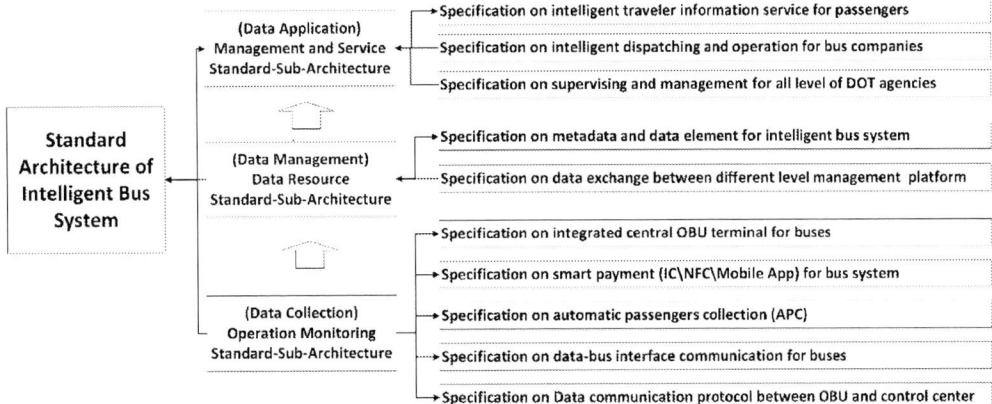

Figure 2. Standard Architecture of Intelligent Bus System

3.2. Data Collection-Operation Monitoring Standard-Sub-Architecture

The Data Collection-Operation Monitoring Standard-Sub-Architecture is focused mainly on sensing and data acquisition by all kinds of sensors to monitoring the status of the passengers, buses, bus-stops and bus-lanes dynamically. This serves as the foundation of the intelligent bus system. Among them, the integrated OBU is the core of this sub-architecture, with this device, the sensing and information transmission network. Figure 3 describes the various service established between vehicles, passengers, traffic (include signal priority), control centre, bus operation environment (bus station and bus lane).

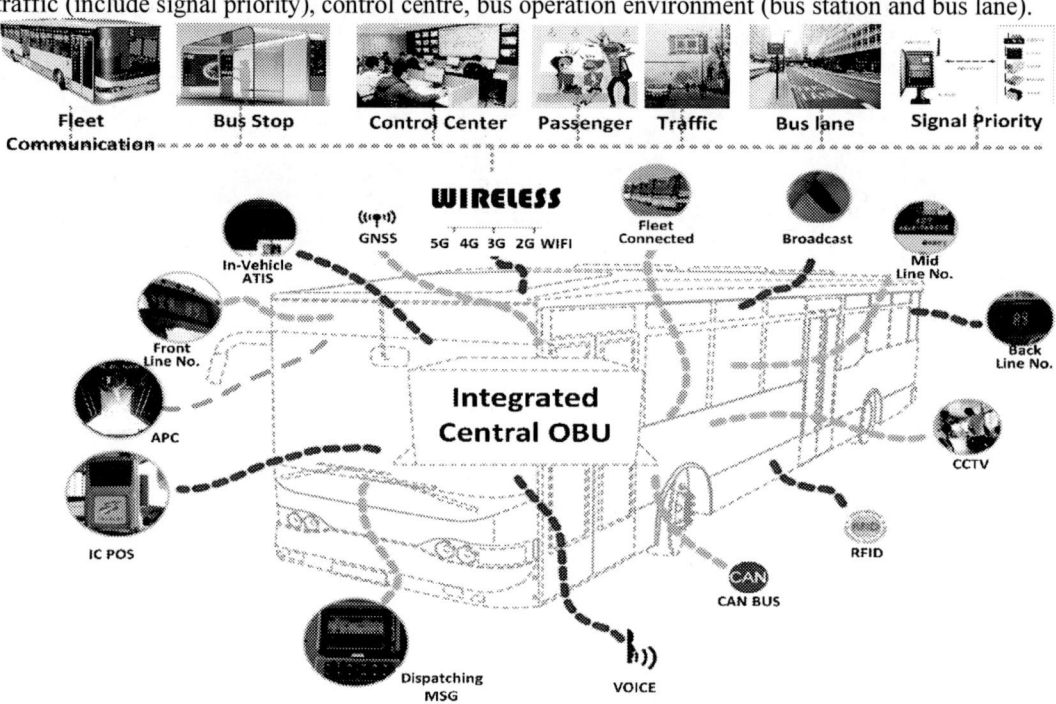

Figure 3. Standard-Sub-Architecture on Operation Monitoring and Data Collection

3.3. Data Management-Data Resource Standard-Sub-Architecture

The data management sub-architecture is focused mainly on data source and data management According to the description of the sub-architecture on data resource. These include the detailed specification on metadata, data exchange between different-level platforms and data quality verification. Figure 4 describes the interactions among the database specification, and the needed data exchange and interface between operating entity levels.

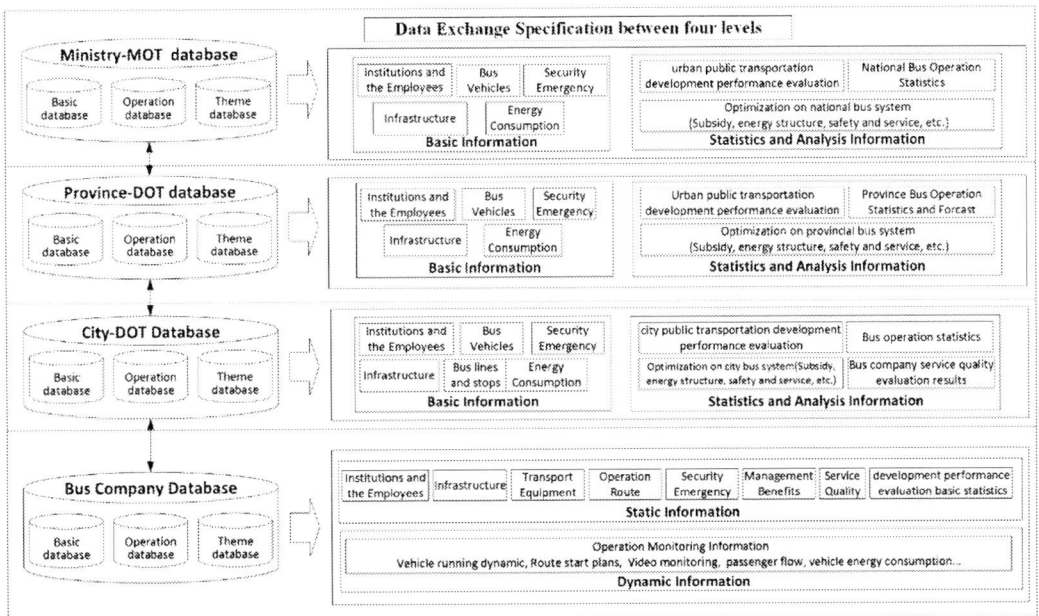

Figure 4. Data exchange between four levels of bus operation and management

3.4. Data Application- Management and Service Sub-Architecture

According to different users of the intelligent bus system, the data application sub-architecture is focused on data application, it consists of three specifications, which are specifications on intelligent dispatching and operation for bus companies, on intelligent supervising and management for all DOT agencies, on intelligent traveller information service system for passengers separately.

3.4.1. Specification on Intelligent Dispatching and Operation for Bus Companies. The specification aims to improve the efficiency for bus companies, focused on the dynamic optimizing allocation of the related resources of bus system such as buses and drivers, considering the real-time traffic situation and spatial-temporal regulation of passengers.

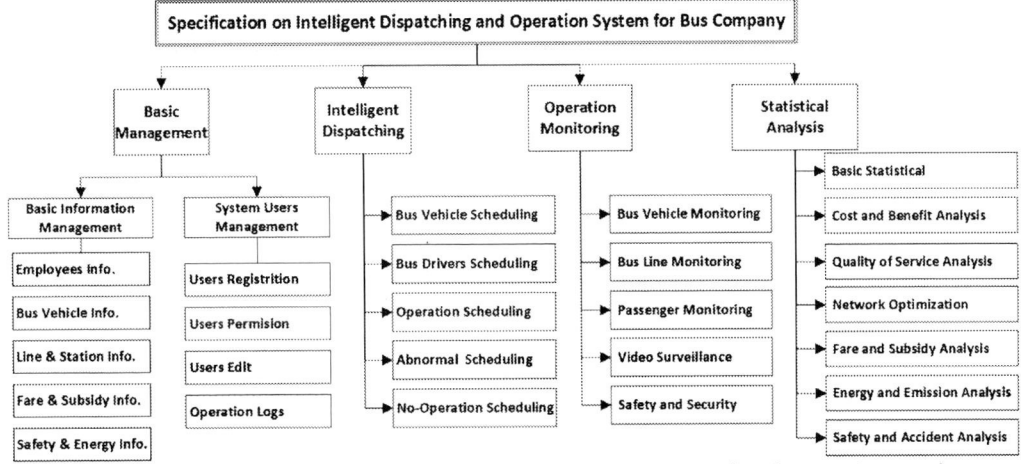

Figure 5. Specification on Intelligent Dispatching and Operation for Bus Companies

3.4.2. Specification on Intelligent Supervising and Management for DOT agencies. The specification has made requirements for the three levels of management agencies separately from comprehensive monitoring, performance evaluation and decision support and optimization, especially, there are much more requirements to the city level compared with other two levels.

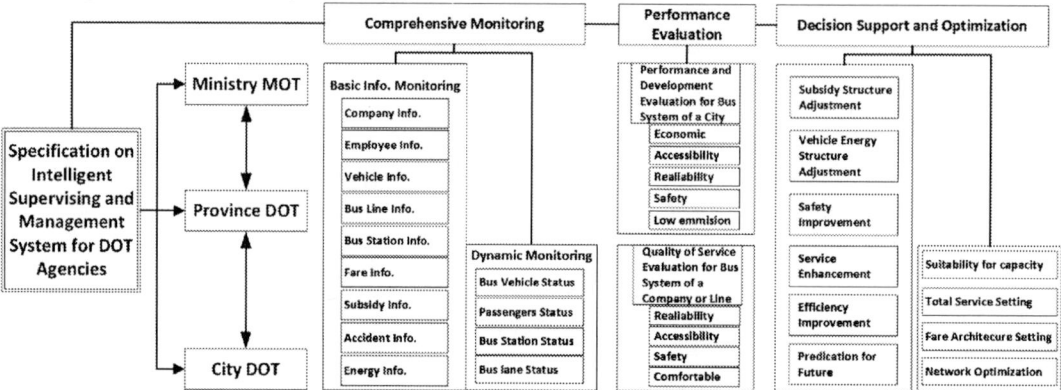

Figure 6. Specification on Intelligent Supervising and Management for DOT agencies

3.4.3. Specification on Intelligent Traveller Information Service System for Passengers. The specification is focused on providing the needed traveller information service to achieve real-time, on-line information delivery to maximize the usability and usefulness of the urban public transit systems. This standard proposed requirements in terms of different stages of travel like the stage of before traveling, bus-stop waiting, inter-bus traveling and transferring.

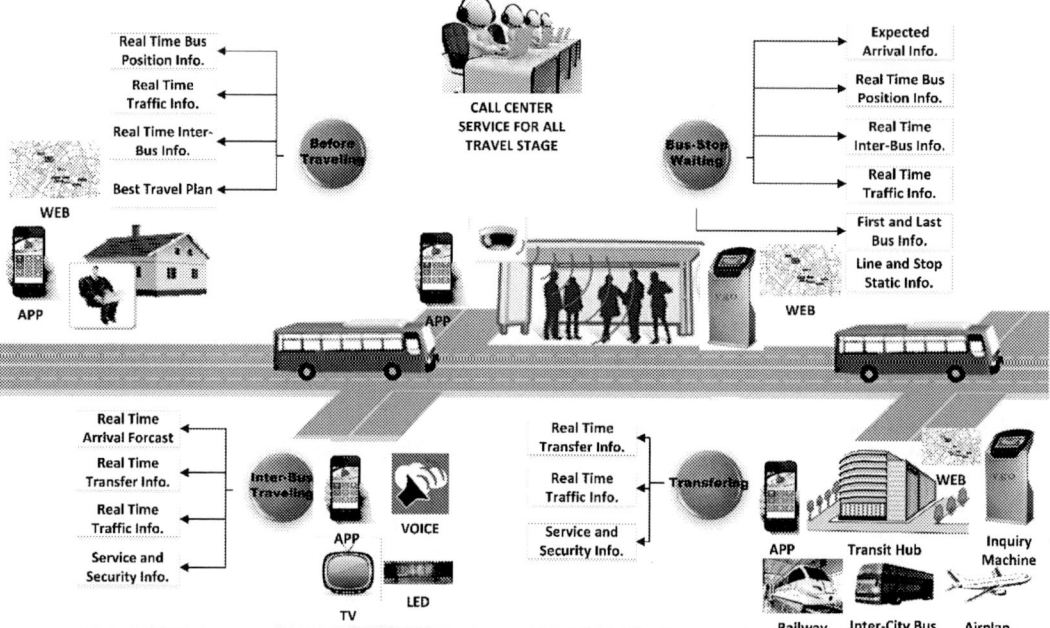

Figure 7. Specification on Intelligent Traveller Information Service System for Passengers

4. Conclusions

This paper summarizes the ITS standards of China, and designs the standard architecture of intelligent bus system, including the technical specifications on devices, communication protocols, data resource, and application software being implemented. Based on the APTS pilot project, 10 national technical specifications already developed and formalized into China National Standards, on different devices,

ISPECE   IOP Publishing

protocols, data resource, and application software being implemented to guide ITS APTS demonstration projects currently ongoing in China.

Continuous improvements in the intelligent bus system are also needed to keep up with the rapid information technology development. Further improvement are also needed to address the multimodal system integration, data mining across operating agencies, and distributed, cloud-based data service that can fully take advantage of the available public and private partnerships.

Acknowledgements

The authors gratefully acknowledge the supports from the Science and Technology Research Foundation for Transportation of China under Grant No. 2015318221020, and the supports from the Special Funds for Volvo Research and Educational Foundations (VREF), and the supports from the National Funds for Nonprofit Research Institutes Basic Research under Grant No. 2015481.

References

[1]   Ahmed El-Geneidy, Jessica Horning, Kevin J. Krizek. (2007). Using Archived ITS Data to Improve Transit Performance and Management. Minnesota Department of Transportation. (Sponsored by the Minnesota Department of Transportation)

[2]   Carol, L, Schweiger, et, al. TCRP synthesis48: Real-Time Bus Arrival Information Systems[R]. America: Transportation Research Board of the National Academies, 2003.

[3]   Federal Transit Administration, (2011). National Transit Database Annual Reporting Manual[R].

[4]   Harrlet R. Smith, Brendon Hemily, and Miomir Ivanovic. Transit Signal Priority (TSP): A planning and Implementation Handbook[M]. ITS America, 2005.

[5]   John, E, (JAY), EVANS, et, al. TCRP report 95: Traveler Response to Transportation System Changes Chapter 9—Transit Scheduling and Frequency[R]. America: Transportation Research Board of the National Academies, 2004.

[6]   Mimi, Hwang, et, al. Advanced Public Transportation Systems: The State of the Art Update 2006[R]. America: Federal Transit Administration, 2006.

[7]   Peter, G, Furth、Brendon, Hemily、Theo, H, J, Muller、James, G, Strathman. TCRP report113: Using Archived AVL-APC Data to Improve Transit Performance and Management[R]. America: Federal Transit Administration, 2006.

[8]   TCRP Report 88. (2002). A guidebook for developing transit performance-measurement system[R]. Transportation Research Board of the National Academics. (Sponsored by the federal transit administration)

[9]   TCRP Report 100. (2003). Transit capacity and service of quality manual: 2nd Edition[R]. Transportation Research Board of the National Academics. (Sponsored by the federal transit administration)

[10]   TCRP Report 113. (2006). Using archived AVL-APC data to improve transit performance and Management[R]. Transportation Research Board of the National Academics. (Sponsored by the federal transit administration

Distributed scalable abstract reasoning based on dl-lite

Bin Xia[1]

[1] 1st School of Computer Science and Technology Wuhan University of Science and Technology Wuhan, China

Abstract. In recent years, researchs have shown that the ontology of a large number of data sets can be inferred effectively by abstract techniques. Distributed environment can further improve the performance of abstract reasoning. Firstly, abstract technology refers to a technology that divides ontology into equivalence classes and uses reasoning tools to reason abstract ontology. Then,The newly generated entities and relationships are added to the original ontology data set through materialization techniques. Finally, in order to ensure the consistency of ontology, the consistency check of ontology is needed later. We use the benchmark to obtain preliminary experimental results. The final results show that this abstract method can deal with a large number of ontologies effectively. Compared with the original reasoning method, the abstract technique proposed in this paper can effectively improve the reasoning efficiency and reduce the memory consumption time.

1. Introduction

With the development of ontology, ontology reasoning has become a hot topic. With the development of research, it is possible to deduce a large number of ontology sets. Some existing reasoning tools such as konclude[1], hermit can deal with a large number of ontologies,but the reasoning efficiency is not very good. This paper introduces an ontology abstract materialization technique called ron-s[2]. The ron-s reasoning's rules include some steps.First, abstracting the original Abox.Then,it divides the same type of data into equivalent classes to achieve the effect of compression (the abstract data may not be smaller than the original data).Finally the ron-s writes the newly generated data back to Abox. Abstract inference technology can greatly reduce the volume of ontologies that need to be inferred. The ron-s technology can also optimize the reasoning efficiency of existing reasoning tools[3]. In the distributed case, the performance of abstract refinement can be further optimized and the fault tolerance of the system can be increased. In this paper, an ontology reasoning optimization method is proposed. The ontology format required for reasoning needs to conform to dl-lite description logic. The whole system runs in a multi-node distributed environment[4]. The experimental results show that the efficiency of the abstract reasoning method proposed in this paper is obvious better than that of reasoning only by means of tools.

2. Abstract Refinement

2.1. Abstract process

In the concept of ontology language, the data of ontology is divided into Abox and Tbox[5]. Tbox is a summary of the concept, and Abox is a concrete implementation of Tbox. For the convenience of understanding and research, Abox and Tbox files were clearly separated in this experiment. ALCHOI ontology is a member of ontology family and is a special ontology[6]. In previous studies, some articles have implemented abstract materialization technology based on ALCHOI ontology. ALCHOI

Content from this work may be used under the terms of the Creative Commons Attribution 3.0 licence. Any further distribution of this work must maintain attribution to the author(s) and the title of the work, journal citation and DOI.
Published under licence by IOP Publishing Ltd

ontology is composed of infinite disjoint countable concept names, noun names, relationship names and individual names[7]. Concept for each assertion D (a), and each principle C ⊆ D, C and D should meet the following syntax definition:

C ::= | ⊥ | A | O | C1 ∩ C2 | C1 ∪ C2 | ∃R.C

D ::= | ⊥ | A | O | D1 ∩ D2 | ∃R.D | ∀R.D | ¬C.

We call the abstract result of Abox abstract B. The abstract process is the division of similar concepts and relationships in Abox into a broad category such as B1,B2 and so on. The result of abstracting B is that B=B1∪B2∪... . We use reasoning tools to reason the abstracted data[8], materialize the ontology to get new concepts and write them back to Abox. The above process may not have inferred all the data, so we need to refine the ontology. The process is to continue calculating the type A and then generate B inference until no new assertion is generated. In general, volume of the abstract Abox is less than original Abox, especially when the number of individual types with large amounts of data is highly repetitive.

2.2. Refinement process
1. First, we get the corresponding abstract B based on input of Abox

2. Then,we use reasoning tools to materialize and abstract B based on the materialization formula B∪ T

3. Adding the new assertion inferred from the previous step to Abox and recalculate the new abstract B

4. Repeat steps 2 and 3 until no new assertion is made,.Then the reasoning ends

The specific process of materialization is as follows: if B is the abstraction of A, there is A corresponding Tbox T for each relational triple, and for each individual A and each concept M, there is the following relationship:

(1) B ∪ T:=M(xtp) can be concluded that A ∪ T:=M(A);

(2) B ∪ T:=M(ytp) and R(a, b) in A can be concluded that A ∪ T:=M(B)

(3) B ∪ T:=M(ztp) and R(c,a) in A can be concluded that A ∪ T:=M(c)

The process of abstract refinement[9] of reasoning is mainly abox in the reasoning ontology and Tbox is ignored. We classify Abox as an equivalent class by type to get an abstract B. Then reason the abstract B. And then we can write the individual mapping back to Abox using the above formula. Iterate until no new assertion is made.

3. Remove the refinement

We have noticed that the above method may be more troublesome in reasoning. Abstract refinement techniques may require many cycles to complete. In real life, the general abstraction refinement process is within 10 cycles. We removed the detailed steps in further research[10]. Ontological reasoning only requires abstract materialization. Since the new technology reduces the reasoning time, the process becomes simpler and the consistency of the reasoning ontology is guaranteed. The idea of new abstraction techniques is suitable for a large number of ontologies[11]. Abstract techniques are used to reduce the volume of ontologies that need to be computed. In general, the more ontologies there are, the better the abstract effect will be. Of course, the ontology still needs to satisfy the dl-lite logical description. Among the various logical descriptions, dl-lite is part of a family of languages designed specifically for ontologies[12]. Dl-lite follows the basic logical description rules shown in fig 1.

3.1. Some Definitions
Definition 1: The abstract B is a union of the concepts and relationships of each type in Abox. Defining Abox triple relationship tp={A,R,R} where the first R represents the left-to-right relationship and the second R represents the right-to-left relationship[13].

Definition 2: The abox and its abstract B are type mappings. Abstract B can be type-mapped to A type and may correspond to multiple instances in Abox.

	Syntax	Semantics
Roles:		
atomic role	R	$R^I \subseteq \Delta^I \times \Delta^I$
inverse role	R^-	$\{\langle e,d \rangle \mid \langle d,e \rangle \in R^I$
Concepts:		
atomic concept	A	$A^I \subseteq \Delta^I$
nominal	O	$O^I \subseteq \Delta^I, \parallel O^I \parallel = 1$
top	T	Δ^I
bottom	$\bot$	$\varnothing$
negation	$\neg C$	$\Delta^I \setminus C^I$
conjunction	$C \cap D$	$C^I \cap D^I$
disjunction	$C \cup D$	$C^I \cup D^I$
existential restriction	$\exists R$	$\{d \mid \exists e \in \Delta^I : \langle d,e \rangle \in R^I \}$
Axioms:		
concept inclusion	$C \sqsubseteq D$	$C^I \sqsubseteq D^I$
role inclusion	$R \sqsubseteq S$	$R^I \sqsubseteq S^I$
concept assertion	$A(a)$	$a^I \in A^I$
role assertion	$R(a,b)$	$\langle a^I, b^I \rangle \in R^I$

Figure 1.Dl-lite syntax

3.2. Abstract Materialization
For an ontology O = A ∪ T

1: Calculating the abstract B of A.

2: Reasoning B based on formula B ∪ T.

3: Mapping the abstract B to Abox. If the corresponding data does not exist in Abox, add the data to Abox.

4: It returns ontology O at last. Finally the reasoning ends.

The ontological reasoning is not over. Because ontology inconsistency may occur in the inferred data. The new Abox still needs to check the ontology consistency.

Ontology consistency check steps:

1: O = A ∪ T

2: Checking whether ontology O meets the consistency requirement. If it meets the requirement, the program returns true directly. Otherwise it returns false and continues to check.

3: Checking whether Abox's abstract B satisfies the consistency requirement. If it satisfies the requirement, the program returns true. If it does not satisfy the requirement, the program return false directly.

In this experiment, the benchmarks[14] such as LUBM and UOBM are used. Experimental tests are carried out by using one, five and ten data formats in LUBM and UOBM. Then, it checks whether the experimental results of ron-s algorithm reasoning and hermit reasoning tool alone are consistent. Finally, experimental results show that the volume of abstract Abox is exponentially lower than the volume of original Abox. In addition, the computational time complexity is significantly reduced. The ron-s algorithm can significantly improve the efficiency of reasoning tools and ensure the consistency of ontology[15].

4. distribute and expriment

4.1. Distributed

To further enhance the ability of reasoning tools to handle a large number of ontologies, we deployed ron-s algorithm to a distributed environment[16]. The ron-s is deployed in multiple nodes on different servers. In addition,we used registry and gateway reverse proxy technology. When a large amount of ontology data is input, the registry can intelligently select an optimized node for processing reasoning. This experiment is built with hermit inference (other inference engines can be built). In addition, *jena* is used to parse RDF, *openAPI* processes ontology files and ontology benchmark library is used as input. The results show that the performance of ron-s algorithm is significantly improved compared with that of the reasoner. If the node is down, the system will automatically switch to a node that is still available for reasoning after a few retries. Distributed virtual IP technology guarantees fault tolerance under distributed environment. And the ontology data is distributed and routed through gateways and eventually flows to different nodes.

4.2. Experiment

We implemented ron-s based on dl-lite. Test data is based on popular standard ontology libraries and evaluated using different methods. The benchmark ontology library includes NPD,DBPedia,IMDB and various versions of LUBM and UOBM. The NPD ontology data set is NPD Fact Pages, which contains information about oil collected on the Norwegian continental shelf. Dbpedia is the data for wikipedia. MDB is data about movies. All three are real data. LUBM is the ontology benchmark of a university. For example, LUBM one represents a university and five represents five universities. UOBM data is also the ontology benchmark of the university. UOBM is an improvement on LUBM which uses ontology generators to generate owl files of different sizes. We extract the ontology that conforms to the dl-lite logical rule. The ontology that does not meet the rule will be removed. We compare the data after abstraction with the data before abstraction. Table 1 records Tbox axioms, atomic concepts, relationships, individuals and inferred assertions. Table 2 records the time required for reasoning by different inference tools. Experimental result suggests that Abox may be 10 percent or less of the original Abox. The time required for the whole reasoning process has also been significantly improved. The experimental environment is as follows: Intel(R) Core(TM) i7-6700hq CPU@2.6GHz, operating memory:16GB, storage memory: 300GB.Ontology data is shown in table 1.

Table 1.Some Information of Ontology

Ontology	axiom	con	Role	Indivi	assert
NPD	400	244	112	804322	1712578
DBPedia	1362	385	705	3236846	21948501
IMDB	133	91	43	6656342	29093482
LUBM 1	80	42	25	1555	5738
LUBM 5	80	42	25	9035	65305
LUBM10	80	42	25	207425	850463
UOBM 1	110	69	35	13654	198363
UOBM 5	110	69	35	11374	937594
UOBM10	110	69	35	242491	1926897

Table 2.Reasoning time

Onto*	ron-s	konclude	hermit	ELK	Pellet
NPD	4	11	579	56	39
DBPedia	98	123	3023	453	856
IMDB	23	142	453	245	234
LU 1	1	2	3	4	4
LU 5	1	2	5	6	5

LU 10	2	4	9	10	9
UO 1	1	2	-	5	4
UO 5	3	10	-	18	15
UO10	8	17	-	23	24

In table 1, with the NPD, Dbpedia and IMDB increasing with the increase of ontology data volume, all data were significantly improved. However, with the increase of ontology data, the axioms, concepts and relationships of LUBM and UOBM have not changed significantly. Because the data they generate is based on the same reasoner. The conceptual materialization time is shown in table 2. In table 2, the ron-s inference algorithm takes the shortest time. The hermit inference engine is embedded in the experiment. For NPD, Dbpedia and IMDB which are based to real-world data. The ron-s comparing reasoning tool has a great optimization. The Performance has also been partially improved in the school ontology LUBM and UOBM.

5. Conclusions

In this paper, an improved reasoning algorithm is presented and experiments are carried out in a distributed environment. The NPD, Dbpedia,IMDB,LUBM and UOBM ontology were used as the benchmark ontology. Experimental results show that ron-s has better performance in distributed environment. In the experiment, we checked the consistency of abstraction and ontology, which can ensure the fairness and integrity of ontology. We mainly focus on the generation of concept assertion in the experiment. The derivation of relationship assertion can be accomplished by horn ontology reasoning refinement mentioned in the front paragraph. The volume of abox after abstraction was significantly reduced exponentially compared with the original abox, and the reasoning time was significantly optimized. In the future, we plan to apply this idea to some new fields and improve the abstract thinking and distributed environment in order to make the speed of ontology reasoning faster and the system more stable.

References

[1] Glimm B, Kazakov Y, Tran T K. Scalable Reasoning by Abstraction Beyond DL-Lite[C]//*International Conference on Web Reasoning and Rule Systems*. Springer, Cham, 2016: 77-93.

[2] Glimm B, Kazakov Y, Liebig T, et al. Abstraction refinement for ontology materialization[C]//International Semantic Web Conference. Springer, Cham, 2014: 180-195.

[3] HermiT O W L. Reasoner[J]. URL: http://www. hermit-reasoner. com, 2016.

[4] Thukral A, Jain A, Aggarwal M, ct al. Semi-automatic Ontology Builder Based on Relation Extraction from Textual Data[M]//Advanced Computational and Communication Paradigms. Springer, Singapore, 2018: 343-350.

[5] Farid H, Haarslev V. Handling Nominals and Inverse Roles using Algebraic Reasoning[J]. arXiv preprint arXiv:1810.00916, 2018.

[6] Glimm B, Kazakov Y, Tran T K. Ontology Materialization by Abstraction Refinement in Horn SHOIF[C]//*AAAI*. 2017: 1114-1120.

[7] Baader F, Borgwardt S, Lippmann M. Query rewriting for DL-Lite with n-ary concrete domains[C]//*Proceedings IJCAI*. 2017.

[8] Zheleznyakov D, Kharlamov E, Horrocks I. Trust-Sensitive Evolution of DL-Lite Knowledge Bases[C]//*AAAI*. 2017: 1266-1273.

[9] Mansour E, Abdelaziz I, Ouzzani M, et al. A demonstration of Lusail: Querying linked data at scale[C]//*Proceedings of the 2017 ACM International Conference on Management of Data*. ACM, 2017: 1603-1606.

[10] Khan S A, Qadir M A, Abbas M A, et al. OWL2 benchmarking for the evaluation of knowledge based systems[J]. *PloS one*, 2017, **12(6)**: e0179578.

[11] Glimm B, Horrocks I, Motik B, et al. HermiT: an OWL 2 reasoner[J]. *Journal of Automated Reasoning*, 2014, **53(3)**: 245-269.

[12] Uschold M. Demystifying OWL for the Enterprise[J]. *Synthesis Lectures on Semantic Web: Theory and Technology*, 2018, **8(1)**: i-237.

[13] Uschold M. Demystifying OWL for the Enterprise[J]. *Synthesis Lectures on Semantic Web: Theory and Technology*, 2018, **8(1)**: i-237.

[14] Sazonau V, Sattler U. TBox Reasoning in the Probabilistic Description Logic SHIQp[C]//*Description Logics*. 2015.

[15] Bak J, Nowak M, Jedrzejek C. RuQAR: Reasoning framework for OWL 2 RL ontologies[C]//*European Semantic Web Conference*. Springer, Cham, 2014: 195-198.

[16] Kim K H, Hwang M S, Jae E Y, et al. A Study on Message Queue Safe Proper Time for Open API Fast Identity Online Fintech Architecture[J]. *International Journal of Software Engineering and Its Applications*, 2016, **10(5)**: 33-44.

Cloud Resource Adaptive Scheduling Framework and Optimization Strategy Based on Swarm Intelligence

HW Zhao[1] , S Zhang[1], Y ruan[1] and XH Jing[1]

[1] Shenyang University, Information College, Shenyang, China

Abstract. Resource scheduling framework is very important for the overall cloud service efficiency and service quality,which is the core module of cloud computing. This paper proposes an adaptive cloud computing resource scheduling framework based on swarm intelligence. The proposed adaptive framework includes resource deployment module, scheduling module, recommendation module, optimization module, and monitoring module.The simulation results show that the proposed framework is more efficient than the traditional framework in terms of task execution time, system load balancing and resource service quality, and can improve the service quality of cloud applications while improving resource utilization.

1. Introduction

Cloud computing is a computing model that provides dynamic and virtual resources by service, including cloud services of various functions.The Scheduling framework of cloud computing mainly includes hardware resource layer, virtual resource layer, cloud system management layer and cloud application layer.The hardware resource layer includes large-scale blades, cabinets and tower servers, centralized and distributed storage, and network infrastructure such as routers and switches.The virtual resource layer mainly integrates various resources by IAAS layer to form a virtual resource pool with super capabilities [1][2]. The cloud system management is mainly responsible for security management, task management, resource management, user management, etc. It is the most important level of cloud computing framework.The cloud system management manages the resource usage of the entire virtual resource layer and the service status of the cloud application layer.The cloud application layer is responsible for integrating virtual resources with different functions for different categories of users, such as integrating application resources and network resources into WEB-based access methods.Cloud resources also provide access to the login interface to ensure that authorized users can use cloud resources[3][4].Therefore, the process of providing the service by the cloud computing service is that the user submits the task to the cloud system through the cloud application layer interface (SAAS/PAAS). After the task is submitted to the cloud system, the cloud application layer is responsible for submitting the task to the cloud system management layer. After the cloud system management layer is processed by the verification process, the resource scheduling component is responsible for allocating resources of the resource pool layer to the task, and finally returns the processing result to the user.

Based on the above research on cloud computing framework and service functions, it can be seen that resource scheduling management is very important for the overall cloud service efficiency and service quality. It is the core module of cloud computing and the key work of this paper. In order to implement the load balancing and improve resource utilization of Cloud Computing system, how to schedule resource becomes a central mechanism in Cloud Computing system.

Content from this work may be used under the terms of the Creative Commons Attribution 3.0 licence. Any further distribution of this work must maintain attribution to the author(s) and the title of the work, journal citation and DOI.
Published under licence by IOP Publishing Ltd

2. Related works

In recent years, with the rapid development of Internet technology, there has emerged various of network-based applications, such as E-mail and microblog, which brings a lot of convenience for people. At the same time that hundreds of millions of users browse information via the Internet, a huge amount of data have been generated continuously, such as electronic trading records, user access logs, etc. Besides, many large enterprises and organizations also generate a mass of data, including the stock information from stock exchanges and the oceanic data from monitoring stations, etc. In the face of such large data sets, how to efficiently process large-scale data sets and dig out valuable information becomes a point of issue concerned by lots of IT enterprises and scholars.

At present,the cloud virtual resource scheduling problem is a major challenge after cloud security issues. Although the work related to the cloud virtual resource scheduling problem has yielded rich research results, these results are often concentrated on an optimization goal, such as application performance priority, energy consumption priority or cost priority[5].

Most of the cloud computing platforms adopt a preset fixed number of physical devices and pre-configure the size of the virtual resources for the application[6][7]. The overall framework is not suitable for the current flexible and on-demand resource characteristics of cloud computing platforms, and the diversity of cloud services and the complexity of cloud computing environments. Requirements, cloud user requirements, and cloud resource prices all have certain dynamics[8].As one of the core problems of cloud computing, the efficiency of scheduling algorithm has a direct impact on the operation capacity of the system. Swarm intelligence algorithm, with good coordination and overall stability, is one kind of swarm intelligence algorithms which imitates swarm intelligence in the process of evolution swarm. These algorithms have characteristics of simple structure and strong searching ability, which can improve the accuracy and efficiency of cloud-computing resource scheduling algorithms. This paper studies on how to use biological group intelligent algorithm of cloud-computing resource scheduling problem. This dissertation proposes three improved artificial intelligence algorithm and an adaptive framework for cloud computing resource scheduling model and implements the application of cloud computing framework[9][10].

3. Adaptive optimization of dynamic scheduling framework(DFAOC)

3.1. Basic functions of DFAOC adaptive dynamic scheduling framework

(1) Cloud computing platform portal

Cloud computing platform portal: mainly through WEB mode, responsible for receiving various tasks submitted by users. The platform portal includes various application services such as IAAS , SAAS [11], and PAAS [12].

(2) Resource deployment module

Resource deployment module: The resource deployment module includes application resource deployment, virtual machine resource deployment, and corresponding deployment strategy. The deployment of the application resource refers to the virtual resource that meets the deployment requirements according to the deployment policy, and the application is deployed on a specific one or several virtual resources. The virtual resource deployment refers to the use of the physical resource. In the case, deploy the requested virtual resources to one or several physical resources according to the deployment strategy. After receiving the deployment task, the deployment policy first determines the deployment type, and then binds the application service and the virtual resource according to the corresponding policy, and binds the virtual resource and the physical resource. The application resource deployment and virtual machine resource deployment can be deployed at the same time or at different times. The deployment strategy of this module mainly uses the physical resource scheduling model and the swarm intelligence algorithm to achieve optimal allocation of resource scheduling and improve the efficiency of using various cloud resources. In addition, the resource deployment module automatically optimizes deployment according to the information that is suitable for optimizing the module, and provides scheduling support for various resources for the scheduling module.

(3) Resource scheduling module

The resource scheduling module needs to ensure the maximization of resource utilization while ensuring the service quality of the cloud computing user. The module is mainly based on the request of the user or the application, and is based on the resource analysis/prediction module, the resource deployment module, and the resource monitoring module. Information, complete the scheduling of cloud resources (virtual scheduling and physical scheduling). This paper mainly uses the virtual resource scheduling model and the swarm intelligence algorithm proposed in the third and fourth chapters.

(4) Adaptive optimization strategy module

The adaptive optimization module includes two main modules: the tuning model and the optimization strategy. The main idea is that, in a time period, the tuning model module obtains the current virtual resource and physical resource load information according to the resource monitoring module, and passes the resource prediction module. Obtain the resource demand information in the next cycle, and perform adaptive comprehensive analysis in combination with various tuning models such as load and energy consumption to determine whether the expected use and demand of various resources exceed the set dynamic threshold. If the expected exceeds the standard, At the same time of initiating the resource warning, the optimization strategy module is used to perform adaptive optimization adjustment such as over-targeting, resource expansion and addition, light load merging, and virtual machine dynamic migration.

(5) Resource analysis and prediction module

The resource analysis and prediction module mainly analyzes and predicts the usage of various resources based on real-time operation optimization and service history information of various resources. The module predicts the use of the PSO-RBF algorithm to predict resource requirements. In addition, the module classifies cloud resources according to resource usage and user dynamic requirements, and sets different indicator weights according to different usage rates of CPU, memory, hard disk, and network.

(6) Resource Monitoring Module

The resource monitoring module mainly performs resource monitoring according to user conditions, resource usage, optimization, and cloud service quality information, including whether the user's QOS and SLA are satisfied, whether the SLO of the application is satisfied, and the virtual threshold is monitored according to the formula. Information such as overload, low load, and failure of resource and physical resource nodes provides support for predictive modules, adaptive optimization modules, and deployment modules.

3.2. Optimization of cloud resource scheduling under DFAOC adaptive process

Based on the DFAOC framework, cloud computing physical resource scheduling optimization problem, each particle represents a physical resource allocation scheme. Currently, typical swarm intelligence algorithms are used to solve continuous optimization problems. Due to the discrete nature of cloud resource scheduling problems, individuals need to be specially coded.

1) Obtain the optimal parameters of the RBF neural network through PSO, and obtain the PSO-RBF optimized neural network;

2) Collect training data and conduct learning training through PSO-RBF neural network;

3) Input the measured data, predict the application resource demand through the PSO-RBF neural network;

4) Initialize the virtual resource pool of the cloud computing system resource according to the prediction result, and dividing the resource pool into the *m* resource subgroup;

5) Use a uniform method to generate an initial particle swarm, and set the initial position and initial velocity;

6) Calculate the fitness value of each particle, and select the optimal particle as the optimal solution of the individual optimal solution, the sub-population optimal solution and the neighbor group;

7) Updating the velocity and position of the particle swarm, and comparing with the individual optimal solution and the optimal group solution, and updating the optimal solution of the individual, the optimal solution of the sub-population and the optimal solution of the neighbor group according to the result;

8) Select the variable particles generated according to formulato generate the mutated particles, and flying backwards;

9) Judge the termination condition. If it is satisfied, the optimal particle position of the group is calculated, and the optimal scheduling scheme of the cloud computing resource is obtained. Otherwise, the iteration is continued.

4. DFAOC framework verification

This experiment verifies that the DFAOC framework is based on the traditional cloud computing resource scheduling model and the cloud resource adaptive model defined in this paper, and uses the adaptive management process of the DFAOC framework as a method. In the DFAOC framework verification process, the adaptive management process and the physical resource scheduling module are mainly used for verification. Other modules adopt the default configuration to test the scheduling efficiency of cloud resources under the DFAOC framework. In the DFAOC framework, the virtual machine initialization resource deployment module, the adaptive optimization module optimization model, and the analysis recommendation module can be dynamically set and changed, with greater flexibility and scalability, and only the DFAOC framework prototype system can be fully implemented. The overall accurate feasibility verification is carried out, but this is a complicated and heavy work. This part of the work will be reserved for the next step of study. Therefore, this paper uses the simulation key module verification method.

This simulation experiment uses the simulation platform of CloudSim, because the CloudSim can simulate the scheduling strategy in the cloud computing environment. This experiment extends the CloudSim platform according to the DFAOC framework, rewrites a series of classes, and defines Cloud resource basic description model, resource load measurement model, dynamic threshold model, multi-attribute weight model, build the program structure of resource deployment, adaptive optimization, analysis and prediction, and simply set the default processing jump. The programming tool is Myeclipse 9.0, verifying the validity of the DFAOC framework.

5. Conclusion

This paper analyzes the specific scheduling framework and scheduling algorithm involved in the cloud computing resource scheduling system. Firstly, the adaptive scheduling framework DFAOC for adaptive optimization of cloud computing resources is designed, and the specific structure and form of the framework system are given. Finally, based on DFAOC cloud resource scheduling framework, PSO optimized RBF neural network algorithm is simulated on CloudSim platform. The simulation results show that the proposed framework can effectively improve the service quality and resource utilization of applications in cloud computing system.

Acknowledgments

The authors are supported financially by International cooperation project(Project No.S2012ZR0191) and the Natural Science Foundation of Liaoning Province (Project No.20170540646).

References

[1] Coello C.,Carlos A.,Lechuga M. S. (2002). MOPSO: a proposal for multiple objective particle swarm optimization, Proceedings of IEEE Congress on Evolutionary Computation. USA, Honolulu: IEEE, 1051-1056.

[2] Cao H. ,Liu Y.(2010) CBFO:The cooperative optimization of Bacterial Foraging, in Computer Application and System Modeling (ICCASM),2010 International Conference on, 2,106-109.

[3] DeviG.,Raju G.,Sridevi P.(2015). Sridevi Application of Genetic Algorithm for Reduction of Sidelobes from Thinned Arrays,AMSE JOURNALS,58(1) 32-52.

[4] Doaa M., Faten H., Ninet M.(2012). Ahmed.A New Control and Design of PEM Fuel Cell System Powered Diffused Air Aeration System. TELKOMNIKA Indonesian Journal of Electrical Engineering,10(4): 291-302.

[5] Guo M.M.,Yang S.L.,Yan.H.,Kan.L.F.,Yang.B.(2014).MOBILE VIDEO ALARM SYSTEM BASED ON CLOUD COMPUTING,REVIEW OF COMPUTER ENGINEERING STUDIES,1(2),5-[6].Hossain S. Infrastructure as a service[J]. Cloud Computing Service & Deployment Models Layers & Management, 2013:22(7),26-49.

[6] Tsafrir D, Schuster A, Benyehuda M, et al. Deconstructing Amazon EC2 Spot Instance Pricing. Third International Conference on Cloud Computing Technology and Science[C], 2012:304-311.

[7] Kippenbrock T, Holloway E, Moore D D. Google Docs[J]. Computers Informatics Nursing Cin, 2010:138-140.72.Karun A K, Chitharanjan K. A review on hadoop — HDFS infrastructure extensions. Information & Communication Technologies[C]. IEEE, 2013:132-137.

[8] Browne, Clarke, Mishra. Automatic Verification of Sequential Circuits Using Temporal Logic[J]. IEEE Transactions on Computers, 2006, 35(12):1035-1044.

[9] Mitra S, Datta S, And T P. Introduction to Physical Polymer Science[J]. Macromolecular Chemistry & Physics, 2010, 207(8):787-787.

[10] Majhi J, Smid M, Smid M. Multi-criteria geometric optimization problems in layered manufacturing.The Fourteenth Symposium on Computational Geometry[C]. ACM, 1998:19-28.

[11] WANG G H, LI Q.H, LIU A.F. Multi-objective optimization cloud workflow scheduling evolutionary genetic algorithm [J]. computer sciences, 2018, (5):31-37.

ISPECE   IOP Publishing

IOP Conf. Series: Journal of Physics: Conf. Series **1187** (2019) 032110   doi:10.1088/1742-6596/1187/3/032110

Research on energy-saving lighting control system of tram station based on traffic and passenger flow information

Binjie Xiao

No. 3447, Dongfang road, Shanghai, China

Binejie Xiao, Shanghai urban construction design group, Senior engineer, Registered electrical engineer&Registered consulting engineer, Engaged in electrical engineering design and research；

Binjiexiao@163.com

Abstract. In order to meet the needs of passengers waiting for the trams, an energy-saving lighting control system of tram station is proposed based on the detection of illumination information, driving in and out of the station information, station passenger flow information. The system can change the lighting control module according to the different conditions of the station, so as to control the opening and brightness of the lighting source and reduce the energy waste caused by the using of the lighting source in the non-necessary period. The system realizes fully automatic lighting energy-saving control of the tram station by making full use of traffic and passenger flow information. The lighting control system can automatically switch according to its internal preset mode, reducing the operation cost of the station.

1. Introduction

In recent years, modern trams have developed rapidly in China and have been put into operation in more than ten cities. Because trams are mostly set along main roads, most of them have no separate lighting, but with the help of street lamps[1,2,4]. The tram platform has certain particularity: sometimes there will be more people waiting for the train at the platform, or even more crowded; there may be very few people waiting for the last bus. If the platforms are not installed energy-saving lighting control system, it may affect the safety of waiting passengers and will cause unnecessary waste when there are few or no people waiting for the bus. Therefore, it is urgent to propose a special lighting energy saving control system, which can automatically adjust the lighting mode according to the situation of waiting passengers and the information of the tram entering and leaving the station. The lighting control system for the tram station can give the balance of lighting and energy efficiency requirements[3~8].

2. Energy-saving lighting control system for tram stations

The energy-saving lighting control system for tram station based on traffic and passenger flow information includes information acquisition module(Tram entering and leaving detection equipment and activity personnel detection equipment), information transmission module, lighting control module, remote lighting control module and communication module set in the control center(See figure 1). The information acquisition module is used to collect the information of passengers entering and leaving the station, illumination information and the information of trams entering and leaving the station. The information transmission module is used to transmit the information collected by the

Content from this work may be used under the terms of the Creative Commons Attribution 3.0 licence. Any further distribution of this work must maintain attribution to the author(s) and the title of the work, journal citation and DOI.

Published under licence by IOP Publishing Ltd

information acquisition module to the lighting control module. The lighting control module controls the brightness of the lighting source according to the station illumination information transmitted by the communication module, the information of tram entering and leaving the station and the information of passengers entering and leaving the station. The illumination control module can be controlled according to the different conditions of the station, so as to control the opening and brightness of the lighting source and reduce the waste of energy caused by the opening of the lighting source in non-essential periods. And the automatic control mode further saves human resources and reduces the operation cost of the station.

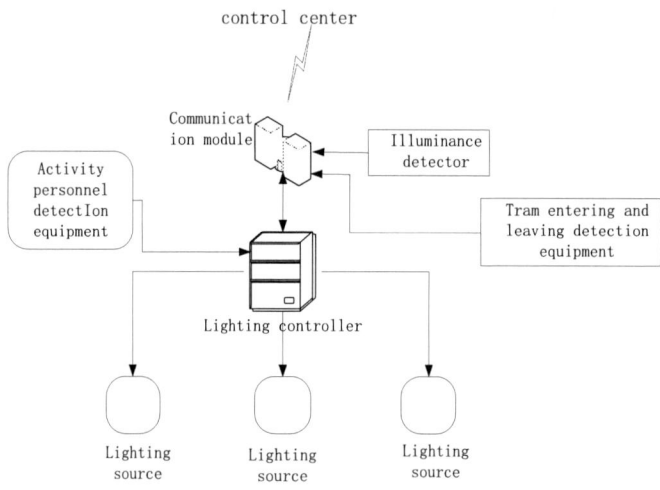

Figure 1. System composition diagram

3. System solutions

The energy-saving lighting control system for tram station based on traffic and passenger flow information is a kind of lighting control system for tram station that can automatically adjust the lighting mode according to the situation of passengers in the tram station and the information of tram driving (entering and leaving the station) and take into account the lighting and energy saving requirements at the same time. In order to achieve the above purposes, the system uses the information acquisition module, information transmission module and lighting control module to collect and transmit the station illumination information, tram entering and leaving station information and passenger entering and leaving station information; and controls the brightness of the lighting source according to the illumination information received, the information of entering and leaving stations of trams and the information of passengers entering and leaving stations; The system also includes a remote lighting control module and a communication module. In case of abnormal situation in the station, the normal transmission between the information transmission module and the lighting control module is interrupted. The remote lighting control module directly controls the lighting control module through the central communication module, and controls the brightness of the lighting source through the lighting control module. The fault information of the lighting source is uploaded to the remote lighting control module through the communication module.

The information acquisition module includes the illumination detector used to detect the illumination information of the station and the detection equipment used to detect the entering and leaving information of the tram. The illumination detector transmits the detected station illumination information to the illumination control module through the station illumination communication module. The information acquisition module also includes a personnel activity detection device for detecting the information of passengers entering and leaving the station. If the lighting control module is in the closed state, the lighting source will be closed; If the lighting control module is opened, the lighting control module controls the brightness of the lighting source according to the information received from passengers entering and leaving the station.

4. Lighting control method and process based on traffic and passenger flow information

The energy-saving lighting control method for tram stations includes the following steps:

Step 1: The illumination detector is used to detect the illumination information of the station and transmitted to the lighting control module through the station illumination communication module. When the station illumination information is in the daytime mode, the lighting control module is closed. The lighting control module controls the brightness of the lighting source based on the information of tram entering and leaving the station. This control mode can ensure the automatic closing of the lighting source in the case of sufficient illumination, which saves energy and labour cost without manually control the opening and closing of the lighting source.

Step 2: start up the detection equipment of tram which is used to detect the information of tram entering and leaving station. The detected information of trams is transmitted to the lighting control module through the detection communication module. When the station illuminance information is in night mode, the lighting control module opens. The lighting control module controls the brightness of the lighting source according to the information of tram entering and leaving the station. If the communication module receives the signal from the trams to the station, the on-board and off-board lighting mode will be turned on, and the lighting source will work at full load to provide 100% illumination, so as to meet passengers' on-board and off-board requirements. When the communication module receives the signal that the tram has left the station, the station illumination is the waiting light mode, providing 30% illumination. In this process, the equipment used to detect the entrance of trams can be based on the current common tram positioning and signal control equipment, as well as a variety of detection equipment such as reading information signs, ring coil detection.

Step 3.start to detect station waiting passengers to the activities of the personnel activity detection device, lighting control module according to the information received 30% of illuminance for waiting passengers. In this detection process, the personnel activity detection equipment used to detect the information of passengers entering and leaving the station can be infrared detection equipment, camera or other equipment.

The remote lighting control module communicate with the center of the communication module in the abnormal situation of the station, such as fire, earthquake, or train fault occurs, because the normal transmission between the information transmission module and the lighting control module is interrupted. The control center directly commands the lighting control module and controls the brightness of the lighting source，for example, turn on the light source and provide 100% illumination. When the lighting source fails, the fault information of the lighting source is uploaded to the remote lighting control module through the central communication module. When the tram is not in operation, the lighting source can also be directly turned off by remote control of the lighting module.

The fault information of lighting source can be uploaded to the remote lighting control module set in the control center through the communication module for information transmission module, so as to timely discover and maintain the fault of lighting source. The brightness of the light source in the waiting mode can be 30%-60%, while the illumination in the boarding mode can be 70%-100%.

The illumination control module can be controlled according to the different conditions of the station due to the illumination detection instrument, the detection equipment for trams entering and leaving the station and the personnel activity detection device. The illumination detection equipment sets day and night modes based on the detection results. The daylight mode is defined as the mode in which natural light can meet the lighting requirements of the station. Under this mode, the lighting source in the station lighting control system is closed. Night mode is defined as a mode in which natural light cannot meet the lighting requirements of the station. In this mode, the lighting source needs to be provided by the station lighting system. In order to further save energy, according to the actual brightness of the lighting source required by the station, the night mode is divided into waiting mode or inbound and outbound mode. Among them, the brightness of the light source in the waiting mode can be 30%, while the lighting mode of getting on and off the bus can be 100%, to avoid accidents caused by insufficient light in the process of getting on and off the bus. In this way, the opening and brightness of the lighting source can be controlled and the waste of energy caused by the opening of the lighting source in non-essential periods can be reduced.

The flowchart of specific lighting control methods for tram stations is shown in figure 2.

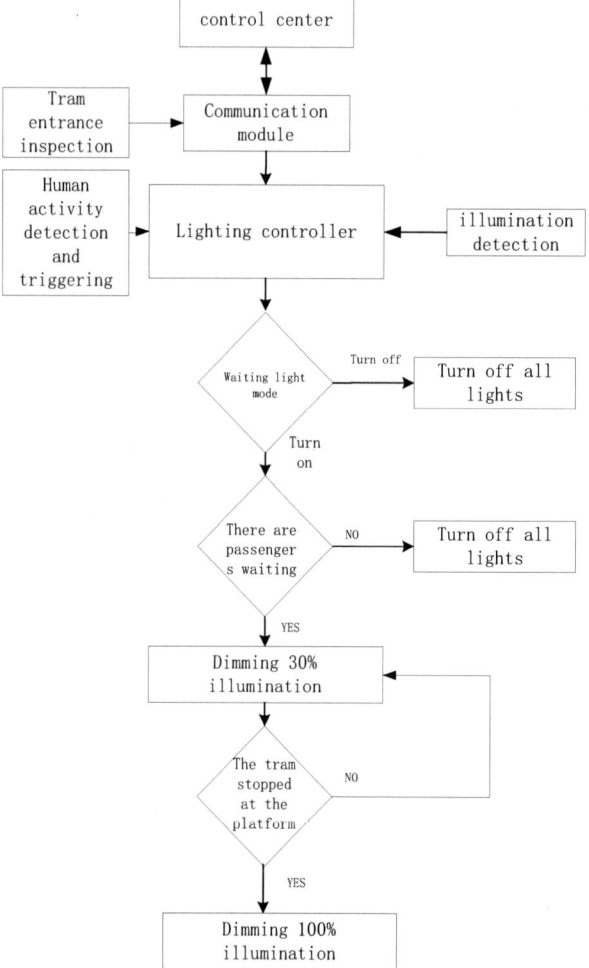

Figure 2. Flowchart of lighting control methods for tram stations

The mentioned above is the cooperation between modules when the lighting control system works normally. Since the lighting control mode of the station is automatic control mode, the lighting control system can automatically switch according to its internal preset mode without manual operation, which saves human costs and reduces the operation cost of the station. The remote lighting control module set in the control center can directly control the lighting control module through the center communication module, adjust the opening and brightness of the lighting source, and provide the ability to control the lighting system of the station in case of emergency. In addition, the fault information of the lighting source can be uploaded to the remote lighting control module through the communication module, so as to avoid the occurrence of the failure of the lighting source.

5. conclusion

The energy-saving lighting control system of tram station can control lighting module based on illumination information, driving in and out of the station information, station passenger flow information. It reduces the unnecessary time lighting source opening by adjusting the brightness or switch the light according to the different situation of the station. The lighting control mode of the station is an automatic control mode by making full use of traffic and passenger flow information, and the lighting control system can automatically switch according to its internal preset mode.

Acknowledgments

This paper are supported by Shanghai talent training and development fund (project number 201324) and Innovation action plan of Shanghai science and technology commission (15DZ1204300).

References

[1] Su jun, su yajun, tang guanghua J 2016 Analysis on energy consumption and energy saving of modern trams in guangzhou. *Research on urban rail transit,* vol 19(12) pp 57-61.

[2] zhao xinmiao D 2014 Calculation of modern tram operation cost and analysis of influencing factors. *Beijing jiaotong university.*

[3] Zhou fang, liu meigen J 2007 Application of intelligent lighting control system in energy saving design of electrical lighting. *Electrical appliances and energy efficiency management technology,* vol.16 pp 20-23.

[4] Xu dianguo, zhang xiangjun, liu xiaosheng, et al. J 2007, Development status and future of lighting electronics technology. *Power electronics technology,* vol. 41(10) pp 2-9.

[5] Li Ming, liu nan, shi junjie. J 2017 Analysis of influencing factors and energy saving measures for tram operation energy consumption. Electric locomotive transmission, No.(2) pp 96-100.

[6] Mu guangyou, li xiaolong, Yin liming, et al. J 2010 Energy consumption analysis and energy-saving countermeasures of subway station lighting system. Urban rail transit research, Vol. 13(8) PP 35-39.

[7] Zhong suyin. J 2008 Study on energy saving of lighting system in guangzhou metro station [J]. Urban fast rail transit, Vol.21(4) pp 81-84.

[8] Wang zhiqiang. J 2013 Design of intelligent lighting control system for subway station [J]. Urban rail transit research, Vol.16(6) pp124-127.

ISPECE   IOP Publishing

The optimization of networking method for the system protection communication networks based on the delay analysis

Yanan Wang[1], Ke Wang[2,1], Ran Zhang[2], Geng Zhang[1] and Yang Wang[1]

[1] China Electric Power Research Institute, Beijing, 100192, China

[2] North China Electric Power University, Beijing, 102206, China

*Corresponding author e-mail: happyandluck_wk@163.com

Abstract: The real-time requirements of the data transmission for the power communication systems are very high, and the system protection demands the communication delay more stringent. The consequences of the malfunction or refusal of the grid control system caused by the communication delay are much more serious than those of other communications. Because the grid stability control system involves the entire system, the malfunction or refusal may lead the entire grid to be disassembled or collapsed. A new networking method is proposed based on the analysis of the existing power communication networking technology in this paper. And the delay of this new networking method is analyzed by analyzing the mechanism of the delay generation. So that the adaptability of the new networking method in the system protection communication networks is determined from the delay level. It is intended to provide the reference and guidance for the construction of a high-real-time system protection communication networks.

1.Introduction

According to the communication requirements of the power system, the maximum transmission distance is considered to be 3,000 kilometers (in the region) and 5,000 kilometers (between the regions). The system protection communication network is required to perform the fault defense control on the important disturbances of the power generation, transmission and distribution in the grid within 300ms. The panoramic state monitoring of the DC system is limited to 60ms, and the acquisition and control of the communication delay should be controlled at 50ms. However, as a matter of fact, a medium-sized regional grid stability control system generally spans a prefecture-level city and the large-scale regional security control systems may even involve multiple substations and power plants in a province or more. Therefore, the grid stability control system is more demanding on the delay of communication. Moreover, one data or command at least should be exchanged within 1.667ms in the two stations of the power communication, and the action must be taken after 3 frames of control commands (5ms) are received continuously. Therefore, the malfunction or the delay action of the grid stability control system caused by the delay may lead the entire grid to be disassembled or collapsed. The communication networking method of the system protection communication network needs to have the characteristics as follows: 1) The reliability requirements of the channel are very high; 2) The speed of the channel should be as fast as possible; 3) The delay of each link of the channel should be as small as possible; 4) The channel should have a certain redundancy and

Content from this work may be used under the terms of the Creative Commons Attribution 3.0 licence. Any further distribution of this work must maintain attribution to the author(s) and the title of the work, journal citation and DOI.

Published under licence by IOP Publishing Ltd

invulnerability; 5) In addition to the normal communication, the channel should be equipped with the corresponding emergency communication method.

2. The networking method of the system protection communication network

2.1. The analysis of the normal networking technology

The power communication system is based on the collection and transmission of the multi-point information. And the strict requirements for the real-time and reliability of the communication depend on the development of the modern communication technology. The favorable opportunity and technical support for the construction of the grid system protection communication network is provided by the wide application and development of the SDH, OTN, PTN and IP switching technology in the various communication fields [1]. The network performance of the normal networking technologies is as shown in table 1.

Table 1. The networking performance of the networking technology

Networking method / Network performance	SDH	OTN	PTN	IP Switching
Multiplexing system	Time division	Wavelength division	Packet division	Packet division
Switching characteristics	Time slot exchange	Time slot exchange	Packet switching	Packet switching
Channel bandwidth	$\geq$10G	$\geq$40G	$\geq$10G	$\geq$40G
Business reliability	Channel protection	Channel protection	Channel protection	Channel protection
Bandwidth particle	2M particles	1G particles	Arbitrary particles	Arbitrary particles
Delay	Satisfied	Satisfied	Partially satisfied	Partially satisfied
Delay consistency	Good	General	General	General
Frequency synchronization	Support	Support	Partial support	Partial support
Time synchronization	Optional	Optional	Support	Support
Multicast	Limited support	Limited support	Support	Support
Safety	High	High	Poor	Poor

It can be seen from the performance comparison of the networking technologies in table 1 that the SDH networking method can meet the reliability requirements of the system protection service, but it also has the disadvantage of poor scalability and the insufficient bandwidth at the branch node and root node [2]. The OTN networking method meets the system protection service requirements, while its cost performance is relatively low based on the delay and uncontrollability which are caused by the system conversion. The PTN networking method also meets the system protection service requirements with the characteristics of small overall delay but uncontrollable, the access network and the transmission network are well connected, but the information security does not meet the requirements. The performance of the IP data network can meet the bandwidth and other requirements, while it has the risk of the uncontrollable delay, and its security also needs a certain protection. In addition to the above normal networking technologies, the dispatching data network of the State Grid, as a dedicated data network for the power dispatching production services, is also the basis for the grid dispatching automation and the management modernization and the important means for ensuring the safe, stable and economic operation of the power grid. However, since there are many services carried on the dispatching data network, the greater risks will be brought if the system protection service is still carried on this network. And the topology of the dispatching data network will also have a greater impact on the delay at the same time.

At present, the existing stable control protection system is mainly carried on the E1/2M dedicated line, and the PMU/WAMS system is mainly carried in the dispatching data network. An effective local communication networking method is proposed in this paper named SDH+EOS networking method by combining the advantages and disadvantages of the above five networking technologies with the requirements of the system protection services to realize the integration of measurement and control of the two mentioned above.

2.2. SDH+EOS networking technology

The normal wide-area stability control is carried on the 2M dedicated line service. And the dedicated line service can be considered to expand to the service of supporting 10M bandwidth to meet the needs of the system protection communication network. Since the substation uses the optical fiber or 100M Ethernet for data transmission, the expansion of the interface bandwidth can be learned from the method of SDH/MSTP. The EOS (Ethernet over SDH) access mode provided by the MSTP equipment is used, and the router or switch adopts the interface of the Ethernet directly. The MSTP is interconnected by the common RJ45 (10/100BaseT) interface which can save the cost of the optical port interconnection greatly, and the end-to-end service management can be realized through the unified network management of MSTP [3]. The MSTP device featuring the EOS (Ethernet Over SDH) technology can form a point-to-point dedicated line based on the transmission equipment, and enter the network in the form of a dedicated "pipe" in different granularities such as STM-1 and STM-4, which conforms the real-time and wide-area local communication scenario. Each layer of the three-layer power communication network is divided into two layers of the main station and substation in this networking method, the status information in the substation acquisition system is transmitted to the main station, and the main station transmits the control commands to the substation. Taking the integration of measurement and control of the system protection communication network into consideration, the internal communication mode of the substation is shown in figure 1.

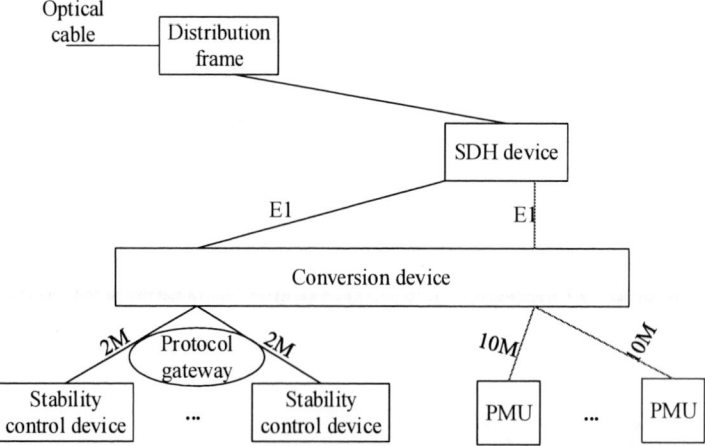

Figure 1. The networking method of the SDH+EOS in the substation

The substation includes a multi-channel PMU device and a multi-channel stability control device in the networking method. The optical fiber signal of the stability control device is mapped to the photoelectric conversion and proprietary protocols by the protocol gateway, and then the data is connected to the conversion router through the Ethernet, while the data of the PMU device is directly connected to the conversion router through the Ethernet.   The conversion router converts the 2M signal of the stability control device into the data of E1, and then combines the Ethernet signal of the PMU to the SDH device, finally the data is packaged into the SDH device through the EOS technology of the MSTP, and transmitted to the main station through the distribution frame.

As is shown in figure 2, there is a corresponding networking method inside the main station.

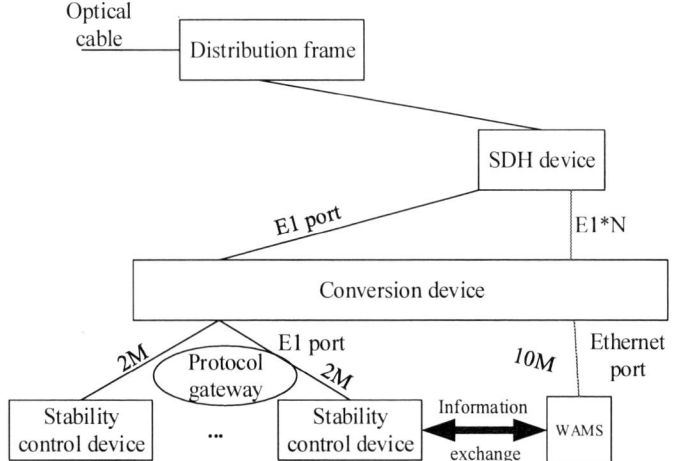

Figure 2. The networking method of the SDH+EOS in the main station

The communication process in the main station is the inverse process of that in the substation. The stability control device and the WAMS system send the control information to each substation and collect the wide-area measurement data from the PMU of them. There is the information interaction between the WAMS system and the stability control device to achieve the real-time decision for the stable control.

The internal communication mode of the substation and the main station are shown above. The wide-area communication networking architecture between the stations should be considered to achieve the wide-area communication. The communication networking method in the wide area is shown in figure 3.

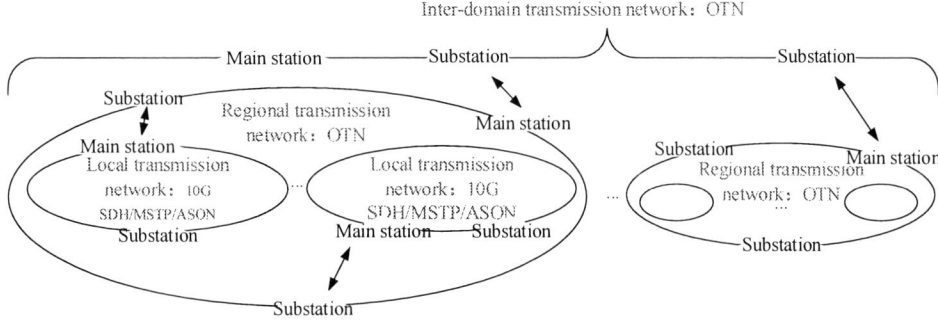

Figure 3. The wide-area networking architecture of SDH+EOS

According to the level of the communication architecture, it can be divided into three layers of the communication network which is in the local area, in the region and between the regions. The figure above shows an inter-domain transmission network consisting of two regional transmission networks, each of which contains two local transmission networks. The main station of the local transmission network communicates with its substation while also interacting with the substation of the regional transmission network. Similarly, the main station of the regional transmission network communicates with its substation and the substation of the inter-domain transmission network meanwhile. The local transmission network mainly uses 10G SDH/MSTP/ASON for networking, and inter-station transmission can select the OTN and other methods. The regional transmission network and the inter-domain transmission network can also be networked by using the OTN.

This networking mode is characterized by the exclusive channel, controllable delay and high security.

The method of the architecture above is feasible. The adaptability of the SDH+EOS networking will be analyzed and verified by using the main factor named delay which affects the network performance in the following.

3. The delay analysis

3.1. The generation mechanism of delay
In the grid communication, the factors which cause the delay are roughly divided into three categories: The delay caused by the transmission channel, the delay generated by the network node and other digital devices and the delay introduced by other factors.

3.1.1. The delay caused by the transmission channel Both the electrical signal and the optical signal are the electromagnetic wave. And the propagation speed of the electromagnetic wave is limited in a certain transmission medium, which is mainly depended on the refractive index of the medium. For example, the transmission delay t_0 of the optical wave signal passing through the optical fiber can be expressed as follows:

$$to = L * n/c \qquad (1)$$

In equation (1), c is the speed of light in vacuum (3×10^5 km/s), L represents the transmission distance (km), the index of refraction of the fiber core is expressed as n with the typical value 1.48. Then it can be calculated that the transmission delay of the optical signal in the optical fiber is approximately 4.93 μs per kilometer.

3.1.2. The delay generated by the network node and other digital devices In a digital connection, the network node devices such as digital switches, digital cross-connect devices, buffers, time slot switching units, and other digital processing devices generate transmission delays in addition to the delay caused by the transmission system. Furthermore, PCM terminals, multiplexers and multiplex converters also produce varying degrees of delay.

3.1.3. The delay introduced by other factors New technologies are emerging in the field of the optical fiber communication, and the introduction of them may decrease or increase the delay which is depended on its working mechanism. For example, the SDH technology can increase the transmission delay by completing the synchronous multiplexing, mapping, positioning, various types of the overhead processing, pointer adjustment and connection processing. In addition, the changes in the network structure, the quality of the installation and construction can also affect the transmission delay [4].

The delay factors caused by the different transmission modes are different in general. Specifically, it is mainly composed of a terminal SDH device, a network node device, and a fiber channel and the like. The typical delay values are shown in table 2

Table 2. The composition of the transmission delay

Device		Terminal SDH/μs	Network node /μs	Optical fiber /μs
Private channel		NO	NO	4.93
Multiplex channel	64K	10~60	20~125	4.93
	2M	10~60	20~125	4.93

3.2 The analysis and calculation of the delay

3.2.1.The calculation formula of the determined value delay Combined with the above analysis of the delay factors, the delay t of the SDH+EOS networking method when transmitting the data can be expressed by equation (2).

$$t = t_{SDH} + nt_i + t_O \tag{2}$$

In equation (2), t_{SDH} is the multiplexing and demultiplexing delay of the SDH device, n represents the total number of the intermediate node device. t_i is the delay of the intermediate node device, such as the forwarding delay of the relay node device, the input and output delay, the router processing delay, the photoelectric conversion delay and the service transmission delay. However, the internal structure of the intermediate node is different from that of the terminal SDH device, so the delay generated is also different.

t_O is the delay generated by the fiber transmission channel, and $t_O = 4.93L(us)$ [5].

3.2.2.The random delay model The delay value of the calculation model listed in equation (2) is calculated based on the determined value. This calculation method is relatively simple and can obtain a typical value that conforms to the actual situation based on a large amount of datas, while the randomness of the delay is not considered to describe the probability of the delay value. The delay of the transmission equipment is a random value in a certain range to meet the certain change trend, not a fixed value in the signal transmission of the actual system protection communication network. While the system protection communication network should anticipate these trends and pre-process the places with too long delays as early as possible. Therefore, some parameters in the above equation should be set as the random variables and the probability model should be established to describe the change of the delay more accurately.

Compared with the deterministic calculations, the random delay model has the following advantages:

(1) The trend of delay can be described more accurately.

(2) The delay value calculated by the random delay model is more general and more consistent with the actual situation.

(3)The probability of the refusal or malfunction of the system protection device can be estimated roughly by establishing a random delay model. On the contrary, the maximum number of the nodes allowed in the network is derived from the national delay standard.

According to the typical values of the device delay and the actual data measured in the paper, it is found that the delay of the protection transmission delay in the system protection communication network has the following characteristics:

(1)The device delay value is concentrated in a certain time interval.

(2)The line delay is relatively fixed, while the device delay is mostly closed to the typical value with a normal distribution.

(3)The impact of the delay of each part of the transmission channel on the total delay is different.

Based on the above reasons, t_{SDH} and t_i are random variables which are represented by T_{SDH} and T_i respectively in equation (2). The above random variables obey the Uniform distribution and the normal distribution respectively, and the distribution parameters are set as shown in table 2. The variance of the random variables is used as a variable to substitute the different values for the comparison calculation in the subsequent calculation [6].

The probability distribution expression is difficult to be derived by the mathematical modeling because of the large number of random variables. Therefore, the MATLAB simulation method is selected to study the random distribution characteristics of the total delay more intuitively.

1000000 random numbers obeying the uniform distribution or the normal distribution were generated by MATLAB based on the parameters listed in table 2, and 1000000 samples were taken as the random delay samples. According to the delay calculation formula, the total delay was calculated by the random delay to obtain a sample space composed of 1000000 total delay values. The probability density function of the total delay can be obtained by the ecdfhist function of MATLAB and the cumulative probability distribution function can be obtained by ecdf.

The degree of dispersion of a data set can be reflected by the standard deviation. The same average does not mean that the standard deviation is the same. The extent to which the delay value deviates from the mean is reflected by the magnitude of the standard deviation when the random variable obeys the normal distribution. As is shown in figure 4, the probability distribution curves with different delays can be got by setting the different standard deviations.

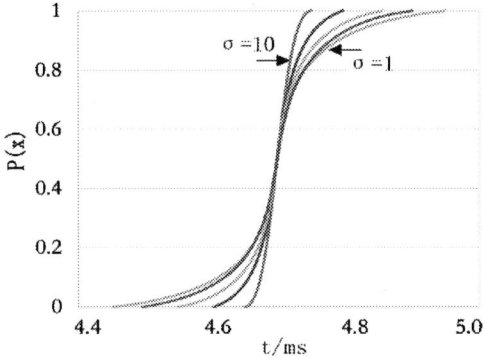

Figure 4.The effect of standard deviation on distribution during the normal distribution

As can be seen from figure 4, the larger the standard deviation is, the steeper the cumulative probability distribution curve is, and the smaller the variation range of the total delay is when the random delay satisfies the normal distribution. The standard deviation in the subsequent simulation delay is taken as 5 to facilitate the study [7].

L is equal to 40, 60, 80, 100, and 120 km, respectively, the number of the network nodes N is 10. The result of the program is shown in figure5

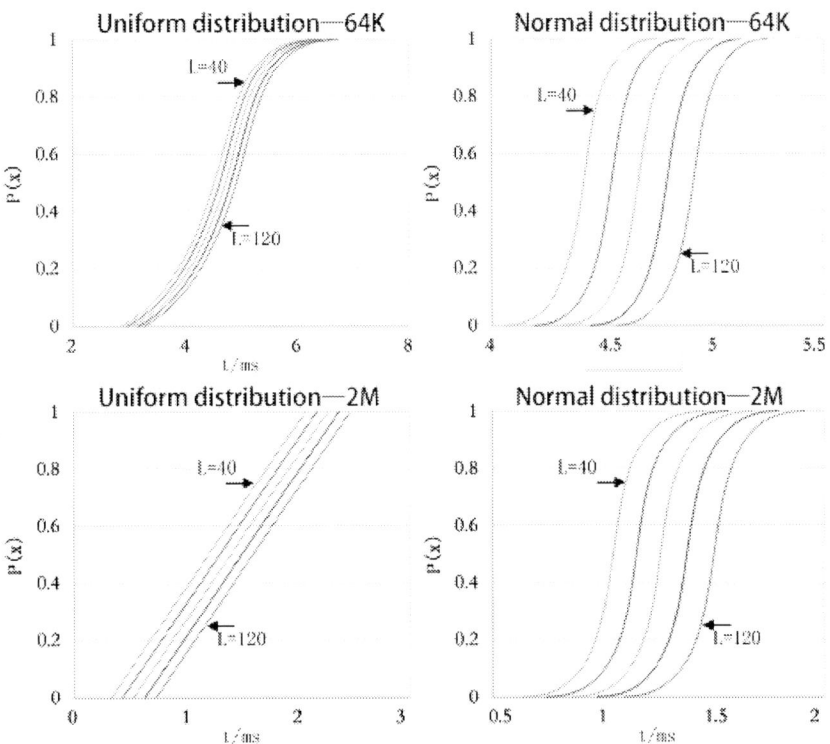

Figure 5. The effect of the change of L on the probability distribution of delay

As can be seen from figure 5, the five curves are the probability cumulative distribution function from left to right respectively for L=40, 60, 80, 100, and 120. The transmission delay gradually increases as the transmission distance increases. When the transmission distance is 120 km, the maximum delay is no more than 7 ms which is much less than the standard 15 ms. When comparing the two types of transmission channels which obey the uniform distribution and the normal distribution respectively, it can be seen that the maximum delay of the random delay which obeys the normal distribution is slightly smaller than that of the uniform distribution. And the change extent of the total delay is greater as the transmission distance increases.

The number N of the network nodes is equal to 10, 15, 20, 25, 30, respectively, L = 100 km. The result of the program is shown in figure 6.

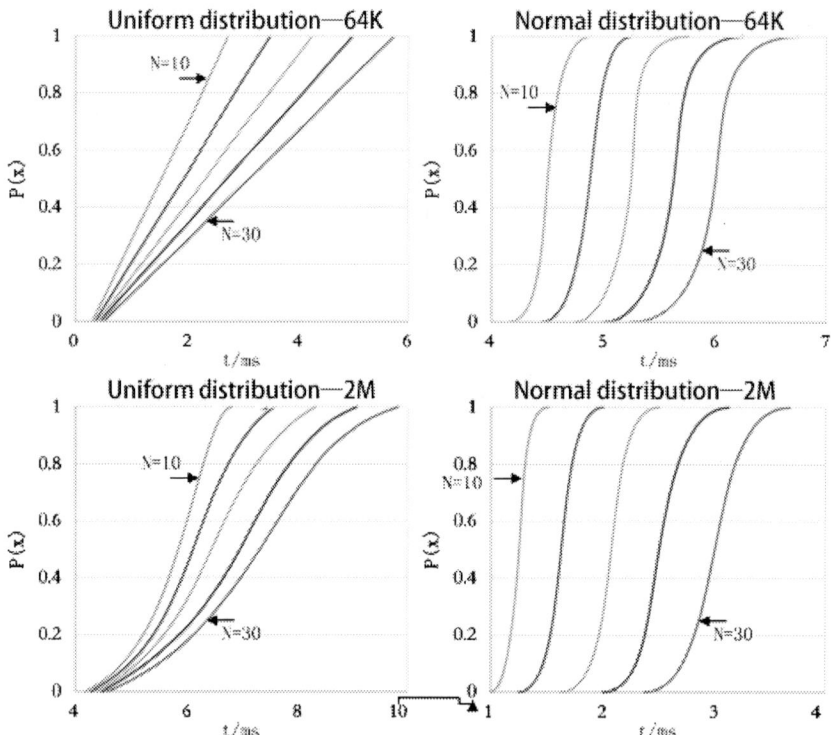

Figure 6.The effect of the change of N on the probability distribution of delay

As can be seen from Figure 6, the five curves are the probability cumulative distribution function from left to right respectively for N=10, 15, 20, 25, and 30. The transmission delay gradually increases as the number of the network nodes increases. When the network node is 30, the maximum delay is no more than 10 ms which is much less than the standard 15 ms. Similar to the case of the change of L, the maximum delay of the random delay which obeys the normal distribution is slightly smaller than that of the uniform distribution.

The following conclusions can be drawn referring to figure 4, figure 5 and figure 6.

(1) The standard deviation has a certain influence on the distribution of the total delay when the random delay satisfies the normal distribution. Therefore, the effect of the variation range of each device delay on the total delay cannot be ignored.

(2) It can be seen from the cumulative distribution curve that the curve is steeper in the case of obeying the normal distribution which shows that the value of the delay is more concentrated under the normal distribution. Meanwhile, the value of the delay under the normal distribution is smaller than that under the uniform distribution, and the maximum delay is smaller [8].

(3) The increase or decrease of the number of network nodes has a greater impact on the total transmission delay compared with the transmission distance. Therefore, the number of the network nodes should be fully considered to avoid the excessive delay due to the excessive network nodes in the actual networking.

(4) Due to the delay generated by the PCM device in the 64k interface accounts for a large proportion of the entire channel delay, the total delay of the 64k interface is greater than that of the 2M interface while regardless of the distribution of the random delay. Therefore, the intermediate link should be avoided to transfer at 64kbps to reduce the total channel delay [9].

4.The SDH+EOS networking technology delay analysis

With the above analysis of the determination of delay and random delay, the delay of the commonly used power communication equipment is investigated. The delay of each equipment is as follows:

(1) The relay node device needs to forward each hop for 7 μs.

(2) The router processing time (including the forwarding time) is 100μs.

(3) The time of the router packages the SDH is 150μs.

(4) The transmission time from the access device to the transmission device is 100μs.

(5) The time for transmitting information from the upper to lower service station is 150 μs.

(6) The photoelectric conversion time of the router is 150μs, FEC + dispersion + transmission delay plus 10% of the time.

The processing delay based on the SDH+EOS networking is $150+2\times(150\times110\%)+n\times[7+2\times(150\times110\%)+100+150+100]+4.93L=150+330+687n=480+687n+4.93L$ by using the equation (2) and the random delay model to assume that there are n hops in the topology.

According to the communication requirements between the regions in the power system, the maximum transmission distance is considered to be 3,000 kilometers (in the region) and 5,000 kilometers (between the regions). It is assumed that the maximum transmission distance in the region is 3000 km, and the transmission delay of the optical fiber on the entire link is about 14790 us by equation (1). And because the communication delay requirement of the wide-area security and stability control system is not higher than 50ms, then $(480+687n)+14790\leq50$. The solution is $n \leq 50$, that is, the maximum hop count allowed in the region is 50 hops. Assuming that the maximum transmission distance between the regions is 5000 km, similarly, the maximum number of the hops allowed between the regions is 36 hops. Therefore, the networking method of SDH+EOS satisfies the high real-time requirement of the system protection communication network as long as the hop count does not exceed the above two values in the case of satisfying the maximum transmission distance.

5. Conclusion

The main functions of the system protection include the AC/DC cooperative control, the disconnect control, the separation control, the precise load shedding and the panoramic state sensing. The current communication network is difficult to meet the high-speed and high-real-time requirements of the above-mentioned acquisition and control information. Therefore, a new networking method is proposed based on the analysis of the existing networking technology of the power system in this paper. Then the mechanism of the communication delay is introduced in detail and the adaptability of the SDH+EOS networking is verified based on the advantages and disadvantages of the delay performance. The final result shows that the new networking method has a better network performance than the existing networking method as long as the number of the nodes (hop count) in the communication network does not exceed a certain limit value (50 hops within the region, 36 hops between the regions) when the system protection communication network is in the case of the maximum transmission distance. It can meet the need for the long distance transmission of the system protection communication network.

Acknowledgments

This paper was supported by the science and technology project from State Grid Corporation of China:"Research on Distributed Simulation and optimization technology of optical transmission network for Electric power communication(5442XX180003-XX71-18-006)"

References

[1] Cao N，LI G，Wang D Q,(2011) Key technologies and construction methods of smart substation, Power System Protection and Control,39:63-68

[2] Zhang Q H,Mu C F,Lu J Y,Liu S,(2016) Research on Construction of Provincial Communication Network for Power Communication, Digital technology and application, 34:45-47

[3] Wang Z Q,Liu W X,Fan Y F,Liu N,(2008) A Reliability Study Framework for Wide Area Measurement System,Automation of Electric Power Systems,32:25-29

[4] Chen W Q,Ma Q F,Zhao J L,(2011) Research on the reliability evaluation of Power optical fiber transmission network, Telecommunications for Electric Power Svstem,32:11-15

[5] Wu Y M,Ding J J,He X X,(2016) Secure consensus control for mufti-agent systems under communication delay, Control Theory&Applications,33:1040-1045

[6] Yang B,Wei L P,Zhan Z B,(2015) Analysis on Characteristics of Communication Delay in Wide Area Measurement System Based on Probability Distribution, Automation of Electric Power Systems, 39:28-55

[7] Liu J,Li D Q,(2013) Distributed Subgradient Method for Multi-Agent Optimization with Communication Delays, Journal of Hefei University of Technology,36:559-565

[8] Liu X L,Xu B G,(2012) Distributed H∞ consensus control for multiple-agent systems with communication delays, Control and Decision,27:494-500

[9] Liu C L,Tian Y P, (2009)Survey on consensus problem of multragent systems with time delays, Control and Decision,24:1601-1609

ISPECE   IOP Publishing

IOP Conf. Series: Journal of Physics: Conf. Series **1187** (2019) 042002   doi:10.1088/1742-6596/1187/4/042002

The research of ship yaw detection method based on virtual navigation channel

Juanjuan Shao, Shu Zhou*, Xin He, Xinzheng Zhang and Xindong Liu

School of Electrical and Information Engineering, Jinan University, Zhuhai 519070, China

*Corresponding author's e-mail: 14104910@qq.com

Abstract. The ship may deviate from the proposed route at any time because it will be influenced by the wind current and other factors during the navigation. Especially near the bridge river area, the river terrain is complex, such as straight and curved navigation channels. If the ship yaws without regulation immediately, this may cause unpredictable consequences such as hitting the bridges. To address the problems above, this paper proposed a ship yaw detection method based on virtual navigation channel. By building the virtual navigation channel from the video stream, the detected distance between the ship and the virtual navigation channel is used to determine whether the ship has the risk of hitting the bridge. The proposed method has been examined in the practical environment. The results show that the method can effectively detect the yawing ship and provide a technical support for the safe driving of the ship in the complex environment.

1. Introduction

With the rapid development of transportation, a large number of cross-river bridges are built. There are many ships passing under these bridges every day. Those ships often carry many heavy cargoes. If ships yaw, they will hit bridges and that will be dangerous for the ships, bridges and people. To address this problem, it is necessary to find a method to avoid this accident. There are two main methods of bridge collision prevention: active collision prevention and passive collision prevention[1]. Active collision prevention is a kind of method to avoid ship collision accidents by intervening in ship navigation management and navigation track[2]. Passive collision prevention is a method to resist ship collision through reinforcing bridge piers or auxiliary anti-collision facilities[3]. To our best knowledge, most researchers focused on passive collision prevention method. In this paper, we proposed an active ship yaw detection method based on virtual navigation channel to avoid the accidents of collision. The method first constructed virtual navigation channel, and then measured the distance between ship and virtual navigation channel[4]. When the distance is larger than the predefined threshold value, the proposed method can send a warning message via alarm and light signals to the people in the ship so that they can control ship to navigate safely. We examined our proposed method in the practical environments. The results show that the ship yaw detection method based on virtual navigation channel can guide the ships to navigate in the river safely and provide a promising strategy for navigation business.

2. Ship yaw detection method based on virtual navigation channel

2.1. Construction of virtual navigation channel

Content from this work may be used under the terms of the Creative Commons Attribution 3.0 licence. Any further distribution of this work must maintain attribution to the author(s) and the title of the work, journal citation and DOI.

Published under licence by IOP Publishing Ltd

The first step of our proposed method is to construct the virtual navigation channel. Since the terrains of the river are different, the shapes of the virtual navigation channels are different. However, the general processes of virtual navigation channel construction are same and listed in figure 1.

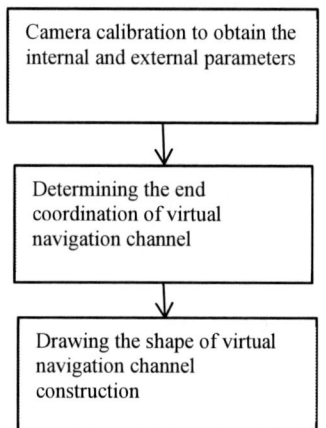

Figure 1. The process of virtual navigation channel construction.

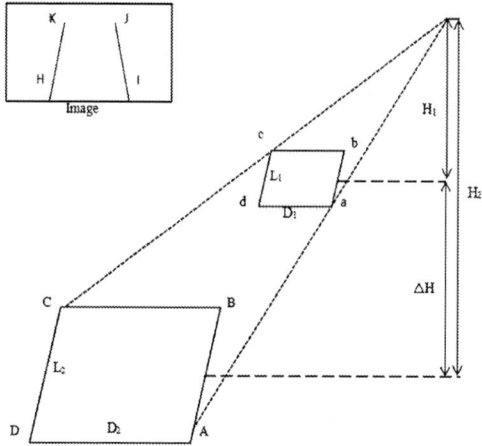

Figure 2. The space structure of camera calibration.

2.1.1. Determination of virtual navigation channel endpoints. Before constructing the virtual navigation channel, the coordinates of the virtual navigation channel's endpoints should be determined so that we can draw the shape of virtual navigation channels with endpoints. The Zhengyou Zhang checkerboard calibration method is used to calibrate the camera [5]. The virtual navigation channel's four endpoints are converted and determined from the world coordinate system to the image coordinate system [6]. As is shown in figure 2, firstly we install the ultrasonic sensor at the same height of the reference plate, and measure the distance ΔH between the reference plate and the water surface. The distance from the camera to the water surface is $H_2 = H_1 + \Delta H$. The navigation width of bridge is L_2 and the distance which can be detected by our method is D_2, Assuming that $Z=0$ and the upper left corner of the zero plane is the origin of the coordinate system, the world coordinates of the four endpoints of the virtual navigation channel on water surface are A $(L_2, D_2, 0)$, B $(L_2, 0, 0)$, C $(0, 0, 0)$, D $(0, D_2, 0)$. According to the similarity theorem of triangle ($\frac{L_1}{L_2} = \frac{D_1}{D_2}$), the length of L_2 and D_2 on the reference plane can be obtained, and the coordinates of A, B, C, D on the reference plate are: a $(L_1, D_1, 0)$, b $(L_1, 0, 0)$, c $(0, 0, 0)$, d $(0, D_1, 0)$. By using the camera calibration parameters and coordinates of a, b, c, d, we calculate the pixel coordinates of the virtual navigation channel endpoints in the image plane, then the virtual navigation channel can be drawn in the image.

2.1.2. Linear virtual navigation channel construction. After calibrating the endpoints of the virtual navigation channel, the moving ship can be detected in the image through using three-frame difference method and mixed Gaussian background difference method [7, 8]. We draw virtual navigation channel in the image by using the straight lines which defined as straight channels l_a and l_b between the four endpoints A, B, C and D, which are shown in figure 3.

2.1.3. Curved virtual navigation channel construction. Due to the navigation environment is complex, sometimes the navigation channel is curved. To address this case, we use lines to construct the straight virtual navigation channel between A and B, which called g, and use Bezier curve[9] to construct the curved virtual navigation channel between C and D, which called h, which is shown in figure 4. As shown in figure 4, to avoid obstacles, we use the second order Bezier curve, which is composed of two

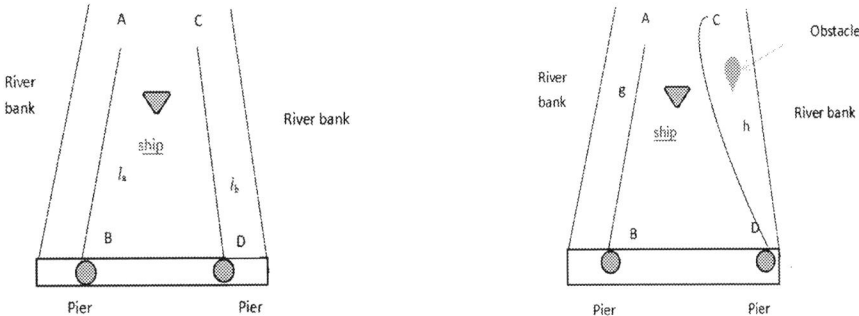

Figure 3. Schematic diagram for the construction of linear virtual navigation channel.   Figure 4. Schematic diagram for the construction of curved virtual navigation channel.

endpoints C and D, and one control point. The mathematical formula is as follows:

$$B(t) = (1-t)^2 P_0 + 2t(1-t) \ P_1 + t^2 P_2, t \in [0,1]$$

(1)

Where P_0 and P_n are endpoints, and P_1 is control point. The shapes of curved virtual navigation channel can be constructed by the second-order Bezier curve, which is shown in figure 5. It is worth noting that the red dot on the left in the figure 5 is the control point, and the blue dash line between the control point and the endpoint is the control line. There are one control point, two endpoints and two control lines in the (a), (b), (c) and (d). The two endpoints have the same coordinates. The concept of the control line is used to help us describe and understand the method easily, it does not exist practically. We can see the y-coordinate of control point in the figure (a) is the same as the figure (b), but the difference of the x-coordinate of control point causes the different curvature of the curve. In contrast, the x-coordinate of control points are the same in the figure (b), (c) and (d), the difference of their y-coordinate cause the bending position of the curves are different. Therefore, if the endpoints are known, we can regulate the shape of the curve by adjusting the coordinate of the control point.

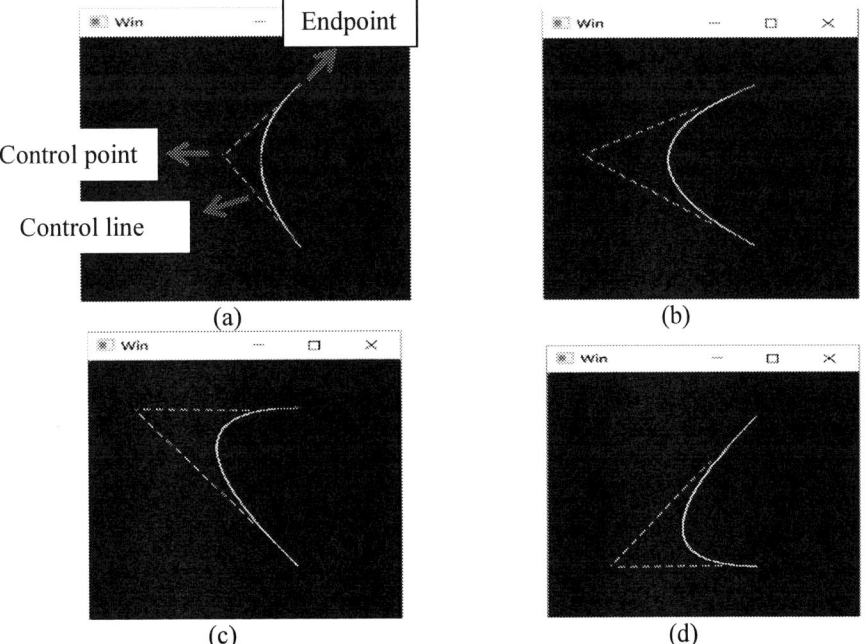

Figure 5. Examples of Bezier curve.

2.2. Ship yaw detection based on virtual navigation channel

On the basis of constructing the virtual navigation channel, we can detect whether the ship yaws by calculating the distance from the ship to the virtual navigation channel. We marked the moving ship by the rectangular frame in the image, and the distance from the rectangular frame vertex to the virtual navigation channel can be regarded as the distance from the ship to the virtual navigation channel.

2.2.1. Ship yaw detection based on linear virtual navigation channel. For the linear virtual navigation channel, since we had known the moving ship's coordinates, the distance from the ship to the linear virtual navigation channel can be obtained by using the formula of the distance between point and line.

As shown in figure 6. The coordinate of vertex A is (x, y), and the coordinate of vertex B is (x+width, y), the width is the width of the rectangular frame. If the four endpoints of the virtual waterway are $P_1(x_1, y_1)$, $P_2(x_2, y_2)$, $P_3(x_3, y_3)$, $P_4(x_4, y_4)$, then the expression of line segment P_1P_2 is:

$$\frac{y - y_1}{x - x_1} = \frac{y_2 - y_1}{x_2 - x_1}$$

(2)

The distance from point A (x, y) to line segment P_1P_2:

$$d_1 = \frac{|(y_2 - y_1)x + (x_1 - x_2)y + (x_2y_1 - x_1y_2)|}{\sqrt{(y_2 - y_1)^2 + (x_1 - x_2)^2}}$$

(3)

Similarly, the distance from point B (x + width, y) to line segment P_3P_4:

$$d_2 = \frac{|(y_4 - y_3)(x + width) + (x_3 - x_4)y + (x_4y_3 - x_3y_4)|}{\sqrt{(y_4 - y_3)^2 + (x_3 - x_4)^2}}$$

(4)

Finally we set a threshold value for the distance from the ship to the virtual navigation channel, when the distance between ship and virtual navigation channel is less than this threshold value, the yaw direction will be indicated.

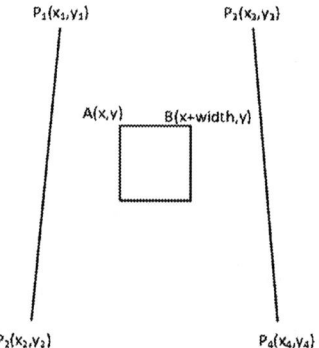

Figure 6. Principle diagram of distance calcu- between ship and linear channel.

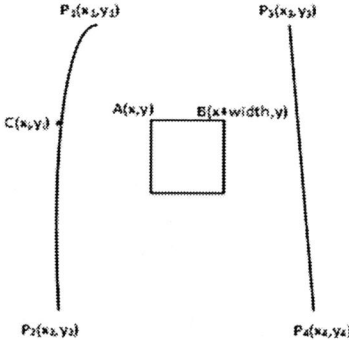

Figure 7. Principle diagram of distance calcu- lation lation between ship and curved channel.

2.2.2. Ship yaw detection based on curved and virtual navigation channel. For curved virtual navigation channel, the coordinates of rectangular frame vertex which marks the moving ship had been known. We extract the coordinate of pixel point on curve, and the distance from the vertex of rectangular frame to the pixel point of curve will be calculated to get the distance from ship to curved virtual navigation channel. As is shown in figure 7, the coordinate of vertex A is (x, y), and the coordinate of vertex B is (x + width, y). In this study, all of curved virtual navigation channels are yellow; their RGB values are (230, 255, 0). All pixel coordinates $B_i(x_i, y_i)$ are extracted from the video frame and then the distance between point A (x, y) and point $B_i(x_i, y_i)$ is calculated. We can detect the ship yaws through the minimal distance which is taken as the reference value. The formula is as follows:

$$d = \min \left\{ \sqrt{(x - x_i)^2 + (y - y_i)^2} \right\} \tag{5}$$

If d is less than the minimal distance, we state that the ship has the risk of yawing.

3. Experimental Results

3.1. Case 1: Linear virtual navigation channel

We examined the proposed method based on linear virtual navigation channel in Sun-Moon Lake of Jinan University, Zhuhai. The figure 8 shows the results. We can see from the figure 8 (a) that the ship doesn't yaw, the white rectangular frame is used to mark the moving ship. Two red linear lines are the virtual navigation channels that we constructed. The ship in the figure 8 (b) yaws and the red rectangular frame is used to mark it. And we use the red arrow to display yaw direction. We further tested our method at the site of Hengqin Bridge of Zhuhai City. The result is shown in figure 9. We can see from the figure 9 that the ship doesn't yaw. There exists a certain distance between the ship and virtual navigation channel. The ship went through the bridge safely, and there was no risk of hitting the bridge.

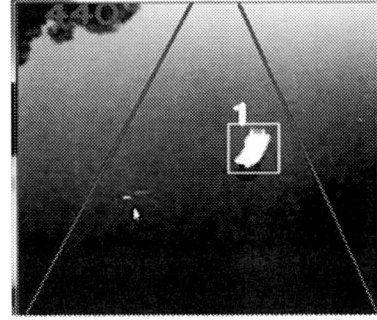

(a)　　The ship doesn't yaw.　　　　　　(b)　　The ship yaws.

Figure 8. The simulation effects of ship yaw detection method of linear virtual navigation channel.

3.2. Case 2: Curved virtual navigation channel

We test the ship yaw detection method based on curved virtual navigation channel at the site of Hengqin Bridge of Zhuhai City. The result is shown in figure 10. In the figure 10, the yellow lines are the virtual navigation channels which we constructed. The moving ship is marked by white rectangular frame, which shows the moving ship doesn't yaw. If the ship yaws, the white rectangular frame will become red and there will be a red arrow to shows the yaw direction of moving ship.

Figure 9. The ship yaw detection of linear virtual navigation channel.

Figure 10. The ship yaw detection of curved virtual navigation channel.

3.3. Analysis

In this section, we will make a detailed analysis of the test results above. From these results, we can see that no matter the ship yaw detection based on the linear or the curved virtual navigation channel, the yaw detection only depends on the distance from the moving ship to the virtual navigation channel. When the distance does not reach the threshold value we set, the ship is marked by the white rectangular frame and safely navigates in the river. Once the ship deviates from the channel, its white rectangular will turn red and a red arrow appears to indicate the direction of its yaw. And we can control the ship's navigation direction depends on the red arrow's direction. Therefore, the key of the method is to construct the correct virtual navigation channel and set the appropriate distance threshold. The above results also indicate that this method has a good practical applicability, for different river environments. We can flexibly use the linear or curved virtual navigation channel to match the actual environment. In the complex bridge river environment, this method can play a good role in early warning.

4. Conclusion

In this paper, we studied the technology of ship yaw detection based on virtual navigation channel. Through constructing the virtual navigation channel and calculating the distance from the ship to virtual navigation channel, we implemented the detection of the ship yaws. The experimental results show that the performance of our proposed method. There are still some points which need to be improved and supplemented in the future work. For example, since where exists strong wind around bridge area, the camera will shake. This is a great impact on the detection of moving ships which affects the calculation of the distance between the ship and the virtual navigation channel. However, considering the methods of ship yaw detection that have exited, this detection method still is a great innovation in past detection method. Compared to other ship yaw detection methods, this ship yaw detection method based on virtual navigation channel can detect the ship's yaw navigation problem in real-time and effectively, and it is a promising method in ship yaw detection research field.

Acknowledgements

This work is supported by Science and Technology Planning Project of Guangdong Province, China under grant number 2017B20218002.

References

[1] Guoyu Chen, Zhengquan Zhang. Two types of Installations and Three Missions for Defending Ships Hitting Bridge [J]. *Urban Road, Bridge and Flood Control*, 2008, (6): 179-179.

[2] Qingwen Deng, Zhongyi Sun, Duan Wang, et al. Design of active collision alarm system for Bridge [J]. *Instruments and Apparatus.*

[3] Bo Geng. Safety Assessment of Bridge ship collision [D]. *Tongji University*, 2007.

[4] Ting An. Research on ship Detection and tracking method based on feature Information extraction [D]. *North China University of Electric Power*, 2015.

[5] Ying Lu, Huiqing Wang, Wei Tong, *et al*. Camera Calibration algorithm based on Harris-Zhang Z. plane Calibration method [J]. *Journal of Xi'an University of Architecture and Technology (Natural Science)*, 2014, 46 (6): 860-864.

[6] Zhang Z. *Camera calibration* [J]. 2014, 28(1-2): 76-77.

[7] Bo-Xuan L I, Shen Y L, Yue H U. New algorithm based on Gaussian mixture model and three frame difference method [J]. *Journal of Engineering of Heilongjiang University*, 2016.

[8] Wang H L, Wang J Q, Ding H F, *et al*. Moving target detection based on the improved Gaussian mixture model background difference method [J]. *Advanced Materials Research*, 2012, 482-484: 569-574.

[9] Wei Yang, Yihong Li. A preliminary study of Bezier Curve based on VC [J]. *Science, Technology and Economics*, 2015, (8).

ISPECE

IOP Publishing

IOP Conf. Series: Journal of Physics: Conf. Series **1187** (2019) 042003   doi:10.1088/1742-6596/1187/4/042003

Configuration Generation of Aircraft Swarm Based on Communication Distance Constraint

Qixi Fu, Xiaolong Liang*, Jiaqiang Zhang, Lyulong He, Chuangchuang Zhu

Air Traffic Control and Navigation College, Air Force Engineering University, Xi'an, Shaanxi, 710051, China

*Corresponding author's e-mail: afeu_fqx@163.com

Abstract. Aiming at the requirements of the aircraft swarm combat task and information consensus, the control method of swarm configuration generation of aircraft swarm based on communication distance constraint is studied. We build the swarm communication relation model based on graph theory combined with the self-organizing obstacle avoidance control strategy and design the control protocol of aircraft swarm space configuration generation. We prove the stability and convergence of the system based on the Lyapunov stability theory and LaSalle invariable principle. Simulation results show that the proposed control protocol can ensure that the aircrafts can stabilize to the desired configuration under the premise of successfully avoiding environmental obstacles.

1. Introduction

After the aircraft swarm is assembled, different configurations have to be formed according to the different tasks. The formation of the configuration of the aircraft swarm is that the individual uses local information to enable all aircraft to implement the assigned task configuration. In the formation of configuration, communication between aircraft has an important impact on the implementation of the corresponding configuration. In the research of aircraft swarm configuration control, the existing literatures usually assume that individuals meet certain communication connectivity conditions without considering the limited communication distance in the actual system[5-7].

Considering aircraft swarm configuration generation communication distance constraint, this paper first constructs the artificial potential field function to achieve network connectivity to maintain communication, the basic idea is: when the adjacent aircraft communication distance tends to radius, the gradient of potential function tends to infinity so as to produce a sufficiently large attraction, the initial time of aircraft in the communication range capable of communicating, always maintain communication network. When carrying out combat tasks, aircraft swarm usually consist of different types and functions of aircraft. The mission is usually initiated by leaders. When other aircraft in the communication range receive instructions, they will cooperate with leaders to generate the specified task configuration. The control protocol which is designed with a leader considering the communication distance aircraft swarm configuration constraint generation control protocol can guarantee the initial communication to meet the constraints of distance of aircraft to avoid obstacles in the environment under the premise of success, and always maintain a spatial configuration within the communication range and stability to the desired.

Content from this work may be used under the terms of the Creative Commons Attribution 3.0 licence. Any further distribution of this work must maintain attribution to the author(s) and the title of the work, journal citation and DOI.

Published under licence by IOP Publishing Ltd

2. Graph theory

Graph theory is an important theoretical analysis tool for the study of aircraft swarm. In the process of research, each individual in the swarm is usually regarded as a node in the graph. The communication connections between nodes are depicted with both directed and undirected edges. Given N state variables $x_i \in R^m, i = 1, 2 \cdots N$, we can draw the following conclusions[8]:

Conclusion 1: for the undirected graph of the adjacency matrix, there is an undirected graph of the adjacency matrix.

$$\sum_i \sum_j a_{ij} x_i^T (x_i - x_j) = \frac{1}{2} \sum_i \sum_j a_{ij} \|x_i - x_j\|^2 \tag{1}$$

If the graph is connected, then:

$$\sum_i \sum_j a_{ij} x_i^T \|x_i - x_j\|^2 = 0 \Leftrightarrow x_i = x_j, \ \forall i, \ j \in v \tag{2}$$

Proof: since $a_{ij} = a_{ji}$, $\forall i, j$, then

$$\sum_i \sum_j a_{ij} x_i^T (x_i - x_j) = \sum_i \sum_j a_{ji} x_j^T (x_j - x_i) = \sum_i \sum_j a_{ij} x_j^T (x_j - x_i)$$

Then

$$\sum_i \sum_j a_{ij} x_i^T (x_i - x_j) = \frac{1}{2} \sum_i \sum_j a_{ij} x_i^T (x_i - x_j) + \frac{1}{2} \sum_i \sum_j a_{ij} x_j^T (x_j - x_i) = \frac{1}{2} \sum_i \sum_j a_{ij} (x_i^T x_i + x_j^T x_j - 2 x_i^T x_j) = \frac{1}{2} \sum_i \sum_j a_{ij} \|x_i - x_j\|^2$$

Where $\sum_i \sum_j a_{ij} \|x_i - x_j\|^2 = 0$, for all $a_{ij} > 0$, then $x_i = x_j$. And when it's a connected graph, you can get $x_i = x_j, \forall i, j$.

3. Problem formulation and some theories

The text of your paper should be formatted as follows:

Theorem 1. Global asymptotic stability theorem[10]. Suppose the system has a equilibrium point $x=0$, if there is a scalar function $V(x)$: $R^n \to R^+$ satisfies the following three condition:

(1) function $V(x)$ is positive in B_{R_0}, that is $V(x) \geq 0$, and if and only $x=0$, $V(x) = 0$ exists;

(2) The derivative $\dot{V}(x)$ in B_{R_0} of the function $V(x)$ is negative semidefinite.

(3) When $x \to \infty$, $V(x) \to \infty$.

Then the equilibrium point $x=0$ is global stable. If the function $\dot{V}(x)$ satisfies in condition (2) is negative, then the equilibrium point of the system is globally asymptotically stable.

Definition 1. The system equation $\dot{x} = f(x)$, if $x(0) \in M \Rightarrow x(t) \in M$ is established for $\forall t \in R$, then the set M is the invariant set of the system, and if $x(0) \in M \Rightarrow x(t) \in M$ is established for $\forall t \geq 0$, the set M is a positive invariant set of the system.

Lemma 1. LaSalle invariance theorem[9]. Let $\dot{x} = f(x)$ be defined on $D \subset R^m$, and $\Omega \subset D$ is a positive invariant set of the system. Let V: $D \to R$: it is a continuous and differentiable function, and there is a $\dot{V}(x) \leq 0$ in Ω. Let E be a set of all points where $\dot{V}(x) = 0$ in medium Ω, and M is the largest invariant set in E, then every solution that starts from Ω tends to a collection M when $t \to \infty$.

Lemma 2. Barbalat lemma. Let x: the first continuous derivable on the $[0, \infty) \to R$, and it has the limit when $t \to \infty$, if $\dot{x} = f(x)$ is consistent, then $\lim_{t \to \infty} \dot{x}(t) = 0$.

Suppose 1. The diagram of the undirected communication is an initial connection, and

$$\forall (i, j) \in E, \ \|p_i(0) - p_j(0)\| < \rho \tag{3}$$

The aim of this paper is to design a distributed control strategy under the suppose 1.

1. communication connectivity:

$$\forall t \geq 0, \forall (i, j) \in E, \|p_i(t) - p_j(t)\| < \rho \tag{4}$$

2. speed consistence:

$$q_i = q^L \tag{5}$$

3. configuration generation:

$$\forall (i,j) \in E, \lim_{t \to \infty} \left\| p_i(t) - p_j(t) - \Delta_{ij} \right\| = 0 \tag{6}$$

4. successful obstacle avoidance:

$$\left\| p_i - o_b \right\| > r \tag{7}$$

Model 1. Individual dynamics model of aircraft swarm.

$$\begin{cases} \dot{p}_i = q_i, \\ \dot{q}_i = u_i, \end{cases} \quad i=1,2,\cdots N \tag{8}$$

Where $p_i \in R^2$, $q_i \in R^2$ represents the position vector and velocity vector of i, and $u_i \in R^2$ is the control input of i. The expected track information can be expressed in the following form:

$$\begin{cases} \dot{p}^L = q^L \\ \dot{q}^L = u^L \end{cases} \tag{9}$$

$p^L \in R^2, q^L \in R^2$ and $u^L \in R^2$ are the position vector, velocity vector and control input of the leader respectively.

4. Design and analysis of control protocol

Considering the communication distance constraints, the generation control problem of leader swarm formation is considered. In this paper, each aircraft is regarded as a point in the potential field, and a smooth potential field function is defined for every two aircraft combination $(i,j) \in E$.

$$\varphi_{ij} = \varphi_{ij}\left(\gamma_{ij}, \beta_{ij}\right) \tag{10}$$

Where γ_{ij} is the control target that the aircraft combination $(i,j) \in E$ is minimization when the desired relative position is reached.

$$\gamma_{ij} = \left\| p_i(t) - p_j(t) - \Delta_{ij} \right\|^2 \tag{11}$$

The function β_{ij} is used to ensure that the aircraft is always in the communication range. It makes the aircraft $j \in N_i^c$, if it is in the communication scope of the aircraft i at the initial time, it will always be in the communication scope of the aircraft i. The function β_{ij} is defined as:

$$\beta_{ij} = \begin{cases} 1 & , p_{ij} \le \bar{\rho}^2 \\ \dfrac{(\rho^2 - p_{ij})(\rho^2 - 2\bar{\rho}^2 + p_{ij})}{(\rho^2 - \bar{\rho}^2)^2} & , p_{ij} > \bar{\rho}^2 \end{cases} \tag{12}$$

where $p_{ij} = \left\| p_i - p_j \right\|^2$, $\bar{\rho} < \rho$ is constant, which satisfies the $\bar{\rho} \ge \max_{i,j \in N_i^c} \left\| \Delta_{ij} \right\|$, and the function is shown in Figure 1.

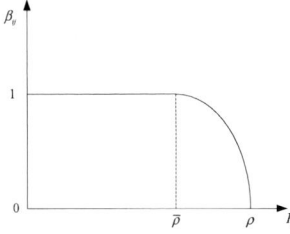

Figure 1. The function of the variable β_{ij}.

On the basis of the potential field function φ_{ij}, and considering the obstacle avoidance in the external environment, the control protocol of the aircraft i is designed.

$$u_i = -\sum_{j \in N_i^c} c_{ij}(q_i - q_j) - \sum_{j \in N_i^c} \frac{\partial \varphi_{ij}}{\partial p_i} - c_i(q_i - q^L) + u^L - \nabla_{p_i} U_{Obsi} \tag{13}$$

The $c_{ij} = c_{ji} > 0$ is the connection weight value of the connection to the aircraft i, j, $c_i \geq 0$. Bring the (13) into the (9):

$$\begin{cases} \dot{p}_i = q_i, \\ \dot{q}_i = -\sum_{j \in N_i^c} c_{ij}(q_i - q_j) - \sum_{j \in N_i^c} \frac{\partial \varphi_{ij}}{\partial p_i} - c_i(q_i - q^L) + u^L - \nabla_{p_i} U_{Obsi}, \end{cases} \quad i = 1, 2, \cdots N \tag{14}$$

Suppose 2. The function φ_{ij} has the following properties:

1. $\varphi_{ij} = \varphi_{ji} \geq 0$; 2. 当 $\beta_{ij} \to 0, \varphi_{ij} \to \infty$; 3. $\dfrac{\partial \varphi_{ij}}{\partial \gamma_{ij}} > 0, \dfrac{\partial \varphi_{ij}}{\partial \beta_{ij}} < 0$。

In order to avoid the obstacle of the external environment, the obstacle avoidance method used in this paper is shown in Figure 2.

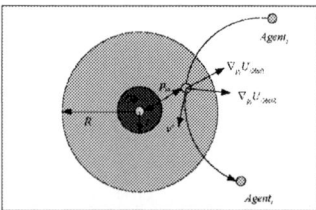

Figure 2. Schematic diagram of obstacle avoidance.

Theorem 2. The system contains a swarm of aviation aircraft components, individual swarm motion model and the leader expected track information respectively (8) and (9) described in suppose 1 and control law (13) under the action, if there is at least one $c_i > 0$, so the aircraft c-an converge to the desired configuration, and is expected to the speed of q^L movement, and can effectively avoid the obstacles and reach the stable swarm.

Lemma 3: in suppose 1, the set $S(p) = \left\{ p \left| \|p_i - p_j\| < \rho, \forall (i, j) \in E \right. \right\}$ is an invariant set of the system (14), so the communication distance constraint condition (4) is always established.

Proof For the system (19), select the following Lyapunov function：

$$V = \frac{1}{2} \sum_i \hat{q}_i^T \hat{q}_i + \frac{1}{2} \sum_i \sum_{j \in N_i^c} \varphi_{ij}(\hat{p}_i, \hat{p}_j) + \sum_i U_{Obsi} \tag{15}$$

Then

$$
\begin{aligned}
\dot{V} &= \sum_i \hat{q}_i^T \left(-\sum_{j \in N_i^c} c_{ij}(\hat{q}_i - \hat{q}_j) - c_i \hat{q}_i - \sum_{j \in N_i^c} \frac{\partial \varphi_{ij}}{\partial p_i} - \nabla_{p_i} U_{Obsi} \right) + \frac{1}{2} \left(\sum_i \sum_{j \in N_i^c} \left(\frac{\partial \varphi_{ij}}{\partial \hat{p}_i} \right)^T \hat{q}_i + \sum_i \sum_{j \in N_i^c} \left(\frac{\partial \varphi_{ij}}{\partial \hat{p}_j} \right)^T \hat{q}_j \right) + \sum_i \hat{q}_i^T \nabla_{p_i} U_{Obsi} \\
&= \sum_i \hat{q}_i^T \left(-\sum_{j \in N_i^c} c_{ij}(\hat{q}_i - \hat{q}_j) - c_i \hat{q}_i \right) - \sum_i \sum_{j \in N_i^c} \hat{q}_i^T \frac{\partial \varphi_{ij}}{\partial p_i} - \sum_i \hat{q}_i^T \nabla_{p_i} U_{Obsi} + \sum_i \sum_{j \in N_i^c} \left(\frac{\partial \varphi_{ij}}{\partial p_i} \right)^T \hat{q}_i + \sum_i \hat{q}_i^T \nabla_{p_i} U_{Obsi} \\
&= -\sum_i \sum_{j \in N_i^c} c_{ij} \hat{q}_i^T (\hat{q}_i - \hat{q}_j) - \sum_i c_i \hat{q}_i^T \hat{q}_i \\
&= -\frac{1}{2} \sum_i \sum_{j \in N_i^c} c_{ij} \|\hat{q}_i - \hat{q}_j\|^2 - \sum_i c_i \|\hat{q}_i\|^2
\end{aligned}
\tag{16}
$$

It is available from the inverse method: if there is at least one aircraft combination $(i, j) \in E$, so that $\beta_{ij} \to 0$ or $p_{ij} \to \rho^2$, then there is $\varphi_{ij} \to \infty$, so that $V \to \infty$ is not consistent with the known conditions. So $p(t) \in S(p)$, $\forall t \geq 0$ can be obtained. Therefore, the lemma 1 is proved that the system can always be kept in the communication range.

According to the LaSalle invariance principle, formula (16) implies that when t tends to infinity, $\dot{V}=0$, $\sum_i \sum_{j\in N_i^c} c_{ij}\|\hat{q}_i-\hat{q}_j\|^2=0$ and $\sum_i c_i\|\hat{q}_i\|^2=0$. $\sum_i \sum_{j\in N_i^c} c_{ij}\|\hat{q}_i-\hat{q}_j\|^2=0$ indicate that all i, j, $\hat{q}_i=\hat{q}_j$ $\sum_i c_i\|\hat{q}_i\|^2=0$ and at least one $c_i>0$ can be at least one $\hat{q}_i=0$. This can be achieved for all i, $\hat{q}_i=0$ or $q_i=q^L$, that is, the follower will agree with the leader to expect speed.

Because every q_i has boundedness limit and $\dot{q}_i$ is uniformly continuous, Barbalat lemma is applied to formula (14). When t tends to infinity, it has $\hat{q}_i=0$ for all i, and $q_i=q^L$ means $i=1,2,\cdots n$ for all.

$$\lim_{t\to\infty}\sum_{j\in N_i^c}\frac{\partial \varphi_{ij}}{\partial p_i}(t)=0 \tag{17}$$

In the global coordinate system, the $\bar{p}_i=p_i-\Delta_i$, $\Delta_{ij}=\Delta_i-\Delta_j$, $\Delta_i\in R^2$ are the position of the aircraft i in the relative global coordinate system when the desired configuration is realized. Construction auxiliary function:

$$F=\sum_i \bar{p}_i^T \sum_{j\in N_i^c}\frac{\partial \varphi_{ij}}{\partial p_i} \tag{18}$$

Then

$$F=\sum_i \sum_{j\in N_i^c} 2\eta_{ij}\bar{p}_i^T(\bar{p}_i-\bar{p}_j)+\sum_i\sum_{j\in N_i^c}2\xi_{ij}\bar{p}_i^T(\bar{p}_i-\bar{p}_j+\Delta_{ij})=\sum_i\sum_{j\in N_i^c}\eta_{ij}\|\bar{p}_i-\bar{p}_j\|^2+\sum_i\sum_{j\in N_i^c}\xi_{ij}(\bar{p}_i-\bar{p}_j)^T(\bar{p}_i-\bar{p}_j+\Delta_{ij})$$

$$=\sum_i\sum_{j\in N_i^c}\eta_{ij}\|\bar{p}_i-\bar{p}_j\|^2+\sum_i\sum_{j\in N_i^c}\xi_{ij}(p_i-p_j-\Delta_{ij})^T(p_i-p_j)=\sum_i\sum_{j\in N_i^c}\eta_{ij}\|\bar{p}_i-\bar{p}_j\|^2+\sum_i\sum_{j\in N_i^c}\xi_{ij}(\|p_i-p_j\|^2-\Delta_{ij}^T(p_i-p_j)) \tag{19}$$

It is clear that $\sum_i\sum_{j\in N_i^c}\eta_{ij}\|\bar{p}_i-\bar{p}_j\|^2\geq 0$ is also able to get $\xi_{ij}\geq 0$ because of

$$\xi_{ij}=\begin{cases}0 & ,p_{ij}\leq\bar{\rho}^2\\ 2\dfrac{\partial\varphi_{ij}}{\partial\beta_{ij}}\dfrac{(\bar{\rho}^2-p_{ij})}{(\rho^2-\bar{\rho}^2)^2} & ,p_{ij}>\bar{\rho}^2\end{cases}$$, and because of $\|p_i-p_j\|>\bar{\rho}\geq\max\|\Delta_{ij}\|$, there are

$$\sum_i\sum_{j\in N_i^c}\xi_{ij}(\|p_i-p_j\|^2-\Delta_{ij}^T(p_i-p_j))\geq\sum_i\sum_{j\in N_i^c}\xi_{ij}(\|p_i-p_j\|^2-\|\Delta_{ij}\|\|p_i-p_j\|)=\sum_i\sum_{j\in N_i^c}\xi_{ij}(\|p_i-p_j\|-\|\Delta_{ij}\|)\|p_i-p_j\|>0 \tag{20}$$

Then

$$F\geq\sum_i\sum_{j\in N_i^c}\eta_{ij}\|\bar{p}_i-\bar{p}_j\|^2+\sum_i\sum_{j\in N_i^c}\xi_{ij}(\|p_i-p_j\|-\|\Delta_{ij}\|)\|p_i-p_j\|>0 \tag{21}$$

The formula (17) is brought into the type (18) to get $\lim_{t\to\infty}F(t)=0$, that is, $\sum_i\sum_{j\in N_i^c}\xi_{ij}(\|p_i-p_j\|^2-\Delta_{ij}^T(p_i-p_j))=0$ and $\sum_i\sum_{j\in N_i^c}\eta_{ij}\|\bar{p}_i-\bar{p}_j\|^2=0$. When the communication graph G is connected, for all i, j, $\bar{p}_i=\bar{p}_j$, is also $\bar{p}_i-\bar{p}_j-\Delta_{ij}=0$, that is, the follower and the leader will maintain the desired configuration movement.

Because of $\dot{V}\leq 0$, it can be concluded that V is bounded, and may as well assume that $V\leq h$. The former formula shows that when $\|p_i-o_b\|\to r$, $U_{Obsi}\to\infty$, and $V\to\infty$ can be derived from the formula (15), it is inconsistent with the hypothesis $V\leq h$, so the aircraft swarm can realize the obstacle avoidance under the effect of the control protocol (13), and theorem 2 can be proved

5. Simulation results and analysis

In this paper, we use the proposed control algorithm[11] to simulate the aircraft swarm cooperative anti stealth configuration in the paper "three rounds" and verify the effectiveness of the proposed method. In the simulation, we take the anti-stealth detection configuration of $R_L = r_T = 52km$. At this time, the three receivers (Rec1, Rec2, Rec3) are distributed in the front of the transmitter in the $[3\pi/20, 17\pi/20]$ range, and are distributed counter-clockwise according to the equal angular distance. The 10000 times the distance of the space configuration is $R_L = r_T = 5.2m$, and the rate of 200 times the contraction ratio is 1m/s. The target matrix of the control configuration is obtained.

$$\begin{bmatrix} 0 & 0 & -4.2 & -3.1 & -5.2 & 0 & -4.2 & 3.1 \\ 4.2 & 3.1 & 0 & 0 & -1 & 3.1 & 0 & 6.2 \\ 5.2 & 0 & 1 & -3.1 & 0 & 0 & 1 & 3.1 \\ 4.2 & -3.1 & 0 & -6.2 & -1 & -3.1 & 0 & 0 \end{bmatrix}$$

The aircraft swarm system contains four aircraft platforms. The initial location is $p_1(0) = [-12.5 \quad 0]^T$, $p_2(0) = [-7.5 \quad 12.5]^T$, $p_3(0) = [-12.5 \quad -5]^T$, $p_4(0) = [-7.5 \quad -12.5]^T$, where p_1 is transmitter, namely leader position, and the remaining three are receiver, follower location.

The location of the obstacle is set to $[50 \quad 0]^T$, the radius is 5m, and the radius of the buffer zone is 30m. The desired speed is set to $q^L = [1 \quad 0]^T$, and the speed is 1m/s. In this paper, the perception radius of aircraft is set to $\rho=20m$, that is to say, that is to say, when the distance between aircraft platforms is $d_{ij} = \|p_i - p_j\| < 20m$, that is, communication between aircraft can be carried out. In simulation, the parameter $\bar{\rho}$ in function φ_{ij} is set to $\bar{\rho}=15m$, and the communication topology of swarm at the initial time is shown in Figure 3.

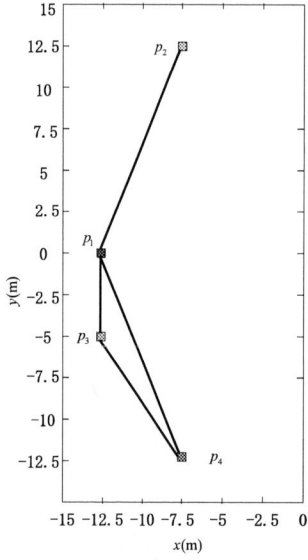

Figure 3. Aircraft swarm initial location and communication topology.

The communication relationship between i and j is $c_{ij} = 2$ when there is no communication relationship $c_{ij} = 0$. Since the leader is known to expect speed information, that is, in the control protocol (12), $c_1 = 4$. In this paper, the potential field function φ_{ij} is defined as $\varphi_{ij} = \lambda_1 \gamma_{ij} - \lambda_2 \ln \beta_{ij}$, in which $\ln(\square)$ is a logarithmic function. When the parameters $\lambda_1 = 3, \lambda_2 = 0, \lambda_1 = 3, \lambda_2 = 2$ are taken, the simulation results are shown as shown in Figure 4 and Figure 5 respectively.

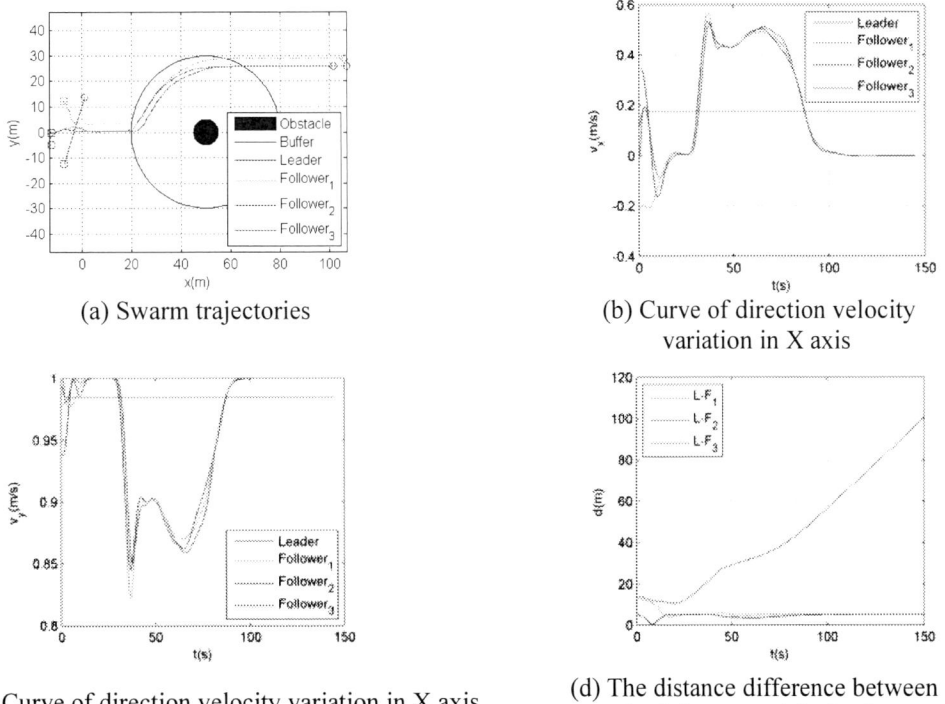

(a) Swarm trajectories

(b) Curve of direction velocity variation in X axis

(c) Curve of direction velocity variation in X axis

(d) The distance difference between the follower and the leader

Figure 4. Simulation results when $\lambda_1 = 3, \lambda_2 = 0$.

From the trajectories of the graph 4 (a) swarm, we can see that the aircraft swarm can not form the desired configuration when the communication maintenance item is not considered (i.e. $\lambda_2 = 0$). This is because the distance between number three follower and other aircraft is increasing gradually, beyond the critical value, making the system communication network no longer connected.

After considering the communication maintenance, the simulation results are shown in Figure 5. From Figure 5 (a), we can see that in the simulation process, the aircraft gradually moved away from the obstacle center in the obstacle buffer area, and successfully achieved the obstacle avoidance. With the passage of time, after the successful obstacle avoidance, the aircraft gradually converged to form a stable configuration. In Figure 5, Figure 5 (b) (c) to adjust the speed of the system oscillation can be seen from the beginning, gradually dynamic stability, in the face of obstacles when the aircraft speed appeared oscillation changes, this is because the obstacle by force, appeared on the direction of adjustment of speed change in obstacle avoidance after the success and gradually converge to the long machine speed. Figure 5 (d) shows that the three followers tend to be stable relative to the long machine, reaching the desired configuration of the 5.2m. aviation.

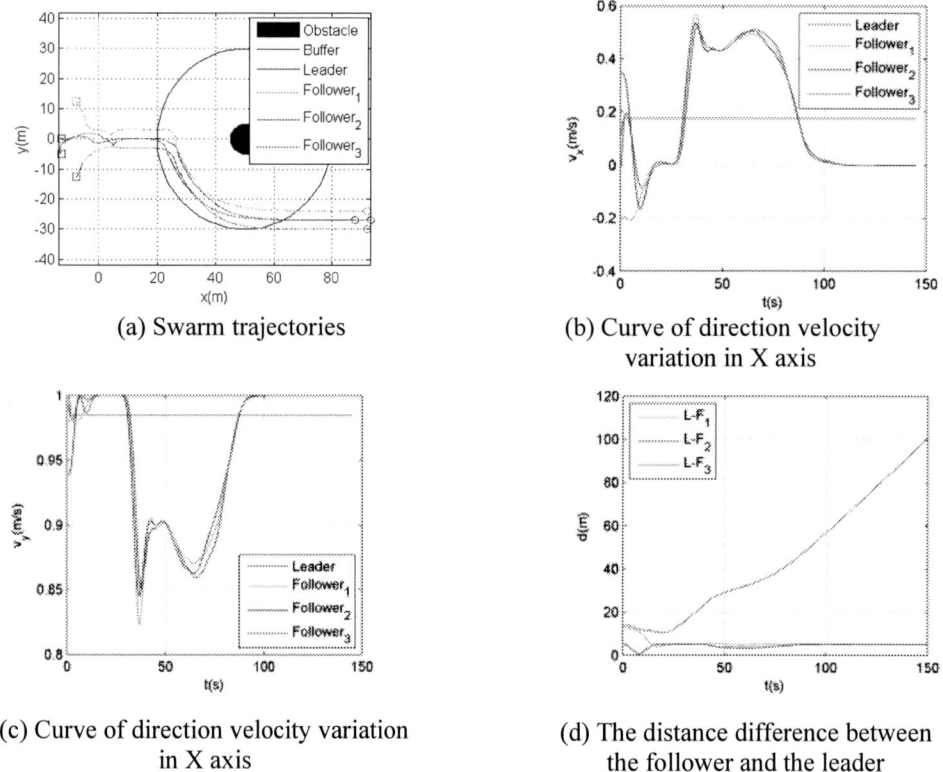

(a) Swarm trajectories

(b) Curve of direction velocity variation in X axis

(c) Curve of direction velocity variation in X axis

(d) The distance difference between the follower and the leader

Figure 5. Simulation results when $\lambda_1 = 3, \lambda_2 = 2$.

To sum up, it can be seen that the communication distance constraint has an important influence on the performance of the control algorithm, and it is a problem that must be considered in the design of the control protocol. The control protocol designed in this paper can realize that the aircraft which always meets the communication distance constraint at the initial time always stays in the communication range, and can successfully avoid obstacles in the environment, and ultimately achieve the gradual stable state and generate the desired configuration.

6. Conclusions

Aiming at the problem of communication distance constraint in the process of aircraft swarm configuration, a distributed control protocol is designed based on consensus theory and artificial potential field method. From the actual operation of the aircraft swarm, the protocol can also realize the speed following and obstacle avoidance for the leader. The use of Lyapunov tools for system stability and convergence are proved. The simulation results show that the initial time in communication within the scope of the aircraft, can always stay connected communication network and can realize the obstacle avoidance and the leader of the speed to follow in the control protocol design; by considering the communication distance constraint and without considering the simulation the communication distance constraint under two kinds of situations, This paper further demonstrated the important role of the communication distance to achieve aircraft swarm configuration.

Acknowledgments

This research was financially supported by the National Natural Science Foundation of China (No. 61472443, No. 61703427) and the Research and Development of Science and Technology Plan Projects in Shanxi Province (No. 2017JQ6035).

References

[1]   Niu Y F., Xiao X J., Ke G Y. (2013) Operation concept and key techniques of unmanned aerial vehicle swarms. Nati onal Defense Science and Technology, 34:37-43.

[2]   Liang X L., Sun Q., Yin Z H. (2013) A study of aviation swarm convoy and transportation mission. Advances in Swarm Intelligence, ICSI, pp. 368-375..

[3]   Liang X L., Sun Q., Yin Z H. (2013) The swarm convoy for transport mission. The Fourth International Conference on Swarm Intelligence, Harbin China, ICSI.

[4]   Liang X L., Li H., Sun Q. (2014) Development trend of air operations and its strategy, Journal of Air Force Engineering University (Military Science Edition), 14: pp. 4-7.

[5]   Ajorlou A., Momeni A., Aghdam A G. (2010) A class of bounded distributed control strategies for connectivity preservation in multi-agent systems. IEEE Transactions on Automatic Control, 55: 2828-2833.

[6]   Dimarogonas D V., Johansson K H. (2010) Bounded control of network connectivity in multi-agent systems. IET Control Theory and Applications, 4: 1330-1338.

[7]   Ji M., Egerstedt M. (2007) Distributed coordination control of multi-agent systems while preserving connectedness. IEEE Transactions on Robotics, 23: 693-703.

[8]   Zhang H., Lewis F L., Qu Z. (2012) Lyapunov, adaptive, and optimal design techniques for cooperative systems on directed communication graphs. IEEE Transactions on Industrial Electronics, 59:3026-3041.

[9]   Huang J. (2015) High order nonlinear multi agents with consistency control. Institute of Technology . Beijing Institute of Technology. Beijing.

[10]   Ho J. (2009) Research on multi agent based cluster motion control method. Nanjing University of Science and Technology. Nanjing.

[11]   Zhu L., Liang X L., Zhang J Q. (2017) Research on the optimal formation configuration of cooperative detection in aeronautical cluster . fire and command control, 42: 69-72.

[12]   P. K. C., Wang, F. Y., Hadaegh. (1998) Optimal formation-reconfiguration for multiple spacecraft, In: AIAA Guidance, Navigation Control Conf., Boston, MA.

[13]   Zhang B., Jia Y. (2013) "Fixed-time consensus protocols for multi-agent systems with linear and nonlinear state measurements," Nonlinear Dynamics, vol.82, 2015, pp. 1683-1690.

[14]   Yang T., Meng Z Y., Dimarogonnas D V. (2014) Global consensus for discrete-time multi-agent systems with input saturation constraints, Automatica, 50: pp. 499-506.

[15]   Han Z M., Lin Z Y., Fu M U. (2015) Distributed coordination in multi-agent systems: a graph laplacian perspective. Frontiers of Information Technology & Electronic Engineering, 16:pp. 429-448.

ISPECE

IOP Publishing

Parameter Estimation Algorithm and Application in Industry Design

Mei Yun, Jiang Haiyang, Lin Ying, Wang Fang, Nie Qimeng

School of mechanical engineering and automation, University of Science and Technology Liaoning, Anshan, Liaoning, 114051, China

e-mail: meiyunliaoning@163.com

Abstract. The paper presents parameter estimation algorithm and application in industry design, it proposes low order autoregressive algorithm, and it presents the parameter estimation model. Experiments show that the proposed algorithm can make parameter estimation effective. The parameter estimation algorithm can be used in more industry design cases.

1. Introduction

One of the most important objectives of statistical inference is to estimate unknown model parameters based on an observed data. There are many applications of industry design involves parameter estimation algorithm [1-6]. Syed Shahnawazuddin [7] has explored the Spectral Moment time-frequency distribution Augmented by features in severe pitch mismatch task. The estimation only can realize the frequency estimation of the stationary process, and the adaptive kernel function is used to improve the algorithm in time-frequency domain, it can realize time-frequency distribution under stable distribution noise environment, and has a certain practical significance.

Parameter estimation is a kind of statistical inference. The process of estimating unknown parameters in the population distribution based on random samples extracted from the population. From the form of estimation, it can be divided into point estimation and interval estimation: from the method of constructing estimators, there are moment estimation, least square estimation, likelihood estimation, Bayesian estimation and so on. There are two problems to be solved: (1) finding the estimators of unknown parameters; (2) pointing out the accuracy of the estimators under certain reliability. Reliability is generally expressed by probability, such as 95% credibility; accuracy is measured by the proximity or error between the estimator and the estimated parameters (or parameters to be estimated).

2. The Parameter Estimation Algorithm of Model

Parameter distribution can use Gaussian distribution, the process variable, and its characteristic function is:

$$\theta(t) = \exp\{n - m[1 + sign(t)\omega(\tau,\alpha)]\} \tag{1}$$

Where, $\omega(\tau,\alpha) = \begin{cases} \tan(\alpha) & if\ \alpha \neq 1 \\ \log|\tau| & if\ \alpha = 1 \end{cases}$, $sign(t) = \begin{cases} 1 & t > 0 \\ 0 & t = 0 \\ -1 & t < 0 \end{cases}$, when $0 < \alpha < 2$.

Content from this work may be used under the terms of the Creative Commons Attribution 3.0 licence. Any further distribution of this work must maintain attribution to the author(s) and the title of the work, journal citation and DOI.
Published under licence by IOP Publishing Ltd

When the system model structure is known, then the process of calculating the parameters of the system model will use the input and output data of the system. At the end of the 18th century, the German mathematician C.F. Gauss first proposed the method of parameter estimation. He used the least square method to calculate the orbit of celestial bodies. In the 1960s, with the popularity of computers, parameter estimation has developed rapidly. There are many methods for parameter estimation, such as moment estimation, maximum likelihood method, uniform minimum variance unbiased estimation, minimum risk estimation, covariant estimation, least squares method, Bayesian estimation, maximum posteriori method, minimum risk method and minimum maximum entropy method. The most basic methods are least square method and maximum likelihood method.

The covariance of distribution does not exist because its variance is not limited. It is similar to covariance of Gaussian random process. In probability theory and statistics, Gaussian process is a stochastic process, it makes every finite collection of those random variables has a multivariate normal distribution. Optimisation software will be used to fit the Gaussian process. The distribution random variable X and distribution random variable Y can be defined as [8]:

$$[X,Y]_\alpha = \int_s xy \ \mu(d) \ , \quad 1 < \alpha \le 2 \tag{2}$$

The least squares method and the maximum likelihood method have recursive forms. In addition, the recursive generalized least squares method, the recursive auxiliary variable method and the recursive extended least squares method are all improved forms of the recursive least squares method.

It can be used to estimate the system with noise interference. In addition, stochastic approximation algorithm, Kalman filter method and Landau recursive estimation are recursive parameter estimation methods from different starting points (see recursive estimation algorithm). The consistency of most recursive parameter estimation algorithms can be proved by martingale convergence, stability of ordinary differential equation, positive and real respectively.

The parameter estimation variable x[n] is defined,

$$x[n] = -\sum_{i=1}^{M} \sum_{l=-L}^{L} a_{i,l} ex[n-i] = -\sum_{i=1}^{M} \sum_{l=-L}^{L} a_{i,l}(x)[n] \tag{3}$$

It also defines parameter estimation variable as

$$X[n] = -\sum_{i=1}^{M} \sum_{l=-L}^{L} a_{i,l} e = -\sum_{i=1}^{M} \sum_{l=-L}^{L} a_{i,l}(x)[n] \tag{4}$$

Where it is stationary distribution process, it define the effective of parameter estimation process as

$$\gamma[n] = \sum_{l=0}^{m} x_l e^{nl} \tag{5}$$

It can get the coefficient $\gamma[n]$ of the parameter estimation. It uses minimum mean square error estimation algorithm for the solution.

The parameter estimation variable $X[n]$ is :

$$T_\alpha(e^f) = \frac{m}{\left| 1 - \sum_{i=1}^{\hat{M}} a_i e^{-f} \right|^\alpha} \tag{6}$$

3. Experiment Results

It will formulate algorithm for comparing the performance of the model method. The model parameters estimation are analysed, the specific simulations are as follows.

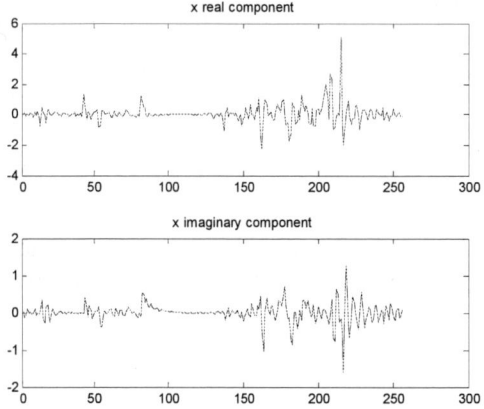

Figure 1. The parameter estimation in time domain

4. Conclusions

The performance of the time-frequency model parameter estimation algorithm and spectrum estimation algorithm degenerate under stable distribution environment. Experiment shows that the proposed parameter estimation algorithm is effective; the proposed algorithm has wider applicability. In future, we will apply the parameter estimation algorithm in industry design cases.

Acknowledgments

This work was financially supported by Liaoning provincial education department project (No.2016HZPY07).

References

[1] Duy C. Huynh ; Bach H. Dinh ; Matthew W. Dunnigan ; Thu A. T. Nguyen ; Nam H. Le. Parameter estimation of a single-phase induction machine using a dynamic particle swarm optimization algorithm[C]. 2011 IEEE Power Engineering and Automation Conference. 8-9 Sept. 2011.Wuhan, China.

[2] Nasar Aldian Ambark Shashoa ; Mohamed A. Hassan ; Abdulmunem Mohammed Almukhtar.Parameter estimation and residual generation for (CARARMA) algorithm model depend on D-RGELS[C]. 2018 Electric Electronics, Computer Science, Biomedical Engineerings' Meeting (EBBT).18-19 April 2018.Istanbul, Turkey.

[3] Shengyu Pei ; Yongquan Zhou ; Qifang Luo.A Hybrid Particle Swarm Algorithm for Nonlinear Parameter Estimation[C].2009 Second International Conference on Intelligent Computation Technology and Automation.10-11 Oct. 2009, Changsha, Hunan, China.

[4] Zhong Lu ; You-chao Sun.Point Estimation Optimization Model of Life Distribution Parameters Based on Genetic Algorithm[C]. 2009 Second International Conference on Information and Computing Science.21-22 May 2009. Manchester, UK.

[5] Xiaolong Li ,Guolong Cui, Wei Yi , Lingjiang Kong.A Fast Maneuvering Target Motion Parameters Estimation Algorithm Based on ACCF[J].IEEE Signal Processing Letters ,Volume: 22 , Issue: 3 , March 2015 ,Page(s): 270 - 274.

[6] Wang Feng, Wang Shaotong.Impact of missing data on parameter estimation algorithm of normal distribution[C]. 2013 2nd International Symposium on Instrumentation and Measurement, Sensor Network and Automation (IMSNA).23-24 Dec. 2013,Toronto, ON, Canada.

[7]   Syed Shahnawazuddin ; Rohit Sinha ; Gayadhar Pradhan.Pitch-Normalized Acoustic Features for Robust Children's Speech Recognition[J]. IEEE Signal Processing Letters ,Volume: 24 , Issue: 8 , Aug. 2017,Page(s): 1128 - 1132.

[8]   LONG Junbo, WANG Haibin, ZHA Daifeng. Fractional Low-order Adaptive Time-frequency Distribution Based on Stable Distribution Noise[J]，Computer Engineering, 2011,37（18）：81-83

A Multi-model Estimation of Distribution Algorithm

Li Hao

School of Electrical Engineering, Southwest Jiaotong University, Chengdu, China
hblh198095@163.com

Abstract. Estimation of distribution algorithm(EDA) is an effective evolutionary algorithm which is working based on the probabilistic model. On the analysis of EDA, we proposed a novel EDA. In this algorithm, multiple probabilistic models are taken to describe the complex problem, the candidate solutions are obtained through sampling the models, each model learned by direct comparison with its own best solution, and the random exchange of evolution targets between models is used to realize information sharing. By setting the upper and lower limit of probability, premature convergence can be avoided. After analyzing this algorithm, we apply it to the knapsack problem, and comparing with common genetic algorithm(CGA) and PBIL, the experimental results illustrate that the proposed algorithm has better performance.

1. Introduction

EDA is an effective evolutionary algorithm. Its conception was first proposed in 1996, and developed rapidly after 2000[1,2]. Different from traditional EAs, EDAs don't use crossover and mutation operators to generate new candidate solutions. Instead, they built a probabilistic model to express the distribution of promising solutions. And according the model, the candidate solutions can be gotten by sampling it. Experiments show that the EDAs often has better performance than other EAs in solving practical problems, so it has been widely used in many fields, including function optimization[3], pattern recognition[4], Intelligent Traffic System[5] and so on[6,7].

Because of its simplicity and efficiency, EDA has become a popular algorithm rapidly. But in practice, it also suffer from premature convergence in the complex problem optimization[8]. In this paper, a multi-model EDA(MEDA) is proposed. In this algorithm, multiple probabilistic models can be taken to describe the solution space, the candidate solutions are obtained through sampling the models, each model evolved by directly compared with its own best solution, premature convergence can be prevented through setting the upper and lower limit of probability, and the random exchange of evolution targets between models is used to realize information sharing. After analyzing this algorithm, we apply it to the knapsack problem, and comparing with CGA and PBIL, the experimental results illustrate that the proposed algorithm has better performance.

The rest of the paper is organized as follows. The EDA framework is introduced in section 2. MEDA is presented in section 3. In section 4, the algorithm analysis is given. And the experimental results are shown in section 5. Finally, conclusion follows in section 6.

2. EDA framework

As an effective evolutionary computation method, EDA has drawn wide attention by scholars all over the world since it was put forward. Meanwhile, it has made great progress in theoretical research and practical application. At present, many different algorithm models have been proposed based on EDA, such as PBIL[9], UMDA[10], MIMIC[11], BOA[12]. Their specific details are different, but their overall structure is similar. In those algorithms, the statistical analysis method is used to analyze the better

Content from this work may be used under the terms of the Creative Commons Attribution 3.0 licence. Any further distribution of this work must maintain attribution to the author(s) and the title of the work, journal citation and DOI.

Published under licence by IOP Publishing Ltd

individuals and a probability model is then generated. By sampling the model, the next generation of solutions is generated. Through repeated modeling and sampling process, the evolution of this algorithm can be implemented. Next, population-based incremental learning(PBIL), as a typical example of EDA, will be introduced.

In PBIL, the probability model representing the spatial distribution of solutions is a probability vector $p(x) = (p(x_1), p(x_2), \cdots, p(x_n))$, where $p(x_i)$ indicates the probability of taking x_i as 1 and $1 - p(x_i)$ is the probability of getting 0. In each generation, M individuals are randomly generated according to the probability vector $p(x)$, then their fitness are determined and the N best individuals ($N<M$) are selected to update $p(x)$. We take $p_t(x)$ to represent the probability vector in the tth generation, $x_t^1, x_t^2, \cdots, x_t^N$ to represent the selected N individuals, the update process is as follows:

$$p_{t+1}(x) = (1-\alpha)p_t(x) + \alpha \frac{1}{N} \sum_{k=1}^{N} x_t^k \qquad (1)$$

where α is the learning rate.

The procedure of PBIL can be described as follows:

1) M individual are generated randomly as the initial population D_t, t=0.

2) The fitness values of M individuals are calculated. If the termination conditions are met, the algorithm is finished, otherwise, it will continue.

3) Select N best individuals as the dominant group D_t^S.

4) A probability model is constructed from the dominant group, and the joint probability distribution is estimated.

5) Based on $p_t(x)$, generating the new generation of population and return to 2).

3. MEDA

In this paper, a multi-model EDA(MEDA) is proposed. In MEDA, multiple probability models are employed, which can be represented as $P = \{p_1, p_2, \cdots p_m\}$, where $p_i = (p_i(x_1), p_i(x_2), \cdots, p_i(x_n))$, m and n are the number of the models and the variables, respectively. In the initialization, let $p_i(x_j)$=0.5, that is, all feasible solutions in the solution space will be obtained at the same probability.

Next, by sampling the probability models, the candidate solutions are obtained. Specifically, for each probability model, each bit of a candidate solution is determined based on the corresponding probability in the model. Thus, we can obtain the population based on the m probability models.

After obtaining the population, each individual is evaluated based on fitness function, and the self best solution and global best solution are saved.

The probability model is updated based on the formula below.

$$p^{t+1}(x) = p^t(x) + \alpha(x_b - p^t(x)) \qquad (2)$$

where $p^t(x)$ is the probability in the tth generation. α is the adjusting factor, satisfying $0 < \alpha < 1$. Through this formula, the probability model is learned from the best solution, and α is used to regulate learning speed. The bigger α, the faster learning speed. Based on (2), the probability model will evolve towards the direction of increasing the probability of obtaining the best solution. The increase is proportional to the difference of the probability value and the corresponding bit of the best solution. When the probability of generating the best solution is small, the difference between x_b and $p^t(x)$ is large, and the model will evolve towards the best solution at a high speed. The maximum range of change is close to α; when the probability of generating the best solution is large, the difference between x_b and $p^t(x)$ is small, and the speed of model evolution is relatively slow, so as to avoid premature convergence. The minimum range of change will tends to 0 as the probability value approximate to the best solution.

To avoid premature convergence, the upper and lower limits are set for all the probabilities in the model, defined

$$p(x) = \begin{cases} \varepsilon & p(x) \le \varepsilon \\ p(x) & \varepsilon \le p(x) \le 1-\varepsilon \\ 1-\varepsilon & p(x) \ge 1-\varepsilon \end{cases} \tag{3}$$

where $0 < \varepsilon \ll 1$. Can be seen, this will make the $p(x)$ always in the area $[\varepsilon, 1-\varepsilon]$, which can avoid the algorithm from losing its search capability because of a probability of 1 or 0.

In the algorithm, each probability model evolves independently, it is very important to share the information about optimal solution among the probability models. In this paper, we choose stochastic information sharing. Specifically, in each generation, we randomly select two probability models to exchange their objective of evolution.

The procedure of MEDA is described as follows:

Begin

Initialize $P(t)$ at $t=0$, in which, all $p_i^0(x_j)$ in $P(0)$ are initialized with 0.5.

make m individuals $X_1^t, X_2^t, \cdots, X_m^t$ by sampling every models.

evaluate all candidate solutions, save the best solutions for every models to $B(t) = \{b_1, b_2, \cdots, b_m\}$, save the global best solution b among $B(t)$

While (not termination condition) do

$t = t+1$

Update $P(t)$ according to (2) and (3)

make m individuals $X_1^t, X_2^t, \cdots, X_m^t$ by sampling $P(t)$

evaluate all candidate solutions, save the best solutions for every probability models to $B(t) = \{b_1, b_2, \cdots, b_m\}$, save the best individual b among $B(t)$

select two individuals randomly to exchange best solution information

end

end

4. Algorithm analysis

EDA is an effective intelligence optimization algorithm. It describes the solution space by a probability model. For most of EDA, the probability model is obtained through learning the best solution or some better solutions. In this situation, the best solution or the average of some better solutions and its adjacent area will become the search focus. But on the other hand, the model is usually just a simple unimodal function, its peak is corresponding to the best solution or the average of the better solutions. Some solutions which are similar to the peak in genes form will have a larger probability to be obtained, and the solutions which are far from it, although which are possible to be obtained according to the model, the obtained probability is relatively small. Due to the complexity of the practical problems, the objective function is often a multi-peaks function. In this situation, if the global best solution is far away from the peak of the model, it is difficult to be found so that the algorithm will be trapped into the local optimum. The single model feature of EDA limits its performance.

However, in multi-model EDA, each model still have one peak like that in other EDA, but many different search focus can exist because of the characteristic of multiple probabilistic models. And besides the focus of each model itself, those overlap areas of more than one models may also be the focus of search. In evolutionary process, Existing of the multiple focuses in the solution space will make the algorithm more likely to find the best solution.

In fact, several EDAs with multiple models have already been proposed, including QEA[13,14], BPSO[15,16], etc[17]. They have been proved to have excellent optimization performance.

ISPECE IOP Publishing

5. Experimental results

To verify the performance of MEDA, the 0-1 knapsack problem are chosen. For the purpose of comparison, CGA and PBIL are also taken.

The 0-1 knapsack problem is a typical combinatorial optimization problem which can be described as: given a set of n items and a knapsack, select a subset of the items so as to maximize the profit:

$$f(X) = \sum_{i=1}^{n} p_i x_i \qquad (6)$$

$$\text{Subject to } \sum_{i=1}^{n} w_i x_i \leq C$$

where $x_i \in \{0,1\}$, $1 \leq i \leq n$, w_i and p_i are the weigh and the profit of the i-th item, respectively. And C is the capacity of the knapsack. $x_i = 1$ if the i-th item is selected, otherwise $x_i = 0$. The knapsack problem belongs to the classical NP-completed problem with $O(2^n)$ complexity. In the experiments, we used the parameter below

$w_i = uniformly\ random\ [1,10]$

$p_i = w_i + 5$

$C = \dfrac{1}{2} \sum_{i=1}^{n} w_i$

Three knapsack problems with 100, 500, 1000 are considered. And $f(X)$ is used as the fitness function. Every solution is repaired with the strategy below

Procedure repair $p\ (t)$

    Begin

        All items in the knapsack are sorted in the order

        Knapsack-overfilled = false

        If $\sum_{i=1}^{n} w_i x_i > C$ then

        Knapsack-overfilled = true

        While (Knapsack-overfilled = true) do

            Select the j-th item which is the last item being chosen

            $x_j = 0$

            If $\sum_{i=1}^{n} w_i x_i < C$ then

            Knapsack-overfilled = false

        End

        While (Knapsack-overfilled = false) do

            Select the j-th item which is the first item being not chosen

            $x_j = 1$

            If $\sum_{i=1}^{n} w_i x_i > C$ then

            Knapsack-overfilled = true

        End

        $x_j = 0$

All problems are optimized 50 times by CGA, PBIL, and MEDA, respectively. In these three algorithms, the population size is 30, and the maximum number of iteration is 500. In MEDA, $r = 0.1$, $\varepsilon = 0.01$. The optimization results of the three algorithms are shown in Table 1, and Figures 1-3.

Table 1. Optimization results with three algorithms.

Items	Algorithm	Best result	Average result	Worst result	Standard deviation
	CGA	623.1686	623.1637	623.1477	0.0050
100	PBIL	623.1687	623.1662	623.1466	0.0035
	MEDA	623.1687	623.1683	623.1670	0.0004
	CGA	3046.9	3041.9	3037	1.2
500	PBIL	3047	3047	3046.9	0
	MEDA	3047	3047	3047	0
	CGA	6148.6	6143.5	6132.6	3.4
1000	PBIL	6158.5	6153.7	6153.6	0.7
	MEDA	6158.6	6154.1	6153.6	1.6

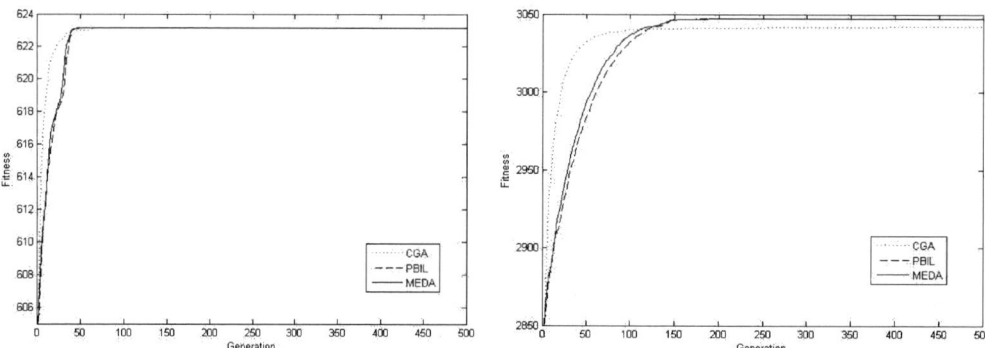

Figure 1. Optimization curves (item 100).　　　Figure 2. Optimization curves (item 500).

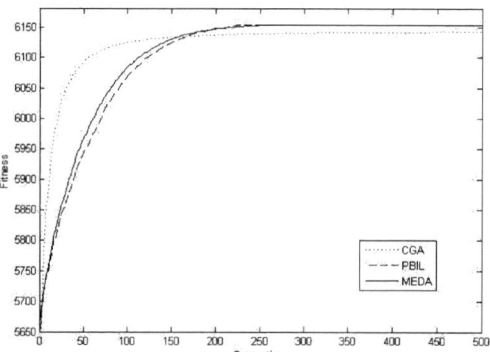

Figure 3. Optimization curves (item 1000).

It can be seen from the table1, MEDA has better optimization result comparing with CGA and PBIL. Figure 1-3 shows that CGA has faster convergence speed in the former period of evolution, but in the later, the premature convergence phenomenon is easy to occur. MEDA can effectively overcome the premature convergence, and its performance is better than CGA. At the same time, it has certain improvement in comparison to PBIL.

6. Conclusion

EDA is an intelligent optimization algorithm based on the probability model. It has good optimization performance. In this paper, a multi-models EDA is carried out. After analyzing it, the 0-1 knapsack problem is taken to verify its performance, and compared with the CGA and PBIL, the results show that the algorithm has better search ability, and its performance is better than the other two algorithms.

References

[1] Zhou Shude, Sun Zengqi. A survey on Estimation of distribution algorithms. ACTA AUTOMATICA SINICA. 2007, 33(2). 113-124.

[2] M Hauschild, M Pelikan. An introduction and survey of estimation of distribution algorithms. Swarm and evolutionary computation. 2011,1(3), 111-128.

[3] Hui Fang, Aimin Zhou, Hu Zhang. Information fusion in offspring generation: A case study in DE and EDA. Swarm and Evolutionary computation(2018). https://doi.org/10.1016/j.swevo. 2018.02.014 .

[4] Cesar R M, Bengoetxea F,Bloch I. Inexact graph matching for model based recognition: Evaluation and comparison of optimization algorithms. Pattern Recognition, 2005,38(11): 2099-2113.

[5] Ying Gao, Waixi Liu. Functional link artificial neural network with cloud estimation of distribution algorithm for traffic flow forcast. 2017 10th International Symposium on Computational Intelligence and Design(ISCID), 194-197.

[6] Wen Shi, Weineng Chen, Ying lin. An Adaptive Estimation of Distribution Algorithm for Multi-Policy Insurance Investment Planning. IEEE Transaction on Evolutionary Computation. 2017, 1-14. DOI:10.1109/TEVC.2017.2782571.

[7] Min Jiang, Zhongqiang Huang, Guiying Jiang. Motion generation of multi-legged robot in complex terrains by using estimation of distribution algorithm. 2017 IEEE Symposium Series on Computational Intelligence(SSCI), 1-6.

[8] S Ivvan Valdez P, Arturo Hernández, Salvador Botello. Repairing normal EDAs with selective repopulation. Applied Mathematics and Computation. 230(2014):65-77.

[9] Baluja S. Population-Based Incremental Learning: A Method for Integrating Genetic Search Based Function Optimization and Competitive Learning. Technical Report CMU-CS-94-163, Pittsburgh, PA: Carnegie Mellon University,1994.

[10] Mühlenbein H, Paass G. From recombination of genes to the estimation of distributions I. Binary parameters. Parallel Problem Solving from Nature - PPSN IV, Berlin, 1996. 178-187.

[11] De Bonet, J S, Isbell C L, Viola P. MIMIC: Finding optima by estimating probability densities. Advances in Neural Information Processing Systems, Cambridge: MIT Press, 1997,9:424-430.

[12] Pelikan M, Goldberg D E, Cantú-Paz E. BOA: The Bayesian optimization algorithm. In: Proceedings of the Genetic and Evolutionary Computation Conference GECCO-99,Orlando,FL: 1999. 525-532.

[13] K.H. Han, J–H Kim. Quantum- inspired Evolutionary Algorithm for a Class of Combinatorial Optimization. IEEE Trans Evolutionary Computation, 2002; 6(6) : 580–593.

[14] Michael D, Stefan S, Nikola K. Quantum- inspired Evolutionary Algorithm: A multi-model EDA. IEEE transaction on evolutionary computation, 2009;13(6):1218-1232.

[15] Kennedy J, Eberhart R, A discrete binary version of the particle swarm algorithm. Proceeding of the world Multiconference on Systemics, Cybernetics and Information. New Jersy: Piscataway, 1997, 4104-410.

[16] Geng Lin, Jian Guan. A hybrid binary particle swarm optimization for the obnoxious p-median problem. Information Sciences. 2018, 425(1): 1-17.

[17] Chugu Wu, Ling Wang. A multi-model estimation of distribution algorithm for energy efficient scheduling under cloud computing system. Journal of parallel and distributed computing. 2018, 117:63-72.

Pruning the deep neural network by similar function

Hanqing Liu[1], Bo Xin[1], Senlin Mu[2], Zhangqing Zhu[1]

[1]School of Management & Engineering, Nanjing University, 210093, China

[2]Nanjing Research Institute for Agricultural Mechanization of National Ministry of Agriculture, 210014, China

xinbo@nju.edu.cn

Abstract. Recent deep neural networks become deeper and deeper, while the demand for low computational cost model will be higher and higher. The exists pruning algorithm usually focus on pruning the network layer by layer, or using the weight sum as important score. However, these methods do not work very well. In this paper, we propose a unified framework to accelerate and compress cumbersome CNN models. We put it into an optimization problem to find a subset of the model which can produce the most comparable outputs. We concentrate on filter level pruning. Experiment shows that our method has surpassed the exists filter level pruning algorithm. Taking the network as a whole is better than pruning it layer by layer. We also have an experiment on the large scale ImageNet dataset. The result shows that we can accelerate the VGG-16 by 3.18× without accuracy drop.

1. Introduction

Deep convolution neural networks (DCNNs) have achieved a great success, especially in areas such as computer vision, e.g., image classification [2, 3, 4], semantic segmentation [8], object detection [9, 10], face recognition [11] and generative models [24, 25]. Modern state-of-the-art neural network architecture becomes more and more deep, and could gain more accuracy on the ImageNet Classification Challenge [1], e.g., AlexNet [2], VGGNet [3], GoogleNet [4], ResNets[5, 6] and DenseNets [7]. However, due to its extreme depth, these neural networks are also suffering a huge amount of calculation burden and usually require powerful computing devices such as GPUs. Since deep learning becomes more and more popular, people become more and more longing to deploy the large DCNN into small devices, such as mobile phones.

Many recent works have been proposed to compress cumbersome large CNNs in many different ways, including but not limited to neural network pruning [12, 14], weight quantization [13], knowledge distillation [15, 16], efficient architecture design [17, 18]. Since recent state-of-the-art network architecture always contains dozens of convolution layers which have a major computation cost, our method is focused on the direction of pruning convolution neural networks' filters by evaluating each filter's important score.

This paper introduces a new approach to pruning neural networks into a slimmer one and can reduce computation greatly. Our method towards to evaluate every filter's importance, and prune out the least important filters. Experiments shows that although pruned a few filters, the networks could regain its accuracy quickly by retrain them as long as the pruned filters are not important. On the contrary, if we prune out the important filters, the networks can hardly regain a high accuracy.

Contemporary researchers usually focused on filters themselves. For example, [14] uses the weight sum to evaluate filters' important score. However, our experiments showed that despite the really

---

Content from this work may be used under the terms of the Creative Commons Attribution 3.0 licence. Any further distribution of this work must maintain attribution to the author(s) and the title of the work, journal citation and DOI.

Published under licence by IOP Publishing Ltd

small filters (close to 0) which contribute little, the rest normal filters do not usually follow this rule, which means that relatively small filters can perform vital roles. Another direction of evaluating filters' importance by considering the difference between feature maps which are produced before and after pruning [12]. However, such criterion is too strict, because even if the feature maps are different from the original networks, the final output could be the same. Moreover, this method only concentrates a single layer which only contains limited information. Different layer has different jobs [26]. Our experiments will show that different layers might have different importance. Thus, focusing on single layers will ignore a lot of information. Unlike [12], our method takes a consideration of the entire network, and the result shows that particular layers could be more important than other layers. And [19] evaluate each neuron by detecting the drop of the accuracy when this neuron is pruned away. However, this criterion is relative relaxed because [15] tells us that the neural networks not only produce the most confident prediction, but also output dark knowledge, which means that a neuron contains a lot of dark knowledge may be seem not important by this criterion, because it might have no influence on the most confident output. To overcome these drawbacks, we proposed a new approach to evaluate the importance of each filter, and we prune out the least important filters. The idea hidden in this method is that we want the network to perform a similar function with the full model.

In a summary, we are looking for a way to find a subset of the original network to perform a similar function with the original full model, and the more similar, the better. We are not checking every intermediate feature maps, instead, we evaluate the final outputs.

In this paper, we introduce a weighting filters' importance method which has a perspective of the entire network, not part of it or filter itself. Our goal is to find a smaller network which can perform a similar function of the original network. As to one particular filter, we weigh its importance by measuring the difference between the functions which are implemented by pruned network and original network. If this difference is large which means that the pruned network's function is very different from the original one leads to a conclusion that this filter is relative important, while if we prune a filter generate a new network whose outputs are similar or even same with the original network's outputs means that this filter is less important. Once we weighed all the filters' importance by using this difference, we could prune out the least important ones. In a summary, we want to find a small convolution neural network which is pruned from the original neural network to perform the most similar function as the original one. To do this, we use the KL divergence to evaluate this difference and use randomly chosen data as input.

Experiments on CIFAR-10 benchmark datasets show that by using our method, we can still compress the VGG-16 [3] by the ratio about 90% with little drop of the accuracy. Compared with the method in [12, 14], our pruned network has better performance.

2. Related Work
In this section, we review some related work in several directions.

Neural Network Architecture Design. Since AlexNet [2] Ignited the interest of deep learning, people began to design neural networks by experiment. From VGGNet [3] to ResNets [5], neural networks became deeper and deeper and had more convolution layer. Recent work began to pay attention to small network architecture in order to run it on mobile devices. By using the idea of sparse connected channels (grouped channels), the state-of-the-art light model such as MobileNets [17], ShuffleNets [18] and MobileNetV2 [20] have much less computation cost compared with cumbersome models but still could maintain a good performance on ImageNet classification benchmark. In our experiment, we found that light models still have redundancy. Our goal is to compress such light model to a more efficient model.

Weight Quantization. Neural networks trained on X86 CPUs or GPUs usually have float32 or float64 weight. Weight quantization always wants to reduce the computation of a network by turning its float weight to more storage and computational efficient type, say, fixed point or even binary numbers. These lower bit numbers such as fixed point numbers can replace the original multiplication by other efficient operation such as XNOR. INQ [13] proposed a method that using a lower bit number than float could produce a comparable accuracy with the original networks. XNOR-Net [21] and Binary-Weighted-Nets [22] forces the weight in neural networks to be binary numbers which means they are restricted to {-1, 1}. Compared with the original networks, the quantized networks are relatively small and more efficient. Moreover, quantized network could be implemented on devices like FPGA. However, these approach usually facing a degradation of accuracy.

Neuron/Filter-level Pruning. A cumbersome network usually requires larger computing resources due to its large number of channels. There are several works aimed to reduce the number of channels in convolutional networks. This idea can trace back to [23]. [23] tries to weigh the importance of weight by measuring the salience of the output with or without these weight, which usually is the cross entropy loss, and uses the Taylor expansion to approximate. [12] measures the salience by detecting the difference of feature maps. [14] uses the so-called magnitude to measure the importance of parameters. [28] also uses the Taylor expansion, however, with first derivative.

Knowledge Distillation. Knowledge distillation tends to transfer knowledge from a teacher model to a student model, and, hopefully, it can bring improvement of precision of the student model. [15] uses a pre-trained cumbersome teacher network to produce so-called soft targets as input for the student network. [15] thinks that a pre-trained teacher model can provide more information than one-hot ground truth label. By taking the advantages of the teacher models, student models can achieve more accuracy. [16] provide a method that training the student network by mimicking the teacher network, and gain the impressive improvement. This method tends to use attention map to complete knowledge transfer.

3. Method
In this section, we will introduce our approach in detail. First we will present our idea behind this method which we called *similar function*. Next we will introduce our framework and algorithm.

3.1 Similar function
Pruning is a popular way to reduce complexity of the cumbersome networks. Our method focuses on filter/neuron-level pruning which means that we prune an entire filter away once we believe that they are unimportant. And our method requires pre-trained models.

We use randomly chosen data, denoted as X, in the dataset which the pre-trained model was trained on as input. Our target is to find a subset of the whole parameters $w \in \theta$, to produce the most similar outputs of the original networks. We use the Kullback–Leibler divergence to calculate this similarity. Thus we also use softmax function to turn the logits which outputted by the network into probability distribution:

$$D(X,\theta) = \phi[f(X,\theta)] \tag{1}$$

Where $\phi(*)$ denotes the softmax function, and $f(*)$ denotes the neural network. And θ represents the corresponding parameters of the network.

And we use $D(X,\theta)$ to calculate the KL divergence:

$$L = KL[D(X,w) \| D(X,\theta)] \tag{2}$$

Thus, we want to find subset of parameters $w \in \theta$ to generate the smallest L:

$$\arg\min_{w \subset \theta} L = KL[D(X,w) \| D(X,\theta)]$$

$$s.t. \ |w| = r \times |\theta|, \ r \in (0,1) \tag{3}$$

Where $r \in (0,1)$ is the pruning ratio which we treat it as a hyperparameter.

Unfortunately, this optimization problem (2) is NP hard, which means that we cannot solve it directly. Thus, we use a greedy algorithm to find an approximate solution.

3.2 Framework

Although we cannot directly find an optimal subset of θ to produce the minimum L, we can find an optimal subset $\hat{\theta}$ which excludes one filter. Thus, it provides an opportunity to solve Eq.3 greedily. After we traverse all the filters in the network by deleting them, we can find the most unimportant ones. After pruning them, the network will usually lose its accuracy. Thus, we are required to retrain the pruned network to regain the accuracy. And we do it again. As illustrated in figure 1, we conclude our framework to this:

1) Each filter's important score is evaluated;
2) The most unimportant filters are pruned;
3) Fine Tuning.

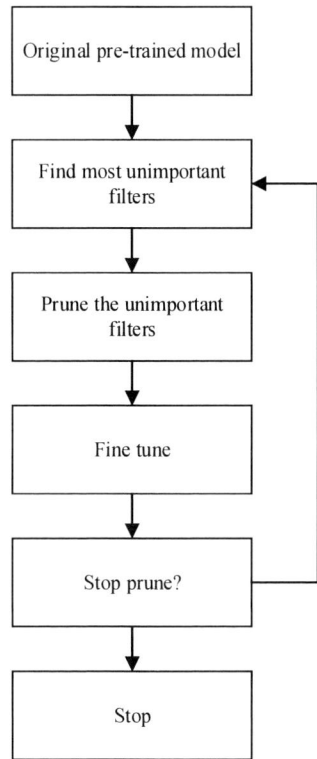

Figure 1. Frame work of pruning algorithm.

3.3 Algorithm

We present a greedy algorithm to find an approximate solution of the optimization problem of eq. 2. We use a triplet $< X, w_i, L_i >$ to denote each filter's process, where $X \in \square^{C \times H \times W}$ is the input tensor which has height H, width W and C channels. And $w_i \in \square^{N_1 \times K \times K}$ is a set of filters in the pre-trained model which has N_1 input channels and with $K \times K$ kernel size. L_i is a scalar, and it represents the difference between outputs generated by the pre-trained model with or without w_i by feeding data X. We use the Kullback–Leibler divergence to calculate L_i:

$$L_i = KL[D(X, \theta) \| D(X, \hat{\theta})] \tag{4}$$

Where $w_i \in \theta, w_i \notin \hat{\theta}$ denotes the parameters of the networks. And the optimal subset:

$$w_{w \subset \theta} = \arg\min \frac{1}{n} \sum_{j=1}^{n} KL[D(X_j, w) \| D(X_j, \theta)] \tag{5}$$

Once we traverse all the filters, we will get the set L_i, $i \in \{1, 2, ..., S\}$. We can find the least important filters. The pseudo code can be found in Algorithm 1.

4. Experiment

We study our algorithm's performance in this section. We will compare with some other several filter level pruning algorithm and show that the performance of our algorithm is far better than others'. Then we will show our result on CIFAR-10 dataset with VGG-16 [3].

Algorithm 1. A greedy algorithm to optimize Eq. 3

Input: Data X; Pre-trained model M; Termination condition r; Pruning filter quantity in one turn a; Training set $\{x, y\}$

Output: Pruned model M';

while not r :
    $S \leftarrow \varnothing$
    for each convolutional layer Λ in M :
        $S \leftarrow S \cup \Lambda.filters$
    $L \leftarrow \varnothing$
    for each i in S :

$$L_i = \frac{1}{n} \sum_{j=1}^{n} KL[D(X_j, w_i) \| D(X_j, \theta)]$$

        $L \leftarrow L \cup L_i$

    $\hat{L} = topa(L)$

    $M' \leftarrow$ Prune $\hat{L}$
    Fine tune M' by $\{x, y\}$
    $M \leftarrow M'$

4.1 Performance comparison

We focus our comparison on the different filter level pruning algorithm as follows:

(1) Random. This is our base line. Because random pruning is the simplest way to prune the network. Without priori knowledge, one could only prune the network randomly. However, we have the pre-trained parameters, which means that we cannot prune the network worse than randomly pruning.

(2) Magnitude [14]. This criterion considers the less important filters are those with smaller parameters. Because those filters produce poorer activation than others'. So each filter's important score is $s_i = \sum |W(i, :, :, :)|$.

(3) Thinet [12]. This criterion is similar to ours. It considers the Euclidean distance between the feature map with and without particular filters as the important scores. And similarly, Thinet solve the optimization problem greedily. The difference between our algorithm and Thinet is that we evaluate every filter in the entire network while Thinet prune the network layer by layer.

To compare these methods, we use the CIFAR-10 benchmark. CIFAR-10 contains 10 classes. Each class has 6000 32×32 pictures. To avoid overfitting, we removed the last fully connected layers in VGG-16 and replaced with a global average pooling layer along with a fully connected layer to classify. Each pruning strategy requires a pre-trained model. We use the same pre-trained model. After pruning, one fine tune epoch is employed. Except the difference of pruning strategy, everything else is same. The pre-trained VGG-16 has 6.7% of top-1 error. The result is presented in figure 2.

Figure 2 shows the result of each pruning strategy's performance. We evaluate four different strategies. In the experiment, we found that the accuracy once we pruned the network without fine tuning could also reveal the performance. And our similar function method could achieve the best accuracy without fine tune, which means that our method can retain the most of information. While random pruning and weight sum strategy is the worst. It seems that magnitude does not possess a strong correlation with weight's importance. A small parameter could have a large impact on the outputs. Thus, the result of magnitude strategy is as poor as the base line is reasonable.

As to the comparison of Thinet and our similar function method, the reason why our method is slightly better is probably that our method takes every filter into account, and prune the network in a whole perspective instead of pruning it layer by layer. The remaining networks show that our method could reveal not only the importance of each layer, but also each filter's. And the result shows that the last layer of each group (conv1-2, conv2-2, conv3-3) is more important than others. The least important layers will remain fewer filters than those important ones. Thus, our method shows better and more robust result.

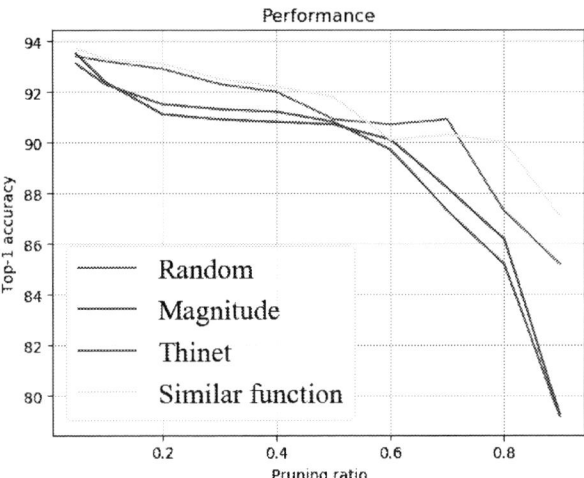

Figure 2. The result of performance comparison with different pruning strategy. Based on benchmark CIFAR-10.

4.2 VGG-16 on ImageNet

In this section, we will evaluate the performance of our similar function method on the large scale ImageNet dataset [1]. ILSCVR-12 is the benchmark of the state-of-the-art networks. It contains about one million images, and these images form 1000 classes. We use 1000 images (each class has one image) to find the importance of each filter via algorithm 1. Also, we use a pre-trained model whose top-1 accuracy is 68.34%. We use the standard training set and validation set to train and validate our models.

During fine tuning, we use SGD algorithm to optimize the model with learning rate from 10^{-3} reduces to 10^{-4}. The mini-batch size is 256. Also, data augmentation is employed. The augmentation contains random crop from 256×256 images and random flip. During the validation, only random crop is used. The pruning ratio is 0.5. We only prune the convolution layers, while the other parameter is same as the original VGG-16 [3]. The result on ImageNet is showed in table 1. Our method pruned fewer filters in the last layer of each group, so it will have more parameters than Thinet. The interesting thing is that after pruning, the network gains a lifting of accuracy. Probably due to discarding the least important filters, the network could somehow avoid the overfitting.

Table 1 Pruning VGG-16 on ImageNet.

Model	Top-1 accuracy(%)	Top-5 accuracy(%)	#parameters(M)	FLOPs(B)
Original VGG	68.34	88.44	138.34	30.94
Similar function	69.62	89.62	131.86	9.73
Thinet	69.11	89.32	131.44	9.58
Training from scratch	66.91	87.32	131.86	9.73

We also train a new model whose architecture is same as the pruned one from scratch. It shows that this network has a lower accuracy than the pruned one, which means that pruning algorithm is better than training scratch. It reveals the fact that although two networks have same architecture, they could have different ability of learning. The only difference of the networks is initialization of parameters. One has part of pre-trained parameters while the other only has randomly initialized parameters. This only difference could bring huge accuracy gap.

We also calculate the FLOPs of each model. When combine with the number of parameters, it shows that convolution layers often have fewer parameters but have huge computation cost. By focusing on pruning convolution layers, one can reduce the inference time dramatically. On the other hand, fully connected layers usually have massive parameter size. One could replace it with global average pooling to reduce the amount of parameters. It can also help alleviating overfitting.

5. Discussion

Cun Y L et al. proposed a method to prune the network by using the information of the second derivative which is also known as Hessian [27]. While [28] proposed an approximate method to determinate which filter is unimportant by utilizing Tylor expansion, because the Hessian is usually unquantifiable. This method, though, does not match the accuracy of the greedy algorithm, save a lot of computing time. Using the first derivative to evaluate each filter's importance is far more efficient than the so-called oracle pruning. The method proposed in this paper is an oracle method indeed. Thus, our method is less efficient than the Tylor expansion approximating. Although the first thing to consider is inference time, the pruning cost still maters. In the future work, we will dedicate to a more economical method, including but not limited to Tylor expansion approximation.

Acknowledgments

This work was supported by the National Key Research and Development Program (NO.2016YFD0702100), the National Natural Science Foundation of China (Nos.61432008and 71732003), the Postgraduate Research and Practice Innovation Program of Jiangsu Province (No.SJCX17-0004).

Reference

[1] Deng J, Dong W, Socher R, et al. Imagenet: A large-scale hierarchical image database[C]//Computer Vision and Pattern Recognition, 2009. CVPR 2009. IEEE Conference on. IEEE, 2009: 248-255.

[2] Krizhevsky A, Sutskever I, Hinton G E. Imagenet classification with deep convolutional neural networks[C]//Advances in neural information processing systems. 2012: 1097-1105.

[3] Simonyan K, Zisserman A. Very deep convolutional networks for large-scale image recognition[J]. arXiv preprint arXiv:1409.1556, 2014.

[4] Szegedy C, Liu W, Jia Y, et al. Going deeper with convolutions[C]. Cvpr, 2015.

[5] He K, Zhang X, Ren S, et al. Deep residual learning for image recognition[C]//Proceedings of the IEEE conference on computer vision and pattern recognition. 2016: 770-778.

[6] He K, Zhang X, Ren S, et al. Identity mappings in deep residual networks[C]//European Conference on Computer Vision. Springer, Cham, 2016: 630-645.

[7] Huang G, Liu Z, Weinberger K Q, et al. Densely connected convolutional networks[C]//Proceedings of the IEEE conference on computer vision and pattern recognition. 2017, 1(2): 3.

[8] Chen L C, Papandreou G, Kokkinos I, et al. Deeplab: Semantic image segmentation with deep convolutional nets, atrous convolution, and fully connected crfs[J]. IEEE transactions on pattern analysis and machine intelligence, 2018, 40(4): 834-848.

[9] Girshick R. Fast r-cnn[J]. arXiv preprint arXiv:1504.08083, 2015.

[10] Liu W, Anguelov D, Erhan D, et al. Ssd: Single shot multibox detector[C]//European conference on computer vision. Springer, Cham, 2016: 21-37.

[11] Taigman Y, Yang M, Ranzato M A, et al. Deepface: Closing the gap to human-level performance in face verification[C]//Proceedings of the IEEE conference on computer vision and pattern recognition. 2014: 1701-1708.

[12] Luo J H, Wu J, Lin W. Thinet: A filter level pruning method for deep neural network compression[J]. arXiv preprint arXiv:1707.06342, 2017.

[13] Zhou A, Yao A, Guo Y, et al. Incremental network quantization: Towards lossless cnns with low-precision weights[J]. arXiv preprint arXiv:1702.03044, 2017.

[14] Li H, Kadav A, Durdanovic I, et al. Pruning filters for efficient convnets[J]. arXiv preprint arXiv:1608.08710, 2016.

[15] Hinton G, Vinyals O, Dean J. Distilling the knowledge in a neural network[J]. arXiv preprint arXiv:1503.02531, 2015.

[16] Romero A, Ballas N, Kahou S E, et al. Fitnets: Hints for thin deep nets[J]. arXiv preprint arXiv:1412.6550, 2014.

[17] Howard A G, Zhu M, Chen B, et al. Mobilenets: Efficient convolutional neural networks for mobile vision applications[J]. arXiv preprint arXiv:1704.04861, 2017.

[18] Zhang X, Zhou X, Lin M, et al. Shufflenet: An extremely efficient convolutional neural network for mobile devices[J]. arXiv preprint arXiv:1707.01083, 2017.

[19] Morcos A S, Barrett D G T, Rabinowitz N C, et al. On the importance of single directions for generalization[J]. arXiv preprint arXiv:1803.06959, 2018.

[20] Sandler M, Howard A, Zhu M, et al. MobileNetV2: Inverted Residuals and Linear Bottlenecks[C]//Proceedings of the IEEE Conference on Computer Vision and Pattern Recognition. 2018: 4510-4520.

[21] Rastegari M, Ordonez V, Redmon J, et al. Xnor-net: Imagenet classification using binary convolutional neural networks[C]//European Conference on Computer Vision. Springer, Cham, 2016: 525-542.

[22] Courbariaux M, Hubara I, Soudry D, et al. Binarized neural networks: Training deep neural networks with weights and activations constrained to+ 1 or-1[J]. arXiv preprint arXiv:1602.02830, 2016.

[23] LeCun Y, Cortes C, Burges C J. MNIST handwritten digit database[J]. AT&T Labs [Online]. Available: http://yann. lecun. com/exdb/mnist, 2010, 2.

[24] Goodfellow I, Pouget-Abadie J, Mirza M, et al. Generative adversarial nets[C]//Advances in neural information processing systems. 2014: 2672-2680.

[25] Kingma D P, Welling M. Auto-encoding variational bayes[J]. arXiv preprint arXiv:1312.6114, 2013.

[26] Zeiler M D, Fergus R. Visualizing and Understanding Convolutional Networks[C]// European Conference on Computer Vision. Springer, Cham, 2014:818-833.

[27] Cun Y L, Denker J S, Solla S A. Optimal brain damage[C]// International Conference on Neural Information Processing Systems. MIT Press, 1989:598-605.

[28] Molchanov P, Tyree S, Karras T, et al. Pruning Convolutional Neural Networks for Resource Efficient Inference[J]. 2016.

ISPECE   IOP Publishing

Application of Internet segmentation research based on Natural Language Processing technology in enterprise public opinion risk monitoring

Di Liu[1, a], Jiangwen Su[2], Lihua Song[2] and Zhen Qiu[1, b]

[1]State Grid Information and Telecommunication Group, Beijing 100000, China;

[2] Fujian Yirong Information Technology CO., Ltd, Fuzhou 350000, China.

[a]liudi@sgitg.sgcc.com.cn, [b]qiuzhen@sgitg.sgcc.com.cn

Abstract: With the advent of the mobile Internet era, the network has become a distribution center of various information such as media, entertainment, sports, economy, politics and so on. A large amount of information is generated and disappeared on the network every day. How to effectively extract and identify the relevant data, and judge and analyze them is an important part of the corporate public opinion control. This paper uses natural language processing technology to study the word segmentation of text information on the network, and applies it to the risk detection of corporate public opinion.

1. current situation of public opinion research

Public opinion analysis is a process of deep thinking, analysis and Research on the public opinion aiming at this problem and getting relevant conclusions according to the needs of specific problems. Public opinion analysis is a subject based on many fields such as language and social communication, news communication and so on. It is different from the early paper media era. After the arrival of the Internet era, the new information modes of Web pages, news portals, short videos, microblogs and other fast-food formats have posed great challenges to public opinion analysis in the past. At the same time, because of mobile interaction. With the advent of the Internet era, the massive information and communication modes have been difficult to cover the previous public opinion analysis models. The data on the network end grow exponentially and erupt. It is very likely that an event will be completed, originated, fermented and erupted in a day or two. All these put forward new choices for public opinion analysis.

1) Definition, origin and development of Internet public opinion in China

Network public opinion refers to the network public opinion which is popular on the Internet and has different views on social issues. It is a form of expression of social public opinion. It is a strong influence and tendentious opinion of the public on some hot and focus issues in real life through the Internet. Network public opinion is a collection of netizens' emotions, attitudes, opinions, opinions, expressions, dissemination and interaction, as well as follow-up influence, with the network as the carrier and events as the core.

Network public opinion refers to the social and political attitudes, beliefs and values that people have towards public issues and social managers through the network around the occurrence, development and changes of intermediary social events in a certain social space. It is the sum of the beliefs, attitudes, opinions and emotions expressed by more people about various phenomena and

Content from this work may be used under the terms of the Creative Commons Attribution 3.0 licence. Any further distribution of this work must maintain attribution to the author(s) and the title of the work, journal citation and DOI.
Published under licence by IOP Publishing Ltd

problems in society. The formation of Internet public opinion is rapid and has a great impact on society. With the rapid development of the Internet in the world, the network media has been recognized as the "fourth media" after newspapers, radio and television, and the network has become one of the main carriers reflecting public opinion.

Internet public opinion is a reflection of social public opinion in the Internet space and a direct reflection of social public opinion. Traditional social public opinion exists in the folk, in the public's ideas and daily comments on the streets and lanes. The former is difficult to capture, while the latter is fleeting. Public opinion can only be obtained through open and secret visits, public opinion surveys and other means. The acquisition efficiency is low, the sample is small and easy to flow biased, and the cost is enormous. Big. With the development of the Internet, the public often express their opinions in the way of informationization. The network public opinion can be easily accessed by means of automatic grasping technology of Turing public opinion network, which is efficient, information fidelity and full coverage.

2) Research methods of public opinion

In recent years, China has made great efforts to use technical means to dig and analyze a large amount of network public opinion information in depth, so as to quickly compile public opinion information, thus replacing the complicated work of manual reading and analysis of network public opinion information. The key technologies related to Internet public opinion are summed up as two types: monomer technology and systematization technology.

(1) Network public opinion collection and extraction technology: network public opinion is mainly formed and disseminated through news, forum/BBS, blog, instant messaging software and other channels. The carriers of these channels are mainly dynamic web pages, which carry loose structured information, making effective extraction of public opinion information very difficult. Mei Xue et al. (2007) realized the extraction and integration of dynamic web page data to a certain extent through the method of automatic generation of web page information extraction Wrapper, which has a certain processing accuracy and extraction efficiency.

(2) Network public opinion topic discovery and tracking technology: Internet users discuss a wide range of topics, covering all aspects of society, how to find hot and sensitive topics from the mass of information, and track its trend change has become a hot research topic. Early research ideas of Alan James, J. Allan, G. Hulten, Qiaozhu Mei and others are based on text clustering, that is, the keywords of text are the characteristics of text. Although this method can aggregate text under a large category of topics, it does not guarantee the readability and accuracy of topics. Duan Jianguo et al. (2007) improved this idea and realized topic discovery and tracking: transforming text clustering into topic feature clustering, and reorganizing and utilizing language text information flow according to events.

(3) Network public opinion tendentiousness analysis technology: through tendentiousness analysis, we can make clear the subjective reflection of the feelings, attitudes, views, positions and intentions of network communicators. For example, Sina's "news mood ranking" divides the mood of users when they read news comments into eight levels as shown in Figure 2-1. The analysis of the tendency of public opinion text is actually an attempt to use computer to achieve the goal of extracting the emotional direction of the author of the text according to the content of the text. Tang Huifeng, Xu Linhong, Li Yanling and others (2007) devoted themselves to the tendentiousness analysis technology of online public opinion texts: by judging the characteristics and types of tendentious feature words in the network environment, and making the identification and annotation of the tone polarity, they constructed an Internet-oriented tendentious mood dictionary and constructed a certain scale of standard data.

(4) Multi-document automatic summarization technology: news, posts, blog posts and other pages contain spam information. Multi-document automatic summarization technology can filter the content of pages and extract summary information to facilitate query and retrieval. To a certain extent, researchers have realized the automatic generation of messages from network public opinion information, and can browse and retrieve information through browsers.r further research on Chinese orientation analysis.

3) the application of computer and machine learning in public opinion.

It is a long-standing pursuit of people to communicate with computers in natural language. Because it not only has obvious practical significance, but also has important theoretical significance: people can use their most accustomed language to use computers, without spending a lot of time and energy to learn various computer languages which are not very natural and habitual; people can also further understand human language ability and intelligence through it. The mechanism.

At the same time, computer is used to analyze and calculate large-scale complex systems, deal with massive text technology, and use statistical and artificial intelligence methods to replace manual text classification and discrimination. At the same time, the analysis and processing based on historical data can quickly complete the judgment and analysis of regional and global hot words in a short time in the future, so as to achieve the role of prevention and control.

2. application of machine learning algorithm in natural language understanding
1) machine learning algorithm text processing technology
Similar to traditional machine learning algorithm, the main modes of text processing technology based on text machine learning algorithm are word segmentation, extraction, discrimination and vectorization.

A Chinese text is a string consisting of Chinese characters (including punctuation marks). Words can form words, phrases can form phrases, phrases can form sentences, and then some sentences can form paragraphs, sections, chapters and chapters. No matter at all levels mentioned above: characters (words), words, phrases, sentences, paragraph,... Or there are ambiguities and polysemy in the transition from the next level to the next level, that is, a string of the same form can be understood as different word strings, phrase strings and so on in different scenarios or contexts, and has different meanings. Generally speaking, most of them can be solved according to the corresponding context and scene. That is to say, in general, there is no ambiguity. This is why we do not feel ambiguity in natural language and can communicate correctly with natural language.

Above all, a Chinese text or a string of Chinese characters (including punctuation marks, etc.) may have multiple meanings. It is the main difficulty and obstacle in natural language understanding. Conversely, an identical or similar meaning can also be represented by multiple Chinese texts or multiple Chinese character strings. And to quantify the data sets.

The process of selecting a subset of related features from a given set of features is called feature selection. Feature selection is from the feature set T={t_1,... In t_s}, we choose a real subset T '={t_1,... T_ (s')} satisfies (s s s). Among them, s is the size of the original feature set, and s^ 'is the size of the selected feature set. The criterion of selection is that feature selection can effectively improve the accuracy of text. Selection does not change the nature of the original feature space, but only chooses some important features from the original feature space to form a new low-dimensional space. Text feature selection can effectively reduce the dimension of text representation.

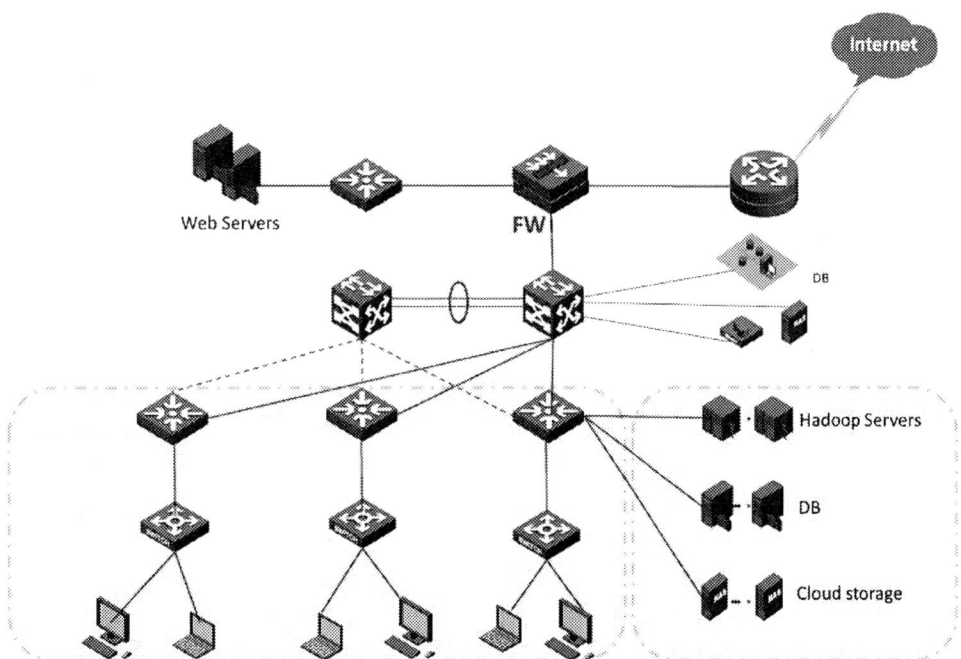

Fig. 1. collection and processing of data information

2) implementation of machine learning algorithm for natural language understanding.

(1) using supervised learning to generate a historical data training model for predicting the type of text.

(2) Mainly delete duplicate data, correct or delete erroneous and invalid data, and check the consistency of data. For example, if the length of text is less than 13, it is meaningless to delete it.

(3) set a set of categories to classify words according to the sample set manually.

(4) using Classified Thesaurus to deal with the processed text information two times.

3) the application and implementation of thesaurus technology.

Word segmentation technology is a technology that search engines use various matching methods to segment keywords according to user's keyword string after query processing for keyword string submitted by users.

The main methods are: string matching method of word segmentation; word segmentation; statistical word segmentation.

Based on these different word segmentation methods, the extracted text constitutes a database and is updated according to the text at any time. The comparative sample database is the word segmentation database. The specific implementation algorithm is not described in detail here, and can refer to the relevant literature at home and abroad.

3. data acquisition method of massive information processing foundation and web crawler method

1) massive information processing technology

The so-called massive information processing means that the amount of data is too large to be solved quickly in a relatively short time and can not be loaded into memory at one time. Based on these problems, many algorithms have been put forward to solve this problem.

In terms of time complexity, we can use ingenious algorithms with appropriate data structures, such as Bloom filter/Hash/bit-map/heap/database or inverted index/trie tree. In terms of spatial complexity, divide and conquer /hash mapping.

The basic methods of massive data processing are summarized as follows:

Divide and conquer /hash mappings + hash statistics + heap / fast / merge sort;

(1) double barrel division;

(2) Bloom filter/Bitmap;

(3) Trie tree / database / inverted index;

(4) external sorting;

(5) Hadoop/Mapreduce for distributed processing.

2) web crawler technology

The object of general search engine is Internet pages. At present, the number of Internet pages has reached 10 billion. So the first problem faced by search engine is how to design an efficient download system to transmit such a large amount of web pages data to the local area and form a mirror backup of Internet pages locally. Web crawler can play such a role to complete this arduous task. It is a key and basic component of search engine system.

Firstly, the crawler system carefully selects part of the web pages from the Internet pages, and takes the link addresses of these pages as seed URLs, and puts these seeds into the waiting URL queue. The crawler reads the waiting URL queue in turn, and parses the URLs through DNS, and converts the link addresses into the corresponding IP addresses of the Web server. Then give the relative path name to the web page downloader, which is responsible for the download of the page. For downloaded local pages, on the one hand, they are stored in the page library, waiting for subsequent processing such as indexing; on the other hand, the URLs of downloaded pages are placed in the crawled queue, which records the URLs of webpages that have been downloaded by the crawler system to avoid duplicate crawling of the system. For the newly downloaded Web pages, extract all the link information and check it in the downloaded URL queue. If it is found that the link has not been crawled, it will be placed at the end of the queue to be crawled, and the corresponding web pages of the URL will be downloaded in the subsequent crawl scheduling. In this way, the formation of a cycle until the URL queue to be grabbed is empty, which means that the crawler system will be able to grab all the pages have been grabbed, at this time completed a complete round of grabbing process.

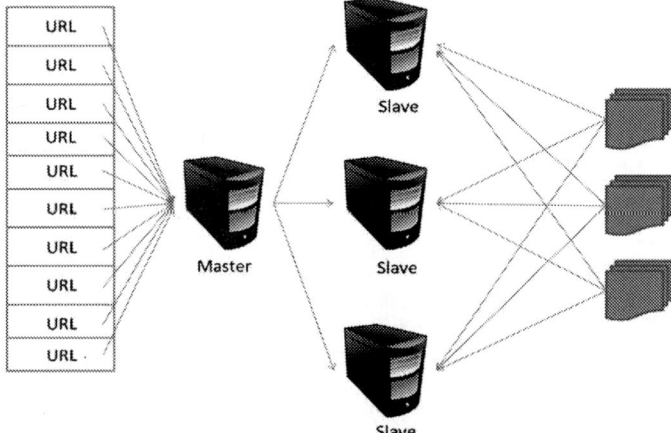

Fig. 2. schematic diagram of web crawler principle

4. the establishment of public opinion monitoring model.

1) public opinion text data analysis

Firstly, according to the latest five-year news reports on the group, two categories of positive news and negative news are selected. At the same time, the text of the group is pre-classified and classified according to the subsequent impact and duration. Finish the text sorting of the thesaurus.

The characteristic text file is segmented and the corresponding sub library is formed.

2) model establishment

By using the established word segmentation library and using the Jieba package of Python language to learn, a model of public opinion discrimination and correction is constructed.

The specific steps are as follows:

[1]. The jieba. cut method accepts three input parameters: a string requiring participle; a cut_all parameter to control whether full mode is used; and an HMM parameter to control whether HMM model is used.

[2]. The jieba.cut_for_search method accepts two parameters: a string requiring participle; and whether to use the HMM model. This method is suitable for search engine to build inverted index segmentation.

[3]. the strings for pending participles can be Unicode or UTF-8 strings and GBK strings. Note: it is not recommended to enter the GBK string directly, it may be unreasonably decoded into UTF-8.

[4]. The structure returned by jieba. cut and jieba. cut_for_search is an iterative generator that can use the for loop to obtain each word (unicode) after segmentation, or use the

[5].jieba.lcut and jieba.lcut_for_search return directly to list

[6]. Jieba. Tokenizer (dictionary = DEFAULT_DICT) creates a new custom word segmenter, which can be used to use different dictionaries at the same time. Jieba.dt is the default participle, and all global participle related functions are the mapping of the word segmentation device.

3) model screening

The initial data and the latest network text are used for preliminary learning, and the selected words in the lexicon are adjusted appropriately according to the results, and the segmentation strategies and corresponding algorithms are adjusted.

5. data learning and adjustment of models

1) public opinion text data collection

Python crawler technology is used to collect data on the main portal websites in our country, and it is textualized and saved as. TXT file format. And classify them according to different data sources.

2) data cleaning and cleaning

The collected data were first screened based on events, space and correlation degree. The obvious irrelevant data and text are eliminated in the first round. The remaining saves are processed in the next step and segmented and vectored.

3) adjustment analysis of learning strategies

According to python's learning curve and effect, the strategy and parameters are adjusted. At the same time, different word segmentation libraries are adopted based on the time limit. The results show that the algorithm convergence achieves the effect of text learning.

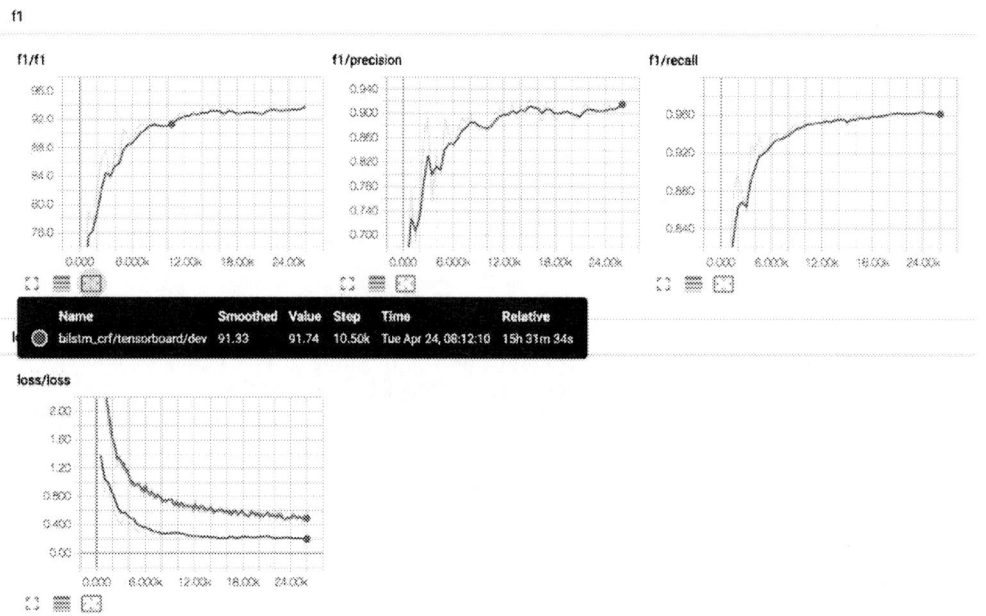

Fig. 3. convergence curve of machine learning algorithm for model

6. summarize the advantages and disadvantages and the main points to be improved.

Using the natural language processing technology of machine learning algorithm to detect and discriminate public opinion system is convenient and fast, and can achieve the purpose of public opinion monitoring. However, the timeliness of the model is relatively short, and the problem of long training time is to be solved. At the same time, sometimes the screening of invalid text is one of the bottlenecks of this technology. So effective segmentation algorithm is the focus of future research.

References
Acknowledgement: This work was supported by State Grid Technical Project (No. 52110418002W).

References:
[1] Zeng run hi. Network public opinion control mechanism research [J]. library and information work, 2009, 53 (18): 79-82.

[2] Zhu Yihua, Zhang Chaoqun, Zheng Dejun, et al. Research on Internet Public Opinion Management from the Perspective of Information Ecology [J]. Information Theory and Practice, 2013, 36 (11): 90-95.

[3] Guo Jianqiang, Zeng Wangfeng. Discussion on the Change of Network Public Opinion Management in the Age of Big Data [J]. Guangxi Social Science, 2015 (8): 145-149.

[4] Zhao Jinping, Zhang Xinyu. Public opinion management strategy for social security incidents [J]. news research guide, 2018 (9).

[5] Lin Yiou, Lei Hang, Li Xiaoyu, et al. Deep Learning in Natural Language Processing: Methods and Applications [J]. Journal of University of Electronic Science and Technology, 2017, 46 (6).

[6] Tian Dong, Zhang Xining. Realization of Weak Supervisory Knowledge Acquisition System Based on Natural Language Processing [J]. Foreign Electronic Measurement Technology, 2017, 36 (3): 60-63.

[7] Bei Chao, Hooper. The influence of language priori knowledge on natural language processing tasks of neural network models [J]. Chinese Journal of Information, 2017, 31 (6).

[8] Ma Yuchun, Song Hantao. Web Chinese text segmentation technology research [J]. computer applications, 2004, 24 (4): 134-135.

[9] Zhang Zhongyao, Ge Wancheng, Wang Liangyou, etc. Research and design of Chinese word segmentation technology based on MMSEG algorithm [J]. Information technology, 2016 (6): 17-20.

[10] Liu Xinliang, Yan Shanshan. Implementation and application of Chinese word segmentation based on Python [J]. Computer and Information Technology, 2008 (11): 85-88.

[11] Xu Xiao, Zhang Weizhe, Zhang Hongli, et al. WAN Distributed Web Crawler [J]. Journal of Software, 2010, 21 (5): 1067-1082.

ISPECE   IOP Publishing

IOP Conf. Series: Journal of Physics: Conf. Series **1187** (2019) 042008   doi:10.1088/1742-6596/1187/4/042008

A Gesture Recognition Algorithm Based on Three-dimensional Projection and Direction Chain Code

Jie Li

School of Electrical and Electronic Engineering, North China Electric Power University, China

Email:jie_li_public@163.com

Abstract.   In this paper, the accelerometer of intelligent terminal is used for gesture recognition. Most of the current motion recognition algorithms in this field have great restrictions on the standardization of motion, and the recognition rate for slightly substandard motion is greatly reduced. Aiming at the existing "$3" algorithm, this paper proposes to project the 3D motion data to the 2D plane for trajectory pattern recognition, and uses the angle and radius to identify the curved area. This method can make up for the people greatly. Physiological habits lead to difficulty in the recognition of gestures, which can greatly improve the recognition accuracy when the movement is not standardized.

1. Introduction

The expression of gestures is the most basic and intuitive behavioral language of human beings. In recent years, gesture-based human-computer interaction patterns have developed rapidly, mainly in two types, one is based on visual image recognition, and the other is based on inertial sensor recognition. Due to the complicated lighting and environmental conditions, image recognition is costly and the accuracy is low. The accelerometer is placed in almost every piece of our electronic products, and we can capture our motion information without external interference. And with the improvement of processing speed and storage capacity of intelligent terminals and the improvement of programmability, we have provided a new platform for our new requirements, which makes us have higher requirements for reliability.

However, the existing gesture recognition is mostly based on the field of heuristic pattern recognition and machine learning, ignoring the study of human physiological habits, which makes the accuracy of the movements not standardized, which greatly limits our freedom.

In order to solve these problems well, Wobbrock et al. proposed the identification method of "without Libraries or Toolkits or Training $1 [1], and others proposed the "Protractor" recognition method and the identification method of $P [2] proposed by Vatavu et al. Later, Kratz et al. simply extended the $1 gesture. In IUI 2010, the $3 [3] gesture recognition algorithm was proposed, which mainly focused on the flat gesture recognition in space, but the user action compatibility is still low, but the action is not qualified. Then the recognition success rate will drop significantly.

This paper proposes an optimization algorithm based on $3, which maps three-dimensional data to three different planes and transforms it into three two-dimensional image data processing methods. The two-dimensional plane is divided according to the circumferential vector to store the database schema.

Content from this work may be used under the terms of the Creative Commons Attribution 3.0 licence. Any further distribution of this work must maintain attribution to the author(s) and the title of the work, journal citation and DOI.
Published under licence by IOP Publishing Ltd

2. Data Collecting and Processing System

This article uses the action information of the Android mobile phone with its own sensor to collect acceleration. The phone comes with a three-axis accelerometer that measures the acceleration in the X, Y, and Z directions. The sensor collects the acceleration state of the current time at any time and converts it into a digital signal through D/A. The three-axis acceleration signal and time signal are acquired by Android Matlab, and transmitted to the PC through Wi-Fi for processing.

The signal of the sensor collected by the mobile phone sensor must be denoised and the active component removed, and then normalized before it can be used. Taking a circular motion as an example, the original data collected is shown in Figure 1. In this paper, the median filter is used to smooth the original signal. The median filter is a non-linear digital filter that uses a movable window of odd-numbered lengths to use the value of the window as a reference to filter out large differences in noise, while others remain. Suppose that the number of sliding window points is N, N data is taken from the input data, and the data is sorted from small to large, and the intermediate value of the filter is taken as the output of the filter. The operation can be expressed as:

$$X = \text{Sort}\{x_1, x_2, \cdots, x_{N-1}, x_N\} = \{x_{u-r}, \cdots, x_u, \cdots, x_{u+r}\} \tag{1}$$

$$X_N = Median\{x_{u-r}, \cdots, x_u, \cdots, x_{u+r}\}_N \qquad (\text{ } u = \frac{N-1}{2} \text{ }) \tag{2}$$

In the formula, Sort represents the sort operation, x_u is the median value of the window, r is the radius of the value, and Median is the operation of taking the median. The result of the processing is shown in Figure 2.

Due to the time error of the test, each piece of information collected will contain invalid information at both ends, causing judgment and interference on the processing of the data, and judging and removing them. The specified starting time condition is: in T_s ($T_s = 100ms$), the absolute value of the maximum value and the minimum value of the acceleration change exceeds the threshold A_S ($A_S = 1m/s^2$) to determine the start of the action; in T_f ($T_f = 400ms$) In the time period, the absolute value of the maximum value and the minimum value of the acceleration change are smaller than the threshold A_f ($A_f = 0.5m/s^2$), and the determination operation ends. The processed data is shown in Figure 3.

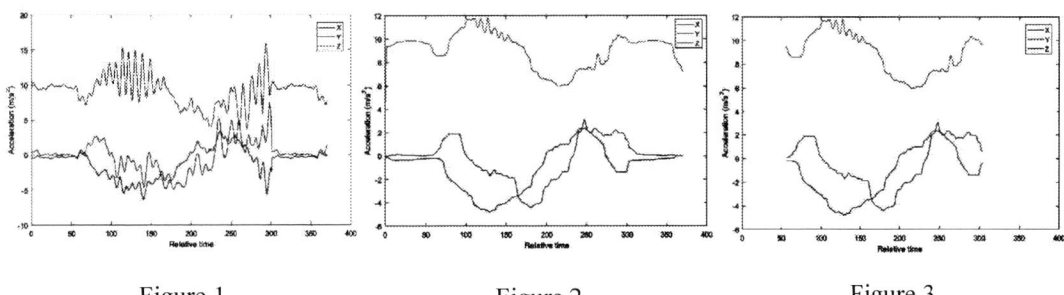

Figure 1   Figure 2   Figure 3

Since the intensity and speed of the action are different for different people, in order to make the final result suitable for most people, it is universal and needs to be standardized. Z-score processing is used here. Obtain the mean and standard deviation of the data separately, and then perform the following operations:

$$\mu = \frac{1}{n}\sum_{i=1}^{n} x_i \tag{3}$$

$$\sigma^2 = \frac{1}{n} \sum_{i=1}^{n} x_i^2 - \mu^2 \tag{4}$$

$$x_i^* = \frac{x_i - \mu}{\sigma}, i = 1,2,3, \cdots, n-1, n \tag{5}$$

μ represents the mean, σ^2 represents the standard deviation, x_i^* represents the value of the acceleration after normalization, and n is the sample size [4].

3. Extraction and Matching of Data Feature Values

3.1. Coordinate projection

Research surface, because each person has great randomness when performing gestures, and does not necessarily move in a plane, it will cause huge errors and difficulty for the plane trying to fit the motion. A lot of valid information will be lost. So I switched to another idea. All three-dimensional motion can be decomposed into two two-dimensional motions that are perpendicular to each other. The measured value of the acceleration sensor includes three vertical directions of acceleration, and each of the two directions constitutes a projection plane. As shown in Figure 4, the storage calculation is performed separately, which can reduce the computational complexity. And each action gesture contains three templates to match, which theoretically reduces the error rate of judgment and improves the accuracy [5].

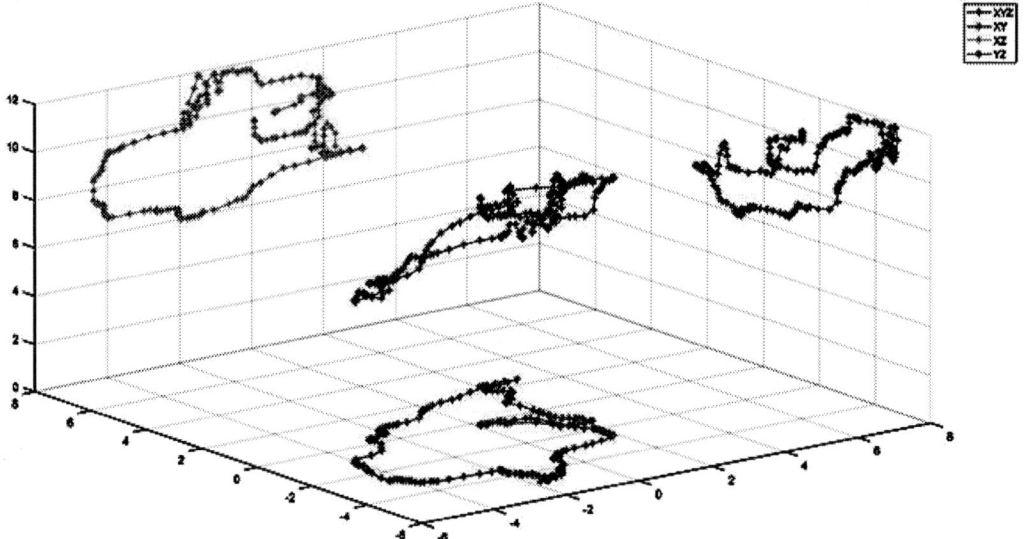

Figure 4

The above preprocessed data is arranged in a matrix. Assuming that a certain gesture is repeated p times, the data collected by the Android mobile phone p times is Q after preprocessing, which can be expressed as:

$$Q_j = \begin{pmatrix} x_{j1}^* & y_{j1}^* & z_{j1}^* \\ x_{j2}^* & y_{j2}^* & z_{j2}^* \\ x_{j3}^* & y_{j3}^* & z_{j3}^* \\ \vdots & \vdots & \vdots \\ x_{jn}^* & y_{jn}^* & z_{jn}^* \end{pmatrix} \quad (j = 1,2,3,\cdots,p) \tag{6}$$

Q is an n*3 matrix, and x_n^*, y_n^*, and z_n^* represent acceleration data in three directions, respectively. Now map to three vertical matrix faces, which can be expressed as:

$$XY_j = \begin{pmatrix} x_{j1}^* & y_{j1}^* \\ x_{j2}^* & y_{j2}^* \\ x_{j3}^* & y_{j3}^* \\ \vdots & \vdots \\ x_{jn}^* & y_{jn}^* \end{pmatrix}, XZ_j = \begin{pmatrix} x_{j1}^* & z_{j1}^* \\ x_{j2}^* & z_{j2}^* \\ x_{j3}^* & z_{j3}^* \\ \vdots & \vdots \\ x_{jn}^* & z_{jn}^* \end{pmatrix}, YZ_j = \begin{pmatrix} y_{j1}^* & z_{j1}^* \\ y_{j2}^* & z_{j2}^* \\ y_{j3}^* & z_{j3}^* \\ \vdots & \vdots \\ y_{jn}^* & z_{jn}^* \end{pmatrix} \tag{7}$$

XY_j, XZ_j, and YZ_j is three mutually perpendicular matrix planes, all of which are n*2.

3.2. Translational scaling of the image

For the sake of simplicity, the algorithm moves the projection center $(\bar{x}, \bar{y})$ of each plane to the origin (0, 0). All other points remain in the same position relative to the origin. The projection center takes the average of the axis data, taking the XY_j plane as an example:

$$\bar{x}_j = \frac{1}{n}\sum_{i=1}^{n} x_{ji} \qquad \bar{y}_j = \frac{1}{n}\sum_{i=1}^{n} y_{ji} \tag{8}$$

Change the plane projection center to $(\bar{x}_j, \bar{y}_j)$.

The distance from the farthest data point from the center of the projection is used as a radius to make a circular area, and the modified plane projection is placed into the circular area. According to the span of the center of the different points, the plane projection is correspondingly scaled.

3.3. Image segmentation and feature acquisition

For existing algorithms, such as the "$1" recognition method and the "Protractor" recognition method, the samples are equally divided. Although these methods have a high degree of recognition for standard actions, they are difficult to recognize for irregular movements due to human arbitrariness. When we perform gestures, the motion trajectory often becomes a closed graph of the curve, so this paper uses a combination of angle and amplitude to obtain the motion trajectory. We divide the circular area into twelve areas according to the equal angle, and then divide the four areas according to the equal radius spacing [6][7][8]. The circular area is divided into 48 areas and named sequentially, as shown in Figure 5.

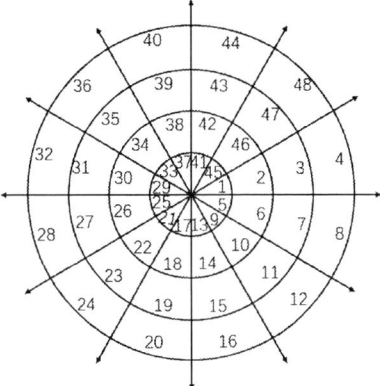

Figure 5

The number of data points in each area is counted, and the probability that the motion trajectory passes through the area is obtained, and finally the trajectory area of the motion can be obtained. Taking the clockwise circular motion as an example, the motion trajectories of the following three projection surfaces are obtained, as shown in Figure 6,7,8.

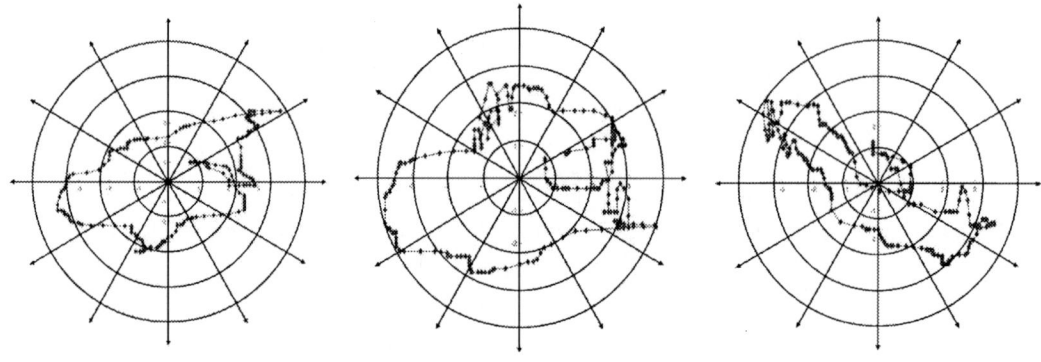

| Figure 6 | Figure 7 | Figure 8 |

In this way, a probability distribution matrix of the track area can be obtained for each projection surface. If assumed that a certain kind of the gesture is repeated p times, change of the p-th selection gesture trajectory region within the same circular area, the number of the selected region is assumed u-q, the probabilities at the region of the track The density is $\varphi_u = q/p$. Finally, the region selection probability matrix of the three projection surfaces of the modified gesture is obtained:

$$\varphi = \begin{pmatrix} \varphi_{xy1} & \varphi_{xy2} & \varphi_{xy3} & \cdots & \varphi_{xy48} \\ \varphi_{xz1} & \varphi_{xz2} & \varphi_{xz3} & \cdots & \varphi_{xz48} \\ \varphi_{yz1} & \varphi_{yz2} & \varphi_{yz3} & \cdots & \varphi_{yz48} \end{pmatrix} \tag{9}$$

3.4. Identification and matching

For the input track to be detected, the area that passes through each projection surface represents 1 and the area that has not passed represents 0, and the matrix γ is established, which is expressed as follows:

$$\gamma = \begin{pmatrix} \gamma_{xy1} & \gamma_{xy2} & \gamma_{xy3} & \cdots & \gamma_{xy48} \\ \gamma_{xz1} & \gamma_{xz2} & \gamma_{xz3} & \cdots & \gamma_{xz48} \\ \gamma_{yz1} & \gamma_{yz2} & \gamma_{yz3} & \cdots & \gamma_{yz48} \end{pmatrix} \quad \gamma_i = \begin{cases} 0 & \text{without i area} \\ 1 & \text{after the i area} \end{cases} \quad (i = 1,2 \cdots, 48) \tag{10}$$

Finally, a weighted summation is performed with each template to obtain a matching degree C between the trajectory and each template:

$$C' = \gamma * \varphi^T = \begin{pmatrix} \gamma\varphi_{11} & \gamma\varphi_{12} & \gamma\varphi_{13} \\ \gamma\varphi_{21} & \gamma\varphi_{22} & \gamma\varphi_{23} \\ \gamma\varphi_{31} & \gamma\varphi_{32} & \gamma\varphi_{33} \end{pmatrix} \tag{11}$$

$$C = \begin{pmatrix} \gamma\varphi_{11} \\ \gamma\varphi_{22} \\ \gamma\varphi_{33} \end{pmatrix} \tag{12}$$

Thereby I select a template with the highest matching degree to the trajectory.

4. Experiment Evaluation

4.1. Establishment of database

In order to verify the gesture recognition method proposed in this paper, the Android Matlab software is installed on the Android platform, and the data transmission is performed with the Matlab on the PC side. The tester's right hand is holding the phone to complete the specified action. The experimenter completed five gestures including clockwise circle, N, W, clockwise square and R in different environments and time, as shown in Table 1.

Table 1 Gesture type.

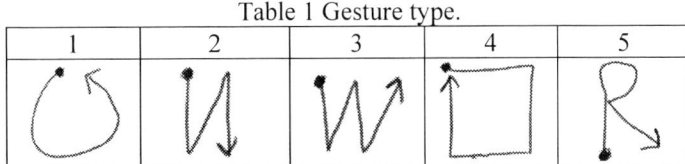

1	2	3	4	5

The black point is the starting point, the arrow position is the end point, and the direction of the arrow is the direction of the path movement. All actions are completed without interruption. Each exercise time is between 3s and 5s. There are 5 experimenters, and each action is repeated 30 times. A total of 750 sets of data are collected and 5 sample templates are created.

4.2. Gesture recognition

Under the premise of not limiting the tester's action normativeness, let 5 testers each perform 10 random and different action inputs, match through Matlab, and simulate the experiment using the "$3" recognition algorithm and "Protractor" identification. The algorithm and the recognition rate of the improved algorithm in this paper, the results are shown in Table 2.

Table 2 Comparison of experimental results.

Gesture number	$3/%	Protractor/%	Algorithm of this paper /%
1	86.4	87.6	95.4
2	88.9	87.2	89.2
3	88.1	89.3	90.4
4	87.9	88.7	92.3
5	89.5	87.4	91.6

The experimental results show that the recognition rate of the "$3" recognition algorithm and the "Protractor" recognition algorithm are relatively low when the tester's action specification is degraded without intentional constraint, and the average recognition rate is 88.16% and 88.04%, respectively. The recognition rate of the algorithm reaches 91.78%, which is higher than the recognition rate of the previous two algorithms. After many tests and observations, it is found that for the closed multi-curve trajectory, the algorithm still has a great degree of recognition under the condition of non-standard behavior, and achieves the expected goal.

5. Conclusion

Aiming at the large uncertainty and unreliable factors of gesture motion, this paper designs and implements a three-dimensional eigenvalue extraction and recognition scheme for gesture acceleration. The scheme is based on the "$3" optimization algorithm, which maps the three-dimensional acceleration data to three mutually perpendicular planes, and uses the angle and radius to partition the trajectories that are reduced to the circular area to construct a sample template. The experimental data is completed by people in a natural state. For different people, there is a different understanding of the arbitrariness of movement, so that each person's movements have a gap, and there is a certain randomness. The experimental results show that under this condition, the design still has a high recognition degree for gestures, which improves the reliability of the system and liberates the user's freedom.

References

[1] Wobbrock, J. O.,Wilson, A. D.,Li, Y.(2007) Gestures without Libraries, Toolkits or Training: A $1 Recognizer for User Interface Prototypes. Assoc Computing Machinery, New York

[2] Vatavu, R. D.,Anthony, L.,Wobbrock, J. O.(2012) Gestures as Point Clouds: A $P Recognizer for User Interface Prototypes. Assoc Computing Machinery, New York

[3]   Popa,M. (2011) Hand gesture recognition based on accelerometer sensors. In: The 7th International Conference on Networked Computing and Advanced Information Management. Spain. pp. 115-120.

[4]   Lichtenauer,J. F., Hendriks, E. A., Reinders, M. J. T.(2008) Sign Language Recognition by Combining Statistical DTW and Independent Classification. IEEE Transactions on Pattern Analysis and Machine Intelligence, 30:2040-2046

[5]   Kratz, S., Rohs, M.(2011) Protractor3D: A closed-form solution to rotation-invariant 3D gestures. In: 15th ACM International Conference on Intelligent User Interfaces. Los Angeles. pp.371-374.

[6]   Li, Y. (2010) Protractor: A Fast and Accurate Gesture Recognizer. Assoc Computing Machinery, New York

[7]   Wang, X. L.,Xie, K. L.(2004) A novel direction chain code-based image retrieval. IEEE Computer Soc. In: The Fourth International Conference on Computer and Information Technology. Wuhan.pp.190-193

[8]   Willems, D.,Niels, R.,van Gerven, M.,Vuurpijl, L.(2009) Iconic and multi-stroke gesture recognition. Pattern Recognition,12: 3303-3312.

Model and Design of High Temperature and Thermal-proof Garment Using Genetic Algorithm

Zhangmaoyi, Sunlele, Jiahaolin.

School of Information and Engineering, China Mining University, Xuzhou 221004

✉Corresponding author, E-mail: kfzqj@126.com

ABSTRACT For question one, We adopt the thought of the "environment-fabric-human body" system; For the second question, we consider from the reverse angle; For question three, we also consider the problem from the perspective of human burns based on the idea of question two. Through the PDE toolbox in MATLAB, a three-dimensional distribution image of the temperature of each layer and air layer over time and thickness of each layer is drawn, and a dual-target optimization model is established by using genetic algorithm to search for the optimal thickness that satisfies the conditions. We use MATLAB to obtain the thickness results of Layer II and Layer IV When the comfort is greatest. Taking into account the body type characteristics of the subject, heat conduction models are established according to different major parts, so as to obtain the optimal thickness of a certain layer of different parts, a greater degree of material saving, and an increase in the number of workers' comfort.

1. Introduction

When working in a warmer environment, people need to wear professional clothes to avoid Burns. Especially in fire welding, steelmaking and other operations, it is very important to wear thermal protective clothing. This article takes the example of square hot clothing made of third-layer fabric material as an example, and takes human safety and comfort as a starting point to reduce R&D costs. A temperature distribution model and a temperature thickness model are established to design suitable clothing for high temperature operations.

For the first question, in the "environmental-fabric-human body" system, the ambient temperature and the initial human body temperature are constant at 75 ° C and 37 ° C, respectively, through the literature[2] It can be seen that the temperature on the outside of the human skin is constant at 48.08 ° C. Assuming that the heat transfer process from the external environment to the human epidermis is carried out in a direction perpendicular to the skin and fabric, the problem can be spatially.

As a one-dimensional model, according to which we can combine one-dimensional unsteady heat conduction models, list the initial conditions and boundary conditions, and then use the finite difference method to obtain the analytical solution of the partial differential equation. The distribution function T(X, T) of temperature T on time T and space X is obtained. Through the PDE toolbox in MATLAB, three-dimensional distribution images of fabric layers and air layer temperature over time and thickness are drawn.

For the second question, we consider from the reverse perspective that given an external ambient temperature of 65 ° C and a four-layer thickness of 5.5 mm, when the dummy test time reaches 60 minutes, the dummy skin external test temperature is just over 44 ° C 5 min, but not More than 47 °

Content from this work may be used under the terms of the Creative Commons Attribution 3.0 licence. Any further distribution of this work must maintain attribution to the author(s) and the title of the work, journal citation and DOI.
Published under licence by IOP Publishing Ltd

C. The optimal thickness of the II layer is L _ min. Based on the heat conduction model established by the problem, we use genetic algorithm to search for the optimal thickness of 12.02 mm.

In response to question 3, we also consider from the perspective of human Burns, that is, when the external ambient temperature is 80 ° C, the outer temperature of the human skin exceeds 44 ° C, and just five minutes, the optimal thickness of the second layer and the fourth layer is obtained., The difference is that in this question, we consider the constraints of human comfort, that is, under the dual constraints of safety and comfort, we solve the optimal solution of two layers of thickness. We consider that the thicker the fabric material, the greater the weight. Even if the suit has a good thermal insulation effect, the weight of the general assembly affects the comfort of the employee and thus affects the efficiency of his or her operations. We define the comfort function $\vartheta = \frac{\vartheta_2 + \vartheta_4}{\sum_{i=1}^{4} \vartheta_i}$, the thickness of Layer II and Layer IV when maximum comfort is achieved under safe conditions, That is, a dual-objective optimization model is established, and the results are 7.4 mm and 6.1 mm using MATLAB, respectively. [1]

2. Restatement of issues

When working in a high temperature environment, people need to wear special clothing to avoid Burns. Dedicated clothing is usually composed of three layers of fabric materials, denoted as layers I, II, and III, where layer I is in contact with the external environment. There is also a gap between layer III and the skin. This gap is recorded as layer IV.

In order to design special clothing, a dummy with a body temperature of 37° C is placed in a laboratory temperature environment and the temperature outside the skin of the dummy is measured. In order to reduce R&D costs and shorten the R&D cycle, please use mathematical models to determine the temperature changes outside the skin of the dummy and solve the following problems:

2.1 Certain parameter values for special clothing materials are derived from[1] Given that the ambient temperature is 75° C, the thickness of the II layer is 6mm, the thickness of the IV layer is 5mm, and the working time is 90 minutes, the temperature outside the skin of the dummy is measured[2] And ... Establish a mathematical model, calculate the temperature distribution, and generate an Excel file with a temperature distribution(file name is problem 1. xlsx).

2.1.1 When the ambient temperature is 65° C and the thickness of layer IV is 5.5 mm, the optimal thickness of layer II is determined to ensure that the outer temperature of the human skin does not exceed 47° C for 60 minutes and does not exceed 44° C for more than 5 minutes.

2.1.2 When the ambient temperature is 80° C, the optimal thickness of Layer II and Layer IV is determined to ensure that at 30 minutes of work, the outer temperature of the human skin does not exceed 47° C and the time exceeding 44° C does not exceed 5 minutes.

2.2 Problem analysis

2.2.1 Analysis of question one

The ambient temperature and the initial temperature of the human body are constant at 75 ° C and 37 ° C, respectively, in the "environmental-textile-human body" system. According to Annex II, the external temperature of the human skin is constant at 48.08 ° C. Assuming that the heat transfer process from the external environment to the human epidermis is carried out in a direction perpendicular to the skin and fabric, the problem can be regarded as a one-dimensional model in space, according to which we can combine one-dimensional unsteady heat conduction models. The initial conditions and boundary conditions are listed, and the differential equation is obtained by the finite difference method. The distribution function T(X, T) of temperature T on time T and space X is obtained.

2.2.2Analysis of question two
From the reverse perspective, given that the external ambient temperature is 65 ° C and the four-layer thickness is 5.5 mm, when the dummy test time reaches 60 minutes, the dummy skin external test temperature is just over 44 ° C 5 min, but not more than 47 ° C. The optimal thickness of the II layer is L _ min.

2.2.3Analysis of question three
We are also based on the idea of question 2, from the perspective of human Burns, that is, when the external ambient temperature is 80 ° C, the outer temperature of the human skin exceeds 44 ° C, and just over five minutes, the optimal thickness of the second layer and the fourth layer is obtained. The difference is that, In this question, we consider the constraints of human comfort, that is, under the dual constraints of safety and comfort, to solve the optimal solution of two layers of thickness. We consider that the thicker the fabric material, the greater its weight, even if the suit has a good thermal insulation effect. However, the weight of the conference affects the comfort of the employees, which in turn affects their operational efficiency. We define the comfort function to solve the thickness of the second layer and the fourth layer When the safety conditions are met and the maximum comfort is achieved,

3. Model assumptions
Assume that the temperature in the "environmental-textile-skin" system is carried out perpendicular to the surface of the skin;
It is assumed that only heat conduction and heat radiation exist in the system, and heat convection is not considered.
Assume that each layer of fabric is isotropic;
It is assumed that the temperature between layers of fabric, between fabric and air, and between air and skin is continuous;
It is assumed that the temperature of fabric and human skin is 37 degrees.
Assume that the properties of density, specific heat and heat conductivity of fabric materials during heat transfer are unchanged;
It is assumed that the temperature distribution is uniform in the texture layer at a certain time and thickness;

4. symbol description
Symbols and Symbolic Meaning see Table 1

Table 1 Symbols and Symbolic Meaning

Symbol	symbol meaning
T	temperature
t	Time
c_i	specific heat capacity
ρ_i	density
a_i	Temperature conductivity
x	Vertical thickness
ϑ_k	Inner Layer Comfort

ϑ_j	Outer comfort
δ, λ	Thermal conductivity

5. Establishment and solution of 4 models

5.1 Construction and Solution of Problem-One Model

5.1.1 Problem analysis
In the "environmental-textile-human system", the ambient temperature and the initial temperature of the human body are constant at 75 ° C and 37 ° C, respectively, and the heat of the external environment is transmitted through the fabric to the air layer, which is a typical heat conduction model. However, there are two types of heat conduction models, steady and unsteady, according to the literature[2] Given the temperature time data, we draw a curve of temperature over time and find that the temperature gradually rises to a constant value over time. Therefore, we judge that this problem is more suitable for non-steady state models. Assuming that heat transfer only considers heat conduction and heat radiation in this problem, that is, heat transfer in the form of heat convection is not considered, and the heat transfer process is perpendicular to the skin, we decided to establish a one-dimensional unsteady state heat transfer model.

5.1.2 Modeling

5.1.2.1 Differential equations and deterministic conditions
For thermal protective clothing, the thickness of each layer of fabric is relatively thin, and the length and width of the area perpendicular to the skin's epidermis are much larger than the thickness of each layer of fabric. There are the following one-dimensional unsteady thermal conduction differential equations:

$$\lambda \frac{\partial^2 T}{\partial^2 x} = \rho C_p \frac{\partial T}{\partial \tau} \tag{1}$$

Or described as

$$\frac{\partial^2 T}{\partial^2 x} = \frac{1}{a} \frac{\partial T}{\partial \tau} \tag{2}$$

The initial condition is:

$$\tau = 0, 0 < x < L (L = \sum_{i=1}^{4} l_i) \tag{3}$$

$$t_1 = t_w = 75, t_2 = t_3 = t_4 = t_n = 37 \tag{4}$$

The initial temperature of the external environment and the human epidermis were respectively. The boundary conditions are:

$$
\begin{aligned}
\tau = 0, x = 0, \frac{\partial T}{\partial x} = 0 \\
\tau > 0, x = l_1, t_1 = t_w \\
\tau > 0, x = l_4, t_4 = t_n
\end{aligned}
\tag{5}
$$

5.1.2.2 Dispersion of the calculated area
In this question, according to the literature[8] we only consider heat transfer along the surface of the human skin, of course, this assumption will cause some error. In order to simplify the model, we establish a one-dimensional unsteady heat conduction model, but "one-dimensional" only means that the heat transfer in space is one-dimensional. Combined with Annex II, it can be seen that the temperature changes over time and eventually tends to a steady state value, which means that the temperature also changes over time. Therefore, in fact, the model we set up solves the problem of "two-

dimensional" and ultimately requires the distribution function T(X, T). However, it should be noted that the time coordinates are one-way, that is, the time progress is not regressive, and the results of the previous moment will have an impact on the results of the latter moment; But the later results will not affect the previous one. The discrete calculation area with X and T as coordinates begins with T = 0 and increases to the J time layer and J +1 layer through one time layer.

5.1.2.3 Discrete of differential equations
For any I node, the differential equation can be written as the following formula at J and J +1:

$$\left(\frac{\partial T}{\partial \tau}\right)_i^J = a\left(\frac{\partial^2 T}{\partial x^2}\right)_i^J$$
$$\left(\frac{\partial T}{\partial \tau}\right)_i^{J+1} = a\left(\frac{\partial^2 T}{\partial x^2}\right)_i^{J+1}$$

(6)

The difference between the left end temperature of the above formula and the partial derivative of time is obtained:

$$\left(\frac{\partial T}{\partial \tau}\right)_i^J = \frac{T_i^{J+1} - T_i^J}{\Delta \tau}$$
$$\left(\frac{\partial T}{\partial \tau}\right)_i^{J+1} = \frac{T_i^{J+1} - T_i^J}{\Delta \tau}$$

(7)

Derivatives of the front end term relative to the I point at time

$$\left(\frac{\partial T}{\partial \tau}\right)_i^J$$

(8)

It's forward differential. We observe that although these two formulas correspond to the right-end difference formula, they have different meanings. In contrast, the right end is the derivative of point I corresponding to the moment's The backward difference.

$$\left(\frac{\partial T}{\partial \tau}\right)_i^{J+1}$$

(9)

By replacing the second-order derivative about the right end of the above equation with the corresponding difference, the following two different difference formats, explicit and implicit, can be obtained:

Explicit:

$$T_i^{J+1} = fT_{i+1}^J + (1-2f)T_i^J + fT_{i-1}^J \quad (J = 0,1,\cdots, i = 2,3,\cdots, N-1)$$

(10)

All implicit:

$$T_i^{J+1} = \frac{1}{1+2f}\left(fT_{i+1}^{J+1} + fT_{i-1}^{J+1} + T_i^J\right) \quad (J = 0,1,\cdots, i = 2,3,\cdots, N-1)$$

(11)

In the above two formulas

$$f = \frac{a\Delta\tau}{\Delta x^2}$$

(12)

It can be clearly seen from the formula that the right end only involves the temperature of the J moment. We calculate from the initial moment(J = 0, T = 0). From the initial conditions listed above, the temperature at the next moment can be obtained, that is, the temperature of the fabric layer and the air layer at J = 1 moment; The temperature obtained by J = 1 is then explicitly calculated to obtain the temperature of each layer at J = 2. By analogy, the temperature values of each moment and each fabric layer and air layer can be directly obtained by explicit formula.

For the following full implicit right end contains the temperature of different nodes at the same time as the left end of the equal sign, so the temperature values of these nodes must be obtained by solving the algebraic equations.

5.1.2.4 Dispersion of boundary conditions
It can be clearly seen from the formula that the right end only involves the temperature of the J moment. We calculate from the initial moment(J = 0, T = 0). From the initial conditions listed above, the temperature at the next moment can be obtained, that is, the temperature of the fabric layer and the air

layer at J = 1 moment; The temperature obtained by J = 1 is then explicitly calculated to obtain the temperature of each layer at J = 2. By analogy, the temperature values of each moment and each fabric layer and air layer can be directly obtained by explicit formula.

For the following full implicit right end contains the temperature of different nodes at the same time as the left end of the equal sign, so the temperature values of these nodes must be obtained by solving the algebraic equations.

$$T_1^{J+1} = T_2^{J+1} \tag{13}$$

$$T_N^{J+1} = \frac{1}{\frac{h\Delta x}{\lambda} + 1}\left(T_{N-1}^{J+1} + \frac{h\Delta x}{\lambda}T_\infty\right) \tag{14}$$

5.1.2.5 Final discrete format
Explicit:

$$T_i = T_0 \ (i = 1,2,\cdots,N) \tag{15}$$

$$T_i^{J+1} = \left(fT_{i+1}^J + fT_{i-1}^J + (1-2f)T_i^J\right) \qquad (i = 2,3,\cdots,N-1) \tag{16}$$

$$T_1^{J+1} = T_2^{J+1} \tag{17}$$

$$T_N^{J+1} = \frac{1}{\frac{h\Delta x}{\lambda} + 1}\left(T_{N-1}^{J+1} + \frac{h\Delta x}{\lambda}T_\infty\right) \tag{18}$$

Among them $J = 0,1,2\cdots$. When the second-order precision element balance method is used for discrete, the corresponding discrete format is:

$$\begin{aligned}
T_1^{J+1} &= T_1^J(1-2f) + 2fT_2^J \\
T_N^{J+1} &= (1-2f-2fBi)T_N^J + 2fT_{N-1}^J + 2fBiT_\infty
\end{aligned} \tag{19}$$

inside $Bi = \frac{h\Delta x}{\lambda}$, Bi For the Biwa number。

Implicit:

$$T_i^0 = T_0 \quad \text{(initial value)} \tag{20}$$

$$T_1^{J+1} = T_2^{J+1} \tag{21}$$

$$T_i^{J+1} = \frac{1}{1+2f}\left(fT_{i+1}^{J+1} + fT_{i-1}^{J+1} + T_i^J\right) \tag{22}$$

$$T_N^{J+1} = \frac{1}{\frac{h\Delta x}{\lambda} + 1}\left(T_{N-1}^{K+1} + \frac{h\Delta x}{\lambda}T_\infty\right) \tag{23}$$

inside $J = 0,1,2\cdots$ When calculating with implicit difference, each time layer needs to iterate to solve the algebraic equations. In each time layer calculation, a temperature field(usually the temperature field of the upper time layer is the initial field of the current time layer) must be assumed first, and then iterative calculation until convergence.

5.1.3 Solution of the model
According to the above method, we find that the final temperature analysis of time and thickness is:

$$T(x,t) = (T_w - T_\infty)\sum_{n=1}^{\infty}\frac{2\sin\zeta_n}{\zeta_n + \sin\zeta_n\cos\zeta_n}\exp(-a^2t)\cos(\zeta_n x) \tag{24}$$

inside ζ_n For the equation $\operatorname{ctg}\zeta = \zeta/B_i$ Root., For each layer of fabric, due to the different material properties, the temperature analysis of time and thickness of the fabric is different, but based on the assumption that each layer of fabric material is isotropic, So it is continuous for the temperature distribution of the entire fabric material and the air layer.

We use MATLAB PDE tool box to draw three dimensional maps of temperature distribution of each layer and three dimensional maps of temperature distribution from thermal protective clothing to human epidermis.

The numerical solution of the four-layer temperature distribution is shown in Figure 1：

By observing the above three-dimensional diagram of the temperature distribution of time and layers of fabric and air, we can get the following conclusions:

From the perspective of time alone, the temperature of the II layer rose fastest in a certain period of time, and the temperature of the IV layer increased by about 10 degrees in 90 minutes.

The temperature of each layer of thermal protective clothing and the temperature of the air layer will eventually rise to a steady state value, with I layer 75 degrees, II layer 63.48 degrees, III layer 52.82 degrees, and IV final 48.08 degrees.

1) The temperature of each layer of fabric material or air layer is transmitted gradually. In this transfer, the temperature is followed by continuous transmission, and there is no unreasonable interruption point in the middle.

2) The temperature of each layer will change over time, further indicating the accuracy of the unsteady state heat conduction model

Based on the above conclusions, we give the following explanation:

According to the physical properties of each layer and air layer of the fabric given in document 1, it is easy to know that the density, specific heat, and heat conductivity of the second layer fabric material are the maximum values in the fourth layer, and it is not difficult to know that the second layer is within a certain period of time., Not only the heat conductivity is the largest, but also the temperature difference of the ascending amount is the largest; On the other hand, the fourth layer, that is, the air layer, is the smallest in terms of both the conduction rate and the rising temperature difference. And in the unsteady state heat conduction model, the temperature eventually rises to a stable value. Numerical solution of temperature distribution in four layers see Fig. 1

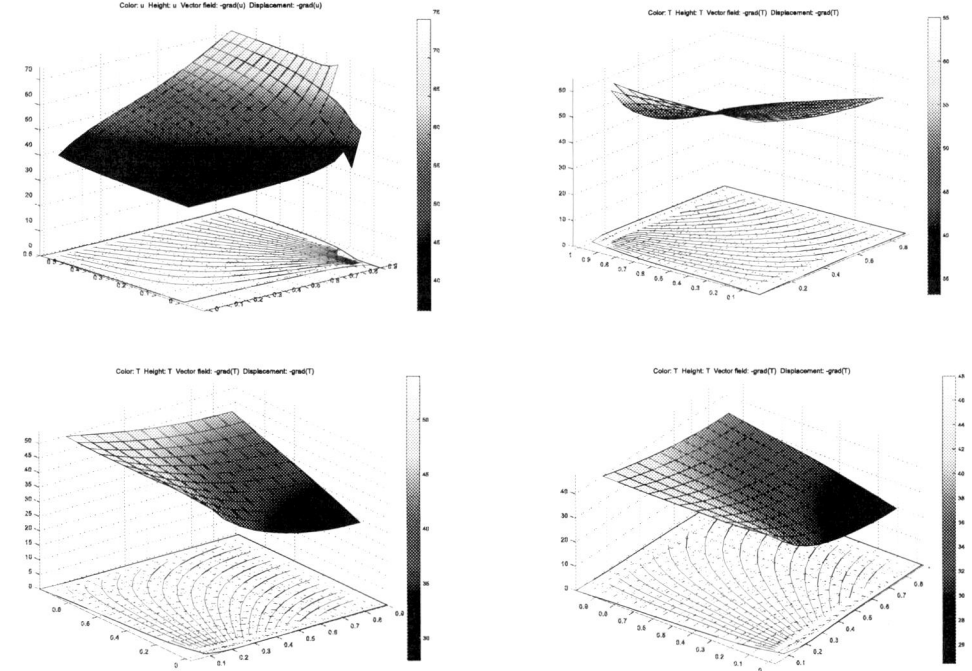

Fig. 1 Numerical solution of temperature distribution in four layers

The following gives the distribution of temperature at certain times: Results of partial temperature distribution see Table 2

Table 2 Results of partial temperature distribution

Time(s)	temperature(° C)	IV lateral	III lateral	II lateral	I lateral

110	39.73	40.09	39.77	42.60	48.00
111	39.77	40.12	39.80	42.70	48.10
1369	48.05	48.29	52.79	63.36	74.93
1370	48.05	48.29	52.79	63.36	74.93
1371	48.05	48.29	52.79	63.36	74.93
2477	48.08	48.29	52.82	63.48	75.00
2478	48.08	48.29	52.82	63.48	75.00
2479	48.08	48.29	52.82	63.48	75.00
3724	48.08	48.29	52.82	63.48	75.00
3725	48.08	48.29	52.82	63.48	75.00
3726	48.08	48.29	52.82	63.48	75.00
3727	48.08	48.29	52.82	63.48	75.00
3728	48.08	48.29	52.82	63.48	75.00

5.2 Establishment and Solution of the Second Problem Model

Based on the analysis and solution results of the issue-one model, it is easy to find that the second problem is actually the inverse process of solving the problem. The first problem is the known thickness of each layer and the ambient temperature conditions. To solve the dynamic distribution of temperature over time at each layer, the second problem is to find the minimum thickness of Layer II given the external temperature and the temperature constraints of the dummy surface. We can use the same method of question one, assuming that the thickness of layer II, according to the Fourier heat conduction principle, lists the general form of the problem two-heat conduction, so as to obtain partial differential equations of the "environmental-textile-human" system, and then increase the boundary constraints. The analytical solution or numerical solution of the differential equation is obtained, and the corresponding critical thickness is found, that is, the thickness of the second layer of the heat service.

5.2.1 Preparation of the model

From the derivation of question one, we can conclude that when the temperature is conducted from the outside to the inside of the human body, the temperature distribution is stepped up. The larger the value of a of the material, the smaller the temperature difference in the cover layer, and the corresponding thermal insulation performance. The worse, We can use this condition as a known premise to constrain the next model, and play a certain guiding role in the solution of the model, establish a one-dimensional coordinate system perpendicular to the inwards of the skin, and the outer side of the I layer is the unit 0 point, The temperature distribution can be expressed as.

$$\begin{cases} \Delta T_1 + \Delta T_2 + \Delta T_3 + \Delta T_4 = T' - T\big|_{x=7.9+x_0} \\ \Delta T_i \propto \dfrac{1}{a_i} = \dfrac{\rho_i c_i}{\lambda_i}, i = 1, 2, 3, 4. \end{cases} \tag{25}$$

In the above equation, T' For external temperature, it is a constant value, a_i is the conductivity of layer I, and the larger the conductivity, the more likely the object is to be in the heat conduction process.

5.2.2 Modeling

Assuming that the thickness of Layer II is x_2, similarly, a partial differential equation is established using the one-dimensional unsteady state heat conduction principle as follows:

$$\frac{\partial T}{\partial t} = a \frac{\partial^2 T}{\partial x^2} + f(x), (0 \le x < 9.7 + x_2) \tag{26}$$

$$\begin{cases} 0 < x < 0.6, a = a_1; \\ 0.6 \le x < 0.6 + x_2, a = a_2; \\ 0.6 + x_2 \le x < 4.2 + x_2, a = a_3; \\ 4.2 + x_2 \le x < 9.7 + x_2. \end{cases} \qquad (27)$$

For this problem, for the four-layer material of the anti-heat suit, there is no heat source in this temperature field, so F(X) = 0. Give the boundary condition:

$$\lambda_1 \frac{\partial T}{\partial x}\bigg|_{x=0} = a_1 (65 - T\big|_{x=0}) \qquad (28)$$

$$T(9.7 + x_0, t) = g(t), 0 < t < 60; \qquad (29)$$

According to the unsteady heat conduction problem with internal heat source, the steady state heat conduction problem with homogeneous boundary conditions and an unsteady homogeneous problem can be decomposed into heat source. The solution of the original equation is:

$$T(x,t) = T_1(x,t) + T_2(x) \qquad (30)$$

5.2.3 Solution of the model

For this problem, we first use the PDE toolbox in MATLAB software to find the analytical solution of the partial differential equation. On the basis of the analytical solution, we use the improved genetic algorithm to recycle and make several iterations to obtain the ideal fabric. Layer II thickness, The specific process is as follows.

Step 1 Determines the region of this partial differential equation

The thickness is from 0 to(9.7 + x_2) and the time is from 0 seconds to 5,400 seconds. Due to the finiteness of the visual graphics in the PDE, we reduce the time to the corresponding scale, such as using 0-1 to represent the change in time. After that, type the boundary value condition and divide the area into the mesh as shown in the figure below to facilitate the solution of the subsequent problem and the drawing of the figure.

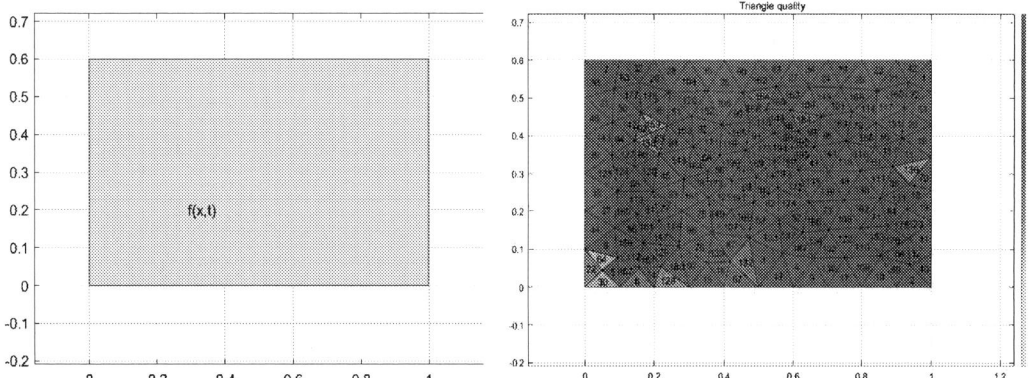

Figure 2 The partial subregions   Figure 3 The breakdown of evaluations.

Step2 analysis of numerical solutions

After solving with the PDE toolbox, the numerical solution of the partial differential equation is first obtained. Figure 4 is an image when x_2 is a certain value between 0.6 and 25. The analysis of the image and model preparation can be found. The thickness of this layer is much thicker than that of other insulation layers so that the two surfaces have enough temperature difference to achieve the insulation effect. The numerical solution of partial differential equations is shown in Figure 4:

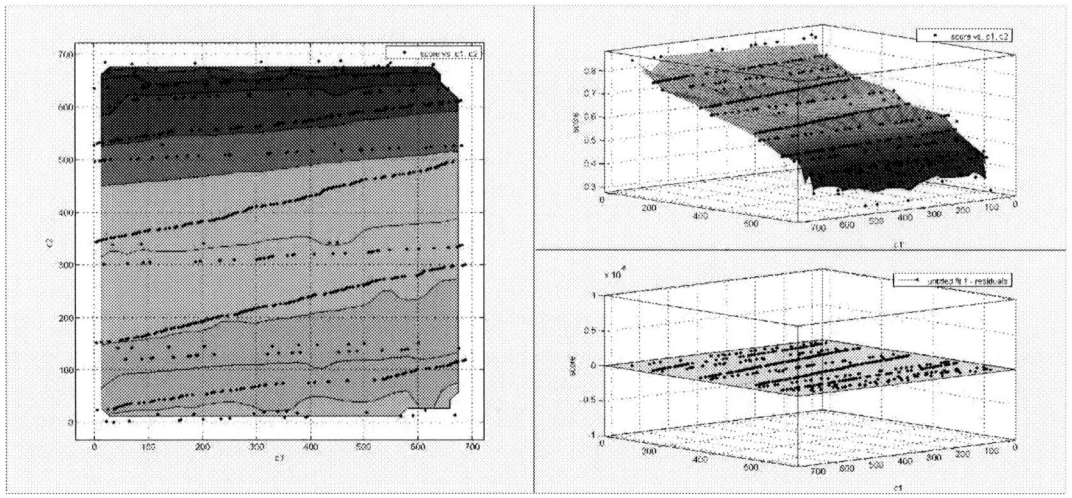

Fig. 4 Numerical solutions of partial differential equations

Step 3 uses genetic algorithm to solve the optimal value of x_2

Based on the initial model assumption, all materials are isotropic materials, and the model is a one-dimensional model. After the establishment of the previous model, a physical problem has been completely abstracted into a mathematical problem, and due to coincidence, This partial differential equation happens to have an analytical solution. The time-varying temperature field function containing the parameter x_2 has been found by the PDE toolbox:

$$T(x,t) = (T_w - T_\infty) \sum_{n=1}^{\infty} \frac{2.56 \sin \zeta_n}{\zeta_n + \sin \frac{1}{2} \zeta_n \cos \zeta_n} x_2 \exp(-a^2 t) \cos(2\zeta_n x) \tag{31}$$

Therefore, we try to use the improved genetic algorithm to solve the minimum value of x_2 based on this function. We use the genetic algorithm to write the principle of programming multiple substitution calculations of all values within the x_2 range. After verifying whether the temperature does not exceed 47 ° C within 60 minutes and the time exceeding 44 ° C does not exceed 5 minutes, each feasible solution is retained, and the previous optimal solution is replaced by a new optimal solution. Finally, the optimal solution with a certain degree of rationality is obtained. That is, the thinnest thickness of Layer II that meets the constraints. The specific flow chart for the calculation of the genetic algorithm is 6, and the procedure is shown in appendix 7. See Figure 5for a map of the optimal solution

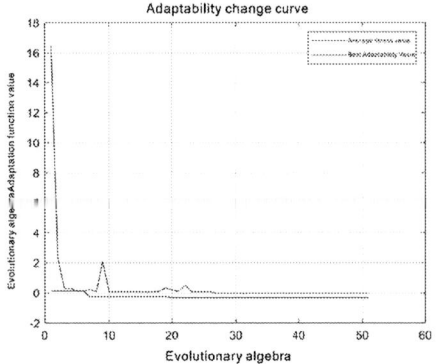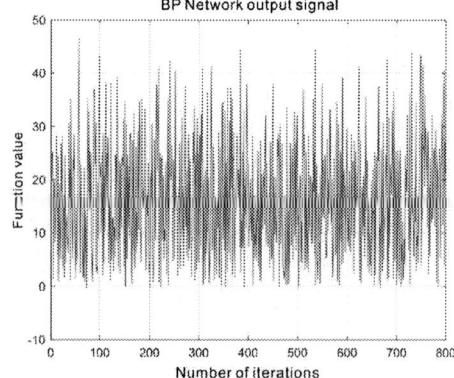

Figure 5 Finding the optimal solution

Figure 6 is the embodiment of the genetic algorithm program solution process. After many iterations, the optimal thickness is found out, and the error evaluation of the solution is given, and the function numerical error evaluation is shown in Table 3.

Table 3 Optimal Thickness Solution Results of Genetic Algorithm

Optimal solution(mm)	Seeking Optimal Solution Error	Find the optimal function numerical error
12.02	0.19748	-0.3122

5.3 The establishment and solution of the third model of the problem

The third problem solution is similar to the second problem solution, but due to security considerations, the working time is reduced, and under the premise of ensuring the thickness of the second layer, the optimization conditions of the fourth layer are added at the same time, so that the thickness of the two layers is achieved. The model needs to consider two conditions at the same time. The distribution function can be obtained by different temperature distribution conditions, and then the optimal value can be obtained by establishing a double-layer programming model. At the same time, due to the actual situation, the human body needs a certain sense of comfort, so the comfort function $\vartheta(L)$ is introduced here to measure.

5.3.1 Modeling

5.3.1.1 Simplification of objective functions

This question needs to consider two thicknesses at the same time to determine the optimal value, and the temperature conditions have changed. This is to simplify the calculation. We simplify the objective function, assuming that the thickness of the second layer and the thickness of the fourth layer are a constant value. Then it is fitted as a parameter to obtain a relatively reasonable surface equation: Temperaturedistributionat8° C as shown in Figure 6。

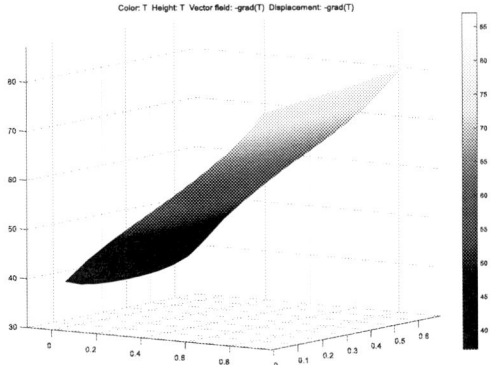

Fig. 6 Temperature distribution at 80 °C

$$T(x,t) = -0.000926 * x * \cos\left(x_2 * t * 4523 * pi\right) - 20.1456 * e^{-(11.2035t * x_4 - 2.0365)^2} + 59.321 \tag{32}$$

5.3.1.2 Establishment of a two-layer planning model

For the second layer of thicknessx_2:

$$\begin{cases} T(x,1800) \le 47 \\ T(x,1500) \le 44 \\ T(0^-) = 80 \\ x_1 = 0.6 \\ x_3 = 3.6 \\ 0.6 \le x_2 \le 25 \end{cases} \tag{33}$$

For the fourth layer thickness x_4:

$$\begin{cases} T(x,1800) \le 47 \\ T(x,1500) \le 44 \\ T(0^-) = 80 \\ x_1 = 0.6 \\ x_3 = 3.6 \\ 0.6 \le x_4 \le 6.4 \end{cases} \tag{34}$$

5.3.1.3 Creation of Comfort Functions

The comfort function represents the range of acceptable temperatures in the human body. It represents a relative value of the part that is in contact with the human body and the whole part, and thus measures the high-temperature resistance performance of the clothing. It is related to the thickness of the second layer of clothing requested and the thickness of the third layer of clothing. Once again, we perform a simple treatment. Set it as a positive function:

Each layer of comfort is $\vartheta_i (i = 1,2,3,4)$, Among them, outer layer comfort $\vartheta_j = \rho_j \times C_j \times \delta_j \times x_j (j = 1,2)$, In ρ, C, δ, x Indicates the density, specific heat, thermal conductivity, and thickness of the clothing; For the inner layer of clothing, there is a certain change

$$\vartheta_k = \frac{C_k \times \delta_k \times x_k}{\rho_k} \ (k = 3,4) \tag{35}$$

The final comfort function is

$$\vartheta = \frac{\vartheta_2 + \vartheta_4}{\sum_{i=1}^4 \vartheta_i} = \frac{\rho_2 \times C_2 \times \delta_2 \times x_2 + \frac{C_4 \times \delta_4 \times x_4}{\rho_4}}{\sum_{j=1}^2 \rho_j \times C_j \times \delta_j \times x_j + \sum_{k=2}^3 \frac{C_k \times \delta_k \times x_k}{\rho_k}} \tag{36}$$

The thickness of the object x_2 is compared with the thickness of the fourth layer of clothing x_4. Considering that the genetic algorithm has different results for each optimization, the obtained value can be taken into the comfort function for testing in order to obtain the relative ideal thickness of the second layer of clothing and the thickness of the fourth layer of clothing.

5.3.2 Model Solving

Considering that question two and question three, although essentially the same, only in the case of the original goal, two additional constraints have been added. If we deal with this separately, we can get the optimal solution. Therefore, this paper still uses genetic algorithm to find the optimal value of the thickness of the second layer of clothing and the fourth layer of clothing, solve it several times, and bring it into the comfort function for testing. The results are as follows: Thickness solution is shown in Table 4:

Table 4 Thickness solution

Layer Number	Optimalsolution（mm）	SeekingOptimal Solution Error	Findtheoptimal unctionnumerical error
II	7.4	0.22132	-0.12147
IV	6.1	0.15409	-0.27148

6.Evaluation and Promotion of 5 Model

6.1 Evaluation of the model

The paper discusses a highly specialized heat conduction problem. This model puts forward too many assumptions for the solution of practical problems, and can only guarantee the rationality of the results to a certain extent or within a certain range. For example, the paper model only analyzes one-dimensional heat conduction models, and does not consider the flow of air in the air layer. In the actual problem, the medium-temperature surface is often not an absolute plane, especially for the human body, and it will certainly have a certain degree of bump. And there will be different shapes of heat transfer temperature field models, the figure of the arm and chest, the rate of heat transfer effect temperature distribution is different.

And the model only simply analyzed the radiation and did not make a thorough discussion. As a major form of heat conduction, it should increase its research weight. Considering the effects of radiation will give reasonable answers to questions that the original model can not explain.

6.2 Extension of the model

We should take into account the body characteristics of the person who is dressed, establish heat conduction models based on different major parts, and then carry out more detailed analysis to obtain the optimal thickness of a certain layer of different parts and a greater degree of material saving. Increase the operator's work Shushichengdu.

7. Conclusion

(1) Conclusion 1 Use genetic algorithm to find the optimal value of the thickness of the second layer of clothing and the fourth layer of clothing, solve it several times, and bring it into the comfort function for inspection. The result is the optimal solution of the second layer of clothing 7.4, and the fourth layer of clothing. The optimal solution 6.1.

(2) Conclusion 2 The paper model only analyzes one-dimensional heat conduction models, and does not consider the flow of air in the air layer. In the actual problem, the medium-temperature surface is often not an absolute plane, especially for the human body. There must be a certain degree of bump. And there will be different shapes of heat transfer temperature field models, the figure of the arm and chest, the rate of heat transfer effect temperature distribution is different.

(3) Conclusion 3 Taking into account the body characteristics of the person wearing the dress, heat conduction models are established according to different major parts, and a more detailed analysis is carried out to obtain the optimal thickness of a certain layer of different parts and a greater degree of clothing saving material.

8. Appendix

8.1 Question 2: Genetic algorithm master procedures:

```
clc
clear all
close all
%%% Set Global Variable
global inputnum hiddennum outputnum net inputn outputn inputps outputps;
%%% First step: raw data processing
% 1.1 Read Data load data input output; %%% Note one: different data formats, different import
commandsinput=input';
output=output';
x1=sort(input(1,:));
x2=sort(input(2,:));
y=x1.^2+x2.^2+7;
figure(1);
plot3(x1,x2,y);
```

```
grid;
 [m,n]=size(output);
%1.2 Decomposition of training data and test data
input_train=input(:,1:n-800);
input_test=input(:,n-800+1:end);
output_train=output(:,1:n-800);
output_test=output(:,n-800+1:end);
%1.3Normalizationofinputandoutputdata
[inputn,inputps]=mapminmax(input_train);%inputn,inputps The normalized data and
structure(including the maximum minimum value average, etc.)
[outputn,outputps]=mapminmax(output_train);
%%% Step 2: BP network algorithm and its mean square error
% 2.1 BP Network Structure Parameters
inputnum=2;
hiddennum=5;
outputnum=1;
%2.2 BP network to get network structure, transfer data
 [BPoutput,BPerror,BPmse]=GB20_BPFitness(input_train,output_train,input_test,output_test);
%%% Step 3: Genetic Algorithm Optimization
% 3.1 Genetic Algorithm Parameters Initializationmaxgen=50;
sizepop=10;
pcross=[0.3];
pmutation=[0.1];
% 3.2 Genetic Algorithm and its Optimal Individuals
[bestchrom,bestfitness,trace] = GB20_GAXunyou(maxgen,sizepop,pcross,pmutation);
% % % 4: drawing display
figure(1)
plot(BPoutput);
grid;
xlabel(Number of iterations);ylabel(' Function value ');
title([' BP network output signal ']);
figure(2)
[r c]=size(trace);
plot([1:r],trace(:,1)','b-',[1:r],trace(:,2)','r-');
grid;
xlabel(' Evolutionary algebra ');ylabel(' Adaptation function value ');
legend(' Average fitness value ','  Best Adaptability Value ');
title([' Adaptability change curve ']);
%%% Step 5: Error Analysis and Recording
Xerror=sqrt((bestchrom(1,1)-0).^2+(bestchrom(1,2)-0).^2);
Yerror=bestfitness-0;
disp([' The optimal solution for BPGA is: ']);
disp([' The optimal solution error for BPGA search is:  ' num2str(Xerror)] );
disp([' BPGA search optimal function numerical error is:  ' num2str(Yerror)] );
```

Author brief introduction

Zhangmaoyi, male, born in 1997, undergraduate student. The main research direction is information and control. Scientific research results and telephone: published many papers, invention patents, utility models. Tel: 18796289172. Email: Kfzqj@126.com

Zhangqijun(communication author), male, was born in 1963 and is a research-level senior worker. His main research direction is engineering machinery and electrical control. Xuzhou Xugong Railway

Equipment Co., Ltd., No. 1 Daishan Road, Jinshanqiao Development Zone, Xuzhou City, Jiangsu Province, Zip Code: 221004; Substitute. Tel: 18796289172, Email: Kfzqj@126.com.

References

[1] Lulinzhen. Thermal transfer model and optimal parameter determination of multi-layer thermal protective clothing[D] .. Zhejiang University of Technology, 2018.

[2] Wan Changfeng, Lin Ji, Hong Yongxing. Boundary node method simulates transient heat conduction and Matlab toolbox development[J] .. Energy and Environmental Protection, 2017(02): 12-17.

[3] Panbin. Mathematical modeling and parameter determination of thermal transfer in thermal protective clothing[D] .. Zhejiang University of Technology, 2017.

[4] Kuangyuyang, Wangtairong.yiwei, two methods for solving the explicit solution of the mixed problem of the heat conduction equation[J] .. Science and Technology Bulletin, 2016, 32(10): 1-4.

[5] Fengqinghua.A New Fractional Projective Riccati Equation Method for Solving Fractional Partial Differential Equations[J].Communications in Theoretical Physics,2014,62(08):167-172.

[6] Yan Laijun,Gao Chunming,Ma Xingchen,Zhao Binxing. Simplified Transmission Photo-thermal Radiometry for Thermal Diffusivity Measurements in Thin Metal Layers[A]. Chinese Institute of Electronics、IEEE Beijing Section.Proceedings of 2013 IEEE 11th International Conference on Electronic Measurement & Instruments VOL.02[C].Chinese Institute of Electronics、IEEE Beijing Section:,2013:4.

[7] Molecular dynamics simulations of the mechanisms of thermal conduction in methane hydrates[J].Science China(Chemistry),2012,55(01):167-174.

[8] Yang Nengbang-Numerical calculation of one-dimensional unsteady heat conduction problems[J] .. Journal of the Qinghai Normal University(Natural Science Edition), 2006(04): 24-26.

[9] Liuhongjie, Wang Wangxiufeng, Wang Zhibao. Use genetic algorithm to search for multiple extreme points[J] .. Journal of Nankai University(Natural Science Edition), 2000(03): 17-22.

[10] Zhao Mingwang. Function optimization mixed numerical algorithm based on genetic algorithm and minimum descent method[J] .. Systems Engineering Theory and Practice, 1997(07): 61-66.

The design of analog signal communication system based on visible light

Z N Zhang[1,2] and H Y Y Hua[1,3]

[1]North China Electric Power University, Beijing 102206, China

[2]825070454@qq.com

[3]18810797692@163.com

Abstract. With the rapid development of science and technology, communication mode has made a great leap forward, and visible optical communication has also been developed around the word. This paper mainly studies the audio signal transmission of LED visible light indoor communication. The system uses audio chip to generate audio signal. We have not only established a channel model, but also received optical signal through BH1750FVI module. In addition, we adopt the direct pulse counting frequency discrimination method for modulation and demodulation, and finally recover the original signal, achieving the purpose of voice signal transmission.

1. Introduction

At present, with the increasing development of LED lighting industry, visible light communication, a means of communication using led light as a media, has been gradually paid attention to and developed globally.

Compared with traditional incandescent lamps, LEDs consume less energy, and are more environmental friendly, so they gradually replace the lighting status of incandescent lamps. In addition, LED modulation performance is good. The communication function can be realized by loading the information to optical signals through modulation technology.

What is more, compared with radio communication, visible light communication is safer and more efficient without electromagnetic interference. Based on this, it can be used for many special occasions, such as hospital, aircraft, military and so on. By controlling the light exposure range, its space reusability is good. Also, basic lighting network is ubiquitous, and the access to communication networks can be realized at low cost.

Therefore, the visible light communication technology based on LEDs has a high research value and application prospect.

2. Model establishment

2.1. Channel model

2.1.1. Characteristics of visible light communication channel. The Figure1 shows the linear base-band transmission model of the indoor visible light communication system, where the pulse response *h(t)* reflects the channel characteristics of the system.

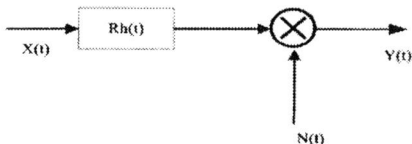

Figure1. A linear base-band transmission
model for indoor visible light communication.

In the indoor visible light communication system, the transmitter light source is white LEDs, and the intensity modulated signal is $X(t)$. The receiver uses a photoelectric sensor, and the photocurrent signal $Y(t)$ received is expressed as

$$Y(t) = RX(t) * h(t) + N(t) \qquad (1)$$

Where R is the photoelectric conversion efficiency of photoelectric sensor; $X(t)$ is the transmitted light power; $h(t)$ is the impulse response of the channel; $N(t)$ represents additive Gaussian white noise.

2.1.2. Calculation of impulse response.
The impulse response algorithm presented in this paper refers to the calculation method proposed by J.R.Barry et al and the improved algorithm proposed by J.B.Carruthers et al.

First of all, we build the model of the light source and receiver. Light source can be generally decided by the position vector r_s, the unit direction vector n_s, the power P_S and radiation mode function $R(\emptyset,\theta)$. $R(\emptyset,\theta)$ is defined as the energy emitted by a light source at a unit solid angle which has an angle of $(\emptyset,\theta)$ with n_s. When the light source of the transmitter adopts Lambert radiation model, the radiation intensity of the light source can be expressed as

$$R(\emptyset,\theta) = \frac{n+1}{2\pi} P_s \cos^n(\emptyset), \emptyset \in \left[-\frac{\pi}{2}, \frac{\pi}{2}\right] \qquad (2)$$

Where n is called the Lambert radiation ordinal number, whose value is related to the half-power intensity angle of the light source, and their specific relationship is

$$n = \frac{(-\ln 2)}{\ln(\cos\theta_{\frac{1}{2}})} \qquad (3)$$

Then the model of reflection surface is made. Assuming that all emission surfaces are ideal Lambert diffuse reflections, the radiation pattern is independent of the incident angle of light.

Finally, we use the impulse response algorithm. As for a specific light source called S and receiver called R, the impulse response can be expressed as follows:

$$h(t; S, R) = \sum_{k=0}^{\infty} h^{(k)}(t; S, R) \qquad (4)$$

Where $h^{(k)}(t)$ is the k time reflection of the response.

We calculate the impulse response of zero degree reflection, which represents the transmission coefficient of light from one point to another without reflected light power. The impulse response of the k time reflection can be iterated by the impulse response of the (k-1) sub-reflection.

2.2. The model of modulation and demodulation
Assume that the modulation signal also called analog signal can be represented by a single frequency signal expressed as

$$u_\Omega(t) = U_\Omega \cos\Omega t \qquad (5)$$

carrier for

$$u_c(t) = U_c \cos\omega_c t \qquad (6)$$

According to the definition of frequency modulation, the instantaneous angular frequency of FM signals is

$$\omega(t) = \omega_c + \Delta\omega(t) = \omega_c + k_f u_\Omega(t) \qquad (7)$$

It is on the basis of ω_c, increased with the frequency offset proportional to $u_\Omega(t)$. Where k_f is the proportionality constant, which is called modulation sensitivity, and its unit is Hz/V. $\varphi(t)$ is the integral of instantaneous angular frequency $\omega(t)$ with respect to time, i.e. :

$$\varphi(t) = \int_0^t \omega(t)dt + \varphi_0 \qquad (8)$$

φ_0 in the formula is the starting angular frequency of the signal. For convenience of analysis, suppose that $\varphi_0 = 0$, so

$$\varphi(t) = \omega_c + m_f \sin\Omega t \qquad (9)$$

Where m_f is the frequency modulation index. Then the expression of FM wave is

$$u_{FM}(t) = U_c \cos[\omega_c t + m_f \sin\Omega t] \qquad (10)$$

The information of a modulated signal is modulated at the frequency of an FM wave. Therefore, in order to obtain the original modulation signal, it is necessary to recover the original modulation signal from the FM wave, namely, frequency discrimination. The frequency of the signal is related to the number of zeros passed in the unit time of voltage. So, the direct pulse counting frequency discrimination method is adopted.

3. Model implementation

3.1. Overall framework
The optical signal communication device can be divided into two parts: optical signal generation system and optical signal receiving system. The overall frame diagram is as shown in Figure 2.

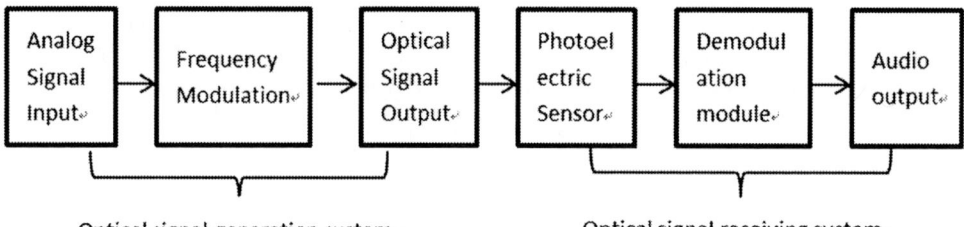

Figure 2. Overall frame diagram.

3.2. Optical signal generation system
Optical signal generation system includes analog signal input module, modulation and optical signal output module.

3.2.1. Analog signal input module. We use the audio module analog output circuit based on audio chip, including microphone and LM386 amplifier. If the input of the circuit is audio signal, voltage signals at different times can be obtained from the output named J3 of the circuit. The internal schematic diagram is as shown in figure 3.

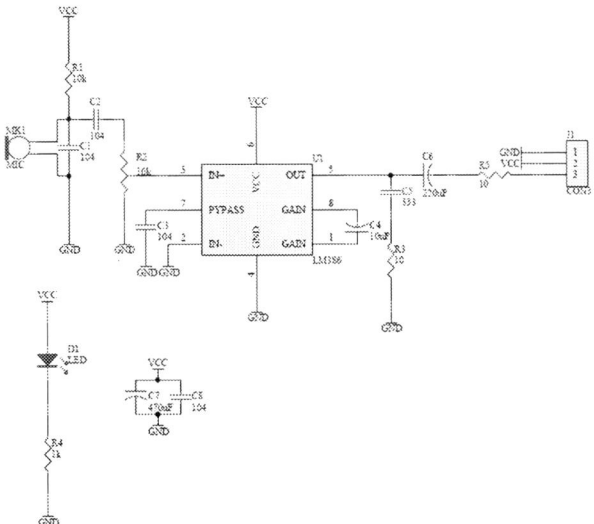

Figure3. Internal schematic diagram.

Sound travels in the form of waves. When encountering obstacles in the propagation path, pressure will be generated on the surface of obstacles, which is sound pressure. The microphone can detect sound pressure. The sound pressure is reflected in the microphone output level. The output level of the microphone is amplified through LM386, which is convenient for the single-chip microcomputer to read data.

3.2.2. Modulation and optical signal output module. The output voltage of the J3 of the sound sensor module is converted into digital voltage by AD conversion module of single-chip microcomputer and read into the single-chip microcomputer for frequency modulation. According to the result of frequency modulation, PWM wave of the corresponding frequency is output by MSP430 to control the flashing frequency of LEDs.

A digital-to-analog conversion is initiated by an ascending edge of a sampled input signal called SHI. After synchronization with ADC10CLK, the sampling timer sets SAMPCON to the height for the selected sampling period. The total sampling time is $t_{sync} + t_{sample}$ synchronization. SAMPCON starts analog-to-digital conversion when it changes from high to low, which requires 13 ADC10CLK cycles. The sampling sequence is shown in the figure.

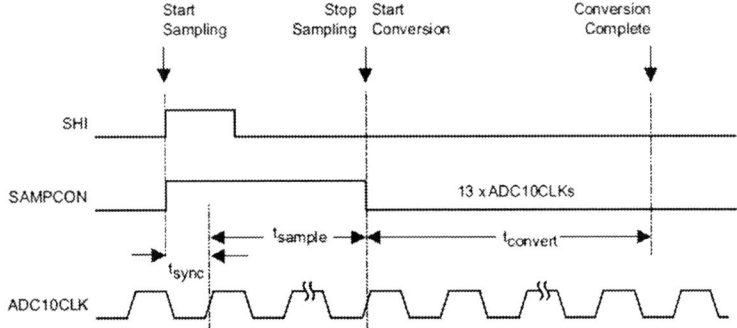

Figure4. Sampling sequence diagram.

After the current analog signal value is obtained by the single-chip microcomputer, the original signal is modulated to the high-frequency carrier by means of frequency modulation, as shown in the figure 5.

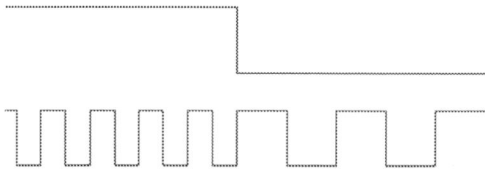

Figure5. Frequency modulation diagram.

Finally, the MSP430 output modulated PWM wave is used as the control signal and connected with the LED drive module to realize the LED flashing according to the timing sequence of the modulated wave.

3.3. Optical signal receiving system

Optical signal receiving system includes the photoelectric sensor, the demodulation module and the audio output module.

3.3.1. Photoelectric sensor. Traditional light sensor mainly uses photosensitive resistances, but its photoelectric characteristic is non-linear, so it is not suitable for detection components. What is worse, photosensitive resistances need to use A/D converter to convert its signals into digital signals, the circuit is complex, high-cost, and its signal acquisition accuracy is not so ideal.

However, the ambient light sensors have the characteristics of low dark current, low illumination response, high sensitivity and linear change of current with the enhancement of illumination. Among them, BH1750FVI module is adopted as the digital photoelectric sensor, which has a wide induction range and is less affected by infrared ray, meeting the accuracy requirements of this paper. It adopts low-cost microcontroller for control, and uses I2C bus interface for data transmission, and can display real-time light intensity measurement values on the LCD.

3.3.2. Frequency demodulation module. The demodulation method adopted in this paper is the direct pulse count frequency discriminator method, and its frequency discriminator diagram is shown in the figure. V_S is the input frequency modulation signal. It realizes the amplification and limiting of the bandwidth through the frequency limiter, and becomes the frequency square wave signal V_1. Then an equal height and shape of the same pulse sequence V_2 is obtained by differential network. Next, the pulse forming circuit is transformed into the corresponding rectangular pulse sequence V_3. Finally, we obtain the output demodulation voltage V_4 by low-pass filter.

The figure 6 is the frequency discriminator composition diagram.

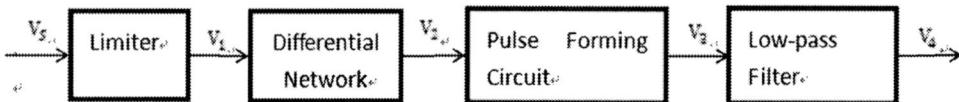

Figure 6. the frequency discriminator composition diagram.

3.3.3. Audio output module. Because the demodulation signal is very small, a small signal amplifier circuit is designed. The audio signal is amplified and eventually played through the speaker.

4. Conclusion

Based on the visible light communication of LED lights, this paper designs a system to simulate the indoor environment by using the semi-closed cubic space of 100cm*100cm*100cm. Through the optical signal generation system and the optical signal receiving system, the audio signal is modulated and demodulated, and finally the signal transmission is realized. Therefore, it can be seen that optical communication not only is safe, efficient, energy saving and environmental friendly, but also has good modulability and no electromagnetic interference. So, it has a good development prospect and research value.

Reference

[1] Chi N 2013 LED visible light communication technology *Beijing: Tsinghua University Press, 2013, 33-35*

[2] Liu ZH, Zhang M, Tang QW, Li JW and Qu YT 2016 Voice transmission based on visible light communication *China New Communications*

[3] Liao BX and Li QH 2018 Design and implementation of visible light digital and analog communication system *Jiangxi Science*

[4] Jovicic A, Li J, Richardson T. Visible light communication: opportunities, challenges and the path to market[J]. *IEEE Communications Magazine*, 2013, 51(12):26-32

[5] Chen T, Liu L, Weiwei H U. Visible Light Communication[J]. *International Journal of Engineering Trends & Technology*, 2013, 4(3):1337-1338

Simulation Study of Dispersion Compensation in Optical Communication Systems Based on Optisystem

Peng Xia[1], ZHANG Li-Hua[2], Yao Lin[1]

[1]School of Electronics and Communication, Anhui Xinhua College,

Hefei 230000,China;

[2]School of Computer and Information, Hefei University of Technology,

Hefei 230000, China

wtt_px@163.com

Abstract. Based on the Optisystem, the simulation model of Optical Communication Systems with Dispersion Compensation is presented. By using FBG, the performance of system in NRZ modulation with M-Z optical external modulator is improved. The parameter setting of the simulation model is presented in detail.The dispersion compensation performances of different pattern is compared through analyzing eye diagrams ,input and output signals.

1.Introduction

Since the advent of fiber Bragg grating(FBG) in the late 1980s, a lot of theoretical analysis and experimental studies have been conducted on their characteristics.In recent years, researches have focused on the improvement of fiber grating on the overall transmission performance of fiber optic system[1]. Relevant research methods include simulation research and experimental simulation. OptiSystem is a optical fiber communication system simulation software released by Optiwave company in Canada, which can help users plan, test and simulate all kinds of situations in optical fiber communication systems. From the physical layer devices to the system level optical communication system design and other conditions can be verified by simulation.

In this paper, optical fiber communication system with dispersion compensation is built based on OptiSystem simulation software, and FBG is used to improve the transmission performance of the system.Precompensation, postcompensation and symmetric compensation are used in simulation, and system parameters, waveform and eye diagram are given in detail.

2.Introduction to FBG compensation technology

Dispersion,loss,and nonlinearity in optical fiber communication system are the main factors affecting transmission performance. Mode dispersion mainly occurs in MMF (multi-mode Fiber), and SMF mainly includes material dispersion and waveguide mode dispersion. Common dispersion compensation technologies mainly include distributed coordination function(DCF),optical fiber compensation, chrip,FBG compensation and other methods[2].In this paper, FBG simulation is used to study the dispersion compensation effect.

The refractive index of FBG can be changed with the spatial distribution of light intensity through certain doping mode, and the refractive index period changes like Bragg grating[3,4]. The dispersion compensation principle of fiber Bragg grating which the specific grating period corresponds to the

---

Content from this work may be used under the terms of the Creative Commons Attribution 3.0 licence. Any further distribution of this work must maintain attribution to the author(s) and the title of the work, journal citation and DOI.

Published under licence by IOP Publishing Ltd

specific light reflection wavelength, and the light reflection position of different wavelengths is different, which is applying to form the time delay.Common FBG include chirp FBG and Uniform FBG(UFBG). UFBG is characterized by narrow reflectance spectrum and is often used in optical filtering.CFBG is characterized by a wide reflection spectrum and can reflect light signals of multiple frequencies, which are mostly used for dispersion compensation[5,6].

At present, the high-speed laser mainly adopts the external modulation mode, which include M-Z (mach-zahnder) waveguide modulator and electric absorption modulator. M-Z modulator is usually made of Lithium Niobate (LiNO3) material, and the Distributed Feedback Laser(DFB) which has good chirp elimination, so M-Z modulator is suitable for long-distance transmission of high-speed systems [7,8].

3.simulation model and parameter setting

The light source used in the simulation system is the continuous wave(CW) laser diode with a wavelength of 193.1THz. The sequence Generator is used to generate the required digital signal sequence, which is converted into electrical pulse signal by the non-return to zero (NRZ) pulse generator. After passing through the M-Z modulator, the electro-optical effect is loaded on the optical wave to become the optical signal when entering the fiber.

The layout of the postposition compensation simulation system is shown in figure 1. The global parameter is set as 10Gb/s and the sequence length is 8bit.1550nm window with a typical value of 0.20db /km.The setting of simulation parameters is shown in figure 2 to 4.

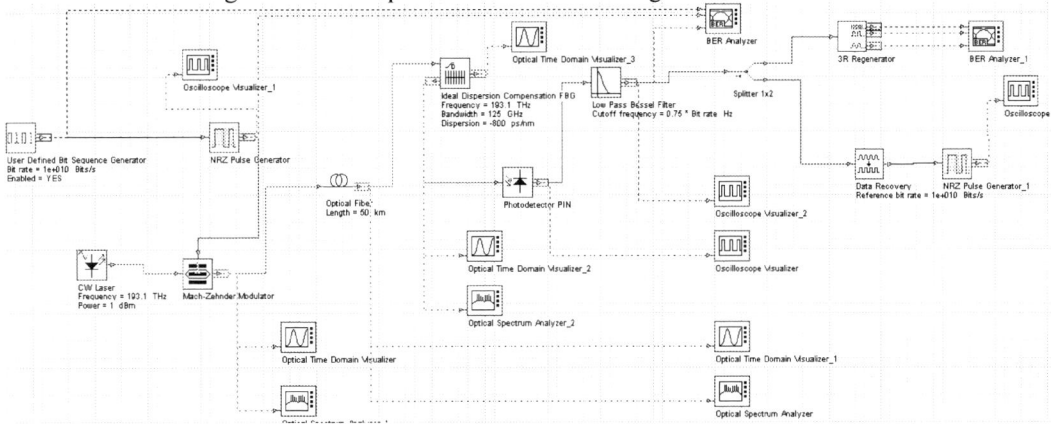

Figure 1.Postposition compensation system

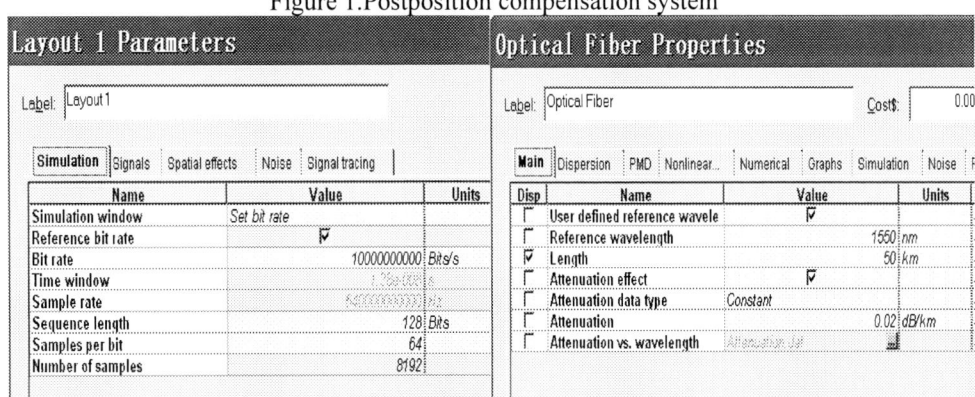

Figure 2.System global parameter setting   Figure 3.Fiber parameter setting

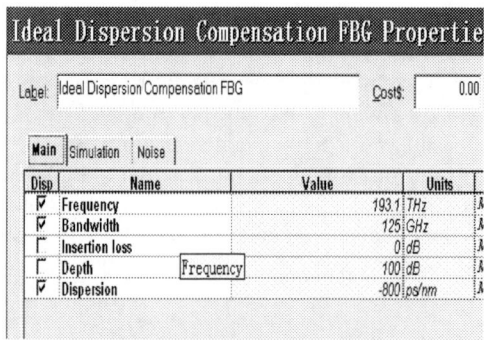

Figure 4.FGB parameter setting

Firstly, the system is simulated without dispersion compensation, waveforms and eye diagram are observed. After that, FBG is set to simulate pre-compensation, post-compensation and symmetric compensation. No FBG system, pre-compensation, post-compensation and symmetric compensation system are shown in figure 5-7.

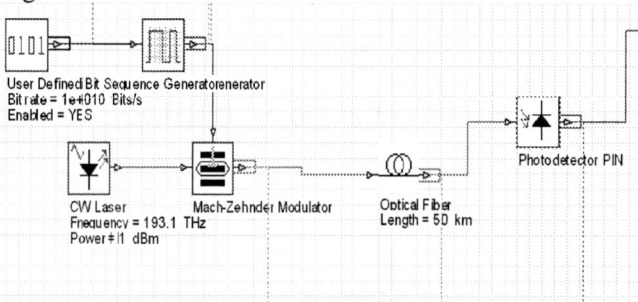

Figure 5.No FBG system

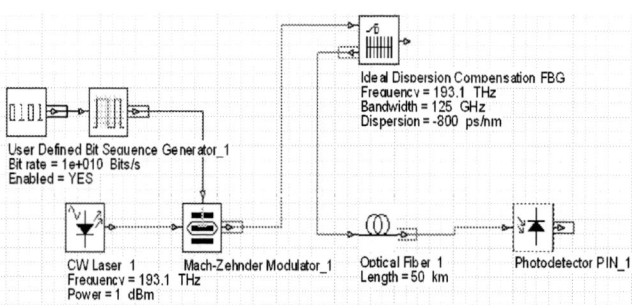

Figure 6.Pre-compensation system

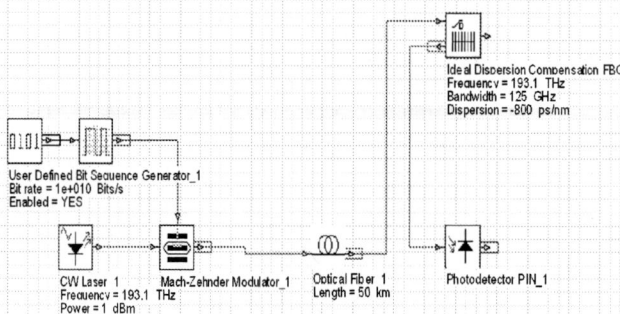

Figure 7.Post-compensationsystem

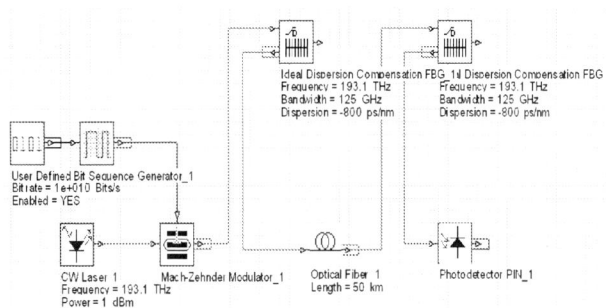

Figure8.Symmetric compensation system

4.Simulation result

The fiber input signal , output signal, eye diagram and spectrum analysis of the system before compensation are shown in figure 9. FBG compensated optical fiber input signal and output signal are shown in figure 10 to 12.They are pre-compensation, post-compensation and symmetric compensation respectively.

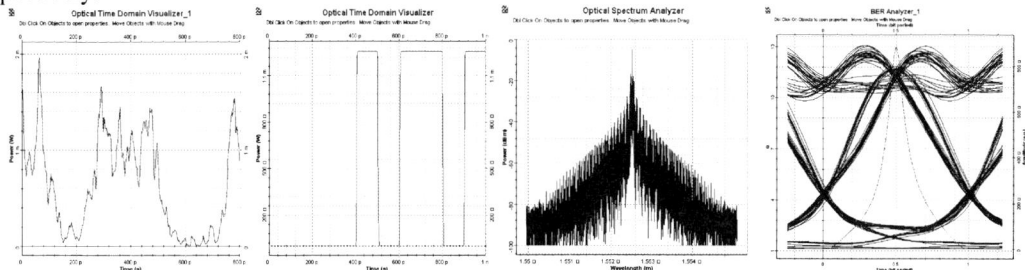

Figure 9. Signals,seye diagram and spectrum analysis before compensation

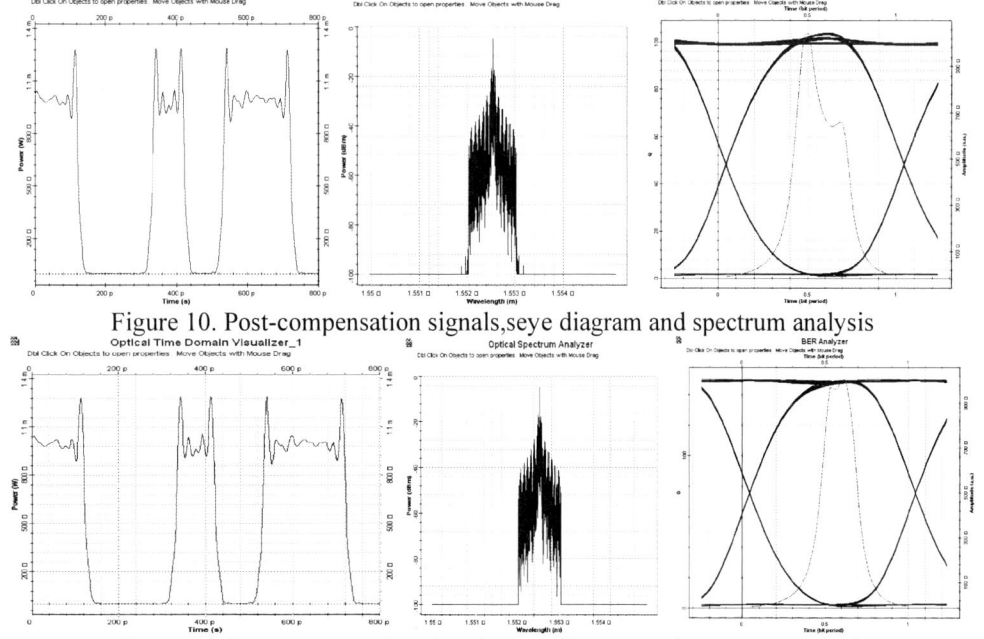

Figure 10. Post-compensation signals,seye diagram and spectrum analysis

Figure 11. Pre-compensation signals,seye diagram and spectrum analysis

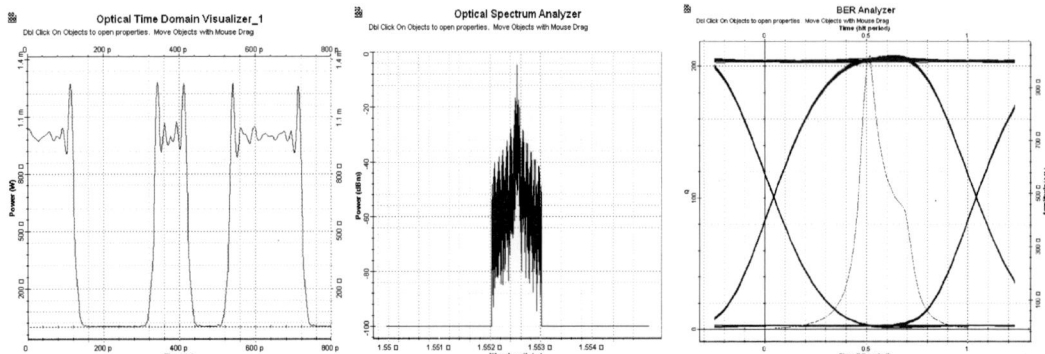

Figure 12. Symmetric compensation signals,seye diagram and spectrum analysis

According to the simulation results, it can be concluded that the uncompensated system has some interference between codes, waveform distortion and unclear eye diagram. It can be observed that signal transmission delay exists in simulation system. After FBG compensation, the eye diagram becomes clear and the waveform distortion is significantly reduced. It can be seen from the spectrum image that FBG has the function of light filtering.Theoretically, when corresponding to a certain input , optical signal to noise ratio (OSNR) transmission performance of pre-compensation, post-compensation and symmetric compensation will be different. In the simulation process, it is also found that under the condition of system setting parameters, the eye diagram opening condition and dispersion compensation effect are analyzed, and the symmetrical compensation performance is the best, followed by the post-compensation.

5.Conclusions

Using M-Z waveguide modulator to construct the transmitter simulation part, the FBG is applied to improve the transmission performance. The simulation adopts the pre-compensation, post-compensation and symmetric supplemen t methods, and gives the system parameters, waveforms and eye diagram in detail during the simulation. When corresponding to the specific fiber input power, the transmission performance of the system with different compensation methods will be different, which is confirmed by the simulation results of the system.The simulation process has some reference value to the system design.

References

[1]WANG Qiuguang,ZHANG Yalin,HU Caiyun,ZHAO Yingqi.Applicationof OpticSystem simulation in experiment teaching of optical fiber communications[J].Laboratory Science 2015,18(1):26-29

[2]Joseph C.Palais. Fiber Optic Communications,Fifth Edition[M].Publishing House of Electronics Industry,2011.

[3]Li Jianzhi,Sun Baochen.Theory analysis of novel fiber Bragg grating temperature compensated method based on thermal stress[J].High Power Laser and Particle Beams,2015, 27(2):76-82.

[4]HUANG Yan-hua. Research on dispersion compensation technology using chirped fiber Bragg grating[J].Optical Communication Technology,2016, 40(11):41-43

[5]Liu Xiaolei,Xiong Xuejuan.Performance analysis of photoelectric dispersion compensation technology based on optisystem[J].Electronic Measurement Technology,2017, 40(11):114-119.

[6]Wang Huiyi,Cao Liankeng,et al.Discussions on optical amplification and dispersion management in 40 Gbit/s fiber-optic communication systems[J].Study on Optical Communications,2014, (1):32-33,44

[7]BI Weihong,LIU Yin.Experiment Research of Multi-wavelength Optical Source Based on the Elect -troptic Intensity Modulator of Mach-Zehnder[J].Opto-Electronic Engineering,2011, 38(7):7-12.

[8]ZHAO Jiamei,SUN Changzheng,XIONG Bing,WANG Jian,LUO YiPackage Design for 4×25 Gb/s Electroabsorption Modulated Laser Array[J].Semiconductor Optoelectronics,2017, 38(1):12-15.

ISPECE

IOP Publishing

The Comparison of Crowd Counting Algorithms based on Computer Vision

Zhaoqing Wang[1,*], Qishu Deng[2,a], Yusheng Zhao[2,b]

[1]School of Information & Communication Engineering, Beijing Information Science& Technology University, Beijing, China
[2]International School, Beijing University of Posts and Telecommunications, Beijing, China

*Corresponding Author E-mail: mj741561@163.com

[a]dengqishu@bupt.edu.cn; [b]zhaoyusheng@bupt.edu.cn

Abstract: This paper aims to compare the three mainstream solutions for today's crowd counting and analyze the highlights of each model. In MCNN, they proposed a multi-column parallel convolutional neural network structure that generates population density maps by adapting crowd changes caused by camera view-points and resolution using filters with different size receptive fields. In Switch-CNN, they added a density classifier to the MCNN to enable the use of local density changes in the crowd. In CSRNet, they abandoned the structure of a multi-column convolutional neural network, using the first ten layers of VGG-16 as the front part and the convolutional neural network as the latter part. From the analysis results, CSRNet shows advanced performance. In addition, we analyzed the comparison results of three convolutional neural networks, and derived the trend of convolutional neural network structure.

1. Introduction

Computer vision began in the late 1960s. Firstly, it was designed to imitate the human visual system as the basic of robots' intelligent behavior [1]. In 1966, to realize the computer vision and let computer present "what it saw", people connect computer to the camera [2][3]. Today, many computer vision algorithms including extracting edges from images, non-polyhedral and polyhedral modeling, labeling lines, representing objects as the interconnection of smaller structures, optical flow and motion estimation are based on the early researches [1]. Then, people use variations of graph cut to solve image segmentation. In a word, computer vision covers several aspects. One of the important applications of computer vision is to accurately estimate the number of people in an image or video.

Crowd counting is a key technology to control crowd and ensure public safety. Monitoring crowds in video or images has important market applications. In addition, information such as the number, density or distribution of participants can be obtained through crowd counting method, which provides effective safety guidance of public places such as stations, shopping malls and plazas. It can also achieve greater economic benefits by improving service quality, analyzing customer behavior, optimizing advertising and resource allocation. Moreover, mature crowd counting technology can be extended to other fields, such as the estimation of vehicle density on traffic roads, microbes counting in microscopic images, and species protection in ecological tribes.

In recent years, with the gradual development of computer vision technology, a large number of crowd counting methods have been proposed. Among them, the core algorithms of crowd counting in the field of deep learning are Multi-CNN, Switch-CNN and CSRNet [4][5][6]. This paper would

Content from this work may be used under the terms of the Creative Commons Attribution 3.0 licence. Any further distribution of this work must maintain attribution to the author(s) and the title of the work, journal citation and DOI.

Published under licence by IOP Publishing Ltd

mainly compare the similarities and differences of those algorithms as well as discuss their advantages and disadvantages. Summarize the trend of those core algorithms of the crowd counting. In all, this paper aims to obtain a clearer understanding on crowd counting algorithm and figure out the future development trends, in order to achieve greater breakthroughs.

2. Literature Review

2.1 Tracked visual features trajectories clustering approaches
Tracked visual features trajectories clustering have been widely used in crowd counting. By using the established pedestrian database to train the classifier, the KLT tracker is used to track the features detected in the pedestrian videos [7] [8]. In addition, the module based on feature tracking and clustering of KLT feature tracker and spatiotemporal clustering module can filter and delete invalid trajectories, calculate the motion features of the target and automatically assign trajectories to each object [7]. However, this is a very computationally intensive method and can only be applied to the crowd counting in the video.

2.2 Feature-based regression approaches
Feature-based regression can be applied to crowd counting of still images. When extracting low-level features, the background can be removed, and various features in the foreground image such as textures can be extracted. Linear, piece-wise linear or neural network regression functions can be established to estimate the number of people [9]. Additionally, feature-based regression approaches can be used to extract and analyze features in image blocks to search objects with specific characteristics. For the crowd with high density and high occlusion, it is more effective to use a regression function that has been trained by low-level features. This approach may be easily affected by sparse and unbalanced situations, causing large changes in feature parameters [10]. There is currently no good solution.

2.3 Deep learning approaches
The CNN-based algorithms are very popular right now, behaving better in accuracy and flexibility in the field of crowd counting. According to the algorithms mentioned above, foreground segmentation is indispensable, but it is difficult to implement. In contrast, the DCNN proposed in [11] does not require foreground segmentation and hand-crafted feature extraction. It can count the total crowd number based on the textures. [4] proposed the MCNN structure. By using different sizes of receptive fields, this algorithm is available to do crowd counting in any static crowd image of any size and resolution. This algorithm can also be transferred to other targets by fine-tuning. [5] Combined with previous studies, Switch-CNN was established, which increased the accuracy of prediction depending on the changes of image density. In addition, researchers are constantly expanding data sets to ensure efficient learning of CNN [12]. Using pure convolutional layer as the core of the structure, a deeper network CSRNet was built, which reduced the computational complexity to some extent [6]. So far, the crowd counting algorithm in deep learning has obtained many breakthroughs and achievements. Therefore, this paper will further compare the three core CNN-based algorithms, discuss their structures and innovation.

3.Comparison
Estimating the number of people in a given population by using the Convolutional Neural Network (CNN) is a relatively engineering study point in the domain of deep learning. There are two main types of solutions. In the first type of network, the input is the image, and the output is the estimated number of people. The input to the second type of network is also an image, but its output is the density map of the crowd, then the number is obtained by integration. However, the output of the second network has more information in the density map than the output of the first one, which gives the distribution of the crowd in space. This paper will use three specific networks as an example to

analyze and compare the network structure using population density counting.

3.1 Multi-CNN

In the case of crowd counting, since the images usually contain heads of very different sizes, it is not possible to detect the characteristics of the crowd density using the filters with reception field of the same size. Hence, they proposed the Multi-Column Convolutional Neural Network (MCNN), which consists of three columns of parallel CNNs with local receptive fields of different sizes. For simplification, in addition to the size and number of filters, each column CNN uses the same network structure. Two 2*2 max pooling layers are used in each column CNN (Because the max pooling layers are used twice, each training sample needs to be down-sampled by 1/4 before the corresponding density map is generated.). At the same time, a rectified linear unit (ReLU) is used as an activation function [13]. To reduce computational complexity, a smaller number of filters are used for CNNs with larger filters. Avoid the distortion of the graphics when stacking the output feature maps of all CNNs and mapping them to the density map. This network uses a 1×1 filter [14] so that the input image can be of any size. Finally, Euclidean distance is used to measure the difference between the estimated density map and the ground truth .The network structure of MCNN is illustrated in Figure 1:

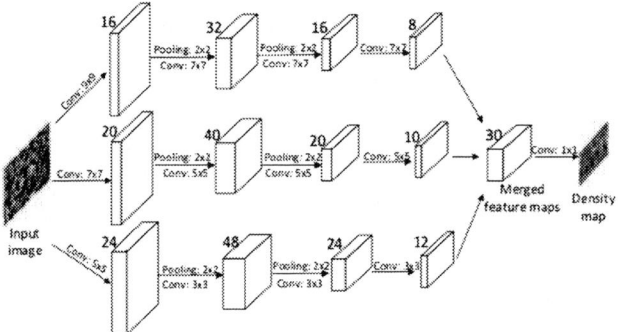

Figure 1: The structure of the proposed multi-column convolutional neural network for crowd density map estimation. [4]

3.2 Switch-CNN

Except for the ability to model large scale variations, the Switch Convolutional Neural Network (Switch-CNN) can also use local density changes in crowd scenarios well. Because the weighted averaging technique used in MCNN will weaken the details in the image, reduce the contrast of the image, and blur the edges in the image to a certain extent. It is difficult to achieve satisfactory fusion effect in most applications, so the ability to leverage local variations in density becomes especially important. The structure of Switch-CNN consists of three CNN regressors with varying receptive fields and the switch that selects the correct regressor for the input patches. This network uses three CNN regressions introduced in it, R1, R2, R3 [4]. R1 is a 9*9 large-size filter that captures advanced features in the scene. R2 and R3 are 7×7 and 5×5 filters respectively to capture features in low scales. The switch network consists of a switch classifier and a switch layer. The switch classifier infers the label of the CNN regressor suitable for the input image patches. The switch layer receives the label and relays the patches to the correct regressor. The switch uses the adaptation of VGG16 network as the switch classifier for three-way classification [15]. The fully-connected layers in VGG-16 are banned by the global average pool (GAP). GAP is followed by a smaller fully-connected layers and a 3-class softmax classifier, corresponding to three CNN regressors in the Switch-CNN. Switch-CNN first divides the image into 3*3 non-overlapping patches based on a certain crowd characteristics, then uses a switch classifier to classify the patches by density standard, and then relays the patches to the independent CNN regressor with different receptive fields and field-of-view. The network structure of

Switch-CNN is illustrated in Figure 2:

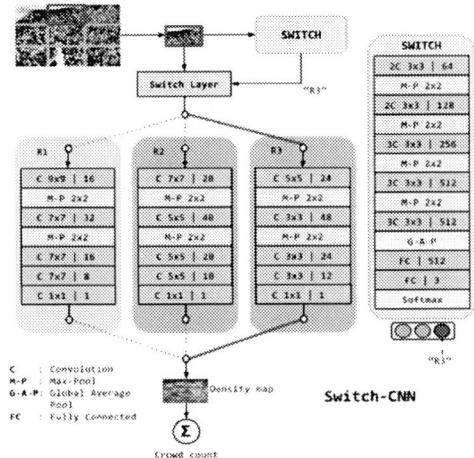

Figure 2. The structure of the proposed switch convolutional neural network for crowd density map estimation. [5]

Configurations of CSRNet			
A	B	C	D
input(unfixed-resolution color image)			
front-end			
(fine-tuned from VGG-16)			
conv3-64-1			
conv3-64-1			
max-pooling			
conv3-128-1			
conv3-128-1			
max-pooling			
conv3-256-1			
conv3-256-1			
conv3-256-1			
max-pooling			
conv3-512-1			
conv3-512-1			
conv3-512-1			
back-end (four different configurations)			
conv3-512-1	conv3-512-2	conv3-512-2	conv3-512-4
conv3-512-1	conv3-512-2	conv3-512-2	conv3-512-4
conv3-512-1	conv3-512-2	conv3-512-2	conv3-512-4
conv3-256-1	conv3-256-2	conv3-256-4	conv3-256-4
conv3-128-1	conv3-128-2	conv3-128-4	conv3-128-4
conv3-64-1	conv3-64-2	conv3-64-4	conv3-64-4
conv1-1-1			

Figure3 The Structure of Dilated Convolutional Neural Networks for crowd density map estimation.[6]

3.3 CSRNet

CSRNet is a network model for high-density population monitoring. The main idea of this network is to deploy deeper CNN. The benefit of this is that you can capture advanced features with larger reception fields while also reducing network complexity. Dilated Convolutional Neural Networks (DCNN) consists primarily of the front-end VGG-16 convolutional layer portion and the back-end dilated convolutional layers. According to similar ideas in [16][5][17], they chose VGG-16 as the front end of CSRNet [6], and then deleted the classification part of VGG-16 fully connected layer, using the convolution layer of VGG-16. Built CSRNet. After making trade-offs between accuracy and the resource overhead (including training time, memory consumption, and the number of parameters), they retained the top ten layers of VGG-16, with three pooling layers. In order to avoid loss of resolution and to extract deeper information of saliency, they proposed dilated convolutional layers. It uses sparse kernels to alternate the pooling layers and convolutional layers. The character expands the receptive field without increasing the number of parameters or the amount of calculation, reducing the computational complexity. In the dilated convolutional layers, the small-size kernel with the a*a filter is expanded to a +(a-1)(b-1) with an expansion step b. Since CSRNet's output density map is only 1/8 of the input size, they choose the bilinear interpolation with a factor of 8 to scale, which ensures that the output has the same resolution as the input image. The dilated convolutional network shows obvious advantages. The first advantage is that the output shares the same dimensions as the input (no pooling and deconvolution). The second advantage is that the output of the classified convolutional contains more detailed information. The network structure of CSRNet is illustrated in Figure 3.

4. Discussion of Application and Function

The ShanghaiTech dataset contains more viewpoints and a larger density of people than most existing datasets. The comparison results for these three models are showed in Table 1. We can clearly see that CSRNet performs better than Switch-CNN and MCNN, achieving the lowest MAE and MSE compared to other methods. This shows that CSRNet perform noticeably well in high-density scenarios and is capable of high-density population density detection. At the same time, the test results of the UCSD dataset characterized by low-density scenes are shown in Table 2. The results show that MCNN performs best and achieves the lowest MAE and MSE. This shows that MCNN is qualified for low-density population testing. Although CSRNet is not the best performer, the effect is considerable. This shows that in addition to being able to perform high-density crowd counting, CSRNet is also suitable for low-density crowd counting and is very versatile.

Table 1.The comparison of three models on ShanghaiTech dataset

Method	Part A		Part B	
	MAE	MSE	MAE	MSE
MCNN	110.2	173.2	26.4	41.3
Switch-CNN	90.4	135.0	21.6	33.4
CSRNet	**68.2**	**115.0**	**10.6**	**16.0**

Table 2.The comparison of three models on UCSD dataset

Method	Part A		Part B	
	MAE	MSE	MAE	MSE
MCNN	110.2	173.2	26.4	41.3
Switch-CNN	90.4	135.0	21.6	33.4
CSRNet	**68.2**	**115.0**	**10.6**	**16.0**

The three columns of parallel CNNs in the MCNN have filters of different size local reception fields, which are significantly modified, compared to the traditional CNN. This can better adapt to the size change of the human head in different scenes. In addition, this network uses the weighted average of the filter of 1*1 to fuse the feature maps from each column CNN, so the density map of any input size image can be generated without distortion. A major change in Switch-CNN is to first divide the input image into 9 patches and then use the switch to take advantage of changes in local population density within the scene. In the switch classifier, the fully connected layer of VGG-16 is replaced by the global average pool (GAP). The advantage of GAP is to regularize the entire network structure to suppress overfitting (random noise of the model overfitting the data set). The highlight of CSRNet is the dilated convolutional layers, which use a sparse kernel to replace pooling layers and convolutional layers to expand the receptive field. This approach brings three distinct benefits: (1) maintaining spatial resolution. (2) Simplified network structure. (3) Reduced computational complexity and number of parameters.

5. Conclusion

In this paper, we compared MCNN, Switch-CNN, and CSRNet from the perspective of network structure and experimental performance. MCNN uses three columns of CNNs with different size receptive fields to achieve adaptation to different sizes of heads. In addition, MCNN also has good adaptability. By using the switch, Switch-CNN uses the factor of local density variation to classify crowd image patches. By using dilated convolutional layers, CSRNet expands the receptive field without reducing spatial resolution, while also enabling density detection for high-density populations. From the overall results, CSRNet is the most condensed and best performing CNN. From MCNN to CSRNet, we have found that the structure of convolutional neural networks has evolved from multiple columns CNN to single columns CNN, and the depth of the network has gradually deepened from shallow to shallow. This is also the future development trend of convolutional neural networks.

Reference

[1]   Szeliski, R. (2010). *Computer vision: algorithms and applications*. Springer Science & Business Media.

[2] Papert, S. A. (1966). The summer vision project.

[3] Margaret, A. (2006). Mind as machine: a history of cognitive science.

[4] Zhang, Y., Zhou, D., Chen, S., Gao, S., & Ma, Y. (2016). Single-image crowd counting via multi-column convolutional neural network. In *Proceedings of the IEEE conference on computer vision and pattern recognition* (pp. 589-597).

[5] Sam, D. B., Surya, S., & Babu, R. V. (2017, July). Switching convolutional neural network for crowd counting. In*Proceedings of the IEEE Conference on Computer Vision and Pattern Recognition* (Vol. 1, No. 3, p. 6).

[6] Li, Y., Zhang, X., & Chen, D. (2018, February). CSRNet: Dilated convolutional neural networks for understanding the highly congested scenes. In *Proceedings of the IEEE Conference on Computer Vision and Pattern Recognition* (pp. 1091-1100).

[7] Rabaud, V., & Belongie, S. (2006, June). Counting crowded moving objects. In *Computer Vision and Pattern Recognition, 2006 IEEE Computer Society Conference on* (Vol. 1, pp. 705-711). IEEE.

[8] Yang, T., Zhang, Y., Shao, D., & Li, Y. (2010). Clustering method for counting passengers getting in a bus with single camera. *Optical Engineering, 49*(3), 037203.

[9] Chan, A. B., Liang, Z. S. J., & Vasconcelos, N. (2008, June). Privacy preserving crowd monitoring: Counting people without people models or tracking. In *Computer Vision and Pattern Recognition, 2008. CVPR 2008. IEEE Conference on* (pp. 1-7). IEEE.

[10] Chen, K., Gong, S., Xiang, T., & Change Loy, C. (2013). Cumulative attribute space for age and crowd density estimation. In *Proceedings of the IEEE conference on computer vision and pattern recognition* (pp. 2467-2474).

[11] Zhang, C., Li, H., Wang, X., & Yang, X. (2015). Cross-scene crowd counting via deep convolutional neural networks. In*Proceedings of the IEEE Conference on Computer Vision and Pattern Recognition* (pp. 833-841).

[12] Liu, X., van de Weijer, J., & Bagdanov, A. D. (2018). Leveraging Unlabeled Data for Crowd Counting by Learning to Rank. *arXiv preprint arXiv:1803.03095*.

[13] Zeiler, M. D., Ranzato, M., Monga, R., Mao, M., Yang, K., Le, Q. V., ... & Hinton, G. E. (2013, May). On rectified linear units for speech processing. In *Acoustics, Speech and Signal Processing (ICASSP), 2013 IEEE International Conference on*(pp. 3517-3521). IEEE.

[14] Long, J., Shelhamer, E., & Darrell, T. (2015). Fully convolutional networks for semantic segmentation. In*Proceedings of the IEEE conference on computer vision and pattern recognition* (pp. 3431-3440).

[15] Simonyan, K., & Zisserman, A. (2014). Very deep convolutional networks for large-scale image recognition. *arXiv preprint arXiv:1409.1556*.

[16] Boominathan, L., Kruthiventi, S. S., & Babu, R. V. (2016, October). Crowdnet: A deep convolutional network for dense crowd counting. In *Proceedings of the 2016 ACM on Multimedia Conference* (pp. 640-644). ACM.

[17] Sindagi, V. A., & Patel, V. M. (2017, October). Generating high-quality crowd density maps using contextual pyramid cnns. In *2017 IEEE International Conference on Computer Vision (ICCV)* (pp. 1879-1888). IEEE.

ISPECE

IOP Publishing

Detector design based on MIMO OTA test

Zhaoqing Wang

School of Information& Communication Engineering, Beijing Information Science& Technology University, Beijing 100101, China

mj741561@163.com

Abstract. The MIMO OTA test is currently one of the more mainstream test methods for multi-antenna terminal performance. However, in the MIMO OTA test, there are still problems such as complicated calculation amount and difficulty in realizing the model. This paper will optimize and improve the traditional MIMO OTA test method for the above problems, mainly using the PAS constraint. Finally, the optimized algorithm in this paper has made great progress in both the amount of calculation and the simplicity of the model.

1. Introduction

The multiple-input and multiple-output (MIMO) technique refers to transmit and receive signals through multiple antennas, which improves the communication quality. MIMO has become an essential element of wireless communication standards such as IEEE 802.11n, HSPA+, WiMAX and LTE 4G. It can make full use of space resources and multiply multiple antennas through multiple antennas. It can multiply the capacity of the system channel (data throughput, Qos and cell coverage) without additional transmit power or bandwidth. The MIMO technology is regarded as the core technology of the next generation mobile communication. Therefore, it is necessary to set a standard test method to test the MIMO device performance. There are many different MIMO test methods. Spatial correlation is a parameter including antenna and propagation characteristics. OTA test can test antenna and propagation effects in one testing time. Therefore, OTA test is a potential test method.

The cluster can be used to model the multipath problem. Each cluster is composed of the same PAS (power azimuth spectrum) scatter. The limited number of probes can be translated to an arbitrary number of clusters. The PAS has two parameters—angle of arrival and angle spread.

The power azimuth spectrum (PAS) plays a very important role in multi-antenna technology. The spatial correlation between the waves impinging on the two antenna elements depends on the

PAS and radiation pattern of the antenna elements, spatial correlation is selected as the main figure of merit to character the channel spatial [1]. The channel will present different PAS when the sampling points are limited, although the spatial correlation is same. Therefore, a new method is proposed to solve the problem. We add constraints on the spatial correlation. The emulated PAS should be close to the target PAS in terms of mean angle of arrival (AoA) and azimuth spread (AS).

2. Background

2.1. 3GPP candidate test methodologies

The seven test methods were proposed to test the performance of MIMO. They can fall into two main type: Anechoic chamber methods and Reverberation chamber methods. The anechoic and reverberation methods are two fundamentally different approaches with same goal—the creation of a spatially diverse

---

Content from this work may be used under the terms of the Creative Commons Attribution 3.0 licence. Any further distribution of this work must maintain attribution to the author(s) and the title of the work, journal citation and DOI.

Published under licence by IOP Publishing Ltd

radio channel. Anechoic chamber method is an effective method. Multiple probes are used to launch signals at the DUT in order to create known angles of arrival, which map onto the required channel spatial model [2]. However, large numbers of probes are required to achieve arbitrary channel model flexibility, which increases the cost and complexity. The reverberation chamber method provides a spatial richness through the natural reflection in the chamber, and further provides a space field by using an oscillating mode agitator, which is close to the isotropic field for a long time. The reverberation chamber can be used to measure the spatial multiplexing gain in the de-correlating antenna [2].

2.2. Configuration of MIMO OTA Setup

The set up of MIMO OTA test are showed in figure 1. The device under test (DUT) is in the center of the multiple probe anechoic chambers. The probes revolve around the DUT and are connected to the channel emulator. The BS emulator measures the signal for testing to the multi- channel emulator. The power amplifiers are required to make up for the path loss in the anechoic chamber [3].

In an emulated channel, the power weights are determined. I use prefaded signal synthesis technique (PFS) to radiate independent fading signals from multiple probes, a single cluster is mapped to multiple OTA probes based on PAS shape and OTA probe angle position to reconstruct PAS, appropriate average power weights should be allocated to the associated OTA probes when reconstructed the spatial characteristics at receiver [1].

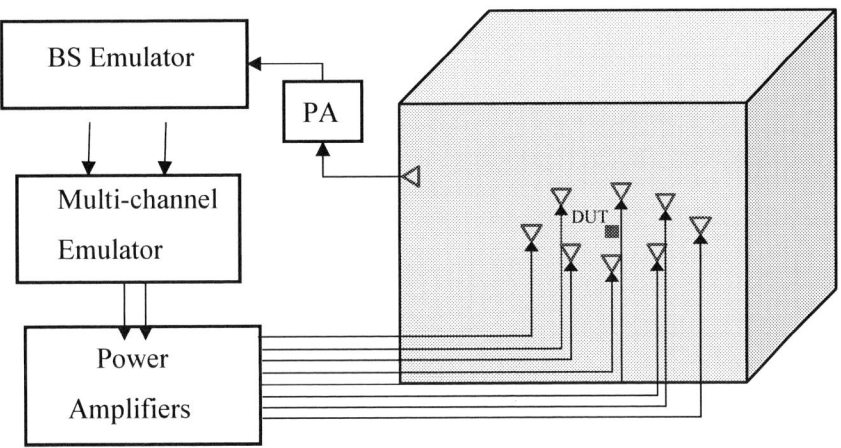

Figure 1 MIMO OTA setup

3. Design and Implementation

In order to consider the influence of antenna correlation on performance, SCME channel model based on SCM model is selected as the target channel model for emulation. Considering the actual situation, the urban macro mode is suitable. In SCME model, a series of fixed power delay and angle parameters are often used to generate the channel model. The simulation process is simple and the channel matrix generates faster.

3.1. The principle of MIMO OTA channel modeling

As description of modeling the MPAC channel in the literature [4], for a NXM linear time- varying MIMO system consisting of N transmitting antennas and M receiving antennas, the transmission matrix between the output and the input can be described by impulse response in the theory of signal and system. The ray-tracing is selected to model the channel. Assuming that there are L roots in the identifiable path in the propagation environment, it has

ISPECE

IOP Publishing

$$H(t,\tau) = \sum_{l}^{L} H_l(t,\tau) = \begin{bmatrix} h_{11}(t,\tau) & h_{1N}(t,\tau) \\ h_{M1}(t,\tau) & h_{MN}(t,\tau) \end{bmatrix} \tag{1}$$

$$H_l(t,\tau) = \iint F_M^T(\phi) h_l(t,\tau,\varphi,\phi) F_N(\varphi) d\phi d\varphi \tag{2}$$

$$h_l(t,\tau,\varphi,\phi) = \begin{bmatrix} a_l^{VV}(t) & a_l^{VH}(t) \\ a_l^{HV}(t) & a_l^{HH}(t) \end{bmatrix} \delta(\tau - \tau_l)\delta(\phi - \phi_l)\delta(\varphi - \varphi_l) \tag{3}$$

H(t, τ) represents the impulse response matrix of the l-th path between the transceivers,

(φ)and **FM**(φ)are the row vector of vector N and vector M, contains the gain information of the antenna array at the transmitter and receiver in the MIMO system. In other words, each element in

(φ)and **FM**(φ) represents the complex gain of each transmit and receive antenna element. This complex gain is a function of the angle and contains amplitude and phase information. φl is the AoD of the l-th path, φl is the AoA of the l-th path. (t, τ, φ, φ) contains the polarization conversion information of the l-th path in the propagation process.

The most important in the channel modeling of MIMO systems is to fully reflect the spread in the spatial domain during signal propagation. It will lead to changes in the spatial correlation between the multiple receiving antennas of the terminal, thus affecting system performance. In the SCME model, the TDL model is introduced. Each mid-path corresponds to different tap power and delay. Multiple sub-paths are superposed to form a mid-path. The mid-path can be close to the Rayleigh distribution .The mid-path can be simulated by classical Gaussian data generator.

3.2. SCME modeling process
Determining the environment
The environment of the MIMO OTA test is suitable for urban macro model in SCME.
Setting parameters
The number of BS terminals and MS terminals is set to 1 and 6. Set the direction of the direct path LOS with respect to BS and MS $\theta BS = \theta MS = 0°$.
Set the delay
In SCME model, we often use fixed power delay and angle parameters to generate a channel model. The delay can be obtained from the parameter configuration table of the channel environment, which is set by 3GPP.

The delays for the six main paths are set as [0 0.3600 0.2527 1.0387 2.7300 4.5977] (us) in Table 2.

The delays for each mid-path can be calculated by $\tau_{n,l} = \tau_n + \triangle_{n,l}$, $\Delta n,l$ is the offset delay of the mid-path. [0 0.0125 0.025 0.3600 0.3725 0.3850 0.2527 0.2652 0.2777 1.0387 1.0512 1.0637 2.7300 2.7425 4.5977 4.6102 4.6227] (us).

$$\tau_{n,l}' = T_s \bullet floor(\frac{\tau_{n,l}}{T_s} + 0.5)$$

Quantize the delay and get:

Determine the relative power Pn,l of each mid-path

Checking Table 2 to get the relative power Pn of each main path. According to the number of sub-path (10, 6, 4) obtained in Table 1, the relative power of each medium diameter Pn,l is obtained, which is 1/2, 3/10, and 1/5 of the main path.

Making normalization depends on the Formula (4):

$$P_n = \frac{P_n'}{\sum_{j=1}^{6} P_j'} \tag{4}$$

1350

5. Determine the AoA for each sub-path $\theta_{n,m,oA} = \theta_{MS} + \delta_{n,AoA} + \Delta_{n,m,AoA}$

Table 1. The parameters of the cluster in SCME (3GPP standard)

Mid-path	Power(number of sub-path)	delay(us) in the main path	Sub-path
1	10/20	0	1,2,3,4,5,6,7,8,19,20
2	6/20	0.0125	9,10,11,12,17,18
3	4/20	0.025	13,14,15,16

Table 2. The parameters of urban macro

Scenario		Urban Macro			
Power P_n and respect delay of each main path	1	0		0	
	2	-2.2204		0.3600	
	3	-1.7184		0.2527	
	4	-5.1896		1.0387	
	5	-9.0516		2.7300	
	6	-12.5013		4.5977	
AS at BS and MS		2		35	
AoA and AoD of each main path	1	65.7489	81.9720	76.4750	-127.2788
	2	45.6454	80.5354	-11.8704	-129.9678
	3	143.1863	79.6210	-14.5707	-136.8071
	4	32.5131	98.6319	17.7089	-96.2155
	5	-91.0551	102.1308	167.76567	-159.5999
	6	-19.1657	107.0643	139.0774	173.1860

4. Result analysis

In SCME, a series of fixed power delay and angle parameters are used to generate the channel model. The simulation process is simpler and the generation speed of channel matrix is faster. However, due to the influence of the mid-path, it takes up more memory. In the process of simulation, the SCME model is selected to model the channel.

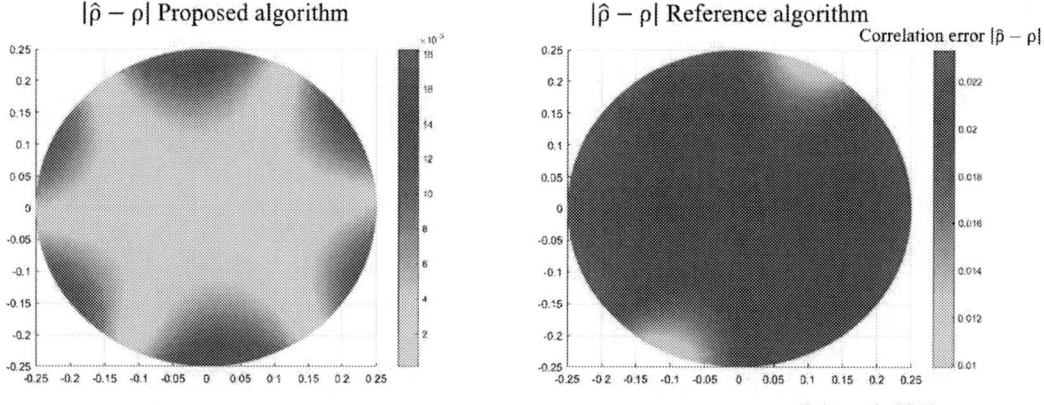

Figure 2 Comparison of correlation for the first cluster of two algorithm under SCME model

The result is showed in figure 6 Correlation errors for the first cluster for the two algorithms are shown in the figure. The reference method is least square algorithm. The reference least square algorithm does not consider the angle, at the area $\Phi a=30°$ and $\Phi a=210°$, the performance is not well. The max deviation (0.024) in reference algorithm is bigger than that in proposed algorithm (0.018). Therefore, the proposed algorithm is better.

5. Conclusion

The paper made a complete design and implement of power allocation for probes in MIMO OTA tests. I finished the main tasks and targets following the schedule. The principle of MIMO OTA was introduced well. The mathematical model for power allocation was established. Spatial correlation was selected to model signal channel. The paper discusses the using of the PAS model, the concept of cluster and prefaded signal synthesis method. After that, paper proposed a formula to calculate the spatial correlation and rewrote the formula as a function of distance and relative position between virtual antennas. The innovation point of the project is introducing constraints on PAS shape in terms of AoA and AS, thus the problem can be converted into a convex problem.

References

[1] Fan, W., Sun, F., & Nielsen, J. Ø. (2013). Emulating spatial characteristics of mimo channels for ota testing. IEEE Transactions on Antennas & Propagation, 61(8), 4306-4314.

[2] Rumney, M., Pirkl, R., Landmann, M. H., & Sanchezhernandez, D. A. (2012). Mimo over-the-air research, development, and testing. International Journal of Antennas & Propagation, 2012(3), 601-617.

[3] Mow, M. A., Niu, B., Schlub, R. W., & Caballero, R. (2014). Tools for design and analysis of over-the-air test systems with channel model emulation capabilities. US, US 8793093 B2.

[4] Sti, P. K., Ms, T. J., & Nuutinen, J. P. (2012). Channel modelling for multiprobe over-the-air mimo testing. International Journal of Antennas & Propagation, 2012.

ISPECE   IOP Publishing

IOP Conf. Series: Journal of Physics: Conf. Series **1187** (2019) 042014   doi:10.1088/1742-6596/1187/4/042014

Research on a fusion gait real-time recognition algorithm

ZHAO Zhi-qiang, LI Meng, DENG Ming-ji, ZHU Zheng and SHI Xin

Chongqing University of Posts and Telecommunications, Chongqing 400065, china

E-mail:lee_meng123@126.com

Abstract. Based on the current market prospects of wearable devices, gait recognition accuracy and real-time market demand, a fusion gait real-time recognition algorithm is designed. The paper introduces the method of limiting the original step signal by using the method of limiting filtering and moving smoothing filtering. At the same time, using dynamic threshold detection and similarity algorithm fusion method, using dynamic window method, the step data is sequentially subjected to dynamic threshold. Only if the two algorithms are simultaneously established, the decision step is established. The fusion gait recognition algorithm is designed to achieve real-time gait recognition. and finally obtain more accurate detection results.

1. Introduction

As a key indicator of human health, the accuracy of the exercise record directly determines the credibility of the data analysis. Therefore, it is necessary to have a better effect in the study of gait recognition algorithms in order to obtain more accurate results. This paper uses STM32 based on M3 core 32-bit processor, and uses ADXL345 three-axis accelerometer to collect three-axis data in motion state, and performs vector synthesis on three-axis acceleration data. We don't consider special situations such as falling and lying on the back in this study.

2. Signal preprocessing

2.1. Raw data description

The original signal collected by the accelerometer contains various interference noises. To avoid noise interference, it is necessary to optimize and improve the hardware design and software algorithms. In terms of hardware design, it is necessary to design a stable power supply and power supply voltage regulator to avoid baseline drift caused by voltage instability. The hardware circuit design also needs to pay attention to high and low frequency separation. Decoupling capacitors are arranged next to the chip power supply pins to remove power frequency interference. Therefore, the design of the experiment is set: the step counter is placed on the waist and abdomen, thereby reducing the large error caused by the conventional device being placed on the wrist due to random swing, so that the gait recognition algorithm is simpler to handle.

Khalil A [1] uses the slope of the adjacent two samples to determine the state of the motion waveform based on the positive and negative slopes. The algorithm roughly determines the peak value. This simple peak judgment method requires the signal to be very filtered. Smooth, no burrs, it is difficult to achieve. Therefore, Mladenov M [2] improved the algorithm of Khalil A. According to the step-by-step characteristics, the window was set, the slope singular point of the fixed window was calculated, the fixed window peak was fixed, and all the peaks and troughs in the window were

Content from this work may be used under the terms of the Creative Commons Attribution 3.0 licence. Any further distribution of this work must maintain attribution to the author(s) and the title of the work, journal citation and DOI.

Published under licence by IOP Publishing Ltd

obtained. The weighted average is used as a condition for judging the threshold step. The peak value is indeed a relatively simple method of judging the number of steps, but this is very complicated for signal processing. The signal must be smoothed to a certain extent before it can be judged by this method. The filtering algorithm is very demanding, and the hardware MCU needs to have Processing speed to guarantee the real-time and effectiveness of the algorithm.

It can be seen from the test data that the acceleration signal of human motion has noise interference in different axial and different frequency bands, and these noises need to be preprocessed. For real-time and effective data processing on the platform, in this paper, the original step signal is preprocessed by limiting filter and moving smoothing.

2.2. Improved limiting filter

Vector synthesized triaxial acceleration data, through MATLAB found that there are some data anomalies, the data value is large, the analysis may be caused by a sudden change in speed during the test, so we need to remove the data first. Limiting filtering can well remove randomly generated interference signals and eliminate the effects of transient instability of the acceleration sensor.

The sampling rate of the original signal is 25 Hz. During the sampling process, the sampling is continuously performed within two sampling intervals, that is, 0.04 s. The values of the adjacent two samples are compared, and the maximum variation is determined within a certain range. The amount of change needs to be generated through experimental experience. The empirical value of this experiment is set to 120. When the actual change amount is greater than the empirical value during the test, the collected signal is considered to be the interference signal, and the previous acquisition data is used instead of the current acquisition, and vice versa. The first data can't be compared, the data error of the first 2s during sampling is large, so it is not used, and the data after 2s is used for testing [3].

First get the vector of the triaxial acceleration xyz and m_n:

$$m_n = \sqrt{x_n^2 + y_n^2 + z_n^2} \tag{1}$$

Amplitude change of vector data:

$$m_n = \begin{cases} m_n, |m_n - m_{n-1}| \le \Delta x \\ m_{n-1}, |m_n - m_{n-1}| > \Delta x \end{cases} \tag{2}$$

The following is the composite waveform, and limited-improved waveform.

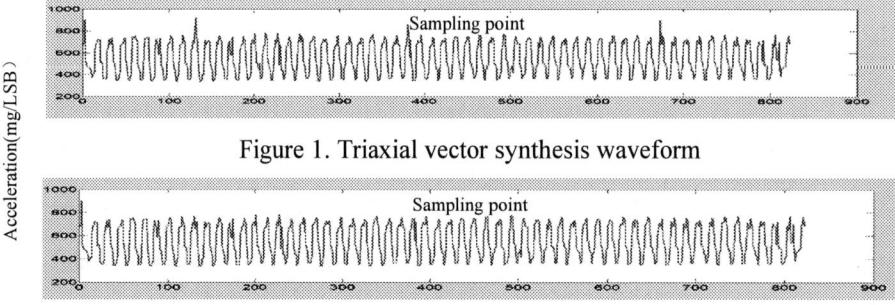

Figure 1. Triaxial vector synthesis waveform

Figure 2. Limiting improved waveform

2.3. Moving smoothing filter

Mobile smoothing filter is often applied to digital signal processing. For time-series discrete data, smooth digital low-pass filter can be used to eliminate glitch and achieve smoothing effect. Smoothing filter is often used in time domain. Signal processing [4].

The smoothing filter processes the continuously sampled w acceleration signals. After one operation is completed, the end data in the array is removed, and the newly sampled acceleration data is inserted into the queue. The remaining w-1 data are moved back, Then, the w data of this queue is calculated, and the input is m(k), the output is y(k), w is the smooth span, and the w/2 elements before and after m(k) are accumulated and averaged. The result is the output y(k), The formula is:

$$y(k) = \frac{1}{w}\sum_{j=k-\frac{w}{2}}^{k+\frac{w}{2}} m(j) \tag{3}$$

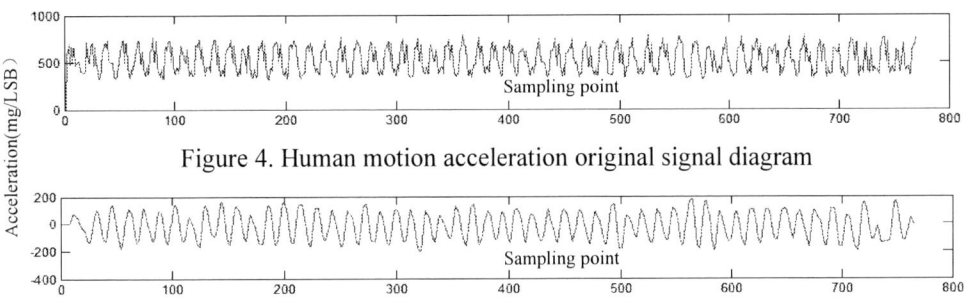

Figure 3. Moving smoothing filter output waveform

According to the gait waveform, first use the limiting filter to remove the singular acceleration signal points in the artificial case, and then use the motion smoothing filter to remove the glitch signal.

The following is the Matlab simulation of the step data. The data is the acceleration raw data of the four-story stairs and the aisle walking for 60s in the first teaching building. The integrated filtering algorithm is used to denoise, and the moving smoothing window length is set to 6. See the front and back contrast waveforms shown in figure 4 and figure 5 below.

Figure 4. Human motion acceleration original signal diagram

Figure 5. Fusion filtered signal diagram

It can be concluded from the simulation results that the use of limiting filtering and moving smoothing filtering can remove signal interference better, and there is almost no glitch signal. The fused filtering algorithm can well realize the preprocessing of the counting signal.

3. Fusion pace decision algorithm design

Through the physical model of human walking, it can be known that after the data preprocessed by the step data, the signal source can be regarded as the processing of the sine wave, and the number of steps is the number of peaks of the sine wave, so the research focus is on Signal feature point extraction, peak search aspect. Combined with multi-party research, the following conclusions can be drawn: the dynamic threshold method alone can be used to determine the number of steps may be too small; the similarity algorithm alone may be too large, depending on the similarity coefficient value. Therefore, this paper adopts the fusion algorithm of dynamic threshold method and similarity algorithm, the advanced similarity degree operation judgment, and then the dynamic threshold judgment. When the two conditions are established at the same time, the step is established, otherwise not counting. This method greatly improves the accuracy of the algorithm.

3.1. Dynamic threshold detection improved algorithm

The dynamic threshold value of the preprocessed signal is determined by step counting. It can be considered that any sine wave has one and only one falling interval, so it is only necessary to detect the number of falling intervals to determine the number of steps. The algorithm diagram is shown in figure 6.

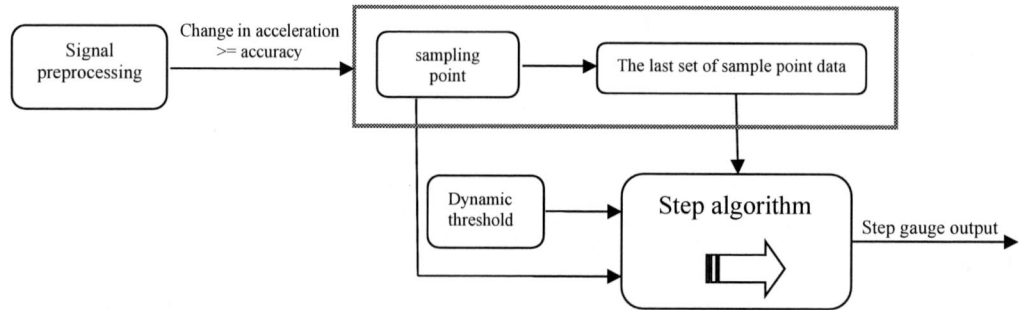

Figure 6. Schematic diagram of step counting dynamic threshold method

The dynamic threshold detection algorithm is to find out S_{max} and S_{min} of the sampled data within a certain time according to the motion frequency of the person, and calculate the average value as the dynamic threshold value, that is, the dynamic threshold $threshold = (s_{max} + s_{min})/2$ Under high frequency interference, the difference between two adjacent sampling points is greater than the dynamic threshold, and the acceleration curve traverses the dynamic threshold curve. From top to bottom, it is considered to be one step [5].

In this paper, the existing dynamic threshold method is improved. Due to the pre-processing of the signal, the step signal will appear like a sinusoidal waveform. For small peaks, only the mean value of the simple signal maximum and minimum values is used. Threshold determination, there will be a step error, so add a coefficient K *(2>K>1)* in the threshold algorithm as follows:

$$threshold = k(s_{max} + s_{min})/2 \qquad (4)$$

The advantage of adding the coefficient K is that the dynamic threshold can be higher than the small peak, and the calculation of the small peak is not performed, and the accuracy of the step determination is improved.

3.2. Waveform similarity algorithm

The step-by-step similarity determination algorithm is to measure the similarity between the processed acceleration signal and the standard signal to obtain the cardinality of the waveform similarity comparison. The similarity degree of the two waveforms is defined, and the two sets of signals are respectively A(i) and B(i), and the multiple k makes k*A(i) approach B(i). The error energy is used to judge the similarity between the two sets of waveforms. Since the processed signal is similar to the cosine signal [6], the cosine signal is used for similarity measurement, and the processed step signal and the standard signal are similarly operated.

Let the original signal be:

$$A = A_a(a_1, a_2, a_3 \cdots a_n) \qquad (5)$$

The signal after preprocessing is:

$$B = B_\beta(\beta_1, \beta_2, \beta_3 \cdots \beta_n) \qquad (6)$$

Here, N is the number of sampling points of the original waveform signal and the processed signal. The similarity of the two discrete signals can be represented by the R_{AB}, and the larger the R_{AB}, the higher the similarity of the two signals.

The similar algorithm coefficient R_{AB} formula is as follows:

$$R_{AB} = \frac{\sum_{i=1}^{N} \alpha_n \times \beta_n}{\sqrt{\sum_{i=1}^{N} \alpha_n^2} \times \sqrt{\sum_{i=1}^{N} \beta_n^2}} \qquad (7)$$

Since the motion waveform is a special periodic waveform, the data is not accurately matched to the original signal and the preprocessed signal. The dynamic window processing data is used. It is assumed that the signal data length is N and there are n waveforms. To improve accuracy, the design of the dynamic window has a step size of 1, and the total sliding distance is m. After experimentation, the similarity is highest when $m = 2N/n(n \geq 3)$.

The following Matlab simulation is the original acceleration data collected by the four testers (A\B\C\D). By comparing the similarity between the filtered data and the sinusoidal signal, the length of the signal segment is divided into 18, and the amplitude of the sinusoidal signal is Comprehensive consideration is based on the acceleration combined vector values. The following is a Matlab diagram of the walking 60-step similarity algorithm.

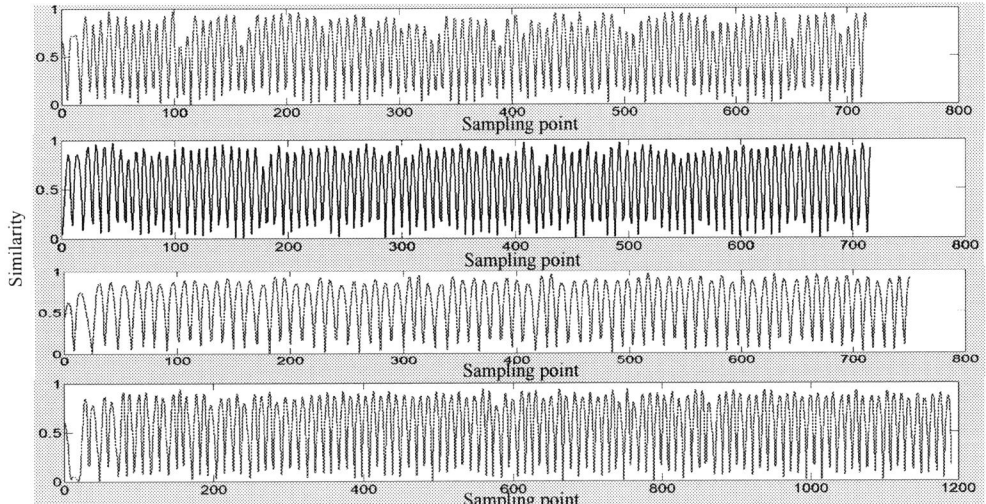

Figure 7. A/B/C/D walking correlation coefficient map

Table 1 is the statistics of the number of similarity signal segments by the ABCD four testers through similar algorithms. It can be seen from the table that the similarity is smaller, the matched signals are met. The more the number of segments, in order to be able to accurately judge the pace, it is necessary to combine other methods.

Table 1. Number of waveforms determined by the similarity coefficient

Tester similarity coefficient	Actual steps	Coefficient of 0.5 Number of signal segments	Coefficient of 0.6 Number of signal segments	Coefficient of 0.7 Number of signal segments	Coefficient of 0.8 Number of signal segments	Coefficient of 0.9 Number of signal segments
A	60	427	364	295	206	87
B	60	432	368	278	175	67
C	60	695	603	502	352	75
D	60	763	639	506	303	28

3.3. Fusion step counting algorithm
In order to improve the accuracy of the step counting, a single dynamic threshold algorithm and a similarity algorithm will interfere with the step determination. This paper uses fusion. The filtering algorithm uses the dynamic window method to sequentially perform the step-by-step data detection by the dynamic threshold method and the similarity algorithm. When both algorithms are determined, the step is established. The gait detection idea of the paper is to process the collected three-axis acceleration data and judge the number of steps. Figure 8 is a flow chart of the fusion gait recognition algorithm.

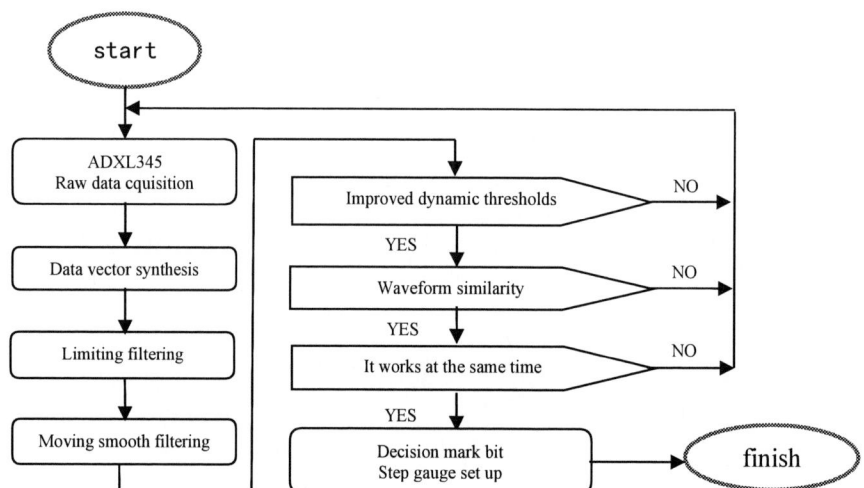

Figure 8. Flow chart of the fusion gait recognition algorithm

4. Experimental results and conclusions

As shown in table 2, the number of steps after testing the comprehensive step determination algorithm for four testers. It can be seen from table 2 that when the similarity is 50%, the fusion algorithm has higher precision. When the algorithm is transplanted, due to the limitation of the hardware platform, there is noise interference of the hardware circuit itself, which is controlled by artificial subjective consciousness. Motion interference increases the difficulty of processing. The 50% fusion determination algorithm is not the most accurate and requires multiple tests. In addition, the fusion algorithm does not consider special postures such as falling and lying, and the judgment of the pace in multiple states needs further study.

Table 2. Fusion step counting algorithm

Tester	Actual steps	Similarity 50% fusion algorithm	Similarity 60% fusion algorithm	Similarity 70% fusion algorithm	Similarity 80% fusion algorithm	Similarity 90% fusion algorithm
A	60	58	56	50	50	35
B	60	57	55	52	42	32
C	60	57	57	55	55	31
D	60	61	59	56	51	18

References

[1] Khalil A, Glal S. SetUp: A Step Counter Mobile Application to Promote Healthy Lifestyle. 2009 International Conference on the Current Trends in Information Technology, 2009:1-5.

[2] Mladenov M, Mock M. A Step Counter Service for Java-Enabled Devices Using a Built-In Accelerometer.Proceedings of the 1st International Workshop on Context-Aware Middlewareand Services, 2015: 1-5.

[3] Fan qiufeng, Hu wanli, Qin changhai. Research on digital filtering in computer control system [J]. Industrial control computer,2016,25(10):39-40.

[4] Huang zheng, Han lixin, Xiao yan. Design of step counting method based on M5 and DTW [J]. Journal of computer science,2016,(39):6-7.

[5] Li yue. Design and research of low power consumption keyless movement wristwatch [D].hangzhou: Zhejiang university,2015.

[6] Pan zhuojin, Wang fang, Zhou zhenhui. Study on waveform similarity measurement in electronic cabin automatic test system of seeker [J]. Computer measurement and control,2014,18(06):1355-1357.

Application research of denoising and super pixel algorithm in image processing

Qian Sun [1,*], Li Xin [1], Hanxu Gao [1], Faliang Chang [2], Zengshun Zhao [1,2,3]

[1]College of Electronics Communications and Phisics, Shandong University of Science and Technology, Qingdao, 266590, China

[2]School of Control Science and Engineering, Shandong University, Jinan, 250061, China

[3]Department of Electrical& Computer Engineering, University of Florida, Gainesville, FL 32611, USA

*Corresponding author e-mail: sunqian940411@163.com

Abstract. With the popularization and development of science and technology, mobile phone, tablet and computer has become the necessities of people, whether work or life, the emergence of science and technology, development and rich brought a whole new world for human civilization, including electronic information in time and space communication provides convenient conditions for people, especially the image processing technology, in the life is very broad. At present, smart phones have become extremely common, and users have a huge demand for images. Every link is inseparable from the formation, acquisition, transmission and acceptance of images. However, in every link, images will be more or less polluted by noise, resulting in users' inability to obtain the desired image effect. However, if the noise is directly optimized or removed, the accuracy of the image will be affected. Therefore, the advanced noise removal technology plays a crucial role in the efficient use of the image. Image superpixel is to gather pixels with similar attributes into a region to represent the image instead of pixels, so as to reduce the order of magnitude of the image atomic structure and further reduce the complexity of the subsequent image processing algorithm, which provides the possibility for the real-time performance of the image processing algorithm.

1. Introduction

Vision is one of the most important ways for human beings to obtain information. It contains more than 70% of all information obtained by human beings. Images are more vivid than words and are the basis of vision. After receiving the image data, the human brain processes, processes and analyzes the data to extract the information it needs. The process of extracting information from an image is actually complicated because the human brain is so exquisitely constructed. With the progress of science and technology and the development of society, human beings have invented computers to help accomplish some tasks. Digital image processing is the process that the computer tries to simulate the human brain to obtain the required information from the

Content from this work may be used under the terms of the Creative Commons Attribution 3.0 licence. Any further distribution of this work must maintain attribution to the author(s) and the title of the work, journal citation and DOI.

Published under licence by IOP Publishing Ltd

image data. In a broad sense, digital image processing includes all image manipulation techniques, which can also be collectively referred to as image engineering [1.2].

Image denoising and super pixel generation are basic problems in the field of image processing, but they are also very important tasks. The purpose of image denoising is to remove noise interference, improve the visual effect of the image, make the image can accurately transmit information and meet the needs of subsequent image processing applications, such as segmentation and super pixel generation, so as to ensure its accuracy. The goal of image superpixel generation is to aggregate pixels into several regions that do not overlap each other and can maintain the original structural characteristics of the image, so as to replace pixel points as the basic atoms to represent the image and reduce the time complexity of subsequent image processing tasks. After de-noising operation and generation of super pixels, the image processing and analysis tasks can not only be protected from noise, but also reduce the complexity of the algorithm, and is conducive to improving the effect of the algorithm. According to figure 1.1 in front of the sample and analysis it can be seen that the existence of the noise generated directly affects the super pixel accuracy, in order to accurate segmentation of noise images, on the one hand to deal with the noise of image, on the other hand can combine image denoising algorithm study design has the noise pixel generation algorithm. Although there are many related literatures on image denoising and super pixel generation, there are still some problems in the algorithm, and the effect still needs to be further improved, more exploration and discovery are needed.

2. Image denoising

2.1. Overview of image denoising

In the process of image acquisition and transmission, it will be affected by equipment and external factors, and noise will be introduced to pollute the image signal. Denoising is a classical and basic problem in image processing and analysis. Common image noises include salt and pepper Noise and Gaussian Noise (AWGN) of Additive White. This paper focuses on salt and pepper noise. Salt-pepper noise, also known as impulse noise, reduces the image quality by randomly changing the color or brightness value of pixels in the image to the maximum or minimum. It is assumed that the level of pepper and salt noise is s,$s \in [0,1]$, pixel x, the probability of 1-S remains unchanged with its true color value $I(x)$, and the probability of s/2 each becomes the maximum value dmax or the minimum value dmin, and its mathematical expression is as follows[3]:

$$I(x) = \begin{cases} d_{min}, s/2 \\ d_{max}, s/2 \\ I(x), (1-s) \end{cases} \tag{1}$$

To generalize the image denoising model, it can be expressed as

$$G = I + N \tag{2}$$

Where I is the clean image matrix without noise, N is the noise matrix, and G is the image matrix with noise observed. The purpose of image denoising is to remove the noise N from the noisy image G and obtain the denoised image I to make i most similar to the real image I..

(a)The original image   (b) Add a Gaussian noise image   (c) Add salt and pepper noise image

Figure 1. Image denoising

2.2. denoising algorithm

Existing image denoising algorithms can be generally divided into two categories, spatial domain method and transform domain method, according to the different basic objects. Image denoising algorithm in spatial domain is to process pixel points directly and separate the real signal from noise. Contrast, transform domain algorithm is to use some method of transform signal transformation to another domain, such as the frequency domain, and then according to the different characteristics of signal and noise in the transform domain to transform domain coefficient, finally, inverse transform to the image spatial domain, and the image denoising after, achieve the goal of denoising. Commonly used transforms include Fourier transform, wavelet transform, etc. [4]

2.3. Common test data of denoising algorithm

Whether the image processing algorithm is designed successfully or not usually needs to be tested by many experiments. Only through comprehensive analysis of the experimental results can a reasonable and credible conclusion be obtained. In the process of algorithm testing, open data sets are generally adopted. Common image denoising test images are shown in figure 2 and figure 3.

Figure 2. Sample grayscale images commonly used in image denoising tests.

Figure 3. Example of color image commonly used in image denoising test.

2.4. Evaluation criteria of denoising algorithm

The evaluation methods of denoising algorithm can be divided into subjective method and objective method. The subjective evaluation method is to perceive the visual effect of the results processed by the algorithm with human eyes, and measure whether the algorithm meets the expectation, removes the noise, causes the blur and so on. Obviously, the subjective evaluation of the results is closely related to the evaluator himself, and the evaluation conclusions obtained by different observers may vary greatly. Therefore, purely subjective evaluation can not be qualitative for algorithm performance[5].

In order to evaluate the denoising algorithm objectively, scholars put forward some quantitative criteria in an attempt to accurately describe the perception effect of vision on image results. Evaluation indexes of common denoising algorithms include Peak signal-to-noise Ratio (PSNR) and Structure Similarity Index Measurement SSIM [6].

3. Image super-pixel generation algorithm

One of the most difficult tasks in the field of image processing is image segmentation. So-called image segmentation is to given an input image, on the basis of all color, gray scale, shape and texture feature information is divided into a number of mutually disjoint and can cover the entire image area, and makes all pixels in each small region with same or similar characteristics, and in any two have the obvious difference between different areas. One of the most important steps from image processing to image analysis is image segmentation. Image over-segmentation is a kind of image segmentation technology. The difference lies in that the image over-segmentation is to divide an input image into more non-overlapping areas of smaller size, and each small area is called super pixel. In other words, the process of image super pixel generation is to combine adjacent pixels together to form a region according to the characteristic information such as grayscale, texture, color and shape, so as to make the features of pixels within the region consistent and the pixels contained in any two different regions have obvious differences. If the input image is set as I and the number of pixels it contains is N, then the hyper-pixel generated by over-segmentation can be expressed as:

$$I = \left\{ S_j \mid S_j \cap S_i = \emptyset, i \neq j \right\}, (i, j = 1, 2 \dots, K) \tag{4}$$

Where K is the number of superpixels, S_j and S_i represent the J_{th} and i_{th} superpixels respectively. One of the most important properties of an ideal hyperpixel segmentation is that it fits well into the image edges.

4. Common test database for super pixel generation algorithm

Research and design of a new algorithm, always through the test before its performance, to determine whether the algorithm is available, effective. For image Segmentation algorithm, this paper used in The test database is also commonly used data sets is by The United States, Berkeley public image Segmentation of The computer vision and Benchmark data sets (The Berkeley Segmentation Dataset and Benchmark, BSD) The test data set is The Berkeley computer vision for image Segmentation and edge detection of The institute to provide The testing image set and its Benchmark. It contains a total of 500 natural images, of which the test set contains 20. , 200 images are contained in training set, and 100 images are contained in val set.

Figure 4 shows an example of an image in BSD. There are two sizes of images in the data set, the first one as shown in the first line of figure 4 with a size of 481 x 321, and the second one as shown in the second line of figure 4 with a size of 321 x 4810.

der to provide a reliable comparison benchmark for algorithm comparison, each image in the database is manually marked by five different individuals on the boundary, and the results are averaged as the boundary benchmark for algorithm evaluation. At the same time, the database also provides a color segmentation benchmark. Each region is displayed with a color. The segmentation benchmark of the image in figure 4 is shown in figure 5. For the boundary, binary images are used to display the database. White pixels represent the marked boundary, and the boundary reference of the image in FIG. 4 is shown in FIG. 6.

Figure 4. BSD centralized image example.

Figure 5.BSD centralized image segmentation region benchmark example (corresponding to the image shown in figure 4).

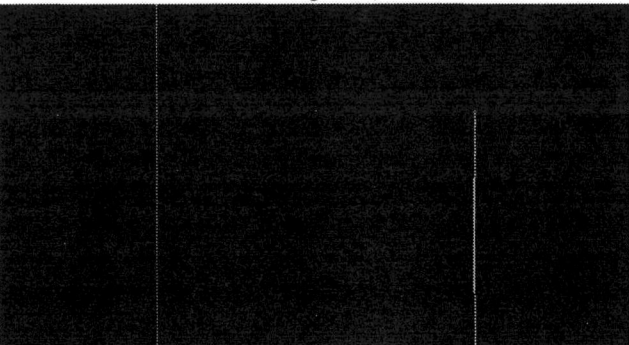

Figure 6. An example of a bsd-focused image boundary benchmark (corresponding to the image shown in figure 4).。

5. Evaluation criteria of super pixel generation algorithm

Image is the main form of visual presentation, so for almost all image processing algorithms, visual effect as a subjective evaluation standard has very important significance. For the same image, different individuals will have different visual perception effects, so the visual evaluation mechanism has strong subjectivity, which generally requires more tests to obtain relatively reliable evaluation results. Therefore, vision can be used as a reference for the evaluation of image processing algorithm results, but it is not all. An objective evaluation standard is needed to give a quantitative evaluation result for the algorithm. In order to objectively evaluate the image superpixel generation algorithm, scholars have proposed several quantitative criteria to accurately describe the visual perception of image results. The

ISPECE

IOP Publishing

commonly used image pixel generation algorithm of evaluation criteria are: Boundary Recall rate (a Boundary Recall BR), Under Segmentation rate (Under SegmentationError USE) Segmentation Accuracy (Achievable Segmentation Accuracy, ASA) and Compactness (Compactness COM).

Acknowledgements
National Natural Science Foundation of China (Grant No. 61403281);
Natural Science Foundation of Shandong Province, China (ZR2014FM002);
China Postdoctoral Science Foundation (2015T80717).

6. Conclusion
In the aspect of image denoising, the characteristics of each method are analyzed from local method and non-local method. The non-local similarity denoising method is relatively effective, but similar block matching is needed, and the computational complexity of this process is relatively high. In the aspect of super pixel generation, several commonly used image segmentation test data sets are also introduced. One of the most important properties of superpixels is to fit the image boundary as well as the compactness of superpixels. However, the opposition between the two makes it necessary to find a balance between them. For the input image and video signal, the noise must be removed before accurate super pixel generation and subsequent image processing.

References
[1]   G Pok, J. C. Liu, A. S. Nair. Selective removal of impulse noise based on homogeneity level information.[J]. IEEE Transactions on Image Processing A Publica-tion of the IEEE Signal Processing Society, 2003, 12(1):85-92.
[2]   Cecile Louchet, Lionel Moisan. Total variation denoising using iterated conditional expectation. In Signal Processing Conference. 2014, 1592-1596.
[3]   David L Donoho, Jain M Johnstone. Ideal spatial adaptation 6y wavelet shrinkage[J]. Biometrika, 1994, 81(3):425-455.
[4]   Xiao Pan, Yuanfeng Zhou, Caiming Zhang, Qian Liu. Flooding based superpixels   generation with color, compactness and smoothness constraints. In Image Pro   cessing (ICIP), 2014 IEEE International Conference on. IEEE, 2014, 4432- 436.
[5]   Zhou Wang, Alan Conrad Bovik, Hamid Rahim Sheikh, Eero P Simoncelli. Image quality assessment: From error visibility to structural similarity[J]. IEEE Transactions on Image Processing, 2004, 13(4):600-612.
[6]   David Martin, Charless Fowlkes, Doron Tal, Jitendra Malik. A database of human segmented natural images and its application to evaluating segmentation algorithms and measuring ecological statistics. In Computer Vision, 2001.ICCV 2001 .Proceedings. Eighth IEEE International Conference on. IEEE, 2001,volume 2, 416-423.

ISPECE

IOP Publishing

Evaluation Method and Experimental Study on Stationarity of High-Precision Linear Motion

Yifan Zhou[1], Xingbao Liu [1], Yangqiu Xia, Shaopeng Cai [1,a]

[1] Institute of Mechanical Manufacturing Technology, China Academy of Engineering Physics, Mianyang City Sichuan Province 621900, China.

[a]Shaopeng Cai: shpenc@163.com

Abstract: In this paper, a method for detecting the high-precision linear motion stability of CNC machine tools is proposed, and the four indexes of maximum fluctuation, rate accuracy, rate stability and rate volatility are used as the evaluation indexes of linear motion stability. Based on the principle of laser interferometry, this method uses Quick View XL™ from Renishaw to collect machine operating data for linear motion stability analysis. The experimental results show that the evaluation method can effectively quantify the key indicators of linear motion stability performance.

1. Introduction

Accurate and stable feed rate is one of the important factors influencing the metal removal rate and surface finish. Especially in the field of ultra-precision machining, the stability of linear motion is particularly important. At present, a lot of research has been carried out at home and abroad on the speed stability of machinery such as turntables and robots. Among them, scholars from Harbin Institute of Technology[1] and Xi'an University of Electronic Science and Technology[2] have conducted a series of research on the speed stability of turntables. Shen J [3], Chwastek S [4]and Hui L[5] et al analyzed the control method of the speed stability of the turntable and feed system. Besides, Yuan J[6] and Choi P J[7] et al studied the motion stability of the robot arm. In theoretical research, Fuller A T et al[8] systematically analyzed the stability of motion. However, for the smoothness of the linear axis, the low-speed inhomogeneity is mainly analysed[9,10], and little research has been done on the high-precision linear motion stability evaluation.

On this basis, this paper proposes a method for detecting the high-precision linear motion stability of CNC machine tools, and uses four indicators as the characterization of the linear motion stability. Based on the principle of laser interferometry, the method collects the data of machine for analysis and calculation, and obtains the quantitative evaluation results of the key indicators of linear motion stability.

2. Evaluation Index

In order to evaluate the smoothness of linear motion, the evaluation index should be first proposed. Through theoretical analysis, four key indicators are proposed and the accuracy and fluctuation of the motion are comprehensively analyzed.

2.1 Maximum fluctuation

Content from this work may be used under the terms of the Creative Commons Attribution 3.0 licence. Any further distribution of this work must maintain attribution to the author(s) and the title of the work, journal citation and DOI.
Published under licence by IOP Publishing Ltd

This indicator intuitively reflects the maximum fluctuation of motion. The formula for the calculation is as follows:

$$\Delta v = v_{max} - v_{min} \qquad (1)$$

Where: v_{max} is the maximum speed; v_{min} is the minimum speed; Δv is the maximum fluctuation.

2.2 Rate accuracy

Rate accuracy refers to the accuracy of the actual rate, indicating the consistency of the actual velocity of the linear motion with the theoretical value. Calculated as follows:

$$v_j = \frac{1}{v_g}\left|\overline{v} - v_g\right| \qquad (2)$$

Where: v_g is the nominal value (theoretical value) of a given rate; $\overline{v}$ is the average (actual value) of the speed of multiple measurements; v_j is the rate accuracy.

At present, the speed measurement method includes two methods, a method of measuring distance at fixed time and a method of measuring time at fixed distance.

2.2.1 Method of measuring distance at fixed time

The machine operates in rate mode at a given rate. When it is stable, the linear axis position increment is read at a time interval t. Based on the speed calculation formula, the displacement increment divided by the time interval t is the machine speed. It should be noted that the selection of the fixed time interval t is related to the given rate command. The mathematical expression of the rate accuracy of the timing ranging method is as follows:

$$v_j = \frac{1}{v_g}\left|\overline{v} - v_g\right| = \frac{t}{L_g}\left|\frac{\overline{L}}{t} - \frac{L_g}{t}\right| = \frac{1}{L_g}\left|\overline{L} - L_g\right| \qquad (3)$$

Where: L_g is the nominal value (theoretical value) of the position increment for the linear axis at a given rate and given time interval. $\overline{L}$ is the average (actual value) of the position increment for multiple measurements.

2.2.2 Method of measuring time at fixed distance

The linear axis operates in the rate mode according to the given rate. The fixed position interval l is selected. After the stable operation, the time taken for the linear motion to travel through the fixed position interval is measured, and the measurement is performed multiple times. The mathematical expression of the rate accuracy of the distance measuring method is as follows:

$$v_j = \frac{1}{v_g}\left|\overline{v} - v_g\right| = \frac{T_g}{l}\left|\frac{l}{\overline{T}} - \frac{l}{T_g}\right| = \frac{1}{T_g}\left|\overline{T} - T_g\right| \qquad (4)$$

Where: T_g is the nominal value (theoretical value) of the time taken by the linear axis at a given rate and a given position interval. $\overline{T}$ is the average of the time increments of multiple measurements (actual value).

2.3 Rate stability

Rate stability refers to the smoothness of speed, in other words, the deviation of speed from its average value. The formula is as follows.

$$v_w = \frac{1}{\bar{v}}\sqrt{\frac{1}{N-1}\sum_{i=1}^{N}\left(v_i - \bar{v}\right)^2} \qquad (5)$$

Where: N is the number of measurements; v_i is the speed of the ith measurement; $\bar{v}$ is the average of the multiple measurements; and v_w is the rate stability factor.

Tip 1 : The choice of the number of measurements N is related to the given rate. In fact, the rate stability is the mean square error of the rate accuracy, which is the average of the distances of the data from the mean. It is used to estimate the fluctuation degree of the actual rate around the mean and quantitatively evaluate the overall volatility of the speed.

In the same way, the rate stability calculation methods of the two methods of timing ranging and distance measurement are analyzed.

2.3.1 Method of measuring distance at fixed time
The formula for calculating the Rate stability of the timing ranging method is as follows.

$$\begin{aligned} v_w &= \frac{1}{\bar{v}}\sqrt{\frac{1}{N-1}\sum_{i=1}^{N}\left(v_i - \bar{v}\right)^2} \\ &= \frac{t}{\bar{L}}\sqrt{\frac{1}{N-1}\sum_{i=1}^{N}\left(\frac{L_i}{t} - \frac{\bar{L}}{t}\right)^2} \\ &= \frac{1}{\bar{L}}\sqrt{\frac{1}{N-1}\sum_{i=1}^{N}\left(L_i - \bar{L}\right)^2} \end{aligned} \qquad (6)$$

Where: N is the number of measurements; L_i is the position increment of the ith measurement; $\bar{L}$ is the average of the position increments of multiple measurements.

2.3.2 Method of measuring time at fixed distance
The formula for calculating the rate stability of the fixed distance measuring method is as follows.

$$\begin{aligned} v_w &= \frac{1}{\bar{v}}\sqrt{\frac{1}{N-1}\sum_{i=1}^{N}\left(v_i - \bar{v}\right)^2} \\ &= \frac{\bar{T}}{l}\sqrt{\frac{1}{N-1}\sum_{i=1}^{N}\left(\frac{l}{T_i} - \frac{l}{\bar{T}}\right)^2} \\ &= \frac{1}{\bar{T}}\sqrt{\frac{1}{N-1}\sum_{i=1}^{N}\left(T_i - \bar{T}\right)^2} \end{aligned} \qquad (7)$$

Where: N is the number of measurements; T_i is the time interval of the ith measurement; $\bar{T}$ is the average of the time intervals of multiple measurements.

2.4 Rate volatility
Rate volatility reflects the fluctuation of motion from another angle. The definition of it is as follows.

$$v_\delta = \frac{\Delta v}{\bar{v}} \times 100\% \qquad (8)$$

Where: Δv is the maximum fluctuation of the rate, $\bar{v}$ is the average speed. If the value of v_δ. is smaller, the velocity fluctuation of linear motion is smaller, and the machine tool runs more smoothly, on the contrary, the operation is more unstable. The initial setting is $v_\delta<0.2$. In principle, it can be considered that the running smoothness is better.

3. Detection method

3.1 Measurement principle
In this paper, the velocity of linear motion is measured based on the principle of laser interferometry. Interferometry is a technology of incremental method based on the principle of light wave interference. It fixedly connects the reflector and the test object; the interferometer mirror is fixed. When the reflector moves with the test object, the difference of the optical path between two beams will change, and the interference fringes will alternate alternately. Finally, the measured distance can be determined by detecting the number of interference fringes in the photoelectric detector. Further, within a specified time interval, the movement distance in that time interval is obtained. On paper, the average value of the moving distance is the average velocity of the object. When the time interval is small enough, it can be approximately considered as instantaneous velocity.

3.2 Test
According to GB/T17421.1, environmental conditions, temperature rise and other test conditions were pretreated. And a standard program is compiled to make the moving parts move along a straight line for a period of time(for example, 5 seconds) and record the actual speed, which is generally a velocity fluctuation curve with burrs.

3.2.1 Test subjects
The Y-axis motion stability of a gantry machining center is tested, the full stroke is 2000mm, and the cutting speed range is(1~7000)mm/min.

3.2.2 Test Instruments
The measurement is performed by the Quick View XL™ provided by Renishaw XL-80). The sampling frequency is 10Hz~50kHz, the speed range is (0~4)m/s and the accuracy is ±0.04%, which meets the require-ment of measurement accuracy. Before the test, the pre-trigger time, post-trigger time, and sampling frequency need to be set. The measuring device is as follows:
- Quick View XL™ software;
- A Renishaw XL80 laser head;
- A Renishaw XC environmental compensation unit;
- A set of measuring optics, including spectroscopes, mirrors and related mounting components;
- A notebook or desktop PC.

The installation of the equipment is shown in Figure 1.

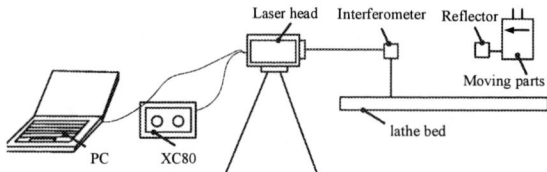

Figure 1. Equipment installation diagram.

4. Results
(1) Test 1: Stationarity of linear motion at different sampling frequencies

The choice of sampling frequency (how many sample points per second) will affect the quality of the test results.

Sampling time is set to 2 seconds, and the detection area is in the middle of the Y axis. The same velocity wave is sampled at different sampling frequencies for data acquisition and stationarity evaluation. When the set speeds are 100mm/min and 1000mm/min respectively, the curve of stationarity of linear motion with sampling frequency is shown in figs. 2 and 3.

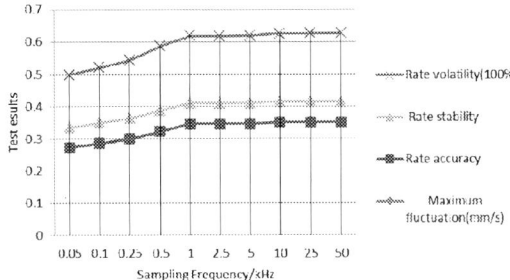

Figure 2. The curve of the stationarity of motion with the sampling frequency when the speed is 100mm/min.

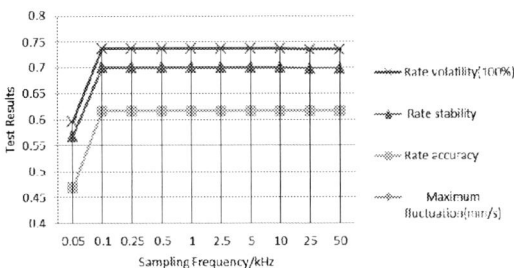

Figure 3. The curve of the stationarity of motion with the sampling frequency when the speed is 1000mm/min.

(2) Test 2: Stationarity of linear motion at different set speeds

Exploring the variation of the linearity of linear motion at different set speeds, the detection parameters are set as follows: sampling time is 2s; the measurement area is the middle of the Y axis; sampling frequency is 50kHz. Figure. 4 is the curve of the stationarity of motion with the setting speed.

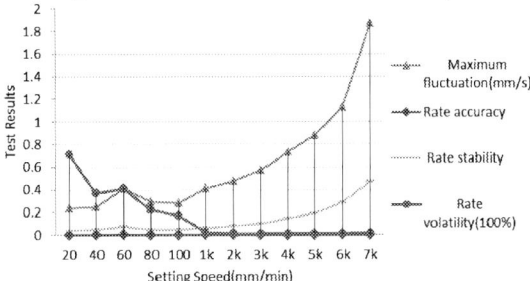

Figure 4.The curve of motion stationarity varying with speed.

(3) Test 3: Stationarity of linear motion of linear axes at different mechanical positions

The detection parameters are set as follows, the sampling time is 2s, the detection position is in the middle of the Y axis, sampling frequency is 50 kHz and the set speed is 100 mm/min.

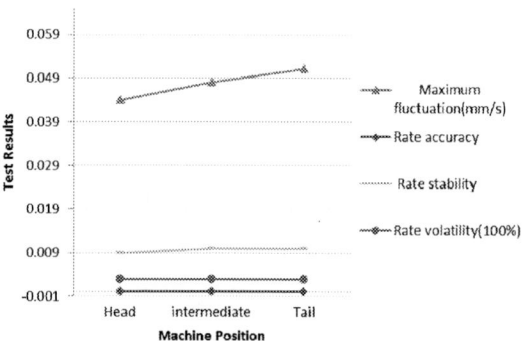

Figure 5. The curve of motion stationarity varying with the detection area

(4) Test 4: Stationarity of linear motion at different sampling times

The detection area is in the middle of the Y axis, the sampling frequency is 50kHz, and the speed is 100 mm/min. The measurement time range is determined by the maximum feed rate and the full stroke of the linear axis. It should be noted that the acceleration/deceleration section should be avoided during the measurement, and the measurement is started after the machine runs smoothly .

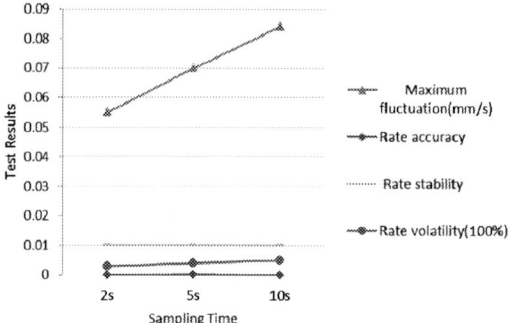

Figure 6. The curve of motion stationarity varying with the sampling time

(5) Test 5: Stationarity of linear motion under different filter time constants

The detection parameters are set as follows: sampling time is 2s, detection area is in the middle of Y axis, sampling frequency is 50kHz and setting speed is 100mm/min.

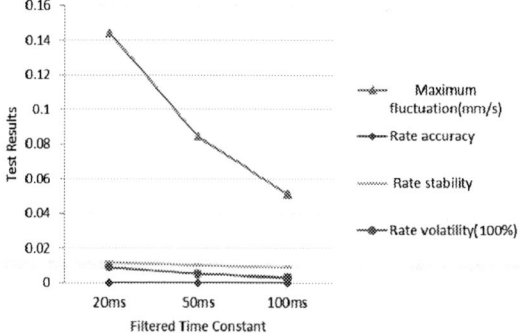

Figure 7. The curve of motion stationarity varying with the filter time constants

5. Discussion

Based on the results, the following conclusions can be drawn:

(1) Test 1, with the increase of sampling frequency, the values of each evaluation index are gradually stable. In figures 2 and figures 3, When the sampling frequency is greater than a certain value, the indicators are basically unchanged. According to the sampling theorem, the lowest sampling frequency must be twice the signal frequency. If a sampling frequency is given, the maximum frequency at which the signal can be correctly displayed without distortion is called the Nyquist frequency. If the signal contains a component with a frequency higher than the Nyquist frequency, the signal will be distorted. The test results are consistent with the theory. The test show that it is recommended to set the sampling frequency to 5~10 times of the highest frequency of the signal. An empirical value 1kHz of sampling frequency is also given.

(2) Test 2, as the speed increases, the value of maximum fluctuation and rate stability increases, and rate volatility and rate accuracy decrease, which is consistent with the actual situation, In fact, when the speed increases, the greater the fluctuation of machine tools;

(3) Test 3, the linear axis has different detection area. The closer to the tail, the maximum speed fluctuation is obviously increased, and other parameters are relatively stable. It is related to the structure of the machine tool itself. The front end of machine tool is commonly used in processing area, and its motion stability is also high.;

(4) Test 4, the longer the sampling time, the larger the maximum fluctuation of the rate is, the trend of the rate fluctuation rate is relatively flat, and the other parameters are relatively stable. Theoretically, the longer the sampling time, the greater the probability of large fluctuations, which is consistent with the actual situation.

(5) Test 5, with the increase of filtering time, the maximum fluctuation, rate stability and rate volatility all show a downward trend, and the rate precision value is relatively stable. The exponential filtering function can be used to reduce the noise caused by electrical noise and vibration or air disturbances. It should be reasonably set to avoid losing valid information. It is recommended to set it up to 63% of the time required for the laser readout change after a step move.

6. Conclusions

Based on the principle of laser interference, a method for detecting the stationarity of linear motion is proposed in this paper. Combined with the experiment, the parameters such as sampling frequency, sampling time, filtering time and motion speed which affect the validity of the detection results are studied. And the stationarity of linear motion is evaluated by four indexes: maximum fluctuation, rate accuracy, rate stability and rate volatility. The experimental results are consistent with the actual situation, which indicates that the detection and evaluation method proposed in this paper can effectively quantify the stationarity of linear motion.

Acknowledgements

This research was financially supported by research on reliability evaluation index system and method of optical components ultra precision manufacturing equipment (Grant No. J043-18-HX).

Reference

[1] Meiling Z. (2013)Research on the stability of turntable speed. Harbin Institute of Technology.
[2] Yachao C. (2014)Research on the control method of turntable speed stability. Xidian University.
[3] Shen J, Xin B, Cui H, et al.,(2017)Speed Control of Single-axis Rotation INS By Tracking Differentiator Based Fuzzy PID. IEEE Transactions on Aerospace & Electronic Systems, PP(99):1-1.
[4] Chwastek S. (2013)Modal Stability Control of a Wheeled Heavy Machine. Applied Mechanics & Materials, 477-478:69-72.
[5] Hui L, Ying H, Huijie Z, et al. (2014) Law of the influence of the change of transmission stiffness on the stability of motion accuracy of CNC machine tool feed system. Journal of Mechanical Engineering, **50** (23): 128-133.

[6] Yuan J, Zhang W, Tao J, et al. (2007)Research on Novel Wire Driving Robot Manipulator for Local Industrial Production Line. In: International Conference on Mechatronics and Automation. IEEE, Harbin, China. 3925-3930.

[7] Choi P J, Rice J A, Cesarone J C. (2010)Kinematics of an infinitely flexible robot arm[J]. Journal of Field Robotics, 10(4):407-425.

[8] Fuller A T, Hostetter G H. (2007)*Stability of Motion*. IEEE Transactions on Systems Man & Cybernetics, 6(12):887-887.

[9] Houari A, Bouabdallah A, Djerioui A, et al. (2017)An Effective Compensation Technique for Speed Smoothness at Low Speed Operation of PMSM Drives. IEEE Transactions on Industry Applications, PP(99):1-1.

[10] Khim G, Park C H, Shamoto E, et al. (2011) Prediction and compensation of motion accuracy in a linear motion bearing table. Precision Engineering, 35(3):393-399.

ISPECE   IOP Publishing

IOP Conf. Series: Journal of Physics: Conf. Series **1187** (2019) 042017   doi:10.1088/1742-6596/1187/4/042017

The Design of Image Depth Information Extraction Algorithm Based on Joint Bilateral Filtering

Lin Yong

Department of Communication, Chongqing College of Electronic Engineering, Chongqing 401331, China

Lin Yong: 65437478@qq.com

Abstract: With the vigorous development of science and technology, specialists and scholars pay their attention to the 2D-to-3D media conversion more and more. As the next generational display, 3D-TV has lots of problems to solve, which must depend on depth estimation. Here it proposes a novel line tracing method and depth refinement filter as core of depth estimation framework. First, edge detection is performed with the downscaled input image. The line tracing algorithm traces strong edge positions to generate an initial staircase depth map. The initial depth map is further improved by a recursive depth refinement filter. It finally presents visual results from depth estimation and stereo image generation. The experimental results show that the algorithm is effective and feasible which leads a good result.

1. Introduction

With the progress of human society and the development of science and technology, people have higher and higher requirements for information processing and information exchange. Traditional machine vision can no longer meet the requirements of three-dimensional object recognition[1,2]. 2D to 3D technology has become an important direction in the development of 3D technology. Depth estimation is one of the key technologies of 3D conversion from 2D to 3D. It can automatically and effectively estimate depth information from one or more images of a scene. The conversion of 2D to 3D has achieved fruitful results overseas. Typical technology providers include ILM, Pass more and other companies. In China, however, there is still no mature technology. This paper is expected to provide a more practical method for depth estimation, improve the image quality of 3D view and users' stereo vision comfort, and further promote the industrialization of this technology in digital television chip applications.

2. Introduction

In order to use depth information in depth estimation framework[3], a linear tracking method of Y. Chang is referred. First, before processing the edge, a grayscale image is processed. Then, linear tracing starts to work, which traces from the leftmost boundary of the edge map to the rightmost boundary. The result of linear tracking produces a line trajectory graph, which starts from the parallel line graph of the initial state, and finally forms the whole line trajectory graph. The initial line trace is composed of the number of parallel lines between the common areas of the line and the line. The number of lines is an important parameter affecting the whole result. The area depth between the bottom trajectories is almost zero, while the area between the top trajectories is the largest. Therefore, a ladder shaped line trace map is obtained.

Content from this work may be used under the terms of the Creative Commons Attribution 3.0 licence. Any further distribution of this work must maintain attribution to the author(s) and the title of the work, journal citation and DOI.
Published under licence by IOP Publishing Ltd

After deep improved filtering, a stereo image is transformed from the depth map to [4]. The horizontal transfer process is based on the depth value of each pixel in the depth map.

3. algorithm description

3.1 Grayscale acquisition and down sampling

Research shows that parallax and 2D video can generate binocular vision. In image processing, YUV space is usually used to represent a frame of 256 color grayscale, that is depth map. Y and UV denote brightness and chromaticity respectively. Each of the 3 components is represented by 8bit, which is represented by 10 in the middle of 0-255. In a gray-scale image with U=V=128, the brightness value Y of each pixel and 2D image form a one-to-one correspondence, and the relative distance between the human eye and the 2D pixel of the corresponding point is expressed by 0 to 255. It is usually agreed that white (255) means the nearest distance, while black (0) means the farthest distance from the human eye. Therefore, the depth information of objects in the 2D diagram is more intuitive. If you want to see a 3D stereo image, you can use a 3D display to process ordinary 2D video and add the corresponding depth map[5].

In this article, we use the following formula to convert the color RGB value to the depth value [5].

$$Gray = (R*38 + G*75 + B*15) \ge 7 \qquad (1)$$

Image down-sampling is an operation to reduce computational complexity. In image super-resolution reconstruction, image resampling is often involved. In all the deduction formulas, image is vectorized, and then the vector is multiplied by a down-sampling matrix D. In this paper, we use matlab to generate the down sampling matrix D of any scale.

For an image with M*N size, the resolution image with (M/s)* (N/s) size can be obtained by S-TIMES down-sampling. So s should be the common number of M and N. If a matrix image is considered, the original image in the s*s window is transformed into a pixel whose value is the average of all the pixels in the window:

$$p_k = \sum_{i \in win(k)} I_i \Big/ s^2 \qquad (2)$$

After the image is vectorized into a vector of 1* (MN), the downsampling process should also have a corresponding matrix whose size is (MN/s^2)* (MN).

Take a 4*4 image as an example: the number represents the pixel location [6].

Table 1. 4*4 Pixel table

0	4	8	12
1	5	9	13
2	6	10	14
3	7	11	15

The vectorization size of the image is 16*1, and the element is:
[0 1 2 3 4 5 6 7 8 9 10 11 12 13 14 15]

The new image after 2 times downsampling has four pixels, which are the mean of 0 145 position, 23 677 position, 8 9 1213 position and 10 11 1415 position respectively.

3.2 Edge detection Sobel operator and its implementation

Compared with other common edge detection operators, Sobel operator has the advantages of fast detection speed, smoothing effect on noise, and some ability to suppress noise. However, because it detects some false edges, it makes the edges rough and reduces the accuracy of detection and positioning. In order to solve these problems, neighborhood or weighted averaging and first-order differential processing are carried out sequentially, and then edge detection is carried out to get edge [7]. Figure 2 shows the matrix window used by the operator (both horizontal and vertical).

1) detect vertical edges in horizontal gradient direction.

2) vertical gradient direction, horizontal edge detection

Figure 1.   Sobel Operator horizontal and vertical template

The operator matrix window (3) and (4) represent the horizontal and vertical convolution operations respectively, and the gradient values are obtained by using the following formulas: $|f_x| + |f_y|$

$$\Delta f_x(x,y) = \{f(x-1,y-1) + 2*f(x-1,y) + f(x-1,y+1)\} - \{f(x+1,y-1) + 2*f(x+1,y) + f(x+1,y+1)\} \quad (3)$$

$$\Delta f_y(x,y) = \{f(x-1,y-1) + 2*f(x,y-1) + f(x+1,y-1)\} - \quad \{f(x-1,y+1) + 2*f(x,y+1) + f(x+1,y+1)\} \quad (4)$$

The algorithm is to make the image two valued, and set the threshold TH to achieve it. The idea of edge detection is that the gradient value of non-edge points is less than the threshold value, on the contrary, the edge points. The main steps of the algorithm are as follows:

1) Move the two direction templates from one pixel to another along the image, and then overlap one pixel position with the center of the pixel.

2) multiplying the coefficients in the template with the corresponding pixel values:

3) add up all the multiplied values.

4) using the values of two convolutions, calculate the gradient value of the place, that is, the new gray value.

5) Select the appropriate threshold TH, if the gray value of the new pixel is greater than or equal to TH, then the pixel is the image edge point.

3.3 Linear tracking

The line tracking algorithm improves an energy function. This energy function is modeled by three constraints. The first one is the constraint condition for tracking the Strong boundary. The second is a smoothing constraint for punning sudden changes in the vertical direction. The third is an elastic constraint on the penalty for significant changes in the vertical direction of the line, so that the vertical position can be avoided from being too far from the initial vertical position.

The following equations describe the three constraints [8]:

Constraint condition 1: edge tracking condition, $E_{lt}(x,y) = \exp(-\text{edge}(x,y)/a)$,   (5)

Constraint condition 2: smoothing constraints:

$$E_s(x,y) = d_s(x,y)/b , \quad (6)$$

Constraint condition 3: elastic constraint conditions: $E_e(x,y) = d_e(x,y)/c$,   (7)

The control parameters a,b,c, depend on the characteristics of the input image and are tentatively, $E_{lt}(x,y)$ defined as edge tracking constraints, a as the control parameters of edge trajectory constraints, $E_s(x,y)$ as the control parameters of smooth constraints, $E_e(x,y)$ as the elastic constraints, $\text{edge}(x,y)$ as the control parameters of elastic constraints, representing the boundary values of points on the edge graph. Represents the vertical distance from the current pixel point to the substitute pixel point. Represents the vertical distance from the starting position of the left boundary on the original line diagram to the point substitute pixel.

3.4 Deep assignment

Line trajectory tracking will track the obvious edges of the edge image from left to right, and get the horizontal non-intersecting line trajectory map[8-10]. Next, we will get the gradient depth map from the bottom to the top according to the line trace. Specific steps are as follows: the distance D of the horizontal trajectory, the position of the initial trajectory and the depth corresponding to each horizontal trajectory are determined by the initial line number n of the line tracing graph and the reference image line number Hi.

The depth assignment of linear graphs must strictly obey the rule of increasing from bottom to top. Each trajectory corresponds to a fixed depth value, which is scanned from the bottom to the top of the column in the assignment. At the beginning, the value is 255. At the time of scanning to line 1, the value is 255-1*d, D is the interval of depth values, until the next trajectory line n is scanned, the value is 255-n*d, and so on, until a row of scanning and assignment is completed and the next column is carried out.

3.5 Joint bilateral filtering

The filter is a mathematical model, through which the image data can be transformed into energy. If the energy is low, the filter can be eliminated. Noise is a low-energy part. If the ideal filter is used, the ringing phenomenon will appear in the image. Using the joint bilateral filter, the system function is smooth, avoiding the ringing phenomenon.

The two nearest key frame is very important for joint bilateral filtering. The key frame is the basis of depth image generation in a certain direction. The nearest backward key frame is the non-key frame. The depth d_{bw} is generated by the inverse bilateral filtering depth recursion algorithm. Next, we use linear interpolation to combine the two methods to get a new depth.

$$d^t = (1-\frac{t}{T})d^t_{fw} + \frac{t}{T}d^t_{bw} \tag{8}$$

The t and T in the formula represent the time distance between the current frame and the forward key frame and the time distance between the two key frames, respectively. However, it can not really reflect the confidence of depth generated from each direction, because the depth weights generated from the front and back end are based on the time distance. The final result is that both the unobscured and the obscured areas may be blurred, even if the averaging reduces part [11].

A simple example of a deep improvement filter will be given by the above equation. Where x and Y represent the coordinates of the image, Z represents the initial depth value calculated by the second part of linear tracking. The first equation describes the task of feature extraction module. The weight of the filter is determined by a series of pixel pairs, and the same quantity is determined by the size of the filter. Here, Y represents pixel brightness while Sigma is a filter parameter. The second equation describes the task of cyclic depth filtering module. Among them, K denotes normalization factor and the ETA represents a set of adjacent pixel positions that describe the filter's length range. This filtering operation runs once or repeatedly until the final depth map is obtained. It relies on the stereograph descriptor. The sample, which is obtained through the use of single-scale loop deep filtering, is shown.

4. Experimental results

In this paper, we use downsampling, Sobel operator for edge detection, linear tracking, depth assignment and joint bilateral filtering to achieve a depth estimation process. The programming tool is MATLAB R2010a. The experiment is based on Win7 Professional platform. The experiment uses ordinary 2D images with resolution of 642*642.

The depth estimation quality of this algorithm is closely related to the parameter setting of Table 2. Generally, the number of line trajectories is 50. When the number of line trajectories is less than 10, discontinuous perceptual depth will be generated due to large depth changes. The control parameters a are generally the average of the pixel values of the edge trajectory, B and C are 1/4 of the image height, the weight a is 0.4, and both beta and_are 0.3. In order to obtain depth information, the algorithm only uses monocular video sequence. Although the accuracy of this method is not as good as that of binocular system, it can be seen from the results that the depth map is clearer after joint bilateral filtering. Although depth blur still exists, it has been greatly reduced. Experiments show that the algorithm is simple and easy to use and can get more accurate depth maps, which provides a good basis for further improvement of depth maps.

Table 2. Parameter values used in depth estimation

a	b	c	α	β	Y	n
0.107	125	125	0.4	0.3	0.3	50

5. Conclusion

Image depth estimation is an important basis of computer graphics based on image feature analysis and extraction. It is widely used in the fields of 3D display, animation, film and television, computer design, stereo photography and so on. This paper explores some of the current mainstream image depth estimation algorithms, and also studies the related technologies involved in depth estimation. Aiming at depth estimation in 3D TV system, a depth estimation algorithm based on relative height cues is proposed. The algorithm uses horizontal line trace to track edge image and template near bottom to achieve depth estimation. The experimental results show that the proposed algorithm can effectively extract the depth information of the image, and the computational complexity is small. It is a method of extracting the depth information of the image that is worth popularizing and applying.

Acknowledgment

Science and technology research project of Chongqing Education Commission(No. KJQN201803104)
Science and technology research project of Chongqing College of Electronic Engineering
(No.XJZK201806)

References

[1] P. Harman, J. Flack, S. Fox and M. Dowley, "Rapid 2D to 3D Conversion[M]," Proceedings of SPIE, (2002)

[2] Y. Chang, C. Fang, L. Ding, S. Chen, and L. Chen,Map Generation   for 2D-to-3D Conversion by Short Motion Assisted Color Segmentation[J],International Conference on Multimedia and Expo, 2007,12(6):1958-1961

[3] J. Ko, M. Kim and C. Kim, "2D-To-3D Stereoscopic Conversion: Depth-Map Estimation in a 2D Single-View Image[J]," Proceedings of SPIE, Vol. 6696, (2007)

[4] S. Battiato, A. Carpa, S. Curti and M. La Cascia, "3D Stereoscopic Image Pairs by Depth-Map Generation[J]," Proceedings of 3DPVT, (2004)

[5] Cheng, C.-C., C.-T. Li, L.-G. Chen. A novel 2Dd-to-3D conversion system using edge information[J]. IEEE Transactions on Consumer Electronics, 2010. 56(3): 1739-1745.

[6] W. Tam, A. Yee, J. Ferreira, S. Tariq and F. Speranza, "Stereoscopic Image Rendering Based on Depth Maps Created From Blur and Edge Information[J]," Proceedings of SPIE, (2005)

[7] Tam, W.J., L. Zhang. 3D-TV content generation: 2D-TO-3D conversion[C]. in 2006 IEEE International Conference on Multimedia and Expo, ICME 2006, July 9, 2006 - July 12, 2006. 2006: 1869-1872.

[8] S. Valencia and R. Dagnino, "Synthesizing Stereo 3D Views from Focus Cues in Monoscopic 2D Images[M]," Proceedings of SPIE, (2003)

[9] LIU Ran, XIE Hui, TAI Guoqin, TAN Yingchun. An Approach to Eliminate Folds Based on view judgment for DIBR. Tongji Daxue Xuebao/Journal of Tongji University, 2013. 41(1):142-147

[10] A. Redert, R.-P. Berretty, C. Varekamp, O. Willemsen, J. Swillens, and H. Driessen, Philips 3D Solutions: From Content Creation to Visualization, The 3rd Int. Symposium on 3D Data Processing, Visualization, and Transmission[J], 2006,14(6):429-431

[11] J. F. Canny, "A Computational Approach To Edge Detection[J]," IEEE Trans. Pattern Analysis and Machine Intelligence, 8, 679-714, (1986)

ISPECE   IOP Publishing

Research on Light-Small Lens Structure Design and Weight Reduction Optimization Based on Neural Network

Shubing An[1]*, Minlong Lian[1], Yiliang Liu [1], Shaofan Tang [1]

[1]Beijing Institute of Space Mechanic & Electrify， Beijing 100094，China

Email of all the authors: 15810135869@163.com, leon810212@163.com,

liuyiliang823@sohu.com, sftang508@sina.com

* Corresponding Author: Shubing An; email: 15810135869@163.com; phone:15810135869

ABSTRACT: In order to meet the design requirements of low weight and high specific stiffness for a small space camera lens assembly, based on the compact coaxial four-mirror optical system, a dynamic optimization design concept is proposed firstly, and then the initial lightweight design of lens structure is realized in an integrated assembly form. And then the BP neural network & genetic algorithm is used to dynamically optimize the lens structure parameters, which reduces the weight of the whole lens while ensuring the high specific stiffness of the lens, and realizes the goal of lens lightening design.

1. INTRODUCTION TO THE RESEARCH

For space load, the reduction of its weight can greatly reduce the launch cost, so there is a weight requirement for the lens assembly of a certain type of small camera. According to the requirements of this model, the weight of the whole lens must be controlled below 1.8 kg, the fundamental frequency must be greater than 180 Hz, and the size of the whole lens should be controlled within the range of φ230 mm x 210 mm. In order to achieve the design goal of miniaturization of lens assembly and meet the requirement of high specific stiffness, this paper proposes a weight reduction idea combining structural design with dynamic optimization. First, the initial lens structure model is designed, and then some size parameters of lens are fine-tuned by dynamic optimization to achieve further weight reduction.

Dynamic optimization design refers to the optimization design problem that includes the dynamic characteristics or dynamic response of the structure in the objective or constraint of the optimization design model. The optimization can realize the weight reduction design and higher stiffness requirement of the structure by changing the size of the parts. The mathematical model is as follows[1]:

Find a set of design variables X=[x1,x2,...xn]T ,Let:

$$f(x) = min$$

$$\text{s.t.} \quad G_i(X) \leq 0 \qquad\qquad (i = 1,2,...L)$$

$$Q_j(X) \leq 0 \qquad\qquad (i = 1,2,...M)$$

$$U_k(X) \leq 0 \qquad\qquad (i = 1,2,...T)$$

Content from this work may be used under the terms of the Creative Commons Attribution 3.0 licence. Any further distribution of this work must maintain attribution to the author(s) and the title of the work, journal citation and DOI.
Published under licence by IOP Publishing Ltd

$$x_r^l \le x_r \le x_r^u \qquad (r = 1,2,\dots N)$$

Where $G_i(X)$is a static constraint function;

$Q_j(X)$is a dynamic constraint function;

$U_k(X)$is another constraint function;

x_r^l, x_r^u are the lower and upper limits of the design variable x_r, respectively

The premise of dynamic optimization is that there is a better structural model. So before optimization, this paper integrates all mirrors into the same connector with the design idea of integrated assembly, reduces the number of connecting elements, and achieves weight reduction in structural design. Then, the finite element model of the lens is established, and some sizes suitable for adjusting in the finite element model of the lens are determined as design variables. By changing the sizes of design variables, the weight and fundamental frequency of the lens in different sizes are obtained respectively. With the input and output data of different sizes of design variables and their weights and fundamental frequencies, the BP neural network model is built, trained and tested. When the test results meet the error conditions, the model of the BP neural network is successful. Taking weight as optimization objective and fundamental frequency as constraint function, genetic algorithm is used to optimize the constructed neural network model, and the optimal solution and variable value are obtained. Modify the value of the variable by integer and so on, and the final parameters are brought into the finite element model to verify the weight and fundamental frequency. Finally, a small lens with light weight and high stiffness is obtained.

2. STRUCTURAL DESIGN OF LIGHT-SMALL LENS

The optical designer of the lens chooses the optical system[2,3] as shown in the Figure 1 according to the relevant requirements, and adds a folding mirror after the quaternary mirror to fold the optical path to make the volume more compact. From primary mirror to quaternary mirror, the aperture are 200 mm, 80 mm, 100 mm and 28 mm, respectively. According to the figure 1, quaternary mirror can be used as field of view diaphragm to shield stray light. This feature can be used in structural design to shorten the length of the external shade and realize weight loss.

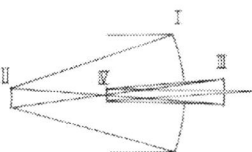

Figure 1. Optical system

Figure 1. Optical layout. I-Primary mirror. II-Secondary mirror. III-Tertiary mirror. IV-Quaternary mirror

In this paper, a lightweight design method is adopted for the design of mirror. As shown in the Figure 2, taking the secondary mirror as an example, its initial thickness is 10mm and its weight is 92g. Because of its small thickness, considering the lightweight degree and the difficulty of processing comprehensively, lightweight holes are excavated from the back, edge thickness is reduced, instead of processing blind holes on the side[4].

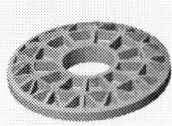

Figure 2. Design of secondary mirror with lightweight form

The primary mirror uses the same form to achieve lightweight design. The distance between the top to the back of the tertiary mirror is small, which makes it inconvenient to design lightweight holes. Therefore, the back edge material is cut accordingly only according to the radian of the mirror. For the quaternary mirror, the aperture and thickness are very small. Considering the difficulty of processing, the lightweight design is not adopted. Through calculation, the lightweight rates of primary, secondary and tertiary mirrors were 73%, 59% and 57%, respectively.

In order to give full play to the advantages of quaternary mirror in eliminating stray light and to realize the weight reduction assembly of mirrors, an integrated extinction and connection element can be designed to suppress stray light more effectively and realize mirror connection. Based on this idea, this paper designs the extinction cylinder as shown in the Figure 3. The relative position of the reflector and the extinction cylinder is marked in the figure(I、II、III、IV Represents the position of the primary mirror to the quaternary mirror, respectively). The light is reflected by the secondary mirror, and it enters the extinction cylinder through the central hole of the quaternary mirror to continue to propagate. By using the aperture function of the extinction cylinder and the mirror, stray light can be suppressed in a compact space. In this way, the lens does not need an external shade, which greatly reduces the weight and volume of the lens.

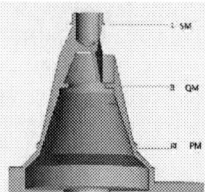

Figure 3. Sectional view extinction cylinder

For coaxial reflective optical systems, in general, the mirror is connected to the inner shade by three-bar support. However, because the diameter of the inner shade is much larger than that of the secondary mirror, the quaternary mirror and the quaternary mirror, if the mirror is connected by three bars, the bars will be relatively long, which will cause more weight gain of the whole machine.

For coaxial reflective optical systems, in general, the mirror is connected to the inner shade by three-bar support. However, because the diameter of the inner shade is much larger than that of the secondary mirror, the quaternary mirror and the quaternary mirror, if the mirror is connected by three rods, the rods will be relatively long, which will cause more weight gain of the whole machine. In order to reduce the weight of connecting elements, this paper makes full use of extinction cylinder as the connecting element of reflector, completes the integration of optical components, realizes the integration of structural forms, better meets the thermal matching of optical elements, greatly reduces the weight of main structure, and meets the development requirements of micro lens. The aperture and thickness of the mirror in this paper are relatively small. The stability of the mirror can be guaranteed by installing the mirror on the wall of the extinction cylinder by gluing. [5]Therefore, the internal diameters of the central opening of the primary and secondary mirrors are respectively connected to the outer diameters of the corresponding positions of the extinction cylinder shown in the figure. The protruding parts of the central hole of the quaternary mirror are glued to the holes at the bottom of the extinction cylinder, and the outer diameters of the quaternary mirror are glued to the inner diameters of the inner hole of the extinction cylinder. At the same time, in the section view shown in Figure 3, the protrusion of the installation position of the primary, secondary and tertiary mirrors in the axial contact with the mirror can limit the displacement of the primary and secondary mirrors in the optical axis direction; the quaternary mirror have small aperture and weight, so this method is not necessary.

For the folding mirror, according to its position in the optical system, it is assembled in the frame and connected to the bottom of the extinction cylinder with three sets of bipod supports.

For all the above components, including the inner shade of the lens, the C/Sic material with high specific stiffness is used in this paper, which ensures the thermal matching of the lens structure while realizing the high specific stiffness of the lens.

The lens and the satellite platform are connected by three groups of flexible bipod support. The form of the support is shown in Figure 4. The support is made of aluminum alloy. When the stress is transferred from the satellite platform to the lens, the flexible support deforms first, thereby releasing the stress and ensuring the stability of the main structure of the lens.

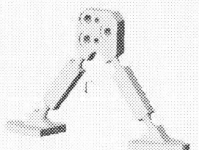

Figure 4. Flexible support connecting lens and satellite platform

After the design of each part is completed, the assembly body is shown as shown in the Figure 5. After calculation, the lens weight is 1.827 kg and the volume is φ220mm * 190mm. The volume meets the requirements of the structural design, and the weight ratio is slightly higher than the required 1.8kg, so the weight reduction design needs to be carried out through subsequent optimization.

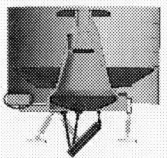

Figure 5. Assembly body of lens

3. RESEARCH ON NEURAL NETWORK MODELING

The finite element model of the lens structure is established[6] and the following parameters are determined as variables in the optimization design: the thickness of the back of the primary mirror （using A representation）, the back of the secondary mirror（using B representation）, the thickness of the back of the refractor（using C representation）, the thickness of the connecting end of the secondary mirror of the extinction support cylinder using A representation（using D representation）, the thickness of the inner shield wall（using E representation）, and the thickness of the connecting end of the shield and the extinction support cylinder （using F representation） . Among these variables, the thickness of primary and secondary mirrors refers to the depth of their lightweight holes. The positions of variables in the lens are shown by 1-6 in Figure 2. The change of the first five parameters can be realized by the change of the thickness of the 2D element of the finite element method. For the connecting end of the shield and the extinction support cylinder, as shown in the red circle part of the Figure 5, if it is modeled by the 2D element, it is impossible to accurately simulate the connection between the part and the inner shade and the support cylinder, thus affecting the accuracy of the model. In this way, a 3D unit is established for this part, and the thickness change is realized by HYPERMORPH function.

After the structural design and finite element modeling are completed, the dynamic optimization simulation of the lens is studied. According to the introduction of the above mathematical model, the optimization model of the lens in this paper is as follows.

(1) Objective function

The objective function f(x) is the weight of the whole machine, and the optimization time is the minimum. That is:

f(x)=min

(2) Design variables

The design variables in this paper are the thickness of five 2D elements and the thickness of the connecting end of the hood.

(3) Constraint conditions

Fundamental frequency constraint:

$f_{max} \geq f_1 \geq f_{min}$

Among them, f_{min} is the minimum required for the fundamental frequency. According to the overall index of the camera, $f_{min}= 180Hz$; f_{max} is the maximum of the fundamental frequency proposed in the structural design. The purpose of this value is to avoid the redundancy of the stiffness and make the structural design more reasonable. This paper takes it as 220 Hz.

In addition, the objective of the optimization is to reduce the weight of the lens. It requires that the weight of the optimized structure should not exceed the original design weight W_0, so there is a self-weight constraint:

$W \leq W_0$

Dynamic optimization design is divided into two processes[7]. First, the dynamic model of the structure is established. Second, a reasonable optimization method is selected to optimize the structure. Its basic flow chart is shown in the Figure 6. In this paper, the dynamic model of structure is established by using BP neural network model.

BP network needs to train input data and output data to establish mapping relationship. In the neural network model, the input data are the above design variables, and the output data are the lens weight and the first-order frequency, respectively, according to the requirements of the objective function and constraints.

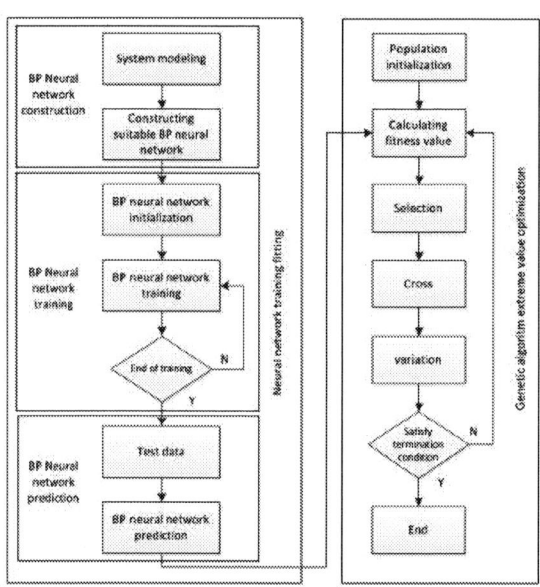

Figure 6. Flow chart for optimization of neural network and genetic algorithm

According to the rule of orthogonal test[8], six design variables are assigned three levels, and then 27 orthogonal test are done to construct the neural network with data of different levels as input. Variables and their levels are shown in Table 1.

Table 1. Variables and their levels

Levels	A(mm)	B(mm)	C(mm)	D(mm)	E(mm)	F(mm)
1	5	8	5	10	2.5	5
2	2	4	2	5	2	2
3	8	12	8	15	4	8

There are 27 sets of input-output data, and it is not necessary to list them all. In this paper, neural network modeling is performed based on these 27 sets of data, and 6 sets of data are listed as schematics. The schematic table is shown in Table 2.

Table 2. The schematic table of input-output data

number	A(mm)	B(mm)	C(mm)	D(mm)	E(mm)	F(mm)	Mass(kg)	Frequency(Hz)
1	5	8	5	10	2.5	5	270.42	1.827

5	5	4	2	5	2	2	164.93	1.646
10	2	8	5	15	2.5	2	190.73	1.718
15	2	4	8	10	4	5	313.84	2.025
20	8	8	8	5	2	5	264.17	1.824
27	8	12	2	10	4	2	252.45	2.215

After the construction of the neural network, the neural network is trained and tested[9]. In this paper, according to the three levels of the six design variables mentioned above, another nine sets of data are selected to test the trained neural network. The predicted value comes from the output of nine sets of test data according to the mapping relationship, which is to be calculated. The expected value is the actual weight and fundamental frequency of nine sets of data in the finite element model, which are known quantities. In order to maintain the accuracy of prediction, this paper requires the root mean square error of weight parameters to be 0.01%, and the root mean square error of fundamental frequency parameters to be 0.5%.

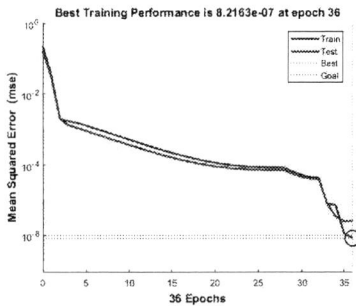

Figure 7. Requirement of training accuracy

Set the training requirement accuracy to $1*10-6$. As shown in the figure 7, the training can achieve the desired effect after 36 iterations. After training, the test results in Figure 8 show that the black circle represents the predicted output of the neural network and the blue circle represents the actual output of the finite element. The root mean square error of weight is 0.076%, the root mean square error of fundamental frequency is 0.43%, which meets the training requirements. It shows that BP network has established a good relationship between input and output.

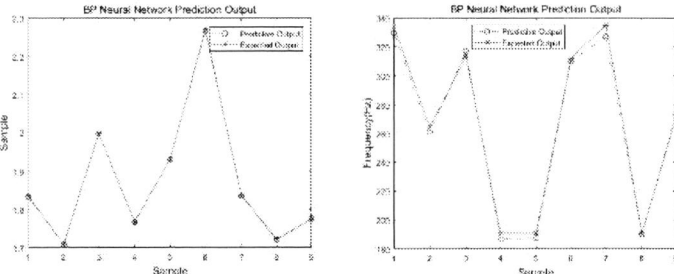

Figure 8. Expected output and actual output (weight on the left and base frequency on the right)

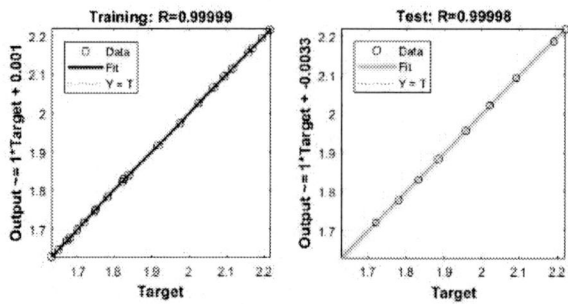

Figure 9. Linear regression results of training sets and test sets

This paper takes weight as an example to further verify the modeling effect of BP neural network. The Figure 9 is the linear regression result of the training set and the test set. From this figure, it can be seen that the correlation between the predicted value and the actual value in the training set is 0.99999, and the correlation between the predicted value and the actual value in the test set is 0.99998. This also shows that the neural network model in this paper is well modeled.

4. THE RESULT OF DYNAMIC OPTIMIZATION BASED ON GENETIC ALGORITHM

After the successful neural network modeling, the lens needs to be weight-reduced. The mapping relationship between the input and output of the BP neural network model can't be expressed by exact functions. Therefore, the traditional mathematical optimization method can't be selected, and the genetic algorithm is used to optimize the neural network. According to the process shown in the figure 6, MATLAB is programmed to minimize the weight as the objective function, and the range of fundamental frequency is constrained from 180 Hz to 230 Hz. The genetic algorithm is used to optimize the neural network. The result of the optimization is shown in the figure 10.

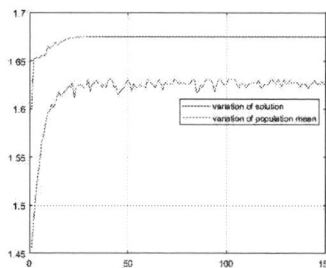

Figure 10. iteration results of the optimization

The figure 10 shows the iteration of the objective function. After 43 iterations, the objective function reaches the optimal solution, which is 1.677kg. At this time, the values of each variable are accurate to 0.001. Considering the processing precision, one decimal point is reserved for each variable. The initial data of each variable and the adjusted data are shown in the Table 3. By substituting the adjusted data into the neural network, the lens weight is 1.682 kg and the fundamental frequency is 189.37 Hz.

Table 3. The schematic table of input-output datas

Variable	A(mm)	B(mm)	C(mm)	D(mm)	E(mm)	F(mm)
Original data	5.562	5.131	3.787	8.324	2.788	4.474
Modified data	5.6	5.1	3.8	8.3	2.8	4.5

5. VERIFICATION OF OPTIMIZATION RESULTS IN FINITE ELEMENT

Through the finite element analysis, the lens weight is 1.687 kg and the fundamental frequency is 189.91 Hz with the same data. Its mode shape is shown in the figure. Compared with the finite element model and the neural network model, the errors of weight and fundamental frequency are 0.30% and 0.29%, respectively. This proves that the optimization results are reliable and also meets the design requirements. At the same time, it shows that the design of this paper has accomplished the goal of lens weight reduction.

6. CONCLUSION
Aiming at the design requirement of low weight and high specific stiffness for a small space camera lens assembly, based on the coaxial four-mirror optical system chosen by optical designers, the initial lightweight design of lens structure is realized in an integrated assembly form. Secondly, taking some parameters of the lens as design variables, the size of the lens is optimized by using the neural network-genetic algorithm, which further reduces the weight. Finally, the finite element method is used to verify that the first-order frequency meets the requirements. It shows that the design goal of low weight and high specific stiffness of lens assembly is realized in this paper. The weight-loss design concept in this paper can be used for reference in the design of other aerospace structures.

References
[1] Feng, Q.H. Dynamic Optimization Design for Box Girder Structure of Bridge Crane Based on Sensitivity Analysis[D]. North University of China,2017. (in Chinese)
[2] Sasian J.M. Flat-field, anastigmatic, four-mirror optical system for large telescopes[J]. Optical Engineering, 1987, 26(12):1197-1199.
[3] Chen, L. Compact Long Focal Length Four Reflector Telescope Objective:CHINA, 107966804[P], 2018.04.27. (in Chinese)
[4] Li, X. Design of Support Structure for Large Mirror of Space Camera[J]. Spacecraft Recovery & Remote Sensing,2016 37(3)：91-99. (in Chinese)
[5] Liu, Q. Calculation and control of adhesive layer in reflector athermal mount.[J].Optics and Precision Engineering.2012 20(10) . (in Chinese)
[6] Zhang, S.L. The Technology of Optimization Design for Structure Based on HyperWorks[M]. BEIJING: China Machine Press,2007. (in Chinese)
[7] Wang, W.Z. Topology and Size Optimization Technologies Applied in Structure Design of Space Camera [J]. Spacecraft Recovery and Remote Sensing, 2012, 33(6): 67-73. (in Chinese)
[8] Dong, Y. Strength Forecasting of Backfilling Materials by BP Neural Network Model Collaborated with Orthogonal Experiment.[J]. Materials Review. 2018,32(06):1032-1036. (in Chinese)
[9] Wang, F. Temperature Sensing of Dislocation Optical Fiber Interference Laser Spectrum Combined with BP Neural Network.[J]. Spectroscopy and Spectral Analysis. 2016,36(11):3732-3736. (in Chinese)

The theoretical development and prospect of two-dimensional topological insulators

Yichen Zhang[1,a]

[1]Kuang Yaming Honors School, Nanjing University, Gulou Campus, 210093, Nanjing, Jiangsu, China

[a]Corresponding author: 151242067@smail.nju.edu.cn

Abstract. Topological insulators can be described as rediscovery of band theory, where numerous secrets on edges and surfaces are unearthed. During the process of rediscovery, the concept of topology into novel theoretical models of topological insulators played a crucial role. This paper aims to give an introduction to the theoretical development of two-dimensional topological insulators in a nutshell, so it is indispensable and efficient to relate topological invariant with the band structure of topological insulators and use specific and vital theoretical models to reveal novel phenomena in those materials. The main focus of the paper is placed on historic landmarks in the theoretical development of two-dimensional topological insulators where fundamental concepts and models have paved the way for further enrichment of family of topological materials and experimental realizations.

1. Introduction

Over the past decades, topological insulator has drawn extensive concern in condensed matter physics due to its unique electronic behavior on its edges (2D) or surfaces (3D). Distinguished from common insulators, conductors and semiconductors, topological insulators (TIs) are insulated inside but have metallic states outside and the most exhilarating characteristic is the topology part of them, which dispels the influence of continuous deformation that does not break the underlying symmetry. Such robustness against continuous deformation is underpinned by two cornerstones. The first is the time reversal symmetry (TRS) where the Hamiltonian function keeps unchanged, while momentum and spin reverse their directions and the other is the spin-orbit coupling (SOC) aspect of TIs. The SOC acts like a magnetic field and induces band overlapping to achieve edge/surface states in TIs. At the beginning of this paper, theoretical development of 2D topological insulators is reviewed in Section 2 by setting out from some basic concepts and the calculation of Hall conductivity. Then in section 3, important theoretical models of 2D TIs are introduced. Finally, conclusions and outlooks are given in section 4.

2. Berry phase, the TKNN invariant and Hall conductivity

One essential question about topological insulator is that why topological insulator having metallic edges/surfaces states is termed as 'topological'. A limpid illustration of topology using common objects is given by considering the relation between a donut and a coffee cup. Imagine that the donut is made of some sort of pliable materials so that one could shape the left half part into a cylinder continuously with the right half donut sticking to the cylinder. Then press the top of the cylinder until the sag is deep enough. In this way, one could smoothly deform a donut into a coffee cup and vice versa, so it can be claimed that these two objects are topologically equivalent. Actually, it is the

---

Content from this work may be used under the terms of the Creative Commons Attribution 3.0 licence. Any further distribution of this work must maintain attribution to the author(s) and the title of the work, journal citation and DOI.

Published under licence by IOP Publishing Ltd

winding number of objects that defines their topological equivalence. For example, a sphere with a different winding number cannot be smoothly reshaped into a coffee cup or a donut unless you cut and open a hole in it. Analogical to this pattern, electronic band dispersion in condensed matter systems also have such topological classification by defining appropriate topological invariant, while the object cited to calculate topological invariant is the wave function of electron instead of being the configuration of objects in real space. In order to identify topologically nontrivial systems, the concept of Berry phase is borrowed as demonstrated below.

Berry phase delineates the adiabatic evolution of a quantum state and can be used to understand the TKNN invariant which identifies the topology of a system. By considering parameter vectors evolving with time slowly enough to describe quantum states and splitting the phase expression into the dynamic phase and the Berry phase, the primitive form of Berry phase is defined in the following form[1].

$$\gamma_n(C) = i\oint_C \langle n(\mathbf{R}) | \nabla_{\mathbf{R}} n(\mathbf{R}) \rangle \cdot d\mathbf{R} \qquad (1)$$

$\mathbf{R}$ is time-dependent and represents a path in the parameter space.

When considering a concrete model of the Chern insulator introduced by Haldane[2], the parameter space becomes the wave vector space and the quantum state is a bulk state $|\psi\rangle$ related to the wave vector $\mathbf{k}$ and spin s. Then defining a vector-potential

$$\mathbf{A}_{\mathbf{k},s} = i \left\langle \psi_{\mathbf{k},s} \left| \frac{\partial}{\partial \mathbf{k}} \right| \psi_{\mathbf{k},s} \right\rangle \qquad (2)$$

and using the Stoke's theorem, the Berry phase can be written in the following form

$$\gamma_S = \int \left(\nabla_{\mathbf{k}} \times \mathbf{A}_{\mathbf{k},s} \right)_z d\mathbf{k}_{x,y} \qquad (3)$$

In order to illustrate the connection between Berry phase integral and the Hall conductivity, an observable quantity revealing topological characteristics of a system, the TKNN formula[3] shall be introduced here. In a two dimensional conductor, the Kubo formula calculating Hall conductivity is written as

$$\sigma_H = \frac{ie^2}{A_O \hbar} \sum_{\varepsilon_\alpha < E_F} \sum_{\varepsilon_\beta < E_F} \frac{\left(\frac{\partial \hat{H}}{\partial k_1} \right)_{\alpha\beta} \left(\frac{\partial \hat{H}}{\partial k_2} \right)_{\beta\alpha} - \left(\frac{\partial \hat{H}}{\partial k_2} \right)_{\alpha\beta} \left(\frac{\partial \hat{H}}{\partial k_1} \right)_{\beta\alpha}}{\left(\varepsilon_\alpha - \varepsilon_\beta \right)^2} \qquad (4)$$

where A_O is the area of the system, $\varepsilon_\alpha, \varepsilon_\beta$ are eigenvalues of the Hamiltonian $\hat{H}$ and E_F is the Fermi level.

Through some basic quantum mechanical calculation, replacing the summation of ε_β with the summation of ε_α and choosing appropriate boundary conditions, the expression for the quantum Hall conductivity in transverse direction can be obtained as:

$$\sigma_{xy} = \frac{ie^2}{\hbar A_O} \sum_{\mathbf{k}} \sum_{\varepsilon_\alpha < E_F} \left(\langle \partial_{k_x} \alpha | \partial_{k_y} \alpha \rangle - \langle \partial_{k_y} \alpha | \partial_{k_x} \alpha \rangle \right) \qquad (5)$$

As is known, when using Kubo formula to calculate conductivity, we only concern about the occupied states, so it is crucial to exclude the contribution of unoccupied states in the process. Then for occupied states, define the vector-potential in the following way

$$A_{\mathbf{k}}^j = i \sum_{\varepsilon_\alpha < E_F} \langle \alpha | \partial_k^j | \alpha \rangle \qquad (6)$$

where j represents x or y direction. From here, the linkage between the integral of Berry curvature over the filled states and the Hall conductivity has been established. The TKNN formula for a two dimensional system can be obtained from the definition in Equation (6):

$$\sigma_{xy} = \frac{e^2}{2\pi h} \int \left(\nabla_{\mathbf{k}} \times \mathbf{A}_{\mathbf{k}} \right)_z dk_x dk_y \qquad (7)$$

Noting that Equation (7) and (3) have similar expression, it is easy to find out mathematical correspondence between the quantum Hall conductivity and the Berry phase curvature. To show the integer characteristic of the Hall conductivity, a transformation shall be exerted on the Brillouin zone(choosing it to be a rectangular for convenience) and consequently, the Brillouin zone is shaped into a torus which has no boundary(see **Figure 1**)[4], thus the domain of integration can be shifted from the whole Brillouin zone (BZ) to the curve C. With curve

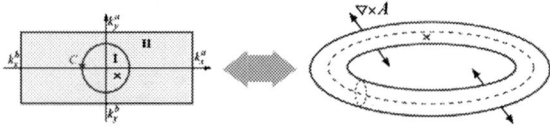

Figure 1. Topologically equivalent transformation of Brillouin zone

C on the right side of area I and on the left side of area II, two parts of the integral have opposite signs. Using the gauge dependent property of vector potential $\mathbf{A}_{k,s}$, which is

$\mathbf{A}_{k,s} \rightarrow \mathbf{A}_{k,s} - \dfrac{\partial \varsigma(\mathbf{k})}{\partial \mathbf{k}}$, we have:

$$\sigma_{xy} = \frac{e^2}{h} \frac{1}{2\pi} \oint_C \nabla_k \varsigma \cdot d\mathbf{k} \qquad (8)$$

The function $\varsigma(\mathbf{k}(t))$ must satisfy the condition that after a closed path, the final value of it and the initial value of it have a difference of the integral multiples of 2π. As a result, the right side of Equation (7) is integral multiples of e^2/h and the integer n, where

$$n = \frac{1}{2\pi} \int \left(\nabla_k \times \mathbf{A}_k \right)_z dk_x dk_y \qquad (9)$$

is the TKNN invariant, also termed as the first Chern number.

On the contrary, for topologically trivial situation, if the Berry potential $\mathbf{A}_k$ is well chosen, the integral over the whole BZ would be zero and the Hall conductance vanishes. So the nonzero Hall conductivity here implies that for topologically nontrivial situation, special singularities must exist in the BZ so that continuous and single-valued gauge selection becomes impossible in a whole BZ.

Conventionally, quantum Hall edge states in 2D systems with a chiral flow along its 1D boundary exhibit such quantized Hall conductivity under the existence of a magnetic field. However, as shwown below, Haldane demonstrated that even without net flux of external magnetic field, the 2D hexagonal model could still harbor nonzero Chern number, i.e., topological nontrivial state.

3. Two-dimensional topological insulators

3.1 Haldane model

In 1988, Haldane introduced a "2D graphite" model[2] with zero external magnetic field where integer Hall conductance arose due to breaking of time reversal invariance (TRI). The diagram of the honeycomb lattice model is shown in **Figure 2**.

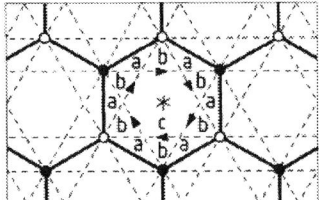

Figure 2. Dots with different colors represent different sub-lattice sites. Black arrows show the directions of positive phase hopping in the state with broken TRI.

The black and white dots have opposite on-site energy. By adding a periodic magnetic flux density, magnetic flux ϕ_a in area a and magnetic flux ϕ_b in area b satisfy the condition that $\phi_a = -\phi_b$ [5], so the total magnetic flux through a unit cell becomes zero, which leads to the consequence that apparently, hopping term t_1 between nearest sites remains unchanged, while hopping term t_2 between second nearest sites is changed due to additional phase term. Setting $\mathbf{a_1}, \mathbf{a_2}, \mathbf{a_3}$ to be the displacement from a white dot to its three nearest black dots and defining that $\mathbf{b_1} = \mathbf{a_2} - \mathbf{a_3}$, $\mathbf{b_2} = \mathbf{a_3} - \mathbf{a_1}$ and $\mathbf{b_3} = \mathbf{a_1} - \mathbf{a_2}$, we can express the Hamiltonian of the system in the following way

$$H(\mathbf{k}) = 2t_2 \cos\phi \left(\sum_{i=1,2,3} \cos(\mathbf{k} \cdot \mathbf{b}_i) \right) \mathbf{I} + t_1 \left(\sum_{i=1,2,3} \left[\cos(\mathbf{k} \cdot \mathbf{a}_i) \sigma_x + \sin(\mathbf{k} \cdot \mathbf{a}_i) \sigma_y \right] \right) + \left[M - 2t_2 \sin\phi \left(\sum_{i=1,2,3} \sin(\mathbf{k} \cdot \mathbf{b}_i) \right) \right] \sigma_z \quad (10)$$

where $\sigma_{x,y,z}$ are Pauli matrices, ϕ is the phase term for t_2 and M is the positive on-site energy.

By expanding the Hamiltonian in a linear way that $\delta\mathbf{k} = \mathbf{k} - \mathbf{k}_\alpha^0$ at two distinct corner zones $\mathbf{k}_\alpha^0$ that is defined as $\mathbf{k}_\alpha^0 \cdot \mathbf{b}_i = 0$ (having assumed that $|t_2/t_1| < 1/3$, which excludes the possibility of bands overlapping), we can express the Hamiltonian of the effective models near the two points as

$$H_\pm = \frac{3t_1 a}{\hbar} \left(\delta k_x \sigma_x - \delta k_y \sigma_y \right) + m_\pm \left(\frac{3t_1 a}{\hbar} \right)^2 \sigma_z \quad (11)$$

where $m_\pm = \pm M - 3\sqrt{3} t_2 \sin\phi$. Obviously, the M term in the Hamiltonian opens a +M and a -M energy gap at two $\mathbf{k}_\alpha^0$ points, respectively. Also we note that it is the $m_\pm$ term that breaks the TRS of the system. The Chern number of the model was calculated by Haldane using the Středa formula[6] and the expression was

$$n = \frac{1}{2} \left[\text{sgn}(m_+) + \text{sgn}(m_-) \right] \quad (12)$$

As a result, the Chern number can be ± 1 or 0. The nonzero Chern numbers happen when $|M/t_2| < 3\sqrt{3} |\sin\phi|$ with finite energy gap between two bands remained.

Haldane model has proved to be quite impossible in experimental realization. However, it is interesting that realization of Haldane model has been achieved in recent research[7] by using ultra-cold fermions where TRS and inversion symmetry of the system was broken.

3.2 Kane and Mele Model and Z_2 Invariant

In 2005, based on Haldane's graphene model, Kane and Mele[8] proposed a quantum spin Hall (QSH) system where spin of electron was taken into account and spin-orbit coupling (SOC) which locked the relative direction between spin and momentum of electron played a crucial role in deriving the helical edge state of the QSH system. In Kane and Mele model, no more magnetic flux exists to break TRS and the two edge states carrying spin currents of electrons are preserved by the TRS. This model, in the absence of a conservation of spin current, is more realistic and assumed to be stable under low temperature. A schematic illustration of the transport behavior of a QSH system versus a QHE system is shown below in order to visualize their spin and momentum configuration.

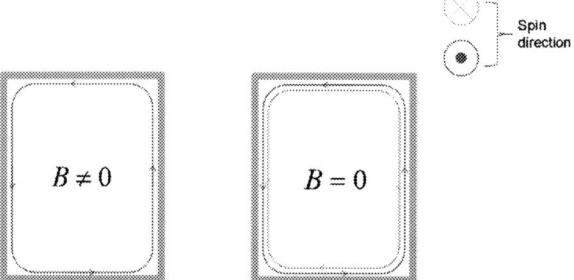

Figure 3. Electronic transport behavior in a quantum Hall insulator (left) and a QSH insulator (right). The direction of spin is locked to the momentum of electron. The spin direction of the red current is out of the plane, while the spin direction of the blue counterpart is into the plane.

The Hamiltonian of the Kane and Mele model can be written in the form that

$$H = t \sum_{\langle ij \rangle} c_i^\dagger c_j + i\lambda_{SO} \sum_{\langle\langle ij \rangle\rangle} v_{ij} c_i^\dagger s^z c_j +$$

$$i\lambda_R \sum_{\langle ij \rangle} c_i^\dagger \left(s \times \hat{\mathbf{d}}_{ij} \right)_z c_j + \lambda_v \sum_i \xi_i c_i^\dagger c_i \qquad (13)$$

The fermion operator in the first term contains the spin index and t is the nearest hopping coefficient. The second term is the SOC term where $v_{ij} = 2/\sqrt{3}\left(\hat{\mathbf{d}}_1 \times \hat{\mathbf{d}}_2\right)_z = \pm 1$ and $\hat{\mathbf{d}}_1$ and $\hat{\mathbf{d}}_2$ are unit vectors along the two bonds the electron traverses going from site j to i. s^z is a Pauli matrix operating on the spin space of electron. The third term and the last term are nearest neighbor Rashba term and sublattice potential($\xi_i = \pm 1$), respectively.

After diagonalization, the Hamiltonian of the system is a four-band one and the Bloch wave function is a four-component spinor, thus the Bloch Hamiltonian must have 16 components. The 16 components consist of the identity matrix, 5 Dirac matrices Γ^a (a=1, 2, 3, 4, 5) and 10 commutators $\Gamma^{ab} = [\Gamma^a, \Gamma^b]/2i$. The 5 Dirac

Table 1. Expressions of coefficients in (14) from Ref[8]

Coefficient	Expression	Coefficient	Expression
d_1	$t(1 + 2\cos x \cos y)$	d_{12}	$-2t\cos x \sin y$
d_2	λ_v	d_{15}	$\lambda_{SO}(2\sin 2x - 4\sin x \cos y)$
d_3	$\lambda_R(1 - \cos x \cos y)$	d_{23}	$-\lambda_R \cos x \sin y$
d_4	$-\sqrt{3}\lambda_R \sin x \sin y$	d_{24}	$\sqrt{3}\lambda_R \sin x \cos y$

matrices are even under time-reversal operator, while the 10 commutators are odd. The Hamiltonian can be expressed in Dirac matrices representation as

$$H(\mathbf{k}) = \sum_{a=1}^{5} d_a(\mathbf{k})\Gamma^a + \sum_{a<b=1}^{5} d_{ab}(\mathbf{k})\Gamma^{ab} \qquad (14)$$

The corresponding nonzero coefficients are shown in Table 1 with $x = k_x a/2$ and $\sqrt{3}k_y a/2$, where $\lambda_{SO}, \lambda_R, \lambda_v$ are coefficients of SOC term, Rashba term and sublattice potential term, respectively.

When $\lambda_R = 0$, $\lambda_v > 3\sqrt{3}\lambda_{SO}$, the system is an insulator. When $\lambda_R = 0$, $\lambda_v < 3\sqrt{3}\lambda_{SO}$, the system is a QSH system where the state can be considered as superposition of two independent quantum Hall states with different spin orientations. The paired states that connect the valence bands and the conduction bands are called edge states. Here, although the total Chern number $n = n_+ + n_-$ becomes zero, yet their difference $n_+ - n_- = \pm 2$ characterizes the special system. However, when $\lambda_R \neq 0$, spin of electrons cannot be separately treated any more, so Kane and Mele introduced Z_2 index from the Bloch wave function. Briefly, the Z_2 index can be obtained by counting the number of pairs of complex zeros of P, where P is defined as

$$P(\mathbf{k}) = Pf\left[\left\langle u_i(\mathbf{k}) \middle| \Theta \middle| u_j(\mathbf{k}) \right\rangle\right] \tag{15}$$

Pf represents the Pfaffian of the matrix, Θ is the time reversal operator and $u(\mathbf{k})$ is the band wave function.

The diagram of energy bands for a 1-dimensional stripe with zigzag boundary condition of both QSH phases and insulating phase is shown in **Figure 3**.

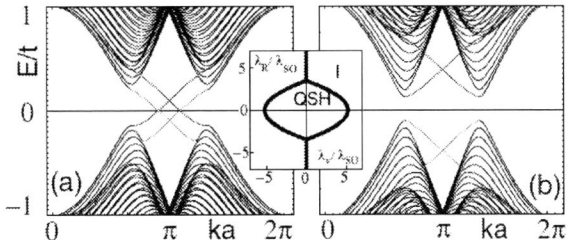

Figure 4. (a) QSH phases with $\lambda_v = 0.1t, \lambda_{SO} = 0.06t, \lambda_R = 0.05t$. (b) Insulating phase with $\lambda_v = 0.4t, \lambda_{SO} = 0.06t, \lambda_R = 0.05t$.

After Kane and Mele model was proposed, Fu and Kane[9] introduced an approach to determine the Z_2 invariant. The Z_2 invariant for 2D situation is given by

$$(-1)^v = \prod_{i=1}^{4} \frac{\sqrt{\det(w(\Gamma_i))}}{Pf(w(\Gamma_i))} \tag{16}$$

where $w(\Gamma_i)$ is a unitary matrix whose components are defined as $w_{mn}(\mathbf{k}) = \left\langle u_m(-\mathbf{k}) \middle| \Theta \middle| u_n(\mathbf{k}) \right\rangle$. Θ is the time-reversal operator satisfying $\Theta^2 = -1$. Methods of calculation for 3D cases and systems with inversion symmetry were also introduced in Fu and Kane's work in 2006[9] and 2007[10].

3.3 HgTe/CdTe quantum wells

The model of mercury telluride-cadmium telluride quantum wells exhibiting the topological property, QSHE, was proposed by Bernevig et.al.[11] in 2006. The transition between topologically trivial state and topologically nontrivial states of the sandwich-like heterostructure of HgTe/CdTe is determined by the critical thickness $d_c (\approx 6.3nm)$ of quantum wells. A schematic of the structure of the quantum well is shown in **Figure 5**.

Figure 5. The sandwich-like structure of HgTe/CdTe quantum wells

In the quantum wells system, an odd number of pairs of helical edge states which are stable and robust against many-body interaction or scattering is protected by the time reversal symmetry where we can write the relationship of two edge states as $\left| -k, \downarrow \right\rangle = \hat{T} \left| +k, \uparrow \right\rangle$. What really helps the realization of the QSH state is actually the unique inverted band gap of HgTe. As seen in **Figure 6**,

CdTe has a normal band pattern where s-type Γ_6 band lies above the p-type Γ_8 band, while HgTe goes just the opposite, so even before the theoretical model was explored, one could get a glimpse of the potential of such a combination.

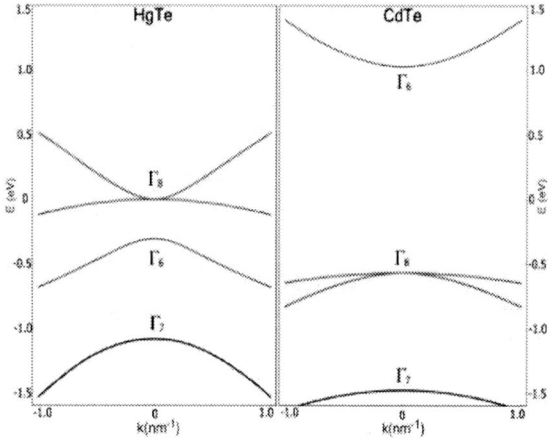

Figure 6. Bulk energy spectrum near the center of BZ

In the deduction of the effective Hamiltonian, Bernevig, Hughes and Zhang took two subbands, E1 and H1, into account and block-diagonalized the Hamiltonian as

$$H_{eff}(k_x,k_y)=\begin{pmatrix} H(k) & 0 \\ 0 & H^*(-k) \end{pmatrix},$$

$$H(k)=\varepsilon(k)+d_i(k)\sigma_i \quad (i=1,2,3) \tag{17}$$

where σ_i are Pauli matrices. Among parameters above, $d_3(k)$ contains a consequential mass parameter M which would change its sign leading to band inversion with the thickness of the HgTe layer varying. The existence of helical edge states can be revealed more explicitly by solving the eigenvalue problem of the effective Hamiltonian, although the form of the effective Hamiltonian and the sign reversal phenomenon have shed light on QSHE.

3.4 Other 2DTIs predicted theoretically

Beyond the models introduced previously, lots of two-dimensional topological insulators have been predicted theoretically in the recent ten years. In 2011, ultrathin Bi(111) films with nontrivial Z_2 index was found theoretically where the topology property of it is independent of the thickness of the material[12]. Besides, atomically thin crystals[13] like silicene, germanene and stanene which possess a stronger SOC than graphene are predicted to be 2DTIs according to first principle calculation[14, 15]. SnX[16], PbX[17], BiX/SbX[18] (X=H, I, Br, Cl, F or OH for SnX, X=H, I, F, Cl, Br for PbX and X=H, F, Cl, Br for BiX/SbX) were also predicted to be 2DTIs, among which BiX/SbX have large band gaps ranging from 0.74 to 1.08eV and are also promising for room temperature application. Another QSH insulator, GaBiCl$_2$, with strong SOC and large energy gap is also predicted to be hopeful candidate for achieving dissipationless transport devices in 2015[19]. Further addition to the family of 2DTIs consists of silicon based chalcogenide[20], functionalized Thallium Antimony films[21], etc. In 2016, VA-VA semiconductors, like β-SbAs, were explored and were considered to be candidates for QSH systems[22]. Among the large family of 2DTIs predicted theoretically, transition metal dichalcogenides (TMD) may be one of the most promising candidates for achieving room temperature QSHE. Since its prediction in 2014[23], extensive work has focused on this family and its structurally distorted phase[24-32]. Notably, these study of QSHE in 2D topological insulators has built up a broad platform for engineering topological insulators by means of strain, tension, electrical and magnetic regulation, optical probe, proximity effect and so on and survival of QSHE has been proved above liquid nitrogen temperatures, which aggrandizes the application value of 2D TIs.

ISPECE

IOP Publishing

IOP Conf. Series: Journal of Physics: Conf. Series **1187** (2019) 042019   doi:10.1088/1742-6596/1187/4/042019

4. Conclusion

To sum up, topological insulators came into public attention through reheating Haldane's work[2] in 1988, which inspired people to consider possibility of topologically nontrivial properties among numerous materials with zero external magnetic flux. Subsequently theoretical and calculative work on searching for topology in seemingly trivial places based on topological band theory sprang up. It turns out that topology hiding in edges and surfaces prepares plentiful surprises for us. Based on the creative work of topological insulators, there emerge a large number of topologically nontrivial condensed matter materials like topological superconductors[33, 34], topological crystalline insulators[35-40], topological Kondo insulators[41], topological Mott insulators[42], some complex oxides[43, 44], Weyl semimetals, etc. The development in this area has greatly enhanced our understanding of different phases like magnetism, superconductivity, and spin of topological materials. However, topological insulator is a discipline in babyhood and still has a long way to go before widespread pragmatic application. In order to get deeper understanding of TIs and other related realms, more profound mathematical tools in topology beyond band theory need implementing and probing. Nevertheless, it is beyond doubt that further probe into the family of topological materials will lead to significant progress in superconductivity, spintronics and quantum computation.

Reference

[1] M. V. Berry, Proc. R. Soc. Lond. A **392**, 45 (1984).

[2] F. D. Haldane, Physical review letters **61**, 2015 (1988).

[3] D. J. Thouless, M. Kohmoto, M. P. Nightingale, and M. den Nijs, Physical review letters **49**, 405 (1982).

[4] G. Tkachov, *Topological insulators: The physics of spin helicity in quantum transport* (Pan Stanford, 2015).

[5] S.-Q. Shen, *Topological Insulators: Dirac Equation in Condensed Matters* (Springer-Verlag Berlin Heidelberg, 2012), Springer Series in Solid-State Sciences, 174.

[6] P. Streda, Journal of Physics C: Solid State Physics **15**, L717 (1982).

[7] G. Jotzu, M. Messer, R. Desbuquois, M. Lebrat, T. Uehlinger, D. Greif, and T. Esslinger, Nature **515**, 237 (2014).

[8] C. L. Kane and E. J. Mele, Physical review letters **95**, 146802 (2005).

[9] L. Fu and C. L. Kane, Physical Review B **74** (2006).

[10] L. Fu, C. L. Kane, and E. J. Mele, Physical review letters **98**, 106803 (2007).

[11] B. A. Bernevig, T. L. Hughes, and S.-C. Zhang, Science **314**, 1757 (2006).

[12] Z. Liu, C. X. Liu, Y. S. Wu, W. H. Duan, F. Liu, and J. Wu, Physical review letters **107**, 136805 (2011).

[13] A. Bansil, H. Lin, and T. Das, Reviews of Modern Physics **88** (2016).

[14] R. Mas-Balleste, C. Gomez-Navarro, J. Gomez-Herrero, and F. Zamora, Nanoscale **3**, 20 (2011).

[15] S. Z. Butler *et al.*, ACS nano **7**, 2898 (2013).

[16] Y. Xu, B. Yan, H.-J. Zhang, J. Wang, G. Xu, P. Tang, W. Duan, and S.-C. Zhang, Physical review letters **111**, 136804 (2013).

[17] C. Si, J. Liu, Y. Xu, J. Wu, B.-L. Gu, and W. Duan, Physical Review B **89**, 115429 (2014).

[18] Z. Song *et al.*, NPG Asia Materials **6**, e147 (2014).

[19] L. Li, X. Zhang, X. Chen, and M. Zhao, Nano letters **15**, 1296 (2015).

[20] R.-w. Zhang, C.-w. Zhang, W.-x. Ji, P. Li, P.-j. Wang, S.-s. Li, and S.-s. Yan, Applied Physics Letters **109**, 182109 (2016).

[21] R.-w. Zhang, C.-w. Zhang, W.-x. Ji, S.-s. Li, S.-s. Yan, P. Li, and P.-j. Wang, Scientific reports **6**, 21351 (2016).

[22] S. Zhang, M. Xie, B. Cai, H. Zhang, Y. Ma, Z. Chen, Z. Zhu, Z. Hu, and H. Zeng, Physical Review B **93** (2016).

[23] X. Qian, J. Liu, L. Fu, and J. Li, Science, 1256815 (2014).

[24] M. N. Ali, L. Schoop, J. Xiong, S. Flynn, Q. Gibson, M. Hirschberger, N. Ong, and R. Cava, EPL

(Europhysics Letters) **110**, 67002 (2015).

[25] F. Zheng *et al.*, Advanced Materials **28**, 4845 (2016).

[26] Z. Fei *et al.*, Nature Physics **13**, 677 (2017).

[27] Z.-Y. Jia *et al.*, Physical Review B **96**, 041108 (2017).

[28] S. Tang *et al.*, Nature Physics **13**, 683 (2017).

[29] V. Fatemi, Q. D. Gibson, K. Watanabe, T. Taniguchi, R. J. Cava, and P. Jarillo-Herrero, Physical Review B **95**, 041410 (2017).

[30] S. Wu, V. Fatemi, Q. D. Gibson, K. Watanabe, T. Taniguchi, R. J. Cava, and P. Jarillo-Herrero, Science **359**, 76 (2018).

[31] Y. Shi *et al.*, arXiv preprint arXiv:1807.09342   (2018).

[32] P. Chen, W. W. Pai, Y.-H. Chan, W.-L. Sun, C.-Z. Xu, D.-S. Lin, M. Chou, A.-V. Fedorov, and T.-C. Chiang, Nature communications **9** (2018).

[33] R. Roy and C. Kallin, Physical Review B **77**, 174513 (2008).

[34] M. Sato and Y. Ando, Reports on Progress in Physics **80**, 076501 (2017).

[35] L. Fu, Physical review letters **106**, 106802 (2011).

[36] T. H. Hsieh, H. Lin, J. Liu, W. Duan, A. Bansil, and L. Fu, Nature communications **3**, 982 (2012).

[37] P. Dziawa *et al.*, Nature materials **11**, 1023 (2012).

[38] J. Liu and L. Fu, Physical Review B **91**, 081407 (2015).

[39] C. Fang, M. J. Gilbert, and B. A. Bernevig, Physical review letters **112**, 046801 (2014).

[40] C. Fang, M. J. Gilbert, and B. A. Bernevig, Physical review letters **112**, 106401 (2014).

[41] M. Dzero, J. Xia, V. Galitski, and P. Coleman, Annual Review of Condensed Matter Physics **7**, 249 (2016).

[42] S. Raghu, X.-L. Qi, C. Honerkamp, and S.-C. Zhang, Physical review letters **100**, 156401 (2008).

[43] H. Jin, S. H. Rhim, J. Im, and A. J. Freeman, Scientific reports **3**, 1651 (2013).

[44] B. Yan, M. Jansen, and C. Felser, Nature Physics **9**, 709 (2013).

ISPECE

IOP Publishing

Calculation Formula of Positioning Error Based on Three Dimensions and Four Datum

Quanli Han[a], Hongqiang Wan and Donchen Han

Xi'An Technological University, School of Mechanical and Electronic Engineering, 710021 Xian, China

[a] Corresponding author: handleel@126.com

Abstract. Calculation of positioning error is difficult for engineers because of its invisibility in forming process, as well as coexistence of various forms. This paper is focused on the calculation of fixture location error from the novel direction to deepen understanding of error origination. In contrast to previous investigations, the authors have developed an equation to compute fixture location error based on three dimensions and four datum. This equation which is given after differential operation allows resource of location error. The findings from the differential equation shows that the fixture location error consists of the change in three misalignments of process datum and locating datum, spatula datum and locating datum, locating datum and stop datum respectively under general condition, and the value of fixture location error is gotten after simple operation of their three components listed above. And in case of locating element of plane or (and) Vee block especially, the fixture location error depends on the change in the distance of process datum and spatula datum. The work brings help to analysis positioning error and its conclusion indicates the method to reduce the value of positioning error quickly.

1. Submitting the manuscript

The fixtures in machining are used to provide accurate and repeatable unique positing of the part and sufficient work-holding to eliminate movement of the part under machining or assembly loads. The costs related to fixture can account for nearly one fifth of the total cost of a manufacturing system [1]. On the other hand, approximately forty percent rejected parts due to dimensioning errors are attributed to poor design of fixtures [2].

Many researchers have investigated the influence of fixture-workpiece system on the aggregate accuracy of machining. The objective of researchers is to improve workpiece location accuracy through optimization during the process of fixture planning, design and manufacture. The researches have been mostly focused on: contact analysis, finite-element analysis (FEA), force analysis, kinematic analysis.

Meyer and Liu brought out a methodology for fixture layout under dynamic machining forces and determination of optimal positions of locating elements and clamping forces [3]. Li and Melkote found an optimal synthesis approach for fixture layout and clamping force considering workpiece dynamics during machining and optimized clamping force for a multiple clamp fixture subjected to a quasi-static force [4, 5]. Amaral et al. developed an algorithm to automatically optimize clamp locations, clamping forces and fixture support in employment of 3-2-1 locating method, to increase machining accuracy by minimizing workpiece deformation [6]. Nee et al. invented a sensor-assisted fixture capable of delivering varying clamping loads, calculated from a quasi-static model, to find minimum of

Content from this work may be used under the terms of the Creative Commons Attribution 3.0 licence. Any further distribution of this work must maintain attribution to the author(s) and the title of the work, journal citation and DOI.

Published under licence by IOP Publishing Ltd

workpiece distortion [7]. Sielenaler and Melkote gave a fixture-workpiece model using FEA to investigate the influence of various parameters on workpiece deformation, including the compliance of the fixture body, contact friction and mesh density [8]. Raghu and Melkote built force-based clamping sequence model to analyze the effects of fixture clamping sequence on workpiece location errors, and they designed an algorithmic procedure to understand how forces and deformations change as the clamp applied sequentially is presented [9]. Tian et al. gave an approach to optically select the locating positions of workpieces and identify feasible clamping regions in terms of the requirements of form-closure principle for fixture layout [10]. Sanchez et al. calculated the contact load and its distribution and valid clamping regions in machining process [11]. Ratchev et al. proposed the behaviour prediction methodology for complex fixture-workpiece during machining processes [12]. Hazarika et al. presented a setup planning methodology for prismatic workpieces by formulating the objective function to minimize the maximum of locating elements reaction forces [13]. Chaari et al. modelled kinematical deviation due to workpiece locating and relocating, and determined dynamic displacements caused by clamping and machining force [14]. Vishnupriyan computed the effect of different layouts and various clamping forces on machining error in terms of system compliance and workpiece dynamics [15]. Selvkumar et al. presented a hybrid scheme to design an optimum fixture layout to reduce the maximum elastic deformation of workpiece from forces while machining [16].

Several important conclusions stem from the analysis of previously discussed investigations. Namely, elastic or plastic workpiece deformations in clamping process can greatly influence final workpiece accuracy. On the other hand, detachment of workpiece from locating elements during machining can lead to less efficient clamping process. Minimization of fixture-workpiece compliance should be paid special attention to fixture layout optimization when considering modern cutting regimes or machining thin-walled and complex-geometry workpieces.

Investigations published so far, basically deal with dynamic behaviour analysis of contact interfaces between fixture elements and workpiece. It is impossible to thoroughly understand the fixture-workpiece interface behaviour because of complex-geometry and various clamping/locating requirements of workpiece machined. The authors maintain that comprehensive analysis of dynamic behaviour and optimization of fixture elements requires, first of all, a dedicated instrument capable of testing of physical models which represent the fixture-workpiece interface behaviour; last but not least, understanding of location error resource and formation of location error. With this in mind, investigations in others of this paper are followed: Four datum and two dimensions formed in locating process are reviewed in section 2, and the computational formulas of process dimensions and dimension error are deduced in section 3, then an example to validate the formula presented is taken in section 4.

2. Review of four datum and two dimensions formed in locating procession
During the use of fixture, the following requirements should be met with, namely, the interface contact between fixture element and certain surface of workpiece should be kept, and the clamping force should be controlled firmly so that the maximum of locating elements reaction forces can be accepted. Once the contact between fixture locating element and surface of workpiece comes into being, the interface on fixture element and workpiece can be determined through stop datum and locating datum respectively. The other two are called as process datum on workpiece and spatula datum on fixture. Based on the above description, the process distance is defined between face to be machined and process datum, while the spatula distance between working face of tool setting element and spatula datum is done. Figure 1 is used as schematic drawing of fixture-workpiece, in which the locating element is central spindle, and workpiece with curve outer surface has central hole in the direction of itself axis, and their two parts are engaged in locating process. Where A- process datum; B- spatula datum; C- locating datum; D- stop datum; M- machining surface; N - interface contact; 1- workpiece; 2- locating element; 3 -fixture; 4- tool; 5- plug gauge; 6- tool setting element.

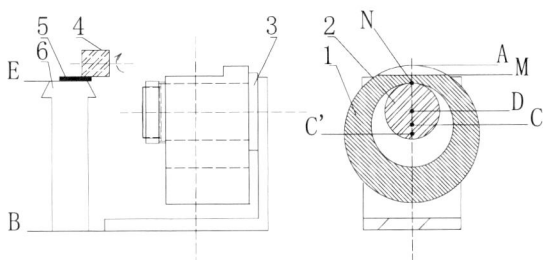

Figure 1. Schematic drawing of fixture-workpiece

3. Computational formulas of process dimension error

The map of dimension geometrical relationship is drawn in Fig.2 to investigate the influence factors of process dimensions error. The outer line on cylindrical workpiece, namely process datum stands for A, and the theoretical center of workpiece is C', then the distance of A and C' is the diameter of workpiece R. On the other hand, the inner circle on the workpiece is the limit surface, which constitutes interface contact with outer surface of locating element. And the theoretical center of inner circle on the workpiece is represented as C. So the distance between C and C' is named as concentricity error. Correspondingly the theoretical center of locating element, that is to say, stop datum is named as D, and then DC stands for the distance between stop datum and locating datum. The spatula datum B usually lays on the worktable of machine, CB is used to express the distance of locating datum and spatula datum, DB is expressed as the distance between stop datum and spatula datum, which is kept constant in one machining operation for one group of workpieces. Finally the distance of stop datum and process datum is stood as CA, E is the working surface of tool aligner, and the distance of E and B is named as tool setting dimension, and the distance between E and M equals to size of the plug gauge, MA stands for process dimension. The following examples in next part will be used to specify carefully.

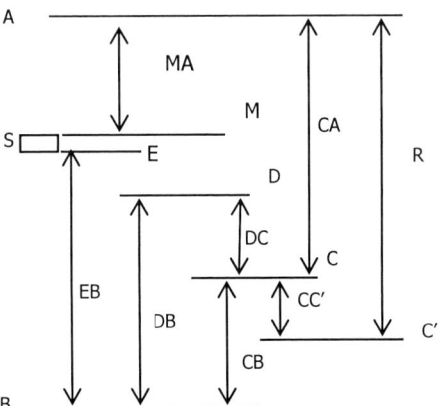

Figure 2. Map of dimension geometrical relationship

From the Fig.2, the following equations are gotten in the terms of geometric diagram.

$$MA + s + EB = R - CC' + CB \qquad (1)$$
$$CA + CC' = R \qquad (2)$$
$$DB = DC + CB \qquad (3)$$

where MA is written as A_1, and EB is done as A_2 whose size is regarded as constant in machining group of workpieces, and CC' is written as e.

Equations 1, 2, 3 are rewritten as followed:

1397

$$A_1 = (R - CC') + CB - (e - A_2) \qquad (4)$$
$$CA = R - e \qquad (5)$$
$$DC + CB - DB = 0 \qquad (6)$$

By substituting Equation 5 into Equation 4,

$$A_1 = CA + CB - (e - A_2) \qquad (7)$$

By adding Equation 6 into Equation 7

$$A_1 = CA + CB - (e - A_2) + 0$$
$$= CA + CB - (e - A_2) + DC + CB - DB$$
$$= CA + 2CB - (e + A_2 + DB)$$
$$= f(A,B,C,D) - k$$
$$f(A,B,C,D) = CA + 2CB$$
$$k = e + A_2 + DB \qquad (8)$$

The following findings are derived from the equation 8.

Process dimension is divided into two subparts, namely, f and k. among them, f is three dimensions which are CA, CB, and DC, they come from the four datum which are the A, B, C and D, and they take direct influence on size of process dimension. Among them, the effect of CB on dimension is bigger than other two factors, which is derived from the coefficient before them.

And k is three dimensions consists of the tool setting dimension, distance between stop datum and spatula datum, tool setting dimension and workpiece's geometrical characteristics such as concentricity error.

After the differential operation on equation 8, the following equation 9 is gotten:

$$d(A_1) = d\{f(A,B,C,D) - k\}$$
$$= d(CA) + 2d(CB) + d(DC) \qquad (9)$$

where d stands for differential operation.

During of machining one group of workpiece, the position of B and D is fixed while that of C an A takes change. And the value of differential operation on k is zero whose components are regarded as constant value in this kind of machining. So the process dimension error stems from change in the three misalignments of A and C, B and C, C and D respectively. And the value of fixture location error can be gotten after simply operation on their three listed above.

Based on the analysis above, the value of location error is naturally the simple operation of change in three misalignments of process datum and locating datum, spatula datum and locating datum, locating datum and stop datum respectively.

Once the misalignments of C and D does not occur, theprocess dimension error stems from two misalignments of A and C, B and C respectively. In this case, the process dimension error depends on the change in the misalignment of BA, and which can be seen in locating procession. The following examples in next part will be used to specify carefully.

4. Example of computer of fixture location error

Described as figure 3, the location of workpiece is done by the combination of plane and half of V-block. The process dimension is h1 and h2, whose process datum is respectively A_1 and A_2. And B is spatula datum; the corresponding dimension of two datum is A1B and A_2B. Their two dimensions are projected into the direction of h_1 and h_2, namely, A_1C and A_2C.

ISPECE   IOP Publishing

IOP Conf. Series: Journal of Physics: Conf. Series **1187** (2019) 042020   doi:10.1088/1742-6596/1187/4/042020

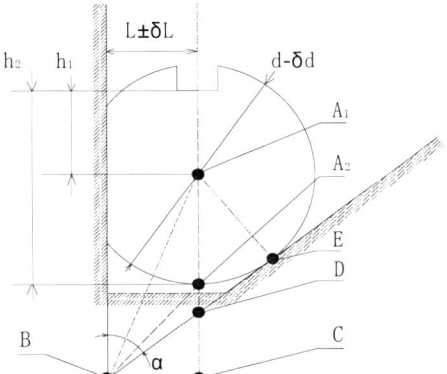

Figure 3. Example of fixture location error used element of plane or V block

Based on geometry in figure 3, the following equation 10 is gotten:

$$\triangle h_1 = \triangle A_1 C = \triangle(A_1 D + DC)$$
$$= (1/\sin\alpha)\delta d/2 + 2(1/\tan\alpha)\delta L \qquad (10)$$
$$\triangle h_2 = \triangle A_2 C = \triangle(A_2 D - A_1 A_2 + DC)$$
$$= (1/\sin\alpha)\delta d/2 - \delta d/2 + 2(1/\tan\alpha)\delta L$$

where, $\triangle h_i (i=1,2)$ is expressed as location error when machining process dimension corresponding. AiC (i=1,2) stands for the component of distance of two datum of A_i (i=1,2) and B in the direction of h_i (i=1,2).

5. Conclusions
After the analysis above, the findings are derived from.

(1). The reason for location error is the misalignments of process datum and locating datum, spatula datum and locating datum, locating datum and stop datum respectively.

(2).When the misalignment of locating datum and stop datum does not occur, the size of location error depends on the change in the distance of process datum and spatula

Acknowledgment
This work supported by Research Program supported by the Shaanxi Provincial Department of Education (Shaanxi key laboratory research project (17JS060), China.

References
[1] Z. M. Bi, W. J. Zhang, J. Prod. **39**, 13(2001).
[2] K. R. Wardak, U. Tasch, P. G. Charambides, J. Msy. **20**,1(2001).
[3] R. T. Meyer, F. W. Liu, J. Prod. **35**, 2 (1997).
[4] B. Li, S. N. Melkote, J. Adv. Manuf. Tech. **17**, 2 (2001).
[5] B. Li, S. N. Melkote, J. Mach. Tool Manuf. **39** ,5(1999).
[6] N. Amaral, J. J. Rencis, Y. Rong, J. Adv. Manuf. Tech. **25** ,8(2005).
[7] A. Y. C. Nee, A. S. Kumar, Z. J. Tao, J. Engine Manuf. **214** , 3 (2000).
[8] S. P. Sielenaler, S. N. Melkote, Comput Ind. **46**, 1 (2006).
[9] A. Raghu, S. N. Melkote, J. Mach. Tool Manuf. **44**, 4 (2004).
[10] S. Tian, Z. Huang, L. Chen, Q. A. Wang, J. Adv. Manuf. Tech. **30** , 1-2 (2006) .
[11] H. Sanchez, M. Estrms, F. Faura, J. Adv. Manuf. Tech. **29**, 5-6 (2006).
[12] S. Ratchev, K. Phuah, S. Liu, J. Matr. Process Tech. **191** ,1-3 (2007).
[13] M. Hazarika, U. S. Sixit, S. A. Deb, J. Adv. Manuf. Tech. **51**, 9-12 (2010)
[14] R. Chaari, M. Abdennadher, J. Louati, M. Haddae, J. Simul. Model, **10**, 2 (2001).

[15] S. Vishnupriyan, Assembly Autom. **32**, 2 (2012).

[16] S. Selvkumar, K. P. Arulshri, K. P. Padmanaban, K. S. K. Sasikumar, J. Adv. Manuf. Tech. **65**, 9-12 (2013).

Application and realization of ray tracing in network planning of wireless private network

Yonghua Zhang[1], Juanjuan Sun[2]

[1]Beijing University of Posts and Telecommunications, Beijing 100876, China;

[2]Beijing University of Posts and Telecommunications, Beijing 100876, China;

zhangyonghua@bupt.edu.cn; sunjj@bupt.edu.cn

Abstract: In order to meet the business requirements of wireless private network in network planning, especially the network requirements of intelligent power private network, such as greater capacity and higher precision. In this paper, deterministic propagation model is used to simulate the network in network planning to improve the accuracy of calculation, and at the same time, must ensure the speed of the planning software. The reverse ray tracing algorithm and mirroring method are used for calculation. Simulation results show that when calculating the direct and low-order reflection paths and ignoring the high-order reflection and diffraction paths that have little effect on the results can ensure a certain calculation rate. And achieve the purpose of balance accuracy and rate.

1. Introduction

Since the beginning of the "Twelfth Five-Year Plan", according to the construction needs of the smart grid, the coverage of the power grid needs to be extended to the residents, which puts higher requirements on the stability and security of the communication network [1]. The coverage advantage of TD-LTE network for mass terminals is obvious. It is necessary to study the application of TD-LTE network in private network.

In wireless network planning, propagation model research is the basis of coverage analysis and capacity simulation. In wireless network planning, the wireless propagation model can accurately simulate the propagation coverage of the network, providing a basis for the design and verification of wireless network planning [2]. At present, the residential users of the power private network are mainly concentrated indoors, where the environment is more complex and the calculation requirements for propagation model are higher. Therefore, it is necessary to study indoor wireless network [3].

2. Propagation model classification

According to the nature of propagation, the propagation models can be divided into two categories: empirical propagation models and deterministic propagation models.[4] Empirical model: An empirical model is a formula derived from statistical analysis of a large number of measured data. The method is simple, and it doesn't require detailed environmental information, which is easy to use, but the method estimates that the path loss is not very accurate.[5] Deterministic model: It is based on electromagnetic theory and is used to solve specific environment. The description of this particular environment is usually done by building and terrain databases. For the deterministic model, the accuracy of the environment description determines the accuracy of the deterministic model. Therefore, this kind of model requires the highest accuracy of environment description. [6].

In the indoor environment, the transmission power of electric wave propagation is relatively small and the coverage distance is relatively short. Its propagation is not related to climate, but related to the size, shape, structure, room layout and human activities of buildings, among which the influence of building materials is the largest [7].

Content from this work may be used under the terms of the Creative Commons Attribution 3.0 licence. Any further distribution of this work must maintain attribution to the author(s) and the title of the work, journal citation and DOI.

Published under licence by IOP Publishing Ltd

For indoor propagation models, deterministic propagation models are usually used to obtain more accurate results. Ray tracing is one of the most widely used deterministic propagation models.

3. Ray tracing model

The ray tracing model is a deterministic propagation model. The ray tracing technique is based on the diffraction theory of geometric optics. The electromagnetic wave is abstracted into a beam of rays, and the object points that make up the object are regarded as geometric points, and the direction of the light represents the direction of propagation of the light energy. The phase, delay and power of each ray are calculated according to the theory of electromagnetic wave propagation, and then the results of coherent synthesis of all rays are calculated.

3.1 Basic principles of ray tracing

The ray tracing technique can approximate the propagation of the ray to the propagation process of the radio wave. There are two general methods for calculating electromagnetic field parameters, analytical methods and numerical methods. For the boundary value problem with less clear boundary, the numerical solution is used to obtain the numerical solution of the electromagnetic field. For the problem with clear boundary conditions and geometric rules, the analytical solution of the electromagnetic field boundary value problem can be used to obtain the analytical solution. The exact solution obtained here for the boundary value condition.

When the received power is calculated by the ray tracing method, the power of each ray is summed. The formula is as follows:

$$P_r = \sum_{i=1}^{n} P_i \qquad (1)$$

Where P_r represents the received power and P_i represents the power carried by one of the rays.

$$P_i = \frac{P_t g_t g_r \lambda^2}{(4\pi d)^2} \left[\prod_j R_j\right]^2 [\prod_k T_k]^2 [\prod_l A_l(s', s) D_l]^2 \qquad (2)$$

Where d is the total length experienced by the ray from the source to the source, P_t is the transmit power, λ is the wavelength, g_t and g_r are the antenna gains at the transmitter and receiver respectively in the direction of the ray; R_j is the reflection coefficient, T_k is the representation Transmission coefficient, D_l represents the edge diffraction coefficient, and $A_l(s', s)$ is the spatial diffusion coefficient used to correct the diffraction coefficient.

Reflection coefficient: For the reflection coefficients R_h and R_v of the horizontally polarized wave and the vertically polarized wave, the calculation formula is as follows:

$$R_h = \frac{\sin\theta - \sqrt{\epsilon_c - cos^2\theta}}{\sin\theta + \sqrt{\epsilon_c - cos^2\theta}} \qquad (3)$$

$$R_v = \frac{\epsilon_c \sin\theta - \sqrt{\epsilon_c - cos^2\theta}}{\epsilon_c \sin\theta + \sqrt{\epsilon_c - cos^2\theta}} \qquad (4)$$

Where ϵ_c is the complex downlink constant of the reflective medium, expressed as:

$$\epsilon_c = \epsilon_r - j60\lambda\sigma \qquad (5)$$

Where ϵ_r is the relative dielectric constant of the reflective medium, σ is the electrical conductivity, and λ is the wavelength.

3.2 Ray tracing algorithm

After the scene modeling, the ray tracing technique needs to simulate the trajectory of the ray. Algorithms for calculating ray trajectories can be divided into two categories: forward ray tracing and reverse ray tracing. The most common method of forward ray tracing is the beam method, which is also commonly called the incident and rebound ray method. This method mainly starts to trace the rays from the source and sequentially traverses all possible rays to complete the forward simulation of the radio wave propagation process. The method needs to traverse all the rays, and needs to judge the receiving range of the receiving point, that is, take the receiving point as a ball to determine whether the ray passes through the ball.。

The reverse ray tracing algorithm starts from the launch point and backtracks each path that can reach the source point according to the principle of geometric optics, including direct reflection, reflection, diffraction, and so on. Abstract different objects in space into plates, columns, cones, and analyze their visible faces. Calculate the arrival power, delay, phase shift, etc. of the ray after reflection, diffraction, etc. Then coherently synthesize the results of all rays. Since the path loss of multiple reflections and diffraction is large, these rays can be selectively ignored, and only the ray paths that have a large influence on the results are calculated. In the case where it is not necessary to calculate all the rays, the degree of computational complexity is reduced by several orders of

ISPECE   IOP Publishing

IOP Conf. Series: Journal of Physics: Conf. Series **1187** (2019) 042021   doi:10.1088/1742-6596/1187/4/042021

magnitude. In addition, proper partitioning of the modeling of the scene can also make the computational complexity smaller.

3.3 Modeling of the indoor environment

For the ray tracing method, it is generally necessary to obtain the specific information of the simulation scene firstly. In this paper, the scene is described mainly by the vector form of the polyhedron model. Specifically, all spatial entities are stored in the form of a polyhedron in the scene, for example, the ground can be represented by a polygon face. In this way, the number of polyhedron, the number of vertices of each polyhedron, and the vertex coordinates of each polyhedron will be stored in the final scene database. Based on this information, information such as the normal vector of the plane can be calculated as parameters for calculating the process of reflection transmission of the ray.

3.4 Tracking of the ray propagation path

In the process of ray tracing, the space can be divided to facilitate the management of the information of the obstacles, and the ray tracing process can be more efficient. The inverse ray tracing method polls the visible faces and sharps from the receiving source to generate a virtual source that can get the receiving source through reflect and diffract, and the virtual source is the starting point of the ray's next propagation path. According to the virtual sources, a virtual source tree is established. As shown in Fig. 1, each child node of the tree represents a virtual source point. In order to ensure the running speed of the planning software, we can ignore the high-order reflection and the diffraction ray to reduce the tree's depth.

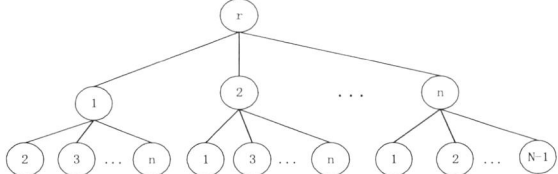

Figure 1 Virtual source of reverse ray tracing algorithm

Establishing a ray tracing model, this paper performs three-dimensional modeling for the indoor environment. The ray will be reflected many times by the ceiling and the floor for the indoor environment is closed. Therefore, in the received power, the proportion of the power of the reflection path will increase, and objects such as furniture in the house will have a large influence to the result. For the forward ray tracing, all rays from the source should be simulated, and the more rays that are tracked, the less likely to miss an effective ray due to environmental factors. In an indoor environment, many rays are reflected or diffracted multiple times to reach the receiving source, so the tracking algorithm is more complex than the outdoor environment. The inverse ray tracing algorithm calculates the ray path that exists in reality, and the calculated magnitude is much smaller than the forward tracking, which ensures the rate of the wireless network planning.

4   Simulation examples and verification

In order to verify the feasibility of the reverse ray tracing method in wireless network planning, this section takes the indoor scene as an example and performs simulation analysis according to the scenario. The specific process is as follows.

4.1 Model and simulation environment

The indoor environment model is shown in Figure 3. According to Recommendation ITU-R P.2040-1, "Effects of Building Materials and Structures on Radio Wave Propagation Above 100 MHz", the relative dielectric constants and conductivity of materials are shown in the table below.

Table 1 Material characteristics

Material	Relative interface constant	Electrical conductivity
Concrete	6.48	0.166
Solid wood	1.64	0.11
Glass desktop	6.2	0.1

4.2 Analysis of simulation results

The transmitting antenna and the receiving antenna are abstracted as points, the coordinates of the transmitting antenna are T_x (1,17,1), the transmitting power is set to 20 dBm, and the coordinates of the receiving antenna

1403

are R_x（15,1,3）. Due to the occlusion of the cabinet, there is no direct path to the propagation path, and tracking is performed using reverse ray tracing. The results obtained are shown in Fig 2.

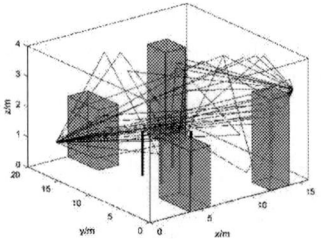

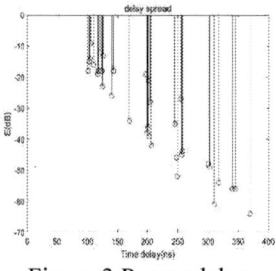

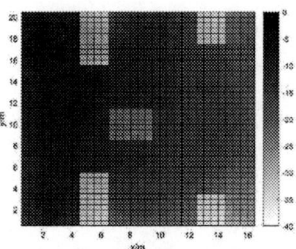

Figure 2 Indoor multipath effect tracking

Figure 3 Power delay distribution

Figure. 4 Planform of power distribution

When the transmitting antenna coordinates are T_x（1,17,1）, the receiving antenna R_x is arbitrary coordinates and the height is 1.5 m, the field strength distribution at different positions is as shown in Fig 4.The depth of the color represents the magnitude of the electric field intensity of the signal. The emission point is T_x（1,17,1）. As the signal is radiating, the electric field intensity of the signal gradually decreases. The lightest area is the location of the obstacle. Since the height of the receiving point is lower than the height of the obstacle, the color of the position of the obstacle is the lightest, that is, the electric field intensity of the signal is the smallest. Consistent with the actual situation the electric field intensity of the signal of the position of the obstacle should be the weakest.

When the coordinates of the transmission point is T_x（1,17,1）, select 10 points R_1（0,2,3）、R_2（5,7,1）、R_3（1,12,2）、R_4（11,3,3.5）、R_5（4,0,1）、R_6（7,18,2）、R_7（7,2,1）、R_8（13,17,3）、R_9（14,18,2）、R_{10}（15,1,3）. In which the first 5 points have a direct path, and the next 5 points do not have a direct path. Table 2 shows the proportion of direct reflection, primary reflection, secondary reflection, tertiary reflection, primary diffraction, higher secondary reflection and diffraction in the received power of the receiving point. Table 3 shows the power ratio of different propagation mechanisms when there is no direct path.

Table 2 Power ratios of different propagation mechanisms when there is a direct path（%）

Receiving point	Direct	Primary reflection	twice reflections	thrice reflecting	Primary diffraction	Higher order reflection and diffraction
1	77.02	17.87	4.65	0.89	0.45	0.12
2	85.13	10.9	2.89	0.65	0.34	0.09
3	93.34	3.91	2.13	0.32	0.24	0.06
4	70.56	21.23	5.75	1.43	0.78	0.25
5	75.06	17.89	5.05	1.14	0.67	0.19
Mean of proportion	80.222	14.16	4.094	0.886	0.496	0.142

Table 3 Power ratio of different propagation mechanisms without direct path（%）

Receiving point	Primary reflection	twice reflections	thrice reflecting	Primary diffraction	Higher order reflection and diffraction
6	56.95	27.79	14.07	0.52	0.67
7	58.54	29.43	10.71	0.54	0.78
8	64.76	22.65	11.2	0.67	0.72
9	48.34	32.34	17.51	0.89	0.92
10	51.54	31.13	15.69	0.75	0.89
Mean of proportion	56.026	28.668	13.836	0.674	0.796

It can be seen from the data in Table 2 that when there is a direct path, the power of the direct path accounts for the highest proportion of the total power. When the distance between the receiving point and the transmitting point is relatively close, the direct ray power ratio can reach more than 90%. The average power of direct power is 80.222%, the average power of primary reflection path is 14.16%. The average power of reflected path that is more than four times and the power of diffracted path that is more than two times accounts for 0.142%. It can be seen that in the indoor environment, the influence of the diffractive power on the total power is very small, but the influence on the rate of the planning software is relatively large, so it can be ignored. The sum of direct ray power and within three (including three) times reflected ray power accounted for more than 99% of the total power.

It can be seen from the data in Table 3, when there is no direct path, the power of the primary reflection path accounts for the highest proportion of the total power. And the power of the reflection within three times (including three times) accounted for more than 98% of the total power. Moreover, the simulation takes 87s when the direct ray and reflected ray within three times (including three times) are calculated., while it takes 1431s when calculate the reflected ray within 12 times and the diffracted ray within 4 times. The speed is 16 times faster. Therefore, it can be concluded that in the wireless network planning, calculating the direct ray power and reflected ray power which setting the maximum number of times of reflection is 3. In this way, the software of network planning can not only meet the requirements of accuracy, but also greatly improve the rate of the software to achieve the purpose of balancing accuracy and rate.

5 Summary

In wireless network planning, how to improve the accuracy of the propagation model, and to increase the operating speed of the software as much as possible is a practical problem that must be faced. Considering the user usage scenarios of the power private network, the simulation of the indoor environment is very important in the network planning of the power private network. Considering the complex environment in the room, the reverse ray tracing method is an accurate method for predicting the indoor radio wave propagation model. The number of reflections that need to be calculated in the simulation can be set according to different requirements for accuracy and rate, thereby achieve the purpose of balancing accuracy and rate.

References

[1] Weiwei Kong，Xiannan Luo，Peng Jia. Research on Planning and Optimization of Power Wireless Special Network [J]. Modern information technology, 2018，2096-4706（2018）12-0062-03.

[2] N. Omaki, T. Imai, K. Kitao, and Y. Okumura, "Improvement of ray tracing in urban street cell environment of non line-of-site (NLOS) with consideration of building corner and its surface roughness," in Proc.EuCAP'16, Davos, Switzerland, Apr. 2016, pp. 1–5.

[3] Li, J., et al.: Measurement-Based Characterizations of Indoor Massive MIMO Channels at 2 GHz, 4GHz, and 6 GHz Frequency Bands. In 83rd Vehicular Technology Conference (VTC Spring), pp. 1–5. IEEE Press, Nanjing (2016)

[4] Wald I, Woop S, Benthin C，et al. Embree：A kernel framework for efficient CPU ray tracing[J]. ACM Transactions on Graphics，2014,33(4):70-79.

[5] Barros F J B, Costa E, Siqueira G L，et al. A Site-Specific Beam Tracing Model of the UWB Indoor Radio Propagation Channel[J]. IEEE Transactions on Antennas & Propagation, 2015, 63(8):3681-3694.

[6] Jundong Tan, Zhuo Su, Yunliang Long. "A Full 3-D GPU-based Beam-Tracing Method for Complex Indoor Environments Propagation Modeling." IEEE Transactions on Antennas and Propagation, pg. 2705-2718, 2015.

[7] Liu Z, Shi D, Gao Y, et al. A new ray tracing acceleration technique in the simulation system of electromagnetic situation[C]// Environmental Electromagnetics. IEEE, 2016:329-333.

Sparse Manifold Learning based on Laplacian Matrix

Wenjing Li

School of informatics and computer science

Indiana University Bloomington

Bloomington, USA

wenjing__li@126.com

ABSTRACT In machine learning, a group of high-dimensional data point set which represents an image set can be looked upon as the point set distributing on a nonlinear manifold. Typically, the manifold dimension is much lower than the dimension of data points. Therefore, it is important to explore the real dimension and the real geometry of high-dimensional data point set. Based on this objective, researchers put forward the concept of manifold learning. Traditional manifold learning can achieve dramatic dimensional reduction on high-dimension points, but there are still some problems of it. In the aspect of processing a large amount of data currently, traditional manifold learning algorithm reveals many weaknesses mainly in time consumption. To improve the deficiency, the paper puts forward a new algorithm, aiming at simplifying the time process of manifold learning algorithm which is also abbreviated as SLEP. In the part of the experiment, the study makes a comparative experiment between the synthetic data set and the real data set. The result shows that the proposed algorithm improves time efficiency.

1. INTRODUCTION

Machine learning is a subset of artificial intelligence based on statistics to train the computers making decisions wisely. The primary methods in implementing data-driven include prediction and predictive analytics. And the essential function of machine learning is finding the pattern by extracting the intrinsic pattern of data from large amount of unstructured data. The major methods, including PCA [6], and Manifold learning [3, 5], and others are divided to supervised methods and unsupervised methods.

In supervised method, the computer is presented with example inputs and their desired outputs, given by a "teacher", and the goal is to learn a general rule that maps inputs to outputs.(wiki) Comparably, unsupervised method are not given with the specific features. As one of the representing method of unsupervised method, Laplacian [1] is local-preserving and has simple core algorithm. And it solved the problem of representing low-dimensional data when data arise from sampling a probability distribution on a manifold.

The one of the deficiency of the current algorithms of manifold is selecting limited neighbors within unqualified high-density data set. In this letter, we explore an approach that solves the problem of constructing adjacency graph by constructing connected set from the constructed adjacency graph. The implementation of SLEP is simple and local-preserving. And the solution reflects the geometrical-intrinsic of the manifold.

2. RELATED WORK

Manifold learning is an important branch of machine learning, which has a great development since the 21st century. In 2000, there were two papers released in Science, Isomap [3] and LLE [5], firstly put forward the concept of manifold learning. The objectives of manifold learning are to reduce the redundant information about high-dimensional data set, and realize dimensional reduction of the data set. Presently, manifold learning algorithm is divided into two types, one of which is global dimensional reduction algorithm like Isomap [3]. The other one is local dimensional reduction algorithm such as LLE [5], LEP [1], LPP [4], LTSA [7], HLLE [2] and so on. The global dimensional reduction algorithm aims at maintaining the global structure during the process of manifold dimension reduction. For example, Isomap [3] algorithm is to maintain the geodesic distance between two arbitrary points of manifold. However, local dimensional reduction algorithm is to maintain local geometric structure of manifold. For example, LLE [5] algorithm is to maintain the linear relationship between local neighborhood point sets of a point while LEP [1] is to maintain the distant structure between local neighborhood points which can keep close relationship between two closer points after dimensional reduction.

The important step of traditional manifold learning algorithm is to divide a data point set into different neighbor sets and the number of neighbor sets should be equal to the number of data points. Then according to different objectives, various geometric structures between data sets can be explored. However, with the social development, there are a lot of large data sets with high dimension and massive data volumes to be solved at present. If researchers still adopt the traditional manifold learning algorithm to reduce dimension of this type of massive data set, it will lead to a complex learning process. Aiming at this circumstance, a new algorithm should be designed to reduce time complexity of the algorithm to a great extent without affecting the result of dimensional reduction. By using the proposed algorithm, the efficiency of processing massive data sets will be improved.

3. SOLUTION OF THE ALGORITHM
Firstly, it needs to explain the symbols used in the algorithm. If the high-dimensional input data point set that we process shows as $\{x_1, x_2, \cdots, x_N\}, x_i \in R^D$, where N refers to the number of data point, and D refers to dimension of data point, the corresponding lower dimension after dimensional reduction shows as $\{y_1, y_2, \cdots, y_N\}, y_i \in R^d$, where d refers to dimension of lower dimensional space. In the following, we first give the method of selecting local minimal neighborhood set, then we describe our proposed sparse manifold learning algorithm (SLEP for short) and in the last we analyze the comparison between traditional LEP algorithm and our SLEP.

3.1 Selection of Local Minimal Neighborhood Set
Firstly, K-nearest neighborhood can be used to solve the local K-nearest neighbor points of x_i at each point, which aims at finding a group of neighborhood subset from the neighborhood set to minimize the number of neighborhood subset, and completely cover all sample points. Moreover, adjacent neighborhood sets are in the relationship of mutual coverage. Therefore, the key of the problem is to find the neighborhood subcovering with finding an algorithm to search the minimal covering subset. The key of the algorithm is to define a set of rules to depict the structural relationship between any two neighborhoods. The existing method is using Greedy Algorithm to search, but it's best to find an algorithm which is more suited to our needs, or improve the existing algorithm based on our objectives. First of all, we can directly construct a k- neighborhood set at any point. However, obviously, there are many redundant sets in this group of neighborhood set. Our objective is to select a subset from this group of neighborhood set. The main idea of this method is referred to [8]. The specific algorithm is shown as following:

Firstly, it needs to construct a K-neighborhood set $\{U_i\}, i = 1, \cdots, N$ at any point x_i, and randomly select a K-neighborhood set U_i as the initial set. Then a neighborhood subset should be selected from the rest neighborhood sets to make this group of the neighborhood subset fully cover all sample point sets and each K-neighborhood is in the relationship of mutual coverage. To make the neighborhoods cover each other and the coverage rate not be too high, we should make an index to measure the coverage rate between two neighborhood sets:

$$|S \cap U| \le (1 - \alpha)|S|$$

Where α refers to the coverage rate between two sets.

3.2 Algorithm Process

The thesis aims at improving the LEP algorithm. The improved algorithm is called SLEP, and the main process is shown as following:

(1) Use K-nearest neighborhood algorithm to solve the K-neighborhood set U_i at x_i of each point.

(2) Use the minimal sub-neighborhood set selection algorithm given in the section 3.1 to select a group of sub-neighborhood set shown as $\{M_j\}_{j=1}^{m}$ from the set$\{U_i\}_{i=1}^{N}$.

(3) Use the reconstructed neighborhood set $\{M_j\}_{j=1}^{m}$ between data points to construct the weight between data points. The constructing algorithm is shown as following:

$$W_{ik} = \begin{cases} e^{-\frac{\|x_i - x_k\|^2}{2t}}, & if \ x_i, x_k \in M_j \\ 0, & if \ x_i, x_k \notin M_j \end{cases}$$

(4) Use LEP to construct the corresponding Laplacian matrix, and then solve the corresponding lower dimension to show $\{y_1, y_2, \cdots, y_N\}$

3.3 The comparison between SLEP and LEP

Different from LEP, the Laplacian matrix constructed by the proposed algorithm is sparse. Because we construct a group of sparse local neighborhood set at a dataset, its number of local neighborhoods is less than the number of data points. Compared with the weight matrix W constructed by LEP, the weight matrix W_s among data points constructed by the minimal neighborhood sets is sparser. The proposed algorithm mainly aims at the circumstances of a huge number of data points, which can lower the time complexity of lower dimension algorithm.

4. EXPERIMENT

In this chapter, a comparative experiment between the proposed algorithm and traditional manifold learning algorithm will be made to illustrate that the proposed algorithm is better than other algorithms from the perspective of time efficiency. In the experiment, the comparative experiment is between synthetic data sets and real data sets. Specifically, synthetic data sets are Swiss Roll data set and Puncture Sphere data set which are generated from the document mani.m while the real data set is USPS data set. During the experiment, five traditional manifold learning algorithms, Isomap [3], LLE [5], LTSA [7], HLLE [2] and PCA [6] are compared with the proposed algorithm, SLEP.

4.1 Synthetic data set

The experiment respectively produces different quantities of Swiss Roll data sets and Puncture Sphere data sets from the document mani.m. These two groups of data sets are three-dimensional data sets and the two-dimensional manifold structure embedding into the three-dimensional space. The objective of manifold learning is to reduce the dimension of these three-dimensional data sets in the two-dimensional space. And the objective of the proposed algorithm is to deduce the time complexity of traditional manifold learning algorithm. In order to verify that the proposed algorithm has an advantage in time consumption, the experiment is based on different quantities of data sets.

During the experiment, researchers respectively select 2000 and 5000 data points to make experiment for each data set. The experiment also takes the advantage of traditional manifold learning algorithm and SLEP to reduce the dimension of these two groups of data sets and then calculate time consumption of each algorithm. The corresponding time consumption is shown in table 1. It can be seen that compared with traditional manifold learning algorithm, time consumption of the proposed algorithm is lower. During the process of processing 5000 data points, the advantage of SLEP in time consumption is more obvious. In results of the experiment, time comparison between SLEP and LEP is much concerned. Table 1 shows that the proposed algorithm SLEP has a great advantage.

In addition to the advantage in time consumption, researchers hope that SLEP can also have an advantage in dimensional reduction, so the experiment is also made in dimensional reduction to verify the hypothesis. The result is shown in graph 1 and 2 which reflect that there is not much difference between the effect of dimensional reduction of the proposed algorithm and LEP.

Table 1. The comparison of time consumption of manifold learning algorithms

	Swiss Roll/5000	Sphere/5000	Swiss Roll/10000	Sphere/10000
PCA [6]	0.6858	0.5886	2.4815	2.3984
Isomap [3]	1071.2383	969.1432	0.6467	0.9261
LLE [5]	1.8241	1.6205	6.5013	4.8359
LEP [1]	2.1818	1.8506	8.6720	7.6406
HLLE [2]	38.3258	34.5965	257.5510	254.9848
LTSA [7]	3.2619	3.6690	18.2465	16.9970
SLEP	0.8965	0.7536	3.1584	2.8156

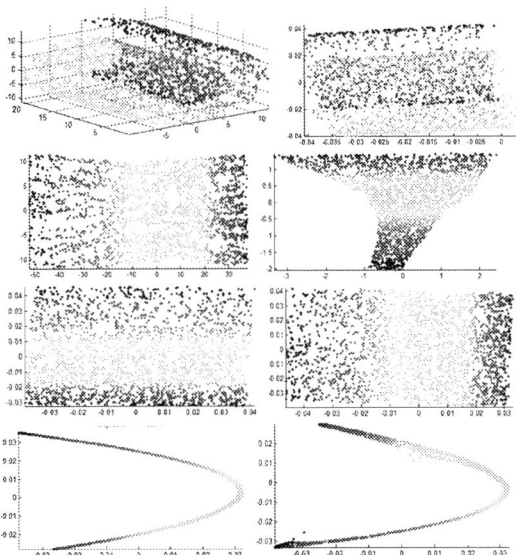

Fig.1.The dimensional reduction of data sets in different manifold learning algorithms, from left to right and from top to bottom: Swiss Roll data set, PCA, Isomap, LLE, HLLE, LTSA, LEP, SLEP

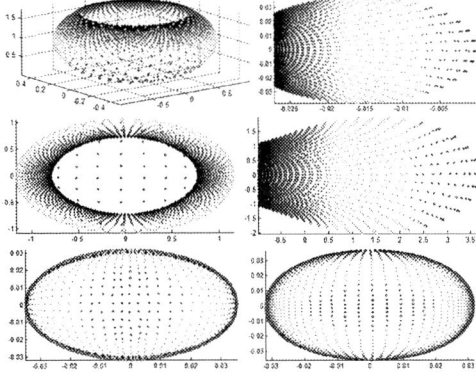

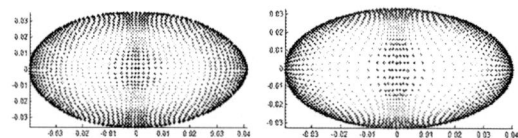

Fig. 2. The dimensional reduction of Puncture Sphere data set in different manifold learning algorithm, from left to right, and from top to bottom: Puncture Sphere data set, PCA, Isomap, LLE, HLLE, LTSA, LEP, SLEP

4.2 Real data set

USPS data set is a group of electronic digital data set, showing the image data set of the script written from 0 to 9, including 9298 images in total. For the pixel of each digital image is 16*16, the dimension of the processed data point is 256 in the experiment. In the aspect of design, there are two steps, including using manifold learning algorithm to reduce dimension of USPS data set and using K-nearest neighbor algorithm to classify and identify data sets in the low-dimensional space. The objective is to compare the image recognition accuracy rate of traditional manifold learning algorithm and SLEP in the real data set and the time consumption during the process of recognition. The experiment selects 400, 500, and 600 data points from different types of data sets as sample sets and other data points are testing data sets. The recognition accuracy rate of different algorithms to USPS data set is shown in table 2. It shows that the recognition accuracy rate of the proposed algorithm is almost the same as traditional manifold learning algorithm. The result indicates that the proposed algorithm does not reduce the accuracy rate of dimensional reduction of data sets.

The final objective of the experiment is to analyze the time consumption of the proposed algorithm, so the experiment compares the time consumption between traditional manifold learning algorithm and the proposed algorithm in different K values. The result is shown in table 3. By the comparison of the experimental result, it can be seen that the time consumption of the proposed algorithm is less than traditional manifold learning algorithm in different K values.

Table 2. Recognition accuracy rate of USPS data set by different algorithms

	USPS-train400	USPS-train500	USPS-train600
PCA [6]	83.95 ± 1.13	84.01 ± 1.26	85.27 ± 1.47
LLE [5]	88.14 ± 1.32	89.54 ± 1.37	90.46 ± 1.25
LEP [1]	90.58 ± 1.25	91.14 ± 1.05	91.69 ± 1.14
SLEP	89.76 ± 1.08	90.81 ± 1.16	91.15 ± 1.02

Table 3. Comparison of time consumption of different dimensional reduction algorithms

	USPS-train400	USPS-train500	USPS-train600
PCA [6]	313.15 s	298.53 s	265.36 s
LLE [5]	388.14 s	425.36 s	509.14 s
LEP [1]	526.31 s	579.12 s	617.43 s
SLEP	**324.36 s**	**368.52 s**	**413.17 s**

5. CONCLUSION

Traditional manifold learning algorithm exists a big problem in time consumption during the process of processing massive data sets. To improve this circumstance, SLEP algorithm is put forward to reduce the time consumption of an algorithm as far as possible without affecting the accuracy of dimensional reduction. A group of the comparative experiment shows that the proposed algorithm has a great improvement in time efficiency.

References

[1] M. Belkin and P. Niyogi, "Laplacian eigenmaps and spectral techniques for embedding and clustering," in NIPS, vol. 14, 2001, pp. 585-591.

[2] D. L. Donoho and C. E. Grimes, "Hessian Eigenmaps: Locally linear embedding techniques for high-dimensional data," in Proceedings of the National Academy of Sciences of the United States of America, vol. 100, 2003, pp. 5591-5596.

[3] J. B. Tenenbaum, V. d. Silva and J. C. Langford, "A global geometric framework for nonlinear dimensionality reduction," Science, vol. 290, 2000, pp. 2319-2323.

[4] X. He and P. Niyogi, "Locality preserving projections," in NIPS, vol. 16, 2003, pp. 153-160.

[5] S. T. Roweis and L. K. Saul, "Nonlinear dimensionality by locally linear embedding," Science, vol. 290, 2000, pp. 2323-2326.

[6] H. Abdi and L. J. Williams, "Principal component analysis," inc. WIREs Comp Stat, vol. 2, 2010, pp. 433-459.

[7] Z. Zhang, H. Zha, Principal manifolds and nonlinear dimension reduction via local tangent space alignment, SIAM Journal of Scientific Computing 26 (2004) 313–338.

[8] Li Yang, "Alignment of overlapping locally scaled patches for multidimensional scaling and dimensionality reduction," IEEE Tran. On Pattern Analysis and Machine Intelligence, vol. 30, 2008, pp. 438-450.

Web advertisement detection using Naive Bayes

Xin Deng[1,a], Lunqing Hou[2,b], Fei Wang[3,c]

[1]College of Computer Science and Software, Shenzhen University Shenzhen, 518060 P.R. China

[2]College of Computer Science and Software, Shenzhen University Shenzhen, 518060 P.R. China

[3]College of Computer Science and Software, Shenzhen University Shenzhen, 518060 P.R. China

[a]1062388964@qq.com, [b]lqhou@outlook.com, [c]wangfei2017@email.szu.edu.cn

ABSTRACT Nowadays, with the development of the internet, more and more people find information on the websites. So the network promoters pretend to write advertising content on websites, advertising content will reduce the user's experiment if the users always find the advertisement instead of the content they need. It will take web managers a lot of time to review whether the content is advertisement or useful content. Even worse, when we daily use message applications, there are always groups full of advertisement messages. It takes a lot of time to manage the users and the content, so many managers just leave it and the groups will be worthless. Basically, when we try to separate useful content from advertisement, we use the keywords. For example, if the content contains keywords like promotion champion, sale, we prefer to define it as advertisement. On the other hand, if the content has keywords like weekend, dinner, we prefer to define it as useful content. This processing can be automated as a text classification problem with the help of machine learning algorithms: k-NN Algorithm, Naive Bayes Algorithm, support vector machine, decision tree, neural network [1][2][3][4][5][6]. This paper applies Naive Bayes to advertisement detection in web content [3].

1. INTRODUCTION

This paper mainly tries to apply Naive Bayes into advertisement problem. Human classify content actually by keywords, when we see content contains keywords like sale, promotion, we prefer to think it is advertisement. This is the same schedule like Naive Bayes. Naive Bayes will count actually what is the probability, when one keyword appears. So, it is convenient for the web managers to delete the advertisement and give the users more useful information. Advertisement content is gathered from a website full of advertisement without any manual management, normal content is gathered from an authoritative news website with strict management.

2. Related work

There are several ways of web spam detection. Some researchers use link based techniques, in detail, they use links and ranking algorithms method like link farm detection to detect spam. They want to detect the web page that attracts search engine referrals [7]. Some researchers try to analysis reviews spam by combining review content and user behavior. They summarize view spam to 3 types: untruthful opinions, reviews on brands only and non-reviews [8][9]. Some researchers use machine

Content from this work may be used under the terms of the Creative Commons Attribution 3.0 licence. Any further distribution of this work must maintain attribution to the author(s) and the title of the work, journal citation and DOI.
Published under licence by IOP Publishing Ltd

ISPECE
IOP Publishing

IOP Conf. Series: Journal of Physics: Conf. Series **1187** (2019) 042023   doi:10.1088/1742-6596/1187/4/042023

learning algorithm to classify the reviews to two classes spam and non-spam. They collect the data by selecting the duplicate review as spam content and other review as normal content [10]. This paper gathers spam data from a website advertisement category and normal data from a news website.

3. Web advertisement detection

3.1 Data source
100000 advertisement articles from a website full of advertisement are labeled as negative dataset. 100000 normal content from an authoritative news website are labeled as positive dataset. We use 80000 positive and 80000 negative data as training set; 20000 positive and 20000 negative data as test set.

3.2 Data processing
At first, the articles contain words that need to be transformed as features that are represented as vectors.

Define articles as D_i, define words as w_i, So:

$D_1 = \{w_{11}, w_{12}, ..., w_{1n}\}$

$D_2 = \{w_{21}, w_{22}, ..., w_{2n}\}$

...

$D_n = \{w_{n1}, w_{n2}, ..., w_{nn}\}$

Let V represents the indexed vocabulary.

$V = \{w_1, w_2, ..., w_k\}$

When the vocabulary is built, every word has a unique id, so the articles can be transferred to vectors. Define article as D_i, define word's id as k_i, so:

$D_1 = \{k_{11}, k_{12}, ..., k_{1n}\}$

$D_2 = \{k_{21}, k_{22}, ..., k_{2n}\}$

...

$D_n = \{k_{n1}, k_{n2}, ..., k_{nn}\}$

3.3 Feature selection
The data is transferred to vectors, and a vocabulary is built. If all the keywords are taken into account, there will be too many features. Not only too many features will make the model too complex with many parameters, but also they may contain noise to the model.

At first the word's frequency need to count, so the word has an id and frequency. The words that have a very high frequency will be removed because they may appear in positive and negative dataset. If the word's frequency is too low, it will be removed from the features, because it is some special content's feature instead of the class's feature.

3.4 Naive Bayes
Naive Bayes is an algorithm that depends on Bayes Rules, the articles $D = \{D_1, D_2, ..., D_n\}$ are represented as vectors d, the classes are represented as $C = \{c_1, c_2\}$ while c_1 is positive and c_2 is negative.

The origin Naive Bayes is:

$$P(c|d) = \frac{P(d|c)P(c)}{P(d)}$$

The class that the content belongs to will be:

$$C_{MAP} = \text{argmax } P(c|d), c \in C$$

$$C_{MAP} = \text{argmax} \frac{P(d|c)P(c)}{P(d)}, c \in C$$

The denominator is the same, so drop the denominator:

$$C_{MAP} = argmax\, P(d|c)P(c), c \in C$$

1413

The document is represented as $(x_1, x_2, \ldots, x_n)$:

$$C_{MAP} = argmax\ P(x_1, x_2, \ldots \ldots, x_n\ |c)P(c), c \in C$$

4. Experiment

4.1 Share words trend

At first the top frequency words are count in both classes. Since they share a same vocabulary of 10000 words. The figure shows that the classes only share a little more than 26% words in top 200, even in top 900 they only share a little more than 34% words. So the two classes will be easy to separate, because they are different enough to each other.

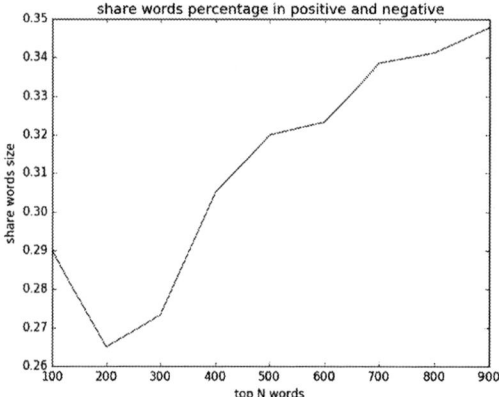

Fig. 1. Share words trend

4.2 Confusion matrix

According to the confusion matrix, the experiment has achieved expected result for spam detection. In 19906 positive samples, only 17 samples classified to negative class by mistake. In 20094 negative samples, only 172 samples classified to positive class by mistake.

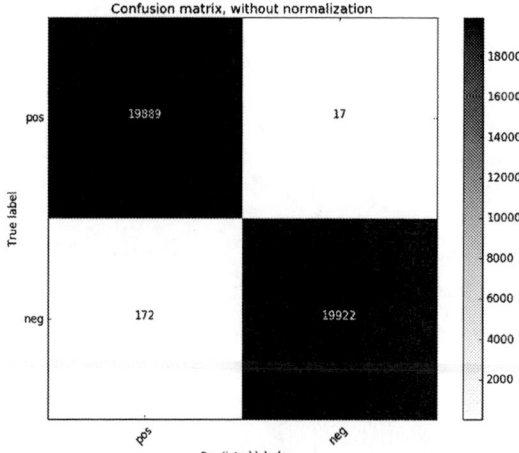

Fig. 2. Confusion matrix

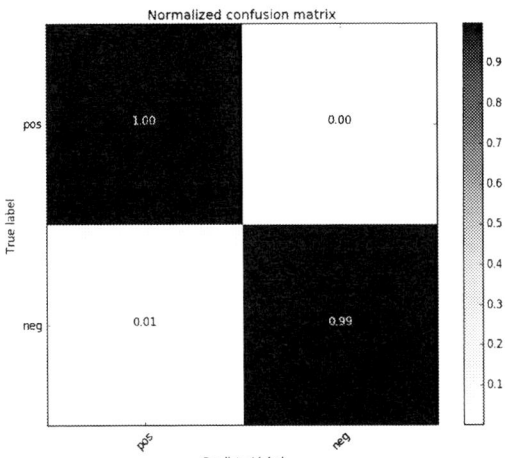

Fig. 3. Normalized confusion matrix

4.3 Analysis Detection percentage of different threshold

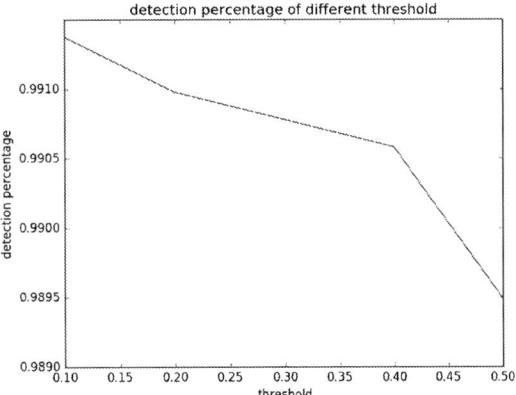

Fig. 4. detection percentage trend

Normally, when the probability is bigger than 50% of a class or the probability of a class is the biggest, the model will predict to this class. To build a sensitive model to the negative content, we use a list of thresholds to show the improvement. It turned out that the thresholds are not useful, because the normal probability has achieved a very high accuracy.

4.4 Precision, Recall, F1 Score

Precision can be used to validate the classifier's ability of not to label a negative sample as positive. Recall can be used to measure the classifier's ability of find all the positive samples [11] [12] [13].
Define：
True Positives (TP) means that the positive article is classified as positive class.
True Negatives (TN) means that the negative article is classified as negative class.
False Positives (FP) means that the negative article is classified as positive class.
False Negatives (FN) means that the positive article is classified as negative class.

$$\text{Accuracy} = \frac{\text{TP} + \text{TN}}{\text{TP} + \text{FP} + \text{FN} + \text{TN}}$$

$$\text{Precision} = \frac{TP}{TP + FP}$$

$$\text{Recall} = \frac{TP}{TP + FN}$$

$$\text{F1 Score} = 2 \cdot \frac{\text{Recall} \cdot \text{Precision}}{\text{Recall} + \text{Precision}}$$

Table1. Precision, Recall, F1 Score

	precision	recall	f1-score	support
pos	0.99	1.00	1.00	19906
neg	1.00	0.99	1.00	20094
avg / total	1.00	1.00	1.00	40000

5. Conclusions

Advertisement detection or spam detection is critically important now days. We use message applications to communicate; websites to find information, shopping applications to buy products. These applications are full of advertisement content or spam. The advertisement content waste a lot of our time and reduce our using experience. The content managers also pay much attention to these content. This paper proposes to use Naive Bayes to solve the problem automatically. At first, we try to analysis the words distribution in the two classes, so we can make sure that they are different enough to separate. The Naive Bayes Algorithm shows good performance in classifying the content to normal and spam classes. We try to figure out whether the threshold can make a contribution to the advertisement detection. It turns out to be helpless, because the model has achieved a very high performance at threshold of 50%.

Advertisement or spam exists everywhere, their words distribution different from each other. In the future work, more advertisement resources will be added to the dataset, so the model can be trained with more data and can detect more type of advertisement or spam. What's more, the advertisers are changing their strategy, so the model has to update with more data added to it.

REFERENCES

[1] Korde, V., & Mahender, C. N. (2012). Text classification and classifiers: A survey. *International Journal of Artificial Intelligence & Applications*, *3*(2), 85.

[2] Bijalwan, V., Kumar, V., Kumari, P., & Pascual, J. (2014). KNN based machine learning approach for text and document mining. *International Journal of Database Theory and Application*, *7*(1), 61-70.

[3] Aghila, G. (2010). A Survey of Na\" ive Bayes Machine Learning approach in Text Document Classification. *arXiv preprint arXiv:1003.1795*.

[4] Tong, S., & Koller, D. (2001). Support vector machine active learning with applications to text classification. *Journal of machine learning research*, *2*(Nov), 45-66.

[5] Johnson, D. E., Oles, F. J., Zhang, T., & Goetz, T. (2002). A decision-tree-based symbolic rule induction system for text categorization. *IBM Systems Journal*, *41*(3), 428-437.

[6] Zhang, X., Zhao, J., & LeCun, Y. (2015). Character-level convolutional networks for text classification. In *Advances in neural information processing systems* (pp. 649-657).

[7] Ghiam, S., & Pour, A. N. (2012). A survey on web spam detection methods: taxonomy. *arXiv preprint arXiv:1210.3131*.

[8] Jindal, N., & Liu, B. (2008, February). Opinion spam and analysis. In *Proceedings of the 2008 international conference on web search and data mining* (pp. 219-230). ACM.

[9] Farooq, S., & Khanday, H. A. Opinion Spam Detection: A Review.

[10] Jindal, N., & Liu, B. (2007, May). Review spam detection. In *Proceedings of the 16th international conference on World Wide Web* (pp. 1189-1190). ACM.

[11] Stehman, S. V. (1997). Selecting and interpreting measures of thematic classification accuracy. *Remote sensing of Environment, 62*(1), 77-89.

[12] Powers, D. M. (2011). Evaluation: from precision, recall and F-measure to ROC, informedness, markedness and correlation.

[13] Pedregosa, F., Varoquaux, G., Gramfort, A., Michel, V., Thirion, B., Grisel, O., ... & Vanderplas, J. (2011). Scikit-learn: Machine learning in Python. *Journal of machine learning research, 12*(Oct), 2825-2830.

Path Planning in Mobile Wireless Sensor Networks

Kezhuang WU [1,a], Junbin LIANG[2,b]

[1]Guangxi Key Laboratory of Multimedia Communications and Network Technology, School of Computer and Electronics Information Guangxi University, Nanning 530004, China

[2]Guangxi Key Laboratory of Multimedia Communications and Network Technology, School of Computer and Electronics Information Guangxi University, Nanning 530004, China

[a]KezhuangWU@st.gxu.edu.cn, [b]liangjb2002@163.com

ABSTRACT Mobile wireless sensor network (MWSN) is a kind of wireless sensor network (WSN) with mobile sink, which is deployed in harsh environment to perform long-term tasks such as data collection and monitoring. Path planning is a fundamental problem in MWSN, which means movement of the mobile nodes in the sensing area should be designed effectively to complete the targets of data collection. However, path planning is a challenging multi-objective optimization issue. Existing schemes of path planning are classified, and their advantages and disadvantages would be compared in this paper. Finally, potential directions of future research are proposed in order to overcome shortcomings of the existing works.

1. INTRODUCTION

With the rapid development of communication technologies and embedded devices nowadays, wireless sensor network (WSN) has been widely used in agriculture, industry and military. The wireless sensor network is a new kind of ad-hoc network. It consists of a large number of nodes with low energy consumption, low computing capability and small storage, which performs special tasks such as environmental monitoring and target tracking. The sensor nodes sense, measure and collect data from the surrounding environment. Each node can not only collect, process, and transmit local data, but also store, calculate, and aggregate data forwarded by other nodes. Sink is a special type of node in WSN, equipped with powerful transmitter, battery, and large storage to collect data across the network and connect to users via the Internet or satellite.

Although applications in WSN are not the same, many of them have a common feature. That is, data of the sensor nodes needs to be collected to the sinks. Therefore, data collection is a very important issue in WSN. In traditional wireless sensor network, there are two main problems in data collection: (1) Because nodes are static, when collecting data, nodes near the sinks need to relay large number of packets forwarded by other nodes. That causes energy consumption of these nodes greatly increasing, which will result in a significant reduction in the lifetime of these nodes. Once these nodes are exhausted, they will not be able to continue relaying data and the lifetime of the WSN will be shortened. (2) Nodes must be densely deployed in the network to ensure network connectivity. However, in some applications, nodes are deployed in separate areas. Some nodes are not able to forward data to the sink directly through the links.

In order to address the problems above, mobile sink is proposed. By moving nearly to nodes that need to forward data, sink can directly collect data from the nodes in WSN or the data could be forwarded with less relays. That would balance the energy consumption of the whole network and prolong its lifetime. At the same time, due to movement of sink, nodes that are not initially connected to the sinks can also transmit data to the sinks when the sinks are within their transmission range.

Path planning is a fundamental issue in data collection based on mobile sinks. Path planning is designing optimal mobile solution of sinks with the goal of minimizing the counts of relays and moving distance of sinks as much as possible. That could reduce not only delay of the whole network, but also the energy consumption of nodes near the sinks, which could prolong lifetime of the network.

Although mobile sinks could prolong lifetime of WSN, it still exists challenging issues. When path length of the mobile sinks becomes longer, energy consumption of nodes would be more balanced. That prolong lifetime of the network. But since the mobile sinks moving too long, the latency of collecting data would also be larger. It means that the path length of the mobile sinks should not be too long. Both energy consumption and delay should be taken into account. It is a challenging multi-objective optimization issue.

In WSN applications, a good path planning algorithm can serve the MAC protocol and routing protocol well. Meeting the demands of data collection and reducing network latency, and prolonging network lifetime are important issues in WSN. Many path planning algorithms have been proposed in recent years, but there are still some problems with these algorithms. This paper divides these existing works into two categories: single-hop data collection path planning and multi-hop data collection path planning.

The rest of this paper is organized as follow: the section 2 and 3 respectively review the single-hop data collection path planning and multi-hop data collection path planning. The section 4 discusses the advantages and disadvantages of single-hop data collection path planning and multi-hop data collection path planning, and point out the future work. Finally, section 5 concludes the whole paper.

2. SINGLE-HOP DATA COLLECTION PATH PLANNING

Clustering to collect data is a common method In WSNs. Divide a large WSN into different clusters, each with a cluster head. Clusters are the basic units of data collection. Cluster member would forward their data to the cluster head. And then the cluster head forwards data to the mobile sink. But in this way, the cluster head needs to relay data from a large number of cluster members, which will lose more energy than the other cluster members. This results in uneven energy consumption, and the short lifetime of the cluster head node makes the lifetime of the network shorter. Based on the considerations above, single-hop data collection path planning comes into being. This type of algorithm requires all nodes to forward data to the mobile sink in single hop.

In reference [1], DkM is proposed. It first uses k-means clustering algorithm to get a lot of rendezvous points (RPs). Then calculate the weight of each potential position of RPs according to the number of nodes in the RPs communication range, the distance of the node, and the average hop distance, so as to remove the redundant RPs according to the weight, It makes least RPs in network to achieve single-hop data collection for all nodes.

The author of reference [2] proposes TSP-DC. It uses straightforward simulated annealing algorithm to get a path. Then the gradient descent is used to optimize the location of the RPs so that the length of the new path is shorter than before. And all nodes are guaranteed to transmit data to the sink in single hop. The paper also proposes TSP-DA, which is an improved solution of TSP-DC. It can dynamically adjust the path based on actual data loads after computing a path by TSP-DC.

Literature [3] proposes a greedy algorithm for single-hop data collection, named Spanning Tree Coverage Algorithm. It calculates all the shortest distance between area covered by a RP and areas covered by the other RPs as cost. Calculates average cost according to the total cost and remaining uncovered nodes. Then searching for the candidate RPs that having the smallest average cost until all the nodes are cover. This paper also designs a solution for multiple mobile sinks for large networks.

In reference [4], considering that obstacles exist in sensing area, a heuristic tourism-planning algorithm is designed. Divide the sensing area into several grids with the same size and treat the obstacles as rectangle. By designing a heuristic algorithm based on spanning graphs, a shortest obstacles-avoidance path could be found.

In reference [5], unmanned aerial vehicles (UAV) is used as mobile sink to collect data. Spiral path planning (SPP) algorithm was designed. The problem that uses UAV to collect data is seen as a traditional traveling salesman problem. The goal of the algorithm is to quickly compute a path with as short as possible. The algorithm divides the sensing area into multiple circles, and the UAV collects data one by one along the circles.

Literature [6] models the Resource Constrained Shortest Path Problem (RCSPP) with UAVs. The model takes into account the energy limitations of UAVs, the path length assignment between the UAVs, and collisions between UAVs. The goal of the RCSPP is to calculate the path and the number of UAVs for meeting demands. The paper proposes a scheme based on the column generation procedure, using path-decomposition formulation to compute a path, and ensure the target constraints.

Table 1 puts together the schemes for paths planning in single-hop data collection to compare their performance.

Table 1.Comparison of single-hop data collection path planning

Path planning strategy	Number of sinks	Path length	Node energy consumption	Delay	Non-uniform network	Obstacles	Scalability	Adaptive	Algorithm complexity
[1]	1	short	low	high	yes	no	high	low	low
[2]	1	short	low	high	yes	no	high	high	high
[3]	≥1	short	low	high	no	no	high	low	high
[4]	1	long	low	high	no	yes	low	low	high
[5]	1	long	low	high	yes	no	low	low	low
[6]	≥1	long	low	high	yes	no	high	high	high

3. MULTI-HOP DATA COLLECTION PATH PLANNING

Single-hop data collection directly collects data generated by each node. It surely can reduce data relay counts of nodes near sinks and prolong lifetime of WSN. However, in this way, since the mobile sink must pass through the communication range of each node when collecting data, the path length becomes longer, resulting in a relatively large transmission delay. This is unacceptable for some delay-sensitive applications. Multi-hop data collection paths planning schemes could make the mobile sink not have to go through the communication range of all nodes, so as to reduce its path length to better solve the problem of transmission delay.

Literature [7] proposes a rendezvous-based solution for delay-constrained data collection with a mobile sink in WSN. This strategy is based on clustering protocol, which prolong lifetime of the entire network by selecting RPs in areas with sufficient energy. When planning a path, use Iterated Local Search (ILS), a heuristic algorithm, to get a shortest path.

Literature [8] proposes WRP, an algorithm based on RPs. The algorithm calculates a weight by size of data that needed to be forwarded and the distance between nodes and the nearest RPs. The nodes with largest weight will be selected as new RPs. These RPs will connect directly to the mobile sinks to forward data, thus avoiding excessive energy consumption.

Work done in [9] considers both obstacle-free network and obstacle network. This work divides the network into multiple regular triangles according to the communication range of sensors. In an obstacle-free network, each node is guaranteed to be covered by at least one RP. At the boundary of the network, the path length can be reduced by adjusting the position of the corresponding RPs. After adjusting the position of the RPs, MCPP, a heuristic algorithm, is used to compute a path. In network with obstacles, the same RPs selection strategy is adopted, and an algorithm for avoiding obstacles is used in the path planning.

Literature [10] proposes a mobile sink path selection (DBMSPS) algorithm with delay constraint. This algorithm first looks for node with the most nodes in its communication range as first RP. Then, by successive iterations, nodes with highest load are selected as RPs until all nodes can forward the data to the nearest RPs with no more than a given hops counts.

In [11], the sensing area is divided into several regular hexagonal regions, and the centers of each region are candidate RPs. The weight of each RPs is calculated according to a cost function. Then iterations are repeated to select the RPs with the highest weight until the delay upper limit is met. Nodes will forward data to the adjacent RPs area if its area does not exist a RP. Then the data would be forwarded to the mobile sink.

Work done in [12] proposes an algorithm called Energy density based trajectory (EDT). This algorithm is based on energy density of each sensor node. Nodes with the lowest energy density will be selected as the RPs until all nodes in the network are covered. The work was improved when considering delay factor, and an improved algorithm, called delay aware energy density based trajectory (DAEDT), is proposed. Based on EDT, the improved algorithm increases the number of nodes selected as RPs until the path length just does not exceed a given delay limit.

Literature [13] proposes an algorithm called NDCMC. The algorithm combines clustered routing and mobile sinks and is designed to easily adjust the balance between energy consumption and latency as needed. The algorithm uses a density-based cluster head selection algorithm. The network is divided into clusters based on the density of nodes and a given cluster radius. Finally, the path is computed by optimal nearest-neighbor algorithm.

The author of the literature [14] proposes ACO-MSPD. This work assumes that the data generated by nodes for each time period is different. And the sizes of data that needs to be forwarded are also different. It results in a certain difference in the load of relay nodes. The ACO-MSPD selects RPs based on the weights calculated based on a given path length of the mobile sink for maximizing network lifetime and minimizing delay. This work also designed a re-selection scheme for RPs, which improved performance of the algorithm.

Literature [15] proposes VRDG algorithm. The algorithm divides the sensing area into multiple data gathering areas (DGA), each containing 1 to 3 circular data gathering units (DGU). Each DGU has a cluster head node (CH). The algorithm calculates position of RPs based on the position of CH in each DGU. Finally, these paths are connected from beginning to end to get a moving path of mobile sink.

Work done in [16] considers the difference in data generation rate and limited storage of nodes. EARTH and eEARTH are proposed. The EARTH algorithm takes the data generation rate, hops counts, and the distance into consideration. Then a spanning tree is formed and RPs are selected by traversing the tree. An improved method eEARTH is subsequently proposed, which further reduces the number of RPs to reduce the path length.

Table 2 puts together the schemes for paths planning in multi-hop data collection to compare their performance.

Table 2. Comparison of multi-hop data collection path planning

Path planning strategy	Number of sinks	Path length	Node energy consumption	Delay	Non-uniform network	Obstacles	Scalability	Adaptive	Algorithm complexity
[7]	1	short	high	low	yes	no	high	high	low
[8]	1	short	low	low	yes	no	high	high	low
[9]	1	long	low	high	no	yes	low	low	low

[10]	1	short	high	low	no	no	high	high	high
[11]	1	long	low	high	no	no	low	high	high
[12]	1	short	high	low	no	no	high	high	high
[13]	1	short	low	low	yes	no	high	high	high
[14]	1	short	low	low	yes	no	high	high	high
[15]	1	long	low	high	no	no	low	low	low
[16]	1	short	high	low	yes	no	high	high	high

4. SUMMARY AND FUTURE RESEARCH

A number of path planning algorithms for data collection have been proposed, which address some problems in path planning. The single-hop data collection path planning algorithm collects data directly by collecting data in single hop, so that the energy consumption of the network is balanced and the lifetime is prolonged. But the path length is long, resulting in a high data collection delay. Multi-hop data collection path planning collect data only need to directly collect data from some nodes in the network, which greatly reduces the path length and reduces the data collection delay. However, this type of algorithm will also cause some nodes to consume more energy because they need to transfer large amounts of data, resulting in a short lifetime of the whole network.

Prolonging network lifetime and reducing data collection latency are two of the most important performance metrics for path planning problems. The main goal of the existing work is to improve the algorithms to obtain better path planning performance. But they still have certain problems, which should be solved in the future work.

a) Multiple mobile sinks. As the network scale continues to expand, a single mobile sink has gradually failed to achieve the required performance. Therefore, using multiple mobile sinks to simultaneously collect data from the sensing area will become a trend. How to divide the sensing area, plan the path, and enable multiple mobile sinks to effectively collect data and make them load balanced will become a focus of future work.

b) Path planning with obstacles. The environment in the sensing area is often harsh or obstructive. In the existing work, obstacles existing in the sensing area are rarely considered. How to solve this problem is a difficult problem in practical application.

c) Dynamic network. The existing work considers static networks, and the mobile elements are only mobile sinks. The movement of a general node causes the topology of the network to change. Implementing path planning in dynamic networks is also a difficult problem.

5. CONCLUSION

This paper summarizes a number of typical path planning algorithms for WSN that are published in recent years. These algorithms are divided into two categories: single-hop data collection path planning and multi-hop data collection path planning, which have their own advantages and disadvantages. The existing works have been analyzed respectively. And their performance has been compared. Finally, the paper points out the advantages and disadvantages of the two types of algorithms and potential directions of future research.

ACKNOWLEDGMENTS

This research is supported by the National Natural Science Foundation of China under Grant Nos: 61562005; the Natural Science Foundation of Guangxi Province (Grant No. 2015GXNSFAA139286), and Thousands of Young and Middle-aged Backbone Teachers Training Program for Guangxi Higher Education (Education Department of Guangxi (2017) No.49).

REFERENCES

[1] Amar Kaswan, Kumar Nitesh, and Prasanta K. Jana. 2017. Energy efficient path selection for mobile sink and data gathering in wireless sensor networks. *AEU-International Journal of Electronics and Communications* 73 (March 2017), 110-118. DOI=https://doi.org/10.1016/j.aeue.2016.12.005

[2] Noralifah Annuar, Neil Bergmann, Raja Jurdak, and Branislav Kusy. 2017. Mobile Data Collection from Sensor Networks with Range-Dependent Data Rates. 2017 *IEEE 42nd Conference on Local Computer Networks Workshops (LCN Workshops)*. IEEE, Singapore, Singapore, 53-60. DOI=10.1109/LCN.Workshops.2017.64.

[3] Ming Ma, Yuanyuan Yang, and Miao Zhao. 2013. Tour Planning for Mobile Data-Gathering Mechanisms in Wireless Sensor Networks. *IEEE Transactions on Vehicular Technology* 62, 4 (May 2013), 1472-1483. DOI=10.1109/TVT.2012.2229309.

[4] Guangqian Xie, Kaoru Ota, Mianxiong Dong, Feng Pan, and Anfeng Liu. 2017. Energy-efficient routing for mobile data collectors in wireless sensor networks with obstacles. *Peer-to-Peer Networking and Applications* 10, 3 (May 2017), 472–483. DOI=https://doi.org/10.1007/s12083-016-0529-1

[5] Wu Yue, Zhu Jiang. 2018. Path Planning for UAV to Collect Sensors Data Based on Spiral Decomposition. *Procedia Computer Science* 131 (2018), 873-879. DOI=https://doi.org/10.1016/j.procs.2018.04.291

[6] Michele Garraffa, Mustapha Bekhti, Lucas Létocart, Nadjib Achir, and Khaled Boussetta. 2018. Drones path planning for WSN data gathering: A column generation heuristic approach. *2018 IEEE Wireless Communications and Networking Conference (WCNC)*. IEEE, Barcelona, Spain, 1-6. DOI=10.1109/WCNC.2018.8377391

[7] Charalampos Konstantopoulos, Grammati Pantziou, Nikolaos Vathis, Vasileios Nakos, and Damianos Gavalas. 2014. Efficient mobile sink-based data gathering in wireless sensor networks with guaranteed delay. *In Proceedings of the 12th ACM international symposium on Mobility management and wireless access (MobiWac '14)*. ACM, New York, NY, USA, 47-54. DOI=http://dx.doi.org/10.1145/2642668.2642674

[8] Hamidreza Salarian, Kwan-Wu Chin, and Fazel Naghdy. 2014. An Energy-Efficient Mobile-Sink Path Selection Strategy for Wireless Sensor Networks. *IEEE Transactions on Vehicular Technology* 63, 5 (Jun 2014), 2407-2419. DOI=10.1109/TVT.2013.2291811

[9] Nimisha Ghosh, Indrajit Banerjee. 2015. An energy-efficient path determination strategy for mobile data collectors in wireless sensor network. *Computers & Electrical Engineering* 48 (November 2015), 417-435. DOI=https://doi.org/10.1016/j.compeleceng.2015.09.004

[10] Amar Kaswan, Kumar Nitesh, and Prasanta K. Jana. 2016. A routing load balanced trajectory design for mobile sink in wireless sensor networks. *2016 International Conference on Advances in Computing, Communications and Informatics (ICACCI)*. IEEE, Jaipur, India, 1669-1673. DOI=10.1109/ICACCI.2016.7732287

[11] Madhvi Mishra, Kumar Nitesh, and Prasanta K. Jana. 2016. A delay-bound efficient path design algorithm for mobile sink in wireless sensor networks. *2016 3rd International Conference on Recent Advances in Information Technology (RAIT)*, IEEE, Dhanbad, India, 72-77. DOI=10.1109/RAIT.2016.7507878

[12] Kumar Nitesh, Amar Kaswan, and Prasanta K. Jana. 2017. Energy density based mobile sink trajectory in wireless sensor networks. *Microsystem Technologies* (2017). DOI=https://doi.org/10.1007/s00542-017-3569-4

[13] Ruonan Zhang, Jianping Pan, Di Xie, and Fubao Wang. 2016. NDCMC: A Hybrid Data Collection Approach for Large-Scale WSNs Using Mobile Element and Hierarchical Clustering. *IEEE Internet of Things Journal* 3, 4 (August 2016), 533-543. DOI=10.1109/JIOT.2015.2490162

[14] Praveen Kumar D., Tarachand Amgoth, and Chandra Sekhara Rao Annavarapu. 2018. ACO-based mobile sink path determination for wireless sensor networks under non-uniform data constraints. *Applied Soft Computing* 69 (August 2018), 528-540. DOI=https://doi.org/10.1016/j.asoc.2018.05.008

[15] Chao Sha, Jian-mei Qiu, Tian-yu Lu, Ting-ting Wang, and Ru-chuan Wang. 2018. Virtual region based data gathering method with mobile sink for sensor networks. *Wireless Networks* 24, 5 (July 2018), 1793–1807. DOI=https://doi.org/10.1007/s11276-016-1431-8

[16] You-Chiun Wang, Kuan-Chung Chen. 2018. Efficient Path Planning for a Mobile Sink to Reliably Gather Data from Sensors with Diverse Sensing Rates and Limited Buffers. *IEEE Transactions on Mobile Computing* (2018). DOI=10.1109/TMC.2018.2863293

Defect detection and recognition based on ADABOOT-SVM integrated model

ZhiKai Liang[1,a], ShaoZhong Cao[2,b], YuKun Tan[3,c]

[1]Beijing Institute of Graphic Communication
Beijing No. two, Xinghua street, China
13051379003, 102600

[2]Beijing Institute of Graphic Communication
Beijing No. two, Xinghua street, China
13520127931, 102600

[3]Beijing Institute of Graphic Communication
Beijing No. two, Xinghua street, China
18801017073, 102600

[a]Liangzhikai0812@gmail.com, [b]chaoshaozhong@gmail.com, [c]tanyukun@gmail.com

ABSTRACT: As the core component of printing machinery, the surface finish and geometric accuracy of printing drum will have an important impact on the quality of printed matter. However, the use of acid ink, alcohol and other chemical raw materials corrode the drum, leading to local collapse or spots. How to effectively identify the types of drum defects has become an important issue. To solve this problem, a defect detection and recognition framework based on adaboot-SVM ensemble learning model is proposed. The framework is composed of two parts: feature extraction and classifier design. The first part is feature extraction from directional gradient histogram (HOG). In the second part, we construct an ensemble of different SVM classifiers to identify defects. The validity of the proposed model is verified by nine different defects. The results show that the integrated model of adaboot SVM is helpful to improve the recognition accuracy of defects.

1. INTRODUCTION

As an important branch of machine vision technology, visual inspection is a hot research direction in the field of product nondestructive testing in China. Data shows that the domestic market in 2015 has reached 350 million US dollars, accounting for 8.3% of the global market, growth rate is 22.2%, ranking first in the world, China has become the world's third largest machine vision market after the United States and Japan. From 2016 to 2020, the growth rate of China's machine vision market is expected to remain above 20%, and will reach a billion dollar market space. Visual sensor has many advantages, such as large amount of information, non-contact with workpiece, high sensitivity and accuracy, strong anti-electromagnetic interference ability, and so on. It is a hot research field. As the core component of printing machinery, the surface finish and geometric accuracy of printing drum will have an important impact on the quality of printed matter. However, the use of acid ink, alcohol and other chemical raw materials corrode the drum, leading to local collapse or spots. How to effectively identify the types of drum defects has become an important issue.

Content from this work may be used under the terms of the Creative Commons Attribution 3.0 licence. Any further distribution of this work must maintain attribution to the author(s) and the title of the work, journal citation and DOI.

Published under licence by IOP Publishing Ltd

Scholars at home and abroad have studied and proposed a variety of new feature extraction algorithms from the analysis of feature extraction, machine learning, depth learning and other methods, and achieved good recognition results. However, there are many kinds of defects, and some defects are very similar. The selection of effective feature combination usually depends on the experience of experts. The accuracy of recognition is difficult to guarantee. Therefore, it is of great significance to find a data-based self-learning method to improve the performance of defect recognition. With the rapid development of artificial intelligence technology in recent years machine learning and its application had a hot topic in the field of artificial intelligence. At present, machine learning has achieved good results in some pattern recognition fields [15-19], such as speech recognition, image classification, natural language processing and so on. At the same time, scholars at home and abroad are committed to introducing machine learning into defect recognition [20-24]. Therefore, this paper uses the gradient histogram of Oriented Gradient proposed in reference [25] to extract the features of defects. At the same time, considering that the idea of ensemble learning can be used to construct an effective combined classifier model, it can be used to replace the soft Max classifier used in most depth learning applications. Based on this, this paper proposes a defect identification method combining HOG and ensemble learning. The specific process of the model is shown in Figure 1. First, noise reduction is done in the preprocessing stage. Then, we create the defect features extracted from HOG images. Finally, in the classifier design stage, a multi-SVM linear combination classifier (MSVMLC) is constructed for classification and recognition.

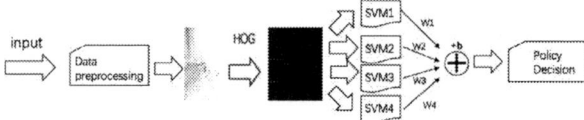

Figure 1. flow chart of defect recognition based on HOG and ensemble learning

2. Histogram of OrientedGradients

Directional gradient histogram (HOG) is an image descriptor for object detection which is widely used in computer vision and image processing. This method uses the histogram of gradient orientation (HOG) feature to express the detected object, extracts the shape information of the detected object, and forms a rich feature set.

Compared with other descriptors, HOG descriptors have some key advantages. Because it runs on a local element, it is invariant to geometric and photometric transformations except for better capturing local shape information, which only occurs in larger spatial regions. Moreover, the HOG is obtained in a densely sampled image block, and the spatial position relationship between the block and the detection window is implicit in the calculated HOG eigenvector.MATEC Web of Conferences 2 (2012) 01001, 2 corresponding to the volume and 01001 to the number of the article (replacing thus the page number).

a) image normalization

The main purpose of normalized image is to improve the detector's robustness to illumination, because the detector must be insensitive to illumination in different occasions when the target is captured.

b) computing gradient of images using first order differential

① Image smoothing

For gray-scale images, in order to remove noise points, we usually use discrete Gaussian smoothing template to smooth: Gaussian function smoothing gray-scale images at different smoothing scales, Dalal and other experiments show that the next, the best human detection effect (that is, do not do Gaussian smoothing), making the error rate reduced by about one. Double. The possible reason for not doing smoothing is that smoothing is based on edges, which reduces the contrast of edge information and thus reduces the signal information in the image.

② Gradient method for image gradient

The first order differential processing generally has a strong response to the step of the gray scale. First derivative:

$$\frac{\partial f}{\partial x} = f(x+1) - f(x)$$

For the function f (x, y), the gradient on its coordinate (x, y) is defined by the following two dimensional column vectors:

The module value of this vector is given in the following form:

$$\nabla f = \begin{bmatrix} G_x \\ G_y \end{bmatrix} = [\frac{\frac{\partial f}{\partial x}}{\frac{\partial f}{\partial y}}]$$

$$\nabla f = ||\nabla f||_2 = [G_x{}^2 + G_y{}^2]^{1/2}$$

Because the computation cost of modulus value is relatively large, it can be roughly solved according to the following formula:

$$\nabla f \approx |G_x| + |G_y|$$

Dalal et al. used many first-order differential templates for gradient approximation, but the results showed that the template [-1,0,1] was the best. Template [-1,0,1] is used as an example to calculate the image gradient and direction. The gradient in horizontal and vertical directions are calculated by gradient template as follows:

$$G_h(x, y) = f(x+1, y) - f(x-1, y) \forall x, y$$
$$G_v(x, y) = f(x, y+1) - f(x, y-1) \forall x, y$$

Which represents the horizontal and vertical gradient values of the pixel respectively. The gradient value (gradient strength) and gradient direction of the pixel are calculated.

$$M(x, y) = \sqrt{G_h(x, y)^2 + G_v(x, y)^2} \approx |G_h(x, y)| + |G_y(x, y)|$$
$$\theta(x, y) = \arctan(G_h(x, y)/G_v(x, y))$$

c) directional weight projection based on gradient magnitude

Generally, there are three kinds of HOG structures: rectangular HOG (R-HOG), circular HOG and center around HOG. Their units are Block (blocks). Dalal's experiment shows that the detection effect of rectangular HOG and circular HOG is basically the same, while the circumferential HOG is relatively poor. Generally, a block is made up of several cells, each of which is made up of pixels. The gradient direction statistics are done independently in each cell so that the gradient direction is the horizontal axis of the histogram. As we mentioned earlier, the gradient direction can be 0 to 180 degrees or 0 to 360 degrees. But the DALAL experiment shows that for human target detection, the direction range of 0 to 180 degrees of neglect can achieve better results. Fruit. The gradient distribution is then averaged into orientation bins, each of which corresponds to a histogram.

d)Normalization of D (HOG) eigenvectors

Normalize the HOG feature vector in block block. The normalization of eigenvectors in block is mainly to make the eigenvector space robust to illumination, shadow and edge changes. The normalization function used in this experiment is L2-norm.

e) get the final eigenvector of HOG.

3. MULTI SVM CLASSIFIER COMBINATION ALGORITHM

SVM is a classification algorithm based on statistical learning theory. It is widely used in many pattern recognition and data mining tasks. SVM classifier not only has strong generalization ability, but also can classify linear non-separable data by mapping data to high-dimensional space through kernel function. Commonly used kernel functions include linear kernel function, Gauss kernel function, polynomial kernel function and so on. At present, the selection of SVM kernel function is still a difficult problem for a given classification task, and multi-SVM combination is an effective method to solve this problem. Therefore, the combined SVM classifier has the following advantages:

(1)according to the above analysis, the combination of SVM classifier can avoid the problem of kernel function selection.

(2)SVM based on different kernel functions has different performance on the same training set. Combining their output results can improve the classification performance.

3.1 Classifier linear combination algorithm

Multi-classifier combination algorithm is a model fusion method, its basic idea is to assume that different classifiers can provide different aspects of the results of information, through the fusion of these information can provide better support for the final decision-making. Reference [27] proposes a linear ensemble method for depth model, which combines the outputs of different DNN models to train a HMM classifier to achieve better results than a single DNN model. In this paper, a linear combination model of multiple classifiers based on posterior probability is constructed on the basis of reference [27]. The model is described as follows:

Consider a schema classification task z, where Z can be assigned to one of the C schema categories, and the object to be classified has C schema categories$\{ \omega_1, \omega_2, \omega_3 \cdots \omega_c \}$。 order $P=[P(x_1), \cdots, P(x_i), \cdots, P(x_N)]T \in R^{N \times c}$ True posterior probabilities for N samples. $p(x_i) \in R^{1 \times c}$ The true posterior probability of a sample is expressed as

$$p(x_i) = [p(\omega_1|x_i), \cdots, p(\omega_c|x_i)]$$

Obviously, x_i is a posteriori probability of belonging to a class is "1" and vice versa "0". Assuming that there are M classifiers, it can be used. $p(x_i) = [p_k(x_1), \cdots, p_k(x_N)]^T \in R^{N \times c}$ Represents the posterior probabilities of K classifiers for N samples. Among them, $P_k(x_i) \in R^{1 \times c}$ Indicates that the posterior probability of the classifier K is output to the sample x_i.

The basic idea of the linear combination strategy of multiple classifiers is that the decision value of the combination model deciding that a sample belongs to a certain pattern category is obtained by linear combination of the output decision values of multiple classifiers, which can be formally described as

$$P(\omega_j|x_i) \approx \alpha_1 P_1(\omega_j|x_i) + \cdots + \alpha_M P_M(\omega_j|x_i) + b_i$$

Furthermore, the linear combination model of multiple classifiers can be described as

$$P_{com}(x_i) = \sum_{k=1}^{M} W_k p_k(x_i)^T + b^T \quad (1)$$

In the formula, $W_k \in R^{c \times c}$ represents the weight matrix of the classifier and $b^T \in R^{c \times 1}$ is the bias vector. Estimating the parameters of Eq. (1) minimizes the difference between the output value and the real value of the combined model, which can be measured by the least square error, i.e.

$$\min_{(W,b)} C = \frac{1}{2N} \Sigma_i ||P_{com}(x_i) - p(x_i)^T||^2 \quad (2)$$

In parameter estimation, regularization term [23] is usually added to the loss function to avoid over fitting. Considering the convenience of solving the model parameters and increasing the L2-regularization term for the loss function, formula (2) is rewritten as

$$\min_{(W,b)} C = \frac{1}{2N} \Sigma_i ||P_{com}(x_i) - p(x_i)^T||^2 + \frac{1}{2N} \Sigma_k \gamma_k ||W_k||^2 \quad (3)$$

In the formula, γ_k is a regularization parameter and can be selected by cross validation. Formula (3) is a multivariate function for extremum problem, which can be solved by least square method. First, the partial derivative of W、b^T is solved.

$$\frac{\partial C}{\partial W} = 0 \text{ and } \frac{\partial C}{\partial b^T} = 0$$

$$
\begin{cases}
\sum_i \left(\left(\sum_{k=1}^{M} W_k p_k(x_i{}^T) + b^T - p(x_i{}^T) \right) p_1(x_i{}^T) \right) \\
\qquad\qquad + \gamma_1 \gamma W_1 = 0 \\
\qquad\qquad\vdots \\
\sum_i \left(\left(\sum_{k=1}^{M} W_k p_k(x_i{}^T) + b^T - p(x_i{}^T) \right) p_M(x_i{}^T) \right) \\
\qquad\qquad + \gamma_M \gamma W_M = 0 \\
\sum_i \left(\sum_{k=1}^{M} W_k p_k(x_i{}^T) + b^T - p(x_i{}^T) \right) = 0
\end{cases}
$$

The solution system can be solved as follows: $[Wb^T]=BA^{-1}$. Among them, the values of A and B are respectively

$$A = \begin{bmatrix} A_1 & A_2 \\ A_3 & N \end{bmatrix}$$

$$A_1 = \begin{bmatrix} P_1{}^T P_1 + \gamma_1 & \cdots & M \\ \vdots & \ddots & \vdots \\ P_M{}^T P_1 I & \cdots & P_M{}^T P_M + \gamma_M \end{bmatrix}$$

$$A_2 = \left[\sum_i P_1{}^T(x_i), \cdots, \sum_i P_M{}^T(x_i) \right]$$

$$A_3 = \left[\sum_i P_1(x_i), \cdots, \sum_i P_M(x_i) \right]$$

$$B = \left[P^T P_1, \cdots, P^T P_M, \sum_i P^T(x_i) \right]$$

In this paper, four different SVM models are selected, which are SVM based on linear kernel (SVM_LK), SVM based on polynomial kernel (SVM_PK), SVM based on Gaussian kernel (SVM_GK) and SVM based on Sigmoid kernel (SVM_SK). From the effect point of view, the combination of multiple SVM classifiers can avoid the problem of SVM kernel function selection to a certain extent. However, the standard SVM can not give the posterior probability output of the sample. In order to use this linear combination algorithm, we need to find a posterior probability output method of SVM.

3.2 A posteriori output of SVM

For pattern classification, the standard SVM can get the decision value of the sample belonging to a certain category, but it can not give the corresponding posterior probability. However, there is a certain correlation between the decision value and the posterior probability. The greater the absolute value of the decision value of the sample is, the higher the reliability of the decision value belonging to a certain category will be. Therefore, the posterior probability of the sample can be obtained by the decision value. A posteriori probability of SVM can be obtained by Platt[29]. The essence of this method is to fit a Sigmoid model by training data, which can map the decision value of SVM to a posterior probability. The Sigmoid model can be expressed as

$$P(y = 1|f) = \frac{1}{1 + \exp(A * f + B)}$$

In the formula, f is the decision value of SVM output, and the parameters A and B can be obtained through training. It should be noted that this method can only be applied to binary classification problems. For multi-classification problems, it can be classified as multiple binary classification problems, and the corresponding posterior probabilities can be obtained respectively and then normalized.

4. EXPERIMENT AND CONCLUSION ANALYSIS

The validity of this method is verified by classifying 9 types of pictures, including traces, scratches, trachoma, fingerprints, stains, water stains, trailing marks, trailing, droplets. Among them, scratches, trachoma, drag marks and tails are drum defects and need to be replaced. Traces, fingerprints, stains, water stains, water droplets are non-defective, only need to take corresponding treatment measures according to different types. This experiment is completed on the platform of Ubuntu 16.04 + Python 3.6. At the same time, the HOG feature extraction model is built by using the scikit-image library. The LibSVM toolbox not only facilitates the construction of SVM model, but also obtains the posterior probability output of SVM using the method proposed in [20].

Experiment 1: In order to verify the effectiveness of the HOG feature extraction model, several common feature extraction algorithms are selected to compare. LBP (Local Binary Pattern) feature extraction and Haar-like feature extraction are used to select SVM based on linear kernel function and SVM based on Gaussian kernel function respectively to achieve the best recognition results. Then, the feature is extracted by HOG and finally classified by Softmax classifier. The average classification accuracy of each method is shown in Table 1.

Table 1. Comparison of classification accuracy.

categories	3	5	7	9
LBP+SVM_LK	67.24	68.02	62.32	53.33
Haar_like+SVM_LKK	73.32	71.03	63.53	60.56
HOG+Softmax	89.02	86.50	83.38	80.63

The results of Table 1 show that the extracted features based on HOG show good performance in classification tasks. And the attenuation rate of accuracy is lower than that of LBP feature extraction and Haar_like feature extraction.

Experiment 2: In order to verify that the multi-SVM combination model can effectively improve the recognition accuracy, this paper extracts the output value of Experiment 1 as the input feature of each comparative classifier (SVM_LK, SVM_PK, SVM_GK, SVM_SK), and combines the four different SVMs according to the algorithm described in Section 2. (MSVMLC), the experimental results are shown in Table 2.

Table 2. Comparison of classification accuracy of different SVM

Classification number	3	5	7	9
HOG+SVM_LK	92.07	91.88	87.32	85.33
HOG+SVM_PK	96.33	91.50	87.53	85.50
HOG++SVM_SK	95.80	91.50	87.20	84.66
HOG+SVM_GK	96.00	90.08	87.34	82.23
HOG+MSVMLC	98.50	93.36	90.38	89.63

From Table 1 and Table 2, we can see that the SVM based on linear kernel function has a certain degree of improvement in classification accuracy compared with the traditional Softmax classifier, and the combined model proposed in this paper can achieve better results than the other four SVM classifiers. Fig. 2 shows the classification accuracy curves of different classification models. It is easy to see that the classification performance of SVM_PK, SVM_PK, SVM_LK and SVM_SK is very similar, and SVM_SK has the best classification performance. At the same time, by comparing the performance curves in Figure 4, we can see that the combination of the four SVM integration model has achieved better classification accuracy.

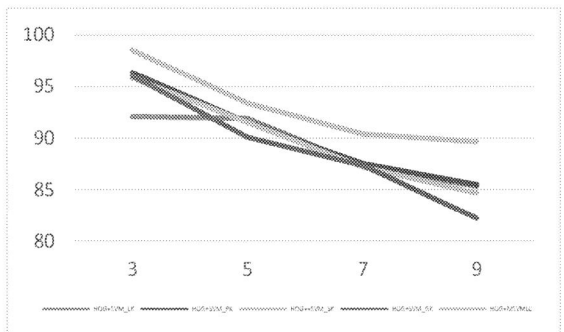

Figure 2. Performance comparison of SVM classifier

Finally, in order to further compare the performance of each classification model, a test set is constructed to test. Table 3 is the classification accuracy of the test set for six different classification models. The combined model can also get better performance than SVM alone.

Table 3. Comparison of classification accuracy rate

Name	HOG+SVM_LK	HOG+SVM_PK	HOG++SVM_SK
accuracy	86.88	84.85	84.98
Name	HOG+SVM_GK	HOG+MSVMLC	HOG+Softmax
accuracy	85.00	87.98	85.25

5. CONCLUSION

In this paper, a fusion model based on HOG and ensemble learning is proposed, which is applied to classification and recognition of printing cylinder defects, and is verified by identifying 9 kinds of common cylinder defects. The results of Experiment 1 show that compared with the features extracted in literature [3], literature [4] and literature [8], the features extracted in this paper based on HOG model have higher accuracy; according to the results of experiment 2, it can be concluded that multi-SVM combined classifier model can further improve the performance of defect recognition.

REFERENCES

[1] Freund, Yoav, Schapire, Robert E. A decision-theoretic generalization of on-line learning and an application to boosting[C]// European Conference on Computational Learning Theory. Springer, Berlin, Heidelberg, 1995:23-37.

[2] Schapire R E, Freund Y, Bartlett P, et al. Boosting the Margin: A New Explanation for the Effectiveness of Voting Methods[J]. The Annals of Statistics, 1998, 26(5):1651-1686.

[3] Lienhart R, Kuranov A, Pisarevsky V. Empirical Analysis of Detection Cascades of Boosted Classifiers for Rapid Object Detection[J]. Dagm, 2003, 2781:297-304.

[4] Sun Xuechun, Jiang Xiaonan, Fu Yao, etc. Camshaft surface defect detection system based on machine vision [J]. Infrared and laser engineering, 2013, 42 (6): 1647-1653.

[5] Yu Ling, Wu Tiejun. Integrated Learning: A Review of Boosting Algorithms [J]. Pattern Recognition and Artificial Intelligence, 2004, 17 (1): 52-59.

[6] Blumberg B, Downie M, Ivanov Y, et al. Integrated learning for interactive synthetic characters[J]. Acm Transactions on Graphics, 2002, 21(3):417-426.

[7] Jamshed M, Parvin S, Akter S. Significant HOG-Histogram of Oriented Gradient Feature Selection for Human Detection[J]. International Journal of Computer Applications, 2015, 132.

[8] Joachims T. Making large-scale SVM Learning Practical[J]. Technical Reports, 1998, 8(3):499-526.

[9] Burges C J C. A Tutorial on Support Vector Machines for Pattern Recognition[M]. Kluwer Academic Publishers, 1998.

[10] Chang C C, Lin C J. LIBSVM: A library for support vector machines[J]. 2011, 2(3):1-27.

[11] Tomasi C, Manduchi R. Bilateral filtering for gray and color images" ICCV[J]. Proc.ieee Inter.conf.computer Vision, 1998:839.

[12] Zhang B, Allebach J P. Adaptive bilateral filter for sharpness enhancement and noise removal.[J]. IEEE Transactions on Image Processing A Publication of the IEEE Signal Processing Society, 2008, 17(5):664-78.

[13] WANG C, WANG J, ZHANG X. Automatic radar waveform recognition based on time-frequency analysis and convolutional neural network[C]// Proc. of the IEEE International Conference on Acoustics, Speech and Signal Processing, 2017:2437-2441.

[14] ZHANG M, DIAO M, GUO L. Convolutional neural networks for automatic cognitive radio waveform recognition[J]. IEEE Access, 2017, 5:11074-11082.

[15] LECUN Y, BENGIO Y, HINTON G. Deep learning[J]. Nature, 2015, 521(7553): 436-444.

[16] BENGIO Y, COURVILLE A, VINCENT P. Representation learning: a review and new perspectives[J]. IEEE Trans. on Pattern Analysis and Machine Intelligence, 2013, 35(8):1798-1828.

[17] ABDEL-HAMID O, MOHAMED A R, JIANG H, et al. Convolutional neural networks for speech recognition[J]. IEEE Trans. on Audio Speech and Language Processing, 2014, 22(10):1533-1545.

[18] SAINATH T N, MOHAMED A R, KINGSBURY B, et al. Deep convolutional neural networks for LVCSR[C]// Proc. of the IEEE International Conference on Acoustics, Speech and Signal Processing, 2013:8614-8618.

[19] HE K, ZHANG X, REN S, et al. Deep residual learning for image recognition[C]// Proc. of the Computer Vision and Pattern Recognition, 2016:770-778.

[20] VINCENT P, LAROCHELLE H, LAJOIE I, et al. Stacked denoising autoencoders: learning useful representations in a deep network with a local denoising criterion[J]. Journal of Machine Learning Research, 2010, 11(12):3371-3408.

[21] MASCI J, MEIER U, CIREŞAN D, et al. Stacked convolutional auto-encoders for hierarchical feature extraction[C]// Proc. of the 21st International Conference on Artificial Neural Networks, 2011: 52-59.

[22] DENG L, PLATT J C. Ensemble deep learning for speech recognition[C]// Proc. of the Annual Conference of the International Speech Communication Association. Singapore, 2014:1915-1919.

[23] NG A Y. Feature selection, L1 vs L2 regularization, and rotational invariance[C]// Proc. of the 21st international conference on Machine learning. Alberta, Canada, 2004: 78.

[24] PLATT J C. Probabilistic Outputs for support vector machines and comparisons to regularized likelihood methods[J]. Advances in Large Margin Classifiers, 2000, 10(4):61-74.

[25] ZADROZNY B, ELKAN C. Transforming classifier scores into accurate multiclass probability estimates[C]// Proc. of the 8th ACM SIGKDD international conference on Knowledge discovery and data mining. Alberta, Canada, 2002: 694-699.

Unmanned Visual Localization Based on Satellite and Image Fusion

Xiaodan Yang

College of Computer Science, Sichuan University

No. 24 South Section 1, Yihuan Road, Chengdu Sichuan610065, China

8613438292728

ykxlxxq@163.com

ABSTRACT Unmanned driving is an important means for future human to achieve locomotion, and it will have broad application prospects. The unmanned vehicle still has the following difficulties in its long distance displacement: Single source sensor cannot meet the requirements of the positioning accuracy of the changeable and complex scenes, and the image analysis accuracy of visual processing in complex interference needs to be improved. In order to solve these problems, an unmanned vision localization algorithm based on multi-sensor fusion is proposed in this paper, by analyzing the positioning perception accuracy of unmanned vehicle, the precision and range of different perception methods at large, medium and small scales are obtained. A vision localization algorithm of multi-source fusion based on pseudo-range equivalence is designed in this paper. In order to reduce the influence of image distortion on localization accuracy, a visual localization algorithm based on image feature matching is proposed. The localization accuracy in complex environment is effectively improved by the multi-source fusion localization algorithm of pseudo-range equivalence. The MATLAB simulation shows that the positioning accuracy of the unmanned driving is improved to a certain extent at different scales.

1. INTRODUCTION

Unmanned vehicle is an automatic displacement technology which uses human-like logic to endow intelligence to machinery. It effectively maps the virtual computing space with the real vehicle and route, and lays a space-time foundation for the diffusion of computational intelligence from small scale to urban scale.

In order to realize group, automatic and precise unmanned driving, we need the cooperation of perception, communication, computing, storage, control, decision-making and feedback, and effectively promote the development of sensor, 5G mobile communication, Internet of Things, large data, cloud computing and other related technologies.

Unmanned driving is an important means to realize human position moving in the future, which has attracted great attention of the world powers. In September 2017, the U.S. federal government formally passed the world's first Driverless Vehicle Act to promote the development of driverless systems and services at the national level. In 2014, the U.S. Department of Transportation and ITS Joint Projects Office jointly proposed ITS Strategic Plan 2015-2019(ITS Strategic Research Plan, 2015-2019), which

---

Content from this work may be used under the terms of the Creative Commons Attribution 3.0 licence. Any further distribution of this work must maintain attribution to the author(s) and the title of the work, journal citation and DOI.

Published under licence by IOP Publishing Ltd

upgraded unmanned technology planning to a dual development strategy of networking and automated control intelligence, and with a focus on breakthroughs in enhancing traffic mobility and supporting information sharing of traffic system and so on. In May 2018, the European Union announced a schedule to strive to enter a fully automatic driving society by 2030, focusing on improving vehicle computing capabilities, and through the accelerated popularization of electric vehicles, in-depth research and development of automatic driving, the integration of automotive computers into digital transport networks to promote the rapid landing of driverless technology. In June 2018, the Japanese government proposed to achieve driverless driving in Tokyo in the 2020 Olympic Games. Regarding China, in May 2015, the "Made in China 2025" put forward by the State Council which explicitly calls for the development of smart vehicles and promoting intelligent vehicles.

According to the latest report from Intel and Strategy Analytics, the technology analysis consulting firm, the global industry related to driverless cars will reach $7 trillion by 2030 and the Chinese market for driverless vehicles will reach $2.5 trillion by 2030.

With the gradual development of driverless technology, there are still some important problems that need to be solved urgently:

1）Single source sensors cannot meet the positioning accuracy requirements of several variable and complex scenes. The displacement process of the carrier is essentially a process of adaptation to the environment. Large-scale displacement needs to undergo a combination of direction, slope, width, and so on. And the number of combinations increases exponentially with the increase of particle size. However, the traditional single-source sensor cannot complete the perception and matching of complex and changeable environment because of its own power and sensing scale limitations.

2）The accuracy of image processing under complex interference should be improved. Vision is the perceptual organ of human intelligence. It needs the effective cooperation of camera, sensor network and computing system to transfer it to AI. Various kinds of interference in various information transmission channels, especially illumination changes, sensor jitter, background clutter, occlusion and scale changes, will cause image distortion in the "brain".

To solve the above problems, this paper carries out unmanned visual localization algorithm of multi-source sensor fusion, through the analysis of positioning perception accuracy in unmanned driving to obtain the precision and range of different perception means in large, medium and small scales. A multi-source fusion vision localization algorithm based on pseudo-range equivalence is designed. In order to reduce the influence of image distortion on localization accuracy, a visual localization algorithm based on image feature matching is proposed. And the localization accuracy in complex environment is effectively improved by multi-source fusion localization algorithm of pseudo-range equivalence. The MATLAB simulation shows that the positioning accuracy of the unmanned driving is improved to a certain extent at different scales.

2. THE BASIC THEORY OF HETEROGENEOUS SENSOR LOCATION

2.1 Basic principle of large scale satellite positioning: distance determination
Satellite positioning is an important method to determine the coordinates of the uncovered targets on the earth's surface by using the geosynchronous orbit reference point and the geometric positioning principle. The basic principles are as follows: The time information of the signal arriving at the receiver is obtained by detecting the signal transmitted by satellite through the ground satellite navigation receiver, and the distance between the satellite and the receiver is obtained by transforming the relationship of time and distance. Finally, the coordinate points of the target are obtained by combining several satellites and performing geometric calculation. The conversion from signal measurement phase to arrival time information is an important means of satellite positioning, as shown in Figure 1. If there is a pseudo range of ρ:

$$\rho = c * \Delta t = c * \tau + c * (\Delta b_u - \Delta b_i) + \Delta D + \Delta \rho_{trop} + \Delta \rho_v (1)$$

In the formula, τ is theoretical arrival time difference, b_u and b_i is satellite clock error at different time points, c is transmission speed of electromagnetic wave, ΔD is ephemeris equivalent distance error,

$\Delta \rho_{trop}$ is ionospheric refraction correction, $\Delta \rho_v$ is receiver noise error. And the following formula can be obtained by joint observation of multiple satellite sources.

$$c * \tau_i = \sqrt{(X^i - X_u)^2 + (Y^i - Y_u)^2 + (Z^i - Z_u)^2} + C \quad (2)$$

In formula (2), τ_i is observed quantity from different satellite sources, X、Y、Z is target coordinates of calculation, C is error source, $C = \Delta D + \Delta \rho_{trop} + \Delta \rho_v$, it can be seen that the errors of satellite positioning mainly come from ephemeris error, ionospheric error and geometric error, as shown in Figure 1.

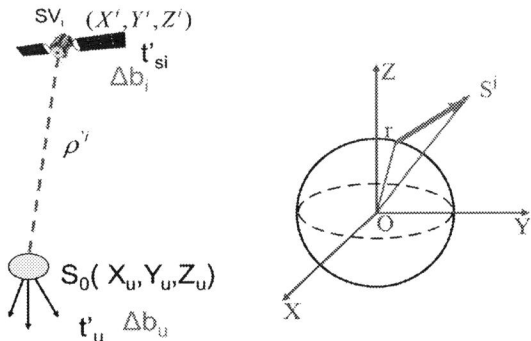

Fig. 1 ranging principle and positioning principle of satellite positioning

Because the variance of large-scale coverage varies little, ephemeris error and ionospheric error can be effectively reduced by ground difference. While the positioning targets are mainly concentrated on the ground, and the statistical proportion of the shaded area is more than that of the open area, which results in the destruction of the basic point distribution of geometric solution as well as the increase of positioning error. The accuracy of general satellite positioning is about 5~20 meters.

2.2 Basic principles of mesoscale wireless networks location: area determination

Wireless sensor network location method is based on a small number of nodes with known location, using its own transmission of the required radio wave signal characteristic parameters to calculate the location of unknown nodes. According to whether using distance measurement or not, it is divided into ranging algorithm and range-free algorithm. Ranging algorithm measures the relative distance or azimuth between nodes to get the actual distance, so that the position of unknown nodes can be calculated. Generally speaking, the positioning accuracy of ranging algorithm is higher than range-free algorithm, but the hardware requirements of the former are higher, and the ability to resist noise is relatively weak.

There are mainly 4 kinds of location methods based on ranging: RSS(Received Signal Strength), TOA(Time of Arrival), TDOA(Time Difference of Arrival) and AOA(Angle of Arrival), different algorithms are adapted to different location scene and accuracy requirements. This paper mainly discusses the positioning results using TDOA as location method. As the name implies, TDOA is a method that uses time difference to achieve positioning, that is, by measuring the time difference of signals arriving at each monitoring station to calculate the location of the signal source.

The node 1 is used as the measuring reference node, and the distance difference of remaining nodes is:

$$\hat{r}_i = (d_i - d_1) + n_{di} = \sqrt{(x - x_i)^2 + (y - y_i)^2} - \sqrt{(x - x_1)^2 + (y - y_1)^2} + n_{di}$$
$$i = 2,3,\cdots,N \quad (3)$$

In the above formula, n_{di} represents measurement noise, and the measured time difference is:

$$\tau_i = \frac{\hat{r}_i}{c}, i = 2,3,\cdots,N \quad (4)$$

Among them, c represents the speed of signal propagation.

If the error in formula (3) is not considered, the TDOA equation can be expressed as:

$$c \cdot \tau_i = \sqrt{(x - x_i)^2 + (y - y_i)^2} - \sqrt{(x - x_1)^2 + (y - y_1)^2} \quad (5)$$

TDOA positioning accuracy mainly depends on the accuracy of time difference measurement. Generally speaking, when the time difference error is about 100 ns, the positioning accuracy of this method can reach about 30 m. However, only using a single ranging algorithm for calculation often cannot meet the requirement of precise positioning, such as TDOA localization method usually needs to be used in conjunction with other methods.

2.3 Basic principle of small scale inertial device location: direction determination
Small-scale inertial device positioning technology is widely used in aviation, aerospace and vehicle positioning and navigation fields due to its strong autonomy, anti-interference and good concealment. It mainly uses the target motion data collected by the inertial sensor terminal to achieve positioning. For example, acceleration sensors and gyroscopes located in the target can measure a series of information such as speed, acceleration, direction of the target. Then, based on the dead reckoning algorithm, the position information of the target is obtained by corresponding calculation, as shown in Figure 2.

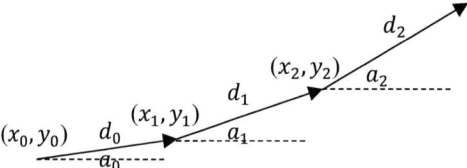

Fig. 2 principle chart of dead reckoning

Based on the premise of short-term navigation, the positioning accuracy of inertial device can reach 15 m at 30 s and will rise to 70-80 m at 60 s. That is because although the inertial navigation positioning based on dead reckoning algorithm has the advantages of data stabilization and no dependence, it also has the characteristics of cumulative error over time, which is the biggest limitation of inertial device localization.

3. VISUAL LOCALIZATION ALGORITHM OF MULTI-SOURCE FUSION BASED ON PSEUDO-RANGE EQUIVALENCE

3.1 Analysis of the problems of heterogeneous sensor location
From start to finish, unmanned driving is a complex process of condition and environment adaptation. In order to accurately analyze the combined errors of unmanned vehicles, it is necessary to subdivide the multiple processes of state change, as shown in Figure 3. Different process divisions are mainly based on the completeness of different number and types of locating sources in each process, thus inferring the problems in improving the locating accuracy.

(1) Satellite state incomplete process: Unmanned large-scale navigation path planning relies on wide coverage of satellite navigation, as in Figure 3, starting point A to starting point B. However, the satellite navigation signal is weak, and it is impossible to obtain enough positioning datum in the urban canyon and under the bridge, that is, the geometric distribution in formula (2) is destroyed. At this time, other positioning means are needed to help improve the accuracy of real-time localization.

(2) Fast baseline collaboration for real-time control: When the unmanned vehicle needs to change the line or avoid collision, the acceleration and direction of driving are determined by the inertial device, and the obstacles are determined by the visual system formed by video and intelligent control. And the inertial device has the problem of long-term accurate drift, but short-term locating accurately, and the positioning accuracy of the visual system is precisely determined by video acquisition. Therefore, in order to ensure the safety of driverless vehicles, rapid positioning means of real-time control is needed.

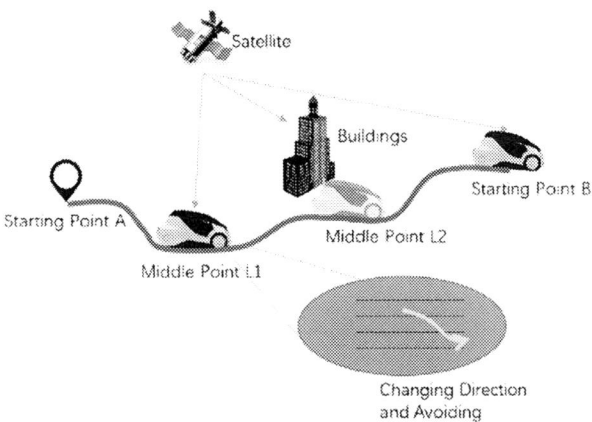

Fig. 3 different location scene in unmanned driving

3.2 Visual location algorithm based on image feature matching

Unmanned vehicles need to respond in milliseconds to avoid obstacles during driving, so they are required to accurately identify the relative position of objects in the scene. The basic principle of visual positioning system is to compare the captured image with the pre-stored image, and the difference between them reflects the distance and angle between them. Here, a characteristic source image on the road is set to be $P_i(x, y)$, and the measured images obtained from different distances and angles are $Q_i(x, y)$, then the corresponding errors and measurements are collected and measured in advance.

$$D(x, y) = \sum_{j=0}^{J-1} \sum_{k=0}^{K-1} [P_i(x + j, y + k) - Q_i(x, y)]^2 \quad (6)$$

In the formula, j and k correspond to distance and angle respectively, and the following formula can be obtained by further changes.

$$DS(x, y) = \sum_{j=0}^{J-1} \sum_{k=0}^{K-1} [P_i(x + j, y + k)]^2 \quad (7)$$

It is the energy of the corresponding area of the template in the source image, which is related to the pixel position (x, y), but the $DS(x, y)$ changes slowly with the pixel position (x, y). At this point, in order to further reduce the error, $DS(x, y)$ is normalized cross-correlation, and gray morphological etching operation is carried out to remove the peak noise of the image, then expansion operation is done to remove the low-valley noise of the image.

$$R(x, y) = \frac{\sum_{j=0}^{J-1} \sum_{k=0}^{K-1} Q_i(x,y) P_i(x+j,y+k)}{\sqrt{\sum_{j=0}^{J-1} \sum_{k=0}^{K-1} P_i(x+j,y+k)} \sqrt{\sum_{j=0}^{J-1} \sum_{k=0}^{K-1} Q_i^2(x,y)}} \quad (8)$$

When x and y change, $Q_i(x, y)$ transforms in the source image area and gets all values of $R(x, y)$. The maximum of $R(x, y)$ indicates the most precise location for matching $Q_i(x, y)$. If a region of the same size as the measured image is extracted from the source image, the matching location information J and K can be obtained.

3.3 Multi-source fusion location algorithm based on pseudo-range equivalence

From the above, ranging, time, field strength, geomagnetism, and so on, are different characteristic quantities, which need to be converted into unified representation quantities when they are solved at the same time. In this paper, pseudo-range is used to solve multi-source fusion localization and it is set to be M_i.

Assuming that the error does not change in time and space, the observation error can be written as

$$\begin{cases} M_1 = \sqrt{(X^i - X_u)^2 + (Y^i - Y_u)^2 + (Z^i - Z_u)^2} + C_1 \\ M_2 = \sqrt{(X^i - X_u)^2 + (Y^i - Y_u)^2 + (Z^i - Z_u)^2} + C_2 \\ M_3 = \sqrt{(X^i - X_u)^2 + (Y^i - Y_u)^2 + (Z^i - Z_u)^2} + C_3 \end{cases} \quad (9)$$

The Kalman filter is used to modify the information so as to get the final correction result of target position. It is assumed that it is the location, $t = (t_x, t_y, t_z)^T$, of different positioning sources aiming at the unified target reference system, then there are

$$M_i = M_k + t \qquad (10)$$

And the different state values $M_{i,k}$ of different locating sources are obtained, then the fusion combination closest to t can be obtained by successive approximation, where the location error is minimum.

4. SIMULATION AND ANALYSIS

4.1 Building simulation environment
In order to prove the validity of the proposed algorithm—unmanned vision positioning of multi-sensor fusion, the performance of the algorithm will be verified by MATLAB. Then setting satellite navigation for large-scale positioning, inertial navigation and vision as a benchmark for small-scale positioning. In the experiment, the sampling period is 1s, the number of sampling points is 3000, and the initial value of the state vector is $[0 \quad 10 \quad 0 \quad 0 \quad 10 \quad 0]$. Monte Carlo method is used as a reference for repeated tests, and collecting the errors of unmanned vehicle in localization and direction selection.

4.2 Positioning accuracy analysis of large scale unmanned vehicle

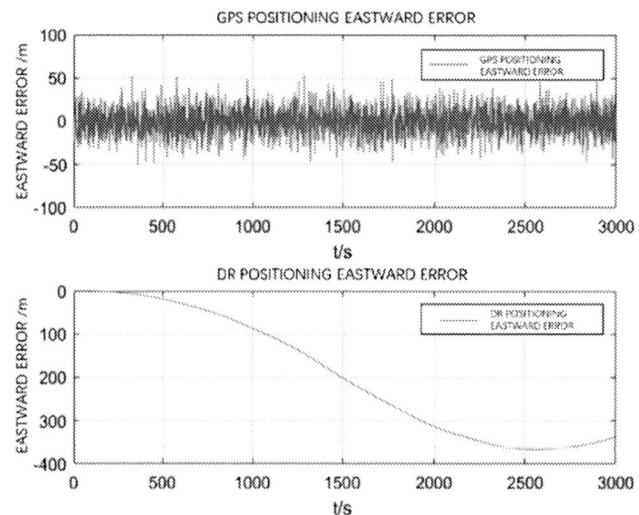

Fig. 4 error of satellite positioning and inertial navigation positioning in unmanned driving

From the graph, we can see that the result of satellite positioning is more stable than that of inertial navigation, of which the error range is basically within 50m. The initial positioning effect of INS is better, but as time goes on, the algorithm has the limitation of error superposition, which leads to rapid divergence of the later positioning results, and the positioning accuracy is getting worse and worse, which cannot guarantee the positioning requirements of high precision. Therefore, it is not advisable to use DR algorithm to target location alone.

4.3 Positioning accuracy analysis of small scale unmanned vehicles
Figure 5 shows the actual effect of using visual methods to distinguish road markings.

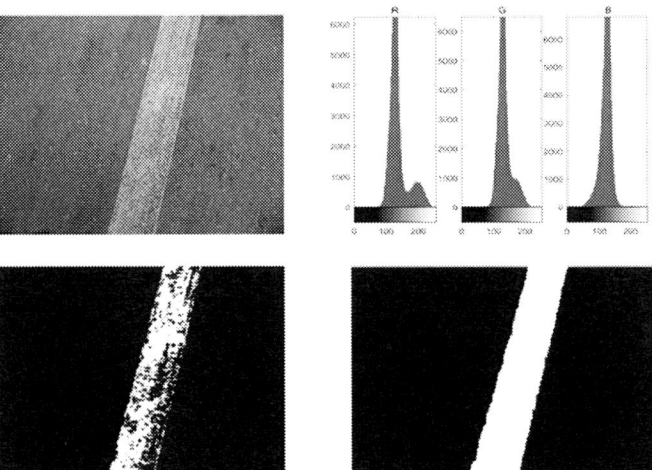

Fig. 5 image matching effect of small scale in unmanned driving—changing line

It shows the original image collected from the video, after RGB feature extraction, and then processed by different color clustering algorithm, and then the resulting image is denoised by morphological closed operation. From the experimental results, it can be seen that the image processed by the algorithm can effectively identify the road situation of original image, and can get a clearer denoising effect map, which can provide support for small-scale behavior of location matching and motion control.

5. CONCLUSION

In this paper, the unmanned vision localization algorithm based on multi-source sensor fusion is developed, and a vision localization algorithm of multi-source fusion based on pseudo-range equivalence is designed. In order to reduce the influence of image distortion for location accuracy, a visual localization algorithm based on image feature matching is proposed, and it is solved by multi-source fusion localization algorithm based on pseudo-range equivalence. Finally, it effectively improves the positioning accuracy under complex environment, and the method has important reference value for practical engineering application in related fields.

REFERENCES

[1] Rulin Huang. Research on Key Technologies of Dynamic Obstacle Avoidance for Autonomous Vehicle[D]. University of Science and Technology of China,2017,05(In Chinese)

[2] WANG Kejun, ZHAO Yandong, XING Xianglei. Deep Learning in Driverless Vehicles[J]. CAAI Transactions on Intelligent Systems, 2018,13(01):55-69(In Chinese)

[3] Kang Junmin. Key Technologies of Positioning System of Unmanned Autonomous Vehicle in Urban Environments[D]. Chang'an University,2016,12(In Chinese)

[4] GREJNER-BRZEZINSKA D A, TOTH C K, MOORE T, et al. Multi-sensor Navigation Systems: A Remedy for GNSS Vulnerabilities[J]Proceedings of the IEEE,2016,104(6):1339–1353.

[5] Bing Hou, Xiaolin Zhang. A Dual-Satellite GNSS Positioning Algorithm of High Accuracy in Incomplete Condition[J]. China Communications,2016,13(10):58-68.

[6] Shafei Wang, Zhongcheng Tian. Electronic Warfare Target Location Methods[M]. Publishing House of Electronics Industry,2014(In Chinese)

[7] Zhongcheng Tian, Congfeng Liu. Passive Locating Technology[M]. National Defense Industry Press, 2015(In Chinese)

[8] YANG Hai, LI Wei, LUO Cheng-ming. Fuzzy adaptive Kalman filter for indoor mobile target positioning with INS/WSN integrated method[J]. Journal of Central South University,2015,22(04):1324-1333.

[9] Yan Zhao, Yuwei Zhai, Eric Dubois, Shigang Wang. Image matching algorithm based on SIFT using color and exposure information[J]. Journal of Systems Engineering and Electronics,2016,27(03): 691-699.

[10] XIA Wei-qiang, FAN Shang-chun, XING Wei-wei, LIU Chang-ting, LI Tian-zhi, WANG Jun-feng. A mathematical morphological approach for region of interest coding of microscopy image compression[J]. Journal of Harbin Institute of Technology,2012,19(3):115-121.

[11] LI Xin, ZHANG Xiaohong, ZENG Qi, PAN Lin, ZHU Feng. The Estimation of BeiDou Satellite-induced Code Bias and Its Impact on the Precise Positioning[J]. Geomatics and Information Science of Wuhan University, 2017,42(10):1461-1467. (In Chinese)

[12] R. E. Kalman. A New Approach to Linear Filtering and Prediction Problems[J]. Journal of Basic Engineering Transactions, 1960, 82(1): 35.

[13] Le DONG, Wenpu DONG, Ning FENG, Mengdie MAO, Long CHEN, Gaipeng KONG. Color space quantization-based clustering for image retrieval[J]. Frontiers of Computer Science, 2017,11(6):1023-1035.

Cooperative Warp of Two Discriminative Features for Skeleton Based Action Recognition

Zheng Sun[1, a)], Xing Guo[1, b)], Wei Li[1, c)] and Zhengyi Liu[1, d)]

[1]School of Computer Science and Technology, Anhui University, Hefei 230601, China

[a)]ahuedu_sz@163.com

[b)]Corresponding author: ahuedu_gx@163.com

[c)]ahuedu_lw@163.com

[d)]ahuedu_lzy@163.com

ABSTRACT The study of human motion recognition has attracted many attentions in recent years. In this paper, a simple but effective feature combination is proposed by us for human action recognition based on skeletal points. The two features that make up this combination are the preprocessed 3D joint positions and the velocities of these 3D joints respectively. Then we use a combination of DTW [1], Fourier transform and SVM to model these actions and classify them. Because in the process of DTW, we use these two features to cooperate with each other to warp action samples, we call it the cooperative warp of two discriminative features (CWTDF). We also considered another scenario that we use these two features to warp their action samples respectively and then combine them before the classification process. We call it the separate warp of two discriminative features (SWTDF). But in this case, the classification performance is not as good as the cooperative warp situation. Although our feature representation is relatively simple, it is so sufficiently discriminative that our classification performance outperforms many state-of-the-art strategies based on skeletal points.

1. INTRODUCTION

Human motion recognition is an ancient research field. It has many applications such as human-computer interaction, security monitoring, elderly care, and entertainment. Because early action recognition methods process the action sequences extracted from RGB cameras [2][3][4][5], it was largely influenced by changes in illumination and subject texture variations. Moreover, the calculation amount of processing RGB image sequences is much larger than that of processing 3D joint position sequences from the depth images. Fortunately, with the invention of Microsoft Kinect and the introduction of a new skeleton tracking algorithm [6] which can extract 3D joint positions in real time, all the problems were solved. Many researchers then built their own action models using these 3D joint positions [7][8]. Vemulapalli et al. [8] used the method of lie group and lie algebra to build the skeletal model.

Inspired but different from [8], We combined two discriminative features which are preprocessed 3D joint position values and 3D joint velocity values on each frame of the action sequence. In the subsequent DTW process, these two features synergized to make the warped features more discriminative. Although the two features are relatively simple, our classification performance

Content from this work may be used under the terms of the Creative Commons Attribution 3.0 licence. Any further distribution of this work must maintain attribution to the author(s) and the title of the work, journal citation and DOI.
Published under licence by IOP Publishing Ltd

outperforms Vemulapalli et al. [8] on the dataset of MSR-Action3D. The duration of our program is much less than theirs. The main contributions of this paper are as follows.

1) A simple but effective feature combination for action recognition is proposed in this paper.

2) We experimentally verified that the cooperative warp of the two discriminative features has higher classification accuracy than the separate warp of these two discriminative features.

Organization: Section 2 is a brief review of action recognition methods based on depth data. Section 3 shows the method of data preprocessing and describes the proposed skeletal representation. Section 4 introduces the method of temporal modeling and classification of the action sequences. Section 5 gives the experimental results. Section 6 concludes the paper.

2. RELATED WORK

Human action recognition approaches based on depth information are divided into three types, which are based on skeletal points, original depth data, and the combination of these two features.

Skeleton-based methods: [9] proposed a feature representation which was called EigenJoints. The feature included static joint positions, the motion between two adjacent frames in an action sequence, and the motion between the initial frame and the current frame. Then the NBNN classifier was employed to classify the actions. Another skeletal joint feature was called HOJ3D [10]. They divided the whole 3d space into n bins and used the Gaussian weight function to associate the skeletal joints with these bins in the 3d space. Then they used a clustering algorithm to get the key postures and modelled the postures using a discrete Hidden Markov Model (HMM). A human action representation which was called SMIJ was proposed by [11]. They calculated the variances or the velocities of the human joint angles over a period to determine which a few angles were the most informative during that period. Then they combined the most informative joint angles of each time period of the entire temporal action sequence as the final action representation. A motion representation using 3d geometric relationships between body parts in each frame of an action sequence was proposed by [8]. Because the geometric relationships of rotation and translation between two body parts are members of the Lie group, they used curves in the Lie group to model the human actions. [22] proposed a descriptor which was called 3DMTG, which combined the 3D Moving Trend feature that captured the dynamic characteristics of the skeletal joints and Geometry feature that captured the offset of the initial frame and the current frame. We used skeleton feature only in this paper.

Original depth data based methods: [12] proposed a depth data based method which was called a bag of 3D points. A static gesture was characterized by some 3D points which were obtained from the original depth data of each frame. And then these static poses made up the whole movement. [13] proposed a method which was called HON4D. The depth sequences were described by a histogram which was obtained by calculating the distribution of the 4D surface normal.

Methods of a combination of these two features: [14] combined the local occupancy pattern feature and the relative 3D positions of skeletal joints. In order to model the action sequences, the method of three levels Fourier time pyramid was employed to represent the two features. [15] proposed an approach to fuse the spatiotemporal features and skeleton joint features. The spatiotemporal features were collected by performing the interest points detection and local feature description. The skeletal joint features were collected by computing the pairwise differences of skeleton joints and the frame difference. Then the random forests method was performed to combine these two features.

3. METHOD OF DATA PREPROCESSING AND THE PROPOSED SKELETAL REPRESENTATION

In this section, we first introduce how to preprocess the original skeletal data from the three datasets, and then introduce how to calculate the action descriptor proposed in this paper.

3.1 Method of Data Preprocessing
All datasets were processed in the following three steps.

First, In order to make the skeletal position data invariant when a person stands in different locations of the scene, we placed the joint position of hip center at the origin of the coordinate system by using every joint position to subtract the hip center position.

Second, To make the skeletal positions invariant to different subject scales, we used the algorithm which was proposed in [16] to process the original data. According to the principle of the algorithm, we took the skeleton of one subject as the reference, and calculated the body part lengths of this skeleton, and normalized all the other skeletons to the reference ensuring that the body parts of all the other skeletons had the same length with the corresponding body parts of the reference skeleton without changing the joint angles of other skeletons. For all the other samples in the datasets, in each frame, we used the breadth-first search (BFS) algorithm, starting from the root joint (hip center joint), moving to the joints associated with this body joint, and successively modified the joint positions without changing the angles of these joints.

Third, when a person is performing an action, different views to the camera will result in different skeletal coordinates. In order to normalize these spatial coordinates, we rotated the x-axis of the original coordinate system to make sure it was parallel to the horizontal plane projection of the left hip to right hip vector, and then we computed the coordinates of the skeletal points in the new coordinate system using the method proposed in [10].

3.2 The Proposed Skeletal Representation

3.2.1 The 3D joint positions feature

Let's define the skeleton joint positions from the action sequences as $P_{ij}=(p_{ijx}, p_{ijy}, p_{ijz})$, where $i \in \{1,...,M\}$ and $j \in \{1,...,N\}$, with M the total number of human body joints and N the total number of frames in an action sequence. P_{ij} is a column vector with length 3. The identifier p_{ijx}, p_{ijy}, and p_{ijz} respectively represent the three-dimensional coordinate value on the x-axis, y-axis, and z-axis of the ith body joint in frame j of an action sequence. So the whole action can be represented as a matrix $A=[P_{ij}]$, where $i \in \{1,...,M\}$ and $j \in \{1,...,N\}$.

Since the length of each action in a dataset is different, for the convenience of later calculation, we interpolate all actions in each dataset into N frames by using the method of cubic spline interpolation [8], where N is the maximum length of the action sequence in each dataset. We denote A' as the action representation after interpolation.

3.2.2 The velocity of 3D joints feature

Let's define the $V_{ij}=(p_{i(j+1)x}-p_{i(j-1)x}, p_{i(j+1)y}-p_{i(j-1)y}, p_{i(j+1)z}-p_{i(j-1)z})$, where $i \in \{1,...,M\}$ and $j \in \{2,...,N-1\}$, with M the total number of human body joints and N the total number of frames of the action sequences (after interpolation). So the velocity feature of the whole action can be represented as a matrix $B=[V_{ij}]$, where $i \in \{1,...,M\}$ and $j \in \{2,...,N-1\}$.

3.2.3 The combination of two features

Let's take the second column to the N-1 column of A' to form a new matrix which we call it A''. So A'' has the same number of columns as B. And then we splice each column of A'' and each column of B into a new column to form a new matrix C. We denote $C=\begin{bmatrix} A'' \\ B \end{bmatrix}$ as the final feature.

4. TEMPORAL MODELING AND CLASSIFICATION APPROACH

When you use a depth camera to collect human movements, the three problems you often face are that different people do the same movements for different durations, different movements have different durations, and some frames in the action sequence will be very noisy, which we call them subject rate variations, action duration variations, and sensor noise.

To handle the subject rate variations, a general idea is to make the actions of the same category as similar as well. The DTW algorithm can resample the action sequence to minimize the sum of the distance between the action sequence and the reference action sequence. So we calculate a nominal action sample as the reference for every action category using the algorithm proposed in [8] and warp the training action samples and test action samples to this nominal sample using DTW. In the MSR-Action3D dataset, for example, the number of action categories is 20. For every category, we calculate a nominal action sample, and then use the nominal sample to warp the training samples and test samples for the entire dataset, so we can warp out altogether 20 sets of data, then after the Fourier temporal pyramid representation of these warped actions which is described in the next paragraph, for each set of data, we use the training data to train out a binary linear SVM classifier, then we can train out altogether 20 linear SVM classifiers.

To handle the action duration variations and sensor noise problems, the warped samples are represented by the representation of Fourier time pyramid which is proposed in [14]. It is robust to noise because it discards the high-frequency Fourier coefficients. It is insensitive to action duration variations because two time series with different durations have the same Fourier coefficient magnitudes. In order to capture more information about the actions, we partitioned the time series in three different ways, which is called the three-layer Fourier temporal pyramid model. The first way is to divide the whole sequence as a partition, the second way is to partition the sequence equally into two parts, and the third way is to partition the sequence equally into four parts, so we get a total of seven time series. For each time series, we apply the Short Fourier Transform [20] and then concatenate these transformed Fourier coefficients as the final features.

Then we use multiple trained one-vs-rest linear SVM classifiers to classify the actions in the test set. The whole process of our approach is shown in Figure 1.

Figure 1. The whole process of our approach. The left half of the figure shows the training process and the right half shows the test process. L represents the number of action categories in a dataset.

5. EXPERIMENTAL RESULTS

In this section, three benchmark datasets are employed to evaluate our skeletal representation. In order to illustrate our proposed feature representation is more discriminative, we did another experiment: we warped the two features separately and represented them by the Fourier temporal pyramid respectively. Finally, we combined them before the classification process.

For all the datasets, we use the same evaluated method with [14] [15], which half subjects are used for training and other half are used for test. We provided 10 different combinations of training data and test data and then averaged their test results to get the final results.

5.1 Comparison with World Famous Methods

We compare our experimental results to many state-of-the-art approaches. The comparison results are shown in Table 1, Table 2 and Table3 respectively.

Table 1. Comparison on MSR-Action3D dataset [12]

Approach	Accuracy (%)
Proposed approach	90.90
Histogram of 3D joints [10]	78.90
EigenJoints [25]	81.40
Actionlet Ensemble [14]	88.20
HON4D [13]	88.36
Lie group [8]	89.48
Skeletal shape trajectories [21]	90.0

Table 2. Comparison on Florence3D-Action dataset [17]

Approach	Accuracy (%)
Proposed approach	88.38
Multi-part bag-of-poses [17]	82.0
Motion Trajectories [19]	87.0

Table 3. Comparison on UTKinect-Action dataset [10]

Approach	Accuracy (%)
Proposed approach	95.38
Random forests [15]	87.90
Grassmann manifold [18]	88.50
Histogram of 3D joints [10]	90.90
Motion Trajectories [19]	91.50

In the three tables above, we can clearly see that the classification accuracy of our method is higher than that of many famous methods in all the three datasets. Especially, our approach outperforms Lie group [8] 1.4 percentage points on MSR-Action3D dataset while we use the same classification method as [8]. This further indicates that although our method is only a combination of two simple features, it is more discriminative than [8] using complex spatial rigid body geometry. The running time of our method is also much less than [8] because our feature dimension is much less than [8].

Figure 2 provides the classification confusion matrix of 20 actions in the MSR-Action3D dataset. In figure 2, we can clearly see that the classification accuracy of only 4 actions is lower than 88% while the classification accuracy of 12 actions is higher than 97%. The classification accuracy of a few actions is low because they are too similar to some movements, such as hammer and forward punch, high arm wave and hand catch, hand catch and side boxing, draw a circle and draw X.

Figure 3 provides the classification confusion matrix of 9 actions in Florence3D-Action dataset. In figure 3, we can see that the more similar two actions are, the more confusing they would be. For example, to answer a phone is likely to be confused with to drink, and to answer a phone is likely to be confused with to read watch.

Figure 4 shows the classification confusion matrix of 10 actions in UTKinect-Action dataset. In figure 4, we can see that the classification accuracy of all movements is higher than 94% except the action throw. The reason is that the action throw is too similar to push.

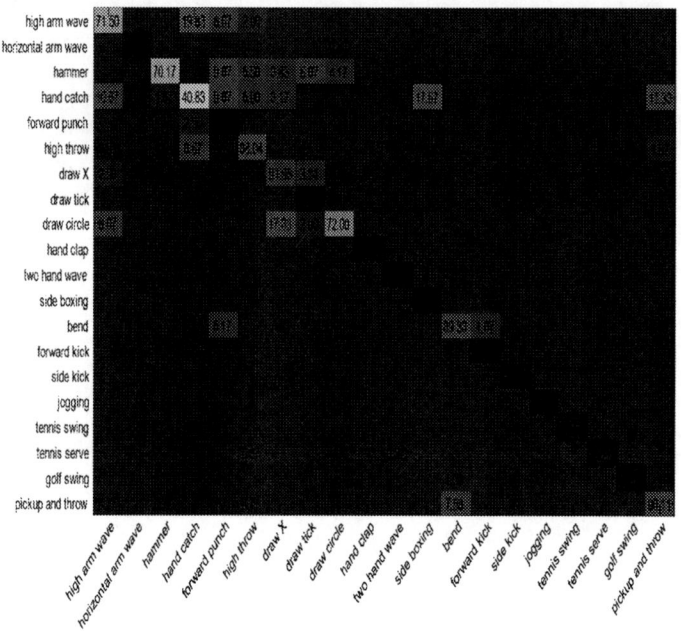

Figure 2. Confusion matrix on the MSR-Action3D dataset.

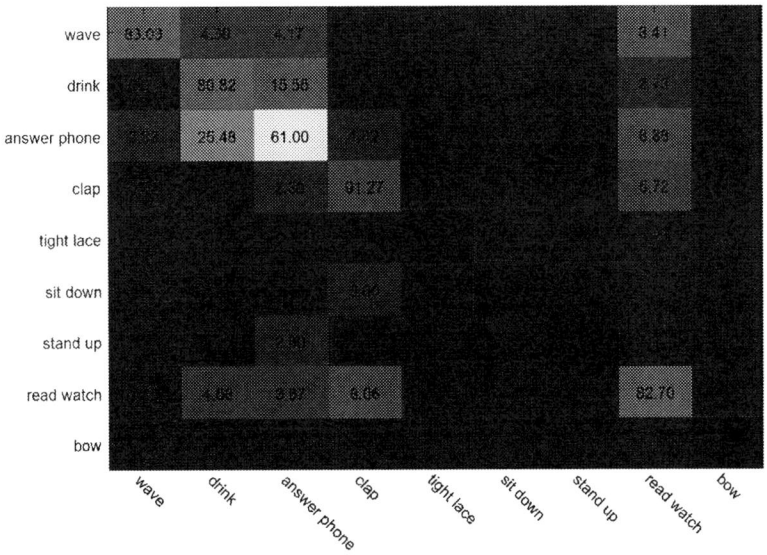

Figure 3. Confusion matrix on the Florence3D-Action dataset.

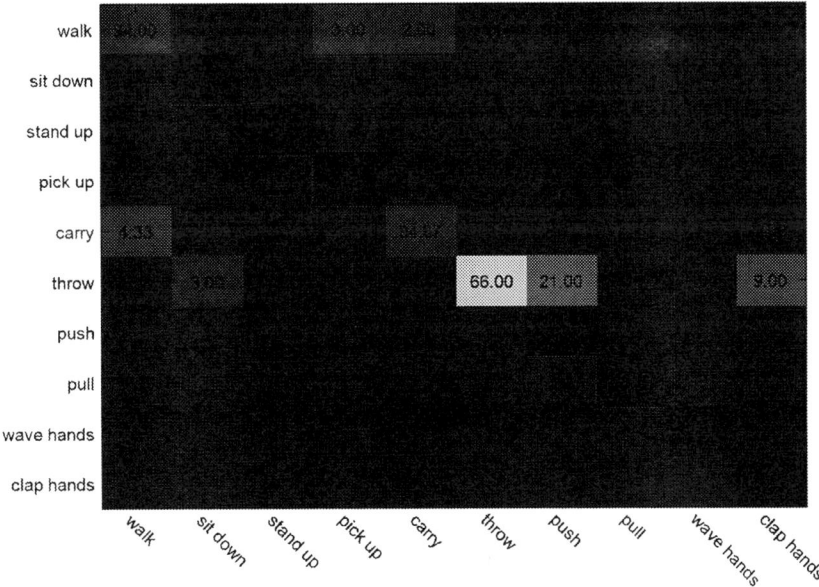

Figure 4. Confusion matrix on the UTKinect-Action dataset.

5.2 Comparison of The Two Methods Proposed in This Paper

Table 4, Table 5, and Table 6 show the comparison result between the two methods proposed in this paper in three datasets respectively. One way is to combine the two features (proposed in 3.2) and then go through the DTW and the Fourier transform. Another way is the two features go through the DTW and the Fourier transform respectively and then being combined. As shown in the following three tables, our experiment proved that the first method is superior to the second one in all three datasets. We analyze the reasons for this result is that two kinds of discriminative information work together to warp more accurate and discriminative information than they work separately.

Table 4. Comparison of the two methods proposed in this paper (MSR-Action3D dataset)

Approach	Accuracy (%)
Cooperative Warp (CWTDF)	90.90
Separate Warp (SWTDF)	90.24

Table 5. Comparison of the two methods proposed in this paper (Florence3D-Action dataset)

Approach	Accuracy (%)
Cooperative Warp (CWTDF)	88.38
Separate Warp (SWTDF)	87.29

Table 6. Comparison of the two methods proposed in this paper (UTKinect-Action dataset)

Approach	Accuracy (%)
Cooperative Warp (CWTDF)	95.38
Separate Warp (SWTDF)	95.18

6. CONCLUSION

A simple but effective feature combination for human motion recognition is proposed in this paper. The performance of our approach outperforms that of many state-of-the-art methods in three benchmark datasets. We also used experiments to prove that cooperative warp is more powerful than separate warp in action recognition.

The limitation of our method is that we need to complete the whole movement to classify it. In the following work, we will study how to classify the movement before it is completed, in order to make it easier for real-time human motion recognition.

ACKNOWLEDGMENTS

Our research was partially supported by the National Key Technology Research and Development Program under Grant 2015BAK24B01, partially by the National Science Foundation of China under Grant 61872002 and partially by the Natural Science Foundation of Anhui Province of China under Grant 1808085MF197.

REFERENCES

[1] Müller M. Information Retrieval for Music and Motion[J]. 2007.
[2] Niebles J C, Feifei N L. A Hierarchical Model of Shape and Appearance for Human Action Classification[C]// Computer Vision and Pattern Recognition, 2007. CVPR '07. IEEE Conference on. IEEE, 2007:1-8.
[3] Niebles J C, Wang H, Li F F. Unsupervised Learning of Human Action Categories Using Spatial-Temporal Words[J]. International Journal of Computer Vision, 2008, 79(3):299-318.
[4] Bobick A F, Davis J W. The Recognition of Human Movement Using Temporal Templates[J]. Pattern Analysis & Machine Intelligence IEEE Transactions on, 2001, 23(3):257-267.
[5] Yu H, Sun G M, Song W X, et al. Human motion recognition based on neural network[C]// International Conference on Communications, Circuits and Systems, 2005. Proceedings. IEEE, 2005:982.

[6] Shotton J, Fitzgibbon A, Cook M, et al. Real-time human pose recognition in parts from single depth images[C]// IEEE Conference on Computer Vision and Pattern Recognition. IEEE Computer Society, 2011:1297-1304.

[7] Wang J, Liu Z, Wu Y, et al. Learning Actionlet Ensemble for 3D Human Action Recognition[C]// IEEE Conference on Computer Vision and Pattern Recognition. IEEE Computer Society, 2012:1290-1297.

[8] Vemulapalli R, Arrate F, Chellappa R. Human Action Recognition by Representing 3D Skeletons as Points in a Lie Group[C]// IEEE Conference on Computer Vision and Pattern Recognition. IEEE Computer Society, 2014:588-595.

[9] Yang X, Tian Y L. EigenJoints-based action recognition using Naïve-Bayes-Nearest-Neighbor[C]// Computer Vision and Pattern Recognition Workshops. IEEE, 2012:14-19.

[10] Xia L, Chen C C, Aggarwal J K. View invariant human action recognition using histograms of 3D joints[C]// Computer Vision and Pattern Recognition Workshops. IEEE, 2012:20-27.

[11] Ofli F, Chaudhry R, Kurillo G, et al. Sequence of the Most Informative Joints (SMIJ): A new representation for human skeletal action recognition[C]// Computer Vision and Pattern Recognition Workshops. IEEE, 2012:24-38.

[12] Li W, Zhang Z, Liu Z. Action recognition based on a bag of 3D points[C]// Computer Vision and Pattern Recognition Workshops. IEEE, 2010:9-14.

[13] Oreifej O, Liu Z. HON4D: Histogram of Oriented 4D Normals for Activity Recognition from Depth Sequences[C]// Computer Vision and Pattern Recognition. IEEE, 2013:716-723.

[14] Wang J, Liu Z, Wu Y, et al. Learning Actionlet Ensemble for 3D Human Action Recognition[C]// IEEE Conference on Computer Vision and Pattern Recognition. IEEE Computer Society, 2012:1290-1297.

[15] Zhu Y, Chen W, Guo G. Fusing Spatiotemporal Features and Joints for 3D Action Recognition[C]// Computer Vision and Pattern Recognition Workshops. IEEE, 2013:486-491.

[16] Zanfir M, Leordeanu M, Sminchisescu C. The Moving Pose: An Efficient 3D Kinematics Descriptor for Low-Latency Action Recognition and Detection[C]// IEEE International Conference on Computer Vision. IEEE Computer Society, 2013:2752-2759.

[17] Seidenari L, Varano V, Berretti S, et al. Recognizing Actions from Depth Cameras as Weakly Aligned Multi-part Bag-of-Poses[J]. 2013, 13(4):479-485.

[18] Slama R, Daoudi M, Daoudi M, et al. Accurate 3D action recognition using learning on the Grassmann manifold[J]. Pattern Recognition, 2015, 48(2):556-567.

[19] Devanne M, Wannous H, Berretti S, et al. 3-D Human Action Recognition by Shape Analysis of Motion Trajectories on Riemannian Manifold[J]. IEEE Transactions on Cybernetics, 2015, 45(7):1340-1352.

[20] A. V. Oppenheim, R. W. Schafer, and J. R. Buck, Discrete Time Signal Processing (Prentice Hall Signal Processing Series). Upper Saddle River, NJ, USA: Prentice Hall, 1999.

[21] Amor B B, Su J, Srivastava A. Action Recognition Using Rate-Invariant Analysis of Skeletal Shape Trajectories[J]. IEEE Trans Pattern Anal Mach Intell, 2016, 38(1):1-13.

[22] Liu B, Yu H, Zhou X, et al. Combining 3D joints Moving Trend and Geometry property for human action recognition[C]// IEEE International Conference on Systems, Man, and Cybernetics. IEEE, 2017:000332-00033

Anomaly Detection Algorithm Based on FCM with improved Krill Herd

Chen Rui[1,a] ,Zhang Fengbin[2,b], Xi Liang[3,c]

[1]Candidate Harbin University of Science and Technology

[2]PhD, professor, PhD supervisor. Senior member of CCF Harbin University of Science and Technology

[3]PhD, associate professor, MS, supervisor Harbin University of Science and Technology

[a]oiaachen@yandex.com, [b]zhangfb@hrbust.edu.cn, [c]xiliang@hrbust.edu.cn

ABSTRACT The anomaly detection algorithm plays an important role in the field of network security. Among them, the most representative method is anomaly detection based on fuzzy C-means (FCM). FCM relies heavily on the initial clustering center and is prone to local extremum. Therefore, the detection effect of the FCM-based anomaly detection algorithm is not ideal in some cases. The herd intelligent optimization technology has a strong global search capability and is widely used in various fields. As a herd intelligence technology, the krill herd algorithm has a relatively simple optimization function structure, which has strong global search ability and is easy to integrate with other optimization strategies. Therefore, a KH algorithm with strong global search capability is introduced, and a hybrid KH-FCM algorithm is proposed. In the hybrid KH-FCM algorithm, the randomly generated initial population will be divided into two subpopulations containing the same number of individuals.

1. Introduction

Cluster analysis is one of the classic methods often used in machine learning. The basic idea is to analyze some invisible relationship which is based on data samples, and then divide these samples into different clusters. This remarkable feature is very suitable for detecting anomalous features [3]. Among them, fuzzy C-means clustering algorithm (FCM) is one of the most classic algorithms in unsupervised machine learning algorithms. Its most notable feature is that it does not need to mark the category information of data records in advance [4] and has been widely used in the field of anomaly detection [5]. However, FCM is highly dependent on the initial value and is prone to fall into local minima. Moreover, it usually requires more iterations [6]. To this end, people have adopted a variety of optimization measures to improve it [7], as the literature [8] combines FCM with genetic algorithm (GA), proposed GA-FCM, the algorithm first selects the best individual and it crosses and mutates the operation, and iteratively generates new optimal individuals until the optimal initial clustering center is generated to improve the FCM's high dependence on the initial value and easy to converge to the local minimum. Xiao Mansheng et al. [9] proposed a spatially correlated FCM, and designed its influence value according to the spatial distribution characteristics of the data set to improve the clustering center, thus reducing the sensitivity to noise.

Chen Haipeng et al. [10] introduced a soft partitioning method, using different clustering numbers to perform multiple clustering analysis. Finally, using the information about membership degree obtained by optimization, the correlation matrix was constructed to obtain the final result.

Content from this work may be used under the terms of the Creative Commons Attribution 3.0 licence. Any further distribution of this work must maintain attribution to the author(s) and the title of the work, journal citation and DOI.
Published under licence by IOP Publishing Ltd

2.RELATED WORK

2.1FCM-based Anomaly Detection Algorithm

Anomaly detection automatically detects and detects various abnormal behaviors by monitoring various events in the system in real time. The basic model is shown in Figure 1 [11]. FCM-based anomaly detection incorporates the idea of clustering. According to the nature of normal and abnormal features, they are divided into different clusters as much as possible and no intersection occurs [17].

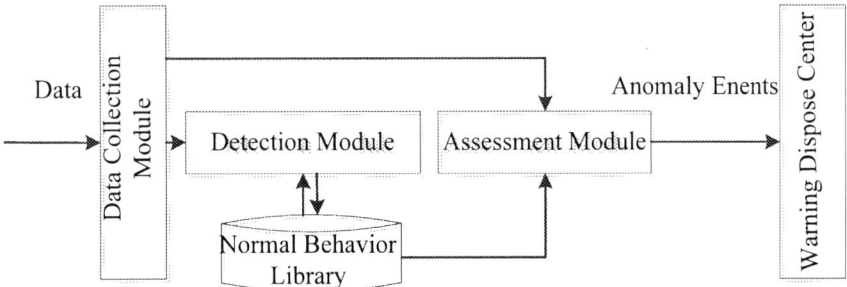

Fig. 1 Basic model of anomaly detection

The FCM-based anomaly detection method is widely used. In the field of medical application, the literature [12] applied the FCM-based anomaly detection to the brain nuclear magnetic resonance image segmentation and corrected the intensity non-uniformity by the FCM algorithm, thus improving the accuracy of the algorithm for tissue segmentation. In [13], FCM-based anomaly detection is applied to brain tumor segmentation nuclear magnetic resonance images, and artificial bee colony algorithm is introduced to reduce the influence of noise and help to identify brain tumors. In the application of power system, the literature [14] introduces the cross-section and cross-over algorithm to optimize the FCM anomaly detection algorithm, effectively compensates for the shortcomings of the single algorithm, and achieves a comprehensive and accurate refinement of the major power customers. Literature [15] combines FCM with adaptive fuzzy reasoning to improve the detection accuracy of the system. The model can quickly detect abnormal conditions and fault level conditions of the distribution network.

1)Mahalanobis Distance

Let A be an n×1 input matrix, which contains n samples，$n_i \in A(1,2,\ldots,n)$. The Mahalanobis distance between n_i and A can be defined as follows:

$$d_M = (\mathrm{n}_i - \overline{\mathrm{n}})^T C^{-1}(\mathrm{n}_i - \overline{\mathrm{n}}). \tag{1}$$

$\overline{n}$ is the sample mean and C is the covariance matrix, expressed as:

$$C = \frac{1}{n}\sum_{i=1}^{n}(\mathrm{n}_i - \overline{\mathrm{n}})(\mathrm{n}_i - \overline{\mathrm{n}})^T. \tag{2}$$

Xiang et al. [22] can adaptively adjust the set distribution of data by using Mahalanobis distance, and apply it to fuzzy clustering, and obtain better results. Therefore, this paper will use the Mahalanobis distance to measure the difference between samples.

2)FCM algorithm based on Mahalanobis distance

Its objective function can be expressed as：

$$J(\mathrm{U},\mathrm{V},\mathrm{C}) = \sum_{i=1}^{c}\sum_{j=1}^{n} u_{ij}{}^m (\mathrm{n}_j - \overline{\mathrm{n}}_i)^T C^{-1}(\mathrm{n}_j - \overline{\mathrm{n}}_i) \tag{3}$$

U is the membership matrix, and V is the cluster center matrix. The goal of the FCM algorithm based on the Mahalanobis distance is to obtain a minimum value for the equation. The constraint is：

$$\sum_{i=1}^{c} u_{ij} = 1, \ u_{ij} \in [0,1] \tag{4}$$

Then use the Lagrange multiplier method to get the following formula:

$$\overline{\mathrm{n}}_i = \frac{\sum_{j=1}^{n} u_{ij}{}^m \mathrm{n}_j}{\sum_{j=1}^{n} u_{ij}{}^m}, \tag{5}$$

$$T_i = \frac{\sum_{j=1}^{n} u_{ij}{}^m (n_j - \overline{n_i})(n_j - \overline{n_i})^T}{\sum_{j=1}^{n} u_{ij}{}^m}, \tag{6}$$

$$d_{ij}{}^2 = (n_j - \overline{n_i})^T [\overline{n_i} \, det(T_i)^{1/l} T_i{}^{-1}](n_j - \overline{n_i}) \tag{7}$$

$$u_{ij} = \frac{1}{\sum_{k=1}^{c} (d_{ij}/d_{kj})^{2/m-1}}. \tag{8}$$

among them, $1 \le i \le c, 1 \le j \le n, c$ is the number of cluster center centers.

FCM algorithm based on Mahalanobis distance requires 3 parameters: Iterative termination error ε and Fuzzy weighted index m. among them, c is given in advance. According to the literature [23], the optimal value interval of m is, in general, the median value of the interval, ε Usually set to 10-5.

KH algorithm is a new heuristic intelligent optimization algorithm, which is mainly based on the simulation study of the survival process of the Antarctic krill herd in the marine environment. For each krill particle, its location update is mainly affected by three factors:

1) induced exercise (induction of surrounding krill);
2) Foraging activities;
3) Random diffusion;

The speed update formula for krill individuals uses the following Lagrangian model:

$$\frac{dx_i}{dt} = N_i + F_i + D_i \tag{9}$$

Among them, N_i, F_i, D_i represent induced movement, foraging movement, and random diffusion, respectively.

The formula for the three factors is constructed as follows:

$$N_i = N_{max}\alpha_i + \omega_n N_{old,i.} \tag{10}$$

$$F_i = v_f \beta_i + \omega_f F_{old,i.} \tag{11}$$

$$D_i = D_{max}\left(1 - \frac{t}{t_{max}}\right)\delta \cdot \tag{12}$$

N_{max}, v_f, D_{max} represent the maximum induction speed, maximum foraging speed and maximum diffusion speed, respectively; $\alpha_i, \beta_i, \delta$ represent induction direction, foraging direction and diffusion direction, respectively; ω_n , ω_f represent the induced weight and the foraging weight, respectively; t , t_{max} are the current number of iterations and the maximum number of iterations. N_{max}=0.01，v_f=0.02，D_{max}=0.005.

The position of the krill individual in the interval t to $(t + \Delta t)$ is updated as follows:

$$x_i(t + \Delta t) = x_i(t) + \left(\frac{dx_i}{dt}\right)(\Delta t) \cdot \tag{13}$$

$$\Delta t = C_t \sum_{j=1}^{N_v}(B_{u,j} - B_{L,j}) \cdot \tag{14}$$

Δt is the scaling factor of the velocity vector; C_t is the step size scaling factor, taking a constant between [0, 2]; N_v represents the number of variables; $B_{u,j}, B_{L,j}$ are the upper and lower bounds of the jth variable, respectively.

To further improve the performance of the algorithm, the genetic operator (crossover or mutation) is executed in the algorithm. After testing, the crossover operator is more effective.

$$x_{i,m} = \begin{cases} x_{r,m}, a_{i,m} < C_r; \\ x_{i,m}, \quad \text{else.} \end{cases} \tag{15}$$

$$x_{i,m} = \begin{cases} x_{g,best,m} + \mu\left(x_{p,m} - x_{q,m}\right), a_{i,m} < M_u, \\ \qquad x_{i,m}, \qquad \text{else.} \end{cases} \tag{16}$$

C_r is the crossover operator; M_u is the genetic operator; a is a uniformly distributed random number on [0,1]; μ is a constant in [0,1].

3.IMPROVED ANOMALY DETECTION ALGORITHMS FOR KRILL HERD FCM
The algorithm mainly includes three parts: the improved krill herd is proposed, and then applied to FCM. Finally, an anomaly detection algorithm based on improved krill herd FCM is proposed.

3.1 Improved Krill Herd Algorithm
According to the standard KH algorithm, since the particle motion is random during the iterative process, when the algorithm moves to a poor position, that is, when the krill herd moves to the harsh environment of the predator, if the krill individual cannot be timely The information transmission makes a dangerous warning, the krill herd is easy to be preyed, and there are a large number of invalid iterations, which makes the algorithm not complete the local search well; Especially when dealing with multi-peak optimization problems, the solution of the algorithm is more likely to fall into local optimum, and the phenomenon of "premature maturity" appears. Based on this, an improved krill herd algorithm based on mutual benefit symbiosis and survival of the fittest is proposed.

3.1.1 Mutual benefit symbiosis strategy of krill herd
Living creatures in nature form a stable ecosystem, and there is a direct or indirect relationship between different species. This relationship can be roughly divided into three categories: mutual benefit and symbiosis, that is, mutual benefit to both parties; symbiosis, only beneficial to one of them, but harmless to the other; parasitic, beneficial to one of them, but harmful to the other. Mutual benefit symbiosis summed up as follows: Individuals of different species live together, and the mutual relationship between the two sides can also refer to the organisms that can survive normally even if they leave each other. Krill will escape from the danger of predators, causing the entire krill herd to move in different directions. In this process, a mutual benefit symbiosis strategy is introduced to enable the krill individuals to communicate with each other and transmit dangerous signals, for other krill warnings. Krill individuals will move toward the krill with good fitness value by comparing the advantages and disadvantages of their respective positions, that is, the difference in the fitness value of the current position, to avoid the predator. Record the current position safety factor level X_k, the global safety factor level X_{best},and the global krill average safety factor level X_{mean}. Use the following formula to indicate the movement of each krill after the hazard warning issued by the world's safest krill to all krills:

$$X'_k = X_k + r_i \times (X_{\text{best}} - \lambda X_{\text{mean}}) \tag{17}$$

$$\lambda = \text{round}[1 + a(0,1)] \tag{18}$$

r_i is a random number between [0,1], λ is an early warning factor, and the value of λ is 1 or 2. If X'_k is better than X_k, then update X_k; Update X_{best} if X'_k is better than X_{best}.

At this point, after receiving the global warning danger signal, the krill herd will transmit dangerous signals to each other. So randomly select two krills p and q, and $X_p \neq X_q$. The mutual symbiosis of this strategy is represented by:

$$X''_p = \begin{cases} X'_p + r_i(X'_p - X'_q), f(X'_p) < f(X'_q); \\ X'_p + r_i(X'_q - X'_p), f(X'_q) < f(X'_p) \end{cases} \tag{19}$$

r_i is a random number between [0,1]; X'_p and X'_q represent the safety factor level of krill p and krill q before exchange, respectively; If X''_p is better than X'_p, update X'_p; If X''_p is better than X'_q, update X'_q.

3.1.2 Survival of the fittest
On the basis of the above improvements, this paper also added the idea of survival of the fittest, and finally obtained the improved krill herd algorithm (MASFKH) based on mutual benefit and survival of the fittest.

The basic idea of survival of the fittest is that the basic idea of survival of the fittest is to maintain the diversity of particles in the population. The specific operation method of the "survival of the fittest" strategy is: after generating a new generation of populations, the newly generated populations will be evaluated for fitness values, that is, sorted according to fitness values, and the worst R particles will be discarded in the algorithm. And randomly generate R new particles in the search space. If the value of R is larger, the newer particles are generated, which is beneficial to maintain the diversity of the population and avoid the algorithm falling into local optimum. However, if the value of R is large, the algorithm tends to search randomly; if the value of R is smaller, it is not conducive to the algorithm to maintain the individual diversity of the particles, and the entire search ability of the algorithm becomes weak. Hence, R here takes NP/10, where NP is the population size.

3.1.3 MASFKH algorithm specific process

Step 1：Determine the size of the population and initialize the settings of the population and related parameters;

Step 2：The objective function value is used as the fitness value to evaluate the fitness of each particle in the population, and the current position and fitness value of the krill herd are stored in the P_{best} of each body, and all P_{best} are optimally stored in G_{best};

Step 3：Calculating the velocity component of the krill individual using equations (2)-(4);

Step 4：Adopting the formulas (9)-(11) to implement the "mutually beneficial symbiosis" strategy for the krill herd;

Step 5：For each body, compare the position it experiences and the objective function, update P_{best} and G_{best};

Step 6：Sort new populations according to fitness values and carry out the strategy of "survival of the fittest", and keep P_{best} and G_{best} unchanged;

Step 7：Determine whether the algorithm satisfies the termination condition. If it is satisfied, output P_{best} and G_{best}; Otherwise, return to Step 3 to continue searching.

3.2 FCM Algorithm Based on Improved Krill Herd

Based on the FCM algorithm of improved krill herd, MASFKH-FCM, the advantage of MASFKH can be well utilized to solve the problem that the FCM algorithm is too dependent on the initial value and is easy to fall into the local extremum. The algorithm idea is to use MASFKH to iteratively calculate the optimal solution of the problem. Then the solution is used as the initial clustering center of FCM, and iterative clustering analysis is carried out to classify different types of samples into different cluster classes. The algorithm flow is shown in Figure 2：

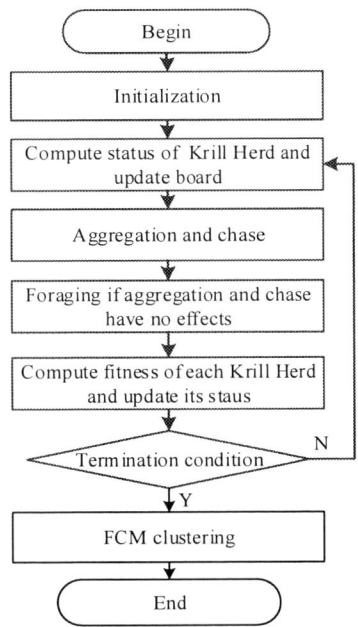

Fig.2 Flow chart of MASFKH-FCM

The algorithm has two key points: problem space coding and fitness function determination.

1) Coding: The algorithm encodes sample data based on the initial clustering center's encoding format, that is, n initial cluster centers form the position information of each krill herd. Let s be the dimension of the problem space, then the position of the krill is the n×s dimension variable:

$$P_n = (c_{11}, c_{12}, \ldots, c_{1s}, \ldots, c_{i1}, \ldots, c_{i2}, \ldots, c_{is}, \ldots, c_{ns}) \quad (20)$$

2) Fitness function: In the FCM algorithm, the minimum value of the objective function J (calculated by equation (3)) is the optimal clustering result. The smaller the value, the more obvious the clustering effect. Therefore, the following formula is set as the fitness function of the algorithm:

$$f = \frac{1}{(J+1)} \quad (21)$$

Specifically, the MASFKH-FCM algorithm steps are as follows:

Step 1. Set the number of krill N, moving Step, fuzzy index m, number of trials TryNumber,δ,number of cluster division c, search range Visual and other parameter values;

Step 2. According to the location of the krill, compare the current results with the records on the bulletin board, select the better value and update the information on the bulletin board;

Step 3. Calculate the initial values of the sample cluster center and the objective function according to equations (6) and (3), and evaluate the fitness;

Step 4. Perform MASFKH induction, foraging, and diffusion behavior;

Step 5. Update the status of the krill and adjust the Visual value adaptively according to equation (15);

Step 6. Compared with the end condition, if the end condition is met, the result is taken as the initial value of the FCM cluster. Go to Step 7 to continue the FCM cluster analysis process; If not, go to Step 2 and continue the MASFKH process.

Step 7. Use the FCM algorithm to perform continuous iterative calculations until the constraints are met and the final result is obtained.

3.3 Anomaly Detection Algorithm Based On MASFKH-FCM

Introducing MASFKH to optimize the FCM algorithm can make it better for anomaly detection. Therefore, the MASFKH-FCM based anomaly detection algorithm is designed. The algorithm flow is as follows：

Step 1. Put the training data into the matrix $X_{n\times l}$；

Step 2. Assign initial values to the krill herd according to the coding rules；

Step 3. Calculated according to MASFKH until the termination condition is met；

Step 4. Take the optimal result from Step 3 as the initial value of FCM；

Step 5. Standard evolutionary processing of test data set $X_{n\times l}$：

$$x_i' = \frac{x_i - min(i)}{max(i) - min(i)}.$$

$max(i)$ and $min(i)$ represent the maximum and minimum values of the attribute, respectively；

Step 6. Iterative calculation based on FCM algorithm. An abnormal sample set is obtained when the termination condition of the algorithm is satisfied.

Wherein, when the FCM algorithm detects the data set, it sets the data to be $X_{n\times l}$ matrix every time a certain amount of data is captured. Warn if there is an exception.

From the time cost analysis of each part above, the time of MASFKH-FCM anomaly detection is O(PN2) +O(nlcf)+O(nl)≈O(PN2) +O(nlcf). This time cost is acceptable.

4. EXPERIMENTAL RESULTS AND ANALYSIS

4.1 MASFKH-FCM Anomaly Detection Experiment

The experiment uses the well-known KDD CUP 1999 dataset in the field. First, data preprocessing is needed to convert the discrete attributes into continuous attributes, such as the protocol attributes. The experimentally set transformation rules are：$TCP \rightarrow 1$, $UDP \rightarrow 1$, $ICMP \rightarrow 1$ etc. Then, randomly select 6 herds of samples for experiment. Set one of them to be the training data set, containing 30,000 records, of which 363 are abnormal records. The remaining 5 herds are used as test data sets, each herd contains 10000 records, of which 120 are abnormal records.

Evaluating the performance of anomaly detection generally requires calculating the detection rate and false-positive rate, its definition is as follows：

$$DR = \frac{Number\ of\ intrusions\ detected}{Total\ intrusion\ in\ the\ data\ set}$$

$$FR = \frac{False\ positive\ data\ for\ intrusion}{Normal\ data\ in\ the\ data\ set}$$

The algorithm related parameter setting process is the same as the previous experiment, and the parameters are set as follows according to the literature [17]：N is 50, P is 200, TryNumber is 50, Step is 0.8, Visual is 4,δ,λ,γ are both 0.5, m is 2 in FCM, and c is 5.

1) In order to verify the effect of the algorithm on different test data sets, and verify the feasibility of each method of capturing a certain amount of data for a centralized processing, we randomly select 6 sets of data from the test data set, each herd contains 1000 data. Make up the $x_{n\times l}$ matrix and conduct experiments. The results are summarized as shown in Figure 3. As can be seen from the figure, each time the test data is selected differently, the algorithm shows different detection performance, but the fluctuation is within the acceptable range. The main reason for the fluctuation is that the algorithm has different cognitive abilities for different anomalies.

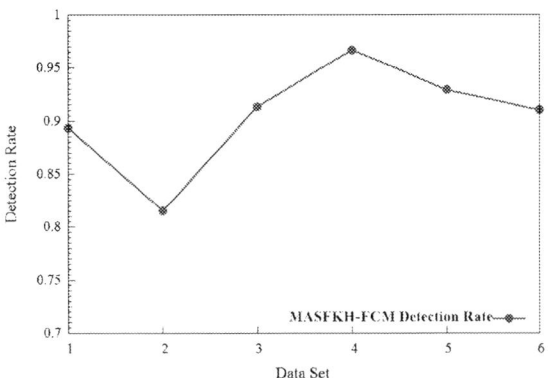

Fig.3 MASFKH-FCM's DR with different test matrixes

2) In order to test the detection stability of the algorithm in different test data sets, we repeat 5 experiments for each data set. The results are summarized in Table 1. The MASFKH-FCM has the same parameters, the mean values of DR and FR of different test sets are different, but the results are satisfactory, and the standard deviation is within the reasonable range of variation. This is because the training set and test set selected in this experiment are randomly sampled, the number is limited, and the characteristics of the included abnormal samples are not the same. Overall, FSF anomaly detection based on MASFKH optimization has better detection performance, especially in terms of stability.

Table 1 MASFKH-FCM's Detection Results with Different Test Set

Data Set	DR		FR	
	Mean	Standard Deviation	Mean	Standard Deviation
1	0.90983	0.00172	0.03477	0.00263
2	0.91885	0.00081	0.01963	0.00067
3	0.93952	0.00077	0.02768	0.00046
4	0.92953	0.00063	0.02173	0.00143
5	0.91988	0.00073	0.04057	0.00167

To verify the performance of FCM anomaly detection after adding MASFKH, the above five sets of test sets were tested using basic FCM, KH-FCM, MASFKH-FCM anomaly detection, and the parameters used were the same. The results are summarized in Table 2 and Figure 8~10.

Table 2 and Figure 4, 5 are the comparison of the detection performance of the final three algorithms. Under different test sets, the DR of MASFKH-FCM anomaly detection is higher than the results of FCM and KH-FCM anomaly detection, and its FR is lower than the FR of the other two algorithms. Overall, the MASFKH-FCM anomaly detection method achieved good results.

Table 2 FCM, KH-FCM and MASFKH-FCM's Detection Results

Data Set	FCM		KH-FCM		MASFKH-FCM	
	DR	FR	DR	FR	DR	FR
1	0.7971	0.0532	0.8673	0.0327	0.9073	0.0312
2	0.8125	0.0417	0.8853	0.0238	0.9187	0.0185
3	0.8491	0.0572	0.9132	0.0316	0.9381	0.0268

| 4 | 0.7736 | 0.0295 | 0.8996 | 0.0249 | 0.9296 | 0.0196 |
| 5 | 0.7894 | 0.0651 | 0.8775 | 0.0561 | 0.9182 | 0.0431 |

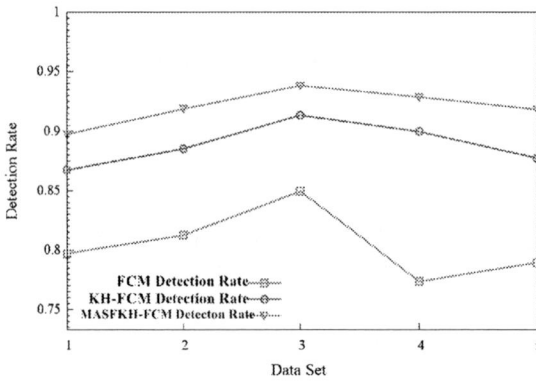

Fig. 4 DR comparison with three algorithms

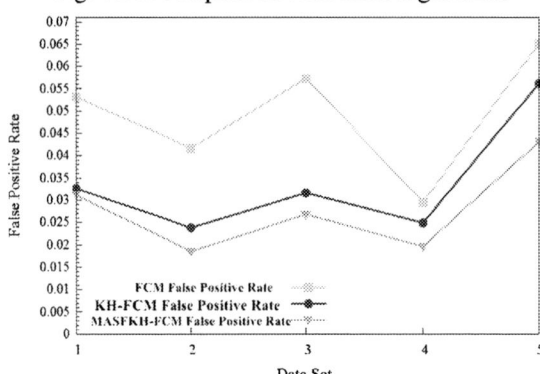

Fig. 5 FR comparison with three algorithms

Figure 6 shows the comparison of detection rates of three algorithms in one of the test sets in different iteration cycles. The MASFKH-FCM anomaly detection can achieve the optimal value when the number of iterations is 85. The other two algorithms require 100~110 iterations to achieve better results, but the results are worse than MASFKH-FCM anomaly detection.

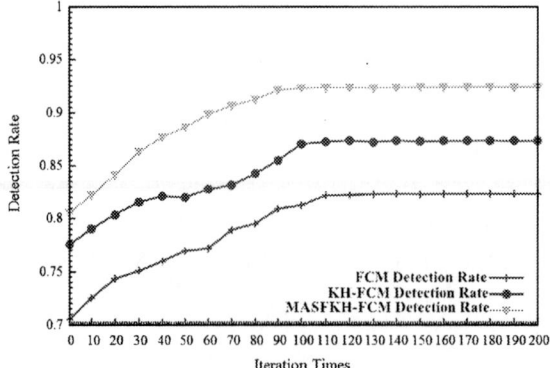

Fig. 6 DR comparison with three algorithms in different iteration cycles

This is because MASFKH optimizes KH, adaptively adjusts the value of Visual, improves the local and global optimization ability of KH, reduces the number of iterations of the algorithm, and makes it difficult to fall into local extremum, thus obtaining the global optimal solution. And this improvement significantly improves the efficiency of the algorithm. Then, combining MASFKH with FCM can solve the problem that FCM algorithm is too dependent on the initial value and is easy to fall into local optimum. Finally, taking full advantage of the advantages of both MASFKH and FCM, the anomaly detection algorithm based on this design can obtain better detection results.

5. CONCLUSIONS
In this paper, the artificial fish swarm algorithm is easy to fall into the local optimal problem when the optimal solution is impending, and the adaptive mechanism is introduced. The value of Visual is adaptively changed with the increase of the number of iterations of the algorithm, thus improving the local and global search ability of MASFKH. Then, using its characteristics of intelligent optimization and better robustness, it is applied to FCM anomaly detection, which solves the drawback that FCM is too dependent on the initial value and is easy to fall into local optimum without getting the optimal solution. Experiments show that the proposed MASFKH-FCM anomaly detection algorithm can obtain better detection performance.

MASFKH-FCM anomaly detection can be adjusted to apply to multiple application areas, such as intrusion detection technology in network security. FCM-based intrusion detection system is one of the important technical branches of intrusion detection. The MASFKH-FCM-based IDS can cluster the network attack sample set to improve the performance of the system to detect various attacks. In actual operation, compared with the traditional FCM algorithm, while the detection performance of the system is effectively improved, the convergence speed and operating efficiency of the algorithm are relatively high, which can better meet the real-time requirements of IDS. How to further balance the accuracy and real-time performance of IDS detection is the focus of future research.

REFERENCES
[1] Mao Jiali, Jin Cheqing, Zhang Zhigang, et al. Anomaly detection for trajectory big data: Advancements and framework [J]. Journal of Software, 2017, 28(1):17-34(in Chinese)
[2] Qian Yanyan, Li Yongzhong, Yu Xiya. Intrusion detection method based on multi-label and semi-supervised learning [J]. Computer Science, 2015, 42(2):134-136 (in Chinese)
[3] Han Jiawei, Kamber M. Data mining: Concepts and techniques [M]. Data Mining Concepts Models Methods & Algorithms Second Edition, 2011, 5(4): 1-18
[4] Qian Sen, Weng Guirong. Medical image segmentation based on FCM and level set algorithm[C]// Proc of 7th IEEE Int Conf on Software Engineering and Service Science. Piscataway, NJ: IEEE, 2017: 225-228
[5] Tang Chenghua, Liu pengcheng, Tang Shensheng, et al. Anomaly intrusion behavior detection based on fuzzy clustering and features selection [J]. Journal of Computer Research and Development, 2015, 52(3): 718-728(in Chinese)
（Xue Xiao, Liu Yian, Kan Yuan, et al. A research of intrusion detection system based on FCM-GRNN clustering [J]. Computer Simulation, 2010, 27(6):151-154 (in Chinese)
[6] Zhang Min, Yu Jian. Fuzzy partitional clustering algorithms [J]. Journal of Software, 2004, 15(6):858-868 (in Chinese)
[7] Jansi S, Subashini P. Modified FCM using genetic algorithm for segmentation of MRI brain images[C]// Proc of 1st IEEE Int Conf on Computational Intelligence and Computing Research. Piscataway, NJ: IEEE, 2015: 150-158
[8] Xiao Mansheng, Xiao Zhe, Wen Zhicheng, et al. Improved FCM clustering algorithm based on spatial correlation and membership smoothing[J]. Journal of Electronics & Information Technology, 2017, 39(5):1123-1129(in Chinese)
[9] Chen Haipeng, Shen Xuanjing, Long Jianwu, et al. Fuzzy clustering algorithm for automatic identification of clusters [J]. Acta Electronica Sinica, 2017, 45(3):687-694 (in Chinese)

[10] Nalluri M S R, Saisujana T, Reddy K H, et al. An efficient feature selection using artificial fish swarm optimization and SVM classifier [C]// Proc of 1st Int Conf on Networks & Advances in Computational Technologies. Piscataway, NJ: IEEE, 2017: 407-411

[11] Manikandan R P S, Kalpana A M. Feature selection using fish swarm optimization in big data[J]. Cluster Computing, 2017, Article in Press

[12] Alobaidi A T S, Hussein S A. An improved artificial fish swarm algorithm to solve flexible job shop[C]// Proc of 2017Annual Conf on New Trends in Information and Communications Technology Applications, Baghdad, Iraq, 2017: 7-12.

[13] Sengottuvelan P, Prasath N. BAFSA: Breeding artificial fish swarm algorithm for optimal cluster head selection in wireless sensor networks [J]. Wireless Personal Communications, 2017, 94(4):1979-1991

[14] Kumar K P, Saravanan B, Swarup K S. Day ahead scheduling of generation and storage sources in a microgrid using artificial fish swarm algorithm[C]// Proc of the 21st International Conference on Century Energy Needs-Materials, Systems and Applications. Piscataway, NJ: IEEE, 2016, Article number: 8052753

[15] Liu Rujuan, Jia Bin, Xin Yang. Network anomaly detection model based on information gain feature selection [J]. Journal of Computer Applications, 2016, 36(s02): 49-53(in Chinese)

[16] Kumari V V, Varma P R K. A semi-supervised intrusion detection system using active learning SVM and fuzzy c-means clustering[C]// Proc of the 1st Int Conf on IoT in Social, Mobile, Analytics and Cloud. Piscataway, NJ: IEEE, 2017:481-485.

ISPECE

IOP Publishing

IOP Conf. Series: Journal of Physics: Conf. Series **1187** (2019) 042029   doi:10.1088/1742-6596/1187/4/042029

An Improved Parallelization of K-means Algorithm based on HADOOP

Yizhuo Guo

Department of Information Engineering, Heilongjiang International University, Harbin 150025, china

610984970@qq.com

ABSTRACT In order to improve the problem that the single-machine serial programming model is not ideal for mass data clustering, we combine big data technology with text clustering related technologies. Implement distributed storage and calculations for text data, parallelization of text vectors and parallel clustering using clustering algorithms based on the Map Reduce programming model. The traditional k-means clustering algorithm is a typical algorithm for solving clustering problems. It has better with good scalability and scalability for processing large data sets, but the initial center of the algorithm is chosen randomly, and the algorithm is unstable every time. To solve the above problems, firstly, based on the idea of density segmentation and sampling thought, the initial clustering center is selected and optimized. Secondly, parallel sampling of the data set to find the best candidate cluster center by referring to the sample maximum and minimum method search and consolidate data objects in parallel. Finally, the selected initial cluster center is replaced by the central point randomly selected by the k-means algorithm, and the clustering algorithm is parallelized. Experiments show that the improved k-means algorithm can effectively reduce the number of iterations.

1. INTRODUCTION

Cloud computing provides an effective solution to deal with massive data, in which the multiple computers can cooperatively work, and then share virtualized resources by Internet. Hadoop which is the basic cloud computing platform can process distributed large data with reliable and scalable. Compared with the existing distributed computing framework, the service components in Hadoop have their own advantages that have parallel computing capability and high availability in storage capacity. Map Reduce is completely open source which provides a concise programming interface. Based on Map Reduce, the relevant algorithm can optimize in various research areas. The data storage mechanism with HDFS is a multi-copy storage strategy. According to the size of the cluster, the number of copies of text datasets are specifies. When the number of nodes storing a copy of the data is reduced by half, the integrity of the stored data can be maintained, the integrity of the stored data can be maintained. The redundant copy strategy ensures high availability of the file system. In this paper, based on Map Reduce parallel sampling and parallel merging data objects, the algorithm is improved for the disadvantage of randomly selecting the center of K-means algorithm. The initial centroid selected by this optimization method is replaced by the original centroid randomly selected by the traditional K-means algorithm. This proposed method can be processed in parallel, effectively reducing the number of iterations and running time of the algorithm. The proposed parallel random sampling is more suitable for large-scale data sets, which can improve the stability of the algorithm and reduce the selection time of initial clustering centers for large data sets. Compared with the effect

Content from this work may be used under the terms of the Creative Commons Attribution 3.0 licence. Any further distribution of this work must maintain attribution to the author(s) and the title of the work, journal citation and DOI.

Published under licence by IOP Publishing Ltd

and quality of traditional K-means algorithm, improved k-means algorithm and canopy-k-means algorithm have improved the effect and quality of text clustering. The experimental results show that the proposed method has superiority in clustering efficiency, clustering quality and parallel performance.

2. The Traditional K-means Algorithm

2.1 The Main Idea of Clustering Algorithm
The main idea of K-means clustering algorithm is to randomly select K objects from the data set as the initial clustering center at first. The second step is to measure the distance from the remaining vectors to the clustering center by distance and merge the remaining vectors with the nearest clustering center into the same cluster and generate K clusters. Next step, the distance formula is used to calculate the distance of the each class. With K cluster centers (cluster mean) as the new clustering center, recalculates the distance between the remaining objects in the data set and K cluster centers, and repeats the above process until the clustering center and the clustering center generated by the previous iteration does not change (the objective function converges). K-means algorithm is easy to understand. Although the algorithm iteration increases the complexity of the algorithm, the whole implementation process is easier and can make the clustering results to be optimal.

2.2 K-means Algorithm Analysis
Compared with other clustering algorithms, K-means clustering algorithm has the following advantages. The high efficiency of large-scale data processing, a wide range of applications, suitable for image processing, information retrieval and other fields, deal with image features and text types of data sets, optimize the clustering effect and get closer to the real clustering effect. Based on the advantages of the above K-means clustering algorithm, this paper selects the algorithm for text clustering. Of course, K-means clustering algorithm also has imperfections whose defects will affect the clustering effect to a certain extent and steps need to be improved in the traditional clustering algorithm. The initial center of traditional K-means clustering algorithm is randomly selected, which will lead to unsatisfactory clustering results. At the present, the following methods are selected for improving the initial cluster centers.

(1) A method based on the maximum and minimum distance. The distance between the initial clustering centers is as far as possible for the data set selection, and two or more objects with similar distances in the space are avoided.

(2) Based on density weighted method. According to density distribution, the objects in the dataset are assigned and compute the average of data objects. This method has good convergence, but the algorithm is of high complexity and low efficiency.

(3) K-means algorithm is executed many times. This method is inefficient.

In this paper, several classical improved methods are fused. We select candidate clustering centers for sampling data by the maximum and minimum distance algorithm, and then candidate clustering centers which are close to each other are merged. The accuracy of selecting the initial clustering centers are guaranteed at the expense of the efficiency of selecting the centers. Therefore, in order to satisfy the K-means clustering algorithm has better efficiency and accuracy, this paper mainly improves the algorithm from three aspects. The parallel random sampling process, parallel merging data object process, data object clustering parallelization process. The improved k-means parallel algorithm is designed and implemented based on Map Reduce.

3. Improved K-means Parallel Algorithm

3.1 Improved K-means Algorithm Parallel Design
The overall process of the improved k-means algorithm based on Map Reduce parallel processing is shown in Figure 1.

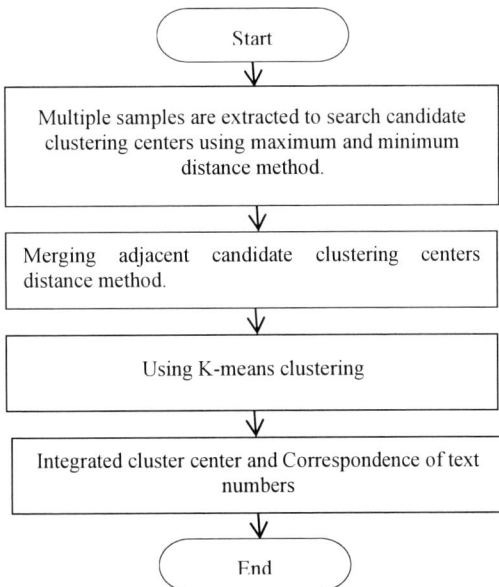

Figure1. The overall process of the improved k-means algorithm based on Map Reduce parallel processing

The improved k-means algorithm is parallelized. Firstly, by two Map Reduce tasks the initial clustering center is optimized based on the parallel design of Map Reduce. Based on Map Reduce, the specific implementation framework with the selection strategy of the initial clustering center is shown in Figure 2.Because the process of computes the cluster center in iteration is independent by K-means algorithm, the locality of K-means clustering algorithm and the distributed computing of Map Reduce distributed are combined. Based on Map Reduce and the initial centroid, the mean of clustering cluster and the computed cluster are divided to several nodes of the cluster and execute at the same time. The specific parallel framework is shown in Figure 3.

ISPECE
IOP Publishing

IOP Conf. Series: Journal of Physics: Conf. Series **1187** (2019) 042029   doi:10.1088/1742-6596/1187/4/042029

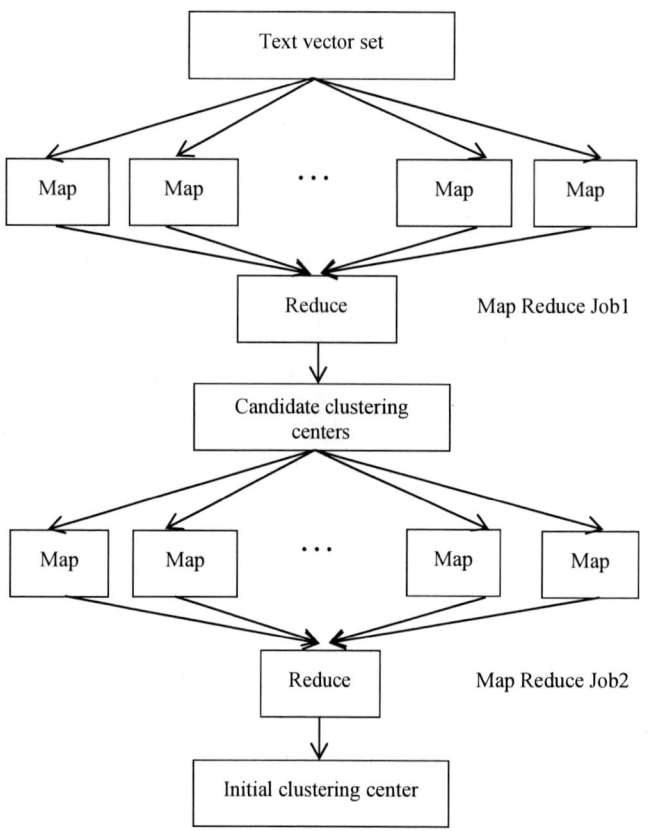

Figure 2. The Map Reduce framework for the initial center point selection strategy

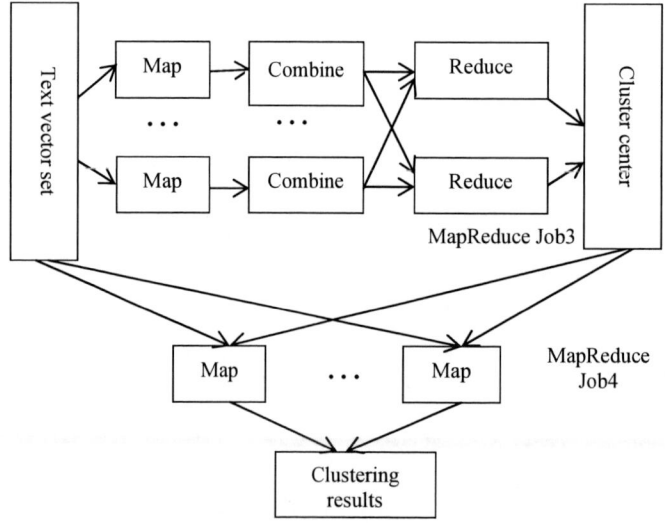

Figure3. MapReduce framework diagram of K-means algorithm

3.2 *Parallel Sampling*

Based on Map Reduce, the design of parallel sampling is divided into two processes. One task of Map Reduce extracts text vectors by several times in parallel. The set of text vectors is divided into samples

1464

which containing the same number of vectors and selects a set of text vectors as candidate clustering centers by the maximum and minimum distance algorithm. Another task of Map Reduce is to evaluate whether candidate clustering centers are adjacent according to density parameters, and vectors which meet adjacent conditions are merged. The distributed dense vectors with points contained in a certain neighborhood are merged. The mean value of each merged class is calculated as the final selection of the initial clustering center.

1. Sample extraction

The most common random sampling methods are traverse sampling and byte offset sampling. The traverse sampling can preserve the original data format of the data set, and it is inefficient to sample large-scale data sets. The byte offset sampling is also inefficient to large-scale data sets sample. In order to improve the efficiency of the above two random sampling methods, we present a random sampling method based on Map Reduce in this paper.

In the first random sampling process Map Reduce Job1, the main task is to sample many times in parallel, the initial cluster center for each sample is selected in the sample set, and the selected text vectors form the candidate cluster center set. In the Mapper stage, the input is in the form of <doc, docVector>. The map function is mainly responsible for extracting S (S = n/m, n is the total number of text in the data set, m is the number of map tasks) from the data set. First, the input text vector is counted from 1 to select the first S vector. When the count value i exceeds S, the selected vector is replaced by the vector used S/i probability. The sample extraction is finished until all the vectors have been accessed. Then, the selected S vectors whose form is <null, case Vector>output. In the Reducer stage, the output of the above step is used as input. The selected vectors are merged into the same Reduce because the all keys of the input are empty. The reduce function takes the selected sample vectors as the input of the maximum and minimum distance algorithm. After the algorithm is executed, the clustering center of the sample is selected to < null, tempVector> form output. In addition, the calculation of radius R is needed to calculate the density parameter of the point when the point is merged near in the second stage. Therefore, in the Reducer stage, the average distance averageDis of candidate cluster centers is calculated and output by<-1, averageDis>. Because the random sampling method used in this paper, the samples are distributed concentration. In this paper, sampled execute many times by above sampling method, that is, Map Reduce Job1 is executed many times.

2. Candidate clustering centers selection

For each sample in the sample set filtrate again to obtain candidate clustering centers. The filtrated method used by the maximum and minimum distance algorithm. The initial clustering center is selected in the sample set which the distance of the data objects is as far as possible. The number of iterations of the algorithm is increased to avoid the near distance of selected objects. This method can be used to firstly filtering of text vector sets.

The idea of the maximum and minimum distance algorithm is to select an object which marks as C_1 from the data set as the first object of the candidate clustering center set. Compared with the distance between the remaining objects in the data set and C_1, using Euclidean distance formula, the object who farthest from C_1 which marks as C_2, C_2 is used as the second object in the candidate clustering center. The remaining objects which are the minimum distance between C_1 and C_2 in the data set marks $\min(d_1, d_2)$. When the object with $\max(\min(d_1, d_2))$ satisfy formula (1), the object is the third candidate clustering center.

$$\max(\min(d_1, d_2)) \geq t(C_2 - C_1) \qquad (1)$$

3.3 Consolidated Data Object Parallelization

Firstly, the distance matrix T is obtained by computing the distance between all objects in candidate clustering centers. The mean distance R of all objects in candidate clustering centers is calculated according to the distance matrix. The density value of all objects in the candidate clustering center is calculated. The density value of the object is the number of objects which is the ball with the radius of R. The density value of all objects is sorted from large to small, and the object with the large density parameter value is selected. The object and the ball with the radius of Rare merged into clusters which are marked as visited points. The objects with the highest density value are selected from the objects that have never been visited, and the merging process of the above data objects is repeated until all the objects have been visited. The mutually disjoint clusters calculate the cluster center. When the candidate cluster center merges operation finish, the final initial cluster center generates.

In the second processing Map Reduce Job2, the candidate clustering centers are merged in parallel. In the Mapper stage, the output of the previous step is input and<null, tempVector > and <-1, averageDis > are input. A branch condition is whether the map function takes -1. At First, it reads the value of the key -1 and calculates the average value R of all candidate clustering centers. The candidate clustering centers whose key is null are read. The cosine distance between candidates clustering centers are calculated to obtain similarity matrix T which will be initialized. For each component of the corresponding matrix T, if its value is greater than the radius R, the corresponding component reset 1, otherwise the matrix reset component is 0. After the initialization matrix is finished, the density parameters of candidate clustering centers are computed. The number of vectors contained in each vector radius R is counted according to the distance matrix T, and the density parameters are sorted from large to small. The above operations are repeated until all vectors are divided into different clusters. Finally, the center of each cluster is calculated as initial vector. The corresponding text number of the cluster center is Cluster ID, which is the initial center of K-means clustering algorithm.

3.4 Cluster Parallelization

The initial centroid selected in parallel from the text vector set replaces the K centers randomly selected by the traditional K-means algorithm. The distance between the remaining vectors in the text data set and the selected cluster centers are computed. The remaining vectors are classified into clusters represented by the initial cluster centers. The average distance of the clusters is recalculated as the cluster center which is the new cluster centers. The above operations are repeated. The improved clustering algorithm finish when the objective function converges.

Map Reduce Job3 in Figure3 mainly calculates the distance between the remaining text vectors in the text dataset and the selected clustering centers. The remaining text vectors are classified into clustering centers which have the smallest distance. The average distance of the cluster clusters are calculated and generate a new cluster center. Map Reduce Job4 mainly generates clustering results based on the final clustering center.

Map Reduce Job1 is the third task processing procedure which can be summarized as follows. In the Mapper stage, the map function is input as<clusterID, initialVector>.The cosine distance from the text vector to the clustering center is calculated and merges the text vector into the cluster represented by the nearest clustering center. The vectors with small degrees become clusters. The each vector label as <clusterID, (1, docVector)> is output. In the Reducer phase, the main task is to calculate the average value of all the vectors in the cluster. First, the number of vectors in each cluster is counted by using the Combine function, and then the average value of all the vectors in the cluster is calculated by using the distance formula. Then the output is in the form of <clusterID, (num, averageVector)>. Num represents the vectors in each cluster. The average Vector represents the cluster mean. Then, reduce function aggregates the local results of each cluster to determine whether the objective function converges or not. When the algorithm finish, the output is the form of <clusterID, centerVector>. The centerVector is a new clustering center generated after iteration by K-means algorithm. If the objective function does not converge, each cluster center is used as a new initial cluster center point. The above process is repeated.

Map Reduce Job4 is the fourth task processing procedure which only needs to perform key-value conversion in the Mapper phase and output as <cluster ID, doc>. The data processing is not request by Reducer.

3.5 The Parallel Implementation of Improved K-means Algorithm

In this paper, the principle of selecting the initial center points is that the distance between the selected initial center points is as far as possible in multidimensional space. To avoid the local optimal solutions caused by the distance between the center points randomly selected by k-means and eliminate the interference of clustering results. The improved K-means algorithm can be summarized as follows.

(1)Sample size and number of samples are estimated in the text vector set $D = \{d_i \mid d_i \in R^p, i = 1, 2, ..., n\}$ generated by MapReduce Job4.Using the method of equal probability sampling, D is sampled many times to get the sample set $S_1, S_2, ..., S_n$.

(2) A text vector d_j is randomly extracted from the sample S_i and used as the first clustering center e_1. The Euclidean distance from the remaining text vector in set D to d_j is computed, and the text vector d_k farthest d is selected as the second clustering center e_2.

(3) For each remaining vector d_i in the text vector set D, the distance d_{i1} and d_{i2} of each vector d_i and clustering center e_1 and e_2 are calculated separately. The minimum values $min(d_{i1}, d_{i2})$ of distance e_1 and e_2 are found.

(4) The larger part of d_{i1} and d_{i2} mark $max(min(d_{i1}, d_{i2}))$, and the corresponding text vector is d_m.

(5) When the n+1 vector in the sample set S_i is selected as the clustering center, the distance between the residual vector in S_i and the selected clustering center is calculated to find the $max(min(d_{i1}, d_{i2}...,d_{in}))$.The corresponding text vector is selected as the n+1 clustering center if the condition of formula (1) is satisfied.

(6) For each sample in the sample set, step (2)-(5) will repeat until it generates a number of candidate clustering centers.

(7) The density parameters of candidate clustering centers are calculated and sorted in ascending order.

(8) The text vector with the largest density parameter is selected. The vectors in the vector radius merge into a cluster. The visited vectors are marked and select the vectors with the largest density parameter from the vector set that has never been visited. Repeat the above steps until all text vectors are accessed. The mean value of clusters with several disjoint clusters is used as the final initial clustering center.

(9) The selected initial clustering center is considered as the initial center. K-means clustering algorithm is executed.

4. Experimental Results and Analysis

The experimental data in this paper are selected from Chinese text corpus. The corpus is collected from news websites including entertainment, education, military, sports, wealth, tourism, finance and so on. Each category in the text data set contains 2000 texts. Before the experiment, the text in the data set is sorted and numbered again from small to large. The storage form is a key which is customized web text numbering. Web text content is the value of key. It is stored in the distributed file system (HDFS) by value of key. In this experiment, two sets of datasets which recorded as S_0 and S_1 respectively from the corpus are used. Five thousand and ten thousand texts of six different categories were extracted.

4.1 Comparison of Iterations Number

The algorithms are applied using Mahout Subproject in Hadoop ecosystem, Canopy-K-means algorithm and the improved K-means parallel algorithm on multi-nodes.

The parameters involved in the calculation process of the improved algorithm are as follows. The threshold frequency of the feature extraction stage is (0.1-0.65). The K value of the K-means algorithm is set to k=6. The sample size is n = 500, n = 700.The data set is extracted five times. The density check parameter t=0.5. Two sets of data sets S_0 and S_1 are constructed based on Map Reduce. The text vector set DS_0 and DS_1 obtained by vectorization of the two sets. The time consumption of clustering and the numbers of iterations of each algorithm are recorded. It is shown in Table1 and Table2.

Table1. Execution of each clustering algorithm in DS_0

Algorithm	Select cluster center time (s)	Number of iterations	Total time (s)
K-means	0	12	678.3
Canopy-k-means	77.3	9	529.3
Improved k-means	98.1	8	478.5

Table2. Execution of each clustering algorithm in DS_1

Algorithm	Select cluster center time (s)	Number of iterations	Total time (s)
K-means	0	25	1518.3
Canopy-k-means	181.5	17	1278.4
Improved k-means	159.2	15	998.5

(1) The execution with test datasets of different sizes is shown in Table1 using three algorithms which are traditional K-means algorithm, improved k-means parallel algorithm, Canopy-k-means clustering algorithm. The experimental results show that the number of iterations is most using K-means algorithm. The number of iterations is significantly reduced using other two algorithms because they optimize the selection of the initial clustering center. From the experimental data, we can see that the improved k-means algorithm proposed in this paper has the least number of iterations and total time consumption. It is verified that the improved k-means clustering algorithm based on the initial center optimization selection is more efficient than the traditional K-means algorithm and Canopy-k-means clustering algorithm.

(2) Because more cosine distances between text vectors need to be computed, more text vectors need to be partitioned multiple times. Using three clustering algorithms on the same cluster, when the number of text in the data increases, the number of iterations of each algorithm increases. When K-means clustering algorithm clustering test data sets are containing 5000 and 10000 texts, the changes number of execution cycles of the algorithm are most. Because the selection of the center point of the algorithm is random, the results of each execution of the algorithm are not consistent and the algorithm is more unstable. Using Canopy-K-means algorithm and the improved K-means algorithm, the number of iterations varies slightly and avoid the situation that the selected center point is not ideal because algorithms use optimization scheme to select the center point. Although the selection of the center point also consumes time, it can effectively reduce the number of iterations and the total time consumption of the algorithm.

4.2 Initial Sampling Rate Comparison

The initial sampling rate compares the operation efficiency of several different sampling methods. Sampling methods for comparison are row-by-row traversal, byte offset, and parallel sampling based on Map Reduce. For multi-group data sets, the sampling time is recorded as timeout when the sampling time is more than one hour. The comparison of sampling time between different random sampling methods for different size data sets is shown in Table 3.

As shown in Table 3, with the increase of the size of the sampled data set, the time consumed by the method of traversing the data line by line is very long. This sampling rate is the lowest. When sampling a small data set, the byte offset sampling method consumes the least time. The sampling time increases linearly with the increase of the amount of data because of limitation of the time complexity. When processing large-scale data sets, this method is not applicable. When small data sets are processed based on Map Reduce parallel random sampling method, the sampling rate is lower than byte offset sampling. Because the start-up of Map Reduce tasks and intermediate steps affect the sampling rate, large-scale data, task processing parallelization and cluster nodes parallel mining, the time consumption tends to be stable and the variation amplitude is small, which verifies that the parallel sampling method designed in this paper can adapt to large-scale data sampling processing. Therefore, parallel random sampling based on Map Reduce in this paper can effectively reduce the selection time of initial clustering centers for large data sets.

Table3. The time of Different sampling methods

The number of elements in the sampled data set				
	400	500	1000	10000
Line by line sampling	1.34.4s	1349.1s	2600.1s	Time out
Byte offset	5.67s	7.32s	467.8s	678.2s
Parallel sampling	76.3s	84.5s	88.5s	93.8s

5. Conclusions

Defects randomly selected for the k-means algorithm center point, further improving the algorithm, parallel sampling and parallel merging data objects based on Map Reduce, and replacing the initial centroid selected by the optimization method with the initial center of random selection of the traditional k-means algorithm. The improved k-means clustering algorithm is based on Map Reduce parallel processing to effectively reduce the number of algorithm iterations and improve the running time of the algorithm. The initial sampling rate of parallel random sampling is compared with other random sampling methods. It is verified that the parallel sampling proposed in this paper is more suitable for large-scale data sets, effectively reducing the selection time consumption of the initial cluster center of large data sets and improving the stability of the algorithm.

REFERENCES

[1] LI Xiaoyu, YU Liying, LEI Hang, TANG Xuefei. Implementation and application of a parallelization of K-means improved algorithm. Journal of University of Electronic Science and Technology of China, 46(01), 61-68 (2017)

[2] Li Guobao, Han Qingju. A MapReduce Parallelization Implementation Based on Improved K-means Clustering Algorithm. Digital Technology and Application, (12),134-136(2016)

[3] Mengya Shao. Transition-based Soft Phase Partition Algorithm of Multiphase Batch Processes Based on K-means Optimal Clustering Proceedings of the 37th China Control Conference, 2018: 6.

[4] Yu Hualong, Han Xuefeng.Bank customer classification algorithm based on improved K-means clustering.Natural Science Journal of Xiangtan University, 40(03),125-128(2018)

[5] [3]Wu Dechao, Liu Xiaohong, Qu Zhijian. Research on Distributed Clustering Algorithm Based on Hadoop. Journal of Shandong University of Technology (Natural Science), 32(04), 25-29(2018)

Remote Sensing Image Building Extraction Based on Deep Convolutional Neural network

Yang qun yi[1,a],Wang wei an[2],Ma xiao dong[3]

[1]Tongji University No. 1239, Siping Road, Shanghai, China18321792670, 86

[2]Tongji University No. 1239, Siping Road, Shanghai, China

[3]Tongji University No. 1239, Siping Road, Shanghai, China

[a]691530291@qq.com

ABSTRACT Segmentation Building extraction in high resolution remote sensing image is difficult due to different object has the same spectral feature. In this paper, we build a convolutional neural network RSBD4 base on theory of deep learning. We also proposes the overlap split method to solve the problem when we split image to speed up compute process and it causes the loss of edge information. We conduct experiments on large area Quick Bird image. Result shows that the proposed method extracts building well and has great practical value

1. INTRODUCTION

The extraction of buildings from remote sensing images is of great significance. The fields of military investigation, urban planning, disaster emergency assessment, etc. need to quickly and accurately extract building targets from remote sensing images [1]. With the continuous improvement of the resolution of remote sensing images, remote sensing images show a lot of new features, such as rich geometry, structure, texture features, spectral refinement, multi-scale of ground objects, etc. [2]. High-resolution remote sensing images bring more details, but also bring a serious of problems, such as increasing of the noise in image, and limit the accuracy of the extraction [3]. Related researchers have been looking for algorithms with higher precision and automation for building extraction.

[4] uses the characteristics of the spectrum and shape of the building, plus the building template as an auxiliary means to extract the buildings in the Quick Bird image. Literature [5] uses the combination of various features of building spectrum, shape and texture features to extract buildings in remote sensing images with an accuracy rate of 72.7%. [6] cites the morphological building index and the shadow index, combined with the object-oriented extraction method, extracts buildings and predicts the height of the building on the 5.8m and 2.1m resolution fused images.

In [7], the deformable component model is used to treat the building as a combination of deformable components. The corresponding parameter template is obtained through training, and the effectiveness of the algorithm is proved in the image of 0.5 m resolution. [8] uses deep convolutional neural network, and the patch-based idea was used. They compare the detection effects of different network structures and extract buildings in high-resolution images.

Based on the principle of deep learning algorithm and the excellent structure of convolutional neural network in computer vision, this paper builds a CNN network for high-resolution remote sensing image building extraction. The strategy of "overlap split" is proposed to optimize the loss of edge information

Content from this work may be used under the terms of the Creative Commons Attribution 3.0 licence. Any further distribution of this work must maintain attribution to the author(s) and the title of the work, journal citation and DOI.
Published under licence by IOP Publishing Ltd

ISPECE   IOP Publishing

IOP Conf. Series: Journal of Physics: Conf. Series **1187** (2019) 042030   doi:10.1088/1742-6596/1187/4/042030

caused by splitting image to parallel computing. The algorithm is tested in a certain area of Shanghai suburbs, which has achieved relatively satisfactory results in accuracy and reliability.

2. Algorithm Flow

2.1 Convolutional neural network structure
The network structure of RSBD4 is shown in Figure 1.

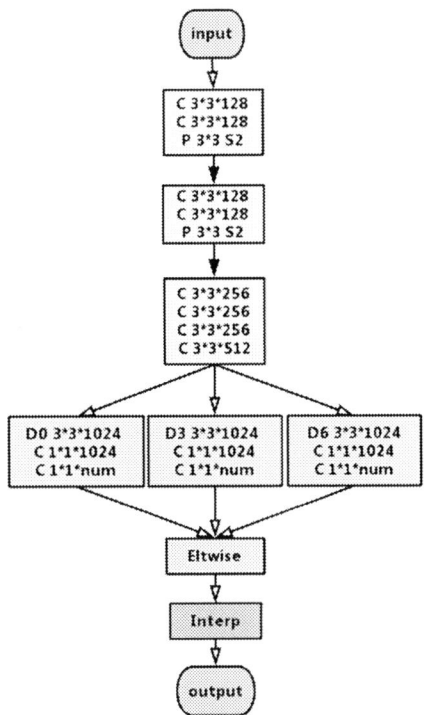

Figure 1. RSBD4 network structure.

In the figure 1, C represents a convolution operation, and C 3*3*128 indicates that the convolution kernel has a size of 3*3 and outputs 128 channels. P is the pooling layer operation. In P 3*3 S2, 3*3 represents the pooling window in size of 3*3, and S2 represents the pooling stride is 2. D is atrous convolution layer, D0 3*3*1024 indicates the hole ratio is 0, convolution kernel size is 3*3, and has output of 1024 channels. Eltwise represents the addition operation. Interp represents an interpolation operation. Num represents the number of extracted categories.

The convolutional layer is the most basic part of a deep convolutional neural network. In convolutional layer, a convolution kernel of a specified size will simultaneously slid over each channel on the input image in a certain step.

In order to bring nonlinear classification to the network, each convolution layer usually has an activation function. The data given in [9] shows that relu can greatly shorten the learning period of the network compared with other activation functions. The network of this paper also adopts relu as the activation function. The formula is as follows:

$$y = \max(0, x)$$

The first two sets of convolutional layers of RSBD4 are followed by a pooling layer. There are two main types of pooling methods, average pooling and max pooling. The average pooling can retain more background information of the image, and the max pooling can retain more texture information. In this

1471

paper, the strong texture features of the building are considered, and the max pooling is selected as the pooling method. Assuming that the window size of a max pool is n×n, the output of the window conforms to the following formula:

$$y = \max(x_1 x_2 x_3 \ldots x_{n^2})$$

The convolutional neural network can extract the features of a image, regardless of which position of the input image the feature point is located, and therefore has spatial position invariance, but the traditional neural network has poor adaptability to the scale change of the input image [10]. The size of buildings on remote sensing images is not fixed, which requires the network to have the ability to identify targets of different scales. In deeplabV2, the author used Atrous convolution layer [11] to solve this ploblem. Atrous convolution layer setting different hole rates, which makes network have the ability to identify targets of different scales. Considering that the size of the building in the remote sensing image is not fixed, RSBD4 use Atrous convolution layers with and the hole ratios are set to 0, 3, and 6, respectively.

The end of the network uses an Eltwise operation to combine the outputs of the three sets of convolutional layers. The Interp operation was used to restore the convolutional layer results to the original image size, and the bilinear interpolation algorithm was used in the experiment.

2.2 Overlap split method

The amount of data faced by remote sensing image processing tasks is usually huge. In order to solve the problem of slow processing speed caused by massive data operations, many remote sensing image processing methods adopt distributed computation [12]. This paper also uses this processing method and try to make full use of the hardware resources of the computer. The detection process is as figure 3:

In order to execute distributed computation, we first split the input image into some blocks, then use multiple compute units to process them, each compute unit will be assigned same number of blocks.

In the experiment, we found that if the block is located in the edge area of the input image, the network is not ideal for the extraction of such block, because some building is cuted into two pieces and is difficult to extract such information-incomplete buildings, so this paper proposes overlap split method in the process, which avoids the loss of precision caused by splitting image.

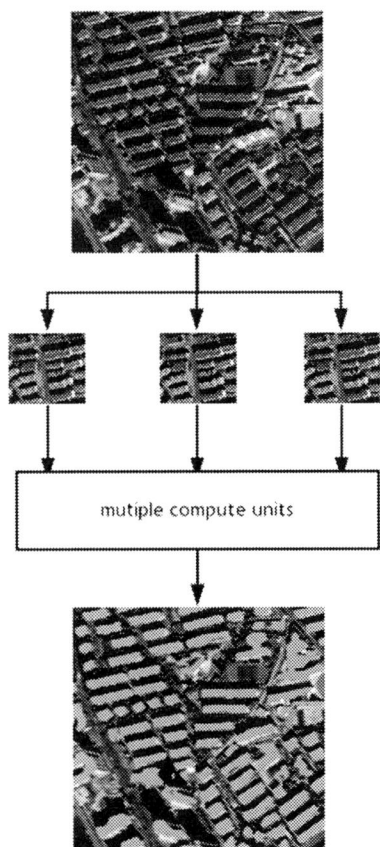

Figure 2. Process procedure.

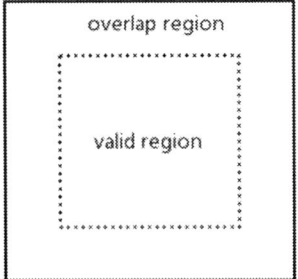

Figure 3. A block of overlap split method.

A block of overlap split method are shown in figure 3. Each block has two kind of region, valid region and overlap region. Although all of the buildings in the block will be extracted, but only extraction result in valid region will be adopted. Overlap region is repeated in two interfacing blocks, which can retain the whole information of buildings in the edge of valid region. Although overlap region is repeated, valid region is unique in different blocks, they can be merged as the final result.

3. Experimental analysis and discussion

In order to verify the effectiveness of the algorithm, we do experiment in Quick Bird high-resolution remote sensing image of Shanghai suburb, and measure results in terms of correctness and completeness.

3.1 Dataset and experiment environment

The experiment is based on the ubuntu16.04 operating system and the open source deep learning framework caffe. We use NVIDIA GTX1080 graphics card for deep learning training, and the 8-core Dell T3600 workstation for various test experiments.

The remote sensing image data used in the experiment is high-resolution remote sensing image acquired by Quick Bird satellite. It has multi-spectral band of 0.6 m resolution and a full-color band of 2.4 m resolution. The training data in the suburbs of Shanghai is about 24 square kilometers. The test data is about 12 square kilometers. The ground truth of the building is manually annotated using ARCGIS 10.1. The test image is shown in figure 4:

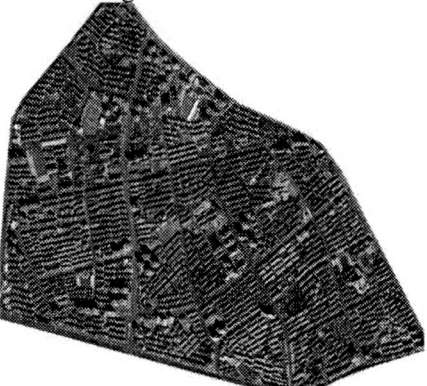

Figure 4.Test image.

3.2 Evaluation method

For remote sensing image extraction buildings, there is currently no uniform accuracy evaluation standard. The literature [17] uses the correctness and completeness to measure the results of the segmentation:

$$correctness = \frac{TP}{TP + FP}$$

$$completeness = \frac{TP}{TP + FN}$$

Where TP represents a positive-class pixel point that is correctly classified, FP is a positive-class pixel point that is misclassified, FN is a negative-class pixel point that is misclassified. The literature [5] has adopted accuracy evaluation standards similar to the correctness rate and completeness rate. Some people also consider the particularity of remote sensing images, and adopt the relaxation accuracy rate and the relaxation recall rate for accuracy evaluation [9] [8], which relaxes the constraint conditions on the conventional accuracy rate and recall rate.

We consider that the extraction of buildings requires the estimation of the area and the extraction of the precise contours in many applications, so we adopt correctness rate and the completeness rate to evaluate experiment.

3.3 result analyze

The test is performed on the above data set by using the network RSBD4 and the overlapping sliding window detection method proposed in this paper. At the same time, in the same data set and environment, the deep learning network segmentation network deeplabV2[10] and the SVM-based object-oriented extraction algorithm and rule-based object-oriented extraction algorithm in ENVI software were selected for comparison experiments. Among them, deeplabV2 selects the best experimental results under the condition of network training convergence. Based on SVM and rule-based object-oriented extraction algorithm, after repeated debugging, the optimal parameters are selected to obtain experimental results. The specific data is shown in Table 1.

Table 1. Comparison between RSBD4 and deeplabv2

Network	correctness	completeness
deeplabV2	78.93%	79.06%
RSBD-4	80.54%	83.18%

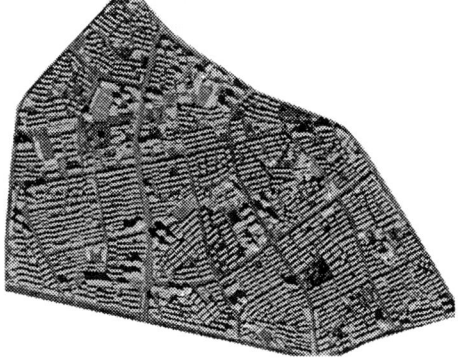

Figure 5. Extraction result

The experimental results show that the correctness rate of RSBD4 reaches 80.54%, and the completeness rate reaches 83.18%. Compared with the deeplabV2, the two have increased by nearly 2 percentage points and 4 percentage points respectively.

By observing the extraction result, we found that compare with deeplabV2, the extracted building edge is more refined. For small buildings, as shown in Figure 6, there is a better extraction result

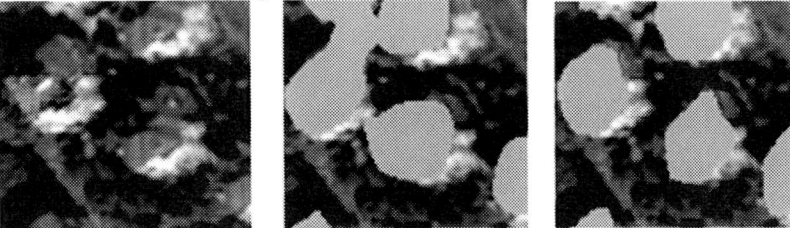

Figure 6. Small building extraction result

4. Conclusion

We build a deep convolutional neural network RSBD4 for the extraction of high-resolution remote sensing image buildings, use the excellent structure of other networks, such as the atrous convolution layer of deeplabV2. We split image into blocks and use distributed computing to accelerate computation. We propose overlap split method to solve the problem of edge information loss caused by splitting the image. The experiment shows that our correctness rate is 80.54% and completeness rate is 83.18%. In next step we can increase the amount of data for training, adjust the network structure, speed up the training time, and improve the efficiency of detection.

REFERENCES

[1] Wang Jun, Qin Qi ming,Ye Xin, et al . A Survey of Building Extraction Methods from Optical High Resolution Remote Sensing Imagery[J].Remote Sensing Technology and Application,2016,31

[2] Li D R, Tong Q X, Li R X, et al. Current issues in high-resolution Earth observation technology. Sci China Earth Sci, 2012, 55: 1043–1051, doi: 10.1007/s11430-012-4445-9

[3] ZHANG Qing yun, ZHAO Dong. Research on Methods of Building Extraction from High Resolution Remote Sensing Images[J] GEOMATICS & SPATIAL INFORMATION TECHNOLOGY, 2015,38(04):74-78

[4] WU Wei, LUO Jiancheng, SHEN Zhanfeng, ZHU Zhiwen. Building Extraction from High Resolution Remote Sensing Imagery Based on Spatial Spectral Method[J] Geomatics and Information Science of Wuhan University, 2012,37(07):800-805.

[5] Xiaoying Jin, Curt H. Davis. Automated Building Extraction from High-Resolution Satellite Imagery in Urban Areas Using Structural, Contextual, and Spectral Information[J] EURASIP Journal on Advances in Signal Processing, 2005, Vol.2005 (14), pp.1-11

[6] Fu Qiankun, Wu Bo, Wang Xiaoqin, et al . Building Extraction and Its Height Estimation over Urban Areas based on Morphological Building Index[J].Remote Sensing Technology and Application,2015,30(1):148 – 154

[7] SHEN Jia jie, PAN Li, HU Xiangyun, Building Detection from High Resolution Remote Sensing Imagery Based on a Deformable Part Model[J] Geomatics and Information Science of Wuhan University 2017,42(09):1285-1291.

[8] Mnih, Volodymyr. "Machine Learning for Aerial Image Labeling." Doctoral (2013).

[9] Xu L, Choy C S, Li Y W. Deep sparse rectifier neural networks for speech denoising[C] IEEE International Workshop on Acoustic Signal Enhancement. IEEE, 2016:1-5.

[10] LIAN Zifeng,,JING Xiaojun, SUN Songlin, HUANG Hai. Multi-Scale Convolutional Neural Network Model with Multilayer Maxout Networks[J] Journal of Beijing University of Posts and Telecommunications. 2016,39(05):1-5+32

[11] Chen L C, Papandreou G, Kokkinos I, et al. DeepLab: Semantic Image Segmentation with Deep Convolutional Nets, Atrous Convolution, and Fully Connected CRFs.[J]. IEEE Transactions on Pattern Analysis & Machine Intelligence, 2016, PP(99):834-848.

[12] YANG Hai ping, SEHN Zhanfeng,LUO Jiancheng and WU Wei, Recent Developments in High Performance GeoComputation for Massive Remote Sensing Data[J] JOURNAL OF GEO - INFORMATION SCIENCE. 2013,15(01):128-136

Speaker identification based on deep learning in FX iDeal system

Yan Ke[1,a], Li Na[2,b], Chen Yutinge[3,c]

[1]Xi'an Jiaotong University No.28, Xianning West Road, Xi'an Shanxi, China +86-15002923953

[2]CFETS Information Technology (Shanghai) Co., Ltd. Building 7,1388 Zhangdong Road, Pudong New Area, Shanghai, China +86-15102117735

[3]CFETS Information Technology (Shanghai) Co., Ltd. Building 7,1388 Zhangdong Road, Pudong New Area, Shanghai, China +86-13564634556

[a]15002923953@163.com [b]lina_zh@chinamoney.com.cn
[c]chenyuting@chinamoney.com.cn

ABSTRACT The FX (foreign exchange) iDeal system, which provides instant foreign exchange information for inter bank traders, with the rapidly increased number of users, there are urgently needs to meet the higher security requirements and improve the compliance of transactions in the inter bank market. In this paper, we make a research on speaker recognition to improve the security of the system. According to the historical sound data sent by the user, we extract the Mel cepstrum features, and then construct the speaker's identification model based on the deep neural network (DNN). Based on the model, we can verify the user. In order to prevent the DNN from over-fitting or falling into the local minimum, we add a dropout mechanism and a dynamic learning rate adjustment algorithm to the network. To maximize the data mining capabilities, we use the restricted Boltzmann machine (RBM) to realize unsupervised training which can make the network initialization parameters are closer to the global minimum, and then optimize the parameters by supervised training. Verification experiment shows that the model can effectively improve the speaker confirmation ability and is more effective than the general BP neural network and GMM model.

1. INTRODUCTION

Nowadays, with the continuous expansion of Chinese bank foreign exchange market, the National Inter bank Funding Center has built a foreign exchange instant messaging system named as FX iDeal, which provides a unified piece of information exchange service for all of bank traders in order to improve the market compliance. As the number of users increases rapidly, the security of the system has received widespread attention, especially in the user chat module which can involve in confidential transaction information interaction. Therefore, ensuring the login user is not confessed is very important. Traditional institutional verification and real-name authentication methods may reveal user information. Human biometrics are non-replicable and voice features are easier to collect. Therefore, the identity authentication based on voice feature has received extensive attention.

The speaker recognition is a biometric identification method which uses voice features, and its implementation is composed of two parts: speaker voice features extraction and the establishment of speaker voice feature model. In terms of voice feature extraction, commonly methods are linear

Content from this work may be used under the terms of the Creative Commons Attribution 3.0 licence. Any further distribution of this work must maintain attribution to the author(s) and the title of the work, journal citation and DOI.
Published under licence by IOP Publishing Ltd

prediction coefficient (LPCC) [1] method, Mel cepstral coefficient (MFCC) [2] method, and traditional methods combined with time-frequency analysis, wavelet analysis, neural network and other extracted acoustic features. In terms of the model building, there are Gaussian mixture models - General Background Model (GMM-UBM) [3], Joint Factor Analysis (JFA) [4] model, and Total Difference Space Factor (i-vector) [5] model. However, with the arrival of big data era, traditional modeling methods are no longer able to obtain information from data effectively. In 2006, Geoffrey Hinton introduced the strategy of "greedy layer-by-layer pre-training" into the neural network [6], which can train larger and deeper networks. From then on, deep learning has received widespread attention from scholars. In 2012, Hinton applied the deep neural network (DNN) to the acoustic model of speech recognition for the first time [7], which verified that DNN had more powerful modeling capabilities and representation features than GMM in large vocabulary continuous speech recognition tasks. DNN becomes the mainstream modeling method for acoustic models [8][9]. However, few scholars use deep learning to construct the speaker recognition model.

In this essay, to improve the security of the FX iDeal system, we aim at the identification problem in the system and research on the speaker recognition technology by considering of stability of the biometric identification. By collecting the voice chat records of the user, the MFCC features can be extracted and the DNN can be built to construct the speaker recognition model. In addition, we use the Boltzmann machine [10] to pre-training the network. To improve the network performance, we add the dropout mechanism [11] and dynamic learning rate adjustment algorithm to the network. To verify the speaker confirmation ability of the DNN training model, we compare it with the GMM-UBM method and the general BP neural network method. In the end, the model will be applied to the FX iDeal system to strengthen the security of the system.

2. User authentication model based on DNN

2.1 The instant messaging system

The FX (foreign exchange) iDeal system is a software platform for various bank traders to provide instant chat services, market display services, regulatory support services, and transaction assistance services. According to the plans of the National Development Department, the associated system of the instant messaging system includes the upstream system, service invocation system, and external user system. If the transaction right is available, users can use the transaction assistance, supervision support and other functions when they log in the system using the unified terminal.

The overall architecture of the system follows three major architectures, including data architecture, application architecture, technical architecture, and six technical layers, including user access layer, access service layer, application service layer, data service layer, technology platform layer and infrastructure layer. All modules of the system include instant messaging module, login authentication module, transaction agent module, field supervision module and interface module. Among them, the instant messaging module provides chat services, including compliance supervision, data analysis, and smart recommendation. User identity authentication in the instant messaging module is one of the important channels to improve the system security. In instant messaging, token authentication is required for each communication. After the verification is successful, the cache is updated, otherwise the communication is rejected. Speaker identification is provided during a chat in the FX iDeal system.

2.2 speaker recognition model based on DNN

The complete speaker recognition system consists of two parts, the first part is the voice feature extraction, and the second part is building the speaker's identification model. Since the MFCC coefficient features are designed in consideration of the principle of human auditory perception, the relationship between the frequency and the actual frequency is shown in equation (1), and has a good characterization ability [12]. This paper extracts MFCC coefficients from user voice data.

$$f_{Mel} = 1127.0\ln(1 + \frac{f}{700.0})$$

(1)

Where f is the physical frequency in Hz.

The deep neural network, used to construct the speaker's identification model, is a multi-layer neural network containing multiple hidden layers, as shown in Figure1. The network input are the voice features. After the hidden layer network, the network output is the verification result of speaker, including two neurons.

At the input layer, the voice features are MFCC coefficient features of one frame or consecutive frames. Taking the 5-frame and 36-dimensional MFCC coefficients as an example, the input layer are 5*36 neurons.

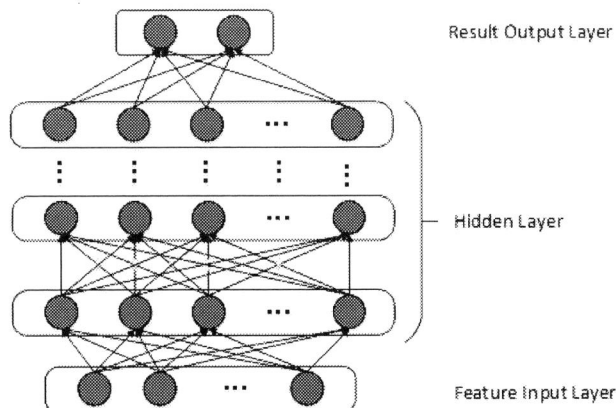

Figure 1 Deep neural network model of speaker identification

In the training phase, first, the network parameter pre-training is performed using a restricted Boltzmann machine, which is an energy model based on the Boltzmann machine. The Boltzmann machine originated from physics, which was simplified and successfully applied to the field of machine learning by Hinton et al. After network pre-training, the network parameters are in a better position. After that, the network training is performed to optimize the parameters. To avoid over-fitting, the dropout mechanism is added which applies a kind of method to set the hidden layer nodes as 0 with a certain probability during training. In addition, in order to prevent network oscillation or slow convergence, a dynamic learning rate adjustment algorithm is added to the network, and the change of the learning rate is determined by the error between the two training sessions in the network, as shown in equation (2).

$$r(i+1) = \begin{cases} \alpha \cdot r(i) & E(i) < E(i\text{-}1) \\ \beta \cdot r(i) & E(i) > E(i\text{-}1) \\ r(i) & other \end{cases} \tag{2}$$

Where, $r(i)$ represents the current learning rate, $r(i+1)$ represents the updated learning rate, $E(i)$ represents the current training error, and $E(i\text{-}1)$ represents the last iteration error.

In the output layer, two neurons are included to characterize the judgment of the confirmation of the speaker. In this paper, the true speaker is represented by outputs 1 and 0, and the false speaker is represented by outputs 0 and 1.

Finally, a feasibility analysis of the network is performed, and the error rate of the model identification is calculated using the formula (3):

$$err = \frac{N}{S} \tag{3}$$

Where N represents the number of output errors, S represents the number of samples.

3. Experimental verification and analysis

The voice test data set used in this paper is the TIMIT voice library established by MIT, and is an authoritative database for speaker recognition. It contains 192 women and 483 male speakers from different parts of the United States. Each speaker has 10 voices and the data set is a total of 6300 voice data. Among them, each voice recording length is about 3-5s, and the acquisition frequency is 16kHz.

In the experiment, n speakers were randomly selected from the speech library, of which 1 was used as the speaker to be recognized, and the remaining n-1 were used as the false speaker. The 8 speech data of each speaker is used as the DNN model test set, and the remaining 2 speech data is used as the DNN model test machine. The 36-dimensional MFCC coefficient feature of each voice is extracted from each speaker's voice, and the voice data is framed by a 256-point Hamming window.

The neural network training takes a frame or consecutive frames of MFCC features as input, and sets the label (0, 1) or (1, 0) to represent the true speaker and the false speaker. In this paper, five consecutive frames are selected, with a total of 5*36=180 features as input; the hidden layer is set to three layers; and the output layer is set to two neurons according to the label. The network structure is 180-100-50-50-2. In this paper, deep neural networks are used to model speaker voice features.The GMM-UBM and the general deep neural network are used to model the speaker, and the results of the three models are compared to verify the validity of the deep neural network model. The error rate is shown in Table 1. From the table we can see that the error rate of deep neural network is lower than GMM-UBM and general deep neural network.

Table1. Error rate based on different model

model	Error rate
GMM-UBM	15.23%
General DNN	13.76%
DNN	8.06%

In addition, in this paper, a dropout strategy is used to deep neural networks to suppress over-fitting of deep neural networks. The experimental results are shown in Table 2. As it can be seen from the table, after adding the dropout strategy, the network has better recognition performance.

Table2. Error rate based on different model

model	Error rate
GMM-UBM	15.23%
General DNN	13.76%
DNN	8.06%
+dropout	6.51%

Theoretically, because the deep neural network is added to the RBM for pre-training, the network parameters can reach an initial optimal position during the initial training of the network. However, compared with the DNN, the general deep neural network parameters are given randomly. Figure 2 shows the iterative process of the first 1800 steps of the two neural networks. It can be seen from the figure that the initial error of the deep neural network added with the RBM is smaller than that of the general deep BP network. Moreover, since the general BP neural network is easy to fall into local minimum points, the network is oscillation. In this experiment, a dynamic learning rate adjustment algorithm is added for the deep neural network, and the degree of oscillation is significantly smaller than that of the BP network. Moreover, the overall error of the deep neural network is small, and the final convergence error is also smaller than the general deep neural network.

The above experimental results show that the deep neural network can effectively reduce the network error rate compared with the traditional speaker recognition models, and the deep neural network has stronger feature representation ability than the general deep neural network. Moreover, the dynamic learning rate adjustment, dropout and other strategies for deep neural networks can effectively optimize network performance.

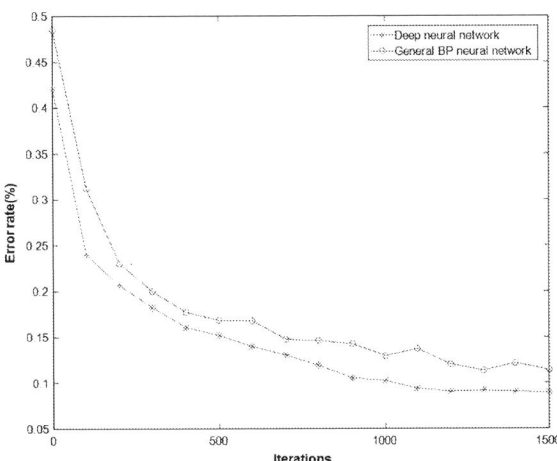

Figure 2 Deep neural network and general BP neural network error iterative process

4. Conclusion

In order to improve the security of the FX iDeal system, this paper uses the deep neural network with dropout strategy and dynamic learning rate algorithm to construct the user voice identification model, and compares it with the general BP neural network and GMM model. The experimental results show that the model has obvious advantages compared with the traditional models, and can exploit the speaker's voice information better. In addition, in the FX iDeal system, a dynamic training strategy can be added to the speaker recognition model by collecting user voice data immediately. Therefore, with the data increases, the recognition performance of the network training model will continuously improve.

REFERENCES

[1]   Atal B S. Automatic recognition of speakers from their voices[J]. Proceedings of the IEEE, 1976, 64(4):460-475.

[2]   Davis S, Mermelstein P. Comparison of Parametric Representations for Monosyllabic Word Recognition in Continuously Spoken Sentences[J]. Readings in Speech Recognition, 1980, 28(4):65-74.

[3]   Reynolds D A, Quatieri T F, Dunn R B. Speaker Verification Using Adapted Gaussian Mixture Models[M]. Academic Press, Inc. 2000.

[4]   Kenny P, Ouellet P, Dehak N, et al. A Study of Interspeaker Variability in Speaker Verification[J]. IEEE Transactions on Audio Speech & Language Processing, 2008, 16(5):980-988.

[5]   Dehak N, Kenny P J, Dehak R, et al. Front-End Factor Analysis for Speaker Verification[J]. IEEE Transactions on Audio Speech & Language Processing, 2011, 19(4):788-798.

[6]   Geoffrey E. Hinton,Simon Osindero,Yee-Whye Teh.A Fast Learning Algorithm for Deep Belief Nets[J].Neural computation,2006,(7):1527-1554.

[7]   Hinton G, Deng L, Yu D, et al. Deep Neural Networks for Acoustic Modeling in Speech Recognition: The Shared Views of Four Research Groups[J]. IEEE Signal Processing Magazine, 2012, 29(6):82-97.

[8]   Richardson F, Reynolds D, Dehak N. A Unified Deep Neural Network for Speaker and Language Recognition[J]. Computer Science, 2015.

[9]   Lei Y, Scheffer N, Ferrer L, et al. A novel scheme for speaker recognition using a phonetically-aware deep neural network[C]// IEEE International Conference on Acoustics, Speech and Signal Processing. IEEE, 2014:1695-1699.

[10] Fischer A, Igel C. An Introduction to Restricted Boltzmann Machines[J]. 2012, 7441:14-36.

[11] Srivastava N, Hinton G, Krizhevsky A, et al. Dropout: a simple way to prevent neural networks from overfitting[J]. Journal of Machine Learning Research, 2014, 15(1):1929-1958.

[12] Wolf J J. Efficient Acoustic Parameters for Speaker Recognition[J]. J.acoust.soc.am, 1972, 51(6).

ISPECE

IOP Publishing

The Improvement of K-NN Classifier with GA-Based Weight-Tunning Method

Wei JIN

State Key Laboratory of Pulsed Power Laser Technology, Electronic Countermeasures Institute, National University of Defense Technology, Hefei, China

kingvee@163.com

ABSTRACT Because the k-Nearest Neighbors(k-NN) algorithm does not take into account different weights values among each feature of data sets, it sometimes achieves low accuracy. In order to increase the k-NN algorithm accuracy, the genetic algorithm-based (GA-based) weight-tunning method is proposed. When the features have been given the same weights values, the k-NN algorithm with EU distance metric is introduced firstly. Secondly, the weighted k-NN algorithm is obtained, when considering different weights values among features. Thirdly, the steps of GA-based weight-tunning method are given in detail. Finally, the experimental tests are running with the real data sets about the students' knowledge status about the subject of Electrical DC Machines. The experimental results show that the weighted k-NN algorithm achieves classification accuracy of 95.2%, whereas the classification accuracy of k-NN algorithm is 86.2%. The GA-based weight-tunning approach for features of data set could play an essential role in improving the k-NN classifier accuracy.

1. INTRODUCTION

k-Nearest Neighbors(k-NN) algorithm is a non-parametric method used for categorization and regression [1]. The basic k-NN algorithm does not take into account differences among each feature of the inputting data set. It may cause low accuracy in some cases. Because in these cases, features have different effects on the classified labels from each other. Weights values are used to measure differences among the features. Therefore, in order to increase the k-NN algorithm accuracy, each feature can be given a weight value.

In this paper, the training and test data sets which are downloaded from the UCI website[2] are the students' knowledge status about the subject of Electrical DC Machines. There are five features and four knowledge levels in the data set. Current challenges are to accurately weight the features of students on their knowledge. The approach that the features are weighted by the weight-tunning method which is proposed in reference [3, 4]. The best weight-values may be obtained by genetic algorithm-based (GA-based) method.

The organization of the paper is as follows: Section 2 presents the details of k-NN algorithm. Section 3 presents the principle and steps for creation of GA-based weight-tunning method. The results of experimental study are presented in Section 4. Section 5 concludes the paper.

2. K-NN ALGORITHM

k-NN algorithm works like this: there is an existing set of example data, also called training set. What labels each sample of training set should fall into are known. It is represented by a Matrix, shown in Eq. (1). There are n features of the training set. In Eq. (1), x_{ij} is the jth feature value of the ith sample

Content from this work may be used under the terms of the Creative Commons Attribution 3.0 licence. Any further distribution of this work must maintain attribution to the author(s) and the title of the work, journal citation and DOI.

Published under licence by IOP Publishing Ltd

of the training set, and l_i is the label value of the ith sample of the training set. A new sample of data with n features and without a label is given, shown in Eq. (2). y is a n-dimensional vector the same as the features number of training set.

$$T = \begin{bmatrix} x_{11} & x_{12} & \cdots & x_{1n} & l_1 \\ x_{21} & x_{22} & \cdots & x_{2n} & l_2 \\ & & \vdots & & \\ x_{m1} & x_{m2} & \cdots & x_{mn} & l_m \end{bmatrix} \tag{1}$$

$$\mathbf{y} = \langle y_1, y_2, \ldots, y_n \rangle \tag{2}$$

The distances among all the samples of training data and the new sample are measured by one mean of distance metric. After that, the sample piece is compared to every sample of training set. The labels of k-number of neighbors are determined if they are the closest to the new sample. For this purpose, the majority-vote method is commonly used to determine the label of the new sample. The three key parameters of the classification process are the integer 'k', training set and distance metric.

There are three metric distance means which are often used including: Eucledion (EU) distance, Manhattan distance and Minkowski distance. However, Eucledion distance is most popular metric distance of the three means. Therefore, EU distance metric is used for measuring the distance between the feature x_i and y, shown in Eq. (3).

$$d(x_i, y) = \sqrt{\sum_{j=1}^{n}(x_{ij} - y_j)^2} \tag{3}$$

The k-NN is used to classify in four steps. The steps are summarized as follows.

i. Determining the k-value (the number of nearest neighbors for the new sample). k is an integer and is usually less than 20.

ii. Calculating the distances between the new sample and the training set with Eq. (3), and saving the distance values $d(x_i, y)$.

iii. Sorting distances $d(x_i, y)$ and determining of nearest neighbors of the new sample. If $k = 3$, the nearest 3 training samples would be obtained.

iv. Making a majority vote among the nearest k training samples in order to determine the label of the new sample y.

3. GA-BASED WEIGHT-TUNNING METHOD

The basic k-NN algorithm has an assumption that every feature is equally important, and reflects the classification results with the same weights values. However, in some cases, the importance of each feature to reflect its relevance for classification is different. The accuracy of k-NN algorithm would be low if each feature has been given the same weights values in these cases. If features are appropriately weighted with an approach, the performance of the k-NN algorithm will not be degraded. The distance metric that is used for the weighted k-NN is a slight variant of the Euclidean metric (Eq. 3), showed in Eq. 4, where w are features weights showed in Eq. 5.

$$d(x_i, y) = \sqrt{\sum_{j=1}^{n} w_j(x_{ij} - y_j)^2} \tag{4}$$

$$\mathbf{w} = \langle w_1, w_2, \ldots, w_n \rangle \tag{5}$$

The genetic algorithm (GA)-based approach for weight-tunning is used to explore the real-valued weights of features. The genetic algorithm is a metaheuristic inspired by the process of natural selection that belongs to the larger class of evolutionary algorithms. Genetic algorithms are commonly used to generate high-quality solutions to optimization and search problems by relying on bio-inspired operators such as mutation, crossover and selection[5].

Two basic steps of GA-based weight-tuning approach coding of individuals in population and updating population are given in following.

i. Coding of individuals: The weight of each feature can be represented in the form of $\langle w_1, w_2, \ldots, w_n \rangle$ with a weight vector. The genes of an individual (chromosome) correspond to the weights and the fitness value of it, showed in Table 1. The last gene (*FV*) is the fitness value of an individual.

Table 1. The genes of chromosome in GA-based weight-tunning method

Gene₁	Gene₂	...	Geneₙ	FV
w_1	w_2	...	w_n	The fitness of individual

ii. Creating and updating of population: The flow diagram that shows creating and updating steps of population is given in Fig. 1.

a. Randomly creating population: h-item individuals are created randomly to constitute the population. The matrix form of population constituting h-item individuals is given in Eq. (6). In Eq. (6), weights values and fitness values (*FV*) are in the interval [0, 1].

$$P = \begin{bmatrix} w_{11} & w_{22} & \cdots & w_{1n} & FV_1 \\ w_{21} & w_{22} & \cdots & w_{2n} & FV_2 \\ & & \vdots & & \\ w_{h1} & w_{h2} & \cdots & w_{hn} & FV_h \end{bmatrix} \tag{6}$$

b. Calculation of fitness value: It assumes that num_e denotes the number of misclassified samples corresponding to any individual in the set of *P*. Fitness value of this individual is calculated as $FV = 1 / num_e$. Fitness values of all individuals in population are calculated similarly.

c.Termination criteria: In this process, the best individual providing the genetic reproduction finalization criteria among individuals in *P* population is explored. Termination criteria might be a definite fitness value or a generation number. After the termination of reproduction, the weights of having the best fitness value individual in *P* will be output as the best weight values of population.

d. Selection of parents: Each individual in population represents a solution for problem. Reproduction operator is used to create new individuals in population.The methods of 'Roulette Wheel' will be implemented to select the parents in this paper.

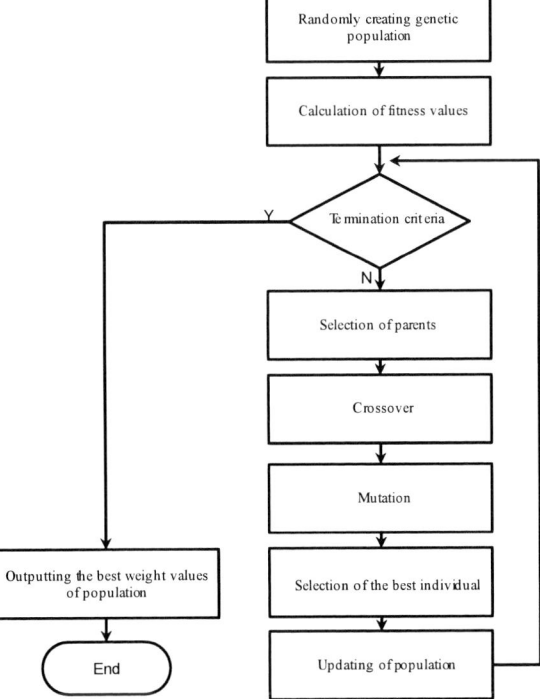

Figure 1. The flow diagram of creating and updating of population.

e. Crossover: It is process that the execution of permutation process vice verse among the genes of children. Thanks to this, the genetic codes of individuals have been changed. In this paper, the crossing methods 'Flip bit' is implemented.

f. Mutation: The genetic codes of child individuals are changed. It is used to maintain genetic diversity in P population. In this study, the random-mixed of mutation methods named 'single point mutation' are used.

g. Selection of the best individual: It is that the individual of the best fitness value through children is selected.

h. Updating of population: It is that the individual of the lowest fitness value in population is replaced by the best individual.

When the best individual has been obtained, the weights values will be used in Eq. (4) to implement the weighted k-NN algorithm.

4. EXPERIMENTS

The training set and testing set are the real data sets about the students' knowledge status about the subject of Electrical DC Machines. The number samples of training set and testing set are 258 and 145 respectively. There are 5 features(STG, SCG, PEG, STR, LPR), and what they mean is shown in Table 2. The labels which have 4 categorical values ('very low', 'low', 'middle', and 'high') represent the knowledge level of user.

Table 2. The data set samples used for experiments

STG	SCG	STR	LPR	PEG	Labels
The degree of study time for goal object materials	The degree of repetition number of user for goal object materials	The degree of study time of user for related objects with goal object	The exam performance of user for related objects with goal object	The exam performance of user for goal objects	The knowledge level of user

The termination criteria of GA-based weight-tunning is set to be the fitness value 0.2 or the generation number 15,000. It means when fitness value of any individual is 0.2 or when the number of generation reaches to 15,000, the weights of individual having the best fitness value will be regarded as the best weights values, and then the updating program of population will be ended. Total number of the created population is 100.

Because the number of features is 5, the number of optimum weights values which may be shown as a vector$\langle w_{STG}, w_{SCG}, w_{STR}, w_{LPR}, w_{PEG}\rangle$ is also 5. The experiments have been performed by the k-NN algorithm under the condition that k value is set to 3. The GA-based weight-tunning has searched the 5 optimum weights values for the 5 features of training set. The curves of searching process is presented in Fig. 2. The maximum fitness of individual in P is about 0.14 when the generation number is updated to over 10,000.

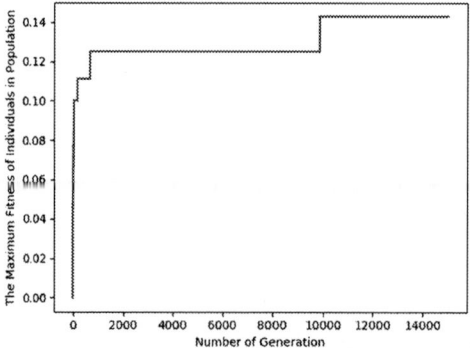

Figure 2. The curve of searching process

Table 3. The best weights values for $k = 3$

W_{STG}	W_{SCG}	W_{STR}	W_{LPR}	W_{PEG}
3.395e-54	2.639e-03	7.064e-60	0.412	0.679

The best weight values are in Table 3. When the values are analyzed, it is seen that W_{PEG} and W_{LPR} have larger weights values than the three others. It shows that the two features of students (PEG, LPR) are considerably more efficient than the other three features (STG, SCG, STR) on the knowledge label of users.

One experimental comparison of the knowledge classifiers has been provided. After training on 258 samples, the average number of misclassified samples, the percentage of average error rates and the average classification accuracy of two knowledge classifiers are measured for 145 testing samples. The weighted k-NN algorithm achieves an average classification accuracy of 95.2% over the validation set, whereas the classification accuracy of k-NN algorithm is 86.2%. Consequently, the weighted k-NN algorithm has produced a considerable improvement: the rate of improvement is 9% for k-NN classifier.

Table 4. Comparing the performance of two classfifers

	k-NN algorithm	Weighted k-NN algorithm
The number of misclassified samples	7	20
The percentage of error rates	4.8%	13.8%
The classification accuracy	95.2%	86.2%

5. CONCLUSION

This paper presents a powerful and efficient genetic algorithm-based weight-tunning approach to improve the accuracy of k-Nearest Neighbors classifier. The experimental studies have shown that when the optimal weights values which are searched by the genetic algorithm-based weight-tunning approach are put into the distance metrics, the classification accuracy of k-Nearest Neighbors will be improved 9 pp. It shows that genetic algorithm-based weight-tunning approach for features of data set could play an essential role in improving the k-NN classifier accuracy.

ACKNOWLEDGMENTS

My thanks to UCI for allowing me to use their data sets freely.

REFERENCES

[1] http://en.wikipedia.org/wiki/K-nearest_neighbors_algorithm.
[2] http://archive.ics.uci.edu/ml/datasets/User+Knowledge+Modeling.
[3] Kelly J D, Davis L. 1991. A hybrid genetic algorithm for classification. *Proceeding of the International Joint Conference on Artificial Intelligence.* (1991): 645-650.
[4] H. T. Kahraman, Sagiroglu, S., Colak, I. 2013. Developing intuitive knowledge classifier and modeling of users' domain dependent data in web. *Knowledge Based Systems.* 37 (2013): 283-295.
[5] Bies, Robert R., Muldoon, Matthew F., Pollock, Bruce G., Manuck Steven, Smith Gwennl, Sale, Mark E. 2006. A genetic algorithm-based, hybrid machine learning approach to model selection. *Journal of Pharmacokinetics and Pharmacodynamics.* (2006): 196–221.

A Route Optimization Model Based on Link State Awareness in SDN

Ran Li[1,a,*], ZhaoWen Lin[2,b], Jie Xu[3,c], Yi Sun[4,d]

[1]Network and Information Center, Institute of Network Technology, Beijing University of Posts and Telecommunications Beijing 100876, China 2 Science and Technology on Information Transmission and Dissemination in Communication Networks Laboratory Shijiazhuang 050081, China 3 National Engineering Laboratory for Mobile Network Security Beijing 100876, China

[2]Network and Information Center, Institute of Network Technology, Beijing University of Posts and Telecommunications Beijing 100876, China 2 Science and Technology on Information Transmission and Dissemination in Communication Networks Laboratory Shijiazhuang 050081, China 3 National Engineering Laboratory for Mobile Network Security Beijing 100876, China

[3]Network and Information Center, Institute of Network Technology, Beijing University of Posts and Telecommunications Beijing 100876, China 2 Science and Technology on Information Transmission and Dissemination in Communication Networks Laboratory Shijiazhuang 050081, China 3 National Engineering Laboratory for Mobile Network Security Beijing 100876, China

[4]Network and Information Center, Institute of Network Technology, Beijing University of Posts and Telecommunications Beijing 100876, China 2 Science and Technology on Information Transmission and Dissemination in Communication Networks Laboratory Shijiazhuang 050081, China 3 National Engineering Laboratory for Mobile Network Security Beijing 100876, China

[a]zendlee@163.com, [b]linzw@bupt.edu.cn, [c]cheer1107@bupt.edu.cn, [d]sybupt@bupt.edu.cn

ABSTRACT In addressing the routing issue in the SDN environment, an improved ant colony algorithm is proposed in this paper, in which the ants tend to balance the path with less pheromone. Besides, the Top-K optimal path algorithm is used to optimize the evaluation set, speed up the evaluation, collect the status of the link in the network, and dynamically adjust the evaporation speed of the pheromone to ensure the real-time accuracy of the routing strategy. The QoS requirements of the service are also guaranteed according to the collected link QoS requirements. In such a scheme, combining the advantages of probabilistic routing and deterministic routing, path optimization is better achieved. The simulation of the traffic shows that the model can significantly improve the throughput of traffic in the SDN network, reduce the packet loss rate, and effectively guarantee load balancing.

Content from this work may be used under the terms of the Creative Commons Attribution 3.0 licence. Any further distribution of this work must maintain attribution to the author(s) and the title of the work, journal citation and DOI.
Published under licence by IOP Publishing Ltd

1. INTRODUCTION

In traditional networks, how to better optimize network forwarding paths has always been a problem for scholars. Software Defined Networking was originally a new network architecture for control and forwarding plane separation proposed by Professor Nick McKeown of Stanford University and others around 2009 [1]. The architecture has the advantages of flexible routing policy control and centralized network topology sensing [2]. Taking advantage of these advantages, we can better calculate and deliver the path forwarding strategy.

In the SDN environment, the default algorithm used by most controllers is based on the Dijkstra algorithm with the shortest hop count. When the controller selects the forwarding path, it always calculates the shortest path from the source host to the destination host in the current network environment. When the load on the network is relatively large, the delay and packet loss rate increase due to congestion on some links on the shortest path, while the link utilization rate on other non-shortest paths is lower. The path is not optimal. In response to this, scholars have proposed many classic routing algorithms, such as ant colony algorithm [3], genetic algorithm [4] and so on. The ant colony algorithm, as a bionic algorithm, has achieved good results in solving the routing problem of SDN networks [5][6]. The adaptive ant colony algorithm proposed by Stefano AD [7] et al. initially implements load balancing by adapting the ant colony algorithm to the SDN network. However, under actual conditions, the current link state and QoS requirements of different traffic are in the path. It should also be taken as a factor in consideration.

Based on the above problems, this paper improves the ant colony algorithm by designing a path optimization model under SDN, and uses the advantages of SDN to filter the Top-K optimal path through QoS requirements and link state to speed up the evaluation of the algorithm. Perceived changes in link state Dynamically adjust the volatilization coefficient of the pheromone, so that the SDN network achieves better results in the QoS requirements and load balancing.

2. ANT COLONY ALGORITHM PRINCIPLE

In the process of foraging, ants express a better path by secreting pheromones of different intensities, and later ants tend to choose more pheromone paths, resulting in more and more ants on the optimal path. The ants on other paths will gradually decrease over time and eventually disappear, thus forming an optimal path. Inspired by this, an ant colony algorithm was proposed.

The transition probability of ants in each node in the traditional ant colony algorithm is:

$$p_{ij}^k = \frac{\tau_{ij}^a \eta_{ij}^\beta}{\sum_{j \in L} \tau_{ij}^a \eta_{ij}^\beta} \quad (2.1)$$

Where k is the label of the individual ant colony, i is the current position of the ant colony, and j is the location that the ant colony can select. p_{ij}^k indicates the probability that the ant k in the i position selects the j position, and L is the set of the next optional position of the ant. τ_{ij}^α indicates the pheromone concentration between i and j. When an ant passes a point or passes all points, it implements an update strategy for the pheromone of the node concerned. The pheromone concentration update strategy can be expressed by the following formula:

$$\tau_{ij}(n+1) = \rho \times \tau(n) + \sum_{k=1}^{m} \Delta \tau_{ij}^k \quad (2.2)$$

ρ represents the evaporation rate of the pheromone on the path. According to the concept of ant colony algorithm, combined with the advantages of SDN, we propose a path optimization model based on link state perception in SDN environment.

3. PATH POTIMIZATION MODEL

In the SDN environment, unnatural ant colony algorithm by improved ant colony algorithm, the data packet is regarded as an ant and the load on the network link is regarded as a pheromone, so that the ant tends to a link with a lower pheromone concentration. At the same time, according to the collected link state, the volatilization speed of the pheromone is adjusted, and the optimization of the path evaluation set is performed by using the Top-K optimal path.

In section 3.1 we describe the architecture of the entire path optimization model. In section 3.2 we describe the unnatural ant colony algorithm, in section 3.3 and section 3.4 we optimize the path model by optimizing the ant colony algorithm's evaluation set and dynamically changing the volatilization velocity of the link pheromone, and in section 3.5 we give a typical route strategy solving process.

3.1 ARCHITECTURE

The path optimization model is divided into the following four modules:

- Pheromone management module (PMM). Collection and update of pheromones on various links in the network
- Network link information collection module (NIM). Collect information about each link in the network, such as delay, packet loss rate, jitter rate, and bandwidth.
- QoS requirement registration module (QRM). Register the QoS requirements of the service traffic to the system as a reference for delivering the corresponding routing strategy.
- Routing strategy formulation and delivery module (RDM). The QoS of the registered service traffic is used as a reference when the corresponding routing strategy is delivered.

The architecture diagram is shown in Figure 1:

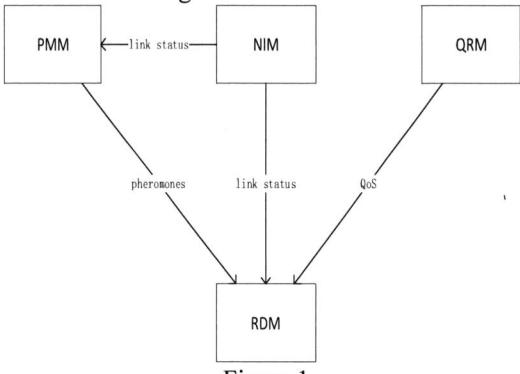

Figure 1

3.2 Unnatural ant colony algorithm

Unlike the standard ant colony algorithm, the unnatural ant colony algorithm is more like an unnatural behavior, which forces the ants to move along all of their covered paths to achieve load balancing. That is, avoid trying the path with the strongest pheromone and exploring the path with the weakest pheromone. We treat packets as ants and treat the load on the network link as pheromones. Optimize network performance in terms of throughput, communication latency and packet loss. Through this strategy, routing algorithms can reduce network congestion, improve throughput and low latency, and ensure even distribution of load across all links.

In the initial phase we put an equal amount of pheromone for each link. Then the probability that the ant selects the line ij at node i is:

$$p_{ij}^k = \prod_{j \in L} \frac{\tau_{ij}^a \eta_{ij}^\beta}{D} (D = \sum_{j \in L} \tau_{ij}^a \eta_{ij}^\beta) \quad (3.1)$$

L represents all candidate links. We reduce the evaluation range of the ant colony algorithm by designing the Top-K optimal path algorithm to determine the set L of optional links.

3.3 Top-K optimal paths algorithm

KC Abbaspour et al. [8] have proposed that pre-processing of pheromone is not applied before the actual operation of the ant colony algorithm, a certain number of paths are found by preprocessing, and then the route is selected according to the pheromone, so that a better path selection can be obtained. effect. So, we use Top-K optimal path algorithm for pre-optimization. First, the path of the first N(N > K) shortest hops is obtained by implementing the Yen algorithm [9], and then by querying the QoS requirements of the service registration, we select the QoS requirements that satisfy the service, and at the same time relatively better. K paths as an evaluation set L for the unnatural ant colony algorithm.

3.4 Pheromone volatilization strategy based on link state

The pheromone volatilization strategy in the ant colony algorithm determines the convergence speed of the algorithm. This paper uses the network status of the NIM to dynamically change the pheromone volatilization speed of the corresponding link. Because the concentration of pheromone in the link in the unnatural ant colony algorithm does not reflect the state of the current link, we set different volatilization speeds depending on the state of the link. The volatilization speed of the path pheromone with a good link state will be faster. After evaluation, this paper formulates the volatility of the pheromone in each link as the formula:

$$\rho = \max\{10 \times \varphi, 10 \times \lambda, \beta\} \quad (3.2)$$

φ indicates the jitter rate of the link, λ indicates the packet loss rate of the link, and β indicates the bandwidth usage of the link.

3.5 Routing strategy solving process

Assume in our SDN network architecture as shown in Figure 2:

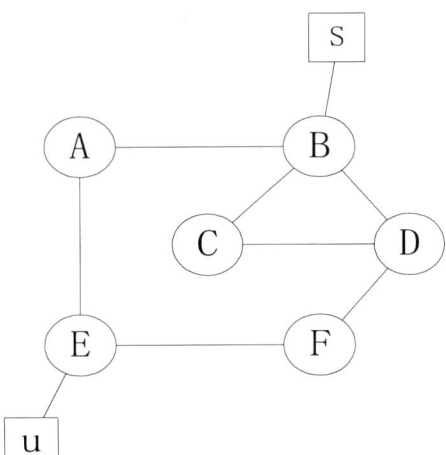

Figure 2

A, B, C, D, E, F are OpenFlow switches. U is the user and S is the service provider. Now we want to import the traffic of user U into service S. The routing strategy solving process is as follows:

(1) QoS requirements for user registration services.

(2) When the first data packet of user U reaches E, there is no matching forwarding rule, and the controller is queried to obtain a forwarding rule. At this point, the controller uses the Top-K optimal path algorithm to select the Top-K optimal path that meets the user's QoS requirements. Because the network size in the graph is small, it is assumed that all reachable paths are selected by us as the optimal path. Then the three candidate paths are E-A-B, E-F-D-B, E-F-D-C-B. Then, L in the unnatural ant colony algorithm is the link in the three paths.

(3) The pheromone concentration of the E-links E-A and E-F of the E-node in the L-set is obtained from the pheromone management module(PMM), and the probabilistic route is selected according to

the formula (3.1). Suppose we have chosen link E-F. In order to prevent the loop from appearing, the link is placed in the taboo table after each link, that is, we put E-F into the taboo table. When the data packet arrives at the switch D, there are two alternative links D-B and D-C in the set L. At this time, we query that B is the destination switch, so, the link D-B is directly selected as the forwarding path.

(4) When the data packet arrives at the switch B, a route from the user U to the network service S is formed in the network.

(5) After that, we use the global update strategy to update the pheromone concentration. The pheromone management module(PMM) detects the number of data packets transmitted through the path, and generates a corresponding backward ant to release the pheromone on the link along the path. At the same time, the pheromone of each link along the way is evaporated according to the volatilization coefficient given in formula (3.2).

4. PERFORMANCE TESTING

Set up the network topology shown in Figure 3 under Mininet and use software named iperf for traffic simulation.

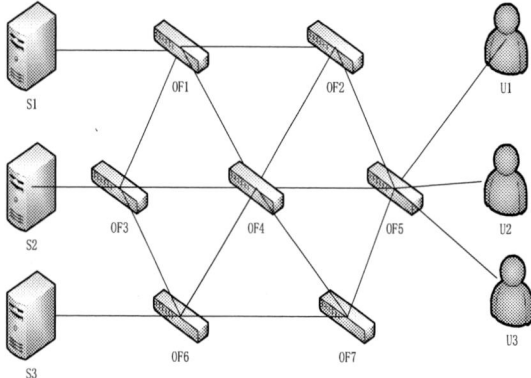

Figure 3

We set the bandwidth of each link to 2Mb/s and continuously increase the load in the link. The path optimization model is compared with the traditional SDN based Dijkstra algorithm with the shortest hop count. The average packet loss rate of the entire network as shown in Figure 4:

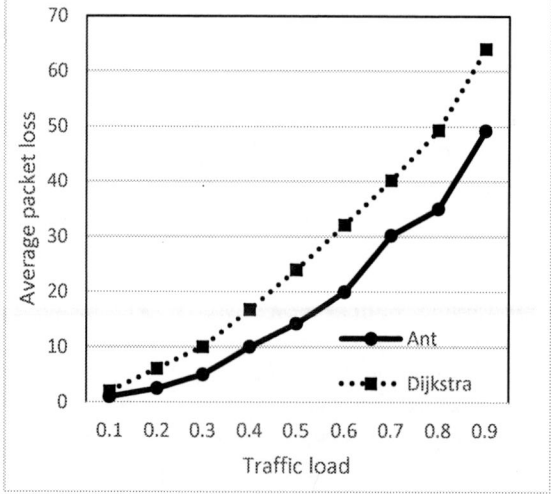

Figure 4

It can be seen that compared with the traditional Dijkstra algorithm based on the shortest hop count, the proposed model can better reduce the packet loss rate in the network and increase the throughput in the network, so that the whole network has a good stability.

At the same time, when we set the bandwidth of a link such as OF1-OF2 to 0.5 Mb/s, the load balancing performance of the path optimization model when the link state is poor is simulated. Because we optimize the pheromone volatilization speed based on the current link state, compared to the model proposed in [7], load balancing can be achieved according to the link state.

The experimental results show that our proposed path state-aware path optimization model achieves our expected results.

5. CONCLUSION
In the scenario of SDN network architecture, this paper proposes a path optimization model based on the unnatural ant colony algorithm to detect the link state of the network so as to load balance the traffic in the network, the QoS requirements of users are taken into account at the same time.

ACKNOWLEDGEMENT
This work is supported by the National Natural Science Foundation of China (Grant no. 61601041), the Fundamental Research Funds for the Central Universities (Grant no. 2018RC55), and the Beijing Talents Foundation (Grant no. 2017000020124G062).

REFERENCES
[1] McKeown N. Software-defined networking[C]. 1NFOCOM keynote talk.2009, 17(2): 30.2.Podani, J. (1994) Multivariate Data Analysis in Ecology and Systematics. SPB Publishing, The Hague.
[2] Monsanto C, Reich J, Foster N, et al. Composing software-defined networks[J]. 2013.
[3] Colorni A. Distributed Optimization of Ant Colonies[J]. Toward A Practice of Autonomous Systems, 1991.
[4] Booker L B, Goldberg D E, Holland J H. Classifier systems and genetic algorithms[J]. Artificial Intelligence, 1989, 40(1):235-282.
[5] Lin W C, Zhang L C. The Load Balancing Research of SDN based on Ant Colony Algorithm with Job Classification[C]. The Workshop on Advanced Research & Technology in Industry Applications. 2016.
[6] Rodr, guez-P, Rez M, et al. An ant colonization routing algorithm to minimize network power consumption[J]. Journal of Network & Computer Applications, 2015, 58(C):217-226.
[7] Stefano A D, Cammarata G, Morana G, et al. A4SDN - Adaptive Alienated Ant Algorithm for Software-Defined Networking[C]. International Conference on P2p, Parallel, Grid, Cloud and Internet Computing. IEEE Computer Society, 2015:344-350.
[8] Abbaspour K C, Schulin R, Genuchten M T V. Estimating unsaturated soil hydraulic parameters using ant colony optimization[J]. Advances in Water Resources, 2001, 24(8):827-841.
[9] Yen J Y. FINDING THE K SHORTEST LOOPLESS PATHS IN A NETWORK[J]. Management Science, 1971, 17(11):712-716.

Numerical simulation of deep learning algorithm for gas explosion in confined space

Li Qizhong[1][2]*, Wang Ye[2], Yangjia[2], Wang Zhongqi[1]

[1] Beijing Institute of Technology, Beijing, China

[2] North China Institute of Science and Technology, East Beijing, China

Email of all the authors:

* Corresponding Author: Li Qizhong; email:30307575@qq.com; phone: 86 10 61590691; fax: 86 10 61590691.

Abstract: With the continuous innovation and development of science and technology, the use of digital Internet and computer and the application of scientific algorithms is becoming more and more extensive. With the development of the computer information technology revolution, various Internet information has been widely spread and developed. According to the data information transmitted above, various intelligent algorithms can be found in the data information network. Based on the depth learning algorithm, the simulation model of the numerical value of gas explosion in restricted space was studied and explored in this paper. The solution of practical problems can be achieved by building a function information model. In this paper, the deep learning algorithm and the various utilization and calculation methods of deep learning algorithm were introduced.

1. INTRODUCTION

Under the impetus of the revolution of science and technology, the development of computer algorithm is getting faster and faster. The computer algorithm can be used to simulate many kinds of experimental forms. The calculation model and the final result can be obtained by the unpractical experiment(Xu Y Let al 2016) [1].In the process of simulation, what we can find is that the computer algorithm has a natural help to build the computing environment. According to the continuous interpretation and research of the algorithm model, the numerical simulation model of gas explosion in restricted space is built for the depth learning algorithm (Kaloudis Eet al 2016) [2].The study of gas explosion in confined space is to study the possibility of gas concentration and gas explosion in the closed space of the mine or underground construction. The possibility of the possible explosion in the closed space is deduced according to the concentration of gas data (Zheng Cet al 2016) [3].On the basis of the morphological simulation of different data models, different airtight spaces produce different simulation results. The gas concentration in the restricted space has a very high correlation performance. According to the precise operation and data matching, the difference between the gas concentration retrieval method in the first space and the actual retrieval method can be found (Sung K Het al 2016) [4].According to the research data, under the confined space, the structure stability is not good and the gas explosion accident is easy to happen. Therefore, in the process of research, we should focus on improving and dealing with the problem of gas concentration and gas explosion. This is the key to our current work (Li Let al 2016) [5].

Content from this work may be used under the terms of the Creative Commons Attribution 3.0 licence. Any further distribution of this work must maintain attribution to the author(s) and the title of the work, journal citation and DOI.

Published under licence by IOP Publishing Ltd

2. STATE OF THE ART

At the end of the last century, a series of intelligent data algorithms have come into being. These intelligent data algorithms are based on the retrieval and processing of some of the factors and the basic content of the data (Yousefi-Lafouraki Bet al 2016) [6].The key global information operation in the data influencing factors is found. According to the model foundation, the characteristics of the data model are found to describe the influence of gas concentration on the possible height of gas explosion in confined space (Loft N J Set al 2016) [7].In observing this part of the search information, for data mining computing algorithms, the realization of the subjective data impact and the exclusion of factors is a problem to be overcome in front of us (Zanotti Oet al 2016) [8].As to the speed of data removal and influencing factors of intelligent algorithms, the operation of intelligent data algorithm is more simple than that of our human beings. However, the depth learning algorithm in it makes global observation and analysis of the research graphics and research data functions. According to the results of observation and analysis, the problem of building the actual model in the learning algorithm is found. In the research process of the depth learning algorithm, the construction steps and the construction process of the neural network algorithm are mainly used to consider the data to eliminate the influence of the interference. In the use of a human brain - like processing and operating system (Feldgun V Ret al 2016)[9].There are many advantages and advantages of the traditional computing algorithm, which can solve the actual factors that face the model and influence (Michael Let al 2016) [10].The foundation of deep learning algorithm is to reduce manual training and use computer algorithms to perform operations.

3. METHODOLOGY

3.1 Operation process analysis of deep learning algorithm and deep learning algorithm

The research direction of deep learning algorithm is actually using deep learning theory to study the construction and integration of arithmetic models. The artificial intelligence algorithm is used to simulate the computing and learning functions of human brain. The computer is given the function of the actual learning and analysis of the problem. In the process of the operation of the depth learning algorithm, the algorithm type and operation order of the depth learning algorithm are considered by using mathematical factors to make a comprehensive comparison and analysis of the data. Combining the theoretical knowledge of the deep learning algorithm, the operation process of the deep learning algorithm can be got. The development of deep learning algorithms originated from Japan. Great breakthrough has been made in technology, and the use of deep learning algorithm in everyday life has begun to be applied in all aspects. The deep learning algorithm can be used to perform comprehensive horizontal operation and result analysis functions. In this paper, a numerical simulation model of gas explosion in confined space is constructed by using the deep learning algorithm. In the process of the actual reference depth learning algorithm, the content of gas concentration in the restricted space is monitored and grasped, and the saturation point of the explosion is the most important aspect of the use of the depth learning algorithm. The successful development of the deep learning algorithm is based on the renewal and perfection of the theory of deep learning algorithm. The guidance and operation experience of the deep learning algorithm have been mastered, which is of great significance to the practical problem of solving the operation and the construction of the data model. In the process of constructing and studying the depth learning algorithm model, the research and analysis of the numerical simulation model of the gas explosion in restricted space based on the depth learning algorithm can be carried out with different function formulas and mathematical calculation models. Considering the problem of data processing, the model and the basic calculation method based on the depth learning algorithm are considered as the steps in the algorithm. In the subsequent improvement steps, the data model and processing problem of the pre designed model algorithm are carried out. The problem of solving the world is regarded as the first problem that the algorithm needs to overcome. This expression can reflect the important characteristics of the depth learning algorithm. The operation feature of the model is as follows.

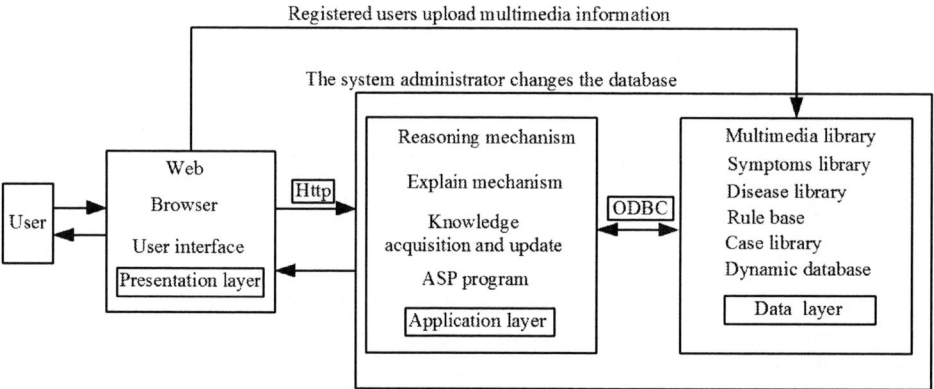

Figure 1. Computer computing model of information delivery network

The depth learning algorithm is an algorithm model that uses a variety of discrimination and understanding measures to make setbacks. According to the operation type of the depth learning algorithm and the classification of data simulation, the difficulty of the overall operation data and operation of the data model algorithm can be reduced. For the operator of the deep learning algorithm, the new algorithm can satisfy the operator's overall operation experience. For the improvement of different operating data, many methods of operation and overall operation test are needed. In the process of developing these deep learning algorithms, the actual operation and test of the algorithm pattern and model are carried out.A variety of physical data need to be used to build the computing process. At the same time, not only the theory of depth learning algorithm and the combination of the data model needs to be updated, but also the solution of the entity model and the updating of the algorithm theory need to be done. Based on the computational data theory of the depth learning algorithm, the analysis and influence process of the operation steps in the depth learning algorithm are carried out. According to the analysis of the overall operation steps in the depth learning algorithm, the steps of the input and output of the algorithm of depth learning algorithm need to carry out the process of continuous operation and process. In the calculation of the depth learning algorithm, the overall data influence and the reference factor of the depth learning algorithm are influenced and analyzed. The decomposition and influence of this part of the algorithm are more important to the influence of the input data of the depth learning algorithm. According to the independent analysis of the depth learning algorithm and the process feedback of the operation, the feedback data factors can be found between the various algorithm types and the calculation results of the algorithm in the depth learning algorithm. Considering the analysis and input process of the feedback data factors and data influence factors, the operation of the model algorithm and function can be carried out according to the various calculation results in the depth learning algorithm. In addition, the deep learning algorithm is decomposed and utilized according to the feedback information of calculation results and calculation factors in the deep learning algorithm. According to the classification of the algorithm model, there is dozens of algorithms for classification. In this paper, because of the limited space, the decomposition process of the depth learning algorithm cannot be analyzed and analyzed one by one. Different algorithms should be used to complete the actual operation process. The calculation frequency and calculation method which consider the depth learning algorithm has different effects on our overall and distribution calculation results. According to the operation and analysis of the information model, the learning structure distribution map of the depth learning algorithm is divided into the following stages.

ISPECE   IOP Publishing

IOP Conf. Series: Journal of Physics: Conf. Series **1187** (2019) 042034   doi:10.1088/1742-6596/1187/4/042034

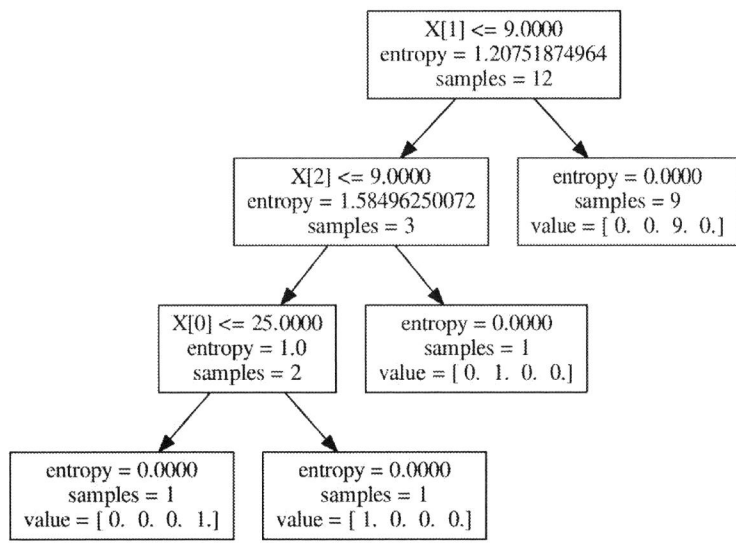

Figure2. Three-layer wavelet packet decomposition tree legend

In the structure diagram of the above learning algorithm, we can learn that the existence of data and influence factors in the depth learning algorithm substitutes and changes the digital meaning of the number and the alphabet representation. According to the different classification numbers and letters, the numbers and letters have different meanings. Under this part of the operation, the impact of the whole data is reflected in the final result. Based on the data model of the algorithm and the result analysis of the operation data degree and influence degree of the alphabet, the problem of the selection of the digital and alphabet influence structure and the data function in the whole calculation process are got. According to the different stratification of selected data functions, the influence degree of different good level data is obtained. The existence of this degree of influence weakens the influence degree data in the depth learning algorithm.The depth learning algorithm becomes more and more complete when the influence degree is more and more perfect. Deep learning algorithm can improve his overall calculation accuracy and data operation ability as much as possible. The research on the use of depth learning algorithms is initially in Japan. Japan has been in the forefront of the world in the intelligent system and intelligent operation model. The content of which includes many aspects, not only the use of data research and use of data and simulation operation.

3.2 Numerical simulation model of gas explosion in confined space based on deep learning algorithm
After the introduction of the deep learning algorithm is completed, an explosion numerical simulation model begins to be built based on the deep learning algorithm. In the process of building the simulation model, the control and selection of simulation model data should be considered, and the selection and control of this part is introduced in detail. The following is the selection and calculation of the simulation data model. After the completion of the computation node, the calculation steps of the depth learning algorithm are carried out according to the actual computing nodes. The detailed steps of the calculation are shown below.

$$m = \sqrt{n+l} + \alpha \qquad (1)$$

m is the number of hidden nodes. n is the number of nodes in the input layer. l is the number of nodes in the output layer. α is a constant, between 0 and 10. However, the general formula can only

1497

be used as a reference and cannot be used directly. Just give us a range of calculations when making calculations.

The calculation formula of the impulse response function is as follows:

$$\psi_{\alpha,\ \tau} = \frac{1}{\sqrt{\alpha}} \psi\left(\frac{t-\tau}{\alpha}\right) = g(t-\tau)e^{-j\alpha t}$$

(2)

In addition, set up $\alpha = \alpha_0^m, \tau = n\tau_0\alpha_0^m$, the upper form can be interpreted as follows:

$$\psi_{m,n}(t) = \alpha_0^{-m/2}\psi\left(\frac{t-n\tau_0\alpha_0^m}{\alpha_0^m}\right)$$

(3)

A neural network algorithm is called the two algorithms. It is usually used to take $\alpha_0 = 2,\quad \tau_0 = 1$ in actual application, which is equivalent to making it discrete on the scale only in the process of filtering continuous neural network algorithm, but its displacement is continuously changed. The two neural network algorithms are shown as the following.

$$\psi_{m,n}(t) = 2^{-m/2}\psi\left(2^{-m}t - n\right)$$

(4)

In addition, the calculation of information decomposition is also required. When calculate, the calculation information of the incoming algorithm is decomposed into two forms of high frequency and low frequency calculation. The calculation formula of the decomposition is as follows:

$$f(t) = f^s(t) + f^d(t)$$

(5)

$f^s(t)$ in the upper form is low frequency. $f^d(t)$ represents high frequency. After a brief analysis of the calculation process of the deep learning algorithm, the complete model of the deep learning algorithm needs to be demonstrated.

4.RESULT ANALYSIS AND DISCUSSION

After introducing the basic calculation process of deep learning algorithm and depth learning algorithm, the simulation and analysis of the operation of the depth learning algorithm and the calculation process of the algorithm are the comprehensive effects of the operation of the algorithm data model of the depth learning algorithm and the overall data and the simulation test environment. According to part of operation and analysis process of the deep learning algorithm, the accuracy and the actual reference degree and influence degree are found in the data model. The numerical simulation model of gas explosion in limited space based on depth learning algorithm is continuously compared and analyzed on the basis of the correctness of the final results of the guaranteed depth learning algorithm. The complete data model is shown as follows.

Serial number	Record
1	A、B、C
2	A、C
3	B、C

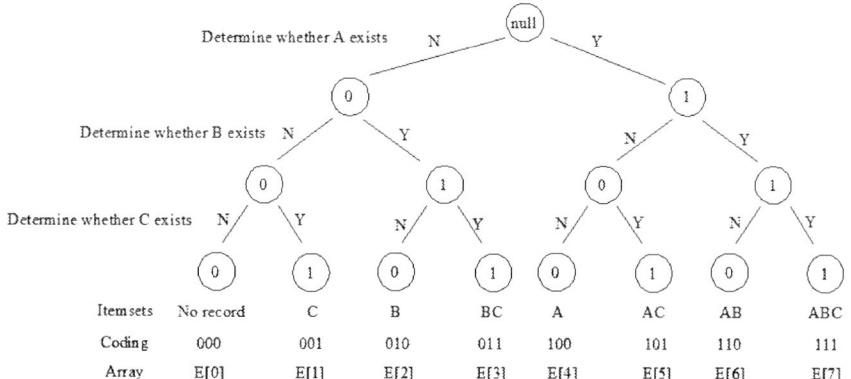

Figure3. Computer computing model of information delivery network

Find the best algorithm model and build the best solution for the centralized test.

Table1. test results expected value table

Status number	Expected output					
	Y0	Y1	Y2	Y3	Y4	Y5
1	0	1	0	1	0	1
2	1	1	1	0	1	0
3	0	0	0	1	1	1
4	1	0	0	0	0	0

In the calculation process of the optimal solution, the following arithmetic calculation and analysis process is made. Under the influence of processing results, part of the data is obtained. This part of the operation data reflects the superiority of the depth learning algorithm that is designed by us, and has a high reference level for the design of the next part of the control test. Use mathematical formula function to compare and select data analysis, and then observe and analyze according to our actual data model. In the process of observation and analysis, we have done a practical research on the overall conversion degree of data. The results of the study are as follows.

Table2. Optimization algorithm test results

Status number	The number of iterations	calculating time	Accuracy
1	5	18	85%
2	6	11	96%

3	5	15	99%
4	3	12	97%

By comparing the data in Table 2 and table 3, the improved algorithm used in this paper greatly reduces the number of iterations we have calculated. The problem of the most common local minimum in the neural network algorithm is solved. The number of iterations is reduced to less than 5 times, and the number of iterations of the traditional algorithm is more than 30 times, which greatly reduces the difficulty of our calculation and reduces the calculation pressure.

Table3. test results of traditional algorithms

Status number	The number of iterations	calculating time	Accuracy
1	32	55	60%
2	41	37	73%
3	37	24	51%
4	51	52	84%

In addition, in the calculation, the local minimum of the sixth sets of data in the traditional algorithm leads to the failure of the calculation, but the optimization algorithm does not have such problems. In the process of similar process, the problem of how to improve the computing speed and the overall calculation efficiency in the operation of neural network algorithm need to be solved. The calculation time of this paper is only about 15 seconds, but for the traditional algorithm, in order to show the computational time contrast of our optimized algorithm more clearly, the contrast of the figure 3 in the form of the lower figure is set up.

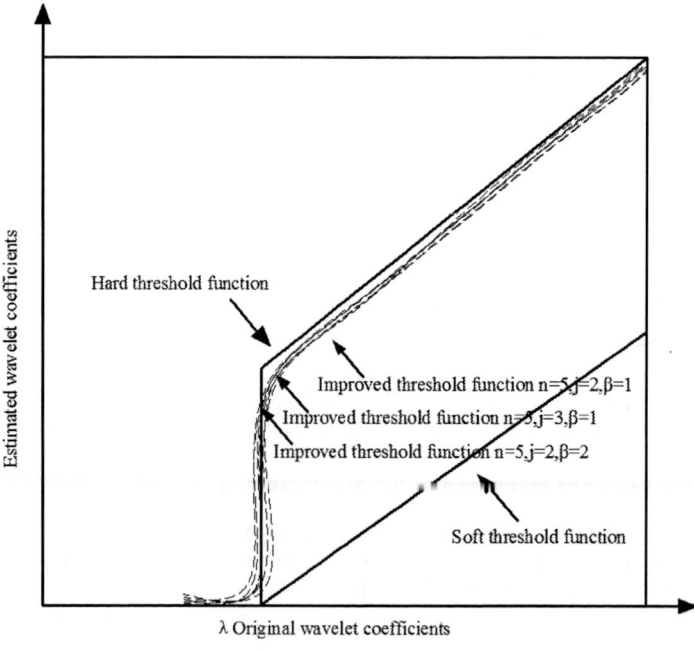

Figure4. Computation of time comparison chart

The information got in the above picture is that after the comparison test of the calculation time, the optimized algorithm reduces the calculation time greatly by reducing the calculation step and reducing the number of iterations, which makes the calculation time of only about 15 seconds under the article. For traditional algorithms, the computation time is generally higher than 40 seconds and the computation time is too long, which show more clearly the computational time contrast of our optimized algorithm.

5.CONCLUSION

With the continuous improvement and perfect of science and technology, the utilization of intelligent algorithm is more and more extensive. In the aspect of model building, the traditional computing algorithms of graphic design and all aspects of life are not suitable for the development of the times. It is necessary to make use of the actual computing algorithm of the Internet to study and deal with the model mode. According to the existing problems and functions in the actual model, the depth learning algorithm we need is studied. The depth learning algorithm proposed in this paper is a bit existing in his ductility and high development ability to meet people's needs and help to provide a lot of technical solutions. The study and improvement process of a depth learning algorithm based on many times of contrast and contrast tests can obtain the accuracy of our calculated and calculate overall data. On the basis of considering the calculation results and calculating the overall accuracy, the target image updating and management of the intelligent data model algorithm is completed. In the research process of deep learning algorithm, the overall accuracy of the algorithm is better than that of the deep learning algorithm. For the study of computer without contrast, the computation time is saved. This is the focus of our research and development in the future. Although there are still some problems in the deep learning algorithm proposed in this paper, we can improve it step by step through continuous improvement.

ACKNOWLEDGEMENTS

A lot of thanks to the fundamental research funds for the Central Universities Fund (3142017066) and the national natural science foundation of China(51874134).

References

[1] Xu Y L, Wang L Y, Yu M G, et al. Study on the characteristics of gas explosion affected by induction charged water mist in confined space[J]. Journal of Loss Prevention in the Process Industries, 2016, 40:227-233.

[2] Kaloudis E, Grigoriadis D G E, Papanicolaou E. Numerical simulations of constant-influx gravity currents in confined spaces: Application to thermal storage tanks [J]. International Journal of Thermal Sciences, 2016, 108:1-16.

[3] Zheng C, Kong X S, Wu W G, et al. The elastic-plastic dynamic response of stiffened plates under confined blast load [J]. International Journal of Impact Engineering, 2016, 95:141-153.

[4] Sung K H, Bang J W, Li L, et al. Effect of crack size on gas leakage characteristics in a confined space[J]. Journal of Mechanical Science & Technology, 2016, 30(7):3411-3419.

[5] Li L, Ji P, Zhang Y. Molecular dynamics simulation of condensation on nanostructured surface in a confined space [J]. Applied Physics A, 2016, 122(5):496.

[6] Yousefi-Lafouraki B, Ramiar A, Ranjbar A A. Numerical Simulation of Two Phase Turbulent Flow of Nanofluids in Confined Slot Impinging Jet[J]. Flow Turbulence & Combustion, 2016, 97(2):571-589.

[7] Loft N J S, Kristensen L B, Thomsen A E, et al. CONAN -- the cruncher of local exchange coefficients for strongly interacting confined systems in one dimension[J]. Computer Physics Communications, 2016, 209:171-182.

[8] Zanotti O, Dumbser M. Efficient conservative ADER schemes based on WENO reconstruction and space-time predictor in primitive variables[J]. Computational Astrophysics & Cosmology, 2016, 3(1):1.

[9] Feldgun V R, Karinski Y S, Edri I, et al. Prediction of the quasi-static pressure in confined and partially confined explosions and its application to blast response simulation of flexible structures[J]. International Journal of Impact Engineering, 2016, 90(2008-05-29):46-60.

[10] Michael L, Nikiforakis N. A hybrid formulation for the numerical simulation of condensed phase explosives [J]. Journal of Computational Physics, 2016, 316:193-217.

Study on Temporal and Spatial Patterns of Brain in Emotional State Based on Steady State Visual Evoked Potentials

Kai Yang[1], Ying Zeng[2,1], Li Tong[1], Bin Liu[1], Xiyu Song[1], Bin Yan[1,*]

[1]China National Digital Switching System Engineering and Technological Research Center

Henan Zhengzhou, 450001;

[2]Key Laboratory for NeuroInformation of Ministry of Education, School of Life Science and Technology, University of Electronic Science and Technology of China, Sichuan Chengdu, 610000;

17638563281 (K.Y.)

ykfer09@163.com (K.Y.); yingzeng@uestc.edu.cn (Y.Z); tttocean@163.com (L.T.); 306043699@qq.com; fsongxiyu@126.com (X.S.); ybspace@hotmail.com (B.Y.).

*Correspondence: ybspace@hotmail.com

ABSTRACT The high signal-to-noise ratio steady state visually evoked potential (SSVEP) signal has been used in many brain computer interface (BCI) experiments and cognition task. In this paper, we used steady-state probe topography (SSPT) to analyze the brain patterns in processing different emotional pictures. We used pleasant, unpleasant and neutral pictures from the International Affective Picture System (IAPS) that were either presented in intact or phase-scrambled form. Pictures were flickering at 10Hz and enabled us to record steady-state visual evoked potentials. Global Power of the electroencephalogram (EEG) signals were computed to yield four time windows, P2, P3, late P3, and slow wave (SW), respectively. SSVEP amplitudes for different emotional pictures were extracted in these windows and submitted to paired-t test. Significantly differences were found between emotional and neutral pictures mainly at both late P3 and SW intervals, as well as at frontal, left parietal, occipital regions as reflected in a significant drop in SSVEP amplitudes. These results revealed the key time windows and brain regions in emotional cognition task.

1. INTRODUCTION

SSVEP and ERP had been used in numerous studies to investigate electrical activities of the brain processing different emotional conditions [1][2]. Studies had reported that frontal, parietal and occipital brain cortices were important in emotion processing [3]. Kemp used pictures from IAPS induced SSVEP, the results indicated that SSVEP amplitudes in frontal and occipital regions decreased at 1462ms [4]. Catherine using SSVEP signals demonstrated that highly arousing emotional pictures consume more processing resources relative to neutral pictures at 400, 700, 1000ms [5]. These studies illustrated SSVEP differences between emotional and neutral conditions at certain time points, but brain responses to emotion was a continuous process [6]. The results at certain time points may can't accurately reflect SSVEP changes in the brain. Meanwhile, Wang used words and faces

Content from this work may be used under the terms of the Creative Commons Attribution 3.0 licence. Any further distribution of this work must maintain attribution to the author(s) and the title of the work, journal citation and DOI.
Published under licence by IOP Publishing Ltd

pictures to elicit ERP, and he summarized ERP-related time course of emotion processing into three stages as early stage (100-200ms), middle stage (200-300ms), late stage (>300ms) [7]. Greg had illustrated the late positive potential (LPP) and P3 components of ERP and the SSVEP were larger for emotional compared to neutral pictures [8]. These findings demonstrated that emotion process was continuous and the changes of SSVEP were related with ERP. In this paper, in order to investigate the spatiotemporal differences, we calculated the Global Power of EEG signals to yield four time windows, SSVEP were averaged in these intervals and submitted to paired-t test.

The main work in this paper were as follows. First, we designed an experiment used pictures from IAPS flicked at 10Hz to elicit SSVEP. Then EEG signals of 30 participants were recorded and the Global Power and SSVEP were extracted. We hypothesized that the SSVEP amplitudes for pleasant, unpleasant and neutral pictures were significantly different in brain regions. Finally, we found that significant differences occurred at P3 and SW intervals and frontal, temporal, left parietal and occipital regions were important in emotion discrimination. The SSPT also showed distinct features among picture categories.

2. MATERIALS AND METHOD

2.1 Participants

All of our subjects were native Chinese undergraduate and graduate students. Beck Anxiety Inventory, Hamilton Anxiety Rating Scale, and Hamilton Rating Scale for Depression tests were administered to exclude individuals with anxiety, depression, or physical abnormalities, as well as those using sedatives and psychotropic drugs. Finally 20 subjects took part in our experiments, including 15 male and 5 female , with a mean age of 22.58 years (range=19-29years ,SD=3.9 years) All participants were right-handed with normal or corrected to normal visual acuity. After the experiment, the subjects received some economic compensation.

2.2 Stimuli

The stimuli including pleasant, unpleasant, neutral and scrambled pictures from the IAPS. The experimental materials were selected based on normative valence and arousal ratings provided by the IAPS set. For each valence category, 20 pictures were selected. The pleasant pictures include pictures of 10 babies and 10 lovely animals; neutral pictures include 10 characters and 10 daily life scenes; unpleasant pictures include 10 fragmentary bodies and 10 animal threats. The valence of each kind of picture was different (pleasant: 7.37, neutral: 5.08, unpleasant: 2.69), and the degree of arousal was also different (pleasant: 5.38, neutral: 3.40, unpleasant 6.24). We adjusted the brightness of all pictures to ensure that there was no significant difference in the brightness of all categories.

In order to yield the scrambled pictures, another 20 pictures were randomly selected from the IAPS set. The phase randomization was performed by a Fourier transform. Firstly, the original phase was replaced by random value and the amplitude kept constant, then inverse Fourier transform was done to rebuild picture [5]. For each picture, phase randomization was done twice to get two scrambled picture, so the number of non-content pictures was 40. Scrambled pictures were presented twice in one trail, one is before the picture appears and the other is after picture end, respectively. With this procedure subjects can observe picture change at any point in time during all experimental conditions.

2.3 Experiment design and procedure

Each picture flicked at a rate of 10 Hz against a black background centrally on a 23-inch computer screen with a frame refresh rate of 60 Hz. The screen was about 80cm in front of the subjects with a visual angle of 10 horizontally and 7 vertically. Each trail lasted 7000ms, including 70 cycles each with 50ms picture on and 50ms black screen. There was a short interval between each trail presenting a fixation cross on the black screen center varying from 6000ms to 8000ms.

Each trial started with a scrambled picture. At a random time point, the picture changed to a normal mode (pleasant, neutral or unpleasant condition) or another scrambled picture of the same origin picture. In order to avoid the experimental response of time point related to picture change, the time points of the picture change were randomly generated and divided into three categories: early (12% trials,100-600ms), medium (70% trials, 800-1300ms) and late (18% trials,1500-2000ms) time window after scrambled picture presented. Trials including early and late time window were regarded as "catch trials", and the data of these trials were not used in final data analysis. The pictures included in these trials were additionally selected from the IAPS set, which were different from those ultimately involved in data analysis. All the pictures included in the final analysis process were randomly presented twice (the same picture was not repeated in three consecutive trials), so there were 160 effective experimental trials (40 trials for each experimental conditions) and 68 "catch trials", a total of 228 trials. The 228 trials were divided into 6 blocks (38 trials each), and there was a short rest between each block (see Figure 1). Before the experiment, there was a practice block (10 trials) so that the subjects were familiar with the experimental process. The pictures used in practice block were different from the experiment pictures that would be included data analysis.

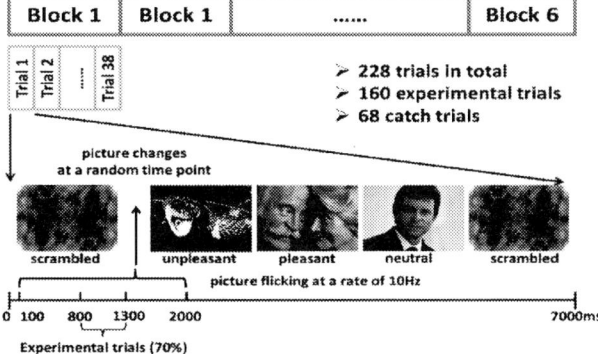

Figure 1. Experimental protocol used in the current study.

After data recording, 60 pictures used in experiment were presented in a randomized order to the subjects again, and the subjects were asked to make a 9 grade Self-Assessment Manikin (SAM) scale for each picture from two dimensions of valence and arousal degree.

EEG signals were recorded by g.tec HIamp System with 62 electrodes located in International 10-20 system, using Fz as a recording reference. All electrodes impedance was lower than 10 KΩ. Online band-pass filter and notch filter of the system were adopted, 0.1-100Hz and 50Hz, respectively. And the data was recorded at a sampling rate of 512 Hz.

2.4 Signal processing
Epochs of 250ms before to 5000ms after pictures change were extracted for each stimulus. First, data was filtered using a low pass filter at a frequency of 40Hz. ERP voltages were baseline corrected by subtracting the 250ms data before pictures presented. Then average reference was done to exclude global artifacts.

In order to measure the global cortical activity and assess the ERP peaks and find their time course, we calculated Global Power $g(t)$ of the ERP. The $g(t)$ was computed as the mean square of the averaged signal of all trials for each category, weighted by the standard deviation of the voltage across trials [9], at each electrode and time point (see equation 1).

$$g(t) = \frac{\sum_{i=1}^{N} X_i^2(t) \cdot s_i^{-1}(t)}{N \cdot \sum_{i=1}^{N} s_i^{-1}(t)} \qquad (1)$$

Thus $X_i^2(t)$ was voltage at electrode i and time t and $s_i^{-1}(t)$ was the standard deviation of the voltage at sensor i and time t across trials.

SSVEP amplitudes were calculated from 10Hz Fourier coefficients (FC) and used a 128-unit Hanning window [10]. The length of the window was greater than two stimulus cycles, so Short Fourier Transform can get all the spectral characteristics of the signal in one cycle [11]. The window shifted one point and magnitude recalculated, and this procedure continued until the window covered the last point of 5 seconds data. Finally SSVEP magnitude was a time series of 4.75 seconds (this was 5 seconds less 0.25 second window length) and all 62 channels data were analyzed. This Fourier analysis can be regarded as a band-pass filter with a central frequency of 10 Hz.

All trials of each category were averaged and then SSVEP amplitudes were normalized. First, for scrambled category, the mean value of time series for SSVEP magnitudes were calculated and yielded 62 values. These 62 values (one for each channel) were then averaged to yield a single Normalized Factor (NF) [3]. SSVEP amplitude time series from each category were divided by NF. This procedure was necessary as there were large inter-subject variations in the SSVEP amplitude. Then the SSVEP segments of each category were averaged between all subjects and the SSVEP segments for scrambled pictures were subtracted as reference to reduce the irrelevant influence of emotional cognition processing.

The whole brain was divided into 8 regions: left (AF3, F1, F3, F5, F7, FP1, AF7) and right (AF4, F2, F4, F6, F8, FP2, AF8) frontal, left (FT7, FC5, T7, C5, TP7, CP5)and right (FC6, FT8, C6, T8, CP6, TP8) temporal, left (FC1, FC3, C1, C3, CP1, CP3)and right (FC2, FC4, C2, C4, CP2, CP4) parietal, left (P1, P3, P2, P4, PO3, PO7, O1) and right (P2, P4, P6, P8, PO4, PO8, O2)occipital. Data of each region was submitted to paired-t test in SPSS19.0 to illustrate the impacts of picture contents.

3. EXPERIMENT RESULTS

As expected, significant differences were found between picture categories in both valence and arousal ratings. For the valence rating scale, pleasant pictures rated more pleasant than neutral pictures (t=19.07, $p<0.001$) as well as neutral pictures rated more pleasant than unpleasant pictures (t=-14.33, $p<0.001$). For the arousal rating scale, both pleasant pictures (t=30.07, $p<0.001$) and unpleasant pictures (t=16.47, $p<0.001$) showed higher arousal ratings than neutral pictures. Unpleasant pictures also more arousing than pleasant pictures (t=25.25, $p<0.001$).

The time series of average differences between ERP for each category were displayed in Figure 2. Four time windows, corresponding roughly to P2 window (210-250ms), P3 window (275-345ms), late P3 window (365-450ms) and slow wave (SW) window (500-670ms) were chosen. These windows were chosen as they contained most voltage peaks. SSVEP data in these time windows were extracted, averaged across subjects and time series to yield 62 values displayed in topographic maps (see Figure 3).

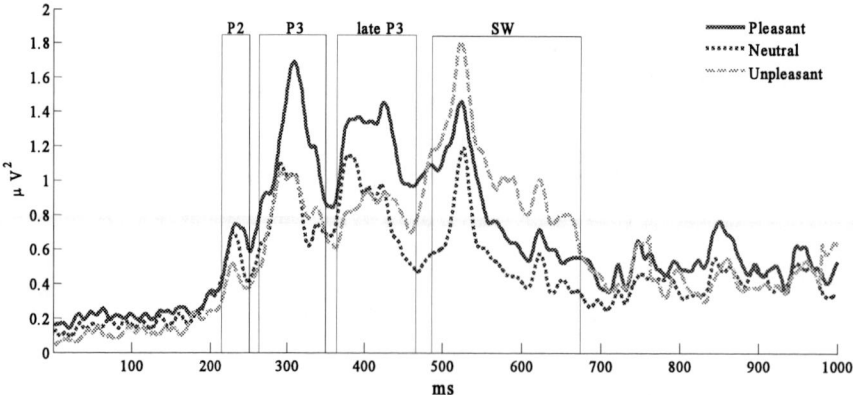

Figure 2. Grand mean (N=20) global power of the voltage obtained at all channels for each category. Red line for pleasant, green line for unpleasant, blue line for neutral.

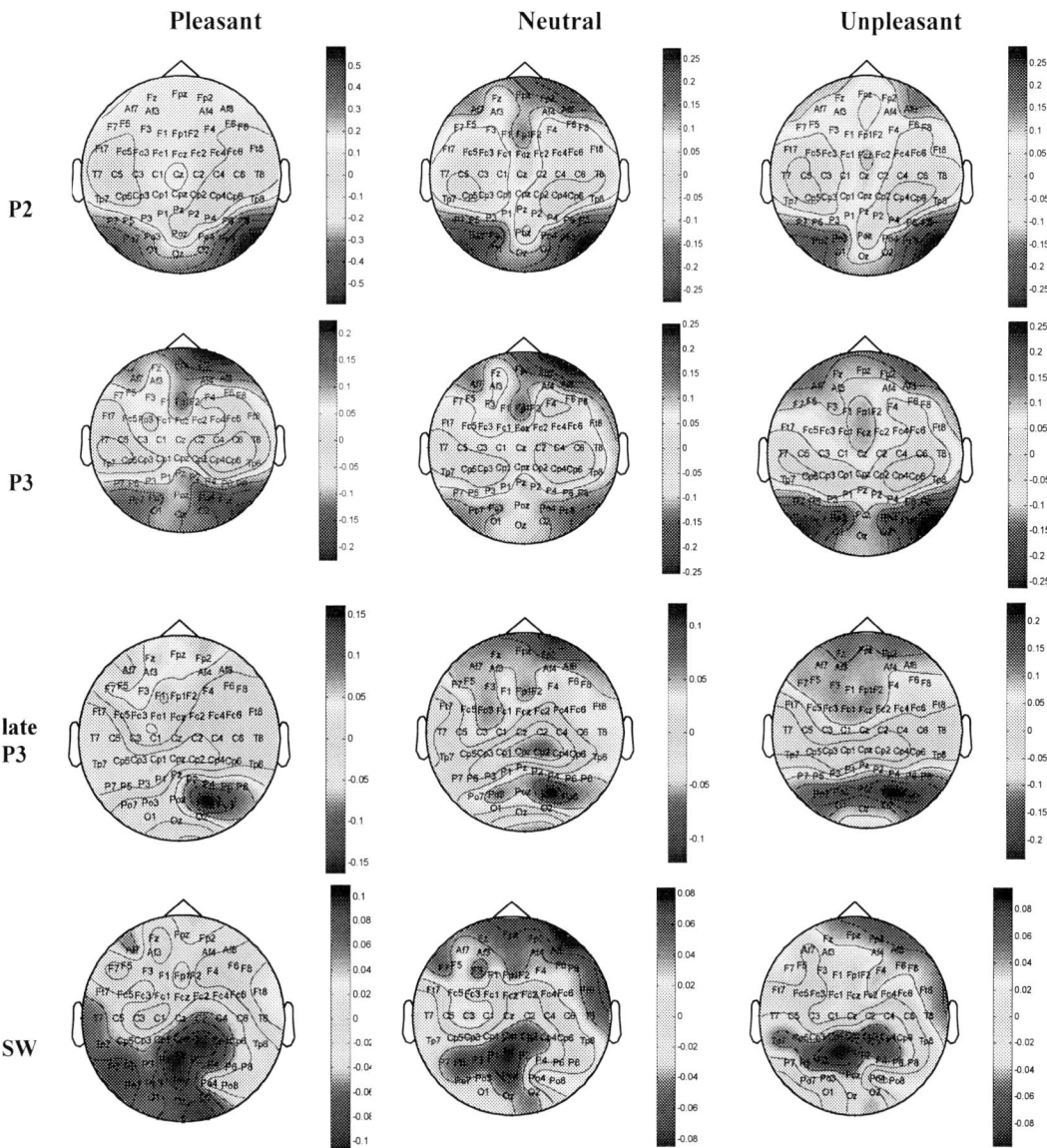

Figure 3. Topography of cross subject averaged normalized magnitude for pleasant，neutral and unpleasant picture contents in P2，P3, late P3 and SW windows.

Picture contents effects were mainly found at left parietal and occipital in P2 window (210-250ms). Positive versus neutral and negative contents were significant at left parietal, $p<0.05$. Differences also existed at occipital, both negative and neutral were statistically significant with positive, $p<0.001$.In P3 window (275-345ms), only at occipital regions, the SSVEP amplitudes of emotional pictures were significantly different with neutral, $p<0.05$. The late P3 window (365-450ms) exhibited a main effect of picture content. At frontal (right and left) and left occipital regions amplitudes of pleasant pictures significantly decreased relative to neutral, $p<0.001$. Similar results were also found at right temporal region, $p<0.05$. Unpleasant relative to neutral at left frontal and occipital regions also showed significantly drop, $p<0.05$. The comparison of SSVEP amplitudes among emotional (pleasant and unpleasant) pictures became significant at right temporal region, p<0.05. Pronounced main effects of picture content were also seen in SW window (500-670ms). The amplitudes of pleasant and unpleasant pictures showed differences at right and left frontal regions, p<0.05, as well as pleasant and neutral categories at right frontal region, p<0.05.

In time series, the SSPT for pleasant and unpleasant relative to neutral pictures showed different tendency. The maps for emotion changed from frontal to occipital as well as from right hemisphere to left zone. On the contrary, the SSPT for neutral changed at occipital first, then the amplitudes for parietal and temporal regions decreased, at the same time it changed from left to right hemisphere. Besides, the amplitudes of emotional decreased from P2 to SW intervals, for neutral conditions the amplitudes increased at P3 window and then reduced in later intervals. At SW window, the SSVEP amplitudes for all categories decreased, particularly at occipital regions.

4. CONCLUSION

The present study aimed at investigating the impact on SSVEP amplitudes for different picture contents on brain regions. First, we compared the SSVEP amplitudes in time series, significant differences were found after P3 intervals as the amplitudes of emotional pictures drop more than neutral. In spatial domain, emotional (pleasant and unpleasant) relative to neutral pictures were significantly different at left parietal, temporal and occipital regions. Among emotional pictures, the SSVEPs were statistically different at frontal, left parietal, occipital and right temporal regions. For SSPT, the amplitudes of pleasant and unpleasant decreased over time as well as the amplitudes of neutral increased in P3 intervals. Interestingly, we found the SSPT for emotion had a different changing trend relative to neutral pictures. Emotional SSPT changed at frontal regions first, at the same time, neutral maps showed discrimination at occipital cortices. Previous studies had given reasons: occipital regions were reported playing an important role in visual task [12][13]. And viewing neutral pictures was mainly a visual task, almost no emotion process included. Brain activities in frontal and temporal regions were related to emotion processing [14][15]. These results might be valuable in revealing the brain mechanism and improving brain computer interaction performance. In the future, we would construct brain-networks to analysis the emotion processing procedure in time series.

ACKNOWLEDGMENTS

The authors would like to thank all subjects who participated in our experiment. This work was supported by the National Key R&D Program of China under grant 2017YFB1002502, the National Natural Science Foundation of China (No. 61701089, No.61601518 and No. 61372172) and the National Defense Science and Technology Innovation Zone Project.

REFERENCES

[1]   Ding, R. **and** Li, P. J. 2017. Emotion Processing by ERP Combined with Development and Plasticity. *Neural Plasticity. Volume* 2017, Article ID 5282670, 1-15. DOI= https://doi.org/10.1155/2017/5282670.

[2]   Zhu, M and Alonso-Prieto. J. 2016. The brain frequency tuning function for facial emotion discrimination: An SSVEP study. *Journal of Vision*, 16(6):12, 1–14, DOI= http://dx.doi.org/10.1167/16.6.12.

[3]   D.A. Camfield. A. Scholey , A. Pipingas , R. Silberstein , M. Kras , K. Nolidin , K. Wesnes , M. Pase , C. Stough, J. 2012. Steady state visually evoked potential (SSVEP) topography changes associated with cocoa flavanol consumption. *Physiology & Behavior* 105 (2012) 948–957. DOI= http://dx.doi.org/10.1016/j.physbeh.2011.11.013.

[4]   A. H. Kemp, M. A. Gray, P. Eide, R. B. Silberstein. J. 2002. Steady-State Visually Evoked Potential Topography during Processing of Emotional Valence in Healthy Subjects. *NeuroImage* 17, (2002) 1684–1692. DOI=

http://dx.doi.org/10.1006/nimg.2002.1298

[5] Catherine H, A, Soren K. Andersen, Matthias M. Müller. J. 2010. Time course of affective bias in visual attention: Convergent evidence from steady-state visual evoked potentials and behavioral data. *NeuroImage* 53 (2010) 1326–1333.DOI= http://dx.doi.org/10.1016/j.neuroimage.2010.06.074.

[6] Mohammad, S. Sadjad, A,E, Yun F, C. 2014 Continuous emotion detection using EEG signals and facial expressions. *2014 IEEE International Conference on Multimedia and Expo (ICME)* (Chengdu China2014).DOI= http://dx.doi.org/10.1109/ICME.2014.6890301.

[7] Wang, Y. Li, X, B. J. 2017. Temporal course of implicit emotion regulation during a Priming-Identify task: an ERP study. *Scientific Reports* 41941 (2017). DOI= http://dx.doi.org/10.1038/srep41941.

[8] Y Greg Hajcak. Annmarie MacNamara. J. 2013. The dynamic allocation of attention to emotion: Simultaneous and independent evidence from the late positive potential and steady state visual evoked potentials. *Biological Psychology* 92 (2013) 447–455. DOI= http://dx.doi.org/ 10.1016/j.biopsycho.2011.11.012.

[9] Andreas keil, Margaret M. Bradley. J. 2002. Large-scale neural correlates of affective picture processing. *Psychophysiology* 39 (2002) 641-649. DOI= http://dx.doi.org/10.1111/1469-8986.3950641.

[10] Silberstein RB, Ciorciari J, Pipingas A Steady-state visually evoked potential topography during the Wisconsin card sorting test. *Electroencephalography and clinical Neurophysiologylogy* 96 (1995): 24–35.DOI= http://dx.doi.org/10.1016/0013-4694 (94) 00189-R.

[11] Wu, Y. Zhang, L. M. *Digital signal processing.* Xi dian university press, 2009.8, 72-76.

[12] Won Heo, June Sic Kim, Chun Kee Chung, Sang Kun Lee. J. 2018. Relationship between cortical resection and visual function after occipital lobe epilepsy surgery. *J Neurosurg* 129 (2018):524–532. DOI= http://dx.doi.org/10.3171/2017.5.

[13] Tetsuya Iidaka, Koichi Yamashita, Kenichi Kashikura, Yoshiharu Yonekura. J. 2004. Spatial frequency of visual image modulates neural responses in the temporo-occipital lobe. An investigation with event-related fMRI. *Cognitive Brain Research* 18 (2004) 196–204. DOI= http://dx.doi.org/10.1016/j.cogbrainres.2003.10.005.

[14] Shota Hori, Koichi Mori, Takehisa Mashimo, Akitoshi Seiyama. J. 2017. Effects of Light and Sound on the Prefrontal Cortex Activation and Emotional Function: A Functional Near-Infrared Spectroscopy Study. *Frontiers in Neuroscience* 11 (2017). DOI= http://dx.doi.org/10.3389/fnins.2017.00321.

[15] J.D. Steele, S.M. Lawrie. J. 2004. Segregation of cognitive and emotional function in the prefrontal cortex: a stereotactic meta-analysis. *NeuroImage* 21 (2004) 868–875. DOI= http://dx.doi.org/10.1016/j.neuroimage.2003.09.066.

Parameter Analysis of Stepped Frequency Pulses Frequency Diverse Array Radar

Ming Tan[1,a],Chun-yang Wang[2,b],Hong-bing Li[3,c],Juan Bai[4,d],Lei Bao[5,e]

[1]Air and Missile Defense College, Air Force Engineering University Xi'an, Shanxi, 710051, China 86 18629679068

[2]Air and Missile Defense College, Air Force Engineering University Xi'an, Shanxi, 710051, China 86 13571041693

[3]Air and Missile Defense College, Air Force Engineering University Xi'an, Shanxi, 710051, China 86 13359226696

[4]Air and Missile Defense College, Air Force Engineering University Xi'an, Shanxi, 710051, China 86 15249042646

[5]Air and Missile Defense College, Air Force Engineering University Xi'an, Shanxi, 710051, China 86 17795834063

[a]tanming_1992@163.com, [b]Wcy_kgd_cn@163.com, [c]84574762@qq.com, [d]b_juan@163.com, [e]2743681405@qq.com

ABSTRACT Owing to its range-angle-dependent transmit beampattern, the frequency diverse array (FDA) can provide the potential capability for novel radar detection, localization and interference suppression. However, the conventional FDA generates range-angle coupled beampattern, which degrades the performance of FDA in radar application. In this paper, stepped frequency pulses frequency diverse array (SFP-FDA) is investigated which can effectively remove the coupling in range-angle dimensions. Influence of parameter selections such as the frequency increment in pulse and element dimension on the proposed scheme are analyzed. The proposed method outperforms the conventional FDA, and it can be effectively utilized through reasonable parameter setting. Numerical results are given to verify the proposed scheme.

1. INTRODUCTION

Unlike phased array radar only angle-dependent, frequency diverse array (FDA) providing a range-angle-dependent beampattern by employing a tiny linearly increasing frequency offset across the array elements [1]-[3]. Due to its promising application potentials, FDA becomes a hotspot and fruitful achievements have been obtained [4]. The multipath characteristics of FDA are presented in [5]. Interference suppression by adaptive frequency offset selection is investigated in [6]. In [7], the derivation of Cramer-Rao lower bounds is made for evaluating the estimation performance in range and angle domains.

However, due to the range-angle-coupling, there might be a group of range-angle pairs hence it is difficult to distinguish the real target from numerous feedback according to the beampattern of FDA. To address this issue, several methods have been proposed recently. Through optimizing the frequency increments with genetic algorithm, a dot-shaped FDA transmit beampattern is presented in [8]. A non-

Content from this work may be used under the terms of the Creative Commons Attribution 3.0 licence. Any further distribution of this work must maintain attribution to the author(s) and the title of the work, journal citation and DOI.
Published under licence by IOP Publishing Ltd

uniform frequency diverse array which can localize the range-dependent targets is proposed in [9]. Based on logistic map, a multi-carrier nonlinear frequency modulation system is presented in [10]. With symmetrical frequency offsets, the beampattern of FDA is discussed in [11]. To attain a required radiation performance, such as the element placements and frequency offsets, the sparse FDA technology based on artificial bee colony optimizer is presented in [12]. We proposed a novel method which called stepped frequency pulses frequency diverse array (SFP-FDA) [13], the localization performances in terms of angle-range dimensions contrast with the double-pulse FDA [14] are analyzed.

In this paper, the impact of parameter selections on SFP-FDA is investigated. By appropriate parameter selections, a narrow pencil beam can be formed which decoupled in range-angle dimensions. The transmit energy can be more concentrated in the desired area than the conventional FDA does. The frequency increment in pulse domain as a key research object is analyzed exhaustively.

The remaining sections are organized as follows. Section 2 proposes mathematical principle of conventional FDA radar. Section 3 provides the derivation of SFP-FDA and the impact analysis of parameter selections. Finally, simulations and results are presented in Section 4, and conclusions are given in Section 5.

2. CONVENTIONAL FREQUENCY DIVERSE ARRAY RADAR

Consider an N element uniform linear array, as shown in Figure 1. The signal operating frequency is f_0, the radiated frequency of the nth element is taken as

$$f_n = f_0 + n\Delta f_a, \quad n = 0,1,K,N-1 \tag{1}$$

where Δf_a is the frequency increment between the adjacent array elements. The inter-element spacing can be represented by

$$d = \frac{\lambda_0}{2} = \frac{c}{2f_0} \tag{2}$$

where c and λ_0 are the light velocity and the basic wavelength, respectively.

The signal transmitted by the nth element can be written as

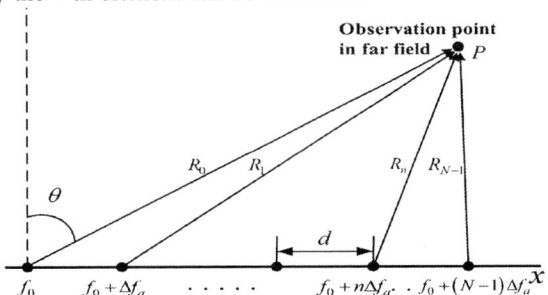

Figure 1. Illustration of conventional FDA.

$$x_n(t) = a_n e^{-j2\pi f_n t} \tag{3}$$

where a_n is the complex weight of the nth signal.

For a far-field point target $P(\theta,r)$, the distance between target and the nth element can be approximately given by

$$R_n \approx R_0 - nd\sin\theta \tag{4}$$

where R_0 is the distance between target and the reference element, note that $R_0 = r$.

The overall signal observed by $P(\theta,r)$ can be derived as

$$x(t,\theta,r) = \sum_{n=0}^{N-1} x_n\left(t - \frac{R_n}{c}\right) = \sum_{n=0}^{N-1} a_n e^{-j2\pi f_n\left(t - \frac{R_n}{c}\right)} \qquad (5)$$

After approximation and arrangement, equation (5) becomes

$$x(t,\theta,r) = e^{-j2\pi f_0\left(t - \frac{r}{c}\right)} \sum_{n=0}^{N-1} a_n e^{-j2\pi n\left(\Delta f_a t - \frac{r\Delta f_a}{c} + \frac{f_0 d\sin\theta}{c}\right)} \qquad (6)$$

Define the element weighting vector as $\mathbf{w} = [a_0 \ a_1 \ L \ a_{N-1}]^T$, where $[g]^T$ denotes the transpose operator. The steering vector of the conventional FDA is

$$\mathbf{a}(t,\theta,r) = \left[1 \ \ e^{-j2\pi\left(\Delta f_a t + \frac{f_0 d\sin\theta}{c} - \frac{r\Delta f_a}{c}\right)} \ \ L \ \ e^{-j2\pi(N-1)\left(\Delta f_a t + \frac{f_0 d\sin\theta}{c} - \frac{r\Delta f_a}{c}\right)}\right]^T \qquad (7)$$

Then, the transmit beampattern can be expressed as

$$B(t,\theta,r) = |x(t,\theta,r)| = |\mathbf{w}^H \mathbf{a}| \qquad (8)$$

where $[g]^H$ is the conjugate transpose operator. If uniform weights are applied, namely $a_0 = a_1 = L = a_{N-1} = 1$, equation (8) becomes

$$B(t,\theta,r) = \left|\frac{\sin\left[N\pi\left(\Delta f_a t + f_0 d\sin\theta/c - r\Delta f_a/c\right)\right]}{\sin\left[\pi\left(\Delta f_a t + f_0 d\sin\theta/c - r\Delta f_a/c\right)\right]}\right| \qquad (9)$$

Note that when $\Delta f_a = 0$, it becomes the beampattern of phased array radar, that is

$$B(\theta) = \left|\frac{\sin\left[N\pi\left(f_0 d\sin\theta/c\right)\right]}{\sin\left[\pi\left(f_0 d\sin\theta/c\right)\right]}\right| \qquad (10)$$

By comparing (9) with (10), it implies that phased array radar is only dependent on angle and the beampattern of FDA radar is not only angle dependent, but also vary as a function of the range and time. However, it is also observed that because of its range-angle coupling, FDA have too many maximums, and it may go against focus energy on the specified location.

3. PROPOSED STEPPED FREQUENCY PULSES FREQUENCY DIVERSE ARRAY RADAR

3.1 Mathematical Principle

Figure 2 shows the radiated frequency versus time in the n th element.

In the first pulse of the SFP-FDA radar, the transmit frequency of each element is the same as that of the conventional FDA. But from pulse-to-pulse, there existing a small frequency increment Δf_t, so the frequency in the m th pulse of the n th element is written as

$$f_{n,m} = f_0 + n\Delta f_a + m\Delta f_t, \quad m = 0,1,K,M-1 \qquad (11)$$

where M is the number of pulses.

Define $a_{n,m}$ as the complex weight of the nth signal in the mth pulse, the pulse width T_p should satisfied with $f_0 T_p = \text{integer}$, the nth element electric field observed at a far field target $P(\theta,r)$ can be expressed as

$$x_n(t,r) = \sum_{m=0}^{M-1} \frac{a_{n,m}}{R_n} \exp\left\{-j2\pi\left[f_{n,m}\left(t - mT_p - \frac{R_n}{c}\right)\right]\right\} \qquad (12)$$

Substituting (4) and (11) into (22) yields

1512

$$x_n(t,r) = \sum_{m=0}^{M-1} \frac{a_{n,m}}{R_n}$$

$$\times \exp\left\{-j2\pi\left[f_0\left(t - \frac{r - nd\sin\theta}{c}\right)\right.\right.$$

$$+ (n\Delta f_a + m\Delta f_t)\left(t - \frac{r}{c}\right)$$

$$- (n\Delta f_a + m\Delta f_t)\frac{nd\sin\theta}{c}$$

$$\left.\left. - (f_0 + n\Delta f_a + m\Delta f_t)(mT_p)\right]\right\} \tag{13}$$

The approximation $R_n \approx r$ is valid in the sense of amplitude, assuming $\Delta f_a = 1/T_p$ and $\Delta f_t = 1/T_p$, due to the fact that the period of function $\exp(g)$ is 2π, so the last term of equation (13) can be ignored. Assuming $\Delta f_t = f_0$, note that $nd\sin\theta = r$ and $\Delta f_a = f_0$, so the third term of equation (13) can also be negligible.

Through approximation and arrangement, (13) can be equivalently reformulated as

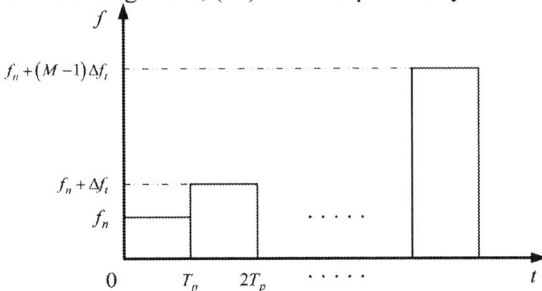

Figure 2. The radiated frequency of the n th element.

$$x_n(t,\theta,r) = \frac{\exp(j\phi_0)}{r} \sum_{m=0}^{M-1} a_{n,m}$$

$$\times \exp\left\{-j2\pi\left(f_0\frac{nd\sin\theta}{c} + n\Delta f_a t + m\Delta f_t t\right.\right.$$

$$\left.\left. - n\Delta f_a \frac{r}{c} - m\Delta f_t \frac{r}{c}\right)\right\} \tag{14}$$

where $\phi_0 = -2\pi f_0(t - r/c)$. Then, the transmit beampattern of SFP-FDA can be expressed as

$$A(t,\theta,r) = \left|\sum_{n=0}^{N-1} x_n(t,\theta,r)\right|$$

$$= \left|\sum_{n=0}^{N-1}\sum_{m=0}^{M-1} a_{n,m} \exp\left\{-j2\pi\left(n\Delta f_a t - n\Delta f_a \frac{r}{c} + f_0\frac{nd\sin\theta}{c}\right)\right\}\right.$$

$$\left.\times \exp\left\{-j2\pi\left(m\Delta f_t t - m\Delta f_t \frac{r}{c}\right)\right\}\right| \tag{15}$$

Provided that the weights in the same order number pulses of each element are equivalent. Define $\mathbf{\alpha}_a(t,\theta,r)$ and $\mathbf{\alpha}_t(t,r)$ as the steering vector of element aspect and signal aspect, respectively, which can be written as

$$\mathbf{\alpha}_a(t,\theta,r) = \left[1 \quad L \quad \exp\left\{-j2\pi(N-1)\left(\Delta f_a t - \Delta f_a \frac{r}{c} + f_0\frac{d\sin\theta}{c}\right)\right\}\right]^{\mathrm{T}} \tag{16}$$

and

$$\mathbf{a}_t(t,r) = \begin{bmatrix} 1 & \mathrm{L} & \exp\left\{-j2\pi(M-1)\left(\Delta f_t t - \Delta f_t \dfrac{r}{c}\right)\right\} \end{bmatrix}^{\mathrm{T}}.$$
(17)

Define $\mathbf{w}_a$ and $\mathbf{w}_t$ as the weighting vector in the element and signal domain, respectively. Rewritten (15) as

$$
\begin{aligned}
A(t,\theta,r) &= \left| \mathbf{w}_a^H \mathbf{a}_a(t,\theta,r) \mathbf{w}_t^H \mathbf{a}_t(t,r) \right| \\
&= \left| \mathbf{w}^H \left[\mathbf{a}_a(t,\theta,r) \otimes \mathbf{a}_t(t,r) \right] \right| \\
&= \left| \mathbf{w}^H \mathbf{a}(t,\theta,r) \right|
\end{aligned}
$$
(18)

where $\otimes$ is the Kronecker product operator, and $\mathbf{w}$ is

$$
\begin{aligned}
\mathbf{w} &= \mathbf{w}_a \otimes \mathbf{w}_t \\
&= \begin{bmatrix} w_{0,0} & \mathrm{L} & w_{0,M-1} & \mathrm{L} & w_{n,m} & \mathrm{L} & w_{N-1,M-1} \end{bmatrix}^{\mathrm{T}}.
\end{aligned}
$$
(19)

It is suggested that the transmit beampattern can be designed by optimizing $\mathbf{w}_a$ and $\mathbf{w}_t$.

3.2 Parameter Selection Analysis

When using uniform weighting, which can be equivalent to $w_{0,0} = w_{0,1} = \mathrm{L} = w_{N-1,M-1} = 1$, equation (18) can be rearranged to

$$
\begin{aligned}
A(t,\theta,r) &= \left| \sum_{n=0}^{N-1} \exp\left\{-j2\pi\left(n\Delta f_a t - n\Delta f_a \dfrac{r}{c} + f_0 \dfrac{nd\sin\theta}{c}\right)\right\} \right. \\
&\quad \times \left. \sum_{m=0}^{M-1} \exp\left\{-j2\pi\left(m\Delta f_t t - m\Delta f_t \dfrac{r}{c}\right)\right\} \right| \\
&= \left| AF_1(t,\theta,r) \mathrm{g} AF_2(t,r) \right|
\end{aligned}
$$
(20)

where

$$
AF_1(t,\theta,r) = \frac{\sin\left[N\pi\left(\Delta f_a t - \Delta f_a \dfrac{r}{c} + f_0 \dfrac{d\sin\theta}{c}\right) \right]}{\sin\left[\pi\left(\Delta f_a t - \Delta f_a \dfrac{r}{c} + f_0 \dfrac{d\sin\theta}{c}\right) \right]}
$$
(21)

and

$$
AF_2(t,r) = \frac{\sin\left[M\pi\left(\Delta f_t t - \Delta f_t \dfrac{r}{c}\right) \right]}{\sin\left[\pi\left(\Delta f_t t - \Delta f_t \dfrac{r}{c}\right) \right]}.
$$
(22)

From (20), it is observed that when we choose $\Delta f_t = 0$, an expression of the conventional FDA is gained, and the phased array beampattern expression is achieved by valuing $\Delta f_a = \Delta f_t = 0$. Therefore, the conventional FDA radar and the phased array radar can be taken for two specific modalities of SFP-FDA radar.

When time t and θ angle are fixed, the cycle of equation (21) and (22) is $T_1 = c/\Delta f_a$ and $T_2 = c/\Delta f_t$, respectively, and both are the function of r. So the cycle of (20) is $T_r = [c/\Delta f_a, c/\Delta f_t]$, where $[c/\Delta f_a, c/\Delta f_t]$ is the least common multiple (LCM) of $c/\Delta f_a$ and $c/\Delta f_t$.

4. SIMULATIONS AND RESULTS

The basic parameters of these numerical results are shown in Table 1.

Table 1. Basic parameters

Parameter	Value	Parameter	Value

N	16	Δf_a	$3kHz$
f_0	$6GHz$	Δf_t	$3kHz$
d	$0.5\lambda_0$	T_p	$2\mu s$
M	16	Target coordinates	$\left(0°,50km\right)$

The additive noise is modeled as complex Gaussian zero-mean spatially and temporally white random sequences with identical variance at each antenna element.

4.1 Comparison with Conventional FDA

The beampatterns of the conventional FDA and the proposed scheme are presented in Figure 3 and Figure 4, respectively. From Figure 3, it is observed that the conventional FDA is range-angle-coupling, and it is difficult to localize the target in accurate position. From Figure 4, we can see that the SFP-FDA can produce decoupled beam points to the target, and it has lower sidelobe and more concentrated energy in the mainlobe area.

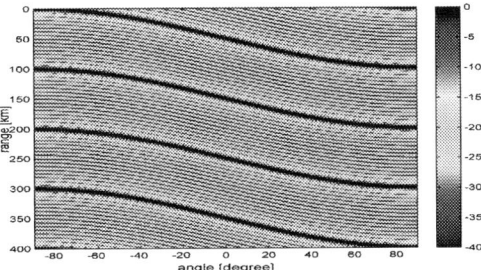

Figure 3. The conventional FDA.

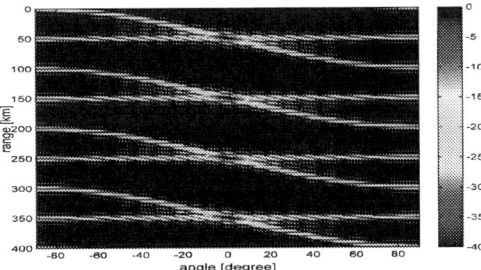

Figure 4. SFP-FDA.

4.2 Frequency Interval Selection

By analyzing the equation (21) and (22), we can find that the parameter Δf_a and Δf_t both influence the cycle and the main lobe width in range domain, and the beampattern in angle domain is chiefly affected by Δf_a, the bigger the Δf_a, the narrower the main lobe width in angle domain. In order to facilitate the analysis, we assumed $\Delta f_t = \delta \Delta f_a$.

Four different δ are selected as shown in Figure 5, for a fixed $\Delta f_a = 3kHz$, $T_{r1} = c/\Delta f_a = 100km$ is gained. The four different cycles corresponding to the different δ can be written as (a) $T_{r2} = 300km$; (b) $T_{r2} = 150km$; (c) $T_{r2} = 75km$; (d) $T_{r2} = 50km$. Owing to the equation (22) and (20), generally speaking, for a narrower and more concentrate beampattern, δ should be as big as possible. But simultaneously, the cycle will shorten, thus the number of the intersection point from (21) and (22) increased sharply, which produce many peak points that contradict with the formation of decoupled beampattern.

Notice that when $\delta > 1$, it is evitable that there are many other peak points which has different position with the desired target, consequently cannot be identified. For a detection distance no more than 100km (according to $c/\Delta f_a$), $\delta > 1$ can be used and another limited condition is $\delta < 2$.

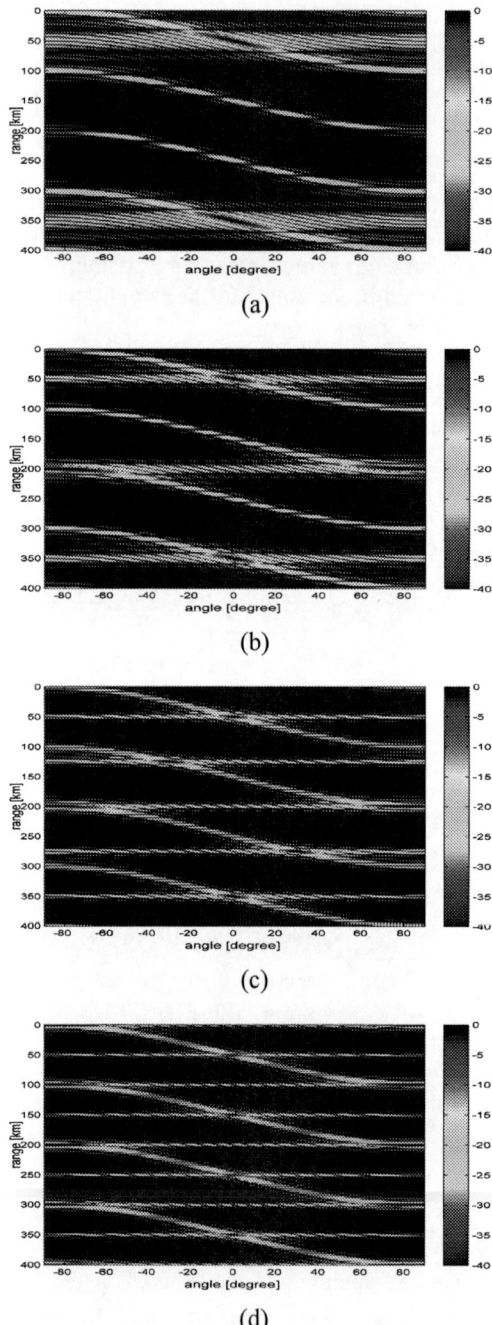

Figure 5. **SFP-FDA with different** δ: (a) $\delta = 1/3$, (b) $\delta = 2/3$, (c) $\delta = 4/3$, (d) $\delta = 2$.

In order to get longer extent of decoupled beampattern, the parameter δ should satisfied with

$$\begin{cases} \delta \leq 1 \\ 1/\delta = \text{integer} \end{cases}$$ (23)

Although the main lobe width of Figure 5. (a) is wider than the other three, the result is within the acceptable limits, meanwhile, the range-angle decoupling is well realized.

4.3 Number of Pulses

By comparing (21) and (22), it is revealed that M to (22) is just like N to (21), which means that the larger the M is, the narrower the main lobe width will be, and the side lobe width also narrower due to the multiple of (21) and (22). The drawback is that the increase of M increases the complexity of computation. Figure 6 is the beampattern that uses parameter values $M = 8$, contrasting with Figure 4, which takes $M = 16$.

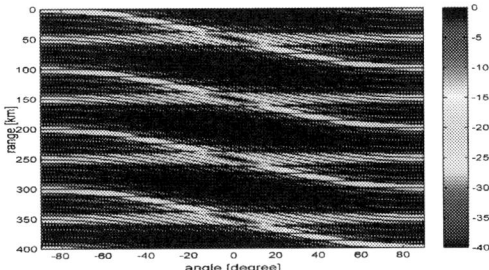

Figure 6. SFP-FDA with $M = 8$.

5. CONCLUSION

We proposed a new method for range-angle decoupled FDA transmit beamforming by stepped frequency pulses synthesizing. The parameter selections matters a lot so it is emphatically analyzed in this paper. The essence of parameter selections is to form range-angle decoupled beampattern without producing redundant peaks which can affect the performance a lot. By reasonable parameter setting, the desired beampattern can be obtained. Simulation results show the effectiveness of the proposed scheme.

ACKNOWLEDGMENTS

This work was supported by the National Science Foundation of China (grant No.61601503).

REFERENCES

[1] Antonik, P., Wicks, M. C., Griffiths, H. D., et al. "Frequency diverse array radars," *in Proc. IEEE Radar Conf.*, Verona, NY, USA, pp.215-217, Apr. 2006.

[2] Wicks, M. C., and Antonik, P. "Frequency diverse array with independent modulation of frequency, amplitude, and phase," U.S. Patent 7319427, Jan. 15, 2008.

[3] Antonik P., and Wicks, M. C. "Method and apparatus for simultaneous synthetic aperture radar and moving target indication," U.S. Patent 20080129584A1, Jun. 5, 2008.

[4] Wang, W. Q. "Overview of frequency diverse array in radar navigation applications," *IET Radar, Sonar Navigat.*, vol. 10, no. 6, pp. 1001-1012, 2016.

[5] Cetinepe, C., and Demir, S. "Multipath characteristics of frequency diverse arrays over a ground plane," *IEEE Trans. Antennas Propag.*, vol. 62, no. 7, pp. 3567-3574, Jul. 2014.

[6] Shao, H. Z., Li, J. C., Chen, H., et al. "Adaptive frequency offset selection in frequency diverse array radar," *IEEE Antennas and Wireless Propagation Letters*, vol. 13, 2014.

[7] Wang, Y. B., Wang, W. Q., and Shao, H. Z. "Frequency diverse array Cramer-Rao lower bounds for estimating direction, range and velocity," *Int. J. Antennas Propag.*, vol. 2014, pp. 1-10, Feb. 2014.

[8] Xiong, J., Wang, W. Q., Shao, H. Z., et al. "Frequency diverse array transmit beampattern optimization with genetic algorithm," *IEEE Antennas and Wireless Propagation Letters*, vol. 16, 2017.

[9] Wang, W. Q., So, H. C., Shao, H. Z. "Nonuniform frequency diverse array for range-angle imaging of targets," *IEEE Sensors J.*, vol. 14, no. 8, pp. 2469-2476, Aug. 2014.

[10] Wang, Z. H., Mu, T. Song, Y. L., et al. "Beamforming of frequency diverse array radar with nonlinear frequency offset based on logistic map," *Progress in Electromagnetics Research M*, vol. 64, pp. 55-63, 2018.

[11] Nusenu, S. Y. "Transmit/received beamforming for frequency diverse array with symmetrical frequency offsets," *Advances in Science, Technology and Engineering Systems Journal*, vol. 2, no. 3, pp. 1-6, 2017.

[12] Yang, Y. Q., Wang, H., Wang, H. Q., et al. "optimization of sparse frequency diverse array with time-invariant spatial-focusing beampattern," *IEEE Antennas and Wireless Propagation Letters*, vol. 17, no. 2, pp. 351-354, 2018.

[13] Tan, M., Wang, C. Y., Li, Z. H., et al. "Stepped frequency pulse frequency diverse array radar for target localization in angle and range domains," *International Journal of Antennas and Propagation*, 2018. DOI=http://doi.org/10.1155/2018/8962048.

[14] Wang, W. Q., and Shao, H. Z. "Range-angle localization of targets by a double-pulse frequency diverse array radar," *IEEE Journal on Selected Topics in Signal Processing*, vol. 8, no. 1, pp. 106-144, 2014.

WebVOS-A WebGIS Application for Volunteer Observation Ships

Honghai Zhu[1,a],YuYu[2,b,*],Shibo Chu[3,c]

[1]Institude of Oceanographic Instrumentation,Qilu University of Technology Miaoling Road 37 Qingdao, China 86-13854214398

[2]National Engineering and Technological Research Center of Marine Monitoring Equipment Miaoling Road 37 Qingdao, China 86-15864749133

[3]Shandong Provincial Key Laboratory of Marine monitoring instrument equipment technology Miaoling Road 37 Qingdao, China 86-13863953223

[a]hyyqyb@163.com, [b]rainertop@126.com, [c]chushibo@126.com

ABSTRACT This paper presents a WebGIS-based application WebVOS which supports management of volunteer observation ships. The design and architecture of WebVOS is fully described, which covers ship equipment, server and client. The server provides ship info, tile map, observation data collection and trajectory data services. Client is responsible for data requesting from server data services and display rendering. WebVOS supports commonly operations of volunteer observation ship management, which increases its availability. A real project is carried on to validate the application and result shows its availability.

1. INTRODUCTION

Volunteer observation ships (VOS) plays an important part in global marine weather observation. It is mainly used to collect various marine hydro meteorological data on the ship navigation route to obtain ocean observation data on the near shore, middle and far ocean. The development of volunteer observation ship can make up for the shortcomings of real-time monitoring capabilities in the far ocean. The valuable information obtained is useful for marine weather forecasting, marine scientific research, marine transportation safety, fishery production safety, aquaculture safety, port terminal safety, national defense military construction [1-3], etc. In [4], a ship monitoring system based on WebGIS is described, in which two critical problems- integrated reading of three format of maps and dynamical display of 3000 ships on the maps is solved. In [5], a navigation aids information distribution system based on network GIS was developed by WEB GIS technique and GeoBeans. In [6], by supporting of Flex and WebGIS technology, a marine meteorological information forecasting and distribution platform has been constructed. In [4], a B/S mode WebGIS application is designed and implemented based on JSP technology. So far as we known, there is no dedicated WebGIS system for volunteer observation ships.

Rest of this paper is organized as follows. The framework architecture of WebVOS is presented. Then, three kernel components of WebVOS are described. After that, an application example is presented. Finally, conclusion and the future work are given.

Content from this work may be used under the terms of the Creative Commons Attribution 3.0 licence. Any further distribution of this work must maintain attribution to the author(s) and the title of the work, journal citation and DOI.
Published under licence by IOP Publishing Ltd

2. WEBVOS SYSTEM ARCHITECTURE

As demonstrated in Fig. 1, the VOS system is based on B/S mode. In the server, services are exported using JSON format. In the client, VOS information is updated by a timer which periodically requests data from the server. The stateless mode between server and client keeps the overall system architecture neat and simply. The kernel services provided by the server are ship info service, tile map service, observation data collection service and trajectory data service. Ship info service provides basic information about the ship, including current position, name and identification, etc. Tile map service serves maps as tiles by splitting the map up into a pyramid of images at multiple zoom levels. Observation data collection service gathers observation data from ship equipment via satellite, mobile and microwave communication. Trajectory data service focuses on ship trajectory management, which provides multi-level spatial indexing and spatial querying. The client uses tile map service to do map rendering. Ship info service is used for ship position and key information display in the client. The client uses trajectory data service for ship trajectory display based on map rendering.

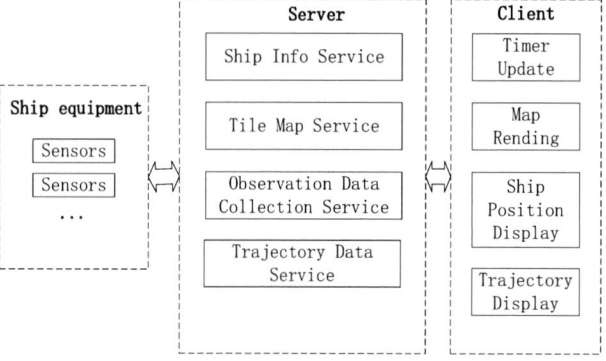

Figure 1. WebVOS system architecture.

3. SHIP EQUIPMENT DESIGN

Ship equipment is installed on the ship which contains a host and other peripherals. As demonstrated in Fig. 2, the host is composed of monitor, keyboard, memory, work indicator, alarm buzzer, AC&DC power supply and acquisition processing system. There are mainly two kinds of peripherals: communication equipment and sensors. Those sensors are responsible for gathering hydro meteorological parameters on the ship navigation routes. Acquisition processing system gathers observation data from sensors. Then, observation data are further processed and delivered to ground station through satellite, mobile communications, etc.

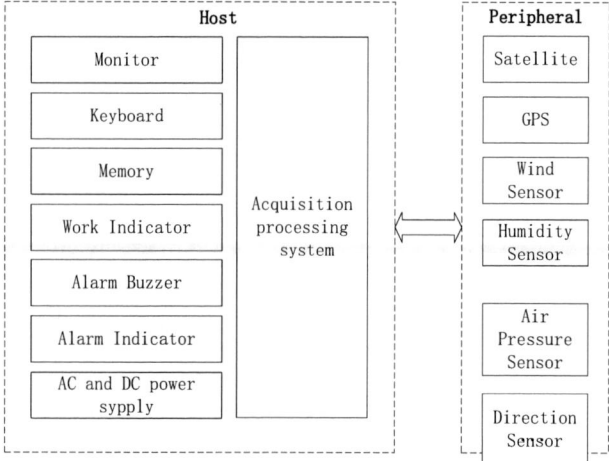

Figure 2. Ship equipment architecture.

4. SERVER DESIGN

The function of the server can be divided into two parts (se Fig.3): web services and data collection. First, data collection server gets observation data from communication equipment and stores those data into database. Then those observation data are presented by web services which are supported by web server. Clusters of web server and database are used to ensure the availability under high concurrency.

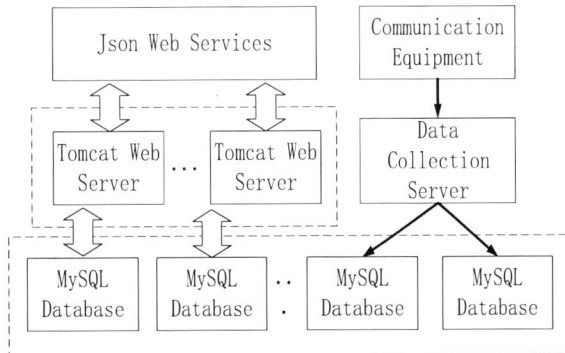

Figure 3. Server architecture.

5. CLIENT DESIGN

A JavaScript open source frame OpenLayer is used for basic map operations. Periodic update is maintained by a timer. In the timer expiring events, data are requested from the server via Json web services and rendered to form map rendering, ship rendering and trajectory rendering (see Fig.4).

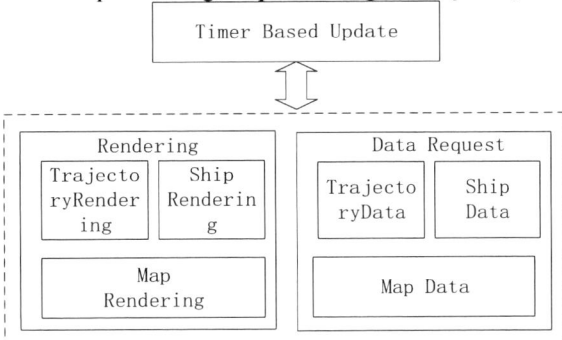

Figure 4. Client architecture.

6. APPLICATION EXAMPLE

In order to validate the Web-GIS application, it has been used in volunteer boat marine observation and management (See Fig. 5 and Fig.6), which is dedicatedly designed for monitoring seawater temperature and salt, wind speed and direction, tidal, rainfall in the ship sailing routes.

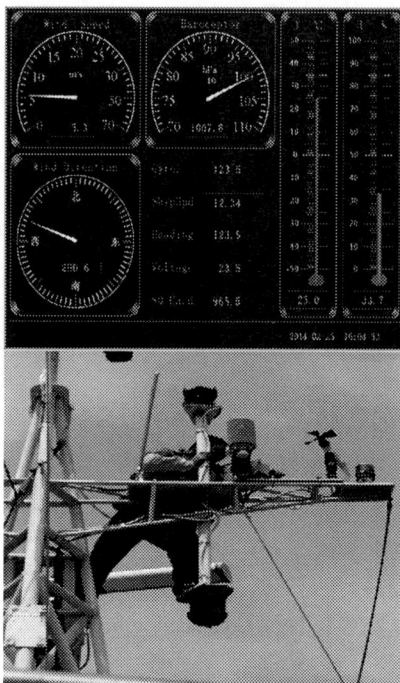

Figure 5. Ship equipment example.

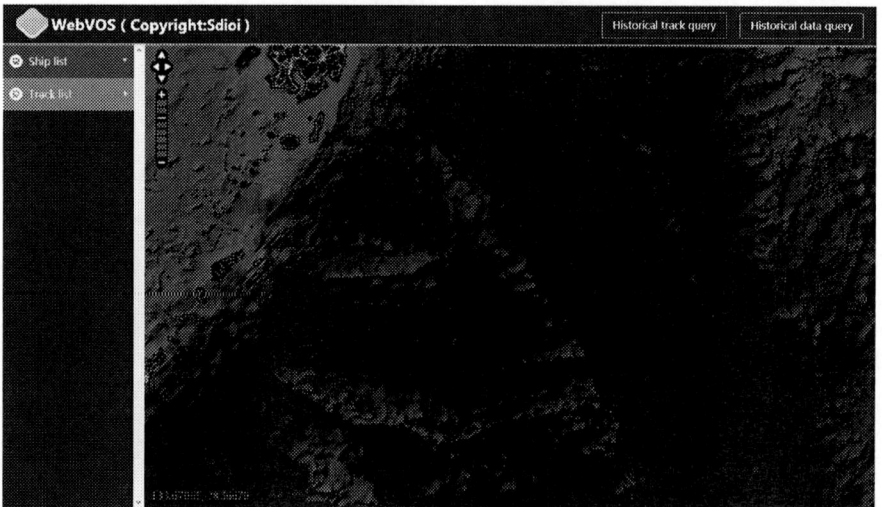

Figure 6. WebVOS application example.

7. CONCLUSION

In WebVOS, the designs and architecture for volunteer observation ship management application are provided. Yet, those components with Java and JavaScript runtime are proved to reliable, stable and easy to use. In future, development and refinement on those designs will be continued and an open source project will be founded.

ACKNOWLEDGMENTS

This work is supported by the National Key R&D Program of China under the Grant 2017YFC1405600, National Science Foundation for Young Scientists of China under the Grant 41706101, Qingdao Applied Fundamental Research Project under the Grant 18-2-2-71-jch, Qingdao City Southern District Science and Technology Development Fund under the Grant 2016-2-012-ZH.

REFERENCES

[1] Owen T, Brewer M, Redmond K, et al. Advances in Web-Based, Near Real-Time Climate Data Ingest For NOAA's Cooperative Volunteer Observation Network[J]. Agu Fall Meeting Abstracts, 2006.

[2] Zhou Y, Jiang Y, Zhu H, et al. Application of WXT520 in volunteer ship observation system[J]. Meteorological Hydrological & Marine Instruments, 2011.

[3] Crimmins T M, Crimmins M A, Gerst K L, et al. USA National Phenology Network's volunteer-contributed observations yield predictive models of phenological transitions[J]. Plos One, 2017, 12(8):e0182919.

[4] Nan X Z, Tao L H. Design of ship monitoring system based on WebGIS[C]// The, International Conference on Information Sciences and Interaction Sciences. IEEE, 2010:6-8.

[5] Peng G J, Zhang X G, Xiang L. Construction of Navigation Aids WEB GIS Information System[J]. Ship & Ocean Engineering, 2007.

[6] Wang H, Sun Y, Min J I, et al. Research of Qingdao Marine Meteorological Services Platform Based on WebGIS and Flex[J]. Geomatics World, 2015.

Improvement of LDA Topic Mining Algorithm and Its Application in Short Text

Kai Li, Chunmei Li*

Department of Computer Technology and Application, Qinghai University Xining 810016, China +8618595841812，+8613897207231

*Corresponding Author

likai614020758@126.com, li_chm0422@sina.com

ABSTRACT In order to quickly provide a large number of short text themes, an improved linear discriminant analysis (LDA) topic mining algorithm is proposed in this paper. Firstly, the acquisition method of obtaining the traditional short text themes is first analyzed. Focusing on the shortcomings of the original algorithm, the improvement process of the improved topic mining algorithm based on LDA is introduced. Finally, through the mining of short text themes by college students as an example, the improved LDA algorithm is improved in the accuracy of short text processing.

1. INTRODUCTION

In today's information age, the development of the Internet can bring a lot of information growth. How to find valuable information in the massive information, organize, manage, and clearly present it to users to help users better use the network information has become an important research topic. In the research of news commentary and similar short-text clustering, some scholars focus on the k-means clustering algorithm' s insignificancy of news comment data, select the initial point by constructing the comment similarity matrix and improve the classification method, and use the cosine distance [1, 2]. In addition, some scholars have analyzed the labels of short texts and used the standard LDA algorithm for cluster analysis [3]. Some scholars have used feature vectors to calculate hot topics in a large number of MicroBlog short texts, and used TD-IDF (term frequency–inverse document frequency) to solve word frequency and word vectors [4]. Although extensive research has been done on the topic mining of traditional texts, traditional text mining algorithms cannot well model special short texts [5].

Extracting assessment topics/objects from the results of topic mining is more suitable for news reviews. This is because the topic mining algorithm based on short text clustering is easy to get the comment topic [6]. Some scholars have proposed preprocessing techniques based on the sparseness, multidimensionality and large-scale nature of MicroBlog information, and used LDA for topic mining [7, 8]. Some scholars have found in the news commentary on integrated LDA and clustering that more noise data in the comments has a great influence on the center point of K-means. Outliers can severely distort the distribution of data, and the squared error function can severely degrade the effects, so k-medoids clustering is used [9,10]. Due to the defects of the LDA algorithm itself, the program overflows and the slow speed caused by the high dimension of the short text matrix, and the LDA is more and more prominent in dealing with the segment text problem. Based on this problem, this paper proposes an improved algorithm based on LDA for a series of MicroBlogging topics, which is

Content from this work may be used under the terms of the Creative Commons Attribution 3.0 licence. Any further distribution of this work must maintain attribution to the author(s) and the title of the work, journal citation and DOI.
Published under licence by IOP Publishing Ltd

combined with the text structure of short text content, and improves the LDA algorithm to find short text keywords of MicroBlog's data.

2. TRADITIONAL LDA THEME MODEL

The LDA topic model is a generation model. Generating a model refers to a given article, determining each word to start with an article, and using a "document-theme" probability distribution to select a topic and selecting a word with a probability distribution based on the topic. Repeat the selection of words with a probability distribution from which the article is generated. Then the probability of the occurrence of each word in the document is calculated as follows:

$$P = \sum P * P$$

LDA is a three-layer Bayesian probability generation model in which two model parameters "word subject" and "topic document" need to be solved. The probability map of the LDA probability topic model is shown in Figure 1 below.

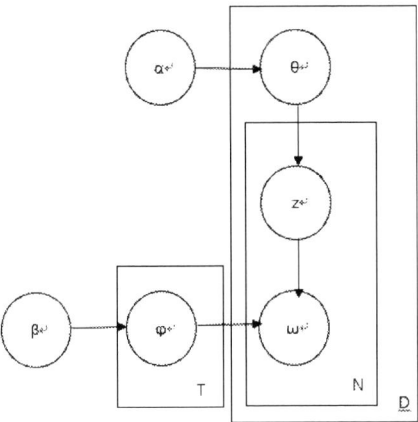

Figure1. LDA probability map model

Φ represents the "subject" probability model, which is a Dirichlet distribution with hyperparameters β. θ represents the "document subject" probability model, which is a Dirichlet distribution with hyperparameters a. T represents the number of topics. D represents the number of documents, and N represents the word length of the document.

By calculating the heat of the word, the corresponding topic heat is calculated, and the topic of the short text is sorted on this basis. The calculation of the topic's hotspot is as follows:

$$H(T) = \sum_d T_w (w \in T, w \in d)$$

3. IMPROVEMENT OF TOPIC MINING ALGORITHM BASED ON LDA

Since text is treated as an unordered word package model in the semantic analysis processing, all short texts of the same time period can be linked together and treated as text processing blocks. In this way, the text space can be reduced, the features of the original text can be expanded, and the similarity of the text can be improved, thereby solving the sparseness and abnormality of the short text to a certain extent. However, it has a high repetition rate. By simplifying the function, the text with the high coincidence rate is changed to the same text as the general text, which is advantageous for the effect of the LDA processing. The algorithm is improved according to the characteristics of the LDA ignoring the word order in the document and the corpus, as follows:

Step 1: the comment of the same paragraph is turned into a text block with a total of m text blocks.

Step 2: the first paragraph is represented by all the characteristic words T. Corresponding word frequency is WF: WF (wf1, wf2, ... wfk). Then express this paragraph with L1 = (log*T1|log*T2|...log*Tk).

Step 3: then repeat step 2 to get L2, ... Lm.

Step 4: Use the short text of a total of m lines as input to the LDA's Single-pass incremental clustering algorithm. The improved LDA's Single-pass algorithm usually selects the first document as the center of the first category, and compares the similarity of the two texts with the next document similarity as the initial cluster. The theme center is recalculated each time when a new topic is generated, or when a topic is added to a new document. The cluster center is the average of the document vectors in the cluster. The improved algorithm flow chart is shown in Figure 2.

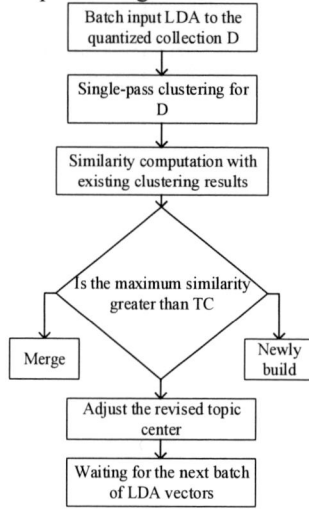

Figure2. Improved Single-pass algorithm flow chart

4. MINING QUALITY TOPICS FROM THE CORPUS

4.1 Application of Improved LDA Topic Mining Algorithm in Short Text

The corresponding number of topics will be automatically mined based on the number of clusters pre-entered manually. These topics are automatically obtained based only on the co-occurrence relationship of words in the corpus. In fact, not all topics consisting of related words can represent a real subject. Table 1 below is the related words of the wedding theme in the corpus, and the topics are found according to the co-occurrence relationship. It can be seen that the top seven words with the highest frequency of the theme are "wedding", "plan", "wedding ceremony" and so on. These words are a good representation of the theme of the wedding, and there is no word " Wedding plan" related to the theme, so this theme is a quality topic.

Table 1.high quality theme (wedding theme)

Key Words	Wedding	Plan	Wedding Ceremony	Ceremony
Probability	0.0918	0.070	0.0310	0.0202
Key Words	Event	Celebration	flower	...
Probability	0.0185	0.0153	0.0144	...

4.2 Mining Noise Themes in the Classification Corpus

Entering the number of topics in the algorithm, here set the number of subjects K to 8. The corpus of different topics are tested, and a non-centered topic with a relatively low frequency of corresponding keywords with a minimum degree of relevance is found in the classified corpus, as shown in Table 2 below. It can be concluded that the words "centre" and "service" are the most frequently used words in this topic. However, common words such as "center" and "service" usually do not represent the subject, and the probability distribution is not uniform. The probability of only "center" is 0.3277. Second, in its high-frequency words, there are two different themes, "shopping" and "Internet cafes." Therefore, the topic is a noise theme that cannot be classified as a topic in the text and should be classified as a non-central theme.

Table 2.noise theme (no central theme)

Key words	Center	Service	Shopping	Internet cafe
Probability	0.3277	0.0503	0.0345	0.0083
Key words	One-stop	Quality	Call	...
Probability	0.0078	0.0052	0.0047	...

4.3 Feature Word Selection

The data source in this test experiment is the 1 million 400,000-dimensional data of Sina Weibo, which will be built in the experiment. We find the probability of the first 200 words under each topic and the probability between 20% and 75%. Can fully represent the corresponding theme. Accordingly, in the screening process of the high quality theme of the LDA model, the first 200 feature words in the ".twords" suffix file output by the LDA model are processed only according to the frequency of each topic.

4.4 Similarity Calculation

The Sim function is defined for independent supervision of each topic. Mainly to calculate the similarity between the two representative words of the subject, that is, the accumulation of the probability product of the co-occurrence words between the two representative words of the subject. In the actual calculation, the first 200 feature words are used to calculate the similarity, and the formula is as follows:

$$Independence(i,m) = \sum_{j=1, j \neq i}^{N} Sim(topic_i, topic_j, m)$$

N is the total number of topics, i.e. the number of artificially predefined LDA clusters; J is the subject of the first document; L is the subject of the second document, and m is the number of texts for the two topics.

4.5 Choice of High Quality Theme

Computation of similarity although clustering operations can be performed, the word frequency distribution of high quality topical representation words are usually unbalanced. A few core words appear more frequently. However, the representation word s of a noise subject are usually composed of some random words, and the correspondence between these random words and high frequency words is small. To calculate the coverage of feature words, that is, the sum of the calculated probabilities, the choice of high quality topics can be achieved. Therefore, it is necessary to set a certain threshold, and some low-probability eigenvalues are deleted by filtering the threshold. If the threshold is already sorted on this basis, the hotspot formula for the topic is as follows:

$$Coverage(i,m) = \sum_{j=1}^{m} wp_{i,j}$$

WPi,j is the probability of the jth character from the topic.

Secondly, considering the probability distribution of feature words, if the volatility is too small, that is, the probability distribution of all feature words is too balanced, most words are composed of random words, lacking the subject core words, such topics can not be used as high quality On the other hand, if the fluctuation is too large, that is, the probability distribution between the feature words is not balanced, this theme is mainly composed of a few general words as high-frequency words, and there is no difference between the topics. For example, non-central themes of high-frequency words, such as "services" and "centers", cannot be considered high-quality topics. So the variance is used to measure the deviation between the random variable and its mathematical expectation, which is used to measure the volatility of a set of data. Therefore, the variance of a high quality theme should fall within a range of fluctuations. The calculation formula is expressed by Equation:

$$\overline{x(1,m)} = \frac{\sum_{j=1}^{m} wp_{i,j}}{m}$$

Entropy was originally a concept in thermodynamics, a physical quantity used to express the degree of disorder of molecular states, and then introduced into information theory to express the order of the system. The lower the information entropy, the less information is contained. The information entropy has an accurate formula, and the corresponding information entropy of the probability sequence is calculated as Equation :

$$H(P_1, P_2,...P_n) = -\sum_{i=1}^{n} P_i * \log(P_i)$$

When choosing each high-quality theme, from an information theory point of view, if all the words appearing are considered to be representation words, they are often stable to the whole system, and the smaller the information entropy of the word sequence is better. Similarly, if only a small number of core feature words are selected to calculate the information entropy, since the probability of the core word is high and the probability distribution between the core words is not balanced, the higher the information entropy of the sequence, the better.

5. ALGORITHM IMPLEMENTATION

5.1 Experiment Setup

In this experiment, the word segmentation tool is used for microblog text, complete word deletion and perform part-of-speech annotation work. Sort the text according to part-of-speech annotation, calculate the heat of microblog, and make preliminary selection according to heat. A Single-Pass and improved LDA algorithm is used for the user to select the subject number K value, the similarity threshold, and the number of microblogs per batch to facilitate topic merging in text clustering.

5.2 Database Design

According to the analysis in the previous chapters, five database tables are designed in this experiment: the microblog data table is used to store the text content and heat value of Weibo, the heat value is composed of forwarding, comment, and praise; the text direction scale table is used for storage. Processing the vector value composed of the text content; the theme table is used to store the keyword used to process the data using the improved algorithm; the user table is used to store the user name, region and gender obtained when the text content is obtained; the log table is used to store the processed data generated Logging.

5.3 Data Sources

The test data source of this experiment data is Sina Weibo. From January 1st, 2018 to November 1st, 2018, all Weibo on the topic of "college students" and no central theme.

5.4 Experimental Results

Use Python's re matching characters to clean out the English letters, numbers, punctuation, and special symbols in the Weibo content. Then the data of the cleaning is performed using the THLAC library of Tsinghua University [11] and the stop word list of Harbin Institute of Technology for word segmentation. Use the IDF method to vectorize the data after the word segmentation. The data after vectorization of the words is calculated separately, one part is the data with the subject, and the other part is the data without the subject. The results of the final processing of the data with the subject are shown in Table 3 below.

Table 3. represents the center theme (college students)

Subject Words	College Student	Game	Work	High School
Probability	0.5718	0.2102	0.1452	0.0311
Subject Words	Youth	Graduation	World Cup	...
Probability	0.0227	0.0100	0.0090	...

Table 3 shows that the two most frequent keywords are the vocabulary of the topic specified in the collection of data, "college student" and "game" Therefore, in terms of the improved algorithm, the accuracy of keyword extraction is higher for short text content with themes. Then, the collected MicroBlog data without the center theme is processed. Can get the following table 4:

Table 4 represents a no-center theme

Subject Words	Victory	Trade	Asian Games	China
Probability	0.4302	0.3281	0.1101	0.0514
Subject Words	USA	Commerce	ZTE	...
Probability	0.0349	0.0677	0.0221	...

Observing Table 4, we can see that the topic-free corpus is about the topic of "Asian Games wins" and "China-US trade" "ZTE". By manually searching for 10 months of topics, these three types of topics are indeed in 10 Among the most discussed topics since the month, the improved LDA algorithm improves accuracy in the processing of short text. Reduce the impact of irrelevant vocabulary

6. CONCLUSION

An improved LDA topic mining algorithm is proposed in this paper. It can improve the shortcomings of the original LDA algorithm applied in the short text data. Aiming at the special text structure of microblog short text, an improved LDA algorithm is proposed to effectively discover the subject words of short text topics. In theory, it is first discovered that the improved LDA subject evolution algorithm can be applied to a series of microblog heat analysis. Secondly, it is found that unlike the evolutionary features of low-dimensional dense vectors, for high-dimensional sparse short text vectors, it needs to be converted into a file structure close to the microblog text. This provides an improved direction for short text LDA topic mining: simplifying a large amount of short text data into general low-dimensional text. In practice, the topic extraction model can be applied to short text heat analysis, such as microblogging, post bar, chat history, and video barrage. Improved algorithms facilitate topic mining due to high precision, which helps managers and decision makers make use of critical information to make decisions.

REFERENCES

[1] Zhao D, He J, Liu J. An improved LDA algorithm for text classification. International Conference on Information Science, Electronics and Electrical Engineering. IEEE, 2014, pp. 217-221.

[2] Zhang C, Yang M. An Improved Collaborative Filtering Algorithm Based on Bhattacharyya Coefficient and LDA Topic Model.International CCF Conference on Artificial Intelligence. Springer, Singapore, 2018, pp. 222-232.

[3] Li Jing, Yin Jian, Liu Shaopeng, et al. Mining of Weibo Topics Based on Topic Tags[J]. Computer Engineering, 2015, 41(4): 30-35.

[4] Li Hui, Wang Liting. Research on Hot Spots of Microblog Based on Term Heats[J]. Information Science, 2018, 36(4): 45-50.

[5] Wang D, Wang S. Improved 2DLDA Algorithm and Its Application in Face Recognition. IEEE, International Conference on Trust, Security and Privacy in Computing and Communications. IEEE, 2015, pp. 707-713.

[6] Qiuqiu L I, Yang H, Feng J, et al. Face recognition algorithm based on 2DPCA+2DLDA and improved LPP. Computer Engineering & Applications, 2015, 90(11), pp. 777-81.

[7] Ma S, Jiang Z, Zhang T. The improved multi-scale Retinex algorithm and its application in face recognition. Control and Decision Conference. IEEE, 2015, pp. 5785-5788.

[8] Hu D, Chen L. A High Efficient Recommendation Algorithm Based on LDA. Human Centered Computing. Springer International Publishing, 2016, pp. 668-675.

[9] Fralenko V P. Localization of text fragments on mixed background: short scientific review. Programmnye Sistemy Teoriya I Prilozheniya, 2014, pp. 33 - 45.

[10] Lee J Y, Dernoncourt F. Sequential Short-Text Classification with Recurrent and Convolutional Neural Networks. 2016, pp. 515-520.

[11] Maosong Sun, Xinxiong Chen, Kaixu Zhang, Zhipeng Guo, Zhiyuan Liu. THULAC: An Efficient Lexical Analyzer for Chinese. 2016.

A security model based on intelligent decision

Shiping Xu, Ying Zhou*, Ronghua Guo, Jiawei Du, Zhe Liu

Luoyang Electronic Equipment Test Center of China, Luoyang, Henan, 471003, China

*Corresponding author's e-mail: zy_jackson@sina.com

Abstract. This paper introduces intelligent decision-making into security model and proposes an adaptive network security model based on intelligent decision-making, based on the analysis of traditional security model. On this basis, the security capability of the model is preliminarily given, and the technical framework of the security protection system is constructed based on the model. Finally, an adaptive security protection process based on the model is given.

1. Introduction

With the rapid development of network and information technology, various security events such as Trojans, worms, DDoS attacks, botnets, and network intrusions have become more frequent, and the complexity and automation of cyber attacks have been increasing, all of which make the traditional security protection methods face a severe challenge. In order to utilize various existing security defense technologies and form a whole security protection system, research on various security models and related technologies has gradually become a hot research topic in the field of network security[1]. Based on the research status of existing PDR security model and its derivative model, this paper proposes a security model based on intelligent decision-making, and discusses the security capability, technical framework and security protection process of the model.

2. Research status of security model

As the foundation of the security protection system, the security model has always been the focus of information and network security research. The classic security models include the PDR model[2] proposed by the US Internet Security Systems Corporation, the P2DR model[3], the PPDRR model[4], and the PDRR model proposed by the US Department of Defense[5]. The dynamic defense model APPDRR model[6], the WPDRRC model proposed by China's 863 information security expert group[7], the adaptive security framework[8] proposed by Gartner for the next generation security system, and the P2OTPDR2 model[9]. The above models are supplemented and improved from the basic PDR model, taking into account the three elements of protection, detection and response, based on the theory of time-based security. The ability and security of different models are measured by time scales, which fully reflects the dynamic thinking of security. However, with the continuous development of cyber-attack technology and the security vulnerabilities that are constantly discovered by the network system itself, the security policy of the network protection system cannot be adjusted in real time according to the security status of the network, resulting in a longer "shutdown" time of the network system. The loss has become larger. Although the existing security model basically uses the security policy as the core to complete the model actions of early warning, protection, detection, response, etc., for the moment, the formulation, generation, delivery, and execution of security policies are based on human experience and When knowledge is implemented, it is often impossible to make security decisions objectively and timely according to the current network security status and provide corresponding security policies to achieve real-time security protection.

3. Security model based on intelligent decision

3.1. Security model design based on intelligent decision

At present, the classical adaptive protection model PDR and its evolution model (such as P2DR, PDRR, P2DRR, etc.) propose a security capability consisting of "protection", "detection", "response", and "recovery" centered on "policy". And the stage is generally recognized. However, systems with the above security capabilities and phases are still passively protected, lacking sufficient ability to detect threats in advance and proactively protect them. With the development of technology ,such as intelligent decision-making, it is possible to use it to predict and dispose of threats before they occur.

Combining the advantages and technical feasibility of the existing adaptive protection models, this paper proposes an adaptive security model for the network system-the ID-P3DR2 model, which will focus on the "security strategy" and include "predicting", "protection", "detection", The five core security capabilities and phases of "response" and "recovery" are shown in Figure 1.

Figure 1 ID-P3DR2 model　　　　　Figure 2 Core capabilities of the security model

Intelligent Decision It is the core of the operation of the entire safety technology system. In the adaptive protection system, the entire security system is constructed and operated by the security policy. As the security state changes, the security policy must be dynamically determined and adjusted to achieve security. In the earlier protection system, security decision-making and policy adjustment mainly relied on people to configure, resulting in inefficient security response. For highly dynamic and complex network systems, security decisions and policy adjustments have failed to meet security requirements[10]. With the development of artificial intelligence technology, intelligent decision making using artificial intelligence technology to achieve automation has become possible. The online model of the network system needs to be centered on intelligent decision-making. With the situational awareness and automated linkage response technology, the adaptive protection of security threats is realized, and the threat response speed is greatly improved.

Security Policy The security policy is a set of rules for guiding security activities based on the inherent security requirements, guidelines and principles of the network system, as well as external constraints such as laws, regulations, and regulatory requirements. It defines how to implement protection of the personnel, equipment, and software in the network system, how to deal with security incidents and other specific requirements. The entire network system security system operates under the guidance of security policies. The essence of adaptive security is that security policies can be adaptively adjusted with the security state of the system itself and the dynamic changes of external threats to guide the security system to effectively protect against threats.

Prediction The core of the forecasting phase/capability is to proactively predict the attack events and potential asset exposures that the network system may encounter by sensing changes in the internal and external environment, and feedback the prediction results to the protection and detection phases to form a closed loop. To avoid potential events, network systems must monitor hackers and other sources of intelligence to predict new attacks. Intranet systems must continuously track changes in assets and the environment and analyze new vulnerabilities.

Protection The key goal of the protection phase/capability is to reduce the attack surface of the network system and prevent the attacker and its attack methods before the attack affects the network system. This phase typically includes a set of strategies, products, and processes designed to prevent the success of the attack, including the ability to strengthen and quarantine systems, vulnerability fixes, transfer attackers, and prevent unauthorized access and activity.

Detection The protection system of the network system must assume that it has been compromised or that there is a possibility of being compromised, so the detection capability is crucial. The goal of detection is to reduce the time it takes to identify threats by identifying and identifying threats. This requires continuous and comprehensive monitoring of the system and the use of increasingly sophisticated analytical tools to provide detection capabilities.

Response The response is to deal with discovered threats or attacks in a timely manner. It is necessary to combine comprehensive investigation and evidence collection and traceability analysis to determine the root cause and scope of the threat, and to develop appropriate strategies for rapid disposal. Changes to security policies and safeguards must then be designed, modeled, and implemented to avoid similar events in the future.

Recovery The main purpose of the recovery phase/capability is to repair the damage caused by the attack in time to ensure the availability and continuity of the services of the network system. It usually includes two aspects of data recovery and business recovery.

The adaptive security protection process as a whole is a security decision based on the perceived security risk, and adjusts the closed loop process of the security policy implementation response. Through the continuous closed-loop operation of the protection process, the safety protection system can follow the dynamic changes of the internal and external security situation and continuously evolve itself.

3.2. Security model security capabilities based on intelligent decision

In order to ensure the security of all levels of the network system and realize the adaptive security protection of the network system, several key capabilities need to be realized in each stage of the model. Figure 2 lists the 14 core security capabilities that a network system's protection system must have and its correspondence with the capability phase of the intelligent decision-based security model.

The three core competencies of "threat knowledge extraction and continuous update", "risk identification and assessment" and "threat prediction" need to be realized in the forecasting stage.

Three major capabilities of "system reinforcement and isolation", "abnormal and attack event protection", and "protection system automatic construction and adjustment" are required in the protection phase.

The "system vulnerability detection", "abnormal and intrusion event detection", "risk confirmation and sequencing" capabilities are required during the detection phase.

Three major capabilities of "event investigation and forensics", "security policy adjustment", and "event linkage response and disposal" are required in the response phase.

The "data backup and recovery" and "business recovery" are required during the recovery phase.

3.3. Intelligent decision-based security model technical framework

The security technology framework of the network system is constructed based on the above security model, as shown in Figure 3. The entire technical framework consists of four parts: security mechanism, security resources, regulatory standards and security management. The security mechanism includes four levels: physical and environmental security, network and communication security, device and computing security, and application and data security. Security resources include security platform, security device and component, and threat knowledge. The basic support; at the same time, the implementation and operation of the technical framework also needs the support of the safety management system.

4. Adaptive security protection based on intelligent decision-making security model

Based on the model proposed in this paper, the security threat adaptive protection process shown in Figure 4 can be implemented.

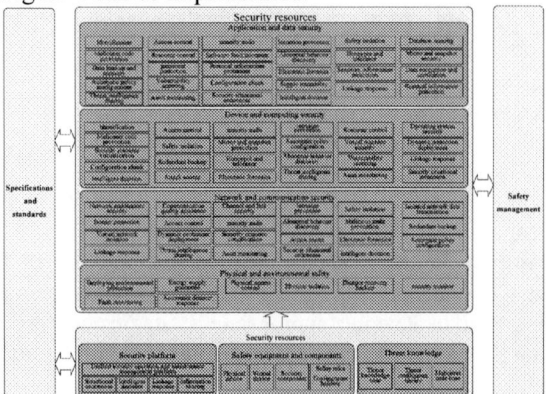
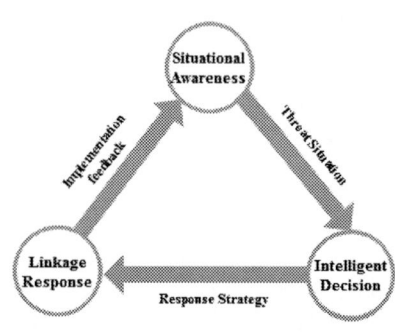

Figure 3 Security Model Technical Framework   Figure 4 Security adaptive protection process

Situational awareness Situational awareness continuously monitors and analyzes the massive data and analysis results of the network system's asset status, system vulnerability, security log, attack events and threat intelligence, knowledge base, etc., and analyzes the context, hazard, scope and development trend of each specific threat, in turn, comprehensively describe the current security situation of the entire network system, predict the next possible attack and its harm.

Intelligent decision Based on the security situation and threat development trend described by situational awareness, intelligent decision-making combines the security objectives, security policies, security capabilities and security knowledge of the network system to comprehensively determine the response strategy for security threats. Depending on the security threat, the response policy can include security policy adjustment, system hardening, defense system adjustment, event handling, backup, and recovery.

Linkage response The linkage response specifically implements the response strategy formulated by the security decision process, and according to the current situation of the security resources, jointly adopts different methods such as technology and management to deal with the security threat, track and verify the response result, and feedback the execution result to the situational awareness for situational awareness. Master the security status of the system in real time and initiate new security decisions and linkage response processes as necessary.

5. CONCLUSION

Based on the analysis of various existing classic security models, this paper introduces intelligent decision-making into the security protection system, and proposes a security model based on intelligent decision-making. Based on the model, the security model should be secure. The capabilities, technical framework and security protection process were initially discussed. How to extract the current network state feature factor, as the input of intelligent decision-making, takes the corresponding deep learning or reinforcement learning algorithm for decision-making, gives the optimal solution strategy to deal with the current network threat, and timely (or even real-time) network responding is the focus of the next step.

References

[1] Jiang Wei. Research on key technologies of active defense based on offensive and defensive game model [D]. Harbin: Harbin Institute of Technology, 2010.

[2] SCHWARTAU W. Time-based security explained: Provable security models and formulas for the practitioner and vendor [J]. Computers & Security, 1998, 17(8): 693-714.

[3] HAN R-S, XU K-Y, ZHAO B. Research and Design of Policy Deployment Model for P2DR Model [J]. Computer Engineering, 2008, 20(0): 69.

[4] Zhou Haigang, Qiu Zhenglun, Xiao Junmo. Network Active Defense Security Model and Architecture[J]. Journal of PLA University of Science and Technology: Natural Science Edition, 2005,6(1):40-3.

[5] Lin Shoumei, Zhang Jiandong, Hu Ruizhi. PDRR Network Security Model [J]. Computer CD Software and Applications, 2010(11):57.

[6] Pan Jie, Liu Aijie. Research on Network Security System Based on APPDRR Model [J]. Telecommunication Engineering Technology and Standardization, 2009(7):27-30.

[7] Yao Chuanjun. Application of WPDRRC Information Security Model in Security Level Protection[J]. Optical Communication Research, 2010(5):27-29.

[8] https://www.csdn.nearticle/2015-06-24/282503.Gartner: Using adaptive security architecture to deal with advanced directed attacks.

[9] Zuo Feng. Research on Information Security System Model[J]. Information Security & Communication Security, 2010(1): 70-72.

[10] HAN Ruisheng, ZHAO Bin, XU Kaiyong. Policy-based Integrated Network Security Management System[J]. Computer Engineering, 2009,35(8):201-204.

A New Clustering Validity Index based on K-means Algorithm

Xiangru Hou*

Department of Information Engineering, Heilongjiang International University, Harbin, 150025, china

*Corresponding author's e-mail: 3306530748@qq.com

Abstract. Although cluster analysis has got great achievements, there are many questions in it. In this paper, the question on determining optimal number of clusters in cluster analysis is studied mainly. KMS (K-means Silhouette) for determining optimal number of clusters in K-means clustering algorithm are proposed. KMS (K-means Silhouette) algorithm improves the way of setting initial clustering centers in K-means clustering algorithm, and uses Silhouette validity index to determine optimal number of clusters. The experimental results on artificial datasets indicate the effectiveness of the proposed algorithms.

1. Introduction

K-means clustering algorithm is one of the most widely used algorithms in clustering analysis. The K-means algorithm, based on the determined clustering number k and the selected initial clustering center, is a clustering algorithm that obtains the minimal sum of the distances (squared) from all samples to the center of the category to which they belong. In practice, it is difficult to accurately define the k value. At present, some function indexes of clustering validity have been proposed.

For the selection methods of K-means algorithm's initial clustering center, k samples were usually randomly selected as the initial clustering center. If this method of randomly selecting the initial clustering center was used to cluster some complex data sets, it might result in unstable and wrong clustering results. At present, researchers have proposed a lot of methods to select the initial clustering center of K-means algorithm. Reference [1] introduced a method of using multiple iterative sampling data sets to obtain initial values, which could solve K-means algorithm's higher dependence on the initial clustering center, Reference [2] used a density-sensitive similarity measurement to calculate the density of samples and to heuristically generate the initial clustering center of samples. Based on max-min distance algorithm, the research on the improvement in the selection of K-means clustering algorithm's initial clustering center was conducted to better determine the optimal clustering number.

2. K-means clustering algorithm

The algorithm selects the k cluster and selects k initial cluster centers, and each sample is assigned to a cluster in the k cluster according to the minimum distance principle. After that, heart cluster and the category of each sample are adjusted continuously, and finally. The sum of the squares of each sample to the center of its category is minimum. The algorithm steps are as follows:

Algorithm K-means clustering algorithm

1. For n samples, k samples are selected as the initial clustering centers $(z_1, z_2, ..., z_k)$.

Content from this work may be used under the terms of the Creative Commons Attribution 3.0 licence. Any further distribution of this work must maintain attribution to the author(s) and the title of the work, journal citation and DOI.
Published under licence by IOP Publishing Ltd

2. For each sample x_i, find the nearest cluster center z_v, and assign it to the cluster u_v marked by z_v.

3. The average method was used to calculate the centers after reclassification

4. Computing $D = \sum_{i=1}^{n} [\min_{r=1...k} d(x_i, z_r)^2]$.

5. If the D value converges, then returns $(z_1, z_2, ..., z_k, U)$, and terminates the algorithm, or goes to the step (2).

3. The method for determining the optimal clustering number based on the improved initial center

3.1. The method for setting initial clustering center

3.1.1. Max-min distance algorithm. Max-min distance algorithm is based on heuristics in the pattern recognition field. Its basic idea is to take the object as far as possible as the clustering center, preventing the initial clustering center from excessive affinity in the initial value selection. This method not only intelligently determines the number of initial clustering centers, but also improves the efficiency of sample partitioning. The data set $S_n = \{x_1, x_2, ..., x_n\}$, select scale factor θ. The algorithm steps are as follows:

Algorithm Max-min distance algorithm

1. Take a sample from S_n as the first clustering center z_1

2. The largest sample of z_1 distance from S_n as the second clustering center z_2

3. The distance between the sample x_i and z_1, z_2 is not calculated as the cluster center, and the minimum value is d

4. If $D_t = \max\{d_i\} > \theta \| z_1 - z_2 \|$, the corresponding sample x_t is used as the third clustering center z_3

5. If there are k cluster centers, a sample that is not taken as a cluster center and the distance of each cluster center d_{ij}, Calculate out $D_r = \max\{\min(d_{i1}, d_{i2}, ..., d_{ik})\}$.

6. Repeat the same treatment, until we can't find a new cluster center that meets the requirement

7. Each sample is divided into various types according to the minimum distance principle

3.1.2. Basic setting method. This paper attempts to combine clustering center initialization with the clustering validity index. In order to determine the optimal clustering number. In the process of searching the clustering space and gradually increasing clustering number by K-means clustering algorithm, when the clustering number is k_{min}, k_{min} samples are selected as the initial clustering center based on the principle of max-min distance algorithm; every time a clustering number is added, add an initial clustering center according to the principle of max-min distance algorithm when maintaining the same center as the last initial clustering, and try to keep the continuity of conditions and the stability of clustering results. In addition, the initial clustering center selected based on max-min distance algorithm is more likely to belong to different clusters, so that better clustering results can be obtained, in order to better determine the optimal clustering number through the validity index. Since the clustering number is unknown in the max-min distance algorithm, the clustering number needs to be obtained by selecting the proportionality coefficient θ as a constraint.

3.1.3. Setting method analysis. The K-means clustering algorithm is used for a data set. Under the known cluster number and the established initial cluster center conditions, the final clustering results are uniquely determined and thus are stable. This can be proved by the operation of replacing the known conditions into the K-means clustering algorithm. In the first iteration of the algorithm, the number and location of the initial cluster center are determined. Each sample is divided into a cluster according to the minimum distance principle, so the classification of each sample is uniquely determined. In the next iteration, taking an average method to calculate the cluster centers after the reclassification. The clustering center and clustering results will change in the two adjacent iterations. However, because each input condition is uniquely determined, the classification algorithm based on the minimum distance principle is stable, so the result of each iteration process is always unique. In this paper, the time complexity of the initial clustering center is $O(ndk)$, so as long as the k-means clustering algorithm is reduced once. The method of this paper is equivalent to the traditional method of generating initial clustering center. If the number of iterations decreases by more than two times, the efficiency of this method will be higher than the traditional method.

3.2. KMS algorithm for determining the optimal number of clusters

The traditional K-means clustering algorithm is used to determine the optimal number of clusters, for each cluster number k, the cluster center should be re initialized. Because the initial cluster centers are different, the clustering results of different cluster numbers have poor comparability of the validity index values, which make the original algorithm to solve the optimal number of clusters unstable. Therefore, this paper improves the existing method of random selection of the initial clustering center method. Initial clustering centers are determined by maximum and minimum distance algorithm. With the increase of cluster number, the original initial cluster centers remain unchanged. Based on the principle of maximum and minimum distance, the initial cluster centers are gradually increased, so that there is an inheritance relationship between the initial cluster centers of different cluster numbers. The upper limit of the search range of the algorithm is changed from the existing $Int(\sqrt{n})$ to the number of clusters k_{AP} generated by the AP algorithm. The Silhouette index was used to analyze the clustering results to determine the best clustering number. The new algorithm to determine the optimal cluster number is recorded as KMS. The algorithm is summarized as follows:

Algorithm KMS algorithm

1. Select the search range of cluster numbers $[k_{min}, k_{max}]$

2. For k= k_{min} to k_{max} .

    (1) Initialize k initial cluster centers z_k

    (2) Using K-means clustering algorithm, update membership matrix U^k and cluster center Z^k

    (3) Check the termination conditions, if not satisfied, turn(2)

    (4) Using clustering results to calculate the Silhouette index value, turn (2).

3. Comparing the Silhouette index value, the k corresponding to the maximum of the index value is the best number of clusters k_{opt}

4. Output the best number of clusters, validity index and clustering results.

4. Experimental Results and Analysis

4.1. k_{max} Simulation experiment and analysis

In this experiment, AP algorithm was used to estimate the k_{max} cluster number for the following 12 datasets. Set the AP algorithm $p = p_m$ .It is shown in Table1. The dataset used in the experiment includes real datasets and artificial datasets, with general numerical data, gene expression data, image data, etc. Kes2 is a two-dimensional two cluster artificial data set, Kes3 is a two-dimensional three cluster of artificial data sets, and the rest of the data set details refer to the relevant references.

Table 1. k_{max} evaluated by Affinity Propagation clustering algorithm

Data set	Sample number	Correct cluster number	AP cluster number	$Int(\sqrt{n})$	Source of data set
Face Image	900	100	106	30	Literature [3]
Random S	500	35	38	20	Literature [4]
fishers_iris	70	6	6	8	Literature [5]
iris	150	3	6	12	Literature [6]
Wine	180	3	9	13	Literature [7]
Model2	100	3	3	10	Literature [8]
Kes2	52	2	5	7	artificial
Kes3	58	3	5	7	artificial

4.2. Experiment and analysis of initial cluster center setting

In this experiment, based on the search range of clustering number and the validity index - Silhouette index, the data sets Kes2 and Kes3 were experimented according to the above initial clustering center setting method 1 and compared with method 2 randomly determining the initial clustering center. The impacts of these two clustering center setting methods on Silhouette index were mainly compared, thus affecting the determination of optimal clustering number. In addition, the random clustering accuracy rate of these two setting methods and efficiency of the clustering algorithm were compared (including running time and number of iterations). While the random clustering accuracy rate (RCAR) was defined as the given correct number of clusters, the clustering algorithm was run for w times on the data set, which was expressed by the percentage of the correct clustering number and w ratio. Here, the running time refers to the initial clustering center's setting time and K-means clustering algorithm running time under the given correct number of clustering. The number of iterations refers to the iteration number of the K-means clustering algorithm based on different setting methods given the correct number of clusters. In order to reduce the error, the algorithm was run 20 times repeatedly, and the running time and the number of iterations took the average of 20 running results.

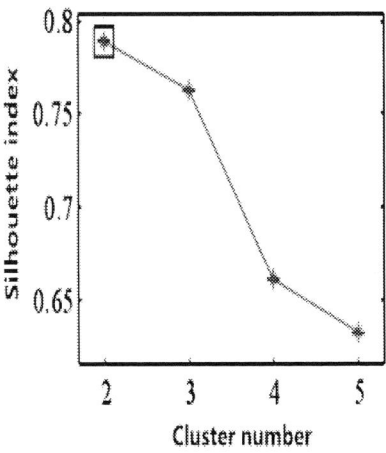 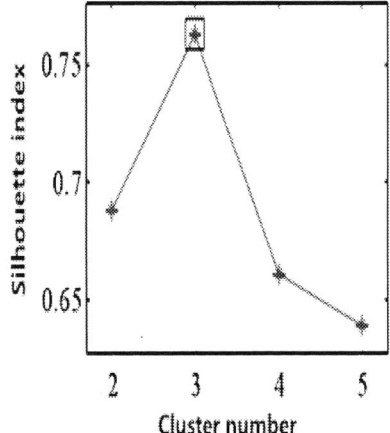

(a)The relation graph of Method1   (b) The relation graph of Method2

Figure1. Clustering numbers-index relationship diagram of Kes2

ISPECE   IOP Publishing

IOP Conf. Series: Journal of Physics: Conf. Series **1187** (2019) 042040   doi:10.1088/1742-6596/1187/4/042040

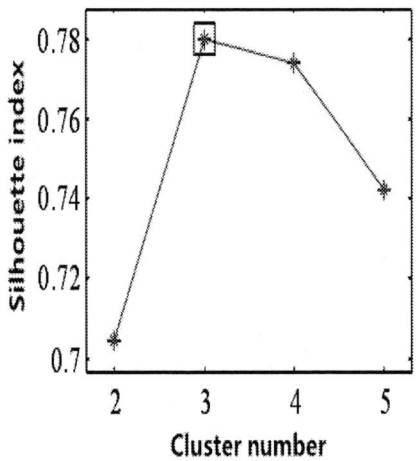 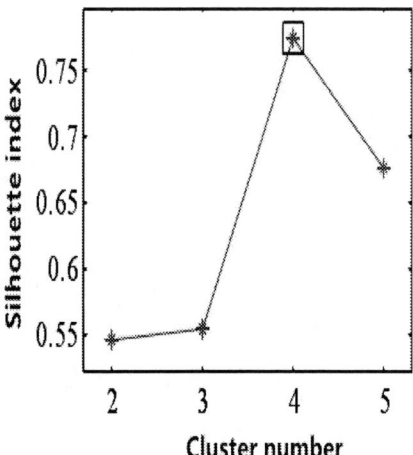

(a)The relation graph of Method1   (b) The relation graph of Method2
Figure2. Clustering numbers-index relationship diagram of Kes3

Figure1 (a) shows the cluster -Silhouette index relation graph with Method1 and Kes2. The optimal clustering number is 2, and the Silhouette index corresponding to the optimal cluster number is 0.7892. Figure 1(b) shows the cluster -Silhouette index relation graph with Method2 and Kes2. The optimal clustering number is 3, and the Silhouette index corresponding to the optimal cluster number is 0.7627. Figure2 (a) shows the cluster -Silhouette index relation graph with Method1 and Kes3. The optimal clustering number is 3, and the Silhouette index corresponding to the optimal cluster number is 0.7797 Figure2 (b) shows the cluster -Silhouette index relation graph with Method2 and Kes3. The optimal clustering number is 4, and the Silhouette index corresponding to the optimal cluster number is 0.7740. Experimental results of determining the optimal number of clusters based on different initial cluster center setting method. It is shown in Table2. From this, by Silhouette index, Method1 can get the correct optimal cluster number. Method2 cannot get the correct optimal number of clusters.

Table2. Optimal clustering numbers based on different setting method

Data set	Correct cluster number	k_{max}	Method1		Method 2	
			Optimal number of clusters	Maximum of Silhouette index	Optimal number of clusters	Maximum of Silhouette index
Kes2	2	5	2	0.789	3	0.76
Kes3	3	5	3	0.771	4	0.77

The correctness and efficiency of the clustering algorithm are obtained based on different initial cluster center setting method. It is shown in Table3. The RCAR value of the Method 1 reflects the quality of clustering. Each clustering result is correct, and it is also proved that the clustering result is uniquely determined under the condition that the initial cluster center is known. Method RCAR values of 2 reflect the data sets Kes2 and Kes3, random initial cluster centers are determined by a random method. The K-means clustering algorithm is unstable and the clustering results are poor. For the Kes2 data set, the running time of the clustering algorithm of method 1 is 0.0067 seconds. The running time of the clustering algorithm of method2 is 0.0128 seconds, which shows that the clustering algorithm of method 1 is significantly less than method 2. Based on Method 1, the number of iterations of the K-means clustering algorithm is 2, and based on method 2, the number of iterations of the K-means clustering algorithm is 6. The number of iterative times of method 1 is obviously less than method 2.

ISPECE

IOP Publishing

IOP Conf. Series: Journal of Physics: Conf. Series **1187** (2019) 042040   doi:10.1088/1742-6596/1187/4/042040

The same conclusion is also found for the Kes3 dataset. Therefore, using method 1, the accuracy of the clustering algorithm and the efficiency of the algorithm are higher than the method 2.

Table3. Accuracy and efficiency of the Clustering algorithm based on different setting method

Data set	Method1			Method2		
	RCAR	Running time	Iteration times	RCAR	Running time	Iteration times
Kes2	100%	0.0067	2	15.41%	0.12	6
Kes3	100%	0.0075	2	14.56%	0.13	6

Using method 1, the clustering effects of data set Kes2 and Kes3 are shown in Figure3 and Figure4, respectively. It can be seen that the clustering of method 1 is very effective.

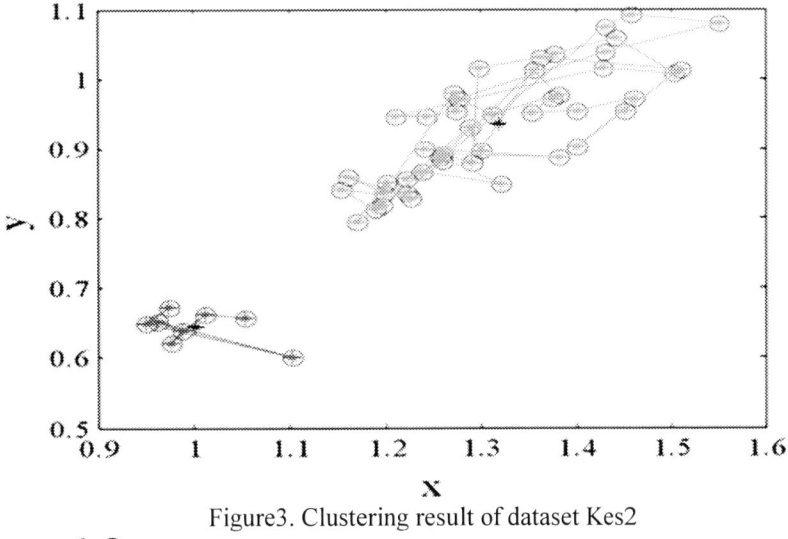

Figure3. Clustering result of dataset Kes2

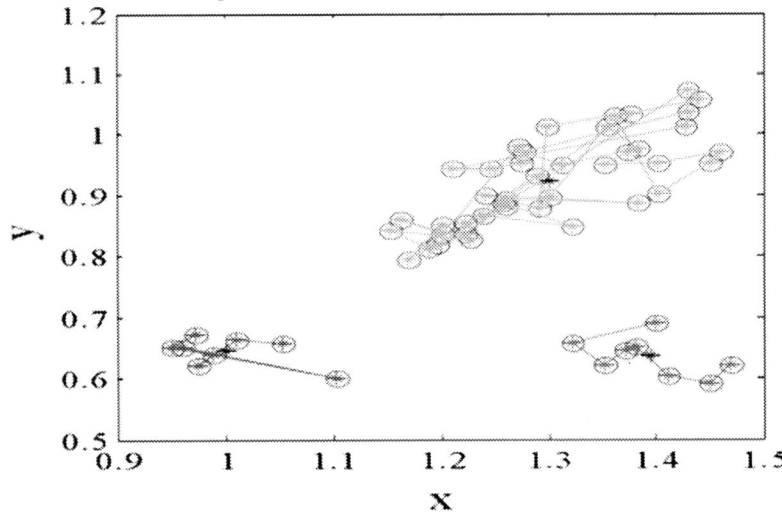

Figure4. Clustering result of dataset Kes3

5. Conclusions

Since k-means clustering algorithm randomly selected k samples as the initial clustering center, the clustering results were greatly influenced by the initial clustering center under the condition that the

1541

clustering structure was not easily identified, the method for solving the optimal clustering number by the K-means clustering algorithm was very unstable. Based on the max-min distance algorithm, a new method for setting the initial clustering center was proposed in this paper, in order to improve the quality of clustering. Algorithm determining the K-means algorithm's optimal clustering number - KMS algorithm was proposed by combining with Silhouette validity index. The theoretical research and experimental results have verified the validity and good performance of the above algorithm.

References

[1]   U.Fayyad, C.Reina, P.S. Bradley. (1998) Initialization of Iterative Refinement Clustering Algorithms, Proceedings of the 4th International Conference on Knowledge Discovering in Databases and Data Mining. New York: AAAI Press,11:194-198

[2]   K. Krishna, M. N. Murty. (1999) Genetic K-means Algorithm, IEEE Transactions on Systems, Man, and Cybernetics-Part B: Cybernetics, 29(3):433-439

[3]   B. Frey, D. Dueck. (2016) Clustering by Passing Messages Between Data Point, Science, 315(5814):972-976

[4]   B. Frey, D. Dueck. (2008) Response to Comment on Clustering by Passing Messages Between Data Points, Science, 319(5864):72-80

[5]   M. Brusco, H. Köhn. (2008) Comment on Clustering by Passing Messages Between Data Points, Science, 319(5864):22-81

[6]   D. Dembélé, P. Kastner. (2013) Fuzzy C-means Method for Clustering Microarray Fata, Bioinformatics, 19(8):973-980

[7]   C. Blake, C. Merz. UCI Repository of Machine Learning Databases, (University of California), Available: Http://mlearn.ics.uci.edu/MLRepository.html.

[8]   D. DAVIES, D. BOULDIN. (1979) A Cluster Separation Measure, IEEE Transactions on Pattern Analysis and Machine Intelligence, 1(2):224-227

Object Detection Based on the Improved Single Shot MultiBox Detector

Songmin Jia[1,2], Chentao Diao[1,2*], Guoliang Zhang[1,2], Ao Dun[1], Yanjun Sun[1,2], Xiuzhi Li[1,2] and Xiangyin Zhang[1,2]

[1]Faculty of Information Technology, Beijing University of Technology, Beijing 100124, China;

[2]Beijing Key Laboratory of Computational Intelligence and Intelligent System.

Email:diaochentao@163.com

Abstract. Aiming at the poor effect of deep learning algorithm on small objects detection, the SSD object detection method based on feature fusion is proposed. The reasons for low detection rate and poor robustness of classical SSD object detection methods are analysed; and through the theoretical analysis and comparative experiments, the characteristic fusion layer was proposed. The shallow layers with high resolution and deep layers with strong semantics are fused with the feature fusion structure; finally, a complete feature fusion structure is designed with the residual block to increase the width and depth of the network. The contrast experiment on the PASCAL VOC dataset was conducted for detection capability and detection accuracy, and experimental result indicates that when the confidence is set to 0.5, the mAP of the SSD method based on feature fusion is 78.04%, which is 0.8% higher than the classical SSD algorithm and 4.8% higher than the Faster RCNN algorithm. Obviously, the proposed algorithm improves the ability of small objects, and verifies the effectiveness of the proposed algorithm.

1. Introduction

In the field of computer vision, object detection is an important research topic. In recent years, various algorithms based on convolutional neural network (CNN) have been applied to object detection tasks, and the detection accuracy and efficiency have been effectively improved [1-5]. However, detecting object at different scales is still a challenging research task. Aiming at the problem of poor performance of multi-scale object detection in current algorithms, the solutions proposed by relevant scholars can be divided into two main categories. One is to extract object features at different scales based on image pyramids to complete multi-scale object detection; the other is to calculate the corresponding feature map of the original image, and then build a feature pyramid on the feature map to complete multi-scale object detection.

Building image feature Pyramid is the basic solution to realize multi-scale object detection at present [6]. These pyramids are scale invariant because the scale changes of objects are canceled out at different levels of the pyramids. Before the deep learning model, manual design of image features Pyramid feature is a popular method [7-9]. This method is particularly important for DPM [10] object detector which relies on a large number of scale sampling. At present, in the task of object recognition, the method of feature extraction using the CNN has replaced the method of artificial design feature

Content from this work may be used under the terms of the Creative Commons Attribution 3.0 licence. Any further distribution of this work must maintain attribution to the author(s) and the title of the work, journal citation and DOI.
Published under licence by IOP Publishing Ltd

[11-12]. In the field of object detection, the CNN has better detection effect than traditional manual design features, but the detection performance still needs to be improved [13, 14].

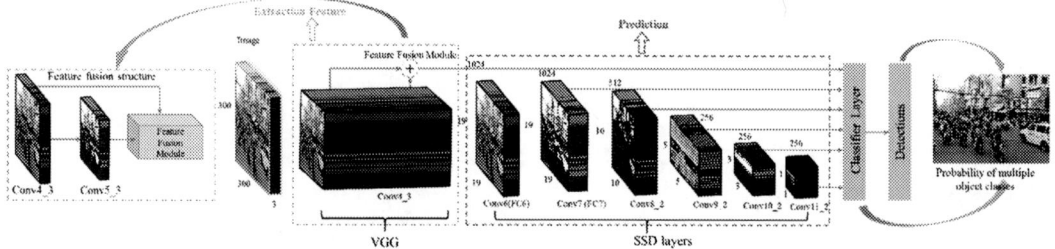

Figure 1. SSD based on feature fusion Network Model

In this paper, a method of SSD object recognition based on feature fusion is proposed (shown in figure 1) for object recognition. In Section 2, the proposed algorithm is introduced about the structure of the method and the processing features. The experiment results and analyses are described in Section 3. Finally, conclusions are drawn in Section 5.

2. Method analysis

The SSD algorithm takes VGG16 as the basic network and several user-defined layers as the functional layers to construct an efficient object detection framework. SSD uses the characteristic pyramid in convolution network to replace the original

Figure 2. The sketch of detect result of SSD for small objects

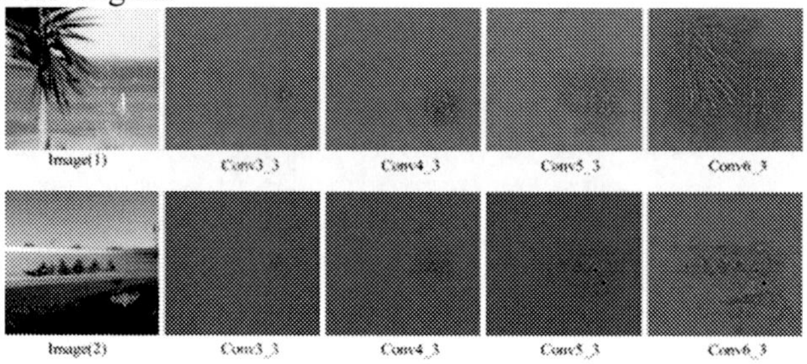

Figure 3. The receptive filed of the partial layer in the SSD

method of multi-scale object detection based on full connection layer. However, SSD algorithm adopts a non-discriminatory approach to different levels of features, which makes it unable to take into account local details and texture features and global semantic features, thus affecting the detection efficiency of the system for small-scale objects. Figure 2 show that the SSD algorithm is less effective in detecting smaller objects (such as people, sheep and cars in figure 2).

It is necessary to determine which level of texture information and semantic information is significantly suitable for feature fusion. Figure 3 shows the receptive field of the SSD algorithm in different layers. Obviously, the receptive field of the object image feature is small when Conv3_3 is used, and the receptive field of Conv6_3 is too large, which leads to a large amount of background noise. In Conv4_3 and Conv5_3, the size of the receptive field is moderate relative to the small-scale object, and the object feature information can be obtained completely.

In the CNN, the receptive field denotes the mapping region of the pixels of the output characteristic image of each layer on the original image. Figure 4 shows the receptive field intention of different layers. It can be seen from figure 4 that the size of the receptive field is related to the size and step size of the convolution nucleus of all the network layers before the layer. Firstly, the sliding step size is calculated. In the neural network, the step value of each layer is the product of the step value of all previous layers, so the step value formula of the first layer is shown in equation (1).

$$strides(i) = \prod_{j=1}^{i-1} stride_j \tag{1}$$

The size of the receptive field is calculated from top to bottom, that is, the deepest receptive field on the first layer is first calculated, and then the receptive field is regressed to the first layer of the network, as shown in equation (2).

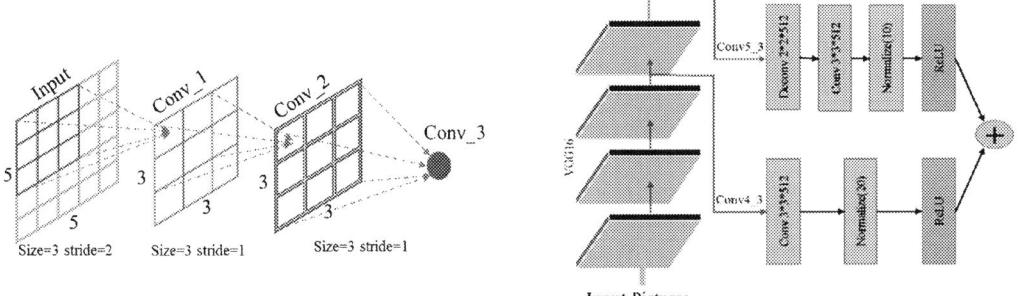

Figure 4. The illustration of the receptive filed   Figure 5. The structure of feature fusion module

$$V'_{rf} = ((V_{rf} - 1) * stride(i)) + Size_{conv} \tag{2}$$

Where V'_{rf} represents the current receptive field, V_{rf} is the current receptive filed in the first layer, $Size_{conv}$ is the size of convolution nucleus.

The receptive field area of Conv4_x to Conv5_x in the SSD network model is calculated by equation (2). The receptive field from Conv4_1 to Conv5_3 are 44, 76, 92, 132, 164 and 196, respectively. Conv4_3 receptive field has a moderate size relative to the small-scale object, and can extract the feature information of the small-scale object completely. In addition, compared with Conv4_3, Conv5_3 has three convolutions and one pooling operation, and the semantic information has been enhanced to avoid the impact of a large number of noises on small-scale object detection. So, this paper proposes a feature fusion framework based on Conv4_3 and Conv5_3.

The aim of feature fusion structure is to design an effective framework to fuse the features of different layers in convolutional neural network, and then send the features of this layer into the detector to complete target detection. The design process is as figure 5.

Assuming that $X_i, i \in Q$ is the extracted depth feature, the fusion architecture should logically satisfy the following equation (3):

$$X_{output} = \Gamma_1 \{ \chi_i (X_i) \} \qquad i \in Q \tag{3}$$

Where χ_i is the feature layer for preprocessing, which is used to fuse all feature maps in the current layer; Γ_1 is the feature fusion framework; Q is the number of channel of the feature map of the current layer. The structure of Γ_1 is shown in figure 5. The feature of Conv4_3 and Conv5_3 is fused by 3*3 convolution layer, normalization layer and activation layer. Considering the size of Conv5_3, it is necessary to be handled by 2*2 deconvolution layer.

Based on the above improvements, the complete object detection framework proposed in this paper can be shown in figure 1. The proposed framework of feature fusion can effectively reduce the possibility of introducing a large amount of background noise into the receptive field to affect small-scale object detection. Meanwhile, fusing feature maps of different layers directly enhances the relevant semantic information, and makes use of the information extracted from each layer of the structure, which makes the proposed features have stronger discriminant performance and effectively improves the object detection accuracy of the proposed algorithm.

3. Experiments

3.1. Experimental setup

In order to evaluate the performance of the algorithm proposed, experiments were carried out on PASCAL VOC 2007 and 2012 datasets. In this section, the test results of SSD algorithm based on feature fusion are compared with those of SSD algorithm and Faster RCNN algorithm. And the performance of the proposed method for small object detection is improved compared with the original algorithm. Finally, the real-time performance of the algorithm is compared and analysed.

The VGG16 network pre-training model used in this paper can be trained based on ImageNet data set. The hyper-parameters are set as follows: batch size = 32, gamma = 0.1, momentum = 0.9, input size 300 x300, optimization type SGD. In addition, the newly added layer is initialized using Xavier [22]. The initial learning rate is set to 10^{-3}, and then adjusted to 10^{-4}, 10^{-5} and 10^{-6} respectively when the number of iterations is 60k, 80K and 100K.

3.2. Performance evaluation algorithm based on dataset
In this paper, the detection model of the proposed algorithm is trained on the dataset of PASCAL VOC2007 and PASCAL VOC2012. The VOC2007 and VOC2012 datasets contain 9963 and 22531 images with 20 targets respectively.

After experiments, the influence of different layer feature fusion on the overall detection performance is analysed on PASCAL VOC2007 test set. The results of experiment are that Conv4_3 and Conv5_3 achieves 78.04% mAP, Conv4_3 and Conv6 achieves 79.95% mAP, and Conv4_3, Conv6 and Conv7 achieves 77.84% mAP. As shown in the result, using Conv4_3 and Conv5_3 for feature fusion can significantly improve the detection accuracy. Conv6 has larger sensing field than Conv5_3, which leads to more background noise being introduced, thus reducing the detection accuracy. In order to evaluate the overall performance of the proposed algorithm, the detection results of 20 kinds of objects are compared and analysed on PASCAL dataset, as shown in table 1. The average accuracy of the proposed algorithm is 78.33%, which is 1.13% and 5.13% higher than SSD and Faster RCNN, respectively. In addition, as shown in figure 6, the detection effect of the first behaviour SSD algorithm and the second behaviour SSD algorithm proposed in this paper are shown. Obviously, the proposed algorithm has been significantly improved than the original SSD algorithm in small-scale object detection. Obviously, the algorithm is superior to the above two methods in object detection accuracy, thus verifying the effectiveness of the proposed algorithm.

4. Conclusion
In this paper, the SSD object detection algorithm based on feature fusion was proposed. By constructing a feature fusion architecture, the object detection performance of the algorithm is effectively improved, and especially for the detection effect of small-scale objects. First, a framework for efficiently integrating image texture features and global feature depth models is designed. Then, the feature information extracted from each layer of the structure is transformed to the detector of the network, and further mining the semantic information of network features. Finally, using the PASCAL VOC dataset, experiments are made for testing and verifying the performance of the proposed algorithm. The experimental results show that the mAP of the proposed SSD object detection algorithm based on feature fusion is 78.0%, which is both higher than that of the SSD and Faster RCNN algorithms, and further verifies the advancement and effectiveness of the algorithm. And the future work will optimize the network parameters, increase the number of training samples, and improve the robustness and adaptability of the model to achieve better detection performance.

Figure 6. Comparison of results between SSD algorithm and feature fusion based on SSD algorithm

Table 1. The test result on PASCAL VOC2007 test dataset (with IOU=0.5)

Methods	MAP (%)	Air	Bike	Bird	Boat	Bottle	Bus	Car	Cat	Chair	Cow
SSD300	77.2	78.5	85.9	75.7	71.0	49.2	85.3	86.5	87.7	60.7	82.3
Faster RCNN	73.2	76.5	79.0	70.9	65.5	52.1	83.1	84.7	86.4	52.0	81.9
Ours	78.0	80.3	86.3	76.8	72.1	51.4	86.1	86.7	88.0	61.4	82.6

Methods	MAP (%)	Desk	Dog	Horse	Mbike	Person	Plant	Sheep	Sofa	Train	TV
SSD300	77.2	76.8	84.3	86.7	84.5	79.1	51.7	77.4	78.8	86.6	76.7
Faster RCNN	73.2	65.7	84.8	84.6	77.5	76.7	38.8	73.6	73.9	83.0	72.6
Ours	78.0	77.1	85.6	87.7	86.4	79.6	53.8	78.9	79.2	87.8	78.1

Acknowledgments

This research is financially supported by the 2017 BJUT United Grand Scientific Research Program on Intelligent Manufacturing (No. 040000546317552) and the National Natural Science Foundation of China (No. 61703012).

References

[1] Redmon, J., & Farhadi, A. (2017). YOLO9000: better, faster, stronger. arXiv preprint.
[2] Ren, S., He, K., Girshick, R., & Sun, J. (2015). Faster r-cnn: Towards real-time object detection with region proposal networks. In Advances in neural information processing systems (pp. 91-99).
[3] Dalal, N., & Triggs, B. (2005, June). Histograms of oriented gradients for human detection. In Computer Vision and Pattern Recognition, 2005. CVPR 2005. IEEE Computer Society Conference on (Vol. 1, pp. 886-893). IEEE.
[4] Lee, K., Choi, J., Jeong, J., & Kwak, N. (2017). Residual features and unified prediction network for single stage detection. arXiv preprint arXiv:1707.05031.
[5] Wang, R. J., Li, X., Ao, S., & Ling, C. X. (2018). Pelee: A Real-Time Object Detection System on Mobile Devices. arXiv preprint arXiv:1804.06882.
[6] Liu, W., Anguelov, D., Reed, S., Fu, C. Y., & Berg, A. C. (2016, October). Ssd: Single shot multibox detector. In European conference on computer vision (pp. 21-37). Springer, Cham.

[7] Redmon, J., Divvala, S., Girshick, R., & Farhadi, A. (2016). You only look once: Unified, real-time object detection. In Proceedings of the IEEE conference on computer vision and pattern recognition (pp. 779-788).

[8] Yang, F., Choi, W., & Lin, Y. (2016). Exploit all the layers: Fast and accurate cnn object detector with scale dependent pooling and cascaded rejection classifiers. In Proceedings of the IEEE conference on computer vision and pattern recognition (pp. 2129-2137).

[9] Bansal, A., Chen, X., Russell, B., Gupta, A., & Ramanan, D. (2016). Pixelnet: Towards a general pixel-level architecture. arXiv preprint arXiv:1609.06694.

[10] Simonyan, K., & Zisserman, A. (2014). Very deep convolutional networks for large-scale image recognition. arXiv preprint arXiv:1409.1556.

[11] Lowe, D. G. (2004). Distinctive image features from scale-invariant keypoints. International journal of computer vision, 60(2), 91-110.

[12] Felzenszwalb, P. F., Girshick, R. B., McAllester, D., & Ramanan, D. (2010). Object detection with discriminatively trained part-based models. IEEE transactions on pattern analysis and machine intelligence, 32(9), 1627-1645.

[13] Girshick, R. (2015). Fast r-cnn. In Proceedings of the IEEE international conference on computer vision (pp. 1440-1448).

[14] LeCun, Y., Boser, B., Denker, J. S., Henderson, D., Howard, R. E., Hubbard, W., & Jackel, L. D. (1989). Backpropagation applied to handwritten zip code recognition. Neural computation, 1(4), 541-551.

Medical image fusion based on statistical modeling

Huaijing Qu[1*], Hengbin Wang[1] and Hongkui Xu[1]

[1]School of Information & Electric Engineering, Shandong Jianzhu University, Jinan, Shandong, 250101, China

*Corresponding author's e-mail: quhuaijing@sdjzu.edu.cn

Abstract. For improving the imaging quality and increasing the clinical applicability, a novel approach to multimodal medical image fusion is proposed based on statistical modeling in contourlet transform domain. Firstly, the coefficients of the approximate subband are modeled as Gaussian mixture distribution, and fused by a new rule using the weighted average of a posterior probability. Then, the coefficients of the detail subbands are modeled by generalized Gaussian distribution, and a selection rule is used for fusion based on the estimated parameters and the matching measure. Finally, an effectively fused image is achieved through inverse contourlet transform. The experimental results show that, compared with existing approaches, the proposed method can make the fused image have better performance, and provide more valuable diagnostic information.

1. Introduction

Due to the practical limitation of medical imaging modality, each source image sensor usually captures different significant information from the same target. In order to make practitioners easily obtain comprehensive and reliable understanding, or meet the requirements of subsequent medical image processing tasks, it is necessary to effectively fuse the related multi-source images. Currently, medical image fusion has been widely applied to medical diagnosis, monitoring and analysis [1,2].

In pixel level context, medical image fusion approaches include that of spatial domain and multiscale transform domain, respectively [3]. Usually, the fusion methods based on transform domain can fully utilize image characteristics of the spatial frequency locality, and obtain better fusion effect [1-3]. Specially, the fusion methods combining multiscale statistical modeling have received a great deal of attention. Burt firstly proposes a statistical fusion algorithm based on saliency measure and matching measure [4]. Achim proposes a fusion method based on generalized Gaussian and alpha-stable modeling[5]. Loza presents a fusion method of multimodal medical image based on non-Gaussian statistical modeling[6]. Howlader proposes a statistical medical image fusion algorithm based on Bayesian maximum posterior probability [7]. Generally speaking, these methods based on wavelet statistical modeling have better effect of image fusion. However, due to the lack of rich directionality, the fusion performance remains to be further improved. In comparison, the contourlet transform can optimally describe the geometric directional information of natural images [8]. Therefore, its statistical subband modeling is used for the image fusion in this paper.

Recently, the multimodal medical image fusion methods based on contourlet transform have aroused extensive attention. Bhatnagar proposes a novel fusion method based on the activity measure and the directive contrast [9]. Yang proposes a fusion scheme based on generalized Gaussian distribution modeling and weight maps [10]. Luo proposes a new fusion method based on hidden Markov model [11]. Considering that any medical image can be regarded as a random distribution

Content from this work may be used under the terms of the Creative Commons Attribution 3.0 licence. Any further distribution of this work must maintain attribution to the author(s) and the title of the work, journal citation and DOI.

Published under licence by IOP Publishing Ltd

implementation, thus the fusion methods using statistical probability modeling may obtain better performance. In recent years, some researchers have made preliminary explorations [10-12].

In this paper, a new fusion method is proposed based on joint modeling in contourlet domain. Specifically, the approximate subband coefficients are modeled as Gaussian mixture distribution (GMD), and fused by using the weighted average of a posterior probability; the detail subband coefficients are modeled by generalized Gaussian distribution (GGD), and the fusion rules of the selection and weighted average are used based on distribution parameters and matching measure.

2. Contourlet transform and Its Approximate Subband Modeling
The contourlet transform includes the multiscale transform and the multidirectional transform. The multiscale transform is achieved by the Laplacian pyramid (LP), and the directional transform of the multiscale detail subbands is implemented by the directional filter bank (DFB). Specifically, the LP firstly decomposes input image into a approximate subband and a detail subband. Then, the multiscale decomposition procedure is iterated after the approximate subband followed by downsampling by 2 in each dimension. Finally, every detail subband image of the LP is further decomposed by a DFB into 2^n directional subband images, where n is the level number of the directional decomposition.

In this paper, the approximate subband coefficient x with the multimodal distribution can be modeled as the GMD, and its mixture probability density function is represented as

$$p(x|\boldsymbol{\theta}) = \sum_{m=1}^{K} \omega_m p(x|\mu_m, \sigma_m^2)$$

(1)

where ω_m, $m = 1, 2 \cdots, K$ denotes the weighted coefficient of the component; K is the number of mixture components, $\sum_{m=1}^{K} \omega_m = 1$, and $K = 5$ is adopted in all the following experiments ; $\boldsymbol{\theta} = \{\omega_i, \mu_i, \sigma_i^2, i = 1, 2, \cdots, K\}$ represents the parameter set of GMD model. Usually, the parameters are estimated according to the expectation maximization (EM) method.

3. Image fusion method based on jointly statistical modeling
The multimodal medical source images are firstly decomposed by contourlet transform. Then, the approximate subband coefficients are modeled as finite GMD and fused according to the weighted average. The choice of the weight adopts our proposed posterior probability of the coefficients. Meanwhile, the detail subband coefficients are modeled as GGD [13], and fused by using the selection and weighted average rule which is based on the refined Burt method. Specifically, the adopted image fusion approach in this paper is described as follows.

Assuming that the source images are A and B, the fused image is C.

1) The source images A and B are respectively decomposed by contourlet transform into the approximate subband and the detail directional subbands.

2) The fusion rule for the approximate subband coefficients adopts the weighted average of a posterior probability. The 5×5 sliding window is used to traverse the approximate subband of the source image. The central pixel coefficient of each window determines the fusion weight according to a posterior probability. Namely, for the approximate subbands of the source images A and B, firstly GMD density functions $p(x_i|\boldsymbol{\theta}_A) = p(x_i|A)$ and $p(y_i|\boldsymbol{\theta}_B) = p(y_i|B)$ are respectively estimated for the central coefficients x_i and y_i, $i = 1, 2, \cdots, N$ in every window, where N is the total number of approximate coefficients; then, the class posterior probabilities $P(A|z_i) = \dfrac{p(x_i|A)P_A(i)}{p(x_i|A)P_A(i) + p(y_i|B)P_B(i)}$ and

$P(B|z_i) = \dfrac{p(y_i|B)P_B(i)}{p(x_i|A)P_A(i) + p(y_i|B)P_B(i)}$ of the corresponding fused central coefficients z_i, $i = 1, 2, \cdots, N$ are respectively computed, where $P_A(i) = x_i/(x_i + y_i)$ and $P_B(i) = y_i/(x_i + y_i)$ are the prior probabilities of the central coefficients in the ith window of the source images A and B, respectively; finally, the central coefficient in every window of the fused image C is obtained by using weighted average, namely

$$z_i = P(A|z_i) \cdot x_i + P(B|z_i) \cdot y_i, \quad i = 1.2. \cdots, N \tag{2}$$

3) The fusion rule of the detail directional subband coefficients uses the selection and weighted average method. Assuming that some detail directional subbands corresponding to A and B are X and Y, respectively. The 11×11 sliding window is adopted. For each of the sliding windows,

(1) the variance parameter of GGD model for the detail directional subband coefficients is estimated according to the refined Newton–Raphson iterative algorithm [13].

(2) determine the salience measure σ_x^2 and σ_y^2, which are equal to the estimated variance of the detail directional subband coefficients, respectively.

(3) compute matching measure $M = 2\sigma_{xy}/(\sigma_x^2 + \sigma_y^2)$, where σ_{xy} denotes the covariance of X and Y.

(4) compute fusion weights $W_{\max}$ and $W_{\min}$ according to the relationship between matching measure M and threshold T; if $M \geq T$, then $W_{\min} = 0.5*(1-(1-M)/(1-T))$ and $W_{\max} = 1 - W_{\min}$, otherwise $W_{\max} = 1$ and $W_{\min} = 0$; in this paper, each threshold selection is carried out based on the optimal fusion performance.

(5) compute and determine the coefficients of the fused detail directional subbands, namely

$$\begin{cases} D_C(m,n) = W_{\max} D_A(m,n) + W_{\min} D_B(m,n) & \sigma_x^2(m,n) \geq \sigma_y^2(m,n) \\ D_C(m,n) = W_{\min} D_A(m,n) + W_{\max} D_B(m,n) & \sigma_x^2(m,n) < \sigma_y^2(m,n) \end{cases} \tag{3}$$

where $D_A(m,n)$, $D_B(m,n)$ and $D_C(m,n)$ represent the coefficients of the source images and fused image in the central position (m,n) of the sliding window, respectively.

4) Reconstruct every fused contourlet subband coefficient, and generate the final fusion image.

4. Experimental Results

The experiments are conducted in two aspects. The modeling accuracy for approximate subband and detail subbands is firstly evaluated through the experimental effect of histogram fitting, then the subjective and objective performance are compared with the existing image fusion methods.

The subjective performance evaluation is based on the visual effect of the fused images. Without requiring a reference image, the objective evaluation adopts the well-known information entropy (E), standard deviation (STD), and Q -index ($Q_{AB/F}$)[1,9]. Among them, the greater the entropy, the more abundant the image information; the larger the STD value, the clearer the fused image; and the larger $Q_{AB/F}$ value shows that the edge details of source images are better preserved.

4.1. GMD modeling for approximate subband coefficients

The contourlet approximate subband can be modeled as finite GMD. In order to compare and verify the modeling accuracy, the estimated GMD probability density curve is used for fitting the approximate subband histogram. In all the experiments below, contourlet transform is decomposed into three levels. The experimental result is shown in Figure 1. Figure 1(a) is the MRI source image, and Figure 1(b) depicts the statistical modeling result. It can be seen from Figure 1(b) that, the approximate subband histogram has a distinct multimodal distribution shape, and its statistical property can be accurately fitted by the estimated GMD.

4.2. GGD modeling for detail directional suband coefficients

Every contourlet detail subband distribution can adopt GGD modeling[13]. The parameter estimation accuracy of GGD model is also evaluated through the effect of histogram fitting. The modeling result for the first directional subband in the first level is also depicted in Figure 1. Figure 1(c) is the CT source image, and Figure 1(d) represents the GGD modeling result. It can be seen from Figure 1(d) that, the histogram of the detail subband presents a single peak mode, and the distribution estimation is also more accurate.

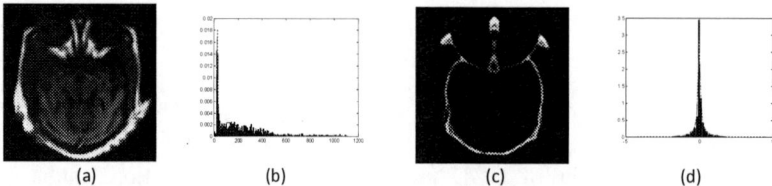

(a) (b) (c) (d)

Figure 1. Statistical modeling of subbands

4.3. Comparisons with existing image fusion methods

Two different modal medical images, namely MRI and CT images with the size of 256×256 are selected as source images [2], as shown in Figure 1(a) and Figure 1(c), respectively.

Firstly, the experimental results are evaluated from the visual effect, as shown in Figure 2(a)-(f). For comparison, we respectively adopt the weighted average method in discrete wavelet transform domain (DWT-W), the weighted average method in contourlet domain (Con-W), the method using the weighted average for approximate subband and GGD modeling for detail subbands in DWT domain (DWT-W-GGD, $T=0.1$), the method using the weighted average for approximate subband and GGD modeling for detail subbands in contourlet domain (Con-W-GGD, $T=0.3$), the method using GMD modeling for approximate subband and GGD modeling for detail subbands in DWT domain (DWT-GMD-GGD, $T=0.1$), and the proposed method using GMD modeling for approximate subband and GGD modeling for detail subbands in contourlet domain (the proposed method, $T=0.3$). It can be seen from Figure 2 that, our proposed approach obtains better visual effect on the contrast, definition and artifact, therefore is more conducive to rational medical diagnosis.

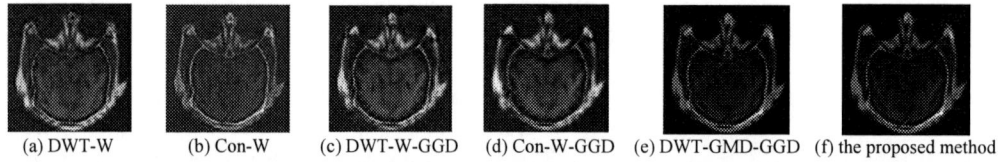

(a) DWT-W (b) Con-W (c) DWT-W-GGD (d) Con-W-GGD (e) DWT-GMD-GGD (f) the proposed method

Figure 2. Visual effect of medical image fusion based on different methods

The objective performance of medical image fusion is respectively evaluated by the information entropy, standard deviation and $Q_{AB/F}$, as shown in Table 1. It can be seen from table data that, the performance achieved by our proposed approach is obviously better than that of the other methods.

Table 1. Objective evaluation for different fusion methods

Methods	E	STD	$Q_{AB/F}$
DWT-W	0.0012	18.7596	0.5768
Con-W	0.0012	18.7600	0.6130
DWT-W-GGD	0.3334	18.7872	0.7130
Con-W-GGD	0.3332	20.9750	0.7133
DWT-GMD-GGD	0.3581	30.7660	0.7379
the propose method	**0.5195**	**31.4354**	**0.7680**

5. Conclusion

In contourlet transform domain, the performance of image fusion can be effectively improved by using statistical modeling. To this end, a new multimodal medical image fusion approach is proposed in this paper based on joint modeling in contourlet domain. Its innovations include that, the approximate subband coefficients are modeled by finite GMD, and fused by using a novel weighted average of a posterior probability; every detail subband coefficients are modeled by GGD, and fused based on distribution parameters and matching measure. The experimental results show that the proposed method effectively improves the performance and visual effect of multimodal medical image fusion.

Acknowledgments
This work is supported by the Shandong Provincial Natural Science Foundation of China (No. ZR2014FM016), and the Shandong Provincial Science and Technology Development Program of China (No. 2014GGX101050)

References
[1]   Du, J., Li, W., Lu, K., Xiao, B. (2016) An overview of multi-modal medical image fusion. Neurocomputing, 215: 3–20.
[2]   James, A.P., Dasarathy, B.V. (2014) Medical image fusion: a survey of the state of the art. Information Fusion, 19: 4-19.
[3]   Li, S., Kang, X., Fang, L., et al. (2017) Pixel-level image fusion: A survey of the state of the art. Information Fusion, 33: 100-112.
[4]   Burt, P.J., Kolczynski, R.J. (1993) Enhanced image capture through fusion. In: Fourth International Conference on Computer Vision. Berlin. pp. 173-182.
[5]   Achim, A., Loza, A., Bull, D., et al. (2008) Statistical modeling for wavelet-domain image fusion. Image Fusion, 2008: 119-138.
[6]   Loza, A., Bull, D., Canagaraiah, N., Achim, A. (2010) A non-Gaussian model-based fusion of noisy image in the wavelet domain. Computer Vision and Image Understanding, 114: 54-56.
[7]   Howlader, T., Jhohura, F.T., Rahman, S.M.M. (2013) A novel statistical image fusion rule for noisy source images. In: 59th ISI World Statistics Congress. Hong Kong. pp. 3630-3635.
[8]   Do, M.N., Vetterli, M. (2005) The Contourlet transform: an efficient directional multiresolution image representation. IEEE Transactions on Image Processing, 14: 2091-2106.
[9]   Bhatnagar, G., Wu, Q., Liu, Z. (2015) A new contrast based multimodal medical image fusion framework. Neurocomputing, 157: 143-152.
[10]   Yang, G., Li, M., Chen, L., et al. (2015) The nonsubsampled contourlet transform based statistical medical image fusion using generalized Gaussian density. Computational and Mathematical Methods in Medicine, 2015: 1-13.
[11]   Luo, X., Zhang, Z., Zhang, B., Wu, X.. (2017) Contextual information driven multi-modal medical image fusion. IETE Technical Review, 34: 598-611.
[12]   Zhang, H., Luo, X., Wu, X., et al. (2014) Statistical modeling of multi-modal medical image fusion method using C-CHMM and M-PCNN. In: 22nd International Conference on Pattern Recognition. Stockholm. pp. 1067-1072.
[13]   Qu, H., Peng, Y., Sun, W. (2007) Texture image retrieval based on Contourlet coefficient modeling with generalized Gaussian distribution. In: the 2nd International Conference on Advances in Computation and Intelligence. Wuhan. pp.493-502.

Automatic integration testing through collaboration diagram and logic contracts

Yi Sun[1], Xiaohua Yang[1], Jie Liu[1], Tonglan Yu[1], Zhuoran Xu[1], Zhiqiang Wu[2] and Zhi Chen[2]

[1]School of Computer, University of South China, Hengyang, Hunan, 421001, China
[2]Key Laboratory for Nuclear Reactor System Design, Nuclear Power Institute of China, Chengdu, Sichuan, 610041, China

E-mail: xiaohua1963@foxmail.com

Abstract. Component-based software development can effectively increase software development efficiency through component reuse. Integration testing is an important technical means to guarantee the quality of software systems composed of components. Aiming at the problem that the scale of software systems is getting larger and larger, and manual design of test cases is costly and difficult, we propose an automatic approach for integration test cases generation based on collaboration diagram and logic contracts. By extracting control flow information from collaboration diagram and combining with contracts as component specification, an intermediate model called execution tree of components is established, then test cases are automatically generated through contract solving technology. This approach not only realizes path coverage, but also completes the automatic integration testing of software system, thereby improving test efficiency and reducing test cost.

1. Introduction

With the increasing scale of software systems, component-based software development technique has been widely accepted in order to reduce development cost and difficulty. Component-based software development builds software system by integrating reusable components, and components collaborate to achieve functions of system. Although the reused components have undergone rigorous unit testing, this does not guarantee that the integrated software system will not have problems, so integration testing is needed to ensure the quality of software.

Unified Modeling Language (UML) is one of the main tools for object-oriented software development, which provides various models for modeling software system. UML models describe static structure and dynamic behaviors of the system, so they can provide guidance for test case generation in software testing. Among these UML models, collaboration diagram is mainly used to describe collaborations and interactions among system objects. The functions of component-based software are completed by interactions among components, so we select collaboration diagram as test model for generating integration test cases.

The main purpose of integration testing is to detect whether interactions among components are implemented correctly. Collaboration diagram points out the components involved in implementing system functions, and interaction relationships among these components can also be found in it. In recent years, researchers [1, 2, 3, 4] proposed some methods to generate test cases from collaboration diagram. These approaches all took advantage of interaction information among objects in collaboration diagram to generate test cases and developed towards automated testing, because it is

Content from this work may be used under the terms of the Creative Commons Attribution 3.0 licence. Any further distribution of this work must maintain attribution to the author(s) and the title of the work, journal citation and DOI.
Published under licence by IOP Publishing Ltd

unrealistic to rely entirely on manual test case generation with the growing complexity of modern software.

An important theory of component quality assurance is contract theory [5], which is widely used as the specification of components. Contracts define interaction rules among components to regulate behaviors of components, which are predicate logic formulas that describe the constraints that must be satisfied before and after the component call. In terms of utilizing contracts to generate test cases, the work in [6, 7] either lacks support for integration testing, or isn't suitable for interactions among components. In addition, the approach presented in [8] does not provide complete system model information when generating test cases and lacks the adequacy of system integration testing.

Therefore, we combine collaboration diagram with logic contracts to automatically generate integration test cases in this paper. By extracting control flow information from collaboration diagram and combining with contracts as component specification, an intermediate model called execution tree of components is established. On the basis of this intermediate tree, contracts of components are collected along path to form path constraint. Solution to path constraint is used as test input, and the postcondition is used as test oracle. Test suite is generated to realize path coverage, thus confirming whether components behave as expected in the system.

The main contribution of this paper is to propose the approach of integrating collaboration diagram and contracts. On this basis, test cases are generated automatically by the technique of contract solving, and path coverage in integration testing of components is realized, so as to solve the problem of integration test case generation of large-scale software system.

The rest of this paper is as follows: Section 2 introduces test model based on UML collaboration diagram. Section 3 presents our proposed approach to generate test cases. Section 4 demonstrates the feasibility of this approach through an example of ATM. Section 5 is a summary of the paper, and future work is drawn.

2. Test model based on UML collaboration diagram

2.1. Component collaboration diagram

UML collaboration diagram (CD) describes collaboration relationships among components. These relationships define the interactions among components during software execution, which is the basis for component integration. Therefore, collaboration diagram can be used as test model for component integration. A CD = <CP, MS>, where CP is a finite set of components and MS is a finite set of messages. Each component in CP is a finite set of methods, and all messages in MS are arranged according to messages' sequence number.

A component is the encapsulation of a series of methods. It only provides limited information about methods to users, including method names, input and output parameters, and contracts. A contract indicates constraints that must be observed when accessing internal methods of component, including precondition and postcondition, representing constraints that must be met before and after the invocation of a method, respectively. Precondition and postcondition are both represented in first-order logic. For a correct method, if the input satisfies precondition, the output must satisfy postcondition.

In collaboration diagram, collaborations start from an external event triggering, then a series of interactions are completed among components, finally the collaborations end when no more messages are generated, resulting in an external output result. In this process, component interactions are achieved through message passing. A sequence of messages represents one execution of the system. Collaboration diagram depicts the execution process of system functions, and each system task corresponds to an execution path in collaboration diagram. Therefore, integration testing of components should be able to reflect the correctness of the system execution process under different tasks, that is, testing each execution path with test data.

Collaboration diagram describes components, which are integrated into software, and interaction messages among these components. A message corresponds to a method of component, and the corresponding method is activated by the message. Message sequence activates corresponding method sequence, which indicates the process of components execution.

In addition to corresponding method, a message also includes other information such as sequence number, condition and type. The sequence number indicates nest relation among messages and the sequential order in which messages are executed. For example, message 1.1 and 1.2 are the first and second nested messages during message 1 execution. The condition represented in first-order logic can be true or false. A message will be executed if condition is true, otherwise not. There are three types of messages: common, conditional and recurrent message. Common message's condition is null. Conditional message is executed when its condition is true, otherwise the message and its nested messages will not be executed. Recurrent message whose nested messages are included in loop body is executed when its condition is true, otherwise not. Loop is restricted to be executed zero times or once to avoid path explosion.

It should be noted that condition of message and contract of method belong to different constraint systems. Condition of message causes different control flows, and contract of method is used to describe the constraints under which method is executed. They are both represented in first-order logic.

2.2. Execution tree of components

In collaboration diagram, the execution path of a system function starts with a message without precursor, then follows a message sequence, finally ends at a message without successor. Meanwhile, methods in components are activated by these messages and a method sequence is formed. This path is called functional execution path.

Condition of message leads to branches which form multiple functional execution paths, thus a tree is created. The nodes of this tree denote messages in collaboration diagram, which will activate corresponding methods in components. The prerequisite of activating a method is to satisfy contract of this method, so contract is also included in the path (i.e., functional execution path) constraint. If path constraint is solvable, it shows the path is reachable; otherwise, the path is unreachable.

In our proposed approach, collaboration diagram with contracts is transformed to an intermediate model called execution tree of components, then this intermediate tree is traversed to collect conditions of messages and contracts of methods to form path constraints. Eventually, test cases used to test each reachable path are acquired with the help of constraint solver.

An execution tree of components (ETC) is a directed tree where ETC = <TN, TE>, TN is a finite set of nodes and TE is a finite set of directed edges. Each node in TN corresponds to a message, and contains constraint information called node constraint. The node constraint is the conjunction of condition of message and contract of method. Moreover, root node represents initial message (i.e., message without precursor) which is triggered by an external event, and leaf node represents terminal message (i.e., message without successor) resulting in external output. A sequence of nodes from root node to leaf node represents a functional execution path, which must satisfy path constraint composed of node constraints.

3. Proposed approach to generate test cases

Collaboration diagram is used as test model to generate integration test cases. Each interaction is processed until all interactions in collaboration diagrams are tested, that is, messages among all component objects in the system are tested at least once. This is adequacy criterion of the proposed approach in the paper.

In the intermediate tree, a sequence of methods activated by messages on a functional execution path represents the functional behavior to be tested, and message passing among component objects on the path describes the necessary interactions among the system functional objects. Therefore, path coverage in execution tree of components meets aforementioned adequacy criterion, so that the problem can be transformed to the analysis and processing of paths in the execution tree of components.

3.1. Constructing an intermediate tree

The way of constructing an intermediate tree ETC is to map messages in collaboration diagram to a set of nodes, which are then built into a tree according to sequence number and condition of message.

In the process of mapping, a mapping table is created to record information about mapping. Because condition of message exists, a message can be mapped to multiple nodes. For example, message 2 is conditional message in message sequence "1→2→3", so functional execution path "1→2→3" is formed when condition of message 2 is true, and functional execution path "1→3" is formed when condition of message 2 is false. In this case, message 3 is mapped to two nodes which are located in two different functional execution paths.

It should be noted that node constraint consists of condition of message which has been processed and contract of method. Still take above message sequence "1→2→3" as an example, condition of message 3 is a part of node constraint when mapping message 3 to node in the functional execution path "1→2→3". In the other functional execution path "1→3", the conjunction of condition of message 3 and negative condition of message 2 is a part of node constraint when mapping message 3 to node. Furthermore, there is a special case: take message sequence "1→2" as an example, since message 2 is conditional message, message 2 is skipped and the following message is mapped to nodes when condition of message 2 is false. However, message 2 is also terminal message, and the following message does not exist, so a virtual node v is created to construct functional execution path "1→v", so as to keep integrality of path. In the process of mapping message to virtual node, the message is null, and node constraint of virtual node consists of negative condition of message 2.

The above describes how to map message to nodes, which is mainly to process condition of message and contract of method so as to form node constraint. According to this operation, we process messages in collaboration diagram in turn.

In the message sequence of collaboration diagram, the first message may be conditional or recurrent message, which results in branches at the beginning and a forest is generated. Therefore, to avoid this situation, a null message is inserted at the head of message sequence, which will be mapped to the virtual root node. Then, remaining messages are processed according to their type.

For common message, it is mapped to nodes directly, whose node constraints are made up of condition of this common message and contract of corresponding method. The number of these nodes is the same as the number of nodes which the precursor of current message is mapped to. Then, use one-to-one correspondence to take these nodes as children of nodes which the precursor of current message is mapped to.

For conditional message, it should be firstly determined whether this message has existed in mapping table. If this message has existed in mapping table, this message will be mapped to these existing nodes when condition of the message is true. On the other hand, take it into consideration that condition of this message is false, this message and its nested messages are skipped, and the following message is mapped to nodes whose number is the same as the number of these existing nodes. Then, use one-to-one correspondence to take nodes, which the following message is mapped to, as brothers of these existing nodes.

After dealing with the above situation, new situation is considered. The conditional message is processed in the same way as common message when condition is true. On the other hand, when condition is false, the conditional message and its nested messages are skipped, and the following message, which hasn't been processed, is mapped to nodes whose number is the same as the number of nodes which the precursor of conditional message is mapped to. Then, use one-to-one correspondence to take these nodes as children of nodes which the precursor of conditional message is mapped to. This operation may cause unprocessed conditional messages to appear in mapping table in advance, this is why it should be firstly determined whether the message has existed in mapping table before processing conditional message.

For recurrent message, it is processed in the same way as conditional message since loop is restricted to be executed zero times or once.

Collaboration diagram with contracts is automatically transformed into execution tree of components through the above operation, which lays a foundation for subsequent test case generation automatically.

3.2. Generating test cases
On the basis of execution tree of components, the automatic generation of test cases can be divided into two steps: the first is to get path constraints, and the second is to solve path constraints.

3.2.1. Getting path constraints. Constraints in collaboration diagram, which include conditions of messages and contracts of methods, have been stored in nodes of tree during the transformation from collaboration diagram to execution tree of components. Consequently, in the first step, the only work is to process constraints in nodes (i.e., node constraints).

As the path is traversed, path constraint, the conjunction of node constraints, is got. As the execution tree of components is traversed, all path constraints are got.

3.2.2. Solving path constraints. A Boolean satisfiability (SAT) problem [9] is a typical constraint solving problem, which is the problem of determining if there exists an assignment that satisfies a given propositional logic formula. The object to be solved of SAT problem is propositional logic formula, whose expressive power is limited to propositional calculus, so researchers extend SAT to satisfiability modulo theories (SMT) [10] by adding theories to SAT to enhance expressive power of SAT. SMT problem is a decision problem for logical formulas with respect to combinations of background theories expressed in first-order logic.

International SMT competitions have been held every year since 2005, and the top several solvers in 2016 and 2017 are Z3, CVC4, Yices2, SMTInterpol, veriT and MathSat5. The analysis results show that the comprehensive solving ability of Z3 solver [11] is the strongest. Therefore, we select Z3 solver to solve path constraints, thereby generating test cases.

A test case is a tuple <ID, path, input, testoracle>, where the elements are defined as follows. ID is a unique identifier used to distinguish different test cases. The path represents path corresponding to this test case. The input represents test input which is solution to path constraint, and it can be got with the help of Z3 solver. The testoracle represents test oracle which is used to check whether actual output is expected. We take postcondition of contract of method which corresponds to leaf node as test oracle.

4. Case study

This paper takes ATM machine as an example to illustrate how to automatically generate integration test cases through collaboration diagram and logic contracts. UML collaboration diagram obtained from the specification is shown in figure 1.

This collaboration diagram describes the process of using ATM machine to realize the withdrawal function. There are five components in it: Client, CardReader, ATMScreen, Account and Withdrawer. Besides, there are also thirteen messages in it, and they are shown as follows.

- 1: isBankCard:=(boolean)insertCard(boolean bankCard)
- 2: isValidCard:=(boolean)readCard(boolean bankCard)
- 3: initializeScreen()
- 4: inputPwd(boolean validPwd)
- 4.1, 5.1: isValidPwd:=(boolean)validatePwd(boolean validPwd)
- 5: *[!isValidPwd && i:=1...3] inputPwd(boolean validPwd)
- 6: displayMenu()
- 7: selectAction(int action)
- 8: inputAmount(int amount)
- 9: [amount<=30000] deductMoney(int amount)
- 9.1: spitSucc:=(boolean)spitMoney()

- 10: opCompleted:=(boolean)ejectCard()

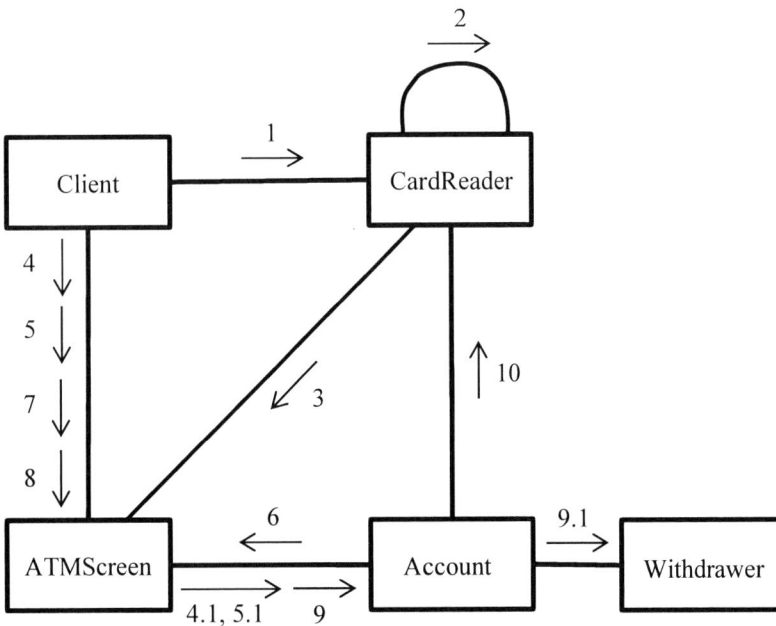

Figure 1. Collaboration diagram of ATM machine.

The initial state of ATM machine is balance(30000), which represents the current account balance is 30000. In addition, for convenience, the corresponding functions of ATM machine are numerically denoted, and the list of functions is as follows: 0 - deposit, 1 - withdrawal, 2 - transfer accounts, 3 - query, 4 - password modification. In this collaboration diagram, message 5 is recurrent message, message 9 is conditional message, and the rest are common messages.

According to section 3.1, the intermediate model generated, execution tree of components, is shown in figure 2.

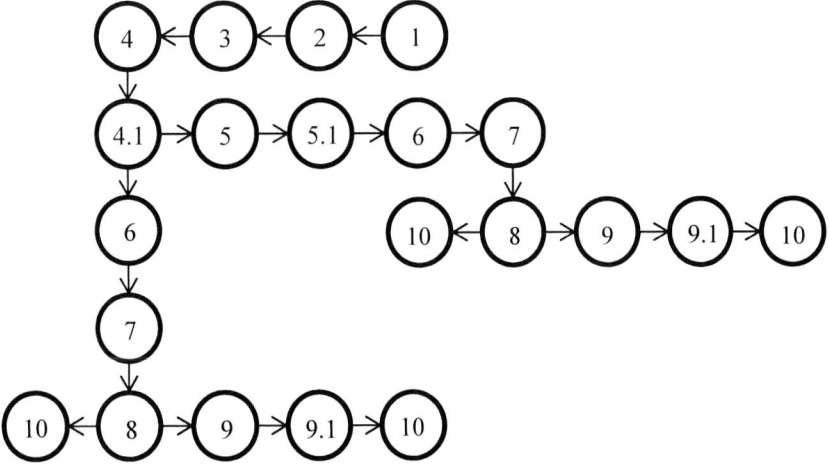

Figure 2. Execution tree of components.

Consider the functional execution path "1→2→3→4→4.1→6→7→8→10", we illustrate test case generation for it. In this path, recurrent message 5 is not executed and condition of conditional message 9 is false, i.e., password entered is valid and amount entered is greater than account balance. Message sequence and processed conditions corresponding to this path are shown in table 1, and the corresponding method sequence and contracts are shown in table 2. Then, according to section 3.2.1, path constraint of the functional execution path generated is shown in table 3. Furthermore, according to section 3.2.2, test case corresponding to this path is generated, and shown in table 4.

Table 1. Message sequence and processed conditions.

message	1	2	3	4	4.1	6	7	8	10
condition of message (processed)	-	-	-	-	-	$\neg$ (!isValidPwd)	-	-	$\neg$ (amount<=30000)

Table 2. Method sequence and contracts.

method	contract	
	precondition	postcondition
isBankCard:=(boolean)insertCard(boolean bankCard)	bankCard == true	isBankCard == true
isValidCard:=(boolean)readCard(boolean bankCard)	bankCard == true	isValidCard == true
initializeScreen()	-	-
inputPwd(boolean validPwd)	validPwd == true	-
isValidPwd:=(boolean)validatePwd(boolean validPwd)	validPwd == true	isValidPwd == true
displayMenu()	-	-
selectAction(int action)	action == 1	-
inputAmount(int amount)	amount > 0 && amount%100 == 0	-
opCompleted:=(boolean)ejectCard()	-	opCompleted == true

Table 3. Path constraint.

path	1→2→3→4→4.1→6→7→8→10
constraint	bankcard == true $\wedge$ validPwd == true $\wedge$ action == 1 $\wedge$ (amount > 0 && amount%100 == 0) $\wedge$ ($\neg$ (amount <= 30000))

Table 4. Test case.

	actual test case
ID	ts1
path	1→2→3→4→4.1→6→7→8→10
input	(true, true, 1, 35000)
testoracle	opCompleted == true

What this test case represents is: a valid bank card is inserted, then valid password is entered and the withdrawal function is selected, but amount entered is greater than account balance, so it fails to withdraw money. Finally, bank card is ejected. "opCompleted == true" represents succeeding in ejecting bank card. The failure to eject bank card means that interactions among components aren't implemented correctly and components doesn't behaves as expected.

5. Conclusions

As an important dynamic model in object-oriented software, UML collaboration diagram is an excellent integration test model. Utilizing component level collaboration diagram with contracts, this paper proposes a new approach for integration test case generation. In this approach, an intermediate model named as execution tree of components is constructed by analyzing messages in collaboration diagram and contracts in components. On this basis, path constraints which correspond to all paths are got and solved to generate test cases. Our proposed approach can automatically generate the intermediate tree by parsing the specification file of collaboration diagram, and then automatically generate test cases through contract solving technology, so as to realize the automation of integration testing. Compared with manual testing, this approach improves test efficiency and reduces the test cost. A case proves this approach is feasible, and our future work is to develop a tool to support it.

Acknowledgments

We acknowledge the support from Nuclear Power Institute of China (grant LRSDT2017304) and Education Department of Hunan (grant 17C1378).

References

[1]   Prasanna M, Chandran K R, Thiruvenkadam K. (2011) Automatic test case generation for UML collaboration diagrams. IETE Journal of Research, 57: 77-81.

[2]   Zeng Y, Liu Q X, Wang C Q, et al. (2013) Generation method of polymorphic path test scenarios with OCL constraints. Computer Engineering, 39: 92-96+102.

[3]   Swain R K, Panthi V, Mohapatra D P, et al. (2014) Prioritizing test scenarios from UML communication and activity diagrams. Innovations in Systems & Software Engineering, 10: 165-180.

[4]   Kaur A, Vig V. (2018) Automatic test case generation through collaboration diagram: a case study. International Journal of System Assurance Engineering & Management, 9: 1-15.

[5]   Meyer B. (1997) Object-Oriented Software Construction. Prentice Hall Publishing, Englewood.

[6]   Zhao Y N, Guo H L. (2014) Component testability research based on contract status checking. Modern Electronics Technique, 37: 83-85+88.

[7]   Zhao Y N, Guo H L. (2014) Component testing research based on contract checking. Journal of Xi'an University of Science and Technology, 34: 290-295.

[8]   Xu D, Xu W, Tu M, et al. (2016) Automated integration testing using logical contracts. IEEE Transactions on Reliability, 65: 1205-1222.

[9]   Guo Y, Zhang C S, Zhang B. (2016) Research advance of SAT solving algorithm. Computer Science, 43: 8-17.

[10]   Jin J W, Ma F F, Zhang J. (2015) Brief introduction to SMT solving. Journal of Frontiers of Computer Science & Technology, 9: 769-780.

[11]   Moura L D, Bjorner N. (2008) Z3: an efficient SMT solver. In: the 14th International Conference on Tools and Algorithms for the Construction and Analysis of Systems. Budapest. pp. 337-340.

Multipurpose IP-Based Space Air-Ground Information Network

Abid Murtaza, Liu Jianwei

Department of electronic and Information Engineering, Beihang University, Beijing, 100191, China

Abid_murtaza47@hotmail.com

Abstract. Integration of space resources (e.g., satellites) in different orbits with the terrestrial/ground network (e.g., internet) is the new evolving era of information technology. This kind of information network can expand the ground services such as internet to the whole world with the help of satellites' broader coverage. On the other hand, it can also provide real-time access to the satellite's useful data (e.g. images of earth), for nearly everyone on the globe. For these reasons, there has been enormous amount of research on this topic, and different network architectures are proposed for the integration of space and ground networks for different applications. However, most of them are either general abstract architecture or for the single dedicated application. Also for such a network, it is desirable that it should be based on the TCP/IP (Transmission control protocol/internet protocol) stack for interoperability with terrestrial networks. However, the biggest challenge for this is the dynamic network topologies due to orbital dynamics of space nodes. In this paper, we have addressed both of these issues by proposing a Space-Air-Ground Information Network (SAGIN) Architecture, which can provide almost all of the services of space and ground networks for all the users around the world according to varying needs of users. Our proposed SAGIN architecture is IP-based to provide interoperability and compatibility with the already deployed ground network. We have also addressed the challenge of dynamic nature of the network topology by designing a simple mechanism which automatically maintains a relatively static IPs of nodes in space (satellites), despite their rapid physical movement in orbits. This SAGIN architecture could be among optimized candidate architecture for future to have for many countries (either individually or by the cooperation of different regional countries). Because, it can provide all the services to nearly everyone in the world, in contrast to traditional individual satellites providing limited services to a limited number of users.

1. Introduction

Over the technology history, Space technologies (e.g., Satellite) and Terrestrial technologies (e.g., internet) have evolved almost independent with each other. Most of the satellites are launched in three well-known orbits; Geostationary Earth Orbit (GEO), Medium Earth orbit (MEO) and Low Earth Orbit (LEO). A vast majority of satellites are totally independent, i.e. not connected with any other satellite in the same orbit or any other orbit, while remaining few are either connected with other satellites in constellation (i.e. satellites within the same layer e.g. LEO or MEO) or connected with satellites in any other orbit [1]. On the one hand, it is established that by connecting satellites in different orbits, the coverage, access, and availability of space resources for users on the ground can be increased significantly [2]. Additionally LASER ISL (inter-satellite link) can provide high bandwidth, robust security, anti-EMI, small antenna size and low power consumption for highly

Content from this work may be used under the terms of the Creative Commons Attribution 3.0 licence. Any further distribution of this work must maintain attribution to the author(s) and the title of the work, journal citation and DOI.

Published under licence by IOP Publishing Ltd

efficient transmission, exchange and access flexibility for multi-service capacity with multiple granularities [3]. Meanwhile, the demand for integrating terrestrial network with space resources has also been increasing over the years for two reasons. First to spread the benefits of terrestrial/ground networks such as internet, to the all the world using global coverage of satellite, which cannot be possible using terrestrial network [4]. Secondly, to provide access to the satellites' data (e.g., images of the earth) to a user through the ground or terrestrial networks such as the internet [5].

For these reason we have seen an enormous amount of research on the same topic of integrating or networking of different space and ground technologies with several different names such as; Space Information Network (SIN), space-based information network (SBIN), Spatial Information Network, Integrated Satellite-Terrestrial network, space–ground heterogeneous network, Interplanetary Internet, Ground Space Merged Architecture etc. More importantly, most of these network architectures are proposed while keeping a particular application in focus, e.g. internet/broadband or mobile satellite communication. Some Authors discussed about different aspects of such integration networks by focusing a more general or abstract architecture containing all possible nodes or sub-networks. Therefore, we have seen a large variety of algorithms, techniques, and protocols for network control, data routing, data transmission, security and other features for different applications of SIN [2][6]. In general, such a heterogeneous network is divided into different sub networks for ease of control and management [7]. We will provide an overview of some of these architectures in the next section.

In this paper, we have proposed a Space-Air-Ground Information Network (SAGIN) architecture that can provide almost all services for different users on the ground according to their varying needs instead of any single application. For an integrated space-ground network, it is desirable that it should be based on the IP (internet protocol), for interoperability with terrestrial networks, as well as to take advantage of synergies with all research and products developed for TCP/IP [8], [9]. However, to achieve this, among the primary challenges, one is the dynamic nature of the network topology. Ground networks usually have relatively static network topologies where the positions of and distances between nodes are relatively fixed. On the other hand nodes (satellites) continuously lose and recover line of sight due to orbital dynamics, and the path lengths between nodes also vary quickly [10]. In this paper we have proposed an IP-based SAGIN, which provide as easy interoperability with the ground network (e.g., internet), devices and protocols, etc. as desired. We have addressed the challenge of dynamic routing by designing a simple network scheme such that all the nodes in space can dynamically update their IPs which will make the nodes virtually static, similar to that on the ground. Our proposed architecture, therefore, can provide a similar to ground availability and access to network nodes in space for users.

This network can provide almost all the civilian, commercial and military services for different users using the same single network. This proposed SAGIN could be among ideal future space architecture for many countries (either individually or in cooperation with many regional countries) which currently having few, dozens or hundreds of independent satellites in space because this can provide remarkable technological and economic benefits for all those countries.

The remainder of this article is arranged as follows; Section 2 provides a quick & brief overview of related work. In Section 3 and 4 we have proposed the architecture of SAGIN and discussed its topologies. In section 5 we have described SAGIN IP scheme. Section 6 describes the dynamic IP updating mechanism. Section 7 provides an overview of services and different applications of SAGIN. Section 8 concludes the paper.

2. Related Work
As mentioned in the introduction, many research for SIN is focused while keeping any particular application in focus. For example, Feihang Dong et al. proposed a novel HAP/satellite architecture to provide emergency communication in emergency scenarios [11]. Similarly, authors in [12] discussed different aspects such as architecture, routing, modelling, and scheduling of complex space information networks from the perspectives of cooperative earth observation application. Mobile satellite communication system is also an example of this application based integration network [13].

There are two approaches towards SIN, first is to divide the whole network into sub networks according to the link characteristics and sub network function and then focus on integration, interfacing and cooperation aspects for these sub-networks. For example, NASA designed the evolvable space communication architecture model to support the Earth sensor web (SW), collaborative observation formation missions, and detailed investigation of planets, moons, and small bodies in the solar system [14], In NASA's space-based network architecture, network elements are divided into sub-networks.

The concept of The system of systems (SOS) was introduced into space-based networks by authors in [15] to analyze independent complex small systems/ networks and enabled them to be developed, integrated, interoperable, and optimal. Based on the SOS structure, availability and capacity of the space-based network can also be analyzed. Lei He et al. proposed a new architecture of the SDN/IP hybrid space information network, where we can use the traditional IP-based system in the ground internet and use centralized SDN in the satellite network.

The alternate approach is to examine the system as a whole single unified network such that nodes either in the ground part or space may be indistinguishable from the user's perspective. This leads the effort towards IP utilization in space part of SIN because it is already widely deployed on earth [9] [5], [16].

A new software-defined architecture for next-generation satellite networks, called Soft-Space, is presented by authors in [17], where they exploited the concepts of network function virtualization, network virtualization, and software-defined radio to facilitate the incorporation of new applications, services, and satellite communication technologies. S. Wang et al. proposed LEO-user-oriented space network architecture, where LEOs are users to the backbone, instead of parts of the backbone. The new architecture simplifies the structure of the backbone network and improves the flexibility and extendibility of it [18]. The Survey of four leading space communication architectures namely OMNI (by NASA), CCSDS, Hi-DSN, and Space VPN, performed by [19], showed that the OMNI architecture provides a sustainable architecture serving future near space exploration demands.

3. SAGIN ARCHITECTURE

Our proposed SAGIN can be divided into three groups of nodes, i.e., space nodes, air nodes and ground nodes as shown in fig. 1. Space nodes include satellites in GEO, MEO, LEO and HAP (High Altitude platform). Air nodes include airplanes, helicopter, UAVs, etc. Ground nodes include internet gateways, fixed or mobile users, terrestrial networks base stations, etc.

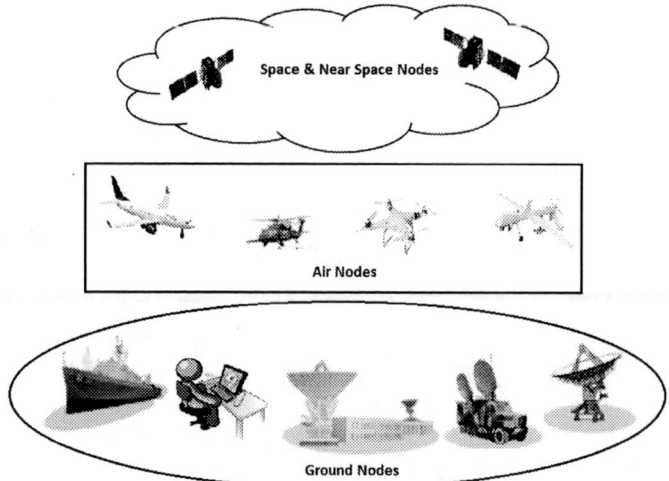

Figure 1. Three Groups of Nodes in SAGIN.

In this paper, we will not focus on ground and air node part of SAGIN for three primary reasons. The first is that many users and nodes on the ground are already connected through the internet which is by now widely deployed. Secondly, our network is IP based so there should not be any compatibility & connectivity issues of space, and air part with ground nodes of SAGIN such as internet gateways. Finally, because of being in the same network the data transmission in SAGIN will be seamless, once a data is entered in the SAGIN with IPs of source and destination, it will flow seamless irrespective of the physical location of the node (either in space air or ground).

Nodes in the air such as Airplanes, Drones, etc. can easily join the SAGIN to connect with the ground nodes because of being a part of the unified IP Network. All that these nodes need to do is to be configured with IPs & able to connect to SAGIN. Similarly, the user on the ground can access to air nodes through SAGIN. We will discuss more about this in section 7.

For designing Space part of our SAGIN architecture, we have considered advantages and disadvantages of different orbits to assign best possible services to satellites in these platforms. This means timing-critical services such as, voice and internet services are assigned to lower platform, i.e., HAP and LEO satellites, while applications which can tolerate little delays such as TV broadcasting and Earth observation are assigned to MEO and GEO satellites.

We have also assumed that next-generation satellites will be capable of serving multiple or at-least dual purposes simultaneously rather than traditional single dedicated purpose. We have already seen some practical demonstration of such multifunctional satellites such as, COMS (communication, ocean, and meteorological satellite), EGNOS (The European Geostationary Navigation Overlay Service), text messages services by Chinese navigation satellites Beidou, S&R payload on Galileo and ADSB receivers on Iridium Next for air traffic monitoring, etc.

The space part of SAGIN is composed of Satellites in four different layers as shown in fig 2;

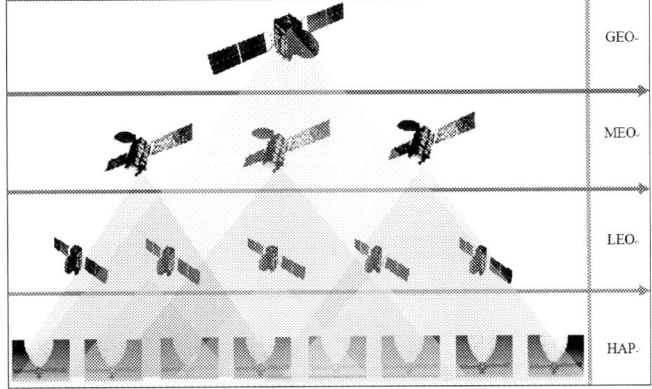

Figure 2. Space Components of SAGIN.

Dual purpose GEO Satellite

In our proposed SAGIN GEO Satellite can provide two primary services; first is broadcasting, as they are typically famous for, and second is network management. In our proposed SAGIN, there may be 6 GEO satellites (3 for TV broadcasting and 3 for real-time relaying of earth observation data to ground). While among 6 GEO satellites, 3 will be responsible for network management while other will be backup for network management.

Dual purpose MEO Satellites

MEO Satellite in our proposed SIN can serve two purposes; first is the navigation service which is the traditional service of MEO satellites, and the second is Earth observation which typically provided by LEO Satellites. However, as we know that now thanks to technology we have already seen GEO Earth observation satellite, so having same or more advanced cameras on MEO satellite will enable us to observe earth in future even comparable to that of traditional LEO resolution.

Dual purpose LEO Satellite

LEO satellite in our proposed SAGIN can serve two services, first is internet "inward" services, and second is Aviation control, text messages, and military communications. The internet service model that we are proposing is slightly different from traditional space-based internet concept, and we will explain that in section 7. However, for now, inward internet data means uni-directional data transfer from internet server towards user.

Dual purpose High Attitude Platform

High altitude platforms are emerging platform which are supposed to be aerostatic platforms or balloons at much lower altitudes than traditional LEO satellites, i.e., from 17-30 km usually. HAP can provide several unique advantages such as larger coverage area than terrestrial towers, compatibility with conventional base station technology and terminal equipment, requires no launch vehicles for deployment like satellites, compare to LEO much shorter signal path and delay. Upgrading, repair, maintenance, and re-deployment are possible, unlike satellites. Low power consumption and low cost are also advantages of HAP.

In our SAGIN, HAP will provide two services first is bidirectional voice communication (e.g. phone calls), second is outward internet service. Means users request to a specific destination server will reach through HAP and similar to LEO of our SAGIN, this is also unidirectional service. HAP may not be fit on the definition of the satellite; however, for ease, we will use term satellite for them throughout in this paper.

4. SAGIN TOPOLOGIES

There are three kinds of data flow through space part in our SAGIN.

- Mission Control Data
- Network Configuration Data
- User Application Data

Mission control data is the telemetry and telecommand data for satellite operators to check satellite's health and for controlling the satellite to keep them in their allowed orbital window or other onboard management on satellite. For mission controlling, the MCC (mission Control center) can access any satellite directly (during its visibility) or indirectly through GEO Satellite anytime.

Network configuration data in our SAGIN is the data that flows within space nodes without the involvement of user and satellite operators to ensure correct network configuration. For the network configuration, we have proposed the tree like topology as shown in figure 3, where data only flow between two immediate layers and no data flow within any layer and we will discuss this in detail in the next two sections.

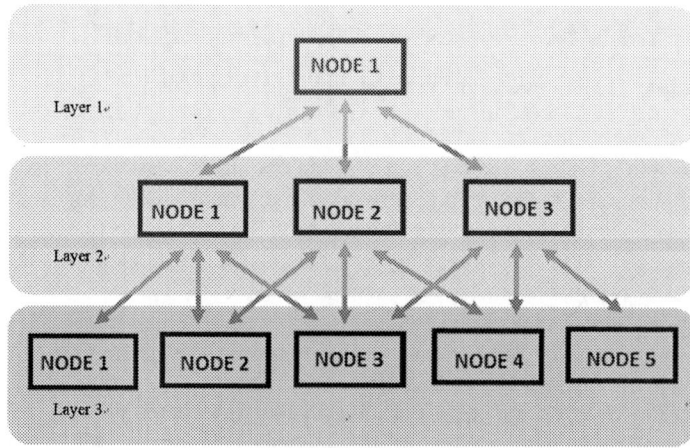

Figure 3. Tree like Topology of SAGIN

User Application data is the data that flows within the networks from terminal nodes (e.g., satellites or server on the ground) to the user and in reverse. For user application data, the topology will vary according to the application. For example for the application of real-time earth observation, images captured by satellites in MEO will be transmitted to GEO satellite that will broadcast them to the ground. So cross-layer topology is required where no data will be shared from one MEO to other MEO. Similarly, for the air traffic monitoring service, the navigation data needs to flow from one MEO to other MEO until it finds the destination nodes (desired Airplane). So for this application, data flow within the MEO layer is desired while no cross data is required. So we can say that for application data flow, the topology in MEO will be hybrid (cross-layer and within a layer). So, all Meo satellites will be connected with its every neighbouring MEO satellite as well as with GEO Satellite above though optical or RF ISL.

Similarly, for internet application and voice communication, there is no cross-layer data flow required in LEO and HAP. So, every LEO satellites will be connected with neighbouring LEO satellites. Similarly every HAP will be connected with neighbouring HAPs. We will discuss more about this in section 7.

5. IP ADDRESSING FOR SAGIN
The ground part of SAGIN is already configured with IPs. For the space part, our proposed network is self-configuring, where nodes will dynamically update their IP address. Therefore, the operator does not need to configure network configuration manually. For the explanation of our IP scheme, we are using IPV4 (internet protocol version 4) addressing. IPv4 addresses are represented in dot-decimal notation, consisting of four decimal numbers or octet (e.g., 192.168.1.5) each. Similarly, nodes (satellites) in our scheme will have IPs as shown below.

www.xxx.yyy.zzz

But, here we are using it with a different interpretation, which is that each octet in our IP scheme is the representation of a node in a particular layer. Which means the first octet "www" (highlighted by green) represents GEO, "XXX" (highlighted with yellow) represents MEO and similarly "YYY" and "ZZZ" represents LEO and HAP respectively. This means by using this IP scheme we can address up to 255 GEO satellites, 255 MEO Satellites, 255 LEO satellites, and 255 HAPs. However if the number of satellites in any layer (e.g., LEO & HAP) is more than 255, IPV6 simply can be used which uses 128 bits for the address.

GEO IPs

As GEO satellite is static for ground, so in our SAGIN, GEO satellites will have static IPs, i.e., IPs of GEO nodes will remain the same. GEO satellites will have following IPs. Assigned IPs (as shown below) are for explanation purpose, and in practice, there may be different IPs.

Geo1-IP=1.0.0.0
Geo2-IP=2.0.0.0
Geo3-IP=3.0.0.0 and so on

MEO IPs

In our SAGIN, Meo Satellites will have following IPs

Meo 1-IP= 1.1.0.0
Meo 2-IP= 1.2.0.0
Meo 3-IP= 1.3.0.0

As mentioned earlier, the highlighted green part of the IPs here is not fixed, instead it represents that MEO satellite is currently under GEO 1 satellite coverage. Similarly the MEO satellites under coverage of GEO3 will have IP as: Meo 9 - IP= 3.9.0.0

LEO IPs

The LEO satellites will have following IPs

LEO 1 –IP = 1.2.1.0
LEO 2 –IP = 1.2.2.0
LEO 45-IP = 1.3.45.0 and so on

The highlighted green octet of above IPs represents that LEO satellite is under GEO 1 coverage. Highlighted yellow part represents that LEO 1 is under coverage of MEO 2 or MEO 3 (LEO 45). So, for example, a LEO satellite (e.g., LEO 50) which is currently under coverage of GEO3 and MEO 8, will have IP 3.8.50.0

HAP IPs

The IPs of HAP will be as follows

$$HAP1 - IP = 1.1.1.1$$
$$HAP2 - IP = 1.1.1.2$$
$$HAP4 - IP = 1.1.1.4 \text{ and So on.}$$

Similar to MEO and LEO, the highlighted green, yellow and pink part of the HAP IP represents the GEO satellite, MEO satellite, and LEO satellites respectively.

One typical scenario will be that at any particular moment one LEO satellite comes under coverage of 2 or more MEO satellites. Similarly, one HAP will come in most cases under coverage of two or more LEO satellites (as shown in fig 1). We will address this common scenario in the next section where we will explain how nodes (satellite) will dynamically update their IPs.

6. DYNAMIC UPDATING OF IP

In this section, we will describe how the IPs of nodes in space will be dynamically updating as they continue moving in the orbit.

MEO Satellites IP Updating

A MEO satellite, on its orbital path will continuously send a beacon message consisting of 56 bits (7 octets) towards the GEO satellite above it, and every GEO satellite will reply to these beacon with its own identity (i.e., 1,2,3 …etc) as shown below

Beacon message from MEO towards GEO = 0.0.0.0.0.0.0
Reply of Beacon from GEO 1 =1.1.1.1.1.1.1
Reply of Beacon from GEO 2 =2.2.2.2.2.2.2
Reply of Beacon from GEO 3 =3.3.3.3.3.3.3

From this reply, MEO satellite can know about identity (IP) of GEO satellite above it. Moving on its path, when MEO satellite will enter the coverage of GEO-2, leaving the GEO 1's coverage, then reply message (i.e., 2.2.2.2.2.2.2) will tell MEO that it is now under the coverage of GEO 2. So, it will automatically change the GEO part of its IP from 1.1.0.0 to 2.1.0.0. Similar will be the case for every MEO satellite. By this method, all the MEO Satellites will dynamically update their IPs according to its current position.

The MEO part of MEO satellite IP (e.g., Highlighted yellow in 2.1.0.0) is also not static; MEO satellite will update this part of its IP based on its current coverage area on earth. Authors in [20] proposed a scheme to exploit geographical location information to make the mobility management independent from handovers. We also used the same concept, which means the whole earth can be divided into X equal parts from EO (earth observation) point of view. Where X is equal to the number of MEO satellite and each part on the ground will be assigned a number from 1 to X. During its orbital path when a MEO satellite will start covering the X part of ground, the MEO satellite will update its MEO part of IP according to that coverage area number as shown in fig. 4.

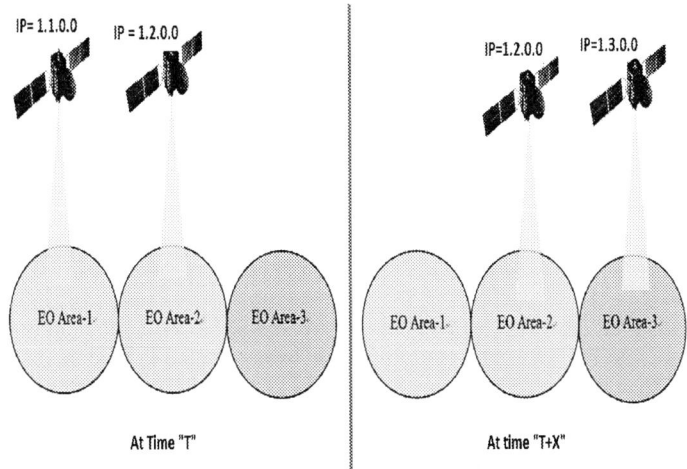

Figure 4. Updating Meo part of IP

Meo can know about its current earth coverage by two ways, either by the real-time acknowledgment from the ground (Lat & Long of ground inside user message) or mathematically as orbital paths can generally be predicted and estimated accurately in usual practice.

LEO Satellite dynamic IPs updating

Similar to MEO, the LEO satellite will also update their IPs dynamically as follows. LEO satellite will continually send Beacon message to MEO satellites above it as follows.

Beacon message from LEO towards MEO= 0.0.0.0.0.0.0

Every MEO will reply to this beacon with its IP as follow

Reply of Beacon from MEO 1 =1.20.1.20.1.20.2
Reply of Beacon from MEO 2 =1.19.1.19.1.19.4

From the reply of MEO 1, LEO can easily deduce that it is under coverage of GEO1 and MEO 20. Similarly, from the reply of MEO 2, LEO can know it is also under coverage of GEO 1 & MEO 19 satellites. The 7th octet in beacon message (highlighted gray) is to tell the traffic load on that particular MEO satellite. The load on the MEO satellite can be divided in to, e.g., Four main categories (1=Low, 2= moderate, 3 is heavy, and 4 is full load). So in response to beacon message, if LEO satellite gets a reply from more than 1 MEO satellites, it can easily decide which MEO satellite it should use for its IP selection, i.e., with lower seventh octet number. This will also do a job of load management to a certain level, similar to DNS (Domain Name Service) on the internet.

Similar to MEO satellite case, the LEO part of LEO satellite IP will be updated based on its current area of coverage. Means earth can be divided into Y parts from LEO coverage view point, and Y is the total number of LEO satellite which requires covering the whole world. Based on the orbital calculations or acknowledgment from ground, LEO can update its LEO part of IP (e.g., 1.4.21.0) accordingly.

HAP dynamic IPs updating

Similar to MEO and LEO, HAP will also update GEO, MEO and LEO part of HAP IP dynamically by sending beacon messages towards LEO satellites above it and according to the reply of LEO, HAP will update its GEO, MEO and LEO part of IP accordingly as shown below.

Beacon message from HAP to LEO = 0.0.0.0.0.0.0
Reply of Beacon from LEO 1 =2.20.5.2.20.5.1
Reply of Beacon from LEO 2 =2.19.6.2.19.6.3

From the reply of LEO 1, HAP can deduce that it is under coverage of GEO 2, MEO 20, and LEO 5. Similarly reply from LEO 2 tells HAP that it is currently under coverage of GEO 2, MEO 19 and

LEO 6. So HAP can update GEO, MEO and LEO part of its IP by considering value in the seventh octet (traffic load).

The HAPS are assumed to be stationary to the ground similar to GEO but with significantly lower coverage area than GEO. Therefore HAP part of HAP IP will be fixed for every HAP.

7. SERVICES

It is important to recall that we are assuming two kinds of users in the world. First are those who already have access to the internet and want to utilize the services of space nodes such as real-time earth images, navigation, TV broadcasting, etc. This kind of user can access all the services of space part of SAGIN through their ISP (internet service providers) who will receive IP data from the satellite using satellite terminal equipment. While others are those users who have no access to the internet, these users can use satellite terminal equipment such as (antenna, modem router, etc.) to access complete SAGIN (space, air and ground part). Our proposed SAGIN architecture can provide many services to the user; some are shown in table 1.

Table 1 Possible services of Proposed SAGIN

Service Name	Layers	Possible User
Internet	LEO & HAP	Everyone
Real-time earth Observation	MEO & GEO	RS, Aviation, Military
Air traffic Monitoring & Control	MEO, GEO & LEO	Airlines/ Airports
TV Broadcasting	GEO	Everyone
Voice and Text	HAP	Everyone
Navigation	MEO	Everyone
Military communication	LEO	Military

It is important to mention that services in table 1could be provided if nodes in space can provide services as we proposed in section 3. However, our SAGIN architecture is designed such that it will not have any impact on the network if satellite provides services other than what we proposed. For example we proposed EO services to be provided through MEO satellite; however, if EO services are provided through LEO satellite, then it will not make any difference in network and service architecture. Only EO users are required to be provided with LEO IPs instead of MEO IPs for EO. Similarly, we have divided internet services into two separate layers (LEO and HAP). However if in practice, only LEO Satellites are used to provide complete internet services it will not have any effect on network configuration. Similar is the case for services of any other satellite in SAGIN.

In this section, we will provide a brief overview of some of the services provided by SAGIN.

Internet

Considering long propagation delays of Geo and MEOs, two kinds of latency-optimized platforms proposed to extend broadband connectivity globally; first is to use HAP for this [21], [22] and second is to use LEOs. Both of these platforms can provide access to the internet for the user in two ways. First is that satellite or HAP platform acts as an access network where they only connect the user to gateways and the user request goes through typical ground internet network until reached destination server. Similarly, the response comes back to the gateway and then through satellite or HAP to the user as shown in fig. 5 (a) [23]. The other way is that space platforms (LEO or HAP) act as a core network as shown in fig. 5 (b). Our proposed SAGIN can be used as an access network, but it is designed and optimized as the core network.

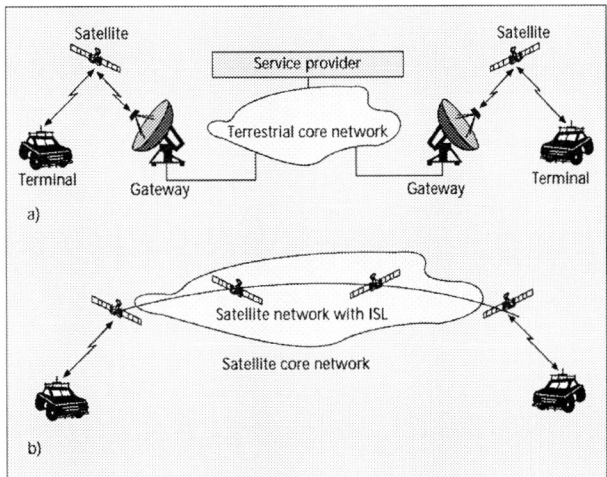

Figure 5. a) Access Network; b) Core network.

In our SAGIN we used LEO satellite as routers, and the user can get the desired data from destination server directly through LEO satellite constellation (logically router network). However, unlike traditional ground routers LEO will only provide inward service. Means data to the user will be delivered from the server using LEO satellites. On the other hand, the data request from the user will reach to servers through HAPs (logical router network), which also provides uni-directional services like LEO but only outward data flow (from user to the server).

Many routing schemes and algorithms are proposed for data flow from source satellite to destination satellite such as [24][25][26][27]. Authors in [12] also discussed various inter satellite routing strategies and protocols. Any of the proposed routing or any further optimized routing scheme can be used in SAGIIN according to requirements of the application.

Through this approach, the overall internet load can be divided into two parts (inward and outward data). At least the outward data load can be removed from LEO by this approach. Additionally, HAPs are placed at altitude very lower than LEOs, so it will reduce at least one-way time delay and hence can improve overall latency than traditional space internet. A further study to compare the performance metrics such as speed and latency of traditional space internet and our proposed internet is essential for better evaluation.

One more important aspect is that we already have projects in line from Space-x, one web, etc. to provide internet access for everyone across the globe through satellite [28]. So, in that case when everyone on the earth will get access to the internet the load on the core ground networks of the internet will be exceptionally increased which could seriously degrade the performance of internet if not completely halts. Damage of optical fiber in some occasional cases also results in disconnection of some regions with others. So considering these aspects, it is a good idea to put all new internet users' load on nodes in space rather than putting already congested ground internet core network.

Voice and Text Communication

Voice communication service through HAP will be similar to that with the terrestrial network. We can imagine that the cellular network communication tower is not on the ground; instead, it will be on HAP. This can provide two fundamental advantages, first is in case of disasters such as an earthquake, flood, Tsunami, and cyclone, etc. on the ground, communication will not be interrupted, and that will help in the rescue operation. Secondly, voice communication will be expanded to nearly all over the world including areas where the terrestrial network is absent such as Sea, Mountain, air, undeveloped areas, etc. Also similar to ground cell network concept for HAP will allow frequency reuse in cells that will help in coping with frequency resources limitation.

Although many internet application can provide voice and text communication, however, these put the load on internet infrastructure (routers and internet links). Using HAP, we can reduce the internet load by separating voice and small text messaging exchange from the main internet load (e.g., web browsing). Also a vast majority of voice and text communication occurs between local users, for that local communication, sending data to such applications server too far (engaging hundreds of routers and links) may be unnecessary, using HAP for voice over IP (VOIP), can provide an efficient alternative to those local communications while it can also support long distance communication. HAPs can also provide unobstructed and better angle of signals than terrestrial networks.

Real-time earth observation

Relaying LEO satellite data through GEO to ground is not a new concept. NASA's TDRS and EUs data relay system are more than two decades old concept [29]. However real-time transmission or broadcasting of earth observation data directly to the users is a new concept, which means seeing the earth from the eyes of satellite right now. In addition to traditional users of remote sensing/earth observation data (for example weather forecasting, agriculture, disaster monitoring, water resource, etc.), this kind of immediate access to earth observation data can open doors of new applications and users. Such as security surveillance, air traffic monitoring, ground traffic monitoring, etc.

In Our SAGIN, each earth observation MEO Satellite will transmit its captured data in IP packets to GEO EO relay satellite. GEO EO Relay Satellite will multiplex these IP packets to form an IP stream, and after encryption, it can broadcast this stream towards the ground. Authorized users on the ground with internet connectivity can access this stream through ISP, while user having no internet connection can receive that data directly from the satellite using terminal hardware.

The next generation IP based satellite communication can transfer images from satellite faster than traditional satellite communication. Saratoga [30] which is a UDP/IP protocol with reliable delivery, can be a suitable candidate protocol for data transfer from MEO to GEO and then from GEO to ground in this case. While to ensure security features, we can use the protocol of [31] for authentication, key exchange, and confidentiality in this real-time earth observation context.

Air traffic monitoring and control

Airlines and aviation authorities can receive the real-time earth observation images data as mentioned above to track the airplanes. Besides, airplanes can continuously send a beacon message to MEO satellite above it to tell MEO that airplane is in its coverage. MEO based on these beacon messages can maintain a list of airplanes under its current coverage. Aviation authorities can then send navigation data request to airplanes which in return can provide their latest navigation position with the help of onboard navigation receivers.

For Aviation control, and military control we have proposed to use LEO satellites because voice communication through MEO will have longer delays than LEO.

TV broadcasting

Similar to real-time earth observation, TV channels can be streamed up independently or through multiplexed service as DTH, and the GEO TV relay satellites will relay them back in the form of IP stream, which could be access through ISP or satellite terminal equipment.

8. CONCLUSION & FUTURE WORK

In this paper, we have proposed a Space air-ground network architecture that could provide access to all the services of satellites to internet users and internet access together with satellite service to users currently not connected with internet. We have proposed our IP-based scheme that will allow interoperability with ground network and internet protocols. We have proposed a mechanism through which each node in space can update its IP virtually static for the ground despite high physical mobility in space. This kind of scheme provides advantages in data routing in space and access to nodes for the user. We finally briefly discussed potential applications such as the internet, real-time earth observation, etc. In our future work, we will focus on applications and their protocols with in-depth details and compatibility and suitability analysis for those protocols in IP based SAGIN architecture environment context.

REFERENCES

[1] ESA, "ESA's Annual Space Environment Report," 2018.

[2] Q. Yu, J. Wang, and L. Bai, "Architecture and critical technologies of space information networks," J. Commun. Inf. Networks, vol. 1, no. 3, pp. 1–9, 2016.

[3] J. Yao, "Microwave photonics," J. Light. Technol., vol. 27, no. 1–4, pp. 314–355, 2009.

[4] B. Evans, M. Werner, E. Lutz, M. Bousquet, G. E. Corazza, G. Maral, R. Rumeau, and E. Ferro, "Integration of satellite and terrestrial systems in future multimedia communications," IEEE Wirel. Commun., vol. 12, no. 5, pp. 72–80, 2005.

[5] D. J. Israel, "Space Network IP Services (SNIS): An architecture for supporting low earth orbiting IP satellite missions," 2005 IEEE Networking, Sens. Control. ICNSC2005 - Proc., vol. 2005, pp. 900–903, 2005.

[6] L. Jianwei, L. Weiran, W. Qianhong, L. Dawei, and C. Shigang, "Survey on key Security technologies for space information networks," Journal of Communication and Information Networks, vol. 1, no. 1. pp. 72–85, 2016.

[7] W. Zhang, D. Bian, Z. Xie, and G. Zhang, "A Novel Space Information Network Architecture Based on Autonomous System," pp. 383–391, 2015.

[8] J. Rash, K. Hogie, and R. Casasanta, "Internet technology for future space missions," Comput. Networks, vol. 47, no. 5, pp. 651–659, 2005.

[9] K. Hogie, E. Criscuolo, and R. Parise, "Using standard Internet Protocols and applications in space," Comput. Networks, vol. 47, no. 5, pp. 603–650, 2005.

[10] D. Selva, A. Golkar, O. Korobova, I. L. i Cruz, P. Collopy, and O. L. de Weck, "Distributed Earth Satellite Systems: What Is Needed to Move Forward?," J. Aerosp. Inf. Syst., vol. 14, no. 8, pp. 412–438, 2017.

[11] F. Dong, H. Li, X. Gong, Q. Liu, and J. Wang, "Energy-efficient transmissions for remote wireless sensor networks: An integrated HAP/satellite architecture for emergency scenarios," Sensors (Switzerland), vol. 15, no. 9, pp. 22266–22290, 2015.

[12] J. Du, C. Jiang, Q. Guo, M. Guizani, and Y. Ren, "Cooperative earth observation through complex space information networks," IEEE Wirel. Commun., vol. 23, no. 2, pp. 136–144, 2016.

[13] G. Comparetto and R. Ramirez, "Trends in mobile satellite technology," Computer (Long. Beach. Calif)., vol. 30, pp. 44–52, 1997.

[14] K. B. and J. L. Hayden, "Space Internet Architectures and Technologies for NASA Enterprises," in Proc. IEEE Aerospace Conf., 2001.

[15] J. M. T. owenes Walker, Murali Tummala and Eachen, "A system of systems study of space based networks utilizing picosatellite formations," in 5th International conference on system of systems Engineering, 2010.

[16] M. Hadjitheodosiou, H. Zeng, A. Nguyen, and B. L. Ellis, "Flexible access for a space communications network with IP functionality," Comput. Networks, vol. 47, no. 5, pp. 679–700, 2005.

[17] S. Xu, X. W. Wang, and M. Huang, "Software-Defined Next-Generation Satellite Networks: Architecture, Challenges, and Solutions," IEEE Access, vol. 6, pp. 4027–4041, 2018.

[18] S. Wang, B. Wu, and B. Wang, "LEO-user-oriented space integrated information network," Proc. 2014 Int. Conf. Cloud Comput. Internet Things, CCIOT 2014, no. Cctot, pp. 166–169, 2014.

[19] O. Y. Tahboub, J. I. Khan, and M. Communications, "Recent Developments in Space Communication Architectures," in AIAA Regional Conference (Region III), Kalamazoo, Michigan, 2008.

[20] H. I. P. M. Management, H. Tsunoda, K. Ohta, N. K. Member, Y. N. Member, N. Kato, and Y. Nemoto, "Supporting IP / LEO Satellite Networks by Handover-Independent IP Mobility Management," in Jsac, 2004, vol. 22, no. 2, pp. 300–307.

[21] D. Grace, Broadband Communications via High Altitude Platforms. willey & sons, 2011.

[22] A. Mohammed, A. Mehmood, F. N. Pavlidou, and M. Mohorcic, "The role of high-altitude platforms (HAPs) in the global wireless connectivity," Proc. IEEE, vol. 99, no. 11, pp. 1939–1953, 2011.

[23] D. J. Bem, T. W. Wieckowski, and R. J. Zielinski, "Broadband satellite systems," Commun. Surv. Tutorials, IEEE, vol. 3, no. 1, pp. 2–15, 2000.

[24] W. Hou, B. Xian, L. Guo, W. Qi, and H. Zhang, "Novel routing algorithms in space information networks based on timeliness-aware data mining and time-space graph," 2015 Int. Conf. Wirel. Commun. Signal Process. WCSP 2015, pp. 0–4, 2015.

[25] G. Yu, C. Zhong, X. Lan, C. Zhang, L. Wei, and Y. Liu, "Research of multi-path routing based on network coding in space information networks," Chinese J. Aeronaut., vol. 27, no. 3, pp. 663–669, 2014.

[26] H. Uzunalioglu and N. Application, "Probabilistic Routing Protocol for Low Earth Orbit Satellite Networks," ICC 200, pp. 89–93, 2000.

[27] S. Cioni and A. Ginesi, "A novel routing design in the IP-based GEO/LEO hybrid satellite networks," Int. J. Satell. Commun. Netw., vol. 28, no. 5–6, pp. 291–315, 2016.

[28] M. Alleven, "From Boeing to SpaceX: 11 companies looking to shake up the satellite space," FierceWireless, 2017.

[29] G. Berretta, A. De Agostini, and A. Dickinson, "The European Data Relay System: Present Concept and Future Evolution," Proc. IEEE, vol. 78, no. 7, pp. 1152–1164, 1990.

[30] L. Wood, "Saratoga: scalable, speedy data delivery for sensor networks," in First CCSR Research Symposium CRS 2011, June 2011, 2011, no. June, p. 2.

[31] A. Murtaza and L. Jianwei, "A Simple, Secure and Efficient Authentication Protocol for Real-time Earth Observation through Satellite," in 2018 15th International Bhurban Conference on Applied Sciences and Technology (IBCAST), 2018, pp. 822–830.

ChanDet: Detection Model for Potential Channel of iOS Applications

GuoMiao Zhou[1], Ming Duan[1*], Qi Xi[1] and Hao Wu[1]

[1]State Key Laboratory of Mathematical Engineering and Advanced Computing, Zhengzhou, Henan, 450000, China

*Corresponding author's e-mail: mdscience@sina.com

Abstract. Despite providing iOS the security, comfortable, powerful mobile operating system, Apple has too many restrictions. Many users prefer to jailbreaking the iOS by using jailbreaking tool, which allows them to do more unavailable things on their devices. This behaviour may cause risks toward applications and many researches have focused on application security in various aspects. We find that a legal application (we call it *potential channel* hereafter) can be hijacked and it acts as a channel between the malware in device and the remote control terminal. In this paper, we introduce a channel model based on five conditions after analysing the entire operation procedure and comparing the similarities and differences of various applications. To approve our argument, we show iOS Messages which meets the five conditions, and demonstrate how to intercept messages, that means, a legal application can be hijacked and become a channel. To eliminate the risks, we propose a solution ChanDet and describe how to test whether an application is a potential channel or not. Finally, we give some protection strategies for applications, and we expect that ChanDet will play a significant role in application security of iOS.

1. Introduction

In order to attack and control a mobile phone, it is necessary to establish a communication channel which is likely through an application. As Apple mobile devices have become a part of our lives, we use them to deal with almost everything in our lives. But, in fact, what we rely on is one application after another. Therefore, how to detect the threats that legal applications are exploited as channels on iOS devices is meaningful.

Previous researches on iOS application security mainly focused on the following aspects. First, literatures [1, 2] have found out application issues. Second, literatures [3, 4, 5] have paid attention to the development of malwares and the use of APIs. Third, literatures [4, 6] have introduced new security mechanisms applying to applications. Fourth, literature [7] has investigated iOS sandbox profile. Fifth, literature [8] has detected vulnerabilities in applications based on mathematical model. Sixth, literature [9] has proposed analysis methods toward applications. Seventh, literature [10] has researched security threats of certificates. However, we make a research on application exploited as a communication channel, which is different from the literature above.

Apple has created and maintained an application distribution platform, App Store, which provides users millions of third-party applications to download. However, many people still jailbreak their devices. Due to the jailbreaking, the device can download and install additional applications, extensions and themes that are not allowed in the official platform. Jailbreaking uses kernel vulnerabilities to implement privilege escalation and remove restrictions applied by security mechanisms.

Content from this work may be used under the terms of the Creative Commons Attribution 3.0 licence. Any further distribution of this work must maintain attribution to the author(s) and the title of the work, journal citation and DOI.

Published under licence by IOP Publishing Ltd

Meanwhile, it provides opportunities for malwares to access system. If a malware exists, a legal application may act as a role which connects malware and remote control terminal. That means, it covertly becomes a channel, receiving commands, notifying the malware and sending out personal data to the remote.

In this paper, an application is authorized which might become a channel. To prevent a legal application from becoming a channel, after an in-depth analysis about the entire operation procedure (as Figure 1 shows) and the similarities and differences of many systems and third-party applications, we propose a solution to test applications which may have possibilities.

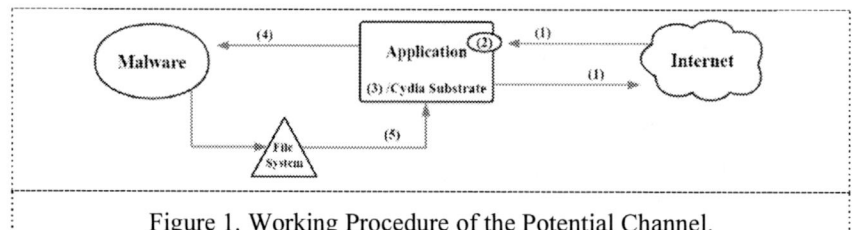

Figure 1. Working Procedure of the Potential Channel.

Five conditions are introduced, and if a legal application meets these five conditions, it has potential to become a channel between the malware and remote. Aiming at detecting an application just mentioned, a solution is put forward. The solution and the method how to test an application are described in detail in the paper. To strength the argument, an attack instance is offered which strongly proves the reality and feasibility. In this example, the mechanism of Messages combining with model is deeply analysed and the messages blocking is implemented. At last, the promotions and limitations of the solution are discussed.

2. Background
iOS is an operating system for iPhone, iPad, and so on and it is well-known by its high security. In this section, some background information covered in this paper is briefly introduced.

2.1. iOS Application and its Programming Languages
iOS applications can be divided into two types, system applications and third-party applications, which offer a great deal of convenience to our lives. App Store was opened on July 10, 2008. As of June 6, 2011, there were 425,000 applications available, which had been downloaded by 200 million iOS users [11]. As of January 2017, the store featured over 2.2 million applications [12].

iOS applications are developed by two programming languages, Objective-C and Swift. Objective-C, which is mainly used for OS X and iOS operating systems is a general-purpose, object-oriented programming language, which first appeared in 1984. Swift is developed by Apple Inc. used for iOS, OS X, etc. introduced at Apple's 2014 Worldwide Developers Conference (WWDC) [13]. Apple aims to boost Swift as one of the most security programming languages.

2.2. iOS Application Reverse Engineering Toolset
Theos. Theos is a jailbreaking development tool package, developed by Dustin Howett and shared on Github. It can develop and deploy iOS software without Xcode. Most tweak developers use Theos and its programming language component is Logos that allows method hooking code to be written easily and clearly.

IDA. The Interactive Disassembler is an interactive, programmable, multi-processor and extensible disassembler for software, written entirely in C++. It is the world's most feature-full disassembler and many software security specialists are familiar with it. It can analyse software statically.

Low Level Debugger(LLDB) and debugserver. LLDB is a dynamic debug tool, which was developed by Apple, and built in Xcode, and debugserver is a console app invoked by Xcode to debug applications on the device. When debugging an application dynamically, LLDB acts as a client sending command on computer and debugserver acts as a server receiving command for remote LLDB debugging on device.

Cydia Substrate. Cydia Substrate (formerly called Mobile Substrate) is a framework that allows third-party developers to provide run-time patches (Cydia Substrate extensions) to system functions, created by Jay Freeman (Saurik). It consists of three major components: MobileHooker, MobileLoader and Safe Mode. We rely on this framework to hook Objective-C method. When installing Substrate, jailbreak device is needed.

3. Potential Channel Model

The *potential channel* is defined for a systematic or third-party application on iOS if this application meets the following five conditions:

(1) The application has network communication capability (i.e. contains communication components or modules).

(2) The communication message of the application can be understood by people (i.e. transmitting plaintext message or encrypted message can be decrypted).

(3) The application process space can load "Cydia Substrate" framework.

(4) The application has inter-process communication (IPC), such as message queue, and notifications.

(5) The application can access the file system.

The potential channel can be exploited to construct a loop for attacks (e.g. content hijacking, content tampering, designated commands and information transmission). Here a hijacking example of the iOS Messages application with potential channel on iOS 9.1 and iOS 10.2 is provided.

4. Channel Verification

First of all, we introduce some applications, processes and plugins related to message. The following analysis is based on iOS 9.1.

Messages (bundle identifier "com.apple.MobileSMS"). This application is what we always see and use on the screen.

imagent (bundle identifier "com.apple.imagent"). This process is always running in the background, waiting for coming messages.

SMS.imservice (bundle identifier "com.apple.imservice.sms"). This plugin is loaded into imagent at running time.

4.1. Messages Analysis

For users, if we want to analyse the iOS Messages mechanism, it is obvious that our target is the Messages application which we touch every time when a message comes, but things are not like that. Have you considered the following question? How can an iPhone receive a message at any time? It is well-known that iOS have strict limitations for processes and a closed application can't run in the background. There must exist a process running in the background. In fact, Messages application is just a client which don't charge of receiving messages, with user interface.

4.2. Imagent Analysis

Imagent is the background process for Messages and iMessage tasks, keeping running when Messages application is closed [14, 15]. It will be launched immediately if it killed. That means, this process is a daemon, which is running in the background forever, which might answer the above question.

After confirming that the imagent is charge of receiving messages, we want to find the first trigger method when a message comes into the mobile. Having found binary file of imagent in a jailbroken device, the class-dump is firstly used to look into its header files and a class called **IMDaemon** is noticed. We scanned all the interface of **IMDaemon** and some of APIs caught our attention.

Then, dynamic and static analysis methods are used to analyse the binary file of imagent, which is dragged into IDA for analysing statically. In IDA, except that the classes and methods have been extracted by class-dump, there are many other methods we can't recognize. According to the methods' name, we can determine that some of methods may related to dealing with message objects. OpenSSH is used to connect Mac and jailbroken device, so that LLDB on Mac and debugserver on mobile could be adopted to dynamically analyse. When some breakpoints are set and these breakpoints are triggered, the process has been interrupted. Looking at the registers' information, we could find the message object (as Figure 2 shows), which consists of many details (e.g. the sender and recipient phone numbers, message content, date and some flags). We can think that message object arriving here after has gone through many methods from the first starting method which we try to find. We tried to find the data source in reverse. Finally, we found that the message object comes from another bundle called SMS.imservice.

```
Apr 25 21:34:36 iPhone imagent[          .dylib)[14830] <Notice>: arg1 = {
    co = "        "; Recipient phone number
    g = "                        ";
    h = "                "; Sender phone number
    k =         (
                {
            data = <313431>; Message content
            type = "text/plain";
        }
    );
    l = 0;
    m = sms;
    n = 460;
    re =        (
    );
    sV = 1;
    w = "2018-04-25 13:34:25 +0000";
}
```

Figure 2. Message Object Structure.

4.3. SMS.imservice Plugin Analysis

SMS.imservice is a plugin. When dragging the binary file of SMS.imservice into IDA, we find that the method names are uniform format (e.g. sub_xxxxx), so we can get some information from the names of the methods. Fortunately, we can get the jump address from SMS.imservice plugin in imagent, so we can continue to reverse it through IDA. We finally find the first trigger method is "**_processReceivedDictionary:**" in **SMSServiceSession** class. Incoming messages firstly processes this method, saves to the database and delivers to other iOS components (e.g. notification). You might find that imagent seems not to use this plugin if disassembling it, because it is being loaded at runtime, which is called "late load".

4.4. iOS Messages Mechanism Overview

Based on all the above analysis, we can describe the whole Messages mechanism of iOS.

When imagent is being started, several plugins (i.e. SMS.imservice, iMessage.imservice and FaceTime.imservice) are loaded into it by using "**-(void) loadServiceBundle**" method in **IMDService** class in **IMDaemonCore** framework, but these plugins are not linked at compile time. When a message is coming in, "**_processReceivedDictionary:**" method in **SMSServiceSession** class of SMS.imservice plugin is being evoked to process the incoming message. SMS.imservice plugin is loaded into imagent, so that imagent completes all the functions for processing a message, such as pushing notifications, storing, and delivering. Then, users will see banners appearing at the top of the screen and Messages application will display the new message.

Figure 3 shows graphically the relations between the various components and the entire working flow.

ISPECE

IOP Publishing

IOP Conf. Series: Journal of Physics: Conf. Series **1187** (2019) 042045   doi:10.1088/1742-6596/1187/4/042045

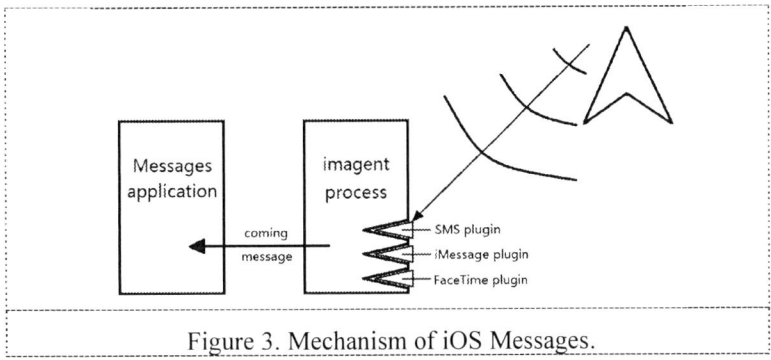

Figure 3. Mechanism of iOS Messages.

4.5. iOS Messages Blocking

Relying on Cydia Substrate, the target method's hook and message blocking on device are implemented, and Theos on Mac is used to implement Tweak development.

Based on all the above analysis, a Tweak is made to block the message and do whatever we want to do.

The code in appendix A is implemented on a jailbroken device of iOS 9.1 and **(id)arg1** (as Figure 2 shows) is a dictionary data type stored message's data and metadata.

After implementing the message blocking on iOS 9.1, we tried to study the message mechanism of newer version iOS 10.2. We find that iOS 10.2 has the same mechanism with iOS 9.1, but it has changed the key method's name from "**_processReceivedDictionary:**" to "**_processReceivedDict-ionary:storageContext:**". The code in appendix B can block messages on a jailbroken device of iOS 10.2 and **(id)arg1** (as Figure 2 shows) is a dictionary data type stored message's data and metadata.

Based on all analysis above, it is confirmed that systematic application Messages is a potential channel and it might be exploited by malware as their communication channels.

Jailbroken devices are used to study Messages mechanism, because it can focus on mechanism research and check debugging information more conveniently for development.

5. Potential Channel Detection

The ChanDet tests the potential channel in five steps. First, for target application, it is confirmed whether it has network communication capability. Second, its communication data and whether it can be understood are intercepted. Third, process space is looked into to check the required framework. Fourth, whether it has IPC is determined. Fifth, whether it can access file system is checked.

The ChanDet flow chart is shown in Figure 4.

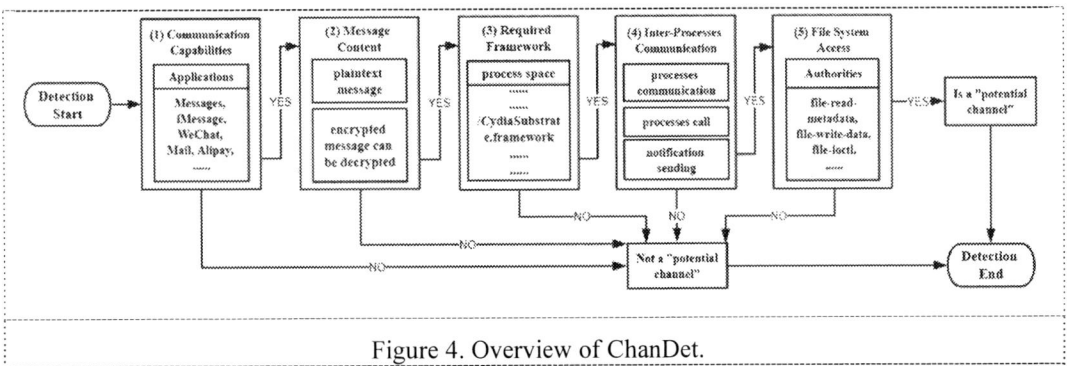

Figure 4. Overview of ChanDet.

5.1. Network Communication Capability and Communication Data

1579

According to life experience, it is easy to know if an application has communication capability, which just needs using. For example, instant messaging and e-mail application classes must have communication capabilities. To look into the application communication data, we aim at the first triggered method of receiving of the application and hook it to intercept the data. By the obtained data, whether the message is plaintext or decrypted could be determined, and an application that has communication capability and could work normally at the same time is confirmed.

For these purposes, the application binary analysed by reverse engineering to catch the first triggered method of receiving data. Because the method is depending on specific application and almost makes sure that the name of the method is different in various applications, only the general approach could be pointed out. By mixing use of dynamic (combination tools of LLDB and debugserver) and static (reverse analysis tool IDA) analysis, we look back to find the original data source of the data object, until the first triggered method.

5.2. Required Framework

Cydia Substrate is a required framework in Tweak's running. In general, it will be installed automatically and immediately after jailbreaking, since it has been a built-in jailbreaking tool. When an application launches, the framework is loaded in, so that in application process space could be found.

LLDB and debugserver are used to check the framework. In dynamic analysis, we always want to find the ASLR offsets of executable files, frameworks, libraries, plugins, etc. and LLDB debugging can list all of them and their offsets. Under LLDB command line, inputting the order "**image list -o -f**", it will output a list containing all in the process space, so it can be found in the output if Cydia Substrate has been loaded in.

5.3. IPC and File System Access

Inter-process communication mechanism is an important component of the iOS philosophy. There are many inter-process communication ways, and only need at least one of them that can work, so the channel can transmit information to aggressive application. Similarly, channel only needs to have read and write permissions in one path, so it can obtain the massive data (it might be user's personal data) from aggressive application.

Therefore, we have an easy way to test the application's IPC and file system access permission. First of all, there is a logic. IPC works on un-jailbreaking iOS, so a jailbreaking iOS should have IPC. Also, jailbreaking has to release sandbox restrictions, that is, file system access restrictions are removed.

However, a verification is needed. Due to the Cydia Substrate framework just checked, we have an easy and effective way in the following description. We can develop a Tweak by Theos, and inject it into application to hook a class. In this Tweak, by using keyword "**%new**", we create a new method which includes process communication testing and file system access codes. If tweak runs, it can be verified based on the running results.

6. Protection

To avoid our applications becoming a channel, some protection measures could be proposed according to the model.

6.1. Do Not Jailbreak

It is known that iOS is considered the most security operating system among the mobile systems. It has built-in many security mechanisms to stop the phone from being invaded, but jailbreaking will damage these security mechanisms, such as removing sandbox restrictions, preventing mandatory access controls, and corrupting code signatures. Without protection of these security mechanisms, iOS can be easily controlled and personal data will face risks. So we advise iOS not to jailbreak.

6.2. Reinforcing Confidentiality

Derived from the second condition of the model, the encryption of the message content (e.g. by using a strong cryptographic algorithm) could be reinforced. If attackers can't understand the message content, they can't hook applications to receive designated commands. Here, an example of reinforcing confidentiality is offered.

When studying the Messages, we also do some preliminary research on iMessage. It has a similar mechanism to Messages. iMessage plugin (bundle identifier "com.apple.imservice.imessage") is loaded when imagent is being started but not linked at compile time. The first evoking method is "**didReceiveMessage:forChat:style:account:**" of **MessageServiceSession** class on both iOS 9.1 and iOS 10.2. When the communication data is intercepted, we can find data has been encrypted and we can't decrypt it easily. iMessage doesn't conform to the model, so it isn't a potential channel.

7. Conclusions

Many people like to jailbreak, removing restrictions and getting more freedom, but Apple don't encourage this, since it might expose devices to more risks due to lacking protection of security mechanisms.

In this paper, we introduce ChanDet, a solution to determine whether a legal application is a potential channel or not. If it is, it has possibilities to become the channel between the malware and remote. After a deep analysis and comparison, a model whose core is five conditions is generalized. Based on the model, ChanDet works on testing these five conditions, and then renders the test result.

The Messages mechanism is deeply analysed and the whole working flow is described. Then the messages blocking to approve the argument is demonstrated, which means that the threat is likely. Additionally, we find that iMessage is not a potential channel due to its confidentiality.

8. Appendices

Appendix A. The code on a jailbroken device of iOS 9.1

```
%group smsLateHook
%hook SMSServiceSession
-(id) _processReceivedDictionary:(id)arg1
{
        %log;
        id r = %orig;
        return r;
}
%end
%end

%hook IMDService
-(void) loadServiceBundle
{
        %orig;
        if ([[[self bundle] bundleIdentifier] isEqualToString:@"com.apple.imservice.sms"] && [[self
bundle] isLoaded])
        {
                NSLog(@"sms plugin is Loaded!");
                %init(smsLateHook);
        }
}
%end
```

Appendix B. The code on a jailbroken device of iOS 10.2

```
%group smsLateHook
%hook SMSServiceSession
-(id) _processReceivedDictionary:(id)arg1 storageContext:(id)arg2
{
        %log;
        id r = %orig;
        return r;
}
%end
```

```
%end

%hook IMDService
-(void) loadServiceBundle
{
        %orig;
        if ([[[self bundle] bundleIdentifier] isEqualToString:@"com.apple.imservice.sms"] && [[self
bundle] isLoaded])
        {
                NSLog(@"sms plugin is Loaded!");
                %init(smsLateHook);
        }
}
%end
```

References

[1] Orikogbo, D., Büchler, M., Egele, M. (2016) CRiOS: Toward Large-Scale iOS Application Analysis. In: Security and Privacy in Smartphones and Mobile Devices. Vienna. pp. 33-42.

[2] Xing, L.Y., Bai, X.L., Li, T.X., Wang, X.F., Chen, K., Liao, X.J., Hu, S.M., Han, X.H. (2015) Cracking App Isolation on Apple: Unauthorized Cross-App Resource Access on MAC OS X and iOS. In: ACM SIGSAC Conference on Computer and Communications Security. Denver. pp. 31-43.

[3] Wang, T.L., Jang, Y., Chen, Y.Z., Chung, S., Lau, B., Lee, W. (2014) On the Feasibility of Large-Scale Infections of iOS Devices. In: USENIX Security Symposium. San Diego.

[4] Bucicoiu, M., Davi, L., Deaconescu, R., Sadeghi, A.R. (2015) XiOS: Extended Application Sandboxing on iOS. In: ACM Symposium on Information, Computer and Communications Security. Singapore. pp. 43-54.

[5] Deng, Z., Saltaformaggio, B., Zhang, X.Y., Xu, D.Y. (2015) iRiS: Vetting Private API Abuse in iOS Applications. In: ACM SIGSAC Conference on Computer and Communications Security. Denver. pp. 44-56.

[6] Werthmann, T., Hund, R., Davi, L., Sadeghi, A.R., Holz, T. (2013) PSiOS: Bring Your Own Privacy & Security to iOS Devices. In: ACM SIGSAC Symposium on Information, Computer and Communications Security. Hangzhou. pp. 13-24.

[7] Deshotels, L., Deaconescu, R., Chiroiu, M., Davi, L., Enck, W., Sadeghi, A.R. (2016) SandScout: Automatic Detection of Flaws in iOS Sandbox Profiles. In: ACM SIGSAC Conference on Computer and Communications Security. Vienna. pp. 704-716.

[8] D'Orazio, J.C., Lu, R.X., Raymond, C.K.K., Vasilakos, V.A. (2017) A Markov adversary model to detect vulnerable iOS devices and vulnerabilities in iOS apps. Applied Mathematics and Computation, 293: 523-544.

[9] Yin Z.Q. (2017) Dynamic Analysis Methods of IOS Application Security. In: International Conference on Information Technology and Management Engineering. BeiJing.

[10] Zheng, M., Xue, H., Zhang, Y.L., Wei, T., Lui, C.S.J. (2015) Enpublic Apps: Security Threats Using iOS Enterprise and Developer Certificates. In: ACM Symposium on Information, Computer and Communications Security. Singapore. pp. 463-474.

[11] WIKIPEDIA. (2018) Mobile app. https://en.wikipedia.org/wiki/Mobile_app#App_Store.

[12] WIKIPEDIA. (2018) App Store (iOS). https://en.wikipedia.org/wiki/App_Store_(iOS).

[13] WIKIPEDIA. (2018) Swift. https://en.wikipedia.org/wiki/Swift_(programming_language).

[14] Stackoverflow. (2013) Block sms on ios6. https://stackoverflow.com/questions/16219799/block-sms-on-ios6/18915532?utm_medimu=organic&utm_source= google_rich_qa&utm_campaign= google_rich_qa.

[15] Stackoverflow. (2014) SMS Interception in Jailbreak iOS 7. https://stackoverflow.com/questions/24525639/sms-interception-in-jailbreak-ios-7.

IOP Conf. Series: Journal of Physics: Conf. Series **1187** (2019) 042046   doi:10.1088/1742-6596/1187/4/042046

Characterizing of Strong Normalization for $\Lambda\mu$-Calculus

Xinxin Shen[1], Kougen Zheng[2*]

[1,2] Department of Computer Science and Technology, Zhejiang University, Hangzhou, Zhejiang, 310027, China

[*]Corresponding author's e-mail: zkg@cs.zju.edu.cn

Abstract. $\lambda\mu$-calculus is introduced by Parigot as an extension isomorphic to an alternative presentation of classical natural deduction. Since then, many properties of it have been studied and, in particular, it does not enjoy the separation property shown by David and Py. $\Lambda\mu$-calculus is proposed by de Groote and developed by Saurin as an extension of $\lambda\mu$-calculus Saurin demonstrates that the separation property holds for the $\Lambda\mu$-calculus. Bakel gives a characterization of strong normalization of $\lambda\mu$-calculus in the view of intersection type. In this paper, we will extend the intersection type assignment system to the $\Lambda\mu$-calculus, and show that it characterizes those terms that are strongly normalizing. The system satisfies the subject reduction and subject expansion properties.

1. Introduction

The intersection type assignment systems, introduced into the lambda calculus in the late 1970s by Coppo and Dezani [1-2] are devised to characterize the set of solvable terms and later extended by Barendregt [3] and Pottinger [4]. For an overview of the various existing systems, please refer to [5]. They extend the simple typed assignment system to include intersections and corresponding rules, allowing for term variables (and terms) to have more than one type. Intersection types have been used in a series of papers for characterizing evaluation properties of λ-terms [6-15]

Meanwhile, the ordinary λ-calculus [16] has been extended in several ways. There is a correspondence, called Curry-Howard correspondence, between simply typed λ-calculus and intuitionistic natural deduction. In order to extend the Curry-Howard correspondence to the classical natural deduction, the $\lambda\mu$-calculus [17] is proposed. Properties of $\lambda\mu$-calculus have been extensively studied both as a typed and an untyped language. Among them, separation property (also called Böhm Theorem) does not hold as shown by [18]. To recover the Böhm Theorem, Saurin [19] develops the $\Lambda\mu$-calculus. The $\Lambda\mu$-calculus can be seen as a stream calculus that enjoys some fundamental properties [20-21].

The full characterisation of strong normalization is a property that is shown for various intersection systems for the λ-calculus. To show that all typeability terms are strongly normalizable, reducibility method introduced by Tait [22] is suggested. The converse of this result is dependent by the subject expansion that can only reliably be shown for left-most outermost reduction [5] or perpetual reduction [23-24]. Bakel [14] characterizes the strongly normalizing $\lambda\mu$-terms and has observed that the intersection type assignment system can be adapted to $\Lambda\mu$-calculus, but whether the characterization result can be extended to $\Lambda\mu$ is unknown. Saurin [20] introduces a simply-typed $\Lambda\mu$-calculus which satisfies the subject reduction and typed terms are strongly normalizable in the type system. However,

---

Content from this work may be used under the terms of the Creative Commons Attribution 3.0 licence. Any further distribution of this work must maintain attribution to the author(s) and the title of the work, journal citation and DOI.

Published under licence by IOP Publishing Ltd

the reverse of this property does not hold for the system. The aim of this paper is to provide a characterization of strongly normalizing $\Lambda\mu$-terms and our approach follows the perpetual reduction.

2. $\Lambda\mu$-calculus

In the grammar, we use the notation $M\alpha$ [24] instead of $[\alpha]M$. The new notation makes explicit the intuition that α represents a potentially infinite stream of terms to which M is applied.

In Parigot's original $\lambda\mu$-calculus, terms of the form $[\alpha]P$, are distinguished as named term from the ordinary terms, and bodies of μ-abstractions are restricted to the named terms. On the other hand, $\Lambda\mu$-calculus considers $P\alpha$ as an ordinary term and any term can be the body of μ-abstraction in the $\Lambda\mu$-calculus.

Definition 1 (Term syntax)

1) The set $\Sigma_{\Lambda\mu}$ of terms are defined inductively by the following grammar
$$M, N ::= x \mid \lambda x.M \mid MN \mid \mu\alpha.M \mid (M)\alpha.$$
where $x \in Var_v$, a set of term variables (denoted by $x, y, z...$), and $\alpha \in Var_S$, a set of stream variables (denoted by $\alpha, \beta, ...$), both denumerable and they are disjoint.

2) A stream $S \in \Sigma^*_{\Lambda\mu}$ is an applicative context of the shape:
$$S := [\,]N_1 \cdots N_m\beta$$
$M, N, ...$ are terms and $M, S, P \cdots$ are streams. Streams are from [19-20] As usual, λ and μ are considered to be binders. The form of $\mu\alpha.M$ is called a μ-abstraction and the form of $M\alpha$ is called a μ-application. Terms shall always be considered up to α-equivalence.

We adopt Barendregt's convention on terms, and will assume that free and bound variables are different. The sets fv(M) and fn(M) of, respectively, *free term variables* and *free stream variables* in a term M are defined in the usual way.

Convention.

1). Application is left-associative.

2). Consecutive abstractions may be collapsed to a single one. eg. $\lambda x_1 x_2 \cdots x_n.M$ denotes $\lambda x_1.(\lambda x_2.(\cdots (\lambda x_n.M) \cdots))$ and similarly for μ.

3). The abstractions extend as far to the right as possible.

Definition 2 (Substitution) There are two forms of substitutions:

- term substitution: $M[x := N]$ is defined as in λ-calculus;
- structural substitution: $M[\alpha \Leftarrow N]$ is defined as the replacement of any subterm $(P)\alpha$ of M with $\alpha \in fn(M)$, by the subterm $(P)N\alpha$. More precisely, $M[\alpha \Leftarrow N]$ is defined by:

$$x[\alpha \Leftarrow N] = x \qquad\qquad (\lambda x.M)[\alpha \Leftarrow N] = \lambda x.M[\alpha \Leftarrow N]$$
$$(M_1 M_2)[\alpha \Leftarrow N] = (M_1[\alpha \Leftarrow N])(M_2[\alpha \Leftarrow N]) \qquad (\mu\beta.M)[\alpha \Leftarrow N] = \mu\beta.M[\alpha \Leftarrow N]$$
$$(M\alpha)[\alpha \Leftarrow N] = (M[\alpha \Leftarrow N])N\alpha \qquad\qquad (M\beta)[\alpha \Leftarrow N] = (M[\alpha \Leftarrow N])\beta \text{ if } \alpha \neq \beta$$

Definition 3 (Axioms) Axioms of the $\Lambda\mu$-calculus are defined by:

$(\beta_T)\quad (\lambda x.M)N \to M[x := N] \qquad\qquad (\beta_S)\quad (\mu\alpha.M)\beta \to M[\alpha := \beta]$

$(\eta_T)\quad \lambda x.(M)x \to M \text{ if } x \notin fv(M) \qquad (\eta_S)\quad \mu\alpha.(M)\alpha \to M \text{ if } \alpha \notin fn(M)$

$(\text{fst})\quad \mu\alpha.M \to \lambda x.\mu\alpha.M[\alpha \Leftarrow x] \text{ if } x \notin fv(M)$

The $M[x := N]$ and $M[\alpha := \beta]$ are the usual capture-avoiding substitutions.

Remark. For $S = [\,]N_1 \cdots N_n\beta$,
$$MS = M\vec{N}\beta$$
$$M[\alpha \Leftarrow S] = M[\alpha := \beta][\beta \Leftarrow \vec{N}]$$
$$M:S = [\,]MN_1 \ldots N_k\beta$$

Definition 4 (Reduction)

The reduction relation $M \to N$ is defined as the compatible closure of (β_T), (β_S), (η_T), (η_S) and (fst).

3. Intersection types for $\Lambda\mu$-calculus

Ugo [25] presents an intersection type for $\Lambda\mu$ to prove the approximation theorem, here we give a modification inspired by [14] and [25].

Definition 5 $\mathcal{T}_T$: δ ::= $\varphi \mid \omega \to \delta \mid \sigma \to \delta \mid \delta \cap \delta$,

$\mathcal{T}_S$: σ ::= $\delta \times \omega \mid \delta \times \sigma \mid \sigma \cap \sigma$ where φ ranges over a denumerable set of type atoms.

$\mathcal{T}_T$ is the set of *term types* and $\mathcal{T}_S$ the set of *stream types*. Comparing with [14], we allow intersection type in the end of functional type. ω cannot be a proper type while all terms are typeable with ω_T or ω_S in the full system of [25].

Notation.

- We use $\rho, \delta, \rho', \delta', \delta_1, \ldots$ to denote term types, $\sigma, \sigma_1, \sigma', \tau, \ldots$ stream types.
- $\cap$ has precedence over $\to$ and $\times$ and $\times$ has precedence over $\to$. $\times$ and $\to$ associate to the right. For example $(\rho \times \sigma \to \tau) \cap \rho \to \sigma \to \tau \equiv (((\rho \times \sigma) \to \tau)) \cap \rho) \to (\sigma \to \tau)$.

A *statement* is an expression of the form M: δ, where M is a term, the *subject* of a statement, and δ is a term type. An *environment for term variables* Γ is a set of statements with distinct term variables as subjects. An *environment for stream variables* Δ is a set of statements with distinct stream variables as subjects. Denote $x \in \Gamma(\alpha \in \Delta)$ if $x(\alpha)$ is a subject of a statement in $\Gamma(\Delta)$ and $\emptyset$ for empty environment. If we write Γ, x: δ $(\Delta, \alpha$: $\sigma)$, we presuppose that $x(\alpha)$ does not occur in $\Gamma(\Delta)$.

A *judgment* is the form of $\Gamma \vdash M$: $\delta \mid \Delta$ where Γ, Δ are environments, M is a term and δ is a term type. It implies that there exists a derivation constructed using the typing rules that has M: δ as its conclusion. We say that a term M is typeable if there exist Γ, Δ and δ such that $\Gamma \vdash M$: $\delta \mid \Delta$.

Definition 6 (Intersection type assignment)

The intersection type system is defined by Figure 1 where the σ in rules (λ) and (app) may be ω.

$$(ax) \quad \frac{(x : \delta) \in \Gamma}{\Gamma \vdash x : \delta \mid \Delta} \qquad (\lambda) \quad \frac{\Gamma, (x : \delta_1) \vdash M : \sigma \to \delta_2 \mid \Delta}{\Gamma \vdash \lambda x.M : \delta_1 \times \sigma \to \delta_2 \mid \Delta}$$

$$(app) \quad \frac{\Gamma \vdash M : \delta_1 \times \sigma \to \delta_2 \mid \Delta \quad \Gamma \vdash N : \delta_1 \mid \Delta}{\Gamma \vdash (MN) : \sigma \to \delta_2 \mid \Delta} \qquad (\mu) \quad \frac{\Gamma \vdash M : \delta \mid \alpha : \sigma, \Delta}{\Gamma \vdash \mu\alpha.M : \sigma \to \delta \mid \Delta}$$

$$(s) \quad \frac{\Gamma \vdash M : \sigma \to \delta \mid \alpha : \sigma, \Delta}{\Gamma \vdash M\alpha : \delta \mid \alpha : \sigma, \Delta} \qquad (\cap\text{-}I) \quad \frac{\Gamma \vdash M : \delta_1 \mid \Delta \quad \Gamma \vdash M : \delta_2 \mid \Delta}{\Gamma \vdash M : \delta_1 \cap \delta_2 \mid \Delta}$$

$$(\leq) \quad \frac{\Gamma \vdash M : \delta_1 \mid \Delta \quad \delta_1 \leq \delta_2}{\Gamma \vdash M : \delta_2 \mid \Delta}$$

Figure 1. The intersection type system

Definition 7 (Preorder)

The relations $\leq_T$ and $\leq_S$ over $\mathcal{T}_T$ and $\mathcal{T}_S$ respectively are the least preorders such that the relations in Table 1 are satisfied.

The preorders on $\mathcal{T}_T$ and $\mathcal{T}_S$ are remarked as $\leq_T$ and $\leq_S$, respectively. The subscripts are normally omitted when there is no confusion. Types may be considered modulo $\sim$. Then $\leq$ becomes a partial order.

Table 1. The Preorder Relation

1. $\delta_1 \leq_T \delta_2 \leq_T \delta_3 \Rightarrow \delta_1 \leq_T \delta_3$; $\sigma_1 \leq_S \sigma_2 \leq_S \sigma_3 \Rightarrow \sigma_1 \leq_S \sigma_3$	2. $\delta_1 \cap \delta_2 \leq_T \delta_i$; $\sigma_1 \cap \sigma_2 \leq_S \sigma_i, i \in \{1,2\}$
3. $(\sigma \to \delta_1) \cap (\sigma \to \delta_2) \leq_T \sigma \to (\delta_1 \cap \delta_2)$	4. $\sigma_2 \leq_S \sigma_1, \delta_1 \leq_T \delta_2 \Rightarrow \sigma_1 \to \delta_1 \leq_T \sigma_2 \to \delta_2$
5. $(\delta_1 \times \sigma_1) \cap (\delta_2 \times \sigma_2) \leq_S (\delta_1 \cap \delta_2) \times (\sigma_1 \cap \sigma_2)$	6. $\delta_1 \leq_T \delta_2, \sigma_1 \leq_S \sigma_2 \Rightarrow \delta_1 \times \sigma_1 \leq_S \delta_2 \times \sigma_2$
7. $\delta \leq_T \delta_1, \delta_2 \Rightarrow \delta \leq_T (\delta_1 \cap \delta_2)$; $\sigma \leq_S \sigma_1, \sigma_2 \Rightarrow \sigma \leq_S (\sigma_1 \cap \sigma_2)$	8. $\delta \leq_T \omega \to \delta \leq_T \delta$
9. $\sigma_1 \to \delta_1 \leq \sigma_2 \to \delta_2 \Rightarrow \delta \times \sigma_1 \to \delta_1 \leq \delta \times \sigma_2 \to \delta_2$	10. $\sim = \leq_A \cap \geq_A$ where A is either T or S.

Definition 8 The preorder on environments is extended as follows.

$$\Gamma \leq \Gamma' \text{ iff } (x : \delta') \in \Gamma' \text{ implies } \exists \delta. (x : \delta) \in \Gamma \text{ and } \delta \leq \delta'.$$

$\Delta \leq \Delta'$ is defined in the same way.

Definition 9 $\Gamma_1 \wedge \Gamma_2 = \{x : \delta_1 \mid x : \delta_1 \in \Gamma_1 \wedge x \notin \Gamma_2\} \cup \{x : \delta_2 \mid x \notin \Gamma_1 \wedge x : \delta_2 \in \Gamma_2\} \cup \{x : \delta_1 \cap \delta_2 \mid x : \delta_1 \in \Gamma_1 \wedge x : \delta_2 \in \Gamma_2\}$. $\Delta_1 \wedge \Delta_2$ is constructed in a similar way.

Proposition 1 $\Gamma_1 \wedge \Gamma_2 \leq \Gamma_i$ and $\Delta_1 \wedge \Delta_2 \leq \Delta_i$ for $i \in \{1, 2\}$.

4. Properties of the type system

In this section, we show some properties on the type system. In particular, the system satisfies the subject reduction property.

Lemma 1 (Weakening and Strengthening)

- (Weakening) $\Gamma \vdash M: \delta | \Delta$ and $\Gamma' \leq \Gamma, \Delta' \leq \Delta, then\ \Gamma' \vdash M: \delta \mid \Delta'$;
- (Strengthening) $\Gamma \vdash M: \delta | \Delta$ and let $\Gamma' = \{x: \delta \in \Gamma \mid x \in fv(M) \}, \Delta' = \{\alpha: \sigma \in \Delta \mid \alpha \in fn(M)\},$ then $\Gamma' \vdash M: \delta \mid \Delta'$.

Proof.

- Prove by induction on the structure of the derivation.
- This is a particular case of the first one.

The above lemma and Proposition 1 lead immediately the following corollary.

Corollary 1 If $\Gamma_1 \vdash M: \delta | \Delta_1$, then for any $\Gamma_2, \Delta_2, \Gamma_1 \wedge \Gamma_2 \vdash M: \delta \mid \Delta_1 \wedge \Delta_2$.

By the definition of relations, we can get the following that will used in some proofs.

Proposition 2 1) $(\sigma_1 \rightarrow \delta_1) \cap (\sigma_2 \rightarrow \delta_2) \leq (\sigma_1 \cap \sigma_2) \rightarrow (\delta_1 \cap \delta_2)$.

    2) $(\delta'_1 \times \sigma_1 \rightarrow \delta_1) \cap (\delta'_2 \times \sigma_2 \rightarrow \delta_2) \leq (\delta'_1 \cap \delta'_2) \times (\sigma_1 \cap \sigma_2) \rightarrow (\delta_1 \cap \delta_2)$.

The next lemma will always be needed if a given type assignment is analyzed.

Lemma 2 (Generation Lemma)

1) $\Gamma \vdash x: \delta | \Delta \Leftrightarrow \exists \ \delta' \sim \Gamma \vdash x: \delta' \mid \Delta \wedge \delta' \leq \delta$.
2) $\Gamma \vdash \lambda x.M: \delta | \Delta \Leftrightarrow \exists\ n, \sigma_i, \delta_i, \delta'_i (i \in \{1, \dots, n\}) \sim \Gamma, x: \delta_i \vdash M: \sigma_i \rightarrow \delta'_i \mid \Delta \wedge \bigcap_i (\delta_i \times \sigma_i \rightarrow \delta'_i) \leq \delta$.
3) $\Gamma \vdash MN: \delta | \Delta \Leftrightarrow \exists\ n, \delta'_i, \delta''_i, \sigma_{i(i \in \{1, \dots, n\})} \Gamma \vdash M: \delta'_i \times \sigma_i \rightarrow \delta''_i | \Delta \wedge \Gamma \vdash N: \delta'_i | \Delta \wedge \bigcap_i (\sigma_i \rightarrow \delta''_i) \leq \delta$.
4) $\mu\alpha.M: \delta \mid \Delta \Leftrightarrow \exists\ n, \delta'_i, \sigma_i\ (i \in \{1, \dots, n\}) \sim \Gamma \vdash M: \delta'_i \mid \alpha: \sigma_i, \Delta \wedge \bigcap_i (\sigma_i \rightarrow \delta'_i) \leq \delta$.
5) $\Gamma \vdash (M)\alpha: \delta \mid \Delta \Leftrightarrow \exists\ \delta', \sigma \sim \Delta(\alpha) = \sigma \wedge \Gamma \vdash M: \sigma \rightarrow \delta' \mid \Delta \wedge \delta' \leq \delta$.

Proof. The right-to-left implications immediately follow from the typing rules.

The converses follow by induction on the structure of derivations.

Lemma 3 (Substitution Lemma)

1) $\Gamma \vdash M[x := N]: \delta \mid \Delta$ with $x \in fv(M)$, iff there exists $\delta' \in \mathcal{T}_{\mathcal{T}}$ such that $\Gamma \vdash N: \delta' \mid \Delta$ and $\Gamma, x: \delta' \vdash M: \delta \mid \Delta$.
2) $\Gamma \vdash M[\alpha \Leftarrow N]: \delta \mid \alpha: \sigma, \Delta$ with $\alpha \in fn(M)$, iff there exists $\delta' \in \mathcal{T}_{\mathcal{T}}$ such that $\Gamma \vdash N: \delta' \mid \Delta$ and $\Gamma \vdash M: \delta \mid \alpha: \delta' \times \sigma, \Delta$.

Proof.

1) The right-to-left is proved by induction on the derivation of $\Gamma, x: \delta' \vdash M: \delta | \Delta$. The left-to-right is proved by induction on the structure of M and using the Lemma 2. We only consider the case $M \equiv M_1 M_2$:

$M[x := N] = M_1[x := N]M_2[x := N]$. By (3) of Lemma 2, there exist $n, \delta_i, \sigma_i, \delta_i'$ such that $\Gamma \vdash M_1[x := N]: \delta_i \times \sigma_i \rightarrow \delta_i' | \Delta, \Gamma \vdash M_2[x := N]: \delta_i | \Delta$ and $\bigcap_i(\sigma_i \rightarrow \delta_i') \leq \delta. x \in fv(M)$, this can be divided into three cases.

   a) $x \in fv(M_1)$ and $x \notin fv(M_2)$. By I.H. there exists δ_i'' such that $\Gamma \vdash N: \delta'' | \Delta$ and $\Gamma, x: \delta'' \vdash M_1: \delta_i \times \sigma_i \rightarrow \delta_i' | \Delta$. $x \notin fv(M_2)$ implies that $M_2[x := N] = M_2$. So $\Gamma \vdash M_2: \delta_i | \Delta$. By Lemma 1, $\Gamma, x: \delta_i'' \vdash M_2: \delta_i | \Delta$. By rule (app), $\Gamma, x: \delta'' \vdash M_1 M_2: \sigma_i \rightarrow \delta_i | \Delta$. Then by rules ($\cap$-I) and ($\leq$).

   b) $x \notin fv(M_1)$ and $x \in fv(M_2)$. $x \notin fv(M_1)$, then $\Gamma \vdash M_1: \delta \times \sigma_i, \rightarrow \delta_i | \Delta$. By I.H., there exists δ'' such that $\Gamma, x: \delta''' \vdash M_2: \delta_i | \Delta$. By Lemma 1, $\Gamma, x: \delta_i'' \vdash M_1: \delta_i \times \sigma_i, \rightarrow \delta_i' | \Delta$. Hence by rules (app), ($\cap$-I) and ($\leq$) the result follows.

   c) $x \in fv(M_1) \cap fv(M_2)$. By I.H., there exist δ_{i1}, δ_{i2} such that $\Gamma \vdash N: \delta_{i1} | \Delta, \Gamma, x: \delta_{i1} \vdash M_1: \delta_i \times \sigma_i \rightarrow \delta_i'' | \Delta$ and $\Gamma \vdash N: \delta_{i2} | \Delta, \Gamma, x: \delta_{i2} \vdash M_2: \delta_i | \Delta$. Then $\Gamma \vdash N: \delta_{i1} \cap \delta_{i2} | \Delta$ and $\Gamma, x: \delta_{i1} \cap \delta_{i2} \vdash M_1: \delta_i \times \delta_{i2} \rightarrow \delta_i' | \Delta$ and $\Gamma, x: \delta_{i1} \cap \delta_{i2} \vdash M_2: \delta_i | \Delta$ by ($\cap$-I) and

Lemma 1. So $\Gamma, x: \delta_{i1} \cap \delta_{i2} \vdash M_1 M_2: \delta_{i2} \to \delta_i \mid \Delta$. Then the result follows by rules $(\cap\text{-I})$ and $(\leq)$.

2) The right-to-left is proved by induction on the derivation of $\Gamma \vdash M: \delta \mid \alpha: \delta' \times \sigma, \Delta$. The other direction is proved by induction on the structure of M and using the Lemma 2. We only consider one case: $M \equiv P\beta$:

   a) $\alpha = \beta . M[\alpha \Leftarrow N] = P[\alpha \Leftarrow N]N\alpha$. That is, $\Gamma \vdash P\ [\alpha \Leftarrow N]\ N\alpha: \delta \mid \alpha: \sigma, \Delta$. By (5) of Lemma 2, $\exists \delta', \sigma', (\alpha: \sigma, \Delta)(\alpha) = \sigma' \ \& \ \Gamma \vdash P[\alpha \Leftarrow N]N: \sigma' \to \delta' \mid \alpha: \sigma, \Delta \ \& \ \delta' \leq \delta$. Then $\sigma' = \sigma$. By (3) of Lemma 2 there exists δ'' such that $\Gamma \vdash P\ [\alpha \Leftarrow N]: \delta'' \times \sigma \to \delta' \mid \alpha: \sigma, \Delta$ and $\Gamma \vdash N: \delta'' \mid \alpha: \sigma, \Delta$.

      i) $\alpha \in fn(P)$: By induction hypothesis, there exists ρ such that $\Gamma \vdash P: \delta'' \times \sigma \to \delta' \mid \alpha: \rho \times \sigma, \Delta$ and $\Gamma \vdash N: \rho \mid \Delta$. Note that $\alpha \notin \Delta$ so that $\alpha \notin fn\ (N)$, hence $\Gamma \vdash N: \delta'' \mid \Delta$. Then we drive $\Gamma \vdash N: \rho \cap \delta'' \mid \Delta$ by rule $(\cap\text{-I})$. On the other hand $\rho \cap \delta'' \leq \rho, \delta''$ implies $\delta'' \times \sigma \to \delta' \leq (\rho \cap \delta'') \times \sigma \to \delta'$ and $(\rho \cap \delta'') \times \sigma \leq \rho \times \sigma$. Therefore, $\Gamma \vdash P: (\rho \cap \delta'') \times \sigma \to \delta' \mid \alpha: (\rho \cap \delta'') \times \sigma, \Delta$ by rule $(\leq)$ and Lemma 1.

      ii) $\alpha \notin fn(P): P[\alpha \Leftarrow N] = P$. Hence $\Gamma \vdash P: \delta'' \times \sigma \to \delta \mid \alpha: \sigma, \Delta$. By Lemma 1, $\Gamma \vdash P: \delta'' \times \sigma \to \delta \mid \Delta$ and then $\Gamma \vdash P: \delta'' \times \sigma \to \delta \mid \alpha: \delta'' \times \sigma, \Delta$. Then by rule (s), $\Gamma \vdash P\alpha: \delta \mid \alpha: \delta'' \times \sigma, \Delta$. Note that by convention $\alpha \notin fn\ (N)$, hence $\Gamma \vdash N: \delta'' \mid \Delta$.

   b) $\alpha \neq \beta$. $M[\alpha \Leftarrow N] = P[\alpha \Leftarrow N]\beta$ and $\alpha \in fn(P)$. By (5) of Lemma 2, $\exists \delta', \sigma', (\Delta, \alpha: \sigma)(\beta) = \sigma' \wedge \Gamma \vdash P\ [\alpha \Leftarrow N]: \sigma' \to \delta' \mid \alpha: \sigma, \Delta \wedge \delta' \leq \delta$. By induction hypothesis, $\exists \delta''. \Gamma \vdash P: \sigma' \to \delta' \mid \alpha: \delta'' \times \sigma, \Delta$ with $\Delta(\beta) = \sigma'$ and $\Gamma \vdash N: \delta'' \mid \Delta$. $\Gamma \vdash N: \delta'' \mid \alpha: \delta'' \times \sigma, \Delta$ by Lemma 1. Then we drive that $\Gamma \vdash P\beta: \delta' \mid \alpha: \delta'' \times \sigma, \Delta$.

Lemma 4 1) $\forall \delta \exists k \geq 1 \exists \delta_1, \ldots, \delta_k, \sigma_1 \ldots \sigma_k \sim \delta \sim (\sigma_1 \to \delta_1) \cap \cdots \cap (\sigma_k \to \delta_k)$. The σ_i may be ω.

   2) $\forall \sigma \exists k \geq 1 \exists \delta_1, \ldots, \delta_k \sim \sigma \sim \delta_1 \times \cdots \times \delta_k \times \omega$

Proof. Induction on the structure of δ and σ.

Theorem 1 (Subject Reduction)

If $\Gamma \vdash M: \delta \mid \Delta$ and $M \to N$, then $\Gamma \vdash N: \delta \mid \Delta$.

Proof. It suffices to check the rules in Definition 3 using Lemma 2 and Lemma 3. We treat *(fst)* only:

$M \equiv \mu\alpha. P$. By (4) of Lemma 2, $\exists n, \delta_i' \sigma_i$ such that $\Gamma \vdash P: \delta_i' \mid \alpha: \sigma_i, \Delta$ and $\bigcap_i (\sigma_i \to \delta_i') \leq \delta$. Then we distinguish two cases:

1) $\alpha \notin fn(P). P[\alpha \Leftarrow x] = P$ for any x. By Lemma 4, there exist ρ_i, σ_i' such that $\sigma_i \sim \rho_i \times \sigma_i'$. We can get $\Gamma, x: \rho_i \vdash P: \delta_i' \mid \alpha: \sigma_i', \Delta$ by Lemma 1 since $\alpha \notin fn(P)$ and $x \notin fv(P)$. Therefore $\Gamma \vdash \lambda x. \mu\alpha. P[\alpha \Leftarrow x]: \rho_i \times \sigma_i' \to \delta_i' \mid \Delta$ by rules (μ) and (λ).

2) $\alpha \in fn(P)$. By Lemma 2, there exist ρ_i, σ_i' such that $\sigma_i \sim \rho_i \times \sigma_i'$. Let $x: \rho \vdash x: \rho \mid \Delta$, then $\Gamma, x: \rho_i \vdash x: \rho_i \mid \alpha: \sigma_i' \Delta$ and $\Gamma, x: \rho_i \vdash P: \delta_i' \mid \alpha: \rho_i \times \sigma_i', \Delta$ by Lemma 1. Therefore $\Gamma, x: \rho_i \vdash P\ [\alpha \Leftarrow x]: \delta_i' \mid \alpha: \sigma_i', \Delta$ by Lemma 3. So $\Gamma \vdash \lambda x. \mu\alpha. P[\alpha \Leftarrow x]: \rho_i \times \sigma_i' \to \delta_i' \mid \Delta$ by rules (μ) and (λ).

Then by the rules $(\cap\text{-I})$ and $(\leq)$, we get $\Gamma \vdash \lambda x. \mu\alpha. P[\alpha \Leftarrow x]: \delta \mid \Delta$.

5. Strong normalization implies typeability

Normal forms are not strictly speaking since a term always has an infinite reduction sequence because of the *(fst)*-rule. Normal forms in this paper are indeed canonical normal forms in [19]. M is strongly normalizable if there no infinite reduction sequence originating in it.

Definition 10 The set of normal forms NF can be defined by the following:

$x \in NF$

$\overrightarrow{M_0}, \cdots, \overrightarrow{M_m} \in NF \Rightarrow \lambda\overrightarrow{x_0}\mu\alpha_1\lambda\overrightarrow{x_1} \cdots \mu\alpha_n\overrightarrow{x_n}. y\overrightarrow{M_0}\beta_1\overrightarrow{M_1} \cdots \beta_m\overrightarrow{M_m} \in NF.$

Lemma 5 For any term $M \in NF$, there exist Γ, Δ and a type $\sigma \to \delta$ such that $\Gamma \vdash M: \sigma \to \delta \mid \Delta$.

Proof. Induction on the shape of M.

If M is a variable, then the statement is immediate.

For simplicity, let $M = \lambda x_0 \mu \alpha_1 \lambda x_1 \cdots \mu \alpha_n \lambda x_n . y M_0 \beta_1 M_1 \cdots \beta_m M_m$ with $M_0, \cdots M_m \in NF$. By induction hypothesis, there exist $\Gamma_i, \sigma_i, \delta_i$ $(0 \leq i \leq m)$ such that $\Gamma_i \vdash M: \delta_i \mid \Delta_i$ (the structure of each δ_i plays no rules). Take $\Gamma' = \Gamma_1 \wedge \cdots \wedge \Gamma_m \wedge y: \delta_0 \times \sigma_1 \times \delta_1 \times \cdots \times \sigma_m \times \delta_m \times \omega \to \delta$, and $\Delta' = \Delta_0 \wedge \cdots \wedge \Delta_m \wedge \beta_i: \sigma_i$ for all $(1 \leq i \leq m)$. By successive application of (app) and (s) rules, $\Gamma' \vdash y M_0 \beta_1 \cdots \beta_m M_m: \omega \to \delta \mid \Delta'$. Let $S = \{\alpha_j \mid \alpha_j = \beta_k \text{ for some } k\}$. If every α_i is different from β_i, the set S is empty. Then let $\Gamma = \Gamma' \wedge x_i: \rho_i$ (for all $0 \leq i \leq n), \Delta = \Delta' - S$, $\tau_j = \tau_j \cap \sigma_k$ where τ_j, σ_k on the right side are the type of α_j, β_k, respectively, By successive application of rules (λ) and (μ) and Lemma 1, we get $\Gamma \vdash \lambda x_0 \mu \alpha_1 \cdots \mu \alpha_n \lambda x_n . y M_0 \beta_1 M_1 \cdots \beta_m M_m: (\rho_0 \times \tau_1) \to \cdots \to (\rho_n \times \omega \to \delta) \mid \Delta$.

We define our perpetual strategy for the $\Lambda \mu$-calculus, using the method in [8]. If a term M is not strongly normalizing, the perpetual path of M is infinite. That is, the perpetual reduction terminates only when the term is strongly normalizing. If M is strongly normalizing, it has a perpetual reduction.

Definition 11 (Perpetual redex)

For any term not in normal form, we define its *perpetual redex* by Figure 2.

The *perpetual strategy* is the strategy that reduces always the perpetual redex. It is denoted by $\rightsquigarrow$.

The perpetual redex of $\lambda x.M$ is:

the perpetual redex of P	if $M \equiv Px$ and $x \notin fv(P)$;
the perpetual redex of M	otherwise.

The perpetual redex of $\mu \alpha.M$ is:

the perpetual redex of P	if $M \equiv P\alpha$ and $x \notin fn(P)$;
the perpetual redex of M	otherwise.

· The perpetual redex of MN is:

$(\lambda x.P)Q$	if $MN \equiv (\lambda x.P)Q$ and $x \in fv(P)$;
the perpetual redex of M	$MN \equiv (\lambda x.P)Q$ and $x \notin fv(P)$ and N is a normal form or M is not a normal form;
the perpetual redex of N	otherwise.

· The perpetual redex of $M\alpha$ is:

$(\mu \alpha.P)\beta$	if $M \equiv (\mu \alpha.P)$ and $\alpha \in fn(P)$
the perpetual redex of M	otherwise.

Figure 2. Perpetual redex

Lemma 6 (Subject Expansion)

Let $M \rightsquigarrow N$ and $\Gamma \vdash N: \delta \mid \Delta$, then $\Gamma \vdash M: \delta \mid \Delta$.

Proof. By induction on the structure of M.

If M is its own perpetual redex and the rule used is β_T or β_S:

1) if $M \equiv (\lambda x.P)Q$ and $x \in fv(P)$, we wish to prove: if $\Gamma \vdash P[x:= Q]: \delta \mid \Delta$, then $\Gamma \vdash (\lambda x.P)Q : \delta \mid \Delta$. By Lemma 3, there exists $\delta' \in \mathcal{T}_T$ such that $\Gamma \vdash Q: \delta' \mid \Delta$ and $\Gamma, x: \delta' \vdash P: \delta \mid \Delta$. By Lemma 4, there exist $\delta_1, \ldots, \delta_k, \sigma_1, \ldots, \sigma_k$ such that $\delta \sim (\sigma_1 \to \delta_1) \cap \cdots \cap (\sigma_k \to \delta_k)$. So $\Gamma, x: \delta' \vdash P: \sigma_i \to \delta_i$. By rule (λ), $\Gamma \vdash \lambda x.P: \delta' \times \sigma_i \to \delta_i \mid \Delta$. Then $\Gamma \vdash (\lambda x.P) Q: \sigma_i \to \delta_i \mid \Delta$. The result follows by rule $(\cap\text{-I})$.

2) $M \equiv (\mu \alpha.P)\beta$. We wish to prove: if $\Gamma \vdash P[\alpha:= \beta]: \delta \mid \Delta$, then $\Gamma \vdash (\mu \alpha.P) \beta: \delta \mid \Delta$. Suppose $\Delta (\beta) = \sigma$. By assumption, $\Gamma \vdash P: \delta \mid \Delta'$ where $\Delta'(\alpha) = \Delta(\beta) = \sigma, \Delta'(\gamma) = \Delta(\gamma)$. Then $\Gamma \vdash (\mu \alpha.M): \sigma \to \delta \mid \Delta' \setminus \alpha$. By Lemma 1, $\Gamma \vdash (\mu \alpha.M): \sigma \to \delta \mid \Delta' \setminus \alpha, \beta: \sigma$. So $\Gamma \vdash (\mu \alpha.M) \beta: \delta \mid \Delta$.

3) $M \equiv \lambda x.Px$ and $x \notin fv(P)$, then $N = P$. By Lemma 4, $\exists k \geq 1 \exists \delta_1, \ldots, \delta_k, \sigma_1, \ldots, \sigma_k$ such that $\delta \sim (\sigma_1 \to \delta_1) \cap \cdots \cap (\sigma_k \to \delta_k)$ and then $\Gamma \vdash P: \sigma_i \to \delta_i$. On the other hand,

$\exists \delta_1' \cdots \delta_n' \sigma_i \sim \delta_1' \times \cdots \times \delta_n' \times \omega$. Therefore, $\Gamma, x: \delta_i' \vdash Px: \delta_2' \times \cdots \times \delta_n' \times \omega \to \delta_i$. So we reason by Lemma 1, rules (λ), (app) to get $\Gamma \vdash \lambda x. Px: \delta_1' \times \delta_2' \times \cdots \times \delta_n' \times \omega \to \delta_i | \Delta$. That is $\Gamma \vdash \lambda x. Px: \sigma_i \to \delta_i | \Delta$. The result follows by rule ($\cap$-I).

4) $M \equiv \mu \alpha. P\alpha$ and $\alpha \notin fn(P)$, then $N = P$. By Lemma 4, $\exists k \geq 1 \exists \delta_1, \dots, \delta_k, \sigma_1, \dots, \sigma_k$ such that $\delta \sim (\sigma_1 \to \delta_1) \cap \cdots \cap (\sigma_k \to \delta_k)$ and then $\Gamma \vdash P: \sigma_i \to \delta_i$. $\alpha \notin fn(P)$, then by Lemma 1, $\Gamma \vdash P: \sigma_i \to \delta_i | \Delta, \alpha: \sigma_i$ So the result follows by rules (s), (μ) and ($\cap$-I).

5) $M \equiv \mu \alpha. P$, then $N = \lambda x. \mu \alpha. P[\alpha \Leftarrow x]$ where $x \notin fv(P)$. Here we suppose the type of N is not an intersection type. We suppose $\lambda x. \mu \alpha. P [\alpha \Leftarrow x]: \sigma \to \delta | \Delta$, then there exist $\sigma_i, \delta_i, \delta_i'$ such that $\Gamma, x: \delta_i \vdash \mu \alpha. P [\alpha \Leftarrow x]: \sigma_i \to \delta_i' | \Delta$ and $\cap (\delta_i \times \sigma_i \to \delta_i') \leq \sigma \to \delta$. So $\Gamma, x: \delta_i \vdash P [\alpha \Leftarrow x]: \delta_i'' | \alpha: \delta_i''', \Delta$ and $\cap (\delta_i''' \to \delta_i'') \leq \sigma_i \to \delta_i'$. If $\alpha \notin fn (P)$, then $\Gamma \vdash P: \delta_i'' | \alpha: \delta_i \times \delta_i''', \Delta$ by Lemma 1 and Lemma 3. Therefore, $\Gamma \vdash \mu \alpha. P \delta_i \times \delta_i''' \to \delta_i'' | \Delta$. Then the result follows by rules ($\cap$-I), ($\leq$) and the preorder. Else, it easily follows by rules and Lemma 1.

$M \equiv M_1 M_2: N = N_1 N_2$. There exist $\delta_i', \delta_i'', \sigma_i$ such that $\Gamma \vdash N_1: \delta_i' \times \sigma_i \to \delta_i'' | \Delta, \Gamma \vdash N_2: \delta_i' | \Delta$ and $\cap_i (\sigma_i \to \delta_i'') \leq \delta$. Then we distinguish two cases:

1) If $MN \equiv (\lambda x. P) Q$ and $x \notin fv (P)$ and N is a normal form or M_1 not a normal form, $N = N_1 M_2$ where $M_1 \rightsquigarrow N_1$. By induction hypothesis, $\Gamma \vdash M_1: \delta_i' \times \sigma_i \to \delta_i'' | \Delta$.

2) Else, $N = M_1 N_2$ where $M_2 \rightsquigarrow N_2$. By induction hypothesis, $\Gamma \vdash M_2: \delta_i' | \Delta$. Hence $\Gamma \vdash M_1 M_2: \delta | \Delta$ by rules (app) and ($\cap$-I).

$M \equiv \lambda x. P: N \equiv \lambda x. P'$ where $P \rightsquigarrow P'$. Application of induction hypothesis and Lemma 2 suffices.

$M \equiv P\alpha$, then $N = P'\alpha$ where $P \rightsquigarrow P'$. Application of induction hypothesis and Lemma 2 suffices.

Theorem 2 M is strong normalizing, then M is typeable.

 Proof. The proof is by induction over the length of the perpetual derivation of M. For the base case we observe that normal forms are typeable by Lemma 5, the induction step follows by Lemma 6.

6. Typeability implies strong normalization

In this section, we use the reducibility method [22] to show that typeable terms are strongly normalizing. The general idea of reducibility method is to interpret types by suitable sets that satisfy certain realizable properties and then to develop semantics in order to obtain the soundness of the type assignment.

Let $\mathcal{SN}$ be the set of strongly normalizable terms and $\mathcal{SN}^*$ the set of streams whose elements are in $\mathcal{SN}$. By the definition of strong normalization, we have the following.

Proposition 3 1) $x S \in \mathcal{SN} \Rightarrow \forall P \in \mathcal{SN}^* x SP \in \mathcal{SN}$.

2) $M [x: = N] S \in \mathcal{SN} \wedge N \in \mathcal{SN} \Rightarrow (\lambda x. M) NS \in \mathcal{SN}$.

3) $M [\alpha \Leftarrow S] \in \mathcal{SN} \wedge S \in \mathcal{SN}^* \Rightarrow (\mu\alpha. M) S \in \mathcal{SN}$.

4) $M \in \mathcal{SN} \Rightarrow \mu\alpha. M\alpha \in \mathcal{SN}$ for $\alpha \notin fn(M)$.

Definition 12 Let $\rho: (Var_v \to \Sigma_{\Lambda\mu}) \cup (Var_s \to \Sigma_{\Lambda\mu}^*)$ be an evaluation. Define the $[\![M]\!]_\rho =$ $M[x_1: = \rho(x_1), \dots, x_n: = \rho(x_n), \alpha_1 \Leftarrow \rho(\alpha_1), \dots, \alpha_m \Leftarrow \rho(\alpha_m)]$ where $fv(M) = \{x_1, \dots, x_n\}$ and $fn(M) = \{\alpha_1, \dots, \alpha_m\}$.

$\rho(x: = N)$ is the valuation ρ' with $\begin{cases} \rho'(x) = N \\ \rho'(y) = \rho(y) \text{ if } y \not\equiv x \end{cases}$ and $\rho(\alpha := S)$ is the valuation ρ' with $\begin{cases} \rho'(\alpha) = S \\ \rho'(\beta) = \rho(\beta) \text{ if } \beta \not\equiv \alpha \end{cases}$

Then by the definition, we have the following proposition that will be used in some proofs.

Proposition 4 1) $[\![x]\!]_\rho = \rho(x)$.

2) $[\![MN]\!]_\rho = [\![M]\!]_\rho [\![N]\!]_\rho$.

3) $[\![\lambda x. M]\!]_\rho = \lambda x. [\![M]\!]_{\rho(x:=x)}$.

4) $\llbracket M \rrbracket_{\rho(x:=N)} = \llbracket M \rrbracket_{\rho(x:=x)}[x:=N]$ if x is not free in the imamage of ρ.

5) $\llbracket \mu\alpha.M \rrbracket_{\rho} = \mu\alpha.\llbracket M \rrbracket_{\rho}$, $\alpha \notin dom(\rho)$.

6) $\llbracket M\alpha \rrbracket_{\rho} = \llbracket M \rrbracket_{\rho}$ if $\alpha \notin dom(\rho)$.

7) $\llbracket M\alpha \rrbracket_{\rho} = \llbracket M \rrbracket_{\rho}S$ if $\rho(\alpha) = S$.

8) $\llbracket M \rrbracket_{\rho(\alpha:=S)} = \llbracket M \rrbracket_{\rho\backslash\alpha}[\alpha \Leftarrow S]$ if α does not occur free in the image of ρ.

Free (term or stream) variable is not free in the image of ρ is important. If it occurs free, only those x free in M are substituted by N when computing $\llbracket M \rrbracket_{\rho}[x:=N]$ while all x are substituted when computing $\llbracket M \rrbracket_{\rho(x:=x)}[x:=N]$. They will not be equal.

Definition 13 (Type Interpretation)
The interpretation of types is defined as follows:

1) We first define $\llbracket \rrbracket$ of types:

$$\llbracket \varphi \rrbracket = \llbracket \omega \to \delta \rrbracket = \mathcal{SN}$$

$$M \in \llbracket \sigma \to \delta \rrbracket \Leftrightarrow \forall S \in \llbracket \sigma \rrbracket \; MS \in \llbracket \delta \rrbracket$$

$$M \in \llbracket \delta \times \omega \rrbracket \Leftrightarrow M \equiv N:S \wedge N \in \llbracket \delta \rrbracket \wedge S \in \mathcal{SN}^*$$

$$M \in \llbracket \delta \times \sigma \rrbracket \Leftrightarrow M \equiv N:S \wedge N \in \llbracket \delta \rrbracket \wedge S \in \llbracket \sigma \rrbracket$$

$$\llbracket \tau_1 \cap \tau_2 \rrbracket = \llbracket \tau_1 \rrbracket \cap \llbracket \tau_2 \rrbracket$$

$$\alpha:\sigma \Leftrightarrow []\alpha \in \llbracket \sigma \rrbracket$$

where $\tau_1, \tau_2 \in \mathcal{T}_T$ or $\tau_1, \tau_2 \in \mathcal{T}_S$.

2) $\rho \vDash M:\delta$ iff $\llbracket M \rrbracket_{\rho} \in \llbracket \delta \rrbracket$;

$\rho \vDash \Gamma, \Delta$ iff $\rho(x) \in \llbracket \delta \rrbracket$ and $\rho(\alpha) \in \llbracket \sigma \rrbracket$ for all $x:\delta \in \Gamma, \alpha:\sigma \in \Delta$.
$\Gamma \vDash M:\delta | \Delta$ iff $\forall \rho \vDash \Gamma, \Delta \; \rho \vDash M:\delta$.
The preorder on types is interpreted as set inclusion.

Lemma 7 For all $\rho, \tau \in \mathcal{T}$, if $\rho \leq \tau$, then $\llbracket \rho \rrbracket \subseteq \llbracket \tau \rrbracket$.
Proof. Induction on the definition of $\leq$.

Lemma 8 1) $\llbracket \delta \rrbracket \subseteq \mathcal{SN}$ and $\llbracket \sigma \rrbracket \subseteq \mathcal{SN}^*$ for all δ, σ.

2) $xS \in \mathcal{SN} \Rightarrow \forall \delta \; xS \in \llbracket \delta \rrbracket$.

3) $S \in \llbracket \sigma \rrbracket$.

Proof. By simultaneous induction on the structure of types. We show some of the cases.

1) We show three cases:

$(\sigma \to \delta)$: $\quad M \in \llbracket \sigma \to \delta \rrbracket \qquad \Rightarrow$ (IH(3))

$\qquad\qquad S \in \llbracket \sigma \rrbracket \wedge M \in \llbracket \sigma \to \delta \rrbracket \quad \Rightarrow$ **Definition 13**

$\qquad\qquad MS \in \llbracket \delta \rrbracket \qquad\qquad \Rightarrow$(IH(1))

$\qquad\qquad MS \in \mathcal{SN} \qquad\qquad \Rightarrow M \in \mathcal{SN}$

$(\delta \times \omega)$: $\quad M \in \llbracket \delta \times \omega \rrbracket \qquad \Rightarrow$**Definition 13**

$\qquad\qquad M \equiv N:S \wedge N \in \llbracket \delta \rrbracket \wedge S \in \mathcal{SN}^* \quad \Rightarrow$ (IH(1))

$\qquad\qquad N \in \mathcal{SN} \qquad\qquad \Rightarrow M = N:S \in \mathcal{SN}^*.$

$(\delta \times \sigma)$: $\quad M \in \llbracket \delta \times \sigma \rrbracket \qquad \Rightarrow$**Definition 13**

$\qquad\qquad M \equiv N:S \wedge N \in \llbracket \delta \rrbracket \wedge S \in \llbracket \sigma \rrbracket \quad \Rightarrow$ (IH(1))

$\qquad\qquad N \in \mathcal{SN} \wedge S \in \mathcal{SN}^* \quad \Rightarrow M = N:S \in \mathcal{SN}^*$

2) We just show the case $\sigma \to \delta$:

$\qquad xS \in \mathcal{SN} \qquad\qquad \Rightarrow$ ((1) of **Proposition 3**

$\qquad \forall P \in \mathcal{SN}^* \; xSP \in \mathcal{SN} \qquad \Rightarrow$ (IH(1))

$$\forall P \in \llbracket \sigma \rrbracket \ xSP \in \mathcal{SN} \qquad \Rightarrow (\mathrm{IH}(2))$$
$$\forall P \in \llbracket \sigma \rrbracket \ xSP \in \llbracket \delta \rrbracket \qquad \Rightarrow \textbf{Definition 13}$$
$$xS \in \llbracket \sigma \to \delta \rrbracket$$

3) Similar to the proof in [14].

From (2) of **Lemma 8**, we immediately get the following corollary.

Corollary 2 For any variable x, $x \in \llbracket \delta \rrbracket$ for all δ.

Lemma 9 1) If $M[x := N]P \in \llbracket \delta \rrbracket$ and $N \in \llbracket \delta' \rrbracket$, then $(\lambda x.M)NP \in \llbracket \delta \rrbracket$.

2) If $M[\alpha \Leftarrow S]P \in \llbracket \delta \rrbracket$ and $S \in \llbracket \sigma \rrbracket$, then $(\mu\alpha.M)SP \in \llbracket \delta \rrbracket$.

Proof. By induction on the structure of types.

1) (φ) and $(\omega \to \delta_1)$: The assumptions become $M[x := N]P \in \mathcal{SN}$ and $N \in \llbracket \delta' \rrbracket$. By **Lemma 7**, $N \in \mathcal{SN}$. Then by **Proposition 3**, $(\lambda x.M)\ N\ P \in \mathcal{SN}$.

$(\sigma \to \delta_1)$: By definition, $\forall L \in \llbracket \sigma \rrbracket, M[x := N]PL \in \llbracket \delta_1 \rrbracket$. By induction hypothesis, $(\lambda x.M)NPL \in \llbracket \delta_1 \rrbracket$. Then $(\lambda x.M)NP \in \llbracket \sigma \to \delta_1 \rrbracket$.

2) (φ) and $(\omega \to \delta_1)$: immediate by **Proposition 3**.

$(\sigma \to \delta_1)$: $\forall L \in \llbracket \sigma \rrbracket \ M[\alpha \Leftarrow S]PL \in \llbracket \delta_1 \rrbracket$. By induction hypothesis, $(\mu\alpha.M)SPL \in \llbracket \delta_1 \rrbracket$. So $(\mu\alpha.M)SP \in \llbracket \delta \rrbracket$.

Lemma 10 (Soundness) $\Gamma \vdash M : \delta \mid \Delta \Rightarrow \Gamma \vDash M : \delta \mid \Delta$.

Proof. Induction on the derivation of $\Gamma \vdash M : \delta \mid \Delta$.

1) (ax): $\Gamma \vdash x : \delta \mid \Delta$ *since* $x : \delta \in \Gamma$. So for any $\rho \vDash \Gamma, \Delta$, $\llbracket x \rrbracket_\rho \in \llbracket \delta \rrbracket$.

2) (λ): $\Gamma \vdash \lambda x.M' : \delta_1 \times \sigma \to \delta_2 \mid \Delta$ since $\Gamma, x : \delta_1 \vdash M' : \sigma \to \delta_2 \mid \Delta$.

Let $\rho \vDash \Gamma, \Delta$. Take $N \in \llbracket \delta_1 \rrbracket$. Since x is bound, x does not occur free in the image of ρ and $\rho[x := N] \vDash (\Gamma, x : \delta_1), \Delta$. By induction hypothesis, $\llbracket M' \rrbracket_{\rho(x := N)} \in \llbracket \sigma \to \delta_2 \rrbracket$. On the other hand $\llbracket M' \rrbracket_{\rho(x := N)} = \llbracket M' \rrbracket_{\rho(x := x)}[x := N]$ by **Proposition 4**. By **Lemma 9**, $(\lambda x.\ \llbracket M' \rrbracket_{\rho(x := x)}N \in \llbracket \sigma \to \delta_2 \rrbracket$. For all $L \in \llbracket \sigma \rrbracket$, $\llbracket (\lambda x.M') \rrbracket_\rho NL \in \llbracket \delta_2 \rrbracket$ by **Lemma 8** and **Proposition 4**. Notice that $N : L \in \llbracket \delta_1 \times \sigma \rrbracket$, so $\llbracket \lambda x.M' \rrbracket_\rho \in \llbracket \delta_1 \times \sigma \to \delta_2 \rrbracket$.

3) (app): $M \equiv PQ$ and $\Gamma \vdash PQ : \sigma \to \delta_2 \mid \Delta$ since $\Gamma \vdash P : \delta_1 \times \sigma \to \delta_2 \mid \Delta$ and $\Gamma \vdash Q : \delta_1 \mid \Delta$. By induction hypothesis, $\Gamma \vDash P : \delta_1 \times \sigma \to \delta_2 \mid \Delta$ and $\Gamma \vDash Q : \delta \mid \Delta$. That is for any $\rho \vDash \Gamma, \Delta$, $\llbracket P \rrbracket_\rho \in \llbracket \delta_1 \times \sigma \to \delta_2 \rrbracket$ and $\llbracket Q \rrbracket_\rho \in \llbracket \delta_1 \rrbracket$. For all $L \in \llbracket \delta_1 \times \sigma \rrbracket$, $\llbracket P \rrbracket_\rho L \in \llbracket \delta_2 \rrbracket$. Notice that $L \equiv N : R, N \in \llbracket \delta_1 \rrbracket$ and $R \in \llbracket \sigma \rrbracket$ by **Definition 13**. *Let* $N = \llbracket Q \rrbracket_\rho$. Hence for any $R \in \llbracket \sigma \rrbracket$, $\llbracket P \rrbracket_\rho \llbracket Q \rrbracket_\rho R \in \llbracket \delta_2 \rrbracket$. Therefore $\llbracket P \rrbracket_\rho \llbracket Q \rrbracket_\rho \in \llbracket \sigma \to \delta_2 \rrbracket$.

4) (μ): $M \equiv \mu\alpha.M'$ and $\Gamma \vdash \mu\alpha.M' : \sigma \to \delta \mid \Delta$ since $\Gamma \vdash M' : \delta \mid \alpha : \sigma, \Delta$. Let $\rho \vDash \Gamma, \Delta$, take any $S \in \llbracket \sigma \rrbracket$, then $\rho(\alpha := S) \vDash \Gamma, (\alpha : \sigma, \Delta)$. By induction hypothesis, $\llbracket M' \rrbracket_{\rho(\alpha := S)} \in \llbracket \delta \rrbracket$. α is bound in M. By Barendregt convention, α does not occur free in the image of ρ. By **Definition 13** $\llbracket M' \rrbracket_{\rho(\alpha := S)} = \llbracket M' \rrbracket_{\rho \backslash \alpha}[\alpha \Leftarrow S]$. So $\llbracket M' \rrbracket_{\rho \backslash \alpha}[\alpha \Leftarrow S] \in \llbracket \delta \rrbracket$, by **Lemma 9**, $\llbracket (\mu\alpha.\ \llbracket M' \rrbracket_{\rho \backslash \alpha}S \in \llbracket \delta \rrbracket$. By **Proposition 4**, $\llbracket \mu\alpha.M' \rrbracket_\rho S \in \llbracket \delta \rrbracket$. Hence by the definition of type interpretation $\llbracket \mu\alpha.M' \rrbracket_\rho \in \llbracket \sigma \to \delta \rrbracket$.

5) (s): $\Gamma \vdash P\alpha : \delta \mid \alpha : \sigma, \Delta$ since $\Gamma \vdash P : \sigma \to \delta \mid \alpha : \sigma, \Delta$. By induction hypothesis, $\Gamma \vDash P : \sigma \to \delta \mid \alpha : \sigma, \Delta$. Let $\rho \vDash \Gamma, (\alpha : \sigma, \Delta)$, then $\rho(\alpha) = S \in \llbracket \sigma \rrbracket$ and $\llbracket P \rrbracket_\rho \in \llbracket \sigma \to \delta \rrbracket$.

$$\llbracket P\alpha \rrbracket_\rho = \llbracket P \rrbracket_\rho S \in \llbracket \delta \rrbracket.$$

6)   (≤): By induction hypothesis and **Lemma 7**.

7)   (∩-I): By induction hypothesis and **Definition 13**.

A consequence of soundness is that every term typeable in the type system belongs to the interpretation of its type. Each type is interpreted as a suitable set satisfying the reduction property considered. Then we obtain the following theorem.

Theorem 3 $\Gamma \vdash M : \delta \mid \Delta$, then M is strongly normalizing.

Let ρ be an evaluation such that $\rho(x) = x$ for $x \in \Gamma$ and $\rho(\alpha) = \alpha$ for $\alpha \in \Delta$. By **Lemma 10**, $\Gamma \vDash M : \delta \mid \Delta$. We observe that $\rho(x) = x \in \llbracket \delta' \rrbracket$ for all $x : \delta' \in \Gamma$ and $\rho(\alpha) = \alpha \in \llbracket \sigma \rrbracket$ for $\alpha : \sigma \in \Delta$ by **Lemma 8** and definition of type interpretation. Hence $\rho \vDash \Gamma, \Delta$ and $\llbracket M \rrbracket_\rho \in \llbracket \delta \rrbracket$, then $M = \llbracket M \rrbracket_\rho \in \mathcal{SN}$.

7. Conclusion

We have shown that the strongly normalizing terms can be characterized by an intersection type assignment system. The system enjoys type preservation under reduction and satisfies the subject expansion property under perpetual reduction.

$\bar{\lambda}\mu\,\tilde{\mu}$-calculus is a term calculus embodying a Curry-Howard propositions-as-types correspondence for classical logic. Characterization of strong normalization in the calculus has been given by Dougherty [12]. We will investigate how to characterize properties of $\bar{\lambda}\mu\,\tilde{\mu}$-calculus with explicit substitution.

References

[1]   Coppo, M., Dezani-Ciancaglin, M. (1978) A new type assignment for lambdaterms. Archiv fur mathematische Logik und Grundlagenforschung, 19:139–156.

[2]   Coppo, M., Dezani-Ciancaglini, M. (1980) An extension of the basic functionality theory for the λ-calculus. Notre Dame J. Formal Logic, 21 (4): 685–693.

[3]   Barendregt, H., Coppo, M., Dezani-Ciancaglini, M. (1983) A Filter Lambda Model and the Completeness of Type Assignment. J. Symbolic Logic 48 (4): 931–940.

[4]   Pottinger, G.. (1980) A type assignment for the strongly normalizable lambda-terms. To H.B.Curry: Essays on Combinatroy Logic, Lambda Calculus and Formalism: 561–578.

[5]   Van Bakel, S., (2011) Strict intersection types for the lambda calculus. ACM Computing Surveys, 43 (3): 1-49.

[6]   Van Bakel, S. (1992) Complete restrictions of the intersection type discipline. Theoret. Comput. Sci., 102 (1) 135–163.

[7]   Dougherty, D., Lescanne, P., (2003) Reductions, intersection types, and explicit substitutions. Math. Structures Comput. Sci. 13 (1): 55–85.

[8]   Lengrand, S., Lescanne, P., Dougherty, D., Dezani-Ciancaglini, M., Van Bakel, S. (2004) Intersection types for explicit substitutions. Information and Computation 189 (1): 17–42.

[9]   Dezani-Ciancaglini, M., Honsell, F., Motohama, Y. (2005) Compositional characterisations of λ-terms using intersection types. Theoret. Comput. Sci. 340 (3): 459–495.

[10] Matthes, R. (2000) Characterizing strongly normalizing terms of λ-calculus with generalized applications via intersection types. In: Goos, G., Hartmanis, J. and van Leeuwen, J. (Eds) CALP. Springer, Geneva, Switzerland. pp.339-354.

[11] Koletsos, G., Stavrinos, G. (2008) Church-Rosser property and intersection types. Australasian Journal of Logic, 6: 37–54.

[12] Daugherty, D.J., Ghilezan, S., Lescanne, P. (2008) Characterizing strong normalization in the Curien-Herbelin symmetric lambda calculus: Extending the coppodezani heritage. Theoret. Comput. Sci., 398: 114–128.

[13] Koletsos, G. (2012) Intersection Types and Termination Properties. Fundamenta Informaticae, 121(1-4): 185–202.

[14] Van Bakel, S., Barbanera, F., De'Liguoro, U. (2012) Characterisation of strongly normalising lambda-mu-terms. In: Proceedings ITRS 2012. Electronic Proceedings in Theoretical Computer Science. Open Publishing Association. pp. 1–17.

[15] Santo, J.E., Ghilezan, S. (2017) Characterization of strong normalizability for a sequent lambda calculus with co-control. In Proceedings of the 19th International Symposium on Principles and Practice of Declarative Programmin. ACM New York, NY, USA, Namur, Belgium. pp. 163–174.

[16] Barendregt, H. (1984) The Lambda Calculus: Its Syntax and Semantics, North-Holland

[17] Parigot, M. (1992) $\lambda\mu$-calculus: an algorithmic interpretation of classical natural deduction. In: Voronkov, A. (Eds) Proceedings of 3rd International Conference on Logic for Programming, Artificial Intelligence, and Reasoning (LPAR). Lecture Notes in Computer Science. Springer, Petersburg, Russia. pp. 190–201.

[18] David, R. Py, W. $\lambda\mu$-calculus and böhm theorem. Journal of Symbolic Logic. 66(1):407-413.

[19] Saurin, A. (2005) Separation with streams in the $\Lambda\mu$-calculus. In: 20th Annual IEEE Symposium on Logic in Computer Science. pp. 356–365.

[20] Saurin, A. (2008) On the Relations between the Syntactic Theories of $\lambda\mu$-Calculi. In: Kaminski, M., Martini, S. (Eds) Computer Science Logic. CSL 2008. Lecture Notes in Computer Science. Springer, Berlin, Heidelberg. Bertinoro, Italy. pp 154-168.

[21] Saurin, A. (2010) Standardization and Bohm Trees for $\Lambda\mu$-Calculus. In: Kaminski, M., Martini, S. (Eds) Computer Science Logic. CSL 2008. Lecture Notes in Computer Science. Springer, Berlin, Heidelberg. Sendai, Japan. pp.134-149.

[22] Tait, W.W. (1967) Intensional interpretations of functionals of finite type i. Journal of Symbolic Logic, 32:198–212.

[23] Dougherty, D., Ghilezan, S., Lescanne, P. (2004) Characterizing strong normalization in a language with control operators. In: Proceedings of the 6th ACM SIGPLAN international conference on Principles and practice of declarative programming. Verona, Italy. pp. 155–166.

[24] Neergaard, P. (2005) Theoretical pearls- a bargain for intersection types- a simple strong normalization proof. Journal of Functional Programming, 15(5): 699-677.

[25] De'Liguoro, U. (2017) The approximation theorem for the $\Lambda\mu$-calculus. Math.Structures Comput. Sci, 27(5):560-580.

Signature handwriting identification based on generative adversarial networks

WANG Siyue and JIA Shijie

College of electrical and information, Dalian Jiaotong University, Dalian 116028, China

jsj@djtu.edu.cn

Abstract. Handwritten signature has been an important identity-verification method since ancient times. Compared with manual handwriting verification, the use of computer image recognition technology for handwriting verification is faster and avoids subjectivity. However, there are still some challenges in traditional image recognition methods, such as feature selection, lack of a standard basis, and low accuracy. For the first time, generative adversarial nets (GAN) technology is adopted to study the task of handwritten signature identification. A special network SIGAN (Signature Identification GAN, SIGAN)is proposed based on the idea of dual learning. The loss value of the trained discriminator of SIGAN is used as the identification threshold. The authenticity of the test handwritten signature is determined by comparing the threshold and loss value of the test image obtained through the network. The experimental data set in this study consists of five hard pen-type signatures, which include some real signatures and some deliberate imitations. The experimental results show that the average accuracy of the SIGAN-based signature identification model is 91.2%, which is 3.6% higher than that of the traditional image classification method.

1. Introduction

Handwriting identification[1] is a special technology to identify the writer's identity according to the characteristics of people's writing habits. Signature identification is an important part of handwriting identification. It is widely used in social life, such as signing contracts, document confirmation, written documents and so on. In criminal investigation, the result of signature identification can be used as an important clue to break the case, and the results of signature identification can also be used as court evidence.

The characteristics of the handwriting are behavioural. For each writer, his handwriting is relatively stable in general, while the local variation of handwriting is the inherent characteristic of each writer's handwriting. The traditional handwriting identification is made by artificial determination of the same characteristics and different features between the handwriting and the sample handwriting. Finally, they make a comprehensive evaluation.such as the comparative test method[2], the auxiliary portable video microscope identification method[3] and so on. These methods are professional, but time-consuming, laborious and costly. The study of handwriting identification using computers began in the Soviet Union in the 60s of the last century. The initial method was to use the different feature extraction methods to extract the features of the handwriting and to classify them with the minimum Euclidean distance and L1 distance. The common feature extraction methods include Fourier transform and autocorrelation[4], range histogram[5], run length analysis method[6], independent component analysis method[7], multi-channel Gabor filter decomposition method[8], etc. These methods

Content from this work may be used under the terms of the Creative Commons Attribution 3.0 licence. Any further distribution of this work must maintain attribution to the author(s) and the title of the work, journal citation and DOI.

Published under licence by IOP Publishing Ltd

need to extract features first, and feature extraction often lacks standard basis, which leads to insufficient accuracy. With the rise of deep learning, a handwriting study based on deep convolution neural network [9][12] has been developed. The deep convolution neural network can automatically learn the characteristics of the sample and help to improve the accuracy of the identification. The above method is an important part of handwriting identification for general handwriting identification. The representative methods include: signature identification based on information entropy[10], signature identification of Ferryman chain code[11], handwriting signature recognition algorithm based on LBP and depth learning[12].

In this paper, the problem of signature handwriting identification is studied by using Generative Adversarial Nets for the first time. By using the idea of dual learning, a special SIGAN network is designed to realize the task of signature identification. The experimental results show that the SIGAN algorithm can achieve better handwriting identification than the traditional method, and is far higher than the average accuracy of the subjective test.

2. SIGAN

2.1. GAN

In 2014, Ian J.Goodfellow proposed an Generative Adversarial Nets[14].The Generative Adversarial Nets is a deep learning model, and its problem is how to learn new samples from the training samples, which is one of the most promising methods of unsupervised learning in the complex distribution in recent years. GAN generally consists of two modules: the data generation model (G) and the discriminant model (D) for estimating the true and false probability of the generated samples. In the training process, the goal of G is to generate realistic samples to deceive D, while the goal of D is to separate the G generated samples from the real samples as much as possible. In this way, G and D form a mutual "game" process, and the optimization of GAN is a minimax game[15] problem, as shown in Figure 1.

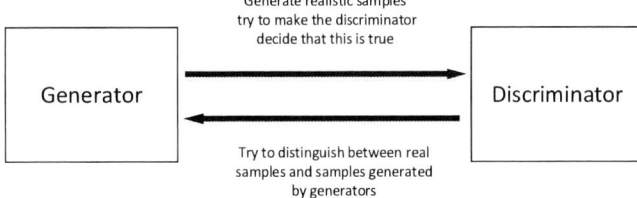

Figure 1. The game process of generator and discriminator.

2.2. The network structure of SIGAN
The overall architecture and data flow of the SIGAN network are shown in Figure 2.

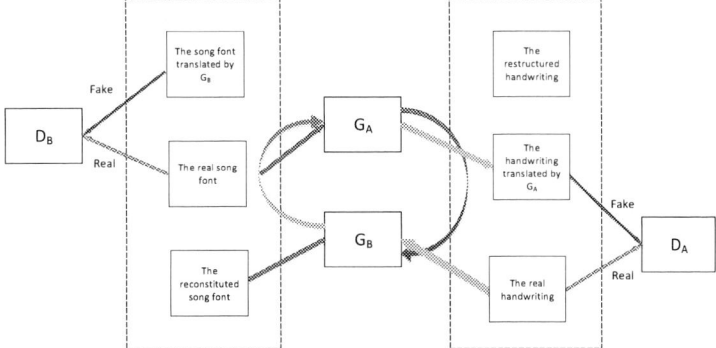

Figure 2. The overall architecture of the network and the flow of data.

1595

(1) The generator G_A translations the standard Chinese character (Song font) image into a handwritten signature image, and its corresponding discriminator D_A is used to distinguish a handwritten signature image from a real image or a image translated by a G_A.

(2) The generator G_B translations the handwritten signature to a standard Chinese character (Song font) name, and its corresponding discriminator D_B is used to distinguish a standard Chinese character (Song font) image as a real image or a image translated by G_B.

(3) The result of the generator G_A translation is sent to the generator G_B. The result is a reconfiguration of the standard Chinese character (Song font) image; the result of the translation of the generator G_B is given to the generator G_A. The result is a reconfiguration of the handwritten signature image. The network minimizes the reconstruction error by continuous iteration.

3. Experiment and analysis of result

3.1. The setting of the experiment

The software and hardware configuration used in this experiment is as follows: the operating system is Ubuntu14.04, the processor is Intel (R) Core (TM) i7, the main frequency is 2.27 GHz, the memory is 16.0GB, the graphics card is GTX1080, and the GPU accelerates. The deep learning platform used in this paper is TensorFlow1.0.0, and the programming language uses Python2.7.6.

The experimental image library contains 640 signed images, of which 320 of the positive samples are written by Liu Yanjiao herself with 5 kinds of pen (neutral pen, ballpoint pen, pencil, blue pen, black pen) and 64 own signatures respectively. The number of negative sample is 320, and the other 4 students in the laboratory imitate Liu Yanjiao with 5 kinds of pen respectively. Each type of pen is modeled on 16, as shown in Figure 3. The ratio of the training set, validation set and test set in this paper is 4:1:5, in which the test set contains 320 signature images, which are composed of 160 random sample signature images and randomly selected 160 negative samples, and all the remaining images form a training set and a validation set. In this paper, the training of SIGAN is used only in the positive samples, while the AlexNet used in the contrast experiment uses all the positive and negative samples in the training center.

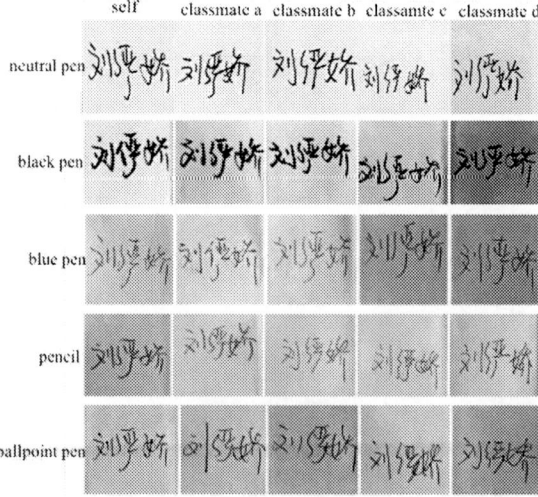

Figure 3. The display of the image's library.

3.2. Experiment

3.2.1. SIGAN. （1）All type of pen.The experimental training set contains all the pen types, and the training set of 160, 80, 40, 20, 10 and 5 are randomly selected for training (the number of each type of pens are the same). Figure 4 gives a comparison of the accuracy of handwriting identification using data enhancement and non data enhancement training. The training time of the mixed pen is shown in Figure 5, and the testing time is 0.27s for each image. Table 2 is the result of models by 160 training set.

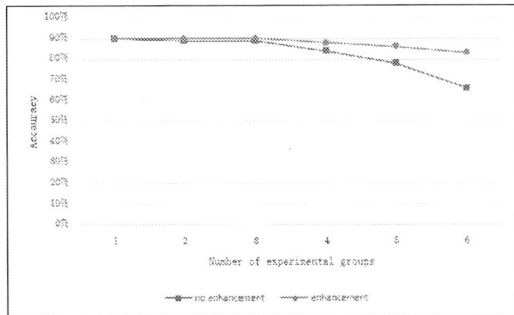

Figure 4. Accuracy of handwriting identification based on mixed pen training.

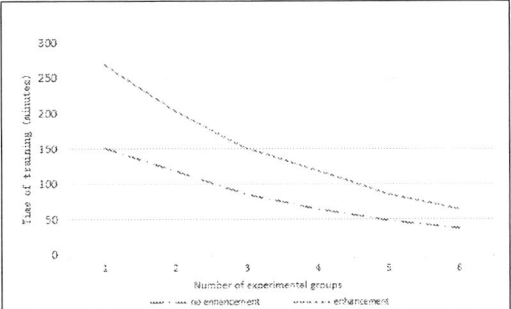

Figure 5. training time of the mixed pen.

Through the above experimental results, we can draw the following conclusions:

1. The accuracy of the model trained with all 160 signature handwriting images is the highest, reaching 90%. However, with the reduction of the number of training sets, the accuracy of the model is gradually reduced. When the number of training sets is less than 10 (no data enhancement), the accuracy drops to below 80%. This indicates that the more the training set is, the better the accuracy of model identification will be.

2. As can be seen from Figure 5, the use of data enhancement can effectively improve the accuracy of the model identification, the less the number of original training samples, the more obvious effect of data enhancement; when the number of original training samples is only 5, the accuracy of the identification can be raised by 17% by the data enhancement.

4. Conclusion

In this paper, the method of signature identification based on SIGAN is studied. By combining the Generative Adversarial Nets with dual learning, the mutual game of the generator and the discriminator is fully utilized to realize the identification task of the signature handwriting. The discriminator identifies the validity of the signature up to 91.2%, which greatly reduces the time and cost of the manual handwriting identification. This paper also needs to further improve the following aspects : (1) How to train handwriting identification model with few training samples; (2) How to improve the performance of SIGAN with the prior knowledge of handwriting identification; (3) In this paper, the image library and algorithm only involve Chinese handwriting in stiff brush. In the future, we will collect samples of brush and various foreign handwriting and test the algorithm.

References
[1]   Zhan Xiang and Yao Jianhong 2015 Test and analysis of handwriting *J. Legality Vision* **2(1)**

53-5

[2] Jia Zhihui 2008 On the judgment of the discrepancy points of the handwriting identification *J. Judicial identification in China* **9(4)** 55-9

[3] Xiong Huaqi and Chen Yaowen 2001 Identification of suspicious handwriting and seal with confocal laser scanning microscope *J. Journal of Laser Biology* **10(4)** 277-80

[4] Kuokuck W 1980 Writer Recognition by Spectral Analysis *J. Int Conf: Security Through Science and Engineering* 5-11

[5] Schomaker M and Bulacu M 2004 Automatic writer identification using connected-component contours and edge-based features of uppercase Western script *J. IEEE Transactions on Pattern Analysis & Machine Intelligence* **26(6)** 787-9

[6] Liu Hong and Tang Sheng 2013 Handwriting identification method based on SVM and texture *J. Journal of computer-aided design & computer graphics* **15(12)** 1479-84

[7] Huang Yaping and Chen Enyi 2003 Handwriting recognition based on independent component *J.* Journal of Chinese Information Processing **17(4)** 566-78

[8] Zhu Yong and Wang Yunhong 2001 Handwriting based identification *J. Automation Journal* **27(2)** 229-34

[9] Cao Jun and Fang Bing 2011 Off-line signature verification based on multi feature and neural network C. Chinese Conference on Pattern Recognition(CCPR) ChongQing 1069-73.

[10] Wang Hongge and Pan Shi 2013 Research on signature recognition of static handwritten Chinese character based on information entropy *J. Computer Applications and Software* **30(1)** 99-102

[11] Nasien D 2014 Freeman chain code (fcc) representation in signature fraud detection based on nearest neighbourand artificial neural network (ann) classifiers *J. Computer Science Journals* **21(1)** 141-9

[12] Ma Xiaoqing and Sang Qinbing 2017 Handwritten signature recognition algorithm based on LBP and deep learning *J. Chinese journal of quantum electronics* **34(1)** 23-31

[13] Krizhevsky A and Sueskeve I 2012 ImageNet classification with deep convolutional neural networks C. International Conference on Neural Information Processing Systems 1097-105

[14] Goodfellow I and Pouge J 2014 Generative adversarial nets C. International Conference on Neural InformationProcessing Systems 2672-80

[15] Blackwell W 2010 An analog of the minimax theorem for vector payoffs *J. Levines Working Paper Archive* **65(1)** 1-8

Object Detection on Underground Low-quality Images

Qi Mu[1], Zhiqiang He[2*] Yankui Liu[3] and Yu Sun[4]

[1] Collage of computer science and technology, Xi'an University of Science and Technology, Xi'an, Shaanxi, 710054, China

[2] Collage of computer science and technology, Xi'an University of Science and Technology, Xi'an, Shaanxi, 710054, China

[3] Collage of computer science and technology, Xi'an University of Science and Technology, Xi'an, Shaanxi, 710054, China

[4] Collage of computer science and technology, Xi'an University of Science and Technology, Xi'an, Shaanxi, 710054, China

*Corresponding author's e-mail: nnkajima@163.com

Abstract. Because of the insufficient illumination and dark environment, the underground image has little difference between the object and the background, and there will be irregular high-light spots, which will have a huge impact on the object detection. The commonly used methods for underground object detection are frame difference method, background difference, etc [1-3]. Both the frame difference method and the background difference method cannot accurately detect the object due to the influence of noise and image quality. This paper uses image enhancement technology to improve the quality of underground image, and proposes a comprehensive object detection method that is more robust to noise, and our algorithm can eliminate the influence of high-light point and improve the detection accuracy.

1. Introduction

Underground object detection is the basis of the current underground video security supervision. At present, the object detection method for underground mainly adopts the frame difference method and the background difference method. The frame difference method performs differential operation on two adjacent frames, and the difference result indicates the amount of gray value change of the corresponding position in the two frames. By setting an empirical threshold, the region with large change of gray value is extracted as the detected object. The frame difference method is sensitive to non-object grayscale changes in the image, such as illumination changes and image noise effects. The background difference method is to extract the object in the object frame by using the difference result between the object frame and the background frame. This method needs to establish a reliable background and consider the selection and update of the background. In the underground video, the difference between the object and the environment is not great, and affected by the underground environment, the image noise changes are obvious, and the illumination changes are not estimable. Therefore, the frame difference method or the background difference method cannot detect the object completely.

We now review existing work on object detection of coal mines. Due to the characteristics of underground video, there are two traditional methods. One is background difference, and another is establishing a more accurate object expression. Zhang Xiehua [4] proposed an adaptive background

---

Content from this work may be used under the terms of the Creative Commons Attribution 3.0 licence. Any further distribution of this work must maintain attribution to the author(s) and the title of the work, journal citation and DOI.

Published under licence by IOP Publishing Ltd

modeling and updating method based on clustering technology, through statistical methods, according to a certain period of time. The change of the pixel value of a point to determine whether the point is the background, this method can adapt to the background change, but it is easy to judge the background pixel as the foreground for the scene with large variation of the underground image noise. Zhang Chen [5] proposed one object detection method based on robust fuzzy kernel clustering. The background and foreground are distinguished by judging the distance between the pixel feature vector and the background subclass. However, this method cannot adapt to sudden spot changes. It is easy to detect an area with a large change in pixel value such as a spot as a foreground object. Cai Limei [6] proposed the idea of checking the position of the person in the well according to the helmet. The helmet is a prominent feature of the miners. This feature can better distinguish the object. And the background, but this method can only find the helmet in the video frame, but not detect the whole body, and the individual helmets are relatively small and easily occluded, and the helmet would not be positioned by the helmet after the occlusion.

Aiming at the particularity of the underground environment, the combination of underground image enhancement and traditional object detection is used to weaken the influence of underground noise on object detection, and multi-threshold segmentation method is used to eliminate the illumination changes around the object to achieve accurate object detection.

2. Proposed approach

Firstly, the frame difference method is used to determine the region of interest, and the interference of illumination variation is reduced. Then the region is compared with the background model. Because the spot is relatively bright relative to the object, the differential image is doubling thresholded to eliminate the interference of the spot. The algorithm flow fig is as follows：

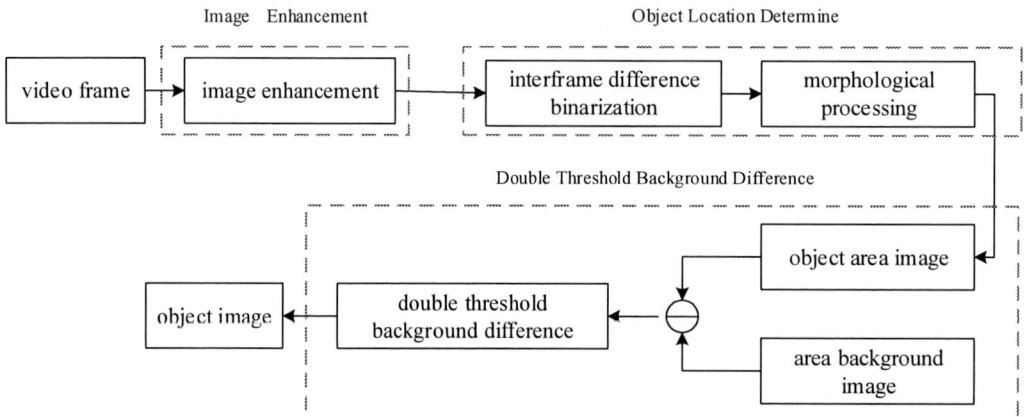

Figure 1. The framework of object detection on underground low-quality images

2.1. Image enhancement

The underground image is affected by the illumination, contains a lot of salt and pepper noise and Gaussian noise, the noise is randomly distributed, and presents an unpredictable change state, so it is impossible to design a targeted filtering method. Wavelet transform has the characteristics of low entropy, multi-resolution, decorrelation and flexibility of selecting wavelet base [7]. It has good localization characteristics in the process of processing images. By using different parameters, it can focus on any detail of the object [8]. Therefore, in the process of processing underground image noise, wavelet transform is used to denoise, which weakens the influence of noise on object detection to some extent.

The artificial lighting method causes the underground image to be extremely uneven and bright. The brightness of the part close to the light source is close to 255, the brightness is close to 0 from the far side, the overall brightness is dark, the contrast is poor, the imaging is uneven, and the details are blurred. Poor image quality results in increased difficulty in mining useful information from images. In order to obtain high-quality images, improve video readability, and improve the accuracy of subsequent processing algorithms, it is necessary to enhance the underground image to highlight its usefulness. In order to improve the difference between light and dark in the image, the histogram equalization is used to evenly distribute the gray value of the image, to brighten the darker region, and to limit the influence of the highlight point with large gray variation on the local region.

2.2. Object location determine

The frame difference method can be used to roughly determine the position of the moving object in the image, which is the first step to achieve accurate object detection. The frame difference method makes the difference between two adjacent frames or multiple frames in the video sequence, the background regions substantially cancel each other, and the regions where the moving objects appear are greatly different, and the position of the moving object can be detected. The basic formula for the two-frame difference is as follows:

$$D_k(x, y) = |I_k(x, y) - I_{k-1}(x, y)| \tag{1}$$

$$T_k(x, y) = \begin{cases} 1, & D_k(x, y) \geq Y \\ 0, & D_k(x, y) < Y \end{cases} \tag{2}$$

T_k is differential foreground image, Y is differential threshold. A reasonable threshold is set according to the amount of change in the pixel value of the image target area, thereby dividing the image foreground object.

The frame difference method yields a binary image that generally contains non-object regions that cause grayscale changes due to environmental factors. These regions are erroneously detected during this process, so it is also important to exclude these erroneous regions. By image etching, a better binary image is created.

2.3. Double threshold background difference

After getting the approximate position of the object, we need to get a more complete representation of the object. In general, the background difference method is a more effective object detection method. This method is simple to calculate, easy to implement, can detect object in real time, and can detect relatively complete moving object.

In the special environment of underground mines, a large amount of artificial illumination is used, and the position and height of the light source are changing. The change of light and dark caused by the change of the light source is difficult to be eliminated. The spot is brighter, the gray value is basically above 240, the underground object is generally dark, and the gray value is lower, so the difference between the spot and the background model is greater than the difference between the object and the background model.

The original image is distinguished from the background image to obtain a difference image, and the difference image is judged using a double threshold. Therefore, the improved background difference method is called a double threshold background difference method. Set two thresholds, the former to distinguish between the object and the background, and the latter to distinguish between the object and the spot. The double threshold method can be used to eliminate the interference of the spot on the object detection, which is expressed by the following formula:

$$W(x, y) = \begin{cases} 1 & T_1 \geq D(x, y) \geq T_0 \\ 0 & Other \end{cases} \tag{3}$$

$D(x, y)$ is background difference image, T_0 and T_1 are threshold, $W(x, y)$ is binary image.

3. Experiments

In this section, we present and discuss the result of our algorithm. We use our algorithm in some underground videos that contain moving objects and spot changes, and we find that our algorithm detection results are more accurate than traditional algorithmic results.

Figure 2. Experimental results

Experimental results are divided into three parts. (a)(b), input image and fixed background. (c)(d), the image enhancement results of the input image. (e)(f), the result of the initial positioning of the object. (g)(h), the final results of traditional algorithms and our algorithms.

4. Conclusion

In this paper, we apply image enhancement algorithms to precise object detection tasks for underground low-quality images, and reduce the impact of image noise and illumination on object detection. During the experiment, it was found that the spot change caused by artificial illumination had a great influence on the object detection. Therefore, the double threshold background difference method was used to eliminate the influence of the spot, and the experimental results performed well on some images. However, since the gray level change of the spot area approximates a continuous process, simply introducing multiple thresholds does not completely eliminate the spot. Therefore, the spot elimination and filling need to explore a better algorithm to further improve the accuracy of the underground object detection algorithm.

Acknowledgments
This work is supported by Scientific and Technology Program Funded by Xi'an City (Program 2017079CG/RC042(XAKD003))

References
[1]   Chen, J.C., Zhang, J.H., Liu, S.J. (2011) Improved object detection algorithm based on background modeling and frame difference. Computer Engineering, S1: 171–173.
[2]   Xue, L.X., Luo, Y.L., Wang, Z.C. (2011) Adaptive moving object detection method based on frame difference. Application Research of Computers, 04 : 1551-1552+.
[3]   Shen, Y., Wang, X.X. (2017) Video moving object detection method based on background subtraction and frame difference method. Automation & Instrumentation, 04 : 122-124.
[4]   Zhang, X.H., Zhao, X.H. (2016) Research on Moving Object Detection in Coal Mine Intelligent Video Surveillance. Industry and Mine Automation, 04: 31-36.
[5]   Zhang, C. (2013) Research on Object Detection and Tracking in Underground Environment. China University of Mining and Technology.
[6]   Cai, L.M. (2010) Research on Human Detection and Tracking in Underground Coal Mine Video. China University of Mining and Technology.
[7]   Zhang, X., Zhang, D.H., Zhang, X.X., Zhang, J.P. (2006) Image Denoising Based on Wavelet Transform. Chinese Journal of Scientific Instrument, S3: 2284-2286.
[8]   Donoho, D.L., Johnstone, I.M. (1994) Ideal spatial adaptation by wavelet shrinkage. Biometrika, 81: 425-455.

ArchiMate Customization and Architecture Repository Management Practices: for a Technology-Intensive Enterprise

Baobao Ding[1], Tong Wu[1*], Yingming Yang[2], Liang Dou[2] and Tiancheng Jin[2]

[1] CFETS Information Technology Co., Ltd, Shanghai, 200120, China

[2] Department of Computer Science and Technology, East China Normal University, Shanghai, 200062, China

*Corresponding author's e-mail: wutong@chinamoney.com.cn

Abstract. Technology-intensive companies need to develop a variety of applications based on the needs of the organizations. Due to the increasing number of projects, a unified Enterprise Architecture specification is required. We propose a situation-specific Enterprise Architecture modeling method for CFETS Information Technology Co., Ltd. By customizing the ArchiMate architecture description language, developing the architectural design modeling tool and establishing the architecture repository management platform, this solution realizes the standardization and unified management of the architectural design model for the enterprise information systems. The evaluation results show that the solution is efficient and easy to use.

1. Introduction

At present, the market environment for enterprises is complex and shifting rapidly. For technology-intensive companies, there is a need to develop a variety of applications based on the requirements of the organizations. With the increasing number of projects, application complexity and circumstances of using software supply chains, outsourcing and collaborative development, the Enterprise Architecture design specifications which model the architectural designs of the various enterprise application projects are required.

China Foreign Exchange Trade System (CFETS) Information Technology Co., Ltd. is a wholly-owned subsidiary of China Foreign Exchange Trade System & National Interbank Funding Center. CFETS provides a trading platform covering interbank foreign exchange market, money market and bond market. The total transaction volume of CFETS in 2017 reached RMB 998 trillion. In order to adapt to the market demand and cope with the rapid growth of data volume, CFETS needs to develop a large number of application projects each year, including some outsourcing projects. In 2014, the delivery cycles of 178 software were less than four months. This number reached 369 in 2015 and 422 in 2016. The iterations of these application projects are very short, the quantity of the projects is increasing, and the maintenance difficulties are rapidly increasing. In order to design standard Enterprise Architecture and build more efficient enterprise information systems, CFETS applies TOGAF (The Open Group Architecture Framework) [1] in the organization, and uses ArchiMate [2], an architecture description language supported by TOGAF, for architecture modeling. However, there are many problems in practice.

Firstly, for technology-intensive companies that pay great attention on technology solutions, the interested technical details need to be described in the architectural design. ArchiMate is a relatively high-level architecture modeling standard [3], which can model the interactions between applications. However, it lacks the ability to describe technical details.

Secondly, the architectural design generated by each project can be treated as an enterprise repository, which needs to be stored in the Architecture Repository [4]. Currently there are no effective and unified management methods to review, maintain and reuse the architectural designs.

To solve the above problems, we propose the following solutions.

- Architecture specification customization: according to the actual needs of CFETS, the default elements and relationships of ArchiMate are streamlined and extended. Furthermore, the concept of architecture view is proposed to show a specific part of the architecture model.
- Architectural design tool implementation: in order to support architecture specification customization, redevelop the open source tool for ArchiMate - Archi.
- Construction of architecture repository management platform: base on the concept that architecture can be regarded as repository [4], an enterprise-wide architecture repository management platform is built to review, maintain and reuse architectural designs conveniently.

The rest of this paper is organized as follows. The section 2 introduces the related work. The section 3 describes the research foundation and the customization of the architecture specifications. The section 4 explains the implementation of the architectural design tool. Section 5 introduces and evaluates the architectural repository management platform.

2. Related work

Sessions [5] made an authoritative comparison and analysis of current mainstream Enterprise Architecture frameworks such as Zachman [6], FEA [7] and TOGAF, which provided an important reference for subsequent research in this field. He pointed out that TOGAF is focusing on practice and provided a practical and feasible Enterprise Architecture method with better reusability and scalability. Lankhorst et al. [8] presented ArchiMate 3.0, an enterprise modeling language that captured the complexity of architectural domains and their relations and allowed the construction of integrated Enterprise Architecture models.

In spite of the potential benefits of Enterprise Architecture, several challenges with Enterprise Architecture adoption still exist, including the inaccurate and unwieldy methods used, as well as ambiguity in terms of goals, concepts and frameworks [9]. Pittl et al. [10] show that the existing enterprise modeling approaches are inappropriate for modeling digital enterprise ecosystems comprehensively. They proposed an idea of how an extension of ArchiMate could be achieved to meet the requirements. Gill et al. [3] evaluated the applicability and integration of six modeling standards including ArchiMate, BPMN, UML, FAML, SoaML and BMM and proposed a hybrid approach for agile Enterprise Architecture modeling.

In order to derive the approaches which consider the specific requirements of organizations and industry, more researches and in-depth knowledge regarding Enterprise Architecture activity in organizations are required [11]. In this paper, we proposed the customized ArchiMate according to the requirements of CFETS. We also developed a set of tool chains, including an architectural design tool and an architecture repository management platform.

3. Architectural specification customization

3.1. ArchiMate

ArchiMate [2] is an Enterprise Architecture modeling standard for TOGAF. It is also a visual Enterprise Architecture description language that can describe the Enterprise Architecture from a high level. Fig. 1 illustrates the basic framework of ArchiMate, showing the levels and attributes of elements and relationships. The basic elements in ArchiMate can be divided into three levels: the business layer, the application layer, and the technology layer, which respectively describe different levels of abstraction in Enterprise Architecture development. The business layer mainly describes the business process. The application layer describes the software architecture of the system. The technical layer describes the underlying technology deployment of the system. The elements can be divided into three categories based on their attributes: objects, behaviors, and subjects, where objects represent the passive elements and subjects represent the active elements in the architecture.

There can be many relationships between elements within and between layers. Relationships are divided into three categories: structured relationships, dynamic relationships, and other relationships. Relationships are used to connect different elements in the architecture. For example, the business object in the business layer is implemented by the data object in the application layer, the technical service of the technical layer serves the application component of the application layer (Serving), and the application interface of the application layer is part of the application component (Composition).

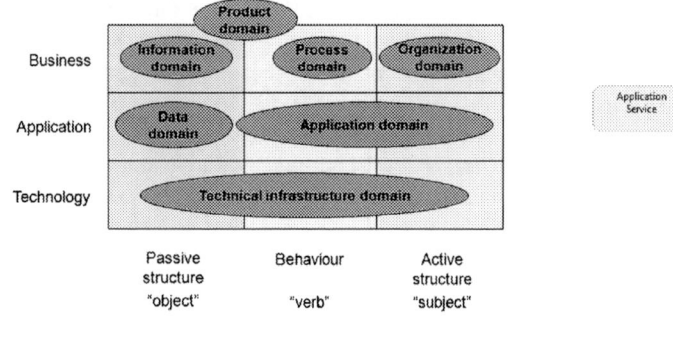

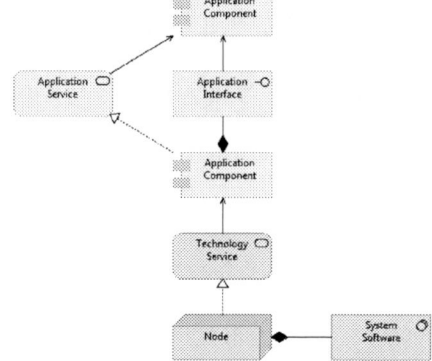

Figure 1. ArchiMate framework.   Figure 2. Basic pattern of customized ArchiMate.

3.2. Elements and relations customization

The ArchiMate 3.0 specification includes the business layer, the application layer, and the technology layer. It involves dozens of elements and more than ten relationships. However, the technology-intensive enterprises focus on the application layer and the technology layer. We streamline and extend the original standard of ArchiMate according to requirement.

Firstly, we remove the entire business layer, leaving only the application layer and technology layer. Secondly, the elements and relationships of the application layer and the technology layer have been streamlined and extended, shown in table 1 and table 2. The basic pattern of designing an architectural model using the customized ArchiMate is shown in fig. 2. Finally, we extend the application layer and technology layer to illustrate more technical details, shown in table 3.

In addition, we propose some basic design rules to provide unified guidance on architectural designing within the company. The specific constraints include the position, the color and the allowable elements for each layer, etc. For example, the overall layout of the architecture model is vertical, and each layer uses different color.

3.3. Architectural view

Architectural view is often defined as part of the description of the Enterprise Architecture that each stakeholder focuses on, and it defines what each stakeholder can see. We summarize the requirements of CFETS and define different architectural views for different stakeholders. Table 4 lists the defined architectural views. For example, the component view only shows the component & interface layer and the reusable artifact layer for developers. The deployment view only shows the component & interface layer and the infrastructure layer for maintenance persons.

3.4. Case study

An Enterprise Architecture modeling case study using our customized ArchiMate is demonstrated here. In order to improve the user experience of the trading system, CFETS has developed the WeChat service system based on the online service mode, so that it can be the supplement to traditional trading mode such as phone service and fax service. The architecture model of the WeChat service system is shown in fig. 3.

Table 1. Streamlined and extended elements.

Layer	Reserved Elements	Deleted Elements	Extended Elements
Application layer	Application component, Application interface, Application service	Application collaboration, Application function, Application interaction, Application process, Application event, Data object	Linux system, Windows system, UNIX system, SQL database, NoSQL database, Web container, Message middleware, etc.
Technology layer	Node, System software, Technology service, Artifact	Device, Technology collaboration, Technology interface, Path, Communication network, Technology function, Technology process, Technology interaction, Technology event, Technology object	

Table 2. Streamlined relationships.

Reserved Relationships	Deleted Relationships
Composition, Aggregation, Assignment, Realization, Serving, Association	Access, Influence, Triggering, Flow, Specialization, Junction

Table 3. Extensions for the application layer and the technology layer.

Original Layer	Extended Layer	Explanation
Application layer	Application service layer	This layer shows the services provided by the software system.
	Component & Interface layer	This layer shows the software architecture that implements application layer services.
	Peripheral system layer	This layer shows the third-party systems.
Technology layer	Infrastructure service layer	This layer shows the services provided by the infrastructure.
	Infrastructure layer	This layer shows the structure and the deployment of the infrastructure.
	Reusable artifact layer	This layer shows the reusable artifacts, which can only be used by application components.

Table 4. Architectural View.

Views	Included Layer	Explanation
Deployment view	Component & Interface layer, Infrastructure layer	Focus on the deployment of the system.
Interface view	Component & Interface layer	Focus on the implementation of the system.
Component view	Component & Interface layer, Reusable	Focus on the use of reusable

Views	Included Layer	Explanation
	artifact layer	artifact.

4. Architectural design tool

4.1. Architectural design tool redevelopment

ArchiMate has a native architectural design tool – Archi. Archi is an Eclipse-based and Java-based open source software for designing architectural models that conform to the ArchiMate specification. The current version is Archi 4.2.

We redevelop Archi to adapt to the customized ArchiMate specification. The original panel and menu is modified to accommodate the extended elements and the added architectural views.

Fig. 4 shows the part of customized panel in the tool. The extended elements including the operating system, Web container, NoSQL database, etc are showed in the red box.

4.2. Model check

When an architect draws the preliminary architecture model, usually there will be errors in the modeling design draft. A verification of architectural specification is needed. When an error occurs, the wrong part will been marked as red by the tool so that the architect will be alerted to correct it.

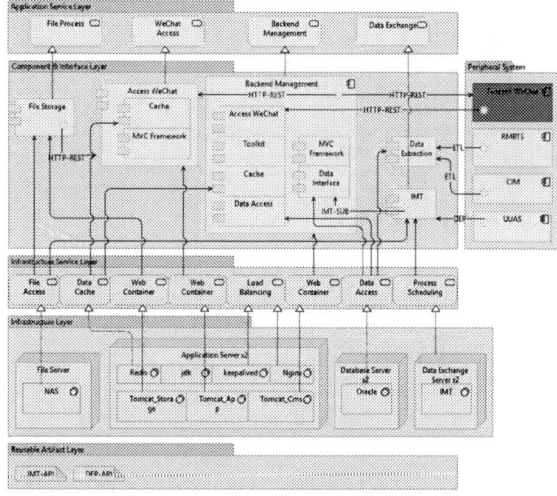

Figure 3. Customized ArchiMate model of the case WeChat service system.

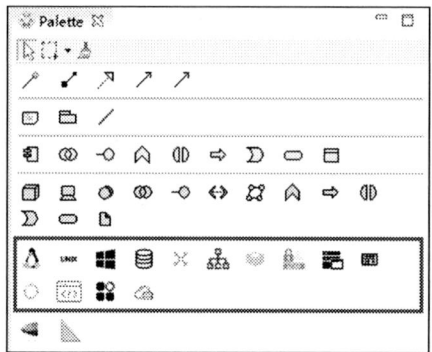

Figure 4. Part of customized panel.

5. Architecture repository management platform

5.1. Overview

In order to implement the concept that architecture is repository, we establish an architecture repository management platform. The architects are the main producer of the architectural design, responsible for designing the architecture models and uploading them to the architecture management platform. The reviewers review the uploaded architecture models. If it does not meet the project requirements, the architecture model is returned to the architect for modifications. The project managers are responsible for the final review of the architecture model. The developers are consumers in the architecture management platform. They can log into the platform and develop systems according to the approved architecture models.

The architecture repository management platform includes three parts: architectural design tool, architecture management system, and architecture repository document library. The framework of the architecture repository management platform is shown in fig. 5.

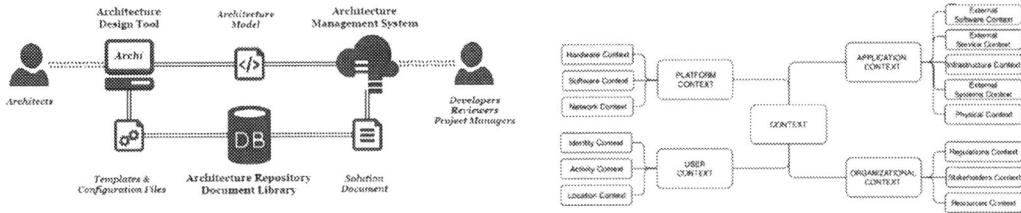

Figure 5. Architecture repository management platform framework.

Figure 6. Context category.

The architecture management system is web-based. The front end of system is developed by HTML and TypeScript, and the back end of system is developed by Java.

The Architecture Repository Document Library is a system that stores all of the architectural design documents of CFETS. The library is developed by C#, and the functionality of SharePoint is packaged as Web API.

5.2. Architecture model context information extraction
The context of a software system comprises the knowledge that architects need to have about the environment in which a system is expected to operate [12]. However, the context information is often overlooked. This leads to software architectural design based on wrong assumptions, which may cause failure of the architectural design.

As shown in fig. 6, Bedjeti [12] summarized the various categories of context models in the software architecture. In our work, the customized ArchiMate specification covers the platform context and application context information. For example, the application context information we need to extract are the service name, function description, component category, component name, and version number, etc.

The context information are extracted during the uploading process of the architecture model, and stored in the cloud. For example, table 5 shows the extracted information of the interfaces between systems.

Table 5. Extracted information of the interfaces between system.

Interface Name	Source System	Target System	Transmission Type	Reliability
i7	User Unified Authentication System(UUAS)	Wechat	DEP	no
i5	RMB Trading System(RMBTS)	Wechat	ETL	no
i6	CFETS Institution Management System (CIM)	Wechat	ETL	yes
i2	Wechat	Tencent WeChat	HTTP-REST	no
i4	Tencent WeChat	Wechat	HTTP-REST	no

5.3. Evaluation

The architecture repository management platform has been put into practice for a period of time. A group of main users have tested the platform, and then a satisfaction survey is conducted for evaluation.

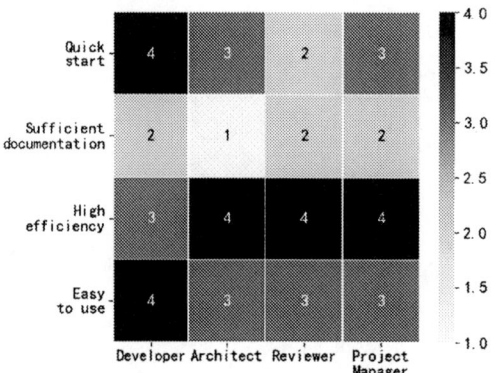

Figure 7. Evaluation result.

The main users of the platform are categorized into developers, architects, reviewers, and project managers. Five people are randomly selected from each stakeholder category, and a questionnaire survey is conducted on them. The content of the questionnaire is prepared in advance. The score ranges from 1 to 4, e.g. for "Easy to use" question, the score 4 indicates that the platform is very easy to use. The average scores of questions are shown in fig. 7.

The results show that most stakeholders think that the platform is easy to get started, however, the platform's documentation is not sufficient enough yet. Most stakeholders believe that the architecture management platform improves the efficiency of architectural design and management as a whole compared to the previous management scheme, and the platform provides more convenient and effective functions. In general, the architecture management platform optimizes and standardizes the process of architectural design and management, which improves the efficiency of the company's architecture repository management, but it also has deficiencies such as lack of documentations, which need to be improved in the future.

6. Conclusion and further research

With the continuous advancement and development of the Internet and informatization, the enterprise's architectural design will inevitably become more important. In the future, we will continue to pay attention to the changes and developments in this field. At the same time, the deployment of Enterprise Architecture methodology in enterprises is also very important. How to guide the relevant personnel to use the architectural design method at low cost that meets the enterprise requirements is also a concern.

Acknowledgments

This research is funded by the Science and Technology Commission of Shanghai Municipality (No. 18511103802).

References

[1] The Open Group. (1995) The open group architecture framework(TOGAF). http://www.opengroup.org/togaf.

[2] The Open Group. (2004) The ArchiMate® Enterprise Architecture Modeling Language. https://www.opengroup.org/archimate-forum/archimate-overview.

[3] Gill, A. Q. (2015). Agile enterprise architecture modelling: Evaluating the applicability and integration of six modelling standards. Information and Software Technology, 67, 196-206.

[4]   The Open Group. (2018) The TOGAF standard, version 9.2, 37.architecture repository. http://pu
      bs.opengroup.org/architecture/togaf9-doc/arch/chap37.html.

[5]   Sessions, R. (2007). A comparison of the top four enterprise-architecture methodologies. Houst
      on: ObjectWatch Inc.

[6]   J. Zachman. (1987) Zachman framework. https://www.zachman.com.

[7]   The U.S. Office of Management and Budget, Office of E-Government and IT. (1999) Federal en
      terprise architecture (FEA). https://obamawhitehouse.archives.gov/omb/e-gov/FEA.

[8]   Lankhorst, M. (2009). Enterprise architecture at work: Modelling, communication and analysis.
      Springer Science & Business Media.

[9]   Barn, B. S., Clark, T., & Loomes, M. (2013, February). Enterprise architecture coherence and th
      e model driven enterprise: is simulation the answer or are we flying kites?. In Proceedings of
      the 6th India Software Engineering Conference (pp. 97-102). ACM.

[10] Pittl, B., & Bork, D. (2017, July). Modeling Digital Enterprise Ecosystems with ArchiMate: A
      Mobility Provision Case Study. In International Conference on Serviceology (pp. 178-189).
      Springer, Cham.

[11] Scholtz, B., Calitz, A., & Connolley, A. (2013, November). An analysis of the adoption and usa
      ge of enterprise architecture. In Enterprise Systems Conference (ES), 2013 (pp. 1-9). IEEE.

[12] Bedjeti, A., Lago, P., Lewis, G. A., De Boer, R. D., & Hilliard, R. (2017, April). Modeling Cont
      ext with an Architecture Viewpoint. In Software Architecture (ICSA), 2017 IEEE Internation
      al Conference on (pp. 117-120). IEEE.

Sea clutter suppression based on sea spikes identification and matrix completion

Zhiyu Shao[1,2*], Jiangheng He[1,2] and Shunshan Feng[1,2]

[1]School of Mechatronical Engineering, Beijing institute of Technology, Beijing, 100081, China

[2]State Key Laboratory of Explosion Science and Technology, Beijing, 100081, China

*Corresponding author's e-mail: shao_zhiyu@163.com

Abstract. The increase of small targets on the sea surface poses a great threat to public security. Sea surface has extremely complex scattering characteristics and sea clutter is mixed with radar target echoes, which seriously affects the detection performance of radar for small targets on the sea surface. Therefore, a method based on sea spikes identification and matrix completion is proposed in this paper. By analyzing the characteristics of sea spikes, the sea spikes in radar echoes are distinguished and eliminated from the original signals. Then, the unqualified data are filled up by matrix completion method, which effectively improves the signal-to-noise ratio of target signals. Finally, the validity of the method is verified by using measured data and simulation data.

1. Introduction
Small targets detection on the sea surface plays a very important role in the field of sea surface rescue and other fields. Sea surface is a constantly changing surface, which has extremely complex scattering characteristics. Sea clutter has become one of the main constraints affecting radar detection performance. How to effectively overcome the interference of sea clutter has been recognized as a difficult point in the field of radar technology.

The rough sea surface is mainly composed of wind wave, gravity wave and capillary wave [1]. When observing the rough sea surface with low grazing angle, the radar echo of sea clutter shows a sea spike, which is not continuous in time and uneven in space, and its probability density function (PDF) curve shows a long tail phenomenon. Because of its short occurrence time, sea clutter changes from steady state to unsteady state and from non-time-varying to time-varying. The Doppler spectrum of sea clutter broadens and deviates from 0 frequency to cover up the weak target echoes, and the radar may distinguish the sea spike as a moving target, resulting in the increase of false alarm probability.

At present, there is no strict physical explanation for the generation of sea spikes. Posner points out that the sea spikes are mainly affected by wind direction, radar view angle and polarization mode [2]. Statistical characteristics of sea spikes are analyzed in [3]. It is considered that amplitude threshold, minimum interval and minimum spike width are three important parameters in describing sea spikes.

With the popularity of compressed sensing theory, matrix completion has attracted more and more attention. The second-order sparsity of matrix is used to restore the target matrix by sampling some elements. It is mainly used in image restoration, network traffic monitoring, and seismic data reconstruction [4].

Content from this work may be used under the terms of the Creative Commons Attribution 3.0 licence. Any further distribution of this work must maintain attribution to the author(s) and the title of the work, journal citation and DOI.

Published under licence by IOP Publishing Ltd

In this paper, a sea clutter suppression method based on sea spikes identification and matrix completion is proposed to improve the signal-to-clutter ratio (SNR) and enhance the detection performance for weak targets.

2. Sea spikes identification

Accurate identification of sea spikes is an important part of this method. As mentioned in [3], radar echo from the same range-gate can be distinguished as sea spikes only if the three conditions are satisfied: (1) The amplitude of the sampling points must exceed a specified threshold; (2) The duration of sampling points exceeding the threshold should not be less than the specified minimum spikes width; (3) The time interval between the sampling points exceeding the threshold should not exceed the specified minimum spikes interval.

$$\begin{cases} |x_i| \geq T \\ W \geq W_{min} \\ I \geq I_{min} \end{cases} \tag{1}$$

where x_i is the i th sampling point, and W and I represent the width and interval of the sea spikes, respectively. The threshold of sea spikes amplitude is usually L times the average power of sea clutter, that is

$$T = \sqrt{\frac{L}{N} \sum_{i=1}^{N} |x(i)|^2} \tag{2}$$

where N is the sequence length.

In this paper, the same parameters as [3] and [5] are used. The minimum spike width is set to 0.1s, the minimum spike interval is set to 0.5s, and the spike amplitude threshold is set to 5 times the average power of sea clutter. The sea clutter data collected by X-band IPIX radar [7] from McMaster University in Canada are analyzed. The main parameters are shown in Table 1.

Table.1 Main parameters of IPIX radar.

Radar parameters	Parameter values
Radar frequency	9.4 GHz
Intermediate frequency	150 MHz
pulse width	20-5000 ns
Transmitting signal wavelength	3 cm
Pulse repetition rate (PRF)	0-2000 Hz
Range resolution	30 m
Ground angle	<1°
Sampling distance	15 m

In this paper, the data #19931118_162658_stareC0000 is selected for analysis. Figure 1 gives the identification results of sea spikes under HH polarization and VV polarization. The red region in the figure represents the sea spikes. It can be seen that under HH polarization, the sea spikes has shorter duration and higher amplitude than VV polarization.

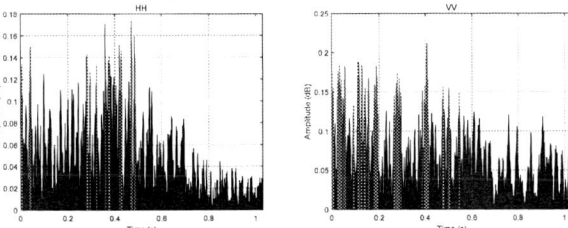

Figure 1. Results of sea spikes identification.

3. Matrix completion

3.1. Algorithm principle

Matrix completion method restores data by using the low rank property of the matrix. At the same time, the method can suppress interference and noise. Suppose that the element $M_{ij} \in \square$ in the given set $(i,j) \in \Omega \subset \{(i,j) : 1 \le i \le m, 1 \le j \le n\}$, Ω is the set of p known elements in X. Than low rank matrix completion can be expressed as a minimizing matrix rank. Its definition is as follows:

$$\min_{X \in m \times n} \ rank(X) \, s.t. \, X_{ij} = M_{ij}, \forall (i,j) \in \Omega \tag{3}$$

The radar echo of a range-azimuth unit can be expressed as

$$r(m) = s(m) + i(m), m = 1, 2, \cdots, M \tag{4}$$

where m represents the pulse sequence, $s(m) = t(m) + c(m)$ represents the target signal and residual clutter, and $i(m)$ is the sea spike. Than we constructe the $P \times L$ dimensional Hankel matrix by using radar echo sequence $r(m), m = 1, 2, \cdots, M$, where $M = P + L - 1$. The rank of Hankel matrix is approximately 3 when the target has low maneuverability [6]. Set $R = H(r)$, $S = H(s)$, $I = H(i)$, the Hankel matrix of the echo signal can be expressed

$$R = S + I + N \tag{5}$$

Hankel matrix S can be expressed as $S = \sum_{m=1}^{M} \alpha_m B_m$ equivalently, where B_m is the base matrix and α_m is the weighted value. According to the theory of matrix completion, the data can be recovered by following optimization problems after the sea spikes are identified and eliminated.

$$\min_{S \in P \times L} \|S\|_*, \, s.t., P_\Omega(R) = P_\Omega(S + N), S = \sum_{m=1}^{M} \alpha_m B_m \tag{6}$$

where $\|*\|_*$ represents the nuclear norm.

3.2. Optimization of cost function

In this section, we use the generalized Lagrange multiplier method to solve the optimization problem. Document [6] extends the classical matrix completion algorithm to the complex domain, and effectively utilizes the phase information. Therefore, we use this method to solve the proposed optimization problem. The Lagrange function in the complex field can be expressed as

$$L(S,Z,\alpha) = \|S\|_* + \Re\left\langle Y_1, R - S - Z \right\rangle + \frac{\mu}{2} \|R - S - Z\|_F^2 + \Re\left\langle Y_2, S - \sum_{m=1}^{M} \alpha_m B_m \right\rangle + \frac{\mu}{2} \left\| S - \sum_{m=1}^{M} \alpha_m B_m \right\|_F^2 \tag{7}$$

where $\Re\langle \bullet \rangle$ denotes the real part of the signal, $\alpha = [\alpha_1, \alpha_2, \cdots \alpha_M]^T$ denotes the complex weighted vector, and $\langle A, B \rangle = tr(A^H B)$ is the inner product of the matrix. Y_1 and Y_2 are Lagrange multiplier matrices, and $\mu > 0$ is the penalty factor. The steps to solve the problem are as follows.

a) Parameter initialization. $Z = 0$, $Y_{1,0} = Y_{2,0} = 0$, $k = 0$, $\rho = 1.3$, $P = 40$, $\eta = 10^{-3}$, $\mu_0 = 10^{-5}$.

b) Updating matrix S.

$$S = U T_{1/(2\mu_k)}(\Sigma) V^H \tag{8}$$

where $E_k = U \sum V^H$ represents the singular value decomposition of the matrix E_k.

c)   Updating matrix Z and vector α.

$$Z = P_{\bar{\Omega}}\left(R - S_{k+1} + \mu_k^{-1} Y_{1,k} \right), \tag{9}$$

$$\alpha_{m,k} = \left\| B_m \right\|_1^{-1} tr\left\{ \left(S_{k+1} - \mu_k^{-1} Y_{1,k} \right)^T B_m \right\} \tag{10}$$

where $\bar{\Omega} = E - \Omega$, E is a matrix whose elements are all 1, and $\alpha_{m,k}$ represents the kth update of the mth element of vector α.

e)   Updating Y_1, Y_2 and μ.

$$\begin{cases} Y_{1,k+1} = Y_{1,k} + \mu_k \left(R - S_{k+1} - Z_{k+1} \right) \\ Y_{2,k+1} = Y_{2,k} + \mu_k \left(S_{k+1} - \sum_{m=1}^{M} \alpha_m B_m \right) \\ \mu_{k+1} = \rho \mu_k \end{cases} \tag{11}$$

f)   Iteration and outputting results.

Turn to the second step until $\left\| S_k - S_{k-1} \right\|_F / \left\| S_k \right\|_F \leq \eta$. Then, the recovery data is given by

$$\hat{r} = \left[\alpha_{1,k}, \alpha_{2,k}, \cdots, \alpha_{M,k} \right]^T \tag{12}$$

4. Performance of the proposed method

The IPIX data #19931118_162658_stareC0000 was used for validation of the proposed method. Figure 2(a) is the time-frequency plot of the original signal. It can be seen from the figure that there are many sea spikes, and it is very easy to cause false alarm for radar detection. Figure 2 (b) is a time-frequency plot with the spikes data set at 0. Although the number of spikes decreases, a large number of gate lobes appear in the spectrum, which will also cause false alarms to a certain degree. Figure 2 (c) is the result of data recovery by compressed sensing after eliminating sea spikes. As we can see from the figure, the sea spikes are recovered due to the sparsity of clutter, and the amplitude is even higher than the original signal. Figure 2 (d) is the result of the proposed method. Compared with figure 2 (b) and figure 2 (c), it can be found that most of the sea spikes have been eliminated, which can greatly improve the detection performance for weak targets in sea clutter.

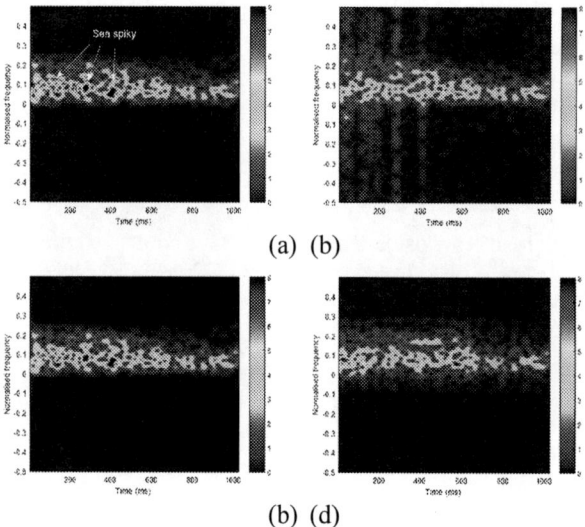

(a) (b)

(b) (d)

Figure 2. The STFT of the sea clutter. (a) Sea clutter before processing. (b) The unqualified data are filled up 0. (c) The unqualified data are filled up by compressed sensing method. (d) The unqualified data are filled up by matrix completion method.

Besides, we add the simulated moving target into the measured sea clutter data to verify the proposed method. The target parameters are as follows: the radar pulse repetition frequency is 1000 Hz, the velocity is about 0.2 m/s. The frequency spectrum and time-frequency plot of the target range-gate are shown in figure 3(a) and (c), respectively. Figure 3(b) and (d) show the frequency spectrum and time-frequency plots of the data after processing by the sea spikes identification and matrix completion algorithm. It can be seen that the sea clutter is obviously weakened and the SNR is increased by about 7.4 dB. Thus, the proposed algorithm can effectively suppress the sea clutter in radar target and improve the SNR.

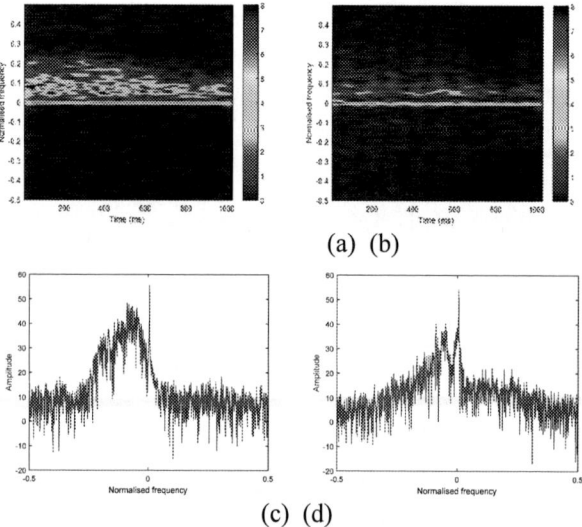

(a) (b)

(c) (d)

Figure 3. (a) Time-frequency plot of sea clutter data with simulated targets. (b) Time-frequency plot of the data after the proposed method. (c) Frequency spectrum plot of sea clutter data with simulated targets. (d) Frequency spectrum plot of the data after the proposed method.

5. Conclusion

In this paper, a method of sea clutter suppression based on the combination of spikes identification and matrix completion is proposed. Firstly, three important parameters are used to identify and eliminate sea spikes. Then, we constructe the Hankel matrix after sea spikes removal, and the unqualified data are filled up by matrix completion method. The Lagrange multiplier method is used to solve the optimization problem. The simulation data and the measured data show that the method can effectively suppress the sea spikes and improve the SNR of weak targets on the sea surface.

References

[1] Ward K, Tough R, Watts S, et al. (2013) Sea Clutter: Scattering, the K-Distribution and Radar Performance. Waves in Random & Complex Media, 17(2):233-234.

[2] Posner F L. (2002) Spikes sea clutter at high range resolutions and very low grazing angles. IEEE Transactions on Aerospace & Electronic Systems, 38(1):58-73.

[3] Greco M, Stinco P, Gini F. (2010) Identification and Analysis of Sea Radar Clutter Spikes. Radar Sonar & Navigation IET, 4(2):239-250.

[4] Chen, L., Chen, S. et al. (2017) A Survey of Matrix Completion Models and Algorithms. Journal of software, 28(6):1547-1564.

[5] Huang, Y., Cheng, X., Guan, J. (2015) Characteristic Analysis and Suppression Method of Measured Sea Spikes. Journal of radar, 4(3):334-342.

[6] Li, M., He, Z. (2015) Transient Interference Suppression Algorithm for Sky Wave Radar Based on Matrix Completion. Journal of electronics and information, 37(5):1031-1037.

[7] Wind H J D, Cilliers J E, Herselman P L. (2010) DataWare: Sea Clutter and Small Boat Radar Reflectivity Databases. IEEE Signal Processing Magazine, 27(2):145-148.

ISPECE   IOP Publishing

Exponential stability for a class of nonlinear singular Markovian jump systems with time-delay

Daixi Liao[1,2*] , Shouming Zhong[1] and Shaohua Long[3]

[1] School of Mathematics Science, University of Electronic Science and Technology of China, Sichuan, China

[2] School of Mathematics, Physics and Energy Engineering, Hunan Institute of Technology, Hengyang , China

[3] School of Science, Chongqing University of Technology, Chongqing, P.R. China

*Corresponding author's e-mail: liaodaixizaici@163.com

Abstract. This paper is concerned with exponential stability criteria of nonlinear singular Markovian jump systems which the time-delay is constant. The Lyapunov method is employed in this paper. A sufficient condition is presented, which ensures that the investigated system is regular, impulse-free and exponentially stable. The obtained result is with the form of linear matrix inequalities(LMIs), which can be solved easily by using Matlab LMI Toolbox. Finally, a numerical example shows the validity of the method.

1. Introduction

During time delay inevitably occurs in such practical systems as networked control system, circuit system and singular Markovian jump system. And the nonlinear system enables a more accurate and general description of the physical system. Therefore, the research on the nonlinear singular Markovian jump system with constant time-delay has been focus of many scholars [1-6]. However, as we know, the exponential stability problem for a class of nonlinear singular Markovian jump systems which the time-delay is constant has not been well studied. For considering the stochastic admissibility of partially unknown transition probabilities systems, there are few research results in the literatures. This inspires this paper's research.

In this work, by means of the Lyapunov functional method, the stochastic admissibility of the system is discussed and the solution is given. Finally, a numerical example shows the validity of the method.

2. Preliminaries

In this paper, we study a class of nonlinear singular Markovian jump systems with invariant time delay as follows:

$$\begin{cases} E(r(t))\dot{y}(t) = A(r(t))y(t) + B(r(t))y(t-h) + C(r(t))f(y(t)), \\ y(\theta) = \phi(\theta), \quad \forall \theta \in [-h,0]. \end{cases} \tag{1}$$

In the above system, we use $y(t) \in R^n$ to represent the state vector, $h > 0$ denotes the constant time-delay. $f(y(t))$ is nonlinear and satisfies $f^T(y(t))f(y(t)) \le \sigma y^T(t)y(t)$, where $\sigma > 0$ is a

---

Content from this work may be used under the terms of the Creative Commons Attribution 3.0 licence. Any further distribution of this work must maintain attribution to the author(s) and the title of the work, journal citation and DOI.
Published under licence by IOP Publishing Ltd

invariant scalar. $\phi(\theta)$ is a continuous initial function. $\{r(t)\}(t \geq 0)$ takes values in a finite set $S^* = \{1, 2, ..., N\}$ and denotes a continuous-time Markov process with right continuous trajectories. The transition probability matrix $\Xi = (\beta_{ij})(i, j \in S^*)$ of the considered Markov process is given by

$$P\{r(t+\Delta) = j \mid r(t) = i\} = \begin{cases} \beta_{ij}\Delta + o(\Delta), & i \neq j, \\ 1 + \beta_{ii}\Delta + o(\Delta), & i = j, \end{cases} \quad (2)$$

where $\Delta > 0$ and $o(\Delta)/\Delta \to 0(\Delta \to 0)$, $\beta_{ij} \geq 0$ is the transition rate from mode i to mode j if $i \neq j$ and such that $\beta_{ii} = -\sum_{j=1, j\neq i}^{N} \beta_{ij}$.

For each $r(t) = i \in S^*$, $E_i \in R^{n \times n}$ satisfies $\det(E_i) = 0$ and we assume that $0 < rank(E_i) = r < n$. The matrices A_i, B_i and C_i are with appropriate dimensions and are known real invariant matrices .

Definition 1. (i) The system (1) is regular if $\det(sE_i - A_i)$ is not identically zero for every $i \in S^*$.

(ii) The system (1) is impulse-free if $\deg(\det(sE_i - A_i)) = r$ for every $i \in S^*$.

(iii) The system (1) is stochastically exponentially stable if initial mode $r(0)$ and for any initial function $\phi(\cdot)$, the situation $y^T(t)y(t) < \alpha e^{-\beta t} \sup_{-h \leq s \leq 0} \|\phi(s)\|$ holds.

3. Main result
Next, the following theorem is presented to derive a solution to the stochastically exponentially stable for the system (1).

Theorem 1. For given a scalar $\eta > 0$, the system (1) is said to be regular, impulse-free and stochastically exponentially stable, if there exist matrices $Q > 0$, $P_i > 0$, T_1, T_2, T_3, Z_1, Z_2, M_1 and S_i such that for each $i \in \bar{S}$,

$$\sum_{5 \times 5}^{i} + \Theta_1 + \Theta_{2i} < 0, \quad (3)$$

$$\Omega_{3 \times 3} > 0, \quad (4)$$

where

$$\Sigma_{11}^i = \eta E_i^T P_i E_i + E_i^T P_i A_i + A_i^T P_i E_i + S_i^T R_i A_i + A_i^T R_i^T S_i + \sum_{j=1}^{N} \pi_{ij} E_j^T P_j E_j + Q + \sigma I,$$

$$\Sigma_{12}^i = E_i^T P_i B_i + S_i^T R_i B_i, \quad \Sigma_{13}^i = 0, \quad \Sigma_{14}^i = hA_i^T M_1, \quad \Sigma_{15}^i = E_i^T P_i C_i + S_i^T R_i C_i,$$

$$\Sigma_{22}^i = -e^{\eta h}Q, \quad \Sigma_{23}^i = 0, \quad \Sigma_{24}^i = hB_i^T M_1, \quad \Sigma_{25}^i = 0, \quad \Sigma_{33}^i = 0, \quad \Sigma_{34}^i = 0, \quad \Sigma_{35}^i = 0,$$

$$\Sigma_{44}^i = -hM_1, \Sigma_{45}^i = hM_1^T C_i^T, \Sigma_{55}^i = -I,$$

$$\Theta_1 = \frac{\tilde{\pi}h^2}{2}[\ell_1, \quad \ell_2, \quad \cdots, \quad \ell_N]diag\{M_1, \quad M_1, \quad \cdots, \quad M_1\}[\ell_1, \quad \ell_2, \quad \cdots, \quad \ell_N]^T,$$

$$\ell_j = [A_j \quad B_j \quad 0 \quad 0 \quad C_j]^T, \quad j = 1, 2, \cdots, N,$$

$$\Theta_2 = [I_{(3n)\times(3n)} \quad 0_{(3n)\times(2n)}]^T e^{-\eta h}F_i^T (hT_1 + \frac{1}{3}hT_2) + Z_1\Delta_1 + Z_2\Delta_2 + \Delta_1^T Z_1^T + \Delta_2^T Z_2^T)F_i[I_{(3n)\times(3n)} \quad 0_{(3n)\times(2n)}],$$

$$\ell_j = [A_j \quad B_j \quad 0 \quad 0 \quad C_j]^T, \quad j = 1, 2, \cdots, N,$$
$$\Omega_{11} = T_1, \quad \Omega_{12} = T_2, \quad \Omega_{22} = T_3, \quad \Omega_{13} = Z_1, \quad \Omega_{23} = Z_2, \quad \Omega_{33} = M_1,$$

$$F_i = diag\{E_i \quad I \quad I\}, \ \Delta_1 = [I \quad -I \quad 0], \Delta_2 = [-I \quad -I \quad 2I],$$
$$\hat{\pi} = \max\{-\beta_{11}, -\beta_{22}, \cdots, -\beta_{NN}\},$$

and $R_i \in R^{n \times n}$ satisfies $R_i E_i = 0$ with full column rank.

Proof : We prove that the system (1) is regular and impulse-free firstly.
Form (4), we can obtain that

$$\Sigma_{11}^i + E_i^T [I \quad 0 \quad 0 \quad 0 \quad 0]\Theta_{2i}[I \quad 0 \quad 0 \quad 0 \quad 0]^T E_i < 0. \tag{5}$$

According to (5), we get

$$E_i^T P_i A_i + A_i^T P_i E_i + S_i^T R_i A_i + A_i^T R_i^T S_i + \pi_{ii} E_i^T P_i A_i + E_i^T [I \quad 0 \quad 0 \quad 0 \quad 0]\Theta_{2i}[I \quad 0 \quad 0 \quad 0 \quad 0]^T E_i < 0 \tag{6}$$

Noting that $rank(E_i) = r < n$, we can find two invertible matrices $\tilde{L}_i$ and $\tilde{G}_i$ such that

$$\overline{E}_i = \tilde{L}_i E_i \tilde{G}_i = \begin{bmatrix} I_r & 0 \\ 0 & 0 \end{bmatrix}. \tag{7}$$

We set that

$$\overline{A}_i = \tilde{L}_i A_i \tilde{G}_i = \begin{bmatrix} \overline{A}_{i1} & \overline{A}_{i2} \\ \overline{A}_{i3} & \overline{A}_{i4} \end{bmatrix}, \ \overline{P}_i = \tilde{L}_i^{-T} P_i \tilde{L}_i^{-1} = \begin{bmatrix} \overline{P}_{i1} & \overline{P}_{i2} \\ \overline{P}_{i3} & \overline{P}_{i4} \end{bmatrix}, \ \tilde{R}_i = R_i L_i^{-1} = \begin{bmatrix} \tilde{R}_{i1} & \tilde{R}_{i2} \\ \tilde{R}_{i3} & \tilde{R}_{i4} \end{bmatrix},$$

$$\tilde{S}_i = S_i \tilde{G}_i = \begin{bmatrix} \tilde{S}_{i1} & \tilde{S}_{i2} \\ \tilde{S}_{i3} & \tilde{S}_{i4} \end{bmatrix},$$

$$\overline{\Theta}_i = \tilde{L}_i^T [I \quad 0 \quad 0 \quad 0 \quad 0]\Theta_{2i}[I \quad 0 \quad 0 \quad 0 \quad 0]^T \tilde{L}_i = \begin{bmatrix} \overline{\Theta}_{i1} & \overline{\Theta}_{i2} \\ \overline{\Theta}_{i3} & \overline{\Theta}_{i4} \end{bmatrix}. \tag{8}$$

Pre- and post-multiplying (6) by $\tilde{G}_i^T$ and $\tilde{G}_i$, respectively, and employing the similar method in [6], we can have that the system (1) is regular and impulse-free.

In the following, we prove the stochastically exponentially stable for the system (1).
Construct a Lyapunov functional as follows:

$$V(t) = V_1(t) + V_2(t) + V_3(t) + V_4(t), \tag{9}$$

where

$$V_1(t) = y^T(t)E^T(r(t))P(r(t))E(r(t))y(t),$$

$$V_2(t) = \int_{t-h}^{t} e^{\eta(\alpha-t)}y^T(\alpha)Qy(\alpha)d\alpha,$$

$$V_3(t) = \int_{-h}^{0}\int_{t+\theta}^{t} e^{\eta(\alpha-t)}\dot{y}^T(\alpha)E^T(r(t))W(r(t))E(r(t))\dot{y}(\alpha)d\alpha d\theta,$$

$$V_4(t) = \sum_{j=1}^{N} \hat{\pi}\int_{-h}^{0}\int_{\theta}^{0}\int_{t+\alpha}^{t} e^{\eta(\xi-t)}\dot{y}^T(\xi)E_j^T W_j E_j \dot{y}(\xi)d\xi d\alpha d\theta.$$

From (9), we have

$$\Lambda V_1(t) = 2y^T(t)E_i^T P_i E_i \dot{y}(t) + \sum_{j=1}^{N} \pi_{ij} y^T(t)E_j^T P_j E_j y(t)$$

$$= 2y^T(t)E_i^T P_i[A_i y(t) + B_i y(t-h(t)) + C_i f(y(t))] + \sum_{j=1}^{N} \pi_{ij} y^T(t)E_j^T P_j E_j y(t), \tag{10}$$

$$\Lambda V_2(t) \leq y^T(t)Qy(t) - e^{-\eta h}y^T(t-h)Qy(t-h) - \eta\int_{t-h}^{t} e^{\eta(\alpha-t)}y^T(\alpha)Qy(\alpha)d\alpha, \tag{11}$$

$$\Lambda V_3(t) = h\dot{y}^T(t)E_i^T M_1 E_i \dot{y}(t) - \int_{t-h}^t e^{\eta(\alpha-t)}\dot{y}^T(\alpha)E_i^T M_1 E_i \dot{y}(\alpha)d\alpha$$

$$+ \sum_{j=1}^N \pi_{ij}\int_{-h}^0\int_{t+\theta}^t e^{\eta(\alpha-t)}\dot{y}^T(\alpha)E_j^T M_1 E_j \dot{y}(\alpha)d\alpha d\theta - \eta\int_{-h}^0\int_{t+\theta}^t e^{\eta(\alpha-t)}\dot{y}^T(\alpha)E_i^T M_1 E_i \dot{y}(\alpha)d\alpha d\theta$$

$$\leq h\dot{y}^T(t)E_i^T M_1[A_i y(t) + B_i y(t-h(t)) + C_i f(y(t))]$$

$$+ e^{-\eta h}g^T(t)F_i^T(hT_1 + \tfrac{1}{3}hT_2) + Z_1 X_1 + X_1^T Z_1^T)F_i g(t)$$

$$+ \sum_{j=1}^N \pi_{ij}\int_{-h}^0\int_{t+\theta}^t e^{\eta(\alpha-t)}\dot{y}^T(\alpha)E_j^T M_1 E_j \dot{y}(\alpha)d\alpha d\theta - \eta\int_{-h}^0\int_{t+\theta}^t e^{\eta(\alpha-t)}\dot{y}^T(\alpha)E_i^T M_1 E_i \dot{y}(\alpha)d\alpha d\theta$$

$$(12)$$

$$\Lambda V_4(t) = \frac{\hat{\pi}h^2}{2}\sum_{j=1}^N \dot{y}^T(t)E_j^T M_1 E_j \dot{y}(t) - \hat{\pi}\sum_{j=1}^N \int_{-h}^0\int_{t+\theta}^t e^{\eta(\alpha-t)}\dot{y}^T(\alpha)E_j^T M_1 E_j \dot{y}(\alpha)d\alpha d\theta$$

$$-\eta\sum_{j=1}^N \int_{-h}^0\int_\theta^0\int_{t+\alpha}^t e^{\eta(\xi-t)}\dot{y}^T(\xi)E_j^T M_1 E_j \dot{y}(\xi)d\xi d\alpha d\theta$$

$$(13)$$

According to $R_i E_i = 0$, we get

$$2[A_i y(t) + B_i y(t-d) + C_i f(y(t))]^T R_i^T S_i y(t) = 0. \tag{14}$$

By (9)-(14), we can see that

$$\Lambda V(t) \leq \varsigma_i^T(t)(\sum_{5\times5}^i + \Theta_1 + \Theta_{2i})\varsigma_i(t) \leq -\tau_i y^T(t)y(t),$$

where

$$\varsigma_i^T(t) = \left[\, y^T(t) \quad y^T(t-h) \quad \int_{t-h}^t y^T(s)ds \quad [E_i\dot{y}(t)]^T \quad f^T(y(t)) \right]^T,$$

$$\tau_i = -\lambda_{\max}(\sum_{5\times5}^i + \Theta_1 + \Theta_{2i}),$$

which guarantees the system (1) is stochastically exponentially stable. This completes the proof.

4. Example
Consider the system (1) with the following parameters:

$$E_1 = \begin{bmatrix} 1 & 0 \\ 0 & 0 \end{bmatrix}, A_1 = \begin{bmatrix} -1.0 & 0.53 \\ 1.74 & 1.56 \end{bmatrix}, B_1 = \begin{bmatrix} -0.11 & 0.06 \\ -0.18 & 0.10 \end{bmatrix}, C_1 = \begin{bmatrix} 0.2 & 0 \\ -0.2 & 0 \end{bmatrix}, E_2 = \begin{bmatrix} 1 & 1 \\ 0 & 0 \end{bmatrix},$$

$$A_2 = \begin{bmatrix} -1.2 & 1.35 \\ -1.02 & -2.38 \end{bmatrix}, B_2 = \begin{bmatrix} 0.12 & 0.08 \\ -0.12 & 0.09 \end{bmatrix}, C_2 = \begin{bmatrix} 0.35 & 0 \\ -1.46 & 0.2 \end{bmatrix},$$

$$\Xi = \begin{bmatrix} -0.5 & 0.5 \\ 0.25 & -0.25 \end{bmatrix}, h = 0.72.$$

In this example, we suppose $R_1 = R_2 = \begin{bmatrix} 0 & 0 \\ 0 & 0.11 \end{bmatrix}$ and $\eta = 0.01, \sigma = 0.1$.

It is clear that that the LMIs of Theorem 1 are feasible on the basis of using the LMI toolbox. Thus, It is obvious that that the considered system is stochastically exponential stable.

5. Conclusion

This paper is concerned with the problem of the stochastically exponentially stable for a class of nonlinear singular Markovian jump systems with time-delay and mode-dependent $E(r(t))$. We derived a solution to the stochastically exponentially stable for the system (1). Finally, we give a numerical example to show the validity of the method.

Acknowledgments

This research was supported by Chongqing Research Program of Basic Research and Frontier Technology (cstc2017jcyjA0923).

References

[1] Long, S.H., Wu,Y.L., Zhong, S.M., Zhang, D. (2018) Stability analysis for a class of neutral type singular systems with time-varying delay. Applied Mathematics and Computation, 339: 113-131.

[2] Wang, Y.Y., Shi, P., Wang, Q.B., Duan, D.P. (2013) Exponential H-infinity filtering for singular Markovian jump systems with mixed mode-dependent time-varying delay. IEEE Transactions on Circuits and Systems—I: Regular Papers, 60: 2440-2452.

[3] Wang, Q., Wu, Z.G., Shi, P., Xue, A.K. (2017) Robust control for switched systems subject to input saturation and parametric uncertainties. Journal of the Franklin Institute, 354: 7266-7279.

[4] Chen, J., Xu, S.Y., Zhang, B.Y., Liu, G. B. (2017) A Note on Relationship Between Two Classes of Integral Inequalities. IEEE Trans. Autom. Control, 62: 4044-4049.

[5] Kao, Y., Xie, J., Wang, C., Karimi, H. (2015) A sliding mode approach to H-infinity non-fragile observer-based control design for uncertain Markovian neutral-type stochastic systems. Automatica, 52: 218-226.

[6] Wang, H.J., Xue, A.K., Lu, R.Q. (2009) Absolute stability criteria for a class of nonlinear singular systems with time delay. Nonlinear Analysis: Theory, Methods & Applications, 70: 621-630.

A New Molecular Encryption Model Based on Microfluidic Techniques

Pengcheng Ma[1] and Heng Sun[1*]

[1] College of Information Science, Jinan University, Guangzhou, Guangdong, 510632, China

*Corresponding author's e-mail: 71831960@qq.com

Abstract. Non-linearity behaviours of fluids have not been fully understood by human beings. In this paper, we present a new molecular encryption model based on microfluidic techniques. The security of molecular cryptography is based on non-linearity of fluids, rather than conventional mathematical hardness assumptions. As there is no mathematical analysis method to predict non-linear behaviours of fluids, it is unaffected by the attack of mathematical analysis. This method provides a new insight to information security and is hopefully to be a new promising direction.

1. Introduction

Molecular cryptography, which is different from conventional cryptography and quantum cryptography, is an emerging promising direction of information security using molecule to encrypt data. A similar research direction of this field is DNA cryptography, which is to encode messages in DNA sequences to conceal the information. With high density of DNA, ultra large information can be stored in a small amount of DNA material [1]. DNA cryptography has provided some new methods to cryptography. Clelland et. al. [2] put forward a DNA-based, doubly steganography technique for sending messages using microdots. This elegant study first showed an implementation of DNA cryptography and the significance fact of using microdots. Several years later, Gehani et. al. [3] proposed two DNA one-time-pad encryption schemes, a substitution method and an XOR scheme. These approaches present a DNA-based cryptographic method rather than number-based methods. Such studies attracted the attention of researchers to study the potential to create a novel cryptography system based on DNA. Plenty of researches are proposed [4, 5] and makes DNA cryptography a main part of molecular cryptography. Lai et. al. [5] proposed an asymmetric encryption method and put forward a new hardness assumption, which was the biological hardness assumption. Using molecules to encrypt messages provides a security different from conventional cryptography which can resist the attack methods developed in conventional cryptography and may even resist quantum computer attack [5]. However, none of these approaches to date hold an efficient transmission method. DNA can only be transported by physical methods, which are not convenient to use in practice. Microfluidics inspired us and we found fluids can be implemented for cryptography.

Recent studies in microfluidics field provided an important explanation of the interaction among droplets in the channel [6]. Microfluidic technology provides a high efficiency which is very suitable for cryptographic requirements. DNA computation and other applications using microfluidic devices have been proposed by early researchers, which show feasibility of using microfluidics in computer applications. Gehani et. al. provided a model for micro-flow based bio-molecular computation device

Content from this work may be used under the terms of the Creative Commons Attribution 3.0 licence. Any further distribution of this work must maintain attribution to the author(s) and the title of the work, journal citation and DOI.

Published under licence by IOP Publishing Ltd

[7]. Grover et. al. developed an integrated microfluidic processor that performs molecular computations using single nucleotide polymorphisms as binary digits [8]. Fuerstman et. al. studied the interaction among droplets in the channel and proposed an encoding/decoding microfluidic network [9]. We design a microfluidic selector to perform molecular encryption methods we proposed. A new hardness assumption, which is based on the nonlinearity of fluids, is provided. This method has not yet explicitly been addressed and in addition included in recently published papers. Lastly, this method is the first try to connect molecular encryption method with modern communication systems, which may provide a new insight of information manipulation and technique.

The paper is divided into two parts. The first part will present an encryption algorithm and the second part will present a designed microfluidic selector to implement this algorithm

2. Methods

2.1. Molecular encryption principle
Several terminologies used in this encryption method, which has never been explained in this field.

1) Molecular code unit. Molecular code is a new encoding scheme which provides a method to encode information in molecular form. The representation of molecular code unit depends on the chosen molecules. For example, we can use DNA triplet to represent each molecular code unit. Multiple molecular code units map a binary number. The coding scheme is the basis of our molecular encryption method. In this paper, DNA molecule is used as an example throughout this paper to explain how to use molecular code unit. Other materials can also be used for molecular code unit representation and the design of each molecular code unit can be different based on the situation.

2) Molecular code pad. The molecular code pad is an integration of all possible molecular code units. The input source can be binary bits, DNA sequences or any other molecular code form. As modern communication systems use binary bits, molecular code can be converted back to its binary form and be transmitted through existed systems. Molecular code units can be implemented using various numbers of DNA bases. For instance, if we use triplets (three DNA bases, ACT, CTG et. al), there will be 64 different triplets. These 64 triplets comprise of a molecular code pad. Then, we choose half triplets to represent 0 and others to represent 1 by using random algorithms. For example, we choose the first triplet denoted as 1 and the second as 0, then the choice continues alternatively. In this way, binary bits are encoded into DNA triplets using previously generated substitution table, which is called 0-1 substitution rule. As we can see, each 1 and 0 have 32 potential DNA triplets which means this substitution is onto, and there are $A_{64 \times 64} \approx 1.2 \times 10^{89}$ molecular code pads in the chosen scheme. Besides, each triplet has the same probability to be chosen as 0 or 1. Thus, there are $C_{64 \times 32} = 1.8 \times 10^{18}$ possibilities for 0-1 substitution. The key space is the multiplication of above, which is $A_{64 \times 64} \times C_{64 \times 32} \approx 2.3 \times 10^{107}$. The security of molecular code pad also lies in the usage of nonlinear characteristic of fluids. As the generation of molecular code pad is nonlinear, it cannot be predicted or analyzed out by mathematical analytical methods. The generation algorithm will be present later.

3) Message string unit. With further research, using only molecular code pad has some special situations, which are insecure. Suppose a sender successfully generates a molecular pad and finished choosing substitution triplets, but when he uses the substitution triplets, each binary bit (0 or 1) is substituted by the same triplet. like all 1 for ACT and 0 for CTG, then there will be only two possible cases for an eavesdropper. To prevent such situations, we create a message string unit which consists of three segments: P segment (position segment), message segment and confusing message segment as shown in Figure 1 (c). Suppose the string unit is of 256 ng. The three segment each other is 3 ng, 144 ng (6 bytes), 109 ng. The position of each three segments are not fixed, which is in order to hide the position of true secret messages. In this way, the eavesdroppers cannot get the secret messages, so they have no knowledge to analyze the statistical frequency of each character. The P segment is used to help the legal receiver to find the secret message segment and decrypt the message. It is selected by the sender and sent with the secret key. The position of message segment is not fixed, but P segment is always before message segment. P segment and message segment is connected and confusing message

segment cannot be in the middle of these two segments. In order to maximize the utility of this structure, we should generate confusing segments according to the probability of each molecular code units in secret messages. Message string unit provides necessary confusion of secret messages.

4) Molecular hardness assumption. Different from early works by other researchers, the hardness of this cryptographic method lies in the non-linearity of fluids. Different viscosity number of different fluid materials can lead to different motion velocities. Based on the non-linearity of viscosity, the output will be different from input sequences, which is the same as diffusion of messages. The condition parameters set is the secret key. The message is meaningless to an eavesdropper who has no knowledge of the condition parameters (key), and hence the correct molecular code pad cannot be generated. Besides, mathematical analysis cannot be applied to such encryption method using molecular hardness assumptions. This molecular hardness assumption is a new hardness assumption for cryptography.

2.2. Cryptographic algorithm

1) Description of algorithm. This algorithm is a one-time-pad stream cipher using molecular code scheme and microfluidic selector as a random pad generator. The input messages can be text, audio or video.

2) Encryption process. In the first step, the molecular code pad can be chosen by the sender. After the sender generates the molecular code pad, he will choose 32 triplets as 0 and others as 1. After converting input messages to molecular codes, message string units are formed according to the principle described. In the last step, the message string units are transformed back into binary format again by simply map A to 11, T to 00, C to 01 and G to 10. Finally, the plaintexts are encrypted and can be transmitted through traditional networks.

3) Decryption process. The process of decryption is essentially the same as that of the encryption like modern cryptography. What we should note is that messages are first converted to molecular form and then use the same generated molecular code pad and 0-1 substitution rule to decrypt the messages. The implementation details of generating molecular code pad are described in the key exchange process.

4) Key exchange process. This algorithm is a secret-key encryption method. The first secret key is the seed and transmitted through a secret channel. After that, the next molecular code pad will use the sequence of code pad generated last time and P segment chosen is transmitted in the secret message.

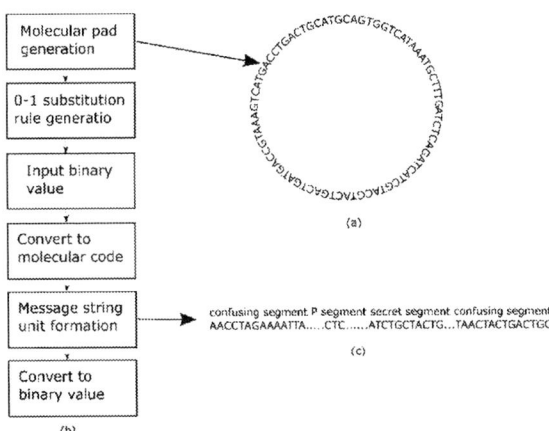

Figure 1. (a) An example of molecular code pad. In this paper, the molecular code units are DNA triplets. The 64 triplets are connected one by one. Each molecular code pad could be different. (b) Flow chart of encryption algorithm. (c) An example of message string unit.

2.3. Microfluidic device design

A key part of our method is the generation of molecular code pad. Fluids at the small length scales that are characteristic of microfluidic systems [6] and viscosity-dominated flows are governed by equation of motion that are non-linear in the velocity of the fluid [9]. With different Reynold number Re and viscosity characteristics, the non-linearity of fluids can be used for random molecular pad generation. The microfluidic platform consists of a droplet generating device as shown in Figure. 2(a) and a micro-selector as shown in Figure. 2(b). The following part shows our design of microfluidic device for encryption.

1) Droplet generating device. The chip comprises three fluid inlets connected to three micropumps to form droplets of different solutions through T-junction. The droplets are moved by a constant force of micropumps. Before pumping into the microfluidic device, each fluid has been heated in desired temperatures given by sender.

2) Microselector. The microselector incorporates three uniquely modified channels. As the resistance force is different due to environment parameters, the velocity of each droplet is different. And due to the length of the channel each droplet pumped in, the time when they are pumped out is different from the original order. Each channel has one droplet in it and another droplet can be pumped in only when the previous drop was pumped out. The process is parallelized by increasing the number of devices. In this way, the upper bound of time complexity is determined by the droplet which consumes the longest time for each device. When a droplet moves through the outlets of microselector, a sensor will distinguish it and record the sequence on computer. After all droplets move out the micro-selector, a molecular code pad is generated. The used droplets can be collected and be prepared to be used next time.

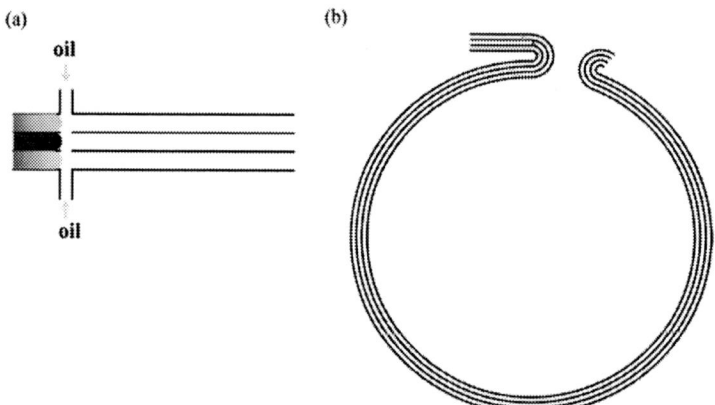

Figure 2. (a) Conceptual design of droplet generating device. The intended triplets are pumped in based on the sequence. (b) Conceptual design of microselector. The droplets move through the channels in different velocities and move out in different sequences from the origin.

3. Results and Discussion

An encryption process has been presented. In the example of this paper, each character contains 8 bits, which is 24 ng, then a secret message segment can consist of 6 characters. As for the security of this scheme, the first question is whether the eavesdropper can get the secret key. If the eavesdropper cannot get the secret key, then the system is secure. Although an eavesdropper by incidence has the correct molecular code pads, he should also know the position of secret messages, otherwise he can only use brute force attack to find the secret message. This proves although the eavesdropper obtains the secret key, he still cannot decrypt the message, which means the system is secure. The total number of key space depends on the number of molecular code pads and the number of 0-1

ISPECE IOP Publishing

substitution rules. In our example, 32 triplets are mapped to 0 and others are mapped to 1. Each triplet has the same probability to be chosen and for each encrypted word of binary length N, so there are 32^N possibilities. What is the most important is that each bit has the same probability 1/32, which satisfy perfect secrecy and One-time-pad condition in Shannon' s theory [10], and if the eavesdropper does not have the correct molecular code pad and 0-1 substitution rule, the secret message cannot be found and then be analyzed by statistical characteristic. To obtain the correct molecular pad, the eavesdropper has to obtain the secret key, which is back to the first question. Another advantage of this method is that there is no worry about that statistical pattern will be revealed as the number of messages increase, because each bit has the same probability using 0-1 substitution. Molecular cryptography has two security folds, mathematical level and physical level. Different from modern cryptography hardness assumptions, molecular cryptography has its own hardness assumptions, which is non-linearity of fluids. Existed DNA encryption methods are like the first case described in Shannon's work [10]. Our method shows how to transmit molecular data by inventing an encoding scheme using currently existed communication systems, like the Internet.

4. Conclusions and Perspectives

In this study, a microfluidic molecular encryption method has been present by using molecular code scheme. Using molecule to encrypt data is different from conventional encryption methods, and hence there is promising future. By converting encrypted messages into binary form, messages can be transmitted through today's communication systems. This encryption scheme is not based on the advantage of the vast storage capacity of DNA, but with non-linearity of fluids, more data can be processed with one molecular code pad generation. Using microfluidic chip to implement this encryption scheme can also provide high-efficiency and be low-cost. With further research of this method and with new discovery in microfluidic field, more molecular characteristics can be used to design encryption methods.

Acknowledgments

This research was supported by the Fundamental Research Funds for the Central Universities (grant No. 21617407), "Climbing Program" Special Funds (grant No. pdjhb0057), and Jinan University Funds for "Challenge Cup" National Undergraduate Curricular Academic Science and Technology Works (grant No. 17112059 &.18113037).

References

[1]   Church, G.M.; Gao, Y.; Kosuri, S. (2012) Next-Generation Digital Information Storage in DNA. Science, 337: 1628.
[2]   Clelland, C.T.; Risca, V.; Bancroft, C. (1999) Hiding Messages in DNA Microdots. Nature, 399: 533-534.
[3]   Gehani, A.; LaBean, T.; Reif, J. (2004) DNA-Based Cryptography. LNCS. 2950: 34-50.
[4]   Chang, W.L.; Guo, M.Y.; Ho, S.H. (2005) Fast Parallel Molecular Algorithms for DNA-Based Computation: Factoring Integers. IEEE T NANOBIOSCI., 4: 149-163.
[5]   Lai, X.J.; Lu, M.X.; Qin, L.; Han, J.S.; Fang, X.W. (2010) Asymmetric Encryption and Signature Method with DNA Technology. Science China Information Science. 53: 506-514.
[6]   Stone, H.A.; Stroock, A.D.; Ajdari, A. (2004) Engineering Flows In Small Devices: Microfluidics Toward a Lab-on-a-chip. Annu. Rev. Fluid. Mech. 36: 381-411.
[7]   Gehani, A.; Reif, J. (1999) Micro flow bio-molecular computation. Bio Systems. 52: 197-216.
[8]   Grover, W.H.; Mathies, R.A. (2005) An integrated microfluidic processor for single nucleotide polymorphism-based DNA computing. Lab Chip. 5: 1033-1040.
[9]   Fuerstman, M.J.; Garstecki, P.; Whitesides, G.M. (2007) Coding/Decoding and Reversibility of Droplet Trains in Microfluidic Networks. Science. 9: 828-832.

[10] Shannon C.E. (2014) Communication Theory of Secrecy Systems. Bell System Technical Journal. 28: 656-751.

ISPECE

IOP Publishing

Multi-peak and power cooperative detection algorithm to detect forwarded spoofing interference signals of BOC modulation receivers

Zhiying Wang[1], Jiawen Wang[1] and Yidong He[2*]

[1] 32021 Troops, Beijing, 100094, China

[2] Satellite Navigation Center, Beijing, 100094, China

*Corresponding author's e-mail: 1012807308 @qq.com

Abstract. In this paper, the multi-peak and power cooperative detection algorithm is proposed to detect forwarded spoofing interference signals of satellite navigation BOC modulation receivers without considering suppression interference. The theoretical value and the simulation value of the ROC curve of the multi-peak and power cooperative detection algorithm can be obtained by the numerical calculation and the experimental simulation. It can be seen from the simulation that with the increase of the number of accumulations after the receiver correlation value and the power intensity of the spoofing signal relative to the real signal, the detection probability of the proposed algorithm is higher. When the power of the spoofing signal is 3 dB greater than that of the real signal, the detection probability can be more than 90% under the condition that the false alarm probability is 5%. While when the power of the spoofing signal is 4 dB greater than that of the real signal, the detection probability can be more than 95% under the condition that the false alarm probability is 0.7%. Given that it only needs to add the module to the software receiver to detect spoofing signals, the algorithm proposed in this paper is of practical significance.

1. Introduction

The Global Navigation Satellite System (GNSS) has become indispensable infrastructures for promoting economic development and maintaining national security. With the continuous development of GNSS technology, binary offset carrier (BOC) modulation technology, which has been seen as a new modulation method, has been used in modernization schemes of GNSS. Moreover, the vulnerability and potential risks of BOC modulation technology have attracted widespread attention. In 2011, Iran captured the US military unmanned reconnaissance aircraft RQ-170 successfully [1], which stimulated the US military to accelerate research on anti-spoofing interference technology greatly. In the GNSS conference organized by the Institute of Navigation (ION) in 2011, there were 18 articles on spoofing and anti-spoofing technology of GNSS, which was the sum of the number of similar research articles in the ION conference in the past 10 years [2]. In 2012, the US Naval Research Office and Rockwell Collins reached the Modern Integrated Spoofing Tracking (MIST) contract to research anti-spoofing technology, which was against attempts by the enemy to disrupt military operations by interfering with GPS signals [3]. Therefore, it can be seen that spoofing and anti-spoofing technology have attracted more and more attention.

Spoofing detection technology is the first step and a key step in the research on anti-spoofing technology. At present, the main anti-spoofing methods include signal encryption authentication

---

Content from this work may be used under the terms of the Creative Commons Attribution 3.0 licence. Any further distribution of this work must maintain attribution to the author(s) and the title of the work, journal citation and DOI.

Published under licence by IOP Publishing Ltd

detection [4], signal arrival time detection [5], signal arrival angle detection [6], signal power detection [7], multi-peak detection [8] and so on. Each detection method has its own shortcomings and applicable scenarios. Meanwhile, there are deficiencies in using only one single detection method. In this paper, the multi-peak and power cooperative detection algorithm is proposed to detect forwarded spoofing interference signals of BOC modulation receivers without considering suppression interference. At the same time, the calculation method of the decision threshold and the corresponding ROC curves are given.

2. Algorithm principle

Forwarded spoofing interference is to receive the real satellite signal through its own antenna, delay appropriately, amplify and transmit to the target receiver to interfere with it. Since the pseudo codes and the navigation messages contained in the signal are all from real navigation satellites, there is no need to know the pseudo code sequence and the navigation message structure of the navigation signal received by the target receiver. Forwarded spoofing interference can deceive many satellite navigation receivers effectively, including military receivers. The principle of forwarded spoofing interference is shown in Figure 1.

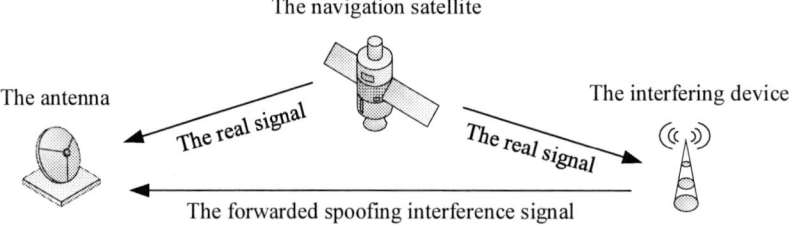

Figure 1. The principle of forwarded spoofing interference.

The multi-peak and power cooperative detection algorithm combines the multi-peak detection method and the signal power detection method to detect forwarded spoofing interference of BOC modulation receivers. It works in the capture phase of the pseudo code of the navigation receiver. If there is no spoofing signal, when the capture threshold is set appropriately, there will be only one correlation peak higher than it typically. Conversely, if there is a spoofing signal, there will be multiple correlation peaks higher than the threshold. It can be seen from the difference between the two cases that when several correlation peaks which are larger than the threshold are detected, at least one spoofing signal exists in the currently received signals. At this time, the spoofing signal can be detected for the power of the spoofing signal is greater than that of the real signal.

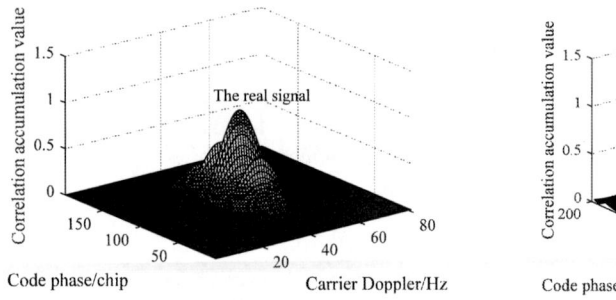

Figure 2. The real signal in the capture phase.

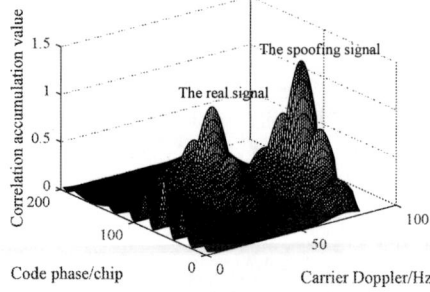

Figure 3. The spoofing signal in the capture phase.

If only one signal was detected during the two-dimensional search including carrier Doppler and code phase, it indicates that the signal should be a real satellite signal, and the receiver should track the signal subsequently, as shown in figure 2. If multiple correlation peaks were detected, there would be a spoofing interference signal in the current signal, as shown in figure 3.

Therefore, the following ternary detection problem can be constructed.

H_0: There is no signal at the current carrier Doppler and code phase search cell.

H_1: There is a real signal at the current carrier Doppler and code phase search cell.

H_2: There is a spoofing signal at the current carrier Doppler and code phase search cell.

By constructing the ternary detection problem mentioned above, it is ensured that the receiver track the real signal normally, thereby preventing forwarded spoofing interference from influencing the final positioning result of the receiver.

3. Detection method

In the process of capturing the satellite signal by the receiver, the received signal is mixed with the sinusoidal replica carrier on the same branch and the cosine replica carrier on the orthogonal branch of the same receiving channel firstly. Then, the mixing result is correlated with the copied C/A code to obtain correlation results named i and q. The correlation results i and q generate I and Q after the coherent integration of time T_c. Finally, I and Q are subjected to non-coherent integration calculation to obtain the amplitude A of the incoherent integration. If there was a forwarded spoofing interference signal, its power would be higher than that of the real one in order to achieve deception. Therefore, the value of A^2 can be used to determine whether there is a forwarded spoofing interference signal. If A^2 was less than the threshold ρ_1, it indicates that the signal would not have been searched yet. If A^2 was greater than the threshold ρ_1 and less than the threshold ρ_2, then the real signal should be in the search cell. If A^2 was greater than the threshold ρ_2, then the forwarded spoofing interference signal should be in the search cell. Figure 4 shows the decision process of the forwarded spoofing interference signal.

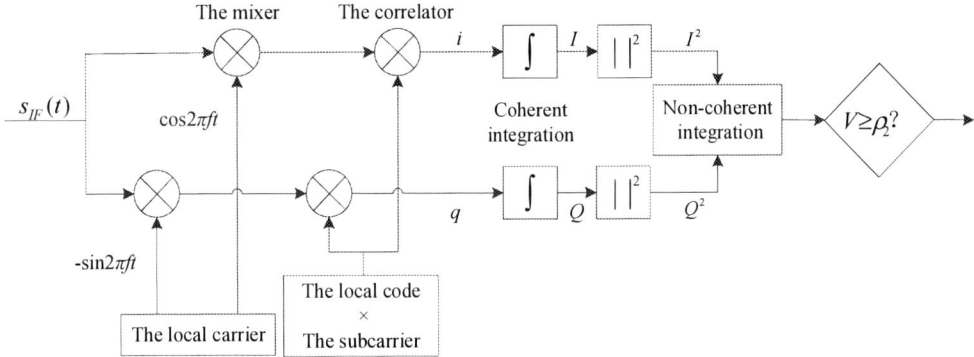

Figure 4. The decision process of the forwarded spoofing interference signal.

It is assumed that the difference of the code phase of the correlation peak between the spoofing signal and the real signal is significant. During the signal capture phase, the receiver despreads each of the carrier Doppler and code phase search cells, and the correlator output is

$$x_n = \frac{1}{T_c} \int_{nT_c}^{(n+1)T_c} s_{IF}(t) s_0(t)^* \, dt \tag{1}$$

In equation (1), $s_{IF}(t)$ is the received signal, $s_0(t)$ is the reference signal of the receiver, $t \in \left[nT_c, (n+1)T_c \right]$.

Then the above three assumptions can be expressed as

$$
\begin{aligned}
H_0 &: x_n = \frac{1}{T_c} \int_{nT_c}^{(n+1)T_c} w(t) s_0(t)^* \, dt \approx w_n \\
H_1 &: x_n = \frac{1}{T_c} \int_{nT_c}^{(n+1)T_c} A^{(1)}(t) \, dt + \frac{1}{T_c} \int_{nT_c}^{(n+1)T_c} w(t) s_0(t)^* \, dt \approx s_n^{(1)} + w_n \\
H_2 &: x_n = \frac{1}{T_c} \int_{nT_c}^{(n+1)T_c} A^{(2)}(t) \, dt + \frac{1}{T_c} \int_{nT_c}^{(n+1)T_c} w(t) s_0(t)^* \, dt \approx s_n^{(2)} + w_n
\end{aligned}
\tag{2}
$$

In equation (2), the normalization condition $\int_{nT_c}^{(n+1)T_c} |s_0(t)|^2\,dt = 1$, $A^{(1)}$ represents the channel gain of the real signal, $A^{(2)}$ represents the channel gain of the forwarded spoofing interference signal, and w_n represents the noise component.

The output of the correlator follows a normal distribution under the above assumptions

$$x_n|H_0 \approx w_n \sim N\left(0,\sigma_n^2\right)$$
$$x_n|H_1 \approx s_n^{(1)} + w_n \sim N\left(s_n^{(1)},\sigma_n^2\right) \qquad (3)$$
$$x_n|H_2 \approx s_n^{(2)} + w_n \sim N\left(s_n^{(2)},\sigma_n^2\right)$$

The forwarded spoofing interference signal power is higher than the real signal. According to the statistical signal theory, in the additive Gaussian noise environment, the sufficient statistics of the above ternary hypothesis test can be expressed as

$$V = h(x) = \mathbf{X}^H \mathbf{X} = \sum_{n=1}^{N} x_n^2 \qquad (4)$$

According to the assumption of the model, under the H_0 condition, the test statistic V follows the centralized Chi square distribution, while under the conditions of H_1 and H_2, the test statistic V follows the decentralized Chi square distribution.

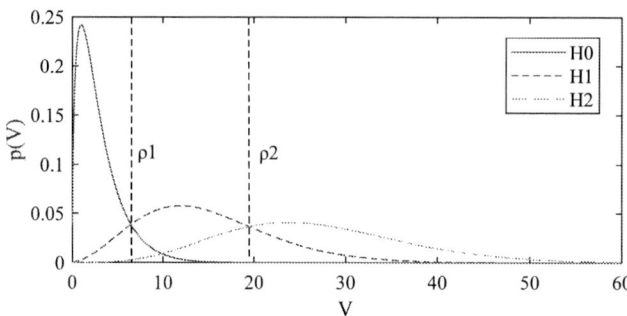

Figure 5. The probability distribution function of the test statistic under different assumptions.

By comparing with the detection thresholds ρ_1 and ρ_2, the decision is made in the ternary hypothesis in equation (7), as shown in figure 5.

If the energy of the correlation value is lower than ρ_1, there should be no signal. If the energy of the correlation value is higher than ρ_2, there should be a forwarded spoofing interference signal. If the energy of the correlation value is between ρ_1 and ρ_2, there should be a real signal.

Therefore, two types of false alarm probabilities are defined. The first type of false alarm is that a cell without a signal is detected as a cell containing a real signal or a spoofing signal. The second type of false alarm is that a cell containing a real signal is detected as a cell containing a spoofing signal.

In order not to lose generality, it is assumed that $\rho_2 > \rho_1$, then the threshold ρ_1 determines the first type of false alarm probability P_{FA1}, which can be expressed as

$$P_{FA1} = P\{V > \rho_1|H_0\} \cup P\{V > \rho_2|H_0\} = \int_{\rho_1}^{\infty} f_{V|H_0}(V)\,dV \qquad (5)$$

So ρ_1 can be expressed as

$$\rho_1 = F_{V|H_0}^{-1}\left(1 - P_{FA1}\right) \qquad (6)$$

The threshold ρ_2 determines the second type of false alarm probability P_{FA2}, which can be expressed as

$$P_{FA2} = P\{V > \rho_2|H_1\} = \int_{\rho_2}^{\infty} f_{V|H_1}(V)\,dV \qquad (7)$$

So ρ_2 can be expressed as

$$\rho_2 = F_{V|H_1}^{-1}\left(1 - P_{FA2}\right) \tag{8}$$

The detection probability of the spoofing signal can be expressed as

$$P_{D2} = \int_{\rho_2}^{\infty} f_{V|H_2}\left(V\right)\mathrm{d}V \tag{9}$$

In equation (5) ~ (9), the $f_{V|H_i}\left(\cdot\right)$ represents the probability density function of the random variable V under the assumption of H_i[9], $F_{V|H_i}\left(\cdot\right)$ represents the cumulative distribution function of the random variable V under the assumption of H_i.

4. Simulation verification

According to the analysis above, the false alarm, the threshold and the detection probability have been expressed. By integrating the probability density function, the cumulative distribution function can be obtained. Substituting the cumulative distribution function of the random variable V under the assumption of H_0 into equation (6) can obtain the threshold ρ_1 for signal detection.

According to the literature, the power of the received satellite navigation signals is about the same when signals reach the ground, while the carrier-to-noise ratio is in the range of 35~55 dB·Hz roughly [10]. Therefore, for a capture process using a coherent integration with an integration step size of 1 ms, the signal-to-noise ratio of the correlator output result is in the range of 5 to 25 dB. Given that the prior information of the power of the real satellite signals and the second type of false alarm probability P_{FA2}, the threshold ρ_2 for spoofing signal detection can be obtained.

Through the above analysis, the probability density function and the cumulative distribution function of the test statistic V under each hypothesis are obtained. The theoretical value and simulation value of the ROC curve of the multi-peak and power cooperative detection algorithm can be obtained by numerical calculation and experimental simulation.

Table 1. The simulation parameter settings.

Parameter	Real signal	Spoofing signal
Pre-detection integration time / ms	1	1
Carrier-to-Noise Ratio / dB·Hz	40	43, 44
Signal to noise ratio / dB	10	13, 14
Degree of freedom (N)	1, 3, 5	1, 3, 5

In order to verify the effectiveness of the proposed algorithm in this paper, 1000 Monte Carlo simulations were performed using MATLAB software. The simulation parameter settings are shown in table 1. Figure 6 and figure 7 show the ROC curves of different spoofing signal-to-noise ratios and post-accumulation times under the condition that the signal-to-noise ratio of the real signal at the output of the receiver correlator is 10 dB.

Figure 6 shows the relationship between the false alarm probability and the detection probability when the signal-to-noise ratio of the real signal is 10 dB, the signal-to-noise ratio of the spoofing signal is 13 dB, and the degrees of freedom are 1, 3, and 5 respectively. It can be seen from figure 6 that in the case of the same false alarm probability, the detection probability of the spoofing signal increases as the number of accumulations after the receiver correlation value increases. When the false alarm probability is 5%, the detection probability is higher than 90%.

Figure 7 shows the relationship between the false alarm probability and the detection probability when the signal-to-noise ratio of the real signal is 10 dB, the signal-to-noise ratio of the spoofing signal is 14 dB, and the degrees of freedom are 1, 3, and 5 respectively. It can be seen from figure 7 that when the false alarm probability is 0.7%, the detection probability is higher than 95%.

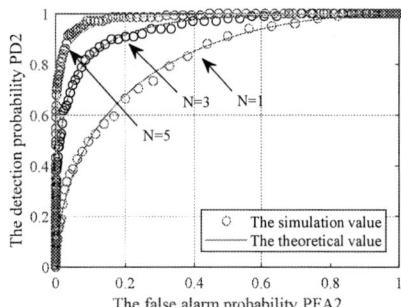

Figure 6. The spoofing detection ROC curve when the spoofing signal-to-noise ratio is 13 dB.

Figure 7. The spoofing detection ROC curve when the spoofing signal-to-noise ratio is 14 dB.

By comparing figure 6 and figure 7, it can be seen that when the number of accumulations after the receiver correlation value and the false alarm probability are the same, the detection probability of the spoofing signal increases as the signal-to-noise ratio of the spoofing signal increases.

In summary, with the increase of the number of accumulations after the receiver correlation value and the power intensity of the spoofing signal relative to the real signal, the detection probability of the forwarded spoofing interference signal detection algorithm proposed in this paper is higher. When the power of the spoofing signal is 3 dB greater than that of the real signal, the detection probability can be more than 90% under the condition that the false alarm probability is 5%. When the power of the spoofing signal is 4 dB greater than that of the real signal, the detection probability can be more than 95% under the condition that the false alarm probability is 0.7%.

5. Conclusion

In this paper, the multi-peak and power cooperative detection algorithm is proposed to detect forwarded spoofing interference signals of BOC modulation receivers without considering suppression interference. When using this algorithm, there is no need to increase the hardware facilities of the receiver. What needs to be done is to add the corresponding module in the signal processing part of the software receiver to detect the spoofing signal, which is of practical significance. However, suppression interference is not considered in this paper. If suppression interference and forwarded spoofing interference were used at the same time, the noise floor of the receiver would be raised and only one peak would appear in the signal capture phase after the RF front-end AGC control. At this time, it is difficult to distinguish whether the peak represents a spoofing signal or a real signal. The above issue needs to be researched in the next step.

Acknowledgments

This work was supported by the National Natural Science Foundation of China (Nos. 61601485).

References

[1] Wullems, C.J. (2012) A spoofing detection method for civilian L1 GPS and the E1-B Galileo safety of life service. IEEE Transactions on Aerospace and Electronic Systems, 48: 2849–2864.

[2] Wesson, K., Shepard, D., Humphreys, T. (2012) Straight talk on anti-spoofing: Securing the future of PNT. GPS World, 23: 32–34.

[3] China News Network. (2012) US company will develop technology for detecting GPS interference and deception for the US Navy. http://www.chinanews.com/mil/2012/10-18/4257529.shtml.

[4]   Humphreys, T.E., Ledvina, B.M., Psiaki, M.L. (2008) Assessing the Spoofing Threat Development of a Portable GPS Civilian Spoofer. In: ION GNSS Conference. Savannah. pp. 165–176.

[5]   Jafarnia, A., Broumandan, A., Nielsen, J. Lachapelle, G. (2012) GPS spoofer countermeasure effectiveness based on signal strength, noise power, and C/N0 measurements. International Journal of Satellite Communications and Networking, 30: 1002–1012.

[6]   Psiaki, M.L., O'Hanlon, B.W., Powell, S.P., Bhatti, J.A., Wesson, K.D., Humphreys, T.E. (2014) GNSS spoofing detection using two-antenna differential carrier phase. Programme & Abstracts the Volcanological Society of Japan, 5: 342–344.

[7]   Humphreys, T.E., Ledvina, B.M., Psiaki, M.L. (2015) Research on GNSS Receiver Spoofing Detection Technology. In: China Satellite Navigation Conference. Xi'an. pp. 1–9.

[8]   Yuan, D., Li, H., Lu, M. (2014) A method for GNSS spoofing detection based on sequential probability ratio test. In: ION Position. Monterey. pp. 351–358.

[9]   Luo, P., Zhang, W. (2014) Basics of Statistical Signal Processing: Estimation and Detection Theory. Publishing House of Electronics Industry, Beijing.

[10] Xie, G. (2009) Principles of GPS and Receiver Design. Publishing House of Electronics Industry, Beijing.

ISPECE   IOP Publishing

IOP Conf. Series: Journal of Physics: Conf. Series **1187** (2019) 042054   doi:10.1088/1742-6596/1187/4/042054

Construction of on-Campus 3D Model based on GIS Technology and OpenGL

Limei Chen*

Department of Information Engineering, Heilongjiang International University, Harbin, 150080, China

*Corresponding author's e-mail:3140025490@qq.com

Abstract. This paper mainly uses professional measurement technology and national standard to carry out actual measurement on the building complex, green vegetation, intra-college roads and other irregular building groups on the campus and realizes data collection and accurate analysis of data. The AutoCAD software is used to draw the map, and the visualization of the campus 3D model is realized through OpenGL. The feasibility of data extraction and related technology is verified.

1. Introduction
In the creation of 3D model of 3D campus based on GIS, the collection of basic data and the accurate analysis of data are very important parts. For any school, the school's floor space is limited [1-3]. In a limited space, both the demand and the humanistic aesthetics must be met. At the same time, the roads on the campus are relatively complex, some buildings are densely populated, and there are a lot of green and landscape belts on the campus [4-6]. Since any element in the space of a limited campus is important, the elements larger than 10 cm on the campus are collected accordingly. Therefore, in practice, this part of the work is also relatively labor-intensive.

2. Measurement and data acquisition

2.1. Plane Control Measurement
(1) The plane control coordinate reference is designed as an independent coordinate system, which requires the vertical axis of the coordinate to be consistent with the direction of the campus center road. The advantage of this approach is that it can make the drawing square and beneficial to the visual effect. (2) The measurement method of the planar control network, uses the Leica (2 seconds, 2mm + 2ppm) total station by measuring the photoelectric distance measuring wire. Both the horizontal angle and the distance are measured in one direction, and one measurement is taken in 1 round [7].

The layout of the plane control points is carried out by means of attaching wires, and the number of attachments is not more than two times. The method of supporting the wires is adopted in places where it is not conducive to the laying of the attached wires. Meanwhile, the control points that are resolved as required shall not be less than those shown in Table 1. In this project, the total area measured on the campus is 0.29 square kilometers, and a total of 26 control points are arranged to meet the requirements in the specification [8].

Content from this work may be used under the terms of the Creative Commons Attribution 3.0 licence. Any further distribution of this work must maintain attribution to the author(s) and the title of the work, journal citation and DOI.
Published under licence by IOP Publishing Ltd
1636

Table 1. Analysis of the Number of Control Points

Mapping Scale	Frame Size (cm)	Analysis of the Control Points (number)
1: 500	50×50	8
1: 1000	50×50	12
1: 2000	50×50	15

Note: ①The points listed in the table refer to all analytical control points available when the map is tested;

②When an electronic speedometer is used to measure the map, the number of control points can be appropriately reduced.

(3) The accuracy requirement of the plane control point is the error in the position relative to the adjacent level control point and should not be greater than 0.1 mm on the map. The main technical requirements for planar conductor measurements are performed in accordance with Table 2. When the length of the attached wire is less than 1/3M, the absolute closing difference should not be greater than 0.3mm on the drawing.

Table 2. Main Technical Requirements for Wire Measurement

Grade	Wire length (km)	Relative closure of the full length of the wire	Side length	Error in angle measurement (")	Error in ranging (mm)	DJ2 Number of Rounds			Azimuth closure difference (")
						1'' class instrument	2'' class instrument	6'' class instrument	
Grade III	14	≤1/55000	3	1.8	20	6	10		$3.6\sqrt{n}$
Grade IV	9	≤1/35000	1.5	2.5	18	4	6		$5\sqrt{n}$
Grade I	4	≤1/15000	0.5	5	15		2	4	$10\sqrt{n}$
Grade II	2.4	≤1/10000	0.25	8	15		1	3	$16\sqrt{n}$
Grade III	1.2	≤1/5000	0.1	12	15		1	2	$24\sqrt{n}$

Note: ①N in the table is the number of stations; ②When the maximum scale of the survey area is 1:1000, the average side length and total length of the first, second and third conductors may be appropriately lengthened, but the maximum length shall not be greater than 2 times the length specified in the table; ③The 1'', 2'', and 6'' instruments of the angle measurement include a total station, an electronic theodolite, and an optical theodolite.

The root of the wire used to determine the detail point should not be more than 25cm; when the length of the wire is less than 1/3 of the height specified in the table, the absolute closure difference is no more than 13cm. When the figure root, wire is laid as a branch wire, the horizontal angle can be

measured by the DJ[6] type theodolite to measure the left and right corners, and the circumferential angle closure difference should not exceed 40''. The relative length error of the side length measurement should not be greater than 1/3000. The average side length and number of sides of the wire shall not exceed the requirements of Table 3.

Table 3. The Average Side Length and Number of Sides of the Tugen Branch Wire

Mapping Scale	Average Side Length (m)	Number of Wire Sides
1: 500	100	2

| 1: 1000 | 150 | 2 |
| 1: 2000 | 250 | 3 |

For the root of the graph, you can choose the angle measurement, side intersection, and internal and external points. When the side and angle intersection method is selected, the intersection angle should be the same as the graph root wire between 30 degrees and 150 degrees. The coordinates calculated by the grouping should not be larger than 0.2mm.

In the actual campus measurement, a total of 4 wires were laid, and the difficult-to-distribute area increased the branch wires. The error of the weakest point of the wire was 5cmm, and the relative error of the full-length was 1/10000, which met the requirements of the specification. Some of the control points of the field are shown in Table 4.

Table 4. Date of Some Control Points

Control point number	x coordinate	y coordinate	z coordinate
A01	100.000	100.000	101.065
A02	216.098	226.687	100.629
A03	222.233	100.544	98.327
A04	108.427	217.432	99.078

2.2. Elevation Control Measurement

(1)The elevation control reference section uses a hypothetical elevation system. (2) The choice of the elevation control measurement method, this part uses the leveling and electromagnetic wave ranging triangle elevation measurement. Among them, the leveling measurement adopts Topology Kang electronic level 1mm/km, and observation should be carried out according to the fourth level. Among them, the electromagnetic wave distance triangulation elevation measurement adopts the triangular elevation measurement with the accuracy of 2 seconds, 2mm+2ppm total station instrument, and the opposite observation method and the vertical angle observation three measurements. (3) The elevation control network is arranged, and the accuracy and reliability of the measurement are usually convenient in the work, and the parts are all set to be attached. (4) For the accuracy requirements of the elevation control network, it is carried out according to the technical indicators of the fourth level measurement, as shown in Table 5.

Table 5.Main Technical Requirements for Fourth-class Leveling

Instrument type	1km height difference error (mm)	Attachment route length (km)	Line of sight length (m)	Number of observations		Attachment or loop closure difference (mm)	
				Joint test with known points	Attach or close route	Flat land	Mountain
DS3	10	≤16	≤80	Round trip once	Go once	$20\sqrt{L}$	$6\sqrt{n}$

Note: The length of the horizontal path (km) of the L attachment or loop. When the level line is laid as a branch line, round-trip observation should be used, and the line length should not exceed 5km. The fourth-order electromagnetic wave ranging triangle elevation, the vertical angle can be used in the

DJ[6]-type total station in the wire method.For the measurement, the difference between the index difference and the vertical angle should not be greater than 10''. The height of the instrument and the height of the station mark should be accurate to 1mm. Adhesion or loop closure difference should not be greater than $20\sqrt{\sum D}$ mm.

Note: D is the length of the electromagnetic wave measuring edge (km). The fourth-level elevation control adopts the level-adjusted route. The error of the height difference per kilometer is 7cm, and the closure or loop closure difference is 15cm, which meets the requirements of the specification.

2.3. Total Station Fragmentation Data Acquisition

The Leica TS50 and TM50 total stations are used to collect the total station data in the field. The advantage is that the performance is stable, and the data collected are reliable.

(1)For the measurement method of the topographic map, a side finder method or a surveying method can be used, and a range finder is used. For all kinds of sketches such as features and features, they should be collected by coding method respectively. For measuring points, horizontal angles and vertical angles, they should be accurate to 1'; zero return inspection is not more than 1.5'. The maximum distance measurement is in accordance with Table 6.

Table 6. Maximum Ranging Length

Scale	Contour Height (m)	Maximum Ranging Length (m)
1:500	0.5	300
1:1000	0.5	500
1:2000	1	700

(2)In this project, computer-aided mapping was used in the industry. When the root of the analytical map failed to meet the measurement, a small number of graphical intersections or line-of-sight fulcrums were added, and the following requirements were made. When plotting the intersection point, select the redundant direction for checking. The diameter of the inscribed circle of the intersection error triangle should be no more than 0.5mm, and the angle of intersection of the adjacent two lines should be between 30 degrees and 150 degrees. The side length of the line of sight fulcrum should be less than or equal to two-thirds of the maximum line-of-sight length of the corresponding scale topographic point, and the method of using the back-and-down line of sight is less than 1/150; the vertical angle of the elevation of the graphical intersection and the line of sight fulcrum uses a round, and the round-trip height difference is less than 1/5 of the contour.

(3)The building groups on the campus and the main ancillary equipment were also mapped. According to the needs of the use of the map, the content and its trade-offs and appropriate integration.

(4) In the actual measurement of the project, the independent objects can be expressed according to the scale, the measured outline, and the symbols are filled; When it cannot be expressed in scale, it accurately indicates its positioning point or positioning line. When the line is dense, the key points have been selected for mapping. Appropriate trade-offs were made when the poles and ancillary facilities were dense. The roads and appendages on the campus, as well as the green plants on the campus, have also been actually mapped.

2.4. Method to Obtain the Data of the Campus Grounds

Thanks to the very wide range of measurement tasks in the application in the total station, it is convenient for working in the field. Three measurements are used in the actual data collected in this project, including height difference measurement, free station measurement and measurement. The measurement procedure flow is (1) input station data and back-sight point data for total station orientation. Sketch the actual scene while recording. The coding method is used for data acquisition. Next, using the resection method, three known points are used to calculate the coordinate data of the measured station through the resection procedure. The result of the final measurement obtained is that the coordinates of the survey site and the position of the survey site obtain the accuracy of the total station orientation to provide a basis for the subsequent accuracy assessment. (2) Get the method of height difference data. This section uses the measurement procedure of the elevation measurement to perform data acquisition and measures the different heights above the building. In the actual measurement, the actual measurement of the intersection method is applied to the height difference that can be hardly measured. The result of this can meet the needs of the collection of different height difference data.

2.5. Total Station Error Calibration

In order to use the accuracy of the collected data, the accuracy index of the total station is usually verified in advance, and the calibration has the measurement of the collimation error and the error measurement of the index difference. The instrument alignment error is due to the error caused by the instrument's horizontal axis and the collimation axis are not perpendicular. The effect of the collimation error on the horizontal angle error increases as the vertical angle increases. The horizontal aiming error and the collimation difference of the horizontal angle are the same. When the line of sight is horizontal, the vertical dial accuracy reading should be 90 degrees. The deviation from this number is called the vertical angle index difference. The process of the measurement is to perform the centering of the total station, aiming at a position of about 150 meters, and the angle of the vertical angle should be less than 5 degrees. Observe by using the face-to-face and right-hand method to compare the data on the face of the face and the right.

3. Analysis of Errors in the Collected Data

(1)Error analysis of plane data of measuring points. The mathematical model of the built-in plane measurement program for the full combat instrument is:

$$X = X_0 + SgCOSa$$
$$Y = Y_0 + SgSINa \tag{1}$$

X and Y represent coordinates; X0 and Y0 represent known point coordinates; S represents a horizontal distance, and α represents an azimuth angle. After applying the error propagation law to the two ends of the above formula, the error propagation law is obtained:

$$MX2 = COS2agMS2 + (SgSINa/\rho)2Ma2$$
$$MY2 = SIN2agMS2 + (SgCOSa/\rho)2Ma2 \tag{2}$$

M indicates the measurement point error

$$M2measuremen = MX2 + MY2 = MS2 + (a/\rho)2Ma2$$

In the formula: the ranging accuracy of the full combat instrument is MS=2mm+2ppm, the maximum ranging length is S≤100m, the azimuth error is Mα≤30 seconds, ρ=206265, so:
$M2measurement = 216.4mm$

The accuracy of the plane position of the measuring point is related to the position of the prism. The plane position error of the prism M cannot be greater than 100 mm, so the average error M of the plane position of the point should be suitable for the specified 200 mm.

$$M2plane = M2measurement + M2square$$

$$Mplane = \pm 101mm$$

(2)Error analysis of elevation data of measuring points. Mathematical model of the built-in elevation measurement program for the full combat instrument:

$$H = H0 + I - V + SgCOS\beta \tag{3}$$

H represents the elevation of the collection point; H0 represents the elevation of the measurement site; I represents the instrument height of the measurement site; V represents the mirror height; S represents the slope distance; and β represents the vertical angle of the zenith distance.

Applying the law of error propagation after differentiating the two ends of the above formula:

$$M2H = M2I + M2V + COS2\beta gM2S + (SgSIN\beta/\rho)2M2\beta H \tag{4}$$

MI and MV should be no greater than 3mm；β should be about 88 degrees；MS=±2mm+2ppm；S≤100m；Mβ≤30 seconds.Substituting the above formula:

$$MH = \pm15.1mm$$

In actual work, the elevation position error M of the prism point is less than 50 mm, and the height error of the elevation position of the point is M:

$$M2\ high = M2H + M2\ high\ square$$
$$M\ high = \pm52mm$$

It can meet the 70mm requirements of the specification.

(3)The mathematical model of the total warhead height difference measurement is:

$$H_{\text{Height difference}} = S(ctg\beta2 - ctg\beta1) + V \tag{5}$$

In formula (1), $H_{\text{Height difference}}$ is the elevation difference of the measurement point; S indicates the horizontal distance from the measurement site to the target point; β1 indicates the vertical angle of the zenith distance when measuring the prism; β2 represents the vertical angle of the zenith distance when measuring the height difference; V represents the height of the prism.

Differentiate the two ends of the above equation, and then apply the law of error propagation to get the formula 2:

$$M2H_{\text{Height difference}} = ctg2\beta2M2S + S2csc2\beta2M2\beta2/\rho2 + ctg2\beta1M2S +$$
$$S2csc2\beta1M2\beta1/\rho2 + M2V \tag{6}$$

Considering $M\beta1 = M\beta2 = M\beta$, it can be concluded:

$$M2H_{\text{Height difference}} = (ctg2\beta1 + ctg2\beta2)M2S + (S/\rho)2(csc2\beta1 + csc2\beta2)M2\beta + M2V \tag{7}$$

β1is usually about 88 degrees, and β2 is usually about 50 degrees；S≤100m；Mβ≤30 seconds；

$$Ms = 2mm + 2ppm$$

Put the value of M_S in formula 3 to obtain after calculation:

$$MH_{\text{Height difference}} = \pm24.2mm$$

Due to the need to consider the height position error of the prism point (should be ≤ 50mm) and the aiming error of the target point (must be ≤ 30mm), the overall error of the target point can be solved is 63mm, in accordance with the requirements of the industry standard (≤ 100mm).

4. Data Diagram

(1) The setting of communication parameters, the total station standard is set to 19200 bits, 8 data bits, no parity, 1 stop bit, carriage return and line feed. The selected baud rate data is a transmission rate of 19,200 bits/second. The transmitted data bits are 8 bits. The stop bit can be set to 1 bit.

(2) The data transmission selects the data output format of the southern cass software and uses the serial port mode to import the data of the total station to the computer, and the transmission data mode is not checked.

(3) In the industry, there are many digital mapping softwares on the mapping market, which are derived from the total station to the data format required by the southern cass. At the same time, the data is imported into the southern cass for internal production and editing, and the effect of the drawing is shown in Figure 1.

ISPECE   IOP Publishing

IOP Conf. Series: Journal of Physics: Conf. Series **1187** (2019) 042054   doi:10.1088/1742-6596/1187/4/042054

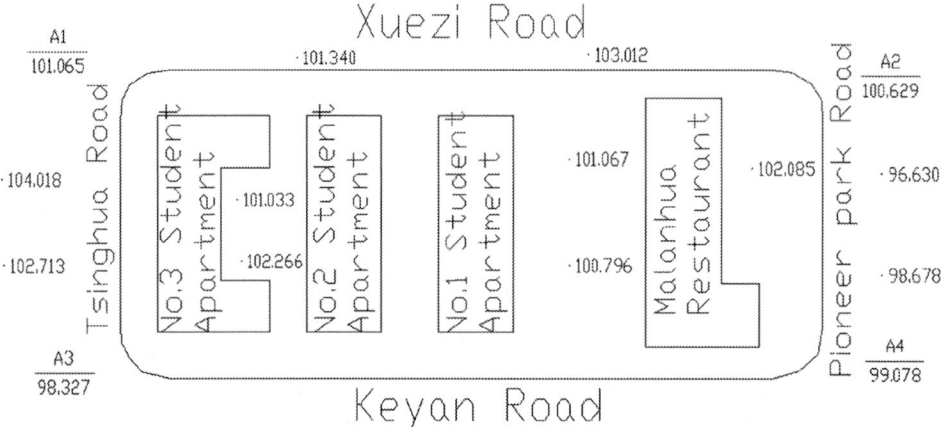

Figure 1. Diagram of Some Buildings of the School' GIS

5. Process Created by OpenGL 3D Model

(1)The OpenGL color mode, color digits, depth digits, drawing style, etc. can be created; 3D models of geometric elements such as points, lines, polygons, images, etc. can be created; the stage set in 3D space can be set, and choose to observe the scene. When creating a 3D object model, OpenGL can set the material and optical properties of the object. For example, you can set the color of the surface of the texture mapping object and increase the conditions for setting lighting and lighting. The mathematical model of the three-dimensional object is converted into pixel information that can be displayed on the computer by effect processing such as scenes and color information.

(2)Accurate coordinate positioning values are very important for the system to create 3D models. In general, the three modes of relative coordinates, absolute coordinates, and polar coordinates can be used, and the actual work needs depend on the situation. When using OpenGL for 3D modeling and drawing graphic images, you can mix and match them. In fact, the UCS coordinate system is often used for OpenGL modeling. UCS can create different coordinate systems. The origin and coordinate axes of the coordinate system can be changed by setting the command UCS\3POINT. The positive direction of the z-axis in the coordinate system is determined following the right-hand principle.

(3)After the user coordinate system is determined, OpenGL can be used to draw the 3D solid model.

Drawing of rule entities. For the two-dimensional large scale of the student dormitory on the campus, the SOLID\EXTRUDE command can be used to stretch the building in a certain direction according to the property of the building and stretch according to the actual situation of different student dormitory. The Extrude surface command can be used for the stretching of the wall; the drawing of the solid with the upper and lower end faces. Normally, the Extrude command is used when drawing entities with upper and lower end faces. This command can be used to stretch any closed object to produce different effects. If it is stretched on a closed circular object, the surface will be created while maintaining an opening above the surface; For the drawing of the form and door in the campus, they can be regarded as the geometric elements of the rule, and the drawing method is relatively simple. You can create a 3D composite by Boolean merging by creating, stretching, and Boolean 2D planar graphics, and then grouping the generated composites to get the desired effect. The fences, fences, etc. in this project are all made by such methods; for the drawing of the roof, on the basis of the closed polylines that have been established, various types of roofs can be painted with the area (REGION). When drawing a feature class, you can use CAD to draw the feature class into a point symbol. In the actual drawing process, Ken will fail to stretch. For example, due to the unconstrained normalization of the closed elements, the position of the vertices may be crossed when the drawing is not closed, or the opening may occur at the intersection of the respective faces. For this reason, it is best to use the corresponding commands to unify the map before stretching.This ensures that the relatively closed area is generated, and the cross lines between the vertices can also be removed. After the stretching of the solid, the height angle and the horizontal angle of the three-dimensional view can be adjusted to obtain a better three-dimensional effect map.

(4)The 3D model map is generated. By rendering the three-bit model, for the geographic information coordinates of the three-dimensional space, OpenGL can be obtained from the CAD interaction. The extraction process is that the points correspond to each other, and the geographic coordinate values are extracted one by one according to the corresponding order. The direction of extraction can be either clockwise or counterclockwise. A three-dimensional model diagram as shown in Figures 2 and 3 obtained by rendering and corresponding program of points.

Figure 2.Front View of 3D Model of Main Teaching Building

Figure 3. Side View of the 3D Model of the Main Teaching Building

6. Conclusion

This paper realizes the creation of a visual 3D model of the campus based on GIS and collects the data of the whole campus. The methods of acquiring various geophysical elements in GIS and the specific accuracy requirements in the process of collecting data are expounded in detail. At the same time, the error analysis of external collection point data is summarized, and the 3D model of campus GIS is realized by OpenGL. The application of the various physical elements of campus GIS is realized, and the creation of a 3D model based on GIS is finally realized, which lays a good foundation for the realization of campus digitalization in the future.

Acknowledgments

This work was supported by the youth special topic of Heilongjiang provincial education department (GBD1317104).

References

[1]   Liu Pingping, Lu Zhaopan, Gao Wuqi. (2018)Research on Visualization of 3D Campus Roaming System Based on OpenGL. Computer Technology and Development,4:18-22

[2]   Xiao Jian, Wei Xiong, Wang Renbo. (2016)Implementation and Significance of 3D Scene Simulation of Large Building Based on OpenGL. Electronic Quality. 11:34-56

[3]   Liu Xiaoying, Zhang Jian. (2014) Design and Implementation of 3D Campus Virtual Reality Platform.. Hubei Agricultural Sciences, (12):56-60

[4]   Jiang Hanqing, Wang Bosheng, Zhang Guofeng, Bao Hujun. (2015)High-quality Texture Mapping for Complex 3D Scenes. Chinese Journal of Computers,(12):88-92

[5]   LAN linlin, QIAN Wei, TIAN Mingyin. (2015)Research and Implementation of 3D Texture Mapping Technology Based on OpenGL. Gansu Science and Technology,(22):90-93

[6]   Tong Lijing, Chen Jing. Research on Application Method of OpenGL Texture Function. Software Guide. 2015(07):111-120

[7]   Li Ying, etc. (2002) OpenGL Function and Example Analysis Manual. National Defense Industry Press,

[8]   Tong Lijing, Chen Jing. Software Guide.(2015) Research on Application Method of OpenGL Texture Function,07:44-48

HTTP Tunnel Trojan Detection Model Based on Deep Learning

Yubo He[1], Yuefei Zhu[1] and Wei Lin[1]

[1] State Key Laboratory of Mathematical Engineering and Advanced Computing, Information Engineering University, Zhengzhou, 450001, China

Abstract: Aiming at reducing the high False Negative rate of the existing Trojan horse detection method based on behavior, this paper utilized the sequence characteristics of tunnel Trojan communication extracted from the transport layer and the bi-directional recurrent neural network in deep learning to build a HTTP tunnel Trojan detection model. The experimental result showed that the deep learning-based detection model reduced the false positive rate of normal network traffic and improved the Trojan detection rate. We also found that the deep learning-based detection model reduced the feature selecting and data cleaning process of generating samples and improved the easy-using of the HTTP tunnel Trojan detection model.

1. Introduce

As one of the key technologies in cyberspace security, trojan detection can guarantee the security of cyberspace. How to detect trojans in a timely and effective way has become a hotspot in the network security researcher's community. Trojan detection is divided into two parts, host-based trojan detection and network-based Trojan detection. The network-based Trojan detection technology distinguishes Trojan traffic from normal traffic by identifying characteristics of network traffic behavior, traffic characteristics, and other traffic transmissions to identify Trojan traffic. In the early days, people used the signature matching method to detect the Trojan traffic, which has achieved high detection rate for the specific Trojans. However, with the appearance of Trojan variant and the new Trojan, the Trojan flow detection based on the signature matching cannot detect unknown Trojans, and Trojan detection based on signature matching may get high false positive rate, which causes false alarm; and a large amount of human and material resources must be consumed to maintain the signature databases of Trojan traffic character features to identify Trojan traffic and to identify variant Trojans. The behavior-based Trojan detection utilizes the unique communication behavior of the Trojan traffic, such as the downlink packet ratio and the heartbeat behavior[1] to detect Trojan traffic, which can identify the variant Trojan and effectively improve applicability the Trojan detection.. With the continuous development of data mining algorithms, data mining algorithms can automatically calculate the behavior pattern of abnormal traffic through input characteristics. The general processing of malware traffic detection based on machine learning is to preprocess the raw traffic, extract the characteristics, input it into the data-mining algorithm and obtain a malware traffic classifier.

Recently, with the maturity of deep learning theory, the research on malware network traffic detection based on deep neural network model has yielded very significant results. Wang[2] proposed a method which converted the raw traffic directly into pictures and used self-encoding network to identify network traffic, and gained high classification accuracy; Wang Yong[3] proposed a way which converted the network traffic characteristics into pictures and put the pictures into the convolutional neural network to classify; Wang Wei[4] cut the original traffic into same length and directly put the raw traffic into the

Content from this work may be used under the terms of the Creative Commons Attribution 3.0 licence. Any further distribution of this work must maintain attribution to the author(s) and the title of the work, journal citation and DOI.
Published under licence by IOP Publishing Ltd

convolutional neural network to learn the raw traffic's character characteristics. However, most of today's deep learning algorithms focus on the character characteristics of traffic, and classifying traffic by self-learning of character features, but compared to normal traffic, Trojan traffic is a time series with very obvious timing characteristics, and Trojan traffic are usually encrypted, which is more difficult to extract the character characteristics. In order to solve these problems, this paper used the recurrent neural network to extract the timing characteristics of Trojan traffic and reduced the false positive rate of deep learning model to identify Trojan traffic.

2. Trojan communication behavior analysis

The HTTP tunnel Trojan is a Trojan that transmits control commands, user data, sensitive data and so on through an HTTP tunnel. Compared with the traditional Trojan, the HTTP tunnel Trojan establishes a connection with the server through the universal port 80. The traditional firewall directly releasees the HTTP data, so the HTTP tunnel Trojan can escape the detection by the firewall, creating a secure Trojan transmission environment.

Since the difference between the HTTP tunnel Trojan and the ordinary Trojan is only the communication channel, there is a big difference between the HTTP tunnel Trojan traffic behavior and the normal traffic behavior, which is the same as the ordinary Trojan. Like the normal application, HTTP tunnel Trojan needs to establish a client-server connection and communication. The HTTP tunnel Trojan communication is usually divided into three phases.

1. Connection establishment. After the console program rooted in the Trojan Controlled Terminal runs, the Trojan sends a DNS request, establishes a TCP connection with the IP address of the controlling terminal obtained by the DNS server and transfers data through an HTTP tunnel with the controlled. While the control terminal is offline, the control terminal receives RST package.

2. Data transmission. The controlled terminal receives the control command information from the control terminal, executes the control command and sends the return result. Since the Trojan controlled terminal is similar to the role that server do in the client/server architecture, the network traffic behavior of the Trojan horse program will be different from the normal program network traffic during data transmission, that is, the Trojan has external control features, and the normal application has internal control features, as shown in Figure 1.

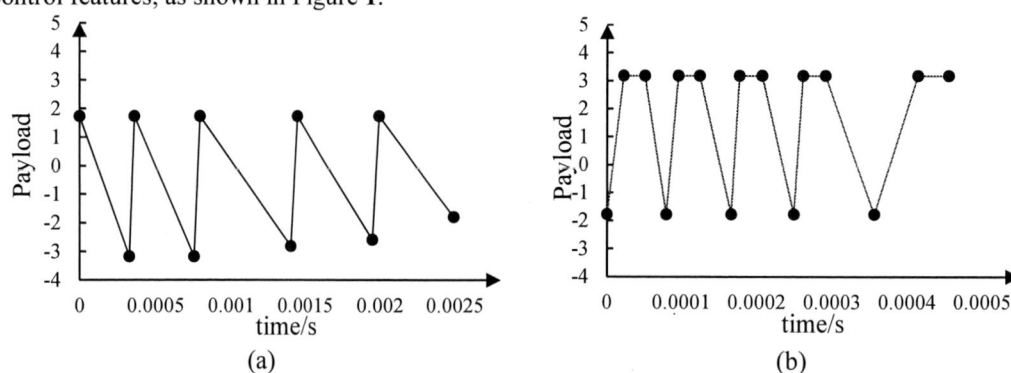

(a)   (b)

Figure 1. Packet Payload size of time. (a) shows the traffic from normal traffic—Edge while (b) shows trojan's traffic, pcshare. A log function is used to the packet payload sizes and out-to-in packet sizes are defined as negative on y-axis.

3. Connection maintenance. In order to maintain the network connection between the controlled terminal and the controlling terminal and probe the network status of the controlled terminal, the attacker introduces a heartbeat mechanism while designing the Trojan. The controlled terminal program will send a data packet containing the host specific information to the controlling terminal through the network connection. The controlling terminal can complete the handshake with the controlled terminal and confirm the network status of the controlled terminal.

As shown in Figure 1, the Trojan has external control features while the normal application has internal control features, therefore during data transmission, the Trojan traffic often begins with the first downstream data packet while the normal application starts with the first upstream data packet. So we can assume that after capturing raw network traffic in the local host and filtering heartbeat packet, if the data transmission behavior is similar to Figure 1(a), it's considered to be a normal program's traffic; if it is similar to Figure 1(b), then it's considered to be a trojan's traffic.

3. HTTP tunnel trojan detection model based on deep learning

In order to verify the validity of the Trojan detection model proposed in this paper, the structure of the Trojan detection model is shown in Figure 2. The raw traffic is preprocessed to obtain the network traffic characteristics, after pre-processing, the network traffic characteristics is put into the long short-term network to learn the behavior feature and constructs a Trojan traffic classifier.

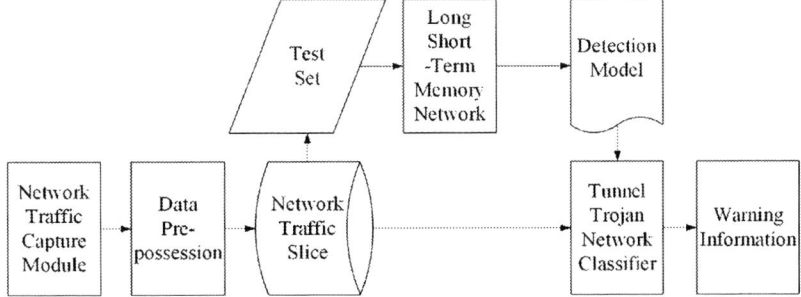

Figure 2. Architecture of HTTP tunnel trojan detection model

3.1. Network traffic capture module

The network traffic capture module is mainly developed by winpcap. The winpcap module captures raw network traffic and saves it in pcap file, and winpcap module can parse and extract communication information such as transmission protocol and communication payload of network data packet. In this experiment, the traffic data of the network was mainly taken from the exit of a laboratory LAN, and the network traffic of the LAN was captured by the mirror port of the switch.

3.2. Data preprocessing

The preprocessing of network traffic mainly includes two modules, traffic clustering module and heartbeat filtering module. The traffic clustering module splits the captured traffic data packets into clusters according to an algorithm; the heartbeat filtering module uses the association rule algorithm to filter the heartbeat packet in each data packet cluster, which gets the pure data transmission traffic. The Data prepossessing process is shown in Figure 3.

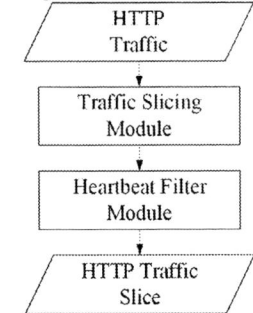

Figure 3. Data Prepossessing process

3.2.1. Traffic slicing module. HTTP traffic has self-correlation and self-similarity properties. But HTTP traffic will appear self-similarity over a long time-span, and HTTP traffic won't show its seconds-level self-correlation characteristics over session scale. Therefore, in this paper we consider splitting the raw IP flow into multiple clusters and begin the correlation analysis of the data packets in a single cluster.

For an IP flow $F = \{P_1, P_2, \ldots, P_n\}$, where P_i represents the i-th packet in stream F, we uses the traffic clustering algorithm to slice the stream F to obtain $F = \{C_1, C_2, \ldots, C_k\}$, where C_j represents the j-th cluster after slicing. According to the literature [5], when slicing network traffic, considering the following factors.

1. Packet arrival time. During the execution of task, the user needs to quickly send the next command after the previous command returning the result. Therefore, we define the user's reaction time t_r, which is the time interval between the return of the result packet of the previous command and the packet of the latter command; At the end of a task the user need to think about the next action, and thus we define the user's thinking time t_t. We assume that the process that sending request and receiving execution results of the request can be considered as a data interaction process. Therefore, it is reasonable to assume that if the time-span of a data slice is greater than the users thinking time t_t, then the next packet can be considered to belong to the next cluster.

2. Time interval scale. Due to the fluctuation of the network bandwidth and the like, the time-span of a packet cluster may be fluctuating, so we define the time interval scale SCALE which reduces the impact of network instability. If the arrival time interval of the adjacent data packet is greater than the SCALE times of the sum of the average intra-cluster time interval and the user's response time t_r, we can assume that the latter data packet may be considered to belong to the next cluster.

3. The maximum arrival time interval. While studying some network traffic we found that if one program contains heartbeat and the heartbeat connection is frequent, especially when the interval is less than the user's thinking time t_t, it's more likely that fault slicing occur because the heartbeat packet will mislead the computer to locate the wrong last packet, so we count the maximum arrival time interval of two data packet, and the maximum arrival time interval is equal to the maximum heartbeat connection interval in this case. In a cluster, if the length of one cluster is greater than the maximum arrival time interval, and If the time interval of the adjacent data packet is greater than the attacker response time t_r, then the next data packet belongs to the next data packet cluster.

3.2.2. Heartbeat filter module. Considering the heartbeat package is control package and will interfere with the Trojan's data transmission process, so we have to filter the heartbeat packet to get the pure IP data transmission flow.

The Trojan's heartbeat can complete the handshake between the controlled terminal and the controlling terminal, so that the Trojan controller can get the network environment status of the controlled terminal, which is beneficial for the attacker to continuing the attack behavior. In general, the Trojan's heartbeat is divided into two types, request response and three-way handshake. The request response type heartbeat means that the Trojan controlling terminal sends a heartbeat packet to the controlled terminal, then the controlled terminal receives the heartbeat packet and responds with a acknowledgement packet. The three-way handshake type is one step longer than the response type, to be more exact, the controlling terminal receives the acknowledgement packet of the acknowledged heartbeat of the controlled terminal, which is similar to the three-way handshake in the TCP connection.

In the Trojan design process, the heartbeat package is sent to characterize the Trojan's network survivability, and the proportion of heartbeat packets in an active Trojan connection is often higher in the entire Trojan IP stream, so in a Trojan stream there will be multiple regular Trojan heartbeat packets with same length, and we can use the association rule algorithm to extract the heartbeat packet rules and filter the heartbeat packets. In this paper, the AprioriAll algorithm [6] is used to extract the heartbeat packet rules and filter the heartbeat packets.

3.3. Traffic classification
Traffic is a time series with strong pre- and post-dependency, so we can use the bi-directional recurrent

neural network to adaptively learn the timing characteristics without manually extracting feature. Thus in this paper, we uses the bi-directional recurrent neural network as a classification model for Trojan traffic.

The Recurrent neural network is a neural network model for processing time series. On the basis of the traditional feed-forward neural network, the recurrent neural network can deal with the time series data by using neurons with self-feedback function in the hidden layer. In order to make the recurrent neural network not only calculate the state information from the previous moment, but also calculate the state information from the future moment, a bidirectional recurrent neural network is introduced, and its structure is shown in Figure 4.

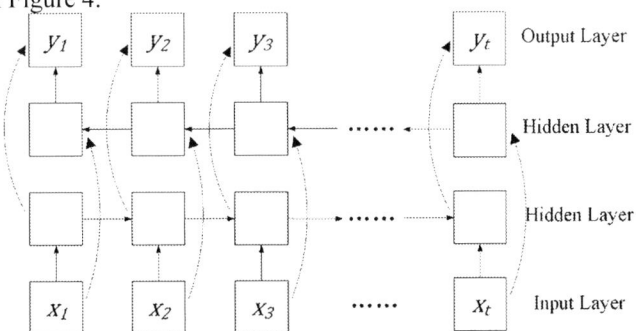

Figure 4. Bidirectional Recurrent neural network structure

The long short-term memory network is a variant of the recurrent neural network model proposed by Hochreiter and Schmidhuber[7] for solving long-term dependence problem. On the basis of the traditional recurrent neural network, a long short-term memory network unit is composed of an input gate, a forgetting gate and an output gate is introduced to replace the hidden layer unit in traditional recurrent neural network. The structure of long short-term memory network unit is shown in Figure **5**.

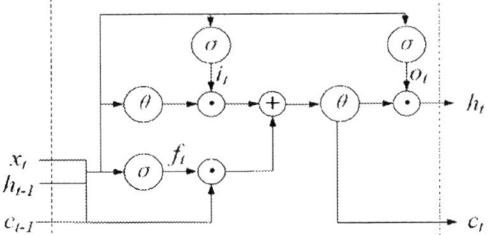

Figure 5. Long short-term memory network Units

The corresponding calculation formula is as follows.

$$i_t = \sigma(W_{xi}x_t + W_{hi}h_{t-1} + W_{ci}c_{t-1} + b_i) \tag{3}$$

$$f_t = \sigma(W_{xf}x_t + W_{hf}h_{t-1} + W_{cf}c_{t-1} + b_f) \tag{4}$$

$$c_t = f_t c_{t-1} + i_t tanh(W_{xc}x_t + W_{hc}h_{t-1} + b_c) \tag{5}$$

$$o_t = \sigma(W_{xo}x_t + W_{ho}h_{t-1} + W_{co}c_{t-1} + b_o) \tag{6}$$

$$h_t = o_t tanh(c_t) \tag{7}$$

Where σ is the sigmoid function, i_t, f_t, o_t, and c_t represent the input gate, the forgetting gate, the output gate, and the unit state at time t. In the long short-term memory network, i_t, f_t, o_t control the information flow. The input flow determines the input ratio. When calculating the unit state, this ratio determines the value of equation (7), and the forget gate decides whether to pass. Passing the value of the previous hidden layer output h_{t-1}, the previous information transfer ratio is calculated by f_t and used to calculate the cell state c_t at time t, and the output gate determines whether the hidden layer unit is passed in the output result. The value of the result c_t. By using long short-term memory network units, we can better deal with the problem of gradient disappearance and gradient explosion in traditional recurrent neural networks.

4. test and result analysis

In order to verify the performance of the HTTP tunnel Trojan detection model based on bi-directional long short-term memory network model, the experimental environment used in this paper is shown in Table **1**.

Table 1. Experimental environment

Type	parameter
Operating system	Windows 10, 64 bit
Processor	Intel Core i7 8700
Memory	16GB DDR4 2133MHz
Graphics Processor Units	Nvidia GTX 1060 6GB
Tensorflow version	Tensorflow-1.8

4.1. Dataset construction

Since there is no standard dataset in the field of HTTP tunnel Trojan detection, we used the mixture of background traffic and manually collected Trojan traffic to construct its own test dataset and valid dataset to evaluate the performance of HTTP tunnel Trojan detection model. However, due to the large difference in the data size of the collection Trojan traffic and the normal Internet traffic, the normal Internet traffic volume will be much larger than the artificial Trojan traffic, which will have a greater impact on the accuracy and false positive rate of the machine learning algorithm, and it is also difficult to guarantee the purity of background traffic. To ensure reliability of system performance and the test, we used some CTU-IDS-2017 dataset[8] as the background traffic of this test, and verified the Trojan traffic detection model proposed in this paper. Two datasets are described in Table **2**. The Trojan traffic of the training dataset was collected in 6 hosts in the lab LAN, 5 of them were infected with HTTP tunnel Trojans and installed Wireshark to capture the trojan traffic(including Pcshare, Gh0st and other HTTP tunnel Trojan randomly collected from the MWCollect[9] and Malfease[10], which are Trojan sample sites) and the rest one of them were used as a controlling terminal. The test dataset's Trojan traffic was collected from 13 hosts in the lab LAN, 10 of which were infected with HTTP tunnel Trojans (including Trojan samples in some test sets and new randomly collected Trojan samples like Bifrost), and 3 were used as controlling terminal.

Table 2. Experimental data samples

Application	Dataset	Time	TCP flow number
sTraining set	Background flow (CTU-IDS-2017)	1day(Monday)	36434
	Trojan flow	×	5431
Test set	Background flow (CTU-IDS-2017)	1day(Tuesday)	25964
	Trojan flow	×	6352

4.2. Experimental evaluation metric

In this experiment, the detection rate of Trojan traffic and the false positive rate of normal traffic are used as the evaluation metric of the Trojan detection algorithm model proposed in this paper. The detection rate evaluates the detection accuracy of the Trojan detection system for Trojan traffic, and the false positive rate evaluates the ratio that Trojan detection model falsely warns user against trojan traffic. The metrics are defined as follows.

The data stream detection rate D is defined as:

$$D = \frac{t_t}{n_t} \tag{8}$$

The data stream false positive rate F is defined as:

$$F = \frac{f_t}{n_n} \tag{9}$$

Where n_t represents the number of all Trojan communication data streams, n_n represents the number of all normal communication traffic data streams; t_t represents the number of properly identified Trojan traffic flows, and f_t represents the number of normal traffic identified as Trojan traffic.

4.3. Recurrent neural network hyperparameter

Hyperparameters are model parameters established before initialization of neural network. Kalus Greff[11] proposed that during the training of neural network the influence of hyperparameters on training results and model convergence is independent after he studied the learning progress of long short-term memory network. Therefore, we adopted variable-controlling approach to fine-tune the hyperparameters like learning rate, the length of the Trojan flow slice sequence, the hidden layer size, the hidden layer number, and the batch size and find a balance point among the model size, the Trojan flow detection rate, and false positive rate. While training neural network, we used 10-fold cross validation to evaluate the capability of the Trojan detection system.

4.3.1. Learning rate. In the experiment, we kept the other variables unchanged, and continuously changed the learning rate. After 10 iterations, the results obtained are shown in Figure 6. As the learning rate decreases, the convergence rate of the Trojan traffic detection rate decreases, but the stability of the detection rate convergence increases. However, when the learning rate is less than 0.05, the stability of the detection rate convergence tends to be stable, but the convergence speed is still reduced. Therefore, the model set the learning rate to 0.05, and the Trojan detection model could be quickly converged and the convergence stability of the Trojan detection rate is maintained at a high level.

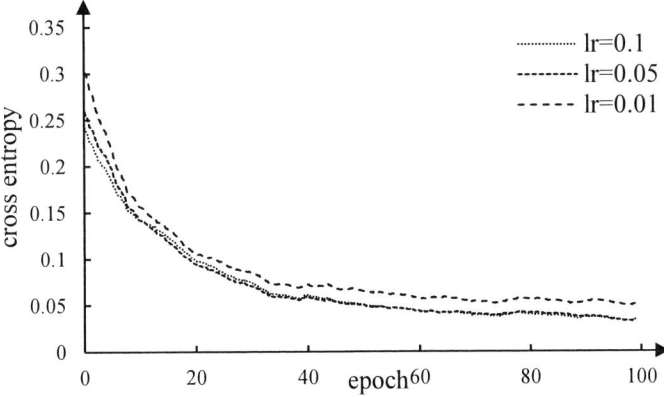

Figure 6. Effect of learning rate on model convergence

4.3.2. Trojan flow slice length. In the experiment, we continuously changed the length of the Trojan slice sequence, and the results were shown in Figure 7. As the length of the slice sequence increases, the detection rate of the Trojan flow increases first, and false positive rate gradually decreases. Then, after the slice length is greater than or equal to 6, the gradual decrease tends to be gentle. This shows that when the sequence length of the traffic session slice is equal to 6, the detection rate of the Trojan is higher, and the subsequent traffic packet length has less influence on the traffic identification capability. Therefore, the model sets the length of the slice sequence of the input traffic session to 6.

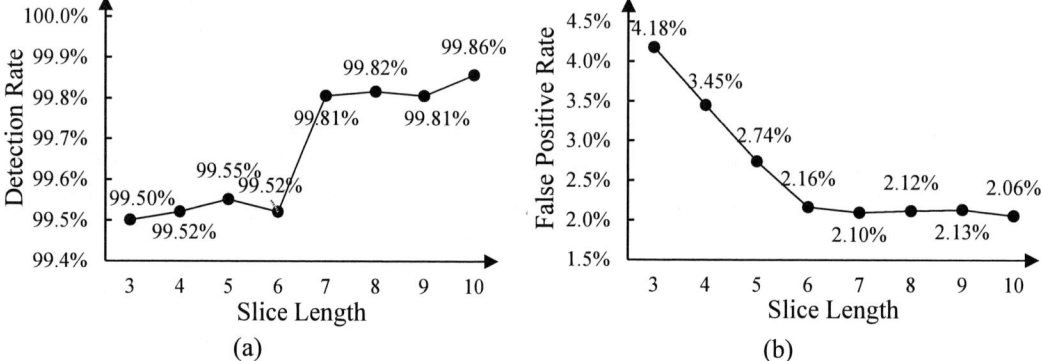

Figure 7. Effect of Trojan slice length on model detection performance

4.3.3. the number of hidden layers. When designing the hidden unit size of each layer, we should consider the relationship between the number of samples and the number of the parameters, because it's widely accepted that the number of parameters should be balanced with the number of samples. So in this experiment the hidden layer sizes of this experiment were 60, 30, 24, 18 with the increasing number of hidden layer number. The relationship between the number of hidden layers and the detection rate and false positive rate is shown in Figure 8. As the number of hidden layers increases, the increasing trend of false positive rate becomes more obvious, and the detection rate of Trojan traffic gradually decreases. Therefore, we set the number of hidden layers to 2.

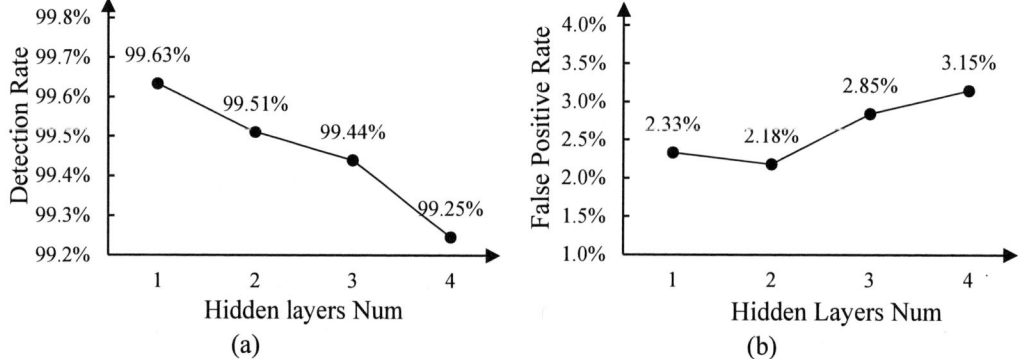

Figure 8. Effect of hidden layer number on model detection performance

4.3.4. hidden layer unit size. The relationship between the size of the hidden layer and the detection rate and false positive rate is shown in Figure 9. In the experiment, we set the hidden layer number to be 1, and we found that when the hidden layer unit is larger than 40, the detection rate of the Trojan flow tends to be gentle, but when the hidden layer unit is in the interval [40, 60], as the hidden layer unit gradually increases, the false positive rate of the model gradually decreases, and then It tends to be flat. Considering too many parameters can cause over-fitting problems in model detection, so we set the hidden layer unit size to 60.

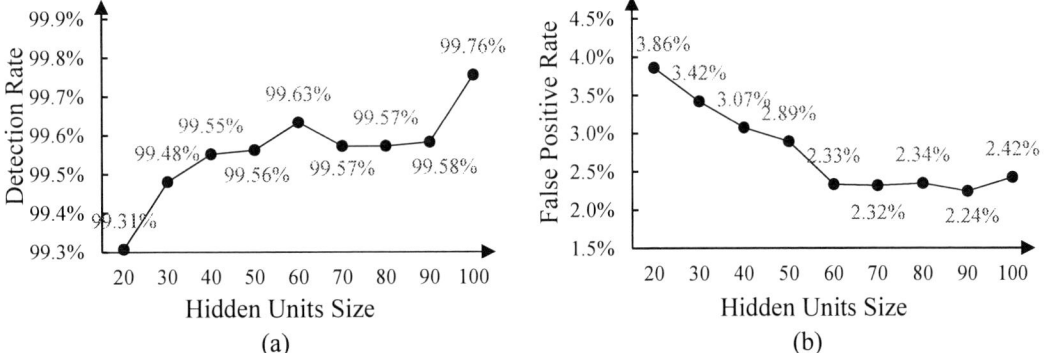

Figure 9. Effect of hidden layer cell size on model detection performance

4.4. Results display

Through the above experiment, we established a bi-directional long short-term memory network model with two hidden layers, each of which contains 30 basic long short-term memory units. The long short-term memory network used softsign function as activation function of the fully-connected layer and used the softmax function as classification function. The loss function optimization algorithm used the Adagrad algorithm.

In the experiment, we used the above bi-directional recurrent neural network model for classification and used the dataset shown in Table **2** as data samples to train and test the Trojan detection model, and the detection performance of the Trojan detection model for some Trojans is shown in Figure **10** & Figure **11**. From the figure we could see that the model proposed in this paper gained high detection rate and low false positive rate and according to Figure **11**, we found that after the introduction of some new Trojans like PainRAT, the detection rate of the new Trojans did not drop significantly. The detection rate of Trojans remained above 99%, and the false positive rate of Trojan traffic remained at around 2%, indicating the Trojan traffic detection model based on sequence slices and bi-directional long short-term memory network proposed in this paper can effectively distinguish abnormal Trojan traffic and normal traffic

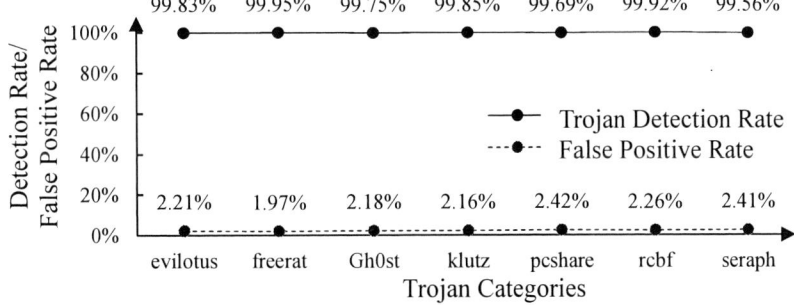

Figure 10. Data test set different categories of Trojan detection rate / false positive rate line chart

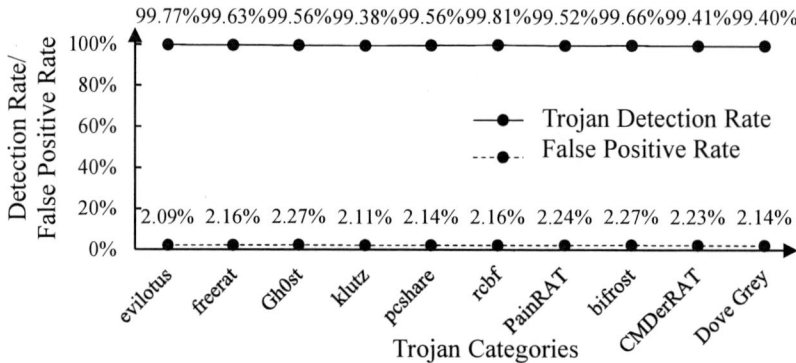

Figure 11. Data evaluation set of different categories of Trojan detection rate / false positive rate line chart

As shown in Table 3, we verified the effectiveness of the proposed method by comparing the proposed model with other network-based HTTP tunnel Trojan detection techniques. From the results, we could see that compared with the original method, the Trojan detection algorithm based on the bidirectional long short-term memory network has a significant improvement on the enhancement of Trojan traffic detection rate and the reduction of normal network traffic false positive rate, which showed that our proposed model could correctly identify the malware network traffic and normal network traffic. Table 3. Comparison of different methods of HTTP tunnel Trojan detection rate and false positive rate

Detection method	detection rate	false positive rate
Literature [13]	91.17%	4.56%
Literature [13]	90.66%	7.28%
This method	99.76%	2.33%

5. Conclusion

Based on the method of Trojan traffic detection utilizing sequence analysis, this paper analyzed the communication behavior of HTTP tunnel Trojan transport layer and used the associated rules algorithm to mine the Trojan heartbeat behavior and filter heartbeat to characterize the data transmission process of the HTTP tunnel Trojan. This paper used the bi-directional long short-term memory network model to obtain the HTTP tunnel Trojan traffic detection model. The results showed that compared with the existing Trojan detection model, the model could improve the detection rate of the HTTP tunnel Trojan and reduce the false positive rate of normal traffic, and during the test of the Validation dataset, we found that the method has certain ability to detect unknown Trojans. The next step we will do is to extend Trojan traffic dataset, improve the network traffic slicing performance, and find other communication behavior characteristics of the HTTP tunnel Trojan to improve the detection performance of the HTTP tunnel Trojan detection model.

Acknowledgment

We thank He Kang, Li Ding and Pan Yan for fruitful discussion. We also thank our teammates for their help with the environment building and test.

References

[1] Sun H., Liu S., Chen J. (2011) Tunnel Trojan Detection Method Based on Operation Behavior. Computer Engineering, 37:123–126.

[2] Wang Z. (2015) The Applications of Deep Learning on Traffic Identification. In: Black Hat. Las Vegas.pp.10.

[3] Wang Y., Zhou H., Feng H., Ye M. (2018) Network traffic classification method basing on CNN.

Journal of Communications, 39:14–23.

[4] Wang W., Zhu M., Zeng X., Ye X., Sheng Y. (2017) Malware traffic classification using convolutional neural network for representation learning. In: International Conference on Information Networking. Da Nang. pp.712–717.

[5] Wu S., Liu S., Lin W., Zhao X., Chen S. (2017) Detecting Remote Access Trojans through External Control at Area Network Borders. In: Symposium on Architectures for Networking & Communications Systems. Beijing. pp. 131–141.

[6] Han, J. (2005). Data Mining: Concepts and Techniques. Morgan Kaufmann Publishers Inc. San Francisco.

[7] Hochreiter S., Schmidhuber J. (1997) Long Short-Term Memory. Neural Computation, 9:1735–1780.

[8] Sharafaldin, I., Lashkari, A.H., Ghorbani, A.A. (2018) Toward Generating a New Intrusion Detection Dataset and Intrusion Traffic Characterization. In: International Conference on Information Systems Security & Privacy. Maderia. pp. 108–116.

[9] Collaborative malware collection and sensing. (2014) Collaborative malware collection and sensing. http://alliance.mwcollect.org. [Accessed: 13-May-2018].

[10] Project malfease. (2014). Project malfease. https://malfease.oarci.net. [Accessed: 13-May-2018].

[11] Greff K., Srivastava R.K., Koutník J., Steunebrink B.R., Schmidhuber J. (2016) LSTM: A Search Space Odyssey. IEEE Transactions on Neural Networks & Learning Systems, 28:2222–2232.

[12] Tegeler, F., Fu, X., Vigna, G., Kruegel, C. (2012) BotFinder: finding bots in network traffic without deep packet inspection. In: International Conference on emerging Networking Experiments and Technologies. Nice. pp. 349-360

[13] Li S., Yun X., Zhang Y. (2012) A Model of Trojan Communication Behavior Detection Based on Hierarchical Clustering Technique. Journal of Computer Researcch and Development, 49: 9–16.

Design and Implementation of connect6 Intelligent Game System

Zengyu Cai[1], Chunfeng Du[1], Xiaoshuang Guo[1] and Jianwei Zhang[2]*

[1] School of Computer and Communication Engineering, Zhengzhou University of Light Industry, Zhengzhou, Henan, 450002, China

[2] Software Engineering College, Zhengzhou University of Light Industry, Zhengzhou, Henan, 450002, China

* zhangjw@zzu.edu.cn

Abstract. As an important part of artificial intelligence, computer game is also considered as one of the most challenging research directions in the field of artificial intelligence. After understanding the research background and development status of the connect6 chess game, the paper analyzes the requirements and feasibility of the system according to relevant theories. This paper describes the intelligent game and the chess shape, and uses the "maximum-minimum" principle, evaluation function and other related technologies to design and implement the development of the connect6 chess game system. After testing the system's functions and intelligence, the system can achieve the desired results.

1. Introduction

With the development of science and technology, artificial intelligence not only is closely related to computer science, but also has a certain relationship with many disciplines such as medicine, psychology and education. At present, artificial intelligence has been booming in many fields such as robotics, machine learning, expert systems, and distributed artificial intelligence [1]. The intelligent chess game not only is a branch of the computer game, but also attracts a lot of attention as an iconic example in the field of artificial intelligence [2]. As ubiquitous online games emerge in an endless stream, chess games are constantly being expanded, such as flying chess, gobang, and chess. At the same time, as a simple and fair chess game, the connect6 chess game is also constantly moving towards intelligent development [3-4]. This paper designs and develops the connect6 intelligent game system, which mainly includes four modules: game search, evaluation function, travel generation and opening library [5]. Each module uses unique techniques and corresponding methods to solve problems that may be encountered in smart games.

2. System function

2.1. Functional Analysis

The overall requirement for system development in this paper is to realize the intelligent game six-game program with certain chess power on the basis of realizing the function of ordinary chess games. The system must also satisfy the human-machine intelligent game on the basis of satisfying the human-human game. The board representation module is used to display the connect6 board, and the board size of the connect6 board is 19×19. The walk generator is used to generate the legal behavior of the party or the other party in the current situation [6-7]. The evaluation function is used to estimate the situation of the move. Two valuation methods are used in this system, based on the valuation of "road" and the valuation based on "chess type" [8]. The search module mainly performs a search for the game tree and strategy. The basic structure model of the system is shown in Figure 1.

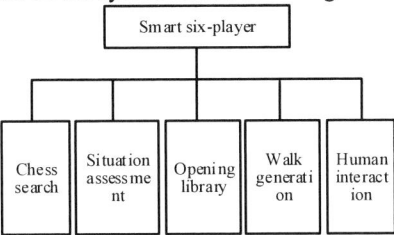

Fig 1 System basic structure model

2.2. Process Analysis

After the game starts, whether it is a mouse click or the computer drops the piece, it is necessary to determine whether the position of the piece is legal. The position of the falling piece will be re-selected, if the position of the falling piece is illegal. Otherwise, update the situation and calculate the next position. Calculate the score to find the chess point with the highest score. Then, the other party plays the chess piece. And the system judges whether it is a victory. The process of the connect6 intelligent game system is shown in Figure 2.

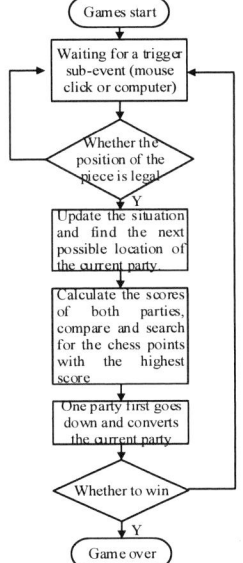

Figure 2. system flow chart

3. Related theories and techniques

3.1. Evaluation Function

In the game, the players are set to *MAX* and *MIN* respectively, and the game assumes that *MAX* goes firstly. Due to the storage space, the number of chess pieces and the search time limit, the complete search of the graph cannot be performed [9]. A static estimation function f is defined to estimate the current state of the game. It is advantageous for MAX when f is a positive number, and MIN is advantageous when f is a nagative number. The following is a step of giving a very small search by taking "Tic Tac Toe" as an example. The principle of "maximal or minimum"value search is shown in Figure 3.

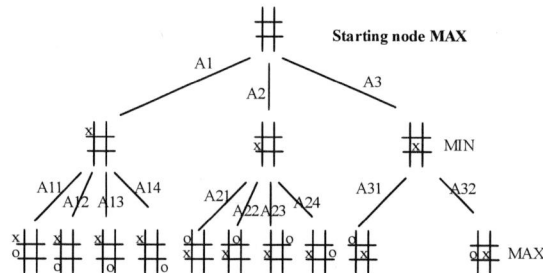

Figure 3. Maximum and very small search

(1) The extended tree node generates a game tree.

(2) Each leaf node is evaluated by the evaluation function f, and the evaluation value of the upper node of the upper layer is obtained.

(3) Use the value of the intermediate node to get the estimate of the middle node of the previous layer. The estimate obtained by the leaf node passing A15, is the smallest [10-11] in A1 walking, and it is advantageous for MIN when the value of f is minus. Therefore, the A1 walking MIN node chooses A15 to walk. The MIN of the same layer A2 and A3 also selects the minimum value to walk.

(4) Repeat step (3) to take the maximum value of its branch at the MAX layer. The MIN layer takes the minimum value of its branch [11] and terminates the operation to the root node.

3.2. Evaluation Function

The evaluation function is a comprehensive evaluation of the current game. According to the current chess situation, Black and White are evaluated separately to judge the current situation of both sides of the game and provide the judgment standard for the search engine [12]. Among the connect6 chess, there are complex chess types. The evaluation of the game is generally done by local scanning. This paper considers that the local scan is not comprehensive enough, so the evaluation function based on the "chess type" global scanning method is adopted. The evaluation function formula based on the "chess type" global scan is as follows:

$$TotalValue = ExternalValue + LocalValue \tag{1}$$

$$LocalValue = MyValue + ThreatValue + BalanceValue \tag{2}$$

$TotalValue$ in equations (1) and (2) represents the total valuation of the situation. $ExternalValue$ represents an estimate outside of the local. $LocalValue$ represents a local estimate. $MyValue$ represents its own local valuation. $ThreatValue$ represents the other party's local estimate. $BalanceValue$ represents the threshold at which the local estimate is unbalanced.

The evaluation function includes a $LocalValue$ and an $ExternalValue$. The external estimate can capture the type and number of moves that exist outside of the local. The external range not is scanned from the checkerboard (0,0) position, but can contain all the pieces of the board and the minimum range outside the partial drop. The local estimate includes the own value, the opponent threat value, and the balance value.

$$MyValue = MyFirstValue + MySecondValue \tag{3}$$

$$ThreatValue = ThreatFirstValue + ThreatSecondValue \tag{4}$$

In the formulas (3) and (4), $MyFirstValue$ represents the valuation of the first piece of the party. $MySecondValue$ represents the valuation of the second piece of the party. $ThreatFirstValue$ represents the threat value that simulates the opponent's first piece. $ThreatSecongValue$ e represents the threat value that simulates the opponent's second piece.

Equation 3 contains a chess analysis of the first piece of the game and a chess analysis of the second piece. And then the sum of the two is used as the own estimate. Equation 4 contains the threat value of the opponent's first piece and the second piece at the same board position. When the difference between the MyValue and the other TheatValue value is large, an imbalance adjustment is required. The adjustment formula is as shown in Equation 5.

$$BalanceValue = W1 + W2 \tag{5}$$

$$W1 = \begin{cases} -\dfrac{min1}{2} & k1 \geq 2.5 \\ 0 & k1 < 2.5 \end{cases} \tag{5}$$

$$W2 = \begin{cases} -\dfrac{min2}{2} & k2 \geq 2.5 \\ 0 & k2 < 2.5 \end{cases} \tag{6}$$

In formulas (5), (6) and (7), $W1$ represents the difference between the valuation of the first piece of the opponent and the valuation of the opponent's first piece. $W2$ represents the difference between the valuation of our second pawn and the valuation of the opponent's second pawn. $K1$ represents the imbalance factor of the first piece estimate, $K1 = MyFirstValue / ThreatFirstValue$. $K2$ represents the imbalance factor of the second piece valuation, $K2 = MySecondValue / hreatSecongValue$. $min1$ represents the smaller value of the first piece of the party and the opponent, and $min2$ represents the smaller value of the first piece of the party and the other party. Adjust the imbalance of the two falling pieces by formulas (6) and (7). When the ratio of the opponent's pawn to the opponent's threat value is greater than 2.5, it is considered to be unbalanced. And then the imbalance is expanded to reduce the value of BalanceValue. When it is less than 2.5, it is considered that no imbalance occurs and the imbalance value is not changed. Through the imbalance adjustment, the way to balance your own and the other party's threats would be choosed to help your own offense and defensive moves.

4. Function test

4.1. Basic Function Test
After the program is started, a human-machine battle or a human-person battle would be selected. At the same time, you can choose to open the background music. If you choose to play Black, the first piece is a black piece. When you choose to play white, the second piece and the third piece will be white pieces. The system includes connect6 rules and system environment related introduction. After testing, basic functions can be achieved.

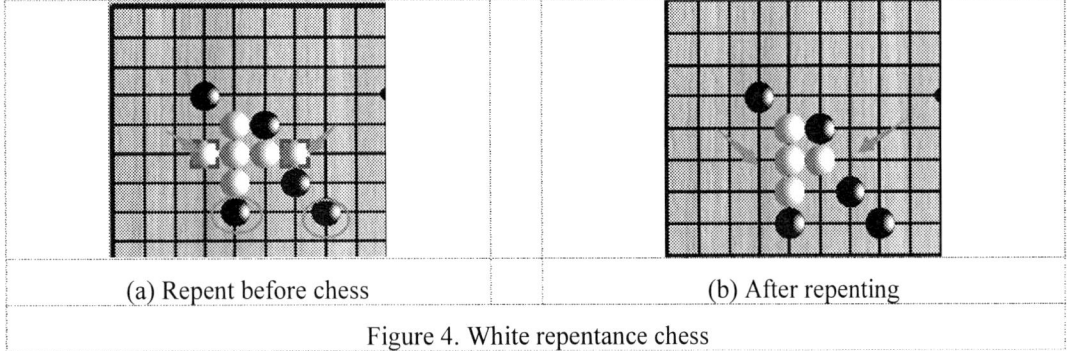

| (a) Repent before chess | (b) After repenting |

Figure 4. White repentance chess

ISPECE

IOP Publishing

IOP Conf. Series: Journal of Physics: Conf. Series **1187** (2019) 042056   doi:10.1088/1742-6596/1187/4/042056

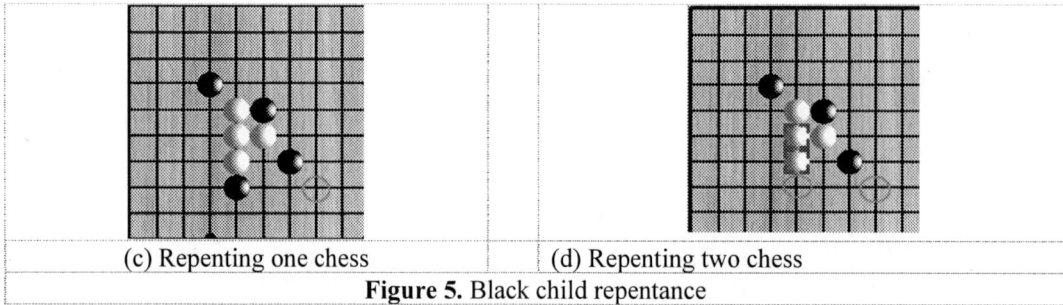

(c) Repenting one chess	(d) Repenting two chess

Figure 5. Black child repentance

Retreat and repentance pieces can be selected during the game. Retreated as shown in Figure 4. Repentance is shown in Figure 5. In Figure 4, since the white color pieces are played two steps at a time, the chess is also retreated at the same time. Figure 5 is a black chess piece. Picture (c) stands for a picture of a chess piece. Picture (d) represents a schematic diagram of repenting two pieces.

4.2. Basic Function Test

The system intelligence test is mainly aimed at the human-machine battle game mode. According to the chess player's falling chess pieces on the chessboard, the computer obtains the best move according to the evaluation function and the game search calculation and analysis, so as to realize the connect6 intelligent game.

Select the player to hold the black color piece firstly, and the computer as the back hand to hold the white color piece. After the chess piece is dropped, the best way to walk is based on the evaluation function evaluation and the game search. There are two cases tested. The chess player's opening game is shown in Figure 6.

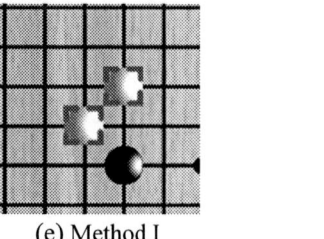

(e) Method I	(f) Method II

Figure 6. Chess player opening game

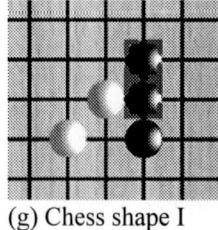

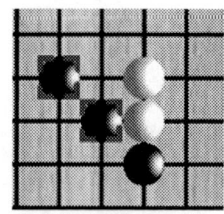

(g) Chess shape I	(h) Chess shape II	(i) Chese shape III

Figure 7. Computer opening game

When the computer first drops the black color piece, the player then chooses to drop the white color piece. After the chess pieces are dropped, the chess pieces that fall by the players can form "living two", "single", "sleeping two" three kinds of chess shapes. In any case, the computer can form a "live three" chess shape in the next game. After testing and analysis, the computer movement is the highest score. The computer opening game is shown in Figure 7.

5. Conclusions

The system designed in this paper has been able to initially complete the man-machine battle, the human-person battle, and the judgment of winning or losing. After testing, the system is stable and can be implemented. In addition to the basic functions, the system also adds the repentance function to make the game more user-friendly. It adds the function of saving the game and opening the game to facilitate the player to record the battle situation. At the same time, it also adds game help to enable the player to quickly understand the game rules and environment. In order to enhance the entertainment of the game, the system added background music. All in all, the connect6 intelligent game system designed and developed in this paper can complete the functions required by the intelligent game system. The research results can contribute to the development of the connect6 chess game in theory and practical application.

Acknowledge
This work is supported by National Natural Science Foundation of China No.61672471, Key Technologies R & D Program of He'nan Province (No.172102210059 and No.172102210060), He'nan Province University science and technology innovation team(No.18IRTSTHN012) and Plan For Scientific Innovation Talent of Henan Province (No.184200510010), Special Program for Fundamental Research Business Fees of Universities Affiliated to Henan Province(16601000020).

References
[1]   Shang W, An P, Wan M, et al. (2017) Research and development overview of intrusion detection technology in industrial control system[J]. Application Research of Computers, 34(2): 328-333.
[2]   Hannaway D B. (2010) Research and Application of Cultivation-Simulation-Optimization Decision Making System for Rapeseed (Brassica napus L.) [C]// Computer and Computing Technologies in Agriculture Iv-- Ifip Tc 12 Conference,ccta 2010 Selected. pp. 441-456.
[3]   Guan-Jun X U. (2012) Research on application and optimization of malfunction alert for computer monitoring system in cascaded control center of Lancang River[J]. Mechanical & Electrical Technique of Hydropower Station. pp. 34(2): 328-333.
[4]   Chen X, Chen Z, Li H, et al. (,2016) Research and implementation of the spectrum analysis system based on FPGA[J]. Electronic Measurement Technology.
[5]   Wang X L. (2014) Search optimization of chess game in Lianzhu mode [D]. Anhui University.
[6]   Lin, Z D. (2014) Intercomparison of the impacts of four summer teleconnections over Eurasia on East Asian rainfall[J]. Advances in Atmospheric Sciences, pp. 31(6):1366-1376.
[7]   Yuan C. (2016) Improvement of Chinese Chess Search Algorithm [D]. Donghua University.
[8]   Huang J P, Zhang D, Miao H. (2009) Research and Implementation of Six-Player Intelligent Game System[J]. Computer Knowledge and Technology, pp. 5(25): 7198-7200.
[9]   Song X L. (2012) Research and Implementation of Chinese Chess Game Tree Search Algorithm[D]. Shenyang University of Technology.
[10] Jiang W J. (2013) Research and application of artificial intelligence in online games [D]. Shanghai Jiaotong University.
[11] Barriga N A, Stanescu M, Buro M. (2017) Game Tree Search Based on Non-Deterministic Action Scripts in Real-Time Strategy Games[J]. IEEE Transactions on Computational Intelligence & Ai in Games, PP (99):1-1.
[12] Takeuchi S, Kaneko T, Yamaguchi K. (2011) Evaluation of Game Tree Search Methods by Game Records[J]. IEEE Transactions on Computational Intelligence & Ai in Games, pp. 2(4):288-302.

Automated Recognition of Retinopathy of Prematurity with Deep Neural Networks

Yifan Wang[1], Yuanyuan Chen[1*]

[1] Machine Intelligence Laboratory, College of Computer Science, Sichuan University, Chengdu, Sichuan, 610065, P.R.China

*Corresponding author's e-mail: chenyuanyuan@scu.edu.cn

Abstract. Retinopathy of Prematurity (ROP) is a blinding disease, which primarily occurs on premature infants whose birth weights is less than 1250 grams or gestation is less than 31 weeks. ROP has become the leading cause of preventable childhood blindness throughout the world. Nowadays, more and more researchers start attempting to develop auto or semi-auto methods based on digital image analysis to diagnose ROP. However, factors like high measurement errors or redundant analysis phrases make traditional analysis methods difficult to assist diagnose ROP perfectly. In this paper, we develop an automated system to analyse premature infants' retinal images using deep neural networks. We primarily try to solve two problems. (1) the existence of ROP, normal or ROP; (2) the severity of ROP, mild-ROP or severe-ROP. Deep neural networks take the advantages of strong representation ability and enable to nonlinear mapping, attain high accuracies and great generalization performances on retinal fundus image datasets.

1. Introduction

Retinopathy of Prematurity (ROP) is a proliferative retinal vascular disease that occurs when abnormal blood vessels grow and spread throughout the retina. Premature infants separate themselves from the maternal environment too early, so the edges of the retina may not get enough oxygen and nutrients. Doctors believe that the periphery of the retina then sends out signals to other areas of the retina for nourishment. As a result, new abnormal vessels begin to grow. These new blood vessels are fragile, weak and can bleed, leading to retinal scarring. When these scars shrink, they pull on the retina, causing it to detach from the back of the eyes.

Nowadays, ROP has become a potentially blinding eye disorder that primarily affects premature infants weighing about 2.75 pounds (1250 grams) or less and being born before 31 weeks of gestation. The shorter gestation of baby, the more likely that baby is to develop ROP.

According to ROP classification guideline [4], 5 stages are used to describe the severity degree of ROP. Stage 1 has a thin but definite structure demarcation line, with recognisable abnormal branching or arcading of vessels. The appearance of Stage 2 is that the ridges arise from the demarcation line and have height and width. Stage 3 demonstrates that neovascularization extends from the ridge into the vitreous. Stage 4 occurs partial retinal detachment. And Stage 5 occurs total Retinal detachment.

The diagnosis of ROP is based on the retinal fundus images from premature infants. A simple screening test and timely treatment by an ophthalmologist can prevent from blindness [8]. The early diagnose of ROP helps to reduce the vision loss of premature infants. However, ROP screening is facing two challenges. Firstly, clinical assessment and grading is difficult, the reason is that the quality of images such as type of lens, focus of image, size of optic nerve, pigmentation and other disease characteristics may influence final clinical assessment badly [1].

Content from this work may be used under the terms of the Creative Commons Attribution 3.0 licence. Any further distribution of this work must maintain attribution to the author(s) and the title of the work, journal citation and DOI.
Published under licence by IOP Publishing Ltd

Due to lack of gold standard, high inter-variability between ophthalmologists causing ROP diagnosis is subjective and prone to low reliability [5]. Secondly, in developing countries like China, India, the large number of population has a conflict with limited medical resources. However, with the rise of birth rate of premature infants, more and more babies cannot gain enough treatment. On the other hand, labor-intensive work is prone to errors occurring.

2. Related Work

Nowadays, due to the excellent results convolutional neural network (CNN) has achieved, more and more researchers start to focus on how to apply CNN on medical image process field. Worrall et al. [12] proposed a novel CNN architecture to diagnose ROP plus disease, which is pre-trained GoogleNet with approximate Bayesian posterior over disease presence. In addition, authors train another CNN to return novel feature map visualizations of pathologies, learned directly from the data. Brown et al. [2] utilize two CNN to diagnose ROP plus disease, the first

CNN is vessel segmentation network, which is used to segment retinal vessel through outputs a probability map whose size is same as input image and the value is between 0 and 1. The second CNN is classification network whose architecture adopts Inception_v1 architecture.

3. Data and Methodology

3.1 Data

Our data comes from Sichuan Provincial Peoples Hospital and Chengdu Women & Children's Central Hospital, which contains 3350 ROP examinations from 2014 to 2017. Every examination consists of 4 to 12 retinal images, which reflects each premature infant's fundus situation.

We adopt samples which have consistent label by different ophthalmologists to construct dataset, and discard samples do not have the same label. Noticeably, the dataset is per-image dataset rather than per-examination dataset. We manually contribute ROP samples to 2 groups, which are mild-ROP and severe-ROP respectively. Last, we divide all samples into three sets, training set, validation set and testing set respectively. Table 1 displays data distribution on every set.

Table 1. The number of samples of different datasets

	Normal	ROP	Mild-ROP	Severe-ROP
Training set	6030	2477	361	814
Validation set	763	465	75	194
Testing set	766	499	93	196
total	7559	3441	529	1204

3.2 Classification Network

3.2.1 CNN architecture. Since 2014, Google proposed inception network architecture continually [7, 9, 10], which improved the best published result on ImageNet again and again. The successful secret of inception network is a module named "inception module" is proposed. Different from traditional convolutional neural network, Inception network contains a series of "inception modules". The inventor of inception network extends the width of "inception module" with using different sizes of kernel to extract different spatial features, which is superior to *traditional hierarchical* convolutional network in performance. Figure 1(a) displays an example of inception module. He et al. [6] proposed a novel CNN architecture named "residual network", which presents a shortcut connection. Residual network achieves the higher classification accuracy on ImageNet. The appearance of these architectures results in the great performance of CNN on image classification field. Figure 1(b) displays the detail of shortcut connection.

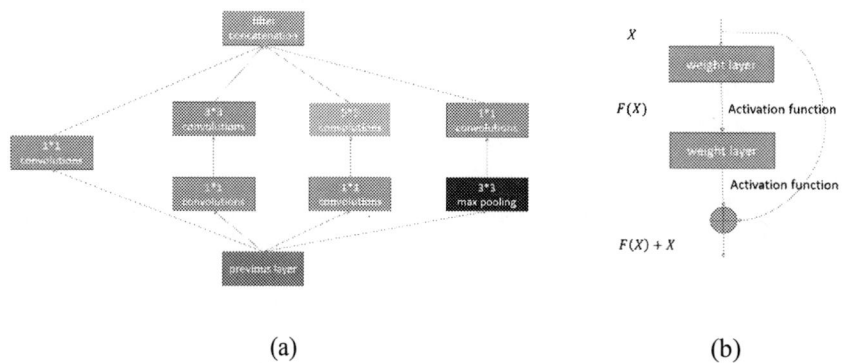

(a)   (b)

Figure 1. Details of different CNN architectures. Fig 1(a) displays an example of inception module. Fig 1(b) displays the details of shortcut connection.

3.2.2 Median Frequency Balancing. Considering that dataset is imbalanced, for example the number of normal samples is much more than ROP samples. We use median frequency balancing on loss function to deal with such a problem. According to median frequency balancing, α_c denotes coordinate of class c while training, which is formulated as:

$$totalloss = \sum_{1}^{n} \alpha_c \cdot loss(c) \qquad where \quad \alpha_c = \frac{medianfreq}{freq(c)}$$

$freq(c)$ denotes the number of class c divided set number, and $medianfreq$ is the median of all frequencies of classes.

3.2.3 Transfer Learning. The primary aim of "Transfer Learning" is proposed to save manual labeling costs with transferring model parameters from labeled dataset to unlabeled dataset. However, considering the difficulties to collect large scale data and the cost of labeling data, more researches gradually to focus on transferring model parameters on two domains, one domain comes from labeled large scale available data, and the other do not have enough data.

4. Experiments and Results

4.1 Normal and ROP

In the experiments of binary classification between normal and ROP. We have a try on pre-trained IncptionV2, InceptionV3 and ResNet-50 neural network. The Result were shown on Table 2. From table 2, we find ResNet-50 architecture demonstrates general best performance on normal and ROP binary classification experiment, and achieves highest accuracy and f1-score. InceptionV2 performs best precision rate on three neural networks. We think that InceptionV2 has the ability to capture more severe ROP disease feature, which is at the expense of reducing recall rate (lowest recall rate in them). However, InceptionV3 perform conversely. InceptionV3 achieves highest recall of 0.9158, but lowest precision rate of 0.8721. In the actual application scenario, doctors would prefer the system not to miss out on any of the ROP cases. Otherwise, it may cause irreversible hurt. However, doctors can tolerate several misdiagnose on negative cases. Through manual screening, they can pick up these false positive samples from positive samples. The process we think is less hard and faster than manual screening on whole of cases. Above of all, InceptionV3 is a good choice for the core of auto-recognition system.

Table 2. Performance of different networks on Normal and ROP binary classification.

Networks	Accuracy	Precision	Recall	F1-Score
InceptionV2(pre-trained)	0.9161	0.9356	0.8453	0.8882
InceptionV3(pre-trained)	0.9138	0.8721	0.9158	0.8935
ResNet-50(pre-trained)	0.9272	0.8998	0.8999	0.8998

4.2 Mild-ROP and Severe-ROP

In the experiment of binary classification between mild-ROP and severe-ROP. Similarly, specific results were shown on table 3. From table 3, we find InceptionV3 achieves best performance among three neural networks on any of evaluation metrics. However, compared with Normal/ROP experiment, almost each evaluation metric is much lower. We think that two primary reasons causing such a result. One is the capacity of mild-ROP/severe-ROP dataset, which is smaller than normal/ROP dataset, and results in easy over-fitting on classification model. Alternatively, there is not enough explicit classification standard on different classes, only depends on the experience of experts.

Table 3. Performance of different networks on mild-ROP and severe-ROP binary classification

Networks	Accuracy	Precision	Recall	F1-Score
InceptionV2(pre-trained)	0.7326	0.7658	0.8718	0.8153
InceptionV3(pre-trained)	0.7847	0.7939	0.9235	0.8538
ResNet-50(pre-trained)	0.7361	0.7913	0.8316	0.8109

5. Conclusion

In the future, we will try to classify the severity of ROP according to different stage and area, and detect position of lesion on premature infants' retinal fundus images. We hope the appearance of the system can help reduce the burden of ophthalmologists and neonatologists for manual screening, and the imbalance of medical resources between developed and less developed regions.

Reference

[1] T. Aslam, B. Fleck, N. Patton, M. Trucco, and H. Azegrouz. Digital image analysis of plus disease in retinopathy of prematurity. Acta Ophthalmologica, 87(5):368–377, 2010. 1, 2

[2] J. M. Brown, J. P. Campbell, A. Beers, K. Chang, S. Ostmo, R. Chan, J. Dy, D. Erdogmus, S. Ioannidis, and J. Kalpathy Cramer. Automated diagnosis of plus disease in retinopathy of prematurity using deep convolutional neural networks. Jama Ophthalmol, 2018. 2

[3] W. M. Fierson. Screening examination of premature infants for retinopathy of prematurity. Pediatrics, 117(2):572, 2006. 1

[4] A. Garner. An international classification of retinopathy of prematurity. Ophthalmology, 92(8):987–994, 1985. 1

[5] Wallace D K , Quinn G E , Freedman S F , et al. Agreement among pediatric ophthalmologists in diagnosing plus and pre-plus disease in retinopathy of prematurity[J]. Journal of AAPOS, 2008, 12(4):352-356.

[6] K. He, X. Zhang, S. Ren, and J. Sun. Deep residual learning for image recognition. CoRR, abs/1512.03385, 2015. 3

[7] S. Ioffe and C. Szegedy. Batch normalization: accelerating deep network training by reducing internal covariate shift. In International Conference on International Conference on Machine Learning, pages 448–456, 2015. 2

[8] P. K. Shah, V. Prabhu, S. S. Karandikar, R. Ranjan, V. Narendran, and N. Kalpana. Retinopathy of prematurity: Past, present and future. World Journal of Clinical Pediatrics, 5(1):35, 2016. 1

[9] C. Szegedy, S. Ioffe, and V. Vanhoucke. Inception-v4, inception-resnet and the impact of residual connections on learning. 2016. 2

[10] C. Szegedy, V. Vanhoucke, S. Ioffe, J. Shlens, and Z. Wojna. Rethinking the inception architecture for computer vision. In Computer Vision and Pattern Recognition, pages 2818–2826, 2016. 2.

[11] D. E. Worrall, C. M. Wilson, and G. J. Brostow. Automated Retinopathy of Prematurity Case Detection with Convolutional Neural Networks. Springer International Publishing, 2016. 2

Research on gait recognition algorithm based on multi-information perception

HuaBin Liu[1]*, ZongMiao Dai[1], YongKang Zheng[2]

[1] Seventh thirteen Institute of China Shipbuilding Industry Corporation, Zhengzhou, Henan, 450015, China

[2] Information Engineering College, Zhengzhou University, Zhengzhou, Henan, 450015, China

*Corresponding author' s e-mail: huabinsky@yeah.net

Abstract: With the collected multi-information, through the information preprocessing, feature vector extraction, analysis and fusion, the recognition model based on BP neural network, genetic algorithm optimization BP neural network, the extreme learning machine and support vector machine are established. We have accurately identified five gait modes: flat walking, upstairs, downstairs, uphill, and downhill. The highest average recognition rate is 96.5%, and the recognition time is 0.156 s. By analyzing the advantages and disadvantages of the four recognition models, the optimal gait recognition algorithm can be obtained.

1. Introduction

In recent years, with the development of human-computer interaction technology, human gait information has been widely studied and applied to the fields of exoskeleton robot control, human motion function pathological detection, and identity recognition. Through the collection and processing of the key data such as human joint motion information, EEG, EMG information, and human-machine force, the current state of motion of the human body is accurately analyzed, which is the premise of sensing the human body' s motion intention, and is also the key technology of the human-computer interaction.

In the 1990s, American scholar Niyogi first proposed the concept of gait recognition, which opened the door for academic research on human gait information. They adopted a computer vision-based approach to gait recognition, analyzed and utilized some of the rules in the gait pattern for the first time [1]. The collected parameters were evaluated by simple pattern analysis of spatio-temporal images, which thus enabled accurate tracking of individuals. In 1997, Richard. F et al. [2] used ultrasonic ranging devices to extract gait information such as pace, stride, gait cycle and peak, and track gait in real time based on changes in gait information. In 2000, Sagawa et al. [3] detected the horizontal and vertical distances of human walking by installing a three-dimensional accelerometer on the toes of the subjects. In 2005, Lee et al. [4] introduced a variety of wearable motion detection sensors to propose a detection method based on spatiotemporal gait parameters. In 2006, the University of Tsukuba in Japan proposed a complete set of human body gesture recognition and prediction theory based on the muscle computer electrical signal and integrated physical sensors [5-6]. In 2010, the University of California at Berkeley established precise mathematical models based on joint motion data and human-machine forces, realizing the multi-pose recognition and control of the human body [7-8]. In 2014, Nogueira et al. of São Paulo University put forward a method of

---

Content from this work may be used under the terms of the Creative Commons Attribution 3.0 licence. Any further distribution of this work must maintain attribution to the author(s) and the title of the work, journal citation and DOI.

Published under licence by IOP Publishing Ltd

estimating the position of lower extremity exoskeleton based on Markov jump linear system using simple exoskeleton data acquisition system [9]. In 2017, Fanello et al. [10] used the sparse representation method for real-time motion recognition and achieved good results.

At present, the gait recognition of bone robots at home and abroad is on the basis of a single sensor. There is no perfect multi-information database. The gait recognition accuracy is low and the real-time performance is not good. In this paper, we collected multivariate information for the needs of gait pattern recognition of new exoskeleton systems, and proposed a multi-person multi-mode gait classification algorithm based on data fusion of multiple physical sensors to provide technical support for human-machine coordinated control of exoskeleton robots.

2. Design of joint information collection system

Human gait information mainly includes plantar pressure, knee and hip joint angle, angular velocity and angular acceleration. In order to accurately obtain these data while eliminating the deviation caused by the sensor installation position, the system introduced a set of exoskeleton models as shown in Figure 1. In this model, the back of the simple upper limb is closely connected with the human body, and the lower limbs are fixed with straps on the thigh and the calf to ensure that the exoskeleton model has good follow abilities to the human body during the experiment. Four nine-axis sensors (MPU6050) are mounted in a coplanar manner on the outside of the thigh and calf to capture angle, angular velocity and acceleration information of the knee and hip joints. When the human body stands, the X axis of the sensor points directly in front of the human body, with the Y axis right, and the Z axis is perpendicular to the earth. The sole is bonded to an independently-developed foot pressure measuring shoe to achieve plantar pressure collection during the subject's walking. In this paper, the multi-point membrane pressure sensor was used to collect the pressure information of the sole. The reasonable distribution of the feature points is beneficial to improve the accuracy of the plantar pressure model. In the experiment, we selected the points with significant plantar pressure changes as the collection points by reference to the plantar pressure distribution of healthy adults. As shown in Figure 3, 15 collection points for each insole can basically satisfy the collection demand of different populations. The plantar pressure collection system is composed of a pressure insole, a connecting wire, and an acquisition module. Joint data and plantar pressure collection frequencies are 40 HZ and 200 HZ, respectively. They can be acquired synchronously when wired

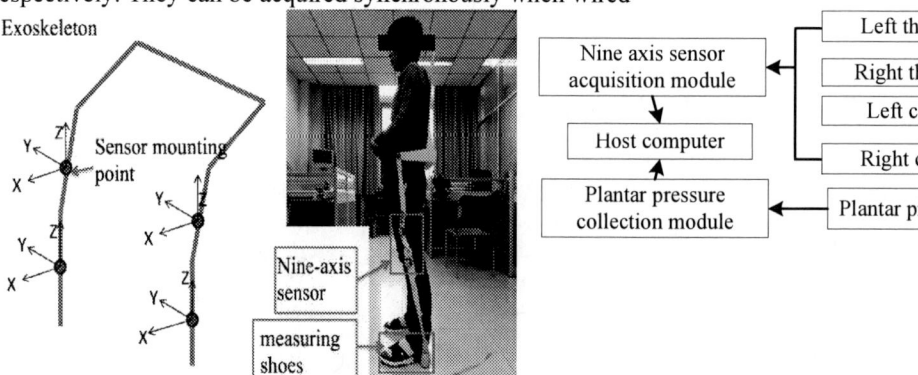

Figure 1. Exoskeleton data acquisition system   Figure 2. System structure diagram

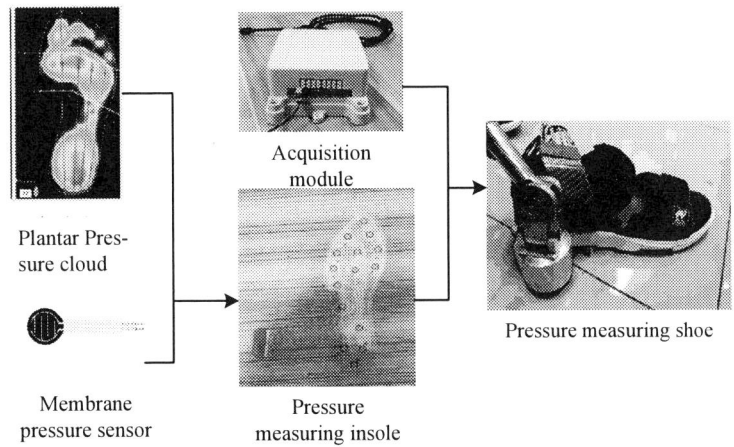

Figure 3. Plantar pressure collection system

3. Data acquisition and preprocessing

3.1. Data collection

The average age of the eight healthy participants is 25.6. All participants realized the experimental procedure and considerations before they participated the experiment. Before doing the experiment, we collected each participant's waist circumference, thigh length and calf length as the basis to adjust the exoskeleton waist width, thigh length and calf length to suit the wearer's needs. Once the experiment began, the participant continued to perform five gait tests: flat walking, uphill, downhill, upstairs and downstairs. Each gait test contained at least 10 gait cycles. During the experiment, there was a flatland between two deferent asynchronous states (about 3-5 meters). There should be a rest for at least 10 minutes during every 5 test periods to ensure the relaxation of the leg muscles and the authenticity and effectiveness of the walking posture data. Every participant was asked to collect ten complete sets of experimental data. The experimental process is shown in Figure 4.

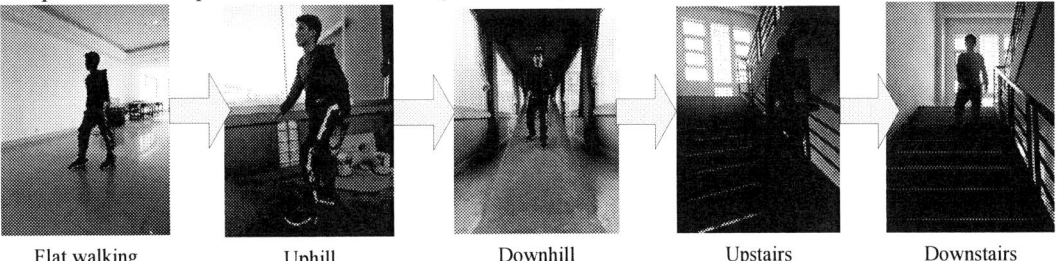

Figure 4. Experimental flow chart

3.2. Data preprocessing

3.2.1. Pretreatment of lower limb motion data

The joint angle is derived from the Kalman filtering, so it does not require denoising. The wavelet transform [11] is used for filtering, and the coif5 wavelet is selected as the wavelet basis function. The decomposition level is N=7. After extracting the high-frequency noise signal, the coefficients of the 1-3 layers are set to 0, and the 4-7 layers use soft threshold filtering. Take the acceleration sensor signal collected by the ground walking as an example. We compared the waveform before and after the wavelet filtering. Figure 5-7 show the hip signal filtering effect.

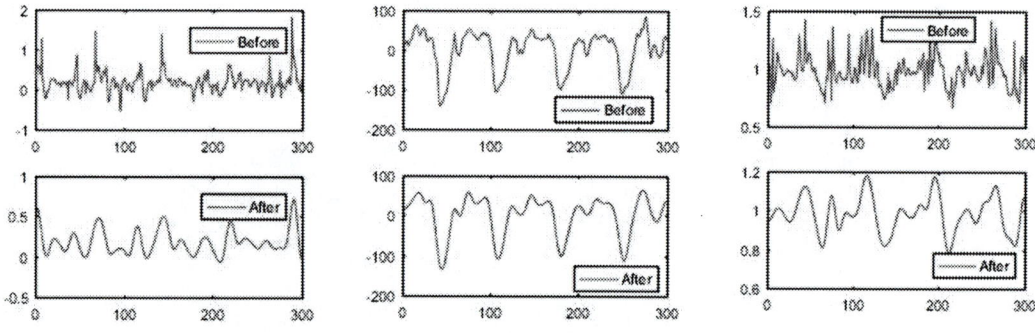

| Figure 5. X-axis acceleration | Figure 6. Y-axis angular velocity | Figure 7 Z-axis acceleration |

It can be seen from the waveforms before and after the signal processing of the nine-axis sensor that the filtering effect is obvious, and the filtered signal exhibits a significant periodicity, which provides data support for the next feature extraction.

3.2.2. Pretreatment of plantar pressure data

We used the butterworth filter [12] to denoise the two typical test points at the forefoot and the heel. The filtering effect is shown in Figure 8.

The data variance after Butterworth filtering is much smaller than the variance of the original data. The filtered curve is smoother and retains valid data in the original signal.

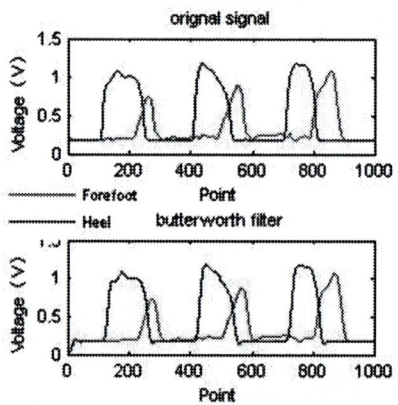

Figure 8. Denoising effect of Butterworth filter

Figure 9. the contribution rate of principal component

4. Feature extraction and fusion

In order to further extract the characteristics in the gait data, we divided a complete gait cycle into the support period and the swing period. The swing period was divided into the pre-swing period, the middle swing period and the late swing period. By calculating the average pressure of the forefoot and the heel, we used threshold analysis to accurately segment the gait phase. We compared features for complete gait cycles.

Common features of the data are generally divided into time domain features, frequency domain features and time-frequency domain features. Since the time domain features of gait information are more suitable for gait recognition, we selected the following features after analyses:

(1) **Mean of angle**, **mean value of the plantar pressure**: reflecting the change of the average amplitude of the signal.

$$\overline{X} = \frac{1}{N}\sum_{i=1}^{n} X_i \tag{1}$$

Where $\overline{X}$ is the average of the angle or foot pressure in a single cycle, and N is the number of samples in a single cycle.

(2) **Mean of angle**: reflecting the fluctuation range of the angle signal

$$S_n^2 = \frac{1}{n}\sum_{i=1}^{n}(X_i - \overline{X})^2 \tag{2}$$

(3) **Variance of angle**: correlation coefficient between acceleration X-axis and Z-axis component

$$r_{xz}^2 = \left[\sum_{i=1}^{n}(X_i - \overline{X})^2\left(Z_i - \overline{Z}\right)\right] \bullet \left\{\sum_{i=1}^{n}(X_i - \overline{X})^2 \bullet \left[\sum_{i=1}^{n}(Z_i - \overline{Z})^2\right]^{\frac{1}{2}}\right\}^{-\frac{1}{2}} \tag{3}$$

Where $\overline{X}$, $\overline{Z}$ are the average values of the X-axis and Z-axis single-cycle signals of the acceleration sensor, respectively, and N is the number of sampling points in a single cycle.

The pairs of different features are shown in Table 1.

Table 1 Comparison of different features

Features	Angle average		Angle variance		Acceleration component correlation coefficient		Plantar pressure		
Gait	Knee	Hip	Knee	Hip	Knee	Hip	Maximum	Varian -ce	Mean
Flat walking	-36.398	-13.408	417.71	129.94	0.6370	-0.0082	1.1032	0.1088	0.5028
Upstairs	-60.529	-24.955	579.19	274.51	0.7254	-0.6040	1.4107	0.0953	0.7102
Downst -airs	-51.281	-20.773	617.45	48.84	0.6850	0.1057	0.7152	0.0421	0.3762
Uphill	-42.059	-16.452	270.63	139.97	0.7628	0.2729	1.2477	0.1648	0.6135
Downhi l-l	-39.204	-15.012	342.57	58.54	0.4790	0.1652	0.8792	0.1244	0.4720

It can be seen from the Table that these features have a high degree of discrimination in the five gaits. Since the sensor can ensure coplanar mounting on the exoskeleton, we can get the angle of each joint by the rotation angle of the Y-axis of the sensor. We selected the correlation coefficient of the X-axis and Z-axis acceleration of the four sensors in the leg, the mean value and variance of the joint angle, the maximum, mean and variance of the plantar pressure to constitute the 15-dimensional matrix about features matrix.

However, not all features have decisive information on the recognition results, many of which are irrelevant and redundant, and it is this unnecessary information that increases the search space of the classifier, thereby reducing its generalization ability [13]. In this paper, we used partial least squares (PLS) to fuse feature vector. Figure 9 shows that the contribution rate of the former 4D principal component of the PLS algorithm to the original feature vector is 93%. We could replace the original feature with the new 4D input, which can effectively shorten the training time of the model and improve the accuracy of the model.

5. Research on gait recognition algorithm

5.1. Algorithm design

This paper used machine learning algorithms to identify five gait modes: flat walking (labeled 1), upstairs (labeled 2), downstairs (labeled 3), uphill (labeled 4), and downhill (labeled 5). We compared the advantages and disadvantages of BP Neural Network, Optimized BP Neural Network by Genetic Algorithm, Extreme Learning Machine (ELM) and Support Vector Machine (SVM) in real time and accuracy. The algorithm design block diagram is shown in Figure 10.

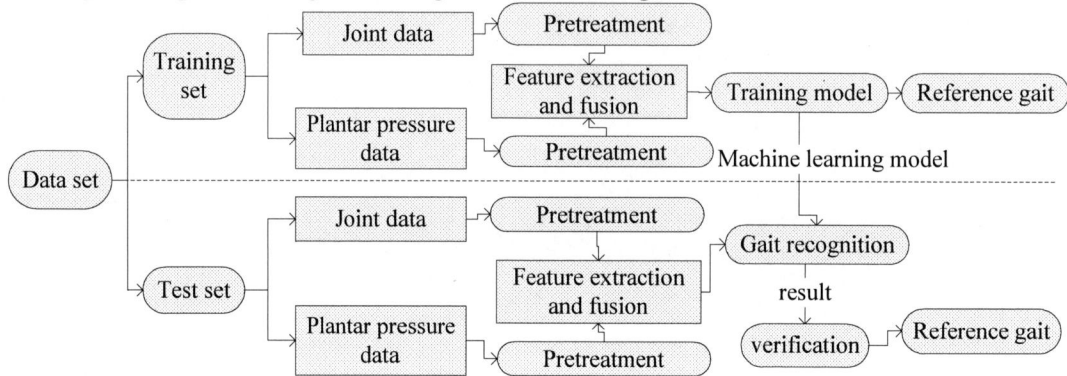

Figure 10. Algorithm flow diagram

We used the eight people's data for modelling and simulation. The fused feature vector in a single gait cycle was used as the algorithm input, and the gait tag was used as the output. We selected 100 samples for each gait, 75 of which were training samples, and 25 were used for testing. The training set totalled 3,000 samples and the test set totalled 1000 samples.

5.2. Calculation and simulation

In order to improve the learning rate of the BP Neural Network and the performance of the recognition model, Genetic Algorithm (GA) was introduced to optimize the initial weights and thresholds of the neural network. In this study, the BP Neural Network adopted a three-layer structure in which the hidden layer was provided with 24 nodes. The initial population of the genetic algorithm was set to 50, and the number of iterations was set to 100. The fitness function was derived from the following formula.

$$SE = \sum_{i=1}^{n}(A_i - A_i^{'})^2 \tag{4}$$

$$val = SE^{-1} \tag{5}$$

Where A_i is the true value of the i th sample, and $A_i^{'}$ is the predicted value of the i th sample. SE is the sum of the squares of the errors, and val is the fitness value of the genetic algorithm. The relationship between the fitness value and the number of iterations is shown in Figure 11. It can be seen that the solution after 50 times is very close to the optimal initial weight and threshold.

The number of hidden layer nodes of ELM [14] has an important influence on the performance of the algorithm model. Figure 12 reflects the influence of the number of nodes on the accuracy of the model. When the number of nodes is 857, the model has the highest accuracy, reaching 93.1%. The Sigmoid function was selected as the activation function of ELM.

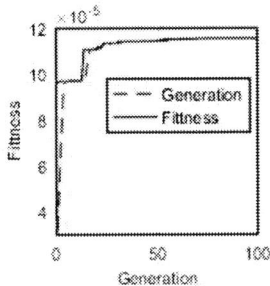 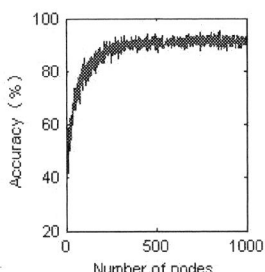

Figure 11. The relationship between the fitness value and the number of iterations

Figure 12. The effect of the number of nodes on the accuracy

In the research, we used the radial basis function kernel. In the SVM [15] training, the kernel function coefficient $g > 0$, and the penalty factor C determined the convergence speed and generalization ability of the SVM. The selection of the C and g parameters directly affected the performance of the SVM. In the study, the gird method was adopted in the parameter selection. Firstly, the range of values of C and g was set. In MATLAB, the for loop was used to calculate the classification accuracy of the training set under the C and g parameters of a single traversal. At the same time, cross validation was used to ensure the validity of the results. Finally, when C =16 and g =0.00097, we got the highest classification accuracy rate. The performance of different algorithms is shown in Table 2.

Table 2. Performance comparison of different algorithms

Algorithm Gait	BP (%)	GA-BP (%)	ELM (%)	SVM (%)
flat walking	89	92	89	100
upstairs	74	87	98	96.5
downstairs	95.5	91	96.5	94.5
uphill	83	86	91	94.5
downhill	67.5	81	91	92.5
average recognition rate (%)	81.8	87.4	93.1	96.5
average time (s)	0.2188	0.125	0.3572	0.0156

5.3. Analysis and comparison of results

In the process of verification, it is found that the BP Neural Network cannot obtain the global optimal solution. This is mainly because the weight used is optimized by the gradient descent method. This optimization process can only optimize it on one point. GA-BP uses genetic algorithms to optimize the initial weights and thresholds of BP Neural Networks, so it exists the same problem. The real time and accuracy and performance of the ELM algorithm are largely affected by the number of neurons in the hidden layer, and the results of each training process are deviations. In general, SVM is better than BP, GA-BP and ELM algorithms. And it has a great improvement in the recognition time. The misidentified data is mainly the uphill and downhill data. This is due to the small slope used in this study, which is similar to the flat walking in some features, leading to the identification error of individual samples.

6. Conclusions

This paper proposes a gait recognition method based on multi-information perception. Firstly, the multi-information of human motion is obtained through the data acquisition experiments, and then the data is smoothed and denoised. The gait cycle is divided according to the plantar pressure and we analyze the gait characteristics in a single cycle, further extract and fuse the effective eigenvalues. Finally, based on the eigenvectors after fusion, the gait recognition model is used to realize the recognition of five kinds of asynchronous states: flat walking, upstairs, downstairs, uphill and downhill. The highest average recognition rate reaches 96.5%. This shows that the SVM algorithm has better generalization performance and recognition accuracy, and the recognition time is short, which provides a theoretical basis for further study of gait information.

References

[1] Niyogi, S. A. , & Adelson, E. H. . (1994). Analyzing gait with spatiotemporal surfaces. IEEE Workshop on Motion of Non-rigid & Articulated Objects. IEEE.

[2] Bae, J. B. J. , Kong, K. K. K. , Byl, N. , & Tomizuka, M. . (2009). A mobile gait monitoring system for gait analysis. IEEE International Conference on Rehabilitation Robotics. IEEE.

[3] Sagawa, K. , Inooka, H. , & Satoh, Y. . (2000). Non-restricted measurement of walking distance. IEEE International Conference on Systems. IEEE.

[4] Lee, C. S., & Elgammal, A. (2005). Towards scalable view-invariant gait recognition: multilinear analysis for gait. International Conference on Audio-& Video-based Biometric Person Authentication.

[5] Kawamoto, H. , & Sankai, Y. . (2004). Power assist method based on phase sequence driven by interaction between human and robot suit. IEEE International Workshop on Robot & Human Interactive Communication. IEEE.

[6] Hayashi, T. , Kawamoto, H. , & Sankai, Y. . (2005). Control method of robot suit HAL working as operator's muscle using biological and dynamical information. IEEE/RSJ International Conference on Intelligent Robots & Systems Iros.

[7] Zoss, A. , Kazerooni, H. , & Chu, A. . (2005). On the mechanical design of the Berkeley Lower Extremity Exoskeleton (BLEEX). 2005 IEEE/RSJ International Conference on Intelligent Robots and Systems. IEEE.

[8] Berkeley Exoskeleton (BLEEX), Berkeley Robotics Laboratory [Online]. Available: http://bleex. me.berkele/edu/bleex.htm.

[9] Nogueira, S. L. , Siqueira, A. A. G. , Inoue, R. S. , & Terra, M. H. . (2014). Markov jump linear systems-based position estimation for lower limb exoskeletons. Sensors, 14(1), 1835-1849.

[10] Fanello, S. R. , Gori, I. , Metta, G. , & Odone, F. . (2017). Keep it simple and sparse: real-time action recognition. Journal of Machine Learning Research, 14(1), 2617-2640.

[11] Wang, D., & Tsui, K. L. (2017). Dynamic bayesian wavelet transform: new methodology for extraction of repetitive transients. Mechanical Systems & Signal Processing, 88, 137-144.

[12] Li, Z. (2008). Electrical design of high order active low-pass filter based on improved butterworth transfer function. Journal of Electronic Measurement & Instrument.

[13] Yang X, Tian Y L. . (2014). Effective 3D action recognition using EigenJoints. Journal of Visual Communication & Image Representation, 25(1): 2-11.

[14] Huang G B, Zhu Q Y, Siew C K. . (2006). Extreme Learning Machine: Theory and Applications. Neurocomputing, 70(3): 489-501.

[15] Chauhan, V. K. , Dahiya, K. , & Sharma, A. . (2018). Problem formulations and solvers in linear svm: a review. Artificial Intelligence Review.

The combination of neural network and "question matching" improves the correct rate of grassland degradation decision

Li Chunmei[1,2], Pi Wei[1], Dong Suo[1] and Li Zhao[1]

[1] Department of Computer Technology and Application, Qinghai University, Xining, Qinghai 810016, China

[2] Li Chunmei's e-mail: li_chm0422@sina.com

Abstract. This paper, analysed the deficiency of grassland degradation by neural network method, and proposed a method combining neural network with "question matching". The neural network evaluation is the main line, supplemented by "inquiry matching" evaluation, and supplemented by "degradation indicating grass species". The combined evaluation of grassland degradation improves the correct rate of grassland degradation decision.

1. Insrtuction

The Sanjiangyuan region is an ecological barrier for the ecological environment security and regional sustainable development of the Yangtze River, the middle and lower reaches of the Yellow River, and Southeast Asian countries. However, under the dual influences of global climate change and human activities in recent decades, grassland vegetation degradation in the Sanjiangyuan region is seriously, the source water conservation capacity has been drastically reduced, and it has directly threatened the ecological security of the Yangtze River, the Yellow River Basin, and even Southeast Asian countries. In 2005, the state invested 7.5 billion yuan to launch the Sanjiangyuan area ecological protection project, and carried out three major constructions: ecological protection, agricultural and herdsmen production and living infrastructure and ecological protection support. Since the implementation of the project in 2005, through the implementation of a series of measures such as returning grazing and returning grass, closing hillsides for afforestation, returning farmland to forests, black soil beach management, rodent control, soil and water conservation and artificial precipitation enhancement projects, the vegetation in the implementation area has been significantly restored, the water source conservation function has been realized initially, the area of the swamp and lake in the source area has expanded to varying degrees [1].

In the long-term research on Sanjiangyuan ecological environment protection, researchers collected a large amount of data on ecology, geography, geology, environmental science, sociology, economics, etc., which condensed several generations of researchers' hard work and sweat, but because of the relationship between the disciplines of scientific research workers, they are often isolated information data, with the characteristics of the discipline and data isolation. It has long relied on expert's experience and manual methods to evaluate grassland degradation. In the long-term research work, we solved the problem of grassland degradation degree determination and decision-making in the Sanjiangyuan region by using computer artificial intelligence technology to develop an expert system for hierarchical decision-making and treatment of alpine meadow grassland in the Sanjiangyuan alpine meadow. On the one hand, it can replace the grassland expert with computer to make expert-level decision-making, saving manpower, material resources and financial resources. On the other hand, the

---

Content from this work may be used under the terms of the Creative Commons Attribution 3.0 licence. Any further distribution of this work must maintain attribution to the author(s) and the title of the work, journal citation and DOI.

Published under licence by IOP Publishing Ltd

expert's knowledge can be systematically summarized, stored in the computer for a long time, and the expert knowledge can be protected and passed down.

2. Neural network decision-making sanjiangyuan grassland degradation decision

For grassland degradation [2, 3, 4], there are many research methods, different angles, different entry points, and different conclusions, such as references 1, 2, and 3. This study carried out detailed research and analysis on the degradation classification of alpine meadow grassland in Sanjiangyuan area, the main factors affecting degradation, and the treatment measures under each degraded condition, and based on the alpine meadow in the Sanjiangyuan area of Qinghai Plateau, Meadow grassland grading for decision making. On the basis of the research data, the data is divided into two categories, and the first type of data is determined by using neural network. At the same time, the second type of data is supplemented by "question matching" method, and the two are organically combined to establish the management of alpine meadow in Sanjiangyuan area, provides technical support and theoretical basis for the Sanjiangyuan ecological protection project.

Neural networks [5, 6, 7, 8] have been widely promoted in many fields because of their parallelism, solving nonlinear relationships, etc. In recent years, methods for improving them have emerged in an endless stream, as in References 5, 6, 7, 8, etc. This research is not based on the improvement of the neural network itself, but on the basis of it, to its deficiencies, uses complementary methods make the neural network more perfect. The use of neural networks to determine grassland degradation has its advantages of speed and intelligence, but it also brings some details of the decision.

2.1. Evaluation of data in the "blind zone"

The title is set 17 point Times Bold, flush left, unjustified. The first letter of the title should be capitalized with the rest in lower case. It should not be indented. Leave 28 mm of space above the title and 10 mm after the title.

2.1.1. Definition of "blind zone". In the study, researchers collected and organized more than 20,000 pieces of data as a training set for neural networks. In the training set, there are five categories of data, namely, non-degraded, mildly degraded, moderately degraded, severely degraded, and extremely degraded. The grass species of each degraded category have obvious characteristics. According to these characteristics, they are summarized into five factors according to the expert's point of view, namely, the smear degree (tbdgd), the edible grass ratio (ksmcbl), and the degraded indicator grass species proportion (thzsczbl), soil organic matter content (tryjzhlbl), rodent damage (shqk). The five input data corresponding to each type of grassland has a distinct feature in distribution, and each type of data is roughly distributed in different regions, as shown in Figure 1 below. Experience has shown that the distribution of the five types of data is regular, as shown. The distribution of the data in the figure is the distribution of the ratio of forage to soil organic matter content. The five categories on the diagonal from the upper left to the lower right are the relation distribution of proportion of edible forage grass and soil organic matter content under the conditions of extreme degraded, severe degraded, moderate degraded, mild degraded and non-degraded respectively. It can be seen from the figure that the input data of the range indicated by the arrow cannot be evaluated, or the test result of the input data between the areas is a category that cannot be judged, "blind zone". Therefore, the data in this area, researchers call it "blind zone" data.

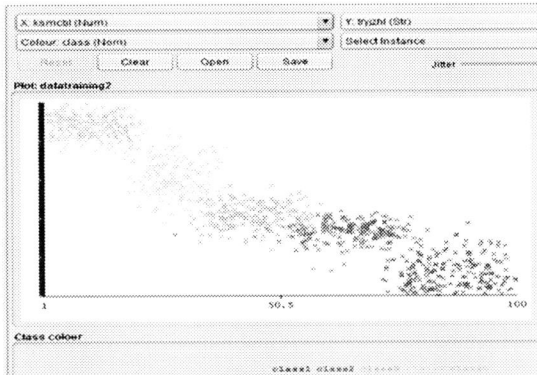

Figure 1. Data distribution of five types of grassland

When "blind zone" data is used as input of neural network, the evaluation result will be wrong. Because the "blind zone" data is not within the range of training data sets and test sets of the neural network, the output of the "blind zone" data as input of the neural network is 100% wrong.

2.1.2. Description of "blind zone" data. The distribution of characteristic data of each type is regular, In soil with a high proportion of edible forage grass, the content of organic matter is naturally high; The organic matter content in soils with high patch coverage and even reduced to black soil paralysis is naturally low.

That is, there is a correlation between the five factors that affect the degree of grassland degradation. As shown in Figure 2, there is a correlation between the proportion of degraded indicator grass species and soil organic matter content. For example, the relationship between people's age and running speed is related. As people get older, their running speed increases; After reaching a certain limit, the older you get, the slower you start to run.

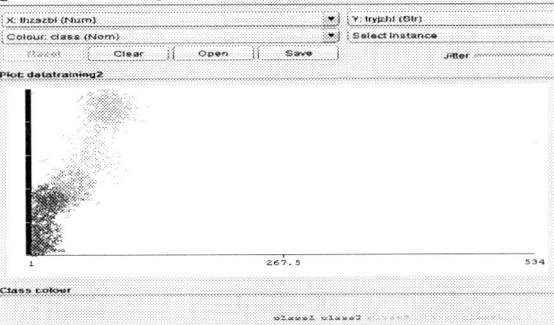

Figure 2. Correlation between the proportion of degraded indicator grass species and soil organic matter content

In this experiment, the relationship between the five grassland types and the proportion of edible forage grass, the proportion of degraded indicator grass species, and the content of soil organic matter were all correlated, as shown in Figure 3, 4, 5 below.

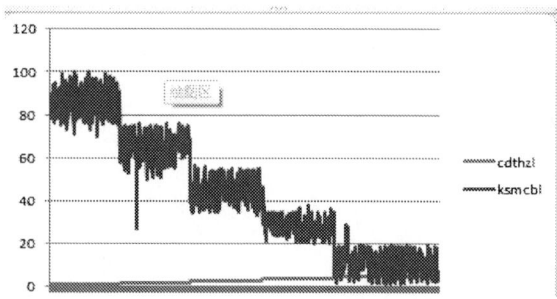

Figure 3. Correlation between grassland degradation species and proportion of edible forage grass.

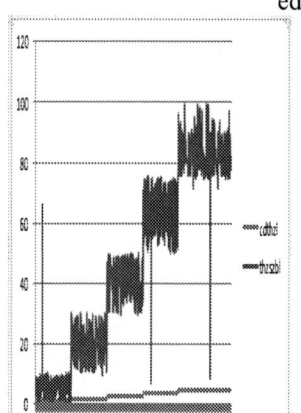

Figure 4. Correlation between grassland degradation species and the proportion of indicator species.

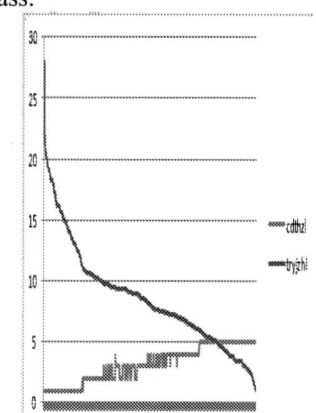

Figure 5. Correlation between grassland degradation species and soil organic matter content.

3. Evaluation of neural network model

3.1. Defects of neural network model evaluation
Due to the defects of the neural network model itself, the testing accuracy of the test data is not 100%, which leads to the accuracy of the neural network model is not 100%, which means that incorrect evaluation results will inevitably appear in the application. Based on this reason, researchers will conduct supplementary evaluation in the following ways to improve the accuracy of grassland evaluation.

3.2. Combination of neural network and "question matching" to improve the decision accuracy
In this study, researchers found that, in addition to the above five important factors, the degradation indicator grass species can be used again to assist in the evaluation of the degradation degree of grassland. Therefore, researchers can further determine whether there is a problem in the previous judgment by indicating whether degraded indicator grass species appears and its proportion. The specific evaluation scheme is shown in Figure 6 below.

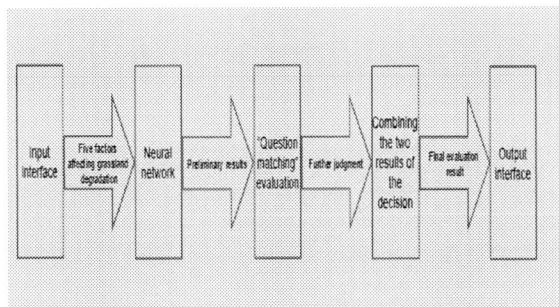

Figure 6. Evaluation method of neural network combined with "question matching".

3.2.1. "Question" whether the input data is in the blind zone to exclude areas that cannot be evaluated. Data in the "blind zone" needs to be excluded. The distribution of the data is within a certain range to fit the turf category. For example, in the non-degraded grassland, the proportion of forage grass is bound to be a large proportion. If all the weeds are poisonous, and the grassland category is non-degraded, this cannot happen. Or, in the non-degraded type of grassland, the organic content of the grassland is very low, and there is a very dense plant community in the barren land, which is not in line with the natural law. Therefore, before entering the neural network for the first evaluation, a judgment should be made. If the data in the "blind zone" does not enter the evaluation process. Directly output, your input data does not match the actual situation, please input the correct data. The evaluation process cannot be entered until the input data is "correct".

3.2.2. "Question" other "knowledge" for further judgment

3.2.2.1. Data at category junction. In the first evaluation result of neural network, there may be a small amount of data input that is wrong, in order to avoid the occurrence of such error, this project adopts the second evaluation method and further "question" [1], the characteristics of each type of grassland can be correctly evaluated if they conform to the characteristics of this type of grassland. Otherwise, compare the "input data" of the first evaluation with the data of the "question matching" of this time, if their common characteristics of " bareland coverage" and "organic content" are similar, if on the basis of some similarity, then "question" dominant grasses and "companion species", if both are matched, the final category can be determined by combining the level of the first evaluation with which category is close. Figure 7, Figure 8 and Figure 9 below show three types of grassland characteristics: non-degraded moderate degraded and extreme degraded.

Suppose, some data input to the neural network is (0.71, 0.19, 0.79, 0.2, 0.7), and the result of the first evaluation is (0, 0, 0, 0.7776, 0.7845), there's no way to tell if it's category 4 or 5. At this point, the input data of the neural network was traced back to the original data (the data before normalization), which was (71.5, 20, 79, 6.9, 0.72) and compared with the criteria for the secondary evaluation, the standard of bareland coverage is slightly greater than extreme degraded, and the content of organic matter is lower than extreme degraded, so again, "question" the companion situation, for example, if the herbage and the arachnids disappear, if they disappear, continue to "question" what the dominant species is, if it is twelve-year-old poisonous weeds, such as Ligularia virgaurea, Ajuga lupulina and other plants, you can interpretation this kind of grass has extreme degraded; If the content of organic matter and the coverage of the plaque land cannot be determined, continue to "question" whether its companion species are mainly containing weeds such as Oxytropis and Lancea tibetica, and if "yes", continue to "question" whether dominant grasses are mainly dominated by stoloniferous weeds, and if so, are severely degraded.

3.2.2.2. Data in category area. The data in the obvious category area, that is, the neural network model evaluates the correct category, generally speaking, will not be wrong. In case of an error, the original data input from last time and this time can be quickly determined by using the same 2.2.2.1 method. Figure 7, 8, 9 are the partial grassland features of the five categories of grassland in this project, such as three categories of grass features of non-degraded, severe degraded and extreme degraded.

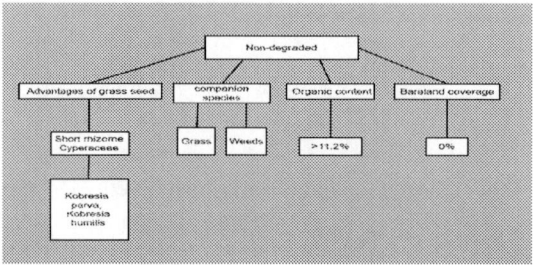

Figure 7. Characteristic of non-degraded grasses.

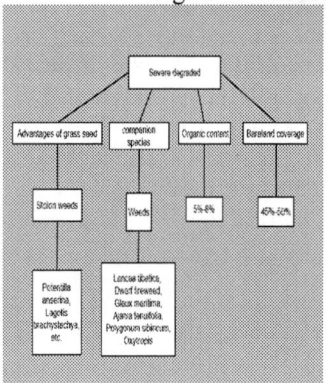

Figure 8. Characteristic of severe degraded grasses.

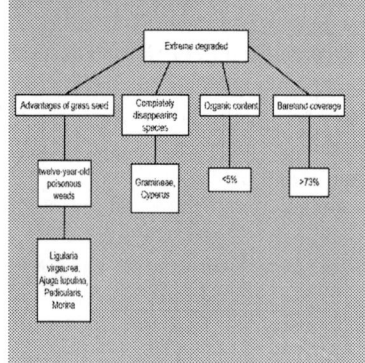

Figure 9. Characteristic of severe extreme grasses.

4. Conclusion

The evaluation of grassland is a complicated process, and the conclusion will be different if the experts study from different perspectives. In this paper, taking the view of some experts as an example, the evaluation process of grassland is completed. The first evaluation of the grassland was carried out by training the neural network model with a large number of experience values from five important factors including the coverage of the patch, proportion of dominant grass species, proportion of edible forage grass, soil organic matter content and rodent damage. After that, from the perspective of organic matter content, patch coverage, companion species and dominant grass species, it was evaluated again with the input data of neural network, and finally the degradation degree of grassland was obtained, which greatly improved the accuracy of grassland decision-making.

References

[1] Li CM, Wang YM, Tian F, Zhou XK, Cui P. (2018) Cluster Computing[j],First Online: https://doi.org/10.1007/s10586-018-1698-x

[2] Liu SL, Tang YH, Zhang FW, Du YG, Lin L, Li YK, Guo XW, Li Q, Cao GM. (2017) Changes of soil organic and inorganic carbon in relation to grassland degradation in Northern Tibet[j], Ecological Research, May 2017, Volume 32, Issue 3, pp 395–404

[3] Su XK, Wu Y, Dong SK, Wen L, Li YY, Wang XX. (2015) Effects of grassland degradation and re-vegetation on carbon and nitrogen storage in the soils of the Headwater Area Nature Reserve on the Qinghai-Tibetan Plateau, China[j], Journal of Mountain Science, May 2015, Volume 12, Issue 3, pp 582–591

[4] Ma L, Yao ZS, Zheng XH, Zhang H, Wang K, Zhu B, Wang R, Zhang W, Liu CY. (2018) Increasing grassland degradation stimulates the non-growing season CO_2 emissions from an alpine meadow on the Qinghai–Tibetan Plateau[j], Environmental Science and Pollution Research, September 2018, Volume 25, Issue 26, pp 26576–26591

[5] Zhou WH, Xiong SQ. (2013) Optimization of BP Neural Network Classifier Using Genetic Algorithm[j], Intelligence Computation and Evolutionary Computation,2013, pp 599-605

[6] Hao G. (2014) Study on Prediction of Urbanization Level Based on GA-BP Neural Network[j], Proceedings of the 21st International Conference on Industrial Engineering and Engineering Management 2014, pp 521-524

[7] He F, Zhang LY. (2018) Mold breakout prediction in slab continuous casting based on combined method of GA-BP neural network and logic rules[j], The International Journal of Advanced Manufacturing Technology,April 2018, Volume 95, Issue 9–12, pp 4081–4089

[8] Cheng Y, Wu R. (2016) The Research of Aviation Dangerous Weather Forecast for Fog and Haze Based on BP Neural Network[j], Proceedings of the 5th International Conference on Electrical Engineering and Automatic Control pp 877-883

The Design and Development of Simulation System for Broad Band Wireless Communication

Lijuan GAO*, Yu HE

School of Space Information, Space Engineering University, Beijing 101416, China

gljhappy@sohu.com   te20100704@sohu.com

* Corresponding Author: Lijuan GAO; gljhappy@sohu.com; 13661324163

Abstract: Based on the analysis of the broad band wireless communication, we design a wireless terrain network with some different subnet of satellite, microwave, LTE, and so on. OPNET provides a virtual network environment, through the modeling of the network. We give the simulation flow of the wireless communication. It designs the station node and satellite node, and builds a simulation system for the broad band wireless communication by OPNET. According the wireless communication topology, it realizes the network simulation system with six zone subnets which contain six single users and two communication vehicles. It contains the communication satellites which use the transparent transponders. The simulation system can display the dynamic route by the simulation system. It also can analyze the performance and efficiency of the wireless system with the parameters, such as: throughput, delay, processing delay, Packet Loss Ratio. It can use "Failure Recovery" module in the simulation system and set the parameters and analyze the communication system influence.

1. Introduction

With the development of the communication technology, wireless communication is used widely and widely with great virtue. We can use network simulation technology in system evaluation, in order to research the wireless communication systems.

Network simulation is widely used in communication network performance analysis [1]. The simulation system of the broad band wireless communication is built by the network simulation tool, which can simulate the wireless communication system [2]. It can build the network equipment, link and protocol model and simulate the transfers of the network flow by network simulation [3, 4]. The simulation system can be used to analyze the performance of the communication system and provide the theory support [5-9]. It can collect the network data for network design and make the network as perfect or effective as possible.

2. Design of the Simulation System

OPNET provides a virtual network environment, through the modeling of the network [6,10]. The user can analyze the network performance more effectively. OPNET can meet the simulation needs of the great complex network with 3 layers models, which can correspond with protocol layer, equipment layer, network layer. OPNET can completely reflect the network characteristics.

　1) Process model

　Process model can simulate the protocol and depict the protocol by finite state machine (FSM).

　2) Node model

Content from this work may be used under the terms of the Creative Commons Attribution 3.0 licence. Any further distribution of this work must maintain attribution to the author(s) and the title of the work, journal citation and DOI.

Published under licence by IOP Publishing Ltd

Node model can simulate the equipment and be made of corresponding protocol model.

3) Network model

Network model can simulate the network and be made of corresponding node model.

OPNET wireless module can provide wireless modeling environment and simulate wireless transmission performance, such as: terrain diffraction, fading, path loss because of atmosphere. It can simulate the influence of the network performance by the node moving and the connection with the other communication system. The simulation flow includes the process of wireless communication protocol modeling, node module modeling, network topology construction, traffic load deployment, statistic selection and simulation analysis and so on, which is showed as figure 1.It can build the wireless communication simulation system by OPNET, which can realize the model modeling, performance evaluation, network layout [6~8].

The broad band wireless communication can analyze the performance of the wireless communication system, and put forward some suggestions for the layout and development of the system.The wireless communication system contains six wireless zone subnets, such as: Beijing, Xi'an, Xichang, Taiyuan, Jiuquan, Xiamen, which can cover the area with 50 kilometers * 50 kilometers. The subnets have some simulation objects, such as: base station, individual user, communication vehicle, which are connected with optical fiber, satellite, microwave, ultrashort wave, wireless Mesh, LTE. It can use the terrain data in the zone subnets' simulation, which can make the simulation system better reflect the true wireless communication and better analyze the performance of the communication.

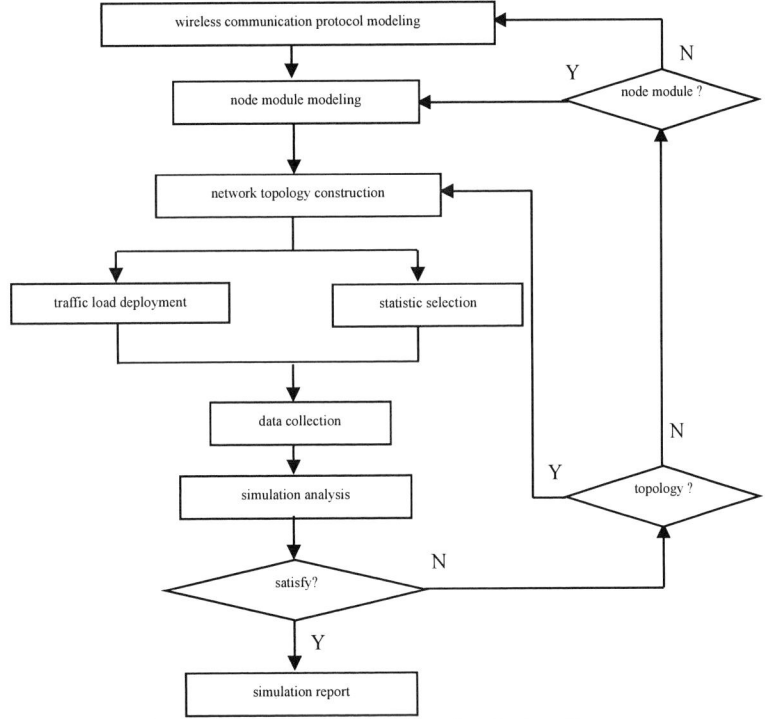

Figure 1. sketch map of the simulation flow

Then it needs to define application and deploy traffic flow. The applications contain FTP, HTTP, Email. It can connect the source node and object node with ip_traffic_flow and deploy the traffic flow with ip_traffic_flow module.

3. Simulation System and Key Node Models

3.1. Simulation System Realization

According the wireless communication topology, it realizes the network simulation system with six zone subnets which contain six single users and two communication vehicles [5], showed as Figure 2.

The terrain model can be used in the simulation system, which can support to load the true terrain data. It can simulate the environment influence, such as terrain factor, the refraction of earth's surface. It can analyze the signal attenuation because of the terrain factors, then well and truly reflect the performance of the wireless communication system. The terrain model contains Free Space Transmission Model, Longley-Rice Model, HATA Model, TIREM Model and so on. It can choose different transmission model and collocate the parameters according to the real terrain in order to reflect the wireless communication system's efficiency in different terrain.

The simulation system can realize the function, such as: model development, terrain simulation, statistic analysis.

1) Model development

The simulation system contains the equipment model from OPNET, such as: router, switch. There are some other equipments, such as: satellite node, ground station, microwave relay node, which need to develop modeling simulation.

2)Terrain simulation

The simulation system can import terrain data by the terrain data interface and use the transmission model in order to well analyze the performance of wireless communication system.

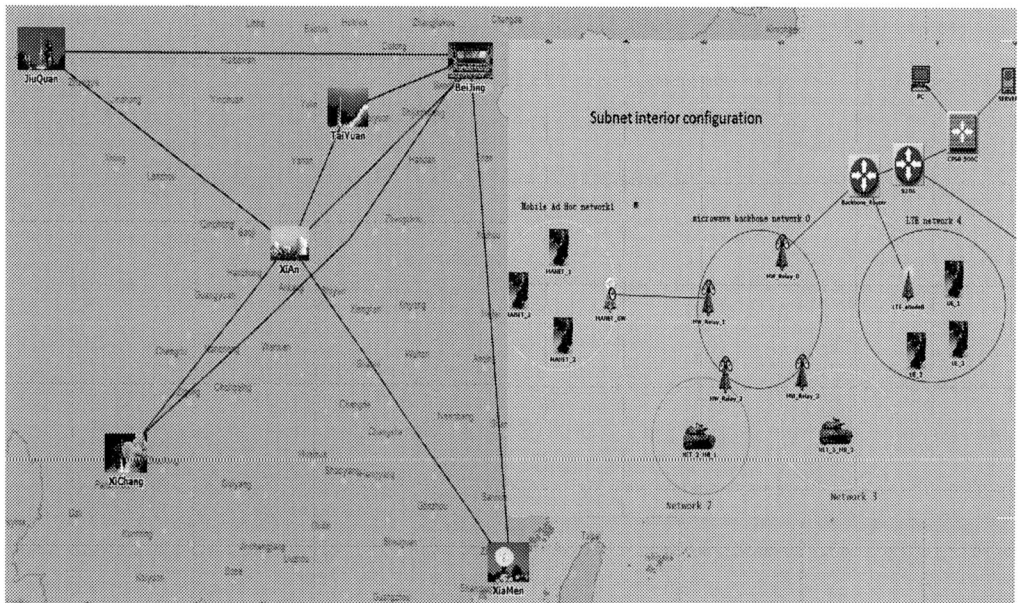

Figure 2. Simulation system and subnet interior configuration

3) Statistic analysis

According to the real evaluation requirement, it can choose the corresponding statistics by Project Editor in the simulation system, such as: point-to-point utilization, point-to-point throughput, in packet jitter, in packet ETE delay, Traffic Receive, Traffic sent, and so on. Then it can collect the simulation data by running the simulation system, which can analyze the system's performance and evaluate the system's efficiency.

3.2. Satellite Model

The satellite nodes in the simulation system need to develop by program. The other nodes can use the models from OPNET, such as: router, switch, microwave node, LTE base station, Which can modify the configuration according to the real equipments.

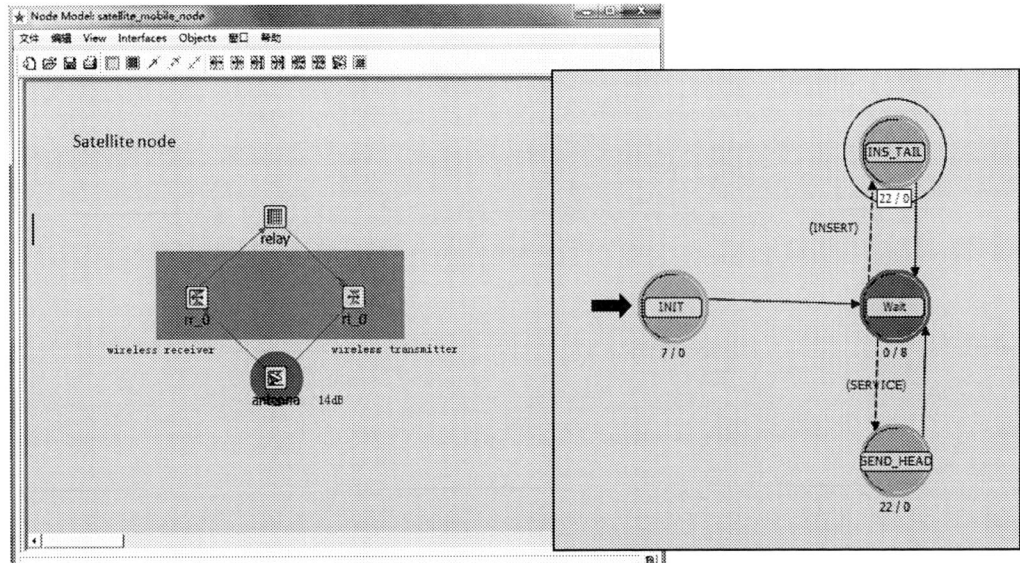

Figure 3. Process model of satellite and relay module

Communication satellites almost are in Geostationary Earth Orbit(GEO), which use the transparent transponders. It can realize the frequency conversion and data transmission after receiving data without the function of onboard switching and onboard processing on satellites. So the satellite node only needs to realize the function of data transmission with the models of relay module, module of Radio Frequency transmitters and receivers, antenna module. The satellite node model and relay module are showed as figure 3. It uses omni-directional antennas with 14dB plus.

The work flow of satellite node is as following:

1) The "antenna" model receives the signal, which can increase 14dB gain;

2) "rr_0" Radio Frequency receiver model receives the signal;

3) "relay" model can choose appropriate frequency to transmit the signal;

4) "rt_0" Radio Frequency transmitter model can transmit the signal.

The "relay" model can realize the repeater's frequency choice as the core of the satellite node. It can realize the function with four process modules, such as: "INIT", "INS_TAIL", "Wait", "SEND_HEAD".

INIT: Process initializes and goes into "Wait" state;

Wait: "Wait" state is ready to be activated by data stream, then turns into "INS_TAIL" satate;

INS_TAIL: It inserts received data packet in queue, then changes the state from "self interrupt" into "SEND_HEAD" state by activation;

SEND_HEAD: It can confirm the suitable frequency in order to transmit satellite signal by "CI" binding in the packet. Then it can acquire the number of the"rt_0" Radio Frequency transmitters and receivers by "confirm_tx_chann_num" function. It can choose suitable channel to transmit the packet by "confirm_transmitter_frequency" function, which can realize the data forwarding.

4. Application of Simulation System

The simulation system of wireless communication can simulate communication process and route choice between different users in different terrain. It also can well display the IP flux route between different users. It can analyze the end-to-end performance, evaluate wireless network efficiency, layout and deployment of the wireless network, the evaluation for network invulnerability [10].

1) End-to-End Performance Analysis

It can set parameters according with real wireless network, such as: network topology, traffic flow, protocol. Then it can collect the simulation data in order to analyze the End-to-End Performance of the wireless communication system.

2) Evaluation of Wireless Network Efficiency and QoS (Quality of Service)

It can evaluate the wireless network efficiency by different evaluation index. It also can compare the different QoS of the network after adding new wireless service.

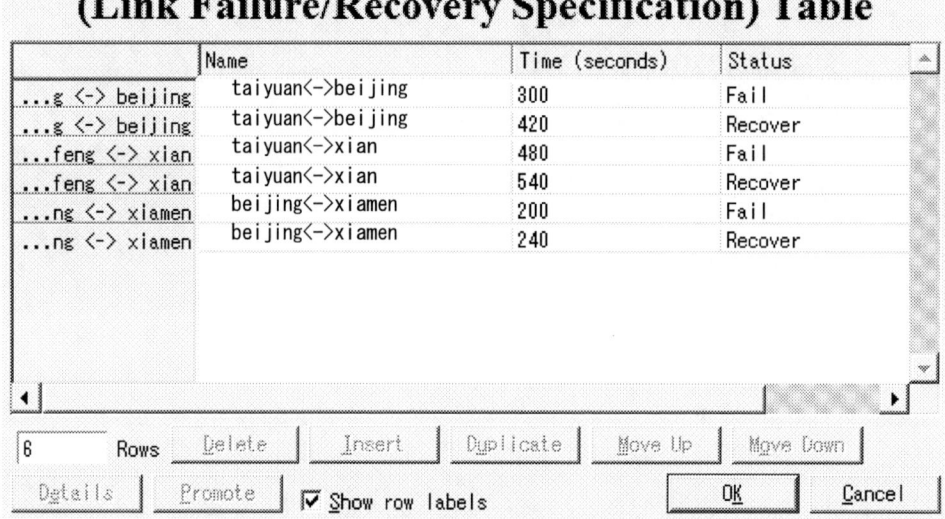

Figure 4. Link Failure and Recovery

3) Layout and Deployment of the Wireless Network

It can evaluate the network efficiency after new layout and deployment, analyze the influence of terrain factors for network deployment.

4) Network Invulnerability Evaluation

It can analyze the influence of the wireless communication system after some key nodes are trouble by changing some simulation parameters. So it can evaluate the network invulnerability viability after being attacked in wartime.

It can use "Failure Recovery" module in the simulation system and set the parameters according to Figure 4. The link between Beijing and Xiamen is failure in 200 second and resume in 240 second. The link between Taiyuan and Beijing is failure in 300 second and resume in 420 second. The link between Taiyuan and Xian is failure in 480 second and resume in 540 second. When the link is failure, the route is interrupt. It needs to choose a new route to continue the data transmission, which can make some packet lost, showed as Figure 5. Once the route is built, it will be work well and have no influence for the communication.

In order to evaluate the system performance, it needs to choose different indices according to different objects. It can choose indices, such as: throughput, delay, processing delay, Packet Loss Ratio for performance analysis of the equipment. It can choose indices, such as: global load, throughput, delay, for performance analysis of the network. It can choose service throughput, delay, delay jitter for performance analysis of the service statistics. It can collect the simulation results by different methods, such as: global statistics results, node statistics results, link statistics results, flux statistics results.

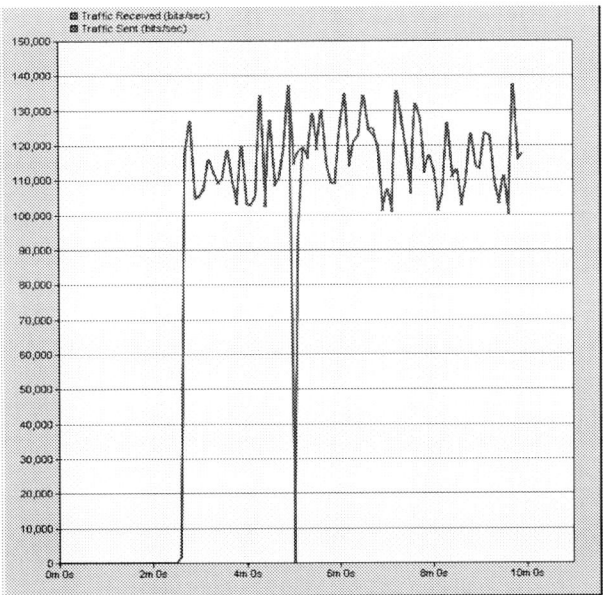

Figure 5. Comparing traffic received with traffic sent

5. Conclusions

It designs the simulation system for the broad band wireless communication, which can simulate the wireless communication network under different environment. The simulation system can well reflect the performance of the wireless communication network and evaluate the system efficiency. It can provide theory support for the system development and ensure the wireless communication network with the performance of security, reliable, high effective, real-time.

References

[1] ZHU Chen, DONG Yin-hu. Network Simulation Technology and Its Application Based on OPNET [J]. Radio Engineering , 2013, 43(3): 12-15,61.

[2] Zhang Shuqiao. OPNET Simulation Test for Multiple Wireless Communication Modes [J]. Digital Communication World, 2014, 10: 63-65.

[3] YE Libang, WANG Manxi, ZHANG Yuling, GENG Hongfeng. Research on multi-connection modes based topology model of wireless communication network [J]. Modern Electronics Technique, 2016, 39(7): 1-4,9.

[4] GUO Pengfei;ZHANG Jie;LV Ming;BO Yuming. Modeling for wireless communication network and fault propagation[J]. Computer Engineering and Applications, 2015, 51(3): 1~5.

[5] XU Heng-jie, LIN Tao, ZHANG Kun, ZHANG Li. Research on Distributed Scene Simulation System Based on WCN [J]. Video Engineering ,2010,34(7): 111~114.

[6] LIU Min, GAO Ming-xia, LU Si-yu, QIAO Hui-dong. Research on OPNET-based simulation of field communication network[J]. Electronic Design Engineering, 2012, 20(4): 120-124.

[7] ZHOU Hua;LIU Zhuang;HAN Wei;HUANG Weifang Design of distributed parallel wireless communication simulation platform based on 5G[J]. Computer Engineering and Applications, 2016, 52(22): 15~21,85.

[8] PEI Xiao-dong, WANG Lin, LIU Wei. Simulation Design of Dynamic Spectrum Access System[J]. Journal of China Academy of Electronics and Information Technology, 2017, 12(4): 414~419.

[9] CHEN Hai-bin. Wireless Network Communication Information Transmission Efficiency Optimization Simulation [J]. Computer Simulation, 2017, 34(10): 159~162.

[10] ZHU Chen, DONG Yin-hu. Network Simulation Technology and Its Application Based on OPNET[J]. Radio Engineering, 2013, 43(3): 12-15,61.

A Deep Learning Approach for Vehicle and Driver Detection on Highway

Peihua Lv[1,2], Yang Zhang[1,2], Xiaobo Lu[1,2,a] and Di Zhou[3]

[1] School of Automation, Southeast University, Nanjing 210096, China

[2] Key Laboratory of Measurement and Control of Complex Systems of Engineering, Ministry of Education, Southeast University, Nanjing 210096, China

[3] Zhejiang Uniview Technologies Co., Ltd, Zhejiang, 310018, China

[a] Corresponding author: xblu2013@126.com

Abstract. The technology of the detection for vehicle and driver is a popular spot in these ten years. In particular, the driver detection is still a troubled question in the study of public security. In our paper, an algorithm based on YOLOv3 and support vector machine (SVM) is proposed for realizing the detection of vehicles on highway, as well as the detection and binary classification of people in the vehicles, so as to achieve the purpose of distinguishing drivers and passengers and form a one-to-one correspondence between vehicles and drivers. The effectiveness of the algorithm is verified under various complicated highway conditions. Compared with other advanced vehicle and driver detection technologies, the model has a good performance and is robust to road blocking, different attitudes and extreme lighting.

1. Introduction

The technology of detection has been widely spread in various kinds of fields, particularly in the field of vehicle and driver detection including the security of the public and the order of traffic. With the rapid development of modern society, various kinds of face detection techniques are being explored. While in practical application, due to the influence of different lighting, different occasions and different attitudes, the performance and effect of detection are not as good as ideal expectation.

The technology of detection has been a hot research area in deep learning in recent years. During 2004, Viola proposed a detection method named cascade, based on the Haar-Like [1] characteristics of AdaBoost [2] to execute the cascade classifier. Unfortunately, subsequent studies [3,4] have shown that it cannot maintain reliable performance in practical applications, because of the affection of people's visual consistency. In the following years, deep CNNs are used for the technology of detection. Yang [5] offered a network with deep neural for the recognition of feature, aiming to obtain a high response rate of regions and generate candidate windows. However, the algorithm is poor in real-time. Subsequently, R-CNN [6] and fast R-CNN [7] and even faster R-CNN [8,9] were generated successively, but the network was too large and the detection speed was not high. The YOLOv3 [10] proposed by Joseph Redmon in 2018 greatly shortens detection time and improves detection efficiency while ensuring accuracy.

At the same time, driver detection has attracted extensive research interest. The research in this field can be done by Support Vector Machine (SVM). By selecting the function, the optimal classification of people can be made to distinguish passengers and driver in the vehicle.

Content from this work may be used under the terms of the Creative Commons Attribution 3.0 licence. Any further distribution of this work must maintain attribution to the author(s) and the title of the work, journal citation and DOI.
Published under licence by IOP Publishing Ltd

In this paper, an improved vehicle and driver detection model based on YOLOv3 and SVM [11] is proposed, which is called IYOLO-SVM to form an adaptive detection model. The model is trained, based on our own database which is composed by pictures of traffic vehicle and driver. These pictures are provided by the Jiangsu Provincial Transportation Department. The result of the IYOLO-SVM model is verified on another database for testing. The finally experimental results show that compared with other advanced methods, IYOLO-SVM model can detect both drivers and the vehicle, and classify people in the vehicle, which has a higher detection rate and a lower error detection rate.

2. Basic Theory

In this article, we have improved and optimized the network structure of YOLOv3 and cascade SVM at the end of the network, which greatly improve the original network. In the next part of the article, we will briefly introduce the structure of the network for detection and the definition of the SVM classifier.

2.1 YOLOv3

The core idea of YOLOv3 is using the picture as a network input, return directly to the bounding box and its dependent categories at the output layer [12]. The full stages of YOLOv3 which is composed of four parts are shown in the following paper.

2.1.1 Bounding Box Prediction

The anchor boxes of YOLOv3 are made by clustering. The values of four coordinates [13] for each of the bounding box prediction (t_x, t_y, t_w, t_h) in predicting cell (a picture into S * S grid cells) based on the left top corner of the picture offset (c_x, c_y), according to the bounding box of p_w, p_h width and height, the bounding box can be predicted as follows:

$$b_x = \sigma(t_x) + c_x \qquad (1)$$
$$b_y = \sigma(t_y) + c_y \qquad (2)$$
$$b_w = p_w e^m \qquad (3)$$
$$b_h = p_h e^n \qquad (4)$$
$$m = t_w, \; n = t_h \qquad (5)$$

The sum of the error loss of square is used to predict the coordinate value, so the error can be calculated rapidly.

YOLOv3 predicts the score of an object for each bounding box by logistic regression. If the prediction of the bounding box and the real border overlap better than that of the other all forecasts, then the value equals 1. If the overlap does not get the value of a threshold (setting 0.5), the prediction of bounding box will be neglected, and is displayed as no loss.

2.1.2 Class Prediction

To classify different kinds of objections, independent logistic classifiers are used instead of a SoftMax. When training, binary cross-entropy loss is used for the class predictions.

2.1.3 Predictions Across Scales

YOLOv3 predicts different boxes in three different scales. YOLOv3 uses FPN (feature pyramid network) to extract feature from scales, and finally predicts a 3-d tensor, containing the bounding box information, object information, and class information.

2.1.4 Feature Extractor

YOLOv3 uses a complex network for performing extraction, which has 53 convolutional layers, called Darknert-53.

2.2 Support Vector Machine

Support Vector Machine (SVM) is a category of generalized linear classifier that classifies data according to supervised learning. The decision boundary is the maximum margin hyperplane for solving learning samples. In order to extend the functionality of our model, we use SVM to classify the two types of objects to determine whether the target is the driver or the passenger, to achieve multiple classification judgments. Experiments show that the connection between SVM and YOLOv3 shows strong ability for detection and classification in different types of environments.

3. Our Improved Method

3.1 The Whole Circuit of IYOLO-SVM

The figure below shows the circuit of the algorithm of IYOLO-SVM. At first, the IYOLO-Net obtains a large number of bounding boxes of cars and people from the input image. Then, the precise regions of vehicle and people are confirmed. After that, the classification model is used to differ whether it is a driver or a passenger and then give labels. Finally, the picture with bounding boxes and labels is expert.

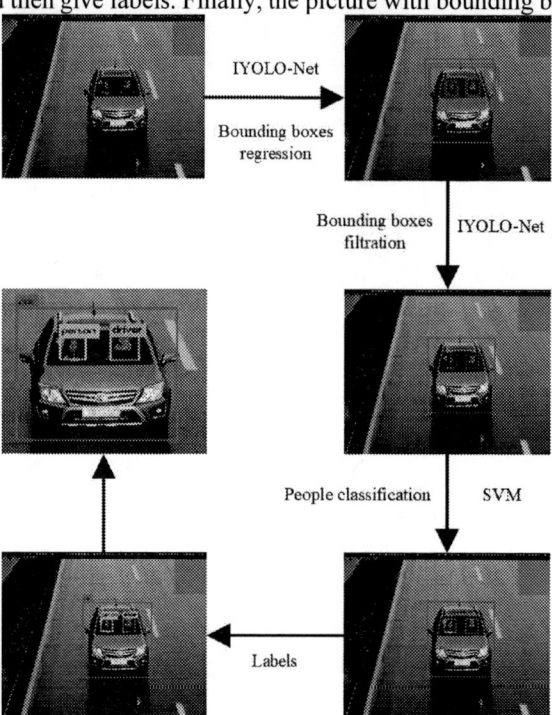

Figure 1. The Whole Circuit of IYOLO-SVM algorithm.

3.2 The Structure of Our New Network

In our IYOLO-SVM algorithm, we have made an obviously improvement on the structure of primary network, which becomes smaller and more efficient (Table.1). The required size for input image is 416*416. There are totally four maxpool layers and twenty-two convolutional layers. In the network [14], the role of the routing layer is to introduce finer-grained granularity features from earlier locations in the network. The reorg layer matches these features to the feature map of the next layer. The size of the end feature map is 13 * 13, the size of the previous feature map is 26 * 26 * 512. The reorganized layer maps the 26 * 26 * 512 feature map to the 13 * 13 * 2048 feature map so that it can be mapped to 13 * 13 resolution function chart. Through this method, high resolution features and low-resolution features are linked together, which can increase the recognition accuracy of small objects such as our people in the vehicle.

Table 1. The Structure of Our Network.

Number	Layer	Filter	Measurement	Out
0	CONV.	32	$3^2/1$	416^2
1	Maxpool		$2^2/2$	208^2
2	CONV.	64	$3^2/1$	208^2
3	Maxpool		$2^2/2$	104^2
4	CONV.	128	$3^2/1$	104^2
5	CONV.	64	$1^2/1$	104^2
6	CONV.	128	$3^2/1$	104^2
7	Maxpool		$2^2/2$	52^2
8	CONV.	256	$3^2/1$	52^2
9	CONV.	128	$1^2/1$	52^2
10	CONV.	256	$3^2/1$	52^2
11	Maxpool		$2^2/2$	26^2
12	CONV.	512	$3^2/1$	26^2
13	CONV.	256	$1^2/1$	26^2
14	CONV.	512	$3^2/1$	26^2
15	CONV.	256	$1^2/1$	26^2
16	CONV.	512	$3^2/1$	26^2
17	Maxpool		$2^2/2$	13^2
18	CONV.	1024	$3^2/1$	13^2
19	CONV.	512	$1^2/1$	13^2
20	CONV.	1024	$3^2/1$	13^2
21	CONV.	512	$1^2/1$	13^2
22	CONV.	1024	$3^2/1$	13^2
23	CONV.	1024	$3^2/1$	13^2
24	CONV.	1024	$3^2/1$	13^2
25	Route			
26	Reorg			
27	Route			
28	CONV.	1024	$3^2/1$	13^2
29	CONV.	40	$1^2/1$	13^2
30	Detection			

3.3 Manual Hard Sample Mining

The traditional method for handling difficult samples is to manually screen out the difficult samples which cannot be classified after self-inspection through the training network. This traditional method [15] is slow and inefficient. This method adopts the method of online difficult sample back propagation. In each mini-batch, the calculated loss is sorted from forward propagation of total samples, then only top seventy percent of the loss is taken as the difficult sample. Then, in the back propagation, only the difficult samples are calculated, and the simple samples are ignored. The online difficult sample back propagation greatly reduces the artificial labour force and improves the training efficiency.

We use the database built by our own to train the IYOLO-SVM model. Similarly, we also use our own testing set for verification. What we need to pay attention to is that, even if we have added online hard sample mining to the network. However, in actual situations, especially in complex highway situations, because of complex lighting changes, attitude changes, object occlusion and so on [16], it is significant to randomly add difficult samples, for the aim to increase the accuracy detection rate and drop the false detection rate in the final experiment. The final results show that this method makes the model improve performance apparently, which is showing in the next section.

4. The Results After Experiments

This following chapter, we compose our own image database through photos provided from the Jiangsu Provincial Transportation Department. It contains about 1,500 photos of the driver's travel in different situations containing their faces and cars. We randomly selected 1200 pictures for training our IYOLO-SVM model and 300 pictures for testing the performance of IYOLO-SVM model. GPU GTX1050ti is applied to train the IYOLO-SVM networks. The convergence graph of loss and results of our model on detection and classification are shown below.

4.1 The Procedure of Training Loss

Figure 2 shows the procedure of training IYOLO-SVM model. The picture indicates that the abscissa stands for the number of times of iterations during training. The ordinate stands for the loss of IYOLO-SVM model. When there are objects in the grid, confidence loss of bounding boxes calculates the weight of the contribution to the total loss for five; When there is no object in the grid, confidence loss of bounding boxes calculates the contribution weight to the total loss for one; The weight of the contribution of category loss to the total loss is calculated for one; The weight of the contribution of bounding boxes' coordinates prediction loss to the total loss is calculated for one.

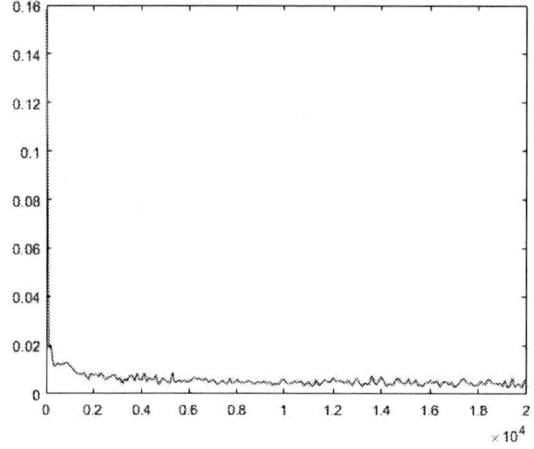

Figure 2. The procedure of training IYOLO-SVM model.

4.2 The Performances of Detection and Classification

The two groups of pictures below reveal the final testing results of the IYOLO-SVM model in different kinds of complex environments. According to the final testing pictures of the IYOLO-SVM model in day and night, the model can be found to be robust under illumination changes. Table 2 is a comparison table comparing the testing results of IYOLO-SVM, YOLOv3 and Cascade CNN. From the experimental testing results, it is obvious to find that the improved method raised by us owns a higher accuracy detection rate while maintaining a lower error detection rate when compared with other methods. In addition, the model also has a good classification effect by using support vector machine. The above phenomena all show that the IYOLO-SVM model greatly improve the accuracy of vehicle and driver detection on complex highway.

(1) The original image

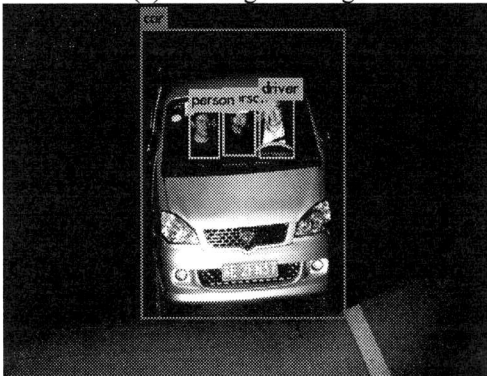

(2) The result picture

Figure 3. The detection and classification consequences in night environments under the IYOLO-SVM model.

(1) The original image

(2) The result picture

Figure 4. The detection and classification consequences in daytime environments under the IYOLO-SVM model.

Table 2. Detecting and Classifying Performance Compared with Our Method and Other Relative Methods.

Detection Method	IYOLO-SVM (OUR)	YOLOv3 [10]	Cascade CNN [17]
Detection Rate	91%	85.6%	78.9%
Error Detection Rate	2.35%	13.28%	19.7%
Classification Rate	94%	81%	68.9%
Error Classification Rate	0.65%	6.2%	15.8%

5. Conclusions

This paper raises a model of vehicle and driver detection based on improved YOLOv3 and SVM. As there might be more than one person in the vehicle, which driver detection through convolutional network in YOLOv3 cannot be realized, it is meaningful to use support vector machine for the judgement of driver and passenger. At the same time, the IYOLO-SVM model also provides a high-accuracy vehicle and driver detection technology in the complex environment on the highway. The final experimental results reveal that our new model raised in this paper is an efficient model with higher detection rate, lower error detection rate, fast detection speed, short training time and strong robustness to complex environments and different illumination on the highway.

However, our model has a bad performance when there is too much object occlusion and the driver's pixel is too small. Therefore, in the future, we need to solve and improve these problems for further study.

Acknowledgments

I would like to thank the National Natural Science Foundation of China (No.61871123), the Jiangsu Provincial Key Research and Development Program (No.BE2016739) and the Jiangsu Provincial University's Priority Academic Project Development Funding Project for our research support. Thanks to the lab's brothers and sisters and teachers for their help in my experiment.

References

[1] Mita, T. , T. Kaneko , and O. Hori . "Joint Haar-like features for face detection." Tenth IEEE International Conference on Computer Vision (2005)

[2] Zhu, Ji, et al. "Multi-class AdaBoost." Statistics & Its Interface 2.3 pp.349-360 (2006)

[3] Q. Zhu, M. C. Yeh, K. T. Cheng, and S. Avidan, "Fast human detection using a cascade of histograms of oriented gradients," in IEEE Computer Conference on Computer Vision and Pattern Recognition, pp.1491-1498 (2006)

[4] M. T. Pham, Y. Gao, V. D. D. Hoang, and T. J. Cham, "Fast polygonal integration and its application in extending haar-like features to improve object detection," in IEEE Conference on Computer Vision and Pattern Recognition, pp. 942-949 (2010)

[5] S. Yang, P. Luo, C. C. Loy, and X. Tang, "From facial parts responses to face detection: A deep learning approach," in IEEE International Conference on Computer Vision, pp. 3676-3684 (2015)

[6] Girshick, Ross. "Fast R-CNN," Computer Science, (2015)

[7] Wang, Xiaolong, A. Shrivastava, and A. Gupta. "A-Fast-RCNN: Hard Positive Generation via Adversary for Object Detection," IEEE Conference on Computer Vision and Pattern Recognition IEEE, pp.3039-3048 (2017)

[8] Ren, Shaoqing, et al. "Faster R-CNN: towards real-time object detection with region proposal networks," International Conference on Neural Information Processing Systems MIT Press, pp.91-99 (2015)

[9] Fan, Quanfu, L. Brown, and J. Smith. "A closer look at Faster R-CNN for vehicle detection," Intelligent Vehicles Symposium IEEE, pp.124-129. (2016)

[10] Redmon, Joseph, and A. Farhadi. "YOLOv3: An Incremental Improvement," (2018)

[11] Suykens J A K, Vandewalle J. "Least Squares Support Vector Machine Classifiers," Neural Processing Letters, 9(3):293-300 (1999)

[12] Liu, Xueliang, et al. "Correction to: On fusing the latent deep CNN feature for image classification." World Wide Web (2018)

[13] Ranjan, Rajeev, V. M. Patel, and R. Chellappa. "HyperFace: A Deep Multi-task Learning Framework for Face Detection, Landmark Localization, Pose Estimation, and Gender Recognition." IEEE Transactions on Pattern Analysis & Machine Intelligence pp.99:1-1(2018)

[14] Foroozandeh Shahraki, Farideh. "Cyclist Detection, Tracking, and Trajectory Analysis in Urban Traffic Video Data.", University of Nevada, Las Vegas, (2018)

[15] Nguyen, D. T., et al. "Combining Deep and Handcrafted Image Features for Presentation Attack Detection in Face Recognition Systems Using Visible-Light Camera Sensors." Sensors 18.3:699. (2018)

[16] Yang, Shuo, et al. "Face Detection through Scale-Friendly Deep Convolutional Networks." (2017)

[17] Yang Zhang, Peihua Lv, Xiaobo Lu. "A Deep Learning Approach for Face Detection and Location on Highway", IOP Conference Series: Materials Science and Engineering, (2018)

ISPECE   IOP Publishing

Robust modeling for fleet assignment problem based on GA-SVR forecast

A J Su[1], W D Yang[2*], C Zhang and M X Kong

College of Civil Aviation, Nanjing University of Aeronautics and Astronautics,
Nanjing, 210016, China

*Corresponding author's e-mail: ywendong@nuaa.edu.cn

Abstract. Fleet assignment is the essential step of overall airline flight cheduling process, and the quality of assignment strategy directly affects economy and safety of air transportation. Market demand forecast is the premise of fleet assignment, and accurate forecast is an important guarantee to reduce passenger overflow and improve aircraft utilization rate. In order to carry out scientific fleet assignment, this paper studies from two aspects: Firstly, support vector machine regression (SVR) is used to forecast flight passenger flow and solve the problem of undeterminable parameters, we present a GA-SVR model with genetic algorithm for parameter optimization. Secondly, from the perspective of flight recovery efficiency, this paper incorporates the concept of fleet robustness, and establishes the robust model for fleet assignment with dual-objective, which maximizes the flight operating profit and minimizes the number of aircraft type in busy airports. Finally, flight network of an airline is analysed to verify the validity of the model and algorithm. It shows that: the MSE mean value of GA-SVR prediction results is between 0.0103 and -0.0031, which is relatively accurate. And the fleet assignment model can significantly improve robustness (16.7%) at the expense of less profit (4.2%).

1. Introduction

Fleet assignment is a process matching suitable aircraft types with various routes and flights to maximize the company's revenue, which is also the basis of aircraft route planning and crew scheduling. It is based on the existing airline network and fleet resources of airline company and Considers factors such as fleet size, cabin capacity, operating cost and potential benefits [1]. Particularly, market size and characteristics is the important factor influencing the efficiency of fleet assignment, the consistency of aircraft size and market demand plays a critical role in creating higher economic benefits [2].

Compared with other modes of transportation, the existing research on the market demand of air passengers is not thorough yet. The econometric model [3-5] is mainly used to predict air passenger flow through analysing the influencing factors, but the factors are difficult to be determined accurately and more subjective. Conversely, machine learning algorithm focuses on analysing the law of data itself and is more objective. Among them, Support Vector Regression (SVR) is a more accurate method, the deficiency is that its results depend on the selection of kernel function parameters. In this regard, this paper proposes a GA-SVR prediction method, which uses Genetic Algorithm (GA) to improve the quality and efficiency of kernel function parameter optimization. For fleet assignment, A robust assignment strategy is conducive to the rapid recovery of irregular flights, which has great significance to the production and operation of airlines and the enhancement of passenger satisfaction.

Content from this work may be used under the terms of the Creative Commons Attribution 3.0 licence. Any further distribution of this work must maintain attribution to the author(s) and the title of the work, journal citation and DOI.
Published under licence by IOP Publishing Ltd

Existing literatures mostly analyse using basic fleet assignment model [6-7]. There are few relevant studies on the robustness of fleet assignment, and the robustness index is relatively single [8], which has no good reference value for the adjustment of irregular flights recovery.

Therefore, this paper improves the prediction method of market demand based on the SVR, and establishes the fleet assignment model with dual-objective: evenue and robustness. So as to better solve the problem of matching between demand and aircraft size.

2. Forecast of airline market demand based on GA-SVR

The airline market demand is a kind of time series, and its value is affected by many factors such as economy, weather, season, holiday, etc. Therefore, the forecast of demand is a complex nonlinear problem. In view of the above characteristics, GA-SVR method is adopted in this paper. In essence, VSR projects a given airline market demand sample set into high-dimensional space through nonlinear mapping, thus transforming nonlinear problems in original space into linear problems in feature space and reducing model complexity [9]. GA is used to optimize the kernel parameters of SVR, which has advantage of good robustness and universality. The specific algorithm flow is shown in the figure below.

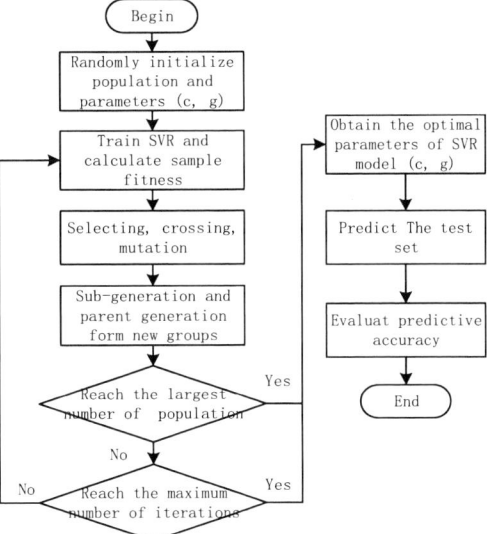

Figure 1. GA-SVR algorithm flow char.

Step1: Selecting. The randomly initialized parameters (c_0, g_0) are brought into the SVR model, the training sample data is input for model training, and the fitness value (mean square error) of sample is calculated according to the equation below. The highly adaptable individuals in the current population were selected to inherit the effective genes to the next generation by means of roulette.

$$MSE = \sqrt{\frac{1}{n}\sum_{i}^{n}(y_i - \hat{y}_i)^2} \tag{1}$$

In the equation, y_i is the ith observation, $\hat{y}_i$ is the ith forecast and n is the quantity of sample. Cross Validation (CV) is used as parameter optimization criteria.

Step2: Crossing. The individuals in the population are randomly paired, the locations of the crossing points are randomly set by the single point intersection method, and some genes between the two individual chromosomes are exchanged by the crossover probability P_c.

Step3: Mutation. Some genes of an individual are changed according to the probability of mutation probability P_m, to adjust the condition that the fitness cannot reach the optimum after the crossing.

Step4: Forming new groups. The fitness of new individuals generated by crossing and mutation is calculated. And then the new generation group is formed together with the parent generation.

Step5: Check whether the maximum number of population or maximum number of iterations which are pre-set can be reached. If so, terminate the operation, and get the optimal parameter (c, g) to predict the test set; otherwise, go back to step2.

3. Robust model assignment model

During the flight operation, it is often affected by weather, flow control and other interference, which can lead to flight delay, and the delay of a single flight will have ripple effects, generating large-scale irregular flights in the hub airport. Therefore, it is of great practical significance to take external disturbances into consideration when planning, and make robust fleet assignment plan. Considering that busy airports are prone to delays, and exchange between aircrafts of the same type is the main and the least loss strategy for flight recovery, this paper proposes a two-objective assignment model with the least type of aircraft in busy airports and the largest profit.

3.1 Profit of fleet assignment

Flight profit is obtained by deducting the cost from the revenue:

$$E_{ij} = P_{ij} - C_{ij} \qquad (2)$$

P_{ij} refers to the revenue of flight i executed by type j. It is calculated by taking the larger value of passenger number pax_i and seat number s_i multiplied by the average passenger ticket price $price_i$: $P_{ij}=max\{pax_i, s_i\} \times price_i$.

C_{ij} refers to the total cost, including the aircraft operation cost OC_{ij} and passenger overflow cost SC_{ij}. $OC_{ij}=a_j+b_jT_is_j$, a_j and b_j are the fixed cost and variable cost coefficients of type j, T_i is the chock time of flight i, and s_j is the seat number of model j. $SC_{ij}=(min\{pax_i, s_i\}-s_i) \times price_i$, which means that take the part of pax_i that exceeds s_i multiplied by the average passenger fare $price_i$.

3.2 Fleet assignment model establishment

A space-time network model $G(N_j, A_j)$ is established. For each aircraft type j, node N_j represents the airport-time two-dimensional point, and arc A_j represents the flight [10]. The following is the specific model parameters: K denotes the aircraft type set, $j \in K$; n_j is the aircraft number, s_j is the seat number; F stands for flight set, $i \in F$; O stands for the set of stop-over flights, $O \subseteq F$, $o \in O$. Most domestic stop-over flights only have one stop, which include two flight segments o_1 and o_2. A denotes the airport set, $a \in A$; m_a refers to the type number of aircraft taking off and landing at airport a; H stands for busy airport set, $H \subseteq A$; M is the space-time network node, $(ak) \in M$, $(ak-)$ is the previous time node right next to (ak), (ak') is the last time node of airport a; $e_{i(ak)}$ is a constant. When flight i lands at the node k of airport a, it equals 1, when it takes off, it equals -1, When neither, it is 0; f_{ia} is a type 0-1 indicator operator, It is equal to 1 when flight i takes off or lands at airport a, otherwise it is 0. $G_{(ak)j}$ is an integer variable, representing the aircraft number of type j at the node k of airport a; x_{ij} is 0-1 decision variable, if flight i is executed by type j, it is 1, otherwise, it is 0; z_{aj} is 0-1 decision variable, which is equal to 1 when type j takes off or lands on airport a, otherwise 0. The model is established as follows:

$$\min \quad w = \sum_{a \in H} m_a \qquad (3)$$

$$\max \quad z = \sum_{i \in F} \sum_{j \in K} (P_{ij} - C_{ij})x_{ij} \qquad (4)$$

$$s.t. \quad \sum_{j \in K} x_{ij} = 1, \forall i \in F \qquad (5)$$

$$G_{(ak^-)j} + \sum_{i \in F} e_{i(ak)}x_{ij} = G_{(ak)j}, \forall (ak) \in M, j \in K \qquad (6)$$

$$\sum_{a \in A} G_{(ak')j} \le n_j, \forall j \in K, (ab) \in M \tag{7}$$

$$f_{ia} x_{ij} \le z_{aj}, \forall i \in F, j \in K, a \in H \tag{8}$$

$$\sum_{j \in K} z_{aj} \le m_a, \forall a \in H \tag{9}$$

$$x_{o_1 j} - x_{o_2 j} = 0, \forall j \in K, o \in O \tag{10}$$

$$x_{ij} = 0,1, \forall i \in F, j \in K \tag{11}$$

$$G_{(ab)j} \in z^+, \forall (ab) \in M, j \in K \tag{12}$$

Equation (3) is the first objective function, with the minimum type number of takeoff and landing in busy airports, and equation (4) is the second, with the maximum profit; Equation (5) is flight coverage constraint; Equation (6) refers to the flow balance. It means that in an airport, the number of planes at a time node is equal to the one at the next time node after the change of takeoff and landing. Equation (7) represents the conservation of aircraft quantity; Equation (8) means that only when the aircraft of type j takes off or lands at airport a, the flight i on airport a can be executed by type j. Equation (9) means that the type number of aircraft taking off or landing at airport a is no more than m_a. Equation (10) specifies that aircrafts assigned for the two sections of a stop-over flight must be of the same type.

This model belongs to dual-objective programming, which is very difficult to solve. Therefore, this paper transforms it into a single-objective programming. Firstly, the single-objective model of profit maximization can be obtained by removing the equation (3), and the result of the model is the maximum profit value E_m. Then, according that proportion of total profit reduction after increasing the robustness is specified to not exceed r, the objective equation (4) is turned into the constraint condition below.

$$\sum_{i=F} \sum_{j=K} (P_{ij} - C_{ij}) x_{ij} \ge (1-r) E_m \tag{13}$$

4. The example analysis

In order to verify the rationality of the model and algorithm, flight network of a large domestic airline company is used for analysis. The case involves a total of 12 aircraft of 3 types, executes 56 flight segments on 21 routes, of which 20 are stop-over flights. All flight numbers, arrival and departure time, take-off and landing airports are read from flight schedules, and the average passenger ticket price is calculated by dividing the annual total revenue of the airline market by the actual passenger volume. The aircraft number, seats and costs of all types are shown in table 1, in which the cost data of different types is obtained by SPSS regression analysis. Market demand forecast uses passenger flow data of 96 months ranging from 2010 to 2017 as the sample.

Table 1. Data of different aircraft type.

Type	Number	Seat	Fixed cost	Variable cost coefficient
A319	6	122	23666.97	3.02
A321	2	177	32583.89	2.10
A325	4	186	36171.47	1.92

4.1 Market demand forecast

The sample data from 2010 to 2016 is taken as the training set and 2017 as the test set. The parameters of the algorithm are set as follows:

Table 2. GA-SVR algorithm parameters.

Algorithm	Parameter	Value
GA	Maximum population	20
	Maximal evolution iterations	200
	Crossing probability	0.9
	Mutation probability	0.09
SVR	Range of penalty coefficient c	$2^{(-2)}$——2^{10}
	Range of kernel parameter g	$2^{(-15)}$——2^{15}

GA-SVR is used for fitting and prediction, and compared with the simple SVR algorithm. 5 flight segments are taken as examples to analyze. Table 3 is the result of VSR parameter optimization, and the optimal parameters obtained by the two methods are significantly different. The results of training set fitting and test set prediction in Chengdu-Wuxi are drawn into broken line graphs, as shown in figure 2 and 3, indicating that the GA-SVR results are relatively accurate.

Table 3. The VSR parameter optimization results of the two algorithms.

Flight segment	Chengdu--Wuxi		Guangzhou--Wuxi		Nanjing--Chengdu		Nanjing--Shenzhen		Nanjing-Xiamen	
Algorithm	GA-SVR	SVR	GA-SVR	SVR	GA-SVR	SVR	GA-SVR	SVR	GA-SVR	SVR
Parameter c	54.98	4.00	32.18	0.44	2.43	1.32	0.28	0.25	125.84	4.00
Parameter g	2937.01	1552.09	2031.07	294.07	402.69	512.00	3978.45	1552.09	2994.17	97.01

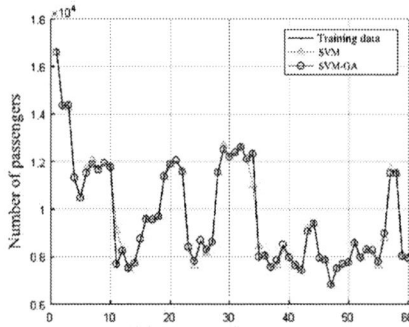

Figure 2. Fitting results of training set. Figure 3. Prediction results of test set.

By analyzing prediction accuracy of the model quantificationally, mean square error (MSE) is compared as shown in table 4. MSE reflects the deviation between the predicted results and the test sample values. The smaller the deviation, the more accurate the prediction. According to the table, the MSE values of GA-SVR are all low in the 12-month prediction of 5 routes, with the mean value between 0.0103 and -0.0031, generally less than that of SVR. This proves again that GA-SVR has a good prediction ability, and the optimization of parameters is better than simple SVR.

Table 4. MSE of the test set prediction results.

Month	Chengdu-Wuxi		Guangzhou-Wuxi		Nanjing-Chengdu		Nanjing-Shenzhen		Nanjing-Xiamen	
	GA-SVR	SVR	GA-SVR	SVR	GA-SVR	SVR	GA-SVR	SVR	GA-SVR	SVR
1	0.0091	0.0090	-0.0091	0.0075	-0.0541	-0.0329	-0.0094	-0.0493	-0.0184	-0.0381
2	-0.0076	-0.0254	0.0091	0.0451	0.1996	0.1756	0.0075	0.0957	0.0215	0.0201
3	0.0105	0.2374	-0.0083	-0.0576	0.1729	0.1363	-0.0070	-0.0068	-0.0179	-0.1286

4	-0.0085	-0.0081	0.0092	0.0090	-0.1980	-0.2088	-0.0004	-0.0067	0.0214	-0.0199
5	0.0081	0.0080	-0.0085	-0.0728	-0.0070	-0.0070	0.0073	0.0150	-0.0231	0.0042
6	-0.0081	-0.0081	0.0095	-0.0090	0.0763	0.0785	-0.0072	-0.0072	0.0299	0.2657
7	0.0082	0.0084	-0.0101	0.0188	-0.0068	-0.0068	0.0017	0.0072	0.0255	0.0028
8	-0.0084	-0.0084	-0.0100	0.0178	0.0071	0.0074	-0.0066	-0.0079	-0.0253	-0.0251
9	0.0081	0.0081	0.0084	-0.0366	0.0040	0.0059	0.0079	0.0087	0.0488	0.9993
10	-0.0078	-0.0391	-0.0079	-0.1394	-0.0709	-0.0697	-0.0067	-0.0397	-0.0215	0.0079
11	0.0113	0.0111	0.0092	0.0092	0.0076	0.0075	0.0066	0.0065	-0.0140	-0.2366
12	-0.0110	-0.0103	-0.0097	-0.0074	-0.0073	-0.0076	-0.0313	-0.0644	0.0173	0.0167
Mean	0.0003	0.0152	-0.0015	-0.0180	0.0103	0.0065	-0.0031	-0.0041	0.0037	0.724

4.2 Robust fleet assignment

According to the characteristics of airline routes and operation bases, the busy airports are set as Nanjing and Shanghai. This paper assumes that the profit reduction should not be higher than 5% after considering the robustness. Firstly, the goal is to maximize profit, and the maximum profit is 177,4697 ￥, with robustness 6. It means that there are 3 types of take-off and landing aircraft in both two busy airports, which is not conducive to the implementation of the replacement strategy. Then, the robust factor is added, solving the dual objective programming can finally obtain the assignment scheme shown in table 5. According to the results, the profit is 1700160￥, and the robustness is 5 . The operating profit decreased by 4.2% after taking into account the robustness, but the robustness increased by significant 16.7%, which can better deal with irregular flights.

Compared with the actual manual assignment plan using in the airlines now and the robust fleet assignment plan of this paper, the number of aircrafts needed for all flights was reduced by 1, and the total aircraft utilization rate increased from 617.92h to 674.09h, which increased by 9.1%, indicating that the model in this paper can make better use of aircraft resources. The number of flights executed by A319 has decreased by 4 compared with the actual number, and the type number and the utilization rate have decreased, while A321 is just the opposite. It indicates that A321 is more economical than A319. Although the number of flights executed by A325 remains unchanged, both the aircraft type and the aircraft utilization rate have increased, its economy is also better. However, it has the largest number of seats, which leading to high cost. Therefore, when the existing passenger volume cannot reach the high passenger rate, it should choose small aircraft to execute, thus saving an aircraft.

Table 5. The results of the robust fleet assignment compared with the actual manual assignment.

Type	Actual manual assignment				Robust fleet assignment			
	Aircraft number	Flight number	Utilization rate of type /h	Utilization rate of aircraft/h	Aircraft number	Flight number	Utilization rate of type /h	Utilization rate of aircraft/h
A319	6	32	3795	632.50	6	28	3515	585.83
A321	2	8	1230	615.00	2	12	1450	725.00
A325	4	16	2390	597.50	3	16	2450	816.67
Total	12	56	7415	617.92	13	56	7415	674.09

5. Conclusion

This paper studies the two important links with closely relation in airline flight scheduling - market demand forecast and fleet assignment. Taking into account the complex nonlinear variation law of air passengers and the problem of limited samples, SVR is used to predict. The SVR parameters are optimized by GA, which greatly improves the efficiency and accuracy. In terms of aircraft fleet assignment, In order to relieve the serious delay in busy hub airports, a dual-objective assignment model with the least number of types in busy airports and the largest profit is established. It can increase the opportunity to exchange aircrafts between the same types when flight recovery, and improve the robustness of assignment. The model and algorithm have important practical significance for alleviating flight delay, and can provide good theoretical basis and effective decision support for airline operation.

Acknowledgment
[1]Fund project: Nanjing university of aeronautics and astronautics graduate innovation base (laboratory) open fund (kfjj20170714, kfjj20170702)
[2]Fund project: Youth science and technology innovation fund (NS2016063)
Fund project: Fundamental research funds for the central universities (NZ2016109)

References
[1]   Zhu X H, Zhu J F and Gao Q 2012 *Science Technology and Engineering*, **12** 1329-33
[2]   Wu D H and Xia H S 2014 *Journal of Transportation Systems Engineering and Information Technology.* **14** 109-37
[3]   Peng T, Zhang Y P and Hao S Q 2016 *Journal of Transport Information and Safety.* **34** 37-44
[4]   Preez J D and Witt S F 2003 *International Journal of Forecasting.* **19** 435-51
[5]   Yao T and Tao J 2015 *Computer Technology And Development.* **25** 147-51
[6]   Zhu B W and Tang F Y 2015 *China Management Informationization.* **18** 135-36
[7]   Tsai M W, Hong T P and Lin W T 2015 *Mathematical Problems in Engineering.* 1-12
[8]   Zhu X H, Zhu J F and Gao Q 2011 *Forecasting.* **30** 71-4
[9]   Chen R, Liang C Y and Lu W X 2014 *Systems Engineering —Theory & Practice.* **34** 1290-6
[10]   Zhe L and Wanpracha Art C 2013 *Transportation Science.* **47** 493-507

An improved SVM web page classification algorithm

Xun-yi Ren, Chen Shi, Dan Zhang and Wen-si Wang

College of Computers, Nanjing University of Posts and Telecommunications, Jiangsu 210003, China

Abstract. This study explored an indepth support vector machine(SVM) classification algorithm suitable for high-dimension computing. An improved SVM algorithm was proposed aiming at two kinds of classical kernel functions of support vector machine. The global kernel function and the local kernel function were weighted into a new mixed kernel function, and the genetic algorithm was used to optimize the function parameters. The improved algorithm avoided the limitation of a single kernel function, taking into account the generalization and learning abilities of the text classification algorithm. Simulation experiments of the collected web pages confirmed the effectiveness of improved SVM algorithm.

1.Introduction

At present, the mainstream text classification algorithms include Naive Bayes algorithm, decision tree algorithm, K-nearest neighbor algorithm, and support vector machine(SVM). The SVM algorithm is more suitable for calculating high dimension, and has better generalization and extension capabilities. Therefore, it is the most widely used algorithm in the era of big data. However, the SVM is associated with the problem of high-dimensional space complexity and how to better solve the quadratic programming problem. This study aimed to improve the SVM algorithm based on the aforementioned two aspects to further enhance the performance of its classification.

Statistical learning theory is based on a small sample of a machine learning domain knowledge. The theory of structural risk minimization principle is used to solve machine learning problem, and the VC dimension theory is used to operate the machine based on the generalization ability of learning [1]. Therefore, before exploring the support vector machine, it is essential to first understand the two concepts of VC dimension and structure risk minimization principle, and then introduce the thought and nature of SVM algorithm to a solid foundation for improving SVM algorithm.

2. SVM algorithm

2.1. VC dimension theory

The basic concept of VC dimension is the set of functions for a test. If the functions in the set can separate k sample elements according to all possible 2k patterns, it is said that the function set can break k sample elements [2]. In general, an n-dimensional space Rn can have n spatial points that are linearly independent. In this case, the VC dimension of the n-dimensional hyperplane is $n + 1$ [3].

For the full function of the test function set, a probability of at least $1-\eta$, $0 \leq \eta < 1$ exists between the test risk $R(\omega)$ and the empirical risk $R_{emp}(\omega)$:

Content from this work may be used under the terms of the Creative Commons Attribution 3.0 licence. Any further distribution of this work must maintain attribution to the author(s) and the title of the work, journal citation and DOI.

Published under licence by IOP Publishing Ltd

$$R(\omega) \le R_{emp}(\omega) + \left(\frac{h\left(\log \frac{2l}{h} + l \right) - \log \frac{\eta}{4}}{l} \right)^{\frac{1}{2}} \tag{1}$$

In the inequality, the number of samples is l, and h is the VC dimension of the test function set. It is evident from the formula that the test risk is determined by the two parts of the inequality on the right-hand side. One is the empirical risk and the second is the confidence risk, expressed by the following formula:

$$\phi\left(\frac{h}{l}\right) = \left(\frac{h\left(\log \frac{2l}{h} + l \right) - \log \frac{\eta}{4}}{l} \right)^{\frac{1}{2}} \tag{2}$$

Therefore, formula 2.1 can be simplified as:

$$R(\omega) \le R_{emp}(\omega) + \phi\left(\frac{h}{l}\right) \tag{3}$$

At this point the relationship is clearer. When the confidence is relatively large, the gap between empirical risk and test risk is widened, leading to the machine learning "over-fitting" situation. Then, it is necessary to control the VC dimension to control the confidence risk in a smaller range and the empirical risk of the minimum value, thereby reducing the test risk and enhancing the generalization performance of the machine learning model.

If the ratio of VC dimension and sample number was appropriate, a better learning model of learning algorithm and neural network could be achieved in accordance with the specific features of the sample itself. This helped in choosing a different VC dimension and minimizing the risk of experience [4]. The subsets were arranged from small to large according to the VC dimension, and then a subset of the function with the same confidence in the risk was selected for the minimum empirical risk [5].

2.2. SVM in the nuclear function
This set of support vectors is a defined set of feature subsets [6]. The SVM algorithm was originally used to solve the problem of linearly separable patterns. The SVM solution can accurately separate two different categories and can also split the plane distance category of the largest spacing, which is the so-called SVM seeking the accurate classification of optimal hyperplane.

The concept of kernel function is as follows [7]. Suppose in a space R^n a function $K(x, y)$ satisfies the Mercer conditions:

$$K(\mathrm{x},y) = \phi(x) \cdot \phi(y) \tag{4}$$

This is called the kernel function. The Mercer condition means that if $g(x) \in L_2(Rn)$, $K(x,y) \in L_2(Rn \times Rn)$, for any, $g(x) \ne 0$ and $\int g(x)^2 dx < \infty$, the following formula is used:

$$\iint K(x,y) g(x) g(y) dx dy \ge 0 \tag{5}$$

For all training elements $x_1, x_2, \cdots x_n \in Rn$, $K(x,y) = \varphi(x) \cdot \varphi(y)$ exists for the positive definite matrix. The commonly used kernel functions have polynomial kernel functions:

$$K(\mathrm{x},y) = \left[(x \cdot y) + c \right]^p, \quad p = 1, 2, \cdots N \tag{6}$$

The Gaussian kernel function is also known as the radial basis function:

$$K(\| x - xc \|) = \exp\left(-\frac{\|x - xc\|^2}{2 \cdot \sigma^2}\right) \tag{7}$$

It is the center of the kernel function. σ is the width of the kernel function parameters, which can be used in another form of the function [8]:

$$K\left(\| x - x_C \|\right) = \exp\left(-\gamma \| x - x_C \|2\right) \tag{8}$$

Sigmoid kernel function:

$$K(x, y) = \tanh(\beta_0 \langle x \cdot y \rangle + \beta_1) \tag{9}$$

where β_0 is a scalar and β_1 is the amount of displacement.

3. SVM algorithm improvement

The kernel function directly affects the ability of SVM algorithm [9]. In the commonly used kernel functions, the polynomial kernel function $K(x, y) = \left[(x \cdot y) + c\right]^p$ is a rotation-invariant kernel function. That is, it can be represented by a function $K(x, y) = f(\langle x, y \rangle)$, where $f: D \rightarrow R$ is a unary real function [10]. Therefore, the polynomial kernel function belongs to the global kernel function, and this typical global kernel function has the advantage of being able to extract the characteristics of the overall sample. The disadvantage is that the interpolation ability is relatively low. The parameters of the polynomial kernel function, that is, homogeneous, exponential parameters of the polynomial function of the kernel function, are shown in figure 1. The test point is 0.3. It can be deduced from the figure that the SVM with polynomial kernel function has a high generalization level. If the number of input samples is different and the difference in sample change is large, the polynomial kernel function still has a strong classification ability.

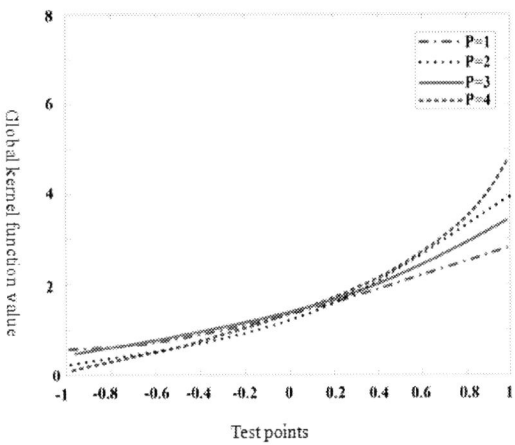

Figure 1. Global kernel function diagram.

The Gaussian kernel function $K(\| x - x_C \|) = \exp(-\dfrac{\|x - x_C\|2}{2 \cdot \sigma 2})$ is a translation invariant kernel function [11]. That is, it can be expressed by a function $K(x, y) = f(x, y)$. Therefore, the Gaussian kernel function is a local kernel function, which is superior in terms of the overall performance for local learning classification.

The value of the parameter σ of the Gaussian kernel function is set to 1, 2, 3, and 4, and the test point is 0. The function of the Gaussian kernel function is shown in figure 2. It is deduced that the Gaussian kernel function is complementary to the polynomial kernel function, which is more conducive to learning the local range of the data. In this case, the Gaussian kernel function is inversely proportional to the radius of the Gaussian kernel function.

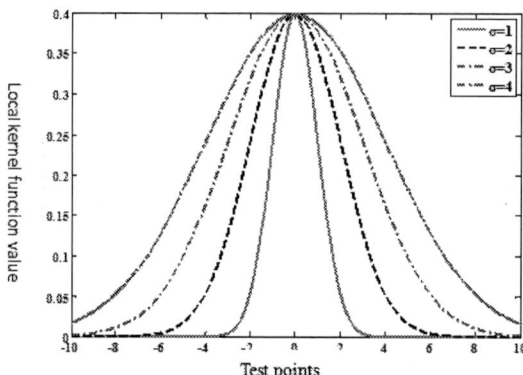

Figure 2. Local kernel function diagram.

In the realization of the web page classification algorithm, the learning and generalization performances of the classifier are mutually restricted. The number and type of test samples are usually unbalanced in a high-dimension space. Satisfactory results could not be achieved in this study based on the aforementioned detailed analysis of the properties and characteristics of polynomial kernel function and Gaussian kernel function, combined with the performance of global kernel function and local kernel function. A new mixed kernel function is constructed:

$$K_{mix} = t\exp\left(-\frac{\|x-y\|^2}{2\sigma^2}\right) + (1-t)\left[(x \cdot y)+c\right]^p \tag{10}$$

The mixed kernel function comprises an effective feasible Gaussian kernel function and the polynomial kernel function. The new mixed kernel function is beneficial to extract the training sample taking into account the global characteristics of the training samples, thereby theoretically establishing a more stable SVM model.

After constructing a new hybrid kernel function, the SVM's ability depends mainly on the penalty factor, parameters of the mixed kernel function, and weight coefficient. These parameters control the search of the VC vector of the feature space after the feature vector is mapped by the SVM algorithm. Therefore, how to make these three parameters achieve the most matching value is the key to improving the SVM algorithm before the need to analyze these parameters.

In the nonlinear SVM the following formula is used in addressing the optimization problem:

$$\begin{cases} \min_{w,b} & \frac{1}{2}\|w\|^2 + C\sum_{i=1}^{n}\xi_i \\ s.t. & y_i\left(\langle w \cdot \phi(x_i)\rangle + b\right) \geq 1-\xi_i, \quad i=1,2,\cdots,n \end{cases} \tag{11}$$

ξ_i is the loss function, and the value of the penalty factor C determines the penalty generated by the empirical error. A lower C value reduces the complexity of SVM, indicating that the empirical risk of SVM is increased. However, a larger C value generates the maximum threshold allowed by the learning feature subspace, indicating that the SVM's generalization and learning abilities can be improved by the fact that the SVM cannot increase the learning complexity. A balance needs to be maintained between the two penalty factors for a better performance of the SVM classifier.

The mixed kernel function includes two parameters. One is the radial radius σ in the Gaussian kernel function, and the other is the order p in the polynomial kernel function. If $\sigma \to 0$, then the training text is all support vector, which helps in improving the accuracy of classification, but at the same time is generated over the fitting situation; if $\sigma \to \infty$, then all the training text completely belongs to a subject class that meets the purpose of generalization, but has low accuracy. For the order p in the polynomial kernel function, if the number of training samples is large, it leads to huge spatial dimension, resulting in the occurrence of dimension disaster.

The new hybrid kernel function constructed in this study also has an important parameter, that is, the weight coefficient t, which ranges from 0 to 1. It is obvious from the formula that if $t \to 1$, then the Gaussian kernel function contributes more to the whole SVM algorithm model, whereas if $t \to 0$, then the overall generalization advantage of the polynomial kernel function is more obvious in the SVM algorithm model. Therefore, it is also important to select the value of the optimal mixed kernel function weight t while improving the SVM algorithm, so that the best performance of the constructed new mixed kernel function is achieved in the classification process.

A key step in the improvement of SVM algorithm is the selection of SVM parameters. The traditional method is the grid search. This study proposes a cross-validation method of genetic algorithm (GA). The essence of the problem is to show the potential of an individual and then use some of the genes encoded by the individual to form a population. The chromosome is the main carrier of genetic material, with a collection of multiple genes [12].

At the beginning of the GA, N initial populations are randomly generated. All the initial populations exist in the data structure of the gene string type, and then the iteration is carried out from the N string populations. Normally, the range of N is $[20,100]$. The next step is to calculate the fitness of each individual in the population, which represents the performance of each individual. The maximum number of evolutionary iterations is set to T, and the current number of evolutionary iterations is t. If $t \le T$, then t is incremented and the following process is continued until $t > T$; the result is the optimal solution. Replication in the algorithmic process is to extract good individuals from the current population to the next generation, while others are passed to the next generation through crossover and mutation operations, where the extraction criteria are the degree of individual adaptation previously calculated. The crossover operation is to randomly transform the chromosomes of all the individuals within the population and then generate new individuals; the newly generated individuals are individuals who have inherited the previous generation of good genes. The last variation of this process is done based on genetic mutations that produce new individuals by mutation.

In this study, GA was used to optimize the SVM hybrid kernel function parameters as follows:

(1) Randomly generate N initial populations and use binary coding, mixed kernel function parameters, weight coefficient, and penalty factor to be encoded, resulting in chromosome gene string.

(2) Encoded gene string is decoded into corresponding parameters and then is inputted into the SVM to train the learning.

(3) The accuracy of training set is determined in the cross-validation as a fitness function to calculate the degree of adaptation of each individual in the group.

(4) The conditions are satisfied to find the best combination of SVM parameters, or the process is continue to the following steps.

(5) Another individual is replaced according to the rules of optimal individual replication.

(6) A crossover probability is selected to cross the generated new individual.

(7) A mutation probability is selected to mutate the new individual and go back to step (2) to continue the loop.

4. Simulation experiments

In this experiment, Matlab was used as simulation tool to introduce LIBSVM open-source software package to build and train the SVM model. The software is used to integrate the kernel function, including Gaussian kernel function and polynomial kernel function. Search optimization algorithm is also integrated. For the newly constructed mixed kernel function, LIBSVM supports custom kernel functions. Therefore, a code can be added to implement the new kernel function in the package, and then the kernel function selection parameter can be set to be customizable. LIBSVM package also integrates the GA function. Therefore, the GA can be used for optimizing new kernel function parameters.

The data used in the simulation experiment was obtained using the WebCrawler; 450 web page texts were selected, including military, scientific and technological information, health, and other eight categories. According to the category of mixed choice, 300 pages of text were used as training samples, and the remaining 150 were test samples. All the text data must first be processed by the data preprocessing module. If the weight difference in the processing of the data is large, it adversely affects the speed and prediction accuracy of the learning and training, making the data normalization indispensable. The TF-IDF algorithm itself is a normalized calculation method. Therefore, all the processed data was normalized in the range of 0 to 1.

In addition, the parameter $c = 0$ of the new mixed kernel function of formula 3.17 was set. That is, the homogeneous polynomial kernel function was taken and then the parameter of the Gaussian kernel function was set to $r = \dfrac{1}{2\sigma^2}$. Consequently, four parameters needed to be optimized: the penalty factor C, Gaussian kernel function parameter r, polynomial order p, and weight coefficient t. Binary coding was used to set the initial population number $N = 50$. The selection operator value was 0.9, the crossover probability was set to 0.45, and the mutation probability was set to 0.1. In this study, the precision value of 10 was chosen as the fitness function of GA. The number of iterations is $T = 50$. The cross-validation of GA was used to select the best parameter change process, as shown in figure 3.

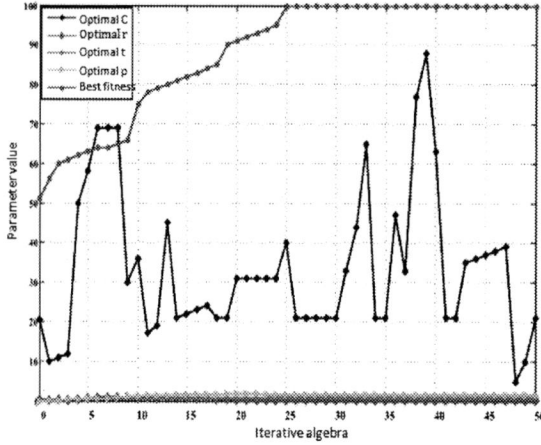

Figure 3. GA is optimized for hybrid kernel parameters.

The figure clearly shows that the number of iterations reached 25 times. Also, the value of the penalty factor did not converge, the remaining three parameters reached convergence, and the best fitness function cross-validation accuracy reached 100%. Therefore, the value of the penalty factor of the 26th iteration was taken for the best parameter value. The result of the optimization was $C = 21.05379$, $r = 0.03562$, $t = 0.81658$, and $p = 1.50175$. All the optimal parameter values were classified into the new kernel function classification model. The correct rate was 98.67%.

The Gaussian kernel function and the polynomial kernel function were integrated in the LIBSVM package. The type of SVM kernel function in LIBSVM was set by the parameter s. The default value was 0, which represented the linear kernel function. In the experiment, the values were set to 1 and 2, each representing a polynomial kernel function and a Gaussian kernel function, respectively. The grid search algorithm function SVMcgForClass was used to optimize the parameters. The comparison of the optimal GA parameters with the correct rate results is shown in table 1.

Table 1. Comparison of the optimal parameters with the correct rate results.

Parameter	Kernel function		
	Gaussian kernel function	Polynomial kernel function	Mixed kernel function
Penalty factor C	3.41328	3.41328	21.05379
δ	1.5374	–	0.03562

t	—	—	0.81658
p	—	3	1.50175
Correct rate	89.33%	88.67%	98.67%

5. Conclusions

This study explored the SVM algorithm that could deal with the large-scale data in the web effectively. After studying and analyzing the kernel function used in the commonly used SVM, a new mixed kernel function was constructed by combining the global kernel function and local kernel function with different characteristics, and then the optimal solution of the parameter was found through the GA. Finally, the influence of the aforementioned two improvements on the classification performance of web pages was verified by experimental simulation. Experiments showed that the proposed method improved the classification performance of web pages in most cases.

References

[1] Vapnik V N 1995 The nature of statistical learning theory *IEEE Trans. Neural Netw.* **8(6)** 1564

[2] Jin-shu Su, Bo-feng Zhang, Xin Xu 2006 Advances in machine learning based text categorization *J. Software* **17(9)** 1848-59

[3] Jian-hui Wang, Hong-wei Wang, Yun-fa Hu 2005 A simple and efficient algorithm to classify a large scale of texts *J. Comput. Res. Develop.* **42(l)** 85-93

[4] Hua Zhao, Tie-jun Zhao, Shu Zhang 2006 Topic detection research based on content analysis *J. Harbin Inst. Technol.* **10(38)** 1740-49

[5] C. Hsu, C. Lin. 2002 A comparison on methods for multi-class support vector machines *IEEE Trans. Neural Netw.* **13** 415-25

[6] Jian-fen Wang, Yuan-da Cao 2001 The application of support vector machine in classifying large number of catalogs *J. Beijing Inst. Technol.* **21(2)** 225-29

[7] X Z Li, J M Kong 2014 Application of GA-SVM method with parameter optimization for landslide development prediction *Nat. Hazards Earth Syst. Sci.* **14(3)** 525-33

[8] Xing-hua Fan, Mao-song Sun 2006 A high performance two-class Chinese text categorization method *Chin. J. Software* **29(l)** 124-31

[9] C H Wu, G H Tzeng, R H Ling 2009 A novel hybrid genetic algorithm for kernel function and parameter optimization in support vector regression *Expert Syst. Appl.* **36(3)** 4725-35

[10] Sebastiani F. 2002 Machine learning in automated text categorization *Acm Comput. Surveys* **34(1)** 1-47

[11] Jian-tao Sun, Dou Shen, Yu-chang Lu 2004 Web document classification techniques *J. Tsinghua University: Sci. Technol.* **44(1)** 65-68

[12] Qing-jie Liu, Gui-ming Chen, Xiao-fang Liu 2012 Genetic algorithm based svm parameter composition optimization *Comput. Appl. Software* **29(4)** 94-1

ISPECE

IOP Publishing

The Application of Convolutional Neural Network in Security Code Recognition

Jingtian Gu[1]

[1]College of Engineering, University of Washington-Seattle, U District, Seattle, WA, United State

Abstract. Security Code or CAPTCHA, which represents Completely Automated Public Turing test to tell Computers and Humans Apart, is used to determine whether the user is human or not. It is widely applied in website management to prevent people from maliciously registering and spreading spam. With the development of machine learning algorithms and artificial intelligence technologies, it is possible to recognize the security codes with machine so that people can have access to the website without registration limits. There are many traditional machine learning algorithms such as Support Vector Machine (SVM) and Random forest are used to recognize security codes, but they have various disadvantages like low efficiency and low learning ability, inability to extract features automatically and difficulty in handling 2-dimension pictures directly. In machine learning area, the convolution neural network (CNN) is famous for its strong learning ability and automatic feature extraction, as well as high learning ability and efficiency, which makes it suitable for 2-dimension image data. Thus, we construct a convolutional neural network for security code recognition. The proposed CNN model is made up with 3 convolutional layers, a flatten layer and a full-connected layer. With the proposed model, we achieve an accuracy of 80.5% in the validation set.

1. Introduction

Security codes have been widely applied in different websites like shopping websites, social media and even search engines. They take advantage of security code for different purpose. For instance, social media use it to filter excessively repetitive registration of advertisement and paid poster accounts to ensure the clean environment. Google gives user security codes when they have unusual behaviors, and ticket websites use these codes to prevent ticket resale. However, some security code recognition models have been built to avoid these codes and make it easier to login or conduct web crawler. Several softwares like automatic login assistant have been published for these purposes. Some even comes into commercial application, for example, 360 has deployed this technology in its software products to handle the security code for its users, which has received a lot of positive comment especially in ticket panic buying.

Most of these security code recognition models are based on template matching [1] algorithm or some traditional machine learning algorithms like SVM [2]. Template matching algorithm matches parts of a processing image with a template image at different levels from pixel level to higher level to judge the texts in the processing image. Two types of matching, feature-based matching and template-based matching, are usually chosen to implement the algorithm. However, the results of matching depend on the template images or features selected, which makes it difficult to apply when the characters are overmuch. When the image has strong features, feature-based matching is considered to be efficient, especially when the image has large resolution. But this method does not view the image as an entirety, so it is highly limited by the image factors. Template-based matching

---

Content from this work may be used under the terms of the Creative Commons Attribution 3.0 licence. Any further distribution of this work must maintain attribution to the author(s) and the title of the work, journal citation and DOI.
Published under licence by IOP Publishing Ltd

partly fills up the inability of feature-based matching since it is used for the image without strong features. To ensure the accuracy, the amount of sampled data point should be very large, thus resolution of the image may be reduced to reduce the required data points. Generally, template matching algorithm has high requirement for the processing image including low noise, high data point quality, suitable size and resolution. As the security codes are becoming more and more complex and irregular, it is difficult for template match algorithm to deal with these codes. Some machine learning algorithms are also used in security code recognition. However, these algorithms are usually based on risk minimization principle and used in classification and regression analysis with 1-dimension data. When it comes to 2-dimension security code images, it is difficult to use these models directly. The images should be converted into gray images and then transformed into black-and-white image with thresholding. Moreover, the features of the image must be extracted to convert the 2-dimension image data into 1-dimension feature data, which can be handled with. Generally speaking, machine learning methods like SVM has a relatively higher accuracy than template matching algorithm when the training samples increase in terms of security code recognition. However, as I mentioned, these traditional machine learning algorithms require feature extraction [3] process. It is difficult to figure out what features of the target images should be picked, and it is also hard to ensure the picked features can represent different types of security codes. Moreover, a lot of useful information may be lost in the process of feature extraction, which may decrease both the efficiency and accuracy of these methods.

Deep learning [4] is a machine learning algorithm based on neural network [5], which imitates the structure of cerebral cortex. The basic element of the neural network is neuron, and each neuron receives input signals from several other neurons. All the input signals are combined together with a given weight and then compared with the bias of current neuron. Finally, it generates the output by activation function. All the neurons connect together by layers to form the neuron network. Throughout training, the weight and bias of each neuron will be adjusted to fit the requirement. Convolution neural network is a type of neural network that works effectively in image processing by extracting feature maps through several convolution layers automatically. With the increase of convolutional layers, the extracted features become more and more complex and suitable for recognition. With the shared weight structure in convolution neural network, the number of parameters to be estimated reduces dramatically reduces and he training expense decreases significantly. Compared with aforementioned image recognition models, convolutional neural network is more suitable for security code recognition, because of the automatic feature extraction and higher learning ability.

In this paper, we constructed a convolutional neural network [6] for security code recognition. The security code images are generated by imitating the mechanism of the security code on the websites, and the image data is used for the training and testing of our proposed CNN model. Each security code image is made up of a combination of four characters or numbers, which are picked randomly. We also add noise lines and points to the image to make it more closed to the real-life security code. Before passing the image data into CNN model, we preprocessed the image by removing the noise point and line with coded filter and cut the image character by character. After preparing the training data set, we built the convolution neural network with three layers. After the three layers, we use a fully connected layer to first flatten the image output from the previous layers and then summarize all the feature maps. A dropout layer is also adopted to avoid overfitting and increase the efficiency. After over 60000 iterations of training, we achieved an accuracy of about 99% for training set. Then we imported the test set and the accuracy is about 80% on the test set.

The rest of the article will discuss the following sections. In section 2, we will introduce how we generate our security code images for training and testing. In section 3, we will briefly talk about how we preprocess our image data before passing it into our CNN model. In section 4, we will discuss how the proposed convolution neural network for security code recognition is built. In section 5, we will go through the process of training our model and how it came out. Finally, we will discuss and evaluate our entire project. The related figures, tables and reference material will be provided.

2. Method/Result

2.1. Security Code Image Generation

To build and evaluate the image recognition model, we needed a large number of images. Although we can get security code images on the websites with the help of web crawler, it is time-consuming to label all the images with the true texts manually. Thus, we decided to imitate the generative mechanism of security code to generate images data, which makes it easier to control the parameters of our image and define the content in the security code images.

Taking the advantage of Python Image Library, it is easy to draw the security code images. A single security code image is made up of four random characters, each of which is either a number in 0-9 or letter in A-Z. For different images, we set different colors and fonts for the characters. Color is picked randomly out of RGB colors. We limited the character codes to be relatively darker in case it is too light to be recognized by human eyes, which is obviously inconsistent with the reality. The font is picked randomly from the system font library. We also added some noise lines and points on the image at random position. To make it closer to reality, we further adjusted the position range of noise line to make sure it can function as noise. The noise point with size of four pixels is set to be different colors randomly, so that it is visible enough. Finally, we affined and enhanced the images and they were ready for next progress.

Figure 1. Example of security code image

2.2. Image preprocessing

In reality, the images are usually preprocessed to reduce the influence of noise such as line noise and point noise. We also preprocessed the image to remove the noise as much as possible to increase the predictability. As we introduced, the main noise involved in our image data is the noise lines and points. For noise points, we directly used a median filter from Python Image Library, which replaced each entry with the median value of neighboring entries. The example image after median filter is shown in the Figure 2, and we see that median filter can remove the noise points effectively.

Figure 2. Image after median filter

Since noise line would cross the characters, a simple filter can't remove it clearly. Thus, we used an algorithm to detect and remove these noise lines. First, we converted the image into binary image and append it into a data matrix to make it easier to process. Our algorithm will detect the two pixels above and two pixels right of each pixel in the image. If they are all black, the selected pixel will be identified as a component of a noise line, and it will be transferred to white. After passing through the filter algorithm, edge enhancement and smoothing will be executed to enhance the image. The outcome showed that most of the noise lines and points can be removed clearly without affecting the characters. The example image after the noise line removal is presented in the Figure 3. In some cases when the noise line nearly vertically crosses the characters, removal of it would cause the character to be separated into two parts. We fixed this drawback later in the character segmentation.

Figure 3. Image after noise line removal

To segment the images by characters, we set two stages: Foundletter and Inletter. Foundletter means the algorithm reaches the starting boundary of a single character, while Inletter means the cursor is currently in the character. Two stages are initially false. When starting to run the algorithm, we similarly pass through every pixel in the image from left to right. Once there is a black pixel and the current Foundletter is false, we change Foundletter to true and mark the position of the pixel as start. For Inletter stage, it is always true as long as the current pixel is black. Then as the algorithm going to right, if Inletter change to false because of white pixel and the Foundletter is still true for now. Foundletter will change to false and the current position will be marked as end meaning that it is the end of the character. Each pair of start and end will be stored in an array. After passing through the entire image, we used the array to identify the boundary of the characters.

As we mentioned, our noise line filter might separate a character into two parts, which would influence our segmentation algorithm. Thus, we added extra module by counting the number of pairs in the start and end array. If there are more or less than 4 pairs, it means that there exists erroneous segmentation. In this case, we used half of mean width of the pair length as reference and compare all the pairs in the array with it. If there are two continuous pairs that are smaller than the reference, they would be identified as the consequence of over segmentation. Then they would be merged together to form the complete pair. On the other hand, if there are fewer than four pairs, although this is less possible, the algorithm would segment the relatively long pair again to separate the joined characters. After segmentation, the preprocessing of the image is finished. The segmented characters will be saved to the related folder according to their content. Totally we have 36 folders for the 26 letters and 10 numbers. The exact code shown in the image is also saved for the subsequent classification in training and test.

2.3. Convolutional Neural Network

Our image recognition algorithm is based on Convolution Neural Network. It consists of three convolutional layers, a flatten layer and a full connection layer. The structure of our proposed CNN model is shown in Figure 4.

Before entering the network, the image would be reshaped to a 32×32 image matrix to unify the size of security code images. The first layer uses filters with the kernel size of 5×5, and there are totally 32 filters applied. For a single image passed into the first layer, there will be 32 feature maps generated from the first convolutional kernel. When executing the convolution, we set padding to be same so that the feature maps will be the same size as the initial image. The stride of convolution is set to be 1. Next step is the pooling of the generated feature maps. The kernel size of pooling is [1, 2, 2, 1] and the strides is [1, 2, 2, 1]. The pooling of the feature maps will finally give us 32 16×16 feature maps which is the output of the first layer.

The filters of the second convolutional layer have the same size and other parameters are the same as those in the first layer. There are only 2 filters generated from the second convolutional kernel. The convolutional kernel in the second layer will increase the number of feature maps from 32 of the first layer to 64. Pooling and padding maintain the same setting as the first layer, which means that the output of the second layer are 64 8×8 feature maps.

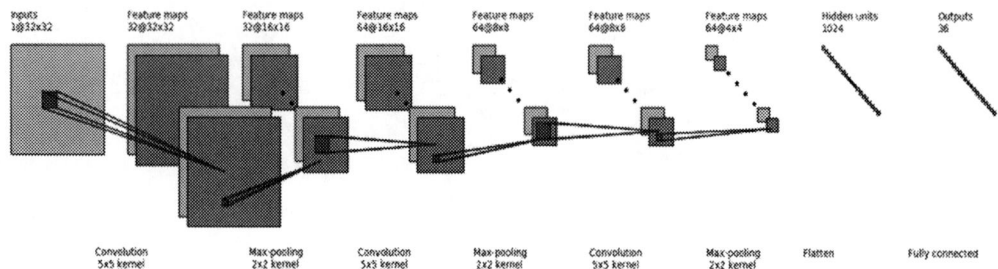

Figure 4. CNN structure diagram

Similarly, the third convolutional layer holds the filter with the same size and parameters. But there is only one filter applied in this layer. Pooling and padding maintain the same setting as the first layer. This means that after convolution and pooling, there are outputs of 64 4×4 feature maps from the third layer, which is the last convolution layer.

After three convolution layers, we built a fully connected layer to summarize all the output from the convolution layers. The first thing this layer does is to flatten the 64 4×4 feature maps from the third layer to vectors with the length of 1024, which in the other way is a 4×4×64 vector. Then all these vectors will be passed into activation function. A dropout layer also exists to randomly drop the neural unit from the CNN with a specific probability of 0.5. In Deep Learning, dropout layer is typically used to limit overfitting, which is a universal phenomenon for machine learning. Then the vectors will be transfer to probability vectors with the length of 36 in the following full-connected layer, and the result will be passed into the last activation function and give us the final output.

In our algorithm, calculation of loss and accuracy is included for us to monitor the training and testing process and determine the performance. For model training, we used TensorFlow [7] from Google as our library. First of all, we organized the input images into batches, each of which had 100 images. Totally, we iterated the training for 100000 times. Every 100 iterations, the current accuracy and loss were calculated so that we could monitor the real-time training result. When the accuracy reached 0.98, the model is potentially applicable, so we saved the existing model every time it iterated. When the accuracy reaches 0.995, it is highly possible that the model has reached its limited and the accuracy hardly increases. In that case, it is meaningless to keep iterating, so the algorithm stopped running, and the last saved model is our final product.

After the model training, we started to import different image set to test it. The test accuracy would be lower than the training accuracy because of overfitting. A validation set a potential adjustment that could be added to more limit overfitting [8].

2.4. Results
To analyze the process and result of training and testing, we used tensorboard [9] to visualize the learning trend, accuracy, loss and other related statistics of the network.

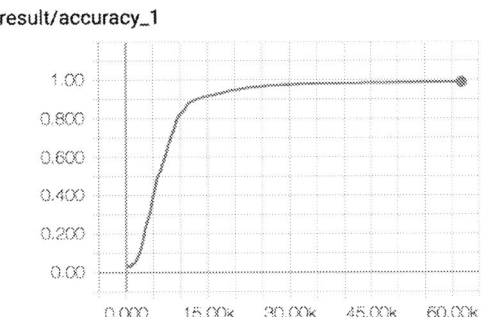

Figure 5. Accuracy plot of training

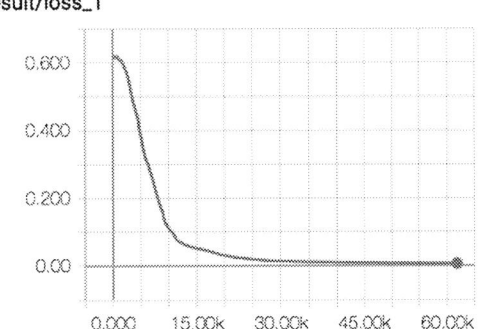

Figure 6. Loss plot of training

Figure 5 and 6 shows the accuracy and loss in the progress of training. We can see that the accuracy increased and the loss decreased rapidly within around first 15000 iterations. Then the change tended to stop over the rest iterations. Finally, the accuracy is extremely closed to 1 and the loss is almost down to 0.

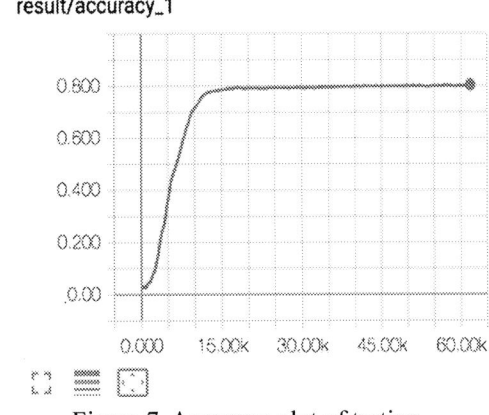

Figure 7. Accuracy plot of testing

ISPECE   IOP Publishing

IOP Conf. Series: Journal of Physics: Conf. Series **1187** (2019) 042064   doi:10.1088/1742-6596/1187/4/042064

result/loss_1

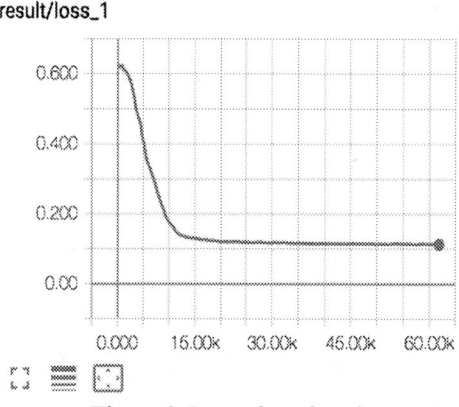

Figure 8. Loss plot of testing

Figure 7 and 8 on the other hand display the accuracy and loss for the testing set. As we can see, the general trend during the testing is similar to that of training. They had a rapid change at the beginning and when it got close to the limit, the change was negligible. What is different is that the upper limit of accuracy is around 0.8 and the floor level of the loss is around 0.1. There clearly exists gap between training and testing. This might because of overfitting in the training set. Overfitting means the model corresponds too exactly to a particular data set, which is our training data.

Other than the analysis of result, we also recorded other detailed statistics of deep learning model, including means, standard deviation, maximum value and minimum value of bias and weights. In the Figure 9 and 10, we show the corresponding statistics for the full-connected layer.

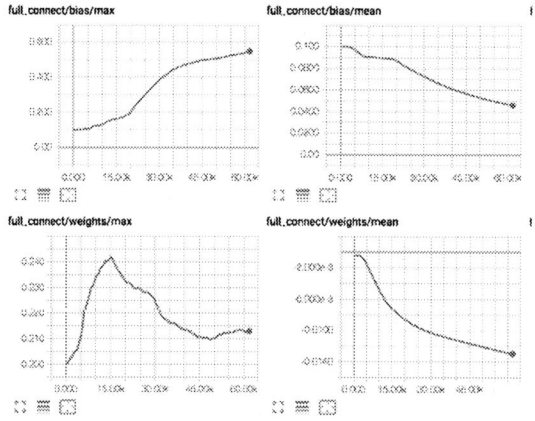

Figure 9. Max and mean value of full connected layer

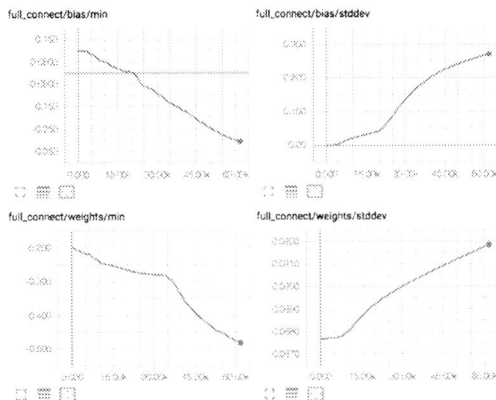

Figure 10. Min and standard deviation statistics of full connected layer

As we can see from the diagrams that the maximum value and standard deviation of bias increased, while mean and minimum value decreased as the training going through. Otherwise, maximum value of weight changed anomalously, and other parameters just share similar patterns with bias.

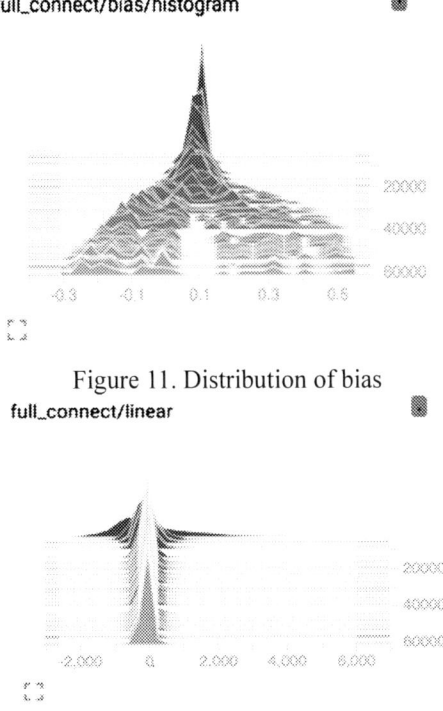

Figure 11. Distribution of bias

Figure 12. Distribution of final output

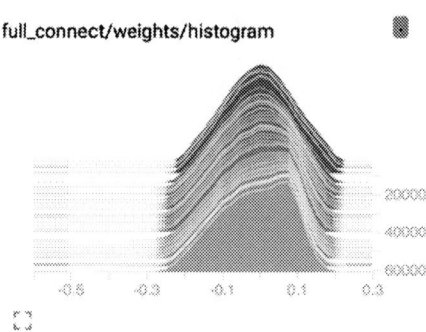

Figure 13. Distribution of weights

In the Figure 11, 12 and 13, we further present the distribution of bias, final output and weights in the full-connected layer. From the histogram, we can directly see the change trend and distribution of all the parameters of the model. Bias scattered over training while weights stayed in a relatively stable distribution. The final output, on the other hand, became more concentrated.

Such diagrams are really helpful to monitor the training process of our model. In most of cases, the neural network is like a black box. We set up how it should learn but we do not know what it finally becomes. These diagrams create a crevice on the black box so that we can peep inside instead of just pass data in.

3. Discussions and Conclusions

The initial goal of this project is to build a security code recognition algorithm that could be used for login and web crawler. Finally, based on Convolution Neural Network, a recognition model was born that can deal with security code images with noise lines and noise points. During the entire process, we did face several problems. For instance, we did not expect the error in segmentation at first because we ignored the potential influence the noise line could bring to segmentation. Thus, we had to add extra methods in the algorithm to fix the mistakes. Another typical negligence happened during training. As is well known, a complex deep learning model requires strong hardware and long enough time to train and the final accuracy it could reach is hardly predictable. We had a high expectation on the accuracy so that we set the break accuracy level to be too high, but we forgot to apply automatic model saving before the accuracy reached our expectation. The result was that the growing speed was tiny before the accuracy got the level, which made the training took an extremely long time to finish and we could not manually stop it because the model would not be saved. Finally, we had to start over to make the condition of the training more reasonable. These problems could be good reminders for deep learning model development. Except for these setbacks, the outcome of our model development turns out to be good. We achieved a training accuracy of 0.9902 and a testing accuracy of 0.805, which are high enough for basic security code recognition. Otherwise, this model still maintains high potential. This model only consists of 3 convolution layers. There are more layers that could be applied to increase the complexity of the network. Other than the structure of the network, some self-adjust functions could be used to change the learning and dropout rates to fit more situations and promote the behavior. The images we feed the model were basic security code images, there are different security code with more twists and noise in many websites. For now, we generated our own images, but web crawler is also an opinion to collect more types of security code images to make the model more comprehensive. Security code recognition is just a small application of image recognition by deep learning. There are so many potentials that we can excavate. Facial recognition, video recognition or even traffic recognition for automatic driving are coming to people's life. The future of this area is bright and hopeful.

References

[1] Hanebeck U D. Template matching using fast normalized cross correlation[J]. Proceeding of Spie

on Optical Pattern Recognition XII, 2001, 4387:95-102.

[2] Schuldt C, Laptev I, Caputo B. Recognizing Human Actions: A Local SVM Approach[C] International Conference on Pattern Recognition. IEEE, 2004:32-36 Vol.3.

[3] Yuille A L, Cohen D S, Hallinan P W. Feature extraction from faces using deformable templates[C] Computer Vision and Pattern Recognition, 1989. Proceedings CVPR '89. IEEE Computer Society Conference on. IEEE, 1989:104-109.

[4] Schmidhuber J, rgen. Deep learning in neural networks[M]. Elsevier Science Ltd. 2015.

[5] Yan W Q, Wang J, Kankanhalli M S. Neural network-based face detection.[J]. IEEE Transactions on Pattern Analysis & Machine Intelligence, 1998, 20(1):23-38.

[6] Li H, Lin Z, Shen X, et al. A convolutional neural network cascade for face detection[C]// Computer Vision and Pattern Recognition. IEEE, 2015:5325-5334.

[7] Abadi M, Barham P, Chen J, et al. TensorFlow: a system for large-scale machine learning[J]. 2016.

[8] Schaffer C. Overfitting avoidance as bias[C]// Sfi Tr. Santa Fe Institute, 1993:267--289.

[9] Gad A F. TensorFlow: A Guide to Build Artificial Neural Networks using Python[J]. 2017.

Depth Enhancement with Improved Inpainting Order and Smoothing Method

Kang Yi*, Yuting Zhao, Yiyan Lei, Jian Pan

School of Information and Communication, National University of Defence Technology, Xi'an, China

*Corresponding author e-mail: penfangs@163.com

Abstract. Holes and noises are two main defects existing in Kinect's depth images, which severely restrict the development of depth-based applications. This paper proposes a novel depth enhancement algorithm with improved propagation order and smoothing methods, based on the classic inpainting ideology. The propagation order is optimized by dividing the structural holes into separate connected regions, which is processed one by one according to definition of the pixel priority with confidence term and directional term, which is calculated with the isophote and the pseudo-normal. Furthermore, the improved smoothing method is presented in two distinctive ways: the linear combination of BF and JBF utilizes the different intensity similarity evaluation, while the joint trilateral filter exploits the properties of Gaussian kernel function, in which the proper combination of the colour similarity, depth similarity and geography distance are considered. Experimental results show that our proposed method obtains outstanding performance over other state-of-the-art methods and stage-tests also demonstrate the advantages of our proposed method over their alternatives. This work can benefit a lot for various depth-based applications, especially for the low cost Kinect's popularity.

1. Introduction

Robot applications, such as map building, path planning and environmental perception, are becoming highly research topics recently. These tasks previously subject to images from conventional camera, however, depth sensors have made a new type of data available. In 2010, Microsoft released an imaging device called Kinect, whose lower price and relatively higher resolution wins great popularity. Although Kinect has made high-resolution real-time depth maps available at a lower cost, there are numerous noises and holes with invalid values. In fact, there are three main reasons leading to the formation of these holes. Firstly, the range is limited, which means objects nearer or farther cannot been measured correctly; secondly, some areas cannot be seen by both the projector and the camera, which is called occlusion in image processing literature; and thirdly, the material properties and reflectivity of the object also make difference. If the surface is too smooth, specular reflection may occur and if the material absorbs all light patterns projected on it by the infrared projector, the camera will capture nothing too. Both conditions will result in holes in the captured depth maps.

As described in [1], the performance of depth-based algorithms can be improved significantly if holes and noises are removed. Hence, the enhancement of depth maps which is aiming at removing noises and filling in the holes, is necessary and practically meaningful.

Researchers have paid great efforts to eliminate these defects and improve quality of depth images captured by state-of-the-art range sensors. Chen et al. [2] detected and removed the pixels with wrong

Content from this work may be used under the terms of the Creative Commons Attribution 3.0 licence. Any further distribution of this work must maintain attribution to the author(s) and the title of the work, journal citation and DOI.
Published under licence by IOP Publishing Ltd

values using a region growing method, with the help of its corresponding RGB image, and filled the holes with a joint bilateral filter. They also proposed an adaptive bilateral filter to effectively reduce the noise. Yang et al. [3] presented another method and adopted an adaptive color-guided auto-regressive (AR) model, which was based on the observation that AR model tightly can fit depth maps of generic scenes for high quality depth recovery from low quality measurements. The task was formulated into a minimization of AR prediction errors subject to measurement consistency, and the predictor for each pixel was constructed according to both the local correlation in the initial depth map and the non-local similarity in the accompanied high quality color image. However, in this method, recovery of an image sized $M \times N$ will have a predictor matrix sized $MN \times MN$, a matrix $\mathbf{P}$ sized $p \times MN$ (p, number of non-zero elements in the original image), and multiple $MN \times MN$ temporary matrices which consumed high computing and storage. Liu et al. [4] formulated the guided depth enhancement problem based on the heat diffusion framework, referred to as GAD. Meanwhile, Le et al. [5] proposed an adaptive directional filters to fill in the holes and suppress the noises in depth maps, whose main contribution lied in the window shapes adaptively adjusted based on the edge direction of the color image.

In our proposed method, conventional inpainting method, which is frequently used in optical images, is adopted for holes filling because of their similarity to a certain degree. We extend the traditional inpainting methods by optimizing its propagation order and introducing the exemplar-based filling method. Filtering with higher color similarity is further applied for noises and artifacts removal.

1.1 Conventional Inpainting Techniques

Inpainting, modification of images in a way that is non-detectable for an observer, can be dated back to the Renaissance, when medieval artwork started to be restored by filling in gaps [6, 7]. Aim of inpainting is to reconstitute the missing or damaged portions of the work, i.e. repairing unknown regions with reference of known regions in an image [8].

The key problems are how to propagate and how to inpaint, namely the inpainting method and its order. Telea A. [9] made a distance transform before inpainting instead of iterative confidence computation, which is based on the FMM (Fast Marching Method). For a region in the image to be inpainted, the algorithm starts from the boundary and goes inside gradually. More weightage is given to those pixels lying near to the point, near to the normal of the boundary and those lying on the boundary contours. Once a pixel is inpainted, it moves to next nearest pixel using FMM. FMM can ensure those pixels near the known are inpainted first, so that it just works like a manual heuristic operation. Another method [10], based on fluid dynamics, utilized partial differential equations, traveled along the edges (meant to be continuous) from known regions to unknown regions and continued isophotes while matching gradient vectors at the boundary of the inpainting region.

Inpainting has been applied to many practical applications with some improvements. Criminisi et al. [11] removed the object and filled in the hole left behind in a visually plausible way, by combining the advantage of texture synthesis and inpainting, which replicated texture and structure information simultaneously and improved computational efficiency by a block-based sampling process. Liu et al. [12] extended the original fast marching method, whose results were limited to only using the information itself, by incorporating another image as guidance to reconstruct the unknown regions or damaged portions. We hence refer to it as a guided FMM, quoted with GFMM later. Furthermore, Gong et al. [13] proposed a new inpainting approach based on the FMM, by extending the inpainting model and the propagation strategy of FMM to incorporate color information for depth inpainting. Zhang et al. [14], inspired by the above-mentioned methods [11, 12], added a level set distance to improve the propagation order and proposed an edge-preserving filtering method to further remove noises and artifacts, which is referred to as ImprovedInpainting later.

1.2 Conventional Inpainting Techniques Noise-removing Strategies

Noise-removing or smoothing has been a much-talked-about topic in image processing literature. Various methods have been proposed and applied to different practical applications.

Simple linear translation-invariant (LTI) filters with explicit kernels, such as the mean, Gaussian, Laplacian and Sobel filters [15] have been widely used in noise-removing community. Low-pass filtering may be the most obvious method, like box filter [16] (also called mean filter), in which the value of i_{th} pixel is replaced by the weighted mean of pixels in the window centered at i_{th} pixel. According to the weighting way, smoothing methods are divided into two categories: isotropic filtering and anisotropic diffusion. Isotropic filtering methods adopt a uniform attitude to all elements, the regions and the edges (including texture and details), which may lead to blurring.

Otherwise, many anisotropic diffusion methods are proposed, like the bilateral filter [17], joint bilateral filter [18], guided filter [19] and so on. They treat regions and edges respectively to preserve details as much as possible and avoid blurring. These efforts have made some difference in the image smoothing community.

2. Material and Methods

In order to remove noises and fill in holes existing in Kinect's captured depth images in a more visually plausible way, thus accelerating development of the depth-based applications, a novel depth enhancement method based on the conventional inpainting method with improved propagation order and smoothing method is proposed in this paper. Figure 1 depicts its framework, which is composed of four stages: the pre-processing, selection of the holes with the highest priority, filling in the holes with Exemplar-based inpainting and smoothing the noises and artifacts caused by inpainting. Holes selection is completed in two steps: drawing the scope of pixels taken into account and then selecting a pixel among them in a certain way. Detailed description of each stage is presented below.

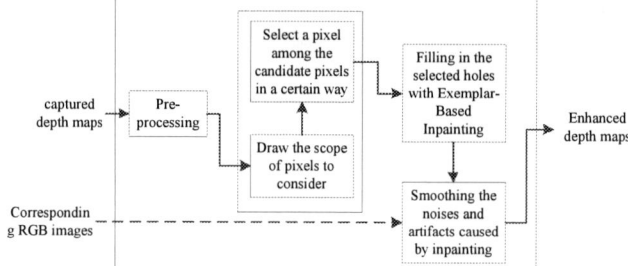

Figure 1. The Framework of the proposed method.

2.1 Pre-processing

As described in Section 1, holes are inevitable in Kinect's captured depth images. Microsoft researchers are also quite clear of this, hence they mark pixels with invalid values using the tag "No-available", which is transferred to zero when the maps are rendered as 8-bit images [20, 21]. Thus, 'zero' does not mean the zero depth, but the absence of a value. Therefore, the holes-removing goal is to replace the pixels value zero with a reasonable value and produce a more visually plausible depth map.

Holes on the captured depth maps can be divided into stochastic holes and structural holes. Stochastic holes consist of discrete single pixels with absent values, while structural holes are made up of regions of holes. As shown in Figure 1, the Kinect's captured depth maps are firstly pre-processed to remove the stochastic holes which may severely influence the inpainting result and the running speed of subsequent processing.

Generally, the mathematical morphological closing operation with structure element size of one pixel is applied to obtain outlier-removed depth maps. Morphological closing operation completed with expansion followed by contraction is efficient to remove the outliers.

Images processed with the above-mentioned method look like Figure 2, in which stochastic holes are removed while remaining structural holes appear black.

2.2 Propagation and Inpainting

Our holes-filling procedure is highly inspired by the object-removal method in [11], which aims at removing the target object and filling in the hole left behind in a visually plausible way. It shares similar challenge on holes filling with the depth enhancement. Exemplar-based texture synthesis contains the essential process required to replicate both texture and structure; the success of structure propagation, however, is highly dependent on the order in which the filling proceeds.

2.2.1 Propagation. As described in Figure 1, selection of the hole to be dealt with is completed in two steps: drawing the scope of pixels to consider of and selecting a hole from the candidate pixels in a certain way. Different methods can be employed to fulfill these tasks.

Drawing the scope of pixels to take into account

Holes in the pre-processed images are comprised by several connected holes regions, shown as Figure 2. In our proposed method, holes in one connected holes region are filled in after another, whose effectiveness has been proved in Section 3.1.1. Due to the fact that a pixel only has relationships with its neighboring known pixels, two separate connected holes regions have little impact on each other. Hence, connected holes regions are processed according to their geometry location, for example, from left-top corner to right-bottom of the image.

To make sure all the connected regions are selected in closed areas and the subsequent processing more easily, the image resulted from pre-processing are expanded with 1 pixel on both sides. For example, the original depth map is denoted with A in matrix form, then the expansion result looks like this: $\begin{pmatrix} 1 & 1 & 1 \\ 1 & A & 1 \\ 1 & 1 & 1 \end{pmatrix}$.

As shown in Figure 2, connected holes regions in pre-processed Art are marked with green closed curves. Image on the left is obtained before expansion and the right-hand one is got after expansion. It's clear that all the holes in Figure 2 (b) is circled in closed green curves, while some holes in Figure 2 (a) is surrounded by unclosed red curves.

(a)Result before expansion   (b) result after expansion

Figure 2. Connected holes regions in pre-processed Art marked with green closed curves.

Picking out the target pixel

Before inpainting, another problem arises after one connected holes region specified, namely the filling order of pixels in the selected region. For ease of description, we adopted notions similar to that used in inpainting literature. The regions of holes, whose values are absent, are denoted as the target regions. Correspondingly, the regions with known values, which will remain fixed and provide reference in the filling process, are defined as source regions.

In our proposed method, each target patch, window size $9*9$ centered at a selected pixel, maintains a pixel priority determined by the priority evaluation function. Two components are combined to model the pixel priority evaluation function, which are the confidence level and a direction term respectively.

The confidence level gives out the reliability of information in the target patch, and is calculated with recursion. The original known pixels, namely pixels in the source regions, are given 100% trust, i.e. the conference value is 1. And the value decreases as we go further into the target regions.

As the same time, the direction term is given by inner product of two vectors, whose directions are respectively the same as the isophote's direction and the gradient of a pseudo-normal. Isophote is defined as the line of the same gray level and is computed by gradient with 90° rotation, for the reason that the gradient is thought to be the fastest changing direction. Meanwhile, pseudo-normal is similar but different to the normal. Directions of lines and textures are not what we care about and noises may make things more complex, therefore, computing the normal direction is not a good choice. What we rarely concern is the direction from the source region pointing into the target region. Therefore, a counterpart of the original depth image, in which pixels located at the corresponding of holes in the original depth image value 0 while others value 1, is created. The pseudo-normal direction is calculated by gradient of the just-created counterpart.

The pixel priority evaluation function is defined as $P(p)$ in Equation (1).

$$P(p) = C(p) * Dir(p) \tag{1}$$

Where $C(p)$ defines confidence level and $Dir(p)$ denotes direction term respectively, which are defined in Equation (2, 3).

$$C(p) = \frac{\sum_{q \in Pt_p \cap \Psi} C(q)}{\left| Pt_p \right|} \tag{2}$$

$$Dir(p) = \frac{\left| \nabla I_p^\perp \cdot n_p' \right|}{\alpha} \tag{3}$$

Where Pt_p is the target patch centered at pixel p, and $\left| Pt_p \right|$ is its area. As shown in the Fig 3, Pt_p is presented with a red rectangular frame with central pixel p. Ψ defines the source regions, which means that q belongs to the intersection of Pt_p and Ψ. $\nabla I_p^\perp$ represents the isophote direction of pixel p and n_p' is its pseudo-normal direction, which is shown in Fig3 with arrow lines. Their inner product gives priority to the pixels whose isophote and pseudo-normal go in the similar direction, namely, when they go in the same direction, the direction term gains its maximal value; otherwise, the value decreases as the angle of isophote and pseudo-normal increases till orthogonal to each other. When the image is 8-bit, α is set 255 to ensure the value of $Dir(p)$ range between $[0,1]$.

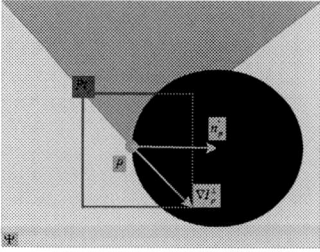

Figure 3. Notations used in definition of priority evaluation function.

2.2.2 Inpainting. The target patch centered at the pixel picked out is filled in with the exemplar-based inpainting method, which is fulfilled in two steps. Firstly, the target patch is used as a sliding window which scans through the whole source regions, to find out its best-match patch. The similarity is measured with the value difference of known pixels in both patches. Then, values of holes in the target patch are replaced with values in the corresponding place of the best-match patch.

The exemplar-based Inpainting method propagates the texture and structure simultaneously and obtains great performance.

2.3 Smoothing

Smoothing is indispensable after holes filled in due to noises existing in the original images and artifacts caused by exemplar-based inpainting algorithm. As described in [22], noises in Kinect's depth maps are a deterministic function of distance added to the random noises present in all systems, modeled as the deterministic noises and random noises. Although the exemplar-based inpainting propagates texture and structure information well, it also brings about artifacts and bumpiness on the surface.

As described in [23, 24], Kinect measures the distance between object and the sensor plane. It's a distance between a set of points and a specified plane, which means that the depth values vary with the geometrical shape of the object. As we all known, any object in the real world varies continuously with no interruption. Therefore, despite the holes, the variation trend is known to us, which is very useful for our depth enhancement.

Thus, the depth value of each pixel in the depth maps has relationship with its neighbors because of the continuity, and only with its neighbors. Therefore, in our proposed method, Gaussian kernel is adopted and non-isotropic-filtering is used to ensure that more details are preserved.

Moreover, pixels are affected by its neighbors, but in structural holes regions, depth information around a hole may be not quite believable. However, information in the corresponding color image may be more believable. Hence, higher weightage is given to the color similarity.

In our proposed method, the smoothing is modeled in two ways, which can obtain similar performance, as shown in Section 3.2.2. Their goal is accordant, though appear different, to combine the influences of three items and enlarge the weight of color similarity at the same time. The first model, defined as Eq.4, is given by a linear combination of the classical bilateral filter [17] and a joint bilateral filter [18]. Meanwhile, the second smoothing model, denoted as Eq.5, is referred to as the joint trilateral filtering.

$$JBF\_BF[I]_p = \frac{1}{\omega_p} \sum_{q \in s} G_{\sigma_s} \left(\|p - q\| \right) \left(\alpha G_{\sigma_{rc}} \left(\left| I_{cp} - I_{cq} \right| \right) + \beta \cdot G_{\sigma_{rd}} \left(\left| I_{dp} - I_{dq} \right| \right) \right) I_q \qquad (4)$$

$$JiontTF[I]_p = \frac{1}{\omega_p} \sum_{q \in s} G_{\sigma_s} \left(\|p - q\| \right) \cdot G_{\sigma_{rc}} \left(\left| I_{cp} - I_{cq} \right| \right) \cdot G_{\sigma_{rd}} \left(\left| I_{dp} - I_{dq} \right| \right) \cdot I_q \qquad (5)$$

Where, $G_{\sigma_s} \left(\|p - q\| \right)$, $G_{\sigma_{rc}} \left(\left| I_{cp} - I_{cq} \right| \right)$ and $G_{\sigma_{rd}} \left(\left| I_{dp} - I_{dq} \right| \right)$ respectively define the weights of distance, color similarity and depth similarity on the central element p , which are defined as Equation (6, 7, 8) respectively. They are all computed between p and q , which belongs to p's neighborhood S . And ω_p is the normalization term, which is used to make sure that range of I is constant.

$$G_{\sigma_s} \left(\|p - q\| \right) = \exp \left(-\frac{\|p - q\|^2}{2\sigma_s^2} \right) \qquad (6)$$

$$G_{\sigma_{rc}} \left(\left| I_{cp} - I_{cq} \right| \right) = \exp \left(-\frac{\left| I_{cp} - I_{cq} \right|^2}{2\sigma_{rc}^2} \right) \qquad (7)$$

$$G_{\sigma_{rd}} \left(\left| I_{dp} - I_{dq} \right| \right) = \exp \left(-\frac{\left| I_{dp} - I_{dq} \right|^2}{2\sigma_{rd}^2} \right) \qquad (8)$$

3. Results And Discussion

We experimented with our depth completion and denoising method on Middlebury stereo dataset[25]. The same as [26], Zero-Mean-White-Gaussian noise with standard deviation 25 and 5 are added to the color and depth images respectively, and around 13% pixels with unknown depth values are added to the original Middlebury stereo dataset.

For ease of comparison, all the experiments are conducted on the 30 images in [26]. One typical example among them is the Aloe, presented in Figure 4. And the control variables are used to demonstrate the advantage of each step with its possible alternatives.

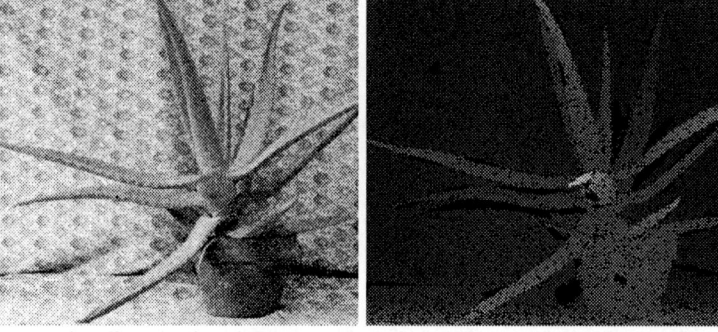

(a) the aligned colour image   (b) the depth map

Figure 4. Example of the depth map and its aligned colour image, the Aloe from the Middlebury Dataset, to be enhanced.

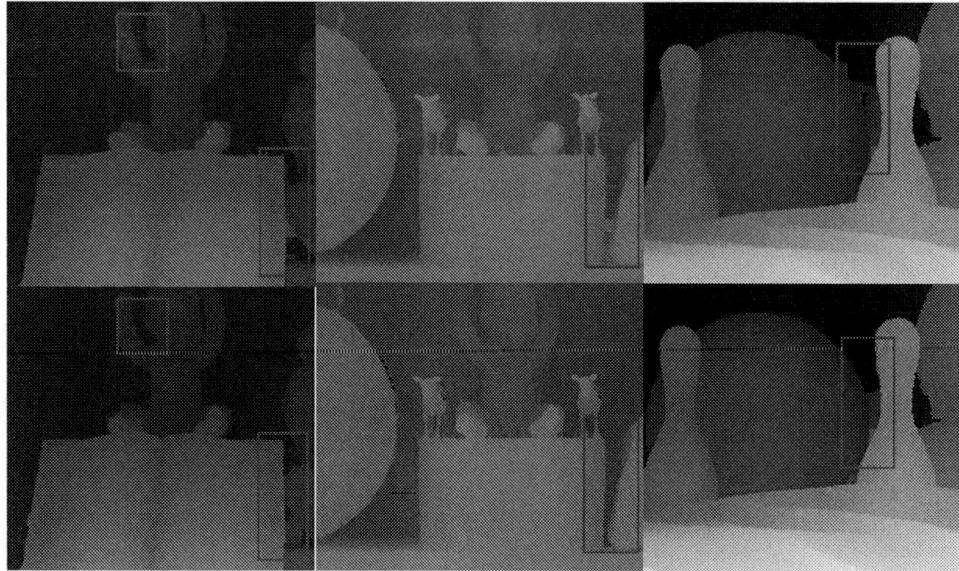

Figure 5. Results obtained with exemplar-based inpainting with selection method ScopeSelectionMethod1 or ScopeSelectionMethod2, Baby2, Baby3 and Bowling1 are given in row (a) and (b).

3.1 Comparison of Propagation methods

As illustrated in Figure 1, the second stage, selecting the hole to be dealt with, is accomplished by two steps: drawing the scope of pixels taken into account and then selecting a pixel among them in a certain way. Possible ways to draw the scopes and select a pixel among them are discussed in the next two sections.

3.1.1 Drawing the scope of pixels to take into account. There are two ways to draw the scope of pixels to take into account: taking all the holes in the pre-processed images into account (referred to as ScopeSelectionMethod1) or dealing with connected holes regions one by one (quoted with ScopeSelectionMethod2). Results of the two scope selection methods are evaluated quantitatively and qualitatively.

As shown in Figure 5, sample results after inpainting, referred to as Baby2, Baby3, and Bowling1, obtained by ScopeSelectionMethod1 lie on the left column and ScopeSelectionMethod2 on the right. Both of them adopt the exemplar-based inpainting. Red rectangles in Figure 5 point out the distinct advantage of ScopeSelectionMethod2 over ScopeSelectionMethod1. It's obvious that the second method works better in structure-preserving and brings in fewer artifacts, especially along the edges.

Given the influence of these two selection methods on the final results, the same subsequent processing methods are adopted. The quantitative results with PSNR and SSIM are presented in Table 1, which obviously tells that ScopeSelectionMethod2 performs better than the other, with 0.3dB higher in PSNR and 0.02 higher in SSIM.

Table 1. Quantization results of average PSNR and SSIM on the testing images.

	ScopeSelection-Method1	ScopeSelection-Method2
PSNR	27.05973	**27.35976**
SSIM	0.92596	**0.92777**

3.1.2 Picking out the target pixel. As described in Section 3.1.1, dealing with pixels in one connected holes region after another works better than taking all the holes into account at once.

Once one connected holes region, referred to as the candidate set, is picked out, there are three methods to determine the order of pixels to be inpainted. Firstly, the pixels can be selected one by one without priority to anyone, namely, all the pixels share the same priority. Secondly, each pixel maintains a priority, determining its subsequent inpainting order, computed by the confidence. Thirdly, the pixel priority is calculated by confidence together with a directional term, as defined in Equation (1).

The variable here is the way selecting some pixels from the specified scope, namely, other procedures are all the same. Comparison results are presented in Table2, which verifies that priority evaluated with confidence and directional term performs relatively better, about 0.09dB and 0.15dB higher in PSNR and 0.00078 and 0.00034 higher in SSIM than the first and second methods.

Table 2. Quantization results of average PSNR and SSIM on the testing images with different pixel priority evaluation methods.

Calculation of priority	no priority	confidence only	confidence and directional term
PSNR	27.26929	27.20078	**27.35976**
SSIM	0.92699	0.92743	**0.92777**

Taking pixels in connected holes regions into account one by one and defining pixel priority with confidence and directional term, adopted in our proposed method, demonstrate the advantages of themselves over their possible alternatives. And all the following experiments are based on the above-proposed pixel selection method combining confidence and directional term.

3.2 Comparison of Smoothing Strategies

As described in Section 2.3, smoothing is an essential procedure after holes filled in Kinect's depth maps. Bilateral filter, referred to as BF, and joint bilateral filter, quoted with JBF, have achieved great popularity in intensity image noise-removing community. But depth enhancement has its own features, invalid pixels in depth maps lead to the BF's inaccuracy and incomplete correspondence of color and depth values disables JBF too. Hence, both methods fail to be suitable perfectly, even though still work to some extent.

In our proposed method, depth similarity and its aligned image's color similarity are both taken into account. Methods to combine the depth similarity, color similarity and geography distance together could be the linear combination of BF and JBF or the joint trilateral filter, defined in Equation (4, 5). As discussed before, in order to suppress the error's accumulation, higher priority is given to the color similarity.

3.2.1 Linear combination of BF and JBF. The method presented in Equation (4), uses the coefficients α and β balancing the influence of BF and JBF on the final results.

As is known to all, the weighting factor in BF is determined by intensity similarity of itself and the geography distance, while in JBF, it's decided by its guidance's intensity similarity and its own geography distance. Correspondingly, in depth enhancement, the image itself is the depth map and the guidance is generally its aligned color image. Therefore, it can be concluded that BF measures the depth similarity, while the geography distance and JBF measures the color similarity and the geography distance in depth enhancement community.

Therefore, if we want to enlarge the influence of color similarity on the final result, only need to make sure that α is bigger than β. Experiments are carried on to compare the results of different ratios, which is presented in Table 3.

Other parameters are set as below. Window size is 7×7, and sigma of distance is 3 while sigmas of color and depth similarity are both set 0.1.

Table 3. Quantization results of average PSNR and SSIM on the testing images with different ratios of BF and JBF.

$\alpha : \beta$	1:1	2:1	3:1	4:1	5:1	6:1	7:1	8:1
PSNR	27.4058	27.4221	**27.4254**	27.4244	27.4214	27.4179	27.4140	27.4100
SSIM	**0.9305**	**0.9305**	0.9302	0.9299	0.9295	0.9292	0.9289	0.9285

More intuitive results are presented in Figure 6, from which we can clearly conclude that when $\alpha : \beta$ equals $3:1$, the results gain the highest average PSNR, while SSIM achieves its maximum at 1 and 2, that's to say, the value of SSIM decreases when color similarity's proportion increases while PSNR enlarges when color similarity's proportion increases.

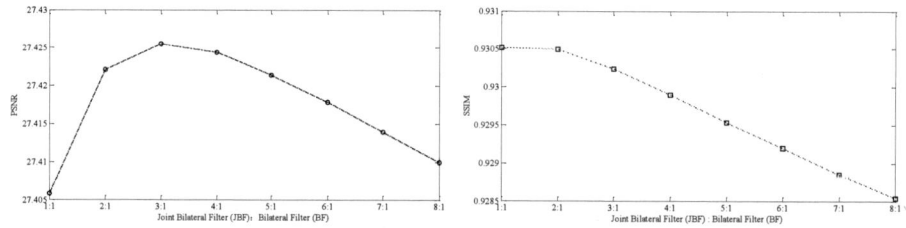

(a)PSNR values with ratio of varies from 1:1 to 8:1   (b) SSIM values with ratio of varies from 1:1 to 8:1

Figure 6. Line Charts of PSNR and SSIM values obtained with different ratios of BF and JBF.

3.2.2 Joint trilateral filter. Another smoothing method is modeled as Equation (5), referred to as the joint trilateral filter. It combined the color similarity, depth similarity and geography distance with multiplication. All the weight factors are calculated with the Gaussian function, which is defined as Eq.9.

$$f(x) = ae^{-\frac{(x-b)^2}{2\sigma^2}} \tag{9}$$

Where, $f(x)$ defines the Gaussian value at x. And a is its peak value, b is the average value, namely, if $x = b$, then $f(x) = a$. All the Gaussian functions used in smoothing kernel share the same average value, i.e. $b = 0$ and the same peak value, i.e. $a = 1$.

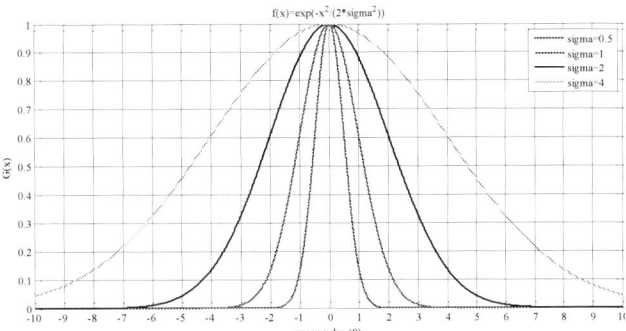

Figure 7. Gaussian curves with different sigmas while average value is 0 and peak value is 1.

As shown in Figure 7, Gaussian curves with the same peak value 1 and average value 0, vary with different σ values. Along with Sigma varying from 0.5 to 4, the curves widen step by step. For example, the Gaussian function values become larger when sigma becomes bigger when mean value is 1.

Therefore, in order to enlarge the influence of color similarity, magnifying the sigma value may take effect. The sigma in color Gaussian function is set n times to depth's, namely if sigma of depth 0.1, then sigma of color n*0.1. Besides, sigma of geography distance is set 3, the same as in Section 3.2.1. Experimental results are presented in Table 4. It can be told directly from the table 4 that when $n = 2$, PSNR gains its highest value 27.4016, while 0.9319 is the best SSIM value at $n = 3$.

Table 4. Quantization results of average PSNR and SSIM on the testing images with different times of color sigma and depth sigma.

n	1	2	3	4	5
PSNR	27.3568	**27.4016**	27.3920	27.3807	27.3721
SSIM	0.9236	0.9317	**0.9319**	0.9315	0.9312

With the same testing data and other former treatment, four smoothing methods, bilateral filter, joint bilateral filter, their linear combination ($\alpha : \beta = 3:1$ is adopted) and the joint trilateral filter ($n = 3$ is adopted), obtain different results respectively. Their quantitative results with PSNR and SSIM are all presented in Table 5.

Table 5. Quantization results of average PSNR and SSIM on the testing images with different smoothing methods.

	BF	JBF	Linear Combination(3:1)	Joint Trilateral Filter(3)
PSNR	27.3462	27.3189	**27.4255**	27.3920
SSIM	0.9298	0.9209	0.9302	**0.9319**
parameters	$\sigma_s = 3$, $\sigma_r = 0.1$	$\sigma_s = 3$, $\sigma_g = 0.1$	$\sigma_s = 3$, $\sigma_{rc} = 0.1$, $\sigma_{rd} = 0.1$	$\sigma_s = 3$, $\sigma_{rc} = 0.3$, $\sigma_{rd} = 0.1$

Figure 8 shows the tendencies of the average PSNR and SSIM on the testing images of the four methods in a more visible way. The proposed method with the two proposed smoothing methods achieve close quality, both better than the conventional BF or JBF on PSNR and SSIM, while BF gains better results than JBF.

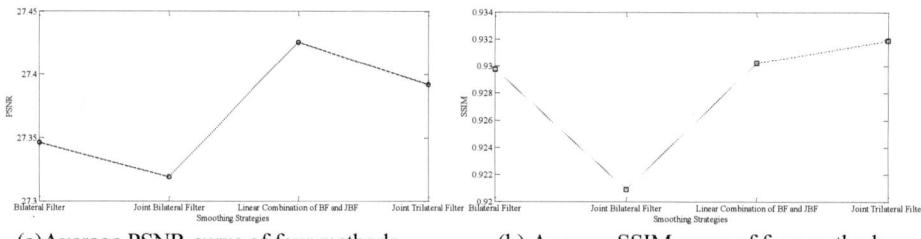

(a)Average PSNR curve of four methods   (b) Average SSIM curve of four methods.

Figure 8. Quantitative results of four smoothing methods, BF, JBF, their linear combination and joint trilateral filter.

3.3 Comparative experiments with other enhancement algorithms

To better demonstrate the superiority of our proposed method, experiments are carried on to compare results with three other state-of-the-art depth enhancement methods, the GAD, GFMM and ImprovedInpainting proposed in [4, 12, 14].

All the three methods share a similar procedure with our proposed method, propagating information from known depth values to unknown regions. In GAD, this is completed with the heat diffusion framework, in which the known depth values are treated as the heat sources. GFMM fulfill this task based on the fast marching method, incorporated with its guidance image. And the ImprovedInpainting defines the priority with the confidence, data term and level set distance factor, and fills in holes with the exemplar-based inpainting method, followed by a certain filter to remove noises.

In our proposed method, similar processing flow is adopted, filling in the holes outside-in. Quantitative results on the testing images with the other three enhancement methods are presented in Table 6, which reveals that our proposed method holds an overwhelming superiority over the other three cutting-edge methods.

Table 6. Quantization results of average PSNR and SSIM on the testing images of GAD, GFMM, ImprovedInpainting and our proposed method with two smoothing methods.

	GAD	GFMM	ImprovedInpainting	Proposed Method(n=3)	Proposed Method(3:1)
PSNR	25.4332	25.6605	26.6498	**27.4255**	27.3920
SSIM	0.8423	0.6974	0.9084	0.9302	**0.9319**

Figure 9, quantization results of PSNR and SSIM on the testing images, further demonstrate the outstanding performance of our proposed method. Meanwhile, the two smoothing strategies in our proposed method share similar results, which are reflected in curves of them almost totally overlapping each other in the line charts. And Figure 10 presents the corresponding qualitative results.

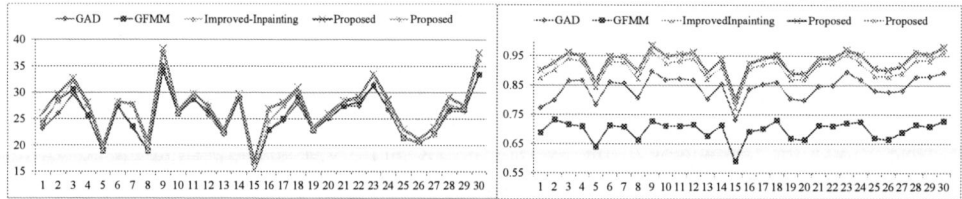

(a)SSIM of the testing images on four methods. (b) PSNR of the testing images on four methods.

Figure 9. Quantitative results of PSNR and SSIM on the testing images of GAD, GFMM, ImprovedInpainting and our proposed method.

Figure 10. Qualitative Comparisons of different algorithms tested on the Aloe, Bowling2, Lampshade1, Monopoly and Teddy are presented above. From left to right, (a) ~ (g) represent the Original Depth images, Corresponding GroundTruth Maps, and results achieved by GAD, GFMM, ImprovedInpainting and our proposed method.(f) shows the results obtained by our proposed method with the linear combination of BF and JBF, while (g) is achieved by the proposed method with joint trilateral filter.

4. Conclusion

In this paper, a novel depth enhancement method for depth maps captured by Kinect is proposed. Our main contributions lie in the selection of the pixel to be filled in and the two smoothing method. Firstly, structural holes in the original depth maps are divided into separate connected holes regions, which are processed one by one. Otherwise, pixels in each connected holes region are filled in based on the priority, determined by the confidence level and directional term, while the directional term is defined with the inner product of isophote and the pseudo-normal. Secondly, two smoothing methods are proposed to enlarge the color similarity, the linear combination of BF and JBF and joint bilateral filter. To enlarge the color similarity, different coefficients are tested and the control variables are used to demonstrate their advantage.

The experimental results demonstrate that our proposed method performs much better than other three state-of-the-art methods, which share similar processing idea. Moreover, experiments on pixel selection present that the method used in our proposed method outperforms its alternatives. And smoothing methods, linear combination of BF and JBF and joint trilateral filter, obtains similar quantative results with proper parameters, both better than the results of single BF or JBF.

Acknowledgments

This work is partially supported by National Natural Science Foundation of China (Grant No. 61305109, No. 61401324), 863 Program (No. 2013AA014601) and Shaanxi Scientific research plan (No. 2014K07-11).

References

[1] Chen L, Wei H, Ferryman J 2013 *Pattern Recognition Letters* **34** 1995-2006

[2] Chen L, Hui L, Shutao L 2012 *21st International Conference on Pattern Recognition* pp 3070-73

[3] Yang J, Xinchen Y, Kun L, Chunping H and Yao W 2014 *IEEE transactions on image processing* **23** 3443-58

[4] Liu J, Gong X 2013 *Advances in Multimedia Information Processing – PCM 2013* Springer International Publishing pp 408-417

[5] Le A V, Jung S W, Won C S 2014 *Sensors* **14** 11362-78.

[6] Langmuir E 1985 *Journal of the Royal Society of Arts* **133** 5351

[7] G. Emile-Male 1976 *The Restorer's Handbook of Easel Painting* (New York: Van Nostrand Reinhold)

[8] Bertalmio, Marcelo, Guillermo S, Vincent C and Coloma Br 2005 *Siggraph* **4** 417-24

[9] Alexandru T 2004 *Journal of Graphics Tools* **9** 23-34

[10] Bertalmío M, Bertozzi A L, Sapiro G 2001 *Proceedings of the 2001 IEEE Computer Society Conference on*, vol 1, pp 355-362

[11] Antonio C, Patrick P, and Kentaro T 2004 *IEEE Transactions on image processing* **13** 1200-12

[12] Liu J, Gong X, Liu J 2012 *In Pattern Recognition (ICPR), 21st International Conference on* IEEE pp 2055-58

[13] Xiaojin G, Junyi, Wenhui Z and Jilin L 2013 *Image and Vision Computing* **31** 695-703

[14] Zhang L, Shen P, Zhang S, Juan S, and Guangming Z. 2016 *IEEE International Conference on Image Processing* IEEE pp 4102-06.

[15] R Gonzalez, R Woods 2002 *Digital Image Processing, 2nd Edition* (Englewood:Prentice Hall)

[16] Mcdonnell M J 1981 *Computer Graphics & Image Processing* **17** 65-70

[17] Tomasi C, Roberto M 1998 *In Computer Vision, 1998. Sixth International Conference on* IEEE pp 839-46

[18] Petschnigg G, Szeliski R, Agrawala M, Cohen M, Hoppe H, Toyama K. 2004 *ACM Transactions on Graphics (TOG)* **23** 664-72.

[19] He K, Sun J, Tang X 2013 *IEEE transactions on pattern analysis and machine intelligence* **35** 1397-1409

[20] Khoshelham K, ISPRS 2011 *ISPRS - International Archives of the Photogrammetry, Remote Sensing and Spatial Information Sciences* **3812** 133-38.

[21] Mallick T, Das P P, Majumdar A K 2014 *Sensors Journal IEEE* **14** 1731-40.

[22] E. Bryan, B P. Velocity Estimation Using an Rgb-D Sensor[EB/OL]. http://www.et.byu.edu/~bmazzeo/ECEn_670_F11/mini_conference_files/Group_9_paper.pdf

[23] "Depth Camera", Microsoft Robotics.User Guide.Robotics Common and Devices, Microsoft Corporation, 2012, https://msdn.microsoft.com/en-us/library/hh438997.aspx.

[24] Depth value of the Kinect and OpenNI, 2011, https://kheresy.wordpress.com/2011/12/27/depth_value_of_kinect_in_openni/.

[25] MiddleburyStereo Dataset: http://vision.middlebury.edu/stereo/data/

[26] Si Lu, Xiaofeng Ren, Feng Liu, Depth Enhancement via Low-rank Matrix Completion(2014), http://web.cecs.pdx.edu/~fliu/project/depth-enhance/.

Recognition of Speed Signs in Uncertain and Dynamic Environments

Zhilong Zhu [1,2,3], Gang Xu[1,2*], Hongmei He[3], Juanjuan Jiang[1,2],Tao Wang[1]

[1] College of Electrical Engineering，Anhui Polytechnic University，Wuhu 241000，China

[2] Anhui Key Laboratory of Detection Technology and Energy Saving Devices，Wuhu 241000，China

[3] Manufacturing Informatics Centre, SATM，Cranfield University，Cranfield, MK43 0AL, UK

Email: zhuzhilong919@ahpu.edu.cn; jiangjuanjuan@ahpu.edu.cn; h.he@cranfield.ac.uk

*Corresponding Author: Gang Xu; email: xugang@ahpu.edu.cn

ABSTRACT: The speed limit signs recognition directly affects the safety of autonomous vehicles. Vehicles are usually running in an uncertain and dynamic environment. The performance of the recognition system is affected by various factors such as the different sizes of pictures, illumination condition and position circumstances, which can lead to misclassification. This makes the speed sign recognition challengeable. To improve the recognition rate of the speed signs in such environments, this work firstly applies the method of the saliency target detection based on the background-absorbing Markov chain, to extract the node in an image, then uses SPP-CNN to classify the extracted nodes with ten-folder validation. The recognition rate is up to 9.32%, higher than that obtained directly by SPP-CNN working on raw dataset.

1. INTRODUCTION

Autonomous vehicles are future trend in automotive sector. Speed sign recognition is important for making good traffic order and ensuring the safety of road users. Traffic signs include text-based and symbolic signs, such as text, numbers, arrows, etc., representing warnings, prohibitions, and other information, using specific colors (e.g. red, yellow, blue, white and black), and specific shapes (e.g. triangles, rectangles, circles and octagonal). Different countries have different traffic signs [1].

The initial road traffic sign recognition (TSR) was based on the color space in the image [2]. Since the first paper on road TSR appeared, various computer vision methods have been developed to detect and identify road traffic signs. The identification of traffic signs is generally divided into three steps: segmentation, detection, recognition. The purpose of segmentation is to obtain the location of traffic signs, and methods include color-based and shape-based. A color-based method is to extract the specific color region; a shape-based method generally relies on edge detection to obtain the shape of traffic signs. Detection determines whether there exists a signpost from the output of segmentation, and the purpose of detection is to reduce the search space to obtain interesting area. Recognition is to classify it into different groups. The speed sign recognition is a critical challenge in the real-time TSR.

However, due to the complexity of the roads and the surrounding environments, speed signs can appear in a variety of scenes, such as the change of sign color due to long-term outdoor environments, dust pollution in the air, the impact of weather conditions, the impact of light, obstruction, view changes, damage, etc. This brings great challenges to automatic speed sign recognition.

Most of the existing TSR systems use the inner region of the signs or the local features such as histograms of oriented gradients, and scale-invariant feature transform for recognition. Although the current technology has made the great progress in the road sign recognition, there is still space to improve the performance of recognition techniques in both effectiveness and efficiency, as well as robustness in dealing with the rotation, illumination, pollution, damage and scale variations in uncertain and dynamic environments.

Deep learning has achieved the success in several areas, such as image processing, natural language processing, etc. Especially convolution neural networks (CNNs) have been widely applied in image–based pattern recognition. In 2014, Girshick [3] first applied R-CNN to categorize the target objects. It calculates the deep CNN to each candidate scheme, and thus generates a lot of information unrelated to the class, increasing the computing and time complexity. In order to improve the efficiency, He et al. [4] developed the Spatial Pyramid Pool Network, and calculated the convolution feature graph of the whole image to produce the feature vector. This accelerated the R-CNN by nearly 100 times. Lee et al. [5] used CNN to recognize the traffic signs based on the boundary estimation and end-to-end training. Zang et al. [6] proposed a framework for the recognition of traffic signs using CNN–based fusion of space and time characteristics. Zhu et al. [7] created a traffic sign dataset, covering illumination and weather conditions, including 100000 examples of 30,000 traffic signs, and they use CNN to detect and classify traffic signs.

In order to improve the recognition rate and real-time performance of Speed Sign Recognition (SSR), we propose a new method that consists of two stages: *Segmentation*, using the Background-Absorbing Markov Chain (BAMC) model [8] [9] to remove background, and *Recognition*, using Spatial Pyramid Pooling based convolution neural networks (SPP-CNN) on the images without background. The experiments will be done on a real-world database, built by the researchers, including various tiled positioned, scaled, and light blurred images.

2. PROPOSED METHOD

2.1. Detection of saliency targets based on BAMC

SLIC segmentation [8] was used to divide the input image into m pixel blocks (nodes). Generally, the pixel blocks at the 4 boundaries (top, bottom, left, right) of an image do not cover the target. Initially k nodes on three selected boundaries are copied as absorption nodes (virtual nodes) in the absorbing Markov Chain. A graph model $G(V,E)$ is established with n nodes, represented by the set of V, $n = k + m$, E is the set of edges, which represent the relations between nodes as the rules: (1) the nodes are correlated to the adjacent pixel blocks; (2) the nodes of all the pixel blocks at all boundaries are correlated with each other; (3) the copied k virtual nodes are not correlated with each other; (4) if node i is a node correlated to a node on the inner boundary of the image, a new node i' is copied, then node i is correlated with node i' and all nodes, correlated with i, are correlated with i'.

Therefore, the correlation matrix A of the nodes in the graph model G (V, E) can be calculated:

$$\alpha_{ij} = \begin{cases} w_{ij} & j \in N(\iota), 1 \le i \le n \\ 1 & i = j \\ 0 & else \end{cases}, \quad w_{ij} = e^{\frac{\|x_i - x_j\|}{\sigma^2}}. \quad (1)$$

where, $N(i)$ is the set of nodes correlated with the node i, the w_{ij} is the weight between correlated nodes i and j, x_i and x_j as a color mean value to the corresponding hyper-pixel block in the color space, σ is a constant. The m pixel blocks in the image are considered as the transfer nodes in the absorption Markov Chain, the k pixel blocks are the absorption nodes. The degree matrix that records the sum of the weights connected to each node is written as $D = diag(\sum_j a_{ij})$. The transition matrix $P = D^{-1} \times A$. With the k-absorbing states and the m-transfer states, P can be reconstructed as

$$P = \begin{bmatrix} Q & R \\ 0 & I \end{bmatrix}, \tag{2}$$

where, $Q \in [0,1]_{m \times m}$, transferring matrix, representing the probabilities between the m transferring nodes; $R \in [0,1]_{m \times k}$, representing the probability between the m transferring nodes and the k absorbing nodes; 0 as a $k \times m$ zero matrix, I is the unit matrix, representing the k absorbing nodes.

For an absorption chain, the basic matrix $N = (J - Q)^{-1}$ can be obtained, n_{ij} is the expected absorption time from the transfer node i to the transfer node j. According to n_{ij}, we can calculate the vector $y = N \times J$, which consists of the absorption time by all transfer nodes, where J is a m-dimension column with all elements are 1. According to the Markov Chain, y is smaller, the corresponding pixel block of the node is more likely to be the background; y is larger, the absorption time is longer, the pixel block may be a significant target. We use the average value of y to represent the significant value s, that is $s(i) = \overline{y}(i), i = 1,2,3, \cdots, m$, the number of transition state nodes in the graph, $\overline{y}$ is the normalized absorption time vector.

Boundary selection [9] [10] is based on the saliency values of the four boundaries. The boundary with largest salience value could include the target; hence, the pixel blocks on the three boundaries are copied as virtual nodes (i.e. absorption nodes) to the set C.

After getting the absorption nodes from the three boundaries, we need to identify more absorption node in the image. The absorption time of all super pixel blocks in M is calculated, and the initial saliency value $S_i^{initial}$ of node i is obtained. If the initial saliency value is less than the threshold T, the node i is more likely background, then node i, as an absorption node, is added to set C.

$$C = C \cup \left\{ i \mid S_i^{initial} \leq T \right\} \tag{3}$$

Algorithm 1 presents the pseudocode of the BAMC process, where, the parameters include: *Im* is an image to be processed; b indicates the size of a $b \times b$ super pixel blocks; M contains all the m super pixel block nodes; C contains all the copied pixel black nodes as background nodes.

The image is first segmented by the SLIC algorithm, and the background pixel block nodes on the three boundaries are identified. For each layer, identify the correlated nodes in the set (M) of super pixel block nodes, and assess if they are belong to background nodes in terms of the defined rules of pixel block relationships and the criteria in Formula (3), move these nodes to the set C, and remove them from M, if they are background nodes. Finally, return the segmented sets of absorption nodes (C) and target nodes (M).

Algorithm 1 BAMC(*Im, b, T, L*)

```
1:  [M]ₘ = SLIC(Im, b, L);
2:  [C]=BoundarySelection(B);
3:  for (i=1...L) do
4:      G=constructGraph(M, C);
5:      A=calMatrix(G);
6:      D = calDegreeMatrix(A);
7:      P=D⁻¹×A
8:      Q = reconstruct(P);
9:      N=(J-Q)^(-1), y=N×J;
10:     for (j=1...m) do
11:         Sᵢⱼ = ȳ;
12:         if (Sᵢⱼ < T) then
13:             C← j;
14:             M ← M \ j;
15:         end if
16      end for
17 end for
18 return (M,C);
```

2.2. Classification with SPP-CNN

ISPECE   IOP Publishing

IOP Conf. Series: Journal of Physics: Conf. Series **1187** (2019) 042066   doi:10.1088/1742-6596/1187/4/042066

The last full connected layer of CNN requires fixed-length vectors [11]. Spatial pyramid pooling can maintain spatial information by pooling in local spatial bins. These spatial bins have sizes proportional to the image size, so the number of bins is fixed regardless of the image size [4] [12]. To make the CNN robust for any size of images, we use spatial pyramid pooling in the deep convolution neural network (SPP-CNN) to recognize the speed signs regardless the size of images produced by the BAMC algorithm. A typical five-layer CNN with the SPP layer as the last pool layer is developed. The CNN passes the features from last convolution layer to the SPP layer. The fixed size of outputs from the SPP layer is passed to the final full connect layer.

3. EXPERIMENT RESULTS AND DISCUSSION

3.1. Experiment setup

Both training and testing of the SPP-CNN were done on a PC with AMD thread ripper 1900x CPU, GTX 1070 Ti Hybrid GPU, 32G memory, the SPP-CNN model is trained on GTX1070 GPU.

The Databases of speed limit sign used in our experiments include 587 samples, which are collected by us on a variety of different roads in the United Kingdom. It includes city streets, countryside, high ways, various luminous environments and positions. Most of samples have a complex background with various target sizes. Two data set were produced as follows: a. the first speed limited sign data set is the original images, denoted as FSLSD as shown in Figure. 1; b. the set of black-and-white traffic speed limit sign extracted from original images with the BAMC algorithm, denoted as the second dataset (SSLSD) in the second line of Figure. 2. From 587 samples, we have chosen 59 samples with different sizes, 52 light-blurred samples, and 46 positions tilted samples for experiment assessment.

Figure 1. FSLSD is obtained by annotate in the original image

Figure 2 SSLSD is obtained based on Background-Absorbing Markov chain

In the segment experiment, the initial number of pixel blocks is set m=250, the threshold value is set T=0.015. As compared effect of the fusion, the number of the layer L = 3. The images in Figure 2 are some results produced by Algorithm 1.

In order to verify how the proposed method improve the performance of the speed limit sign recognition, we conducted the experiments as follows: a. For each dataset, we use the five-layer

SPP-CNN, using ten-folders crossing validation. Namely, for each run, 10% samples are randomly picked as test set, and the rest 90% of images as the training set. b. take the 59 samples with different sizes, 52 light-blurred samples and 46 positions tilted samples as the test set respectively, and calculate the recognition rate.

3.2. Experiment results
Figure 3 is the mean loss of training set results to the two datasets; Table 1 is the results of ten-folders crossing validation; Table 2 is the test results of severe circumstances samples with proposed method

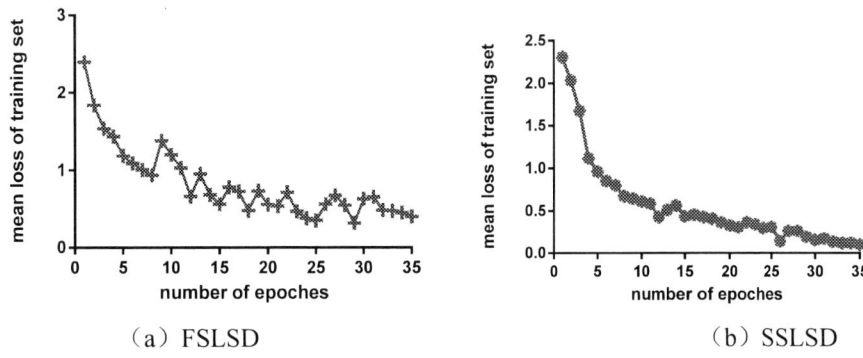

(a) FSLSD   (b) SSLSD

Figure 3 The mean loss curve of training set for ten-folder training

Table 1 Ten-folder crossing validation test results of the two Datasets

Test turns	1	2	3	4	5	6	7	8	9	10	mean
FSLSD(%)	86.44	93.22	88.14	91.53	93.22	86.44	86.44	89.83	89.83	91.53	89.66
SSLSD(%)	98.30	100	100	98.30	96.61	100	98.30	100	100	98.30	98.98

Table 2 Special circumstances sample test results based on proposed method

Image conditions	Different size	Light-blurred	Positions tilted
Recgonition results	59/59	50/52	45/46
Accuracy(%)	100	96.15	97.83

3.3. Discussion
It can be seen from Figure 3 (a) and (b) that the baseline model used in this paper is convergent during the training process with the iteration epochs increases, and the convergence trend of figure (b) is better than that of figure (a), it illustrated that the black-and-white traffic speed limited sign, which is obtained by proposed method in this paper has better effect to neural network learning.

From Table 1, it can be obviously seen that the recognition rate of SSLSD is better than that of FSLSD. This is because that the traffic speed limit signs as a saliency targets are ensured by the transfer nodes, then the recognition of a speed limit sign is completed by the SPP-CNN.

From 587 original images, some with different sizes, blurred lights or titled positions in uncertain environments are selected, respectively. The proposed approach is applied on these subsets. Table 2 shows the results. With the proposed approach, BAMC-SPP-CNN, the recognition rate on the 59 images with different sizes is 100%, the recognition rate on the 52 image with blurred lights is 96.15%, and the recognition rate on the 46 images with tiled positions is 97.83%. Due to the introduction of SPP, the CNN is robust for different sizes of speed signs.

4. CONCLUSIONS

The contribution of this paper is summarized as follows: (1) for salient target segmentation based on Markov chain, first three boundary nodes are used as the absorption nodes, the absorption time of all the super-pixel blocks in an image is calculated, and the foreground is distinguished from the background based on the specified threshold. (2) a new database of images without background is produced to be the input of CNN for classification, and the ten-folder crossing validation is used for the performance assessment. (3) The recognition rate of SPP-CNN, fed with SSLSD is up to 9.32% better than that of the SPP-CNN, fed with FSLSD.

Acknowledgements

This research was financially supported by the Key project of natural science in universities in Anhui Province (KJ2018A0111). Anhui Key Laboratory open project of Detection Technology and Energy Saving Devices, Anhui Polytechnic University, China (2017070503B026-A01), (1506c085002). Anhui Polytechnic University student Research Project (KC00418025).

References

[1] M.-Y. Fu, Y.-S. Huang. (2010) A survey of traffic sign recognition. in Proceedings of International Conference on Wavelet Analysis and Pattern Recognition, July, 119–124.

[2] H. Fleyeh, M. Dougherty. (2005) Road and traffic sign detection and recognition," in Proceedings 10th EWGT Meet/ 16th Mini-EURO Conference, pp.644 – 653.

[3] Girshick, R., Donahue, J., Darrell, T., et al. (2014) Rich feature hierarchies for accurate object detection and semantic segmentation. Proceedings of the IEEE Computer Society Conference on Computer Vision and Pattern Recognition. 580–587.

[4] He, K., Zhang, X., Ren, S., et al. (2014) Spatial Pyramid Pooling in Deep Convolutional Networks for Visual Recognition. IEEE Transactions on Pattern Analysis and Machine Intelligence 1–14.

[5] Lee, H. S., Kim, K. (2018) Simultaneous Traffic Sign Detection and Boundary Estimation using Convolutional Neural Network. Transactions on Intelligent Transportation Systems. 1–12.

[6] Zang, D., Wei, Z., Bao, M., et al. (2018) Deep learning–based traffic sign recognition for unmanned autonomous vehicles. Proceedings of the Institution of Mechanical Engineers, Journal of Systems and Control Engineering. 232(5) 497–505.

[7] Zhu, Z., Liang, D., Zhang, S., et al. (2016) Traffic-Sign Detection and Classification in the Wild. 2016 IEEE Conference on Computer Vision and Pattern Recognition (CVPR), 2110–2118.

[8] Jiang B W, Zhang L H, Lu H C, et al. (2013) Saliency detection via absorbing Markov chain. Proceedings of IEEE International Conference on Computer Vision. Sydney, NSW, Australia. 1665-1672.

[9] Jiang F L, Zhang H T, Yang J, et al. (2018) Image saliency detection based on background absorbing Markov chain [J]. Journal of Image and Graphics. 23(6) 0857-0865.

[10] C. Liu, F. Chang, Z. Chen, et al. (2016) Fast traffic sign recognition via high-contrast region extraction and extended sparse representation. IEEE Transactions on Intelligent Transportation Systems. 17(1) 79–92.

[11] T. Chen, S. Lu. (2016) Accurate and efficient traffic sign detection using discriminative AdaBoost and support vector regression. IEEE Transactions on Vehicular Technology. 65(6) 4006–4015.

[12] Zhao, H., Shi, J., Qi, X., et al. (2017) Pyramid scene parsing network. Proceedings - 30th IEEE Conference on Computer Vision and Pattern Recognition, CVPR 2017, 2017–January, 6230–6239.

ISPECE

IOP Publishing

IOP Conf. Series: Journal of Physics: Conf. Series **1187** (2019) 042067   doi:10.1088/1742-6596/1187/4/042067

Structure Analysis and Generation of X.509 Digital Certificate Based on National Secret

Hua Jiang[1], Gang Zhang[1,a], Jinpo Fan[1]

[1]Communication Engineering Department, Beijing Electronics Science and Technology Institute, No. 7, Fu Feng Road, Beijing, China

[a] Corresponding Author: email: 18810968002@163.com

Abstract: The X.509 format certificate is a widely used digital certificate standard and is a series of data used to identify the identity information of parties to a communication. The design is based on the national secret algorithm SM2, SM3, replacing the ECDSA algorithm and the SHA-256 algorithm in the strongswan source code with the SM2 and SM3 algorithms respectively, Replace the ECDSA-SHA256 certificate with the SM2-SM3 certificate and use the X.509 certificate format for the encryption product.

1.Preface

At present, the RSA signature algorithm has always occupied the dominant position of digital certificate authentication, and the SM2 algorithm researched independently in China is a new domestic asymmetric algorithm, which is more advanced than the RSA algorithm. Based on the national commercial password security and other reasons, the national password management department is being widely promoted nationwide, and the digital certificate based on the national secret algorithm should be further studied and promoted.

2.basic knowledge

2.1 X.509 digital certificate

X.509 is the format standard for public key certificates in cryptography. X.509 certificate have been used in many Internet protocols including TLS/SSL (the cornerstone of WWW World Wide Web Safe Browsing). They are also used in many non-online applications, such as electronic signature services. The X.509 certificate contains the public key, identity information (such as the network host name, organization name or individual name, etc.) and signature information (which can be the signature of the certificate issuing authority CA or self-signed). For a certificate signed by a trusted certificate issuing authority or otherwise verifiable, the certificate owner can use the certificate and the corresponding private key to create a secure communication and digitally sign the document. X.509 is a set of certificate standards defined by the ITU-T standardization department based on their previous ASN.1 definition. The digital certificate based on SM2-SM3 is an X.509 digital certificate.

2.2 National secret algorithm

Like the RSA algorithm, the SM2 algorithm belongs to the asymmetric algorithm system and belongs to the Elliptic Curve Cryptography (ECC) algorithm. However, unlike the RSA algorithm, the RSA algorithm is based on the mathematical problem of large integer decomposition. The SM2 algorithm is

Content from this work may be used under the terms of the Creative Commons Attribution 3.0 licence. Any further distribution of this work must maintain attribution to the author(s) and the title of the work, journal citation and DOI.

Published under licence by IOP Publishing Ltd

based on the discrete logarithm problem of the point group on the elliptic curve. Since the algorithm is based on ECC, its signature speed and key generation speed are faster than RSA. The ECC 256 bit (the SM2 uses one of the ECC 256 bits) has a higher security strength than the RSA 2048 bit, but the operation speed is faster than the RSA.

SM3 is mainly used for digital signature and verification, message authentication code generation and verification, random number generation, etc. Its security and efficiency are comparable to SHA-256.

3. Analysis of X.509 digital certificate structure

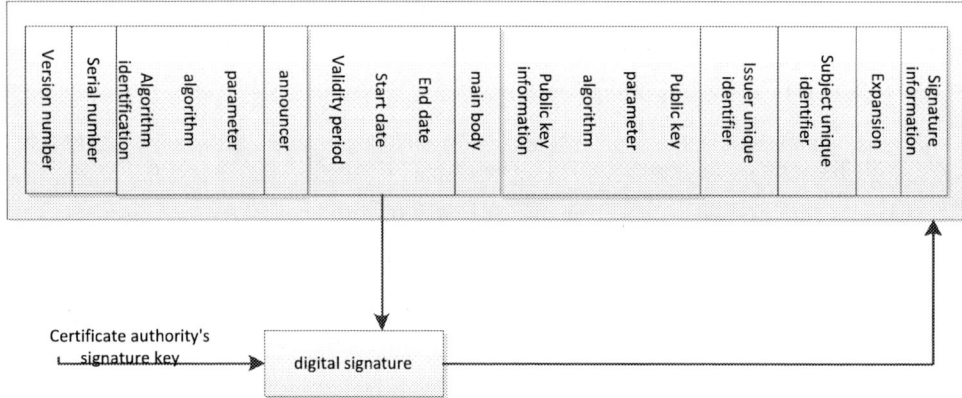

Figure 1. X.509 certificate

3.1 the basic part of the X.509 certificate

3.1.1 Version number
Identifies the version of the certificate (version 1, version 2, or version 3).

3.1.2 Serial number
A unique integer identifying the certificate. A unique identifier for this certificate assigned by the certificate issuer.

3.1.3 Signature
The algorithm identifier used for the visa book, consisting of the object identifier plus the relevant parameters, is used to describe the digital signature algorithm used in this certificate. For example, the object identifiers for SHA-1 and RSA are used to indicate that the digital signature uses SHAA to hash SHA-1 hashes.

3.1.4 Issuer
The distinguished name (DN) of the certificate issuer.

3.1.5 Validity period
The period of time during which the certificate is valid. This field consists of "Not Before" and "Not After", which are represented by UTC time or normal time, respectively.

3.1.6 Subject
The distinguished name of the certificate owner. This field must be non-null unless you have an alias in the certificate extension.

3.1.7 Subject public key information

The public key of the principal (and the algorithm identifier).

3.1.8 Issuer unique identifier
Identifier—The unique identifier of the issuer of the certificate, which is optional.

3.1.9 Subject unique identifier
The unique identifier of the certificate owner, which is optional.

3.1.10 X.509 certificate extension
Optional standards and specialized extensions, the elements of the extension have this structure:
Extension ::= SEQUENCE {
extnID OBJECT IDENTIFIER,
Critical BOOLEAN DEFAULT FALSE,
extnValue OCTET STRING }
extnID: indicates the OID of an extension element
Critical: indicates whether this extension element is extremely important
extnValue: indicates the value of this extension element, the string type.

4.software design
The research results are mainly used in the design of strongswan-based IPSec VPN gateway, written in Python language, using ECDSA-SHA-256 certificate as a template to modify and replace and generate a new SM2-SM3 certificate.

4.1 SM2 algorithm replacement
In strongswan, the ECDSA (Elliptic Curve Digital Signature Algorithm) in openssl is called by default to implement the signature verification function. Therefore, the key pair, signature, and checksum of SM2 are implemented by modifying the source code in ECDSA. Mainly change the parameters of the SECP256k1 curve in the ECDSA source code to the parameters of SM2, and add sm2verifies(), sm2sign(), sm2verify_digest(), sm2sign_digest(), sm2sign_number() and other functions to the source code. Finally, through the national standard algorithm test V1.3.3 software for SM2 to generate key pairs, signatures, check and comparison, fully compatible with each other, fully interoperable, indicating correct rewriting.

4.2 SM3 algorithm replacement
The hash algorithm that is loaded by default in strongswan is sha-1, the digest value of sha-1 is 96 bits or 160 bits, and the length of SM3 algorithm is 256 bits. The digest value of the sha-256 algorithm is 256 bits long, which corresponds to SM3. Write SM3.py with python3.5, the specific code will not go into details.

4.3 Interface UI Design
This interface is developed using eric6 software. Eric6 is an IDE program for the Python programming language. It is powerful and does not lose any IDE program under the Python platform. It takes up less memory and runs faster. The most important thing is that it is seamless with PyQt5. PyQt5 is based on the Python programming language. The external GUI development language, its solid underlying foundation and powerful visual interface design make PyQt5 a leader in Python language GUI development. The update speed is fast, and the speed of developing GUI programs is fast. It can be said that other GUI development languages are beyond the reach.

The Qt designer provided by eric6 software is used to design the overall interface framework, which is designed with the public key, the serial number of the certificate, the certificate issuer, the effective start and end date of the certificate, and the user of the certificate.

The power of eric6 software lies in the separation of interface and logic. The .py file here is compiled from .ui file, so when the .ui file changes, the corresponding .py file will also change. The .py file compiled from the .ui file is called the interface file. Since the interface file is initialized each time it is compiled, you need to create a new .py file to call the interface file. This new file is called a logical file or a business file. In the logical file, write the code, extract the certificate body in the digital certificate, replace the public key in the main body with the public key provided by the external to obtain the new subject, SM3 hashes the new subject, and then uses SM2 to sign. Get a 64-bit new signature value, and finally convert the decimal data entered in the interface box into a hexadecimal certificate format by Ascill encoding. Finally, the data is integrated according to the X.509 digital certificate encoding method and displayed in the box of the new certificate. Finally, the certificate data can be saved in the desired format. Figure 2 below shows the interface for generating a CA certificate.

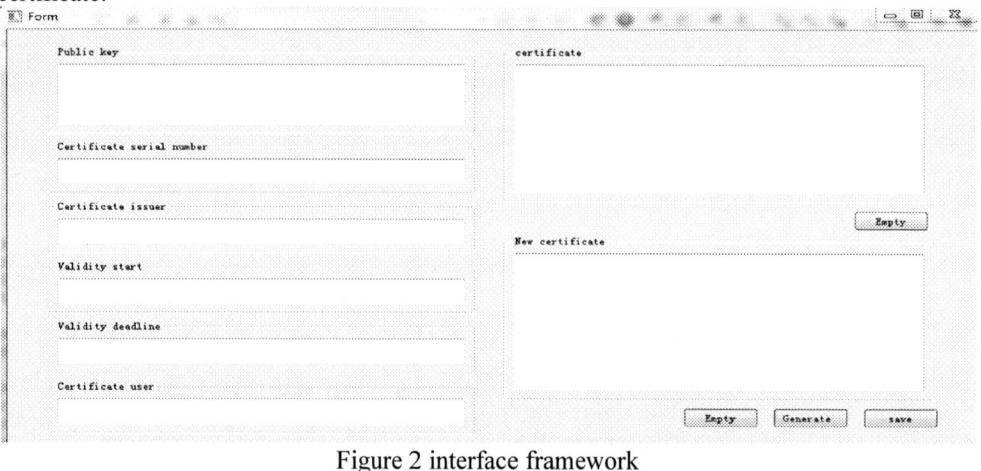

Figure 2 interface framework

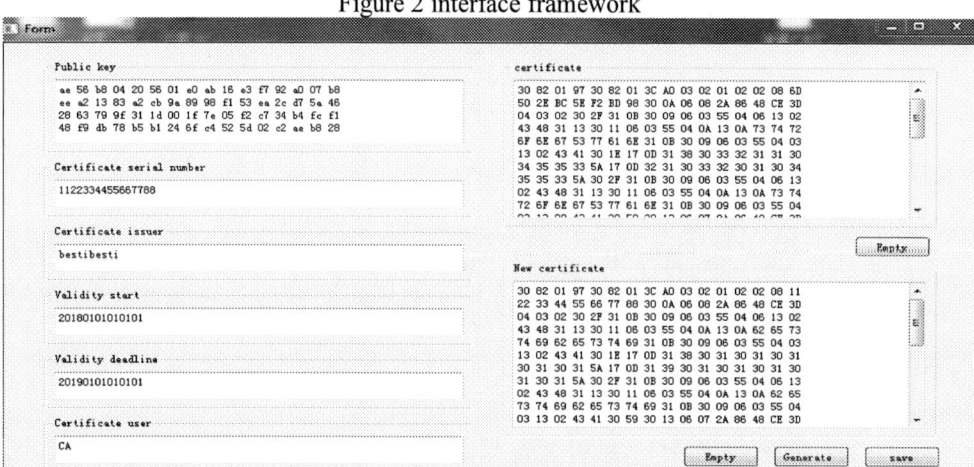

Figure 3 generates a certificate

5. Analysis of certificate structure examples

ISPECE   IOP Publishing

IOP Conf. Series: Journal of Physics: Conf. Series **1187** (2019) 042067   doi:10.1088/1742-6596/1187/4/042067

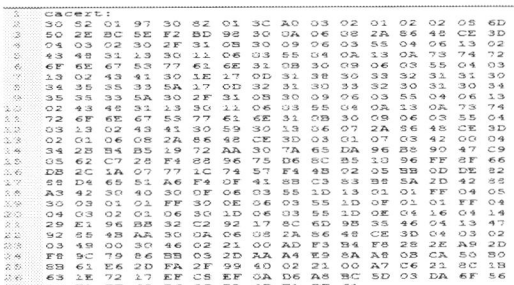

Figure 4 CA certificate instance

Certificate analysis：

address	content	significance
0000	30 82 01 97	Certificate header, SEQUENCE type (30), data block length byte is 2 (82), length is 407 (01 97)
0004	30 82 01 3C	The certificate body header, SEQUENCE type (30), data block length byte is 2 (82), length is 316 (01 3C)
0008	A0 03 02 01 02	Special Content - Certificate Version (A3), Length 3, Integer Type (02), Length 1, Version 3 (2)
0013	02 08 6D 50 2E BC 5E F2 BD 98	Integer type (02), length 8 (08), certificate serial number (6D 50 2E BC 5E F2 BD 98)
0023	30 0A 06 08 2A 86 48 CE 3D 04 03 02	SEQUENCE type (30), length 10 (0A), OBJECT IDENTIFIER type (06), length 8, algorithm (2A 86 48 CE 3D 04 03 02)
0035	30 2F 31 0B 30 09 06 03 55 04 06 13 02 43 48 31 13 30 11 06 03 55 04 0A 13 0A 73 74 72 6F 6E 67 53 77 61 6E 31 0B 30 09 06 03 55 04 03 13 02 43 41	Certificate issuer: SEQUENCE type (30), length 47 (2F) SET starts a collection with a length of 11 (0B), id-at-countryName (55 04 06), corresponding to CH (43 48) SET starts a collection with a length of 19 (13), Id-at-organizationUnitName (05 04 0A), corresponding to strongSwan (73 74 72 6F 6E 67 53 77 61 6E) SET starts a collection with a length of 11 (0B), Id-at-commonName (55 04 03), corresponding to CA (43 41)
0084	30 1E 17 0D 31 38 30 33 32 31 31 30 34 35 35 33 5A 17 0D 32 31 30 33 32 30 31 30 34 35 35 33 5A	SEQUENCE type (30), length 30 (2F) Notbefore: UTCTime type (23), length 13, UTCTime '180321104553Z' Notafter: UTCTime type (23), length 13, UTCTime '210320104553Z'
0116	30 2F 31 0B 30 09 06 03 55 04 06 13 02 43 48 31 13 30 11 06 03 55 04 0A 13 0A 73 74 72 6F 6E 67 53 77 61 6E 31 0B 30 09 06 03 55 04 03 13 02 43 41	The following data block represents the body information, length 47 (2F), SET starts a collection with a length of 11 (0B), id-at-countryName (55 04 06), corresponding to CH (43 48) SET starts a collection with a length of 19 (13), Id-at-organizationUnitName (05 04 0A), corresponding to strongSwan (73 74 72 6F 6E 67 53 77 61 6E) SET starts a collection with a length of 11 (0B), Id-at-commonName (55 04 03), corresponding to CA (43 41)
0165	30 59 30 13 06 07 2A 86 48 CE 3D 02 01 06 08 2A 86 48 CE 3D 03 01 07 03	SEQUENCE type (30), length 89 (59)Key parameter (06 08 2A 86 48 CE 3D 03 01 07) public key value, expressed as integer

1743

	42 00 04 34 2B B4 B5 19 72 AA 30 7A 65 DA 96 B8 90 47 C9 05 62 C7 28 F4 88 96 75 D6 8C B5 10 96 FF 8F 66 DB 2C 1A 07 77 1C 74 57 F4 4B 02 05 BB 0D DE 82 88 D4 65 51 A6 F4 0F 41 8B C3 83 B8 5A 2D 42 88	type, length 64, public key（34 2B B4 B5 19 72 AA 30 7A 65 DA 96 B8 90 47 C9 05 62 C7 28 F4 88 96 75 D6 8C B5 10 96 FF 8F 66 DB 2C 1A 07 77 1C 74 57 F4 4B 02 05 BB 0D DE 82 88 D4 65 51 A6 F4 0F 41 8B C3 83 B8 5A 2D 42 88）
0256	A3 42 30 40 30 0F 06 03 55 1D 13 01 01 FF 04 05 30 03 01 01 FF 30 0E 06 03 55 1D 0F 01 01 FF 04 04 03 02 01 06 30 1D 06 03 55 1D 0E 04 16 04 14 29 E1 96 BB 32 C2 92 17 8C 6D 9B 35 46 04 13 47 92 85 4B AA	Extended value Subject Key Identifier (55 1D 13) Key usage (55 1D 0F) Certificate Strategy (55 1D 0E)
0324	30 0A 06 08 2A 86 48 CE 3D 04 03 02 03 49 00 30 46 02 21 00 AD F3 B4 F8 28 2E A9 2D F8 9C 79 86 BB 03 2D AA A4 E9 8A A8 0B CA 50 B0 8B 61 E6 2D FA 2F 99 40 02 21 00 A7 C6 21 8C 1B 63 1E 72 17 EF C8 EF 0A D6 A8 BC 5D 03 DA 6F 56 B0 E1 EE 69 76 3C DA 4D E1 DE 61	Certificate signature algorithm identifier（30 0A 06 08 2A 86 48 CE 3D 04 03 02 03 49 00） Certificate signature value（30 46 02 21 00 AD F3 B4 F8 28 2E A9 2D F8 9C 79 86 BB 03 2D AA A4 E9 8A A8 0B CA 50 B0 8B 61 E6 2D FA 2F 99 40 02 21 00 A7 C6 21 8C 1B 63 1E 72 17 EF C8 EF 0A D6 A8 BC 5D 03 DA 6F 56 B0 E1 EE 69 76 3C DA 4D E1 DE 61）

6. Conclusion

In this paper, the generation of digital certificates based on national secret algorithm has been realized. The python source code library for ECDSA be amended as SM2 algorithm. Support national secret algorithms to meet the needs of national security and economic development. This paper also analyzes the format of the X.509 digital certificate and elaborates on the meaning of each part of the digital certificate. You can freely set the contents of the certificate to generate and save. However, there are also some problems that need to be solved. For example, the generated certificate is transmitted to the encryption board of the IPSec VPN to facilitate the reading of the encryption card.

References

[1] Mukkamala R, Balusani S. Active Certificates: A New Paradigm in Digital Certificate Management[C]// International Conference on Parallel Processing Workshops. IEEE Computer Society, 2002:30.

[2] Zhang J, Hu N, Raja M K. Digital certificate management: Optimal pricing and CRL releasing strategies[J]. Decision Support Systems, 2014, 58(58):74-78.

[3] Maes S H, Sedivy J. Portable information and transaction processing system and method utilizing biometric authorization and digital certificate security[J]. Lancet, 1999, 1(7793):32.

[4] Maes S H, Sedivy J. Portable information and transaction processing system and method utilizing biometric authorization and digital certificate security[J]. Lancet, 1999, 1(7793):32.

[5] Han S L, Min M A, Wang T. Design and Implemtntation of Digital Certificate Application System[J]. Netinfo Security, 2012.

[6] Haiwen O U, Wang Y, Ouyang C, et al. SM2-based digital certificate parsing and validity verification[J]. Journal of Computer Applications, 2016.

[7] Harwani B M. Introduction to Python Programming and Developing GUI Applications with PyQT[J]. Chromatographia, 2011, 32(12):1088.

[8] Bouda, Peter, Bouda, et al. Pyqt Und Pyside: Gui- Und Anwendungsentwicklung Mit Python Und Qt[J]. Open Source Press, 2015.

Research on parking detecting analysis based on projection transformation and Hough transform

Xuemei Yu and Yaojie Sun

School of Electronic Information Engineering, Hebei University of Technology, Tianjin, China

yuxuemei2015@163.com

Abstract. In order to improve the utilization of effective parking spaces in the parking lot, real-time monitoring is one of the effective methods for detecting the number of parking spaces in the parking lot. Since the single camera has a dead angle in the process of taking pictures, the use of a dual camera for image acquisition at the same time eliminates dead ends. Positioning of the parking space in the image is an important part of monitoring the parking space with the camera. In this paper, the position of the camera is analyzed by using Hough transform line detection technology, and the positioning can be performed according to the analysis result. By merging the two images after the detection and positioning, the picture without dead space can be accurately obtained.

1. Research background

The parking space condition detection system is an important part of the intelligent transportation system ITS, and its application scope has covered many fields of daily life of residents. At present, the parking space detection methods are mainly divided into two categories based on image and non-image detection. Non-image-based parking space detection includes ultrasonic detection, geomagnetic detection and so on, which has the advantages of being reliable and inexpensive. But its disadvantages are damage to the road surface during construction, interference with traffic, and high maintenance cost in the later period; and image-based parking space detection effectively avoids the above disadvantages. In the actual parking lot, different combinations of single camera, dual camera or multiple cameras can be used according to the actual situation to eliminate the shooting dead angle of the collecting surface. Therefore, this paper mainly studies the method and implementation of parking space detection in the parking lot based on image detection. The image is collected by the dual camera on the parking space of the parking lot, so that there is no dead angle in the parking lot. The specific flow chart is shown in Figure 1.

Content from this work may be used under the terms of the Creative Commons Attribution 3.0 licence. Any further distribution of this work must maintain attribution to the author(s) and the title of the work, journal citation and DOI.

Published under licence by IOP Publishing Ltd

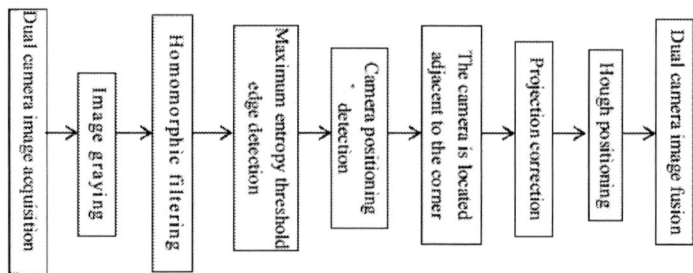

Figure 1 system flow chart

2. Pre-processing and positioning detection of parking space positioning

The image detected by the camera is sequentially subjected to gradation processing, homomorphic filtering, maximum entropy threshold edge detection, and first angle correction for image correction and positioning.

The image captured by the camera is an RGB color image, and the gray image is a special color image with the same components of R, G, and B. The variation range of one pixel is 255, and the variation range of one pixel of the color image is more than 16 million ($255 \times 255 \times 255$), so grayscale it in the pre-processing process. In this paper, homomorphic filtering is used to process the grayscale image, which reduces interference and improves image contrast. The maximum entropy threshold segmentation method is used in edge detection to retain more parking space information.

The camera positioning detection uses the Hough line detection algorithm to calculate the angle θ between the longest line and the horizontal line by extracting the longest line in the picture (recording this angle for subsequent camera position classification); according to the angle θ rotates the picture clockwise to achieve picture correction. The position of the camera is judged based on the magnitude of the line angle θ detected by the Hough line. There are two ways to place two cameras in the parking lot. One method is that the camera is located on the left and right sides of the parking space, that is, $20^o \leq \theta \leq 80^o$,as shown in Figure 2, after image correction the two pictures can be directly fused to detect the number of parking spaces in the parking lot. However, the disadvantage of this installation position is that the wide angle of the camera is required to be large enough, so it is rarely used. Another method is that the camera is located at an adjacent corner of the parking lot, and then the image is subjected to projection transformation and Hough positioning processing, as shown in Figure 3.

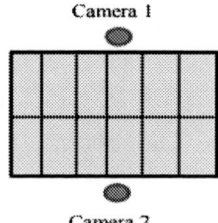

Figure 2 The camera is located on both sides of the parking space (top view)

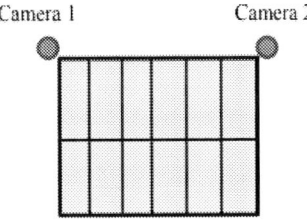

Figure 3 The camera is located at an adjacent corner of the parking lot (top view)

3. projection transformation

Projection transformation is a process of transforming the coordinates of a map projection point into the coordinates of another map projection point. In the parking space positioning, the projection transformation is used to realize the malformation correction of the image.

Set the point under the camera coordinate system (x, y, h) to the pixel plane coordinate system (w, z), and set the world coordinate system to (X, Y, Z), and the corresponding relationship is shown in Figure 4.

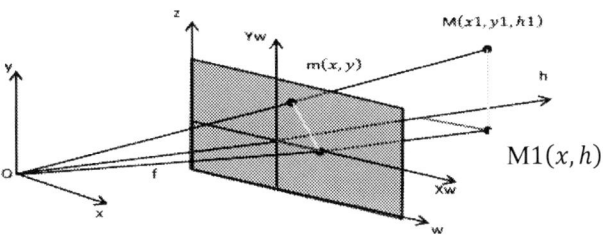

Figure 4 Relationship between each coordinate system in the projection transformation

Set a point M(x, y, h) in the camera coordinate system, and project the M point onto the xoh plane to get the M1 point. According to the triangle similarity principle, the yellow line in the figure is proportional to the red line, according to this. The principle can convert the point M in the camera coordinate system into a point m in the pixel coordinate system to realize image projection correction. Therefore, it is only necessary to find the corresponding projection matrix T to achieve projection correction.

$$[w, z, 1] = T^{-1} \times [x, y, h] \qquad (1)$$

Set the picture size to w1×h1, select four points on the picture edge, and select the four positions of the parking space with the upper left, upper right, lower left and lower right as (x1, y1), (x2, y2), (x3, y3), (x4, y4), operation matrix A.

$$A = \begin{bmatrix} 1 & 1 & 1 & 0 & 0 & 0 & -x1 & -x1 & -x1 \\ 0 & 0 & 0 & 1 & 1 & 1 & -y1 & -y1 & -y1 \\ w1 & 1 & 1 & 0 & 0 & 0 & -w1 \times x2 & -x2 & -x2 \\ 0 & 0 & 0 & w1 & 1 & 1 & -w1 \times x2 & -y2 & -y2 \\ 1 & h1 & 1 & 0 & 0 & 0 & -x3 & -h1 \times x3 & -x3 \\ 0 & 0 & 0 & 1 & h1 & 1 & -y3 & -h1 \times x3 & -y3 \\ w1 & h1 & 1 & 0 & 0 & 0 & -w1 \times x4 & -h1 \times x4 & -x4 \\ 0 & 0 & 0 & w1 & h1 & 1 & -w1 \times x4 & -h1 \times x4 & -y4 \end{bmatrix} \qquad (2)$$

Let U be 8×8 matrix, V be 8×9 matrix, T1 be 9×9 matrix, and T1 can be calculated according to singular matrix decomposition formula 3.

$$A = U \times V \times T1^T \qquad (3)$$

Let H be a 9×1 matrix, then: $\qquad H(i) = T1(i, 9)/T1(9, 9) \quad (1 \le i \le 9) \qquad (4)$

The projection matrix T can be obtained by arranging the data in H in 3×3.

4. Image positioning under Hough transform

After the above image correction, the obtained image still contains the non-target area, so the image location based on Hough line detection is used for screening.

The traditional Hough line detection is performed on the image after the projection correction, and the coordinates of the detected starting point of the line are respectively saved in M_point1 and M_point2. To reduce noise interference, calculate the length of all detected lines and remove all lines that are less than 2/3 of the image width in M_point1 and M_point2. The original image captured by the camera is shown in Figure 6.

The projection corrected image is sequentially subjected to grayscale, homomorphic filtering, maximum entropy threshold edge detection and etching to obtain a binary image image1. In image1, search from top to bottom and left to right. Find the first white point and record it as A. This white point must be the top left boundary point. At least one of its right, bottom right, bottom, and bottom left neighbors is the boundary point, denoted as B. Start B finds the boundary point C in the adjacent point in the counterclockwise direction with dir=3 (4 connected domain). If C is the next vertex of point A, and the previous vertex of C is the A vertex, it means that it has been rotated, and the program ends; otherwise, it continues to search from point C until it finds A. In the process of searching, all the found points are stored in the structure D, and all the points are sequentially connected in the order of the search to form the boundary contour of the parking space target, and the boundary tracking can be completed.

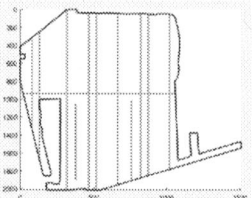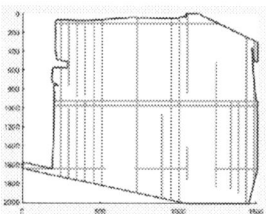

Figure 5 Hough line detection and boundary tracking diagram

In Figure 5, we can see that there are interference lines after the line detection and boundary tracking solutions. Therefore, the straight line obtained in the Hough line detection is processed in accordance with the angle with the horizontal direction, and all the straight lines in the horizontal direction and the vertical direction are retained, and the horizontal straight line starting point is stored in the structural body A_lines, and the vertical direction is straight. The starting point is stored in the structure B_lines.

Based on the boundary information obtained in the boundary tracking, points equal to or equal to $\leq$ 50 in the structure D are sequentially searched for in B_lines and A_lines and recorded in the structure C. These points are the effective points of the parking space boundary, and the coordinates of the points in all the structures C are retrieved, and the maximum value Xmax to the X-axis, the minimum value Xmin, the Y-axis maximum value Ymax, and the minimum value Ymin are compared. Returning to the image after projection correction, the image is intercepted by four points (Xmin, Ymin), (Xmin, Ymax), (Xmax, Ymin) and (Xmax, Ymax), and an accurate target parking space picture can be obtained, such as Figure 7 shows.

5. parking spaces

In order to visually display the parking space situation, the pictures after the Hough positioning process are merged. According to Figure 8, the picture contains complete parking space information. Therefore, according to the number of parking spaces, the specific parking spaces of the two pictures can be simultaneously divided and marked, and the marking order is marked from left to right and top to bottom. Classify pictures with the same number. Positions 1, 2, 7, and 8 are based on the classification result of the right image. Positions 5, 6, 11, and 12 are subject to the left image classification result. Other locations can be determined by any set of images (determined by the image on the right in this article). All the pictures marked last are arranged in order, and the car position fusion can be completed. The picture after the car position is merged as shown in Figure 8.

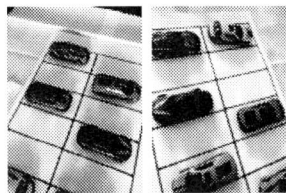 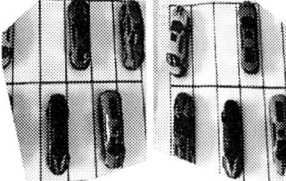 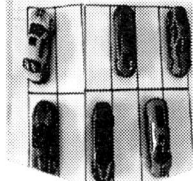

Figure 6 original image   Figure 7 After positioning   Figure 8 After fusion

6. Conclusions

The parking space detection system can be divided into the following two main research directions: multiple acquisition surfaces are integrated with each other, and multiple imaging surfaces of a single acquisition surface are combined. In this paper, the image is acquired by multi-angle of single-collector and dual-camera. The camera is positioned and detected by the captured image to determine the correction angle. The camera position is classified according to different angle ranges. If the angle detected by the camera positioning is greater than 20 degrees and less than 80 degrees, the combination of projection transformation and Hough positioning is used to locate the parking space. Then, the position number detection and the parking space fusion of the positioned picture can be obtained, and accurate parking space information can be obtained. If the camera in the parking lot is located at an adjacent corner, and the angle detected in the camera positioning detection is greater than 20 degrees and less than 80 degrees, the combination of Hough line detection and projection transformation can accurately detect the parking space.

References

[1]   Richard E and Rafael C.2013. *MATLAB Implementation of Digital Image Processing [M]*. (Beijing: Tsinghua University).

[2]   Liu C L.2017.*MATLAB Image Processing [M]*. (Beijing: Tsinghua University).

[3]   Wen Z and Sun K H.2017.*MATLAB Intelligent Algorithm [M]*. (Beijing: Tsinghua University).

[4]   Zhang Z and Li T.2018. (16). *Image recognition in the field of intelligent transportation   [J]*. (Unlimited internet technology) pp 139–143.

[5]   Li L.2018. *Research on Automatic Correction Algorithm Based on Single Wide-angle Distortion Image [D]*. (Beijing: Beijing Jiaotong University).

[6]   Liu L L and Liang X H.2017, 25(19). *QR code image correction based on improved canny operator and Hough transform [J]*. (Electronic design engineering) pp 183–186.

Research on the efficiency of Beijing-Tianjin-Hebei airport group based on system dynamics

C Y Wang[1,3], W W Wu[1,2,4] and J Zhang[1]

[1]College of Civil Aviation, Nanjing University of Aeronautics and Astronautics, 2111 00 Nanjing, China

[2]Corresponding author email: nhwei@nuaa.edu.cn

[3] Fund project: Nanjing University of Aeronautics and Astronautics Graduate Innovation Base (Lab) Open Fund[grant number kfjj20170705]

[4] Fund project: National Natural Science Foundation of China [grant number 71201081, 71731001]

Abstract. The Beijing-Tianjin-Hebei Airport Group is a nonlinear complex system with dynamic changes and feedback mechanisms. It is difficult for a general model to simulate an airport group. This paper adopts the system dynamics method to establish the airport group efficiency evaluation model, adjust the variable parameters, and verify the policies implemented in each airport in this airport group. The study found that the overall efficiency of the Beijing-Tianjin-Hebei Airport Group has increased year by year. However, the Beijing Capital International Airport has experienced over-utilization of resources in the past three years. The efficiency of Tianjin Binhai International Airport has been rising, changing from the excessive resource redundancy in 2010 to the nearly full use of airport resources in 2017. Although the efficiency of Shijiazhuang Zhengding Airport has increased, the airport resources have been still wasteful. Through the simulation of the airport group, we can understand the efficiency of the airport group and each airport, and fully understand the integrity of the airport group and the mutual impact mechanism between the airports.

1. Introduction

In 2017, a notice was issued, proposing to accelerate the coordinated development of civil aviation in Beijing, Tianjin, Hebei, and strive to build a world-class aviation airport group. The Beijing-Tianjin-Hebei airport group currently has nine airports, including Beijing Capital International Airport (PEK), Tianjin Binhai International Airport (TSN) and six airports in Hebei province including Shijiazhuang Zhengding International Airport (SJW). In 2017, the ratio of the passenger throughput of each airport in this airport group is shown in Figure 1. This airport group presents a single-level agglomeration of PEK. The passenger demand concentrated in PEK leads to a shortage of flights, exceeding the capacity the airport can guarantee, causing flight delays. TSN and SJW have less air transportation demand and airport resources are idle. The oversaturation of PEK and the redundancy of the other two airports have affected the overall development of the airport group. Therefore, research on the overall efficiency of airport groups and airport development policies is urgently needed.

At present, there are many research methods for efficiency, including Data Envelopment Analysis (DEA). DEA is very mature in the evaluation of airport efficiency, and can evaluate the efficiency of multi-input and multi-output, which is not affected by the data units, but is susceptible to extreme values and random disturbances. Fernandes, Mustafa Lsa Dogan, He Yan, Jia Pinrong [1]-[4] studied the operational efficiency of airports using different models of DEA. Currently the study of airport group efficiency is little. Zhang Weina [5] used DEA to evaluate the coordination efficiency between the two airports. It evaluated the correlation between the input of one airport and the output of another airport, but did not reflect the characteristics of the airport group.

The airport group is composed of airport individuals that distinguish each other and interact with each other and work together on the development of the airport group. It is a complex system under the combined effect of the factors including regional economy, population, policies, airports, other airport groups and so on. The impact of regional economy, passenger demand, resource supply, etc. on the airport group is systematic. The development of the regional economy has a positive impact on the overall passenger demand of the airport group. When selecting an airport, passengers who choose an airport in the airport group may not choose another airport. To avoid homogenization when airlines establish bases or develop routes, only one airport within the airport group will be selected. Therefore, the airport group emphasizes the common development of the individual and the overall.

System dynamics (SD) is precisely a method that emphasizes the combination of macro and micro, and studies the external influence and internal structure of the system. It can deal with complex system problems and can carry out policy simulation and predict policy effects. At present, there is no study of the airport group efficiency with SD, but the method has certain applications in the civil aviation field. Chen Yaqing [6] established a model for air traffic forecasting. Ioanna E. Manataki [7] assessed the performance of the Athens International Airport under different needs and resource allocations. In addition, there are many achievements in the study of urban transportation complex system using SD. Among them, the more famous researches are the urban dynamics model of Professor Forrester [8] and the travel generation model of Professor Shirazian. Hossein Haghshenas, Shiyong Liu, Xu Tianyou, and Yang Haoxiong [9]-[12] also used SD to establish urban traffic models, study urban traffic development, manage traffic congestion, and propose development policies.

Compared with the data envelopment method, the use of SD to study the efficiency of the airport group can not only better reflect the systematic, dynamics and complexity of the airport group, but also reflect the characteristics of the airport group, and simulate the impact and external influences between airports in the airport group. Therefore, based on SD, this paper analyses the dynamic structure and feedback mechanism of the airport group, and uses the Beijing-Tianjin-Hebei airport group as the research object to construct the Beijing-Tianjin-Hebei airport group efficiency model. Then this paper uses Vensim software to simulate and analyses the simulation results, and verifies the impact of the policies adopted by the airports on the airport group and other airports in this group.

2. Model building

2.1. Model hypothesis

Hypothesis 1: The passengers of Beijing-Tianjin-Hebei airport group are mainly concentrated in PEK, TSN, SJW. So this article assumes that this airport group consists of these three airports, and the other six airports are temporarily not considered.

Hypothesis 2: In general, airport efficiency is used to measure the extent to which resources are used. Therefore, it is assumed in this paper that the efficiency of each airport is the ratio of passenger throughput per airport to the capacity of each airport terminal. If the efficiency value is 1, the airport has the best allocation of resources and the highest efficiency. If the efficiency value is less than 1, the airport resources are not fully used and the airport efficiency is not high. If the efficiency value is greater than 1, meaning the excessive use of resources in the airport which can lead to congestion and declining service levels, so airport efficiency is not high too.

Hypothesis 3: The impact of each airport in the group is different, so the airport group efficiency is assumed to be the weighted average of the airport efficiency, and the weight is the ratio of the passenger throughput per airport to the total passenger throughput.

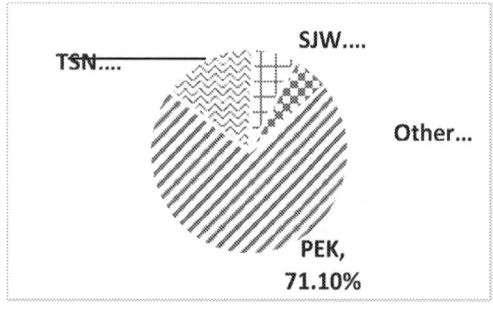

Figure 1. The ratio of the passenger throughput of airports.

Figure 2. The causal relationship diagram.

2.2. Variable meaning

Table 1. Model variables and meaning.

Variable name	Meaning	Variable name	Meaning
GDP	Regional GDP	TLCC	Low-cost base carriers of SJW
AGDP	Regional per capita GDP	TBSP	Branch shipping points of SJW
RP	Regional permanent resident population	FE	Efficiency of PEK
TSPPV	Regional transportation storage postal production value	SE	Efficiency of TSN
RTP	Regional total passenger throughput	TE	Efficiency of SJW
AFP	Passenger throughput of PEK	FC	Capacity of PEK
AFPI	Passenger throughput increment of PEK	SC	Capacity of TSN
ASP	Passenger throughput of TSN	TC	Capacity of SJW
ASPI	Passenger throughput increment of TSN	FW	Efficiency Weight of PEK
ATP	Passenger throughput of SJW	SW	Efficiency Weight of TSN
ATPI	Passenger throughput increment of SJW	TW	Efficiency Weight of SJW
FIPR	International passenger ratio of PEK	E	Total efficiency of the airport group
SBSP	Branch shipping points of TSN		

2.3. Causality diagram, feedback loop

The causal diagram can make complex problems simple and systematic, and provide scientific and clear ideas for analysing problems. After preliminary analysis, the system dynamics model of the Beijing-Tianjin-Hebei airport group proposed in this paper mainly includes three subsystems: PEK, TSN and SJW. The causal relationship diagram of the airport group is shown in Figure 2. The passenger throughput of three airports constitutes the total passenger throughput in this airport group, and the regional passenger throughput affects the productive value of the transportation, storage and post in the region, which in turn affects the regional overall GDP. The three airport subsystems are linked by economic factors to form the entire system of the airport group. There are three loops in the system dynamics model of this paper, as shown in Figure 3, reflecting the relationship between airport passenger throughput and economy. The increase in regional GDP leads to an increase in per capita GDP. The rise in per capita GDP stimulates people's travel needs, including air travel demand. As a result, the increase of the passenger throughput at PEK, TSN and SJW can bring the growth of airport benefit. The airport benefit can drive the development of the local economy to a certain extent. Therefore, the increase in transportation, storage and post production value directly increases GDP.

2.4. Flow diagram
The causality diagram of the Beijing-Tianjin-Hebei airport group does not distinguish between state variables and rate variables, and can only reflect the influence between various variables. However, in the process of further modelling, the model needs to be quantified to distinguish different types of variables. Therefore, the system flow diagram is drawn on the basis of the causal graph, as shown in Figure 4. The passenger throughput at each airport is a state variable, and accordingly, the passenger throughput increment is increased as a rate variable.

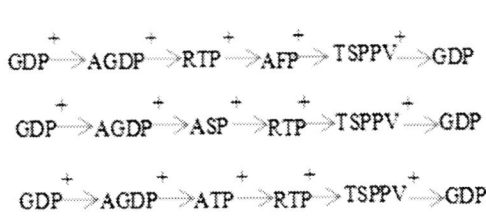

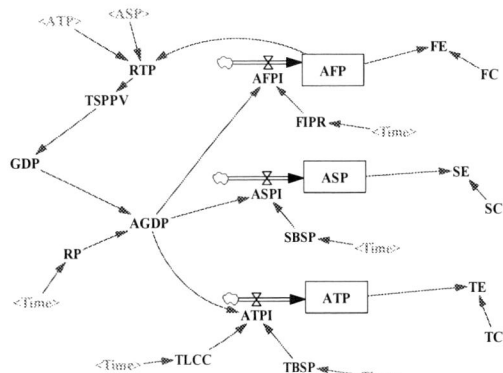

Figure 3. The feesback loop.   Figure 4. The flow diagram.

3. Case study
In this paper, the VEMSIM PLE system dynamics software is used to simulate the Beijing-Tianjin-Hebei airport group efficiency model. The data mainly comes from the 2010-2017 Civil Aviation Airport Production Bulletin and the National Economic and Social Development Bulletin. The historical data is eight years long and the forecast period is five years from 2018 to 2022, with a time step of one year. It mainly includes the following data:

GDP (2010~ 2017);

Regional permanent resident population (2010~2017);

Passenger throughput in three airports (2010~2017).

3.1. Structural equation

$$L\ AFP.K = AFP.J + AFPI.JK \tag{1}$$

$$N\ AFP = 7394.81 \tag{2}$$

$$N\ ASP = 727.71 \tag{3}$$

$$N\ ATP = 272.36 \tag{4}$$

$$R\ AFPI.JK = 27.401 * FIPR.J + 33.221 * AGDP.J + 34.875 \tag{5}$$

$$R\ ASPI.JK = 22.078 * *SBSP.J + 180.635 * AGDP.J - 47.425 \tag{6}$$

$$R\ ATPI.JK = 24.422 * TLCC.J + 4.3 * TBSP.J + 18.162 * AGDP.J - 80.119 \tag{7}$$

$$A\ RTP.K = AFP.K + ASP.K + ATP.K \tag{8}$$

$$A\ TSPPV.K = 0.284 * RTP.K + 1003.05 \tag{9}$$

$$A\ GDP.K = 27.578 * TSPPV.K - 44973.9 \tag{10}$$

$$A\ AGDP.K = GDP.K/RP.K \tag{11}$$

1753

$$A\,FE.K = AFP.K/FC \tag{12}$$
$$C\,FC = 8550 \tag{13}$$
$$C\,SC = 2500 \tag{14}$$
$$C\,TC = 2000 \tag{15}$$
$$A\,FIPR = TABLE(TFIP\,R, TIME, 2010, 2022, 1) \tag{16}$$
$$FW.K = AFP.K/(AFP.K + ASP.K + ATP.K) \tag{17}$$
$$E.K = FW.K * FE.K + SW.K * SE.K + TW.K * TE.K \tag{18}$$

In the structural equation, L, R, A, N, C, J and K represents the state equation, the rate equation, the auxiliary equation, the initial value, the constant equation, the previous moment, the current moment respectively. The equations of ASP, ATP are similar with that of AFP. The equations of SE, TE are similar with that of FE. The equations of SW, TW are similar with that of FW. TFIPR indicates the value of FIPR variable over time, and the equations of SBSP, TLCC, TBSP, and RP are similar with that of FIPR.

3.2. Model checking

In order to ensure that the constructed model is consistent with the actual situation, the model can reflect the characteristics of the real system, the model results need to be compared with the historical data. If the error is within the acceptable range, the model can be used to simulate the reality and the subsequent prediction results are reasonable. In this paper, the GDP and PEK passenger throughput data from 2010 to 2017 are selected as the test objects, as shown in Table 2. The errors between the predictive value of GDP, the passenger throughput of PEK and the actual data are all within 10%, indicating that the model is constructed reasonably and can be used for the next policy analysis.

Table 2. The comparison of predictive values and actual values.

Year	GDP actual value	Predictive value	Error absolute value	passenger actual value	Predictive value	Error absolute value
2010	43732.3	45911.4	4.98%	7394.8114	7394.8114	0.00%
2011	52074.97	50907.9	2.24%	7867.4513	7636.31	2.94%
2012	57348.29	53909.4	6.00%	8192.9352	7884.61	3.76%
2013	62685.77	57656.2	8.02%	8371.2355	8142.24	2.74%
2014	66478.91	62249.5	6.36%	8612.8313	8411.78	2.33%
2015	69358.89	68130.8	1.77%	8993.9049	8696.65	3.31%
2016	75624.97	75134.5	0.65%	9439.3454	9002.58	4.63%
2017	82559.73	83708.2	1.39%	9578.6296	9334.37	2.55%

3.3. Simulation results

The system simulation results are shown in Table 3. From 2012 to 2014, the efficiency value of PEK was in the range of [0.9, 1], indicating that the resource usage in this airport was about to reach its limit. In 2015, the efficiency value exceeded 1, indicating that PEK was overloaded. The efficiency value of TSN from 2010 to 2014 was less than 0.5, indicating that the airport's resources were redundant and not fully utilized. In 2017, the airport efficiency value tended to 1, so airport resource usage was close to saturation. The efficiency value of SJW in 2010-2017 has been below 0.5, it means that the waste of resources was serious. Since 2010, the overall efficiency of the Beijing-Tianjin-Hebei airport group has been continuously improved. In 2015 and 2016, although the efficiency of TSN and SJW has increased rapidly, the impact on overall efficiency was not as large as that of PEK due to the low weight. The efficiency of PEK has exceeded the limit in nearly three years, so the overall efficiency was near saturation. This also shows that for PEK, measures should be taken to divert or increase capacity, and TSN and SJW should take a different approach to increase the flow.

Table 3. The efficiency value of the airport group.

Year	FE	SE	TE	FW	SW	TW	E	Year	FE	SE	TE	FW	SW	TW	E
2010	0.86	0.29	0.14	0.88	0.09	0.03	0.79	2017	1.09	0.93	0.49	0.74	0.18	0.08	1.02
2011	0.89	0.31	0.15	0.88	0.09	0.03	0.82	2018	1.13	1.32	0.62	0.68	0.23	0.09	1.13
2012	0.92	0.34	0.18	0.87	0.09	0.04	0.84	2019	1.18	1.69	0.75	0.64	0.27	0.09	1.28
2013	0.95	0.40	0.21	0.85	0.10	0.04	0.86	2020	1.23	2.13	0.90	0.60	0.30	0.10	1.47
2014	0.98	0.48	0.27	0.83	0.12	0.05	0.89	2021	1.29	2.67	1.07	0.56	0.34	0.11	1.73
2015	1.02	0.62	0.34	0.80	0.14	0.06	0.92	2022	1.35	3.32	1.26	0.52	0.37	0.11	2.07
2016	1.05	0.79	0.41	0.76	0.17	0.07	0.96								

3.4. Policy verification

For PEK, the policy that can be implemented is to expand the airport capacity. In the model, the passenger capacity of PEK is adjusted to simulate the expansion policy. The efficiency comparison before and after is shown in Figure 5. After the expansion, the efficiency value is less than 1, no more overuse. From the appearance, the efficiency value is reduced, but in fact, the congestion of PEK is reduced, the airport resources are fully utilized. So the airport efficiency is improved. The policy of expanding airport capacity is also in line with the completion of the Beijing New Airport and its operation in 2019. By then, the overall capacity of Beijing airports will be larger, which will meet the needs of passengers in Beijing.

In addition, the proportion of international passengers in PEK has increased year by year, and the flight schedule is allocated to international routes. Also, PEK tries to cancel low-efficiency flights and flights operated by 100 or less seats, or transfer them to TSN and SJW, so the positioning is more biased towards international routes. This policy is embodied in the model as the variable "International passenger ratio of PEK (FIPR)". Although the increase in efficiency value is small after increasing the value of this variable , the change in international passenger numbers can increase the efficiency of PEK, TSN and SJW under the premise of the airport's unsaturated resources, so that the efficiency of the airport group improves. It reflects the behaviour of an airport can influence other airports and the whole airport group. Therefore, the policy of PEK towards international routes is reasonable and effective.

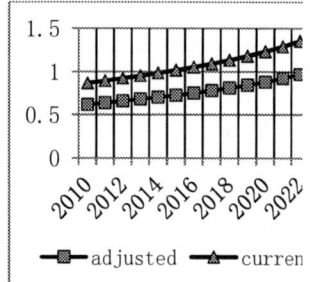 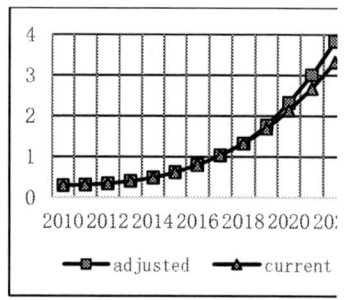 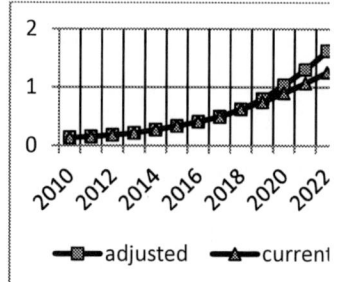

Figure 5. The impact of the expansion policy of PEK on the efficiency of PEK.

Figure 6. The impact of increasing branch points of TSN on the efficiency of TSN.

Figure 7. The impact of increasing low-cost carriers of SJW on the efficiency of SJW.

TSN is a supplement to Beijing Airport in terms of trunk flights, but its competitiveness is far less than that of Beijing. TSN adapts to the needs of the Beijing-Tianjin-Hebei airport group and focuses on the development of the branch network to form a route network combining trunk and branch lines. In the model, the policy is embodied as the variable "Branch shipping points of TSN (SBSP)". Increasing this variable value, the comparison of the efficiency of TSN before and after the change is shown in Figure 6. After the adjustment, the efficiency value of TSN has increased clearly, but as the predicted passenger demand is far greater than the capacity of TSN, it may be necessary to expand capacity in the later period. The efficiency of PEK and SJW has increased slightly. The increase makes the efficiency of the airport group raise. The impact reflects the interconnectedness of the airports in the airport group. It also shows that this policy is feasible.

Increasing the number of low-cost base airlines in SJW in the model, the comparison of the efficiency of SJW before and after the change is shown in Figure 7. After the adjustment, the efficiency value of SJW has increased significantly, the impact on PEK and TSN is relatively small due to the small weights of SJW. The change make the efficiency of airport group improve overall. Changing the number of branch shipping points in SJW and changing that at TSN have similar effects on these three airports, so it will not be repeated here. Therefore, SJW needs to be positioned at low cost and on the branch lines continuously. SJW can encourage the establishment of low-cost base airlines, open low-cost routes, increase branch routes, and actively expand regional flights continuously.

4. Conclusion

The airport group is a complex system influenced by many factors such as regional economy, population, and various airports. The study of airport group efficiency needs to reflect the dynamic changes and integrity of the airport group. Compared with other methods, system dynamics (SD) can better reflect the characteristics of airport groups and simulate the internal and external impacts of airport groups. Therefore, this paper uses the system dynamics to simulate the Beijing-Tianjin-Hebei airport group and establish the efficiency model. The results show that the model is consistent with the development trend of the airport group. The overall efficiency of the airport group is increasing year by year. However, since 2015, PEK has an efficiency value higher than 1, indicating that the airport resources were overused and the airport efficiency was reduced. The efficiency of TSN has increased year by year. In 2017, the efficiency value reached 0.93, and the resources tended to be fully utilized. Similarly, the efficiency of SJW has increased, but it still has not reached the efficiency value of 0.5, and the waste of resources was serious. Then, the paper verifies the internationalization of PEK, the combination of TSN's trunk-branch lines, the low-cost and branch positioning of SJW, and finally proves that these policies are reasonable and feasible. At the same time, the policy proposed to expand the capacity of PEK in this paper coincides with the construction of the new airport in Beijing. Through the subjective regulation of these policies, it is possible to promote the relatively balanced service of the airport group, and avoid the situation of excessive accumulation of aviation resources in PEK. Research on the efficiency of airport groups based on system dynamics fully reflects the inseparable links between airports in the airport group and the integrity of the airport group.

This paper is a preliminary exploration of the efficiency of the airport group. Further studies on it and policy analysis can be more in-depth, including:

(1) The research can introduce other airports into the SD model of airport group efficiency;

(2) Establish a SD model with more factors;

(3) This paper uses the flow capacity ratio to define efficiency, and how to define efficiency more rigorously needs further study.

References

[1] Fernandes E, Pacheco R R. *Transportation Research Part A*, 2002, **36**(3):225-238.

[2] MUSTAFA L D，ASIR G. *Transportation Policy*，2016：92-104.

[3] He Yan, Zhang Yu. *Journal of Logistics Science*, 2011, **34**(5): 4-7.

[4] Jia Pinrong. *Chinese Journal of Management Science*, 2015, (S1): 496-503.

[5] Zhang Weina. *Nanjing University of Aeronautics and Astronautics*, 2014.

[6] Chen Yaqing, Han Yunxiang. *Journal of Traffic Information and Safety*, 2009, **27**(5): 146-148.

[7] Ioanna E. Manataki*, Konstantinos G. Zografos. *Journal of Air Transport Management* 16 (2010) 86–93.

[8] J.W.Forrester, *Urban Dynamics*, Mass., Cambridge:M.I.T. Press, 1969.

[9] Haghshenas H, Vaziri M, Gholamialam A. *Cities*, 2015, **45**:104-115.

[10] Liu S, Triantis K P, Sarangi S. *Transportation Research Part A Policy & Practice*, 2010, **44**(8):596-608.

[11] Xu Tianyou. *Beijing Jiaotong University*, 2014.

[12] Yang Haoxiong, Li Jindan, Zhang Hao, et al. *Systems Engineering - Theory & Practice*, 2014, **34**(8): 2135-2143.

ISPECE

IOP Publishing

IOP Conf. Series: Journal of Physics: Conf. Series **1187** (2019) 042070   doi:10.1088/1742-6596/1187/4/042070

The Application of Alternating Direction Method of Multipliers on l_1-norms Problems

Yanchen He[1] [a]

[1]School of statistics and mathematics, Central University of Finance and Economics

[a] Yanchen He: yanchenhe@foxmail.com

Abstract. l_1 regularization is commonly used for variable or model selection, and it is often incorporated into the penalized likelihood framework for the sparse solutions. Lasso, group lasso and graphical lasso are three representative methods among them. The objective functions of these methods all contain a l_1 penalized term. Several algorithms have been developed to solve those problems, such as least angle regression for lasso, group nonnegative garrote for group lasso and block coordinate descent for graphical lasso. However, they are designed for the solution of a specific problem and none of them can solve l_1-norm problems generally. A convex optimization algorithm called ADMM(Alternating Direction method of multipliers)splits targets problems into two distinct parts and handles them separately, which makes it a natural fit for l_1-norm problems consisting of a likelihood term and a penalty term.

Based on the methodology of ADMM, we derived a framework for the solution of lasso, group lasso and graphical lasso. The proposed framework outperforms the existing methods in simulations. Finally, we apply our method to analyze a birth weight dataset *birthwt* which could be found in *R* package *MASS*.

1. Introduction

We are now embracing an era of big data. With the rapid growth of the data generated from internet statistical models tend to be more and more complex. Such complexity could be caused by, for example, the huge increase of the explanatory variables for a response variable. However sometimes people only seek for these variables which are "the most relevant to the response variable" in a statistical model. Here are some reasons for that:

1: One may seek for simpler way to explain the data.

2: One may try to reduce the complexity of model by selecting a reasonable subset of explanatory variables because of the need to reduce the cost of prediction.

3: Redundant explanatory variables may cause colinearity. To avoid colinearity one could apply variable selection.

In practice there are a cluster of methods for model selection. Among them a popular approach is to add a l_1 penalty term to the model. The geometric property of l_1 penalty ensures its better performance in variable selection since it enforces the solution to be sparse (For example, l_1 penalty performs better than l_2 penalty because of their different constraints boundaries[1]. The shape of the boundaries of l_1 penalty makes some variables shrink to 0 exactly.), thus l_1 penalty is applied to a wide range of statistical model like LASSO[1] for variable selection, group lasso for grouped variables selection[2], graphical lasso for determining the direct interaction in Gaussian graph problem[3].

Content from this work may be used under the terms of the Creative Commons Attribution 3.0 licence. Any further distribution of this work must maintain attribution to the author(s) and the title of the work, journal citation and DOI.

Published under licence by IOP Publishing Ltd

The LASSO(Least Absolute Shrinkage and Selection Operator), proposed firstly by R.Tibshirani in 1996 [1], adds a l_1 penalty term to the least square loss in order to select the key factors and increase the prediction accuracy. . Let A and b be the design matrix of explanatory variables and vector of observations respectively, and denote x the vector of coefficients. The aim of LASSO is to minimize:

$$\frac{1}{2}\|Ax-b\|_2^2 + \lambda\|x\|_1 \qquad (1.1)$$

where λ is the tuning parameter. To solve the LASSO problem, several methods have been proposed. B.Efron et al. developed LARS(Least-Angle Regression) [4],which gets the solutions by adjusting the coefficients in the direction that make all the correlations with the residual the same. The coordinate descent method, which in each step minimizes the function in coordinate direction, is also applicable to LASSO. Early introductions of such application can be seen on Fu[5].

Group lasso, introduced by M.Yuan and Y.Lin[3],was motivated by the idea that in a linear model one may have several groups of variables but only a few groups of them are useful. Let the coefficients of vector be grouped as follows:

$$x = \begin{pmatrix} x_1^T & x_2^T & \cdots & x_N^T \end{pmatrix}^T$$

The group lasso problem is to minimize the following objective function:

$$\frac{1}{2}\|Ax-b\|_2^2 + \lambda\sum_{i=1}^N \|x_i\|_2 \qquad (1.2)$$

In the paper of M.Yuan and Y.Lin, they developed Group LARS for solving group lasso problem, which is a generalized version of LARS. They also extended the coordinate descent method to block coordinate descent method, in which every step focuses on optimizing the objective function with respect to one group instead of one coordinate.

In Gaussian graphical model, for a dataset generated from a multivariate Gaussian distribution with mean μ and covariance Σ (the covariance is assumed to be positive definite so that the precision matrix $X=\Sigma^{-1}$ is well-defined), it has been proved that(see, for example, [6])two components x_i and x_j of the random variable are conditional independent if and only if $X_{ij} = X_{ji} = 0$, thus estimating the inverse matrix of the covariance matrix Σ^{-1} is crucial in the detection of relation between factors in the model. Banerjee[7] ,Yuan and Lin[8] proposed graphical lasso, a l_1-penalized model to solve the estimating problem. Given the empirical covariance matrix S, the estimation problem requires to minimize:

$$\operatorname{tr}(SX) - \log\det(X) + \lambda\|X\|_1 \qquad (1.3)$$

where S and X are symmetric positive definite matrices and $\|\cdot\|_1$ is elementwise l_1 norm, that is, $\|X\|_1 = \sum_{i,j}|X_{ij}|$.

For the algorithm solving graphical lasso, in the original paper of Yuan and Lin they solve it using interior point algorithm. Jerome H.Friedman, Trevor Hastie and Robert Tibshirani handle the problem by repeatedly solving LASSO to determine a sub-block of the matrix.

Although some algorithms have been designed to deal with the problems above, it should be noted that all of them are designed to solve some specific problems. It's natural for us to seek for an algorithm that can deal with all of the problems we mentioned, or, to speak more generally, all the optimization problems with l_1 penalty.

Alternative Direction Method of Multiplier (ADMM), originally proposed by Glowinski and Gabay[9] and reviewed by Boyd et al.[10], is an algorithm suitable for decomposable problems. For a constrained optimization with a decomposable objective function, ADMM separate the problem into several independent parts (here the term "independent" means each part relies on only one coordinate) and deal with each one separately. Such feature of ADMM indicates that it is suitable for l_1 penalized problem since l_1 -penalized problem consists of a likelihood part and a l_1 penalty part, and by introducing new variable we can make the problem decomposable. For specific problems, especially l_1

-penalized problem, ADMM has been proved to be efficient with high convergence speed[10]. Due to ADMM's good performance in solving l_1-penalized problem, in this article, we use ADMM algorithm to solve the three typical l_1-penalized problem: LASSO, Group lasso and graphical lasso.

The rest of the article is organized as follows. In section 3 we introduce the general procedure of ADMM algorithm and the background of three classical l_1-penalized problems, including lasso, group lasso and graphical lasso. The framework for the solution of these problems is derived. In section 4, several simulations are conducted and we compare the accuracy and efficiency of our proposed method with other existing algorithms. Finally in section 5,we demonstrate our methods by an application to the birth weight data served in R package $MASS$.

2. ADMM approach to solve l_1 problem

2.1. General framework of ADMM

ADMM solves problems in the form:

$$\min f(x) + g(z)$$
$$\text{s.t. } Ax + Bz = c \qquad (2.1)$$

, where f and g are convex functions.

Firstly, we write the augmented Lagrange function of the objective function as

$$L_p(x,z,y) = f(x) + g(z) + y^T(Ax + Bz - c) + (\frac{\rho}{2})\|Ax + Bz - c\|_2^2 \qquad (2.2)$$

Let $u = (1/\rho)y$, and we may simplify it as:

$$L_p(x,z,y) = f(x) + g(z) + (\frac{\rho}{2})\|Ax + Bz - c + u\|_2^2 - (\frac{\rho}{2})\|u\|_2^2 \qquad (2.3)$$

ADMM repeats the following steps until convergence:

$$x^{k+1} = \arg\min_x \left(f(x) + (\frac{\rho}{2})\|Ax + Bz^k - c + u^k\|_2^2 \right) \qquad (2.4)$$

$$z^{k+1} = \arg\min_z \left(f(x) + (\frac{\rho}{2})\|Ax^{k+1} + Bz - c + u^k\|_2^2 \right) \qquad (2.5)$$

$$u^{k+1} = u^k + Ax^{k+1} + Bz^{k+1} - c \qquad (2.6)$$

In (2.4), we first solve for x that minimizes the objective function with z and u fixed, then we look for z that minimize with x and u fixed in (2.5), and finally the dual variable u is updated.

Now we try to solve l_1 penalized problem via ADMM. Problems with l_1 regularization often take the below form:

$$\min l(x) + \lambda\|x\|_1 \qquad (2.7)$$

, where $l(x)$ is a convex loss function. To make the application of ADMM possible, we shall state problem (2.7) in another way. By introducing a new variable z we rewrite the problem in an equivalent form

$$\min l(x) + \lambda\|z\|_1$$
$$\text{s.t } x = z \qquad (2.8)$$

The ADMM algorithm for l_1 problem is

$$x^{k+1} = \arg\min_x \left(l(x) + (\frac{\rho}{2})\|x - z^k + u^k\|_2^2 \right) \qquad (2.9)$$

$$z^{k+1} = S_{\lambda/\rho}\left(x^{k+1} + u^k\right) \qquad (2.10)$$

$$u^{k+1} = u^k + x^{k+1} - z^{k+1} \qquad (2.11)$$

, where $S_\theta(x) = \left(1 - \theta/\|x\|_2^2\right)_+ x$ is the soft thresholding operator.

2.2. Three specific problems

2.2.1. LASSO

In 1996, Tibshiranit proposed a new method, called LASSO, to solve linear regression problem. He added a l_1 penalty to the sum of squared residuals[1]. The benefit brought by l_1 penalty is that the geometric property of l_1 norm ensures that estimated values of some parameters can be shrunk to 0 precisely, thus increase the accuracy and interpretability of the linear model. LASSO has a wide range of applications. For example, in video concept detection, an adjusted form of LASSO(parallel lasso) was proposed to build the detectors of visual features [11].

Recall that the LASSO problem is to minimize

$$\frac{1}{2}\|Ax-b\|_2^2 + \lambda\|x\|_1 \qquad (1.1)$$

Applying the ADMM procedure (2.9), (2.10) and (2.11) for l_1 problem, we obtain following algorithm:

$$x^{k+1} = \left(A^T A + \rho I\right)^{-1}\left(A^T b + \rho\left(z^k - u^k\right)\right) \qquad (2.12)$$

$$z^{k+1} = S_{\lambda/\rho}\left(x^{k+1} + u^k\right) \qquad (2.13)$$

$$u^{k+1} = u^k + x^{k+1} - z^{k+1} \qquad (2.14)$$

The updating step for x is obtained by replacing the loss-function term in (2.9) with $\frac{1}{2}\|Ax-b\|_2^2$, compute the gradient of (2.9) and let the gradient be 0.

2.2.2. Group lasso

The group lasso, firstly proposed by Ming Yuan and Yi Lin[2], stresses the group structure in parameters. It replaces the penalty term inside classical LASSO with groupwise l_2 penalties. The group structure occurs in many circumstances. For example, in model of categorical variables, a factor with several levels could be represented by several dummy variables. And it's reasonable that these dummy variables should be considered together.

Recall that the group lasso problem is to minimize

$$\frac{1}{2}\|Ax-b\|_2^2 + \lambda\sum_{i=1}^{N}\|x_i\|_2 \qquad (2.15)$$

, where $x = \left(x_1^T \quad x_2^T \quad ...... \quad x_N^T\right)^T$

This is not a classical l_1 problem. However ADMM approach is still applicable. Rewrite this problem:

$$\min \frac{1}{2}\|Ax-b\|_2^2 + \lambda\sum_{i=1}^{N}\|z_i\|_2$$

$$\text{s.t } x = z = \left(z_1^T \quad z_2^T \quad ...... \quad z_N^T\right)^T \qquad (2.16)$$

Now form the augmented Lagrangian with the scaled dual variable like what we do in (3.3):

$$L_p(x,z,y) = \frac{1}{2}\|Ax-b\|_2^2 + \lambda\sum_{i=1}^{N}\|z_i\|_2 + (\frac{\rho}{2})\|x-z+u\|_2^2 - (\frac{\rho}{2})\|u\|_2^2 \quad (2.17)$$

The step of updating x is the same as the case of LASSO since they share the same parts involving x. For the step of updating z, we first group z, c and u in the same way as x:

$$z = \left(z_1^T \quad z_2^T \quad ...... \quad z_N^T\right)^T, \ c = \left(c_1^T \quad c_2^T \quad ...... \quad c_N^T\right)^T, \ u = \left(u_1^T \quad u_2^T \quad ...... \quad u_N^T\right)^T$$

Then it is easy to obtain the following equation:

$$(\frac{\rho}{2})\|x-z+u\|_2^2 = \sum_{i=1}^{N}(\frac{\rho}{2})\|x_i - z_i + u_i\|_2^2 \qquad (2.18)$$

It means that we can update each z_i separately. Plug (2.17) into (2.18) we get:

$$L_p(x,z,y) = \frac{1}{2}\|Ax-b\|_2^2 + \sum_{i=1}^{N}\left(\lambda\|z_i\|_2 + (\frac{\rho}{2})\|x_i - z_i + u_i\|_2^2\right) - (\frac{\rho}{2})\|u\|_2^2 \tag{2.19}$$

So for each z_i the task is to find z_i that minimize:

$$\lambda\|z_i\|_2 + (\frac{\rho}{2})\|x_i^{k+1} - z_i^k + u_i\|_2^2 \tag{2.20}$$

The solution to that problem is:

$$z_i^{k+1} = S_{\lambda/\rho}\left(x_i^{k+1} + u_i^k\right) \tag{2.21}$$

Now we can summarize the 3 steps in one iteration:

$$x^{k+1} = \left(A^T A + \rho I\right)^{-1}\left(A^T b + \rho\left(z^k - u^k\right)\right) \tag{2.22}$$

$$z_i^{k+1} = S_{\lambda/\rho}\left(x_i^{k+1} + u_i^k\right) \text{ for each sub-vector } z_i \tag{2.23}$$

$$u^{k+1} = u^k + x^{k+1} - z^{k+1} \tag{2.24}$$

2.2.3. Graphical lasso

As is stated before, in the Gaussian graphical model determining the sparse pattern of Σ^{-1} is important since $\left(\Sigma^{-1}\right)_{ij} = 0$ if and only if the i-th component and j-th component of the variable are not connected.

To estimate Σ^{-1} we often minimize the negative likelihood of the distribution. Such problem is called sparse inverse covariance estimation. A typical way to solve the problem is to apply graphical lasso, which add a l_1 penalty to the negative log-likelihood of Σ^{-1}. Recall that the sparse inverse covariance estimation problem is to minimize

$$\text{tr}(SX) - \log\det(X) + \lambda\|X\|_1 \tag{2.25}$$

, where S and X are symmetric positive definite matrices and $\|\cdot\|_1$ is elementwise l_1 norm, namely,

$$\|X\|_1 = \sum_{i,j}|X_{ij}|.$$

Convert it into ADMM form:

$$\min \text{tr}(SX) - \log\det(X) + \lambda\|Z\|_1$$

$$\text{s.t. } X = Z \tag{2.26}$$

Now deduce the augmented Lagrangian form of (2.26)

$$L_p(x,z,y) = \text{tr}(SX) - \log\det(X) + \lambda\|Z\|_1 + \text{tr}(X^T X) + (\frac{\rho}{2})\|X - Z\|_F^2 \tag{2.27}$$

The updating step for X is computed by taking the gradient of X and force it to vanish.

$$X^{k+1} = Q\tilde{X}Q^T \tag{2.28}$$

where Q is obtained by spectral decomposition:

$$\rho\left(Z^k - U^k\right) - S = Q\begin{bmatrix} \lambda_1 & 0 & 0 & 0 \\ 0 & \lambda_2 & 0 & 0 \\ 0 & 0 & ... & 0 \\ 0 & 0 & 0 & \lambda_n \end{bmatrix}Q^T$$

$\tilde{X}$ is a diagonal matrix and $\tilde{X}_{ii} = \dfrac{\lambda_i + \sqrt{\lambda_i^2 + 4\rho}}{2\rho}$.

The updating step for Z is elementwise optimization:

$$Z_{ij}^{k+1} = S_{\lambda/\rho}\left(X_{ij}^{k+1} + U_{ij}^k\right) \tag{2.29}$$

3. Simulation

In this part we simulate some examples related to the three models (LASSO, group lasso and graphical lasso) and apply our proposed methods to solve them. The purpose is to evaluate the efficiency and accuracy of three ADMM algorithms proposed in the last section. We compare the efficiency of ADMM with other commonly used methods such as coordinate descent (CD) and block coordinate gradient descent (BCGD)[13]. When comparing the time cost, we run each algorithm for 100 times and compute the average of the total time for each method. The results are shown roughly in figures and exact numerical results are in appendix. In all the figures in the rest of the article the black bars indicate the time cost of ADMM while the white bars indicate the time cost of function to compare with. For the test of numerical accuracy we first compute the difference between ADMM's result and the result of other commonly used algorithm corresponded to the problem, and then check the l_1 norm of the difference. The norm will be listed in a series of tables.

All the observations represented by the rows of the matrix X are generated from a multivariate Gaussian distribution $N(0,\Sigma)$. Here we consider two cases: The explanatory variables are independent or dependent because in some circumstances part of the covariate may correlate. For the independent case, we set $\Sigma = I$. For the dependent case, we set $\Sigma_{ij} = 0.5^{|i-j|}$. To test the performance of our proposed algorithms under different situations, we set the dimension p to be 20, 50 or 100, and the number of observations n to be 100, 200 or 400. We consider three different tuning parameters $\lambda = 0.1, 1$ or 10.

3.1. Simulation for LASSO

We constructed a classical linear model where the vector of coefficients β is $(1,-1,2,-3,1,2,0.......,0)^T$ (the first index represents the intercept in this linear model).We used the *glmnet* function embedded in *R* package *glmnet* to compare with our algorithm. This function use coordinate descent method which is stated by Friedman, Hastie and Tibshirani[12].

The l_1 norm of the difference between the results of two algorithms are shown in Table 1-6. The results show that ADMM can obtain ideal numerical convergence.

Table 1.The l_1 norm of the difference between the results of ADMM and *glmnet*
($\lambda = 0.1$, independent case)

	n=100	n=200	n=400
p=20	0.001413	0.000208	≤ 0.0001
p=50	0.05676	0.004496	0.000751
p=100	1.438052	0.087495	0.010916

Table 2. The l_1 norm of the difference between the results of ADMM and *glmnet*
($\lambda = 1$, independent case)

	n=100	n=200	n=400
p=20	0.000389	0.000277	≤ 0.0001
p=50	0.035364	0.009722	0.000729
p=100	0.214013	0.03883	0.009497

Table 3. The l_1 norm of the difference between the results of ADMM and *glmnet*
($\lambda = 10$, independent case)

	n=100	n=200	n=400
p=20	≤ 0.0001	0.000107	≤ 0.0001
p=50	≤ 0.0001	≤ 0.0001	0.000126
p=100	0.001315	0.000191	0.000323

Table 4. The l_1 norm of the difference between the results of ADMM and *glmnet* ($\lambda = 0.1$,dependent case)

	n=100	n=200	n=400
p=20	0.009231	0.004715	0.003296
p=50	0.07291	0.01915	0.012644
p=100	2.653374	0.135962	0.028054

Table 5. The l_1 norm of the difference between the results of ADMM and *glmnet* ($\lambda = 1$,dependent case)

	n=100	n=200	n=400
p=20	0.00498	0.003228	0.001594
p=50	0.014189	0.0118	0.004164
p=100	0.24706	0.055075	0.017698

Table 6. The l_1 norm of the difference between the results of ADMM and *glmnet* ($\lambda = 10$,dependent case)

	n=100	n=200	n=400
p=20	0.00087	0.00052	0.000958
p=50	0.00039	0.000518	0.001317
p=100	0.004306	0.001983	0.001537

The comparison tables of time cost are listed in Figure 1-6, where the black strips indicate the time cost of ADMM and the grey strips represent the time used by *glmnet*. The figures above imply that ADMM's time cost is less than that of *glmnet*. These figures imply that ADMM's time cost is less than that of *glmnet* in almost all the cases. Also, for fixed tuning parameter, as the number of coefficients and observations increase, the time cost of ADMM increase more slowly than *glmnet*.

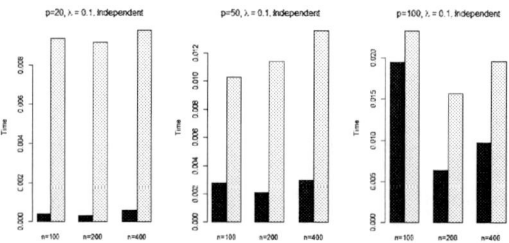

Figure 1.The time cost of ADMM and *glmnet* ($\lambda = 0.1$, independent case)

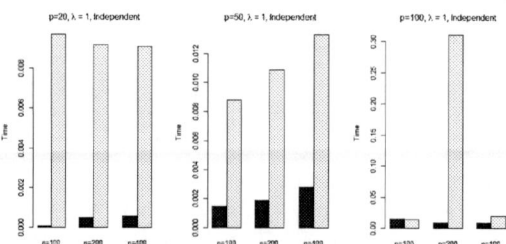

Figure 2.The time cost of ADMM and *glmnet* ($\lambda = 1$, independent case)

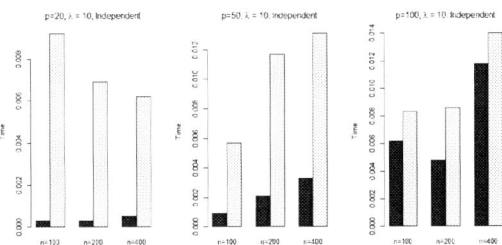

Figure 3.The time cost of ADMM and *glmnet* ($\lambda = 10$, independent case)

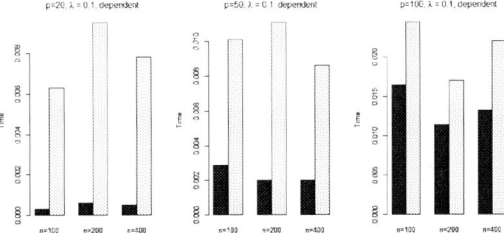

Figure 4.The time cost of ADMM and *glmnet* ($\lambda = 0.1$, dependent case)

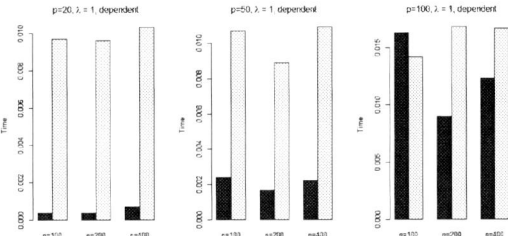

Figure 4.The time cost of ADMM and *glmnet* ($\lambda = 1$, dependent case)

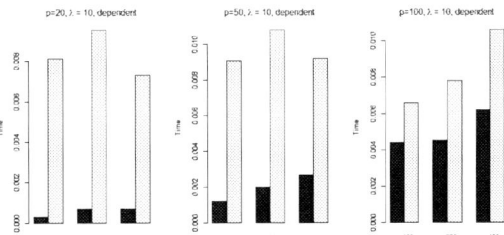

Figure 6.The time cost of ADMM and *glmnet* ($\lambda = 10$, dependent case)

3.2. Simulation for Group lasso

We use the same procedure as in LASSO to generate the dataset. However in this case we divide the covariates into some groups:

$$\beta = (1,-1,2,-3,1,2,0.......,0)^T = (1,\beta_1,\beta_2,\beta_3,\beta_4,......)^T$$

$\beta_1 = (-1,2,-3), \beta_2 = (1,2), \beta_3 = (0,0,0), \beta_4 = (0,0), \beta_m = (0)$ for $m \geq 5$.

l_2 penalty will be applied separately to those groups. Note that actually only the first two groups are nonzero vectors. The procedure of constructing the observations is the same as in LASSO, as stated above. The function *grplasso* in *R* package *grplasso* is applied. For the details of the algorithm, please refer to [13].

The l_1 norm of the difference between the results of two algorithms are shown below:

Table 7.The l_1 norm of the difference between the results of ADMM and *grplasso* ($\lambda = 0.1$, independent case)

	n=100	n=200	n=400
p=20	0.009978	0.004202	0.001662
p=50	0.107016	0.016152	0.006910
p=100	4.663680	0.078458	0.016133

Table 8.The l_1 norm of the difference between the results of ADMM and *grplasso* ($\lambda = 1$, independent case)

	n=100	n=200	n=400
p=20	0.074356	0.041666	0.014318
p=50	0.585938	0.145715	0.066060
p=100	2.255765	0.542820	0.156378

Table 9.The l_1 norm of the difference between the results of ADMM and *grplasso* ($\lambda = 10$, independent case)

	n=100	n=200	n=400
p=20	0.170725	0.222433	0.094887
p=50	0.277922	0.359937	0.357325
p=100	0.868632	0.766472	0.562727

Table 10.The l_1 norm of the difference between the results of ADMM and *grplasso* ($\lambda = 0.1$, dependent case)

	n=100	n=200	n=400
p=20	0.025212	0.006882	0.002378
p=50	0.149601	0.025561	0.012626
p=100	4.337529	0.139232	0.034336

Table 11.The l_1 norm of the difference between the results of ADMM and *grplasso* ($\lambda = 1$, dependent case)

	n=100	n=200	n=400
p=20	0.110155	0.064208	0.022173
p=50	0.521091	0.210995	0.113988
p=100	2.140328	0.696216	0.290701

Table 12.The l_1 norm of the difference between the results of ADMM and *grplasso* ($\lambda = 10$, dependent case)

	n=100	n=200	n=400
p=20	0.106464	0.214301	0.007447
p=50	0.244576	0.379445	0.393245
p=100	0.690592	0.742223	0.688781

The comparison tables of time cost are listed in Figure 7-12. Exact numerical values are in the appendix as in 3.1. In the table below the black bars still indicate the time cost of ADMM while the white bars still indicate the time cost of function *grplasso*. As is shown by these figures, ADMM wins against *grplasso* if we consider the time cost and the growing speed of time cost as the parameters increase.

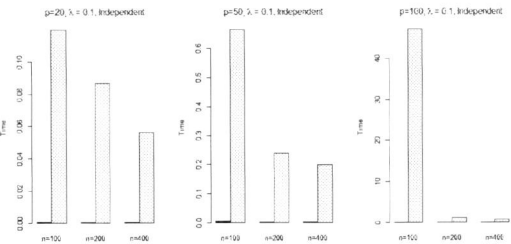

Figure 7.The time cost of ADMM and *grplasso* ($\lambda = 0.1$, independent case)

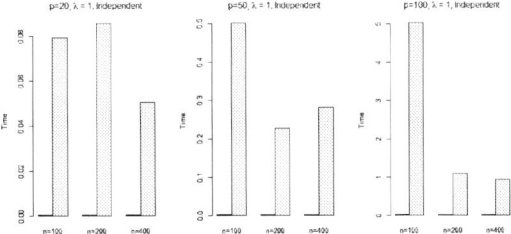

Figure 8.The time cost of ADMM and *grplasso* ($\lambda = 1$, independent case)

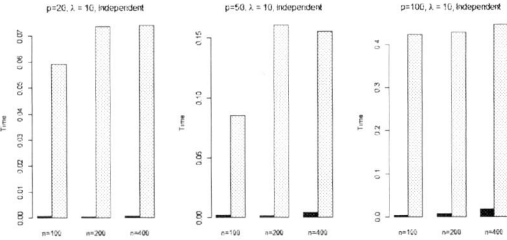

Figure 9.The time cost of ADMM and *grplasso* ($\lambda = 10$, independent case)

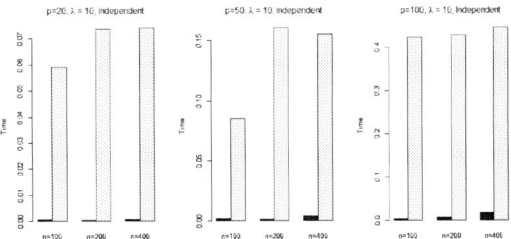

Figure 10.The time cost of ADMM and *grplasso* ($\lambda = 0.1$, dependent case)

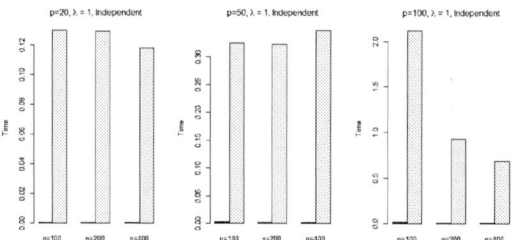

Figure 11.The time cost of ADMM and *grplasso* ($\lambda=1$, dependent case)

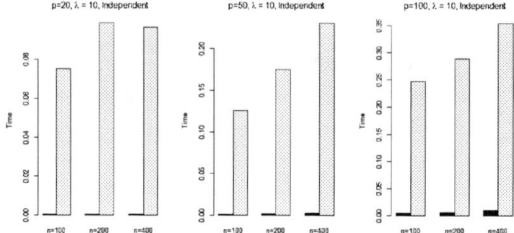

Figure 12.The time cost of ADMM and *grplasso* ($\lambda=10$, dependent case)

3.3. Simulation for Sparse inverse covariance estimation

Given a set of samples generated from multivariate Gaussian distribution $N(0,\Sigma)$, our goal is to estimate the precision matrix Σ^{-1}. In this example we let

$$\Sigma_{ij} = \begin{cases} 2.0\,(i-j=0) \\ 0.6\,(|i-j|=1) \\ 0.4\,(|i-j|=2) \\ 0\,(|i-j|\geq 3) \end{cases}$$

Note that in this section we do not need to consider the case that every component of the covariate is independent.

The function *glasso* in *R* package *glasso* is applied. This procedure is based on the paper of Friedman, Hastie and Tibshirani[3]. For the measurement of numerical accuracy we compute the difference matrix of two matrices obtained by ADMM and *glasso* and then check the elementwise l_1 norm of the difference matrix. From the tables below we could see ADMM achieves enough numerical accuracy.

Table 13.The l_1 norm of the difference between the results of ADMM and *glasso* ($\lambda=0.1$)

	n=100	n=200	n=400
p=20	0.001509	0.001247	0.001146
p=50	0.009142	0.004697	0.003215
p=100	0.034976	0.014704	0.008182

Table 14.The l_1 norm of the difference between the results of ADMM and *glasso* ($\lambda=1$)

	n=100	n=200	n=400
p=20	0.000049	0.000067	0.000055
p=50	0.000124	0.000132	0.000157
p=100	0.000194	0.000267	0.000290

Table 15. The l_1 norm of the difference between the results of ADMM and *glasso* ($\lambda = 10$)

	n=100	n=200	n=400
$p=20$	0.000018	0.000020	0.000016
$p=50$	0.000049	0.000035	0.000045
$p=100$	0.000066	0.000078	0.000094

However, the simulation indicates that ADMM may cost more time than *glasso* to reach the ideal result. The time cost of matrix decomposition (see 2.28) may leads to low deficiency of ADMM.

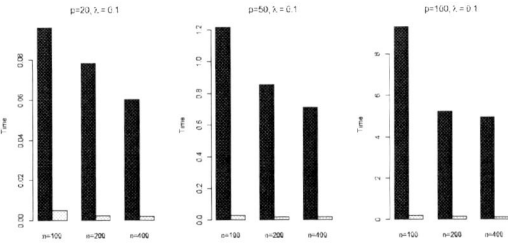

Figure 13. The time cost of ADMM and *glasso* ($\lambda = 0.1$)

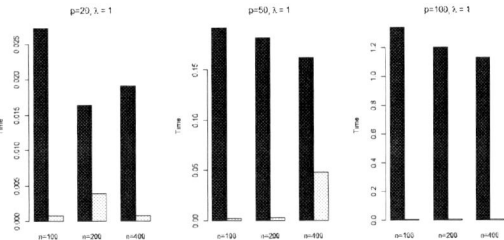

Figure 14. The time cost of ADMM and *glasso* ($\lambda = 1$)

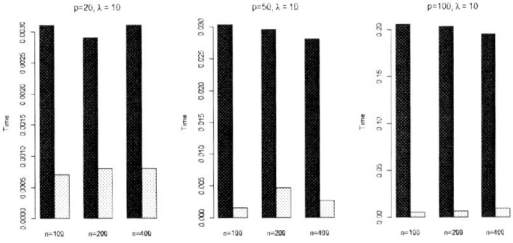

Figure 15. The time cost of ADMM and *glasso* ($\lambda = 10$)

4. Real-world data analysis

In this part we apply ADMM to deal with data from real world. The dataset we choose is *birthwt* in *R* package *MASS*. This dataset was collected during 1986 by Baystate Medical Center, Springfield[14]. The variables recorded in the dataset are presented in Table 16.

Table 16. Variables in *birthwt* dataset

Variable name	Meaning
age	mother's age in years.
lwt	mother's weight in

	pounds at last menstrual period.
race	mother's race (1= white,2= black,3= other).
smoke	smoking status during pregnancy.
ptl	number of previous premature labours.
ht	history of hypertension.
ui	presence of uterine irritability.
ftv	number of physician visits during the first trimester.
bwt	birth weight in grams.

4.1. Linear regression analysis on birthwt with LASSO

We first applied lasso to find out the factors related to the birth weight of infants, and the ADMM method was used for the numerical solutions. We used 10-fold cross validation to determine the tuning parameter, and the *R* function *cv.glmnet* was implemented for this. For this dataset the tuning parameter was set as 39.3249.

We implement ADMM to solve the LASSO problem with tuning parameter $\lambda = 39.3249$,and the result are shown in Table 17.To check if the result is precise we also implement *glmnet*. As is indicated in table 17, the error is acceptable.

Table 17. Results for dataset *birthwt* via ADMM and *glmnet*, the results of *glmnet* are in the brackets.

Variable	Results of ADMM	Results of *glmnet*	Numerical Error
intercept	2375.853	2375.844	0.009
age	2.990352	2.990529	0.000177
lwt	3.794469	3.794497	0.000028
smoke	-111.2471	-111.2424	0.0047
ptl	$\leq 10^{-8}$	0	$\leq 10^{-8}$
ht	$\leq 10^{-8}$	0	$\leq 10^{-8}$
ui	-136.2305	-136.2310	0.0005
ftv	$\leq 10^{-8}$	0	$\leq 10^{-8}$
*race*_white	138.6107	138.6087	0.0020
*race*_black	$\leq 10^{-8}$	0	$\leq 10^{-8}$

Under LASSO the factors selected are *age*, *lwt*, *smoke*, *ui*, and *race*. So we can conclude that the age of mother, mother's weight in last menstrual period, the presence of uterine irritability and the race of mother are the key factors.

4.2. Polynomial regression analysis on birthwt with group lasso

As all continuous functions can be approximated by polynomials, regression with polynomials may capture the nonlinear feature. We now analyze the dataset with polynomial regression. We use polynomials of 3 degree to approximate the feature of mother's age and mother's weight at last menstrual time. To do this we set $agei = age^i$ and $lwti = lwt^i$ $(i = 1, 2, 3)$ .And we select 3 variables of the form $agei$ and 3 variables of the form $lwti$ seperately as 2 groups. The discrete variable ptl and ftv are grouped separately with their dummy variables. In conclusion variables are grouped as below: $\{age1, age2, age3\}, \{lwt1, lwt2, lwt3\}, \{white, black\}, \{smoke\}, \{ptl1, ptl2m\}, \{ht\}, \{ui\}, \{ftv1, ftv2, ftv3m\}$.

Then, we used group lasso to determine the key factors instead of simple LASSO.

Similarly, a 10-fold cross validation is implemented and the tuning parameter is set as $\lambda = 11.64447$. The result is in the table 18.

Table 18.Results for group lasso via ADMM

Variables	ADMM
intercept	3.02225961
$age1$	0.11423533
$age2$	0.41561593
$age3$	0.24677664
$lwt1$	0.52619988
$lwt2$	-0.13975513
$lwt3$	0.41530045
$white$	0.15388875
$black$	-0.04713589
$smoke$	-0.16275857
$ptl1$	-0.14279620
$ptl2m$	0.03469140
ht	-0.23939033
ui	-0.36043124
$ftv1$	0
$ftv2$	0
$ftv3m$	0

Age was discovered to have nonlinear effects on the birth weight of the infant. Also, the number of physician visits during the first trimester is not a crucial factor, just as same as LASSO implies.

4.3. Detection of interaction between factors in birthwt with graphical lasso

In this section we try to detect the interaction between some factors that effect the infants' weight usin g graphical lasso. The factors we select are: *age, lwt, smoke, ptl ,ht ,ui ,ftv* and *race*. As race takes only 3 values(1=white, 2=black, 3=others) we replace it withrace_white and race_black as dummy variable s. We apply ADMM to estimate the pr-cision matrix with the tuning parameter set as $\lambda = 0.1$.

The obtained estimation of precision matrix is as follows:

Table 19. Estimation of precision matrix via ADMM. In this matrix elements whose absolute value is over 0.1 are marked bold and italic(except the diagonal elements).

	age	lwt	smoke	ptl	ht	ui	ftv	race_white	race_black
age	0.93	-0.07	0	0	0	0	-0.09	-0.09	0.01

lwt	-0.07	0.94	0	0.03	*-0.11*	0.04	-0.03	0	*-0.11*
smoke	0	0	0.95	-0.07	0	0	0	*-0.18*	0
ptl	0	0.03	-0.07	0.93	0	*-0.11*	0	0	0
ht	0	*-0.11*	0	0	0.92	0	0	0	0
ui	0	0.04	0	*-0.11*	0	0.92	0	0	0
ftv	-0.09	-0.03	0	0	0	0	0.92	0	0
race_ white	-0.09	0	*-0.18*	0	0	0	0	1.03	*0.27*
race_b lack	0.01	*-0.11*	0	0	0	0	0	*0.27*	1

The matrix indicates some underlying interaction between mother's weight at last menstrual time and history of hypertension. Potential relationship is also found between 2 pairs: number of previous premature labors with presence of uterine irritability and smoking status during pregnancy with mother's race. Note that the interaction between people of black race and white race is also implied in the matrix, but it is somewhat obvious because they are dummy variables related to the race of mother. Figure 16 provides a clearer view of the structure.

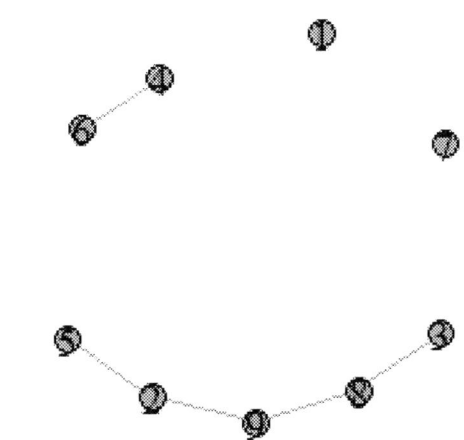

Figure 16: Undirect graph from graphical lasso. The nodes are:
1:*age*, 2:*lwt*, 3:*smoke*, 4:*ptl*, 5:*ht*, 6:*ui*,7:*ftv*, 8:*race_white*, 9:*race_black*

5. Conclusion

In this paper we state that ADMM is an appropriate algorithm for l_1 penalized problem. By simulation we show that ADMM is competitive among other commonly used algorithm we regard to numerical accuracy and time cost. Although ADMM may not be the best choice in every situation (as is shown in 3.3), there are many situations such that ADMM performs ideally. Finally we apply ADMM to analyze the real-world dataset *birthwt*, which is about birth weight of infant. LASSO, group lasso and graphical lasso, associated with ADMM to solve themselves, are implemented to capture the linear and nonlinear feature inside the model and the potential interaction between some factors.

References

[1] Tibshirani, R. (2011). Regression shrinkage and selection via the lasso: a retrospective. *Journal of The Royal Statistical Society Series B-statistical Methodology*, 73(3), 273-282.

[2] Yuan, M., & Lin, Y. (2006). Model selection and estimation in regression with grouped variables. *Journal of The Royal Statistical Society Series B-statistical Methodology*, 68(1), 49-67.

[3] Yuan, M., & Lin, Y. (2006). Model selection and estimation in regression with grouped variables. *Journal of The Royal Statistical Society Series B-statistical Methodology*, 68(1), 49-67.

[4] Efron, B., Hastie, T., Johnstone, I. M., Tibshirani, R., Ishwaran, H., Knight, K., ... & Weisberg, S. Least angle regression. *Annals of Statistics*, 32(2), 407-499. EPL, **84** (2008)

[5] Fu, W. J. (1998). Penalized Regressions: The Bridge versus the Lasso.*Journal of Computational and Graphical Statistics*, 7(3), 397-416.

[6] Lauritzen, S. (2004).*Graphical models*. Oxford: Clarendon Press.

[7] Banerjee, O., Ghaoui, L. E., & Daspremont, A. (2008). Model Selection Through Sparse Maximum Likelihood Estimation for Multivariate Gaussian or Binary Data. *Journal of Machine Learning Research*,, 485-516.

[8] Yuan, M., & Lin, Y. (2007). Model selection and estimation in the Gaussian graphical model. *Biometrika*, 94(1), 19-35.

[9] Gabay, D., & Mercier, B. (1976). A dual algorithm for the solution of nonlinear variational problems via finite element approximation. *Computers & Mathematics with Applications, 2*(1), 17-40.

[10] Boyd, S., Parikh, N., Chu, E., Peleato, B., & Eckstein, J. (2011). Distributed optimization and statistical learning via the alternating direction method of multipliers. *Foundations & Trends in Machine Learning, 3*(1), 1-122.

[11] Bo Geng, Yangxi Li, Dacheng Tao, Meng Wang, Zheng-Jun Zha, & Chao Xu. (2012). Parallel lasso for large-scale video concept detection. *IEEE Transactions on Multimedia, 14*(1), 55-65.

[12] Friedman, J., Hastie, T., & Tibshirani, R. (2010). Regularization paths for generalized linear models via coordinate descent. *Journal of Statistical Software, 33*(01), 1-22.

[13] Lukas Meier, Sara van de Geer, & Peter Bühlmann. (2010). The group lasso for logistic regression. *Journal of the Royal Statistical Society: Series B Statistical Methodology, 70*(1), 53-71.

[14] Hosmer, D. W., & Lemeshow, S. (2000). *Applied logistic regression.* J. Wiley.

ISPECE   IOP Publishing

IOP Conf. Series: Journal of Physics: Conf. Series **1187** (2019) 042071   doi:10.1088/1742-6596/1187/4/042071

Pre-flight rerouting combining A* algorithm and AHP under severe weather

Ding Wencan, Sui Dong

College of Civil Aviation, Nanjing University of Aeronautics and Astronautics, Nanjing 211106, China

samathadwc@nuaa.edu.cn

Abstract. Delays due to bad weather have become common in recent years. When the delay occurs, in most cases, flights will choose ground-holding which results in the waste of airspace resources and reduces the operating efficiency of the entire air route network. In order to deepen the reform of the ATC operations, local rerouting strategy before flight is imperative. In this paper, based on the influence degree and scope of severe weather, historical meteorological data and flight plan data, the damage situation of the air route is comprehensively analyzed, and the segment that pilots need to avoid in route selection is defined. Taking the priority of different city pairs into consideration, the OD pair allocation order model based on AHP (analytic hierarchy process) was established. In addition, this paper establishes the evaluation index of navigation modification combined with the actual operation, and makes a comparative analysis with an example.

1. Introduction

Developed countries such as Europe and the United States have already begun to study how to arrange safe, orderly and efficient flights under the influence of severe weather, and put forward rerouting strategies. According to the difference of implementation time, relevant literatures are classified into pre-flight route change planning and real-time route change planning. About the rerouting before flight: in 1993, Dixon and Weidner established the navigation modification strategy under the grid navigation modification environment [1]. Sarah Stock Patterson established a relatively perfect mathematical model for static navigation strategy, and used Lagrange relaxation algorithm to solve the model [2]. In 1999, Krozel considered reducing the workload of pilots and ATC controllers, and increased the limit on the number of turning points [3]. Sridhar and Chatterji proposed a rerouting planning method based on polygon[4]. In 2018, Mayara Condeet al. studied the cooperative rerouting strategy in ATFM and proposed a route and airspace time slot resource allocation scheme, which minimized the disutility cost of flight delay and rerouting by airline operators [5].About real-time route change planning: In 2008, Kees van Balen and Cees Bil[6], based on the previous studies, constructed the free flight reraneling software under the circumstance of partial airspace closure and verified the feasibility.

In 2009, Mark Hansen[7] studied the characteristics of route change and established a static ground waiting and dynamic route change model, in which the priority of flights was added into the model, and CPLEX9.1 was employed to solve the problem on SunfireV250.The model is suitable for flights with short flight times or when weather conditions are relatively stable. In 2017, Zhang and Mahadeva[8] studied a method based on simulation to optimize aircraft rerouting process by considering multiple uncertain sources.

Content from this work may be used under the terms of the Creative Commons Attribution 3.0 licence. Any further distribution of this work must maintain attribution to the author(s) and the title of the work, journal citation and DOI.

Published under licence by IOP Publishing Ltd

Motivated by the analysis above, this paper puts forward the local pre-flight rerouting strategy combining A* algorithm and AHP. This method divides the complex rerouting problem into several sub-problems.

2. Rerouting problem description

For flight safety, rerouting in the event of a thunderstorm is necessary to bypass weather-affected areas. It is assumed that all levels are disabled in the weather-affected area. The area is represented by a flat, two-dimensional polygon, and no aircraft is allowed to fly through it. The fixed route of a flight is composed of airway points. It is assumed that the fixed route of a flight is F and the airway point is $f_1, f_2, ... f_N$.

The rectangular coordinate system as shown in the figure is established for these flights that need to be diverted, where the magnetic north is in the positive direction of y axis and the magnetic north is 90 degrees east of x axis. Assuming that each vertex of the polygon in the flight limit area is $p_i(x_i, y_i), i = 1, 2, ..., n$, the specific method to determine the rerouting point is described below:

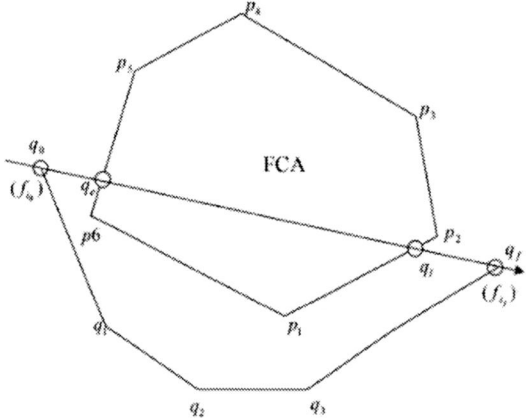

Figure 1 Diagram of rerouting model under adverse weather conditions

Step1: For the affected flights, first determine the flight segment within the flight restriction zone.

The initial influence point is the point where it intersects with the flight limit area at first along the route direction, $p_i(x_i, y_i), i = 1, 2, ..., n$; similarly, the final influence point is the point where it intersects with the flight limit area at last along the route direction, $q_l(x_l, y_l)$.

Step2: Determine the starting and ending points of rerouting flight segment. The airway points with the shortest selection distance to q_i and not within the range affected by severe weather are defined as $q_f, q_f = f_{i_f}$.

Step3: Determine the middle point of the rerouted section, and use A* algorithm to search the shortest path from the starting point and the ending point respectively to ensure that the rerouting part does not cross the FCA. What needs to be explained here is that the figure shows only some routes affected by severe weather, and some routes not affected will not be rerouted. This paper studies local rerouting affected by severe weather before flight.

3. Establishing the mathematical model

3.1 the mathematical model

$$\min D = \sum_r \sum_s \sum_k c_k^{rs} \qquad (1)$$

$s.t.$	$\sum_k f_k^{rs} = q_{rs}$	Traffic volume conservation	(2)
	$f_k^{rs} > 0 \;\forall (r,s) \in W$	Non-negative flow condition	(3)
	$k \le B$	Number of paths, $B=2$	(4)
	$\sum_r \sum_s f_a^{rs} \left| \beta_i^a \right| \le 80\% C_i$	Capacity constraint	(5)
	$\sum_r \sum_s f_a^{rs} \beta_i^a = 0$	Node flow balance	(6)

r	All the starting points involved in an OD pair
s	All the ending points involved in an OD pair
x_a	The flow on segment a
d_a	The distance of segment a
f_k^{rs}	Traffic volume on kth path between OD pair(r-s)
$\delta_{a,k}^{rs}$	Represents the relationship between segments and paths,0-1 variable
W_{rs}	Set of all paths between OD pair (r-s)
C_k^{rs}	The distance of kth path between OD pair(r-s)
q_{rs}	Traffic demand between OD pairs (r-s) extracted from the flight plan
i	Airway point $i \in U$,U represent set of all nodes
β_i^a	Represents the relationship between nodes and flight segments,,0-1 variable

3.2 OD priority research

3.2.1 Possible Influencing factors
Various factors will affect the importance of OD pairs, such as OD volumes, affected segment number under serious weather, the following segment number, etc.

3.2.2 OD priority research
OD priority means the sequence of allocating traffic volumes. To make it specific, OD pair with more traffic volumes should have the priority to be rerouted. Three factors are considered, OD volumes, the following segment numbers and affected segment numbers. The flight data comes from August 2,2016 and the meteorological information comes from 8:42 at that day. According to the vulnerable segment searching method mentioned, we can find the affected OD pair and the corresponding affected segments.

With the data mentioned above, AHP(Analytic hierarchy process) is applied into OD priority research. In this part, we put forward an alternative OD ordering method using AHP.The hierarchical model for OD priority is as follows.

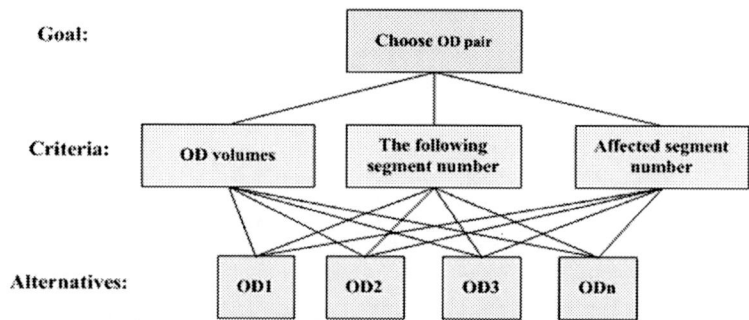

Figure 2 The hierarchical model for OD priority

As is shown in the Figure 2, we have three levels of the hierarchical model for OD priority, the goal is to determine the sequence of OD pairs, the criteria level includes three factors mentioned and the alternative level are the OD pairs we should reroute.

Table 1 Part of OD and 3 factors' grades

Departure	Arrival	OD volumes	The following segment	Affected segment	Grade 1	Grade 2	Grade 3	Total grade
ZBAA	ZSSS	40	54	6	9	5	8	7.8519
ZBAA	ZSHC	24	43	5	6	4	8	5.6912
ZSSS	ZGSZ	29	16	2	7	1	9	5.6492
ZBAA	ZGGG	23	29	4	6	3	9	5.5368
ZBAA	ZGSZ	24	29	6	6	3	8	5.4307
ZUCK	ZBAA	23	16	2	6	1	9	5.0158
ZSSS	ZGGG	23	10	2	6	1	9	5.0158
ZSSS	ZSAM	20	34	2	5	3	9	4.9034
ZBAA	VHHH	14	29	6	4	3	8	4.1639
ZBAA	ZSPD	12	54	6	3	5	8	4.0515
ZBAA	ZGHA	14	18	4	4	2	9	4.0095
ZSPD	LEMD	1	116	7	1	9	8	3.8267

According to the procedure of AHP, after constructing a judgment matrix and hierarchical single ordering and consistency checking, the data of three factors and the corresponding grades of three grades are listed. Considering the weight of three factors, each OD pair has a final score. The larger is the value, the higher is the priority of this OD.

3.3 Rerouting based on A* algorithm

3.3.1 Improved A Star Algorithm

In solving the problem of known starting and ending nodes and multiple target nodes, the improved A* algorithm takes A* algorithm as the basis, retains its original path-finding idea and increases the function of handling the traffic flow of air route nodes.

Due to the constant change of node capacity, when considering the connectivity between two points, it is necessary to judge according to the residual capacity of the node. Only when the residual capacity of the node is not zero, can the path pass through.

When the flight passes through the node, the remaining capacity of the node decreases by 1. In this way, capacity is regarded as a shared resource. After a flight decides the route, the capacity of certain nodes will be changed, thus affecting subsequent flights.

3.3.2 Searching process based on OD priority
- Select the affected OD pairs at the peak period for a test
- For the affected flights, we keep the departure point and the arrival points the same, start searching from the affected node.
- At first, find 2 nearest points outside the affected part fore and aft ends, keep finding the shortest path from both sides using A-star algorithm.
- When finding the departure point or the arrival point, the single-direction searching is over. When the departure point and the arrival point are both found, the searching is over.

4. Case Studies
The rerouting strategy in this study refers to the static local rerouting before flight. Our purpose is to make a prearranged plan before departure. The approach we are going to take is a combination of polygonal rerouting (to identify the affected sections) and rerouting based on existing route points (to find available routes on some other sections affected by severe weather).

4.1 Data collection
(1)Flight data
This study is based on the meteorological radar data and flight plan data from June to August in 2016.

The data involved in the rerouting model include meteorological radar data, FPL (Flight Plan) data, traffic capacity data, thunderstorm statistics data, etc. Among them, we extracted FPL data on August 2, 2016. The information includes flight number, latitude and longitude, departure and landing airport, route, etc.

(2)Meteorological radar data

We get the reflectivity data based on the Doppler meteorological radar in the captured aviation weather data. Radar reflectivity is used to represent the intensity of a meteorological target. According to U.S. meteorological service center radar echo and dangerous weather grade[7][8],the location of whose reflectivity data is more than 41 DBZ will be considered as a dangerous unit. Thus, we extract the affected flight and the corresponding vulnerable segments.

We collect the flight data on August 2, 2016 and sort the OD volumes according to different time. The overall distribution of the number of flights in a day increases first and then decreases. The flights are mainly focused on 9 AM-9 PM. Therefore, we take that period as the peak period for testing.

(3)Capacity data

According to the research of air route network node capacity[9],based on different configurations, we figure out the capacity value of each node.

4.2 Discussions of Results
Safety, economy and the difference between route and flight plan are the basic factors to evaluate a route.In this paper, the rerouting results that does not cross the region affected by thunderstorm are regarded as meeting the requirements of safety. Combined with the common evaluation indexes of air route network and the requirements of actual operation, this section evaluates the rerouting results through the following indexes: non-linear coefficient, increased distance,increased time and operation cost.

(1)Non-linear coefficient

Non-linear coefficient is an important indicator in road network layout planning. The non-linear coefficient between two nodes of the network is defined as the ratio of the actual distance between the two nodes to the straight line distance between two points.

$$R_{ij} = \frac{L_{ij}}{D_{ij}} \tag{7}$$

D_{ij} is the straight line distance between OD-pair i and j

L_{ij} is the actual distance between OD-pair i and j

(2)Increased distance

The difference between the total length of the rerouting route and the total length of the original route is the increased distance.

$$l_{ij} = D_{ij} - d_{ij} \tag{8}$$

D_{ij} is the length of the rerouting route between OD-pair i and j

d_{ij} is the length of the original route between OD-pair i and j

(3)Increased time

Increased time reflects the delay cost. Assuming that the average flight speed is 800km/h, and the increased time is the ratio of increased distance to average speed.

$$T_{ij} = \frac{l_{ij}}{800} \tag{9}$$

l_{ij} is the increased distance after rerouting between OD-pair i and j

(4)Operation cost

The cost reflects the total flight mileage of rerouting route. Under the premise that ignores differences of aircraft types, operation cost can be expressed through total mileage.

$$TOC = \sum_{i=1}^{n} f_i a_i \tag{10}$$

a_i is the length of segment i

f_i is the traffic flow on segment i

Table 2 Weight priority data with AHP data

	Non-linear coefficient of rerouting routes	Increased distance(km)	Increased time(min)	Operation cost(km·times)
AHP	1.18	99.67	7.48	855282

5. Conclusion

In this paper, through the polygonal region division of the impact range of severe weather, the static local rerouting under the condition of partial ARN damage is discussed, and the pre-flight local rerouting model is presented combining the Wardrop principle and the air traffic control program and aircraft performance are considered.

In addition, this paper defines the concept of OD priority, which can meet the requirements of different cities. The effectiveness and practicability of the local rerouting method combined with the analytic hierarchy process proposed are analyzed by examples. Meanwhile, relevant evaluation indexes are proposed based on the requirements of practical operation.

References

[1] M Dixon, G Weiner. Automated aircraft routing through weather-impacted airspace[C]. Fifth international Conference on Aviation Weather Systems,Vienna, VA, 1993:295~298.

[2] Krozel J.Estimating time of arrival in heavy weather conditions.[C]//AIAA Guidance Navigation and Control Conference. New Orleans:AIAA,1999:1481-1490.

[3] Hauf Thomas，Hupe Patrick，Sauer Manuela． Aircraft route forecasting Under adverse Weather Conditions[J]．Meteorologische Zeitschrift，2016，26(2)：189-206.

[4] Mayara Condé, Rocha Murça.Collaborative air traffic flow management: Incorporating airline preferences in rerouting decisions[J].2018(71):91-107.

[5] Chiang Y J, Klosowski J T,Lee C, et al. Geometric algorithms for conflict detection/resolution in air traffic management[C]. 36th IEEE Conference on Decision and Control,San Diego,Cal,1997:1835~1840.

[6] Menon P K, Sweriduk G D, Sridhar B. Optimal strategies for free flight air traffic conflict resolution[J]. Journal of Guidance, Control and Dynamics,1998,22(2):202~211.

[7] Avijit Mukherjee, Mark Hansen, A dynamic rerouting model for air traffic flow management.Transportation Research, 2009, B43: 159-171.

[8]]Xiaoge Zhang, Sankaran Mahadevan.Aircraft re-routing optimization and performance assessment under uncertainty. 2017(96)：67-82

[9] Hoffman B, Krozel J, Jakobavits R. Potential Benefits of Fix-Based Ground Delay Programs to Address Weather Constraints[C]. AIAA Guidance, Navigation, and Control Conference, Providence, RI, 2004: 1

[10] Research on capacity of air route network node based on different configurations.[A] Journal of Harbin University of Commerce(Natural Sciences Edition).2017

ISPECE

IOP Publishing

Attacking Intel UEFI by Using Cache Poisoning

Dong Wang[a] Wei Yu Dong

State Key Laboratory of Mathematical Engineering And Advanced Computing
Zhengzhou, China

[a]18844195710@163.com

Abstract—The Unified Extensible Firmware Interface (UEFI) is a software interface between an operating system and platform firmware designed to replace a traditional BIOS. In this paper, we evaluated the security mechanisms used to protected SPI Flash, and then analyzed the attack surface presented by those security mechanisms. Intel provides several registers in its chipset relevant to locking down the SPI Flash chip that contains the UEFI in order to prevent arbitrary writes. Since these registers implement their functions through the system management mode, the main attack surface is concentrated in the system management mode. In this paper, we propose an attack vector for the system management mode, which uses the method of cache poisoning to attack the system management mode and destroy the protection mechanism of SPI Flash. This method can overcome the limitations for the traditional attacks. Experimental results proved that this kind of attack can arbitrarily write to the UEFI.

1. INTRODUCTION

The BIOS is the first code to execute on a platform during power on. It is written in assembly language and is difficulty to extend due to its difficulty in writing. The BIOS is the interface between the operating system and the hardware. However, there is no uniform standard for this interface, which makes the products of each BIOS manufacturer to be different from each other. Moreover, the traditional BIOS has many problems in hardware compatibility and security. In 2003, Intel designed and proposed a new generation of firmware standards: Unified Extensible Firmware Interface (UEFI)[1]. Compared with traditional BIOS, UEFI provides a well interface specification for drivers and applications with favorable scalability.

UEFI's responsibilities include configuring the platform, initializing critical platform components, and locating and transferring control to an operating system[2]. UEFI is also responsible for configuring and instantiating System Management Mode (SMM), a highly privileged mode of execution on the x86 platform. Thus any malware that controls the UEFI is able to place arbitrary code into SMM, which may generate a huge impact on the system. The UEFI's residence on an SPI Flash chip means it will survive operating system reinstallations. These properties make the UEFI a desirable residence for malware. Therefore, an examination of the security of UEFI is necessary

Based on the UEFI open source specification, this paper analysis the protection mechanism of the SPI Flash where UEFI storage. And then, this paper presents an attack method, which attacks the SMM by means of cache poisoning. This attack can bypass the protection mechanism of SPI Flash and reflash the UEFI

Content from this work may be used under the terms of the Creative Commons Attribution 3.0 licence. Any further distribution of this work must maintain attribution to the author(s) and the title of the work, journal citation and DOI.
Published under licence by IOP Publishing Ltd

2. UEFI TECHNOLOGY

2.1. UEFI Architecture

UEFI is a new software interface between an operating system and platform firmware designed to replace a traditional BIOS[1]. In general, UEFI has many technical advantages over BIOS such as pre-OS environment, boot services and runtime services, CPU-independent drivers and powerful security architecture for OS kernel.

UEFI uses modular technology to divide its functions into two parts[3]: Hardware Control and Operating System Software Management. Figure 1 shows the architecture of UEFI.

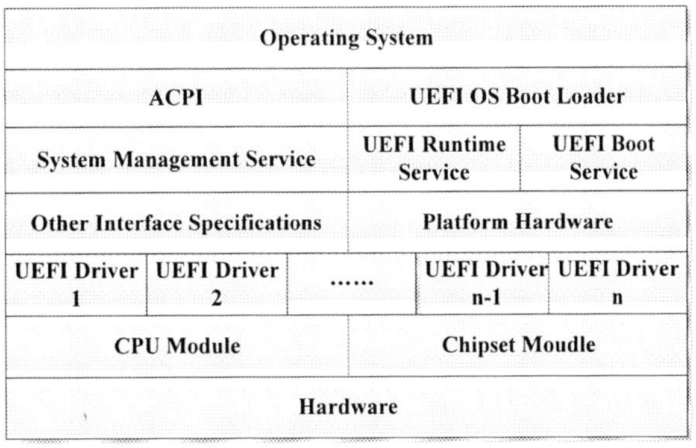

Operating System			
ACPI		UEFI OS Boot Loader	
System Management Service		UEFI Runtime Service	UEFI Boot Service
Other Interface Specifications		Platform Hardware	
UEFI Driver 1	UEFI Driver 2	...... UEFI Driver n-1	UEFI Driver n
CPU Module		Chipset Moudle	
Hardware			

Figure 1. UEFI Overall Framework Diagram

The hierarchical relationship between UEFI firmware, operating system and operating system bootloader can be seen in the figure. After the system power on, UEFI is executed first. It verifies the integrity of other firmware and is the root of trusted of firmware execution. After that, UEFI initializes and tests the underlying hardware, loads and executes additional firmware modules. Finally, UEFI choose the boot device and pass control to bootloader to authenticated OS kernel.

2.2. UEFI Security Boot

UEFI boot process pass through several stages[5], each stage has its own role and may potentially pose some security risks, the startup process of UEFI is shown in Figure 2.

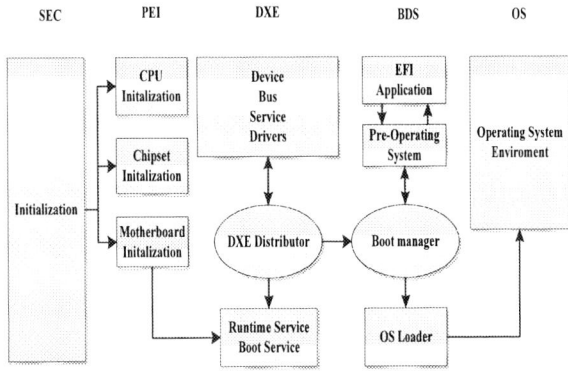

Figure2. UEFI Framework Execution Flow Chart

- SEC. The SEC phase is the first phase executed by the system. It is the root of system security and provides a control node for firmware startup. The SEC stage handles the platform restart event, which uses the processor's cache as a temporary memory area to perform security checks on the initial code of the platform. After that, the SEC forwards the pointer to the temporary memory and the temporary stack, and transfers control to the next phase.
- PEI. The PEI (Pre-EFI Initialization) phase, which is written in C language, and the c execution environment needs to be established during execution. The PEI phase is responsible for initializing the CPU and main memory while providing initial configuration space for components such as processors, chipsets, and system mother boards. The PEI phase passes the status information and control to the next phase through the HOBs list.
- DXE. The DXE (Driver Execution Environment) phase is where the majority of the system initialization takes place. The DXE phase queries the platform related information through the HOBs and builds a complete running environment for the driver. The DXE core produces a set of Boot and Runtime Services. The DXE enumerates DXE drivers and external device drivers through the boot service, allocating memory and loading executable images. UEFI allows for a variety of sources of driver. In order to ensure the integrity and security of the platform, UEFI must ensure the legitimacy of the firmware by hash encryption and encryption authentication for the firmware. After that, The DXE transfers control to the next phase.
- BDS. The Boot Device Selection phase is the final phase in which UEFI has firmware control, and it manages all possible boot setting through the boot manager. The BDS cooperates with the DXE, provide a user launch control platform, which is responsible for discovering the possible boot devices, selecting one to boot from, loading the Boot Loader and executing it.

3. UEFI SECURITY

The UEFI is the first code to execute on a platform during the startup process. UEFI completes the initialization operation of the system and the loading of the system. Therefore, ensuring the security of the UEFI is very important. This chapter analyzes the UEFI protection mechanism to lay the foundation for follow-up attacks

3.1. SPI Flash Protection Mechanisms

UEFI code is stored in SPI Flash. Once an attacker can maliciously write to SPI Flash, UEFI security will be greatly affected. Intel's ICH documentation[4] provides a range of SPI Flash protection mechanisms, the most important mechanisms are the BIOS_CNTL register and the Protected Range Register. At boot time, both registers are specified by the ICH [4]and configured by the UEFI.

The protection mechanism of the BIOS_CNTL register is shown in Figure3.

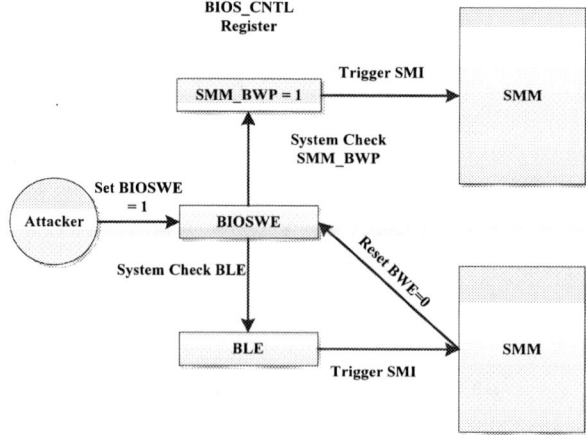

Figure 3. BIOS_CNTL Register Protection Mechanism

The BIOS_CNTL register contains 3 important bits in this regard, BWE, BLE and SMM_BWP[6][8]. The BWE is the write enable bit, if BWE is set to 1, the user can write anything to SPI Flash. If BWE is set to 0, SPI Flash will be locked. The BLE is the lock enable bit, which is used to ensure the security of BWE. Once BLE is set to 1, any modification to BWE will trigger a System Management Interrupt (SMI) and enter SMM. In SMM, the CPU will reset the BWE. The SMM_BWP is the write protection bit in SMM. When SMM_BWP is set to 1, The BWE will only be modified in SMM

Intel specifies a number of Protected Range (PR) registers that can also protect the flash chip against writes. These 32bit registers specify Protected Range Base and Protected Range Limit fields that sets the relevant regions of the flash chip for the write protection enable and read protection enable bits. When the write protection enable bit is set, the region of the flash chip defined by the Base and Limit fields is protected against writes. Similarly, when the Read Protection Enable bit is set., that same region is protected against read attempts. To ensure that the PR registers are not modified, Intel provides HSFS, FlOCKDN to protects the PR register. Once this bit is set, the PR register will not be modified. Although the PR register has a good protection effect, in practice, some motherboard manufacturers are not correctly configuring PR registers, because of the upgrade and maintenance. After investigation, there were (4779/5197) 92% percent of systems did not bother to implement the PR register.

3.2. System Management Mode
System Management Mode is the x86 architecture operating mode of the CPU[9][12], mainly used for motherboard control and power management. SMM is a special mode of operation, it does not depend on the specific operating system and is completely controlled by firmware. The SMM can only be entered via an SMI and can only be exited via the RSM instruction. The Northbridge chipset will generate an SMI signal by writing to Advanced Management Power Control.

The SMM entry is located in a memory called SMRAM, which is located in RAM. An SMI handler is executed from SMRAM, whenever the SMI handler executes an RSM assembly instruction, the CPU state is restored from the map saved in memory. The operating system does not even notice when it is being interrupted by management software running in SMM. The address of the SMRAM is specified by the SMBASE register of the CPU. In practice, the SMRAM address is usually 0xa0000

SMM code runs with full privileges on the platform, even more privilege than operating system kernels. There is a need to prevent access to the SMRAM when the system is not in SMM so that only the SMI Handler can modify the content of the SMRAM. The main mechanism to prevent modification of the SMI handler is the D_LCK bit. If the D_LCK bit is set, configuration bits for the SMRAM in the chipset become read only.

In summary, this decision entangles the security of the UEFI with the security of SMM. Any vulnerabilities that can be exploited to gain access to SMM can be leveraged into an arbitrary reflash of the UEFI. So, for the SMM, this paper proposes a UEFI attack method.

4. CACHE POISONING
At the CanSecWest conference in 2006, a privilege escalation scheme was proposed, which is an actual privilege escalation scheme. The 2008 Black Hat Conference Sparks hide the keylogger in the SMI Handler[11][14]. However, these rootkits have strong limitations and will not exist once the platform restarted. Most rootkits can only be designed for a specific platform, making it difficultly to design a generic SMM rootkit. Also, the biggest limitation is that the D_LCK bit cannot be bypassed. This paper adopts the method of cache poisoning, which can overcome the limitations of traditional attacks

For x86 processors, CPU cache strategy can be divided into Write Back, Write Through and not cacheable[8]. We assume that the memory area of the SMRAM is cached as WB. Once the SMI handler is executed, it will be cached to the CPU's data and instruction buffer. When the SMI handler passes control to the operating system, then the SMI handler will be reside in the CPU cache for a while.

Based on the above principles, we can use this method set BWE bit as 1, so that we can arbitrarily erase the UEFI. The attack process is shown in Figure 4.

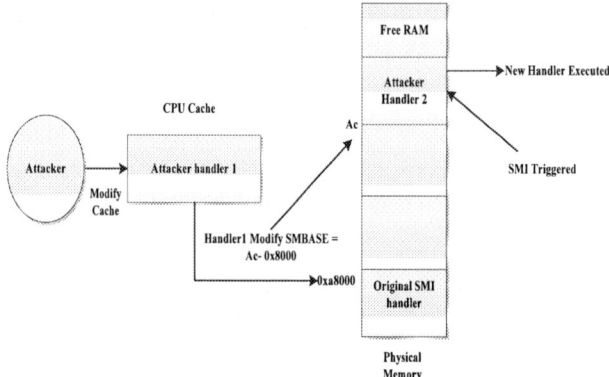

Figure 4. Cache Poisoning Attack

Assume that the rootkit wants to hide some malware in SMRAM. Because of the D_LCK, we cannot access to SMRAM unless in SMM. We can modify the SMRAM cache strategy as Write Back. Then the SMM handler will be temporarily parked in the cache. Then we put an attack SMM handler into a free RAM, we call this area C and the address of this area is Ac. Modify the SMI handler, the SMI handler should modify the SMBASE value in the save state of the CPU so that SMBASE = Ac-0x8000. And then trigger an SMI and the modified SMI handler will executed. The value of the SMBASE register will be set to Ac-0x8000. When the next SMI is triggered, the CPU will determine that the new SMRAM location is SMBASE+0x8000=Ac, and the attack SMI handler will be executed from memory. After that, the attack vector will reside in memory and be located and executed. At the same time, the memory will no longer be protected by D_LCK, so even the D_LCK is set, the attacker can still modify the SMI handler. At this point, we can create an SMI handler which can set BWE = 1. Now, SPI Flash write protection is turned on, UEFI can be arbitrarily erase.

5. CONCLUSIONS
UEFI is proposed to overcome the limitations of the traditional BIOS. It is a new boot platform with many advantages and great development space. However, due to its high scalability, UEFI also has many security risks. The method of cache poisoning in this paper overcomes the limitations of the traditional SMM rootkit and can successfully bypass the D_LCK protection, so that the rootkit code can reside in memory for a long time. Successfully implement the attack of UEFI. The next work has the following two points:

♦   This method needs to modify the CPU's cache strategy, so it only applies to kernel-level rootkits.

♦   On a multicore architecture system, each CPU use a different SMRAM, so the SMI handler may be different. When an SMI occurs, SMI handler cannot be specified.

REFERENCES
[1] Intel Unified Extensible Firmware Interface Specification: Version2.3.1[EB/OL]. 2011.http://www.uefi.org/home/.

[2] Intel Corporation. DQ57TM Development Kit BIOS [EB/OL].2014. http://uefidk.com/develop/development-kit

[3] Intel Corporation. Intel 7 Series Family Platform Controller hub Datasheet[EB/OL]. http://www.intel.com/content/www/us/en/chisets//7-series-chipset-pch-datasheet.html.

[4] Intel Corporation. Intel I/O Controller Hub 10 (ICH10) Family Datasheet.[EB/OL]. http://www.intel.com/content/www/us/en/io-controller-hub-10-family-datasheet.pdf.

[5] C.Kallenberg, J.Butterworth, X.Knovah, C.Cornwell , "Defeating Signed BIOS Enforcement," In EkoParty. Buenos Aires.2013.

[6] Corey Kallenberg, Sam Cornwell, Xeno Kovah "Setup For Failure: More Ways To Defeat SecureBoot".In Hack In The Box Amsterdam ,Amsterdam. 2014.

[7] R.Wojtczuk, C.Kallenberg, "Attacking UEFI Boot Script." In Chaos Communicating Congress. Humburg; Germany. 2014.

[8] Corey Kallenberg, Rafal Wojtczuk. Exploiting an Intel Flash Protection Rase Condition. 2013.

[9] Sherri Sparks, Shawn Embleton, "SMM Rootkits: A New Breed Of OS Independent Malware," Lasveagas. 2008.

[10] Rod Smith. "Managing EFI Boot Loaders For Linux: Dealing With Secure Boot.". http://www.rodsbooks.com/efi-bootloaders/secureboot.html.

[11] Cve-2013-3582. http://www.kb.cert.org/vuls/id/912156, 2013-01-10.

[12] J. Brossard, Hardware backdooring is practical. In BlcakHat ,Las Vegas,USA,2012.

[13] Y. Bulygin. Evil maid just got angrier. In CanSecWest,Vancouver, Canda, 2013.

[14] Y.Bulygin, A Furtak, O. Bazhaniuk. A tale of one software bypass of windows 8 secure boot. In Black Hat, Las Vegas, USA,2013.

[15] J. Butterworth, C. Kallenberg, X. Kovah. Bios chronomancy: Fixing the core root of trust for meansurement. In BlackHat ,Las Vegas, USA,2013.

ISPECE   IOP Publishing

IOP Conf. Series: Journal of Physics: Conf. Series **1187** (2019) 042073   doi:10.1088/1742-6596/1187/4/042073

A systematic review: Road infrastructure requirement for Connected and Autonomous Vehicles (CAVs)

Yuyan Liu[1,2], Miles Tight[2], Quanxin Sun[1] and Ruiyu Kang[3]

[1]School of Traffic and Transportation, Beijing Jiaotong University, Beijing 100044, China

[2]School of Engineering, University of Birmingham, Birmingham B15 2TT, United Kingdom

[3]Drycoolers Inc, Oxford MI 48371, United State

E-mail: 15531189909@163.com

Abstract. There is currently a significant worldwide interest in Connected and Autonomous Vehicles (CAVs), not least for the reason that their realisation and implementation would transform the nature of transportation and provide new impetus for social and economic change. However, the road to support CAVs has not been well prepared, at the risk of leaving potential barriers to CAVs deployment. This paper, therefore, focuses on the gap between current status and future requirements of CAV-compliant road infrastructures, summarizes the possibilities of upgrades, and proposes a three-phase road infrastructure upgrade plan that evolves over time. The first phase is maintenance, followed by a segregated-infrastructure expansion phase leading to phase three which involves the application of simplified standard. The paper is based on an extensive literature review and evidence synthesis and is intended as a stimulus for future study and further debate. For objectiveness, the proposal is general but would need to be refined, when put into practice, to cater for specific implementation contexts.

1. Introduction

Connected and Autonomous Vehicles (CAVs) use a variety of devices including radar, lidar, cameras as sensors, to perceive their surroundings [1]. However, current shortcomings in the perception process may need to be overcome through well-prepared road infrastructure [2]. Otherwise, the realisation of a future with increased mobility for disabled people, a reduction of road accidents caused by human error and the mitigation of pollution as well as congestion could remain but a distant aspiration.

Currently, the research into road infrastructure to support CAVs is still in its infancy. Even though some reports and white papers containing 'scattered views' are slowly being released along with today's CAVs testing projects (e.g. TSC in the UK), the planning and guidance documents used by the highway authorities (e.g. DMRB and WebTAG in the UK) have not taken CAVs into account yet [3]. Therefore, the aim of this paper is to assess the requirement for existing road infrastructure to support CAVs. and the review question is proposed as:

- How should the road infrastructure be upgraded to accommodate and support CAVs?

Following on from the Introduction, the paper is laid out as follows: Section 2 provides essential background information of both CAVs and the road. Section 3 is the detailed literature review process, followed by Section 4, which summarises the literature findings into two categories, namely the aspect

Content from this work may be used under the terms of the Creative Commons Attribution 3.0 licence. Any further distribution of this work must maintain attribution to the author(s) and the title of the work, journal citation and DOI.
Published under licence by IOP Publishing Ltd

of infrastructure considered to be upgraded and the upgrade standard for different scenarios. A discussion is conducted in Section 5 to give an upgrade plan, whilst Section 6 concludes the paper.

2. Background

2.1. Taxonomy of driving automation

To know how roads should be upgraded, it is firstly needed to clarify the difference between different levels of AVs and the conventional cars. The SAE International [4] published a six-level definition for automation, see in table1. With each higher level of automation, the responsibility for performing driving task will increasingly shift from the driver to the vehicle.

Table 1. Taxonomy of driving automation

No	Level Name	Driving task		Respo-nse to failure	Operat-ion area
		Lateral & longitudi-nal control	Monitor surroun-dings		
Driver perform part or all of the driving task					
0	No driving automation	Driver	Driver	Driver	N/A
1	Driver assistance	Driver & System	Driver	Driver	Limited
2	Partial driving automation	System	Driver	Driver	Limited
System perform the entire driving task					
3	Conditional dri-ving automation	System	System	System &Driver	Limited
4	high driving automation	System	System	System	Limited
5	Full driving automation	System	System	System	Un-limited

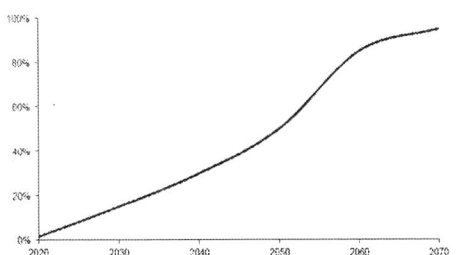

Figure 1. Prediction of AVs implementation timeline.

2.2.CAVs implementation timeline

The time when different levels of CAVs will be implemented in fleet is another essential topic when considering upgrades. There are a number of factors that impact the prediction of the implementation timeline [5~6].

The Automation level2 is widely proved available now by many automobile manufactors, including: Maserati MY 2018, Nissan Leaf 2018 and Audi A8 2019. But when the vehicle manufacturers will produce higher levels of AVs is less certain. For example, Ford is planning to jump-over Level 3 and directly produce Level 4 vehicles in 2021 [7]. The same year, 2021, for producing level 4 vehicles is also mentioned by BMW, who also predicts that the fully AVs will be available between 2025 and 2030 [8]. However, even after full automation is proved to be reliable, it does not mean that people are happy with it and ready to buy. Additional time will also be needed for regulatory approval by the government.

Although the narrative about AVs fleet proportion is uncertain. However, an S-curve theory or Gal's Insight [4], which was usually used as innovation deployment prediction can be adopted, see in figure1. The prediction shows that around 2030s, the AVs fleet proportion will reach 20%, while in 2050s, it will account 50% of vehicles on the road. Similar ideas were mentioned by KPMG [9], which predicted that in 2035, AVs in the fleet will reach 25%, and in 2050, the figure will exceed 50%.

2.3. Classification of the road infrastructure assets
To avoid any missing or making fragmented review, the road assets classes needs to be defined to ensure the upgrades are considered in a systematic level. This paper referred the Network Management Manual published by the Highway Agency [10] and defined five classes of road infrastructure assets: 1) Communications; 2) Road pavement; 3) Structures; 4) Geotechnical and 5) Drainage.

3. Methodology
A literature review can be a good starting point for an academic article that investigates relatively new fields of academic study [11]. This paper generally followed the methodological framework of a systematic review, as developed by Gough et al [12].

During key terms searching, synonyms of the keywords and Boolean operator are applied, see in table2. In order to identify existing articles potentially answering the review questions, 5 bibliography databases, 19 organizational databases were used, see in table3. Among the extensive databases, 7252 studies match the initial search. By filtering out those not written in English, published before 2012, and not transportation related papers, 1831 results were left. Then, based on a quick review of title and abstract, 57 results remained. After removing duplicates and no full text available, 29 papers made it through the first-round screening. The author then applied a Quality of Evidence (QoE) standard to the first-round results for evidence appraising, see in table4. 7 literatures are defined as "high quality" results thus were directly obtained, while the references list from the rests were reviewed for the second-round, and 6 articles were adopted, which gives a total amount of 13 results for the final synthesis of evidence. Figure2 shows a flowchart of the acquisition process.

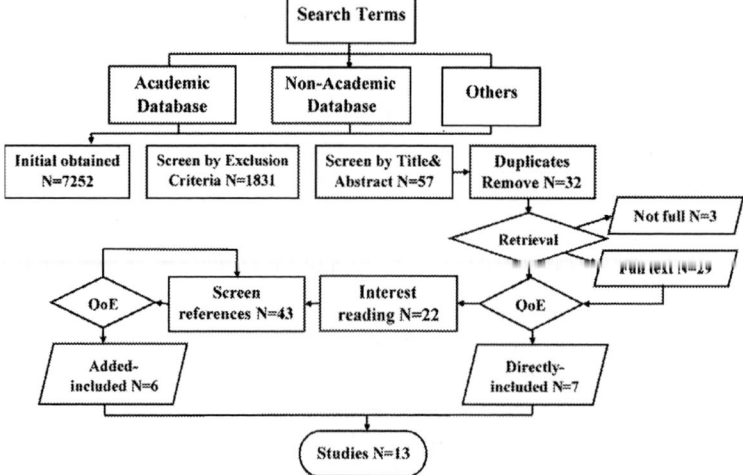

Figure 2. Overview of literature acquisition process.

Table 2. Key searching terms.

Key words	Search strings
CAVs	(Connect* OR Autonomous OR automated OR self-driv* OR auto-pilot OR driverless) AND (Vehicle* OR car* OR automobile* OR driv*)
Upgrade	AND (upgrad* OR chang* OR design OR maintain* OR construct*)
Road infrastructure	AND (Road* OR facilit* OR infrastructure)

Table 3. Academic and non-academic databases.

Types	Name
Bibliography Database	TRID/ Scopus/ Web of Science/ GeoBase/ Compendex
Organization Database	TRL/ OECD/ CAR/ CACC/ DfT/ KPMG/ SAE International/ TSC/ ITF/ Atkins/ RAND Corporation/ GATEway/ USDOT/ UKAutodrive/ Infrastructure Partnership Australia/ iRAP/ MEARGE Greenwich/ RAC Foundation/ VENTURER/ Victoria Transport Policy Institute

Table 4. The Quality of Evidence (QoE).

Standard	Description
Soundness of the studies	The study itself have explicit and effective evidence with reliable conclusion.
Relevance of review scope	Studies focus both on the field of AVs and the likely road infrastructure changes.
Appropriateness for the review needs	Provides at least one aspect of upgrade suggestions from either the problems emerged in real situation or the critical thinking through engineering experience.

Notes: Any literature meets at least two of the standards will be defined as "high quality" studies

The acquisition results are shown in table5. The evidence was synthesized and summarized into 15 infrastructure relevant topics, which consists of 12 topics of infrastructure aspects (belongs to the five classes of road assets), and 3 topics of road upgrade standards.

Table 5. Summary of literature acquisition results.

Ref	Author(s)	Year	Outlet type	Upgrade aspect	Upgrade standard	location
[13]	Johnson	2017	Report	TR/DC/PS/RS/P/S/R/B/G/DS	M/SE/SI	UK
[14]	Lyon et al.	2017	Report	TR/IRC/DC/P/B/G	M	Australia

[15]	Lawson	2018	report	TR/R/DS	M/SE	UK
[16]	Shladover & Bishop	2015	White paper	TR/DC/P/B	M/SE/SI	EU-US
[3]	TSC	2017a	report	TR/IRC/DC/PS/P/S/SHA/R/B	M/SE/SI	UK
[17]	McCarthy	2016	report	TR/DC/P/B	M/SI	UK
[18]	UKAutodrive	2018	White paper	TR/DC/P/R	N/A	UK
[19]	KPMG	2012	Report	TR/DC/P	SI	N/A
[20]	Begg	2014	Report	DC/P	M/SE/SI	UK
[21]	Chen et al.	2016	Journal	P/G	M	Sweden
[22]	CAVita	2017	White paper	DC/P/G	M	USA
[23]	Godsmark et al.	2015	report	TR/DC/P/S/R/B/G	SI	Canada
[24]	Kuutti et al.	2017	Journal	TR/DC	N/A	UK

Notes: Upgrade aspects under five infrastructure classes: 1) Communications: TR/IRC/DC; 2) Road pavement: PS/RS; 3) Structures: P/S/SHA/R/B; 4) Geotechnical: G; 5) Drainage: DS. Where TR=Traffic signs and Road markings; IRC=Incident and Roadwork Communications; DC=Digital Communications; PS=Pavement Structure; RS=Road Surface; P=Parking; S=Service station; SHA=Safe Harbour Area; R=Roundabout; B=Bridge; G=Geotechnical; D=Drainage System;
Upgrade standards: M=Maintenance; SE=Segregated infrastructure; SI=simplified-standards.

4. Summary of findings

4.1. Aspect of infrastructure considered to be upgraded

4.1.1. Traffic signs and road markings. CAVs require highly visible road edges, curves, speed limit and other signages to complete the task of locating [24], navigating [13], and parking [3]. However, the design and maintain of signs and markings are far away from accommodating CAVs. For example, faded road markings have discontinuous issues in North America [25]; low maintenance priority (e.g. dirty) cause signs unreadable in the UK [26]; non-standard road signs confused users in the EU [27].

In order to adopt CAVs, governments and organizations should call for uniform road marking across EU members, or even suggest global standardization [3], and also improve maintenance criteria and establish monitoring systems [3,14]. To complement the "readability" maintenance of roads, upgrades could also enhance the brightness of street lights [16] and frequently control roadside vegetation.

4.1.2. Incident and roadworks communications. Incidents and roadworks will change the road layout, which require CAVs to interpret real-time changes, such as merging-lanes suggestions provided by temporary signs and cones. Currently, two websites provide real-time information on traffic accidents and roadworks in the UK (roadworks.org and www.waze.com). Even though, both websites provide the cause and time of accidents, the lack of accurate on-site layout imaging still makes it difficult for CAVs to distinguish real-time changes from historical maps, thus hard to navigate.

Two suggestions for the future: establish uniform CAV-readable emergency signs, barriers or cones [3]; establish digital roadside communications to replace static facilities to provide real-time data [14].

4.1.3.Digital communications. The successful deployment of CAV requires not only vehicle-mounted instruments, but also road-side digital infrastructure to meet various forms of connectivity requirements (V2X communication) [1]. Sensors and Internet connectivity are required here for applying digital infrastructures:

• Substantial amounts of sensors including in-roadway sensors (such as loop detectors and magnetic detectors), and over-roadway sensors (cameras, radars and ultrasonic etc.) [14]. These sensors could help the road management center monitoring traffic demands and allocating transportation resource through I2V communications [1] to achieve smoother traffic flow.

• Internet connectivity can be implemented either via mobile network (4G/5G) or wifi-based facilities (ITS-G5) [3]. Some articles also mentioned fitting new roads and major junctions with the fiber-optic-cable beforehand to make future digital infrastructure easier to realize in the future [3,13].

4.1.4.Pavement structure. CAVs are more likely to operate continuously in the middle of the carriageway by using Lane-Keeping-system (LKS) [28], which will accelerate the appearance of rutting and other pavement deteriorations (e.g. potholes, cracking) [21]. So, certain areas beneath the CAV operation track need to be strengthened consequently [13].

In AV-dedicated lanes, higher stiffness and more deformation-resistance materials (e.g. asphalt concrete) can be considered for the surface layer. The maintenance frequency for the remaining pavement layers can also be increased in the future.

4.1.5.Road surface. CAVs can adjust vehicle speed more predictively through V2V and V2I communications to avoid sharp braking [13] thus reducing the stopping distance design standard. The requirement for the coefficient of friction (represented by the PSV and the texture depth) can therefore be relaxed, so that less skid-resistance materials are possible for use in the surface course in the future.

4.1.6.Parking. Multiple researchers have foreseen that Automated-Valet-Parking (AVP) can significantly reduce future parking demands [14,19,29], especially in the context of shared CAVs — shared cars can serve different customers at different times, eliminating the needs for long-term parking [30]. Once arrived the parking lot, the AVP will enable them parking more closer with each other to save more space. But current carparks are not yet able to support self-parking, as most parking spaces are made of concrete and are located underground where GPS signals are not strong, causing difficulties in navigation. Aiming at this, some researchers suggested solutions by digitally fitting those areas with Bluetooth and near field communications [18]. However, it is still uncertain when the retrofitting of existing carparks and the building of new ones will start [31].

It is expected that more flexible design approaches are obtained to meet the parking needs during transition phase [3]. A case study was provided by an architecture design firm from Boston, US [32], whose design could support both AVs and conventional vehicles in transition, see in figure3.

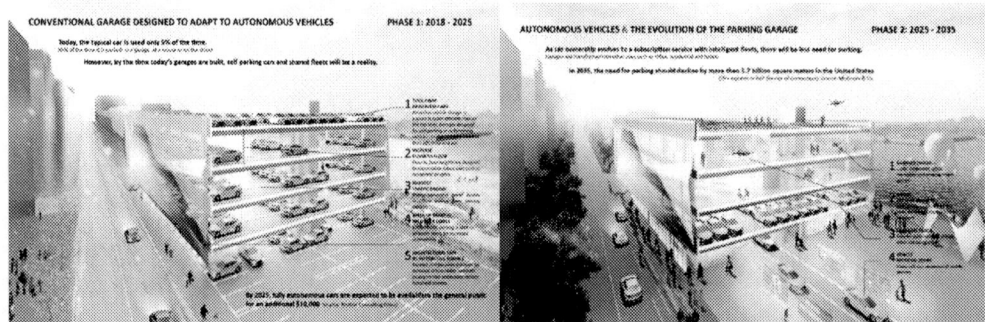

Figure 3. Garage adapts to AVs (left) and to full AVs and new uses (right).

The top floor of the building was originally designed as only a CAV-compatible parking for efficiency, whilst the remaining floors were retained as traditional parking spaces. As the amount of CAV increases, the lower floors are expected to be converted into CAV parking or even charging spaces, and the top layer then becomes a recreation area to avoid construction-waste.

4.1.7.Service stations. Traditional service stations providing drivers places to stop, to refresh, and to refuel. With the British government's announcement that pure petrol/diesel-powered vehicles will be banned from 2040 [33], and the ultimate goal is making Electric Vehicles (EVs) account for at least 9% of the total fleet by 2020 [18], it can be expected that future service stations are likely to provide more chargers [3,23]. A bolder idea can even be replacing charging points completely by a more efficient charging road [34]. In fact, the first electrified road in Sweden has already been put into use so that vehicles on this road can be charged while driving [35].

4.1.8.Safe harbour area. Safe harbour area is essential for AVs, since the level 3-4 vehicles require safe places to stop or restart in an emergency or when reaching the exit of an automated operation areas [4].

The traditional safe harbour area in the UK is a hard shoulder along the motorway but a number of these have recently been converted into running lanes. The former definition of safe harbour referred to an emergency refuge area (ERA), see in figure4, which is 100m long and located at intervals of 2.5km. These were sometimes misused by drivers, leading to serious consequences (for instance, if the ERA is occupied by other cars, a driver is forced to stop in a busy lane when broken down).

Figure 4. Example of hard shoulder (left) and ERA (right).

In the future, the frequency and the design standard of safe-harbours needs to be reconsidered to accommodate CAVs and may also need the introduction of regulations to avoid misuse [3].

4.1.9.Roundabout. There are two opposite considerations about the future of roundabouts:
•At present, roundabouts essentially reduce the severity of accidents by transforming dangerous side-collisions into passing-collisions. However, with the advent of AVs, signalized intersections may be safer because of their more predictable elements [15].
•In terms of achieving smoother traffic flow, roundabouts may be preferred due to shorter delays and queuing times [36]. This benefit is predicted to extend as the merging action is easier to achieve with the AV's computer programmers [23].

4.1.10.Bridges. As the amount of CAV increases, the segregated infrastructures, including bridges [23], tunnels and underpasses [13] are also needed to be provided to reduce conflicts between different traffic flows. The current bridge designs, particularly long-span bridges, has not catered for the 'platooning' needs of heavy trucks. Therefore, the inadequate design standard needs to be updated [3].

4.1.11.Geotechnical. With the technique of LKS, many researchers advocate the possibility of narrower carriageways [14], extra lane(s) based on existing road width [22], and tighter corners [13] in the future. Besides, reduced road gradients with more cuts and fills, embankments and tunnels are also preferred in response to the needs of platooning, since the vehicle speed can be easily-controlled on flat slopes [13].

4.1.12.Drainage. According to the Annual Assessment of Highways England's performance [37], 69% of road network drainage data in the UK is unknown. Since severe surface water accumulation may cause the CAV systems becomes 'paralysed' [15], improving the design and maintain of drainage infrastructure (e.g. culvert. channels and gullies) should be given higher priority [13].

4.2.Upgrade standards for different scenarios
Instead of simply combining the upgrade aspects together to develop a retrofit plan for the uncertain future, it is better to determine more targeted plans for different scenarios. A comprehensive review of three different upgrade standards is given below, considering different CAV proportions on the road.

4.2.1.Maintenance. No matter how rapidly the CAVs evolves, it is better to update the infrastructure through maintenance rather than being radically changed at an early stage [17]. This is not only to meet the needs of CAV development [13], but also to meet the safety requirements of conventional vehicles [15]. Maintenance work can be considered from:
- Establishing enhanced standards and shorten maintain intervals of communication facilities.
- Increase the frequency of pavement maintenance (e.g. grading, sealing and patching) [21].
- A broader maintenance range of drainage system is also worth considering.

4.2.2.Segregation. As the automation level evolves and the market penetration of CAV increases, the road may be shared by different levels of AVs, which requires not only the preservation of the current facilities for lower-level AVs [13], but also the construction of additional infrastructure for full AVs. However, the co-existing infrastructure will undoubtedly bring about road right issues, which prompts further consideration of the expansion needs of the AV-segregated facilities [16]. Including:
- Specific routes [17,38,39] with reinforced surface materials.
- Safe harbour areas with devices that prevent head-on-collisions (e.g. central barriers) [15].
- Other structures like shared-CAVs gathering stations, service stations or even charging roads.

4.2.3.Simplification. Some existing road infrastructure may become redundant when full automation is proven efficiency. Therefore, new standards for different classes of infrastructure are expected to be developed to simplify road facilities [13].
- In terms of communications: Use simplified road markings [16] in AV-specific areas; Replace the static roadway-signs with digital in-vehicle devices [17] or establish single communication beacons to replace several traffic signals [13]; Remove the speed cameras since CAVs are programmed to obey traffic rules [20].
- In terms of structures: Roundabouts and some crash countermeasures (e.g. speed bumps) can be reduced [20]; The layout of carparks can be unified into a more concise standard; Service stations can be considered as a proper safe harbour areas for faulty AVs [3].
- Road geometry can be simplified, making future roads less confusing and more attractive [13].

5. Discussion

5.1.A general description of the upgrade plan for the transition phases

Based on the timeline prediction in section2.2 and the upgrade standard in section4.2, a three-phase transition plan is proposed. For each phase, it assigns different road infrastructure upgrade work consider the status of each stakeholder (manufacture, road users, and the government), shown in table6.

Table 6. The upgrade plan for transition phases.

No.	Start time	CAVs proportion	Automation level	Manufacture (techniques) status	Road users (market) status	Government (policy) status	Upgrade standard
1	Now	<20%	Level 0	Mature	Dominant	Release CAVs testing standards	Main-tenance
			Level 1-2	Prove to be efficiency	Early adopter		
			Level 3-4	Prove to be reliable	Turn-out		
			Level 5	Testing	Rarely seen		
2	2030s	20~50%	Level 0	Mature	Gradually eliminate	Road right division	Segre-gation
			Level 1-2	Mature	Dominant		
			Level 3-4	Prove to be efficiency	Faithful adopter		
			Level 5	Prove to be reliable	Turn-out		
3	2050s	>50%	Level 0	Mature	Nearly disappear	Ownership change; Energy substitution	Simplif-ication
			Level 1-2	Mature	Gradually eliminate		
			Level 3-4	Mature	Dominant		
			Level 5	Prove to be efficiency	Faithful adopter		

5.1.1.Phase 1.
Phase 1 is essentially underway and should continue until sometime in the 2030s when level 2 automation will be dominant. Current, techniques are established, the market is dominant, and all road policy is suitable for automation level 0, which is no driving automation. However, researchers and vehicle OEMs are trying to prove the efficiency of Level 1-2 and reliability of Level 3-4, and testing Level 5. Therefore, the government along with other highway agencies need to publish a certain level of new road standards for supporting the testing of AV technologies.

The current upgrade standard includes: 1) Maintaining the existing road infrastructure (including certain types of communications facilities, pavement structure and drainage systems) to ensure the safe operation of conventional vehicles. 2) Building certain types of digital communication facilities for vehicles with partial automation functions at the same time.

5.1.2.Phase 2.
Phase 2 is the key and the most complicated stage. Although Level 2 vehicles account for vehicles on the road, the market share of Level 3-4 vehicles can no longer be ignored. Also, Level 5 vehicles may become available on the market as well. This mixed-level operation will certainly be a problem for road-rights campaigners. Therefore, new laws and regulations must be released by the government to highlight the separation of responsibilities, for example, in accidents that involve AV.

Furthermore, the transportation sector needs consequently adjust the upgrade standard to build some road facilities that are able to reduce conflicts. Investing in AV-segregated infrastructure like AV-specific routes, gatherings and service stations are all possible ways for further investigation. Flexible design, such as building CAV-compatible parking lots, is also needed. In addition, since lower-level vehicles are still available on the market, the existing facilities such as road signs and traffic signals should not be abandoned in this phase immediately [15,39].

5.1.3.Phase 3. Phase 3 will not start until Level 3-4 vehicles take the majority of market share and the automation level 5 are proved to be efficient and attracting customers. For phase 3, many articles envisaged the 'ideal future' with full AVs, for example, 'middle-class' families will be able to afford level 5 AVs which allow them to sleep or work during daily commute.

Also, during phase 3, the government could spare more attention to social or environmental issues that generated from phase 2 (e.g. the increasingly serious transportation pollution). New solutions can therefore be considered, such as ownership-model changes and the substitution of energy options. Two highly possible products that may be seen are unmanned buses and shared driverless taxis, which will take a significant number of privately-owned vehicles off the road [18]. EVs or other clean-energy-powered vehicles will grow rapidly. Also, road infrastructure could and will become simpler, making the road less complex and more attractive. For example, multiple road markings and traffic signals could be replaced by a single digital communication beacon, very few or none parking spaces appeared in the city center, most service stations replaced by charging stations, etc.

5.2.Assessment of impacts of the upgrade plan
It is worth noting that there may be gaps or overlaps among the 3 phases, as it is not an outcome of accurate prediction or calculations. However, a potential cost-benefit assessment is still needed to help estimate the worthiness of the upgrade.

To avoid any lengthy and fragmented analysis for the unconfirmed future, table7 provides a weighted framework to qualitatively assess the costs-benefits impact on the upgrade plans. The measuring indicators were developed from the Road Investment Strategy published by DfT [40].

Table 7. Assessment of cost and benefits for the upgrade choices.

Indicators of cost and benefits	a	b	c	d	e	f	g	Ave
Maintain-oriented								
Unified design and enhanced maintain for road signs & markings	2	5	3	5	5	3	3	3.7
Setup readable incident roadwork signage	2	5	3	5	5	3	3	3.7
Enhanced maintenance standards for pavement structure	2	4	3	5	4	2	3	3.2
Enhanced maintenance standards for drainage system	2	4	3	5	4	3	3	3.4
Segregated-expansion								
Setup specific lanes with more bridges, tunnels & underpasses	1	5	3	5	5	1	5	3.5
Setup CAV-compatible parking lot	1	5	5	5	5	1	5	3.8
Setup safe harbour area with safe defence	1	5	3	5	4	1	3	3.4
Setup charging machine for service station	1	5	3	4	5	5	3	3.7
Simplified-standard								
Setup digital roadside communications	1	5	5	5	5	1	5	3.8
Reduce the number of roundabouts	5	1	3	1	3	5	5	3.2
Relaxed skid-resistance standards for road surface	5	1	3	1	3	5	3	3.0
Simplified geometry design standards	3	2	3	2	4	5	5	3.4

Notes: Scores: 1=disadvantage; 2=possible disadvantage; 3=neutral; 4=possible advantage; 5=advantage.

Indicators: a=Construction & maintenance cost; b=Needs of extra CAVs techniques cost; c= Easy for future change; d= Improved road safety; e=Improved road efficiency; f=Less energy consumption and pollution; g= Better land use; ave=average score

In table7, the average score is calculated under the assumption that each evaluating indicator has the same weight, and a score of more than 3 suggests that the upgrade option is cost-effective when applied appropriately.

Based on this assumption, most upgrade points are shown as very cost-effective, especially the "Setup digital roadside communications" and "Setup CAV-compatible parking lot". (Both scored 3.8). However, "Relaxed skid-resistance standards for road surface" with score of 3.0 shows that it is not very desirable.

6. Conclusion

Through a wide-ranging literature review and in-depth evidence synthesis, this paper proposes a 3-phase road infrastructure upgrade plan in answer to the research question "How should the road infrastructure be upgraded to accommodate and support CAVs?" Phase 1 mainly comprises maintenance work, phase 2 involves the construction of CAV-segregation infrastructure, and phase 3 concerns a simplified design standard for constructing road infrastructure. Although the 3-phase plan has limitations largely associated with the immaturity of the CAVs market and the lack of collectable data, it should be regarded as a starting point, which can be used to stimulate further research in the field of road facilities upgrades during the CAVs transition time.

Future considerations, however, could be more accurately proposed based on, say more sophisticated data analysis. Appraising software (e.g. the HDM4) could be used for qualitative assessment. A limitation of the research is that the 3-phase upgrade plan did not consider the effects from different regions, countries or under different cultures. Future study could also create specific plans by considering different city sizes, populations, policies and even energy allocation constraints.

References

[1] Guanetti J, Kim Y and Borrelli F 2018 Control of connected and automated vehicles: State of the art and future Challenges. *Annual reviews in control.* Vol **45**, pp. 18-40.

[2] Pendleton S D et al 2017 Perception, Planning, Control, and Coordination for Autonomous Vehicles. *MDPI, Machines,* Vol.**5**, Issue.1. DOI:10.3390.

[3] TSC 2017a *Future Proofing Infrastructure for Connected and Autonomous Vehicles.* Available : https://s3-eu-west-1.amazonaws.com/media.ts.catapult/wp-content/uploads/2017/04/251153 13/ATS40-Future-Proofing-Infrastructure-for-CAVs.pdf.

[4] SAE 2018 *Taxonomy and Definitions for Terms Related to Driving Automation Systems for On-Road Motor Vehicles.* J3016_201806.

[5] Litman T 2018 *Autonomous Vehicle Implementation Predictions: Implications for transport planning.* Victoria Transport Policy Institute.

[6] TSC 2017b Market Forecast: for Connected and Autonomous Vehicles. Available: https://assets .publishing.service.gov.uk/government/uploads/system/uploads/attachment_data/file/642813 /15780_TSC_Market_Forecast_for_CAV_Report_FINAL.pdf.

[7] Walker J 2018 *The Self-Driving Car Timeline– Predictions from the Top 11 Global Automakers.* Available: https://www.techemergence.com/self-driving-car-timeline-themselves-top-11-aut omakers/.

[8] BMW 2018 *Autonomous Driving: What You Need to Know in 2018.* Available: https://www.yo utube.com/watch?v=xsQvq4WlUYU&t=126s.

[9] KPMG 2017 *Impact of Autonomous Vehicles on Public Transport Sector.* Available: https://ho me.kpmg.com/ie/en/home/insights/2017/07/impact-autonomous-vehicles-public-transport-se ctor.html.

[10] HA 2009 Network Management Maunal: Standards for Highway. Available: http://www.standar dsforhighways.co.uk/ha/standards/nmm_rwsc/docs/nmm_part_0.pdf.

[11] Wee B V and Banister D 2015 'How to Write a Literature Review Paper?' *Transport Reviews*. **36:2**, pp:278-288, DOI: 10.1080/01441647.2015.1065456

[12] Gough D, Oliver S and Thomas J (ed.) 2017 *An introduction to systematic reviews*. Los Angeles: SAGE. 2nd edn.

[13] Johnson C 2017 *Readiness of the road network for connected and autonomous vehicle*. London: RAC Foundation.

[14] Lyon B, Hudson N, Twycross M, Finn D, Porter S, Maklary Z and Waller T 2017 *Automated vehicles: Do we know which road to take?* Infrastructure Partnerships Australia.

[15] Lawson S 2018 *Roads that Cars Can Read: Report III – Tackling the Transition to Automated Vehicles*. Available: http://resources.irap.org/Report/2018_05_30_Roads%20that%20cars%2 0can%20read_FINAL.PDF.

[16] Shladover S E and Bishop R 2015 Road transport automation as a Public-Private Enterprise. *Third EU-U.S. Transportation Research Symposium.* Washington D.C. 14-15 May 2015 Washington, D.C: Transportation Research Board.

[17] McCarthy J, Bradburn J, Willians D, Piechocki R, and Hermans K 2016 *Connected and Autonomous Vehicle: Introducing the Future of Mobility.* London: Atkins Ltd.

[18] UK Autodrive 2018 *Paving the Way: Building the Road Infrastructure of the Future for the Connected and Autonomous Vehicles.* Available: http://www.ukautodrive.com/downloads/.

[19] KPMG 2012 *Self-driving Cars: the Next Revolution.* Available: https://faculty.washington.edu/j bs/itrans/self_driving_cars[1].pdf.

[20] Begg D 2014 *A 2050 Vision for London: What Are the Implications of Driverless Transport.* London: Transport Times. Available: https://www.transporttimes.co.uk/Admin/uploads/6416 5-transport-times_a-2050-vision-for-london_aw-web-ready.pdf.

[21] Chen F, Baileu R and Kringos N 2016 'Potential influences on long-term service performance of road infrastructure by Automated Vehicles'. *Transportation Research Board.* pp. 72–79. DOI: 10.3141/2550-10.

[22] CAVita 2017 *Connected and Automated Technologies and Transportation Infrastructure Readiness.* National Cooperative Highway Research Program Project 20-24(111).

[23] Godsmark P, Kirk B and Flemming B 2015 *Automated vehicle: The Coming of the Next Disruptive Technology.* The Conference Board of Canada.

[24] Kuutti S, Fallah S, Katsaros K, Dianati M, Mccullough F and Mouzakitis A 2018 'A survey of the state-of-the-art localisation techniques and their potentials for Autonomous Vehicle applications'. *IEEE Internet of Things Journal.* DOI 10.1109/JIOT.2018.2812300,

[25] Sage A 2016 *Where's the Lane? Self-driving Cars Confused by shabby US roadways.* Available: https://www.reuters.com/article/us-autos-autonomous-infrastructure-insig/wheres-the-lane-s e lf-drivingcars-confused-by-shabby-u-s-roadways-idUSKCN0WX131.

[26] DMRB 2015 *Design Manual for Roads and Bridges. Part 2*. TD 25/15. Vol **8** Traffic signs and road lighting, Sect 2. Traffic signs and road markings.

[27] EuroRAP 2013 *Roads that Cars Can Read: A quality standard for road markings and traffic signs on major rural roads.* Available : www.eurorap.org/wp-content/uploads/2015/03/roads_that_ cars_can_read_2_spread1.pdf.

[28] Vine S L and Polak J 2014 *Automated cars: A smooth Ride Ahead?* Independent Transport Commission.

[29] West D M 2016 *Moving forward: Self-driving vehicles in China, Europe, Japan, Korea, and the United States.* Center for technology innovation at Brookings.

[30] Cavoli C, Phillips B, Cohen T and Jones P 2017 *Social and Behavioural Questions Associated with Automated Vehicles, A Literature Review.* London: Department of Transport.

[31] Wagner J, Baker T, Goodin G and Maddox J 2014 *Automated Vehicles: Policy Implications Scoping Study.* Texas A&M Transportation Institute. 600451-00029-1.

[32] Arrowstreet 2016 Available at: http://www.arrowstreet.com/portfolio/autonomous-vehicles/.

[33] BBC News 2017 *New Diesel and Petrol Vehicles to Be Banned from 2040 in UK.* Available: https://www.bbc.co.uk/news/uk-40723581.

[34] Kenny S 2017 *Road Charging for Cars: What the European Commission Should Do?* Freight & Rail Policy Officer. Transport & Environment.

[35] Boffey D 2018 *World's First Electrified Road for Charging Vehicles Opens in Sweden. Available*: https://www.theguardian.com/environment/2018/apr/12/worlds-first-electrified-road-for-charging-vehicles-opens-in-sweden.

[36] DMRB 2007 *Deign Manual for Road and Bridges. Part 3.* TD 16/07. Vol **6**. Road geometry. Sec 2. Junctions.

[37] ORR 2017 *Annual Assessment of Highway England's Performance.* ISBN 9781474145565.

[38] Bierstedt J, Gooze A, Gray C, Peterman J, Raykin L and Walters J 2014 *Effects of Next-generation Vehicles on Travel Demand and Highway Capacity.* Available: http://orfe.princeton.edu/~alaink/Papers/FP_NextGenVehicleWhitePaper012414.pdf.

[39] Glancy D J, Peterson R W and Graham K F 2016 A Look at the Legal Environment for Driverless Vehicles. Washington, D.C: *Transport research board.* DOI 10.17226/23453.

[40] DfT 2014 *Road investment strategy. Summary leaflet.* Available : https://assets.publishing.service.gov.uk/government/uploads/system/uploads/attachment_data/file/410877/ris-leaflet-accessible.pdf.

ISPECE   IOP Publishing

IOP Conf. Series: Journal of Physics: Conf. Series **1187** (2019) 042074   doi:10.1088/1742-6596/1187/4/042074

Ship Detection and Tracking in Nighttime Video Images Based on the Method of LSDT

L Liu[1,2], G Liu[3], X M Chu[1], Z L Jiang[1], M Y Zhang[1] and J Ye[3]

[1] National Engineering Research Center for Water Transport Safety, Wuhan University of Technology, Wuhan, Hubei, China

[2] School of Energy and Power Engineering, Wuhan University of Technology, Wuhan, Hubei, China

[3] School of Computer Science and Technology, Wuhan University of Technology, Wuhan, Hubei, China

Abstract: Ship detection and tracking has been recognized as a challenging task in the maritime administration. The method of LSDT (Light Spot Detection and Tracking) has been applied to achieve ship detection and tracking based on the nighttime surveillance video. Firstly, the light spots in the video images are detected through LOG and invalid spots are filtered by the gray threshold. Multiple targets are subsequently tracked by Kalman filtering and light spots are marked to determine properties in order to add and delete spots. A case study has been performed in the middle reach (Wuhan) of the Yangtze River. Nighttime surveillance video has been obtained and further utilized for method verification. The results show that the ship can be detected and tracked by LSDT effectively. The present study provides an alternative for ship detection at night, thus improving the maritime surveillance efficiency.

1. Introduction
Ship detection and tracking technologies are of significance for ensuring safe navigation and reducing maritime accidents. It could be achieved by a number of different ways, such as radar, AIS and video surveillance. However, shortage have been noted by using aforementioned methods in practical circumstances. For example, visibility is low and the scene is blurred at night. Although the radar can help navigation, the radar targets are not clear enough to determine what targets are [1]. The AIS signal is easily lost and affected by shipborne equipment [2]. Maritime accidents happen frequently at night. To improve ship navigation safety, it is essential to explore robust ship detection and tracking method at night.

The traditional ship detection at night can be divided into three categories. First of all, the infrared camera is utilized to perform ship detection because of its adaption to the light at night and severe weather, such as the foggy day. Liu et al [3] built a platform on the pier and set up an infrared camera to construct a system which can detect ship during day and night to avoid collisions. Liu et al [4] used the infrared camera for ship monitoring to realize abnormal behaviour detection in port and reduce the possibility of theft of ships, but the infrared camera is more expensive. Secondly, synthetic aperture radar SAR is chosen for ship detection. Tello et al [5] and Marino [6] used wavelet transform and notch filter for ship detection by SAR image respectively, but the energy consumption of SAR is higher so that it is difficult to ensure continuous monitoring. And this method is mainly used for fishery monitoring and oil spill management. Finally, satellite imagery is used for fishing vessel detection,

Content from this work may be used under the terms of the Creative Commons Attribution 3.0 licence. Any further distribution of this work must maintain attribution to the author(s) and the title of the work, journal citation and DOI.
Published under licence by IOP Publishing Ltd

mainly detecting ship by boat lighting. The earliest operation system is a linear scanning system and then developed into visible infrared imaging radiation system VIIRS to collects DNB data [7]. D. Elvidge et al. designed the DNB data ship detection system. Yamaguchi et al. [8] proposed the ship traffic flow density assessment algorithm based on DNB data and BT3.7. However, this type of method is more suitable for fishing vessel detection while less applied in other scenarios.

In summary, there are limited researches on ship detection at night by an ordinary camera. Considering real-time traffic signal detection method [9], when ship sails at night, ship lights would be turned on. Given the ship lights can be detected and tracked, it is thus possible to realize ship detection and tracking by this way. On the basis of this, the ship detection and tracking method LSDT was studied by collecting ship navigation video at night in Wuhan Section of the Yangtze River. Specifically, it mainly consists of two steps: the first step is ship light spots detection by LOG [10]; then the Kalman filtering is utilized to perform light spots tracking [11]. Through the LSDT proposed in this paper, ship detection and tracking at night by an ordinary camera can be accomplished. The content of this paper is organized as follow: the current ship detection and tracking methods at night are introduced firstly; ship light spots detection based on LOG and multiple targets tracking based on the Kalman filtering are thus illustrated; the effectiveness verification of LSDT is further demonstrated through a case study; some preliminary conclusions are drawn at the end.

2. Methodology

It is always difficult to identify ship directly at night, especially in the inland river environment. Due to the background in the video, features of the ship image cannot be extracted effectively. The navigation scenery at night near the Wuhan Tianxingzhou Bridge in Wuhan Section is shown in figure 1. The ship could be hardly detected in the nighttime video images. However, while the ship is sailing, ship lights are turned on. Thus ship detection and tracking could be accomplished through the method of light spot detection and tracking (LSDT).

Figure 1. Nighttime video image of ship navigation at Wuhan reach, Yangtze River

The ship light spot detection and tracking method mainly includes the following steps, as shown in figure 2.

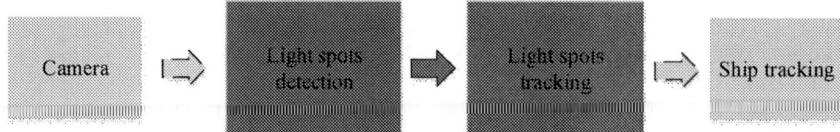

Figure 2. Flowchart of ship detection and tracking

In the first step, all visible light spots are detected in the grayscale image of night surveillance video based on LOG. The image collected in the inland river at night has other light sources. As shown in figure 1, the visible light spots include the bridge and building lights in addition to ship lights. Therefore, considering the influence of other lights, after spot detection is implemented, the spots should be filtered according to the gray value of spots and the ship light spots should be kept as much as possible.

In the second step, light spots are tracked. Since one ship has multiple ship lights, multiple target tracking is required. In this paper, the Kalman filtering tracking method is adopted to track multiple moving objects.

Finally, the method is verified by the surveillance video at night near Wuhan Tianxingzhou Bridge.

3. Light spot detection
The ship detection is actually light spot detection which means to find the bright area surrounded by the low gray pixel in the gray image. Therefore, the spot detection can be converted into blob detection. In this paper, the LOG is used for blob detection [12].

3.1. LOG
Marr and Hildreth combined Gaussian filtering and Laplacian to form the LOG (Laplacian of Gaussian) blob detection [7]. The basic thought of LOG is to filter the image through a Gaussian filter firstly and then perform a Laplacian operator on the filtered image to obtain the edge. This filter makes full use of the Gaussian function to reduce the noise influence, and the probability of detecting the false edge is reduced by Laplacian template.

The two-dimensional Gaussian kernel function is written as:

$$G_{\sigma}(x,y) = \frac{1}{\sqrt{2\pi\sigma^2}} \exp\left(-\frac{x^2+y^2}{2\sigma^2}\right) \quad (1)$$

The Laplacian operator is expressed as follows:

$$\nabla^2 f = \frac{\partial^2 f}{\partial x^2} + \frac{\partial^2 f}{\partial y^2} \quad (2)$$

Apply a Laplacian operator to a two-dimensional Gaussian function:

$$\nabla^2 G = \frac{\partial^2 G}{\partial x^2} + \frac{\partial^2 G}{\partial y^2} = \frac{-2\sigma^2 + x^2 + y^2}{2\pi\sigma^6} \exp\left(-\frac{x^2+y^2}{2\sigma^2}\right) \quad (3)$$

Thus, LOG is obtained:

$$LOG = \sigma^2 \nabla^2 G \quad (4)$$

In fact, when LOG operator is for image processing, Gaussian convolution can be used for smoothing firstly, and then the Laplacian operator is used for Laplacian convolution.

3.2. Algorithm flow of spot detection
The main process of spot detection includes image gray processing and blob detection through LOG, then maximum value filtering and background detection to detect image peaks in order to get the center position and radius of spots, and finally spot overlap determination according to the proportion of overlap area occupied itself area. The algorithm processing flow of light spot detection is shown in figure 3.

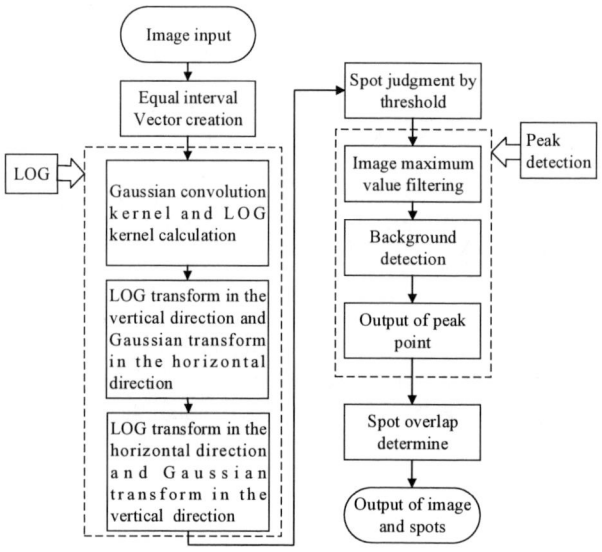

Figure 3. Flowchart of spot detection algorithm

3.2.1. Image input and scale transformation. The original image captured by the camera is the RGB image and it needs to be converted to the grayscale image. Since the spot size is not definite, the image is performed by multiple-scale LOG kernel to ensure detection effect. Therefore, the equal interval vector is constructed for the generation of different scale filters which depend on the convolution kernel parameter σ.

3.2.2. Detection through LOG. In order to simplify the calculation, Gaussian convolution and LOG convolution are performed in the horizontal direction and the vertical direction respectively and then the results in the two directions are summed. Taking spot detection with $\sigma=1$ as an example, the specific flow is as follow: firstly, LOG operator is performed on the vertical direction and Gaussian convolution is performed on the horizontal direction, the result is shown in figure 4(a); secondly, Gaussian convolution is performed on the vertical direction and LOG operator is performed on the horizontal direction, the result is shown in figure 4(b); finally, summation operation is performed and the result is shown in figure 4(c).

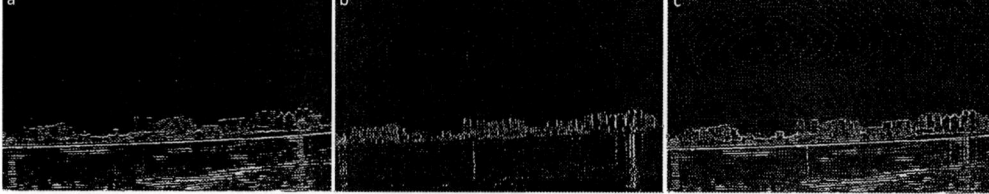

Figure 4. LOG convolution and Gaussian convolution results in the vertical direction and the horizontal direction

3.2.3. Spot determination by threshold. Since the nighttime inland video image contains background lights such as architectural lights, it is necessary to filter the detected non-ship spots. The gray threshold is set to make it. If the threshold is too low, the detected spots contain other light spots. And as the amount of spots increases, the tracking difficulty increases. On the other hand, if the threshold is too high, ship detection and tracking cannot be realized because the lights of ship cannot be detected. In this paper, the threshold is determined by the experiment.

3.2.4.Peak detection. Peak detection consists of three steps: maximum filtering, background detection and image erosion. Since the size of detected spots is small, the maximum filtering is utilized to increase it. Then according to the comparison of the image before and after maximum filtering, the background of the image is determined and eroded. Finally, the result of the maximum value filtering is compared with the background image after erosion and the peak is obtained which means the center point of the spot in the image are obtained.

3.2.5.Spot overlap determination. The flow of spot overlap determination is as follow: first of all, calculate the sum r_{sum} and the difference absolute value r_{diff} of the two spots' radius, calculate the distance d between the center points of the two spots. If $d <= r_{diff}$, it is indicated that the smaller spot is located in the larger spot so that the smeller spot can be disregarded; if $r_{sum} <= d$, it means there is no overlap. Otherwise, the overlapping area between the two spots need be calculated. The calculation of the overlapping area is shown in figure 5, where r_1, r_2 and d indicate the radius of two spots and the distance between the center points respectively. Based on this, the ratio of the overlapping area occupied each spot is calculated separately. If the ratio is greater than the threshold (0.1 is taken here), the spot is considered to overlap. The pseudo code of the overlapping area calculation is written as follows.

input：$x_1, y_1, r_1, x_2, y_2, r_2$, supposed $r_1 \geq r$

1： $d = \sqrt{(x_1 - x_2)^\wedge 2 + (y_1 - y_2)^\wedge 2}, r_{sum} = r_1 + r_2, r_{diff} = r_1 - r_2$

2： if $d \leq r_{diff}$, *then*

3： area$=\pi * r_2 * r_2$

4: else if $d \geq r_{sum}$

5： area=0

6: else

7： $\theta_1 = a\cos * ((d^2 + r_1^2 - r_2^2)/2 * r_1 * d)$

8： $\theta_2 = a\cos * ((d^2 + r_2^2 - r_1^2)/2 * r_2 * d)$

9： area$=2 * (0.5 * r_1^\wedge 2 * \theta_1 + 0.5 * r_2^\wedge 2 * \theta_2 - 0.25 * \sqrt{(d + r_1 + r_2) * (r_1 + r_2 - d) * (r_1 + d - r_2) * (r_2 + d - r_1)})$

10: end if

return area

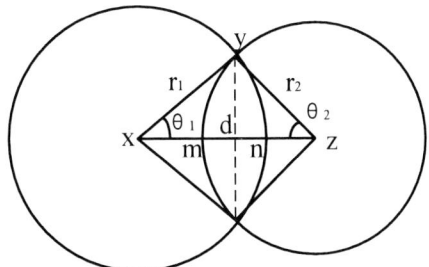

Figure 5. Schematic diagram of spot overlapping area calculation

4. Light spot tracking

After the light spot detection is implemented, the Kalman filtering is used to track light spots. Since the number of light spots detected in the image is more than one, it is necessary to accomplish multiple target tracking [13]. For the spots detected in a single frame image, Kalman filtering single target tracking is sequentially cycled, and then Kalman filtering multiple target tracking can be realized.

4.1. Kalman filtering

Taking a single spot tracking as an example, assuming x_k and y_k represent the position of the center of the spot respectively, v_x and v_y represent the speed in the direction x and y respectively, a_x and a_y represent the acceleration in the direction x and y. The vector X_k defined to describe the state of moving spot is as follow:

$$X_k = \left[x_k, y_k, v_x, v_y \right]^T \tag{5}$$

The observation vector Z_k is:

$$Z_k = \left[x_k, y_k \right]^T \tag{6}$$

On the basis of this, the state transition matrix, the controlling input matrix, the control value at the time K and the system measurement matrix are determined as follow:

$$A = \begin{bmatrix} 1 & 0 & dt & 0 \\ 0 & 1 & 0 & dt \\ 0 & 0 & 1 & 0 \\ 0 & 0 & 0 & 1 \end{bmatrix}, B = \begin{bmatrix} dt^2/2 & 0 \\ 0 & dt^2/2 \\ dt & 0 \\ 0 & dt \end{bmatrix}, U_k = \begin{bmatrix} a_x \\ a_y \end{bmatrix}, H = \begin{bmatrix} 1 & 0 & 0 & 0 \\ 0 & 1 & 0 & 0 \end{bmatrix} \tag{7}$$

After Kalman filtering parameters are obtained, the process of target tracking by Kalman filtering is to predict the state of the next moment and calculate the covariance matrix firstly, as shown in equation (8).

$$X_{k+1}' = A * X_k + B * U_k$$
$$P_{k+1} = A * P_k * A^T + Q \tag{8}$$

The state variable value matrix and the covariance matrix are obtained after the correction value is added by the update step, as shown in equation (9).

$$X_{k+1} = X_{k+1}' + K_{k+1} * (Z_{k+1} - H * X_{k+1}')$$
$$P_{k+1} = P_{k+1}' - K_{k+1} * H * P_{k+1}' \tag{9}$$

The Kalman gain coefficient calculation is shown in equation (10).

$$K_{k+1} = P_{k+1}' * H^T * \left(HP_{k+1}' H^T + R \right)^{-1} \tag{10}$$

After predicting and updating, as shown in equations (8) and (9), the correction value at the next moment is obtained. And the target tracking through Kalman filtering is implemented by cycling the two above steps continuously.

4.2. Multiple target tracking process based on Kalman filtering

Some major issues have to be resolved for multiple target tracking of the ship at night. Firstly, because the number of light spots detected is more than one, it is necessary to match them in each frame image. Secondly, due to the problem of flashing of lights, the spot detection is unstable, for example, the same spot may not be detected all the time in the adjacent frame images. At last, when the ship leaves the video surveillance area, the corresponding spots need to be deleted. For the first problem, the Hungarian algorithm [14] is adopted to match spots. For the second problem, Kalman filtering is performed on the tracking spots of the previous frame image, and then the observation result is matched with the measurement result through the Hungarian algorithm. According to the matching results, whether the spots are retained is determined. For the last problem, the new detected spots unmatched is recognized as new spots and the spots whose number of unmatched time reaches the threshold is set as spots that have left the video monitoring area and have to be deleted. The tracking process is illustrated in figure 6.

ISPECE   IOP Publishing

IOP Conf. Series: Journal of Physics: Conf. Series **1187** (2019) 042074   doi:10.1088/1742-6596/1187/4/042074

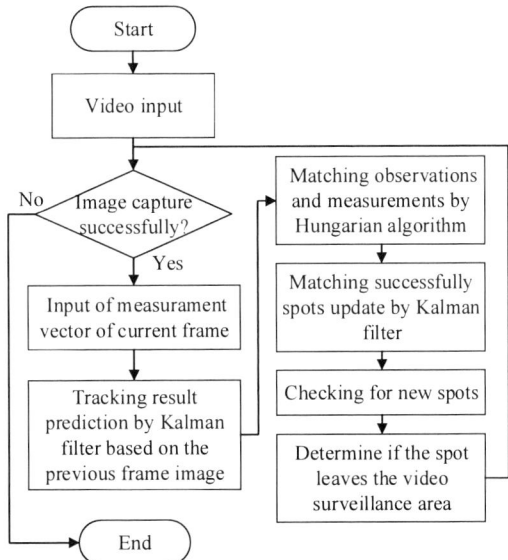

Figure 6. Flowchart of spots tracking algorithm

4.2.1. Prediction by Kalman filtering. The spot detected in the image of the previous frame are sequentially predicted according to the parameters of the Kalman filtering described above, and the covariance matrix and the Kalman gain coefficient are calculated.

4.2.2. Data match. The corresponding Euclidean distance between the measurement points and observation points is calculated, the Hungarian algorithm is used to match the two point sets according to the distance matrix. The distance threshold is set and the distance between the matching points is compared with the threshold to determine whether it is a valid match. If the measurement point is far from the observation point, the pairing is considered invalid.

4.2.3. Update by Kalman filtering. The effective matching points of observation based on the measurement points and the covariance matrix are updated.

4.2.4. Spot addition and deletion. Considering the possibility that new ships enter the video surveillance area, new spots need to be tracked. In the data matching process, in addition to the invalid match, the measurement points that are not matched are included and added to the current spot tracking data as new spots. At the same time, the unmatched points in the observation points are marked. If the time number of marking reaches the threshold, the spots may have left the video monitoring area and the observation points are deleted.

Spots tracking is achieved by iterating the above steps until the ship tracking is accomplished.

5. Case study and discussions

The VS2015 and OpenCV programming environment are chosen to realize ship detection and tracking in the nighttime surveillance video. Taking the ship surveillance video around 19 o'clock on March 22, 2018, as a case study of LSDT method verification, the image is 1920*1080, as shown in figure 1. The current time and camera label are marked on the upper left and lower right of the video screen due to the equipment, so the image is cropped before ship detection and tracking. The cropped image is shown in figure 7.

1807

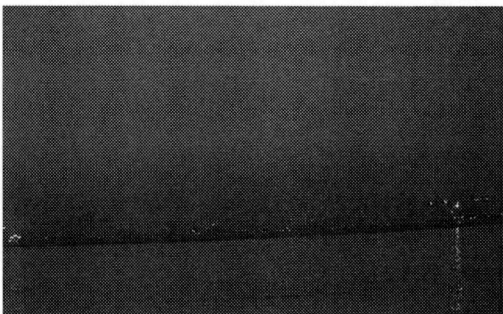

Figure 7. Image after cropping

5.1. Light spot detection

5.1.1.Spot detection results. According to the method flow, the spots are detected firstly. In order to make the ship tracking effect obvious, the images are taken every 30 frames in the video for light spot detection and tracking. The images are taken every 30 frames from the first frame to the 991st frame in the video, and the number of spot detections is as shown in table 1.

Table 1. The number of detected spots

Frame	Number	Frame	Number	Frame	Number	Frame	Number
1	9	271	7	541	7	811	9
31	8	301	7	571	8	841	7
61	9	331	7	601	10	871	9
91	8	361	8	631	9	901	8
121	8	391	7	661	8	931	8
151	10	421	12	691	8	961	7
181	10	451	6	721	8	991	10
211	9	481	10	751	9		
241	8	511	8	781	11		

The results of spot detection in 1st, 181st, 361st, 541st, 721st, and 901st frame are shown in figure 8. It can be seen that the number of detected spots is relatively stable. The number of ship light spots is 3 to 4 and other spots are coastal architectural light spots. It can be concluded that the stability of spot detection by this way is satisfactory.

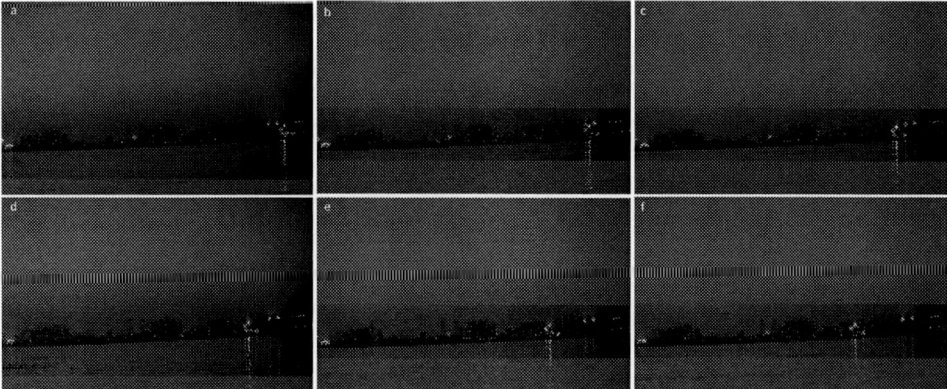

Figure 8. Spot detection results

5.1.2. Gray threshold setting. When ships sail in the inland river, the influence of coastal lights and reflected lights from the river surface increase the difficulties of ship light spot detection. In order to filter other spots, after the LOG convolution, the spots are filtered by the gray threshold. As shown in figure 9, when the threshold is set 60/70/80/90/100/110, the number of detected spots is 32/24/13/9/5/2. It is found through experiments that when the threshold is between 80 and 100, the spot detection result is more satisfactory; when the threshold is set to 100, the ship spots in some images are not detected. Therefore, 90 is taken as the gray threshold.

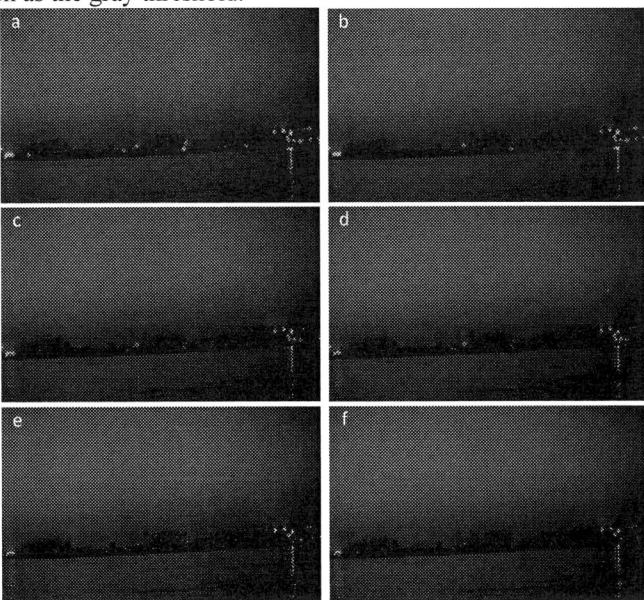

Figure 9. Spots detection results under the different gray threshold

5.2. Light spot tracking

5.2.1. Spot tracking results. According to the algorithm flow of multiple target tracking by Kalman filtering, the spots are tracked and the tracking results corresponding to figure 9 are shown in figure 10 (the distance threshold 10). The same spots in the adjacent frame images are connected by colored polylines.

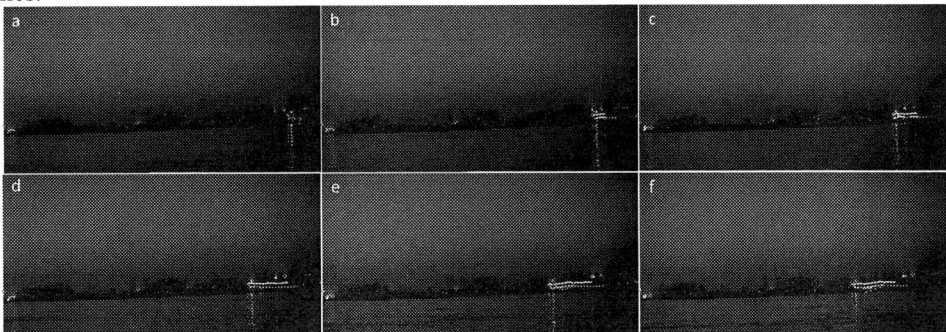

Figure 10. Spots tracking results

It can be seen that the method based the on LSDT can realize the goal of ship detection and tracking. The image tracking visual effect has been affected to some extent due to the fact that the number of spots detected on the same ship is not unique and the lights are flickering. However, it is feasible to determine ships and its running state at night through an ordinary camera by detecting the spots and tracking them.

5.2.2. Distance threshold setting. After matching the observation points with the tracking points by Hungarian algorithm, it is necessary to determine the results of match. If the distance between the matching points is greater than the distance threshold, it is considered to be an invalid match. In this paper, the distance threshold is obtained experimentally. The tracking result in 421st frame with distance threshold 5 is shown in figure 11(a) and the corresponding result in 871st frame with distance threshold 15 is shown in figure 11(b). It can be seen that when the distance threshold is 5, the same spots in adjacent frames is disconnected which affects the tracking result; when the distance threshold is 15, some unrelated spots are connected. Therefore, the distance threshold in this paper is set 10 and the corresponding tracking result is shown in figure 10.

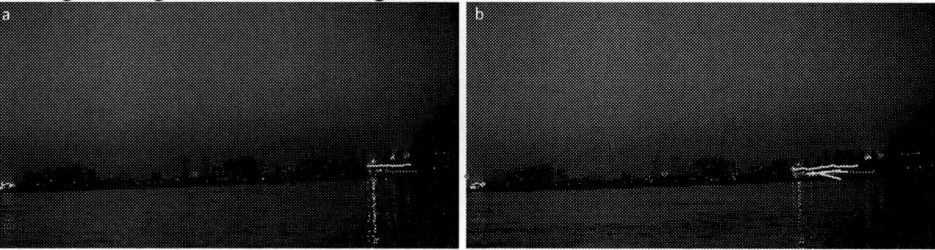

Figure 11. Distance threshold experiment results

6. Conclusions
In this paper, a method of LSDT for nighttime ship detection and tracking is proposed, and the experiment based on the actual monitoring video verifies the effectiveness of the method. So far, the previous researches in ship detection at night are mostly based on infrared camera and SAR. By using LSDT, we are capable of detecting and tracking light spots of inland ship with an ordinary camera. Further researches are still necessary for improving the proposed method, such as: how to filter the lights of background and how to convert multiple target tracking into single target tracking for multiple spots of the same ship to improve the tracking effect.

Acknowledgment
This paper is supported by the National Natural Science Foundation of China (No. 51709220 and No. 51479155) and Science and Technology Plan Project of Wuhan (No. 2017010201010132).

Reference
[1] Fefilatyev S and Goldgof D. Detection and tracking of marine vehicles in video. *International Conference on Pattern Recognition.* IEEE, 2012:1-4.
[2] Liu L, Liu X, Chu X, et al. Coverage effectiveness analysis of AIS base station: A case study in Yangtze River. *International Conference on Transportation Information and Safety.* IEEE, 2017: 178-183.
[3] Liu J, Wei H, Huang X Y, et al. A bridge-ship collision avoidance system based on FLIR image sequences. *Advances in Electrical Engineering and Computational Science.* Springer Netherlands, 2009: 123-133.
[4] Liu L, Zhang Y J, Liu J, et al. Application of infrared binocular vision for monitoring moving target in port. *Applied Mechanics & Materials,* 2014, 678: 155-161.
[5] Tello, M., C. Lopez-Martinez and J.J. Mallorqui. A novel algorithm for ship detection in SAR imagery based on the Wavelet Transform. *IEEE Geoscience and Remote Sensing Letters,* 2005. **2**(2): p. 201-205.
[6] Marino, A. A notch filter for ship detection with Polarimetric SAR data. *IEEE Journal of Selected Topics in Applied Earth Observations and Remote Sensing,* 2013, **6**(3): 1219-1232.
[7] Elvidge C, Zhizhin M, Baugh K, et al. Automatic boat identification system for VIIRS low light imaging data. *Remote Sensing,* 2015, **7**(3):3020-3036.

[8] Yamaguchi T, Asanuma I, Park J G, et al. Temporal monitoring of vessels activity using day/night band in Suomi NPP on South China Sea. SPIE. 2017:101860K.

[9] Charette R D and Nashashibi F. Real time visual traffic lights recognition based on Spot Light Detection and adaptive traffic lights templates. *Intelligent Vehicles Symposium. IEEE*, 2009: 358-363.

[10] He Q and Yan L. Edge detection algorithm based on LOG and canny operator. *Computer Engineering*, 2011(3): 210-212.

[11] Cai D, Duan X H and Gao H Z. Detection and tracking of water sports vessels. *Computer & Digital Engineering*, 2017, **45**(7):1313-1317.

[12] Chen J, Chen GH, Shi LH, et al. Edge Detection technology in image tracking. *China Optics*, 2009, **02**(1): 46-53.

[13] Zhao GH, Zhuo S and Xu X L. Multi-target tracking method based on Kalman filter. *Computer Science*, 2018(8).

[14] Zhu M, Liang D, Fan Y Z, et al. Image matching algorithm based on spectral features. *Journal of South China University of Technology (Natural Science Edition)*, 2015(9): 60-66.

ISPECE IOP Publishing

A novel approach to multi-resolution technique for fast pattern recognition

Xiaochun Liu[1], Xiaohua Ding[2], Xiang Zhou[2], Wen Cui[3]

[1] Department of Military Aerospace, College of Aerospace Science and Engineering, National University of Defense Technology, Changsha, 410073, China

[2] Shenzhen Eagle Eye Online Electronic Technology Co. Ltd, Shenzhen, 518100, China

[3] 96656 Troops of Rocket Army, Beijing, 100000, China

Abstract. Multi-resolution is a commonly used acceleration algorithm for pattern search. However, there still exist two unsolved questions: "how many levels of image pyramid should be built" and "how to propagate the badly degraded truth patterns to the next pyramid level". We address these two questions through a novel distinction analysis and propagation strategy based on multimodality analysis. Experimental results show that this method has good practicability and advantage.

1. Introduction

By reducing the area to be searched and the number of pixels used in pattern matching algorithm, multi-resolution technique is able to speed up the computational efficiency of pattern matching dramatically. Multi-resolution technique has been commonly used and also played an important role in many real-time pattern matching applications. Important applications of multi-resolution include image matching [1]-[4], pattern recognition [5]-[8], motion estimation [9]-[11], frame interpolation [12], image registration [13] and Fractal decoding [14]. A multi-resolution technique usually works as follows: Upon building the image pyramids for both the pattern template and searching image, the first search is conducted with the most compressed pattern template and searching image.

The resulting location provides a coarse location of the truth pattern template to the next level of the searching image. Therefore, instead of performing a complete search in the next level, one requires to only search a close vicinity of the coarse location obtained from the previous search. This sequence is iterated until the search in the original image is completed. One disadvantage of building image pyramid is the loss of detail information. In some cases, the truth and false corresponding templates only differ in relation to details, which may render the template unrecognizable for template matching algorithms when running at the most compressed level of image pyramid. Thus, a novel distinction analysis is presented to measure the self-distinction between the template and its neighboring templates, and then determine whether to continue building image pyramid or not based on the value of self-distinction. In practical use, the truth corresponding point maybe degrade to the second best matching point (or even worse) because of the loss of detail information, which means the conventional method that propagates only one candidate point may not find the truth point. To address this problem, we present a multimodality detection that can provide a set of rather than only one candidate points to the next level.

Content from this work may be used under the terms of the Creative Commons Attribution 3.0 licence. Any further distribution of this work must maintain attribution to the author(s) and the title of the work, journal citation and DOI.
Published under licence by IOP Publishing Ltd

The paper is structured as follows. In section 2, the definition and procedures of distinction analysis are described, and some experimental results are also shown to demonstrate our idea. The propagation strategy based on multimodality detection is described, and comparison experiments are also given in section 3. Finally, concluding remarks are provided in section 4.

2. Distinction Analysis

Because of image filtering and scale reduction when building image pyramid, the detail information of pattern template will be inevitably lost, which may render the template too ambiguous to recognize. To answer the question "how many levels should be used to construct the image pyramid to keep the template recognizable", a novel method that determines the proper number of level according to the characteristics of template as a heuristic manner is described as follows (as shown in Fig.1): (1) conduct self-distinction analysis using the template and its neighboring templates whose centers are equally spaced on a circle of radius R as shown in Fig.2; (2) If the template on the current level is still recognizable, the self-distinction value will be relatively large, and then we can decide to continue building the next image pyramid. Otherwise, end the iteration.

Inspired by the work in the literatures [16, 17], we define the distinction measurement using the template and its neighboring templates. The definition of distinction measurement on the 1st pyramid level goes as follows. (1) As shown in Fig.2, select the center of template as the center of circle, and select other neighboring templates whose centers are equally spaced on the circle of radius R. (2) In this paper, we employ the zero-mean normalized cross-correlation (ZNCC) as the similarity measure function which is a contrast invariant variation. Let f and g denote the template and its neighboring template respectively. The definition of ZNCC can be expressed as follow:

$$ZNCC(u,v) = \frac{\sum_{i=1}^{m}\sum_{j=1}^{n}\left(f(x+i,y+j)-\overline{f}\right)\cdot\left(g(u+i,v+j)-\overline{g}\right)}{\sqrt{\sum_{i=1}^{m}\sum_{j=1}^{n}\left(f(x+i,y+j)-\overline{f}\right)^2}\sqrt{\sum_{i=1}^{m}\sum_{j=1}^{n}\left(g(u+i,v+j)-\overline{g}\right)^2}}, \quad (1)$$

where: $\overline{f}$ and $\overline{g}$ denote the mean value of f and g, m×n denotes the size of template, (x, y) and (u, v) are the beginning points of f and g respectively. (3) Using Eq.1, get the ZNCCs between the pattern template and its neighboring templates. The definition of Distinction Measurement (DM) can be expressed as Eq.2,

$$DM = \left(1 - \max\left(ZNCC\right)\right)/2. \quad (2)$$

ISPECE   IOP Publishing

IOP Conf. Series: Journal of Physics: Conf. Series **1187** (2019) 042075   doi:10.1088/1742-6596/1187/4/042075

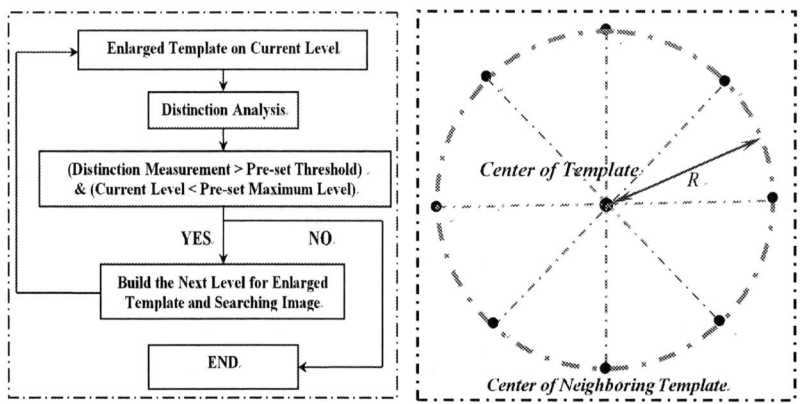

Fig.1. Determination of image
pyramid based on distinction

Fig.2. Calculation of distinction
measurement

If the pattern template is completely distinctive, the value of DM is equal to 1.If we define R as the radius of the circle on the 1st level, the radius on the kth level can be defined as Eq.3,

$$R_k = \frac{R}{2^{k-1}}.$$ (3)

From Eq.3, although each level is given a different value of R, the equivalent radius keeps invariant for all levels when converting to the 1st level. To save more computational time, the enlarged template instead of the whole image is adopted to calculate DM.

2.1. Experiments of Distinction Analysis

Because Gaussian pyramid is able to efficiently suppress the high-frequency noise caused by pixel interpolation while constructing pyramid and provide much of the information in the coarser scales, in this paper we adopt the Gaussian pyramid rather than four points mean pyramid as multi-resolution image representation [15]. (Our computation is slightly more than four points mean pyramid.)

The results of distinction analysis are shown in Fig.3-4 and Table 1. The pre-set threshold of distinction measurement is given a value of 1/3. How many levels should be used for image pyramid is decided according to the characteristic of the pattern template. For example: the first template including a person is more distinctive than the second template, thus it can afford more levels of image pyramid, and has larger value of distinction (as shown in Tab.1). From Table 1, we can see that the distinction values of the template 1 and 2 all decrease as the increase of pyramid level, which agree with our theoretical analysis above.

3. Propagation Strategy Based On Multimodality Detection

3.1. Theoretical Analysis

The pattern template normally becomes increasingly hard to recognize as the pyramid level increases. In some cases, the resulting location propagated from the most compressed level may be wrong, which will lead the total matching fail because of cascade mode. To address this problem, a multimodality-based propagation strategy is proposed to improve the conventional algorithm to handle multimodality situations that are quite common in practical applications. Actually, the conventional multi-resolution search relies on the "unimodality assumption". As a result, its applicability is severely limited to the situations when the pattern templates are highly unique. However, it is common that a template has multiple similar candidate templates that only differ in relation to detail information in the most compressed search area, which will produce a correlation matrix with multiple peaks. In this situation,

1814

the truth location may be corresponding to any one of peaks of correlation matrix, which obviously violate "unimodality assumption". The traditional propagation strategy, which propagates only one candidate location corresponding to the first highest peak, will inevitably cause wrong matching.

As shown in Fig.5, we first choose a template from a visible image, and then conduct the search in an infrared search area. The size of template and search area are 101×101(pixels) and 201×201(pixels) respectively. The correlation matrix is calculated according to ZNCC. The template matching is conducted on different pyramid level successively. From the images of Fig.5, we can see on the 1st level, the truth location corresponds to the first highest peak of correlation matrix since relatively rich detail information. However, as the loss of detail information when constructing pyramid, the truth location begins to degrade (the second highest peak on 2nd level and the third highest peak on 3rd level). To solve the problem, one can propagate those candidate locations whose values of ZNCC are larger than a pre-set value, and then conduct the search at the vicinity of those locations on the finer pyramid level [5]. This approach may have better performance as more candidate locations have been provided. However, there exist two main problems in this solution. If the first highest peak that may correspond to false location is relatively strong, the candidate locations will gather at its vicinity, which may lead the truth location fail to be propagated. In addition, the approach often provides too many candidate locations, and increases the computational time too much.

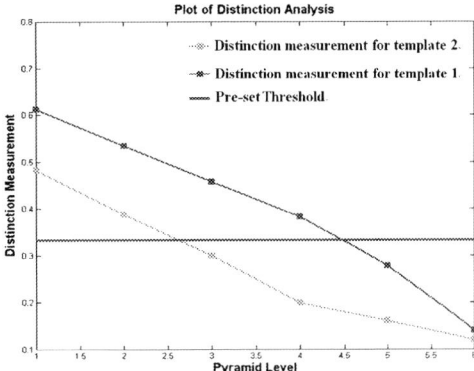

Fig. 3. Distinction analysis for the pattern template1 and pattern template 2

In this work, a new propagation strategy based on "multimodality assumption" is presented to find the truth location that may degrade. Our method begins with a reasonable assumption that the truth template is still one of templates similar to the template no matter how it has degraded. Because there normally exists a peak that is formed by the similar template in the correlation matrix, in order to propagate truth locations, we just need find all locations of peaks. Hence, the multimodality detection is presented to detect peaks and find their locations in the correlation matrix. The value of peak is normally higher than its neighbors, which means it corresponds to a local maximum value. To find those peaks, each value of correlation matrix is compared to its neighbors within a circle of radius R in the correlation matrix. It is selected only if it is larger than all of these neighbors. If we define the ZNCC(i, j) and ZNCC(i+m, j+n) are the value of location(i, j) and (i+m, j+n), the multimodality detection can be expressed as follow:

$$ZNCC(i, j) \geq ZNCC(i + m, j + n), \quad (\sqrt{m^2 + n^2} \leq R). \tag{4}$$

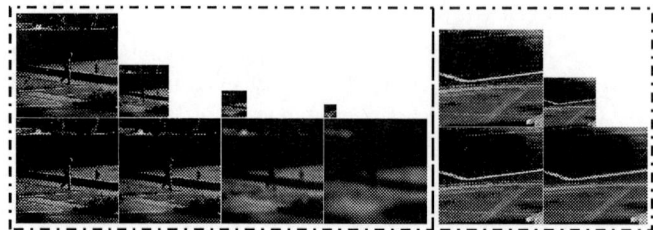

Fig. 4. Examples for constructing image pyramid based on distinction analysis. The images in the left and right rectangle are the results for the first and second pattern template respectively. For each group, upper row is the Gaussian pyramid for each level. Bottom row is the Gaussian pyramid expanded to the size of the original image.

To save computational time, one can choose a pre-set threshold (CT) to constrain the number of candidate locations. Let Cmax denote the maximum value of correlation matrix, PVR denote the Peak Value Ratio, then a threshold namely CT can be defined as follow:

$$C_T = C_{max} \times PVR. \tag{5}$$

One doesn't need to reset CT when the template and/or search area changes, because the value of the CT can be adjusted according to the characteristic of correlation matrix. One can also set a threshold (NT) to constrain the number of candidate locations. We first need to sort the value of ZNCC at the candidate locations as a descending order, and then a location is selected only if its value of ZNCC ranks among the top NT. Using this Method, we can make sure the candidate locations and the peaks are one-to-one correspondences, thus we can avoid those candidate locations gathering at vicinity of some strong peak. In this work, the radius R, the Peak Value Ratio PVR and the threshold of the number are experimentally given a value of 5, 0.75, and 3 respectively.

3.2. Experiments for Multimodality Detection

We have evaluated the performance of the proposed algorithm using three multi-sensor image pairs and many pattern matching experiments. Because in our experiments all the truth locations are not corresponding to the highest peak of correlation matrix on the most compressed level, the traditional method, which propagates only the highest peak, cannot handle this situation. Therefore, we only need to compare the results obtained from the proposed algorithm with those from previous method presented in paper [5]. In order to justify the improved propagation strategy based on the multimodality detection, the two methods were both used to get the number and locations of similar templates for each experiment with the predetermined PVR or NT respectively. The experimental results are shown in the table 2 and 3. The matching images were well registered before the template matching. The original size of template is 101×101(pixels).

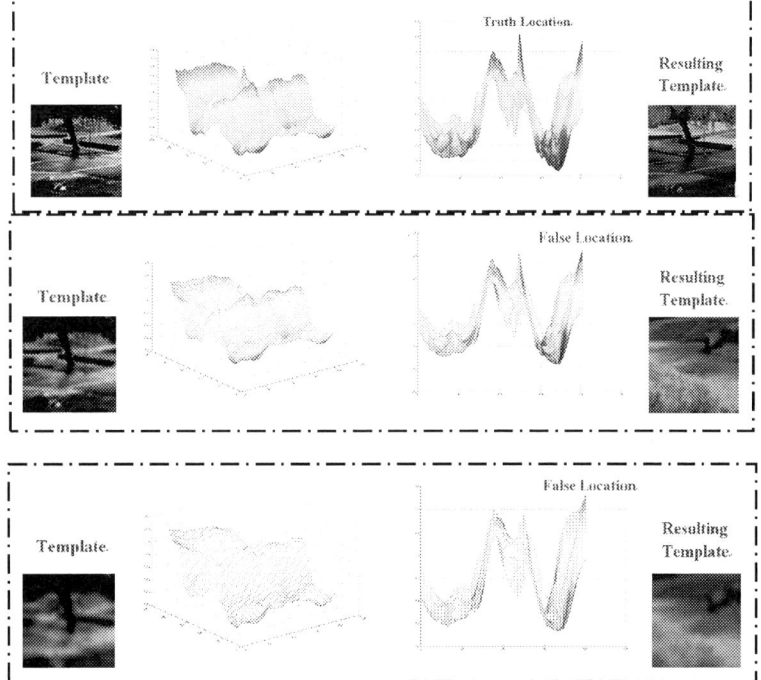

Fig. 5. The results of template matching on different levels of Gaussian pyramid. The top row is the visible pattern image (left) and the infared search image (right), and the resulting locations for different levels are denoted in the search image. The results of different pyramid levels are shown from the second row to the fourth row. In each row, from left to right: (1) pattern template which are expanded to the original resolution, (2) the three dimensional correlation matrix, (3) the correlation matrix viewed from X to Y, (4) the resulting template.

From table 2, we can get the conclusion that: for the given pre-set threshold NT, the candidate locations obtained from the previous method gather at the vicinity of some false but strong peaks, the truth location is almost not propagated; however, because the candidate locations obtained from the proposed method are a one-to-one correspondence with the peaks of correlation matrix, it can propagate the truth location robustly. From table 3, we can get the conclusion that: for the given pre-set threshold PVR, both methods are able to propagate the truth location, but the number of candidate locations obtained from our method is much less than the previous method. The mean number of our method is 3.4, whereas the number of the previous method is 155.7. Apparently, our method can save more computational time. Note that: only one failure occurred because the truth location has degraded to the sixth highest peak. If NT is given a value of 6, we will also propagate the truth location correctly.

Table 2. Comparisons between our method and previous method when NT = 3.

Image pair	Candidate number	True/False	Candidate number	True/False
1	3	True	3	False
1	3	True	3	False
2	3	False	3	False
2	3	True	3	False
2	3	True	3	False
3	3	True	3	True
3	3	True	3	False

Table 3. Comparisons between our method and previous method when PVR = 0.75.

Image ID	Candidate number	True/false	Number (TMRT)	True/false
1	4	True	658	True
1	4	True	92	True
2	6	True	71	True
2	2	True	14	True
2	4	True	158	True
3	2	True	57	True
3	2	True	40	True

4. Conclusion

A novel approach to multi-resolution technique for fast template matching is proposed in this paper. There are two main contributions in this work. (1) We present a distinction analysis to decide how many levels should be built in the image pyramid according the characteristics of the pattern template. (2) We present a propagation strategy based on multimodality detection. Using this strategy, we can propagate the truth location even it has degraded badly. In addition, through two thresholding scheme, we can constrain the number of candidate locations. The experimental results demonstrate that our method can propagate the truth location efficiently and robustly even in the challenging situation, and has a significant advantage over the previous methods.

References

[1] Sabyasachi Pal, Titas De, Mainak Sen, Arnab Chatterjee: Grey-Scale Template Image Matching by Area Based Matching Techniques. In IECON 2011 organized in collaboration with IEEE on 5th&6th, (2011).

[2] E H Adelson, C H Anderson, J R Bergen, P J Burt, and J M Ogden: Pyramid methods in image Processing. RCA Engineer, vol. 41, pp. 29-33, (1984).

[3] Ryo Takei: A New Grey-Scale Template Image Matching Algorithm Using the Cross-Sectional Histogram Correlation Method. (2003).

[4] Lin, S.J., Chung, K.L., Chang, L.C.: An improved search algorithm for vector quantization using mean pyramid structure. Pattern Recognition Letter. 22 (3/4), pp.373–379(2001).

[5] Jingge Wu, Jordan Kuhn: Pattern matching method using a pyramid structure search, image detecting device, and computer-readable medium. United States Patent, Patent No.: 7215816B2 (2007).

[6] Sheikholeslami G., Chatterjee S., Zhang A. Wave Cluster: A Multi-Resolution Clustering Approach for Very Large Spatial Databases. In Proc. 24th Int. Conf. on Very Large Data Bases, New York, NY, pp. 428 – 439, (1998).

[7] W.J. MacLean and J.K.T. Tsotsos: Fast pattern recognition using gradient-descent search in an image pyramid. In Proceedings of the 15thInternational Conference on Pattern Recognition, pp. 877-881(2000).

[8] Lee, C.H., Chen, L.H.: A fast search algorithm for vector quantization using mean pyramids of code word. IEEE Transactions on Consumer Electronics, vol. 43 (2/3/4), pp.1697–1702(1995).

[9] Jong-Nam Kim and Tae-Sun Choi: Adaptive matching scan algorithm based on gradient magnitude for fast full search in motion estimation. IEEE Transactions on Consumer Electronics, vol. 45(3), pp. 762 – 772(1999).

[10] G. de Haan: Progress in motion estimation for video format conversion," IEEE Transactions on Consumer Electronics, vol. 46, pp. 449–459(2000).

[11] B Jahne, H Haussecker, and P Geissler: Bayesian Multi-Scale Differential Optical Flow. Handbook of Computer Vision and Applications, vol. 2, chapter 14, pp. 297-422, Academic Press (1999).

[12] B.W. Jeon, G.I. Lee, S.H. Lee and R.H. Park: Coarse-To-fine Frame Interpolation for Frame Rate Up-Conversion Using Pyramid Structure. IEEE Transactions on Consumer Electronics, vol. 49(3), pp. 499-508 (1997).

[13] Philippe Th´evenaz, Urs E. Ruttimann, and Michael Unser: A Pyramid Approach to Subpixel Registration Based on Intensity. IEEE Transactions on Image Processing, vol. 7(1), pp. 27-41(1998).

[14] Hyun-Soo Kang, Seong-Dae Kim: Fractal decoding algorithm for fast convergence. Optical Engineering, vol. 35(11), pp.3191-3198(1996).

[15] Burt and Adelson: The Laplacian Pyramid as a Compact Image Code. IEEE Transactions on Communications, vol. COM-31, no. 4, pp. 532-540(1983).

[16] John E.Hutchinson: Fractals and Self Similarity. Indiana University Mathematics Journal, Vol.30 (5), pp. 713-747(1981).

[17] Mark E. Crovella and Azer Bestavros: Self-Similarity in World Wide Web Traffic: Evidence and Possible Causes. IEEE/ACM Transactions on Networking, vol. 5(6), pp. 835-846 (1997).

ISPECE

IOP Publishing

Research on PLC Information Model Based on UML Class Diagram

Xu Ran Yu, Yun Xiao Zu*, Wei Hai Li

School of Electronic Engineering, Beijing University of Posts and Telecommunications, Beijing 100876, China

zuyx@bupt.edu.cn

Abstract. In the industrial control system, there are a variety of PLC, which are made of different manufacturers and with various types, working in the operation line for different enterprises. Due to the different communication protocols and information exchange standards, it is difficult to achieve information sharing and unified management, and easily leading to "information island". In this paper, the PLC information model is put forward to design the information standard, and further to realize the standardization of data collection and industrial information, system integration and information sharing, and then set up the industrial control information standard. The modeling and drawing principle of PLC information model based on UML is introduced, and the modeling and drawing method of PLC information model based on EA UML modeling tool is presented.

1. Introduction

PLC has high reliability and strong anti-interference ability, and has been widely used in industrial control field, having become one of the three pillars of modern industrial automation [1]. In the industrial control sys-tem, there may be cases where various types of PLCs produced by various manufacturers operating in production lines of different enterprises. Since various PLCs adopt different communication protocols and information exchange standards, and the compatibility is poor, it is difficult to achieve information sharing and unified management, and it is easy to form an "information island" [2]. The current problem of non-opening, non-interworking, and difficult interconnection of PLC has become a paradox or bottleneck restricting the development of smart manufacturing. To this end, it is necessary to follow the unified standards and build a unified industrial control information platform for the development of sensing, measurement, control, communication, network, cloud computing, and big data in PLC and even industrial control fields. The compilation of information standardization standards in the industrial control field includes information standards, interface specifications, information security rules, and so on. The PLC information model proposed in this paper is to design information standards, standardize data collection, standardize industrial control information, and facilitate system integration and information sharing.

The existing industrial control information related standards mainly describe the system structure system, but lack the information system architecture, information modeling and communication interface standards. Take the national standard "GB/T 15969 Programmable Controller" as an example. This standard only reflects the basic elements and usage methods of the equipment, but does not explain the information representation between PLC and equipment, PLC and network, PLC and MES system. Methods, information exchange methods, information security and other issues.

Content from this work may be used under the terms of the Creative Commons Attribution 3.0 licence. Any further distribution of this work must maintain attribution to the author(s) and the title of the work, journal citation and DOI.
Published under licence by IOP Publishing Ltd

2. Material and Methods

UML provides a set of graphical symbols for modeling object-oriented systems [3]. Class Diagram is the most common and important diagram in object-oriented system modeling. It is the basis for defining other graphs. It is mainly used to display the classes, interfaces and static structures and relationships between them. a static model. The relationships between classes are generalized, related, dependent, and implemented.

2.1 Several relationships between classes

2.1.1. Generalization. Generalization is an inheritance relationship that represents a general and special relationship, and the sub-class inherits all the details of the parent class. In the class diagram, a solid line with a hollow arrow is used, and the arrow points from the subclass to the parent class.

2.1.2. Implementation (Realization). Implementation is a relationship between a class and an interface, indicating that the class is the implementation of all the features and behaviors of the interface. In the class diagram, the dotted line with a hollow arrow is indicated, and the arrow points to the interface.

2.1.3. Association. An association is a connection between a class and a class. It makes a class know the properties and methods of another class and indicates the connection between the objects of the thing. The class diagram is represented by a solid line with a common arrow. The two-way association can have two arrows or no arrows, and the one-way association has an arrow.

The aggregation relationship is a relationship between the whole and the part. Even if there is no whole, the part can exist separately. The class diagram is represented by a hollow diamond, and the diamond points from the local to the whole.

Combinations are whole and part, but parts cannot exist separately and exist separately. The class diagram is represented by a solid diamond, and the diamond points from the local to the whole.

Multiplicity is often used in associations, aggregations, and combinations to represent how many associated objects exist. Expressed by "number: asterisk (number)".

2.1.4. Dependency. A dependency is a dependency between two or more classes. The class diagram is represented by a dashed line with an arrow pointing to the dependent element.

2.2 PLC information model

In UML modeling, class diagrams can be used to represent how different entities are related to each other. UML modeling can make it easier to abstract the relationship between classes and classes by using forms such as class diagrams. The information model proposed in this paper is based on the analysis of PLC data collection requirements. Based on the object-oriented thinking, seven classes are abstracted from the basic properties of PLC, and the relationship between these classes is analyzed and determined.

Based on the object-oriented technology, the UML class diagram is used to build the PLC information mod-el. This paper proposes the information model modeling flow chart shown in Figure 1, and divides the UML class diagram modeling into six steps. The first four steps are detailed as follows.

(1) Analyze the problem domain and determine the requirements:

The standard is formulated with reference to the IEC information model, which describes the information model in an object-oriented manner and abstracts the information through classes. The basic components of the PLC [4] include the central processing unit (CPU), memory, input/output interfaces (I/O, including input interfaces, output interfaces, external device interfaces, expansion interfaces, etc.), external device programmers, and power modules. The internal components of the PLC are connected by the power bus, the control bus, the address bus and the data bus, and the external device is configured according to the actual control object to form the PLC control system [5].

The demand is mainly to obtain the classes required for modeling, the parameters of the class, the attributes, methods, and the relationships between the classes from the basic components of the PLC and the basic attribute information. Then, verify the information needed for modeling according to the existing standard specifications, and make the necessary supplements, and finally classify the information into 7 standard classes.

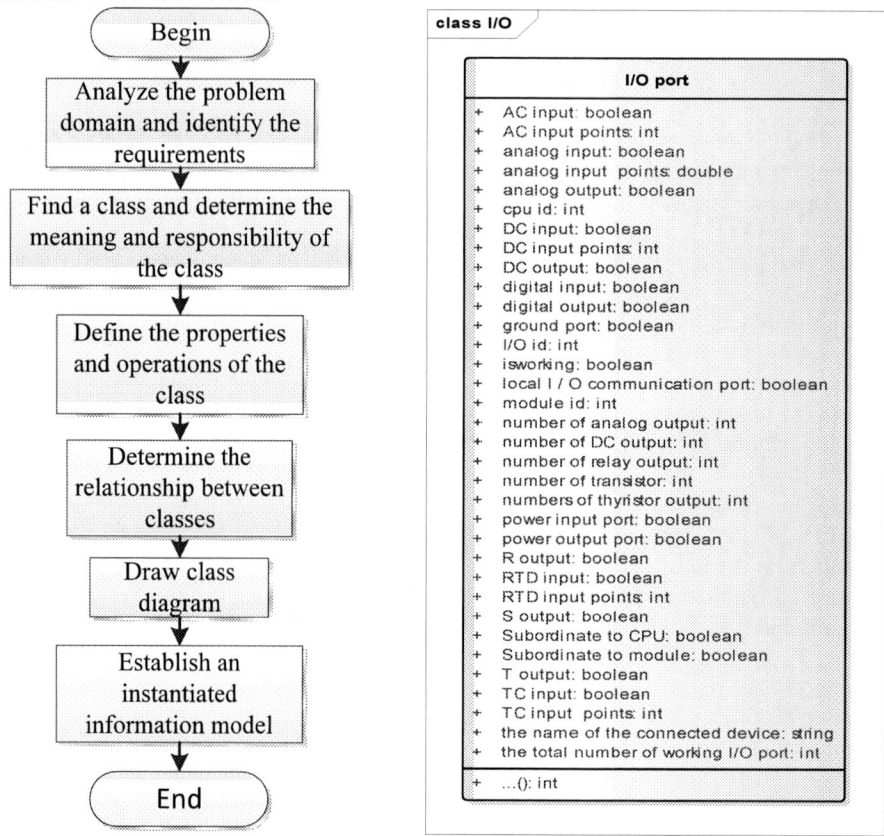

Figure 1. Information model modeling flow chart.

Figure 2. The properties and operation of I/O extended module.

(2) Find the class and determine the meaning and responsibilities of the class:

Based on the functions of the system, these functions are analyzed and summarized, and the relationship between the functions is clarified. The basic attributes of the PLC are divided into seven categories, namely CPU module, temperature control module, I/O port module, I/O expansion module, register module, power module, and communication module.

(3) Define the properties and operations of the class:

The I/O port module mainly realizes the input interface to convert various control signals input from the field control or detection component to the PLC into signals that the CPU can receive and process, and output interface circuit converts the weak electric control signal sent by the CPU into a strong electric signal required in the field [6]. Its properties and operations are shown in Figure 2.

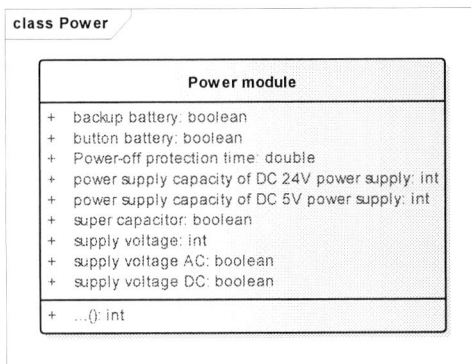

Figure 3. The properties and operation of power module.

Figure 4. The properties and operation of communication module.

The power module mainly converts the externally supplied AC power into DC power required for the CPU, memory, etc., and its attributes and operations are shown in Figure 3.

The communication module [7] is used for program and data exchange between the programmable controller and other programmable controllers or any devices in an external device or automation system. It provides functions such as device verification, data acquisition, alarm reporting, program execution and I/O control, application transfer, and connection management between external devices and PLC signal processing units. Its properties and operations are shown in Figure 4.

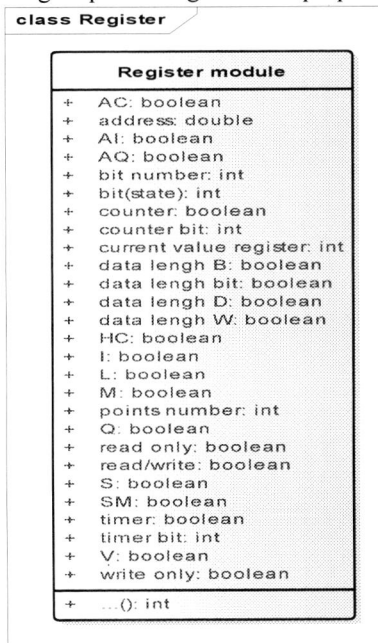

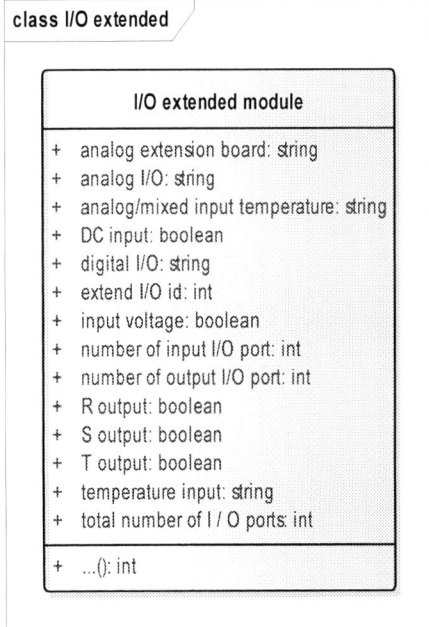

Figure 5. The properties and operation of register module.

Figure 6. The properties and operation of I/O port module.

Programmable controllers require a variety of logic devices and computing devices with different functions in software design, called programming components. These programming components are used to complete the logic operations, arithmetic operations, timing, counting functions given by the program, and the name of the relay [8] is used and named according to their functions, such as input relays, output relays, timers, counters. Wait. However, these devices are not real physical devices, but

ISPECE   IOP Publishing

IOP Conf. Series: Journal of Physics: Conf. Series **1187** (2019) 042076   doi:10.1088/1742-6596/1187/4/042076

are composed of electronic circuits and memories. Each programming device corresponds to a memory cell of the component image register in the PLC memory, such as timer T, counter C, auxiliary relay M, The state device S and the like are all composed of memories. When the PLC is working, the data in the register is read by changing the state of the register. Its properties and operations are shown in Figure 5.

The I/O expansion module is used to connect the expansion module to the basic unit, so that the configuration of the PLC is flexible to meet the needs of different control systems. When each function module is connected to the PLC host, it can be simply plugged in, which is flexible and convenient. When the host I/O point does not meet the requirements of the control system, various I/O modules can be expanded as needed. Its properties and operations are shown in Figure 6.

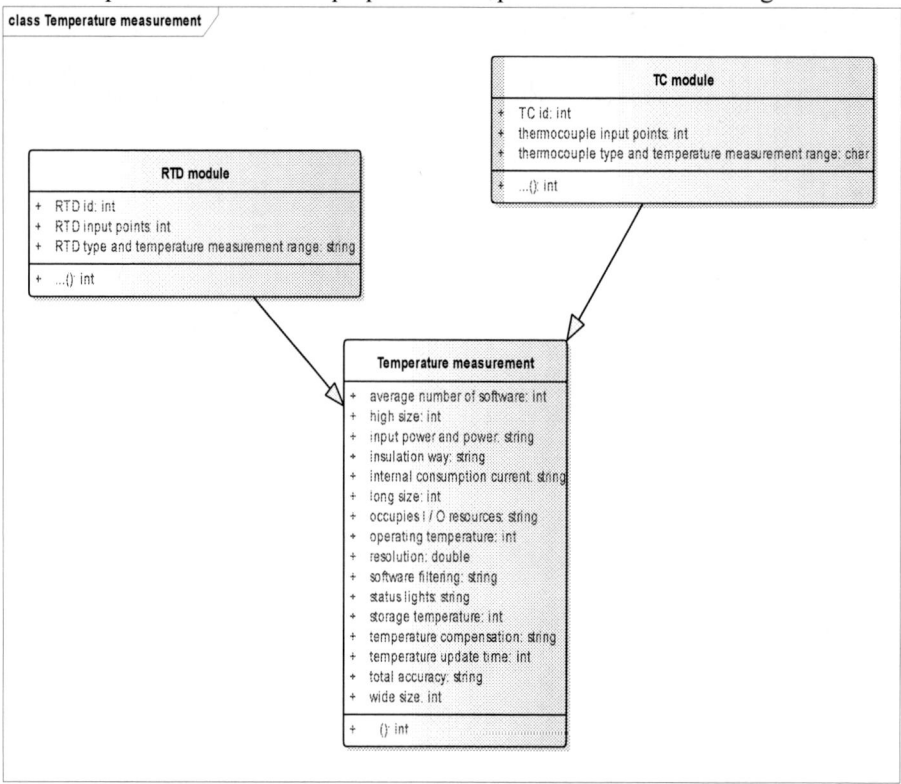

Figure 7. The properties and operation of temperature module.

The temperature measurement module is used to measure and display the temperature of the process control, including the thermocouple temperature measurement module TC and the platinum resistance temperature measurement module RTD, and the temperature measurement input signal from the process control is converted into a digital quantity of a certain number of bits for PLC operation and processing. Its properties and operations are shown in Figure 7.

(4) Determine the relationship between the classes

The CPU module is the core class of the whole model. The power supply, I/O port, register module and CPU module are combined. The temperature control, communication, I/O expansion module and CPU module are related, thermocouple, platinum resistance temperature control module. It is a generalization relationship with the total temperature control module.

3. Results

An example of a UML class diagram of the PLC information model is shown in Figure 8. The CPU module is the core class of the entire model, which processes the signals acquired from the sensors and the internal data memory according to the application, generating signals to the actuator and the internal data memory.

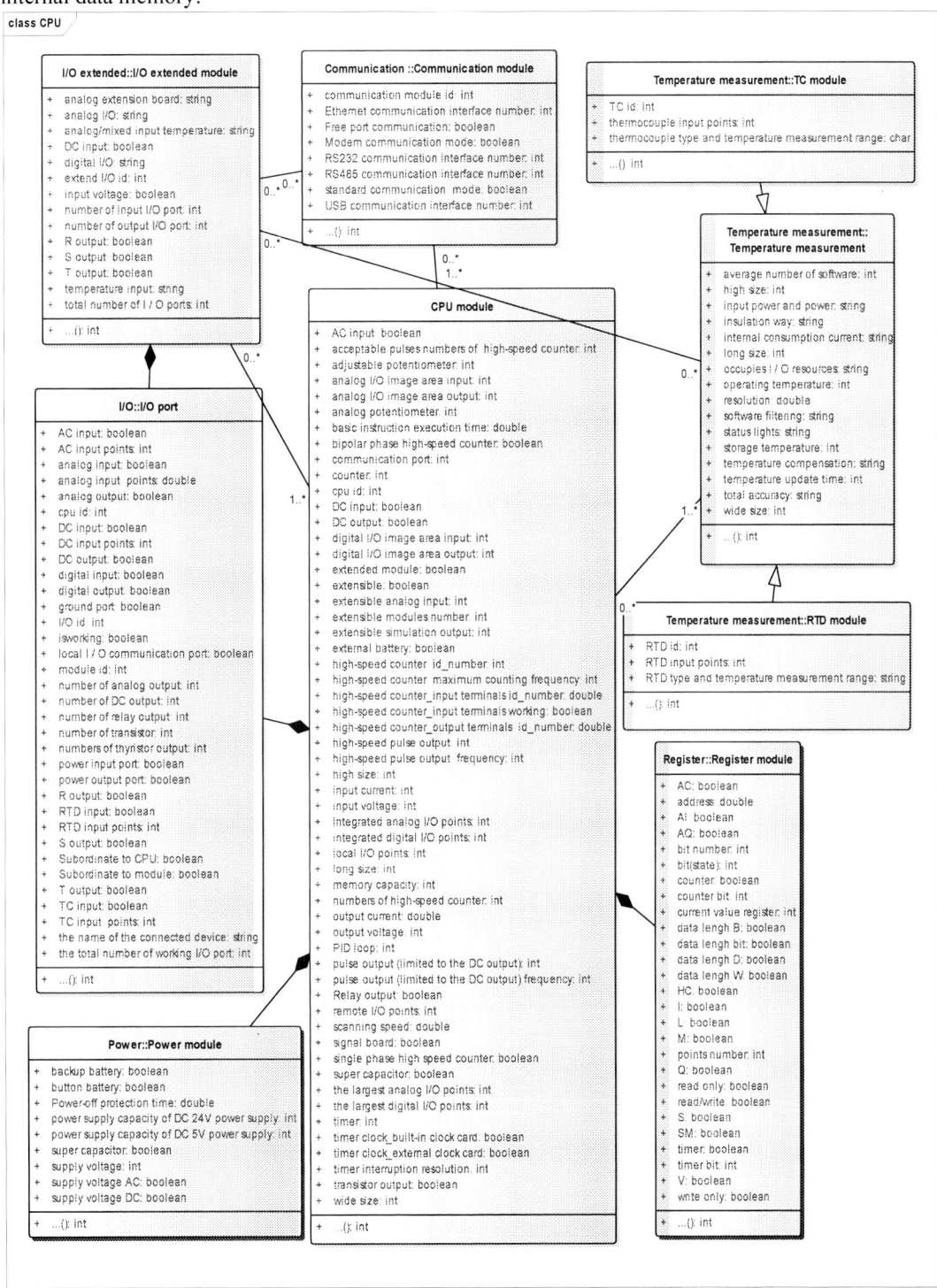

Figure 8. UML Class Diagram Example of PLC Information Model.

4. Conclusions

The PLC information model proposed in this paper describes the information model in an object-oriented manner, and abstracts the information through the class to complete the design of the information standard in the industrial cloud platform. The unified information representation mode in the constructed industrial control information platform is compatible with various PLCs, which is convenient for industrial system integration and information sharing. Based on this, the definition of information interface between modules, the design inter-face specification can be further clarified, the PLC information model can be realized, and the standardization of information exchange can be realized, which provides basic conditions for interoperability and information fusion of industrial control cloud terminals in the future.

Acknowledgments

This paper is supported by Beijing Key Laboratory of Work Safety Intelligent Monitoring (Beijing University of Posts and Telecommunications).

References

[1] Shi, L., Hua, B., Zhu, X., & Wu, M. (2011). Real-time communication between pc and siemens plc based on opc. *Marine Electric & Electronic Engineering.*

[2] Pfrommer, J. (2016). Semantic interoperability at big-data scale with the open62541 opc ua implementation. 173-185.

[3] Standards S. Programmable Controllers - Part 1: General Information[J].

[4] Rumpe, B. (2016). Modeling with UML: Language, Concepts, Methods. Springer Publishing Company, Incorpo-rated.

[5] Bayindir, R., & Cetinceviz, Y. (2011). A water pumping control system with a programmable logic controller (plc) and industrial wireless modules for industrial plants--an experimental setup. *Isa Transactions*, 50(2), 321-328.

[6] Cristian, B., Constantin, O., Zoltan, E., Adina, P. V., & Florica, P. (2014). The control of an industrial process with PLC. International Conference on Applied and Theoretical Electricity (pp.1-4). IEEE.

[7] Carmel, Carmel Zhu Yongqiang, & Wang Wenshan. (2015). *PLC Industrial Control : Programmable logic controllers: industrial control.* Mechanical Industry Press.

[8] Basile, F., Chiacchio, P., & Gerbasio, D. (2013). On the implementation of industrial automation systems based on plc. *IEEE Transactions on Automation Science & Engineering*, 10(4), 990-1003.

[9] Zhang, X. (2013). Research and design of temperature monitoring system in industry applications based on plc. *Applied Mechanics & Materials*, 345, 364-367.

A New Improved Simplified Particle Swarm Optimization Algorithm

LIU Haikuan, YUE Dachao, ZHANG Lei, LI Zhiyuan, JIANG Dawei

Jiangsu Normal University of Electrical Engineering and Automation, Xuzhou 221116, China

Email: jsnu_paper@163.com

Abstract. A new simplified particle swarm optimization algorithm is proposed for the question that the basic particle swarm optimization algorithm is easy to fall into the local optima. The algorithm introduces new parameters on the basis of simplifying particle swarm, adjusts parameters by adaptive method, and coordinates the relationship between various parameters, which increases the use of particle information and ensures the difference between particles. Through the test function of eight, the improved algorithm can effectively avoid the premature convergence and greatly improve the convergence speed and convergence precision.

1. Introduction

Particle swarm optimization (PSO) is an optimization algorithm proposed by Kenndy and Eberhart et al in 1995. The algorithm simulates the bird's foraging behavior and considers the solution of the optimization problem as a bird in the search space called particles. Particles have three properties: position, velocity and fitness of objective function. Firstly, the algorithm initializes a group of particles randomly within the search range, and then iterates continuously. In the process of iteration, two parameters are updated, one is the optimal position of the particle itself and another is the optimal position of the population, until the terminating condition is reached.

The particle swarm algorithm method still has many problems, such as convergence speed and local optimality. In order to correct the shortcomings, many scholars have proposed improved methods and achieved certain results. At present, the refined method can be summarized as five types [1]:

- The deformation of standard PSO algorithm [2], such as adjustment of inertia weight and learning factor, add convergence factor, constraint factor, etc.
- Mixture of particle swarm algorithm and other algorithms [3-4], such as quantum particle swarm optimization, immune particle swarm optimization, chaotic particle swarm optimization, etc.;
- Binary particle swarm optimization also known as discrete particle swarm optimization.
- Cooperative particle swarm optimization [5]. For example, mixed three groups of collaborative particle swarm optimization algorithm.
- Simplified particle swarm algorithm [6-8].

The traditional particle swarm algorithm(PSO) usually only uses the particle's information of individual history optimality and global optimality, but tends to ignore the contemporary optimal particle generated in each iteration process. In this paper, a new simplified particle swarm algorithm is proposed. On the basis of simplified particle swarm algorithm, the new algorithm introduces a new parameter that is contemporary social guidance, namely contemporary optimal particle. In addition,

Content from this work may be used under the terms of the Creative Commons Attribution 3.0 licence. Any further distribution of this work must maintain attribution to the author(s) and the title of the work, journal citation and DOI.

Published under licence by IOP Publishing Ltd

the new algorithm works as much as possible to coordinate the relationship between the various influencing factors, improves the use of particle information, ensures the differences between particles, improves the premature phenomenon, improves global convergence, and improves the performance of the algorithm.

2. Particle swarm optimization

Particle swarm optimization is an intelligent optimization algorithm. The algorithm initializes some population particles first, and then iterates the optimal solution. Supposing that in the D dimensional search space, S particles are set and iterate T times. The i-th particle is represented by X_{id} at the d-dimensional position, the velocity is expressed in V_{id}, the best position of the particle is recorded with $pBest_{id}$, and the best location of the population is recorded by $gBest_d$. After obtaining the above two optimal values, the position and velocity are updated according to

$$V_{id}^{t+1} = \omega V_{id}^{t} + c_1 r_1 \left(pBest_{id} - X_{id} \right) + c_2 r_2 \left(gBest_d - X_{id} \right) \tag{1}$$

$$X_{id}^{t+1} = X_{id}^{t} + V_{id}^{t+1} \tag{2}$$

In (1),(2), $t = 1, 2, ..., T$, $i = 1, 2, ..., S$, $d = 1, 2, ..., D$. r1 and r2 are random numbers that follow the U(0,1) distribution; c1 and c2 are learning factors; W is a weighting coefficient (between 0.1 and 0.9), and the algorithm terminates when it reaches the maximum number of iterations or target accuracy.

3. Improvement of simplified particle swarm optimization

3.1. Simplified Particle Swarm Optimization

Hu Wang et al simplified the basic particle swarm algorithm to:

$$X_{id}^{t+1} = \omega X_{id}^{t} + c_1 r_1 (pBest_{id} - X_{id}^{t}) + c_2 r_2 (gBest_d - X_{id}^{t}) \tag{3}$$

In (3), the first item on the right is particle inertia, which regulates the influence of individual history to the present, uses the ω to adjust. The second is self-cognition, which is the particle's thinking on itself. c_1 and c_2 are the learning factor. The third is the social guidance, which indicates the particle's imitation of the best particle in history. r_1 and r_2 are random numbers that submit the U(0,1) distribution. The algorithm is updated through the position of particles, and the equation is reduced from second order to first order. The validity and superiority of the method is available by theory and experiment. This improvement makes the whole process more simple.

3.2. New Simplified Particle Swarm Optimization

In the simplified particle swarm algorithm(SPSO), each round of particle update is to make the particle individuals greatly approach the extremum of the individual and the entire population, and each particle is evolved with the same iterative formula. The difference between the particles is feeble. It is easy to fall into a local optimal solution, so that individuals of the entire population are surrounded by a certain point, and a phenomenon of premature occurs. In order to improve the phenomenon of premature and maintain the diversity of the population, in the improved algorithm (N-SPSO), the concept of contemporary social guidance is introduced. That is to select the best particles of each round and use them as impact factors to join. In the new update strategy, compared to the simplified particle swarm algorithm, the particle information can be used more fully to ensure the diversity of the particle population, and the performance of the algorithm can be better. The new update strategy is:

$$X_{id}^{t+1} = \begin{cases} \omega^t X_{id}^t + c_1 r_1 (pBest_{id} - X_{id}^t) \\ + c_2 r_2 (gBest_d - X_{id}^t) \qquad rand() \geq \eta \\ \omega^t X_{id}^t + c_1 r_1 (pBest_{id} - X_{id}^t) \\ + c_2 r_2 (gBest_d - X_{id}^t) \qquad rand() < \eta \\ + c_3 r_3 (nBest_{id}^t - X_{id}^t) \end{cases} \qquad (4)$$

In each round of evolution, with certain probability η, the best contemporary particle is used to update the particle position. In (4), $nBest_{id}^t$ is the optimal particle of the present and c_3 is a learning factor.

In the SPSO, the three parameters of ω, c_1 and c_2 are completely independent, but this will have an adverse effect on some extent. If the individual's self-recognition and social guidance are overused, it will cause the global search to fail. For example, if c_1 is large, the particles will linger in their own range, and the search speed will be slow. If c_2 is large, the particles will converge to the local optimum at the initial stage and fall into the local optimum. Conversely, if the parameter values are small, the particle information is not fully utilized. It is also possible that the algorithm can not obtain the global optimal value, and the particles may be too slow to move out of the local optimal value [9-11].

Therefore, the better parametric adjustment strategy is that in the early stages of evolution, there should be greater individual inertia influence factors, and smaller social guidance capabilities, which will preserve the diversity of particles, and the ability of search throughout the space. In the late stages of evolution, it should have small inertia and large social guidance capability, which will help the algorithm converge to a global optimal solution, thereby improving the algorithm convergence speed and accuracy.

In summary, this paper develops a dynamic evolutionary strategy of "large-scale small search, small-scale large search". That is, the search space is large, in the early stage of the algorithm, and the particle needs to be searched in a small area around itself so that the particles have a strong self-Cognition. This is used to improve the ability of the algorithm to search globally and prevent it from falling into a local optimum. At the later stage of the algorithm, the search space becomes smaller and the particles need to have stronger social cognition to perfect the local fine search ability of the algorithm. A certain potential correlation has been established between particle inertia, individual cognition, social guidance, and contemporary social guidance, as follows:

$$\omega^t = \begin{cases} \omega_{max} & X_{id}^t > X_{mean} \\ \omega_{min} - \dfrac{(\omega_{max} - \omega_{min}) * (X_{id}^t - X_{min})}{(X_{mean} - X_{min})} & X_{id}^t \leq X_{mean} \end{cases} \qquad (5)$$

$$c1 = 0.5 + 2 * \cos\left(\frac{\pi * (t-1)}{2 * (T-1)}\right) \qquad (6)$$

$$c2 = 0.5 + 2 * \sin\left(\frac{\pi * (t-1)}{2 * (T-1)}\right) \qquad (7)$$

$$c3 = c1 \qquad (8)$$

The adaptive inertia weight tactic is used. The range is $\omega \in [\omega_{min}, \omega_{max}]$. The introduction of trigonometric functions in learning factors is to avoid large mutations. It can be seen that the new algorithm pays more attention to the individual in the early stage, and then highlights the population intelligence of the whole population at a later stage. The newly introduced contemporary social guidance parameter takes the same position as individuals.

The flows of the new algorithm are as follows:

1) Initialize N particle positions, $pBest_{id}$, $gBest_d$, and $nBest_{id}^t$ within the search scope.

2) Perform the following operations on all particles in the population:

2.1) According to (4), (5), (6), (7), calculating c1, c2, c3, ω.

2.2) Calculating particle fitness and reset $nBest_{id}^t$.

2.3) If the current particle's fitness is better than its own history $pBest_{id}$, update $pBest_{id}$ to the current position.

2.4) If the current particle's fitness is better than the global particle $gBest_d$, then the $gBest_d$ is updated to the current position.

2.5) If the current particle's fitness is better than all contemporary particle records $nBest_{id}^t$, update $nBest_{id}^t$ to the current position.

3) To determine whether the algorithm satisfies the accuracy or reaches the maximum number of iterations. If yes, go to step 4, otherwise go to step 2.

4) Outputting $gBest_d$, get the corresponding value, the algorithm ends.

4. Experiment analysis

In order to test the effectiveness of the new algorithm proposed in this paper, five algorithms are selected for comparative analysis. The five algorithms are

- Adjust particle swarm optimization (APSO) [12].
- Extreme simplified particle swarm optimization (ESPSO).
- Adjust weight particle swarm optimization (AWPSO).
- Proposed simplified particle swarm optimization (PSPSO) [13].
- Simplified particle swarm optimization (SPSO) [14].

Eight classical test functions are used to test the performance of the algorithm. These test functions are

Table 1. Test functions

Function name	Formula	Function name	Formula
1. Sphere	$f(x) = \sum x_i^2$	5. High conditioned elliptic	$f(x) = \sum (x_i + 10^{6\frac{d-1}{D-1}} x_i^2$
2. Rastrigin	$f(x) - \sum \left(x_i^2 - 10\cos(2\pi x_i) \right.$	6. Quartic	$f(x) = \sum i x_i^4 + rand()$
3. Sum of different power	$f(x) = \sum (x_i + |x_i|^{d+1})$	7. Schwefel 1.2	$f(x) = \sum_i (\sum_j x_j)^2$
4. Alpine	$f(x) = \sum (x_i + |x_i \sin(x_i) + 0.1 x_i|)$	8. XinSheYang01	$f(x) = \sum (x_i + |x_i|^d * rand$

Each set of tests used the same population size of 40 particles, $x_i \in [-100, 100]$, a particle dimension of 10, a cycle of 30 times, and an average of 100 generations per evolution. The results obtained are as follows

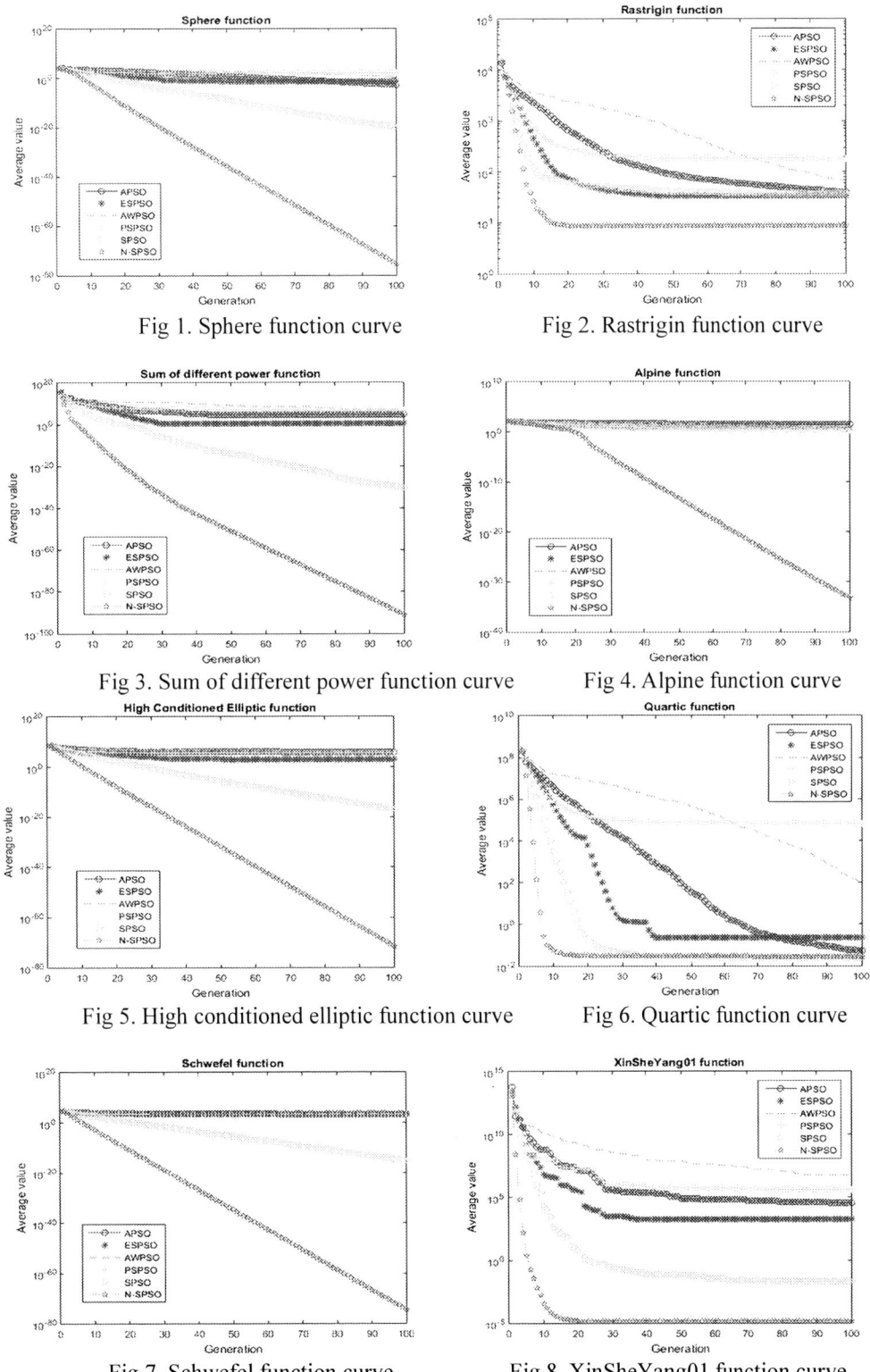

Fig 1. Sphere function curve

Fig 2. Rastrigin function curve

Fig 3. Sum of different power function curve

Fig 4. Alpine function curve

Fig 5. High conditioned elliptic function curve

Fig 6. Quartic function curve

Fig 7. Schwefel function curve

Fig 8. XinSheYang01 function curve

Table 2. Comparison of function results A

function algorithm	Sphere		Rastrigin		Sumofdifferentpow		Alpine	
	mean	optimum	mean	optimum	mean	optimum	mean	optimum
APSO	0.0014	3.62e-5	4.01e1	8.9130	1.37e7	4.90e-8	6.76e1	3.8363
ESPSO	8.37e3	2.30e-18	2.16e4	7.3e3	6.9e14	1.30e-38	1.56e2	0.0014
AWPSO	3.0650	0.2172	7.948e1	3.62e1	3.37e8	2.89e2	1.96e1	3.6626
PSPSO	50.852	1.2062	1.783e2	3.61e1	1.58e7	5.41e1	1.0485	0.0676
SPSO	5.9e-20	5.5e-24	1.132e2	2.21e-8	7.1e-33	8.9e-46	3.2410	4.3e-10
N-SPSO	3.5e-76	1.14e-78	2.414e1	1.35e-9	9.2e-93	5.1e-100	1.5e-36	2.43e-4

Table 3. Comparison of function results B

function algorithm	HighConditionedE		Quartic		Schwefel 1.2		XinSheYang01	
	mean	optimum	mean	optimum	mean	optimum	mean	optimum
APSO	9.31e5	0.0814	0.129	0.013	2.11e3	6.19e2	4.44e5	0.0244
ESPSO	3.80e7	9.9e-17	5.21e8	1.24e6	1.01e5	0.0022	2.3e4	8.3e6
AWPSO	1.87e6	3.87e2	108.9	1.557	3.45e3	7.36e2	4.9e7	3.95e1
PSPSO	3.31e5	1.65e3	6.57e4	1.13e2	1.51e3	2.66e2	1.95e6	2.61e1
SPSO	1.7e-17	1.7e-21	0.675	0.0345	8.3e-16	2.5e-16	3.49e-1	1.7e-8
N-SPSO	2.6e-72	1.7e-73	0.071	0.0053	2.8e-75	4.71e-76	3.12e-5	1.51e-30

Fig. 1 to Fig. 8, Table 2 to Table 3 show the results of testing. It can be seen that N-SPSO has obvious advantages compared to the other five algorithms. Both the evolutionary speed and the convergence speed are very fast, and the convergence accuracy is also high, that is, N-SPSO has a better optimization performance.

5.Conclusion

Aiming at the shortcomings of the original particle swarm optimization algorithm, this paper proposes an improved scheme: a new simplified particle swarm optimization (N-SPSO). The algorithm introduces fresh variables. This can regulate and balance the relationship among the parameters adaptively, thus enhancing the utilization of particle information. The experiments of eight classical test functions show that, compared with the other particle swarm optimization algorithms mentioned above, N-SPSO has the advantages of strong global search ability, fast convergence speed, high precision, and can effectively avoid precocity and so on. The promoted scheme strengthens the performance of the algorithm. However, there are many factors that affect the N-SPSO algorithm, such as how to adjust the relationship between various parameters, how to determine the optimal minimum number of iterations, etc. these are all the future research directions.

References

[1] Huang Taian, Sheng Jia gen, Xu Hongyang, et al. Improved Simplified Particle Swarm Optimization[J]. computer simulation, 2013,30 (02): 327-330+335.

[2] Zhao Yuandong. Particle swarm optimization with weight function's learning factor [J]. computer application, 2013,33 (08): 2265-2268.

[3] Pan Dazhi, Liu Zhibin. Realization of improved Quantum- behaved Particle Swarm Optimization algorithm [J]. computer engineering and application, 2013,49 (10): 25-27.

[4] Ceng Yanyang, Feng Yunxia, Zhao Wentao. Adaptive Mutative Scale Chaotic Particle Swarm Optimization Based on Logistic Mapping [J]. system simulation journal, 2017,29 (10): 2241-2246.

[5] Gao Yunlong, Yan Peng. Unified optimization algorithm based on multi-swarm PSO algorithm and cuckoo search algorithm [J]. control and decision, 2016,31 (04): 601-608.

[6] Zhou Dan, Nan Jing Chang, Gao Mingming. Fuzzy neural network for modeling based on improved simplified particle swarm optimization [J]. computer application research, 2015,32 (04): 1000-1003.

[7] Chen Qunlin, Gao Yue Lin, Guo Xiang. Particle swarm optimization algorithm based on equal replacement and random opposition[J]. computer application research,2017,34 (08): 2364-2367.

[8] Xiong Zhong Wang, Luo Ke. Clustering algorithm based on improved simplified particle swarm optimization [J]. computer application research, 2014,31 (12): 3550-3552.

[9] Sun Zhenlong, Li Xiaoye, Wang Ying. Improved Simple Particle Swarm Optimization Algorithm [J]. computer science, 2015,42 (S2): 86-88.

[10] Zhao Jiaxin, Gao Yue Lin. A Modified Adaptive Particle Swarm Optimization [J]. Journal of Ningxia University (NATURAL SCIENCE EDITION), 2016,37 (02): 125-130+134.

[11] Ma Guoqing, Li Ruifeng, Liu Li. Particle swarm optimization algorithm of learning factors and time factor adjusting to weights [J]. computer application research, 2014,31 (11): 3291-3294.

[12] Zhou Haotian, Wu Zhiyong, Tian Yubo. Research on improving Simple Particle Swarm Optimization [J]. computer engineering and application, 2012,48 (24): 41-44.

[13] Mustafa Servet Kiran. Particle swarm optimization with a new update mechanism[J]. Applied Soft Computing, 2017,60.

[14] Hu Wang, Li Zhishu. A simpler and More Effective Particle Swarm Optimization Algorithm [J]. software journal, 2007,18 (04): 861-868.

AUTHOR INDEX

Ailing, Qi 1974
An, Shubing 1378
An, Wang W. 1470
An-Wen, Ying 479
Bai, Juan 1121, 1510
Bai, Xiaoye 252
Bao, Lei 1121, 1510
Bao, Wenxia 1853
Bao, Yiting 780
Bi, Mingkai 1939
Bin, Xu 2166
Bo, Yang 434
Bo, Zhang 2315
Bouchart, Vanessa 732
Boxing, Zhang 1063
Cai, Biao-Hua 751
Cai, Guoliang 1209
Cai, Peng 409
Cai, Shaopeng 1365
Cai, Sun 214
Cai, Xiaoyu 2283
Cai, Zengyu 1656
Cao, Shaozhong 1425
Cao, Shukun 640
Cao, Xinli 87
Cao, Yan 276
Cao, Yichao 2360
Cao, Yundong 239
Cen, Tao 2278
Chang, Faliang 1359
Chang, Rui 1130
Chang, Wen 2638
Changhui, Ma 268
Chao, Xiang 2478
Che, Renfei 366, 386
Chen, B. W. 1932
Chen, Baiyu 2602
Chen, Chen 2720
Chen, Chunlong 331
Chen, H. 2089
Chen, Hai-Quan 808
Chen, Hemu 1908, 1987
Chen, J. J. 1932
Chen, Jianjun 108
Chen, Jie 135
Chen, Jing 2272
Chen, Jinqiang 2506
Chen, Liang 772

Chen, Lili 2494
Chen, Limei 1636
Chen, Ming 2232
Chen, Qiaoling 1209
Chen, Quan 323
Chen, Shanji 2109
Chen, Shuyu 2488
Chen, Ting 2042
Chen, Wei 200
Chen, Xiangzhou 447, 1859
Chen, Xiaolin 1108
Chen, Xiaoxiao 568, 589
Chen, Xueli 1869
Chen, Yanfang 693
Chen, Yazhen 2720
Chen, Yong 1869
Chen, Yuanyuan 1662
Chen, Yueyue 108
Chen, Zhi 1554
Chen, Zhonghe 553
Cheng, D. S. 1932
Cheng, Si 2323
Cheng, Wang 701, 738
Cheng, Yang 510
Chenglin, Zhang 144
Chevrier, Pierre 732
Chi, Zhang 562
Chong, Gao 168
Chu, Qianqian 2278
Chu, Shibo 1519
Chu, X. M. 1801
Chuan, Wang 716
Chuanli, Wang 701, 738
Chunmei, Li 1675
Cong, Wang 2252
Cui, Fang Y. 894
Cui, Jing Gang 581
Cui, Wen 1812
Cuicui, Liu 168
Dai, Jian 2551
Dai, Zongmiao 1667
Dandan, Sun 888
Danyang, Li 2211
Deng, Fanyi 93
Deng, Ming-Ji 1353
Deng, Qinghong 882
Deng, Qishu 1342
Deng, Shaoxiang 214

Deng, Xin	1412	Gan, Baiqiang	2592
Di, Xiaofeng	2494	Gan, Hua	2304
Diao, Chentao	1543	Gan, Ping	1979
Ding, Baobao	1604	Gang, Wang	434
Ding, Fu-Jun	2371	Gao, Chao	1994
Ding, Hailan	2466, 2559	Gao, Dongliang	409
Ding, Hongyan	732	Gao, Fei	135
Ding, Huixia	1859	Gao, Fei-Fei	181
Ding, Jie	1000	Gao, Hanxu	1359
Ding, Lili	984	Gao, He	1994
Ding, Xiaohua	1812	Gao, Hong	1171
Ding, Yanfeng	1209	Gao, Jian	2368
Dong, Ma X.	1470	Gao, Lijuan	1682
Dong, Sui	1775	Gao, Qingshui	208
Dong, Wei Y.	1782	Gao, Zhenxing	2113
Dong, Xianlei	2421, 2429	Gao, Zihan	1006
Dong, Xiaoming	323, 337, 359	Geng, Lei	1049
Dou, Liang	1604	Gong, Chunwei	2466, 2559
Du, Chunfeng	1656	Gong, Taorong	296
Du, Jiawei	1531	Gong, Xinman	833
Du, Jinyang	453	Gou, Yating	114, 162
Du, Wen	366, 386	Gu, Bochuan	2176
Duan, Lijin	276	Gu, Jingtian	1710
Duan, Lunqin	955	Guan, Denggao	346
Duan, Ming	1575	Guan, Shilei	57
Duan, T.	1834, 2609	Guan, Wanlin	35
Dun, Ao	1166, 1543	Guangyao, Jia	461
Fan, Dandan	2232	Guanhui, Wang	2216
Fan, Jie	102	Gui, Xinyue	2183
Fan, Jinpo	1739	Guiqiang, Liu	888
Fan, Mingqi	2649	Guizhong, Wang	27
Fan, Xingyuan	453	Guo, Gongde	2565
Fang, Wang	1282	Guo, Hejia	640
Fang, Zhuo	168	Guo, Hongwei	102
Feng, Hao B. W.	2624	Guo, Kai-Feng	1190
Feng, Lansheng	258	Guo, Ming-Xuan	808
Feng, Ruzhi	524	Guo, Ronghua	1531
Feng, Shanqiang	2176	Guo, Runqiu	258
Feng, Shunshan	1612	Guo, Shaobing	917
Feng, Wang	168	Guo, Sheng H.	2712
Feng, Xiao	1063	Guo, Siyuan	822
Feng, Xiaoche	2257	Guo, Song	751
Fengbin, Zhang	1450	Guo, Wei	530
Fu, Da	2602	Guo, Xiaoshuang	1656
Fu, Hongyong	2308	Guo, Xing	1111
Fu, Jie	2453	Guo, Xueqi	346
Fu, Jun	2297, 2329	Guo, Yajie	93
Fu, Qixi	1273	Guo, Yingjun	93
Fu, Yifan	882	Guo, Yizhuo	1461
Fu, Yuyang	2036	Guo, Ziteng	574
Fusheng, Chen	144	Haibo, Tan	596
Gai, Lujie	882	Haichao, Chen	607

Haijun, Lei 1974
Haijun, Peng 596
Hai-Lan, Ding 2446
Haiyang, Jiang 1282
Han, Donchen 1395
Han, Jun 1939
Han, Liang 848
Han, Qianru 2021
Han, Quanli 1395
Han, Tao 640
Han, Tongxin 1057
Han, Wang 2572, 2578
Han, Xueshan 331, 372
Hanyan, Wang 2221
Hao, Chen 144, 596
Hao, Cheng 2203
Hao, Chuxue 18
Hao, Jinshun 221
Hao, Junjie 1919
Hao, Li 1286
Hao, Weiguang 263, 787, 795, 802
Hao, Yun 553
Haolin, Jia 1315
He, Fangzheng 1
He, Hongmei 1733
He, Jiangheng 1612
He, Juntao 366, 386
He, Lyulong 1273
He, Ming 1115
He, Renke 2278
He, Shiwei 1939
He, Shuming 2368
He, Xin 1267
He, Yanchen 1758
He, Yidong 1629
He, Yu 1682
He, Yubo 1645
He, Zhiqiang 1599
Hengjie, Li 416
Hong'Bo, Chen 911
Hong-Zhi, Yu 2446
Hou, Aijun 2176
Hou, Lunqing 1412
Hou, Xiangru 1536
Hou, Yan 2746
Hou, Yueqi 2071
Hu, Beibei 2421, 2429
Hu, Chang'An 745
Hu, Dehao 346
Hu, P. C. 1834, 2609
Hu, Shi-Cheng 990
Hu, Yue 228

Hu, Yulan 263, 787, 795, 802
Hu, Yunpeng 2001
Hua, Dengxin 894
Hua, H. Y. Y. 1330
Hua, Wang 434
Hua-Jie, Chen 668
Huang, Bihui 485
Huang, Jingzhi 1228
Huang, Li 2629
Huang, Lin 1979
Huang, Min 2065
Huang, Qiuzi 2488
Huang, R. 2089
Huang, Wei 1029
Huang, Wenhan 693
Huang, Xiaoping 2117
Huang, Xulong 1919
Huang, Yangfan 1979
Huang, Yizhi 693
Huang, Yuwei 1029
Hui, Baofeng 2109
Huimin, Fan 2211
Huimin, Sun 1036
Huitao, Wang 2014
Huiying, Song 461
Ji, Ke 46
Ji, Weiyan 1228
Ji, Yang 808, 2079
Jia, Guangyao 427
Jia, Guoqing 2109
Jia, Hongwei 2706
Jia, Qiang 510
Jia, Shanjie 1859
Jia, Shijie 1594
Jia, Songmin 1166, 1543
Jia, Wenbo 35
Jia, Yafang 2401
Jia, Yunfei 188, 662
Jia, Zhigang 2638
Jiachen, Tian 168
Jiajia, Han 2315
Jian, Wang 875
Jian, Zhou 518
Jianbo, Yin 401
Jiang, Cheng 2323
Jiang, Dawei 1827
Jiang, Hua 1739
Jiang, Juanjuan 1733
Jiang, Xiaoying 2565
Jiang, Z. L. 1801
Jiang, Zhanjun 2283
Jiang, Zhe 372

Jianhui, Zhou 168
Jianwei, Liu 1562
Jianzheng, Liu 63
Jianzhi, Tuo 634
Jianzhong, Yang 401
Jiaojiao, Xi 2544
Jiaxin, Liu 1215, 1222
Jia-Zhi, Yang 479
Jie, Cheng 738
Jie, Huang 768
Jie, Ren 2519
Jiefeng, Mou 1036
Jikang, Wang 63
Jikun, Guo 542
Jin, Fei 78
Jin, Ge 214
Jin, Jian 518
Jin, Li 2544
Jin, Tao 323
Jin, Tiancheng 1604
Jin, Wei 1483
Jin, Weiqi 1955
Jing, Jing 1147
Jing, X. H. 1246
Jing, Zhang 168
Jing, Zhu 1215, 1222
Jingshi, He 2347
Jinjie, Shan 518
Jinliang, Qiu 461
Jinxiu, Wang 888
Jiyao, Tian 1967
Jun, Wei 434
Junning, Qin 2315
Kang, Ruiyu 1788
Kang, Yang 135
Ke, Yan 1477
Ke, Zhang 634
Kong, Juan 1919
Kong, M. X. 1696
Kong, Weizheng 130
Kong, Xiangzeng 2565
Kou, Xu-Peng 2674
Kuaia, Tengfei 905
Kui, Zhang Yong 875
Kun-Yu, Qi 2446
Lai, Ming-Ming 2669
Lan, Ru 2638
Lan, Yunsheng 2048
Le, Guigao 905
Lei, Chu 27
Lei, Lei 1839, 1846
Lei, Min 87

Lei, Wang 195, 2221
Lei, Xiang 1967
Lei, Yiyan 1720
Lei, Zhipeng 379
Lele, Sun 1315
Li, Bin 751
Li, C. 2436
Li, Chunmei 1524, 1889, 1898
Li, Cui 911
Li, Dezhi 296
Li, Guanghui 870
Li, Guanyu 2466, 2515, 2559
Li, Guoqiang 78
Li, Haifeng 323
Li, Hong-Bing 1121, 1510
Li, Huanran 1883
Li, Hui 1883
Li, Huizhi 156
Li, Jiahao 1166
Li, Jian 613, 2152
Li, Jie 1308
Li, Jing 239
Li, Jinping 2706
Li, Jiping 1092
Li, Kai 1524, 1898
Li, Li 822
Li, Lulu 372
Li, Maohua 2662
Li, Meng 1353, 2304
Li, Ming 1084
Li, Mingchao 2001
Li, Minwei 1013
Li, Nianlian 732
Li, Pengyang 894
Li, Ran 1488
Li, Shuangxi 1883
Li, Sicong 35
Li, Tao 2692
Li, Tong 2506
Li, Wang 2460
Li, Wanze 745
Li, Wei H. 1820
Li, Wei 1441
Li, Weichao 78, 1199
Li, Wenbo 331, 372
Li, Wenjing 1406
Li, X. L. 2436
Li, Xiao 2602
Li, Xingxing 745
Li, Xiuzhi 1166, 1543
Li, Xu 2701
Li, Xuefei 135

Li, Yan 894
Li, Yanyun 1979
Li, Ye 2071
Li, Yingqi 833
Li, Yulong 188
Li, Zhi L. 1839, 1846
Li, Zhifei 2036
Li, Zhiming 258
Li, Zhiyuan 1827
Lian, Minlong 1378
Liang, Bin 673
Liang, Dong 1853
Liang, Gang 2304
Liang, Junbin 1418
Liang, Li 738, 1036
Liang, Ning 114, 162
Liang, Shutian 18
Liang, Xi 1450
Liang, Xiaolong 1273, 2071
Liang, Yi 1029
Liang, Ying 57
Liang, Yuqing 1919
Liang, Zhikai 1425
Liao, Daixi 1618
Liao, Minfu 882
Liao, Xiaoming 693
Liao, Zitian 1138
Liling, Liu 2355
Liman, Shen 144
Limei, Zhao 2335
Lin, Dansheng 2267
Lin, Doudou 2429
Lin, Jinghui 346
Lin, Shaofu 2393, 2401
Lin, Sheng 1883
Lin, Wei 668, 1645
Lin, Yao 1336
Lin, Zhang 2079
Lin, Zhaowen 1130, 1488
Ling, Liu 1142
Li-Qing, He 668
Liu, Bin 984, 1503
Liu, Can 870
Liu, Chang 290
Liu, Changli 967
Liu, Cuicui 114, 162
Liu, Di 1300
Liu, Fuyang 123
Liu, G. 263, 787, 795, 802, 1801
Liu, Haikuan 1827
Liu, Hanqing 1292
Liu, Haojie 315

Liu, Huabin 1667
Liu, Jianwei 2048
Liu, Jiawei 653
Liu, Jiaxin 188, 662
Liu, Jie 1554, 2140
Liu, Jingli 78
Liu, Jun 2030, 2058
Liu, Ke Cheng 581
Liu, Kun 2342
Liu, L. 195, 1801
Liu, M. 2089
Liu, Qianru 2283
Liu, Renzhang 78
Liu, Shi 208
Liu, Shuxin 239
Liu, Tingxiang 200
Liu, Wei 955
Liu, Wenchang 276
Liu, Wenda 18
Liu, X. 2089
Liu, Xianglong 1234
Liu, Xiaochun 1812
Liu, Xiaoliang 78
Liu, Xiaoqian 2734
Liu, Xindong 1267
Liu, Xingbao 1365
Liu, Xueyan 2551
Liu, Yajie 263, 787, 795, 802
Liu, Yang 346
Liu, Yangyang 2408
Liu, Yankui 1599
Liu, Ye 2267
Liu, Yi 625, 629
Liu, Yiliang 1378
Liu, Yonggang 315
Liu, Yongxia 2048
Liu, Yu 188, 2679
Liu, Yuting 296
Liu, Yuyan 1788
Liu, Zefeng 1013
Liu, Zhe 1531
Liu, Zhengyi 1441
Liu, Zhenzhen 2629
Liu, Zhizhen 276
Liu, Zilin 263, 787, 795, 802
Liyu, Xia 2572, 2578
Long, Luo 602
Long, Pan 856
Long, Shaohua 1618
Long, Wang 701, 738
Long, Wu 416
Lu, Jiangang 1228

Lu, Jun	409
Lu, Ligen	258
Lu, Shikun	894
Lu, Xiaobo	1688, 2360
Luai, Almadhehagi	427
Luhua, Xing	268
Luo, Lisai	2466
Luo, Shihui	723, 1183
Luo, Taorui	409
Luo, Wanbo	2065
Luo, Zhen	123
Lv, Peihua	1688
Ma, Hongfeng	379
Ma, Jianwei	680, 687
Ma, Kun	2342
Ma, Lulu	305
Ma, Panwei	510
Ma, Pengcheng	1623
Ma, Shiwei	917
Ma, Te	2384, 2388
Ma, Xiaodong	2009
Ma, Xinling	759
Ma, Zhi-Run	2686
Mao, Wanfeng	2649
Mao, Yanrong	427
Mao, Yazhou	315
Maotao, Yang	144, 596
Maoyi, Zhang	1315
Meng, Xiaocheng	366, 386
Meng, Yuting	2001
Mi, Yongsheng	10
Miao, Feng	524, 530
Miao, Lanfang	2036
Min, Huang	2741
Min, Liu	768
Mou, Pengbo	625, 629
Mouhai, Liu	144, 596
Mu, Jiong	581
Mu, Qi	1599
Mu, Senlin	1292
Mu, Xihui	156
Murtaza, Abid	1562
Na, Li	1477
Nannan, Liu	1036
Ni, Xue	228
Nie, Li	1153
Ouyang, Chengtian	2042
Ouyang, L.	2089
Pai, Liu	888
Pan, Fangyu	1153
Pan, Jian	1720
Pan, Qiao	2262

Pang, Xun	673
Peng, Du	634
Peng, Fengzhi	152
Peng, Jianjun	1092
Peng, Lin	2669, 2674, 2686, 2701
Peng, Luxi	2499
Peng, Wang	174
Peng, Yanfei	1092
Penghou, Liu	607
Ping, Wang	888
Qi, Yingchuan	510
Qiang, Li X.	2166
Qiang, Li	2166
Qiang, Lin	152
Qiang, Wu	888
Qiao, Yulong	1962
Qimeng, Nie	1282
Qin, Hua	46
Qin, Luxing	2360
Qing, Wu	2323
Qingjun, Guo	995
Qinyuan, Li	2315
Qiu, Mengyue	2232
Qiu, Zhen	1300
Qiu, Zhiwen	693
Qiuqiu, Wang	1159
Qizhong, Li	1494
Qu, Huaijing	1549
Ran, Jilin	967
Ran, Li	2472
Ren, Jie	905
Ren, Jiyuan	123
Ren, Xun-Yi	1703
Ren, Zhong	814
Ren, Zongjin	305
Riaz, Umair	870
Rong, Tang	2079
Ru, Cong	934, 941, 948
Ru, Zhang	1914
Ruan, Y.	1246
Ruan, Zhenzhen	1979
Rui, Chen	1450
Rui, Zhang	542
Run-Dong, Wang	2519
Ruopeng, Yang	2014
Saeed, Muhammad J.	870
Sang, L. Z.	1834, 2609
Shan, Gao	888
Shao, Bao-Zhu	181
Shao, Juanjuan	1267
Shao, Xuebin	2408
Shao, Zhiyu	1612

She, Jintao 491
Shen, Gao Q. 1839, 1846
Shen, Guiquan 1228
Shen, Wei 2638
Shen, Wuqiang 1228
Shen, Xinxin 1583
Sheng, Jiayue 841
Sheng, Tingran 1057
Sheng, Xuanyu 841
Shengdongt, An 401
Shi, Changkai 57
Shi, Chen 1703
Shi, Haoqiang 2283
Shi, Kai 1883
Shi, Peiji 2408
Shi, Xin 1353
Shi, Zhao-Cun 751
Shi, Zhe 93
Shixu, Li 1036
Shuai, Chen 2245
Shudong, Wang 461
Shuifeng, Zhang 2245
Shushuang, Liang 2656
Shuzhen, Yang 710
Sicong, Li 2472
Situ, Shuwei 1
Song, Deyu 379
Song, Huiying 427
Song, Lihua 130, 1300
Song, Min 2649
Song, Ping 568, 589
Song, Q. H. 2436
Song, Xing 2368
Song, Xiyu 1503
Song, Yuqin 2030
Song, Zilong 1939
Songze, Lei 1063
Su, A. J. 1696
Su, Hongsheng 51
Su, Jiangwen 1300
Su, Tongdan 1115
Su, Y. 2089
Sui, Wei 394
Sun, Chenzhe 2615
Sun, Chuanmin 346
Sun, Cong 305
Sun, Feng 181
Sun, Heng 1623
Sun, Hexu 93
Sun, Hua 337, 359
Sun, Jiabin 2127
Sun, Jianyong 1108

Sun, Juanjuan 1401
Sun, Lin 2291
Sun, Liying 252
Sun, Qian 1359
Sun, Qibo 2429
Sun, Quanxin 1788
Sun, Ruifeng 1013
Sun, Xianhai 768
Sun, Xiao 1949
Sun, Xiaoxiao 848
Sun, Xin 1199
Sun, Yanjun 1166, 1543
Sun, Yao 346
Sun, Yaojie 1745
Sun, Yi 613, 1130, 1488, 1554
Sun, Ying 453
Sun, Yu 1599
Sun, Zhen P. 613
Sun, Zheng 1441
Sun, Zhijie 2297, 2329
Suo, Dong 1675
Suo, Shuangfu 221
Tan, Guangyu 870
Tan, Jinjun 2267
Tan, Ming 1121, 1510
Tan, Yukun 1425
Tang, B. M. 2436
Tang, Guoshen 276
Tang, Jinjin 2528
Tang, Jun 1908
Tang, Shaofan 1378
Tang, Xiao 290
Tang, Xinhuai 2291
Tang, Yanqun 2001
Tang, Ying 2537
Tao, Kepeng 35
Tao, Yu 710, 716
Tao, Zhengping 2528
Teng, Xiaofei 1071
Tian, Bin 502
Tian, Feng 2453
Tian, Jiachen 114, 162
Tian, Jin 2021
Tian, Jing-Jing 2371
Tian, Zhengbing 252
Tianfang, Wu 2245
Tight, Miles 1788
Ting, Ding 461
Tingting, Liu 1142
Tong, Fei 1908, 1987
Tong, Guan 888
Tong, Li 1503

Tu, Jingzhe	323
Tu, Jinlong	10
Tu, Yaqing	1084
Tu, Yongcheng	1908, 1987
Wan, Hongqiang	1395
Wan, Xing	2065
Wanbo, Luo	2741
Wang, Bo	1029
Wang, C. Y.	1750
Wang, Caishen	290
Wang, Chang'An	2734
Wang, Chao	2297, 2499
Wang, Chaochao	337
Wang, Chunlin	1108
Wang, Chun-Yang	1121, 1510
Wang, Denggui	759, 826
Wang, Dong	510, 1782, 2597
Wang, Endong	1877
Wang, Enshi	485
Wang, Fei	1412
Wang, Feng	114, 162
Wang, Guanhong	296
Wang, Guirong	833
Wang, Haibin	2323
Wang, He	917
Wang, Heng	640
Wang, Hengbin	1549
Wang, Hu	808
Wang, Hui	135
Wang, Jiawen	1629
Wang, Jing	252
Wang, Jiong	2071
Wang, Juan	2453
Wang, Ke	282, 447, 1256, 1859
Wang, Kun	200
Wang, L.	1834
Wang, Lanwen	841
Wang, Li	2297, 2329
Wang, Lihua	228
Wang, Lingxia	826
Wang, Lingxue	1955
Wang, Lingyu	2421
Wang, Liquan	934, 941, 948
Wang, Minghao	620
Wang, Nian	1908, 1987
Wang, Ping	2140
Wang, Qi	2238, 2323
Wang, Qingjia	2342
Wang, Qiuling	130
Wang, Runjiao	955
Wang, Shiyu	2262
Wang, Shudong	427

Wang, Shuyuan	1100
Wang, Siyue	1594
Wang, Song	2515
Wang, Tao	1733
Wang, Ting	1100
Wang, Wen	1049
Wang, Wenjie	221
Wang, Wen-Si	1703
Wang, Wentao	46
Wang, X. S.	1932
Wang, Xi	2042
Wang, Xiangpei	1919
Wang, Xianli	524, 530
Wang, Xiaogang	1153
Wang, Xiaolan	200
Wang, Xiaoming	2304
Wang, Xingong	46
Wang, Xuewei	548
Wang, Yan	346, 1043
Wang, Yanan	282, 447, 1256, 1859
Wang, Yang	221, 282, 1256, 1979
Wang, Yanyan	955
Wang, Yaokun	57
Wang, Yifan	1662
Wang, Yijing	1100
Wang, Yingjing	662
Wang, Yisheng	2140
Wang, Yuanmin	1919
Wang, Yudong	409
Wang, Yujiang	1029
Wang, Yuqiao	2692
Wang, Zhao	2408
Wang, Zhaoqing	1342, 1348, 2134
Wang, Zhe	123
Wang, Zheng	653
Wang, Zhiping	290
Wang, Zhiying	1629
Wang, Zhi-Yuan	990
Wannian, Zhu	1967
Wei, Caisheng	574
Wei, Chen	416
Wei, Liu	1927
Wei, Pi	1675
Wei, Qianwen	2393
Wei, Shicheng	1029
Wei, Zhang	634
Wei, Zhengxian	2649
Wei, Zheyu	46
Weidong, Xu	1967
Weihai, Li	1927
Weihua, Ma	723, 1183
Wei-Jun, Pan	2519

Weiwei, Qi	1142	Xie, Dong-Fan	2371
Wen, Chang	2679	Xie, Feng	2329
Wen, Chao	826	Xie, Lingling	1
Wen, Guangqi	1889	Xie, Minzhen	1078
Wen, Junhao	2488	Xie, Yong-Jun	1190
Wenbo, Li	268, 352	Xin, Bo	1292
Wencan, Ding	1775	Xin, Li	1359
Wensheng, Yin	562	Xin, Ma	2203
Wenwen, Jiao	2472	Xin, Peizhe	409
Wenxue, Liu	268	Xin, Sun	2315
Wu, Chunshang	2377	Xin, Xiaoyu	2478
Wu, Hao	1575	Xin, Yan	875
Wu, Haobo	2048	Xin, Yang	1967
Wu, Hong	1962	Xincheng, Ren	471
Wu, Hongmei	1919	Xing, Wan	2741
Wu, Jianhong	2499	Xinliang, Cao	471
Wu, Jinbo	822	Xin-Zhe, Yin	479
Wu, Jun	1049	Xu, Binshi	1029
Wu, Jun-Jie	808	Xu, Feng	1919
Wu, Kezhuang	1418	Xu, Gang	1733
Wu, Qinqin	2267	Xu, Guangping	1883
Wu, Qiong	1006, 2123	Xu, Guanli	346
Wu, Ran	2123	Xu, Hongkui	1549
Wu, Tong	1604	Xu, Jie	1130, 1488
Wu, W. W.	1750	Xu, Peiyuan	2727
Wu, Xiaoquan	2267	Xu, Sanchuan	536
Wu, Xuehui	2360	Xu, Shiping	1531
Wu, Yingying	502	Xu, Wanjin	1108
Wu, Yusi	2140	Xu, Wei	2706
Wu, Zhiqiang	1554	Xu, Wenjing	315
Wufan	2014	Xu, Xiangqian	640
Xi, Hongyan	2140	Xu, Xiaoshen	2232
Xi, Qi	1575	Xu, Xin	2297, 2329
Xia, Bin	1240	Xu, Yugong	2203
Xia, Peng	1336	Xu, Zhuoran	1554
Xia, Rongzhen	379	Xue, Qiao	447, 1859
Xia, Sibin	2393, 2401	Xue-Chao, Liao	646
Xia, Yangqiu	1365	Xueshan, Han	352
Xianfang, Tang	1914	Xuli, Zhu	1036
Xiangbin, Liu	144	Xuxiang, Huang	352
Xiangguo, Su	2252	Yachao, Jia	1914
Xiangzhou, Chen	282	Yan, Bin	1503
Xiao, Binjie	1251	Yan, Chunyu	917
Xiao, Zhitao	1049	Yan, Haotian	2134
Xiaofei, Zou	2014	Yan, Kedi	73
Xiaokun, Wang	2221	Yan, Qianghu	87
Xiaomei, Hu	710, 716	Yang, Guang	2009
Xiaoping, Li	1063	Yang, Guohui	924
Xiao-Shu, Wang	2098	Yang, Huiyue	1084
Xiaotie, Ma	2335	Yang, Jian	46
Xiaoyan, Zhang	1159	Yang, Jianxi	315
Xie, Cheng	108	Yang, Jiebin	745

Yang, Jun-You	181
Yang, Kai	1503
Yang, Li	995
Yang, Lin-Nan	2674, 2701
Yang, Lu	2629
Yang, Ning	2559
Yang, Qunyi	2009
Yang, W. D.	1696
Yang, Wan	668
Yang, Wei	2712
Yang, Weijun	2291
Yang, Wentai	2304
Yang, Xiaodan	1433
Yang, Xiaohua	1554
Yang, Yi	208
Yang, Yingming	1604
Yang, Yiyong	221
Yang, Yongxi	924
Yang, Yu	1177
Yang, Yuansheng	1171
Yang, Yuanyuan	2602
Yang, Ziwei	1147
Yangjia	1494
Yanhong, Wang	1063
Yanhong, Zuo	861
Yanrong, Mao	461
Yanwei, Shang	152
Yanzhe, Du	607
Yao, Jianchun	394
Yao, Jianyu	826
Yao, Jiawei	2195
Yao, Ling	2117
Yao, Zheng	394
Yating, Gou	168
Ye, J.	1801
Ye, Wang	1494
Ye, Xuanyu	2720
Yi, Jun	323
Yi, Kang	1720
Yi, Wang	63
Yi, Yang Q.	1470
Yi, Zhijun	2494
Yin, Aiping	2113
Yin, Ningxia	870
Yin, Yanan	1171
Yin, Zhiqin	1115
Ying, Lin	1282
Yinzheng, Zheng	973
Yong, Lin	1373
Yonggang, Yue	401
Yonggang, Zhu	861
Yongqiang, Fan	401

You, Fucheng	548
You, Zhou	646
Youzi, Wang	2572, 2578
Yu, Fengyun	917
Yu, Jiujiu	2160
Yu, Ling	1092
Yu, Lu	553
Yu, Nan	2478
Yu, Shida	2189
Yu, Tianbiao	780
Yu, Tonglan	1554
Yu, Xiaochen	394
Yu, Xin	394
Yu, Xu R.	1820
Yu, Xuemei	1745
Yu, Yu	1519
Yuan, Bo	246
Yuan, Ziyan	2746
Yue, Chen	27
Yue, Dachao	1827
Yun, Mei	1282
Yunxiao, Zu	1927
Yutinge, Chen	1477
Zéman, Zoltán	2662
Zeng, Hanghang	51
Zeng, Jijun	2267
Zeng, Tianlong	2058
Zeng, W. D.	2584
Zeng, Ying	1503
Zeng, Yue	2103
Zhai, Xiujun	2615
Zhai, Yayu	568, 589
Zhan, Hong-Yuan	1190
Zhan, Y. C.	422
Zhang, Bin	625
Zhang, Bo	1199
Zhang, C.	1696
Zhang, Chenglin	2089
Zhang, Chengning	924
Zhang, Chu	208
Zhang, Chun J.	2712
Zhang, Cunlin	1949, 1962
Zhang, Dan	1703
Zhang, Dewen	35
Zhang, Dong	1877
Zhang, Fang	1049
Zhang, Feng	2272
Zhang, G. R.	2584
Zhang, Gang	1739
Zhang, Geng	282, 447, 1256
Zhang, Guan-Feng	181
Zhang, Guanglei	2615

Zhang, Guoliang	1543	Zhang, Yichen	1386
Zhang, Hanhua	2692	Zhang, Yidu	1006
Zhang, Hao	366, 386, 882	Zhang, Yonghua	1401
Zhang, Haoxue	882	Zhang, Yue	2103
Zhang, Hongda	35	Zhang, Z. N.	1330
Zhang, Hua	2117	Zhang, Zhi	290
Zhang, Huanping	2662	Zhang, Zhongshi	2238
Zhang, J.	1750	Zhangkang	2113
Zhang, Jian	826	Zhanjun, Wang	195
Zhang, Jianwei	1656	Zhao, Changfang	905
Zhang, Jiaqiang	1273, 2071	Zhao, Chengqiang	934, 941, 948
Zhang, Jincheng	258	Zhao, Enmin	239
Zhang, Jinwei	745	Zhao, Guorong	1994
Zhang, Jun	305	Zhao, H. W.	1246
Zhang, Junhao	2262	Zhao, Hui	2488
Zhang, Lan	934, 941, 948	Zhao, Kai	305
Zhang, Lei	1827	Zhao, Li	1675
Zhang, Liang	239, 2368	Zhao, Lili	258
Zhang, Li-Hua	1336	Zhao, Liujun	130
Zhang, Lin	2304	Zhao, Qian	984
Zhang, M. Y.	1801	Zhao, Qing	2238
Zhang, Min	723, 1183	Zhao, Qing-Song	181
Zhang, Na	1177	Zhao, Yangze	548
Zhang, Nan	2127	Zhao, Ying	453
Zhang, Peng	2140	Zhao, Yuanmeng	1949, 1955, 1962
Zhang, Qiang	2551	Zhao, Yuejin	1949, 1962
Zhang, Qingqing	2551	Zhao, Yufeng	1199
Zhang, Qiurong	693	Zhao, Yusheng	1342, 2134
Zhang, Ran	447, 1256, 1859	Zhao, Yuting	1720
Zhang, Rui	228	Zhao, Zengshun	1359
Zhang, Ruiqi	359	Zhao, Zhi-Qiang	1353
Zhang, Ruiqiu	772	Zhaoren, Pan	710
Zhang, S.	1246	Zhen, Qiao	416
Zhang, Shoushou	822	Zheng, Jinxin	78
Zhang, Shuo	924	Zheng, Kougen	1583
Zhang, Tieshan	814	Zheng, Wei	379
Zhang, Tinglei	1877	Zheng, Yongkang	1667
Zhang, Xiangyin	1543	Zheng, Yufu	1147
Zhang, Xianmin	772	Zhen-Huan, Chen	646
Zhang, Xiaotong	2238	Zhenwei, Zhang	174
Zhang, Xiaoying	200	Zhichao, Guo	63
Zhang, Xin	2669, 2686	Zhiyong, Wu	596
Zhang, Xincheng	276	Zhizhong, Guo	27
Zhang, Xinyu	2048	Zhong, Shouming	1618
Zhang, Xinzheng	1267	Zhong, Zhongzhi	1853
Zhang, Xuangong	156	Zhongqi, Wang	1494
Zhang, Xuchong	772	Zhou, Chun	453
Zhang, Xudong	102	Zhou, Di	1688
Zhang, Xujuan	2551	Zhou, Guomiao	1575
Zhang, Yanan	2408	Zhou, Jian	629
Zhang, Yang	1688	Zhou, Qinqin	2734
Zhang, Yanjun	239	Zhou, Rundong	346

Zhou, Shu 1267
Zhou, Xiang 87, 1812
Zhou, Yifan 1365
Zhou, Ying 1531
Zhou, Yong 1908
Zhou, Zeyu 653
Zhu, Chuangchuang 1273
Zhu, Haoming 296
Zhu, Honghai 1519
Zhu, Hongwei 2048
Zhu, Jingli 984
Zhu, Junjie 693
Zhu, Leiye 2323
Zhu, Ming 1853
Zhu, Weijun 822
Zhu, Xingxiong 2597
Zhu, Yu 581
Zhu, Yuan 2147
Zhu, Yuancheng 379
Zhu, Yuefei 1645
Zhu, Zhangqing 1292
Zhu, Zheng 1353
Zhu, Zhengbin 934, 941, 948
Zhu, Zhilong 1733
Zhuanga, Duoduo 2113
Zhuo, Fang 114, 162
Ziqiang, Lou 2252
Zonghua, Xie 634
Zou, Yuan 102
Zu, Yun X 1820

International Symposium on Power Electronics and Control Engineering (ISPECE 2018)

Journal of Physics: Conference Series Volume 1187

Xi'an, China
28-30 December 2018

Part 3 of 3

ISBN: 978-1-5108-8673-5
ISSN: 1742-6588

Printed from e-media with permission by:

Curran Associates, Inc.
57 Morehouse Lane
Red Hook, NY 12571

Some format issues inherent in the e-media version may also appear in this print version.

This work is licensed under a Creative Commons Attribution 3.0 International Licence.
Licence details: http://creativecommons.org/licenses/by/3.0/.

No changes have been made to the content of these proceedings. There may be changes to pagination and minor adjustments for aesthetics.

Printed with permission by Curran Associates, Inc. (2026)

For permission requests, please contact the Institute of Physics
at the address below.

Institute of Physics
Dirac House, Temple Back
Bristol BS1 6BE UK

Phone: 44 1 17 929 7481
Fax: 44 1 17 920 0979

techtracking@iop.org

Additional copies of this publication are available from:

Curran Associates, Inc.
57 Morehouse Lane
Red Hook, NY 12571 USA
Phone: 845-758-0400
Fax: 845-758-2633
Email: curran@proceedings.com
Web: www.proceedings.com

TABLE OF CONTENTS

VOLUME 1

Preface

Peer Review Statement

POWER ELECTRONIC EQUIPMENT AND SYSTEM

Study of the Mechanism of Tangent Bifurcation in Voltage Mode Controlled DCM Buck Converter 1
Lingling Xie, Fangzheng He, Shuwei Situ

Research on Transformer Fast OLTC System 10
Jinlong Tu, Yongsheng Mi

Research on the Design of the Fuzzy Control System of Full Bridge DC Converter 18
Wenda Liu, Shutian Liang, Chuxue Hao

Measurement of Inrush Current in Transformer Based on Optical Current Transducer 27
Chu Lei, Guo Zhizhong, Chen Yue, Wang Guizhong

Study of the Standard Sine Wave Frequency Conversion Power Supply Based on Analog and
Digital Integrated Control 35
Wenbo Jia, Hongda Zhang, Dewen Zhang, Kepeng Tao, Sicong Li, Wanlin Guan

Optimal Installation of Distributed Generators Based on an Enhanced Harmony Search Algorithm 46
Ke Ji, Wentao Wang, Xingong Wang, Zheyu Wei, Jian Yang, Hua Qin

Self-Adaptive Control of Rotor Inertia for Virtual Synchronous Generator in an Isolated Microgrid 51
Hanghang Zeng, Hongsheng Su

Software Consistency Checking Method for Distribution Terminal Based on Chaotic Map 57
Yaokun Wang, Ying Liang, Changkai Shi, Shilei Guan

Three-Level Generalized Discontinuous Pulse-Width Modulation Strategy Considering Neutral
Point Potential Balance 63
Wang Jikang, Liu Jianzheng, Wang Yi, Guo Zhichao

Research on Intelligent Charging System Technology of Automobile Group 73
Kedi Yan

Coordination Between Converter-Based Wind Turbines and Synchronous Generators During Inertia
Control 78
Fei Jin, Xiaoliang Liu, Guoqiang Li, Jingli Liu, Weichao Li, Jinxin Zheng, Renzhang Liu

Research of a High Voltage and High Value Resistors Standard Device 87
Xiang Zhou, Xinli Cao, Qianghu Yan, Min Lei

Modeling and Control of DC Microgrid System Based on Hydrogen Production Load 93
Yingjun Guo, Zhe Shi, Yajie Guo, Fanyi Deng, Hexu Sun

A Novel State of Health Estimation Method for Lithium-Ion Battery in Electric Vehicles 102
Jie Fan, Yuan Zou, Xudong Zhang, Hongwei Guo

The Key Design Technology of Successive Approximation Analog-to-Digital Converter to Improve Efficient and Precision 108

Cheng Xie, Yueyue Chen, Jianjun Chen

Research on Reliability Assessment of Thyristor in HVDC Converter Valve 114

Ning Liang, Jiachen Tian, Cuicui Liu, Yating Gou, Fang Zhuo, Feng Wang

Smart Grid and Electric Power Informatization 123

Jiyuan Ren, Zhe Wang, Zhen Luo, Fuyang Liu

Construction of Power Industry Corpus Based on Data Mining and Machine Learning Intelligent Algorithm 130

Liujun Zhao, Weizheng Kong, Qiuling Wang, Lihua Song

Online Identification Method of Induction Motor Parameters Based on Rotor Flux Linkage 135

Xuefei Li, Yang Kang, Hui Wang, Jie Chen, Fei Gao

Design of Power Consumption Tester for HPLC Power Line Carrier Communication Module 144

Liu Mouhai, Zhang Chenglin, Liu Xiangbin, Chen Hao, Yang Maotao, Chen Fusheng, Shen Liman

Research on Power Enterprise Network Security Solution 152

Shang Yanwei, Lin Qiang, Fengzhi Peng

Morphological Analysis of Optocoupler Accelerated Degradation Test Data 156

Xuangong Zhang, Xihui Mu, Huizhi Li

Research on Overvoltage Distribution of HVDC Converter Valve in Special Environment 162

Ning Liang, Cuicui Liu, Yating Gou, Jiachen Tian, Fang Zhuo, Feng Wang

Calculation of Electrical Stress Distribution and Influencing Factors Analysis of HVDC Converter Valve in Special EMP Environment 168

Gao Chong, Zhou Jianhui, Zhang Jing, Gou Yating, Liu Cuicui, Tian Jiachen, Zhuo Fang, Wang Feng

Numerical Simulation of Internal Flow in Direct Burning Coal-Fired Hot Flue Gas Furnace 174

Zhang Zhenwei, Wang Peng

Virtual Synchronous Generator Grid Connected Control Method Based on Virtual Impedance 181

Bao-Zhu Shao, Guan-Feng Zhang, Jun-You Yang, Fei-Fei Gao, Feng Sun, Qing-Song Zhao

Design of Memory Test System for Measuring Transmission Lines Galloping 188

Yulong Li, Jiaxin Liu, Yunfei Jia, Yu Liu

Application of Hybrid Conjugate Gradient Algorithms in Inverse Problems of Electromagnetic Tomography 195

Li Liu, Wang Lei, Wang Zhanjun

Probabilistic Modeling of Output Characteristics Based on ECM Algorithm for Wind Farms 200

Tingxiang Liu, Xiaoying Zhang, Kun Wang, Wei Chen, Xiaolan Wang

Study on the Influence of Insulator on the Coupling Effect of Transmission Tower-Line System 208

Qingshui Gao, Shi Liu, Yi Yang, Chu Zhang

The Impact Research of Delay Time in Steam Turbine DEH on Power Grid 214

Ge Jin, Sun Cai, Shaoxiang Deng

Investigation on the Relationship Between Winding Wire Size and Total Loss of BLDC 221
Jinshun Hao, Shuangfu Suo, Yiyong Yang, Yang Wang, Wenjie Wang

Chaos Control of Bi-Directional DC-DC Converter by Resonant Parametric Perturbation Method in
a DC Microgrid 228
Lihua Wang, Xue Ni, Yue Hu, Rui Zhang

Application Research of Multi-Source Information Fusion Technology in Power Network Fault
Diagnosis 239
Shuxin Liu, Enmin Zhao, Yanjun Zhang, Jing Li, Liang Zhang, Yundong Cao

Research on Power Quality Acquisition and Reconstruction Method Based on Compressed Sensing 246
Bo Yuan

A Study of Simulation on Relationship Between Young's Modulus of Cable Joints and Interface
Pressure Based on Finite Element Method 252
Zhengbing Tian, Liying Sun, Jing Wang, Xiaoye Bai

Effect of Heat Shield on the Heating Efficiency in MOCVD Chamber by Resistive Heating 258
Lili Zhao, Zhiming Li, Jincheng Zhang, Runqiu Guo, Ligen Lu, Lansheng Feng

Study on Basic Experiment and Optimization Prediction Model of Orthogonal Electrolytic
Machining of Film Cooling Hole in High Temperature Nickel-Based Alloy Blades 263
Yulan Hu, Weiguang Hao, Guoqiang Liu, Yajie Liu, Zilin Liu

Multi-Objective Reactive Power Optimization of Hybrid AC/DC Power System Considering Power
System Uncertainty 268
Liu Wenxue, Xing Luhua, Ma Changhui, Li Wenbo

Optimization of Charging Method for Scaled EVs 276
Xincheng Zhang, Zhizhen Liu, Yan Cao, Lijin Duan, Guoshen Tang, Wenchang Liu

Operation Quality Evaluation of Power Communication Network Based on Business QOS
Indicators 282
Geng Zhang, Ke Wang, Yang Wang, Yanan Wang, Chen Xiangzhou

A Stator Flux Calculation Method for Permanent Magnet Synchronous Motor in 60° Coordinate
System 290
Xiao Tang, Zhi Zhang, Chang Liu, Caishen Wang, Zhiping Wang

Analysis of Marketing Strategy of Electricity Selling Companies in the New Situation 296
Taorong Gong, Dezhi Li, Yuting Liu, Guanhong Wang, Haoming Zhu

Study on Zero Drift of Charge Amplifier Based on MOSFET 3N165 and OPA LF356N 305
Zongjin Ren, Cong Sun, Jun Zhang, Lulu Ma, Kai Zhao

Study on Radial Vibration of Circular Piezoelectric Ceramic 315
Yazhou Mao, Jianxi Yang, Haojie Liu, Yonggang Liu, Wenjing Xu

A Method of Power Flow Calculation Considering New FACTS and HVDC 323
Quan Chen, Xiaoming Dong, Haifeng Li, Tao Jin, Jun Yi, Jingzhe Tu

Combined Heat and Power Optimal Dispatch Considering Wind Power Uncertainty 331
Chunlong Chen, Xueshan Han, Wenbo Li

Analysis of Power System Vulnerability Considering Multiple Disturbances Corresponding to Information and Physics 337
Chaochao Wang, Hua Sun, Xiaoming Dong

Influence of Ni-Cu-La-B-Coated Glass Fiber on Conductivity and Electromagnetic Shielding Performance of Coatings 346
Denggao Guan, Yang Liu, Dehao Hu, Rundong Zhou, Xueqi Guo, Yan Wang, Yao Sun, Guanli Xu, Jinghui Lin, Chuanmin Sun

A Renewable Energy Assessment Model Considering the Effect of Frequency Regulation 352
Huang Xuxiang, Han Xueshan, Li Wenbo

The Comparison of Thermal Characteristics of AC Cable and DC Cable 359
Ruiqi Zhang, Hua Sun, Xiaoming Dong

Research on the Protection Range of Bird Droppings of 110kV Transmission Line Based on ANSYS Maxwell 366
Hao Zhang, Renfei Che, Wen Du, Xiaocheng Meng, Juntao He

A Novel Aggregation Method for Doubly Fed Wind Farm 372
Lulu Li, Xueshan Han, Wenbo Li, Zhe Jiang

Research on Substation Perimeter Isolation Based on Phased Array Radar and Multi-Video Fusion Technology 379
Yuancheng Zhu, Zhipeng Lei, Wei Zheng, Hongfeng Ma, Rongzhen Xia, Deyu Song

Fault Causes Identification for Transmission Lines Based on HHT and PNN 386
Wen Du, Renfei Che, Hao Zhang, Xiaocheng Meng, Juntao He

Transmission Line Insulator Fault Detection Based on Ultrasonic Technology 394
Zheng Yao, Xin Yu, Jianchun Yao, Wei Sui, Xiaochen Yu

Insulation Defect Detection of Solid Insulating Material Based on Nanosecond Pulse Voltage 401
Yang Jianzhong, An Shengdongt, Fan Yongqiang, Yin Jianbo, Yue Yonggang

Dynamic Variance Equalization Planning Optimization Method for Power Grid System Protection Communication Network 409
Dongliang Gao, Taorui Luo, Peizhe Xin, Yudong Wang, Jun Lu, Peng Cai

System for Real-Time Transmission Control of AC Motor Temperature Data Based on Linux System 416
Li Hengjie, Wu Long, Chen Wei, Qiao Zhen

Comparison Between UofC Model and Ionosphere-Free Combination Model in PPP 422
Y. C. Zhan

Dynamic Reactive Power Compensation and Harmonic Suppression of Optical Storage Microgrid Control in Natural Coordinates 427
Guangyao Jia, Shudong Wang, Huiying Song, Yanrong Mao, Almadhehagi Luai

Comparison and Analysis of X86 Server and Minicomputer Application in Power Enterprises 434
Yang Bo, Wei Jun, Wang Hua, Wang Gang

Risk Assessment of Power Communication Network Based on LM-BP Neural Network 447
Yanan Wang, Ke Wang, Ran Zhang, Qiao Xue, Xiangzhou Chen, Geng Zhang

Research on Remote Meter Reading Scheme and IoT Smart Energy Meter Based on NB-IoT Technology 453
Xingyuan Fan, Chun Zhou, Ying Sun, Jinyang Du, Ying Zhao

Optimal Configuration of Optical Storage Microgrid Under Demand-Side Response Based on Cooperative Game 461
Wang Shudong, Mao Yanrong, Jia Guangyao, Song Huiying, Qiu Jinliang, Ding Ting

Permittivity Model for GNSS-R Telemetry Wetlands 471
Cao Xinliang, Ren Xincheng

Design of Continuous Automatic Wire-Feeding Device Based on Electric Explosive Wire 479
Ying An-Wen, Yang Jia-Zhi, Yin Xin-Zhe

Calculation and Verification of Voltage Drop When Starting Tunnel Axial-Flow Fan 485
Enshi Wang, Bihui Huang

A Novel Searching Method of Fault Chains for Power System Cascading Outages Based on Quantitative Analysis of Dynamic Interaction Between System and Components 491
Jintao She

INTELLIGENT CONTROL SYSTEM AND MECHANICAL DESIGN

Warehouse Design Model for Shuttle Based Storage and Retrieve System 502
Bin Tian, Yingying Wu

Research on Control Strategy of a Buck-Type Harmonic Injection Three-Phase Rectifier 510
Qiang Jia, Panwei Ma, Yingchuan Qi, Dong Wang, Yang Cheng

Research and Analysis of MRC and IRC Algorithm Based on L TE System 518
Jian Jin, Shan Jinjie, Zhou Jian

Rotor-Mechanical Coupled Fault Feature Extraction Based on Second-Order Blind Identification 524
Feng Miao, Ruzhi Feng, Xianli Wang

Separating for Nonlinear Mixed Rotor Fault Signals Based on Adaptive Particle Swarm Optimization 530
Feng Miao, Xianli Wang, Wei Guo

A Survey of Knowledge-Based Intelligent Fault Diagnosis Techniques 536
Sanchuan Xu

Structure and Simulation of Roadway Disaster Simulation Control System for High Temperature Smoke Drill 542
Guo Jikun, Zhang Rui

Combination of CNN with GRU for Plate Recognition 548
Fucheng You, Yangze Zhao, Xuewei Wang

Methods to Solve Salt &Pepper Noise, and Frame Dropping of Timed Address Event Representation Vision Sensor 553
Lu Yu, Zhonghe Chen, Yun Hao

A Double-Channel Iterative NFXLMS Algorithm Used in Horizontal Vibration Isolation 562
Zhang Chi, Yin Wensheng

Projectile Velocity Measurement System Based on PVDF and Data Processing Method 568
Xiaoxiao Chen, Ping Song, Yayu Zhai

Some New Results on the Finite-Time Control and Its Application to a Chemical Reactor System 574
Ziteng Guo, Caisheng Wei

Dynamic Weighing System Based on Internet of Things Technologies 581
Jing Gang Cui, Jiong Mu, Ke Cheng Liu, Yu Zhu

Application of Weighted Fusion Algorithm in Air Tightness Detection Device 589
Xiaoxiao Chen, Ping Song, Yayu Zhai

Design of Intelligent Commutation Switch System Based on HPLC Carrier Scheme 596
Liu Mouhai, Tan Haibo, Yang Maotao, Chen Hao, Wu Zhiyong, Peng Haijun

Research on Four Axis Manipulator Trajectory Tracking Based on RBF Neural Network Algorithm 602
Luo Long

Design of Multifunctional Intelligent Security Robot Based on Single Chip Microcomputer 607
Liu Penghou, Chen Haichao, Du Yanzhe

Multi-Lane Detection using CNNs and a Novel Region-Grow Algorithm 613
Yi Sun, Jian Li, Zhen P. Sun

A Survey of Cloud Computing Access Control Technology 620
Minghao Wang

The Torsion Bars System Reliability Analysis with Failure Mode in Crawler Vehicle 625
Yi Liu, Pengbo Mou, Bin Zhang

Simulation and Life Prediction of Gear Meshing Process of Gearbox of a Crawler Vehicle 629
Pengbo Mou, Yi Liu, Jian Zhou

Study on Cargo-Swing Reduction of General Gantry Crane using Hybrid Optimal Input Shaper 634
Tuo Jianzhi, Du Peng, Zhang Ke, Zhang Wei, Xie Zonghua

Design and Analysis of the Leveling Hydraulic System of the Combine Harvester 640
Heng Wang, Shukun Cao, Xiangqian Xu, Tao Han, Hejia Guo

Research on Fast Self-Learning Improvement of ADRC Control Algorithm for Film Thickness
Control System 646
Liao Xue-Chao, Zhou You, Chen Zhen-Huan

Research and Analysis of Intelligent RGV Based on Dynamic Scheduling Optimization Model 653
Zheng Wang, Zeyu Zhou, Jiawei Liu

Research on Intelligent Near-Power Early Warning System for Mechanical Vehicles 662
Yingjing Wang, Jiaxin Liu, Yunfei Jia

Design and Research of a Aero Engine Operating Status Monitoring System 668
Wei Lin, He Li-Qing, Wan Yang, Chen Hua-Jie

Motion Recognition Based on Kinect for Human-Computer Intelligent Interaction 673
Xun Pang, Bin Liang

Parameter Optimal Design and Simulation of Power System of Electric Vehicle Based on AVL-CRUISE 680
Jianwei Ma

Parameter Design and Simulation Analysis of Power System in Plug-In Hybrid Vehicle 687
Jianwei Ma

Effect of Injection Compression Process Parameters on Residual Stress of Products Based on Numerical Simulation 693
Junjie Zhu, Yanfang Chen, Wenhan Huang, Qiurong Zhang, Xiaoming Liao, Yizhi Huang, Zhiwen Qiu

Multiphysics Modelling of Warm Shot Peening of AISI 4140 Steel 701
Wang Cheng, Wang Long, Wang Chuanli

Study on the Control System of Agaricus Bisporus Picking Robot 710
Hu Xiaomei, Pan Zhaoren, Yang Shuzhen, Yu Tao

Design and Application of Visual System in the Agaricus Bisporus Picking Robot 716
Hu Xiaomei, Wang Chuan, Yu Tao

Application of Levitation Frame with Mid-Set Air Spring on Maglev Vehicles 723
Min Zhang, Ma Weihua, Shihui Luo

Study on Preparation Methods of Copper-Based Composites 732
Nianlian Li, Vanessa Bouchart, Pierre Chevrier, Hongyan Ding

Study on the Performance of the Wind Turbine Airfoil with Icing 738
Wang Long, Wang Cheng, Cheng Jie, Wang Chuanli, Li Liang

Research on Straightness Error Detection and Quality Control of Multi-Crankshaft Bores for Large Medium Speed Engine Block 745
Jiebin Yang, Xingxing Li, Chang'An Hu, Wanze Li, Jinwei Zhang

Study on Pressure Pulsation Suppression of Reciprocating Pump 751
Bin Li, Song Guo, Biao-Hua Cai, Zhao-Cun Shi

Design of Automated Guided Vehicle for Conveying Objects 759
Denggui Wang, Xinling Ma

A Hydraulic Fault Diagnosis Method Based on IMF Entropy Feature Fusion 768
Liu Min, Huang Jie, Xianhai Sun

Natural Gesture Control of a Delta Robot using Leap Motion 772
Xuchong Zhang, Ruiqiu Zhang, Liang Chen, Xianmin Zhang

Research on Manufacturing Technology of Thin-Walled Parts of Fe105 Metal Based on Laser Cladding 780
Tianbiao Yu, Yiting Bao

Research on Electrolyte Jet Assisted Laser Micromachining Technology 787
Yulan Hu, Weiguang Hao, Guoqiang Liu, Yajie Liu, Zilin Liu

Thermal Barrier Coating Processing Based on Improved Ant Colony Algorithm Process Optimization and Verification 795
Yulan Hu, Guoqiang Liu, Weiguang Hao, Yajie Liu, Zilin Liu

Experimental Study on Regression Model of Ultraviolet Laser Processing Thermal Barrier Coating Based on Response Surface Method 802
Yulan Hu, Guoqiang Liu, Weiguang Hao, Yajie Liu, Zilin Liu

Shape Optimization of Hook for Marine Crane.. 808
Yang Ji, Hu Wang, Hai-Quan Chen, Ming-Xuan Guo, Jun-Jie Wu

Design and Experimental Study of Vibration Reducing Experimental Device for Magneto-Rheological Elastomer.. 814
Tieshan Zhang, Zhong Ren

A Simple Safety Control Method for PSS Critical Gain Test 822
Siyuan Guo, Shoushou Zhang, Weijun Zhu, Li Li, Jinbo Wu

Design of Seat Clamping Device for Automobile DOF Shaker 826
Jian Zhang, Jianyu Yao, Chao Wen, Denggui Wang, Lingxia Wang

Contour Error Control of X-Y Platform Based on Nominal Model in Polar Coordinate System 833
Guirong Wang, Xinman Gong, Yingqi Li

Analysis of the Pressure Expansion of Bridge Plug Tools and Packers by Equivalent Material Method 841
Lanwen Wang, Xuanyu Sheng, Jiayue Sheng

A New Numerical Force Analysis Method of CBR Reducer with Tooth Modification.................................. 848
Xiaoxiao Sun, Liang Han

Influence of Electropulsing Treatment on Residual Stresses and Tensile Strength of As-Quenched Medium Carbon Steel.. 856
Pan Long

High Accuracy Numerical Simulation on 3D Weld-Pool Shape of Large Parts 861
Zhu Yonggang, Zuo Yanhong

Experimental Study on Cutting Force Comparison Between Inner Cooling and Outer Cooling in Zig-Zag Milling... 870
Umair Riaz, Can Liu, Guangyu Tan, Guanghui Li, Ningxia Yin, Muhammad J. Saeed

Study on Vibration Reduction of Crane Monitoring System 875
Wang Jian, Zhang Yong Kui, Yan Xin

Research on Intelligent Communication System for Circuit Breaker Condition Monitoring........................... 882
Qinghong Deng, Hao Zhang, Minfu Liao, Haoxue Zhang, Yifan Fu, Lujie Gai

Research on Energy Saving and Consumption Reduction Technology of Underground Gas Storage Compressor.. 888
Guan Tong, Liu Guiqiang, Wang Jinxiu, Wang Ping, Sun Dandan, Gao Shan, Liu Pai, Wu Qiang

Load and Stress Distribution of Thread Pair and Analysis of Influence Factors 894
Shikun Lu, Dengxin Hua, Yan Li, Fang Y. Cui, Pengyang Li

Study on the Effect of Temperature on Dynamic Characteristics of Rotor System with Straight Crack 905
Tengfei Kuaia, Changfang Zhao, Jie Ren, Guigao Le

Mechanical Productivity Design and Mechanical Process Analysis Framework Construction.........................911
Cui Li, Chen Hong' Bo

VOLUME 2

Simulation and Experimental Research on the Influence of Tool Geometries on the Cutting Force of High Temperature Alloy 917
Fengyun Yu, Chunyu Yan, Shaobing Guo, Shiwei Ma, He Wang

Research on Extraction and Analysis of Characteristic Conditions of Hub Motor for Electric Vehicles 924
Guohui Yang, Shuo Zhang, Chengning Zhang, Yongxi Yang

Design and Experimental Research of Expansion-Anchorage Device in Deepwater Pipeline........................ 934
Lan Zhang, Chengqiang Zhao, Liquan Wang, Cong Ru, Zhengbin Zhu

Study on Wear Mechanism of Diamond Particles in the Cutting of Pipeline Steel 941
Lan Zhang, Zhengbin Zhu, Liquan Wang, Chengqiang Zhao, Cong Ru

Analysis of Impact Characteristics of Diamond-Beaded Rope and Its Influence on Cutting Efficiency and Life 948
Lan Zhang, Cong Ru, Liquan Wang, Zhengbin Zhu, Chengqiang Zhao

Analysis of Mesh Stiffness of Herringbone Gear Considering Modification... 955
Wei Liu, Lunqin Duan, Runjiao Wang, Yanyan Wang

Modeling of the Stiffness of Corrugated Cardboard Considering Material Non-Linear Effect....................... 967
Jilin Ran, Changli Liu

Numerical Analysis on Fluid-Solid Coupling Cooling of Minimal Surface Lattice Structure 973
Zheng Yinzheng

Preparation and Characterization of Composite Resin Containing Anion Powder.. 984
Lili Ding, Jingli Zhu, Qian Zhao, Bin Liu

Research on Fuzzy PID Control of Forearm of Tunnel Steel Arch Mounting Machine 990
Zhi-Yuan Wang, Shi-Cheng Hu

Early Fault Diagnosis of Rolling Bearing Based on Lyapunov Exponent... 995
Guo Qingjun, Li Yang

The Effect of Longitudinal Shock Absorber on the Vibration Response of Train-Bridge Coupling System in the Articulated Train 1000
Jie Ding

Simulation and Experiment of Passive Orbit Disconnected Support 1006
Zihan Gao, Yidu Zhang, Qiong Wu

Research on Trajectory Tracking and Vibration Suppression of a Smart Flexible-Joint-and-Link Space Manipulator.. 1013
Zefeng Liu, Ruifeng Sun, Minwei Li

Effect of Ultrasonic Treatment on Morphology and Microwave Absorption Performance of ZnO Spheres 1029
Wei Huang, Shicheng Wei, Yi Liang, Bo Wang, Yuwei Huang, Yujiang Wang, Binshi Xu

Effect of Different Volume Fraction Magnetorheological Fluids on Its Shear Properties 1036
Sun Huimin, Zhu Xuli, Liu Nannan, Mou Jiefeng, Li Liang, Li Shixu

Research on Butt Joint of Ultrafine Grained Steel of Manual Arc Welding 1043
Yan Wang

Fiber Diameter Measuring Method of Textile Materials Based on Phase Information 1049
Wen Wang, Fang Zhang, Zhitao Xiao, Lei Geng, Jun Wu

Mechanism Analysis of Ferromagnetic Resonance of Electromagnetic Voltage Transformer in
Neutral Ungrounded System 1057
Tingran Sheng, Tongxin Han

Fast Aerial UAV Detection using Improved Inter-Frame Difference and SVM 1063
Li Xiaoping, Lei Songze, Zhang Boxing, Wang Yanhong, Xiao Feng

Discussion About Artificial Intelligence's Advantages and Disadvantages Compete with Natural
Intelligence 1071
Xiaofei Teng

Development of Artificial Intelligence and Effects on Financial System 1078
Minzhen Xie

A Variable Step-Size Adaptive Notch Filter for Frequency Estimation using Combined Gradient
Algorithm 1084
Huiyue Yang, Yaqing Tu, Ming Li

Smart Home System Based on Deep Learning Algorithm 1092
Yanfei Peng, Jianjun Peng, Jiping Li, Ling Yu

Bounded Noises Estimation Based on Cognitive Radio in Distributed Fusion System 1100
Shuyuan Wang, Ting Wang, Yijing Wang

Datacentre TCP Protocol of Centralized Window Control 1108
Chunlin Wang, Jianyong Sun, Wanjin Xu, Xiaolin Chen

Gas Packaging Container Based on ANSYS Finite Element Analysis and Structural Optimization
Design 1115
Zhiqin Yin, Tongdan Su, Ming He

Influence of Double Stealth Aircraft Approach Forward Support Cooperative Jamming on Radar
Detection Performance 1121
Lei Bao, Chun-Yang Wang, Hong-Bing Li, Juan Bai, Ming Tan

MD-UCON: A Multi-Domain Access Control Model for SDN Northbound Interfaces 1130
Rui Chang, Zhaowen Lin, Yi Sun, Jie Xu

Design Research on Information Coding System Under the Concept of Agile Manufacturing 1138
Zitian Liao

Research on Milling Force Prediction Model Based on Improved Particle Swarm Optimization
Algorithm 1142
Liu Ling, Qi Weiwei, Liu Tingting

Optimization of Adaptive Handover Algorithm Based on Distributed Antenna in LTE-R 1147
Ziwei Yang, Yufu Zheng, Jing Jing

A Game-Theory Approach Based on Genetic Algorithm for Flexible Job Shop Scheduling Problem............1153
Li Nie, Xiaogang Wang, Fangyu Pan

An Improved Hybrid Structure Multi-Classification Support Vector Machine 1159
Zhang Xiaoyan, Wang Qiuqiu

Hand-Eye Calibration for Flexible Manipulator...1166
Jiahao Li, Xiuzhi Li, Ao Dun, Songmin Jia, Yanjun Sun

Bounds on the Total Signed Domination Number of Generalized Petersen Graphs $P(n,3)$...........................1171
Hong Gao, Yanan Yin, Yuansheng Yang

Change Impact Analysis of Complex Mechanical Product Based on Complex Network Theory 1177
Na Zhang, Yu Yang

Influence of Variable Slip Frequency Control Strategy on Tractiv E Performance 1183
Min Zhang, Ma Weihua, Shihui Luo

Study on Detection System of Grooved Rail Based on Inertial Measurement - Laser Triangulation
Comprehensive Algorithm...1190
Hong-Yuan Zhan, Kai-Feng Guo, Yong-Jun Xie

Mimic Defense Structured Information System Threat Identification and Centralized Control 1199
Bo Zhang, Weichao Li, Xin Sun, Yufeng Zhao

SMC Chaos Control of a Novel Hyperchaotic Finance System using a New Chatter Free Sliding
Mode Control 1209
Guoliang Cai, Yanfeng Ding, Qiaoling Chen

Project Evaluation and Analysis of Metrological Verification Regulation Based on Fuzzy
Comprehensive Analysis Method.. 1215
Zhu Jing, Liu Jiaxin

Safety Adaptability of Engine Retarder (Jacobs) on Long Downhill of Expressways 1222
Liu Jiaxin, Zhu Jing

Design of Information System Vulnerability Governance Platform Based on Distributed Asset
Acquisition and Vulnerability Verification Radar... 1228
Guiquan Shen, Jiangang Lu, Wuqiang Shen, Jingzhi Huang, Weiyan Ji

Standard Architecture of China Intelligent Bus Systems.. 1234
Xianglong Liu

Distributed Scalable Abstract Reasoning Based on Dl-Lite 1240
Bin Xia

Cloud Resource Adaptive Scheduling Framework and Optimization Strategy Based on Swarm
Intelligence 1246
H. W. Zhao, S. Zhang, Y. Ruan, X. H. Jing

Research on Energy-Saving Lighting Control System of Tram Station Based on Traffic and
Passenger Flow Information.. 1251
Binjie Xiao

APPLICATION OF COMPUTER NETWORK AND INFORMATION TECHNOLOGY

The Optimization of Networking Method for the System Protection Communication Networks
Based on the Delay Analysis 1256
    Yanan Wang, Ke Wang, Ran Zhang, Geng Zhang, Yang Wang

The Research of Ship Yaw Detection Method Based on Virtual Navigation Channel 1267
    Juanjuan Shao, Shu Zhou, Xin He, Xinzheng Zhang, Xindong Liu

Configuration Generation of Aircraft Swarm Based on Communication Distance Constraint 1273
    Qixi Fu, Xiaolong Liang, Jiaqiang Zhang, Lyulong He, Chuangchuang Zhu

Parameter Estimation Algorithm and Application in Industry Design 1282
    Mei Yun, Jiang Haiyang, Lin Ying, Wang Fang, Nie Qimeng

A Multi-Model Estimation of Distribution Algorithm 1286
    Li Hao

Pruning the Deep Neural Network by Similar Function 1292
    Hanqing Liu, Bo Xin, Senlin Mu, Zhangqing Zhu

Application of Internet Segmentation Research Based on Natural Language Processing Technology
in Enterprise Public Opinion Risk Monitoring 1300
    Di Liu, Jiangwen Su, Lihua Song, Zhen Qiu

A Gesture Recognition Algorithm Based on Threedimensional Projection and Direction Chain
Code 1308
    Jie Li

Model and Design of High Temperature and Thermal-Proof Garment using Genetic Algorithm 1315
    Zhang Maoyi, Sun Lele, Jia Haolin

The Design of Analog Signal Communication System Based on Visible Light 1330
    Z. N. Zhang, H. Y. Y. Hua

Simulation Study of Dispersion Compensation in Optical Communication Systems Based on
Optisystem 1336
    Peng Xia, Li-Hua Zhang, Yao Lin

The Comparison of Crowd Counting Algorithms Based on Computer Vision 1342
    Zhaoqing Wang, Qishu Deng, Yusheng Zhao

Detector Design Based on MIMO OTA Test 1348
    Zhaoqing Wang

Research on a Fusion Gait Real-Time Recognition Algorithm 1353
    Zhi-Qiang Zhao, Meng Li, Ming-Ji Deng, Zheng Zhu, Xin Shi

Application Research of Denoising and Super Pixel Algorithm in Image Processing 1359
    Qian Sun, Li Xin, Hanxu Gao, Faliang Chang, Zengshun Zhao

Evaluation Method and Experimental Study on Stationarity of High-Precision Linear Motion 1365
    Yifan Zhou, Xingbao Liu, Yangqiu Xia, Shaopeng Cai

The Design of Image Depth Information Extraction Algorithm Based on Joint Bilateral Filtering 1373
    Lin Yong

Research on Light-Small Lens Structure Design and Weight Reduction Optimization Based on Neural Network 1378
Shubing An, Minlong Lian, Yiliang Liu, Shaofan Tang

The Theoretical Development and Prospect of Two-Dimensional Topological Insulators............................ 1386
Yichen Zhang

Calculation Formula of Positioning Error Based on Three Dimensions and Four Datum............................ 1395
Quanli Han, Hongqiang Wan, Donchen Han

Application and Realization of Ray Tracing in Network Planning of Wireless Private Network 1401
Yonghua Zhang, Juanjuan Sun

Sparse Manifold Learning Based on Laplacian Matrix 1406
Wenjing Li

Web Advertisement Detection using Naive Bayes 1412
Xin Deng, Lunqing Hou, Fei Wang

Path Planning in Mobile Wireless Sensor Networks 1418
Kezhuang Wu, Junbin Liang

Defect Detection and Recognition Based on ADABOOT-SVM Integrated Model.. 1425
Zhikai Liang, Shaozhong Cao, Yukun Tan

Unmanned Visual Localization Based on Satellite and Image Fusion 1433
Xiaodan Yang

Cooperative Warp of Two Discriminative Features for Skeleton Based Action Recognition........................ 1441
Zheng Sun, Xing Guo, Wei Li, Zhengyi Liu

Anomaly Detection Algorithm Based on FCM with Improved Krill Herd 1450
Chen Rui, Zhang Fengbin, Xi Liang

An Improved Parallelization of K-Means Algorithm Based on HADOOP 1461
Yizhuo Guo

Remote Sensing Image Building Extraction Based on Deep Convolutional Neural Network 1470
Yang Q. Yi, Wang W. An, Ma X. Dong

Speaker Identification Based on Deep Learning in FX iDeal System 1477
Yan Ke, Li Na, Chen Yutinge

The Improvement of K-NN Classifier with GA-Based Weight-Tunning Method... 1483
Wei Jin

A Route Optimization Model Based on Link State Awareness in SDN 1488
Ran Li, Zhaowen Lin, Jie Xu, Yi Sun

Numerical Simulation of Deep Learning Algorithm for Gas Explosion in Confined Space 1494
Li Qizhong, Wang Ye, Yangjia, Wang Zhongqi

Study on Temporal and Spatial Patterns of Brain in Emotional State Based on Steady State Visual Evoked Potentials 1503
Kai Yang, Ying Zeng, Li Tong, Bin Liu, Xiyu Song, Bin Yan

Parameter Analysis of Stepped Frequency Pulses Frequency Diverse Array Radar....................................... 1510
  Ming Tan, Chun-Yang Wang, Hong-Bing Li, Juan Bai, Lei Bao

WebVOS-A WebGIS Application for Volunteer Observation Ships... 1519
  Honghai Zhu, Yu Yu, Shibo Chu

Improvement of LDA Topic Mining Algorithm and Its Application in Short Text.. 1524
  Kai Li, Chunmei Li

A Security Model Based on Intelligent Decision.. 1531
  Shiping Xu, Ying Zhou, Ronghua Guo, Jiawei Du, Zhe Liu

A New Clustering Validity Index Based on K-Means Algorithm 1536
  Xiangru Hou

Object Detection Based on the Improved Single Shot MultiBox Detector 1543
  *Songmin Jia, Chentao Diao, Guoliang Zhang, Ao Dun, Yanjun Sun, Xiuzhi Li, Xiangyin
  Zhang*

Medical Image Fusion Based on Statistical Modeling 1549
  Huaijing Qu, Hengbin Wang, Hongkui Xu

Automatic Integration Testing Through Collaboration Diagram and Logic Contracts................................ 1554
  Yi Sun, Xiaohua Yang, Jie Liu, Tonglan Yu, Zhuoran Xu, Zhiqiang Wu, Zhi Chen

Multipurpose IP-Based Space Air-Ground Information Network 1562
  Abid Murtaza, Liu Jianwei

ChanDet: Detection Model for Potential Channel of iOS Applications 1575
  Guomiao Zhou, Ming Duan, Qi Xi, Hao Wu

Characterizing of Strong Normalization for Λμ-Calculus... 1583
  Xinxin Shen, Kougen Zheng

Signature Handwriting Identification Based on Generative Adversarial Networks 1594
  Siyue Wang, Shijie Jia

Object Detection on Underground Low-Quality Images... 1599
  Qi Mu, Zhiqiang He, Yankui Liu, Yu Sun

ArchiMate Customization and Architecture Repository Management Practices: For a Technology-
Intensive Enterprise 1604
  Baobao Ding, Tong Wu, Yingming Yang, Liang Dou, Tiancheng Jin

Sea Clutter Suppression Based on Sea Spikes Identification and Matrix Completion 1612
  Zhiyu Shao, Jiangheng He, Shunshan Feng

Exponential Stability for a Class of Nonlinear Singular Markovian Jump Systems with Time-Delay 1618
  Daixi Liao, Shouming Zhong, Shaohua Long

A New Molecular Encryption Model Based on Microfluidic Techniques.. 1623
  Pengcheng Ma, Heng Sun

Multi-Peak and Power Cooperative Detection Algorithm to Detect Forwarded Spoofing
Interference Signals of BOC Modulation Receivers 1629
  Zhiying Wang, Jiawen Wang, Yidong He

Construction of on-Campus 3D Model Based on GIS Technology and OpenGL 1636
    Limei Chen

HTTP Tunnel Trojan Detection Model Based on Deep Learning 1645
    Yubo He, Yuefei Zhu, Wei Lin

Design and Implementation of Connect6 Intelligent Game System 1656
    Zengyu Cai, Chunfeng Du, Xiaoshuang Guo, Jianwei Zhang

Automated Recognition of Retinopathy of Prematurity with Deep Neural Networks 1662
    Yifan Wang, Yuanyuan Chen

Research on Gait Recognition Algorithm Based on Multiinformation Perception 1667
    Huabin Liu, Zongmiao Dai, Yongkang Zheng

The Combination of Neural Network and "question Matching" Improves the Correct Rate of
Grassland Degradation Decision 1675
    Li Chunmei, Pi Wei, Dong Suo, Li Zhao

The Design and Development of Simulation System for Broad Band Wireless Communication ... 1682
    Lijuan Gao, Yu He

A Deep Learning Approach for Vehicle and Driver Detection on Highway 1688
    Peihua Lv, Yang Zhang, Xiaobo Lu, Di Zhou

Robust Modeling for Fleet Assignment Problem Based on GASVR Forecast 1696
    A. J. Su, W. D. Yang, C. Zhang, M. X. Kong

An Improved SVM Web Page Classification Algorithm 1703
    Xun-Yi Ren, Chen Shi, Dan Zhang, Wen-Si Wang

The Application of Convolutional Neural Network in Security Code Recognition 1710
    Jingtian Gu

Depth Enhancement with Improved Inpainting Order and Smoothing Method 1720
    Kang Yi, Yuting Zhao, Yiyan Lei, Jian Pan

Recognition of Speed Signs in Uncertain and Dynamic Environments 1733
    Zhilong Zhu, Gang Xu, Hongmei He, Juanjuan Jiang, Tao Wang

Structure Analysis and Generation of X.509 Digital Certificate Based on National Secret 1739
    Hua Jiang, Gang Zhang, Jinpo Fan

Research on Parking Detecting Analysis Based on Projection Transformation and Hough Transform 1745
    Xuemei Yu, Yaojie Sun

Research on the Efficiency of Beijing-Tianjin-Hebei Airport Group Based on System Dynamics 1750
    C. Y. Wang, W. W. Wu, J. Zhang

The Application of Alternating Direction Method of Multipliers on l_1-Norms Problems 1758
    Yanchen He

Pre-Flight Rerouting Combining A* Algorithm and AHP Under Severe Weather 1775
    Ding Wencan, Sui Dong

Attacking Intel UEFI by using Cache Poisoning 1782
    Dong Wang, Wei Y. Dong

A Systematic Review: Road Infrastructure Requirement for Connected and Autonomous Vehicles (CAVs) 1788
Yuyan Liu, Miles Tight, Quanxin Sun, Ruiyu Kang

Ship Detection and Tracking in Nighttime Video Images Based on the Method of LSDT 1801
L. Liu, G. Liu, X. M. Chu, Z. L. Jiang, M. Y. Zhang, J. Ye

A Novel Approach to Multi-Resolution Technique for Fast Pattern Recognition 1812
Xiaochun Liu, Xiaohua Ding, Xiang Zhou, Wen Cui

Research on PLC Information Model Based on UML Class Diagram 1820
Xu R. Yu, Yun X. Zu, Wei H. Li

A New Improved Simplified Particle Swarm Optimization Algorithm 1827
Haikuan Liu, Dachao Yue, Lei Zhang, Zhiyuan Li, Dawei Jiang

VOLUME 3

Research on Precise Maintenance Method for Green Belt of Municipal Road Based on UAV Image Sequence 1834
T. Duan, P. C. Hu, L. Z. Sang, L. Wang

TPO-MAC: Traffic-Priority-Based Opportunistic MAC Protocol for Multi-Channel Cognitive Radio Networks 1839
Gao Q. Shen, Lei Lei, Zhi L. Li

Opportunistic Routing with Available Bandwidth Assurance for High Dynamic UAV Swarms 1846
Zhi L. Li, Lei Lei, Gao Q. Shen

Image Flame Recognition Algorithm Based on M-DTCWT 1853
Wenxia Bao, Zhongzhi Zhong, Ming Zhu, Dong Liang

Defect Prediction Model for Object Oriented Software Based on Particle Swarm Optimized SVM 1859
Yanan Wang, Ran Zhang, Xiangzhou Chen, Shanjie Jia, Huixia Ding, Qiao Xue, Ke Wang

Research on Knowledge Graph Application Technology 1869
Yong Chen, Xueli Chen

Predicting Failures in Hard Drivers Based on Isolation Forest Algorithm using Sliding Window 1877
Tinglei Zhang, Endong Wang, Dong Zhang

A New Crossover Algebra of GA for Solving the Degree Constrained Minimum Spanning Tree Problems 1883
Hui Li, Kai Shi, Huanran Li, Sheng Lin, Guangping Xu, Shuangxi Li

Research on Hybrid Recommendation Model Based on PersonRank Algorithm and TensorFlow Platform 1889
Guangqi Wen, Chunmei Li

Improvement of LDA Algorithm Based on Microblog Short Text Hotspot Analysis 1898
Kai Li, Chunmei Li

Extraction of Cerebral Hemorrhage and Calculation of Its Volume on CT Image using Automatic Segmentation Algorithm 1908
Nian Wang, Fei Tong, Yongcheng Tu, Hemu Chen, Yong Zhou, Jun Tang

Thinking and Exploration on the Teaching of the Course of "College Computer Foundation" Under the Mode of "Internet Plus" 1914
Tang Xianfang, Jia Yachao, Zhang Ru

Prediction of the Anti-Inflammatory Mechanism of Clematis Chinensis Based on Network Pharmacology 1919
Xulong Huang, Junjie Hao, Yuqing Liang, Yuanmin Wang, Juan Kong, Xiangpei Wang, Feng Xu, Hongmei Wu

Streaming Information Transmission Based on OPC UA 1927
Liu Wei, Zu Yunxiao, Li Weihai

A New Point-Weighting Finite-Difference Modelling for the Frequency-Domain Wave Equation 1932
D. S. Cheng, B. W. Chen, J. J. Chen, X. S. Wang

Research on Layout Optimization of Express Parcel Transportation Network Distribution Center Based on Node Operation Process 1939
Jun Han, Shiwei He, Mingkai Bi, Zilong Song

Using Markov Constraint and Constrained Least Square Filter to Develop a Novel Method of Passive Terahertz Image Restoration 1949
Yuanmeng Zhao, Xiao Sun, Cunlin Zhang, Yuejin Zhao

A Novel Dual-Band Video Fusion Algorithm using Fast Lookup-Tables: Toward Naturalistic Color 1955
Yuanmeng Zhao, Weiqi Jin, Lingxue Wang

Terahertz / Visible Dual-Band Image Fusion Based on Hybrid Principal Component Analysis 1962
Yuanmeng Zhao, Yulong Qiao, Cunlin Zhang, Yuejin Zhao, Hong Wu

A Camouflage Effect Detection Model for Fixed Targets 1967
Yang Xin, Xu Weidong, Xiang Lei, Zhu Wannian, Tian Jiyao

Welding Defect Signal Extraction Technology Based on OMP Algorithm 1974
Qi Ailing, Lei Haijun

Calculation Method of Cross Section Area of Collapsing Dangerous Rock Based on Parallel Binocular Vision 1979
Ping Gan, Zhenzhen Ruan, Yang Wang, Lin Huang, Yanyun Li, Yangfan Huang

Automatic Measurement Algorithm of Scoliosis Cobb Angle Based on Deep Learning 1987
Yongcheng Tu, Nian Wang, Fei Tong, Hemu Chen

A Robust $\mathcal{H}_\infty$ Approach of In-Flight Calibration for UAVs with Low-Cost IMU 1994
He Gao, Chao Gao, Guorong Zhao

Invisible Information Transmission System of Visible Light Based on Interleaved Code 2001
Yuting Meng, Yunpeng Hu, Mingchao Li, Yanqun Tang

Application of Deep Convolution Neural Network in Automatic Classification of Land Use 2009
Xiaodong Ma, Guang Yang, Qunyi Yang

Research on IP Address Allocation of Tactical Communication Network 2014
Wang Huitao, Yang Ruopeng, Wufan, Zou Xiaofei

The Priority Assignment of Messages Effects on Delay Performance in VANET 2021
Jin Tian, Qianru Han

An Improved Adaptive Weighted Median Filter Algorithm 2030
Yuqin Song, Jun Liu

Research on Long-Distance Hand Recognition Based on Depth Information 2036
Yuyang Fu, Lanfang Miao, Zhifei Li

A Fault Repair Method for Workstation Cluster Based on Probabilistic Model Checking 2042
Xi Wang, Ting Chen, Chengtian Ouyang

Anomaly Detection for Time Series using Temporal Convolutional Networks and Gaussian Mixture
Model 2048
Jianwei Liu, Hongwei Zhu, Yongxia Liu, Haobo Wu, Yunsheng Lan, Xinyu Zhang

Automatic Detection of Follicle Ultrasound Images Based on Improved Faster R-CNN 2058
Tianlong Zeng, Jun Liu

Research on Network Security of Campus Network 2065
Min Huang, Wanbo Luo, Xing Wan

DATA MINING AND ANALYSIS

Online Route Planning for Cooperative Area Coverage Search of Aircraft Swarm 2071
Yueqi Hou, Xiaolong Liang, Jiaqiang Zhang, Ye Li, Jiong Wang

AdaptiveSLA: A Two-Stage Scheduling Framework for SLA Profit Maximization in Multi-Tenant
Database 2079
Yang Ji, Zhang Lin, Tang Rong

Research on Reporting Scheme of Grading Stop-and-Recharge Event of Low-Voltage Acquisition
Terminal.. 2089
R. Huang, Chenglin Zhang, L. Ouyang, M. Liu, X. Liu, Y. Su, H. Chen

Progress and Trends in Mobile Cloud Computing Research 2098
Wang Xiao-Shu

Review of Research on Blockchain Application Development Method... 2103
Yue Zeng, Yue Zhang

Research on the Random Corresponding of Privacy Data Mining in the Association Rules of Cloud
Computing... 2109
Baofeng Hui, Guoqing Jia, Shanji Chen

Discussion on Supplier Selection in the Selection of Large Civil Passenger Aircraft..................................2113
Aiping Yin, Zhangkang, Zhenxing Gao, Duoduo Zhuanga

Analysis on Error Compensation for Integrated Navigation Based on Forgotten Kalman Filter 2117
Ling Yao, Hua Zhang, Xiaoping Huang

Construction and Analysis of Campus Knowledge Payment Platform Under the Wave of Big Data 2123
Ran Wu, Qiong Wu

The Mobile Payment Based on Public-Key Security Technology... 2127
Jiabin Sun, Nan Zhang

The Advisable Technology of Key-Point Detection and Expression Recognition for an Intelligent Class System.. 2134
Yusheng Zhao, Haotian Yan, Zhaoqing Wang

Research on the Interaction Between Language and Economy.. 2140
Yusi Wu, Jie Liu, Ping Wang, Yisheng Wang, Peng Zhang, Hongyan Xi

An Expertise-Enhanced Collaborative Filtering Method for Keywords Recommendation in Searching Engine Marketing 2147
Yuan Zhu

Analyse the Data Tendency in the Public Opinion Monitoring System 2152
Jian Li

Analysis and Design of Course Website for Software Testing Based on SPOC............................ 2160
Jiujiu Yu

Product Information Modeling Based on Polychromatic Sets and Scheme Optimum Selection for Conceptual Design 2166
Li Qiang, Xu Bin, Li X. Qiang

Research of a Method for Synchronized Phasor Data Transmission Based on IEC61850 2176
Shanqiang Feng, Aijun Hou, Bochuan Gu

The Design and Creation of an Interactive E-Book: "Book of Answer" 2183
Xinyue Gui

Encrypted Tag Design for RFID Systems.. 2189
Shida Yu

Automated Sentiment Analysis of Text Data with NLTK 2195
Jiawei Yao

A Study of Speech Feature Extraction Based on Manifold Learning 2203
Cheng Hao, Ma Xin, Yugong Xu

Research and Improvement of CHI Feature Selection in Sentiment Analysis............................... 2211
Li Danyang, Fan Huimin

A Research on the Innovation and Development of Micro and Small Sci-Tech Enterprises.......... 2216
Wang Guanhui

Mode of Freight Rolling and Collecting and Distributing Based on Cloud Logistics Platform 2221
Wang Xiaokun, Wang Lei, Wang Hanyan

Research on the Integration Demonstration System of Space Information Network Based on TDRSS 2232
Dandan Fan, Mengyue Qiu, Xiaoshen Xu, Ming Chen

The Resource Aggregation and Integration Platform for Shared Development of the Direct Bank.............. 2238
Qi Wang, Zhongshi Zhang, Xiaotong Zhang, Qing Zhao

Implementation of Interactive Classroom Design Based on WI-FI Service................................... 2245
Chen Shuai, Zhang Shuifeng, Wu Tianfang

Research on the Challenges and Innovation Models of Public Management in the Age of Big Data 2252
Wang Cong, Su Xiangguo, Lou Ziqiang

Applications Research of Cluster Analysis in Chinese Acupuncture Therapy 2257
    Xiaoche Feng

Prediction of Alzheimer's Disease Based on Bidirectional LSTM 2262
    Qiao Pan, Shiyu Wang, Junhao Zhang

A Survey of the Key Technology of Software Vulnerability Mining 2267
    Dansheng Lin, Xiaoquan Wu, Qinqin Wu, Ye Liu, Jijun Zeng, Jinjun Tan

Research on Security Evaluation of Government Cloud Platform Based on Fuzzy Analytic
Hierarchy Process 2272
    Jing Chen, Feng Zhang

Big Data Mining for Investor Sentiment 2278
    Tao Cen, Qianqian Chu, Renke He

Research on Frequency Hopping Synchronization Strategies Based on TOPSIS Method 2283
    Xiaoyu Cai, Zhanjun Jiang, Qianru Liu, Haoqiang Shi

An Event Detection Method Based on Association Link Network 2291
    Lin Sun, Weijun Yang, Xinhuai Tang

A Comparative Study of Customer Complaint Prediction Model of Time Series, Multiple Linear
Regression and BP Neural Network 2297
    Xin Xu, Zhijie Sun, Li Wang, Jun Fu, Chao Wang

Anomaly Detection Based on PMF Encoding and Adversarially Learned Inference 2304
    Lin Zhang, Wentai Yang, Hua Gan, Meng Li, Xiaoming Wang, Gang Liang

Mimic Defense System Security Analysis Model 2315
    Li Qinyuan, Han Jiajia, Sun Xin, Qin Junning, Zhang Bo

The Present of Education Big Data Research in China: Base on the Bibliometric Analysis and
Knowledge Mapping 2323
    Cheng Jiang, Qi Wang, Wu Qing, Leiye Zhu, Si Cheng, Haibin Wang

Method for Creating, Updating and Maintaining a Case Library of Service Business Guidance 2329
    Xin Xu, Zhijie Sun, Jun Fu, Li Wang, Feng Xie

Design and Implementation of Body Quality Index App Based on Android 2335
    Zhao Limei, Ma Xiaotie

Emotional Analysis of Public Opinions in Colleges and Universities:Based on Naive Bayesian
Classification Method 2342
    Qingjia Wang, Kun Liu, Kun Ma

Dual-Channel Supply Chain Sale Strategies with Return Guarantee 2347
    He Jingshi

Summary of Recommendation System Development 2355
    Liu Liling

An Effective Method for Forest Fire Smoke Detection 2360
    Luxing Qin, Xuehui Wu, Yichao Cao, Xiaobo Lu

Demand Analysis of Material Reserve Optimization 2368
    Jian Gao, Shuming He, Xing Song, Liang Zhang

Mining Spatiotemporal Characteristics of Car-Sharing Demand 2371
Jing-Jing Tian, Dong-Fan Xie, Fu-Jun Ding

An Empirical Study on Regional Logistics Competitiveness in Guangdong 2377
Chunshang Wu

Evaluation Effect of Internet Word of Mouth and Application of Big Data 2384
Te Ma

Purchase Decision Under Network Environment and Application of Big Data 2388
Te Ma

Research on the Evaluation Index System of Intelligent Railway Passenger Station 2393
Shaofu Lin, Qianwen Wei, Sibin Xia

Research and Analysis on the Top Design of Smart Railway 2401
Shaofu Lin, Yafang Jia, Sibin Xia

Study on Vulnerability Rating of the Intelligent and Connected Vehicle's Cybersecurity 2408
Yangyang Liu, Zhao Wang, Yanan Zhang, Peiji Shi, Xuebin Shao

Analysis of Spatio-Temporal Distribution Characteristics of Passenger Travel Behaviour Based on
Online Ride-Sharing Trajectory Data 2421
Xianlei Dong, Lingyu Wang, Beibei Hu

Research on Road Traffic Flow Status Based on Survival Analysis 2429
Beibei Hu, Doudou Lin, Qibo Sun, Xianlei Dong

Research on the Management and Maintenance of Infrastructures in Fog Section of Motorway
Based on the MOT Model 2436
X. L. Li, Q. H. Song, B. M. Tang, C. Li

Research on Semantic Prediction Analysis of Tibetan Text Based on Word2Vec 2446
Ding Hai-Lan, Yu Hong-Zhi, Qi Kun-Yu

Research on the Characteristics of Bitcoin Price Fluctuations Based on ARCH Effect 2453
Juan Wang, Feng Tian, Jie Fu

Design and Realization of Scenes of 3D Virtual Digital Library 2460
Wang Li

End-to-End Speech Synthesis for Tibetan Lhasa Dialect 2466
Lisai Luo, Guanyu Li, Chunwei Gong, Hailan Ding

Research on China's Transport Connectivity Along Corridors of the Belt and Road Initiative 2472
Jiao Wenwen, Li Ran, Li Sicong

A Study on Location of Logistics Hubs of Hub-and-Spoke Network in Beijing-Tianjin-Hebei
Region 2478
Xiaoyu Xin, Nan Yu, Xiang Chao

Blockchain-Based Intelligent Hospital Security and Data Privacy Construction 2488
Qiuzi Huang, Shuyu Chen, Hui Zhao, Junhao Wen

Research and Analysis on the Transformation of Road Passenger Transport Industry 2494
Lili Chen, Zhijun Yi, Xiaofeng Di

Study on the Operation Mode of Suburban Railway at Home and Abroad and the Inspiration to Beijing 2499
Luxi Peng, Chao Wang, Jianhong Wu

Application Scenarios Based on SDN: An Overview 2506
Tong Li, Jinqiang Chen, Hongyong Fu

Overview of End-to-End Speech Recognition 2515
Song Wang, Guanyu Li

Air Traffic Management Process Quality Assessment Model Based on Improved Fuzzy Matter Element Analysis 2519
Pan Wei-Jun, Ren Jie, Wang Run-Dong

Real-Time Estimation of Urban Rail Transit Passenger Flow Status Based on Multi-Source Data 2528
Zhengping Tao, Jinjin Tang

Comprehensive Assessment of Green Development Level for Urban Rail Transit Enterprises Based on ANP and Entropy Weight Method 2537
Ying Tang

Simulation Study on Emergency Evacuation of Metro Stations in Fire Degradation Mode 2544
Xi Jiaojiao, Li Jin

The Evolutionary Game Analysis of Incentive Mechanism for Crowd Sensing of Public Environment 2551
Qiang Zhang, Qingqing Zhang, Xueyan Liu, Jian Dai, Xujuan Zhang

Study on Tibetan Word Vector Based on Word2vec 2559
Ning Yang, Guanyu Li, Hailan Ding, Chunwei Gong

Analysis of Seismic Anomalies of the Jiuzhaigou Earthquake 2565
Xiaoying Jiang, Xiangzeng Kong, Gongde Guo

Research on the Development Orientation and Thinking of New Media in State-Owned Enterprises 2572
Wang Han, Wang Youzi, Xia Liyu

Analysis of the Situation Faced by New Media Propaganda in State-Owned Enterprises 2578
Wang Han, Xia Liyu, Wang Youzi

A New Linguistic Decision Making Method—FLM-VIKOR 2584
G. R. Zhang, W. D. Zeng

Design and Application Research of VR/AR Teaching Experience System 2592
Baiqiang Gan

Application of Blockchain in Document Certification, Asset Trading and Payment Reconciliation 2597
Xingxiong Zhu, Dong Wang

Study on the Relationship Between the Sharing Rate of Vehicle Exhaust Pollution and the Quantity of Possession 2602
Baiyu Chen, Da Fu, Yuanyuan Yang, Xiao Li

Research on Route Planning of Aerial Photography of UAV in Highway Greening Monitoring 2609
T. Duan, P. C. Hu, L. Z. Sang

Research on Specific Eye Movement Mode of Qualified Railway Driver 2615
Chenzhe Sun, Guanglei Zhang, Xiujun Zhai

Research on Big Data Decision-Making Support Platform of New Energy Bus... 2624
Hao B. W. Feng

Linked Data Crowdsourcing Quality Assessment Based on Domain Professionalism 2629
Lu Yang, Li Huang, Zhenzhen Liu

Research on Quantitative Risk Assessment Method of Packaged Cargoes Carried by Ship Based on
Online Dynamic Big Data Fusion Technology 2638
Ru Lan, Wen Chang, Wei Shen, Zhigang Jia

An Analysis Method for Error Propagation Reachability of Component-Based Software............................ 2649
Mingqi Fan, Min Song, Zhengxian Wei, Wanfeng Mao

Thoughts on the Orientation of Mathematics Education in Colleges and Universities 2656
Liang Shushuang

Study on Chinese Technical Economy and Global Social Responsibility 2662
Huanping Zhang, Maohua Li, Zoltán Zéman

Research on Semantic Information Retrieval Model of Bamboo & Rattan Domain Based on Query
Extension 2669
Lin Peng, Ming-Ming Lai, Xin Zhang

Research on Semantic Information Retrieval Model of Bamboo Rattan Domain Based on Semantic
Relevance 2674
Lin Peng, Xu-Peng Kou, Lin-Nan Yang

Emergency Event Matching using Hierarchical Blocking Method 2679
Chang Wen, Yu Liu

Research and Construction of Yunnan Plant Vertical Retrieval System 2686
Lin Peng, Zhi-Run Ma, Xin Zhang

A Social Network Water Army Detection Model Based on Artificial Immunity... 2692
Hanhua Zhang, Tao Li, Yuqiao Wang

Design and Implementation of Ontology Knowledge Base of Endemic Genera of Seed Plants in
Yunnan Province.. 2701
Lin Peng, Xu Li, Lin-Nan Yang

The Applications of the Edge Detection on Medical Diagnosis of Lungs 2706
Wei Xu, Jinping Li, Hongwei Jia

A Novel Rating Style Mining Method to Improve Collaborative Filtering Algorithm.................................. 2712
Wei Yang, Sheng H. Guo, Chun J. Zhang

Study on Traffic Organization and Work-Zone Optimization of Four-Lane Freeway Reconstruction
and Expansion 2720
Xuanyu Ye, Yazhen Chen, Chen Chen

Review on Studies of Machine Learning Algorithms.. 2727
Peiyuan Xu

Spatial Spillover Effects of the Real Estate Industry on Economic Development -From Destocking Perspective... 2734
    Xiaoqian Liu, Qinqin Zhou, Chang'An Wang

Application of Big Data in Forecasting Traffic Flow 2741
    Luo Wanbo, Wan Xing, Huang Min

Financial Risk Analysis and Early Warning Research Based on Data Mining Technology 2746
    Yan Hou, Ziyan Yuan

Author Index

ISPECE

IOP Publishing

Research on precise maintenance method for green belt of municipal road based on UAV image sequence

T Duan[1,2], P C Hu[1,2], L Z Sang[1], L Wang[1]

[1]China Transport Telecommunications & Information Center, Beijing, 100011

[2]National Engineering Laboratory for Transportation Safety and Emergency Informatics, Beijing, 100094

duantaohao@126.com

Abstract. Urban road greening refers to the planting of trees, flowers and grass and road protection forests on both sides of roads and separation belts to achieve the purpose of noise isolation, air purification and environment beautification. It will bring a huge amount of work to the maintenance of the road green belt, because of the conflict between the diversity of plant growth and the standardization of road greening. However, with the rapid development of high technology, it is of great significance to develop a high-throughput inspection and monitoring platform for the qualitative and quantitative evaluation of road greening projects. The high-throughput unmanned aerial vehicle (UAV) platform can be used for monitoring the plant canopy structure of road greening belt, which has the advantages of high efficiency, high reduction degree, low cost and so on. In the future, it will become a feasible technical means of road greening project maintenance. Based on the acquisition of plant growth status information of road green belt, the application prospect of high and new technology in the monitoring of functional indexes of road green belt is explored. Then, it can be a technical support for road safety, green belt design, structure optimization and maintenance management.

1. Introduction

1.1. The significance of road greening
With the rapid development of the national economy, the urban construction has shifted from the previous only function to the combination of function, landscape and ecology, gradually. As the artery of urban infrastructure, road greening takes up a relatively heavy weight. Road greening can not only play the role of ecological protection, but also form the green isolation belt and linear green landscape of urban[1]. There are three main functions for road greening, environmental protection, landscape effect and traffic safety. Environmental protection can be achieved through air purification, noise reduction and water and soil conservation. The landscape effect can be achieved through the combination of a variety of plants and pruning of the canopy structure of plants, then improving the stiff monotonous driving environment[2]. Road central green belt is an important part of highway construction to realize the traffic safety function by the anti-glare effect, the guidance to the driver's vision and the alleviation of mental fatigue. It can alleviate the driver's visual fatigue in the daytime and prevent glaring at night effectively by planting green plants with a recommended height and canopy width in the edge of the road and the central separation zone, so as to guarantee the driving safety[3].

Content from this work may be used under the terms of the Creative Commons Attribution 3.0 licence. Any further distribution of this work must maintain attribution to the author(s) and the title of the work, journal citation and DOI.

Published under licence by IOP Publishing Ltd

1.2. Road greening maintenance status

There always are a variety of landscape plants planted in the green belt of municipal road, whose growth conditions vary from one to the other. It can affect the normal driving of vehicles when plants grow too much, that will encroach on the driveway and requires inspection and pruning regular [4]. Dwarfness or death from disease or traffic accident will affect the landscape of municipal roads, so the treatment or retransplantation was required regular. At the present stage, as most of the maintenance work of road greening projects still relies on sampling test in China, which is mainly based on the inspection by workers and vehicles. It can be a lot of work but not very accurate. At the same time, due to the particularity of the location of road green belts, the personal safety of maintenance workers cannot be guaranteed in the high-risk working environment.

1.3. Advantages of UAV in road greening application

UAV has the characteristics of remote operation, real-time acquisition of high-resolution data and low cost. So that it can solve the difficulties in maintaining heavy workload, labor cost and work safety of road green engineering by applying UAV to the daily maintenance and monitoring of vegetation in municipal road green belt at present [5]. Meanwhile, UAV aerial survey has included in the recommended application method of road greening engineering detection in the newly revised highway engineering quality inspection and evaluation standard of 2018 (JTG F80/1-2017) [6], which will also provide policy support for the application and promotion of UAV technology in road greening maintenance.

1.4. The foreground of UAV in road greening application

With the development of computer technology, a comprehensive evaluation method combining qualitative characteristics with quantitative analysis of the highway greening way is proposed based on the development and application of efficient intelligence testing platform. It can directly reflect the function, status and maintenance level of road greening, and provide reference basis for rational utilization, management and improvement of road greening. At present, only qualitative description and evaluation of highway greening is carried out, lack of quantitative parameter evaluation system. Based on the maintenance needs of the greening project and the quantitative analysis of aerial image data of UAV [7], a quantitative monitoring platform was founded for road greening in the paper. The efficient and non-destructive technology based on UAV platform is explored to meet the quantitative monitoring of highway greening function index in the transportation field.

2. Materials and methods

In this study, there were 6 basic processes for each aerial photography experiment (figure 1). It includes setting of aerial photography equipment, sensor parameters and flight parameters of aircraft, design and pre-flight simulation of specific flight missions according to aerial photography requirements, planning of autonomous flight routes after adjustment; calibration of original aerial images and reference of geographic information; the reconstruction of ortho-mosaic, three-dimensional point cloud and digital surface model (DSM) for the whole monitoring scene based on the preprocessing of the calibrated aerial image sequence, and target region identification and feature parameter extraction according to different research purposes.

ISPECE   IOP Publishing

IOP Conf. Series: Journal of Physics: Conf. Series **1187** (2019) 042078   doi:10.1088/1742-6596/1187/4/042078

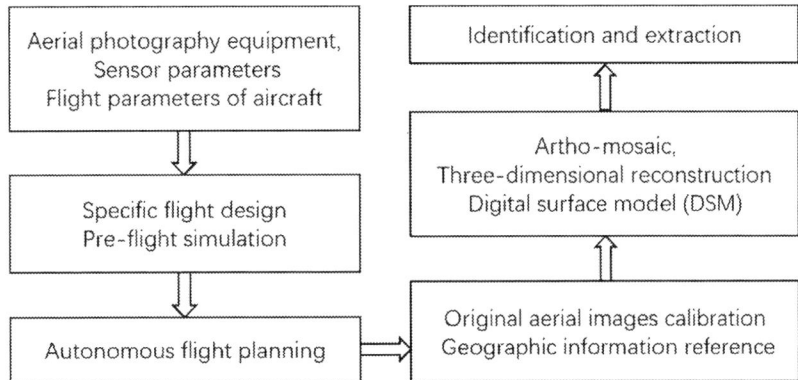

Figure 1. Processes of aerial photography experiment

2.1. UAV aerial photograph system and route planning

A four-rotor UAV monitoring platform was selected for the experiment (Figure 2C), the main hardware components including the UAV (Inspire 1 Pro), holder, and the aerial photography sensor (Chan Si X5 camera). The process of UAV aerial photography is as follows: locate the target aerial photography area on the satellite map before aerial photography, plan the flight path, set UAV and camera parameters. The open source software Mission planner for UAV was used for planning the route. The flight speed could be calculated and generated in the software after the flight height, image overlap and shooting interval are set, automatically (Figure 2A and B). The overlapping degree of the adjacent images setting was about 80%, and the camera's shooting interval was set to 2 s. In order to ensure the aerial photography area cover the target area fully, the possible deviation between satellite map and actual area was taken into account in the route planning.

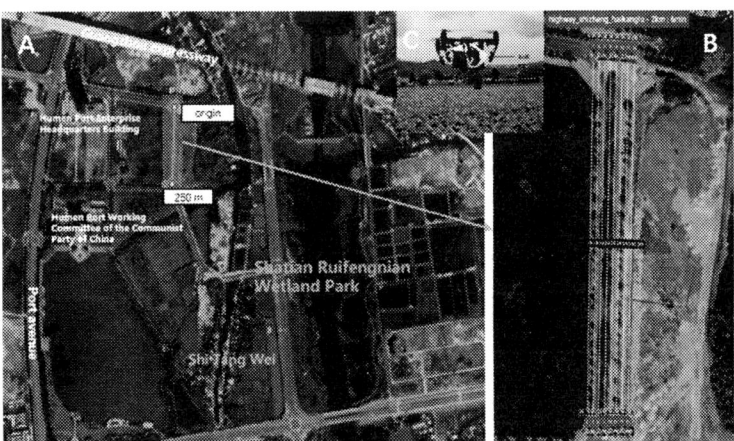

Figure 2. Route planning of UAV aerial photography for target monitoring section

2.2. Aerial image sequence processing

After exporting aerial images from drone storage devices, ortho-mosaic, 3D point clouds and DSM were generated using Agisoft PhotoScan software. The GPS positioning system, which is carried by the drone, can get geographic information of the image sequence. Firstly, feature matching points between adjacent image pairs were found in the original image sequence, and dense 3D point cloud of the whole monitoring area was generated based on the key matching points. The generated 2D ortho-mosaic and DSM carried the color and elevation values of each reconstruction point, respectively. Both of the reconstructed 3D point clouds carried the color and height value, and 2D ortho-mosaic covered the entire scene within the flight area.

3. Results

3.1. Reconstruction and visualization of road target section scene based on UAV platform

After obtaining the image sequence based the UAV near-sensing platform, the image data can be processed to obtain different forms of panoramic reconstruction models. The aerial shot monitors the target section with an actual length of 252 m, and adopts the flight height of 30 m with a total of 8 routes, and the flight duration is 6 min. A total of 184 effective original graphs are obtained. Thus, the 2D ortho-mosaic, 3D point clouds reconstruction and DSM of the target monitoring scene can be obtained (Figure 3). The reconstructed 3D dense point cloud is composed of about 17,300,000 points, and each point carries the spatial position coordinates and RGB information. The corresponding pixel of reconstructed DSM carries its relative elevation value, which can be used to represent the elevation change of target monitored section.

3.2. Identification of problem areas

The original image size of the single image obtained by the camera on the UAV platform is 4000 × 2250, and when the flight height is 30 m, the corresponding GSD is 0.67 cm. This resolution can be used to identify individual plants with the naked eye, and the morphological differences between neighboring plants can be compared by visual discrimination. The volume of single-row trees on both sides of the road can be extracted through pixel classification, and the growth difference between adjacent trees on both sides of the road can be compared and analyzed, so as to identify unhealthy plants effectively and accurately (Figure 3 red box marked). It can identify the area where excessive foliage needs to trim in the central green belt accurately (Figure 3 yellow box marked), through the comparison of area of central green belt in the early stage of road planning and the actual growth obtained from the reconstruction based on UAV.

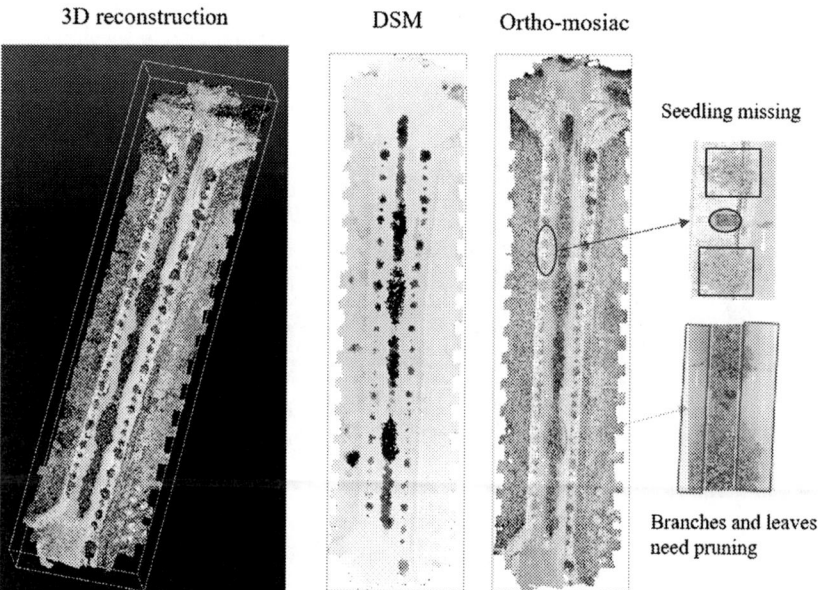

Figure 3. Reconstruction, visualization and accurate recognition of road target section scene based on UAV platform

4. Conclusions

According to the actual requirements of road green belt maintenance work, the functional indexes of the greening project are qualitatively described in this paper, systematically. Then, a precise monitoring method of municipal road green belt based on the UAV near-sensing telemetry platform is proposed, combined with the requirements of quantitative analysis technology in the later maintenance work. The feasibility in the research and application of UAV aerial image in road green belt vegetation is further determined through the overall study of the method for the plant structure characteristic parameters obtaining based on image data at the present stage. Through the geographic information carried by aerial images and image information interpretation technology, the location of the green belt areas that need to be pruned or replanted can be found accurately. A lot of road green belt parameters can be extract automatically, such as the quantification of greenbelt richness, plant morphology and health status, based on the UAV precise maintenance platform of road green belt, which is equipped with multi-spectral sensor and high-resolution camera to obtain multi-source image set. Combined with the evaluation of the design parameters of road green belt, the application direction and prospect of aerial image of UAV in the precise maintenance of road green belt can be further clarified. In the future, the dynamic growth changes of structural characteristic parameters of road green belt will be realized based on the high-throughput UAV platform, which can monitor the target road section periodically, providing theoretical and technical support for the design, tree species selection, structural optimization and pruning management of road green belt.

Acknowledgements

We would like to acknowledge Bangyou ZHENG (CSIRO) for his generous help with the photography and software used for 3D reconstruction, Wei GUO (The University of Tokyo) for his help in image interpretation algorithm, and Yan GUO (China Agricultural University) for his hardware support of UAV field experiments and image data processing. This study is supported by the National Key Research and Development Program of China (2017YFC0804904) and National Natural Science Foundation of China (31771678).

References

[1] Yuan L, Lu J, Xiang Q J, Song G S 2005. Research on the indexes and evaluation method of freeway green landscap. *Journal of Transportation Engineering and Information.* **3**, 39–43.

[2] Xiao D, Wen Z, Xu X, Šarić Ž 2016. Simulation and analysis of the buffer function of freeway greening on out-of-control vehicles. *Promet-Traffic Transp.* **28**, 257–265.

[3] Bi R Z, Zhao Y H, Liu H X 2011. Ergonomic analysis of anti - dazzle tree planting in highway central partition zone. *Highway.* **5**, 188–192.

[4] Marcucci D J, Jordan L M 2013. Benefits and challenges of linking green infrastructure and highway planning in the United States. *Environmental Management.* **51**, 182–197.

[5] Chen Y, Song Q, Li Z, Shi L 2017. Design of highway monitoring system based on four-rotor UAV. *China Science a   nd Technology Information.* **16**, 73–74.

[6] Research Institute of Highway Ministry of Transport 2018. Highway engineering quality inspection and evaluation standards: Civil engineering (Volume I). *China Communications Press.*

[7] Guo W, Rage U K, Ninomiya, S 2013. Illumination invariant segmentation of vegetation for time series wheat images based on decision tree model. *Computers Electronics in Agriculture.* **96**, 58–66.

TPO-MAC: Traffic-Priority-based Opportunistic MAC Protocol for Multi-Channel Cognitive Radio Networks

Gao Qing SHEN[1,2], Lei LEI[1], Zhi Lin LI[1,2]

[1]College of Electronic and Information Engineering Nanjing University of Aeronautics and Astronautics, Nanjing, 211106, China.

[2]Science and Technology on Communication Networks Laboratory, Shijiazhuang, 050081, China.

leilei@nuaa.edu.cn

Abstract. Opportunistic medium access control (MAC) protocol design is a challenging issue for multi-channel cognitive radio networks (CRN). This paper provides a novel traffic-priority-based opportunistic MAC (TPO-MAC) protocol in multi-channel CRN. An important mechanism in TPO-MAC is that whether the primary users (PUs) can access the licensed channels immediately depends on their traffic priority, which alleviates the collision between secondary users (SUs) and PUs. In order to investigate the performance of TPO-MAC, a Markov chain model is further proposed to derive the saturation throughput of TPO-MAC. We validate our model by extensive simulations and compare the performance of TPO-MAC with the results from other literature. The results show that TPO-MAC performs better than the existing opportunistic MAC protocols.

1. Introduction

In the past decade, the demand for spectrum has increased dramatically due to the wildly application of wireless networks and the electronic devices. However, according to the studies conducted by Federal Communication Commission (FCC), the utilization of spectrum is extremely uneven. The unlicensed portion of the spectrum (2.4-GHz and 5-GHz bands) is nearly exhausted while the licensed portion of the spectrum is only utilized between 15% and 85%. Cognitive radio networks (CRN) as an effective way to improve the utilization of licensed spectrum have attracted significant attention in recent years. In CRN, the unlicensed users (or secondary users, SUs) opportunistically use the licensed spectrum which is unused by licensed users (or primary users, PUs). The key problem in CRN is to allocate the resources (spectrum) among the users. Thus, spectrum allocation strategy, i.e., medium access control (MAC) protocol design becomes an essential problem in CRN.

Within this context, researchers have proposed plenty of MAC protocols for CRN. Zikria et al. [1] and Dappuri et al. [2] designed an opportunistic MAC protocol based on 802.11 DCF under the condition of multi-channel scenario and proposed an analytical model to analyse the performance of their proposed MAC protocols, respectively. Thilina et al. [3] proposed a dynamic common-control-channel-based MAC protocol which eliminates the requirement of a dedicated channel for information exchange. However, all of them assume that the SUs have to vacate the licensed channel whenever the PUs return while the traffic priority of PUs is not always high.

In this paper, we propose a traffic-priority-based opportunistic MAC (TPO-MAC) protocol for multi-channel CRN. There are two types of PUs' traffic priority in TPO-MAC: high and low. The

---

Content from this work may be used under the terms of the Creative Commons Attribution 3.0 licence. Any further distribution of this work must maintain attribution to the author(s) and the title of the work, journal citation and DOI.

Published under licence by IOP Publishing Ltd

biggest difference between our protocol and others is that if the traffic priority is low, PUs sense the channel state first before their transmission and delay their transmission when the channel is occupied by SUs. Meanwhile, we further put forward a two-dimensional Markov chain model with pseudo state for saturation throughput analysis of TPO-MAC. Simulated results validate the accuracy of our model. We also compare the throughput obtained by TPO-MAC with other existing protocols for CRN and the results show that TPO-MAC performs better than the existing ones.

2. TPO-MAC protocol for CRN

2.1. Model assumptions

Consider there are N_s SUs and M licensed channels in a cognitive radio networks whose range is limited to a wireless local area network (WLAN). On one hand, the SUs contend to use the vacant licensed channels which are not occupied by PUs. An unlicensed channel is used as the common control channel (CCC) for contention and information exchange. On the other hand, the traffic priority of PUs can be divided into two kinds of high and low, and the proportions of them are p_h and p_l, respectively. If the traffic priority of PUs is high, the PUs can occupy the licensed channels at any time, irrespective of the SUs' transmission. However, if the traffic priority of PUs is low, they need to sense the channel state first before their transmission and delay their transmission when the channel has been occupied by SUs.

The PUs use the licensed channel in a slot-by-slot way with a fixed length T_p, and each slot is further divided into several mini-slots for SUs opportunistically access the vacant channel unoccupied by PUs. The SUs have two independent transceivers [4] and they can work simultaneously. One is used for the contention of licensed channel over the CCC while the other is worked for tuning, sensing and transmitting over any one of the licensed channel.

2.2. Details of TPO-MAC

Assume that SUs work in saturated condition, i.e., SUs always have a packet available for transmission. Each SU attempts to access the channel on the basis of carrier sense multiple access with collision avoidance (CSMA/CA). Similar to IEEE 802.11, TPO-MAC also adopts the binary exponential backoff (BEB) scheme to avoid collision. As shown in figure 1, each SU needs to wait a random number of backoff slots which is uniformly chosen in the range [0, CW-1] before their transmission. The value of CW is called contention window size and it is initialized to a constant value CW_{min}. Once collisions occur, the contention window size is doubled unless CW_{max} is reached. Slot time (δ) is the unit time of the contention window, it accounts for the propagation delay, and the time needed to switch from receiving to transmitting state. If the CCC is idle for a distributed interframe space (DIFS) time, then the backoff time counter decreases as long as the CCC is sensed idle, freezes when the CCC is sensed busy, and restarts when the CCC is sensed idle for a DIFS time again.

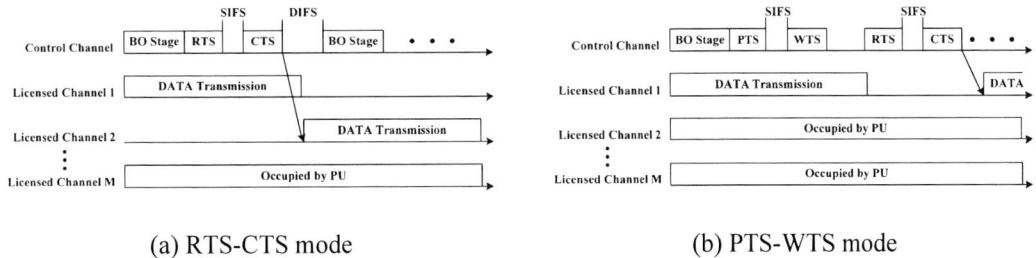

(a) RTS-CTS mode   (b) PTS-WTS mode

Figure 1. Example of TPO-MAC

When the backoff time counter reaches zero, as shown in figure 1(a), the SU sends a control frame called request-to-send (RTS) if there is at least one vacant licensed channel available. When the receiver detects the RTS frame, it responds a clear-to-send (CTS) frame after a short interframe space (SIFS) time. With a successful RTS-CTS interaction, sender and receiver acknowledges the licensed channel that will be used in the data transmission stage. If there is no available licensed channel when the backoff time counter reaches zero, as shown in figure (b), the SU sends a control frame called prepare-to-send (PTS) to reserve the priority of transmission. The receiver responds a wait-to-send (WTS) frame after a SIFS time. After the PTS-WTS interaction, all SUs suspend their backoff timer and sense the licensed channels state persistently. Once any one of the licensed channels is idle, the sender and receiver which have reserved the priority of transmission will restart the process of the RTS-CTS interaction. The rest of SUs resume the backoff timer till they receiver the CTS frame.

After a successful RTS-CTS interaction, the sender and receiver start data transmission stage on the appointed licensed channel. The receiver responds acknowledgement (ACK) upon successful reception of the packet. If there is any interference by PUs during data transmission stage, it is regarded as a failure transmission.

3. Performance analysis of TPO-MAC

3.1. Transmission probability

Consider a WLAN with N_s SUs, one control channel, M licensed channels and M PUs. We use a two-dimensional Markov chain $\{s(t), b(t)\}$ to analyse the process of backoff of SUs, where t and $t+1$ correspond to the beginning of two consecutive backoff time slots. Let $s(t)$ be the stochastic process representing the backoff stage $(0,...,m)$ of SUs. Let $b(t)$ be the stochastic process representing the backoff time counter. The value of $b(t)$ is chosen in the range of $[0, w_i-1]$, where $w_i = \min(2^i \text{CW}_{\min}, \text{CW}_{\max})$ for $i=(0,...,m)$. The w_i will be reset to $\text{CW}_{\min}$ if a packet is still not successfully transmitted until the retransmission counter achieve its maximum value. In addition, the pseudo state S_i represents the transmission stage in licensed channel. In contrast to Bianchi's model [5], other nodes which are not involved in the transmission will restart their backoff process after receiving the CTS frame in TPO-MAC. For the purpose of consistency of the model, we use a virtual backoff process $\{S_i(t), c(t)\}$ to analyse the transmission stage of SUs. As shown in figure 3, $S_i(t)$ represents the pseudo state of SUs at the backoff stage i, and $c(t)$ represents the virtual backoff time counter at time t. The value range of $c(t)$ is in $[0, T_L]$ wherein T_{L1} and T_{L2} represent the time of the successful transmission and the failed transmission, respectively.

We define the transition probability from state "a" to state "b" as $P\{b|a\}$. Let p_C be the collision probability of SUs in control channel and p_D be the collision probability of SUs in data transmission stage. The non-null one-step transition probabilities of the model are

$$\begin{cases} P\{i,j \mid i,j+1\} = 1 & i \in [0,m] \quad j \in [0, W_i - 2] \\ P\{i,j \mid i-1,0\} = \dfrac{p_C}{W_i} & i \in [1,m] \quad j \in [0, W_i - 1] \\ P\{0,j \mid S_i\} = \dfrac{1}{W_i} & i = 0, m \quad j \in [0, W_i - 1] \\ P\{0,j \mid S_i\} = \dfrac{1-p_D}{W_0} & i \in [1, m-1] \quad j \in [0, W_i - 1]. \\ P\{i,j \mid S_i\} = \dfrac{p_D}{W_i} & i \in [1, m-1] \quad j \in [0, W_i - 1] \\ P\{S_i \mid i,0\} = 1 - p_C & i \in [0,m] \\ P\{0,j \mid m,0\} = \dfrac{p_C}{W_0} & j \in [0, W_0 - 1] \end{cases} \quad (1)$$

Let $b_{i,j} = \lim_{t \to \infty} P\{s(t)=i, b(t)=j\}$, $i \in [0,m]$, $j \in [0, W_i - 1]$ be the stationary distribution of the Markov chain. First, note that

$$
\begin{aligned}
b_{i,W_i-1} &= \frac{b_{i-1,0}\,p_{\mathrm{C}} + P_{S_i}\,p_{\mathrm{D}}}{W_i} \\
&= \frac{b_{i-1,0}\,p_{\mathrm{C}} + b_{i,0}(1-p_{\mathrm{C}})p_{\mathrm{D}}}{W_i} \quad 0 < i \le m.
\end{aligned}
\tag{2}
$$

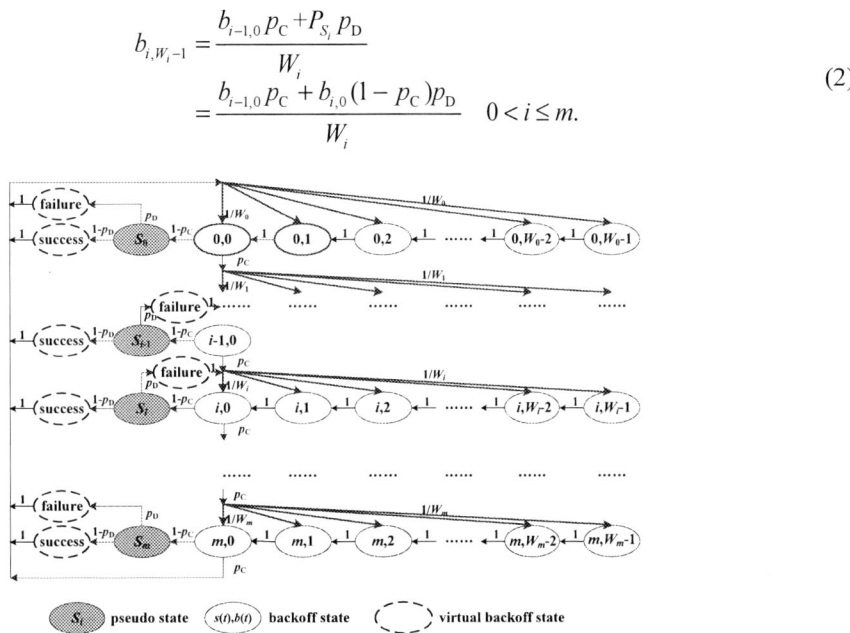

Figure 2. Markov chain model with pseudo state for TPO-MAC.

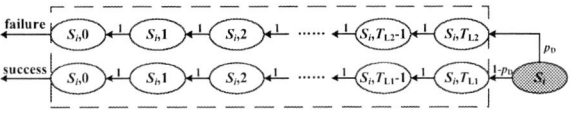

Figure 3. Virtual backoff process for transmission stage of Pus.

Owing to the regularities of the Markov chain, we have

$$
\begin{aligned}
b_{i,0} &= W_i b_{i,W_i-1} \\
&= W_i \frac{b_{i-1,0}\,p_{\mathrm{C}} + b_{i,0}(1-p_{\mathrm{C}})p_{\mathrm{D}}}{W_i} \\
&= b_{i-1,0}\,p_{\mathrm{C}} + b_{i,0}(1-p_{\mathrm{C}})p_{\mathrm{D}}.
\end{aligned}
\tag{3}
$$

Simplify equation (3), we get

$$
b_{i,0} = \frac{b_{i-1,0}\,p_{\mathrm{C}}}{1-(1-p_{\mathrm{C}})p_{\mathrm{D}}}.
\tag{4}
$$

For writing convenience, we define $\lambda = p_{\mathrm{C}}/(1-(1-p_{\mathrm{C}})p_{\mathrm{D}})$. Based on the recurrence relationship of the chain, we derive that

$$
\begin{cases}
b_{i,0} = b_{0,0}\lambda^i & 0 < i \le m. \\
b_{i,j} = \dfrac{W_i - j}{W_i} b_{0,0}\lambda^i & 0 \le i \le m, 0 \le j \le W_i - 1
\end{cases}
\tag{5}
$$

By imposing the normalization condition:

$$1 = \sum_{i=0}^{m} b_{i,0} + \sum_{i=1}^{m}\sum_{j=1}^{W_i-1} b_{i,j} + \sum_{j=1}^{W_0-1} b_{0,j} + (1-p_C)\left[(1-p_D)\sum_{i=0}^{m}\sum_{j=0}^{T_{L1}-1} b_{i,0} + p_D\sum_{i=0}^{m}\sum_{j=0}^{T_{L2}-1} b_{i,0}\right]$$
$$= b_{0,0}\left\{\left\{(1-p_C)\left[T_{L1}(1-p_D)+T_{L2}p_D\right]+1\right\}\frac{(1-\lambda^{m+1})}{1-\lambda} + \sum_{i=1}^{m}\frac{2^i W_0-1}{2}\lambda^i + \frac{(W_0-1)}{2}\right\}. \tag{6}$$

$b_{0,0}$ is finally determined as

$$b_{0,0} = \left\{\left\{(1-p_C)\left[T_{L1}(1-p_D)+T_{L2}p_D\right]+1\right\}\frac{(1-\lambda^{m+1})}{1-\lambda} + \sum_{i=1}^{m}\frac{2^i W_0-1}{2}\lambda^i + \frac{(W_0-1)}{2}\right\}^{-1}, \tag{7}$$

where

$$\begin{cases} p_C = 1-(1-\tau_C)^{N_s-1} \\ p_D = p_h \dfrac{T_L}{T_P} \end{cases}. \tag{8}$$

From (5) and (7), the probability τ_C that a SU transmits in a random slot time in control channel can be written as

$$\tau_C = \sum_{i=0}^{m} b_{i,0}$$
$$= b_{0,0}\frac{1-\lambda^{m+1}}{1-\lambda}. \tag{9}$$

3.2. Saturation throughput
Let S be the normalized system throughput, defined as the payload information transmitted per unit time. Let P_i denote the probability that the control channel is idle in a randomly given slot, which can be written as

$$P_i = (1-\tau_C)^{N_s}. \tag{10}$$

The probability of a successful RTS-CTS interaction in control channel can be written as

$$P_s = N_s\tau_C(1-\tau_C)^{N_s-1}. \tag{11}$$

Now, we can express S as

$$S = \frac{N_s\tau_C(1-p_C)(1-p_D)E[P]}{P_i\delta + P_s T_s + (1-P_i-P_s)T_f}. \tag{12}$$

Here $E[P]$ is average packet payload size, T_s is the average time that the control channel is occupied by SUs due to the RTS-CTS or PTS-CTS interaction, T_f is the average time that the control channel is sensed busy by a collision. Determining the value of T_s and T_f is not a difficult problem and we omit the detailed discussion of them.

4. Model validation and numerical results
In this section, we adopt the network simulator QualNet 4.5 to verify the proposed model. The values of the parameters we used in the simulations are listed in table 1.

Table 1. Simulation Parameters.

Parameters	Value	Parameters	Value
Propagation model	Two-Ray	SIFS	10μs
Transmission rate	11Mbps	RTS	20bits

Slot time	20μs	CTS	14bits
CW_{min}	32	PTS	20bits
CW_{max}	1024	WTS	14bits
p_h	0.5	T_p	5000δ

The simulation and numerical saturation throughput results of single SU against number of SUs for different number of licensed channels M=1, 2, 3 and 4 are compared in figure 4. On one hand, we find that the throughput of single SU is sensitive to the value of the number of SUs. It decreases when the number of SUs increases. The main reason accounts for this phenomenon is that the increasing of the number of SUs increases the collision due to contention among SUs. On the other hand, we observe that the more licensed channels, the better performance of throughput. It is easy to understand that more licensed channels bring more opportunities for transmitting.

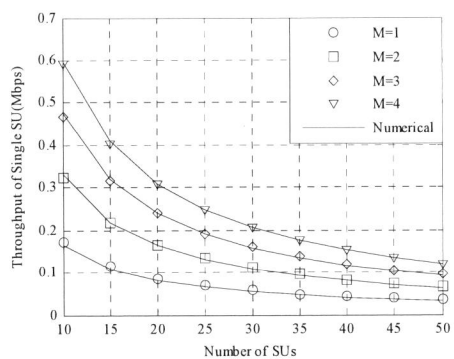

Figure. 4 Saturation throughput: numerical and versus simulation.

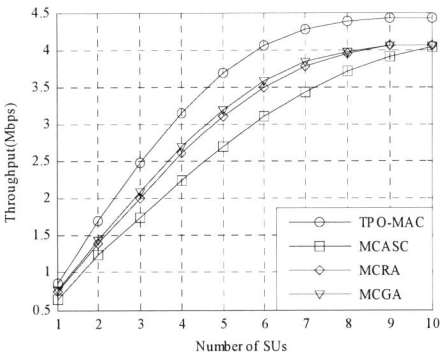

Figure. 5 Comparison of TPO-MAC with other existing protocols.

In figure 5, we compare the performance of TPO-MAC with the MAC protocols proposed in [6]. The number of licensed channels is fixed as 10. It can be seen that the throughput obtained by TPO-MAC performs better than MCASC, MCRA and MCGA. It benefits from the priority classification of PUs' traffic in TPO-MAC. Whether the PUs can access the licensed channels immediately depends on the priority of their traffic. It alleviates the conflict between PUs and SUs. Thus, there is a marginal promotion of the saturation throughput in TPO-MAC compared to the existing ones.

5. Conclusions
In this paper, we have proposed a simple but effective traffic-priority-based opportunistic MAC protocol for multi-channel CRN. The traffic priority classification of PUs decreases the collision between PUs and SUs. Meanwhile, we further propose a two-dimensional Markov chain model to analyse the saturation throughput of TPO-MAC. The accuracy of our model is validated by extensive simulations. The results show that TPO-MAC performs better than other existing protocols for CRN.

Acknowledgments
This work was supported by the National Natural Science Foundation of China (No. 61572254), the Natural Science Foundation of Jiangsu Province of China (No. BK20161488), the Aeronautical Science Foundation of China (No. 2016ZC52029).

References
[1] Y. B. Zikria, F. Ishmanov, M. K. Afzal, S. W. Kim, S. Y. Nam, and H. Yu, "Opportunistic channel selection MAC protocol for cognitive radio ad hoc sensor networks in the internet of things," Sustainable Computing: Informatics and Systems, vol. 18, pp. 112-120, 2018/06/01/ 2018.
[2] B. Dappuri and T. Venkatesh, "Design and Performance Analysis of Multichannel MAC protocol for Cognitive WLAN," IEEE Transactions on Vehicular Technology, 2018.

[3] K. G. M. Thilina, E. Hossain, and D. I. Kim, "DCCC-MAC: A dynamic common-control-channel-based MAC protocol for cellular cognitive radio networks," IEEE Transactions on Vehicular Technology, vol. 65, pp. 3597-3613, 2016.

[4] X. Zhang and H. Su, "CREAM-MAC: Cognitive radio-enabled multi-channel MAC protocol over dynamic spectrum access networks," IEEE Journal of Selected Topics in Signal Processing, vol. 5, pp. 110-123, 2011.

[5] G. Bianchi, "Performance analysis of the IEEE 802.11 distributed coordination function," IEEE Journal on selected areas in communications, vol. 18, pp. 535-547, 2000.

[6] J. Lai, E. Dutkiewicz, R. P. Liu, and R. Vesilo, "Opportunistic spectrum access with two channel sensing in cognitive radio networks," IEEE Transactions on Mobile Computing, vol. 14, pp. 126-138, 2015.

Opportunistic Routing with Available Bandwidth Assurance for High Dynamic UAV Swarms

Zhi Lin LI[1,2], Lei LEI[1], Gao Qing SHEN[1,2]

[1]College of Electronic and Information Engineering, Nanjing University of Aeronautics and Astronautics, Nanjing, 211106, China

[2]Science and Technology on Communication Networks Laboratory, Shijiazhuang, 050081, China

leilei@nuaa.edu.cn

Abstract. The reconstruction of routing schemes constitutes an important quality of service (QoS) support for UAV swarms to keep their applications and services stable and active. In this paper, we develop a forwarding distance and available bandwidth (AB) estimation based opportunistic routing (FD-ABOR) protocol for high dynamic UAV swarms. We first improve the AB estimation algorithm for multi-hop UAV ad hoc networks by taking full account of the disparity between the sending and receiving ability and reconsidering the bandwidth consumption induced by hidden nodes. Secondly, we propose the scheme that the sender broadcast the RTS frame piggybacked with available bandwidth information for route request. Based on available bandwidth and forwarding distance, neighbour node decides it to be the candidate forwarder and computes the forwarding priority, then competes with each other to forward data packets. The simulation results show that our method outperforms AODV and DSR in terms of throughput and packet delivery ratio.

1. Introduction

Recently, the great advances made in electronic, sensor and communication technologies enable the development of Unmanned Aerial Vehicles (UAVs) worldwide [1]. The UAV swarms, which is formed by a large number of mini-UAVs, is actually a function distributed and intelligent combat system. The UAV swarms have some advantages over single-UAV systems in cost, scalability, survivability and speed-up. It has been investigated increasingly due to their promising applications in warfare, such as cooperative reconnaissance, wide area surveillance, and saturation attack, etc.

Communication and networking are crucial technologies for UAV swarms to exchange information with each other and with the control station in an autonomous and collaborative manner [2]. Flying Ad Hoc Networks (FANETs) make it possible for UAV swarms to interconnect UAVs dynamically by multi-hop ad hoc networks [1]. However, several networking challenges, such as the high mobility, uneven UAVs distributions, and the fast changing topology, have made the implementation of communication for UAV swarms quite difficult [3]. One of the most prominent design problems lies on medium access control (MAC). In contrast to synchronized time division multiple access (STDMA) based MAC protocols, utilizing the carrier sense multiple access with collision avoidance (CSMA/CA) based MAC protocols in UAV swarms can obtain a satisfactory performance by taking advantage of their flexibility, robustness and reliability. Nevertheless, quality of service (QoS) provisioning is still an essential issue in CSMA/CA ad hoc networks. Available bandwidth (AB), that the maximum MAC

Content from this work may be used under the terms of the Creative Commons Attribution 3.0 licence. Any further distribution of this work must maintain attribution to the author(s) and the title of the work, journal citation and DOI.

Published under licence by IOP Publishing Ltd

throughput that can be obtained between two adjacent nodes without disrupting the existing flows, has been regarded as the fundamental information which can contribute to the design of QoS solutions [4].

On the other hand, the reconstruction of routing schemes constitutes an important QoS support for UAV swarms. Many factors make opportunistic routing (OR) the best routing decision for UAV swarms in FANETs. During the past few years, an important number of opportunistic routing protocols with various strategies are proposed to exploit its benefits [5]. The available bandwidth is regarded as a routing metric because it approximates the residual data relaying capacity of a wireless channel, and it implicitly considers the wireless channel condition [6].

In this paper, we propose a forwarding distance and available bandwidth estimation based opportunistic routing protocol (FD-ABOR) for high dynamic UAV swarms. Specifically, an available bandwidth estimation algorithm for multi-hop UAV ad hoc networks is presented firstly. In contrast to previous efforts, we take full account of the difference between the node sending and receiving capacity and reconsider the bandwidth consumption induced by hidden nodes. In our design, the sender broadcast the RTS (Request-To-Send) frame piggybacked with the available bandwidth information to its neighbors for route request. Based on available bandwidth and forwarding distance, the neighbor decides it to be the candidate forwarder and computes the forwarding priority, then competes with each other to forward data packets. Finally, we evaluate the performance of our scheme and compare it with AODV and DSR.

2. Available bandwidth estimation for FD-ABOR
We first briefly introduce the available bandwidth estimation algorithm and then give the detailed account of FD-ABOR, which aims to discover a real-time transmission path with better link quality and fewer transmission hops. A similar available bandwidth estimation algorithm can be found in our previous work [7]. We also divide the estimation algorithm into three parts: maximum available bandwidth, preliminary estimation and refined estimation.

2.1. Maximum available bandwidth
The maximum available bandwidth denoted by AB_{max} is the maximum MAC throughput a link can achieve regardless of interfering traffic. We define a transmission cycle t to be the time taken for a successful packet transmission. Based on the specification of MAC protocol, we derive AB_{max} as

$$AB_{max} = \frac{L_{DATA}}{t}. \tag{1}$$

Here L_{DATA} is the size of a DATA frame. Note that the time duration for RTS/CTS handshaking includes the maximum wait time (t_{wmax}) for forwarders to reply with CTS.

2.2. Preliminary estimation of available bandwidth
To calculate the available time of a link (T_L), the node employs its carrier sensing capability to measure the local available transmission/reception duration in an estimation period of T.

Considering the disparity between sender and recipient, the basis for identifying the available transmission time is whether the sensed signal power is less than the carrier sensing threshold and the idle interval is longer than the length of a distributed interframe space (t_{DIFS}). While the sensed signal power is less than the collision threshold for available reception time. Lastly, we obtain the total available transmission time of sender S as $T_S(S)$, and the total available reception time of recipient R as $T_R(R)$.

We define CES as the combination event that node S can serve as the sender but node R cannot be the recipient, while CER represents that node R can serve as the recipient but node S cannot be the sender. The probabilities of the above two combination events are estimated as $P_{CES}(S,R)$ and $P_{CER}(S,R)$ in [7]. Consequently, the total available time of link (S, R) can be calculated as

$$T_L(S,R) = \min\{[1 - P_{CES}(S,R)] \cdot T_S(S), [1 - P_{CER}(S,R)] \cdot T_R(R)\}. \tag{2}$$

The recipient of the link locally computes the available time, thus derives the preliminary estimation of the AB by

$$AB_{\text{pre}}(S,R) = \frac{T_{\text{L}}(S,R)}{T} \cdot AB_{\text{max}}.$$ (3)

2.3. Refined estimation of available bandwidth

We next reconsider the transmission failures induced by hidden nodes and improve the accuracy of the preliminary estimation. To elaborate the transmission failures, we introduce a representative scenario with related situations to refine the preliminary estimation in figure 1. The scenario is divided into 9 zones by the carrier sensing range (R_{cs}), the transmission range (R_{tx}) and the collision range (R_{co}) of node S and node R.

Similarly, we deduce the probability of RTS collisions and DATA collisions due to interference from hidden nodes that locate in A_7-zone area as p_{RTS} and p_{DATA} respectively. Besides, we evaluate the probability of the inability to respond with CTS due to interference from the hidden nodes within A_8-zone area as p_C. Ultimately, by deducting the bandwidth consumption induced by the above situations, the final refined estimation of the AB of a link can be expressed as

$$AB_{\text{ref}}(S,R) = AB_{\text{pre}}(S,R) \cdot (1 - p_C) \cdot (1 - p_{\text{RTS}}) \cdot (1 - p_{\text{DATA}}).$$ (4)

3. Forwarding distance and available bandwidth estimation based opportunistic routing protocol

3.1. Route discovery

During the operation of the network, each node executes the available bandwidth estimation algorithm and then acquires and stores the time duration continuously for the calculation of available bandwidth. When sending the data packet, the sender does not specify the next hop forwarder in advance and broadcast the RTS frame piggybacked with available bandwidth information to its neighbours for route request.

The RTS frame has the function of route discovery like the RREQ packet in AODV so that the necessary routing information contained in RREQ is added to RTS frame.

3.2. Candidates selection and prioritization

3.2.1. Forwarding distance

The neighbor node of the sender receives RTS frame successfully then calculates the forwarding distance according to the geographic location between itself and the destination. We denote by N_S the sender (may be the source or relay node), N_D the destination, and by N_i a forwarding node. The distance between the source and the destination is denoted by $Dist(N_S,N_D)$. Thus the forwarding distance of node N_i (denoted by $DF(N_i,N_D)$) is computed by

$$DF(N_i, N_D) = Dist(N_S, N_D) - Dist(N_i, N_D).$$ (5)

It is the Euclidian distance between the sender and the destination minus the Euclidian distance between the neighbor and the destination. We state that $DF(N_i,N_D) \in (0,R_{\text{tx}}]$, where R_{tx} is the transmission range. The primary candidate node satisfies the following three conditions: The candidate node must be the neighbor of its sender; the next hop forwarder is located between the source and the destination; the current candidate is not the neighbor of the last hop forwarder.

We define DF_{min} ($0<DF_{\text{min}}<R_{\text{tx}}$) the minimum value of forwarding distance, hence $DF(N_i,N_D) \in [DF_{\text{min}}, R_{\text{tx}}]$. Therefore, we compute the forward progress of node N_i as $f_F(N_i, N_D)$ by

$$f_F(N_i, N_D) = \frac{DF(N_i, N_D) - DF_{\text{min}}}{R_{\text{tx}} - DF_{\text{min}}}.$$ (6)

From (6), we know $f_F(N_i,N_D) \in [0,1]$. If the forward progress is greater than or equal to zero, the neighbor node can be selected as the primary candidate and added to the primary candidate relay set $\{C_i\}$.

3.2.2. Link available bandwidth

The primary candidate nodes analyse the available transmission duration of the sender ($T_S(N_S)$) from the received RTS frame. Then these nodes calculate their available bandwidth of the link consist of the sender and the candidate node as $AB(N_S,N_i)$, according to the available bandwidth estimation algorithm introduced formerly. Considering the required bandwidth of flow (RBF) in actual communication, the admission of traffic flow is controlled and the *RBF* is set to the lower limit of available bandwidth. The primary candidate can act as a qualified candidate forwarder only if it meets the minimum bandwidth requirement, i.e., $AB(N_S,N_i) \in [RBF, AB_{max}]$. We define $f_{AB}(N_S,N_i)$ the bandwidth factor and calculate it as

$$f_{AB}(N_S, N_i) = \frac{AB(N_S, N_i) - RBF}{AB_{max} - RBF} \quad . \tag{7}$$

The bandwidth factor $f_{AB}(N_S,N_i)$ is chosen as [0,1]. On the premise of satisfying the requirement of forwarding distance, the candidate relay set $\{F_i\}$ can be obtained by screening the primary candidate node under the condition that the bandwidth factor is greater than zero.

3.2.3. Forwarding priority

Based on forwarding distance and available bandwidth, the forwarding priority of candidate forwarder denoted by $pri(N_S,N_i)$ can be calculated as

$$pri(N_S, N_i) = \begin{cases} \alpha \cdot f_{AB}(N_S, N_i) + (1-\alpha) \cdot f_F(N_i, N_D), N_i \neq N_D \\ 1 \qquad\qquad\qquad\qquad\quad ,N_i = N_D \end{cases} . \tag{8}$$

Here α is a weight coefficient, and $\alpha \in (0,1)$, $pri(N_S,N_i) \in [0,1]$. The specific value of α should be decided on the network performance requirements. The priority of the destination is directly set to 1 if the candidate relay set contains the destination.

3.3. Back-off and compete to forward

The candidate forwarders set back-off time on basis of forwarding priority and compete to respond with CTS frame constructed by RTS frame. The back-off time denoted by t_{BFCTS} is set to

$$t_{BFCTS} = t_{SIFS} + [1 - pri(N_S, N_i)] \cdot t_{wmax} . \tag{9}$$

Among that t_{wmax} is the maximum waiting time and it can be valued as a time slot length (t_{slot}). The candidate nodes must wait for SIFS earlier and continues to wait for a certain time related to priority. The time duration of the sender to wait for CTS is

$$t_{WFCTS} = t_{SIFS} + t_{wmax} + N_{en} \cdot t_{CTS} . \tag{10}$$

Here N_{en} is the number of nodes enabled to reply with CTS. Meanwhile, t_{WFCTS} is considered as the maximum waiting time in the back-off process. After waiting for the long time of t_{WFCTS}, the candidate forwarder will give up the response if it has not reply with CTS.

3.4. Packet forwarding and acknowledgment

If the sender has received CTS successfully, it transmits DATA to the responding node. Otherwise it will employ the retransmission mechanism to attempt to broadcast RTS again.

If the forwarder received the desired data packet within the waiting time, it replies with ACK and prepares to forward the data packet. When other candidate forwarders receive the data frame in the process of back-off to reply with CTS, they will cancel the back-off and exit the competitive forward.

4. Performance evaluation

4.1. Methodology and parameters

We have implemented the FD-ABOR protocol in Exata3.0 simulator and evaluate the performance through simulations. The global routing algorithm, AODV and DSR, are also used in the comparison. The main typical simulation parameters are listed in table 1.

Table 1. Simulation parameters.

Parameters	Value	Parameters	Value
Transmission power	15dbm	Traffic model	CBR flow
$SINR_{th}$	10dB	Packet size	1000bytes
Transmission range	250m	RTS size	52bytes
Carrier sensing range	550m	CTS size	30bytes
Transmission rate	2Mbps	ACK size	30bytes
Synchronization time	192μs	Slot time	20μs
DIFS	50μs	SIFS	10μs

To ensure the rationality of the number of candidates in candidate relay set, we design a network topology composed of 73 UAV nodes in figure 2. The entire topology consists of adjacent regular triangles and the UAVs are placed at the intersections with 200m distance. We assign four traffic flows (f1, f2, f3 and f4) with different required bandwidth (500K, 400K, 300K, and 100K) and establish time (0s, 5s, 10s, and 15s). The estimation period is set to 0.1s, and the routing lifetime of AODV and DSR is 5s. The simulation time is 20 seconds. We compare our FD-ABOR with AODV and DSR protocols in terms of throughput, packet delivery ratio and average end-to-end delay. The three performance metrics are counted up to the computing time.

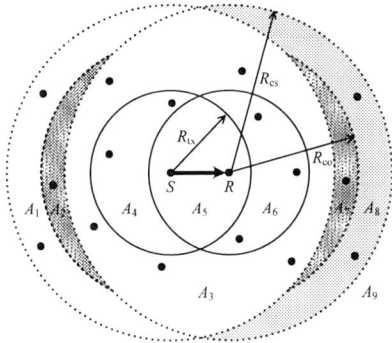

Figure 1. Scenario for refined AB.

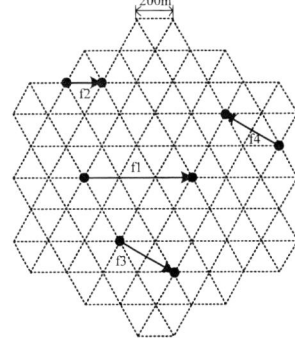

Figure 2. Simulation topology.

4.2. Simulation results

The results of throughput and packet delivery ratio of f1 over simulation time are shown in figure 3(a) and figure 3(b).

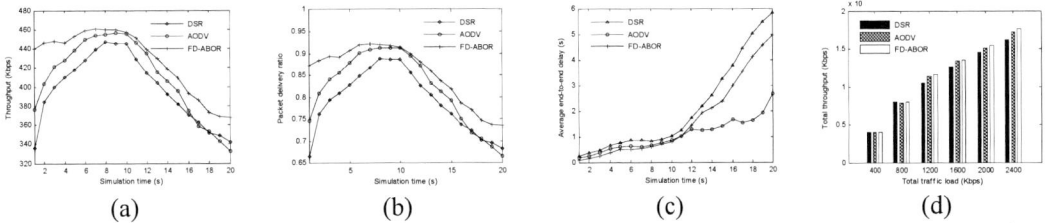

Figure 3. Simulation results. (a) Throughput of f1. (b) Packet delivery ratio of f1. (c) Average end-to-end delay of f1. (d) Total throughput.

ISPECE

IOP Publishing

Compared with the other two protocols, our approach has the best performance of throughput because it relies on the candidate relay set to forward data packets. The candidate forwarder is endowed a priority with larger available bandwidth thus can achieve higher throughput. Owing to initiating the route discovery process, AODV and DSR have lower throughput in the initialization phase. Yet FD-ABOR discovers the route by broadcasting RTS frame every time. However, the throughput declines sharply because of the significant interfere induced by flow 3, and it is aggravated due to the impact of flow 4. Predictably, the trend of packet delivery ratio is consistent with that of throughput.

The simulation results in figure 3(c) depict how average end-to-end delay of f1 varies over time. The admission of other flows aggravates competition and leads to data collisions, which makes the end-to-end delay of f1 increase gradually. Although AODV has the least delay, the throughput and delivery ratio are worse than FD-ABOR. Our method selects the relay node by considering not only forwarding distance, but also link available bandwidth. Nevertheless, it also determines that FD-ABOR does not always select the shortest path. Our scheme increases the end-to-end delay appropriately, but guarantees the QoS more importantly.

Afterwards, the four flows access the channel at the same time while the traffic load of each flow increases from 100Kbps to 600Kbps. The simulation results of total throughput shown in figure 3(d) further demonstrate the superiority of our scheme.

5. Conclusions

The design of routing schemes becomes essential and mandatory to better assist the transmission of packets for UAV swarms. In this work, we develop a forwarding distance and available bandwidth estimation based opportunistic routing protocol for high dynamic UAV swarms. We first improve the AB estimation algorithm for multi-hop UAV ad hoc networks. Based on available bandwidth and forwarding distance, the neighbor node that received the broadcast RTS decides it to be the candidate forwarder and competes to forward data packets. The simulation results show that our scheme outperforms AODV and DSR in throughput and packet delivery ratio.

In the future, combining available bandwidth and geographic location with other QoS metrics, such as delay and energy, is also an interesting attempt.

Acknowledgments

This work was supported by the National Natural Science Foundation of China (No. 61572254), the Natural Science Foundation of Jiangsu Province of China (No. BK20161488), the Aeronautical Science Foundation of China (No. 2016ZC52029).

References

[1] İ. Bekmezci, O. K. Sahingoz, and Ş. Temel, "Flying Ad-Hoc Networks (FANETs): A survey," *Ad Hoc Networks*, vol. 11, pp. 1254-1270, 2013.

[2] Y. Cai, F. R. Yu, J. Li, Y. Zhou, and L. Lamont, "Medium Access Control for Unmanned Aerial Vehicle (UAV) Ad-Hoc Networks With Full-Duplex Radios and Multipacket Reception Capability," *IEEE Transactions on Vehicular Technology*, vol. 62, pp. 390-394, 2013.

[3] O. S. Oubbati, A. Lakas, F. Zhou, M. Güneş, and M. B. Yagoubi, "A survey on position-based routing protocols for Flying Ad hoc Networks (FANETs)," *Vehicular Communications*, vol. 10, pp. 29-56, 2017.

[4] Y. Sarikaya, I. C. Atalay, O. Gurbuz, O. Ercetin, and A. Ulusoy, "Estimating the channel capacity of multi-hop IEEE 802.11 wireless networks," *Ad Hoc Networks*, vol. 10, pp. 1058-1075, 2012.

[5] N. Chakchouk, "A Survey on Opportunistic Routing in Wireless Communication Networks," *IEEE Communications Surveys & Tutorials*, vol. 17, pp. 2214-2241, 2015.

[6] M. O. Farooq, T. Kunz, C. J. Sreenan, and K. N. Brown, "Evaluation of available bandwidth as a routing metric for delay-sensitive IEEE 802.15.4-based ad-hoc networks," *Ad Hoc Networks*, vol. 37, pp. 526-542, 2016.

[7] L. Lei, T. Zhang, L. Zhou, and X. Chen, "Estimating the Available Medium Access Bandwidth of IEEE 802.11 Ad Hoc Networks With Concurrent Transmissions," *IEEE Transactions on Vehicular Technology*, vol. 64, pp. 689-701, 2015.

ISPECE   IOP Publishing

IOP Conf. Series: Journal of Physics: Conf. Series **1187** (2019) 042081   doi:10.1088/1742-6596/1187/4/042081

Image Flame Recognition Algorithm Based on M-DTCWT

Wenxia Bao[1], Zhongzhi Zhong[2] , Ming Zhu[1], Dong Liang[1]

[1]National Engineering Research Center for Agro-Ecological Big Data Analysis & Application, Anhui University, Hefei 230039, China，   Supported by the National Natural Science Foundation of China (61401001,61501003,61672032)

[2]College of Electronic Information Engineering, Anhui University, Hefei 230039, China

Corresponding author e-mail: bwxia@ahu.edu.cn

Abstract. In order to improve the effectiveness of flame image feature extraction, we propose a M-DTCWT (multidirectional dual-tree complex wavelet transform) complex frequency domain feature extraction method, combined with multi-feature fusion to achieve flame recognition in multiple scenes.First, the suspected flame region is detected by the RGB-HSI mixed color space. Secondly, Combining the filter bank in the M-DTCWT with the hourglass filter bank to construct more M-DTCWT in the diagonal direction, M-DTCWT decomposition on the suspected flame region image, and extracting the improved LBP (Local Binary Patterns)texture feature and circularity feature in the low frequency coefficients.Finally, through feature fusion, the SVM(Support Vector Machine) using the cross grid search method identifies the flame.A large number of experimental results verify the effectiveness of the algorithm.

1. Introduction

Horng et al. [1] selected the HSI color space and extracted the flame region using a fixed threshold segmentation method, which can roughly segment the flame region. Jiang B [2] proposed to a method of combining the global color features of the LAB histogram with the local SURF texture features to identify the flame, which has a certain improvement in the recognition rate compared to the single feature or local feature, but cannot distinguish between real flame and flame-like objects. Foggia, P [3] et al. proposed a fire flame detection method based on YCbCr [4] color, shape change feature in the spatial domain and flame motion information feature. Kuang-Pen Chou et al. [5] used pixel block-based analysis of local features, where local features include the color of the flame and the immobility of the flame, and the flame was further identified by the LBP feature. This method improves the flame recognition rate in a simple scene to some extent. Sam G Benjamin et al. [6] used the HSV-YCbCr color space and the five texture features of gray-level co-occurrence matrix (GLCM) to identify the flame and achieve a higher recognition rate, but the GLCM feature is a statistical texture feature, so this method has limitations in the classification of textures at the pixel level.

The above image-based flame detection and recognition method basically extracts the flame feature in the time domain or the spatial domain. Since the spatial frequency of the actual image is complex and contains a lot of redundant information, the flame feature is not used efficiently. To solve this problem, this paper proposes a flame recognition algorithm combining RGB-HSI and M-DTCWT complex frequency feature fusion.The flame image is transformed by the M-DTCWT complex frequency domain, which not only reduces the redundant information of the ILBP feature and the

Content from this work may be used under the terms of the Creative Commons Attribution 3.0 licence. Any further distribution of this work must maintain attribution to the author(s) and the title of the work, journal citation and DOI.
Published under licence by IOP Publishing Ltd

circularity feature, but also more effectively describes the geometric features of the multidirectional edges and textures of the flame image, thereby improving the recognition accuracy of the flame.

2. Suspected flame region extraction

Color is a typical feature of flame. Commonly used color spaces are HSV, RGB, HSI, YCbCr, etc. Different color spaces have different characteristics [7]. According to the characteristics of flame color, the RGB and HSI mixed color space is used to analyze the flame. The candidate flame region obtained by RGB and HSI mixed color space is then subjected to the median filtering and the closed operation by the morphological processing method, and finally the filling operation .This series of operations can better eliminate the noise and fill in small holes. This method not only eliminates some false fire areas, but also obtains a relatively complete flame region.

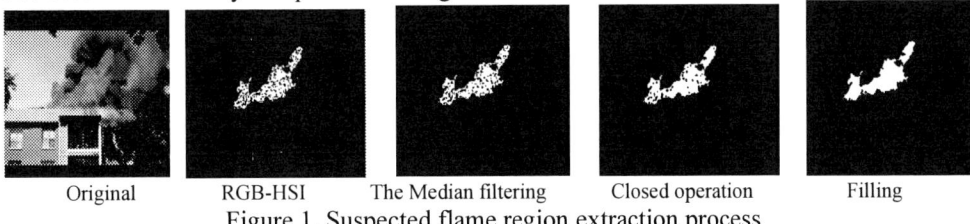

Original RGB-HSI The Median filtering Closed operation Filling

Figure 1. Suspected flame region extraction process

3. Extraction of complex frequency domain features of flame region

3.1. M-DTCWT(Multidirectional dual-tree complex wavelet transform)

In order to further increase the directional selectivity of DTCWT (Dual-Tree Complex Wavelet Transform) [8], an hourglass decomposition filter [9] combined reconstruction filter bank is added on the basis of DTCWT filter bank to construct an M-DTCWT filter bank. The input image is transformed by M-DTCWT.

If the complex scaling function is equal to
$\Phi_{i,j}(n_1,n_2)=\Phi^{re}{}_{i,j}(n_1,n_2)+\sqrt{-1}\Phi^{im}{}_{i,j}(n_1,n_2)$, the complex wavelet function can be expressed as
$\Psi_{i,j,\,k}(n_1,n_2)=\Psi^{re}{}_{i,j,k}(n_1,n_2)+\sqrt{-1}\Psi^{im}{}_{i,j,k}(n_1,n_2)$. Here n_1 and n_2 are the corresponding pixels in the image, j and k are the scaling and translation index, i are the number of directional subbands, and re and im are the decomposed real and imaginary parts, respectively.Then the two-dimensional image decomposed by M-DTCWT can be represented by a series of complex scaling functions and complex wavelet functions.

$$f(n_1,n_2)=\sum_{k\in Z^2}c_{jr,k}\Phi_{jr,k}(n_1,n_2)+\sum_{i=1}^{8}\sum_{j\geq jr}\sum_{k\in Z^2}d^{(i)}{}_{i,j,k}\Psi^{(i)}{}_{i,j,k}(n_1,n_2) \qquad (1)$$

Where Z is a natural set, $c_{j,k}=\left\langle MS^{(b)}(n_1,n_2),\Phi_{j,k}(n_1,n_2)\right\rangle$ is the scale coefficient , and $d^{(i)}{}_{j,k}=\left\langle MS^{(b)}(n_1,n_2),\Psi_{i,j,k}(n_1,n_2)\right\rangle$ is the complex wavelet coefficient in the i direction.

The multi-scale and multidirectional decomposition of the image of fire flame region extracted from RGB-HSI color space is carried out to obtain the decomposed high and low frequency subband coefficients, and then the subband coefficients are divided into four low frequency subblocks and eight high frequency subblocks.Average low-frequency subblock are obtained by averaging 4 low-frequency subblocks and average high-frequency subblock are obtained by averaging 12 high-frequency subblocks.

 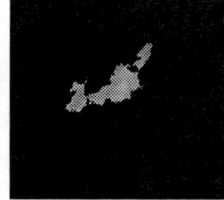  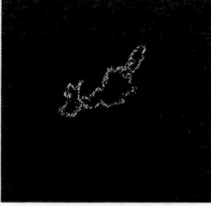

Original   Fire flame region   Average low frequency block Average high frequency block

Figure 2 .M-DTCWT decomposition process of flame image

As can be seen from Fig. 2, the flame region is decomposed into low frequency blocks and high frequency blocks by M-DTCWT, the low frequency blocks represent the detail information of the flame and the high frequency blocks represent the contour information of the flame. Therefore, the low frequency block can be used to extract the ILBP feature and the circularity feature of the flame, and the high frequency block can be used to extract the edge feature[10] of the flame.

3.2. The improved LBP texture feature in complex frequency domain

In order to make LBP [11] [12] multidirectional selectivity, we consider using the local features of the image to assign a direction to the central pixels in each domain. On the basis of the traditional LBP algorithm [11], the centroid of the circular region centered on the pixel is determined by the first and the zeroth moment of the image, and the main direction of the local texture of the region is determined by the centroid. The multidirectional LBP feature, which is very helpful for the feature similarity calculation, also improves the accuracy of fire classification.

After obtaining the central pixel of a certain field, the image moments $G(x, y)$ is calculated, and the calculation method is as follows:

$$m_{p,q} = \sum_{x,y} x^p y^q G(x, y) \qquad (2)$$

According to the image moment theory, the first moment and the zeroth moment of the image are used to calculate the centroid C of the central pixel field. The calculation formula is as follows:

$$C = \left(\frac{m_{10}}{m_{00}}, \frac{m_{01}}{m_{00}} \right) \qquad (3)$$

According to the imaging principle of the camera, the distribution of the pixels in the image area is not uniform, so the centroid C and the geometric center O of the image moment $G(x, y)$ are not in the same position. Then a vector $\overrightarrow{OC}$ can be constructed. The angle between $\overrightarrow{OC}$ and X is defined as the main direction of the central pixel in a certain field. The calculation formula of the angle θ is as follows:

$$\theta = \tan^{-1}\left(\frac{m_{01}}{m_{10}} \right) = \tan^{-1}\left(\frac{\sum\limits_{x,y} y G(x, y)}{\sum\limits_{x,y} x G(x, y)} \right) \qquad (4)$$

Through the above process, combined with the method of calculating the LBP features of rotation invariant mode [12], we can finally get the LBP features of multiple directions. Based on M-DTCWT processing of fire flame region, the low frequency coefficient features are obtained. The size of four low frequency blocks and one average low frequency block are 320*240, and each low frequency block is divided into 8*8 cells. In this paper, we use ILBP feature extraction of gray scale invariant + rotation invariant + uniform pattern, so there are nine patterns in each cell histogram. All cells of each low frequency block are combined to form a feature vector 9*64 dimensions, and the other four low frequency blocks are fused into a feature vector group 5*9*64 dimensions to describe the ILBP feature of this flame image.

3.3. The circularity feature of the complex frequency domain

The circularity is a measure of the degree of circular similarity. According to the circumference and the area of the circle, the formula for calculating the circularity is defined :

$$C = \frac{L^2}{4\pi S} \qquad (5)$$

Where L is the perimeter of the flame boundary and S is the area of the flame. The acquired flame image is transformed into binary image, 1 is fire flame region (white region) and 0 is non-fire flame region (black region). The perimeter can be obtained by the boundary algorithm. The vertical or horizontal distance between two adjacent flame pixels is unity 1. The distance between the other two points is obtained by using the Pythagorean theorem. The maximum circularity of the circular object is 1, and the more complex the object is, the smaller the value is.

Based on the M-DTCWT processing fire flame region to obtain the low frequency coefficient feature, the circularity feature of the five low frequency coefficient blocks is extracted by the above algorithm and merged into one feature vector group. The circularity feature and the ILBP feature set are optimally weighted [13], and then the circularity feature and the ILBP feature are serially fused into a set of (5*9*64+5) dimensional feature vector groups.

4. Experimental results and analysis

The algorithm flow of this paper is as follows: input training sample image for image preprocessing, then extract the suspected flame region through the color space of RGB-HSI, then process the extracted region with M-DTCWT to obtain the low frequency image, then realize ILBP feature and circularity feature extraction on low frequency images and the fusion of ILBP feature and circularity feature, and finally the optimal SVM is obtained by the cross-grid search method. The test samples are put into the optimal SVM for flame recognition.

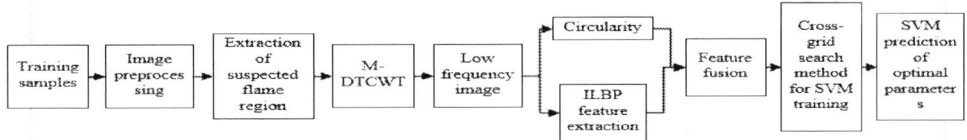

Figure 3. Algorithm flow chart

The experimental environment of this paper is as follows:Window7 system, Intel(R) Core(TM) i5-6200U CUP @2.30GHz,2.40GHz,4GB RAM. There is currently no authoritative and complete library of photo and video detection samples for fire flame detection. Part of the experimental samples comes from self-recorded video images, part of which comes from the Internet. There are 7858 images in total, and the image size is 640*480. It comes from different occasions, including night indoor and outdoor fire flame video images, night indoor and outdoor non-fire flame video images, and daytime Indoor and outdoor fire flame video images, indoor and outdoor non-fire flame video images during the day.

In the experimental data set, a total of 5800 flame images and 2400 non-flame images were selected as training samples, and the remaining 2058 were used as test samples. The SVM kernel function is selected as the radial basis kernel function and the optimal parameter kernel parameter g and the penalty parameter c [14] are obtained.The optimal parameters of RGB-HSI+LBP+circularity algorithm(Algorithm 1)is c=4, g=0.25. The optimal parameters of the algorithm are c=5.6569, g=0.25. The parameters of SVM of this algorithm and RGB-HSI+LBP+ circularity algorithm were changed to the optimal parameters, and then the test images were tested. The following is the recognition rate of the four algorithms.

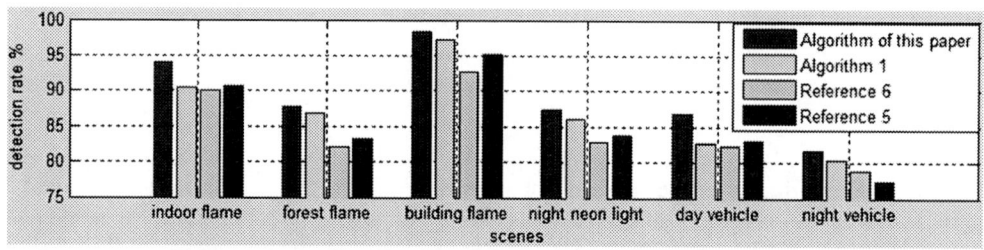

Figure 4.SVM classifier for flame detection results

The image of indoor fire and building fire is relatively clear, and the flame features are more obvious. The difference between indoor flame and surrounding environment is large, and the interference factors are small. Therefore, the recognition rate is relatively high. Forest fire flames are greatly affected by the environment such as sunlight and wind. The flame feature may be unclear, and the noise is large, so the recognition rate will decrease. For night vehicles, the recognition rate is not very high due to the influence of street lights and some police cars. The neon lights flashing at night are similar in color to the fire flame and their circularity are similar. The probability of false detection is large, which will reduce the detection rate.

The fire flame image is processed by M-DTCWT, and the circularity and texture features are extracted in the low frequency image to detect the fire flame. It can be seen from the experimental data in Table 1 that the recognition rate of the algorithm in the five scenarios is obviously higher than that of Algorithm 1, indicating that the image flame recognition algorithm based on M-DTCWT can improve the efficiency of circularity and ILBP feature extraction, and the fire recognition rate has been improved on the basis of the original, so it has a good effect on fire flame image recognition.

Table 1.Total average recognition rate

Image description	Method		Average recognition rate
Indoor flame scene	YCbCr-HSV+GLCM	Reference 6	90.01%
	Local Feature (YCbCr+Fire source anlysis)+LBP	Reference 5	90.66%
	RGB-HSI+LBP+circularity	algorithm 1	90.36%
	RGB-HSI+M-DTCWT+ILBP+circularity	this paper	93.97%
Outdoor flame scene	YCbCr-HSV+GLCM	Reference 6	87.45%
	Local Feature (YCbCr+Fire source anlysis)+LBP	Reference 5	89.1%
	RGB-HSI+LBP+circularity	algorithm 1	90.34%
	RGB-HSI+M-DTCWT+ILBP+circularity	this paper	93.08%

It can be clearly seen from Table 2 that compared with the reference [6], the reference [5]and the algorithm 1, the proposed method combining the RGB-HSI color space and the M-DTCWT complex frequency domain feature fusion method has higher and more stable recognition rate.

5. Conclusion

The suspected flame region is segmented by RGB-HSI color space, and the ILBP and circularity feature are extracted from the suspected flame region by W-DTCWT at low frequencies for flame recognition. Experiments show that the flame recognition algorithm has a high recognition rate and can adapt to many complex scenes and has good robustness. The main innovations of the experiment are: Combined with W-DTCWT, the complex frequency domain feature extraction of images can not only reduce the problem of inefficient use of features due to a large amount of redundant information, but also achieve more stable ILBP features and circularity features in the frequency domain and the high accuracy of flame recognition. On the other hand, W-DTCWT decomposes the image in multiple directions, which increases the more effective ILBP and circularity features and improves the flame recognition rate.

References

[1]Wen-Bing Horng, Jian-Wen Peng, A New Image-Based Real-Time Flame Detection Method Using Color Analysis. Proceedings of 2005 IEEE International Conference on Networking, Sending and Control,March 2005 , pp:100-105.

[2]Jiang B, Lu Yi, Towards a solid solution of real-time fire and flame detection. Multimedia Tools and Applications, 2015,74(3):698-705.

[3]Foggia, P., Saggese, A. Vento, M.Real-time fire detection for video surveillance applications using a combination of experts based on color,shape, and motion. IEEE Trans Circuits Syst Video Technol, 2015, pp:1545–1556.

[4]Premal C E, Vinsley S, Image processing based forest fire detection using YCbCr colour model[C]. International Conference on Circuit, Power and Computing Technologies. IEEE, 2014.

[5]Kuang-Pen Chou, Mukesh Prasad,Deepak Gupta, Bloack-Based Feature Extraction Model for Early Fire Detection. IEEE,August 2017.

[6]Sam G Benjamin ,Radhakrishnan B, Nidhin T G, Extraction of Fire Region From Forest Fire Images Using Color Rules and Texture Analysis. 2016 International Conference on Emerging Technological Trends [ICETT] , 2016.

[7]WU Xiyin, YAN Yunyang, DU Jing, Fire detection based on fusion of multiple features[J]. CAAI Transactions on Intelligent Systems, 2015, pp:240-247.

[8]I.W.Selesinick, R.G.Baraniuk, N.G, Kingsbury, The dual-tree complex wavelet transform. EEE Signal Processing Magezine, 2005,pp:123-151.

[9]Yue M.Lu, Minh N.Do, A Mapping-Based Design for Nonsubsampled Hourglass Filter Banks in Arbitrary Dimensions. IEEE Transactions on Signal Proceesing,2008,pp:1466-1478.

[10]OUYANG Ji-neng, BU Le-ping, YANG Zhi-kai, WANG Teng, An Early Flame Identification Method Based on Edge Gradient Feature . IEEE, 2018 , pp:642-646.

[11]Mehta R, Egiazarian K, Dominant Rotated Local Binary Patterns(DRLBP) for texture classification[J]. Pattern Recognition Letters, 2016, pp:16-22.

[12]M.Pietikainen et al, Local Binary Patterns for Still Images. Computational Imaging and Vision 40, DOI 10.1007/978-0-85729-748-8_2, 2011, pp13-47.

[13]Bo Sun, Song Shi-Ji, Cheng Wu, A new algorithm of support vector machine based on weighted feature. Proceedings of the 2009 International Conference on Machine Learning and Cybernetics, 2009, pp: 1616-1620.

[14]LUO Xiaoyan, CHEN Huiming, LU Xiaojiang, XIONG Yang, Forecast of SVM mill load based on grid search and cross validation. China measurement&Test,2017, pp: 132-136.

Defect Prediction Model for Object Oriented Software Based on Particle Swarm Optimized SVM

Yanan Wang[1], Ran Zhang[2, 1, *], Xiangzhou Chen[1], Shanjie Jia[3], Huixia Ding[1], Qiao Xue[1] and Ke Wang[2]

[1]China Electric Power Research Institute, Beijing 100192, China

[2]North China Electric Power University, Beijing 102206, China

[3]State Grid Shandong Electric Power Company Economic and Technological Research Institute, Shandong 250000, China

* Corresponding Author: Ran Zhang; *email: zrzhangran111@163.com*

Abstract. In terms of the security problem of power information system, this paper analysed the importance of the software defect prediction method in object-oriented software development, and proposed a software prediction model based on particle swarm optimized Support Vector Machine (SVM) corresponding to the features of object-oriented software. The model mainly consists of three parts: the first is the pre-processing module which normalizes the original data and selects feature, then the second is adaptive inertia weight particle swarm module which optimizes the parameters of SVM with the prediction accuracy as the fitness. Finally, the last SVM classification module predicts categories of reduced-dimension data using the optimal parameters from the second module. Experimental results show that the accuracy of the proposed model is 8.2%−12.2% higher than the comparative model, and 9.9%, 5.6% and 7.7% higher on the precision, recall and F value, which proves the validity of the proposed model.

1. Introduction

The power information system software was developed using object-oriented design techniques in the long-term development process. Object-oriented system design is the process of searching the power software architecture and the solutions of power software functional model in essence. The object-oriented power system software gradually evolves from a single function to a comprehensive development for the reason that the comprehensiveness and reliability of the functions. While the more complex of the structure and model of the software itself and the larger scale of the object, which lead to more serious security problems faced by the software of the power information system. The corresponding solutions are proposed including the allocation of resources such as human funds and the control and arrangement of development progress by determining whether the module to be tested and the object have errors which are predicted by the software defect prediction in the early stage of software development. Therefore, software defects and loopholes should be discovered and resolved as soon as possible in the process of software development to provide a necessary protection for national production and normal market operation and offer an important approach for reducing the cost and cycle of testing and improving the quality of the software in the grid [1].

Content from this work may be used under the terms of the Creative Commons Attribution 3.0 licence. Any further distribution of this work must maintain attribution to the author(s) and the title of the work, journal citation and DOI.
Published under licence by IOP Publishing Ltd

The software defect prediction technology can be divided into static and dynamic defect prediction techniques. The dynamic defect prediction technology focuses on the defect distribution and the variation of quantity over time in the whole life cycle or test phase of the software to predict the future defect distribution of the software. While the static prediction technology which is more commonly used pays more attention to the metrics of the software defect and takes defect prediction based on metrics of different defect-related attributes. It can provide corresponding prediction functions in the software analysis and design stage and in the early stage of development. The existing static defect prediction techniques are basically based on different machine learning algorithms, such as decision tree, random forest, naive Bayes, BP neural network and artificial immune system. The module attributes are obtained according to the software metrics that describes the software characteristics and then classified (defective or not) [2]. These methods have a certain degree of defect prediction ability, while imply some problems more or less. Taking some methods for example. The decision tree ignores the correlation between feature attributes due to overfitting; The naive Bayes has higher requirements for attribute independence which needs to know the prior probability. The neural network is prone to the problem of local optimization or insufficient fitting. It is necessary to obtain the factors related to the defects according to the expert experience like the Bayesian model, while with the low calculation efficiency; The Support Vector Machine has capabilities of good learning and expansion without a unified and efficient method for setting the optimal parameters. Moreover, various algorithms need to use a large number of class and object feature attributes to measure the software in dealing with the object-oriented software which lead to the "dimensionality disaster". And the predictive model is less practical with the long detection time.

The software metrics method, as a standardized method for obtaining the software architecture and module attributes, analyzes the attributes of the software entities and describes the characteristics of the software quantitatively according to the corresponding metrics essentially, which can provide the necessary data sources for the work of the software quality assessment and the metric-based software defect prediction. Traditional structured software metrics include the McCabe structural complexity metrics, LOC statement line metrics and Halstead software science metrics, etc. The complexity software metrics are aimed at the attributes within the software module, while the attributes between modules should also be taken into account because of the significance of the interaction between modules with the development of the object-oriented software technology. The attributes specific to the object-oriented software such as data abstraction, encapsulation, inheritance, polymorphism, information hiding, cohesion and coupling are unable to be extracted by the previous structured metrics in the process of the object-oriented analysis and object-oriented design [3]. The specific metrics models need to be proposed for the object-oriented software because the feature attributes of the software obtained based on the traditional metrics are not sufficient to fully represent the intrinsic characteristics of the object-oriented software. The most representative object-oriented software metrics models are the MOOD model and the CK metric proposed by Chidamber and Kemerer [4]. The CK metric contains six metric attributes that are based on a rigorous metric theory to characterize the scale and complexity of the object-oriented software design. These metric indicators include as follows: Weighted method of class WMC, depth of inheritance tree DIT, number of children in the class NOC, coupling between objects CBO, response for class RFC and lack cohesion of method LCOM. They not only contain most features of the object-oriented software, but also expand their attribute sets correspondingly in each data set to cope with the complexity of the software. The CK metric which is flexible is used for the software defect detection to test the effectiveness of our proposed model in this paper [5].

Defect prediction model for object-oriented software based on particle swarm optimized SVM (Support Vector Machine) is proposed in this paper in order to optimize the performance of the software defect prediction methods. That is, the feature selection is carried out by the relief algorithm firstly, and then the defect prediction of the model is taken to obtain the optimal prediction results, in which the model is combined the strong generalization ability of SVM with the high efficiency of optimization of the PSO (Particle Swarm Optimization) algorithm which updates the inertia weight dynamically with the degree of particle aggregation.

2. Problem description and related algorithms

2.1. Problem description
The defect detection problem of the software in the power information system is researched in this paper. The generation of the software defect is inevitable in the implementation of programming and has a significant impact on the software quality. Therefore, we need to utilize the statistical learning technology to predict the number and type of defects in software systems based on the historical data, existing fault data sets or the software metric data such as defects that have been discovered, and count the number of defects that are not found but may still exist to determine whether the system can be delivered for use [6]. In the process of using the statistical learning technology, it is necessary to reduce the dimensionality of the feature attributes of the classification set and normalize the attribute values to avoid the "dimensionality disaster" and prevent the excessive processing cost firstly. Secondly, the appropriate parameters should be set for the selected classification algorithm. While the parameters set by the expert experience are unable to show the characteristics of the new prediction problem accurately. Therefore, a suitable optimization algorithm is needed to find the corresponding parameters of these classification algorithms to obtain the optimal defect prediction results finally.

2.2. Support Vector Machine
SVM is a general-purpose feedforward neural network that can be used for pattern classification and non-linear regression [7]. The SVM learning algorithm comes from the statistical theory with the core of the empirical risk minimization principle and the basic idea of solving the kernel function and quadratic programming problem. The reason why SVM is chosen in this paper is that it is more suitable to obtain the classification optimal solution from the finite sample. And the introduction of the kernel function enables the SVM model to deal with the non-linear problem of software defect detection effectively meanwhile. The kernel function maps the data to the high-dimensional feature space which makes the sample set linearly separable in it. And the most commonly used kernel function is the Radial Basis Function which preserves the ability to map to the infinite dimensional space while ensuring the positive semi definite of the kernel matrix with the wider scope of the application and the stronger learning ability.

The training sample set is $T = \{(x_1, y_1), (x_2, y_2), \ldots, (x_n, y_n)\}, y_i \in \{-1, +1\}$ with the number of samples n in it. And the dimension of each sample feature attribute is d, that is $x_i = (x_i^1, x_i^2, \ldots, x_i^d)$. The purpose of SVM classification is to find a partition hyperplane $w^T x + b = 0$ that maximizes the interval between the two heterogeneous support vectors in the feature space. Separating samples by class produces the most stable classification results and the strongest processing capacity for the future samples. It is necessary to use the method of Lagrange multipliers to obtain the dual problem for the basic model of the SVM convex quadratic programming to find the partition hyperplane with the largest interval. The basic Lagrangian function is as shown in equation (1):

$$L(w, b, k) = \sum_{i=1}^{d} k_i \left(1 - y_i(w^T x_i + b)\right) + \frac{1}{2}||w||^2 \tag{1}$$

The dual problem as shown in equation (2) can be obtained according to the equation above.

$$\max_{k} P = \sum_{i=1}^{d} k_i - \frac{1}{2} \sum_{i=1}^{d} \sum_{j=1}^{d} k_i k_j y_i y_j x_i^T x_j$$

$$\text{s.t.} \quad \sum_{i=1}^{d} k_i y_i = 0 \tag{2}$$

x is mapped to the high-dimensional feature space to solve the problem of non-linear partitioning. The original model is converted to $f(x) = w^T \varphi(x) + b$, in which $\varphi(x)$ represents the eigenvector after mapping x to the high-dimensional feature space. The dual problem is also converted to the following equation (3).

$$\max_{k} P = \sum_{i=1}^{d} k_i - \frac{1}{2} \sum_{i=1}^{d} \sum_{j=1}^{d} k_i k_j y_i y_j \varphi(\boldsymbol{x}_i)^T \varphi(\boldsymbol{x}_j)$$

s.t. $\sum_{i=1}^{d} k_i y_i = 0; \ k_i \leq 0$ (3)

The kernel function $\varphi(\boldsymbol{x}_i)^T \varphi(\boldsymbol{x}_j)$ is the inner product of the high-dimensional feature space, which is defined as the result of the function $\varphi(\boldsymbol{x}_i, y_i)$ in the original sample feature space to define the feature space of the map implicitly [8].

When using the above model for the sample classification, there may be a "hard margin" problem which means that each sample is required to meet the constraint requirements but with the high cost and it is difficult to ensure that the classification is not derived from overfitting. So we relax the constraints of the sample and try to get as few unqualified samples as possible while maximizing the interval to alleviate this problem. And then the alternative loss function hinge is added to the optimization goal and a slack variable is introduced to obtain a new optimization goal.

$$\min_{w,b,\zeta} \frac{1}{2} ||\boldsymbol{w}||^2 + C \sum_{i=1}^{d} \zeta_i$$

s.t. $y_i(\boldsymbol{w}^T \boldsymbol{x}_i + b) + \zeta_i \geq 1; \quad \zeta_i \geq 0$ (4)

In equation (4), the former term indicates the interval of the support vector called "structural risk" which describes the property of the model after regularization; The latter term represents the error called "experience risk" which shows the matching degree of the training data set and the current model; The parameter penalty factor C is used to balance the weights of the two terms to obtain the model that meets people's requirements with the principle of minimizing experience. The Lagrangian function which is an optimized objective function is also calculated by the method of Lagrange multipliers in equation (5).

$$L(\boldsymbol{w}, \boldsymbol{\kappa}, \boldsymbol{\lambda}, \boldsymbol{\zeta}, b) = \frac{1}{2} ||\boldsymbol{w}||^2 + C \sum_{i=1}^{d} \zeta_i + \sum_{i=1}^{d} k_i \left(1 - y_i(\boldsymbol{w}^T \boldsymbol{x}_i + b) - \zeta_i\right) - \sum_{i=1}^{d} \lambda_i \zeta_i \quad (5)$$

The dual problem in equation (6) can be obtained according to the equation above.

$$\max_{k} P = \sum_{i=1}^{d} k_i - \frac{1}{2} \sum_{i=1}^{d} \sum_{j=1}^{d} k_i k_j y_i y_j \varphi(\boldsymbol{x}_i, \boldsymbol{x}_j)$$

s.t. $\sum_{i=1}^{d} k_i y_i = 0; 0 \leq k_i \leq C$ (6)

The support vector expansion $f(x) = \sum_{i=1}^{d} k_i y_i \varphi(\boldsymbol{x}_i, \boldsymbol{x}_j) + b$ can be obtained after calculating the equations above. Thus the classification function is as shown in equation (7).

$$classify(\boldsymbol{x}) = \text{sgn}(\sum_{i=1}^{d} k_i y_i \varphi(\boldsymbol{x}, \boldsymbol{x}_i) + b) \quad (7)$$

2.3. Particle Swarm Optimization

An evolutionary technique called the Particle Swarm Optimization (PSO) was put forward based on the behavioural research on the predation of birds proposed by Eberhart and Kennedy [9]. The PSO algorithm initializes a set of stochastic populations firstly which is also called particle swarms. Each particle passes through the solution space at the rate of initialization, in which the velocity is a function of the historical behaviour (speed, position, fitness) of the particle and others in the swarms and changes in each iteration. The position attribute of each particle corresponds to a potential solution. And the pros and cons of the current solution can be evaluated by calculating the fitness of the current particle. If it fails to meet the requirements of the solution, the particle velocity and position will be updated and the next iteration process will be entered until we find the optimal solution or the maximum number of iterations that meets the requirement.

3. Software defect prediction based on particle swarm optimized SVM

3.1. Software defect prediction model

The software defect prediction model proposed in this paper is shown in figure 1. The training samples are normalized and preprocessed numerically and the relief algorithm is used for feature selection at first to solve the problem of low performance and fitness of the large sample data dealt by the PSO-SVM model. Then the parameter penalty factor C of the SVM model and the bandwidth σ of the Gaussian kernel are optimized by the PSO algorithm in order to find out the parameters with the highest accuracy of the classification model. The trained model is used to predict the defects of the test data and future data at last [10].

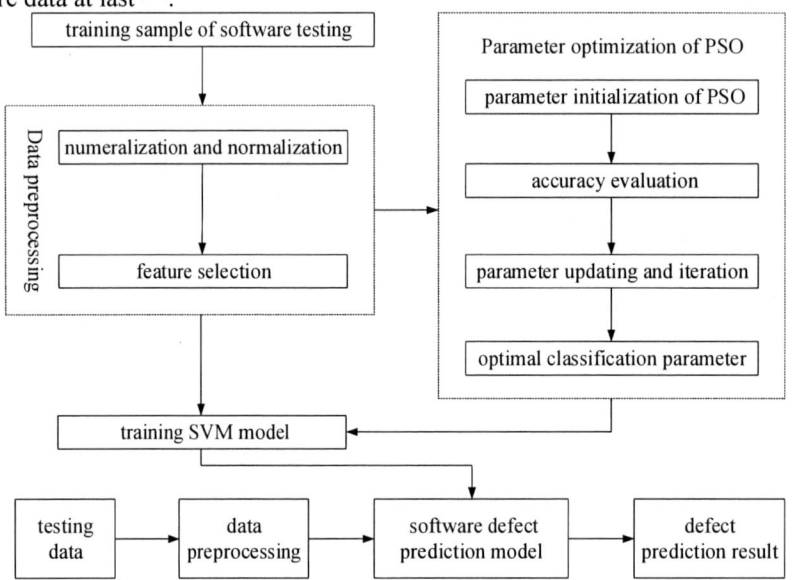

Figure 1. Software defect prediction model based on particle swarm optimized SVM.

3.2. Feature attribute pre-processing

Each data set to be tested contains a large number of sample feature attributes in the actual project, which brings an extremely heavy computational burden to the defect prediction model. Moreover, it is impossible to delete or add some attributes arbitrarily due to the different contribution of each attribute in the classification. Therefore we need the feature selection algorithm called relief algorithm to solve this problem. The relief algorithm is designed for the classification problem of the two kinds of data, which can calculate the weight and ranking of each feature attribute for the software defect prediction problem, and ensure the accuracy of the classification model with the high operation efficiency after completing the feature selection. The algorithm is as follows [11].

The sample instance space is $T = \{T_1, T_2, \ldots, T_n\}$, in which $T_i = (t_1, t_2, \ldots, t_m)$.

Step 1: Assigning the value 0 to the weight of each feature attribute of the sample, quantifying the attribute represented by the non-numeric value, and then normalizing all the values according to the method of maximum and minimum normalization.

Step 2: Selecting a sample T_i randomly and choosing the nearest neighbour samples T_{hit} and T_{miss} from its samples of the same and different types.

Step 3: Updating the weight of each attribute t_k as follows in equation (8):

$$weight(t_k) = weight(t_k) + \frac{1}{n} \times \frac{D(T_i, T_{miss}, t_k)}{\max(D(t_k))} - \frac{1}{n} \times \frac{D(T_i, T_{hit})}{\max(D(t_k))} \qquad (8)$$

$D(T_i, T_j, t_k)$ represents the Euclid distance of T_i and T_j over the attribute t_k. $\max(D(t_k))$ means the maximum Euclid distance of all samples on the attribute t_k.

Step 4: Iterating the process for n times from the begin of step 2, and calculating the average weight of each feature attribute at last, according to which the feature attributes are sorted, and then compared with the feature attribute threshold or the final retained feature attribute set is obtained according to the preset quantity for the training of the SVM model.

3.3. Optimization of software defect prediction model parameters

The model of SVM has been discussed in Chapter 2. The kernel function named the Radius Basis Function (RBF) is selected to improve the generalization ability of the software defect prediction model in equation (9), that is:

$$\varphi(x_i, x_j) = \exp(-||x_i^2 - x_j^2||/2\sigma^2) \tag{9}$$

Therefore, the correlation parameters that need to be optimized by the particle swarm optimization are the penalty factor C and the bandwidth σ of the Gaussian kernel in the software defect prediction model proposed in this paper. The algorithm is as follows [12].

Step 1: Initializing the speed interval, learning factors c_1 and c_2, inertia weight w, the number of iterations and the particle swarm $S = \{(s_{1\_c}, s_{1\_\sigma}), (s_{2\_c}, s_{2\_\sigma}), ..., (s_{num\_c}, s_{num\_\sigma})\}$, in which num is the swarm quantity, including the position of each particle (s_{i_c}, s_{i_σ}) and the velocity (v_{i_c}, v_{i_σ}).

Step 2: Calculating the fitness $fitness(S_i^k)$ of this iteration of the particle for the position of each particle. The prediction accuracy of the SVM model obtained by the current penalty factor s_{i_c} and the bandwidth s_{i_σ} of the Gaussian kernel is used as the return value of the fitness function in this paper, as is shown in equation (10), that is:

$$fitness(S_i^k) = accuracy_{SVM}(s_{i\_c}, s_{i\_\sigma}) \tag{10}$$

Step 3: The individual extremum of the particle will be updated by $fitness(S_i^k)$ if $fitness(S_i^k)$ is better than the individual extremum $PBest_i$. And the swarm extremum $GBest^k$ will be updated by $fitness(S_i^k)$ in this iteration when $fitness(S_i^k)$ is superior to the individual extremum of all other particles and the swarm extremum $GBest^{k-1}$ in the previous iteration meanwhile.

Step 4: The iteration will be quit and the swarm extremum $GBest^k$ will be output as the optimal parameter for training the SVM model if the maximum number of iteration is reached or the current swarm extremum $GBest^k$ satisfies the accuracy requirement. Otherwise, according to the following equation (11):

$$\begin{cases} V_i^{k+1} = w \times V_i^k + c_1 \times rand1() \times (PBest_i^k - S_i^k) + c_2 \times rand2() \times (GBest^k - S_i^k) \\ S_i^{k+1} = S_i^k + V_i^{k+1} \end{cases} \tag{11}$$

Updating the speed and position of each particle and starting to the next iteration from step 2. c_1 and c_2 mainly affect the balance between the individual and swarm memory of particles. And we set c_1 = 1.6, c_2 = 1.5 according to the experience.

The inertia weight *w* mainly influences the balance between the historical memory of the particle and the current state: When the particle is close to the optimal solution with an excessive value, it ignores the effect of the local search in order to pay attention to the result of the global search without falling into the local optimization which results in the overriding of the optimal solution. Otherwise, the speed of the movement is too slow to approach the optimal solution as quickly as possible [13]. Therefore, a dynamic inertia weight particle swarm optimization algorithm is proposed in this paper, which makes the swarm gradually reduce the speed of movement in the process of concentrating near the optimal solution to make each particle search the fitness of the surrounding space more accurately and enhance the overall performance of the standard PSO algorithm. A variable *close* is defined to indicate the degree of swarm aggregation in this paper based on the analysis above. The swarm aggregation of each iteration *k* is as follows.

$$close_k = \frac{1}{num \times |S_{max} - S_{min}|} \times \sum_{i=1}^{num} D(S_i^k, \overline{S_i^k}) \tag{12}$$

In equation (12), $close \in (0,1)$. $D(S_i^k, \overline{S_i^k})$ represents the Euclid distance of the average position (center of gravity) between each particle and the swarm. $|S_{max} - S_{min}|$ means the maximum diameter length of the solution space. *close* describes the case of approaching the optimal solution space of the

ISPECE   IOP Publishing

IOP Conf. Series: Journal of Physics: Conf. Series **1187** (2019) 042082   doi:10.1088/1742-6596/1187/4/042082

particle swarm after each iteration. And the larger the value, the more dispersed the swarm, on the contrary, the more concentrated it is.

It is necessary to reduce the value of the inertia weight w gradually after the particles are aggregated, otherwise, the inertia weight should be increased. They can be mapped to the solution space of the inertia weights by quantifying the degree of aggregation of particles to obtain the values of inertia weights at different concentration levels. Using the definition (13) to calculate the value of w to achieve the above purpose.

$$w_k = \frac{1}{1+\exp(-10(close_k - 0.38))} \times (w_{max} - w_{min}) + w_{min} \qquad (13)$$

It optimizes the local optimization after quickly closing to the optimal space, and w_{min} and w_{max} are the lower and upper bounds of w which are set to 0.8 and 1.2 by experience.

4. Software prediction model experiment based on PSO-SVM

4.1. Experimental data set
The model proposed in the paper is implemented based on Matlab and compared with LE-SVM and LE-KNN in order to verify the performance of the defect prediction model for object-oriented software. Four experimental data sets that conform to the CK metric in this paper are used to verify the validity of the defect prediction model. The first one is the Class-level data for KC1 provided by National Aeronautics and Space Administration (NASA), which contains 145 samples for a total of 89 feature attributes, 60 defect-free samples and 85 defective samples [14]. The second one is the eclipse2.0 data set based on the real data of open source eclipse [15]. There are 6728 different samples, including 975 defective samples and 5753 defect-free samples. The next one is eclipse3.0 data set containing 9470 samples with 1522 defective samples, and the last one named the ant-1.7 data set which has 745 samples with a total of 166 defect-free samples [16]. Since whether there is a defect is represented by the number of bugs for the eclipse and ant data sets, we need to update them to logical variables 1 and 0 that indicate whether there is a defect firstly. Meanwhile, 700 samples are randomly selected from the last three data sets and divided into two groups with equal numbers as the training sets and test sets respectively because the manifold learning algorithm will have the problem of data point loss in the process of high dimensional dimensionality reduction.

The swarm size of the PSO algorithm is set to 50 and the number of evolutions is set to 100, at the same time, the SVM model is trained by the ten-fold cross-test method to obtain the defect detection accuracy of the SVM on the test set in this paper. The SVM model is implemented on Matlab R2015a based on libsvm-3.21 [17]. The output feature dimension in the LE algorithm is 10. The penalty factor of the SVM model of the LE-SVM algorithm on the KC1 data set is $C = 1$, and its Gaussian kernel bandwidth $\sigma = 0.01176$. While the penalty factor $C = 1$ with the Gaussian kernel bandwidth $\sigma = 0.04176$ on other data sets. The feature dimension retained by relief is 15, the search space of the penalty factor C of the SVM model is (0.1, 100.0) and that of the Gaussian kernel parameter is (0.01, 100.0) in the algorithm proposed in this paper.

Table 1. Cross matrix of actual defect situations and prediction results.

Actual defect situation	Defect prediction result	
	Defective module	Defect-free module
Defective module	correct prediction, defective-$N1$	error prediction, defect-free-$N2$
Defect-free module	error prediction, defective-$N3$	correct prediction, defect-free-$N4$

In table 1,The total number of test samples is $N = N1 + N2 + N3 + N4$, in which the correct number of samples predicted is $N1 + N4$ and the total number of samples predicted incorrectly is $N2 + N3$. The performance of the model defect prediction is analyzed according to the general evaluation index in this paper [18].

The accuracy represents the ratio of the number of samples with the correct prediction to the total number of samples being predicted, in which the correct prediction means that the defective modules

1865

were detected as defective accurately, and the non-defective modules were not misjudged. The calculation equation is as follows in equation (14).

$$accuracy = (N1 + N4)/N \qquad (14)$$

The precision indicates the ratio of the number of samples that are actually defective and predicted to be defective to the number of all defective samples being predicted, which can be expressed as follows in equation (15).

$$precision = N1/(N1 + N4) \qquad (15)$$

The recall shows the ratio of the number of samples that are actually defective and predicted to be defective to all actual defective samples. The calculation equation is as follows in equation (16).

$$recall = N1/(N1 + N2) \qquad (16)$$

The value of F is the harmonic mean of the accuracy and recall, and it can be expressed as follows in equation (17).

$$F = \frac{2}{1/recall + 1/precision} \qquad (17)$$

4.2. Result analysis

The accuracy can reflect the performance difference between the software defect prediction model proposed in this paper and the comparative model more objectively because the training accuracy of the SVM model is taken as the fitness of the PSO algorithm. Figure 2 shows the accuracy of the proposed model and the comparative model on the four data sets. It can be seen from that the proposed model leads the comparative algorithm on the four data sets which is mainly reflected in the following aspects [19]. The attributes that are unfavourable to the classification such as the attribute with too small numerical differences can be removed by the proposed model through the relief algorithm to make the prediction result more accurate. At the same time, the penalty factor and Gaussian kernel bandwidth that optimize the performance of the SVM training model are obtained through PSO to further improve the accuracy of the prediction results. While the parameters of the comparative algorithm can only be obtained through the experience with the lack of the process of optimization, which leads that the result has a certain distance from the optimal solution and the prediction result has a certain gap with the proposed model. The accuracy of the proposed model on the four data sets is higher than the comparative model 8.2% ~ 12.2%.

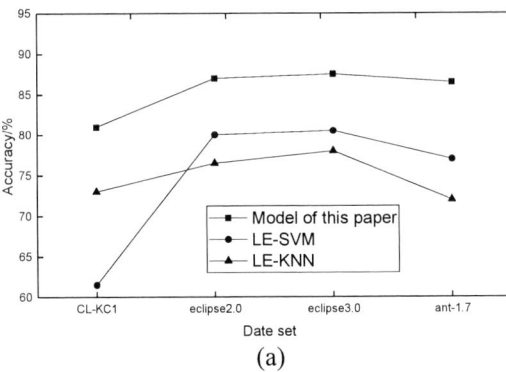

(a)

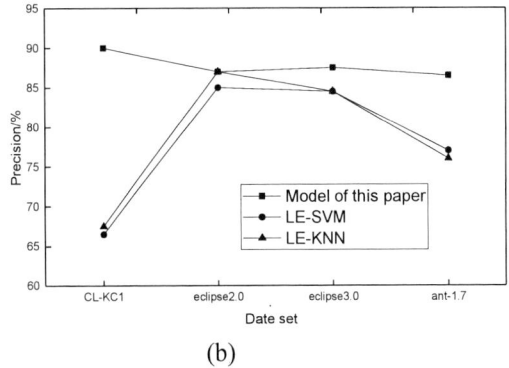

(b)

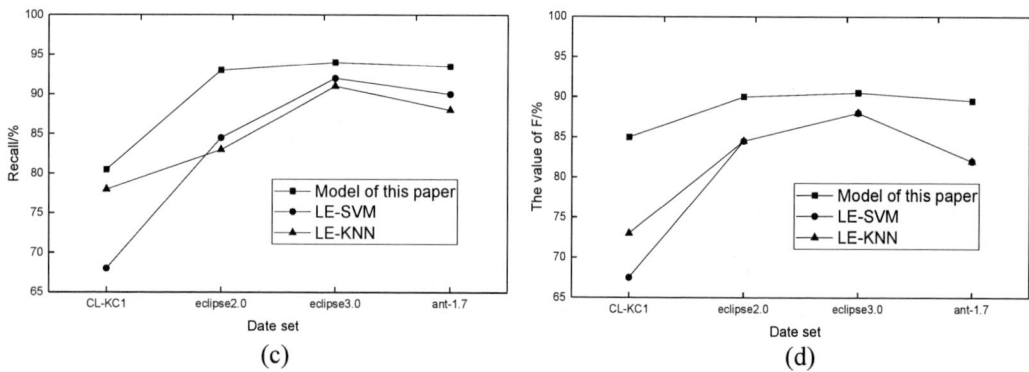

Figure 2. Performance on each data set.

On the basis of ensuring the overall performance of the model which means making sure that the prediction model has a sufficient good defect prediction accuracy, it is necessary to improve the ability of the defect prediction model to identify the defective modules in the test set, that is, to obtain the higher prediction accuracy and recall. These two indicators respectively illustrate the proportion of the defective results in the predicted defective modules and the proportion of predicted defective results in the defective modules. The former reflects the need to detect overhead on the defect-free modules, and the latter reflects the additional overhead that needs to be re-predicted on the defect-free prediction modules. Therefore, the accuracy and recall should be improved to ensure the lower additional overhead.

The value of F is the harmonic mean of the accuracy and recall which is a balance and comprehensive consideration of the two indicators. The reason for evaluating this indicator is to avoid the problem that the extra cost cannot be reduced led by the case that the single indicator is very high while the other one is poor.

The accuracy, recall and the value of F on the four data sets of the proposed and comparative models are shown in figure 2 (a) ~ (d) respectively. It can be seen that the three indicators of the LE-SVM model and the LE-KNN model are similar in the last three data sets, and the gap of those in CL-KC1 is mainly caused by the recall. This phenomenon mainly occurs because the main information in the original data set is saved in the newly generated low-dimensional data set instead of retaining some of the original feature attributes after the manifold learning for feature dimension reduction. The impact on the prediction results by using the prediction model will be reduced for such data. Similarly, the difference in the accuracy between the two models in figure 2 is relatively small, which can also illustrate the problem [20]. The relief algorithm which is efficient and extremely suitable for the two-class problem is used in the dimension reduction processing for the prediction model proposed in this paper. The problem of the fixed convergence rate and the local optimization are avoided to the utmost by the improved dynamic inertia weight PSO algorithm because of the optimization process of the penalty factor and the Gaussian kernel bandwidth solution space. So that the algorithm has the corresponding optimal model parameters in the face of different test sets to obtain the optimal results of the defect prediction. It can be obtained that the proposed model has a 9.9%, 5.6% and 7.7% lead in accuracy, recall and the value of F respectively over the LE-SVM model with better performance by calculating the average of the indicators of the three models on the four data sets.

5. Conclusion

A corresponding defect prediction model is proposed in this paper for the problem of the defect prediction for object-oriented software in the power information system. The basic idea is that the relief algorithm performs the feature selection to avoid the problem with the excessive feature dimension. And then the optimal SVM training model is obtained by the PSO algorithm which updates the inertia weight dynamically according to the degree of the particle aggregation, so that it is used to predict the defect for the software module of CK metric. The experimental result shows that the model can get a good prediction result with a good performance on the four general indicators. While the model cannot cope with the prediction problem in the software running process as a static analysis model. Moreover, the time performance of the model needs further optimization due to the iterative loop process of the PSO algorithm. The next step is to optimize the model prediction results and the algorithm runtime to cope with future defect prediction.

Acknowledgments

This paper was financially supported by the Science and Technology Project from State Grid Corporation of China: "Research on Distributed Simulation and optimization technology of optical transmission network for Electric power communication (5442XX180003-XX71-18-006)".

References

[1] Gang LI, Yinsheng SU and Chen CHEN 2001 *J. Journal of Computer Applications.* **21**(9): 78-80.
[2] ABAEI G and SELAMAT A 2014 *J. Vietnam Journal of Computer Science.* **1**(2): 79-95.
[3] Yao ZHANG, Zhihai YUAN and Haiyan JIANG 2010 *J. Computer Technology and Development.* **20**(8): 37-40.
[4] Tong YI 2011 *J. Application Research of Computers.* **28**(2): 427-34.
[5] Lukui SHI, Chunjuan MA and Jingxin WANG 2014 *J. Computer Engineering and Design.* **35**(11): 3859-63.
[6] Ramandeep Kaur and Harpreet Kaur 2017 *J.* Volume-3, Issue-1.
[7] CORTES C and VAPNIK V 1995 *J. Machine Learning.* **20**(3): 273-97.
[8] CRISTIANINI N and SCHOLKOPF B 2002 *J. AI Magazine.* **23**(3): 31-41.
[9] KENNEDY J 2011 *Encyclopedia of Machine Learning* (Berlin: Springer) p 760-66.
[10] Zhang H G, Zhang S and Yin Y X. 2017 *Chin J Eng.* **39** (1): 39.
[11] KIRA K and RENDELL L A 1992 *Proceedings of the Ninth International Workshop on Machine Learning* (San Francisco: Morgan Kaufmann Publishers) p 249-56.
[12] Wenli SHANG, Shengshan ZHANG and Ming WAN 2014 *J. Electronic Journal.* **42**(11): 2314-20.
[13] HUANG C L and DUN J F 2008 *J. Applied Soft Computing.* **8**(4): 1381-91.
[14] BOETTICHER G, MENZIES T and OSTRAND T (2007-01-01) [2013-03-17] *DB/OL. http: // promise data. org / repository.*
[15] ZIMMERMANN T, PREMRAJ R and ZELLER A 2007 *Proceedings of the 3rd International Workshop on Predictor Models in Software Engineering* (Washington, DC: IEEE Computer Society) No. 9.
[16] JURECZKO M and MADEYSKI L 2010 *Proceedings of the 6th International Conference on Predictive Models in Software Engineering* (New York: ACM) Article No. 9.
[17] CHANG C C and LIN C J. LIBSVM 2010 *J. ACM Transactions on Intelligent Systems and Technology.* **2**(3): Article No. 27.
[18] Huiyan JIANG, Mao ZONG and Xiangying LIU 2011 *J. Chinese Journal of Computers.* **34**(6): 1148-54.
[19] Yan B Y, Sheng Z F and Li G 2016 *J. Solid Rocket Technol.* **39**(1): 106.
[20] Nanshuai WANG, Jingfeng XUE and Changzhen HU 2015 *J. Chinese scientific papers.* **10**(2):159-163.

Research on Knowledge Graph Application Technology

Yong Chen[1, a] Xueli Chen[2, b]

[1]Nanjing Normal University
Jiangsu Province of China
15605199883, incl. China

[2]Nanjing Normal University
Jiangsu Province of China

[a]harry.cy1994@gmail.com [b]61021@njnu.edu.com

ABSTRACT. The development of the Internet has mainly gone through three stages. The first stage can be defined as the era of document interconnection. At this time, the main task of the Internet is to provide corresponding layout content for readers to read.

The second stage can be defined as the era of data interconnection. Recently, the main task of the Internet provides an interaction that makes the user not only the reader of the web content, but also the maker of the web content [26].

The third stage can define the era of semantic interconnection, which pays more attention to the Internet. The creator and editor of network knowledge, the network truly becomes the user's understanding and provider of needs. Knowledge graph's ability is to understand its own strong semantics, as well as the openness and interconnectivity of the knowledge graph, makes the vision of the semantic connected era with circumstance as close as possible. The knowledge graph was first proposed in the project released by Google on May 17, 2012.Knowledge graph is not a new technical concept. As early as in 2006, [20], the concept of Semantic Web was mentioned. Nowadays, scholars began to call on the use of ontology model in data to express data implicit semantics. At present, intelligent information service applications are increasingly appearing in all aspects of technology and life, such as intelligent search, intelligent question and answer, personalized recommendation, and companies that have appeared in various fields and are in contact with consumers and users.

1. Introduction

The second part of the article first introduces the development of the Internet and the future trends, and leads to the application scenarios of the current knowledge graph. The concept of knowledge graph is traced back to the source. And some basic information and concepts are described.

The third part of the article describes in detail some related knowledge concepts of knowledge graphs, and systematically describes the existing technical concept application fields and research methods. The development and connection of data cleaning, data mining, knowledge extraction, knowledge acquisition, knowledge fusion, knowledge reasoning and other technologies are introduced. The fourth part of the article gives a detailed introduction to the application of knowledge graphs, and describes the different methods applied in different periods. Introduce the application of knowledge graphs in search systems, question and answer systems, recommendation systems, etc.

Content from this work may be used under the terms of the Creative Commons Attribution 3.0 licence. Any further distribution of this work must maintain attribution to the author(s) and the title of the work, journal citation and DOI.
Published under licence by IOP Publishing Ltd

2. Definition and structure of knowledge graph

2.1. Definition of knowledge map
The initial definition stems from a description of the Knowledge Graph project which published by Google is defined as a bridge between Google's traditional search and smart search. In essence, the knowledge graph is a description of objective things. So far, knowledge graphs have been widely used in various large-scale knowledge bases.

2.2. Architecture of Knowledge Graph
Knowledge graph is a kind of structural data to describe objective things. Objective things are generally represented by conceptual abstractions of concepts, entities, attributes, etc. In the process of constructing knowledge graphs, common representations are triples, three. Tuples represent the two most common representations such as (entity 1 - relationship - entity 2) and (entity - attribute - attribute value). The construction of knowledge graphs is generally divided into two types. One is to extract concepts related to entity attributes from the structured data of various encyclopedia websites, and add them to the knowledge base; the other is to collect data from publicly. The resource model is extracted and added to the knowledge base through a series of screenings such as manual review. Therefore, the knowledge graph mainly involves such knowledge as data cleaning, knowledge mining, knowledge acquisition, knowledge representation, knowledge fusion, and knowledge reasoning.

3. Architecture of Knowledge Graph

3.1. Knowledge mining
Knowledge mining is the process of extracting implicit, previously unknown and potentially valuable information from a database.

In order to determine how data mining technology (DMT) and its applications are developed, Liao, S. H. et al. surveyed and classified the literature from 2000 to 2011, reviewing data mining techniques and their applications and developments. The author retrieves the required data from previous journal articles and other literatures, and analyzes the three types of knowledge, analysis, and architectures, depending on the survey and classification, and also analyzes them in different studies and Application in the field of practice.

In 2016, Banuqitah, HU et al. proposed an agent-based two-level self-supervised relationship extraction system based on the UMLS unified medical language system knowledge base [2]. The model uses a self-supervised method for relation extraction (RE). The method is to construct an enhanced training example using UMLS information with mixed text features, and the model combines Apache Spark and HBase BD technology with multi-data mining and machine learning techniques, and also with a multi-agent system (MAS).

3.2. Knowledge extraction
The KNOWITALL system is designed to automate the lengthy process of extracting a large number of facts (such as the names of scientists or politicians) from the Web in an unsupervised, domain-independent and scalable manner [8], which outlines the new architecture of KNOWITALL and design principles, and highlight its unique ability to extract information without any manual-tagged training examples.

The article raises a challenge: how can we improve the recall and extraction rate of KNOWITALL without sacrificing accuracy? Organize the full text, and then propose three different methods to deal with this challenge and evaluate their performance. Pattern learning learns domain-specific extraction rules that support additional extraction. Subclass extraction can automatically identify subclasses to facilitate recall (eg, "chemists" and "biologists" are identified as subclasses of "scientists"). The list extracts a list of positioning class instances, learning a "wrapper" for each list. Since each method is guided from KNOWITALL's domain-independent method, these methods also exclude manual training examples. At the end of the article, some experiments were also reported, focusing on the establishment of a list of named entities to measure the relative effectiveness of each method and to demonstrate their synergy.

Some scholars have proposed an application of a kernel method for extracting relationships from unstructured natural language sources [24] and introduced the kernel defined on the shallow analytical representation of the text and designed an efficient algorithm for computing the kernel. The author combines the support vector machine and the voting perception learning algorithm to extract the human-dependent relationship and the organization-position relationship from the text, and evaluates the proposed method, and compares it with the feature-based learning algorithm.

3.3. Knowledge Reasoning

Knowledge reasoning is mainly divided into two categories, one is graph-based reasoning and the other is logic-based reasoning. In a 2013 article, we introduced the use of neural tensor networks to reason about entities and relationships [16]. The article mainly introduces a neural tensor network of expressions, which is used to reason the relationship between two entities. Previously we always represented an entity as some discrete atomic unit, or a single entity vector. We have experimentally shown that performance can be improved when the entity representation constitutes the average of the word vectors. Finally, the authors also demonstrate that the performance of all models is improved when these word vectors are initialized with word vectors learned from unsupervised large corpora. The author evaluates the real relationship between the entities of a given subset of the knowledge base of the model by considering the prediction of additional problems. It is concluded that this model is superior to the previous model, and can classify the invisible relationships in WordNet and the accuracy of FreeBase is 86.2% and 90.0%, respectively.

Some scholars have also considered the reasoning method based on logical combination [6]. The purpose is to solve the dilemma of existing knowledge reasoning in the knowledge base of similar fields. They extend the representation of the logic of the description to generate a combination of description logic, using the similarity of concepts to relate the concepts of similar fields, thus giving the logical grammar and semantics of the combined description. The article finally gives The Tableau algorithm is used to implement this concept, and finally the feasibility of logical reasoning is confirmed by an example.

4. Application of knowledge graph

4.1. Intelligent Search System

The most important feature of the intelligent search system is that it can truly understand the needs of human beings, and can use various aspects of knowledge to analyze the needs of users and give the search answers that best meet the needs of users. The form of network ontology language was originally transformed from SHIQ and RDF to OWL. As early as 2003, scholars proposed a process of network ontology language [11]. The article introduces the necessity of OWL ontology language as a new language and also comprehensive summary the OWL's DL axioms and facts. The article also introduces its advantages and unstoppable development trend as a new language, as well as some comparisons in its development the stage. The traditional information retrieval system expresses documents and queries through keyword collections, but such search queries are not very efficient and performance-oriented.

Scholars in related fields have proposed a semantic text search method based on ontology query extension [13]. The main purpose is to use the ontology feature of the named entity and its potential related entities to perform semantic interpretation of documents and queries and to provide different advantages for semantic annotation and search with different ontology combinations.

One of the biggest problems facing intelligent search systems is the query problem. Some scholars have proposed a method to enable semantic association query in the Semantic Web. The main idea is to represent the concept of semantic association as a complex relationship between resources and entities, called ρ-Queries [1]. The main idea is to formalize the RDF data model by introducing the concept of attribute sequences as types, and to provide a formal framework for supporting complex queries such as semantic associations. After an open source semantic retrieval system YaSemIR[5] was proposed by Davide Buscaldi et al., it also attracted great interest from experts in related fields. The query is mainly based on one or more ontology of OWL format normalized into the ontology of SKOS format for semantic query indexing out a collection of text.

4.2. Question and answer system

The question-and-answer system is similar to smart search, but it is very different. The main function of the callback system is to be able to talk to the user in the form of question and answer, and to give feedback to the answers raised by the user. Some scholars use the word embedding model to construct the question and answer system in a supervised mode [3]. They use auto-learning data and Wikipedia data for model training, and the answers to the questions are expressed in a set of triples. The author also designed a new model of word embedding to be used in the evaluation of the correctness of the triplet answer. Finally, it is confirmed by examples that the performance of the model has been greatly improved, but it uses a lot of manual tag data in the research process.

Some scholars have done an in-depth study of the question and answer system from the perspective of the figure. The study of the question and answer system based on subgraph embedding is one of them [4]. Using fewer manual attribute tags is a big advantage for them. The author is committed to learning how to answer questions under the broad subject from the knowledge base, and they use low-dimensional words and knowledge base embedding to score the answers to the answers. Finally, experiments proved that the effectiveness of the system shows good performance. Some scholars are based on the multi-column convolutional neural network question-and-answer system [7]. The author uses convolutional neural networks to learn their distribution and representation from three different aspects: answer type, answer path and answer context. A low-dimensional approach is used in entities and relationships, combined with a problem vocabulary score calculation method designed by them to assess the core vocabulary of the problem. Finally, a lot of experiments were carried out on Freebase, which shows that our system has great advantages in performance compared with other conventional systems.

Anthony Fader et al. at the University of Washington describe a new learning algorithm for open-ended question-and-answer systems [3]. The author introduces a template for a scalable learning algorithm that can lead to general problems, as well as lexical changes in entities and relationships. The biggest advantage of this algorithm is that it does not require manual annotation and can be applied to large, triad knowledge bases with no data. With the increase of the number of knowledge bases, the question and answer system of multiple databases has also emerged.

Among them, Yuanzhe Zhang et al. proposed a multi-knowledge question-and-answer system model [25]. The biggest problem we face in the process of constructing multiple knowledge bases is the alignment between the index libraries. Different alignments also fundamentally affect the interaction process. The author has an integer linearity based on this problem. The new joint model of planning unifies the two processes of alignment construction and query construction into a unified framework, which ultimately proves to be effective and can improve the performance of alignment constructs and query constructs. As shown in the figure 1, it is a simple intelligent question-and-answer system.

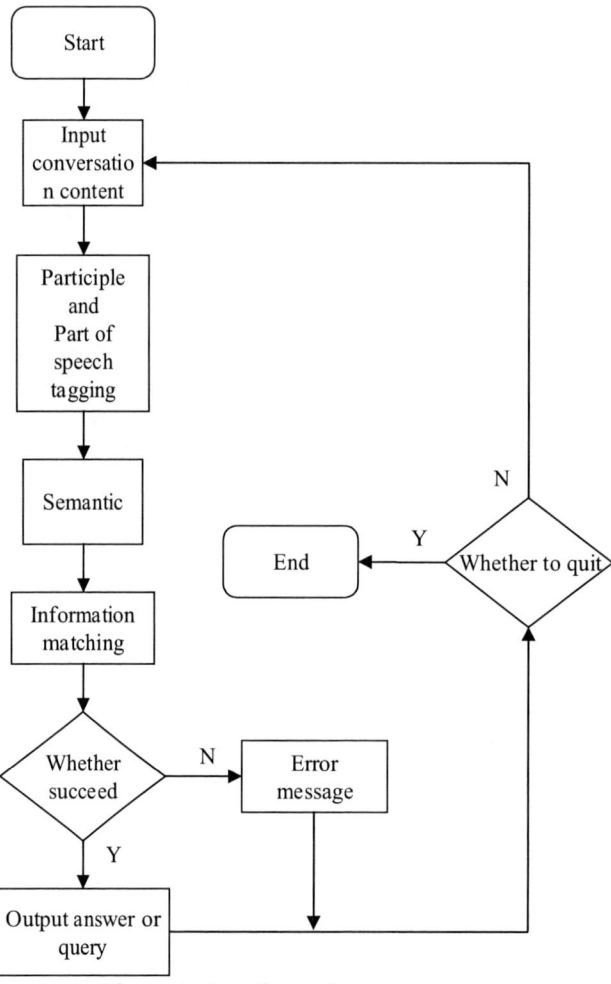

Figure 1. Question and answer system.

4.3. recommendation system

The recommendation system is the machine's knowledge of using big data, knowledge graphs, etc., comprehensively analyzing the potential needs of users, analyzing, giving conclusions, and recommending relevant content to users, thus forming a benign cycle. User-like content will be recommended. The system's data set, the recommendation system in turn recommends relevant content to the user. The key technology in the recommendation system is the construction of recommendation algorithms, and there are mainly collaborative filtering recommendation algorithms. There are three main types of recommendation systems, namely user-user (UU matrix), user-item (UV matrix), and articles. - Item (VV matrix).

Jennifer Nguyen et al. proposed a matrix decomposition algorithm to improve the accuracy of the proposed algorithm [15]. The author first introduces the matrix decomposition method (MF) of collaborative filtering. The MF method is to observe the statistical user's preference through proof decomposition. The author's enhanced matrix decomposition algorithm mainly focuses on the content information directly collaborate into the matrix decomposition method, and filter together. Use forced alignment penalty to narrow the regression constraint and force potential project features to become content attributes. In addition to the matrix decomposition, there is a more popular method is the Markov chain (MC). The MC method is to predict the next action by learning the user's recent actions. Some scholars have proposed a combination of the two methods.

A graph of personality was based on Markov chain [18]. The main features of the method are not the same as usual. Instead of using the same transformation matrix for all users, a separate transformation matrix is used for each user, which is generally a transformation cube. And the decomposition model is introduced so that each conversion can receive the effects of similar users, projects, and similar transformations.

In the field of recommendation systems, some scholars are using cyclic neural networks to enhance the construction of recommendation systems for long-term user history [10]. The main idea is to provide more accurate suggestions by modeling the entire callback, and for classics. The ranking loss function has been modified to make it more suitable for a particular problem. Labels are widely used in the web2.0 era.

At this stage, label optimization based on recommendation systems has also attracted the interest of scholars in related fields. Some scholars have extended Bayesian personalized recommendation criteria to task tag recommendation [17], and combined with stochastic gradient learning algorithm and bootstrap sampling learning algorithm to provide a linear pairwise interaction tensor decomposition model (PITF).It has a linear operation time and experimentally proves that the algorithm runs much better than the best quality method (RTF-TD) when the running time is reduced from O(k3) to O(k).

As shown in the figure 2, it is the main application of knowledge mapping.

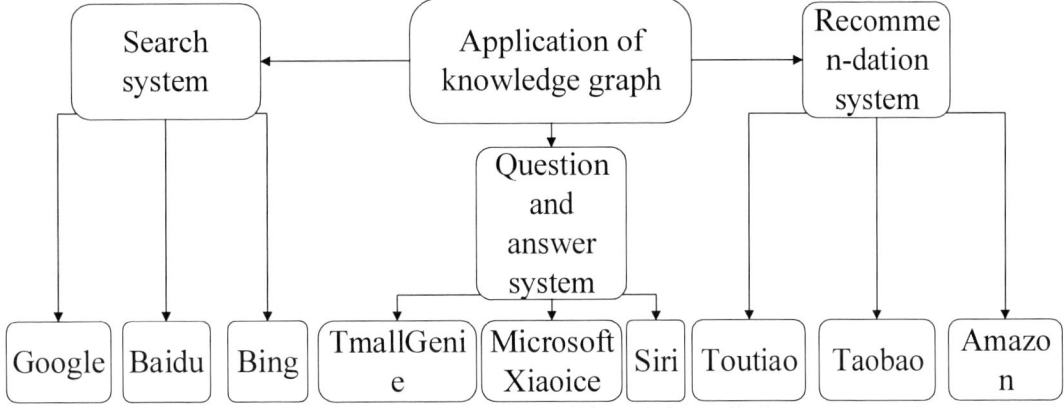

Figure 2. Knowledge graph is mainly applied.

5. Conclusion

With the continuous development of knowledge graph, the application of future knowledge graph will become more diversified and multi-scenarios. Applications in various industries will be continuously developed and supplemented. It can be expected that in the future, you can truly realize human-machine barrier-free communication with intelligent machines. The future question-and-answer system will also be diversified and will pay more attention to the interaction of multiple rounds of dialogue. All of this can be applied to the application scenarios of the knowledge map. In the future, deep knowledge reasoning with knowledge graph and improved computational efficiency of large-scale knowledge graph require people to continually explore user needs, explore more important application scenarios, and propose more practical application algorithms. Therefore, the development of future knowledge graph requires both a rich accumulation of knowledge map technology, a keen perception of human needs, and the right application. I believe that in the future, life and work with the participation of knowledge graph will be smarter, simpler and better.

References

[1] ANYANWU., K. and SHETH., A., 2003. ρ-Queries: Enabling Querying for Semantic Associations on the Semantic Web. In Proceedings of the 12th International Conference on World Wide Web (2003), 690-699.

[2] BORDES, A., WESTON, J., and USUNIER, N., 2014. Open Question Answering with Weakly Supervised Embedding Models. Machine Learning and Knowledge Discovery in Databases, 165-180.

[3] BORDES., A., CHOPRA., S., and JASONWESTON., 2014. Question Answering with Subgraph Embeddings. Proceedings of EMNLP.

[4] BUSCALDI., D. and ZARGAYOUNA., H., 2013. YaSemIR: Yet another Semantic Information Retrieval System. In Proceedings of the 6th International Workshop on Exploiting Semantic Annotations in Information Retrieval, 13-16.

[5] DAOSHE, L. and SHIHAN, Y., 2012. Interdisciplinary reasoning on description logic. Application Research of Computers, 4503-4511.

[6] DONG., L., FURUWEI., ZHOU., M., and XU., K., 2015. Question Answering over Freebase with Multi-Column Convolutional Neural Networks. In Proceedings of the Proceedings of the 53rd Annual Meeting of the Association for Computational Linguistics and the 7th International Joint Conference on Natural Language Processing (2015), 260-269.

[7] ETZIONI, O., CAFARELLA, M., DOWNEY, D., POPESCU, A.M., SHAKED, T., SODERLAND, S., WELD, D.S., and YATES, A., 2005. Unsupervised named-entity extraction from the Web: An experimental study. Artificial Intelligence 165, 1 (Jun), 91-134. DOI= http://dx.doi.org/10.1016/j.artint.2005.03.001.

[8] HIDASI, B., KARATZOGLOU, A., BALTRUNAS, L., and TIKK, D., 2015. Session-based Recommendations with Recurrent Neural Networks. Computer Science.

[9] HORROCKS, I., PATEL-SCHNEIDER, P.F., and VAN HARMELEN, F., 2003. From SHIQ and RDF to OWL: the making of a Web Ontology Language. Web Semantics: Science, Services and Agents on the World Wide Web 1, 1, 7-26. DOI= http://dx.doi.org/10.1016/j.websem.2003.07.001.

[10] M., V., NGO., and CAO., T.H., 2010. Ontology-Based Query Expansion with Latently Related Named Entities for Semantic Text Search. In Proceedings of the 2nd Asian Conference on Intelligent Information and Database Systems (2010), 41-45.

[11] NGUYEN, J. and ZHU, M., 2014. Content-boosted Matrix Factorization Techniques for Recommender Systems.

[12] R. SOCHER, CHEN, D., MANNING, C.D., and NG, A., 2013. Reasoning with Neural Tensor Networks for Knowledge Base Completion. In Proceedings of the Neural Information Processing Systems 26. Red Hook (NY, USA2013), 926–934.

[13] RENDLE, S. and SCHMIDT-THIEME, L., 2010. Pairwise Interaction Tensor Factorization for Personalized Tag Recommendation. Association for Computing Machinery, 81–90.

[14] RENDLE., S., FREUDENTHALER., C., and SCHMIDT-THIEME., L., 2010. Factorizing Personalized Markov Chains for Next-Basket Recommendation, 811-820.

[15] SHADBOLT, N., HALL, W., and BERNERS-LEE, T., 2006. The Semantic Web revisited. IEEE Intelligent Systems 21, 3 (May-Jun), 96-101. DOI= http://dx.doi.org/Doi 10.1109/Mis.2006.62.

[16] ZELENKO, D., AONE, C., and RICHARDELLA, A., 2003. Kernel methods for relation extraction. Journal of Machine Learning Research 3, 6 (Aug 15), 1083-1106. DOI= http://dx.doi.org/Doi 10.1162/153244303322533205.

[17] ZHANG., Y., HE., S., LIU., K., and ZHAO., J., 2016. A Joint Model for Question Answering over Multiple Knowledge Bases. In Proceedings of the American Association for Artificial Intelligence (February, 12-17. 2016).

[18] ZHAO, C.L., LI, X.Y., XU, C.Y., and WANG, F., 2009. Informal learning based on web2.0. Icaie 2009: Proceedings of the 2009 International Conference on Artificial Intelligence and Education, Vols 1 And 2, 393-397.

ISPECE

IOP Publishing

Predicting failures in hard drivers based on isolation forest algorithm using sliding window

Tinglei Zhang[1], Endong Wang[1] and Dong Zhang[2]

[1]School of the Zhengzhou's Smarter City, Zhengzhou University, Henan 450000, China

[2]System Software R&D Department, Inspur Electronic Information Industry Co., Ltd, China

zhangtinglei_zzu@sina.com; wangend@inspur.com; zhangdong@inspur.com

Abstract. With the surge in the development of cloud computing and big data, the scale of the data center is constantly expanding. The data missing caused by the damage of large data center disk has become a problem that cannot be ignored. Predicting the failures of the hard drivers can effectively improve the reliability and reduce the maintenance cost of the data center. The SMART data is volume and multi-dimensional, as well as the high dynamic characteristics of data center hard disk accessing, so we propose an algorithm based on isolation forest using sliding window. The experiments show that this algorithm can effectively save the time of modeling and ensure high prediction accuracy.

1. Introduction

With the surge in the development of cloud computing and big data, the scale of the data center is constantly expanding, and data center downtime costs are increasing significantly. In 2016, Emerson Network Power released "Cost of Data Center Outages in 2016", which shows that the cost per minute of data center failure is close to $9,000 in the U.S [1]. IT equipment failure is the main cause of data center downtime, one of them is hard driver failure. It has been estimated that 78% of the hardware replacements are caused by hard driver failures [2]. Predicting the failures of the hard drivers can effectively improve the reliability and reduce the maintenance cost of the data center.

Hard disk failure prediction methods including threshold method, the statistical methods, the Bayesian methods, the Support Vector Machine (SVM), decision tree, and the Hidden Markov Model (HMM) [3]. Most of them may not get good results, on the one hand, the data sets are small , which are far from the data of data center in real word, on the other hand some methods were used in many years ago, which were not quite effective [12].

In recent years, many researchers have proposed some improved algorithms. Zhao et al. [4] regard SMART attribute value as time series data, and used Hidden Markov Model (HMM) and Hiddle semi-Markov Model (HSMM) to the failures of the hard drivers, which obtains zero false alarm rate and 52% prediction accuracy in Hughes data set. Song et al. [5] propose to use SMART data to predict hard disk failure based on the Support Vector Machine (SVM) Classification using lOcal clusterinG with Over-Sampling (COG-OS) framework. Compared with the traditional SVM algorithm, it improved the recall ratio of failure class prediction, while leads to accuracy of failure class prediction and performance of normal class prediction decline. The authors of [6] propose an attribute selection method based on neural network weight matrix, combining with three kind of non-parametric

---

Content from this work may be used under the terms of the Creative Commons Attribution 3.0 licence. Any further distribution of this work must maintain attribution to the author(s) and the title of the work, journal citation and DOI.
Published under licence by IOP Publishing Ltd

statistical methods (the Rank Sum test, the RAT reverse arrangement test, Z-Score) to select useful attribute for building prediction model based on two kinds of machine learning methods (CART decision tree and BP neural network), which obtain very good performance, the experimental results also show that the accuracy may decline when applying this method to unbalanced data.

The SMART data set of data center has the characteristics of huge amount and multi-dimension. According to the research of BACKBLAZE data centre, there are about 8.95 million pieces of data in the data set of the second quarter of 2018 (SMART Attribute values are read per day for each disk) [7]. So the prediction algorithms must have the ability of dealing with volume and multi-dimensional data set. Compared with other prediction algorithms, the isolation forest algorithm predicts the failure through compute the path length of data in iTree, which reduces the calculation cost, has a linear time complexity with low constant and a low memory requirement. So this algorithm has the capacity to scale up to handle extremely large data size and high-dimensional problems.

The access to the hard disk in the data center is highly dynamic, resulting in large fluctuations in some SMART properties. Therefore, it is difficult to accurately determine the failure of the hard disk through SMART data at a single point in time series. The algorithm based on sliding window comprehensively considers the prediction results of multiple time points in the window and reduces the false alarm rate of single point prediction.

In this paper ,Aiming at the characteristics of disk driver predictions of large data centers, we proposes an isolation forest algorithm based on sliding window, which can make best use of the advantages of isolation forest algorithm and sliding window algorithm. We also apply it to the SMART data of a real large data center for failure prediction, to verify the availability and efficiency of the algorithm.

The rest of this paper is organized as follows: Section 2 briefly introduces the related work. The algorithm based on isolation forest using sliding window is described in Section 3. In Section 4, the experimental results and analysis are presented. Finally, Section 5 concludes the whole paper.

2. Related Work
In this section ,the brief introduction of SMART, the basic principle of the isolation forest algorithm and the general framework of sliding window will be presented.

2.1. Introduction of SMART
The SMART (self - monitoring, analysis and reporting technology) detect hard driver failure by monitoring the hard driver heads, platters, motor, circuit operation data (such as temperature, read data error rate, start-stop times, temperature, etc.) [8]. Each drive manufacturer defines a set of attributes, and sets threshold values beyond which attributes should not pass under normal operation, if exceed those threshold, the system can automatically send warning information. Although the accuracy of the SMART technology is low and the SMART warning-time goal is 24 hours before driver failure, the monitoring data of SMART technology provides the basis for other prediction algorithms.

2.2. Principle of isolation Forest
The core idea of the isolation forest algorithm is that the number of abnormal points is usually small, and there are significant differences between normal points in attributes. Therefore, when the points are partitioned by iTree, the path length of normal points is greater than the path length of abnormal points [9]. As shown in Figure 1, the Figure 1(a) contains 7 normal points and 1 abnormal point. After a randomly generated iTree(as shown in Figure 1(b)) partition, the path length of the abnormal point in iTree is significantly smaller than that of other normal points.

2.3. General Framework of sliding window
In this paper, the algorithm based on sliding window comprehensively considers the prediction results of multiple time points in the window. The general form of SMART data can be described as follows:
$V = \{v(1), v(2),..., v(t), v(t+1),...\}$

where $v(t) \in R^N$ for $t \geq 1$.

The prediction window at time t can be described as follows:

$$p_{window}(t) = \{v(t-1), v(t-2), \ldots, v(t-k+1)\}$$

where k denotes the size of sliding window.

For example, the SMART data of the hard driver at time 10 is $v(10)$, the time interval is 1, and the window size is 3, then the prediction window of the hard driver at time 10 is

$$p_{window}(10) = \{v(10), v(9), v(8)\}$$

,

and the prediction window at the next moment is

$$p_{window}(11) = \{v(11), v(10), v(9)\}$$

.

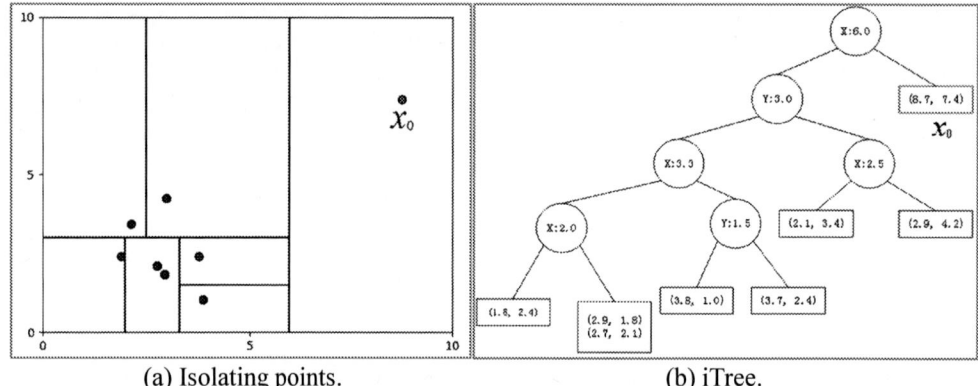

(a) Isolating points.   (b) iTree.

Figure 1. abnormal points partitioned by iTree.

3. The Algorithm Based on isolation Forest using Sliding Window

In the training, isolation forest is composed of a certain number of isolation trees. In the process of constructing an isolation tree, first, we extract N pieces of data from the training data set by uniform sampling, it can be used as the training data of the isolation tree. Second, we randomly select an attribute and a value within the range of all values of this attribute (between the minimum value and the maximum value), then divide the data by selected attribute and value, if the value of data's attribute less than the selected value, it will be divided into the left of the node, contrarily, it will be divided into the right of the node. Finally, we can obtain a split condition and two depending cubes, and then repeats the partition process on two depending cubes, until only one record is in the data set or the height of the tree is reached. The algorithm can reference to [9].

In the predicting, we assume that the prediction algorithm is $f(v_t)$, The general form of hard driver's state can be described as follows:

$$\sum_{i=0}^{i=k-1} f(v_{t-i}) > \sigma$$, the state of hard driver is failure.

$$\sum_{i=0}^{i=k-1} f(v_{t-i}) \leq \sigma$$, the state of hard driver is failure.

where $f(v_t) = 1$ denotes that the predicted state is abnormal, $f(v_t) = 0$ denotes that the predicted state is normal, σ is the abnormal threshold . The algorithm is shown in Figure 2.

```
Algorithm: predict
Inputs: V-dataset of window, F-iForest, $l_{max}$ -the path
length of normal points , $\sigma$ -threshold
outPuts:  the state of hard driver
predict(V, F, $l_{max}$, $\sigma$){
    abnormal=0        //the number of abnormal points
    for(int i=0;i<V.size;i++){
        sumPL=0;
        for(int j=0;j<F.size;j++){
            sumPL+=pathLength(V[i],F.get(j),0,1);
        }
        If(sumPL/F.size<$l_{max}$){
            abnormal++;
        }
    }
    if(abnormal>$\sigma$){
        return 1;       //return failuer
    }else{
        return 0;       //return good
    }
}
```

Figure 2. The algorithm of prediction.

4. Result And Discussion

All samples [11] were collected from an enterprise-class disk model of Seagate named
ST31000524NS. There are samples from 23,395 disks in the data set. Each disk was labeled good or
failed, with only 433 disks in the failed class and the rest of disks (22,962) in the good class. Each
SMART data contains 10 SMART attributes. In the experiment, we construct 11 training data sets and
1 test data set , as shown in Table 1 .

Table 1. The date sets of experiment.

Name	Number of Hard Driver		Name	Number of Hard Driver	
	good	failed		good	failed
Train_Data_1	1000	100	**Train_Data_2**	2000	100
Train_Data_3	3000	100	**Train_Data_4**	4000	100
Train_Data_5	5000	100	**Test_Data**	1000	333

In the first experiment, we chose 6 data sets (Train_Data_1, Train_Data_2, Train_Data_3,
Train_Data_4, Train_Data_5) to train prediction model.Then, we use the test data set (Test_Data) to
test the model.We use accuracy and false alarm rate to evaluate the classifier's performance, the Table
2 shows the result.

Table 2. The accuracy and false alarm rate of algorithm.

Name	Accuracy	false alarm rate
Train_Data_1	93.79%	5.48%
Train_Data_2	93.92%	5.40%
Train_Data_3	94.22%	5.25%
Train_Data_4	95.05%	4.28%
Train_Data_5	94.52%	4.72%

In the second experiment, we use the open-source data mining tool WEKA 3.8.3, and the test
environment is Intel(R) Core(TM) i3-6100 CPU @3.70GHZ with 4GB of RAM. we chose Random
Forest, Bayesian network, BP neural network, LibSVM and isolation Forest to train prediction model

in 5 data sets (Train_Data_1, Train_Data_2, Train_Data_3, Train_Data_4, Train_Data_5) .The training times of five kinds of algorithm in each data set are shown in Table 3.

Table 3. The training time of five kinds of algorithm in each data set.

Algorithm	Train Time 1	Train Time 2	Train Time 3	Train Time 4	Train Time 5
Random Forest	61.05 S	131.06 S	211.25 S	281.02 S	394.72 S
Bayesian network	1.33 S	2.97 S	5.00 S	6.22 S	9.93 S
BP neural network	154.40 S	281.59 S	403.56 S	515.53 S	667.74 S
LibSVM	1129.21 S	1325.21 S	>1500 S	>1500 S	>1500 S
isolation Forest	0.68 S	1.81 S	3.11 S	4.48 S	5.89 S

The first experiment shows that the average accuracy of the algorithm is 94.30% , and the average false alarm rate of the algorithm is 5.03% . The algorithm has higher accuracy and lower false alarm rate.The Second experiment shows that the training time of isolation forest algorithm in each data set is least, while the algorithm of LibSVM costs most time. The Figure 2 shows the relationship between train time and the size of data set. As the number of data increases, the training time increases linearly.

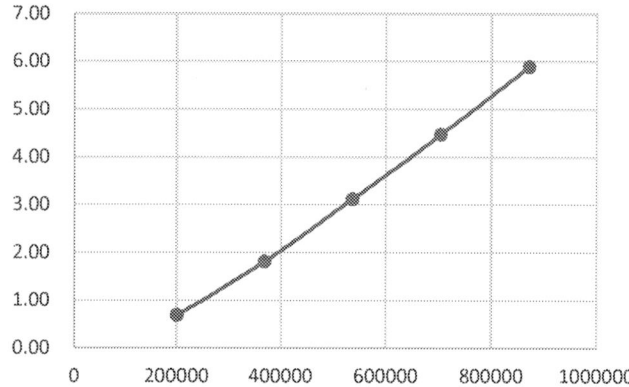

Figure 2. The relationship between train time and the size of data set.

Compared with other prediction algorithms, the algorithm based on isolation forest using sliding window reduces the calculation cost, has a linear time complexity with low constant and a low memory requirement. So this algorithm has the capacity to handle extremely large data size and high-dimensional problems.

5. Conclusion

In this paper, we propose an algorithm based on isolation forest using sliding window for predicting the failures of the hard drivers. We also applies it to the SMART data of a real large data center for failure prediction, to verify the availability and efficiency of the algorithm. The experiments show that this algorithm can effectively save the time of modeling and ensure high prediction accuracy.

Acknowledgments

This work was supported by China National Key Research and Development Plan "Cloud Computing and Big Data Key Projects" under the grants NO. 2017YFB1001700.

References

[1] Emerson Network Power releases "Cost of Data Center Outages in 2016". *Computer World*, Apr 2016(008).

[2] K. V . Vishwanath and N.Nagappan. Characterizing cloud computing hardware reliability. *ACM Symposium on Cloud Computing DBLP*, 2010.

[3] Teerat Pitakrat, Lars Grunske et al. A comparison of machine learning algorithms for proactive hard disk drive failure detection. *International ACM Sigsoft Symposium on Architecting*

Critical Systems, 2013.

[4] Zhao Y, Liu X, et al. Predicting disk failures with HMM- and HSMM-based approaches. *Industrial Conference on Advances in Data Mining: Applications & Theoretical Aspects Springer-Verlag*, pp 390-404, 2010.

[5] Song Yunhua, Bai Wenyang and Zhou Qi. Prediction on hard disk failure of cloud computing framework by using SMART on COG-OS framework. *Journal of Computer Applications*, pp 31-35, 2014.

[6] Jia Yuhan, Li Jing, Jia Runying, et al. Hard disk failure prediction model validation in large data center environment. *Journal of Computer Research and Development*, pp 54-61, 2015.

[7] Hard Drive Stats for Q2 2018. https://www.backblaze.com/blog/hard-drive-stats-for-q2-2018/.

[8] S.M.A.R.T. https://en.wikipedia.org/wiki/S.M.A.R.T..

[9] Fei Tony Liu, K. M. Ting and Z. H. Zhou. Isolation-Based Anomaly Detection. *ACM Transactions on Knowledge Discovery from Data*, pp 1-39, 2015.

[10] Wenjun Yang, Dianming Hu and Yuliang Liu. Hard drive failure prediction using big data. *2015 IEEE 34th Symposium on Reliable Distributed Systems Workshop (SRDSW) IEEE Computer Society*, 2015.

[11] Bingpeng Zhu, Gang Wang, et al. Proactive drive failure prediction for large scale storage systems. *Mass Storage Systems & Technologies IEEE*, pp 1-5, 2013.

[12] Li Jing, Wang Gang, et al. Review of reliability prediction for storage system. *Journal of Frontiers of Computer Science and Technology*, 2017.

ISPECE

IOP Publishing

A New Crossover Algebra of GA for Solving the Degree Constrained Minimum Spanning Tree Problems

Hui Li, Kai Shi, Huanran Li, Sheng Lin, Guangping Xu, Shuangxi Li

Tianjin University of Technology

2966788181@qq.com, shikai0229@163.com,
164940886@qq.com,{lins,gpxu,sxli}@tjut.edu.cn

Abstract. The degree constrained minimum spanning tree (DCMST) problem seeks a spanning tree of a connected graph with the smallest weight. It is of crucial importance to the design of communication networks. There is no polynomial algorithm for solving DCMST problem, heuristic algorithm is usually used to find an approximate solution. The heuristic algorithms, such as genetic algorithm (GA), face the problems on tree heritage, which means the offspring may not be a feasible solution. To address this problem, we propose a novel crossover algebra which considers both the edge learning and structure learning, and can guarantee the offspring be a feasible solution. Experiment results validate the effectiveness of the proposed algorithm.

1. Introduction

Degree constrained minimum spanning tree (DCMST) is the problem about generating a minimum spanning tree with degree constraint. It's closely related to the problem about network design and network optimization. DCMST is widely used in computer network, communication network, transportation network and other network related problems. For example, with the increasing cost of infrastructure and equipment, the construction of today's communication network and transport network impose more requirements and restrictions on the structure or function of nodes. Most of the nodes have limited storage and transfer ability, so it can connect to a finite number of nodes. This is the application of degree constrained minimum spanning tree in real life.

It has been proved that DCMST is NP-Hard, a heuristic algorithm is usually used to search for the best solution within acceptable computational cost [1]. The traditional heuristic algorithm can find the minimum spanning tree quickly when the edge weight is taken into account in the search process. But it is easy to fall into the local optima. So there is no guarantee to find an effective solution for DCMST [2]. Tradition heuristic algorithm faces the problems of tree representation and new solution generation which degrades its performance for finding good results.

A good tree representation method should have the following properties: it can represent all possible trees; it is unbiased in the sense that all trees are equally; it can represent only trees; it can be easy to go back and forth between the trees and codes; it should be localized in this sense that small changes in the representation make small changes in the tree. Tree representation should cover the entire space of solutions, produces only feasible offspring, and possesses locality, and all these necessary characteristics, in order to use genetic algorithm effectively [4]. However, for DCMST, the tree represents still faces some problems: The offspring are not necessarily represented in the form of trees, and the gap between them is too large. Aiming at addressing this problem, in this paper, we propose a novel crossover algebra to improve the feasibility and locality of GA algorithm, thus obtaining better results for DCMST problem.

Content from this work may be used under the terms of the Creative Commons Attribution 3.0 licence. Any further distribution of this work must maintain attribution to the author(s) and the title of the work, journal citation and DOI.

Published under licence by IOP Publishing Ltd

2. Related work

The problem of the degree constrained minimum spanning tree is a well known problem. There are many algorithms proposed to solve the problem. The following will briefly introduce the related research in recent years.

The first, a branch and bound algorithm is proposed by Narula and Ho [5]. The step to select next node is rigidly and blindly, and it gives no priority to node that may quickly retrieve the answer node. The minimum spanning tree problem with degree constraint can be expressed as a linear 0-1 integer programming problem, which is solved by the construction method based on the original heuristic and dual heuristic and the branch and bound algorithm [6].

Secondly, a new representation method and an appropriate mutation operator can be used to solve the degree constrained minimum spanning tree problem. As the representation of candidate solutions and mutation operators are the basic design choices of evolutionary algorithms, this method has better locality and efficiency performance [7].

In addition, Prufer number can also be used as the chromosome in the current study to improve the crossover and mutation process of the existing genetic algorithm, so as to obtain high locality, heritability and self-adaptation performance [8].

In summary, most of the optimization improvements are based on crossover process. So they have the problem for generating a feasible solution in the neighborhood. Therefore, in this paper, we will propose a new crossover algorithm to solve these problems.

3. Crossover operator

As described above, the crossover operator, which is crucial to the performance of GA, faces some problems on generating new solutions. This means that the new solution may lose the character of feasible and locality. In this part, we will analysis these problems further and then design a new crossover operator to solve these problems.

3.1. Problems of traditional crossover algorithm

3.1.1. Crossover algorithm. In the original crossover algorithm, a pair of parents is selected for generating new children. In this process, a cross point k will be set at first, then the two parents exchange the genes from the cross point. Let the chromosomes of the parents be a($a_1, a_2 \cdots\cdots a_n$), and b($b_1, b_2 \cdots\cdots b_n$), and the cross point is k. After the crossover process, the new children are a($a_1, a_2 \cdots\cdots a_{k-1}, b_k, b_{k+1} \cdots\cdots b_n$) and b($b_1, b_2 \cdots\cdots b_{k-1}, a_k, a_{k+1} \cdots\cdots a_n$) respectively.

3.1.2. Problems. In this part, we will show the problems of traditional crossover algorithms under different tree representation method.

- Unfeasibility. As described above, one new generated child after the crossover process is ($a_1, a_2 \cdots\cdots a_{k-1}, b_k, b_{k+1} \cdots\cdots b_n$). However, in some representation methods, the new child may not be a tree. Although the Prufer number representation method can guarantee that the new child is a tree, it may violated the node degree constraint.

- Locality. The new generation should keep the similarity of the apparent. However, the Prufer number method, which has the advantage of feasibility, will lost the character of locality at the same time.

3.2. A Novel Crossover operator

3.2.1. Basic idea. The traditional genetic algorithm only considers the edge heritage, which means the child inherits one half edges of the parent. In our mechanism, we propose a novel idea of structure heritage. Structure heritage means that the child will inherit the structure character of the parents. So in the new crossover mechanism, the children will inherit both the edge and structure character. It means that a sub-tree of a child is isomorphic to one of its parents. Based on such idea, we will design our new crossover method in detail which incorporates both the edge heritage and structure heritage. The new algorithm comprises two parts: determining the cross point and generating new child trees.

3.2.2. Determining the cross point. The first step is finding two sub-trees in parents A and B respectively which has the same size of K. The size of K is about n/2. Let the nodes of the sub-trees be $(a_{i1}, a_{i2} \cdots\cdots a_{ik})$ of Parent A and $(b_{i1}, b_{i2} \cdots\cdots b_{ik})$ in Parent B.

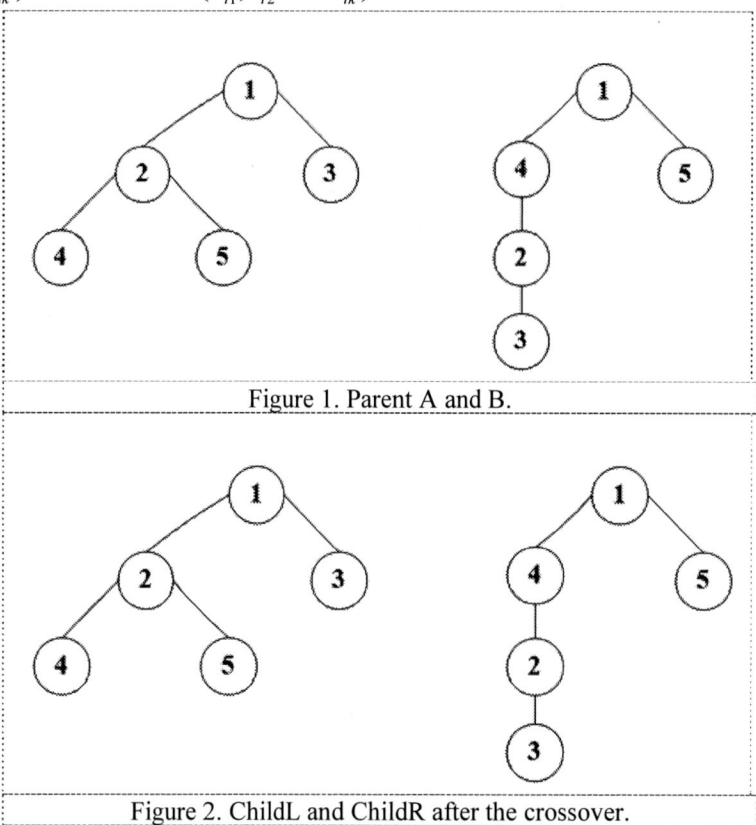

Figure 1. Parent A and B.

Figure 2. ChildL and ChildR after the crossover.

3.2.3. Generating new children trees. Let the two parents be Parent A and Parent B and the two children be ChildL and ChildR. We will introduce the method to generate ChildL at first. The new generated ChildL comprises two sub-trees. One of which is from Parent A excludes the nodes $(a_{i1}, a_{i2} \cdots\cdots a_{ik})$, the other is a sub-tree with nodes $(a_{i1}, a_{i2} \cdots\cdots a_{ik})$ constructed based on the structure of sub-tree $(b_{i1}, b_{i2} \cdots\cdots b_{ik})$. The construct process can be describe as follows:

- If there is an edge between b_{ip} and b_{iq} (p, q={1...k}) in Parent B, then construct an edge between a_x and a_y for the ChildL. Otherwise, delete the edge between a_{ip} and a_{iq}.

- Connect the two sub-trees, then forms the new ChildL.

- Similarly, we can generate ChildR which comprise one sub-tree from B excludes $(B_{i1}, B_{i2} \cdots\cdots B_{ik})$ and the other sub-tree $(B_{i1}, B_{i2} \cdots\cdots B_{ik})$ which is with isoconstructural to $(a_{i1}, a_{i2} \cdots\cdots a_{ik})$. Fig. 1 and Fig.2 shows an example the crossover process.

In short, we can conclude that the generating process is keeping one half of the structure of a parent unchanged, while updating the other half by exchanging the structure with another parent.

4. Genetic algorithm

Genetic algorithm (GA), also known as evolution algorithm, is a heuristic search algorithm, which is inspired by Darwin's theory of evolution and used for reference in the process of biological evolution. Understanding genetic algorithms requires understanding the concepts of populations, individuals, genes, chromosomes, heredity and mutation, and the concept of natural selection and survival of the fittest.

Population: the evolution of organisms takes place in the form of groups. Such a group is called a population. Individual: a single organism that makes up a population. Gene: a genetic factor. Chromosome: it contains a set of genes. Heredity and mutation: the new individual inherits the genes from each part of the parent, and at the same time has a certain probability of genetic variation. Natural selection and survival of the fittest: individuals who are highly adaptive to the environment are more likely to participate in reproduction, so more and more offspring will be born. In contrast, low-adaptability individuals are less likely to participate in reproduction, and fewer offspring [9].

There are three basic steps in genetic algorithm: selection, crossover and mutation. By genetic algorithm, the solution with low fitness function value is eliminated and the solution with high fitness function value is increased, so that the excellent gene can be preserved. In this way, it is possible to evolve individuals with high fitness function after the evolution of N generation.

5. Performance Evaluation

Each figure should have a brief caption describing it and, if necessary, a key to interpret the various lines and symbols on the figure. We have described the algorithms in detail before, and now we need to use bench mark data sets (CRD) to evaluate their performance.

In CRD data-set, each node corresponds to the horizontal and vertical coordinates of a point. Taking a 30-nodes data set as an example, the distance of each two nodes (the edge weight) needs to be calculated according to the Euclidean distance of the coordinates [10]. Therefore, we can get a 30 ×30 weight graph after the calculation, the same diagonal value of the horizontal coordinate is infinitely large, which means the node can not be connected with itself. The parameters of GA algorithm are set as Table1.

Table 1. Parameters setting

The crossover probability	0.2
Population	200
Genetic generation	4000
Degree-constrained	3

Therefore, at a certain number of nodes, the crossover probability of 0.2, 200 population, genetic generation 4000, degree of 3 is limited. The results are shown in Table2 and Table3. From the table, we can see that the proposed mechanism can improve the performance by reducing the weight about 20 percent.

Table 2. Target value of different algorithms at 30 nodes

30 nodes	the original GA algorithm	the proposed algorithm
CRD300	10584	9798
CRD301	11227	8990
CRD302	11868	9487
CRD303	12421	9887

CRD304	13039	10289
CRD305	11745	10267
CRD306	12413	9386
CRD307	13570	10086
CRD308	11754	9154
CRD309	11542	10384

Table 3. Target value of different algorithms at 50 nodes

50 nodes	the original GA algorithm	the proposed algorithm
CRD500	22530	18919
CRD501	17875	17368
CRD502	21536	18130
CRD503	19484	18058
CRD504	20608	17448
CRD505	20721	17703
CRD506	21946	18146
CRD507	21023	18062
CRD508	21125	17486
CRD509	21222	18063

6. Conclusion

In this paper, we investigated the DCMST problem using genetic algorithm. A new crossover algebra was designed for improving the feasibility and locality performance. Experiments results on CRD data-set showed that the proposed method can get better performance. In our future work, we will try to give a theoretical analysis of the new crossover algorithm.

Acknowledgment

This work was partly supported by an NSFC project under Grant No.11426163, Tianjin Nature Science Foundation Projects under Grant Nos.15JCQNJC00500, 17JCYBJC15600, a Tianjin Major Scientific and Technological Research Plan and Program under Grant No.16ZXHLSF00160.

References

[1] Narula, S.C. and C.A. Ho. (1980). "Degree Constrained Minimum Spanning Tree." *Computers and Operations Research* 7, 239–249.

[2] Collins, N., R. Eglese, and B. Golden. (1998). "Simulated Annealing: An Annotated Bibliography.' *American Journal of Mathematical and Management Sciences* 8(3–4), 209–307.

[3] G. Zhou, M. Gen, "Approach to the degree-constrained minimum spanning tree problem using genetic algorithms", *Engineering Design and Automation*, vol. 3, no. 2, pp. 156-165, 1997.

[4] G. Zhou, M. Gen, "A note on genetic algorithm approach to the degree-constrained spanning tree problems", *Networks*, vol. 30, pp. 105-109, 1997.

[5] J. W. Moon. (1967). "Various proofs of cayley's formula for counting trees." A Seminar on Graph Theory, pages 70.

[6] Palmer, C.E. and A. Kershenbaum. (1994). "Representing Trees in Genetic Algorithms." Technical report, IBM T.J. Watson Research Center, P.O. Box 704, Yorktown Heights, NY 10598, 1994.

[7] Gen, M., G. Zhou, and M. Takayama. (1998). "A Comparative Study of Tree Encodings on Spanning Tree Problems." *IEEE Conference on Evolutionary Computation.*

[8] T. C. Hu. (1974). "Optimum communication spanning trees." *SIAM Journal on Computing*, 3(3), 188-195.

[9] Gabow, H.N. (1978). "Good Algorithm for Smallest Spanning Trees with a Degree Constraint," *Networks* 8(3), 201–208.

[10] Boldon, B., N. Deo, and N. Kumar. (1996). "Minimum-Weight Degree-Constrained Spanning Tree Problem: Heuristics and Implementation on an Simd Parallel Machine." *Parallel Computing* 22(3), 369–382.

Research on Hybrid Recommendation Model Based on PersonRank Algorithm and TensorFlow Platform

Guangqi Wen[1,a], Chunmei Li[1,b,*]

[1] Department of Computer Technology and Application, Qinghai University Xining 810016, China

[a]595800468@qq.com

[b]li_chm0422@sina.com

[*] Corresponding Author

Abstract. At present, the borrowing data of university libraries in western China generally has the problem of large sparseness and inaccurate recommendation results. However, the traditional recommendation algorithm does not solve the problem. Therefore, solving the data sparsity problem and improving the recommendation accuracy has become a very important issue. Through research, the author proposes a hybrid recommendation model that combines the PersonRank algorithm with the neural network trained by Google's artificial intelligence framework TensorFlow in the context of excessive data sparseness in western college libraries. First, using the existing borrowing records to model through the bipartite graph probability random walk method, the author obtains a list of books that the reader may be interested in and calculates the probability value of interest. After that, the author converts the probability value into a pseudo-scoring table, and uses TensorFlow to calculate by pseudo-scoring, and supplements the data training set and book classification information to obtain the recommendation result. The model effectively solves the problem of data sparsity in the recommendation system. Finally, it compares with the traditional recommendation algorithm and an improved recommendation algorithm. The accuracy of the model is evaluated by AUC index, and the accuracy of the proposed model is higher.

1. Introduction

With the advent of the era of big data, it has become increasingly difficult for people to obtain the resources they need or are interested in from massive amounts of data. The emergence of this problem has brought enormous challenges to the related work of the library, especially for readers, who cannot quickly and accurately find the collection resources they need.

The concept of "recommended system" was proposed by Robert Armstrong, a professor at CMU University, in the American Association of Artificial Intelligence (AAAI) in 1995. The more famous recommendation system applications are: Amazon and Taobao's e-commerce recommendation system, Netflix and MovieLens' movie recommendation system, Youtube's video recommendation system, Douban and Last.fm's music recommendation system, Google's news recommendation system, and Facebook and Twitter's friend recommendation system. Among them, the collaborative filtering recommendation algorithm as a representative of the traditional recommendation algorithm has been the focus of many scholars. These two problems are the data sparsity problem and the cold start problem. The collaborative filtering algorithm recommends very good results when the amount of data

Content from this work may be used under the terms of the Creative Commons Attribution 3.0 licence. Any further distribution of this work must maintain attribution to the author(s) and the title of the work, journal citation and DOI.

Published under licence by IOP Publishing Ltd

is sufficient and a certain user base is recommended. But when the data is too sparse, its accuracy will drop dramatically.

There are also many well-known traditional recommendation algorithms, such as topic-based model-based recommendations. These traditional recommendation algorithms usually use the similarity between computing readers or books to cluster. Similarly, these traditional algorithms have good recommendations when the amount of data is sufficient and the data size is small. However, with the explosive growth of book resources, these traditional recommendation algorithms gradually reveal their drawbacks. With the explosive growth of data volume, they cannot evolve in time, and the sparseness of data becomes more and more prominent, which directly leads to the failure of traditional recommendation algorithms to make accurate recommendations.

Zhou[1] et al. proposed a network-based resource allocation model in 2007, which achieved better results than the traditional collaborative filtering algorithm. *Wang Qian*[5] et al. proposed an improved recommendation algorithm based on the bipartite graph, introducing the ratio Θ of the sum of the project degree and the weight of the project, which improves the recommendation accuracy and diversity.

At the same time, the emergence and popularity of machine learning artificial intelligence has also brought huge development space for the recommendation work. Google's YouTube recommendation uses the TensorFlow framework to recommend video to users and has achieved great success. Some scholars have introduced the FFM model into the recommended field, and the recommendation results are generated by predicting the scores of the items. Although this solves the sparseness problem of the traditional book recommendation algorithm, the efficiency is low and the recommendation effect is not obvious.

2. The source of the data

2.1. Data collection
In the research process of this paper, the author collected the data of Qinghai University Library. The collected data table has more than 90,000 pieces of reader information table, more than 300,000 pieces of data in the borrowing record table, and a book information table and a book category table. Simple processing of each data table to remove useless data. Keep valid information such as reader name, job number (student number), borrowed books, and book categories.

2.2. Organize data
All the data tables are associated, and all the information is organized into one table. The data in the final data table is the reader barcode (work number)-name-reader level-unit-grade-professional-title.

Table 1. data				
Reader's barcode	Reader's name	Reader's level	Reader's unit	Borrowing books
201100502	Shen Hui	student	Medical school	Introduction to Organic Chemistry
201113080	Zhang Dejing	student	School of Chemical Engineering	Principle of inorganic chemistry
209021404	Bian Pengguo	student	material science	Gas turbine (1)
209021404	Bian Pengguo	student	material science	Gas turbine (2)

201135280	Zhang Shou	teacher	Agricultural and pastoral college	Chemistry
209025945	Li Aimin	student	Surveying Engineering	Engineering Mathematics
210014367	Wang Huifeng	student	School of Chemical Engineering	Material mechanics
209019658	Xu Zhengjie	student	School of Mechanical and Electronic Engineering	Material mechanics

The above is some of the data listed. After the statistics are completed, a total of 11137 rows of data are obtained.

2.3. Data cleaning

There are many problems in the data set after sorting, and the data needs to be cleaned.

- In the process of building a library management system, testers often test the system and generate a large amount of test data. These are often not real borrowing data and need to be manually deleted.

- The library has undergone many changes in the development process, and each change will take a different form of record. For example, the book type, the book number format corresponding to a group of books is similar to K249.07; but there is also a batch of books corresponding to the classification number is similar to the J993/997 format. The research in this paper has dealt with this kind of data uniformly and unified the format.

- The research data table format in this article uses the csv file format. The csv file format uses a comma to separate the data. In many book names, there are commas, which have a great impact on the reading of the data. Therefore, using python to write a script to traverse the data set, mark out the problematic data bar, and manually adjust the title of the book to adjust the problem.

- In the process of importing datasets into the model, the variety of reader barcode formats always leads to the unrecognizable model. Therefore, each reader is mapped and mapped to the list of 0-Personmax. Personmax is the total number of readers. After the mapping is completed, the mapped data table is trained as a model input part.

3. Mixed model

3.1. PersonRank

In the research process of this paper, the author collected the data of Qinghai University Library. The collected data table has more than 90,000 pieces of reader information table, more than 300,000 pieces of data in the borrowing record table, and a book information table and a book category table. Simple processing of each data table to remove useless data. Keep valid information such as reader name, job number (student number), borrowed books, and book categories.

3.1.1. PersonRank algorithm

The idea of the PersonRank algorithm is to calculate the importance of each node by means of material diffusion. The algorithm transforms the reader's borrowed data into a bipartite graph of the image. For a borrowed record dataset, the reader and the book are divided into two parts. If a reader has read and borrowed a book, the reader and the book are connected by a line. In this way, all the data is connected, which constitutes the reader-book. We define the nodes, the user collection is {A, B, ..., W, ... Z}, the book collection is {a, b, c, ...}, and R(n) represents the importance of the n-node.

The PersonRank algorithm is for a certain reader. Here, reader W is used as an example to walk around the established bipartite graph by random walk, and the process is iterated until the data converges. Finally, the degree of importance to all other nodes of the reader W is obtained, from which the book object nodes and the probability values of interest are selected.

3.1.2. PersonRank algorithm calculation process

Select the reader W to recommend, and initialize the two parts:

$$R(A) = 0, R(B) = 0, \ldots \ldots, R(W) = 1, \ldots \ldots, R(Z) = 0$$

After initialization, the two parts are swam, and the starting point of the walk must be a point where the R value is not zero. After starting the walk, the probability of going to the next point is α, and the probability of staying at the current node is 1-α. Let the current node be n and the node of the reader W be N. Then the importance value of a node is calculated as follows:

$$R(n) = \begin{cases} \alpha \times \sum_{n' \in in(n)} \dfrac{R(n')}{|out(n')|} & n \neq N \\ (1 - \alpha) + \alpha \times \sum_{n' \in in(n)} \dfrac{R(n')}{|out(n')|} & n = N \end{cases}$$

calculation process:

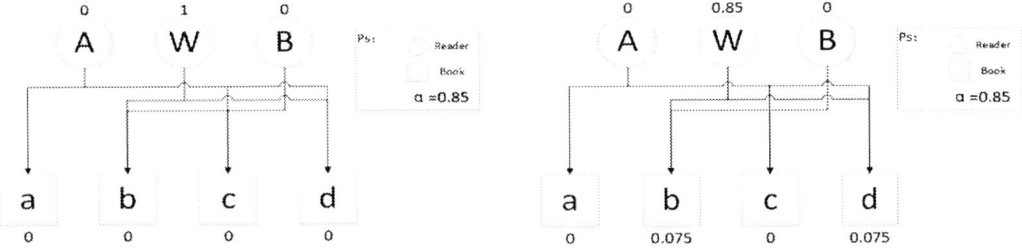

Figure 1. Initialization diagram   Figure 2. Take a walk

After a finite number of iterations, the probability values gradually stabilize and eventually reach stability. Screen out the book nodes that are more important to the reader and scale them up. Since the value of importance is a probability value, which is a value between 0 and 1, the magnitude is increased by two values that reach the range of 1-100. Calculate all readers and obtain a pseudo-recommendation list for each reader and the reader's level of interest in each book.

3.1.3. Pseudo-scoring table

In many cases, there is no reader's rating of books in the library borrowing data. Without the score sheet, we can't grasp the reader's personal preferences, and we can't make accurate predictions. So we will ask for books that are of interest to each reader. The conversion of probability values into pseudo-score tables can effectively solve this problem. For example:

The books and probabilities calculated by a reader are as follows (only five books are listed):

Table 2. Probability value of interest

	"mathematical model"	"SPSS Data Statistics and Analysis Application Tutorial"	"Wake up and still love you"	"MATLAB Scientific Computing and Visual Simulation Collection"	"computer network"
Probability value	0.1828	0.1342	0.0344	0.0220	0.0205

The probability values are all in the range of 0-1. The current work is to convert it into a value within 0-100, which is a pseudo-score. Multiply all probability values by 100 and the result of the calculation is a pseudo-score. Calculate the pseudo-scores of all readers and organize them into pseudo-scores.

3.2. Tensorflow

3.2.1. TensorFlow platform introduction
TensorFlow is Google Brain's second-generation open source machine learning system, and the versatility of the formal system makes it widely used in other computing fields.

The core of TensorFlow is the data flow graph, which uses a directed graph of "nodes" and "lines" to represent a large number of mathematical calculations. "Node" generally refers to some mathematical operations, sometimes also to the beginning or end of the system; "line" represents a relationship between points. As shown below:

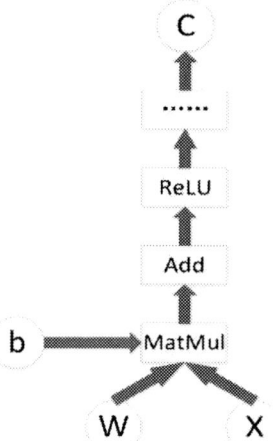

Figure 3. TensorFlow workflow

In this paper, we use TensorFlow to predict pseudo-scoring tables and reader eigenvalues that have been collated using the PersonRank algorithm. The model used is a content-based recommendation model.

3.2.2. TensorFlow training on data sets
Through the secondary cleaning and finishing of the data, each reader's rating information for each book and the category of each book are obtained. The training data set must contain the user's preference information as well as the rating information, so we divide the data into a single set to facilitate the training of TensorFlow.

Content-based recommendations are divided into two parts. First, the data is compiled to calculate the similarity between readers and books, and simple clustering between similar objects is performed. The second is to recommend the reader through the results of similarity.

Calculate the loss value of the training model:

$$\text{loss1} = \frac{1}{2}\sum_{p=1}^{n}\sum_{Wr(W,p)=1}\left((\theta^{(p)})^T \times x^W - y^{(W,p)}\right)^2$$

n represents the number of readers; r(W, p) represents a pseudo-scoring table; $(\theta^{(p)})$ represents the preference of the p-user; x^W represents the feature of the W-book; and N represents the number of features.

$$\text{loss2} = \frac{\lambda}{2}\sum_{p=1}^{n}\sum_{k=1}^{N}\left(\theta_k^{(p)}\right)^2$$

Loss2 represents a regularization term, and the role of regularization is to prevent the training data from eventually over-fitting. There are usually two ways to solve the over-fitting phenomenon. One is to minimize the number of selected variables, and the other is to use regularization. The regularization method automatically weakens unimportant feature variables and automatically "extracts" important feature variables from many feature variables, reducing the order of magnitude. The introduction of regularization is very helpful to this model, which balances empirical risk and structural risk.

$$lose = lose1 + lose2$$

The loss value loss of the model is obtained, and further calculations and optimizations are required. The algorithm is optimized, the calculated loss value is calculated again, the loss gradient with respect to the model parameters is calculated, and then the variable is updated by the calculated loss gradient to achieve a better training effect. The loss value can be used to measure the degree of error in the prediction, the loss gradient

Next, the data set is trained by the CPU, and the training result is stored in the file after one training, so that the next direct acquisition can reduce the calculation amount and time consumption.

3.3. Feature-incremental hybrid model

At present, the accuracy of the traditional, single recommendation model is getting lower and lower, resulting in unsatisfactory final recommendation results. Hybrid models often compensate for certain deficiencies in a single model, and the results are complemented by complementary advantages between models.

In this paper, the feature-incremental hybrid model is used to calculate the former calculation result as part of the input of another algorithm. The pseudo-scoring table calculated by the PersonRank algorithm in this paper is used to calculate the TensorFlow data set, which is a standard feature-incremental hybrid model. The pseudo-scoring table calculated by the PersonRank algorithm effectively solves the sparseness problem of the data and provides the potential interest of the reader. The content-based recommendation algorithm makes up for the inaccuracy caused by the PersonRank algorithm ignoring the edge weight.

4. Experimental result

4.1. AUC indicator evaluation

AUC is a model evaluation index for the evaluation of the two-category model. The recommended result in this paper is a two-category model, which divides the book object into two categories: recommendation and non-recommendation. To calculate the AUC, the ROC curve is first constructed. The ROC curve is drawn based on the real category and prediction probability of the sample. Building the ROC curve requires predicting the true category and predicted category of the object. By calculating the AUC we can know the accuracy of the prediction. When the value of the AUC indicator is higher than 0.5, the prediction result is good. If it is lower than 0.5, the prediction result is not ideal, and the randomness is relatively strong.

4.2. Pseudo-scoring table generation and detection

By collecting the borrowing data of Qinghai University Library and calculating it, the data of Qinghai University Library obviously has the problem of excessive data sparsity, and there is no book score. The calculation of the model used in this paper is very suitable. The following is a pseudo-scoring table calculated by the PersonRank algorithm by borrowing data from a teacher. (only the first five pieces of data are taken):

Table 3. PersonRank algorithm calculation result

	"JAVA from entry to mastery"	"MATLAB Scientific Computing and Visual Simulation Collection"	"PHP, MYSQL and JAVASCRIPT study manual"	"PYTHON scientific calculation"	"Data Mining: Practical Case Analysis"
Pseudo score	7.67	5.49	5.33	5.33	4.48

4.3. TensorFlow training model results analysis

The pseudo-scoring table is made into a scoring matrix and is calculated as part of the model input. The results obtained are as follows (only the first five data are taken):

Table 4. TensorFlow training results

	"JAVA from entry to mastery"	"MATLAB Scientific Computing and Visual Simulation Collection"	"PHP, MYSQL and JAVASCRIPT study manual"	"Data Mining: Practical Case Analysis "	"Algorithms in the Age of Big Data: Machine Learning, Artificial Intelligence, and Typical Examples"
Final rating	8.23	5.11	5.08	5.03	4.63

The calculated book score results are all converted into probability values of 0-1, which are used to represent the user's degree of interest value. The recommended results (ten recommendations) are manually labeled, that is, the recommended object selects the recommended results and selects the book of interest. Finally, two lists are obtained, namely a 0-1 list (1 is of interest, 0 is not of interest) and a probability table. Draw the ROC curve:

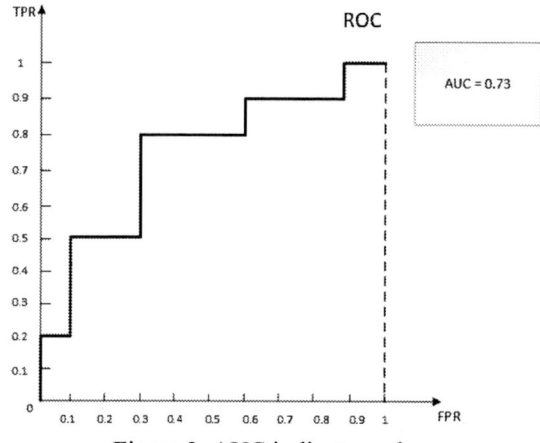

Figure 3. AUC indicator value

The X-axis represents the false positive rate, and the Y-axis represents the true positive rate. By calculating the area under the ROC curve, the value of the AUC index is 0.73, the AUC index is greater than 0.5, and the model prediction result is good.

4.4. Comparative analysis

The results calculated by looking at the model presented in this article are far from explaining the problem. It is necessary to compare the results of some of the recommended models that are currently used extensively, and then to observe the recommended effects of the model in this paper.

Here, the average AUC index value of each model is obtained by performing a large number of calculation comparisons with the collaborative filtering recommendation model and the weighted bipartite graph-based recommendation model under the same data set:

The text of your paper should be formatted as follows:

Table 4. Model comparison

	Collaborative filtering model	Recommendation model based on weighted bipartite graph	Model studied in this paper
Average AUC indicator value	0.68	0.72	0.73

The comparison chart is as follows:

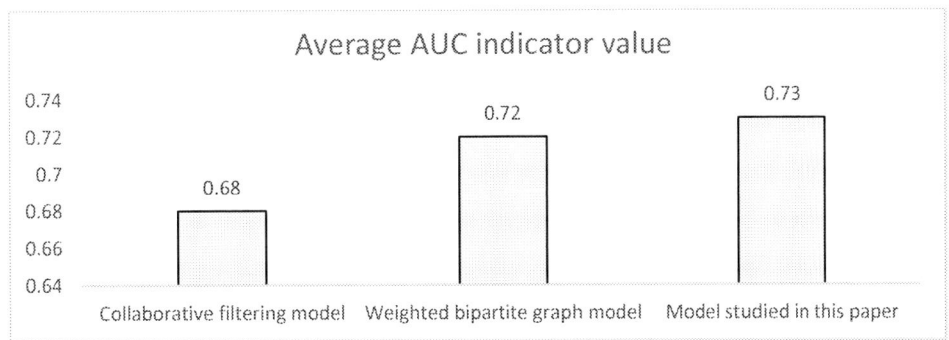

Figure 4. Comparison of average AUC indicator values

The collaborative filtering model is the representative of the older recommendation model. The recommendation model based on the weighted bipartite graph is the latest recommendation model. By comparison, we can see that the proposed model has better recommendation results.

5. Conclusion

Aiming at the data set with too large data sparseness and no scoring information in the recommended field, this paper proposes a hybrid recommendation model based on the combination of PersonRank algorithm and TensorFlow. The pseudo-score table is calculated by the PersonRank algorithm and then the TensorFlow platform is used to perform data set training on the content-based recommendation algorithm to obtain the recommended result. Finally, through comparison with the traditional recommendation algorithm, the hybrid recommendation model proposed in this paper has higher accuracy. However, for data sets with large data volume and comprehensive information, the proposed algorithm is less accurate, because ecause the recommended model in this paper is only applicable to data sets with large data sparsity.

References

[1] Zhou T, Ren J, Medo M, et al. Bipartite network projection and personal recommendation[J]. Physical Review E, 2007, 76(4): 046115.

[2] Zhang Jun, Li Xin. Handwritten Character Recognition under TensorFlow Platform[J]. Computer Knowledge and Technology: Academic Exchange, 2016 (6): 199-201.

[3] Zhang Shuang, Huang Peng. An Empirical Analysis of the Sparse Problem in the Book Recommendation System of University Libraries[J]. Journal of Academic Libraries, 2014 (6): 47-53.

[4] Zhou Lingyuan, Duan Longzhen. Design and Implementation of Personalized Book Recommendation System——Taking Nanchang Aviation University Library as an Example[J]. Library Theory and Practice, 2014 (12): 106-109.

[5] Wang Wei, Duan Shuangyan. An improved recommendation algorithm based on bipartite graph network structure [D]. , 2013.

[6]   Li Shuqing, Xu Xia, Xu Minjia. Book Qualification Measurement Method and Personalized Book Recommendation Service Based on Reader Borrowing Two-Division Network[J]. Journal of Library Science In China, 2013 (3): 83-95.

[7]   Xia Jianxun, Wu Fei, Xie Changsheng. Application of Data Filling to Alleviate Sparse Problems and Realize Personalized Recommendation[J]. Computer Engineering and Science, 2013, 35(5): 15-19.

[8]   ZHANG Xinmeng, JIANG Shengyi. Personalized recommendation algorithm based on weighted bipartite graph[J]. Journal of Computer Applications, 2012, 32(03): 654-657.

[9]   Wang Lianxi. A Personalized Book Recommendation System for University Libraries[J]. Modern Information, 2015, 35(12): 41-46.

[10] YANG Wu, TANG Rui, LU Ling. News recommendation method based on content-based recommendation and collaborative filtering[J]. Journal of Computer Applications, 2016, 36(2): 414-418.

[11] An Dezhi, Liu Guangming, Zhang Heng. Book recommendation model based on collaborative filtering[J]. Library and Information Service, 2011, 54(01): 35-38.

[12] Jiang Zhoufeng, Yang Jun, E Haihong. Content-based recommendation algorithm combined with social labeling[J]. Software, 2015 (2015 01): 1-5.

[13] ZHANG Yiwen, WANG Wei, CHENG Jiaxing. Improved recommendation method for bipartite random walk based on user interest[J]. Journal of Computer Applications and Software, 2015, 32(6): 76-79.

[14] Zheng Xiangyun, Chen Zhigang, Huang Rui, et al. Personalized book recommendation algorithm based on topic model[J]. Journal of Computer Applications, 2015, 35(9): 2569-2573.

ISPECE   IOP Publishing

Improvement of LDA Algorithm Based on Microblog Short Text Hotspot Analysis

Kai Li[1,a], Chunmei Li[1,b,*]

[1] Department of Computer Technology and Application, Qinghai University Xining 810016, China

[a]likai614020758@126.com

[b]li_chm0422@sina.com

[*] Corresponding Author

Abstract. In view of the high sparsity of microblog short text, the simple and random text content and the number of typos, and the use of standard LDA algorithm to calculate the result deviation of the keyword of short text in microblog is insufficient. In this paper, an improved LH-LDA model of microblog short text based on LDA algorithm is proposed. The model simultaneously links the heat of short text labels and short text. Through the relationship between the two and the text content, the keywords of a large amount of text are mined. At the same time, the correctness of the model is derived by Gibbs sampling, and the keywords of Weibo content are obtained. This article was adopted Python web crawler microblogging content, comment, forwarding, and praise. The results show that the traditional LDA algorithm is subject to more interference factors and the processing results are not accurate enough. From the results of the improved LH-LDA model processing in this short text, the keywords of this type of short text can be better found.

1. Introduction

With the development of modern society, the network has developed rapidly as a product of the new era, and many network-related industries have emerged. As one of the types, Weibo has developed rapidly. As a contemporary mainstream entertainment social segment, Weibo is popular in the world due to the convenience and speed of Weibo. Weibo users can log in a variety of ways or post their own content to share the content of others. At the same time, the number of Weibo users has reached 400 million, and the number of microblog readings in a single month has reached more than 25 in the field of 10 billion. Due to the huge number of users and the amount of information content, it is more important to mine and analyze the hot topics of microblog short text.

The hot topic of Weibo is that the number of people and attentions participating in a topic in Weibo reaches a certain level in a certain period of time. Therefore, whether the topic is a hot spot is usually determined by the number of participants and the degree of attention. Because the content of Weibo is simple, highly sparse, the text is randomly written, and the sentence structure is not fixed, the information expressed by the text vocabulary is very limited. Therefore, these factors determine that using the standard LDA algorithm to select the topic of the microblog text will cause a large error. At the same time, if the K-Means clustering algorithm is used to calculate the keywords of the short text, the clustering algorithm will produce a large deviation because the similarity between the texts ignores the internal relations of the text.

Content from this work may be used under the terms of the Creative Commons Attribution 3.0 licence. Any further distribution of this work must maintain attribution to the author(s) and the title of the work, journal citation and DOI.

Published under licence by IOP Publishing Ltd

Combining the characteristics of essay, an LDA algorithm based on label and heat is proposed to improve the standard LDA algorithm. By collecting labels and heat, you can improve the LDA standard and increase the weight value. Secondly, sparse short texts with short texts with heat will make a measure of the importance of each short text. Helps to find keyword headlines for short text and increase accuracy.

2. Overview

In the discovery of the topic of short texts such as Weibo and social media, scholars at home and abroad have done a lot of research. Zhou Fuxing et al. proposed a method for mining hot topics on microblogs using a hybrid approach [1]. Although Weibo is labeled with and without tags, Weibo is a short text with a heat value. The heat of a text can represent the degree of attention of the topic; Huang Chang et al. proposed a BBTM improved model to mine the hot topic of microblogging [2], using the BBTM model to take into account the contextual meaning of short text, and integrated micro The popularity of the blog text, but did not take into account the loss of some hot topics after the word segmentation. Wang Hui used a model based on the POSTTLA model for microblog hot topic discovery [3], and used PSO to optimize the bp neural network to train the prediction model composed of time series. This method only considers the influence of time on the topic heat, and does not consider the attention of Weibo content and the influence of the tag of Weibo content on topic discovery; some scholars improve the LDA algorithm [4, 5], improve the essay The accuracy of the extraction of subject matter in this content. At the same time, some scholars analyze vocabulary [6,7,8,9], and mine microblogs and Twitter hot topics through eigenvalues such as vocabulary heat, label and time.

It is found that Weibo content related to hot topic keywords often has hotspot tags, and the tags are usually created by the official or individual and have a large number of people to discuss. Considering the attention of Weibo content is usually determined by forwarding, comments and praise. This paper proposes a short text topic analysis model based on label and attention. The core content of this model is the potential Dirichlet algorithm. In this paper, the hot topic of microblog short text is analyzed by the optimization of the algorithm.

3. Microblog short text topic analysis model based on tag and heat

3.1. Traditional LDA algorithm model

First, the LDA algorithm model is based on the PLSA algorithm model. The PLSA algorithm model is a word bag method. The process of generating a document using the PLSA algorithm model is to select the topic of the article first, and then determine the topic to generate the word. The difference from the LDA algorithm model is that the samples of the PLSA algorithm model appear randomly, and the parameters are unknown but fixed. The LDA algorithm sample is fixed and the parameters are unknown, but the parameters are not fixed and are a random variable. The steps to generate a document for the PLSA model are given below:

- The first step: select a random document d_j according to a certain probability $P(d_j)$.
- The second step: Select the subject z_i of the document according to the probability $P(z_i|d_j)$ in the document d_j.
- The third step: selecting a word W_k of the subject according to the word frequency $P(W_k|z_i)$ in the already selected topic z_i.

Therefore, in the use of the PLSA algorithm model to determine the topic of the article is the selected word, backwards to derive the topic to which the word belongs, and finally determine the topic of the article. The probability plot of the PLSA algorithm model is shown in Figure 1.

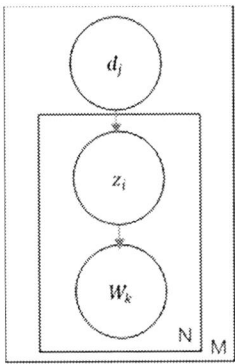

Figure 1. PLSA algorithm model probability map

Where N is the number of words under the topic distribution and M is the number of articles.

LDA is a topic model, which is a Bayesianization of the PLSA algorithm model. It adds a priori function when calculating the topic distribution and word distribution. The a priori function a priori distribution of the subject distribution a, and a word distribution of the a priori function β. It can give the subject words of the document in probabilistic form, so that by analyzing some documents to extract their themes (distribution), topic clustering or text categorization can be performed according to the theme (distribution). Then the probability formula for each word appearing in the article is as shown in Equation 1.

$$P(a_i, w_i) = P(w_i, a_i) * P(a_i) \tag{1}$$

Therefore, the LDA model judges the subject of this article by counting the probability of each word appearing. Therefore, the LDA algorithm model probability map can be obtained, as shown in Figure 2.

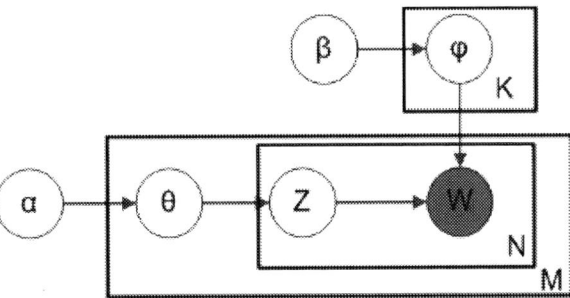

Figure 2. Probability plot of the LDA algorithm model

Where φ represents the probability of the subject word of the article, θ represents the probability of the topic of the article, z is the topic distribution, N is the number of words under the topic distribution, and M is the number of articles.

3.2. Label hotspot model based on LDA algorithm

Since the content of Weibo is different from the normal document, the short text content of Weibo has its own characteristic values, such as the microblog tag "#", forwarding, comment, praise, and so on. Therefore, a topic analysis algorithm LH-LDA (label hotspot LDA algorithm) for microblog short text is proposed. The algorithm is a comprehensive algorithm model based on LDA algorithm, which is suitable for the mining of microblog topic keywords. The Bayesian probability distribution graph is shown in Figure 3. The micro-blog hotspot tag and attention degree are used to add additional feature values to the short text content of each microblog, and the theme distribution of the content is calculated according to the microblog content and its attached feature values. The Weibo content containing the Weibo hotspot label is classified into one category, and the Weibo content that does not contain the Weibo hotspot label is classified into one category. Then calculate the theme distribution using different calculation methods for different categories. The following is a definition of Weibo with Weibo hotspot tags and no hotspot tags:

1) Hot topic Weibo: refers to the microblog that contains two "#" hot topic tags in the Weibo content. This kind of Weibo can respond well to the subject matter of Weibo.

2) General topic Weibo: refers to the microblog that does not contain hot topic tags in the Weibo content. The Weibo may be a microblog that is original or forwarded by the user. This kind of Weibo content is often diverse and needs to be converted into a word vector to analyze its subject word distribution.

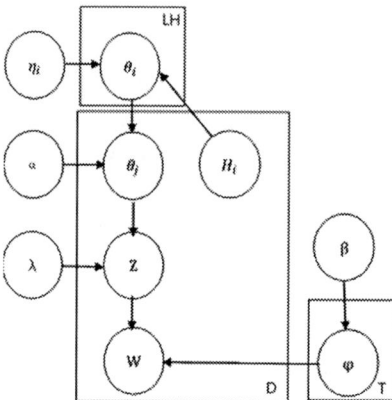

Figure 3. Bayesian probability plot of the LH-LDA algorithm model

The process of the algorithm is introduced by the Bayesian probability graph of the calculation model.

First, the first step is to determine whether the short text of the microblog has a hotspot label, and if the microblog has a hotspot label, that is, a hot topic microblog. The relationship θ_i between the hot topic H_i and the different topics is extracted from the distribution of the attention parameter η_i, and θ_i is assigned to θ_j. If the Weibo does not have a hot topic tag, that is, the general topic Weibo, the relationship θ_i between the microblog and different topics is extracted directly from the subject distribution of the alpha parameter. In the entire algorithm model, the probability distribution formula of θ is shown in Equation 2 below.

$$P(\theta|\eta,\alpha,H)=\frac{P(\theta,\eta,\alpha,H)}{\sum_{T=1}^{T} P(\theta,\eta,\alpha,H)} \tag{2}$$

The formula for calculating the degree of interest η [10], as shown in Equation 3:

$$\eta=\log_{10}(a)\times0.1283+\sqrt{c}\times0.4082+F\times0.6827 \tag{3}$$

Where a represents the number of microblogs; c represents the number of Weibo comments; and F represents the number of microblog forwardings. Multiplying the value at the end of the expression is the result of a survey of 10,000 Weibo users. The heat value of Weibo can be approximated by the attention parameter η. At the same time, for the Weibo content, we use the word bag model without considering the relationship between the upper and lower semantics of the vocabulary content, so when the microblog text has m words You can use Equation 4 to calculate the heat value of each word:

$$\eta(W)=\frac{\eta(w)}{m} \tag{4}$$

In the second step, the probability distribution formula of θ is solved. For the hot topic microblog, in the different topics included in the microblog, the theme Z to which θ belongs and the word W to which the theme Z belongs are determined by the Dirichlet distribution of the parameter λ. For the general topic microblog, the generated word W is extracted by the word polynomial distribution parameter φ. Finally, the joint distribution of vocabulary and subject results in Equation 5:

$$P(Z,W|\theta,\lambda,\beta)=P(\lambda)P(Z|\theta)P(W|Z,\theta) \tag{5}$$

The process of generating microblogs by LH-LDA algorithm model is used to inversely derive the distribution of microblog topic keywords. Therefore, the microblog hot topic analysis process diagram based on the LH-LDA algorithm model is shown in Figure 4:

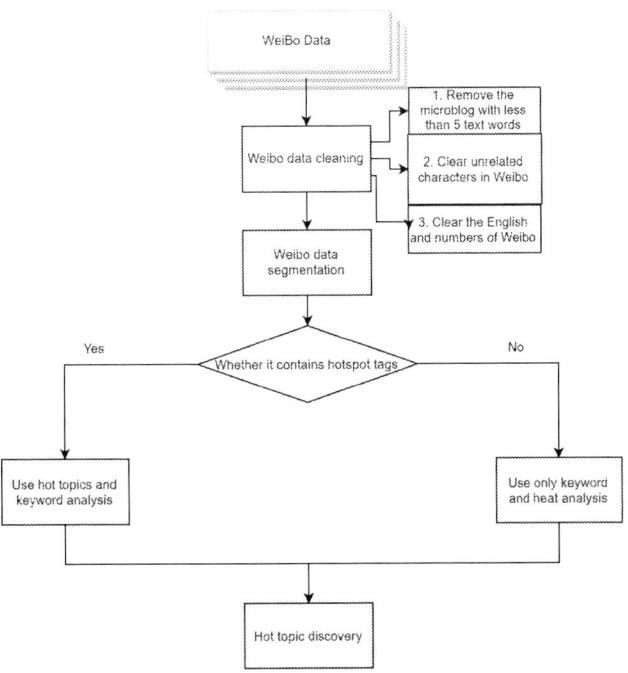

Figure 4. Microblogging hot topic analysis process diagram

3.3. Application of LH-LDA algorithm in short text subject mining

The traditional LDA algorithm model solves the word distribution and can be based on the distribution Z from the subject and the priori Dirichlet parameter β. Find the word distribution ψ Formula 6:

$$P(w|z,\psi) = \prod_{i=1}^{W} P(w_i|z_i) \qquad (6)$$

However, the proposed LH-LDA algorithm model estimates the unknown parameters differently from the traditional LDA variable EM algorithm to estimate the unknown parameters. The LH-LDA algorithm model derives unknown parameters using Gibbs Sampling, which is Gibbs sampling. Gibbs sampling [11] is a method for observing unknown parameters in a series of specified multidimensional probability distributions (joint probability distributions of multiple random variables) in Markov chain Monte Carlo theory (MCMC). For a document D_i in the LH-LDA algorithm model, the attention parameter η, the Dirichlet distribution α, the Dirichlet distribution β and the hotspot label parameter H are empirically obtained a priori parameters, while other variables The parameter word topic distribution parameter Z, the topic distribution parameter θ, and the word distribution parameter φ are unknown variable parameters. The relational expression 7 of the word topic distribution parameter Z under the vocabulary W is derived by the word topic joint distribution formula, where M is the number of words of the text:

$$P(Z_i = T|Z_{\neg i}, W_i) \propto P(Z_i = T, W_i = M|Z_{\neg i}, W_{\neg i}) \qquad (7)$$

Finally, the formula 8 of the distribution parameter θ of the microblog in the subject relationship and the formula 9 of the distribution parameter φ of the subject in the vocabulary are obtained by stepwise iterative sampling:

$$\varphi = \frac{M_T^{(t)} + \beta_t + H_t + \eta_t}{\sum_{v=1}^{V} M_T^{(t)} + V\beta_t + H_t + \eta_t} \qquad (8)$$

$$\theta = \frac{M_D^{(k)} + \alpha_k + H_k + \eta_k}{\sum_{k=1}^{K} M_D^{(k)} + K\alpha_k + H_k + \eta_k} \qquad (9)$$

The improved LDA algorithm can better calculate the microblog with hotspot tags and heat values, and can find the keyword that best represents the content of the text in the short text of Weibo.

In addition, the LH-LDA algorithm can be applied not only to the extraction of keyword texts in short texts, but also to the filtering of spam messages and emails. Through the analysis of SMS and email, find the keywords that best represent SMS and email, and filter SMS and email according to the keyword.

4. Experiment and result analysis

4.1. Short text data set and data preprocessing

The short text dataset collected in this article is a microblog text content obtained by simulating web browsing using python's selenium library and using Xpath. The microblog short text data set collects 229,900 microblog contents from January 1, 2018 to November 1, 2018. The key words of the data set are mined using the LH-LDA algorithm.

The data preprocessing is first to clean the data collected by the microblog short text. Data cleaning is the deletion of characters and text that are not related to the research content. The data collected in this article is the text collected by the crawler, which may contain URLs, emoticons, numbers, English, Japanese, Korean, etc., so the regular expression is used to clean the collected microblog text. Keep only the text content. At the same time, in the process of cleaning the data, the microblog text with characters less than 5 characters is deleted, because the text is too short to display the theme content well.

Then, the cleaned data is classified. Contains the characters "#....#", that is, hot topic labels, which are classified into one category, as shown in Table 1; the Weibo content that does not contain the characters is divided into one category, as shown in Table 2. Then, the text contained in the middle of the character "#....#" is taken as a hot topic by a specific character extraction, as shown in Table 3.

Table 1. Weibo with hot topic tags

#Wang Junkai# 18 years of desire for the college entrance examination smoothly and like people to test an ideal university together, and he is not far from the two or three friends to see a relatively stable life, see Wang Junkai
#Pure art# #draw# Wimbledon's painting alumni @YUCHU's rich little expressions, very playful oh#London Art University UAL#
Colorful Garden # On the celebration party, Miss Chu Yuxi's vibrating video was praised 80W a day, and videos from universities such as Tsinghua University and the People's Public Security University were rated as excellent campus videos.
#new skill get# How to pass the College English Level 4 and College English Level 6 exams in one month? Need to take the test of the students of the 4th and 6th grade
Renmin University of China # # Rookie return plan # A campus environmental awareness campaign advocating recycling and reuse of waste packagings was launched at Renmin University of China in Beijing recently.

Table 2. Weibo without hot topic tags

The New Year's Day is more than a year. I have to go to the University for the New Year's Day. I have to review the exams.
Today, I am very happy at noon. I have a good time at the grandmother's house. I have a very good grapefruit super sweet afternoon. I'm playing a mobile game with my brother. You're still like a junior high school.
Hard work, small fresh meat arrived early in the morning to pick up the university friend Xixi in the light valley small gathering
Have your friends recently indulged in a game called jumping?
Let's let the woods that have passed through the battle to tell everyone how to get the high scores.

Table 3. Hot topics extracted

Tsinghua University
Campus things
Warm video
I am a scientist
a moment of ease

Secondly, the Python library of Tsinghua University's word segmentation tool THULAC [12] is used to perform word segmentation on the preprocessed data set. Then use Harbin Institute of Technology's stop word list to remove the stop words from the data set after the word segmentation. The appearance of stop words has a greater impact on the short text of Weibo, because the number of words in Weibo content does not exceed 140 words, and the number of stop words occupies a certain amount, they do not help the topic analysis.

4.2. Vectorization of words

The tool for the word vector used in this article is Word2vec. Word2vec is a toolkit for Google to acquire word vectors for open source in 2013. Due to the high sparsity of Weibo content, and the LH-LDA algorithm model is a word bag model, the word vector used to construct Weibo texts with Word2vec is more accurate. Using Word2vec to perform word frequency statistics on the preprocessed data set, the word frequency is formed into a word frequency matrix, and the vectorization operation of the data set is completed. The result is shown in Table 4, which is the word frequency vector of each word in all documents.

Table 4. Data set conversion to word frequency vector results

...
(230125, 54857)
(230125, 42018)
(230125, 143516)
(230125, 45527)
(230125, 4599)
(230125, 70529)
(230125, 151658)

4.3. Microblogging hot topic analysis experiment of LH-LDA algorithm model

The experimental environment used in this article is the Intel Core i7-8700@3.20Ghz CPU, 32GB of RAM PC. The experimental tool used is python 3.7.2.

First, the a priori Dirichlet parameters α and β are determined according to the LH-LDA algorithm model derived above. For the traditional LDA algorithm model, set α = 1 / K, β = 1 / K. For the LH-LDA algorithm model of this paper, α=50/K and β=0.01 are set according to the empirical value, where K represents the number of topics. Here we have a default number of topics of 10. The traditional LDA algorithm and the LH-LDA algorithm model have a default iteration of 1000 times. Table 5 lists the Top 10 topics after the LDA algorithm and the LH-LDA algorithm process the data set.

Table 5. Results for dataset LDA and LH-LDA algorithms

Algorithm model	Top10 topic
LDA	1.Education 2. Graduation 3. Play 4. Weibo 5. Games 6. University 7. Student 8. Campus 9. World Cup 10. Teacher
LH-LDA	1.Game 2. Work 3.High school 4. Youth 5. Admissions 6. Education 7. Graduation 8. Campus 9. Dormitory 10. World Cup

Through the experimental results, it can be found that the LDA and LH-LDA algorithms are based on the theme of "university students" in general theme types, which proves that the analysis of microblog topic topics using LDA algorithm is correct. The calculation formula 10 of uncertainty is introduced below, where u represents an uncertain subject vocabulary, and topic represents the total number of vocabulary words obtained:

$$\text{Uncertain(topic)} = \frac{\sum_{n=1}^{n} u}{\sum_{k=1}^{K} topic} \qquad (10)$$

From the results of the traditional LDA algorithm model calculation data set, the words "Weibo", "Student" and "Teacher" show the "university" theme, but its uncertainty is relatively large, so it is not Can more accurately derive the keywords in the dataset. The Top10 keywords obtained by the LH-LDA algorithm are subject words related to college students, and the results calculated using the LH-LDA algorithm are less uncertain, and the obtained keywords are more accurate. Table 6 shows the uncertainty of the LDA and LH-LDA algorithms.

Table 6. Uncertainty of LDA, LH-LDA algorithm

Algorithm model	uncertainty
LDA	0.4
LH-LDA	0.2

4.4. Analysis of results for qualified topic content

This paper analyzes the accuracy of different algorithms from a given limited theme "Xi'an Smoking". Collected 8,000 Weibo data through python crawlers. Firstly, the microblog content is manually labeled, and then the data is processed by the LDA and LH-LDA algorithms respectively. The final result is shown in Table 7.

Table 7. Qualified topic LDA, LH-LDA algorithm solution results.

Calculation method	Topic Words
Manual labeling	Xi'an, Non-smoking, Comprehensive, Public Places, Health, Smoking, Smoke-free, Control, Management
LDA	Method, Xi'an, Non-smoking, Tourist, Public places ,Smoking, caveat, fine, control ,country
LH-LDA	Method, Xi'an, Non-smoking, Tourist, Public places, Smoking, caveat, fine, control, country, The strictest

The results of the processing of different algorithms are compared with the keywords that are manually labeled. It can be found that the traditional LDA algorithm selects a large number of vocabulary words, but does not consider whether the vocabulary itself belongs to the hot tag content, and does not consider the heat value of the vocabulary, so there will be some irrelevant subject vocabulary, such as " "The tourists", "the most strict" and other factors affecting the subject matter. LH-LDA processes a large amount of microblog data, and labels and heats the vocabulary of the word segmentation so that each vocabulary has two attributes. Therefore, from the experimental results, the difference between LH-LDA and manual labeling is no more than 30%. Through the above analysis of the microblog texts of the undefined subject and the microblog text analysis of the limited topics, it can be seen that the annotation of the vocabulary by the LH-LDA algorithm to calculate the keywords of a large number of microblog short texts can make up for the lack of context semantics of the microblog content. High sparsity and other characteristics, better mining of microblog short text content.

5. Conclusion

Due to the particularity of microblog short text, that is, the sparsity is high, the text content is simple, and the context semantics are lacking, the traditional standard LDA is not very effective for the discovery of keyword words in microblog short text content. Aiming at this problem, this paper proposes an improvement of LDA algorithm based on microblog hotspot label and heat value. Combining Word2vec algorithm to convert vocabulary into word frequency vector, the microblog hotspot label and heat value are given different labels to microblog vocabulary, effective the keywords of the microblog short text are calculated, and the algorithm can be used for various types of short text topic analysis. Second, the algorithm converts a large amount of high-sparse short text content into low-sparse text content. The LH-LDA algorithm can be applied to a variety of short text heat analysis and keyword analysis, such as SMS, email, Weibo text, video barrage, etc. for subject extraction analysis. Future research content will continue to optimize short text content processing and LH-LDA algorithm optimization. Also consider adding parallelization operations in the data processing process to speed up the short text processing.

References

[1] Zhou Fuxing, Chen Xiuzhen, Ma Jin, Li Shenghong. A new method for mining hot topic topics based on tag semantics[J]. Computer Engineering, 2018: 1-11.

[2] Huang Chang, Guo Wenzhong, Guo Kun. Research on Improved BBTM Model for Hot Spots of Microblogs[J]. Computer Science and Exploration, 2018: 1-14.

[3]   Ma Xiaoning, Wang Hui. Topic trend prediction based on BP neural network optimized by PSO[J]. Computer Engineering and Design, 2018, 9: 036

[4]   Wu Wankun, Wu Qinglie, Gu Jinjiang. Hot topic discovery of e-commerce microblog based on EM-LDA synthesis model[J]. Data Analysis and Knowledge Discovery, 2016, 31(11): 33-40.

[5]   Zhang Chenyi, Sun Jianbiao, Ding Yiqun. Mining of Weibo Theme Based on MB-LDA Model[J]. Journal of Computer Research and Development, 2011, 48(10): 1795-1802.

[6]   Li Hui, Wang Liting. Research on Hot Spots of Microblog Based on Term Heats[J]. Information Science, 2018, 36(4): 45-50.

[7]   Li Jing, Yin Jian, Liu Shaopeng, et al. Mining of Weibo Theme Based on Topic Tag[J]. Computer Engineering, 2015, 41(4): 30-35.

[8]   Shang Qingxia. Chinese microblog hot event detection and sentiment analysis based on TH-LDA model [D]. Southwest University, 2017.

[9]   Ishikawa S, Arakawa Y, Tagashira S, et al. Hot topic detection in local areas using Twitter and Wikipedia[C]//ARCS Workshops (ARCS), 2012. IEEE, 2012: 1-5.

[10]   Yan Yueming. Evolution analysis of hot topics on Weibo based on topic model [D]. Xidian University, 2017.

[11]   HAN Dong, WANG Chunhua, XIAO Min. Text Classification Method Combining Semi-supervised Learning and LDA Model[J]. Computer Engineering and Design, 2018, 39(10): 3265-3271.

[12]   Maosong Sun, Xinxiong Chen, Kaixu Zhang, Zhipeng Guo, Zhiyuan Liu. THULAC: An Efficient Lexical Analyzer for Chinese. 2016.

ISPECE   IOP Publishing

Extraction of Cerebral Hemorrhage and Calculation of Its Volume on CT Image Using Automatic Segmentation Algorithm

Nian Wang, Fei Tong, Yongcheng Tu, Hemu Chen, Yong Zhou, Jun Tang

Electronic and Information Engineering, Anhui University, Hefei, 230039, China
The First Affiliated Hospital of Anhui Medical University , Hefei, 230022,China
Supported by the Natural Science Foundation of China (61672032)
Corresponding author e-mail: wn_xlb@ahu.edu.cn

Abstract. Cerebral hemorrhage is caused by a number of factors leading to cerebral vascular rupture, which causes blood to flow into the brain tissue and then clumps together to form hematoma. If the treatment is improper for patients with cerebral hemorrhage, their lives will be in great danger. Rapid, accurate and repeatable estimation of cerebral hemorrhage is considerably important for medical diagnosis and treatments. In this paper, an automatic system is proposed to segment the regions of cerebral hemorrhage from CT images. Then the volume of cerebral hemorrhage can be calculated by extracting the regions of cerebral hemorrhage. The segmentation method proposed in this paper is based on the Otsu algorithm, Chicken swarm optimization algorithm and local recursion algorithm. Meanwhile, the results of the proposed algorithm are compared with those given by the experts.

1. Introduction

Stroke is an acute cerebrovascular disease that causes damage to brain tissue due to sudden rupture of cerebral blood vessels or obstruction of blood vessels. [1] Stroke has the characteristics of high incidence, mortality and disability rate . At the same time, the investigation shows that the death rate from stroke has risen to the highest in China. [2] Stroke apoplexy can be divided into two types: ischemic stroke and hemorrhagic stroke. [2] Accurate estimation of the volume of cerebral hemorrhage is of great significance in the treatment of hemorrhagic stroke and has important clinical value.

At present, manual and automatic segmentation are commonly applied at home and abroad to measure the cerebral hemorrhagic volume. [3-4] Manual segmentation is extremely wasteful and difficult to implement, even with non-ignorable disadvantages which are low in efficiency and high in repeatability. As far as manual segmentation is concerned, automatic segmentation is not only fast in segmentation, but also high in accuracy. However, due to the complex shape and appearance of medical images, the automatic segmentation also faces the following technical challenges. [5-7] (1)There is a gray overlap between parts with and without cerebral hemorrhage. (2) In the regions of the cerebral hemorrhage, the substantial gray level changes. As a result, how to design a robust, fast and effective segmentation algorithm for extracting the regions of the cerebral hemorrhage and calculating its volume in order to describe the degree of the cerebral hemorrhage has become an intense and difficult topic for scientific researchers . [2, 8]

Several types of method have been proposed by different researchers. Sumijan et al. proposed a method of combined thresholds to segment the regions of the cerebral hemorrhage, which uses image

Content from this work may be used under the terms of the Creative Commons Attribution 3.0 licence. Any further distribution of this work must maintain attribution to the author(s) and the title of the work, journal citation and DOI.
Published under licence by IOP Publishing Ltd

features and Otsu algorithm to extract the regions of the cerebral hemorrhage. [9]Soroushmehr, S. Mohamad R., et al. developed a method of using Gaussian mixture model to obtain the region of the cerebral hemorrhage. Firstly, the algorithm used linear contrast stretching, Gray Matter Subtraction (GMS) and Total Variation (TV) to process the images of cerebral hemorrhage, and then GMM model is used to segment the images of cerebral hemorrhage. Finally, Intensity Constraint, Area Constraints and Midline Pattern Matching are used for post processing of the image. [10] Sun, Mingjie, et al. developed a method of using fuzzy clustering and threshold method. The coarse segmentation area was reconstructed in three dimensions, and then hyper volume segmentation was carried out. [11] Prakash, KN Bhanu, et al. proposed a modified distance regularized level set evolution (MDRLSE) algorithms for segmentation of the region of cerebral hemorrhage. [12] Moreover, another method is proposed in[13], which segments the regions of the cerebral hemorrhage by using Random Forests which adopts the standardized-to-template and neighborhood mean to reduce complexity and computation time.

2. The Skull Removal

The skull was removed in order to eliminate a part of the brain image, which is unfavorable to extract the region of cerebral hemorrhage. After removal, it became easier to make research and analysis. Threshold method and the left and right scanning algorithm are applied to the skull removal in this paper. Each slice was divided into the regions of bone and none-bone. The skull and some tissues outside the skull were removed by the left and right scanning algorithm. Then some noise spots appeared on the image of the skull removal, which can be eliminated by the median filter.

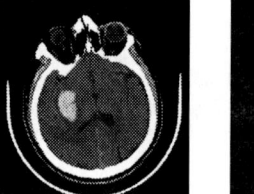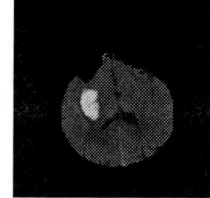

Figure 1. (a) Original Image, (b)After the skull was removed.

3. Enhanced Segmentation Algorithm

3.1. Otsu Segmentation Algorithm

Otsu segmentation algorithm uses threshold to divide an image into two parts: the target C1 and the background C2. The best threshold will be obtained when their inter class variance reaches its largest value. The discriminant is shown as follow.

$$t = \underset{1 \leq t \leq L}{\arg\max}\left\{\sigma_B^2(t)\middle|\sigma_B^2\right\} = \omega_1\left(u_1 - u_\tau\right)^2 + \omega_2\left(u_2 - u_\tau\right)^2 \quad (1)$$

$$\omega_1 = \sum_{i=1}^{t} p_i, \quad \omega_2 = \sum_{i=t+1}^{l} p_i, \quad u_1 = \sum_{i=1}^{t} i\, p_i / \omega_1, \quad u_2 = \sum_{i=t+1}^{l} i\, p_i / \omega_2 \quad (2)$$

Where p_i refers to the probability of gray level i occurrence, $\omega_1, \omega_2, u_1, u_2, u_\tau$ are respectively represented as the accumulative probability of C1,the accumulative probability of C2, the average gray level of C1, the average gray level of C2, the average gray level of the whole image.

3.2. Chicken Swarm Optimization Algorithm

Chicken swarm optimization (CSO) algorithm is a random optimization algorithm that imitates the foraging behavior of natural chickens with hierarchical structure. In the chicken swarm optimization algorithm, the chicken swarm is divided into three grades, which are roosters, hens and chickens depending on the individual's ability in the process of foraging,which is full of competition and cooperation. As time changes, the role of individual also changes, and the chicks grow into big chickens, while the adaptive range and fitness of the algorithm are improved. The algorithm is characterized as follows:

1.The chicken swarm is divided into several subgroups, each of which contains a rooster, several hens and chicks.

2.The allocation of individual roles in the chicken swarm is determined by the individual's fitness. Several individuals with high fitness are assigned as roosters and become the heads of each subgroups, while a few less adaptable individuals are assigned to chicks. For the rest of those individuals, they are assigned to hens. Hens live randomly in any chicken swarms, and the mother-child relationship between hens and chicks is determined at random.

3.Even if dominant relationship and mother-child relationship in chicken swarm have been established, these relationships will remain unchanged as long as the updated conditions are not satisfied. If the updated conditions are met, they should be reestablished.

4.In the chicken swarm, each grade has its own way of foraging. The individual follows the rooster in search for food, and the chicks are fed around the hens. The roosters have the greatest advantage in the competition of the food, the hens have the second, and the chicks have the least advantage .

When searching for the optimal solution for the optimization problem, each individual in the chicken swarm is a solution of the optimization problem. Suppose NR, NH, NC and NM represent the numbers of roosters, hens, chicks, and mother hens, respectively. In the whole chicken swarm, all the individuals are assumed to be N. The position of each individual $x_{i,j}(t)$ denotes the j dimension of the i body in the t

iteration value. The whole chicken swarm is divided into three types of chicken, and the location updated formula for each type of the chicken swarm is not the same.

Roosters correspond to the best fitness individuals in the chicken population. Position update equations of the roosters is as follows:

$$x_{i,j}(t+1) = x_{i,j}(t) \times (1 + Randn(0, \sigma^2)) \quad i \in (1,...,pop), j \in (1,...,\dim) \quad (3)$$

Where Randn(0, σ^2)denote a Gaussian distribution that the mean value are 0,and the standard deviation is σ^2, ε are 1, a very small constant in order to prevent zero division error. Position update equations of hens are as follows:

$$x_{i,j}(t+1) = x_{i,j}(t) + c_1 \times Rand \times (x_{r1,j}(t) - x_{i,j}(t)) + c_2 \times Rand \times (x_{r2,j}(t) - x_{i,j}(t))$$

$$c_1 = \exp((f_i - f_j) \Big/ abs(f_i) + \varepsilon) \quad\quad (4)$$

$$c_2 = \exp((f_{r2} - f_i))$$

Where Rand is a uniformly distributed random number between 0 and 1, r1 is the rooster in the chicken swarm of the ith hen, r2 is a randomly selected chicken from the chicken swarm which is different from the ith hen.

Position update equations of the chicks is as follow.

$$x_{i,j}(t+1) = x_{i,j}(t) + F_L \times (x_{m,j}(t) - x_{i,j}(t)) \quad (5)$$

Where m is the hen corresponding to the ith chick, $F_L \in (0,2)$ is the follow coefficient that chicks follow the chick mothers to looking for food.

3.3. Local Recursive Algorithm For Chicken Swarm Optimization Based On Otsu

Local recursive algorithm can be implemented by following steps. Firstly, the whole image is segmented using the chicken swarm optimization algorithm based on Otsu, and the threshold t is generated. The pixels larger than the threshold t keeps the original gray level, and the smaller ones are 0. Then, the threshold t is determined whether it meets the recursive termination condition or not, if that condition is not met, the image will continue to be segmented by chicken swarm optimization algorithm based on Otsu(CSOO).The flow chart of this algorithm is as follows.

a)Input an image that shows the skull has been removed.

b)The fitness function of the chicken swarm optimization algorithm is constructed by using the Otsu model, see formula (1).

c)Parameters of the chicken swarm optimization algorithm are set,including the maximum number of iterations M, chicken size pop, the solution space dimension dims, the number of chicken renewal G, the proportion of roosters in the whole chicken swarm scale r Percent, the proportion of hens in the scale of the whole chicken swarm h Percent, and the proportion of hens that can raise chickens the proportion of hen scale m Percent. Since the gray image pixels are between 0 and 255, the interval is regarded as the initial solution space of the function, and the initialization of the chicken swarm is generated in this interval.

d)The first threshold t approaches the optimal segmentation threshold by cooperating with roosters, hens and chickens in different subgroups.

e)The threshold t will be judged whether satisfies the local recursive termination condition. Because the pixel values of the hematoma area are higher than that of the normal brain tissue, the upper limit of threshold is regarded as the local recursive termination condition in this paper. In this paper, the initial value of upper limite of threshold is 255 gray level, and every 5 gray level decreases gradually. When the upper limit of threshold is 130 gray level, the experimental results of this algorithm are the best. So the local recursive condition of this paper is that threshold t is less than 130 gray level. If the condition is satisfied, the steps of c and d will be executed, and the solution space in the step of c is updated to between t and 255, otherwise f is executed.

f)The image will be segmented using the optimal threshold obtained from the step of e.

g)The post-processing method is used to process the segmented image and get the final segmentation image.

3.4. Post-Processing

In this paper, a post-processing method is proposed to remove the noise which contain the normal brain tissues of small amount of blood from the segmented images.

Step1: The noise from the segmented images will be filtered out by median filter that has a large parameter. At the same time, the location information that is the top, the bottom,the far left and the far right pixel points of bleeding regions of filtered image is obtained.

Step2: The post-processing method in this paper uses the position information of bleeding regions only to digging out the pixels information from the bleeding regions and nearby regions of the segmented image to generate a new image.

Step3: A median filter that has small parameter is applied to filter the noise near the bleeding regions of the new image and obtain the final segmentation result image.

3.5. Volume calculation of cerebral hemorrhage

After segmenting the regions of cerebral hemorrhage, the volume of cerebral hemorrhage is calculated. The formula for calculating the volume of cerebral hemorrhage is as following.

$$V = \sum_{i=1}^{N} s_i \Delta h \quad (6)$$

Where N represents the number of images of the bleeding slice; s_i denotes the area of i th slice image, Δh represents the thickness of the slice. First, the area of brain bleeding on each slice is calculated .The area of bleeding regions on each slice is obtained by multiplying the area of pixel with the number of pixels. In this paper, the row and column spacing of the pixel in the bleeding regions are 0.468mm. The number of pixels in the bleeding regions can be obtained by counting the number of pixels, whose gray value are not 0. Then, the volume of cerebral hemorrhage on each slice is obtained by multiplying the area of the cerebral hemorrhage and the layer thickness. In the end, the volume of cerebral hemorrhage on each slice accumulates the total volume of cerebral hemorrhage in the patients.

4. Experimental results and analysis

There are several CT images of 6 mm slice thicknesses from 30 patients suffered from cerebral hemorrhage shown below. Take one for example which provided by First Affiliated Hospital of Medical University Of An Hui provided the images. The performance of the experimental results in this paper and the corresponding segmentation results are shown in Figure 2.

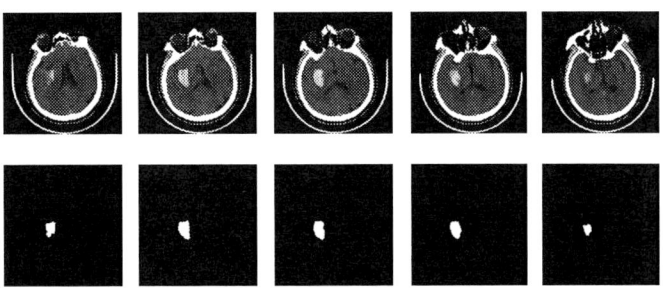

Figure 2. Segmentation results of 8 bleeding slices of a patient

Table 1. Comparison of segmentation results between three segmentation algorithms.

Slices	Slice 1	Slice 2	Slice 3	Slice4	Slice5	The average
Kmean(DSC%)	72.43	93.98	93.14	90.99	64.41	82.99
CSOO(DSC%)	64.48	90.66	86.60	76.83	52.96	74.31
Our Method (DSC%)	95.60	98.33	97.62	97.86	90.38	95.96

As shown in Table 1, the experimental results in this paper are compared with the Ground Truth, according to the results of calculation, Ours yields a higher dice similarity coefficient of 95.96%.The more accurate the segmentation of bleeding regions is, the more accurate the measurement of volume of cerebral hemorrhage would be.

Table 2. The Volume Of Cerebral Hemorrhage To 5 Slices Of The Patient

Slices	Slice 1	Slice 2	Slice 3	Slice4	Slice5
Ground Truth (Number of Pixel)	2188	3140	3269	2823	1590
Our Method (Number of Pixel)	2040	3037	3117	2738	1311
Slice bleeding volume Cm3	2.68	3.99	4.10	3.60	1.72
Total bleeding volume Cm3			16.09		

Table 2 shows the volume of cerebral hemorrhage of each slice and the total volume of cerebral hemorrhage of the patient. The total volume of cerebral hemorrhage of this patient is 16.09cm3. Ground truth is derived from the result of manual segmentation by doctors.Accurate calculation of the volume of cerebral hemorrhage facilitates the doctor to diagnose and treat the patient properly.

5. Summary

It is of great significance to assisting doctors in diagnosing hematoma patients to accurate segmentation of hematoma area and measurement of hematoma volume. In this paper, an automatic segmentation algorithm for cerebral hematoma in CT images is proposed. The dice similarity coefficient of this algorithm has reached to 95.96% . The algorithm proposed in paper is not only of high convenience, but also of good robustness. Only the volume of the bleeding regions was calculated in this paper, while it is critical to obtain the location and the shape of the bleeding regions for the diagnosis of cerebral hemorrhage, in which the algorithm needs to be improved.

References

[1] Sumijan, S., Yuhandri, Y., & Boy, W. Detection and Extraction of Brain Hemorrhage on the CT-Scan Image using Hybrid Thresholding Method.Turk J Elec Eng & Comp Sci 2016, Volume 1 , 1:10-19.

[2] Valadka AB, Gopinath SP, Contant CF, Uzura M, Robertson. Relationship of brain tissue PO2 to outcome after severe head injury. Critical care medicine. 1998; Volume 26.9:1576-1581.

[3] Roy, S., Wilkes, S., Diaz-Arrastia, R., Butman, J. A., & Pham, D. L. (2015, March). Intraparenchymal hemorrhage segmentation from clinical head CT of patients with traumatic brain injury. In Medical Imaging 2015: Image Processing; Volume 9413, International Society for Optics and Photonics.

[4] Shahangian B , Pourghassem H. Automatic brain hemorrhage segmentation and classification algorithm based on weighted grayscale histogram feature in a hierarchical classification structure. Biocybernetics and Biomedical Engineering 2016 . Volume 36.

[5] Bhadauria, H. S., Dewal, M. L. Intracranial hemorrhage detection using spatial fuzzy c-mean and region-based active contour on brain CT imaging. Signal, Image and Video Processing 2014; Volume 8 . 2: 357-364.

[6] Bhadauria, N. S., Bist, M. S., Patel, R. B. (2015, March). Performance evaluation of segmentation methods for brain CT images based hemorrhage detection. In Computing for Sustainable Global Development (INDIACom), 2015 2nd International Conference on (pp. 1955-1959). IEEE.

[7] Rudin, L. I., Osher, S., & Fatemi, E.Nonlinear total variation based noise removal algorithms. Physica D: nonlinear phenomena 1992; Volume 60, 9: 259-268.

[8] Karuna, M., & Joshi, A. Automatic Detection and Severity analysis of brain tumors using GUI in matlab. International Journal of Research in Engineering and Technology 2013 , Volume 2, 10: 586-594.

[9] Paulson, O. B. Cerebral apoplexy (stroke): Pathogenesis, pathophysiology and therapy as illustrated by regional blood flow measurements in the brain 1971; Stroke, Volume 2, 4: 327-360.

[10] Soroushmehr, S. M. R., Bafna, A., Schlosser, S., Ward, K., Derksen, H., & Najarian, K. (2015, August). CT image segmentation in traumatic brain injury. In Engineering in Medicine and Biology Society (EMBC), 2015 37th Annual International Conference of the IEEE (pp. 2973-2976). IEEE.

[11] Sun, M., Hu, R., Yu, H., Zhao, B., & Ren, H. (2015, October). Intracranial hemorrhage detection by 3D voxel segmentation on brain CT images. In Wireless Communications & Signal Processing (WCSP), 2015 International Conference on (pp. 1-5). IEEE.

[12] Prakash, K. B., Zhou, S., Morgan, T. C., Hanley, D. F., & Nowinski, W. L.Segmentation and quantification of intra-ventricular/cerebral hemorrhage in CT scans by modified distance regularized level set evolution technique. International journal of computer assisted radiology and surgery 2012;Volume 7, 5:785-798.

[13] Muschelli, J., Sweeney, E. M., Ullman, N. L., Vespa, P., Hanley, D. F., & Crainiceanu , C. M.PItcHPERFeCT: Primary intracranial hemorrhage probability estimation using random forests on CT. NeuroImage: Clinical 2017; Volume 14, 4:379-390.

Thinking and Exploration on the Teaching of the Course of "College Computer Foundation" under the Mode of "Internet Plus"

Tang Xianfang   Jia Yachao Zhang Ru

Northwestern Polytechnical University Mingde College **710124**

Absrtact: At present, information technology is in the rapid development stage.The development of information technology has made the Internet technology widely used, has also been applied in the network teaching platform, and has gradually formed the "Internet Plus" mode. The "Internet Plus" mode has been applied to the teaching of the course of "College Computer Foundation",at this stage, some teachers have made a full study of the "Internet Plus" mode,it is hoped that this mode can be better applied to the teaching of the course of "College Computer Foundation", so that the teaching quality and teaching efficiency of "College Computer Foundation" can be improved significantly.

1.   Analysis of the Problems in the Teaching of the Course of "College Computer Foundation"

There are lots of students in universities. These students choose different majors, and their knowledge levels are also different. Some students have already mastered some basic knowledge of computer before entering university. But some students did not contact with the basic knowledge of computer before entering college. In this case, it is difficult to ensure that the students can master the basic knowledge of computer effectively if the teachers use a unified way to teach. However, at present,in the course of the development of the course of "College Computer Foundation" , the teaching mode and teaching materials used by teachers are unified, and the time of practical operation of the students set up is relatively short, so the different needs of students are difficult to meet, and the learning progress of students is not consistent.;In addition, the time students use the computer to carry out practical operations is short, resulting in students' poor actual operation ability. The teaching mode of teachers is relatively single, the large screen PPT and blackboard teaching methods used in the past are difficult to meet the learning needs of college students, and it is difficult to fully mobilize the students' learning enthusiasm. Teachers fail to teach according to their aptitude, which makes it difficult to improve the teaching quality of the course and to cultivate talents who meet the needs of the society.

2.   the Meaning of the "Internet Plus" Mode

In terms of "Internet Plus" mode, this mode combine all kinds of industries effectively by information and communication technology and the Internet platform,so that it creates a new ecology in a certain field.To put it simply, the "Internet Plus" mode is to combine the Internet with various industries effectively, that is, the Internet and the various industries have a deep integration through the Internet platform information and communication technology . The "Internet Plus" mode is a representative of the new social formation.,it not only makes full use of the social resources of the Internet,but also optimizes the resources of the Internet further , and the Internet can fully serve all kinds of industries, and then can benefit people by the continuous development of the "Internet Plus" mode. In the case of

Content from this work may be used under the terms of the Creative Commons Attribution 3.0 licence. Any further distribution of this work must maintain attribution to the author(s) and the title of the work, journal citation and DOI.

Published under licence by IOP Publishing Ltd

university education, it should keep pace with the development of the times and constantly applies the "Internet Plus" mode to the university education, so that the "Internet Plus" mode can be effectively combined with the teaching mode,and a new model of education that meets the needs of students and lays a favorable foundation for further development of students will be creates.

3. the Way to Combine the "Internet Plus" Mode with the "College Computer Foundation" Course Teaching Effectively

3.1. Reform in Education
In the case of the "College Computer Foundation" course, the teaching materials used in the course correspond to the situation in the various institutions, these materials contain basic knowledge of computer science, in addition to teaching materials for computer majors.Other professional textbooks are written based on the use of literature and history, science and technology, and this kind of knowledge is not highly targeted, coupled with the continuous development of computer technology and information technology,the new technology and new knowledge of computer are emerging constantly, so some teaching materials will be difficult to fit in with the new knowledge in teaching, which makes it difficult for some teaching materials to meet the needs of the students.

In order to meet the needs of college students for computer operation skills and basic knowledge and the development of computer technology, each university should make a detailed analysis of the needs and characteristics of students majoring in computer science,in view of the concrete situation, each university compiles the relevant contents in the teaching material of "College Computer Foundation", and then the computer knowledge in the textbook can be combined effectively with the knowledge of each major. Teachers can make full use of the "Internet Plus" mode, the relevant teaching website , in which teaching cases and courseware can be shared and displayed can be set up by the mode. This not only realizes the full use of network teaching resources, but also makes students study what they need, it also plays an auxiliary role in the further development of students.

3.2. Reform of Examination Mode
As far as the course of "College Computer Foundation" is concerned, it is a course that pays attention to the training of students' computer practical operation skills and theoretical knowledge,the course

requires the students' assessment standards to be comprehensive, scientific and objective. However, it is difficult for the traditional assessment mode and thought to realize this point, which makes it difficult for the traditional examination mode to coincide with the examination and teaching requirements under the new situation. Therefore, in order to judge the students' academic achievement more scientifically, teachers should reform the assessment method. Teachers should constantly innovate their teaching ideas and make a more comprehensive, objective and fair examination according to the teaching level of the students and teachers in the school,the new mode can make the students' learning effect be judged scientifically, and the comprehensive application ability of the students can be improved through this kind of examination mode.

Schools and teachers can separate education from assessment and apply diversified assessment programs when schools and teachers reform the examination mode. Teachers can not carry out teaching according to the assessment goals, nor can they make the assessment goals according to the teaching contents,the assessment should be effectively combined with the characteristics of the curriculum, and then the comprehensive evaluation should be carried out from multiple levels and various directions. Based on this, teachers can make full use of the network platform to develop and set up an online examination system, and use the system to replace the traditional paper form of assessment. In this process, the test system extracts questions from the question bank, these questions are effectively combined with the questions in the exams. This system can evaluate the scores of students scientificly after the exams,it can not only improve the efficiency of teachers' examination paper, but also ensure the fairness of examination paper evaluation. In addition, to achieve diversified assessment,the examination system should continue to expand and update the question bank; the proportion of the examination should be 60% on the computer examination, 20% on the computer operation and 20% on the daily learning performance. According to practice, it is shown that the use of diversified assessment methods, which is separated from teaching and assessment, can not only comprehensively and objectively evaluate the students' learning effect, but also can stimulate the students' initiative and enthusiasm for learning .

3.3. Pay Attention to Differentiated Teaching

The learning level of school students in each class has great difference, and the level of students' absorbing knowledge is also different,if teachers want to improve teaching quality better, they need to adopt different teaching methods. When teachers develop differentiated teaching, they need to master the differences of middle school students, and then make targeted teaching programs according to the students' individual learning needs, so as to ensure that students can get better education on the basis of their original education. "College Computer Foundation" is a compulsory course for students, students from all over the world in the college, students in different regions, some students live in poor areas, some students live in developed areas ,this leads to great differences in the degree of teaching and the acceptance of computers. In view of this situation, if the computer basic teaching methods used by teachers are not targeted, this will lead to polarization,on the one hand, students with a weaker foundation will find it difficult to digest the knowledge explained by teachers,the students' ability of computer application is difficult to improve effectively, which is a great obstacle to the students' further study;On the other hand, students with a good foundation can absorb all the knowledge taught by the teacher, but they need to absorb more computer knowledge. Therefore, teachers should master the students' knowledge level completely and adopt different teaching methods according to the students' knowledge level, which can make common progress for the students with poor foundation and good foundation. Teachers can make full use of the "Internet Plus" mode, set up the corresponding learning platform through the network, let students acquire more knowledge in the network platform, and choose the knowledge they need to learn according to their own situation,the needs of differentiated teaching can be met by this way.

3.4. Developing Practical Teaching

The application of practical teaching mode in the teaching of the course of "College Computer

Foundation" can make teachers' teaching more compatible with their daily life, enable students to practice what they have learned, and make students who has learned be useful, students' learning interest and enthusiasm can be fully stimulated by the use of practical teaching,which is an effective way to improve students' learning initiative. For example, when teaching about computer hardware, a teacher can use the online shopping guide platform to design a corresponding teaching theme,the teaching can be formulated as "choose and purchase the computer of one's own heart" , how to select computer,performance of various parts can be passed on to the students. Teachers can combine with students' professional characteristics when teaching related knowledge of office software, so that students can produce the article of picture and text mixed cutting according to the function of office software. The teachers can let the students make the score books and summarize and analyse the scores by using Excel form in class.According to the content of the teaching material and the practical teaching of the students' life, teachers can accord with the students' professional characteristics, and improve the students' subjective initiative and learning enthusiasm simultaneously in the process of teaching.

3.5. Put into Effect a Diversified Teaching Mode
At the present stage, diversified teaching modes are widely used in various fields of education. Among them, task-based teaching and case teaching are all relatively good teaching modes,these teaching models can play a good role in promoting basic computer teaching. With the continuous development of network technology, teachers can make full use of the network information platform in the teaching process of computer basic knowledge, and make full use of the new and diversified teaching mode under the "Internet Plus" mode, such as micro-class,flip class,admiration class etc.the teaching quality and efficiency of "College Computer Foundation" can be significantly improved by the way.

3.5.1. Application of the Micro-Class and Admiration Class.
In recent years, the teaching modes of admiration and micro-class have been widely used,these educational modes belong to online education,the development of these educational modes has laid a foundation for the formation of the educational form of "Internet Plus". In the teaching of the course "College Computer Foundation", the teachers can input the key knowledge of the teaching into the micro-class, and then put the recorded micro-lesson into the network platform, and communicate and discuss these important and difficult knowledge with the students in the platform. With the deepening of teaching, teachers can set up a group of micro-courses in the course of "College Computer Foundation", and then gradually form the teaching method of "admiration class".

According to the corresponding practice, it is shown that the multi- teaching mode, such as micro-class and admiration class, can effectively combine the task-driven teaching and case teaching, and then stimulate the students' subjective initiative sufficiently, and realize the aim of improving teaching quality and the teaching effect synchronously . Teachers should choose the corresponding teaching methods according to the teaching content in the course of College Computer Foundation", and the teaching mode should be effectively combined with the characteristics of the students, and then the targeted teaching will be realized.

3.5.2. Application of Flipping Class
Flipping class teaching mode is a teacher creates a video in the course of teaching and puts the video into the corresponding learning website, and students can watch the videos after class or at home,students communicate what they learn from video in class, which is similar to the form of homework. the main role of the students can be brought into full play, the students can be guided to study independently, and the students should be self-evaluated, and the teachers will give corresponding guidance in the process of the students' learning evaluation by the application of the flipping classroom teaching mode .

Teachers can not only provide the teaching videos to the students, but also help the students to learn after class by using the flipped class in the process of teaching. In the process of teaching, teachers can select the corresponding teaching topics according to the teaching contents, and then carry out targeted teaching to the students. The school can divide the teachers into many teaching and research groups and let the teachers carry on the thorough analysis and research to the related subjects, and then design the subjects which are more suitable for the students' needs, so that the student's study becomes more pertinence. After a period of project teaching, the teachers can discuss and communicate the students' learning results, and let the students learn what they have learned,the teacher will extend or supplement the knowledge according to the students' knowledge. In the process of teaching, teachers should give full play to the role of participants and organizers, and effectively use the teaching mode of flipping the class to improve the students' thinking ability. At the same time, the students' autonomous learning ability is improved, and then the teaching goal will be realized on the basis of improving the teaching effect.

4. Conclusion:

With the continuous development of the Internet era, the field of education faces many challenges and opportunities, "Internet Plus" is the product of the development of the Internet era,it is not only the representative of the advanced productive forces, but also can promote the further evolution of the economic form,and it provides more advanced technology for the development, reform and innovation of various industries. When it comes to the teaching of the course "College Computer Foundation", it should make full use of the "Internet Plus" mode, combine this model with the diversified teaching mode, and gradually establish a diversified examination mode in the course of teaching,this mode not only improves and develops the teaching of "College Computer Foundation" ,but also enhances the students' practical ability and comprehensive ability gradually.

References :

[1] Liu Pingping, Lu Zhaopan, Wang Jianguo, Xu Liangchen. Exploration and practice of teaching reform of 《College Computer Foundation 》 in the environment of "Internet Plus" [J]. Educational Teaching Forum, 2018 (29): 132-133.

[2] Song Hui, Liu Xiaoqiang, Du Ming, Wang Zhijun. Computer basic teaching system for the cultivation of innovative application ability of "Internet Plus" [J]. Computer Education, 2018 (05): 130-132.

[3] Yang Hui. Construction and practice of basic courses in University computer based on "Internet Plus"[J]. Education and Teaching Forum, 2016 (33): 64-65.

[4] Tang Kunjian, Du Guangzhou. Thinking and exploring the teaching of the basic course of University computer under the mode of "Internet Plus"[J]. Asia Pacific Education, 2016 (01): 101-102.

Prediction of the Anti-inflammatory Mechanism of Clematis chinensis based on Network Pharmacology

Xulong HUANG, Junjie HAO, Yuqing LIANG, Yuanmin WANG, Juan KONG, Xiangpei WANG, Feng XU *, Hongmei WU *

Department of Pharmacy, Guiyang University of Chinese Medicine, Guiyang City, Guizhou Province, Guiyang 550002, PR China

* Corresponding author. Department of Pharmacy, Guiyang University of Chinese Medicine, Guiyang City, Guizhou Province, Guiyang 550002, PR China. E-mail: 810230331@qq.com(Feng XU), whm0425@126.com（Hongmei WU）

Abstract. Clematis chinensis is a traditional Chinese medicine with good anti-inflammatory effects, but the underlying molecular mechanism is unclear. In this work, the potential mechanism of Clematis chinensis with anti-inflammatory effect was explored using the network pharmacology. The active chemical constituents of Clematis chinensis were screened according to the pharmacodynamic activities and ADME parameters, then the potential anti-inflammatory targets of the active chemical constituents in Clematis chinensis were screened with the databases of TCMSP, PubChem, TCMID, etc., and the target data sets were established through the BATMAN-TCM database, traditional Chinese medicine target database, HIT database, and TTD database. The targets related to inflammation were screened by using OMIM database, and the interactive target between clematis chinensis and inflammation were constructed by using PPI database. The complex network map of "drug composition-target-disease" was constructed by Cytoscape software, and GO (gene function) analysis was carried out by using ClueGO plug-in. Protein interaction analysis, target gene function enrichment analysis, and signal pathway analysis were carried out by STRING database, biological information annotation database (DAVID), and KEGG Pathway database. 39 out of 458 chemical constituents of Clematis chinensis had good drug-likeness and activity. 35 key targets were screened out by Degree, Betweenness Centrality, Closeness Centrality, and other network topology characteristics, and these targets were involved in 25 signaling pathways related to the anti-inflammatory action including TNF signaling pathway, MAPK signaling pathway, NF-kappa B signaling pathway, etc. In addition, it was also related to the positive and negative regulation of many biological processes such as inflammatory response, positive regulation of I-kappaB, kinase/NF-kappaB signaling, positive regulation of NF-kappaB, etc. The TNF signaling pathway, MAPK signaling pathway, and NF-kappa B signaling pathway were closely related to the anti-inflammatory activity of Clematis chinensis. Network pharmacology could be used to study the anti-inflammatory molecular mechanism of Clematis chinensis, which is beneficial to the development and utilization of Clematis chinensis.

1. Introduction

There are three sources of Clematis chinensis, which are *Clematis chinensis* Osbeck, *Clematis hexapetala* Pall., and *Clematis manshurica* Rupr. The dried roots and rhizomes are used as Clematis chinensis herbs, and it has many traditional functions like dispelling rheumatism, dredging meridians, collaterals, etc. Clematis chinensis is mainly used to treatment of rheumatic arthralgia, limb numbness, muscle and pulse contracture, flexion, and extension adverse [1]. Modern studies have shown that Clematis chinensis mainly contains saponins, flavonoids, lignans, phenols, alkaloids, and other components [2-6]. Interestingly, these components have anti-inflammatory, analgesic, anti-tumor, anti-bacterial, and other pharmacological activities [7-12]. At present, many studies have reported that Clematis chinensis has good anti-inflammatory activity, but the underlying mechanism is still unclear. However, the network pharmacology with holistic and systematic characteristics provide a new idea and method for revealing the anti-inflammatory mechanism of Clematis chinensis, which coincides with the "multi-component, multi-pathway and multi-target" synergetic characteristics of traditional Chinese medicine [13-15]. Therefore, in this paper, network pharmacology was used to analyse the possible targets and signalling pathways of the active chemical constituents for preventing and treating inflammation in Clematis chinensis, and the results could provide a reference for the further study of Clematis chinensis.

2. Method

2.1. The molecule structure and target protein prediction
The chemical components were obtained from the relevant literature (likely CNKI database, PubMed database), PubChem database, TCMSP database, etc. Then, the potential components were screened out using Pharmacological activity and ADME parameters. Based on the comprehensive analysis of BATMAN-TCM database and traditional Chinese medicine target database, Score cutoff (> 35 score) and P-value (p< 0.05) were used as indexes to screen the active constituents and predict the targets of Clematis chinensis. According to the similarity comparison, the active component database of Chinese herbal medicine (HIT, http://lifecenter.sgst.cn/hit/) and the control target database (TTD, http://bidd.nus.edu.sg/group/cjttd/) were used to screen the active constituents and its target data set was constructed. Human gene and gene phenotype database (OMIM, http://www.omim.org/) was used to screen the anti-inflammatory genes and protein targets of Clematis chinensis, and the data set of anti-inflammatory targets was established. Human target connexins were obtained through an interactive protein database (http://dip.doe-mbi.ucla.edu). Finally, all the selected targets were transformed into UniProt ID format by UniProt database query.

2.2. Network construction and topological profile analysis
Through PPI (http://www.genome.jp/kegg/) analysis, the active components, corresponding targets, inflammatory targets and interactive protein targets were connected into the "component-target-disease" network. The above network was visualized using Cytoscape 3.6.1 software. Among the topological parameters, degree and betweenness centrality were used as crucial factors to describe the most influential nodes in network. Thus, the node values higher than the median value of all nodes were chosen as the hubs.

2.3. Enrichment Analysis of Biological process and Pathway Diagram
The STRING database (https://string-db.org/) was used to analyze the protein-protein interaction of the screened targets. Finally, DAVID (https://david.ncifcrf.gov/) database) was used to analyze the KEGG pathway and the biological process of GO (Gene Ontology).

3. Result

3.1. Network construction
Through the literature, from PubChem database and TCMSP database, a total of 458 chemical constituents were obtained, of which 39 were the potential constituents. In addition, 685 interactive proteins were screened by Cytoscape software. The anti-inflammatory interactive network of Clematis chinensis was constructed by network pharmacology. After visualized with different colors and shapes, the network relationship between active components and disease targets could be directly seen in figure 1. Among them, the yellow square represented the common target for drugs and diseases, and it was

also the most important protein target of Clematis chinensis for its anti-inflammatory activity. The yellow ellipse represented the direct target of inflammation, the red triangle represented the predicted active compound in Clematis chinensis, the blue ellipse represented the direct target of active components of Clematis chinensis, and the purple ellipse represented the interacting protein banded to the components of Clematis chinensis to the disease targets.

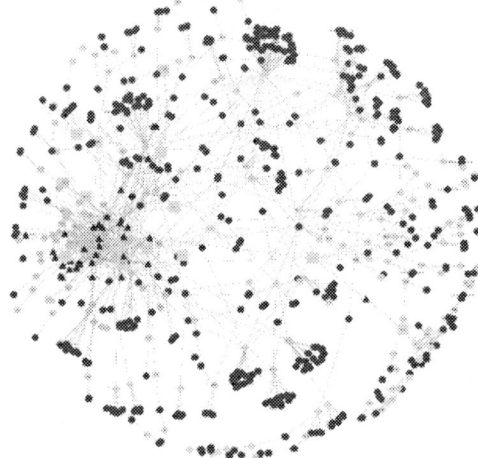

Fig. 1 "Component-Target-Disease" Interactive Network for Anti-inflammatory Action of the active components in Clematis chinensis

3.2. Topological profile analysis

The protein targets associated with the active components of Clematis chinensis were obtained, and the topological parameters of these protein targets were calculated. The median value of the three topological parameters (Degree, Betweenness centrality and Closeness centrality) was 6, 0.005, and 0.173, respectively. The selected targets with three topology parameters of each target greater than the median values were selected as the important target proteins in the anti-inflammatory activity of Clematis chinensis. Finally, 35 targets were screened out, and the results were shown in Table 1. The protein interaction of the selected targets were shown in Fig. 2. The protein interaction showed that the active components of Clematis chinensis are interaction with each other and regulation each other.

Table. 1 Topological parameters related to the Target of Anti-inflammatory effects of active components of Clematis chinensis

Uniprot ID	Protein names	Gene names	Closeness Centrality	Degree	Betweenness Centrality
P05067	Amyloid-beta A4 protein	APP	0.193	12	0.033
P31749	RAC-alpha serine/threonine-protein kinase	AKT1	0.234	13	0.029
P41182	B-cell lymphoma 6 protein	BCL6	0.217	8	0.019
Q13936	Voltage-dependent L-type calcium channel subunit alpha-1C	CACNA1C	0.176	8	0.019
Q14790	Caspase-8	CASP8	0.193	6	0.009
P24385	G1/S-specific cyclin-D1	CCND1	0.247	10	0.041
P00533	Epidermal growth factor receptor	EGFR	0.220	19	0.040
Q13158	FAS-associated death domain protein	FADD	0.189	7	0.011
P04150	Glucocorticoid receptor	NR3C1	0.237	17	0.035
Q16665	Hypoxia-inducible factor 1-alpha	HIF1A	0.191	12	0.025
P25963	NF-kappa-B inhibitor alpha	NFKBIA	0.219	18	0.036

O15111	Inhibitor of nuclear factor kappa-B kinase subunit alpha	CHUK	0.211	18	0.021
O14920	Inhibitor of nuclear factor kappa-B kinase subunit beta	IKBKB	0.225	15	0.018
P05231	Interleukin-6	IL6	0.186	6	0.014
P05412	Transcription factor AP-1	JUN	0.237	15	0.056
Q99558	Mitogen-activated protein kinase kinase kinase 14	MAP3K14	0.209	13	0.020
Q99759	Mitogen-activated protein kinase kinase kinase 3	MAP3K3	0.200	10	0.022
Q9Y6K9	NF-kappa-B essential modulator	IKBKG	0.226	23	0.046
P19838	Nuclear factor NF-kappa-B p105 subunit	NFKB1	0.191	20	0.017
P35228	Nitric oxide synthase, inducible	NOS2	0.265	12	0.070
P04637	Cellular tumor antigen p53	TP53	0.263	60	0.201
O15350	Tumor protein p73	TP73	0.202	9	0.006
P09874	Poly [ADP-ribose] polymerase 1	PARP1	0.212	11	0.031
P23219	Prostaglandin G/H synthase 1	PTGS1	0.237	14	0.010
P35354	Prostaglandin G/H synthase 2	PTGS2	0.241	18	0.025
P37231	Peroxisome proliferator-activated receptor gamma	PPARG	0.240	17	0.029
Q13546	Receptor-interacting serine/threonine-protein kinase 1	RIPK1	0.228	7	0.064
Q13586	Stromal interaction molecule 1	STIM1	0.187	7	0.011
Q86WV6	Stimulator of interferon genes protein	TMEM173	0.213	8	0.010
Q04206	Transcription factor p65	RELA	0.195	19	0.022
P02786	Transferrin receptor protein 1	TFRC	0.174	7	0.008
P01375	Tumor necrosis factor	TNF	0.215	10	0.158
P19438	Tumor necrosis factor receptor superfamily member 1A	TNFRSF1A	0.202	10	0.030
P25445	Tumor necrosis factor receptor superfamily member 6	FAS	0.217	11	0.096
Q9Y4K3	TNF receptor-associated factor 6	TRAF6	0.215	22	0.044

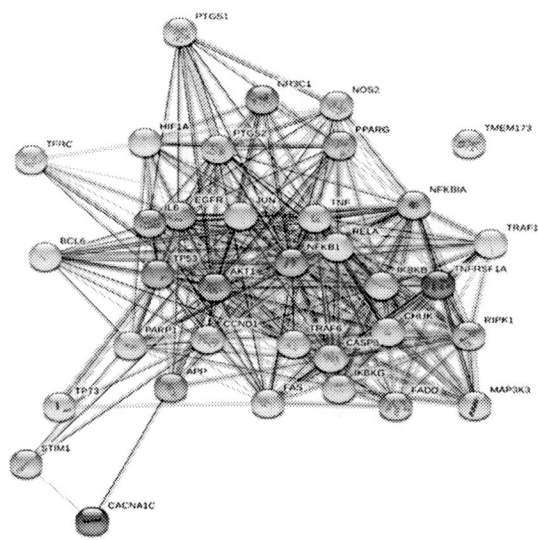

Fig. 2 Protein-Target interaction Diagram of Anti-inflammatory effects of active components of Clematis chinensis

3.3. Biological function analysis of GO

25 potential targets were mapped to the DAVID database for bio-functional enrichment and systematic analysis of their biological processes. 296 biological processes were enriched, 20 of which are P≤0.01 and Count≥10, and the results were shown in Table 2. The results showed that these targets are involved in a variety of biological processes including inflammatory response, positive regulation of I-kappaB, kinase/NF-kappaB signaling, positive regulation of NF-kappaB, etc. These biological processes were closely related to the occurrence and development of inflammation. It indicates that the pathogenesis of inflammation is involved in the abnormality of many biological processes in vivo, and these screened biological processes might be related to the anti-inflammatory effects of Clematis chinensis.

Table. 2 GO of bioaccumulation Analysis of Anti-inflammatory effect of Clematis chinensis

Category	Term	Count	Count (%)	P-Value
GOTERM_BP_DIRECT	positive regulation of transcription from RNA polymerase II promoter	23	65.7	1.6E-19
GOTERM_BP_DIRECT	inflammatory response	13	37.1	5.2E-12
GOTERM_BP_DIRECT	positive regulation of I-kappaB kinase/NF-kappaB signaling	12	34.3	1.1E-14
GOTERM_BP_DIRECT	positive regulation of apoptotic process	11	31.4	2.6E-10
GOTERM_BP_DIRECT	apoptotic process	11	31.4	1.1E-07
GOTERM_BP_DIRECT	positive regulation of NF-kappaB transcription factor activity	10	28.6	4.1E-12
GOTERM_BP_DIRECT	negative regulation of apoptotic process	10	28.6	2.1E-07
GOTERM_BP_DIRECT	positive regulation of transcription, DNA-templated	10	28.6	5.9E-07
GOTERM_BP_DIRECT	negative regulation of transcription from RNA polymerase II promoter	10	28.6	9.3E-06

GOTERM_CC_DIRECT	cytoplasm	26	74.3	7.6E-08
GOTERM_CC_DIRECT	cytosol	22	62.9	2.2E-08
GOTERM_CC_DIRECT	nucleus	21	60	0.00039
GOTERM_CC_DIRECT	nucleoplasm	14	40	0.00094
GOTERM_MF_DIRECT	protein binding	34	97.1	7.1E-09
GOTERM_MF_DIRECT	identical protein binding	21	60	5.1E-19
GOTERM_MF_DIRECT	transcription factor binding	11	31.4	1.4E-10
GOTERM_MF_DIRECT	enzyme binding	11	31.4	6.7E-10
GOTERM_MF_DIRECT	ubiquitin protein ligase binding	10	28.6	3.8E-09
GOTERM_MF_DIRECT	protein kinase binding	10	28.6	3.9E-08
GOTERM_MF_DIRECT	protein homodimerization activity	10	28.6	0.00001

3.4. KEGG pathway analysis

The results showed that 25 signaling pathways ($P \leq 0.01$ and Count$\geq$10) related to the anti-inflammatory effect of Clematis chinensis were screened out by KEGG pathway enrichment analysis (Table 3). These pathways have shown a direct or indirect role in anti-inflammatory, such as TNF signaling pathway, MAPK signaling pathway, and NF-kappa B signaling pathway.

Table. 3 KEGG Pathway enrichment Analysis of Anti-inflammatory effect of Clematis chinensis

Category	Term	Count	Count (%)	P-Value
KEGG_PATHWAY	Pathways in cancer	20	57.1	1.4E-15
KEGG_PATHWAY	TNF signaling pathway	17	48.6	6.4E-21
KEGG_PATHWAY	Chagas disease (American trypanosomiasis)	16	45.7	2.5E-19
KEGG_PATHWAY	MAPK signaling pathway	16	45.7	2E-13
KEGG_PATHWAY	Apoptosis	15	42.9	5.8E-21
KEGG_PATHWAY	Hepatitis B	15	42.9	1.7E-15
KEGG_PATHWAY	Herpes simplex infection	15	42.9	4.7E-14
KEGG_PATHWAY	HTLV-I infection	15	42.9	4.4E-12
KEGG_PATHWAY	Toll-like receptor signaling pathway	14	40	9E-16
KEGG_PATHWAY	Osteoclast differentiation	13	37.1	5.3E-13

KEGG_PATHWAY	Hepatitis C	13	37.1	6.3E-13
KEGG_PATHWAY	RIG-I-like receptor signaling pathway	12	34.3	1.3E-14
KEGG_PATHWAY	Small cell lung cancer	12	34.3	1.2E-13
KEGG_PATHWAY	NF-kappa B signaling pathway	12	34.3	1.6E-13
KEGG_PATHWAY	Toxoplasmosis	12	34.3	2.2E-12
KEGG_PATHWAY	Epstein-Barr virus infection	12	34.3	7E-12
KEGG_PATHWAY	Measles	11	31.4	4.6E-10
KEGG_PATHWAY	NOD-like receptor signaling pathway	10	28.6	3.6E-12
KEGG_PATHWAY	Prostate cancer	10	28.6	2.4E-10
KEGG_PATHWAY	T cell receptor signaling pathway	10	28.6	7.8E-10
KEGG_PATHWAY	Neurotrophin signaling pathway	10	28.6	4E-09
KEGG_PATHWAY	Non-alcoholic fatty liver disease (NAFLD)	10	28.6	3.1E-08
KEGG_PATHWAY	Influenza A	10	28.6	1E-07
KEGG_PATHWAY	Tuberculosis	10	28.6	1.2E-07
KEGG_PATHWAY	PI3K-Akt signaling pathway	10	28.6	0.000031

4. Discussions

In this paper, the network pharmacology methods were used to predict the anti-inflammatory targets and pathways of the active chemical constituents from Clematis chinensis, and 39 active chemical constituents were screened out. The interactive network diagram of "component-target-disease" was constructed. The results showed that 25 pathways and 20 biological processes are directly or indirectly involved in the anti-inflammatory effects of Clematis chinensis through pathway enrichment analysis, and the main signaling pathways included TNF signaling pathway, MAPK signaling pathway, NF-kappaB signaling pathway and the biological processes were inflammatory response, positive regulation of I-kappaB, kinase/NF-kappaB signaling, positive regulation of NF-kappaB, and so on. Finally, the research for anti-inflammatory mechanism of Clematis chinensis biological effects with multi-component, multi-target, and multi-pathway were achieved. The predicted results suggest that there are three potential anti-inflammatory signaling pathways involved in the anti-inflammatory effects of Clematis chinensis, and the next step is to select the NF-kappa B signaling pathway, which is most closely related to the pathogenesis of inflammation, to be verified. The aim of this work is to provide the reference for the in-depth study of Clematis chinensis.

Acknowledgments

This work acknowledges the funding from Open Laboratory of Key Laboratory of Miao Medicine in Guizhou, China [Qian Miao medicine K (2017) 026], the Guizhou domestic first-class construction project [(Chinese Materia Medica) (GNYL [2017] 008)], and Guiyang university of Chinese medicine first-class professional da chuang he zi (2017), No.158. The authors thank the government of China for their financial support.

References

[1] National Pharmacopoeia Commission. Pharmacopoeia of the people's Republic of China. A [M]. Beijing: China Medical Science and Technology Press, 2015: 250-251

[2] Shi S, Jiang D, Dong C, *et al*. Triterpene saponins from clematis mandshurica.[J]. Journal of Natural Products, 2006, 69(11):1591-1595.

[3] Gong Y X , Hua H M , Xu Y N , *et al*. Triterpene Saponins from Clematis mandshurica and Their Antiproliferative Activity[J]. Planta Medica, 2013, 79(11):987-994.

[4] Dong C X, Shi S P, Wu K S, *et al*. Studies on chemical constituents from root of Clematis hexapetala[J]. China Journal of Chinese Materia Medica, 2006, 31(20):1696.

[5] Zhang W K , Yang G E , Li Q , *et al*. Studies on chemical constituents of Euphorbia sororia.[J]. China Journal of Chinese Materia Medica, 2006, 31(20):1694.

[6] Liu J Y, Zhou N, Gong Y X, *et al*. Isolation and identification of chemical constituents from Clematis mandshurica[J]. Chinese Journal of Medicinal Chemistry, 2016,26(01):56-60.

[7] Fu Q , Zan K , Zhao M , *et al*. Triterpene saponins from Clematis chinensis and their potential anti-inflammatory activity.[J]. Journal of Natural Products, 2010, 73(7):1234.

[8] Liu Y., Shi R.b., Liu B., *etc*. Study on HPLC fingerprint of Chemical constituents of Salvia miltiorrhiza Bunge [J]. Journal of Beijing University of traditional Chinese Medicine, 2006, 29 (3): 188-192.

[9] Zhao Ying, Yu Chunfan, Zhang Guiying, *et al*. Antitumor effect of total saponins of Clematis chinensis and its effect on proliferation cycle of cancer cells [J]. Shizhen Medical Chinese Medicine, 2010, 21 (08): 1908-1909.

[10] Yun-Yi Z , Hong-Wei Z , Pei-Feng L I , *et al*. Antispasmodic Anti-inflammation and Analgesic Effect of Clematis[J]. Chinese Traditional Patent Medicine, 2001（11）：30-33.

[11] Xin-Rong Z , Chang-Jiang W , Xiao-Qin W . Effect of Extracts from Clematidis Chinensis Radix et Rhizoma on Diabetic Nephropathy in Rats[J]. Chinese Journal of Experimental Traditional Medical Formulae, 2015,21(16):152-156.

[12] Chen F Y, Xiang Jie G U, Zhong M K. Effect of sixpetal clematis root parenteral solution on interlukin-1 level in osteoarthritis joint fluid and chondrocyte culture supernatant[J]. Orthopedic Journal of China, 2004(07):43-45.

[13] Hartog A, Hougee S, Faber J, *et al*. The multicomponent phytopharmaceutical SKI306X inhibits in vitro cartilage degradation and the production of inflammatory mediators[J]. Phytomedicine International Journal of Phytotherapy & Phytopharmacology, 2008, 15(5):313-320.

[14] Yan D, Xiao X. [Investigation on pattern of quality control for Chinese materia medica based on famous-region drug and bioassay--the work reference].[J]. China Journal of Chinese Materia Medica, 2011, 36(9):1249-1252.

[15] Liang X, Li H, Li S. A novel network pharmacology approach to analyse traditional herbal formulae: the Liu-Wei-Di-Huang pill as a case study[J]. Molecular Biosystems, 2014, 10(5):1014-1022.

Streaming Information Transmission Based on OPC UA

Liu Wei, Zu Yunxiao*, Li Weihai

School of Electronic Engineering, Beijing University of Posts andTelecommunications, Beijing 100876, China

zuyx@bupt.edu.cn

Abstract. OPC UA is a new specification developed on the basis of classic OPC. It overcomes the limitations of classic OPC, has the advantages of platform independence and scalability, and is widely used in industrial fields. Node-Red is a visual programming tool based on node.js that allows you to combine modules to write applications and make programming simpler. This paper designs a Node-Red to create a stream as an OPC UA client by connecting OPC UA related node modules, connect to the OPC UA server, and call services such as reading and writing, subscription, etc., to realize real-time monitoring and display of the workshop temperature, and start the fan when the temperature is too high. In addition, Node-Red has a display module that provides a more diverse selection of information.

1. Introduction

With the concept of Germany's "Industry 4.0", smart manufacturing has become a research hotspot in the global manufacturing industry. For this reason, China has formulated the "Made in China 2025" project with information technology and industrialization as the main line. OPC UA (Unified Architecture) technology is a new technology proposed by the OPC Foundation and is widely used in industrial control. It provides a secure, reliable and vendor-independent communication interface that enables the transfer of raw data and preprocessed information from the manufacturing level to the production planning or ERP level [1].

The OPC UA client connects to the server, invokes the server's services, completes the task of reading and writing content on the address space, and subscribes to variables or events through monitoring items. The real-time monitoring of temperature has an important significance in the industry to ensure smooth production and safety.

1.1. OPC UA

OPC unified architecture (OPC UA) is a new specification developed by the OPC foundation on the classic OPC technology. It not only eliminates the limitations of the classic OPC, but also enriches the functions of the classic OPC, solves the existing problems, with the advantages of platform independence, scalability, safety and reliability, OPC UA is widely used in data exchange between industrial automation systems in manufacturing and process industries. Based on an advanced platform-independent transport protocol, OPC UA enables OPC UA applications to run on smart devices and controllers, DCS and scada systems, and even on MES and ERP systems.

The most common architectural mode of OPC UA is the client/server (C/S) model, which is also the basic OPC UA communication model. The server encapsulates the source of the process information so that the information can be accessed through the interface, after the client connects to the server, it can

Content from this work may be used under the terms of the Creative Commons Attribution 3.0 licence. Any further distribution of this work must maintain attribution to the author(s) and the title of the work, journal citation and DOI.
Published under licence by IOP Publishing Ltd

access and use the data the server provides. The server provides the service, the client sends a request message to the server, and the server responds to the request.

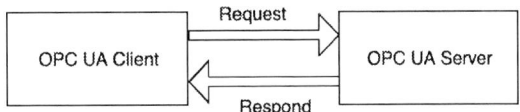

Figure 1. Client/Server model.

OPC UA is based on Service Oriented Architecture (SOA) [2], making OPC UA easier to use than traditional OPC. A service is an interface between a server and a client. It is independent of the transport protocol and development environment. It is defined in an abstract way, using a transport mechanism to exchange information between the client and the server.

The basis of OPC UA is data transmission and information modeling. Traditional OPC can only provide pure data, such as temperature values measured by temperature sensors, or pressure values measured by pressure sensors. OPC UA provides a more efficient way of presenting data semantics, in addition to the data provided by traditional OPC, it also allows the display of measurement results to be provided by a certain type of sensing device and allows for the display of the type hierarchy supported by that device.

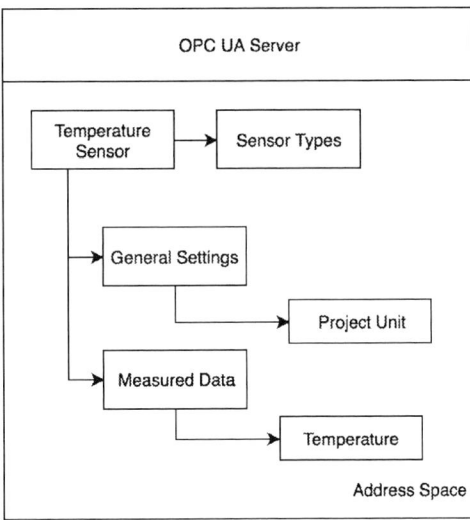

Figure 2. Temperature sensor data provided in the OPC UA Server.

1.2. Node-Red

Node-Red is a visual programming tool developed by IBM that allows programmers to write applications by combining components. These components can be hardware devices Web APIs, functions, or online services.

Node-Red provides a web-based programming environment. Programming is done by dragging and dropping the defined node to the workspace and creating a data stream with a wire connection node. The programmer saves and executes it with one click by clicking the deploy button. The program is saved in the format of JSON string, which is convenient for users to share and modify.

Node-Red is based on Node.js, and its execution model is the same as Node.js, which is also event-driven non-blocking. In theory, all modules of Node.js can be encapsulated into one or several nodes of Node-Red. As a visual programming tool, Node-Red simplifies the tedium of programming and makes the interface more concise.

Node-Red is a graphical programming method based on node.js. This article designs the OPC UA client based on the existing nodes of the Node-Red library.

2. OPC UA client design

OPC UA usually uses a client/server communication mode, and the client gets the service by sending a request to the server. The OPC UA client application uses the application program interface (API) to invoke the services provided by the server. The OPC UA communication stack converts the client's application program interface (API) calls into messages, which can send the underlying communication entity to the server [3].

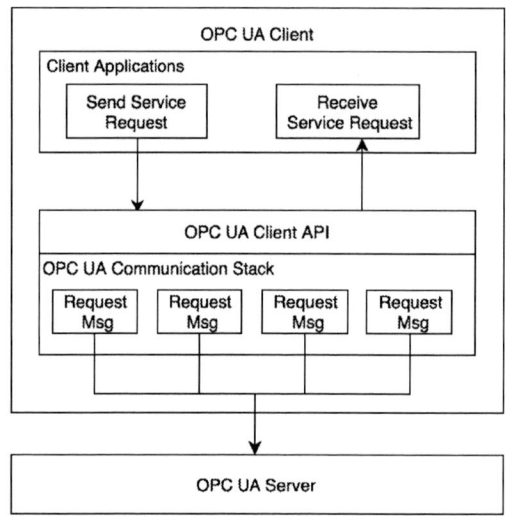

Figure 3. OPC UA Client Architecture.

2.1. Nodes

There are sets of OPC UA nodes in Node-Red, and the OPC UA nodes can also be customized, which encapsulates various types of section objects and necessary functions. To build the OPC UA client with Node-Red, connect the required modules by dragging and dropping, and set the end node in the corresponding module to connect the client to the server. Just set the Id of the object node to implement the read and write function, avoiding large-scale programming and simplifying the workflow.

There are also database module interfaces in Node-Red, which support databases such as mysql and mongodb. The read data can be stored in the database, which is convenient for querying at any time, and can also directly operate the database in Node-Red, which is convenient and quick. The Dashboard module is used as a display module in Node-Red to assume display functionality.

2.2. The function to be realized

This paper designs a factory-based workshop client to achieve real-time monitoring and display of the workshop temperature, and can start the fan program when the temperature is too high. The temperature sensor contains a status variable to indicate if it is running. There are also configuration variables that the client can read, and the configuration can be changed by writing variables [4]. The client can call methods to start or stop the fan. In addition, the client can subscribe to real-time temperature data measured by the temperature sensor and display it.

（1）Connecting to the server

（2）reading data

（3）Subscription data

（4）calling method

（5）Chart display

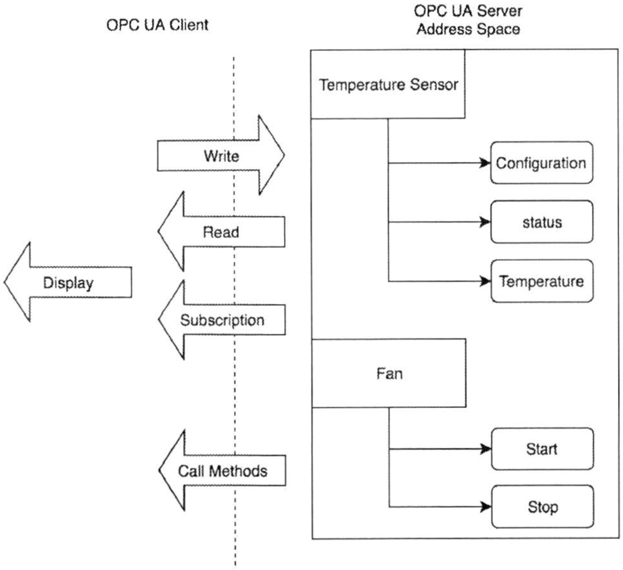

Figure 4. OPC UA Client Functions.

3. Implementation

The server is uniquely identified by the endpoint and connected to the server [5]. The read and write service can not only read and write the value of the variable, but also read and write the attributes of the node. By directly defining the NodeId of the target node in the input node, the content in the node can be read or written.

Subscriptions are groups of information sources that are managed by a monitoring item. There are two common types of monitoring items. The first is monitoring data and the second is monitoring events. Monitoring items usually select some common settings, such as subscription interval, monitoring mode, etc [6].

By inputting the nodeId and the subscription interval of the monitoring object in the input node, real-time monitoring of the data stored in the node at the time interval can be realized.

The chart node can output the data as a chart. By using the function node, the format of the node data is adjusted so that the abscissa of the graph is time and the ordinate is temperature [7].

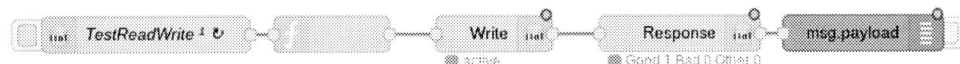

Figure 5. OPC UA Client in Node-Red.

The Switch node determines the temperature of the subscription. When the temperature is too high, the method is called to start the fan. When the temperature is normal, the fan is stopped.

4. Conclusion

In the context of "Smart Manufacturing 2025", this paper designs an OPC UA client based on OPC UA. In addition to basic read and write services, real-time monitoring and display of the shop floor temperature can be realized, and the temperature can be too high. When the fan is started, centralized management of the production workshop is realized, and the workload of the on-site maintenance personnel is reduced [8]. Temperature monitoring is widely used in modern industrial production and is of great significance to the safety of industrial production.

Acknowledgments

This paper is supported by Beijing Key Laboratory of Work Safety Intelligent Monitoring (Beijing University of Posts and Telecommunications).

References

[1] Thomas Hadlich 2006 Providing device integration with OPC UA Industrial Informatics IEEE International Conference

[2] Stefan-Helmut Leitner and Wolfgang Mahnke 2007 OPC UA-service-oriented architecture for industrial applications

[3] Wang Hua and Liu Feng 2008 OPC UA technology and application Industrial Control Computer 21 (12)

[4] Lu Huiming and Yan Zhifeng 2010 Research on key technology of the address space for OPC UA server Advanced Computer Control (ICACC) 2010 2nd International Conference

[5] Zhang Lizhan, Jin Qibing and Zhao Dali 2008 Integration system of management and control based on OPC UA specification [J] Industrial Control Computer 21(9):26-27

[6] IEC 62541 -1:OPC Unified Architecture Specification -Part 1:Overview and Concepts[S].IEC, 2008

[7] Nguyen Thi Thanh Tu and Huynh Quyet Thang 2013 design and development of the air conditioning system by using OPC UA specifications and modbus protocol IEEE 8th Conference on Industrial Electronics and Applications (ICIEA) p105

[8] Daoqu Geng, Yi Wen and Qi Gao 2018 research on semantic extension method of real-time data for OPC UA 3rd International Conference on Materials Science, Machinery and Energy Engineering (MSMEE 2018) p104

ISPECE

A new point-weighting finite-difference modelling for the frequency-domain wave equation

D S Cheng[1], B W Chen[1,3], J J Chen[2] , X S Wang[1]

[1]School of Software Engineering, Shenzhen Institute of Information Technology, Shenzhen 518172, PR China

[2]School of Mathematics and Statistical Science, Guangxi Teachers Education University, Guangxi 530299, PR China

[3]Author to whom any correspondence should be addressed.

Email address: chenbw@sziit.edu.cn

Abstract. In this paper, we propose a new 21 point finite-difference scheme for modelling the frequency-domain wave equation in 2-dimensional domain. To discretize the second derivative term, we develop a modified central difference scheme in which each of the gridpoints is replaced with a linear combination of it and its neighbouring girdpoints along one direction. For the discretization of the zeroth-order term, we use the weighted average of all the 21 points. The combination coefficients and the weight parameters are determined by minimizing the numerical dispersion. In comparison with the scheme of a derivative-weighting one, the new scheme has a much better performance in reducing the numerical dispersion when the step sizes are not equal in different directions. Numerical experiments are presented to illustrate the effectiveness of the new scheme.

1. Introduction
The frequency-domain wave equation is extensively applied in many fields of science and engineering, for instance, geophysics, marine technology, petroleum exploration. Frequency-domain modelling has many advantages over the time-domain modelling. For example, it is very convenient to manipulate a single frequency, and easy to implement attenuation. Moreover, as each frequency can be computed independently, it is very favourable for parallel computing.

For the modeling frequency-domain wave equation, the finite difference method is preferred, since it is easily implemented and has less computational complexity. For the classical central difference, it leads to a poor numerical dispersion, which is not able to be eliminated [1]. To reduce numerical dispersion, the rotated 9-point difference scheme was proposed in [2], which combined the classical Cartesian coordinate system and its rotated systems to discretize the equation. In [3], the rotated difference scheme was extended to the 25-point formula. However, the rotated difference scheme is not robust. On the one hand, it fails when directional sampling intervals are different, that is, the step sizes are not equal in different directions. On the other hand, it is not pointwise consistent, which is an important property for the convergence. To improve the robustness, [4] proposed an average-derivative 9-point scheme and [5] developed a point-weighting 9-point scheme. To obtain a consistent scheme, [6] and [7] constructed two types of derivative-weighting schemes, in which a weighted derivative and linear combination of the gridpoints are employed to discretize the Laplacian and the zeroth-order term respectively. Nevertheless, the derivative-weighting schemes still cannot handle the

Content from this work may be used under the terms of the Creative Commons Attribution 3.0 licence. Any further distribution of this work must maintain attribution to the author(s) and the title of the work, journal citation and DOI.

Published under licence by IOP Publishing Ltd

ISPECE

IOP Conf. Series: Journal of Physics: Conf. Series **1187** (2019) 042092   doi:10.1088/1742-6596/1187/4/042092

problem of non-equidistant sampling. High order schemes [8-10] are also proposed to improve the accuracy, however, they are demanding with the smoothness of the right-hand side.

In this paper, we propose a new 21 point-weighting finite-difference scheme, which is pointwise consistent and suitable for the situation that the step sizes are not equal in different direction. To discretize the Laplacian, the new scheme employs a modified central difference scheme in which each of the gridpoints is replaced with a linear combination of it and its neighbouring girdpoints along one direction. The discretization of the zeroth-order term adopts weighted averages of all the 21 points. The combination coefficients and the weight parameters are obtained by minimizing the numerical dispersion. The new 21 point-weighting scheme is more flexible compared to the derivative-weighting scheme. Furthermore, it also outperforms the 9-point schemes in reducing the numerical dispersion. Finally, numerical simulations are given to demonstrate the efficiency of the new scheme.

2. A new point-weight finite-difference scheme

In this section, we present the construction of the new finite-difference scheme. Consider the 2-dimensional Helmholtz equation

$$-\frac{\partial^2 u}{\partial x^2} - \frac{\partial^2 u}{\partial^2 y} - k^2 u = 0, \tag{1}$$

where u is the displacement, $k = \frac{2\pi f}{v}$ is the wavenumber with f and v being the frequency and velocity respectively. To discretizing equation (1), we ues the 21-point finite difference stencils which are presented in figure 1 with $(0,0)$ representing the central point, while the others denoting the neighboring points of $(0,0)$. Let $u|_{m,n} := u(x_0 + m\Delta x, y_0 + n\Delta y)$ denote the discretization of u at location $(x_0 + m\Delta x, y_0 + n\Delta y)$, where $\Delta x, \Delta y$ are the step sizes in the horizontal and vetical directions respectively, and (x_0, y_0) is a given inital point . Here, let $\Delta x = h$, $\Delta y = rh$ with r being a positve number. Then, $x_m = x_0 + mh$, $y_n = y_0 + nrh$.

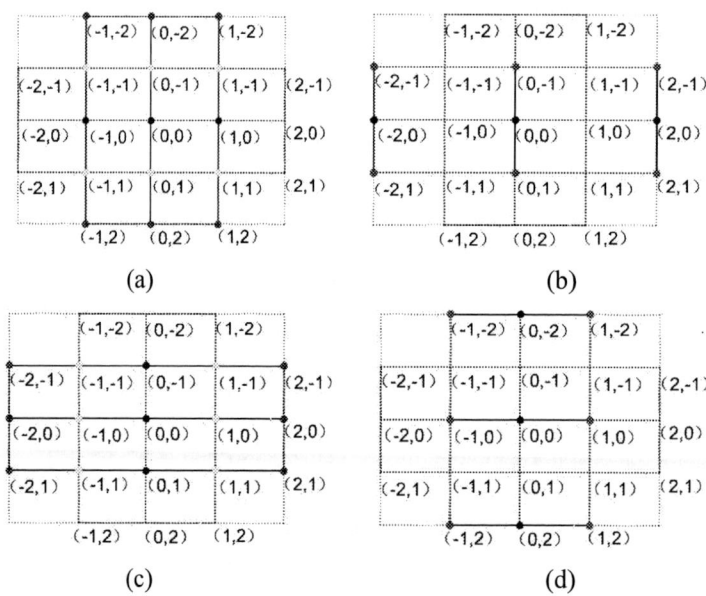

Figure 1. Finite difference stencils. (a),(b) are used for discretizing $\frac{\partial^2 u}{\partial x^2}$, and (c),(d) are for $\frac{\partial^2 u}{\partial y^2}$.

Define $\mathrm{T}_{h,x} u|_{m,n}$, $\mathrm{T}_{2h,x} u|_{m,n}$ as follows

1933

$$\mathrm{T}_{h,x}u\,|_{m,n} := \frac{1}{h^2}\left(\tilde{u}_{m-1,n} - 2\tilde{u}_{m,n} + \tilde{u}_{m+1,n}\right), \tag{2}$$

$$\mathrm{T}_{2h,x}u\,|_{m,n} := \frac{1}{(2h)^2}\left(\hat{u}_{m-2,n} - 2\hat{u}_{m,n} + \hat{u}_{m+1,n}\right), \tag{3}$$

where

$$\tilde{u}_{m-1,n} = a_1 u_{m-1,n} + \frac{a_2}{2}\left(u_{m-1,n-1} + u_{m-1,n+1}\right) + \frac{a_3}{2}\left(u_{m-1,n-2} + u_{m-1,n+2}\right),$$

$$\tilde{u}_{m,n} = a_1 u_{m,n} + \frac{a_2}{2}\left(u_{m,n-1} + u_{m,n+1}\right) + \frac{a_3}{2}\left(u_{m,n-2} + u_{m,n+2}\right),$$

$$\tilde{u}_{m+1,n} = a_1 u_{m+1,n} + \frac{a_2}{2}\left(u_{m+1,n-1} + u_{m+1,n+1}\right) + \frac{a_3}{2}\left(u_{m+1,n-2} + u_{m+1,n+2}\right),$$

$$\hat{u}_{m-2,n} = a_4 u_{m-2,n} + \frac{a_5}{2}\left(u_{m-2,n-1} + u_{m-2,n+1}\right),$$

$$\hat{u}_{m,n} = a_4 u_{m,n} + \frac{a_5}{2}\left(u_{m,n-1} + u_{m,n+1}\right),$$

$$\hat{u}_{m+2,n} = a_4 u_{m+2,n} + \frac{a_5}{2}\left(u_{m+2,n-1} + u_{m+2,n+1}\right).$$

Here, parameters $a_1, a_2, \ldots, a_5$ satisfy $\sum_{j=1}^{5} a_j = 1$, which guaratees that the approximations of $\frac{\partial^2 u}{\partial x^2}$ is second order in accuracy. Similarly, define $\mathrm{T}_{h,y}u\,|_{m,n}$, $\mathrm{T}_{2h,y}u\,|_{m,n}$ as follows

$$\mathrm{T}_{rh,y}u\,|_{m,n} := \frac{1}{r^2 h^2}\left(\tilde{u}_{m,n-1} - 2\tilde{u}_{m,n} + \tilde{u}_{m,n+1}\right), \tag{4}$$

$$\mathrm{T}_{2rh,y}u\,|_{m,n} := \frac{1}{(2rh)^2}\left(\hat{u}_{m,n-2} - 2\hat{u}_{m,n} + \hat{u}_{m,n+2}\right), \tag{5}$$

with

$$\tilde{u}_{m,n-1} = a_1 u_{m,n-1} + \frac{a_2}{2}\left(u_{m-1,n-1} + u_{m+1,n-1}\right) + \frac{a_3}{2}\left(u_{m-2,n-1} + u_{m+2,n-1}\right),$$

$$\tilde{u}_{m,n} = a_1 u_{m,n} + \frac{a_2}{2}\left(u_{m-1,n} + u_{m+1,n}\right) + \frac{a_3}{2}\left(u_{m-2,n} + u_{m+2,n}\right),$$

$$\tilde{u}_{m,n+1} = a_1 u_{m,n+1} + \frac{a_2}{2}\left(u_{m-1,n+1} + u_{m+1,n+1}\right) + \frac{a_3}{2}\left(u_{m-2,n+1} + u_{m+2,n+1}\right),$$

$$\hat{u}_{m,n-2} = a_4 u_{m,n-2} + \frac{a_5}{2}\left(u_{m-1,n-2} + u_{m+1,n-2}\right),$$

$$\hat{u}_{m,n} = a_4 u_{m,n} + \frac{a_5}{2}\left(u_{m-1,n} + u_{m+1,n}\right),$$

$$\hat{u}_{m,n+2} = a_4 u_{m,n+2} + \frac{a_5}{2}\left(u_{m-1,n+2} + u_{m+1,n+2}\right).$$

Then, the second partial derivatives $\frac{\partial^2 u}{\partial x^2}$ and $\frac{\partial^2 u}{\partial y^2}$ at (m,n) are approximated respectively by

$$\mathrm{T}_x u\,|_{m,n} := \mathrm{T}_{h,x}u\,|_{m,n} + \mathrm{T}_{2h,x}u\,|_{m,n}, \quad \mathrm{T}_y u\,|_{m,n} := \mathrm{T}_{rh,y}u\,|_{m,n} + \mathrm{T}_{2rh,y}u\,|_{m,n}.$$

ISPECE   IOP Publishing

IOP Conf. Series: Journal of Physics: Conf. Series **1187** (2019) 042092   doi:10.1088/1742-6596/1187/4/042092

Moreover, the zeroth term k^2u is approximated by weighted averages of all the 21 points as follows.

$$\mathrm{I}(k^2u)\big|_{m,n} := b_1(k^2u)\big|_{m,n} + \frac{1}{4}\sum_{j=1}^{5} b_{j+1}\mathrm{I}_j(k^2u)\big|_{m,n}, \tag{6}$$

where

$$\mathrm{I}_1(k^2u)\big|_{m,n} := (k^2u)\big|_{m,n+1} + (k^2u)\big|_{m,n-1} + (k^2u)\big|_{m+1,n} + (k^2u)\big|_{m-1,n}),$$

$$\mathrm{I}_2(k^2u)\big|_{m,n} := (k^2u)\big|_{m,n+2} + (k^2u)\big|_{m,n-2} + (k^2u)\big|_{m+2,n} + (k^2u)\big|_{m-2,n},$$

$$\mathrm{I}_3(k^2u)\big|_{m,n} := (k^2u)\big|_{m-1,n-1} + (k^2u)\big|_{m-1,n+1} + (k^2u)\big|_{m+1,m-1} + (k^2u)\big|_{m+1,m+1},$$

$$\mathrm{I}_4(k^2u)\big|_{m,n} := (k^2u)\big|_{m-2,n+1} + (k^2u)\big|_{m+2,n-1} + (k^2u)\big|_{m-1,n-2} + (k^2u)\big|_{m+1,n+2},$$

$$\mathrm{I}_5(k^2u)\big|_{m,n} := (u\big|_{m-2,n-1} + u\big|_{m+2,n+1} + u\big|_{m-1,n+2} + u\big|_{m+1,n-2}).$$

Here, parameters $b_1, b_2, \ldots, b_6$ satisfy $\sum_{j=1}^{6} b_j = 1$, which guarantees the approximation of k^2u is second order in accuracy. Finally, we obtain the new difference scheme for equation (1) as follow

$$-\mathrm{T}_x u\big|_{m,n} - \mathrm{T}_y u\big|_{m,n} - \mathrm{I}(k^2u)\big|_{m,n} = 0. \tag{7}$$

Substituting equation (2)-(6) into (7), then the 21 points difference scheme is given by

$$T_1 U_{m-1,n-2} + T_2 U_{m,n-2} + T_3 U_{m+1,n-2} + T_3 U_{m-2,n-1} + T_4 U_{m-1,n-1} + T_5 U_{m,n-1} + T_4 U_{m+1,n-1} +$$

$$T_1 U_{m+2,n-1} + T_2 U_{m-2,n} + T_5 U_{m-1,n} + T_6 U_{m,n} + T_5 U_{m+1,n} + T_2 U_{m+2,n} + T_1 U_{m-2,n+1} + T_4 U_{m-1,n+1} \tag{8}$$

$$+ T_5 U_{m,n+1} + T_4 U_{m+1,n+1} + T_3 U_{m+2,n+1} + T_3 U_{m-1,n+2} + T_2 U_{m,n+2} + T_1 U_{m+1,n+2} = 0,$$

where $U_{m-j,n-l}$ ($j,l = -2,-1,0,1,2$) denote the unknowns, and

$$T_1 = \frac{a_3}{2h^2} + \frac{a_5}{8r^2h^2} + \frac{b_5}{4}k^2, \quad T_2 = \frac{-a_3}{h^2} + \frac{a_4}{4r^2h^2} + \frac{b_4}{4}k^2, \quad T_3 = \frac{a_3}{2h^2} + \frac{a_5}{8r^2h^2} + \frac{b_6}{4}k^2,$$

$$T_4 = \frac{a_5}{8h^2} + \frac{a_3}{2r^2h^2} + \frac{b_6}{4}k^2, \quad T_5 = \frac{a_2}{2h^2} + \frac{a_2}{2r^2h^2} + \frac{b_3}{4}k^2, \quad T_6 = \frac{-a_2}{h^2} - \frac{a_5}{4h^2} + \frac{a_1}{r^2h^2} + \frac{b_2}{4}k^2,$$

$$T_7 = \frac{a_5}{8h^2} + \frac{-a_3}{2r^2h^2} + \frac{b_5}{4}k^2, \quad T_8 = \frac{a_4}{4h^2} + \frac{-a_3}{r^2h^2} + \frac{b_4}{4}k^2, \quad T_9 = \frac{a_1}{h^2} - \frac{a_2}{r^2h^2} - \frac{a_5}{4r^2h^2} + \frac{b_2}{4}k^2,$$

$$T_{10} = \frac{-2a_2}{h^2} - \frac{a_4}{2h^2} - \frac{2a_1}{r^2h^2} - \frac{a_4}{2r^2h^2} + b_1 k^2.$$

3. Numerical dispersion analysis and determination of the weight parameters
In this section, we perform the classical dispersion analysis, and obtain the weight parameters by minimizing the numerical dispersion.

For equation (1), the classical plane-wave solution is $U(x,y) := e^{-ik(x\cos\theta + y\sin\theta)}$, where θ is the propagation angle from the y-axis, and the wavenumber $k = \frac{2\pi f}{v}$ is assumed to be a constant. Let λ be the wavelength and G be the number of gridpoints per wavelength, that is, $G = \frac{\lambda}{h}$. Since $\lambda = \frac{v}{f}$, we have $kh = \frac{2\pi}{G}$. Then, substituting the discrete plane-wave solution $U_{m+j,n+l} := e^{-ik(x_{m+j}\cos\theta + y_{n+l}\sin\theta)}$ into equation (8) and applying the Euler formula $e^{-ix} = \cos x + i\sin x$, we obtain

1935

$$T_1R_1 + T_2P_2 + T_3S_2 + T_4R_2 + 2T_5P_1Q_1 + T_6P_1 + T_7S_1 + T_8Q_2 + T_9Q_1 + \tfrac{1}{2}T_{10} = 0, \qquad (9)$$

where

$$p_1 = \cos\left(\frac{2\pi}{G}r\sin\theta\right), Q_1 = \cos\left(\frac{2\pi}{G}\cos\theta\right), S_1 = \cos\left[\frac{2\pi}{G}\left(2\cos\theta - r\sin\theta\right)\right], R_1 = \cos\left[\frac{2\pi}{G}\left(\cos\theta + 2r\sin\theta\right)\right],$$

$$p_2 = \cos\left(\frac{4\pi}{G}r\sin\theta\right), Q_2 = \cos\left(\frac{4\pi}{G}\cos\theta\right), S_2 = \cos\left[\frac{2\pi}{G}\left(\cos\theta - 2r\sin\theta\right)\right], R_2 = \cos\left[\frac{2\pi}{G}\left(2\cos\theta + r\sin\theta\right)\right].$$

Let k_N denote the numerical wavenumber. Then, replacing k with k_N in T_j ($j = 1, 2, \ldots, 10$) yields

$$k_N^2 h^2 L = R, \qquad (10)$$

with

$$L = \frac{1}{2}\left(b_1 + \frac{b_2}{2}\left(P_1 + Q_1\right) + b_3 P_1 Q_1 + \frac{b_4}{2}\left(P_2 + Q_2\right) + \frac{b_5}{2}\left(R_1 + S_1\right) + \frac{b_6}{2}\left(R_2 + S_2\right)\right),$$

$$R = a_1\left[\frac{P_1}{r^2} + Q_1 - \left(1 + \frac{1}{r^2}\right)\right] + a_2\left[\left(1 + \frac{1}{r^2}\right)P_1 Q_1 - \frac{Q_1}{r^2} - P_1\right] + a_3\left[\frac{R_1}{2} - P_2 + \frac{S_2}{2} + \frac{R_2}{2r^2} + \frac{S_2}{2r^2} - \frac{Q_2}{r^2}\right]$$

$$+ a_4\left[\frac{P_2}{4r^2} + \frac{Q_2}{4} - \frac{1}{4}\left(1 + \frac{1}{r^2}\right)\right] + a_5\left[\frac{R_1 + S_2}{8r^2} + \frac{R_2 + S_1}{8} - \frac{P_1}{4} - \frac{Q_1}{4r^2}\right].$$

It follows from $kh = \frac{2\pi}{G}$ that

$$\frac{k^N}{k} = \frac{G}{2\pi}\sqrt{\frac{R}{L}}. \qquad (11)$$

Equation (11) is the so-called dispersion relation formula. To obtain a good difference scheme with small numerical dispersion, it requires that value of equation (11) should be close to 1, that is, k_N is close to k. Consequently, parameters a_j, b_l ($j = 1, 2, \cdots, 5$, $l = 1, 2, \cdots, 6$) can be determined by

$$(a_1, \cdots, a_5, b_1, \cdots, b_6) = \arg\min\left\{\left\|J\left(a_1, \cdots, a_5, b_1, \cdots, b_6\right)\right\|_{\infty, I_G, I_\theta}\right\}, \qquad (12)$$

where

$$J\left(a_1, \cdots, a_5, b_1, \cdots, b_6\right) := \frac{k_N}{k} - 1 = \frac{G}{2\pi}\sqrt{\frac{R}{L}} - 1,$$

with $\sum_{j=1}^{5} a_j = 1$, $\sum_{j=1}^{6} b_j = 1$, $\theta \in \left[0, \frac{\pi}{2}\right]$, $G \in \left[G_{min}, G_{max}\right]$. The Nyqusit sampling limit requies $G_{min} \geq 2$.

We solve numerically the optimalization problem (12) by the least-squares method [5-6], and finally the parameters a_j, b_l ($j = 1, 2, \cdots, 5$, $l = 1, 2, \cdots, 6$) are obtained.

4. Simulation experiments

In this section, we present simulation examples to demonstrate the efficiency of the new difference scheme. Firstly, we use the least-squares method to solve the optimalization problem (12) with setting $G_{min} = 4$ and $G_{max} = 400$. Then, the parameters obtained are $a_1 = 0.2065$, $a_2 = 0.2000$, $a_3 = 0.0164$, $a_4 = 0.3871$, $a_5 = 0.1900$, $b_1 = 0.3398$, $b_2 = 0.4792$, $b_3 = 0.1490$, $b_4 = 0.0266$, $b_5 = 0.0027$.

We next plot the normalized phase velocity curves for new difference scheme in Figure 2, where the horizontal coordinate is G^{-1}, and the vertical coordinate is $k_N k^{-1}$. The propagation angles are chosen to be 0°, 15°, 30°, 45^0. As a comparison, we also present normalized phase velocity curves for the derivative-weighting scheme in [6]. As is observed in Figure 2, the new scheme is efficiency in reducing the numerical dispersion, and performs much better than the scheme in [6], especially for the situations of non-equidistant sampling. Specifically, in Figure 2(c) and (d), set $r=0.5$, that is, $\Delta x=h$, $\Delta y=rh=0.5h$. It is seen that the new scheme performs robustly, however, the scheme in [6] fails. For $r=2$ in Figure 2 (e) and (f), we have the similar results. Finally, we use the new difference scheme to model the frequency-domain wave equation (1) with a point source placed at the center of the 2D computational domain $(0,1)\times(0,1)$. In Figure 3, we present the real part of the simulation result (numerical solution) for $f=5,20$ respectively, which confirms the effect of the new point-weighting difference scheme.

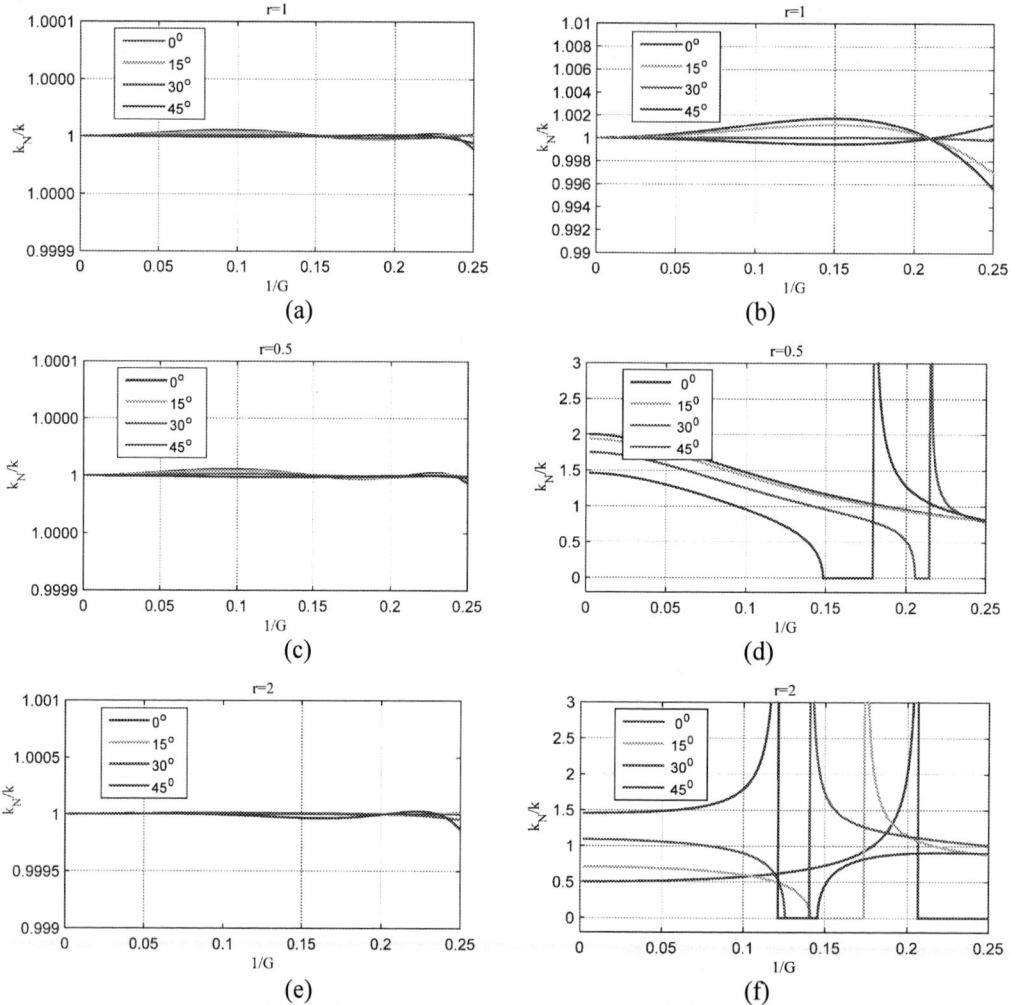

Figure 2. Normalized phase velocity curves for the new point-weighting difference scheme and the derivative-weighting scheme. (a),(c),(e): new point-weighting scheme with $r=1,0.5,2$. (b), (d), (f): derivative-weighting scheme with $r=1,0.5,2$.

ISPECE   IOP Publishing

IOP Conf. Series: Journal of Physics: Conf. Series **1187** (2019) 042092   doi:10.1088/1742-6596/1187/4/042092

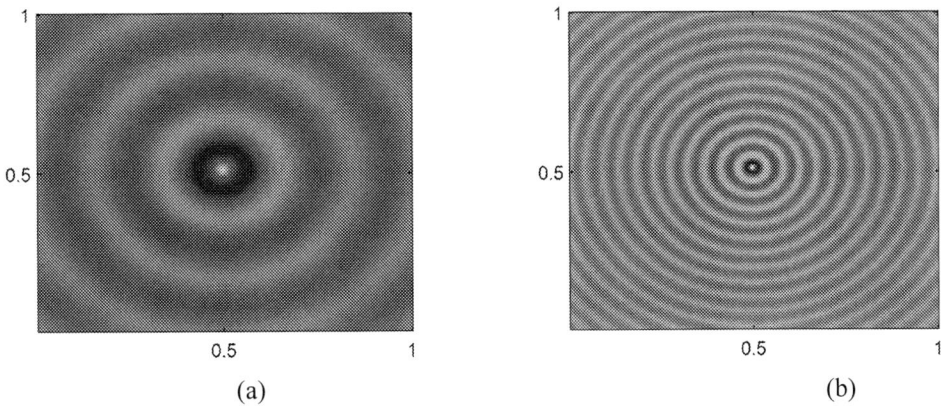

(a)   (b)

Figure 3. Real parts of the numerical solutions for different frequencies. (a) $f = 5$, (b) $f = 20$.

5. Conclusion

In this paper, a new point-weighting finite-difference scheme is proposed for modeling the frequency-domain wave equation in 2-dimensional domain. The new scheme uses a modified central difference formula to discretize the second derivative term. Specifically, in the modified central difference formula, each of the gridpoints is replaced with a linear combination of it and its neighbouring girdpoints along one direction. Additionally, the weighted average of all the 21 points is used to discretize the zeroth-order term. The new scheme is second order in accuracy and is pointwise consistent. Simulation experiments illustrate the efficiency of the new scheme. It outperforms the 9-point derivative-weighting scheme not only in reducing the numerical dispersion, but also in tackling the problems of non-equidistant samplings.

Acknowledgement

This research was financially supported by the Technology Innovation Commission of Shenzhen under grants JCYJ20160527102119211, and the Research Projects of Shenzhen Institute of Information Technology under grants QN201710 and ZY201714.

References

[1]   Babuška I, Sauter S 2000 *SIAM Rev.* **42** 451-84.
[2]   Jo C, Shin C and Suh J 1996 *Geophys.* **61** 529-37.
[3]   Shin C, Sohn H 1998 *Geophys.* **63** 289-96.
[4]   Chen J 2012 *Geophys.* **77** 201-10.
[5]   Cheng D, Tan X and Zeng T 2017 *Comput. Math. Appl.* **73** 2345-59.
[6]   Chen Z, Cheng D, Feng W and Wu T 2013 *Int. J. Numer. Anal. Model.* **10** 389-410.
[7]   Cheng D, Liu Z, Wu T 2015 *Math. Comput. Simulat.* **117** 54-67.
[8]   Stolk C 2016 *J. Comput. Phys.* **314** 618-46.
[9]   Wu T, Xu R 2018 Comput. Math. Appl. 75 2520-37.
[10] Turkel E, Gordon G, Gordon R and Tsynkov S 2013 J. Comput. Phys. 232 272-87.

Research on layout optimization of express parcel transportation network distribution center based on node operation process

Jun Han[1], Shiwei He[2*], Mingkai Bi[3] and Zilong Song[4]

[1,2,3,4] Key Laboratory of Transport Industry of Big Data Application Technologies for Comprehensive Transport, Ministry of Transport, Beijing Jiaotong University, Beijing, 100044, China

[2]School of Transportation, Beijing Jiaotong University, Beijing, 100044, China

*Corresponding author's e-mail address: shwhe@bjtu.edu.cn

Abstract. Through refining the ability of distribution center node and combining the national political factors, this paper proposes an improved P-median model, which considers the constraints of the import operation scale, export operation scale and intermediate operation capability of each candidate distribution center node and mandatory nodes. Finally, this paper takes a large express logistics enterprise parcel transportation network as an example to analyse the optimization result of case and parameter sensitivity of node service scope. We find that the optimized layout of the distribution center breaks the administrative division constraint, realizes the nearest transit of parcel, fully considers the scale of each operation of the node, and meets the requirements of mid-to-long term development trend of the parcel express transportation network, which has reference significance for the development of the parcel transportation business and the research on site selection model.

1. Introduction

In recent years, China's express delivery business has continued to develop. Under the influence of logistics demand in various industries, the competition of domestic express logistics enterprises has become increasingly fierce. However, the layout of the existing parcel transportation network distribution center is difficult to support the rapid development of express delivery business, which is reflected in the following aspects: (1) Some provinces use the provincial capital as a network hub connecting the whole country, but they are not the geographic center of the province and cannot meet the growing needs of express delivery parcels. (2) In some areas, the capacity of distribution center is limited, and the future development need is difficult to meet. (3) In some provinces, there are few channels for parcel distribution, hence the pressure of the express network parcel flow will gather in the provincial capital in the peak season. (4) The distribution level of some distribution centers is not clear. In the northwest, northeast, southwest and other regions, there is a lack of transit hubs to serve remote areas.

In order to improve the transportation efficiency of express logistics industry, seize the central market, and better meet the customer's small batch size, multi-category and time-sensitive distribution requirements, logistics enterprises need to optimize the layout of the parcel transportation network in addition to improving staff productivity and infrastructure construction.

The parcel transportation network distribution center is also called the transshipment center and the processing center. It refers to the production operation unit responsible for parcel sorting, shipping and transfer operations in the parcel transportation network, which constitutes the node of the parcel transportation network[1].

The layout optimization of the distribution center of the parcel express transportation network belongs to the location problem. In the existing research, experts and scholars have carried out sufficient and in-depth research on site selection theory, which has played a good guiding role in production operations[2]. Specifically, in foreign countries, literature[3] has reviewed the research of hierarchical location model; literature[4~5] proposed a hospital location model based on capacity limitation; literature[6~8] proposed hierarchical selection models with capacity limitation. Domestically, Wu Xing et al[9] built a railway logistics center hierarchical location model with the goal of maximizing total demand and minimizing logistics transportation costs; Liu Qiang et al[10] proposed the two-level planning model of regional integrated transportation hub layout; Li Daikun[11] classified the transportation hubs, and then used the median model to optimize the layout of each level of hubs; Li Tingting et al[12] optimized the layout of regional passenger hubs using a hierarchical location model; Ding Jinxue et al[13] optimized and analyzed the layout of China's transportation hubs.

In summary, the current research on the construction of location model mainly considers the node capacity limitation and the demand stratification. According to the actual production situation, which plays a guiding role of the model construction, the current research mainly has two problems. First, most studies only consider the overall operational capability of the candidate nodes but ignore the differences between distribution centers in the import and export capabilities which may have an impact on the selection of the distribution centers; second, the analysis and constraints on the factors influencing the selection of the distribution center ignore the fact that the political factors will have a decisive influence on the choice of the distribution centers.

In view of the limitations of the current research, this paper proposes an improved optimization model of distribution center, which refines the operational capabilities of the alternative distribution centers and their associated relationships, fully considers the import operations scale, the export operations scale and the constraints of intermediate operations of each candidate node, and also integrates political factors in order to provide decision-making reference for the layout optimization of the parcel express transportation network and the construction of the relevant location model.

2. Analysis of Factors Affecting Site Selection of Distribution Center

In the parcel express transportation network, the distribution center is the node of each transportation line. Other aspects of parcel delivery, except transportation, including loading, unloading, sorting, distribution, and sending, occur most at the distribution center. Therefore, the distribution center is an important part of the entire express transportation network, and the rationality of its layout determines the transportation efficiency of the entire parcel transportation system. The selection of the parcel express delivery center needs to start from the internal operations of each candidate node, refine the operation capability, and pay attention to the import and export operation scale and intermediate operation processing capability of each candidate node.

The import operation means that the distribution center receives parcels shipped from various places. The export operation means that the distribution center sends the sorted parcels to their destinations. The scale of import and export operations is mainly affected by the number of goods stacks, the handling capacity of goods stacks and the efficiency of the operators. The intermediate operation means that the distribution center sorts the imported parcels, and performs the corresponding operations such as circulation, classification, and collection of the parcels according to the destination. The operation process of the parcel express distribution center is shown in Figure1.

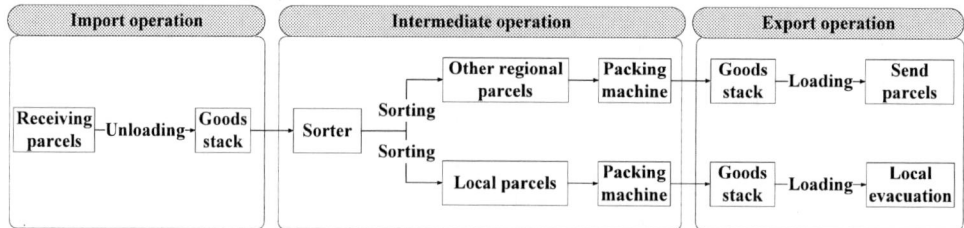

Figure 1. The summary operation process of parcel express transportation network distribution center

The scale of import and export operations reflects the parcel transportation demand of the distribution center in the express transportation network, that is, if the demand reaches a certain standard, the candidate node should be selected as the distribution center. The intermediate operation capacity limits the total operation of the distribution center. If only one overall capability value is used for constraint, the data with smaller import and export operations will become the constraint value, so that other capabilities may have surplus, which in turn causes waste of a large amount of transportation resources. This paper fully considers the difference in the scale of the import and export of alternative nodes, and constrains the import scale and export scale of the alternative nodes separately. Specifically, this paper calculates the import and export scale of each candidate node separately and stipulates the scale standard of import and export operations.

In addition, the location of the distribution center should be conducive to the delivery and transportation of the parcels in the choice of the express delivery network distribution center, so the administrative center or traffic center should be selected as much as possible[14]. Therefore, it is necessary to set some mandatory distribution centers according to the political status and economic scale of the alternative distribution center. Furthermore, for some remote areas, the parcel transportation volume is relatively small and the coverage is vast, so it is also necessary to set some mandatory distribution centers in order to balance the transportation network and increase the accessibility of the transportation network.

In summary, the layout optimization problem of the parcel express delivery network distribution center studied in this paper is a mixed integer linear programming problem with minimum weighted distance for full network demand transportation, which takes into account the scale of the import operations, the scale of export operations, the constraints of intermediate operations of each candidate node, and the constraints of political factors. In this problem, the distance matrix between each candidate node, the coverage service distance of each node, the minimum number of distribution centers that can cover all the requirements, the mandatory distribution centers, the import and export volume of the candidate nodes, and the total business volume are all known.

3. Improved parcel express delivery network distribution center layout optimization model

3.1. Model hypothesis
(1) The number of the distribution center is known and unchanged.

(2) The size and structure of the parcel express transportation network is stable, and the parameters related to nodes and arcs are known.

(3) The mandatory node information limited by political conditions is known.

3.2. Related symbol definition
The collections are defined as follows: set V be a collection of distribution centers, indexed by i, j; set Z_j be a collection of mandatory distribution centers, in general, provincial capitals and municipalities are mandatory distribution centers.

The parameters are defined as follows: set P as the selection number of distribution centers; d_{ij} as the transportation distance between nodes; D_j as node coverage service distance; h_i as the total amount of transportation at the hub node, also the demand weight of the hub node; o_i as node's export volume;

d_i as node's import volume; C_1 as import operation scale standard; C_2 as total operation scale standard (export operation volume add import operation volume); C_3 as operation capacity of the distribution center; M as a positive integer large enough.

The decision variables are defined as follows: x_j --- whether j is selected as the distribution center, if it is, the value is 1, otherwise is 0; y_{ij} --- whether node i is served by node j, if it is, the value is 1, otherwise is 0.

3.3. Improved parcel express delivery network distribution center layout optimization model

$$\min \sum_{i \in V} \sum_{j \in V} h_i d_{ij} y_{ij} \tag{1}$$

s.t.

$$\sum_{j \in V} x_j = P \tag{2}$$

$$\sum_{j \in V} y_{ij} = 1 \quad \forall i \in V \tag{3}$$

$$y_{ij} - x_j \le 0 \quad \forall i \in V, j \in V \tag{4}$$

$$\sum_{i \in V} y_{ij} o_i - M x_j \le C_1 \quad \forall j \in V \tag{5}$$

$$\sum_{i \in V} y_{ij} (o_i + d_i) - M x_j \le C_2 \quad \forall j \in V \tag{6}$$

$$\sum_{i \in V} y_{ij} (o_i + d_i) \le C_3 \quad \forall j \in V \tag{7}$$

$$d_{ij} \le D_j + M \cdot (1 - y_{ij}) \quad \forall i, j \in V \tag{8}$$

$$x_j \ge z_j \quad \forall j \in V \tag{9}$$

$$x_j, y_{ij} \in \{0,1\} \quad \forall i \in V, j \in V \tag{10}$$

Among them, formula (1) is the objective function that minimizes the total transportation distance for demand weighting; formula (2) is a limit on the number of choices for the express parcel transportation network distribution center; formula (3) represents that for demand node i, it can only be served by one distribution center j; formula (4) is a logical constraint, which means that if the demand point i is served by the distribution center j, the candidate distribution center j must be selected; formula (5) is a logical constraint, that is, if the import scale of candidate node reaches C_1, it must be set as the distribution center; formula (6) is a logical constraint, that is, the sum of import and export scale of the candidate node reaches C_2, it must be set as the distribution center; formula (7) is the intermediate operational ability constraint of the distribution center; formula (8) is the node coverage constraint; formula (9) is a mandatory node constraint; formula (10) is a decision variable 0-1 constraint.

3.4. Model solving

The model is a 0-1 mixed integer linear programming model. It can be concluded from the model that if V contains n values, the whole model has $(n+n^2)$ decision variables and $(2n^2+5n+1)$ constraint expressions. So it can be seen that the solution scale of the model is $O(n^2)$, that is, as n increases, the problem solving complexity changes as the number of square polynomials.

ILOG CPLEX software has significant advantages in solving such problems. It embeds various precise algorithms such as Branch and Bound, Lagrangian Relaxation and Column Generation. The most suitable algorithm is determined according to the complexity of the problem, the efficiency of the solution and the quality of the solution[15]. Based on the above analysis, this paper uses CPLEX optimization software to solve the problem. In addition, this paper writes VBA code to sort out the running results of CPLEX, and obtains the optimized distribution center layout plan.

4. Case study

This paper takes a large express logistics enterprise express parcel delivery network as an example. The large express logistics enterprise parcel express delivery network now contains 338 parcel nodes. In order to break the influence of the regional administrative division on the layout plan, all the nodes are both the distribution center candidate nodes and the service demand nodes, so that the demand parcel of each node can be transferred nearby. The distance matrix between 338 nodes and the OD parcel volume matrix between nodes are known.

In order to speed up the efficiency of parcel transportation, the service distance interval is determined to be [300, 350] km according to the current next day delivery business of the large express logistics enterprise. Combining the differences in natural geography conditions in China, the service distance of each node is divided into the eastern, central and western regions according to the location of the nodes. The service distance of nodes in the eastern region is appropriately reduced and the service distance of the western node is expanded according to the situation that the eastern region has developed economy and high demand, and the western region is sparsely populated and the parcel amount is small.

The scale of import and export operations and the restriction of intermediate operation capacity are shown in Table 1. The provincial capitals and municipalities directly under the Central Government are designated as the mandatory distribution centers. Under the coverage mode of 150/300/600 km, the basic coverage set model can be used to obtain the minimum number of distribution centers, which is 85.

Table 1. Import and export operation scale and intermediate operation capability of each node

Import volume	Total business volume (sum of import volume and export volume)	The intermediate operation capacity
20 (ten thousand pieces)	35（ten thousand pieces）	120（ten thousand pieces）

4.1. Solution result

In the Windows7 operating environment, this paper uses the optimization software CPLEX to solve the model, each run time is less than 4s, then uses Excel software to write the VBA code to sort out the calculation results. The layout plan of the express delivery network distribution center under the coverage distance mode of 150/300/600 km is shown in Table 2. Table 2 only shows part of the layout and coverage plan of the distribution center.

Table 2. Optimization result of layout plan of distribution center under the coverage distance mode of 150/300/600 km

No.	Distribution Center	Service City
1	Jingdezhen	Jingdezhen
2	Lanzhou	Lanzhou, Baiyin, Jinchang, Tianshui, Wuwei, Dingxi, Pingliang, Linxia, Guyuan, Weinan, Gannan, Huangnan
3	Beijing	Beijing, Langfang
4	Tianjin	Tianjin, Zhangzhou, Tangshan
5	Chaoyang	Chaoyang, Qinhuangdao, Chengde, Chifeng, Jinzhou, Fuxin, Huludao
6	Zhangjiakou	Zhangjiakou
...	...	...
85	Haixi State	Haixi State, Guoluo State

According to all the data in Table 2, the layout of the 85 distribution center is shown in Figure 2. The node name is not marked in the figure due to too many nodes.

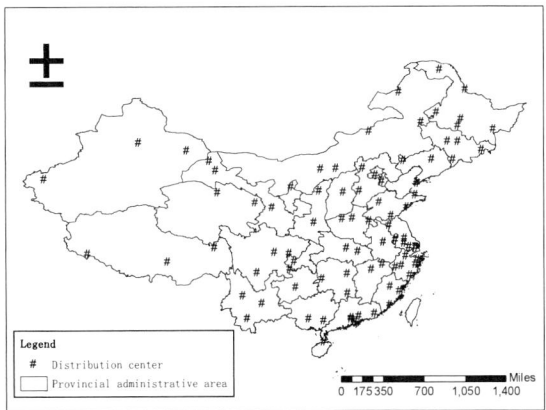

Figure 2. The layout plan diagram of 85 distribution centers

4.2. Result analysis

At present, the large express logistics enterprise has set up 75 distribution centers, and implements the processing mode of "Relatively concentrated exports and stratification of imports". This model has planned 85 distribution center nodes and their specific coverage plans for the large express logistics enterprise package transportation network. Compared with the existing parcel transportation network, the optimized distribution center layout plan changes greatly. Take the two provinces of Henan and Hubei as examples, as can be seen from Table 3, Table 4 and Figure 3, the city which is underlined means that the city served is a non-provincial city.

Table 3. The layout plan of distribution center in Henan and Hubei provinces before optimization

Province	Distribution Center	Service City	The Number of Service City
Henan	Zhengzhou	Zhengzhou, Shangqiu, Anyang, Hebi, Jiaozuo, Kaifeng, Luoyang, Luohe, Pingdingshan, Fuyang, Xinxiang, Xuchang, Zhoukou, Sanmenxia	14
	Xinyang	Xinyang, Nanyang, Zhumadian	3
Hubei	Wuhan	Wuhan, Ezhou, Huanggang, Huangshi, Suizhou, Xianning, Xiaogan, Jingmen, Jingzhou, Yichang, Enshi	11
	Xiangyang	Fuyang, Shiyan	2

Table 4. The layout plan of distribution center in Henan and Hubei provinces after optimization

Province	Distribution Center	Service City	The number of Service City
Henan	Zhengzhou	Zhengzhou, Handan, Kaifeng, Anyang, Hebi, Xinxiang, Xuchang, Zhumadian	8
	Luoyang	Luoyang, Jincheng (Shanxi), Yuncheng (Shanxi), Changzhi (Shanxi), Pingdingshan, Jiaozuo, Sanmenxia, Nanyang	8
	Shangqiu	Shangqiu, Bengbu (Anhui), Huaibei (Anhui), Fuyang (Anhui), Jining (Shandong), Heze (Shandong), Fuyang, Luohe, Zhoukou, Bozhou (Anhui)	10

1944

Hubei	Wuhan	Wuhan, <u>Xinyang (Henan)</u>, Huangshi, Xiaogan, Xianning, Huanggang	6
	Jinmen	Jingmen, Shiyan, Yichang, Ezhou, Suizhou, Jingzhou, Xiangyang	7

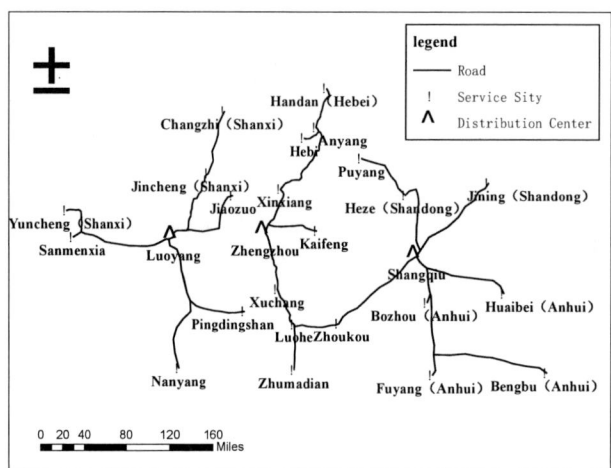

Figure 3. The layout plan of distribution center in Henan provinces after optimization

As can be seen from Table 3, Table 4 and Figure 3:

(1)The service targets of some distribution centers in Henan and Hubei provinces are not limited to the demand points in the province. Take Shangqiu City as an example, it is located in the handover area of Henan, Fujian and Shandong provinces, so it undertakes a large number of tasks outside the province because of its relatively important hub status. In the same way, Luoyang and Wuhan also undertake the demand for some nodes in Shanxi Province and Henan Province. The node layout is more uniform, and it is more in line with the actual road network construction needs. It verifies the effectiveness of the planning principle of breaking the administrative division and realizes the nearest transit of parcel delivery.

(2)In the optimized layout plan of the distribution centers of Henan and Hubei provinces, some service cities of Zhengzhou and Wuhan were transferred to non-provincial-capital distribution centers such as Luoyang, Shangqiu and Jingmen, which effectively reduces the pressure on the transfer operations of provincial capitals and also prevents the parcel flow from accumulating excessively in the provincial center. The optimized layout plan effectively solved the problem of imbalance in the amount of parcels of distribution centers in some provinces.

In addition, the optimized layout's level is more distinct and the positioning is clearer. The optimized layout appropriately increases the transit hubs to remote areas such as the southwest and northwest, which enhances the efficiency of parcel express distribution in the case of poor transportation conditions in remote areas. The corresponding optimized layout is shown in Table 5.

Table 5. The layout plan of distribution center in some southwest and northwestern provinces

Province	Distribution Center	Service City
Yunnan	Kunming	Kunming, Qujing, Chuxiong, Yuxi, Wenshan, Qianxinan, Honghe
	Dali	Dali, Baoshan, Dehong, Lijiang, Nujiang
	Puer	Pu'er, Linyi, Xishuangbanna
Xinjiang	Karamay	Karamay, Altay, Bozhou, Tacheng, Yili

	Hami	Hami
	Kashi	Kashi, Aksu, Hetian, Kezhou
Xizang	Changdu	Changdu, Diqing, Yushu Prefecture
	Lasa	Lhasa, Shigatse, Linzhi, Naqu, Shannan

In summary, the optimized parcel delivery network layout is more reasonable, which will provide decision-making reference for further transportation route planning of the large express logistics enterprise.

4.3. Node coverage sensitivity analysis

In response to the call of the country's "Rise of Central China" strategy, the state has clearly emphasized the optimization of transportation resource allocation, and required the acceleration of the construction of railway networks in the central provinces and the improvement of the transportation capacity of the highway trunk lines. With the further improvement of the transportation network and road network structure, the timeliness of parcel transportation will inevitably be greatly improved. In this paper, the coverage parameters of the nodes are further considered. The coverage distance of the nodes in the central provinces is adjusted from 300km to 150km, which is consistent with that in the east. The model is solved again and the solution results are shown in Table 6.

Table 6. The optimization results under different coverage distance

Hub Coverage Distance（km）			Number of Distribution Center	Objective Function Value（km）
East	Central	West		
150	300	600	85	2 514 958 620
150	150	600	124	1 814 880 930

According to the result of re-solving, the layout plan of the optimized distribution center under the coverage distance mode of 150/150/600 km is shown in Fig.4. Since the number of distribution centers is too large, the node name is not marked in the figure.

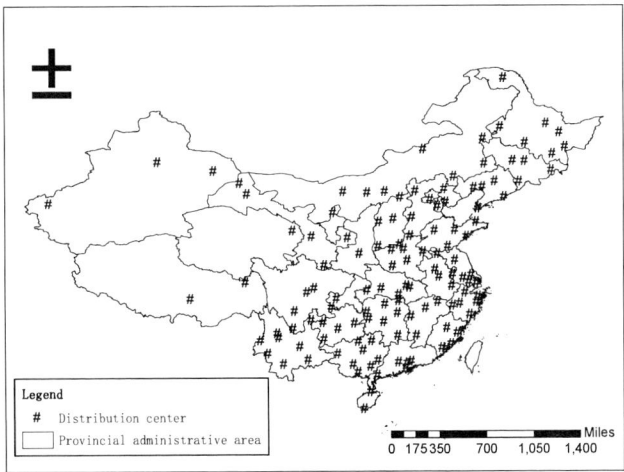

Figure 4. The layout plan diagram of 124 distribution centers

Combining Table 6 and Figure 4, the number of distribution centers has increased from 85 to 124. The re-solving layout plan increases a certain number of distribution centers especially in the central provinces. This is in line with the gradual improvement of the transportation network in the central region, and the trend that the demand for parcel transportation is increasing and the demand for the number of distribution centers is further increasing. However, the objective function value, that is, the demand-weighted transportation distance, is reduced by nearly 30%. This is because the increase of the distribution center makes the distribution center of the entire parcel express transportation network more densely distributed, and the accessibility between nodes is further increased, thereby the demand weighted distance decreases. This also shows that the coverage distance of the node has a direct impact on the layout of the parcel express delivery network distribution center.

With the development of China's economic level and the further improvement of the structure of the parcel express transportation network, especially with the implementation of the strategy of "Rise of Central China", the demand for parcel transportation and distribution centers will increase day by day. The layout of parcel express transportation network needs to have certain forward-looking and moderately ahead of the development of the corresponding business. By comparison, it can be obtained that the optimization plan under the coverage distance mode of 150km in the east, 150km in the middle and 600km in the west can provide important decision-making reference for the long-term planning of the parcel express transportation network.

5. Conclusion

This paper analyzes the process and impact mechanism of the distribution center operation, innovatively decomposes the scale of the import and export operations of the nodes, constrains the logistics capability of the nodes from the aspects of import volume, export volume and intermediate operation capacity, integrates the political factors, and proposes an improved distribution center layout optimization model in the optimization problem of the parcel transportation network node. Then this paper takes a large express delivery logistics enterprise parcel express transportation network as an example. According to the service distance model of 150km in the east, 300km in the middle and 600km in the west, 85 layout centers and their coverage plans are solved. Compared with the existing package express delivery network distribution center layout, the optimized layout fully considers the various operational capabilities of the node, breaks the administrative division limit, realizes the nearest transit of parcel delivery, and appropriately increases the distribution center to remote areas. The optimization plan enhances accessibility of the entire parcel express delivery network, which also verifies the feasibility of the model proposed in this paper. Finally, this paper further considers the node coverage distance sensitivity and obtains the layout and coverage optimization plan of the 124 distribution centers of the parcel express transportation network, which satisfies the trend of the increasing volume of parcels and the increasing demand for distribution centers, increases the forward-looking nature of the package express delivery network layout, and also provides a decision-making reference for the medium and long-term planning of the parcel express transportation network.

Acknowledgments

The authors wish to acknowledge the financial support by National Key R&D Program of China(2018YFB1201402) and the guidance of professor He.

References

[1] Luo Xianli. Process Design and Simulation Research of Mail Processing Center [D]. Beijing University of Posts and Telecommunications, 2010.

[2] Li Tingting, Song Rui. Robust optimization model for hierarchical layout of integrated passenger hubs at the national level[J]. Journal of Southeast University (Natural Science Edition), 2015, 45(01): 189-195.

[3] Guvenç Şahin, Haldun Süral. A review of hierarchical facility location models[J]. Computers & Operations Research, 2007, 34(8):2310-2331.

[4] Galvao R D, Espejo L G A, Boffey B, et al. Load balancing and capacity constraints in a hierarchical location model[J]. European Journal of Operational Research, 2006, 172(2):631-646.

[5] Mestre A M, Oliveira M D, Barbosa-Póvoa A. Organizing hospitals into networks: a hierarchical and multiservice model to define location, supply and referrals in planned hospital systems[J]. Or Spectrum, 2012, 34(2):319-348.

[6] Teixeira J C. A hierarchical location model for public facility planning[J]. European Journal of

Operational Research, 2008, 185(1):92-104.

[7]Farahani R Z, Hekmatfar M, Fahimnia B, et al. Survey: Hierarchical facility location problem: Models, classifications, techniques, and applications[J]. Computers & Industrial Engineering, 2014, 68(1):104-117.

[8]Smith H K, Harper P R, Potts C N. Bicriteria efficiency/equity hierarchical location models for public service application[J]. Journal of the Operational Research Society, 2013, 64(4):500-512.

[9]WU Xing, SONG Rui, YAN Guowei. Research on Location-Distribution of Railway Logistics Center[J]. Journal of Dalian Jiaotong University, 2017, 38(01): 7-11

[10]LIU Qiang, LU Huapu, WANG Qingyun. Bi-level programming model for regional integrated transportation hub layout[J]. Journal of Southeast University (Natural Science Edition), 2010, 40(6): 1358-1363.

[11]Li Daikun, He Shiwei, Shen Yongsheng, et al. Research on Urban Node Node Layout Optimization of Integrated Transportation Hubs[J]. Journal of Traffic Information and Safety, 2012, 30(4): 52-55.

[12]Li Tingting, Song Rui, He Shiwei, et al. Research on Layout Optimization of Regional Passenger Hub Based on Hierarchical Location Model[J]. Journal of Transportation Systems Engineering and, 2014, 14(6): 36-41.

[13]Ding Jinxue, Jin Fengjun, Wang Chengjin, et al. Evaluation, Optimization and Simulation of Spatial Distribution of China's Transportation Hubs[J]. Acta Geographica Sinica, 2011, 66(4): 504-514.

[14]Wu Chengmao, Fan Xiangyu, Wen Xiaozheng. A New Method for Location Selection of Distribution Centers in Postal District[J]. Journal of Xi'an University of Posts and Telecommunications, 2002, 7(1): 5-7.

[15]Tang Luohao. Location problem, model and algorithm considering facility failure [D]. National University of Defense Technology, 2016.

ISPECE   IOP Publishing

Using Markov constraint and constrained least square filter to develop a novel method of passive terahertz image restoration

Yuanmeng Zhao[1], Xiao Sun[1,2], Cunlin Zhang[1] and Yuejin Zhao[2]

[1] Key laboratory of Terahertz Optoelectronics, Ministry of Education of China, Beijing Key Laboratory for Terahertz Spectroscopy and Imaging, Department of Physics, Capital Normal University, Beijing 100048, China

[2] School of Optoelectronics, Beijing Institute of Technology, Beijing 100081, China

zhaoym@vip.163.com

Abstract. In recent years, passive terahertz imaging has gained significant attention in both research and practice. One big challenge with passive terahertz imaging is its low-quality images with high level of noise. State-of-the art image restoration methods have been developed for image denoising, such as methods based on Markov constraint and regular filter methods. Building upon these two methods, this paper develops a novel method for passive terahertz image restoration which preserves well both high frequency and low frequency information of the images. Performance of our method is evaluated using two common image criteria of the image sharpness, i.e. edge intensity and definition. Experimental results showed our method outperform state-of-the art methods for passive terahertz image restoration.

1. Introduction

Passive terahertz imaging is a newly introduced technology that shows good potential in security check field and has drawn much attention in research in recent years [1-3]. One of the major advantages of passive terahertz imaging is its good transmission capability, which can be used to detect dangerous objects concealed in personnel's clothes. Moreover, terahertz wave is radiation free to human body, which meets the safety concern of security screening [4-5].

Although passive terahertz imaging permits many advantages over traditional security check technology, its' application in real-world scenarios suffers from its relatively low noise ratio image. Marcin Kowalski et al. adopted threshold to image de-noising and contrasted various edge detection methods [6-7]. Dai etc. put forward a terahertz image average denoising method using suitable level Gaussian noise and adaptive color block-matching 3D filtering. Their method improved the visual effect and removed interference [8]. However, it lacks an appropriate image restoration method for terahertz security check images.

In this study, we developed a novel method based on Markov random field theory to improve passive terahertz imaging quality. In the following sections, we will report our method in detail. In particularly, in section 2, we will describe the image capture and pre-process. In section 3, we will present our method of passive terahertz image restoration based on Markov Random Field theory. Lastly, we will conclude our study findings and algorithm performance.

Content from this work may be used under the terms of the Creative Commons Attribution 3.0 licence. Any further distribution of this work must maintain attribution to the author(s) and the title of the work, journal citation and DOI.
Published under licence by IOP Publishing Ltd

ISPECE   IOP Publishing

IOP Conf. Series: Journal of Physics: Conf. Series **1187** (2019) 042094   doi:10.1088/1742-6596/1187/4/042094

2. Pre-processing of passive terahertz image

Passive terahertz imaging mainly relies on detecting the thermal radiation of targets and backgrounds. In this study, we used a terahertz point detector, a terahertz scanner, and terahertz lenses to capture the image. The terahertz scanner was composed of flapping mirror and trihedral mirror. The trihedral mirror was used for vertical scanning. Each surface of trihedral mirror was ellipsoid. During each rotation of the trihedral mirror, the detector captured three lines of the imaging target. The horizontal scanning was completed with the flapping mirror. Then the terahertz signal entered the terahertz detector and was converted to electronic signal. After processed by a computer program, a passive terahertz image was generated.

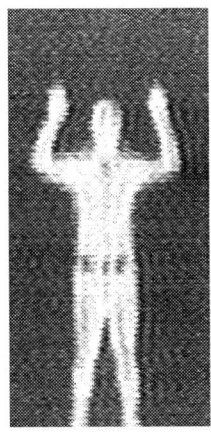 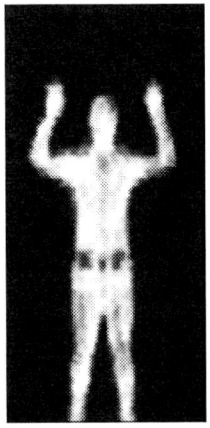

(a) Original image   (b) Pre-processed image

Figure 1. Passive Terahertz image and pre-processed image

The original image is shown in Figure 1 (a). It is quite blurred with the influence of diffraction limit of the imaging system. There are a lot of radiation noises of background, random noises from detector and electromagnetic noise from the scanning device. To solve the problem of original terahertz images, it is necessary to pre-process the images. A filter in frequency domain and adaptive median filter is adopted to remove the noises. After the de-noising processing, linear contrast stretch is used to enhance the image. The result image is shown in Figure 1 (b).

3. Passive terahertz image restoration

According to Fourier optics, optical imaging system is considered as a low-pass filter. The goal of this step is to recover a high-quality image $f(x,y)$ from its degraded observation $g(x,y)$.

$$g(x, y) = f(x, y) * h + n \tag{1}$$

In equation (1), g is a degraded image, and f represents the high-quality idea image. h is point spread function (PSF), a known degradation matrix and n is additive noise. g can be treated as an ideal image influenced by optical transfer function H and additive noises n.

3.1. Terahertz image restoration based on Markov constraint

The grayscale of pixel in passive terahertz images is only influenced by its nearby pixels, and this matches with the characteristic of the Markov random field. We considered image model follow Poisson distribution and then we applied Markov constraint to the image model. Equation (2) shows the iteration process.

$$\hat{f}_{n+1} = \hat{f}_n \exp\left\{\left(\frac{g_{ij}}{(\hat{f}_n * h)_{ij}} - 1\right) \oplus h_{ij} - \alpha \frac{\partial}{\partial f} U(\hat{f}_n)\right\} \tag{2}$$

h is the PSF of imaging system. n is the number of iterations. $U(f)$ is called energy function, and in this study, we chose to use Geman model. Equation (3) shows the complete $U(f)$ function. α is regularization parameter, which used to balance the weight of the Markov constraint. In doing so, we can achieve good results of preserving the objects' edge in the image.

$$U(f) = \sum_C \frac{(D_c(F)/\gamma)^2}{1 + (D_c(F)/\gamma)^2} \tag{3}$$

In equation (3), c is the associated clique. We defined c using the 2nd neighborhood system in this study. Dc(F) is the difference of the image. The algorithm has good capability to extrapolate the frequency spectrum of images, but it will damage the low frequency information of the images.

3.2. Improved Terahertz image restoration method using Markov constraint and constrained least square filter

To address the shortcomings of the image restoration based on Markov constraint, we adopted a regular filter to improve the image quality.

Regular filter is also called constrained least square filter, whose core idea is to find an estimation of original image $\hat{f}$ to minimum Euclidean norm of noise $\|g - H\hat{f}\|$. It is described in frequency domain as follow:

$$\hat{F}(u,v) = \left[\frac{H^*(u,v)}{|H(u,v)|^2 + \gamma|P(u,v)|^2}\right] G(u,v) \tag{4}$$

$\hat{F}$ and G are the frequency spectrum of estimated and degraded images respectively. γ is Lagrange coefficient, and $P(u,v)$ is Laplacian operator. The regular filter algorithm permits good retaining of the low frequency information despite of the poor ability to extrapolate the frequency spectrum of images. The low frequency information of regular filter and the high frequency information of the original method are combined in this improved method.

3.3. Experimental Results

In this section, we compared restored images using the above mentioned three different methods, namely original Terahertz image restoration based on Markov constraint, regular filter Terahertz image restoration, and the improved Terahertz image restoration method using Markov constraint and constrained least square filter. In particularly, we attained both restored images and their according low-pass filtered images to highlights the characteristics of each method.

The resulting image restoration using different methods are shown in Figure 2. In particularly,

- The left image of Figure 2 (a) is the regular filter restoration image and the right image of Figure 2 (a) is its spectrum image.
- The left image of Figure 2 (b) is the low–pass filtering image of the left image of Figure 2 (a) and right image of Figure 2 (b) is its spectrum image.
- The left image of Figure 2(c) is the restored image using original Terahertz image restoration based on Markov constraint and the right image of Figure 2(c) is its spectrum image.

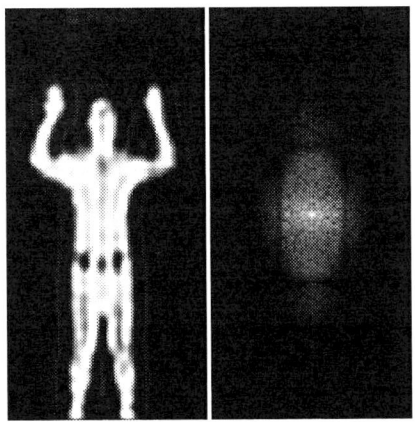 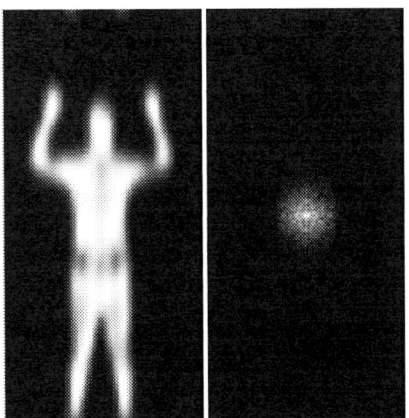

(a) Image restored by regular filter and its frequency spectrum

(b)The result image of (a) low-pass filtered and its frequency spectrum

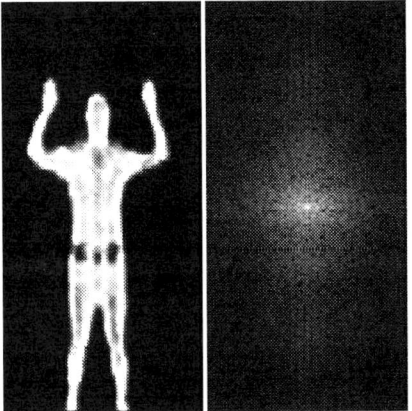 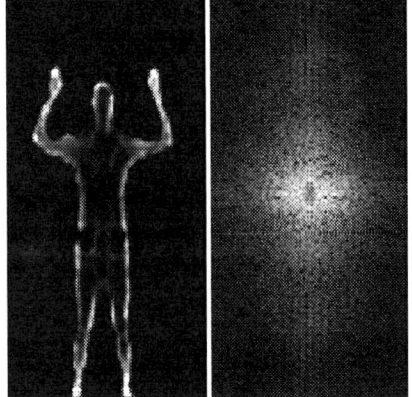

(c) Image restored by MAP restoration based on Markov constraint and its frequency spectrum

(d) The result image of (c) low-pass filtered and its frequency spectrum

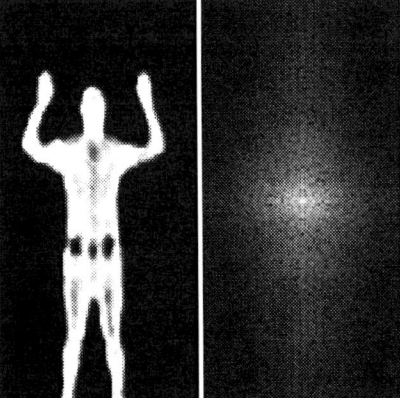

(e) Image processed by algorithm in this article and its frequency spectrum

Figure 2. Results of different image restoration algorithm

- The left image of Figure 2(d) is the low–pass filtering image of the left image of Figure 2 (c) and right image of Figure 2 (d) is its spectrum image.
- The left image of figure 2(e) is the restored image using improved Terahertz image restoration using Markov constraint and constrained least square filter, and the right image of figure 2(e) is its spectrum image.

The images clearly showed that the original method has good capability of preserving the high frequency information of the images. However, the regular filtering method has good capability of preserving the low frequency information of the images. By combing both the original method and the regular filtering method, the improved method did show good capability of preserving both high frequency and low frequency information.

3.4. Experimental Results

Edge intensity (*CV*) and image definition (*Definition*) are adopted to evaluate the quality of terahertz restoration images using the three above methods. Equation (5) and (6) show the formula:

$$CV = \frac{\sum_{i=1}^{N_1} \sum_{i=1}^{N_2} |f * La|}{N_1 \times N_2} \tag{5}$$

$$Definition = \frac{\sum_{i=1}^{N_1-1} \sum_{i=1}^{N_2-1} \sqrt{[f(i+1,j)-f(i,j)]^2 - [f(i,j+1)-f(i,j)]^2}}{N_1 \times N_2} \tag{6}$$

f is the image needs to be evaluated, La is a Laplacian operator, $N_1 \times N_2$ is the size of the image. *CV* and *Definition* are both used to describe the sharpness of the images. The degree of the image quality is determined by the value of both of the *CV* and *Definition*. The results of image quality evaluation are shown in Table 1.

Table 1. *CV*, *Definition* of different restoration images.

	Regular filter	Original method	Improved method
CV	0.0387	0.0402	0.0414
Definition	0.0548	0.0602	0.0639

4. Conclusion

In this study, we presented a novel method for the quality restoration of terahertz images with noises. This method exploits the disadvantage of original image restoration based on Markov constraint and the advantage of regular filter image restoration method. By combining these two methods, the improved terahertz image restoration method preserves good high frequency and low frequency information of the images. Thus, the resulting images achieve good quality.

The performance of our method was evaluated on two important image quality criteria, edge intensity and definition. Results showed the improved method has the highest *CV* and *definition* value among the three methods. Therefore, the experimental results demonstrate good performance, i.e. shaper, of our method in terms of terahertz image restoration.

Acknowledgments

Supported by the National Natural Science Foundation of China (61875140) and Beijing Natural Science Foundation (4181001).

References

[1] Chan W L, Deibel J A and Mittleman D M 2007 Imaging with terahertz radiation *Reports on Progress in Physics* **70**(8) pp 1325-1379

[2] Sun Q, He Y, Liu K, Fan S, Parrott E P and Pickwell-MacPherson E 2017 Recent advances in

terahertz technology for biomedical applications *Quantitative imaging in medicine and surgery* 7(3) 345

[3] Mittleman D M 2018 Twenty years of terahertz imaging [Invited] *Optics Express* **26**(8) pp 9417-9431

[4] Federici J F, Schulkin B, Huang F, Gary D E, Barat R, Oliveira F and Zimdars D 2005 THz imaging and sensing for security applications—explosives, weapons and drugs. *Semiconductor Science and Technology* **20**(7)

[5] Morozov D, Doyle S M, Banerjee A, Brien T L, Hemakumara D, Thayne I G, Hadfield R H and et al 2018 Design and characterisation of titanium nitride subarrays of kinetic inductance detectors for passive terahertz imaging *Journal of Low Temperature Physics* **193**(3-4) pp 196-202

[6] Kowalski M, Piszczek M, Palka N and Szustakowski M 2012 October Improvement of passive THz camera images. *In Millimetre Wave and Terahertz Sensors and Technology* V (Vol 8544 p 85440N) International Society for Optics and Photonics

[7] Kowalski M, Kastek M, Walczakowski M, Palka N and Szustakowski M 2015 Passive imaging of concealed objects in terahertz and long-wavelength infrared *Applied Optics* **54**(13) pp 3826-3833

[8] Dai L, Zhang Y, Li Y and Wang H 2014 MMW and THz images denoising based on adaptive CBM3D *International Conference On Digital Image Processing*

ISPECE   IOP Publishing

IOP Conf. Series: Journal of Physics: Conf. Series **1187** (2019) 042095   doi:10.1088/1742-6596/1187/4/042095

A Novel Dual-band Video Fusion Algorithm Using Fast Lookup-Tables: Toward Naturalistic Color

Yuanmeng Zhao[1], Weiqi Jin[2] and Lingxue Wang[2]

[1] Department of Physics, Capital Normal University, Beijing 100048, China

[2] School of Optoelectronics, Beijing Institute of Technology, Beijing 100081, China

zhaoym@vip.163.com

Abstract. This paper presents an algorithm for fast dual-band video fusion with good naturalistic color using lookup tables. We establish a color lookup table based on the methods of frame extraction, pseudo-color mapping and color transfer; then we fill up the missing values of the lookup table using the Euclidean distance among its elements. From this lookup table, the luminance value and chromatic values of each pixel in the output frames are retrieved, and consequently the final color fused video is attained. Experiments show that the algorithm is a fast and practical approach to fuse dual-channel videos with stable color appearance and perceived high naturalness.

1. Introduction

The technique of image fusion makes it possible to take advantage from different sensors, not a single sensor. Color fusion images are considered more efficient than gray fusion images in helping people for tasks especially based on human eye observation. This is because human eyes are very much more (hundreds of times) sensitive to the levels of color scale than to the levels of gray scale. To put it another way, human is more likely to remember the details within a color image than within a gray-scale image. Considering humans' physiological characteristics, color images, compared to the gray scale images, have the advantage of enabling better memory of the scenes, faster response to the scenes, and higher impression of the scene by human beings. Many researchers have studied multi-band natural color fusion imaging algorithms. Waxman et al. [1-3] enhanced the in-band image contrast and the inter-band color contrast based on the biological vision characteristics, and continuously proposed many low-light level and infrared image fusion methods. Toet et al. [4-6] put forward many natural color fusion algorithms, such as by using the local minimum operators, by drawing on the photo color cast correction processing method introduced by Reinhard et al. [7], and so on. Toet's algorithms transferred color from natural daytime images to the fusion images with high contrast but less naturalistic colors, and improved fusion images' color naturalness. Other color fusion algorithms based on color transfer were also developed later [8-10].

However, most of the prior color fusion algorithms for attaining naturalistic color requires very large amount of computation, which is not time efficient and not applicable to the real tasks. In recent years we developed the multi-band and the single-band image fusion algorithms to attain perceived naturalistic colors based on color transfer. The purpose of this study is to explore a new method for fast processing of video fusion with enhanced naturalness.

Content from this work may be used under the terms of the Creative Commons Attribution 3.0 licence. Any further distribution of this work must maintain attribution to the author(s) and the title of the work, journal citation and DOI.

Published under licence by IOP Publishing Ltd

2. Dual-Band Image Color Fusion

We captured optical registered infrared image (denoted by I_{IR}) and grayscale television image (denoted by I_{vis}), and then these images were mapped to RGB color space ($I_{IR} \rightarrow R$ channel, $I_{vis} \rightarrow G$ channel, $0 \rightarrow B$ channel), so we got a source image (denoted by I_{source}), as is shown in Figure.1. Figure. 1. showed that I_{source} had low color saturation and poor color hues, negatively affecting target's recognition capability and scene's perceived depth. We transformed I_{IR} and I_{vis} from most commonly used color space RGB to opponent color space $l\alpha\beta$.

Three natural daytime color images with different scenes were chosen as the reference images (denoted by I_{ref}, as is shown in Figure.1(a)~1(c)). In $l\alpha\beta$ space, we transferred color effects from I_{ref} to I_{source} using color space channels' mean and standard deviation which is shown in equation (1):

$$\mu_C = \frac{1}{mn}\sum_{i=0}^{m-1}\sum_{j=0}^{n-1} C(i,j), \qquad \sigma_C = \sqrt{\frac{1}{mn}\sum_{i=0}^{m-1}\sum_{j=0}^{n-1}\left[C(i,j)-\mu_C\right]^2} \tag{1}$$

where C represented l channel, α channel or β channel. m and n denoted the numbers of image rows and columns. Images' origin coordinates were (0, 0). Means and standard deviations of I_{ref}'s three channels were separately transferred to I_{source} so that I_{source} gets similar color effects as I_{ref}, by operating

$$C_{trans} = \mu_C^{ref} + \frac{\sigma_C^{ref}}{\sigma_C^{source}}\left(C_{source} - \mu_C^{source}\right) \tag{2}$$

where C_{trans} represented l, α or β channel of resulting image of color transfer. μ_C^{source}, μ_C^{ref}, σ_C^{source} and σ_C^{ref} denoted means and standard deviations of source image and reference image. After this operation, mean and standard deviation of C_{trans} were the same as those of reference images. However, the operation in $l\alpha\beta$ space led to some loss of image details. I_{trans} was a 24-bit image, but the RGB values of some of the pixels were likely to overflow the range of [0, 255] when doing color space reverse conversion. Therefore, when doing data type conversion, we compressed the values of the high and low ends of each channel while preserving the range of [26, 229]:

$$C''_{trans} = \begin{cases} \text{round}\left(25\left[C'_{trans}-\min(C'_{trans})\right]/\left[25-\min(C'_{trans})\right]\right), & C'_{trans} \in \left[\min(C'_{trans}),25\right] \\ \text{round}\left(230+25\left[C'_{trans}-230\right]/\left[\max(C'_{trans})-230\right]\right), & C'_{trans} \in \left[230,\max(C'_{trans})\right] \end{cases} \tag{3}$$

where C'_{trans} and C''_{trans} were the data of I_{trans}'s R, G or B channel before and after the compression.

Ref images' colors were separately transferred into I_{source} to get the result images (denoted by I_{trans}, as shown in Figure.1(d)~1(f)), which were more colorful than I_{source}. The natural sense and color saturation of result images were remarkably improved.

Because of the individual differences in terms of the physiological and psychological vision, people's color effect evaluations on three result images are not fully consistent. We recruited 8 trained subjects to subjectively evaluate the color naturalness of Figure.1(d)~1(f). The result (see Table 1) showed that most of the subjects (6 out of 8) ranked Figure.1(f) as the best one, and there was no obvious pattern for the ranks of the other two figures (Figure.1(d), Figure.1(e)). This result provided us with the basis for choosing reference images in connection with specific scenes; meanwhile, the diversity of the ranking orders from the subjects showed the need to provide observers with different reference images.

Long wave infrared image (I_{IR})   Grayscale visible image (I_{vis})   Source image (I_{source})

(a)   (b)   (c)

Reference images (I_{ref})

(d)   (e)   (f)

Color transfer results (I_{trans})

Figure 1. Improved pseudo-color fusion image referring to images with natural color

Hardware implementation of video color transfer needs to store only the means and standard deviations of three color-channels (6 scalars in total) of reference image selected by observers not the whole reference image. Therefore, a lot of storage space can be saved. However, each individual source video frame performs division operations in addition to calculating the mean and standard deviation, so the amount of computation is still large.

Table 1. Ranking order of the color naturalness on three color transfer results (Figure.1(d)~1(f))

Observer Number	1	2	3	4	5	6	7	8
Order of color naturalness	f>d>e	f>e>d	d>f>e	f>e>d	f>d>e	f>e>d	d>f>e	f>d>e

3. A Novel Dual-band Video Fusion Algorithm Using Fast Lookup-Tables

3.1. Two-dimensional color look-up table

The color look-up table (LUT) is a commonly used data structure to store and display color images among the multi-media technologies. We colorized dual-band images I_{IR} and I_{vis} by color space mapping and color transfer to obtain a resulting image I_{trans}. Then we established a two-dimensional color look-up table L_0. The method is to set the grayscale values of each pixel in I_{IR} and I_{vis} to be the horizontal and vertical coordinates in L_0 and set the chromatic values (RGB values) of I_{trans}'s pixel, whose locations is the same as the pixels in I_{IR} and I_{vis}, to be the value of L_0 (see Figure.2(a)). The expressions were as follows:

$$\begin{cases} L_0[\mathbf{I}_{IR}(x,y),\mathbf{I}_{vis}(x,y),1]=\mathbf{I}_{trans}(x,y,1) \\ L_0[\mathbf{I}_{IR}(x,y),\mathbf{I}_{vis}(x,y),2]=\mathbf{I}_{trans}(x,y,2) \\ L_0[\mathbf{I}_{IR}(x,y),\mathbf{I}_{vis}(x,y),3]=\mathbf{I}_{trans}(x,y,3) \end{cases} \tag{4}$$

where $x\in\{0, 1, 2, \ldots, m\text{-}1\}$, $y\in\{0, 1, 2, \ldots, n\text{-}1\}$. m and n denoted the numbers of image rows and columns respectively. The value ranges of image I_{IR}, I_{vis}, and each color channels in I_{trans} were all [0, 255]. Accordingly, the color look-up table L_0's horizontal and vertical coordinates' ranges were [0, 255], and L_0 can index up to 65536 colors.

Because the actual dual-channel video frames do not commonly comprise all the grayscale values ([0, 255]), we cannot get a complete LUT. Put it more clearly, on one hand, the sets of grayscale values from I_{IR} and I_{vis} are included in the sets of coordinate values in L_0, namely

$$\left\{\left[\mathbf{I}_{IR}(x,y),\quad \mathbf{I}_{vis}(x,y)\right]\right\}\subseteq\left\{(x_0,y_0)\,|\,x_0,y_0\in\{0,1,2,\cdots,255\}\right\} \tag{5}$$

On the other hand, the data provided by the color fusion image I_{source} cannot cover the whole color look-up table. Therefore, in the look-up table L_0 (see Figure.2(a)), there are a certain number of vertical lines and break points. Therefore, L_0 only provides part of the mapping relationships between grayscale and chromatic values. If this is directly applied to video fusion, part of the pixels of some frames cannot be colorized.

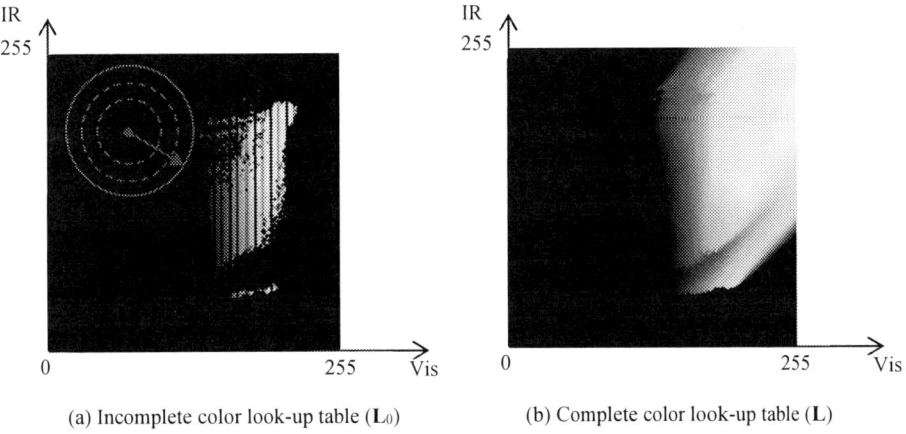

(a) Incomplete color look-up table (L_0)   (b) Complete color look-up table (L)

Figure 2. Set up two-dimensional color look-up table

So, we assigned the chromatic values of the nearest pixels to the pixels without color in L_0, according to the principle that the closer the pixels' Euclidean distances is, the closer the chromatic values will be. Thus, complete color look-up table is established. The specific implementation method is to take the pixel whose R + G + B =0 as center, and draw a smallest circle containing chromatic pixels. Take the mean of chromatic values on the circle as the chromatic values of the center (see Figure.2(a)). After traversing over the whole image of L0, we obtained the complete color look-up table L (see Figure.2(b)).

3.2. Fast algorithm based on two-dimensional color look-up table

The flowchart shown in Figure.3 described the procedure we followed to get the color look-up tables base on dual-band color fusion and color transfer. The fusion effect of an actual scene was shown in Figure.4. In the next, the algorithm processing procedure will be explained via the color transfer procedure of these images.

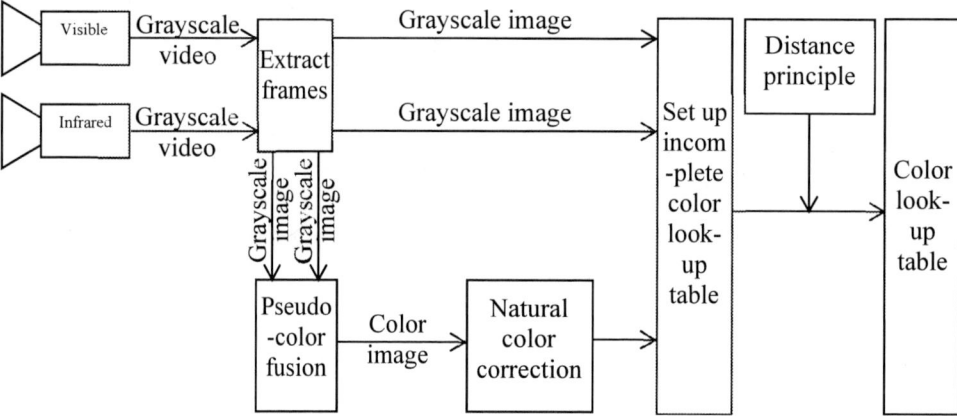

Figure 3. The flowchart for getting color look-up table base on dual-band color fusion and color transfer

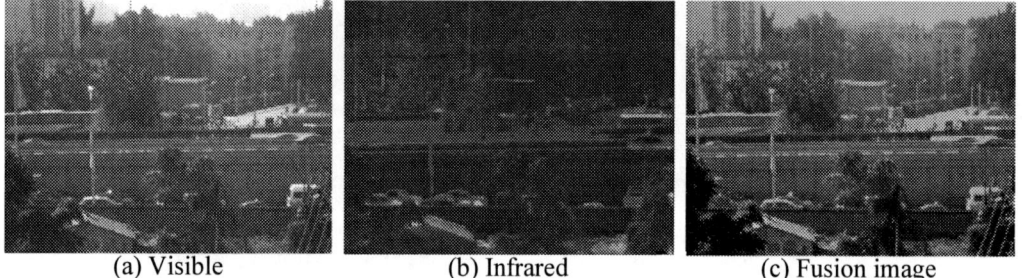

(a) Visible   (b) Infrared   (c) Fusion image

Figure 4. The pseudo-color fusion image derived from single-band grayscale images

Multiple images were used as reference images to transfer color to source image separately, so we obtained different resulting images accordingly. Only the mean and standard deviation of each channel of de-correlated opponent color space lαβ were used when we transferred reference image's color (as table 2), so we can just pre-store these statistics in the system hardware. Thus, a certain number of selected reference images (such as hills and woods, shrubs, deserts, towns, sea and blue sky, white clouds, etc.) don't take up much storage space, guaranteeing the application requirements in different environments. Figure.5 shows 6 result images obtained by transferring the color of 15 reference images in Table 1 to Figure.4(c). As Figure.5 showed, the resulting images have different naturalistic color. According to observer's visual preference, one of them was chosen to establish two-dimensional color look-up table following the methods in the previous section. Figure.6 showed the two-dimensional color look-up table corresponding to Figure.5(d).

Inputted dual-channel videos can be colorized directly with the help of two-dimensional color look-up table. That is to determine corresponding pixel's chromatic values after natural color fusion from indexing the grayscale values, which belong to pixels with the same coordinates of simultaneous frames extracted from dual-channel videos, in two-dimensional color look-up table. Figure.7 showed 4 frames of the actual resulting video transferred. Fusion images have good color stability and avoid the same object's inter-frame color changing with the camera's movement. In traditional methods, camera's movement may cause the mean and standard deviation of source image channels changing. As this two-dimensional color-lookup-table based method saves a lot of procedures for processing each frame, such as color space transformation, $l\alpha\beta$ channels' mean/standard deviation computation, and color transfer operation, the overall computational amount is significantly reduced. This advantage makes it possible to naturally color fuse dual-channel videos in real-time.

Table 2. Reference image data pre-stored in processing board

Reference image number	μ_l	μ_a	μ_β	σ_l	σ_a	σ_β
1	3.0279	-0.1221	-0.0455	0.6649	0.2004	0.0396
2	3.7828	-0.0085	-0.0252	0.3800	0.2805	0.0442
3	2.8324	0.2030	0.0210	0.9783	0.2865	0.0376
4	3.2456	0.1491	-0.0378	0.4725	0.1488	0.0249
5	3.0740	0.2515	0.0504	0.7478	0.2267	0.0447
......			......			
15	3.2625	0.2693	0.0019	0.5952	0.1928	0.0231

Figure 5. Get different color appearances according to a set of reference image

4. Conclusion

This paper proposed an algorithm which establishes complete two-dimensional color look-up tables via color space mapping and color transfer and implements fast processing dual-channel video fusion using the look-up tables. Experiment results showed that this algorithm can significantly improve the speed of image fusion with good naturalistic color, which needs very complicated calculation. The obtained color video had good natural senses, rich color hues, and high color saturation. Meanwhile, the details and features of each video channel were well maintained. The color of look-up table in accordance with certain observer's subjective judgment and selection was of benefit to current observer's perception of scene. This operation mode was in line with the further practical application. The same set of grayscale values inputted to color look-up table will index to the same chromatic values. So, if a certain object's reflection and radiation remain unchanged in different frames, its color will maintain stability, excluding the impact of image sensor's automatic gain. The application of look-up tables provides a new way of thinking for other complex image processing methods.

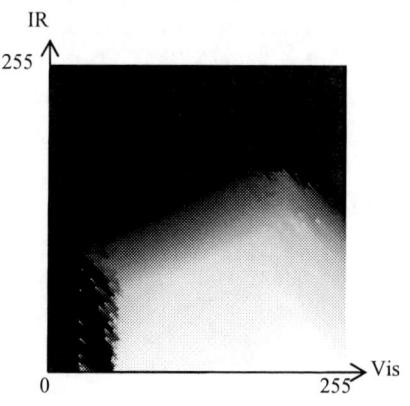

Figure 6. Set up two-dimensional color look-up table according to observer's choice

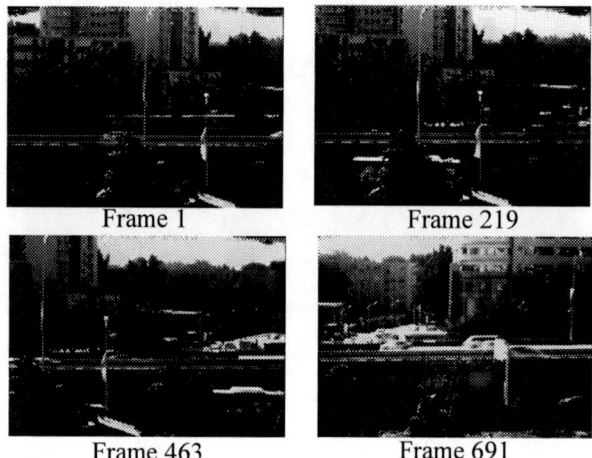

Figure7. Frames of fused video using natural color look-up table

References

[1] A M, Waxman A N, Gove M C, Seibert, et al 1996 Progress on color night vision: visible/IR fusion, perception & search, and low-light CCD imaging *SPIE Enhanced and Synthetic Vision* pp 96-107

[2] A M, Waxman M, Aguilar D A, Fay, et al 1998 Solid-state color night vision: fusion of low-light visible and thermal infrared imagery *Lincoln Laboratory Journal* **11**(1) pp 41-60

[3] Waxman A M , Aguilar M , Baxter R.A, Fay D A , Ireland D B , Racamoto JP and Ross W D 2017 Opponent-Color Fusion of Multi-Sensor Imagery: Visible, IR and SAR *Available online: http://www.dtic.mil/docs/citations/ADA400557* (accessed on 28 August)

[4] Toet A and Walraven J 1996 New false color mapping for image fusion *Optical Engineering* **35**(3) pp 650-658

[5] Toet A and Hogervorst M A 2008 Portable real-time color night vision *SPIE Multisensor, Multisource Information Fusion: Architectures, Algorithms, and Applications* pp 697402-1-697402-12

[6] Maarten H and Toet A 2017 Improved Color Mapping Methods for Multiband Nighttime Image Fusion *Journal of Imaging* 3.3: 36

[7] Reinhard E, Adhikhmin M, Gooch B, et al 2001 Color transfer between images *IEEE Computer Graphics and Applications* **21**(5) pp 34-41

[8] Rabin J, Ferradans S and Papadakis N 2014, October Adaptive color transfer with relaxed optimal transport. *In Image Processing (ICIP) IEEE International Conference* pp 4852-4856

[9] Yu X, Ren J, Chen Q and Sui, X 2014 A false color image fusion method based on multi-resolution color transfer in normalization YCBCR space *Optik-International Journal for Light and Electron Optics* **125**(20) pp 6010-6016

[10] Ancuti C, Ancuti C O, De Vleeschouwer C and Bovik A C 2016 September Night-time dehazing by fusion. *In Image Processing (ICIP), 2016 IEEE International Conference* pp 2256-2260

Terahertz /Visible Dual-band Image Fusion Based on Hybrid Principal Component Analysis

Yuanmeng Zhao[1], Yulong Qiao[1,2], Cunlin Zhang[1], Yuejin Zhao[2] and Hong Wu[2]

[1] Key laboratory of Terahertz Optoelectronics, Ministry of Education of China, Beijing Key Laboratory for Terahertz Spectroscopy and Imaging, Department of Physics, Capital Normal University, Beijing 100048, China

[2] School of Optoelectronics, Beijing Institute of Technology, Beijing 100081, China

zhaoym@vip.163.com

Abstract. This study puts forward a kind of passive terahertz image and visible image fusion method. Our method takes two steps of image fusion. First step involves a pre-processing of the passive terahertz image and image fusion using principal component analysis. Registered terahertz image is segmented and pseudo-color encoded. Taking the characteristics of terahertz image and visible image into account, principal component analysis is used to fuse the dual-band images. HIS method is used for the second step of image fusion. The experiment results show that this algorithm can effectively fuse passive terahertz and visible image and is helpful to detect and locate dangerous objects concealed under the clothes of the target subjects to strengthen the practicability of terahertz imaging system for security screening.

1. Introduction

Terahertz has unique advantages as a security screening technology for its high penetrating capability through plastic, cloths, leather and other materials. In addition, terahertz is found to have relatively very low radiation to human body, and therefore, is suggested to be a no-harm security check technique [1-3]. Recent years have observed a significant amount of research devoted into analyzing security check by using terahertz technique [4,5]. However, terahertz images only give incomplete information about subjects under checked. One major drawback with using terahertz for security check is lacking detailed information about security check scene and the subjects need to be checked. Visible images give these kinds of detailed information but cannot present the hidden objects carried by the subjects under checked. The technique of image fusion makes it possible to take advantage from different sensors, not a single sensor [6-8]. It is needed to fuse these two bands images to advance the terahertz security screening technique. In this paper, we propose a novel dual-band terahertz and visible image fusion method based on hybrid principal component analysis.

2. Terahertz image acquisition

We developed terahertz scanning system using single terahertz detector and this system adopts total reflection working mode to reduce terahertz radiation loss after penetrating through terahertz lenses. Figure 1(a) presents the imaging system. The system is composed of five parts, including the subjects to be checked, the frame scan plane mirror, trihedral focusing scanning mirror, reflecting mirror, and the terahertz imaging front-end. Best imaging distance of the system is 1.7m. The system has imaging

---

Content from this work may be used under the terms of the Creative Commons Attribution 3.0 licence. Any further distribution of this work must maintain attribution to the author(s) and the title of the work, journal citation and DOI.

Published under licence by IOP Publishing Ltd

speed as fast as 2s, and its best resolving ability is 4cm@0.1THz. Figure 1(b) shows the image it captures.

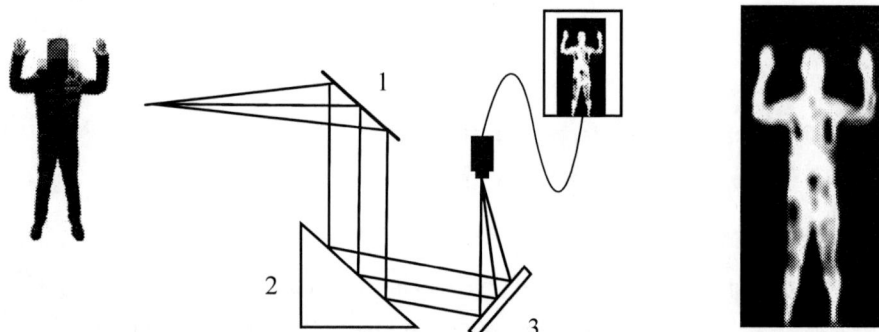

(a) terahertz imaging system 1. the frame scanning plane mirror 2. focusing scanning mirror 3. reflecting mirror 4. imaging front-end   (b) terahertz image

Figure 1. terahertz imaging system and the image captured

3. Passive Terahertz/Visible image registration and pre-processing

Image registration is the condition for image fusion. Because passive terahertz imaging system and the visible imaging system gather images at different frame rate, using different equipments, at different angles, therefore, one of the two bands of images must be transformed in order to register dual-band images. It is challenging to extract characteristic points of passive terahertz image. This study adopts mutual information as the similarity measure. Genetic algorithm combined with Powell algorithm yield to the hybrid algorithm which is adopted in this study for dual-band image registration.

In particularly, adopting mutual information means that the algorithm uses the statistical correlation between two band image grey scale values. This approach does not require extracting characteristic points and has its advantage of high precision degree, but the disadvantage is it will be very likely to arrive at local extremum. GA's advantage is in its searching capability, but low precision. Powell algorithm can reach high precision but is very strict at selecting the initial points. Therefore, we combine these three approaches to have good precision, and global searching capability.

In the GA algorithm, initial population is 100, selection operator is Roulette selection method, variation function is Gaussian function. Stagnation band number is 100. Before the image registration, to easy the calculation of the mutual information between the two-band images, we pre-processed the visible image by using morphological processing. Figure 2(a) shows the original visible image, figure 2(b) shows the image after morphological processing. Figure 2(c) shows the original passive terahertz image, figure 2(d) shows the image after GA registration, and Figure 2(e) gives the image after processed with the hybrid method. The mutual information of Figure 2(c), Figure 2(d), and Figure 2(e) is 0.2076、0.5793、0.9908. Apparently, images processed using hybrid approach have the best quality.

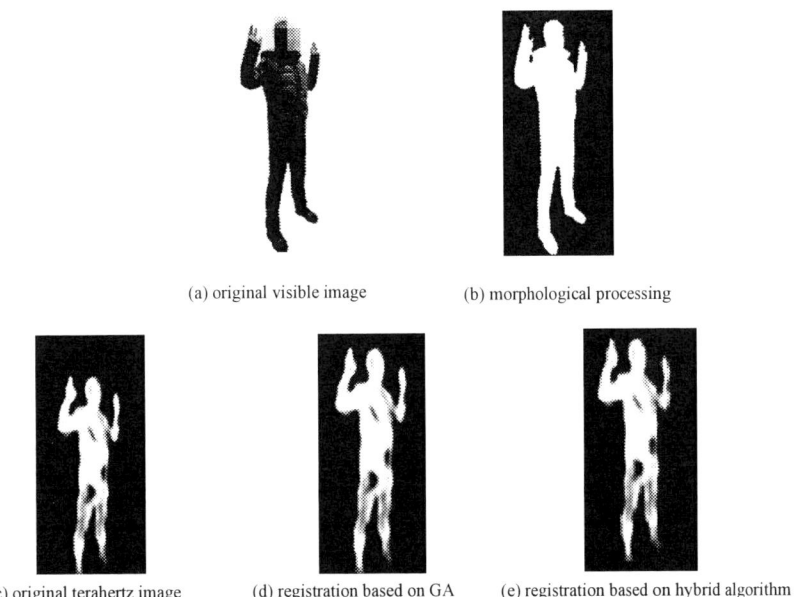

(a) original visible image   (b) morphological processing

(c) original terahertz image   (d) registration based on GA   (e) registration based on hybrid algorithm

Figure 2. visible image and terahertz image after registration

With the objective of making the targets more identifiable, we pre-processed the terahertz image after image registration. We took three steps. The first step was to perform image segmentation using region-growing method. Secondly, we assign values to the target region using the normalized mutual correlation algorithm. Lastly, we enhanced the color of the image by using pseudo-color method. Figure 3 shows the experimental results.

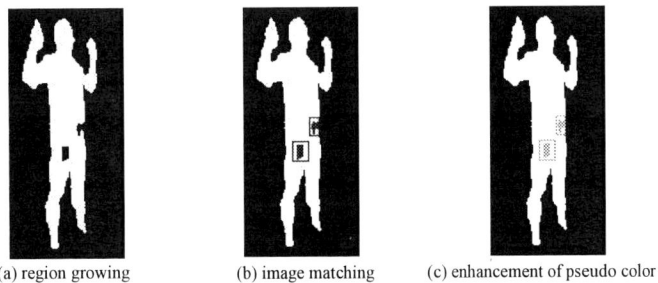

(a) region growing   (b) image matching   (c) enhancement of pseudo color

Figure 3. pre-processing

4. Passive Terahertz/Visible dual-band image fusion

PCA image fusion method has spatial information preserved from image with high spatial resolution and spectral information from image with high spectral resolution. We analysed the image captured through our system. Visible image has high spatial resolution and big volume of spectral information. Passive terahertz image's resolution is low, and the image is grayscale. Fusing these dual-band images can achieve both of the target identification and the high spectral information and high resolution. The following describes the steps of the PCA image fusion method: 1) compute the principle component transformation matrix and the respective feature vectors of the visible image; 2) compute principle component using equation (1):

$$pc_k = \sum_{j=1}^{n} x_j \varphi_{jk} \tag{1}$$

3) perform the histogram matching between the terahertz image and visible image's first principle component and replace first principle component with terahertz image. Then we perform the inverse principle component transformation with the replaced image to get the fused image. Figure 4 shows the fused resulting image by using PCA method.

By only using PCA methods to perform image fusion, we can obtain image with high resolution but lost spectral information. Therefore, we combine HIS transformation with the PCA method by following these steps. 1) compute the visible image's principle component transformation matrix and its according feature vector, and then we perform the histogram matching between the terahertz image and first principle component of visible image. The first principle component was replaced by the terahertz image and after performing the inverse principle component transformation on the replaced image, we get the PCA fused image. 2) we convert the R, G, B three channels of visible image to the HIS space and get I, H, S channels. Then we use histogram matching to guide the image fusion. We fused the PCA fused image and the I channel of visible image to replace the luminance component. We put together H, S channels and finally get the fused image by HIS inverse transformation back to RGB space. Figure 5 shows the image fusion results using our hybrid principal component analysis method. Results shows that fused image has both of the visible image's high resolution and spectral information and the passive terahertz image's identifiable targets.

Figure 4. fused image based on PCA method   Figure 5. fused image based on hybrid method

Table 1. objective evaluation of terahertz/visible image fusion

	spatial frequency	standard deviation	entropy	average gradient	correlation	distortion degree
Visible image	31.0693	94.0404	3.1557	7.5264		
Terahertz image	55.7501	111.9837	0.8282	8.9718		
PCA fused image	32.1748	93.6275	3.1349	7.9654	0.9876	5.5203
HIS fused image	32.9408	93.0246	3.1015	7.6221	0.9863	5.8334
hybrid method fused image	32.9889	93.0239	3.2404	8.0923	0.9991	1.1565

5. Method Objective Evaluation

We rely on spectral and spatial information to evaluate the quality of our proposed image fusion method. Specifically, we used various statistics parameters to evaluate the image. Details of spatial information, such as spatial frequency, standard deviation, entropy, and average gradient, were used.

Meanwhile, details of spectral information, such as correlation and distortion degree. Table 1 shows the results. Results showed our method outperformed the other methods.

6. Conclusion

In this study, we proposed a novel method for visible and passive terahertz images fusion using PCA method combined with HIS method. The improved method can quickly locate the targets carried by human objects from the background scene. This method has the advantage of preserving the visible images' high resolution and spectral information. In order to speed up the security check speed and reduce the errors of the target identification, future study can continuously improve the correctness and efficiency of the image fusion method.

Acknowledgments

Supported by the National Natural Science Foundation of China (61875140) and Beijing Natural Science Foundation (4181001).

References

[1] Hu B B and Nuss M C 1995 Imaging with terahertz waves *Optics Letters* **20**(16) pp 1716-1718
[2] Lee Y S 2009 Principles of terahertz science and technology *Springer Science & Business Media* **170**
[3] Hangyo M 2015 Development and future prospects of terahertz technology *Japanese Journal of Applied Physics* **54**(12)
[4] Federici J F, Gary D, Barat R and Zimdars D 2005 May THz standoff detection and imaging of explosives and weapons *In Optics and Photonics in Global Homeland Security International Society for Optics and Photonics* **5781** pp 75-85
[5] Knipper R, Brahm A, Heinz E, May T, Notni G, Meyer H G, Popp J, et al 2015 THz absorption in fabric and its impact on body scanning for security application *IEEE Transactions on Terahertz Science and Technology* **5**(6) pp 999-1004
[6] Tu T, Su S, Shyu H and Huang P S 2001 A new look at IHS-like image fusion methods *Information Fusion* **2**(3) pp 177-186
[7] Wang Z, Ziou D, Armenakis C, Li D and Li Q 2005 A comparative analysis of image fusion methods *IEEE Transactions on Geoscience and Remote Sensing* **43**(6) pp 1391-1402
[8] Zhang Q, Liu Y, Blum R S, Han J and Tao D 2018 Sparse representation based multi-sensor image fusion for multi-focus and multi-modality images: A review *Information Fusion* **40** pp 57-75

ISPECE   IOP Publishing

A Camouflage Effect Detection Model for Fixed Targets

YANG Xin[1], XU Weidong[1]*, XIANG Lei[2], ZHU Wannian[3], TIAN Jiyao[1]

[1]Key Laboratory of Science and Technology for National Defense, Amy Engineering University, Nanjing, Jiangsu, 210007, China

[2]61912, Beijing, 100089, China

[3]Teaching and Research Office of Camouflage in Training Center, Army Engineering University, Xuzhou, Jiangsu, 221004, China

Corresponding Author: XU Weidong; email: 1435227062@qq.com

Abstract: The traditional camouflage effect detection mainly implements the evaluation process for a single image, and cannot effectively reflect the statistical characteristics of the target. In order to better simulate the dynamic interpretation process of reconnaissance personnel on the target, a dynamic and statistically characteristic camouflage effect evaluation model is proposed for the problem of camouflage effect detection of fixed targets. Combined with the Mean shift target tracking technology, the model statistically correlates the target with the background eight-link domain and establishes a normalized joint Gaussian distribution. The target's camouflage effect is then evaluated using the distribution of probability density. The experiment performs complete camouflage, partial camouflage and non-disguise on the exit target of a simulated cavern, and calculates the logarithmic amplification probability and statistics of the curve after collecting the data. According to the 3σ criterion, the mean value is compared with a preset threshold value, and corresponds to the original camouflage state, the better camouflage state, and the invalid camouflage state, respectively. The experimental results show that the model can clearly distinguish the different camouflage states.

1. Introduction

In engineering practice, the camouflage problem of a fixed type of target is relatively common. Compared to general objectives, fixed targets have long-term and fixed characteristics, such as defense engineering, cave mouths, and work objectives. Fixed targets are usually located in remote areas. Personnel and ground detection equipment are difficult to reach. Usually, the data is obtained by using drones or satellite reconnaissance means. Therefore, the acquired data is redundant and diverse. For the problem of camouflage detection, the traditional camouflage effect detection application personnel interpret the discovery probability of the target, which requires a lot of manpower and material resources[1]. In recent years, some scholars have proposed many methods for detecting camouflage effects. Some neural network processing based on image features, some based on the establishment of psychology-based stimulation functions, and some based on the distance of background spots[2-5]. These methods can quickly and objectively evaluate the camouflage effect of the target through a large number of calculations. However, the actual reconnaissance process is often dynamic, multi-view joint and statistical, and these methods for evaluating a single image cannot effectively solve this problem.

The research ideas in this paper are as follows. Aiming at the problem of camouflage effect evaluation of fixed targets, and using its characteristics reasonably, a fixed target camouflage effect detection model is established, which can effectively overcome the shortcomings of traditional evaluation methods. The

---

Content from this work may be used under the terms of the Creative Commons Attribution 3.0 licence. Any further distribution of this work must maintain attribution to the author(s) and the title of the work, journal citation and DOI.

Published under licence by IOP Publishing Ltd

model adopts the anomaly detection model in machine learning. By quantifying the target feature data and the feature parameters in the disguised state, the camouflage state evaluation process of the target is realized.

2. Model Establishment

2.1. Target and background area division
The eight-way domain method is used to divide the target and background. The method takes the target in the image as a template, and the eight-way domain with the target center as the origin as the direct background of the target. Considering only the background near the target area, not all the image areas are classified as background areas, which is in line with the law of the human eye to read the camouflage effect. Regardless of the proportion of the target in the image during imaging, the target only melts into the background near it, regardless of the background of the far region. In addition, the segmentation method in citation 6 is irregular, and such segmentation requires a large amount of manual segmentation work, which is also disadvantageous for computer processing. So this paper applies the method of rectangular segmentation, and divides the target and background with the smallest adjacency rectangle of the target, as shown in Figure 1. The rectangular segmentation divides a part of the background into the target area. In the state of good camouflage, the target and the background are well integrated, and can be ignored.

Back 1	Back 2	Back 3
Back 8	Target	Back 4
Back7	Back 6	Back 5

Figure 1. Eight-way domain partition map of target and background area

Use $A_{ij}^{(k)}$ to represent continuous image data frames. Where k represents the number of frames, and i, j respectively determine the coordinate position of the pixel. The background area number is numbered 1-8 in the order from the upper left corner and the counterclockwise rotation. The target area of the kth frame is represented by $A_0^{(k)}$, $A_1^{(k)}$, $A_2^{(k)}$, ..., $A_8^{(k)}$ respectively represent 1-8 of the kth frame. Let a, b, c, and d be the target segmentation template abscissa, ordinate, length and width, respectively, then the target area is expressed as:

$$A_0^{(k)} = \left\{ A_{i_0 j_0}^{(k)} \mid a \leq i_0 \leq a + c, b \leq j_0 \leq b + d \ i_0, j_0 \in Z \right\} \tag{1}$$

Similarly, the eight-way background area can be represented by a similar formula, and each background area has a clear relationship with the target area. The background area No. 3 of the kth frame is expressed as formula 2:

$$A_3^{(k)} = \left\{ A_{i_3 j_3}^{(k)} \mid a + c \leq i_3 \leq a + 2c, b \leq j_3 \leq b + d \ i_3, j_3 \in Z \right\} \tag{2}$$

It may be easy to indicate that the horizontal and vertical coordinates of the target area between the adjacent two frames are shifted to the right by Δa and Δb.

$$i^k = i^{k+1} + \Delta a \tag{3}$$
$$j^k = j^{k+1} + \Delta b \tag{4}$$

Therefore, the numerical relationship between each target and the background between successive image data frames can be clearly calculated by the above formula 3 and formula 4.

2.2. Target Tracking
Knowing the target coordinate position of each frame is a prerequisite for calculating the similarity

feature data. The target area $A_0^{(1)}$ of the start frame can be obtained by manual labeling or global query matching (in the case of a known target structure). If the rest of the frames are acquired in the same way, the work efficiency will be greatly reduced. Therefore, it is necessary to use the target tracking algorithm to obtain the values of Δa and Δb. Target tracking algorithms can be divided into four categories: active contour based tracking, feature based tracking, region based tracking, and model based tracking [7]. Among them, the feature-based tracking algorithm has an advantage of being insensitive to changes in scale, deformation, and brightness of moving objects[8]. This paper selects the Mean shift tracking algorithm based on the gray histogram feature. The algorithm was proposed by Fukunaga in 1975 [9], which uses the non-parametric estimation of density gradient to achieve fast tracking of moving target regions.

The target area has n pixels ($n = i_0 \times j_0$), its absolute position is represented by z_i, and z_i^* indicates its relative position, then the target area model q_u is:

$$q_u = C \sum_{i=1}^{n} K_E(\|z_i^*\|^2)\, \delta[b(z_i - u)] \tag{5}$$

The δ and b functions determine whether the color value at z_i belongs to u. The normalized parameters C, δ function and z_i^* are calculated as follows:

$$C = 1/\sum_{i=1}^{n} K_E(\|z_i^*\|^2) \tag{6}$$

$$z_i^* = \left(\frac{(x_i - x_0)^2 + (y_i - y_0)^2}{x_0^2 + y_0^2}\right)^{0.5} \tag{7}$$

$$\delta(x) = \begin{cases} 1 & x = 0 \\ 0 & x \neq 0 \end{cases} \tag{8}$$

When establishing the target model, the Epanechikov kernel function is used, which is expressed as follows:

$$K_E(x) = \begin{cases} c(1 - \|x\|^2) & \|x\| < 1 \\ 0 & others \end{cases} \tag{9}$$

When the k-1 frame is set, the center area where the target is located is f_0, the candidate target center area is f, and the probability of the candidate area is:

$$p_u(f) = C \sum_{i=1}^{n} K_E\left(\left\|\frac{f - z_i}{h}\right\|^2\right) \delta[b(z_i - u)] \tag{10}$$

Where h is the kernel function window size.

In this paper, the Bhattacharyya coefficient is used to establish the probability density function in the similarity measure, which is described as formula 11:

$$\rho(p, q) = \sum_{u=1}^{m} \sqrt{p_u(f) q_u} \tag{11}$$

After the Taylor expansion is performed on the probability density function, the approximate expression can be obtained:

$$\rho(p, q) \approx \frac{1}{2} \sum_{u=1}^{m} \sqrt{p_u(f_0) q_u} + \frac{C}{2} \sum_{i=1}^{n} w_i K_E\left(\left\|\frac{f - z_i}{h}\right\|^2\right) \tag{12}$$

$$w_i = \frac{1}{2} \sum_{u=1}^{m} \sqrt{\frac{q^u}{p_u(f)}} \delta[b(z_i) - u] \tag{13}$$

Deriving the formula 12, we can get it after finishing:

$$f_{t+1} = f_t + \frac{\sum_{i=1}^{n} w_i(f_t - z_i) K_E\left(\left\|\frac{f_t - z_i}{h}\right\|^2\right)}{\sum_{i=1}^{n} w_i K_E\left(\left\|\frac{f_t - z_i}{h}\right\|^2\right)} \tag{14}$$

Finally, the iterative process ends by limiting the number of iterations or setting an iteration change threshold. The coordinate change value can be calculated as shown in formula 15:

$$\Delta f = f_{end} - f_0 = (\Delta a, \Delta b) \tag{15}$$

2.3. Feature Extraction

The feature extraction in the existing camouflage effect analysis model is mainly considered from the aspects of image color, texture, and spot shape. These features are selected from the perspective of human eye observation. However, the calculation of these features does not have a good analytical theory to guide the evaluation of camouflage effects. Based on this, this paper considers the selection of features from the following three aspects. The first aspect is to look at the pixel space of the target and background, rather than the feature space of the image. The second aspect uses prior information to evaluate existing features, thereby avoiding the extraction of complex image information. In the third aspect, in order to avoid losing the structural information on the space and increasing the calculation speed, the rotation transformation and the contraction transformation of the image are not considered, otherwise the calculation amount will be greatly increased, and the real-time requirement cannot be achieved.

Considering the relationship between image features and camouflage effects, the following six image features are selected to characterize the similarity of eight backgrounds and targets.

- H-histogram feature relationship in HSV space:

$$\rho_1 = \frac{H \cdot H'^T}{\|H\|\|H'\|} \tag{16}$$

- S-histogram feature relationship in HSV space:

$$\rho_2 = \frac{S \cdot S'^T}{\|S\|\|S'\|} \tag{17}$$

- Image gray histogram feature relationship:

$$\rho_3 = \frac{G \cdot G'^T}{\|G\|\|G'\|} \tag{18}$$

- Peak signal to noise ratio characteristic relationship[10-12]:

$$\rho_4 = 10 log_{10} \left[\frac{(2^n-1)^2}{MSE} \right] \tag{19}$$

- Perceived hash feature relationships. To make the calculation faster, after reducing the image to an 8×8 gray matrix, calculate[13-15]:

$$\mu_x = \frac{1}{64} \sum_{i=1}^{64} x_i \tag{20}$$

$$\mu_y = \frac{1}{64} \sum_{i=1}^{64} y_i \tag{21}$$

$$\rho_5 = \sum_{i=1}^{64} b(x_i, \mu_x) \oplus b(y_i, \mu_y) \tag{22}$$

Where $\oplus$ denotes an exclusive OR operation and function b is calculated as follows:

$$b(x,y) = \begin{cases} 1 & x \geq y \\ 0 & y \geq x \end{cases} \tag{23}$$

- Autocorrelation model feature relationship[16,17]:

$$\rho_6 = \frac{\sum_m \sum_n (A_{mn} - \bar{A})(B_{mn} - \bar{B})}{\sqrt{\sum_m \sum_n (A_{mn} - \bar{A})^2 \sum_m \sum_n (B_{mn} - \bar{B})^2}} \tag{24}$$

2.4. Parameter Estimation and Effect Evaluation

It can be known from Lyapunov's theorem that the superposition of multiple random variables approaches a normal distribution. The distribution of image feature samples is affected by various complex factors and can be approximated as obeying a normal distribution. Therefore, as long as the parameters of the normal distribution are estimated, the feature distribution parameters in the case where the entire camouflage state is good can be obtained. The parameters are estimated using a first-order estimation method. Let the sample feature set be $P = \{p_u^f\}$, where u=1,...,m, f=0,...,8. p_u^f represents the fth background of the uth frame data in the training sample. When f=0, it represents the target. Then the feature representation of the sample can be represented by a three-dimensional matrix A:

$$A = \{a_{ijk}\} \tag{25}$$

$$a_{ijk} = \rho_i(p_k^0, p_k^j) \tag{26}$$

Therefore, the parameters of the sample distribution can be estimated:

$$\mu_{ij} = \frac{1}{m} \sum_{k=1}^{m} a_{ijk} \tag{27}$$

$$\sigma_{ij} = \frac{1}{m} \sum_{k=1}^{m} (a_{ijk} - \mu_{ij})^2 \tag{28}$$

According to the above model, the flow chart of the feature parameter extraction of the algorithm is shown in Figure 2. After reading the training data, the target area is selected to obtain the background eight-way domain (if the background is in the edge area, the complete eight-way domain cannot be obtained, then it is lost.). Calculate the feature to obtain the mean variance while tracking and establish a normalized joint distribution. The algorithm flow chart of the effect evaluation process is shown in Figure 4. The data to be detected is read and the eight-way domain is acquired, and the feature is substituted into the normalized joint distribution to calculate the joint probability density. Since the value fluctuation range is $0\sim6.94\times10^{-20}$, it is not conducive to the setting of the threshold. In order to enhance the sensitivity of the density value, the result is first magnified 10 times and then the logarithm is obtained. The process is as follows:

$$r = \sum_{j=1}^{8}\sum_{i=1}^{6} ln\ 10p_N\left(\frac{a_{ij}^{P_x}-\mu_{ij}}{\sigma_{ij}}\right) \tag{29}$$

By the logarithmic amplification probability limit, the value varies from $-\infty\sim66.41$. It can be seen that the range of variation of the data is significantly larger, which can effectively reduce the calculation error. In order to effectively distinguish different camouflage effect states, and considering the 3σ criterion of the Gaussian distribution, the result is divided into three-level camouflage states. They are the original camouflage state ($r\geq0$), the better camouflage state ($0>r\geq-1000$), and the failed camouflage state ($r<-1000$).

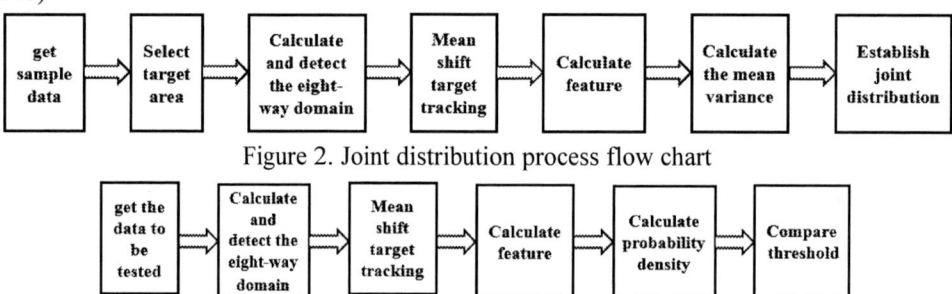

Figure 2. Joint distribution process flow chart

Figure 3. Flow chart of camouflage effect evaluation process

3. Experiment and Result Analysis

The experimental process was carried out in a southern suburb of Nanjing in 2018, and continuous aerial imaging acquisition was performed on a certain simulate export target. The altitude of the aircraft is about 50 meters, and the flight conditions are selected in the morning, noon, afternoon, sunny, rainy days and other time periods. The data collected after the masquerading is completed is used as a training sample, as shown in Figure 4. The image in the full camouflage state is used as the test data of the original camouflage state; the image in the partial camouflage state is used as the test data of the better camouflage state; the image in the non-camouflage state is used as the test data of the failed camouflage state, and the three types of data are separately collected 25 frames.

    (a) Completely camouflage   (b) Partially camouflage   (c) Not camouflage

Figure 4. Three camouflage states in the experimental data

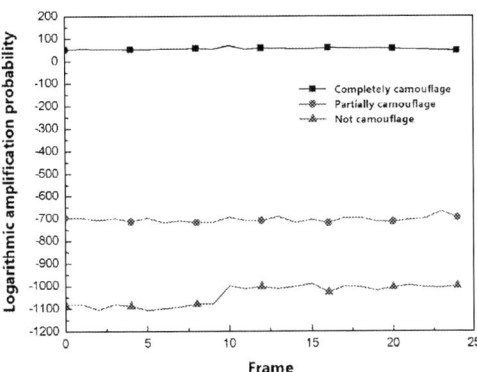

Figure 5. Probability density plot of three camouflage states

The probability density graph shown in Figure 5 is calculated by separately calculating the three types of data to be detected shown in Figure 4. The abscissa represents the number of frames of the data to be detected, and the ordinate represents the logarithmic amplification probability density value. Mathematical statistics on these three curves yield the results shown in Table 1. It can be concluded from the figure and the table that in the state of complete camouflage, the mean value of the curve is 58.2364, which is in accordance with the original camouflage state; in the state of partial camouflage, the mean value of the curve is -727.6583, which is in accordance with the better camouflage state; in the state of not camouflage, the curve The average value is -1005.2298, which is in compliance with the failure camouflage state. The mean data of the curve reflects its camouflage state very well. The fluctuation of the curve is relatively stable, and the variance and the extreme difference are relatively stable, indicating that the model can run smoothly during the dynamic detection process. From the full camouflage state to the partial camouflage state, the arrangement orientation of the camouflage net is changed and the ornaments arranged above are removed, but the curve is already close to the threshold of the failed camouflage state. This indicates that the sample data is not sufficiently expressed for the data space, and the next step should be to supplement the sample data reasonably. Overall, the model provides an accurate reflection of changes in the target camouflage state.

Table 1. Statistical data of three camouflage state probability density curves

Three camouflage states	Statistical data		
	Mean	Variance	range
Completely camou-flage	58.2364	5.8953	21.5698
Partially camouflage	-727.6583	6.9555	29.0336
Not camouflage	-1005.2298	8.1569	22.8956

4. Conclusion

Based on the target tracking and anomaly detection techniques, this paper proposes a dynamic camouflage effect evaluation method. The model can extract features from the well-prepared state to establish a normalized joint Gaussian distribution, and evaluate the target from multiple frames of images and multiple angles. Different from the static evaluation method of single image, the model establishment process uses the feature data when the camouflage state is good, which can reflect the target state more objectively. Six correlation characteristics were selected. From the histogram and the mean variance table of the experimental data, it can be concluded that when the camouflage state is stable, the data dispersion range is relatively regular, which basically conforms to the trend of Gaussian distribution. This shows that using the Gaussian distribution to simulate the variation range of the camouflage state, it can well handle the influence of factors such as illumination, weather, and shooting angle. By increasing the number of samples, the distribution of feature data can be more comprehensively and correctly reflected. In the experimental process, a certain simulate export target was completely camouflaged, partially camouflaged and not camouflaged. The aerial imaging acquired multi-frame data and calculated its logarithmic amplification probability density and the mean value of the statistical curve. The results were 58.2364, -727.6583 and -1005.2298, respectively. Compared with the set threshold range, it corresponds to the original camouflage state, the better camouflage state and the invalid camouflage state. The experimental results show that the model is effective and the calculation results can better reflect the camouflage state of the target.

References

[1] XU Wei-dong, WANG Xiang-wei. (2015) Camouflage detection and evaluation theory and technology. Beijing: National Defense University Press.

[2] LU Xu-liang, LIN Wei, etc. (2005) Applying the Fuzzy Clustering Analysis of ISODATA to the Classification of Camouflage Effectiveness. Acta Armamentarii, 26(05):681-684.

[3] LIN Wei, CHEN Yu-hua, etc. (2013) Camouflage Assessment Method Based on Image Features and Psychological Perception Quantity. Acta Armamentarii, 34(04):412-417.

[4] YU Jin, ZHU Li-fan, etc. (2009) Evaluation Model of Optical Camouflage Effect Based on BP Neural Network. Shipboard Electronic Countermeasure, 32(06):55-57.

[5] Jianfei Qin, Liyong Qu, etc. (2016) Optical Camouflage Effect Objective Evaluation Method Research Under The Condition of Complex Backgrounds. MATEC Web of Conference, 61(10):1-4.

[6] CUI Bao-sheng, XUE Shi-qiang, etc. (2010) Camouflage effectiveness evaluation based on image feature. Infrared and Laser Engineering, 39(06):1178-1183.

[7] X. L. Lv, Q. Jia, etc. (2016) Research on Camouflage Assessment Psychological Impact of Different Pattern Paintings Based on The Identification Probability Model. Basic & Clinical Pharmacology & Toxicology, 118(36):30-40.

[8] Haritaoglu I, Harwood D, LS David. (2013) Real-time Surveillance of People and Their Activities. IEEE Trans. on PAMI, 22(8):809-830.

[9] K. FUKUNAGE, L. D. HOSTETLER. (1975) The estimation of the gradient of a density function with application in pattern recognition. IEEE Trans. on Information Theory, 21(1):32-40.

[10] Chiuhsiang Joe Lin, Chi-Chan Chang, etc. (2014) Developing and Evaluating a Target-Background Similarity Metric for Camouflage Detection. Plot One, 9(2):1-1.

[11] Xue Feng, Wu Fan, etc. (2018) Camouflage texture design based on its camouflage performance evaluation. Neurocomputing, 274(24):106-114.

[12] YANG Jun-tang, XU Wei-dong, etc. (2017) A Surendra-based Improved Detection Method of Moving Target Camouflage Effect. Acta Armamentarii, 38(01):190-194.

[13] Xu Li. (2013) Moving object detection using LAB color space. Journal of Huangzhong University of Science and Technology, 41(1):220-224.

[14] Pan B, Xie H M, Wang Z Y. (2012) Equivalence of digital image correlation criteria for pattern matching. Applied Optics, 49(28):234-238.

[15] Li J, Dang J W, Bu F. (2014) Analysis and improvement on bacterial foraging optimization algorithm. Journal of Computer Science and Engineering, 3(1):1-7.

[16] Cao X, Wang H, Shi Z. (2015) The research of image matching algorithm in visual inspection system. Journal of Electronic Technology, 4(2):91-94.

[17] Sun Q, Zhou X H. (2010) Application of Matlab in camouflage effectiveness evaluation of protective construction. Computer & Digital Engineering, 4(7):134-138.

Welding defect signal extraction technology based on OMP algorithm

Qi Ailing, Lei Haijun

School of Computer Science and Technology, Xi'an University of Science and Technology, Xi'an, China

1302077899@qq.com, 2500091461@qq.com

Abstract. Flip-chip technology has been rapidly developed and widely used in the field of microelectronic packaging, and defect detection has also received increasing attention. In the experiment, ultrasonic testing was used to detect defects on flip chip. Aiming at the problem that the interference noise of the ultrasonic detection defect signal seriously affects the location of the defects and the signal extraction, we use the algorithm based on orthogonal matching pursuit (OMP) to extract the defect signal and adopt the Gabor atom library which is optimally matched with the ultrasonic signal to achieve matching the ultrasonic echo signals adaptively and greatly reducing the complexity of the sparse decomposition algorithm. The simulated and actual ultrasonic defect signals were tested separately and compared with the matching pursuit algorithm. The result shows that OMP can extract defect signals more effectively in the noise background.

1. Introduction

As a fast-developing and widely used microelectronic packaging technology, flip-chip soldering technology has many advantages such as high alignment precision, short interconnect line, high input and output density, etc. It is an important means to reduce package size and increase package density[1-2]. When ultrasonic testing is used to detect welding defects, the system is usually accompanied by interference noise, which will pollute the signal ultrasound. Even the signal will be annihilated which will bring difficulties to the subsequent processing of the signal and defect identification under severe cases. Therefore, the issues of signal extraction and noise suppression are important.

In 1993, Mallat and Zhang proposed an idea of sparse decomposition based on overcomplete dictionary[3], by which the signal is represented as a linear combination of several atoms in an overcomplete dictionary to make the signal characteristics more precise and achieve the purpose of signal extraction. This idea has been greatly developed in various fields of signal processing, such as signal processing[4-6], compression[7] and feature extraction[8]. Among many sparse decomposition algorithms, the MP algorithm is almost the fastest, but because of the larger size of atomic library, the computational cost in searching for atoms is too large.

In this paper, OMP algorithm which is matching pursuit introducing orthogonalization is used to extract the welding defect signal. It can accelerate the convergence speed, improve the performance of the algorithm, and effectively extract the defect signal.

Content from this work may be used under the terms of the Creative Commons Attribution 3.0 licence. Any further distribution of this work must maintain attribution to the author(s) and the title of the work, journal citation and DOI.

Published under licence by IOP Publishing Ltd

2. Matching pursuit algorithm

The idea of the matching pursuit algorithm is that it thinks the input signal has a certain correlation with the atom in the dictionary library. This correlation is represented by the inner product of the signal and the atom in the library, that is, the larger the inner product, the greater the correlation of this atom and the atom in the library, thus the atom can be used to approximate this signal.

The iterative calculation is used to find the atom (also called the best atom) that best matches the original signal from the overcomplete dictionary, and the signal is sparsely decomposed. The basic process is:

1）Calculate the inner product of the signal y and each column (atoms) in the dictionary matrix, and select the one with the largest absolute value, it is the best atom. To meet the condition

$$| < y, x_{r_0} > | = \max_{i \in (1,...,k)} | < y, x_i > | \qquad (1)$$

r_0 represents the column index of a dictionary matrix.

2）Signal y is decomposed into the vertical projection component and the residual value of the most matching atom x_{r_0}:

$$y = < y, x_{r0} > x_{r0} + R_1 f. \qquad (2)$$

3）For residual value $R_1 f$, the same decomposition is did in step 1), then step K can be obtained:

$$R_k f = < R_k f, x_{r_{k+1}} > x_{r_{k+1}} + R_{k+1} f. \qquad (3)$$

$x_{r_{k+1}}$ satisfies:

$$| < R_k f, x_{r_{k+1}} > | = \max_{i \in (1,...,k)} | < R_k f, x_i > |. \qquad (4)$$

4）After the K-step decomposition, the signal y is decomposed into:

$$y = \sum_{n=0}^{k} < R_n f, x_{r_n} > R_n f + R_{k+1} f. \qquad (5)$$

among them $R_0 f = y$.

3. Denoising of welding defect signals based on OMP algorithm

3.1. Principle of OMP algorithm

The OMP algorithm[9] is optimized on the basis of the MP algorithm. The atom selection method is unchanged, and orthogonalization processing is added to the selected atoms in the decomposition process. For the atom x_i selected in formula (4), using Gram-Schmidt orthogonalization method:

$$U_{k+1} = x_{r_{k+1}} - \sum_{m=0}^{k} < x_{r_{k+1}}, U_m > U_m. \qquad (6)$$

This makes the OMP algorithm converge faster than the MP algorithm with the same accuracy requirements.

3.2. Principle of OMP algorithm

A noisy signal is a signal synthesized by a noiseless (original) signal and noise, noiseless signal is considered to be sparse, that is, it can be represented by a finite number of atoms. The noise is random and non-sparse, that is, it cannot be represented by a finite number of atoms. So the sparse components of the signal can be extracted by noisy signal, and used to reconstruct the signal. In this process, the noise is processed as the residual between the noisy signal and the reconstructed signal. The residuals are discarded during the process to achieve the denoising effect.

In order to get an accurate representation of the signal, it is important to design an appropriate dictionary. The literature[10] shows a good characterization of the ultrasonic detection signal by the Gabor dictionary. According to the original signal characteristics, a compatible Gabor dictionary library is built, and the signal is sparsely decomposed according to the OMP algorithm step. Finally, the noise is removed to achieve the purpose of defect extraction.

4. Experimental results and analysis

In order to compare the signal extraction effect of the OMP algorithm, it is compared with the MP algorithm. Figure 1(a) shows the simulated defect signal, figure 1(b) shows the noise-added signal after adding Gaussian white noise, figure 1(c) shows the denoised approximation signal of the five atomic reconstructions searched by the OMP algorithm, and figure 1(d) shows the denoised approximation signal of the five atomic reconstructions searched by the MP algorithm. It can be seen from the figure that the OMP algorithm can accurately extract the defect signal and is closer to the original signal. Though the MP algorithm can detect the defect, the approximated effect is not as good as the OMP algorithm.

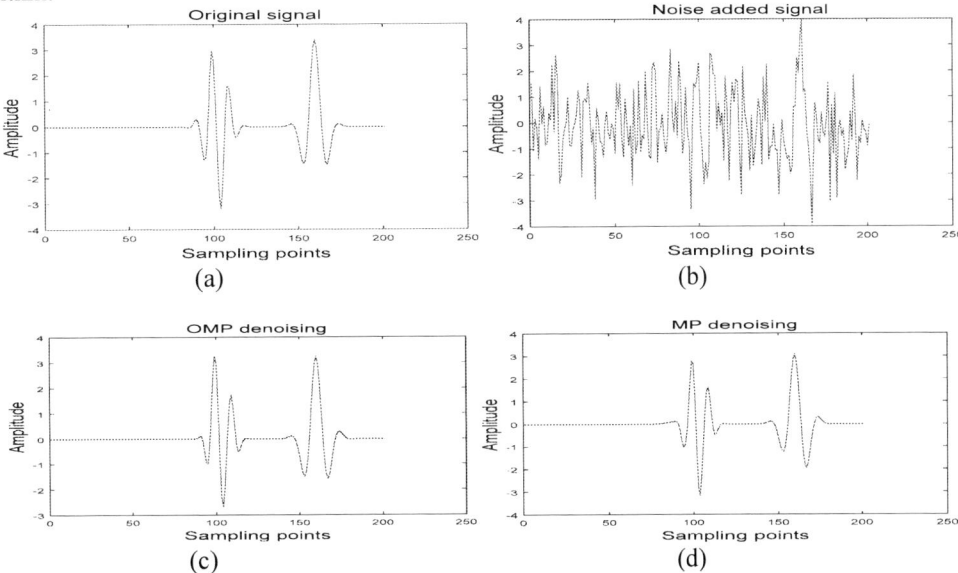

Figure 1. Signal extraction effect of two detection methods

In order to observe the difference between the two algorithms more intuitively, the signal-to-noise ratio SNR is used to evaluate the denoised effect. Figure 2 shows the change in signal-to-noise ratio between the two algorithms in each iteration. It can be seen from the figure that the denoising effect of the OMP algorithm is better than the MP algorithm.

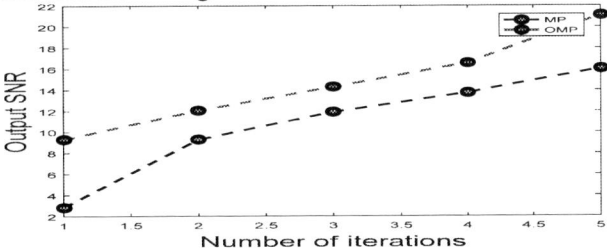

Figure 2. Comparison of signal-to-noise ratios of two methods in the iterative process

The data acquisition was performed by an atomic force sound microscope manufactured by Sonoscan, and the waveform was obtained using the A-scan method at a sampling frequency of 230 MHz. Figure 3(a) is a waveform diagram of the acquired defect signal; figure 3(b) is the result of the OMP algorithm denoised process; figure 3(c) shows the result of adding noise to the measured signal; figure 3(d) shows the result of the OMP algorithm after noise is added. It can be seen that the original defect signal can be well matched using the OMP algorithm.

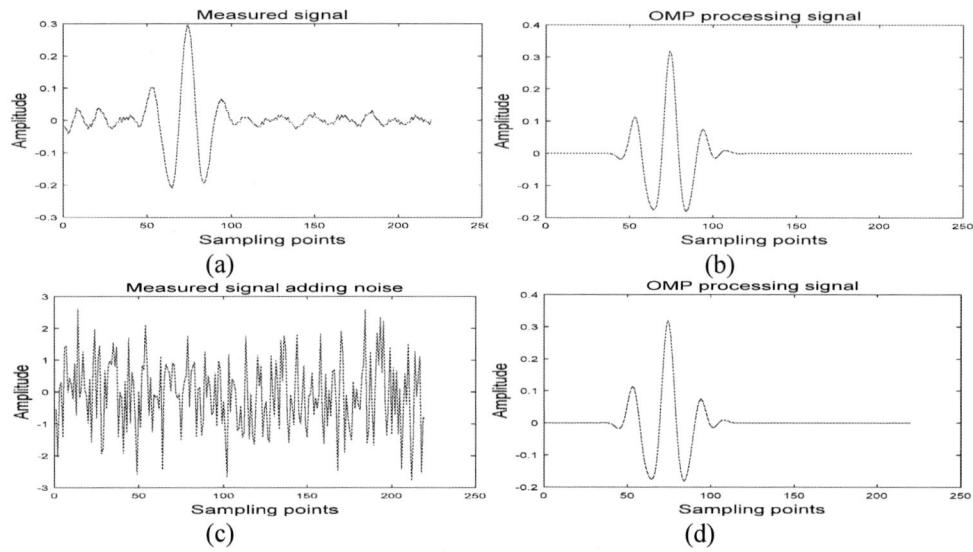

Figure 3. Measured signal denoising results

5. Conclusions

In this paper, the OMP algorithm is used to denoise the noisy welding defect signal, and the Gabor dictionary which is best matching the ultrasonic detection signal characteristics is adopted. Compared with the MP algorithm, the result can better match the signal characteristics. Through the simulation of the simulated signal and the measured defect signal, it is proved that the OMP algorithm can extract the defect signal more effectively, and the signal-to-noise after denoising is relatively large.

Acknowledgment

Fund project: the National Natural Science Foundation of China (61674121)

References

[1] Lei Yongwei 2016 Advantages and Applications of Microelectronic Packaging Technology[J]. *Technology and Market*, 23(11) p 99

[2] Mallik D, Radhakrisham K and He J Q, et al. 2005 Advanced package technologies for high—performance systems[J]. *Intel Technology Journal*, 9(4) pp 259—271

[3] Mallat S G, Zhang Z 1993 Matching pursuits with time-frequency dictionaries[J]. *IEEE Trans on Signal Processing*, 41(12) pp 3397 - 3415

[4] Fotiadou K, Tsagkatakis G and Tsakalides P 2014 Low Light Image Enhancement via Sparse Representations[C]// *International Conference Image Analysis & Recognition.* Springer, Cham

[5] Sahoo S K, Makur A 2015 Signal Recovery from Random Measurements via Extended Orthogonal Matching Pursuit[J]. *IEEE Transactions on Signal Processing*, 63(10) pp 2572-2581

[6] Yin Zhongke, Wang Jianying and Shao Jun 2005 Signal Sparse Decomposition Based on Structural Properties of Atomic Library[J]. *Journal of Southwest Jiaotong University*, 40(2) pp 173-178

[7] Fang Y, Chen L, Wu J, et al 2011 GPU Implementation of Orthogonal Matching Pursuit for Compressive Sensing[C]// *IEEE International Conference on Parallel & Distributed Systems. IEEE Computer Society*

[8] Wu Yunxia, Tian Yimin 2016 Feature extraction and recognition method of coal and rock images based on dictionary learning[J]. *Journal of China Coal Society*, 41(12) pp 3190-3196

[9] Pati Y C, Rezaiifar R and Krishnaprsad P S 1993 Orthogonal Matching Pursuit: Recursive Function Approximation with Application to Wavelet Decomposition[J]. *Proc. 27 th Annu. Asilomar Conf. Signals Syst. Comput*, 1 pp 40-44

[10] Zhang G M, Harvey D M and Braden D R 2006 Advanced Acoustic Microimaging Using Sparse Signal Representation for the Evaluation of Microelectronic Packages[J]. *IEEE Transactions on Advanced Packaging*, 29(2) pp 271-283

ISPECE   IOP Publishing

Calculation method of Cross Section area of collapsing dangerous Rock based on parallel Binocular Vision

Ping Gan[1], Zhenzhen Ruan[1], Yang Wang[1], Lin Huang[1] Yanyun Li[1] and Yangfan Huang[1]

[1] School of Microelectronics and Communication Engineering, Chongqing University, Chongqing, China

vigor_gp@163.com, r1041942883@163.com, wy20124448@163.com, 1244358767@qq.com, 815183417@qq.com, hyf@cqu.edu.cn

Abstract. Dangerous rock collapse may cause traffic accidents and other potential traffic safety hazards in mountainous areas. In this paper, aiming at the problem that dangerous rock collapse disaster is difficult to monitor and judge, a method of measuring the cross section area of collapse rock is proposed based on parallel binocular vision, which is convenient for the relevant departments to deal with the emergency ahead of time. This paper first calibrates the binocular camera on the MATLAB platform, and then corrects and matches the collected images on the OpenCV platform. The feasibility of measuring area by pixel method and spatial coordinate method is verified by using regular triangular and rectangular shapes. Then the pixel method is applied to the measurement of cross sectional area of rock collapse, which provides a method for predicting the disaster grade of collapse of dangerous rock.

1. Introduction

The disaster of dangerous rock collapse in mountain area mostly occurs in the remote mountainous area. Because of its complex environment, remoteness, sudden occurrence, and difficult to determine the disaster grade, it is particularly difficult to monitor and warn the disaster of dangerous rock collapse. With the development of computer vision technology, the methods to infer various three-dimensional spatial problems through two-dimensional image information have been widely used in various industries and become one of the hot research fields, but in computer vision, binocular vision has the advantages of high efficiency, suitable precision and low cost. It is suitable for non-contact measurement on the site of highway rock collapse. In this paper, the binocular vision system is applied to the measurement of the cross sectional area of the collapse rock, which facilitates the relevant management departments to release the information in advance, to deal with the emergency and to grasp the field situation in real time. It helps to reduce threats to people's personal and property security.

The realization process of the pixel method in this paper is as follows: Firstly, the distance parameters and 3D coordinates of the object are obtained by binocular vision, and then the regular triangular and rectangular images are preprocessed. Then the object is segmented by the method of inter-class variance, then the pixel area of the object in the image is obtained, and finally the actual area is obtained. The method is applied to measure the cross-sectional area of collapse rock.

2. Binocular stereo vision principle

According to the principle of binocular stereo vision, two cameras are used to obtain the information

Content from this work may be used under the terms of the Creative Commons Attribution 3.0 licence. Any further distribution of this work must maintain attribution to the author(s) and the title of the work, journal citation and DOI.

Published under licence by IOP Publishing Ltd

of the same object in the three-dimensional world.

2.1.Binocular stereovision model

Compared with the camera arrangement of the intersecting binocular optical axis, In stereo vision, if the left and right cameras meet the conditions: the central line of the left and right cameras is their common X axis and the optical axes of the two cameras are parallel [1], the two cameras are called parallel alignment state.

This paper is based on the parallel optical axis model. The principle of a typical binocular visual imaging system without distortion[2] is shown in figure 1 below:

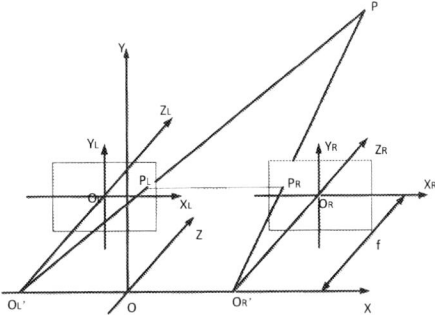

Figure 1. Parallel binocular visual model.

The imaging points of the space point P in the image plane of the left and right camera are respectively P_L, P_R. Through the imaging model of the camera, the relationship between the three-dimensional position of the target in space and the imaging point in the image can be determined. The geometric model parameters of camera imaging can be solved by transforming the pixel coordinate system to the world coordinate system [3].

The pixel coordinate system represents the number of rows and columns of a pixel point in the image. The corresponding point of space point p on the imaging plane is (u, v) in the pixel coordinate system, and the world coordinate of the space point P is (x_w, y_w, z_w), the transformation relationship between the world coordinate system and the pixel coordinate system can be obtained as follows: (here A is the inner parameter matrix, [R t] is called the outer parameter matrix, R denotes the rotation of the camera from the origin of the world coordinate, and the translation of t means the translation.)

$$
s \begin{bmatrix} u \\ v \\ 1 \end{bmatrix} = \begin{bmatrix} \alpha & \gamma & u_0 \\ 0 & \beta & v_0 \\ 0 & 0 & 1 \end{bmatrix} [R \quad t] \begin{bmatrix} x_w \\ y_w \\ z_w \\ 1 \end{bmatrix} = A[R \quad t] \begin{bmatrix} x_w \\ y_w \\ z_w \\ 1 \end{bmatrix} \tag{4}
$$

2.2.Binocular camera calibration

In the calibration process of binocular stereo vision system, the main task is to obtain camera internal parameters, external parameters and distortion parameter vectors. The inner parameters are only related to the camera itself, while the outer parameters (rotation matrix and translation vector) are determined by the relative position and attitude of the camera and the world coordinate system.

In this paper, MATLAB version of the toolbox_Calib(Camera Calibration Toolbox based on MATLAB) calibration toolbox is used to calibrate, and then the subsequent image correction and matching processing are carried out on the Open CV platform. The calibration board is calibrated with a 12 × 9 black and white checkerboard with a high accuracy of 5 × 5 mm for each cell [6]. In the process of collecting the image of calibration board, we try to make the calibration board in different position in the image, different angle, the top and bottom of the screen are covered to a certain extent, the number is about 15-20 groups is more suitable. The 20 groups of pictures taken by the camera in this paper are shown in figure 1.2 below.

Figure 2. Calibration images.

2.3.Image correction

After calibrating and getting the inner and outer parameter matrix of the camera, it is necessary to correct the images taken by the left and right cameras.

The plane of the left and right images is coplanar and the line pair is punctual, so that the stereo parallax can be calculated. The correction consists of two steps: one is to eliminate the distortion caused by the lens to the image [7][8][9], which is a transformation for a single camera. Two is binocular parallel correction, is for binocular camera operation. The distortion is divided into radial distortion and tangential distortion. The radial distortion is caused by the technological problems of the lens itself, which is usually much larger than the tangential distortion. Therefore, the radial distortion is only considered in this paper, but the tangential distortion is ignored.

The following (a) is a calibration board image collected directly from the camera, and the edge of the image is obviously distorted, (b) as the corrected image of the calibration board. It can be clearly compared that the edges of the original image with obvious curved edges are well corrected.

Figure 3. Image before Rectify.   Figure 4. Rectified image.

In this paper, the stereoRectify() function of Open CV platform is selected for image correction:

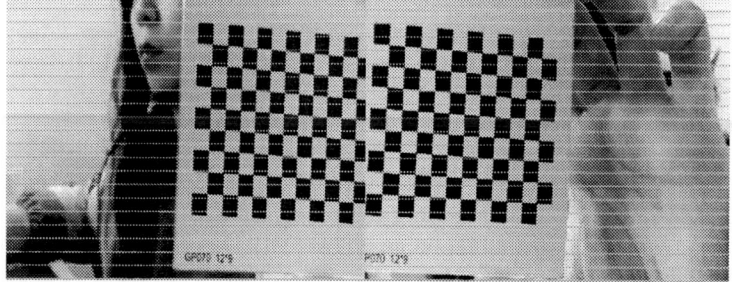

Figure 5. Rectified image.

It can be seen that the result of image correction is ideal and the right and left image matching points after correction are at the same horizontal line.

3. image preprocessing

It is necessary to preprocess the images collected by the camera before the area measurement, which mainly includes image denoising, image segmentation and so on.

3.1.image denoising

In image processing, before further processing such as image segmentation, the first step is to reduce the noise of the image to a certain extent. In this paper, the median filtering method is used. Its principle is to set the gray value of each pixel point to the median value of all pixel points in a neighborhood window of that point, so as to eliminate the isolated noise points.

The following (a) is the gray image of the original rectangular image, and the figure (b) is the image after the median filter. It can be seen that the noise effect has been filtered out in a certain part.

Figure 6. Original rectangular image.

Figure 7. Image after median filtering.

3.2.image segmentation

In order to further process the target, it is necessary to separate the object from the background first. In this paper, the method of inter-class variance is used to segment the image. The method divides the image into two parts: target and background according to the gray characteristic of the image. It can obtain satisfactory results when the image quality is better and the background changes steadily. The following (a) is a gray image after denoising, and the graph (b) is a binary image which is segmented by the method of inter-class variance.

Figure 8. Image after median filtering.

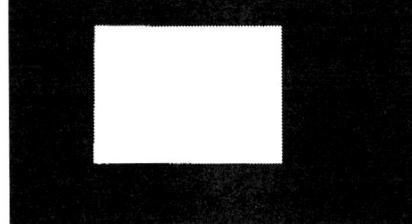

Figure 9. The binary image after segmentation.

4. Area calculation method

4.1.Pixel method

Suppose that the widths of a single pixel in the x direction and y direction in the image plane are x_p, y_p, respectively, and the actual widths in the x direction and y direction in the plane with depth of field L are X_p, Y_p [10], respectively. The focal length of the camera is f. Based on the proportional relation of the pinhole imaging model (only the x direction model is drawn, the y direction is the same):

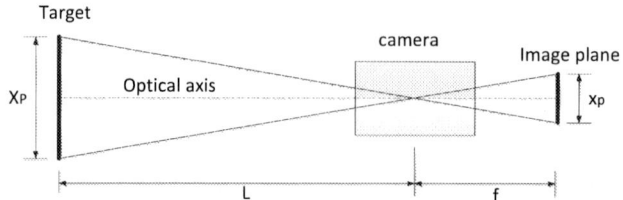

Figure 10. Pinhole imaging model.

$$\begin{cases} \dfrac{x_p}{X_p} = \dfrac{f_x}{L} \\ \dfrac{y_p}{Y_p} = \dfrac{f_y}{L} \end{cases} \tag{6}$$

Then the actual area of a single pixel on a plane with a distance of L is [5]:

$$a = X_p Y_p \tag{7}$$

Then calculate the number of pixels of the object after image preprocessing, that is, the pixel area is A, we can calculate the actual area of the object according to the formula:

$$S = A * a = A X_p Y_p \tag{8}$$

After image correction, the SGBM matching algorithm is implemented on the platform of OpenCV. The parallax map can be obtained by matching the left and right images. Through the parallax image, the three-dimensional information of image pixels can be obtained in the command window. By reading out the value of z axis in its 3D coordinate, the depth of field distance data L is obtained, and the actual area of the object is calculated.

4.2. Spatial coordinate method

After matching according to binocular stereo vision, according to parallax, the information of 3D world obtained is as follows:

$$\begin{cases} x_w = \dfrac{BX_{left}}{d} \\ y_w = \dfrac{BY}{d} \\ Z_w = \dfrac{Bf}{d} \end{cases} \tag{9}$$

Where d is the parallax of the space point in the X direction of the left and right image plane, B is the baseline distance of the binocular camera, and f is the focal length. In this paper, the reprojectToImage() function of OpenCV platform is used to calculate the three-dimensional coordinates. For regular graphs such as triangles and rectangles, we can calculate the side length according to the spatial coordinates of each vertex, and then calculate the area according to the formula [10] of each area of the regular image.

5. experimental result

5.1. measure the area of a regular pattern

1) Pixel method

We use the above method to calculate the area of triangle and rectangle respectively. The experimental data are shown in the following table:

Table 1. Pixel method.

	triangle	rectangle
Number of pixels	118616	250321
distance(mm)	350.912mm	350.912mm
area	147.68cm^2	311.66cm^2

The actual area of the triangle is 137.38 cm^2 and the actual area of the rectangle is 288.86 cm^2, so the error of the triangle area measured by the pixel method is as follows:

$$e_1 = \left|\frac{S_1' - S_1}{S_1}\right| \times 100\% = 7.5\% \tag{10}$$

The error of rectangular area measured by pixel method is as follows:

$$e_2 = \left|\frac{S_2' - S_2}{S_2}\right| \times 100\% = 7.89\% \tag{11}$$

2) Space coordinate method

According to the reprojectToImage() function based on the OpenCV platform in this paper, the spatial coordinates of each vertex of the rectangle are obtained respectively (-34.8004,-89.3379,350.912), (170.375,-88.4611,350.912), (-35.9695,52.999,350.912). (170.375, 52.4144,350.912). Because there will be some errors, the difference between the left and right coordinates or the average values of the upper and lower coordinates will be taken to calculate the area of the side length. The area of the rectangle can be calculated as 291.37 cm^2 by the rectangular area formula. On the basis of the experimental data, the area of the triangle is obtained as follows:

Table 2. Space coordinate method.

	triangle	rectangle
area	146.34cm^2	291.37 cm^2

The actual area of the triangle used in the measurement experiment is 137.38 cm^2, the actual area of the rectangle is 288.86 cm^2, so the error between the triangle area and the actual area measured by the spatial coordinate method is 6.5%. The area error of the rectangle is 0.87%. From the experimental data above, we can see that the error of the spatial coordinate method is smaller than that of the pixel method.

5.2. Measurement of Cross Section area for Slope collapse

Based on the above experimental results, the pixel method calculates the area error within an acceptable range, and the collapsed slope rock is an irregular object, which cannot be calculated by the existing area formula using the space coordinate method. Therefore, this paper applies the pixel method to the measurement experiment of the cross-sectional area of the slope.

Since the cross section of the camera can also be an irregular pattern, taking the image of the collapsed rock collected by the left camera as an example, the gray image of the original image and the image corrected by binocular are as follows:

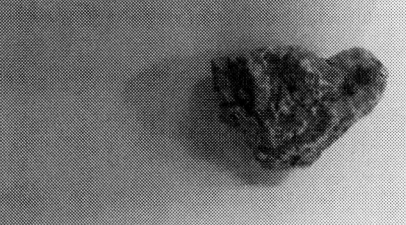 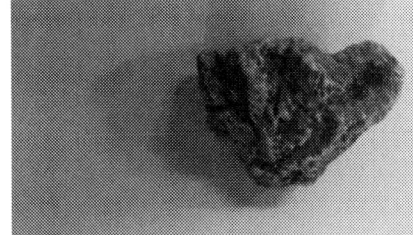

Figure 11. Grayscale image of original image.   Figure 12. Rectified image.

Since the collapsed rock is a three-dimensional object, the depth value of each point calculated by OpenCV is different. In this paper, the average depth of the rock mass surface is used as the actual distance of the rock. The pixel method is then used to calculate the actual cross-sectional area of the collapsed rock.

Before seeking the average depth value of the entire collapsed dangerous rock, we first need to distinguish the depth of the entire collapsed rock from the background[11]. According to the specific experimental environment, we set the depth threshold to [100,400]. The depth values of the pixels within the threshold range are averaged. The following image is a depth map of the collapsed rock and a binary image:

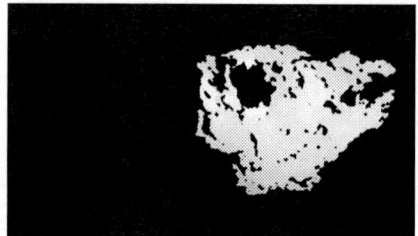 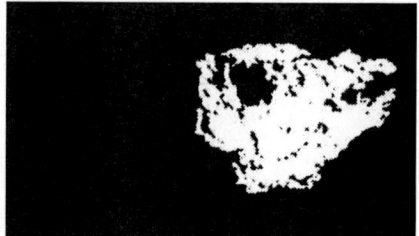

Figure 13. depth map.   Figure 14. binary image.

According to the pixel method, the results of the calculation are shown in the following table:

Table 3. cross-sectional area of the slope based on pixel method.

	Number of pixels	distance	area
data	176923	191.91mm	67.936 cm^2

At the actual site where the dangerous rock collapses, the level of the disaster can be estimated according to the actual measured cross-sectional area of the collapsed rock, and the relevant departments are given an alarm to classify the relevant measures in advance.

6. conclusion

When applied to the actual slope, because the background environment of the site is more complicated, it is necessary to improve the performance of the equipment and the accuracy of image preprocessing. The error of the experiment will be smaller and the prediction accuracy will be better.

References

[1] Ding-Cai, C. , Ding-Cheng, W. , & Jin-Shui, Z. . (2006). Research on measurement of realistic planting leaf area based on machine vision. *Journal of Computer Applications, 26*(5), 1226-1228.

[2] Song, L. M. , Wang, M. P. , Lu, L. , & Jing Huan, H. . (2007). High precision camera calibration in vision measurement. *Optics & Laser Technology, 39*(7), 1413-1420.

[3] Gherardi, R. , Farenzena, M. , & Fusiello, A. . (2010). Improving the efficiency of hierarchical structure-and-motion. *Computer Vision & Pattern Recognition.* IEEE.

[4] Okutomi, M., & Kanade, T. (2002). A multiple-baseline stereo. *IEEE Computer Society Conference on Computer Vision & Pattern Recognition, Cvpr.*

[5] Yakimovsky, Y. , & Cunningham, R. . (1978). A system for extracting three-dimensional measurements from a stereo pair of tv cameras. *Computer Graphics & Image Processing, 7*(2), 195-210.

[6] Bleyer, M. , & Gelautz, M. . (2007). Graph-cut-based stereo matching using image segmentation with symmetrical treatment of occlusions. *Signal Processing: Image Communication, 22*(2), 127-143.

[7] Shih, S. W. , Hung, Y. P. , & Lin, W. S. . (1992). Efficient and accurate camera calibration technique for 3-d computer vision. *Proceedings of SPIE - The International Society for Optical Engineering, 1614*, 133-145.

[8] Lenz, R. K. , & Tsai, R. Y. . (2003). Techniques for calibration of the scale factor and image center for high accuracy 3D machine vision metrology. *IEEE International Conference on Robotics & Automation.* IEEE.

[9] Wei, G. Q. , & Ma, S. D. . (2002). Implicit and explicit camera calibration: theory and experiments. *IEEE Transactions on Pattern Analysis and Machine Intelligence, 16*(5), 469-480.

[10] Geiger, D. , Ladendorf, B. , & Yuille, A. . (1995). Occlusions and binocular stereo. *International Journal of Computer Vision, 14*(3), 211-226.

[11] Se, S. , Lowe, D. , & Little, J. . (2001). Vision-based Mobile Robot Localization And Mapping using Scale-Invariant Features. *IEEE International Conference on Robotics & Automation.* IEEE.

ISPECE

IOP Publishing

Automatic measurement algorithm of scoliosis Cobb angle based on deep learning

Yongcheng Tu[1], Nian Wang[1], Fei Tong[1], Hemu Chen[2]

[1] Electronic and Information Engineering, Anhui University, Hefei, 230039, China
[2] The First Affiliated Hospital of Anhui Medical University, Hefei, 230022, China

Corresponding author e-mail: wn_xlb@ahu.edu.cn

Abstract. Aiming at the subjective experience of the physician, the high measurement error in the Cobb angle measurement of scoliosis X-ray images and the X-ray image of spine is difficult to segment. A deep learning based scoliosis Cobb angle measurement algorithm which can automatically calculate Cobb angle without the physician's manual definition is proposed. A DU-Net detection and segmentation network is proposed in this paper to remove the unrelated regions and to segment the spine contour in the spine X-ray image. The aggregated channel features in pedestrian detection algorithm is introduced to scoliosis image to realize the spine region detection. And the DU-Net network is training to segment spine contour. Therefore, the spine curve can be fitted by the spine contour and the Cobb angle can be automatically measured by the tangent line of spine curve. As a result, the Cobb angle measure methods yields an average error of 2.9° to reference Cobb angle which are measured manually by special orthopaedist. The detection algorithm in this paper yields an average precision of 98.5% and a recall of 99.5%. Moreover, the DU-Net reach an average Dice coefficient to reference segmentation of 90.28%, an IOU of 82.29% and a precision of 86.30%.

1. Introduction

The spine is the backbone of the body, located in the middle of the back, composed of the vertebrae and the intervertebral disc. Scoliosis is a three-dimensional spinal disorder, generally characterized by the lateral deviation of spine, which is accompanied by an angle of spine curvature in coronal plane larger than 10°. The Cobb angle which can be seen in Figure 1 have become a quantitative standard for doctors to diagnose or observe the symptoms of scoliosis patients[2].

The radiologists always measure Cobb angle by using protractor after the end-vertebrae was selected manually. Therefore, the accuracy of Cobb angle measurement was mainly depended on the subjective experience of radiologists[2]. Some researchers had investigated the deviations of manually measured Cobb angle under different end-vertebrae selection or operation methods, and they reported that the maximum measurement error could be up to 11.8°[3]. The error was so high that it would affect the diagnosis and treatment of scoliosis patients. In addition, tedious and time-consuming operations for scoliosis increase the possibility of operation mistakes. Therefore, computer-aided methods to measure Cobb angle are in urgent need due to the less rely on prior-knowledge and personal operation.

Zhang et al. [4] proposed a computer-aided method to measure Cobb angle on the basis of Hough-transform, which can calculate Cobb angle automatically after manually selected the end-vertebrae region of interest (ROI) and adjusted the brightness and contrast of the images. Samuvel et al. [5] proposed a segmentation algorithm to measure Cobb angle by putting mask on images. However, the

Content from this work may be used under the terms of the Creative Commons Attribution 3.0 licence. Any further distribution of this work must maintain attribution to the author(s) and the title of the work, journal citation and DOI.
Published under licence by IOP Publishing Ltd

accuracy of Cobb angle measurement mainly depended on the place of mask. Zhang [6] proposed a Cobb-angle computer-aided measurement algorithm based on deep neural network, which can automatically estimate the slope of the spine after manually selecting the block of interest in the upper and lower vertebrae, and automatically measure the Cobb angle. Moreover, some computer-aided and mobile-aided soft-wares were designed to measure Cobb angle for the purpose of improving the efficiency of radiologists[7][8]. They indeed improved the efficiency of measuring Cobb angle, however, upper and lower end-vertebrae had to be manually selected, which was time-consuming and subjective. With the continuous development of computer vision, image processing algorithms such as machine learning target detection algorithms[1] and medical image automatic segmentation algorithms [9] are constantly improved. The computer-aided diagnosis methods for medical images are gradually proposed and improved.

In this study, we propose an automatic Cobb angle measurement algorithm based on deep learning. A DU-Net detection and segmentation network is proposed to segment spine contour. The aggregated channel features is used to construct the spine detection model while the U-Net is used as the spine image segmentation model. In the end, the spine curve can be fitted by the spine contour and the Cobb angle can be automatically measured by the tangent line of spine curve. The 6th polynomial can better characterize the curvature of the spine [10][11], so the Cobb angle can be automatically measured.

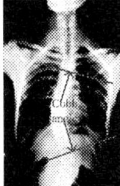

Figure 1. Cobb angle measurement

2. Spine detection

2.1. Aggregated channel features

The Aggregated channel features(ACF) proposed by Pitor Dollar[1] for the pedestrian detection algorithm, which combines features such as LUV, gradient magnitude and gradient histogram. The feature is introduced in our study to realize the spine region detection. The characteristics of each channel of the spine image are shown in Figure 2.

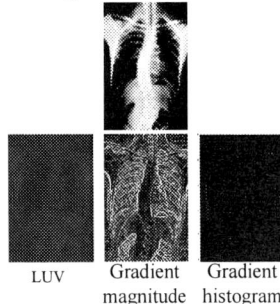

LUV   Gradient   Gradient
magnitude   histogram

Figure 2. Aggregated channel features

The spine ACF are trained and cascaded by the weak classifier. The classification algorithm used in this paper is Adaptive Boost (Adaboost)[15].

2.2. Spine detection algorithm

Due to the strong expression ability of the ACF, it can effectively describe the contours of the target. Therefore, this paper introduced the ACF to detect the spine, extracts the features of the multi-scale features from the spine image, and use the Adaboost algorithm to train the cascade classifier. The detection steps and detection flow chart are shown in Figure 3.

Step1:Label the area where the foreground spine is located, extract the features of the front and background ACF in input image, form the feature vector and send the corresponding background label to the weak classifier for classification training, iteratively obtain the current optimal weak classifier;

Step2:constructing a spine image feature pyramid, sliding window detection on different scale feature images, and using the above-mentioned trained classifier to determine the front result image;

Step3: Use the non-maximum suppression algorithm to obtain the final detection result.

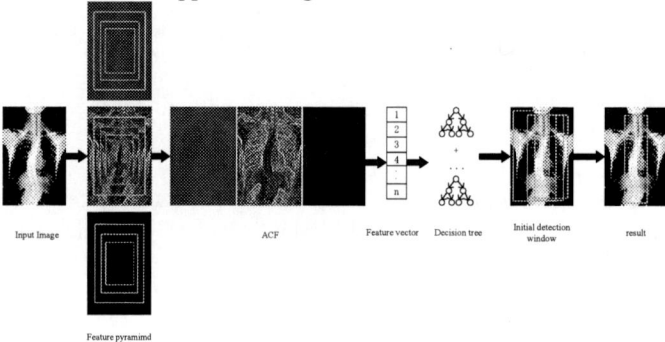

Figure 3. Detection algorithm flow chart

3. Cobb angle automatic measurement algorithm

In this paper, the DU-Net detection segmentation network is proposed. The spine image detection algorithm is used to detect the spine region, and the region where the spine is located is obtained. The DU-Net segmentation network can automatic segment the spine to obtain the spine contour. The spine contour central point can be extracted after segmentation, and the curve fitting is performed using a 6th order polynomial curve fitting algorithm. The scoliosis Cobb angle can be automatically calculated by calculating the slope of the curve.

3.1. DU-Net segmentation algorithm

The algorithm flow of the DU-Net detection segmentation network proposed in this paper is shown in Figure 4. The ACF of the X-ray spine image is extracted, and the Adaboost algorithm is used to train the cascade detection classifier. The labelme is used to label the spine contours and use the spine detection area to train the DU-Net segmentation network.

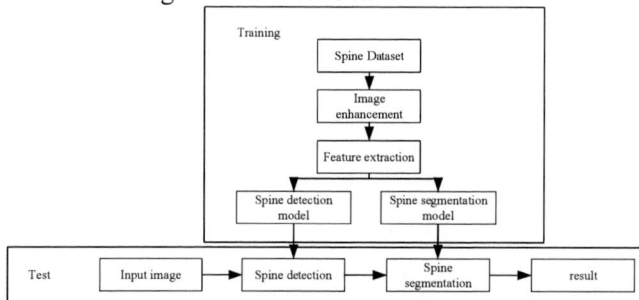

Figure 4. Segmentation algorithm flow chart

*3.1.1. U-Net.*U-Net [9] is a semantic segmentation network model based on FCN[14] structure proposed by Olaf Ronneberger et al. in 2015. It was originally applied to medical cell image segmentation and is suitable for medical image segmentation. Therefore, U-Net is introduced to implement the spine segmentation module in DU-Net.

*3.1.2. Training.*Using a deep convolutional neural network based on the PyTorch framework, implemented in Python and accelerated with the GPU. PyTorch is a deep learning neural network framework that Facebook has open source, which is flexible and efficient to use. The Loss function in our paper can be seen in below:

$$Loss(x) = SoftmaxLoss(x)+1\text{-}Dice(x) \qquad (1)$$

As shown in the above formula, Loss function combines SoftmaxLoss and Dice coefficients. The Dice coefficient is used to measure the difference between the prediction results and the labels during training. The network uses the RMSprop optimizer to update the parameters. The model training time is about 8 hours, and the convergence speed is faster. Part of the feature map during training are shown below.

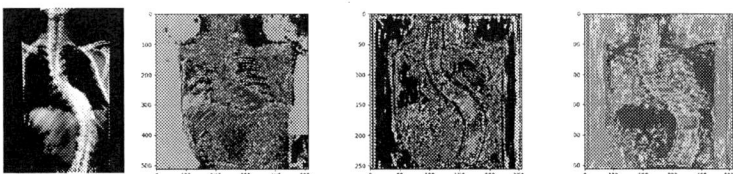

Figure 5. Feature map

3.2. Spine curve fitting

The spine contour has been obtained in above. And the central point can easily extracted by the spine contour. Therefore, the spine curve can be fitted by a set of points in the center of the spine contour, reflecting the shape and orientation of the central set of points in the spine and obtaining the curve equation of the spine.

3.3. Automatic Cobb Angle Calculate

Cobb angle is defined as the angle between the lines drawn along the most tilted vertebrae on the spine image, which is actually the angle between the tangent vector to the curve at the corresponding central point [13]. The definition of Cobb angle in this work is illustrated in Figure 6.

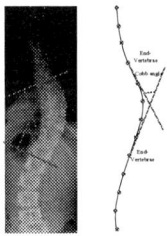

Figure 6. Cobb angle

4. Experimental results and analysis

4.1. Dataset

The self-built dataset of spinal image training, verification and testing was obtained from the Department of Radiology of the First Affiliated Hospital of Anhui Medical University and the NIH [12] public chest X-ray dataset. The dataset have a total of 800 X-ray images of the spine, including 600 training sets and 200 test sets. A total of 100 X-ray images of the scoliosis were concentrated in the training, and a total of 10 X-ray images of the scoliosis were measured. In order to carry out network training, testing and algorithm verification, labelme is used to segment the mask of the image to obtain the corresponding segmentation label. The results obtained by the Cobb angle automatic measurement algorithm in this paper are verified by the orthopedic experts of the Affiliated Hospital of Anhui Medical University. The image and label in the dataset are shown in Figure 7.

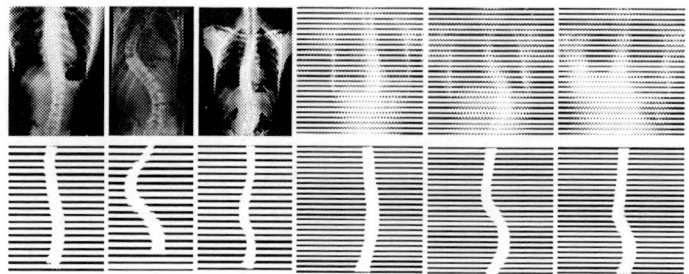

Figure 7. Training data

4.2. Result and analysis

In this section, the DU-Net and U-Net segmentation results are compared. Figure 8 shows the comparison between the DU-Net and the U-Net segmentation result.

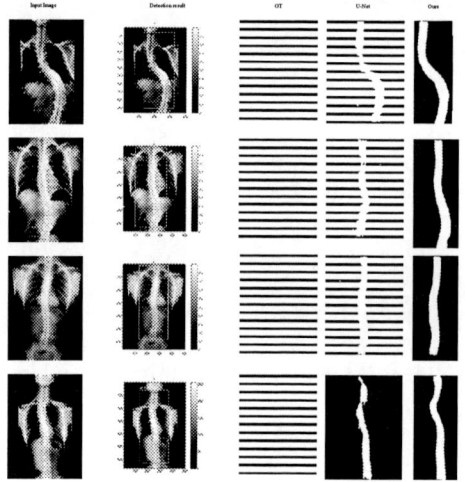

Figure 8. Segmentation result

In order to quantify segmentation results among DU-Net and U-Net, four measures of Dice coefficient, IOU, Precision and Recall[16] are selected. Moreover, the higher values of the segmentation results, the better the segmentation result will be. Our method is superior to U-Net algorithm in the segmentation of spine for X-ray images. ours yields a higher Dice coefficient of 90.28% and an IOU of 82.29%. The more accurate the spine segmentation is, the more accurate the Cobb angle measurement would be.

Table1 Segmentation indicator

Method	Dice	IOU	Precision	Recall
U-Net	0.7127	0.5700	0.7036	0.7457
DU-Net	**0.9028**	**0.8229**	**0.8630**	**0.9493**

The Cobb angle can be automatically calculated by the fitted curve with the tangents and the comparison result of ours and the reference Cobb angle which are measured manually by special orthopaedist are as follows:

Table2 Cobb angle measure result

Test	ours	reference	bias
1	49.2°	47.8°	1.4°
2	39.1°	40.1°	-1.0°
3	43.8°	49.2°	-5.4°
4	23.5°	25.2°	-1.7°
5	24.2°	29.3°	-5.1°

6	12.2°	15.5°	-3.3°
7	54.8°	51.5°	3.3°
8	31.5°	33.8°	-2.3°
9	32.1°	28.1°	4.0°
10	16.9°	15.2°	1.7°

5. Conclusion

Aiming at the subjective experience of the physician, the high measurement error in the Cobb angle measurement of scoliosis X-ray images and the X-ray image of spine is difficult to segment. A deep learning based scoliosis Cobb angle measurement algorithm which can automatically calculate Cobb angle without the physician's manual definition is proposed. A combination of target detection algorithm and image segmentation algorithm is introduced to form a DU-Net network structure and apply it to the spine image segmentation task. The results of this work have showed a good consistency between the automatic and reference measurements of Cobb angle. Our work can not only improve the accuracy of spine X-ray image segmentation but also simplifies the Cobb angle measurement step. The future work will optimize the segmentation network and to further improve the accuracy of spine X-ray image segmentation.

Acknowledgements

This work is supported by the Natural Science Foundation of China under Grant 61672032, and the Key Projects of Outstanding Youth Talent Support Program of Anhui Provincial Universities under Grant gxyqZD2016012.

References

[1] Dollar P, Appel R, Belongie S, et al. Fast Feature Pyramids for Object Detection[J]. IEEE Trans Pattern Anal Mach Intell, 2014, 36(8):1532-1545.

[2] Zhou G Q, Jiang W W, Lai KL, et al.Automatic Measurement of Spine Curvature on 3-D Ultrasound Volume Projection Image With Phase Features[J]. IEEE Transactions on Medical Imaging, 2017, 36(6):1250-1262.

[3] Pruijs J E H , Hageman MAPE, Keessen W, et al. Variation in Cobb angle measurements in scoliosis[J].Skeletal Radiology, 1994, 23(7):517-520.

[4] Zhang J, Lou E, Hill D L, et al. Computer-aided assessment of scoliosis on posteroanterior radiographs[J]. Medical & Biological Engineering & Computing, 2010, 48(2):185-195.

[5] Samuvel B , Thomas V , Mini M G , et al. A Mask Based Segmentation Algorithm for Automatic Measurement of Cobb Angle from Scoliosis X-Ray Image[C]// International Conference on Advances in Computing & Communications,IEEE,2012,110-113.

[6] Zhang J , Li H , Lv L , et al. Computer-Aided Cobb Measurement Based on Automatic Detection of Vertebral Slopes Using Deep Neural Network[J]. International Journal of Biomedical Imaging, 2017, 2017:1-6.

[7]Wu W. Reliability and reproducibility analysis of the Cobb angle and assessing sagittal plane by computer-assisted and manual measurement tools[J]. Bmc Musculoskeletal Disorders, 2014, 15(1):33-33.

[8] Shaw M , Adam C J , Izatt M T , et al. Use of the iPhone for Cobb angle measurement in scoliosis[J]. European Spine Journal, 2012, 21(6):1062-1068.

[9]Ronneberger O , Fischer P , Brox T . U-Net: Convolutional Networks for Biomedical Image Segmentation[J]. 2015.

[10] Zhou G Q , Jiang W W , Lai K L , et al. Automatic Measurement of Spine Curvature on 3-D Ultrasound Volume Projection Image With Phase Features[J]. IEEE Transactions on Medical Imaging, 2017, 36(6):1250-1262.

[11] Krejci J , Gallo J , Stepanik P , et al. Optimization of the examination posture in spinal curvature assessment[J]. Scoliosis, 2012, 7(1):10.

[12] Wang X , Peng Y , Lu L , et al. ChestX-Ray8: Hospital-Scale Chest X-Ray Database and Benchmarks on Weakly-Supervised Classification and Localization of Common Thorax Diseases[C]// CVPR 2017,3462-3471.

[13] Tomaž Vrtovec, Franjo Pernuš, Boštjan Likar. A review of methods for quantitative evaluation of spinal curvature[J]. European Spine Journal, 2009, 18(5):593-607.

[14] Long J , Shelhamer E , Darrell T . Fully Convolutional Networks for Semantic Segmentation[J]. IEEE Transactions on Pattern Analysis & Machine Intelligence, 2014, 39(4):640-651.

[15] Schapire R E, Singer Y. Improved Boosting Algorithms Using Confidence-rated Predictions[M]. 1999.

[16] Chang H H , Zhuang A H , Valentino D J , et al. Performance measure characterization for evaluating neuroimage segmentation algorithms.[J]. Neuroimage, 2009, 47(1):122-135.

A Robust $\mathcal{H}_\infty$ Approach of In-flight Calibration for UAVs with Low-cost IMU

He Gao[1,2], Chao Gao[2,3,*] and Guorong Zhao[2]

[1] Naval Research Academy, Beijing 100011, P.R. China

[2] Naval Aeronautical University, Yantai 264001, P.R. China

[3] The 92664th unit of PLA, Qingdao 266000, P.R. China

*Corresponding author: gaochao.shd@163.com

Abstract. In this paper, the on-line calibration problem for the low-cost IMU of UAV has been studied. A robust H_∞ estimation scheme is designed such that, in the presence of modelling error, asynchronous sampling issue, and filtering bias, the overall estimation dynamics is exponentially stable in the mean square; at the same time, this scheme is applicable to the on-line calibration with asynchronous sampling time. More specifically, a novel attitude angle matching calibration scheme is proposed. Hereinto, the measurement model is deduced in detail in the scheme, and the influences of the amplitude and frequency on the estimation of the misalignment angle is analyzed in detail. A numerical simulation example has been used to demonstrate the potential of the proposed approach to robust estimation in terms of both performances and computational tractability.

1. Introduction

Unmanned aerial vehicle (UAV) is one of the emerging methods to implement unmanned logistics transport and aerial photography. And thus, how to complete the online calibration of UAV airborne inertial navigation system (INS) quickly and accurately is the key to improve localization accuracy and route planning efficiency. However, INS has the weakness of low observability, especially in the case of static pedestals, which will lead to the least observability [1]. Besides, during the state estimation of the navigation system of UAV, the observability of the INS will affects the convergence speed and convergence accuracy of Kalman-like estimator. Meanwhile, during the calibration process of the moving base UAV, the movement of the pedestal can improve the observability of the airborne INS. Therefore, the accuracy and speed of the INS online calibration can be improved by the purposeful maneuvering of the pedestal under certain conditions [2, 3].

However, the online calibration problem of the UAV airborne INS has the following two characteristics: (i) the flight environment is complex: the navigation process in the urban environment requires rapid response, and the flight environment is intensive, if the calibration time is too long It may reduce the safety of flight due to insufficient navigation accuracy; (ii) UAV usually equipped with low-cost IMU; (iii) the security requirement of UAV needs the low-cost IMU has high accuracy [4-6]. And thus, the traditional velocity matching calibration scheme cannot meet the special requirement of the UAV.

Literature [7] proposed an angular velocity matching scheme, which shows that it can get better results with the swaying wing; however, this scheme is only applicable to the strapdown to strapdown

Content from this work may be used under the terms of the Creative Commons Attribution 3.0 licence. Any further distribution of this work must maintain attribution to the author(s) and the title of the work, journal citation and DOI.

Published under licence by IOP Publishing Ltd

system. Literature [8] proposes a combination matching scheme with velocity and attitude angle; however, this approach didn't give the specific Kalman-like filtering model. In this paper, a novel attitude angle matching calibration scheme is proposed. Hereinto, the measurement model is deduced in detail in the scheme, and the influence the amplitude and frequency the wing on the estimation of the misalignment angle is analyzed in detail.

2. Mathematical description of robust H_∞ filter

The key feature of H_∞ filtering approach is applied the H_∞ norm to the filter design, in order to solve the various uncertainties of the target system. Besides, H_∞ filtering approach regards the noise and uncertain inputs as the random signals with limited energy, making the H_∞ norm of the closed-loop transfer function from system disturb to estimation error less than a given positive value γ.

In this paper, let the dynamic equation of the generalized controlled object be

$$
\begin{cases}
\dot{x} = Ax + Bw \\
y = Cx + Dv \\
z = Lx
\end{cases}
\tag{1}
$$

where x is the state vector, y is the observation vector, z is the estimated vector, w and v denote the system noise and observation noise, respectively; A, B, C, D, and L denote the constant matrices with corresponding dimensions.

The system is supposed to satisfy the following constraints: (i) (A,B) is controllable, while (A,C) is observable; (ii) $DD^T = I$. More specifically, constraints (i) is the prerequisite for the existence of the estimator, while constraints (ii) is equivalent to the performance index in the general quadratic form.

Let $\hat{z}$ be the estimation of z, $\hat{z} = L\hat{x}$, and then $\Delta z = z - \hat{z} = Tw$, where T denotes the transfer function matrix from w to Δz. For a linear steady-state system (1) with finite energy interference w, the optimal estimated performance index can be written as

$$
J = \|T\|_\infty = \sup_{0 \neq w \in L_2(0,\infty)} \left\{ \frac{\|z - \hat{z}\|_2}{\|w\|_2} \right\} < \gamma
\tag{2}
$$

In order to make the estimation result of H_∞ estimator unbiased and have a predictive correction structure, the state estimator should satisfy the structure of the observer as follows

$$
\dot{\hat{x}} = A\hat{x} + K(y - C\hat{x}), \quad \hat{z} = L\hat{x}, \quad \hat{x}(0) = 0
\tag{3}
$$

where, denotes the feedback gain matrix of the state estimator. Let $e = x - \hat{x}$, and then

$$
\begin{cases}
\dot{e} = (A - KC)e + Gw - KDv \\
\Delta z = z - \hat{z} = Le
\end{cases}
\tag{4}
$$

The transfer function from interference w to estimation error Δz can be expressed as follows

$$
T = \begin{bmatrix} A - KC & G & -KD \\ L & 0 & 0 \end{bmatrix}.
$$

For the system model (1), which considers the disturbance effects and satisfies the constraints (i) and (ii), given a positive number $\gamma > 0$, the necessary and sufficient condition for equation (2) to be established is that, there is a unique positive definite symmetric matrix $P(P = P^T > 0)$ that satisfies the following algebraic Riccati equation

$$AP + PA^T + P(\gamma^{-1}L^T L - C^T C)P + BB^T = 0 \qquad (5)$$

where the feedback gain matrix to make the estimator (3) stable is $K = PC^T$.

An thus, we can conclude from the above analysis that, the existence of the H_∞ filter depends not only on its own structural parameters, but also on the vector to be estimated, i.e., z. γ is required to be as small as possible in the design of the H_∞ filter, and thus the value of γ should choose the minimum value of the solution of equation (5).

3. Error Models of online calibration for airborne low-cost IMU

Coordinate frames used in the case study of on-line calibration of the low-cost IMU can be referred in [3,7]; hereinto, in the earth fixed frame (e-frame), we can obtain $\Omega = [0 \quad \omega_{ie}\cos L \quad \omega_{ie}\sin L] = [0 \quad \Omega_N \quad \Omega_U]$, where ω_{ie} is the rotational angular velocity of the earth, L denotes the local latitude. The IMU error differential equations in the navigation frame mechanization are given as follows:

$$\begin{cases} \dot{X} = AX + \eta \\ Y = DX + \zeta \end{cases} \qquad (6)$$

where

$$X = [\phi_e \ \phi_n \ \phi_u \ \delta V_e \ \delta V_n \ \delta L \ \delta\lambda \ \nabla_e \ \nabla_n]^T$$

hereinto, ϕ_e, ϕ_n, ϕ_u denote the misalignment errors in east angle, north angle and azimuth angle, respectively, $\delta V_e, \delta V_n$ denote the errors in east velocity and north velocity, δL denotes latitude error, $\delta\lambda$ denotes longitude error, ∇_e, ∇_n denote the accelerometer bias errors, η denote the system white noise vector, $E(\eta_k \eta_j) = Q\delta_{kj}$.

$$A = \begin{bmatrix} \overline{A} & \overline{C} \\ 0_{5\times 7} & 0_{5\times 5} \end{bmatrix}$$

$$\overline{A} = \begin{bmatrix} 0 & \omega_{ie}\sin L + V_e\tan L/R_e & -(\omega_{ie}\cos L + V_e/R_e) & 0 & -1/R_e & 0 & 0 \\ -\omega_{ie}\sin L - V_e\tan L/R_e & 0 & -V_n/R & 1/R_e & 0 & \omega_{ie}\sin L & 0 \\ \omega_{ie}\cos L + V_e/R_e & V_n/R_e & 0 & \tan L/R_e & 0 & \omega_{ie}\cos L + V_e\sec L/R_e & 0 \\ 0 & -f_u & f_n & V_n\tan L & 2\omega_{ie}\sin L + V_e\tan L/R_e & 2\omega_{ie}\cos L \ V_n + V_eV_n\sec L/R_e & 0 \\ f_u & 0 & -f_e & 2\omega_{ie}\sin L + 2V_e\tan L/R_e & 0 & -(2\omega_{ie}\cos L + V_e\sec L/R)\ V_e & 0 \\ 0 & 0 & 0 & 0 & 1/R_e & 0 & 0 \\ 0 & 0 & 0 & \sec L/R_e & 0 & 0 & 0 \end{bmatrix}$$

where f_e, f_n, f_u denote the east, north and celestial component of the specific force, respectively, R_e denotes the radius of the earth.

$$\overline{C} = \begin{bmatrix} C_{11} & C_{12} & C_{13} & 0 & 0 \\ C_{21} & C_{22} & C_{23} & 0 & 0 \\ C_{31} & C_{32} & C_{33} & 0 & 0 \\ 0 & 0 & 0 & C_{11} & C_{12} \\ 0 & 0 & 0 & C_{21} & C_{22} \\ 0 & 0 & 0 & 0 & 0 \\ 0 & 0 & 0 & 0 & 0 \end{bmatrix}$$

where, C_{ij} is an element in the attitude transformation matrix from sensor coordinate system to e-frame. The horizontal velocity error from IMU/GPS output is used as observation equation as follows,

$$Y = V_{INS} - V_{GPS} = DX + \zeta \tag{7}$$

and

$$D = \begin{bmatrix} 0 & 0 & 0 & 1 & 0 & 0 & 0 & 0 & 0 & 0 & 0 & 0 \\ 0 & 0 & 0 & 0 & 1 & 0 & 0 & 0 & 0 & 0 & 0 & 0 \end{bmatrix}$$

where, ζ denotes observation white noise vector, $E(\eta_k \eta_j) = R\delta_{kj}$.

4. Numerical Simulation and Analysis

4.1. Simulation configurations and metrics

In this section, we present a numerical simulation example on the low-cost IMU of the UAV to demonstrate the proposed estimator design method. The primary sensor system used is the low-cost IMU that generates position, velocity and attitude information. Velocity error formed from the difference between the IMU and the GPS output yield observation to the robust H_∞ estimator. Here, the IMU computations are carried out at 20Hz; while the robust H_∞ estimator update interval is 1Hz. This test has been carried out for 475s.

In the experimentation, the designed values of matrices for the process noise covariance and the measurement noise covariance are as follows:

$$Q = diag\{(50\mu g)^2, (50\mu g)^2, (0.25°/h)^2, (0.25°/h)^2, (0.25°/h)^2, 0,0,0,0,0\}$$

$$R = diag\{(0.01m/s)^2, (0.01m/s)^2\}$$

where the initial horizontal and azimuth misalignment angles are 1°, initial velocity error is 0.1m/s, the constant error of accelerometer is 100μg, the constant bias of gyroscope is 0.3°/h, the random bias is 0.25°/h, the measurement noise of system velocity is 0.01m/s, and the initial localization error is $(10^{-4})°$.

The weighted matrix of the proposed robust H_∞ estimator is

$$S = diag\{(1°/3600)^2, (1°/3600)^2, (1°/3600)^2, (10^{-4})^2, (10^{-4})^2, (10^{-6})^2, (10^{-6})^2, 0,0,0,0,0\}$$

$$W = Q/3, \ V = R/5, \ \gamma = 3$$

4.2. Simulation results and analysis

In this subsection, several key properties of the proposed robust H_∞ estimation approach are tested on the UAV with the dynamics models in Subsection 4.1. During the simulation evaluation, the motion of UAV is divided into two stages, i.e., (i) uniformly accelerated motion with acceleration value 0.04m/s², and initial velocity 5m/s, and (ii) uniform circular motion with acceleration value 0.04m/s².

For comparison, a Kalman-like linear estimator in [7] is chosen to compare with the newly proposed robust H_∞ estimation algorithm in our simulation. The estimation results, i.e., misalignment error estimation of UAV and error estimation of UAV velocity, are listed in Figures 1 and 2, respectively.

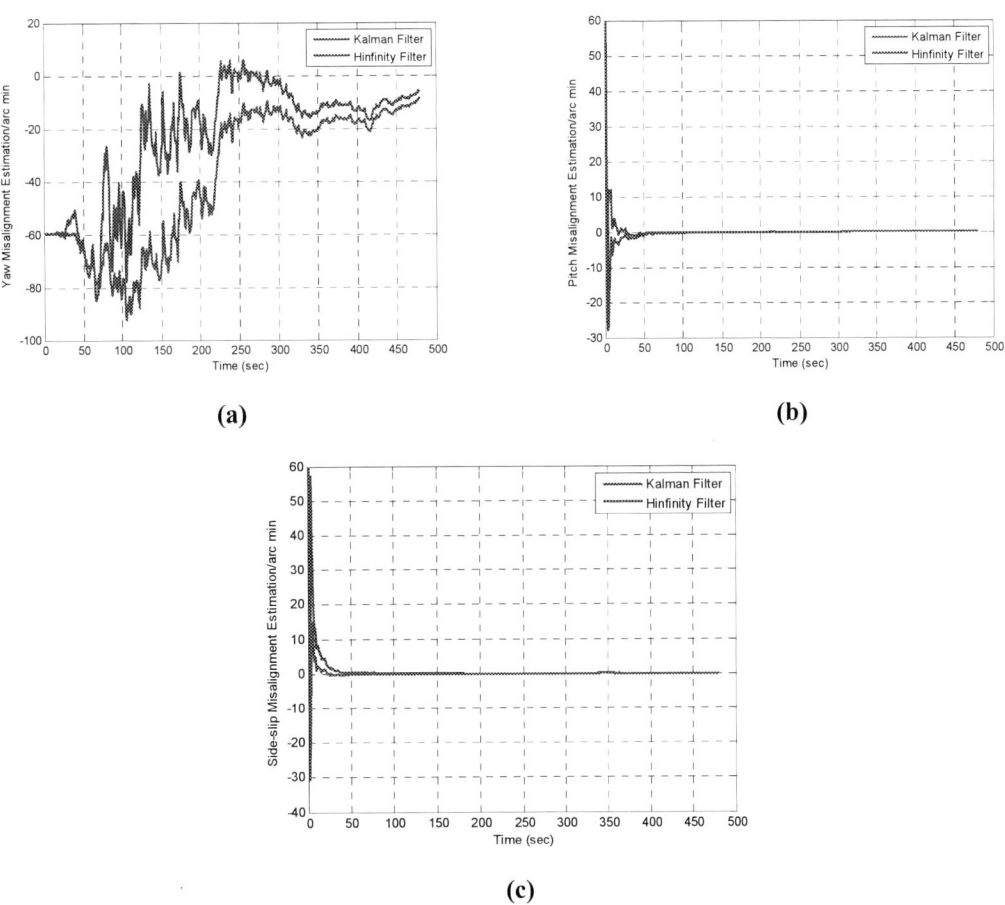

(a)

(b)

(c)

Figure 1. Misalignment error estimation of UAV.
(a) Estimation error of heading angle; (b) Estimation error of pitch angle; (c) Estimation error of yaw angle.

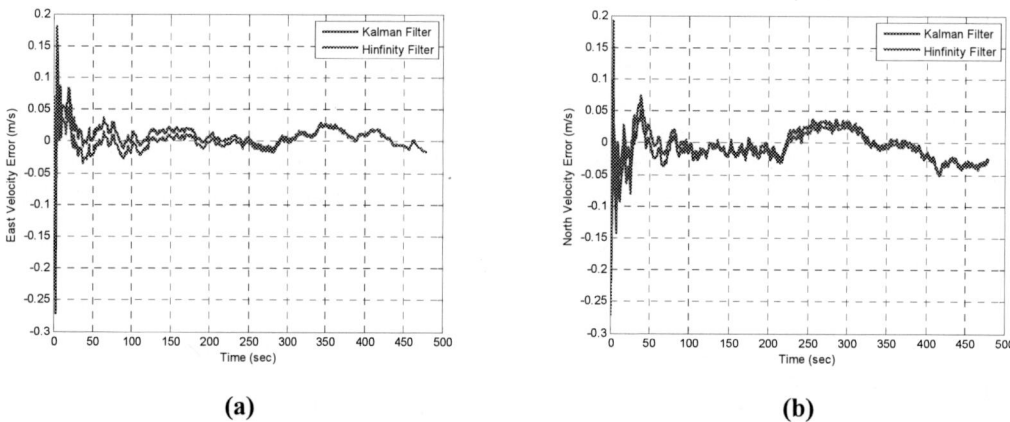

(a)   **(b)**

Figure 2. Error estimation of UAV velocity.
(a) East velocity estimation error; (b) North velocity estimation error.

Furthermore, in Table 1, we analyse the calibration accuracy of Kalman-like estimator and robust H_∞ estimator in the form of numerical results. It is clearly from the results that the calibration accuracy of robust H_∞ estimator is higher than Kalman-like estimator, especially when the heading angle of the UAV is changing constantly. Besides, the calibration velocity of the proposed H_∞ estimator is faster than the Kalman-like estimator.

Table 1. Simulation comparison between Kalman filter and robust H_∞ filter

	Algorithm	ϕ_e /"	ϕ_n /"	ϕ_u /'
Accelerated motion	Kalman	14.457	14.365	7.978
	H_∞	12.675	13.354	5.786
Uniform Circular Motion	Kalman	11.564	11.567	7.464
	H_∞	9.786	9.452	5.243

5. Conclusions

In this paper, the on-line calibration problem for the low-cost IMU of UAV has been studied. A robust H_∞ estimation scheme has been designed such that, in the presence of modelling error, asynchronous sampling issue, and filtering bias, the overall estimation dynamics is exponentially stable in the mean square and, at the same time, this scheme is applicable to the on-line calibration with asynchronous sampling time. A numerical simulation example has been used to demonstrate the potential of the proposed approach to robust estimation in terms of both performances and computational tractability.

Acknowledgments

This work was supported by the National Natural Science Foundation of China under Grant No. 61473306.

References

[1] John Baziw, Cornelius T. Leondes, Initial rapid alignment/calibration of a marine inertial navigation system, IEEE Transactions on Aerospace and Electronic Systems, 1971. (4), pp.450-465.

[2] Nebot E., Durrant-Whyte H., Initial Calibration and Alignment of Low Cost Inertial Navigation Units for Land Vehicle Applications, Journal of Robotics Systems, 1999, 16 (2), pp. 81-92.

[3] Li Chan, Cao Yuan, Zhang Shi-feng, New calibration method for shipboard platform inertial navigation system, 34th Chinese Control Conference, 2015, pp.5342-5347.

[4] Gökçen Aslan Aydemir, Afsar Saranlı, Characterization and calibration of MEMS inertial sensors for state and parameter estimation applications, Measurement, 45 (2012) 1210–1225

[5] Chao Gao, Guorong Zhao, Jianhua Lu, Shuang Pan, Decentralised Moving-Horizon State Estimation for a Class of Networked Spatial-Navigation Systems with Random Parametric Uncertainties and Communication Link Failures, IET Control Theory and Applications (2015), vol. 9, iss. 18, pp. 2666-2677.

[6] Chao Gao, Jianhua Lu, Guorong Zhao, Shuang Pan, Decentralized Navigational State Estimation for Networked Navigation Systems with Finite Channel Capacity and Randomly Switching Topologies, Proc IMechE Part G: Journal of Aerospace Engineering (2018), 232 (2), pp. 201-214.

[7] Chao Gao, Guorong Zhao, Jianhua Lu, Shuang Pan, Decentralized State Estimation for Networked Spatial-Navigation Systems with Mixed Time-Delays and Quantised Complementary Measurements: The Moving Horizon Case, Proc IMechE Part G: Journal of Aerospace Engineering (2018), 232 (11), pp. 2160-2177.

[8] Xinlong Wang, Fast alignment and calibration algorithms for inertial navigation system, Aerospace Science and Technology, 2009, 13(4), pp. 204-209.

ISPECE

IOP Publishing

Invisible Information Transmission System of Visible Light Based on Interleaved Code

Yuting Meng[1], Yunpeng Hu[1], MingchaoLi[1], Yanqun Tang[1]

[1]Information System Engineering Institute, Information Engineering University of PLA, Zhengzhou, Henan, China

myt940227@163.com

Abstract. In this paper, based on the characteristics of display-camera link and the secondary imaging mixing phenomenon caused by the hardware structure in invisible information transmission system, we propose a reasonable display-camera link model. Then by analyzing the application environment and the methods to realize of Interleaved Code(IC), we propose an invisible information transmission system design of visible light based on Interleaved Code under the characteristics of the display-camera link, which uses the complementary frame modulation. This system can well resist the secondary imaging mixing problem. The experiment results are shown to validate the robustness and effectiveness of the proposed designs.

1. Introduction

As a new emerging technology, visible light communication (VLC) [1] has great advantages over traditional radio frequency (RF) communication in terms of communication security, green environmental protection and broadband high speed, attracting more and more attention. In recent years, with the popularization of smartphones equipped with high definition cameras and displays with high resolution, a new technology named optical camera communication (OCC) [2-10] comes into being. Operating in the visible light spectrum and utilizing light emitting diodes (LEDs) or displaying as transmitters and cameras as receivers, OCC provides a convenient communication mode interaction with environment without increasing complexity of the existing hardware. The most common application of display-camera link is using a mobile phone to scan a quick response (QR) code for achieving information. In order to avoid the QR code occupying a certain space on the display screen and disturbing the look and feel of the images or videos, the area occupied by the QR code in the application is small, which also causes a problem of reduced throughput and short communication distance. Visible light implicit imaging communication as a communication method that can solve the above problems and realize information transmission under the condition of no human vision is attracted by more and more researchers [2, 3, 6].

To ensure the integrity of the received signal, the camera frame rate is usually set to twice as much as that of the display refresh rate [2, 6, 8]. In order to achieve the implicit effect, the display refresh rate should not be less than 60Hz, which means that the camera frame rate should be up to 120Fps. In addition, the inherent characteristics of display-camera link can cause a series of unsynchronized problems resulting in the reduction of transmission reliability [11-14]. In [11-12], the authors conduct the preliminary research, coming to conclusion that the link characteristics can cause frame losing and frame mixing problems. In [11], an unsynchronized display-camera system is established. It

Content from this work may be used under the terms of the Creative Commons Attribution 3.0 licence. Any further distribution of this work must maintain attribution to the author(s) and the title of the work, journal citation and DOI.

Published under licence by IOP Publishing Ltd

compensates frame synchronization by using in-frame color tracking and linear erasure code. In [12], the mixed frame is modeled as a linear superposition of two adjacent frames, and a mixed frame seam detection algorithm is proposed. In [13, 14], the preliminary model and frame synchronization compensation algorithms for display-camera link is proposed. In our previous research, the hardware characteristics of the four imaging directions in the display-camera link are studied, and the maximum likelihood estimation of the mixing factor is derived, and the trend is analyzed. Although the existing literatures has studied and analyzed the secondary imaging mixing phenomenon appearing in the display-camera link, its performance against this phenomenon is poor, and the paper does not solve the impact of this phenomenon on information transmission. In this paper, a system design based on Interlaced Code (IC) is proposed for the secondary imaging mixing phenomenon in the display-camera link. It solves the interference problem caused by the secondary imaging mixing phenomenon at the receiver and improve the reliability of the system for the identification of implicit information detection.

2. System Model

The implicit VLC communication is as shown in Figure 1. Signal is modulated into the carrier video, and then transmitted through optical line-of-sight channel. Meanwhile, the modulated signal visually transparent doesn't impair user-viewing experience, while can be captured by cameras and recovered by subsequent processing. In this paper, we leverage the complementary frame design and embed signal into the B-channel (Red, Green and Blue channel of a pixel) [7] to ensure better implicit effect.

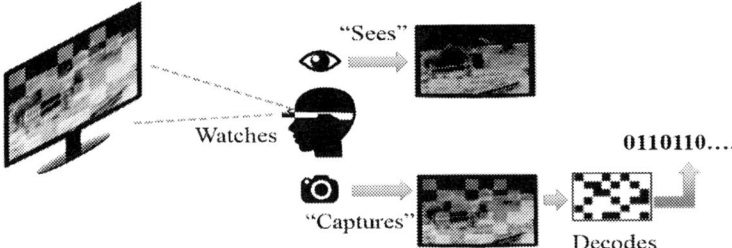

Figure 1. Demonstration of the unobtrusive VLC communication [[5]].

In the display-camera link, due to the rolling shutter characteristics of the CMOS camera and the line refreshing characteristics of the display, the display can able to respectively refresh the pixels of two adjacent frames when exposing a certain pixel. Then, the pixel information actually received by the camera is a mixture between the two adjacent frame pixels, so a new phenomenon called secondary imaging mixture [11,12,15,16] appears. In order to facilitate the analysis, display-camera link is modeled under the assumption of the optical axes of the transmitter and the receiver aligned. So, the received frame $\tilde{y}_k(i,j)$ can be written as,

$$\tilde{y}_k(i,j) = \sum_{l=1}^{L} G\delta(i,j,d,\theta)\frac{T_{k,l}(i,j)}{T_e}y'_{k,l,\tau}(i,j) + n_k(i,j)$$

$$y'_{k,l,\tau}(i,j) = \frac{1}{M'}y_{k,l,\tau}\left(\frac{i'}{M'},\frac{j'}{M'}\right)$$

$$\sum_{l=1}^{L}T_{k,l}(i,j) = T_e$$

$$y_{k,l,\tau}(i',j') = y_{(k,l)+\tau}(i',j')$$

(1)

where $\tilde{y}_k(i,j)$ is the received frame mixed by L adjacent original frames, the number L is related to the transceiver refresh rate and the transceiver relative refresh direction, and G is the lens responsiveness. Attenuation factor $0 \le \delta(i,j,d,\theta) \le 1$ represents attenuation of signal, which is a function of spatial location (i,j) and capturing distance d and angle θ, is the exposure duration of CMOS, $0 \le T_{k,l}(i,j) \le T_e$ is the exposure time for pixel (i,j) in the l-th frame among L adjacent original frames. τ is time-delay factor introduced by frame losing, $y_{k,l,\tau}(i',j')$ is the l-th frame in L adjacent original frames with τ frames lost before. $y'_{k,l,\tau}(i,j)$ is the image of the transmitted image $y_{k,l,\tau}(i',j')$ on the imager plane due solely to geometric optics. M' denotes the magnification predicted by geometrical optics of the imaging system from the transmitter plane to the imager plane [17]. $n_k(i,j)$ is modeled as zero-mean additive Gaussian white noise with variance of σ^2.

As mentioned before, the existing commercial displays mostly adopt the synchronization technology promising the frame losing probability extremely low, thus the impact caused by frame losing can be overlooked. Assuming the image on the imager plane and the transmitted image are the same in size, M' equals to 1. So, the display-camera link model can be further simplified as,

$$\tilde{y}_k(i,j) = \sum_{l=1}^{L} G\delta(i,j,d,\theta) \frac{T_{k,l}(i,j)}{T_e} y_{k,l}(i,j) + n_k(i,j) \tag{2}$$

Since the refresh direction of the display and the camera are the same, the received frame can be considered as a mixture of two adjacent transmitted frames. Therefore, the link model in this case can be expressed as,

$$\tilde{y}_k(i,j) = \sum_{l=1}^{2} G\delta(i,j,d,\theta) \frac{T_{k,l}(i,j)}{T_e} y_{k,l}(i,j) + n_k(i,j)$$
$$= G\delta(i,j,d,\theta)\left\{\left[1 - \frac{T_{k,2}(i,j)}{T_e}\right] y_{k,1}(i,j) + \frac{T_{k,2}(i,j)}{T_e} y_{k,2}(i,j)\right\} + n_k(i,j) \tag{3}$$

Defining mixing factor $\lambda_k(i,j) = \dfrac{T_{k,2}(i,j)}{T_e}$, the link model can be rewritten as,

$$\tilde{y}_k(i,j) = G\delta(i,j,d,\theta)\left\{\left[1 - \lambda_k(i,j)\right] y_{k,1}(i,j) + \lambda_k(i,j) y_{k,2}(i,j)\right\} + n_k(i,j) \tag{4}$$

3. System Design
In this section, due to its simplicity and practicability, the complementary frame design has been widely used in the existing research. So we also discuss our work based on complementary frame design [3, 5-8, 10].

3.1. Complementary Frame Design IC-Based System Design
Complementary frame design to visually hide embedded data in display-camera VLC is based on the following simple idea. A two-dimensional (2D) data pattern is added to one frame and then subtracted from the next frame. So the basic complementary frame design is as follows,

$$\begin{cases} y_k(i,j) = V_k(i,j) + d_k(i,j) \\[2mm] d_k(i,j) = (-1)^k q_{\left\lceil \frac{i}{H/Q_1} \right\rceil \left\lceil \frac{j}{L/Q_2} \right\rceil} \Delta \end{cases} \tag{5}$$

where $V_k(i,j)$ is the k-th original frame, $d_k(i,j)$ is the embedded signal, $y_k(i,j)$ is the displayed frame, H and L represent the frame resolution, and the coordinate range of the pixel of the transmitting end is $1 \le i \le H$, $1 \le j \le L$. The information factor q is the corresponding element in the information matrix $Q_{\left\lceil \frac{k}{T} \right\rceil} = (q_{mn})_{Q_1 Q_2}$. The elements in the information matrix Q are all 1 or -1, and they are mapped by a sequence of 0, 1 bit of length $Q_1 \times Q_2$, 0 and 1 are respectively mapped to -1 and 1. $\lceil \bullet \rceil$ is round up operation, Δ represents signal embedding intensity in B-channel.

In receiver, the embedded signal is recovered from the difference between adjacent camera-captured frames (difference frame decision) under perfect frame synchronization condition with no frame losing and no frame mixing,

$$
\begin{aligned}
\hat{d}_k(i,j) &= \frac{1}{2}\left(\bar{y}_{2k-1}(i,j) - \bar{y}_{2k}(i,j)\right) \\
&\approx \frac{1}{2}\left(y_{2k-1}(i,j) - y_{2k}(i,j)\right) \\
&= d_k(i,j)
\end{aligned}
\tag{6}
$$

3.2. IC-Based System Design

In the display-camera link when information is transmitted, there are about 2.5 consecutive seconds in each 15 seconds, which is a serious mixture and the mixing factor is around 0.5, so that the information at the receiver cannot be recovered well, and the error performance of the system is reduced. We can think of this secondary imaging mixing phenomenon as a kind of periodic deep fading, which can lead to long burst errors. The most common coding technique for solving burst errors is Interleaving Code, which can separate such long burst errors in time. Therefore, we introduce interleaved coding technology into implicit imaging communication to improve system performance.

3.2.1. Interleaving Code Technique. Interleaved code is a channel modification technology, which is essentially a time diversity technique that can correct long burst errors or multiple burst errors. Through the interleaving technique, the burst error channel can be converted into a statistically independent error channel, so that long burst errors are broken up in time, and long burst errors are evenly distributed over the entire time axis. Therefore, the long burst error is converted into a short burst error and there is less error code in one codeword. The burst error can be solved by further using common channel coding techniques.

The principle of the interleaved code is as shown in Figure 2, assuming that the original transmission information of the channel is a codeword A', and $A' = (a_{11}, a_{12}, \cdots, a_{1n}, a_{21}, a_{22}, \cdots, a_{2n}, \cdots, a_{m1}, a_{m2}, \cdots, a_{mn})$. The interleaver is designed as an array memory which is written in row-wise and read out in column-wise. The interleaver is used to read the code array into the interleaver first by row, and then read into the channel for transmission by column. The receiver first needs to deinterleave the information transmitted by the channel, and the deinterleaver works exactly the opposite of the interleaver. The deinterleaver is still an array memory, which is written in column-wise and read out in row-wise. After deinterleaving, we can decode the transmitted information.

Common interleaved code has packet interleaved code and convolitonal interleaved code. In this paper, we use packet interleaved code to design the system. The packet interleaved code arranges the encoded data in a matrix of m (interleaver of degree) row and n (length of one codeword) columns, and the interleaved code matrix A' of the codeword A can be expressed as

$$
A = \begin{matrix}
a_{11} & a_{12} & \cdots & a_{1n} \\
a_{21} & a_{22} & \cdots & a_{2n} \\
\vdots & \vdots & & \vdots \\
a_{m1} & a_{m2} & \cdots & a_{mn}
\end{matrix}
\tag{7}
$$

Each line codeword is a row code or sub code of the interleaved code matrix, and is transmitted in columns in order from left to right in the transmission.

3.2.2. System Design. RS code has strong ability to correct burst errors, and RS code has matured and is widely used in practice. Therefore, in our system design, we adopt reed-solomon (RS) code and interleaved code to solve long burst errors introduced by secondary imaging mixing.

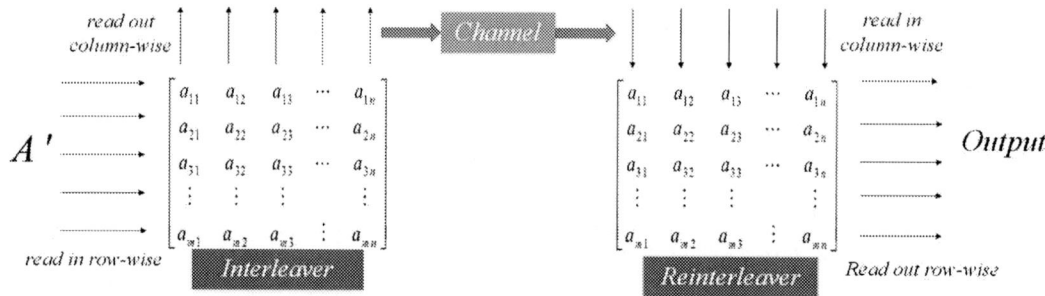

Figure 2. The principle of the Interleaved Code.

Interleaver of degree is a very important parameter in interleaved code, so it needs to be determined in our system design. If a row code (n,k) that can handle burst errors of length $b \leq \dfrac{n-k}{2}$, then it can be combined with an interleaver of degree m to create an interleaved (mn, mk) block code that can handle bursts of length l, which can be expressed as

$$l = mb \tag{8}$$

So, the interleaver of degree can be calculated by,

$$m \geq \left\lceil \frac{r \times F_s \times t}{K \times b} \right\rceil \tag{9}$$

where t is the continuous error time, r is the display refresh rate, F_s is the probability of continuous long error bits, K is the parameter in complementary design.

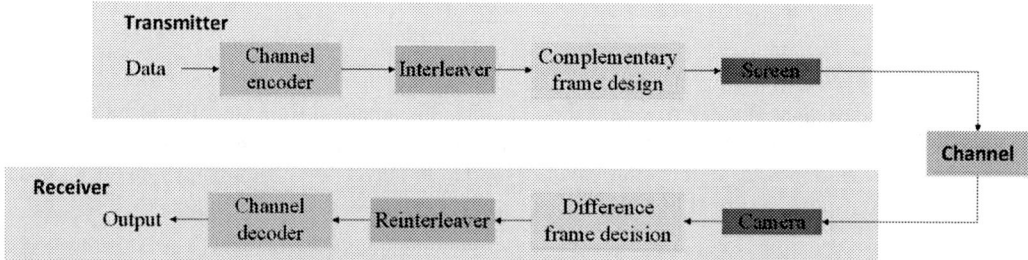

Figure 3. System block diagram of IC-based design.

Our system design based on interleaved code is shown as Figure 3. Firstly, at the transmitter, the data is RS encoded by the channel encoder. Then the RS-encoded data is interleaved by the interleaver and subjected to the complementary frame modulation. After that, the modulated frames are pushed by the display and transmitted through the optical channel. At the receiver, the received frames are captured by the camera and then subjected to different frame decision demodulation. Then deinterleave the data demodulated by different frame decision, and finally decode the data by channel decoder, and the transmitted information can be obtained.

4. Experiment Results and Analysis
In order to test the performance of the proposed systems, prototype systems are set up under laboratory conditions to carry out experiments.

4.1. Experiment Parameters Setup

In our experiments, we leverage an off-the-shelf 27-inch liquid crystal display (LCD) monitor (PHILIPS-272G5DYEB) with resolution of 1920*1080 as transmitter. At receiver, we capture videos of resolution 1920*1080 by using an Apple Iphone6s smartphone. The display refresh rate and camera frame rate are both set to 60. The hardware configurations are shown in the Table 1.

Table 1. Hardware configurations.

Transceiver	Type	Resolution	Frame rate/FPS
Transmitter	PHILIPS-272G5DYEB(27inch)	1920*1080	60
Receiver	Apple Iphone6s	1920*1080	60

Experiment conditions are shown in the Table 2. In the IC-based system design, the RS (15, 9) code is adopted by taking the decoding complexity, coding redundancy and error correction performance into account. The parameters in equation (9) are shown in Table 3, so the interleaver of degree is set to 10 in our experiments.

Table 2. Experiment conditions.

Parameter	Value
Embedding intensity	{1, 2, 3, 4, 5, 6, 7, 8}
Diversity-multiplexing (Block number)	24*40
Distance/cm	110
Angle/degree	0
Illumination intensity/lux	100-200

Table 3. Parameters in equation (9).

Parameter	Value
t/s	2.5
r	0.4
K	2

4.2. Analysis of Experiment Results

In the same experiment environment, the invisible information transmission system based on interleaved code and the system without interleaved code are tested respectively, and the error performance of the two systems under the same information embedding strength is compared. Experiment results is shown as Figure 4.

It can be seen from Figure 4 that the error performance of the invisible information transmission system based on interleaved code increases with the increase of the embedding strength, and the error performance of this system based on interleaved code is better than that without interleaved code. And when the embedding intensity is about 8, the BER is close to 10^{-4}, which indicates that the invisible information transmission system designed in this paper increases the complexity of the system, but the reliability of the system is improved and the information can be transmitted effectively.

5. Conclusions

In this paper, we firstly presented a reasonable model of the display-camera link based on its characteristics. From our analysis, we can see the characteristics of the display-camera link can lead to secondary imaging mixing phenomena, and we further simplified the display-camera link model with the same refresh rate for both the display and the camera. Next, based on the characteristics of the secondary imaging mixing phenomenon in the display-camera link model, we proposed a system design of the invisible information transmission based on interleaved code, which can improve the reliability and validity of the system. Finally, we constructed the system under experiment conditions

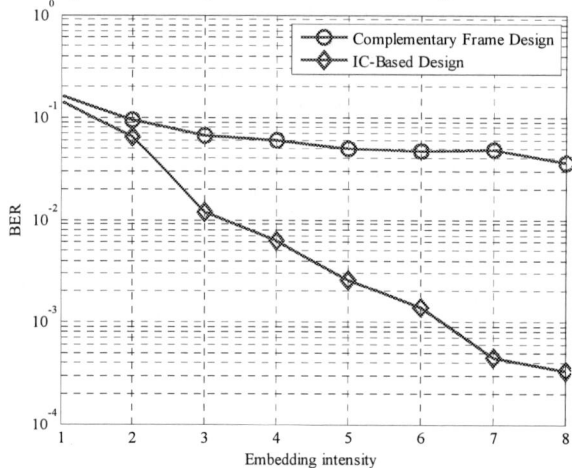

Figure 4. The BER of IC-based system design with different embedding intensity Δ .

and test performance of the system. The experiment results show that the proposed system design can effectively reduce the impact of secondary imaging mixing phenomena and greatly improve the BER performance of the system.

Acknowledgments

This research is supported by National Natural Science Fund of China (No.61601516).

References

[1] Grobe, L., Paraskevopoulos, A., Hilt, J., et al. (2013) High-speed visible light communication systems[J]. IEEE Communications Magazine, 51(12): 60-66.

[2] Yuan, W., Dana, K., Ashok, A., et al. (2012) Dynamic and invisible messaging for visual MIMO[C]. In: IEEE Workshop on Applications of Computer Vision. Breckenridge. pp. 345-352.

[3] Woo, G., Lippman, A., Raskar, R. (2012) Vrcodes: unobtrusive and active visual codes for interaction by exploiting rolling shutter[C]. In: IEEE International Symposium on Mixed and Augmented Reality. Munich. pp. 59-64.

[4] Li, T., An, C., Campbell, A. T., et al. (2014) Hilight: hiding bits in pixel translucency changes[J]. ACM SIGMOBILE Mobile Computing and Communications Review, 18(3): 383-386.

[5] Wang, A., Peng, C., Zhang, O., et al. (2014) Inframe: multiflexing full-frame visible communication channel for humans and devices[C]. In: ACM Workshop on Hot Topics in Networks. Los Angeles. pp. 1-7.

[6] Shi, S., Chen L., Hu, W., et al. (2015) Reading between lines: high-rate, non-intrusive visual codes within regular videos via implicitcode[C]. In: ACM International Joint Conference on Pervasive and Ubiquitous Computing. Osaka. pp. 157-168.

[7] Izz, M., Li, Z., Liu, H., et al. (2016) Uber-in-light: Unobtrusive visible light communication leveraging complementary color channel[C]. In: IEEE International Conference on Computer Communications. San Francisco. pp. 1-9.

[8] Wang, J., Huang W., Xu,Z. (2017) Demonstration of a covert camera-screen communication system[C]. In: International Wireless Communications and Mobile Computing Conference. Guilin. pp. 910-915.

[9] Kim, B.W., Kim, H.C., Jung, S.Y. (2015) Display Field Communication: Fundamental Design and Performance Analysis[J]. Journal of Lightwave Technology, 33(24): 1-1.

[10] Kays, R., Brauers, C., Klein, J. (2017) Modulation concepts for high-rate display-camera data transmission[C]. In: IEEE International Conference on Communications. Paris. pp. 1-6.

[11] Hu, W., Gu, H., Pu, Q. (2013) LightSync: unsynchronized visual communication over screen-camera links[C]. In: ACM International Conference on Mobile computing and networking. Miami. pp. 15–26.

[12] Shu, X., Wu, X. (2016) Frame untangling for unobtrusive display-camera visible light communication[C]. In: ACM Multimedia Conference. Amsterdam. pp. 650-654.

[13] Li, M.C., Hu, Y.P., Tang, Y.Q., et al. (2017) Frame loss detecting for unobtrusive display-camera visible light communication[C]. In: IEEE International Conference on Signal Processing, Communications and Computing. Qingdao. pp. 1-6.

[14] Li, M.C., Hu, Y.P., Yao, X.W., et al. (2018) Frame synchronization compensation for visible light implicit imaging communication[J]. Acta Optica Sinica, 38(1): 0106002.

[15] Do, T.H., Yoo, M. (2016) Performance analysis of visible light communication using CMOS sensors[J]. Sensors, 16(3): 309.

[16] Lee, H.Y., Lin, H.M., Wei, Y.L., et al. (2015) RollingLight: enabling line-of-sight light-to-camera communications[C]. In: International Conference on Mobile System, Application, and Services. Florence. pp. 167-180.

[17] Hranilovic, S., Kschischang, F.R. (2006) A pixelated MIMO wireless optical communication system[J]. IEEE Journal of Selected Topics in Quantum Electronics, 12(4): 859-874.

Application of Deep Convolution Neural Network in Automatic Classification of Land Use

Xiaodong Ma[1], Guang Yang[1], Qunyi Yang[1]

[1]College of Surveying and Geo-informatics,Tongji University, Shanghai, 200092, China

* E-mail: maxiaodong@tongji.edu.cn

Abstract. Automatic classification of land use has always been a topic of concern for remote sensing and land science. It plays an important role in the field of land survey and land management and is the basis for the country to carry out land use planning. In last few years, with more and more high resolution remote sensing platforms is becoming usable, it is possible to update and evaluate land use classification quickly with the advantage of huge volume of data and more frequent of the image data updating. At the same time, we are facing more and more challenges of the big data in practice. With the rapid development and achievements of deep learning in the field of image recognition, this paper introduces a deep convolutional neural network to classify and evaluate the existing land use information, and conduct experiments and demonstrations through the self-constructed convolutional neural network. The test results show that the method has a good effect in the determination of houses, factories, greenhouses, waters and woodlands. Due to the small number of samples and the inconspicuous features, the site is confused with other land features, resulting in lower classification accuracy. The method of this paper can realize the automatic classification of land use types and the evaluation of classification effects.[1]

1. Introduction

Land use type has complex natural and social attributes, making it a hot and difficult problem in the field of land resource management to meet the needs of users for effective classification of land use[2]. Nowadays, with the continuous development of sensor technology, the sources of remote sensing images are becoming more and more abundant, making the solution of the above problems possible[3].

Automatic classification of land use has always been one of the key directions of remote sensing image classification research. The essence lies in summarizing and summarizing various types of land types, so as to obtain the unique characteristics of the category image and other land types, and automatically match the predicted land class to verify the probability of belonging to a certain land class, thereby realizing automatic classification.

Automatic land use classification plays an important role in land surveys, annual change surveys, and natural resource surveys. It can standardize and simplify the workflow of image interpretation, reduce the difficulty of working for internal staff, and improve the production efficiency of image data. Especially with the extensive use of deep learning in the field of image recognition and the continuous development of algorithms in recent years, as well as the improvement and excellent performance of various mature algorithm frameworks, it shows that deep learning is portable in the field of remote sensing image recognition.Based on this point, this paper explores the application of deep learning in land use classification.

Content from this work may be used under the terms of the Creative Commons Attribution 3.0 licence. Any further distribution of this work must maintain attribution to the author(s) and the title of the work, journal citation and DOI.
Published under licence by IOP Publishing Ltd

2. Introduction to data and research method

2.1. Research progress on current land use classification methods

The main methods of current land use classification include field investigation and automatic interpretation of images. This article focuses on automatic interpretation of images. The methods of image classification mainly include supervised classification, unsupervised classification, and semi-supervised classification. Among the commonly used classification algorithms, support vector machine (SVM) classification, decision tree (DT) classification, random forest (RF) classification, and semantic modeling based classification algorithms encounter many difficulties in dealing with practical engineering problems (such as high resolution). Remote sensing image classification), the theoretically optimal method is difficult to obtain satisfactory results in practice. These are all automatic classification algorithms for shallow structure models. They are characterized by the fact that the input signals are often processed using only a small number of linear or nonlinear methods, and it is often difficult to achieve satisfactory results for complex signals. [4]

Deep learning is a deep-rooted machine learning method. It is the latest development trend of artificial neural networks. By extracting more abstract features from the lower layer to the higher layer, the input data is formed into a network weight structure that best fits the required features. Accuracy. Hinton et al. use the deep learning model to classify the data, and conclude that the deep neural network structure can learn more abstract features than the existing methods, and has stronger classification ability and good generalization ability. Dang Yu et al. used the AlexNet convolutional neural network to classify and evaluate the surface coverings, and achieved good classification results [1]. Men Jilin et al. used multi-structured convolutional neural networks for classification verification of high-resolution image land use. The research indicates that deep learning has certain feasibility in land use classification. [5]

2.2. Image and experimental area introduction

The data sources used in this study include: aerial imagery collected by the surveying and mapping department from June to September 2017. The image is RGB full-color band with a resolution of 1 m. The supporting data includes the administrative division maps of various levels in Shanghai, which are used to determine the scope of research, the current situation of land use in Pudong New Area (field survey, visual interpretation of images, and information provided by relevant departments) as a comparison standard for classification and classification results.

The research area of this experiment is located in the southern part of Chuansha New Town, which is a typical suburban area. There are a total of 496 land use maps, and the land use types are divided into six categories: buildings, farmland, greenhouses, woodland, waters and construction sites.

2.3. Make samples

This experiment uses land parcels from other regions as sample set. The features selection is done by the GIS masking tool. According to the purpose of classification, the samples are divided into waters, farm land, greenhouses, construction sites, buildings and forests. The feature sample selection shows in table 1.

Table 1. Features selection summary.

Features / Number	Factory	Building	Farmland	Waters	construction	Greenhouse	Forest
	101	107	183	80	60	106	50

2.4. Introduction of the experimental environment

The experiment uses the Tensorflow open source framework and CUDA-GPU acceleration scheme under the Ubuntu operating system, and the NVIDIA GeForce 1080Ti (11 G memory) used for the graphics card for GPU acceleration. The other main hardware is Intel Xeon E5-2620 eight-core processor, 64G memory, PCI. -E X8 interface, 120 G solid state drive, etc.

2.5. Introduction of experimental methods

This experiment uses a multi-layer convolutional neural network built by itself, which is divided into six layers, which are two-layer convolution, two-layer pooling, and two full-connection layers. The first convolutional layer uses 96 convolution kernels (3*3*3) to filter the 96*96*3 image with a single sliding step. After being processed by the convolutional layer, the data is subjected to ReLu activation and normalization transformation, and then pooled, and passed as an output to the next layer.

The first fully linked layer is stacked after the last convolution and pooling layer, and the number of signatures is reduced by half after the final pooling layer. The last full link layer output fuses the softmax result of the tag. There are 6 nodes in this layer, which correspond to 6 categories of farmland, forest land, water body, house, construction site and greenhouse. After being processed by the fully connected layer, the network accurately distinguishes the six land features and accurately classifies the imported house maps. As follows:

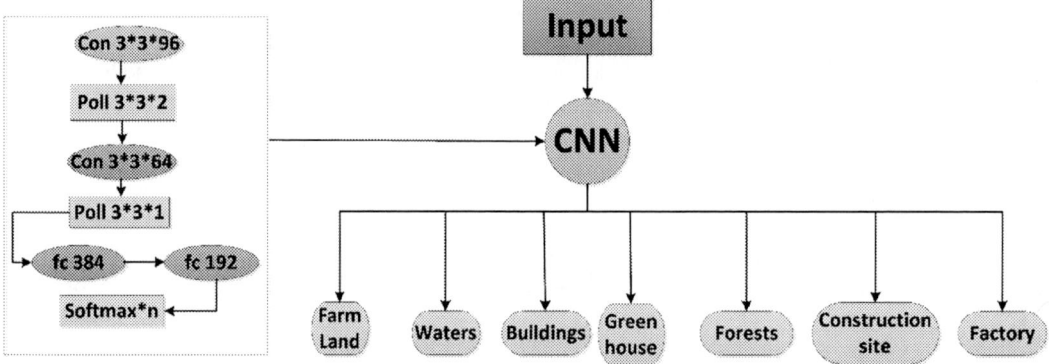

Figure 1. CNN structure diagram

2.6. Experimental process

According to the method described in 2.2, the sample characteristics are determined and the sample set is selected. The sample set is used as input to carry out model training. According to the training result, the convolution layer, the pooling layer and the fully connected layer are fine-tuned to obtain a suitable prediction network model. The experimental process shows in figure.2.

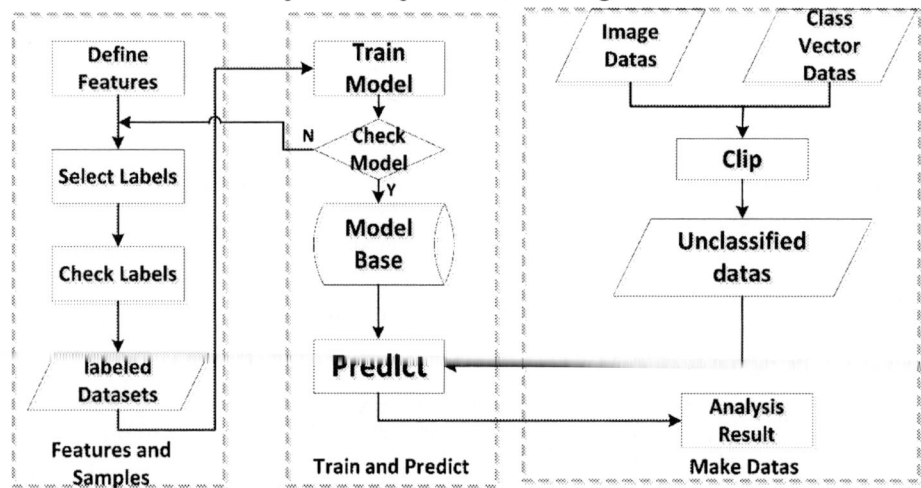

Figure 2. Classification work flow diagram

The parameters of the full connection layer were trained with the selected sample set. The parameters were set to a learning rate of 0.01, a batch size of 100, a weight attenuation rate of 0.002, and a training set of 687. Through the training loss value and the development trend of the verification loss value, the degree of fine adjustment and over-fitting phenomenon are judged[6]. The figure.3 shows the loss and classification accuracy of the training set after each iteration.

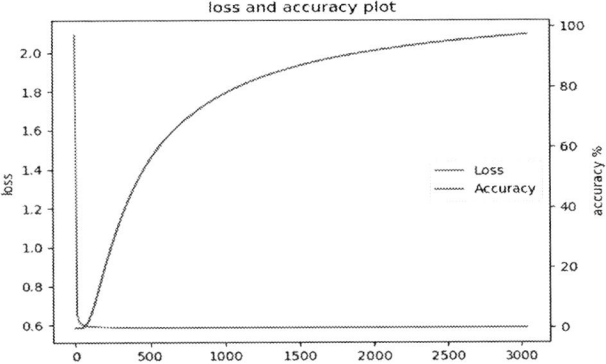

Figure 3. Loss and accurancy curve diagram

It shows that in 100 iterations, the loss rate drops rapidly but the accuracy does not change much. After 3000 trainings, the model's loss rate and accuracy tend to be stable, so the 3000th training model is selected as the prediction model.

3. Experimental analysis

The prediction data set is taken as input, and the training model derived from 2.6 is used as the prediction model for classification prediction. The classification result is compared with the land use status map to obtain the classification result confusion matrix. The result shows in table2:

Table 2. Classification result confusion matrix.

Predict category	Actual category							Total	User Accuracy/%
	Factory	Building	Farmland	Waters	Constru-ction	Green-house	Forest		
Factory	29	6	0	0	0	0	0	35	82.86
Building	2	90	2	0	6	4	0	104	86.54
Farmland	0	7	151	1	7	5	5	176	85.80
Waters	5	3	0	35	0	3	1	47	74.47
Construction	0	5	2	0	3	1	0	11	27.27
Greenhouse	2	10	2	0	2	89	0	105	84.76
Forest	0	0	2	0	0	2	14	18	77.78
Total	38	121	159	36	18	104	20		
Cartographic Accuracy/%	76.32	74.38	94.97	97.22	16.67	85.58	70.00		

Overall classification accuracy=82.86%   **Kappa coefficient=0.7792**

As the results shows the overall classification accuracy is 82.86%, Kappa coefficient is 77.92%, the consistency test results are highly consistent, and the classification results are highly reliable.

Among them, the classification accuracy of users in farmland, buildings, factories and greenhouses is higher than 80%, the classification accuracy of water and forest land is above 70%, and the classification accuracy of construction is low,which is 27.27%. The cartographic accuracy in farmland

and waters is high, about90%; The classification accuracy of greenhouse is 85.58%; The cartographic precision of plant, building and forest land is above 70%, and the classification accuracy of construction site is low, which is 16.67%. This confusion matrix shows most of the construction lands are classified into buildings and farmland for the reason that the characteristics are overlaped among them, which leads to ambiguity of features. The number of prediction sets of the construction features is small, which is one of the reasons for the high misclassification rate. Among them, the comprehensive classification accuracy of farmland and water area is the highest, and the classification effect of greenhouses and buildings is better. Because forests' feature is close to farmland and greenhouse, there is also a certain degree of misclassification.

In summary, it is applicable for using the convolutional neural network of deep learning in land classification.

4.Conclusion

This paper mainly studies the application feasibility of deep learning convolutional neural network in land use classification, and takes the aerial image of a certain area in Shanghai as an example for experimental demonstration. The experimental results show that deep learning is applicable in land use classification.

In land use surveys, land use monitoring, and land change surveys, deep learning can also be used to automatically classify features. At the same time, the classification prediction model can be continuously improved, so that it can adapt to the classification of different seasons and different regions, and improve the prediction accuracy of classification. The current depth learning in the field of pattern recognition has reached 90% of the overall recognition accuracy of objects. Above, of course, the land use type has certain characteristics such as high complexity and homogenization, which makes the classification accuracy more challenging.

(1) Feature extraction automatically. The extraction of current features also requires manual intervention, which requires automatic segmentation of remotely sensed images, thereby reducing manual workload. Achieve complete automatic classification.

(2) The classification accuracy of subdivisions needs to be improved. At present, it is only possible to achieve higher classification accuracy between the first-level land types and the land types with obvious differences between the classes. How to improve the classification of features with small differences between classes needs to be discussed in the future.

References

[1]Dang Yu, Zhang Jixian, Deng Kazhong,et al. 2017.Study on the evaluation of land cover classification using remote sensing images based on AlexNet[J]. Journal of Geo-information Science, 19(11):1530-1537.

[2] Gao Xianjun, Zheng Xuedong, Shen Dajiang, et al.Automatic Building Extrac tion Based on Shadow Analysis from High Resolution Images in Suburb Areas[J]. Geomatics and Information Science of Wuhan University, 2017, 42(10):1350-1357.

[3] Huang Wei, Li Yonggang, Wang Yi, et al. Spatial Cooccurrence Kernel Based Aerial Image Classification[J]. Geomatics and Information Science of Wuhan University, 2017, 42(7):884 - 889.

[4] Cao Linlin,Li Haitao,Han Yanshun,,et al.Application of convolutional neural networks in classification of high resolution remote sensing imagery[J].Science of Surveying and Mapping.2016,41(9);170-175.

[5] Men Jilin,Li Yueyan,Zhang Bin,Zhou Fan. (2018) Land Use Classification Based on Multi-Structure Convolution Neural Network Features Cascading. https://doi.org/10.13203/j.whugis20180137.

[6]Ge Yun, Jiang Shunliang, Ye Famao, et al. Remote Sensing Image Retrieval Using Pre -trained Convolutional Neural Networks Based on ImageNet[J].Geomatics and Information Science of Wuhan University.

Research on IP Address Allocation of Tactical Communication Network

Wang Huitao[1], Yang Ruopeng[1], Wufan[1]and Zou Xiaofei[1]

[1]College of Information and Communication, National University of Defense Technology, Wuhan 430010, China

81445785@qq.com

Abstract: According to the IP address allocation problem of tactical communication network, the structure and IP address requirements of tactical communication network are analyzed in paper. A tactical communication network IP allocation method based on Classless Inter-Domain Routing (CIDR) addressing technology and without network-wide routing configuration is proposed. The allocation principle, addressing method and allocation process of the method are analyzed in detail. Based on this, the algorithm and model of IP address allocation in tactical communication network are given. Finally, the model and algorithm are analyzed and verified with examples. .

1. Introduction

The tactical communication network is a layer-by-layer communication network that uses wireless communication as the main means to interconnect various communication user terminals and channels in the network, and realizes interconnection and intercommunication between units at all levels to meet the needs of information and communication support of troops. Due to the adoption of IP interconnection technology, various types of communication terminals in a tactical communication network are required to uniformly allocate and address their IP addresses to meet the needs of the communication network for IP addresses. The IP address allocation of the existing tactical communication network is mainly carried out by network planning software. However, from the perspective of distribution, the following problems still exist: First, manual allocation is dominant, randomness is easy, and error is easy. Second, each unit is separately allocated and lacks systematic distribution strategies. The third is the separation of communication and allegations, lacking integrated design and distribution. In view of the above problems, this paper proposes a tactical communication network IP address allocation model and algorithm, and carries out verification analysis to improve the tactical communication network IP address utilization and network opening time.

2. Tactical communication network structure [1][2]

The tactical communication network is mainly composed of a variety of routers, switches, computers, communication terminals, etc., and interconnects numerous devices through a tactical communication platform to improve stable and reliable information and communication support for the troops. Its network structure is shown in Figure 1.

Content from this work may be used under the terms of the Creative Commons Attribution 3.0 licence. Any further distribution of this work must maintain attribution to the author(s) and the title of the work, journal citation and DOI.

Published under licence by IOP Publishing Ltd

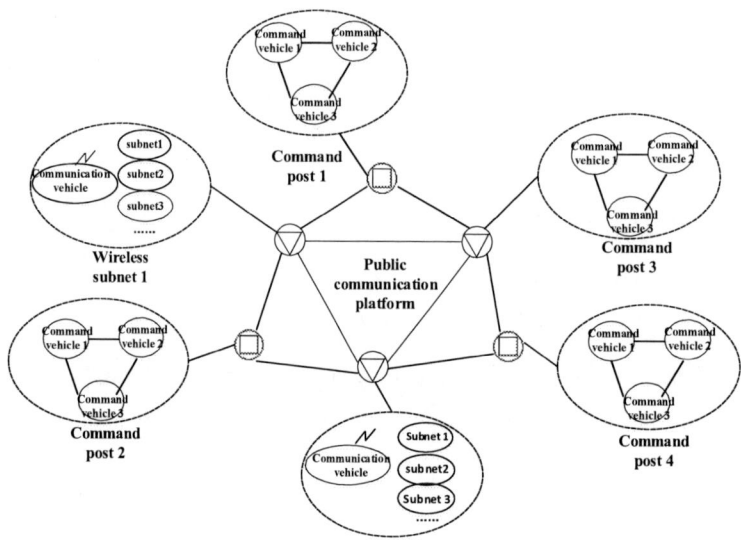

Figure 1.Schematic diagram of tactical communication network structure

The tactical communication network mainly consists of a public communication platform, a command post network and various wireless subnets, and each network is interconnected by an IP interconnection protocol. In each network, the used IP address resources mainly include an IP service address and an IP management address. The IP service address mainly includes the switch core address, the switch port address, various mobile communication device addresses, computer terminals, and the like. The IP management address mainly includes an IP address used for device management, such as a node management device, a communication device, and a computer terminal device, etc.

2.1 Command post network

The command post network is mainly used to ensure the communication of the command post. It is mainly based on various types of communication and accusation vehicles. The IP address used by each communication vehicle mainly includes the terminal address of the command post user, device management address, and Interconnection address and reserved address of the communication platform within the command post. Among them, the terminals of the command post's user at all levels are mainly connected by Ethernet interfaces. Device management addresses are mainly network switch interfaces; the interconnection addresses of the communication platforms of the command post are mainly switches, communication terminals and communication interconnection devices.

2.2 Public communication platform

The public communication platform is mainly used to build a public communication network covering the entire operational area, and to ensure access to users at all levels of the command post. The IP addresses required for each public platform communication vehicle mainly include the switch address, the in-vehicle device management address, the communication platform address, and the user terminal address.

2.3 Wireless subnet

The wireless subnet is mainly used for the communication between various users to access the superior command post or the command post of the same level.

It is mainly composed of various VHF Radio stations to build a hierarchical radio subnet, whose IP address is mainly the tactical radio port address.

3. IP address allocation and addressing

3.1 Distribution principle
According to the actual application and technical requirements of the tactical communication network, the following principles must be followed when assigning IP addresses:

- *The principle of uniqueness.* The uniqueness of the IP address is the basis for the tactical communication network user terminal to correctly carry the service. To avoid the IP address conflict of the entire network, the IP address of each device should be unique to the entire network.
- *The principle of aggregation.* Due to the large scale of tactical communication networks, the large number of equipments, and the existence of a large number of heterogeneous networks, in order to reduce the number of router entries, the principle of router aggregation should be considered in planning.
- *The principle of autonomy.* For tactical communication networks, they are usually divided into different autonomous regions according to the force preparation. Each autonomous region can be divided into different small autonomous regions, and each autonomous region is relatively independent, which can effectively reduce the difficulty of IP allocation.

3.2 Addressing method [3][4]
The traditional IP address allocation mainly takes the IP address into a fixed class. The two-level addressing method is adopted. The IP address is in the form of "Network number + Host number". This addressing method can effectively save the IP address space. It is advantageous to divide the subnets. However, as the number of networks increases, the number of routing increases greatly, which reduces the forwarding speed of IP packets and is not flexible enough entries h. In order to improve the efficiency of IP address utilization, reduce the number of routing entries, improve the speed of network forwarding, and reflect the relationship between hierarchical management and unified planning, this paper uses Classless Inter-Domain Routing (CIDR) addressing technology for subnetting, A "CIDR address block" is composed of continuous IP addresses with the same network prefix, and the addressing method is "Network prefix + Host number". The IP address is divided into subnets with different network prefixes and assigned to each node. Each node divides the next subnet into the next subnet according to the assigned subnet and assigns it to the next subnet, and so on, until the IP address is assigned to each terminal device and user in the network. Because the prefix of the network is variable, the subnets of different lengths can be divided according to different situations, which fully meets the requirements of flexible and multi-level hierarchical division of battlefield network.

3.3 Assignment process
First, determine the total IP address allocation resources according to the operational requirements, network structure, and user scale of the tactical communication network. Take the bottom-up, level-by-level summary, and analyze the IP address requirements of each network layer by layer. When determining the requirements of each network, the network IP address segment requirements are determined according to the number of nodes and interconnections of each network platform, and then the seat address segment and the IP address segment of the internal and external subnet of the command post are clearly defined according to the number of seats in each command post and the internal and external network interconnection relationship of the command post. The IP address segment of each subnet is determined according to the number of wireless packet subnets and the number of users in the subnet.

The second is to analyze and determine the IP requirements. A unified planning and distribution method is adopted. According to the hierarchical division method, the CIDR technology is used for subnetting, and the bottom-up and level-by-level aggregation methods are adopted to analyze the IP address requirements of each node layer by layer. Continuous address block allocation and hierarchical partition are adopted to implement route aggregation, ensure the relative independence of IP addresses between networks, and improve network stability. The basic idea is shown in Figure 2.

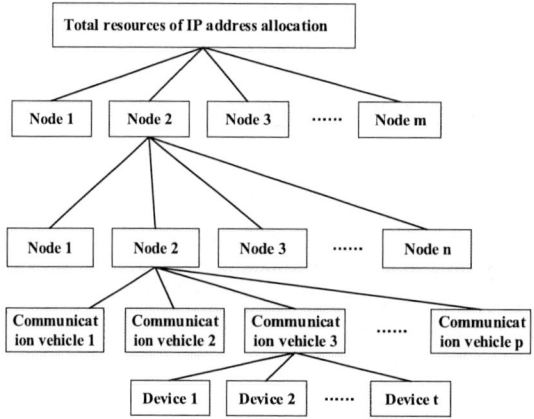

Figure 2.IP layering

The third is to specifically assign an IP address. According to the IP address requirements, IP addresses are allocated to each device and user in a segmented aggregation manner.

The fourth is to verify the allocation of IP addresses. It mainly checks the route aggregation and conflicts of IP address allocation.

4. Models and algorithms [5-8]

4.1 Assumptions

- The IP address has the uniqueness of the whole network;
- The IP address adopts a layered manner, and each command post and public platform network adopts an independent IP address segment;
- The radio station interconnected by the Internet controller in the subnet uses the core IP address controlled by the Internet, and no longer allocates the address separately;
- The communication and accusation equipment IP addresses are uniformly planned, and the address segments are not separately divided.

4.2 Assignment models and algorithms

4.2.1 Assignment algorithm

According to the CIDR addressing rules, CIDR usually uses a slash notation to indicate an IP address. If the number of bits occupied by the network prefix is **m**, the IP address can be expressed as: IP address/**m** according to the CIDR rule. It represents that the prefix bit in the IP address is **m** bits, the default mask of the class A address can be represented by **/8**, the class B address can be represented by **/16**, and the class C address can be represented by **/24**.

Hypothesis $\mathbf{m_1}$——The number of network prefix bits before the subnet is divided. $\mathbf{m_2}$——The number of network prefix bits after the subnet is divided.

The number of subnets in a tactical communication network is $\mathbf{N_2}$, the required number of subnets is$\mathbf{n_1}$, then meet: $\mathbf{2^{n_2} - 2 \geq N_2}$, thus the number of required subnets is:$\mathbf{n_2 \geq \log_2(N_2 + 2)}$.

Similarly, the number of IPs used by devices in a subnet in a tactical communication network is $\mathbf{N_2}$, the required number of users is $\mathbf{n_3}$, then meet:$\mathbf{2^{n_3} - 2 \geq N_3}$, The required number of users is: $n_3 \geq \log_2(N_3 + 2)$.

According to the above formula, the number of network prefix bits before the subnetting is known:

$m_1 = 32 - n_2 - n_3$.The number of network prefix bits after subnetting is $m_2 = 32 - n_2$, according to the number of network prefixes s_2 ,The corresponding subnet mask is available.

The above is a general allocation algorithm for IP address allocation. Next, the route aggregation algorithm is discussed.

First, find out the key bytes that are aggregated in the IP address, that is, the bytes with different values in each subnet address. Secondly, analyze the number of users in the subnet, analyze the number of bits used in selecting the subnet address, so that the number of subnet addresses is greater than or equal to the maximum keyword section minus the minimum key section, and find the appropriate subnet address block by calculation, so that: **Keyword Section/Address Block Size = Maximum Keyword Section/Address Block Size**.

Again, the address block size is calculated, and it can be concluded that the user borrows **n** bits from the prefix for subnet aggregation.

Finally, the network routing address after aggregation = **Keyword Section - Keyword Section / Address Block Size**.

4.2.2 IP address calculation
(a) Calculation of the IP address of the command post

The command bases of tactical communication networks are mainly based on communication vehicles and are interconnected according to certain network topologies. Depending on the level of each command post, it is also different for the required IP address devices. For Class 1 command posts, which mainly include communication vehicles and command vehicles, the maximum IP address on each vehicle is no more than **64**, and the mask is an address segment of **/25**.For the type 2 command organization, mainly the command vehicle, the maximum IP equipment is usually no more than **32**, and its mask is **/26** address segment. For the type 3 command terminal, the maximum IP equipment is no more than **16**, and its mask is **/27** address segment. The specific requirements are shown in the table.

Table 1. *the IP address demand table of the command post*

Device type	Class 1 command agency command vehicle IP address requirements	Class 2 command agency command vehicle IP address requirements	Class 3 command agency command terminal IP address requirements
User terminal	24	16	4
device management	16	8	4
communication device	8	4	4
Reserved	16	8	4

(b) Calculation of IP address of public communication platform

The public communication platform is mainly constructed according to various communication vehicles. The maximum number of IP addresses used by each communication vehicle is **64**, and the mask is an address segment of **/25**.

(c) Calculation of IP address of Wireless packet subnet

The packet subnet address segment is mainly used for the VHF Radio network. Generally, the maximum number of subnet members is not more than 16, and the mask is /27.In order to facilitate route aggregation, the terminal IP address should be obtained from the reserved address of the troop (unit) of the nearest access point.

5. Examples

Assume that the IP address resource assigned by the superior to a tactical communication network type 1 command organization is a Class B address XXX.XXX.0.0/16, which is mainly used to secure the Class 1 command post network, the public platform network, and the wireless packet subnet. IP address is used. According to the tactical communication network allocation model and algorithm, the

IP address allocation scheme of the type 1 command organization is shown in Table 2.The following takes the command post network as an example, and assigns an IP address to it. The allocation is shown in Table 3.

Table 2. IP address allocation scheme for tactical communication

	X17 X18 X19	X20 X21	X22 X23 X24 X25	X26 X27 X28	X29 X30 X31 X32
Network Address Resource Pool	Command post network (max=8)	0: User terminal	Number of node communication vehicles max=16	Number of user terminals per vehicle max=64	
		1: Device Management	Number of node communication vehicles max=16	Each vehicle equipment max=16	IP address of each device max=16
		2: Communication platform	Device type max=8	IP address of each device	
		3: Reserved			
	Public communication platform (max=8)	0: User terminal	Number of node communication vehicles max=16	Number of user terminals per vehicle max=64	
		1: Device Management	Number of node communication vehicles max=16	Each vehicle equipment max=16	IP address of each device max=16
		2: Communication platform	Device type max=8	IP address of each device	
		3: Reserved			
	Wireless subnet (max=32)		Number of node subnets max=8	Each vehicle equipment max=4	Number of members per subnet max=16
	Reserved				

Table 3. IP address allocation table of command post network

Command post network (XXX.XXX.0.0/16 ~ XXX.XXX.255.255/16)		Command organization 1 (XXX.XXX.0.0/19 ~ XXX.XXX.31.255/19)		User terminal (XXX.XXX.0.0/21~XXX.XXX.7.255/21)		......
Command organization 1	XXX.XXX.0.0/19	User terminal Equipment management	XXX.XXX.0.0/21	Car 1	XXX.XXX.0.128/25	
Command agency 2	XXX.XXX.32.0/19		XXX.XXX.8.0/21	Car 2	XXX.XXX.1.0/25	
Command organization 3	XXX.XXX.64.0/19	Communication platform	XXX.XXX.16.0/21	Car 3	XXX.XXX.1.128/25	
Command organization 4	XXX.XXX.96.0/19	Reserved	XXX.XXX.24.0/21	Car 4	XXX.XXX.2.0/25	
Command agency 5	XXX.XXX.128.0/19			Car 5	XXX.XXX.3.128/25	
......	......			...	......	

6. Conclusions

The IP address planning problem of tactical communication network is an important part of tactical communication network management planning, which plays an important role in ensuring the normal operation of the communication network. Based on the analysis of the structure of tactical communication network and the requirement of IP address, this paper proposes a tactical communication network IP allocation method based on Classless Inter-Domain Routing (CIDR) addressing technology and without whole network routing configuration, which achieves fast and automatic IP address allocation and configuration of communication devices of tactical communication network. The actual use effect shows that the allocation method and the allocation process can ensure the correctness of the IP address allocation, shorten the network startup time, and reduce the difficulty of network startup.

References

[1] Yu Jingdong, Zeng Renjie. Overview of IP Address Allocation Technology in Ad Hoc Network[J].Communication Technology,2007(12):130-133

[2] Dai Fei, Zhong Lianjiong, Zhang Kunao. IP address management of tactical communication network [J]. Technology and Innovation Management, 2010.01: 115-118

[3] Xie Xiren, Computer Network (seventh edition) [M]. Beijing: Publishing House of Electronics Industry, 2017.1

[4] Qiu Xiurong. Automatic allocation method of IP address in wireless mobile communication network [J]. Science and Technology Bulletin, 2018(7): 23-35.

[5] LI Haitao, Long Yixiang, Meng Fanxin. A Method for IP Address Assignment in Mobile Communication Network[J].Communication Technology,2012.07:95-96

[6] Fan Xianxue, Jin Xinghua. Networking of Mobile Command and Control System Based on Mobile IP Technology[J]. Command Information System and Technology, 2016.06: 32-36

[7] ABDELMALED A, FEHAM M, TALEB AHMED A. On Recent Security Enhancements to Auto configuration Protocals for MANETs Real Threats and Requirements[J].International Journal of Computer Science and Network Security, 2009.9(4):401-407

[8] Xie Shuizhen. Research on Subnet Partitioning Method [J], Communication Technology, 2011.09: 60-62

ISPECE

IOP Publishing

The priority assignment of messages effects on delay performance in VANET

Jin Tian[1], Qianru Han[2]

[1]Engineering School of Networks and Telecommunications,

Jinling Institute of Technology
Nanjing, China

[2]College of Internet of Things Engineering,

Hohai University

Changzhou, China

Jim.tian@jit.edu.cn   171619020003@hhu.edu.cn

Abstract. Low delay performance between neighbor vehicles in VANT need to an indepth research. In this paper, we conceived a novel priority assignment plan to reduce transmission delay. A priority was assigned to each message based on static and dynamic factors and size of the message. Dynamic scheduling sequence was carried out based on the messages priority assigned. The performance of proposed method was analysed in highway scenario on the average delay and waiting delay in queue. Simulation results are consistent with theoretical deriving. Therefore, low latency environments can be provided in Vehicular Ad-hoc Network using the novel plan.

1. Introduction

Vehicular Ad-hoc Network (VANET) has special characteristics such as high rate of topology change, high mobility of nodes, high nodes density, sharing the wireless channels, and frequently broken rout. Those special characteristics in VANET give rise to some challenges in data transferring and scheduling [1-4]. When the channels are saturated due to the increasing number of vehicles, congestion happens in the networks. In other words, when the vehicles send messages simultaneously in high density situations, the shared channels are easily congested. Congestion indeed leads to overload the Medium Access Control (MAC) channels, increases the packet loss and delay, and consequently decreases the performance of VANET. In this situation, safety messages (especially emergency messages) cannot be properly transmitted due to deficiency in the messages scheduling. It should be also noted that the scheduling in VANET is faced to some challenges because of sharing wireless communication channel and employing multi-channel technology with single-radio transceivers. Therefore, congestion should be controlled for enhancing the reliability of VANET [5-8].

In VANET, data traffic initiated from vehicles is expected to be random and bursty in nature. Vehicle-to-vehicle (V2V) communication is generally used in a multi-hop fashion in order to allow vehicles to connect to the out-of-transmission range road-side unit (RSU) [9]. In this way, bursty data flows are delivered to a remote vehicle through multi-hop transmission, which facilitates to support a broad range of applications including, e.g., car accident alert, road condition warning, video streaming,

Content from this work may be used under the terms of the Creative Commons Attribution 3.0 licence. Any further distribution of this work must maintain attribution to the author(s) and the title of the work, journal citation and DOI.
Published under licence by IOP Publishing Ltd

web browsing, and file sharing [10]. Although multi-hop V2V communication can significantly improve the connectivity probability, the increases of end-to-end backlog and delay are difficult to avoid, which impact the quality of service (QoS) of the vehicular communication and may lead to inevitable network resources waste. From passengers' perspective, end-to-end backlog and delay are the most critical factors and are arguably the most important service requirements that directly affect their experiences and satisfaction levels. Thus, it is necessary to have an indepth understanding of the multi-hop end-to-end queueing performance of different types of data services. However, how to analyze the multi-hop end-to-end backlog queueing bound still remains as a challenging issue under the circumstances of the bursty data traffic characteristics and highly dynamic channel environment.

The multi-hop delay [11] of the safety related messages is derived by multiplying the one-hop delay and the average hop count. One-hop delay is a period of time under which a packet is delivered from source to destination. The delay is composed of the processing delay (D_{proc}), queuing delay (D_{queue}), transmission delay (D_{trans}), and propagation delay (D_{prob}) [12]. The processing delay is the time needed for extracting header of packets and executing various algorithms (e.g., routing algorithms, congestion control algorithms, so on). The queuing delay is the waiting time of a packet in a queue before transferring. The transmission delay is the required time for transferring. The propagation delay is the required time for propagation of the packet. Thus:

$$Delay = D_{proc} + D_{queue} + D_{trans} + D_{prop} \qquad . \tag{1}$$

The rest of the paper is organized as follows. Section II researches delay element. Section III proposes how to prioritize the messages in order to reduce delay time of packets transmission. Section IV analyses delay performance of transmission delay and queueing delay. Section V simulates the proposed strategy in the highway and discusses the obtained results.

2. Motivated Element

A delay control strategy in VANET is prioritizing messages in MAC layer. When the messages need access the channel, we can first define priority for the messages and then send them into different communication channels [13], [14]. This method maybe serve more requests, reduce download delay, packet loss, and so on [15]. Bouassida et al. [16] first introduced a congestion control strategy that the priorities were assigned to messages based on utility and validity of messages, and speed of senders and receivers. Suthaputchakun et al. [17] proposed a priority-based strategy using Enhanced Distributed Channel Access (EDCA) mechanism to increase safety in highway environments. EDCA can classify the message into four categories according to message urgency condition.

Generally, there are two types of congestion control mechanisms in networks: 1) open-loop mechanism without information feed and 2) closed-loop mechanism by information feed [18].

Congestion control strategies in VANET include prioritizing and scheduling the messages in communication channels [19]. Generally, the prioritizing and scheduling the messages is a very common open-loop congestion control strategy in communication channels [20]. Some performance metrics are considered to increase efficiency of message transmission in VANET [21]. In this paper, a delay control strategy is presented to reduce message delay. The paper mainly changes a few metrics such as weather condition and geography position.

3. Priority Assignment for the Message

In the priority assignment unit, priorities are assigned to the messages generated by applications in the vehicle or received from the other vehicles. Then, what time the message is transmitted is determined based on its priority assigned. The priority of each message is defined based on static and dynamic factors as well as size of message:

$$\mathrm{Pr}\,iority_{Message} = \frac{Static_{Factor} \times Dynamic_{Factor}}{Message_{size}} \tag{2}$$

$Priority_{Message}$ is directly proportional to $Static_{Factor}$ and $Dynamic_{Factor}$. However, because the emergency and high priority safety messages have smaller size compared to the other messages, $Priority_{Message}$ is opposite proportional to $Message_{Size}$.

The $Static_{Factor}$ is defined based on the content of messages and type of applications. $Static_{Factor}$ for a message is considered to be 1, 3, 4, or 5 if the message belongs to $Priority_{Service}$, $Priority_{Safety-Low}$, $Priority_{Beacon}$, or $Priority_{Emergency}$ category, respectively [16]. $Message_{size}$ is considered to 1 or 100 if the message size is less than 500 byte or bigger than 500 byte. In the following, each category is detailedly defined as [22], [23].

In contrast of static factor defined based on the content of messages and type of applications, dynamic factor is defined based on circumstances of VANET. The metrics considered for calculating the dynamic factor are velocity of vehicles, usefulness of messages, validity of messages, directions of sender and receiver vehicles, weather condition, and geographic position. In the following, these metrics are described in details.

1.**Velocity metric (Vel)**: This metric represents the relative speed of message sender that is defined based on the total coverage area of a vehicle traveling with velocity v during time dt (Figure 1) [24]:

$$Vel = \frac{\pi \times R^2 + 2 \times R \times v \times dt}{\pi \times R^2} \tag{3}$$

where R is communication range, and v is average speed of vehicle in time dt. A higher priority should be assigned to the message with higher Vel metric.

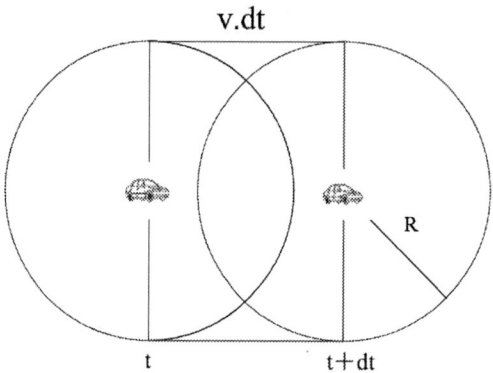

Fig. 1. Velocity metric (Vel).

2. **Usefulness metric (Use)**: This metric is defined according to the probability of message retransmissions by the neighbor vehicles. The usefulness is determined by the inverse of retransmissions:

$$Use = \frac{1}{retransmissions} \tag{4}$$

3. **Validity metric (Val)**: Validity metric is defined as the remaining time to the message deadline in real-time applications. The validity can be given by Equation (5) [24]:

$$Val = \frac{\text{Re}\, maining\ Time\ to\ the\ Deadline}{Transferring\ Time} \tag{5}$$

Transferring Time in this equation, which is used for normalization, shows an estimated time to transfer message between sender and receiver vehicles.

4. Distance metric (Dis): This metric is considered as a relative distance between message sender and receiver [24].

5. **Direction metric (Dir)**: direction metric shows that two vehicles (sender and receiver) are driving closer to each other (*Dir*=0) or they are driving away from each other (*Dir*=1) [24].

6. **Weather Conditions metric (WC)**: messages generated in severer weather condition should first be transferred.

7. **Geographic Position metric (GP)**: the priority of messages produced in rapid position is higher than that in ordinary position.

By combining Equations (3) to (5), the dynamic factor is calculated by Equation (6):

$$Dynamic_{Factor} = \begin{cases} \dfrac{Vel \times Use}{(Val+1) \times Dis \times WC \times GP} & dir = 0 \\ \dfrac{Vel \times Use \times Dis}{(Val+1) \times WC \times GP} & dir = 1 \end{cases} \tag{6}$$

Based on Equations (6) and (2), dynamic factor and consequently message priority are directly proportional to *Vel* and *Use* metrics. However, dynamic factor and message priority are opposite proportional to *Val*, *WC*, and *GP* metric. In this equation, *Val* metric is added to 1 to avoid ambiguous result when the validity is equal to zero. Equation (6) shows that dynamic factor is opposite proportional to *Dis* metric when *Dir* is equal to 0. However, dynamic factor is directly proportional to *Dis* when *Dir* is equal to 1. EDCA above mentioned is the default strategy of prioritizing in VANET [25], [26].

4. Delay Performance Analysis

Processing delay (nanosecond) is smaller than other delay factors (millisecond). So, the processing delay can be omitted in our computations. The propagation delay equals dividing distance between sender and receiver by the light speed. The propagation delay also can be omitted in our computations since the distance is much smaller than the light speed.

A. Transmission delay Analysis

Theoretically, transmission delay is decided by the average number of backoff slot and successful transmission time after using priority assignment. Transmission delay is calculated by Equation (7):

$$D_{trans} = T_B + T_F + T_s \tag{7}$$

where T_B is back-off period, T_F is freezing back-off period, and T_s is successful transmission period.

If EDCA mechanism is applied to contest, $(T_B + T_F)$ can be computed on reference [27]. The reference analysed delay performance of IEEE 802.11p in saturation condition. To transmit a frame successfully, the average number of backoff slot that a station needs is

$$E(X_i) = \frac{W-1}{2} + d_i \tag{8}$$

where d_i indicates the remaining frozen time before the backoff counter is reacted for states (k, l) with $k \geq 1$. The variable X_i ($i = 0, 1, 2, 3$) represents the total number of backoff slots, which a frame encounters without considering the case when the counter freezes.

If we just consider the successful transmissions, the total number of slots which a frame encounters when the counter freezes is represented the random variable B_i ($i = 0, 1, 2, 3$). The average frozen slot can get from

$$E(B_i) = \frac{E(X_i)}{1-p_i} p_i \tag{9}$$

where the probability p_i means that a station in backoff stage for the priority i class senses the channel busy.

Let p_b denote the probability that the channel is busy. p_s denotes the probability that a successful transmission occur in a slot time. The frame transmission delay of the priority i class is average time of δ, $(p_s/p_b) T_s$ and T_s for an idle slot at state (k, l) $(k>0)$, a busy slot at states (k, l) $(k>0)$ and a successful transmission at states $(k, 0)$, respectively. D_{trans} can be gotten from

$$D_{trans} = E(X_i)\delta + E(B_i)\frac{p_s}{p_b}T_s + T_s \tag{10}$$

The second term in the formula (10) affects D_{trans}. When priority assignment unity is used, the term can be decreased until zero.

B. Queuing Delay Analysis

Link delay consists of queuing delay and channel contest delay. To reduce queuing delay, high priority messages will be immediately sent to the head of line before contest the channels. For this purpose, the message scheduling is conducted in two steps of static and dynamic scheduling.

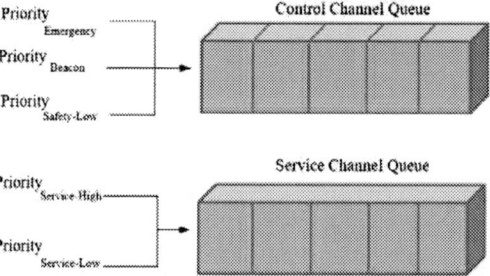

Fig. 2. Static scheduling process.

In the static scheduling step, the messages are transferred to either control channel queue or service channel queue based on static factor. The method is similarly to EDCA mode. The messages with Priority$_{Emergency}$, Priority$_{Beacon}$, and Priority$_{Safety-Low}$ priorities are transferred to control channel queue, and the messages with Priority$_{Service-High}$ and Priority$_{Service-Low}$ priorities are transferred to service channel queue. Figure 3 shows the static scheduling process in the message scheduling unit. In dynamic scheduling step, the packets in each queue are reordered based on their priorities when a new packet is entered to the queue. High priority message will be going to the head of line to waiting to contest the channel. This method is referred as "DySch". The method by a long way reduces the time to queue because the message needn't queue nearly.

The queuing delay is calculated by Equation (11):

$$D_{queue} = \frac{1}{\mu - \lambda} - \frac{1}{\lambda} \cdot \frac{Q_L \rho^{Q_L}}{1 - \rho^{Q_L}} \tag{11}$$

where, ρ is utilization which is equal to $\dfrac{\lambda}{\mu}$, where λ and μ are packet arrival rate and packet service rate, respectively. Q_L shows maximum queue length. If reschedule is used on priority assignment unity, D_{queue} will be reduced substantially.

Since the priority was added on a message, the messages don't queue and immediately will be sent to the head of line. Therefore, it does not take time for queuing to the head of line.

5. Simulation Result Discussions

For evaluating the performance of the proposed scheduling strategies in VANET, network simulators should be employed. In this paper, Network Simulator (NS) version 2.35 [28] were used for network simulation. Table 1 and Table 2 show the parameters used in the simulations of highway scenario. IEEE 802.11p was considered as the communication protocol. CSMA/CA strategy was also used as transmission strategy in MAC layer. TwoRayGround was employed to model the propagation in highway scenario. The Poisson distribution was also used for generating the data traffic. A table-driven routing protocol like Destination-Sequenced Distance-Vector (DSDV) is assumed in simulations.

Table 1. Configuration parameters for simulation of the highway

Parameters	Value
Transmission rate	6 Mbps
Bandwidth	6 MHz
Emergency message size	500 Bytes
MAC type	IEEE 802.11p
Propagation model	TwoRayGround
Routing protocol	DSDV
Simulation time	200 s
Simulation runs	20
Back-off time slot length	13 μs

Table 2. EDCA parameters used in the CCH of WAVE

Access category	CWmin	AIFSN
AC_0	15	9
AC_1	7	6
AC_2	3	3
AC_3	3	2

For more evaluation of the proposed strategy, the variation of the average delay with simulation time is investigated in Fig. 3. EDCA in IEEE 802.11p is the traditional channel contest method. So, average delay of DySch strategy compares with that of EDCA strategy. Here, the number of vehicles is assumed to be 50. Fig. 3 illustrates that by advancing the simulation time, the average delay of the packet transmission decreases for EDCA strategy and DySch strategy. However, using DySch strategy, the amount of reduction of average delay is higher than the EDCA strategy. In Fig. 3, it can be also seen that the average delay using DySch at simulation time 50 s is much lower than the EDCA strategy. It means using the proposed strategy, congestion is controlled before it occurs. Here, it should be emphasized that DySch strategy is open-loop strategy.

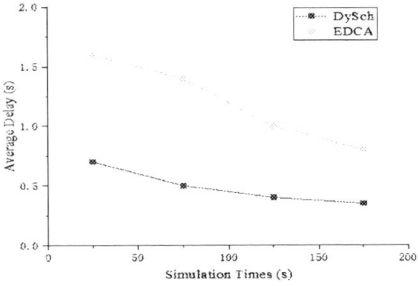

Fig. 3. Variation of the average delay with simulation time.

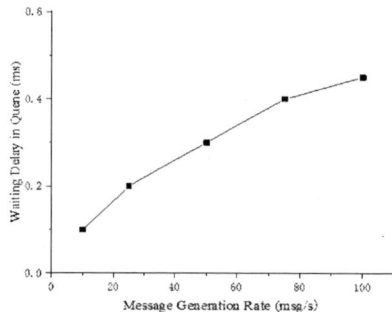

Fig. 4. Impact of the message generation rate on waiting delay in queue.

The impact of the message generation rate on average waiting delay in queue is evaluated for safety while congestion control is conducted using DySch. Fig. 4 illustrates that the average waiting delays in queue for safety messages are much low to 0.1~0.5 ms. A negligible delay for safety messages can be seen in Fig. 4. This result show that DySch transfers the safety messages in VANET without any significant waiting delay in queue.

6. Conclusion

In this paper, we proposed an improved DySch delay control strategy. DySch is distributed strategy. So, each vehicle independently prioritized and scheduled all the messages on its specification. The proposed strategy operated through two units: 1) priority assignment unit, and 2) message scheduling unit. In priority assignment unit, first, static and dynamic factors were calculated based on the content of messages and situation of vehicles, respectively. Then, a priority was assigned to each message based on static and dynamic factors and size of the message. The paper calculates the priority of messages by using new metric such as weather condition and geography position.

The performance of DySch strategy was analysed in highway scenario on the average delay and waiting delay in queue. Simulation results are consistent with theoretical deriving. Application of DySch strategy improved the performance of VANET by reducing the average delay. The average delay may highly be decreased on formula (12). Also, the queue delay can be nought since a packet can immediately arrive at head of line. So, the result showed that the applications of the strategy led to the lower waiting delay in queue. Therefore, more safe and reliable environments can be provided in VANET using DySch strategy.

In the future, some issues will be further researched. QoS model for VANET must be set up. QoS route and QoS MAC for VANET need further modified by tradition and new machine learning theories based on the model.

Acknowledgment

This work was support in part by Grant of Jinling Institute of Technology (jit-n-201304) and the Science & Technology Projects of Nanjing (2012ZD003)

References

[1] Zeadally, S., Hunt, R., Chen, Y.-S., Irwin, A., Hassan, A.: Vehicular ad hoc networks (VANETS): status, results, and challenges. Telecommun. Syst. 50, 217--241 (2012)

[2] Guerrero-Ibáñez, J., Flores-Cortés, C., Zeadally, S.: Vehicular Ad-hoc Networks (VANETs): architecture, protocols and applications, in: Next-Generation Wireless Technologies, Springer, 49--70 (2013)

[3] Karagiannis, G., Altintas, O., Ekici, E., Heijenk, G., Jarupan, B., Lin, K., et al.: Vehicular networking: A survey and tutorial on requirements, architectures, challenges, standards and solutions, Commun. Surv. Tut. IEEE 13, 584--616 (2011)

[4] Ghosh, T., Mitra, S.: Congestion control by dynamic sharing of bandwidth among vehicles in VANET, in:12th International Conference on Intelligent Systems Design and Applications (ISDA), 291--296 (2012)

[5] Jabbarpour Sattari, M., Md Noor, R.: Dynamic D-FPAV congestion control for algorithm for VANETs to rescue human lives, Archives Des. Sci., vol. 65, 2012

[6] Sepulcre, M., Mittag, J., Santi, P., Hartenstein, H., Gozalvez, J.: Congestion and awareness control in cooperative vehicular systems, in: Proceedings of the IEEE, 99, pp. 1260--1279 (2011)

[7] Singh, R.K., Tyagi, N.: Challenges of routing in vehicular ad hoc networks: a survey, IJECCE 3, 126--132 (2012)

[8] Campolo, C., Vinel, A., Molinaro, A., Koucheryavy, Y.: Modeling broadcasting in IEEE 802.11p/WAVE vehicular networks, IEEE Commun. Lett. 15, 199--211 (2011)

[9] Tornell, S.M., Calafate, C.T., Cano, J.C., Manzoni, P.: DTN protocols for vehicular networks: an application oriented overview, IEEE Communications Surveys & Tutorials, vol. 17, no. 2, pp. 868-887, Second Quater (2015)

[10] Zang, Y., Stibor, L., Cheng, X., Reumerman, H.-J., Paruzel, A. and Barroso, A.: Congestion control in wireless networks for vehicular safety applications, Proceedings of the 8th European Wireless Conference (2007)

[11] Li, X., Hu, B.J., Chen, H., Li, B., Teng, H. and Cui, M.: Multi-hop delay reduction for safety-related message broadcasting in Vehicular Ad Hoc Networks, IEEE International Conference on Communication, Networks and Satellite (ComNetSat), 44--49 (2012)

[12] Li, J. and Chigan, C.: Delay-aware transmission range control for VANETs, IEEE Global Telecommunications Conference (GLOBECOM 2010), 1--6(2010)

[13] Sattari, M.R.J., Noor, R.M. and Keshavarz, H.: A taxonomy for congestion control algorithms in Vehicular Ad Hoc Networks, IEEE International Conference on Communication, Networks and Satellite (ComNetSat), 44--49 (2012)

[14] Chen, J., Lee, V. and Ng, J.K.: Scheduling real-time multi-item requests in on-demand broadcast,14th IEEE International Conference on Embedded and Real-Time Computing Systems and Applications, RTCSA'08, pp. 207-216, 2008

[15] Bouassida, M.S. and Shawky, M.: On the congestion control within VANET, Wireless Days, 2008. WD'08. 1st IFIP, 1--5 (2008)

[16] Suthaputchakun, C.: Priority-based inter-vehicle communication for highway safety messaging using IEEE 802.11e, International journal of vehicular technology, vol. 2009, (2009)

[17] Tanenbaum, A.S.: Computer Networks, 5-th Edition, Prentice Hall, Englewood Cliffs (NY), (2010)

[18] Sepulcre, M., Gozalvez, J., Harri, J., Hartenstein, H.: Contextual communications congestion control for cooperative vehicular networks, IEEE Trans. Wireless Commun. 10, 385--389 (2011)

[19] Gui, Y. and Chan, E.: Data Scheduling for Multi-item Requests in Vehicle-Roadside Data Access with Motion Prediction Based Workload Transfer, 26th International Conference on Advanced Information Networking and Applications Workshops (WAINA), 569--574 (2012)

[20] Kumar, V. and Chand, N.: Data Scheduling in VANETs: A Review, International Journal of Computer Science and Communication, vol. 1, pp. 399-403 (2010)

[21] Pesel, R. and Maslouh, O.: Vehicular Ad Hoc Networks (VANET) applied to Intelligent Transportation Systems (ITS), *Universite de Limoges, France*, 149 (2011)

[22] Kargl, F.: Vehicular communications and VANETs, Talks 23rd Chaos Communication Congress (2006)

[23] Taherkhani, N., Pierre, S.: Prioritizing and scheduling messages for congestion control in vehicular ad hoc networks, Computer Networks. 108, 15--28 (2016)

[24] Rawat, D.B., Popescu, D.C., Yan, G. and Olariu, S.: Enhancing VANET performance by joint adaptation of transmission power and contention window size, IEEE Transactions on Parallel and Distributed Systems, vol. 22, pp. 1528-1535 (2011)

[25] Torrent-Moreno, M., Jiang, D. and Hartenstein, H.: Broadcast reception rates and effects of priority access in 802.11-based vehicular ad-hoc networks, Proceedings of the 1st ACM international workshop on Vehicular ad hoc networks, 10--18 (2004)

[26] Hartenstein H. and Laberteaux, K.: VANET vehicular applications and inter-networking technologies, vol. 1: John Wiley & Sons (2009)

[27] Tian, J., Liu, R.P., Zhang, X.: Saturation delay performance analysis by Markov chain for WAVE, International Journal of Advanced Information Science and Technology. 45, 93--97, January 2016

[28] Issariyakul, T. and Hossain, E.: Introduction to network simulator NS2, Springer Science & Business Media(2012)

ISPECE

IOP Publishing

An improved adaptive weighted median filter algorithm

Yuqin Song [1, 2, 3], Jun Liu [1, 2]

[1]School of Computer Science and Technology, Wuhan University of Science and Technology, Wuhan 430081, P. R. China

[2]Hubei Province Key Laboratory of Intelligent Information Processing and Real-time Industrial System, Wuhan 430081, P. R. China

[3]Correspondence: 986027187@qq.com

Abstract. An improved adaptive weighted median filtering method is proposed to deal with the interference noise of ultrasonic RF signal. Firstly, edge pixel points are determined to be filtered by the method of extending edge points; secondly, mean value is used to replace the median value which considered to be noise points; finally, weighted smoothing processing is carried out. The final experimental results in this paper show that the proposed method has better effect on RF signal processing.

1. Introduction

The RF signal required by ultrasound elastography technology [1] will be affected by the instrument in the process of producing, causing noise interference, and even submerging the effective signal. Therefore, the elimination of noise interference has become an important point in ultrasound elastography technology. Filtering plays an important role in signal processing.

In the process of signal processing, the size and shape of the traditional median filter window are defined, and edge details will be lost while filtering. In order to solve the contradiction between denoising and retaining details at the same time, researchers propose many improved median filtering algorithms. Literature [2] is a standard adaptive median filter algorithm, which uses the median value in the window as the response for pixel points containing noise, but is not applicable for high-density noise images. In order to improve the adaptive median filtering algorithm, literature [3] solves the problem of repeated operation of pixel points in the process of window iteration, but it is easy to misjudge the extreme points as noise points for filtering. Literature [4] is a center-weighted median filtering algorithm. By weighting, the proportion of the center pixel in the window increases, but it is easily affected by noise points. Literature [5] is an image adaptive median filtering algorithm, combining one-dimensional and two-dimensional median filtering to propose adaptive median filtering, but there are certain limitations in window selection.

Based on the above, this paper proposes an improved adaptive weighted median filtering algorithm, which can alleviate the loss of edge details while reducing noise.

2. Another section of your paper

2.1. Median filtering

Median filter (MF) [6] is a nonlinear signal processing technique based on sorting statistics theory that can effectively suppress noise. It is a neighborhood operation that sorts pixels in the neighborhood according to gray level. And the intermediate value of the group is then selected as the output pixel

---

Content from this work may be used under the terms of the Creative Commons Attribution 3.0 licence. Any further distribution of this work must maintain attribution to the author(s) and the title of the work, journal citation and DOI.

Published under licence by IOP Publishing Ltd

value. In 1971, JWTukey proposed the concept of median filter on time series analysis. The advantage of this filter is that it is simple and fast, and it shows excellent performance in filtering out superimposed white noise and long tail superimposed noise.

The idea of median filtering is to compare the size of pixels in a certain domain, and take out the median of this field as the new value of the center pixel of this field. The standard median filtering algorithm relies on a fast sorting algorithm. It is a nonlinear filtering method with less edge blur. It can not only remove or reduce random noise and pulse interference, but also preserve the information at the edge of the image.

The standard median filter (MF) is defined as:

$$\hat{f}(i,j) = Median\{g(s,t), (s,t) \in S_{ij}\} \tag{1}$$

In equation(1), $g(s,t)$ is noise, the median filtering method is to sort the pixels in the sliding filter window, and the output pixel value $\hat{f}(i,j)$ of the filtering result is the median value of the sequence.

2.2. Adaptive median filtering
According to the basic nature of median filtering, the size of the sliding window plays a crucial role in signal filtering performance. The classic median filter sliding window will remain unchanged, the smaller window can protect the signal well, but cannot effectively remove the noise; the larger sliding window can better suppress the noise, but at the same time the edge of the signal is blurred or even lost effective information, which is also an important drawback of median filtering.

In order to compensate for this defect, adaptive median filtering (AMF) [7] is used to adaptively adjust the sliding window length, which can effectively denoise and protect the effective signal details.

Assuming that the image size is $M*N$, (x,y) is the coordinates of the image signal point p , and S_{xy} is the domain window centered on (x,y). Which defines:

Z_{min} represents the smallest gray value in S_{xy}, Z_{max} is the largest gray value, Z_{med} is the median of all gray values in the representation, Z_{xy} is the gray value of the x row and the y column, S_{max} is the maximum window size allowed by S_{xy}.

The adaptive median filtering algorithm can be represented as two processes, processes A and B:
Processes A:

$$\begin{cases} A_1 = Z_{med} - Z_{min} \\ A_2 = Z_{med} - Z_{max} \end{cases} \tag{2}$$

For process A, if $A_1 > 0$ and then $A_2 < 0$, go to process B, otherwise judge the window suspected noise point, increase the window size; if the window size $n <= S_{max}$, repeat process A, otherwise output $Z_{xy} = Z_{med}$, think that the window has noise points.

Processes B:

$$\begin{cases} B_1 = Z_{xy} - Z_{min} \\ B = Z_{xy} - Z_{max} \end{cases} \tag{3}$$

For process B, if $B_1 > 0$ and $B_2 < 0$, it is judged that there is no noise point in the window, output $Z_{xy} = Z_{xy}$, otherwise output $Z_{xy} = Z_{med}$.

2.3. Improved adaptive weighted median filter (IAWMF)
The traditional adaptive median filtering is based on the extreme point of the adaptive window as the basis for determining the noise point. There are three defects: 1) the adaptive window takes the point of the image as the center point and filtering step-by-step using a template of $N*N(N \geq 3)$, and the pixel points at the edge of the signal are ignored, which affects the overall filtering effect. 2) When the

detected suspected noise point is judged beyond the filter window size, the median value of the output may be a noise point. 3) For the window containing the noise point, the gray value of the pixel of the region in which it is located is taken as the processing result of the filtering, and the magnitude of the median has a large effect on the filtering effect of the noise.

In response to the above defects, this paper proposes the following solutions:

Extended edge: the adaptive median filter takes (x, y) as its center point in the process of implementation, and ignores its edge value in the process of filtering, which affects the experimental results. For a matrix of size 5*5, in the process of taking the median, it is equivalent to convolving with a matrix of size 3*3. In this process, only the region of size 3*3 that is not the edge has an impact, and the edge remains the same. In order to make up for this defect, this paper adopts the extended adjacent edge method of ad =1, so that all pixel points can be detected by noise in the filtering window. After expanding the image, the image becomes a 7*7 size matrix, and the convolution operation of the non-edge 5*5 matrix (corresponding to the original 5*5 matrix).

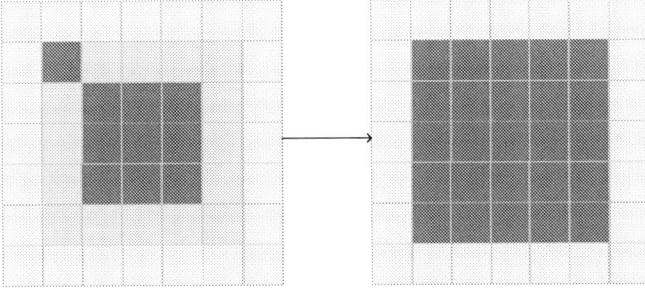

Figure 1.   The Process of Extending Edge to Median Value

Combined mean filtering: Mean filtering [8] is one of the classic algorithms for image denoising. The noise-free gray value is estimated by the average gray value in the neighborhood of each pixel of the noise image. Mean filtering is a typical linear filtering algorithm, which refers to giving a template which includes neighboring pixels around it to a target pixel on an image, and then replaces the original pixel value with the average of all the pixels in the template.

In the traditional median filtering algorithm, it is considered that the median point Z_{med} when $Z_{min} < Z_{med} < Z_{max}$ is satisfied is not a noise point. And if $Z_{med} = Z_{min}$ or $Z_{med} = Z_{max}$ is satisfied, it is a noise point. If the window size is not satisfied, the median point of the output is a noise point. Therefore, a way to combine the mean filtering is proposed, and the mean value Z_{mean} of the domain is used instead of the median.

$$Z_{xy} = Z_{mean} \qquad (4)$$

Weighted filtering: The weighted median filtering is to multiply each pixel in the window by a corresponding weight, and then statistically sort, taking the median instead of the noise value. The traditional median filtering can be seen as a weighted median filter with a weight of 1 for each pixel.

$$W = \frac{1}{n*n} one(n,n) \qquad (5)$$

In this paper, the weight matrix of equation (5) is selected where $n \geq 3$, then convolution operation on the signal after adaptive median filtering, to obtain a new weighted adaptive median filtered signal which replacing the original signal value.

3. Experiment and result analysis
The experimental data in this paper were derived from the University of Michigan's agar-graphite tissue mimicking phantoms and ex vivo kidneys. The size and dimensions of the experimental data cannot be easily estimated which is a 1536*128*225 three-dimensional matrix.

To further verify the filtering effect, the signal-to-noise ratio and the root mean square error of the signal are compared, and the definitions are respectively:

$$SNR = 10\log[\sum_{i=1}^{N} S^2(i) \Big/ \sum_{i-1}^{N} (S(i) - \hat{S}(i))^2]$$ (6)

$$RMSE = \sqrt{\frac{1}{N}\sum_{i=1}^{N}(S(i) - \hat{S}(i))^2}$$ (7)

Where $S(i)$ represents the original signal, $\hat{S}(i)$ representing the filtered signal. The results are shown in Table 1.

Table 1 Filtering vs. SNR and RESE of experimental data

	SNR	RMSE	Time(s)
MF	0.96	931.82	0.096
AMF	6.56	479.32	1.31
MAMF	6.85	472.75	1.62
AWMF	6.85	472.06	1.93
IAWMF	7.61	433.76	3.46

It can be seen from Table 1 that for large-scale experimental data, Mean Adaptive Filtering (MAMF) and Adaptive Weighted Median Filtering (AWMF) have almost the same filtering effect, but the latter consumes longer running time than the former. The SNR of this method is the highest, and the RMSE is the smallest, indicating that the filtering result is the best, but the RMSE error value is very large for this experimental data.

In order to further verify the experimental method, another set of 500*200 data is selected for testing while a noise signal is added, and the experimental results are compared (the selected filtering window size $S_{max} = 7$, the added noise is uniform noise) as shown in Table 2.

Table 2   Filtered SNR and RMSE comparison

	SNR	RMSE	Time(s)
MF	6.87	3.24	0.011
AMF	13.86	0.145	0.048
MAMF	15.0	0.127	0.081
AWMF	23.68	0.0425	0.1179
IAWMF	41.36	0.0039	0.1532

It is concluded from Table 2 that the improved method has much higher SNR than other methods and the RMSE is the smallest which indicating that the signal processing effect is better.

Trying to change the filter window during the experiment and finding that the filtering effect will change accordingly. In order to further verify the conjecture, the experiment chooses to change the window size S_{max} and noise types to experiment:

(1) Select the added noise as uniform noise, and change the value of S_{max}, $S_{max} = 3i + 2(i = 0,1,2 ...)$.

(2) Select $S_{max} = 7$ to change the type of noise added (uniform noise, Gaussian noise, Rayleigh noise, salt and pepper noise).

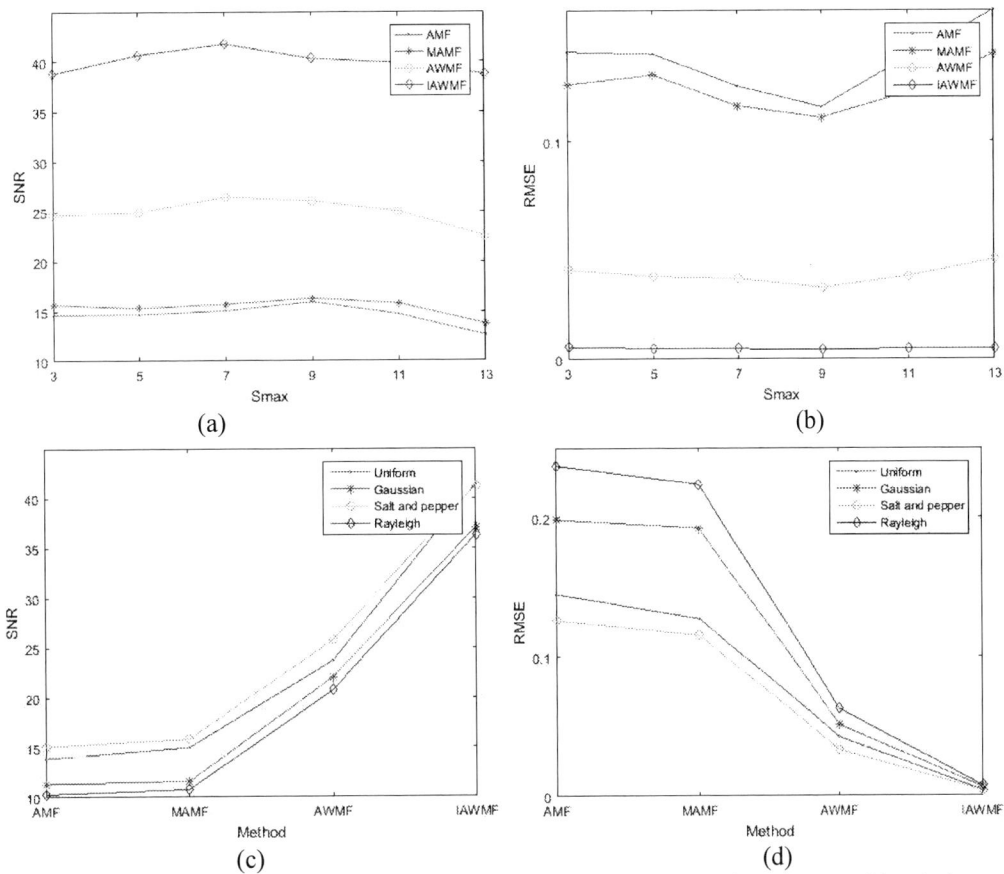

Figure 2. (a)and (b) respectively represent SNR and RMSE that change with window size;(c)and (d) respectively represent SNR and RMSE that change with different noise types.

Combining the above experimental data, the following conclusions are drawn:

(1) The method presented in this paper has a good filtering effect. As shown in figure 2(a), its SNR is much higher than that of other methods, and RMSE of root mean square error is close to 0 in figure 2(b). For different filter window sizes, RMSE of IAWMF remains almost unchanged, indicating that IAWMF has a certain stability.

(2) As the size of the filter window increases to a certain extent, the filtering effect is getting better and better. As shown in Figure 2(a), the larger the SNR value, the smaller the corresponding RMSE. For AWMF and IAWMF, when $S_{max} \leq 7$, the SNR increases with the increase of S_{max}. When $S_{max} \geq 7$, the value of SNR gradually decreases, indicating that the filter window size $S_{max} = 7$ of weighted filtering is the most suitable. The same result compares AMF and MAMF, the filter window size $S_{max} = 9$ is the best.

(3) The type of noise will have a certain impact on the experimental results. The filtering effect on the addition of salt and pepper noise is the best, and the processing of Rayleigh noise is the worst. The method in this paper has a very close effect on the processing of various noises which indicating that this method is almost immune to the type of noise.

In summary, the method IAWMF has better processing effect on noise, and has better filtering performance than other methods.

4. Discussion

In this paper, on the basis of adaptive median filtering, an improved adaptive weighted median filter is proposed. The median value of the suspected noise point is replaced by means of mean value, and the weight matrix convolution operation of the filter window after adaptive median filtering is performed. The final results are better than other algorithms, and the feasibility of this method is illustrated. It laid the foundation for the subsequent ultrasonic elastography experiment.

References

[1] G. Cortela, L. Leija. Elastograms of the diabetic foot by ultrasonic impulse elastography [J]. IEEE Transactions on Image Processing, 2016.

[2] Wang D D, Wang F M. Research on digital image noise removal technology based on Matlab [J]. Mechanical engineering and automation, 2015(2): 98-99.

[3] Liu pengyu, ha rui, jia kebin. Improved adaptive median filtering algorithm and its application [J]. Journal of Beijing university of technology, 2017, 43(4):581-586.

[4] Ko S J, Lee Y H. Center weighted median filters and their applications to image enhancemen [J]. IEEE Transactions on Circuits and Systems, 1991, 38(9):984-993.

[5] Liu hai. An image adaptive median filtering algorithm [J]. Software guide, 2018(5).

[6] Gong S R, Liu C P, Wang Q. Digital image processing and analysis [M]. Beijing: Tsinghua University press, 2006.

[7] Department of ECE, Shri JJT University, et al. A Novel Algorithm for Image Denoising using Modified Adaptive Median Filter [J]. Research Journal of Applied Sciences, Engineering and Technology, 2015, Vol.10 (4):373-375.

[8] Mehdi Mafi, Hoda Rajaei, Mercedes Cabrerizo. A Robust Edge Detection Approach in the Presence of High Impulse Noise Intensity through Switching Adaptive Median and Fixed Weighted Mean Filtering[J]. IEEE Transactions on Image Processing, 2018.

ISPECE   IOP Publishing

IOP Conf. Series: Journal of Physics: Conf. Series **1187** (2019) 042108   doi:10.1088/1742-6596/1187/4/042108

Research on long-distance hand recognition based on depth information

Yuyang Fu, Lanfang Miao, Zhifei Li

College of Mathematics and Computer Science, Zhejiang Normal University, 668 Yingbin Avenue, Jinhua, Zhejiang, 321004, China

197361794@qq.com

Abstract. This paper proposes a long distance gesture recognition algorithm based on Kinect. First, we use Kinect to capture human skeleton and depth information, track and extract hand information. For the characteristics of the depth image determine that, the experimental results will be not affected by the background, light, skin color and clothing. Then the initial obtained data of hand shape is denoised and smoothed, and the contour and skeleton of hand shape are extracted. When the hand is at long distance, the accuracy of Kinect is not sufficient to get the detail hand shape-information. So we combine the hand depth information with the color information to get hand shape. Finally, we use the Hu moment of the hand shape contour binary image and the hand skeleton binary image as data feature, and utilize SVM to train and identify hand gesture. The experimental results show that the Hu moment of the hand skeleton binary image is more advantageous than the Hu moment of the hand contour binary image, and the proposed long-distance hand recognition algorithm also has the recognition accuracy similar to the close-distance.

1. Introduction

Gesture recognition has always been an important topic in the field of computer vision. Due to the flexibility of the hand, it is more important for human to express their inner thoughts and interaction with the external environment.

Lin et al.[1] proposed that Kinect offers a variety of opportunities for both new and old applications, and it can be used as a reference in gesture recognition, human activity recognition, human biological measurement assessment, 3D surface reconstruction, and healthcare applications. Thanh et al.[2] proposed an approach to extract the discriminative patterns for efficient human action recognition. And they consider each action is consisted of a sequence of unit actions, each of which is represented by a pattern. In addition, they first automatically extract the key-frames from a skeleton sequence and categorize them into different patterns.

Zhao et al.[3] presentsed an improved strategy for hand segmentation using the randomized decision forest framework based on depth images. In the proposed method, a new depth feature derived from the central point of hand structure is induced to strengthen the ability of generalization of depth feature as well as reduce the requirement for training dataset, while not sacrifice the accuracy of hand segmentation. Compared to traditional images, depth images can also avoid self-occlusion of the hand based on depth information[4]. With the development of deep learning, a hand segmentation method based on complete convolution network has been proposed[5].

Gesture recognition requires the extracting features of the hand data. It is usually divided into static gesture recognition and dynamic gesture recognition. Dynamic gesture recognition is mainly based on

Content from this work may be used under the terms of the Creative Commons Attribution 3.0 licence. Any further distribution of this work must maintain attribution to the author(s) and the title of the work, journal citation and DOI.

Published under licence by IOP Publishing Ltd

the direction and speed of hand movement. Static gesture recognition is mainly based on hand shape judgment and it will be studied in this paper. The methods of classification recognition mainly include template matching, support vector machine (SVM)[6], neural network and so on.

In this paper, we will use Kinect to extract depth and skeleton information of hand. The Hu moment is selected as the feature of the static gesture, and the SVM classifier is used for gesture recognition. And the recognition effects between the hand contour Hu moment and the hand skeleton Hu moment of the same static gesture at close and long distances are compared by experiment.

The rest of the paper is organized as follows. In Section II, the close distance and long distance hand extraction algorithms are described in detail. In Section III, we introduce the features and tools used in gesture recognition. Section IV shows the experimental results. Concluding remarks are drawn in Section V.

2. Hand Preprocess

In this section, we will introduce the process and related algorithms for extracting hand from close distance and long distance. The flow chart of our method is shown in Figure 1. As shown in Figure 1, our method is divided into two parts: (1) Extract the hand from a close distance (within 2m); (2) Extract the hand from a long distance (beyond 2m).

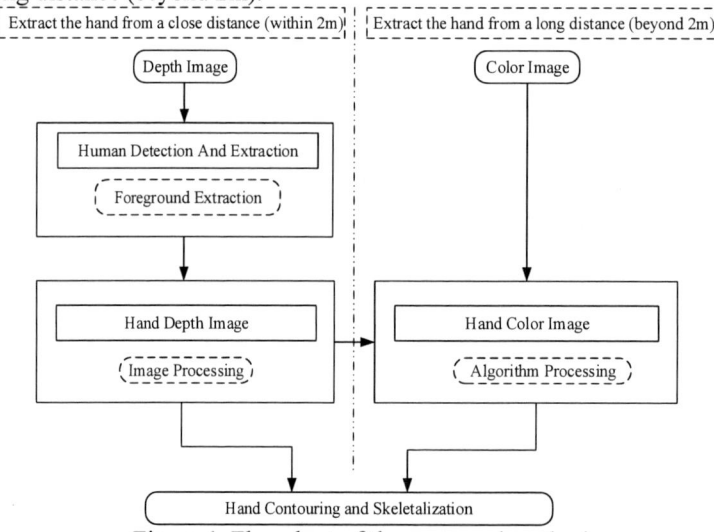

Figure 1. Flowchart of the proposed method

2.1. Extract hand shape at a close distance (within 2m)

With the Kinect device, we can get the depth and color data streams. The pixel information in the depth image represents the calibrated distance information in the scene. In order to get the human body from the image, we can use the foreground segmentation function provided by Kinect. The main idea of this function is to use thresholds to filter the depth of the background.

In this paper, we use Kinect to extract the foreground extraction of depth image, that is, obtain the corresponding mask depth indirectly from the index number which is randomly assigned for each object by Kinect, as shown in Figure 2(b). Kinect can also track 20 joint points of the human body, so we can get the depth map and color map of the hand according to the hand joint. Since Kinect uses the principle of laser speckle to obtain hand depth image, we can find the edge of the image is rough, especially in the target edge there are many holes and noise points. If the rough image is processed directly, some errors may be caused when extracting the features later. We use the filtering method to repair defects on the image. Although this method cannot eliminate all the noise, they have considerable results[7].

ISPECE   IOP Publishing

IOP Conf. Series: Journal of Physics: Conf. Series **1187** (2019) 042108   doi:10.1088/1742-6596/1187/4/042108

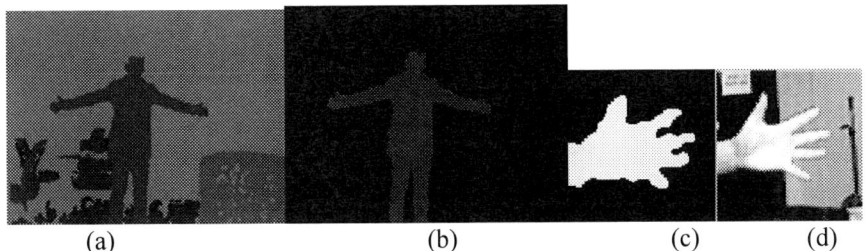

| (a) | (b) | (c) | (d) |

Figure 2. (a) Depth image from Kinect; (b) Human body depth image; (c) Binary image of human hand; (d) Color image of human hand

2.2. Extract hand shape at a long distance (beyond 2m)

When the human body is far away from the Kinect camera, the hand area becomes a group in the depth image, and we can hardly distinguish the movement of the finger, so the hand recognition cannot be performed, as shown in Figure 3(a). In order to recognize the finger at a long distance, we propose a new hand extraction method based on the color image.

First，because the color camera and depth camera of Kinect are not in the same position, we need to use its conversion function to align the color image with the depth image. Then, based on the bone tracking function, we can get the range of hand areas in depth and color images. Although the hand area in the depth image cannot obtain detailed finger information, it can roughly obtain the contour of the hand. We can get the intersection of the depth image and the hand area of the color image to get Figure 3(b). According to the depth map of the hand, we can get the contour of the hand, and the centre point is the point on the hand. Therefor we use the pixel value of the centre point as a reference to recalculate the pixels in the contour to obtain Figure 3(c).

After obtaining Figure 3(c), we should calculate the value of threshold to divide the hand part (area A) and the other part (area B). We assume there are 255 values and the sum of the remaining blocks is N, the number of blocks whose depth value is i is n_i , and the probability of appearance of each depth value is p, then obviously:

$$N = \sum_{i=0}^{255} n_i \,, p_i = {n_i}/{N} \tag{1}$$

The appearance probability of the area A and the area B satisfies the following equation

$$p_A = \sum_{i=0}^{x} p_i \,, p_B = \sum_{i=x+1}^{255} p_i = 1 - p_A \tag{2}$$

Where x is the selected threshold.

The inter-class variance of the area A and the area B satisfies the following equation:

$$w_A = \sum_{i=0}^{x} {ip_i}/{p_A} \,, w_B = \sum_{i=x+1}^{255} {ip_i}/{p_B} \tag{3}$$

$$w_0 = p_A w_A + p_B w_B = \sum_{i=0}^{255} ip_i \tag{4}$$

$$\sigma^2(x) = p_A(w_A - w_0)^2 + p_B(w_B - w_0)^2 \tag{5}$$

Where w_A and w_B represent the average pixel values of area A and area B respectively, w_0 represents the average pixel values of the global image and $\sigma^2(x)$ represents the inter-class variance of the two areas.

The main idea of this process is to choose the value of threshold according to the pixel characteristic so as to maximize the variance between the area A and the area B. Variance is a measure of the pixel distribution uniformity. The greater inner-class variance between area A and area B, indicates that the greater the difference between these two areas. Therefore, segmentation with the highest variance between classes means that the probability of misclassification is the smallest. The final optimal threshold x is:

$$x = \max_{0 \leq i \leq 255} \{\sigma^2(x)\} \tag{6}$$

After calculating optimal threshold of the above equation, we can extract the gesture area as shown in Figure 3(d).

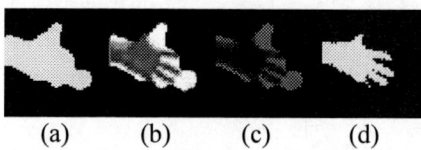

(a)   (b)   (c)   (d)

Figure 3. (a) Original binary image of human hand; (b) Original image intersect with gray image; (c) Difference calculation; (d) Improve binary image of human hand

2.3. Hand Skeleton Extraction

Human skeleton extraction can use image thinning algorithms, which have been used for pattern recognition and image analysis for a long time. The thinning algorithm reduces the binary number pattern to obtain a unit width skeleton that maintains geometric and topological properties.

Zhang et al.[8] proposed a fast parallel thinning algorithm. It consists of two sub-items: one to remove the southeast boundary point and the northwest corner point, and the other to delete the northwest boundary point and the southeast corner point. The connectivity and insensitivity to boundary noise are very good in this method. Tarabek et al.[9] proposed a robust parallel thinning algorithm based on ZS algorithm. It maintains the good performance of the ZS algorithm and overcomes the shortcomings of not producing a unit width skeleton by removing the post-processing steps of redundant pixels by combining the additional conditions for identifying critical patterns. This paper uses this algorithm for human skeleton extraction.

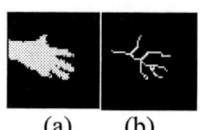

(a)   (b)

Figure 4. (a) Improve binary image of human hand; (b) Skeleton binary image of human hand

3. Gesture Recognition

3.1. Feature Extraction

Some people used the hand image Hu moment as the static gesture data feature, while Liu et al.[10] used the hand contour Hu moment as the static gesture data feature, they all used the SVM classifier to recognize the gesture. In order to explore the data characteristics with the maximum recognition rate, this paper chooses the hand contour Hu moment and the hand skeleton Hu moment as the data features of the same static gesture. Then we use the SVM classifier to identify and compare the recognition effects.

The Hu moment mainly represents the geometric features of the image region, and has the characteristics of rotation, translation, and scale invariance. The Hu moment is characterized by seven moment invariants, which are composed of a linear combination of the second and third order central moments of the image[11].

3.2. Classification and Identification tools

This paper uses SVM as a tool for gesture recognition. Support Vector Machine (SVM) is a machine learning method developed by statistics. The main idea is to map the sample space into a high-dimensional feature space through a nonlinear mapping, so that the problem of nonlinear separability in the original sample space is transformed into a linearly separable problem in the feature space.

The experiment uses the opencv-2.4.3 library to calculate the Hu moment of the image, and the SVM module inside it used to train the model.

4. Experiment Results

4.1. Gesture Recognition Experiment
The purpose of the experiment is to classify and recognize the five static gestures shown in Figure 5. The experimental development environment is Visual Studio 2013 and opencv-2.4.3 library, the programming language uses C++, and the hardware device uses Kinect for Windows 1.8.

In order to complete the experiment, the programmed program performs the following functions in order:

- Different people use five static gestures within 2m and beyond 2m respectively. We use Kinect to record depth gesture map and color gesture map separately, and process them to get corresponding hand contour map and hand skeleton image.
- After remove some of the blurred images in the sample due to frame skipping, we select 20 images of each hand contour and hand skeleton image within 2m as training samples. And the other image are test samples. We use the OpenCV to calculated Hu moment and use SVM to train model to get the rate of hand recognition.
- Using the same method, we do experiment with the images beyond 2m. Because Kinect has been unable to extract hand images beyond 2m, we use our hand extraction method to compare hand contour and hand skeleton recognition rate.

Figure 5. Five kinds of static images of hand gesture

4.2. Experimental Results and Analysis
Table 1 is the experimental result of the hand recognition rate. The total recognition rate of the hand contour image within 2m which Hu moment is used as the data feature is Accuracy = 51.73%. The total recognition rate of the data feature using hand skeleton map Hu moment is Accuracy = 79.34%.

Beyond 2m, Kinect cannot be used to recognize gestures accurately, so we use the hand recognition algorithm proposed in this paper. For the hand contour image using Hu moment as data feature, the total recognition rate is Accuracy = 68.98%. And for the hand skeleton image, the total recognition rate is Accuracy = 78.25%.

Table 1. Hand gesture recognition rate

Recognition rate of hand gesture within 2m				Recognition rate of hand gesture beyond 2m			
Contour	Rate/%	Skeleton	Rate/%	Contour	Rate/%	Skeleton	Rate/%
	73.93		88.69		100		86.25
	31.50		78.74		78.95		78.95
	21.01		30.25		69.39		38.78
	36.99		94.52		24.14		71.93
	66.29		97.75		66.36		93.46

The above experimental results show that the recognition rate of the data feature of the hand skeleton image Hu moment is higher than that of the hand contour image Hu moment. The hand extraction method proposed in this paper effectively increased the hand recognition distance.

5. Conclusion and Future

In this paper, a method for extracting hand based on depth information is proposed. The Hu moment of hand contour and hand skeleton is used as data feature. We use SVM classifier to recognize five types of hand static gesture. It is found that the hand skeleton recognition rate is higher than the hand contour recognition rate.

Although the gesture recognition method implemented in the paper has a good recognition rate, the number of recognition gesture is still relatively less. At the next step, we will combine with dynamic gesture recognition or limb skeleton as the type of recognition, to realize a functionally rich somatosensory interaction system.

References

[1]   Lin, Yen Yu , et al. "Depth and Skeleton Associated Action Recognition without Online Accessible RGB-D Cameras." IEEE Conference on Computer Vision & Pattern Recognition IEEE Computer Society, 2014.

[2]   T.T.Thanh, F.Chen, B.Le, and B.Le,"Extraction of Discriminative Patterns from Skeleton Sequences for Accurate Action Recognition," Fundamenta Informaticae, vol. 130, no.2, pp.247-261, 2014.

[3]   Zhao, Mengyi , and Q. Jia . "Hand Segmentation Using Randomized Decision Forest Based on Depth Images." International Conference on Virtual Reality & Visualization IEEE, 2017.

[4]   Kuang, Hailan , et al. "An Effective Skeleton Extraction Method Based on Kinect Depth Image." International Conference on Measuring Technology & Mechatronics Automation IEEE, 2018.

[5]   Long, Jonathan , E. Shelhamer , and T. Darrell . "Fully Convolutional Networks for Semantic Segmentation." IEEE Transactions on Pattern Analysis & Machine Intelligence 39.4(2014):640-651.

[6]   Rossi, Matteo , et al. "Hybrid EMG classifier based on HMM and SVM for hand gesture recognition in prosthetics." (2015).

[7]   Shen, Yujie , et al. "A Novel Human Detection Approach Based on Depth Map via Kinect." Computer Vision and Pattern Recognition Workshops (CVPRW), 2013 IEEE Conference on IEEE, 2013.

[8]   Zhang, T. Y , and C. Y. Suen . "A fast parallel algorithm for thinning digital patterns." Comm Acm 27.3(1984):236-239.

[9]   Tarabek, Peter . A robust parallel thinning algorithm for pattern recognition. 2012.

[10]   Liu, Yun , Y. Yin , and S. Zhang . "Hand Gesture Recognition Based on HU Moments in Interaction of Virtual Reality." International Conference on Intelligent Human-machine Systems & Cybernetics IEEE, 2012.

[11]   Huang, Zhihu Huang Zhihu , and J. L. J. Leng . "Analysis of Hu's moment invariants on image scaling and rotation." International Conference on Computer Engineering & Technology IEEE, 2010.

A Fault Repair Method for Workstation Cluster Based on Probabilistic Model Checking

Xi Wang[1,*], Ting Chen[1] and Chengtian OuYang[1]

[1]School of Information Engineering, Jiangxi University of Science and Technology, Jiangxi, China

*Corresponding author e-mail: wang_xi_happy@163.com, dustinchan123@163.com, oyct@163.com.

Abstract. To analyze the component modules and maintenance unit modules in the workstation cluster, a fault repair method for workstation cluster based on probabilistic model checking is proposed. In the proposed method, the queue model is introduced when workstations are waiting for repairing, and different priorities are assigned according to the importance of the component functions. The formal model of the system is established by an extended continuous time Markov chain, the attributes of the system are described by continuous stochastic logic, and the fault repair module is verified by PRISM. The experimental results show that the proposed method can greatly reduce the time required for the maintenance process, and improve the maintenance efficiency and the fault tolerance of the system.

1. Introduction

Cluster technology is characterized by reliability, availability and expansibility [1]. High reliability cluster systems require higher standards of reliability and disaster defense. When a node's computing or service fails or the system fails, the cluster system needs to spread the tasks running on the failed node to other normal nodes, continue to complete the current service or calculation, and can be repaired in a short time.

Model checking technology has been used in the analysis and verification of computer hardware, communication protocols, system's safety design [2-3]. Probabilistic model checking usually builds a probabilistic model for the system, then uses the steady-state and instantaneous probability calculations to complete the quantitative verification [4, 13]. The model checking tool PRISM is usually adopted [5, 11]. Probabilistic model checking is also used to detect the correctness and performance of various stochastic distributed algorithms [6], service quality in communication networks and protocols [7-8], and performance analysis in areas such as power management systems [9].

Based on the research [10], this paper studies the fault repair module in high reliability workstation cluster, and proposes a fault repair method based on probabilistic model checking. In this method, component module and maintenance unit module in workstation cluster are analyzed. And they are formally modeled and analyzed by using model checking tool PRISM. Additionally sub algorithm FCR and sub algorithm REP are proposed. During the repair process, multiple faulty workstations waiting for repairing are put into the proposed queue model, then different priorities are assigned

---

Content from this work may be used under the terms of the Creative Commons Attribution 3.0 licence. Any further distribution of this work must maintain attribution to the author(s) and the title of the work, journal citation and DOI.
Published under licence by IOP Publishing Ltd

according to the importance of the function for the component. Finally, scheduling strategy of the maintenance unit is adjusted. The formal model of the system is established by an extended continuous time Markov chain. The attributes of the system are described by continuous stochastic logic, and the fault repair module is verified by PRISM. The experimental results show that the proposed method not only shortens the repair time, but also makes the system faster recover to normal working state. So the method provided in this paper is an available way to improve the maintenance efficiency and the ability of system's fault tolerance.

2. Algorithm description

In this paper, the fault repair module of workstation cluster system is mainly composed of component module and repair module. Correspondingly, the proposed algorithm is mainly composed of sub algorithm FCR and sub algorithm REP.

2.1. Sub algorithm FCR

The continuous time Markov chain (CTMC for short) model [12] is extended, then extended continuous-time Markov chain (ECTMC for short) is used for modeling, and the properties of the workstation cluster are described by continuous stochastic logic (CSL for short). The definitions of CTMC and ECTMC are as follows.

Definition 1 CTMC is a tuple $C = (S, s_0, R, AP, L)$, where S is a finite set of states. s_0 is the initial state. $R:S \times S \rightarrow R_{\geq 0}$ is the migration rate function. AP indicates an atomic proposition. $L:S \rightarrow 2^{AP}$ is a state identification function.

Definition 2 ECTMC is a tuple $ECTMC = (W, S, s_0, R, Act, status, T)$, where S, s_0, R are the same as Definition 1. W represents the set of components in the system. $Act = \{work, repair\}$, work indicates the normal working behavior of the system, and repair indicates the maintenance behavior of the system. $status$: $S \rightarrow 2^{Act}$ represents the status identification of the components. T represents time.

In the workstation cluster system, the number of components is n, and an ECTMC model is used to describe the process of failure, repairing, and reoperation of each component. In the ECTMC model, $w(1, 2, \ldots n)$ is the component queue, where $w(i) \in W$ is the ith component. $S = \{run, failure\}$ indicates that there are two kinds of operation states of components. s_0 is the initial state of the component. Rate $\lambda_i \in R$ is the rate for state migration. State identification $status_i\{work, repair\} \in status$ represents whether the current component $w(i)$ is working properly or needs to be maintained. $t \in T$ represents the time. The input and output of the sub algorithm FCR are as follows.

Input: $w(i)$, $status_i\{work, repair\}$, T.

Output: S, $status(i)$.

Sub algorithm FCR starts from the current state of the component $w(i)$. With the passage of time t, $w(i)$ fails at the migration rate λ_i. $status(i)$ and S are changed. The maintenance module runs after the failure occurs, if $w(i)$ is repaired, then $status(i)$ and S are changed again. Specific algorithmic description for FCR is shown in Figure 1.

2.2. Sub algorithm REP

Because there is only one maintenance unit, improving the efficiency of the maintenance unit can greatly improve the performance of the system. If multiple components fail at the same time, a reasonable scheduling strategy is needed to enable the system to return to normal state in a very short time. Therefore, this paper introduces queue model to maximize the efficiency of maintenance unit when multiple components fail. According to the functional distinction of system components, it can be divided into three priority levels:

1. workstation. The workstation sub cluster is consisted of multiple workstations.

2. switch. Switch components connect workstations.

3. backbone network. When a standby sub cluster is required, the backbone network connects the two sub cluster.

Sub-algorithm FCR.	Sub-algorithm REP

Defined variables: w(i): 1≤i≤n; S; work, repair: Boolean; t;
Initialization : w(i) ←w(1), $\bar{s}$ ←run, work← true , repair← false , t←0;
While t <=T do
 If(w(i)=1) // enter component w(1)
   if(work=true) //working well
      λ_i : work ←false; //w(1) fails at λ_i,
   end
   if(repair=false & work=false)
     repair ←true; // w(1) needs repair
   end
  if(repair=true &work=false)
     repair ←false; work ←true;
     // w(1) returns to normal
   end
end
……… // other w(i) is the same as w(1)
	Initialization: q←0 ; r←0 ; t←0 ;
	While(t <= T) do
	if(w(i)=1) // enter component w(1)
	 if(q<n) // queue q is not full
	  q←q+1 ;// failed w(1) enter the queue q
	 end
	 if(r=0&repair=false)
	  rate(1): r←1 ; // repairing status of w(1)
	 end
	 if(r=1&q>0) //repairing for w(1) is finished
	  μ(1):r←0 ; q←q-1 ;
	 end
	end
	if (w(i)=2) // enter component w(2)
	 if(r=0&repair=false)
	  rate(2):r←2 ; // repairing status of w(2)
	 end
	 if(r=2)
	  μ(2): r←0 ;// repairing for w(2) is finished
	 end
	end
	…// enter other components w(i), such as workstation is the same as w(1), switch and backbone is the same as w(2).

Figure 1. Sub algorithm FCR.   Figure 2. Sub algorithm REP.

As a key component connecting two sub clusters and having only one, the backbone network has the highest priority of 3. Secondly, the priority of the two switches connected to the workstation is 2. There are usually a good many failures in the workstation due to more liable to failure. When a single workstation fails, it will not affect the operation of the whole system, so its priority is 1. When multiple workstations fail, they enter the queue to wait for repair. When they encounter higher priority component obstacles, the repair unit needs to repair the components with higher priority immediately after the current component has been repaired. The maintenance unit also has two states: idle and working. According to the different repaired components, the working state can be divided into four different kinds of states, which can be distinguished by different value of variable r. r=0, 1, 2, 3 indicates idle, repairing workstation, repairing switch and repairing backbone network respectively. The queue length is represented by variable q. The value of variable q cannot exceed the total number of workstations.

Input: q, r, where the migration process is {*arrival, start(i), repair(i)*}; T represents time, it is synchronized with the time of component module in sub algorithm FCR. t is a time variable.

Output: Real-time output of the current repair module status r and queue length q.

Starting from t=0, if the workstation components fail, the status of the maintenance module changes to the mode of repairing workstation, and the faulty workstation enters the queue and leaves the queue after repairing. If other components fail, the status of maintenance module changes to the mode of repairing other components, after the repair is completed, the next faulty component is repaired till $t = T$. The specific algorithm is shown in Figure 2.

3. Experiment

This paper takes the scenario of workstation cluster [9] as a case study, and its scenario schematic diagram is shown in Figure 3.

The system is a highly reliable workstation cluster, which is connected by two sub clusters through the backbone network. The sub cluster is composed of several sub workstations, which are connected by star topology. The switch serves as a bridge between the sub cluster and the backbone network. Each component is likely to collapse, but only one repair unit is responsible for repairing each component. Based on the characteristics of high reliability cluster, in order to minimize the time of service interruption and ensure the uninterrupted demand for computing services, it is necessary to maintain the cluster system at a relatively high level of quality of service (QoS for short).

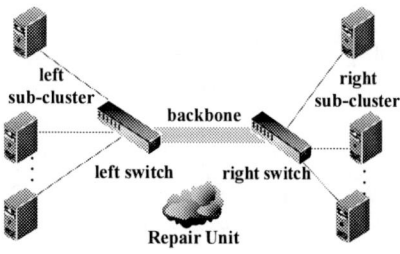

Figure 3. Scenario of workstation cluster.

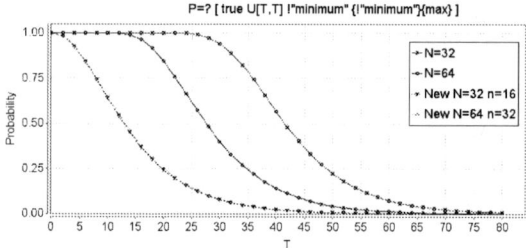

Figure 4. Probability of encountering the same problem.

Therefore, the following constraints should be followed: i) in order to provide the lowest quality of service (QoS$_{min}$ for short), at least k ($k \geq (3/4)*N$) workstations are required to maintain normal working, and k workstations are connected. If the current number of sub cluster workstations is less than k, it is necessary to connect another sub cluster through the backbone network to ensure that the number of workstations working normally is larger than k. Otherwise, lower than QoS$_{min}$ cannot ensure the normal operation of the whole system. The whole system can be restarted only when the number of workstations is higher than that of QoS$_{min}$ after repairing the faults by the maintenance unit; ii) in order to provide high quality of service (QoS$_{max}$ for short), at least h($k < h \leq N$) workstations are required to keep working normally.

The system can be divided into six main components: two sub clusters, two switches, backbone network and maintenance unit. Workstation module, switch module, backbone module and maintenance module can be modeled in PRISM, according to the different functions of components. These modules can interact with each other, and the parallel combination of each element constitutes the whole system.

According to the sub algorithm FCR and sub algorithm REP proposed in this paper, each component of the system is modeled separately. The attributes of the system are verified by PRSIM.

Table 1. Experimental parameter table.

Parameter	Meaning	Parameter value
N	Preset value of workstation number	actual quantity
left_mx	Number of workstations in the left sub-cluster	N
right_mx	Number of workstations in the right sub-cluster	N
n	Capacity of queue	(0.5~2)*N
k	QoS$_{min}$	(3/4)*N
h	QoS$_{max}$	N
ws_fail	Average failure probability of single workstation	1/500
switch_fail	Average failure probability of switch	1/4000
line_fail	Average failure probability of backbone network	1/5000

If the value of QoS$_{min}$ is (3/4)*N, neither of the workstation sub clusters is working properly, then the minimum value of queue length is the total number of workstations that failed in the whole system, i.e. (1/2)*N, the maximum value of queue length is the total number of workstations. Consequently, capacity of queue is (1/2)*N≤n≤2*N.

In the experiment, the probabilistic model checking tool PRISM is used to verify and analyze the attributes of the system, and the results are compared with those in literature [10]. The experimental process and results are as follows. In Figure 4-6, the solid line represents the experimental results in literature [10], and the dotted line represents the experimental results in this paper.

(1) Attribute 1: **P=?[true U[T,T] !"minimum" {!"minimum"}{max}]**

Attribute 1 describes the maximum probability for encountering the same problem again after T time units when the number of workstations working normally in the system is less than QoS_{min}, as shown in Figure 4.

In Figure 4, the green and blue solid lines represent respectively the different number of workstations, and the probability changes with time. Two dotted lines represent the probability trend after queue introduction under the same conditions. It is found that the probability of encountering the same fault again decreases greatly in the same time after queue introduction. Therefore, the introduction of queue can effectively reduce the probability of encountering the same fault.

(2) Attribute 2: **P=? [true U<=T "premium" {"minimum"}{min}]**

Attribute 2 describes the probability of the system from QoS_{min} to QoS_{max} within T time units. In attribute 2, the label "premium" represents that the formula satisfies QoS_{max}, and the label "minimum" represents that the formula satisfies QoS_{min}. The experimental results are shown in Figure 5.

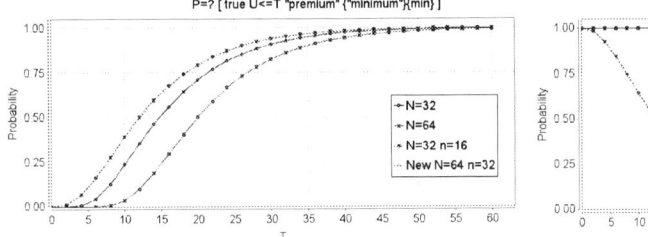 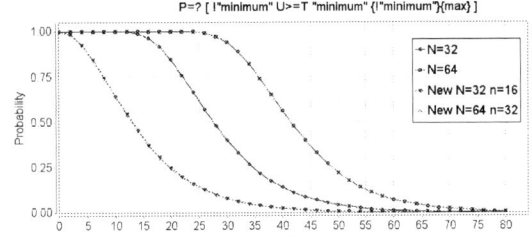

Figure 5. Probabilistic Trend Chart from QoS_{min} to QoS_{max} in T time unit.

Figure 6. Probabilistic Trend Chart of Over T Unit Time for Restoring from Lower than QoS_{min} to Normal.

From Figure 5, the solid line shows the trend of probability change of different workstation number N over time. With the increasing of workstation number, the more time it takes to reach the same probability P, that is, the number of workstations will affect the time when low-level QoS is restored to high-level QoS. Under the same experimental conditions, the dotted line represents the change of probability after the queue is introduced. Thus, after introducing queue model, the time required for the system to recover from the lowest QoS to the high-level QoS is greatly reduced under the same probability P. That is to say, the introduction of queue in this paper improves the efficiency of maintenance, reduces the interaction between component modules and repair unit, and improves the disaster resistance of the whole system. Even if there is a fault, it can recover quickly and return to normal working condition.

(3) Attribute 3: **P=? [!"minimum" U>=T "minimum" {!"minimum"}{max}]**

Attribute 3 describes the probability that it takes more than T time units to recover to normal operation when the QoS of system is below standard. The experimental results are shown in Figure 6.

In Figure 6, the solid line reflects that the time required for different number of workstations returning to normal state increases with the number of workstations increasing. The dotted line indicates that the time required after queue being introduced is much less than that before. From the experimental analysis of Figure 2-4, it can be seen that when the number of fault workstations is large, due to the introduction of queue model in the algorithm proposed in this paper, the maintenance scheduling of fault components becomes more orderly, and the data interaction among components is reduced, so that maintenance units do not need to spend more time to detect other components. It improves the efficiency of maintenance unit, thus improving the reliability and disaster resistance of the system.

4. Conclusion

Taking the scenario of workstation cluster as a case study, this paper proposes a workstation cluster fault repair method based on probabilistic model checking. By defining the extended continuous-time Markov chain model and the proposed sub algorithm FCR and sub algorithm REP, the operation scenario of workstation cluster is formalized as the ECTMC model. The attributes of the system are described by CSL, and the attributes of the system are verified and analyzed by the probabilistic model checking tool PRISM. The experimental results show that in the interaction between component modules and maintenance unit, the queue model is used in the proposed method. When the number of workstations with faults is large, they enter the queue to wait for repair. According to the importance of component function, different priorities are divided and the scheduling strategy of repair unit is adjusted. This method reduces the interaction time between each functional component and repair unit, which greatly cuts down the repair time, further improves the maintenance efficiency. In addition, the proposed method enables the whole system to return to normal work faster, and improves the capability for fault tolerance and disaster resistance of the system. Compared with the existing algorithms, the proposed method has some advantages and provides a better solution for improving system performance and reliability.

Acknowledgment

This work has been supported by the national natural science foundation of China (No. 61462034, No. 61561024), and Scientific Research Fund of Jiangxi Provincial Education Department, China (No. GJJ160632, No. GJJ170517).

References

[1]   Zhang X, Hu Z, Zheng J and Tang S 2003 *Comput. Eng.* **4** 26
[2]   Clarke E M, Grumberg O, Hiraishi H, Jha S, Long D E, McMillan K L and Ness L A 1993 *Proc. of Chdl'93 (Ottawa)* vol A (Amsterdam: Elsevier Science Publishers) pp 15-30
[3]   Lahtinen J, Valkonen J, Björkman K, Frits J, Niemela I and Heljanko K 2012 *Reliab. Eng. Syst. Safe.* **105** 104
[4]   Ji M 2014 *Research on the Model Checking Method for Complex Stochastic Systems* (Harbin: Harbin University of Engineering)
[5]   Kwiatkowska M, Norman G and Parker D 2011 *Proc. of CAV 2011 (Snowbird)* vol 6806 (Berlin: Springer) pp 585-591
[6]   Feng C, Zhang H, Yan S, Fu Y and Bao X 2017 *Proc. of ICRSE(Beijing)* vol 1 (Piscataway, N J: IEEE) pp 1-6
[7]   Guo X, Yang Z 2016 *Proc. of ICSESS(Beijing)* vol 1 (Piscataway, N J: IEEE) pp 564-567
[8]   Sesic A, Dautovic S and Malbasa V 2008 *IEEE.T.Comput.Aid.D.* **27** 403
[9]   Haverkort B R, Hermanns H and Katoen J P 2000 *Proc. of SRDS'00(Erlangen)* vol 1 (Piscataway, N J: IEEE) p 228
[10]   Workstation Cluster on http://www.prismmodelchecker.org/casestudies/cluster.php
[11]   Liu Y,Li X,Ma Y and Wang L 2015 *Chinese Journal of Computers* **11** 2145
[12]   Zhou C 2014 *Theory and Application of Random Model Checking*(Beijing: Science Press)
[13]   Gu Y, Zhu F and Tang S 2012 *Journal of Jiangxi University of Science and Technology* **33** 51

Anomaly detection for time series using temporal convolutional networks and Gaussian mixture model

Jianwei Liu[1], Hongwei Zhu[1], Yongxia Liu[2], Haobo Wu[3], Yunsheng Lan[4] and Xinyu Zhang[4]

[1] College of Information Science & Electronic Engineering, Zhejiang University, Hangzhou, Zhejiang, China

[2] College of Electrical Engineering, Zhejiang University, Hangzhou, Zhejiang, China

[3] School of Information Science and Engineering, Lanzhou University, Lanzhou, Gansu, China

[4] Gacia Electrical Appliance Co., Ltd., Wenzhou, Zhejiang, China

Email: ljw0608.mail@gmail.com

Abstract. Anomaly detection, as an important research field in the analysis of time series, has practical and significant applications in many occasions, such as network security, medical health, Internet of Things (IoT), fault diagnosis and so on. However, due to the inherent characteristics of time series, such as tremendous data volumes, the imbalance of normal data and abnormal data, additional constraints and challenges are added for anomaly detection for time series. We present a novel anomaly detection framework, which applies temporal convolutional networks to extract features of time series and combined Gaussian mixture model with Bayesian inference to detect anomalies of systems. In order to evaluate the effectiveness of our approach, experiments are carried out on two typical time series datasets including EEG dataset and current dataset of electrical equipment. The experiments indicate that temporal convolutional network can contribute to extracting salient features of time series and Gaussian mixture model with Bayesian inference has good generalization and reliability for anomaly detection. Meanwhile, the designed architecture and analysis approach of anomaly detection reveal the method's effectiveness and generalization in the feature extraction and anomaly detection for other time series.

1. Introduction

A time series is a sequence of data points changing over time, which is widely available in real life. Anomaly detection, as an important research field in the analysis of time series, has attracted the attention of many researchers. In order to ensure stable and safe operation of the system, anomaly detection plays a crucial role in many applications, such as network security, medical health, Internet of Things and system fault diagnosis [1].

Anomaly detection aims at detecting abnormal behaviors in time series, then triggering alerts or performing specific actions for abnormal behavior. Anomalies are usually classified into three categories: (1) point anomalies, which are the simplest type of anomalies. A single data point is abnormal relative to others in the sequence regardless of other factors. (2) contextual anomalies, which rely on the context, meaning that other factors can affect whether the value is abnormal or not. (3) collective anomalies, in which a single data point is not abnormal but the set of consecutive values deviates from a set of normal consecutive values [2]. As shown in figure 1, (a) is the EEG signal of the epileptic patient transition from seizure free to seizure, and (b) is the current waveform of the electrical equipment transition from normal operation to arc fault. Collective anomalies are usually difficult to detect in real data streams, but they can provide a lot of useful information and can be used as early warning of potential problems in practical applications. This paper will focus on collective anomaly detection.

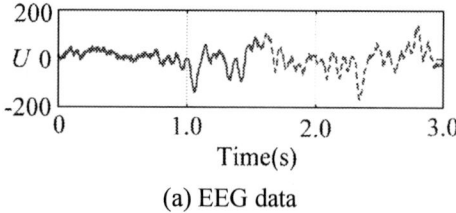
(a) EEG data

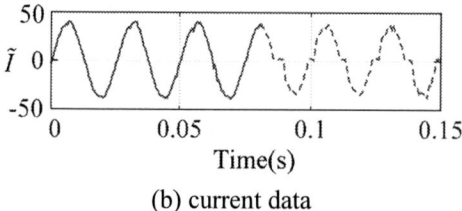
(b) current data

Figure 1. Examples of collective anomalies in time series. Blue lines (solid lines) and red lines (dotted lines) respectively denote normal and abnormal time series.

Anomaly detection is an important research topic in the field of data science and machine learning, and has attracted the attention of many researchers. The existing anomaly detection methods can be divided into two categories: supervised and unsupervised methods. The former adopts supervised learning algorithms such as Support Vector Machine (SVM) and Artificial Neural Network (ANN) to anomaly detection [3]. The output of the model is generally defined in two types. The first is the predicted value of the current state. Whether the current state is abnormal depends on the differences between the observed value and the predicted value. The other way of model output is a label that represents whether the current state is abnormal. Despite of the effectiveness of supervised methods, there are still two problems in anomaly detection: (1) class imbalance in datasets, which means that the number of abnormal samples is far lower than that of normal samples in the dataset. (2) Obtaining high-quality labels can be costly.

In contrast, unsupervised methods can be promising for anomaly detection, as they can be modeled using unlabeled training data. Unsupervised anomaly detection methods can be further divided into two categories, namely, probabilistic methods and unsupervised learning methods. Assuming that the normal data follows an underlying distribution (e.g., Gaussian distribution), statistical method can estimate the probability density of the current state. Yamanishi et al. [4] proposed an anomaly detection engine based on statistical theory to score the current state by the probability model using online learning. The higher the score, the more likely it is to be an anomaly. Recently, unsupervised learning methods have achieved remarkable performance in handling large and complicated datasets. Among them, clustering analysis is widely used in anomaly detection. Clustering analysis groups normal observations into one or several clusters, and each cluster has a cluster center. Some observations that do not belong to any cluster or are far away from all of cluster centers are defined as anomaly. Moreover, advanced clustering algorithms are well-designed for anomaly detection, for example, Gaussian mixture model for epilepsy detection, which proves the effectiveness of automatic epilepsy diagnosis [5]. Jacobs et al. [6] make use of the Gaussian mixture model for fault detection and location of gas turbines.

However, the properties of time series add additional constraints and limitations to the machine learning. Firstly, because time series generate quickly, and the amount of data volume is huge, it is impossible to label abnormal data manually and store all of data. Furthermore, anomalies rarely occur, and the time series contains a large number of normal data and few of abnormal data, which results in class imbalance in datasets, and it is difficult to acquire a satisfying anomaly detection model by supervised learning. Therefore, an unsupervised algorithm is needed to detect anomaly for time series.

What is more, collective anomaly detection aims at detecting anomaly sub-sequences in a time series. In general, sub-sequences can be mapped into another feature space in which anomaly detection is performed. Therefore, feature extraction will be critical in anomaly detection. Traditionally, feature extraction for anomaly detection utilizes signal analysis tools, such as Fourier transform (FFT), Discrete wavelet transform (DWT) or chaos analysis to manually extract features. However, studies have shown that traditional methods of feature extraction are subjective in a certain sense, which limits the expression ability of original sequence [7]. In recent years, deep learning has achieved state of the art in many applications, such as face recognition [8], person re-identification [9], image semantic segmentation [10] and natural language processing [11]. It is well-known that the essence of deep learning is feature representation learning [12], which is a basic modeling tool for acquiring, representing and compressing large-scale signals. Convolutional neural network (CNN) [13] has been used for sequence modeling for several decades. Nowadays, CNN has been widely applied to machine translation [14], audio synthesis [15], and language modeling [16]. Therefore, in order to improve the performance and applicability of anomaly detection, it is necessary to apply the representation learning of CNNs to extract feature of time series.

In this paper, a novel framework is proposed to detect anomaly for time series. Temporal convolutional networks (TCN) are applied to extract feature of time series, mapping high-dimensional time series into low-dimensional feature space, in which anomaly detection are performed. Furthermore, based on the Gaussian mixture model (GMM), the Bayesian inference is introduced to effectively detect anomaly with low computational complexity, and it can be applied to online anomaly detection for time series.

2. Architecture

Considering the particularity of collective anomaly detection in time series, we propose a novel anomaly detection framework, which firstly maps the high-dimensional signals to low-dimensional feature space to extract salient features of time series. Inspired by Van et al. [15], this paper optimizes temporal convolutional network to make it more suitable for feature extraction of time series. Secondly, Gaussian mixture model and Bayesian inference together are used to detect anomalies. The details about anomaly detection framework is as follows.

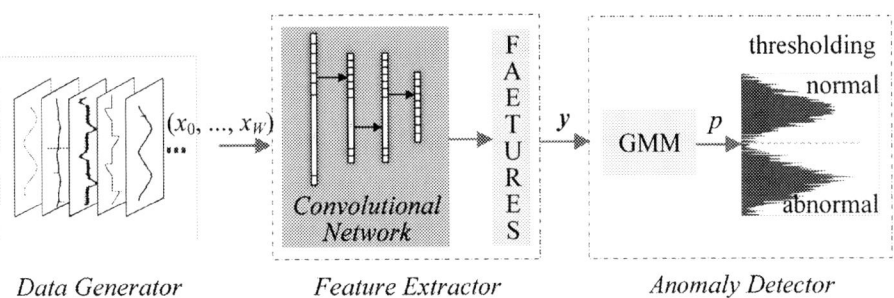

Figure 2. The architecture of anomaly detection for time series.

As shown in figure 2, the basic architecture of the anomaly detection proposed in this paper consists of the following parts:

- *Data Generator*: this module mainly extracts input series from data stream and feed it to the next stage. In the analysis of time series, a time window is commonly used to extract sample by a fixed length (i.e., window width), and the window moves a fixed length (i.e., a stride size) over time. Assuming that the width of the time window is W and the stride size is S.

- *Feature Extractor*: The module consists of a well-designed stack of temporal convolutional networks to extract features of time series. The input series $(x_0, x_1, ..., x_W)$ passes through the feature extractor to obtain a low-dimensional feature vector y.

- *Anomaly Detector*: The module utilizes the Gaussian mixture model to detect whether the current system state is abnormal and output the detection result. GMM can calculates the confidence probability p of current system belonging to the abnormal state, and compares and analyzes with the given confidence threshold T to obtain the detection result.

Therefore, the anomaly detection framework designed in this paper can output an operational alarm signal that can be used in systems such as automatic fault diagnosis or medical diagnosis. Next, we will demonstrate the details about the anomaly detection framework.

3. Temporal convolutional networks

Temporal Convolutional Networks (TCN) based on a 1D convolutional network, is a generic network structure for sequence modeling. The following network structure is an optimized version of structure proposed by Van [15]. This revised network structure improves its performance to extract the salient features of the sequence.

3.1. Casual convolutions

TCN structure is mainly based on two principles: 1) For every layer, the numbers of inputs and outputs are equal; 2) The casual constraints, which means that the output y_t at time t only depends on the previous inputs $x_0, x_1, \ldots, x_t$ but not depends on the future inputs $x_{t+1}, x_{t+2}, \ldots, x_T$. According to the first principle, TCN uses a 1D fully convolutional networks structure. To satisfy the second principle, TCN utilizes casual convolutions, in which the output at time t is convolved with elements of convolutional kernel and the current and previous inputs.

However, this basic network has an obvious drawback. In order to ensure the effectiveness of inputs in the very long history, very deep network or very large convolutional kernel is needed. In fact, experiments have revealed that neither of the methods is feasible. Therefore, modern convolutional network is supposed to be elaborated to optimize the performance of TCN.

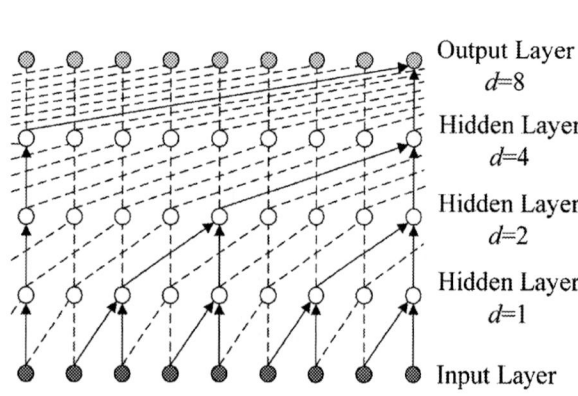

Figure 3. Architectural elements in a dilated-causal convolution network.

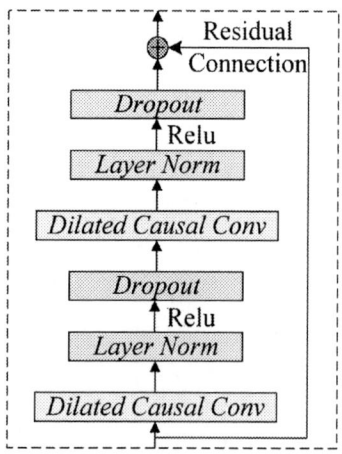

Figure 4. Architecture elements of TCN.

3.2. Dilated convolutions

As is stated above, receptive field of TCN is limited if only casual convolutions are included, leading to the fact that output at time t can't receive the inputs in the long history. Thus, dilated convolutions are necessary to be introduced to TCN to improve its receptive field without extra computation cost. For input sequence $x = (x_0, x_1, \ldots, x_t)$ and k-sized convolutional kernel f, the s-th output of dilated convolutions is shown in equation (1).

$$F(s) = (\boldsymbol{x} * f)(s) = \sum_{i=0}^{k-1} f(i) \cdot x_{s-d \cdot i} \qquad (1)$$

where d denotes dilation rate. Obviously, dilated convolution degrades to ordinary convolution when $d=1$. And larger dilation rate means larger input range of top layer in the network, thus increasing the receptive field.

Figure 3 illustrates the architecture of dilated-causal convolutions with dilation rate $d=1, 2, 3, 4$. It can be seen that receptive field are enlarged layer by layer, as the dilation rate increase exponentially with the linear increase of network depth.

3.3. Residual connections
It can be seen from figure 3 that increasing the network depth or extending its width can improve the performance of network. However, with the increasing of network depth, it is challenging to train the network, and the performance of network begins to deteriorating gradually. Thus, residual networks can be applied to simplify the training progress. This method makes the optimization of deep network easier and network deeper, improving the network performance apparently.

Receptive field of TCN structure depends on the network depth n and the size k of convolution kernel, so the stabilization of network is very significant. As is illustrated in figure 4, residual connections are included in the dilated-causal convolutional network. Besides, dropout layer and layer normalization are added following the dilated-causal convolutional layer to improve the generalization performance of network.

To sum up, TCN structure has been optimized to adapt to time series. The network proposed in this paper contains input layer, three TCN blocks, a fully-connected layer and a softmax layer. After fully-connected layer, L2 normalization are added as feature embedding layer. We train the TCN model in the form of classification to improve the ability of feature representation.

3.4. Squeeze-and-excitation block
Convolutional network, based on the convolution operation, extracts features through integrating spatial information and channel information locally. Fortunately, the dependence between channels can be modeled explicitly by a new network called squeeze-excitation (SE) block, which can improve the feature extraction ability of TCN. SE block is shown in figure 5, F_{tr} is traditional convolutional operation, while $\mathbf{X}$ and $\mathbf{U}$ are its input and output respectively. In the SE block, global average pooling follows the input $\mathbf{U}$, outputting $1 \times C$ data, which is a procedure called squeeze ($\mathbf{F}_{sq}$). Then, the data go through two fully-connected layers, which is a procedure called excitation ($\mathbf{F}_{ex}$). The mechanism of SE block automatically obtains the significance of each channel, enhancing the major features and weakening the minor features to improve the discrimination of extracted feature.

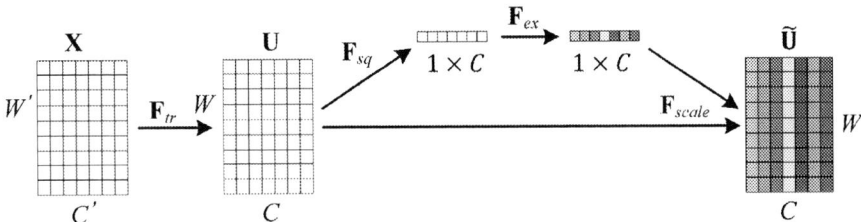

Figure 5. Architecture of SE Block.

3.5. Loss function

3.5.1. Focal loss. Due to the class imbalance in datasets, trained TCN model by cross-entropy loss function will be more inclined to the class with more samples, which results in an unsatisfying model.

To solve this problem, focal loss revises cross entropy loss function by introducing adjustment factor $(1-p_t)^\gamma$ and balance factor α_t to improve the classification performance of the model.

$$p_t = \begin{cases} p, & if \ y = 1 \\ 1-p, & otherwise \end{cases} \tag{2}$$

$$FL(p_t) = -\alpha_t(1-p_t)^\gamma \log(p_t) \tag{3}$$

where p denotes the possibility that the sample x belonging to the positive class, and λ is focal parameter.

3.5.2. Center loss. The features extracted by the trained TCN networks with focal loss are separated but not discriminated. To solve this problem, center loss function is supposed to be added to train the TCN network. The basic mechanism of center loss function is that increasing variance among different classes and minimizing variance within the same class can extract discriminated features. The center loss function is shown in equation (4).

$$L_c(x) = \frac{1}{2}\sum_{i=1}^m \| x_i - c_{y_i} \|_2^2 \tag{4}$$

where x denotes input feature vector and c_y represents the center of x corresponding to class y.

For time series classification, there are only two centers representing positive and negative samples respectively. During the training process, center loss and focal loss are supposed to be combined, and the centers c_y are updated in each iteration. Focal loss can increase the feature variance between different classes while center loss can decrease the feature variance within the same class. Thus, the training method combines focal loss and center loss can help obtain the more discriminated features.

4. Gaussian mixture model for anomaly detection

The Gaussian mixture model (GMM) is a weighted linear combination of multiple Gaussian components, which is often used to solve the problem that data contains multiple different distributions. GMM can smoothly approximate the density distribution of arbitrary shapes and can be applied to complicated object modeling.

Let $x \in \square^D$ be a D-dimension sample from time series, its probability density function can be formulated as following:

$$\Pr(x) = \sum_{k=1}^K \omega_k \psi(x; \mu_k \Sigma_k) \tag{5}$$

$$\psi(x; \mu_k \Sigma_k) = \frac{1}{(2\pi)^{D/2} |\Sigma_k|^{1/2}} \exp\left[-\frac{1}{2}(x-\mu_k)^T \Sigma_k^{-1}(x-\mu_k) \right] \tag{6}$$

where ω_k is the weight of the k-th Gaussian component, $\sum_{k=1}^K \omega_k = 1, \ 0 \le \omega_k \le 1$. $\mu_k \in \square^D$, $\Sigma_k \in \square^{D\times D}$ respectively represents the mean vector and covariance matrix of the k-th Gaussian component. We can utilize expectation maximum (EM) algorithm to estimate the parameters $\theta = (\omega, \mu, \Sigma)$ of GMM.

Based on GMM, the Bayesian inference probability (BIP) index [17] is introduced to detect anomaly. The BIP is formulated as follows:

$$BIP = \sum_{k=1}^K P(C_k | x_j) \cdot P_L^{(k)}(x_j) \tag{7}$$

$$P_L^{(k)}(x_j) = \Pr\left\{ D(x, C_k) | x \in C_k \le D(x_j, C_k) | x_j \in C_k \right\} \tag{8}$$

$$D((x_j, C_k) | x_j \in C_k) = (x_j - \mu_k)^T \Sigma_k^{-1}(x_j - \mu_k) \tag{9}$$

where $P_L^{(k)}(x_j)$ indicates the local Mahalonobis distance-based probability index of testing sample x_j.

$D(x_j, C_k)$ is the Mahalanobis distance of testing sample x_j to the center of k-th Gaussian component.

Given the threshold T, the BIP index is calculated by equation (7) and compared with the given the threshold T to determine whether the current state is abnormal.

5. Experiment
In this section, experiments are carried out on two typical time series datasets to evaluate the anomaly detection for time series proposed in this paper, and the experimental results are analysed.

5.1. Performance metrics
In order to make the experiment results readable, some objective evaluation metrics are introduced to evaluate the model performance. Formally, *TP*, *FP*, *TN* and *FN* denote respectively the number of normal time series correctly detected as normal (True Positives), the number of anomalies wrongly detected as normal (False Positives), the number of anomalies correctly detected as abnormal (True Negatives), and the number of normal time series wrongly detected as abnormal (False Negatives). Accuracy, recall, FPR (False Positive Rate) and F1-score are defined as following:

$$Accuracy = \frac{TP + TN}{TP + FP + TN + FN} \tag{10}$$

$$recall = \frac{TP}{TP + FN} \tag{11}$$

$$FPR = \frac{FP}{TN + FP} \tag{12}$$

$$F1 = \frac{2 \times Precision \times Recall}{Precision + Recall} \tag{13}$$

where *Precision=TP/(TP+FP)*.

5.2. Implement details
The samples of datasets are collected from time series by using a sliding window, and the dataset is divided into training dataset and testing dataset by the ratio of 1:4. Then we use focal loss and center loss to conduct supervised training for TCN. During the process of training, a simple cross-validation method is used to verify the performance of the model.

Then the trained TCN model was applied to extract the features of normal time series. Determined the number of Gaussian components K, the GMM parameters $\theta = (\omega, \mu, \Sigma)$ were iteratively optimized by EM algorithm. Given a testing sample, the corresponding BIP index was calculated by equation (7). If the BIP index exceeds the given confidence threshold T, the observation is considered to be abnormal, otherwise it is marked as a normal sample.

5.3. Results and analysis
This section conducts experiments on the available EEG datasets and private current datasets of electrical equipment to analyse and evaluate the performance of the anomaly detection techniques proposed in this paper.

5.3.1. Epilepsy signal detection. This experiment carried out on the available EEG dataset, whose detailed information can be found in [18]. The EEG dataset contains five subsets, each of which contains 100 fragments from single channel EEG signal. Subsets A and B were acquired from patients with open and closed eyes, and subsets C and D were collected from patients during epilepsy free intervals. Subset E was collected from patience during epilepsy. The length of all EEG signal fragments was 4096, and the sampling rate was 173.61Hz. As shown in figure 6 (a), EEG data without epilepsy were taken as normal samples while the EEG data during epilepsy as abnormal samples for experiments.

In this paper, the sliding window size W of EEG data was set to 150. The TCN model was trained on the EEG dataset, and the trained TCN was used as feature extractor. As shown in figure 7 (a), 1000 samples were mapped into 2D feature space, it can be seen that the abnormal and the normal samples are respectively clustered in different spatial regions. This distribution characteristic reflects that the TCN model can help to extract salient and discriminated features of the time series.

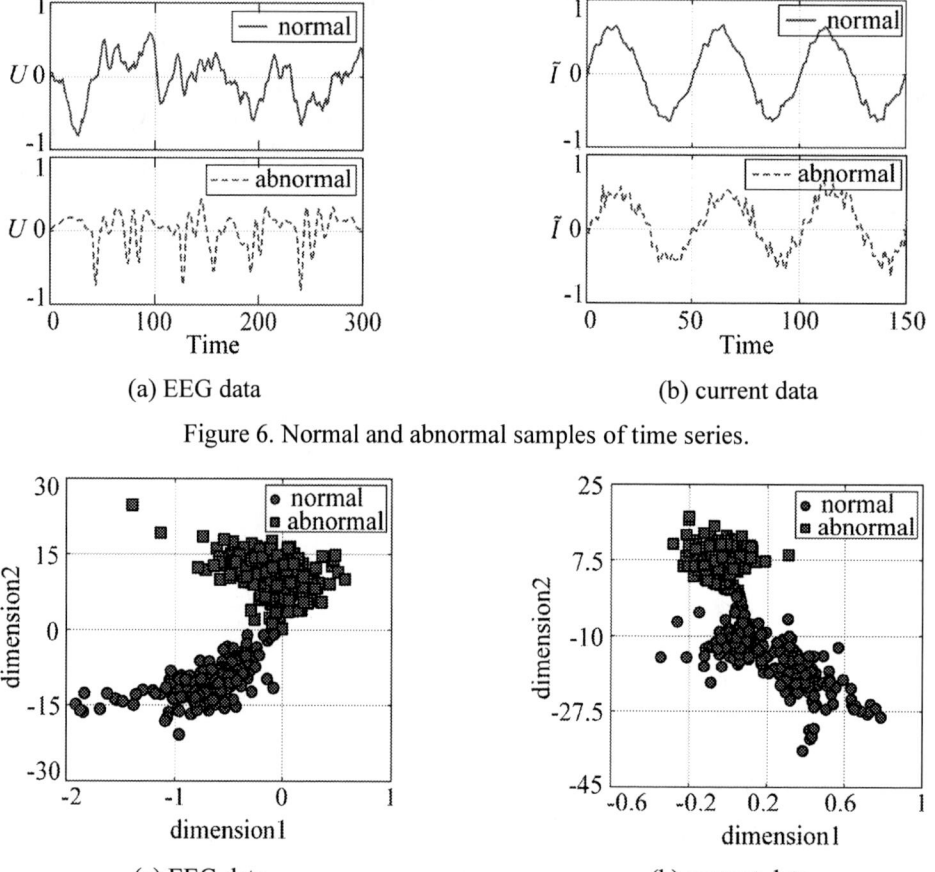

(a) EEG data   (b) current data

Figure 6. Normal and abnormal samples of time series.

(a) EEG data   (b) current data

Figure 7. The distribution map of high-dimension time series mapping into 2D feature space.

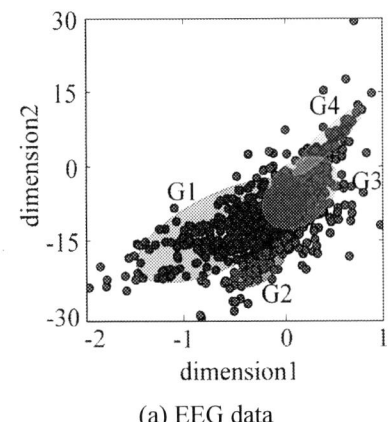

(a) EEG data

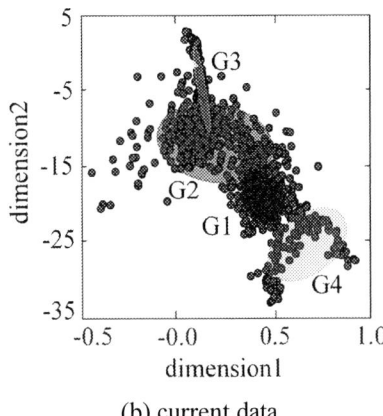

(b) current data

Figure 8. Diagram of the modeling results for normal data using GMM.

We extracted the features of normal EEG data by trained TCN model, and optimized Gaussian mixture model by EM algorithm. As is shown in the figure 8 (a), the number of Gauss components K = 4, they were marked as G1, G2, G3 and G4 respectively, and each ellipse represented a Gaussian component.

The 10-fold cross validation was utilized to evaluated the performance of the GMM model on testing dataset, and the evaluation results of GMM on EEG dataset was shown in table 1. Specifically, when the FPR was 1‰, the recall could reach 93.52%.

Table 1. The evaluation results of GMM.

Datasets	Accuracy	recall	FPR	F1
EEG	99.45%	99.67%	0.796%	99.54%
Arc Fault	97.10%	97.096%	0.498%	95.66%

5.3.2. Arc Fault detection. The current dataset was collected from the AC current signals of the electrical equipment. In the experiment, we collected the current signal of the air conditioner during normal operation and the arc fault at a sampling rate of 96 Hz. As shown in figure 6 (b), the current signals of the electrical equipment working normally were taken as normal samples, while the current signals during arc fault were taken as abnormal samples. The size W of sliding window 96, and we trained the TCN model on the current dataset, and extracted the features of current series. As shown in figure 7(b), the high-dimensional time series were mapped into the 2D feature space. Obviously, the normal and abnormal samples were respectively clustered in different spatial regions, which further showed that the TCN model can help to extract the salient and discriminated features of the time series.

Similarly, the feature vectors of normal time series were extracted using TCN model, and optimized Gaussian mixture model by EM algorithm. As shown in figure 8 (b), the number of Gaussian components K = 4, they were marked as G1, G2, G3 and G4 respectively, and each ellipse represented a Gaussian component. Table 1 also showed the evaluation results of GMM on arc fault datasets.

To summarize, we can draw the following conclusions through the analysis of the experiments: (1) the TCN model can help to extract the salient and discriminated features of the time series. (2) In the case of class imbalance in datasets, GMM with Bayesian inference for anomaly detection can obtain high recall rate while keeping a low FPR, and it can be applied to detect anomaly for time series.

6. Conclusion

This paper proposed an online anomaly detection framework for time series based on temporal convolution networks and Gaussian mixture model, which can map high-dimensional time series into low-dimensional feature space for anomaly detection. For the sequence modeling, temporal convolutional networks were proposed for feature extraction of time series. Based on the GMM, Bayesian inference was introduced in this paper to obtain better performance in anomaly detection.

Experiments were conducted to train and evaluate the TCN and GMM on EEG dataset and arc fault dataset. The experiment results indicate the optimized temporal convolutional network can help to extract the salient and discriminated features of the time series, and the Gaussian mixture model has good generalization and reliability in anomaly detection. At the same time, the design architecture and analysis method of anomaly detection proposed in this paper reveal its effectiveness and generalization in feature extraction and anomaly detection of other time series.

References

[1] Chandola V, Banerjee A and Kumar V 2009 *CSUR* **41** 15
[2] Saarinen I 2017 *Adaptive real-time anomaly detection for multi-dimensional streaming data* (Espoo : Aalto University)
[3] Batta P, Singh M, Li Z, Ding Q and Trajkovic L 2018 *Proc. Of the IEEE Int. Symposium on Circuits and Systems (Florence)* pp 1-4
[4] Yamanishi K, Takeuchi J I, Williams G and Milne P 2004 *Data Mining and Knowledge Discovery* **8** 275-300
[5] Li Y, Cui W, Luo M, Li K and Wang L 2018 *International Journal of Neural Systems* **28** 7
[6] Jacobs W R, Edwards H, Li P, Kadirkamanathan V and Mills A R 2018 *International Journal of Prognostics and Health Management* **9** 26
[7] Wang Y, Zhang F and Zhang S 2018 *IEEE Transactions on Instrumentation and Measurement* **67** 2526-37
[8] Schroff F, Kalenichenko D and Philbin J 2015 *Proc. of the IEEE Conf. on Computer Vision and Pattern Recognition (Boston)* pp 815-823
[9] Chen S Z, Guo C C and Lai J H 2016 *IEEE Transactions on Image Processing* **25** 2353-67
[10] Girshick R, Donahue J, Darrell T and Malik J 2014 *Proc. of the IEEE Conf. on Computer Vision and Pattern Recognition (Columbus)* pp 580-587
[11] Zhang Y, Lease M, Wallace B C 2017 *31st AAAI Conference on Artificial Intelligence (San Francisco)* pp 3386-3392
[12] Bengio Y, Courville A and Vincent P 2013 *IEEE Transactions on Pattern Analysis and Machine Intelligence* **35** 1798-1828
[13] LeCun Y, Boser B, Denker J S and Henderson D 1989 *Neural Computation* **1** 541-551
[14] Kalchbrenner N, Espeholt L, Simonyan K, Oord A V D, Graves A and Kavukcuoglu K 2016 Neural machine translation in linear time *preprint* arXiv/1610.10099
[15] Van Den Oord A, Dieleman S, Zen H, Simonyan, K, Vinyals O, Graves A, Kalchbrenner N, Senior A and Kavukcuoglu K 2016 Wavenet: A generative model for raw audio *preprint* arXiv/1609.03499
[16] Dauphin Y N, Fan A, Auli M and Grangier D 2016 Language modeling with gated convolutional networks *preprint* arXiv/1612.08083
[17] Yu J and Qin S J 2008 *AIChE Journal* **54** 1811-29
[18] Andrzejak R G, Lehnertz K, Mormann F, Rieke C, David P and Elger C E 2001 *Phys. Rev. E* **64** 061907

ISPECE

IOP Publishing

Automatic detection of follicle ultrasound images based on improved Faster R-CNN

Tianlong Zeng[1,2,3] , Jun Liu [1,2]

[1] School of Computer Science and Technology, Wuhan University of Science and Technology, Wuhan 430081, P. R. China

[2] Hubei Province Key Laboratory of Intelligent Information Processing and Real-time Industrial System, Wuhan 430081, P. R. China

[3] Correspondence: 942636327@qq.com

Abstract. Follicle Ultrasonic images detection technology plays an important role in the monitoring of bull-follicle. Because of follicle ultrasound image containing lots of speckle noise and fuzzy edges, the traditional image detection algorithm is difficult to get better detection results on the ultrasonic image, and the traditional image detection algorithm needs to carry out sample feature extraction for each image, which is time-consuming and time-consuming and labor-intensive. According to the characteristics of the cattle follicle ultrasound image sets, this paper proposes a model of image detection based on improved deep learning Faster R - CNN to automatically detect cattle ovarian follicles, through joint VGG-16 different network layer characteristic figure to replace single deepest characteristic figure, retain the deep semantic characteristics at the same time, also keep the shallow characterization information. The experimental results show that this method has a better effect on the ultrasonic image detection of bovine follicle.

1. Introduction

Nanyang Yellow Cattle is a famous yellow cattle breed in China with good working capacity, meat performance and adaptability. In the field of monitoring follicles of yellow cattle, ultrasound technology has been widely used[1][2]. The traditional yellow cattle breeding industry relies mainly on artificial farming experience to arrange breeding and conception, which leads to high breeding costs and low reproductive rates. Ultrasound imaging technology, as a new technology, is widely used in cattle breeding and embryo production because of its non-invasive, non-radioactive, real-time dynamic display and low price. It is an effective means to improve the breeding of yellow cattle. Ultrasound image detection technology plays an important role in the monitoring of yellow cattle follicles. By monitoring the follicles of yellow cattle, finding the best time and conception for breeding can effectively improve the reproductive capacity of yellow cattle. Follicular detection is one of the important steps in follicular ultrasound imaging. Accurate positioning of the target area using image object detection technology has a decisive influence on the accuracy of subsequent follicle segmentation, quantitative calculation and measurement. Foreign scholars Pierson and Ginthers used ultrasound imaging to test and verify the hypothesis proposed by Rajakoski that the cattle would have two follicular developmental waves in one estrous cycle[3]. In China, Nong et al. used ultrasound to observe the follicular development and ovulation of Leiqiong yellow cattle under different estrus conditions[4]. In the field of ultrasonic image detection, Potocnik summarized three methods for automatic detection of ultrasound images of follicles:

Content from this work may be used under the terms of the Creative Commons Attribution 3.0 licence. Any further distribution of this work must maintain attribution to the author(s) and the title of the work, journal citation and DOI.

Published under licence by IOP Publishing Ltd

cellular neural network algorithm, region growing algorithm and predictive correction algorithm[5]. In 2010, PS Hirenmath proposed three kinds of follicle detection algorithms: edge detection[6], geometric features[7], and active contourlet waves[8]. Since 2012, the University of Toronto Alex proposed AlexNet[9] and achieved the best results of ImageNet in 2012. After that, various better and deeper networks have been proposed. As a branch of machine learning, deep learning has developed rapidly in recent years and has won all the champions of the years. In 2014, Ross B.Girshick used the region proposal combined with CNN (Convolutional Neural Network) to replace the sliding window and manual design features used in traditional object detection, and designed the R-CNN framework[10], which made a huge breakthrough in object detection and opened a craze of object detection based on deep learning. In the same year, Kaiming He et al. proposed the SPP (Spatial Pyramid Pooling) net[11], which greatly improved the speed of R-CNN. In 2015, Ross B. Girshick improved R-CNN to Fast R-CNN[12] and the final Faster R-CNN[13].

2. Method

Faster R-CNN is Ross Girshick's end-to-end neural network that combines object detection and classification. It has two earlier versions, R-CNN and Fast R-CNN, as well as Ross's early research result.

2.1. R-CNN

R-CNN, the Region with CNN, is a milestone in the application of convolutional neural networks to object detection. CNN has good performance for feature extraction and classification, and the Region Proposal method is used to achieve object detection. At that time, Ross used CNN for feature extraction. From the experience-driven artificial feature paradigm HOG and SIFT to the data-driven representation learning paradigm, the feature representation ability of the sample was improved. The extraction of Region Proposal was based on the traditional Selective Search method. The extracted feature is to train and classify with SVM. In the training process, there are supervised pre-training under large samples and fine-tuning under small samples to solve the problem that small samples are difficult to train or even over-fitting.

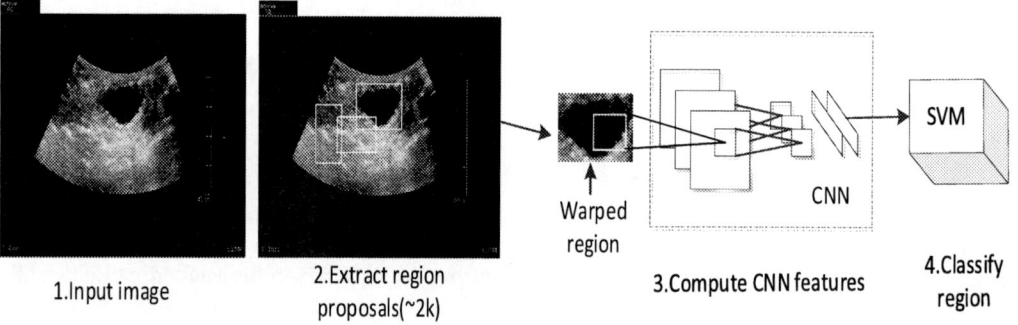

Figure 1 Schematic diagram of the R-CNN model

2.2. Faster R-CNN

Fast R-CNN has improved R-CNN by first replacing the SVM classifier with Softmax, and replacing the last max pooling layer with the ROI pooling layer to map region proposals of different sizes to the same dimension, and the ROI pooling layer can back propagate, thus there are only once feature extraction of the entire image area avoiding redundant feature extraction operations in R-CNN; the Fast R-CNN network uses parallel different fully connected layers at the end, which can simultaneously output classification results and the box regression results ,and uses SVD to decompose it, which reduces the computational complexity, speeds up the detection, implements end-to-end multitasking training, and does not require additional feature storage space.

The Figure 2 shows the schematic diagram of Fast R-CNN.

The RoI pooling layer is the main reason why the Fast R-CNN is significantly faster than the R-CNN. It is simply the special-case of the spatial pyramid pooling layer used in SPP nets in which there is only one pyramid level. Each RoI is defined by a four-tuple (r, c, h, w) that specifies its top-left corner (r, c) and its height and width (h, w). There are two main functions, one is to locate the RoI in the image to the corresponding patch in the feature map, and the other is to down sample(max pooling)the feature map patch into a fixed-size feature with a single-layer SPP layer and pass it to the fully-connected layer FC.

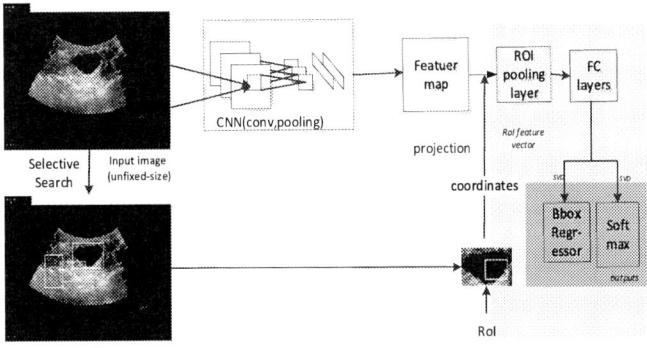

Figure 2 Schematic diagram of Fast R-CNN

2.3. Faster R-CNN

Following Fast R-CNN, the region proposal algorithm Selective Search (2s/image) and EdgeBoxes (0.2s/image) implemented on the CPU become the biggest bottleneck in object detection speed improvement. The Faster R-CNN replaces the Selective Search with the RPN layer, thus achieving a complete end-to-end of the network.

RPN, Region Proposal Networks, draws on the ideas of SPPnet and RoI, first maps the feature map obtained by CNN to the original image, and designs boxes of different sizes on the original image. The box is called an anchor. Then calculate the IOU between these anchors and ground truth, and divide the anchor into positive and negative according to the calculated IOU value. The positive class indicates that the value of IOU is greater than 0.7, and the negative class indicates that the value of IOU is less than 0.3. Finally, a certain amount of positive and negative samples are selected according to the numerical value for training, so that the boundary of the boxes is regressed and corrected.

The RPN structure is shown as Figure 3:

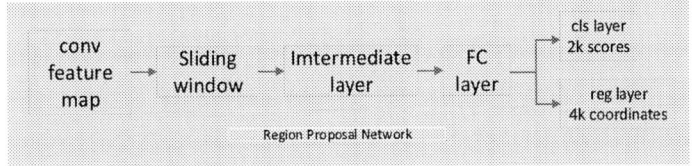

Figure 3 Schematic diagram of RPN

In a certain amount of anchors as the positive and negative samples obtained by the RPN, each anchor will output two scores first, one indicating the probability of the object, another indicating the probability of not being the object, and uniformly represented by p_i; then, outputting 4 coordinate values are used to represent the position coordinates of the anchor. Therefore, the loss function of the RPN layer is a multi-task loss function, which consists of two parts: one is the SoftmaxLoss of the object probability, and the other is the smooth L1 Loss of the coordinate position between the anchor and the ground truth. The specific definition of the Loss function is as follows:

$$L(\{p_i\}\{t_i\}) = \frac{1}{N_{cls}}\sum_i L_{cls}(p_i, p_i^*) + \lambda\frac{1}{N_{reg}}\sum_i p_i^* L_{reg}(t_i, t_i^*) \qquad (1)$$

Here, i represents the i-th anchor in a mini-batch and p_i is the predicted probability of anchor i being an object. The ground-truth label p_i^* is 1 if the anchor is positive, and is 0 if the anchor is negative. When the label of the anchor is positive, the label is negative when it is negative. t_i is a vector representing the 4 parameterized coordinates of the predicted bounding box, and t_i^* is that of the ground-truth box associated with a positive anchor. Assume (x, y, w, h) denote the box's center coordinates and its width and height, then variables (x, x_a, x^*) are for the predicted box, anchor box, and ground-truth box respectively (likewise for y, w, h). So the task of learning is to make the values of the two similar. as the figure 4 shows:

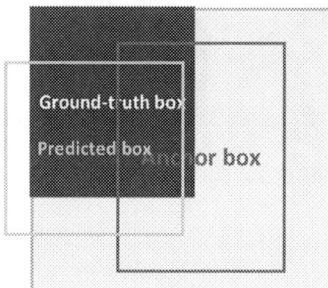

Figure 4 the predicted box, anchor box, and ground-truth box

Including:

$$
\begin{aligned}
t_x &= (x - x_a)/w_a, t_y = (y - y_a)/h_a, \\
t_w &= \log(w/w_a), t_h = \log(h/h_a), \\
t_x^* &= (x^* - x_a)/w_a, t_y^* = (y^* - y_a)/h_a, \\
t_w^* &= \log(w^*/w_a), t_h^* = \log(h^*/h_a)
\end{aligned}
\qquad (2)
$$

In formula (2), L_{cls} is the SoftmaxLoss for two categories, and the formula is:

$$L_{cls}(p_i, p_i^*) = -\log[p_i^* p_i + (1 - p_i^*)(1 - p_i)] \qquad (3)$$

L_{reg} is the smooth L1 Loss for two offsets, the formula is:

$$L_{reg}(t_i, t_i^*) = R(t_i - t_i^*) \qquad (4)$$

in which the function R is defined as follows:

$$R(x) = smooth_{L_1}(x) = \begin{cases} 0.5x^2 & |x| \le 1 \\ |x| - 0.5 & otherwise \end{cases} \qquad (5)$$

The hyper-parameter λ in formula (2) controls the balance between the classification loss and the regression loss. N_{cls} and N_{reg} are used to normalize the classification loss L_{cls} and the regression loss L_{reg}, respectively.

For the entire network, nearly half of the time is in the full connection layer operation. Therefore, the truncated SVD is used to accelerate the FC calculation.

The Figure 5 shows the schematic diagram of Faster R CNN.

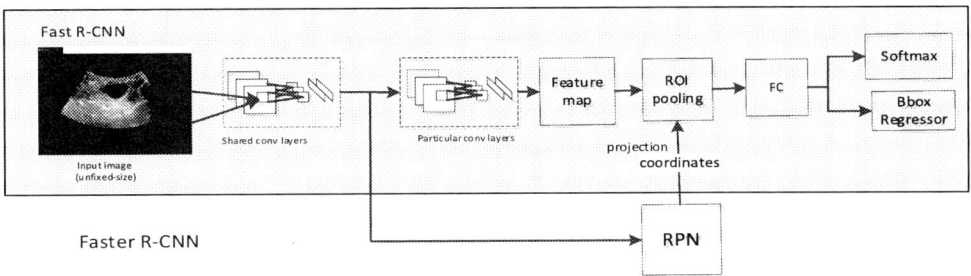

Figure 5 Schematic diagram of Faster R-CNN

2.4. Improved Faster R-CNN

The improved faster R-CNN borrowed the thought of FCN which connected the conv3, conv4 and conv5 layers of vgg_16 network together to form a Hyper Feature map instead of the original conv5_3 for RPN and Fast R-CNN. Because of the small target of follicular ultrasound image, in order to maintain high accuracy, the deep semantic features and the shallow characterization information are retained at the same time.

The Figure 6 shows the schematic diagram of Hyper Feature map.

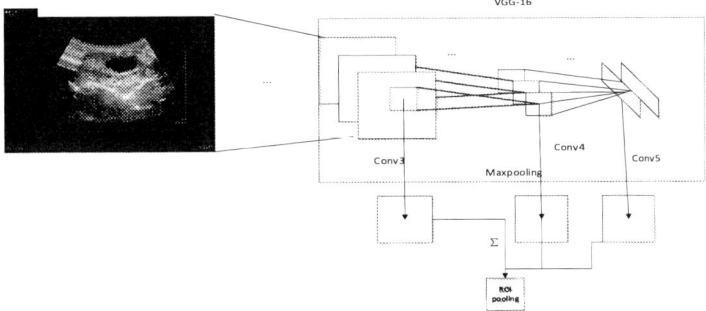

Figure 6 Schematic diagram of Hyper Feature maps

3. Implementation Detail

The ultrasound yellow cattle follicle image set of this experiment is derived from the National Natural Science Foundation of China, "Key Techniques for Monitoring Dynamic Changes of Follicles in Nanyang Yellow Cattle Based on 3D Ultrasound Imaging". The data set contained 6000 section images from 115 yellow cattle follicles three-dimensional ultrasound images.

3.1. Data process

The raw data of this experiment is a three-dimensional ultrasound image of the mvl file. Before training, the 3D ultrasound images in the file are saved as bmp files in frames by View3DXI software. Then, through the MATLAB software, the obtained bmp file is image data enhanced by using image flipping, rotation, random cutting, etc., and is saved as a jpg file and numbered. Finally, the image is labled by ImageLabel in github, and finally the data set is saved in the form of a VOC 2007 data set.

3.2. Training

4-Step Alternating Training [13]is used during training. During training, the data set is randomly divided into two sets of Trainval and test, each accounting for 50% ,3000 images, and Train_val is divided into two parts of train and val, which still account for 50%, that is,1500 image. Pre-training was performed using VGG-16 network during training.

Table 1 The results of the experiment

model	Proposals	mAP(%)
Faster R_CNN	300	70.1
Faster R_CNN	2000	75.4
improved Faster R_CNN	300	73.9
improved Faster R_CNN	2000	78.3

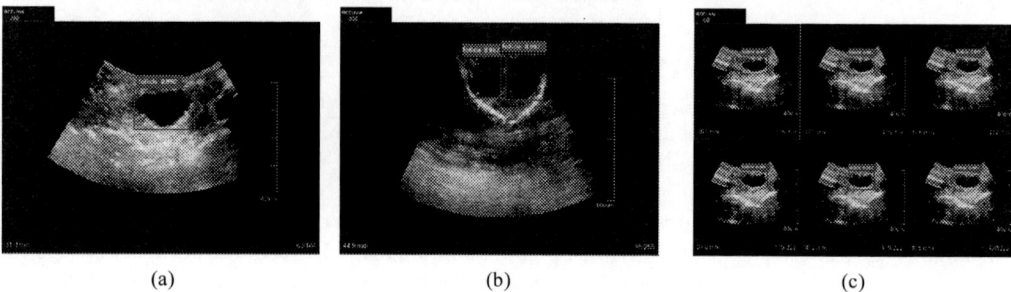

(a)   (b)   (c)

Figure 7 The results of the experiment

3.3. Main result

The improved Faster R_CNN mAP experimental results are shown in the table 1.

The results of the experiment on the ultrasound images of various types of follicles are shown in the figure 7.

Experiments show that the deep learning method Faster R-CNN has higher recognition accuracy and recognition precision and faster recognition speed for follicular ultrasound images, compared with the traditional image processing method. Moreover, the multi-follicle ultrasound image and the multi-image combined follicle ultrasound image have good recognition ability, which is not possible by the traditional image processing method.

4. Conclusion

This paper introduces the research background, significance and research status of follicle ultrasound images and neural networks. Then, using the improved Faster R-CNN model to detect and locate the ultrasound image of follicle, the speed and precision are greatly improved, especially for multi-follicle images. However, there are still many shortcomings in this experiment, and there is a lot of room for improvement in the recognition accuracy and recognition speed of the follicle super-god image. And generally a follicle will have 30-60 slices. Therefore, it is possible to consider the continuous relationship between the upper and lower layers when detecting, which greatly reduces the calculation amount of the network and greatly improves the recognition speed of the ultrasound image of the follicle.

References

[1] G. Cortela, L. Leija. Elastograms of the diabetic foot by ultrasonic impulse elastography [J]. IEEE Transactions on Image Processing, 2016.

[2] Filteau V. DesCôteaux L, Predictive values of early pregnancy diagnosis by ultrasonography in dairy cattle[C]//American Association of Bovine Practitioners. Conference (USA). 1998.

[3] Fricke P M. Scanning the future—Ultrasonography as a reproductive management tool for dairy cattle[J]. Journal of Dairy Science, 2002, 85(8): 1918-1926.

[4] Nong H J. Study on Dynamic Changes of Follicle Development in Leiqiong Yellow Cattle based on B-ultrasound [D]. Guangxi University, 2007.

[5] Potocnik B, Cigale B, Zazula D. The XUltra project - Automated Analysis of Ovarian Ultrasound Images[C]// Computer-Based Medical Systems. IEEE, 2002:262-267.

[6] Hiremath P S, Tegnoor J R. 2010 "Automatic Detection of Follicles in Ultrasound Images of

Ovaries using Edge Based Method[J]. International Journal of Computer Applications Special Issue on Recent Trends in Image Processing & Pattern Recognition, 2011, RTIPPR(3):120-125.

[7] Hiremath P S, Tegnoor J R. Recognition of follicles in ultrasound images of ovaries using geometric features[C]// International Conference on Biomedical and Pharmaceutical Engineering. IEEE, 2010:1-8.

[8] Hiremath P S, Tegnoor J R. Follicle detection in ultrasound images of ovaries using active contours method[C]// International Conference on Signal and Image Processing. IEEE, 2010:286-291.

[9] Krizhevsky A, Sutskever I, Hinton G E. ImageNet classification with deep convolutional neural networks[C]// International Conference on Neural Information Processing Systems. Curran Associates Inc. 2012:1097-1105.

[10] R. Girshick, J. Donahue, T. Darrell, and J. Malik. Rich feature hierarchies for accurate object detection and semantic segmentation. In CVPR, 2014.

[11] He K, Zhang X, Ren S, et al. Spatial Pyramid Pooling in Deep Convolutional Networks for Visual Recognition[J]. IEEE Trans Pattern Anal Mach Intell, 2015, 37(9):1904-1916.

[12] Girshick R. Fast R-CNN[J]. Computer Science, 2015.

[13] Ren S, He K, Girshick R, et al. Faster R-CNN: Towards Real-Time Object Detection with Region Proposal Networks.[J]. IEEE Trans Pattern Anal Mach Intell, 2017, 39(6):1137-114

Research on Network Security of Campus Network

Min Huang,Wanbo Luo and Xing Wan

Leshan Vocational & Technicai college,Sichuan Leshan,614000

Corresponding author Xing Wan Krantson@163.com

Abstract: Nowadays, network security incidents often occur, such as network worms, denial of service attacks, network fraud and so on. Nowadays, the methods of network security incidents are constantly innovating, resulting in data leakage, destruction and other incidents. Campus network is the basis of daily teaching, scientific research and normal operation of schools. It is of great significance to strengthen the construction of campus network security. Based on this, this paper studies the campus network security.

1. Introduction
With the rapid development of network nowadays, the security of campus network has been paid more and more attention by us. Although our network security technology means are mature now, for the campus network, which is a special network carrier, there are many problems in its management because of the large number of users and the wide coverage of the network. Study and solve.

2. Campus Network Security
According to the scale of the usual campus network design, we can see the campus network topology in Figure 1, which is divided into two parts: server and user. The Internet accessed from outside the campus will pass through the campus firewall, then through the core switch, through the access layer switch, and then connect to the various terminals of the school for use.

Content from this work may be used under the terms of the Creative Commons Attribution 3.0 licence. Any further distribution of this work must maintain attribution to the author(s) and the title of the work, journal citation and DOI.
Published under licence by IOP Publishing Ltd

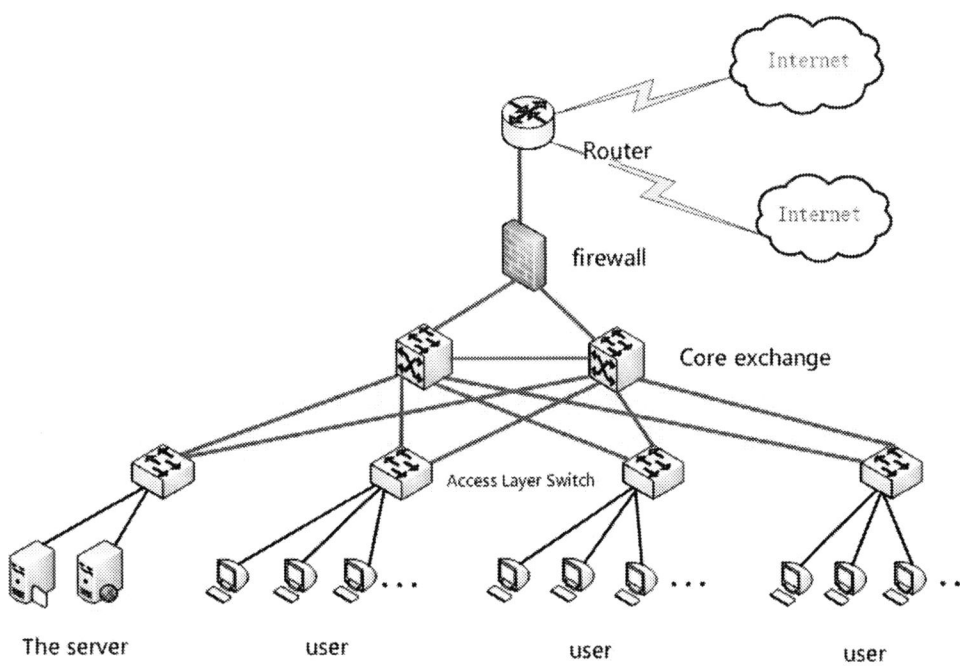

Figure 1 - Campus Network Topology

3. Current Problems of Campus Network Security

In many universities nowadays, the scale of campus network is very large. How to ensure the normal and safe operation of campus network is an urgent problem for colleges and universities. There are mainly the following problems:

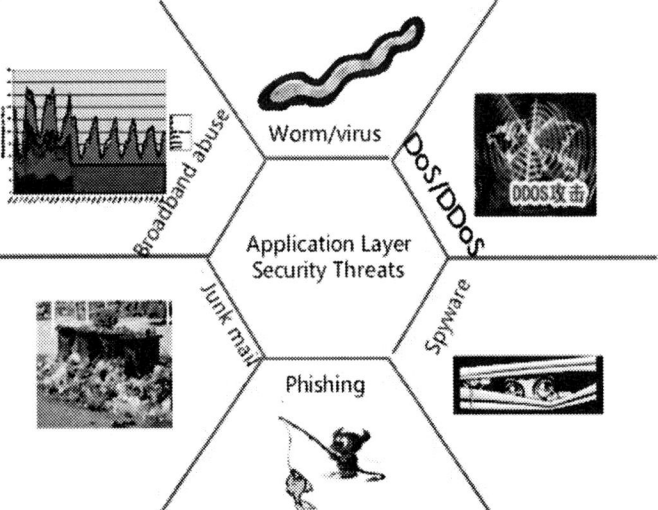

Figure 2 - Campus Network Security Threats

3.1 Computer Virus Invasion

Computer virus is the main threat to the network security of campus network. It can spread through various forms. The harm of computer virus is also enormous. It will affect the transmission of campus network, paralyse the operation of campus network and so on. In many cases, a computer in the campus network will be infected by virus and other computers in the network will also be infected with virus.

3.2 DoS、DDoS attack

Campus network is connected with the external Internet. While using the Internet conveniently, it is very vulnerable to hackers'attacks. DoS and DDoS are used to attack the campus network and servers, which makes the traffic in the campus network increase sharply. This will make the campus network and services not be used normally and eventually lead to the collapse of the entire campus network[1].

3.3 Spyware

Some malicious attacks can make use of the "fool-like" spy attack software flooded on the network. Illegal users can carry out unscrupulous attacks on the campus network without any computer technology. This has caused a huge impact on the security of the campus network.

3.4 Phishing

Many illegal people set up some counterfeit websites on the network to deceive users and maliciously extract bank accounts and passwords, which will cause irreparable losses to teachers and students of schools. These phishing websites are often set up very similar to regular websites, making it difficult for people to identify them.

3.5 Junk mail

Some people on the network, through mass spam, to defraud school teachers and students of clicks, on the one hand, will make teachers and students use computers infected with related viruses, on the other hand, also defraud teachers and students of personal information, and then use these personal information illegally[2].

3.6 Broadband abuse

BT, Xunlei and other download software are widely used in the campus network, which occupies the bandwidth resources of the campus network, making the network run slowly and other teachers and students can not use it normally. While users download network resources by downloading software, there are also hidden dangers of introducing external network virus into campus network, which has a great impact on the safe operation of campus network[3].

4. Preventive Measures for Network Security of Campus Network

4.1 Reasonable Configuration of Firewall

Based on the research of firewall technology and existing conditions, this paper designs an improved scheme of campus network firewall system, as shown in Figure 3.

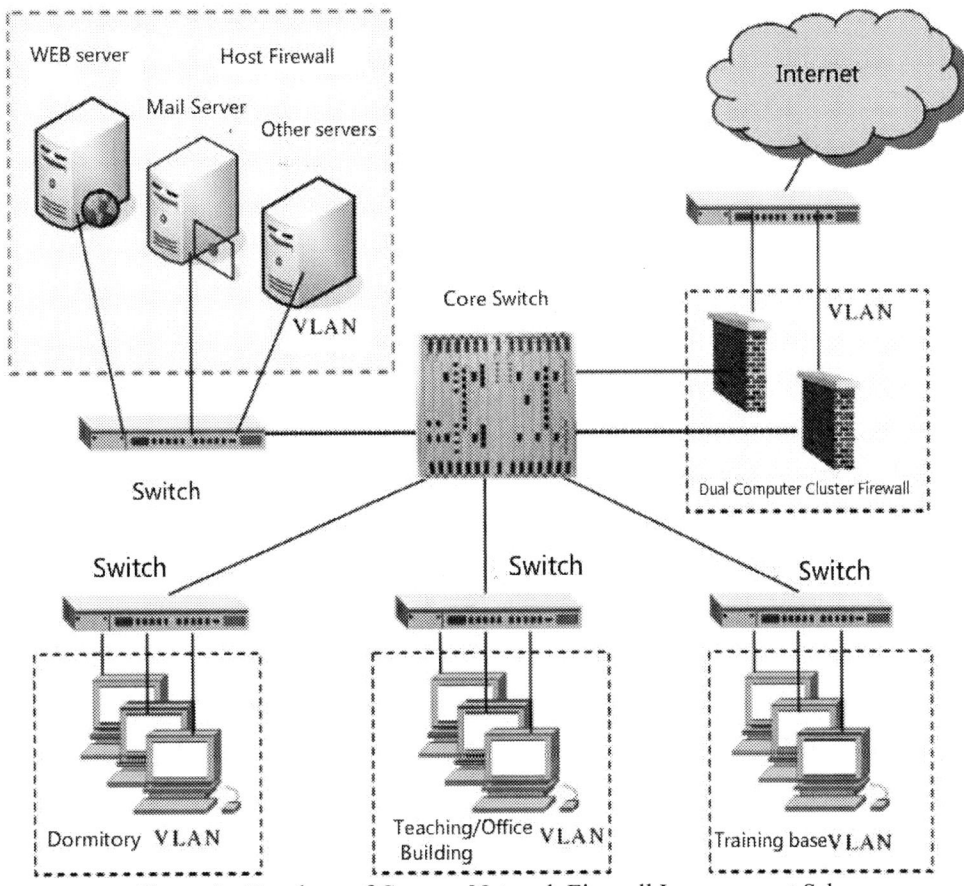

Figure 3 - Topology of Campus Network Firewall Improvement Scheme

(1) According to the situation that the firewall of the campus network may have single-point function, using the design of the dual-computer cluster firewall, its principle is that if the main firewall is ineffective due to the failure, then the backup firewall can be timely repaired and undertake the corresponding work. The two firewalls can easily replace the work, which guarantees the uninterrupted network security protection of the campus network[4].

(2) In the internal network server, according to the lack of security capability, the server host firewall is set up. Its main responsibility is to refuse unauthorized access. Server host firewall can not be felt by users. It realizes precise access control, guarantees the security of the server and fully escorts the campus internal network server.

(3) According to the different situations of internal users, different VLANs are set up to effectively control unauthorized access of authorized users.

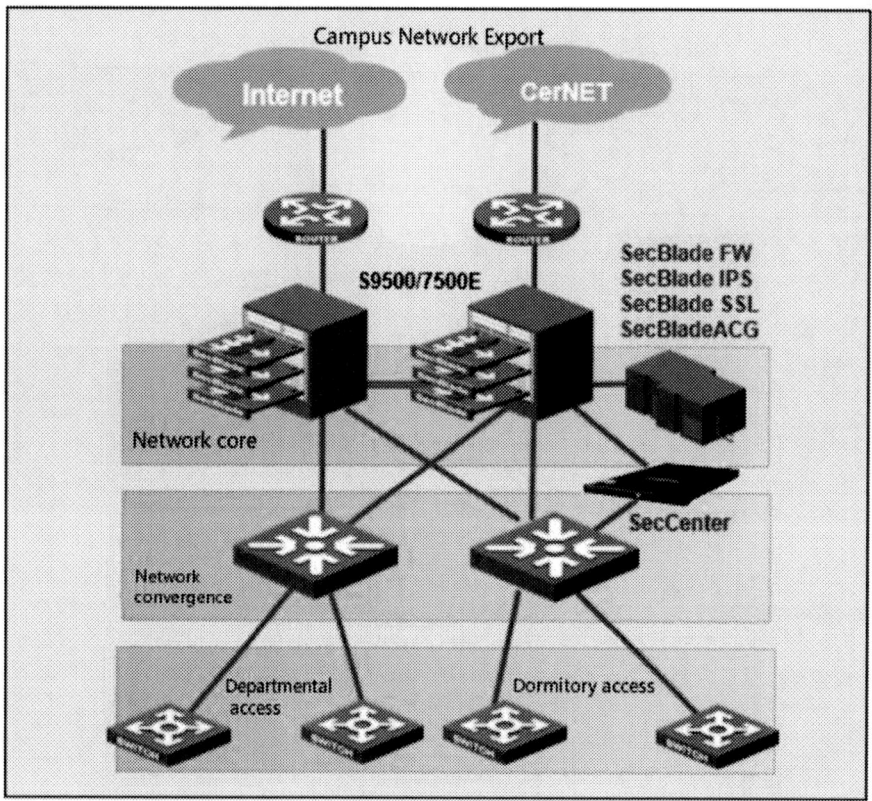

Figure 4 - Campus Network Backbone Network Protection Solution

This is a kind of solution for the backbone network protection of campus network. It mainly includes:

Remote access: SSL scheme completes remote security access of campus network.

In-process control: firewalls and IPS find attacks in parallel.

Post-audit: SecCenter records and outputs reports.

4.2 Intrusion Detection System

We actively introduce Intrusion Detection System (IDS) in the network security of campus network. Its main functions are two aspects. The first is to detect the behavior of unauthorized users and intrusive systems, and the second is to detect the illegal operation of system resources by authorized objects. Intrusion detection is an active means of security protection technology. It can do a good job of accurate protection for internal and external attacks or misoperation. When the network security of campus network is infringed, it can real-time monitor and make corresponding prompts to fully guarantee the network security of campus network.

4.3 Encryption of core data

Network security incidents often occur in campus network, such as obtaining, attacking and tampering with various kinds of information, such as examination results, examination papers, school card funds information and so on. Faced with this situation, we can adopt the means of data encryption, which usually means to change the way information is expressed, and its purpose is to protect the real information. This will make it impossible for unauthorized users to access protected information, we

There are always criminals trying to get information that he should not get through various means. There are many similar information in the campus network, such as the result information of the educational administration network, the examination information of the educational administration, the examination papers, and even the fund information of the meal card campus card. Public-key and private-key cryptosystems are widely used, that is, private-key cryptosystems for encryption and decryption. Many new technologies have been applied to information encryption. These new technologies will provide the strongest backing force for information security. Figure 4 is the schematic diagram of Guiyang University campus network security solution.

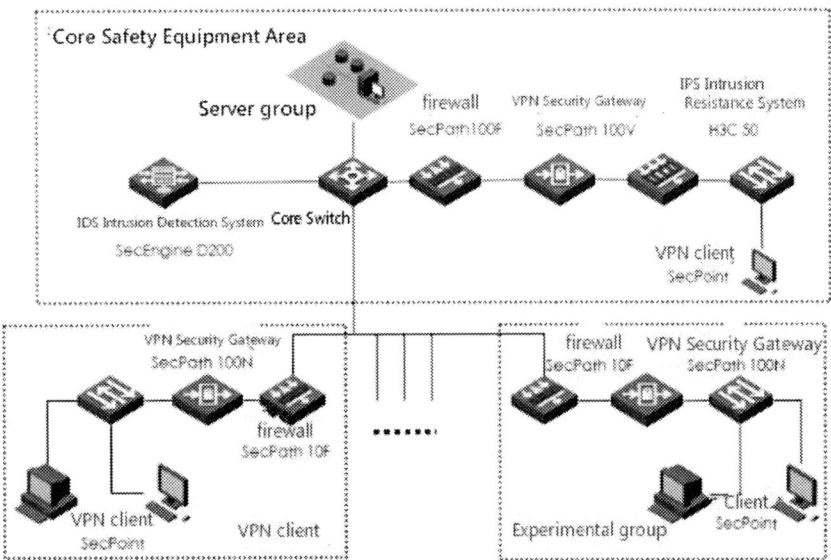

Figure 4 - Guiyang University Campus Network Security Solution

5. Summary

In the network environment of campus network, it is very vulnerable to external attacks, and the means and technology of network destroyers are constantly changing. The network security scheme of campus network designed by us is not unchangeable. On the one hand, we should make some scientific and reasonable adjustments according to the actual situation, on the other hand, we should adapt to the changes of the external environment. When adjusting our campus network security design, only in this way can we ensure the safety and reliability of the campus network.

Reference

[1] Kong Linghao. Brief discussion on the analysis and Countermeasures of campus network security [J]. Network security technology and application, 2018 (05): 56-57.

[2] Li Xiufeng. Research and analysis of network security mechanism in campus network construction [J]. Network security technology and application, 2017 (10): 93-94.

[3] Meng Jingxin. Campus Network Planning and Design [J]. Information and Computer (Theoretical Edition), 2017 (06): 171-173.

[4] He Shuyi. Application of Network Security Technology in Campus Network [J]. Communication World, 2017 (02): 72-73.

Online Route Planning for Cooperative Area Coverage Search of Aircraft Swarm

Yueqi Hou[1], Xiaolong Liang[1*], Jiaqiang Zhang[1], Ye Li[1] and Jiong Wang[2]

[1] Air Traffic Control and Navigation College, Air Force Engineering University, Xi'an, Shaanxi, 710051, China

[2]Air Force Xi'An Flight Academy ,Xi'an， Shaanxi, 710306, China

*Corresponding author's e-mail: afeu_hyq@163.com

Abstract. Aiming at the problem of cooperative search for aircraft swarm in an unknown environment without prior information, a cooperative search algorithm with the coverage rate as the optimization objective is proposed, based on the Model Predictive Control theory and the Differential Evolution algorithm. Firstly, the Area Coverage Map (ACM) is established to describe the mission area, and a rapid method of updating the ACM based on Hadamard product is given. Then, the search effect is measured by coverage rate calculated based on the ACM. We regard the aircraft swarm as a control system, and establish a systematic prediction model. The maximum coverage rate in the predicted period is defined as the optimization goal, and the DE is used to solve this problem. Finally, the simulation results verify the effectiveness of the proposed method.

1. Introduction

Aircraft swarm is a flight system composed of a number of manned and unmanned aerial vehicles with different functions, which can cooperate with each other and have the ability to emerge as a whole[1-4]. Carrying out regional search missions using aircraft swarm has important value of application, such as searching the distribution of targets in the enemy area or searching and rescuing in the desert region.

In the traditional way, in [5-7], the mission area was divided in several parts and the search routes in each sub-area was designed. These methods are usually pre-calculation, offline planning, and the flight route is fixed. Due to the uncertainty of the environment and the unexpected situation, the cooperative area search is a dynamic process[8]. Online route planning is needed according to real-time information and aircraft state in search process.

Ref. [9] classified the environment based on a hexagonal grid, using information entropy to describe the environmental uncertainty. However, the establishment of search probability maps depends on prior information. In [10], the environment was described using Cartesian grids, and each grid was given a value to represent the uncertainty of the target distribution. The search reward function and the no-fly zone avoidance strategy were designed. Based on [10], Ref. [11] considered communication constraints and analyzed the influence of different communication constraints on the efficiency of cooperative search. Although both of them consider the maneuvering limitations of the UAV, it is assumed that the flight of the UAV between the grids resulting in more turnovers and larger turn angles. Ref. [12] effectively reduce the size of the solution of cooperative search decision-making problems based on distributed model predictive control framework, using the algorithm based on the combination of Nash

---

Content from this work may be used under the terms of the Creative Commons Attribution 3.0 licence. Any further distribution of this work must maintain attribution to the author(s) and the title of the work, journal citation and DOI.

Published under licence by IOP Publishing Ltd

optimization and Particle Swarm Optimization(PSO). In general, the above method still has some limitations in practical application:

a) Relying on a priori information to establish environmental search map;
b) Aircraft moves across the grid, and every turn angle is fixed;
c) Coverage rate of the area is still to be improved.

In view of the above problems, this paper mainly studies: how to control the aircraft swarm effectively implementing online route planning in the unknown environment without any prior information, and search the mission area with the largest coverage rate.

2. Model of the problem

Without the priori information of the targets and their location within the region, aircraft swarm carrying communication equipments and sensors to search for a specific mission area, as shown in Fig. 1. In order to search all the targets in the area as soon as possible, the maximum coverage rate of the mission area needs to be achieved. Once an aircraft being shot down by enemy fire [8], traditional offline planning methods are not robust enough to perform adaptive search missions.

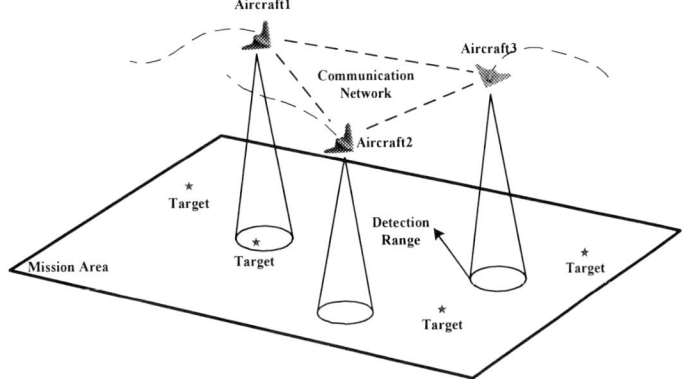

Figure 1. Mission area

Let the mission area Ω be a rectangular area of $L_x \times L_y$, rasterize the area into $M_e \times N_e$ grids and construct the Area Coverage Map (ACM). As the search progresses, the ACM updated and shared among the aircraft swarm in real time, providing decision information for each aircraft. Each grid is given a value $\mu(i,j,k)$, which is used to describe whether or not the grid (i,j) has been searched at time k, as shown in Fig. 2.

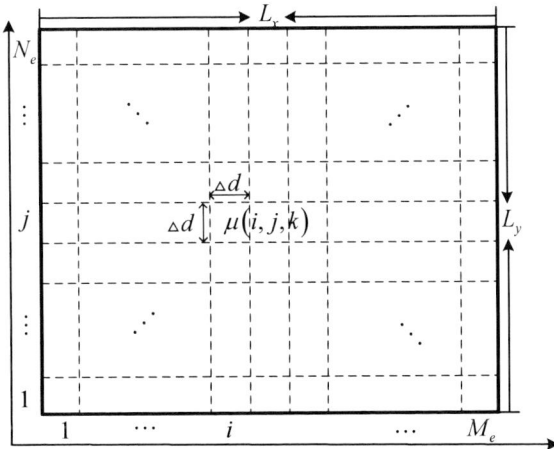

Figure 2. Mission area rasterization

For the sake of research, we made the following assumption: Once the grid (i, j) is within the scope of the aircraft's sensor detection, it is considered that the grid has been searched and the target distrubution within the grid is absolutely known. Based on this assumption, the mission area Ω can be divided into two areas: searched area $\Omega_c(k)$ and unsearched area $\Omega_{nc}(k)$. $\mu(i, j, k)$ can be expressed as

$$\mu(i,j,k) = \begin{cases} 0, & (i,j) \in \Omega_c(k) \\ 1, & (i,j) \in \Omega_{nc}(k) \end{cases} \tag{1}$$

The ACM can be defined as

$$\text{ACM} = \left\{ \mu(i,j,k) \,\middle|\, (i,j) \in \Omega \right\} \tag{2}$$

Represent the ACM as a matrix and establish an environment matrix

$$C_e(k) = \left[\mu(i,j,k) \right]_{N_e \times M_e} \tag{3}$$

We assumed that the aircraft is flying at a constant altitude over the mission area and that the aircraft is considered as a mass of motion in a two-dimensional space whose motion equation is

$$\begin{cases} x_i(k+1) = x_i(k) + v_0 \Delta t \cos \varphi_i(k+1) \\ y_i(k+1) = y_i(k) + v_0 \Delta t \sin \varphi_i(k+1) \\ \varphi_i(k+1) = \varphi_i(k) + \Delta\varphi_i(k) \\ \Delta\varphi_i(k) \in \left[-\varphi_{max}, \varphi_{max} \right] \end{cases} \tag{4}$$

Where, $[x_i(k), y_i(k)]$ is aircraft position; v_0 is flying speed; Δt is time step; φ_i is heading angle; $\Delta\varphi_i$ is heading angle increment for the decision input; φ_{max} is the maximum mobility under the turning angle. The state of the aircraft at time k is $p_i(k) = (x_i(k), y_i(k), \varphi_i(k))$. Aircraft swarm consisting of n aircrafts can be regarded as a control system, the state of the system is $P(k) = (p_1(k), p_2(k), \cdots, p_n(k))$, the equation of state is

$$P(k+1) = f\left(P(k), u(k) \right) \tag{5}$$

Among them, $u(k) = (\Delta\varphi_1(k), \Delta\varphi_2(k), \cdots \Delta\varphi_n(k))$ is the input of the system; $f(\bullet)$ is the state transfer function of the system, determined by the equation of motion of the aircraft.

From the moment k, the position of all aircraft can be predicted, given multiple predictive control inputs. By optimizing the control inputs, it leads the aircraft swarm to cover as many unsearchable areas as possible. According to the thought above and system state equation, a systematic prediction model is established

$$\begin{aligned} P(k+j+1 \,|\, k) &= f\left(P(k+j \,|\, k), u(k+j \,|\, k) \right) \\ j &= 0, 1, 2, \cdots H-1 \end{aligned} \tag{6}$$

Where H is the prediction period and $P(k+j+1 \,|\, k)$ is the system state at $k+j+1$ based on the prediction of the system state at time k, the value of which depends on the system state $P(k+j \,|\, k)$ and the control input $u(k+j \,|\, k)$.

3. ACM Modeling and updating
Coverage rate is used to measure the pros and cons of search methods as an evaluation index of search mission. During the mission, the increase of coverage rate from k to $k+1$ can evaluate the merits and demerits of decision-making course at time k. Therefore, how to calculate coverage rate quickly and accurately is of great significance to mission evaluation and real-time decision-making.

In this paper, the environment matrix is updated rapidly by using the Hadamard product of the environmental sub-matrix and the detection matrix, so as to update the ACM. Through the perception of the ACM, the aircraft can draw the current and incremental coverage rate of the next decision-making, which can provide basis for the decision-making.

In this paper, the detection range of the sensor is simplified to a circular area with R_s radius centered on the location of the aircraft. The circumscribed square area of the detection area is denoted by SQ and the area SQ is rasterized by $M_u \times M_u$ grids as shown in Fig. 3. M_u can be expressed as

$$M_u = 2 \times \left\lceil \frac{R_s}{\triangle d} \right\rceil + 1 \tag{7}$$

Where, R_s is sensor detection radius; $\triangle d$ is raster fixed interval; $\lceil \circ \rceil$ is rounded up function.

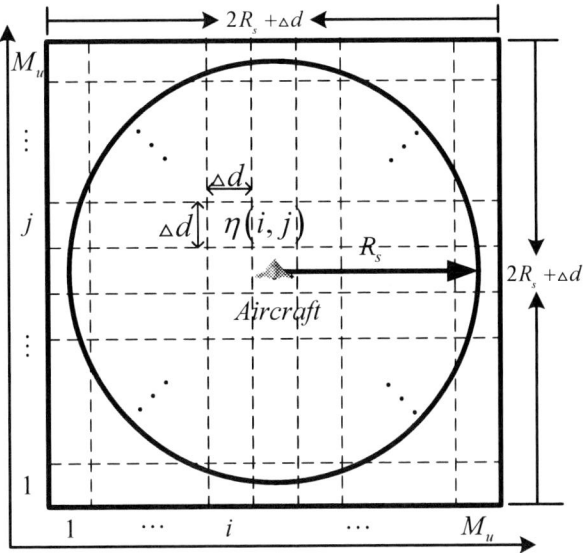

Figure 3. Detection area rasterization

Each grid is assigned a value of $\eta(i,j)$, which describes whether the grid sensor can be detected. Assuming that the area SQ is divided into two areas: detectable area SQ_c and undetectable area SQ_{nc}. If the grid (i,j) is within the scope of the aircraft sensor detection, $(i,j) \in SQ_c$, conversely, $(i,j) \in SQ_{nc}$. $\eta(i,j)$ can be expressed as:

$$\eta(i,j) = \begin{cases} 0, & (i,j) \in SQ_c \\ 1, & (i,j) \in SQ_{nc} \end{cases} \tag{8}$$

Establishing the detection matrix $\boldsymbol{D}_u = [\eta(i,j)]_{M_u \times M_u}$, which is the capabilities of aircraft searching for adjacent grids. When the aircraft is orbiting at a fixed altitude, the sensor detection radius R_s is constant. Therefore, the detection matrix $\boldsymbol{D}_u$ does not change with time k and is uniquely determined.

Define environment sub-matrix $\boldsymbol{C}_e^{(m_0,n_0)}(k)$: In the environmental matrix $\boldsymbol{C}_e(k)$, the sub-blocks with the element $\mu(m_0,n_0,k)$ as the central element and the dimension $M_u \times M_u$ are called the environmental sub-matrix $\boldsymbol{C}_e^{(m_0,n_0)}(k) = \left[\mu^{(m_0,n_0)}(i,j,k) \right]_{M_u \times M_u}$, $(m_0,n_0) \in \Omega$, $(i,j=1,2,\cdots M_u)$. The relationship between the element $\mu^{(m_0,n_0)}(i,j,k)$ in $\boldsymbol{C}_e^{(m_0,n_0)}(k)$ and the element $\mu(i,j,k)$ in $\boldsymbol{C}_e(k)$ can be expressed as

$$\mu^{(m_0,n_0)}(i,j,k) = \mu(i+m_0-M_u-1, j+n_0-M_u-1, k) \tag{9}$$

When the aircraft is within grid (m_0, n_0), it is approximately assumed that the aircraft is in the center of the grid. In this case, the environment matrix $\boldsymbol{C}_e^{(m_0, n_0)}(k)$ overlaps with the detection matrix $\boldsymbol{D}_u$, and the dimensions are equal. Hadamard product of the above two matrices

$$\boldsymbol{C}_e^{(m_0, n_0)}(k+1) = \boldsymbol{C}_e^{(m_0, n_0)}(k) \circ \boldsymbol{D}_u = \left[\mu^{(m_0, n_0)}(i, j, k) \times \eta(i, j) \right]_{M_u \times M_u} \quad (10)$$

Where, $\circ$ is Hadamard product. The Hadamard product of $\boldsymbol{C}_e^{(m_0, n_0)}(k)$ and $\boldsymbol{D}_u$ is multiplied by the corresponding elements $\mu^{(m_0, n_0)}$ and η. The updated environment sub-matrix $\boldsymbol{C}_e^{(m_0, n_0)}(k+1)$ is replaced to the corresponding sub-block of the environment matrix $\boldsymbol{C}_e(k)$, that is, the update of the ACM is implemented.

The above method is extended to the case of multiple aircraft movements: Assuming that the detection period of the sensor is T_s, the linear movement of n aircraft from time k to time $k+1$ is dispersed as a point trace with T_s as the time interval. For each point on the trace of doing the above operation, you can achieve time $k+1$ ACM updating.

4. Objective function and algorithm

The key to realize cooperative search is to design a search objective function to evaluate the pros and cons of each step decision. Therefore, in the route planning, each step of the decision mainly considers the increment of coverage rate as the search efficiency, and considers the boundary distance and turning angle as the search cost.

Based on the above considerations, the objective function can be expressed as

$$\max J\left(\boldsymbol{P}(k), \boldsymbol{u}(k) \right) = \omega_1 \gamma J_1(k) - \omega_2 J_2(k) - \omega_3 J_3(k) \quad (11)$$

Where, J_1 is the increment of coverage rate of all aircraft performing one step; J_2, J_3 is the cost function of turning angle and boundary distance; $\omega_1, \omega_2, \omega_3$ is the corresponding weight; γ is the importance factor, $\gamma \geq 1$; Since the above revenue and cost functions have different dimensions, therefore, we need to standardize them.

Coverage rate at time k is the area percentage of $\Omega_c(k)$ to Ω, approximately equal to the ratio of the number of searched grids to the total number of grids, which can be expressed as

$$Cover(k) = \frac{M_e \times N_e - \sum\limits_{i=1}^{M_e} \sum\limits_{j=1}^{N_e} \mu(i, j, k)}{M_e \times N_e} \quad (12)$$

The increment of coverage rate at moment k to moment $k+1$ is the difference between $Cover(k+1)$ and $Cover(k)$. Search earnings can be expressed as

$$\begin{aligned} J_1(k) &= Cover(k+1) - Cover(k) \\ &= \frac{\sum\limits_{i=1}^{M_e} \sum\limits_{j=1}^{N_e} \left[\mu(i, j, k) - \mu(i, j, k+1) \right]}{M_e \times N_e} \end{aligned} \quad (13)$$

If the turning angle is too large, fuel consumption will increase, which will affect the battery life. Therefore, a cost function is designed to minimize the turning angle of each aircraft and reduce the cost of fuel consumption caused by turning. The cost function of turning angle can be expressed as

$$J_2(k) = \frac{1}{n} \sum\limits_{i=1}^{n} \frac{|\Delta \varphi_i(k)|}{\varphi_{max}} \quad (14)$$

2075

In the search process, the closer to the boundary, the less effective area can be covered by the sensor. Drawing on the idea of virtual potential function, a cost function is designed. The aircraft will be subjected to the virtual "repulsion" of the boundary. The closer to the boundary, the larger the "repulsion" will be. Therefore, the cost function of the boundary distance can be expressed as

$$J_3(k) = \frac{1}{n} \sum_{i=1}^{n} \left(\frac{1}{x_i(k)} + \frac{1}{L_x - x_i(k)} + \frac{1}{y_i(k)} + \frac{1}{L_y - y_i(k)} \right) \tag{15}$$

Model Predictive Control (MPC) is a method to design the optimal control input of the system in the predictioncycle. The core idea is rolling optimization. In this paper, we combine MPC and Differential Evolution (DE), and design a reasonable cooperative search algorithm.

Assuming that the system state at time k is $P(k)$ and the input of H-step predictive control is $U(k) = \left(u(k), u(k+1), \cdots, u(k+H-1) \right)$, the search performance of system after H-step prediction is recorded as

$$J^{(H)} \left(P(k), U(k) \right) = \sum_{j=0}^{H-1} J \left(P(k+j \,|\, k), u(k+j \,|\, k) \right) \tag{16}$$

Then, the optimal model for solving system optimal control input at time k can be expressed as

$$U^*(k) = \arg\max J^{(H)} \left(P(k), U(k) \right)$$
$$s.t. \begin{cases} P(k+1+j \,|\, k) = f \left(P(k+j \,|\, k), u(k+j \,|\, k) \right) \\ \qquad\qquad j = 0, 1, 2, \cdots H-1 \\ P(k \,|\, k) = P(k) \end{cases} \tag{17}$$

Where: $U^*(k)$ is the system optimal predictive control input. The first item $u^*(k \,|\, k)$ of $U^*(k)$ is a control input in time k, that $u(k) = u^*(k \,|\, k)$. At $k+1$, the aircraft repeats the above optimization process based on current environmental information and system state. Different from the greedy search algorithm, which only considers the current search revenue, the MPC-based rolling optimization method predicts the long-term search revenue and improves the overall search efficiency.

5. Simulation results and analysis

The cooperative search algorithm based on coverage rate is given in the previous section. In order to verify its feasibility, this section simulates it. Simulation environment for I7-4960, clocked at 2.60GHz, 16G memory, based on Matlab 2014a platform for simulation. The mission area for the rectangular area of 8000m*8000m, each grid size of 20m*20m, the implementation of the search mission of the four aircraft takeoff coordinates were (200,0), (2700,0), (5200,0) , (7800,0). Some other parameters are respectively set as $v_0 = 30m/s$, $R_s = 200m$, $\varphi_{max} = 60°$, $\triangle t=10s$ and $H=2$.

The coverage rate is greater than 90% for the termination conditions, the simulation results as shown in Fig. 4.

Observing the Fig. 4(a), we can intuitively see that the four aircraft keep the detection range of the sensor as non-overlapping as possible in the initial search phase, and achieve a higher coverage rate increment. Through simulation, the proposed method can make the coverage rate increase at a higher rate in the early stage of the search. The angle and the boundary cost function in the objective function better achieve the smoothing of the route and the boundary constraints. From the Fig. 4(b), we can see that the four aircrafts achieve a better cooperative search, the overlap area of the sensor detection area is less, and finally the area coverage rate of more than 90% is reached.

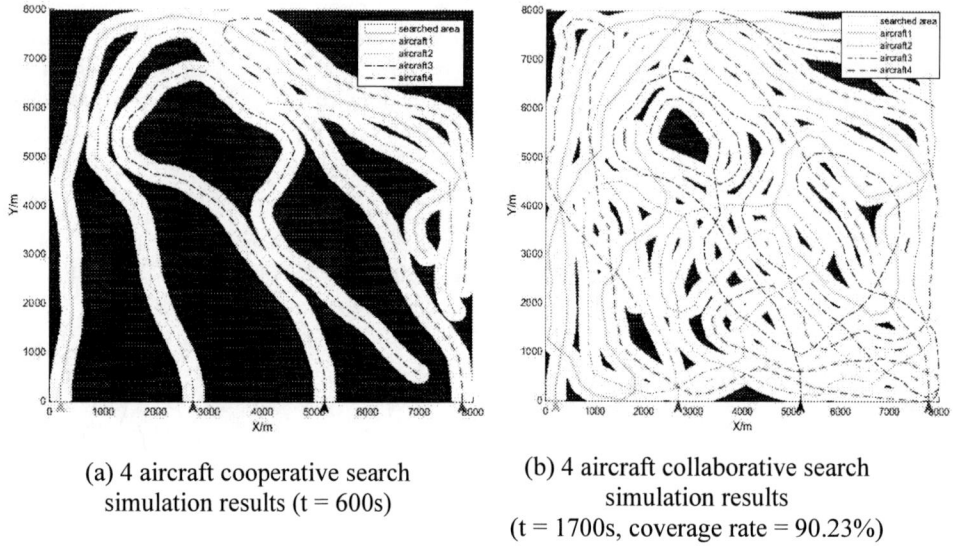

(a) 4 aircraft cooperative search simulation results (t = 600s)

(b) 4 aircraft collaborative search simulation results (t = 1700s, coverage rate = 90.23%)

Figure 4. Coperative search simulation results

In order to further verify the advantages of this method in improving coverage rate, 100 simulations are performed. The curves of the coverage rate obtained with time are shown in Fig. 5.

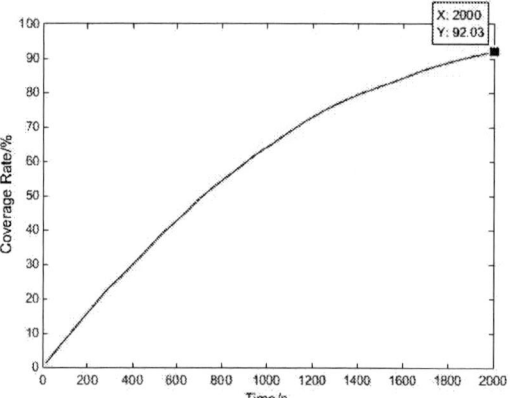

Figure 5. Coverage rate changing curve

It can be seen from the Fig. 5 that the proposed method has a good performance of cooperative search, and the coverage rate has maintained a relatively high growth rate in the initial search period. The search capability has declined in the middle and late stages due to the decentralization of unsearchable regions. When the search time t=2000s, the coverage rate of 92% or more, achieving the desired search purposes.

6. Conclusions

In order to improve the regional coverage rate of the aircraft swarm when performing unknown environment search mission, this paper proposes a fast ACM updating method based on Hadamard product, and designs the search objective function with the coverage rate as the main objective.Draw lessons from MPC idea, take DE algorithm to solve. Simulation results show that the cooperative search algorithm proposed in this paper can implement coverage search for unknown environments and maximize the search capability of each aircraft. In the case of a failure of some aircraft, the online planning method proposed in this paper can continue to achieve collaboration based on the current environment. Compared with other online planning methods, this method can achieve higher coverage rate.

Acknowledgments

This research was financially supported by the National Natural Science Foundation of China (No. 61472443, No. 61703427) and the Research and Development of Science and Technology Plan Projects in Shanxi Province (No. 2017JQ6035).

References

[1] Niu Y F, Xiao X J, and Ke G Y, (2013)Operation concept and key techniques of unmanned aerial vehicle swarms. National Defense Science and Technology, 34: 37-43.

[2] Liang X L, Sun Q, and Yin Z H, (2013)A study of aviation swarm convoy and transportation mission. Advances in Swarm Intelligence, 368-375.

[3] Liang X L, Sun Q, and Yin Z H, (2013)The swarm convoy and transport mission. In: The Fourth International Conference on Swarm Intelligence. Harbin China. pp. 46-56

[4] Liang X L, Li H, and Sun Q, (2014)Development trend of air operations and its strategy. Journal of Air Force Engineering University (Military Science Edition), 14: 4-7.

[5] Peng H, Shen C L, and Huo X H, (2007)Research on multiple UAV cooperative area coverage searching. Journal of System Simulation, 19: 2472-2476.

[6] Chen H, Wang X M, Jiao Y S, and Li Y, (2010)An algorithm of coverage flight path planning for UAVs in convex polygon areas. ACTA Aeronautica et Astronautica Sinica, 31: 1802-1808.

[7] Yu S N, Zhou R, Xia J, and Che J, (2015)Decomposition and coverage of multi-UAV cooperative search area. Journal of Beijing University of Aeronautics and Astronautics, 41: 167-173.

[8] P. Vincent and I. Rubin, (2004)A framwork and analysis for cooperative search using UAV swarms. ACM Symposium on Applied Computing,: 79-86.

[9] Tian J, Chen Y, and Shen C L, (2007)Cooperative search algorithm for multi-UAVs in uncertainty environment. Journal of Electronics & Information Technology, 29: 2325-2328.

[10] Wu W C, Huang C Q, Song L, Tang S Q, and Bai R C, (2011)Cooperative search and path planing of multi-unmanned air vehicles in uncertain environment. ACTA Armamentraii, 32: 1337-1342.

[11] Fu X W, Wei G W, and Gao X G, (2016)Cooperative area search algorithm for multi-UAVs in uncertainty environment. System Engineering and Electronics, 38: 821-827.

[12] Peng H, Shen C L, and Zhu H Y, (2010)Multiple UAV cooperative area search based on distributed model predictive control. ACTA Aeronautica et Astronautica Sinica, 31: 593-601.

ISPECE   IOP Publishing

AdaptiveSLA: A Two-Stage Scheduling Framework for SLA Profit Maximization in Multi-tenant Database

YANG Ji, ZHANG Lin, TANG Rong

State Grid Chongqing Electric Power Company, Chongqing,400014

Zhkwang21@163.com

Abstract. The requirement of multi-tenant applications is continuously increasing in cloud computing epoch, and they usually have larger data volume compared with traditional applications. Considering the quality of managing these data, SLAs (Service Level Agreement) are usually defined between multi-tenant database service providers and tenants. The providers will get revenues if SLAs are met, otherwise they will pay penalties. In order to maximize the profit, this paper presents AdaptiveSLA, a two-stage scheduling policy for multi-tenant database. In the first stage, AdaptiveSLA detects performance crises and leverages sliding window algorithm to mitigate crises in distributed multitenant database. The crises may derive from heterogeneity of requests or performance degradation of nodes. In the second stage, AdaptiveSLA makes requests' execution sequence. This stage aims at completing requests as many as possible. The executing process is constrained by SLA, which is modeled as slack time. Extensive experimental results demonstrate the effectiveness and efficiency of AdaptiveSLA scheduling policy.

1. Introduction

Database service provident is critical to efficient work of SaaS application. With the rapid expansion of tenant scale, database service provider maintains larger amounts of data and faces more unpredictable requests such as queries and transactions sent by application servers. SLA - an agreement signed by tenant and service provider - guarantees the effectiveness of multi-tenant database service [1]. More specifically, response time is in the service provider's best interest to directly optimize the profit [2]. Hence, it is very important to make profit-oriented schedule for database service providers according to request's response time. In fact, the problem requires both monitoring resources and making execution sequence. (1) Resources monitoring. When a node detects performance crises, target nodes should be chosen timely to execute its subsequent workloads. Crises may derive from heterogeneity of requests or performance degradation of the nodes, and they can extend request's response time if not eliminated in time. (2)Execution sequence making. Requests' execution sequence has directly correlation with the final profit, The more requests providers finished before deadline, the more possible they maximize the holistic profit.

Taking the problems and challenges into consideration, a research presents Delphi to mitigate crises using hill-climbing algorithm[3]; while another study designed SmartSLA, addressing the issue of how to intelligently manage the resources among different tenants to maximize its own profits[4]. However, their polices have several weak points. Firstly, they focused on system level metrics such as database throughput, average query response time. But the granularity they used is too rough to leverage the resources efficiently. Second, they neglected the significance of making execution sequence online, the cloud computing model has to operate in a fast changing environment to provide

Content from this work may be used under the terms of the Creative Commons Attribution 3.0 licence. Any further distribution of this work must maintain attribution to the author(s) and the title of the work, journal citation and DOI.
Published under licence by IOP Publishing Ltd

services to unpredictably diverse set of tenants, providers should develop lightweight and real-time decision policy.

In order to maximize the profit, this paper presents AdaptiveSLA, a fine-grained two-stage scheduling framework aiming at dispatching analysis queries. It detects and mitigates performance crises in the first stage to make good use of the cluster's resources (e.g., CPU, Mem, and I/O, this paper don't take I/O bandwidth into consideration for the reason of connection pooling in LAN). Compared with other methods AdaptiveSLA just migrate the following requests to nodes who have the same tenant data. Then the second stage makes profit-effective execution sequence for admitted requests, the constrain of scheduling is represented by slack time which describes the SLA constrain. The major contributions are as following:

A self-managing peer-to-peer multi-tenant DBMS architecture. This paper design and implement AdaptiveSLA, which eliminates the bottleneck of master node. Every node of it can execute requests or dispatch requests to other nodes without directions of a master.

Crisis migration mechanism. AdaptiveSLA uses sliding window algorithm to help find targets when nodes detect crises. The principle not only considers the status of free resources of target nodes, but also thinks about their resource consumption rate in case of causing new crises in target nodes.

Requests execution sequence. When making execution sequence of requests, decisions are made in a holistic fashion by synthetically considering the profit and slack time. The slack time model expresses the time a request can be further postponed without violating SLA constraint.

The rest of the paper are organized as follows. Section 2 provides background of database multi-tenancy and overview of AdaptiveSLA. Sections 3 presents the detailed implementation of AdaptiveSLA and explains how it works. At last, this paper presents the experiment results in Section 4.

2. Database Multi-tenancy and Overview of AdaptiveSLA

In this section, this paper first introduces the background of database multi-tenancy, describing how data is organized in database. Then it provides an overview of AdaptiveSLA, including several models used when making the schedule. At the end of the section the slack time model is explained in detail.

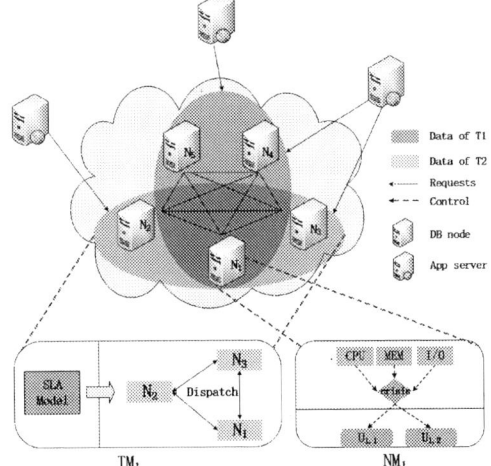

Figure 1. Overview of AdaptiveSLA architecture

2.1 Database Multi-tenancy

In multi-tenant applications, data of one tenant is not so much compared with the overall tenants, so it is of big advantage to manage data in tenant fashion. Many companies have studied a lot in designing multi-tenant database. For instance, Frederick Chong identified three distinct approaches for creating data architectures, such as separating database, sharing database but separating schema and sharing database and schema architecture [6]. Stefan Aulbach compared the three approaches, and proved that numerous tables can intensify the competition in node resources [11], which increased the risk of degrading of the performance of the node or even crashing it. Therefore, nowadays sharing table attracts more and more attention when maintaining data for several tenants. Every request submitted by application server were marked with tenant property. In response to the solution, Force.com's optimized metadata-driven architecture for multi-tenant applications, however metadata node had a latent visit bottleneck. AdaptiveSLA eliminated these limitations by developing the system with peer-to-peer (P2P) architecture, then focused on maximizing the whole SLA profit.

2.2 Overview of the AdaptiveSLA Models

2.2.1 Node Model. AdaptiveSLA introduces node model NM_i to model behaviors of node N_i. In practice, it has two functions:(1) Execute requests. It introduces a concept to define the request execution unit at a node model in this paper.

Definition 1(Execution Unit): Unit $U_{i,j}$ means a request execute unit at NM_i that can handle requests q_k^j about T_j .

As can be seen from Figure 1, NM_1 maintains data of T_1 and T_2, so it can execute requests about T_1 and T_2 using $U_{1;1}$ and $U_{1;2}$ separately. Besides, units coexist at a node may have different resource demands and enjoy respective SLA profits, node model allocates different CPU, Mem, I/O resources to different units according to their tenant resource consumption and SLA profit [5].

(2) Mitigate crises. To provide well resource provision, node models must mitigate performance crises in time. In order to show the process this paper introduces a vector to express resource state and define a concept of crisis.

Definition 2(Resource Vector): The resource vector V = [CPU; Mem; I/O] is the resource state of CPU, Mem and I/O. Respectively, V_i expresses the free resource of NM_i and $V_{i,j}$ expresses the resources NM_i needs to hold on $U_{i,j}$.

Definition 3(Crisis): For each dimension $D \in \{CPU, MEM, I/O\}$ a crisis is detected to $U_{i,j}$ at NM_i, if $V_{i,j}[D] <= \eta * V_i[D]$, (η is a coefficient of relaxation generated by experiments).

The crisis of $U_{i,j}$ means the resources demand for $U_{i,j}$ exceeds the supply ability of NM_i, then should look for targets to relieve its workload pressure, which will be described in more detail latter.

2.2.2 Tenant Model. One tenant's data is replicated to several nodes, they compose a tenant model to collaborate for the tenant's requests. Tenant model $TM_j = \{NM_k \mid \forall NM_k \to T_j\}$ means a set of node models who have the same data replica of T_j , and can execute requests about T_j .

Definition 4(Collaborative Node): Node models in a tenant model maintain the same tenant data replica and can provide the same database service, we call them collaborative node for each other.

Table1. Profile in NM1

	NM_2	NM_3	NM_4	NM_5
$T M_1$	1	1	0	0
$T M_2$	0	0	1	1

Review the example of Figure 1, tenant models $TM_j = \{NM_k \mid \forall NM_k \to T_j\}$; NM_2 and NM_3 are collaborative nodes in TM1. Here are two reasons why to design this device. For one reason, tenant model evaluates economic results of executing requests. Requests that have the same tenant property share the same SLA profit model assigned by the tenant model, e.g. the profit out-coming of executing requests of T_1 is decided by TM_1. For another, tenant model records how the tenant data distributed in system, it represents dispatching candidates when a node model of it detects performance crises. Every node model likes NM_1 is configured with a profile as show in Table 1 in order of distinguishing collaborative nodes, value 1 means collaborative relation and value 0 doesn't, e.g., *profile*[$T M_1$; NM_2] = 1 means if NM_1 detects crises to $U_{1,1}$, NM_1 can dispatch the coming requests about T_1 to NM_2, it is tenant model who responses for finding targets by using sliding window algorithm.

2.2.3 Slack Time Model. It is a key problem to make a rational execution sequence for requests in maximizing the profit. To make an accurate decision, this paper adopts a concept of the slack time.

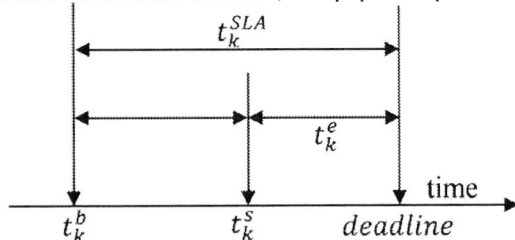

Figure 2. Slack Time Model

Definition 5(Slack Time): For each submitted request q_k^j at a node model, AdaptiveSLA defines four parameters for it: t_k^b, t_k^s, t_k^e and t^{SLA}. t_k^b is arrival time and t^{SLA} is the response time under SLA constrain. t_k^e is the execution time for q_k^j, the slack time t_k^s can be calculated by $t_k^s = t^{SLA} + t_k^b - t_k^e$.

As illustrated in Figure 2. Slack t_k^s is the time that q_k^j can be further postponed without introducing additional SLA penalties, the scheduling policy is made based on this constrain. The fact is that, to calculate t_k^s of q_k^j, we just need to know t_k^e, which is predicted by using machine learning technology. Both TYPE (a version of Gatekeeper implemented by Tozer et al. with load shedding added) and Q-Cop approaches start by predicting the execution time of a query using the number of currently running queries as the feature in their model for each query type. This paper uses the same method with boosting approach method for it iteratively obtains weak learners (namely, learners that do not necessarily have good performance individually) and combines the weak learners to create a strong learner which reduces the prediction error. The features for each query is the number of predicates and cardinality for each predicate. Algorithms used in this paper are all from the off-the-shelf machine learning package WEKA.

2.3 SLA Profit Model

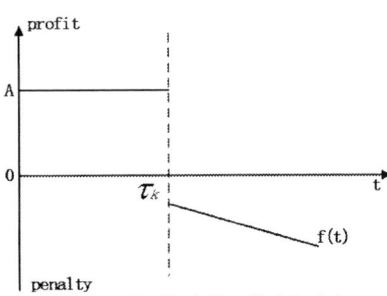

Figure 3. SLA Profit Model

In multi-tenant database, we assume there is an associated SLA profit model for each tenant, requests of one tenant enjoy the same profit model. Different studies have explored many profit modes in various shapes, AdaptiveSLA believes that a segmented function as shown in Figure 2 is a more commonly choice used in the real-world. The figure denotes that if the response time t of request q_k^j is shorter than τ, the service provider obtains a revenue A, otherwise, pays a penalty back to the tenant according to the final response time. In addition, what has not been illustrated is if a request rejected up-front, the service provider has to pay a constant penalty P_j according to its tenant. The profit formulation is defined as:

$$
\text{profit}(q) = \begin{cases} A, & \text{if } t <= \tau \\ f(t), & \text{if } t > \tau \\ -P, & \text{if } q \text{ is rejected} \end{cases} \tag{1}
$$

Once a request finished at a node model, the tenant model will report its economic result to the service provider by using Equation (1).

3. The Two-Stage Scheduling Policy

AdaptiveSLA designed to have no single point of failure, nodes in this system have equal role of providing database service. To maximize SLA profit, this paper decomposes the policy into two stages as shown in Figure 4. Firstly, AdaptiveSLA mitigates performance crises using sliding windows algorithm to provide effective resource provisioning, secondly, it makes SLA-oriented request's execution order for every node model according to request's slack time.

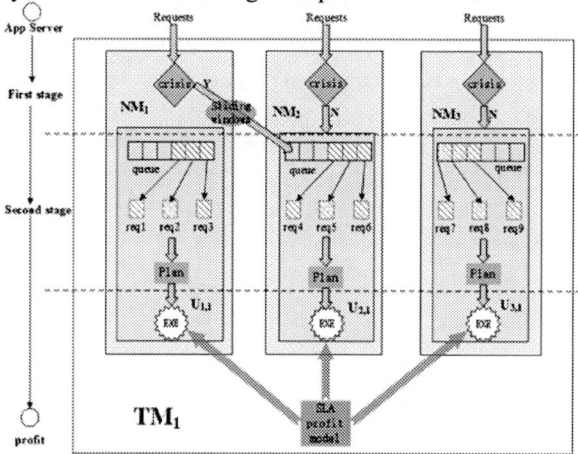

Figure 4. The Two-Stage Scheduling Policy

3.1 The First Stage: Dispatching Requests

Every node model can execute requests or dispatch them to other node models, this stage determines in which node model requests will be executed.

3.1.1 Crisis Detection.
At first, application server submits analysis queries to AdaptiveSLA, and they will be routed to the nearest node model which maintains their tenants' data. Then, the node model periodically checks its resource status using crisis detecting mechanism, if the current resource state is sufficient to support every request execute unit of it, which means no crises are detected, so requests will be sent to the queue of relative execute unit according to the tenant property. Otherwise, the current resource state can't meet the need of executing requests of some request execute unit, therefore, this node model should search for target node models to help mitigate crises by dispatching the following requests of crisis unit to other node models.

$$crisis = \exists V_{more}^{j} > V_{left} \left(\forall T_j \rightarrow NM_i \right) \tag{2}$$

The goal of Equation 2 is to find a node who has the most sufficient resources to execute these requests among all the collaborative nodes of NM1, the measurement is defined as the distance of resource vector V_i' and $V_{i,j}$, V_i' shows the current resource status of target nodes and $V_{i,j}$ means how much $U_{i,j}$ need to run its requests. Besides, it has two constraint conditions. The first one restrains the resource consumption ratio of the target node, one resource vector V_i is recorded by NM$_1$, which means the nearest resource status of targets, another vector V_i' is got by a communication mechanism, which means the current resource status of targets, every resource vector V is marked with a timestamp T. The threshold value is also an empirical vector gotten by several experiments, which evaluates that the target node has little possibility to detect crises. The second constrain condition indicates that the CPU, Mem and I/O resources are all enough to handle requests dispatched by $U_{i,j}$.

3.1.2 Crisis Mitigation. When crises are detected to a node model, AdaptiveSLA uses sliding window here rather than using the last resource status, providing a more confident view about shifts in behaviour. AdaptiveSLA maintains a per-node sliding window of last two resource vectors, one records the latest history resource status, another shows the current resource status, each V is marked with timestamp. For instance, assume NM_i detects crises on $U_{i,j}$. NM_i records resource status V_k of all NM_k ($NM_k \in TM_j$) except itself at the beginning. NM_i then fetches new resource status vector V_k' to fill the second window. AdaptiveSLA will choose those who have abundant resource and low consumption rate by comparing all collaborative nodes' sliding results. After choosing, V_k' takes place of V_k and the window slides forwards.

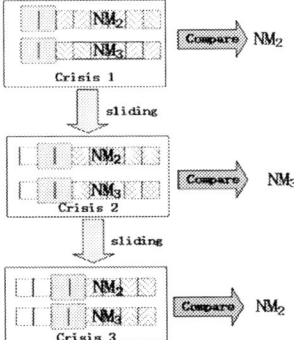

Figure 5. Overview of Sliding Window Algorithm

Take Figure 5 as example, at first, NM_1 detected a crisis on $U_{1,1}$, and the following requests about TM_1 can be dispatched to NM_2 or NM_3 by scanning the configurable table of NM_1. NM_1 recorded their initial resource status before and then send message to get their current resource status. The sliding window algorithm used the two windows (resource vector) of every candidate to calculate its potential and chose the best one. After choosing the window of every candidate, each window slides forward to wait for next crisis. Conclusively, the compare process can be formulated as:

$$target = \arg\max_{i=2,3} (\|V_i - V_{1,1}\|)$$

$$\text{s.t.} \quad \frac{V_i' - V_i}{T_i' - T_i} < \bar{\theta} \,, \ (\bar{\theta} \text{ is an empirical vector}) \tag{3}$$

$$V_{1,1}[D] < V_i[D] \times \eta, \ D \in \{CPU, MEM, I/O\}$$

The first stage eliminates crises automatically, aiming at utilizing resources rationally. If this stage can't find a target for some node model, it means the current cluster is unable to withstand the workload pressure, new servers should be added in, AdaptiveSLA handles this event for capacity planning by a human administrator.

3.2 The Second Stage: Planning Execution Order

The second stage makes execution sequence for requests waiting to be executed. Requests in a queue share a common SLA profit model, to maximize the profit, requests should be finished as many as possible under their SLA constrains, the constrain used here is modeled as slack time defined above, which means the latest time a request can be delayed. Suppose a request q_k^j about tenant T_1 is admitted by NM_1 at a moment t_k^b, We can also get its SLA constrain time t_k^{SLA} according to its tenant property from TM_1, besides, the execution time t_k^e of q_k^j can be predicted by ML technology, so its slack time can be calculated by $t_k^s = t_k^{SLA} + t_k^b - t_k^e$.

3.2.1 Machine Learning Technology. As execution time is so important in our decision, this paper take the model with the least prediction error. Here uses a boosting approach called additive regression in

WEKA package and also uses the regression trees and linear regression as the weak learners. Then evaluates two potential feature vectors: one based on the SQL text of the query such as number of nested subqueries, total number of selection predicates, number of equality selection predicates, number of non-equality selection predicates, the other describes the relationship between the input parameters (the normalized CPU share, memory size, number of database replicas, request rate). Experiments show that the model prediction error is further reduced with the boosting.

3.2.2 Sequence Making Algorithm. The Algorithm 1 is a re-constructive job sequencing problem, it shows how a request be arranged to execute. A candidate can be appended to execution queue Q if its slack time is longer than total execution time of the execute queue, otherwise reverse the execute queue and replace the first request whose execution time is longer than it. A request will be rejected if it misses the deadline or can't be added in the execution queue.

Algorithm 1 :Enqueue Algorithm

Input:

    q_k; #a new request
    Q; #a request queue waiting to be executed
    T (Q); #total execution time of requests in Q

1: if $t_k^s > T (Q)$ then

2:    $Q \rightarrow$ append(q_k);

3:    $T (Q) := T (Q) + t_k^e$;

4: else

5:    if $t_{current} > t_k^{SLA} + t_k^b$ then

6:       for all $q_p :=$ reverse(Q) do

7:          if $t_p^e ; > t_k^e$; then

8:            swap(q_p; q_k);#exchange the place of q_p; q_k and enqueue q_k

9:            $T (Q) := T (Q) + t_k^e$;

10:         end if

11:       end for

12:    end if

13:    reject(q_k);

14: end if

15: return Q;

The Algorithm 2 shows that the execution unit just executes requests of the execution queue Q in turn.

Algorithm 2 :Dequeue Algorithm

Input:

    Q; #a request queue waiting to be executed
    T (Q); #total execution time of requests in Q

1:    if $Q \neq \phi$ then

2:          $q := Q \rightarrow$ pop();

3:          T (Q):= T (Q)- ;

4: end if

5: return q;

To prove the correctness of this algorithm, first suppose requests of a tenant are ordered by deadline considering their arrival orders, which is concluded by the formulation deadline $= t_k^{SLA} + t_k^b$, requests of a tenant have the same SLA constrain, and different begin time. For a coming request q_k, if $t_k^s > T(Q)$, executing this request after executing all requests before will not violate its SLA constrain, it can work normally if it is appended to the tail of the queue. Otherwise, replace it with q_p whose execution time is longer than it. That is if $t_p^e > t_k^e$, and deadline(t_k) > deadline(t_p), so $t_k^s > t_p^s$, q_k can work well without violation after replacing and the total execution time of execution requests in Q reduced, which not only has no bad impact on executing other requests in Q but also benefits to arrange more requests to execution.

4. Experiment

In this section, this paper deployed and evaluated a system on a cluster of 10 nodes dedicated to database processed. For the test bed, the database server run MySQL v5.0 with InnoDB storage engine on CentOS generated by openstack, each is configured with an Intel Xeon E312xx processor, 4G memory, and 20GB Disk. The following section describes data set and benchmark used for workload generation, and a detailed evaluation of AdaptiveSLA.

4.1 Benchmark Description

Existing benchmarks provide little support for evaluating the effects of multiple tenants' database, such as the TPC suite, focusing on testing the performance and limits of a single high performance DBMS dedicated either for transaction processing (TPC-C) or for data analysis (TPC-H) [3]. To adapt to multiple tenant environment, AdaptiveSLA changed TPC-C, systematically generating workloads and data set for 20 tenants. Tenant's data size varies from 100MB to 2GB.

This benchmark is centered on the principal activities (transactions) of an order-entry environment. The workloads generator maintains a configuration file describing how tenants' data distributed in the system, it generated tenants' workloads with different rate to simulate heterogeneous tenant pressure. In the training phase, we executed requests in an idle node with maximum possible resources, then calculated the average response time as the SLA standard, besides, we experimented the resource consumption of every execute unit. During the working phase of the system, the arrival rate of requests followed a Poisson distribution with the rate set in each test, for the reason that it is widely used to model independent arrival requests to a website [4]. Every node model gathered CPU and Mem status using Linux tool *top* and gathered the I/O usage with the tool *iostat* per minute. If crises were detected, node model sent requests to others by TCP pipelines, otherwise, the node model used relative execute unit to execute requests. After finishing a request, the node model reported its profit. This paper measured the total resource consumption and the final profit under 5 random parameters of the workload setting.

4.2 Experiment Results

ISPECE

IOP Publishing

IOP Conf. Series: Journal of Physics: Conf. Series **1187** (2019) 052002   doi:10.1088/1742-6596/1187/5/052002

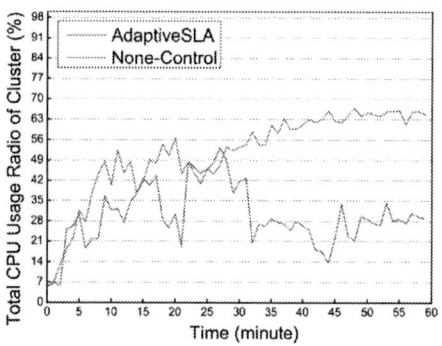
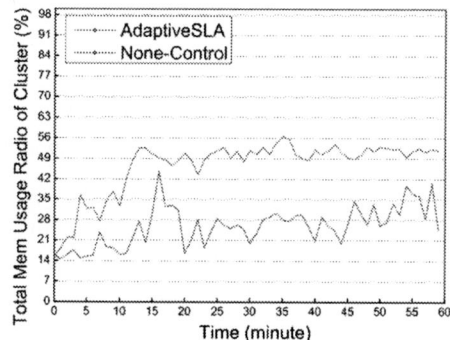

Figure 6. Consumption of CPU and Memory

We compared AdaptiveSLA's performance with the previous works, to view the results, we plot the distribution of the outcoming with average statistical result. We firstly proved that our system can make good use of the resource by mitigating crises. Take the example of CPU and Mem consumption, Figure 6 shows their utilization ratio of the whole cluster, the data is summarize with several experiments. In each sub figure, the x-axis represents the system running time, during which we kept sending requests to nodes with different rate, and the y-axis represents the resource utilization ratio, it indicates the load-balancing capacity of the system. From the plot of Figure 6, we can see clear distinguish of resource usage ratio between AdaptiveSLA and a none-control schedule that doesn't use any schedule policy. With the workload increasing, the resource consumption of none-control fashion is substantially worse than AdaptiveSLA, the experiment result shows AdaptiveSLA improves 30% to 35% in CPU usage, and 15% to 20% in Mem usage compared with none-control method. For one reason that heterogeneous workloads will cause wasting of resource, but AdaptiveSLA balances workloads to free nodes if some nodes detect crises, for another, AdaptiveSLA can arrange more requests to execute, so its resource consumption is higher. What's more, the resource usage in none-control fashion is more susceptible to the skew of requests. However, as can be seen, the rate of change of Mem is not as obvious as CPU, it may because that the Mem resource is more abundant than CPU in our system. Besides, the max consumption of AdaptiveSLA is confined such as the max CPU using ratio is near 70%, it is because the date set is fixed before experiment, sometimes, workloads from a tenant is too heavy, but they can just be dispatched to limited nodes who have the same tenant data, we cannot leverage the resource of other free nodes.

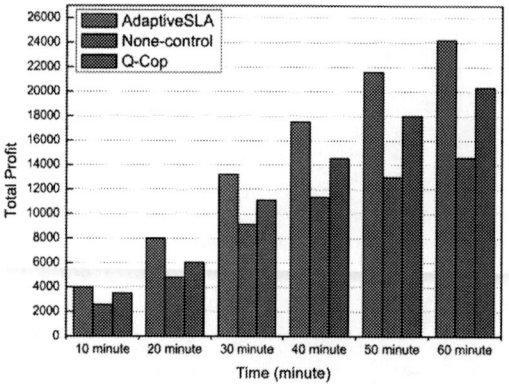

Figure 7. The Total SLA Profit

We then compared the final profit gained by different methods in out experiment. Figure 7 demonstrates the profit gained by AdaptiveSLA, Q-Cop and a none-control schedule. As can be seen,

2087

(1) At first, the workload isn't heavy, so all the three methods gained the similar profit because these systems are underloaded. (2) With the workload pressure increase, some nodes have performance crises, the outcoming is different. The profit increasing rate of AdaptieSLA is the highest, while none-control fashion becomes slow. The final statistic shows AdaptiveSLA achieved 35% to 40 % more profit than none-control method, for one reason we have proved above, AdaptiveSLA used the total resources effectively when facing the same workloads, for another, the second stage of AdaptiveSLA can arrange more requests to execute under their SLA constrains. AdaptiveSLA also achieved 20% to 25% more profit compared with Q-Cop. The result shows AdaptiveSLA makes more reasonable schedule policy. (3) We carefully analyze this phenomenon and conclude that our schedule policy does well in maximizing the profit. In summary, the experiment results in this section demonstrate that AdaptiveSLA, using a two-stage schedule policy, can make good use of resources of the cluster, and make a well profit-oriented execution plan, the final profit of service provider is improved.

Acknowledgments

First and foremost, I would like to show my deepest gratitude to my supervisor, a respectable, responsible and resourceful scholar, who has provided me with valuable guidance in every stage of the writing of this thesis. Without his enlightening instruction, impressive kindness and patience, I could not have completed my thesis. My sincere appreciation also goes to the colleagues from the State Grid Chongqing Electric Power Company, who participated this study with great cooperation. Last but not least, I'd like to thank all my friends, for their encouragement and support.

References

[1] Lehner W, Sattler K U, Sattler K U. (2013) 'Web-scale data management for the cloud', pp.137-145

[2] Abadi, D. J., Madden, S., Hachem, N. (2008) Columnstores vs. row-stores: how different are they really. Sigmod

[3] Elmore A J, Das S, Pucher A, et al. (2013) Characterizing tenant behavior for placement and crisis mitigation in multitenant dbmss. SIGMOD

[4] Xiong P, Chi Y, Zhu S. (2011). Intelligent management of virtualized resources for database systems in cloud environment ICDE

[5] Popovici F I, Wilkes J. (2005) Profitable services in an uncertain world.Proceedings of the 2005 ACM/IEEE conference on Supercomputing. IEEE Computer Society

[6] Azeez A, Perera S, Gamage D, et al. (2010) Multi-tenant SOA middleware for cloud computing.

Research on reporting scheme of grading stop-and-recharge event of low-voltage acquisition terminal

R Huang[1,2], Chenglin Zhang[3], L Ouyang[1,2], M Liu[1,2], X Liu[1,2], Y Su[1,2], H Chen[1,2]

[1]State Grid Hunan Power Supply Service Center, Changsha, 410004, China.

[2]Hunan Province Key Laboratory of Intelligent Electrical Measurement and Application Technology, Changsha, 410004, China.

[3] State Grid Hunan Technical Training Center, Changsha 410004, China.

424379318@qq.com

Abstract. For a long time, power supply companies mainly rely on the power supply service command platform for organization and dispatch of work orders and work orders for low-voltage network power outages. However, the analysis of power outage location, power outage scale, requires personnel strength and repair plan is mainly based on the prejudgment of the phone entered after the user loses power, which is prone to misjudgment of workload or recovery time, low processing efficiency and high cost. There are many security risks. This paper analyzes the problem of stop-and-recharge events, single-user and meter-box stop-and-recharge events, and proposes a reporting plan for the three-level stop-and-return event to solve the problem of immediate reporting of events after power outages. Make full use of the acquisition advantages of the electricity information collection system, and comprehensively improve the customer service response speed and distribution network operation and management level.

1. Introduction

For a long time, power supply companies mainly rely on the power supply service command platform for the troubleshooting of low-voltage network, including organization and dispatch and work orders [1-2] for power outages. However, the analysis of power outage location, power outage scale, requires personnel strength and repair plan is mainly based on the prejudgment of the phone entered after the user loses power, which is prone to cause misjudgment of workload or recovery time, low processing efficiency, high cost and many security risks [3-5] for follow-up processing on site.

The electricity information acquisition system is connected with the power supply service command platform. By pushing power failure and recovery events of the low-voltage household meter, acquisition unit, main meter of the substation area or concentrator in real time, the electricity information acquisition system is able to assist the power supply service command platform in analyzing and judging the authenticity, location, cause, property and scope [6-8] of a fault comprehensively. This accelerates the troubleshooting response time, reduces the cost and improves the customer satisfaction and system operation index [9-10].

This paper analyzes the problem of power failure and recovery events of the substation area, of monitoring points for key branch and users and of single user and meter box, and proposes a reporting plan for the three-level stop-and-return event. On the basis of the original concentrator, electric energy meter and acquisition unit in the substation area, this scheme is assisted with monitoring devices for

Content from this work may be used under the terms of the Creative Commons Attribution 3.0 licence. Any further distribution of this work must maintain attribution to the author(s) and the title of the work, journal citation and DOI.
Published under licence by IOP Publishing Ltd

monitoring points and combines the active reporting function of a power outage event with the research and judgment system of the power outage fault acquisition in the main substation to realize the three-level online monitoring of the power supply status from the substation area to the household meter. This solves the problem of immediate reporting of events after power outages.

2. Scheme overview

Currently, there is a relevant metering acquisition device terminal (including a Type I concentrator, Type II concentrator, monitoring and metering terminal for distribution and transformation, load management terminal and data acquisition terminal of station electric energy), electric energy meter (three-phase and single phase electric energy meters) and acquisition unit (Type I and II acquisition units) in the low-voltage network. The application scene can be divided into the following types based on different data acquisition modes:

(1) The wireless public network is used between the main substation and the terminal for data communication. RS485 is used between the terminal and the electric energy meter for data communication;

(2) The wireless public network is used between the main substation and the terminal for data communication. The low-voltage power line carrier/micropower wireless is used between the terminal and the electric energy meter for data communication;

(3) The wireless public network is used between the main substation and the terminal for data communication. The low-voltage power line carrier/micropower wireless is used between the terminal and the acquisition unit for data communication. RS485 is used between the acquisition unit and the electric energy meter for data communication;

(4) The wireless public network is used between the main substation and the electric energy meter for data communication;

According to the classification of devices with power failure and recovery events, the events include electric energy meter, acquisition unit and terminal events.

According to the scope of power failure and recovery events, events are classified as follows:

(1) Event of singe device (such as single electric energy meter, acquisition unit or terminal);

(2) Event of multiple devices (such as multiple electric energy meters, acquisition units or terminals);

(3) Event of all devices (such as all electric energy meters, acquisition units or terminals);

For the way to inform the main substation in time based on aforementioned different device and scope events, this scheme is confirmed to include the following four aspects:

(1) Event sensing. The way for the device to sense a power failure and recovery event correctly;

(2)Event reporting. The way for the communication network to support the reporting of a power failure and recovery event;

(3) Event researching and judging. The way for the substation to analyze and judge a power failure and recovery event correctly.

(4) Event processing. The way for the substation to configure a follow-up processing flow for a power failure and recovery event.

Currently, the relevant event reporting of an electric energy meter is designed based on utilizing the power failure and recovery event before reporting the operation status under the live status. And state grid power companies have complete acquisition schemes for power failure events after the power-on status of a terminal and an electric energy meter. Therefore, this scheme only involves the way to report a power failure event rapidly.

In the communication network, if a node senses a device at the user side has a power failure event, corresponding reporting mechanisms shall be designed to support reporting of events in different coverage scopes.

(1) Active reporting mechanism

Reporting of single node event in the communication network requires the active reporting mechanism. After receiving a power failure event (such as communication address of a node with a power failure) from the node with a power failure, adjacent nodes without a power failure transmit and report the power failure event to the local communication module of the terminal by grade under the data acquisition or free status of the current terminal and the local communication module of the terminal reports it to the terminal before the terminal reporting it to the main substation ultimately.

(2)Conflict detection/collision avoidance mechanism

Reporting of a multiple-node event in the communication network requires the conflict detection/collision avoidance mechanism. Limited by the bandwidth of the communication channel, it is inevitable to lose an event reporting if plenty of nodes conduct the event reporting simultaneously. Adding the conflict detection/collision avoidance mechanism can reduce the probability of mutual conflicts among event reporting signals and improve the success rate of the event reporting.

3. Reporting scheme of three-level power failure and recovery events

According to different power failure scenes, power failure events are divided into three major categories: power failure and recovery of the substation area, of monitoring points for key branch and users and of single user and meter box.

3.1. Power failure and recovery of substation area

Power failure of the entire substation area caused by the distribution and transformation and line faults influences the production, livelihood and personal safety greatly. Occupying the top grade in reporting of the three-level power failure and recovery event, it is the most urgent power failure status for repairing. Realizing the reporting of a power failure and recovery event of the substation area will turn "passive repair" to "active operation and maintenance", which is crucial for enhancing the power supply reliability, reducing user complaints and improving the power supply service quality.

According to its causes, power failure and recovery of the substation area can be divided into the following types.

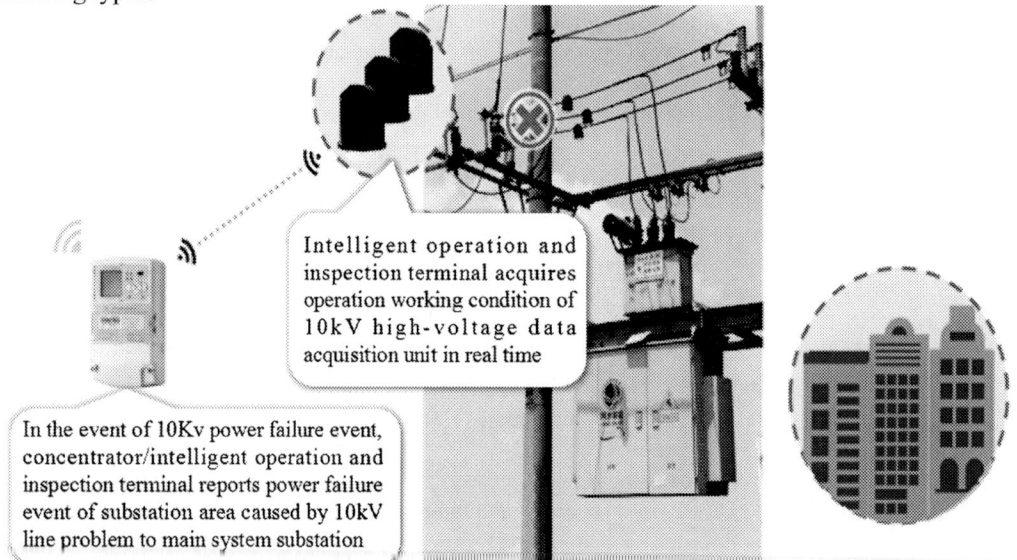

Figure 1. Power failure and recovery of substation area caused by 10kV line fault

3.1.1. Power failure and recovery of substation area caused by 10kV line fault.
The high-voltage data acquisition unit is installed to 10kV high-voltage line. The concentrator/ intelligent operation terminal is used to acquire the operation working condition of the acquisition unit in real time via the wireless channel. After sending 10kV power failure event, the concentrator/ intelligent operation terminal reports the power failure event of substation area to the main system substation as shown in Figure 1.

3.1.2. Power failure and recovery of substation area caused by low voltage branch switch abnormality. The branch monitoring device with the super capacitor is installed at the output end of the low-voltage branch switch in the power distribution room to acquire the status and trip causes of the low-voltage switch in real time. In the event of power failure, it is reported to the concentrator/ intelligent operation and inspection terminal via the carrier/wireless channel and the concentrator/ intelligent operation terminal reports it to the main system substation as shown in Figure 2.

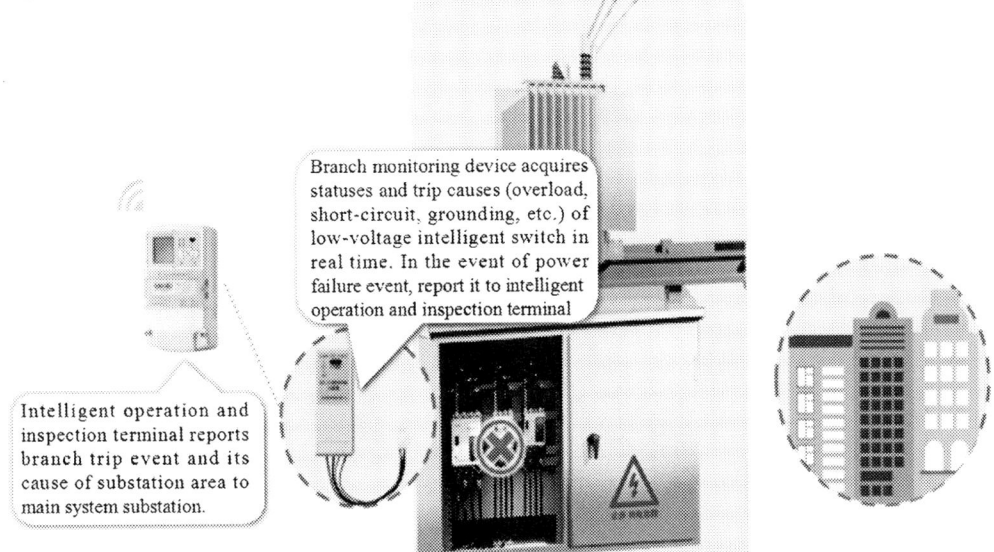

Figure 2. Power failure and recovery of substation area caused by low voltage branch switch abnormality

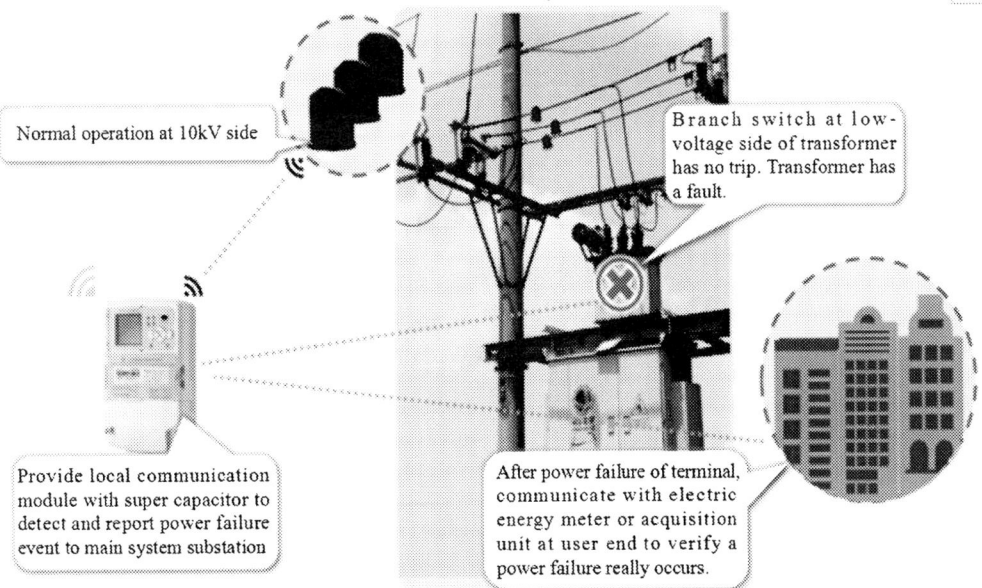

Figure 3. Power failure and recovery of substation area caused by transformer fault

3.1.3. Power failure and recovery of substation area caused by transformer fault. Under the circumstance of a normal operation at 10kV side and trip-free branch switch at the transformer side, a transformer fault can also cause a power failure of the substation area. In this case, the concentrator/

intelligent operation monitoring terminal will have a power failure. Therefore, the local communication module with the super capacitor shall be provided. In the event of a power failure, communication between the local channel and the electric energy meter at the user end or the acquisition unit is used for verifying a real power failure event before reporting it to the main system substation as shown in Figure 3.

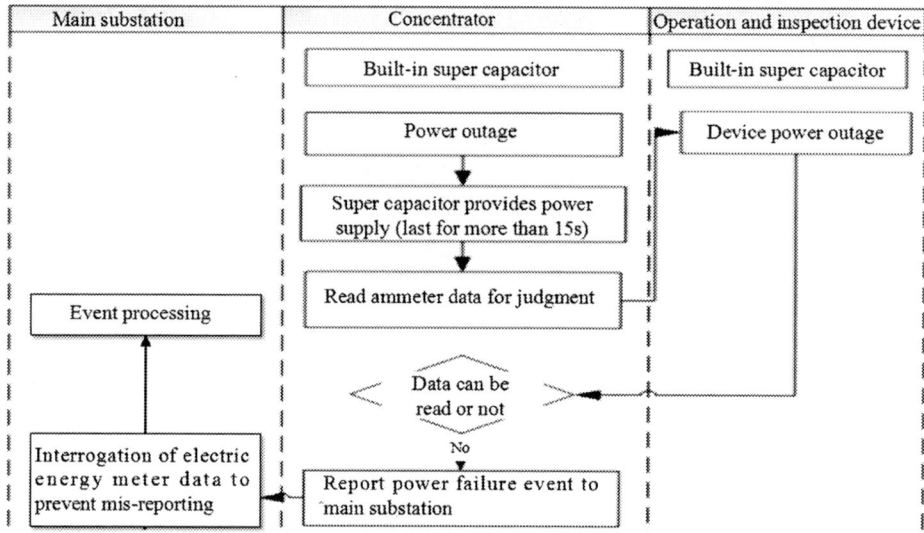

Figure 4. System structure

3.1.4. Realization flow. Conduct software upgrading of the concentrator and the main substation after adding the super capacitor or operation and inspection device to the concentrator communication module and adding the auxiliary device for power failure operation and inspection, built-in super capacitor and built-in communication module with carrier/ micropower channels to the main branch of the substation area. The flow in Figure 4 is followed for reporting.

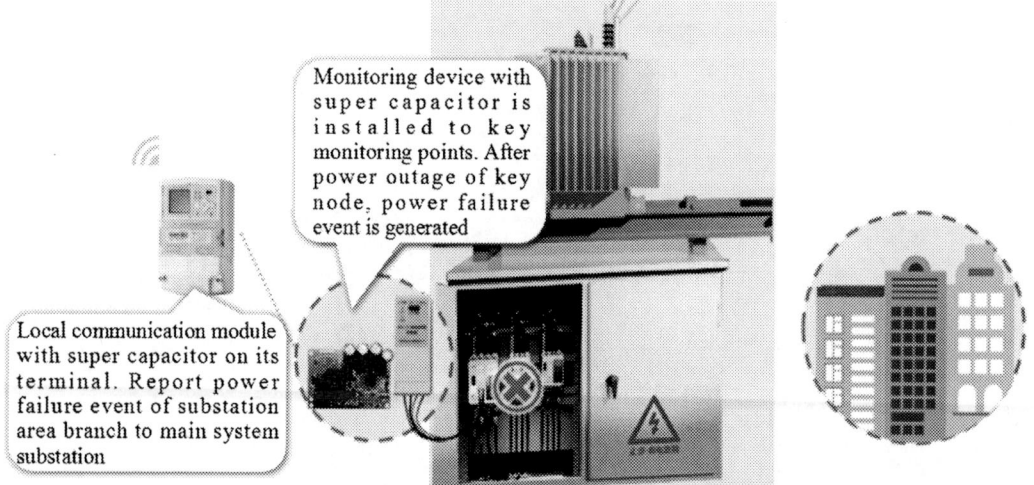

Figure 5. Power failure and recovery of monitoring points for key branch and users

3.2. Power failure and recovery of monitoring points for key branch and users
The monitoring device with the super capacitor is installed for key monitoring points, which can be an acquisition unit, an electric energy meter or a special device to generate a power failure event after a

power failure of the monitoring point. The local communication module of the terminal is configured with the super capacitor. After a power failure, it conducts communication with the monitoring device to obtain and report the power failure event to the main system substation as shown in Fig. 5.

3.3. Power failure and recovery of single user and meter box

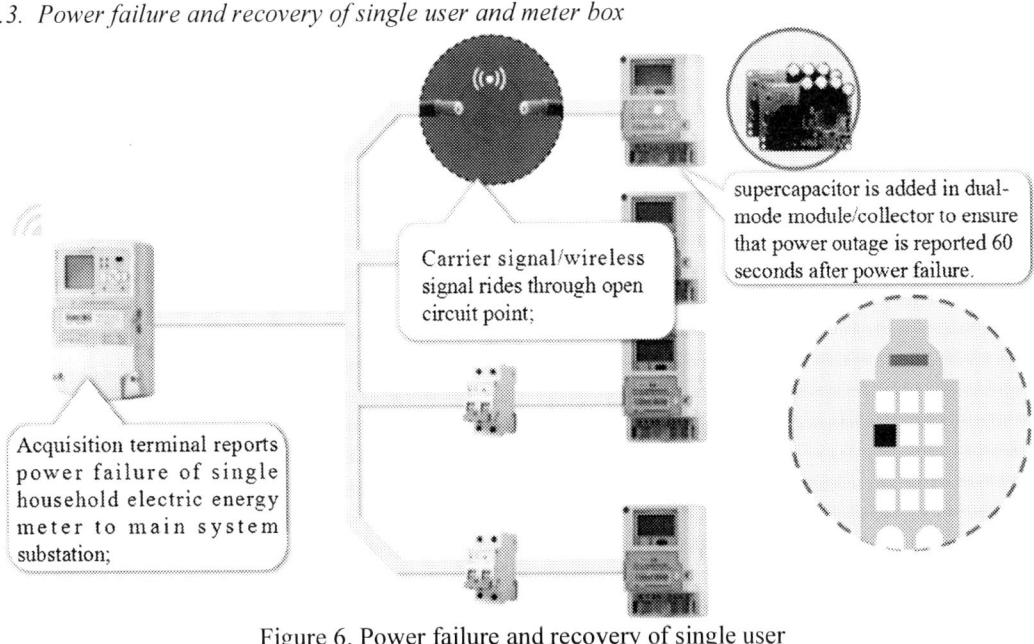

Figure 6. Power failure and recovery of single user

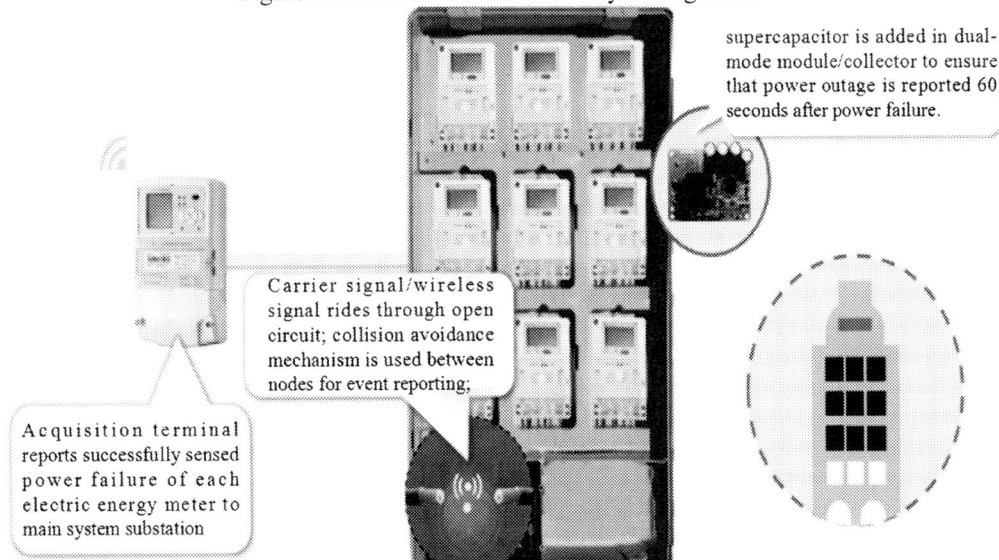

Figure 7. Power failure and recovery of meter box

(1) The carrier, micropower dual-channel and super capacitor are equipped in the concentrator carrier module and electric energy meter communication module. After a power failure, it can maintain the communication power for 60s.

(2) According to the fact whether the zero-cross signal loses judging a power failure and recovery, the electric energy meter communication module judges the occurrence of a power failure and

recovery event via the communication between the electrical level of a module ELV pin and an electric energy meter.

(3) The local communication channel has the collision avoidance mechanism. The electric energy meter communication module reports a power failure and recovery event to the concentrator directly or via adjacent nodes.

(4) The concentrator reports a power failure and recovery event to the main acquisition substation via GPRS.

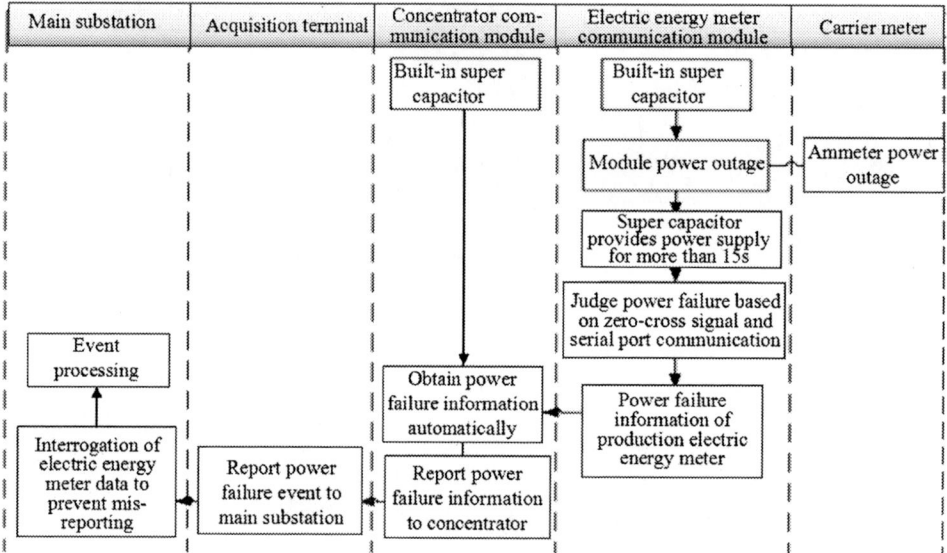

Figure 8. Scheme 1 on power failure of carrier meter

3.3.1. Power failure of carrier meter. Modification scheme 1: as shown in Figure 8, add the super capacitor and conduct software upgrading for the communication module; conduct software upgrading for the concentrator communication module; conduct software upgrading for the main substation.

Modification scheme 2: as shown in Figure 9, add the wireless dual-channel module with carrier-micropower; add the super capacitor design to the communication module; conduct software upgrading of the communication module; conduct software upgrading of the concentrator communication module; conduct software upgrading of the main substation.

3.3.2. Power failure of RS485 meter under acquisition unit. Modification scheme: Type I acquisition unit communication module/Type II acquisition unit are changed to three-phase power supply and equipped with the super capacitor; software upgrading of acquisition unit communication chip, acquisition unit, concentrator communication module, concentrator and main substation.

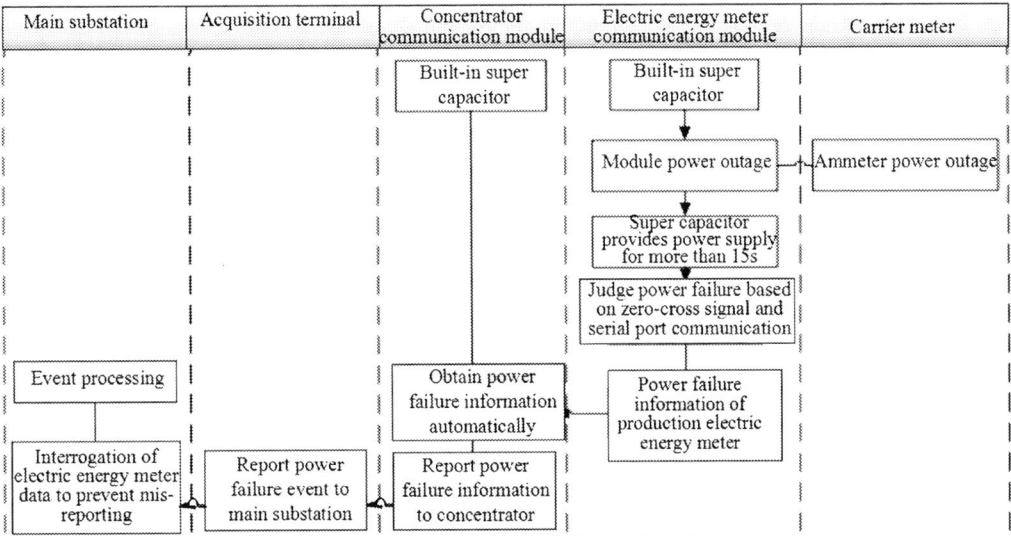

Figure 9. Scheme 1 on power failure of carrier meter

3.3.3. Power failure of RS485 meter under concentrator. At the moment of power outage of the electric energy meter, only basic operations such as display and button are maintained and its own power failure status cannot be reported automatically. Therefore, only the concentrator for electric energy meter polling can be relied on to diagnose a power failure. But this mode may have a mis-reporting caused by 485 line fault. As a result, an error-free reporting of a power failure event can be realized by installing the communication module to the electric energy meter or adding the device which can conduct communication under power outage to the meter end.

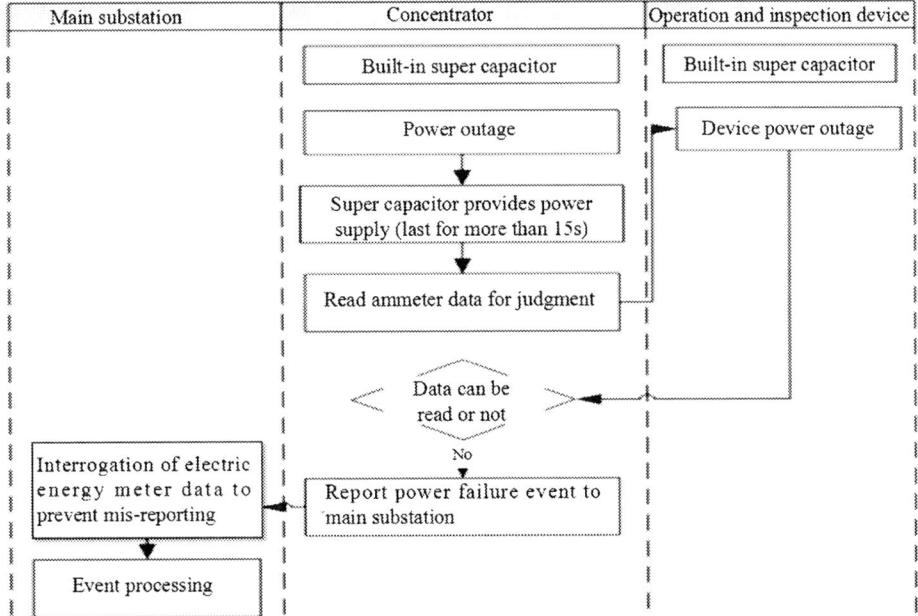

Figure 10. Report on power failure of RS485 meter for concentrator

Modification scheme: add the super capacitor to the concentrator communication module or add the operation and inspection device; install the auxiliary operation and inspection device for a power failure to each ammeter phase; the operation and inspection device has the built-in super capacitor and the built-in communication module with the carrier/micropower channels; software upgrading of the concentrator and main substation.

4. Conclusion
This paper proposes a reporting plan for the three-level power failure and recovery event. On the basis of the original concentrator, electric energy meter and acquisition unit in the substation area, this scheme is assisted with monitoring devices for monitoring points and combines the active reporting function of a power outage event with the research and judgment system of power outage fault acquisition in the main substation to realize the three-level online monitoring for the power supply status from the substation area to the household meter and reporting a power failure and recovery event. This solves the problem of immediate reporting of events after power outages. It makes full use of the acquisition advantages of the electricity information acquisition system and combines with the power supply service command platform deeply to turn the "passive repair" to "active operation and maintenance" to further solves the problem of "the last one kilometer" of the power supply service. It comprehensively improves the customer service response speed and distribution network operation and management level.

References
[1]   Lishan B, Chen L, Bo F, et al 2016 J Function Design and Application of Fault Repair and Command Platform for Intelligent Distribution Network *Distribution & Utilization* 33(12) pp 29-33.
[2]   ChangguoZ, Xiaoshu H, Jianbin Y, et al 2016 J Marketing-distribution-dispatch Management Model Optimization for Country Grid *Automation of Electric Power Systems* 33(12) pp 29-33.
[3]   Canming Z, Zhuhong L, Lei T, et al 2016 J Fault localization for electric power communication network based on fault propagation model and supervised learning *Journal of Computer Application* 36(4) pp 905-908.
[4]   Weikang R, Jihua H, Zhihao L, et al 2012 J Research on Intelligent Fault Identification System in Low Voltage Power Distribution Network *Low Voltage Apparatus*12 pp 37-39.
[5]   Jiangyi H, Engo Z, Xingang D ,et al 2014 J Application Status and Development Trend of Power Consumption Information Collection System *Automation of Electric Power Systems* pp 131-135.
[6]   Chuangxin G, Haibo L, Bin Y, et al 2013 J A Survey of Research on Security Risk Assessment of Secondary System *Power System Technology*37(1) pp 112-118.
[7]   Bo L, Quanyi B, Minghai Y, et al 2015 J Analysis and Coping Strategy of Low Voltage Remote Cost-based Control on the Electricity Service Interruption/Recovery Problem *Distribution & Utilization*32(12) pp 61-67.
[8]   Haifan H 2012 J Research of Distribution Network Dispatching Fault Rapid Restoration Support System *China Electric Power (Technology Edition)* pp 22-24.
[9]   Yin Z J Study on Event Records Function of Smart Meters *Electrical Measurement & Instrumentation.*
[10]   Yunlong H, Shusheng Z, Wen C,et al 2016 J Study on Testing Method for Power Fails of Smart Electricity Meter and Outage/Power-on Events of Concentrator 44 (3) pp 71-76.

Progress and Trends in Mobile Cloud Computing Research

Wang Xiao-Shu

Dalian Vocational Technical College   116035

Abstract: This paper briefly introduces the definition and characteristics of mobile cloud computing, expounds the research progress of mobile cloud computing, and conducts in-depth research and analysis on the development trend of mobile cloud computing. It has published some suggestions and hopes to play a certain reference the application and development of cloud computing and help to better meet people's needs in mobile cloud computing, improve the effectiveness and extensiveness of mobile cloud computing applications, and lay a good foundation for the sustainable and stable development of China's social economy.

1. Introduction

In the rapid development of the Internet, mobile applications are becoming more abundant, and mobile terminals have gradually become an important component of people's work and life. With the realization of mobile games, mobile payment, etc., mobile applications are becoming more complex, and more stringent requirements are imposed on the computing power, security, and storage capabilities of mobile terminals. However, due to factors such as heat dissipation, size, and weight, the resource storage capacity and computing power of the mobile terminal are still very different from those of the non-mobile device, especially the battery capacity limitation, which has a very serious impact on the user experience. In order to break the limitations of computing, storage and battery, and enrich the application experience of mobile users, it is of great value and significance to introduce cloud computing in the mobile environment. Based on this, this paper has carried out research and analysis.

2. Definition and Characteristics of Mobile Cloud Computing

Mobile cloud computing mainly utilizes the advantages of cloud in storage and computing, breaks the impact and limitations of mobile terminals in terms of resources, enriches mobile user applications, and brings better experience to users [1]. In terms of definition, mobile cloud computing can be expressed as a usage mode for acquiring resources or information such as software and platforms from the cloud through a mobile terminal, using a wireless network, and adopting an easy extension.

Content from this work may be used under the terms of the Creative Commons Attribution 3.0 licence. Any further distribution of this work must maintain attribution to the author(s) and the title of the work, journal citation and DOI.
Published under licence by IOP Publishing Ltd

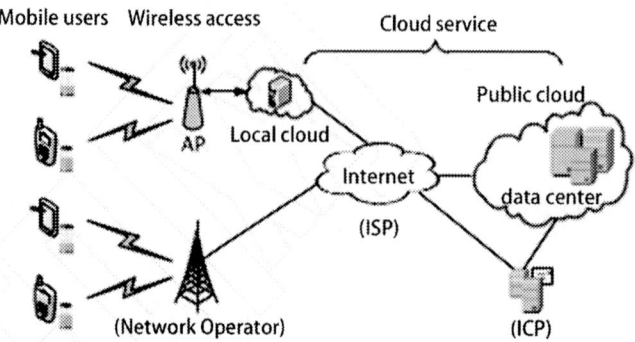

Figure 1 Mobile Cloud Computing System Framework

Figure 1 above shows the system framework of mobile cloud computing. Through wireless accounting and network operators, the effective connection between users and the Internet is realized. The data center can provide services such as storage and computing for users. Internet content providers can publish games, videos, and news resources in appropriate data centers to enrich users' content services [2]. If the user has high requirements in terms of security, power, and network delay, you can also use the LAN to connect with the local micro cloud to obtain the corresponding cloud service, which is characterized by scalability. The local micro cloud and the public cloud can be connected through the Internet to expand and optimize the storage and computing capabilities, and enrich user resources.

Based on cloud computing, mobile cloud computing has the advantages of cloud computing multi-service integration, user sharing, and resource expansion. At the same time, it also has the characteristics of user mobility and wireless network security vulnerability.

3. Research Progress in Mobile Cloud Computing

Mobile cloud computing involves many fields such as wireless networks, cloud computing, and mobile computing. In terms of technology, it includes computing migration technology, mobile cloud-based location services, mobile terminal energy-saving technologies, and data security and privacy protection [3]. Currently, mobile cloud computing is mainly used in mobile cloud storage, micro cloud applications, mobile cloud games, and mobile intelligent group applications, but it is affected by factors such as computing storage resource limitations, user mobility characteristics, limited battery energy, and illegal intrusion. There are still very large limitations in applications.

3.1 Computational Migration Technique

The computing migration technology mainly refers to the use of mobile terminal computing, storage, etc. to migrate tasks to servers with rich resources nearby. In this way, the requirements of storage, computing and energy of mobile terminals are minimized. In the development process of cloud computing, computing migration is applied in the cloud environment and other aspects, playing a very important value and role in mobile cloud computing. In terms of calculating the overall migration goal, it includes reducing service delay, saving mobile terminal energy consumption, and expanding CPU capacity.

Computational migration includes steps such as discovery, awareness, partitioning tasks, scheduling, and execution control. Execution control is the core component, including reliable remote proxy connection, information execution, and return calculation results. If the mobile app has migration requirements, the app sends a pause request and the local agent migrates the agent to the cloud agent after reading the state. The remote agent returns to the mobile terminal after the result processing.

3.2 Mobile Cloud Based Location Service

Location services are a very important technical content in mobile cloud computing and are currently receiving a lot of attention. The traditional positioning technology based on GPS has a very wide range of services, and the technology development is relatively mature. It has been widely used in military, transportation, construction, agriculture, etc., but because of the disadvantages of large energy consumption and weak penetration, it is already very inconvenient to user action recognition and precise indoor positioning and other mobile user needs, such as shopping guide. Introducing the mobile cloud computing model in the new location service construction can effectively solve this problem.

3.3 Mobile Terminal Energy Saving Technology

Mobile terminals are not very ideal in terms of battery capacity growth. In particular, mobile applications have become more abundant in recent years, and there are very big conflicts and contradictions between them and limited mobile terminal power. It has become a major influencing factor for user experience, the attention is getting higher, and it is developing towards energy saving in data transmission and energy saving in positioning services.

In terms of data transmission and energy saving: the current mobile terminal wirelessly transmits more data, and the energy consumption in the data transmission process occupies a very large proportion in the mobile terminal energy consumption. For example, the WiFi network is the most widely used wireless transmission technology. The energy consumption caused by the mobile terminal WiFi network is mainly concentrated in the idle listening state of the CSMA mechanism, and the function loss and data transmission generated by the mobile terminal in idle listening state. In the energy-saving scheme, the power consumption generated by the webpage loading is dynamically adjusted by the sensing network state mode, and the downloading strategy and the screen rendering mode are dynamically adjusted, thereby saving 24.4% of the energy consumption when the user experience is satisfied. In addition, the use of video tail traffic or channel dynamic allocation, etc., can save 29%-61% of video transmission energy consumption in WiFi networks [4].

In terms of location service energy conservation: In the mobile cloud computing environment, there are many applications that need to use location services, and the positioning process needs to generate very large computing and communication energy consumption. Positioning service energy saving is also an important development direction of mobile terminal energy-saving technology. Its energy-saving includes two aspects based on mobile terminal optimization and cloud-based optimization. The optimization of the energy consumption of the mobile terminal needs to predict and change the data source selection. The cloud-based optimization mainly achieves the corresponding energy-saving effect by migrating the positioning calculation to the cloud or sharing the cloud positioning data.

3.4 Data Security and Privacy Protection

As cloud computing services acquired by mobile users become more abundant, security threats such as data leakage and privacy exposure are becoming more serious.

First, in the cloud data security, in the mobile cloud computing environment, user data is migrated to the cloud data center using the wireless network, and services such as multi-user sharing and online query are also required. In this operation, not only is it easy to be attacked by an external attacker, but also it is easy to cause data leakage or loss due to improper operation or the like. These problems can be controlled and solved through a series of technical measures such as identity authentication and access control. Identity authentication belongs to an important technical content of cloud data security storage. Because mobile terminals calculate their own mobility and resource limitation characteristics, they design a cloud authentication platform, which combines implicit authentication with TrustedCube, implicit authentication is not taken. Traditional biometric information and user stored data are used for user authentication. Based on the mathematical statistical model, the user's legality is used to authenticate the user's legality. TrustedCube belongs to a cloud authentication infrastructure. A variety of different authentication methods had been thought in order to ensure data security, cipher text is often used to store data in the cloud, which greatly increases data access. The user attribute can be used as the public key to associate the ciphertext with the user private key to achieve flexible representation of the access control policy, and the system has good scalability, which is an ideal control scheme.

Secondly, in terms of user privacy protection, mobile cloud computing should not only pay attention to the security of user data, but also need to pay sufficient attention to user privacy protection. In mobile cloud computing, privacy-related personal information, home address, activity location, etc., in terms of protection technology, include the following three types: First, identity privacy protection, multi-user data sharing, public the key encryption mechanism is based on the protection of user privacy and avoids the theft of user identity information in the process of verifying user identity. Second, location privacy protection, such as fuzzy location mechanism which is achieved by reducing the accuracy of user location information to effectively protect the user privacy; third, behavioral privacy protection, using security index and virtual query to protect query privacy.

Finally, mobile terminal security, mobile cloud computing not only faces the security threats which is faced by cloud computing, but also faces problems caused by malicious code and operating system vulnerabilities. The best way to detect and defend against these security threats is to install security software such as online updates, remote locks. The function and operation effect of the current security software are limited by the energy and processing capabilities of the mobile terminal. Based on the cloud-based security software platform, the user only needs to install a lightweight agent, and can use the cloud server to implement virus protection and effectively detect attacks, it is not only complete functions, but also save about 30% energy consumption.

4. Trend of Mobile Cloud Computing

4.1 Service Quality Assurance for Mobile Cloud Computing

On the one hand, efficient and continuous service, the current communication technology transmission rate on the market is as high as 100 megabytes, which can lay a good foundation for users to enjoy higher quality mobile cloud computing services. However, there are still major conflicts and contradictions in the explosive growth of mobile traffic and the availability of broadband resources. With radio technology, an effective enhancement of bandwidth utilization can be achieved. In addition, Cellular network traffic migrates towards the WiFi network, which can also effectively alleviate the problem of air interface resources [5].

On the other hand, the cloud data consistency guarantee and the wireless environment are more complex. By controlling the consistency of the user terminal and the cloud data, the mobile cloud computing service quality can be effectively guaranteed. The terminal data is transmitted by means of redundant backup, etc., and can meet the requirements of data consistency between the cloud and the multi-terminal.

4.2 Mobile Cloud Computing Enhancements

First, to calculate the efficient environment awareness and decision-making in the migration, in the process of popularization and development of mobile cloud computing applications, in order to improve the application of computing migration technology, it is necessary to accurately estimate the local execution and remote execution costs, and collect environmental information and systems, and fill and pay attention to the protection of user data security and privacy.

Secondly, mobile cloud computing services based on motion recognition and precise positioning, such as indoor positioning technology, have significantly improved in terms of positioning accuracy and system performance. From the previous positioning accuracy optimization is moving toward the direction of movement and motion recognition, and is expected to provide users with efficient and convenient and personalized cloud services.

Finally, mobile cloud computing services based on new communication technologies and network architectures are enhanced. For example, 5G communication is almost the same, network access speed is significantly improved, and indoor positioning, micro base stations, etc. can be better supported, and the characteristics of 5G will be improved. And the characters and advantages in the mobile cloud environment have become a major research and development direction.

4.3 Secure and Available Mobile Cloud Services
On the one hand, cloud-based mobile terminal security protection technology, in view of the security threats to current mobile terminals, requires more in-depth research on intrusion detection and virus protection functions based on limited power and processing capabilities. The migration of password mechanisms and security protection functions in the cloud has gradually become a major development trend in the current mobile cloud security field. In addition, the solution to the contradiction between the encryption mechanism and the redundancy conflict needs to be fully considered in the research of mobile cloud computing security.

On the one hand, cloud-based mobile terminal security protection technology, in view of the security threats to current mobile terminals, requires more in-depth research on intrusion detection and virus protection functions based on limited power and processing capabilities. The migration of password mechanisms and security protection functions in the cloud has gradually become a major development trend in the current mobile cloud security field. In addition, the solution to the contradiction between the encryption mechanism and the redundancy conflict needs to be fully considered in the research of mobile cloud computing security.

5. Conclusion
Currently, mobile Internet and wireless data communication applications are becoming more widespread, and people are paying more attention to mobile cloud computing. With the emergence of applications such as smart homes and virtual reality, mobile cloud computing is also facing a severe test in terms of reliability, security and efficiency. In the future development of mobile cloud computing, we must first pay attention to the service quality assurance of mobile cloud computing. Secondly, we must pay attention to the enhancement of mobile cloud computing. Finally, we must pay attention to the security and availability of mobile cloud services. Through these methods, we can better support mobile cloud computing development to meet people's needs in mobile cloud computing.

References
[1] Xie Yang. Analysis of the application of mobile cloud computing in mobile learning [J]. Information Technology and Informatization, 2018 (07): 128-130.
[2] Chen Siying, Guo Yue, Hu Huanbo. Execution Scheduling of Random Task Sequences in Mobile Cloud Computing[J]. Computer Knowledge and Technology, 2018, 14(21): 36-39.
[3] Luo Xiaozhe. Energy-saving countermeasures for mobile cloud computing under 5G network [J]. China New Communications, 2018, 20 (13): 142.
[4] Liu Haitao, Yang Qiong. Research on Security Access Control Technology of Office Resources in Mobile Cloud Service Environment (English) [J]. Machine Tool & Hydraulics, 2018, 46(12): 128-132+170.
[5] Bi Yan. Mobile cloud computing data security issues [J]. Electronic Technology and Software Engineering, 2018 (12): 229.

Review of research on blockchain application development method

Yue Zeng, Yue Zhang*

Shanghai Key Laboratory for Trustworthy Computing, Software/Hardware Co-design Engineering Research Center of MOE, East China Normal University Shanghai, China

51174500152@stu.ecnu.edu.cn, yzhang@sei.ecnu.edu.cn

*Corresponding author: Yue ZHANG, yzhang@sei.ecnu.edu.cn

Abstract. The blockchain technology is complex and involves a wide range of fields, which leads to the lack of uniform specifications for the development of blockchain applications. Although blockchain used to be divided according to the decentralized degree by some organizations, which are difficult to give developers specific guidance and results. To this end, this paper proposed a feature-oriented classification method of blockchain applications based on the analysis and comparison of current typical blockchain applications and frameworks, including digital currency blockchain, development platform blockchain, decentralized application and extended blockchain, helping developer create blockchain applications of a targeted manner and clarifying the functional architecture of different types of above various blockchain applications. Finally, the above classification method has been verified by the analysis of an extended blockchain.

1. Introduction

Blockchain is an innovative application model of computer technology in the era of Internet in distributed storage, point-to-point transmission, consensus mechanism and encryption algorithm[1]. As a distributed "ledger" with decentralized, immutable and transparent features, blockchain has captured world's attention. China Blockchain Industry Conference was held in Beijing with a series of related policies formulated on August 21, 2016[2]. Two months later, the Ministry of Industry and Information Technology released the white paper about blockchain technology in china at the first time[3].

The development of blockchain technology has not only brought blockchain applications into digital currency, but also in the fields of credit, energy, logistics, finance and so forth. As a complex network-based software connector, blockchain has many components and configurations, in which its internal structure will lead to some changes[4]. Up to now, the academia still lacks a unified standard for the classification of blockchain applications, and most of the known blockchain classification methods are based on the level of conceptual or logic, lacking comparison of specific functions. Ingo Weber classified the architecture of the blockchain according to the quality and performance of the blockchain system[5]. Blockchain were divided into two dimensions by Hitoshi Okada based on whether there exists on market incentives and authority control[6]. This paper investigated several mainstream blockchains in the market, and then a classification method based on the functional feature of blockchain was proposed which divided the blockchain applications into digital currency, development platform, decentralized application(DAPP) and extension class.

Content from this work may be used under the terms of the Creative Commons Attribution 3.0 licence. Any further distribution of this work must maintain attribution to the author(s) and the title of the work, journal citation and DOI.
Published under licence by IOP Publishing Ltd

The next of paper is arranged as following: The second section mainly introduces the architecture of the mainstream blockchains. The third section comes up with a feature-oriented classification method for blockchain development. The last section verifies the above classification method with a specific application example.

2. Application architecture of Blockchain

Blockchain applications are widely used for which different features is not the same. The article will introduce architectures of several typical blockchain applications.

2.1. Bitcoin

Bitcoin has the most basic blockchain architecture whose structure is shown in Figure 1. The infrastructure layer is mainly used to implement data storage, transaction transmit and network security. The kernel layer includes POW(Proof of Work) consensus, transaction scripts, token rewards, mining and information verification[7]. They can support efficient transactions, token release and data consistency between nodes. The basic function of Bitcoin is to realize the decentralized transaction of currency, so the scope of application is relatively limited. Also, users can manage their accounts and view transaction information in real time through client wallet.

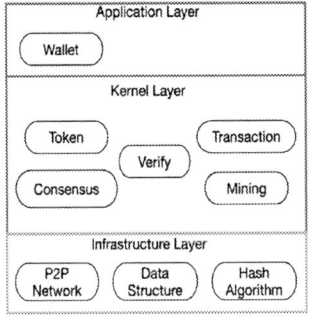

Figure 1. The architecture of Bitcoin

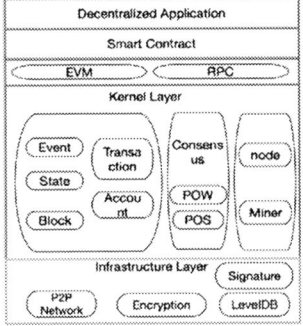

Figure 2. The architecture of Ethereum

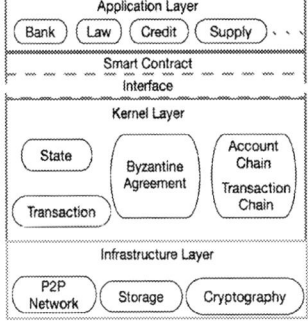

Figure 3. The architecture of Beihang Chain

2.2. Ethereum

The basic layer of Ethereum is similar to Bitcoin, but the core layers are much different. The architecture of Ethereum is shown in Figure 2. The transaction in Ethereum is essentially a process of state transition. The user can trade through the account or trigger the relevant smart contract, then the miner will pack the block to the chain after verifying the transaction information. The event helps the client read the smart contract and store the contract information. Of course, the EVM (Ethereum Virtual Machine) is the core component of Ethereum[8]. Since EVM is Turing-Complete, developers can use high-level languages to compile complex operations such as solidity.

2.3. Beihang Chain

Beihang Chain is the representative of applied blockchain, and the functional architecture of that is shown in Figure 3. The chain aims at scalability, introducing double-chain services based on the common blockchain platform. ABC (account chain) creates account indexes to speed up the query rate, and TBC (transaction chain) aims at chain transactions. The consensus agreement adopts Byzantine voting in which the master node elected by the duplicate node can verify and submit, improving the blockchain transaction rate greatly[9].

The architecture of typical blockchain application has limitations and is relative to specific products. To this end, many organizations began to develop uniform development standards for blockchain applications. The Ministry of Industry and Information Technology issued "The Functional Architecture of Blockchain"[10] in May 2017 (Figure 4). It depicts the abstraction of technical details in blockchain, but it ignores the diversity of blockchain applications and does not describe the characteristics of blockchain application.

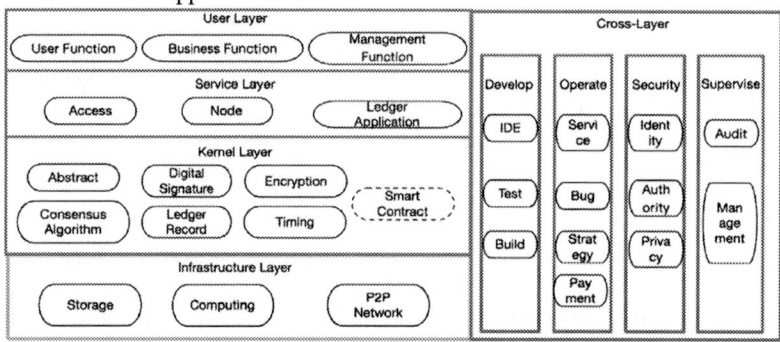

Figure 4. The functional architecture of Blockchain

3. Feature-oriented classification method for blockchain applications

It is essential to understand the type of blockchain in the development of blockchain applications. The mainstream classification idea is to divide the blockchain into public chain, consortium chain and private chain according to the degree of participation of nodes. That mainstream method neglects the connection between different blockchains and its own functional characteristics. Therefore, we propose a feature-oriented classification method for blockchain applications after analyzing the functional characteristics of typical blockchain. The classification principle is shown in Table 1:

Table 1. Classification principle of blockchain.

	Only Involves Upper Interface	Only Currency Function	Smart Contract	Cross-Chain
Digital currency	no	yes	no	no
Development platform	no	no	yes	no
Decentralized application	yes	no	yes	no
Extended blockchain	no	no	yes	yes

3.1. Digital currency blockchain

Digital currency blockchain aims at creating a digital currency that can achieve value storage, transfer, and liquidation. The best-known blockchain representative is Bitcoin. According to the functional characteristics of the digital currency blockchain, its structure can be summarized as shown in Figure 5.

The infrastructure layer includes block structure, hash algorithms and P2P Network. The block data is used to store the users' transaction information, and the hash algorithms can solve anonymity and the lightweight storage of information. P2P Network ensures the distributed characteristics of the system. The kernel layer includes consensus algorithms, encryption, digital signature, timestamps and script. Among them, the guarantee of data consistency is provided by consensus algorithms, and digital signature make transaction more secure and reliable. In addition, time stamp record when the block is created, and the script defines the rule of transaction. The application layer includes trading websites, wallets and API, which can provide user-friendly interface.

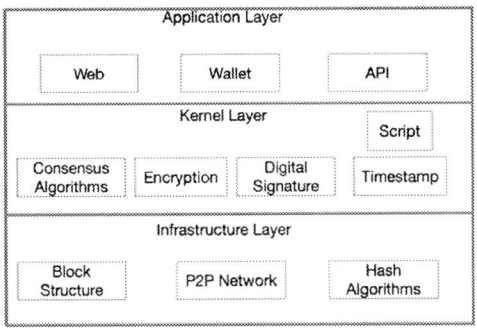

Figure 5. The architecture of digital currency blockchain

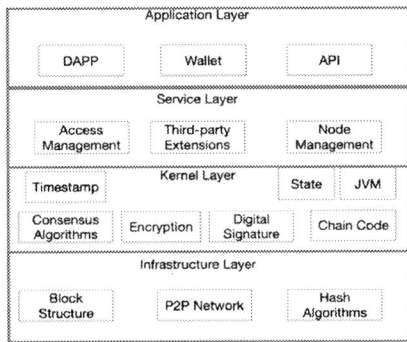

Figure 6. The architecture of development platform blockchain

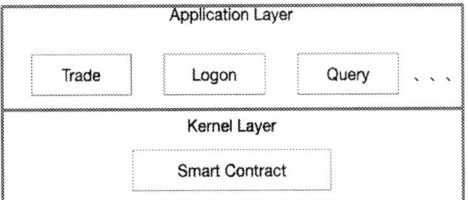

Figure 7. The architecture of decentralized application

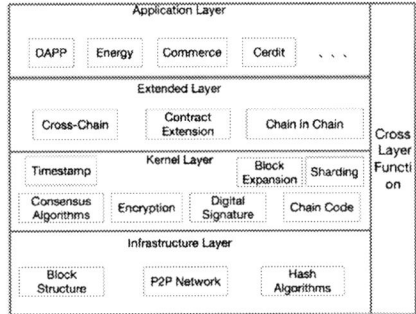

Figure 8. The architecture of extended application

3.2. Development platform blockchain

The development platform blockchain mainly provides developers with a platform to develop decentralized applications rapidly. Its application architecture is shown in Figure 6. The infrastructure layer is similar to the digital currency blockchain, providing core technical support for the upper layer. The kernel layer adds smart contracts and virtual machine technology. Among them, smart contracts can accept and store the value, and virtual machines are the basic components on which high-level languages written by developers need to be compiled and executed. The service layer helps the blockchain platform to provide management services, including user node access, third-party extension development and node authority management. The application layer focus on the interaction with developers such as DAPP.

3.3. Decentralized application

Decentralized application (DAPP) is a distributed application running on a blockchain platform. It has the following characteristics: running on a distributed network, storing information safely, and greatly protecting privacy. Similar to mobile apps, decentralized applications usually contain a user-friendly interface. The difference is that they are decentralized, and data is immutable. The functional structure of the decentralized application is shown in Figure 7. The application layer only provides user registration, transaction, query and other normal function. The core layer consists of smart contracts written by developers.

3.4. Extended blockchain

There are remarkable functions in extended blockchain of which the core layer incorporates many supererogatory technologies including sharding network, multi-channel technologies and block expansion, helping to greatly improve transaction speed and throughput. For example, the extended blockchain ZEPPELIN used sharding techniques to increase the throughput of blockchain up to million-second transaction speeds[11].The extended layer introduces new technologies such as cross-chain technology, contract extension, and chain-chain to improve system performance and help implement the practical of blockchain. For example, Plasma autonomously expands smart contracts to improve system robustness[12]. Fabric Token adds multi-chain structure to support high-performance smart contract interaction[13]. The architecture of extended application is shown as Figure 8.

4. Analysis of blockchain application

This paper takes the P2P trading system on the Power Ledger platform as an example to analyze how to develop blockchain applications.

First, we have analysis of the P2P trading system to reveals three roles in the power trading system: consumers, producers and holders. Among them, consumers corresponding with individual users purely need electric energy, and producers produce renewable energy (such as solar energy) by themselves and can trade electrical energy with other consumers through blockchain technology. The holders are responsible for purchasing the POWR token and redeem it to a Spakz token for distributing[14].

According to the user role, we can initially summarize the functions of the system including as follows:

- Rely on smart contracts to control user transactions and currency exchange.
- Provide smart meters to protect users' power information at any time.
- Double-chain structure for cross-regional token clearing.
- Control node permissions to implement user access management.

Based on our analysis, the extended blockchain is selected as the architecture prototype. The functional structure of Power Ledger as shown in Figure 9.

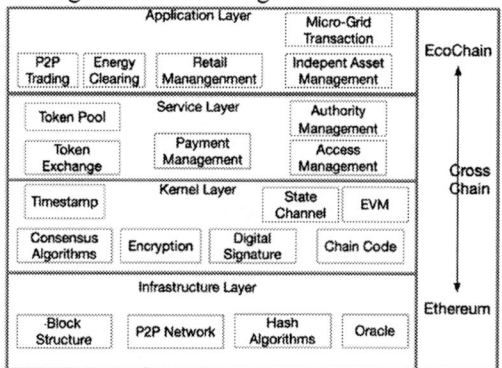

Figure 9. The structure of Power Ledger

The infrastructure layer design is similar to Ethereum, and the difference is that the system uses oracle to store data for data management in order to effectively store transaction history information. The kernel layer uses POS as consensus protocol to minimize the consumption of internal resource. In addition, Power Ledger uses the self-developed EcoChainTM (EcoChain) as the upper blockchain to form a double-chain structure.

The service layer is mainly provided by EcoChain Service and mainly includes the following components: token exchanging, smart bond, token hosting pool and node authority management. The token exchanging provides exchange of Sparkz tokens and POWR tokens. Sparkz is a medium for maintaining exchange rates between tokens and local currencies. Smart Bonds ensure that transactions between users are carried out efficiently, otherwise the POWR tokens are blocked. The token hosting pool hosts the tokens of the power beneficiaries and provides the user with token rewards.

5. Summary
Starting from the technical characteristics of current mainstream blockchain applications such as Bitcoin, Ethereum and Beihang Chain, this paper discusses the functional features and architectural components of different types of blockchain applications. Based on that, a feature-oriented classification method for blockchain applications is proposed. This paper innovatively divides the blockchain application into digital currency blockchain, development platform blockchain, decentralized application and extended blockchain, giving the functional architecture of the corresponding type blockchain application. At last, combined with the example of Power Ledger, we validate the effectiveness of our classification method.

Acknowledgments
This paper is finished under the support of Key Program of the National Nature Science Foundation of China(No.61532019, No.61772034); Shanghai Committee of Science and Technology, China (No.15511104700, No.16DZ1100600); National Defense Basic Scientific Research Program of China (JCKY2016212B004-2); Open Project Fund of Shanghai Key Lab for Trustworthy Computing (No.07dz22304201607); Natural Science Foundation of Anhui Province(1808085MF172).

References
[1]   Swan M. Blockchain: Blueprint for a New Economy[M]. USA: O'Reilly Media In, 2015:12-14.
[2]   Zhu Jianming And Fu Yonggui. Progress in the application of blockchain [N]. Science and Technology Review, 2017.35 (13): 71-73
[3]   White Paper on China's Blockchain Technology and Application Development (2016). China Ministry of Industry and Information Technology. 2016.10
[4]   C. Pautasso, L.Zhu AndV. Gramoli, et al. The blockchain as a software connector[C]. 2016 13th Working IEEE/IFIP Conference on Software Architecture (WICSA). Venice: IEEE Press, 2016: 182-191
[5]   Xiwei XU, Ingo WEBER And Mark STAPLES. A Taxonomy of Blockchain-Based Systems for Architecture Design[C]. 2017 IEEE International Conference on Software Architecture (ICSA). Gothenburg: IEEE Press, 2017:243-251
[6]   Hitoshi Okada, et al. Proposed Classification of Blockchains Based on Authority and Incentive Dimensions[C]. 2017 19th International Conference on Advanced Communication Technology(ICACT). Phoenix Park: ICACT-TACT 2017:593-597
[7]   https://bitcoin.org/en/developer-guide
[8]   Dr. GAVIN Wood. Ethereum: A decentralised transaction ledger[C]. 2014:1-10
[9]   Cai Weide, Yu Lian, Wang Rong et al. Research on Application System Development Method Based on Blockchain[J]. Journal of Software, 2017(6): 1474-1487
[10]   Blockchain Reference Architecture (release) [S]. China Blockchain Technology and Industry Development Forum Standard. 2017. 5
[11]   The Zilliqa Team. The Zilliqa Project: A Secure, Scalable Blockchain Platform. 2018. 5.
[12]   Joseph Poon, Vitalik Buterin. Plasma: Scalable Autonomous Smart Contracts[R]. 2017: 1-8
[13]   NIKOLAY Nikov, MARIN Ivanov, DONCHO Karaivanov, et al. The Fabric Token (FT) Ecosystem High-Level Development & Management of Smart Contracts[R]. 2017: 18-22
[14]   Power Ledger Platform Token Interactions[OL]. https://powerledger.io/media/Power-Ledger-Platform-Token-Interactions.pdf

ISPECE   IOP Publishing

Research on the Random Corresponding of Privacy Data Mining in the Association Rules of Cloud Computing

Baofeng Hui, Guoqing Jia, Shanji Chen

Qinghai University Nationalities, Xining Qinghai, 810007, China

Abstract: To strengthen data privacy protection, and improve data mining accuracy, random response mode is adopted to design privacy protection mining method based on association rules. Granular computing method and technology are applied to mining fields of association rules in data mining, and mining to association rules is researched in more extensive way from another perspective in this paper. Firstly, partial concealing mode is adopted to conceal and transform original privacy data and improve data security; secondly, associated frequent item set is utilized to construct simple and efficient privacy protection mining algorithm; finally, algorithm proposed is verified to have higher privacy and accuracy through theoretical analysis and experimental verification. After classical association rules mining algorithm is analyzed and researched in detail with its characteristics and restrictions summarized through examples in this paper, association rules mining model based on granular computing is proposed, which makes theoretical preparation for proposal and construction of association rules pick-up algorithm based on granular computing. Experimental result shows that association rules mining method based on granular computing is feasible and effective.

1. Introduction

In recent years, data mining technology has been widely concerned by information industry circle, which is the inexorable outcome of paradoxical movement between rapidly increasing data quantity and increasingly poor information amount. Systematic and intensive research to data mining technology is objective requirement of global information-based development. Data mining technology includes many research fields, of which association rule is an important research direction, having vitally important application value in business decision. This topic mainly makes related research to association rules mining. Traditional association rules mining algorithm, such as Apriori algorithm and its improved algorithm etc., makes mining to certain and accurate concept, and it is difficult to mine non-accurate or blurry concept. Through experiment, it can be found that main calculation to search frequent item set lies in frequent 2-item set generation, and frequent 2-item set generation process is Apriori algorithm mining bottleneck, and therefore, a kind of new association rules mining algorithm based on fuzzy sets is proposed in this paper, fuzzy set theory and semantic association rule concept are introduced in the algorithm, reasonable and non-accurate semantic translation is made to numerical attribute of database, and algorithm efficiency is improved by improving size of item set of pruning part scanned, which avoids exponential growth tendency of length of set scanned. Because core problem of Apriori algorithm is to find the maximum item set, the process to find the maximum item set is global search process, and genetic algorithm is a kind of global optimization algorithm, and avoids local optimization in search process. Therefore, truly useful rules can be found by applying genetic algorithm to rule finding and extraction. Therefore, a kind of association rules mining algorithm based on genetic algorithm is proposed in this paper, which mainly

Content from this work may be used under the terms of the Creative Commons Attribution 3.0 licence. Any further distribution of this work must maintain attribution to the author(s) and the title of the work, journal citation and DOI.
Published under licence by IOP Publishing Ltd

mines quantitative association rules, and algorithm mainly includes association rules coding method design, fitness function construction and genetic operator improvement etc. According to 2 kinds of association rules mining algorithm that are proposed and designed in this paper and that are based on computational intelligence, we extract association rules respectively by taking medical database and student database as mining prototype, and make experimental analysis; experimental result verifies effectiveness of 2 kinds of algorithm and also illustrates wide application prospect of association rules mining.

2. PFP algorithm

PFP algorithm is a kind of parallel FP-Growth algorithm that was proposed by Li et al. of Google Beijing Research Institute in 2008 and that is based on MapReduce frame. Because of high expansibility and high fault tolerance of MapReduce frame, algorithm can process big data in relatively good way.

Basic thought of algorithm is to transform transaction database into a new "intra-group dependency transaction" database and distribute it to corresponding Reducer, and in the process of recursive construction of Reducer to FP-Tree, local FP-Tree generated through different "intra-group dependency transaction" is mutually independent.

3. PFP-P algorithm

Basic starting point of PFP-P algorithm is to replace mining to all frequent item sets with mining to closed frequent item set. Different from mining to all frequent item sets, mining to closed frequent item set does require "intra-group dependency transaction" of all items in transaction data, but "intra-group dependency transaction" of partial suffix items to reduce data transmission quantity. In addition, another advantage of mining to closed frequent item set is that mining result data can be significantly reduced under the premise that information completeness is guaranteed, which is convenient for storage and further processing.

3.1 Suffix item list

For the convenience of discussion, this paper maps transaction data as transaction mode, which means that transaction data ranked according to sequence of items in F-List after non-frequent items in transaction data are deleted is expressed as T; item in transaction T is expressed as I; support degree of mode T and submode is expressed as sup(), and the minimum support degree meeting frequent mode is expressed as sup_min.

Definition 2.1 Closed frequent mode. Assumed that submode $X=I_1I_2...I_k$, $sup(X)=u \geqslant sup\_min$; if $\forall I_p$, where $p \geqslant k+1$, $sup(I_1I_2...I_k...I_p)<u$, then X is called as closed frequent mode of item I_k, and I_k is called as suffix of X. Closed frequent mode corresponds to closed frequent item set one by one.

Definition 2.2 Suffix item. For submode $X=I_1I_2...I_k$ of T, if $sup(I_1)= sup(I_2)=...= sup(I_k)=u$ is met, then:

Definition 2.2.1 If $\forall I_p$, where $p \geqslant k+1$ and $sup(I_p)<u$, then I_k is called as u support degree suffix item of T.

Definition 2.2.2 For any item I_i and any suffix item I in X ($I_i \neq I$), if $sup(I_iI)<u$, then I_i is called as u support degree suffix item of T.

Theorem 2.1 Suffix of closed frequent mode must be suffix item.

Prove. Prove through proof by contradiction. Assumed that submode $X=I_1I_2...I_k$ of T is closed frequent mode, $sup(X)=u$ and I_k is not suffix item, then according to definition 2.2.1, item I_p must exist in T, where $p \geqslant k+1$ and $sup(I_p) \geqslant u$, because support degree of item in T is monotonous and does not increase, $sup(I_p) =u$; according to definition 2.2.2, $sup(I_kI_p)=u$; therefore, a submode $X'=I_1I_2...I_kI_p$ must exist in T, to make $sup(X')=u$; according to definition 2.1, X is not closed frequent mode, which conflicts with assumption, so original conclusion is verified.

According to theorem 2.1, when Mapper distributes data to Reducer, it just need to distribute data to suffix item to obtain enough information to construct closed frequent item set, and it does not need to distribute data to non-suffix item. List consisting of all suffix items in transaction data is called as suffix item list, and it is a subset of F-List.

3.2 PFP-P algorithm analysis

This section analyzes data communication complexity of PFP-P. Quantity of data transmitted in step 1 of algorithm is the same with that of PFP, being M_{DB}. Quantity of data needing to be transmitted in step 2 to construct P-List is also M_{DB}.

Data transmission quantity of step 3 is:

$$M_{Dup} = M_{DB} + \sum_{i=1}^{n}\sum_{j=1}^{l_i-1} jf(I_j I_{l_i})g(I_j) \tag{1}$$

Where function g(I) is defined as follows:

$$g(I) = \begin{cases} 0 & I \text{ is non-suffix item.} \\ 1 & I \text{ is suffix item.} \end{cases} \tag{2}$$

Add data transmission quantity of 3 steps together to obtain data transmission quantity of algorithm in the whole process:

$$M = M_{DB} + M_{Dup} = 3M_{DB} + \sum_{i=1}^{n}\sum_{j=1}^{l_i-1} jf(I_j I_{l_i})g(I_j) \tag{3}$$

For the convenience of discussion, assumed that average length of transaction data is l, the maximum value of grouping is obtained, and average total data transmission quantity of PFP and PFP-P can be obtained according to formula 1-4 and 2-3:

$$\overline{M_1} = (3/2)nl + (1/2)nl^2 \tag{4}$$

$$\overline{M_2} = 3nl + n\sum_{j=1}^{l-1} jg(I_j) \tag{5}$$

According to formula 2-4 and formula 2-5:

$$\overline{M_2} - \overline{M_1} = n\sum_{j=1}^{l-1} jg(I_j) - n(1/2)l(1-l) + nl \tag{6}$$

When all g(I) is 1, which means that all items of transaction data are suffix items, value of formula 2-6 is nl of the third item, which means that PFP-P transmits one more M_{DB} data than PFP. When g(I) is item of 0, which means that sum of coefficient j of non-suffix item is greater than average length l of transaction, data transmission quantity of PFP-P will be lower than that of PFP. When average length l increases, effect of nl of the third item in formula 2-6 on total transmission quantity will decrease rapidly, and what plays a decisive role will be former 2 items in the formula. When l is relatively great, only few non-suffix items are required to make sum of coefficient exceed 1.

According to suffix item list construction algorithm, assumed that the number of item meeting support degree sup is k_{sup}, then probability of non-suffix item in item with support degree being sup can be expressed as:

$$P(\exists I \notin PList) = Min\left(1, \frac{k_{sup} - 1}{n(n-1)(n-2)...(n-sup+1)}\right) \tag{7}$$

Assumed that the minimum support degree of frequent item in transaction data is minsup, and the maximum support degree of item is maxsup, then

$$M_{DB} = \sum_{i=1}^{n} l_i = \sum_{i=\min sup}^{\max sup} i k_i \tag{8}$$

Assumed that n and support degree of each item are kept unchanged in transaction data, according to formula 2-8, k will increase with increase of l; according to formula 2-7, probability that item in transaction data is non-suffix item also increases with increase of it. Therefore, for data with relatively great average transaction length, PFP-P can reduce average transmission quantity of data effectively.

Taking data $\{T1=(a_1,a_2,...a_{50}), T2=(a_1,a_2,...a_{100})\}$ as example, assumed that the minimum support degree threshold is 1 with the maximum grouping adopted, then in PFP, data shall be divided into 100 groups, and total transmission quantity of data is 150*(3/2)+(1/2)*(50*50+100*100)=6475. But in

PFP-P, a_{50} is suffix item of support degree 2, while a_{100} is suffix item of support degree 1, and other items are non-suffix items with division of 2 groups (a_{50} and a_{100}), and total transmission quantity of data is 150*3+49+99=589.

4. Conclusion

Based on PFP algorithm, a kind of parallel closed frequent item set mining algorithm PFP-P based on suffix item list is proposed in this paper. This algorithm replaces mining to all frequent items in original algorithm with mining to closed frequent item set to improve mining efficiency; aimed at features of closed frequent item set, suffix item list is introduced in mining process to reduce transmission quantity of grouped data in mining process, and lower internal consumption of system. Experiment shows that the algorithm is superior to original algorithm in average performance, and in decrease of the minimum support degree threshold, it can shorten mining time effectively; the algorithm can lower consumption of communication between nodes effectively with good speedup quality; compared with processing to low-dimension dataset, the algorithm has more advantages in processing high-dimension dataset, and therefore, the algorithm is more applicable to mining task of massive high-dimension data.

Acknowledgement

This work was supported by the International Science & Technology Cooperation Project of Qinghai (2013-H-811, 2014-HZ-821) and Application Basic Research Project of Qinghai(2015-ZJ-721).

Reference

[1] Saad A. Abdelhameed, Sherin M. Moussa, Mohamed E. Khalifa. Restricted Sensitive Attributes-based Sequential Anonymization (RSA-SA) approach for privacy-preserving data stream publishing[J]. Knowledge-Based Systems,2018.

[2] Lou Ann Scarton, Liqin Wang, Halil Kilicoglu, Margaret Jahries, Guilherme Del Fiol. Expanding vocabularies for complementary and alternative medicine therapies[J]. International Journal of Medical Informatics,2018.

[3] Rouzbeh Razavi, Amin Gharipour, Martin Fleury, Ikpe Justice Akpan. Occupancy detection of residential buildings using smart meter data: A large-scale study[J]. Energy & Buildings,2019,183.

[4] Peter J. Carew, Qixin Lu, Larry Stapleton. Perceived Risk Factors for Cloud-Based Data Storage and Control Systems: A Cross-Cultural Comparative Study of Irish and Chinese Companies[J]. IFAC PapersOnLine,2018,51(30).

[5] Wafa Qadadeh, Sherief Abdallah. Customers Segmentation in the Insurance Company (TIC) Dataset[J]. Procedia Computer Science,2018,144.

[6] Md. Zahid Hasan, K.M. Zubair Hasan, Abdus Sattar. Burst Header Packet Flood Detection in Optical Burst Switching Network Using Deep Learning Model[J]. Procedia Computer Science,2018,143.

ISPECE   IOP Publishing

Discussion on Supplier Selection in the Selection of Large Civil Passenger Aircraft

Aiping Yin[a], Zhangkang[a], Zhenxing Gao[b], Duoduo Zhuang[a]

[a] Shanghai Aircraft Design and Research Institute,

No. 5188 JinKe Road, PuDong New District, Shanghai 201210, China

[b] *Nanjing* University of Aeronautics and Astronautics,

No. 29 JiangJun Road, JiangNing District, Nanjing   211106, China

yinaiping@comac.cc

Abstract: At present, the competition in the air transport industry is becoming increasingly fierce. It is increasingly being paid attention to by airlines to explore cost reduction and efficiency gains in aircraft procurement and selection. Taking the comprehensive and scientific evaluation of suppliers as the starting point, the analytic hierarchy process was used to study the choice of suppliers in aircraft selection. Taking the aero-engine as an example, the method of assigning different weights to different evaluators is used to compare the indicators with the principle of analytic hierarchy process, so as to obtain the weight distribution value of each evaluation index, that is, the complete evaluation index system. In the scoring process, in order to avoid the impact of different scoring personnel on the final scoring results due to differences in professional knowledge and work experience, the scores of different categories of personnel were weighted and averaged differently.

1. INTRODUCTION
Aircraft selection refers to the process of determining the use requirements of the aircraft purchased and reasonably selecting according to the airworthiness department or the manufacturer's technical specifications to meet the needs of the airline. Aircraft selection is a complex system engineering involving disciplines and aircraft parameters, engine performance, communication and navigation technology, flight dynamics, meteorology, art design, financial management, and corporate strategic planning. Since the price of the aircraft is as high as 100 million US dollars and the service life is about 20 years, whether the aircraft is suitable for selection, economical and reasonable choice of technical parameters will directly affect the development of the airline in the next 20 years. profit. This shows that aircraft selection is an extremely important task for airlines. Supplier selection is an important part of the aircraft selection program. This article takes the aero engine as an example to study the choice of suppliers in aircraft selection. In order to find scientific and reasonable secondary indicators, this paper uses brainstorming and causal analysis to find and analyze the supplier evaluation index system that needs to be considered in aircraft selection. In the specific implementation process, we organized the procurement department, the engineering department, and the leadership decision-making level to participate in brainstorming, and applied the causal analysis method to organize the items that everyone discussed and focused into a complete system. The steps of the analytic hierarchy process are shown in Figure 1.

Content from this work may be used under the terms of the Creative Commons Attribution 3.0 licence. Any further distribution of this work must maintain attribution to the author(s) and the title of the work, journal citation and DOI.
Published under licence by IOP Publishing Ltd

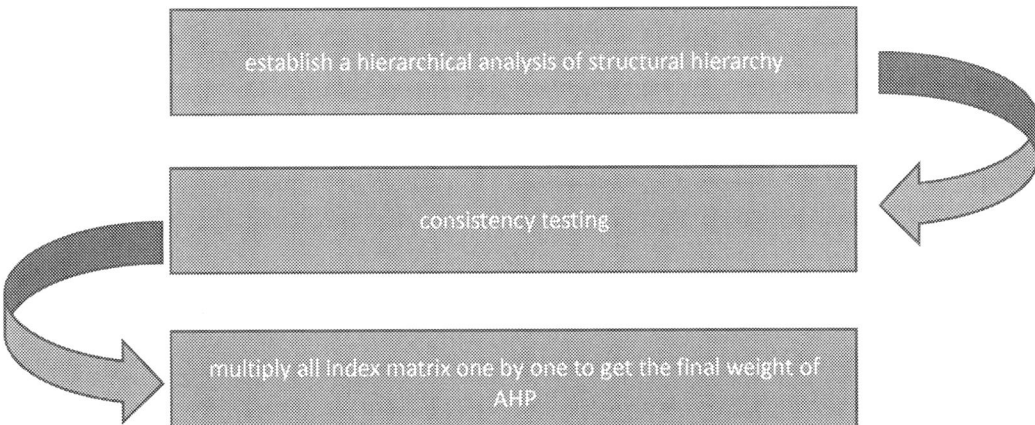

Figure 1:   Steps of the analytic hierarchy process

2. ENGINE SUPPLIER EVALUATION INDEX WEIGHT

Analytic Hierarchy Process (AHP) is a multi-objective decision analysis method that combines elements related to task decision-making into goals, criteria, and programs. Based on this, qualitative and quantitative analysis is combined. This method is a hierarchical weight decision analysis method proposed by American operations researcher Pittsburgh University professor T. L. Satty in the early 1970s. It is characterized by the in-depth analysis of the influencing factors and intrinsic relationships of complex problems that require decision-making, using a small amount of quantitative data to mathematically make the decision-making process, thus making complex decisions with multiple criteria and no obvious structural characteristics. The problem provides an easy way to make decisions. Analytic Hierarchy Process (AHP) is an effective method for multi-objective decision-making. The basic operation steps are as follows: firstly establish a hierarchical target and sub-objective system framework, then the experts will compare the objectives to form a comparison matrix, and finally conduct a consistency test on the comparison matrix. The weight vector is obtained, and the final decision value is calculated from the product of the weight vector and the target assignment vector. The decision-making problem of civil engine optimization selection can be attributed to the scope of operations research. At present, fuzzy judgment method, cluster analysis method and analytic hierarchy process are used at home and abroad for calculation decision. Among them, the analytic hierarchy process has the smallest error, and it is the most widely used in large-scale decision-making problems because of its practicability and effectiveness in dealing with complex decision-making problems. The biggest advantage is that it can deal with the combination of qualitative and quantitative problems, and can introduce the decision maker's subjective judgment and policy experience into the model and quantify it. AHP is used to comprehensively evaluate candidate civil aircraft engine suppliers[1].

The noise and emissions of the engine directly affect the airworthiness and comfort of the passenger aircraft. The fuel consumption and reliability of the engine greatly affect the economy and safety of the aircraft. In the early stage of civil passenger aircraft design, the evaluation of the candidate civil passenger aircraft engine must be based on a set of scientific and credible evaluation selection methods. The technical selection of civil passenger aircraft engines is a complex system engineering involving disciplines and contents such as aircraft performance, engine performance, aerodynamic drag, quality dimensions, safety, environmental protection, and supplier capability experience. And because the price of civilian passenger aircraft is as high as 100 million US dollars, the service life is about 20 a. Therefore, whether the selection of civil engine technical parameters is reasonable and whether the selection is appropriate will directly affect the development and profitability of the airline in the next 20 years. This paper makes use of the combination of qualitative and quantitative methods of analytic hierarchy process to conduct comprehensive and objective evaluation and technical selection of candidate aviation civil engines, and considers that the engine needs to meet various requirements of aircraft requirements, and establishes a set of technical evaluation index system. The weighting factors of various influencing factors are obtained. Through the evaluation and analysis of the various indicators affecting the civil engine, the candidate engine is evaluated comprehensively, and the most suitable engine that meets the requirements of the civil passenger aircraft is selected. It is of great practical significance to provide a practical and feasible quantitative comprehensive evaluation method for the overall comprehensive evaluation of candidate civil engines[2].

Taking the first-level indicator in the indicator system as an example, the weight values of the six first-level evaluation indicators such as cost C1, quality C2, use characteristic C3, after-sales service C4, supply capacity C5, and market share C6 are analyzed. When comparing the weights of these six primary evaluation indicators in decision-making, we use the 1-9 score scale method. "9" means very important, and "1" means equal importance. In order to judge the importance of the first-level indicators, we invited the purchasing manager P1, the engineer P2 and the company management personnel P3 to evaluate the above indicators, and assign different weights to the evaluation results according to different division of labor. The purchasing manager P1 gives 0.3, the engineer P2 gives 0.2, and the company manager P3 gives 0.5. After the expert evaluation, the final first-level indicator weight vector can be obtained by using the corresponding weight weighting[3]. The weight of each relevant factor is shown in Table 1.

Table 1: Training impact weights

Index	Purchasing Manager	Engineer	Company Manager
Weight	0.3	0.2	0.5

For the optimization model of civil passenger aircraft engine, the civil engine selection and evaluation problem should be organized and hierarchical, and a structural model of hierarchical analysis should be constructed. In the analysis model, the complex problem of the selection is decomposed into various components (called elements), and then the elements are divided into groups according to the attributes. Using the same group of elements as a criterion, the elements within the group are dominant, and at the same time they are dominated by the target layer of the same level. The highest level is the target layer, which is the predetermined goal that needs to solve the problem. The middle layer groups are the criteria involved in and the objectives to be achieved. The bottom layer is the various measures and technical indicators that may be adopted to achieve the goal. The analytic hierarchy process is used to establish an engine selection model consisting of four levels: the first level target layer is the final preferred civil passenger engine; the second level criterion layer is the evaluation aspect of the engine selection (criteria); the third layer is under each criterion layer. Corresponding specific engine index parameters; the fourth floor plan layer is an alternative civil engine. The judgment matrix structure, the weight coefficient calculation, the judgment matrix consistency test, the weight calculation of the target layer, and the like are conventional methods, and are not described in detail herein[4].

3. ENGINE SUPPLIER EVALUATION
When evaluating quantitative indicators, it is only necessary to standardize the corresponding assignments. However, when evaluating qualitative indicators, these qualitative indicators are somewhat biased towards engineering services, and some are more focused on procurement. Therefore, in the actual score, it is necessary to evaluate the evaluation according to different evaluators. The indicator types are given different weights. The civil passenger engine is an extremely complicated

system, which has many influencing factors on the passenger aircraft. Therefore, the objective technical evaluation of the candidate civil engine is a difficult and complicated task, and the selection of the passenger aircraft engine directly affects the operation of the civil passenger aircraft. Performance and R&D costs. This paper introduces the analytic hierarchy process into the technical evaluation and selection of candidate engines for civil passenger aircraft, establishes its comprehensive evaluation index system and selection evaluation method, and establishes a selection evaluation model to realize an effective, scientific and objective comprehensive evaluation of candidate engines. select. From the analysis and evaluation results of the above application examples, the method is feasible and in line with objective reality. The various influencing factors of the candidate engine are expressed in numerical form, and the contrast effect is intuitive, which helps to select a suitable civil passenger engine, improve the competitiveness of the passenger aircraft in the future market, and provides a certain theory for the comprehensive evaluation of civil engines. Basis has very important practical significance[5].

4. CONCLUSION
In summary, the analytic hierarchy process used in this study can make the supplier selection in aircraft selection more economical, comprehensive and rapid. The specific results have the following two points: comprehensive consideration of various departments' evaluation of different aspects of suppliers, optimization of supplier selection results, making the final selection results more scientific, comprehensive and accurate; shortening the choice of suppliers in aircraft selection the cycle ensures the progress of the project.

References
[1] Naseem Ahmadpour,Jean-Marc Robert,Gitte Lindgaard. Aircraft passenger comfort experience: Underlying factors and differentiation from discomfort[J]. Applied Ergonomics,2016,52.
[2] Katrin Kölker,Peter Bießlich,Klaus Lütjens. From passenger growth to aircraft movements[J]. Journal of Air Transport Management,2016,56.
[3] Naseem Ahmadpour,Jean-Marc Robert,Gitte Lindgaard. Aircraft passenger comfort experience: Underlying factors and differentiation from discomfort[J]. Applied Ergonomics,2016,52.
[4] Zhaosong Fang,Hong Liu,Baizhan Li,Andrew Baldwin,Jian Wang,Kechao Xia. Experimental investigation of personal air supply nozzle use in aircraft cabins[J]. Applied Ergonomics,2015,47.
[5] Kong Xiangjun,Le Ningning,Li Chunsheng,Xiao Xianbo,Lin Bo,Xu Bing. Landing Gear Ground Load Measurement and Verification Test for a Large Passenger Jet[J]. Procedia Engineering,2015,99.

Analysis on Error Compensation for Integrated Navigation based on forgotten Kalman Filter

Ling Yao[1*], Hua Zhang[1] and Xiaoping Huang[1]

Anhui Xinhua University, Hefei, 230088, China

*yaoling214@163.com

Abstract. Aiming at the error caused by accelerometer bias and gyro drift in SINS/GPS integrated navigation system, an error compensation algorithm based on forgetting Kalman filter is proposed. This algorithm can not only effectively estimate the constant error of inertial devices, but also compensate the error to obtain the optimal estimation output of integrated navigation system. Experiments show that for integrated navigation system the algorithm can achieve relative ideal navigation accuracy, verify the correctness of the integrated system and is superior to the subsystem in reliability. It has a very good application value.

1. Introduction

With the development of computer technology and modern control technology, various integrated navigation systems are widely used in aviation, aerospace, navigation and other fields. Strap-down Inertial Navigation System (SINS) and Global Position System (GPS) are the most commonly used integrated navigation systems. SINS/GPS integrated navigation system realizes the complementary advantages of the two independent navigation systems and improves the reliability as well as the adaptability of the system. Adaptive Kalman Filter (KF) is usually used in integrated navigation system. According to the dynamic characteristics of SINS/GPS integrated system, KF outputs the optimal error estimation of system state variables, compensates the system error so as to improve the accuracy of integrated navigation system.

2. Mathematical Model of SINS/GPS Integrated Navigation

Completely SINS relies on the motion carrier equipment to complete the navigation task, without the limitation of meteorological conditions, and can provide relatively complete navigation parameters in a short period. However, the accuracy of SINS system mainly depends on inertial measurement elements. The errors of navigation parameters accumulate with time. It can not meet the requirements of long-distance, high-precision navigation and rapid response. GPS can provide high-precision three-dimensional position, velocity and time information in real time and continuously all over the world without accumulating systematic errors. However, GPS system is easy to be vulnerable to external environmental impact, limited bandwidth, difficult to extract carrier attitude information and low data update rate. SINS/GPS integrated navigation system integrates the advantages of each subsystem, improves the system performance and environmental adaptability.

SINS/GPS integrated navigation system is divided into tight combination and loose combination. The tight combination implementation model is widely used since it is simple and accurate. The system establishes state equation with indirect method and the state equation is built with 15-dimensional state parameters. The Northeast Sky (ENU), i.e. the geographical coordinate system, is

Content from this work may be used under the terms of the Creative Commons Attribution 3.0 licence. Any further distribution of this work must maintain attribution to the author(s) and the title of the work, journal citation and DOI.
Published under licence by IOP Publishing Ltd

selected as the navigation coordinate system. Its state space is described as:

$$\begin{cases} X = F\dot{X} + GW^b \\ Z = HX + V \end{cases} \tag{1}$$

Where:

$$X = [\Phi \ \delta V \ \delta P \ \varepsilon \ \nabla] \tag{2}$$

Φ represents three misalignment angles in the northeastern sky direction, δV velocity error, δP position error, ε denotes gyro drift, ∇ accelerometer bias.

3.Adaptive Kalman Filter

3.1 Standard Kalman Filterin

Error Compensation is an important part of integrated navigation system. The accuracy and speed of Error Compensation determine the performance of the system. Kalman Filter (KF), based on prediction and correction, is the most commonly used optimal estimation algorithm in parameter processing of integrated navigation system. KF uses state variables to establish statistical mathematical models to describe the dynamic characteristics of the system, so as to estimate and compensate the attitude parameters and related error sources in real time.

The standard KF estimates the optimal state of the system by combining the system state equation with the measurement equation. It can be divided into two processes: prior prediction and posterior correction. It is assumed that the state equation and measurement equation of the discrete linear system are as follows：

$$\begin{cases} X_k = \Phi_{k/k-1}X_{k/k-1} + \Gamma_{k/k-1}W_{k-1} \\ Z_k = H_k X_k + V_k \end{cases} \tag{3}$$

where X_k is the state vector and Z_k is the measurement vector. W_{k-1} is the system noise vector and V_k is the measurement noise vector. The two are unrelated zero-mean Gauss white noise vector series with variances of Q_k and R_k, respectively.$\Phi_{k/k-1}$, $\Gamma_{k/k-1}$ and H_k are known structural parameters of the system.

KF error estimation is divided into five steps as follows.

One-step state prediction.

$$\hat{X}_{k/k-1} = \Phi_{k/k-1}\hat{X}_{k-1} \tag{4}$$

Mean square error of one-step stateprediction.

$$P_{k/k-1} = \Phi_{k/k-1}P_{k-1}\Phi_{k/k-1}^T + \Gamma_{k-1}Q_{k-1}\Gamma_{k/k-1}^T \tag{5}$$

Filtering gain.

$$K_k = P_{k/k-1}H_k^T(H_k P_{k/k-1}H_k^T + R_k)^{-1} \tag{6}$$

State estimation.

$$\hat{X}_k = \hat{X}_{k/k-1} + K_k(Z_k - H_k\hat{X}_{k/k-1}) \tag{7}$$

Mean square error of state estimation.

$$P_k = (I - K_k H_k)P_{k/k-1} \tag{8}$$

SINS/GPS integrated navigation system has abrupt noise change and poor system observability, which affects the error estimation performance of KF and even leads to filtering divergence. Adaptive Kalman Filter (AKF) is usually used. AKF uses the results of the filter parameters obtained at the previous time point and then automatically adjust the filter parameters at the current time point in order to adapt to the statistical characteristics of the signal and unknown noise changing with time, so as to achieve the optimal filtering.

3.2 AKF algorithm based on forgetting factor estimation

The commonly used AKF algorithms are AKF algorithm based on forgotten factor estimation, AKF algorithm based on noise statistical characteristics estimation and AKF algorithm based on filter gain matrix estimation. In order to alleviate the over convergence and make the error estimation better adapt to the new measurement changes, an AKF based on forgotten factor estimation is proposed.

When there is a deviation between the selected mathematical model and the actual system, the weights of system noise and measurement noise are modified in the filtering process in order to reduce the weights of historical observation data and improve the weights of new information, which can better reflect the actual situation and achieve the purpose of reducing the inertia of the filter. Therefore, AKF based on forgotten factor estimation has stronger robustness to process parameters and better estimation accuracy than standard KF by utilizing the effective information in "new information".

Variable forgotten factors are used to adjust the mean square deviation matrix of state estimation errors in real time and the past data are gradually eliminated. The mean square error matrix of the state estimation error is as follows:

$$P_{k/k-1}^* = \Phi_{k/k-1}(sP_{k-1}^*)\Phi_{k/k-1}^T + \Gamma_{k/k-1}Q_{k-1}\Gamma_{k/k-1}^T \tag{9}$$

where s is a forgotten factor or a fading memory factor.

It can be observed that the mean square deviation matrix is formally independent of the current time N and it only needs to multiply the P_{k-1}^* of the previous time state mean square deviation matrix by the forgetting factor s, which is equivalent to expanding the uncertainty of the state prediction and forgetting the previous estimates.

If the forgetting factor s > 1, the state vector is as follows

$$X_k^* = (I - K_k^* H_k)X_{k/k-1}^* + K_k^* Z_k \tag{10}$$

The formula above reflects that AKF based on forgotten factor estimation enhances the weight of measurement Z_k and reduces the weight of state prediction $X_{k/k-1}^*$ correspondingly, which reduces the impact of historical measurement data.

3.3 Determination of Forgotten Factor

The determination of forgotten factor is one of the key points of AKF algorithm based on the estimation of forgotten factor.

By introducing forgotten factor, the statistical characteristics of noise can be continuously corrected and predicted in real-time so that the algorithm achieves adaptive effect. However, in this data tracking algorithm the estimated value will adapt faster if the forgotten factor H is too small. But the deviation trend of tracking data will be larger. If the forgotten factor s is selected too big, it will reduce the adaptive speed of the estimated value greatly. So it is very important to select the appropriate forgotten factor for this algorithm. The determination of forgotten factor is one of the key points of AKF algorithm based on the estimation of forgotten factor. It is assumed that the initial values of Q_k、 R_k and P are positive definite symmetric matrices and the matrix H_k is full rank. Then the optimal forgotten factor can be determined as follows:

$$s = \max\{1, tr[N_k]/tr[M_k]\} \tag{11}$$

$$\text{where} \begin{cases} S_k = [Z_k - H_kX_{k/k-1}][Z_k - H_kX_{k/k-1}]^T \\ M_k = \Phi_{k/k-1}P_{k-1}H_k^T \\ N_k = S_k - H_kQ_kH_k^T - R_k \end{cases} \tag{12}$$

4. Simulation Analyses

The white noise of the gyro is 0.01°/h, the white noise of the accelerometer is 10-4g, and the speed error is 0.1m/s. The test was carried out 300 s in total with a sampling time of 0.1 s. The update period of the adaptive Kalman filter is T=0.1s, and the forgetting factor is 1.001. At this time, the height channel of the system is considered. The attitude error curve of the system after Kalman filtering is shown in Figure 1. It can be seen from the figure that the error results no longer diverge with time accumulation. Although the initial period is fluctuating, the northeast sky three-way attitude error is bounded overall and the curve is smooth.

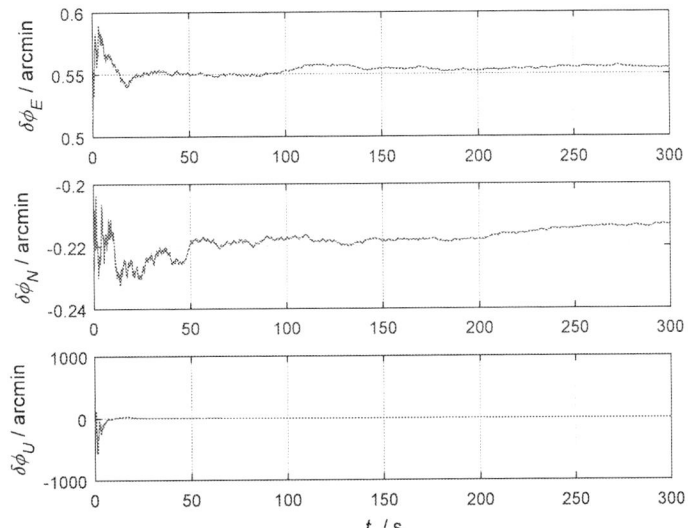

Figure.1 Adaptive Kalman Filtering posture error

In order to compare the performance of the SINS/GPS integrated navigation system, the SINS navigation system been tested statically at first. The navigation sampling time is 0.1s, and the three subsample rotation vector optimization algorithm is used. The test is performed for 3600s. A graph of attitude error, velocity error, and position error for the SINS navigation system is shown in Figure 2.

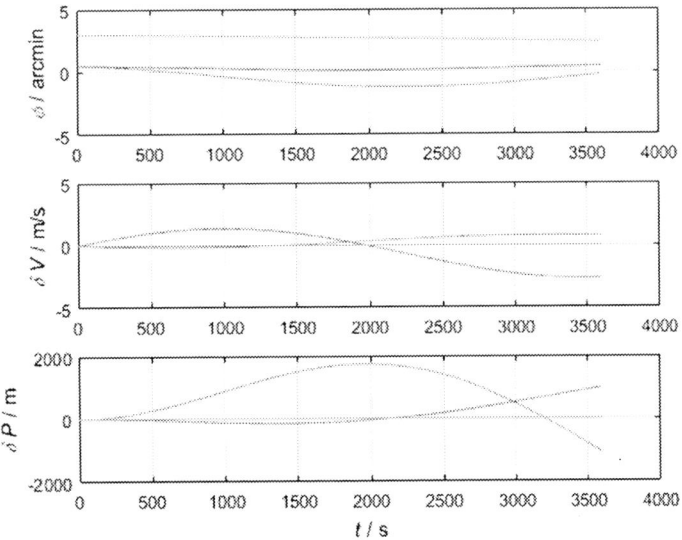

Figure 2.SINS navigation system attitude error, speed error, position error

The integrated navigation system uses the navigation parameter error of SINS/GPS was inputted as the KF input based on the forgetting factor, the error of various combined navigation parameters was estimated by filters,the parameters such as the filtered output attitude of the combined navigation was to correct the navigation parameters, Thereby, the optimal estimated output of the attitude calculation result of the integrated navigation system could be executed. The three misalignment angles, speed error, position error, gyro drift and accelerometer bias curve of the SINS/GPS integrated navigation system in the northeast direction was showed in Figure 3 and Figure 4. The test is carried out for

3600s in total.

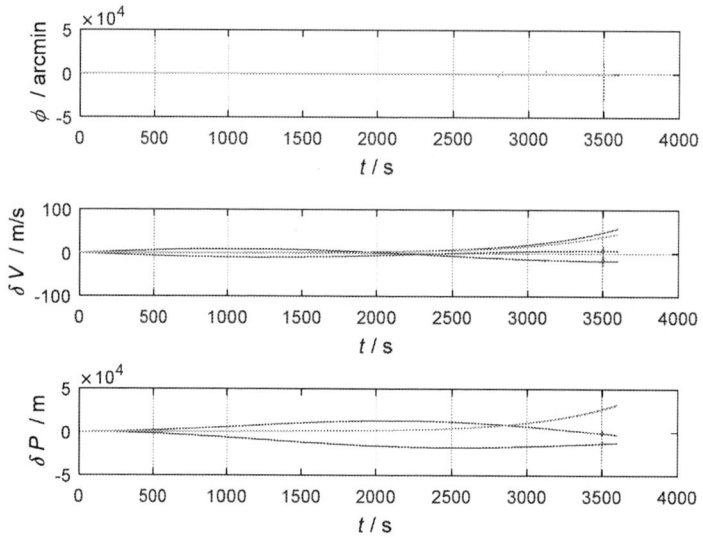

Figure 3. SINS/GPS integrated navigation system attitude error, speed error, position error

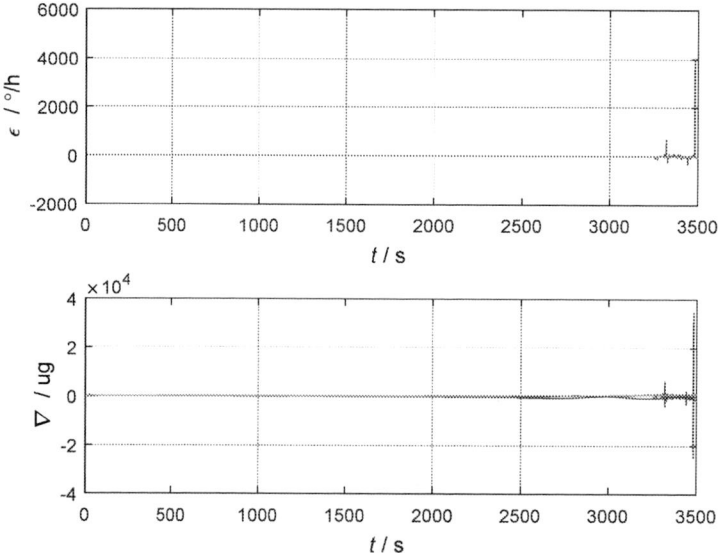

Figure 4.SINS/GPS integrated navigation system gyro random constant drift, accelerometer random constant bias

The error of the SINS/GPS integrated navigation system has been reduced according to time range, and the accumulated error in the previous period has few influence on the system operation, the accuracy of the system are improving hereby. However a certain degree of distortion occurs in the initial stage. The main reason of the initial value is selected in the KF process, but the curve tends to converge with time.

5. Conclusion

It can be seen from test data that the SINS/GPS integrated navigation system has the advantages of high precision and good convergence. The advantages of GPS and INS was given full play which be designed by the SINS/GPS integrated navigation system in this paper , and Navigation parameter error was inputted in process to estimate navigation parameter based on forgotten Kalman filter to input signal which improved the navigation accuracy and suppressed the issue of Filter divergence. Overall the SINS/GPS integrated navigation system has high value of application on precision, reliability, wide application range.

Acknowledgments

This work was supported by the Scientific Research Project of Anhui Xinhua College No. 2018zr012,department of Anhui province University quality engineering projects No.2017jyxm0531 and No.2017sxzx37.

References

[1] Khan B,Qin Y Y .Tightly coupled integration of a low cost MEMS-INS/GPS system using adaptive Kalman filtering[J].International Journal of Control and Automation,2016,9(2):179-190.

[2] Han H Z,Xu T H,Wang J.Tightly coupied integration of GPS ambiguity fixed precise point positioning and mems-ins through a troposphere-constrained adaptive Kalman filter[J].Sensors ,2016,16(7);1057.

[3] Werries A,Doian Hohn M.Adaptive Kalman filtering methods for iow-cost GPS/INS iocalization For autono-mous vehiles[R]Carnegie Mellon University,2016.

[4] Meng X Y, Wang Y Y. An improved adaptive extended kalman filter algorithm for SINS/GPS compact integrated navigation system [J]. Journal of Beijing institute of technology, 2008,38(06):625-630+636.

[5] CaoJ. Research on SINS/GPS integrated navigation adaptive filtering algorithm [D]. East China normal university,2018.

Construction and analysis of campus knowledge payment platform under the wave of big data

Ran Wu; Qiong Wu

School of Business Management, Hohai University, Changzhou, 213022.China.

Abstract: With the increasing popularity of the network information age, "big data" emerging at the historic moment, has become the hottest IT industry vocabulary. In this context, the application of big data technology plays an important role in the construction of campus knowledge payment platform. Therefore, this paper is dedicated to studying the technical support for the construction of campus knowledge payment platform and the influence of big data on the platform construction, so as to promote the platform technology upgrade and optimize the platform management.

1. Introduction

With the rise of cloud computing, social media, mobile Internet, a large amount of fragmented data emerge. Network sharing and openness make everyone can obtain and store information on the Internet, but because there is no quality control and management mechanism, information resources, the good and bad are intermingled, formed a complicated world of information. How to find suitable information in a variety of resources has become a problem most people are distressed about. Because of its paid characteristics, knowledge content on the knowledge payment platform is more professional and personalized, providing targeted services for customers.

In addition, at the current rate of data generation, 2.5 terabytes of data are generated every day. A large amount of data requires the platform to continuously improve the "data capability", that is, the ability to process a large amount of data at high speed, responsibly and sustainably, so as to give users an extremely fast, efficient and accurate user experience. Big data technology is the ability to quickly obtain valuable information from various types of data, which is exactly what the campus knowledge payment platform needs.

Domestic research on the "knowledge payment" platform is mainly based on the well-known platforms such as "Dedao" and "Himalaya" to discuss its development mode, existing problems and future trends, etc., and there is no payment platform that takes college students as knowledge creators. However, under the traditional mode of knowledge sharing, there are often problems such as resource dispersion, low efficiency, low information quality and lack of effective supervision mechanism. With the application of big data technology, the platform can well grasp the needs of users, integrate platform resources, accurately locate and provide satisfactory services.

2. Research status at home and abroad

2.1 domestic research status

Wang chuanzhen proposed that "Fenda" and "Dedao" are knowledge dissemination models that rely on open content communities, screen out more valuable information, and provide online consultation, online courses, information sharing and other content services to the Internet by individuals on the basis of payment [1]. Payment model based on knowledge in certain principles of economics. Zhang chunxiao pointed out that the knowledge payment platform connects the knowledge sharer and the knowledge demander, and the platform improves the efficiency for bilateral users. Knowledge consumers are willing to pay when the amount of payment is less than the cost of time (energy) to obtain specific knowledge, and knowledge sharers can benefit from getting access to the platform to share knowledge [2]. Sun xiaozhen came to the conclusion that college students generally hold an attitude of recognition towards knowledge payment and have a wide demand for paid knowledge, and that paying will become an important means for college students to acquire knowledge [3].

2.2 foreign research status

Foreign concept of "knowledge paid" appear earlier. In the field of education, foreign knowledge sharing is embodied in MOOC (Massive Open Online Course) online course mode in colleges and universities. In addition, well-known paid professional skill sharing platforms include Coursera and Skillshare, etc. Quora has also added a paid video question-and-answer function, which integrates social and knowledge payment, and is relatively complete. However, it is quite different from the large-scale purchase of high-quality knowledge resources under the background of network payment pioneered in China.

3. Technical support of big data knowledge payment platform

Big data technology is used to improve the management mode of platform data and the quality of platform data, so as to achieve more economic benefits.

3.1 database integration technology

Data integration technology can integrate data into unrelated data systems through unified application standards and data structures, so that each system or different users can effectively acquire data. Therefore, data integration technology can be used to unify data platform and interact with various heterogeneous database data.

3.2 graphical user interface

Compared with the character command language interface based on symbols, the graphical interface based on visual perception has certain cultural and language independence, and can improve the efficiency of visual target search [4]. In the visual stage, computer graphics technology is used to give full play to the potential of image perception and image thinking, which improves the efficiency of information transmission [4]. Graphical user interface can improve the comfort of users, increase the artistic and ornamental value of the platform.

3.3 advanced server structure

This structure is composed of data server, application server and client, which has obvious advantages. It can make the sharing of platform resources convenient, save cost and improve system performance [5].

3.4 object-oriented technology and programming language

Platform construction is mainly based on Java language, DreamWeaver platform and MySQL database.

4. Function module design of big data knowledge payment platform

Campus "knowledge payment" platform is mainly divided into three modules

ISPECE

IOP Publishing

IOP Conf. Series: Journal of Physics: Conf. Series **1187** (2019) 052009   doi:10.1088/1742-6596/1187/5/052009

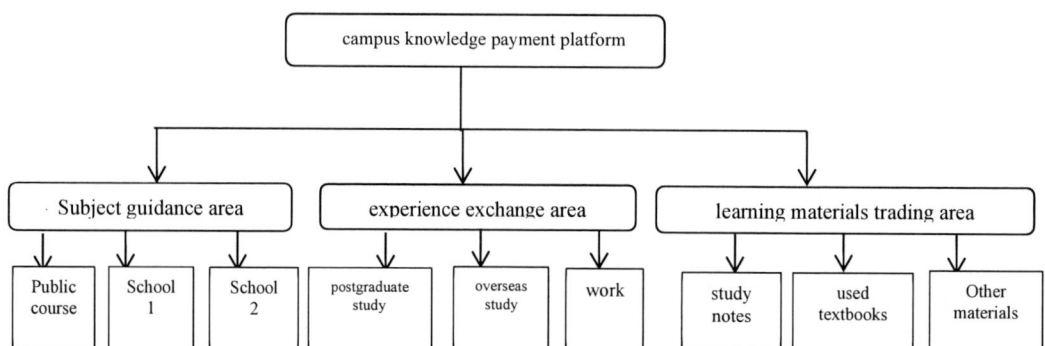

Figure 1 function module design of big data campus knowledge payment platform

4.1 subject guidance area
It is divided into different colleges, different majors, and different disciplines. It integrates the information of students who have excellent academic performance and counseling ability in different disciplines on the platform and uploads it to the database for users to search.

4.2 experience exchange area
There are four sections: postgraduate entrance examination, postgraduate recommendation, overseas study and work. There are excellent postgraduate admission institutions, overseas study institutions and job-hunting units. Information profiles of college students with relevant experience (including only publicly available information) are uploaded to the database for users to search and contact.

4.3 learning materials trading area
Learning materials include study notes, used textbooks and other learning materials. Upload the version and pricing information of the materials to the database for users to select transactions.

5. Implementation ideas of big data knowledge payment platform

5.1 Big data technology deals with massive data resources
Big data processing and analysis is becoming the node of the fusion application of the new generation of information technology. Platform users will constantly generate big data when uploading resources, searching information and consulting services. Cloud computing provides storage and computing platforms for these massive and diversified big data. Through the management, processing, analysis and optimization of data from different sources, the results are fed back to the platform, which will create huge economic and social value.

5.2 Big data technology excavates users' needs
Analyze the massive data of the platform, including the resources provided by knowledge producers and the types and contents of knowledge sought by knowledge consumers, so as to find the laws in the data, explore the needs of both parties, adjust the operation mode of the platform, and improve the repurchase rate and satisfaction of the knowledge payment platform.

5.3 Big data technology becomes the core competitiveness of the platform
The application of big data will become a key factor to improve the core competitiveness of the platform. Decisions in all walks of life are changing from "business-driven" to "data-driven", and knowledge payment platforms are no exception. The analysis of big data can enable the platform to grasp the market dynamics in real time and respond quickly, making more accurate and effective

2125

marketing strategies. It can help the platform provide more timely and personalized services for knowledge consumers.

6.Summary

Through the application of big data knowledge payment platform, it can provide convenient services for knowledge producers and consumers. It is also advantageous for the platform side to manage massive data, reduce time cost and improve efficiency. Then, it can meet the needs of platform users, and enhance the competitiveness of the platform.

References:

[1] Wang chuanzhen. Singularity and future of knowledge payment [J]. Internet economy,2017(Z1):68-73.

[2] Zhang chunxiao. Research on pricing mechanism of knowledge payment platform based on bilateral market theory [J]. Journal of jilin university of business and technology,2017,33(04):32-35.

[3] Sun xiaozhen, Dong xiao, Li ying, Zhao tong. Survey and analysis of college students' knowledge payment [J]. Knowledge economy,2017(19):162+164.

[4] Zhou Hui, Xiao-min tian. Design and implementation of billiard game based on OpenGL [J]. The design and implementation of information technology, 2013, (2): 130-134139. The DOI: 10.3969 / j.i SSN. 1009-2552.2013.02.038.

[5] Li bing, Wang wei, Wang feng. Analysis on countermeasures for the construction of big data knowledge service platform [J]. Electronic technology and software engineering,2016(17):173.

The Mobile payment based on public-key security technology

Jiabin Sun[a], Nan Zhang[b]

International School, Beijing University of Posts and Telecommunications, Beijing 100876, China

[a]sunjiabin@bupt.edu.cn

[b]2015212939@bupt.edu.cn

Abstract This paper mainly concentrates on data security technology applied on the mobile payment application in China. The widely used technology of public-key encryption will be introduced in this paper and it also includes the mobile payment system mode applied in Chinese mobile payment application. And it aims at how the technology serves for different business to meet the users' requirement, and how the technology adjusts to different market requirement.

1.Introduction

With great convenience and low cost, mobile payment has become an essential part of e-commerce. And therefore, it is particularly important to ensure the information security during the paying process. As a result, it is very necessary to conduct a research on efficient security mechanisms for mobile payment, and on the relationship between the security mechanisms and the e-commerce business mode. With these considerations, this paper will demonstrate the issue from the following aspects:

First, a general theoretic research is conducted for the Mobile Payment System (MPS), including operation mode, and structure. Second, security mechanism of MPS is studied, including encryption technology, the comparison of symmetric and asymmetric encryption, and authentication method. Then the paper will focus on one of the most commonly used MPS in China, Alipay, and also the popular digital currency. By analyzing the differences of the two application, this research will try to find out how the technology adjusts to the market requirement.

2.Mobile payment system

2.1.Operation Mode

There are three main operation mode, distinguished by the different business operation entity: 1) mobile operators 2) banks 3) third-party service provider [2].

When the mobile operator acts as the main body of the mobile payment platform, the mobile operator will use the mobile phone number as the account, the user's mobile payment transaction costs the department deducts from the user's credit account. The mobile payment service with the mobile operator as the main body has the characteristic that there is no need for bank participation, and it is impossible to invoice the non-calling business.

When bank plays the main role as the main body of operation, it interconnects with the mobile communication network through a dedicated line, and binds the bank account to the mobile account. Banks provide users with trading platforms and payment channels, while mobile operators only provide information channels for banks and users, and do not participate in the payment process. At present,

Content from this work may be used under the terms of the Creative Commons Attribution 3.0 licence. Any further distribution of this work must maintain attribution to the author(s) and the title of the work, journal citation and DOI.
Published under licence by IOP Publishing Ltd

most of Chinese business bank provides mobile banking, and operates their own mobile payment platforms.

The mobile payment service with the bank as the main operation has the following characteristics: each bank can only provide the use services of its own bank, and the mobile payment services cannot be interconnected between banks [2].

Payment service provider or mobile payment platform operator is the third independent of banks and mobile operator and it is also a bridge and link between mobile operators, banks and merchants. Through the operator of the trading platform, users can easily realize the mobile payment service across banks. The third party service provider is now playing a role of merging users and banks as users can access to multiple bank accounts on third party service provider, such as Alipay and Wechat [2].

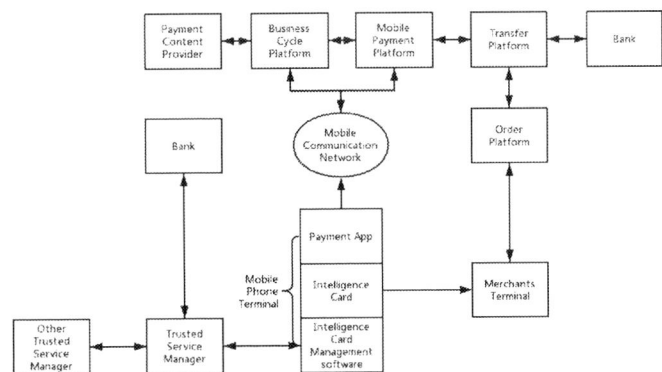

Figure 1. Flow chart of the mobile payment process [1]

2.2.Structure

The Mobile payment system (MPS) is consists of two parts: PA (Mobile Payment Agent) and MPP (Mobile Payment Platform) and they are connected by digital data network [2]. MPA is responsible for the management, fee-counting, and communication in nation-wide scale, connected with national business operation support system (BOSS); as MPP is responsible for the management, fee-counting, and communication in the local scale, connected with service providers. The UASS (Unified Account Service System) in BOSS provides unified management of the user accounts nation-wide. Users can attach their mobile phone number to their mobile payment account and also to their bank account [2].

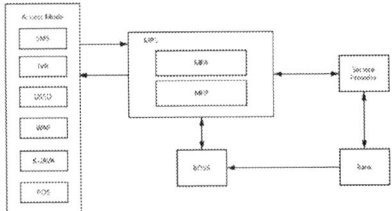

Figure 2. Structure of MPS [2]

3.Security mechanism of mobile payment system

3.1.Encryption technology

3.1.1.Symmetric encryption

Symmetric encryption uses a shared key (symmetric) to encrypt. That means the sender and the receiver hold the same key to encrypt and decrypt the message in the communication.

Key should be exchanged between the communicating entities before the transmission of data [3]. For the encryption process, exchanging keys is important. Weak and short keys can be easily attacked as compared to longer keys which are more difficult to break [3]. Symmetric key encryption algorithms are still widely used as powerful techniques in insecure communication channel [3].

The DES (Data Encryption Standard) is a commonly used symmetric encryption. DES was the first block cipher which was designed by IBM and it was adopted by national bureau of standard in 1977. DES algorithms take 64 bits plaintext as an input and transform it into 64 bits cipher text as output. The key length of DES is also 64 bits. DES is called a complex block cipher as it has 16 blocks of complex round ciphers and each block itself has a complex function [3].

3.1.2. Public-key encryption

Public-key encryption is asymmetric encryption because the sender and receiver are not using the same key to encrypt and decrypt. There are two types of keys used in public-key encryption, the public key and the private key.

The RSA algorithm is the one of the most commonly used public-key encryption. The algorithm uses two keys, the private key and the public key for encryption and decryption. The sender encrypts the message with the receiver's public key, which is known by everyone, and then the receiver decrypt the message with the private key which is only known by the receiver. In this way, the stand-in attack will be efficiently prevented, that even the attacker captures the message in the middle of the traffic, there is no way for the attacker to decrypt the message since the private of the receiver is needed.

The encryption process can be described as followed:1) Select two large prime numbers p, q; 2)Calculate n=p*q; 3) Calculate $\phi(n)=(n-1)(q-1)$; 4)Select integer e, such that $gcd(\phi(n),e)=1$ and $1<e<\phi(n)$; 5)Let $d=k*e^{-1}mod\ \phi(n)$ (k=1,2,3,...); 6)Public key is $k_u=\{e,n\}$, and private is $k_r=\{d,n\}$

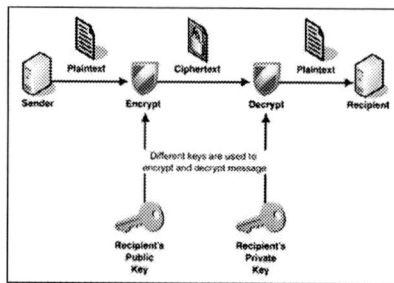

Figure 3. Process model of asymmetric key cryptography [3]

3.1.3. Comparison between symmetric encryption and public-key encryption

In general, there several differences between symmetric and public-key encryption. Symmetric encryption requires only one shared key while public-key encryption requires two secret keys, the public key and the private key. And as for only one key is required, symmetric encryption has better operation speed than public-key encryption but lower security level. Although public-key encryption has the advantage of security, the calculation of large prime numbers also increase complexity of the algorithm.

Table 1. Comparing the two different encryption method

Features	Symmetric encryption	Public-key encryption
Number of keys	1	2
Speed	Fast	Slow
Security	Low	High
Complexity	Low	High

3.2 Authentication

3.2.1 Digital Signature

Digital signature is an alphabetic string obtained through processing the transmitted text by a Hash, with the purpose of verify the source of text and confirm whether the text is undergoing changes [4].

To ensure the availability of digital signature, PKI (Public Key Infrastructure) is often used. It follows the standard public key encryption technology and offers a full set of security assurance infrastructure for sectors like e-commerce, e-government, e-banking and on-line banking securities [4].

Data signature technology can ensure that: information cannot be known by other sides except the senders and receivers; information during transmission will not be tampered with; the recipient is able to confirm the identity of the sender; sender information for their own cannot be denied. Digital signature uses public key cryptosystem, that is, it uses a pair of matching asymmetric keys to achieve the encryption and decryption, signature and verification at the same time [5].

The digital signature for a message is generated in two steps:1) generating a message digest which is a summary of the message using hash algorithms 2) encrypt the message digest with the sender's private key.

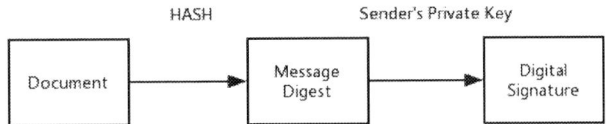

Figure 4. Digital Signature [5]

The digital signature is attached to the message, and sent to the receiver. The receiver then does the following: 1) Using the sender's public key, decrypt the digital signature to obtain the message digest generated by the sender. 2) Uses the same message digest algorithm used by the sender to generate a message digest of the received message. 3) Compares both message digests (the one sent by the sender as a digital signature, and the one generated by the receiver) [5]. If they are not exactly the same, the message has been tampered with by a third party. It is sure that the digital signature was sent by the sender because the message digest is encrypted by the sender's private key, so the receiver can decrypt it with the sender's public key, and also with the private key, the receiver is certain about the sender's identity.

3.2.2 Certificate Authority

CA (Certificate Authority) is a trusted authority in a network that issues and manages security credentials and public keys for message encryption. As part of a PKI, a CA checks with a registration authority to verify information provided by the requester of a digital certificate. If the RA (Registration Authorities) verifies the requestor's information, the CA can issue a digital certificate. Indeed, the CA is responsible for the distribution and revocation of the certificate. Depending on the PKI implementation, the certificate might include the owner's public key, the expiration date of the certificate, the owner's name, and other information about the public key owner [6].

3.3 Firewall

A firewall protects a local system/network from network-based security threats and at the same time allows access to the outside world. In most cases, firewall is required since it is difficult to equip every single device with strong security features. Usually, the firewall is inserted between the premises network and the Internet and it establishes a controlled link and security wall between the premises and the Internet.

4.Application of public-key technology in different business mode

4.1.Scanning-code payment

4.1.1.Security mechanism of QR Code

QR Code is an array of bits that can be used to store information. Information embedded in QR Code can be secured using various methods such as TTJSA algorithm, SD-EQR, hash function, reversible data hiding, steganography, histogram, symmetric encryption, asymmetric encryption, Reed-Solomon method, Signed QR Code (SQR Code), etc. [7].QR Code has its core structure (shown in Fig. 5). The encrypting and decrypting of the QR code relies on the encoding region of the code [8].

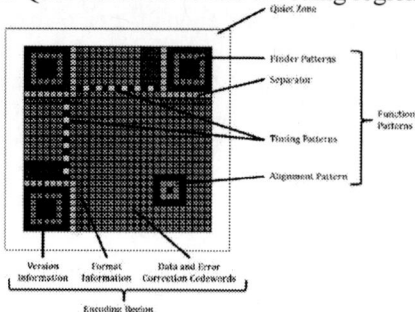

Figure 5. QR Code Structure [8]

If QR Code is dirty or damage, it does not matter because QR Code has error correction capability [7]. The QR code achieves powerful error-correction capability by using Reed-Solomon codes, a widely used mathematical error-correction method,and four levels of error correction(L-7%, M-15%, Q-25%, H-30%) are available[8]. Data can be restored even if the symbol is partially dirty or damaged. As long as the Finder patterns are not masked and no more than 30% of the code is hided, then the data can be restored [7].

4.1.2.QR Code in mobile payment

The QR Code encryption through PKI in mobile payment system can be divided into the following seven steps: 1)the user gives user ID; 2) the merchant receive user ID;3) the merchant request public key to third party; 4)the third party generate public key and private key based ID;5) the third party authenticate and receive private key; 6)the merchant generates QR code uses public key; 7) the user scan QR code ,decrypt using private key, and verify received message[7].

In the case of Alipay in China, there are three modes of QR code that are commonly used in real business scenario: 1) the identity code 2) the receiving code 3) the paying code. As for the identity code, what is encoded in the code is a URL with an encrypted identity message, and this information is unchanged and cannot be seen. For the receiving code, what is encoded in the QR code is also a URL (a different API) with an encrypted message, and this information is also unchanged and cannot be seen. For the paying code, it is not a URL that is encoded in the QR code, instead, a string of numbers is encoded, and this string of number changes every minute. From the QR code model below it is clear that the string of number that encoded in the QR code is the public key that is generated by the third party used in every transaction. This could explain why merchants using Alipay can have a permanent QR code for receiving money, because the information encoded in the QR code is unchanged, and the code itself does not need to be changed. For different payment application, the difference between their QR codes is the URLs and APIs that encoded in the QR code.

For the users of Alipay, what they require is both safety and convenience for their payment, and therefore, Alipay uses QR code technology through PKI to encode and encrypt the users' data to ensure the payment safety and enhance the user experience.

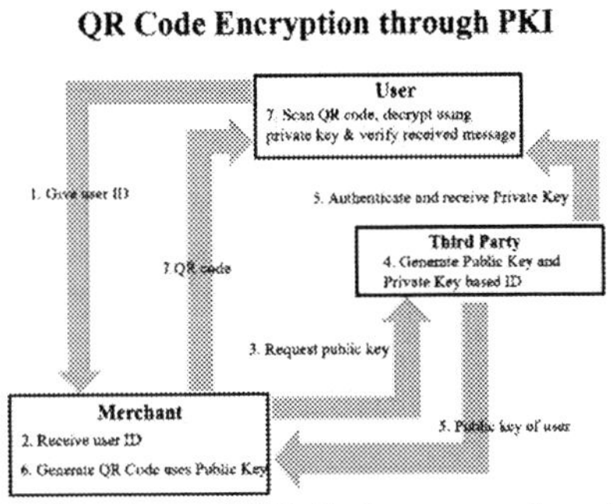

Figure 6.Structure of QR Code payment system [7]

4.2.Digital currency

Bitcoin is one of the most famous digital currencies, and it was first put forward by Satoshi Nakamoto in 2008. In the transaction of Bitcoin, a third party financial institution is not needed. This is then added to the continuous chain of hashes known as the 'Blockchain' which is a permanent record/ledger of all events witnessed on the network and can be considered the backbone of the Bitcoin network [9]. The blockchain utilizes complex mathematical algorithms to construct a chain of SHA-256 cryptographic data that cannot be replicated or altered. All transactions are added to the Blockchain and must be signed using a private key held by the owner of that particular Bitcoin which prevents third parties tampering with the transaction and Blockchain [9].

A transaction using Bitcoins requires two items; a private key and a Bitcoin address. The Bitcoin address is a random sequence of numbers and letters that makes up the Bitcoin public key and combined with a private key makes up the asymmetric key pair [9]. The sending of Bitcoins between users requires the private key of the sender for signature and the public key of the receiver. A number of confirmations are required for this transaction in order to secure the integrity of Bitcoin against tampering or hacking. The miners produce Bitcoins in the network and the blockchain ensure the safety of the trade.

5.Conclusion

This paper mainly discuss the security technology used in mobile payment system, and describes two popular MPS applications, Alipay and Bitcoin. For the users of Alipay, both security and convenience is important. And therefore, QR code technology is used in mobile app. As for the Bitcoin miners, the security requirement is even higher than Alipay users, so that blockchain is used to ensure the absolute security.

References

[1] L.You. Security Technology Analysis of Mobile Payment [J].Telecom World, 2017(06):35-37.

[2] D.Fangming. Research on Security Mechanism of Mobile Payment System [D].XiDian University, 2006.

[3] P. Chaudhury *et al.*, "ACAFP: Asymmetric key based cryptographic algorithm using four prime numbers to secure message communication. A review on RSA algorithm," *2017 8th Annual Industrial Automation and Electromechanical Engineering Conference (IEMECON)*, Bangkok, 2017, pp. 332-337.

[4] Junling Zhang, "A study on application of digital signature technology," *2010 International Conference on Networking and Digital Society*, Wenzhou, 2010, pp. 498-501.

[5] J. Zhu, "Study on the e-commerce security model based on PKI," *2010 International Conference on Computer Application and System Modeling (ICCASM 2010)*, Taiyuan, 2010, pp. V4-6-V4-9.

[6] S. F. Al-Janabi and A. K. Obaid, "Development of Certificate Authority services for web applications," *2012 International Conference on Future Communication Networks*, Baghdad, 2012, pp. 135-140.

[7] A. T. Purnomo, Y. S. Gondokaryono and C. Kim, "Mutual authentication in securing mobile payment system using encrypted QR code based on Public Key Infrastructure," *2016 6th International Conference on System Engineering and Technology (ICSET)*, Bandung, 2016, pp. 194-198.

[8] S. Tiwari, "An Introduction to QR Code Technology," *2016 International Conference on Information Technology (ICIT)*, Bhubaneswar, 2016, pp. 39-44.

[9] J. G. Fraser and A. Bouridane, "Have the security flaws surrounding BITCOIN effected the currency's value?," *2017 Seventh International Conference on Emerging Security Technologies (EST)*, Canterbury, 2017, pp. 50-55.

The Advisable Technology of Key-Point Detection and Expression Recognition for an Intelligent Class System

Yusheng Zhao[1, a], Haotian Yan[1, b], Zhaoqing Wang[2, *]

[1]International School, Beijing University of Posts and Telecommunications Beijing, China

[2]School of Information& Communication Engineering, Beijing Information Science& Technology University, Beijing, China

*Corresponding author email: mj741561@163.com
[a]zhaoyusheng@bupt.edu.cn; [b]yanhaotian@bupt.edu.cn

Abstract The intelligent classroom system has a very wide range of application scenarios in modern education. With the continuous breakthrough of artificial intelligence technology, a real-time intelligent system that can judge students' performance in class has a large demand market. In this paper, the real-time detection technology needed by the smart classroom system is studied from the aspects of behavior detection and expression recognition. In terms of behavior detection, this paper combines the two key point detection technologies CPM (Convolutional Pose Machines) and CMU (Carnegie Mellon University) OPENPOSE [1] [2]. In terms of expression recognition, in order to detect the position of the face in real time, this paper studies several popular target detection algorithms and improves the traditional CNN network, and proposes a real-time image classification network architecture [8].

1. Introduction

In the key point detection part, this paper analyzes the popular CPM (Convolutional Pose Machines) and CMU (Carnegie Mellon University) OPENPOSE algorithms, and describes and compares the main structures of the two algorithms. Finally, combined with the key point detection for the significance of this application scenario, we chose the CMU OPENSE algorithm for further exploration. In the expression detection section, based on the idea of the target detection algorithm RCNN [10]. In this paper, we use the deep learning method Single Shot MultiBox Detector (SSD) algorithm to select the bounding box, and then use our modified CNN to classify the selected faces [14]. Figure 1 shows the recognition results of combining the two algorithms. This article will elaborate on these two parts.

Content from this work may be used under the terms of the Creative Commons Attribution 3.0 licence. Any further distribution of this work must maintain attribution to the author(s) and the title of the work, journal citation and DOI.
Published under licence by IOP Publishing Ltd

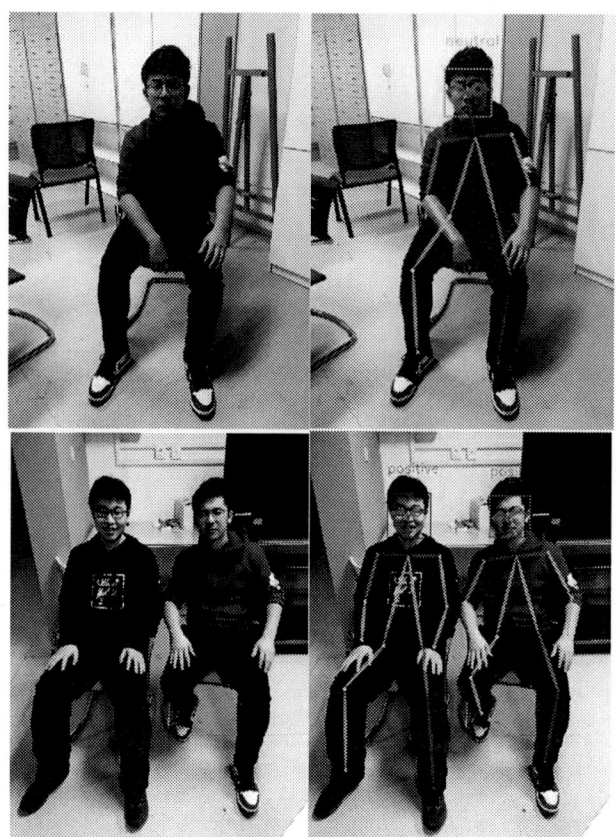

In Figure 1, the left column is the original image, and the right column is the result of the recognition. The output includes the structure of each human key point and its expression. The upper and lower lines correspond to the single-person detection and multi-person detection of the algorithm.

2. Literature review

In the key detection part, there are two main directions, one is top-down and the other is bottom-up. The top-up human bone key point localization algorithm mainly consists of two parts, human detection and the human body key point detection. That is, firstly, each person is detected by the target detection algorithm, and then the human skeleton key point detection is performed for a single person based on the detection frame, wherein the representative algorithms are RMPE, Mask R-CNN, etc. [5][6]. The bottom-up approach also consists of two parts, key point detection and key point clustering. That is, first all the key points in the picture need to be detected, and then all the key points are determined by the relevant strategy clustering into different individuals. The representative algorithms for modeling the relationship between key points is Associative Embedding [7]. The algorithms CPM and CMU OPENSE used in this paper are all bottom-up methods.

In 2016, Google proposed a new CNN, Xception, which greatly reduced the amount of computation of the CNN convolutional layer [9]. In 2014, RGB proposed the target detection algorithm RCNN. Its idea is to extract the region proposal first, and then use CNN to classify the regional proposal. Based on this idea, a faster target detection algorithm such as Fast RCNN and Faster RCNN is generated [11][12]. In 2016, RGB proposed YOLO with faster response, but its monitoring accuracy for small objects is not high [13]. In 2017, Weiliu et al. proposed a new target detection algorithm based on regression theory, Single shot detector, which has very fast response and high monitoring accuracy for small objects [14].

3. Key-point detection

3.1 Convolutional Pose Machines

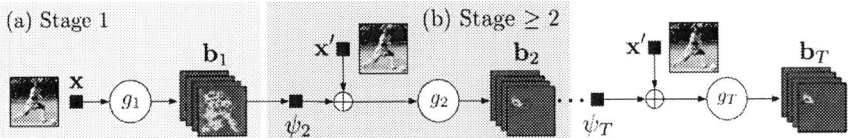

Figure 2.The framework of Pose Machines [1]

CPM takes advantage of both the deep convolutional network and the spatial modeling of the Pose Machine framework [3]. The basic structure of the Pose Machine is shown in Figure 2.It was proposed by Shih-En Wei in 2016[1].

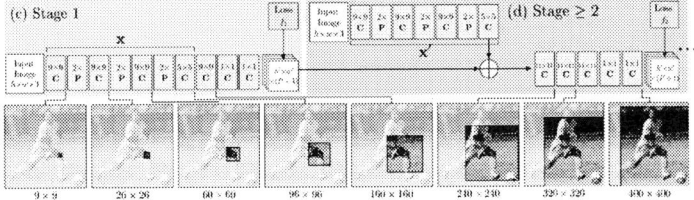

Figure 3.The framework of Convolutional Pose Machines[1]

CPM uses Pose Machine framework to replace the original classifier with a full convolution layer, as shown in Figure 3[4].

When stage $t = 1$, CPM predicts joint points according to local image evidence. Using local information local means that the acceptance field of the network is constrained to the local part of the output pixel value. Picture block. The input image passes through the full convolution network and outputs the predicted results of the joint points.

When stage $t = 2$, the input to the classifier includes: the original picture feature, the previous stage's beliefs for each joint point, and the Gaussian center constraint of the generated center (used to bring the response back to the image center, handling multiple characters in the image The situation, to achieve multi-objective attitude estimation).

When stage $t > 2$, the input of the classifier no longer includes the original picture feature, but is replaced by the convolution result of the previous layer. The other inputs are the same as t=2. It is also three inputs.

Finally, through multiple iterations, the estimation of the relevant nodes is realized. However, the connection between the joint points cannot be achieved.

3.2 CMU OPENPOSE

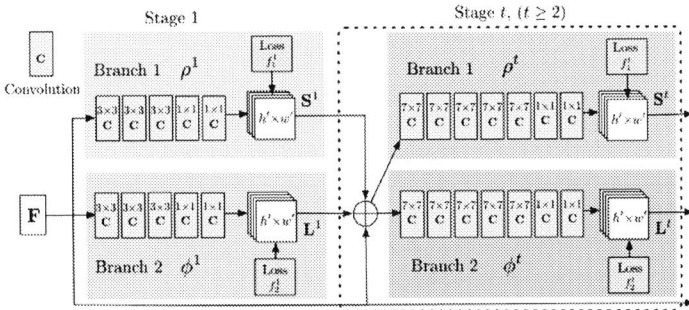

Figure 4.The framework of CMU OPENPOSE[2]

The main idea of CMU OPENPOSE is to convolve separately from two branches, as shown in Figure 4.

Branch1 (Part Confidence Maps): Find all joint points, there are two parts. The first part is stage t=1, which accepts the input as the original image, and then outputs confidence maps to each joint point according to the classifier. The second part is stage t ≥ 2, the input it accepts is all the confidence maps and the original image obtained by the previous stage, and its output is still confidence maps. looping until convergence.

Branch2 (Part Affinity Fields): Its structure is basically the same as the first branch. The only difference is that each of the confidence maps of the output contains a certain type of connection (which can be simply understood as a bone).

The core of this algorithm is to successfully represent these connections as joints of joints, thus realizing the basic structure of the human body. As shown in Figure 5.

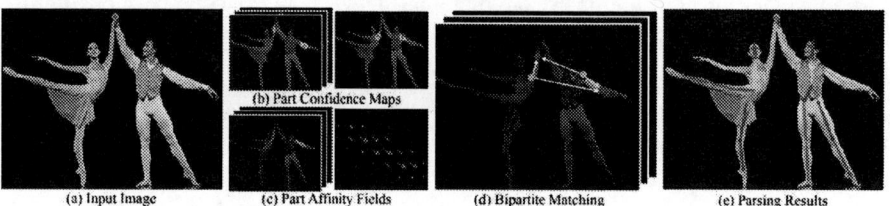

Figure 5.The output of CMU OPENPOSE [2]

4. A real-time expression detection system with a few parameters

4.1.A Deep learning method based on regression for face detection

RCNN, an object detection method, has achieved an unprecedented accuracy, but because it does not meet the real-time characteristics requirements, we use the SSD method to detect human faces [10][14]. The SSD is applied to a neural network and generates a series of bounding boxes for each feature map cell on different levels of feature maps, and then uses the IoU algorithm to find the bound closest to ground truth. The response of SSD is very fast, which can achieve the mAP of 73.1% with a processing speed of 58 frames per second [14]. The efficiency of SSD is ample for real-time detection.

4.2. Modified CNN for real-time classification

The traditional CNN network generates many parameters during training. Due to the increasing in computation, CNN has a serious delay in classifying expressions. Reducing the parameters is one of the main ideas of accelerating the network, aiming to recognize facial expressions in real time on devices without high performance. Because the fully-connected layer produces many parameters, the fully-connected layer of the CNN should be considered for removal. This paper uses a Depth-wise Separable Convolution (DSC) combined with the residue parts of CNN (Figure 6). The DSC is divided into two phases. Suppose the convolution kernel's size is Dn · Dn and the input is M feature maps. In the depth-wise convolution phase, M convolution kernels perform one-to-one convolution with M feature maps to generate M results. In the point-wise convolution phase, convolution operations are performed with N convolution kernels whose size is 1 · 1 and M results from the last phrase, so N results are generated, namely, the input of M channels are converted into the outputs of N channels [9]. Fig6 shows the whole network structure, and through the comparison we found that DSC greatly reduces parameters and computation.

The improved network (Figure 7) consists of four layers of DSCs, with a batch normalization and a ReLU function sequenced after each convolutional layer. Additionally, an average pooling layer and a softmax function are placed at the end of the modified network. Finally, there are almost 60,000 parameters included in the model, which is 80 times smaller than the original CNN.

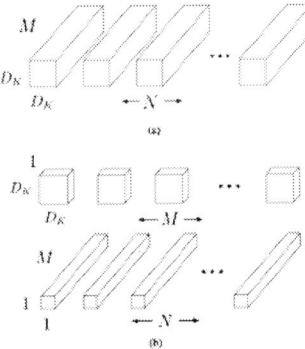

Figure 6. The number of convolution operations of CNN is DK · DK · M · N · DF · DF, and DSC is DK · DK · M · DF · DF + M · N · DF · DF. The CNN convolution kernel parameter is DK · DK · N · M, and the convolution kernel parameter is DK · DK · M+N · M.

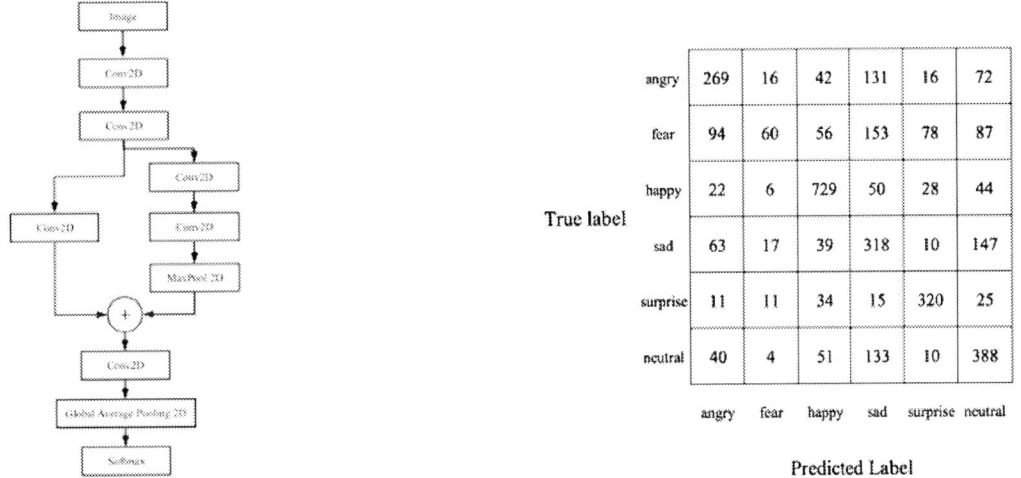

True label						
angry	269	16	42	131	16	72
fear	94	60	56	153	78	87
happy	22	6	729	50	28	44
sad	63	17	39	318	10	147
surprise	11	11	34	15	320	25
neutral	40	4	51	133	10	388
	angry	fear	happy	sad	surprise	neutral

Predicted Label

Figure 7. The improved network   Figure 8. This matrix shows the accuracy of the classification

5.conclusion

This paper studies two important techniques applied to the smart class system, key point detection and expression recognition. We compared two key point detection algorithms, CPM and CMU OPENSE, because CMU OPENSE is more suitable for multi-person scenes, and can detect key points of hands, CMU OPENSE is more suitable for building the behavior detection module of smart class system. For the traditional CNN, based on the idea of reducing the parameters and the computation time, we rebuild a network that can recognize the expression in real time by deleting the full connection layer, using depth-seperable convolution and detecting the face areas with SSD.

Reference

[1] Wei, S. E., Ramakrishna, V., Kanade, T., & Sheikh, Y. (2016). Convolutional pose machines. In Proceedings of the IEEE Conference on Computer Vision and Pattern Recognition (pp. 4724-4732).

[2] Cao, Z., Simon, T., Wei, S. E., & Sheikh, Y. (2016). Realtime multi-person 2d pose estimation using part affinity fields. arXiv preprint arXiv:1611.08050.

[3] Ramakrishna, V., Munoz, D., Hebert, M., Bagnell, J. A., & Sheikh, Y. (2014, September). Pose machines: Articulated pose estimation via inference machines. In European Conference on

Computer Vision (pp. 33-47). Springer, Cham.

[4] Long, J., Shelhamer, E., & Darrell, T. (2015). Fully convolutional networks for semantic segmentation. In Proceedings of the IEEE conference on computer vision and pattern recognition (pp. 3431-3440).

[5] Fang, H., Xie, S., Tai, Y. W., & Lu, C. (2017, October). Rmpe: Regional multi-person pose estimation. In The IEEE International Conference on Computer Vision (ICCV) (Vol. 2).

[6] He, K., Gkioxari, G., Dollár, P., & Girshick, R. (2017, October). Mask r-cnn. In Computer Vision (ICCV), 2017 IEEE International Conference on (pp. 2980-2988). IEEE.

[7] Newell, A., Huang, Z., & Deng, J. (2017). Associative embedding: End-to-end learning for joint detection and grouping. In Advances in Neural Information Processing Systems (pp. 2277-2287).

[8] Krizhevsky, A., Sutskever, I., & Hinton, G. E. (2012). Imagenet classification with deep convolutional neural networks. In Advances in neural information processing systems (pp. 1097-1105).

[9] Chollet, F. (2017). Xception: Deep learning with depthwise separable convolutions. arXiv preprint, 1610-02357.

[10] Girshick, R., Donahue, J., Darrell, T., & Malik, J. (2014). Rich feature hierarchies for accurate object detection and semantic segmentation. In Proceedings of the IEEE conference on computer vision and pattern recognition (pp. 580-587).

[11] Girshick, R. (2015). Fast r-cnn. In Proceedings of the IEEE international conference on computer vision (pp. 1440-1448).

[12] Ren, S., He, K., Girshick, R., & Sun, J. (2015). Faster r-cnn: Towards real-time object detection with region proposal networks. In Advances in neural information processing systems (pp. 91-99).

[13] Redmon, J., Divvala, S., Girshick, R., & Farhadi, A. (2016). You only look once: Unified, real-time object detection. In Proceedings of the IEEE conference on computer vision and pattern recognition (pp. 779-788).

[14] Liu, W., Anguelov, D., Erhan, D., Szegedy, C., Reed, S., Fu, C. Y., & Berg, A. C. (2016, October). Ssd: Single shot multibox detector. In European conference on computer vision (pp. 21-37). Springer, Cham.

Research on the interaction between language and economy

Yusi Wu[1,a], Jie Liu[2,b], Ping Wang[2,c], Yisheng Wang[1,d], Peng Zhang[2,e], Hongyan Xi[1,f]

[1]School of Business Administration Hohai University Changzhou, Jiangsu 213022, China.
[2]College of Mechanical and Electrical Engineering Hohai University Changzhou, Jiangsu 213022, China.

[a]1294054460@qq.com; [b]1643549712@qq.com; [c]137584860@qq.com; [d]861956245@qq.com; [e]1135401844@qq.com; [f]2385975321@qq.com.

Abstract: With the increasing economic globalization, the language spoken by people has also undergone great changes. This paper analyzes the changes in language use as the economy develops and the impact of regional language on economic globalization. Firstly, we use reliability analysis to analyze the reliability of the collected data, and the results show that the data used in this paper has high reliability. Then, we define the Economic Attraction (EA) to measure the motivation of people learning the language of the country or region, which will give people an idea of the future usage of each language. The results show that the number of users in English and Japanese will increase in the future. Then, selecting a multinational company as a typical example of economic globalization for analysis, the TOPSIS algorithm was used to analyze the location of a multinational company's office in the global region, respectively, Tokyo, London, Dubai, Paris, Madrid, Singapore. And the language of the office staff is Japanese, English, Arabic, French, Spanish, Chinese and English.

1. Introduction

There are currently about 6,900 languages spoken on Earth. About half of the world's population claim, one of the following ten languages (in order of most speakers) as a native language: Mandarin (incl. Standard Chinese), Spanish, English, Hindi, Arabic, Bengali, Portuguese, Russian, Punjabi, and Japanese. [1][2] However, much of the world's population also speaks a second language. [3]When considering total numbers of speakers of a particular language (native speakers plus second or third, etc. language speakers), the languages and their order change from the native language list provided.[4] The total number of speakers of a language may increase or decrease over time because of a variety of influences to include,[5] but not limited to, the language(s) used and/or promoted by the government in a country, the language(s) used in schools, social pressures, migration and assimilation of cultural groups, and immigration and emigration with countries that speak other languages. But the economy has an increasing influence on people's daily lives, the economy of other countries has become the biggest factor in learning the language of the country.

With the development of economic globalization, multinational service companies have gradually set up offices overseas to handle daily affairs in order to expand their business. The language learning of office personnel will be affected by the language of the region.

Content from this work may be used under the terms of the Creative Commons Attribution 3.0 licence. Any further distribution of this work must maintain attribution to the author(s) and the title of the work, journal citation and DOI.
Published under licence by IOP Publishing Ltd

ISPECE

2. Data collection and verification
First, we collect the number of native speakers in the world's 10 major languages and the number of second and third language users for later analysis. The use of these 10 languages is shown in Table 1.

Table 1: ten languages (in order of most speakers) in the word

Native Language Rank	Native Language	Native Speakers	Second(or 3rd,etc)Language Speakers
1	Mandarin Chinses	897 million	193 million
2	Spanish	436 million	91 million
3	English	371 million	611 million
4	Hindustani	329 million	215 million
5	Arabic	290 million	132 million
6	Bengali	242 million	19 million
7	Portuguese	218 million	11 million
8	Russian	153 million	113 million
9	Punjabi	148 million	1 million
10	Japanese	128 million	1 million

Due to the limitations of data collection, we need to analyze the reliability of the data we are looking for.

Here we use α-reliability coefficient method to test the reliability of the data. α-reliability coefficient method is the most commonly used reliability coefficient. The value of the Cronbach α coefficient usually is between 0 and 1. If the alpha coefficient does not exceed 0.6, it is generally considered that the internal consistency is insufficient; when it reaches 0.7-0.8, the scale has considerable reliability, and when it reaches 0.8-0.9, the reliability of the scale is very good. An important characteristic of the Cronbach α coefficient is that their values increase with the increase of the scale item. Therefore, the Cronbach α coefficient may be artificially and unreasonably increased due to the inclusion of redundant measurement items in the scale. There is also a coefficient that can be used simultaneously with the Cronbach α coefficient. The coefficients can help in the evaluation of whether the calculation of the mean number masks some unrelated measurement items in the calculation of the Cronbach α coefficient. Different researchers have different views on the threshold value of the reliability coefficient. Some scholars believe that the Cronbach α coefficient should be at least 0.8 in the basic research, and the Cronbach α coefficient should be at least 0.7 in the exploration study. In the practice study, the Cronbach α coefficient only needs to reach 0.6.

Then we use SPSS to analyze the reliability of the data. The results are shown in Table 2.

Table 2: Reliability Statistics

Cronbach's Alpha	N of Items
0.87	10

We can get 0.87. In a reasonable range, the data is a high degree of credibility. So we can use the data in the following analysis.

3. Language Learning

3.1 Economic Attraction
Since there are more than two hundred countries and regions, we divide the world into five continents, namely America, Asia, Oceania, Africa, Europe. And for every continent, we selected a typical country. Where we introduce a parameter called Economic Attraction (EA) to show the economy of a region. We assume that the people who want to learn language focus on the economy of the region. We normalize the per capita GDP coefficient and the growth rate of GDP coefficient to value the economic attraction[6].

$$GDP_P^* = \frac{GDP - GDP_{\min}}{GDP_{\max} - GDP_{\min}} \qquad (1)$$

$$GDP_{gr}^* = \frac{GDP_{gr} - GDP_{gr\min}}{GDP_{gr\max} - GDP_{gr\min}} \qquad (2)$$

$$EA = GDP_P^* + GDP_{gr}^* \qquad (3)$$

Where GDP_P^* the country is's per capita GDP coefficient, GDP_{gr}^* is the growth rate of GDP coefficient of a country, GDP_{gr} and is the growth rate of GDP.

We also fully attach great importance to the GDP attraction while choosing the typical country for every continent. They are America for America, Japan for Asia, Australia for Oceania, South Africa for Africa, Spain for Europe. Among them, South Africa is not one of the top 3 countries of GDP in Africa, but its comprehensive national strength can't be ignored. Compared with the countries that hold higher GDP but only specialize in tourism or resources, we carry out the GDP attraction of South Africa is more stable and reliable. Similarly, Spain's GDP doesn't rank first in Europe, but Spanish is one of the top ten influential languages given by the topic. Due to this phenomenon, we exclude the countries that hold higher GDP but only specialize in tourism or resources.

3.2 Language Learning Trend
We apply the five countries per capita GDP in 2013, 2014, 2015, 2016, 2017.The growth rate of the country GDP are listed below in Figure 1 and Table 3

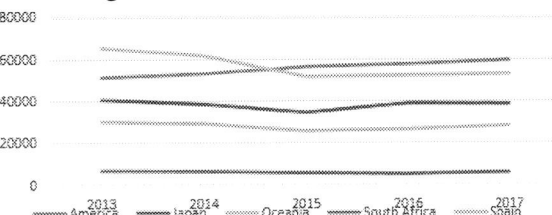

Figure 1: 2013-2017 GDP per capita of representative countries in five continents

Table 3: 2013-2017 growth rate of the five countries' GDP

Nation	America	Japanese	Australia	Africa	Spain
Rate	6.9	6.09	-7.12	-8	-0.16

We normalize the per capita GDP coefficient and the growth rate of GDP coefficient to value the economic attraction from 2013 to 2017. We combine the weight of the GDP with the EA in a country in Figure 2. And it can be seen that English and Japanese are highly attractive languages for language learners. Therefore, the number of users in English and Japanese will increase in the future.

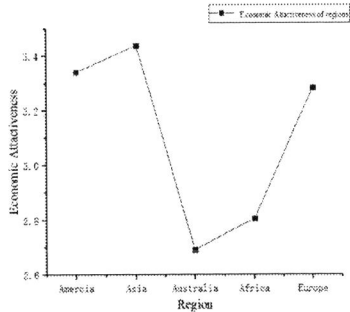

Figure 2: Economic attraction of different regions

We use the total EA of four continents as a denominator, each EA of the four continents as a molecular. So we learn the proportion of Asia immigrants in each continent. Similarly, we know the distributions of the other four continents. There are five continents distributions as listed:

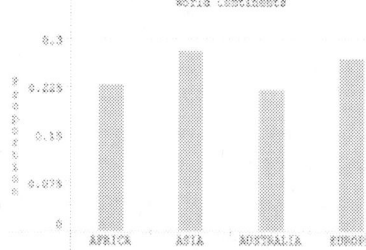

Figure 3: Distribution of American immigrant population

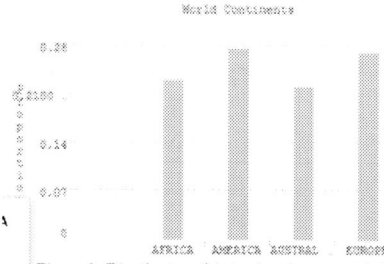

Figure 4: Distribution of Asian immigrate population

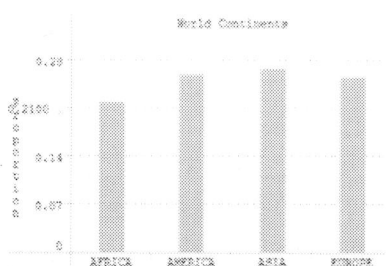

Figure 5: Distribution of Australian immigrant population

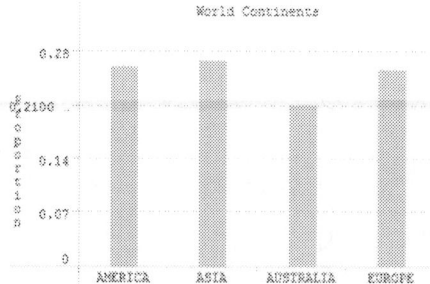

Figure 6: Distribution of African immigrant population

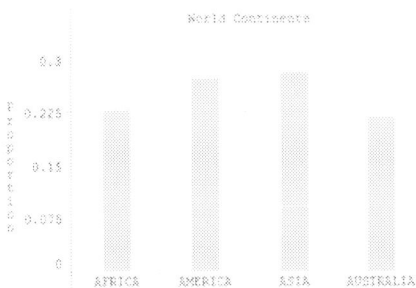

Figure 7: Distribution of European immigrant population

4. International Offices Options of Multinational Service Company

This paper selects a multinational company as a typical example of economic globalization for analysis. The multinational company with offices in New York City in the United States and Shanghai in China, is continuing to expand to become truly international. On one hand, the company formulate regional strategies to build inter-regional webs, improve the distribution of client resources. On the other hand, a company seeks to make the most efficient use of client resources in the different areas.[7] In order to do business better, the staff at the office need to speak the local language.

4.1 Model Building

We use TOPSIS to choose the location of the office. TOPSIS is an effective multi-index evaluation method. [8]This method evaluates positive ideal solution and negative ideal solution. We sort the optimal solution by computing the relative relevance of each ideal solution to the best solution. The steps are as follows:

① Use the method of vector programming to get the canonical decision matrix. Supposing the decision matrix of multi-attribute decision makes problem $A = (a_{ij})_{m \times n}$, standardized decision matrix $B = (b_{ij})_{m \times n}$, a technique for order preference by Similarity to an Ideal Solution can be specified as follows:

$$b_{ij} = a_{ij} / \sqrt{\sum_{i=1}^{m} a^2_{ij}} \ , i = 1, 2, \cdots, m; j = 1, 2, \cdots, n \tag{4}$$

② Constitute a matrix of weighted norms $C = (c_{ij})_{m \times n}$. Suppos that the decision vector given by each attribute is $w = [w_1, w_2, \cdots, w_n]^T$, then

$$c_{ij} = w_j \times b_{ij}, i = 1, 2, \cdots, m; j = 1, 2, \cdots, n \tag{5}$$

③ Determine positive ideal solution C^* and negative ideal solution C^0. Assume that the j property value of positive ideal solution C^* is c_j^*, the j property value of negative ideal solution C^0 is c_j^0, then

$$c_j^* = \begin{cases} \max c_{ij}, j: \ \text{benefit}, \\ \\ \min c_{ij}, j: \text{cost}, \end{cases} \quad j = 1, 2, \cdots, n, \tag{6}$$

$$c_j^0 = \begin{cases} \min c_{ij}, j: \text{benefit}, \\ \\ \max c_{ij}, j: \text{cost}, \end{cases} \quad j = 1, 2, \cdots, n \tag{7}$$

③ Calculate the distance between each solution and the ideal solution. The distance d_i between the alternative and the ideal solution is

$$s_i^* = \sqrt{\sum_{j=1}^{n}(c_{ij} - c_j^*)^2}, i = 1, 2, \cdots, m; \tag{8}$$

④ The distance d_i between the alternative and the negative ideal solution is

$$s_i^0 = \sqrt{\sum_{j=1}^{n}(c_{ij} - c_j^0)^2}, i = 1, 2, \cdots, m. \tag{9}$$

⑤ Calculate the number of queuing indicators for each program is

$$f_i^* = s_i^0 / (s_i^0 + s_i^*), i = 1, 2, , m. \tag{10}$$

According to f_i^* descending order of the arrangement of programs.

4.2 Model Solution

As we see from United Nations Industrial Development Organization ,[9]there are 6 factors about the site selection of the international office: political environment, infrastructure conditions, geographic conditions, policies and institutional environment, modern service industry agglomeration, human resources. We assume the weight of six factors is the same and the chosen cities' political environment is almost the same.

Considering the 5 factors, we use the city GDP to measuring the level of infrastructure conditions and geographic conditions, per capita GDP measuring the level of the level of modern service industry agglomeration, the number of famous universities measures the human resources.

Based on The World Bank, we know some cities: London, Singapore, Hong Kong, Paris, Tokyo, Dubai, Sydney, Bombay, Moscow, Madrid. Since these cities above attract multinational company, we have these as the locations of national office. Normalize the GDP per capita, the city GDP, the number of famous universities Use the Weighted Average method to get the weights of the GDP per capita, the city GDP, the number of famous universities respectively 0.25,0.5,0.25.

We apply MATLAB to calculate the advantage order of the ten cities and the results are in Table 4.

Table 4: Advantage order of the ten cities

City	Tokyo	London	Dubai	Paris	Madrid
Index	0.8725	0.7086	0.6017	0.5825	0.5623
City	Singapore	Sydney	Hong Kong	Moscow	Bombay
Index	0.5537	0.4412	0.4012	0.3527	0.3524

Based on the data above, we choose Tokyo, London, Dubai, Paris, Madrid, Singapore for a large multinational service company to open the additional international offices in Figure 8. And the language of the office staff is Japanese, English, Arabic, French, Spanish, Chinese or English.

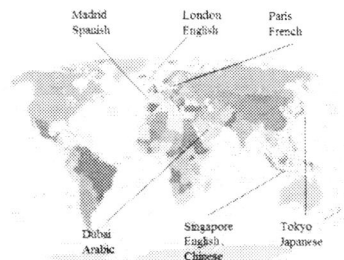

Figure 8: The location of the six new offices and the language used by their employees

5. Conclusions

This paper analyzes the interaction between economy and language. First, we use reliability analysis to analyze the reliability of the collected data, and the results show that the data used in this paper has high reliability. the economic attractiveness is defined to estimate the changes in future language use under the influence of the economy. The results show that the number of users in English and Japanese will increase in the future. Then we choose multinational companies as a typical example of economic globalization, use TOPSIS to select the location of the office and the language usage of their employees. At last, these offices are Tokyo, London, Dubai, Paris, Madrid, Singapore.

References

[1] Tao Chen. Distribution and abundance of Indo-Pacific humpback dolphins (Sousa chinensis) in the Moyang River Estuary: the western boundary of the world's largest population of humpback dolphins [A]. The International Society of Zoological Sciences (ISZS)、The Northwest Institute of Plateau Biology (NWIPB), CAS、Zoological Society of Qinghai Province. Proceedings of the 9th International Symposium of Integrative Zoology[C].The International Society of Zoological Sciences (ISZS)、The Northwest Institute of Plateau Biology (NWIPB), CAS、Zoological Society of Qinghai Province:,2017:1.

[2] List of Languages by Total Numbers of Speakers https://en.wikipedia.org/wiki/List of languages by total number of speakers January17, 2018

[3] Min Yang. Cohesive Features in English as a Second Language Students' Argumentative Writing——Taking Kunming University as an Example[A]. INTI University & Colleges, Tunku Abdul Rahman University College. Proceedings of the 2018 International Seminar on Education Research and Social Science(ISERSS 2018)（Advances in Social Science, Education and Humanities Research(ASSEHR),VOL.195）[C].INTI University & Colleges, Tunku Abdul Rahman University College:,2018:5.

[4] Geng Liu. A Forecast Model for Language under the Influence of Immigration [A].Proceedings of 2018 3rd International Symposium on Mathematics and Computer Science（ISMCS2018）[C]. 2018:9.

[5] KHALID AHMED. Pakistani ESL Learners' Pragmatic Competence: Motivation and Development[D]. Huazhong Normal University,2016.

[6] https://data.worldbank.org.cn/indicator/SE.XPD.TOTL.GD.ZS

[7] Odhiambo Odera,Albert Scott,Jeff Gow. An examination of the quality of social and environmental disclosures by Nigerian oil companies [J]. Corporate Governance,2016,16(2).

[8] Andrzej Kobryń,Joanna Prystrom. A Data Preprocessing Model for the Topsis Method[J]. Folia Oeconomica Stetinensia, 2016, 16(2).

[9] Christine Jasch. Governmental initiatives: the UNIDO (United Nations Industrial Development Organization) TEST approach[J]. Journal of Cleaner Production,2015,108.

ISPECE

IOP Publishing

An Expertise-enhanced Collaborative Filtering Method for Keywords Recommendation in Searching Engine Marketing

Yuan Zhu

School of Trade and Economics, Taizhou Vocational and Technical College, Zhejiang, China

*Corresponding author e-mail: tzzy246@163.com

Abstract. Nowadays, with the development of information technology, people like to use searching engine (SE) for help. Thus stores and businessmen start to realize the necessity of marketing through searching engine. Searching engine marketing (SEM) becomes a kind of new marketing tools, which can also be called advertisement campaign. SE business users select keywords in the SE system to attract customers searching of these. How to select the right keyword in SEM is a necessary problem needed to be solved recently for SE business users. Traditional keywords recommendation only recommend semantic similar keywords to business users which still makes the result set too large to choose. The new influential factor - expertise value combined with click rate will be integrated into the traditional collaborative filtering to minimize the keywords set waiting to be selected. The final recommendation is based on our expertise-enhanced collaborative filtering combined with semantic similarity.

1. Introduction

With the repaid growth of Web 2.0, customers are used to shopping online and searching the Internet for suggestions instead of going to the offline shopping store in person. Thus the online information has become very important for guiding customers to the right destinations they want. More guidance or references appears online to meet their increasing needs, which has caused information explosion. The exploded data have increased the difficulty for customers to find the right commodity, which made them more and more confused.

Searching Engine (SE) was developed to solve this problem by recommending information or commodities related to the customer searching words. By pressing the enter bottom online of Google Searching Engine, related web links will be provided immediately based on the keywords being typed on the searching bar. As the results of this searching engine can easily affect the shopping decisions of customers, Searching Engine Marketing (SEM) [1] was proposed for businessman to broadcast their products online by paying the SE company on the user clinks searched from the keywords they set. The click-paid method had soon attracted the attention of many retailers. However, as more and more retailers participate in the SEM plan, more keywords appear in the SE database. Which keywords are the most suitable for these retailers start to become a difficult problem. Current SEM system can recommend some keywords to retailers. Yet these keywords are only based on their similar commodities, which lack the consideration of other retailers. Thus we propose a keyword recommendation method, which considers the influence of both retailers and commodities. We classifies the retailer into different levels according to their expertise index extracted and calculated from their own profile to make our recommendation more proper for retailers to succeed in advertise campaign.

Content from this work may be used under the terms of the Creative Commons Attribution 3.0 licence. Any further distribution of this work must maintain attribution to the author(s) and the title of the work, journal citation and DOI.
Published under licence by IOP Publishing Ltd

2. Related work

Recommendation system has proven to be an effective response to the information overload currently in numerous areas [2], providing suggestions for all kinds of users. Traditional methods of recommendation can be classified into two kinds:

1) Content-based recommendation.
2) Collaborative filtering (CF).

The former one relies on the completeness of user/item profile. The recommendation process cannot be carry on if the system lack user's content, which was defined as a cold start problem.

The latter one is more used in modern business. For example, Amazon has used collaborative filtering method to recommend books considering other users' choice as references. The predicted recommendation is based on ranking the ratings calculated from its nearest users'/items'. Nevertheless, in the case of rating scarcity, the CF can hardly execute because of rating prediction problem.

For the calculation of CF, usually we use the Pearson correlation coefficient or the cosine measure [3] to acquire the similarities between different users/items, shown as follows [3],

$$Cos(A,B) = \frac{\sum_{i=1}^{n} R_{A,i} \times R_{B,i}}{\sqrt{\sum_{i=1}^{n}(R_{A,i})^2} \times \sqrt{\sum_{i=1}^{n}(R_{B,i})^2}}$$

(1)

where n denotes the number of co-rated items by both user A and user B, $R_{A,i}$ and $R_{B,i}$ are the ratings for the ith item from the n co-rated items rated by user A and user B respectively.

Neither of these two methods can solve the above problem. Recently some methods [4] have been proposed which combine the two methods together to make the recommendation more suitable in different extreme cases. New elements such as trust [5], demographic factors [6], etc. calculated through different algorithms and models have been integrated into traditional recommendation method, which can influence the prediction precision. Parvin, Moradi and Esmaeili [5] have used ant colony optimization to calculate the user trust to enhance the traditional CF method.

Recently, expertise has been discussed by many anthers [7-9] in recommendation systems as a key point for rating prediction, which can measure the professionalism of the reference users. In the process of words recommendation, the suggestions of professional ones can be considered more valuable than other reference. Recommended results based on domain experts will be more trustable than those simply based on semantic similarity [7] or rating similarity. Thus we propose an expertise-based personalized keywords recommendation method in SEM to differ from others with the redefinition of experts adapting to the new area.

3. An enhanced expertise-based recommendation method for SEM

Fig.1 below shows the architecture of our expertise-enhanced collaborative filtering recommendation method for SEM, which can be separated into three parts:

1) Data extraction. Keywords and user contents that have attended the former advertisement campaign will be extracted from the underlying data forming SE keywords database and SE user log database.

2) Rating computation. In this section, we calculate the similarity between keywords and users through the data extracted from the database. Besides, the expertise level of the users is also computed to acquire the expertise-based words rating similarity for the latter recommendation. The process of computation will be elaborated in details later on.

3) Recommendation. The top-ranking keywords will be recommended by the calculated expertise-enhanced words rating.

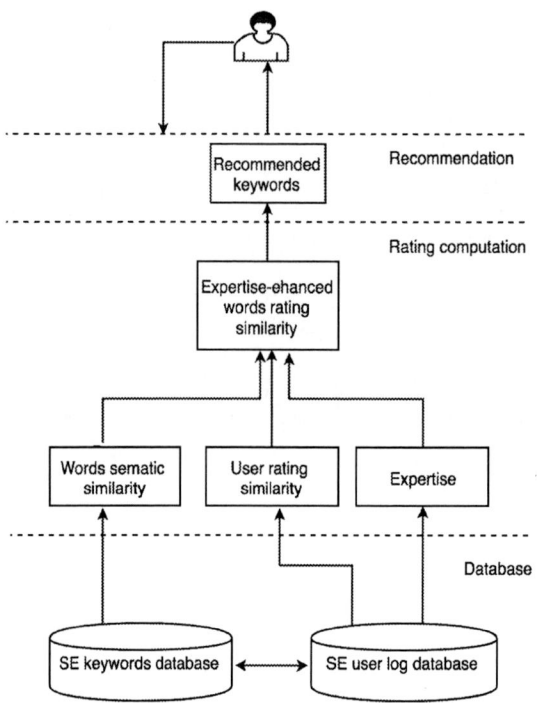

Figure 1. The architecture of our expertise-enhanced CF for SEM

3.1. Words semantic similarity

Ontology-based semantic similarity has long been researched by literature [11-12], which is proved to be accurate and quick responsible. Cai [13] has proposed a semantic similarity method in the area of manufacturing services with his manufacturing ontology. In our work, we would like to take Cai [13]'s method as a reference to compute keyword semantic similarity based on the Abdeljaber [14]'s proposed ontology.

The semantic similarity between keyword KW_i and keyword KW_j within Abdeljaber [14]'s proposed ontology O can be calculated as the equation below [2],

$$Sim(KW_i, KW_j, O) = \alpha \frac{n\left(C_{super}(KW_i, O) \quad C_{super}(KW_j, O)\right)}{n\left(C_{super}(KW_i, O) \quad C_{super}(KW_j, O)\right)} + \beta \frac{n\left(C_{sub}(KW_i, O) \quad C_{sub}(KW_j, O)\right)}{n\left(C_{sub}(KW_i, O) \quad C_{sub}(KW_j, O)\right)}$$

(2)

where $C_{super}(KW_i, O)$ and $C_{super}(KW_j, O)$ represent the super classes subsuming KW_i and KW_j within ontology O separately while $C_{sub}(KW_i, O)$ and $C_{sub}(KW_j, O)$ denote the sub classes subsumed by KW_i and KW_j respectively within O. $n\left(C_{super}(KW_i, O) \quad C_{super}(KW_j, O)\right)$ is the number of super classes subsuming both KW_i and KW_j while $n\left(C_{super}(KW_i, O) \quad C_{super}(KW_j, O)\right)$ is the number of super classes subsuming either KW_i or KW_j. Besides, $n\left(C_{sub}(KW_i, O) \quad C_{sub}(KW_j, O)\right)$ is the number of sub classes subsumed both KW_i and KW_j while $n\left(C_{sub}(KW_i, O) \quad C_{sub}(KW_j, O)\right)$ is the number of sub classes subsumed either KW_i or KW_j. α and β are the weighting factors and we define $\alpha + \beta = 1$ to restrict the value of $Sim(KW_i, KW_j, O)$ to be not too large.

The semantic similarity between user U_i and U_j can be computed as equation [3] likewise,

$$Sim(U_i, U_j, O) = \alpha \frac{n\left(C_{super}(U_i, O) \quad C_{super}(U_j, O)\right)}{n\left(C_{super}(U_i, O) \quad C_{super}(U_j, O)\right)} + \beta \frac{n\left(C_{sub}(U_i, O) \quad C_{sub}(U_j, O)\right)}{n\left(C_{sub}(U_i, O) \quad C_{sub}(U_j, O)\right)}$$

(3)

where $C_{super}(U_i, O)$ and $C_{super}(U_j, O)$ represent the super classes subsuming U_i and U_j within ontology O separately while $C_{sub}(U_i, O)$ and $C_{sub}(U_j, O)$ denote the sub classes subsumed by U_i and U_j respectively within O.

3.2. Expertise computation

In this paper, the expertise user of a searching engine database can be defined as ones whose former keywords being selected to participating in the advertisement campaign have reaped lots of customer user clicks. The more the customer clicks the link related to the keywords, the more professional the user is. Finally it will increase the expertise value of the user.

Suppose a user U_i has selected N keywords to participate in the advertisement campaign and one keyword $KW(U_i, n)$ selected by him has accepted $C_{KW(U_i, n)}$ clicks, thus the total clicks received of user U_i bought by his keywords can be calculated as $\sum_{n=1}^{N} C_{KW(U_i, n)}$.

Thus the expertise value of a user U_i can be computed through the following equation,

$$EV(U_i) = \frac{\sum_{n=1}^{N} C_{KW(U_i, n)}}{\sum_{i=1}^{M} \sum_{n=1}^{N} C_{KW(U_i, n)}}$$

(4)

where $\sum_{i=1}^{M} EV(U_i) = 1$ and $EV(U_i)$ is the expertise value of user U_i and $\sum_{i=1}^{M} \sum_{n=1}^{N} C_{KW(U_i, n)}$ is the total clicks within the search engine database of M total users taking part in the former advertisement campaign.

This way of computation for expertise value is to normalize the result, which make the comparisons between two different users easier.

3.3. Expertise-enhanced collaborative filtering

Thus the expertise-enhanced predicted click rate of a new keyword $KW(U_j)$ by user U_j can be computed through the equation below (6),

$$PC\left(KW(U_j)\right) = Sim_U \times Sim_{KW} \times C_{KW(U_i)} \times EV(U_i)$$

(5)

where

$$Sim_U = Sim(U_i, U_j, O)$$

(6)

$$Sim_{KW} = Sim\left(KW(U_i), KW(U_j), O\right)$$

(7)

$Sim_U = Sim(U_i, U_j, O)$ means the semantic similarity between user U_i and U_j while $Sim_{KW} = Sim\left(KW(U_i), KW(U_j), O\right)$ means the semantic similarity between keywords KW_i and KW_j. $C_{KW(U_i)}$ is the click rate of user U_i and $EV(U_i)$ is the expertise value of U_i.

The predicted click rate $PC\left(KW(U_j)\right)$ of user U_j's keyword $KW(U_j)$ will be ranked according to the final results and the keywords with the top highest predicted click rate will be recommended to the user for them to select as the advertisement campaign reference.

4. Conclusion

In this paper, an expertise-enhanced collaborative filtering method has been proposed to address the keywords selection problem in searching engine market (SEM) for advertisement campaign. We use ontology-based semantic method to compute the user similarity and keywords similarity to firstly filter the dissimilar keywords with less semantic similarity. Secondly, we give a new expert definition for the users in SE system and calculate the expertise value of the users through keywords click rates. The final predicted click rate for a new coming keyword will be computed considering the influence of both user/keywords semantic similarity and user expertise. However, there are still some restrictions in the process of rating prediction. The method in our work does not consider the various domains of different keywords. Besides, experts in different fields may have different expertise value. In this case, more studies should be conducted for further improvement and perfection.

Acknowledgments

This work was supported by Normal Program of Provincial Education Department of Zhejiang (No.Y201636388).

References

[1] Drèze, Xavier, and François-Xavier Hussherr. 2003. "Internet Advertising: Is Anybody Watching?" Journal of Interactive Marketing. Vol. 17, no. 4, pp. 8–23.

[2] M. I. Matin-Vicente, A. Gil-Solla, M. Ramos-Cabrer, Y. Blanco-Fernandez, M. Lopez-Nores,"Semantic inference of user's reputation and expertise to improve collaborative recommendations", Expert System with Applications, vol. 39, pp. 8248-5258, 2012.

[3] L. Mekouar, Y. Iraqi, R. Boutaba, "An analysis of peer similarity for recommendations in P2P systems", Multimedia Tools and Applications, vol. 60, no. 2, pp. 277-303, 2012.

[4] Christou, I. T., Amolochitis, E., & Tan, Z.-H.. AMORE: design and implementation of a commercial-strength parallel hybrid movie recommendation engine. Knowledge and Information Systems, Vol. 47, No. 3, pp. 671–696, 2015.

[5] Parvin, H ; Moradi, P; Esmaeili, S . TCFACO: Trust-aware collaborative filtering method based on ant colony optimization, EXPERT SYSTEMS WITH APPLICATIONS. Vol.118, pp. 152-168, 2019.

[6] B. Yapriady, A. L. Uitdenbogerd, "Combining Demographic Data with Collaborative Filtering for Automatic Music Recommendation", Knowledge-Based Intelligent Information and Engineering Systems, vol. 3684, pp. 201-207, 2005.

[7] Y. H. Xu, X. T. Guo, J. X. Hao, J. Ma, R. Y. K. Lau, W. Xu, "Combining social network and semantic concept analysis for personalized academic researcher recommendation", Decision Support Systems, vol. 54, no. 1, pp. 564-573, 2012.

[8] K. Kwon, J. Cho, Y. Park, "Multidimensional credibility model for neighbor selection in collaborative recommendation", Expert Systems with Application, vol. 36, no. 3, pp. 7114-7122, 2009.

[9] Z. X. Huang, X. D. Lu, H. L. Duan, C. H. Zhao, "Collaboration-based medical knowledge recommendation", Artificial Intelligence in Medicine, vol. 55, no. 1, pp. 13-24, 2012.

[10] Chen Y, Lin SP. Technology of semantic search based on ontology. Comput Eng Appl. 2006;S1:78-80 [SEP]

[11] Bhogal J, MacFarlane A, Smith P. A review of ontology based query expansion. Inf Process Manag. Vol. 43, no. 4, pp. 866-886, 2007. [SEP]

[12] Yong LI, Zhang ZG. Semantic retrieval research based on ontology. Comput Eng Sci. vol. 30, no. 4, pp. 17-18, 2008. [SEP]

[13] M, Cai, W, Y, Zhang, K. Zhang, "ManuHub: a Semantic Web system for ontology-based service management in distributed manufacturing environments", IEEE Transactions on Systems, Man, and Cybernetics Part A: Systems and Humans, vol. 41, no. 3, pp. 574-582, 2011.

[14] Hikmat A. M. Abdeljaber, "Profile-Based Semantic Method using Heuristics for Web Search Personalization" International Journal of Advanced Computer Science and Applications(ijacsa), vol. 9, no. 9, pp. 191-198, 2018

Analyse the data tendency in the public opinion monitoring system

Jian Li*

Luoyang Campus of the PLA Information Engineering University, Luoyang, China

*Corresponding author e-mail: maomaotfntfn@163.com

Abstract. With the rise of the network era of big data, many of a colour view of public opinion data emerge in endlessly, network also drives the various views in the different public opinion direction, so the tendency of public opinion data analysis is becoming more and more attention by many scholars and government officials, public opinion is most critical text content, data is mixed of text content. This paper analyses the tendency of network comment data, introduces language processing, subject extraction and other related knowledge, and achieves the purpose of improving the accuracy of tendency classification.

1. Introduction

With the rapid development of Internet technology, people are more inclined to express their opinions on the Internet, such as blogs, BBS, WeChat, live broadcast, twitter, etc., or spread events and participate in comments through news, current events, social channels and so on. So public opinion information has been widely spread on the Internet in various media, many media, business organizations, such as e-commerce platform will use public opinion to implement the regulation, promotion, marketing and other purposes, for example, when consumer is buying a product or service, will refer to the comment below has the user information, to achieve direct understanding of the product or service. Merchants collect these reviews, optimize customer products and after-sales services, and implement better production and marketing strategies. The national government analyzes public opinion data, understands social needs, provides better services for the people, analyzes information sources for emergencies, finds the most critical and popular problems, reasonably guides and solves them, and maintains social stability.

Data trend analysis has an important position in the field of public opinion monitoring, big data in the Internet environment, using the analysis of mining technology to judge subjective emotional color text, text data to mixed results, a process that involves the different level of word, word, sentence discourse data content, and produces different tendency analysis method, including the data mining technology, machine learning techniques [1], [2] language processing technology, information retrieval technology [3] and other related domain knowledge, and analysis of existing complexity. In addition, the application of these technologies in the field of trend analysis has expanded the scope of public opinion analysis system, new technical problems have been emerging, and numerous researchers have been solving new problems, which has also made the subject of trend analysis increasingly perfect in the field of public opinion monitoring.

Nowadays at home and abroad, this paper study tendency analysis method, summarized based on machine learning method based on semantic judgment method and mode of different characteristics, focus on the solution produced by different population level view differences of misjudgment effects on the fact that, in accordance with point of view of people's difference and the formation process of the

---

Content from this work may be used under the terms of the Creative Commons Attribution 3.0 licence. Any further distribution of this work must maintain attribution to the author(s) and the title of the work, journal citation and DOI.

Published under licence by IOP Publishing Ltd

theme idea, adopting reasonable subject building model and feature selection method to extraction point topic. In view of the massive network public opinion data, a reasonable solution is found. By building a combinatorial classifier model on the distributed platform, the trend analysis results of public opinion data can be obtained quickly and accurately. The research in this paper can reduce the impact of subjective crowd thinking differences on the accuracy of tendency classification, and achieve a rapid and efficient analysis of public opinion tendency.

2. Data tendency classification method

2.1. Feature selection technology
Feature selection is an important part of data tendency classification. The feature item should fully reflect the information carried by the text data. Therefore, the more reasonable the feature item is selected, the more accurate the classification result will be.

2.1.1 word frequency method, document frequency method
The word frequency method (TF) counts the frequency of characteristic words appearing in the text, and involves the technical content of some statistics. The size of the threshold is used to determine the topic words, and the statistical results and the range of the threshold are compared, which will exceed the threshold interval. The vocabulary outside the scope is eliminated, which not only extracts the high-frequency vocabulary, but also reduces the text dimension, which provides convenience for the subsequent classification by the classifier.

The document frequency method (DF) is developed based on the word frequency method. First, a feature item is defined. The feature item is placed in a specified database corpus, and the number of the feature items in the data set is counted and calculated. It accounts for the ratio of the total dataset corpus. The formula is as follows [4]

$$DF(t_i) = \frac{N_{ti}}{N_{all}} \qquad (1)$$

In formula (1), N_{ti} is a feature item, N_{all} is the number of documents contained in feature item i, and N_all is the number of documents in all data sets. The ratio of the feature items in the corpus is obtained by the method, so that the theme is extracted for the feature item, that is, the word frequency method is used again. Use the previously specified threshold range to filter the topic, exclude words that do not meet the range of the threshold, or use the size of the DF value to judge. If the value is small, the frequency of a document appears to be small, then the document matches. The feature item does not adequately reflect the subject matter of the document. Conversely, when the value is large, it indicates that the document appears more frequently, and the matching feature item has clear content recognition ability. The center of the document word frequency method is reflected in the vocabulary labeling, matching documents, and judging the devaluation. It is an early widely used topic extraction method, but there are many uncertainties depending on the number of words.

2.1.2 Mutual information
Mutual information is used to analyze the relationship between two different random variables, that is, to measure whether the variables are related. It embodies the correlation between feature items and classifiers from the side, so that we can select the appropriate classification system based on this relationship. Different from the word frequency method, the feature item does not depend on the vocabulary, but is related to the tightness between the variables. The closer the relationship between the two random variables, the larger the feature item, the looser the association degree and the smaller the feature item. The essence is the degree of mutual correlation of feature items, and the expression of mutual information between category C and feature t is as follows

$$MI(t_i, c_i) = \log \frac{p(t_i, c_i)}{p(t_i)p(c_i)} \qquad (2)$$

In formula (2), t_i is the feature item, c_i is the category to which the feature item belongs, MI is the relationship between a feature item and the belonging category, and p is the probability that the item appears in the category. It can be seen that when Ml is 0, it indicates that the feature item has no relationship with the category, that is, two random events are independent of each other. If the value of Ml is larger, it indicates that the feature item is closely related to the category. By extracting topics from mutual information, the steps are cumbersome. It is necessary to label a certain category or a feature item first, so we must first use the classifier to do a classification operation, but this method directly categorizes the feature items and reduces the classifier. The influence between the topic extraction and the theme extraction is conducive to the subsequent classification model construction, and has a high degree of precision.

3. optimization tendency analysis method

3.1. Feature selection technology optimization method

When preprocessing the text data to be analyzed, if the vector feature is established directly by the word segmentation technique or the statistical word frequency to obtain the text feature, the text vector will have a huge dimension, the workload is very large, and the classifier classification is greatly reduced. Accuracy, so to ensure the core information of the text, simplify the workload, improve the classification accuracy, you need to adopt the most representative feature items, which is the focus of research feature selection technology. Common feature selection methods are Document Frequency Method (DF) and Information Gain Method (CIG). These methods all have their own selection criteria, but they also have their own shortcomings. This chapter analyzes the advantages and disadvantages of these feature selection methods, and improves the inadequacies. It adds word frequency parameters to reduce the interference factor and improves the accuracy of feature extraction.

3.1.1 Document Frequency Method Improvement

The document frequency method is simple to calculate. It only needs to consider the number of feature documents contained in the entire document set. It is necessary to calculate the weight value of each word in the data set and compare the set threshold value selection feature items. Feature items that are within the range of the threshold are retained, and feature items that are outside the range of the threshold are filtered out. This method is suitable for feature screening of large-scale document sets. However, the disadvantage is that there are many characteristic words with few occurrences, but these words contain important classification information. If it is directly filtered, it will have a great influence on the classification accuracy.

3.1.2 Improvement of information gain method

The information gain is used to measure the difference between the feature item and the classification information. If a feature item can make the amount of information generated by the classifier larger, the more important the feature item is. Suppose there is a variable Y, and its value has n possibilities, namely $Y_L \ldots .. Y_N$ and each value probability is $p_1 \ldots \ldots p_n$, then the information expression of the variable Y is :

$$H(y) = -\sum_{i=1}^{n} p_i \times log_2 p_i \qquad (3)$$

It can be seen from formula (3) that the richer the number and type of variables, the larger the amount of information, indicating that the amount of information is related to the number of values and the probability of each number, and the value of the variable itself. The value does not matter much.

3.2. Subjective extraction technology optimization method

Among the various news and commentary topics on the Internet, the topic has a wide range of topics, which may be aimed at different social events, government decisions, facts, or opinions. Some comments contain repetitive and diverse views, ie, subject matter. Fuzzy, contradictory views, which will affect the screening of features, and finally cause the problem of reduced classification accuracy. Therefore, an excellent topic recognition method can extract information from a wide variety of topics, and achieve the most attention-grabbing content in mining online public opinion. Topic recognition is divided into two aspects, perspective topic recognition and text topic recognition. There are connections between the two and there are big differences. The subject of opinion refers to the object of subjective emotion published in the text content. The text topic refers to the object expressed by the text content, that is, by whom to express and what to express. In a variety of online news or comments, most researchers only analyze text topics, and ignore the topic, which leads to overlap between the two, affecting the accuracy of text topic extraction. Therefore, for both text and perspective, it is necessary to find a reasonable topic recognition technology to improve the accuracy of topic recognition.

4. Based on the concept of machine learning to build a public opinion classification model

4.1. Target requirements
The majority of netizens use various online platforms to comment on news reports or social dynamics in the society, express their own opinions, and form word of mouth or public opinion. The social public opinion monitoring department and the majority of social survey scholars have obtained different emotional tendencies, derogatory or derogatory meanings of the masses through different methods of inclination analysis. Through different degrees of analysis, we can measure the weight of public opinion and judge the possible impact of public opinion events. However, the perspectives of different social groups are different, and the opinions expressed are different. They may violate the original truth of the facts and form a wrong thinking guide. Therefore, public opinion will have deviations in the process of communication, leading to the accuracy of the tendency to discriminate. Sexual decline, so in order to maintain a high degree of accuracy in the final analysis, it is necessary to consider how to use an efficient method to solve the impact of crowd differences.

4.2. Subjective thinking differences
The difference in subjective thinking is manifested in two aspects: the complexity of Chinese expression and the diversity of people in different fields.

(1) Part of speech ambiguity: Different emotional words have different meanings under different parts of speech, and thus the tendency of emotions is different, such as "you are a person who is easy to be proud" and "I am proud of him", in two Under the word, "pride" has different tendencies, adjectives mean derogatory, and nouns mean derogatory.

(2) Collocation ambiguity: In the process of text expression, some words have no emotional tendency, but when used in conjunction with other words, they have emotional tendencies, such as "high quality" and "high price". The former means derogatory, The latter represents derogatory. "Learning prominent" and "lumbar prominence", the former derogatory and the latter derogatory.

(3) Statement Ambiguity: The same emotional words show different emotional polarities under different contexts. For example, some textual content will use the derogatory words: "True is OK" and "You are extremely excellent". "Eye" and other expressions convey a concept of irony and accusation.

(4) Ambiguity of the evaluation object: Generally, the content of the news report will describe and comment on the object of an evaluation in different language expressions. However, due to the complexity of the Chinese context, the evaluation object needs us to make a clear judgment. Inaccurate, it may have the opposite emotional tendency, such as "a good cook made a bad meal", "a rotten chef made a delicious dish", evaluation chefs or evaluation dishes, we need to determine the object of evaluation, otherwise it will Misleading emotional tendency results. It can be seen that the semantic complexity of Chinese is restricted by part of speech, collocation, text environment, etc. Therefore, it is necessary to do a lot of optimization processing when judging the tendency of text, helping us to eliminate the influence of these potential factors and make the structure of sentiment analysis more To be accurate.

4.3. Solution
(1) The tendency analysis of social public opinion topics is a large-scale data set analysis. In order to take into account the emotional differences caused by different fields and different groups of people, a

single classifier cannot meet the demand. Therefore, it is necessary to build a framework and adopt The pattern of multiple classifications requires reasonable selection of text extraction methods and classifiers to ensure that different classifiers can be combined reasonably to obtain accurate data trend analysis results.

(2) Extract the theme first, and then tend to classify it as the main idea. It aims to resolve the differences in thinking between different groups of people and to weaken the interaction between different thinking in communication. After extracting the theme, it is necessary to construct a classifier, fully consider the classification method, the classification object, the association and influence of the classification steps, realize the distinction between the population, and reduce the influence of subjective thinking on the objective lyric facts.

Finally, a large amount of data is processed, and the results of the classification are obtained, and the accuracy of the results is verified.

(3) Because the concept of propensity analysis of data content is closely related to the technology of data mining and machine learning related fields [4], the related concepts of data mining and machine learning are introduced, and reasonable mining methods and learning modes are analyzed and selected. And on the basis of this, improve the construction of the classifier, and provide reliable and efficient text content support for our subsequent text data trend analysis operation.

4.4. Machine learning and data mining concepts

Data mining refers to searching in a large way in a large number of data. This process of searching is divided into automatic or semi-automatic [5]. For relatively large-scale data groups, automatic or manual supervision can be used for mining, but artificially assisted. The semi-automatic method is extremely labor intensive, so it is necessary to implement the concept of automated mining by means of machine learning, and to find hidden or unknown potential information in a large amount of data. Machine learning refers to the use or design of some analysis algorithms, according to specific rules, allowing the computer to automatically complete the analysis of the text content, and to achieve regular judgment of the content of the unknown text, and obtain the prediction results. Machine learning is the core part of data mining. It is mainly divided into supervised and unsupervised learning [6], as shown in Figure 1:

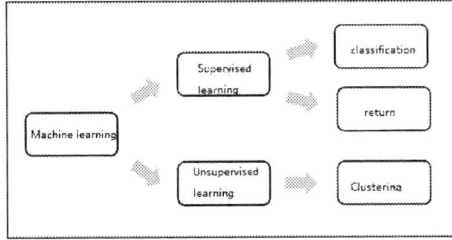

Figure 1: Data mining and machine learning flow chart

Figure 1 shows the implementation process of machine learning and related algorithms, which are mainly divided into supervised learning and unsupervised learning:

1) Supervised learning process

Supervised learning is also called predictive learning. In the case of text-oriented classification, it uses the classified text data set as a training sample, the predicted text data as a test sample, the design learning algorithm to achieve classification, and the result and sample set. The content is compared and the algorithm is optimized. Including classification and prediction algorithms.

2) Unsupervised learning process

Unsupervised learning refers to the analysis of text data directly through the algorithm under the condition of the determined sample data set, realizing the prediction of unknown text data, mining valuable information, no comparison of training sets, and completely based on the prediction set itself. The algorithm includes two modes: cluster analysis and association analysis. The clustering algorithm implements the subdivision of samples, finds the similarities of sample features and classifies them into one class. Association analysis finds the connection between things and is used to optimize the analysis results.

It can be seen that the unsupervised learning method mainly relies on data mining to achieve, but there is no clear measurement standard for the specific algorithm optimization concept, that is, it only realizes mining, so this paper uses supervised learning to measure and update the algorithm.

5. public opinion analysis experiment

5.1. Environment Construction

The experimental background uses the eclipse development environment, the graphics editor Origin, to complete the text data preprocessing work. Feature extraction and classifier construction are implemented using the weka platform to verify the feature selection optimization effect. The combined classifier was constructed using the Hadoop distributed environment [7] to verify the accuracy of the combined classifier.

5.2. Evaluation indicators

The classification evaluation index in this paper is also an important reference for measuring feature selection and theme refinement, including: precision rate, recall rate, and F value as evaluation indicators:

(1) Precision and recall rate:

Precision: Precision: is the ratio of the number of texts that are to be classified to $N_{w \to r}$, and the number of texts with correct classification results from $N_{r \to r}$, indicating whether the classification result is accurate, such as Formula (4):

$$p = \frac{N_{r \to r}}{N_{r \to r} + N_{w \to r}} \tag{4}$$

Recall: The full rate refers to $N_{r \to w}$ in all texts belonging to a category. This class is correctly judged. The proportion of the number from $N_{r \to w}$ reflects whether the classification result is complete, as shown in formula (5):

$$R = \frac{N_{r \to r}}{N_{r \to w} + N_{r \to w}} \tag{5}$$

(2) F value:

F is a comprehensive evaluation of the recall rate and precision rate:

$$F = \frac{2 \times P \times R}{P + R} \tag{6}$$

However, the precision ratio P and the recall rate R are inverse relations. The change of the precision and the recall rate show a reverse trend. It is necessary to use the F value to balance the direct relationship between the two, but in the complex network public opinion environment, accurate It is very difficult to calculate the recall rate. Therefore, under the premise of introducing the F value, this paper mainly evaluates the precision.

5.3. Feature Optimization Effect

According to the weka visualization interface, we can calculate the trend of the curve after optimization by different feature selection algorithms, and analyze the optimization effect through the graph.

(1) The TF optimization effect map reflects the curve trend of the F value as the data volume changes, as shown in Figure 2:

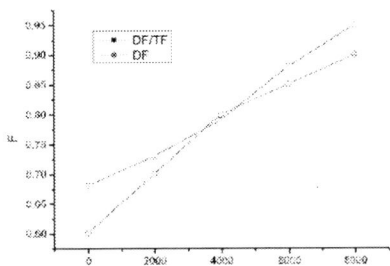

Fig.2 TF curve feature extraction optimization effect

It can be seen from Fig. 2 that the extraction effect of DF and DF/TF is analyzed, and the result curve is reflected by F value: DF/TF shows better accuracy as the number of feature items increases, and the F value gradually increases. High, indicating that the number of feature items does not affect the optimization method of adding TF, but the optimization effect is not obvious, and there is no optimization effect when the value is 4000 items.

(2) The IF optimization effect map reflects the trend of the F value as a function of the amount of data, as shown in Figure 3:

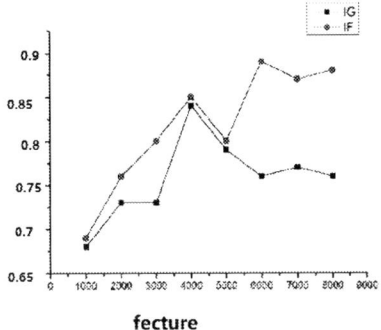

Fig.3 IF feature extraction optimization effect

It can be seen from Fig. 5-7 that the extraction effect of IG and IF is analyzed, and the curve result is expressed by F value: it can be seen that IF has more obvious effect, but with the increase of characteristic phase, F value shows a downward trend. It shows that IG is not affected by TF, and feature items are more dependent on classification results. When the feature item reaches about 4500, the effect is best.

3) The MI optimization effect reflects the trend of the F value as the amount of data changes, as shown in Figure 4:

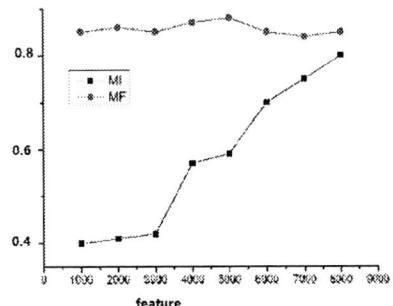

Fig.4 MI feature extraction optimization effect

It can be seen from Fig. 4 that the mutual information M optimization effect is the most obvious among the three. When the TF is not considered, the F value gradually increases with the number of feature items, and when the TF factor is added, the F value is not affected by the feature. The number of items is affected.

6. Summary

In the context of massive public opinion data, this paper takes the tendency classification of the paper data as the main research content, and aims to find a text tendency classification method with high precision and fast recognition speed. Achieve accurate trend classification results for Internet news events or product reviews. This paper mainly considers the influence of subjective emotional color on classification accuracy, improves feature extraction algorithm, introduces topic recognition mode to optimize classification effect, and combines machine learning features to construct combined classifier to realize text data classification.

(1) The feature selection algorithm is optimized in the feature selection stage. The TF word frequency correction parameters are added to the DF, MI, and IG feature extraction algorithms, aiming to optimize the classifier performance through a better feature selection algorithm.

(2) proposes a solution to the subjective crowd's emotional color. Through the topic perspective extraction technology, and using the LDA topic extraction model and the derivative terminology, the classifier performance is optimized based on the feature extraction.

(3) Based on the machine learning concept, this paper uses the integrated learning method to construct a combined classifier, and combines the clustering algorithm with the classification algorithm. Based on the optimal topic results, the classifier is used to derive the propensity results of different texts.

From the final point of view of the optimization tendency classification results, this paper adopts a reasonable optimization method for each step of the inclination analysis, which not only effectively reduces the influence of emotional deviations at different levels of the population, but also makes the text data tend to analyze.

The research on public opinion analysis and emotional judgment can't stay at the research stage, but it should be applied to many actual social statistics, such as government, enterprises, social organizations, etc., so that the statistical information is more accurate and practical. The reference value, although the public opinion monitoring can not reach the high level of foreign countries, but it must be in line with China's national conditions, and take advantage of the advantages of foreign monitoring methods, and steadily advance in the social environment.

References

[1]   Dev S, Wen B, Lee Y H, et al. A Tutorial on Machine-Learning Techniques and Applications[J]. IEEE Geoscience&Remote Sensing Magazine, 2016, 4(2):79-93.

[2]   Cui-Xia LI. Modern Computer Intelligent Recognition Technology of Natural Language Processing Research Progress and Application[J]. Science Technology and Engineering, 2012,12(7):51-53.

[3]   Hammett I M. Study on the Cross-Language Information Retrieval Technology and Application[J]. Journal of Information, 2007, 34(1):85-110.

[4]   Yang K F, Zhang Y K, Yan L I. Feature Selection Method Based on Document Frequency[J]. Computer Engineering, 2010, 36(17):33-35.

[5]   Zhang W M, Jiang W U, Yuan X J. K-means teat clustering algorithm based on density and nearestneighbor[J]. Journal of Computer Applications, 2010, 30(7):1933-1935.

[6]   Wang F, Lei B. Modeling and Analysis of Hadoop Distributed File System[J]. Tele communications Science, 2010:18(3),32-39.

[7]   Davis J, Goadrich M. The relationship between Precision-Recall and ROC curves [C].International Conference on Machine Learning. ACM, 2006:233-240.

Analysis and Design of Course Website for Software Testing Based on SPOC

Jiujiu Yu

College of Computer Engineering, Anhui SanLian University, Hefei 230601, China

Corresponding author: yjjyjL@163.com

Abstract. Technologies of the architecture on SSH(Struts-Spring-Hibernate), database interface access, and web data mining are used to analyse and design of a lightweight course website for software testing which is based on SPOC(Small Private Online Course) for blended teaching on campus. Functions of learning topics releasing on the course, organizing and coordinating learners' discussion to promote the learning process can be achieved through the website for teachers, and the website can also provide learners with selecting some social learning tools(BBS, QQ, WeChat) to complete online communication for the course learning. Finally, further work is expected in the paper. Firstly, the targeted and distinctive mobile learning resources and smart education mode should be constructed which is based on SPOC. Secondly, the visualized, diversified third-party platforms and tools in the course website should be integrated effectively which are enable learners to complete the process of online exercising or examining for some types of subjective testing items. Thirdly, how to ensure the reliability, toughness, running performance, safety, stability for the course website which is formed by cutting out some non-essential functions from existed MOOC(Massive Open Online Course) platforms, and improve the functions of multiple data statistics with web data mining technology and motivational observation for the website development.

1. Introduction

SPOC (Small Private Online Course), as the blended learning model for online and offline which is emerged in post-MOOC (Massive Open Online Course) era, and is the product of MOOC's in-depth integration with classroom teaching [1]. Compared with MOOC, SPOC has some advantages obviously as the following: it is aimed at small-scale specific learning groups, redefines the role of teachers, improves the effect of blended learning, serves the process on college teaching easier, and so on. SPOC activities on campus in some well-known universities in China and some new local universities have been carried out actively in recent years.

However, blended learning has higher requirements for students 'autonomous learning in post MOOC era. The online learning in blended learning is mainly depended on the many well-known MOOC platform that are existed whatever in domestic or abroad to complete the autonomous learning process, and it tends to be disconnect SPOC with actual teaching [2]. Additionally, some of the existing MOOC platforms still have a series of difficulties in completing qualifications for learners, performance evaluation, and how to connect and recognize credits for similar courses in the universities [2-4]. Through some teaching platforms or course website are developed in most of the universities, they are just the simple pile of all kinds of static course resources, lacking of the functions of comments, feedback, interaction, and so on. The development of the SPOC will be limited too. Therefore, as a case, we try to analysis and design a lightweight course website with the architecture

Content from this work may be used under the terms of the Creative Commons Attribution 3.0 licence. Any further distribution of this work must maintain attribution to the author(s) and the title of the work, journal citation and DOI.

Published under licence by IOP Publishing Ltd

on SSH(Struts-Spring-Hibernate) for the course of software testing which is based on SPOC, and apply to the actual teaching process for local learners on campus. In this paper, the structure is organized as follows: some key technologies of the development on this course website are analysised in Section 2. Section 3 presents the requirement for users and the architecture on the course website. Section 4 describes the realization on functions for some core modules of the course website. Section 5 points out further work in this paper.

2. Key technologies

The designing of the website is based on the construction of SPOC for the course of software testing on local campus, and some key technologies and their descriptions on development are shown as Table 1.

Table 1. Key technologies.

Technology	Brief description
SSH	SSH (Struts-Spring-Hibernate), as the three layers on architecture based on J2EE for web server.
Database interface access	JDBC, which encapsulates some of the common operations that used by website designing into a different collection class in a namespace, and the data is presented in the form of tables by Gird View.
Web data mining	Technology of web data mining is used to analyse learner's access behaviour on learning frequency, resident time, and so on [5-6].

3. Requirement analysis and architecture design of the website

3.1 Requirement for users

The requirement for users is focused on the functions of resource sharing and teaching interaction of website, user roles are divided into three categories: teachers, learners (students), and system administrators. Different users have different access authorities. The website provides teachers with releasing course unit and activities, organizing and coordinating between learners' discussion to promote the learning process, it also provides learners with selecting BBS, QQ, WeChat and other social learning tools for online communication and interaction and study to achieve purpose [7]. The use case diagram for the requirement is shown as Figure 1.

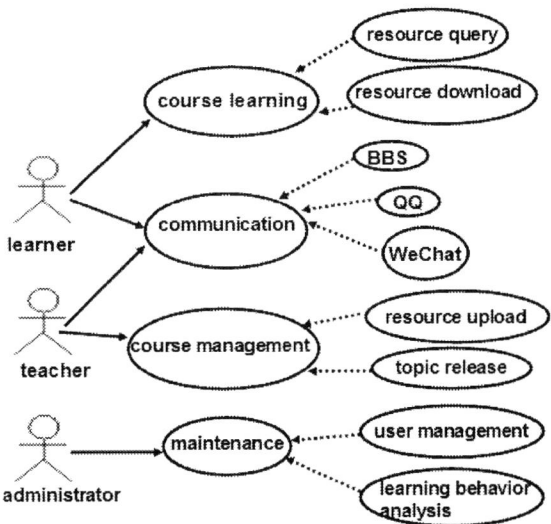

Figure 1. Use case diagram for requirement.

3.2 Design of the physical architecture

The physical architecture of the site is shown in Figure 2. Web server(Microsoft IIS 8.0) uses the SSH architecture and is separated from the database server(SQL-Server 2015) in physical structure, which can reduce the amount of data transmission, improving data access speed, and ensure data security. Client users use the website through web browser in PC.

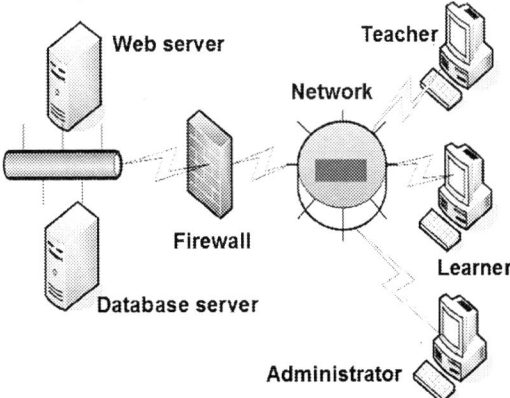

Figure 2. Physical architecture on the website.

By the way, the data Map/Reduce method is used to solve the problem of rapid increase in data scale by us in order to handle a large number of concurrent requests [8].To ensure that the website can be quickly responded to the requests using for user's learning resource.

3.3 Analysis on the database

According to the characteristics on SPOC, some new data tables and their E-R relationships are added in the website, such as course forum, discussion topic, post bar, learning behaviour, and so on.

4. Realization on core modules

4.1 Course center

The course center is the main body of the website. Teachers could upload relevant information on software testing course (such as course introduction, course guide, learning plan, various learning resources, etc.) through the course center, and release learning topics for students learning instruction. Teachers can carefully organize and design a learning topic according to the hierarchy of "learning themes-learning courseware or expanding resources-organizing learning activities" and design one or more related learning activity sequences to interpret a certain knowledge point [9]. Here, an example on learning topic on software testing process of software testing course is shown as Table 2.

By the way, teachers need to provide expanded resources that related to learning content (course-related E-books or auxiliary information materials, etc.) to design learning topics, and release online communication to organize students 'online or offline discussions [7].

When learners register and log on to the course center successfully, some operations such as query, browse, and download files for various electronic resources of the "Software Testing" course could be done. But for the unregistered learners, they don't have permission to download files.

Table 2. Learning topic on software testing process.

Topic		Software testing process	
No.	Learning content	Teaching strategy	Learners behaviour
1	Unit testing	Case study	Watching video
2	Integration testing	Case study	Watching video
3	Validation testing	Case study	Watching video
4	System testing	Case study	Watching video
5	Acceptance testing	Case study	Watching video
Learning resource	Course video	Testing process.MPG4	
	Extended resource	1.Releationship on unit testing and integration testing.pdf 2.Regression testing.ppt	
Learning activity	Course forum	Optimization on software testing process	
	Communication online	Relationship on software development process and testing process	
	Resource sharing	http://www.51testing.com	
	Blog writing	Improvement of testing process.	
	Exercise	Single-choice question 1. 2. 3. 4. ...... Multiple-choice question 1. 2. 3. 4.	
	Thinking	1. Test strategies on integration testing? 2. Basis on system testing? 3. Agile testing?	

4.2 Communication and discussion

The module of communication and discussion mainly integrates some social learning tools such as QQ, WeChat, blog, and so on, to provide the learning environment for learners to participate in course discussion.

In this module, on one hand, both registered learners and anonymous learners can make fully discussions through the course forum real-time. Such as posting and replying for course topics which are released by teachers. On the other hand, learners can share the learning topics and content on discussion by their personal communication tools of QQ and WeChat, and maintain real-time contact with teachers and other online learners to achieve positive interaction[2][7].

Of course, learners can also share their learning experience to their blogs quickly and easily through the website.

4.3 learning behaviour management

The most function of this module is to help teachers to analysing, tracking and recording learner's learning state on the course by using web data mining technology, which is reflected by the learner's relevant learning behavior data.

Through this module, teachers can see some statistics on students' learning course, some important data for learning behaviour which are included in learning topics of each course chapter, such as time duration of learning, learning frequency, resident time, the number of students' submissions, watching videos, reading all sorts of E-learning materials, replying posts, blogs writing and sharing, rates of average completion for exercises, and so on [8].

Teachers can query a certain type of learning data according to actual teaching. Administrators can export various types of learning data from the backstage system of the website in the form of visual reports and provide them to teachers, so teachers could understand the relevant learning behaviours on students in a more realistic and objective form. Using web data mining technology to analyse learner's access behaviour

5. Further work

Although we analysis and design of the course website for the course of software testing which is based on SPOC above, in my opinion, some further work will be researched in the future for development and application on blended teaching and learning actually.

Firstly, with the influence of rapid development of knowledge updating and the concept of life-long learning, the ubiquitous learning that based on mobile terminal devices, such as smart phones and tablet PCs will be further developed [8][10]. The construction of targeted and distinctive mobile learning resources and smart education mode which is based on SPOC is an important direction for development [11].

Secondly, the process of exercising or examining should be completed on line by peer assessment system in SPOC.

Now, the implementation of this process can be realized basically in most of the famous MOOC or SPOC platforms, and the application of choosing question algorithm is relatively mature in examining system. However, discrepancy will be existed by students peer assessment and lack of impartiality [11]. So, in further work, how to according to the features on the course and cognitive ability of small-scale learning group, integrating the visualized, diversified third-party platforms and tools in the course website effectively which enable learners to complete the process of exercising or examining online for some types of subjective items more conveniently, and develop standardized, scientific, and systematic scoring standards to enable students to evaluate each other relatively fair [11].

Thirdly, as a lightweight course website with the characteristics of good expansibility, usable and flexible, some functions which are auxiliary and non-essential should be cut out from the website for a certain local students group on campus.

It is also the future work in this paper that how to ensure the course website which is formed by some existed MOOC platform functions cutting out but without affecting the reliability, toughness, running performance, safety, stability, and so on.

And the fourth, the diversification on requirements of learning behaviour for different learners in different learning phase will be achieved in big data era [12]. Functions of multiple data statistics with web data mining technology and motivational observation in development of this course website should be improved by course teachers and website developers together. It is also the future work.

Acknowledgments

This work was supported by Foundation of "Massive Open Online Course on Software Engineering" of Anhui Province under Grant No. 2015mooc104, "Software Engineering-Excellent Resource Sharing Course" of Anhui Province under Grant No. 2016gxk048, "Research on Software Exploratory Testing and Its Key Technology" of Anhui SanLian University under Grant No.KJZD2017008, and

"Teaching Team of Software Engineering Course" of Anhui SanLian University under Grant No. 15zlgc029.

At last, as the corresponding author, I would like to express my heartfelt gratitude to my colleagues and all the authors of the references which are listed at the end of this paper.

References

[1] H. Zhang, Analysis of the Blended Teaching Mode of English SPOC in Higher Vocational Industry in the "Post-MOOC" Era, Journal of Chinese Vocational and Technical Education. Issue 23 (2018), p 63-66.

[2] J.J. Yu, Design and Application on Self-learning System Based on SPOC, Journal of Xichang College-Natural Science Edition. Issue 1 (2017), p 57-61.

[3] D.M. Zhao, Y. Yin, Study on the Teaching Practice of Blackboard-based B-learning Model, Journal of Modern Educational Technology. Issue 9 (2012), p 41-44.

[4] X.S. Zhai, J. Yuan, Development Dilemma and Countermeasures of MOOC in Higher Education in China, Journal of E-education Research. Issue 10 (2014), p 97-101.

[5] X.W. Sun, Online learning behaviour analysis based on SPOC platform log data and influencing factors, Journal of Shenyang Normal University：Nat Sci Ed. Issue 1 (2017), p 103-107.

[6] M. Huang, X. Liang and X.L. Gu, Introduction on Massive Open Online Course (Publishing House of Electronics Industry, Beijing, 2015)

[7] J.J. Yu, Design and Implementation on Autonomous Learning System for Software Engineering Based on MOOC, Journal of Xichang College-Natural Science Edition. Issue 4 (2016), p 39-44.

[8] H.D. Yin, Design of SPOC Teaching Platform in the Post-MOOC Period, Journal of Jiangsu Radio & Television University. Issue 4 (2015), p 44-50, 90.

[9] Z. Zhang, Study on Design of Massive Open Online Course, MASTER'S THESIS of Central China Normal University, 2014.

[10] H.D. Yin, Design of SPOC Platform for Teacher Teaching Skills Based on Educational Ecology, Journal of Chongqing RTV University. Issue 2 (2017), p 17-24.

[11] J.J. Yu, Research on Key Development Technologies for SPOC Platform, MATEC Web of Conferences. (to be published)

[12] H.D. Yin, Exploration of Blended Teaching Model based on Fanya SPOC Platform during the Post-MOOC Period, Journal of Modern Educational Technology. Issue 11 (2015), p 53-59.

ISPECE

IOP Publishing

Product Information Modeling Based on Polychromatic Sets and Scheme Optimum Selection for Conceptual Design

Li Qiang[1], Xu Bin[2] and Li Xiao Qiang[3]

[1]Safety and Production Dept, State Power Investment Co. Ltd Tibet Branch， Lhasa 850000, Tibet China

[2]Fen Yi power plant of Jiangxi Province

[3]Dispatching control center, State Grid Tibet Electric Power Company Limited3Dispatching control center, State Grid Tibet Electric Power Company Limited

[a]Li Qiang: 15446640@163.com

Abstract. In this paper a conceptual design product information model is established. The polychromatic sets theory was applied to this new model which was aimed at the deficiency of approach for solving product conceptual design presently. This method helps to realize the deduction from requirements to different product scheme combinations by the boolean matrixes which obtains unified colors and individual colors, the constraint relationship in the product model. To solve the conceptual design scheme evaluation problem, a decision-making method was put forward based on fuzzy analytic hierarchy process (AHP) and genetic algorithm (GA) . This approach was designed to convert fuzzy information into numerical value by combining the principles of fuzzy mathematic and information axiom, and use AHP to determine the evaluation index weights . An example of vibration isolation system conceptual design was illustrated the feasibility of this method.

1. Introduction

Conceptual design is the forefront of the overall design process of the product, which has a direct impact on the subsequent design results and costs of the product. It is the focus of product design theory research. Researching on conceptual design is mainly divided into the construction of product information model to describe the conceptual design process and the reasoning based on product information model and the selection of optimal design schemes.

For conceptual design product modelling problems, Gao[1] established a conceptual design process model based on axiom design, and used the Bill Of Material (BOM) view to formalize the product concept space. Johannes et al [2] used the axiom design of the polyline mapping principle to establish a conceptual design product knowledge expression model. The above research studies the formal expression of conceptual design product information from different angles, but the built-in model has insufficient integrated mapping ability for different design information in complex systems, which is not conducive to the integrated research of conceptual design and subsequent detailed design process.

For the optimization of conceptual design schemes, Tijana et al. [3] solved the optimization problem of press concept design scheme by using improved ant colony algorithm. Lorenzo et al. [4] used fuzzy analytic hierarchy process to evaluate the scheme of product concept design. The above

Content from this work may be used under the terms of the Creative Commons Attribution 3.0 licence. Any further distribution of this work must maintain attribution to the author(s) and the title of the work, journal citation and DOI.

Published under licence by IOP Publishing Ltd

research has achieved the optimization of the scheme from the qualitative or quantitative perspective, but it has not considered the conceptual design product modeling and the scheme reasoning and the evaluation of the scheme together, which is not conducive to the research of the integration of different stages of product design.

Based on the polychromatic sets set theory, this paper establishes a conceptual design product information model and uses the polychromatic sets surrounding matrix to realize the reasoning process of the conceptual design scheme. When the concept design feasible scheme is preferred, the fuzzy analytic hierarchy process is used to determine the evaluation index weight and fuzzy evaluation matrix, and the conceptual design scheme optimization model is established. Finally, the genetic algorithm is used to obtain the optimal scheme, which provides a new idea for product conceptual design modeling and optimal integration research.

2. Product conceptual design process model

It is generally believed that conceptual design refers to the workflow involved in a system that requires input as a design and output as an optimal solution. Usually, the conceptual design input function requires an output structure scheme, so it is a function-to-structure conversion process [5].

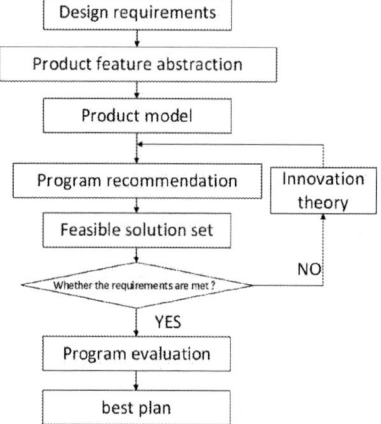

Fig.1 Mechanical product concept design process

Figure 1 depicts the workflow for product conceptual design, which consists of two basic processes, synthesis and evaluation. In the comprehensive part, the total function of the system is decomposed into relatively independent sub-function units through the functional analysis, and then all the program elements that can realize the function are found. Finally, these program elements are combined to form a total program set. The evaluation part is to quantitatively evaluate the plan according to the functional indicators of the product in the obtained program set, and select the optimal plan. The synthesis part is a set of divergence generated by the design requirement reasoning, which is a divergent process; the evaluation is to select the optimal solution from the program set, which is a convergence process [6].

3. Conceptual design product modelling and scheme reasoning based on polychromatic sets

The polychromatic sets set hierarchy model is decomposed from top to bottom, and nodes are given different colors to describe their properties, parameters and other attributes. The connection relationship between nodes represents different constraint relationships, corresponding to different color edges, each edge corresponding to an inference matrix. The characteristics of the polychromatic sets set hierarchy model can describe the specific process of conceptual design products from total function decomposition to solution element implementation. The specific steps are as follows:

1)Uniform color and personal color determination. According to the product design requirements, the function decom -position is completed according to the product working process or the principle of action until it is decomposed to the lowest level function or has mature parts. Detailed definitions of the uniform color $F_i(A)$ and personal color $F_i(a)$ of each layer are in the hierarchical model. The function of the product in the n layer is represented by $F^i(A)$, and the implementation method or the specific scheme is represented by $F^i(a)$, corresponding to the function $F^{i+1}(A)$ to be implemented in the next layer.

2)Constraint relationship classification and edge coloring. According to the positional relationship between functions and methods in the hierarchical decomposition structure model, the constraints are classified: $R = \{R_1, R_2, R_3, R_4\}$. Using the processing method for the constraint relationship in the polychromatic sets set, the set of edges C, the personal color set $F(c)$ of the edges, and the shading matrix $[C \times F(c)]$ are established.

3)The establishment of a product information model was shown in picture 2.

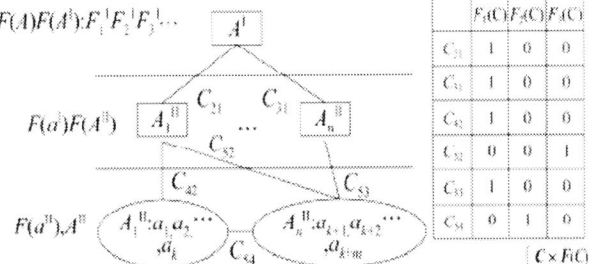

Fig.2 Example of conceptual design product information model based on polychromatic sets

In the polychromatic sets conceptual design process model, the connections between nodes correspond to different "color" attributes, and the coloring of edges is represented by $F_i(c)$, which corresponds to different inference or constraint matrices. ①if $F_1(c)=1$,it means that there is a direct reasoning relationship R_1 between the two vertices, which is described by matrix $[F(a) \times F(A)]$; ② if $F_2(c)=1$,it means that there is a direct reasoning relationship R_2 or R_5 between the two vertices, which is described by matrix $[F(A) \times F(A)]$ or matrix $[F(a) \times F(a)]$; ③if $F_3(c)=1$,it means that there is a direct reasoning relationship R_3 between the two vertices, which is described by matrix $[F(a) \times F(A)]$. The relationship between set $F(A)$ and personal color $F(a)$ of all elements is represented by matrix $[F(A) \times F(a)]$:

$$\|c_{i(j)}\|_{F(a),F(A)} = [F(a) \times F(A)]$$

$$=
\begin{matrix}
& F_1 & \cdots & F_2 & \cdots & F_m & \\
\begin{bmatrix}
c_{1(1)} & \cdots & c_{1(j)} & \cdots & c_{1(m)} \\
\vdots & \vdots & \vdots & \vdots & \vdots \\
c_{i(1)} & \cdots & c_{i(j)} & \cdots & c_{i(m)} \\
\vdots & \vdots & \vdots & \vdots & \vdots \\
c_{n(1)} & \cdots & c_{n(j)} & \cdots & c_{n(m)}
\end{bmatrix}
&
\begin{matrix}
f_1 \\
\vdots \\
f_i \\
\vdots \\
f_n
\end{matrix}
\end{matrix}
\qquad (1)$$

In the matrix, if $f_j \in F(a_i)$, then $c_{i(j)} = 1$, else $c_{i(j)} = 0$. Further, the constraint relation- ship between the same layers is represented by the autocorrelation matrix $[F(a) \times F(a)]$ or $[F(A) \times F(A)]$. In the process of product conceptual design, the scheme reasoning is to implement the search process from the upper-level requirement function to the lower-level function to the lower-level scheme element, and then search the constraint matrix to exclude the constraint, and finally get the product conceptual design feasible solutions sets.

4. Evaluation of conceptual design

The solution process of product concept design can be divided into two stages. The first stage is to obtain the feasible program set of the product through the analysis and reasoning of the problem, and the second stage is to evaluate the series of feasible solutions obtained.

4.1 Scheme evaluation based on fuzzy theory

In the conceptual design stage, the quantitative information of product parameters, structure, etc. are all fuzzy, so it is difficult to make evaluation decisions. The fuzzy theory is used to quantify the fuzzy information by using set and fuzzy mathematics, and the program evaluation and optimization are based on the membership degree of the product plan. High and low determination schemes have excellent grades, and preferred schemes with high grades; when scheme grades are the same, preference is given to schemes with high degree of membership.

Assume that the conceptual design to be evaluated is evaluated according to m levels, and the corresponding mathematical expression is $V = \{v_1, v_2, v_3, \cdots, v_m\}$, The performance of the n indicators is evaluated, and the corresponding indicator set is mathematically expressed as: $U = \{u_1, u_2, \cdots, u_n\}$. The multi-index multi-level evaluation matrix corresponding to the design scheme is as follows:

$$R = \begin{bmatrix} R_1 \\ R_2 \\ \vdots \\ R_i \\ \vdots \\ R_n \end{bmatrix} = \begin{bmatrix} r_{11} & r_{12} & \cdots & r_{ij} & \cdots & r_{1m} \\ r_{21} & r_{22} & \cdots & r_{2j} & \cdots & r_{2m} \\ \vdots & \vdots & & \vdots & & \vdots \\ r_{i1} & r_{i2} & \cdots & r_{ij} & \cdots & r_{im} \\ \vdots & \vdots & & \vdots & & \vdots \\ r_{n1} & r_{n2} & \vdots & r_{nj} & \cdots & r_{nm} \end{bmatrix} \quad (2)$$

V The elements in V correspond to different evaluation levels, such as {very good, good, normal, bad,...}; r_{ij} indicates the degree of subordination of the product plan for the evaluation index u_i with respect to level v_j.

Conceptual design scheme evaluation index weight $W = \{w_1, w_2, \cdots w_i \cdots, w_n\}$, among them, $0 < w_i < 1$, $\sum_{i=1}^{n} w_i = 1$. Then the weighted fuzzy comprehensive evaluation vector is as follows

$$B = W \times R = (b_1, b_2, \cdots, b_j, \cdots, b_m) \quad (3)$$

Among them, $b_j = \sum_{i=1}^{n} w_i r_{ij} \quad j \in [1, m]$ is the subordination of the j comment in the fuzzy comprehensive evaluation.

When the fuzzy theory is used to evaluate and optimize the scheme, the fuzzy evaluation of the design scheme is used to determine the superior level of the membership degree. The scheme with high grade is preferred. When the scheme level is the same, the membership degree is high.

4.2 Establish program evaluation model

Firstly, the weight of the evaluation index is determined based on the AHP method:

①Establish a judgment matrix based on the relative importance of each design factor in product design: $A = [a_{ij}]_{n \times n}$

② Calculate the maximum eigenvalue λ_{max} of A and the corresponding eigenvector $W = (w_1, w_2, \cdots, w_n)$

③Calculate CI and CR, $CI = (\lambda_{max} - n)/(n-1)$; $CR = CI / RI$

Quantify the program evaluation level $V = \{v_1, v_2, v_3, \cdots, v_m\}$ according to the specific situation, convert it to comment vector $Q(q_1, q_2, \cdots, q_m)$. Each of the program elements in a series of feasible solutions obtained through the program reasoning is assigned to the expert evaluation to determine the respective comprehensive evaluation vectors R. Calculate the comprehensive evaluation vector B of each component according to the weight vector W of each evaluation index [7]. Then the comprehensive evaluation score for each component is as follows:

$$c = <Q \cdot B> = \sum_{j=1}^{m} q_j \cdot b_j \qquad (4)$$

The choice of the conceptual design is transformed into the component combination of the constituents, and the objective function is determined as:

$$\max y = \sum_{i=1}^{m} \sum_{j=1}^{mi} c_{ij} u_{ij} \qquad (5)$$

c_{ij} is the evaluation score of the selected component. u_{ij} indicates the j component of the i module, if $u_{ij} = 1$, it includes the component, $u_{ij} = 0$ indicates that it is not included.

4.3 Solution of Scheme Evaluation Model Based on Genetic Algorithm

(1) Determination of chromosome coding method

The design of the coding scheme is based on the idea of gene expression, using a binary multi-parameter cascade coding method. When the product can achieve a certain function or have a certain structure, it is represented by gene 1; otherwise, it is represented by 0. When the product concept is designed, it is broken down into different modules according to functions. The structure of each module is a gene segment. The chromosome length of each module is the number of components that meet the requirements. Each module can only select one mechanism. Therefore, each segment can only have one gene value of 1, and all other gene values are 0.

(2) Determination of fitness function

Determine the fitness function of the algorithm according to the objective function:

$$f_i = \frac{y_i}{\sum_{i=1}^{N} y_i} \qquad (6)$$

Among them, N is the size of the initial population, y_i is the fitness value of the i individual.

(3) Determination of algorithm control parameters

Determine the population size of the algorithm N, the number of iterations of the algorithm M, the probability of crossover P_c, and the probability of mutation P_m.

The flow of using the fuzzy genetic algorithm to optimize the design scheme is shown in Figure 3.

ISPECE   IOP Publishing

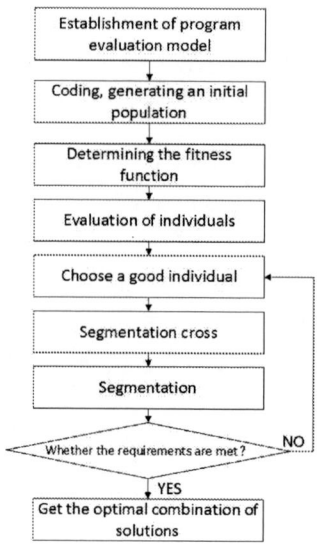

Fig. 3 Algorithm implementation flow chart

5. Instance application

Take the conceptual design of the vibration isolation system as an example. The known user requirements are: passive vibration isolation system, complex structure, high requirements for high-frequency vibra- tion isolation performance, low equipment centroid position, minimum vibration frequency required to be isolated, high temperature resistance and oil resistance are not required.

5.1 polychromatic sets modeling of vibration isolation system

Figure 4 is a functional decomposition hierarchical structure model of the vibration isolation system design, in which the dotted line indicates the selection constraint of the vibration isolation component by the support mode, and the thick solid line indicates the selection constraint of the support mode component element to the vibration isolation component scheme element.

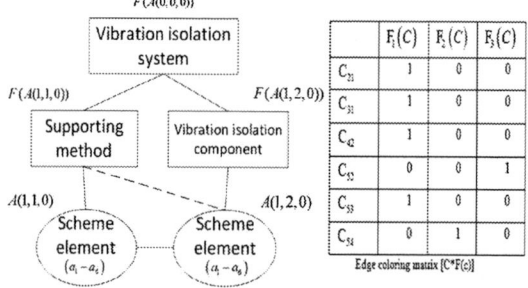

Fig.4 Vibration isolation system conceptual design model

In Fig.4, The 0th layer represents the user's design requirements for the vibration isolation product $F(A(0,0,0)) = F_1^0 F_2^0 \cdots F_{14}^0$; Functional characteristics of the first layer of vibration isolation products $F(A(1,1,0)) = F_1^1 F_2^1 F_3^1 F_4^1$, $F(A(1,2,0)) = F_5^1 F_6^1 \cdots F_{10}^1$; The method set corresponding to the second layer support mode is $A(1,1,0) = a_1 a_2 a_3 a_4 a_5 a_6$, The method set corresponding to the vibration isolating component is $A(1,2,0) = a_7 a_8 \cdots a_{21}$.

5.2 Vibration isolation system scheme reasoning

There are six edges in the hierarchical structure model of the vibration isolation system, which correspond to three kinds of constraint relationships. Each edge has a corresponding reasoning or

2171

constraint matrix. For the first type of direct reasoning relationship R_1, Corresponding edge color $F_1(c)=1$, As $[F(a) \times F(A)]$ and $[F(A) \times F(A)]$ establish an inference matrix, as follows:

$$T_1 = [F(A(1,1,0)) \times < F(A(0,0,0)); A(0,0,0) >]$$
$$T_2 = [F(A(1,2,0)) \times < F(A(0,0,0)); A(0,0,0) >]$$
$$T_3 = [A(1,1,0) \times F(A(1,1,0))]$$
$$T_4 = [A(1,2,0) \times F(A(1,2,0))]$$

As can be seen from Fig5, $A(1,1,0)$ and $A(1,2,0)$ contain the second type of constraint R_2 represented by matrix $F(a) \times F(a)$. $F(A(1,1,0))$ and $A(1,2,0)$ contain the second type of constraint R_3 represented by matrix $F(a) \times F(A)$ as follows:

$$S_1 = [A(1,2,0) \times A(1,1,0)]$$
$$S_2 = [A(1,2,0) \times F(A(1,1,0))]$$

Fig.5 is a polychromatic sets surrounding matrix that characterizes the constraint relationship in the design of vibration isolation system.

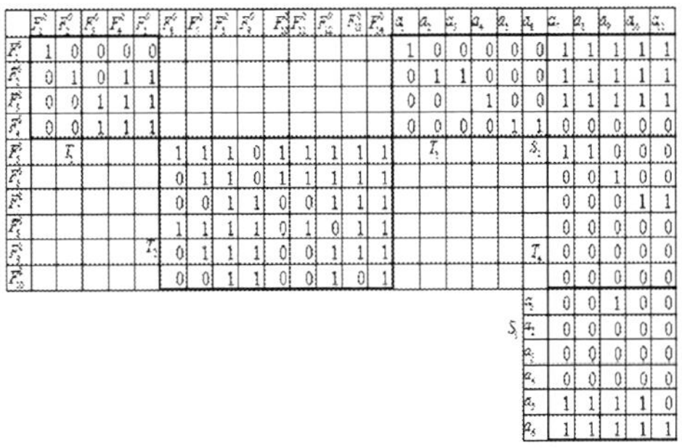

Fig.5 Vibration isolation system polychromatic sets inference and constrained surrounding matrix

The user's requirements for the design of the vibration isolation product are expressed as follows:

$$F_1^0 \quad F_2^0 \quad F_3^0 \quad F_4^0 \quad F_5^0 \quad F_6^0 \quad F_7^0 \quad F_8^0 \quad F_9^0 \quad F_{10}^0 \quad F_{11}^0$$

$$0 \quad 1 \quad 0 \quad 0 \quad 0 \quad 0 \quad 1 \quad 0 \quad 1 \quad 0 \quad 0$$

According to the reasoning matrix T_1, T_2, the following two design schemes are obtained, which are represented by polychromatic sets lanes as follows:

	F_1^1	F_2^1	F_3^1	F_4^1	F_5^1	F_6^1	F_7^1	F_8^1	F_9^1	F_{10}^1
Scheme 1	0	1	0	0	0	0	0	1	0	0
Scheme 2	0	1	0	0	0	0	0	0	1	0

For scheme 1, according to the reasoning matrix T_3, T_4, four design scheme combinations can be obtained: (a_2, a_{14})、(a_2, a_{15})、(a_3, a_{14})、(a_3, a_{15}); For scheme 2, 6 design combinations can be got: (a_2, a_{16})、(a_2, a_{17})、(a_2, a_{18})、(a_3, a_{16})、(a_3, a_{17})、(a_3, a_{18}). According to the constraint matrix S_1, S_2, it can be seen that the constraints of the scheme (a_2, a_{16}) and the scheme (a_2, a_{17}) cannot be realized in reality, and finally eight feasible schemes are obtained: (a_2, a_{14})、(a_2, a_{15})、(a_3, a_{14})、(a_3, a_{15})、(a_2, a_{18})、(a_3, a_{16})、(a_3, a_{17})、(a_3, a_{18}).

5.3 Vibration isolation system evaluation

(1) Determine weight

Firstly, comprehensively evaluate the design factors of the vibration isolation system considering mechanical performance, economy, system maintainability and versatility, and score the design factors importance judgment matrix by experts.

$$E = \begin{bmatrix} 1 & 2 & 2 & 1 \\ 1/2 & 1 & 1 & 1/2 \\ 1/2 & 1 & 1 & 1/2 \\ 1 & 2 & 2 & 1 \end{bmatrix} \qquad (7)$$

Then calculate the maximum eigenvalue of E as $\lambda_{\max} = 4.001$, $CI = \dfrac{\lambda_{\max} - n}{n-1} = 0.0007$, look up the table $RI = 0.9$, so that: $CI / RI = 0.0007 < 0.1$, Compliance with conformance testing standards. Finally, the weight vector of the evaluation index of the conceptual design of the vibration isolation system is obtained by normalizing the feature vector:

$$W = (0.333, 0.167, 0.167, 0.333)$$

(2) Fuzzy comprehensive evaluation of scheme elements

For vibration isolation system, comment set: $V = \{v_1, v_2, v_3, \cdots, v_m\} = \{very\,good, good, normal, bad\}$ Establish each component evaluation matrix R based on the membership of the four components of mechanical performance, economy, system maintainability and product versatility:

$$R = \begin{bmatrix} R_1 \\ R_2 \\ R_3 \\ R_4 \end{bmatrix} = \begin{bmatrix} r_{11} & r_{12} & r_{13} & r_{14} \\ r_{21} & r_{22} & r_{23} & r_{24} \\ r_{31} & r_{32} & r_{33} & r_{34} \\ r_{41} & r_{42} & r_{43} & r_{44} \end{bmatrix} \qquad (8)$$

The weighted evaluation indicator vector becomes:

$$B = W \cdot R = (b_1\ b_2\ b_3\ b_4) \qquad (9)$$

Table1 shows the membership of the 21 components of the vibration isolation system to the above four evaluation indicators:

Table 1 Comprehensive evaluation table for each component

	a1	a2	a3	a4	a5	a6	a7	a8	a9	a10
b1	0.39	0.52	0.15	0.44	0.38	0.23	0.24	0.11	0.29	0.50
b2	0.10	0.30	0.32	0.14	0.13	0.41	0.24	0.30	0.19	0.18
b3	0.15	0.13	0.17	0.23	0.19	0.14	0.31	0.30	0.27	0.17
b4	0.36	0.05	0.36	0.19	0.30	0.22	0.21	0.29	0.25	0.15

	a11	a12	a13	a14	a15	a16	a17	a18	a19	a20	a21
b1	0.32	0.41	0.37	0.44	0.37	0.34	0.32	0.23	0.36	0.27	0.34
b2	0.21	0.18	0.24	0.27	0.24	0.28	0.26	0.27	0.18	0.28	0.19
b3	0.19	0.13	0.26	0.20	0.26	0.21	0.17	0.16	0.27	0.16	0.27
b4	0.28	0.28	0.13	0.09	0.13	0.17	0.25	0.34	0.19	0.29	0.20

(3) Scheme optimization based on genetic algorithm

The design of the vibration isolation system is quantified, and the corresponding vector after quantization is as follows:

$$Q = (q_1, q_2, q_3, q_4) = (10, 7, 5, 2) \qquad (10)$$

The vibration isolation system is composed of two parts: the supporting method and the corresponding component of the vibration isolating component. The corresponding conceptual design optimization model is as follows:

$$y = \max \sum_{j=1}^{4} q_j x_{ij} + \max \sum_{j=1}^{4} q_j x_{kj} \ (i \in [1,6]; j \in [7,21])$$
$$s.t (x_{ij} \neq 0) \bigcup (x_{kj} \neq 0) \qquad (11)$$

Among them, x_{ij} is the degree of membership of evaluation component j for vibration isolation scheme component i. The fitness function of the algorithm is shown in Equation (6).

The parameters of the algorithm are set as follows: binary encoding, The population size is 10 and the maximum number of iterations is 100, $p_c = 0.8$, $p_m = 0.1$. The optimal combination scheme calculated by calculation is (a_2, a_{14}), Its corresponding target value is 15.52. The optimal value curve and the average fitness convergence curve during the running of the algorithm are shown in Fig. 6.

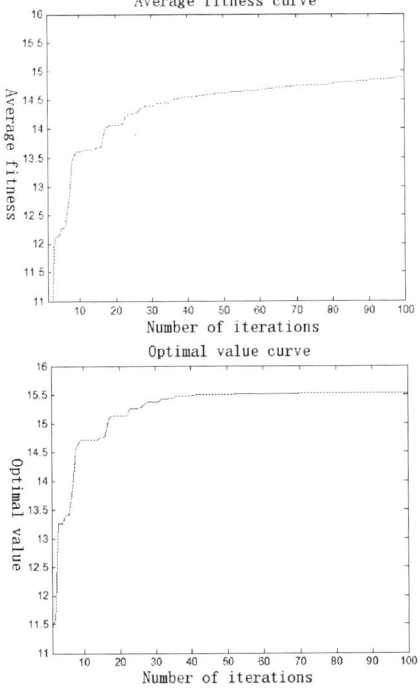

Fig. 6 Convergence curve of genetic algorithm

It can be seen that the first group of schemes is the optimal design scheme, and the high-level support 1 is used as the support mode. The spring and the rubber are connected in parallel as the support scheme element. The vibration isolation scheme has good design mechanical performance. The high support mode 1 is nested in the external mass base through the inner mass, which can realize the lateral limit brake well, prevent lateral vibration slip and improve stability. The spring and the rubber are arranged in parallel, and the vibration isolation effect is better. The specific structure diagram is shown in Fig. 7.

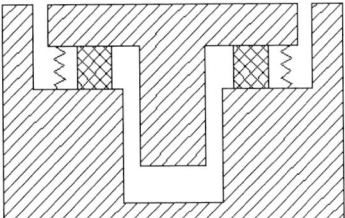

Fig. 7 Vibration scheme system conceptual design optimal scheme structure diagram

6. Conclusion

In this paper, the polychromatic sets set theory is used to establish the product conceptual design information model. The concept of unified color and individual color in polychromatic sets set is used to establish the inference and constraint matrix to describe the complex constraint relationship in the model. The model is more conducive to computer expression. For the solution of the conceptual design scheme, firstly, the feasible design set of product design is obtained by searching the polychromatic sets set surrounding matrix from top to bottom, and the solution space is reduced. The fuzzy evaluation matrix is established by using fuzzy theory, and the weight of the evaluation index of the scheme is determined by AHP method. The scheme evaluation model is established and the optimal scheme combination is solved by genetic algorithm. Taking the conceptual design of vibration isolation system as an example, the product information model of vibration isolation system and the corresponding reasoning and constraint matrix are established. By searching the surrounding matrix, the feasible scheme of vibration isolation system design is obtained. The fuzzy theory, analytic hierarchy process and genetic algorithm are used to obtain the optimal combination of vibration isolation system, which shows that the conceptual design idea proposed in the paper is feasible.

References

[1] Xinqin Gao, Zongbin Li. Computer Aided Conceptual Design of Mechanical Product Using Polychromatic Sets[P]. Mechatronics and Automation, Proceedings of the 2006 IEEE International Conference on,2006.

[2] Johannes Bräuer, Reinhold Plösch, Matthias Saft et al. Measuring object-oriented design principles: The results of focus group-based research[J]. The Journal of Systems & Software, 2018, 140.

[3] Tijana Vuletic, Alex Duffy, Laura Hay et al. The challenges in computer supported conceptual engineering design[J]. Computers in Industry, 2018, 95.

[4] Lorenzo Fiorineschi, Francesco Saverio Frillici, Federico Rotini et al. Exploiting TRIZ Tools for enhancing systematic conceptual design activities[J]. Journal of Engineering Design, 2018, 29(6).

[5] Thomas Jusselme, Emmanuel Rey, Marilyne Andersen. An integrative approach for embodied energy: Towards an LCA -based data-driven design method[J]. Renewable and Sustainable Energy Reviews, 2018, 88.

[6] Syuhaimi Kassim , Fareq Malek. EM/Circuit Co-Simulation: A Highly Accurate Method for Microwave Amplifier Design[J]. Universal Journal of Computer Science and Engineering Technology, 2010, 1(2).

[7] Edwards Phil. Questionnaires in clinical trials: guidelines for optimal design and administration[J]. Trials, 2010, 11(1).

ISPECE

IOP Publishing

Research of a method for Synchronized Phasor Data transmission Based on IEC61850

Shanqiang Feng[1,a], Aijun Hou[1,2] and Bochuan Gu[1]

[1]Electric Power Research Institute of Guangdong Power Grid Corporation, 510080Guangzhou, China

[2]China Southern Power Grid Key Laboratory of Power Grid Automation Laboratory, 510080Guangzhou, China

[a]Corresponding author: fengshanqiang@139.com

Abstract. At present, the communication services for the Wide Area Measurement system in intelligent substation don't use the IEC-61850 system yet, which causes a lot of inconvenience and problems. In order to solve these problems, this paper proposes a method of Synchronized Phasor Data transmission based on IEC-61850. In this method, the communication services of Wide Area Measurement system are converted to IEC-61850 system, and find a way that can map the existing functions to the IEC-61850 system.Firstly, a method for high-speed data transmission and fast control is proposed.Then functional services that cannot be fully implemented by IEC-61850 services are expanded to fill the gaps in a prescribed way. Finally, from the perspective of unified substation information platform, the prospects of this scheme are analyzed.

1. Introduction

Intelligent substation is a more advanced application based on digital substation[1].In order to make an intelligent substation "smarter" with some prominent features, multiple information systems representing the characteristics of "Information Isolated Island [2]" within the substation shall be integrated so as to integrate all the information on a single unified platform [3], avoid waste of resources due to information data repeatability and achieve information interaction among different information systems, thus providing a much "smarter" advanced application.

After years of development and construction, power system has increasingly high requirements on power quality and stability control. Application of the wide area measurement system also becomes more widespread, which is vital for monitoring the dynamic characteristics of the power system which cannot be obtained from the traditional real-time monitoring system [4-6]. At present, equipment and systems with different functions within the intelligent substation have gradually completed resource integration and information unification. Multiple information systems such as the real-time monitoring system, the information protection slave station, the fault recording system, the on-line monitoring system of primary equipment and the power energy collection system have already achieved an information communication system based on IEC-61850. Therefore, it is imperative for conversion of wide area measurement system based on synchronized phasor data acquisition system to IEC-61850 information communication system.

Although 61850 modeling of synchronized phasor data is described in Q/GDW 1844-2012 Technical Specification for Synchronized Phasor Measurement Units for Intelligent substations[7] in China, the description only contains the outline of modeling and no clear definition of information service is made. Therefore, the traditional transmission protocol is still used in practical application, making the synchronized phasor data acquisition system still isolate from other information systems.

Content from this work may be used under the terms of the Creative Commons Attribution 3.0 licence. Any further distribution of this work must maintain attribution to the author(s) and the title of the work, journal citation and DOI.

Published under licence by IOP Publishing Ltd

In order to realize an information system based on IEC-61850, this paper proposes a method of synchronized phasor data transmission based on IEC-61850. In this method, the functions of synchronized phasor dynamic data transmission, on-line management and off-line data query will be completed based on IEC-61850 communication protocol so that the information of synchronized phasor data acquisition system is integrated into the information system of the intelligent substation in a standard interaction method.

2. Existing Synchronized Phasor Transmission Method

At present, the 1344 protocol is used for transmission between synchronized phasor units and WARMS master station. Two common versions of the protocol are *Q/GDW 131-2006 Technology Guidance of Power System Real Time Dynamic Monitoring System* [8] and *GB/T 26865 Real-time Dynamic Monitoring System of Power System* [9]. The major difference between the two versions is the definition of the server and that of the client in the data channel, where the former defines the master station as the server and the latter defines the slave station as the server. The functional service is completely the same and the fields generated in the message are slightly different. Seeing from functional services, there is no difference between the two versions.

2.1 Overview of existing Synchronized Phasor transmission mode

At present, synchronized phasor data are collected and calculated by each acquisition unit first, and a self-describing file CFG1 of the acquisition unit is generated to describe its own functional configurations simultaneously while the phasor data are generated. Afterwards, the self-describing file of each acquisition unit will be integrated by data concentrator to produce a complete self-describing file CFG1 of all the information within the substation and summarize synchronized phasor measurement data of all the acquisition units, realizing data interaction with the master station in the form of complete self-describing information and data.

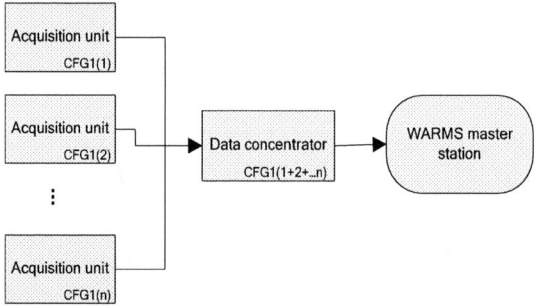

Fig. 1 Structure for Synchronized Phasor Data transmission in intelligent substation at present

2.2 Significance of changing existing transmission mode

Since an exclusive protocol is employed by the synchronized phasor data acquisition system at present, it is necessary to establish a separate network system physically in the construction of intelligent substation, as shown in the figure below:

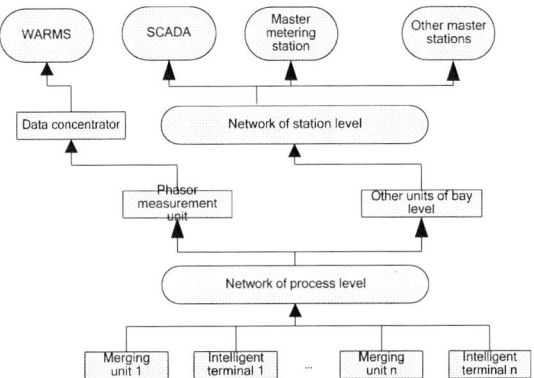

Fig. 2 Network structure of intelligent substation at present

At present, sampled data within the intelligent substation are completed by the same unit of the process layer in a unified manner, providing a basis that the measurement of synchronized phasor, real-time measurement value and electric energy can be finished in the same unit. Although the data network of the process layer has completed unification, the transmission of synchronized phasor data to the upper layer has not been unified into IEC-61850 system. The measurement unit must provide a separate physical network card to form a separate network to the upper layer, making the synchronized phasor data acquisition system substantially separate from other information systems.

Since both the process level and the bay level have finished unification, the integration to IEC-61850 system for the transmission mode of synchronized phasor data to the master station is significant. Once the transmission of synchronized phasor data is converted to IEC-61850 system, a network structure of information system will be directly reduced physically. Meanwhile, synchronized phasor data are no longer separate data and they will be integrated into the network of station level together with other information within the substation, making it possible to analyze problems through combined use of synchronized phasor data and other data information.

3. Overview of Synchronized Phasor Data Transmission based on 61850

After IEC-61850 system is employed by synchronized phasor data acquisition system, the synchronized phasor data will form a unified data service interface for the station level. Master stations of any level or with any functions may only need to access to the network of the station level to simultaneously access to synchronized phasor data service and services of other information systems with the standard and uniform access method, as shown in the figure below:

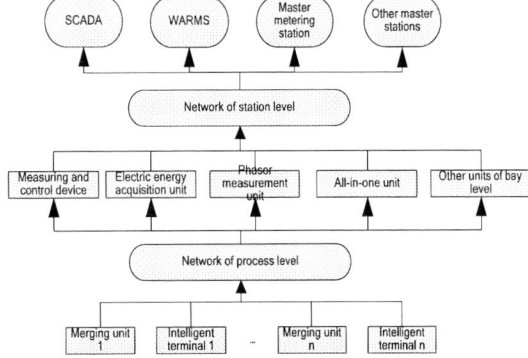

Fig. 3 Network structure of intelligent substation after communication service unified

As shown in Fig.3, if synchronized phasor data acquisition system completes the conversion to IEC-61850 system, data of various units of the bay level will be uniformly integrated into the network of the station level. The system network will be clearly divided into two layers. If other information systems are required to obtain synchronized phasor data for advanced applications, they can obtain data from the station level. On the other hand, if WARMS master station needs information except for synchronized phasor, such as various emergency signal and device alarm, it can also obtain relevant data from other information systems via the network of the station level.

4. Research of a Method for Synchronized Phasor Data Transmission Based on 61850

Main functions of the synchronized phasor acquisition system include:

1.Dynamic data transmission
2.Off-line data query
3.Management

Now we will conduct a research on how to realize these functions in IEC-61850 system.

4.1. Dynamic data transmission

Dynamic data transmission includes the transmission of phasor, analog and state.

(1) The application of phasor and analog requires high density data to form time section. Therefore, a very high transmission speed is necessary. It is not feasible to use MMS message with a low speed and transmission has to be carried out by SV message at the station level. The IEC 61850-90-5 part has defined the R-MSVCB data control block[10], which is used for transmission of SV message at the station level.

(2) In order to realize the quick control launched by the master station and that triggered by the received state signal, it is necessary to provide a state transmission service with high transmission speed and fast response, which cannot be satisfied by low-speed MMS of the station level. Then the IEC 61850-90-5 part has defined the R-GOCB data control block for transmission of GOOSE message at the station level to the master station so as to realize the demand on high-speed transmission of state and application of quick control of the master station.

Process of interaction for dynamic data is as follows:

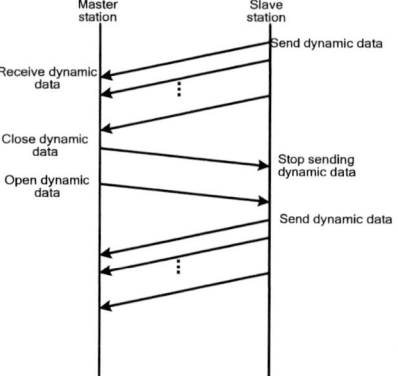

Fig. 4 Process of transmission for Dynamic data

4.2 Off-line data query

Due to lots of synchronized phasor data, it is necessary for the local measurement unit to store all the dynamic data to form dynamic data files. Meanwhile, in case of system disturbance, the synchronized phasor measurement unit will generate recording files with local storage. The master station needs to query the dynamic data files and recording files with local storage when analyzing disturbance.

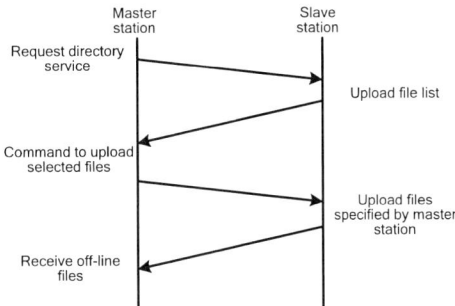

Fig. 5 Process of querying off-line files

4.3 Data management function

4.3.1. Open or close dynamic data

Opening or closing dynamic data is completed by writing services of the characteristic value SvEna of R-MSVCB control block. When the characteristic value is written as True, dynamic data transmission is opened; when it is written as False, dynamic data transmission is closed.

4.3.2. Change dynamic data transmission speed

Similar to sampled data of the process level, dynamic data are also measured values with high density and equal time intervals. Therefore, the dynamic data here can be regarded as sampled data, with the transmission speed as the sampling rate. In this way, transmission speed can be set based on the characteristic value SmpRate of R-MSVCB control block.

4.3.3. Upload self-describing file CFG1

In IEC-61850 system, the self-describing file is ICD file. The master station obtains the ICD files of slave stations with two methods: the first one is off-line mode, which means the off-line production of SCD file containing ICD of slave stations. The master station may obtain ICD of slave stations by reading SCD files. The second one is on-line mode, which means direct on-line reading of ICD information of slave stations through directory services. This mode is applicable to slave stations with less information. If ICD of slave stations contains much information, this mode will be very time consuming.

4.3.4. Download CFG2 file

Since dynamic data covers all the data collected in the whole station, data in real-time transmission by the master station at ordinary times are only a part of all the data. It is necessary to provide the master station with the function to transmit the selected data from all the data of the whole station.

Here, it may be realized by custom data set of IEC-61850. As loss of custom data will occur when power is down, an interaction process is defined here to ensure the correctness of phasor transmission, as shown in the figure below:

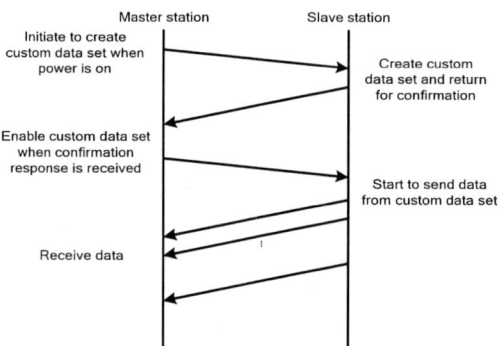

Fig. 6 Process of data enable when power is on

5. Conclusion

It is feasible and significant to convert synchronized phasor acquisition system to IEC-61850 system. This method integrates the synchronized phasor acquisition system and other information systems of the substation on a single unified platform, closely connecting the originally isolated information with the information of other information systems, which provides a basis for conjoint analysis of information. In the future, with integration of various information systems, simplification of repeated data acquisition, unification of repeated data transmission and integrated application of various system data, the means of conjoint analysis will become more and more mature, making it possible to conduct unified analysis and process of information from multiple substations within a region in order to complete macro overall control.

Acknowledgments

The work described in this paper was fully supported by a
grant from China Southern Power Grid scientific research project : Research and application of on-line centralized maintenance and control technology for substation automation equipment (No.GDKJXM20161913). Thanks are due to Yi Wang for assistance with the experiments and to Yanxu Zhang for valuable discussion.

References

[1] LI Mengchao, WANG Yunping, LI Xianwei, et al. Smart substation and technical characteristics analysis[J]. Power System Protection and Control, 2010, 38(18): 59-62.
[2] Q/GDW 383-2009 Technical guide for smart substation[S]. Beijing: State Grid Corporation of China, 2009.
[3] WAND Dongqing, LI Gang, HE Feiyue. Design of integrative information platform for smart substation [J]. Power System Technology, 2010, 34(10):20-25.
[4] CHENG Yun-feng, ZHANG Xin-ran and LU Chao. Research progress of the application of wide area measurement technology in power system[J].Power System Protection and Control,2014,V42(4):145-153
[5] GAN Lei, KANG Hewen, HE Min. Dynamic monitoring of voltage stability based on wide area measurement system[J]. Power System Protection and Control, 2010,V38(21).132-135,161.
[6] CHEN Jing, LIU Dichen, WANG Baohua. Research of wide area backup protection based on limit PMU [J]. Power System Protection and Control, 2012, V40 (17):67-71, 77.
[7] Science and Technology Department of State Grid Corporation of China. Q/GDW 1844-2012 Technical specification for synchronized phasor measurement units for smart substations [S].2012.
[8] National Power Dispatching and Communication Center. Q/GDW 131-2006 Technology guidance of power system real time dynamic monitoring system[S].2006.

[9] Power Systems Management and Associated Information Exchange. GB/T 26865.2-2011 Real-time dynamic monitoring system of power system [S].2011.

[10] Draft IEC 61850-90-5 TR Ed 1. Use of iIEC 61850 to transmit synchrophasor information according to IEEE C37.118[S].Geneva: IEC, 2010.

[11] ZHU Yongli, HUANG Min, LIU Peipei, et al. Research on transmission of power telecontrol information based on IEC61850 and network[J]. Relay, 2005, 33(11): 45-49.

[12] ZHU Yongli, WANG Dewen, WANG Yan. A telecontrol communication modeling method based on IEC61850 [J]. Automation of Electric Power Systems, 2009, 33(21): 72-76.

[13] ZHANG Daojie, WANG Guangmin, et al. Research and application of integration of network protection dispatching and control[J]. Power System Protection and Control, 2013, 41(14): 149-153.

[14] HUANG Kai, YANG Ji, GU Quan. A source-based maintenance technology for integrated intelligent operation system of power grids[J]. Automation of Electric Power Systems, 2014, 38(15): 71-75.

The Design and Creation of an Interactive E-Book: "Book of Answer"

Xinyue Gui

Department of Electrical Engineering, University of Bridgeport, 221

University Avenue, Bridgeport, Connecticut 06604, USA

+86 13971621290

2114224755@qq.com

ABSTRACT Interactive E-book is a digital book which the reader can direct the storyline and interact with it. 'The book of answer' is a kind of popular book that reader can interact with. In this project, we aim to develop an PVE game to simulating the process that the reader interact with the computer with the content of the book saved in. We use Python to realize the interactive dialogue process in GUI (Graphical User Interface).

1. Background

E-book refers to the type of book that is published in a digital form. Like the ebooks that we can find on Amazon and Kindle, ebook has changed the concept of books from heavy to light and portable. Besides, normally, E-book is cheaper than traditional paper book. Also, E-books are convenient to carry on. Those features make E-book popular now.

Interactive E-book, literally, is a E-book that reader can interact with. In charging with the E-book is a wonderful user experience while reading. Users can control the page choosing, video playing and context reading by themselves

'Book of Answer' (Carol Bolt,2016) is a popular book with difference words in each page. Once people want to find an answer, they just need to consider about a question before they page to a randomly number with an answer on the blank. The book is very popular because of its interesting, unexpected, unpredictable answer and its randomness. Lots of people found their answer in the book and treat it as a tool to soothe the mind when they face any trouble.

In terms of the book 'The book of Answer' is a paperback 640 pages heavy book. It is impractical to carry the book during the journey, working or other situation. Considering the book's attributes of interesting and high interaction, it would be a wonderful material for an interactive E-book. Therefore, The research proposes to develop a interactive E-book system with the materials come from 'Book of Answer'. Occupy the 'Book of answer' in interactive E-book would build a 'Game-like' reading experience. The E-book will turn the page and get the answer dependent on the reader.

2. Literature Review

Nowadays, the latest interactive E-book product such as 'Campbell Biology', 'Running in the Gauntlet ' and there also are some software to make interactive book released like Moglue and SnapAPP. Besides there are some provoking essay showing very significant information and method. There is a successful sample of interactive E-book, which is, 'Campbell Biology' (Jane B. Reece .et al, 2010). The 1600 pages book can be displayed by a pad following the instruction by readers to perform

---

Content from this work may be used under the terms of the Creative Commons Attribution 3.0 licence. Any further distribution of this work must maintain attribution to the author(s) and the title of the work, journal citation and DOI.

Published under licence by IOP Publishing Ltd

the video, audio, hot spot including in the E-book, The essay 'Designing and Building an Interactive E-Book' described.(Maria Bartoszewicki, 2013) There are some buttons and mouse sensing area designed to make a chance for reader to interact during the reading in order to construct an imitation of reality users experience.(Yuanyuan Min, 2010)

Applying E-book as an interactive role into higher education can make a great difference in study. Mohammed Mohammed Ahmed Ebied has done his experiment for 60 students and gave the readers an unexcepted answer that:

There is a significant differences between the study groups in academic achievement favor to experimental group students whose study with e-book. (Mohammed,2015)

Of cause, iPad or iPhone can play the same role in graduating student study. But less attention has been paid to children. (Huang, Yueh Min，2012) Clare claimed that the advantages of interactive E-book that the storyline can be developed by user and content displaying in more interesting way like 3D diagrams and other tricky. It also shows the successful example of interactive E-book, such as 'Running in the Gauntlet' (Jeffrey Hayzlett,2012), which used Snap Tags to attract many users. (Clare mcdermott, 2012) For the method to create a interactive E-book, a very basic, easy-learned and functional software named MoglueBuilder has been introduced to build a interactive E-book. This software designed for ordinary people, thus it does not need the ability of programming. User control the storyline by drag and pulling in the UI. According to the essay, it is obviously that the interactive E-book will became more and more popular and common. (Annie, 2012)

Based on the articles gathered, a summery can be drawn that the interactive Ebook introduce a enjoyment way of reading. Because of the popularization of those easy-mastered software and promising developing future, one day it will develop as the main-string reading method.

Except for increasing the interestingness of reading, the development of interactive E-book also can be manipulated learining difficult and unpredictable things, such as natural catastrophe and disaster. The outcome turned out as a optimistic view:

The results reveal that the attention of users can be improved and the image can also be improved during the interaction and operation process. Finally, this study proves that the interactive electronic book makes learning easier and faster than the traditional book. (Shih,2013)

Not only in this field, the E-learning can be built as a template to do the study in order to save the labor and time. Hatipoglu has explained his research and method to do the E-book learning problems and shared his steps. (Hatipoglu,2012)

3. Proposal

This project aims at designing and creating an interactive E-book of 'Book of Answer' with python. This programme simulates the process of 'choosing page' and 'showing answer' actions via programming in python. Thus there are two windows needed in this project. One for asking and one for answering. The demo created in the project can work like the interactive dialogue system as figure 1 shown. The ebook would allow us to interact with users, allow users to choose pages, and our system would provide the answers for them. Therefore, we need to develop the programme that can accept the data produced by users and read the Ebook document. Finally get the right answer and display it to customers.

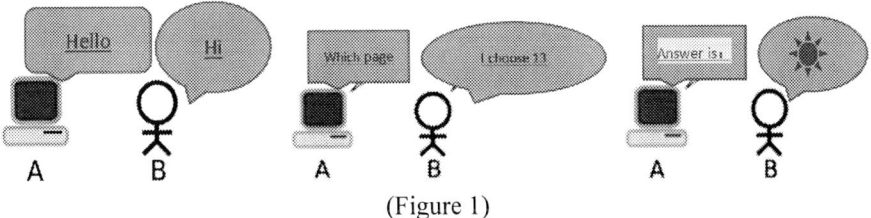

(Figure 1)

4. Framework:

Our interactive E-book of "Book of Answers" is fulfilled by several models in Python, such as GUI, Pandas, Tkinter, Dataframe and easygui. GUI (Graphical User Interface) can create an enviroment that the customers can chat with computer by graphical icons. Pandas (The data analysis library) is a package that can build a data structure and do the data analysis. In the project, It is used to process the data in DataFrame. Tkinter is a class in the file tkinter, which is a TK GUI toolkit for python binding. In the programming, the import of the module tkinter is the method chosen to create a dialogue window. DataFrame is the data structure for processing data table in Pandas. In the project, It is used as a data source for the computer to read,research as find the right answer. Easygui is also a module for GUI programming. It can be used to create another window to bring out the answer and can be used independently. Just import the module and use it after the function Insert number triggered. It is a relatively simple and easy to use module in python, which is good for who new to python.

Due to the copyright of the book, the research only use the document of 'Book of Answer' with 25 pages and 25 answers and it is saved as a csv form with the index: Page number. One page number only correspond to a very answer. Once customers input a very page number, the computer will get the corresponding answer based on the saved document. In the programming, the vital parts are constructing the function, setting the parameter of each label, button on the canvas and combining the two canvas together with the button which is used to command the function. While running the programme, the canvas will be shown on the window to ask 'Which page would you like to choose', after the customer inputs a certain number and clicks the 'Enter' button, then Another canvas will show up to give the answer.

The development has 3 steps. Firstly, A data table to be established to store the page number and answers then save it as a csv mode. That is the source where the computer can get the data. Secondly, A function will be built to combine the data table and the customer's instruction. Thirdly, A window will be constructed for customers to input and polish the dialogue canvas. Here is a diagram shown below: (Figure 2)

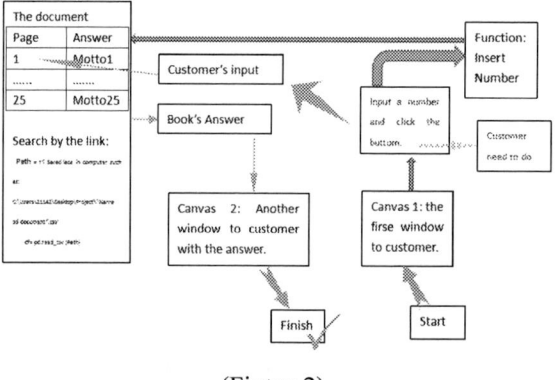

(Figure 2)

5. Pseudo-code

The code will be shown as three parts which corresponding the threes steps which has been mentioned above:

5.1 The definition of function 'insert number'

```
def insert_number():

    global x1

    x1 = int(entry1.get())

    f = r'C:\Users\21142\Desktop\Project\DataFrame.csv'
```

df= pd.read_csv (f)

easygui.msgbox('Here is your Answer:'+ df['Here is your Answer：'][x1])

Firstly en entry will be made as x1 and give a path f to guide the programme read the document. The key word [x1] in the last row ensure the only right answer to print out. The method easygui is to used to build the second window for presenting the answer.

5.2 The definition of a canvas

canvas1 = tk.Canvas(root, width = 350, height = 200,bg='violet')

canvas1.pack()

I=tk.Label(root,bg='green',width=50,height=5,text='Welcome to the "Book of Answer"')

I.pack()

n=tk.Label(root,bg='yellow',width=50,height=5,text='Which page would you like to choose')

Pack

m=tk.Label(root,bg='blue',width=50,height=3,text='Thank you')

m.pack()

Establish four packs for the first window. Build canvas1 and three labels on the canvas. At last determine the parameter for it, such as the length, width, background color and the content of the label. The order of the code determines the place located on the canvas.

5.3 The definition of the Button

button1 = tk.Button (root, text='Enter ',command=insert_number)

canvas1.create_window(250, 100, window=button1)

The button for customers to click is a command to trigger the function 'insert number'. Also some parameter will be set to determine its place on the canvas1.

6. Solved improvement

During the programming, the easiest way to solve it is that listing all the answer in the code script and make it as a select button mode. Such as:

def print_selection():

    I.config(text='Here is your answer: '+var.get())

r1=tk.button(window,text='Page1',variable=var,value='Thank you',command=print_selection)

r1.pack()

But the number of page of book always should be higher than one hundred. It will be unrealistic to show all the possibilities in the script. Thus the research use the Pandas method to store the large information in the memory and only call them when the command inert button is triggered.

7. The process of running the programme showing below

Figure 3 shows the first window to ask the question.

(Customers inset a number and enter the number. In the project, It is set the page number is from 1 to 24)

Figure 4 shows the second window to answer the question.

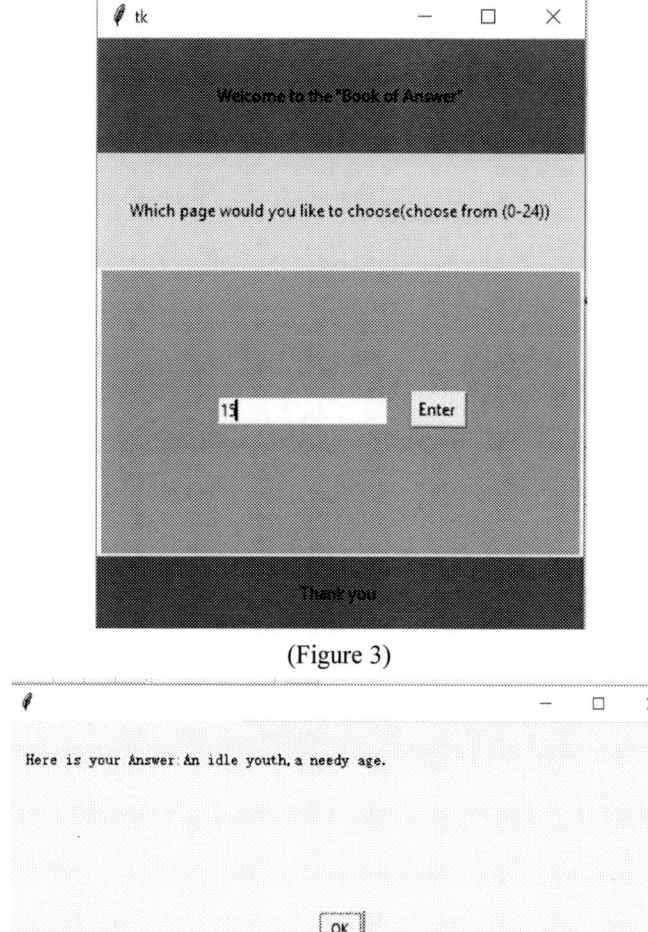

(Figure 3)

Here is your Answer: An idle youth, a needy age.

OK

(Figure 4)

8. Conclusion:

The reason to develop the program is that the book can soothe the emotion when people facing trouble. But the book about hundreds of page will be tough to carry on anytime. If people can play the game easily by phone, it will become convenience and portable.

The overcomed trouble including the encoding problem, which is to do with the UTF-8 converting, and the function choosing. The tkinter has lots of function parts including Entry, Index and Get method. Also In the project, it is set the page number is from 1 to 24. If one customer input number 0 page, the answer will be none. If the customers input a number over then the largest page number of the customer does not input a rational page number, the programme can give a instruction. (let n=the largest number of page) Such as:

```
(x=int(x)
    if 0<=x<n
      break
    else:
      print('Please choose another page')
  except:
    print('Please input a Number').)
```

The demo made in the project is comparable to a interactive dialogue between the computer and user. The interaction between them present by the 'asking' and 'Answering' process. The program can do the debug process to minimize the error. By developing this project, people can interact with the book 'The book of answer' anytime as long as they carried phone. The programme would be improved in the video interface by using bootstrap to get into a website area. Or it also can be develops as a really PVE game in App program. In addition, the content of 'Book of answer' can be changed not only in word, but in audio can video to make the interactive E-book more attractive to customers.

Reference:

[1] Huang, Yueh Min, et al. "Empowering personalized learning with an interactive e-book learning system for elementary school students." Educational Technology Research & Development 60.4(2012):703-722.

[2] Shih, et al. "How to manipulate interactive E-book on learning natural catastrophe-An;example of structural mechanics using power machine." Natural Hazards 65.3(2013):1637-1652.

[3] Ebied, Mohammed Mohammed Ahmed, and S. A. A. Rahman. "The Effect of Interactive e-Book on Students' Achievement at Najran University in Computer in Education Course. " Journal of Education & Practice 6(2015).

[4] Hatipoglu, Nuh, and N. Tosun. "The Design Of Renewable And Interactive E-Book Template For E-Learning Environments." Online Journal of Communication & Media Technologies (2012).

[5] Carol Bolt. "The book of Answer." Foreign language teaching and research press. Jane B.Reece/ lisa A.Urry/ Michael L.Cain/ Steven A.Wasserman/ Peter V.Minorsky/ Robert (2016).

[6] B.Jackson. "Campbell Biology". Benjamin Cummings.(2010).

[7] Bartoszewicki, M. (2013). Designing and Building an Interactive ebook. [online] Aptaracorp.com. Available at: https://www.aptaracorp.com/sites/default/files/designing-and-building-interactive-ebook.pdf

[8] Yuanyuan Min. (2010). Designing and implementation of science interactive E-book. Available at: https://max.book118.com/html/2017/0622/117371742.shtm

[9] Clare mcdermott. (2012). what Content Marketers Need to Know About Interactive E-books. Available at: http://contentmarketinginstitude.com/2012/09/ interactive-ebook.

[10] Annie. Moglue. (2012). The software making interactive ebook without programming. Available at :https://m.leiphone.com/news/201406/0518-annie-moglue.html .

Encrypted Tag Design for RFID Systems

Shida Yu

Department of Electrical Engineering and Automation
Luoyang Institute of Science and Technology
Luoyang, China
8617603872355

857940118@163.com

ABSTRACT: The wireless implantable RFID tag combines wireless communication technology, medical sensor technology, and data conversion technology to achieve continuous detection of body indicators. The implantable tag system includes a small signal detection circuit, a readout circuit to complete sensor data acquisition and conversion, and a digital baseband and RF front end to perform communication with an external reader. The digital baseband processor handles communication protocols, controls communication flow, configures the RF front end, and controls the sampling of human body data by the ADC, which is a key component of the implantable tag system.

Based on the 13.56MHz IS015693 radio frequency identification protocol, this paper designs and implements an encrypted tag baseband. Through the analysis of the protocol and the application requirements of the tag, the system architecture of the digital baseband is formulated. Subsequently, the functional logic was verified using Cadence's Incisive EDA tool and synthesized using Synopsys' EDA synthesis tool Design Compiler. The simulation results show that the logic function is correct. The synthesis tool report shows that under the 0.13um process, timing closures. The setup time and hold time meet the requirements.

1. INTRDUCTION

1.1 RFID system security issues.
RFID systems are composed of three key components:
• The RFID tag, or transponder, carries object identifying data.
• The RFID tag reader, or transceiver, reads and writes tag data.
• The back-end database stores records associated with tag contents

1.1.1 Tags
Every object to be identified in an RFID system is physically labeled with a tag. Tags are typically composed of a microchip for storage and computation, and a coupling element, such as an antenna coil for communication. Tags may also contain a contact pad, as found in smart cards. Tag memory may be read-only, write-once read-many or fully rewritable.

1.1.2 Readers
Tag readers interrogate tags for their data through an RF interface. To provide additional functionality, readers may contain internal storage, processing power or connections to back-end databases. Computations, such as cryptographic calculations, may be carried out by the reader on behalf of a tag.

Content from this work may be used under the terms of the Creative Commons Attribution 3.0 licence. Any further distribution of this work must maintain attribution to the author(s) and the title of the work, journal citation and DOI.
Published under licence by IOP Publishing Ltd

1.1.3 Back-End Database

Readers may use tag contents as a look-up key into a back-end database. The back-end database may associate product information, tracking logs or key management information with a particular tag. Independent databases may be built by anyone with access to tag.[1]

As you can see from the above introduction, each electronic tag usually contains an integrated circuit, which is essentially a microchip with memory. The security of data on electronic tags and the security of data on computers are also threatened. When an unauthorized party enters an authorized reader, it still sets a reader to communicate with a particular electronic tag, and the data of the tag is attacked. In this case, an unauthorized user can read the data on the electronic tag like a legitimate reader. On writable tags, data may even be modified or even deleted by illegal users.

1.2 Development status of RFID technology at home and abroad

The development of RFID technology abroad is also relatively early. Especially in the United States, the United Kingdom, Germany, Sweden, Switzerland, Japan, and South Africa, there are relatively mature and advanced RFID systems. According to different operating frequencies, RFID tags can be classified into low frequency (LF), high frequency (HF), ultra-high frequency (UHF) and ultra-high frequency (SHF). Among them, the low-frequency close-range RFID system is mainly concentrated in the 125KHz and 13.56MHz systems; the high-frequency long-distance RFID system is mainly concentrated in the UHF band (902MHz-928MHz) 915MHz, 2.45GHz, 5.8GHz. Long-range RFID systems in the UHF band have been well developed in North America; European applications have gained more applications with active 2.45 GHz systems. The 5.8 GHz system has mature active RFID systems in Japan and Europe[2].

China's research on RFID technology has also developed rapidly, and market cultivation has begun to bear fruit. More typical is the introduction of a remote automatic identification system with independent intellectual property rights in the construction of automatic identification system for railway numbers in China. After years of on-site operation test, the railway number automatic identification system project was fully invested in construction in 1999. After two years of construction and trial operation, the current railway number automatic identification system project has played a system design function, which has fulfilled the dream of railway people, and its role in radiation and penetration into other applications has become increasingly apparent. In the field of close-range RFID applications, many cities have implemented bus RF cards as prepaid electronic tickets, prepaid electronic rice cards and other applications. In terms of RFID technology research and product development, China has already developed the technical capabilities and system integration capabilities of self-developed low-frequency, high-frequency and microwave RFID electronic tags and readers. The gap between advanced RFID technologies and foreign RFID technology is mainly reflected in the technology of RFID chips. Despite this, in the design and development of tag chips, there have been many successful low-frequency RFID system tag chips available in China. However, there is still much room for development based on the function and research of the baseband control circuit of the RFID chip.

1.3 The research of this paper

This paper mainly deals with the tag of high frequency passive HF RFID system based on ISO15693 protocol. A new set of encrypted 13.56MHz high frequency RFID tag chip system is proposed for the security problem of RFID tag system. The system supports the tag part of the RFID system with encryption function based on the ISO 15693 international standard protocol, using advanced RF circuit design and embedded microcontroller, combined with efficient data processing technology[3].

2. DESIGN INTRODUCTION

The tag portion of the RFID system designed in this paper mainly includes a baseband module and an encryption module.

2.1 Baseband Module

The digital baseband is mainly responsible for the processing of digital signals, and its functions are summarized into three functions: decoding, instruction processing and encoding. Introduced below:

2.1.1 Decoding

After receiving the instruction from the reader (electromagnetic wave signal), the tag converts the demodulated signal into a digital signal and transmits it to the baseband through the analog front end. At this time, the signal is recognized by the tag, that is, it is determined what instruction it is, What kind of operation is performed, what data is included, etc., and decoding is required first. Therefore, the signal first enters the decoding module when it enters the digital baseband.

2.1.2 CRC Check

The signal received by the tag from the reader may be erroneous due to various factors during the transmission. In order to verify the correctness of the instruction, the decoded data in the baseband is checked for correctness.

2.1.3 Instruction Analysis

After the CRC check is correct, the baseband will analyze it. First, according to the first byte of the instruction, the relevant format of the tag to return instruction is analyzed, such as single/double load wave, data transfer rate and the like. Then, according to the second byte of the instruction, the type of the instruction is resolved, and the corresponding operation module is entered according to different instruction types.

2.1.4 Anti-collision Handling

When a reader communicates with an electronic tag, it is prone to a situation where multiple tags enter the active field of the reader at the same time, when the reader sends a query command to the tag, if multiple tags return to the reader at the same time The unique identifier (UID), due to the superposition and interference of the electromagnetic wave signal, the reader cannot get the correct return signal. The anti-collision module in the tag is to solve the problem that when the reader interrogates multiple tags, the tag also returns signals at the same time. The anti-collision module is responsible for calculating the order in which the tags of different UIDs are returned to the reader, so that the plurality of tags can sequentially return signals to the reader in an orderly manner, so that the reader receives and processes the signals one by one.[4]

2.1.5 Access EEPROM

After the baseband analyzes the instructions, different operations are performed according to different types of instructions, such as reading data blocks, writing data blocks, etc., essentially performing read/write operations on the EEPROM. According to the instruction, after the corresponding read/write operation of the EEPROM is successful, the corresponding information will be returned.

2.1.6 State transition

The baseband switches the tag to the corresponding working state according to different types of state transition instructions.

2.1.7 Coding

After the instruction in the baseband is executed, the corresponding information will be returned to the reader, and the completion or error information of the reader instruction will be fed back, thereby realizing the interaction of information once. In the encoding module, the baseband encodes the instructions that return the tag to the reader, and then passes it to the analog front end, which is sent through the antenna. Each tag needs to store certain information to be recognized as a specific object to be recognized by the reader. Currently, more non-volatile memory EEPROMs are used. It not only stores the user's information, such as the origin, price, etc. of a certain product, but also stores the information of the tag itself, such as the unique identifier (UID) of the tag. When the corresponding read/write instruction arrives, the tag baseband reads and writes the EEPROM as needed.

2.2 Cryptographic Module

In order to prevent the tag from being read by an illegal user, we added the operating mode of this module. In this mode, tags cannot make predictable responses to queries from unauthorized users. The implementation of this module is divided into hash lock and hash unlock[5].

2.2.1 Hash lock protocol:

A tag owner locks tags by first selecting a key at random, then computing the hash value of the key. The hash output is designated as the metaID.

The tag owner will then store the metaID on the tag and toggle it into a locked state. Writing the metaID may occur either over the RF interface or over a physical contact channel for added security. Upon receipt of a metaID value, the tag enters its locked state. While locked, a tag responds to all queries with only its metaID and offers no other functionality.

Finally, the tag owner will store the key and metaID in a back-end database, indexed on the metaID.

2.2.2 Hash unlock protocol:

To unlock a tag, the owner first queries the metaID from the tag and uses this value to look up the key in a back-end database. The owner transmits this key value to the tag, which hashes the received value and compares it to the stored metaID. If the values match, then the tag unlocks itself and offers its full functionality to any nearby readers.

To prevent hijacking of unlocked tags, they should only be unlocked briefly to perform a function before being locked again

This unlocking protocol is summarized and illustrated in Figure 1.

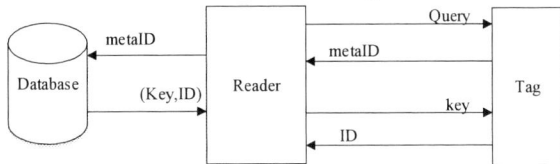

Figure 1 A reader unlocks a hash-locked tag

3. SIMULATION AND DESIGN COMPILE

3.1 Simulation

Anti-collision handling is an essential feature of RFID systems. When there are multiple tags entering the effective field of the reader at the same time, and the reader sends a query command to the tag, if multiple tags return a unique identifier (UID) to the reader at the same time, due to the superposition and interference of the electromagnetic wave signals, the reader cannot get the correct return signal. It is a key feature so that we must verify it. So in our testbench, we instantiate five tag modules to ensure the necessary conditions for conflicts. At the same time, in our test and test platform, a Reader was also instantiated. It randomly reads one of the five tags

The hierarchy of the testbench is as follow:

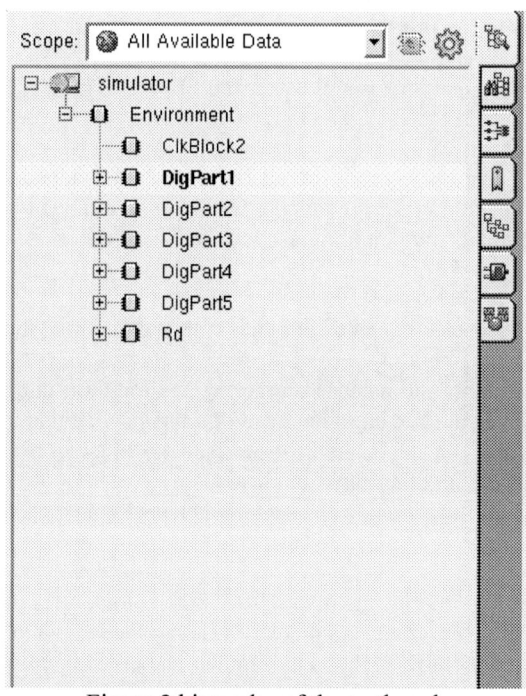

Figure 2 hierarchy of the testbench

The wave of the simulation is as follow:

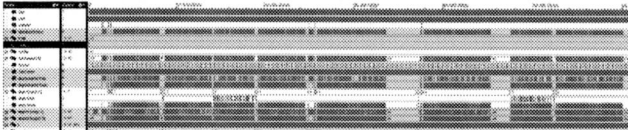

Figure 3 the wave of the simulation

The dbgGlobalOutData was written in to a text file, and the contents is as follow:

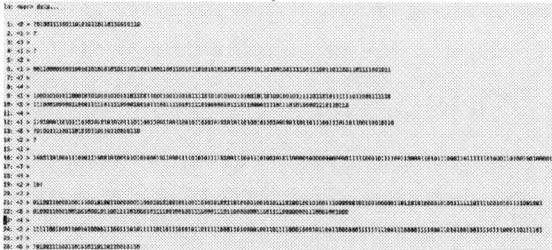

Figure 4 the output text

In the text file, the "?" means that there is a collision. But it only appears in the head or tail and the data is not interrupted. According to the above analysis, the collision has been avoided.

3.2 Design Compile

After we finished the RTL, we use the Design Compiler of Synopsys to compile our RTL. The tag in this paper is composed with eight blocks: ClkBlock, InBlcok, MainBlcok, MemBlock, RndBlock, HashBlock, RWMBlock and OutputBlock. The MainBlock is the control block, which provide all the control of the tag. And RndBlock is the randomize block, which make up the cryptographic module with the HashBlock.

4. CONCLUSION AND FUTURE WORK

In this paper, we designed an encrypted tag design for RFID systems according to the hash-lock protocol. The tag design is compatible with IS015693. After finishing the RTL, we use the EDA tools to simulate the function and compile the design to check out that the logic function is correct and the timing is closure.

In the future work, we will optimize the design, which includes less area and lower power. In addition, we will add some feature. For example, we will make our tag design be able to control the ADC to simple voltage, which can be used in the human implantable chip.

References

[1] OHTA H, IZUMI S, YOSHIMOTO M. A more acceptable endoluminal implantation for remotely monitoring ingestible sensors anchored to the stomach wall[J]. Conference proceedings: ... Annual International Conference of the IEEE Engineering in Medicine and Biology Society. IEEE Engineering in Medicine and Biology Society. Annual Conference, 2015, 2015: 4089–4092.

[2] XIAO Z, TAN X, CHEN X et. al. An Implantable RFID Sensor Tag toward Continuous Glucose Monitoring[J]. IEEE journal of biomedical and health informatics, 2015, 19(3): 910–919.

[3] DEHENNIS A D, MAILAND M, GRICE D et.al. A near-field-communication (NFC) enabled wireless fluorimeter for fully implantable biosensing applications[C]//2013 IEEE International Solid-State Circuits Conference Digest of Technical Papers. 2013: 298–299.

[4] DEHENNIS A, GETZLAFF S, GRICE D et.al. An NFC-Enabled CMOS IC for a Wireless Fully Implantable Glucose Sensor[J]. IEEE journal of biomedical and health informatics, 2016, 20(1): 18–28.

[5] WEIS S A. Security and Privacy in Radio-Frequency Identification Devices[J]. : 79.

Automated Sentiment Analysis of Text Data with NLTK

Jiawei Yao

Department of Communication, Beijing University of Posts and Telecommunications, China

chelsea_bean@yeah.net

ABSTRACT At present, most of researches in natural language processing primarily are focused on deep learning way to solve accuracy problems. This paper discusses a branch of natural language processing, sentiment analysis. In order to implement the function of sentiment analysis efficiently, we built an algorithm by calling the NLTK library firstly. As a result, this method can roughly obtain the emotional scores of different sentences and compare them with the scores given by human being who create these sentences. The results are presented by correlation calculation and images such as boxplot. The analysis of the images demonstrates the deficiency of the method. In addition, for the results, we found the direction of learning to solve the accuracy problems and made reasonable arrangements and planning.

1. BACKGROUND

1.1 Natural Language Processing

Natural language processing (NLP) is an area of computer science and artificial intelligence concerned with the interactions between computers and human (natural) languages, in particular how to program computers to process and analyze large amounts of natural language data. Natural language processing is a model for studying language ability and language application. A computer (algorithm) framework is built to implement such a language model, and it is perfected, evaluated, and finally used to design various practical systems. Its branches are Automatic Speech Recognition (ASR), Named entity recognition (NER), Optical character recognition (OCR), Sentiment analysis and so on.

1.2 Sentiment Analysis

Sentiment analysis (sometimes known as opinion mining or emotion AI) refers to the use of natural language processing, text analysis, computational linguistics, and biometrics to systematically identify, extract, quantify, and study subjective preferences and affective states. Generally speaking, sentiment analysis aims to determine the attitude of a speaker, writer, or other subject with respect to some topic or the overall contextual polarity or emotional reaction to a document, interaction, or event.

1.3 Comma-separated Values File

In computing, a comma-separated values (CSV) file is a delimited text file that uses a comma to separate values. A CSV file stores tabular data (numbers and text) in plain text. Each line of the file is a data record. Each record consists of one or more fields, separated by commas. The use of the comma as a field separator is the source of the name for this file format. The CSV file format is not fully standardized. The basic idea of separating fields with a comma is clear, but that idea gets complicated when the field data may also contain commas or even embedded line-breaks. CSV implementations may not handle

Content from this work may be used under the terms of the Creative Commons Attribution 3.0 licence. Any further distribution of this work must maintain attribution to the author(s) and the title of the work, journal citation and DOI.

Published under licence by IOP Publishing Ltd

such field data, or they may use quotation marks to surround the field. Quotation does not solve everything: some fields may need embedded quotation marks, so a CSV implementation may include escape characters or escape sequences.

In addition, the term "CSV" also denotes some closely related delimiter-separated formats that use different field delimiters. These include tab-separated values and space-separated values. A delimiter that is not present in the field data (such as tab) keeps the format parsing simple. These alternate delimiter-separated files are often even given a .csv extension despite the use of a non-comma field separator. This loose terminology can cause problems in data exchange. Many applications that accept CSV files have options to select the delimiter character and the quotation character.

1.4 The Natural Language Toolkit

The Natural Language Toolkit, or more commonly NLTK, is a suite of libraries and programs for symbolic and statistical natural language processing (NLP) for English written in the Python programming language. It was developed by Steven Bird and Edward Loper in the Department of Computer and Information Science at the University of Pennsylvania. NLTK includes graphical demonstrations and sample data. It is accompanied by a book that explains the underlying concepts behind the language processing tasks supported by the toolkit, plus a cookbook.

NLTK is intended to support research and teaching in NLP or closely related areas, including empirical linguistics, cognitive science, artificial intelligence, information retrieval, and machine learning. NLTK has been used successfully as a teaching tool, as an individual study tool, and as a platform for prototyping and building research systems. There are 32 universities in the US and 25 countries using NLTK in their courses. NLTK supports classification, tokenization, stemming, tagging, parsing, and semantic reasoning functionalities.

1.5 Random matrix

In mathematics, random matrices (also called probability matrices, transition matrices, substitution matrices, or Markov matrices) are the matrices used to describe the transformation of a Markov chain. Each of its terms is a non-negative real number that represents the probability. It applies to probability theory, statistics and linear algebra and is used in computer science and population genetics.

The random matrix describes the Markov chain in a finite state space S. If the probability of moving from i to j within a time step is $Pr(j|i)=pi,j$, the i-th row and j-th column elements of the random matrix P are given by pi,j.

Since the sum of the probabilities from state i to the next state must be 1, this matrix is a right random matrix, so we can get $\Sigma pi,=1j$. The probability of a two-step transition from i to j is given by the (i,j) number element of the given square matrix of P: $(p2)i,j$ [4].

2. DESIGN AND IMPLEMENTATION

2.1 Targets

For each dataset, we need to complete analysis as follows
- Import modules
- Open the input file (csv) using the csv module and read content (texts and ratings).
- Run the sentiment analysis function to each text review and retrieve a score.
- Collect all the scores from the entire dataset.
- Evaluate correlation between user-generated ratings and NLTK-generated scores.
- Visualize the result using R-language.

The basis of these tasks is to extract texts and read content (texts and ratings). Now let me introduce the relevant codes and methods that I used.

IOP Conf. Series: Journal of Physics: Conf. Series **1187** (2019) 052020 doi:10.1088/1742-6596/1187/5/052020

2.2 Import Modules

In the Python script, I need to import the modules I need, which contains the functions I am going to use. The codes and libraries as follow:

```
from nltk.sentiment.vader import SentimentIntensityAnalyzer
from nltk import tokenize
from numpy import *
import numpy
import sys
import os
import csv
```

Take sys for example, first, we use the import statement to enter the sys module. Basically, this statement tells Python that we want to use this module. The sys module contains functions related to the Python interpreter and its environment. When Python executes the import sys statement, it looks for the sys.py module in the directory listed in the sys.path variable. If the file is found, the statement in the main block of the module will be run, and then the module will be available to you. Note that the initialization is only done when we first input the module. In addition, "sys" is short for "system."

2.3 Datasets

In this project, we will use the NLTK's sentiment analysis function to analyze text sentiment using three datasets: 1) Amazon product review, 2) beer review, and 3) movie review. Each dataset provides a list of pairs of a review content and a numeric rating like figure 1. For instance,

- Text: "I like this move"
- Rating: 5

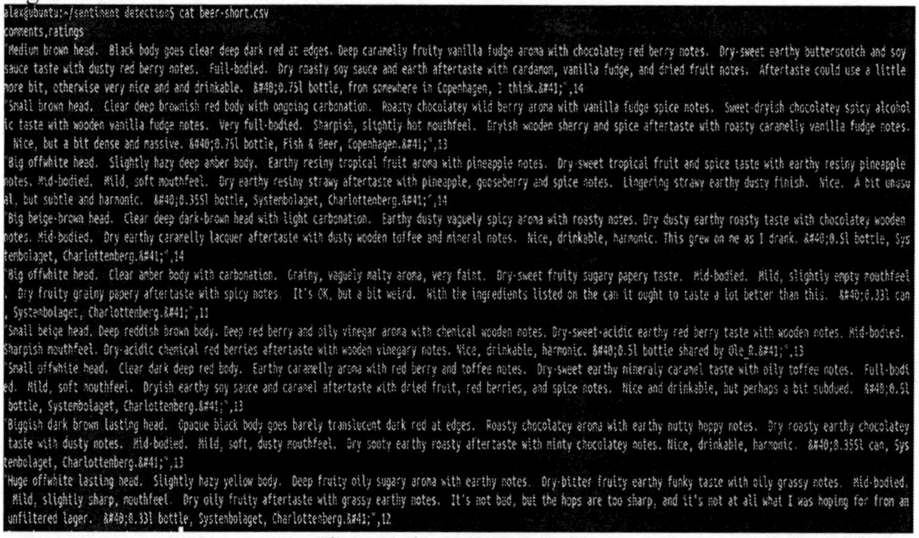

```
alex@ubuntu:~/sentiment detection$ cat beer-short.csv
comments,ratings
"Medium brown head.  Black body goes clear deep dark red at edges. Deep caramelly fruity vanilla fudge aroma with chocolatey red berry notes.  Dry-sweet earthy butterscotch and soy
sauce taste with dusty red berry notes.  Full-bodied.  Dry roasty soy sauce and earth aftertaste with cardamon, vanilla fudge, and dried fruit notes.  Aftertaste could use a little
more bit, otherwise very nice and and drinkable.  &#40;0.75l bottle, from somewhere in Copenhagen, I think.&#41;",14
"Small brown head.  Clear deep brownish red body with ongoing carbonation.  Roasty chocolatey wild berry aroma with vanilla fudge spice notes.  Sweet-dryish chocolatey spicy alcohol
ic taste with wooden vanilla fudge notes.  Very full-bodied.  Sharpish, slightly hot mouthfeel.  Dryish wooden sherry and spice aftertaste with roasty caramelly vanilla fudge notes.
 Nice, but a bit dense and massive. &#40;0.75l bottle, Fish & Beer, Copenhagen.&#41;",13
"Big offwhite head.  Slightly hazy deep amber body.  Earthy resiny tropical fruit aroma with pineapple notes.  Dry-sweet tropical fruit and spice taste with earthy resiny pineapple
notes. Mid-bodied.  Mild, soft mouthfeel.  Dry earthy resiny strawy aftertaste with pineapple, gooseberry and spice notes.  Lingering strawy earthy dusty finish.  Nice.  A bit unusu
al, but subtle and harmonic.  &#40;0.355l bottle, Systembolaget, Charlottenberg.&#41;",14
"Big beige-brown head.  Clear deep dark-brown head with light carbonation.  Earthy dusty vaguely spicy aroma with roasty notes. Dry dusty earthy roasty taste with chocolatey wooden
notes. Mid-bodied.  Dry earthy caramelly lacquer aftertaste with dusty wooden toffee and mineral notes.  Nice, drinkable, harmonic. This grew on me as I drank. &#40;0.5l bottle, Sys
tembolaget, Charlottenberg.&#41;",14
"Big offwhite head.  Clear amber body with carbonation.  Grainy, vaguely malty aroma, very faint.  Dry-sweet fruity sugary papery taste.  Mid-bodied.  Mild, slightly empty mouthfeel
.  Dry fruity grainy papery aftertaste with spicy notes.  It's OK, but a bit weird.  With the ingredients listed on the can it ought to taste a lot better than this.  &#40;0.33l can
, Systembolaget, Charlottenberg.&#41;",11
"Small beige head. Deep reddish brown body. Deep red berry and oily vinegar aroma with chemical wooden notes. Dry-sweet-acidic earthy red berry taste with wooden notes. Mid-bodied.
Sharpish mouthfeel. Dry-acidic chemical red berries aftertaste with wooden vinegary notes. Nice, drinkable, harmonic. &#40;0.5l bottle shared by Ole_R.&#41;",13
"Small offwhite head.  Clear dark deep red body.  Earthy caramelly aroma with red berry and toffee notes.  Dry-sweet earthy mineraly caramel taste with oily toffee notes.  Full-bodi
ed.  Mild, soft mouthfeel.  Dryish earthy soy sauce and caramel aftertaste with dried fruit, red berries, and spice notes.  Nice and drinkable, but perhaps a bit subdued.  &#40;0.5l
 bottle, Systembolaget, Charlottenberg.&#41;",13
"Biggish dark brown lasting head.  Opaque black body goes barely translucent dark red at edges.  Roasty chocolatey aroma with earthy nutty hoppy notes.  Dry roasty earthy chocolatey
 taste with dusty notes.  Mid-bodied.  Mild, soft, dusty mouthfeel.  Dry sooty earthy roasty aftertaste with minty chocolatey notes. Nice, drinkable, harmonic.  &#40;0.355l can, Sys
tembolaget, Charlottenberg.&#41;",13
"Huge offwhite lasting head.  Slightly hazy yellow body.  Deep fruity oily sugary aroma with earthy notes.  Dry-bitter fruity earthy funky taste with oily grassy notes.  Mid-bodied.
 Mild, slightly sharp, mouthfeel.  Dry oily fruity aftertaste with grassy earthy notes.  It's not bad, but the hops are too sharp, and it's not at all what I was hoping for from an
 unfiltered lager.  &#40;0.33l bottle, Systembolaget, Charlottenberg.&#41;",12
```

Figure 1. beer-short.csv file

2.4 Content Extraction

Read content (texts and ratings) and store them in two lists. Part of codes for implementing these features is as follows:

```
with open(filename,'rt') as csvfile:
    reader = csv.reader(csvfile)
    listx = [row[1] for row in reader]
    del listx[0]
```

```
floatlistx = [float(_s) for _s in listx]

for index1 in range(0,len(floatlistx)):
    floatstrx.append("%.4f" %(floatlistx[int(index1)]/20))
print (floatstrx)
```
Open the CSV file and use a loop to read the second list in the file, meanwhile listx can store all elements about the second list. We use the same method to extract sentences and store all elements about the first list in col.

2.5 NLTK-generated Scores

```
sid = SentimentIntensityAnalyzer()
for sentence in col:
    ss = sid.polarity_scores(sentence)
    listy.append(ss["compound"])

floatlisty = [float(_s) for _s in listy]
```
Using the function polarity_scores(x) [1] to calculate floats for sentiment strength based on the input text which is a single string text data. The output is a dictionary that has four fields, {'compound', 'neg', 'neu', 'pos'}. We can select any one of the properties for the next study.

2.6 Evaluate Correlation

Evaluate correlation between user-generated ratings and NLTK-generated scores. Moreover, the correlation is Pearson correlation coefficient. Pearson correlation coefficient, used to measure the correlation (linear correlation) between two variables, is between [-1, +1].

```
cor = numpy.corrcoef(floatlistx,floatlisty)
print "The correlation is:",'\n',cor
```
Two lists containing numbers. The shape of floatlistx and floatlisty should be same. We can get the return of the correlation coefficient matrix of the variables. For instance:

```
Import numpy
numpy.corrcoef([1,2,3], [1,3,2])
> array([[1. , 0.5],
         [0.5, 1. ]])
```

2.7 Collect All the Scores

Create a newfile to store all outputs about sentence, ratings and sentiment.

```
newfile = file(outputname,'wb')
writers = csv.writer(newfile)
writers.writerow(['sentence','ratings','sentiment'])

for index in range(len(floatlisty)):
    data=[col[int(index)],floatlistx[int(index)],
    floatlisty[int(index)]]
    writers.writerow(data)

newfile.close()
```

csv.writer(csvfile, dialect='excel', **fmtparams)

Return a writer object responsible for converting the user's data into delimited strings on the given file-like object. csvfile can be any object with a write() method. If csvfile is a file object, it must be opened with the 'b' flag on platforms where that makes a difference. An optional dialect parameter can be given which is used to define a set of parameters specific to a particular CSV dialect. It may be an instance of a subclass of the Dialect class or one of the strings returned by the list_dialects() function.

The other optional fmtparams keyword arguments can be given to override individual formatting parameters in the current dialect. For full details about the dialect and formatting parameters, see section Dialects and Formatting Parameters. To make it as easy as possible to interface with modules which implement the DB API, the value None is written as the empty string. While this isn't a reversible transformation, it makes it easier to dump SQL NULL data values to CSV files without preprocessing the data returned from a cursor.fetch* call. Floats are stringified with repr() before being written. All other non-string data are stringified with str() before being written. The output.csv file's screenshot as follows: Figure 2.

	A	B	C	D	E
1	sentence	ratings	sentiment		
2	This is a sr	5	0.9933		
3	The movie	5	0.9442		
4	A fine exa	4	0.9905		
5	I love this	5	0.8763		
6	Break out	5	0.9729		
7	As owners	5	0.6705		
8	"Not Real	1	-0.9653		
9	I acquired	4	0.7579		
10	So I expec	3	0.9891		
11	This is the	5	0.9528		
12	This is one	5	0.9614		
13	This film w	2	-0.6331		
14	It's 1964 a	5	0.886		
15	...just exac	1	-0.6705		
16	"Mary Pop	5	0.9991		
17	Movie car	4	0.4404		
18	It is, witho	5	0.9949		
19	An interes	5	0.6901		
20	Why is this	2	0.7467		
21	We are tru	5	0.8852		
22	I agree wi	5	0.7345		
23	The movie	4	0.7003		
24	This is quit	5	0.8258		
25	This is a m	2	0.868		
26	the movie	1	0.474		

Figure 2. Outputs.csv file

2.8 Markov Matrix

Suppose M is a map from time series $X(X = \{x(t) | t \in N, x(t) \in R\})$ to a network $g \in G(g = \{N, A\}$ is a set which have nodes N and arcs A). We can assume the Q quantiles, and then M assigns each quantile q_i to a node $n_i \in N$ in the relevant network with weight w_{ij}^k as long as two values x(t) and x(t + k) belong to quantiles q_i and q_j, with t = 1, 2, ..., T and the time differences k = 1,..., $k_{max} <$ T.

Weights w_{ij}^k are simply given by the number of times a value in quantile q_i at time t is followed by a point in quantile q_i at time t + k, normalized by the total number of transitions. Repeated transitions through the same pair increase the value of the corresponding weight. With proper normalization, the weighted adjacency matrix becomes a Markov transition matrix W_k, with $\sum_j^Q w_{ij}^k = 1$[2].

We can use an example to learn this algorithm, see Figure 3. When k = 1, Q = 4 quantiles and X have T = 60 time points (three coloured lines mean four quantiles) [5].

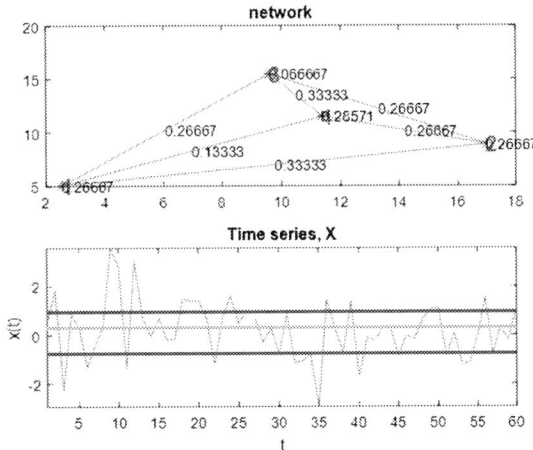

Figure 3. Time series, X and Network, g

3. RESULTS AND DISCUSSION

3.1 Boxplot and Histogram-R language

The American statistician John Tukey invented the box map in 1977. It consists of five numerical points: minimum (min), lower quartile (Q1), median, upper quartile (Q3), and maximum (max). You can also add averages (mean) to the box diagrams, as shown above. The next quartile, median and upper quartile make up a "box with compartments." An extension line is established between the upper quartile and the maximum value. This extension line becomes a "whisker" [6].

Since there are always "dirty data" and "outliers" in real-life data, these outliers are remitted separately so as not to shift the overall characteristics due to these few outliers. The two levels of whiskers in the box diagram were modified to the minimum and maximum observations. The maximum (minimum) observation value is set an experience here to 1.5 IQR distances from the quartile values.

When analysing the data through the box diagram, the box diagram can help us to identify the characteristics of the data effectively:

Visually identify outliers in the data set (see outliers).

Determine the degree of data dispersion and bias in the data set (observe the length of the box, the shape of the upper and lower compartments, and the length of the beard).

In addition, a histogram is an accurate representation of the distribution of numerical data.

Related codes:
#boxplot
library(ggplot2)
acs = read.csv("C:/Users/46025/Desktop/SentimentDetection.csv",header=T)
ggplot(acs,aes(x = ratings,y = sentiment)) + geom_boxplot()
#histogram
library(ggplot2)
acs=read.csv("C:/Users/46025/Desktop/SentimentDetection.csv",header=T)
ggplot(acs, aes(x = ratings)) + geom_histogram(binwidth = 1, fill = "lightblue", colour = "black")

The ggplot2 is a data visualization package for the statistical programming language R. Geom represents a geometric object, which is an important layer control object in ggplot because it is responsible for the type of graphics rendering. Therefore, we can figure the boxplot and histogram by using this package easily.

ISPECE   IOP Publishing

IOP Conf. Series: Journal of Physics: Conf. Series **1187** (2019) 052020   doi:10.1088/1742-6596/1187/5/052020

3.2 Boxplot and Histogram-display

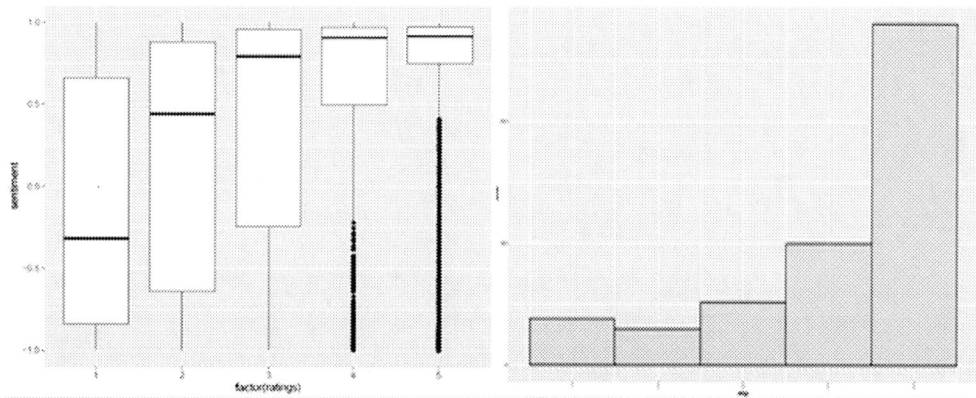

Figure 4. boxplot and histogram

Figure 4 is the boxplot and histogram results I generated using the dataset movie.csv.

According to the boxplot, we can see rating 4 and 5, the number of outliers are very much. In addition, we can know that the number of rating 4 and 5 in the histogram is the most, from which we can know that our method is limited to a small amount of text. If we handle a large number of sentences by using this method, it can reduce accuracy and get many of the wrong scores. Meanwhile, the calculation of correlation also proves this conclusion. For instance, we calculated the correlation of beer.csv and beer-short.csv, the results as follow:

```
[[1.         0.51653668]
 [0.51653668 1.        ]]
```

Figure 5. beer.csv correlation

```
[[1.         0.12516795]
 [0.12516795 1.        ]]
```

Figure 6. beer-short.csv correlation

We can get the same conclusion from figure 5 and figure 6, because beer-short.csv just have 10 sentences, beer.csv have 5000 sentences. Compared with the correlations, we can know if we handle a large number of sentences by using this method, it can reduce accuracy and get many of the wrong scores.

4. CONCLUSION AND FURTHER WORK

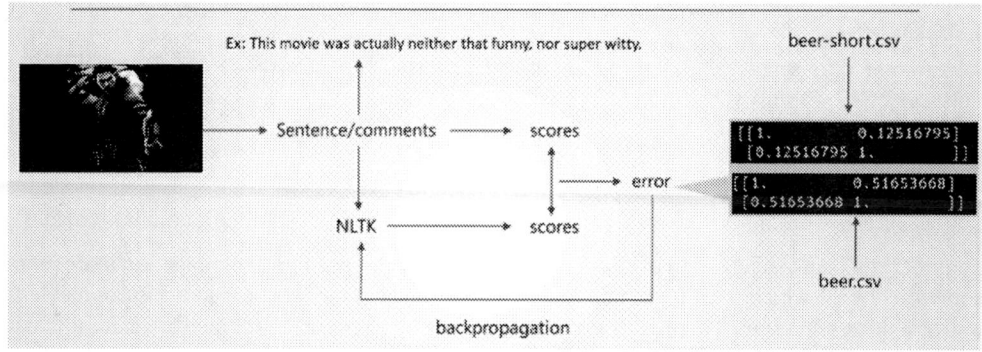

Figure 7 Flow chart

The sentiment prediction system we used in this paper works just by looking at words in isolation, giving positive points for positive words and negative points for negative words and then summing up these points. Through evaluating correlation between user-generated ratings and NLTK-generated scores and figuring the boxplot and histogram, we can know the order of words is ignored and important information is lost, ex: This movie was actually neither that funny, nor super witty. It caused the results maybe have lower accuracy with the number of sentences increasing. In contrast, the fact that we want to build a new deep learning model actually builds up a representation of whole sentences based on the sentence structure. It computes the sentiment based on how words compose the meaning of longer phrases. We can get scores that are more accurate by constantly training data sets and adjusting weights like figure 7. This way, the model is not as easily fooled as previous models.

The next step I am going to do is to understand TensorFlow and TFlearn, which is a modular and transparent deep learning library built on top of Tensorflow. Based on these fundamental knowledges, I will build a new deep learning model which have higher performance in Sentiment Analysis [3].

REFERENCES

[1] Hutto, C.J. & Gilbert, E.E. (2014). VADER: A Parsimonious Rule-based Model for Sentiment Analysis of Social Media Text. Eighth International Conference on Weblogs and Social Media (ICWSM-14). Ann Arbor, MI, June 2014.

[2] A. Campanharo, E. Doescher, F.Ramos, "Automated EEG signals analysis using quantile graphs," 14th Int. Work-Conf. Artif. Neural Netw. IWANN, Proceed. 2017.

[3] M. Baroni and A. Lenci. 2010. Distributional memory: A general frame work for corpus-based semantics. Computational Linguistics, 36(4):673–721.

[4] TAO Wenbing, JIN Hai. A novel based on graph theory image segmentation method [J]. Chinese Journal of computers, 2007, 30(1): 110-119.

[5] MathWorks, (1994-2018). Documentation - Quantile-quantile plot. [online]Available from: http://cn.mathworks.com/help/stats/qqplot.html?s_tid=gn_loc_drop [(Accessed: 5/1/2018)].

[6] John Hunter, Darren Dale, Eric Firing, Michael Droettboom and the Matplotlib development team; 2012 - 2018 The Matplotlib development team. Last updated on Aug 11, 2018. Created using Sphinx 1.7.6. Doc version v2.2.3-1-gd47e15e7a.

A Study of Speech Feature Extraction Based on Manifold Learning

Cheng Hao[1,a], Ma Xin[2,b]*, Yugong Xu[3,c]

[1]School of Information Science and Engineering Shandong University Qingdao China

[2]School of Information Science and Engineering Shandong University Qingdao China

[3]School of Information Science and Engineering Shandong University Qingdao China

[a]cheng92hao@163.com, [b]max@sdu.edu.cn, [c]1069202299@qq.com

ABSTRACT Manifold learning is a nonlinear data dimension reduction method. It can look for the essence of things from the observed phenomena, and find the inherent law of data. Traditional MFCC feature will lead a slower learning speed on account of it has high dimension and useless noise. Therefore, a speech feature extraction method based on manifold learning is proposed. Firstly, we use the manifold learning dimension reduction algorithm for the dimension reduction of Mel features and then for vowels classification. In order to further demonstrate the effectiveness of manifold learning feature in speech recognition, we propose a fusion speech feature extraction method and apply it to the identification of Chinese isolated words. Experiments prove that the fusion feature extraction method has achieved a better result than that of traditional MFCC feature extraction method.

1. INTRODUCTION

There are abundant feature parameters in speech signal, and different feature vectors represent different physical and acoustic meanings.The choice of characteristic parameters is of great importance to the success or failure of the speech recognition system.If a good feature parameter is selected, it will help to improve the recognition rate. Feature extraction is to extract or reduce the information in the speech signal which has nothing to do with the recognition, reduce the amount of data to be processed during the subsequent identification stage, and generate characteristic parameters that represent the speaker information carried in the speech signal. By far, the most commonly used speech characteristics are Linear Prediction Cepstrum Coefficient (LPCC) [1] and MEL Cepstrum coefficient (MFCC) [2]. All of them have achieved good recognition effect in speech recognition.

Due to physiological limits, speech organs only have a few degrees of freedom, people can only produce a limited range of sounds, which occupy a restricted area of acoustic space. Therefore, the speech signal data can be regarded as a low-dimensional manifold embedded in the high-dimensional space of all possible waveform. The purpose of feature extraction is to represent the most important essential rule of speech signal with as few data dimensions as possible.

Manifold learning is a nonlinear data dimension reduction method, which was first proposed in the famous science magazine Science in 2000 [3, 4]. In 2005, Jansen et al. discovered that voice data is data embedded in high dimensional acoustic space, and in 2006, the Fourier transform of voice signal was used to further prove the essence of the existence of manifold in voice signal [5]. Errity Andrew et al. [4] studied the speech signal manifold and the possibility of the use of manifold learning method for speech signal analysis. They believe that manifold learning can be used to explore the hidden low-

Content from this work may be used under the terms of the Creative Commons Attribution 3.0 licence. Any further distribution of this work must maintain attribution to the author(s) and the title of the work, journal citation and DOI.
Published under licence by IOP Publishing Ltd

dimensional manifold structure in speech signal, and the manifold feature extracted by manifold learning algorithm is beneficial to improve the accuracy of ASR system. Wang li et al. [7] put the MFCC feature into the lower dimension feature space by using the method of manifold dimension reduction, and extracted the manifold features, and presented a research idea of using the manifold learning method to extract the characteristic parameters of speech signal.

The manifold learning algorithm can discover the geometric structure contained in the high dimensional data space, and the essential connection between sample data, and realize data dimension reduction [8, 9]. In this paper, manifold learning algorithm is applied to feature extraction, which can overcome dimension disaster, obtain essential features, save storage space, and eliminate useless noise.

2. MANIFOLD LEARNING ALGORITHM

Isometric Feature Mapping (ISOMAP) is a nonlinear data dimension reduction algorithm proposed by Tenenbau et al. [3] based on MDS algorithm. Assuming that the data is distributed on a manifold space, in order to keep inherent geometry relationship in high-dimensional space, the relative distance between each sample point in a high dimensional sample set usually needs to remain constant. In Isomap algorithm, the author introduced Geodesic Distance to replace the traditional Euclidean Distance, which can better maintain high dimensional sample space geometric relationship between each sample, and then obtain the optimal structure of dimension reduction. Tenenbau gives a method to approximate the geodesic distance by using the shortest path between two points in the data neighborhood graph in [3]. Bernstein et al. [8, 9] have proved that the geodetic distance can be approximated by the shortest distance when there are enough data samples and under uniform sampling conditions.

Classic ISOMAP algorithm has three steps:

(1) Establish neighborhood relationship diagram $G(V, E)$: For each $x_i (i = 1, \cdots, N)$, according to certain criteria to determine K phase velocity point and calculate its K neighbor. To point x_i for vertex. Euclidean distance $d_o(x_i, x_j)$ construction diagram $G(V, E)$.

(2) According to the graph set good path to calculate the shortest path between any two points use them as myopia geodesic distance $d_l(x_i, x_j)$ to construct the geodesic distance matrix $D[d_l(x_i, x_j)]_{N_x N}$.

(3) For geodesic distance matrix $D[d_l(x_i, x_j)]_{N_x N}$ use traditional MDS algorithm, looking for a low dimensional data $Y = (y_1, y_2 \ldots y_n)$.

Laplacian Eigenmaps (LE) [8, 12] is a nonlinear dimension-reduction algorithm for semi-supervised learning, which is also an algorithm to maintain the local relationship between samples. LE algorithm base on spectral graph theory, through the calculation of adjacency matrix of sample points in higher dimensional sample space and feature decomposition of the laplacian matrix of the adjacency matrix, and then calculate the nonlinear dimension reduction result of the high-dimensional sample data.

Locality Preserving Projection (LPP) [9] was proposed by He Xiaofei and Partha Niyogi in 2003. LPP algorithm is a linear approximation of LE [8, 12] algorithm. LPP algorithm calculates the transformation matrix by constructing the adjacency graph of the data set and using the Laplace matrix of the adjacency graph. By using this transformation matrix, data points can be mapped to the low-dimensional subspace.

3. FEATURE EXTRACTION OF VOICE SIGNAL

3.1 MFCC Feature

MFCC [2] feature is a kind of speech characteristic parameter based on human auditory mechanism and is an important characteristic parameter in the field of speech signal processing. According to human auditory mechanism, the division of subjective perceptual frequency domain is not linear. Long-term studies have shown that the relationship between perceived frequency and actual frequency is

$$F_{mel}(f) = 1125 \times \ln(1 + \frac{f}{700}), \qquad (1)$$

Where, f is the actual frequency, and the unit is Hz, is the perceived frequency, and the unit is Mel. By analyzing the speech signal in the perceptual frequency domain, we can get a better approach to the auditory processing process and get a better result.According to the critical band theory, a series of frequency groups are divided on the frequency to form several triangular bandpass filter groups, namely Mel filter groups. The extraction and calculating process of MFCC features as show in Figure 1.

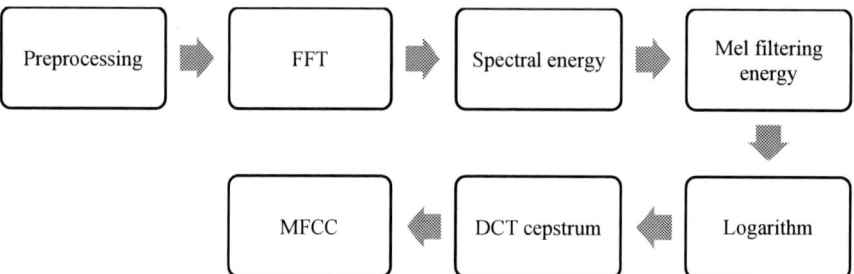

Figure 1. MFCC feature extraction process

3.2 Manifold Learning Feature

In most applications, the number of triangular filters in the Mel filter group is generally 26, so the dimension D of the initial sample parameter vector of speech signal is 26.However, relevant studies have shown that the degree of freedom of speech signal is generally 5 to 6, and the dimension of initial sample parameters can be reduced by using manifold learning method, so as to obtain the eigenfeature parameters of speech signal.Therefore, a new method of extracting speech feature parameters is proposed in this paper. The extraction process of phonetic characteristic parameters based on manifold learning as shown in Figure 2.

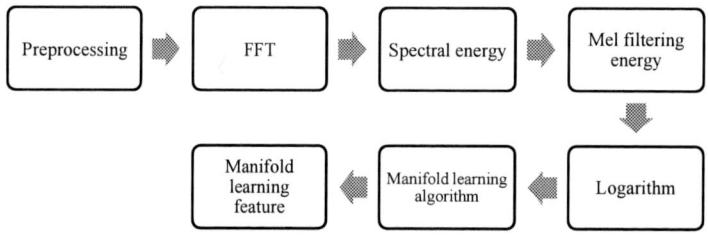

Figure 2 Manifold learning feature extraction process

3.3 Feature Fusion

On the basis of vowel classification, in order to further prove the viewpoint proposed in this paper in the identification of Chinese numeral isolated words, the manifold learning features and the MFCC features are fused as new features for speech recognition. In this way, we can not only use the advantages of MFCC features based on human auditory mechanism, but also use the advantages of manifold learning features. The feature fusion method adopts series fusion, namely, the MFCC feature matrix is directly spliced behind the manifold learning feature matrix. The extraction process of fusion features is shown in Figure 3.

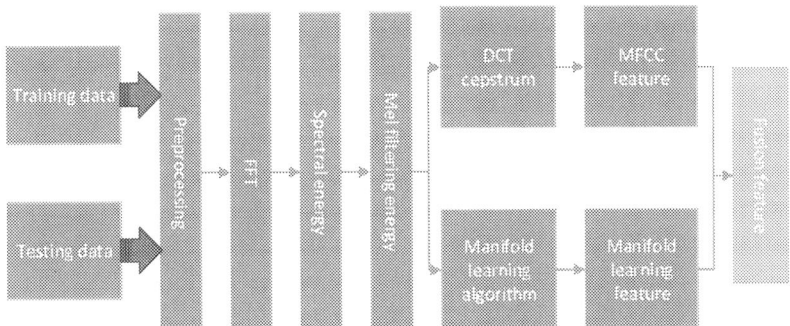

Figure 3. Fusion feature extraction process

4. EXPERIMENTS

4.1 Vowel Separability Analysis

According to the vowel phoneme map given by IPA, 5 vowels, such as aa, ae, uw, iy and eh, are selected, and 400 samples are extracted from the TIMIT [12] corpus for each vowel, and then we calculate the MFCC features and manifold learning features of the 2000 samples respectively. The manifold learning algorithm used in this paper is ISOMAP, LE and LPP, and the parameter k (the number of nearest neighbor samples) used by the algorithm is 12 by default. In order to investigate the performance of each algorithm in different dimensions, the 2-15 dimensional features of these vowel samples are calculated. FIG. 4 shows the distribution of 2-dimensional manifold feature sample points of the four vowels aa, ae, uw and iy. It should be noted that because ae and eh sound alike, the separation degree of each algorithm is not very good, so only four vowels are used for visualization. It can be seen from the figure that the manifold learning algorithm is effective in the visualization of vowel clustering analysis, while the traditional MFCC feature is not effective.

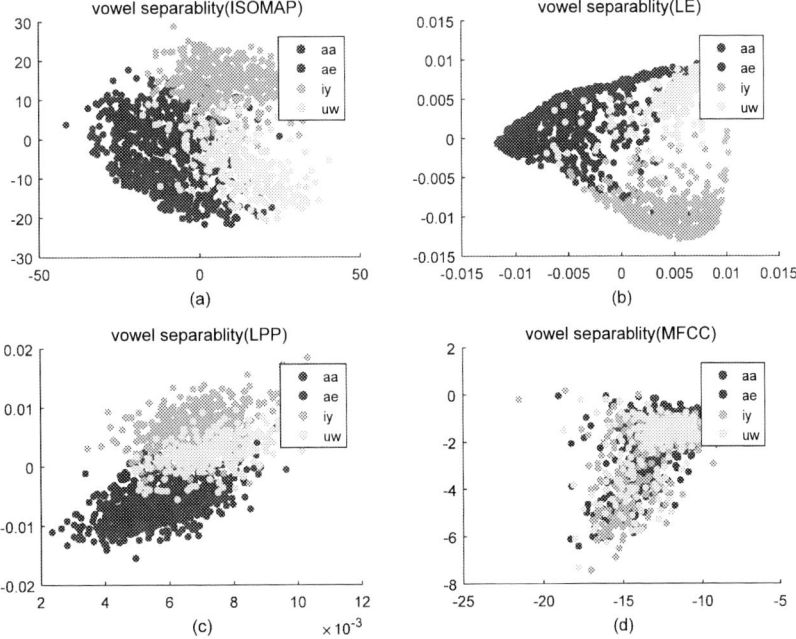

Figure 4. Distribution of manifold features in two-dimensional space

The performance of each algorithm can be measured by calculating the vowel separability in low dimensional space. In this paper, Bhattacharyya distance [6, 14] is used to measure the separability of two kinds of samples. The mean value of the two samples is M_1 and M_2, and the variance is Σ_1 and Σ_2. The calculation method of Bhattacharyya distance is as follows

$$D_{Bhattacharyya} = \frac{1}{8}(M_2 - M_1)^T [\frac{\Sigma_1 + \Sigma_2}{2}]^{-1}(M_2 - M_1) + \frac{1}{2}\ln\frac{|\frac{\Sigma_1 + \Sigma_2}{2}|}{\sqrt{|\Sigma_1|}\sqrt{|\Sigma_2|}} \quad , \quad (2)$$

The Bhattacharyya distance between each of the five vowels in 2-15 dimensional feature space is calculated. By comparison, the separation degree of the vowels in the manifold feature space is significantly better than that of the MFCC feature space. FIG. 5 shows the results in the 3-dimensional feature space, which is obviously better than the results in [6].

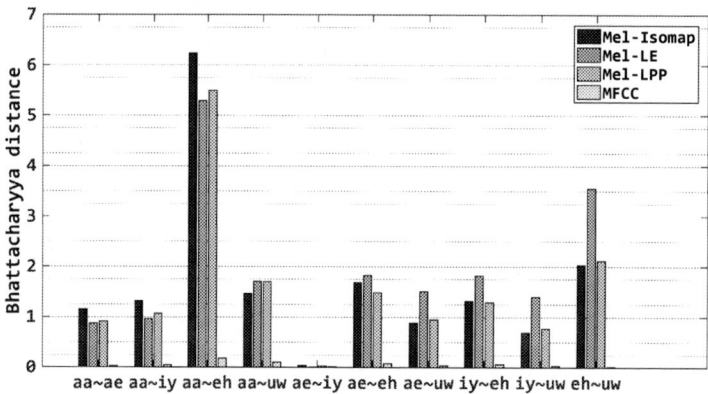

Figure 5. Vowel separability test

4.2 Vowel Classification Analysis

In order to test the usability of manifold learning algorithm, this section conducts a classification analysis of these five vowels. In order to investigate the performance of each algorithm in different dimensions, the 2-15 dimensional features of these five vowel samples are calculated. In the classification analysis, 300 samples of each vowel are taken and marked to form the training set, and the remaining 100 samples were used for classification test. KNN (k-nearest Neighbour) [15] classifier was used for the classification test, and the parameter k=5. FIG. 6 shows the results of kNN classification experiment. It can be seen from the figure that the classification results obtained by using the manifold features are obviously better than the traditional MFCC features. The features based on manifold learning can achieve better classification result at low dimension, and can reflect the manifold structure of speech signal. With the increase of dimension, the error rate does not decrease significantly, which means that LE and LPP algorithm cannot extract more information from local linear space.

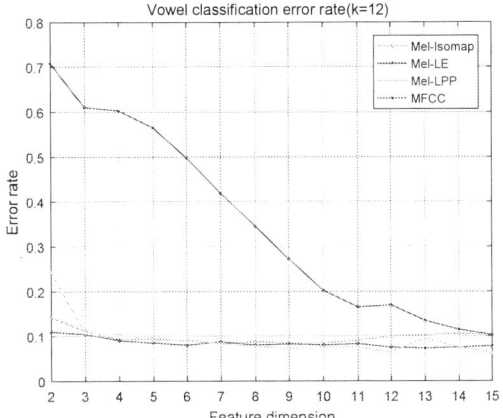

Figure 6. Vowel classification test

In the above experiment, the default nearest neighbor parameter k of the manifold learning dimension-reduction algorithm we used is 12, which is also one of the most important parameters of the dimension-reduction algorithm and has a great influence on the classification results of vowels. It can be seen from Figure 7 that the parameter k has a significant influence on the accuracy of vowel classification. When k=8 or k=16, the accuracy is not very obvious with the change of dimension. When k=10, k=12 or k=14, the accuracy is increased with the increase of feature dimension. When k=12 and the feature dimension is 8, the accuracy of vowel classification reaches up to 92.25%.

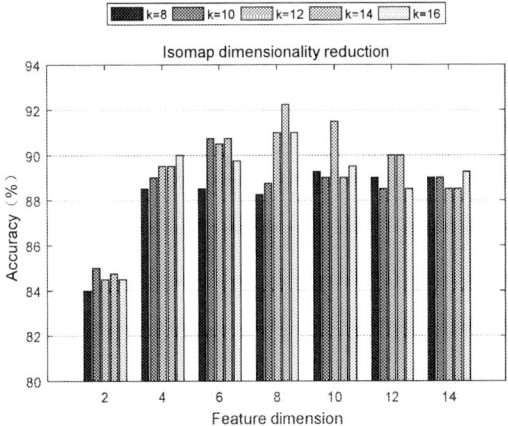

Figure 7. Effect of manifold learning parameter k on vowel classification

4.3 Speech Recognition Analysis

In order to test the effect of manifold learning algorithm in continuous speech recognition, this paper uses the HMM-GMM to conduct the experiment of isolated words recognition in Chinese pronunciation. The training set uses 10 Arabic numerals from 0 to 9, 200 samples in total, 20 samples for each number, and the sampling rate is 16KHz. There are 50 samples in the test set, 5 samples for each number. In the experiment, in order to integrate the advantages of the two algorithms, the manifold learning features and the MFCC features are fused together to obtain the new fusion features for the recognition of Chinese isolated words. In this experiment, ISOMAP manifold learning algorithm is adopted, and the nearest neighbor parameter k is 12. FIG. 8 is the result of the experiment, which compares the average recognition rate of fusion feature extraction method and MFCC feature extraction method in different dimensions. It can be seen from the figure that the average recognition rate of ISOMAP fusion feature is much higher than that of traditional MFCC feature in the case of low dimension or even some high dimension. And we can see from Table 1 that the average recognition rate of MFCC and MFCC-ISOMAP are 42.7% and 52% respectively. The MFCC-ISOMAP fusion feature extraction method has better performance in Chinese isolated words recognition.

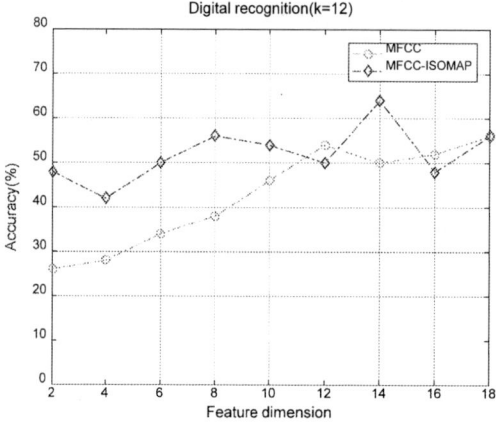

Figure 8. 0-9 Chinese number recognition

Table 1. Average recognition rate of different feature extraction method (%)

Feature dimension	MFCC	MFCC-ISOMAP
2	26	48
4	28	42
6	34	50
8	38	56
10	46	54
12	54	50
14	50	64
16	52	48
18	56	56
Average recognition rate	42.7	52

5. CONCLUSION

In this paper, we apply the manifold learning algorithm to reduce the dimension of vowel features, and then use KNN classifier to test the manifold feature extraction method. The results show that the vowel classification error rate of manifold learning feature extraction method is much lower than that of traditional MFCC feature extraction method, which means the manifold learning feature extraction method is effective in vowel classification. In order to further demonstrate the effectiveness of manifold learning algorithm in speech recognition, the ISOMAP fusion feature extraction method is used to recognize Chinese isolated words. The HMM-GMM is used to model the speech signal. The experimental results show that the recognition rate of ISOMAP fusion feature extraction method is also higher than that of traditional MFCC feature extraction method, which indicates that the ISOMAP fusion feature extraction method is helpful to improve the recognition rate of Chinese isolated words.

Whether other manifold learning algorithms are also effective, and whether the manifold learning algorithm can be combined with CNN, RNN and other deep learning models for speech recognition is the next step of this paper. In any case, manifold learning as the latest machine learning algorithm, has obtained the certain achievement in the face recognition, which will become an important research direction to promote the breakthrough of speech recognition.

ACKNOWLEDGEMENT

This work was supported by the National Natural Science Foundation of China under Grant No.62171453. The authors would like to thank Xu Yugong for his helpful works and comments.

REFERENCES

[1] Gupta H, Gupta D. LPC and LPCC method of feature extraction in Speech Recognition System[C]// Cloud System and Big Data Engineering. IEEE, 2016:498-502.

[2] Davis S, Mermelstein P. Comparison of parametric representations for monosyllabic word recognition in continuously spoken sentences [J]. Readings in Speech Recognition, 1990, 28(4):65-74.

[3] Tenenbaum J B, Silva V D, Langford J C. A Global Geometric Framework for Nonlinear Dimensionality Reduction [J]. Science, 2000, 290(5500):2319-2323.

[4] Roweis S T, Saul L K. Nonlinear dimensionality reduction by locally linear embedding. [J]. Science, 2000, 290(5500):2323-2326.

[5] Jansen A, Niyogi P. Intrinsic fourier analysis on the manifold of speech sounds [J]. Science China, 2006, 56(3):1-11.

[6] Errity A, Mckenna J. An investigation of manifold learning for speech analysis[C]// INTERSPEECH 2006 - Icslp, Ninth International Conference on Spoken Language Processing, Pittsburgh, Pa, Usa, September. DBLP, 2006.

[7] Zhang P, Wang L. The application research of speech feature extraction based on the manifold learning [J]. ICCSEE-13, 2013.

[8] Belkin M, Niyogi P. Laplacian eigenmaps and spectral techniques for embedding and clustering [J]. Advances in Neural Information Processing Systems, 2009, 14(6):585-591.

[9] He X. Locality preserving projections [J]. Advances in Neural Information Processing Systems, 2003, 16(1):186-197.

[10] Balasubramanian M, Schwartz E L. The isomap algorithm and topological stability. [J]. Science, 2002, 295(5552):7.

[11] Bernstein M, Silva V D, Langford J C, et al. Graph Approximations to Geodesics on Embedded Manifolds [J]. Stanford University, 2000, 24(9):153--158.

[12] Belkin M, Niyogi P. Laplacian Eigenmaps for dimensionality reduction and data representation [M]. MIT Press, 2003.

[13] Garofolo J S, Lamel L F, Fisher W M, et al. TIMIT Acoustic-phonetic Continuous Speech Corpus[C]// Linguistic Data Consortium. 1993.

[14] Fukunaga, Keinosuke. Introduction to statistical pattern recognition (2nd ed.)[J]. 1990, 60(12-1):2133-2143.

[15] Zhang M L, Zhou Z H. ML-KNN: A lazy learning approach to multi-label learning [J]. Pattern Recognition, 2007, 40(7):2038-2048.

Research and Improvement of CHI Feature Selection in Sentiment Analysis

Li Danyang[1,a], Fan Huimin[2,b]

[1]School of Computer Science and Engineering Xi'an Technological University
86-15229895010, 710021

[2]School of Computer Science and Engineering Xi'an Technological University
86-13891982323, 710021

[a]821563942@qq.com,[b]492896361@qq.com

ABSTRACT Feature selection is a very important step in sentiment classification based on machine learning method. This paper will focus on the better CHI feature selection method and make improvements to overcome the shortcomings of the traditional method. The experimental results show that the improved IM-CHI feature selection method improves the F1 value of emotion classification by 4.9% in different feature dimensions, 4.1% in different classifiers, and 2.0% in data sets of different fields. It proves the effectiveness of this method in emotion classification.

1. INTRODUCTION

With the rapid development of the Internet, people freely express their opinions and opinions on the Internet, which contain a lot of valuable information. One of the most representative is e-commerce, people in the major e-commerce websites on the merchandise comments to a large extent affect consumer purchasing decisions [1].

Sentiment analysis, also known as opinion mining, refers to mining the emotional tendencies expressed by reviewers through the analysis of text content. The task of text emotion analysis mainly includes Sentiment classification, emotion information extraction and emotion information retrieval and induction [2]. Machine learning is a common method of text Sentiment classification. Sentiment analysis method based on machine learning [6-8] includes data preprocessing, text representation, feature selection, training classifier and prediction, and feature selection plays an extremely important role in it. Features are part of the characteristics of classified objects and important basis for classification. At the beginning, the selected features will reach tens of thousands or even hundreds of thousands of dimensions, which not only makes the computer operation time very long, but also reduces the accuracy. Feature selection is the use of a certain feature selection algorithm from the beginning of the high-dimensional feature selection part of the information-rich features as a classifier classification features, reduce noise, improve classification accuracy. Reasonable feature selection not only reduces the classification time, but also removes redundant information and improves the classification accuracy. Therefore, feature selection is very important for text emotional classification [3]. At present, the commonly used feature selection algorithms are document frequency (DF), information gain (IG), mutual information (MI), chi-square statistic (CHI), expected cross-tropy (ECE). After analyzing and comparing the feature selection methods such as IG, DF, MI and CHI, Yang draws a conclusion that the classification effect of CHI method is relatively good [4]. But the

Content from this work may be used under the terms of the Creative Commons Attribution 3.0 licence. Any further distribution of this work must maintain attribution to the author(s) and the title of the work, journal citation and DOI.
Published under licence by IOP Publishing Ltd

algorithm does not involve any form of word frequency information, there is a large word frequency defect, which leads to the accuracy of the method is reduced.

Aiming at the shortcomings of traditional CHI, this paper proposes an improved IM-CHI feature selection method in Sentiment classification, which combines the characteristics of sentiment analysis and overcomes the defect of word frequency. Experiments show that the algorithm proposed in this paper is superior to the traditional CHI algorithm in all aspects for sentiment classification.

2. THE TRADITIONAL CHI ALGORITHM

CHI, proposed by British statistician Pearson, is used to test the independence or relevance of class variables. When a feature is selected using the CHI algorithm, the larger the CHI value of the feature word in the class, the greater the representation of the word for the class. First, suppose the feature word w and category ci are independent. If N denotes the total number of documents in the document set, A denotes the number of documents containing the feature w and belonging to category c_i, B denotes the number of documents containing the feature w but not the category c_i, C denotes the number of documents that do not contain the feature w but belong to the category c_i, and D denotes the number of documents that do not contain feature w nor belong to category c_i. Then, the formula of CHI is shown in formula (1).

$$CHI(w,c_i) = \frac{N \times (AD - CB)^2}{(A+C)(B+D)(A+B)(C+D)} \qquad (1)$$

this paper study deeply the traditional CHI algorithm based on sentiment classification, and analyze the following disadvantages:

(1) Word frequency defect. The traditional CHI algorithm only considers whether a feature word appears in a document or not, but does not consider the number of occurrences of the word. If a feature word appears frequently only in a small number of documents of a certain kind, the calculated CHI value is relatively small. This makes the algorithm more likely to choose low-frequency words and reduce the accuracy.

(2) It does not reflect the particularity of sentiment classification. For sentiment classification, the words used to express emotions should be preferred when choosing features. If only a small number of emotional words appear, the calculated CHI value is very small, and it is likely to be excluded in feature selection, which leads to the classification accuracy will be extremely inaccurate.

3. IMPROVED IM-CHI ALGORITHM

Aiming at the two shortcomings of traditional CHI algorithm, this paper adds word frequency information calculation and increases the amount of emotional word information calculation, and improves the algorithm successfully. The improved feature selection algorithm is named IM-CHI algorithm.

In order to solve the problem of word frequency defect in traditional algorithm, this paper proposes the computational complexity CHI_freq, which is used for reference to the traditional Chi-Square statistics, as shown in Formula (2).

$$CHI\_freq(w,c_i) = \frac{N}{C \times C'} \times \frac{(F \times C' - F' \times C)^2}{WF \times (N - WF)} \qquad (2)$$

In formula(2), w is the feature word, c_i is the category, N is the number of the feature words contained in all categories, C is the number of the feature words contained in class c_i, C' is the number of the feature words contained in other categories, F is the frequency of the feature word w in class c_i, F' is the frequency of the feature word w in other categories, and WF indicates that the word w is the word frequency in all categories.

For certain datasets, N, C, and C' correspond to constants, so from the right half of formula (2), it can be concluded that the more frequent a feature word w is in a certain category, the greater the value of the molecular part, and the greater the calculated value of CHI_freq.

For the problem that traditional algorithms do not embody the particularity of affective classification problem, this paper proposes the affective parameter λ to increase the computational value of affective words. The value of the emotional parameter λ is shown in Formula (3).

$$\lambda(w) = \begin{cases} 1, & w \text{ is an emotional word} \\ 0, & w \text{ is not an emotional word} \end{cases} \tag{3}$$

By querying the sentiment dictionary, if the characteristic word w is the emotional word, the value of λ is 1, otherwise, the value of λ is 0. Therefore, the improved *IM-CHI* algorithm is shown in Formula(4) by combining the computational complexity of *CHI_freq* and affective parameter λ.

$$IM-CHI(w,c_i) = CHI \times CHI\_freq + \lambda(w) \times CHI\_freq \tag{4}$$

CHI is the traditional Chi square statistic value. The improved *IM-CHI* algorithm adds the calculation of the frequency of the feature words to the traditional chi-square value, which overcomes the shortcoming of the traditional algorithm which tends to select the low-frequency words, and the affective parameter λ increases the calculation value of the affective words. Formula (4) shows that when a feature word is an affective word, the value of lambda is 1, then the chi-square value of the affective word is more than the chi-square value of the non-affective word. This makes the possibility of preferential choice of affective words in feature selection more likely, and has better effect in affective classification.

4. EXPERIMENTAL RESULTS AND ANALYSIS

In order to verify the effectiveness of the improved algorithm, this paper compares the algorithm from several different perspectives, such as data set domain, feature dimension and classifier. The experiment uses the weighted average F1 value of accuracy and recall rate as the evaluation index. The classifier obtained by using CHI algorithm and IM-CHI algorithm respectively in the training process will be used in the test process. Therefore, two classification results will be obtained in the test process. The validity of the improved algorithm is verified by comparing the F1 values of the two results.

The data set used in this experiment is a total of 16000 internet shopping reviews datasets including Chinese emotion mining corpus provided by Dr. Tan Songbo of the Chinese Academy of Sciences, covering six areas: hotels, computers, books, mobile phones, milk, water heaters. The positive and negative reviews of the data set were randomly selected 20% and randomly sorted as test sets.

4.1 Comparative Experiments Under Different Characteristic Dimensions
In order to find the appropriate feature dimension and verify the change of classification F1 value of CHI algorithm and IM-CHI algorithm under different feature dimension, SVM [5] with better classification effect is used to carry out the experiment. The experimental results are shown in figure 1.

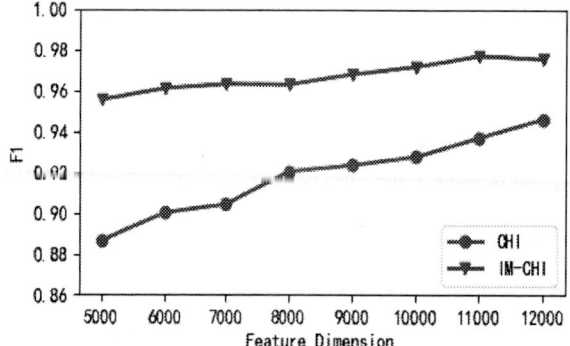

Figure 1. Comparison test results of different feature dimensions under SVM
Experiments show that the F1 value of the improved IM-CHI algorithm is higher than that of CHI

algorithm in different dimensions.

Analysis of figure 1 shows that when the dimensions are small, the F1 values of the two algorithms are relatively low, and with the increase of the dimensions, the F1 values show an upward trend. After calculation, the improved IM-CHI algorithm improves the F1 value by 4.9% on average in different feature dimensions. In addition, it can be seen from the graph that the F1 value is higher when the feature dimension is about 11000, and then the F1 value tends to be stable.

4.2 Comparative Experiments Under Different Classifiers

In order to verify the effectiveness of IM-CHI algorithm under different classifiers, this experiment selects Naive Bayes, Logical Regression, Support Vector Machine, Decision Tree, K-Nearest Neighbor classifier, and carries on the contrast experiment under the same environment of the feature dimension 11000. The experimental results are shown in Figure 2.

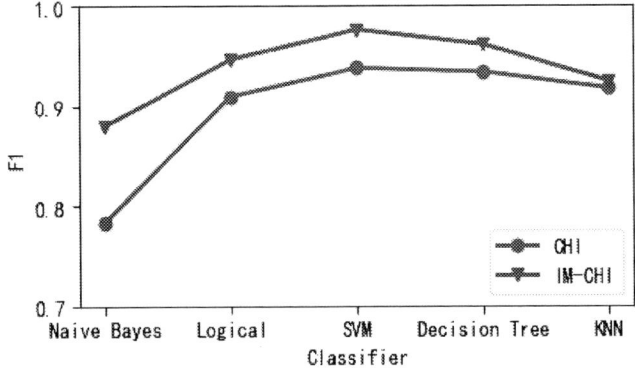

Figure 2. Comparison results of CHI and IM-CHI under different classifiers

Experiments show that the improved IM-CHI algorithm has higher F1 value than CHI algorithm even if different classifiers are used.

Analysis of figure 2 shows that Bayesian classifier has the worst classification effect, but the F1 value of IM-CHI algorithm has been greatly improved, reaching 9.6%. It can also be seen that among all the classifiers, the F1 value of SVM is the highest and the classification effect is the best. This result also corresponds to the results of Pang [5].

4.3 Comparative experiments under different data sets

To verify the effectiveness of the improved IM-CHI algorithm in different data domains, the data sets are divided into six domains: hotel, computer, book, mobile phone, milk and water heater. The data sets of six domains are shown in Table 1 below. Data sets in each field are balanced corpus.

Table 1. Size distribution of datasets in six fields

Dataset domain	Hotel	computer	book
Comment number	3500	3780	3712
Dataset domain	mobile phone	milk	water heater
Comment number	2316	2012	682

In the environment of SVM classifier and feature dimension 11000, the experiments are carried out by using CHI algorithm and IM-CHI algorithm respectively. The experimental results are shown in Figure 3.

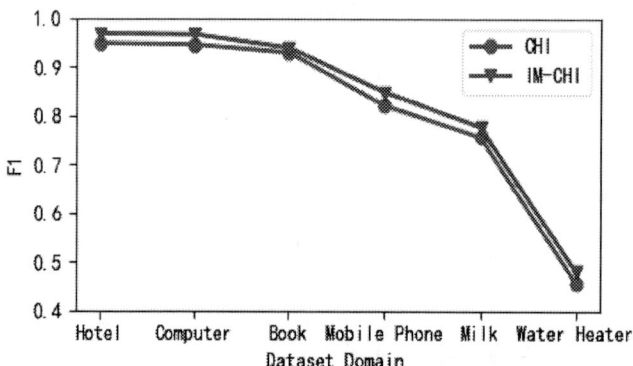

Figure 3. Comparison results of CHI and IM-CHI under different data sets

Experiments show that the F1 value of the improved IM-CHI algorithm is higher than that of CHI algorithm under different datasets.

Figure 3 shows that the F1 value decreases with the decrease of the training dataset size, so the training dataset must be above a certain size to ensure the reliability of classification. At the same time, it can be concluded that the improved algorithm's F1 value is very stable, compared with the CHI algorithm's F1 value increased by an average of 2.0%.

5. CONCLUSION

(1) In this paper, an improved IM-CHI feature selection algorithm is designed to overcome the shortcomings of traditional CHI algorithm.

(2) Compared with the traditional CHI algorithm, the improved IM-CHI algorithm improves the classification F1 value by 4.9% on average in different feature dimensions; under different classifiers, the improved algorithm improves the classification F1 value by 4.1%; under different data sets, the improved algorithm improves the classification F1 value by 2.0% on average.

(3) Experiments show that the IM-CHI algorithm proposed in this paper is superior to the traditional CHI algorithm, but the adaptability of the algorithm to data sets in different fields is not stable, so the performance of the algorithm needs to be further improved.

REFERENCES

[1] Li Piji, Ma Jun, Zhang Dongmei. 2012. Label extraction and sorting in user reviews. Journal of Chinese Information Processing, 26: 14-19,45.

[2] Zhao Yanyan, Qin Bing, Liu Ting. 2010. Text sentiment analysis. Journal of Software, 21: 1834-1845.

[3] Yan Lei. 2017. Sentiment text classification based on improved CHI feature selection. Transducer and Microsystem Technologies, 36: 47-51.

[4] Yang Y, Pedersen J O. 1997. A comparative study on feature selection in text categorization. Proceedings of the Fourteenth Internatio-nal Conference on Machine Learning, Morgan Kaufmann Publi-shers Inc, 412-420.

[5] B. Pang,L. Lee,S. Vaithyanathan. 2002. Thumbs up Sentiment Classification Using Machine Learning Techniques. Proceedings of Conference on Empirical Methods in Natural Language Processing (EMNLP). Philadelphia: ACL, 79-86.

[6] Song Jingjing. 2013. Sentiment analysis tendency analysis of Chinese short text. Chongqing University of Technology.

[7] Xu Libo. 2012. Extraction of the collocation relationship between product evaluation objects and emotional words. Beijing University of Posts and Telecommunications.

[8] Hu Cheng. 2016. Design and implementation of emotion analysis system based on dependency syntax. South China University of Technology.

A Research on the Innovation and Development of Micro and Small Sci-tech Enterprises

Wang Guanhui

School of International business, Tianjin Foreign Studies University
Tianjin, China

e-mail: newhop@126.com

Abstract: For micro and small sci-tech enterprises with insufficient resource endowment and management experience, it is of great significance to carry out innovation to improve their level of management. Based on resource-based theory and dynamic capability theory, the paper explores the influences from the angle of ecological entity, populations and communities. The internal influences include resources and capability, and the external influences include the population and development environment. Based on the above, the model of growth mechanism is suitable for technology-based small firms to show the developing process of technology-based small firms. In view of the development status of technology-based small firms, this paper researches on the growth mechanism of technology-based small firms to find the micro and macro issues for improvement. All the specific comments and suggestions can offer reference for the leaders to make decision.

1. Introduction

Micro and small enterprises are an indispensable part in China's economy. They have created job opportunities for urban and rural residents and eased the employment pressure. Especially after Premier Li Keqiang issued the call for "mass entrepreneurship and innovation" at the Summer Davos Forum in September 2014, local governments immediately formulated preferential policies to facilitate the development of micro and small enterprises. Micro and small sci-tech enterprises have become a center of concern, because they are created by high-tech talents and have played a very important role in providing employment and enhancing local economic vitality and improving industrial structure. Studying the growth mechanism of micro and small sci-tech enterprises can help us in identifying the difficulties and obstacles of their development from different perspectives, deepen our understanding of the root causes of the obstacles, and put forward effective suggestions and countermeasures, thus providing decision-making reference for the development of micro and small sci-tech enterprises.

2. Related concepts and theoretical foundations

The definition of micro and small sci-tech enterprises varies from region to region. Here it is defined as those enterprises with core technology, less capital investment, high proportion of sales income from technical products, small number of employees but with high average educational level. In such enterprises, the ownership and management are usually not divided, so the owners are usually the managers. And as the name suggests, such enterprises are far from monopolizing the industry.

The theory of organizational ecology focuses on the relationship between organizational structure and its impact on the environment. Enterprises are compared to organic organisms, and are studied

Content from this work may be used under the terms of the Creative Commons Attribution 3.0 licence. Any further distribution of this work must maintain attribution to the author(s) and the title of the work, journal citation and DOI.
Published under licence by IOP Publishing Ltd

from the angles of individual enterprises, enterprise populations and the external environment of enterprises. They can adapt to the changes of the environment through inheritance and innovation in operation mode. Correspondingly, the adaption of enterprises and the exchange of resources between enterprises will also have impact on the environment. The organic characteristics, population and community characteristics and growth environment of micro and small sci-tech enterprises conform to constructing the growth mechanism from the perspective of organizational ecology .

From the individual point of view, the ecological characteristics of micro and small sci-tech enterprises are as follows: a single organism composed of capital, technology and other elements; exchange and interaction with the external environment for purchasing and selling; a certain life cycle and organizational genes. From the point of view of the population of micro and small sci-tech enterprises, it is an organic entity composed of interconnected micro and small sci-tech enterprises, which has some similarities with the biological population.

Biological communities interact with each other through energy and material exchange to form an ecosystem. Similarly, micro and small sci-tech enterprise communities also interact through transaction and operation, thus constituting their organizational ecosystem. The individuals, populations, communities and external environment of micro and small sci-tech enterprises interact with each other to form their special ecosystem.

Organizational ecosystems are influenced not only by community relationships, but also by the surrounding environment, specifically, by resource limitations. Different types of enterprises result in different ecosystems, and micro and small sci-tech enterprise form their unique organizational ecosystems under the influence of individual, cluster and external environment.

The development of micro and small sci-tech enterprises follow the general law of life cycle, but the size of enterprises has a profound impact on the length of their life. Take the small and medium-sized enterprises in the United States as an example. The chance of survival within one year for small and micro enterprises is lower than that of large enterprises (250 people and above) by 23%, and is also lower than enterprises with 10 to 20 people by 8%. When the number of employees exceeds 20, the relationship between enterprise size and enterprise survival rate can be relatively stable, showing a positive correlation. According to the White Paper on Human Resource Management of SMEs in China, the life of SMEs in China is about 2.5 years, and the short life indirectly reflects the low survival rate.

3. An analysis of the growth of micro and small sci-tech enterprises from the perspectives of individuals, populations and environmental factors

3.1 Individual factors

According to the endogenous growth theory, the individual factors of the growth of micro and small sci-tech enterprises can be divided into two groups: resources and capabilities. Resources play an important role in the survival and development of micro and small sci-tech enterprises. (1) capital, as the lifeblood of an enterprise, supports all business operations of an enterprise. (2) Human resources are the most indispensable force for the development of enterprises, especially in micro and small sci-tech enterprises where the number of employees is small, and the technical level, R&D capability and teamwork of employees are highly demanded. (3) Technologies are the first and foremost condition to set up micro and small sci-tech enterprises, and technological innovation ability is an important factor affecting the long-term development of enterprises. Technology not only needs to be transformed into productivity, but also needs continuous improvement to meet the growing needs of customers. The failure of most micro and small enterprises is due to incompatibility between technology and market demand, which makes it difficult to achieve effective application. (4) Before and after the start-up of micro and small sci-tech enterprises, it is necessary to have comprehensive relevant information regarding supply and demand situation, competition degree, government policies and regulations, etc. Only by expanding the sources of information and grasping the latest information can they win in the competition.

Micro and small sci-tech enterprises are mostly led and managed by the founders, so the founders' ability play an important role in the growth and development of enterprises, mainly in the following aspects: (1) Comprehensive management ability. The growth and development of enterprises require leaders to have the ability not only to make full use of resources, but also to integrate and allocate resources. (2) Marketing ability. China's micro and small sci-tech enterprises mostly rely on the market niche for further development. If mistakes are made in the judgment on the market, then the goods will not find any ready market. (3) Human resource management ability. How to recruit, train and employ talents is one of the important strategies related to the development of enterprises. The work efficiency, mobility and incentive policies of technical backbones are especially crucial. (4) Learning ability. It refers to the timely grasp and feedback of market information and new technology. The exchange of technological information between micro and small sci-tech enterprises can give full play to the advantages of clusters and enhance their overall competitiveness.

3.2 Population factors

The growth of organisms is not only influenced by themselves, but also influenced by their relationship with other organisms and the environment in which they live. Similar to the social agglomeration of organisms, there exists a relationship of competition and cooperation between micro and small sci-tech enterprises in the same region, thus forming a population. The population of micro and small sci-tech enterprises and their suppliers, distributors and other surrounding populations constitute the organizational community of the region. Micro and small sci-tech enterprises do not hold a dominant position in the market, the probability of survival alone is very low. Only by effectively integrating into the population and community and being aided by the advantages of the cluster can they achieve mutual cooperation and development. Micro and small sci-tech enterprises compete with each other to for market shares, but they also need to cooperate with each other to reach a win-win situation. In the process of cooperation and competition, they may find the opportunity to expand, but also may face the risk of being eliminated. Generally speaking, the population factors of micro and small sci-tech enterprises are similar to the family environment of organisms. They are crucial to the growth and development of enterprises. Without them, enterprises are hard to survive.

Micro and small sci-tech enterprise clusters are mostly populations networks intertwined by related or supportive industries because of the production-marketing relationship and sharing of public facilities and equipment and others. They strive to secure more sensitive, more clustered, more affordable factors of production, shorten the time to obtain market information, and enjoy a higher level of social service system. Compared with the independent development of micro and small sci-tech enterprises, the population can inject more vitality into the enterprises within the population and improve the survival rate.

3.3 Environmental factors

In the ecosystem, the survival of organisms is affected by the natural environment; similarly, the growth of micro and small sci-tech enterprises is also affected by the environment.The environment here refers to policy environment, market environment, cultural environment and so on.

In this paper, the environmental factors in the growth of micro and small sci-tech enterprises are divided into four aspects. (1) The impact of the market environment. One aspect of the market environment is the external market environment, including supply and demand, access threshold, competition, etc.. It provides an important reference for enterprises to make decisions concerning production and sales. The market environment plays a decisive role for micro and small sci-tech enterprises to stand firm and seek development in the competition. The other aspect is financing environment. Developed capital market has a relatively perfect investment and financing system, which can provide a good trading platform for enterprises to obtain capital support. (2) The impact of government policy and legal environment. The tendency of relevant government policies and regulations, the implementation of relevant supportive policies, the market access mechanism, the stability and fairness of relevant laws and regulations play an important guiding role in the development of micro and small sci-tech enterprises. (3) Social services and infrastructure construction, including related conditions for enterprise clusters, such as infrastructure, logistics level, innovation environment, degree of sharing and intellectual support. A good social service environment is conducive to the sound growth of enterprise clusters. (4) Cultural environment, including the concept of family business and the mainstream values of society. Many micro and small sci-tech enterprises are family businesses. They attach great importance to kinship but neglect the importance of contract. It is easy to bring credit risk because of inbreeding. Under this value background, the management and reform of micro and small sci-tech enterprises will be hindered correspondingly. Combining with the mainstream values of society, flexible, innovative, open and inclusive corporate culture is more conducive to the growth and long-term development of micro and small sci-tech enterprises.

4. Conclusions and recommendations

At present, the growth of micro and small sci-tech enterprises in China is still facing internal difficulties, such as obsolete technology and equipment, unreasonable management, financing difficulties, lack of talents and external obstacles, such as inadequate national policies and laws, imperfect social service system and so on. It is necessary to further overcome the internal and external problems in order to achieve growth and give full play to their social and economic functions.

The author believes that micro and small sci-tech enterprises should exert themselves in 4 aspects. First, they should accurately position themselves in the market and concentrate on seizing the niche market in which they have competitive edges. Second, they should adopt appropriate imitation strategies to avoid unnecessary risks and innovate under the premise of imitation. Third, they should strengthen external cooperation and cooperate with others in marketing to achieve win-win situation. Fourth, they should enhance collaborative innovation and cooperative reciprocity, improve collective bargaining ability and exert agglomeration effect while utilizing agglomeration advantages. At the same time, the government should further improve relevant policies, laws and regulations, build a complete financial system, improve the level of social services to facilitate the development of micro and small sci-tech enterprises.

The growth system model of micro and small sci-tech enterprises reflects the influence exerted on their growth by individual, population and environment of their own kind in a certain region. Based on the theory of organizational ecology, this paper studies micro and small sci-tech enterprises, which is conducive to in-depth study on the basis of enterprise ecosystem. The exploration of the population and cluster of micro and small sci-tech enterprises and the analysis of the competition and cooperation between enterprises are to be deepened so that the growth system theory of micro and small sci-tech enterprises is more comprehensive, enriching theoretical research while providing reference suggestions for the growth of micro and small sci-tech enterprises.

Acknowledgement

The research is supported by Tianjin Science and Technology Project *A Research on the Linkage Between the Innovative Development of Tianjin Sci-tech Enterprises and the Tech-powered Region* (No. 17ZLZXZF01040), and the 13th Five-Year Plan Project for Education Science of Tianjin *A Research on the Interaction between the Development and Opening up of Tianjin Binhai New Area and the Cluster of Higher Education* (No.HE3088).

References

[1] XUE Jie, *Research on the joint effects of technological capabilities，market capabilities and design capabilities on innovation activities in TSMEs[J].* Studies in Science of Science, 2017,(9): 1409-1421.

[2] JIANG Zhi-peng, *The long-term and short-term effect of technological potential difference on enterprises, technological capacity: Empirical evidence from industry-university-research patents[J],* Studies in Science of Science, 2018,(1): 123-139.

[3] Tucci C, Massa L, *Business model innovation[J].* Oxford Handbook of Innovation Management, 2013, 50(3): 500-504.

[4] Mu J. *Marketing capability, organizational adaptation and new product development performance[J].* Industrial Marketing Management, 2015, (49): 151-166.

[5] Kollmann T, Stockmann C. *Filling the entrepreneurial orientation-performance gap: The mediating effects of exploratory and exploitative innovations[J],* Entrepreneurship Theory and Practice, 2014, 38 (5), 1001-1026.

[6] Abebe M, Angriawan A. *Organizational and competitive influences of exploration and exploitation activities in small firms[J].* Journal of Business Research, 2014, 67(3): 339-345.

[7] Homburg C, Schwemmle M, Kuehnl C. *New product design: Concept, measurement, and consequences[J].* Journal of Marketing, 2015, (79): 41-56.

[8] Eisenman M. *Understanding aesthetic innovation in the context of technological evolution[J].* Academy of Man agement Review, 2013, 38, (3): 332-351.

[9] Jindal R P, Sarangee K R, Echambadi R, et al. *De signed to succeed: Dimensions of product design and their impact on market share[J],* Journal of Marketing, 2016, 80(4): 72-89.

[10] Mugge R, Schoormans J P L. *Newer is better The influence of a novel appearance on the perceived performance quality of products[J].* Journal of Engineering Design, 2012, 23 (6): 469-84.

[11] Noble C H, Kumar M. Exploring the appeal of product design: A grounded, value-based model of key design elements and relationships[J]. Journal of Product Innovation Management, 2010, 27: 640-657.

Mode of Freight Rolling and Collecting and Distributing Based on Cloud Logistics Platform

Wang Xiaokun[1,a], Wang Lei[2,b], Wang Hanyan[3,c]
[1]Dalian Jiaotong University
Dalian
China
86+13840814266

[2]Dalian Dazhan technology development co. LTD
Dalian
China
86+13614090504

[3]Dalian Jiaotong University
Dalian
China
86+18840843323

[a]24014902@qq.com, [b]171950486@qq.com, [c]790210812@qq.com

ABSTRACT: With the application of Internet and electronic information technologies , cloud logistics platform, as a means of operation of new information flow, can effectively solve the problem of goods rolling, collecting and dispatching by utilizing its efficient information transmission, timely response, optimization and integration. This paper summarizes the development status and main problems of cargo rolling transport, and proposes the preliminary planning framework of the cargo rolling and collecting logistics network platform, including the construction requirements, system architecture and functional system design. In addition, we also propose the rolling and collecting mode and the operation mode based on the cloud logistics platform.

1. INTRODUCTION

Ro-Ro transport is a multi-modal transport mode that realizes seamless connection, improves cabin utilization and ensures transport safety in the organization mode of land and sea transport. It is an important link to promote the circulation and commercial development of goods on both sides of the Straits, the Inland Sea or the river. Ro-Ro transport is a transportation organization mode. The whole process is as follows: first the goods on the trailor are moved by the tractor to the ro-ro ship; after the trailer and the cargo are smashed down, the tractor is disembarked; and then the cargo and the trailer are bundled fixed, and shipped to the destination port; finally the tractor waiting for the port tows the trailer to disembark and transports the goods to the destination.

In the literature, many studies have focused on research goods collection and distribution. Jin [1] studied the organization method and countermeasures of the hanging transport in the ro-ro transport; Liu [2] investigated the seamless transport system of the land and sea in the Bohai Sea area; Fan [3] examined the scheduling optimization of the mounted and transported tractors; Qu [4] studied the ro-

ro and hoisting transport of Luluo Luhai. Li [5] carried out the analysis of the transportation mode of land, sea and cargo rolling and its countermeasures. Chu [6] took Guojiayu port area as an example to study the optimization analysis of the port-based transportation and distribution system based on ro-ro transportation; Li et al.[7] studied the transportation mode and promotion strategy of the cargo around the Bohai Bay. In the research of "cloud computing" and "cloud logistics", Liu [8] proposed the goal, principle, implementation process, promotion mode and development trend of building smart cloud logistics; Cao [9] summarized relevant theories and technologies of cloud computing and logistics public information platform and analyzed the issues of the system architecture, system design and function realization of the logistics information platform. Han [10] studied in detail the role of the Internet of Things technology, cloud computing technology and SOA architecture in the construction of each layer of the platform. Kong [11] proposed a task allocation method and task assignment model with cloud logistics characteristics from the perspective of cloud logistics platform. Gong [12] proposed a method of using cloud computing and cloud logistics to form a logistics demand information integration platform to realize information's Exchange, process, transfer, integrate logistics resources, and maximize logistics benefits.

The contribution of this work is to propose the construction requirements, system architecture and functional design of the logistics and logistics platform for cargo rolling through analyzing the current situation and problems of freight-rolling and hoisting transportation. Moreover, it develops the organization mode of cargo collection and distribution based on cloud logistics platform

2. THE DEVELOPMENT STATUS AND PROBLEMS OF GOODS ROLLING AND HANGING TRANSPORTATION

2.1 Development Situation of Goods Rolling and Hanging Transportation

China's ro-ro transportation has formed four major ro-ro transport regional markets, including the Qiongzhou Strait, the Zhoushan Islands, the upper reaches of the Yangtze River and the Bohai Bay. Due to the late start of China's ro-ro transportation development, it has the low efficiency of cargo collection, the poor information circulation, low sharing, and the high logistics cost. At present, it is still in the process of transition from the traditional mode to the information model. There are many transportation companies. However, most are the traditional business management mode with small scale and low informationization, service level and service awareness. At the same time, the transportation organization mode of cargo rolling and slinging is still in the pilot operation, and has not been widely used. It has not yet formed a platform for cargo collection and distribution information service.

The reasonable combination of ro-ro transport and drop and pull transport can solve the problems of safety and standardization in ro-ro transport, and can greatly improve the transport efficiency. Therefore, after the informatization development reaches a certain level, it has been widely applied in Europe.. Today, the ro-ro-hanging transportation in developed countries and regions in Europe and America has entered a mature stage. The information management of ro-ro-hanging transportation is high. A regional logistics information service platform has been set up. The ro-ro transportation terminal management system has obtained the comprehensive construction and improvement. Roll-to-roll transportation in developed countries in Europe and America has begun to develop in the direction of intelligent, cross-regional services, large-scale ro-ro ship and high-speed transportation

2.2 The problem of traditional goods rolling and transporting

China's ro-ro-hanging transportation enterprises have a low informatization . Most enterprises are still in the traditional ro-ro cargo transportation stage. Information transmission is mainly based on telephone, fax, paper documents and manual entry. Low cooperation results in independent logistics, slow information flow, long response time and low level of logistics services. The traditional cargo rolling and collecting organization mode is shown in Figure 1.

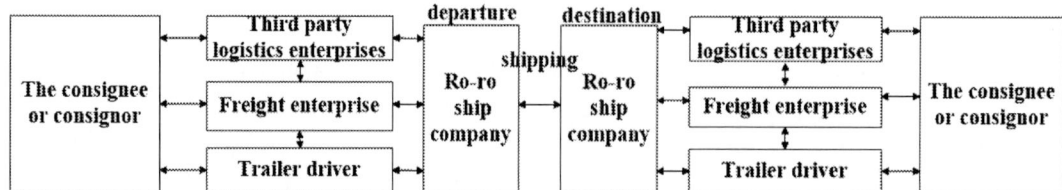

Figure 1. Schematic diagram of the traditional organization mode of the goods

2.3 The problems with traditional cargo rolling and shipping

(1) The types of domestic trailers are numerous and complicated. In the process of hanging and hanging, there are often cases. For example, the trailer and trailer types do not match; the pairing is unsuccessful; the goods cannot be transported out of the ro-ro terminal in time.

(2) The shipping company has fewer sources of supply, less supply and instability, and difficulties in collecting goods, resulting in a low loading rate of ro-ro ships.

(3) The cargo source information distribution channel and the carrier information acquisition channel are blocked. It is difficult to find a suitable carrier driver and carrier shipping company. The goods cannot be shipped out to the destination in time, the storage and transportation costs are high.

(4) The few sources of the source information of the trailer driver, the high cost of the source information channel, and the high short-selling rate, result in low driver income.

(5) Due to the information asymmetry between the loading and unloading of the trailer, vehicles return without goods, which results in high empty driving

3. CARGO ROLLING COLLECTION AND DISTRIBUTION CLOUD LOGISTICS PLATFORM PLANNING

3.1 Construction needs

The users of the logistics and logistics platform include the logistics service provider, the logistics service demander, the government management department and the platform operation organization. The information needs of users are as follows:

(1) Logistics service provider

The logistics service providers mainly include ro-ro ship companies, freight forwarding companies, third-party logistics companies, and transportation drivers. Ro-Ro shipping companies need to release information on Ro-Ro ships, cabins, freight rates, etc. on the platform; freight forwarding companies need to publish and find information on sources, shipping resources, and transport drivers on the platform, especially in advance booking and matching, hanging, transportation and distribution; third-party logistics companies need system support in the collection of scattered goods, dynamic tracking of trailers, transportation organization, vehicle scheduling, etc.; transportation drivers need the information support and guidancein supply, transportation fleet, ship location etc.

(2) Logistics service demand side

The logistics service demand side, that is, the cargo owner, needs to find the information of the logistics enterprise, vehicle and ship quickly; it needs to understand the basic information of the logistics enterprise, credit record, qualification information of the employees and operating vehicles, etc. in order to choose a good reputation and high service quality. Logistics enterprises with security guarantees carry out logistics transaction activities; real-time access to shipping schedules and ro-ro ship transportation status information is required in order to arrange delivery plans and intermodal handovers in advance.

(3) Government management department

Government management departments need to provide the information resources to the market, and obtain market feedback information or data to improve the supervision and decision-making ability of the goods in the market. Meanwhile, they are obliged to release the relevant policies, regulations, and procedures regarding goods transportation ,increase the channels for propaganda of laws and regulations, and reduce the cost of approval for suppliers and demanders of logistics services.

(4) Platform operators

In addition to the daily management needs, the platform operators can also provide support for platform management through the mining and processing of historical data and information. On the other hand, it provides value-added services for logistics consulting for users with special needs. The operational service level provides objective evaluation and analysis, and proposes appropriate improvement and upgrade plans in the maintenance and update of the platform software and hardware.

3.2 Architecture

(1) Operating system

It establishes a cloud logistics platform for customer service with information integration and efficient intelligence. Focusing on the integrated logistics needs of customers, it uses the Internet technology to turn ro-ro shipping companies, freight forwarding companies, third-party logistics companies, all kinds of information of transport drivers, cargo owners and other institutions together to realize logistics information sharing, provide operational process support and intelligent management services for the whole process of collecting and distributing, and realize the seamless docking of land and sea transport in two areas of ro-ro transport. .

(2) Architecture

On the basis of cloud services, the architecture of the logistics system is mainly composed of three types of cloud services: infrastructure as a service (IaaS), platform as a service (PaaS), and software as a service (SaaS). The logistics and logistics platform is located between the terminal users and the physical infrastructure, including the application software layer, the platform service support layer and the infrastructure service layer. The physical infrastructure consists of physical hardware facilities such as servers, computer rooms, and network equipment. It is the foundation of the cloud logistics platform and provides physical support for the platform. The infrastructure service layer is based on the basic service layer built on the logistics infrastructure. The application software layer and platform service support layer provide cloud storage, cloud computing and cloud transmission capabilities; the platform service support layer is the provider of applications and common components required by the cloud logistics platform, through which users can register and manage their personal information and security settings; the application software layer provides various services for terminal users, including platform information services, platform interface services, application system services, and software services.

(3)Functional system design

According to the platform architecture design and the needs of customers, it develops the logistics platform based on the platform portal, and the platform of the user service window and the comprehensive logistics information display site. These platforms can be expanded through the portal content, service system and operation ways. With logistics service providers, logistics service demanders, government management departments and platform operators as the entry point, a comprehensive information web site on the integration of goods, information, regional influence Portal is developed. According to the information needs of customer service, qualification certification and credit service, business service, e-commerce application service, collaborative operation service, vehicle and ship dispatch service and dynamic tracking service, the company develop the cloud logistics platform with the service purposes of "trust one time". ", the bill of lading "one in the end", and "one-time collection" . The platform service function system includes: "one-unit" business service function, e-commerce application service function, qualification certification and credit service function, transportation organization collaborative operation service function, vehicle and ship dispatch service function, cargo vehicle and ship dynamic tracking service function, "one The "stationary" customer service function and operational decision support service functions, as shown in Figure 2.

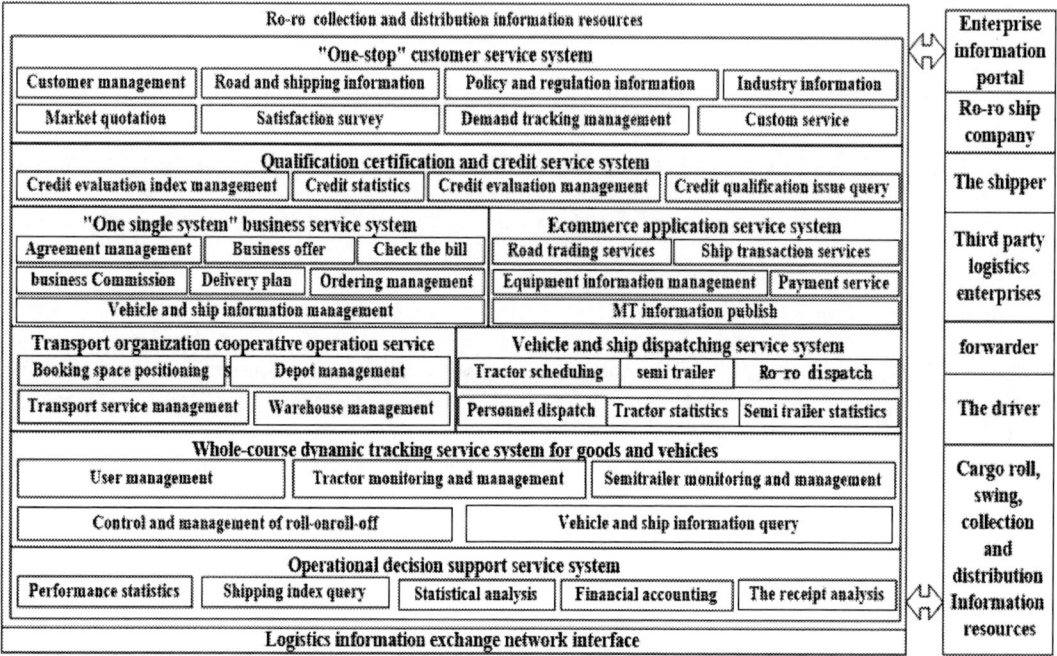

Figure 2. Functional system diagram of the cargo rolling and collecting and distributing cloud logistics platform

4. THE CARGO ROLLING AND COLLECTING MODE BASED ON THE CLOUD LOGISTICS PLATFORM

4.1 The goods roll, the collection and distribution organization mode under cloud logistics platform
Taking the cloud logistics platform as the core, the platform's information mining, information processing and information distribution functions are used to effectively integrate the logistics resources of Ro-Ro ship companies, cargo owners, drivers, freight forwarders and third-party logistics companies onto the platform. The processing technology is used to allocate the data reasonably and effectively, and finally the processed information is distributed to the logistics entities of the goods collection and distribution organization. Through making advantages of each logistics entity and effectively matching them, the platform adopts an optimal way to integrate decentralized logistics resources, so as to maximize the utilization of logistics resources and to form an efficient organization model for goods collection and distribution.

The Ro-Ro shipping company can realize the rapid arrival of goods and the distribution of goods by port, so as to achieve rapid collection of goods; freight forwarding companies can better cooperate and share information with ro-ro ship companies and transport drivers. They timely find the transportation of goods tools and transport the company's foods to the destination as soon as possible, while using the platform for online booking business; third-party logistics companies can better cooperate and share information with the Ro-Ro ship company, get more sources and better arrange transportation fleet transportation and distribution; the trailer driver can transport the goods back and forth between the two places on time and can load the goods; the cargo owner can quickly find the carrier and timely load the goods out of the warehouse, transport to the destination, and realize real-time monitoring of the whole process. The mode of transporting cargo collection and storage is shown in Figure 3:

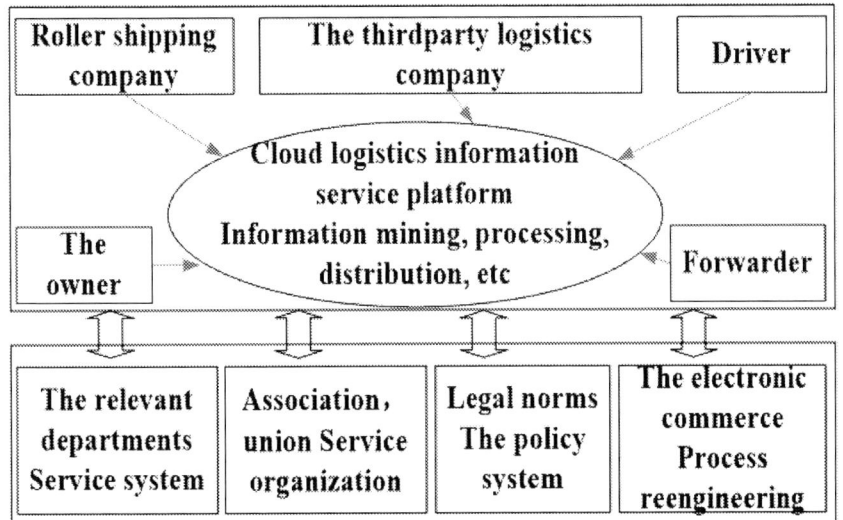

Figure 3. The organization of cargo collection and shipping under the cloud logistics platform

4.2 Cloud logistics platform-based the operation mode of the goods collection and distribution service

4.2.1 Ro-Ro ship company operating mode

There are two operating modes of the Ro-Ro ship company: one is the passive cargo collection and operation mode; the other is the active cargo collection and operation mode.

The passive cargo collection and operation mode can release the ro-ro transportation information through the platform, such as ro-ro ship cabin, navigation route, ro-ro transportation information on ports, departure times and ro-ro shipping prices; online booking services are also available. Cargo owners, freight forwarders and third-party logistics companies in the port area can carry out ro-ro ship information inquiry and space reservation through the platform, and deliver the goods to the port within the time specified by the Ro-Ro ship-company, and carry out detailed information such as cargo and trailer. After the registration is filed, the trailer enters the cabin and is shipped. After the ro-ro ship departing from ports, the shipping company will release the information on the loading cargo, the trailer model, the arrival time of the ro-ro ship, the port of call, etc. on the platform. The driver of the trailer in the destination port will inquire through the platform and pick up the transportation task wait at the port for the specified time required by the Ro-Ro shipping company. After the ro-ro ship is docked, the trailer is loaded onto the trailer. After registration, the trailer is towed away from the ro-ro ship and sent to the destination to complete the cargo transportation task，as shown in Figure 4.

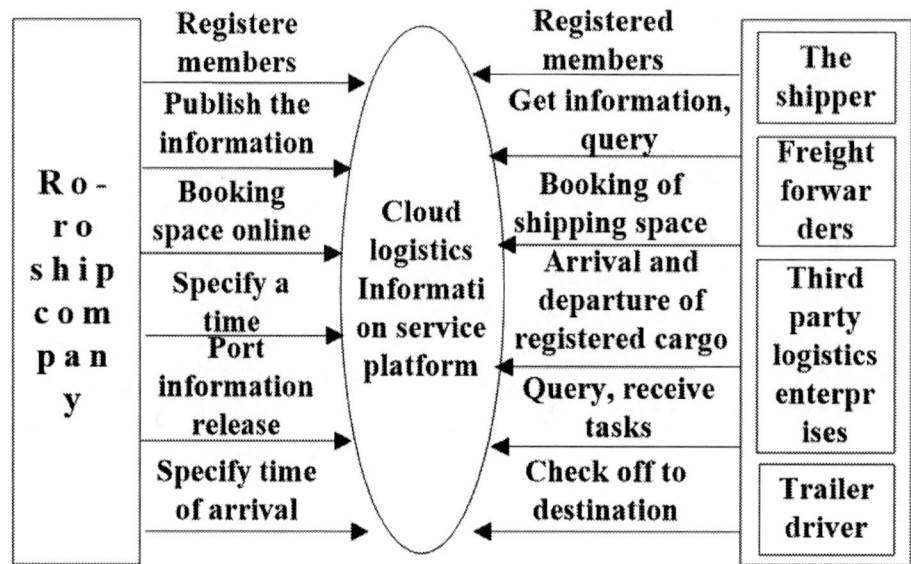

Figure 4. The passive cargo collection and operation mode of the Ro-Ro ship company

The active cargo collection and operation mode is that the Ro-Ro shipping company can release the information of the Ro-Ro transportation through the platform. It can strengthen the communication with the freight forwarding enterprises and the third-party logistics enterprises through the platform, promote the cooperation agreement on the Ro-Ro ship transportation and signed a cooperation contract. When the scheduled transportation volume is reached within the time limit stipulated in the contract, the freight forwarding company or the third-party logistics enterprise that cooperates with the Ro-Ro shipping company can preferentially book the cabin and obtain certain preferential offers for mutual benefit.

In addition, the Ro-Ro shipping company can take advantage of its rich logistics resources information and cooperate with the transportation drivers to form a stern transport fleet led and operated by the Ro-Ro ship company, responsible for the transportation and delivery of goods arriving at ports. Through the cloud logistics platform, the Ro-Ro shipping company provides logistics resource information to the transportation drivers. After the ship arrives at the port, the organization organizes the transportation fleet to deliver the goods to the destination at the fastest speed. At the same time, the Ro-Ro shipping company through cooperating with freight forwarders and third-party logistics enterprises can collect the goods at the destination and transport them to the port, so as to realize the seamless connection of land and sea transportation, as shown in Figure 5.

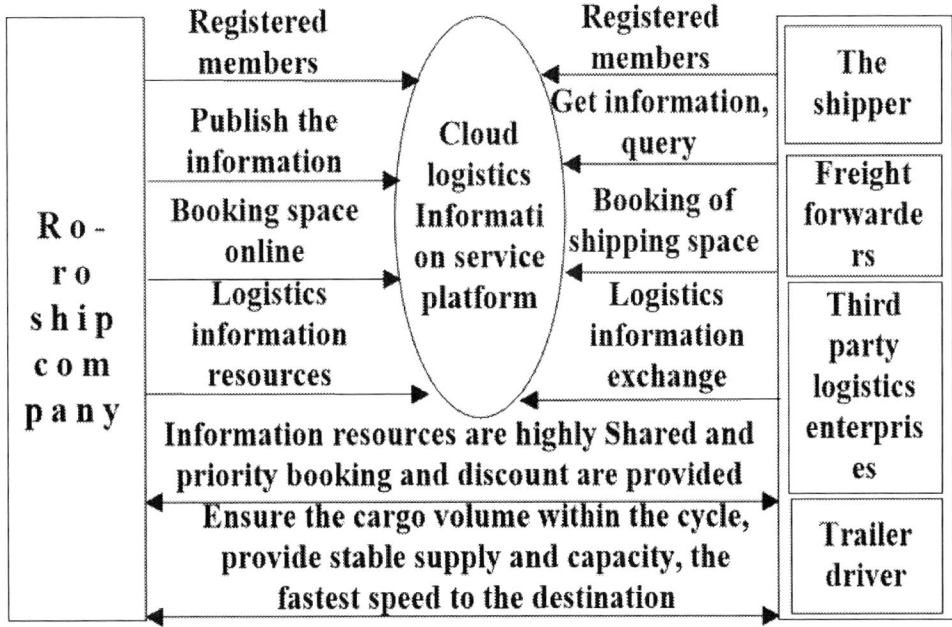

Figure 5. Rolling and shipping company active cargo collection and operation mode

4.2.2 Freight Forwarding Enterprise Operation Mode

The freight forwarding enterprise can query the source information through the platform, negotiate with the owner through the platform, and finally reach the cargo shipping agreement. The freight forwarder can also release the source information on the platform. This provides cargo information for drivers and third-party logistics companies. The freight forwarding enterprises can use the platform to make cargo transportation agreements with drivers and third-party logistics companies to complete transportation tasks. Freight forwarders can also complete cabin reservations through platforms and reach cargo transportation cooperation with Ro-Ro shipping companies. The agreement provides supply to the Ro-Ro ship company for transportation cooperation, as shown in Figure 6.

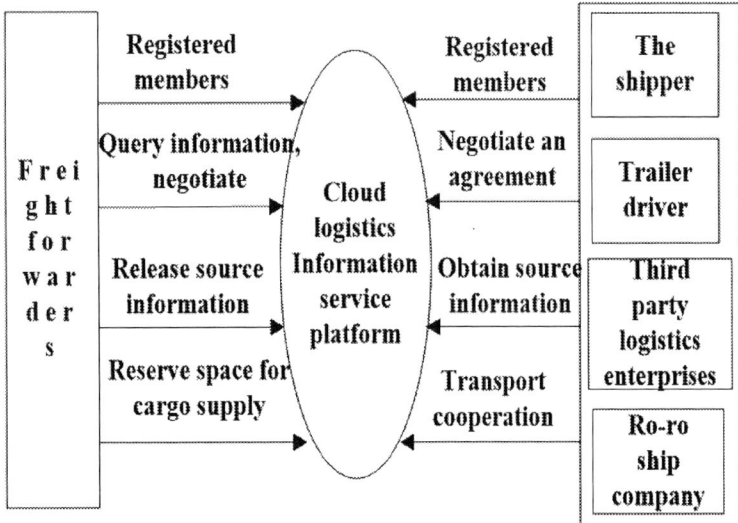

Figure 6. Freight agent business operation mode

4.2.3 Third Party Logistics Enterprise Operation Mode

The third-party logistics enterprise becomes a member of the platform. It can release the fleet transportation capacity, warehousing and other information through the platform, and at the same time find the source information published by the cargo owner and freight forwarding company. After the enterprise reaches the carrier agreement, the goods are loaded and transported to the warehouse and freight center of the logistics enterprise, and then the less-than-truckload goods are packed and carpooled. Then the company fleet transports the goods to the terminal for ro-ro transportation; after goods reaching the destination terminal, the owners with less-than-truckload cargo are provided by devanning services and carpooling. That is, unpacking separate delivery services, ultimately transports the goods to their destination. Third-party logistics companies can also complete the booking of the space through the platform, and reach a cargo transportation cooperation agreement with the Ro-Ro shipping company to provide the Ro-Ro shipping company with a transport fleet to carry out cargo transportation cooperation, as shown in Figure 7.

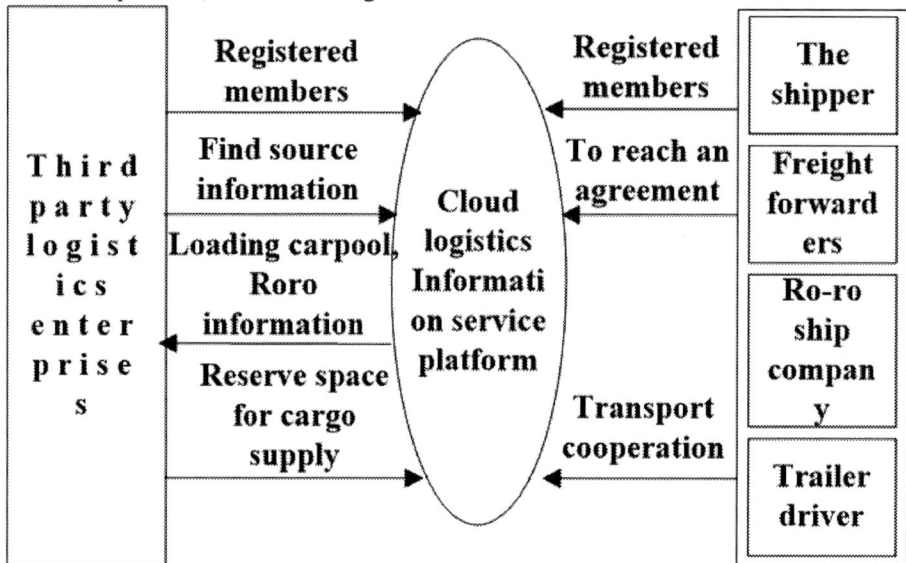

Figure 7. Third-party logistics business operation mode

4.2.4 Transport driver transport mode

The transport drivers can publish the vehicle source information on the platform, waiting for the owner to contact and reach the carrier agreement to complete the carrier task. They also can find the source information through the platform and contact the owner, freight forwarding companies or ro-ro shipping companies to pick up transportation tasks. After drivers making a shipping agreement, they safely deliver them to destinations for completing transportation tasks. Drivers can build transport fleets through platforms in order to receive more transportation tasks and improve the efficiency of transportation, as shown in Figure 8.

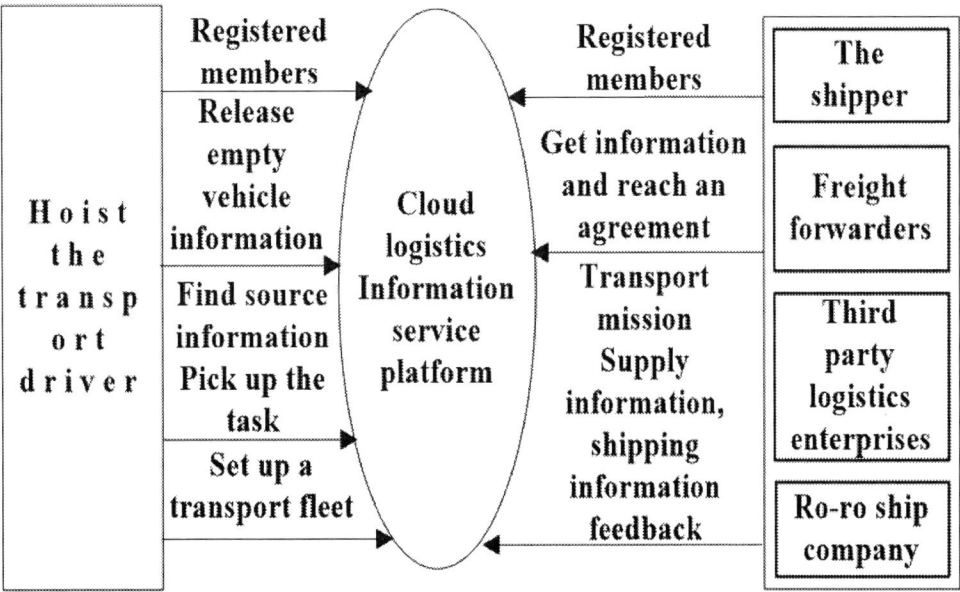

Figure 8. the transportation mode of the transport driver

4.2.5 Shipper Delivery Mode

Cargo owners can release goods information on the platform, waiting for the transportation driver or the third-party logistics enterprise to pick up the transportation task and transport the goods to the destination. The cargo owners can also inquire the transportation ability of the driver or third-party logistics enterprise through the platform and finally reach the carrier agreement to complete the delivery task, as shown in Figure 9.

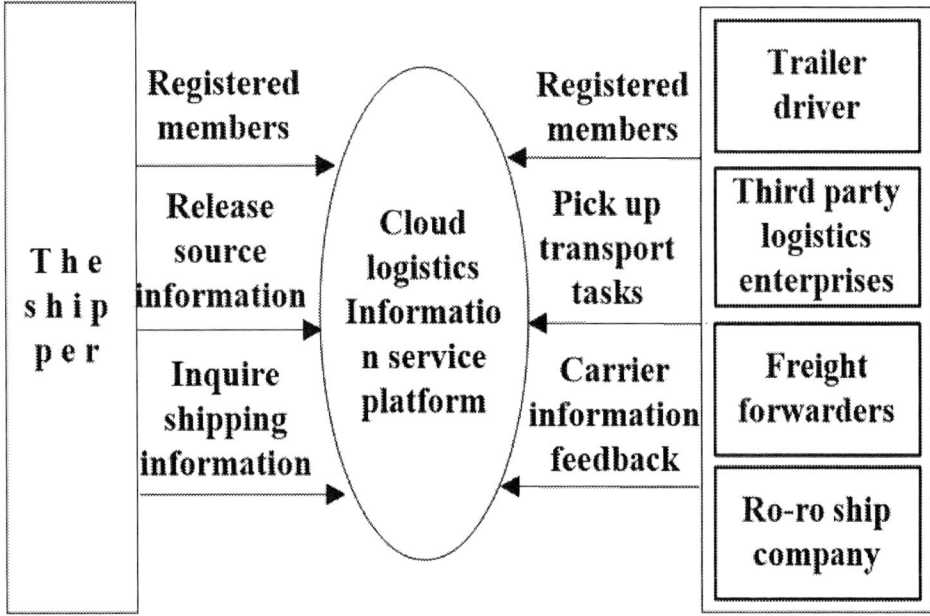

Figure 9. Shipper delivery mode

5. CONCLUSION

Based on the cloud logistics platform, the roll-and-roll collection and distribution mode will integrate and optimize all elements related to goods organization and transportation consisting of the rolling company, cargo owner, transportation driver, freight forwarding company, third-party logistics enterprise and ro-ro terminal enterprise. Through the cloud logistics platform, efforts are made in this work to solve problems such as imperfect information systems, poor flow of resource information, and low level of information sharing. The cloud logistics platform should be the inevitable trend of the future development of the ro-ro transport industry. It still needs the relevant departments to promote it. It is widely recognized in the industry and forms a coalition organization to jointly formulate relevant rules and regulations and jointly organize operation management.

REFERENCES

[1] Jin Zhihong. Analysis and countermeasures of hanging transport organization in ro-ro transport. *Technology and Methods*, 2013: 59-61.DOI=10.3963/j.ISSN 1674-4861.2010.02.015

[2] Liu Tao. Research on the seamless transportation system of land and sea passenger and cargo in the Bohai Sea area. *Dalian: Dalian Maritime University*, 2008.

[3] Fan Ningning.Study on the optimization of the scheduling of the towing and hoisting transport tractors in Yantai.*Dalian: Dalian Maritime University*.2012.06.DOI=10.3963/j.issn.2095-3844.2013.04.047

[4] Qu Wenqiang.Study on Rolling and Hanging Transportation of Luliao Land and Sea.*Research and Discussion,* 2015,09(4).011.DOI=10.3969/j.issn.1005-152X.2014.12.142

[5] Li Zuona. Rolling land and sea cargo transport mode rejection hanging Analysis and Counter measures. *World Shipping*, 2014.06.016.

[6] Chu Pengjie.*Optimization Analysis of Port Collection and Distributing System Based on Ro-Ro Transportation—Taking Guojiatun Port Area as an Example.* Chongqing: Chongqing Jiaotong University.2013.

[7] Li Hongqi,Chang Xinyu,Gao Hongtao.The Transportation Mode of Rolling and Hanging in the Bohai Bay and Its Promotion Strategy. *Transportation Management,* 2014.11:73-79.DOI=10.3969/j.issn.1673-0194.2016.20.040

[8] Liu Qin,Liu Gang.Building Smart Cloud Logistics. *Enterprise Information Management,*2013.10:116-118.

[9] Cao Liming. Architecture and Design of Logistics Public Information Platform Based on Cloud Computing.*Price Monthly,* 2014.01:53-57.DOI=10.3963/j.issn.1000-8969.2017.01.001

[10] Han Haiwen. Research on Architecture of Port Logistics Integrated Service Platform Based on Cloud Computing and Internet of Things Technology. *Computer Science*, 2013.06: 232-235.

[11] Kong Lingqi.Research on cloud logistics operation mode and task allocation.*Qinhuangdao:Yanshan University*.2015.05.

[12] Gong Xianglin,Yang Rong.Application of "Cloud Computing" and "Cloud Logistics" in Logistics.*China Circulation Economy*,2012(10):29-33.

[13] Yongsheng Liu, Yu Li, Juan Chen. Research on Modes of Cargo Ro-Ro,Drop and Pull Transport in Land-Sea Transportation Channel between Shandong and Liaoning. *Journal of Service Science and Management*, 2015, 8, 229-235.

Research on the Integration Demonstration System of Space Information Network Based on TDRSS

Dandan Fan[1,a], Mengyue Qiu[1,a], XiaoshenXu[1,a], Ming Chen[2,b]

[1]Beijing Space Information Relay and Transmission Technology Research Center, China

[2]Beijing No.1906 Maibox, China

[a]vandd82@sina.com, [b]jerkey_cn@sina.com

ABSTRACT Aiming at demonstration of space information network, this paper proposed the integration demonstration system based on TDRSS, which can make up for the difference of transport protocols, operation mode and data distribution between space information network and TDRSS. The system expanded the topology structure of TDRSS in order to be compatible with space information network service and supply the theoretical and technical base for the improvement of space information network.

1. INTRODUCTION

Space information network is the network system that can acquire, transmit and process space information using space platform as carrier, such as satellites, near-space balloons, planes and so on. Now space information network is the global research hotspot [1] [2]. As the depth and range of theoretical research on space information network is developing sustainably, we need an open demonstration platform for the theory and technology research in order to establish quantitative knowledge of space information network and improve the developing of research. As a result, the integration demonstration system of space information network is a significant research area now [3].

As an integrate system in orbit, TDRSS is similar to space information network to a certain extend on the structure, function and service objects, which can provide feasible method to the demonstration of space information network [4]. The design proposal of demonstration system based on TDRSS was proposed in this paper, in which the open and flexibility of TDRSS was utilized to illustrate space information network. The proposal can supply a restructurable and expandable demonstration platform to accelerate the development of space information network in the future.

2. STRUCTURE OF SPACE INFORMATIO-N NETWORK

According to the transmission requirement of space information network and the main characteristic of transmission platforms, the space information network has two-layered architecture consisting of backbone layer and regional access layer [1], as shown in Fig.1. As the core of the space information network, the backbone layer should satisfy the requirements such as global overage, stable structure, wideband carrier capacity, convenient access, supporting multiple types of service access and interconnection of heterogeneous networks. The service enhancement layer is composed of the systems which have the relative simple functions, but can provide regional coverage. The service enhancement layer mainly aims at the non-spacecraft users and improving the regional information transmission capability. It achieves the rapid information transmission among different users in the same region by establishing the regional network. As the access terminals of space information

Content from this work may be used under the terms of the Creative Commons Attribution 3.0 licence. Any further distribution of this work must maintain attribution to the author(s) and the title of the work, journal citation and DOI.

Published under licence by IOP Publishing Ltd

network, users are the originator of transmission services, and the end to end transmission can be performed through the transmission channel provided by space information network. Users could be the information acquisition and application platforms, such as the spacecrafts, planes, ships and the relevant ground application centers. Users can implement the convenient access to space information network through the backbone layer and service enhancement layer.

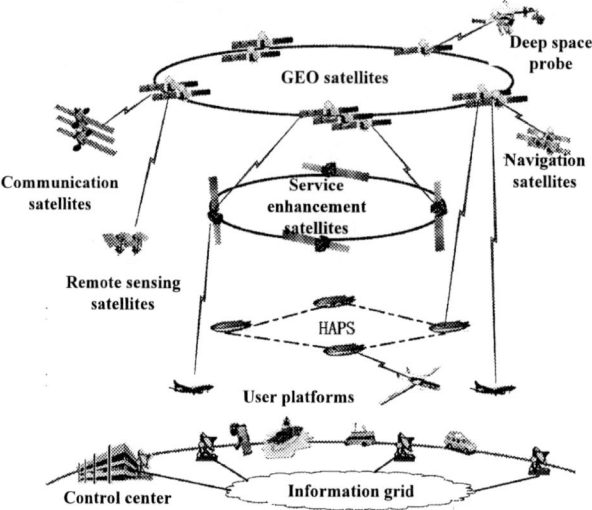

Figure 1. Space information network structure

The two-layered space information network has an open structure, which is compatible with different types of access subnet. The backbone layer is consisted of GEO satellites and relevant ground system, which has the advantage of stable structure. There are different types of service provided by the space information network, such as wideband data transmission service of single user, narrowband data transmission service of single user, trunking communication of multi-user in a group, data transmission service between groups. Divided by the movement characteristic of users, there are predictable trajectory services and unpredictable trajectory service. Divided by relay requirement, there are real-time service and non-realtime service. Divided by the access way, there are planned service and burst service.

3. DEMONSTRATION SYSTEM OF SPACE INFORMATION NETWORK

As an enormous system, the space information network can provide different data transmission services for different users. But there are many key techniques to be solved, such as networking protocols, transmission techniques and so on. This paper proposed an integrated demonstration system based on TDRSS, which can provide the proof platform for key techniques.

3.1 Introduction of TDRSS

TDRSS is a geostationary satellite system, which can provide the space-based measurements and data transmission for the global LEO satellites and can get nearly Gbps transmission rate. We can see from the structure that three TDRS can get global coverage. Through the ISL and satellite-to-earth link, TDRSS can be connected with LEO satellites and ground stations. The connectivity between different LEO satellites can be realized through TDRS and ground stations. Thus the structure has partly the characteristic of network, which is shown in Fig.2.

On the other side, the characteristics of TDRSS such as high dynamic tracking and bending transmission are convenient for the reengineer of network protocols and technique proof, which is very helpful for the realization and demonstration of system opening, expansibility and reconstruction.

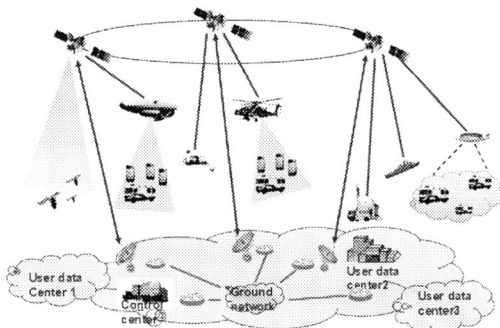

Figure 2. System structure of TDRSS

As a conclusion, TDRSS is similar to the space information network in the system structure, system function and service objects. And the successful application experience of TDRSS on different user platforms can provide feasible method for the comprehensive integration demonstration of key techniques of space information network. But compared with the service feature of space information network, there are some differences between TDRSS and space information network, which are given below.

3.1.1 Protocol difference

The transmission protocol of TDRSS is special designed for it, which can not provide interconnection between heterogeneous networks. In the transmission protocol of TDRSS, the data transmission is operated on the data link layer, while the ground station of TDRSS and the user center must pack, extract and distribute the forward data and reward data on the application layer. According to the transmission protocol, the data from space to ground can not be routed and switched directly. As a result, the space target can not be connected with the ground system directly on the network layer and the protocol conversion of data will increase the processing delay and complexity of system.

3.1.2 Operation difference

TDRSS strictly operates as the operation plans which are planned several days ago. Usually, the users of TDRSS must submit the request a week ago. And the control center of TDRSS will accomplish the resource allocation and decide the operation plans of TDRSS. Since the operation plans are decided, there is no much room for the users to adjust working plans. This kind of strictly planned operation mode can not be used to the quick access of the space information network.

3.1.3 Control difference

TDRSS distributes data from a storage point in the control center, which is inflexible to the users. The data of LEO satellites transmitted through TDRS to the ground station must be received all by the control center of TDRSS and retransmitted to the users after protocol conversion. The data distribution method has too many segments. As a result, when the number of users and the amount of data increases, the control center of TDRSS as the storage and retransmit point will bear more and more pressure on the reliability and data distribution capability.

3.2 Design of demonstration system

Aiming at the difference of transmitting protocols, operation mode and data distribution method between the space information network and TDRSS, the design of demonstration system was proposed in this paper.

3.2.1 Transporting protocols

According to the IP connection demand of space and ground, a system composed of LEO user satellites, TDRS, ground station, control center and the user data centers was proposed, which is illustrated in Fig.3.

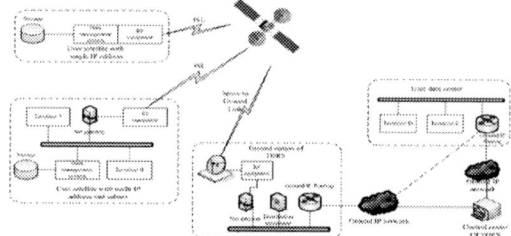

Figure 3. Data Relay Transmitting system structure based on IP connection

3.2.1.1 LEO user satellites with multi IP address

As for the LEO user satellites, there is subnet in the users. In order to transmit the subnet data, a gateway is designed to complete the protocol conversion from data link layer protocol of subnet to the data link layer protocol of space channel, for example the CCSDS AOS protocol. After the data is transmitted to ground through the TDRS channel, a gateway in the ground station is also designed to extract IP message from the data link layer protocol of space channel and accomplish the protocol conversion from the data link layer of space channel to the data link layer of ground network. After the protocol conversion, the standard routing equipment will distribute the data to the user data centers according to the destination address in the IP message of the space subnet. Thus the routing path can go through the control center or not as needed.

3.2.1.2 LEO user satellites with single IP address

The user satellites with single IP address has no subnet, so the link layer protocol conversion inside the user satellites is not necessary. We can directly package the IP message into the data link layer protocol of space channel and transmit through the space channel. The other processes of transmission are the same with user satellites with multi IP address.

According to the lately research result of space data transmission protocols, the protocols which combined the IP and CCSDS protocols were adopted in our system, as shown in Fig.4. The data link layer protocols of ground network and subnet of user satellites adopted the data link layer protocols of Ethernet and 802.11 wireless network. In the space channel, the SDLPs of CCSDS were adopted. Because the data link layer protocols of ground and space are different, the protocols conversion must be accomplished before transmission through ISL channel and SGL channel. In the network layer, we adopted the standard IP protocols. Aiming at the reliability and efficiency of transmission, the transport layer not only adopted the standard TCP/UDP protocols but also adopted the segmented techniques of TCP to improve the performance of transport layer. The segmented techniques of TCP can overcome the long link delay of transmission, the high error rate of network and unsymmetrical link bandwidth and so on. The application layer can adopt many standard protocols such as HTTP, SMTP, FTP, Telnet and so on, or the users can design special application layer protocols according to their requirements.

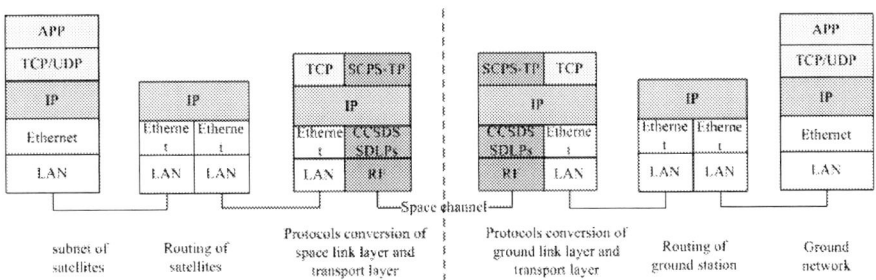

Figure 4. Protocols stack of space information network based on TDRSS

3.2.2 Operation mode driven by tasks

In order to improve the flexibility of resource application and adjustment of TDRSS, an operation mode driven by tasks was proposed which can dynamically adjust the resource of TDRSS according to the tasks requirements of users [5]. After the resource applications were received by the control center of TDRSS, the system can adjust the resource assignment rapidly and automatically and the difference between new allocation plan and old allocation will be minimized in order to minimize the disturbing to other users. In the operation mode driven by tasks, the flow of resource adjustment of TDRSS was given in Fig.5. First the insert resource requirement was preprocessed by the dynamic adjustment algorithm, which can calculate the TDRSS resource that can supply the demand.

3.2.3 Networking receiving and sharing system on ground

As the transmission pivotal of all kinds of different user platforms, a networking receiving and sharing system was designed in this paper. In the system, the different kinds of user data can be stored and shared in order as needed, which can be the base of the massive data sharing of space information network in future [6]. The system adopted the structure of universal platforms and special plug-ins, which was composed of data receiving, data reprocessing and resource sharing service, as shown in Fig.6. The data receiving system can receive and store user data in real time. The data reprocessing system can decrypt, analysis and decompress the user data according to the user demands. The special processing plug-ins can be added to the system to do further processing such as multi-frame imaging, radiometric correction, geometry correction and so on. The resource sharing service system can store, organize and manage the data universally and supply data catalog search and download to users.

4. CONCLUSION

Aiming at the spaces information network structure, this paper designs the demonstration system based on TDRSS, which is composed of data relay transmission system through IP connection between space and ground, operation mode driven by tasks and networking receiving and sharing system on ground. The demonstration system can provide the verify platform for the research of key techniques in space information network and improve the development of theories and techniques research.

REFERENCES

[1] Zhang Wei, Zhang Gengxin, Gou Liang, J. 2016. Satellite Constellation Design in Space Information Network In *ZTE Technology Journal*. 19-23,45.

[2] Liao Yong, Fan zhuochen, Zhao ming, J. 2017. Survey on Security Protocol of Space Information Networks In *Computer Science*. 202-204.

[3] Zhou hongbin,Yin bo, J. 2016. Research on Spatial Information Network System Validation Technology In *Radio Communications Technology*. 16-19,47.

[4] Ji Wenlong, Yu Xiaohong, J. 2014. Analysis of Development and Application of Foreign Data-relay Satellite Systems In *Journal of Sichuan Ordance*. 118-120,124.

[5] He chuan,Li Yajing, Qiu Zhen, J. 2017. Task programming models and algorithms of tracking and data relay satellite in application-on-demand In *Chinese Space Science and Technology*. 46-55.

[6] CHEN Hao ZHAI Zheng'an GAO Shenghua ZHANG Bin, J. 2016. Design of Ground Cloud Storage Structure for the Tracking and Relay Satellite System In *Chinese Space Science and Technology*. 392-399.

The Resource Aggregation and Integration Platform for Shared Development of the Direct Bank

Qi Wang[1,a], ZhongShi Zhang[2,b], Xiaotong Zhang[3,c], Qing Zhao[4,d]

[1]Central University of Finance and Economics
No. 39 South College Road, Haidian
District, Beijing, China

[2]University of South Florida
4202 E. Fowler Avenue, Tampa,
FL, USA

[3]Binghamton University-SUNY
4400 Vestal Parkway East
Binghamton, NY, USA

[4]Beijing University of Technology
No.100, Pingleyuan, Chaoyang District,
Beijing, China
+86 18611329377

[a]Jr107g@163.com, [b]542845510@qq.com, [c]Regina-tong1211@163.com,
[d]zhaoqing1025@emails.bjut.edu.cn

ABSTRACT: Direct Banks is critical to provide online users a novel investment path without the intermediation of traditional banks and financial institutions. However, a massive resources are underutilized due to dearth of sharing concepts. The objective of this study was to propose a Direct Bank resource integration platform (DBRIP) by addressing the problem of non-circulating user resources and data resources, and enhance the resource utilization rate among the Direct Banks, including three main layers: first, data layer, multiple data resources is aggregated in this layer; second, function layer, it's the core of DBRIP architecture that provides a lot of necessary functions for Direct Banks and customers; finally, service layer that provides a series of service functions and ensures customers and Direct Banks have a good user experience. We believe this study provides valuable insights for the development of Direct Banks in era of sharing economy.

1. INTRODUCTION

Sharing economy can be defined as Internet mediated based on Internet platform which can share, swapping, trade, rent products or service, and it disrupts traditional business mode, generates new economy activity and potentially leads to environmental and social benefits. In the early stage, sharing products mainly are fixed assets, such as a house and a car [1]. Nowadays, the sharing economy involves multiple areas, such as medical treatment, education, finance, resource, travel and accommodation.

The advancement of IT (information technology) has resulted in a rapid development of electronic economy markets. The key feature of these electronic economy markets is reducing and eliminating the traditional middlemen, so it's very convenient for customers to connect the product providers and

Content from this work may be used under the terms of the Creative Commons Attribution 3.0 licence. Any further distribution of this work must maintain attribution to the author(s) and the title of the work, journal citation and DOI.
Published under licence by IOP Publishing Ltd

product service directly [2]. A typical example is online P2P (peer-to-peer) lending (e.g., Kiva) and crowdfunding services (e.g., Kickstarter) [3-4] which is a novel financing model for the Internet users. P2P based activity including giving, obtaining and sharing the permission for products and services, coordinated through Internet online services. As the development of Internet finance, traditional banks borrowed the experience from other countries, and tried to develop the Internet finance by building Direct Banks.

Direct Banks are built based on the traditional banks that provide the financial products and services through online platforms, such as E-mail, telephone bank, mobile software, etc., without the intermediation of any financial institutions. However, the results produced by Direct Banks are not satisfied compared with P2P lending, because of resource limitation that is produced by the lack of Shared concepts. The development of IT alongside the growth of web 2.0 has enabled the online platforms that promote user sharing and collaboration [5]. As new application of IT entered in financial fields, Direct Banks has been able to reduce costs and provide all functions effectively to complete transactions like traditional banks.

The main contribution of this paper is as follows: (1)the essential characteristic of sharing economy is analyzed, and then, we introduce the relationship between sharing economy and Direct Banks; (2)we analyze the problems caused by dearth of understanding shared concept for Direct Banks; (3)system architecture of Direct Bank resource integration platform we proposed and illustrate the characteristics and advantages of it.

2. THEORETICAL CONTEXT

2.1 The Essential Characteristic of Sharing Economy

Sharing economy is a novel economic mode, the essential of which is sharing the permission of idle resources through the third-party Internet information technology platform. This economy mode has been expected to alleviate societal problems, such as pollution and hyper-consumption, it also can increase the utilization of resources and optimize the allocation of resources. The following we will describe more detail. The main characteristics of sharing economy includes three parts:

(1) Base on third-party Internet platform: sharing economy emerges from a number of information technological developments that have simplified sharing products and services through the availability of multiple information systems on the Internet. In the traditional society, instant messaging is difficult and expensive, through information technology platforms, communication cost will be significantly reduced and solving the problem of low utilization of resources. The business models of sharing economy are often platform-based to match demand and supply. As shown in Figure 1. As increasing use of the Internet, the possibilities of Internet enable online platforms cheaper and easier to access.

Figure 1. Structure of business model of sharing economy

(2) Reducing and eliminating the traditional middlemen: before the advent of the sharing economy, transaction costs might be quit high because direct interaction is costly, people need to find resources in your vicinity or through specific intermediary [6]. The Internet, laptop and mobile devices or other new technologies overcome parts of this problem, resources can be concentrated and integrated by sharing platforms. Especially information costs and decision costs are usually dramatically reduced compared with face to face interaction. Consumer can view and select the products in a uniform interface, and trading by themselves. Sharing platforms will greatly reduce the transaction costs for consumer. Table 1 demonstrates this using an example of sharing-based Direct Bank, comparing traditional bank with Internet-enable sharing bank.

Table 1. Advantages of Direct Banks in the sharing economy.

Activities in Direct Bank from the consumer point of view, for example.

	Traditional Bank	**Direct Bank**
Search financial information costs	Finding the financial instructions in your vicinity	Finding the Internet platform that supply the financial products
Trading and decision costs	Negotiating with a sales personnel and conditions of the deal individually	Checking the price and conditions specified by the platform Selecting and trading by self
Enforcement costs	Organizing payment method and payment	Payment via the platform

(3) Emphasize sharing rather than occupies: people used to regard possessions as symbols of wealth and social status. But now, with the improvement of the living standard, consumers are more prefer to enjoy the use of permission of the products than to possess them. Thus, the living costs can be reduced by resource sharing. The resources people like to share include car, house, money, knowledge and financial products.

2.2 The Relationship between Sharing Economy and Direct Banks
Direct Bank is a new Internet technology of selling efforts by banks, and the result of that in two interactive developments, (1) the computer technologies is changing; (2) the demand for personal selling efforts is changing. The personal contact between products seller and the customers in financial institutions is replaced by online communication between an online banking customer service and the customers. Hence, the high employee wage costs and the capital of the traditional branch are substituted by lower costs.

There is growing interest in Internet bank in the societies, due to traditional banks are time consuming and labor consuming, The advantages of Direct Banks, from the banks' perspective: (1) the building expenses are lower due to the bank does not need to established many branches located in a density populated area; (2) the wage costs are lower because both a reduction in the number of employees and the wage rate of the staff [7]. From the customers' perspective: (1) Direct Banks have the online financial experts who can give advice about the products for customers; (2) enable customers to search and purchase financial products anytime.

In Germany and America, the Direct Banking system has matured and widely spread in other countries [8]. Up to now, Direct Banks have reached 114 in China. Compare with P2P, Direct Banks have the stronger capability of risk resistance and the higher credibility. However, the low utilization of resources will be existence because of each Direct Bank develops independently. There are two mainly negative effects is described as: (1) user data without circulation, extremely easy to cause the waste of user resources; (2) product data without circulation, it will lead to inaccurate market positioning, high homogenization of financial products, coupled with make Direct Banks trapped in a vicious circle of price competition. Therefore, in order to improve resource utilization, Direct Banks need to borrow the concepts of sharing economy. The optimal allocation of user resources and data resources can be realized by promoting the circulation of resources with the resource integration platform. The ecological environment of Direct Banks is described in Figure 2.

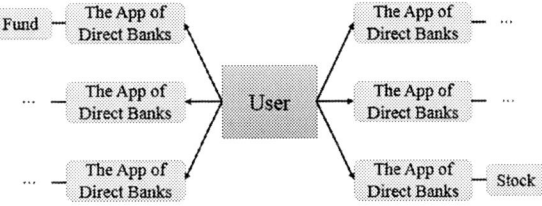

Figure 2. The ecological environment of Direct Banks

3. THE DEVELOPMENT OF DBRIP WITH SHARING ECONOMY

Direct Banks in the sharing economy evolve round a resource integration platform base on sharing economy business that compete with other traditional banks or financial institutions. Therefore, this chapter first characteristics and advantages of Direct Bank resource integration platform is illustrated. In a second step, the system architecture of Direct Bank in the sharing economy in general is explained more detail.

3.1 The Advantages of DBRIP in Sharing Economy

The DBRIP is produced by the concepts of sharing economy business. The relationship between resource integration platform, Direct Bank and users is shown in Figure 3. The main function of the resource integration platform is aggregate data resources of a number of Direct Banks and integrate user history information. This platform fulfill supply demand between Direct Banks and customers as well as select demand between customers and financial products, and improving the utilization of customer resources and data resources.

The DBRIP breaks the service mode of single line connection between Direct Banks and customers. Through DBRIP users can invest or purchase financial products more freely and Banks can also offer more suitable products for customers. There are many opening functions are provided by resource integration platform for Direct Bank, such as, user management function, uniform data interface, data security, privacy protection, products recommendation, etc. We believe these functions could help Direct Bank to connect resource integration platform safely and efficiently, enable users to select financial products securely.

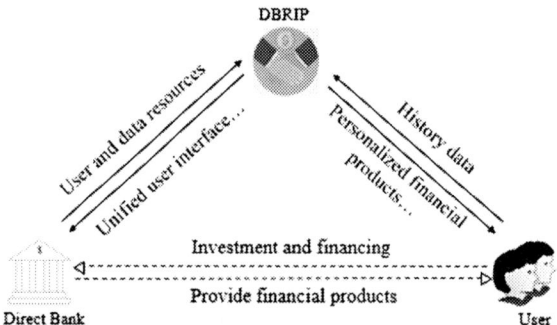

Figure 3. The relationship between information sharing platform, Direct Bank and users

The business model of the resource integration platform are basically virtual networks that connects user resources and data resources of a number of Direct Banks. Due to DBRIP is formed based on network and sharing economy, so we can borrow the characteristics of network and sharing economy [9] to describe the DBRIP, includes compatibility, consistency, regulations and economies of scale.

(1) Compatibility: in software engineering, compatibility means that a piece of software can work steadily on several different operating systems. In DBRIP, The product resources and user resources of a lot of Direct Banks can be integrated on the platform.

(2) Consistency: user purchase demand and product resource supply need to be consistent for a DBRIP. For example, a platform will not recommend futures when a user want to buy financial products with low risk.

(3) Regulation: Another essential aspect of DBRIP is regulations. Generally speaking, agree on regulations within a transaction is need to coordination. In the sharing economy, a platform set the regulations for the transactions, such as payment, terms of privacy, communication and terms of business.

(4) Economy of scale: economy of scale is a typical characteristic of network [10]. This also holds for sharing economy. DBRIP is able to attract a large number of users through integrate financial

products of multiple banks. In consequence, it is fairly cheap for DBRIP to reach a large number of suppliers and customers. For example, a bank needs to possess a significant number of capital and financial products to attract users. But in DBRIP, this threshold is basically nonexistent, because a large number of resources is easily reached through integrate banks and sharing information. Most important of all, the platform need not to purchase any money to provide this supply. This is the reason why sharing economy companies is quite easy to competition with the incumbents.

The advantages of DBRIP is described as: (1)DBRIP provides a place for customers and Direct Banks to deal anywhere and anytime; (2)DBRIP will save a large amount of manpower and material resources for banks, because they do not need to build an internet bank respectively; (3)User-friendly design, customers need not to download App of each bank, because all banks can be included in DBRIP and they can view and purchase products of each bank on DBRIP.

3.2 System Architecture of DBRIP

As shown in Figure 4, the components of the DBRIP contains three modules: data layer, function layer and service layer. In the following, we will introduce each of these modules in detail.

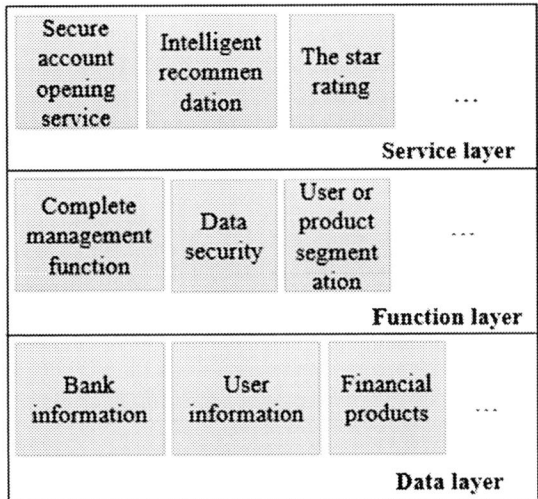

Figure 4. The architecture of DBRIP

Data layer

The multiple data resources are integrated in this module. These resources mainly consist of financial products, bank resources and user resources. That is clear these resources will produce a vast amounts of data. In order to offer better service to users, cloud computing is a good solution to storage and computing problem of big data. In the numerous cloud computing and storage research, a famous distributed file system HDFS (Hadoop Distributed File System) [11] is a standard model. HDFS provides high fault tolerance and high performance which ensure file access efficiently and securely.

The architecture of HDFS cluster consist of two parts: a single master node called Namenode that maintains and manages all files and a plurality nodes called Datanodes. The function of HDFS cluster is data storing, this is a critical part of the whole system, and it provides reliable and persistent storage capabilities. As illustrated in Figure 5. the function of Client is the entry of the system which provides the interface to upload, down load, and browse file. HDFS client is mainly responsible for data processing, the file system is accessed through HDFS clients, which first contact the Namenode for data location, and the second, that transfer data from the specified Datanode to another Datanode.

2242

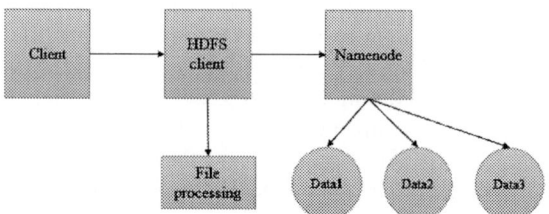

Figure 5. The architecture of HDFS system

Function layer

This module contains a series of public functions that ensure the comfortable services for Direct Banks and customers. (1)Complete management function and nice operation interface, which enable users have a good user experience and increase customer stickiness. (2)User and product segmentation, the customers should be classified according to income, age and education. Many results shows the customer who prefer Internet bank tend to be younger, wealthier and high education than traditional customers [12]. Products also need to be segmented in capital preservation and venture investment. NB (Naïve Bayes) [13] is a classical probabilistic classifier in machine learning, and it can predict class membership probabilities. (3)A criteria for star rating and growth factor ranking, the evaluation indicator of Direct Banks include: the background, operation situation, trading volume, earning rate and social influence. These evaluation results will help banks develop products as well as users select purchase products. (4)Multi-source semantic fusion, this function can process the complex and changeable multi-dimensional data, which is helpful to take advantage of the data and produce more reliable data analysis results. (5)Unified data access standards can utilize data resources efficiently. (6)Data security and privacy protection, the aim of this function is protects the information security of users and Direct Banks on the information sharing platform. Compare to traditional data, big data faces higher risk due to multiple sources and a large amount of data. In order to solve this problem effectively, there are some techniques might be useful, such as, k-anonymity, l-diversity and t-closeness [14]. These approach can withstand probabilistic inference attacks mistake identify and mistake attributes. These functions together guarantee the safe and effective operation of resource integration platform.

Service layer

The module provides a series of services for Direct Banks and users. (1)Secure account opening service, this function enable users transaction with an easy mind on the platform. (2)Intelligent recommendation of financial products, that recommend products through user history data and prediction with the similarity between user profiles. Collaborative filtering (CF) [15] is the one of the most popular recommendation algorithm, which recommend products to customers that other customers with similar tastes have liked in the past. The aim of which is reduce the time cost for user selecting products, and help banks deeply understanding user' preferences. (3)The star rating and the growth factor ranking of Direct Banks, this function reflects the overall strength of Direct Banks and the prosperity degree of industry. The ranking result will help user conduct a comprehensive risk assessment, reduce investment risk and develop an effective operational strategies for Direct Banks.

4. CONCLUSION

We conclude that DBRIP is the result of two interactive developments, the concept of sharing economy and application of Internet technological in financial filed. The incentive to propose this platform depends on the expectation of cost reduction and the expectation of handling the problem of resource underutilized. That is under the market competition premise, a large number of banks have more advantages than one. Nevertheless, there is an incentive for Direct Banks to establish DBRIP in order to attract more customers and reap long run profits through information sharing. But it does not imply that traditional bank is obsolete. DBRIP is not comfortable for uninformed customers and a customer who prefer manual services of traditional banks.

Overall speaking, the main motive to use DBRIP is lower time costs and browse or purchase anytime. Direct Bank is a booming industry that needs to be improved continuously and it also need to strengthen the cooperation with information platform. In our future studies, further efforts will be requested to functions and enhancing the performance of DBRIP.

REFERENCES

[1] Allen, Darcy / Berg, Chris, The sharing economy, How over-regulation could destroy an economic revolution, Institute of Public Affairs, Melbourne, 2014.

[2] Patsuris, P., Cut out the middleman, Forbes, 1998.

[3] Dervojeda, Kristina Verzijl, Diederik Nagtegaal, Laurent, The Sharing Economy, Accessibility-Based Business Models for Peer-to-Peer Markets, Case study no. 12, European Commission, Directorate-General for Enterprise and Industry, 2013.

[4] C De Roure , L Pelizzon, P Tasca, How Does P2P Lending Fit into the Consumer Credit Market?, Social Science Electronic Publishing, 2016.

[5] Kaplan, A.M., & Haenlein, M., Users of the world, unite! The challenges and opportunities of Social Media. Business Horizons, 2010, 53(1), 59–68.

[6] Dahlman, Carl J., The Problem of Externality, in: Journal of Law and Economics, 1979, Vol. 22 (1), p. 141–162.

[7] Ehrlich, Isaac and Lawrence Fisher, The Derived Demand for Advertising: A Theoretical and Empirical Investigation, American Economic Review, 1982, 72, 366-388.

[8] Steltzner, Holger, Bankgeschäfte auf der Datenautobahn, Frankfurter Allgemeine Zeitung No. 210, 9.9.1996.

[9] Shapiro, Carl Varian, Hal R., Information Rules, A strategic guide to the network economy, Boston, Massachusetts, 1999.

[10] Shy, Oz, The economics of network industries, Cambridge, UK, 2001.

[11] AK Karun, K Chitharanjan, A review on hadoop — HDFS infrastructure extensions, Information & Communication Technologies, 2013:132-137.

[12] Pischulti, Helmut, Direktbank - Bank der Zukunft, in: Allgemeine Deutsche Direktbank AG (ed.): Direktbanken. Die moderne Bankverbindung, Frankfurt, 1995, a.M., 4-11.

[13] J. W. Han and M. Kamber. Data mining concepts and techniques, The 2nd edition, Morgan Kaufmann Publishers, San Francisco, CA, 2006.

[14] N Li, T Li, S Venkatasubramanian, t-Closeness: Privacy Beyond k-Anonymity and l-Diversity, IEEE International Conference on Data Engineering, 2007 :106-115.

[15] S Liao, Y Chen, An Improved Collaborative Filtering Recommendation Algorithm, International Conference on Networking & Distribu, 2017, 32 (9) :204-208.

Implementation of Interactive Classroom Design based on WI-FI Service

CHEN Shuai[1,a], ZHANG Shuifeng[2,b], WU Tianfang[3,c]

[1]Huainan normal university
school of electronic engineering,
Huainan, China
0086-0554-6863698

[2]Huainan normal university
school of electronic engineering,
Huainan, China

[3]Huainan normal university
school of electronic engineering,
Huainan, China

[a]chen232001@126.com , [b]1565458403@qq.com, [c]1433624596@qq.com

ABSTRACT: In order to realize the teaching interactive classroom in the local area network, the WI-FI router is used to set up the virtual server, and the web service system is established by using the Apache TOMCAT service software. The chat room was designed using Java Server Pages (JSP). Through the chat room, teachers and students can interact. First the WIFI network architecture was introduced, then the interactive classroom chat room design was proposed, finally the experiment was carried out. The results show that the interactive classroom design based on Wi-Fi service can achieve the effect of teaching interaction. The system can save network traffic without accessing the Internet, while increasing classroom interaction.

1. INTRODUCTION

According to the survey [1], the mobile phone dependence in the classroom is more serious, which has affected the normal classroom learning for the current college students. College students who rely on mobile phones may have different levels of health problems, suggesting that correct guidance should be given to reduce the dependence of college students on mobile phones [2]. Colleges and universities have set off a wave of activities to promote "no mobile phone classroom"[3]. Some schools have launched a series of study styles such as "no mobile class" and "fingerprint attendance", aiming to prevent students from playing mobile phones and skipping classes [4]. This classroom activity is contrary to the modern educational concept and the development of the times. Although it is considered that mobile phone dependence has become a pathological [5], if you take advantage of the benefits of mobile phones, you can help improve teaching functions, increase classroom atmosphere, and make the classroom more and more intelligent.

Content from this work may be used under the terms of the Creative Commons Attribution 3.0 licence. Any further distribution of this work must maintain attribution to the author(s) and the title of the work, journal citation and DOI.
Published under licence by IOP Publishing Ltd

2. RELATED WORK

The paper [6] applies the same screen display technology and the teacher's personal smart phone, builds a mobile teaching platform, solves the shortcomings of the multimedia system itself, and exerts the functions of the smart phone, the interaction is strong, and the mobile is convenient, but the special screen device and its supporting device and software are needed. The article [7] in the ordinary classroom, through the wireless network and teaching interaction system, builds a smart electronic classroom based on student smart phone, realizes the information interaction between the teacher computer and the student mobile phone, but the system needs to use the mobile Internet and generate traffic cost. Reference [8] suggests an immersive, simulated learning environment built using virtual reality technology, but at a high cost. The intelligent classroom information platform [9] is a classroom information environment consisting of three parts: "cloud", "net" and "end", and the cloud platform provides cloud infrastructure, support platform, resource service, data processing, teaching services, etc., such as building a complete teaching resource management platform. In view of the practical problems of college students playing mobile phones, sleeping, and less interaction between teachers and students, this project[10] is based on the Android platform, using client/server mode, based on the campus wireless Wi-Fi network, developed a university intelligent classroom interactive teaching system.

For the sake of cost, most teachers and schools need cheap smart interactive courses. In order to realize low-cost interactive courses, this paper studies a smart classroom interactive implementation that is cheap and convenient and does not require the Internet, as long as a router and a service computer are used.

3. WI-FI NETWORK ARCHITECTURE

The network architecture is shown in Figure 1. The hardware consists of a web server and a wireless router.

The web server accesses the router, and the web server is provided by a virtual server configured by the router. The terminal device such as the user's mobile phone can access the web server content by wirelessly accessing the WIFI and connecting to the router by the router SSID and password.

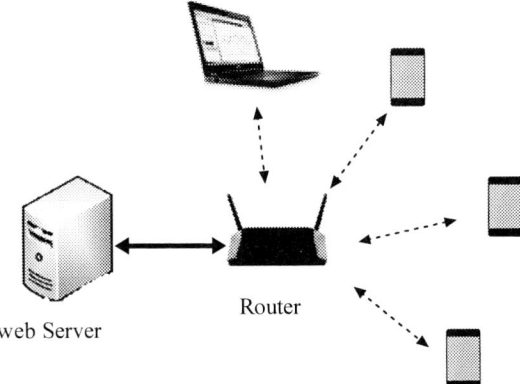

web Server   Router

Figure 1. Service network architecture.

3.1 Router settings

Log in to the router through the router IP address (for example, 192.168.8.1) and password.

Assume that the IP address assigned by the router to the web server (PC) is 192.168.8.101. Set the virtual server in the router. The common server selects HTTP, the port selects 80, the IP address is set to the IP address 192.168.8.101 assigned by the router to the web server (PC), and the protocol type selects TCP. As shown in table 1.

Table 1. Virtual server in the router settings

Serve r	External Port	Inner Port	IP Address	Protocol Type
HTTP	80	80	192.168.1.1 01	TCP

3.2 WEB server software

Install Apache TOMCAT on the web server, and modify the TOMCAT port configuration to 80. Start the TOMCAT service, and then open the web server of port 80, as shown in Figure 2.

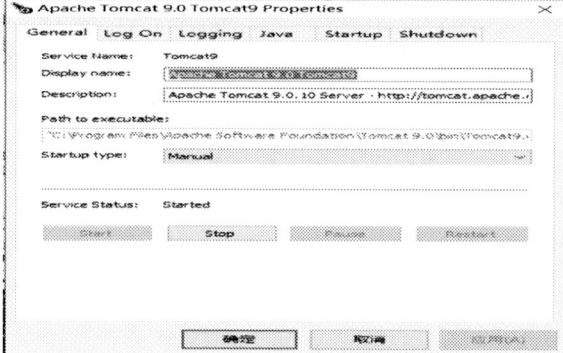

Figure 2. Start the HTTP service.

3.3 Connect to the virtual server and test

Enter the IP address and port through the browser of the terminal that accesses the router. That is, enter: http://192.168.8.101/

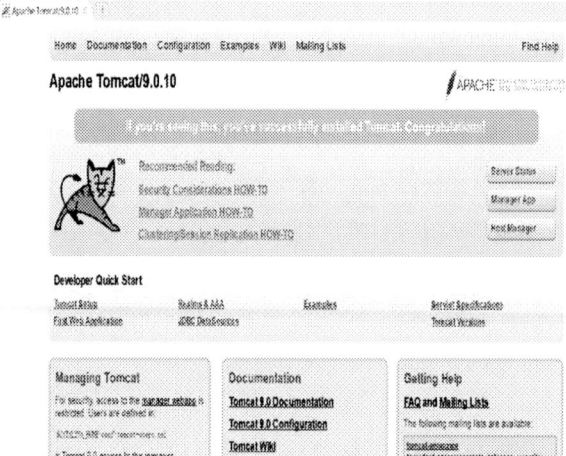

Figure 3. Http service content accessed via WI-FI.

Then you can access the http service content set by the web server (Personal computer). As shown in Figure 3

Please use a 9-point Times Roman font, or other Roman font with serifs, as close as possible in appearance to Times Roman in which these guidelines have been set.

4. INTERACTIVE CHAT ROOM DESIGN

A simple chat room was designed for the wireless communication between the teacher and the students. The chat room consists of three parts: an interactive content area, a personal speaking area, and a speaking button. As shown in Figure 4.

The interactive content area displays the date, time, and IP address of each registered speaker who is speaking in the local area network, and has already spoken, and can only output the display. The personal speaking area is the conversation content that a single user is prepared to issue and needs to be entered. The speaking button is used to control the content of the individual speaking area, and when the button is clicked, the content of the speaking is sent.

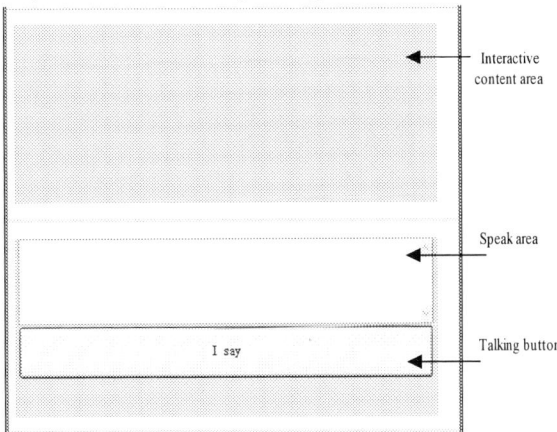

Figure 4. Chat room interface.

4.1 Indexi.jsp file

A web document consists of three files.

The contents of the Indexi.jsp file are as follows：

```
<%@ page language="java" import="java.util.*" pageEncoding="UTF-8"%>
<%@ page language="java" contentType="text/html;charset=GB2312"%>
<html>
 <!-- frameset rows="*,150"  -->
  <frameset rows="20,20">
    <frame src="show.jsp" >
    <frame src="send.jsp">
 </frameset>
</html>
```

4.2 Show.jsp file

The contents of the show.jsp file are as follows:

```
<%@ page language="java" import="java.util.*" pageEncoding="UTF-8"%>
<%@ page language="java" contentType="text/html;charset=GB2312"%>
<html>
<meta http-equiv="refresh" content="3;url=show.jsp">
<body>
```

```
<div style="text-align: left;background-color:#F8D8A9;width:100%;height:100%">
   <%if (application.getAttribute("words") != null)
        out.println(application.getAttribute("words"));
   %>
   </div>
</body>
</html>
```

4.3 Send.jsp file
The contents of the send.jsp file are as follows:

```
<%@ page language="java" import="java.util.*" pageEncoding="UTF-8"%>
<%@ page language="java" contentType="text/html;charset=GB2312"%>
<html>
<body>
   <%
   try
    {
      request.setCharacterEncoding("GB2312");
      String mywords = request.getParameter("message");
      String t = "";
      if (application.getAttribute("words") == null
         && mywords != null) {
       t = "<b>" + (String) request.getRemoteAddr() + ":</b>"
          + mywords + "<br/>";
       application.setAttribute("words", (Object) t);
      }
     else if (mywords != null) {
       t = (String) application.getAttribute("words");
       t += "<b>" + (String) request.getRemoteAddr() + ":</b>"
          + mywords + "<br/>";
       application.setAttribute("words", (Object) t);
        }
    }
  catch (Exception e) {} %>
 <div style="text-align: center; background-color: #F8D8A9;width:100%;height:100%">
 <form method="post" action="send.jsp">
 <textarea name="message" style="width: 500px; height: 80px">
 </textarea><br>
  <input type="submit" value="I say" style="width: 500px; height: 50px" />
</form>
</div>
</body>
</html>
```

5. EXPERIMENT
Assume that the IP address of Apache TOMCAT is: 192.168.1.102. Two terminals connected to the same local area network, such as a mobile phone or a PC, have IP addresses of 192.168.1.100 and 192.168.1.102, respectively. The two terminals enter the web server address: 192.168.1.102 through the browser, and the chat interface of FIG. 4 appears. In this way, the two terminals can chat and teach interaction. The experiment result is shown in figure 5.

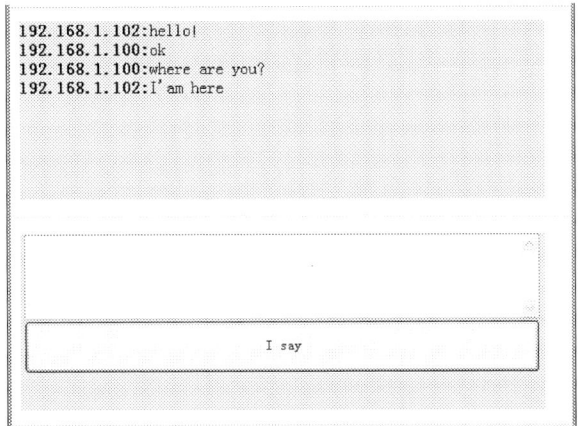

Figure 5. Experiment result.

6. ANALYSIS

Table 2. Comparison of several methods

No.	method	description	cost	flexibility
1	the same screen display technology	Dedicated equipment, single interaction, mobile phone with the same screen	High	Low
2	the wireless network	Private server	High	High
3	virtual reality technology,	VR device	High	Low
4	the cloud platform	Cloud Server	High	High
5	the campus wireless Wi Fi network	Campus wireless network, arranged in advance	High	High
6	Method of this paper	Low cost, flexible at any time	Low	High

The performance of several interactive classrooms is shown in Table 2. The several other methods are expensive due to the need for dedicated equipment and the use of billing traffic. The method proposed in this paper is inexpensive because it uses a WI-FI router and a personal computer and uses a local area network. In terms of flexibility of use, the wireless network method, the cloud classroom method, the campus wireless Wi-Fi network method, and the method herein are highly flexible.

7. CONCLUSIONS

Wireless interactive classrooms are implemented in the LAN using Wi-Fi and Apache web services. The teacher and the student can realize wireless conversation through the mobile computer, and carry out class discussion and answering questions. The simplicity of this design is that no network traffic is required. The hardware only needs one ordinary router and a PC (laptop) with network function.

The next step is to improve the web page, add functionality, and increase the flexibility of teaching interaction. By adding a database server, one-click sign-on can be achieved, and the attendance situation can be quickly counted; the quiz test supports single-choice, multi-selection, and subjective question-limited answers, and the answer status is displayed in real time, and the test score is automatically saved [11].

Through classroom data analysis, you can grasp the student's learning situation and teaching interaction emotions. Online testing is also available through this interactive system.

ACKNOWLEDGMENTS

This study was supported by the major fund of nature science for colleges and universities in Anhui province of China (KJ2014A239), the fund of nature science for colleges and universities in Anhui province of China (KJ2011Z342), Scientific research project of HUAINAN normal university (2018xj34, 2016xj01zd,2016xj46);Teaching research project of HUAINAN normal university(2017hsjyxm51); Provincial College Students Innovation and Entrepreneurship Training Program of ANHUI province (201710381087).

REFERENCES

[1] LUO Shao-ye,ZHUANG Mei-lian. The Status Quo and Cause Analysis of College Students' Classroom Mobile Phone Dependency[J]. Journal of Zunyi Normal College, 2017,19(1):146-150.

[2] YANG Li. Research on the Status Quo of College Students' Mobile Phone Dependence and Its Impact on Health[J]. Western quality education, 2018(16): 78-79,81.

[3] LI li. Thoughts on the "No Mobile Classroom" Activities Tried by Colleges and Universities[J], Journal of Campus Life & Mental Health ,2017,15(5):383-385.

[4] Qi Yuanli, Chen Jianquan. "Fancy punch card" lived the classroom learning style[N]. China Education News, 3-28-2016.

[5] LIU Guo-sheng. Mobile phone dependence is also a disease[J]. Everyone's health, 2018(17):20-21.

[6] XUE Shenglan. The Construction and Application Base on Smart Phone's Teaching Support Platform[J]. China's electrification education, 2017,(3):127-131.

[7] XUE Shenglan. Research on the Application of Smartphones Integrating into Classroom Teaching[J]. China's electrification education, 2018,(1):86-91.

[8] WANG Yu-bo.Application and Design of Virtual Reality in Smart Classroom[J]. Electronic world, 2018(14):155.

[9] http://js.news.163.com/yz/16/1021/16/C3TR45EB04041DP8.html.2018-10-10.

[10] ZHU Jian,DU Xuan,MENG Qing-hui,et al. Constructing the Wisdom Class in the University——to Effectively Improve the Teaching Effect[J]. Computer knowledge and technology, 2016,12(15):187-189.

[11] http://www.caigou.com.cn/news/2017112715.shtml.2017-11-27.

Research on the Challenges and Innovation Models of Public Management in the Age of Big Data

Wang Cong, Su Xiangguo, Lou Ziqiang

Qingdao Institute of Technology, Qingdao, 266300, China

Abstract: In recent years, China's computer level has been further developed, and the computing power of big data technology has been further improved, which has also caused a great impact on China's public management work. Through the application of big data technology in the public management process, it can further improve its decision-making level and service capability, and promote the social risk management ability to be further improved to meet the individual needs of the people. The challenges of public management and the innovation model in the era of big data are explored and analyzed.

1. Introduction

The arrival of the era of big data has had a relatively large impact on people's daily life, production and thinking patterns. The continuous development and wide application of big data technology has also led to the application of more information methods in the field of public management, and the time and space situation of public management has undergone major changes. It can be said that big data technology can provide key technical support for the innovation and optimization of the public management service model. It also requires the active application of big data thinking and means at all levels of government, as well as various modes such as institutional and organizational innovation can be used to improve their public management level to meet people's needs for public management work.

2. The Connotation and Characteristics of The Era of Big Data

Big data is not a simple evaluation of the amount or amount of data, it is the information technology which captures useful data by screening large amounts of data. China's life will also change with the arrival of the era of big data. Through the application of a series of big data means, it can also accelerate the development of public management and public services, and can achieve good public management from the three levels of data, analysis and processing and industry integration by achieving secure collaborative governance.

At present, there is no uniform definition of the meaning of big data in the academic world, but it is certain that big data is an information explosion effect formed by the revolutionary advancement of information technology, and can obtain great value products and services [1] through a model for effective analysis of massive data. Big data technology also has three features: large, high speed and diverse. With the continuous development of computer technology in China, the current society is also developing towards the era of innovative Internet, and the data generated by people in daily life and work has also been greatly improved. Through the application of big data technology, effective information can be filtered in massive data to meet the individual needs of the public. Therefore, various government departments also need to actively apply big data technology to continuously optimize and innovate the existing public management model, thereby obtaining good innovation

---

Content from this work may be used under the terms of the Creative Commons Attribution 3.0 licence. Any further distribution of this work must maintain attribution to the author(s) and the title of the work, journal citation and DOI.

Published under licence by IOP Publishing Ltd

management effects.

3. Changes in The Field of Public Management in The Era of Big Data

The arrival of the era of big data can first promote the service quality and service efficiency of public services, and can also obtain good government behavior network supervision effects, and the quality of public services has been effectively improved. Through the application of big data technology, relevant tracking and evaluation work can also be completed, and the quality of public service provision and the problems in the service process can be effectively analyzed, and the service effect is further improved.

Through the application of big data technology, the public opinion crisis ability of public administration can be promoted. There are still many complicated factors in the process of social development in our country, and there are many emergencies. The unpredictability and uncertainty brought by emergencies also make the handling of public opinion ability be an important indicator to measure social governance ability [2]. The application of big data technology can reasonably predict the uncertain factors in the process of social development, and promote the further improvement of the handling of public opinion events.

Finally, the application of big data technology can also effectively meet the needs of the public. With the advent of the era of big data and the continuous development of society, individuals have begun to present multi-dimensional and heterogeneous features, and the independence and subjectivity of individuals have become more significant. Through the application of big data technology, more detailed data mining can be achieved, which can achieve good public management effects and meet the various needs of the public with sufficient technical support.

4. Opportunities for Public Management Work in the Era of Big Data

4.1 Improve Decision-Making Science

In order to improve the decision-making ability in public management work, big data technology can also be applied to effectively integrate and analyze various information resources, and then obtain key information in massive data to provide good support and decision basis for subsequent public management work. With the continuous development of information in a timely manner, people can participate in the discussion of public decision-making through various modes such as Weibo and WeChat, so that relevant departments can fully analyze the actual needs and hotspots of the people. Guarantee the effectiveness and scientificity of public management decisions. In addition, in the process of applying big data technology, people's participation, initiative and responsiveness can be effectively enhanced. Therefore, public management decisions in the era of big data need to rely on rich and detailed data to be carried out, which can also promote the scientific and democratic decision-making of public decision-making, has a certain positive significance and further develop the public management work in China.

4.2 Big Data Technology Can Enhance The Risk of The Management Ability of Society

In recent years, the incidence of various risk events in China has been greatly improved, like the natural disasters, financial crises, and public health crises have also caused great threats to people's lives and production. With the application of big data technology, it can also directly convert the passive response after the original risk into the active prediction between the risk occurrence, and promote the initiative, effectiveness and forward-looking of risk management. The application of big data technology can also effectively predict the occurrence of social crisis and risk, so as to further improve the controllability of risk. For example, the tsunami warning in Japan after the 3.11 earthquake was also an important example of the use of big data, and effectively reduced the casualties and economic losses [3].

5. Public Management Work Optimization Measures in The Era of Big Data

5.1 Challenges of Public Management in the Age of Big Data

①The feasibility is not strong and the means are too single. Through the application of modern public management technology, it has good forward-looking characteristics for some public events, so it has been widely used in the public government management departments of our country. However, some emerging management technology goals still have problems with low application significance. These public management technologies and platforms have become one of the indicators of government performance, and have caused great obstacles to the further development of public management work in China. In addition, some public administration departments still have a single and rigid management model in the current management work, and they cannot play a good role in promoting the optimization of public management. In recent years, the stampede in Shanghai, the restrictions on the purchase of milk powder in Hong Kong, and the law enforcement of urban management violence have led to problems such as the simple and rude public management model at all levels of government in China, so that the effect of good public management can be obtained. But it will lead the question like the management ability of the government department is too rude, and will have many adverse effects on the subsequent development of public management work.

②Management lacks big data awareness. China's big data technology started relatively late and has developed rapidly in recent years. However, the public's awareness of big data is limited, and some management departments have problems with lack of data awareness in their work. In the current analysis of mass incidents in China, many incidents are caused by lack of public opinion ability and poor data collection ability, which makes some public opinions unable to be solved in time, and can only respond passively to public crises and leads public event processing effect. Analysis of the stampede accident on the Shanghai Bund, it seems to be an accident, but still can be judged by the flow of mobile phones. However, due to the lack of big data awareness, the local management department did not strictly control the density of personnel, and did not do the relevant prevention work, which led to the occurrence of tragedy [4].

③Data distortion is more serious. In the current performance appraisal system, some government personnel will also make access to promotion opportunities by fabricating false data models. For example, when the GDP increase is determined, because the evaluation indicators are too single or their economic resources are limited, This has led some government personnel to be less satisfied with the actual data, and to obtain the opportunity for promotion through the model of renewal data. For example, the total industrial output value of Henglan Town in Guangdong in 2012 was only 2.22 billion yuan, but the total industrial output value in the local annual report was as high as 8.51 billion yuan, which means that the total industrial output value of the town is increased nearly 6 billion yuan, it is impossible to fully display the actual development of the local area. In addition, in the era of big data, personal privacy and daily behavior will be monitored in many cases, and national security is subject to relatively large threats. Therefore, there are certain application risks in the application process of big data technology.

5.2 Innovative Mode of Public Management in The Era of Big Data

①Issue relevant laws and regulations to maintain the security of big data. Compared with some developed countries such as Europe and the United States, there are still problems in China's personal data privacy legislation. Although the arrival of the era of big data has given certain convenience to public management, it will infringe on the legitimate rights and interests of the public to a certain extent, and it shows the gaps in government data security maintenance and data legislation. Therefore, the relevant government units also need to continuously optimize and improve the existing data security system, and also actively introduce relevant laws and regulations to fully play the guiding role of legal norms. In addition, there is a need to make reasonable provisions for the use of organizational or personal data, and to regulate the type of data. In the process of data collection and processing, a standardized management system is also needed to fully utilize the performance of big data technology.

In addition, in the process of data security legislation, it is necessary to define the data storage time limit, and then perform storage restriction on data positioning based on combining different types and purposes. In addition, the "forgotten right" of personal data needs to be fully protected to avoid problems such as user information and privacy disclosure, and to protect citizens' normal rights and interests from being harmed.

②Do a good job in public opinion analysis. In order to allow public opinion supervision and analysis functions to be fully utilized, public administration departments are required to continuously improve and strengthen the information collection channels. Through the powerful public opinion analysis platform, it can also effectively cover many functions such as publicity, evaluation and public opinion report. With the help of the public opinion analysis function in big data technology, it can effectively collect and analyze various Internet information through information collection channels, and can also have good tracking effect for some hot topics, and can timely analyze the public opinion events. In addition, it is necessary to do a good job of public opinion guidance through a variety of modes, and some negative information needs to be processed in a timely manner, and useful information can also be disseminated in the first time [5]. Finally, it is necessary to adopt the public opinion evaluation and the public opinion processing function, and also need to provide feedback on the handling situation in the first time to enhance the public's conviction.

③Strengthen the construction of social public management data platform. Only on the basis of the social public management data platform and the main body construction work can we meet the needs of public health services. For example, through the transformation of street light micro base stations, construction of data acquisition platforms and other modes, the application of big data technology can be fully exerted. In the process of constructing the social public management data platform and the main body, taking the construction of smart city as an example, it is necessary to construct a new social public management model based on data value-added services and intelligent technologies to make its own public management service level. And the quality of management services has been effectively improved. Government departments need to strengthen the construction of big data management products and actively develop related technologies. The public can access the government website through network technology, and through the establishment of the relevant push mechanism, the use value of a large amount of original data can be fully utilized, thereby obtaining the optimization and utilization effect of the data resource.

6. Conclusion:

As an important basis for ensuring the stable development of the society, public management work needs to strengthen the attention of relevant management departments, and it is necessary to continuously optimize the existing public management mode which is based on the actual development of the society. The arrival of the era of big data has also had a very big impact on China's public management work. This requires government departments to have a full understanding of the connotation and characteristics of big data technology, and to be able to change the public management in the era of big data, at the same time the challenges faced were fully recognized, and subsequent measures were taken to improve the use of big data technologies, thereby promoting the improvement of local public management and providing the public with better public management services.

Acknowledgments

1. A Study of the Current Scientific Research Situation and Countermeasures in Private Colleges and Universities under the New Normal Situation of Education (Project Number: J17RA232);

2. The Research on the Promotion Strategies of Teaching Ability of Young and Middle-aged Teachers in Private Colleges and Universities (Project Number: YC2017013);

3. The Research on the Connotation Construction Path of Private Undergraduate Colleges and Universities in New Situation (Project Number :18-ZC-GL-02)

References:

[1] Zhou Fangjian, He Zhen. Challenges and Countermeasures of Urban Public Security Emergency Management in the Age of Big Data[J]. Journal of Yunnan University for Nationalities (Philosophy and Social Sciences Edition), 2018, (1): 117-123.

[2] Wang Songgao. Public Management Innovation in the Age of Big Data [J]. Management Observation, 2018, (24): 62-63.

[3] Hu Yangming, Li Tao. Problems and Countermeasures in the Construction of Rural Public Crisis Management Informationization in the Age of Big Data[J]. Science and Technology Management Research, 2016, (12): 170-177.

[4] Gao Qunan. The Challenges and Countermeasures of Local Government Public Management in the Age of Big Data[J]. Chi Zi,2018, (8):178.

[5] Chen Deli. The Enlightenment of Public Management in the Age of Big Data[J]. Modern Economic Information, 2018, (4): 112.

ISPECE

IOP Publishing

Applications Research of Cluster Analysis in Chinese Acupuncture Therapy

Xiaoche Feng[1]

[1]The First Affiliated Hospital of Guangxi University of Chinese Medicine, Nanning, 530000, China

Abstract: It is significant to improve the clinical therapeutic effect of Chinese acupuncture for an acupuncturist. Cluster analysis can classify the patient samples and help the acupuncturist to innovate the traditional acupuncture therapy under the concept of personalized treatment. This paper points out the concrete steps of the cluster analysis application in the improvement of the treatment effect of Chinese acupuncture, which incudes indexes selection, data collection and clustering analysis. Then, the paper provides a concrete case of cluster analysis application for thirty patients of lumbar intervertebral disc protrusion. We divided the patients into three categories and adopt different types of patients with different acupuncture treatment to achieve good treatment results.

1. Introduction

Chinese acupuncture therapy is a medical discipline that studies meridians, acupoints, acupuncture techniques, and the use of acupuncture and moxibustion therapy to prevent and treat diseases. For thousands of years, acupuncture has made great contributions to the health of the Chinese people and the growth of China's population. Today, even with the rapid development of medical science, it is still regarded as an important part of China's cultural heritage. We should vigorously promote standardized clinical research and other aspects, to innovate acupuncture theory, promote the modernization and internationalization of acupuncture and moxibustion process for China's health and social development. Computer technology has begun to have a more extensive application in the field of acupuncture and moxibustion in the clinical, teaching and scientific research and other fields. Medical statistics, including medical data analysis, is an important part of medical science. The application of the logic of data analysis method to medical research and management can play an important role in assisting decision-making. Cluster analysis can find interesting patterns in potential data, and it is a very active research field of data mining technology. Clustering is an unsupervised learning method. The main difference between clustering and classification is that it divides all data points into several classes or clusters without knowing how many classes there are in the database beforehand, and ensures the data in each cluster by combining some criteria such as similarity measure. This paper intends to use the cluster analysis to enhance the curative effect of acupuncture.

2. Concept and methods of cluster analysis

2.1 Concept of cluster analysis

Clustering is the process of dividing data sets into multiple sisters or clusters through specific methods. The objects in clusters are highly similar, but the similarity of objects between different clusters is very poor. In machine learning, classification is called supervised learning as the given class labelling information, that is, learning is supervised. Clustering is called unsupervised learning because it does not provide the initial category of objects. Therefore, we use observation learning to cluster, rather than sample learning. Generally, the difference between data objects is measured by their own attribute values, usually according to some distance between data objects, and the calculation of distance is different according to the different types of data variables. The usual variable types are discrete variables, continuous variables, or have these types of attributes. Clustering analysis algorithm

Content from this work may be used under the terms of the Creative Commons Attribution 3.0 licence. Any further distribution of this work must maintain attribution to the author(s) and the title of the work, journal citation and DOI.
Published under licence by IOP Publishing Ltd

is used in many data mining processes, because it can be used as a pre-processing of related mining methods, and clustering technology itself is a data mining method. At the same time, clustering analysis technology can be used to deal with the noise in the data and provide pre-processing results for other data mining methods.

2.2 Partition cluster method

The idea of partitioning algorithm is to divide the data objects to be mined into k groups (k < N, N represents the number of objects in the data set), and each group represents a cluster. And to satisfy that any data object can only belong to one cluster, each cluster has at least one data object. This algorithm usually requires the parameter k to determine the number of clusters after the algorithm starts. The algorithm establishes an initial grouping according to the parameter k. Later, the algorithm uses iterative relocation technology to reallocate the data objects in each cluster repeatedly, and then obtains the final relatively satisfactory clustering results. It is a good clustering analysis algorithm to minimize the gap between data objects within clusters and to maximize the gap between data objects among clusters.

$$E = \sum_{i=1}^{k} \sum_{p \in C_i} |p - m_i|^2$$

Among the above formula, p represents an arbitrary data object, E represents the sum of square errors of all data objects, and m_i is the mean value of the c_i of the data p cluster. At the beginning of the algorithm, k data objects are randomly selected as the initial centre of k clusters in the data set, and the remaining data objects are allocated to the nearest cluster centre according to the distance between them and each cluster centre. Then the average value of all data objects in each cluster is recalculated, and the result is used as the new cluster centre. The above process is repeated gradually until the objective function converges.

2.3 Hierarchical cluster method

The idea of hierarchical clustering algorithm is to decompose a given data set hierarchically until the relevant termination conditions meet the stop. According to the idea of the algorithm, hierarchical clustering method can be divided into two categories, one is bottom-up clustering and the other is top-down decomposition hierarchical clustering. The idea of clustering algorithm is to take each point as a single cluster at the beginning, and then merge two closest data objects or clusters each time, all objects are divided into a cluster, or to achieve the relevant termination conditions to stop. Cluster proximity needs to be defined here; decomposition hierarchical clustering starts with the point containing all data pairs as a cluster, then gradually divides into smaller clusters until each data object is separated in a cluster, or some relevant termination condition is reached. Hierarchical clustering algorithm has the advantages of clear hierarchical structure, simple algorithm ideas, and can dynamically adjust the number of clusters according to the needs of users. But the disadvantages of hierarchical clustering are also obvious. We need to evaluate many clusters or objects before making the decision of clustering or splitting, which leads to the poor scalability of the algorithm in dealing with clusters of different sizes. In the hierarchical clustering process, once the data objects are split or condensed, the processing already done cannot be changed. This leads to the fact that if the choice of merger or split is unreasonable, it cannot be further amended.

3. Application steps of cluster analysis in Chinese acupuncture therapy

To improve the curative effect of acupuncture and moxibustion of traditional Chinese medicine, we should cluster the effects of acupuncture and moxibustion of traditional Chinese medicine. We divided the patients with better acupuncture effect into one group, the patients with moderate acupuncture effect into one group, and the patients with worse acupuncture effect into one group. Different ways can enhance the efficacy of acupuncture and moxibustion.

Specifically, we use acupuncture rehabilitation as an example to treat lumbar disc herniation. Protrusion of lumbar intervertebral disc is a clinical disease caused by lumbar strain, strain or senility leading to intervertebral disc lesions, fibre ring damage or stimulation of nerve roots and blood vessels. With the accelerated pace of life and the increase in the number of people engaged in mental work, lumbar disc herniation has become the primary cause of lumbar and leg pain, directly affecting people's normal life. Acupuncture and moxibustion of traditional Chinese medicine, as a treasure in Chinese history and culture, has obvious effect on the treatment of this disease. It can effectively relieve the lumbar and leg pain caused by lumbar disc herniation, with high efficiency, safety, convenience and other characteristics, can improve the patient's self-care ability and quality of life. The specific steps are as follows:

The first step is to select indexes. To objectively evaluate the therapeutic effect of acupuncture on lumbar disc herniation, the JOA score, McGill pain score, inflammatory factors and LFR were selected to evaluate the results. The second step is to collect data. Collect the indicators needed to analyse acupuncture efficacy. These indicators need to be obtained by expert scoring and real-time detection. Taking LFR as an example, the shilling patient stands upright, measures the vertical distance from the middle fingertip to the ground, then lets the patient body bend forward as far as possible, measures the vertical distance between the middle fingertip and the ground currently, and finally takes the difference of the vertical distance of the middle finger tip between the states of upright and forward flexion as LFR. Step three: cluster analysis. Cluster analysis can be done easily by using and other professional software, and a genealogical clustering diagram is generated. Step four: results analysis. The results of clustering are analysed, and different methods are adopted to improve the efficacy of Chineses acupuncture.

4. Empirical analysis of the effect improvement of Chinese acupuncture therapy

Thirty patients with lumbar disc herniation were selected from The First Affiliated Hospital of Guangxi University of Chinese Medicine from September 2017 to August 2018. We use clustering algorithm to cluster patients into relatively bad efficacy, moderate efficacy and relatively good efficacy of three categories.

4.1 Step 1: indexes selection

According to the analysis of the third part of the paper, the author chooses the following four indicators as basic variables. Their names and meanings are shown in the following table:

Table 1. Index selection of curative effect of lumbar intervertebral disc protrusion

Index	Connotation
JOA score I_1	Include subjective symptoms, clinical signs and limitation of daily activities
McGill pain score I_2	Include pain rating index, existing pain intensity and visual analogue scale
Inflammatory factors I_3	Tumor necrosis factor (TXB2)
LFR I_4	The difference of the vertical distance of the middle finger tip between the states of upright and forward flexion

4.2 Step 2: data collection

The data of the selected indexes of 30 patients with lumbar disc herniation obtained by expert scoring and real-time detection is shown as Table 2.

Table 2. Data of the selected indexes of 30 patients of lumbar intervertebral disc protrusion

Index	I_1	I_2	I_3	I_4
Patient 1	11	6	36.8	41.0
Patient 2	5	32	37.4	43.1
Patient 3	16	22	32.9	42.8
Patient 4	17	45	30.3	51.9
Patient 5	15	22	31.0	49.4
Patient 6	13	38	38.3	55.4
Patient 7	11	17	35.5	51.5
Patient 8	19	18	36.6	50.8
Patient 9	10	40	30.9	40.8
Patient 10	16	9	31.3	47.2
Patient 11	14	22	30.7	54.6
Patient 12	24	42	31.5	51.4

Patient 13	24	8	37.6	50.9
Patient 14	14	23	30.2	51.6
Patient 15	17	27	30.9	48.7
Patient 16	23	5	36.5	45.8
Patient 17	15	12	34.4	55.3
Patient 18	21	7	30.8	40.0
Patient 19	11	10	30.1	43.1
Patient 20	15	33	30.8	52.1
Patient 21	7	19	34.0	52.2
Patient 22	15	33	36.1	52.4
Patient 23	18	42	35.9	46.9
Patient 24	14	45	30.1	55.1
Patient 25	12	21	33.9	43.9
Patient 26	5	13	40.6	43.7
Patient 27	19	33	34.2	52.1
Patient 28	18	45	40.7	55.1
Patient 29	16	26	36.3	46.5
Patient 30	18	11	32.4	46.0

4.3 Step 3: clustering analysis
We use the system clustering method to do the clustering analysis with the help of the software of SPSS 24.0. The result of systematic clustering analysis is shown in Table 3.

Table 3. Clustering result of thirty patients

Case	Category	Case	Category	Case	Category
Patient 1	B	Patient 11	B	Patient 21	B
Patient 2	B	Patient 12	B	Patient 22	B
Patient 3	A	Patient 13	B	Patient 23	C
Patient 4	B	Patient 14	B	Patient 24	C
Patient 5	B	Patient 15	B	Patient 25	B
Patient 6	C	Patient 16	B	Patient 26	B
Patient 7	A	Patient 17	B	Patient 27	B
Patient 8	B	Patient 18	A	Patient 28	C
Patient 9	B	Patient 19	B	Patient 29	B
Patient 10	A	Patient 20	B	Patient 30	C

From the data index features, Patient 3, Patient 7, Patient 10 and Patient 18 are the patients with relatively good efficacy; Patient 6, Patient 23, Patient 24, Patient 28 and Patient30 the patients with relatively bad efficacy. The Chinese acupuncture therapy of other patients are moderate.

4.4 Step 4: result analysis
For patients with type A, the original treatment can be maintained. Patients of type B have a general therapeutic effect. They are confident in Chinese acupuncture, but they may not be able to guarantee their devotion to time and energy. The potential of patients with type B is enormous. Acupuncturists can adopt the "task driven method"

when improving the treatment level of patients of type B. The acupuncturist worked out a detailed and reasonable acupuncture program for the short board of patients with type B. Most of the patients with class C treatment lack confidence in the treatment of diseases, and they unconsciously do not believe that acupuncture can really play a substantial role in the treatment. In this regard, acupuncturists should actively improve the cognitive level of patients to help patients establish confidence in the treatment of disease. At the same time, acupuncturists can adopt the combination of acupuncture and medicine to enhance the efficacy.

5. Conclusions
Acupuncture and moxibustion can relieve the pain of patients with lumbar intervertebral disc herniation to improve their motor function and quality of life. This paper discusses the feasibility of cluster analysis to classify the patients with lumbar intervertebral disc herniation. The concrete conclusions are as follows:

(1) The process of using clustering analysis to classify the patients includes four steps of indexes selection, data collection, clustering analysis and result analysis.

(2) The indexes of the treatment effect of Chinese acupuncture for treating lumbar intervertebral disc herniation can be selected as the follows: JOA score I_1, McGill pain score I_2, Inflammatory factors I_3 and LFR I_4.

(3) We can use cluster analysis to classify patients into the categories A, B and C. For Class A patients, acupuncturists should follow the previous acupuncture methods; for Class B patients, acupuncturists should adopt a project-based approach, focusing on the specific acupuncture points; for Class C patients, acupuncturists should greatly enhance the cognitive level of patients and use a variety of treatment methods to enhance efficacy comprehensively.

References
[1] Wang Y, Liu Z, Wu Y, et al. Efficacy of acupuncture is noninferior to nicotine replacement therapy for tobacco cessation: results of a prospective, randomized, active-controlled open-label trial[J]. Chest, 2018, 153(3): 680-688.

[2] Tao W W, Jiang H, Tao X M, et al. Effects of acupuncture, tuina, tai chi, Qigong, and traditional Chinese medicine five-element music therapy on symptom management and quality of life for cancer patients: a meta-analysis[J]. Journal of pain and symptom management, 2016, 51(4): 728-747.

[3] Liao J, Shen P. Observations on the Efficacy of Mind-regulating and Kidney-reinforcing Acupuncture in Treating Attention Deficit Hyperactivity Disorder[J]. Shanghai Journal of Acupuncture and Moxibustion, 2017, 36(1): 30-33.

[4] Wang C, Liu B, Liu Y, et al. Analysis on the concepts related to adverse events and adverse reactions of acupuncture[J]. Zhongguo zhen jiu= Chinese acupuncture & moxibustion, 2018, 38(1): 87-90.

[5] Qian X, Zhou X, You Y, et al. Traditional chinese acupuncture for poststroke depression: A single-blind double-simulated randomized controlled trial[J]. The Journal of Alternative and Complementary Medicine, 2015, 21(12): 748-753.

[6] Du Y. Essential characteristics and clinical treatment regularity of acupuncture therapy[J]. Zhongguo zhen jiu= Chinese acupuncture & moxibustion, 2018, 38(6): 650-654.

[7] Zhou L, Xia Y, Ma X, et al. Effects of" menstrual cycle-based acupuncture therapy" on IVF-ET in patients with decline in ovarian reserve[J]. Zhongguo zhen jiu= Chinese acupuncture & moxibustion, 2016, 36(1): 25-28.

[8] Wang J, Pei J, Cui X, et al. Interactive dynamic scalp acupuncture combined with occupational therapy for upper limb motor impairment in stroke: A randomized controlled trial[J]. Zhongguo zhen jiu= Chinese acupuncture & moxibustion, 2015, 35(10): 983-989.

[9] Chang C C, Chen T L, Chiu H E, et al. Outcomes after stroke in patients receiving adjuvant therapy with traditional Chinese medicine: A nationwide matched interventional cohort study[J]. Journal of ethnopharmacology, 2016, 177: 46-52.

[10] Liu Y Z, Zhang B H, Luo Y, et al. Curative Effect of YU Yun Pulse-feeling-based Acupuncture Therapy for Treatment of Middle-late Liver Cancer[J]. Journal of Guangzhou University of Traditional Chinese Medicine, 2018, 35(1): 66-69.

[11] Shi Y, Quan R, Li C, et al. The study of traditional Chinese medical elongated-needle therapy promoting neurological recovery mechanism after spinal cord injury in rats[J]. Journal of ethnopharmacology, 2016, 187: 28-41.

[12] He J, Zhang X, Qu Y, et al. Effect of combined manual acupuncture and massage on body weight and body mass index reduction in obese and overweight women: a randomized, short-term clinical trial[J]. Journal of acupuncture and meridian studies, 2015, 8(2): 61-65.

Prediction of Alzheimer's Disease Based on Bidirectional LSTM

Qiao Pan[1], Shiyu Wang[2] and Junhao Zhang [1*]

[1]Computer science and technology, Donghua University, Shanghai, 201620, China

[2]Computer science and technology, Donghua University, Shanghai, 201620, China

[1*]Computer science and technology, Donghua University, Shanghai, 201620, China

[*]Shiyu Wang e-mail: 1248965945@qq.com

Abstract. Alzheimer's disease (AD) is a common disease in the elderly. It affects human life seriously and is difficult to cure, so if you can predict the occurrence of the disease and the development trend in advance, you can prevent or treat Alzheimer's disease as soon as possible. Mild cognitive impairment (MCI) is a syndrome that occurs in the preclinical phase of Alzheimer's disease (AD). It is a transitional state between normal aging and early AD and may be an early sign of AD. This article uses the basic information of the patient's neuropsychological test scale data, genetic data and tomographic data in first, six and twelve months as input data and bidirectional LSTM plus Attention mechanism as a model to obtain a three-dimensional model. The output of the vector is divided into normal (NL), mild cognitive impairment (MCI) and Alzheimer's disease (AD). The experimental results can predict the development of Alzheimer's disease (AD), and determined the model has a good performance.

1. Introduction

Alzheimer disease (AD), commonly known as senile dementia, is a slow progress and incurable course. It starts with memory impairment, gradually develops to loss of multiple cognitive abilities, changes in personality and behavior, and eventually loses basic bodily functions until death. Although the pathological changes in AD begin to appear very early, their typical clinical symptoms do not appear until later. But when the patient shows clinical symptoms, the condition has missed the best treatment opportunity. Therefore, if you can conduct an in-depth study of MCI, which is an early signal of the onset of AD, based on the patient's early diagnosis, it is hoped that the high-risk population of AD will be discovered and screened, and the best time window for AD treatment will be provided to prevent or delay the occurrence of AD.

This paper uses a bidirectional LSTM plus Attention mechanism as a model. In addition, the patient's basic information, genetic data and three time points of neuropsychology scale were used as input to predict the development trend of the patient's condition, which is normal (NL), mild cognitive impairment (MCI) or Alzheimer's disease (AD).

2. Related work

Medically, clinical/cognitive measures have been established to assess the cognitive status of patients and to refer to these scales as an important criterion for clinical diagnosis of Alzheimer's disease. For example, the mini mental state examination (MMSE) and the Alzheimer's disease assessment scale-cognitive subscale (ADAS-Cog) [1].

Content from this work may be used under the terms of the Creative Commons Attribution 3.0 licence. Any further distribution of this work must maintain attribution to the author(s) and the title of the work, journal citation and DOI.

Published under licence by IOP Publishing Ltd

Duchesne et al. [2] studied the relationship between magnetic resonance image (MRI) and cognitive function changes, and used the component analysis (PCA) to perform the MRI image of the obtained nuclear magnetic resonance image. After dimension reduction, a linear regression model was used to predict the trend of MMSE changes in patients with Alzheimer's disease for one year. Stonnington et al. [3] used the correlation vector regression (RVR) method to study the relationship between MRI scans and related clinical scores in patients with Alzheimer's disease, on two independent medical data sets (Mayo). After analyzing the data in the Clinic and the Alzheimer's Disease Neuroimaging Initiative, a continuous model of predicting the patient's clinical score based on individual MRI scans of Alzheimer's patients was established.

All of the above methods are based on clinical data at a time point to predict clinical scores related to the condition of Alzheimer's disease. To improve predictive performance, Zhou et al. [4] proposed a simultaneous use of multiple time points.

During the collection process, there are often cases where some entries are missing, resulting in incomplete patient records. To solve these problems, Xiang et al. [5] proposed a unified two-layer efficient optimization model based on complete multi-source data. Another method is to use the multi-core learning (MKL) method [6] for data fusion. The disadvantage of this method is that only the source data layer is analyzed, without considering the feature layer and the data source layer, the sub-optimal solution is produced when the source is high-dimensional data.

In view of the above problems, this paper proposes a method to predict the development of AD by using bidirectional LSTM plus Attention mechanism as a model. Compared with the traditional method, it has the following advantages: firstly, this paper starts from the acquired neuropsychological data, analyzes the time availability of the data, and refers to the time series problem in deep learning to apply the key attribute of time to the model. This reduces the performance problems associated with making predictions at a single point in time. Secondly, this paper also makes use of the special methods in deep learning to attach weights to different attributes (chapter 3), so that the attribute values have their respective weights.

3. Model

This section introduces a model of bi-directional LSTM attentional mechanism used to predict the patient's condition. Figure 1 shows the overall structure of the model. After the data is inputted, it is processed by dropout and transmitted to the hidden layer. In the hidden layer, the data processed by BI-LSTM is divided into two parts. One part is used for attention calculation of attention_vec and the other part is used directly for standby in attention_mul. The two parts are calculated into a unified matrix and output through the output layer.

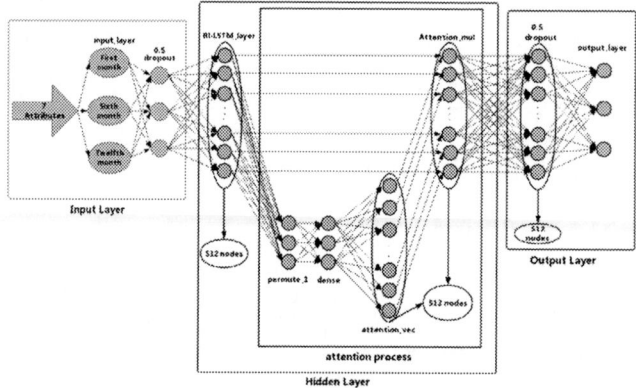

Figure 1. Structure diagram.

3.1. Bidirectional long-term and short-term memory (BI-LSTM)

RNN has shown extraordinary ability to handle sequence problems. However, RNN will eventually cause the gradient disappearance. In order to solve this problem, scholars proposed RNN variant structure based on gate control mechanism, namely LSTM. [7]LSTM added three gates on the basis of RNN: input gate, forget gate, output gate, and a core memory module (cell).

Figure 2 shows the operation mode of the three gates. In the formula, w and h are weight matrix and a number of 0-1, indicating whether the t-th cell needs to be forgotten, 1 means complete retention, 0 means complete abandonment, and b is the value of paranoia.

The input gate is used to control how much of the current input can be transferred to the core memory module. The mathematical formula is as (1). The output gate controls how much of the current core memory module can be output to the hidden node (2). The Forgotten Gate is used to properly forget some things to improve the learning speed of the model when the context is low (3).

$$i_t = \sigma(w_{xi}x_t + w_{hi}h_{t-1} + w_{ci}c_{t-1} + b_i) \tag{1}$$

$$o_t = \sigma(w_{xo}x_t + w_{ho}h_{t-1} + w_{co}c_t + b_o) \tag{2}$$

$$f_t = \sigma(w_{xf}x_t + w_{hg}h_{t-1} + w_{cf}c_t + b_f) \tag{3}$$

The basic idea of Bi-LSTM is to propose two forward and backward LSTMs for each training sequence, and both are connected to the output layer. This structure provides complete past and future context information for each point in the input sequence of the output layer.

For the hidden layer of bi-lstm, forward calculation is the same as backward LSTM. But for the two hidden layers, the input sequence goes in the opposite direction. The output layer will wait until both hidden layers have passed before the output information is updated. Figure 3 shows the standard form of bidirectional LSTM. X_i is the input at different times and Y_i is the corresponding output, A is the forward LSTM cell, A′ is the reverse LSTM cell, s_0 to s_1 is the change of the forward information transfer matrix, s_0' to s_1' is the change of the reverse information transfer matrix.

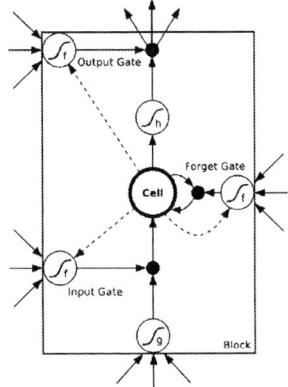

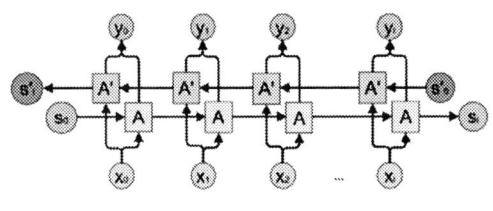

Figure 2. LSTM structure.   Figure 3. Bidirectional RNN.

3.2. Attention

Currently, the Encode-Decode model is very popular with the public because it has achieved better results in many areas than other models. The usual practice is to encode an input sentence into a fixed-size vector and then as the initial state of the decode. Such a fixed encoding method causes some long sentences to lose features, and may also allow some shorter sentences to be mixed noise. In response to this problem, the attention mechanism is introduced. At this time, the decoder makes the input at each moment different according to the time. Attention mechanism also has its shortcomings, which leads to an increase in the computational complexity of the model and a longer training time for the model.

4. Experiment

The data in this paper is from the Alzheimer's Disease Neuroimaging Initiative (ADNI public data set), which mainly uses the data of ADNImerge.

In order to improve the practicability of the model, model uses the random method to separate test data and training data, and allocate them according to the proportion of the total ratio of 8:2. The dataset has 1405 patients and 4218 data. The number of patients with different symptoms is shown in figure 4. Figure 4 shows the number of patients with different symptoms, with 0 indicating normal, 1 indicating MCI and 2 indicating AD. There were 794 men and 612 women in the data. Age is also a closely related property of Alzheimer's disease.

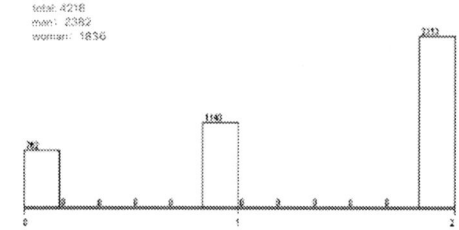

Figure 4. Number of patients with different symptoms.

4.1. Subsections

Java package of Weka and data of 25 dimensions was used as input. Weka used patient symptoms as labels, and set poolsize in CFS and lookuolocksize in BF to 1, searchtermination to 5, and then ran the code. The algorithm screened 7 common AD diagnostic attributes as shown in table 1.

Table 1. Attributes filtered by the algorithm.

Serial number	ATTRIBUTES	Serial number	ATTRIBUTES
1	AGE	5	RAVLT.forgetting
2	PTGENDER	6	FAQ
3	FDG	7	CDRSB.bl
4	MMSE		

4.2. Baseline setting

In order to show the excellent performance of model, different experiments were used for comparison:

- A simple RNN model based on time series uses the attribute values mentioned above and combines the data of three time nodes to predict the last patient condition.
- The LSTM model, which is excellent in timing, also uses the same input data and functions as BI-LSTM model.
- The GRU model simplified by LSTM has the same input data and the same model function.

4.3. Experimental result

For different models, the parameters of the model are set to the same, the epoch (that is, the number of iterations of data per round when training the neural network) was set to 20, the dropout (training percentage) was set to 0.3, and the batch_size (that is, the amount of data input per round by the neural network) was set to 16.

Table 2 shows the training data set accuracy, test data set accuracy, and the model score provided by keras.

Table 2. Model accuracy and score (keras).

Model	Train-ACC	Test-ACC	Score
Simple RNN	0.45	0.50	1.03

GRU	0.46	0.51	1.06
LSTM	0.51	0.56	1.03
BL-LSTM attention	0.84	0.87	0.41

The results show that the accuracy of the proposed model has been greatly improved. In the later stage, after further adjusting the parameters and changing the initialization mode inside the model, the accuracy was improved to over 90%.

While the accuracy of the model has been greatly improved, each round of epoch changed from 3 seconds to 15 seconds in the model training. Attention mechanism actually increased the model's training time.

5. Conclusion

In this paper, the deep learning method was used to predict the development of the patient's condition from the neuroimaging scale, and at the same time the attention in BI-LSTM was added to let the attributes get their own weight. The patient's condition was successfully judged without using MRI.

Acknowledgments

This work was supported by the Science and Technology Development Foundation of Shanghai(16JC1400802, 16JC1400803, 18511102703),the Special Fund of Shanghai Municipal Commission of Economy and Informatization (RX-RJJC-08-16-0483,2017-RGZN-01004, XX-XXFZ-02-18-2666，XX-XXFZ-01-18-2604).

References

[1] Petrella, J. R., Coleman, R. E., & Doraiswamy, P. M. (2003). Neuroimaging and early diagnosis of alzheimer disease: a look to the future. Radiology, 226(2), 315-36.

[2] Duchesne, S., Caroli, A., Geroldi, C., Collins, D. L., & Frisoni, G. B. (2009). Relating one-year cognitive change in mild cognitive impairment to baseline mri features. Neuroimage, 47(4), 1363-1370.

[3] Stonnington, C. M., Chu, C., Klöppel, S., Jr, J. C., Ashburner, J., & Frackowiak, R. S. (2010). Predicting clinical scores from magnetic resonance scans in Alzheimer's disease. Neuroimage, 51(4), 1405-1413.

[4] Zhou, J., Yuan, L., Liu, J., & Ye, J. (2011). A multi-task learning formulation for predicting disease progression. ACM SIGKDD International Conference on Knowledge Discovery and Data Mining, San Diego, Ca, Usa, August (Vol.3, pp.814-822). DBLP.

[5] Xiang, S., Yuan, L., Fan, W., Wang, Y., Thompson, P. M., & Ye, J. (2013). Multi-source learning with block-wise missing data for Alzheimer's disease prediction. Paper presented at the Proceedings of the 19th ACM SIGKDD international conference on Knowledge discovery and data mining, Chicago, Illinois, USA.

[6] Lanckriet GR, De Bie T, Cristianini N, Jordan MI, & Noble WS. (2004). A statistical framework for genomic data fusion. Bioinformatics, 20(16), 2626-35.

[7] Bahdanau, D., Cho, K., & Bengio, Y. (2014). Neural machine translation by jointly learning to align and translate. Computer Science.

ISPECE   IOP Publishing

A Survey of the Key Technology of Software Vulnerability Mining

Dansheng Lin[1], Xiaoquan Wu[1], Qinqin Wu[1*], Ye Liu[1], Jijun Zeng[1], Jinjun Tan[2]

[1]Information Center, Guangdong Power Grid Company Limited, Guangzhou, Guangdong,China

[2]Guangdong Information Technology Security Evaluation Center, Guangzhou, Guangdong, China

[*]Corresponding author's e-mail: 149679600@qq.com

Abstract. In the current cyberspace, vulnerability has attracted the widespread attention. This paper briefly introduces the typical software vulnerability mining techniques. Based on the existing research work, this paper puts forward the whole idea of the research on the vulnerability mining technology of software in the future, and then makes some research on some key points and key technologies respectively. In this paper, a brief study of the software vulnerability mining technology is given, which is helpful to carry out the follow-up research work.

1. Introduction

Software vulnerabilities manifest in many forms, including race conditions, buffer overflows, integer overflows, dangling pointers, poor input validation (e.g., SQL injection, cross-site scripting), information leakage, violation of least privilege and other access control errors, use of weak random numbers in cryptography, protocol errors, and insufficient authentication[1-2]. Discovering vulnerabilities has become an extremely widespread activity and new threats are constantly being faced. Vulnerabilities in software represent a serious risk for information systems. According to the National Vulnerability Database (NVD), which maintains records of all acknowledged vulnerabilities of software products on the market, more than 90,000 vulnerabilities have been discovered since 1997. The number of vulnerabilities is increasing rapidly due to the development of new hacking techniques. Between 2010 and 2015, around 80,000 vulnerabilities were newly registered in the major database known as the CVE (Common Vulnerability Enumeration)[3].

In recent years, the scope of security threats has also been expanded. As cataloged in the National Vulnerability Database, prominent recent examples of serious software vulnerabilities include Heartbleed (CVE-2014-016), Shellshock (CVE-2014-6271), the glibc buffer overflow (CVE-2015-7547), VENOM (CVE-2015-3456), and Microsoft Malware Protection Engine vulnerability (CVE-2017-0290). According to the website CVE Details, there have been 937 vulnerabilities between January and August 2017 to date. Ten percent of those vulnerabilities were in the "critical" (highest severity) category with assigned Common Vulnerability Scoring System (CVSS) scores of 9 or 10.

There are many kinds of vulnerability mining techniques. It is very difficult to complete the analysis work by using only one vulnerability mining technique. Generally, several vulnerability mining techniques are optimized to find a balance between efficiency and quality

Content from this work may be used under the terms of the Creative Commons Attribution 3.0 licence. Any further distribution of this work must maintain attribution to the author(s) and the title of the work, journal citation and DOI.
Published under licence by IOP Publishing Ltd

2. Vulnerability Mining Techniques

2.1. Manual Analysis

Manual analysis is a grey box analysis technique. Aiming at the target program being analyzed, special input conditions are manually constructed, output and target state changes are observed, and vulnerability analysis techniques are obtained. Input includes valid and invalid inputs, including normal output and abnormal output. Abnormal output is the premise of a loophole or a loophole in the target program. The change of abnormal target state is also a harbinger of vulnerabilities and the direction of deep excavation. Manual analysis is highly dependent on the experience and skills of the analyst. Manual analysis is mostly used for target programs with human-computer interaction interface. Manual analysis is mostly used in Web vulnerability mining [4-5].

2.2. Fuzzing

Fuzzing technology[6-7] is an automatic software testing technology based on defect injection. It uses black box analysis technology, uses a large number of semi-effective data as the input of the application program, and uses the abnormality of the program as a sign to discover the possible security vulnerabilities in the application program. Semi-valid data means that the necessary identification part and most of the data of the target program under test are valid, and the intentionally constructed data part is invalid. When the application processes the data, it may have errors, which may lead to the collapse of the application or trigger the corresponding security vulnerabilities.

According to the characteristics of the target, Fuzzing can be divided into three categories: 1) Dynamic Web page Fuzzing, for ASP, PHP, Java, Perl and other web page programs, also includes B/S architecture applications built using this technology, the typical application software is HTTP Fuzz; 2) File format Fuzzing, for various document formats, typical application software is PDF Fuzz.

The construction method of Fuzzer software input is similar to that of black-box testing software. Boundary values, strings, additional strings of file head and end can be used as basic construction conditions. Fuzzer software can be used to detect a variety of security vulnerabilities, including buffer overflow vulnerabilities, integer overflow vulnerabilities, formatted string and special character vulnerabilities, competition conditions and deadlock vulnerabilities, SQL injection, cross-site scripting, RPC vulnerability attacks, file system attacks, information leakage, etc.

Compared with other technologies, Fuzzing technology has the advantages of simple thought, easy to understand, easy to find vulnerabilities, easy to reproduce vulnerabilities, and no false alarm. At the same time, it also has all the shortcomings of black box analysis, and has some problems such as non-universal, long construction test cycle and so on. Common Fuzzer software includes SPIKE Proxy, Peach Fuzzer Framework, HTTP Fuzzer of Acunetix Web Vulnerability Scanner, OWASP JBroFuzz, WebScarab, etc.

2.3. Patch Comparison

Patch matching techniques[8-9] is mainly used by hackers or competitors to find bugs that have been corrected but not yet disclosed by software publishers. It is a common technical means used by hackers before exploiting bugs. Security announcements or patch release instructions generally do not specify the exact location and cause of vulnerabilities, and it is difficult for hackers to exploit vulnerabilities only according to the announcement. Hackers can determine the location of vulnerabilities by comparing binary files before and after patching, and then combine with other vulnerability mining technologies to understand the details of vulnerabilities. Finally, they can get the attack code of vulnerability exploitation.

Simple comparison methods include binary byte and string comparison, and the comparison after reverse engineering of the target program. The first method is suitable for the comparison of minor changes before and after patching, and is commonly used for the analysis of vulnerabilities caused by string changes and boundary value changes. The second method is applicable to the analysis of vulnerabilities caused by the change of function parameters according to decompilation. Neither of these two methods is suitable for document modification. Complicated comparison methods include Tobb Sabin's graphical comparison based on instruction similarity and Halvar Flake's structured binary comparison. Some unstructured changes in files, such as changes in buffer size, can be found and displayed graphically.

The commonly used patch comparison tools are Beyond Compare, IDACompare, Binary Differing Suite (EBDS), BinDiff, NIPC Binary Differ (NBD). In addition, a large number of advanced text editing tools have similar functions, such as Ultra Edit, HexEdit and so on. These patch comparison tools are based on string comparison or binary comparison techniques.

2.4. Static Analysis

Static analysis technology[10-11] is a typical white-box analysis technology, which analyses and detects the source program of the target being analyzed and finds the security vulnerabilities or hidden dangers in the program. Its methods mainly include static string search and context search. Static analysis process is mainly to find incorrect function calls and return states, especially function calls that may not have boundary checks or incorrect boundary checks, which may cause buffer overflow functions, external call functions, shared memory functions and function pointers.

For open source programs, security flaws in programs can be found by detecting file structures, naming rules, functions and stack pointers that do not conform to security rules. When the target is not accompanied by the source program, it is necessary to reverse-engineer the program, obtain reverse-engineering code similar to the source code, and then search. Using a method similar to the source code, vulnerabilities can also be found in programs. This kind of static analysis method is called disassembly scanning. Because of the use of the underlying assembly language for vulnerability analysis, in theory, all computer vulnerabilities can be found. For programs that do not open source code, it is often the most effective way to find security vulnerabilities. However, this method also has great limitations. The expanding feature library or dictionary will result in large result set and high false alarm rate. At the same time, this method focuses on the analysis of the features of the code, but does not care about the function of the program, and will not have the analysis and inspection of the function and program structure.

2.5. Dynamic Analysis

Dynamic analysis technology[12-13] originates from software debugging technology. It uses debugger as a dynamic analysis tool. But unlike software debugging technology, it often deals with the analyzed program without source code or the analyzed program which has been reverse-engineered.

Dynamic analysis needs to run the target program in the debugger, and discover the vulnerabilities by observing the running state of the program, memory usage and register values during execution. The general analysis process is divided into code flow analysis and data flow analysis. Code flow analysis mainly traces the target program code flow dynamically by setting breakpoints to detect defective function calls and their parameters. Data flow analysis is to trigger potential errors by constructing special data.

In the process of dynamic analysis, dynamic code replacement technology can be used to destroy the process of program operation, replace function entrance and function parameters, which is equivalent to constructing semi-effective data, so as to find the hidden defects in the system. Common dynamic analysis tools include SoftIce, OllyDbg, WinDbg and so on.

3. Typical Technique Applications

3.1. Acunetix Web Vulnerability Scanner

We can use Acunetix Web Vulnerability Scanner to mine vulnerabilities, this software provides some predefined parameters Library of Fuzz operation, which can be easily used by beginners and analysts. The process is as follows:

1) Define the HTTP request (Request), that is, define the web page URL needed to access.

2) Define operation parameters (Add generator), that is, define string expressions that may cause vulnerabilities, such as: $password, $passwd, $token;

3) Insert in request, which binds defined parameters to a search strategy;

4) Define Fuzzer Filters, which binds parameters to HTTP requests;

5) Scan (Start);

6) Waiting for software to return matching items, which are possible vulnerabilities.

Through above steps, a possible loophole in a web page is discovered.

3.2. Patch Comparison Technology Application

In October 23, 2008, Microsoft released the patch of MS08-067, which was listed as serious. The security update solves a secret report loophole in the server service. If a user receives a special RPC request on the affected system, the vulnerability may allow remote code execution. The process of vulnerability mining using patch comparison technology is as following: First, keep an original file, then install a new patch, extract the same and new files, you can use the software to compare. After comparison, 3 functions that have been modified are found. The comparison software lists three function names, 0.25, 0.67 and 0.94 respectively, and lists the similarities before and after patching. By comparing the results, we can construct parameters and observe the behavior before and after patching. Finally, we find that two of the three functions are directly related to vulnerabilities.

4. Conclusion

Vulnerability mining technology is derived from software testing theory and software development debugging technology, which can greatly improve the security of software. Third-party organizations and technology enthusiasts in the field of network security also use this technology to find various software vulnerabilities and release them to the public in time, contributing to improving the overall level of information security, but vulnerability mining is also a double-edged sword, which has become the mainstream technology for hackers to crack software. The development of vulnerability mining technology has broad prospects. With more and more attention to information security, software development technology is more and more advanced, and new analysis methods will emerge.

Acknowledgments

This work was financially supported by GDKJXM20162130(037800KK52160003).

References

[1] Khalili A, Sami A, Azimi M, et al. Employing secure coding practices into industrial applications: a case study[J]. Empirical Software Engineering, 2014, 21(1):1-13.

[2] Pham N H, Nguyen T T, Nguyen H A, et al. Detection of recurring software vulnerabilities[C]// Ieee/acm International Conference on Automated Software Engineering. ACM, 2010:447-456.

[3] NVD, https://nvd.nist.gov/.

[4] Zhao M, Grossklags J, Liu P. An Empirical Study of Web Vulnerability Discovery Ecosystems[C]// ACM Sigsac Conference on Computer and Communications Security. ACM, 2015:1105-1117.

[5] Holm H, Ekstedt M, Sommestad T. Effort Estimates on Web Application Vulnerability Discovery[C]// Hawaii International Conference on System Sciences. IEEE, 2013:5029-5038.

[6] Takanen A, Demott J, Miller C. Fuzzing for software security testing and quality assurance[M]. Artech House, 2008.

[7] Zhi-Yong W U, Wang H C, Sun L C, et al. Survey on Fuzzing[J]. Application Research of Computers, 2010, 27(3):829-832.

[8] Guo H, Wang Y Y, Pan Z L, et al. Research on Detecting Windows Vulnerabilities Based on Security Patch Comparison[C]// International Conference on Instrumentation & Measurement. IEEE, 2016:366-369.

[9] Hua X L, Long S Y, Feng G, et al. Security patch comparison techniques based on graph isomorphism theory[J]. Journal of Computer Applications, 2006, 26(7):1623-1622.

[10] Xia Y M. Security Vulnerability Detection Study Based on Static Analysis[J]. Computer Science, 2006, 33(10):279-282.

[11] Li P, Cui B. A comparative study on software vulnerability static analysis techniques and tools[C]// IEEE International Conference on Information Theory and Information Security. IEEE, 2011:521-524.

[12] Tang H, Huang S, Li Y, et al. Dynamic taint analysis for vulnerability exploits detection[C]// International Conference on Computer Engineering and Technology. IEEE, 2010:V2-215 - V2-218.

[13] Kim S, Kim R Y C, Park Y B. Software Vulnerability Detection Methodology Combined with Static and Dynamic Analysis[J]. Wireless Personal Communications, 2016, 89(3):777-793.

Research on Security Evaluation of Government Cloud Platform Based on Fuzzy Analytic Hierarchy Process

Jing Chen, Feng Zhang*

Department of information engineering and art design, Shandong Labor Vocational And Technical Collage, Jinan, Shandong, 250000, China

*Corresponding author's e-mail:zhangfengx@sdlvtc.cn

Abstract. E-government cloud, combined with the emerging technology of cloud computing, streamlines, optimizes and integrates government management and service functions. It effectively solves the shortcomings of traditional e-government platforms, and opens up a new path for China's e-government construction. The application of cloud computing has greatly promoted the efficiency of the government system, but the security of cloud computing has seriously affected the development of e-government cloud. This paper analyzes the security problems faced by e-government cloud, summarizes the coping strategies, and proposes a method for e-government cloud platform security assessment, which provides reference for the construction of e-government cloud.

1. Introduction

Electronic cloud government is a general term for government organizations, and it includes new public service models and governance models in the cloud computing environment. In the traditional e-government information system, the platforms are independent of each other, resources cannot be shared, and utilization is low, thus forming an "information island", which causes great waste of resources and financial resources. Through server virtualization, network virtualization, and storage virtualization, the cloud computing technology realizes the flexible architecture of resources provides on-demand services, improves resource utilization, and solves the problem of low utilization and information of traditional e-government Platforms Island and other issues [1].

E-government cloud also faces many security risks brought by new cloud computing technologies. The network of government work itself and the security of information networks are a contradiction [2]. "Government" involves all aspects of the national economy and the people's livelihood. There are many state secrets, and the information content is highly sensitive and confidential. The network environment has become a "hotbed" for many unsafe factors because of its unique openness and public nature. Therefore, information security has become a core issue to ensure the smooth implementation of e-government.

2. Government Cloud Platform Security Assessment

The security risks of the e-government cloud platform are also of great concern to the government. The hard part of the assessment of security risk is the uncertainty and multi-cause. Government cloud security is the fundamental guarantee for the steady operation of e-government under the cloud environment. The cloud government platform is evaluated for security capabilities to ensure the safe operation of the government cloud system. Taking one government department in Shandong as an example, the fuzzy analytic hierarchy process is used to explore the security assessment process of cloud

Content from this work may be used under the terms of the Creative Commons Attribution 3.0 licence. Any further distribution of this work must maintain attribution to the author(s) and the title of the work, journal citation and DOI.

Published under licence by IOP Publishing Ltd

computing platforms.

2.1 Fuzzy comprehensive evaluation method

Fuzzy comprehensive evaluation is a combination of qualitative and quantitative methods [3]. Use the principle of fuzzy relational synthesis to quantify some factors that are not easily quantified or whose boundaries are unclear. In a more complex system, the factors affecting the result are often multi-layered, thus forming a discriminate tree structure. The main steps are as follows:

1. Establish a fuzzy comprehensive evaluation index system

Before the assessment work begins, it is necessary to construct a fuzzy comprehensive safety assessment indicator system, which is the basis and key to the overall assessment. Constructing a comprehensive evaluation indicator system is to find out the objective factors that affect safety.

2. Determine indicator weights by analytic hierarchy process

This paper mainly uses the analytic hierarchy process to determine the weight of the evaluation indicators. Analytic hierarchy process can effectively solve the problem between qualitative and quantitative indicators. This method first determines the fuzzy evaluation index system, establishes a fuzzy evaluation matrix through the system, and then gives a quantitative representation of the relative importance of each level of indicators according to the fuzzy judgment, finally uses the mathematical method to solve the weight value.

When constructing the pair wise comparison judgment matrix, we compare the importance of the two indicators to the evaluation of the superior indicators, and assign them accordingly. Usually we use the form of 1 to 9 and the reciprocal, as shown in Table 1.

Table 1. Definition of importance

1	with the same degree of importance
3	one is slightly more important than the other
5	One is more important than the other
7	One is much more important than the other
9	One is more and more important than the other
2 .4. 6.8	The importance is between the median of the two adjacent judgments above
reciprocal	If the ratio of the factor i to the importance of the factor j is a_{ij}, Then the ratio of factor j to factor i is 1 / a_{ij}.

According to the above method of comparing the index factors, we can draw the evaluation matrix A of each level.

Step1 Summing. Summing each column of the evaluation matrix A;

Step2 Building a standardized matrix. A standardized matrix $\overline{A}$ is obtained by dividing each element by the sum of its corresponding columns;

Step3 Calculating weight vector. For the normalized matrix A, find the average of each row to get a vector;

$$A = \begin{bmatrix} a_{11} & a_{12} & \cdots & a_{1n} \\ a_{21} & a_{22} & \cdots & a_{2n} \\ \vdots & \vdots & & \vdots \\ a_{n1} & a_{n1} & \cdots & a_{nn} \end{bmatrix} \qquad \overline{A} = \begin{bmatrix} \overline{a_{11}} & \overline{a_{12}} & \cdots & \overline{a_{1n}} \\ \overline{a_{21}} & \overline{a_{22}} & \cdots & \overline{a_{2n}} \\ \vdots & \vdots & & \vdots \\ \overline{a_{n1}} & \overline{a_{n1}} & \cdots & \overline{a_{nn}} \end{bmatrix}$$

Step4 Calculating the maximum eigenvalue of weight. The largest eigenvalue of the discriminated matrix A obtained is calculated according to the weight vector $\lambda_{max} = \sum_{i=1}^{n} \frac{(AW_i)}{n(W_i)}$.

Step5 Testing the judgment matrix consistency. Based on the maximum eigenvalue of the matrix, the consistency index is $CI = \frac{\lambda_{max} - n}{n-1}$ If = 0, the discriminated matrix is identical; and the greater the CI, the worse the consistency. Discriminate matrices of order 1 or 2 are usually consistent. For the discriminated matrix with higher order, in order to make it consistent, it is necessary to compare the consistency index CI with the average random consistency index RI , as $CR = \frac{CI}{RI}$. The values of RI is as shown in Table 2.

Table 2. Average random consistency indicator RI

Order	1	2	3	4	5	6	7	8	9
RI	0	0	0. 58	0.94	1.12	1.24	1.32	1.41	1.45

If $CR = 0.1$ or $CR < 0.1$, it is considered that the consistency is better. But if $CR > 0.1$, it means that the discriminated matrix needs to be re-adjusted to redistribute the weight coefficients. After satisfying the judgment matrix consistency, the calculated W is regarded as the weight vector of the hierarchy.

3. Construct an evaluation matrix and seek security membership

The comment set is a collection of elements of the various evaluation results that the evaluator may make to the evaluated object. Generally, five rating levels are indicated in the safety assessment, as a commentary domain $V = [V1,\ V2,...,\ V5]$. In this paper, {much better, better, good, average, bad} are used. According to the expert survey to judge the score results and the corresponding membership function, it can be determined that each factor index is rated as the membership degree of each V_i, and a fuzzy evaluation

matrix R_k is established: $R_k = \begin{bmatrix} r_{11} & r_{12} & \cdots & r_{1n} \\ r_{21} & r_{22} & \cdots & r_{2n} \\ \vdots & \vdots & & \vdots \\ r_{m1} & r_{m1} & \cdots & r_{mn} \end{bmatrix}$.

In this fuzzy evaluation matrix, r_{ij} represents V_j in the comment set V of the indicators u_i in the indicator domain U. This paper uses the method of percentile scoring to find each membership degree. For the five reviews in the comment sct, thc five scores of 90, 80, 70, 60, and 50 are the boundaries. According to the calculation of the following membership function formula, each membership degree can be obtained, and the fuzzy evaluation matrix is obtained.

Substitute the expert's score into 5 membership functions so that each indicator will get 5 different memberships. They respectively represent the membership of the five security levels, and each indicator will get a 1 × 5 security membership matrix.

4. Evaluate matrix and weight synthesis

According to the formula $B_k = W_k \bigcirc R_k$, the comprehensive evaluation result $B = [B_1 B_2 \cdots B_m]^T$ of each layer of evaluation indicators is calculated. If $\sum B_1 \neq 1$, you need to normalize it. Where: "$\bigcirc$" represents the synthesis operator.

$$u_{v1}(u_i) = \begin{cases} 1 & u_i \geq 90 \\ \frac{(u_i - 80)}{10} & 80 \leq u_i \leq 90 \\ 0 & u_i < 80 \end{cases} \quad (1)$$

$$u_{v2}(u_i) = \begin{cases} \frac{(u_i - 70)}{10} & 70 \leq u_i < 80 \\ \frac{(90 - u_i)}{10} & 80 \leq u_i < 90 \\ 0 & u_i < 70 \ \text{or} \ u_i \geq 90 \end{cases} \quad (2)$$

$$u_{v3}(u_i) = \begin{cases} \frac{(u_i - 60)}{10} & 60 \leq u_i < 70 \\ \frac{(80 - u_i)}{10} & 70 \leq u_i < 80 \\ 0 & u_i < 60 \ \text{or} \ u_i \geq 80 \end{cases} \quad (3)$$

$$u_{v4}(u_i) = \begin{cases} \frac{(u_i - 50)}{10} & 50 \leq u_i < 60 \\ \frac{(70 - u_i)}{10} & 60 \leq u_i < 70 \\ 0 & u_i < 50 \ \text{or} \ u_i \geq 70 \end{cases} \quad (4)$$

$$u_{v5}(u_i) = \begin{cases} 0 & u_i \geq 60 \\ \frac{(60 - u_i)}{10} & 50 \leq u_i \leq 60 \\ 1 & u_i < 50 \end{cases} \quad (5)$$

5. Multi-level fuzzy comprehensive evaluation

In this paper, the indicator system hierarchy is divided into three layers. In the multi-level fuzzy comprehensive evaluation, the evaluation result of the next level $B = [B_1 B_2 \cdots B_m]^T$ is taken as the evaluation matrix of the upper level. Then synthesize it with the weight vector W obtained at this level, and obtain the overall comprehensive evaluation result as $S = W \bigcirc B$. Finally, judge the results of the comprehensive evaluation based on the principle of maximum membership.

2.2 The Security Assessment of Government Cloud Platform
Through in-depth investigation and consultation with experts in related fields, the evaluation matrix is first established by the analytic hierarchy process. The weights of the security assessment indicators at all levels of the government cloud system are obtained by comparison, and any evaluation object of the protocol layer is randomly selected. The specific process is as follows:

1. Establish a judging matrix to determine the weight of the indicator

The evaluation object is selected as five evaluation indicators under hardware security [4]: sufficient Server backup, regularly replaced hardware, good physical isolation, secure storage media and prohibited illegal physical access. Compare these five factor indicators in pairs and compare them with a scale of 1 to 9 scales. The results are shown in Table 3.

After comparing the judgment matrices in the table, we obtain the comparison results and construct the discriminated matrix A. By normalizing the matrix A according to the canonical column averaging method, a normalized matrix by column can be obtained.

Table 3. Comparison results of indicators at the indicator level

hardware security	sufficient Server backup	regularly replaced hardware	good physical isolation	secure storage media	prohibited illegal physical access
sufficient Server backup	1	2	3	3	1 /3
regularly replaced hardware	1 /2	1	2	2	1 /4
good physical isolation	1 /3	1 /2	1	1 /2	1 /5
secure storage media	1 /3	1 /2	2	1	1 /3
prohibited illegal physical access	3	4	5	3	1

Then calculate the eigenvector of matrix A: $W = [0.23\ 0.14\ 0.07\ 0.09\ 0.45]$.

$$
AW = \begin{bmatrix}
1 & 2 & 3 & 3 & \frac{1}{3} \\
\frac{1}{2} & 1 & 2 & 2 & \frac{1}{4} \\
\frac{1}{3} & \frac{1}{2} & 1 & \frac{1}{2} & \frac{1}{5} \\
\frac{1}{3} & \frac{1}{2} & 2 & 1 & \frac{1}{3} \\
3 & 4 & 5 & 3 & 1
\end{bmatrix}
\begin{bmatrix}
0.230 \\ 0.142 \\ 0.068 \\ 0.088 \\ 0.450
\end{bmatrix}
=
\begin{bmatrix}
1.13 \\ 0.68 \\ 0.35 \\ 0.52 \\ 2.31
\end{bmatrix}
$$

After finding the feature vector, verify the consistency of the judgment matrix of this indicator layer. Calculate the maximum eigenvalue according to the formula of the largest eigenvalue:

$$
\lambda_{\max} = \sum_{i=1}^{n} \frac{(AW_i)}{n(W_i)} = \frac{1}{5}\left(\frac{1.13}{0.23} + \frac{0.68}{0.14} + \frac{0.35}{0.07} + \frac{0.52}{0.09} + \frac{2.31}{0.45}\right) = 5.13
$$

Consistency indicator $CI = \frac{\lambda_{\max} - n}{n-1} = 0.03$, $CR = \frac{CI}{RI} = \frac{0.03}{1.12} = 0.027 < 0.1$, It can be seen that the discriminated matrix satisfies the consistency index, so $W = [0.23\ 0.14\ 0.07\ 0.09\ 0.45]$, that is, the index weight of the indicator layer hardware security .

2. Construct an evaluation matrix and find the security membership of the indicator

The analog experts are also judged by the indicator of the hardware security of the indicator layer. Its first factor is "full server backup" u_1, Assume that the scores given by the ten experts are "89, 85, 93, 88, 85, 90, 85, 86, 87, 88". Then the scores are respectively brought into the membership function, and the factor server backup is fully subordinate to the comment set V1, V2, V3, V4, and V5 are respectively named $rV_1(u_1), rV_2(u_1), rV_3(u_1), rV_4(u_1), rV_5(u_1)$:

$$rV_1(u_1) = \frac{1}{10}\left[\begin{array}{l} u_1(89) + u_1(85) + u_1(93) + u_1(88) + u_1(85) + u_1(90) + u_1(85) + u_1(86) + u_1(87) \\ + u_1(88) \end{array} \right]$$

$$= \frac{1}{10}[0.9 + 0.5 + 1 + 0.8 + 0.5 + 1 + 0.5 + 0.6 + 0.7 + 0.8] = 0.73$$

And as the same reason, $rV_3(u_1)=0$, $rV_4(u_1)=0$, $rV_5(u_1)=0$。

Therefore, the membership r_1 of the factor u_1 is (0. 73, 0. 27, 0, 0, 0). Similarly, the hardware is periodically replaced, the physical isolation is good, the storage medium is safe, and the illegal physical connection is prohibited. The security memberships are (0.75, 0.25, 0, 0, 0), (0.59, 0.41, 0, 0, 0), (0.63, 0.37, 0, 0, 0), (0.70, 0.30, 0, 0, 0).

Comprehensive assessment results:

$$S = W \bigcirc R = [0.23\ 0.14\ 0.07\ 0.09\ 0.45]\begin{bmatrix} 0.73 & 0.27 & 0 & 0 & 0 \\ 0.75 & 0.25 & 0 & 0 & 0 \\ 0.59 & 0.41 & 0 & 0 & 0 \\ 0.63 & 0.37 & 0 & 0 & 0 \\ 0.70 & 0.30 & 0 & 0 & 0 \end{bmatrix} = [0.69\ 0.31\ 0\ 0\ 0]$$

According to the principle of maximum membership degree, the security rating level is 0.69 and it indicates that the hardware security index is very good, thus we complete the evaluation. And this can also be called unit evaluation. Since the index is divided into three layers of tree structure in this study, the hardware security membership degree is saved and the second layer index weight is matrixed to obtain the security membership degree of the second layer indicators, and then we save the index security membership degree of the second layer, and then the second layer index is compared with each other to obtain the weight, and then the membership degree is calculated to obtain the first layer of 1×5 security membership matrix, and the most affiliated comment is selected. Levels are the result of a security assessment of the entire government cloud system, which is a multi-level fuzzy security assessment.

3. Conclusion

This paper analyzes the security problems faced by e-government cloud, summarizes the coping strategies, proposes a method for e-government cloud platform security assessment, and proposes a fuzzy comprehensive evaluation method to apply fuzzy mathematics theory to security assessment, combined with analytic hierarchy process. In the paper we determine the weights, use the multi-level assessment method to conduct a practical assessment of the security issues in the cloud platform, and obtain the security assessment results of the government cloud system.

Acknowledgement

This paper is supported by Shandong Social Science Planning Project, 18CZZJ07, Research on Innovative Government Service under the Vision of "Internet +" ;And Shandong Province Vocational Education Teaching Reform Research Project,2017205,The Construction of Professional Group Curriculum System under the Background of Credit System in Higher Vocational Colleges.

References

[1] Dimitrios Zissis，Dimitrios Lekkas. "Securing e—Government and e—Voting with an open cloud computing architecture",Government Information Quarterly,2011,28(2) .

[2] Aliyun: "General Administration of Customs chooses big data cloud solution".https: //www. aliyun.com/customer/detail/haiguan? Spm.

[3] Xiao Yang, Li Yang, Zichun Yang. Research And Application of Industrial SCDA Security Risk Assessment method based on Fuzzy AHP[J]. Computer Application and Software, 2017, 34(5):54-60.

[4]   Baoguang Chen.Research on the Cloud Evalutation System of Government[J]. Application and Pratice of New Technology, 2017(10):229-230.

Big Data Mining for Investor Sentiment

Tao Cen[1*], Qianqian Chu[1] and Renke He[1]

[1] Business School, Qingdao University, Qingdao, Shandong, 266000, China

*Corresponding author's e-mail: centao007@126.com

Abstract. Investor sentiment is a key factor affecting asset volatility, but it is hard to quantify. Previously investor sentiment is mainly measured from questionnaires and market exchange data. With rapidly developing of big data, the way of knowledge exploration has been changed. Various types of data are produced and reserved every minute, particularly unstructured Internet big data which can reflect investor behaviour and investor sentiment directly. Investor sentiment will be better quantified through big data. In this paper, we review new data sources and analytic methods for quantifying investor sentiment, and discuss the future of big data in behavioral finance.

1. Introduction

Investor sentiment influences asset price, liquidity and volatility [1]. Although it is important to investors decision-making, investor sentiment is hard to measure. Traditionally, indicators of investor sentiment include subjective and objective indicator. Subjective indicator is based on questionnaires. But due to low survey coverage, it has sample representativeness and reliability problems. Also, survey is very cumbersome and takes a long time. Objective indicator is based on stock exchange data, but it tends to be quite simple, and proxies such as turnover rate and trade volume lack micro-foundation. Therefore, new proxies of investor sentiment should be developed. With rapid development of Internet and Mobile Internet, a large amount of investor related big data is generated online every minute. Big data mining provides new methods for tracking investor behaviour and measuring investor sentiment, and is becoming more and more extensive. Internet textual sentiment is a supplement of traditional sentiment. This paper reviews big data sources and methods for measuring investor sentiment, then summarizes processes of disposing chaos big data, and finally points out advantage and challenges of big data mining for investor sentiment.

2. Understanding big data

2.1. Definition of big data

With development of information technology, the big data era is coming [2]. Yet big data has not a clear definition, it refers to big data set that is difficult to store, read, process and analyse by traditional ways. Big data can be well described by 5V, including: (1) Volume, referring to a large amount of data, in unit of PB, EB or even ZB; (2) Variety, involving to various data types, including structured, semi-structured and unstructured data such as HTML, audio, video, pictures, GPS information and so on; (3) Value, meaning that big data is valuable although relatively low value density; (4) Velocity, relating to high growth speed, and high standard for processing data; (5) Veracity, emphasizing the authenticity and objective record of big data, but different sourced data have different credibility. In a word, big data is big but useful, and has become means of production.

Content from this work may be used under the terms of the Creative Commons Attribution 3.0 licence. Any further distribution of this work must maintain attribution to the author(s) and the title of the work, journal citation and DOI.

Published under licence by IOP Publishing Ltd

2.2. Big data transform scientific research paradigm

Hidden rules and laws can be better understood with big data analysis. Big data will transform humans' live, work and think in the following three aspects [3]. First of all, big data use overall data, avoiding drawback of sample survey. Secondly, big data is willing to accept diverse and complex data, and no longer limited to precise data. The more data dimensions, the better. Simple algorithm of big data is more effective than complex algorithm of small data. Finally, big data emphasizes on correlating and predicting, and weakens casualty effect. 'What is that' is more important than 'why is that'. Big data mining bases on massive data induction and statistics, promoting novel relationship exploration and new empiricism formation. Knowledge derives from bottom-up data mining without explicit theory. In the big data era, everything can be quantified [3]. Data can be obtained from the most unlikely places, and then correlations can be found directly through data mining. That has transformed scientific research paradigm to the fourth paradigm — data intensive paradigm. The paradigm can digitize events that were invisible and unmeasurable in the past, and discover new knowledge through big data mining. Therefore, big data has significant impact on the human thinking, cognitive models, social-economic life and value orientation. In particular, Internet big data is a large data sets related to investor sentiment and investor behaviour.

3. Big data sources for mining investor sentiment

Big data is no longer limited to officially published data, because personal and institutional activities are digitalized, recorded and stored. Big data includes structural and unstructured data, the latter includes text, audio, video and image. Because of huge number of Internet users, Internet big data is large volume and continue growing exponentially. Internet textual data has been widely used in sentiment analysis [4]. Because of their relatively easy accessibility and operability, search engine data, social media, stock forum and Internet news have been widely studied for forecasting stock market. We review four main sources as follow.

3.1. Search engine

Investors get information through search engine at anytime and anywhere. Investor attention and investor information demand are directly reflected in searching keywords. Well-known search engines data include Google trend and Baidu Index. Researchers use these indexes extensively to study relationship among investor attention, stock returns, price volatility, market volatility, and market efficiency. Google search volume, as a proxy of investor attention, was used to predicts Russell 3000 Index stocks, showing a positive correlation between the volume and stock prices in the next two weeks, but price reversal eventually within the year. And it also predicts the large first-day return and long-term underperformance of IPO stocks [5]. FEARS index bases on search queries from words like 'recession', 'unemployment' and 'bankruptcy' and so on, and the index predicts return reversals, temporary market volatility and money flow from equity funds to bond funds [6]. Google search volume closely correlate to liquidity of S&P500 stocks as well [7].

3.2. Social media

Social media has become the primary medium for investors to get information. Social media network has two characteristics. First, message and information are spread rapidly on the social network, providing a suitable platform for studying investor sentiment and market response. Investors exchange investment ideas through social media as well. On the one hand, it is profitable to be an influential speaker because people tend to overestimate others' opinions. Participants in social media are willing to get information from others, especially from the most influential people. On the other hand, once people form their own opinions, they tend to spread these opinions to other individuals. Second, social media records a large number of investors' emotion, which is useful to study heterogeneity risks and market efficiency. During information spreading in social networks, because of investors having different geography position, investors receive information at different speeds. Therefore, individuals form beliefs or emotions at different speeds. The negative sentiment contained in Twitter messages is negatively correlate with stock index, and the more negative sentiment will follow with the higher volatility of stock indexes next day [8]. More than 200 million pieces of posts on Twitter about 30 NASDAQ stocks are mining to construct investor sentiment index; this index can predict the stock market movement with an accurate rate up to 70 percent [9].

3.3. Stock forum
Stock forum is an online platform for investors to track company news and exchange investment ideas. Investors get public information, forecast information, speculative information, and personal comments from stock forums. Stock forums have some advantages in analysing investor behaviour and sentiment. First, forum posts reflect investor concerns and emotions. Second, posts on stock forum contain disagreements and emotional differences among investors. Third, posts on stock forum contain non-public information, which is useful for predicting stock returns. Finally, the financial online forum topics are rather professional, and posters and readers have knowledge of financial market, thus decreasing noise trading. Textual sentiment, mining from Yahoo! Finance and Raging Bull Message Boards by Naive Bayes algorithm, is positively correlated with stock return. And the post number is negatively correlated with the yield next day [10]. Das and Chen (2007) collect messages from Yahoo Stock Forum, and label each message to bullish, bearish or flat view, then measure small investor sentiment of high-tech stocks, suggesting that the index is closely related to stock volatility [11].

3.4. Internet news
Internet news refers to news released by the media on the Internet, including politic, economic, affairs, company dynamics, stocks analysis and so on. Internet news is large number, high timeliness and diversity, and has been used by many scholars to study the relationship between investor sentiment and asset prices. Measuring investor sentiment based on Internet news is indirect. Online news cannot directly represent investor attention. Only when investors receive such information will they pay attention to the securities. Internet news can be divided into objective news and subjective news. Objective news describes the event, and subjective news refers to biased individual report. Therefore, even for a same event, medias publish different news or even the opposite news, which will convey positive or negative emotions to the public. Tetlock et al (2008) built investor sentiment index based on Wall Street Journal stock news, and find that negative sentiment led to stock price fallen [12].

4. Big data mining process for investor sentiment
Big data contains quantitative and qualitative information. Quantitative data includes article number, reading number, forwarding number, search volume and collection number and so on, reflecting the demand and attention of investors. Qualitative data directly reflect emotions of market participants, such as article views and commentary information. Several processes for quantifying investor sentiment are as follows. The first step is to get bid data. Using web crawlers, website APIs to get big data sets. The second step is storing and cleaning data. Distributed management system is adopted to ensure standardized access of big data based on data types, formats, update periods. Then, qualitatively and quantitatively analyse text information, for example, marking article with positive, negative, or neutral, then counting article number, reading number and so on. Finally, combining the qualitative and quantitative information to calculate the investor sentiment. Qualitative analysis is the most difficult process and has become the main content of sentiment analysis [13].

Sentiment analysis is mainly based on machine learning and lexicon method. Machine learning can be divided into supervised learning and unsupervised learning. Machine learning classifier algorithm includes K-mean algorithm, maximum entropy, neural network, support vector machine and so on. Lexicon method regards article as any combination of words (i.e., bag of words), ignoring article structure, word order, grammar, and syntax. Then, every word in article is labelled as negative or positive base on predefined dictionary. The highest proportion of the marked mood determines the text mood. The higher proportion of positive vocabulary, the more optimistic article is. Lexicon method face two problems: word coverage and the weight of each word. Higher coverage in the dictionary result in more accurate classification. Both machine learning and lexicon method have advantages and disadvantages. Machine learning is easy to use and accurate. But it uses more computing resources, and needs to manually mark the training set. It is necessary to ensure the accuracy of manual marking. So, machine learning relies on the accuracy of manual tagging in some sense. In contrast, the lexicon method does not require manual marking of text, and has a good classification given high vocabulary coverage. However, a vocabulary does not always have the same meaning in different contexts. For example, in the general GI dictionary and Harvard dictionary, 73.8% words are negative, but they are positive or neutral in the financial context [14].

5. Advantages for big data mining on capital market
Measuring investor sentiment and behaviour is hard, as well as mechanism of changing from individual behaviour to overall behaviour. Individual preferences and information disseminations behind Internet big data provide good material for studying investor behaviour and investor phycology. Containing a lot of company valuation information, big data is significant to capital market. Internet big data is large volume and time-sensitive, directly reflecting investor sentiment, and more effective than the previous proxy variables. In addition, Internet big data spread rapidly and interact easily. Owing to the wisdom of crowd, fake news and inferior information are more likely to be eliminated by investors, preserving superior information, and thus market will be more efficient.

Big data is a new type of production material. Investor sentiment mining is just one aspect of big data application in the capital market. Capital market participants, if they have big data resource and data mining technologies, can better understand market movement based on big data mining. Big data mining platform has become an important driving force for institutions competition, expanding business scale and creating value. Big data promotes capital market participation institutions, such as financial institutions, stock exchanges and other mechanisms to create value. Market participant face challenges of how to grasp investment timing and opportunities in a rapidly changing information age. Big data mining improves the reaction speed of investors and is conductive to decision-making. Based on big data, institutional investors can reduce information sharing costs, and improve products and operational efficiency. Big data sentiment analysis may effectively recognize the current market conditions and development trends, improving risk management ability.

6. Challenges and future
Rapid development of information technology provides not only opportunities but also challenges for financial research. First of all, data dimension is quite simple. Many factors affect investor sentiment. There are still many helpful data types such as Online Shopping Platforms, Mobile Payment, GPS Information, Night Light Intensity, and Sensor Data. These data have been used widely in the macro-economic research, but rarely used in researching investor sentiment. More unstructured data are expected to be used. In this context, huge and multi-dimensional big data means that the amount of data is extremely large, and data types and sources are diverse. Because humans' mood is affected by physical conditions, such as temperature, humidity, air pollutant, noise and so on. Sensor data related to physical condition will be useful to measure investor sentiment. Also, individual sport data along with medical big data should not be neglected. Muti-dimensional data facilitate cross-validation with more robust result. Although big data contains great value, data are not knowledge before algorithms are executed. In asset pricing area, the gap between chaos data and knowledge is big data mining.

Secondly, more suitable algorithms should be developed and applied in quantifying investor sentiment. Machine learning methods have inherent defects such as over-fitting and slow convergence, and rely too much on artificial designing [15]. Market data is complex, traditional artificial neural networks are also difficult to accurately measure investor sentiment, and easy to be influenced. If data quality of Internet big data is low, the model fitting results are hard to be satisfactory. Furthermore, for ambiguous sentences, correct attitude classification by machine learning is less than 30 percent [15].

Last but not least, research paradigm in finance discipline is to be changed. Under traditional research paradigms, it requires so many assumptions that it is difficult to adapt to a rapidly changing market, resulting in lost a lot of information. And traditional paradigm is not suitable for analysing complex, high-dimensional and high-noise financial data. There is still a gap between financial research and investment application. In contrast, the fourth paradigm is based on big data mining digitalized event and behaviour, and could be mainstream in the near future.

7. Conclusion

Investor sentiment is a factor for asset pricing, but it is difficult to measure. With the widespread use of big data in various fields, data-intensive science has become a new research paradigm, providing new data and method for investor sentiment measurement. The most popular big data source is Internet big data, including textual data from social media, search engine and online forum. These data directly reflect investor sentiment by data mining. But big data mining for investor sentiment still have some flaws, such as low data dimension, and more suitable algorithms are need. And future research paradigm in finance is expected to be changed.

Acknowledgments

This work is supported by the Qingdao Postdoctoral Applied Research Funding (2016044).

References

[1] De Long, J.B., Shleifer, A., Summers, L.H., Robert, J.W. (1993) Positive Feedback Investment Strategies and Destabilizing Rational Speculation. The Journal of Finance., 2: 379-395.

[2] McKinsey Global Institute. (2011) Big data: The next frontier for innovation, competition, and productivity. https://www.mckinsey.com/business-functions/digital-mckinsey/our-insights/big-data-the-next-frontier-for-innovation

[3] Mayer-Schonberger, V., Cukier, K. (2013) Big data—A Revolution That Will Transform How We Live, Work, And Think. Houghton Mifflin Harcourt Publishing, New York.

[4] Teoh, S.H. (2018) The Promise and Challenges of New Datasets for Accounting Research. Accounting, Organization and Society., 40-69: 109-117.

[5] Da, Z., Engelberg, J., Gao, P.J. (2011) In Search of Attention. The Journal of Finance., 66 (5): 1461-1499.

[6] Da, Z., Engelberg, J., Gao, P.J. (2015) The Sum of All FEARS Investor Sentiment and Asset Prices. The Review of Financial Studies., 28 (1): 1-32.

[7] Ding, R., Hou, W.X. (2015) Retail Investor Attention and Stock Liquidity. Journal of International Financial Markets, Institutions & Money., 37: 12-26.

[8] Bollen, J., Mao, H.N., Zeng, X.J. (2011) Twitter mood predicts the stock market. Journal of Computational Science., 2 (1): 1-8.

[9] Li, B., Chan, K.C.C., Ou, C., Sun, E.F. (2017) Discovering public sentiment in social media for predicting stock movement of publicly listed companies. Information Systems., 69: 81-92.

[10] Antweiler, W., Frank, M.X. (2004) Is All That Talk Just Noise? The Information Content of Internet Stock Message Boards. The Journal of Finance., 3: 1259-1294.

[11] Das, S.R., Chen, M.Y. (2007) Yahoo! for Amazon: Sentiment Extraction from Small Talk on the Web. Management Science., 53(9): 1375-1388.

[12] Tetlock, P.C., Saar-Tsechansky, M., Macskassy, S. (2007) More Than Words: Quantifying Language to Measure Firms' Fundamentals. The Journal of Finance., 63(3):1437-1467.

[13] Bukovina, J. (2016) Social media big data and capital markets—An overview. Journal of Behavioural and Experimental Finance., 11: 18-26.

[14] Loughran, T., Mcdonald, B. (2011) When Is a Liability Not a Liability? Textual Analysis, Dictionaries, and 10-Ks. The Journal of Finance., 66 (1): 35-65.

[15] Su, Z., Lu, M., Li, D. (2017) Deep Learning in Financial Empirical Applications: Dynamics，Contributions and Prospects (in Chinese). Journal of Financial Research., 5:111-126.

Research on Frequency Hopping Synchronization Strategies based on TOPSIS Method

Xiaoyu Cai[1], Zhanjun Jiang[2], Qianru Liu[3] and Haoqiang Shi[4]

[1,2,3,4] School of Electronic and Information Engineering, Lanzhou Jiaotong University, Lanzhou, Ganshu, 730070, China

*Corresponding author e-mail: 59444069@qq.com

Abstract. The synchronization of hopping frequency is a key step in frequency hopping communication systems. During system synchronization, the contradiction emerges between synchronization duration and credibility in the capture confirmation process after frequency acquisition, which make it difficult for the system to establish synchronization quickly and steadily. Thus, a solution is presented based on the classic Technique for Order Preference by Similarity to an Ideal Solution (TOPSIS) to solve this problem. Firstly, a mathematic model for recognition process is built by Markov chains. Secondly, TOPSIS is employed to make analysis and suggestions to the recognition strategies. Results show that the presented mothed can shorten the synchronization time, ensure the reliability of the synchronization systems, and achieve an optimal effect in frequency hopping capture confirmation process.

1. Introduction

Frequency-hopping communication is an important type of spread spectrum communication system. As an effective means of anti-jamming and anti-interception in modern communication field, it has been widely used in military and civil communication fields. The essential premise of frequency hopping communication system is that the system can establish and realize accurate frequency hopping synchronization. Therefore, the synchronization process is the key step of frequency hopping system. The speed of synchronization and the reliability of synchronization system will directly affect the performance of the whole frequency hopping communication system [1-2].

Frequency hopping synchronization consists of three processes: initial acquisition, capture validation and tracking [3]. Initial acquisition is the detection of a potential synchronization through correlation, acquisition confirmation needs to determine whether the synchronization acquisition is caused by interference or real acquisition, usually through successive detection to determine[4]; The final tracking phase is to complete the fine synchronization on the system frequency[5]. The core of acquisition and confirmation is to compare the frequency hopping between the two ends of the receiver and transceiver. At present, if the same number of frequency points exceed the predetermined threshold value, usually set to half of the total number of hops, the symbol system has completed the frequency hopping synchronization and can enter the frequency hopping system. Data transmission phase; conversely, re-enter the frequency capture phase [6-9]. In theory, the more frequency points are detected in this stage, the higher the reliability of frequency hopping synchronization is at the receiving end. However, in the actual situation, the system often has very strict requirements for the time required for frequency hopping synchronization. Therefore, the detection time and reliability of the verification process are a set of contradictions, which need to be considered in practice in order to ensure the comprehensive performance of the synchronization phase of the system.

Content from this work may be used under the terms of the Creative Commons Attribution 3.0 licence. Any further distribution of this work must maintain attribution to the author(s) and the title of the work, journal citation and DOI.
Published under licence by IOP Publishing Ltd

In this paper, the mathematical model of frequency capture and confirmation process is established. The contradiction between detection time and reliability is regarded as a multi-attribute comprehensive decision making problem, which is called "multi-attribute comprehensive evaluation and decision method" in mathematical modeling. There are a large number of algorithms suitable for different application scenarios in this field. The most classical and widely used algorithms are the TOPSIS method and the grey correlation analysis. The idea of TOPSIS method is to calculate the Euclidean distance between each index of the to-be-assessed method and the ideal index, then make them have a weighed summation [10], whereas gray correlation analysis method regards the value of each index as a point on a curve and compares the curve of the to-be-assessed method with the ideal value curve. Therefore, when the evaluation index is relatively less, the evaluation results of the TOPSIS method is more accurate, and when the number of evaluation index is large, there are enough indexes to fit the curve accurately, so the evaluation result of the gray correlation analysis method is more accurate. In this study, four evaluation indexes are adopted, the number of which is small, so the TOPSIS method is used to further analyze, and through TOPSIS analysis, the optimal frequency capture confirmation scheme is selected from many options.

2. Mathematical Modelling of the Frequency Capture Confirmation Process

The frequency hopping frequency capture confirmation process can be mathematically abstracted into a mutual transformation problem in which a plurality of discrete states satisfy a certain condition. At the same time, when the system frequency hopping detection reaches a certain state, the state of the next moment depends only on the current state and is independent of the previous state. Stochastic processes for this feature are mathematically typically modeled using homogeneous Markov chain [11].

In homogeneous Markov chain, $P_{ij}(n)$ indicates the n-step transition probability, and $\mathbf{P}(n) = \left(P_{ij}(n) \right)$ indicates n-step transition probability matrix. The sum of the row elements of this matrix is equal to 1, that is: $\sum_{j=1}^{+\infty} P_{ij}(n) = 1$. When n is a value of 1, it is a one-step state transition probability:

$$P_{ij} = P_{ij}(1) = P\left\{ S_{m+1} = a_j \,\middle|\, S_m = a_i \right\} \tag{1}$$

And the one-step transition probability matrix $\mathbf{P} = \mathbf{P}(1) = \left(P_{ij} \right)$. In the homogeneous Markov chain situation, the n-step transition probability is completely determined by the one-step transition probability:

$$\mathbf{P}(n) = \left[\mathbf{P}(1) \right]^n = \mathbf{P}^n \tag{2}$$

The Markov chain is used to mathematically model the frequency hopping synchronization confirmation process. The continuous acquisition confirmation mode is adopted, that is, the continuous detection of the L hops frequency in the K hops detection is determined to be passed. As shown in the following figure 1:

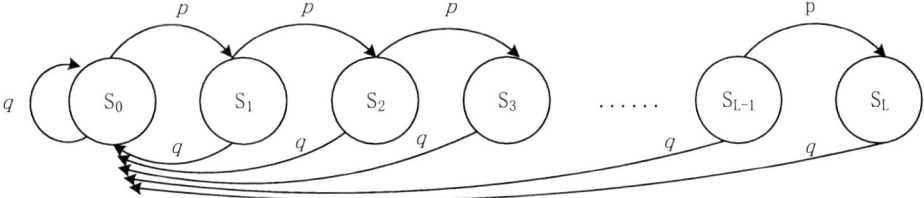

Figure 1. Continuous capture mode.

In figure 3, q is the false detection rate of detecting a whole hop in the system, $p = 1 - q$. For a homogeneous discrete Markov chain, the one-step state transition probability matrix can describe the whole states of Markov chain [12]. For the continuous detection method, the matrix is denoted as $\boldsymbol{\pi}$:

$$\boldsymbol{\pi} = \begin{bmatrix} q & 1-q & 0 & 0 & K & 0 \\ q & 0 & 1-q & 0 & K & 0 \\ q & 0 & 0 & 1-q & K & 0 \\ K & K & L & K & K & 0 \\ q & 0 & 0 & K & K & 1-q \\ 0 & 0 & 0 & K & 0 & 1 \end{bmatrix} \tag{3}$$

Especially,

$$\mathbf{P}(n) = \mathbf{P}(0) * \boldsymbol{\pi}^{n} \tag{4}$$

The probability distribution $\mathbf{P}(0)$ of the initial stage of the continuous detection scheme is $(1, 0, 0, 0, 0L\ L)$, and the transition probability of the Markov chain at any time can be obtained according to the above formula.

According to the above analysis, it is necessary to use the Markov chain to solve the frequency acquisition confirmation probability. After the K hops, the frequency hopping receiver can successfully reach the probability of detecting the detection threshold L, that is, the Lth value in the $\mathbf{P}(K)$ probability vector, which is recorded as P_L. In addition to the K hop and the detection threshold L, the other one involved in calculating the P_L is the parameter q in the state transition probability matrix.

Through the mathematics software Mathematica operation, the relationship between P_L and the matrix parameter q in the case of different total hops K and detection threshold L is simulated as shown in figure 2 and figure 3.

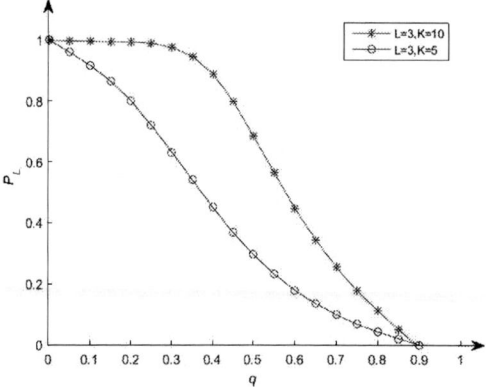

Figure 2. The probability of continuously detect 3 hops in 5 hops and 10 hops.

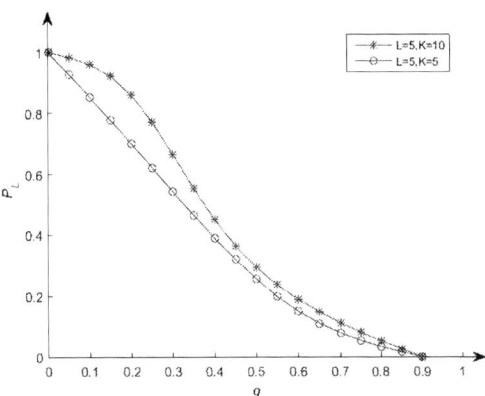

Figure 3. The probability of continuously detect 5 hops in 5 hops and 10 hops.

It can be seen from the analysis results that when the total hop count K and the detection threshold L are given, the probability of successful detection decreases as the system false detection rate q increases. When L is constant, as the total number of hops K increases, the influence of the system false detection rate q on the successful pass detection decreases. In extreme cases, when the total number of hops is infinite, even if the false positive rate is very high, the system can finally reach the required threshold. For the same K and L, $L = 5$ and $L = 3$, the probability of successful detection is small. Obviously, continuous detection of 3 hops is simpler than continuous detection of 5 hops. At the same time, the simulation results are in line with the normal logical judgment, which fully proves the rationality of this mathematical model. The following further analysis will continue to adopt the mathematical model.

3. Strategy Analysis of the Frequency Capture Confirmation Process

3.1. Indicator Selection

After completing the mathematical modelling of the detection process, it is necessary to consider how to choose a better detection scheme which is to select the optimal combination of K and L values. Two issues should be considered here the first is to select which evaluation index, the second is how to use an algorithm to integrate these evaluation indexes.

Selection of frequency hopping evaluation indicators to be evaluated: the detection time t, the capture probability p_{bh}, the normalized average length B and the credibility index T. The parameters used in the calculation of the above indicators include: the hopping speed v, the signature length m, the signature threshold n, the error rate of baseband signal p_b, the total detected hop count K and the detection threshold L.

The concrete calculating idea is:

Detection time t is the total time required for the frequency hopping capture confirmation process:

$$t = v \cdot K \tag{5}$$

The acquisition probability p_{bh} is the probability that the capture confirmation process can be successfully completed after the detection time t, and is calculated by the Markov chain state transition matrix under different detection conditions.

The normalized average length B refers to the average step size from the initial state to the captured state in the Markov chain:

$$B = \frac{1-p^L}{qp^L} \tag{6}$$

The credibility index T represents the reliability of the results of the program to be evaluated:

$$T = \frac{\dfrac{n}{16}\dfrac{K}{K_{max}}}{\dfrac{K-L}{K}} = \frac{nK^2}{16K_{max}(K-L)} \tag{7}$$

3.2. Frequency Acquisition Confirmation Scheme using TOPSIS Method

The TOPSIS method is a more popular multi-index evaluation method. This method will favor ideal solution and negative ideal solution of all evaluation indexes, called the optimal solution and the worst solution. Then it will calculate the relative similarity degree of each method away from the ideal method that is the degree of each solution closed to the optimal solution and away from the worst solution. Finally, we'll get a sort of all solutions, and choose the best one.

The concrete algorithm steps of the TOPSIS method are shown below [13]:

- The vector decision method is used to find the canonical decision matrix, and the decision matrix for setting the multi-attribute decision problem is $\mathbf{A} = (a_{ij})_{m \times n}$, where a_{ij} is the attribute value of each scheme, and the normalized decision matrix is $\mathbf{B} = (b_{ij})_{m \times n}$, where $b_{ij} = a_{ij} / \sqrt{\sum_{i=1}^{m} a_{ij}^2}$.

- Constitute a weighted normative matrix $\mathbf{C} = (c_{ij})_{m \times n}$. The decision maker set the weight vector for each attribute $\mathbf{w} = [w_1, w_2, w_3, \text{K K}, w_n]^T$. And $c_{ij} = w_j b_{ij}$, $i = 1, 2, 3, \text{K}, m$, $j = 1, 2, 3, \text{L}, n$.

- Determine the positive ideal solution c^* and the negative ideal solution c^0.

- Calculate the Euclidean distance of each solution between it and the positive ideal solution and negative ideal solution.

$$s_i^* = \sqrt{\sum_{j=1}^{n}\left(c_{ij} - c_j^*\right)^2} \qquad i = 1, 2, 3, \text{K}, m \tag{8}$$

$$s_i^0 = \sqrt{\sum_{j=1}^{n}\left(c_{ij} - c_j^0\right)^2} \qquad i = 1, 2, 3, \text{K}, m \tag{9}$$

- Calculate the queued index value of each method (called the comprehensive evaluation index)

$$f_i^* = \frac{s_i^0}{s_i^* + s_i^0} \qquad i = 1, 2, 3, \text{K}, m \tag{10}$$

- Arrange f_i^* from large to small.

The following table 1 selects 10 groups of schemes and different combinations of K and L values, representing different hopping frequency synchronization confirmation schemes.

Table 1. Evaluation plan data sheet.

Scheme Number	Signature Threshold	Total Hop Count	Detection Threshold
SCH-1	10	10	8

SCH-2	10	8	6
SCH-3	10	8	4
SCH-4	11	7	6
SCH-5	11	7	4
SCH-6	12	6	4
SCH-7	12	5	3
SCH-8	12	4	3
SCH-9	14	4	3
SCH-10	14	3	3

According to the frequency hopping parameters of the 10 groups of schemes, 10 sets of indicators to be evaluated are calculated by the index calculation formula, and then normalized, as shown in Table 2:

Table 2. Normalized indicator value.

Scheme Number	Detection Time	Capture Probability	Normalized Average Length	Credibility Index
SCH-1	0.4834	0.3437	0.5075	0.5118
SCH-2	0.3867	0.3400	0.3768	0.3276
SCH-3	0.3867	0.3447	0.2511	0.1638
SCH-4	0.3384	0.3402	0.3805	0.5517
SCH-5	0.3384	0.3436	0.2528	0.1839
SCH-6	0.2900	0.3384	0.2618	0.2211
SCH-7	0.2417	0.3386	0.1947	0.1536
SCH-8	0.1933	0.3330	0.1947	0.1965
SCH-9	0.1933	0.2055	0.3076	0.2293
SCH-10	0.1450	0.1694	0.3076	0.3194

In MATLAB, the above data is brought into the TOPSIS algorithm for comprehensive evaluation to obtain the ranking results of the 10 groups of programs. The best to worst solutions are from left to right:

Table 3. The best to worst solutions

Scheme Number	SCH-4	SCH-8	SCH-7	SCH-6	SCH-2	SCH-7	SCH-5	SCH-10	SCH-3	SCH-9
Weight	1	2	3	4	5	6	7	8	9	10

Take the top three schemes and re-list:

Table 4. Top three schemes.

Scheme Number	Signature Threshold	Total Hop Count	Detection Threshold
SCH-4	11	7	6
SCH-8	12	4	3
SCH-7	13	5	3

According to the analysis, from the perspective of detection credibility, although the No. 4 scheme has lower requirements on the signature threshold than the No. 8 and No. 7 schemes, the requirements for detecting the hop count are more demanding, and it is necessary to continuously detect 6 hops in 7 hops in order to successfully capture the confirmation process. The No. 8 and No. 7 schemes increase the signature threshold, but reduce the total number of detected hops. It can be seen that the three preferred schemes are more focused on the credibility of the capture confirmation process, except that the No. 4 scheme focuses more on the credibility of the multi-hop combination test results, while the No. 8 and No. 7 schemes are more focused on 1 hop detection credibility (1 hop corresponds to a complete signature). At the same time, the total number of detected hops of these three preferred schemes is basically at the intermediate level of 10 schemes, thus ensuring a fast capture confirmation. Therefore, the earlier ranking scheme considers the two factors of the length and reliability required for simultaneous confirmation.

Then take out the scheme in the last few places and re-list:

Table 5. Lower ranking scheme.

Scheme Number	Signature Threshold	Total Hop Count	Detection Threshold
SCH-10	14	3	3
SCH-3	10	8	4
SCH-9	14	4	3

It is analyzed that the requirements of the 10th and 9th schemes are very demanding, and the signature threshold is set to 14 hops in the 16-hop signature, and 3 hops must be detected in the detection of 3 hops and 4 hops respectively to pass the capture confirmation process. Such a strict setting will result in a greatly reduced probability that the receiving end will successfully capture the confirmation. The system will not be able to pass the confirmation process and then need to re-enter the initial capture phase, which will take longer synchronization time, so the solution ranks lower. And No. 3 The scheme is at the other extreme, the low requirements result in a much lower confidence in the validation process.

From the brief analysis of the evaluation results of the scheme, it can be clearly seen that in the frequency hopping synchronization acquisition confirmation process, the optimal scheme should ensure the high reliability of each hop detection result, and at the same time reduce the total detection hop count as much as possible. The strategy for ensuring the frequency hopping frequency capture confirmation process takes into account the reliability and synchronization duration.

4. Conclusion
In this paper, we use the Markov chain to mathematically model the capture confirmation process, then select 4 sets of indicators and use the TOPSIS algorithm for comprehensive evaluation of 10 sets of programs. It is then pointed out that in a particular frequency hopping communication system, time and credibility as evaluation indicator are more important than other indicators. Therefore, with the application of frequency hopping technology in military communications, mobile communications, and personal communication systems, this technology should receive more attention.

Acknowledgments
This work is supported by Foundation of A Hundred Youth Talent Training Program of Lanzhou Jiaotong University.

References
[1] OU, C. X., Wu, Z. J. Synchronization Scheme for Frequency Hopping Communication System Based on TOD and PN Code Synchronization [J]. Modern Defence Technology, 2018, 46(02):93-98.
[2] Li, Y. Y. Research and implementation of frequency hopping synchronization based on TOD [D]. Southeast University, 2017.
[3] Li, F. L., Li, Z. Q., Lou, D. K. Analysis and research of synchronization technique for requency-hopping communication systems [C]. International Conference on Computer Science and Network Technology. IEEE, 2011: 1968-1972.
[4] Jiang, E. G. Jiang. Frequency hopping communication system design and research of synchronization acquisition [D]. Hangzhou Dianzi University, 2013.
[5] Jiao, L. Research of synchronization method for shortwave frequency hopping communication system [D]. University of Electronic Science and Technology of China, 2013.

[6]   Chen, B., Li, X., Du, X. L. Synchronization technology research based on Fast-Slow frequency hopping [D]. Computer Simulation, 2014, 31(03):226-229+238.

[7]   Qin, J. Research of key technology for frequency hopping synchronization systems [D]. University of Electronic Science and Technology of China, 2016.

[8]   Chen, Z. J. Research on frequency hopping radio and synchronization technology of short-wave radio[D]. Xidian University, 2012.

[9]   Li, L. Research on de-hopping and synchronization technology of high speed frequency hopping system [D]. Hangzhou Dianzi University, 2014.

[10]   Zhang, X., Wu, Q. L. Research on the personalized recommendation based on TOPSIS method [J]. Journal of Intelligence,2009,28(12):127-130.

[11]   Shi, R. J. Markov chain foundation and application[M]. Xi'an University of Electronic Science and Technology Press,1992.

[12]   Wang, Z. K. Birth and death process and Markov chain[M]. Science Press,1980.

[13]   Zong, P., Zeng, F. Z. TOPSIS method for customer satisfaction[J]. Business research,2006(19):19-51.

ISPECE

IOP Publishing

An Event Detection Method Based on Association Link Network

Lin Sun[1], Weijun Yang[2] and Xinhuai Tang[1]*

[1] School of Software, Shanghai Jiao Tong University, Shanghai, China

[2] School of Software, Shanghai Jiao Tong University, Shanghai, China

*tang-xh@cs.sjtu.edu.cn

Abstract. With the explosive growth of network data, it becomes more and more difficult to acquire and understand information quickly and accurately. Therefore, events discovery from a large amount of document is a very challenging problem. In this paper, we propose a method to detect events, a semantic community network is constructed by ALN (Association Link Network) firstly, and then we update the word node cluster based on WCC metric, and finally merge the communities by word co-occurrence similarity. The experimental results show that this method can complete the event detection well, and it can ensure that every document in an event cluster is describing the same topic.

1. Introduction

With the progress of network infrastructure technology, the total amount of information in the network also shows explosive growth. It becomes very difficult to detect events quickly and accurately from the huge and scattered information. This paper focuses on documents clustering to represent the event semantic community. At present, there are many and mature researches on clustering methods, such as spectral analysis and information theory. However, in the field of text clustering, the applicability and effectiveness of these algorithms need to be verified experimentally. Because most text-oriented clustering methods are expanded from the representation of spatial features of word sparse distribution,but has poor clustering effect when texts at a large number. The semantics of an event is distributed in the chronologically ordered documents, and traditional approaches pay less attention on the word association links, which have a higher ability to organize semantically related words together presenting an event's semantic information [1].

Our method focuses on news documents, which are simple and stable in structure, clear in meaning, and often contains specific phrases. Therefore, words used in the documents describing the same event are basically the same, therefore, in this paper, we construct association link network based on the frequency of co-occurrence words between documents. Based on the modified WCC metric, nodes updating strategy was proposed to divide the semantic network into communities. At the same time, community similarity is applied to implement integration between communities. Finally, documents are mapped to the community to enrich semantic network. And this network can provide support for the event tracking according to the similarity of the document and the semantic network.In the experimental part, we collect news documents about actual events from news websites through web crawlers, measure the effect of the event detection method by comparing the rate of event recognition and the degree of cohesion within the event community.

Content from this work may be used under the terms of the Creative Commons Attribution 3.0 licence. Any further distribution of this work must maintain attribution to the author(s) and the title of the work, journal citation and DOI.

Published under licence by IOP Publishing Ltd

2. Related work

Several event detection methods have been proposed based on clustering algorithms. Li et al. [2] develop a topic detection prototype system based on K-means algorithm. Garcia et al. [3] proposed two clustering algorithms aiming to construct a cluster hierarchy, dealing with dynamic data sets. Chen et al. [4] present an effective Fuzzy Frequent Itemset-Based Hierarchical Clustering approach, which uses fuzzy association rule mining algorithm. Tu and Seng [5] propose a more precise set of prediction indices based on time, volume, frequency for emerging topic detection.

The common problem with these methods is that they ignore the association link between words, which can largely represent the semantic information of events. Therefore, this paper constructs ALN (Association Link Network) network through co-occurrence relationship to emphasize semantic relationship between words, so as to improve the effect of event detection.

Association Link Network [6] is a kind of semantic link network to organize the associated resources (e.g., keywords and web pages) loosely distributed in the Web, aiming at extending the loosely connected network without semantics to an association-rich network. ALN can be donated by:

$$ALN = \langle N, L \rangle \tag{1}$$

where N is a set of web resources, L is a set of weighted semantic links.

Many researchers proposed clustering algorithms based on the semantic network. Yang Liu et al. [1] proposed ALN-based event discovery approach by label propagation. But nodes in the network cannot be shared by different communities, which may lead to different event communities being divided together, or some communities miss some key words. Arnau et al. [7] propose a metric called WCC guarantees communities with structure and cohesion. WCC calculate only the number of triangles formed by current node and all other nodes in that community network, so by this metric every edge between word is the same without using weight of edges. To solve the above problems, we introduce a modified metric based on WCC which calculate perimeter of the triangle, and a node update strategy to update community partition.

3. Event detection

3.1. Preprocessing

The preprocessing part refers to the operation of word segmentation and filter. In the news text, some words in the statement are enough to clearly express the event semantics. Building semantic network with word segmentation results will lead to a large scale of the network and increase the complexity of subsequent operations. Therefore, filtering the text participle results before constructing the ALN semantic network is necessary. Filtering aims to retain key semantics in a statement (e.g., nouns, gerunds, proper nouns, and phrases, etc.). This kind of word is often found in the same event and can clearly describe the characteristics of the event.

3.2. Building ALN

For the news text, we take the word after preprocessing as the node, and use the co-occurrence frequency between words as the edge to construct the relational semantic network. Weight calculation formula of edges between nodes is calculated as follows:

$$w(a,b) = Co(a,b) \big/ (DF(a) \cdot DF(b))^{1/2} \tag{2}$$

where $Co(a,b)$ donates the times that word a and word b appear together in the same text, $DF(a)$ donates the number of texts containing the word a.

At this point, we have the original semantic network of ALN. In this network, the word nodes in the same text are all joined. Different texts that describe the same event will contain the same word combination to a large extent, so the full join graph of different texts that describe the same event will overlap, with a high degree of connection probability between the words and a high edge weight, thus forming a highly cohesive community. Words that describe different events are less likely to connect and have lower edge weights, which can be dispersed among different communities. This is the main reason why ALN can be used for event clustering.

In order to make full use of the highly cohesive feature of ALN semantic network, after the construction of the original network is completed, an edge reduction is carried out to reduce the impact of the insignificant edge on the community division.

3.3. Node update strategy

In this paper, we mainly consider the perimeter of the triangle formed by current node and all nodes in that community network since the weight of edge is introduced. The difference is that WCC calculate only the number of triangles formed by current node and all other nodes in that community network. Given a semantic network $G = (V, E)$, the problem is to divide nodes of the graph into disjoint cohesive sets. We apply improved metric based on WCC to initiate community segmentation of nodes in the community. In term of a node, the degree of cohesion between the node and the community can be donated by:

$$CC(a) = \sum (w(a,b) + w(a,c)) / N_a \cdot \overline{Weight_a} \qquad (3)$$

where N_a donates the number of triangles formed by node a and all other nodes in that community network, $\overline{Weight_a}$ donates the average weight of all the edges derived from a.

For a community C, the improved WCC metric are calculated as follows:

$$\text{wcc}_p(x,C) = \begin{cases} \dfrac{p(x,C)}{p(x,V)} \cdot \dfrac{vt(x,V)}{|C-x| + vt(x,V-C)} & , if\ p(x,C) > 0 \\ 0 & , if\ p(x,C) = 0 \end{cases} \qquad (4)$$

where $p(x,C)$ donates the perimeter of the triangle formed by current node and all nodes in the community network, $vt(x,V)$ donates the number of nodes in V that can form triangle with node x.

For a partition of V donated by $P = \{C_1,\dots,C_n\}$ which is a pairwise disjoint subsets of V. For a partition P, the improved WCC metric are calculated as follows:

$$WCC_p(P) = \sum_{1}^{n} \sum_{x \in C} WCC_p(x,C_i) \qquad (5)$$

Thus, in the subsequent iterative update process, the updating strategy of community clustering is determined according to the numerical changes of the metric. In this paper, we propose four corresponding updating strategies for nodes.

Keep. Node x still belongs to the current community, and the segmentation results of ALN semantic network remain unchanged.

$$\Delta_K = 0 \qquad (6)$$

Detachment. Node detach from current community C_k. Current partition of V is $P = \{C_1,\dots,C_k,\dots,C_n\}$, the partition after detachment is $P' = \{C_1,\dots,C_k',\dots,C_n,\{x\}\}$, and $C_k = C_k' \cup \{x\}$,

$$\Delta_D = WCC_p(P') - WCC_p(P) \qquad (7)$$

Transfer. Node x transfer from the community C_p to which it belongs to another community C_q.

$$\Delta_T = \Delta_D(x,C_p) - \Delta_D(x,C_q) \qquad (8)$$

Join. Node x joins a new community but is not detached from the current community.

$$\Delta_J = -\Delta_D(x,C_p) \qquad (9)$$

In each iteration calculation of the community partition update, the strategy with the maximum value is selected to operate on the node. Among them, the last strategy is the key to solve the problem that the node is shared by multiple communities, and it is also the basis for subsequent community integration

3.4. Community integration
Node update strategy indicates that the number of communities does not decline significantly in each update iteration, nodes only alternate between different communities. However, the existence of the last strategy means that the overlap between communities keeps increasing. When the overlap reaches a certain level, we can consider that the events described by these two communities are the same, so the two communities can be merged.

Based on this idea, this paper calculates the weight proportion of common nodes in two communities respectively and then calculates their cosine similarity. Two communities of similar proportions can be merged into one. It is donated by:

$$sim(C_p, C_q) = \sum_{i=1}^{k} w(n_i, C_p) w(n_i, C_q) \Big/ (\sum_{i=1}^{k} w(n_i, C_p)^2 \cdot \sum_{i=1}^{k} w(n_i, C_q)^2)^{1/2} \tag{10}$$

Since the similarity judgment in this paper only considers the common nodes of two communities, small number of common nodes which proportion is similar will lead to the integration of less relevant communities. Therefore, our method introduces a threshold to exclude the effect of this case by calculating the ratio of the number of common nodes to the node number of communities with fewer nodes.

3.5. Map documents to communities
For the communities that have been clustered, our method maps the documents to the semantic community that is consistent with the same event, and then rebuild the semantic community by combining the existing community and documents to enrich the ALN-based semantic network. Our method applies mutual information to measure the semantically association strength between document d and community C_p :

$$I(d, C_p) = p(d, C_p) \cdot \log(p(d, C_p) \big/ p(d) p(C_p)) \tag{11}$$

where $p(d)$ donates the probability that document d appears. $p(C_p)$ donates the probability that community C_p is selected is the ratio of the total weight of the internal edge of community C_p to the sum of the total weight of the internal edge of the community to be selected. $p(d, C_p)$ is the joint probability between document d and community C_p, which is composed of similarity and correlation.

4. Experiments

4.1. Data set
The clustering method proposed in this paper aims to describe the same event in each clustering text, so the experimental data set should be clustered based on the event. Therefore, we crawl news about 10 events from the web page of Tencent news website as the data set of this experiment. In the meantime, our method applies word segmentation tool to divide crawled documents and then filter out words that are not descriptive and noun.

4.2. Evaluation criteria
In this paper, recognition rate and purity are used to quantify the effect of our clustering method. The recognition rate refers to the ratio of the number of events divided by our method (donated by $count_c$)

to the number of actual events (donated by $count_e$), and the similarity between the two values indicates whether the result of our method's event recognition is similar to the actual situation.

$$recognition\_rate = count_c/count_e \qquad (12)$$

Purity is used to measure whether word nodes in a community network describe the same event, and high purity means that the event semantic community is highly cohesive:

$$purity = \sum_{C_i \in P} |C_i| \cdot Max(F\_Measure(C_i, Ev_j))/P \qquad (13)$$

4.3. Experimental Results Analysis
In this paper, two data sets of different quantities are selected as input for experiment and comparison with the label propagation method based on ALN mentioned above.

Table 1. Experimental results.

Method	6 actual events		10 actual events	
	Recognition rate	Purity	Recognition rate	Purity
ALN with label propagation	0.83	0.78	0.8	0.73
Our method	1.33	0.85	1.3	0.83

As shown in the above table, the number of recognized events by label propagation method is closer to the actual number of events than the clustering algorithm proposed in this paper, but the clustering results are generally less than the actual number, leading to the decline in the purity of the clustering results. Since our method introduces weight on the basis of WCC and emphasizes more on the correlation between semantic nodes in community network, so the performance on purity is much better than that of the former. As the words in the same document are fully connected in the ALN semantics network, the degree of cohesion of subnets is high in the processing procedure, which leads to the less possibility of selecting *join* policy in the nodes updating iteration. In addition, other methods, such as the adjusted cosine similarity and the cosine similarity of both communities, are also considered in the community integration, however, the co-occurrence word cosine similarity used in this paper is more effective. Although the method proposed in this paper divide the event semantic community in slightly larger numbers than it actually is, it ensures the quality of the community in terms of purity. In other words, an event is likely to be divided into more specific communities, which is acceptable in engineering and can be corrected by other means.

5. Conclusion
In this paper we propose an event detection algorithm based on Association Link Network. This algorithm firstly constructs ALN network according to the co-occurrence relationship of words in the documents, and then adopts corresponding nodes update strategy based on WCC metric for each node, and finally coordinates with the community integration and text mapping method to achieve the event detection. The experimental results show that this method can complete the event detection well, and it can ensure that every document in an event cluster is describing the same topic. In the future studies, time dimension can be considered to be added for word nodes, and the calculation of inter-community correlation in the community integration mechanism can further improve the effect of event detection of this method.

References
[1] Liu Y, Luo X, Xuan J. Online hot event discovery based on Association Link Network[J]. Concurrency & Computation Practice & Experience, 2015, 27(15):4001-4014.

[2] Li S, Lv X, Wang T, et al. The key technology of topic detection based on K-means[C]// International Conference on Future Information Technology & Management Engineering. 2010.

[3] Reynaldo Gil-García, Pons-Porrata A. Dynamic hierarchical algorithms for document clustering[J]. Pattern Recognition Letters, 2010, 31(6):469-477.

[4] Chen C L, Tseng F S C, Liang T. Mining fuzzy frequent itemsets for hierarchical document clustering[J]. Information Processing & Management, 2010, 46(2):193-211.

[5] Yang H C, Lee C H. A novel self-organizing map algorithm for text mining[C]// International Conference on System Science & Engineering. IEEE, 2010.

[6] Luo X, Xu Z, Yu J, et al. Building Association Link Network for Semantic Link on Web Resources[J]. IEEE Transactions on Automation Science and Engineering, 2011, 8(3):482-494.

[7] PratPérez, Arnau, Dominguezsal D, Brunat J M, et al. Shaping Communities out of Triangles[J]. Computer Science, 2012:1677-1681

ISPECE

IOP Publishing

A Comparative Study of Customer Complaint Prediction Model of Time Series, Multiple Linear Regression and BP Neural Network

Xin Xu[1,2,3], Zhijie Sun[1,2,3], Li Wang[1,2,3], Jun Fu[1,2,3],Chao Wang[1]

[1] State Grid Jibei Electric Power Company Limited.

[2] State Grid Jibei Electric Power Company Limited Electric Power Research Institute

[3] Country Huadian Electric Power Research Institute Co., Ltd.

Corresponding author's e-mail: 676269930@qq.com

Abstract. The current prediction algorithm is mainly the "time series", the "multiple linear regression" and the "BP neural network". This article studies and compares the three algorithms in the field of customer complaint prediction. By using SPSS, this article predicts taking "Jibei Electric Power Customer Complaints" as the target. By comparing and analyzing the actual prediction results, the "BP neural network" algorithm is described as the most suitable complaint prediction for customer.

1. Introduction

The customer complaint is a mirror of the quality of the enterprise. How to prevent complaints is an unavoidable problem for enterprises with customer demand-oriented today. The prevention is better than the disaster relief. The enterprise should pay attention to the prevention of the complaint. The enterprise is required to eliminate the customer's dissatisfaction in the initial stage, so as to avoid the deterioration of problem and the cost of the enterprise investment [1]. The premise of prevention is to be predicted. At present, the prediction of customer complaints constructs prediction model often based on "ARIMA (Autoregressive Integrated Moving Average Model) Time Series" [2], "Multiple Linear Regression" [3], and "BP (Back Propagation) Neural Network" [4]. The three algorithms each have their own characteristics and disadvantages. However, when constructing the complaint prediction model, the selection of the appropriate core algorithm will essentially determine the success of the complaint prediction model. In this article, a comparative study of three kinds of algorithm prediction is carried out based on the "Number of complaints" of the power customers in the Jibei Region as the prediction target to explore the algorithms that apply to the customer's complaint prediction model.

2. The principle of algorithm

2.1. Arima time series algorithm
The time series refers to a set of statistical data in the order of time. The ARIMA time series algorithm includes a self-regression process AR, a moving average process MA, and a differential process $DX = \text{diff}(y, i)$.

The concrete implementation process is divided into three steps.

Content from this work may be used under the terms of the Creative Commons Attribution 3.0 licence. Any further distribution of this work must maintain attribution to the author(s) and the title of the work, journal citation and DOI.
Published under licence by IOP Publishing Ltd

(1) Time series differential/stationary processing.

The stability of the sequence is checked by a scatter diagram, an auto-correlation function and a partial auto-correlation function, and it is determined whether to perform the differential processing and the differential order according to the stationarity characteristics.

(2) Model parameter order recognition.

The parameter p in the ARIMA (p, d, q) model is a self-regression term, the parameter q is the number of moving average items, and the parameter d is the number of times that the time series becomes stationary. [5] ARIMA (p, d, q) model parameters are set according to the truncated and trailing characteristics of the autocorrelation function and the offset function of the data sequence.

(3) Model test

The residual sequence white noise test is performed by constructing the correction statistic Q of the box-pierce.

$$Q = (N - D - \max(p, q)) \sum_{k=1}^{m} p_k^2(\hat{a}).$$

(1)

2.2. Multiple linear regression

Regression is a study of the relationship between the variable and the independent variable, and the relationship between the independent variable and the dependent variable is expressed by means of the regression equation. The multivariate linear regression model is a correlation between a variable and a plurality of independent variables.

$$y = \beta_0 + \beta_1 x_1 + \ldots + \beta_p x_p + \varepsilon$$

(2)

The concrete implementation process is divided into three steps:

(1) Independent variable selection

At the time of modeling, the selection of the independent variable is first made. The method is commonly used as stepwise regression and grey relational degree.

(2) Coefficient estimation of multivariate linear regression model

The estimation of the parameter vector B is performed using the least square method. The residual error is E:

$$E = Y - \hat{Y}$$
$$\hat{Y} = XB$$

The least square method is used:

$$E'E = (Y - \hat{Y})'(Y - \hat{Y}) = (Y - XB)'(Y - XB) = \min$$

(3)

According to the extremum principle, the above formula is for B and B=0:

$$\frac{\partial E'E}{\partial B} = \frac{\partial (Y - XB)'(Y - XB)}{\partial B} = -2(Y'X)' + 2(X'X)B = 0$$

(4)

It is therefore available:

$$\hat{B} = (X'X)^{-1} X'Y$$

(3) Model test of multiple linear regression

Commonly used as the complex correlation coefficient test (R), and F test

$$R = \sqrt{1 - \frac{\sum (y_i - \hat{y}_i)^2}{\sum (y_i - \overline{y})^2}}$$

(5)

R describes the degree of linear correlation between the independent variable and the dependent variable.

$$F = \frac{\sum (\hat{y}_i - \bar{y})^2 / (m - 1)}{\sum (y_i - \hat{\bar{y}})^2 / (n - m)}$$

(6)

F test is to verify hypothesis whether $\beta_1 = \beta_2 = \beta_3 \ldots = \beta_m = 0$ is right. [6]

2.3. BP neural netword

The full name of BP neural network is Back Propagation Neural Network, which was proposed by a group of scientists led by Rumelhart and McCelland in 1986. The BP neural network is a kind of "multi-layer feedforward neural network", which uses the "error inverse propagation algorithm" to train the neural network. It is also one of the widely used neural network algorithms. The neurons of the BP neural network have three basic functions, namely, "modified weight value", "summation" and "transfer". [7]

The implementation steps are as follows:

(1) Neural network initialization.

Randomly give the random number of each neuron connection weight value (Wji) and a threshold value (Bj) valuation (-1, 1).

(2) Calculate the output of the input layer.

Randomly select an input vector $X_p = (\begin{array}{cccc} x_1 & x_2 & \cdots & x_n \end{array})$, and calculate expected output vectors $Y_p = (\begin{array}{cccc} y_1 & y_2 & \cdots & y_n \end{array})$

(3) Calculate the hidden layer output.

$$O_{pi} = f_j(net_{pi}) = f_j(\sum W_{ji} O_{pi})$$

(7)

(4) The error of each node is calculated in the reverse direction, and the weight of each neuron is corrected.

$$W_{ji}(t + 1) = W_{ji}(t + 1) + \eta \delta_{pj} Q_{pj}$$

(8)

3. Test procedure

3.1. Data description

In this article, the predicted target is a total amount of single-week of customer complaint of the State Grid Jibei Electric Power Company. The used date time interval is from July 2015 to September 2016. The ARIMA time series prediction only needs the statistical data of the "Complaints" category.

The multiple linear regressions and the BP neural network will use the total amount of the single-week complaints as the dependent variable. Through the data screening, 25 secondary work orders from the "information query", "service consultation", "fault report", "report", "opinions", "suggestions", "praise" and "service application" as the independent variable of the prediction model. The time interval is also from July 2015 to September 2016 and the data is about 1444248.

3.2. ARIMA time series algorithm

Because the predicted target is total number of complaints in one week, the study needs to divide the original data into 7 groups of time series according to the period of 7. 7 ARIMA time series prediction is respectively carried out to obtain all the predicted values. Then the original time series is divided into seven different time series, and the time series prediction model is established. The model parameters and model evaluation are shown in Table 1 below:

Table 1 ARIMA Model Evaluation Form

Model	Model parameters	Model fitting statistic		Ljung-Box Q(18)			Outlier
		Smooth R -square	R Square	Statistic	DF	Sig	
Complaint model _ Monday	(2,0,1)	.659	.680	17.175	16	.376	0
Complaint model _ Tuesday	(1,0,0)	.624	.633	17.384	17	.429	0
Complaint Model _ Wednesday	(1,1,0)	.633	.643	13.240	17	.720	0
Complaint Model _ Thursday	(1,0,0)	.639	.628	12.244	17	.785	0
Complaint Model _ Friday	(1,1,0)	.552	.655	13.201	17	.723	0
Complaint Model _ Saturday	(1,0,3)	.645	.674	14.181	16	.585	0
Complaint model _ Sunday	(1,0,3)	.626	.662	14.410	16	.568	0

The prediction value is evaluated as shown in Table 2 below:

Table 2 ARIMA Predicted Value Evaluation Form

The relative error is less than 1 0%	The relative error is less than 2 0%	The relative error is less than 3 0%	Relative error is less than 4 0%
22.48%	43.56%	60.21%	71.12%

3.3. Multiple linear regression

The "multiple linear regression" uses the "complaint amount/week" as the dependent variable. When the complaint amount/ week of t day is to be predicted, the t-1 day other secondary work order other than the complaint work order will be used as the argument. And the variable selection is carried out by adopting the stepwise regression method.

The multivariate linear expression is as follows:

$$y = 0.113x_1 + 0.88x_2 + 0.207x_3 + 0.183x_4 - 1.804x_5 + 4.890x_6 - 0.81x_7$$
$$- 10.336x_8 - 0.179x_9 + 0.718x_{10} - 1.394 \tag{9}$$

Table 3 Parameter Schematic Form of the Multiple Linear Regression Equation

System parameters	Business order class
y	Complaint amount/week
x_1	User information
x_2	Electric energy quality fault
x_3	Customer internal fault
x_4	Charge of electricity and electricity
x_5	Metering device
x_6	Urgency complaints
x_7	Call-up failure report
x_8	Change of power utilization information

x_9	Demand for electric service
x_{10}	Request for the service of the call-up service
To be abandoned	High-voltage fault
To be abandoned	...

Model evaluation:

Table 4 Evaluation Form for Multivariate Linear Regression Model

R	R Square	R square after adjustment
0.705	0.498	0.486

Predicted value evaluation:

Table 5 Evaluation Form for Multivariate Linear Regression Prediction

The relative error is less than 10%	The relative error is less than 20%	The relative error is less than 30%	Relative error is less than 40%
21.67%	36.57%	55.53%	68.62%

3.4. Neural network algorithm

Because the data volume is small, the training quantity and the calculation amount are not small, a lower learning rate is adopted to improve the matching degree of the model. Using the IBM SPSS Statistics as a tool to set up a neural network model, the relevant configuration information is shown in Table 6 below.

Table 6 Parameters of the Configuration of the Neural Network.

Configuration item	Parameter
Interval	70% training 30% test 0% detection
Network system	Single layer hidden layer
Initial learning rate	0.01
Lower boundary of learning rate	0.001
Time-history learning rate reduction	10
Kinetic energy	0.9
Interval center point	0
Interval offset	±0.5
Minimum relative change in training errors	0.0001
Minimum relative change in training error rate	0.001
Hide layer activation function	Hyperbolic tangent function
Output layer activation function	Identity function

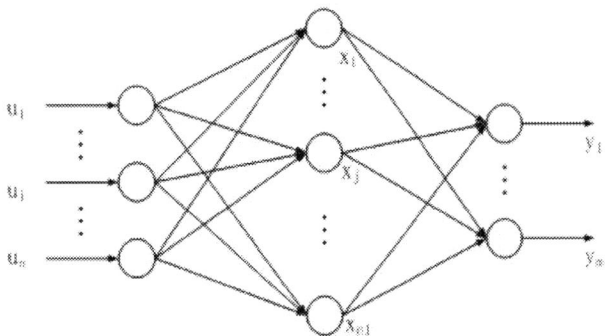

Figure1 Schematic Diagram of the Neural Network

Table 7 Evaluation Form of Neural Network Predicted Value

The relative error is less than 10%	The relative error is less than 20%	The relative error is less than 30%	Relative error is less than 40%
31.38%	58.47%	76.52%	90.29%

4. Conclusion

Table 8 Comprehensive Prediction Value Evaluation Form

Algorithm	The relative error is less than 10%	The relative error is less than 20%	The relative error is less than 30%	Relative error is less than 40%
Multiple linear regression	24.83%	45.15%	62.75%	74.27%
ARIMA time series	22.48%	43.56%	60.21%	71.12%
Neural network algorithm	31.38%	58.47%	76.52%	90.29%

It can be concluded from the above table that the prediction model constructed by the neural network algorithm is much higher than the other two algorithms. From the essence of the algorithm and the prediction results, the essence of the time series is the prediction of the future value through the process of self-regression and moving average. Its prediction is based on the identification of the "long-term trend", the "seasonal variation", and the "cyclic variation", and is poor for the "irregular variation". The problem of multiple linear regression is not a linear relationship between the "Complaints" and the "Other customer claims" in the reality, and the way of regression fitting does not really embody the mapping relation between the independent variable and the dependent variable. Similarly, in the process of constructing the model, the choice of the independent variable is a difficult problem. Gray correlation in a large number of independent variables is not much different from each other. A large number of independent variables can not be well fitted, and a small number of selected independent variables can not be accurately predicted. The neural network algorithm is based on the non-linear model between the independent variable and the dependent variable, which reflects the more and more complex mapping relationships between independent variables and dependent variable through multi-layer network structure. At the same time, the error inverse propagation algorithm of the BP neural network gives it a better ability to study and train.

By the full text, the BP neural network is the optimal algorithm for constructing the customer complaint prediction model from the perspective of the accuracy of numerical example prediction, and the degree of fitting between the independent variable and the dependent variable.

References

[1] Liu Yingying, Tian Xue, Yao Jia, and so on. Customer complaint management for e-commerce[J]. E-commerce, 2015 (1): 54-54.

[2] Yan Wei, Cheng Chao, Xue Bin, and so on. Monthly sales forecast method combining X12 multi plication model and ARIMA model[J]. Power system and its automation journal, 2016, 28 (5): 74-80.

[3] Fu Qianrao. Study on the Method of the Prediction of the Haze Based on the Multiple Linear Reg ression[J]. Computer Science, 2016, 43 (S1): 526-528.

[4] Chen Xinyun, Jiang Yongkang, Li Muyuan, and so on. Research on the Demand for Single-Site Dispatch of Public Bicycle based on BP Neural Network[J]. Traffic Standardization, 2016, 2 (3): 30-35.

[5] Han Chao, Song Su and Wang Cheng-hong. Real-time Adaptive Prediction of Short-term Traffic Flow based on ARIMA Model[J]. Journal of System Simulation, 2004, 16 (7): 1530-1532.

[6] Chen Yongsheng and Song Lixin. Multiple Linear Regression Modeling and the Application of S PSS Software[J]. Journal of Tonghua Normal University, 2007, 28 (12): 8-9.

[7] Liu Xiaotong. BP Neural Network Input Layer Data Normalization Research[J]. Mechanical engi neering and automation, 2010 (3): 122-123.

ISPECE   IOP Publishing

IOP Conf. Series: Journal of Physics: Conf. Series **1187** (2019) 052037   doi:10.1088/1742-6596/1187/5/052037

Anomaly Detection Based on PMF Encoding and Adversarially Learned Inference

Lin Zhang[1], Wentai Yang[2], Hua Gan[1], Meng Li[1], Xiaoming Wang[1], Gang Liang[3]

[1] Chengdu City Electric Power Engineering Design Company, Chengdu, Sichuan, 610065, China

[2] College of Computer Science, Sichuan University, Chengdu, Sichuan, 610065, China

[3] College of Cyber-security, Sichuan University, Chengdu, Sichuan, 610065, China

[*]Corresponding author's e-mail: 2274504924@qq.com

Abstract. In order to solve the problem of increasing the dimension and sparse feature space caused by the categorization coding method in the existing abnormal traffic detection problem, a coding method based on Probability Mass Function (PMF) is proposed. Secondly, in order to improve the ability of abnormal traffic detection algorithms to identify unknown attack type data and improve detection efficiency, we use Adversarially Learned Inference as the basic detection algorithm. The comparison experiments on the standard dataset show that the proposed method has improved the accuracy and detection efficiency greatly compared with the existing anomaly detection methods.

1. Introduction
The network has now gone deep into life. The internet connects the world, and the intranet connects whole organization, both of which have greatly improved the efficiency of work and brought convenience to people's lives from time to time. This is undoubtedly the welfare brought by the advancement of information technology. However, it should not be overlooked that cyberspace is not a pure land. In fact, cyber security issues are becoming more and more serious. In recent years, malicious network intrusion events, cyber fraud incidents, and network information disclosure incidents with the network as the carrier and implementation path have emerged in an endless stream, which make individuals, enterprises and countries suffer economic and reputation losses. Frequent network security incidents have made network users pay attention to network security, and people are gradually realizing the importance of network security to the network age. The frequent cyber security incidents, and subjective demand of cyber security, give great importance to network security related works.

Network security field can be generally divided into sub-fields such as intrusion detection, intrusion prevention, information confidentiality, security auditing and etc. Among many network security works, network abnormal traffic detection belonging to intrusion detection is of great significance. The main reason is that it is often used as a means of finding problems. For network security workers, the ability to find problems is very important. If it is impossible to accurately determine whether an intrusion event occurs on the network, it is impossible to make a timely response, so that subsequent operations such as intrusion event implementation response and system state recovery cannot be performed. Network abnormal traffic detection is a means to discover abnormal traffic based on the normal use state of the network, and thus provide more warning information for intrusion detection.

Content from this work may be used under the terms of the Creative Commons Attribution 3.0 licence. Any further distribution of this work must maintain attribution to the author(s) and the title of the work, journal citation and DOI.
Published under licence by IOP Publishing Ltd

According to the adopted detection ideas, the abnormal traffic detection methods can be divided into four classes, which are respectively based on statistics, information theory, classification and clustering [1]. 1) Probability and statistics theory has an abnormal point detection algorithm. For example, in the case of one-dimensional data and assumption of normal distribution, if the difference between the mean and the value to be detected is greater than 3 times of standard deviations, it can be marked as an abnormal point with high probability. For multidimensional data, in addition to the extended one-dimensional anomaly detection method, there are correspondingly multivariate Gaussian distribution method, chi-square method, PCA method, etc; 2) Information Entropy in information theory is often used for abnormal traffic detection, which monitors the overall traffic rate. Information Entropy-based method assumes that Information Entropy will change correspondingly when anomaly occurs. So that abnormal traffic detection depends on monitoring of Information Entropy of whole traffic. Such methods have strong applicability to attack types with significant changes of traffic information entropy (eg DDOS); 3) cluster-based methods use unsupervised machine learning clustering algorithms as model algorithms, such as K-Means. Clustering-based detection methods have innate advantage of detection in the context where attack types are unknown; 4) classification-based method is mainly based on machine learning classification method to detect abnormal traffic, involved algorithms are: Support Vector Machine (SVM)), Decision Tree (DT), Random Forest (RF), Neural Network (NN), and etc. Most of these methods use supervised machine learning where annotated data is necessary, whereas their detection efficiency and accuracy are relatively high.

The main work of this paper is as follows: The first part introduces the significance and background of abnormal traffic detection. The second part analyses and summarizes related network abnormal traffic detection works. The third part introduces our contributions, including Probabilistic Quality Function (PMF) based encoding method and abnormal traffic detection algorithm based on the Adversarially Learned Inference (ALI). The fourth part describes experiment on the NSL-KDD dataset, and analysis and discussion are given. At last, we summarize our contribution

2. Related works
In this section, we first introduce related existing works of four types mentioned above in section 1 briefly, and then we summarize the problems in them.

2.1. Anomaly Traffic Detection Methods
Statistical-based methods appear earlier, its basic idea of anomaly detection is to determine whether observations fall within some confidence interval. Applied methods include Chi-Square Distribution, Wavelet Transform, Smooth Regression, Principal Component Analysis (PCA), and etc. Qian YK [2] proposed an anomaly detection algorithm that anomaly was detected by comparing Chi-Square statistic of both test data and normal data. W LU [3] applied the Wavelet Transform to anomaly detection and verified its algorithm performance on DARPA dataset, where dataset was organized in a time series style. HZ Moayedi [4] applied Autoregressive Integrated Moving Average model (ARIMA) to model time series data, comparing with model of normal data. M-L Shyu [5] detected abnormal traffic based on PCA, which was used to extract better features for classifier.

Main idea of Information Entropy-based methods is that Information Entropy of abnormal traffic is special, so such methods detect anomaly by monitoring Information Entropy of stream data. Information Entropy based methods perform well in the case of attack type are unknown, this is because that no specific label is applied. Lakhina [6] applied the Information Entropy based on feature distribution to abnormal traffic detection task for the first time. Zheng LM [7] studied the Information Entropy based traffic classification problem in the multi-dimensional case.

The principle of cluster-based abnormal traffic detection methods is that similar traffic is classified into one class according to certain traffic similarity threshold. While traffic that is not classified into any cluster is considered to be abnormal. Similarly, cluster-based methods are also applicable to the case where attack type is unknown. L PORTNOY [8] used clustering algorithm to realize abnormal traffic detection for the first time, which gave abnormal traffic detection the ability to detect unknown attack types. Zuo J [9] improved selection method of the initial clustering centers in K-means clustering algorithm, avoiding the situation where outlier points were selected as the initial clustering centers, thereby reducing the number of iterations, and applying K-means clustering algorithm more efficiently to abnormal traffic detection task.

Classification-based anomaly traffic detection methods use annotated data. Conventional machine learning algorithms have good classification performance and are widely used in abnormal traffic detection tasks. Generally, detection efficiency and accuracy of these methods are better. Zhu Yingwu [10] regarded abnormal traffic detection problem as binary classification problem based on Information Entropy and SVM. H SAXENA [11] used Information Gain to pre-selected features that were beneficial to classification, then the model was validated on the KDD-99 dataset and SVM algorithm. Li Q [12] used C4.5 algorithm to detect abnormal traffic, and optimal features were found while model was constructing. In addition to conventional machine learning algorithms, Neural Network (NN) based anomaly traffic detection methods have also been studied. In abnormal traffic detection task, NN is usually used as dimension reduction method as well as detection algorithm. 1) Because of its strong representation of dataset, NN is often used as a pre-training part for constructing hybrid classification model, which plays the role of dimension reduction and feature extraction. J YANG [13] combined Restricted Boltzmann Machine (RBM) with SVM to construct a hybrid abnormal traffic detection model, when Spark was leveraged to accelerate training. F LIU [14] used Deep Belief Networks (DBN) as feature extraction and dimension reduction method to construct a hybrid APT detection model with Support Vector Data Description (SVDD). 2) NN can be also used as detection algorithm. U FIORE [15] proposed a discriminative restricted Boltzmann machine (DRBM), which gives RBM the ability for classification, and finally achieving semi-supervised abnormal traffic detection. D WULSIN [16] used DBN as semi-supervised classification method to verify model performance on clinical medical image dataset. For the first time, J AN [17] imported Variational Auto-Encoder (VAE) into anomaly detection task, and then compared the model with Auto-Encoder(AE) and PCA on MNIST dataset and KDD-99 dataset. For the first time, T SCHLEGL [18] applied Generative Adversarial Network (GAN) in anomaly detection, and proposed a new measurement for classifier based on Feature Matching (FM) beyond existing Cross entropy based measurement. The work trained the model based on both generator loss and classifier loss, and performance of the model was tested on image dataset.

2.2. Challenge of Existing Methods

The above related works involve nearly all aspects of the field of abnormal traffic detection. Except non-negligible contributions, there are also challenges they are faced with. We conclude as bellow:

1) Existing methods can hardly satisfy the requirements of both identifying unknown attack and high detection efficiency. Among existing methods, there are three kinds of algorithms for identifying unknown attack types: one class classification-based methods, clustering-based methods and generation model based methods. Among them, the accuracy of clustering-based methods and one class classification-based methods are low relatively. While generative model based methods have higher accuracy, whereas their detection efficiency is too low to meet practical demands.

2) Existing encoding methods for categorical feature bring about dimensions expansion problem. These encoding methods are One-Hot Encoding, Dummy Encoding, and Label Encoding. The first two are similar with each other in principle and form. They both are based on all possible values of categorical features. Due to the expansion of original data dimension after encoding, the feature space becomes very sparse and harmful to model training. Therefore, One-Hot Encoding or Dummy Encoding is often applied together with dimensionality reduction technology such as PCA. Label Encoding assigns specific number to each possible value of categorical features. Although this encoding method can avoid dimension expansion, unnecessary bias may be drawn into model because that encoding numbers have little correlation with true distribution of features.

3. Our model

Abnormal Traffic Detection model work on a series of input data, and specify prediction label of current input. Actual network traffic data usually are multi-dimensional. the i th input data can be denoted as vector form $T_i = \{f_1, f_2, \cdots, f_n\}$. The abnormal traffic detection model $g(T_i)$ outputs the label of T_i, so the detection process can be illustrated as formula (1):

$$C_i = g(T_i) \tag{1}$$

If the true label of current input is denoted as $G_i = \{T, F\}$, then the training goal of the model is to minimize error rate of the model on the training dataset. And minimizing the model error rate is equivalent to minimizing the following loss function:

$$S(D) = \sum_{i=1}^{m} \begin{cases} 0, G_i = C_i \\ 1, G_i \neq C_i \end{cases} \tag{2}$$

Where D denotes the input dataset, m denotes the size of this dataset.

There are many factors that affect the performance of abnormal traffic detection model, such as data preprocessing methods, data dimensions and classification algorithms. This paper mainly works on encoding methods for categorical features and anomaly detection algorithms adopted in abnormal traffic detection. We propose a new abnormal traffic detection method based on Probabilistic Quality Function (PMF) and Adversarially Learned Inference (ALI).

3.1. PMF encoding
Most of existing related works rely on vectorization and normalization for input features. For non-continuous features, also known as categorical features, specific encoding is needed to convert them to continuous features firstly. Conventional encoding methods for categorical features include Label Encoding, One-Hot Encoding, Dummy Encoding.

Label Encoding designates encoding value for categorical features based on an encoding dictionary built manually. For example, for feature "gender", the encoding result may be 1 for male and 0 for female. The problem with Label Encoding is that the encoding value is meaningless and can hardly reflect true distribution. In the above example, it is difficult to explain why the value of male is larger than the value of female, and actually 0 for male and 1 for female is also okay in Label Encoding. However, different encoding results lead to different classification performance, and the detection performance of classifier may be hurt when encoding result has nothing to do with true data distribution.

One-Hot Encoding method overcome the shortcomings of Label Encoding by generating encoding value from feature distribution instead of manual labor. It expands dimension according to number of whole possible values of categorical features, and every dimension represents a possible value of current feature. Due to that one feature can only have one value for current input data, there is only one dimension getting non-zero encoding value while the other dimensions get zero value. Dummy Encoding is very similar to One-Hot Encoding method. The only difference is that Dummy Encoding represent a value of the feature by zero for all encoding dimensions. As a consequence, Dummy Encoding could save one encoding dimension. The problem with One-Hot Encoding and Dummy Encoding is dimension expansion, which reduces model efficiency. And dimension expansion also causes sparse and high-dimensional feature space. In extreme cases, it may cause Curse of Dimensionality, which makes model hard to train.

In this paper, we proposed a new encoding method for categorical feature based on PMF. PMF denotes probability of Discrete Type Random Variable in statistics. It has the following three important characteristics: 1) each value is between 0 and 1; 2) the sum of the values of equals 1; 3) each value denotes probability of a value of current feature. If the value of the PMF in a certain category is large, it indicates that the sample of the category has a large proportion in the dataset. Thus PMF indeed reflects the true distribution of feature.

The aforementioned three characteristics of PMF make PMF not only satisfy the requirements of vectorization and normalization but also be meaningful to reflect origin information of dataset. Compared with Label Encoding, PMF based encoding wouldn't disturb classifier. Moreover, PMF is a one to one encoding method, which means that the dimension stays constant and there is no dimension expansion problem like in One-Hot Encoding and Dummy Encoding.

Figure 1 illustrates PMF based encoding by the "weather" example. Suppose all possible values of feature "weather" are rainy, sunny, cloudy, and the values above bars are PMF values of corresponding weather in current dataset. For example, 0.3 of "rainy" means there are 30 percent rainy data of all data. And we can also find the total PMF of all possible weather (including rainy, sunny and cloudy) equals 1, which means PMF based encoding normalizes data at the same time.

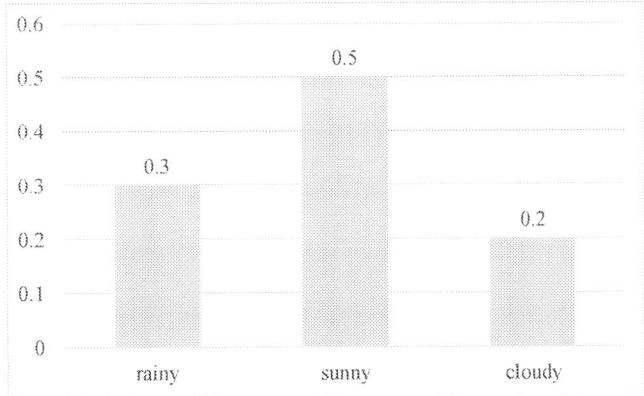

Figure 1. PMF Based encoding for the weather

For an actual dataset D, the probability is represent by frequency. So the PMF encoding value $p_{c,k}$ for feature c and its categorical value k can be computed by formula (3):

$$p_{c,k} = \frac{\sum_{i=1}^{|D|} I(i,c,k)}{|D|} \tag{3}$$

$$I(i,c,k) = \begin{cases} 0, i_c \neq k \\ 1, i_c = k \end{cases} \tag{4}$$

$|D|$ represents the size of the dataset, and i_c represents the value of feature c of sample data i.

Compared with One-Hot Encoding and Label Encoding, PMF based encoding shows obvious dimensionality reduction effect. Giving the number of features N and the number of categorical features C, the ratio of dimensionality reduction of the PMF based encoding can be calculated by formula (5):

$$R_{C \to \infty} = \frac{N + \sum_{i=1}^{C} x_i}{N + C} \geq 1 + \frac{C}{N + C} \approx 2 \tag{5}$$

x_i denotes the number of value of feature i. Because that x_i is greater than 2 at least, so the lower bound of R is always greater than 2 when. C .approaches infinity. Namely, we can at least get 2 times of dimensionality reduction of One-Hot Encoding and Dummy Encoding when categorical features are the main part of all features.

3.2. Anomaly Detection Based on ALI
According to Section 2.2, existing abnormal traffic detection related works cannot guarantee both of high detection performance and the detection ability for unknown attack types. To solve this problem, we investigate the latest research progress about abnormal traffic detection and deep generative model, and we propose a new detection method based on Adversarially Learned Inference (ALI).

Adversarially Learned Inference (ALI) was first proposed by V DUMOULIN [20] based on both of Variational Auto-Encoder (VAE) and Generative Adversarial Network (GAN) to overcome existing problems with which VAE and GAN were faced. While the model was evaluated on MNIST dataset. In this paper, we apply ALI to abnormal traffic detection task.

GAN, VAE and ALI are deep generative models, which means that the methods based on these algorithms have ability to identify unknown attack because of their unsupervised characteristic. Moreover, because of strong representation of deep networks, these methods may get higher accuracy than clustering based methods and classification based methods such as One-class SVM

3.2.1. Methods based on VAE and GAN

VAE is a deep generative network composed of an encoder and a decoder. The principle of generating data is as follows: firstly, the hidden variable z is obtained by sampling from the conditional distribution $p(z \mid x, \theta)$, and then new data is generated by sampling from the conditional distribution $p(z \mid x, \theta)$. Encoder is used to train and get $p(z \mid x, \theta)$ and decoder is used to train and to get $p(x \mid z, \theta)$. When training is successful, generated data will be similar to origin input data. The basic detection principle of VAE based methods is to compare differences between generated data and origin data. For test data, if the difference is small, the test data and training data may belong to the same class with a high probability, which means the test data is normal. If the difference is large, the test data may belong to the opposite class of training data, in this case, the test data is likely to be anomaly data. Figure 2 shows the structure of the three-layer VAE.

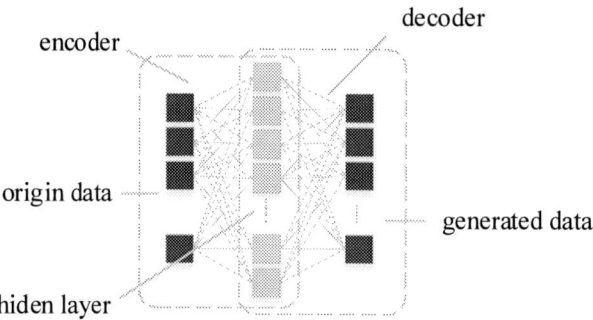

Figure 2 Variational Auto-Encoder example

The loss function of VAE is based on Cross Entropy. Equation (6) shows how to compute Cross Entropy. Equation (7) shows the loss function of VAE.

$$H(p,q) = -\sum_x p(x) \log q(x) \tag{6}$$

$$\zeta(q) = E_{z \sim q(z|x)} \log p_{model(z,x)} + H(q(z \mid x)) \tag{7}$$

$$H(q(z \mid x)) = E_{z \sim q(z|x)} \log p_{model}(x \mid z)$$
$$- D_{KL}(q(z \mid x) \| p_{model}(z)) \tag{8}$$

GAN is a deep generative model based on game theory. It consists of generative network G and discriminant network D. G tries to forge fake data that can be judged as true by D, and D tries to distinguish the generated forged data from G. G and D are trained by turn. According to the zero-sum rule in game theory, training ends up when it is unable to obtain more benefits for either G and D. Therefore, the model of GAN is trained by formula (9).

$$g^* = \arg \min_g \max_d v(g,d) \tag{9}$$

$$v(g,d) = E_{X \sim p_{data}} \log d(x) + E_{X \sim p_{model}} \log(1 - d(x)) \tag{10}$$

p_{data} represents the real distribution of input data, and p_{model} represents the distribution of generated data. The training process of GAN is shown in Figure 3.

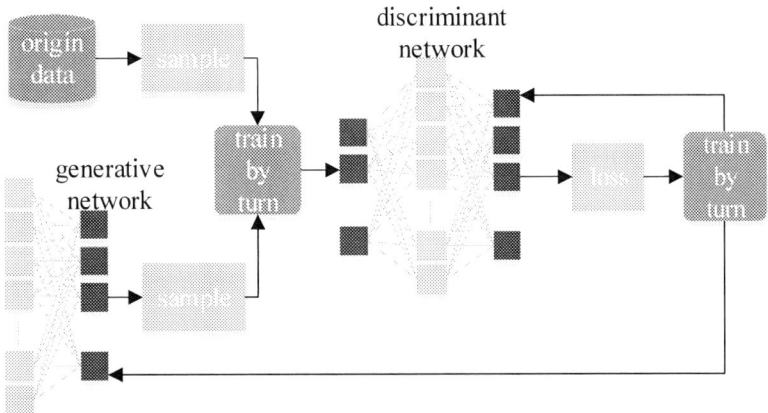

Figure 3 Training process of GAN

Compared with VAE, GAN has lower error[20], but since GAN doesn't have inference network which maps the input data to the hidden layer variables, it takes a lot of time to recover the hidden layer representation in the test phase in order to calculate the overall loss, so the test efficiency of GAN based methods is much lower than the VAE based method.

3.2.2. Anomaly detection based on ALI

ALI is a deep generative network that have complementary advantages of VAE and GAN. Specifically, ALI adds an inference network to GAN (similar to the generator in VAE), and ALI includes generative network, discriminant network and inference network, thus ALI needs only a little time to restore hidden distribution of test data, which makes calculation of overall test error more efficient.

Unlike GAN, the discriminant network in ALI receives a vector pair at the same time, and there are two cases for this vector pair: 1) real data and its encoding; 2) generated data and generated hidden variables in the network. The goal of the discriminator is to determine if they match with each other. ALI is trained by equation (11).

$$g^* = \arg\min_{g} \max_{d} V(d,g) \tag{11}$$

$$V(d,g) = E_{q(x)}(\log(d(x,g_z(x)))) + E_{p(z)}(\log(1 - d(g_x(z),z))) \tag{12}$$

$q(x)$ represents the mapping of the actual feature vector to the hidden layer vector, and $p(x)$ represents the mapping of the hidden layer vector to the actual feature vector. Figure 4 shows the training process of ALI.

The principle of ALI for anomaly detection is similar to VAE and GAN. Firstly, normal data is used to train ALI, and then the test data including the abnormal data is used to test the model. Namely, the abnormal traffic is detected by monitoring the difference of generated data and test data. If the difference of generated data and test data is found to be big, the current test data can be determined as abnormal data.

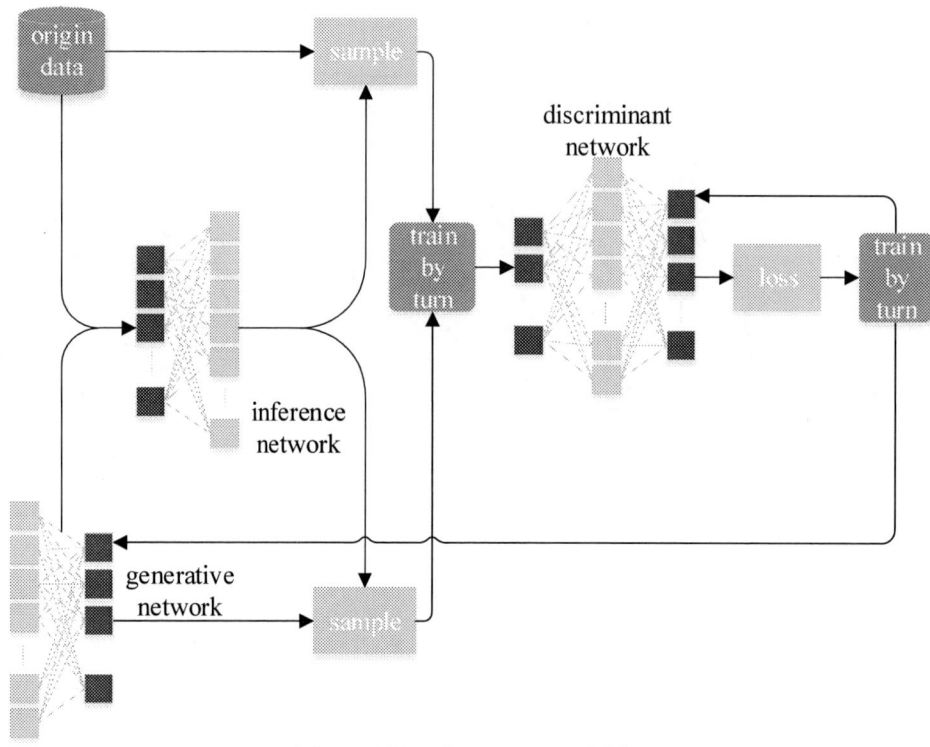

Figure 4 Training process of ALI

The abnormal traffic detection task has the requirements of algorithm detection accuracy, detection efficiency and the ability to recognize unknown attacks. Compared with VAE and GAN, ALI algorithm has higher detection accuracy and higher detection efficiency[20]. ALI also belongs to unsupervised generation model, which means that it has the ability to identify unknown attack. Therefore, we apply ALI into abnormal traffic detection task.

4. Experiment

4.1. Dataset

Table 1 Categorical features in KDD-99 dataset

Features	Number of possible values
protocal type	3
service	70
flag	11
land	2
logged_in	2
is_host_login	2
is_guest_login	2

We evaluate our model based on PMF encoding and ALI with others on real abnormal traffic detection dataset KDD-99. This article used 10% of the KDD-99 dataset as experimental dataset. According to the PMF based encoding method proposed in section 3.1, there are seven categorical features are encoded. Table 2 describes these seven categorical features. The feature dimension based on PMF encoding is 41, and the feature dimension based on One-Hot encoding is 121. Therefore, our encoding method for categorical features reduces the feature dimension by 66%.

In this paper, the features except above seven features are processed by the maximum and minimum normalization method (Maxmin), and the Maxmin value of feature x for sample i can be calculated by formula (13).

$$f(x_i) = (x_i - \min(x)) / (\max(x) - \min(x)) \tag{13}$$

4.2. Environment and evaluation indicators

4.2.1. Experimental environment.
Ubuntu 16.04 LTS operating system, Inter(R) Core(TM) i7-6700 CPU @3.40GHz 3.41GHz processor, 12.0GB RAM.

4.2.2. Evaluation indicator
a)   Detection accuracy $F_1 - measure$, the higher value indicates the higher accuracy;
b)   Train time $train - t$ and test time $test - t$, the shorter time indicates the higher efficiency.

4.3. Training
The experiment dataset is divided randomly into training datasets and test datasets in a ratio of $1:1$. For the training dataset, only the normal data is retained, and the test dataset is not further processed.

The setting of hyper-parameters in ALI model refers to existing related work[21]. In addition, batch size is 50, initial learning rate and initial weight are 0.1. On our dataset, it takes 16 minutes and 20 seconds to finish model training.

4.4. Results and Discussion
We compare our methods with SVM based on binary classification [11] (SVM), standard One-Class SVM (OC-SVM), One-Class SVM with 3 features selected in advance (OC-SVM-3), and GAN based method [18] (GAN).

Table 2 Experiment results of five methods

Methods	F$_1$-measure	train-t	test-t
SVM	0.8318	37.9	1.4
OC-SVM	0.0078	940.5	257.5
OC-SVM-3	0.9691	9.9	1.8
GAN	0.9247	1223	7291
ALI	0.9602	980.3	6.8

The settings of SVM based method refers to the work of H SAXENA [11]. the setting of GAN based method refers to the work of T SCHLEGL [18]. The only difference between OC-SVM and OC-SVM-3 is that OC-SCM-3 use 3 features selected in advance based on experience instead of all origin features. We implement the above models with TensorFlow and Scikit-Learn.

Table 2 gives $F_1 - measure$, $train - t$, and $test - t$ (in seconds) of these five methods on our dataset. Figure 5 shows the confusion matrix of the detection results of ALI method in the KDD test dataset. Among the total of 24,701 test data, 243,107 data is classified correctly.

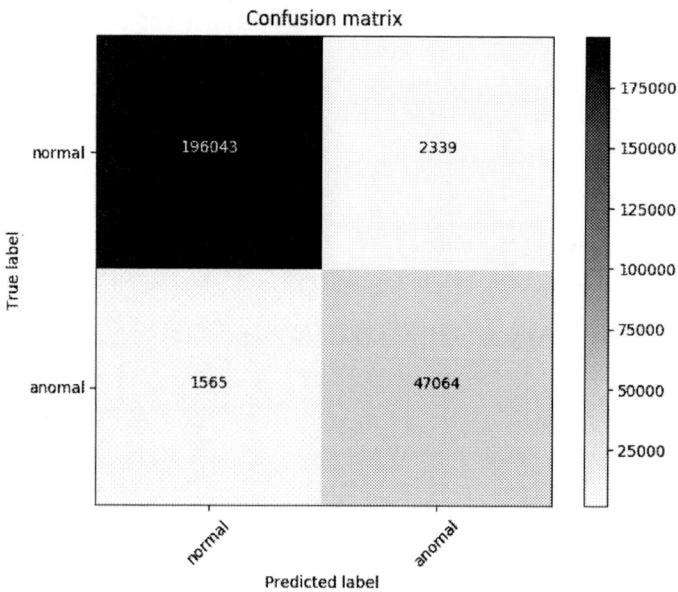

Figure 5 Confusion matrix of ALI based method

We can draw from Table 2 that our method gets higher $F_1 - measure$, $train - t$ and $test - t$ than the other 4 methods. It is worth noting that the test time of our method is 1072 times shorter than GAN based method. The accuracy and efficiency of SVM based method are high relatively, but the fatal disadvantage of this method is its supervised characteristic, which makes it hard to recognize unknown attacks. The accuracy of OC-SVM-3 is higher than OC-SVM, but the pre-selected features depend on manual labor, which makes its generalization ability is limited. Based on above analysis, our method has better performance on accuracy and efficiency than existing methods, with the ability to detect unknown attacks and without manual feature selection, and our method shows better practical application ability

5. Conclusions

Based on the investigation of background, significance and related works of abnormal traffic detection, we propose a new encoding method based on PMF towards categorical features, we also propose a new detection method based on ALI. We evaluate our method on KDD-99 dataset. The experiment results show that our method has the advantages of high detection accuracy and high training efficiency, and has the ability to detect unknown attacks. In the future, we will continue to pay attention to latest generative models and to study their application in the cyberspace security.

Acknowledgments

We would like to express our gratitude to the corresponding author of this paper, Yang Wentai, a respectable, responsible and resourceful scholar, who has provided us with valuable advice in every stage of the writing of this article.

This work is partially supported by Sichuan Science and Technology Department Application Foundation Project (2018JY0193,2018GZDZX0010), the Sichuan Provincial Department of Education focuses on funding research projects (61370065, 61502040), Chengdu Chengdian Power Engineering Design Co., Ltd. and Sichuan University Cooperation Project (17H0781).

References

[1] Tang LJ, Li HY. Survey of Network Anomaly Traffic Detection Based on Large Data[J]. Computer Knowledge and Technology, 2017,13:19-21

[2] Qian YK, Chen M, Ye LX, Liu FR, Zhu SW, Zhang H. Network-Wide anomaly detection method based on multiscale principal component analysis. Journal of Software, 2012,23(2):361−377

[3] LU W, GHORBANI A A. Network anomaly detection based on wavelet analysis[J]. EURASIP J. Adv. Signal Process, 2009, 2009: 1-16.

[4] MOAYEDI H Z, MASNADI-SHIRAZI M. Arima model for network traffic prediction and anomaly detection[C]. Information Technology, 2008. ITSim 2008. International Symposium on, 2008 : 1-6.

[5] SHYU M-L, CHEN S-C, SARINNAPAKORN K, et al. A novel anomaly detection scheme based on principal component classifier[R]. : MIAMI UNIV CORAL GABLES FL DEPT OF ELECTRICAL AND COMPUTER ENGINEERING, 2003.

[6] LAKHINA A, CROVELLA M, DIOT C. Mining anomalies using traffic feature distributions[C]. ACM SIGCOMM Computer Communication Review, 2005 : 217-228.

[7] Zheng LM, Zou P, Han WH, Li AP, J Y. Traffic anomaly detection using multi-dimensional entropy classification in backbone network[J]. Journal of Computer Research and Development, 2012,49(9):1972-1981

[8] PORTNOY L, ESKIN E, STOLFO S. Intrusion detection with unlabeled data using clustering[C]. In Proceedings of ACM CSS Workshop on Data Mining Applied to Security (DMSA-2001, 2001 : .

[9] Zuo J, Chen ZM. Anomaly detection algorithm based on improved K-means clustering[J]. Computer Science, 2016,43(8):258-261

[10] Zhu YW, Yang JM, Zhang JX. Anomaly detection based on traffic information structure[J]. Journal of Software, 2010,21(10):2573-2583

[11] SAXENA H, RICHARIYA V. Intrusion detection in KDD99 dataset using SVM-PSO and feature reduction with information gain[J]. International Journal of Computer Applications, 2014, 98(6).

[12] Li Q, Yan CH, Zhu Y. Analysis and detection of network traffic anomaly based on decision tree[J]. Computer Engineering, 2012,38(5):92-95

[13] YANG J, DENG J, LI S, et al. Improved traffic detection with support vector machine based on restricted Boltzmann machine[J]. Soft Computing, 2017, 21(11): 3101-3112.

[14] LIU F, LI Y, XIA F, et al. A Method of APT Attack Detection Based on DBN-SVDD[J], 2017, : .

[15] FIORE U, PALMIERI F, CASTIGLIONE A, et al. Network anomaly detection with the restricted Boltzmann machine[J]. Neurocomputing, 2013, 122: 13-23.

[16] WULSIN D, BLANCO J, MANI R, et al. Semi-supervised anomaly detection for EEG waveforms using deep belief nets[C]. Machine Learning and Applications (ICMLA), 2010 Ninth International Conference on, 2010 : 436-441.

[17] AN J, CHO S. Variational autoencoder based anomaly detection using reconstruction probability[J]. Special Lecture on IE, 2015, 2: 1-18.

[18] SCHLEGL T, SEEBöCK P, WALDSTEIN S M, et al. Unsupervised anomaly detection with generative adversarial networks to guide marker discovery[C]. International Conference on Information Processing in Medical Imaging, 2017 : 146-157.

[19] FRIEDMAN J H. On bias, variance, 0/1—loss, and the curse-of-dimensionality[J]. Data mining and knowledge discovery, 1997, 1(1): 55-77.

[20] DUMOULIN V, BELGHAZI I, POOLE B, et al. Adversarially learned inference[J]. arXiv preprint arXiv:1606.00704, 2016, : .

[21] [ZENATI H, FOO C S, LECOUAT B, et al. Efficient GAN-based anomaly detection[J]. arXiv preprint arXiv:1802.06222, 2018, : .

Mimic Defense System Security Analysis Model

Li Qinyuan [1], Han Jiajia [1], Sun Xin [1], Qin Junning [2], and Zhang Bo[3]*

[1] State Grid Zhejiang Electric Power Research Institue, Hangzhou, Zhejiang, 310014, China

[2] State Grid Zhejiang Electric Power CO.,LTD, Hangzhou, Zhejiang, 310007, China

[3] Nanjing University of Science & Technology, Nanjing, Jiangsu, 210094, China

*Corresponding author's e-mail: zhangbo3@epri.sgcc.com.cn

Abstract. In this article, the mimic defense technology based on dynamic heterogeneous redundancy architecture is studied, and the mimic defense system is described formally to improve the proactive defense capability of information systems and key equipment in the network. A method integrating dynamic feature, heterogeneous feature and redundant feature that carries out security analysis of mimic defense system by probabilistic analysis is proposed for the lack of effective security assessment method about mimic defense system. Vali-dated by simulation experiments simulating the attacker and the mimic defense system, this model can calculate the safety of the mimic defense system according to the attack factor, heterogeneity degree, dynamic transformation and other related factors. According to the research results, the proposed model is of certain guiding significance for helping designers to construct mimic defense system.

1. Introduction

During the development plan of the "10th Five-Year Plan" period, it was clearly pointed out that all directions of rapid development in China cannot be separated from the support of the information industry. Information technology plays a role as a booster for the development of various fields such as people's livelihood, economy, and national defense. The information construction can be used to promote the development of new industries. Immediately after the "Thirteenth Five-Year Plan", it was pointed out that while pursuing the rapid development of information technology, we must pay more attention to the construction of network security and ensure the healthy development of information technology. While Internet technology is rapidly promoting the rapid development of power, finance, transportation and other industries, the continuous disclosure of security incidents at home and abroad has fully proved that the security of Internet technology is an important challenge in the current Internet development. Traditional cybersecurity defense techniques have increased the attacker's threshold of intrusion and blocked most types of attacks through a large number of rules. However, the traditional network security defense technology is often static before the attacker successfully invades, which allows the attacker to have sufficient time to study the attack target before the attack. Once obtaining the permission of target system, the attacker can continuously infiltrate the attack. In short, the traditional network security defense technology does not actively change over time, and the attacker's attack capability continues to increase with the increase of the amount of information and the intrusion technology, which makes the defender always in a passive defense position. Therefore, it has become an important research direction in the field of network security to break the situation of network attack and

Content from this work may be used under the terms of the Creative Commons Attribution 3.0 licence. Any further distribution of this work must maintain attribution to the author(s) and the title of the work, journal citation and DOI.
Published under licence by IOP Publishing Ltd

defense where it is easy to attack but hard to defend, and break through the situation that network security defense relies on prior knowledge of network attacks, and construct a network security proactive defense mechanism[1].

Based on a "non-similar redundancy" structure with high availability and high reliability, the mimic defense technology[2] takes calculations or service components with equivalent function and different structure as element, cooperates with multi-mode voting mechanism without the basis on rules and features, and disturbs the attacker's judgment by nonlinear transformation of the external features of the system.According to the dynamic heterogeneous redundancy structure, using the harsh conditions that an attacker cannot construct an attack that satisfies all heterogeneous components at the same time, a dynamic scheduling policy that avoids coordinated attacks is introduced to make it difficult for an attacker to maintain the attack chain through a voting mechanism, and increase the difficulty of attackers to detect and scan.

In recent years, more and more research work and research results have focused on developing and advancing mimic defense technology. Proposed in Literature[3]，a DHR (Dynamic Heterogeneous Redundancy) model with "dynamic heterogeneous redundancy" provides security architecture of guiding significance for the construction of mimic defense system; according to the DHR model, the literature [4] uses the multi-level structure of the web server to design the "dynamic heterogeneous redundancy" basis in the operating system layer and the server software layer, and realize the establishment of the mimic defense system in the field of web servers; in the literature [5] , according to the DHR model, seven sets of heterogeneous redundant routing kits, such as ZTE ZXR10, FiberHome Fengine S5800, and Maipu MP3900, are applied in the router application layer to realize the establishment of mimic defense system in the router field. However, very little research has focused on the safety analysis methods of the mimic defense system.

Established in this paper based on the network attack chain and the dynamic heterogeneous redundancy characteristics of the mimic defense system, a security analysis model can be used to calculate the possibility that each key node in the mimic defense system can be successfully attacked, so that different parameters can be used to analyze the security defense effectiveness of the mimic defense system, which makes it easy for system designers to better understand how to use mimic defense technology to improve the security of key information systems.

2. Construction of mimic defense system security analysis model

2.1. Construction of mimic defense
The basic model of the dynamic heterogeneous redundancy model is mainly iPo model [6]. As shown in Figure 1, when entering the system, the submitted request input is first copied into n parts to be transmitted to the execution set through the input proxy unit. The execution set contains n non-similar redundant executables (P1, P2, P3, ..., Pn), where P1, P2, P3, ..., Pn are executive bodies with the same functions but different implementation methods; each executive body accepts a copy of the request and processes it; the response of processing result of each executive body is outputted after voting by the voter. Taking advantage of the cyber attack's dependence on the environment, an attack against a specific vulnerability cannot be effectively played in the heterogeneous executive body (P1, P2, P3, ..., Pn), thus achieving the defense against the vulnerability attack. At present, systems with mimic defense structure such as the mimic defense construction router, the mimic defense construction distributed storage system and mimic defense construction web server have been formed based on the mimic defense technology.

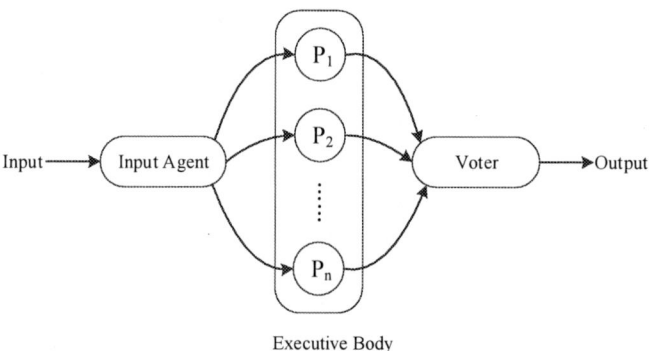

Fig. 1 iPo model

On this basis, the dynamic heterogeneous redundancy (DHR) model [7] adds heterogeneous component sets, dynamic scheduling algorithms [8] and heterogeneous element pools. The heterogeneous element pool provides the diversity design of components at various levels, which can form a heterogeneous set of heterogeneous components to improve the security of the system. When the executive body is attacked by the executive body set, the system selects the component in the heterogeneous component set to replace the executed body that is attacked in the execution body set according to the dynamic scheduling algorithm. The environment necessary for attack triggering is eliminated to make it difficult for the same attack to occur continuously. On the other hand, the existence of a dynamic scheduling algorithm causes the system to present different system attributes during the period change, which disturbs the attacker's judgment and increases the attacker's scanning detection difficulty. If the system has a DHR model, it is said to conform to the structure of mimic defense.

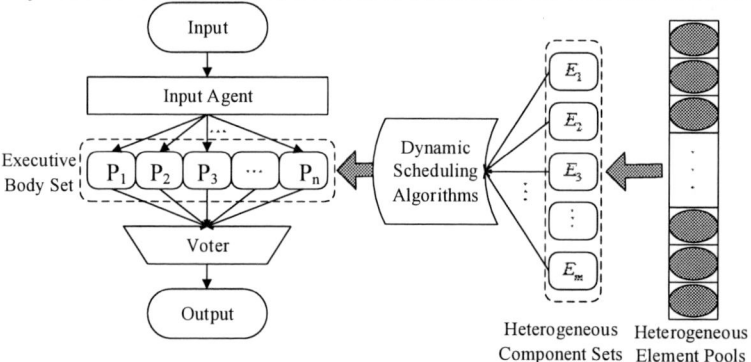

Fig. 2 dynamic heterogeneous redundancy model

2.2. Model introduction

The mimic defense system security analysis model is designed to deeply understand how key features in the mimic defense system affect the security defense capabilities of the system and guide the design and implementation of the mimic defense system. Therefore, the model should have good computational efficiency and scalability, and clearly demonstrate the security defense capabilities of the mimic defense system. The key challenge in designing a security analysis model for mimic defense systems is the non-monotonic nature of the mimic defense system. Because the heterogeneous nature of the mimic defense system and the multi-mode arbitration mechanism can block the information chain between the attacker and the attack target, and the multidimensional dynamic performance of the mimic defense system can disrupt the judgment of the attacker, the typical assumption in the security analysis model—the attacker has enough time to discover, infiltrate, exploit the vulnerability, can not be applied to the mimic defense system.

If the mimic defense system is modeled by Markov chain [9], the explosive growth state space of the model will appear with the increase of network scale, which makes the difficulty of analysis increased. In addition, the non-monotonicity of the mimic defense system does not conform to the assumption that the current node state in the Markov chain depends only on the state of its previous node.

Based on the above situation, the model is made in this paper according to the state of the node, that the attacker's location is transferred from the current node to its next node is taken as the successful invasion of current node by attacker; that the attacker's location is transferred from the current node to its previous node is considered as and the control loss of current node by attacker. This situation in the mimic defense system is usually expressed as the heterogeneous dynamic transformation of the node or an abnormality in the voting output of the voter. In order to avoid the difficulty of model analysis increased by the forward and backward transfer between a large number of nodes with the increase of network scale, the model only focuses on the attacker's invasion of the next node from the current node and staying at the current node, and the transfer situation of the attacker to the previous node is indirectly analyzed to reduce the complexity of the transfer between nodes.

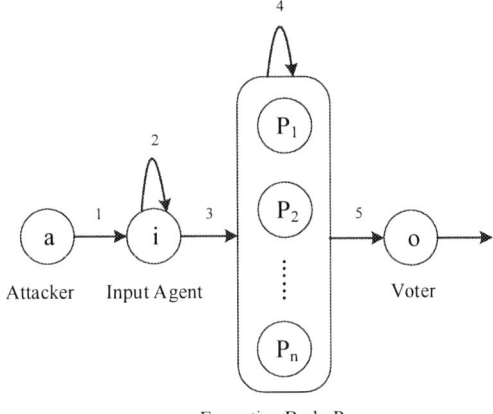

Fig. 3 The structure diagram of security analysis model

The model structure is shown in Figure 3, where the node a represents the attacker; the node i represents the input proxy module in the mimic defense system; the logical P node is represented by the executive body set in the mimic defense system, where P1, P2, ..., Pn represents specific executive body nodes; the o node is represented as the voter in the mimic defense system. These two nodes are the mimic defense boundary of the system without heterogeneous redundancy features. Therefore, the dynamic defense technology is used to prevent the attacker from using the input agent as a springboard to continuously attack the executable body of P1, P2 ... Pn and the hijacking voter to tamper with the correct output of the system.

The 1, 2, 3, 4, and 5 processes in the model represent the transfer process between nodes, where 1, 3, and 5 represent the process of the attacker invading the next node from the current node; 2, 4 represents the process of attacker staying at the current node.

3. Model security analysis

3.1. Model assumptions and input parameters
It is assumed when this model is used to evaluate the security of the mimic defense system that there is sufficient heterogeneous executive body to construct the mimic defense structure for any kind of attack,

which is not subject to the diversity of hardware and software. At the same time, the model has the following input parameters:

$T_{dynamic}$: It is the time period of the dynamic transformation when proxy node, the executive body node, and the voter node are input in the model, which reflects the dynamic characteristics of the mimic defense structure. $T_{dynamic}$ can be fixed values and random values. In this paper, only fixed values of $T_{dynamic}$ are considered;

T_{attack} : It is the time required for the attacker to successfully invade its next node in the model, which reflects the complexity of the attacker successfully implementing an attack;

p_h: It is the probability that the heterogeneous attributes are expressed for a certain attack among the execution body nodes in the model. That is, the probability that two executive body nodes produce different results in a certain attack is p_h, which reflects the heterogeneous characteristics in the mimic defense structure;

$p_{(i,j)}$: It is the probability that an attacker successfully invades from node i to the next node j in a static system without heterogeneous characteristics and dynamics, which reflects the difficulty of the attacker successfully implementing the attack.

3.2. Model analysis

In order to analyze the security of the model, p_1、p_2、p_3、p_4、p_5 are respectively used to indicate that the possibilities of the invasion of node a into the input agent node i, continuing to stay at the i node, the invasion from i node to logic P node, continuing to stay at a logic P node, and invasion of logic P-node to voter o node. The derivation process of p_1 is as follows: node i is near any time node that dynamically changes, the probability that the specific attack initiated by the attacker is heterogeneous is p_h. Therefore, the probability that node i does not affect the continuous implementation of attack at any time of dynamical change is $1 - p_h$. Node i can undergo up to $T_{attack} / T_{dynamic}$ dynamic transformations during a period of successful attack implementation T_{attack}. Therefore, the probability that dynamic transformation does not affect the attack for the node i within the unit time required to complete the intrusion attack is $(1 - p_h)^{T_{attack}/T_{dynamic}}$. Based on the above analysis, the possibility that an attacker successfully invades node i from node a can be expressed as:

$$p_1 = p_{(a,i)} \times (1 - p_h)^{T_{attack}/T_{dynamic}} \tag{1}$$

After successfully invading the node i, the attacker can initiate $T_{dynamic} / T_{attack}$ times of attacks to the executive body node P in each dynamic transformation period. Then the possibility that all the intrusion attacks from node i to node P fail is $(1 - p_{(i,P)})^{T_{dynamic}/T_{attack}}$. Therefore, the possibility of successful intrusion from node i to node P during the dynamic transformation period $T_{dynamic}$ is:

$$1 - (1 - p_{(i,P)})^{T_{dynamic}/T_{attack}} \tag{2}$$

The attacker will continue to stay at node i with T_{attack} if one of the following two situations occur: First, the penetration attack initiated by the attacker from the node i to the node P fails, and the dynamic transformation of the node i does not affect the attack initiated by the attacker. In this case, the probability that the attacker continues to stay at the node i is:

$$(1 - p_{(i,P)})^{T_{dynamic}/T_{attack}} \times (1 - p_h)^{T_{attack}/T_{dynamic}} \tag{3}$$

Second, the penetration attack initiated by the attacker from node i to node P is successful, and the dynamic transformation of node i does not affect the attack initiated by the attacker. But the dynamic transformation of node P affects the effective implementation of the attack. In this case, the probability that the attacker continues to stay at node i is:

$$
\begin{aligned}
&(1-(1-p_{(i,P)})^{T_{dynamic}/T_{attack}}) \times \\
&(1-p_h)^{T_{attack}/T_{dynamic}} \times (1-(1-p_h)^{T_{attack}/T_{dynamic}})
\end{aligned}
\tag{4}
$$

Combining the above two cases, the ultimate possibility that the attacker continues to stay at node i is expressed as:

$$
\begin{aligned}
p_2 &= (1-p_{(i,P)})^{T_{dynamic}/T_{attack}} \times (1-p_h)^{T_{attack}/T_{dynamic}} + \\
&\quad (1-(1-p_{(i,P)})^{T_{dynamic}/T_{attack}}) \times (1-p_h)^{T_{attack}/T_{dynamic}} \times \\
&\quad (1-(1-p_h)^{T_{attack}/T_{dynamic}}) \\
&= (1-p_h)^{T_{attack}/T_{dynamic}} - (1-(1-p_{(i,P)})^{T_{dynamic}/T_{attack}}) \times \\
&\quad (1-p_h)^{2T_{attack}/T_{dynamic}}
\end{aligned}
\tag{5}
$$

Similarly, under the situation that dynamical changes initiated by the attacker at the node i and the node P do not affect the attack initiated by the attacker, the probability of successful invasion from the node i to the node P is $(1-(1-p_{(i,P)})^{T_{dynamic}/T_{attack}})$, and thus p_3 can be expressed as:

$$
p_3 = (1-(1-p_{(i,P)})^{T_{dynamic}/T_{attack}}) \times (1-p_h)^{2T_{attack}/T_{dynamic}}
\tag{6}
$$

Through the expression of p_1, p_2, p_3, the possibility p_P that the attacker successfully invades the node P from the node a can be calculated as:

$$
\begin{aligned}
p_P &= p_1 \times (p_2^{0} + p_2^{1} + \ldots + p_2^{n}) \times p_3 \\
&= \frac{1}{1-p_2} \times p_{(a,i)} \times (1-(1-p_{(i,P)})^{T_{dynamic}/T_{attack}}) \times \\
&\quad (1-p_h)^{3T_{attack}/T_{dynamic}}
\end{aligned}
\tag{7}
$$

Next, the possibility of successful invasion from node a to node o is calculated. First, p_4 and p_5 should be calculated, which represents the possibility that the attacker continues to stay at the node P, the possibility of successful invasion from the node P to the node o, and the possibility of continuously staying at the node o. According to process representation and analysis method of p_1, p_2 and p_3, the expressions of p_4 and p_5 are calculated as follows:

$$
\begin{aligned}
p_4 &= (1-p_h)^{T_{attack}/T_{dynamic}} - (1-(1-p_{(P,o)})^{T_{dynamic}/T_{attack}}) \\
&\quad \times (1-p_h)^{2T_{attack}/T_{dynamic}}
\end{aligned}
\tag{8}
$$

$$
p_5 = (1-(1-p_{(P,o)})^{T_{dynamic}/T_{attack}}) \times (1-p_h)^{2T_{attack}/T_{dynamic}}
\tag{9}
$$

Through the expressions of p_1, p_2, p_3, p_4, p_5, the possibility p_o that the attacker successfully invades the node o by the node a is calculated as:

$$
\begin{aligned}
p_o &= p_1 \times (p_2{}^0 + p_2{}^1 + \ldots + p_2{}^n) \times p_3 \times \\
&\quad (p_4{}^0 + p_4{}^1 + \ldots + p_4{}^n) \times p_5 \\
&= \frac{1}{1-p_2} \times p_{(a,i)} \times \left(1 - (1 - p_{(i,P)})^{T_{dynamic}/T_{attack}}\right) \times \\
&\quad (1 - p_h)^{3T_{attack}/T_{dynamic}} \times \frac{1}{1-p_4} \times \\
&\quad \left(1 - (1 - p_{(P,o)})^{T_{dynamic}/T_{attack}}\right) \times (1 - p_h)^{2T_{attack}/T_{dynamic}} \\
&= \frac{1}{1-p_2} \times \frac{1}{1-p_4} \times p_{(a,i)} \times \left(1 - (1 - p_{(i,P)})^{T_{dynamic}/T_{attack}}\right) \\
&\quad \times \left(1 - (1 - p_{(P,o)})^{T_{dynamic}/T_{attack}}\right) \times (1 - p_h)^{5T_{attack}/T_{dynamic}}
\end{aligned}
\tag{10}
$$

4. Conclusion

This paper proposes a model integrating dynamic feature, heterogeneous feature and redundant feature that performs the safety analysis of the mimic defense system using the probabilistic analysis method, which provides innovative ideas to solve the problem of the lack of effective safety assessment methods for the mimic defense system. However, some simplifications and assumptions have been made during the construction of the model. For example, the attacker's attack ability and the heterogeneity of the system are expressed by probability, which is impossible to comprehensively consider different attack types and mimic defense systems. In order to improve the accuracy and applicability of the model, key factors such as how to assess the attacker's attack ability and the heterogeneity of the system will be studied in the future.

Acknowledgments

Supported by the science and technology project of State Grid Zhejiang Electric Power CO.,LTD (Grand No. SGGR0000XTJS1800310).

References

[1] Yang lin, Yu Quan. Dynamically-enabled Cyber Defense [M]. Posts & telecom Press, 2016.
[2] Wu Jiangxing. Meaning and Vision of Mimic Computing and Mimic Security Defense [J]. Telecommunications Science, 2014, 30(7): 1-7.
[3] Wu Jiangxing. Research on Cyber Mimic Defense [J]. Journal of Cyber Security, 2016, 1(4): 1-10.
[4] Tong Qing, Zhang Zheng, Wu Jiangxing . The Active Defense Technology Based on the Software/Hardware Diversity [J]. Journal of Cyber Security, 2017, 2(1): 1-12.
[5] Ma Hailong, Yi Peng, Jiang Yiming, He Lei. Dynamic Heterogeneous Redundancy based Router Architecture with Mimic Defenses [J]. Journal of Cyber Security, 2017, 2(1): 29-42.
[6] Luo Xingguo, Tong Qing, Zhang Zheng. Mimic Defense Technology [J]. Engineering Science, 2016, 18(6): 69-73.
[7] Hu Hongchao, Chen Fucai, Wang Zhenpeng. Performance Evaluations on DHR for Cyberspace Mimic Defense [J]. Journal of Cyber Security, 2016, 1(4): 40-51.
[8] Ma B, Zhang Z. Security Research of Redundancy in Mimic Defense System [C]. Advanced Information Management, Communicates, Electronic and Automation Control Conference, 2017: 1447-1451.
[9] Xie G. RELIABILITY THEORY OF THE GENERALIZED GAUSS-MARKOFF MODEL AND ITS SIMPLE APPLICATION[J], 1989.
[10] Ma B, Zhang Z, Zhu Y. A formalization research on web server and scheduling strategy for

heterogeneity[C]. Advanced Information Management, Communicates, Electronic and Automation Control Conference, 2017: 1447-1451.

ISPECE   IOP Publishing

IOP Conf. Series: Journal of Physics: Conf. Series **1187** (2019) 052039   doi:10.1088/1742-6596/1187/5/052039

The Present of Education Big Data Research in China: Base on the Bibliometric Analysis and Knowledge Mapping

Cheng Jiang[1], Qi Wang[2], Wu Qing [2, *], LeiYe Zhu[1], Si Cheng[1], HaiBin Wang[3,1, *]

[1] School of Educational Science, Huangshan University, Huangshan, 245041, China

[2] Student Office, Huangshan University, Huangshan, 245041, China

[3] School of Business Administration, Zhejiang Gongshang University, Hangzhou 310018, China

*Corresponding author. Wu Qing, qingwu@hsu.edu.cn or HaiBin Wang, wanghaibin@hsu.edu.cn

Abstract. Big data is a term used to refer to data sets that are too large or complex for traditional data-processing application software to adequately deal with. With the deep integration between information technology and education, researchers at home and abroad have made great progress in the field of big data in education. Nevertheless, a comprehensive quantitative analysis of the emerging research trends and topics has not yet been discovered. To reveal the research characteristics and current status of educational big data, 2052 related papers from the China National Knowledge Infrastructure (CNKI) were analyzed by CiteSpace V software. The results display that: (1) the domain of education big data was launched in 2013 and has experienced a rapid growth in the past five years, reaching an all-time high in 2017. XianMing Yang and Central China Normal University are at the forefront of contributing authors and organization respectively. (2) "Education in ideology and politics" is the most frequently-used keywords and "electronic schoolbag", "educational reform" and "MOOC" have the longest outbreak range. "Big data era", "educational innovation", "big data", "educational evaluation", "educational management", "educational application", "higher education" and "educational informatization" are the top eight largest clusters. Those are the topical research topics in the field of educational big data.

1. Introduction

Big data is a term used to refer to data sets that are too large or complex for traditional data-processing application software to adequately deal with, which has the characteristics of sheer volume, wide variety, high velocity and extreme veracity [1]. With the deep integration between information technology and education, great progress has been made by researchers at home and abroad, in areas of big data in education. Thereafter, western researchers from different disciplines began to the study of big data in education, and obtained important achievements, including the construction of education big data technology system, the educational application of big data, the teaching innovation with big data, and so on [2-3].

The 13th Five-Year Plan for the Development of National Education issued by The State Council in 2017, clearly proposed "accelerating the construction of education big data and open sharing" [4]. Although the focus on education big data starts relatively late in China [5], China's educational big data research has achieved a rapid development in the context of the country's emphasis on education big data. Chinese researchers pay more attention to the technology and application research of

Content from this work may be used under the terms of the Creative Commons Attribution 3.0 licence. Any further distribution of this work must maintain attribution to the author(s) and the title of the work, journal citation and DOI.
Published under licence by IOP Publishing Ltd

education big data among these achievements. In the technology research of education big data, researchers explore the techniques of collecting, mining and analyzing educational big data, such as with the adoption of the Python [6] and a map-based visual analysis method for big data in education [7]. In the application research of education big data, researchers start with the significance and function of the education big data construction [8-9], and the educational reform by education big data [10]. As has been shown, Chinese researchers have acquired rich achievement in the area of big data in education in the past 5 years. Nevertheless, a comprehensive quantitative analysis to objective and effective authors and organizations have not been completed, along with the burgeoning research trends and topics. For the sake of better promoting the development of Chinese education big data, this paper has conducted a quantitative review of education big data research in China based on the bibliometric analysis and knowledge mapping with the support of the CiteSpace V.

2. Methodology

2.1. data sources
The source and reference analytical papers come from the China National Knowledge Infrastructure (CKNI). There are 2052 papers of all types published conducting in the CNKI collection, under the headings "big data" and "education" or "educational big data". The data sources contain 1991 journal articles and 61 Dissertations.

2.2 data processing
2052 papers were saved in RIS format which includes the crucial information, such as titles, authors' names and affiliations, abstract and keywords. The bibliometric analyses and knowledge mapping were conducted with Java CiteSpace V software [11], which supports the construction and visualization of bibliographic record networks.

3. Results

3.1 Bibliometric analysis results

3.1.1 Chronology statistics. In view of annual analysis, the number of annual studies from 2013 to 2018 in China is revealed in Fig. 1. Chinese research of education big data began in 2013 and showed a year-on-year growth trend, reaching an all-time high in 2017.

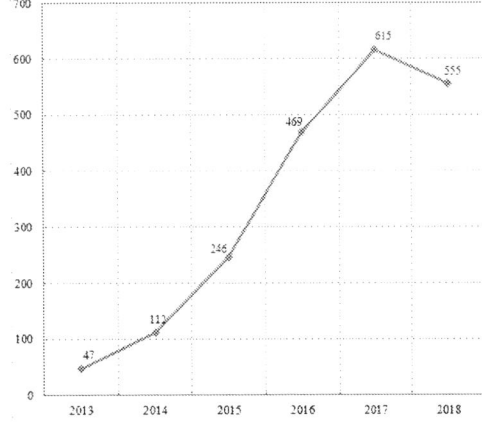

Fig 1. The annual research of education big data published in China.

3.1.2. Author statistics. Table 1 shows the Chinese top 10 outstanding authors and the number of their published papers. As shown, XianMing Yang tops the list with a total of eleven articles, followed by KeYun Zhao and Hai Zhang, each of which has published six articles.

Table 1. Top 10 contributing authors and number of published papers.

Rank	Author	Frequency
1	XianMing Yang	11
2	KeYun Zhao	6
3	Hai Zhang	6
4	PengGao Zhang	4
5	NanZhong Wu	4
6	HongHai Yu	3
7	YanNan Zhang	3
8	FuXiang Hu	3
9	HuaiJie Li	3
10	FanGang Hu	3

3.1.3. Affiliation statistics. Table 2 depicts the Chinese top 10 organizations facilitating the studies on education big data. As indicated, Central China Normal University has the best performing, followed by East China Normal University and Southwest University. Not surprising, nine of the top 10 outstanding organizations are normal university, proving that normal universities play a totally important role in education big data research.

Table 2. Top 10 contributing organizations.

Rank	Author	Frequency
1	Central China Normal University	30
2	East China Normal University	29
3	Southwest University	29
4	Northeast Normal University	23
5	Beijing Normal University	23
6	Qufu Normal University	18
7	Jiangsu Normal University	17
8	University of Electronic Science and Technology of China	16
9	Shaanxi Normal University	16
10	Fujian Normal University	14

3.2. Knowledge mapping results

3.2.1. High frequency keywords. Table 3 lists the top 10 keywords of studies on Chinese education big data, apart from their frequencies. Obviously, the most frequently cited keywords are "education in ideology and politics" in addition to the subject keywords "big data". In addition, "colleges and universities", "innovation", "undergraduate" are positioned in the 2[ed] 3[rd]and 4th places respectively, except the subject keywords. It indicates that the current researchers focus on the educational innovation with the help of the big data technology.

Table 3. Top 10 keywords.

Rank	Keywords	Freque
1	big data	2161
2	big data era	732
3	education in ideology and politics	563
4	big data in education	264
5	colleges and universities	234
6	innovation	177
7	undergraduate	152
8	university's ideological and political education	144
9	education	111
10	higher education	111

3.2.2. Popular research trends. To seek the popular research trends, the keywords with citation burst are analyzed in this paper. Fig 2 presents the top ninth keywords with the strongest citation. As showed, "electronic schoolbag", "educational reform" and "MOOC" have the longest outbreak range, each of which exceeds more than two years. Furthermore, the topic of "data processing" is one of the latest research frontiers, indicating that the processing technology of education big data can be the popular research trends in future.

Top 9 Keywords with the Strongest Citation Bursts

Keywords	Year	Strength	Begin	End	2013 - 2018
educational evaluation	2013	3.2246	**2013**	2014	
efficient path	2013	2.149	**2013**	2014	
electronic schoolbag	2013	4.6733	**2013**	2015	
educational reform	2013	10.2203	**2013**	2015	
data mining technology	2013	2.149	**2013**	2014	
flipped classroom	2013	5.3783	**2013**	2014	
evaluation system	2013	2.149	**2013**	2014	
MOOC	2013	6.2178	**2014**	2016	
data processing	2013	5.8623	**2015**	2016	

Fig 2. Top ninth keywords with the strongest citation bursts.

3.2.3. Popular research topics. In order to explore popular research topics, this paper analyzes the keywords with clustering. With running CiteSpace, it presents 39 clusters whose Modularity Q is 0.898 and the Mean Silhouette is 0.901. Fig 3 presents the top eight largest clusters. They were big data era, educational innovation, big data, educational evaluation, educational management, educational application, higher education and educational informatization. As presents, big data era (#0) is the largest cluster. It contains 32 articles and its Silhouette is 0.972. The cluster's high-frequency keywords include "big data era", "information innovation" and "digital campus". Educational innovation is the second largest cluster (#1), which includes 27 articles and its Silhouette is 0.864. The high frequency keywords of this cluster contain "innovation", "big data technology", "education mode" and "educational innovation". Big data is the third largest cluster (#2), which comprises 25 articles and its Silhouette is 0.983. The high-frequency keywords in this cluster contain "big data", "ideological and political education", "undergraduate" and "data mining".

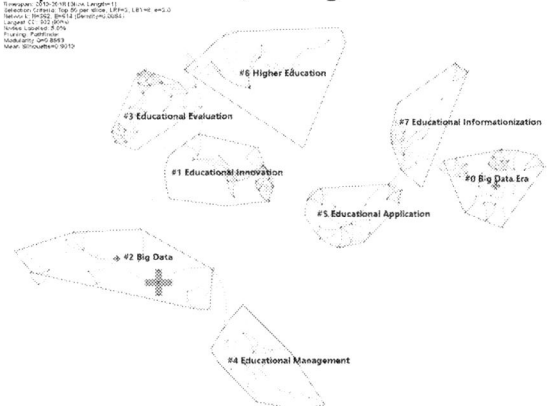

Fig 3. Clusters of the current researches.

4. Conclusions and further research

In the past five years, there are a large number of papers on education big data having published in China. Nevertheless, a comprehensive quantitative analysis to objectively identifying influential researchers and organizations as well as the emerging research trends and topics has not yet been accomplished. To order to fill this vacancy, with the help of the CiteSpace V software, this paper quantitatively reviews China's education big data research on account of the bibliometric analysis and knowledge mapping. It was presented the chronology, author and affiliation statistics in the bibliometric analysis section. The results show that research of education big data in China started since 2013 then showed the year-on-year growth trend since 2013, and reached its highest point in history in 2017 with the annual number 615. Xianing Yang tops the list with a total of eleven articles and Central China Normal University has the best performing among the contributing organizations. Furthermore, it is discovered that nine of the top 10 outstanding organizations are normal university, proving that normal universities play a totally important role in education big data research.

In the section of Knowledge mapping analysis, the high frequency keywords are obtained. As shown in table 3, the most frequently cited keywords were "education in ideology and politics", "colleges and universities", "innovation", "undergraduate", "university's ideological and political education" and "higher education" besides the subject keywords "big data" or "education", illustrating that Chinese researchers pay more attention to the application and educational innovation with the help of the big data technology, especially the education in ideology and politics among undergraduates. It is particularly important to discover the major popular research trends and topics in the field of Chinese education big data through keywords bursting and clustering. Fig 2 shows the top nine keywords with the strongest citation bursts. "Electronic schoolbag", "educational reform" and "MOOC" have the longest outbreak range, illustrating that the three topics are the primary popular research trends in last years. In addition, the topic of "data processing" is one of the typical frontiers of recent research, indicating that the processing technology of education big data can be the fashionable research trends in future. The top eight largest clusters are depicted in Fig 3. Big data era (#0) is the largest cluster, which includes 32 articles. And "big data era", "information innovation" and "digital campus" were the high frequency keywords. Educational innovation (#1) is the second largest cluster, which includes 27 articles. And the high frequency keywords were "innovation", "big data technology", "education mode" and "educational innovation". Big data (#2) is the third largest cluster, which contains 25 articles, and the high frequency keywords were "big data", "ideological and political education", "undergraduate" and "data mining". It implies that the current study focuses on the education application and reform with the support of the big data technology.

Obviously, there are still some limitations in this research. The database used is only from the China National Knowledge Infrastructure (CKNI). As a matter of fact, there are generous excellent articles about education big data in abroad. Future research needs to further expand the scope of the search and to further explore the current status quo of the education big data research.

Acknowledgement

This research was supported by the Key Project of Humanities and Social Sciences in Anhui Province of China (Grant NO. SK2014A390 and SK2016A0882) and the Teaching research project of Anhui Province in China (Grant NO. 2017zhkt414 and 2017jyxm0445).

References

[1] Breur, T. (2016) Statistical power analysis and the contemporary "crisis" in social sciences. Journal of Marketing Analytics, 4:1-5.

[2] ShiJing, L., Rong, Z. (2018) A comparative analysis of research hot spots and trend on educational big data at home and abroad. China medical education technology, 32: 240-245.

[3] RenFeng, H., Yao, L. (2018) Analysis on the hotspots of educational big data research. Journal of Yanbian Institute of Education, 32: 58-60

[4] The State Council. (2017) Notice of the State Council on Printing and Distributing the 13th Five-Year Plan for the Development of National Education. http://www.moe.gov.cn/jyb_xxgk/moe_1777/moe_1778/201701/t20170119_295319.html

[5] XiaoJun, W. (2013) The future of education big data. Primary and secondary school information technology education, 2: 23.

[6] ShiChun, W., XinHua, X., HongChun, Z., JiaChen, H. (2018) Research on the influence of

individual differences on python crawler's access to educational big data. China Education Informationization, 9:79-81.

[7] SanNvYa, L., DongBo, Z., Hao, L., JianWen, S., Jie, Y. (2018) A study of map-based visual analysis method for big data in education. Audio-visual education research, 7:49-56.

[8] XiaoMin, L. (2017) Analysis of the current situation of the construction of educational big data in China and the initial establishment of the framework. China's electrification education, 8:128-131.

[9] XiaoFeng, Y. (2016) Education in the era of big data: prospect and action. Higher education research, 37:7-12.

[10] Ying, P. (2018) Problems and challenges in the study of education big data in China. Modern trade industry, 39:58-59.

[11] Chen, C. (2017) Expert Review. Science Mapping: A Systematic Review of the Literature. Journal of Data & Information Science, 2: 1-40.

Method for Creating, Updating and Maintaining a Case Library of Service Business Guidance

Xin Xu[1,2], Zhijie Sun[1,2], Jun Fu[1,2], Li Wang[1,2], Feng Xie[1,2]

[1]State Grid Jibei Electric Power Company Limited Electric PowerResearch Institute

[2]CountryHuadian Electric Power Research Institute Co., Ltd.

Corresponding author's e-mail: 676269930@qq.com

Abstract. This article provides a method for creating, updating and maintaining a case library of service business guidance. On the one hand, it uses genetic algorithm to filter and summarize cases that have been processed in the service business area to form a guidance case library. Therefore,service business staff are able to query and refer to similar existing cases when dealing with business in a follow-up; on the other hand, in the follow-up continuous use of new cases, new business in accordance with the new policy to update the guidance case base, the guidance case library always has the latest and most comprehensive cases to meet the needs of various service business.

1. Introduction

95598 is a customer-oriented service window for power supply enterprises. It is a direct channel for communication with the public and service objects. It is also a vital part of building a harmonious power supply environment. With the enhancement of social service supervision and customer service demands, the service quality of 95598 service also has higher requirements.

Prior to the operation of 95598, in order to provide business guidance to the service staff of the power supply service, the case summary of various business types is generally classified into a guidance case by the manual, which is completely dependent on the manual completion. The efficiency is low, and the maintenance cost is high.After the operation of 95598, in order to ensure that the service personnel of the power supply service of the provincial and municipal remote workstations can be familiar with the processing flow of various business work orders, the provincial power companies need to carry out pre-job system training and operation exercises for the service personnel of the power supply service. However, due to the change of the working mode, the manual summary guidance case is more complex, which brings great difficulty to the training work.

In addition, with the introduction of the new policy, there will be a new type of business appearance, which will provide the service personnel of the power supply service with a good command of the process, and can bring a lot of difficulties to the accurate strain in the case of various types of business cases.The case keywords can be selected by the genetic algorithm[1-8], and the case library can be created according to the keywords, and the case database can be updated and maintained by the latest policies and new cases, so as to form the latest and comprehensive guidance case library, and provide pre-job training for the service business personnel and business processing.

2. Creation method of case libraryof service business guidance

2.1. Creation method of case libraryof service business guidance

Depending on the characteristics of the service business, the keyword "9""10" needs to be extracted in each case to identify a number of alternative keywords. When implemented, a number of words reflecting the business characteristics of this type of service are identified as alternative keywordsdepending on the type of service business to be applied.For example, for the power supply service business, the following alternative keywords can be identified: power outage, re-power, ladder electricity price, electric vehicle, photovoltaic power generation, meter reading, low voltage, power theft, home appliance damage, arrears, default electricity, electricity invoices, fault repair …

Using genetic algorithm to screen the optimal keyword from alternative keywords, the method comprises the following steps: establishing an initial population by using a plurality of alternative keywords as an individual, and calculating a short-term ratio and a total proportion of each individual in the current population. The near-term ratio refers to the proportion of the case with the individual in the recent case, and the total proportion refers to the proportion of the case with the individual in the total case source.For example, in the field of power supply services, the "total case source" includes all business cases received over the past two to three years as of the creation of the service business guidance case base. "The source of the recent case" includes all business cases received in the last quarter as of the creation of the service business guidance case library.

All individuals in the current population are sorted in ascending order of the recent share. Using the following fitness functions to calculate the fitness of each individual in the current population:

$$ft(i) = \sum_{n=1}^{i} CR(n) \qquad (1)$$

$$CR(i) = \begin{cases} 1, & |PO(i) - AO(i)| \leq T \\ 0, & |PO(i) - AO(i)| > T \end{cases}$$

Where i is the ranking ordinal of an individual in the current population, $PO(i)$ is the recent share of an individual with serial number I, $AO(i)$ is the total share of an individual with a serial number i, and $ft(i)$ is the fitness of an individual with an ordinal number i, and$T$ is the threshold value.

Since $ft(i) = CR(1) + CR(2) + CR(3) \ldots\ldots + CR(i-1) + CR(i)$, therefore, the larger the number of individuals, the more matching coefficients included in their fitness calculation; the smaller the number of individuals, the less the matching coefficient included in their fitness calculation; When an individual with an ordinal number i, the matching coefficient $CR(i) = 1$, the fitness of the individual is greater than that of any individual in front of it; When an individual with an ordinal number i, the matching coefficient $CR(i)$=0, the fitness of the individual is not less than the fitness of any individual in front of it; In other words, individuals with larger serial numbers are likely to be more adaptable and have a higher probability of being retained in the population.

The average of the fitness of all individuals in the parent population of the current population is calculated.For each individual in the current population, it is determined whether the fitness of the individual is less than the average of the fitness of all individuals in the parent population, and if so, the individual is determined to be a non-conforming individual.All non-qualified individuals are discarded from the current population.A genetic operation is performed on the remaining individuals in the current population to produce a new generation of populations.

In which, for the initial population, the initial population is considered to be the average of the fitness of all individuals as a result of their absence of a parent population.

For example, the average value of the fitness of all individuals in the parent population is 2, the case of the current population is shown in Table 1 below, in which the non-conforming individuals are "Web service", the remaining eligible individuals are "test list", "paid service", and "frequent power failure", and the remaining individuals are copied, crossed. The genetic manipulation, such as mutation, is used to breed the population.

Table 1. Individual fitness of the current population

Individual	Web service	Test list	Paid service	Frequent power failure
Recent share PO	3%	5%	12%	17%
Serial No.	1	2	3	4
Total Share ratio AO	1%	4%	1%	13%
Matching coefficient CR	1	1	0	1
Degree of adaptability ft	1	2	2	3

(1) Copy

After the non-qualified individuals are deleted from the current population, the remaining individuals in the current population are copied so that the remaining individuals are all in the next generation population.

(2) Crossing

Pre-setting the crossing condition, and after the non-qualified individuals are deleted from the current population, cross-combining any two remaining individuals in the current population according to a preset crossing condition to generate a cross-individual;the fitness function of the cross-individual is calculated by the fitness function of the formula 1;And judging whether the fitness of the cross-individual is greater than or equal to the average value of the fitness of all the individuals in the parent population, and if so, allowing the cross-individual to enter into a new generation population.The cross-conditions set in advance are as follows: a cross probability is set in advance (the value range is generally 0.5 to 0.95), and a floating point number between 0 and 1 is generated for the remaining individuals in the current population before crossing, and the floating point number is smaller than that of the cross-probability.

For example, in the current population, the remaining individuals that meet the cross-conditions are "Web service" and "first-aid repair quality", then the two individuals are subjected to cross-processing, and the resulting cross-individuals are "Site Quality" and "emergency repair service".The fitness of the crossover individual is then calculated according to Formula 1, as shown in Table 2, the fitness of the "emergency repair service" is equal to the average 2 of the fitness of all individuals in the parent population, retained, and entered into a new generation population.

Table 2. Individual fitness after cross-combination

Cross-individual	Site Quality	Emergency repair service
Recent share PO	0	3%
Serial No.	1	2
Total Share ratio AO	0	4%
Matching coefficient CR	1	1
Degree of adaptability ft	1	2

(3) Variation

Pre-setting the variation condition; after the non-qualified individuals are deleted from the current population, each of the remaining individuals in the current population meets the preset variation condition is subjected to the mutation operation to generate a variant;the fitness function is used to calculate the fitness of the variation individual; And judging whether the fitness of the variation individual is greater than or equal to the average value of the fitness of all the individuals in the parent population, and if so, allowing the variant to enter into a new generation population.Wherein the pre-set variation condition is as follows: a mutation probability is set in advance (the value range is generally 0.01 to 0.1), a floating point number between 0 and 1 is generated for the remaining individuals in the current population before the mutation is carried out, and the floating point number is smaller than that of the variation probability, The specific gene position of the mutation is determined by generating a random number.

For example, in the current population, the remaining individuals in accordance with the variation conditions are "test list", and the individual is subjected to a mutation treatment, and the resulting variation is an "electroscope", a "look-up table". The fitness of the variant is then calculated according to the formula 1, as shown in Table 3, the fitness of the "look-up table" is equal to the average value 2 of the fitness of all individuals in the parent population, retained, and entered into a new generation population.

Table 3. Individual fitness after variant operation

Crossover individuals	Power inspection	Table checking
Recent share PO	1%	2%
Serial No.	1	2
Total Share ratio AO	0	4%
Matching coefficient CR	1	1
Degree of adaptability ft	1	2

According to the above algorithm, the genetic operation is carried out until the preset termination condition is met, and the "pre-set termination condition" can be designed as a population update algebra to reach a certain preset value, for example, the termination condition is to stop when the fifth generation population is generated; In addition, the termination conditions can also be designed as the fitness average of all individuals in the population to a certain preset value. An individual in the current population is selected as an optimal keyword when the preset termination condition is met, and an optimal keyword corresponding case is selected, and all cases are stored in the classification of each keyword to form a service business guidance case library.

3. Update and maintenance method of service business guidance case base

3.1. The case library itself is purified and streamlined.
Because the optimal keywords represent a lot of and centralized demands on the customer, it is necessary to include some duplicate case data in each keyword case sub-set, so that the case library is too large to cause the information explosion and influence the typical case search. The call and learning speed, in order to ensure that the case in the case database is typical in all kinds of problems, it is necessary to perform self-purification and reduction for each keyword case sub-set to eliminate the duplicate redundant data.

The main steps of the purification and reduction work are to carry out the accurate matching of the case service classification and the power supply unit area attributes within a certain keyword case subset, and for the service of the power supply service, the business classification includes consulting, failure, complaints, reports, opinions, suggestions, praise, and the like; Take Jibei Electric Power Co., Ltd. as an example, the power supply unit includes Tangshan Power Supply Company, Zhangjiakou Power Supply Company, Qinhuangdao Power Supply Company, Chengde Power Supply Company and Langfang Power Supply Company. If a case of a certain keyword case contains a large number of the same business classification and the case of the power supply unit, only the most recent case is reserved as a typical case of such a case, and the other cases are deleted as redundant data, and the self-purification of the sub-set of each keyword case can be obtained to obtain a typical case base.

3.2. Learning and maintenance of case base

With the development of the society and the development of the economy, the electric power industry has changed constantly, and in order to change the relevant rules and regulations and the policy documents, the processing method of the same case is changing constantly due to the change of the related policies. Although the service business staff deal with the case according to the new processing method, the original processing method needs to be reserved as the basis of the case processing effect comparison. In order to change the case, the policy Aisle shall be set in the case database maintenance system. The newly issued related policies enter the case database through the policy Aisle, and the case keyword information contained in the policy is extracted.

When a new case enters a typical case library, the keyword information in the new case is first matched with the keyword information in the recent publishing policy, and if the matching key can be matched, the matching key is label, and the new case enters the case database; If the new case keyword does not match the keyword in the policy file, the new case directly enters the case library. The new case is matched to the information in the case library. If the same key is not retrieved, the new case is included in the typical case library to form a new typical keyword case. If the same keyword is searched, the new case is added to the corresponding keyword case sub-set, the new case is exactly matched with all the cases in the keyword case sub-set, the power supply service is taken as an example, the accurate matching of the two attributes of the service classification and the power supply unit can be carried out, if all the above two attributes match successfully, and the new case key is the label key, the new case is saved, the old case in the case sub-set corresponding to the new case is set to the read-only mode, only the call is allowed, and the update is no longer carried out, and is stored in a temporary library as an expired case for use. If the new case key is an unlabeled key, delete the old case corresponding to the new case, and keep the new case. If one of the above two attributes fails to match, the new case is saved as a new typical case in the keyword case collection.

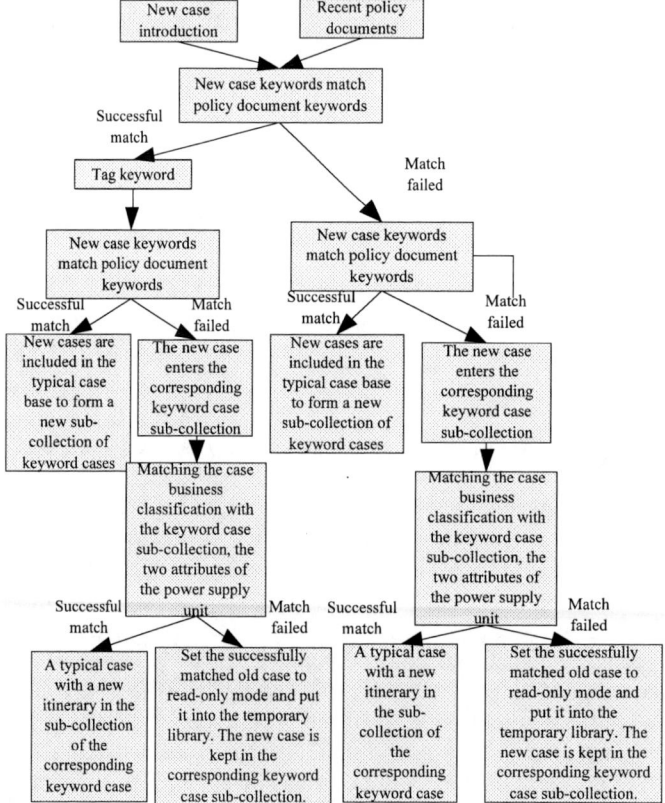

Figure 1. Case Library learning and maintenance flowchart

4. Conclusion:

This article presents a method for creating and updating a service business guidance case base, which is to filter the keywords in the case by the genetic algorithm, to set up a typical case base according to the key words, and to update and maintain the typical case base according to the new policy, and the case library is kept real-time and comprehensive. Due to the fact that a specific fitness function is set, the optimal keywords that are finally screened can not only reflect the frequently occurring cases in a long period of time, but also reflect the frequently occurring cases in a short period of time. The screening method fully considers the common types of the service business, so that the created guidance case library has good practicability and can bring very practical guidance value to the service business staff. When the service business people deal with the case, the case sub-set in the typical case base can be called according to the case key words, the accurate meaning of the case information can be learned, a plurality of response schemes in the client demands can be given in real time, or a plurality of elimination schemes for determining the fault can be provided. It is possible to infer the future evolution of a customer or a work process based on past and present conditions of the case operation information, in addition to the past and present circumstances of the case operation information. The typical case library as a content covers a wide range of data sets with typical experience in various fields, and provides offline knowledge reference for each discipline, and becomes an important source for the extraction of relevant knowledge organize teaching and training. This article introduces the service of power supply service in the power industry as an example, and the method can also be applied to the service industries such as water supply and gas supply.

References

[1] Qu Zhijian, Zhang Xianwei, Cao Yanfeng, Liu Xiaohong, Feng Xiaohua. Genetic Algorithm Research Based on Adaptive Mechanism[J]. Computer Application Research. 2015 (11).

[2] Chen Chao. Improved research and application of adaptive genetic algorithm[D]. Institutes of Technology of South China. 2011.

[3] Sun Guoqiang. Multi-center mass data layout research based on genetic algorithm[J]. Software guidance. 2015 (1).

[4] Zhuang Jian, Wang Sun 'an. Research on the Self-adjusting Genetic Algorithm[J]. Journal of System Simulation. 2003 (2).

[5] Tian Yanshuo. Research and Application of Genetic Algorithm[D]. Electronic Science and Technology University. 2004.

[6] Ma Yongjie, Yun Wenxia. Research progress of genetic algorithm[J]. Computer application research. 2012 (4).

[7] Dunwei Gong,Guangsong Guo,Li Lu,Hongmei Ma.Adaptive interactive genetic algorithms with individual interval fitness[J].Progress in Natural Science.2008(03).

[8] Guangmin Wang,xianjia Wang,Zhongping Wan,shihui Jia. An adaptive genetic algorithm for solving bilevel linear programming problem[J]. Applied Mathematics and Mechanics (English Edition).2007(12).

[9] Gao Junbo, Luan Cuiji, Wang Xiaofeng. New keyword extraction algorithm research[J]. Computer engineering and design. 2008 (3).

[10] Research on keyword extraction strategy based on word network[D]. Southwest University. 2008.

Design and Implementation of Body Quality Index App Based on Android

Zhao Limei[1]*, **Ma Xiaotie[2]**

[1] College of Information Engineering, Beijing Institute of Fashion Technology, Beijing, 100029, China

[2] College of Information Engineering, Beijing Institute of Fashion Technology, Beijing, 100029, China

*Corresponding author's e-mail: 1067403637@qq.com

Abstract. The body quality index is a commonly used standard for measuring the fatness and health of the human body at home and abroad, referred to as the BMI index. The design is based on the android software development platform, which has developed an app that can help people measure their body quality index. The design module consists of height and weight input module, calculation module, query result module and BMI common sense module. In the development process, with the help of Android studio software, the interface of the app is laid out and designed through the use of xml layout language. The java programming language is used to show users the five functions of body mass index, body weight status, standard weight, healthy weight range and BMI common sense.

1. Introduction

At present, with the diversification of diet and the increase of contemporary work pressure, the per capita weight in China is rising, the proportion of obese people is increasing year by year, and the phenomenon of sudden disease has become a major concern. Now everyone pays special attention to their physical health. To determine whether a person has reached the standard weight, the BMI index of mobile phone software has played a useful role in solving this problem. Whether in domestic or foreign, the BMI index is an important indicator of the health of a person based on the relationship between height and weight. Through the development of this app, users can query their BMI index and health status at any time.

2. App development design

In this design, the body quality index software has two interfaces, the main interface and the query interface. The main interface is to enter the height, weight and gender selection, and the user can edit this. The query interface is calculated by the relevant algorithm, so that the user's BMI index value, physical health status, standard weight, weight range under health, and related knowledge about the BMI index are reflected in the interface. The overall functional block diagram of the App is shown in figure 1.

Content from this work may be used under the terms of the Creative Commons Attribution 3.0 licence. Any further distribution of this work must maintain attribution to the author(s) and the title of the work, journal citation and DOI.
Published under licence by IOP Publishing Ltd

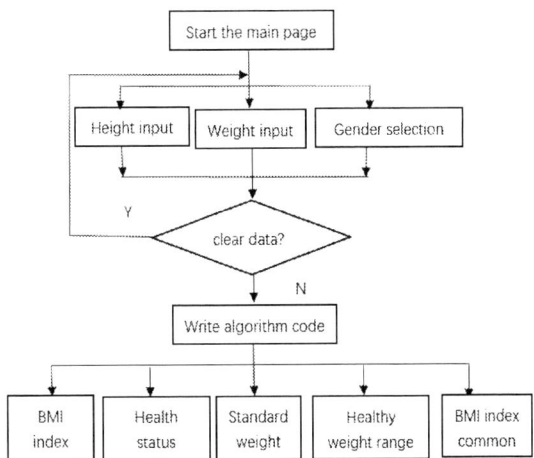

Figure 1. The overall block diagram.

3. App interface setting

3.1. Layout method

This article mainly applies linear layout. Linear layout is the more common layout method in Android, using the LinearLayout tab. The layout is divided into a vertical layout and a horizontal layout. And there are two types of controls: vertical and horizontal. The vertical layout has only one control per line and the horizontal layout has only one row, and all the controls are arranged from left to right.

3.2. Main page layout

In the created Bmi project, the main layout file is activity_main.xml under the app/res/layout folder. The parent layout of the page adopts the vertical layout in the linear layout, which is realized by setting android: orientation= "vertical". The input of height, weight and gender in the submodule adopts horizontal layout and is realized by setting android: orientation= "horizontal". The page background is implemented by setting android: background= "@drawable/ background". PaddingLeft, paddingRight and paddingTop are the distances of the internal text from the left border, the right border and the top [1].

　　<LinearLayout
　　android : layout width="match _ parent"
　　android : layout height="match_ parent"
　　</LinearLayout>
　The layout_ width and layout_ height properties are used to set the width and height respectively.

3.2.1. Layout control. Each control of the UI layout control has its corresponding properties. By selecting different properties and giving them values, different effects can be achieved. The layout of the article mainly uses three major controls.

　　1.Text representation control. The text presentation control is what presents to the user. In the main page, such as the text of height, weight and gender, their function is the content displayed on the mobile phone, users can not edit.

　　<TextView/>

　　2.Edit text control. The function is to let the user input the content required by the text, which needs to be edited by the user. Taking the height text as an example, the layout xml statement is as follows.

　　<EditText

android ：id="@+id/height"

android ：hint=" Please enter the height "/>

By setting the android: id="@+id/height" attribute, an identifier named "height" is created, which adds a new id to the control. This id is automatically generated in the R.java file. It is convenient to use the "R.id.height" to refer to the height object in the main program. And through the android: hint property setting to get the "Please enter the height" prompt [2].

3.Button control. Button controls are divided into two main controls: radio buttons and command buttons.

Radio button controls are applied to the gender selection module. This is because men and women are mutually exclusive, and only one of the radio buttons in the same group can be selected. Radio button controls must be nested under the RadioGroup container.

Among them, the selected state of the radio button control, in the xml file can be set using android: checked="", set to true if selected, set to false if not selected. This design sets the male to true, that is, after the page is launched, the gender selection defaults to male.

Command button controls are used to respond to the user's click operation. When receiving the response command, it jumps to the next page for corresponding processing. Buttons can be represented in text or image. In this paper, the calculation button and the clear button are designed in text form, and the BMI value, health status, standard weight and healthy weight range are obtained by calculating the button. The trigger event of the calculation button is implemented by setting the android: onClick="calculator" property.

The clear button is a one-key clear when the height and weight values are entered incorrectly, allowing the user to re-enter the value. The android: text="clear" is used to set the button text name, and the trigger event of the clear button is implemented by android: onClick="clear".

3.3. Result page layout

The activity_second.xml layout file is created under the app/res/layout folder to display the app's query results. The parent layout of the page adopts the vertical layout, the same as the main page. There are five modules on the results page, which are weight status, BMI index, standard weight, healthy weight range, and related BMI index.

Among them, weight status, BMI, standard weight and healthy weight range were all laid out in the same way. And they all adopted the horizontal layout in the linear layout. And set the property android: layout_ gravity to left, so that the position displayed by the control is at the leftmost end of the interface.

In the results page, the layout control is applied to the text representation control, which is used to render the calculated result. Take the weight status as an example, set the android: id="@+id/weightcondition" attribute in the TextView. It is convenient to use the "R.id.weightcondition" to refer to the object in the Activity program [3]. The BMI index, standard weight and healthy weight range are the same as above and add the new android: id="@+id/myBMI" attribute, android: id="@+id/standard_ weight" attribute and android: id="@+id/health_ weight" attribute in the text representation control to add new the id.

The final introduction to the BMI index common sense is implemented by the android: text=" " attribute in the text representation control.

4. Main program design

MainActivity.java is the main program file. The onCreate method has been automatically called by the system when the activity is created. It is the beginning of the activity life cycle. The programming uses the java programming language and uses the BMI algorithm to calculate the BMI value. The BMI value is compared with the index reference standard to determine the weight status. Finally, the standard weight and the range of healthy weight is calculated according to the relevant algorithm and the derived inverse formula. The specific flow chart is shown in figure 2.

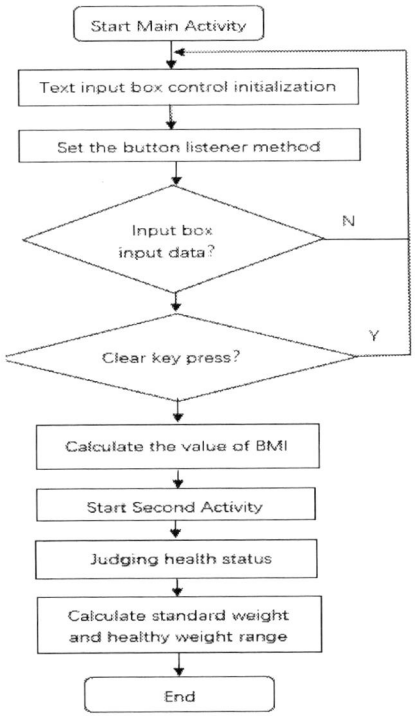

Figure 2. Specific flow chart of main program design.

4.1. Initialization settings

First, declare the user height and user weight in the MainActivity, rewrite the parent class through @override, and use the onCreated method to complete the initialization work when the activity starts.

The main program needs to continue adding the findViewById method in the default onCreate method to initialize the height and weight input box control. Among them, R.id.height indicates that the id of the height input box is height, which is mentioned in the height layout file in 3.2.1.

4.2. Button listener implementation

Implementing a button listener event can be done by declaring the button's id in the activity_ main file and setting the setOnClickListener click event for it in MainActivity.java. Or declare the button property setting in the activity_ main file, and set its method in MainActivity.java [4]. This article uses the second method, does not require id and initialization, directly in the layout file by setting android: onClick="calculator" and android: onClick="clear" attribute, to achieve the triggering event of the calculation and clear button.

4.3. BMI index calculation and display

4.3.1. BMI index calculation. First, the algorithm needs to obtain the user's height and weight values. And use the getText () method to convert the user's height text content into a string and assign it to the weight, and convert the user's weight text content into a string and assign it to the weight.

Second, convert the height and weight variables of String type into double type and represent them with h and w variables respectively. Convert by applying the Double.parseDouble method.

Finally, the BMI index is equal to the weight divided by the square of the height. The algorithm formula is BMI index = weight (kg) / (height * height (m)). In the case where the height h is not zero, the calculation result is assigned to the result variable.

double result=w/(h*h)

4.3.2. Starting SecondActivity. First, generate an intent object in MainActivity, call the setClass method to set the SecondActivity to be started, call the startActivity method to start the Activity. After that, the activity needs to add the passed data, use bmi to receive the value of result, and height to receive the input height value.

Finally, register the SecondActivity in the app/manifests/AndroidManifest.xml manifest file.

4.3.3. BMI index display. In the activity_second.xml layout file, the text indicates that the control has set the android: id= "@+id/myBMI" attribute, and the myBMI identifier is created. This requires that the application in secondactivity.java file use R.id. myBMI to get the interface component.

In SecondActivity.java, the interface of the Activity is displayed using the activity_second.xml layout file, which is implemented by the code setContentView(R.layout.activity_second). The findViewById method is used to initialize the text representation control that outputs BMI value, which is implemented by myBMI = (TextView) findViewById(R.id. myBMI). After that, BMI index data is extracted by getDoubleExtra method. Finally, the calculated BMI value is displayed by the setText method. And the BMI value retains two decimal places.

int v=(int)(value*100+0.5);

myBMI.setText(v/100.0+ "(21-22) is the best");

In the process of BMI calculation, two cases need to be paid attention to. One is to determine whether the edited text control has numeric input. If the text of one of the controls does not input a value, then an error message will appear, and get a message prompt for height or weight please input data. The second is to determine whether the input height value is zero. If it is zero, the text content is cleared by the userheight,setText("") code, and the user needs to re-enter the height.

4.4. Weight status judgment and display

The weight status is classified according to the size of the BMI index. The smaller the index, the lighter the weight, the bigger the index and the higher the weight. The specific classification criteria in this article is shown in Table 1.

Table 1. Classification criteria.

Guideline	BMI classification	Health status
BMI<18.5	Too light	Need nutrition
18.5<=BMI<24	normal	Continue to keep
24<=BMI<28	overweight	Healthy weight loss
28<=BMI<30	Grade 1 obesity	Exercise
30<=BMI<40	Grade 2 obesity	Exercise
BMI>=40	Grade 3 obesity	Exercise

First, the output of the weightcondition text represents the control initialization, implemented by the code weightCondition = (TextView) findViewById (R. id. weightcondition). According to the BMI classification of Table 1, the following code is written to realize the situation judgment and displayed in the text representation control with the id weightCondition.

4.5. Standard weight and healthy weight calculation and display

The standard weight calculation is only related to the user's height value. When the user's height value is obtained, the code is calculated according to the summarized promotion formula to calculate the standard weight. Standard weight (kg) = [height (m) - 1] * 0.9, the formula is universal, and the results of the calculation can be applied to all Asians.

    double height=intent.getDoubleExtra("height",-1);
    double w=(int) ((height-1) * 0.9);
    standard_ weight.setText(w*100+ " ");

The healthy weight range refers to the state of weight under normal conditions, that is, the BMI value is between 18.5 and 24. By inversely deriving the formula for calculating the BMI index, it is concluded that when the BMI value is 18.5, the body weight is at a normal minimum. Min=18.5* height* height. When the BMI value is 24, it is the normal maximum body weight, Max = 24 * height * height. The definition of a healthy weight range is between the minimum and maximum values.

    double min=(int) （（18.5*height*height）*100+0.5）/100;
    double max=(int) （（24*height*height）*100+0.5）/100;
    height_ weight. setText(min+ " " +max);

4.6. App physical map

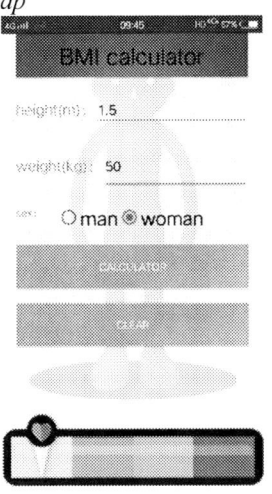
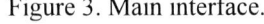
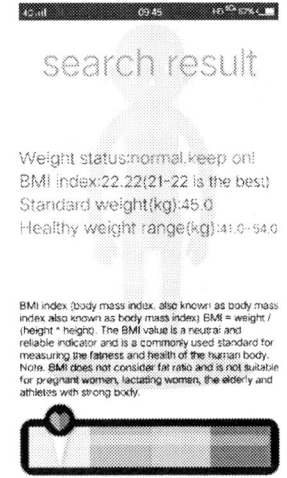

Figure 3. Main interface. Figure 4. Query interface.

5. Conclusion

The design of this App is mainly accomplished through two parts: interface setting and main program implementation. Through the familiarity of the xml layout language and the java language, an App with five functions of displaying weight status, BMI index, standard weight, healthy weight range and BMI common sense is realized. The advantage of the design is in the development process, through the derivation of formula calculation, constantly modify the source code and repeatedly run debugging, so as to complete the function of the test. The disadvantage is that it only realizes the basic functions of BMI, and it still needs further study and optimization of the code to achieve more functions.

References

[1] Wang Xianghui, Zhang Guoyin, Shen Jie. (2010) Android application development. Tsinghua university press, Beijing.

[2] Dong Minghua, li Hongwei. (2017) Android entry program development-BMI calculator. Computer knowledge and technology, 163: 51–59.

[3] Yin Mengzheng. (2016) Overview of android-based APP development platform. Communications power technology, 33:154-155+213.

[4] Yao Haozhe, Liu Lijuan. (2016) Construction of an App for personal health promotion based on Android platform. Computer era, 11: 28-31.

Emotional Analysis of Public Opinions in Colleges and Universities:Based on Naive Bayesian Classification Method

Qingjia Wang[1], Kun Liu[1*] and Kun Ma[1]

[1] School of Information Science and Engineering, University of Jinan, Jinan, Shandong, 250022, PR China

* liukun@ujn.edu.cn

Abstract. In this paper, we use the emotion dictionary to process and express the public opinion texts of colleges and universities. We construct the public opinion texts of colleges and universities sentiment emotional classifier based on Naive Bayesian theory, and then judge the emotional tendency of public opinion texts of colleges and universities. Experiments show that the method has the characteristics of fast classification and high accuracy, and is suitable for the public opinion system of colleges and universities. The method facilitates the university management's interpretation and mastery of the public opinions related to the school in the network, promotes and guides the positive energy in the communication network, corrects the vulgar and biased content, prevents the forwarding and dissemination of inappropriate speech, and establishes a positive campus environment among the teachers and students of the whole school.

1. Introduction

The Internet has become the main way of information dissemination in the 21st century, but at the same time it has two sides, it can spread positive energy and positive content, and also spread some negative and misleading speech. Colleges and universities are the initial place for higher education, and it plays a pivotal role. As college students are gradually becoming the main force and target of online public opinions, more and more colleges and universities have become the high-risk places for online public opinions emergencies. Therefore, the emotional analysis of college public opinions is of great significance. Through the emotional analysis of college public opinions, it guides and controls the public opinions of colleges and universities, so that school leaders and student workers can timely understand students' views and attitudes toward the event, keep an eye on the development of events and monitor the development trend of public opinions in colleges and universities, provide early warning of speeches that endanger school safety, guide social opinion orientation and provide guarantee for the healthy and stable development of society.

2. Research

2.1. Research on public opinions in colleges and universities

The psychological status of college students in a period of time and space is reflected by the enthusiasm of colleges and universities. At present, many researchers have made certain research results in colleges public opinions, which has led to the rapid development of college public opinion platforms. The amount of data generated on campus every day is in PB, which puts some pressure on the students' mental state and the organization of college administrators. So Wang J F used the idea of

---

Content from this work may be used under the terms of the Creative Commons Attribution 3.0 licence. Any further distribution of this work must maintain attribution to the author(s) and the title of the work, journal citation and DOI.

Published under licence by IOP Publishing Ltd

big data to construct a college public opinion platform in 2017[1]. Jumadi et al. conducted research on public opinions data mining and sentiment classification based on SVM [2]. Zhao Y Q proposed a complex proxy network model for college public opinion systems [3]. Through complex adaptive system theory and complex network analysis methods, the impact of future public opinions can be evolved in this model to help managers to better guide the direction of public opinions and the way students think about problems in colleges and universities, and provide powerful practical guidance for human progress. Many well-known scholars at home and abroad have carried out certain research on the public opinions of colleges and universities. Through the mining and preliminary analysis of the public opinions data of colleges and universities, many guiding analysis models for the network public opinions of colleges and universities have been put forward. This includes building a data analysis platform or using relevant theories to monitor public opinions in colleges and universities. These are just excavations and preliminary analysis of university public opinions data.

2.2. Research on text sentiment analysis

Text sentiment analysis is a key algorithm in the field of data mining and machine learning, which is one of the research hotspots in the field of natural language processing (NLP). In recent years, with the continuous deepening of people's emotional analysis of online texts, the research scope of text emotions is expanding and gradually occupying an important position in the academic world, and many domestic and foreign scholars have long studied this aspect. The text orientation analysis method based on machine learning produces text data that satisfies a certain format by labeling the tendency of some texts, then uses these marked text as training data, constructs an emotion classifier by using the classification method of machine learning and data mining. Finally, the method uses the classifier to classify the test data, and calculates the emotional tendency and emotional intensity of the test data. Pang et al. used Native Bayesian Algorithm, Support Vector Machines Algorithm,and Maximum Entropy Algorithm for experimental verifications and comparative experiments on text emotions as early as 2002. The end result is that the effects of the three algorithms are not much different [4]. Mullen et al. used the Support Vector Machines method to establish a text classification model and perform sentiment analysis on texts [5]. There are still many scholars who choose different features and machine learning algorithms to study, and the sources of the research texts are more diversified. Gammon et al. choose customer feedback texts as the research object [6], and Li et al. choose sentence context and emotion transfer words as the research object[7]. In addition, some scholars have proposed to measure the semantic tendency of emotional words in the text, obtain the average value of the semantic directional value of all words and finally obtain the tendency metric of the overall text.

3. Research method

The college sensation sentiment analysis method proposed in this paper is mainly to construct a Naive Bayesian sentiment emotional classifier. First, web crawlers are used to obtain public opinion data of universities. After obtaining the data, the text corpus preprocessing is performed, including word segmentation, deleting the stop words, reading the sentiment dictionary to obtain the emotional words in this article, and calculating the feature weights. After the text corpus preprocessing is completed, the text classifier is constructed by the Naive Bayesian algorithm. The basic idea of the construction is based on the Bayesian formula.First, the training corpus is trained to obtain the prior probability of the text category and the posterior conditional probability of the feature attribute. Then, the Bayesian formula is used to calculate the probability that the text to be classified belongs to all categories, and finally the text to be classified is divided into classes with higher probability values.

Naive Bayesian is a probability-based learning algorithm that is based on the prior probability of hypotheses. The probability of observing different features is given under the given assumptions. To the sentimental tendency of the public opinion text $d=\{w_1, w_2,...,w_n\}$ of the university to be classified belongs to $C=\{C_{Positive}, C_{Negative}, C_{Neutral}\}$, considering the weight of the feature words, in the case where the features are independent of each other, the classification algorithm formula is shown in formula 1.

$$category = \underset{c_i \in C}{\arg\max}\{P(c_i) \prod_{k=1}^{n} P(w_k, c_i)^{W_t(W_k)}\} \tag{1}$$

Where $P(c_i)$ is the prior probability of the category c_i, and the pre-estimate based on training expectations that have been manually labeled correctly, that is the ratio of positive text, negative text, and neutral text to the total text, respectively, is calculated as formula 2:

$$P(c_i) = \frac{d(c_i)}{\sum_{c_i \in C} d(c_i)} \tag{2}$$

Where $d(c_i)$ is the number of texts belonging to c_i.

$P(w_k, c_i)$ in the formula 1 is a posterior probability that the feature word w_k appears in the category c_i, and $w_t(w_k)$ is the weight of the word w_k in the text to be classified. This paper uses the sum of the weights of the feature word w_k in the text belonging to the category c_i divided by the weights of all the words of the category c_i. If a feature word that does not appear in the training corpus appears in the text to be classified, $P(w_k, c_i)$ will be 0. In order to avoid this situation, using Laplace transform, $P(w_k, c_i)$ is calculated as shown in formula 3:

$$P(w_k, c_i) = \frac{weight(w_k, c_i) + 1}{\sum_{k=1}^{n} weight(w_k, c_i) + V}$$

$$(V = \sum_{c_i \in C} \sum_{k=1}^{n} weight(w_k, c_i)) \tag{3}$$

4. Experiment and analysis

4.1. Data set
The data set contains pieces of 1000 data, including positive and negative college public opinion data 500 each, each experiment selects 50 positive and negative data as test corpus, and the remaining 900 data as training corpus. Experiments were conducted by two feature selection methods, one is to use emotion dictionary to select emotional words, and the other is TF-IDF.

4.2. Standards for experimental evaluation
The experiment used the current widely used accuracy rate, recall rate and F_1 measure to evaluate the experimental results. TP indicates the number of positive texts classified to positive categories, FN indicates the number of positive texts classified to negative categories, FP indicates the number of negative texts classified to positive categories, and TN indicates the number of negative texts classified to negative categories. The three indicators are calculated as follows:

$$P_{pos}=TP/(TP+FP) \tag{4}$$
$$R_{pos}=TP/(TP+FN) \tag{5}$$
$$P_{neg}=TN/(FN+TN) \tag{6}$$
$$R_{neg}=TN/(FP+TN) \tag{7}$$
$$F_1=2PR/(P+R) \tag{8}$$

P_{pos} and P_{neg} are the accuracy rates, and R_{pos} and R_{neg} are the recall rates. P_{pos} reflects the proportion of positive corpus classified by the classifier into positive corpus, and P_{neg} reflects the proportion of negative corpus classified by the classifier into negative corpus. R_{pos} reflects the proportion of positive corpus that is correctly classified to the total positive corpus, and R_{neg} reflects the proportion of negative corpus that is correctly classified to the total negative corpus.

4.3. Results and analysis

Each evaluation criterion was used to evaluate the classification effect. Tables 1 and 2 respectively list the classification accuracy rate, recall rate, F_1 measure and their arithmetic mean results of the positive corpus and negative corpus in four experiments..

Table 1. Accuracy, recall and F_1 measure of classification methods characterized by emotional words

Data set	P_{pos}	R_{pos}	F_1	P_{neg}	R_{neg}	F_1
C1	0.98	0.98	0.98	0.98	0.98	0.98
C2	0.98	0.98	0.98	0.98	0.98	0.98
C3	0.83	0.98	0.90	0.98	0.80	0.88
C4	0.86	0.84	0.85	0.84	0.86	0.85
Average	0.91	0.95	0.93	0.95	0.91	0.92

Table 2. Accuracy, recall and F_1 measure of classification methods characterized by TF-IDF

Data set	P_{pos}	R_{pos}	F_1	P_{neg}	R_{neg}	F_1
C1	0.93	0.86	0.89	0.94	0.94	0.90
C2	0.80	0.86	0.83	0.78	0.78	0.81
C3	0.71	0.90	0.79	0.64	0.64	0.73
C4	0.81	0.52	0.63	0.88	0.88	0.75
Average	0.81	0.79	0.79	0.81	0.81	0.80

As can be seen from Tables 1 and 2, using emotional words as feature selection, the F_1 measurement on all corpora is improved compared with TF-IDF. This shows that the Naive Bayesian classification method using emotional dictionary as feature selection can improve the emotional classification effect of college public opinion information.

5. Conclusion

Experiments show that the sentiment analysis of college public opinions can be effectively realized through the combination of sentiment dictionary and Naive Bayesian algorithm, but this paper does not distinguish application fields. In the field of colleges and universities, the characteristics of each emotional word are different from other fields, so it is necessary to establish a comprehensive emotional dictionary based on the university field. It is necessary to expand the number of training corpora and improve the quality of training corpus. Based on the sentiment dictionary, the Naive Bayesian probability model of the university field is constructed and continuously revised.

References

[1] Wang, J.F. (2017) Research on Campus Network Public Opinion Mining Based on Big Data.Value Engineering, 26:196-198.

[2] Jumadi, Maylawati, D.S., Subaeki, B., et al. (2016) Opinion mining on Twitter microblogging using Support Vector Machine: Public opinion about State Islamic University of Bandung.In: International Conference on Cyber and IT Service Management. Bandung.

[3] Zhao, Y.Q. (2016) Research on Evolving Model of University Public Opinion Based on Complex Agent Network. Information Science, 34:130-134.

[4]   Bo, P.,Lillian, L.,and Shivakumar, V.(2002) Thumbs up? Sentiment Classfication using Machine Learning Techniques.In: Empirical Methods in Natural Language Processing. Philadelphia.pp: 79-86.

[5]   Mullen, T., Collier, N.(2004) Sentiment Analysis using Support Vector Machines with Diverse Information Sources. In: Empirical Methods in Natural Language Processing. Barcelona.pp: 412-418.

[6]   Gamon, M. (2004) Sentiment classification on customer feedback data:noisy data, large feature vectors,and the role of linguistic analysis. In: International Conference on Computational Linguistics. Switzerland. pp: 831.

[7]   Li, S.S., Lee, S., Chen, Y., et al. (2010) Sentiment classification and polarity shifting. In: International Conference on Computational Linguistics. Beijing. pp: 635-643.

Dual-channel Supply Chain Sale Strategies with Return Guarantee

Jingshi HE

Logistic department, Dong Guan Polytechnic, Dong Guan, Guangdong,523808

Corresponding author's e-mail: hejings@163.com

Abstract. This paper explores the optimal pricing strategies under the centralized and decentralized dual-channel supply chain concerning customer returns and return guarantee. The influence of parameters on the optimal value such as optimal price, return policies, return guarantee, market share was analyzed. It shows that the profit of the enterprise is reduced with the increase of the return rates, and the return guarantee price is positively related to the profit of the enterprise. Enterprises can actively use the means of return guarantee and increase its price in order to obtain more supply chain profit, and increasing the size of the network direct sales is also effective to supply chain.

1. Introduction

With the development of online retailers, more and more manufacturing enterprises sell their products by combining traditional retail channels with online direct marketing, and formed the dual channel supply chain. Retail enterprises such as GOME and SUNING also adopted the method of combining traditional retail channels with online marketing. International famous brands opened e-commerce marketing while selling goods in traditional retail channels such as Estee Laud, Nike, Dell, Cisco. Dual-channel sales not only keep the consumers who like shopping in traditional channels but also attract new consumers and those consumers whom traditional sales difficult to cover, so it can expand the market scope, control sales prices, and improve supply chain's profits[1].

In order to maintain a high degree of customer satisfaction and attract more consumers in the market full of fierce competition, more and more companies introduced a loose return policy, and false failure returns become the popular way of the market. Returning products without giving reasons within seven days for customers are not only popular in online shopping, but also prevalent in physical store. False failure returns will attract more consumers, but it also causes product re-inspection, repackaging, and cost incensement of reverse logistics. Baiman[2] showed that false failure returns increase consumer satisfaction, but it will bring unnecessary returns. Partial returns policy makes it difficult for consumers to be satisfied, and reduce consumer loyalty in his research. So some businesses use cost hassle to replace partial of the return policy such as fixed term return, self-freight, holding invoices, etc[3]. Return guarantee is a new model of network sales in recent years. Customers who purchase return guarantees are provided with a full refund. Elong and other booking site agents provide insurance of cancel order, and TMALL also provide insurance of return shipping. If customers purchase their freight fee insurance, then return freight fee will be paid by the third party, and that is to say, customers enjoy full return policy. Insurance of cancel order and insurance of return shipping can be regarded as a kind of return guarantee.

Return guarantee has just started in the practice, and the theoretical research in this field is rare. This paper build the supply chain pricing model considering the return guarantee, the influence of

Content from this work may be used under the terms of the Creative Commons Attribution 3.0 licence. Any further distribution of this work must maintain attribution to the author(s) and the title of the work, journal citation and DOI.
Published under licence by IOP Publishing Ltd

supply chain pricing strategies and return strategies under return guarantee. ZHENG Chun-dong[3]using questionnaire method show that insurance of return shipping can significantly reduce the perceived risk of consumers, improve the level of trust, and then have a positive impact on the purchase intention of consumers. Mukhopadhyay[4] analyzed the optimal price and return policy in the online marketing from the perspective of reverse logistics. Yan[5] studied the influence of the retail service on pricing and profit distribution under dual-channel supply chain. CHEN Chong-ping[6] built the model of refund and pricing based on return guarantee in online sale environment, and optimal decision was analyzed in two cases of return and no return option, where goods price, return policies and return guarantee were seen as variables.

2. The Model Description

This paper considering manufacturers produce a single product to sale in the dual-channel supply chain (figure 1).The notations and the parameters are defined as follows:

p_d the online direct channel price.

p_r offline traditional retail channels price.

c_d manufacturer's product cost.

c_r offline retailer selling cost.

ω the wholesale price of manufacturer to the retailer.

θ the market share of retailers, $0 < \theta < 1$.

$1 - \theta$ the market share of online direct marketing,

where α is the potential market scale, $\alpha_r = \theta\alpha$ represents the basic market demand of retail channels, $\alpha_d = (1 - \theta)\alpha$ represents the basic market demand of online direct channels and $\alpha_r > 0, \alpha_d > 0$[7].

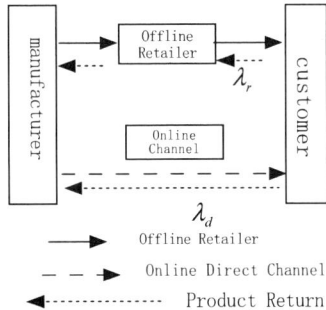

Figure 1 The model of dual-channel supply chain

Considering the impact of price and cross price effects on traditional retail and online direct selling in the market, it is assumed that $b_1 > b_2 > 0$, b_1 is the price elasticity of demand, and b_2 is the cross-price elastic,

According to the above descriptions, the sales function of offline traditional retail (D_r) and sales function of online sale(D_d)can be given as follows:

$$D_r = \theta\alpha - b_1 p_r + b_2 p_d; \quad D_d = (1 - \theta)\alpha - b_1 p_d + b_2 p_r \quad (1)$$

Customer returns have become one of the effective channels to attract customers either offline traditional retail or online direct marketing channels, we assume that return function $R = \lambda X_i (i = d, r)$, where λ represents basic return rate. The return amount is φp when customer do not purchase return guarantee, and return factor accord with $0 < \varphi \leq 1$.

Generally speaking, the offline traditional retailers return rate is less than the rate of the network direct sales channel, i.e. $\lambda_r \leq \lambda_d$, since the store customers can touch the production in entity stores.

We assume that g represents the return guarantee price. Customer enjoy a full refund when their purchase return guarantee, and there is $\varphi = 1, g = 0$ represents the customers don't purchase the return guarantee, so they enjoy partial return refund.

The self pay freights in actual business when not purchase return guarantee also be viewed as partial return refund. Where ϕ represents the proportion amount which purchase return guarantee accounted for the market sales.

Profit function of offline traditional sale channels is follows:

$$\pi_r = (p_r - \omega - c_r)D_r + D_r\phi g - \lambda_r\varphi p_r D_r - \lambda_r\phi p_r D_r = (\theta\alpha - b_1 p_r + b_2 p_d)[\phi g - \omega - c_r - (\lambda_r\varphi + \lambda_r\phi - 1)p_r]$$

(2)

Where $D_r(p_r - \omega - c_r)$ is the sales revenues, $\phi D_r g$ is the revenues of purchase return guarantee, $\lambda_r\phi D_r p_r$ is the amount return back to consumers who purchase return guarantee (full refund when purchase the return guarantee). $\lambda_r D_r\varphi p_r$ is the amount return back to consumers who don't purchase the return guarantee.

Profit function of manufacturer direct sale online as follows:

$$\pi_d = D_r(\omega - c_d) + D_d(p_d - c_d + \phi g - \lambda_d\varphi p_d - \lambda_d\phi p_d)$$
$$= (\theta\alpha - b_1 p_r + b_2 p_d) * (\omega - c_d) + [(1 - \theta)\alpha - b_1 p_d + b_2 p_r][\phi g - c_d - (\lambda_d\varphi + \lambda_d\phi - 1)p_d]$$

(3)

We assume the $k_d = \lambda_d\phi + \lambda_d\varphi - 1 < 0, k_r = \lambda_r\phi + \lambda_r\varphi - 1$, so we get the supply chain profit function:

$$\pi_{sc} = (\theta\alpha - b_1 p_r + b_2 p_d) * [\phi g - c_r - c_d - (\lambda_r\phi + \lambda_r\varphi - 1) p_r] + [(1 - \theta)\alpha - b_1 p_d + b_2 p_r] * [\phi g - c_d - (\lambda_d\phi + \lambda_d\varphi - 1) p_d]$$
$$= (\theta\alpha - b_1 p_r + b_2 p_d) * (\phi g - c_r - c_d - k_r p_r) + [(1 - \theta)\alpha - b_1 p_d + b_2 p_r] * (\phi g - c_d - k_d p_d)$$

(4)

3. Analysis of the Centralized Supply Chain Decision

We get the first derivative of price(p) with respect to supply chain profit function(π_{sc}):

$$\frac{\partial\pi_{sc}}{\partial p_r} = 2b_1 k_r p_r - b_2(k_r + k_d)p_d - \theta\alpha k_r + b_1 c_r - (b_1 - b_2)(\phi g - c_d)$$

(5)

$$\frac{\partial\pi_{sc}}{\partial p_d} = 2b_1 k_d p_d - b_2(k_r + k_d)p_r - (1 - \theta)\alpha k_d - b_2 c_r - (b_1 - b_2)(\phi g - c_d)$$

(6)

Then second derivative of π_{sc} get follows: $\frac{\partial^2\pi_{sc}}{\partial p_r^2} = 2b_1 k_r; \frac{\partial^2\pi_{sc}}{\partial p_d^2} = 2b_1 k_d$.

We assume that $k_d = \lambda_d\phi + \lambda_d\varphi - 1 < 0, k_r = \lambda_r\phi + \lambda_r\varphi - 1$, So the supply chain profit function is a convex function of p_r, p_d, and the supply chain profit function has the maximum value.

Proposition 1: There is an optimal price p_r, p_d to maximize the profit of the supply chain. That is when the selling price is less the optimal price, the supply chain profit increases with the rise selling prices both offline retail and online direct. When the price is greater than the optimal price, the supply chain profit is decreases as the price.

We assume the first derivative (5) (6) equal to zero and get the optimal supply chain price.

$$p_{dsc}^* = \frac{(k_r - k_d)b_1 b_2 c_r + \theta\alpha b_2(k_r + k_d)k_r + 2b_1(1 - \theta)\alpha k_d k_r + (2b_1 k_r + b_2 k_r + b_2 k_d)(\phi g - c_d)(b_1 - b_2)}{4b_1^2 k_r k_d - b_2^2(k_r + k_d)^2}$$

(7)

$$p_{rsc}^* = \frac{(b_2^2 - 2b_1^2)c_r k_d + b_2^2 c_r k_r + (1 - \theta)\alpha b_2 k_d^2 + (2b_1\theta + b_2 - \theta b_2)\alpha k_d k_r + (2b_1 k_d + b_2 k_r + b_2 k_d)(\phi g - c_d)(b_1 - b_2)}{4b_1^2 k_r k_d - b_2^2(k_r + k_d)^2}$$

(8)

According to the previous research it can be known that $b_1 > b_2, \lambda_r \le \lambda_d$, so $k_r < k_d$, we get the follows:

$$\frac{4b_1^2 k_r k_d}{b_2^2 (k_r + k_d)^2} > \frac{4b_2^2 k_r k_d}{b_2^2 (k_r + k_d)^2} = \frac{4 k_r k_d}{(k_r + k_d)^2} > \frac{k_r^2}{k_r^2} = 1 \quad , \quad 4b_1^2 k_r k_d - b_2^2 (k_r + k_d)^2 > 0$$

The first derivative of θ, g (market share of retailers and return guarantee) with respect to optimal price of offline traditional sale and online sale get results as:

$$\frac{\partial p_{rsc}^*}{\partial \theta} = \frac{\alpha k_d [(2b_1 - b_2)k_r - b_2 k_d]}{4b_1^2 k_r k_d - b_2^2 (k_r + k_d)^2} > 0 \quad (9)$$

$$\frac{\partial p_{dsc}^*}{\partial \theta} = \frac{\alpha k_r [(b_2 - 2b_1)k_d + b_2 k_r]}{4b_1^2 k_r k_d - b_2^2 (k_r + k_d)^2} < 0 \quad (10)$$

Proved as follows:

$$\alpha k_d [(2b_1 - b_2)k_r - b_2 k_d] > \alpha k_d [(2b_1 - b_2)k_d - b_2 k_d] = \alpha k_d^2 (2b_1 - 2b_2) > 0$$

$$\alpha k_r [(b_2 - 2b_1)k_d + b_2 k_r] < \alpha k_r [(b_2 - 2b_1)k_d + b_2 k_d] = \alpha k_r k_d [(b_2 - 2b_1) + b_2] = 0$$

Proposition 2: Retailers' optimal prices increase with the increase of the offline retailer`s market share, where the optimal price of the online selling channel decreases with the increase of the retailer's market share under centralized supply chain decision.

With the increase of the offline retailer's market share, retailers will get more supply chain power and then making higher retail prices to gain more profits. On the contrary, with the increase in the proportion of online direct sales, the seller of online can also set higher prices to obtain greater profits.

$$\frac{\partial p_{dsc}^*}{\partial g} = \frac{(2b_1 k_r + b_2 k_r + b_2 k_d)\phi}{4b_1^2 k_r k_d - b_2^2 (k_r + k_d)^2} < 0 \quad (11)$$

$$\frac{\partial p_{rsc}^*}{\partial g} = \frac{(2b_1 k_d + b_2 k_r + b_2 k_d)\phi}{4b_1^2 k_r k_d - b_2^2 (k_r + k_d)^2} < 0 \quad (12)$$

$$\frac{\partial \pi_{sc}}{\partial g} = D_r \phi + D_d \phi > 0 \quad (13)$$

Proposition 3: The optimal price in online direct channel and offline traditional retail channels are inversely proportional with return guarantee. The profit under the centralized supply chain is directly proportion to return guarantee revenue.

The result of take partial derivative of λ, ϕ with respect to the supply chain profit function(π_{sc})as:

$$\frac{\partial \pi_{sc}}{\partial \lambda_r} = -(\phi + \varphi)p_r(\theta\alpha - b_1 p_r + b_2 p_d) = -(\phi + \varphi)p_r D_r < 0 \quad (14)$$

$$\frac{\partial \pi_{sc}}{\partial \lambda_d} = -(\phi + \varphi)p_d[(1 - \theta)\alpha - b_1 p_d + b_2 p_r] = -(\phi + \varphi)p_d D_d < 0 \quad (15)$$

Proposition 4: Supply chain profit decreases with the increase of the return rate. Therefore, enterprises should reduce the return rate can thorough the following measures: strengthen the analysis of the reasons for customer returns, improve product quality, strengthen the pre-sale and after-sales service, provide a more detailed description of the goods, etc.

Take partial derivative of ϕ with respect to optimal price and supply chain profit π_{sc}^*, p^* under the centralized supply chain decision

$$\frac{\partial p_{dsc}^*}{\partial \phi} = \frac{(2b_1 k_r + b_2 k_r + b_2 k_d)g(b_1 - b_2)}{4b_1^2 k_r k_d - b_2^2 (k_r + k_d)^2} < 0 \quad (16)$$

$$\frac{\partial p_{rsc}^*}{\partial \phi} = \frac{(2b_1 k_d + b_2 k_r + b_2 k_d)g(b_1 - b_2)}{4b_1^2 k_r k_d - b_2^2 (k_r + k_d)^2} < 0 \quad (17)$$

$$\frac{\partial \pi_{sc}^*}{\partial \phi} = D_r * g + D_d * g > 0 \quad (18)$$

Proposition 5: The profit of supply chain is proportional to scales of purchase return guarantee under centralized decision, the optimal price of the online direct selling channel and the offline retail channel is inversely proportional to the scales of purchase return guarantee. Therefore, in order to obtain a higher profit enterprise should to strengthen the sale of return security options.

4. Analysis of the Decentralized Supply Chain Decision

According to the profit of traditional retail function, we get follows:

$$\frac{\partial \pi_r}{\partial p_r} = 2b_1 k_r p_r - b_2 k_r p_d - b_1(\phi g - \omega - c_r) - \theta \alpha k_r, \quad \frac{\partial^2 \pi_r}{\partial p^2_r} = 2b_1 k_r$$

According to the previous reasoning we know that $k_r < 0$, so $\frac{\partial^2 \pi_r}{\partial p^2_r} < 0$ .we deduct that the profit of offline traditional retail function is a convex function of p_r , and it has the maximum value.

According to the profit of manufacturer`s online sale:

$$\frac{\partial \pi_d}{\partial p_d} = 2b_1 k_d p_d - b_2 k_d p_r + b_2(\omega - c_d) - b_1(\phi g - c_d) - (1 - \theta)\alpha k_d \quad \frac{\partial^2 \pi_d}{\partial p^2_d} = 2b_1 k_d$$

According to the previous we know that $k_d < 0$, so there is $\frac{\partial^2 \pi_d}{\partial p^2_d} < 0$, we deduction that the profit of offline traditional retail function is a convex function of p_d, and it has the maximum value.

According to $\frac{\partial \pi_d}{\partial p_d} = \frac{\partial \pi_r}{\partial p_r} = 0$, we get follows results:

$$p_d^* = \frac{b_1 b_2(\phi g - \omega - c_r)k_d + (2b_1\alpha - 2\theta\alpha b_1 + \theta\alpha b_2)k_r k_d + 2b_1^2(\phi g - c_d)k_r - 2b_1 b_2(\omega - c_d)k_r}{(4b_1^2 - b_2^2)\ k_d k_r} \quad (19)$$

$$p_r^* = \frac{2b_1^2(\phi g - \omega - c_r)k_d + (2\theta\alpha b_1 + \alpha b_2 - \theta\alpha b_2)k_r k_d + b_1 b_2(\phi g - c_d)k_r - b_2^2(\omega - c_d)k_r}{(4b_1^2 - b_2^2)k_d k_r} \quad (20)$$

Take partial derivative of θ, g, ϕ with respect to optimal price of offline traditional retail and online seals, the result follows:

$$\frac{\partial p_d^*}{\partial \theta} = \frac{-\alpha(2b_1 - b_2)}{4b_1^2 - b_2^2} < 0 \quad (21)$$

$$\frac{\partial p_r^*}{\partial \theta} = \frac{\alpha(2b_1 - b_2)}{4b_1^2 - b_2^2} > 0 \quad (22)$$

$$\frac{\partial p_d^*}{\partial g} = \frac{b_1 b_2 \phi k_d + 2b_1^2 \phi k_r}{(4b_1^2 - b_2^2)k_d k_r} < 0 \quad (23)$$

$$\frac{\partial p_r^*}{\partial g} = \frac{b_1 b_2 \phi k_r + 2b_1^2 \phi k_d}{(4b_1^2 - b_2^2)k_d k_r} < 0 \quad (24)$$

$$\frac{\partial \pi_d}{\partial g} = D_d \phi > 0 \quad \frac{\partial \pi_r}{\partial g} = D_r \phi > 0 \quad (25)$$

$$\frac{\partial p_d^*}{\partial \phi} = \frac{b_1 b_2 g k_d + 2b_1^2 g k_r}{(4b_1^2 - b_2^2)k_d k_r} < 0 \quad (26)$$

$$\frac{\partial p_r^*}{\partial \phi} = \frac{2b_1^2 g k_d + b_1 b_2 g k_r}{(4b_1^2 - b_2^2)k_d k_r} < 0 \quad (27)$$

$$\frac{\partial \pi_r}{\partial g} = D_r \phi > 0; \frac{\partial \pi_d}{\partial g} = D_d \phi > 0 \quad (28)$$

Proposition 6: Retailers' optimal prices increase with the increase of the offline retailer's market share, where the optimal price of the online selling channel decreases with the increase of the retailer's market share under decentralized supply chain decision.

Proposition 7: The profit of supply chain is proportional to scales of purchase return guarantee under decentralized decision, the optimal price of the online direct selling channel and the retail channel is inversely proportional to the scales of purchase return guarantee.

5. Numerical Studies

Numerical analysis will be carried out to verify the previous theoretical analysis. We use the following numbers as the base values of the parameters and take MATLAB as the tool. $\alpha = 200, \ \theta = 0.5, \ b_1 = 3, b_2 = 2, c_r = 6, c_d = 8,$
$\omega = 10, \phi = 0.2, \varphi = 0.95, \lambda_d = \lambda_r = 0.2$

Figure 2 indicates the relationship between supply chain profit and price of online direct selling or offline retail channel. It shows that supply chain profit function is a convex function of p_r, p_d, and it has the maximum value. The optimal price under centralized supply chain decision is higher than the price under decentralized supply chain decision, and the supply chain profit under centralized decision is also greater than decentralized decision.

With the increase of return rate, the profit of retail and online direct model is decreased, the return rate reach one point and the profit is zero as figure 3 indicates. Enterprises should reduce the return rate by the following measures: strengthening the analysis of the reasons for customer returns, improving product quality, strengthening the pre-sale and after-sales service, or providing a more detailed description of the goods, etc.

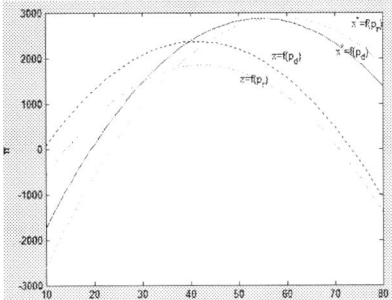

Figure 2 Effect of price on supply chain profit   Figure 3 Effect of return rate on profits

As the figure 4 indicates, with the prices of return guaranteed increased, the supply chain profit increased slowly. So it is beneficial to increase the profit of the supply chain and the offline traditional retail and the online direct sales when raise the price of return guarantee.

As figure 5 indicates that the retailer's price increases with the raise of offline market share, while decreases with the online direct selling share whether it is centralized decision or decentralized decision. This shows that with the offline market share increase, the seller has a greater right to develop market prices.

But the profit of supply chain decreases with the offline market share increase according figure 6.In order to achieve greater profits, manufactures can reduce the offline traditional retail market share and expand the market share of online direct sales under the dual channel sales.

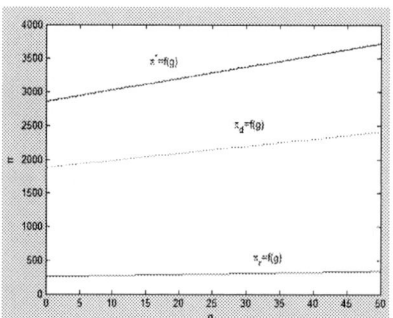

Figure 4 Effect of return guarantee on profit

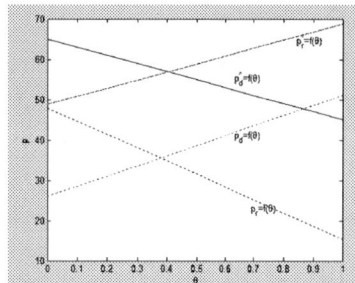

Figure 5 Effect of retailer's market share on price

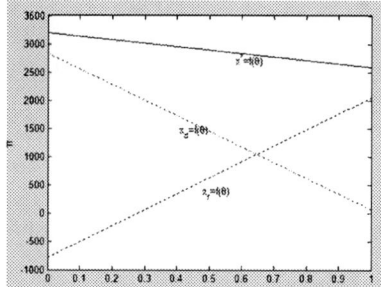

Figure 6 Effect of retailer's market share on profit

6. Conclusions

Return guarantee is a new model of network sales in recent years, such as insurance of cancel order, insurance of return shipping, etc. Return guarantee can ensure the buyer obtain a full return service at a lower cost, which has been applied in practice and has been gradually extended.

It can be expected that return guarantee will be put in more further applications in the future no matter in the network store or entity stores. By now it has not been studied deeply in theoretical research.

The influence of parameters on the optimal value such as optimal price, return policies, return guarantee, market share was analyzed. We show that the profit of the enterprise is reduced with the increase of the return rates, and the return guarantee price is positively related to the profit of the enterprise. In order to obtain more supply chain profit. Enterprises can actively use the means of return guarantee and increase its price, Increasing the size of the network direct sales is also effective to supply chain.

Further research maybe done considering service level and return policy is different under the dual channel, and the risk aversion`s supply chain strategy.

Acknowledgements

This research was financially supported by the team of Logistics management research and service innovation. (Grant No. CXTD201803).

REFERENCES

[1] Ofek E，Katona Z，Sarvary M．(2011)"Bricks and clicks": the impact of product returns on the strategies of multichannel retailers．Marketing Science， 30 : 42-60.

[2] Baiman s, Fischer P E, Rajan M v. (2000)Information contracting, and quality costs.Management Science,46:776-289.

[3] Cachon G P, Swinney R. (2009)Purchasing, pricing, and quick response in the presence of strategic consumers. Management Science,55:497-511.

[4] ZHENG Chun-dong，LIU Yi-fan，ZOU Meng. (2016)Influence of insurance of return-of-goods freight on purchase intention of online consumers. Journal of Shenyang University of Technology (Social Science Edition),4:150-156.

[5] Mukhopadhyay S K, Setoputro R. (2007)A dynamic model for optimal design quality and return policies.European Journal of OperationalResearch,180:1144-1154.

[6] Yanr R，Pei Z. (2011)Information asymmetry，pricing strategy and firm`s performance in the retailer-multi-channel manufacturer supply chain. Journal of Business Research，64:377-384.

[7] CHEN Chong-ping, CHEN Zhi-xiang. (2016)Return Guarantee and Pricing Decision in Online Sales[J].Chinese Journal of Management Science ,6:52-60.

[8] LIU Yong-mei，LIAO Pan，HU Jun-hua，CHEN Xiao-hong. (2015)Pricing Strategies in Dual-channel Supply Chain with Retail Services and Customer returns. Operations Research and Management Science,6:79-87.

Summary of recommendation system development

LIU Liling[1]

[1] Beijing, Haidian District, Zhong Guan Cun South Street, No.5 Yard, Beijing Institute of Technology, China

Abstract. A recommendation system is an information service system that connects users and projects: for one thing, it helps users discover potential projects of interest; for another thing, it helps project providers to deliver projects to users who are interested in it. The recommendation system is a powerful system that can add value to the company or business. In the future, it will continue to be researched and developed to bring a better experience to users.

1. INTRODUCTION

A recommendation system is a tool that actively finds information that may be of interest to a user from a large amount of information. Building a system that supports online user decisions, recommending a personalized, highly-matched product or project is a core issue in the recommended system area. It can be traced back to cognitive science, approximation theory, information retrieval, prediction theory, management science, and customer selection models in the market [1]. In view of the theoretical and practical application value of the recommendation system, this paper reviews the research progress of the recommendation system, and attempts to lay a foundation for further research on the recommendation system theory and the expansion of its application field.

2. RELATED WORK

2.1. Recommendation system and algorithm

The recommendation system is based on recommended techniques. The recommended technique, also known as personalized information filtering, is used to predict whether a given user will like a particular project (predictive problem) or to identify a set of N items of interest to a given user (top-N recommendation) problem). The recommendation system proactively provides users with items that may be of interest, essentially by linking users and projects in a certain way. Figure 1 shows the working principle of the recommendation system. From left to right, the input data source is followed by a recommendation algorithm to generate recommendation results for personalized recommendation. Different recommendation systems use different recommendation algorithms, so the core of the recommendation system is to use different recommendation algorithms according to different data sources.

Content from this work may be used under the terms of the Creative Commons Attribution 3.0 licence. Any further distribution of this work must maintain attribution to the author(s) and the title of the work, journal citation and DOI.
Published under licence by IOP Publishing Ltd

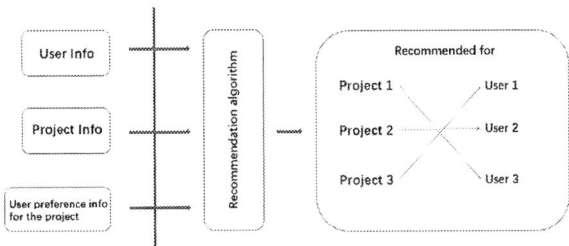

Fig.1 Operating principle of recommender system

The recommended algorithms are endless, and different classification results can be obtained according to different classification criteria. The mainstream recommendation algorithms are divided into three categories [1]: content-based recommendations, collaborative filtering recommendations, and hybrid recommendations. With the deepening and development of the recommendation system research, more and more algorithms and models have emerged. According to the model, it is divided into nearest neighbor model, hidden factor model and graph model. It can also be divided into e-commerce domain recommendation, social network domain recommendation, multimedia domain recommendation, mobile application domain recommendation, cross-domain recommendation [2], etc., depending on the application domain. The recommended system classification framework [2] is shown in Figure 2. The left branch of the frame diagram is the application area of the recommendation system, such as books, texts, pictures, movies, music, shopping, TV programs and others. The right branch is used by the recommendation system. Data mining techniques such as association rules, clustering, decision trees, K-nearest neighbors, link analysis, neural networks, regression, and heuristics.

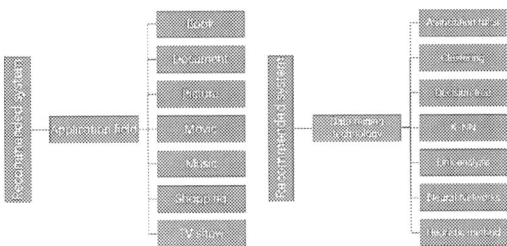

Fig.2 Classification framework of recommender system

2.2. Deep learning technology

As more and more data on the Internet can be acquired, multi-source heterogeneous data including images, texts, and tags contain rich user behavior information and personalized demand information, and a mixture of multi-source heterogeneous auxiliary information is combined. The recommended method is more and more important because it can alleviate the problem of data sparseness and cold start in the traditional recommendation system, but the auxiliary information often has complex features such as multi-modality, data heterogeneity, large-scale, data sparseness and uneven distribution. The research on hybrid recommendation methods combining multi-source heterogeneous data still faces serious challenges.

In recent years, deep learning has made breakthroughs in the fields of image processing, natural language understanding and speech recognition [3], which has become a boom of artificial intelligence and brings new opportunities for the research of recommendation systems. On the one hand, deep learning can characterize the massive data of users and projects by learning a deep nonlinear network structure. It has a powerful ability to learn the essential features of data sets from samples, and can obtain deep feature representations of users and projects. On the other hand, deep learning can achieve the unified representation of data by mapping different data into a single hidden space by performing automatic feature learning from multi-source heterogeneous data [4]. Based on this, the traditional recommendation method is combined. It is recommended to effectively utilize multi-source heterogeneous data to alleviate the problem of data sparseness and cold start in traditional recommendation systems.

Covington et al. [5] proposed a deep neural network model (see Figure 3) for YouTube video recommendation by using multi-source heterogeneous data such as user information, context information, historical behavior data, and project feature information.

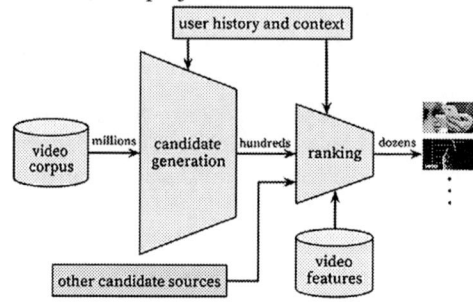

Fig.3 Covington's Recommendation system

2.3. Knowledge map

On May 17, 2012, Google officially proposed the term "knowledge map" [7]. The knowledge map is intended to describe the various entities or concepts that exist in the real world, and the relationships between them. Each entity or concept is identified by a globally unique ID, each attribute-value pair is used to characterize the intrinsic properties of the entity, and the relationship is used to connect the two entities to characterize the association between them. Knowledge maps can be combined with a variety of data sources to enrich data semantic information, and can provide services to users in combination with inferred implicit information. With the development requirements of information retrieval, smart city and other application fields, applying knowledge maps to these areas to improve user experience and system performance has become a hot topic in academic and industrial circles.

In the field of recommendation systems, people tend to focus on the connection between users and projects, and lack of considerations for the interconnection between users and users, projects and projects. The knowledge map-based recommendation system enhances the semantic information of the data by connecting users and users, users and projects, and projects and projects to further improve the accuracy of recommendation. It has important research significance and practical value, and gradually becomes a recommendation system, which nowadays is one of the most active branches of research.

3. OVERVIEW OF TYPICAL RECOMMENDATION ALGORITHMS

3.1. Demographic-Based Recommendation

The basic assumption of this method is that "a user may like an item similar to a user similar to it." When we need to personalize a User recommendation, we use the User Profile to calculate the similarity between other users, and then select the top K users that are most similar to them, and then use those users' purchase and score information to recommend. A simple and common recommendation is to use the Items covered by these users as a list of recommendations, sort by the average of the scores of Item on these Users, and supply the sorted list of recommendations to the user.

3.2.Content-Based Recommendation

The first step in a typical Content-Based approach is to build a User Profile. A simpler construction method is to consider all the items that the User has ever scored, and make a weighted average of the Item Profiles of these items as the UserProfile of the User. Obviously, the strategy for building a User Profile can be complex. For example, we can consider the time factor and calculate the Profile of User in different time periods to understand the changes in User's preference on historical data. With the User Profile, we can start recommending. The simplest recommendation strategy is to calculate the similarity between all the items that the user has not tried and the user's User Profile, in order of similarity. A list of recommendations is generated and output as a result. Besides, the recommendation strategy can also be very complicated, such as considering the real-time interaction data collected during the user interaction process on the data source to determine the ordering, using the decision tree and artificial neural network on the model, but the core of these methods The links are all calculated using the similarity between the User Profile and the Item Profile.

3.3.Collaborative Filtering-Based Recommendation

Collaborative Filtering-Based Recommendation refers to collecting the past behavior of the user to obtain explicit or implicit information about the product, that is, according to the user's preference for the item or information, discovering the relevance of the item or the content itself, or the relevance of the user, and then Based on these associations, recommendations are made. According to the foregoing, recommendations based on collaborative filtering can be based on User-based Recommendations, based on Item-based Recommendations, and based on subclasses such as Model-based Recommendation. The user's preference or scoring matrix for the item is often a large sparse matrix. In order to reduce the amount of calculation, Clustering items for Collaborative Filtering can be used.

4. RECOMMENDED SYSTEM POTENTIAL DEVELOPMENT DIRECTION

4.1.Variety

The data available in the recommendation system is complex and complex. For example, information in the social network, location location information, and other context-aware information are taken into account. Not only does the amount of data increase, but the computational complexity also increases exponentially. In addition, the recommendation system research involves privacy protection. How to ensure personalized recommendation and protect user's privacy is a confrontational problem, which brings great challenges to researchers and developers.[8]

4.2 Interpretable

As an important product in the field of artificial intelligence, the recommendation system is widely accepted and applied. The core of the recommendation is the rationality of the high recommendation result, which also requires the recommendation result to be well interpretable, although this has long been known. It is realized, but the special research on interpretability is still lacking [9]. In the current research, the interpretability discussion of the recommendation algorithm is generally the selection process after the algorithm evaluation. With the high demands of users, the research of "recommendation reasons" has received more and more attention in industry and academia.

5. CONCLUSION

A recommendation system is an information service system that connects users and projects: on the one hand, it helps users discover potential projects of interest; on the other hand, it helps project providers to deliver projects to users who are interested in it. The recommendation system is a powerful system that can add value to the company or business. In the future, it will continue to be researched and developed to bring a better experience to users.

References

[1] Adomavicius G, Tuzhilin A. Toward the next generation of recommender systems: a survey of the state-of-the-art and possible extensions[J]. IEEE Transactions on Knowledge and Data Engineering, 17(6): 734-749, 2005

[2] Park D H, Kim H K, Choi I Y, et al. A literature review andclassification of recommender systems research[J]. ExpertSystems with Applications, 39(11): 10059-10072, 2012

[3] Lecuny, Bengio Y, Hinton G. Deeplearning. Nature, 521(7553): 436-444, 2015

[4] Yu-xin, PENG, Wen-wu, et al. Cross-media analysis and reasoning: advances and directions[J]. Frontiers of Information Technology & Electronic Engineering, 18(1):44-57, 2017

[5] Covington P, Adams J, Sargin E. Deep Neural Networks for YouTube Recommendations[C]// ACM Conference on Recommender Systems. ACM, 2016:191-198.

[6] Park D H, Kim H K, Choi I Y, et al. A literature review andclassification of recommender systems research[J]. ExpertSystems with Applications, 39(11): 10059-10072, 2012

[7] Jing L. Automatic Image Annotation Based on Concept Indexing[J]. Journal of Computer Research & Development, 44(3):452, 2007,

[8] Aciar S, Zhang D, Simoff S, et al. Recommender System Based on Consumer Product Reviews[C]// IEEE/WIC/ACM International Conference on Web Intelligence. IEEE Computer Society, 2007:719-723.

[9] Tintarev N, Masthoff J. A Survey of Explanations in Recommender Systems[C]// IEEE, International Conference on Data Engineering Workshop. IEEE, 2007:801-810.

ISPECE   IOP Publishing

IOP Conf. Series: Journal of Physics: Conf. Series **1187** (2019) 052045   doi:10.1088/1742-6596/1187/5/052045

An Effective Method for Forest Fire Smoke Detection

Luxing Qin [1,2], Xuehui Wu [1,2], Yichao Cao[1,2] and Xiaobo Lu [1,2,a]

[1]School of Automation, Southeast University, Nanjing 210096, China

[2]Key Laboratory of Measurement and Control of Complex Systems of Engineering, Ministry of Education, Southeast University, Nanjing 210096, China

[a]Corresponding author: xblu2013@126.com

Abstract. This paper focuses on smoke detection in forest environments. In this paper, dark channel prior and OTSU based multi-threshold are used to find the disturbances such as sky and haze. These regions are blocked to reduce false alarm rate. A motion detection method based on frame difference is adapted to find the motion objects. Color moments, HOG and LBP are chosen as features of smoke and SVM is used as the classification. To reduce more false alarms, the motion regions are classified for several consecutive frames. They won't be regarded as smoke regions unless M frames of them are classified as smoke ones. Experiment results showed that the proposed method can detect smoke in video effectively and work real-timely.

1. Introduction

Forest fire happens every year around the world, which causes serious damage to ecological environment, animals and humans. Traditional smoke detecting systems are made up of sensors. These systems either have high false alarms rate or cost too much as plenty of sensors are required. Besides, sensors have many other limitations, such as communication and control. In order to detect the forest fire as soon as possible, many different forest fire detection algorithms were proposed.

Forest fire smoke detection algorithm can be divided into several categories. Wavelet transform and other signal processing algorithms to a single image is an approach [1]. These methods don't need training before detection, but perform badly on change of lightness and other disturbances. Methods based on feature extraction are widely adapted. Usually, a background model of the scene should be built to find the motion area and avoid unnecessary calculation. The features are then put into a classifier such as support vector machine (SVM). There are lots of features which have been proved to be useful to describe smoke. Static features such as color feature, LBP, HOG and compactness perform well. Dynamic features include movement directions, area change and so on. Cai et al. [2] combined static features and dynamic features and got better results. In recent years, deep learning makes a great success in computer vision. Convolutional neural network (CNN) gets much higher accuracy rate on images classification and recurrent neural network (RNN) can reinforce the knowledge for a set of sequential frames. Tao et al. [3] showed that CNN works better on smoke detection. A.F. et al. [4] applied CNN and RNN to video sequences. Though deep learning performs better, the speed of training and working is not satisfactory. In practice, it's difficult to work real-time without a high performance GPU.

The real forest environment can be complicated. The features of smoke are changeable and disturbances such as haze, sky and change of lightness can also be captured by the motion detection algorithm. All of them increase the false positive rate significantly. In this paper, Dark channel prior and multi-threshold segmentation are used to remove the disturbances such as white house, haze and sky before detection. After motion object segmentation, the regions which are moving but not blocked are

Content from this work may be used under the terms of the Creative Commons Attribution 3.0 licence. Any further distribution of this work must maintain attribution to the author(s) and the title of the work, journal citation and DOI.

Published under licence by IOP Publishing Ltd

activated. Finally, features of motion cells are extracted and support vector machine is used to classify the motion cells.

2. Method

The proposed method consists of three major steps:

(1) Extracting mask regions with dark channel prior and multi-threshold segmentation.
(2) Motion detection.
(3) Extracting features and classifying using SVM.

2.1. Dark channel prior

The dark channel prior theory [5] is a method based on statistics on a huge amount of outdoor pictures. It's found that in most non-sky areas, there is always at least one of the three color channels with a low value. This channel with the minimum value is called dark channel.

For an input image J, the dark channel can be expressed as:

$$J^{dark}(x) = \min(\min(J^C(y))_{C \in \{r,g,b\}, y \in \Omega(x)}) \tag{1}$$

Where J^C is a color channel of input image J, and $\Omega(x)$ is a local patch centered at x. In practical calculation, we usually find the minimum value of the three channels of every pixel and store them in a grayscale image whose size is the same as the original picture. Then a filter is adapted to smooth the grayscale image. The radius is a changeable parameter. Usually the window size is $2*Radius+1$. According to the dark channel theory, $J^{dark}(x)$ should be small for close object but relatively large for sky and haze. As is shown in Figure 1, the dark channel values of the image pixels at different distances differ greatly.

In computer vision, the graph formation is used widely:

$$I(x) = J(x)t(x) + A(1-t(x)) \tag{2}$$

Where $I(x)$ is the original image that is captured by the camera, $J(x)$ is the result picture, A is the global atmosphere light component and $t(x)$ is the transmission. The aim is to calculate $J(x)$ with $I(x)$ already known.

In practice, we select the top 0.1% pixels with high brightness in the dark channel which is the most opaque. Among these pixels, the highest intensity pixel mean in the input image is selected as A. Assuming that $t(x)$ is a constant value in each window, the estimated value of transmission can be derived:

$$\tilde{t}(x) = 1 - \min(min_C(\frac{t(y)}{A_C}))_{y \in \Omega(x)} \tag{3}$$

In actual, the atmosphere is not absolutely free of any particles. For a camera used in forest fire detection, it's required to be installed on a high place to cover a large area, even several kilometers. The existence of haze is the basis of human perception of the depth of the scene, while it greatly increases the difficulty of smoke detection. It's hard to distinguish fire smoke and haze at a far distance.

2.2. Multi-threshold segmentation

After the calculation of the air transmission of each pixel of the image, we get a grayscale image. In order to find the area that we want to mask, a threshold segmentation method should be used.

The OTSU algorithm [6] is an adaptive method to get the proper threshold and segment the image. Suppose the grayscale space of an image is represented by L. For each pixel, i is the grayscale value in the grayscale space. The number of each grayscale value in the grayscale histogram of the image is represented by n_i, and the total number of pixel of the image is denoted by N, then $N = n_1 + n_2 + ... + n_{L-1}$. If the pixels of the image are divided into 2 classes, the class probability $\omega_{0,1}(t)$ is computed from the histogram:

$$\omega_0(t) = \sum_{i=0}^{t-1} p(i) \tag{4}$$

$$\omega_1(t) = \sum_{i=t}^{L-1} p(i) \tag{5}$$

Where the probability of each grayscale value is represented by $p(i)$ and t is the single threshold of the segmentation.

The total mean value of the image is μ_T, and the mean values of each of the 2 classes in the image are μ_0 and μ_1 respectively:

$$\mu_0(t) = \frac{\sum_{i=0}^{t-1} ip(i)}{\omega_0(t)} \tag{6}$$

$$\mu_1(t) = \frac{\sum_{i=t}^{L-1} ip(i)}{\omega_1(t)} \tag{7}$$

$$\mu_T(t) = \sum_{i=0}^{L-1} ip(i) \tag{8}$$

The inter-class variance σ_b^2 and the intra-class variance σ_w^2 of the two classes can described:

$$\sigma_b^2(t) = \sigma^2 - \sigma_\omega^2(t) \tag{9}$$

$$\sigma_w^2 = \omega_0(t)\sigma_0^2(t) + \omega_1(t)\sigma_1^2(t) \tag{10}$$

When using the OTSU method to find the threshold, the optical threshold can be calculated by maximizing the inter-class variance or minimizing the intra-class variance iteratively:

$$t_{opt} = \substack{Max \\ 0 \le t \le L-1} \{\sigma_b^2(t)\} = \substack{Min \\ 0 \le t \le L-1} \{\sigma_w^2(t)\} \tag{11}$$

As it's shown in Figure 2 , the forest scene usually can be made up of three or more regions, so it's necessary to extend to multi-threshold method [7]. The grayscale space is still $[0, L-1]$ with $N = n_1 + n_2 + \cdots + n_{L-1}$. For multi-threshold segmentation, the image should be segmented into n classes, so there are $n-1$ thresholds $t_0, t_1, \ldots, t_{n-1}$. The total mean value of the image is still μ_T, and the mean values of the regions corresponding to n classes are $u_0, u_1, \ldots, u_{n-1}$. Similar to the single threshold method, the optical threshold for multi-threshold algorithm should follow the principle of maximizing inter-class variance or minimizing intra-class variance.

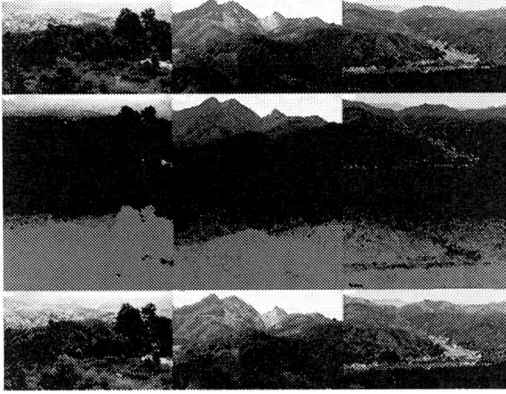

Figure 1. (a) Source image (b) Dark channel image (c) Multi-threshold segmentation results (d) Mask regions of the images

2.3. Motion detection

Motion detection is an important part of video smoke detection. A good motion detection algorithm can locate the moving objects accurately and avoid unnecessary calculation. Usually, a background model should be built and updated every frame. Then the foreground can be separated from the difference between current frame and the background model.

Gaussian mixture model (GMM) [8] and ViBe [9] have been widely used as motion detection methods before smoke detection. GMM needs several frames to build the background model and requests various calculations. Compared to GMM, ViBe needs less modeling process and costs less time. Recently, RPCA has been used in background modeling and performs better [10]. However, this approach requests a large amount of memory space and complicated calculations, which is hard to apply to 720P and 1080P videos.

In forest environments, the field of a camera vision is wide. Therefore, the motion detection method should be highly sensitive to catch every possible motion. Besides, it should be fast enough to satisfy the need for real time.

In this paper, we use an improved method based on frame difference, which is showed in Figure 2:

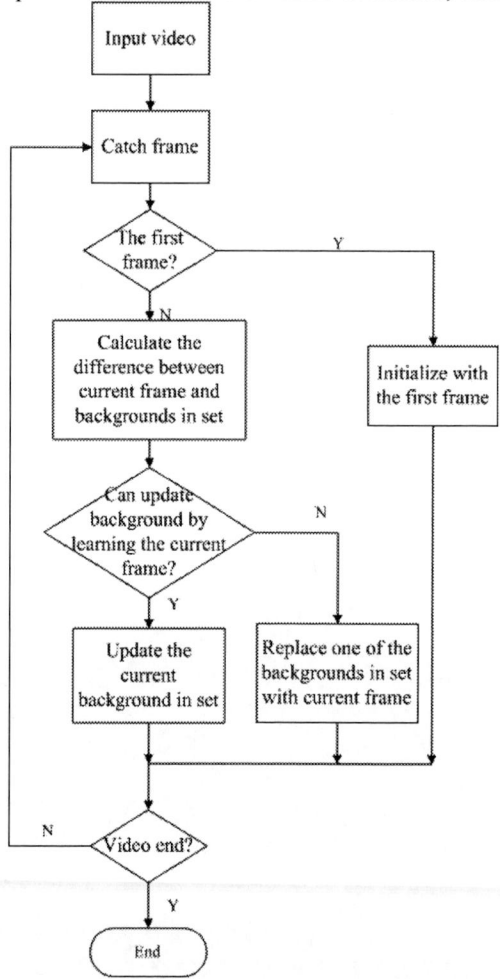

Figure 2. Proposed background modeling flow chart

Experiments were carried out to test the efficiency of the proposed method, GMM and ViBe. Figure 3 shows that our method locates the motion regions effectively. Table 1 shows that our method is faster than both of the others.

Figure 3. Motion detection with ViBe, GMM and proposed

Table 1. Speed of GMM, ViBe and proposed

Method	Size of video	Average time per frame
GMM	1280×720	33.7863ms
ViBe	1280×720	31.2695ms
Proposed	1280×720	24.2788ms

2.4. Feature extraction and smoke detection

As a widely used feature, color feature doesn't need much calculations. For smoke detection, color is a very useful feature. Color moments can measure color distribution of an image. It is usually that the first three color moments are used as features. Color histogram counts the color information of the entire image and quantifies the proportion of different colors in the entire image. In this paper, color moments in HSV color space is chosen to describe the color characteristic of the smoke.

Local binary patterns (LBP) is a type of visual descriptor used for classification in computer vision. There are many improvements of original LBP methods. In this paper, original circle LBP is adapted as a characteristic of the smoke.

The histogram of oriented gradients (HOG) is also a feature descriptor used in computer vision and image processing for the purpose of object detection. As a statistical feature, it counts occurrences of gradient orientation in localized portions of an image.

As smoke moves slowly and won't disappear in a short time, the moving regions won't be regarded as a potential smoke region if it can't be detected by the motion detection in consecutive N frames. The mask regions calculated from dark channel image and multi-threshold segmentation will not be detected, for it's useless to detect these regions. Color moments, LBP and HOG are adapted as features for image classification. In practice, we classify the motion cells for consecutive 50 frames. If M of them are regarded as smoke, this region will be classified as smoke regions.

3. Experiment results

Figure 4. The result images of different videos

The judgment of the performance of video smoke detecting algorithm is not easy to define. Smoke fire detection is not the same as car or face. Actually, a cloud of smoke can be divided into several cells, and a correct detection of one of the cells is enough. From this point, we can judge the performance of the algorithm by whether detecting smoke from the video. However, it will be more reliable if improving the recognition rate of every cell.

The experiment was carried out on a PC with an Intel Core i3, 3.40GHz processor using the method proposed in this paper. Figure 3 shows images of smoke detection. Red box regions are smoke cells and regions circled by red contours are mask regions.

Twenty 720P videos which are captured by a practical camera used for forest fire detection have been tested in this experiment with a trained SVM classifier. A half of them are with smoke while others are not but can easily cause error detection. The classifier is trained with about 5000 picture samples in total. Color moments, LBP and HOG are adapted as the features to train the classifier and detect smoke. To prove the efficiency of the proposed method, Cai's method [2] is tested on the same videos, too. The experiment results of non-smoky videos and smoky videos with mask or not and Cai's method are showed in Table 2 and Table 3 respectively.

In forest smoke detection, a single false alarm region will cause a false alarm video. As is showed in the table, the false positive rate of regions and videos decreases significantly, while the true positive rate only decreases a little. Actually, all of the smoke videos do detect smoke. Besides, the speed of the method can reach about 18fps. As for Cai's method, some of the smoke regions are detected but the false alarm rate is higher. Besides, for some videos whose smoke regions are small and far away from the camera, smoke is not detected. As is showed in Table 4, in order to use dynamic features, multi-object tracking costs too much calculation, which makes it much slower than static ones.

Table 2. Comparison of non-smoky videos

Method	Motion regions	Detected regions	False alarm rate of regions	Smoky videos

Classifi-cation only	1271	11	0.87%	3
With mask regions	1259	1	0.08%	1
Cai's method	540	68	12.6%	3

Table 3. Comparison of smoky videos

Method	Motion regions	Actual regions	Detected regions	Detection rate	Smoky videos
Classifi-cation only	1736	88	70	79.5%	10
With mask regions	1711	88	61	69.3%	10
Cai's method	2118	121	84	69.4%	6

Table 4. Comparison of speed

Method	Size of video	Time per frame
Classification only	720P	57.49ms
With mask regions	720P	57.50ms
Cai's method	720P	257.69ms

4. Conclusion

In this paper, a method for forest smoke detection is proposed. In terms of the specific forest scene, Dark channel prior and multi-threshold segmentation are used to mask the region that may easily cause error detection. The results show that the false positive rate decreases significantly while the true positive rate almost keep the same. The method can be real-time on a PC with no GPU.

Acknowledgment

This work was supported by the National Natural Science Foundation of China (No.61871123), Key Research and Development Program in Jiangsu Province (No.BE2016739) and a Project Funded by the Priority Academic Program Development of Jiangsu Higher Education Institutions.

References

[1]   G.R. et al.. Wavelet-Based Smoke Detection in Outdoor Video Sequences [C]. MWSCAS, pp.383-387(2010)

[2] M. Cai, et al.. Intelligent Video Analysis-based Forest Fires Smoke Detection Algorithms [C]. ICNC-FSKD, pp.1504-1508(2016)

[3] C. Tao, et al.. Smoke detection based on deep convolutional neural networks [C]. ICIICII, pp.150–15(2016)

[4] A.F., L.K., K. Jo. Smoke Detection on Video Sequences Using Convolutional and Recurrent Neural Networks. University of Ulsan, Ulsan, Republic of Korea(2017)

[5] He K, Sun J, Tang X. Single Image Haze Removal Using Dark Channel Prior [J]. IEEE Trans Pattern Anal Mach Intell, 33(12):2341-2353(2016)

[6] Otsu N. A Threshold Selection Method from Gray-Level Histograms [J]. IEEE Trans.syst.man.& Cybern, 9(1):62-66(2007)

[7] Y. Cao et al.. Single Image Depth Level Estimation Using Dark Channel Prior [C]. (to be published)

[8] T. B. et al.. Background Modeling using Mixture of Gaussians for Foreground Detection – A Survey [J]. Recent Patents on Computer Science, pp.219-237(2008)

[9] Olivier Barnich, Marc Van Droogenbroeck. ViBe-a universal background subtraction algorithm for video sequences [J]. IEEE Trans Image Processing, 20(6):1709-1724(2011)

[10] M. Wu, Y. Sun et al.. Multi-component group sparse RPCA model for motion object detection under complex dynamic background [J]. Neurocomputing, pp.120-131(2018)

Demand analysis of material reserve optimization

Jian Gao[1,2], Shuming He[3], Xing Song[1] and Liang Zhang[2]

[1] Department of Equipment Command and Administration, Army Engineering University, Shijiazhuang, China

[2] Unit 65153 PLA, Chaoyang, China

[3] Unit 32134 PLA, Tianjing, China

863872761@qq.com

Abstract. Material reserve is an important task in the nation construction and military development. On the basis of analyzing the significance of modern logistics reserve and the present situation of the storage industry, this paper mainly combs and summarizes the problems existing in the storage industry of our country, puts forward the pertinent optimization measures and strategies, and points out the optimal demand for material reserve. The paper will lay a solid foundation for the next stage of material reserve construction and development.

1. Importance of optimizing material reserves

1.1. Reduce transportation and production costs

Although material reserve will increase the cost of warehousing products, it can also improve the efficiency of transportation and production, relatively reduces the cost of both. If the market demand is uncertain, unpredictable circumstances, the reserve of a certain amount of products can effectively prevent the production of shortage costs, and also to a certain extent to ensure production. The rhythm of operation makes the production plan well implemented and the production cost can be reduced. At the same time, warehousing can concentrate and integrate the small batch and scattered product transportation tasks, which is conducive to forming the overall optimization of the whole transportation and transportation routes, thus reducing the transportation cost[1].

1.2. Regulating supply and demand

The significance of warehousing for modern logistics has been mentioned before. For example, the production of certain products has seasonal characteristics due to raw materials and other reasons, but the demand for products is continuous. So warehousing here helps to adjust the contradiction between supply and demand. For example, when the price of a raw material needed for the production of a product is lower at a certain time, the manufacturer of the product can purchase a certain amount of raw material in advance and store it for future production.

1.3. Production needs

Storage is not only needed in the process of product circulation, but also exists in the process of product production. For example, in the production logistics, we often mention the temporary storage of products in process, the storage of raw materials and so on.

Content from this work may be used under the terms of the Creative Commons Attribution 3.0 licence. Any further distribution of this work must maintain attribution to the author(s) and the title of the work, journal citation and DOI.
Published under licence by IOP Publishing Ltd

1.4. Marketing needs

In order to win consumers and obtain long-term loyalty of consumers, modern enterprises usually adopt the strategy of rapid customer response, and the implementation of this strategy must rely on the role of warehousing. Storing products close to customers can effectively prevent out-of-stock and shorten the delivery time, thus effectively improving the quality of customer service[2].

2. The status quo of material reserves

2.1. Management system based on departmental management

The long-term formation of this division of departments, regional division, their own warehouse for their own use, closed to each other, the situation of repeated construction has not yet completely changed. The low degree of socialization of the warehouse industry, as a result of this decentralized and self-governed management system, resulting in decentralized funds, backward management, outdated equipment, low warehouse utilization. It is understood that the average utilization rate of warehouse area in China is less than 40%. Some warehouses are idle for a long time, but some are not enough to continue to invest in building new warehouses. Because of blindness and disorderly construction of warehouses, market competition is excessive and warehouse prices are in disorder[3].

2.2. Not suited to the requirements of market economy development

Existing warehouses are bungalows, whose function is simply to store products. Coupled with the current situation, warehousing market is not standardized, unequal competition. The vast majority of state-owned warehouses economic benefits are not good, have many long-term losses, not only lacks stamina, and even have survival problems.

2.3. Warehouse equipment is old and backward

Many of the warehouse equipments are still in the primitive state of manual work. Even though people lift their shoulders, their work efficiency is still very low. Many warehouses goods can not go in and out smoothly. Goods in the warehouse stay too long, or improper storage and damage, mildew, serious losses, increased logistics costs. With the rapid development of China's economy and the deepening of the reform of logistics system, as well as the continuous expansion of opening up and the development of foreign trade, logistics should have a rapid development, especially the warehousing industry, to meet the needs of this new situation[4].

3. Measures to optimize material reserves

3.1. Strengthen storage infrastructure construction

Investments should be intensified to upgrade the infrastructure of existing warehouses, constantly renovate obsolete and aging warehouses, and update and use modern warehousing equipment. We should learn from the advanced experience and technology both at home and abroad, but also combined with the actual situation of various regions, can not be greedy, to form a scientific and reasonable network of storage facilities.

3.2. Introducing competition mechanism and establishing market system

In order to ensure the healthy development of the warehousing industry, it is necessary to standardize the market order, speed up the introduction of competition mechanism, and establish a unified open, fair competition, standardized and orderly modern warehousing system. Abolish all kinds of relevant provisions that do not conform to the provisions of national laws and regulations, and create a relaxed external environment for the operation and development of warehousing enterprises.

3.3. Strengthening resource integration and improving storage standards

The storage and transportation facilities of warehousing enterprises in different industries in China can not be shared, which affects the ability of enterprises to rationally co-ordinate the storage resources. In order to meet the requirements of modern logistics, we must strengthen the integration of resources and establish a storage network. Storage standardization is not only to achieve close cooperation between storage links and other links, but also an effective means to improve operational efficiency within the warehouse. Therefore, warehousing enterprises should constantly improve their standardization system[5].

3.4. Building information platform to realize storage informatization

To improve the utilization rate of warehouse and realize effective inventory control, it is necessary to establish an effective information network, realize the sharing of warehouse information, and actively promote the informationization of enterprise warehouse management. Using modern information technology to build a public information platform, to achieve effective combination of public information network and storage network, improve the level of enterprise warehousing information.

3.5. Cultivating warehousing talents and improving training system

Talents are important resources of enterprises. To develop warehousing enterprises, we must have technical and managerial talents. Warehousing enterprises should make full use of all kinds of resources, actively introduce relevant talents from colleges and universities, and strengthen the on-the-job training of employees in logistics enterprises, and strengthen the training and training of warehousing professionals.

3.6. Establishing warehouse management laws and regulations

Establishing and improving the rules and regulations with the responsibility system as the core is a basic work of warehousing management. At present, China lacks a relatively complete legal and policy system, so it is necessary to speed up the formulation and improvement of warehouse management laws and regulations, standardize the industrial competition order, adjust the policies to help enterprises better management.

4. Summary

The ultimate goal of logistics activities is to shorten the time of commodity circulation, to reach the destination quickly, to ensure the quality of commodities intact, to reduce the cost of commodities in circulation and to achieve social benefits. To sum up, one sentence is to spare no effort to use less mileage, less links, less inventory, less time and less cost to achieve the goal of "timely, accurate, economic and safe" and to provide customers with satisfactory quality of service.

References

[1] Yao Kay, GuoShizhen, Fu Xiaozhong, and so on. The scale control model of ammunition physical reserve construction [J]. equipment college journal, 2016, 27 (1): 52-56.

[2] Summer Changchun. Elementary introduction to the direction of the development of light weapons technology [J]. scientists, 2017 (3): 18-19.

[3] Jiang Da. Military logistics system model and application of [M]. Chinese material press, 2006.

[4] He Jianfeng, Li Cui. New high, future war weapons and equipment production capacity on the construction of [J]. automation, 2014 (10): 4-6.

[5] Hu Bin, evergreen. Practice analysis and Research on model modeling of military concept model [J]. system simulation, 2008, 20 (12): 3085-3088.

Mining Spatiotemporal Characteristics of Car-sharing Demand

Jing-Jing Tian[1], Dong-Fan Xie[1] and Fu-Jun Ding[2]

[1]MOE Key Laboratory for Urban Transportation Complex Systems Theory and Technology,Beijing Jiaotong University, Beijing 100044, P.R. China.

[2]Gansuyixiangxing New Energy Developments Ltd, Lanzhou 730000, China

E-mail address: 17120756@bjtu.edu.cn (Tian) , dfxie@bjtu.edu.cn (Xie)

Abstract. Since car-sharing demand plays an essential role on the management and service of car-sharing system, this paper attempts to analyze the temporal and spatial characteristics of the car-sharing demand, aiming to discover spatial and temporal patterns and association rules. Based on a clustering algorithm (i.e., DBSCAN), the spatiotemporal characteristics of car-sharing demand are studied. On the basis, the demand is divided into various clusters in space. Furthermore, the correlations between any two clusters are studied. The results show that car-sharing demand is high on Friday, Saturday and Sunday, and has strong correlation with time and space. It is expected the results can support the management of car-sharing system, and further promote the service level for passengers.

1. Introduction

With the intensification of traffic problems such as traffic congestion, car-sharing is considered to be one of the effective ways to relieve these problems. Under the strong support of government policies, car-sharing have achieved rapid development in China. Correspondingly, research on car-sharing is gradually becoming a hot topic. There are a lot of researches on the demand forecasting and location depot of car-sharing, while the analysis of spatial-temporal characteristics of demand is not enough.

As for studies based on the trajectory, most existing works focus on clustering and spatial-temporal association rules. The former based on segmentation of the moving trajectories, such as segment extraction and clustering of the moving trajectories (Ma X et al. 2013, Arthur et al. 2014,Almannaa et al. 2018). The latter mainly taking into account the spatial and temporal constraints simultaneously, and focusing on the improvement of efficiency (XL Zhao et al. 2010, Y Xia et al. 2011, MQ Yan et al. 2017).There is also a method that main uses heat maps(Chang Yu et a!, 2017) to analyze features of demand .

In order to have a deeper understanding of the car-sharing demand characteristics, this paper mainly explores the temporal and spatial characteristics of car-sharing demand through clustering and spatiotemporal association algorithm. The rest of the paper is organized as follows. In Section 2, we describe the dataset used in this paper and make some basic data analysis. In Section 3, we introduce the clustering algorithm named DBSCAN and analyze the results. In Section 4, we expand the apriori association algorithm and analyze the spatiotemporal relevance of car-sharing demand. In Section 5, we summarize the conclusion of this paper.

Content from this work may be used under the terms of the Creative Commons Attribution 3.0 licence. Any further distribution of this work must maintain attribution to the author(s) and the title of the work, journal citation and DOI.

Published under licence by IOP Publishing Ltd

2. Data analysis

In this study, the data is obtained from a car-sharing company, including the rental data and the trajectory data of the car-sharing in Beijing. The rental data contains 81455 rental records from December 17, 2016 to October 16, 2017(278days). The car-trajectory data contains 123864309 records of trips from January 1, 2017and October 31, 2017(293days). Preprocess raw data, and the datasets used in this study are given in tables 1 and 2 for rental records and GPS data, respectively.

Table 1 Car-sharing Rental Data

Fields	Description
Rental number	Identity of an cars-haring use transaction
Pick-up station	Name of car picking-up station
Return station	Name of car return station
Time	Time of picking up car
Duration	Total travel time of an transaction
Distance	Total mileage of an transaction
License Number	Identity of a car

Tab 2 Trajectory Data of car-sharing

Fields	Description
Car number	Identity of a car
Time	the time of recording car's position
Longitude	Longitude of car position
Latitude	Latitude of car position
Speed	Instantaneous speed of car

2.1.Rental Data

Car-sharing system can be divided into three categories based on operating patterns: two-way (or round-trip) systems, one-way system and free floating. For the first, the car must be returned to the station where it was picked up. While for the second, it allows users to return a rented car to any of the designated stations. And the third allows the car to park in any place where it can be parked. It is clearly that the company applies the one-way car-sharing pattern, which is more prevalent. Therefore, station imbalance will occur.

In this study, the datasets contain 174 car-sharing stations and 960 cars. Based on the dataset of 278 days, the difference between the total numbers of borrowed and returned cars is used to indicate the degree of imbalance in the stations, as shown in Figure 1. The greater the absolute value of the difference, the higher the imbalance of the station. It can also be found that the station with larger amount of rental volume has a higher degree of imbalance. Few stations are in balance states.

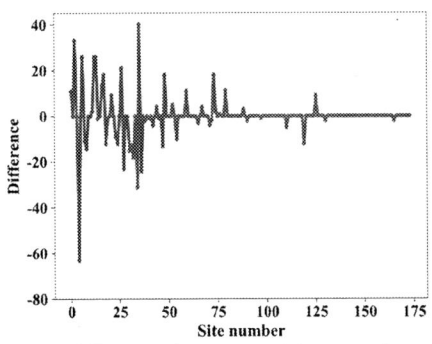

Fig 1 Difference between pick-up and return

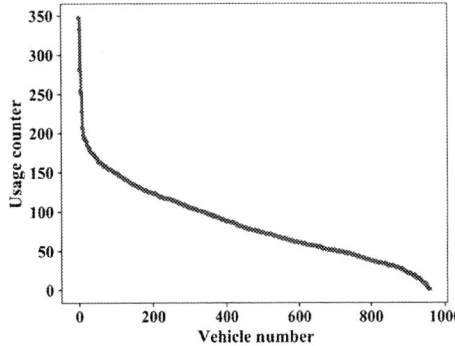

Fig2 Car usage frequency

2372

In addition, the car-sharing are sorted in terms of the rental amount, as shown in Figure 2. One can see that there are two turning points on the curve. The turning point divides the cars into three categories according to rental amount of each car.

For the type of cars with a high rental amount, it is found that almost all the cars are picked up and returned at the same site, and these stations are more balanced. We can infer that even if there is not much rental data at these sites, the utilization rate of cars is higher.

2.2. Trajectory Data

The trajectory data has total 123,864,309 track points, including 1071 cars. The trajectory of a car consists of multiple trips by many users, and the trajectory data implies the departure point and destination of the user, which should be split for the following analysis. In this study, for the sake of simplicity, we refer to the related literature, the point where the stay time exceeds ten minutes is used as the travel stop point, and the stop point is used as the end point of the last trip and the starting point of the next trip. The OD demand is extracted from the trajectory data, concluding 194 working days, and 99 non-working days. The resulting dataset is shown in table 1.

Table3 OD travel information

Field	Description
Otime	Time of origin
Olon	Longitude of origin
Olat	Latitude of origin
Dtime	Time of destination
Dlon	Longitude of destination
Dlat	Latitude of destination
Distance	Distance between origin and destination
Time	Duration between origin and destination

3. Clustering analysis of car-sharing demand

Clustering analysis is a branch of data mining. Based on similarity, there is more similarity between patterns in one cluster than patterns in different clusters. The algorithm of cluster analysis can be divided into five classes: Partitioning Methods, Hierarchical Methods, density-based methods, grid-based methods and Model-Based Methods.

3.1. Clustering Algorithm

This study applied DBSCAN algorithm for clustering. DBSCAN (Ester et al. 1996) is a density-based method with two parameters: eps and minpts, corresponding to the maximum radius of the neighborhood and the minimum number of neighbors for a core point. A cluster is a maximal set of density-connected points.

3.2. Clustering Analysis

After OD trips were extracted from trajectory data, the DBSCAN clustering algorithm is used to analysis the start and end points of the OD trip. By continuously adjusting the parameters, the clustering results are given in Figures 3 and 4.

It can be seen that the spatial distribution of each cluster of origin and destination is consistent. Therefore, only the characteristics of the origin point will be studied in the following. Figure 5 presents the demand of all clusters. The number refers to the proportion of non-working day (including holiday). It is clear that cluster 5 has the most demand, and cluster 6 has the most demand of non-working day.

Figures 6, 7 and 8 reveal the demand from month, hour, and day of the week, respectively. The demand of cluster 5 has increased significantly in August, while other clusters have seen a significant downward trend. The distribution of each cluster is basically the same in the day, and there are three peak hours in a day: 8-10am, 12-14pm, and 16-18pm. Figure 8 show the distribution of each cluster on the day of the week. For all clusters, the proportion of Friday, Saturday and Sunday is larger, and the difference between other working days is small.

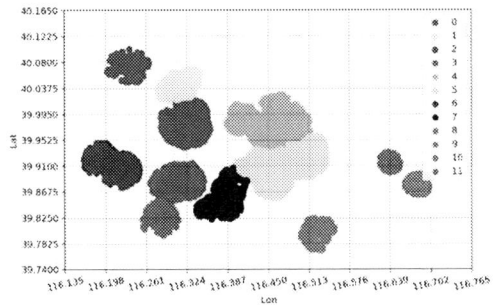

Fig 3 Clustering result of origin   Fig 4 Clustering result of destination

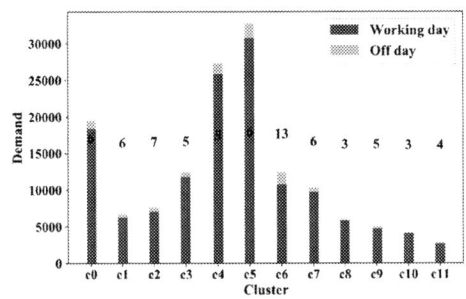
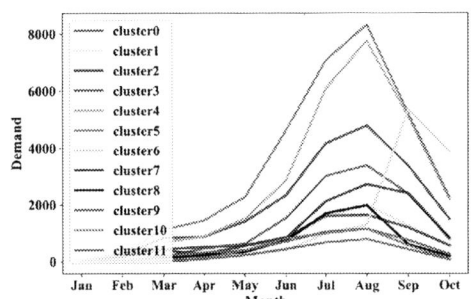

Figure 5 Distribution on weekdays and off days   Fig 6 Distribution in each month

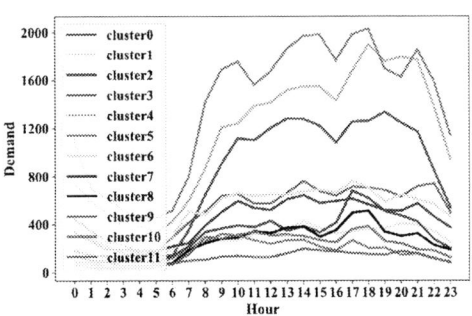
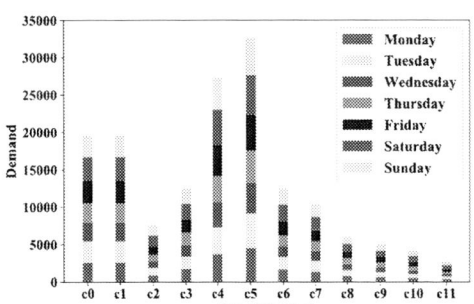

Fig 7 Distribution of a day   Fig 8 Distribution on day of the week

The extracted origin-destination trips also include duration and distance. In order to analyze their characteristics, we make their own cumulative distribution curves, as shown in Figures 9 and 10. We can see that 90% of the origin-destination travel distance is within 30km, and 95% of the duration is within 90 minutes.

2374

ISPECE   IOP Publishing

IOP Conf. Series: Journal of Physics: Conf. Series **1187** (2019) 052047   doi:10.1088/1742-6596/1187/5/052047

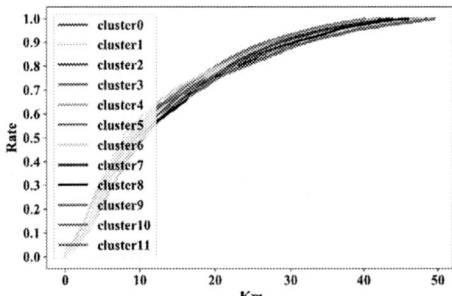

Fig 9 Cumulative distribution of distance   Fig 10 Cumulative distribution of duration

4. Association analysis of car-sharing demand

The study of association rules is another topic of data mining. It refers to discover the connections between things from the data. The association rules were originally proposed for the Market Basket Analysis problem in order to discover the correlation between different commodities in the transaction database.

4.1. Apriori association Algorithm

The Apriori algorithm (Agrawal, 1994) is the most commonly used algorithm for mining association rules. The main idea of the algorithm is to discover frequent item-sets gradually by increasing the number of elements in the item-set.

Before mining association rules, we need to divide data of each attributes into different levels according to certain rules. In this paper, we divide the study area into grids, and a day into eight time periods. Then, we count the demand for each time period of every grid. In order to make the results more accurate and reliable, this study divides the dataset into two categories: workday (excluding holidays) and non-working days (including holidays). After the initial dataset is established, set minimum support and confidence and get strong association rules.

4.2. Apriori Analysis Results

Perform the correlation analysis according to the method in 4.1, and visualize the obtained strong association rules. Fig 11 and 12 show the association rules for weekdays and non-working day, respectively. Blue, green and red represent the three levels of low demand, medium demand and high demand. The flat formed by the horizontal grid number and the vertical grid number represents the geographic plane of the study areas, and the vertical axis represents the time period.

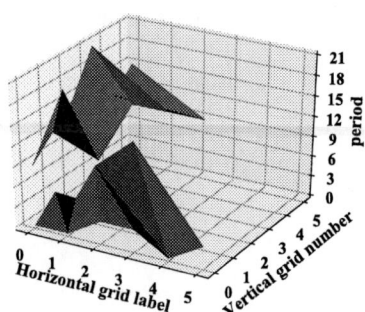   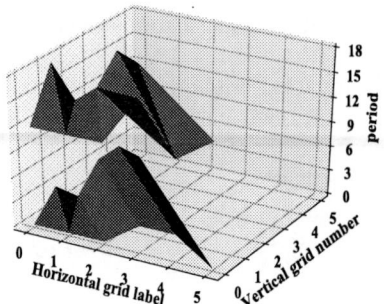

Fig 11 Association rules for weekdays   Fig 12 Association rules for non-working day

2375

The association rules can be described as: area A in the period B ⇒ demand (sparse, medium and dense). Whether it is a working day or a non-working day, the area covered by the association rules is similar, indicating that this part has a distinctive feature. And it's clear that the demand on non-working day is higher than weekdays.

On working days, it can be seen that the area covered by blue is concentrated at 0-6 points, that is, the demand is low, and the part with medium demand is wide distributed. On non-working days, the distribution of low demand is relatively uniform, and the spatiotemporal range covered by the high demand part is widely distributed.

5. Conclusion

With the development of car-sharing, it is increasingly necessary to study its mechanism. In order to conduct more reliable researches on the location depot and car relocation of car-sharing, it is important to conduct a further analysis of the car-sharing demand.

Through the clustering analysis and spatiotemporal correlation analysis, the following effective conclusions can be obtained. These conclusions can be applied to subsequent demand forecasting and site depot researches:

(1) The total amount of the site with high degree of imbalance is higher, while the site with high balance is lower, but the utilization rate of the car is quite high.

(2) Users mainly for short-term travel and short or medium distance.

(3) Different clusters have different characteristics, and should be considered separately when forecasting car-sharing demand.

(4) The strong association rules are mainly reflected in the relationship between spatiotemporal and demand, and the gap between working days and non-working days is large.

Acknowledgments

This work is supported by the project from Gansuyixiangxing New Energy Developments Ltd, Lanzhou, China.

References

[1] Ma X, Wu Y J, Wang Y, et al. 2013 Mining smart card data for transit riders' travel patterns *Transportation Research Part C Emerging Technologies.* **36** 1-12

[2] Arthur, Shaw and Gopalan 2014 Finding frequent trajectories by clustering and sequential pattern mining *Journal of Traffic and Transportation Engineering.* **1** 393-403

[3] Almannaa M, Elhenawy M and Rakha H 2018 A Novel Clustering Algorithm for Traffic Operational Analysis *Transportation Research Board*

[4] Zhao X L and Xu W X 2010 Mining spatio-temporal association rules in bus IC card databases *International Conference on Power Electronics and Intelligent Transportation System.* 125-128

[5] Xia Y and Chengdu 2011 Spatio-temporal Association Rule Mining Algorithm and its Application in Intelligent Transportation System *Computer Science.* **38** 173-176

[6] Yan M Q, Guo Z Y and Ren Z H 2017 Spatio-temporal analysis of bus pickpocketing using association rules based on clustering *Journal of East China Normal University*

[7] Yu C and He Z C 2017 Analysing the spatial-temporal characteristics of bus travel demand using the heat map *Journal of Transport Geography.* **58** 247-255

[8] Ester M, Kriegel H P and Xu X 1996 A density-based algorithm for discovering clusters a density-based algorithm for discovering clusters in large spatial databases with noise *International Conference on Knowledge Discovery and Data Mining.* 226-231

[9] Agrawal R and Srikant R 1994 Fast Algorithms for Mining Association Rules in Large Databases *International Conference on Very Large Data Bases.* 487-499

ISPECE   IOP Publishing

An Empirical Study on Regional Logistics Competitiveness in Guangdong

Chunshang Wu

Heyuan Polytechnic, Heyuan city , Guangdong Province, 517000,China

City University of Macau,china

Corresponding Author: Chunshang Wu; email: wuchunshang1023@163.com; phone:+8613794737728.

wuchunshang1023@163.com

Abstract: Through the construction of factor analysis of regional logistics competitiveness evaluation model, we objectively evaluate the regional logistics competitiveness of Guangdong Province. Then we find out the main influencing factors of regional logistics competitiveness. According to the short-term of each city, we propose some suggestions to improve regional logistics competitiveness.

1. Introduction

With industrial upgrading and transforming of economic development mode, the logistics industry has a deeper impact on economic development and its position in the national economy has become more important . So, the governments have introduced many measures to accelerate the development of the logistics industry. However, due to the lack of scientific evaluation of competitiveness of logistics development in the region, some cities have blindly copied the development experience of others, which seriously affected regional logistics efficiency and logistics competition.

With the rapid development of logistics industry, regional logistics has become a research hotspot. For regional logistics competitiveness, scholars conduct qualitative and quantitative research. Shun Yao (2007) [1] defines the concept and connotation of regional logistics competitiveness, and constructs a "network diamond model" to measure regional logistics competitiveness. Xiuli Gao (2010) [2] ,Cheng Zhang (2012) [3] , Peihua Zhao (2013) [4] Zhang Cheng (2014) [5] Separately analysed regional logistics competitiveness in Guangdong, six central provinces, Henan and Jiangxi provinces ,by using factor analysis, cluster analysis and other quantitative methods. However, the current research mainly focuses on the static study of logistics competitiveness, but the regional logistics competitiveness is dynamic change. We will introduce competitive potential to dynamically evaluate the regional logistics competition.

2. Evaluation model of regional logistics competitiveness in Guangdong

2.1. Construction of evaluation model

Based on the references, it is found that the factor analysis model is more suitable for regional logistics competitiveness evaluation, because it is not affected by different indicators and different dimensions,

Content from this work may be used under the terms of the Creative Commons Attribution 3.0 licence. Any further distribution of this work must maintain attribution to the author(s) and the title of the work, journal citation and DOI.
Published under licence by IOP Publishing Ltd

and it can find significant impact factors. Therefore, we use factor analysis model to analyze the comprehensive logistics capacity of Guangdong province.

The form of factor analysis model:

$$\begin{cases} X_1 = a_{11}F_1 + a_{12}F_2 + \cdots + a_{1m}F_m + \varepsilon_1 \\ X_2 = a_{21}F_1 + a_{22}F_2 + \cdots + a_{2m}F_m + \varepsilon_2 \\ \quad\vdots \\ X_p = a_{p1}F_1 + a_{p2}F_2 + \cdots + a_{pm}F_m + \varepsilon_p \end{cases} \qquad (1)$$

X= $(X_1, X_2, \ldots, X_p)$ are Observable random variables, E(X)=0.

F=$(F_1, F_2, \ldots, F_m)$ (m<=p) are unknown vectores., E(F)=0.

Cov (F, ε) =0.   D(F)=I.

After extracting the common factor, the common factor can be represented by the linear combination of the common factor variables. Formula is (2).

$$F_j = \beta_{j1}X_1 + \cdots + \beta_{jp}X_p \qquad j=1,\cdots,m \qquad (2)$$

2.2. Selection of evaluation index

Regional logistics competitiveness refers to the comparative advantages of logistics industry, including regional logistics foundation, logistics demand, logistics supply, logistics environmentRegional logistics. Eompetitiveness is dynamic development, so regional logistics development potential index is increased . Freight volume growth rate reflects the scale potential of logistics development, while Per capita GDP growth rate reflects the regional logistics demand potential. In order to avoid the sudden change of indicators, the two indicators have chosen the average growth rate in the past five years.The evaluation indicators are shown in Table 1.

Table 1. Evaluation index system of regional logistics competitiveness in Guangdong Province

First level index	Second level index
Regional social development economic environment	Per capita GDP（X1）
	Per Capita Disposable Income of Urban Permanent Households（X2）
	GDP of primary industry（X3）
	GDP of second industry（X4）
	GDP of tertiary industry（X5）
	Total Retail Sales of Consumer Goods（X6）
Development scale of regional logistics	Freight Traffic（X7）
	Freight Ton-kilometers（X8）
Regional logistics foundation support	Length of Highways （X9）
	Freight Vehicles (X10)
	Business Volume of Postal and Telecommunication Services（X11）
	Number of Higher Education and Secondary Schools（X12）
Regional logistics development potential	Freight volume growth rate（X13）
	Per capita GDP growth rate（X14）

3. Evaluation of regional logistics competitiveness in Guangdong

The data are from the Guangdong statistical yearbook 2017 and the statistical yearbook of various cities in 2017.

3.1. Data standardization

Because of the different dimensions of the 14 indicators, standardization of index data is needed. Using SPSS23.0.the normalized data matrix is obtained by averaging the original data .

3.2. Applicability test

Using SPSS23.0, we get KMO and Bartlett spherical test results (Table 2). KMO 0.640>0.5,sig.=0, Through testing, it is suitable for factor analysis.

Table 2. KMO and Bartlett's Test

Kaiser-Meyer-Olkin Measure of Sampling		0.640
Bartlett's Test of Sphericity	Approx. Chi-square	528.906
	df	91.000
	Sig.	0.000

3.3. Factor extraction and common factor determination

The principal component is used to extract the factors. The results are shown in Table 3. As is shown in Table 3，The first three factors can explain 88.395% of the variance. After the fourth factor, the curve gradually flattened and the explanation ability was not strong. Therefore, the first three common factors can better represent the competitiveness of Guangdong's cities.

Table 3. Total Variance of Explained

Factor	Initial Eigenvalues			Ctraction Sum of Squared Loadings			Rotations Sum of Squared Loadings		
	Total	Variance %	Cumulative %	Total	Variance %	Cumulative %	Total	Variance %	Cumulative %
1	8.548	61.057	61.057	8.548	61.057	61.057	7.219	51.564	51.564
2	2.779	19.849	80.907	2.779	19.849	80.907	2.805	20.037	71.601
3	1.048	7.488	88.395	1.048	7.488	88.395	2.351	16.795	88.395

3.4 Calculation of factor matrix and rotated factort matrix

In order to make the common factor have more obvious meaning, when the explanation meaning of the initial factor is not very clear, it is necessary to take the rotation of the row factor to get a more satisfactory principal factor, and enhance the explanatory meaning of the factor. Table 4 is the factor matrix and rotated factort matrix.

As it is showed from table 4, Common factor F1 has a large load in X1, X2, X4, X5, X6, X7, X8, X10, X11, X12. These indicators mainly reflect the regional economic strength and logistics development scale. Common factor F2 has a large load in X3 and X9, which indicates that F2 is related to the development of primary industry and regional traffic conditions. Common factor F3 has a great load on X13 and X14, which mainly reflects the development scale potential and demand potential of regional logistics.The common factor F1 and F2 together constitute the regional logistics competitiveness, expressed in F4.

Table 4. Component Matrix and Rotated Component Matrix

Index	Component			Rotated Component		
	Factor			Factor		
	1	2	3	1	2	3
Zscore(X1)	0.864	-0.364	-0.081	0.596	0.500	0.530
Zscore(X2)	0.850	-0.413	0.031	0.587	0.592	0.447
Zscore(X3)	-0.209	0.750	-0.476	0.028	-0.911	0.045
Zscore(X4)	0.918	-0.201	-0.123	0.703	0.359	0.525
Zscore(X5)	0.978	0.078	-0.042	0.889	0.184	0.374
Zscore(X6)	0.976	0.175	0.038	0.944	0.142	0.270

Zscore(X7)	0.876	0.390	0.137	0.966	-0.015	0.068
Zscore(X8)	0.807	0.499	0.217	0.968	-0.086	-0.066
Zscore(X9)	-0.377	0.713	-0.318	-0.103	-0.849	-0.139
Zscore(X10)	0.947	-0.111	-0.110	0.768	0.299	0.490
Zscore(X11)	0.959	-0.031	-0.060	0.823	0.261	0.423
Zscore(X12)	0.788	0.505	0.209	0.951	-0.100	-0.069
Zscore(X13)	-0.079	0.796	0.448	0.355	-0.463	-0.707
Zscore(X14)	-0.529	-0.161	0.608	-0.409	0.284	-0.654

ZXi is Xi after standardization.

3.5. Calculation factor score

In order to facilitate the comparative analysis of regional logistics development differences in Guangdong Province, the regression method was used to calculate the factor scores of each city through SPSS23.0. The results are shown in Table 5.

Table 5. Factor Score Component Matrix

Index	Factor		
	1	2	3
ZX1	0.019	0.099	0.153
ZX2	0.032	0.165	0.067
ZX3	-0.004	-0.452	0.275
ZX4	0.040	0.032	0.167
ZX5	0.105	-0.011	0.066
ZX6	0.135	-0.003	-0.011
ZX7	0.176	-0.025	-0.124
ZX8	0.202	-0.022	-0.207
ZX9	0.005	-0.372	0.145
ZX10	0.059	0.013	0.145
ZX11	0.083	0.012	0.095
ZX12	0.199	-0.028	-0.202
ZX13	0.200	-0.032	-0.473
ZX14	0.042	0.314	-0.493

The score of factor 5 of Table 2 is substituted by (2), and the scores of common factors F1, F2 and F3 are obtained. The three common factors all reflect the regional logistics competitiveness of Guangdong Province from different angles, but the single factor can not reflect the overall competitiveness of regional logistics, Therefore, we use the variance contribution rate of Table 3 to weigh F1, F2 and F3, and get the score of comprehensive factor F. Calculation formula is shown in formula (3).

$$F=0.6106F1+0.1985F2+0.0849 F3 \quad (3)$$

The results are shown in table 6.

Table 6 Factor scores and rankings

City	F1	F1 rank	F2	F2 rank	F3	F3 rank	F4	F1 rank	F	F1 rank
Guangzhou	4.08	1	-0.52	15	-0.70	17	3.56	1	2.34	1
Shenzhen	0.95	2	1.13	4	2.97	1	2.08	2	1.02	2
Zhuhai	-0.38	17	1.32	2	0.26	7	0.94	5	0.05	6

City										
Shantou	-0.29	11	0.82	7	-0.84	19	0.53	8	-0.08	8
Foshan	0.32	3	0.99	5	0.51	4	1.30	4	0.43	3
Shaoguan	-0.31	14	-0.77	17	0.11	9	-1.07	16	-0.33	14
Heyuan	-0.21	9	0.07	10	-1.47	20	-0.13	10	-0.22	11
Meizhou	-0.32	15	-0.68	16	-0.40	16	-1.00	15	-0.36	16
Huizhou	-0.02	5	0.10	9	-0.37	14	0.08	9	-0.02	7
Shanwei	-0.58	20	0.18	8	-0.37	15	-0.40	13	-0.34	15
Dongguan	0.06	4	1.62	1	0.06	10	1.69	3	0.37	4
Zhongshang	-0.36	16	1.25	3	0.41	6	0.89	6	0.06	5
Jiangmen	-0.26	10	-0.01	12	-0.06	11	-0.27	11	-0.17	10
Yangjiang	-0.85	21	-0.39	14	1.07	3	-1.24	18	-0.51	21
Zhanjiang	-0.05	6	-1.88	21	0.21	8	-1.92	21	-0.38	18
Maoming	-0.20	8	-1.57	20	0.42	5	-1.77	19	-0.40	19
Zaoqing	-0.44	19	-1.37	19	1.16	2	-1.81	20	-0.45	20
Qingyuan	-0.29	13	-0.93	18	-0.22	13	-1.22	17	-0.38	17
Chaozhou	-0.19	7	0.87	6	-1.86	21	0.69	7	-0.08	9
Jieyang	-0.29	12	-0.29	13	-0.13	12	-0.58	14	-0.24	12
Yunfu	-0.40	18	0.07	11	-0.77	18	-0.34	12	-0.29	13

Note: F4 is the competitive strength factor of regional logistics, which is weighted by F1 and F2.

4. Discussion on the results of regional logistics competitiveness in Guangdong

As is showed in table 6, we find that the top three of F4 are respectively won by Guangzhou, Shenzhen and Zhuhai, and the top six are all cities in the Pearl River Delta except Shantou. This shows that the regional logistics competitiveness in Guangdong Province is most affected by economic strength. The only exception to the Pearl River Delta is Zhaoqing, ranked 15th, mainly because Zhaoqing's freight volume and highway mileage rankings are behind. The ranking of F3 is quite different from that of F4. Foshan Foshan ranks first, and Jiangmen, Qingyuan and Maoming have great potential for development. Guangzhou and Zhuhai are lagging behind other Pearl River Delta cities.

Based on the Boston matrix analysis method, we classifies the regional logistics competitiveness of 21 l cities in Guangdong Province by using the two indexes of regional logistics competitiveness and regional logistics competitiveness potential, as shown in Figure 1.

5. Conclusions

The logistics competitiveness of each city in Guangdong Province shows a serious imbalance, and the situation and characteristics of logistics competitiveness vary from place to place. Therefore, targeted development proposals should be put forward according to the characteristics of logistics competitiveness in different regions.

Area A: Regions in this area have good performance both in competitiveness and in competitiveness potential. These regions are in the leading position in the regional logistics competitiveness of Guangdong Province, as long as these regions continue to give full play to their advantages.

Area B: Although the region in this Area have strong competitive strength, but have logistics competitiveness potential. The typical city is Guangzhou. Although the economy in these area are developed, but growth momentum are lack.The most important thing for these areas is making full use of the advantages of economic strength and taping the potential of regional logistics growth.

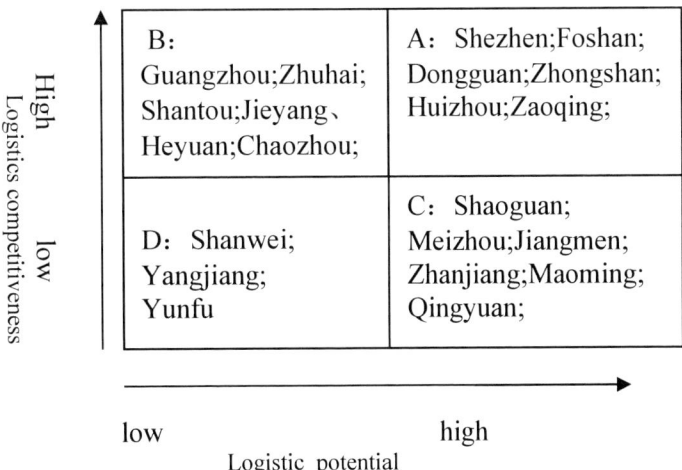

Area C: In this area, the logistics competitiveness is poor, but the logistics competitiveness potential is huge. The typical city is Shanwei . For these regions, on one hand, they should continue to keep their growth momentum,on the other hand they should enhance their competitiveness. And the key to enhance the competitiveness of these areas lies in accelerating economic development by undertaking industrial transfer , and improving the regional traffic environment and personnel training conditions.

Area D: In this area, regional logistics competitiveness and competitive potential are both poor. In order to change this situation, these regions should develop economic innovate the logistics development model through adopting the unconventional development model. For example, they can vigorously develop e-commerce logistics, reverse logistics to enhance the level of regional logistics competitiveness.

Acknowledgements

This research was financially supported by Guangdong Provincial Education Department 2017 key platform and research projects（Grant No.2017GWTSCX044）research projects of China Society of Logistics（Grant No.2018CSLKT3-172）.

References

[1] Shun Yao. (2007) Analysis of Regional Logistics Competitiveness Model and Construction of Reticulate-diamond Model. In:Northeast Normal University.

[2] GAO Xiu-li, WANG Ai-hu (2010) Integrated Evaluation System and Empirical Research of the Regional Logistics Competitiveness, In:Industrial Engineering and M anagemnt. pp. 41-45.

[3] Zhang Chen,Chen Xiao-meng (2012) Empirical Study on Logistics Competitivity of Six Provinces of Central China, In:Logistics Technology. pp. 257-259.

[4]   Zhao Peihua (2013) Study on Regional Logistics Competitivity of Henan Based on Factor and Clustering Analysis. In:Logistics Technology. pp. 151-153.

[5]   Zhang Cheng, Zhang Yuan, Zhang Zhijian (2014) Evaluation and Cluster Analysis of Regional Logistics Competitivity of Jiangxi In:Logistics Technology. pp. 147-150.

Evaluation effect of Internet word of mouth and application of big data

Te Ma

School of tourism, Dalian University, Dalian 116024, China

*Corresponding author's e-mail: 717126748@qq.com

Abstract. The role of online word-of-mouth has been gradually recognized by enterprises, but enterprises are more concerned about the marketing effect of online word-of-mouth and how to improve the effect of online word-of-mouth. Based on the comparison of advertising, online word-of-mouth and traditional word-of-mouth, this paper introduces the evaluation methods of online word-of-mouth effect from the perspectives of enterprises, consumers and third-party websites, and expounds the significance of online word-of-mouth effect evaluation to the development of word-of-mouth theory and enterprise practice.

1. Introduction

In summary, there are two kinds of non-personnel promotion, one is mass media promotion, mainly for advertising; the other is interpersonal promotion, that is word-of-mouth. With the popularity of the Internet, especially the emergence of interactive marketing based on Web 2.0 technology, many enterprises have realized the important promotional role of online word-of-mouth. But, what is the effect of Internet word of mouth? _How to evaluate and manage the effect of online word-of-mouth? These are the issues that enterprises are more concerned about in the process of online word-of-mouth marketing, and also the problems that need to be solved urgently in the field of online word-of-mouth marketing.

2. Research status of Internet word of mouth effect

Through the statistics of more than 60 publications (including top marketing publications) in EBSCO and Emerald database from 2005 to 2017, it is found that more than 100 articles have been published during this period, two-thirds of which have been published since 2000. Generally speaking, the research results of online word-of-mouth mainly focus on the motivation of communicators and receivers; and the influencing factors of the effect of online word-of-mouth communication. The research on online word-of-mouth effect evaluation is very limited. For the need of theoretical development and enterprise network marketing practice, the related research on online word-of-mouth effect evaluation needs to be carried out in depth.

2.1. Internet word of mouth communication process

Buttle and Goldsmith call this kind of information online word-of-mouth. The spread of online word-of-mouth differs from traditional word-of-mouth mainly in the following aspects: The basis of non-acquaintance communication is the weak relationship between website and netizens. Second, because of the weakness of network virtuality, anonymity and trust, online word-of-mouth information has a greater risk to consumers. Third, consumers can operate the Internet by clicking, browsing,

Content from this work may be used under the terms of the Creative Commons Attribution 3.0 licence. Any further distribution of this work must maintain attribution to the author(s) and the title of the work, journal citation and DOI.

Published under licence by IOP Publishing Ltd

replying and registering on the Internet. With the ordering and so on, the enterprise may track the record through the "click stream" software, and carried on the analysis, thus appeared "the customer behavior management". Compared with the Internet, consumers have a great degree of freedom of search and expression, and the cost of reproducing online word-of-mouth is very low. This is an obvious feature of online word-of-mouth which is different from advertising and traditional word-of-mouth. Because of anonymity, corporate marketers may be involved in the dissemination or employment of network push hands.

2.2. Characteristics of Internet word of mouth
Previous studies have called word-of-mouth advertising , based on the fact that word-of-mouth is generally the same as the audience of advertising and produces similar effects. However, Alexander argues that advertising is distinguished from word-of-mouth by formal, paid and non-interpersonal communication. Word-of-mouth and advertising, as two main ways of information diffusion, are different in disseminators, channels and information dissemination. For Internet word of mouth, it preserves some basic characteristics of traditional word-of-mouth, but also has some characteristics of advertising.

2.3. The role of Internet word of mouth.
Word of mouth can affect people's cognition, emotion, expectation, attitude, behavior intention and behavior. Comparing with traditional word-of-mouth, online word-of-mouth has a wider scope, a faster speed and a greater influence on the community. For enterprises, its role lies in: First, Promote products, reduce marketing costs and promote sales. Second, cultivate corporate brand and enhance corporate reputation. Improving products and services. Support customer information and relationship management. Through its own network or third-party network, enterprises can provide consumers with convenient and inexpensive channels of word-of-mouth communication. They can also track the process of word-of-mouth communication, collect and analyze the content of word-of-mouth communication, thus more effectively grasp consumer psychology, guide and control word-of-mouth communication, and achieve B C C marketing.

2.4. The effect of Internet word of mouth.
Effect is generally the result of the operation and function of things. In sociology, effect refers to the effective result of human behavior. In communication science, effect has two meanings: (1) narrow sense, communication effect refers to the change of psychology, attitude and behavior caused by communication behavior on the object of communication; (2) broad sense, communication effect refers to the change of audience and social property. The above two meanings constitute two basic aspects of the study of communication effects: one is the micro-process that produces effects on individuals; the other is the macro-process that produces effects on society.

The effect of online word-of-mouth refers to the psychological and behavioral reactions of consumers caused by the spread and acceptance of online word-of-mouth and the degree of its impact on enterprises. Buttle (1998) classifies word-of-mouth effects into pre-purchase effects and post-purchase effects. Pre-purchase effects are mainly the cognitive responses of consumers caused by word-of-mouth, such as awareness and understanding, purchase and re-dissemination intention. The post-purchase effect mainly refers to the re-purchase and re-dissemination behavior of consumers after purchase experience influenced by word-of-mouth. Research by Bickart and Schindler (2001) suggests that since online reviews can exist in a network system for a long time, they not only have a short-term promotional effect, but also have a lasting impact.

3. Evaluation of Internet word of mouth effect
The effect of online word-of-mouth is a practical problem that enterprises are more concerned about, and it is also the weakness in the current study of online word-of-mouth. The evaluation of online word-of-mouth can be carried out from three perspectives: enterprises, consumers and third parties, in

which consumer evaluation is the basis for the study of the effect of word-of-mouth, third party evaluation is the practical need, and enterprise evaluation is the ultimate destination.

3.1. Enterprise perspective evaluation

Similar to the evaluation of advertising effect, the effect of online word-of-mouth can be measured by comparing the input cost and sales revenue of enterprises. The problem is that online word-of-mouth input is generally difficult to be as specific as advertising input and can be clearly reflected in the financial statements. However, with the emergence of third-party online word-of-mouth service organizations, online word-of-mouth investment can not be as specific as advertising input. Income can be calculated gradually, such as network search ranking fees and third-party community network word-of-mouth promotion service fees, and can be included in the accounting of marketing costs, or the addition of network word-of-mouth promotional special subjects.

For sales revenue, enterprises can take the increase of sales in a certain period of time after the input of online word-of-mouth promotion as the performance of online word-of-mouth in this period, the time can be determined according to the accounting cycle, but also according to the cycle of sales promotion effect. One of the most prominent problems is how to ensure that sales growth is due to the input of online word-of-mouth, which proves that online word-of-mouth is the cause of sales growth in this stage may be a difficult problem. Therefore, the evaluation of the effect of online word-of-mouth from the perspective of enterprises needs to be repeated and verified, and the influence share of other causes needs to be excluded through experimental analysis.

3.2. Consumer behavior perspective assessment

There are many factors leading to the realization of sales in enterprise practice. It is difficult to determine which sales are produced by word of mouth. However, consumers are generally certain about what factors trigger their own purchase behavior, so the evaluation of word-of-mouth effect is more reasonable for consumers. Network word-of-mouth has both some characteristics of traditional word-of-mouth and network advertisement. The evaluation of network word-of-mouth effect can draw lessons from the evaluation method of advertising effect. Lavidge put forward the basic way to evaluate the effect of advertisement in the ladder-step model: it can evaluate the effect of advertisement by measuring consumers' cognition, emotion and behavior. According to TRA theory, if from the perspective of consumers, online word-of-mouth effect measurement dimension can also increase consumer behavior intention.

3.3. Third party perspective assessment

At present, some websites and consulting firms are engaged in the evaluation of the third-party online word-of-mouth effect. The main variable is "attention", which represents the degree of cognitive preference of the subject in a certain period of time. It reflects the influence of specific brands. The main research institutes in China are:

(1) Baidu Index is a data analysis service based on Baidu web search and Baidu news search. It reflects the impact of online word-of-mouth with different keywords "user attention" and "media attention" over the past period of time. User attention is based on the amount of search done by netizens in Baidu. The weighted sum of search frequencies of each key word in Baidu web search is calculated and displayed in the form of a graph. Media attention is based on the number of news related to the keyword in Baidu News Search in the past. The final data is obtained by scientific weighting calculation and displayed in the form of a graph.

Word-of-mouth index (http://www.koubei.com) is the Word-of-mouth network according to the majority of members on the business of the key indicators of evaluation, with 5-star system, the higher the score, the higher the Word-of-mouth index. The accuracy of grading depends largely on the number of comments.

Zhongguancun online attention index (http://zdc.zol.com.cn), according to daily user visits, statistics of product and brand attention, and through the product attention, brand exposure and channel coverage of these three aspects of weight analysis, calculate the ZDC concern index. This index is a weighted composite index, which mainly reflects the degree of attention and brand influence of brands and products in the market.

4. Conclusions

Network word-of-mouth is not only a promotion strategy, but also a good way for enterprises to cultivate brand, establish credibility and disseminate corporate culture. As online word-of-mouth is a C_C-based marketing, it may be an effective way to evaluate the effect of online word-of-mouth by evaluating consumers' attitudes and behaviors, and integrating controllable data of sales and network management, based on previous theories of communication effects and characteristics of online word-of-mouth. In the specific evaluation process, enterprises can choose the evaluation method according to the brand maturity and product characteristics. The effect of different products through the network word-of-mouth communication will be significantly different. Factors that affect the network word-of-mouth communication can be analyzed according to the specific situation, such as the network channel may be more suitable for the promotion of digital products.

Complicated products may be difficult to understand. At this time, we should increase the network experience and expand the network channel of group communication because of experience. Strategy generally promotes cognition, and the developed communication network promotes people to get information. If it is emotional problems, such as low consumer preferences for products, this can enhance product personalization. If it is the process of behavior intention behavior conversion, it may be that the way consumers get products is too narrow, the price is too high, then the problem is mainly reflected in the pricing and channel problems, so it is better to adjust prices and expand product channels.

References

[1] Hogan John E., Lemon,Katherinen. Libai Barak, Quantifying the Ripple: Word-of-Mouth and Advertising Effectiveness, September 2004 ,Journal of Advertising Research.

[2] Robert C,Brooks Jr，"Word-of-Mouth" Advertising In Selling New Products[J]，Journal of Marketing，October 1957.

[3] Alexander, Ralph S, Marketing Definitions, Chicago: American Marketing Association.

[4] Barbara Bickart，Rutgers Robert M. Schindler，Rutgers ，Expanding the Scope of Word of Mouth: Customer-to-Customer information on the Internet， Advances in Consumer Research Volume 29，2002.

[5] Te Ma, Feng Ding. Research on the dynamic effect of the intelligent urban experience to the tourists' two-way internet word-of-mouth,November 2017,International Journal of Communication Systems.

Purchase decision under network environment and application of big data

Te Ma

School of tourism, Dalian University, Dalian 116024, China

*Corresponding author's e-mail: 717126748@qq.com

Abstract: According the theory of traditional economics, consumers will be more rational when they make decision in the context of Internet, because they could get lots of information through the web that advance the symmetry of information. But, it is not the fact. Why would the contradiction happen? In the present paper, by the theory and literature review, we found that in the context of Internet versus tradition.Consumers are more likely to be impatient in decision-making. Consumers have more freedom and irrationality in decision-making. Therefore, consumers' network behavior is more difficult to predict.

1. Introduction

If consumers want to understand a product, they just need to type in a few keywords lightly on the web search engine, and a lot of relevant product information can be presented in front of them, which we might not have imagined 10 years ago. According to traditional economics, the more fully consumers acquire information and the more symmetrical the relative information, the more rational their behavior and behavioral tendencies should be. But the fact is far from simple. According to the data of the Internet Consumer Behavior Survey in August 2008 in the New Heng Management Research, 32.3% of Internet information browsers have no pre-purchase plan. Or under the condition of demand, affected by network information and purchasing intention. It is possible for consumers to purchase irrational products under the network environment. How to explain this contradiction theoretically will be answered from the perspective of the characteristics of rational and irrational decision-making of consumers.

2. "Rational economic man" and "limited rational man"

Traditional economic theory assumes that consumers are rational economic people. Consumers can know exactly what they need, obtain, process and use information adequately, and make rational choices according to their goals, and make rational decisions under the condition of symmetrical information. The concept of "rational economic man" is based on ideal environment and conditions, but it lacks attention to the psychological characteristics, social environment and situation of real people. In the 1980s, Richard Taylor and others, based on Simon's theory of "bounded rationality", drew inspiration from evolutionary psychology that most people were neither completely rational nor completely selfish, but related to the social environment in which they lived. In 2002, Nobel Prize winner Karniman put forward the theory of human decision-making judgment behavior and cognitive bias under uncertain conditions, which greatly advanced the theory of bounded rationality. Overall, consumers are actually more and more difficult to grasp the future information, and more and more difficult to control the situational factors. Therefore, consumers only have moderate rationality under certain conditions.

Content from this work may be used under the terms of the Creative Commons Attribution 3.0 licence. Any further distribution of this work must maintain attribution to the author(s) and the title of the work, journal citation and DOI.

Published under licence by IOP Publishing Ltd

Consumer decision making process is also a process of information searching and processing. The basic goal of consumer purchase decision is still utility maximization and cost minimization. The early view holds that utility maximization is the primary goal of purchasing decision, while the recent limited rationality view holds that people's limited information processing ability restricts the realization of the goal. Because of bounded rationality, consumers often do not have the ability to evaluate all available alternatives in a complex decision-making environment. As an alternative, consumers tend to use a two-stage approach to reduce decision-making efforts. In the first stage, consumers face many kinds of products and form a large set of related products according to their beliefs and preferences, and then choose the most feasible alternative subset; in the second stage, consumers make a more in-depth evaluation of alternative subset, compare the important attributes of different products, and make a comparison. Purchase decisions.

3. The main theory of consumer decision-making process

In general, the information search process of consumers begins with internal search. Most of the information in consumer's memory comes from previous experience in products. When the information extracted from the memory is sufficient, consumers no longer search for information. When the information extracted from the internal information resources is insufficient or there is a conflict between the information in the memory, consumers search for relevant information from the outside (Li Dongjin, 2000). Internal search is carried out passively without time and cost, while external search is carried out actively in order to obtain greater search benefits under time and cost conditions.

External search is mainly through two ways: one is the source of non-marketing control information, not the source of business information. Jiang Lin (2002) divides this kind of information into Personal experience and Word of mouth information. Personal experience information, refers to the information which the consumer obtains personally, such as the information which the consumer inspects, operates or observes the product and actually uses the product obtains. Word of mouth information refers to the product related information of consumers through personal relationships. (3) Third-party organization information refers to information provided by third-party organizations other than transactions, such as inspection reports of government agencies and other non-profit organizations, professional comments or professional knowledge lectures. Second, the source of marketing control information, also known as business information sources, such information often has corporate profit intentions, such as various media advertising, exhibitions, online reviews.

The theory of consumer information search, processing and purchase decision mainly includes the following categories:

3.1. EKB model

The EKB model was proposed by Engel, Kollat and Blackwell in 1968 and revised in 1984 and 1993. The model considers that the consumption process consists of two parts: purchase decision and purchase behavior. Purchasing decision refers to the psychological activities and behavior tendency of consumers before purchasing products, which belongs to the formation process of consumers'attitude; purchasing behavior is the implementation process of purchasing decision. In real consumption activities, these two parts penetrate and influence each other and constitute a complete consumer behavior. The EKB model divides the consumer purchase decision-making process into three stages: information reception (search) information processing decision-making processing.

The information receiving stage mainly refers to the consumer gains information through the enterprise's advertising, promotion, public relations and other marketing strategies. Information processing stage refers to the processing process of consumers after obtaining information, which includes five stages: contact, attention, understanding, acceptance and retention. Consumers will selectively screen information and store it in the memory system. The decision processing stage is the core part of EKB model, which describes the whole process of consumer's purchase decision-making. It includes five stages: problem confirmation, information search, scheme evaluation, purchase choice and decision-making.

3.2. The basic mode of S-O-R theory

The stimulus response (S-R) theory was proposed by Watson in 1913. Reynolds in 1974 proposed the S-O-R theory based on the concept of psychology. S: Stimulus denotes stimulus; O: Organism denotes organism or reactant; R: Response denotes the response caused by stimulus. Based on SOR theory and consumer decision-making process, Nicosia divides the consumer behavior model into the following processes: receiving enterprise information stimulus information collection and scheme evaluation purchase decision and behavior information feedback. The Nicosia mode emphasizes the continuity of consumer purchase decisions, as shown in Figure 1.

Howard-S Heth consumer purchase behavior model was proposed by Howard and S Heth in 1969 on the basis of S-O-R theory. The model considers that behavior originates from psychological motivation and its composition includes input factors, intrinsic variables and output results. Input factors include three different stimulus factors: first, product entity stimulus factors, such as product quality, price and service; second, commercial media information, such as brand, trademark and packaging product symbol stimulus; third, consumer social environment. Intrinsic variables refer to the consumer's purchase intention, including perceptual variables and learning variables, after receiving external information or stimulation, to form product impression in the mind and through self-learning. The output result is that consumers produce certain reactions after stimulation. These reactions include arousing attention, increasing understanding, forming attitudes, generating purchasing intentions, and even triggering purchasing behavior.

3.3. ELM:Elaboration Likelihood Model.

The Elaboration Likelihood Model (ELM) proposed by Petty and Cacioppo (1986) assumes that consumers will process information along both core and peripheral routes. Route choice depends on the degree of participation and attention of consumers. If consumers have a very rational understanding of information and the ability to process information, then consumers may process information along the core line; if consumers are not rational to deal with information, do not have the ability to process information, and there is a peripheral route.

When consumers process information along the core line, their attitudes are greatly influenced by information. The core line contains three elements of attitudes: cognition, emotion and behavioral intention, but more emphasis is placed on cognition. The attitudes formed by consumers through a very rational evaluation process are likely to lead consumers to take corresponding actions. The information processing process of consumer's choice of peripheral route is more focused on the emotional aspect of attitude. Consumer's processing of information relies on those peripheral factors which have no direct connection with information content, and establishes irrational connection between these peripheral factors and brand. At this point, consumers began to exchange information, express their ways and communicate with each other. Factors such as degree of relationship will become an important factor for consumers to judge the credibility and value of information. In the state of rational core route, consumers' attitude is more likely to decide behavior. Behavior-influenced attitudes are more likely to occur in peripheral routes, such as irrational purchases and consumer attitudes towards products in the context of interest-oriented and immediate expected value.

4. Influencing factors of consumer decision making process

There are two factors influencing consumer decision-making, for example, EKB model concludes that the internal factors influencing consumer purchase decision-making include trust, risk, utility, knowledge, attitude, personality, values, motivation, lifestyle and purchase intervention, and the external factors include family, culture, social class, interpersonal influence.

On the basis of the two-factor theory, the three-factor theory adds the influencing factors of "enterprise marketing". Kotler (2001) attributes the influencing factors of consumer decision-making to four aspects: social, cultural, personal characteristics, personal psychology. In addition, product categories and characteristics have a direct impact on consumer information search and processing. According to the degree of understanding and the way consumers understand the characteristics of the products, Nelson (1974) divides the products into search products and experience products : (1) search products: the consumers easily to obtain products information, such as soap and toothpaste and other daily necessities; (2) experience products: consumers need to be able to make judgments after they are used, such as tourism and dining entertainment. (3) Credence product: The general consumer can not obtain the relevant information about the characteristics of the product even after use, so it is impossible to verify the quality of the product, and can only give trust, such as medical services and counseling. This classification method is applied more widely in the network environment.

5. Conclusions

Compared with the traditional environment, the network is not only a tool, but also an important context for consumer decision-making. Especially when consumers combine the search and processing of network information with online shopping, the impact of network situation on consumer behavior is more significant, which may be an important aspect ignored by previous studies. There are three advantages for consumers to obtain purchase decision-making information in the network environment: greater freedom. In the network environment, netizens have more network information available, more kinds of products for choice, anonymous virtual communities and blogs appear, so that people can fully share information, and the development of online stores, consumers can stay at home, free. Choice. More convenient, because of the rapid update of network technology and equipment, the Internet has been widely used and has become one of the most important channels of information access. With the development of 3G mobile phone and network interconnection, the advantages of network information transmission will be more obvious. Search engine and network instant messaging tools make people's daily life almost soak in the network. Lower costs, the use of network technology to process information, retrieval and classification work has been simplified, information processing costs, time and energy input reduced, and the speed of unprecedented increase.

Through the summary of the advantages of network information acquisition, combined with the above-mentioned theory, the overall view: In the network situation where consumers are, there are a large number of commercial and non-commercial information such as online advertising, online word-of-mouth, and so on, and frequent stimulation, consumers are in a continuous and high-intensity stimulation state, the possibility of external information to guide consumption Increase; Consumers search and process information in the network context, relatively more active and free, less constraints, decision-making behavior may increase arbitrariness; For search costs, can almost be ignored, regardless of the cost of the state, consumers may be less from self-memory to extract information, but more dependent on the network context. Provide information resources, so that the probability of using personal experience information based on rationality will be reduced; Consumers face more sufficient information, a variety of products, a variety of prices can be selected, consumer personality choice space ahead of the expansion, while network consumers in most cases for a separate "go online", often independent decision-making, rarely with Others consult, lack of "collective decision-making" or "reference decision-making", which is undoubtedly a little less rational; The entertainment function of the network is extremely powerful, consumers in the process of surfing the Internet are mostly in a relaxed state of enjoyment, taking irrational peripheral route to search and process information is likely to increase. Therefore, compared with the traditional situation, consumers are more likely to make irrational purchasing decisions in the network situation.

Reference

[1] Te Ma, Feng Ding. Research on the dynamic effect of the intelligent urban experience to the tourists' two-way internet word-of-mouth,November 2017,International Journal of Communication Systems.

[2] Bettman,J.R. Constructive consumer choice processes. Journal of consumer research, Dec 1998, Vol.25, Issue 3,pp.187-217.

[3] Nelson,Advertising as Information,Journal of Political Economy 1974,82(July/August):729-54.

[4] Feng Ding, Te Ma.Research on the dynamic relationship of homogeneity tourist destination based on tourist data,IEEE Access,2018.05.

Research on the evaluation index system of intelligent railway passenger station

Shaofu Lin[1,2]*, Qianwen Wei[2], Sibin Xia[2]

[1]Beijing Advanced Innovation Center for Future Internet Technology, Beijing, 100124, China

[2]College of Software Engineering, Beijing University of Technology, Beijing, 100124, China

*Corresponding author, e-mail: linshaofu@bjut.edu.cn

Abstract. The evaluation index system of intelligent railway passenger station is a set of scientific and systematic evaluation index system, a method standard for scientific evaluation and quantitative calculation of passenger station construction, and a concrete embodiment of evaluation results of intelligent railway passenger station construction, playing a leading, guiding and testing role. This paper analyses the basic connotation and definition of intelligent railway passenger station and reference to the relevant field index system design experience, extracting the basic framework of intelligent railway passenger station evaluation index system, taking "Beijing-Zhangjiakou High Speed Railway" as the application case and combining with the actual needs to propose relevant suggestions and countermeasures, thus providing a basis for the planning and management of intelligent passenger station.

1. Introduction

The passenger station is the main window for the interaction between the railway and the passengers. It is responsible for the whole process of passengers entering the station-waiting-outbound. With the development of informatization and the innovative development mode of the transportation industry, the passenger transportation service is intelligent, the informatization of passenger transportation services, diversified functions and the business district of the station have become an important direction for the development of stations. Smart stations use modern technology such as big data, cloud computing, and Internet of Things to effectively reduce station operating costs, improve service levels and quality, and reduce environmental pollution in a business, problem-demanding, and technology-driven manner. However, in the construction of intelligent railways at home and abroad, a complete and standardized intelligent passenger station evaluation index system has not been developed to assist the state in the intelligent management and dispatch of intelligent railways. The intelligent passenger station evaluation index system is a method system for quantitative calculation and scientific evaluation of intelligent railway construction achievements. It is a concrete manifestation of testing the results of intelligent railway construction, and can also promote the construction process of intelligent transportation efficiently. It is essential and necessary to carry out the research work on the evaluation index system of the intelligent passenger station, which has important scientific research value and social significance.

Content from this work may be used under the terms of the Creative Commons Attribution 3.0 licence. Any further distribution of this work must maintain attribution to the author(s) and the title of the work, journal citation and DOI.
Published under licence by IOP Publishing Ltd

2. The concept and development status of intelligent railway passenger station

2.1. Basic Concepts and Definitions of Intelligent Railway Passenger Station

Intelligent railway passenger as the subsystem of the intelligent traffic system in the intelligent railway, new generations of information technology such as big data, cloud computing, Internet of Things, artificial intelligence, next-generation communication, Beidou satellite navigation, BIM, and robots are widely integrated in the planning, construction, operation, service and management of passenger stations, with the goal of convenient travel by passengers, warm service of stations, efficient production organization, safe real-time protection, energy saving and environmental protection of equipment. It comprehensively utilizes all the mobile, fixed, space, time, manpower and other resources of the railway to realize the planning, construction, organization and transportation of the railway passenger station, and the high degree of informatization, automation and intelligence of the whole life cycle. It is an important support for realizing the intelligent transportation infrastructure and promoting the sustainable development of the transportation industry. It is also an important part of smart city construction.

2.2. Research and development of relevant industry evaluation indicators system at home and abroad

The evaluation index system is the basis for the ranking and optimization of the evaluation objects, and is the key support for improving the practice process and optimizing management measures. For example, internationally, the IBM Urban Intelligence Assessment White Paper states that assessing a city's core systems and activities is the most fundamental content in the smart city strategic planning phase. In addition, China has continuously explored the research topics of the "smart city" evaluation index system. For example, the Nanjing Information Centre published the "Study on the Evaluation Index System of "Smart City" in 2010, the "Smart City Construction Evaluation Index System 2.0" issued by Shanghai Pudong New Area, etc. It provides the main basis for the development of smart city evaluation in China, and is an important means to guide the healthy development of smart cities across China. In addition, "Evaluation standard of Zhejiang local standard town " comprehensively use the basic theories of urban and rural planning, urban economics, environmental science, ecology and systems science to reflect the comprehensive level of development of characteristic towns. An overview of the research on the evaluation index system of related fields, as shown in Table 1, the analysis of the commonality and personality points of the first-level indicators in the relevant field evaluation index system, as shown in Table 2.

Table 1. Overview of research on evaluation index system in related fields

Index name	index system	index principle
EU medium-sized cities smart ranking evaluation indicators	6 first-level indicators, 31 second-level indicators, and 74 third-level indicators.	--
IBM Smart City Assessment Criteria and Elements	7 first-level indicators and 28 secondary indicators.	tailored, holistic, comprehensive, comparable
Nanjing City Information Centre Smart City Evaluation Index System	4 first-level indicators and 23 second-level indicators.	collectability, additivity, cognition
Pudong New Area Smart City Indicator System 2.0	6 first-level indicators, 18 second-level indicators, and 37 third-level indicators.	collectability, representativeness, comparability
Evaluation standard of Zhejiang local standard town	evaluation of the combination of commonality and personality evaluation indicators	Scientific, systematic, operability, etc.

Table 2. Analysis of commonality and personality point of the first-level index

Index name	Common first level index	Personality first level index
EU medium-sized cities smart ranking evaluation indicators	smart city infrastructure construction,	Smart People and Mobility
IBM Smart City Assessment Criteria and Elements	smart city public management and service,	Smart building
Nanjing City Information Centre Smart City Evaluation Index System	smart environment construction, intelligent humanities construction,	Internet and smart industry
Pudong New Area Smart City Indicator System 2.0	smart government construction	Wisdom city perception of democracy
		Smart city soft environment construction
Evaluation standard of Zhejiang local standard town	--	Function "gather together"、Form is "small and beautiful"、System is "new and live"

Although China has formed a complete evaluation index and application guideline for the standard system in the smart city, it lacks in the construction of the urban intelligent traffic evaluation system, especially in the construction of the evaluation index system of the intelligent railway passenger station. The perfect intelligent railway passenger station evaluation index system can be used to guide the public travel demand information, and provide strong support for the pre-control, coordination and resilience of transportation operations.

3. Basic ideas for establishing evaluation index system of intelligent railway passenger station

3.1. Design basis

3.1.1. Policy basis. The evaluation of the intelligent passenger station evaluation index system is based on the "13th Five-Year Plan for the Development of Modern Integrated Transportation System" (Guo Fa [2017] No. 11), "Smarter Transportation Makes More Easy Action Plan (2017-2020)" and the spirit of the report of the 19th National Congress is the policy basis.

3.1.2. Theoretical. Basis theories of information management, passenger station management, operation management, public service management, and information resource management at home and abroad are the main theoretical basis for the development of the evaluation index system for intelligent passenger stations.

3.1.3. realistic basis. When designing the evaluation index system of intelligent passenger station, it is necessary to fully consider the basic requirements of the construction of intelligent passenger vehicles in the political, economic, cultural and social macro-development environments, as well as the status quo of China's smart railway construction and the demand characteristics of different types of urban railway construction，and then put forward relevant index items and evaluation criteria.

3.2. Design Principles

This paper constructs the evaluation index system of intelligent railway passenger station. Based on the design principles of evaluation indicators in relevant fields, the comprehensive consideration of the development goals of smart railway makes the evaluation of intelligent passenger station objective, comprehensive and scientific.

3.2.1. Guiding principle. Taking the construction target of the intelligent passenger station as the standard, the evaluation index system will be guided by the national policy spirit, the railway development plan and the actual needs. It is necessary to reflect the construction requirements of the railway station for the intelligent station construction, and to reflect the development focus of the intelligent passenger station in the index element project and weight arrangement.

3.2.2. Systematic principles. The evaluation index system of the intelligent passenger station should be a well-defined whole. The index items of different dimensions should be at different levels, forming a certain order, and there is an inherent logical relationship between the index layer and the index layer and the indicators at the same level.

3.2.3. Scientific principles. Science is the soul of the indicator system. The preparation process of the intelligent passenger station indicator system should be as objective as possible, and the influence of subjective consciousness should be excluded as much as possible, with quantitative indicators as the main factor and qualitative indicators as the supplement. In addition, the indicator system should be able to scientifically and accurately reflect the existing strength and future potential of smart stations.

3.2.4. Principle of operability. The evaluation index of the intelligent passenger station should be an indicator that is easy to quantify in the actual operation, so as to facilitate quantitative evaluation and comparison of the intelligent passenger station. In addition, data or information reflecting the assessment indicators should be easily collected to fully reflect their actual application value and operational possibilities.

3.2.5. The principle of scalability. The intelligent passenger station is a dynamic development process. The evaluation index system of the intelligent station should also be based on the evolution of the "smart station" development path, and at the same time weigh the current development of the railway and timely supplement and adjust it dynamically.

3.3. The basic framework of the evaluation index system for intelligent railway passenger station
According to the definition, development status and construction goals of the intelligent railway passenger station. The indicator system framework is divided into three levels, including 4 first-level indicators, 19 second-level indicators, as shown in Figure 1.

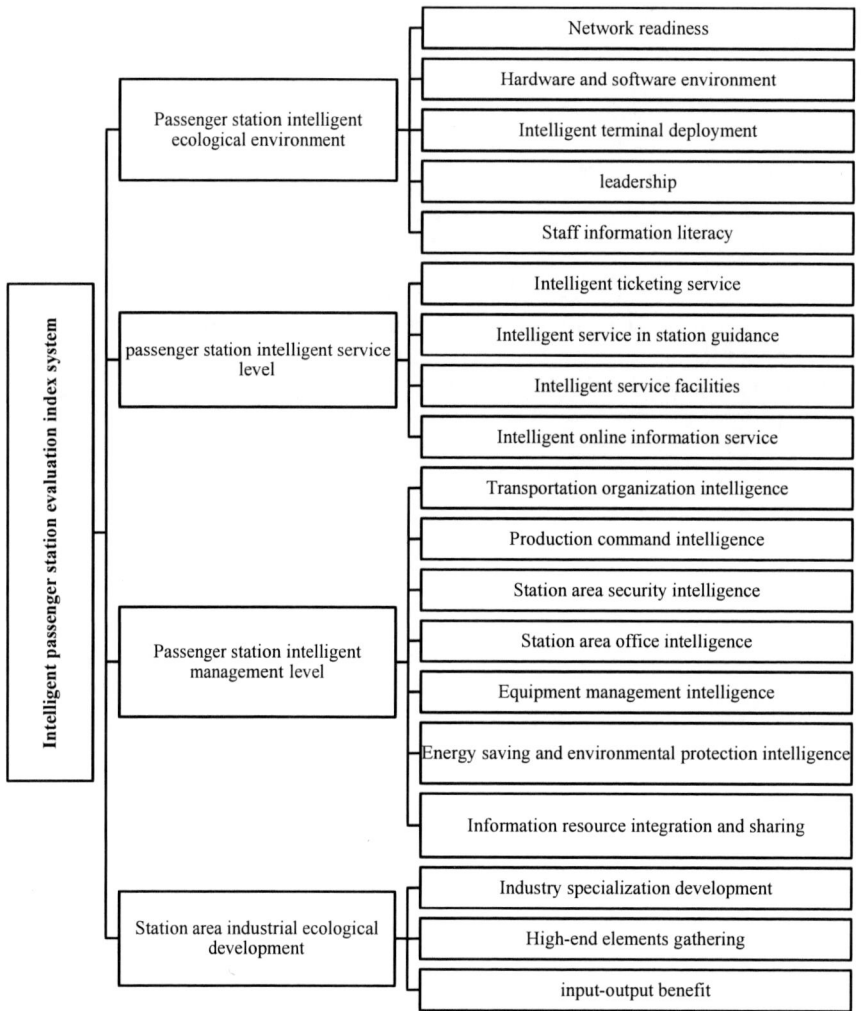

Figure 1. Intelligent Railway Passenger Station Evaluation Index System Framework.

4. Application demonstration of indicator system in Beijing-Zhangjiakou High-speed Railway

The Beijing-Zhangjiakou High-speed Railway is the world's first high-speed railway built with intelligent technology. According to the above-mentioned evaluation, the overall framework of the system is designed, and the application scenarios are designed and the implementation plan is formulated.

4.1. Application scenario design

4.1.1. Benchmark evaluation. The benchmarking assessment comprehensively evaluates intelligent passenger stations nationwide from the aspects of intelligent infrastructure, intelligent management, and intelligent services, facilitating horizontal comparison of stations between cities. According to the evaluation results of the benchmarking, the station can find out its own gaps and deficiencies, determine the construction direction and construction focus of the next stage, and thus improve the competitiveness of the station. In addition, through the evaluation of the target, we can understand the current status of the construction of smart passenger stations in various places, and provide the basis for decision-making by government departments, so as to formulate corresponding guiding policies or opinions.

4.1.2. Diagnostic assessment. The diagnostic evaluation of the indicator system should cover the whole process of planning, construction and operation of the intelligent passenger station in combination with the actual situation of urban development and the urban strategic objectives. Among them, the diagnostic assessment of the construction process includes the fit of key projects and urban needs, and the standardization and controllability of the construction process. The diagnostic assessment of the operational process includes whether the operational model is coordinated with the economic and social development of the city, and how effective the operation is. According to the results of the diagnosis and evaluation, not only can the problems existing in the passenger station be solved in time, the solutions can be formulated in a targeted manner, and the advantageous resources can be used to make up and improve.

4.1.3. Performance appraisal. With performance as the driving force, it can effectively improve the operational effects of various service management departments at the station. First, the evaluation indicators should extract the commonalities of different service management areas, grasp the performance level of different areas as a whole, and carry out transformation according to the evaluation results. Secondly, in different places, different fields, even different departments in the same field or the same place, there are significant differences in business scope and service items. The evaluation indicators should be deepened within the field, combined with the business characteristics of each department to refine the details of service management. Finally, for cross-disciplinary and cross-sectoral businesses and services to be assessed separately, the focus should be on resource sharing and coordination.

4.2. Application management workflow
Establishing an application management system framework in the system is conducive to accelerating the development of regulations, policies, technical standards, management systems and other regulations for the construction of intelligent passenger stations. The main processes of application management work are as follows:

4.2.1. Evaluation preparation. The evaluation management team promotes the formation of the evaluation expert group and the implementation group, and publishes the evaluation plan for the intelligent passenger station, which clarifies the evaluation scope, region, and time. According to the objectives and management needs of the development and construction of the intelligent passenger station, the implementation group will set the index weights, form the corresponding data collection list, questionnaire, clear evaluation indicators, evaluation criteria and filling requirements, etc., and submit them to the expert group for review and report to the development management group.

4.2.2. data collection. The review implementation team will publish a data collection list and a questionnaire, and the organization will be reported by the organization. The receiving unit shall report the internal audit of the data, and report it to the management team and the implementation team. The implementation team supervised and guided the whole process of data collection and coordinated communication problems in a timely manner.

4.2.3. Evaluation execution. The evaluation of the intelligent passenger station is carried out in two stages of initial evaluation and re-evaluation. The evaluation implementation group summarizes and counts the indicator data reported by the evaluated units according to the established rules, and organizes the level of the evaluation indicators of the initial evaluation of the expert group. The expert group reviews the initial evaluation results with reference to the level characteristics of each level, and gives specific scores. For those who do not have obvious characteristics, adjust the corresponding level. The leader of the expert group will summarize the scores of each evaluation index, calculate the total score, and determine the evaluation level of the evaluated unit.

4.2.4. Evaluation release. The evaluation implementation team prepares an evaluation report based on the evaluation results. The evaluation report shall include the scores of the evaluated units, the corresponding evaluation level, and the sub-items of each evaluation index. The evaluation report is submitted to the expert group for approval, and is submitted to the review management team for approval, and is released by the management team.

4.3. Implementation plan and recommendations

4.3.1. Improve the institutional mechanism and standard specification of application management of evaluation index system. The framework of the application management system mechanism should be refined according to the actual needs, clarify the division of responsible units and responsibilities at all levels, and evaluate the operational guarantee mechanism, and finally form an application management plan for the evaluation system of the intelligent passenger station to ensure the orderly evaluation. There are plans to proceed.

4.3.2. The weight of the evaluation index system and the basis for scoring. On the basis of the pilot application of the evaluation index system, the proportion of the importance of each evaluation index in the whole system should be divided, that is, the weight of the individual evaluation indicators. In addition, according to the top-level design and technical route of the intelligent passenger station, combined with the objective and technical system of the intelligent station and information construction of the passenger station and the actual construction of the station in the pilot application, the classification index of the evaluation index system is formed.

4.3.3. Evaluation and rating training work of the intelligent passenger station. The government department took the lead and the local authorities actively cooperated to carry out the application training of the standard system and evaluation index system of the intelligent passenger station, and selected a group of local standardization personnel to participate in the development of national standards and international standards for intelligent passenger stations. At the same time, through regular organization of publicity training courses, seminars and other activities, increase the publicity and development of intelligent passenger transport station development standards and evaluation indicators, improve the impact of the evaluation of intelligent passenger station evaluation, expand the training and introduction of intelligent passenger station evaluation work Talent.

4.3.4. Research and build a unified mobile cloud station real-time data monitoring unified cloud platform. The construction of intelligent passenger station will use a large amount of information technology. In order to facilitate the unified management of station data, it is necessary to build a unified research on real-time data monitoring and unified cloud platform to realize real-time data collection, aggregation, collation, statistics, release and visualization of national passenger stations. Function, constantly introducing cutting-edge science and technology, providing program and decision support for unified scheduling of station resources, global disaster prevention early warning, and inter-station business coordination.

5. Conclusions

Taking the demonstration application of Beijing-Zhangjiakou High-Speed Railway as the research object, this paper analyses and explores the evaluation index system of intelligent passenger station, and proposes relevant development ideas to ensure the passengers' full-service, integration, self-help and personalized travel service quality. However, the preparation of the indicator system is not a one-step process. The process itself is a dynamic process that is constantly revised and improved. In the future, it is necessary to dynamically adjust the indicator system according to the deepening of the understanding of intelligent passenger stations and the practice of building intelligent passenger stations. Timely adjustment and improvement of the indicator system will provide strong institutional guarantee and scientific guidance for the construction and development of intelligent railway passenger stations in the future.

References

[1] De-Dao G U , Wen Q . Study on the Construction of Evaluation Index System of China's Smart City[J]. Future & Development, 2012.

[2] Ning B , Tang T , Gao Z , et al. Intelligent railway systems in China[J]. IEEE Intelligent Systems, 2006, 21(5):80-83.

[3] Jia L, Nie A, Wang F. Railway Intelligent Transportation System-State of Art, Challenges and Development[J]. Communication & Transportati0n Systems Engineering & Information, 2001, 1(3):207-211.

[4] Khekare G S , Sakhare A V . A smart city framework for intelligent traffic system using VANET[C]// International Multi-conference on Automation. IEEE, 2013.

[5] Liu Y , Luo X . A Study on Planning of Large Urban Road Passenger Station[C]// International Conference on Intelligent Computation Technology & Automation. IEEE Computer Society, 2008.

[6] Lanke N , Koul S , Lanke N , et al. Smart Traffic Management System[J]. International Journal of Computer Applications, 2014, 75(7):19-22.

[7] Yong Q, Jia L, Yuan Z. Railway intelligent transportation system and its applications[J]. Engineering Sciences, 2011, 09(1):53-59.

[8] Li P, Jia L M, Nie A X. Study on railway intelligent transportation system architecture[C]// Intelligent Transportation Systems, 2003. Proceedings. IEEE, 2003:1478-1481 vol.2.

Research and Analysis on the Top Design of Smart Railway

Shaofu Lin[1,2]*, Yafang Jia[2], Sibin Xia[2]

[1]Beijing Advanced Innovation Center for Future Internet Technology, Beijing, 100124,China

[2]College of Software, Beijing University of Technology, Beijing, 100124,China

*Corresponding author, e-mail: linshaofu@bjut.edu.cn

Abstract. Deepening the informatization and intelligent construction of railways has become an inevitable choice to promote the innovation and development of railways and enhance their core competitiveness.At present,China has in-depth research in smart cities,smart transportation and other fields, but the research on smart railways is still in its infancy, and it is urgent to make plans for the development of smart railways to provide guiding suggestions for the development of railway informationization.Based on the research results of smart cities and smart transportation in related fields at home and abroad, combined with the application trends of internet technology and big data technology in railway informatization, this paper attempts to give a clear definition of smart railway from the perspective of smart city development. It also proposes the overall structure of the top-level design of the smart railway, and the application of the smart railway in combination with the development needs of the construction of the Jing-Zhang high-speed railway.

1. Introduction

The smart railway is a new stage of the development of intelligent transportation informationization, and it is the total integration and comprehensive embodiment of the railway informationization public service system .In 2010, the leadership of the Ministry of Railways, in conjunction with the need to change the way of railway development, proposed the development direction for the smart railway. It is expected to improve the overall capacity of the railway through the intelligent development of the railway, accelerated the transformation of the railway development mode, and realized the sustainable development of the railway.In recent years, China has been working hard in the direction of railway informationization. However, as of now, there is still no standardized norm in the field of smart railways. People's understanding of smart railways is rather vague. In the research of top-level design of smart railways, it is necessary to establish a conceptual model of smart railways, clearly define the connotation and characteristics of smart railways, and build a model of a smart railway architecture system, which will not only help deepen the understanding of the smart railway, but also play a guiding role in the development planning and construction of the smart railway.

2. Research and development of intelligent railways at home and abroad

Japan RTRI introduced the railroad system for the 21st century in 2000 - Cyber Rail (Digital Network Interconnected Railway). In 2005-2010, the EU invested in supporting the Inte GRail project to meet the development needs of the European railway system integration. In 2014, the EU proposed the Shift2Rail program, which is the first large-scale railway research and innovation project in the history of the European Union. In 2007, British rail operator Network Rail proposed the Intelligent

Content from this work may be used under the terms of the Creative Commons Attribution 3.0 licence. Any further distribution of this work must maintain attribution to the author(s) and the title of the work, journal citation and DOI.

Published under licence by IOP Publishing Ltd

Infrastructure strategy, which changed the strategy of detecting faults in the infrastructure maintenance process to predictive avoidance, and finally avoided by design. In 2009, IBM proposed the Smarter Rail development concept based on the wisdom of the Earth strategy. It outlines the needs of the development of smart railways and the three key intelligent features that smart railways should have: more thorough perception, more comprehensive interconnection, and deeper intelligence.In 2017, China Railway Corporation set up a smart station forum at China Railway International Equipment Exhibition to discuss the application and development of smart stations at home and abroad.In summary, the smart railway is the development direction of railway informationization, the basis for realizing the sustainable development of green railways, and the overall embodiment of the railway modernization level.Therefore, the development of a clear intelligent railway architecture model is an effective way to solve the problems faced by China's railway informatization process, and will have a great significance for scientifically guiding the development and planning of smart railways.

3. Smart Railway Concept and Model

3.1. The concept of smart railway

Under the background of accelerating the construction of smart cities in China, railways are an important transportation channel between cities, and their information construction should also rely on emerging technologies. It should move from the traditional concept of "railway" to the "smart railway". According to the current research status of railway intellectualization, the definition of smart railway is proposed in this paper:

Smart Railway refers to a railway transportation subsystem of intelligent traffic system in smart cities.

It mainly uses new generation information technology such as Internet of Things, cloud computing, big data, satellite positioning and navigation, geospatial information, and artificial intelligence.It is a new system and new ecology that fully integrates with railway transportation planning to support, promote and guide the intelligent development of railway transportation. At the same time, it is also a comprehensive service platform and mobile information physics space on the railroad for mobile leisure, office, learning and consumption.

3.2. The overall architecture model of the top design of the smart railway

Smart Railway integrates new generations of information technology such as cloud computing and big data with railway management to build an intelligent information railway, and realize various services such as accessing various intelligent terminals online at anytime and anywhere in the railway operation. The main features are network ubiquitous interconnection, intelligent sensing, data sharing, business collaboration, and intelligent services.

The overall structure model of the smart railway is shown in Figure 1. It mainly includes the intelligent sensing layer, intelligent transmission layer, information resource layer, application support layer, application platform layer, standard specification management system and information security system.

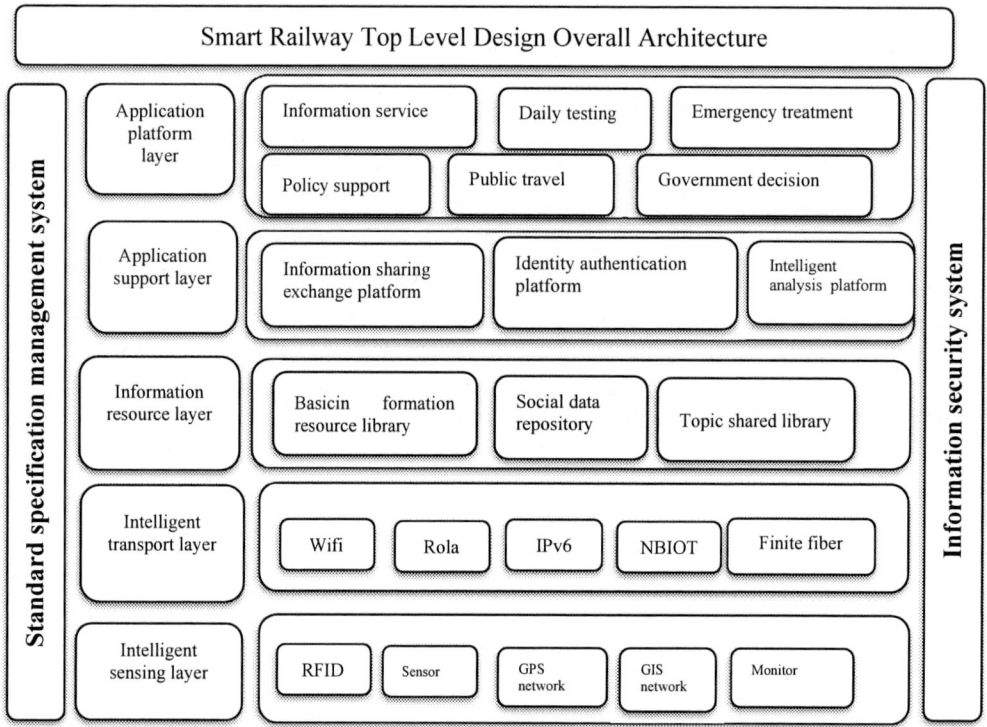

Figure 1.Smart Railway Top Level Design Overall Architecture

4. The idea of building a smart railway

4.1. general idea

The construction of smart railways must adhere to the development concept of "innovation, coordination, green, openness, and sharing", adhere to the people-centered development thinking[10], implement the development requirements of the "Thirteenth Five-Year Development Plan for Railways", and focus on the three core functions of smart railway transportation, operation services, and promoting industrial development, closely linked to the construction of smart railways, vehicles, stations, people "one main line".At the same time, it is necessary to work hard to promote the construction of a new generation of information infrastructure, enhance the smart development capability of the railway, and improve the railway service system to enhance the passenger service experience and strengthen security measures. In addition, it also need to complete the "five major tasks" of improving the information management and security support capabilities of smart railways, promoting the sustainable development of railway, promoting the construction of high-end cultural brands, and enhancing international influence.

4.2. Construction principles

4.2.1. Improve basic and safe development. It is necessary to adhere to the use of independent innovation technologies, products and services, accelerate the construction of railway safety and controllable information infrastructure, strive to improve the intelligent railway information governance and security support capabilities, improve the safety service guarantees of railway operation processes and emergency response capabilities of major events to promote the safe development of the railway.

4.2.2. Green low carbon, energy saving and environmental protection. In the process of railway planning and construction, Firstly,we must save resources such as land, line, passage, and promote the comprehensive development and utilization of railway stations and surrounding land. Secondly, it is necessary to strengthen railway environmental protection management, establish and improve railway environmental protection technical standards, assessment and evaluation systems and product certification systems. In addition, we should actively promote the application of new environmental technologies, new materials and new technologies, increase investment in environmental protection and the renovation and renovation of existing environmental protection facilities. Finally, comprehensive measures should be taken to effectively prevent noise and vibration along the railway.

*4.2.3. People-oriented, service-oriented.*It should adapt to the needs of integrated and high-quality travel services, revise and improve the railway passenger transport service quality standard system, strengthen the coordination of various transport modes, and actively carry out passenger transport services. At the same time, it is necessary to optimize the ticketing organization and services, further improve the 12306 network ticket sales, actively adopt Internet ticketing and mobile APP to purchase tickets, etc..Thereby improving the passenger train punctuality rate, improving the service quality of motor trains, improving the service level of ordinary passenger trains, and providing passengers with more good travel experience.

*4.2.4. Transforming and upgrading, improving quality and efficiency.*It is necessary to speed up the transformation of the railway development mode, accurately match the changes in transportation demand, and comprehensively improve service quality and operational efficiency. It will strengthen the integration with other modes of transportation and modern logistics, extend the service chain, tap the resources for revitalization, and enhance the capacity for sustainable development of the railway.

4.3. The goal of building
Through in-depth study and implementation of the spirit of the 19th Party Congress and the implementation of the "Thirteenth Five-Year Development Plan for Railways", this paper aims to improve the level of passenger transport services, firmly establish the awareness of safe production red lines, improve the level of rule of law in railway safety management, and save intensive use of resources. It is the construction goal to ensure the maximum energy saving and consumption reduction.

4.4. Main task

4.4.1. Building a new generation of information infrastructure to enhance the ability of railway intelligence development .It is necessary to build high-speed rail network, speed up railway fiber optic network coverage, with expansion capability gigabit access, to achieve a minimum bandwidth of 50 megabytes user subscription, the popularity of 100 megabytes.There will provide free broadband wireless access to users throughout the railway through the government's purchase of services, and gradually improve the IOT facilities for intelligent operation of railways.In addition, it will strengthen the integrated management of the object management object and the sensing device, and promote the sharing and integration of the IOT data.

4.4.2. Strengthen security measures to improve smart railway governance and security capabilities.
It is necessary to strictly implement relevant standards and acceptance requirements to ensure the quality and safety of high-speed railway construction and comprehensively improve the safety management level of high-speed railways.In addition, it will establish a safety production situation analysis and early warning and emergency response plan mechanism to improve the ability to detect and prevent safety accidents, strengthen safety measures, and enhance the wisdom of railway management and safety.

*4.4.3. Improve the service system and enhance the passenger service experience .*It should give full play to the advantages of efficient, fast access and wide coverage of the railway transportation network, optimize organization and dispatch, deepen the potential of passenger transportation, and improve the network passenger transportation capacity.In order to meet the needs of integrated and high-quality travel services, it will revise and improve the railway passenger transport service quality standard system, strengthen the transport mode and cooperation of various modes of transport, and actively carry out passenger transport services to improve the passenger service level.

*4.4.4. Promote the sustainable development of railway green .*In the process of building a smart railway, it should give play to the comparative advantages of railways, advocate green travel modes, give full play to the role of railway backbone transportation, and accelerate the construction of a composite logistics channel and an energy-saving comprehensive transportation system with green railways as the backbone. In order to strengthen railway environmental protection management, it is necessary to establish and improve railway environmental protection technical standards, assessment and evaluation systems and product certification systems.Moreover,it will strengthen energy management, promote intelligent energy conservation control, and improve energy comprehensive utilization. Optimize transportation organization, improve transportation efficiency, and further reduce railway transportation energy consumption.

*4.4.5. Promote the construction of high-end cultural brands and enhance international influence .*It should make full use of domestic and foreign resources, and actively promote all-round external cooperation in China's technical consultation, construction, equipment manufacturing, transportation management, personnel training and technical standards, and promote the development of railway "going out" to the high-end direction of the industrial chain and value chain.We should actively follow the development trend of international railway standards, carry out comparative analysis of Chinese and foreign standards research, actively transform international standards suitable for China's national conditions, to build China's smart railway cultural brand and enhance international influence.

5. Application demonstration in Jing-Zhang high-speed railway

In order to build a smart Jing-zhang high-speed rail, the railway department actively adopts advanced technologies such as cloud computing, big data, Internet of Things, mobile Internet, artificial intelligence, BIM, etc., carefully designing, organizing and constructing, focusing on building quality projects to break through intelligence. Key technologies such as construction, intelligent equipment and intelligent operation will further improve the modernization level of railway safety production, operation management and passenger transportation services. The above design concept is consistent with the design idea of the smart railway in this paper. Therefore, the conceptual model designed in this paper is demonstrated in the Jing-Zhang high-speed railway, which has the meaning of reality and operability.

5.1. Application scenario design

*5.1.1. Smart train .*Smart trains refer to the intelligentization of railway mobile transport equipment and trains. Through the data transmission and communication platform, the smart train realizes the real-time information transmission between the on-board sensors and the reliable mass information transmission between the vehicle and the ground-to-vehicle. The train status information is perceived and monitored through the real-time dynamic digital platform and the operating environment digital platform. The model is to realize the modern detection of high-speed trains and the operation control of trains, and then to connect digitalization of power supply, line engineering and digitalization of geographical environment, and to establish a smart train customer service platform to achieve high-quality services during passenger travel.

*5.1.2. Smart infrastructure.*Smart infrastructure is the intelligentization of railway fixed infrastructure, including road networks, signal equipment and traction power supply equipment. It has automatic equipment such as station interlocking, line blocking control, signal control, traction power transmission control. It can more accurately and reliably support the realization of automatic identification of vehicle number, disaster prevention safety monitoring, traffic safety monitoring, emergency command and rescue, transportation resource management and other systems.

*5.1.3. Smart transportation guarantee.*Smart transportation security is the intelligentization of the railway transportation guarantee process, assisting in the automation of information processing and safety inspection and monitoring of railway equipment maintenance and repair operations, ensuring the reliable provision of transportation production resources and ensuring quality, and ensuring the efficiency and safety of the transportation production process. This article will make full use of communication technology, information technology, decision-making technology, geographic information and virtual reality technology to establish an intelligent emergency platform, timely and accurately grasp the situation, scientifically deploy various types of rescue resources, and create a multi-party collaborative meeting model to achieve complete, unified, efficient emergency information management and command coordination, effectively implement emergency rescue and disposal, and improve the ability of the railway to respond to emergencies.

*5.1.4. Smart management.*Smart management is data-centric, and comprehensively acquires information on transportation production, transportation market demand changes, and business management business process status through various components of smart railways. Among them, various decision support systems can predict the demand for passenger and freight transportation, analyze the current transportation plan's ability to adapt to transportation needs, infrastructure resources, operational efficiency of transportation resources, and completion of business plans, so as to adjust transportation production plans in a timely manner, and continue to optimize the line plan, adjust the allocation of capacity resources.

*5.1.5. Smart passenger service.*Smart passenger service supports a full range of passenger travel services, providing convenient ticket purchase methods, digitizing passenger identity information, providing passengers with information navigation, personalized information services and cross-modal passenger services. The intelligent user navigation service provides passengers with information on train schedules, travel and consignment, and assistance decision support. The smart passenger transport service timely releases railway transportation product information and various service contents, station and agent point distribution through the Internet. At the same time, the intelligent passenger transport service takes the passenger's demand as the optimization goal, and takes the train time, fare and train grade as the constraints, and provides the auxiliary optimization decision for the passenger travel planning through the intelligent optimization technology.

5.2. Safety measures

*5.2.1. Strengthen organizational leadership .*The national railway bureau should give full play to the organizational leadership and overall coordination role of industry standards work, and strengthen planning guidance. At the same time, the relevant departments and units will be guided to put the standard work in the strategic position of priority development, and will be included in the relevant department and unit work plan for key deployment and arrangement.

*5.2.2. Improve organizational management mechanism.*The construction and operation of smart railways adhere to the methods of politics, production, learning, research and media alliance. In light of the current situation of the railway, relevant departments set up operational entities with government, industry and academic research as the core, and promoted the construction and operation activities of different modes with industrial funds as the guarantee. At the same time, government agencies should establish an efficient operation organization, continuously improve the standard operation mechanism and set up a special organization or committee to support the overall planning and design team and key project construction management team to carry out corresponding work.

5.2.3. Strengthen the construction of professional talents. It is necessary to give full play to the resource advantages of scientific research institutes in universities, encourage joint research and development of production, education and research, strengthen the discipline construction and talent cultivation in the field of smart railways, and focus on strengthening the cultivation of complex talents such as cloud computing and big data. The development of smart railway application demonstration and publicity can improve the information literacy, knowledge learning ability and information consumption level of the whole people. It is also necessary to improve the high-end informationized talent service system and introduce talents from key technologies and industries in the domestic and international smart railways and related fields.

6. Conclusion

This paper takes the application of Jing-Zhang high-speed railway as the research object, analyzes and studies the top-level design of the smart railway, clarifies the connotation and definition of the smart railway, deepens people's understanding and understanding of the smart railway, and proposes the overall design architecture model. It has a certain guiding role for the development of railway informationization in China. However, this paper stands in the perspective of the overall development of the smart railway. There are still many shortcomings in the preliminary exploration and research of the Smart Railway. The top-level design is a macro-plan that affects the subsequent development of the smart railway. In the future, it should be updated and improved in real time according to the development of the smart railway, providing powerful guidance for the information development of the railway.

References

[1] Khekare G S , Sakhare A V . A smart city framework for intelligent traffic system using VANET[C]// International Multi-conference on Automation. IEEE, 2013.

[2] COOPER Dave E. Intelligent transportation systems for smart cities:a progress review[J]. Science China(Information Sciences), 2012, 55(12):2908-2914.

[3] Stefansson G, Lumsden K. Performance issues of Smart Transportation Management systems[J]. International Journal of Productivity & Performance Management, 2009, 58(1):55-70.

[4] Huang X. Smart Antennas for Intelligent Transportation Systems[C]// International Conference on ITS Telecommunications Proceedings. IEEE, 2006:426-429.

[5] Li X, Song J. The Top Design Methodology of Smart City in China[C]// International Conference on Intelligent Computation Technology and Automation. IEEE, 2014:861-864.

[6] Jianbo, Cheng, Peng. Top-Level Design of Smart City Based on "Integration of Four Plans"[J]. ZTE Communications, 2015, 13(4):34-39.

[7] Lanke N , Koul S , Lanke N , et al. Smart Traffic Management System[J]. International Journal of Computer Applications, 2014, 75(7):19-22.

[8] Bouhedda M, Bellatreche S, Ahmed-Serier R. Smart traffic signal controller design and hardware implementation based ant colony system[C]// International Conference on Modelling, Identification and Control. IEEE, 2017:1110-1116.

ISPECE   IOP Publishing

IOP Conf. Series: Journal of Physics: Conf. Series **1187** (2019) 052054   doi:10.1088/1742-6596/1187/5/052054

Study on Vulnerability Rating of the Intelligent and Connected Vehicle's Cybersecurity

Yangyang Liu[1,3], Zhao Wang[2], Yanan Zhang[1], Peiji Shi[1] and Xuebin Shao[1]

[1] Automotive Data Center, China Automotive Technology and Research Center Co., Ltd., Tianjin, China.

[2] Auto Standardization Research Institute, China Automotive Technology and Research Center Co., Ltd., Tianjin, China.

[3] Supported by program of scientific and technological activities for the returned Chinese who study abroad in Tianjin.

Abstract. Based on the automobile vulnerability data, the rating system of automobile cybersecurity vulnerabilities was constructed by referring to CIA, HEAVENS et al. First of all, basic rules such as naming, coding, classification and description of vulnerabilities are developed to ensure the uniqueness, normalization and compatibility of vulnerabilities. Next, the vulnerability rating system was developed by taking the three elements of scene, threat and impact and the 13 sub-elements of subdivision as evaluation indexes. The evaluation of the existing vulnerabilities and the comparison with the professional evaluation by using the established rating system show that the system is reasonable and effective.

1. Foreword

Since the beginning of the 21th century, automobiles have kept developing toward a direction of electrification, intellectualization, net connection and sharing, in which background intelligent connected vehicles emerge at the right moment. As a development direction of automobile industry, intelligent connected vehicle will again impact the pattern of the automobile industry and our trip mode, but also bring us corresponding threats on cybersecurity. Some hackers take use of automobile vulnerabilities to send out malicious codes to attack the intelligent connected vehicle, or even endanger personal safety and social security. To build a comprehensive and efficient intelligent automobile cybersecurity system, the promotion of automobile cybersecurity vulnerability level system is imperative [1]. There are vulnerability databases operating at home and broad already [2-3], but there is no relevant standard or specification grading of automobile vulnerabilities. Moreover, the grading of vulnerabilities is critical to vulnerability management, risk assessment and emergency response. Therefore, the study on automobile cybersecurity vulnerability level evaluation system is extremely urgent [4].

2. Significance of the establishment of automobile vulnerability level system

Along with the application of emerging technologies such as Internet, Artificial Intelligence, wireless network, cloud computing, big data and so on, degree of intellectualization and net connection of automobile has kept on growing. Automobile has become a veritable intelligent terminal device in the Internet Era and the cybersecurity of automobile will always be a problem requiring great attention in intelligent connected vehicles. Automobile cybersecurity now is still in a relatively preliminary state

Content from this work may be used under the terms of the Creative Commons Attribution 3.0 licence. Any further distribution of this work must maintain attribution to the author(s) and the title of the work, journal citation and DOI.
Published under licence by IOP Publishing Ltd

which is easily to be broken through. Numerous works are still required, among which the threat analysis for cybersecurity is more complicate and important. For the whole vehicle and various subnet components, there are too many unknown vulnerabilities and attacking modes. We need to put ourselves on attackers' position to analyze with non-traditional vulnerability analysis, osmotic and statistical techniques, which is relatively difficult for traditional auto workers. Thus, a vulnerability grading system shall be developed to provide reference for whole vehicle cybersecurity evaluation, emergency response level constitution and so on, so as to take actions beforehand to prevent malicious use of vehicle vulnerabilities which may cause problems such as economic loss, driving mistake, stealing of privacy and damage on functional security.

From the view of automobile emergency response, when cybersecurity event occurs on intelligent connected vehicle, we need a set of feasible auto emergency response scheme to reduce response time as much as possible, so that we can respond quickly to mitigate or eliminate impact caused by the security event. According to the vulnerability level system, different repair modes are constituted for vulnerabilities in different levels, studies on emergency response mechanism are carried out and loss caused by improper vulnerability handling is reduced. Finally, whether the vulnerability is repaired is verified, and lessons learned are concluded and analyzed, which provide reference for vulnerability emergency response and security development in future.

From the view of auto vulnerability management, risk assessment can be carried out on vulnerabilities to evaluate the probability for intelligent connected vehicles to suffer threats and influence on road participants after their occurrence, as well as output a series of threats with corresponding risk levels. It can test the severity of vulnerabilities, rate and divide levels of security vulnerabilities preliminarily, as well as help applicators to determine the urgency and importance of response required by vulnerabilities.

3. Vulnerability level system

3.1. Basic rules of vulnerability level system
This part introduces basic rules of auto vulnerabilities including coding, naming, classification, description and so on.

3.1.1. Coding of auto vulnerabilities
To facilitate the identification and reflection of uniqueness, normativeness and compatibility of auto vulnerability data, unique codes are distributed to auto vulnerabilities indexed uniformly. The following principles shall be observed: 1) uniqueness: code for each auto vulnerability data shall be unique; 2) chronological order: it shall be easy to identify time of auto vulnerability indexing and define timeliness of vulnerabilities; 3) confidentiality: it shall be avoided to disclose business information in codes. According to the above principles, the following grammatical rule for coding is constituted: QCLD+YYYYMMDD+form code of any figure, i.e. "QCLD" is a fixed 4-bit code; "YYYY" is a four-bit sequenced code; "MM" is a two-bit sequenced code; "DD" is a two-bit sequenced code; "any figure" is a three-bit code ordered according to the sequence of generation of auto vulnerability, of which the quantity can be expanded when necessary, e.g.: QCLD-20180801-001, QCLD-20180801-1000.

3.1.2. Naming of auto vulnerabilities
Based on the coding of auto vulnerabilities, auto vulnerability data shall be named to facilitate retrieval and consultation [4]. Therefore, the following principles are constituted: 1) readability: it shall cover information related to the auto vulnerability data such as auto OEM, component manufacturer, auto service provider and auto model, sub-type of vulnerability and so on; 2) confidentiality: it shall be avoided to disclose any detail of the vulnerability technique. According to the above principles, the following naming method is proposed: "auto manufacturer" + "auto model name" + "vulnerability position". Besides, "vulnerability" or "defect" is postfixed at the end. Wherein,

name of manufacturer refers to short name of the manufacturer of the vehicle affected by the vulnerability. If the manufacturer usually uses Chinese name or if it is a foreign company without Chinese name, English name of the company will be adopted; vehicle name refers to the official name of the vehicle model affected by the vulnerability; vulnerability position refers to the specific position of the vehicle affected by the vulnerability. For example: XX company XX vehicle steering lamp defect.

3.1.3. Classification of auto vulnerabilities

The utilization of auto vulnerability is generally not possible by single component only. It requires the use of a series of combined defects on different components. Therefore, the classification of auto vulnerabilities based on defective object module may cause unclear classification [5-7]. According to concepts of threat and property in cybersecurity, two parameters are designed - vulnerability component and affected components.

Vulnerability component refers to vehicle components containing defects. It is a kind of threat. Attackers attack vehicle by using vulnerabilities in vulnerability components;

Affected component refers to components endangered after successful use of vehicle vulnerabilities. It is a kind of property. Affected component can be the vulnerability component itself, or other software, hardware or network components in the vehicle.

This auto cybersecurity vulnerability system is based on vulnerability components. It classifies auto vulnerabilities into seven categories (APP, cloud platform, IVI, T-BOX, radio, ECU, bus and so on) according to threat analysis on cybersecurity of intelligent connected vehicles.

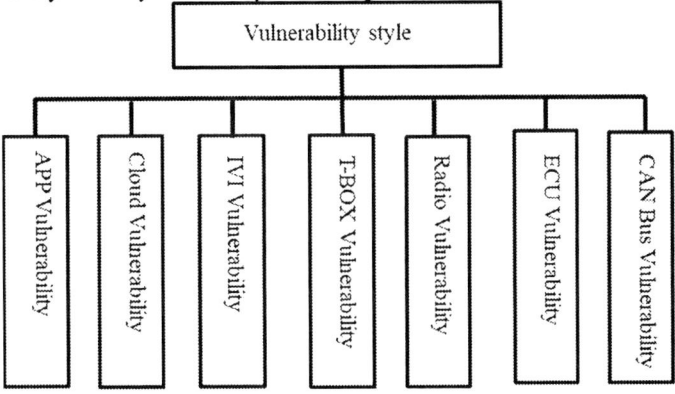

Figure 1 Vulnerability types

3.1.4. Description of auto vulnerabilities

For vulnerability data confirmed to be indexed, text description shall be adopted to describe cause, existing position, affected range and so on of the vulnerability in a uniform format. According to the above requirements, the following principles are constituted: 1) simplicity: describe the vulnerability simply, clearly and straightforward; 2) authenticity: describe the vulnerability truly and objectively; 3) confidentiality: avoid to disclose excessive technical details of the vulnerability which may cause secondary hazard. According to the above principles, method for content description is constituted:

The method for auto vulnerability data description includes "vulnerability body introduction" and "vulnerability content introduction". In the "vulnerability body introduction", it is required to sketch basic information of the system or product where the vulnerability exists, which includes "name of the vulnerability body", "name of the vehicle related company", "definition of the vulnerability body", "functional overview", e.g.: XXX is a XXX product (software, solution, system, tool and so on) of XXX (English) Company in XXX (country). The product possesses function of XXX; in the "vulnerability content introduction", it is required to sketch type, cause, attacking mode and impact of the vulnerability, including "type of vulnerability", "cause", "use manner", "impact and consequence". E.g.: this car has XX (type) type vulnerability which is caused by XX of XX (name of the product or system). Attackers can fulfill XX attack via XX (manner) to cause XX danger or result.

3.2. Vulnerability level evaluation system

With reference to cybersecurity CIA model, HEAVENS model, CVSS model and real vehicle tests, 3 categories of parameters and 13 types of sub-parameters are summarized and stated in a tree structure[8-10]. Among sub-factors, different to the static grading made by traditional Internet on vulnerabilities, dynamic factors are added (e.g. danger time) to reduce uncertainty brought to risk assessment, emergency response and so on.

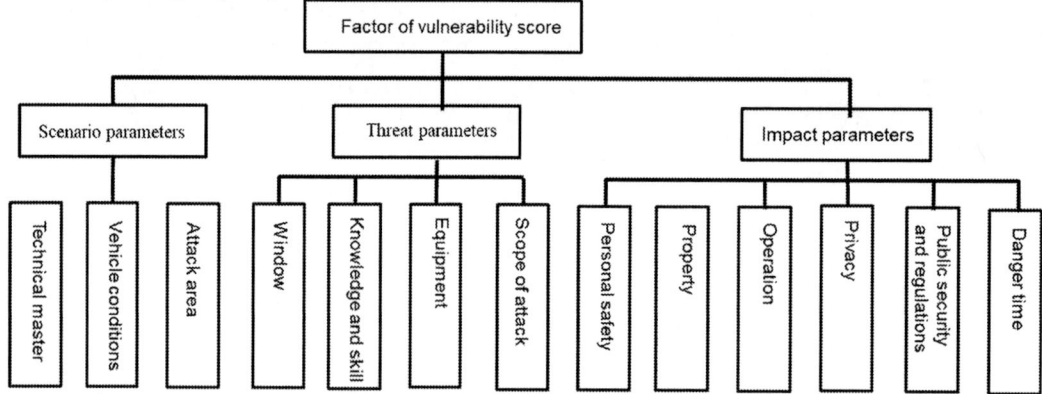

Figure 2 Tree structure for vulnerability rating parameters

3.2.1. Scene parameters (SP)

The evaluation factor "scene parameter" refers to the correlation factors describing state and range of the affected vehicle by using relevant technical mastery situation when the vulnerability is attacking the vehicle. Three kinds of sub-factors are set: technical mastery, vehicle condition and scope of attack:

1) Technical mastery refers to the comprehensive consideration on the understanding on the attacking vulnerability used and mastery degree on technology;

2) Vehicle condition refers to the state of vehicle when attacked, distinguished by driving speed;

3) Scope of attack refers to the comprehensive consideration on quantities and varieties of auto targets attacked;

Levels, corresponding definitions and benchmark value of sub-factors are as following:

Table 1 Scene Parameters

3.2.1. Parameter name	Level	Description	Score
Technical mastery (TM)	Low	Dimly aware position of vulnerability; master attack flow including attacking tools, attack steps and so on	1
	Medium	Can acquire vulnerability material and information in details; know relevant principles of the vulnerability in a "low" level	3
	High	Be proficient on the work mode of vulnerability triggering in a "medium" level; mature principles have been accumulated	10
Vehicle	Still	The car is parking or idling, i.e. the car speed is	1

conditions (VC)		0	
	Low speed	The car is driving in a speed of 0-15km/h	2
	Medium speed	The car is driving in a speed of 16km/h to 25km/h	4
	Relatively high speed	The car is driving in a speed of 26km/h to 50km/h	7
	High speed	The car is driving in a speed above 50km/h	10
Attack area (AA)	Single vehicle attack	Launch vulnerability attack on one vehicle only	1
	Single model attack	Can use the vulnerability to attack same vehicle model	7
	Multi model attack	Can use the vulnerability to attack more than one vehicle models	10

Take the mined vulnerability data as study sample and use neural network modification formula coefficient to obtain the formula for calculation score for scene parameter:

$$SP = 0.952 * TM + 1.905 * VC + 1.429 * AA \qquad (1)$$

Wherein: SP refers to the score of scene parameter; TM refers to the value taken for technical mastery parameter; VC refers to the value taken for vehicle condition parameter; AA refers to the value taken for the attacking area parameter.

3.2.2. Threat parameter

The evaluation factor "threat parameter" refers to the correlation factors successfully triggering the vulnerability when an attack is launched for the vulnerability to be evaluated. It can be divided into four kinds of factors: window, knowledge and skill, equipment, scope of attack and so on:

1）   Window refers to the attack vector adopted when using the vulnerability for attack;

2）   Knowledge and skill refer to comprehensive consideration using basic principle, method, audience of knowledge groups and so on when an attack is launched;

3）   Equipment refers to the advanced level of equipment used when mining and using vulnerabilities, as well as launching attacks;

4）   Scope of attack refers to the quantity of component targets attacked by using the vulnerability when an attack is launched;

Table 2 Threat parameters

Parameter name	Level	Description	Score
Window (WD)	Remote	The attacker can launch attacks to automobile by using the vulnerability via the Internet, e.g. 4G, 3G and so on	10
	Close range	The attacker can use the vulnerability to attack automobiles by shared physical or logics, e.g.: bluetooth, Wi-Fi, IEEE 802.11, local IP subnet and so on	5

	Local	The attacker can use the vulnerability by reading/writing operations or running applications/tools, i.e. local participation is required for the use of vulnerability, for example: the attacker shall login locally; users shall download, accept the malicious contents	3
	Physical contact	The attacker shall contact the automobile physically to launch an attack, for example, launch the attack on automobile bus by OBD II	1
Knowledge and skill (KS)	Amateur	The attacker takes use of existing attack to execute simple command to launch attack but will not improve the attack method and attack tool	10
	Skillful operator	The attacker has certain knowledge on safety field or auto field, is able to carry out relevant businesses, knows simple and popular attack flow and is able to improve the attack tools used	5
	Automobile safety expert	Be familiar with algorithm, protocol, hardware, architecture of key component bottom such as ECU and so on or attack techniques and tools lately defined in the safety field; possess solid knowledge of cryptography, classical attack method	2
	Multi-field expert	The attacker launches attacks by using vulnerabilities. Knowledge of different professional field is required in different attack steps	1
Equipment (EM)	Open hardware equipment and software	The equipment used to launch attack has been publicly available, relatively common in traditional safety field, such as protocol analyzer, downloader, common IT equipment - notebook computer and so on	10
	Open hardware equipment and software for specific use	The equipment is not easy to be obtained by attackers; but its script can be attacked by purchasing or development; for example: on-board communication equipment, vehicle spy, CANoe, USRP and so on	7
	Customized or specific hardware equipment and software	The equipment is specifically produced or especially complicate software; the equipment or software is under control, or very expensive, for example Unpacker ExeCryptor exclusively used by some national organizations	3
	Multiple customized or specific hardware equipment and software	Equipment or software needs to be customized by different disciplines for different attack steps when the attacker launches attacks	1
Scope of	Single	Launch attack on unique component in the attacked vehicle	1

attack (SA)	Multiple	Launch attack on more than one components in the attacked vehicle but not all components	7
	Almost all	Against all electronically-controlled components in the vehicle	10

Take the mined vulnerability data as study sample and use neural network modification formula coefficient to obtain the formula for calculation score for threat parameter:

$$TP=1.905*WD+0.952*KS+0.952*EM+1.905*SA \tag{2}$$

Wherein: TP refers to the score of threat parameter; WD refers to the value taken for window parameter; KS refers to the value taken for knowledge and skill parameter; EM refers to the value taken for equipment parameter; SA refers to the value taken for the scope of attack parameter.

3.2.3. Impact parameter

The evaluation factor "impact parameter" refers to the correlation factors related to dangers caused by attack on vehicles. They can be divided into six categories: personal safety, property, operation, privacy, public security and regulations, danger time and so on:

1) Personal safety refers to the severity of safety injury on people in the automobile;

2) Property refers to the total property consideration on direct and indirect losses caused to the automobile manufacturer, component manufacturer and individuals;

3) Operation refers to unexpected losses on automobile function after attack;

4) Privacy refers to losses caused by infringement on personal privacy after attack;

5) Public security and regulations refer to comprehensive consideration on losses caused by harm on surrounding public security and damage on laws and regulations;

6) Danger time refers to the duration of suffering by the vehicle after attack;

Table 3 Impact parameters

Parameter name	Level	Description	Score
Personal safety (PS)	NA	No personal injury	0
	Minor injury	The driver and passengers are slightly injured but able to move freely	3
	Serious injury	The driver and passengers are not able to move freely	7
	Life threat	Life crisis on the driver and passengers; or many people are injured	10
Property (PP)	NA	No property loss	0
	Low	Property loss on single vehicle	2
	Medium	Property loss on multiple vehicles	6
	High	Huge property loss to the OEM or component manufacturer or even huge property loss to the national auto industry	10
Operation (OA)	NA	No operational impact	0
	Low	Impact operation of entertainment system only	2
	Medium	Impact operation of body system	4
	High	Impact operation of power control system	10
Privacy	NA	No loss of privacy data	0

(PA)	Low	Infringe privacy data such as account, password, address book and so on of a single person	1
	Medium	Infringe privacy data such as account, password, address book and so on of multiple persons	6
	High	User privacy data of the whole vehicle model, whole OEM or even all vehicle manufacturers	10
Public security and regulations (PR)	NA	No loss on public security and regulations	0
	Low	Will not cause social harm; cause slight damage to laws and regulations	2
	Medium	Cause slight social harm	6
	High	Cause serious social harm; cause serious damage to laws and regulations	10
Danger time (DT)	NA	Able to retrieve immediately after attack	0
	Short time	Able to retrieve within several hours after attack	2
	Long time	Able to retrieve one day or above after attack	5
	Not able to retrieve	Not able to retrieve without interference from professionals after attack	10

Take the mined vulnerability data as study sample and use neural network modification formula coefficient to obtain the formula for calculation score for impact parameter:

$$IP = 3.333 * PS + 1.429 * PP + 0.952 * OA + 0.953 * PA + 2.381 * PR + 0.952 * DT \qquad (3)$$

Wherein: IP refers to the score of impact parameters; PS refers to value taken for personal safety parameter; PP refers to value taken for property parameter; OA refers to value taken for operation parameter; PA refers to value taken for privacy parameter; PR refers to value taken for public security and regulations parameter; DT refers to value taken for danger time parameter.

3.3. Vulnerability level

3.3.1. Attack level (AL)
Attack level is determined by the superposition of positive effects of scene parameters and threat parameters.

$$AL = SP + TP \qquad (4)$$

Wherein: AL refers to the value taken for attack level parameter.

Table 4 Attack level evaluation form

Total of attack level parameter (AL)	Attack level	Score for attack level
0-15	Low	1
16-40	Medium	2
41-70	High	3
Above 70	Extremely high	4

3.3.2. Impact level (IL)
Impact level is determined by the function of impact parameters.

$$IL = IP \qquad (5)$$

Wherein: IL refers to the value taken for impact level parameter.

Table 5 Impact level evaluation form

Total of impact level parameter (IL)	Impact level	Score of impact level
0-15	Low	1
16-40	Medium	2
41-70	High	3
Above 70	Extremely high	4

3.3.3. Vulnerability level (VL)

Vulnerability level is determined according to the 2D diagram formed by impact level and attack level.

Table 6 Vulnerability level rating form

Vulnerability level (VL)	Impact level (IL)			
Attack level (AL)		1	2	3	4
	1	Low risk	Low risk	Low risk	Medium risk
	2	Low risk	Medium risk	Medium risk	High risk
	3	Low risk	Medium risk	High risk	High risk
	4	Medium risk	High risk	High risk	Serious

Wherein: VL refers to vulnerability level, including four levels: low risk, medium risk, high risk and serious.

4. Application of vulnerability level system

Famous hackers Charlie Miller and Chris Valasek in America fulfilled remote attack without physical contact on 2014 Jeep Cherokee produced by Chrysler. When the attack was launched, the hackers took use of a series of combined defects, including relevant vulnerabilities of OMAP-DM3730 system, V850 controller embedded in the on-board information interaction system, cellular network service defect of operator Sprint, open port D-Bus (6667) and so on. Finally, the hackers fulfilled the control over comfort system (such as air conditioning system), information and entertainment system (such as radio, IVI) and power control system (such as acceleration, steering) of the vehicle [11,12]. For the "Chrysler Jeep Cherokee auto power braking system remote controlled vulnerability", the intelligent connected vehicle cybersecurity vulnerability level system was applied:

4.1. Scene parameter analysis

For "technical mastery" parameter: in relevant information available, it was only dimly known that the position of vulnerability of that vehicle was on the OMAP-DM3730 system and V850 controller embedded in the on-board information interaction system without knowing technical details on firmware level, therefore basis system was defined as "low"; for "vehicle condition" parameter: sudden braking and driving in a speed above 50km/h were realizable under a speed of 30km/h, as well as sudden loss of power for the vehicle and failure of accelerator. To sum up, it could be defined as "relatively high speed"; for "attack area" parameter, the vulnerability was not only accessible on 2014 Jeep Cherokee, but also on 2014 dodge Durango, thus was defined as "multi-model attack".

According to formula (1), score of impact parameter is calculated as:

$$SP = 0.952 * 1 + 1.905 * 7 + 1.429 * 10 = 28.577 \qquad (6)$$

4.2. Threat parameter analysis

For "window" parameter: hackers took use of the vulnerability via Sprint cellular network, thus it was defined as "remote"; for "knowledge and skill" parameter: mining and use of this vulnerability required knowledge on multiple professional fields including software engineering, communication, cryptology, hardware and so on, thus it was defined as "multi-field expert"; for "equipment" parameter: when the hackers launched an attack, they used ordinary notebook computer; but in firmware study, firmware extraction and professional programmer required were required, thus it was defined as "open hardware equipment and software for specific use"; for "scope of attack" factor: the vulnerability not only enabled attack on electrically-controlled components related to power control system, but also brought impact to all electrically-controlled components on Jeep Cherokee cars, thus it was defined "almost all".

According to formula (2), score of threat parameter is calculated as:
$$TP = 1.905 * 10 + 0.952 * 1 + 0.952 * 3 + 1.905 * 10 = 41.908$$
$$(7)$$
According to formula (4), score of attack level is calculated as:
$$AL = SP + TP = 70.485 \tag{8}$$

4.3. Impact parameter analysis

For "personal safety" parameter: when the power braking system was attacked, sudden braking might seriously threaten lives of the driver and passengers, thus it was defined as "life threat"; for "property" parameter: if the vulnerability was not repaired timely, Chrysler might suffer huge property loss, thus it was defined as "high"; for "operation" parameter: after the attack was launched, operation of power control system was affected, thus it was defined as "high"; for "privacy" parameter: the vulnerability didn't involve privacy issues, thus it was defined as "NA"; for "public security and regulation" parameter: the vulnerability was applicable to multiple vehicle models and attack might cause sudden brake of multiple vehicles, causing serious social danger, thus it was defined as "high"; for "danger time" parameter: when the attack was launched, only malicious code was input into the on-board terminal. Once stop transmitting, the vehicle would retrieve. Thus it was defined "NA".

According to formula (3), score of impact parameter is calculated as:
$$IP = 3.333 * 10 + 1.429 * 10 + 0.952 * 10 + 0.953 * 0 + 2.381 * 10 + 0.952 * 0 = 80.95 \tag{9}$$
According to formula (5), score of impact level is calculated as:
$$IL = IP = 80.95 \tag{10}$$

4.4. Vulnerability level evaluation

Score of attack level was 70.485, above 70, thus was defined "extremely high"; score of impact level was 80.950, above 70, thus was defined "extremely high"; according to the vulnerability level-rating form, the vulnerability was graded as "serious" level.

4.5. Application of vulnerability description

According to instance analysis, the weak component in this vulnerability was IVI system. According to the classification rule, the vulnerability was IVI type; according to coding rule, the vulnerability was coded as QCLD-20180728-001; according to the naming rule, the vulnerability was named "Vulnerability that Chrysler Jeep Cherokee auto power braking system was controlled remotely"; in the vulnerability description, for description on impacted components, key components involved in this vulnerability mainly included OMAP-DM3730 and V850 controller. OMAP-DM3730 is a processor based on open multi-media application platform OMAP produced by Texas Instruments (TI) in America. It is composed by 1GHz ARM Cortex-A8 Core and 800MHz TMS320C64x+DSP Core, and possesses functions such as 3D graphic processing and video acceleration; V850 is a micro-controller supporting real time operating system produced by Renesas (NEC) in Japan. For the description on vulnerability contents, the vehicle contained IVI type vulnerability which was caused by that the firmware updating of V850 micro-controller was not protected by signature mechanism and attackers could adopt method such as flash of V850 firmware, insertion of malicious code and so on. Moreover, V850 under the high level could awake the OMAP, which fulfilled the conditions of remote control on acceleration, braking and steering system of the vehicle, and could cause serious personal injury and property loss. The collection is as shown in the following figure:

Item		Description of Content	
Classification		The weak component in this vulnerability was IVI system，so the vulnerability was IVI type.	
Rating		According to scores and the vulnerability level-rating form, the vulnerability was graded as "serious".	
	Coding	CAVD-20180228-001	
	Naming	Vulnerability that Chrysler Jeep Cherokee auto power braking system was controlled remotely	
Vulnerability Collection	Description	Description of Impacted Component	Key components involved in this vulnerability mainly included OMAP-DM3730 and V850 controller. OMAP-DM3730 is a processor based on open multi-media application platform OMAP produced by Texas Instruments (TI) in America. It is composed by 1GHz ARM Cortex-A8 Core and 800MHz TMS320C64x+DSP Core, and possesses functions such as 3D graphic processing and video acceleration; V850 is a micro-controller supporting real time operating system produced by Renesas (NEC) in Japan, which can send CAN signal.
		Description of the Vulnerability's Content	For the description on vulnerability contents, the vehicle contained IVI type vulnerability which was caused by that the firmware updating of V850 micro-controller was not protected by signature mechanism and attackers could adopt method such as flash of V850 firmware, insertion of malicious code and so on. Moreover, V850 under the high level could awake the OMAP, which fulfilled the conditions of remote control on acceleration, braking and steering system of the vehicle, and could cause serious personal injury and property loss.

Figure 3 Vulnerability Collection

5. Verification of Vulnerability Rating System

Comparison between grading of partial auto vulnerabilities by expert evaluation and the results from the vulnerability grading system is shown below:

Table 7 Vulnerability system verification

NO.	Name of vulnerability	Expert evaluation	System evaluation
1	XX1 In-car network Car flameout vulnerability caused by OBD	Low risk	Low risk
2	XX2 In-car network causes car flameout vulnerability by OBD	Low risk	Low risk
3	XX2 Vulnerability of diagnosis network controlling I/O identifiers	Low risk	Low risk
4	XX2 Wireless key signal hijacking vulnerability	Medium risk	High risk
5	XX2 IVI contains external opening port vulnerability	Medium risk	Medium risk
6	XX2 Cell phone APP installation package non-encrypted data vulnerability	Medium risk	Medium risk
7	XX2 Auto cloud platform contains sensitive information vulnerability	High risk	Medium risk
8	XX3 Cell phone APP account key data non-encryption vulnerability	Medium risk	Medium risk
9	XX5 In-car network triggers backup camera vulnerability by OBD	Medium risk	Medium risk
10	XX5 Vulnerability of diagnosis network reading	Low risk	Low risk

	data by identifiers		
11	XX5 Cell phone APP password transmission non-encryption vulnerability	Medium risk	Medium risk
12	XX5 Cell phone APP contains XSS cross-site script attack vulnerability	Medium risk	Medium risk
13	XX6 Cloud platform contains input vulnerability	High risk	Medium risk
14	XX6 APP contains replay attack vulnerability	Medium risk	Medium risk
15	XX3 APP contains replay attack vulnerability	Medium risk	Medium risk
16	XX4 In-car network controls acceleration by OBD	Medium risk	Medium risk
17	XX1 Vulnerability of in-car network out of control via OBD	Medium risk	Medium risk
18	XX7 vulnerability of permanent flashing via OBD	High risk	High risk

Among the 18 vulnerabilities verified, only the "XX2 auto wireless key signal hijacking vulnerability", "XX2 auto cloud platform contains sensitive information vulnerability", and "XX6 auto cloud platform contains injection vulnerability" are different. Results by expert evaluation are medium risk, high risk, high risk; while the results from the vulnerability system are high risk, medium risk, medium risk. Some parameter setting requires improvement, for example: vehicle condition. To sum up, the intelligent connected vehicle cybersecurity evaluation system can rationally and scientifically evaluate auto vulnerabilities.

6. Conclusions

This paper builds a set of intelligent connected vehicle cybersecurity vulnerability level system: 1. Fundamental rules: this part contains rules such as coding, naming, classification, description and so on. It can facilitate management of auto vulnerabilities; 2. Auto vulnerability level evaluation system: this part is based on models such as CIA, HEVENS and CVSS system. It constitutes an auto vulnerability level evaluation method containing three categories of factors (scene, threat and impact), totally 13 types of sub-factors. This system has broken the situation of no specification and standard for reference in grading of intelligent connected vehicle cybersecurity vulnerabilities. It provides a practical method for evaluation of auto vulnerabilities and certain theoretical supporting for construction of vulnerability database for use in autos, as well as guarantees cybersecurity of intelligent connected vehicles in certain degree.

References

[1] LIU Qi-xu, ZHANG Chong-bin, ZHANG Yu-qing, et al. Research on key technology of vulnerability threat classification[A]. Journal on Communications, 2012, 33(Z1): 79-87.
[2] National vulnerability Database. http:nvd.nist.gov/.
[3] Common Vulnerability Scoring System Version 3.0 Calculator. https://www.first.org/cvss/calculator/3.0.
[4] Common Vulnerabilities and Exposures (CVE). https://cve.mitrc.org.·
[5] The Open Source vulnerability Database. http:osvdb.org/.
[6] US-CERT vulnerability Notes Field Descriptions. http:www.kb.cert.org/vuls/html/fieldhelp/#help.
[7] SANS Critical vulnerability Analysis. http:www.sans.org/newsletters/cva/.

[8] WANG L, JAJODIA S, SINGHAL A, et al. K-zero-day safety: measuring the security risk of networks against unknown attacks[A]. 15th European Symposium on Research in Computer Security[C]. Athens, Greece, 2010. 573-587.

[9] T. Becsi, S. Aradi and P. Gaspar, "Security issues and vulnerabilities in connected car systems", in Proc. 2015 International Conference on Models and Technologies for Intelligent Transportation Systems (MT-ITS 2015), pp. 477-482, 2015.

[10] FRUHWIRTH C, MANNISTO T. Improving CVSS-based vulnerability prioritization and response with context information[A]. 2009 3rd International Symposium on Empirical Software Engineering and Measurement[C]. Lake Buena Vista, FL, USA, 2009. 535-544.

[11] SAE International J3061. "Cybersecurity Guidebook for Cyber-Physical Vehicle Systems," 2016.

[12] A. Bouard, M. Graf and B. Burgkhardt, "Middleware-based security and privacy for in-car integration of third-party applications", 401:17-32,2013.

Analysis of spatio-temporal distribution characteristics of passenger travel behaviour based on online ride-sharing trajectory data

Xianlei Dong, Lingyu Wang, Beibei Hu *

School of Business, Shandong Normal University, Jinan, China. * Corresponding author tel: +86 13280027089

E-mail: sddongxianlei@163.com,1311009466@qq.com, *xizixinxiang@163.com

Abstract: With the rapid development of information technology, the travel mode of " Internet + Transportation " has brought convenience to residents while causing a series of urban traffic problems. Based on GPS data from Didi Chuxing, this paper relates the residents' travel behaviour to their daily activities, and studies the regularity of the pick-up and drop-off activities of residents from the dimensions of time and space. The results are as follows: Firstly, the daily variation trend of residents' travel is unbalanced. At the same time, the travel volume of residents is at a higher level during night. Secondly, the travel of urban residents is closely related to urban areas and urban functional structures. On the administrative scale, the distribution of residents' travel is concentric, and the farther the distance from the downtown area is, the smaller the number of orders is. On the functional scale, the hot functional areas of residents' travel are related to residents' needs. At last, mastering the distribution of residents' travel hot spots and time will help to understand the residents' travel mode and the current situation of urban traffic, and provide new ideas for solving traffic problems in big cities.

1. Introduction

At present, the related research on the characteristics of residents' travel behavior mainly starts from the following three aspects. Firstly, researches are based on traditional questionnaires. For example, small sample surveys use travel time, travel mode and commute distance to study the urban commuting behavior and reflect the home-work separation [1]. It constructs a model of resident travel to explore the intrinsic relationship between the spatial and temporal distribution characteristics of residents and their daily activities by focusing on the number of residents' travel trips, purposes, methods and time consumption [2].

Secondly, based on the taxi trajectory data, the passengers' pick-up and drop-off activities are identified. And some papers search for the regularity of spatiotemporal distribution of the pick-up and drop-off activities. For example, it analyzes the key factors that influence the spatiotemporal distribution of the pick-up and drop-off activities, and studies the relationship between the spatial and temporal dynamics of passengers and the urban facilities from the perspective of urban spatial dynamics [3]. Based on the pick-up and drop-off activities, scholars have also found the gathering roads and regions of residents' travel in different time periods to understand the residents' travel demand and traffic status in real time [4]. This paper validates the model of intra-urban human mobility by Monte Carlo simulation, which shows the geographical heterogeneity and distance decay effect improve each other when

Content from this work may be used under the terms of the Creative Commons Attribution 3.0 licence. Any further distribution of this work must maintain attribution to the author(s) and the title of the work, journal citation and DOI.
Published under licence by IOP Publishing Ltd

influencing human mobility models [5, 6]. This paper obtains the regularity of spatial and temporal distribution of passengers by clustering the trajectory data, which can help to reduce the cruise rate, and then alleviate the urban road congestion [7]. Thirdly, based on the geographical statistics method, the pick-up and drop-off activities of taxi passengers can be accurately matched with the geographical location to explore the spatial and temporal characteristics of residents' travel. Current researches visualize the aggregate region, and use POI data to reflect the difference of residents' travel characteristics in different functional areas [8]. Based on the spatial interactivity between each functional area from the aspects of time and distance, it shows that the time elasticity of commuting is the greatest. That is, time is more important to alleviate traffic problems.

In summary, previous studies have focuses on the behavior characteristics of residents' travel from using traditional questionnaires, or using traffic big data to study the relevance between the behavior characteristics of people's travel and urban functional areas, and so on. However, from the perspective of ride-hailing cars, few papers have targeted the correlation study of distinguishing the characteristics and regularities of residents' travel behaviors from time and space dimensions. Based on the ride-hailing cars and taxi trajectory data from the online ride-hailing platform, this paper researches the regularity of the pick-up and drop-off activities of residents from the dimensions of time and space, and reveals the flowing and traveling mechanism of residents within the urban area from a new perspective.

2. Data Source

We obtain the order trajectory data from Beijing online ride-hailing platform (Didi Chuxing). After data filtrating, 439,742 valid orders were obtained, including 66,186 taxi orders and 373,556 ride-hailing orders. Each order data contains 10 fields: Order ID, Driver ID, Vehicle ID (cryptographic), City Name, Order Start Latitude, Order End Latitude, Order Start Time, Order Finished Time, Product Line (Taxi or Ride-hailing), Itinerary ID (for the ride-hailing carpooling list, the Itinerary ID is the same.).

Geographically match the original order data. Firstly, the geographic information system (Arcgis) is used to divide the administrative division of the order's pick-up and drop-off activities of passengers. The order data is accurately matched to the vector map and the order data in each administrative area is counted. The original order data is geographically matched by spatial statistics method of Arcgis. Then, the order data is accurately matched to obtain the origin and destination address of orders. Finally, the address-matched order's origin and destination data are classified into the first-level industry of Baidu Map POI, that is, all the data will be categorized into 12 functional areas including shopping, traffic, education, financial, hotel, residential, catering, life service, culture, leisure, medical and government functional areas.

3. Results

3.1. Spatial Distribution of Passengers' Travel

Differences in geographic location and regional functions among different administrative districts lead to different spatial distribution of passenger travel. The distribution of orders (including the pick-up and drop-off activities of passengers) in 16 administrative districts of Beijing is shown in Figure 1. And one order includes a pick-up activity and a drop-off activity. We consider the distribution of average daily pick-up activities as the distribution of average daily orders.

ISPECE   IOP Publishing

IOP Conf. Series: Journal of Physics: Conf. Series **1187** (2019) 052055   doi:10.1088/1742-6596/1187/5/052055

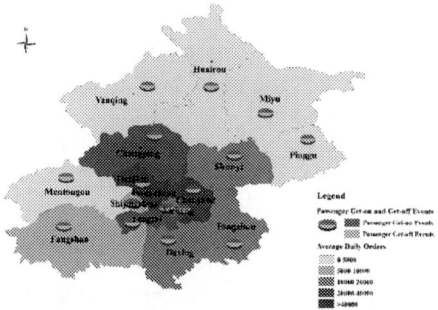

Figure.1 The average daily orders of taxis and ride-hailing and the proportion of the pick-up and drop-off activities of passengers in each administrative district (The proportion of the pick-up activities to drop-off activities is equal to one in any administrative district.).

From the distribution of the pick-up activities, the daily average order demand of various administrative districts in Beijing is quite different, as shown in Figure 1. Chaoyang District has the largest order volume. The daily average order volume accounts for 29.46% of the total order volume, respectively. Located in the central area of Beijing, Chaoyang District has regional advantages, relatively developed economy, a large number of permanent residents and large population density. which leads to the increasing demand for ride-hailing service. Yanqing County, Pinggu District, Fangshan District, Huairou District, Miyun County, and Mentougou District have relatively small daily orders,which are located in the remote suburbs of Beijing with underdeveloped economy, weak talent attraction and small density. As a result, the demand for ride-hailing service is small.

The distribution of the pick-up and drop-off activities of passengers is unbalanced. From Figure 1, we can know that the average number of the pick-up activities per day is significantly higher than that of the average number of the drop-off activities in Chaoyang District, Haidian District and Xicheng District. While in Changping District, Fengtai District, Shunyi District, Dongcheng District, the situation is to the contrary. The imbalance between the pick-up and drop-off activities in different administrative districts reflects the phenomenon of cross-regional mobility of passengers, as shown in Figure 2. From the proportion of cross-region orders to the total number of orders in this district, Dongcheng District has the highest proportion of cross-region orders on taxi(ride-hailing), and followed by Xicheng District, which are 86.09%(73.51%) and 79.88%(63.46%), respectively. The population policies of Dongcheng District and Xicheng District in recent years have led to the migration of the people to the surroundings and peripheral areas, resulting in residents' cross-district travels.

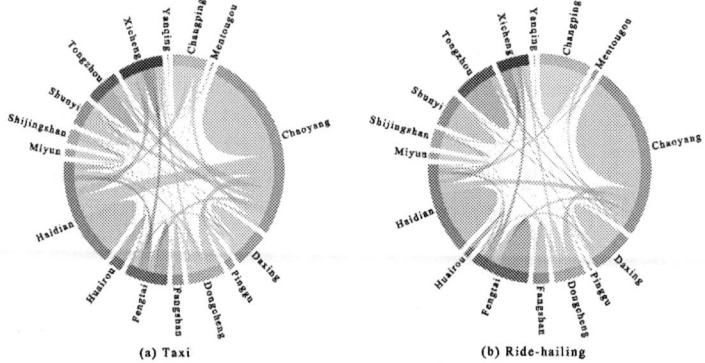

Figure. 2 Order flow status in different districts

From the flow of cross-regional orders, the cross-regional orders (including cross-regional inflow orders and outflow orders) in Chaoyang District are the most. The proportion of cross-regional outflow orders on the taxi(ride-hailing) to total cross-regional outflow orders is 29.53% (24.53%). And the proportion of cross-regional inflow orders is 22.97% (22.39%). Chaoyang District gathers wealthy

2423

scientific and technological innovation resources, attracting people's travel, employment and even life through industrial agglomeration. Under the background of relaxing the core functions of the non-capital, Chaoyang District's radiation-driven role is obvious. And the radiation diffusion of high-tech industries and their achievements has strengthened people's inter-regional mobility.

It is noteworthy that passengers' travel behavior in the six districts of Beijing are closely related, mainly in the case that the cross-regional orders from Dongcheng District or Xicheng District to the other five administrative districts account for more than 92% of the total cross-regional orders in the district. The number of cross-regional orders from Chaoyang District, Haidian District or Fengtai District to the other five administrative districts accounted for more than 68% of their total cross-regional orders. First of all, Geographically, the City Six District is located in the center of Beijing, with closed geographical location. Secondly, the City Six District, a central carrying area of Beijing's capital function, is a center for national politics, culture, international exchanges and scientific and technological innovations of the whole country, and the exchange among the regions is frequent, which makes the travel behavior of residents closely linked in City Six District.

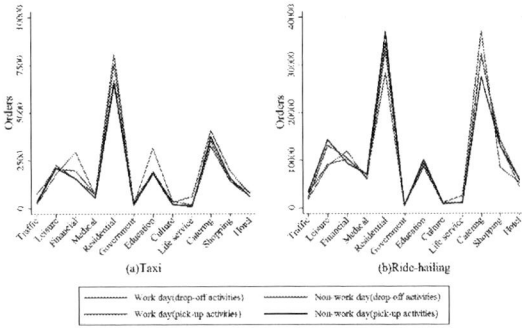

Figure3. The daily average quantity of the pick-up and drop-off activities on taxi and ride-hailing for each function area.

The specific geographic locations of the pick-up and drop-off activities are classified into 12 functional areas to further analyze the characteristics of residents' travel, as shown in Figure 3. The number of the pick-up and drop-off activities in the residential and food function areas is relatively high, respectively accounting for 34.17% (26.41%) and 17.19% (25.50%) of the daily average orders. These two types of functional areas are densely populated and have a large flow of people. As hot spots for orders (including the pick-up and drop-off activities), they reflect the daily demand of residents to a certain extent. The number of orders in cultural and life service functional areas are relatively small, which account for 4.63% (0.70%) and 1.14% (1.10%) of the daily average orders. On the one hand, these functional areas are relatively less distributed in cities. On the other hand, they are also affected by geographic locations, such as culture functional areas. Most of the culture functional areas are located in the city center, and the roads are crowded. People mostly choose to travel by public transportation.

3.2. Time Distribution of Passenger's Travel
In the following, we compare the time distribution of the pick-up and drop-off activities of passengers for taxi and ride-hailing orders, and then study the rule of time distribution of passengers' travel behavior. Figure 4 shows the daily changes of the proportion of taxi and ride-hailing orders per hour to daily average orders at different time.

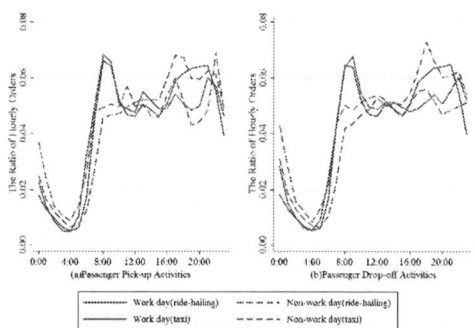

Figure4. The proportion of the pick-up and drop-off activities of taxi and ride-hailing passengers' orders to daily average orders at different time.

Overall, the proportion of daily changes in the pick-up and drop-off activities on the taxi and ride-hailing are roughly the same. On work days, the morning and evening rush hours of the pick-up and drop-off activities of taxi and ride-hailing passengers' orders respectively appear at 7:00-9:00 and 16:00-18:00. And the morning rush hours on non-work days is postponed. Specifically, on non-work days, the proportion of the pick-up and drop-off activities on the taxi per hour is 6.92% and 4.98% at 22:00.

From the perspective of taxis and ride-hailing, the order proportion is roughly the same in the morning rush hours. While in the evening rush hours, the proportion on ride-hailing per hour is higher than that on the taxi. In the evening rush hours, the travel demand of passengers increases, which led to the supply less than demand for the taxi. And compared to taxi, ride-hailing has a price advantage. After 22:00, the proportion of orders on the taxi per hour is higher than that of ride-hailing. This indicates that residents have higher trust in taxi than on ride-hailing at night to a certain extent. And some ride-hailing drivers are part-time drivers so they offer fewer night-time services.

3.3. Time Distribution of Passengers Traveling in Different Functional Areas

Figure 5 shows the changes of the pick-up and drop-off activities of passengers on the taxi and ride-hailing to daily average orders in different functional areas. During the period of 0: 00-6: 00, the proportion of orders per hour in 12 functional areas is at a low level, and the lowest proportion appears at around 4:00. In other time periods, due to the difference in functional area positioning and passenger demand, the orders of different functional areas have a distinct difference in daily changes.

The daily changes in the proportion of the pick-up and drop-off activities of passengers in the leisure, shopping, food and hotel areas are roughly the same on work days. During the morning rush hours from 7:00 to 9:00, the proportion of the pick-up activities on the taxi and ride-hailing is higher than that in the evening rush hours. And in the evening rush hours from 17:00 to 19:00, the proportion is opposite to the morning rush hours. The proportion of passengers on the taxi and ride-hailing is comparatively high from 20:00 to 22:00. In leisure areas, the proportion of the pick-up activities of taxis' passengers is 21.57% on non-work days, which is 8.51% higher than in the morning rush hours. It shows that residents have a lot of leisure activities at night, while the reduction of public transportation makes the demand for taxis and ride-hailing increase.

Government, medical and financial areas are mainly business oriented. The proportion of the drop-off activities on ride-hailing is larger than that on the taxi, and the proportion of the pick-up activities of passengers is opposite during morning rush hours. In the government area, the total proportion of the drop-off activities on ride-hailing is 25.53% during the morning rush hours on work days, while the proportion of taxi is 16.56%. The cost performance of the ride-hailing is higher than that of taxi, as a result, residents are more willing to choose the ride-hailing. Compared with government departments, the overtime work in the financial and medical functional areas is serious, which leads that the government area's evening rush hours are more obvious than the other two areas.

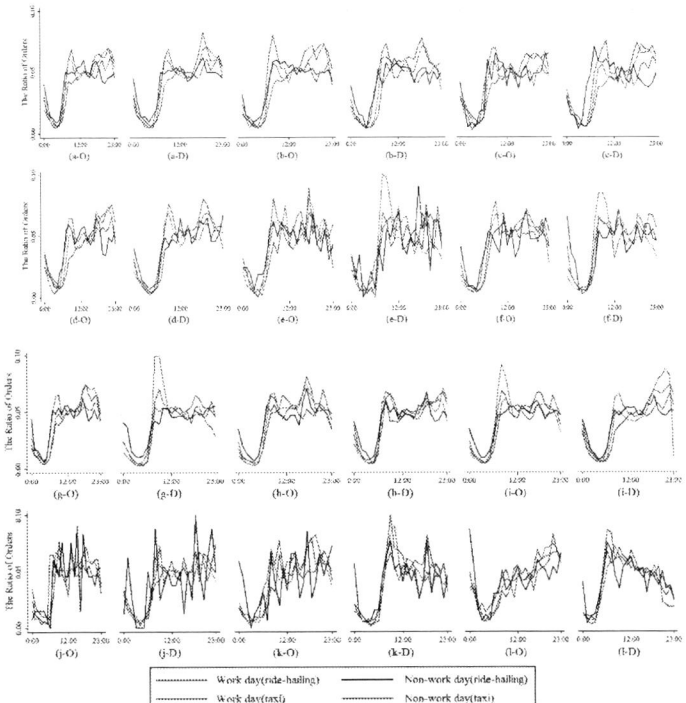

Figure5. Daily changes of the proportion of the pick-up and drop-off activities per hour to daily average orders in different functional areas: (a-O)-(l-O) and (a-D)-(l-D) show the situation of passenger pick-up and drop-off activities on catering, shopping, hotel, leisure, government, medical, financial, education, residential, life service, traffic functional areas.

The daily changes of the proportion of the pick-up and drop-off activities in the education area are similar to that in the residential area. From the difference in the proportion of the pick-up and drop-off activities of passengers, the proportion of the pick-up activities in the residential area is large. Take one as an example; the proportion of the pick-up activities on ride-hailing in the residential area is as high as 9.17%, while the education functional area is only 6.44%. From 10:00 to 16:00, the changes of the pick-up and drop-off activities are stable, and the proportion of the pick-up and drop-off activities of passengers is roughly the same, and the ratio was stable at around 5%. During this period, most of the residents go out to work or study, and the flow of people is small.

In life service areas, due to the diverse purposes of residents' travel, the proportion of the pick-up and drop-off of passengers on the taxi fluctuates greatly on non-working days. For example, the proportion of the pick-up activities on the taxi fluctuates within the range of 1.52%-8.33% during the period from 14:00 to 18:00. The proportion of the pick-up activities in the culture area show a steady growth trend after 6:00, and the daily changes are relatively random. To take one example, on non-work days, the proportion of the pick-up activities on the taxi is 0.66% at 7:00, while the proportion surges up to 6.56% at 8:00. Compared with other functional areas, the travel demand of residents in the culture area is not an essential requirement, so the daily changes are random and not stable.

From the perspective of the proportion of the pick-up and drop-off activities, except for the period from 0:00 to 8:00, the proportion of the pick-up activities shows an increasing trend and the proportion of the drop-off activities is on a downward trend. The proportion of the pick-up activities of in the morning rush hours is higher than that of the drop-off activities. The proportion of the drop-off activities is higher in the evening rush hours. During the rush hours in Beijing, the roads are crowded. Most of the residents go to work by subway or other public transport modes. However, due to the long commuting distance and time, residents often transfer in the transportation facilities, using combined transportation.

4. Conclusions

In this paper, based on the trajectory data, we analyze the travel characteristics of residents in different districts and different functional areas of Beijing from the perspective of time and space by using the spatial statistics method of GIS and address matching algorithm. Our conclusions are as follows:

From the perspective of the administrative district, the spatial distribution of residents' travel is concentric, mainly showing that the farther away from the city center, the smaller the number of orders. Driven by the city's functional positioning in different administrative regions, there is a spatial stratified heterogeneity of distribution of orders, which leads to the phenomenon of cross-regional travel. Cross-regional mobility mainly occurs in the City Six District. Similar to the administrative district, the distribution of orders among functional areas are unbalanced because of the various travel purposes of residents. Relevant departments should further adopt reasonable allocation of transportation resources, advocate green travel concepts or other measures to alleviate traffic congestion.

On the whole, daily changes of orders are unbalanced and exist at the obvious morning and evening rush hours. The quantity of order is at a high level at night, and the variation of the proportion of the drop-off activities lags slightly behind the pick-up activities. First of all, the distribution of orders in the morning rush hours on work days is more obvious than that on non-work days, while the distribution of orders in the evening rush hours is the opposite. Secondly, the orders in different functional areas show different time distribution characteristics. For example, in the morning rush hours on non-work days, the proportion of orders per hour in education and residential functional areas are larger than leisure, food, shopping and hotel areas. Researching the distribution of residents' travel in different functional areas is an important reference for solving traffic congestion problems during rush hours and achieving the fine traffic management.

Due to the deficiency of data and methods, this paper does contain some shortcomings needing to be improved. On the one hand, only using the GPS data can't completely analyze the characteristics of the residents' travel behavior. On the other hand, we have not taken into consideration the situation of mixed land. In future research, we will use the multi-source data to analyze the characteristics of residents' travel behavior, and adopt a more sophisticated quantitative method to study the characteristics of residents' travel behavior.

Acknowledgements

This study was supported by the programs of National Social Science Foundation of China (Grant: 16CJY056).

References

[1] Liu D, Tongyan Q I, Ke Z, et al. J 2009 Beijing Residents' Travel Time Survey in Small Samples. *Journal of Transportation Systems Engineering & Information Technology*, 9(2):23-26.

[2] ZHOU Suhong, DENG Lifang. J 2010 Spatio-temporal Pattern of Residents' Daily Activities Based on T-GIS: A Case Study in Guangzhou, China.*Acta Geographica Sinica*, 65(12): 1454-1463.

[3] Jiansheng W U, Bo L I, Huang X. J 2017 Spatio-temporal Dynamics and Driving Mechanisms of Resident Trip in Small Cities. *Journal of Geo-Information Science*, 19(2):176-184 .

[4] Han Y, Wang W.J 2018 On the Spatial and Temporal Distribution of Resident Trip Based on Taxi GPS Data. *Geomatics & Spatial Information Technology*, (2):87-89.

[5] Liu Y, Kang C, Gao S, et al. J 2012 Understanding intra-urban trip patterns from taxi trajectory data.*Journal of Geographical Systems*, 14(4):463-483.

[6] Liu X. J 2016 Inferring trip purposes and uncovering travel patterns from taxi trajectory data. *Cartography & Geographic Information Science*, 43(2):103-114.

[7] Qin K, Zhou Q, Wu T, et al. J 2017 Hotspots Detection from Trajectory Data Based on Spatiotemporal Data Field Clustering . *ISPRS - International Archives of the Photogrammetry, Remote Sensing and Spatial Information Sciences*, XLII-2/W7:1319-1325.

[8] Duan Y M, Liu Y, Liu X H, et al. J 2018 Identification of Polycentric Urban Structure of Central

Chongqing Using Points of Interest Big Data. *Journal of Natural Resources*, 33(5):788-800.

Research on road traffic flow status based on survival analysis

Beibei Hu, Doudou Lin, Qibo Sun, Xianlei Dong[*]

School of Business, Shandong Normal University, Jinan, China * Corresponding author tel: +86 13280029639

E-mail:xizixinxiang@163.com, 1193535083@qq.com, 1003667231@qq.com, [*]sddongxianlei@163.com

Abstract: Based on the data of bayonet and floating car, this paper studies the traffic flow state of the main roads in Shenzhen, analyses the relationship between speed and flow, and establishes a speed-flow model. It studies the characteristics of traffic flow in Shenzhen and the overall distribution of congestion duration. The results demonstrate that there is a strong correlation between flow and speed, which have obvious characteristics of morning and evening peak within a day, and the flow and speed show significant differences between work days and non-work days. As for the congestion duration of each road, they are generally under 16 minutes. Specifically, when the congestion duration is less than 8 minutes, there is a high probability that the congestion state ends; while the congestion lasts more than 10 minutes, the congestion status is less likely to end in a short time. In view of the above analysis, it is recommended that relevant departments strengthen monitoring and control of real-time road traffic so that the traffic environment in Shenzhen can be improved.

1. Introduction

Along with the advancement of urbanization, traffic problems have become one of the major problems of all cities around the world. By the end of 2017, China's motor vehicle ownership reached 310 million. The rapid growth of vehicle ownership has improved people's travel efficiency and happiness as well as causes the traffic congestion and environmental pollution [1], which increases people's travel costs [2], and blocks the coordinated operation of urban functions. Therefore, real-time monitoring of road traffic flow and research on the distribution of road congestion are of great significance for traffic management and urban function realization.

The urban transportation systems are complex and non-linear. In addition to the traffic flow and road traffic capacity, the traffic conditions of urban roads are also affected by pedestrian crossings, traffic lights and vehicle lane changes, etc. With the development of network information technology, the intelligent transportation system makes real-time monitoring of road traffic status come true, and it is possible to obtain real-time road condition information and predict road status in real time [3]. In recent years, domestic and foreign scholars have done a lot of valuable research on road traffic flow, which mainly includes the theoretical description of road traffic flow state and model construction, as well as prediction of congestion status. For example, based on the traffic flow measurement data, they divide the free flow, steady flow and crowded flow, then establish flow-speed model, and analyze the relationship between speed, flow and density [4]; They employ vehicle speed and flow data to build based on the time-space flow-speed model and traffic flow prediction model, and study the relationship between highway traffic flow parameters [5]; they use radial basis function neural network method to collect traffic characteristics such as pedestrian characteristics, traffic flow, traffic

Content from this work may be used under the terms of the Creative Commons Attribution 3.0 licence. Any further distribution of this work must maintain attribution to the author(s) and the title of the work, journal citation and DOI.
Published under licence by IOP Publishing Ltd

facilities, traffic situation, traffic sociality, etc. Based on the urban congestion theory model of traffic 5S elements, they evaluate urban traffic congestion [6]; on the basis of the overall change law of urban road traffic congestion index, they establish a K-nearest urban road traffic congestion index prediction model to achieve short and medium term prediction of traffic congestion index [7]; a system dynamics model of urban road congestion is established to seek solutions to traffic congestion problems from the perspective of residents travel and urban logistics [8].

In summary, the previous researches on the state of road traffic flow focus more on the construction of traffic flow model and traffic congestion prediction model, while there is a small number of studies which use traffic flow big data to analyze the traffic flow state and the overall distribution of congestion law in specific urban roads. In this paper, taking specific roads as an example, we use urban road data to construct a speed-flow model. and explore the distribution law of each road congestion duration. Then we provide theoretical and methodological support for traffic management and planning.

2. Material and Methods

2.1. Data Source and Pretreatment
This research obtains the bayonet data and floating car data of 4 roads in Futian District, Shenzhen from March 25 to March 31, 2018. The bayonet data come from the four monitoring points of the Xinzhou interchange on Beihuan Road, Fulong Road'Henglongshan Entrance, Xiangmihu Road'Municipal Party School, Jingtian Road Jingmi Village Park. Floating car data are mainly collected from GPS trajectory data of vehicles such as taxis, dump truck, heavy-duty semi-trailers and coach car, etc. After the invalid data and the duplicate data are deleted, we get a total of 1,528,672 bayonet data and 1,834,340 floating car data. The floating car data includes six fields, which include the timestamp corresponding to the trajectory, the license plate number after desensitization, the longitude coordinates, the latitude coordinates, the instantaneous speed and satellite speed.

Next, it is about the further pretreatment of the raw data. Firstly, the timestamp field in the floating car trajectory data is converted. And then the Java program is employed to call the Aliyun Map to match the latitude and longitude of the floating car data, for obtaining the road segment where the vehicle is located. Finally, the Beihuan Road, Fulong Road, Xiangmihu Road and Jingtian Road are screened out from floating car data as the main research objects for empirical analysis.

2.2. Methodology
Based on the traffic measured data, we use the survival analysis method to study the distribution law of road congestion duration. The specific measurement steps are as follows:

（1）Computing survival time. The survival time of road congestion refers to the duration of the congestion. The floating car data are divided into two minutes as a time unit per day. According to the congestion status threshold of different types of roads, we make a judgment on the congestion status of each road in each time unit. The time unit of continuous congestion state is recorded as a congestion sample. And the sample numbers under different survival time are recorded.

（2）Computing survival function. $S(t)$ denotes the survival function, which indicates the probability that the duration of the road congestion state is longer than t. As $S(t)$ increases, the possibility of the road continuing to congested is greater. In this paper, we use the non-parametric method of kaplan-meier to estimate the survival function of a congestion, as shown in Equation (1):

$$S(t) = P(T > t) = \prod_{t_i \leq t} \frac{n_i - d_i}{n_i} \tag{1}$$

Here, T is the actual congestion duration of the road, n denotes the number of time samples for all congestion, d_i refers to the number of samples that end congestion in t_i. n_i is the number of samples that are still in a congested state at t_i ($t_1 < t_2 < \cdots < t_k, i = 1, 2, 3 \cdots k$).

(3) Computing risk function. The risk function refers to the probability that the congestion state will end instantaneously within a short time Δt after the time t has elapsed. The larger the function value is, the greater possibility the congestion ends, and conversely, the congestion continues. The risk function $h(t)$ is shown in Equation (2):

$$h(t) = \lim_{\Delta t \to 0} \frac{P(t \leq T < t + \Delta t \mid T \geq t)}{\Delta t} \tag{2}$$

In which, T denotes the road congestion duration, and P is the probability that the congestion sample will end the congestion state instantly within a short time Δt after the time t has elapsed.

3. Results

The traffic flow status of roads is mainly affected by three parameters: flow, speed and density. This paper uses Greenhills' speed-flow model [9] $Q=VK$ to analyze the relationship between speed and flow, wherein Q denotes the flow, V denotes the speed, and K refers to the density. Greenhills' theory believes that the vehicle speed is higher when the flow of the road is not saturated, the mutual interference among vehicles is slight, and the traffic flow is in free flow. While the road is in the peak period, the traffic flow increases sharply, the mutual interference increases, and the vehicle speed decreases gradually. Finally, the free flow state will be destroyed. Even in severe cases, the vehicle speed will drop to zero.

3.1. the Research on Speed and Flow Characteristics

The daily traffic flow on work days and non-work days on the four roads, such as Beihuan Road, Fulong Road, Xiangmihu Road and Jingtian Road, is shown in Figure 1. It is noticeable that on the work days and non-work days, the daily traffic flow distribution of four roads is almost same, which demonstrates "U" curve during the period from 0:00 to 8:00, and the minimum value of traffic flow appears at around 4:00. Besides, the morning peak is mainly at 7:00-9:00, the evening peak at 17:00-19:00. Compared with the work days, the peak period on the non-work days is delayed one hour. For example, the morning peak of the work days on Beihuan Road appears at 7:00, while the non-work days appears at 8:00.

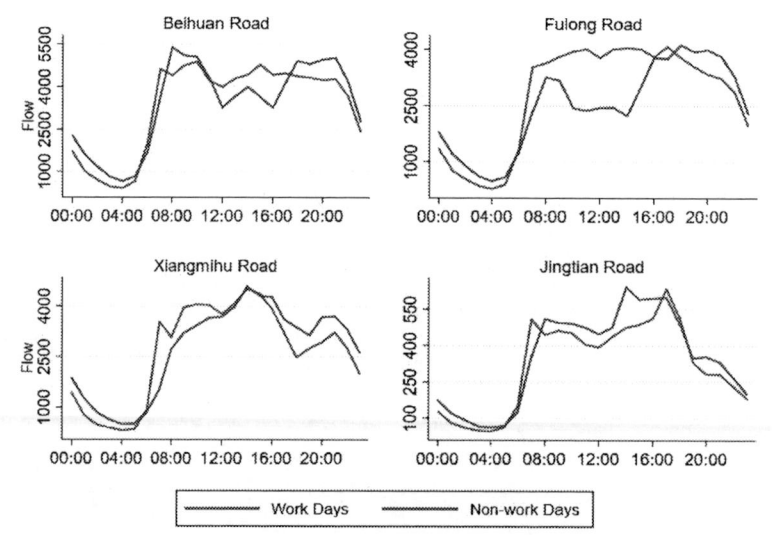

Figure 1 the Daily Traffic Flow of Each Road on Work day and Non-work day.

In Figure 1, the largest and least traffic volume is Beihuan Road and Jingtian Road respectively. The Beihuan Road, as a city expressway, undertakes the functionality of the city's large capacity, long distance and fast traffic, and the traffic is smooth and the traffic volume is relatively high.

Correspondingly, the Jingtian Road is a subordinate road, connecting with main roads and branches, and the flow is relatively low. In addition, the amount of traffic flow at the morning and evening peak of the four roads between work days and non-work days is different. For example, in the morning and evening peak on Beihuan Road, the traffic flow on non-work days is 7.1% higher than that on work days. The reasons are that the number of tourists on non-work days increases, and the limited vehicle on the work days results in relatively high traffic on non-work days.

Next, the speed of the four roads in one week is shown in Figure 2. On a whole, the speed curves of the four roads all show a "W" shape. The period of low speed is mainly in the morning and evening peaks(7:00-9:00,17:00-19:00), while the high speed is mainly from 0:00 to 6:00.

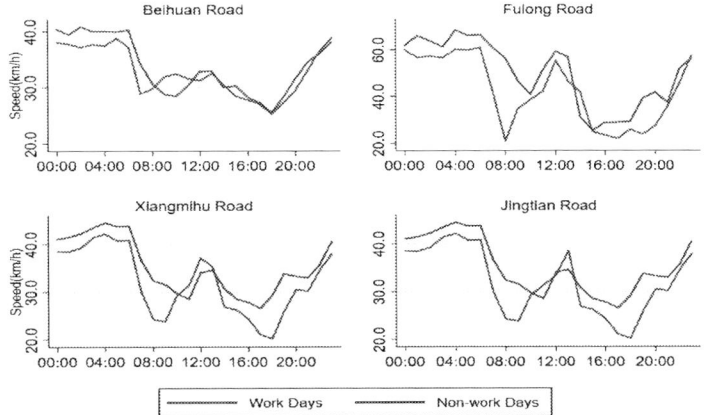

Figure 2 the Speed of Each Road on Work days and Non-work days.

As shown in Figure 2, due to the decreasing commuting vehicles during non-work days, the speed of each road on non-work days is higher than work days, especially in the peak period on Fulong Road, Xiangmihu Road and Jingtian Road. Generally, the better the traffic conditions, the higher speed is. On Beihuan Road, the difference of speed between work days and non-work days is the smallest, and speed curves on work days and non-work days are roughly consistent.

3.2. the Speed-Flow Model

Next, we fit the traffic flow and vehicle speed data of four roads and construct the speed-flow model. The formula is shown in Equation (3):

$$V = aQ^2 + bQ + c \tag{3}$$

The fitting effect of each road is shown in Table 1.

Table 1 the Speed-Flow Expression and R-squared of Four Roads.

Road Name	the Speed-Flow Expression	R-squared
Beihuan Road	$V = -0.249Q^2 - 3.458Q + 33.42$	0.4816
Fulong Road	$V = -5.702Q^2 - 12.25Q + 50.21$	0.4377
Xiangmihu Road	$V = 1.427Q^2 - 4.24Q + 31.63$	0.4262
Jingtian Road	$V = -0.5829Q^2 - 5.41Q + 33.72$	0.5447

The fitting curves of each road are shown in Figure 3.

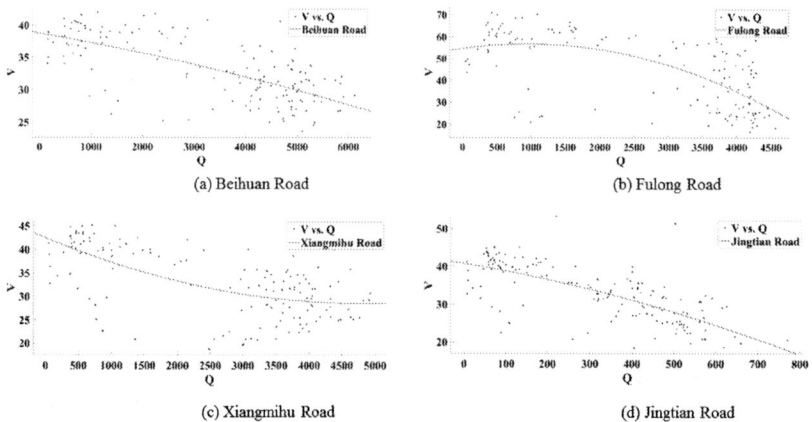

(a) Beihuan Road   (b) Fulong Road

(c) Xiangmihu Road   (d) Jingtian Road

Figure 3 the Relationship Between Speed and Flow of Each Road.

According to Figure 3, as the flow increases, the speed shows a declining trend, which indicates a negative correlation, except the speed-flow curve of Fulong Road on the early stage, which is positively correlated (as shown in Fig3(b)). On the early stage of Fulong Road, as traffic flow increases, the speed rising slowly, and there is a short period of a flat peak which indicates that the mutual interference between vehicle speed and flow is weak. The tendency of Beihuan Road, Jingtian Road and the state of the remaining part of Fulong Road are roughly same, and illustrate that at the primary step the traffic flow increases, the speed of the vehicle decreases steadily; when the flow continues to increase, the vehicle speed begins to drop sharply. By contrast, the speed-flow fitting curve of Xiangmihu Road is in a downward convex state, which illustrates as traffic flow increase gradually, the speed of the vehicle always decreases slowly and reaches a stable state. At this moment, the correlation between speed and flow is obscure.

3.3. the Distribution of Road Congestion Duration

According to the construction standard of Shenzhen Road Traffic Operation Index, congestion status is defined that the expressway speed (such as Beihuan Road, Fulong Road and Xiangmihu Road) is less than 31.5km/h and the subordinate road speed (such as Jingtian Road) is less than 25km/h. In this paper, we take two minutes as the shortest time for congestion persistence, and analyze the changes in the congestion status of the four roads in a week, and count the number of samples under different congestion durations, as shown in Table 2.

Table 2 the Number and Proportion of Samples for Congestion Duration.

Road Name	the Number of Samples With Congestion Duration <16 minutes	Sample Ratio	the Number of Samples With Congestion Duration >16 minutes	Ratio
Beihuan Road	302	87.5%	43	12.5%
Fulong Road	149	77.6%	43	22.3%
Xiangmihu Road	312	100%	0	0
Jingtian Road	142	97.9%	2	2.1%

It can be seen from Table 2 that the congestion duration is mostly under 16 minutes. To make the research more universal, only the samples of congestion duration less than 16 minutes are discussed, thus the survival function and risk function of the congestion are estimated, as shown in Figure 4. Overall, when the congestion occurs, as the duration of congestion increases, the survival curve shows a downward trend. Among them, the survival curve has a large decline in the interval [0,8], and then shows a gentle trend, which indicates that the probability of ending the congestion state is relatively large. The overall trend of the risk curve firstly rises and then falls. The specific performance is that when the survival time is within two minutes, the risk rate of congestion increases sharply, and the probability of the end of the congestion state is greater. Two minutes later, the risk curves of four roads show a downward trend, and at the point of 10 minutes gradually tend to be gentle, indicating that the probability of ending congestion is gradually decreasing.

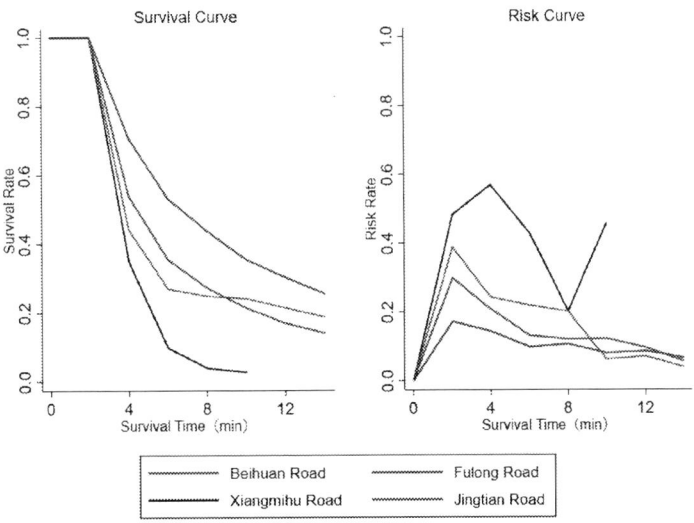

Figure 4 Survival Curve and Risk Curve of Each Road.

For the survival curve, the congestion survival rate of Fulong Road is obviously higher than the other three roads. Specifically, the survival rate of Fulong Road is the highest, and the survival rate of Xiangmihu Road is the lowest, indicating that Fulong Road has the highest probability of continuing to remain congested. Namely, Fulong Road has more serious congestion state. However, there is no significant difference in the congestion status between the Beihuan Road and Jingtian Road. For the risk curve, the risk curve of Xiangmihu Road fluctuates greatly, and the risk of congestion remains at a high state. It shows that the congestion state of Xiangmihu Road has a higher probability to end under different congestion time, and the road has better ability to dissipate congestion. In addition, the rate of risk on Beihuan Road, Fulong Road and Jingtian Road fluctuate significantly, and the difference among the three roads is unobvious, and the probability of the end of the congestion state is low.

4. Conclusions

Taking Shenzhen as an example, this paper uses the bayonet passing data and floating car data of four main roads in Futian District to establish a speed-flow model, and analyzes the state of traffic flow in Shenzhen, and uses survival analysis method to study the distribution of congestion duration. The conclusions are as follows:

The traffic flow and speed of the four main roads in Futian District of Shenzhen City have obvious morning and evening peak characteristics, whose extent between work days and non-work days is different. Firstly, the flow of each road maintain at a high level in the morning peak, and then the downward trend occurs till the end of the evening peak, and the total daily traffic flow on non-work days is higher than work days. Then, the speed has a "W" fluctuation in one day, and the morning and evening peak characteristics are very prominent, which in non-work days is obviously higher than the work days. Next, the speed-flow model shows that there is a strong correlation between speed and flow, while due to different road type and function, as the flow increases, there is a different change in speed, and the condition of flow switch between different states.

For the congestion phenomenon, under different congestion survival times, the probability of four roads continue to maintain and end the congestion state is different. The survival curve indicates that when the congestion time increases, the probability of the road maintaining the congestion state is small. Fulong Road has the highest survival rate, which means the possibility of continuing to maintain congestion is the highest, and the ability to relieve congestion is poor. While the risk rate curve is hump-like, and the risk rate fluctuates significantly under different congestion time. When the congestion time is long, the probability of ending the congestion state is low.

The study of the traffic flow state is of great significance for road construction and traffic control. In order to improve the traffic situation in Shenzhen, it is recommended that relevant departments strengthen the monitoring of road conditions and enhance road reconciliation guidance and actively adjust the restrictive policies to control traffic flow during the morning and evening peak periods. It is also advised that they should improve urban intelligent transportation systems and strengthen the collection of real-time data to provide more accurate travel recommendations for residents, such as choosing the right time to go out, and from the perspective of prevention and control to alleviate road pressure and improve urban traffic, and so on.

Acknowledgments
This study was supported by the programs of National Social Science Foundation of China (Grant: 16CJY056).

References
[1]   Huang W, Fan H, Qiu Y, et al. 2016 Causation mechanism analysis for haze pollution related to vehicle emission in Guangzhou, China by employing the fault tree approach. J. *Chemosphere*, 151:9-16.
[2]   Yang Z W, Liu X M. 2012 The Effect of Urban Travel Cost on Travel Structure. J. *Journal of Transportation Systems Engineering & Information Technology*, 12(2):21-26.
[3]   Chmiel W, Dziech A, Ernst S, et al. 2016 INSIGMA: an intelligent transportation system for urban mobility enhancement. J. *Multimedia Tools & Applications*, 75(17):10529-10560.
[4]   Xu C, Chen X M, Coltd T T. 2014 Analysis of Characteristics of Traffic Flow Speed-density Relation Model. J. *Journal of Highway & Transportation Research & Development*, 31(2):114-120.
[5]   Lv Y, Duan Y, Kang W, et al. 2015 Traffic Flow Prediction With Big Data: A Deep Learning Approach. J. *IEEE Transactions on Intelligent Transportation Systems*, 16(2):865-873.
[6]   Xiong L, Yang S F, Zhang Y, et al. 2018 Research on the Urban Traffic Congestion Evaluation Model Based on 5S Theory. J. *Operations Research & Management Science*, (1):117-124.
[7]   Wei Q B, He Z C, Zheng X S, et al. 2017 Prediction of Urban Traffic Performance Index Considering Multiple Factors. J. *Journal of Transportation Systems Engineering & Information Technology*, 17(1):74-81.
[8]   Yang H X, Li J D, Zhang H, et al. 2014 Research on the governance of urban traffic jam based on system dynamics. J. *Systems Engineering-Theory & Practice*, 34(8):2135-2143.
[9]   Chang Y L, Wang X T, Zhang P. 2016 Cosine Function Speed-flow Relationship Model for Road Segment Basesectiond on Upstream Arrival Traffic Flow. J. *Science Technology & Engineering*, 16(11):257-261.

Research on the Management and Maintenance of Infrastructures in Fog Section of Motorway based on the MOT Model

LI X. L.[1,3], SONG Q. H.[1,2]*, TANG B.M.[3], LI C.[2]

[1]Chongqing Institute of Geology and Mineral Resources, Chongqing, 400042, China

[2]Army Logistics University, Chongqing Key Laboratory of Geomechanics & Geoenvironment Protection, Chongqing, 401311, China

[3]School of Civil Engineering, Chongqing Jiaotong University, Chongqing, 400074, China;

*Corresponding author: songbook@163.com

Abstract. Nowadays, a crop of prominent problems like road aging, rebuilding or extending, equipment maintenance and technology upgrading are faced in the management and maintenance of Chinese motorways. With multiple advantages, the infrastructure management and maintenance based on public-private partnership (PPP) model is extensively used at home and abroad. From the perspective of road transportation safety management, this paper discusses the maintenance of infrastructures in state-run motorways with maintain, operate and transfer (MOT) model, analyses its features, and finds problems that should be noticed in implementation plan, implementation steps, and the operating process of MOT project.

1. Introduction

Being a kind of special section in motorways, there is heavy fog in fog section in some periods of time because of elements like topography, climate and road environment, which may even cause road traffic. Road traffic safety involves people, vehicles and traffic environment. Among which, traffic environment refers to the traffic and transport environment which drivers, passengers and vehicles are in, including the management and maintenance of road. The management and maintenance of traffic environment have large influence on driving safety and the service efficiency of motorways. Let alone the management and maintenance of fog section motorways. In fog environment, the visual range of drivers is limited. If necessary road facilities are damaged at this moment, the coupled superposition may increase the probability of traffic accident. Always locating in intensive sections like mountainous areas, bridges and tunnels, fog sections require higher maintenance cost and quality requirement on facilities because of complex road condition and easily and frequently damaged facilities. With high maintenance requirement and difficult maintenance technology, road sections with high-tech facilities like road meteorological watch system and fog section inducible system need to be maintained by professional companies. Therefore, it is necessary to make in-depth research on maintenance of facilities in fog sections.

The PPP (Public-Private-Partnership-a cooperation model of private enterprises and state-owned enterprises) model has generalized and narrowed meanings[1]. The generalized meaning is the form of cooperation between government and (profit or non- profit) organizations on certain project, including BOT (Build, Operate and Transfer) model, TOT (Transfer, Operate and Transfer) model, BOO (Build, Own and Operate) model, BLT (Build, Lease and Transfer) model and so on. The narrowed meaning is the model that government and private departments create SPV, which is a kind of special institution, built to introduce social capital, do exploitation, and assumes risks jointly.

In 1992, England firstly applied PPP model to manage public facilities (such as: highway, railway, airport, port, hospital, and jail etc.) and other public immovable properties. There are three stages during the development and perfection process, which are privatization stage, contract signing and contracting stage, and private investment encouraging stage. The research on PPP in China starts a little later. However, certain progress has been got. For example, Chen [2] analysed the connotation and matters needing attention during application; Sha [3] objectively analysed opportunities and challenges of PPP model in China. Deng et al. [4]analysed the PPP model risk sharing principle and the flow of this principle. Lu[5]made deep analysis on PPP model and the development of urban public utilities in China.

There are different attitudes on PPP model in educational circles: Bloomfield et al. [6], Hall[7], Hodge [8], Gaffney et al. [9], Walker [10], Jean et al. [11] keep critical attitude on PPP model, thinking that PPP model does not achieve the effect on public infrastructure required by government in the capital saving. Among which, Hall believes private capital neither eases government's pressure in financial budget nor provides more infrastructure construction. However, Pollitt [12] keep positive attitude on PPP model. He studied the experience of PPP model in England, pointed out that PPP model in England was able to manage and perform technical improvement of engineering project effectively, and effectively transfer risks of project. The joint management scheme with tradable credits and road tolls for PPP networks was proposed [13]. A transformed first-price sealed-bid auction with independent private values to determine the equilibrium royalties and subsidies in PPP was presented [14].

However, as a whole, most government departments support PPP model. Among which, to confirm the positive and significant function of PFI, England launched PFI: Strengthening Long-term Partnerships in 2006. In 1990 a, countries like Chile, Portugal and Brazil applied this model, which decreased government's financial burden, in large scale public infrastructure successively. Therefore, there is much research and application value of PPP model in large scale governmental infrastructure.

Since 2013, China has reinforced the construction of PPP project, inspiring a new round of PPP construction trend in China. PPP project is performed from rail traffic and road traffic construction to maintenance and management in later period. However, the operation procedures of PPP project are different in road traffic construction period and management and maintenance period. Huang et al. [15] studied PPP project in aspects like financing, investment, stock rights, and project management from economic and management perspectives. From highway traffic safety and traffic service quality perspectives, there are relative few researches on management and maintenance of motorways in China with PPP operation model.

It is not difficult to find that, there are multiple PPP project models with different applicability. The cooperation features of transactions with different models are quite different. Therefore, based on narrowed PPP model, this paper comes up to manage and maintain fog section in motorways with MOT model (maintain, operate and transfer), and discusses the implementation scheme and steps of this model.

2. Main Problems in Management and Maintenance

Main problems in management and maintenance of motorways in current China:

(1) Different maintenance requirements and standards. During the 15 years from 2000, the motorway mileage increased 33.2% every year. With different aging and damaging degree of infrastructures, every motorway has different maintenance requirements and standards. As the increasing of mileage and increasing of construction technology and technological level, old motorways and new ones are maintained by quite different methods, which results in different

maintenance expenditure.

(2) A part of motorways is badly in need of being remoulded and upgraded. Motorways develop along with the development of economy and technology. Along with China's economic system reform and the speeding up of urbanization, with a series prominent problems like aging road surface, scrapped facilities and unreasonable or unscientific design, many motorways which are built from 1970s to 1980s cannot satisfy current traffic flow and need. For example, some road sections are designed without considering the influence of atrocious weather like heavy fog, ice, strong wind, and high temperature. Traffic safety problems, which are caused by unreasonable designs in long downgrade, tunnel, truck escape ramp, ramp and curve, become more serious along with the increasing of motor vehicle holding quantity in China. The maintenance and upgrade of infrastructure in motorways become more and more urgent.

(3) The pressure of managing and maintaining motorways becomes larger and larger. Motorways in China can be divided into two types according to right of management: state-operated one and private one. The previous type occupies the main portion. According to literature, on the premise of conforming to relevant regulations, China encourages domestic and foreign economic organizations to invest road construction. After regulated period, all Chinese private motorways will be transferred to state-operated ones finally. How can administrative departments ensure the increasing but not decreasing of technical level of motorways when mileages increase constantly? How to maintain infrastructure of highway in time when the vehicle amount keeps rising in China? Obviously, these problems cannot be solved only by traffic management department no matter from financial resources, manpower, material resources, and technical level. At present, there is no research on special maintenance of facilities in fog section of motorways in China.

(4) As shown in Figure 1, the change trend chart of China fog day from 1950 to 2011 shows that the number of fog days in China is increasing. With the increase of fog days, the visibility of expressway is significantly reduced. According to the literature [16], visibility is reduced by about half.

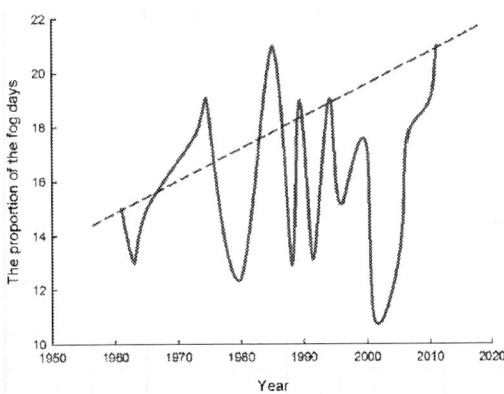

Figure 1   China fog daily variation trend chart

3. Analysis on the Feasibility and Necessity of MOT Model

3.1. MOT model

Belonging to narrowed PPP model, compared with BOT, BOO and BLT models, MOT model lacks construction links in projects, and is mainly used in maintenance, operation and management of fog section in motorways in later period. The project of management and maintenance of fog section in motorways with MOT model is called FRMOT (Fog Region Maintain Operate Transfer) for short, and is called MOT project in this paper. Essentially, MOT project is a PPP project. Except for characteristics owned by PPP project, it also has following characteristics:

(1) Compared with regular PPP project, MOT project requires lesser prior-period investment causes lower financing difficulty to private sectors, and is mainly adopted in the management and maintenance of facilities in fog section in motorways which have been constructed or operated for a period of time.

(2) As the major management and maintenance subject in later period, private sectors' objective estimation on fog section in existing motorways with current technology is in favour of making scientific and reasonable plan on the management and maintenance. However, the technical disclosure effect with prior-period road construction greatly influences the operation effect of MOT project, and increases operation risks of private sectors. Fog sections and sections without fog in the same road network should be effectively divided. The special fog section in which MOT project will be adopted should be estimated scientifically.

(3) With the major implementation risk of long project cycle which may last for twenty to thirty years, the sustainability of project, stability of support from public sectors, and credit of public sectors are very important. The appearance of fog is greatly influenced by weather. Fog appears in different time and place without discipline in short time under the influence of many factors like time and weather. This requires different demand on periodical maintenance of facilities in fog sections, and increases the difficulty to estimate MOT project in early stage.

(4) Being different from regular PPP project, it is difficult to define MOT project and enact clauses when signing contract. This requires public sectors to supervise MOT projects in later period. However, problems like how to ensure status of public and private department should be solved.

3.2. Difference between MOT Model and Other PPP Models

Except for the features of regular PPP models, compared with frequently used PPP models like TOT and BOO, MOT model has following features. Please refer to table 1. Compared with PPP model, MOT model has low financing difficulty, project operation risk, project risk estimation difficulty, project cost overrun risk, and project market competition risk. However, these projects should be done by enterprises with high technological capacity and high economic strength because of its long operation period.

Table 1   Comparison between MOT and conventional PPP mode

Content	MOT	Conventional PPP Mode
Financing difficulty	Slightly Lower	High
Project operation risk	Slightly Lower	High
Project risk assessment	Easy	Difficult
Project operation cycle	Long	
Demand for private enterprises	High	
Risk of cost overruns	Low	High
Market competitive abilities and Market risk	Low	High

3.3. Analysis of feasibility

According to literature [17], the PPP model is most suitable to be used in highway field, which owns relative mature pricing model and process. It is showed by the investigation result that the emerging PPP market has a good prospect and powerful development strength. To assist the development of PPP model, relevant national departments have issued some policies. In the 18th National Congress of the Communist Party of China, "let market play the decisive effect in resource allocation" is proposed. This provides a political and economic environment for the operation of PPP project, upgrading its promotion and application from simple financing cooperation to the level of state economy reform.

PPP projects have been implemented in China from 1990 a. According to data from Bank of Asia, PPP model was used in 1018 projects up to 2011. Please see figure 2 for the specific trend chart.

In China, PPP project can be roughly divided into three stages: a) Foreign companies were in the dominant position in 1990s; b) State-owned enterprises were in the dominant position from 2000 to 2012; c) China encourages social capitals to participate in the construction of infrastructure from 2013 to now, and raises a new round of PPP trend in the whole China. According to statistical data from the World Bank, China had run 1065 PPP projects (narrowed franchise PPP project) from 1990 to 2012, and had achieved many successes in the construction of public projects with PPP model in fields like motorways, bridges, tunnels, water supply, sewage treatment and power plants. For example: Projects like Beijing No. 4 Subway Project, Quanzhou-Citong Bridge, South Extension Project of Nanjing No. 1 Subway, Shanghai Shenfengjin Motorway Project, and 3rd Nanjing Changjiang River Bridge, have got expected effect to a large extent. With good leading and reference value, Beijing No. 4 Subway Project and Bird Nest Project are the representatives of PPP project up to now. In recent years, Chinese scholars, likeYuan et al. [18] have made in-depth summary and analysis on failed PPP projects or those with problems in China, and found main risks, causes, and inherent laws, providing certain reference value for the successful operation of PPP projects in the future. Domestic scholars including Hao et al. [19],Lu and Ma [20], Yao [21] analysed Beijing No. M Subway Line, which is the first large scale successful PPP project in China, and went depth into its successful experience, providing detailed basis for PPP projects in China in the future.

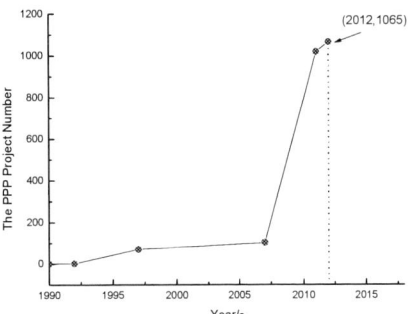

Figure 2 The changing trend of the PPP projects in
China

It can be seen, along with the increasing of mileages in China, the pressure in motorway traffic grows every day. It is a good way to effectively ease government pressure and increase management pressure of road traffic through applying PPP model in road traffic safety combining with its features.

3.4. Analysis of the Necessity
China reached the goal of road construction, which costs developed countries 50 years, with only 20 years. However, because of rapid development of vehicle industry in China, the increasing of vehicle holding quantity and traffic volume result in coexistence of people and vehicles, traffic jam, frequent occurring of traffic accidents, as well as insufficient maintenance in manpower, financial and material aspects for road infrastructure. The infrastructure which relates to traffic safety becomes a decoration and even the hidden danger.

With construction of motorways being emphasized, the maintenance and upgrade in later period are of more importance. It becomes more unrealistic for government departments to assume this task along with the geometric increasing of mileages of motorways. Governments now maintain and upgrade infrastructures with PPP model as it becoming more mature, which does not only solve the problem of insufficient manpower, financial resources and energy, and drives the reformation of maintenance and management of infrastructure and public facilities [22].

4. Research on Management and Maintenance of Motorways with MOT Model
Combining with the operation process of PPP project and features of maintenance of fog sections, the implementation process of MOT projects can be showed in the Figure 3. These projects include 8

steps of project estimation, project biding, constitution of SPC project company with special targets, SPC financing, maintenance and operation, user usage, user payment, and project transferring. The specific implementation and operation process of every step is showed as follows:

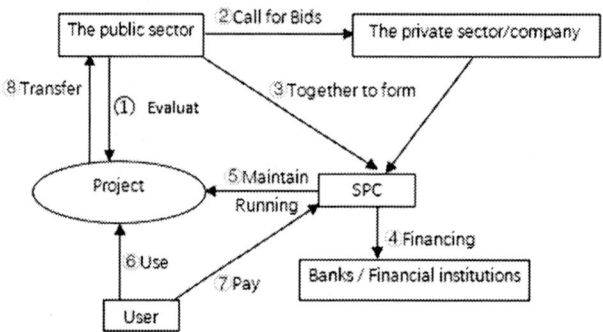

Figure 3   The MOT project implementation process

4.1. Bidding of MOT project
With long periods, some even last for several decades, PPP projects have high qualification requirement to private departments or enterprises. Being started late in China, there is no specialized qualification requirement documents of private sectors or enterprises about PPP projects required. Therefore, we may use for reference of experience from England [23 24]. The EC Merger Regulation issued by England government makes clear regulation on financial capacity, company scale, market share, and turnover of private sectors or enterprises [25]. To the MOT projects, the development of private sectors is more important than financial capacity. Company development better shows vitality of enterprises and ensures the survival time of MOT projects.

4.2. Construction of SPC
SPC (Special Purpose Company) refers to public sectors or private sectors/enterprises which operate PPP projects including specific matters like financing and organizing of project operation institutions through organizing specialized agencies. The specific leading organs of SPC might be different according to real demand of MOT project. The mission of SPC terminates when MOT project is completed. During the construction of SPC, the requirements on MOT projects from public sectors should be defined to ensure the quality and quantity of MOT project. To not influence the function of SPC, public sector should not participate in too much during the operation of MOT project. When maintaining infrastructure in fog section in motorways, SPC needs to ensure timely and effective maintenance, efficient supervision, and avoid unnecessary intervention from administrative departments in technological innovation and renewal of infrastructure. At the same time, fog sections always locate in high mountains, gorges and network of rivers where there are many bridges, tunnels, and complex roads, which require high qualification requirements on SPC companies because of difficult technical maintenance and frequent maintenance.

Most MOT projects have long time of duration. As to the maintenance of infrastructure in motorways, SPC might survive until the scrapping of motorways when private sectors gain considerable benefits and develop well. Therefore, public and private departments should regulate obligations and rights of SPC, public sectors and private sectors in contract form at the beginning of preparing SPC, trying to regulate all issues like personnel and capital during operation in contract. Anything may happen during the implementation of contract because of long periods (some even last for several decades) of MOT project. It is believed by some scholars that there should be a mid-term contract to modify contract when it is needed.

4.3. Financing of MOT projects

One of the largest advantages of PPP model is to spread risks effectively [26] and solve government's shortage of funds in the construction of large scale public infrastructure. In 2014 a, Wang Baoan, the vice-minister of Ministry of Finance pointed out that "the expired repaying debts of local governments occupy 21.89% of overall debts". At present, the urbanization rate in China is 53.6%, and this rate is predicted to be 60% in 2020, which will bring investment demand of RMB 4.2 billion. According to operation experience of foreign PPP projects, capital problem is the decisive problem in MOT projects after SPC being organized by public or private departments.

4.4. Existing problems and advices during the implementation of MOT

The reasons for which infrastructure of fog section in motorways being damaged are divided into: man-made destruction and non-artificial destruction. As shown in Figure 4, the proportion of damage to equipment caused by different environmental factors is shown. Man-made destruction contains: deliberate stealing, smashing, road accident, jerry-building during maintaining, as well as unqualified infrastructure or those cannot be used because of not operating according to procedure. Non-artificial destruction contains: natural depreciation and abrasion, corrosion because of light and air, as well as other non-artificial factors, environmental climates. Different damaging ways determine different maintenance methods and costs.

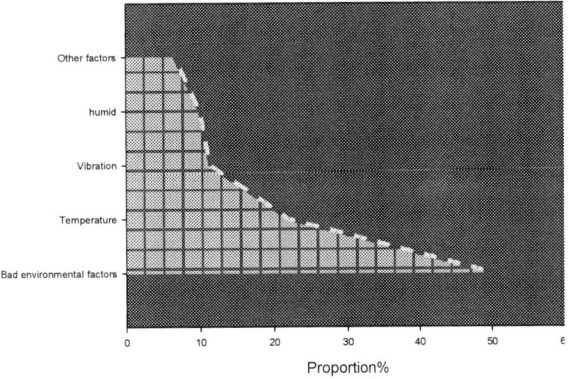

Figure 4   The proportion of equipment damaged by
different environmental factors

The aging degrees of infrastructures in fog sections are different because of different time, period, traffic flow, and external environment. MOT projects should be estimated according to different fog sections and using conditions of facilities, and equipped with different maintenance schemes and grades. The returns to scale and range benefits should be considered when estimating MOT project for fog section. Before maintaining infrastructures in motorways, public sectors should firstly make professional subsection and subitem estimation on facilities in fog sections, confirm their service life period, depreciation condition, the natural environment and traffic flow condition, record road facility conditions, make scheme and expense budget for maintenance or upgrade schemes on this basis, and lay foundation for bidding in later period. Following problems should be solved during the implementing of MOT projects:

(1) How can public sectors and private departments (or enterprises) cooperate effectively and complement each other's advantages. Table 2 is the simple comparison of public sectors and private sectors that make construction of infrastructures in motorways on 7 aspects like national policies, information resources and capitals. It can be seen from Table 2 that public and private sectors are able to complement each other's advantages. However, it is worth thinking deeply about how to do this during the implementation process. The managing and delegating powers to lower levels and playing positive functions of government during implementing process of MOT projects are the key to the

success or failure.

Table 2 The contrast to the Characteristics of the Public Sector and the Private Enterprise

Sector	National Policy of Information Sources	Fund	Managerial Experience	Management Efficiency	Initiative and Innovation	Risk Appetite	Stability of Decision making
Public sector	Fast	Shortage	Shortage	Low	Low	Low	Low
Private sector	Slow	Adequate	Adequate	High	Low	Low	Low

(2) Project risk impacts the investment activity of private sectors.

The non-uniform revenue time, standards and acceptance standards of MOT project increase investment risk Zheng (2015). Meanwhile, as the authorities in China, public sectors cannot cooperate with private sectors (or enterprises) equally or effectively because of unequal administrative status. Private sectors (or enterprises) are always in the weak position and lack enough right of speech. On the other hand, because of long period, the personnel change of both parties may decrease the constant stability of projects and increase risk. MOT project also has problems in PPP projects.

The Quanzhou-Citong Bridge Project is a typical sample [27]. This is the first BOT project for construction of infrastructure which adopts private capital in China. It was constructed in trial in 1993 a, and was open to traffic in 1996 a. It was said by the person in charge that the invested capital has not been returned until 2014 a mainly because there is no effective and substantial agreement between the enterprise and local government. Most investment risks are assumed by the enterprise. The failure of this project influences domestic PPP projects, and makes a part of enterprises dare not to cooperation with governments on these projects later. Therefore, the research team of Li and Shen [28] puts forward 9 principles about risk distribution based on long term research. At the same time, PPP projects develop rapidly in recent years in China because of political support from governments.

(3) Imperfect policies and systems

From 2013 to now, the government of China has issued a series of documents and regulations on PPP projects, while they are still insufficient in operation because of its complexity. For example, there is still no standard model for how to effectively standardize contracts signed between public and private sectors based on PPP projects. During actual operation, public sectors, like the municipal government of Chongqing, call up specialists to discuss and try to refine contract terms. This operation avoids contract risks to the largest extent. However, risk brought by imperfect contract is still very large. The selection standards of private sectors, such as qualification of assuming PPP projects, are not required by specific documents. In academic circles, scholars are divided into two parties. One side recommends governments to strictly examine qualification of private sectors, rigorously enforce market access, and require enterprises who be responsible PPP projects to have high-growth and powerful economic strength. Another side keeps the opposite opinion, believing that governments should decrease market access for private sectors and let more enterprises to participate in PPP projects.

This MOT project for maintenance of infrastructure in fog section in motorways discussed in this paper costs a large number of capital and long period. As to enterprises with poor anti-risk capability, the operation cost and risk will be increased if these enterprises close down or have capital problem during the contract period. Most PPP projects have similar demands with MOT project. Therefore, the government of China should improve technical specification and legislative guarantee of MOT projects to provide laws to be abided by and based on.

(4) Unstable decisions of local governments

From implementation condition of PPP projects in recent years, the condition of "one side wishful thinking and another side little reaction" exists between public and private sectors. Governments want more private companies to participate in the bidding of PPP projects, while private enterprises give little reaction. The reasons are the long implementation period and variable and random decisions made by local governments. For example, the phenomenon of "making frequent changes in policies" and "employees change when leaders change" exist in some local governments, which make many MOT projects going broke. Withal, it is pointed by Wang and Feng [29] that "government credit" is one of major reasons which influence the success or failure of PPP project. This requires government to protect projects from executive ability perspective to ensure projects not being influenced by change of the term of government or personnel and projects being driven constantly by specialized institutions.

5. Conclusions

From the perspective of traffic safety management, with MOT model, this paper analyses the feasibility and necessity to maintain infrastructure in motorways when state-owned motorway quantity increases, and problems like aging, reconstructing (expanding), facility maintaining, maintaining, and technical upgrading of a part of motorways, especially fog section (including state-run ones or ones transferred from private ones) become more serious. It comes up with following conclusions:

(1) The traditional PPP model analyses from perspectives of project management, investment and financing, while MOT model adopted in motorways analyses from perspectives of fog section in motorways and traffic safety management.

(2) Being compared with traditional PPP project, MOT model has features like low financing difficulty in early stage, clear implementation target, and risk sharing.

(3) MOT model has advantages like professional technology and detailed project subdivision which are not owned by traditional PPP model. However, being the same with traditional PPP model, it has the same problems like large risks in later period and government credit. Therefore, a good environment in governmental level should be created.

Acknowledgments

This research was substantially supported by the Natural Science Foundation of Chongqing, China (cstc2017shmsA0581), the Science and Technology Plan Project of Chongqing Land Resources and Housing Administration (KJ-2017019) and the Special funding for postdoctoral research projects in Chongqing (Xm2017006).

References

[1]   Zhang Ping, *The application of PPP model in Hunan and the practice research,* Journal of Management,(23), pp.123, 2014.

[2] Chen Liuqin,*The financing model of PPP new public-private partnership*, Construction Economy, **269**, pp.76-80, 2005.

[3] Sha Ji, *The research on the application of PPP in China's infrastructure projects*,Nanjing: Southeast University, Master Dissertation, pp.58, 2004.

[4] Deng Xiaopeng, Li Qiming, Wang Wenxiong, et al.,*The PPP mode risk sharing principle of review and application*, Construction Economy,(9),pp.32-35, 2008.

[5] Lu Qingcheng, *Public private partnership (PPP) mode and the development and study of our urban public utilities*, Huazhong science and Technology University, Ph. D Dissertation, pp. 124, 2008.

[6] Bloomfield, P., Westerling, D., Carey, R, *Innovation and risk in a Public-Private Partnership: financing and construction of a capital project in Massachusetts*, Public Productivity & Management Review, **21**(4), pp.460-471, 1998.

[7] Hall, J, *Private Opportunity, Public Benefit?* Fiscal Studies, **19**(2), pp.121-140 , 1998.

[8] Hodge, G.A, *Risks in Public-Private Partnerships: shifting, sharing or shirking?* Asia Pacific Journal of Public Administration, **26**(2),pp.155-179, 2004.

[9] Gaffney, D., Fellow, R., Pollock, A.A.M., et al., *The private finance initiative: the politics of the private finance initiative and the new NHS.* BMJ Clinical Research, **319**(7204),pp. 249-253,

1999.

[10] Walker, R.G, *Public-private partnerships: form over substance?* Australian Accounting Review, **13**(30), pp.54-59, 2003.

[11] Jean Shaoul, *Railpolitik: The financial realities of operating Britain's national railways*, Public Money & Management, **24**(1), pp.27-36, 2004.

[12] Pollitt, M.G. *The declining role of the state in infrastructure investments in the UK// Faculty of Economics,* University of Cambridge , 2000.

[13] Wang Hua, Zhang Xiaoning. *Joint implementation of tradable credit and road pricing in PPP networks considering mixed equilibrium behaviours.* Transportation Research Part E: Logistics & Transportation Review. **94**, pp. 158-170, 2016.

[14] Kang, C.-C., Lee, T.-S., Huang, S.-C, *Royalty bargaining in Public–Private Partnership projects: insights from a theoretic three-stage game auction model,* Transportation Research Part E: Logistics & Transportation Review, **59**(11), pp.1–14, 2013.

[15] Huang Teng, Ke Yong jian, Li Zhanzhan, et al, *A china-foreign comparative analysis on government management in the PPP mode,* Project Management Technology, **7**(1),pp.9-13, 2009.

[16] Ding Y H, Liu Y J, *Analysis of long-term variations of fog and haze in China in recent 50 years and their relations with atmospheric humidity,* Science China: Earth Sciences, **57**,pp.36–46, 2014.

[17] Wu Liyang, *Government subsidies in the theory of PPP project reasonable,* Journal of Engineering Consultation in China, (8),pp.22-24, 2016.

[18] Yuan Xia, Ke Yongjian, Wang Shouqing, et al., *Analysis of major risk factors of PPP project in China based on case study,* Soft Science in China, **1**, pp.107–113, 2009.

[19] Hao Yawei, Wang Yingying, Ding Huiping, *Core elements of public-private partnership in urban rail transit case of Beijing metro line M,* Journal of Civil Engineering, **45**(10),pp. 176–180, 2012.

[20] Lu Qingcheng, Ma Jian, *Urban rail transit project adopts the model of public-private financing problems,* Urban Mass Transit, **11**(4), pp.5-8, 2008.

[21] Yao Pengcheng, Wang Song jiang, *Survey on pricing theory of public private partnership highway project,* Science and Technology Management Research, **31**(9),pp. 180-184, 2011.

[22] Li Keqiang, *Li Keqiang chaired a state council executive meeting,* http://www.gov.cn/guowuyuan/2014-04/23/content_2665259.htm, 2014.

[23] Sparrowe, R.T., Liden, R.C., Wayne, S.J., et al, *Social networks and the performance of individuals and groups,* Academy of Management Journal, **44**(2),pp. 316-326, 2001.

[24] Jone Felicity AE, Daniel PH Knights, Sinclair Vita FE, Paula Baraitser, *Do health partnerships with organisations in lower income countries benefit the UK partner?* A review of the literature, Globalization and Health. http://globalizationandhealth.biomedcentral.com/articles/10.1186/1744-8603-9-38, 2013.

[25] Perry-Smith, J.E, *Social yet creative: The role of social relationships in facilitating individual creativity,* Academy of Management Journal, **49**(1),pp. 85-101, 2006.

[26] Zheng Zhiquan, *The PPP mode social investors of risk management research,* Fujian Agriculture and Forestry University, Master Dissertation, pp.58, 2015.

[27] Song Jinbo, Song Danrong, Wang Dongbo, *Quanzhoutung Bridge in BOT project of operational risk,* Management case study and review, **2**(3), pp.196-201, 2009.

[28] Li Qiming, Shen Liyin, *Infrastructure BOT project concession period decision model,* Journal of management engineering, **14** (1), pp.43-46, 2000.

[29] Wang Shouqing, Feng Ke, *PPP rush to calm thinking,* Journal of Urban and Rural Construction, (8): pp.38-40, 2015.

ISPECE

IOP Publishing

IOP Conf. Series: Journal of Physics: Conf. Series **1187** (2019) 052058   doi:10.1088/1742-6596/1187/5/052058

Research on Semantic Prediction Analysis of Tibetan Text Based on Word2Vec

Ding Hai-lan[1a], Yu hong-zhi[2b], Qi kun-yu[3c]

[1]Northwest Minzu University, China National Institute of Information Technology，
Gansu Lanzhou 730030,

[2]Northwest Minzu University ,China National Institute of Information Technology，
Gansu Lanzhou 730030，

[3]Northwest Minzu University ,China National Institute of Information Technology，
Gansu Lanzhou 730030，

[a]e-mail: www.dinghailan@qq.com

[b]e-mail: www.zwbgzdh@126.com

[c]e-mail: www. 2630103542@qq.com

Abstract. This article uses Google's open source Word2Vec tool to input the corpus of the Tibetan text "*Sage wedding*" after the word segmentation. The words are mapped to a K-dimensional space in the text and transformed into word vectors by using the context information of the vocabulary. The Word2Vec tool then learns to get a vector model, each of which is represented by a unique word vector. A vocabulary is constructed through training text data and then the vector is represented by learning the words. Word vectors capture the laws of many languages, which results the similarity of the distance between words and words. The experimental results show that the accuracy and recall rate based on the Word2Vec training model are very high.

1.   Introduction

In 2013, Tomáš Mikolov and his colleagues at Google released Word2Vec, which is an unsupervised tool for computing continuous distributed representations of words. The tool provides a vector representation of the continuous word bag model CBOW and Skip-Gram models for calculating words. The Word2Vec tool requires a text corpus as input and produces a word vector as output. Firstly, it constructs a vocabulary based on the training text data and then learns the vector representation of the word. The generated word vector file can be used as a function in many natural language processing and machine learning applications. This tool finds the closest word for a user-specified word. For example, if you enter "China", the distance will display the most similar words and their related words that are closest to the "China" distance. In the article, the Tibetan text is taken as an example of the " Sage wedding " because it is a profound Tibetan historical work. Its writing is exquisite, including Tibet's history, religious history, cultural development and the economy of ancient Tibet. It has a unique style historic value. In this paper, Word2vec's CBOW training model is used to express words into word vectors quickly and efficiently. The word vectors capture the semantic features between words in natural language, and then output according to distance and distance to output the most

Content from this work may be used under the terms of the Creative Commons Attribution 3.0 licence. Any further distribution of this work must maintain attribution to the author(s) and the title of the work, journal citation and DOI.

Published under licence by IOP Publishing Ltd

similar or related to high frequency words. Vocabulary, these words serve as important information to predict the approximate semantics of text.

2. Word2 Vector

The computer can only recognize 1 and 0 for natural language processing, then if you want the computer to process the text, you must convert the text into a language that the computer can recognize. The most direct way is to convert the word into a word vector. The vectorization of words refers to the mathematical representation of words, mainly including one-hot representation, distributed representation and word2vec model training word vector. The first way is to use a very long vector to represent a word with a component of 1, and the rest are all 0, 1. The shortcoming is that it cannot provide semantic information. The second way was first proposed by Hinton, who mapped the words into a low-dimensional and dense 100-200-sized real vector space, so that the closer the words are, the closer the word distance is. The third way is to use the NNLM model (Nin Network Language Model) proposed by Bengio and the Log-Linear model and the Mikolov model of Hinton to propose the language model of Word2Vec. Word2vec can effectively train the word vector at high speed. [1]

3. Word2 Vec Tool

Word2Vec is a vector model representation tool. The tool turns words into vectors, and calculates the similarity by calculating the cosine between the two word vectors. The Word2Vec tool takes the corpus after the word segmentation as input and learns to get a vector model. It first constructs a vocabulary from the training text data and then learns the vector representation of the word. Word vectors can capture the laws of many languages, such as the operation of a vector: vector ("ri bo"), vector ("ri mthon po"), vector ("ri   khul"), vector ("ri rtse")etc, translated into "mountain, mountain, mountain, mountain top" in order. Through the context operation, the semantic distance between the vector and the vector is very similar in the results obtained by the above examples. Therefore, we can understand Word2Vec as a distance tool. When you input a word, it will display all words close to the word according to the distance.

3.1. Two training models of Word2Vec

The word vector is actually a vector that maps words to a semantic space. Word2Vec is implemented by means of neural network. Considering the context of text, Word2Vec has two models, CBOW and Skip-gram. These two models are similar in the training process. The Skip-gram model uses a word as input to predict the context around it. The CBOW model takes the context of a word as input to predict the word itself.

3.1.1. Skip-gram training model

If you use a word as input to predict the context around it, the model is called the Skip-gram model. First determine the window size Window, generate 2*window training samples for each word, (i, i-window), (i, i-window+1),..., (i, i+window-1), (i, i+window). Immediately after determining batch_size, note that the size of batch_size must be an integer multiple of 2*window, which ensures that each batch contains all samples corresponding to one vocabulary. There are two training algorithms: Hierarchical Softmax and Negative Sampling. [2] Finally, the neural network is iterated and trained for a certain number of times, and the parameter matrix of the input layer to the hidden layer is obtained. The transposition of each row in the matrix is the word vector of the corresponding word. The specific model is shown in Figure 1:

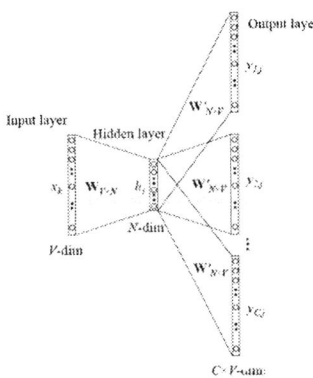

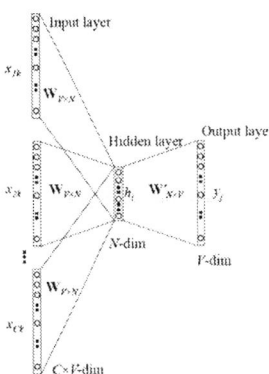

Figure 1 Skip-gram2 model   Figure 2 CBOW model

3.1.2. CBOW training model

The CBOW (Bag-of-words model) model takes the context of a word as input to predict the word itself. Firstly,it determines the window size window, generate 2*window training samples for each word, (i-window, i), (i-window+1, i),..., (i+window-1, i), (i+window, i). Secondly,it determines the batch_size. Note that the size of the batch_size must be an integer multiple of 2*window. This ensures that each batch contains all the samples corresponding to one vocabulary. There are two training algorithms: Hierarchical Softmax and Negative Sampling. Finally, the neural network is iterated and trained for a certain number of times, and the parameter matrix of the input layer to the hidden layer is obtained. The transposition of each row in the matrix is the word vector of the corresponding word. The specific model is shown in Figure 2:

In the two training models of Word2Vec's Skip-gram and CBOW, it is recommended to use the Skip-gram training model to train when there are more corpora of training. When the corpus is relatively small, it is recommended to use the CBOW training model to train. In general, Word2Vec can use the trained word vector model, convert the input words into word vectors and then train the model, and finally output the words according to the distance, and turn these words into a synonym set.

4. Experimental process and results analysis

First, we select the Tibetan text corpus "Sage wedding" (mkhsa p'i dg'a ston) which size is 2.37MB and import the text into the Tibetan WordSegmentEx3.0 Tibetan word segmentation software, and get the text after the word segmentation and the text sorted according to the word frequency. Next, we install Anaconda3, an open source Python distribution, which is an open source package management system and environment management system. After the installation is successful, use the Conda to build a development environment in the terminal.

4.1. Experimental steps

After the terminal builds the environment variables, in the Spyder development environment of Anaconda3, the word vector test code is written in the python programming language. We call the Word2Vec CBOW model algorithm in the Gensim toolkit to train. The size of the trained word vector is 50, the training window is 5, and the minimum word frequency is 5. Firstly, we calculate the similarity of two words, then calculate the related words of a word, and finally output the set of words that are closest to the semantic distance of the two words. The core code for a word vector test written in the python programming language is as follows:

```
# genism modules
from genism.models import Word2Vec
from genism.models.word2vec import Text8Corpus
import os.path
```

```
import sys
import numpy as np
```
The corpus of training is after the word segmentation, the batch processing has 33244 sentences for each sentence. Then, all the part-of-speech annotations in the corpus are removed and the tokens are replaced by spaces. Each sentence retains the ending symbol in the Tibetan sentence, that is, the single suffix, which is the unit of the sentence. Finally, we get a list of features for a text. In the word bag model (CBOW), the characteristic of a document is the word it contains. The specific steps are shown in Figure 3:

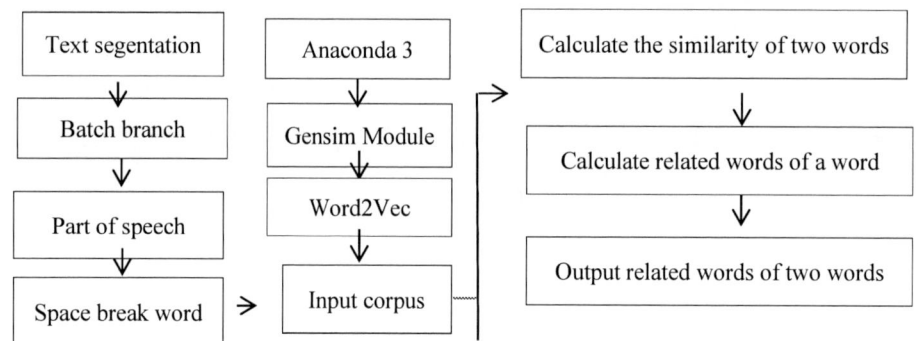

Figure 3 Word2Vec experimental steps

After extracting the feature table, the high-frequency vocabulary is selected for training. When the high-frequency vocabulary is selected, the helper words [3] (*rnma dbye*), conjunctions (*dnga sgra*), numerals (*grngs p'I tshig*) , decorative set words (*rgyna sdud*), pronouns (*tshba tshig*), to-be-reported words (*lhga bcsa*), negative words (*dgga sgra*), refers to human nouns suffix (*bdga sgra*) , adjectives (*khyda chos kyi ming*) and single vertical symbols (*shda*) in the word frequency table are removed[4]. After removing these high-frequency vocabulary and symbols, select 10 sets of high-frequency vocabulary for training. The specific word frequency is shown in Table 1:

Table 1 text " Sage wedding " high-frequency vocabulary

High frequency word	Word frequency	High frequency word	Word frequency	High frequency word	Word frequency	High frequency word	Word frequency
Rgyla po	1035	*chos*	917	*mdada*	747	*tshe*	740
byung	731	*lo*	650	*bysa*	640	*lha*	602
Sngsa rgysa	474	*bod*	470	*byon*	410	*byed*	409
Bstna pa	406	*shing*	399	*dus*	395	*bu*	383
bya	380	*gnsa*	374	*Ston pa*	353	*jul*	346

10 groups of high frequency words are used as the training target to enter the Word2Vec model for training. The training results are shown in Table 2, Table 3, Table 4 and Table 5:

Table 2 Word2Vec model training related words of two words

rgyla po	chos	mdzda	tshe	byung	lo
spel	0.8170989751815796	*bynga phyoks*	0.6811710596084595	*dgung lo*	0.8219879865646362
gdms ngga	0.8086523413658142	*phying*	0.6716954112052917	*lon pa*	0.7987777590751648
sgrub	0.8076239228248596	*shing rta*	0.6692179441452026	*rgyla sa*	0.789211094379425
bsgrubs	0.8062039017677307	*zhla snga*	0.6634161472320557	*bzhes pa*	0.7790915966033936
thugs	0.8038085699081421	*khye'u*	0.6567350625991821	*bskos*	0.7712806463241577
dge b'i bshes gnyen	0.7964973449707031	*mdog*	0.6561226844787598	*druk bcu*	0.7650943994522095

2449

rje 'bngsa		dbus ma		lng bcu	
	0.7910181879997253		0.6552908420562744		0.7649886012077332
gnda		phok pa		bcu	
	0.79095458984375		0.6549668908119202		0.7566709518432617
bon		rjessu bzung		brmkyd bcu	
	0.790745735168457		0.6528903841972351		0.7561909556388855
mna ngga		smin mtshma		rtsa	
	0.7862948179244995		0.6524414420127869		0.7513672709465027

In the development environment of the Word2Vec model, when the two words of the highest word frequency "*Rgyla po chos*""(king, religion) are input in the text training, the words that are closest to and related to the two words are obtained in turn "*Spel,gdms ngga,sgrub,bsgrubs,thugs,dge b'i bshes gnyen,rje 'bngsa,gnda,bon,mna ngga*" (development, teaching, practice, achievement, spirit, *gexi*, monarch, elite, Bon and teaching). When you enter the two terms of "*Mdad tshe*" (career, time), you will get the words that are closest to and related to the two words "*bynga phyoks,phying,shing rta,zhla snga,khye'u,mdog,dbus ma,phok pa,rjes su bzung,smin mtshma*" (North, Backward, Big Car, Pre-driver, Junior, Color, Center, Teaching, Support and Eyebrow). From the above high-frequency vocabulary set, we can see that the relevant words obtained by training and the close-range words are all related to the Buddhist language, the development of Buddhism, and the historical part of the evolution of the word, so that we can roughly predict the semantics of the text is about religion which is a work of history.

Table 3 Word2Vec model training related words of two words

sngsa rgysa	bod	bysa	lha	byon	byed
rgya gra		'phrul	0.7499148845672607	Ci zhig	0.7952665686607361
	0.8007855415344238				
dra ba		Zhu ba	0.7326218485832214	brjed	0.7951421141624451
	0.7687827944755554				
mu stegs		Brgya byin	0.7192088961601257	mtshon	0.7947472333908081
	0.764392614364624				
rkyla khmsa		Khri btsun	0.7027950286865234	Rtga tu	0.7946202158927917
	0.7596821784973145				
jo bo rje		thogs	0.7001242637634277	Rgyun du	0.7859824895858765
	0.7370045185089111				
gdong dmra		Rol mo	0.6994699835777283	zhen	0.7819713354110718
	0.729033887386322				
srol		Phba pa	0.697623610496521	lhung	0.7802630662918091
	0.7245499491691589				
bon		sgra	0.6975480318069458	skyes bu	0.7793267369270325
	0.7234969139099121				
shra phyogs		bus	0.6924548149108887	De dga	0.7787083983421326
	0.7211496233940125				
bla		Lha mo	0.6915058493614197	bzod	0.7785230875015259
	0.7209500074386597				

In the development environment of the Word2Vec model, when the two words of the word frequency"*sngsa rgysa bod*" (Buddha, Tibetan) are input in the text, the words that are closest to and related to the two words are obtained in turn"*rgya gra, dra ba,mu stegs,rkyla khmsa,jo bo rje,gdong dmra,srol,bon,shra phyogs,bla*"(India, development, foreign roads, national land, Jue Wojie, red face, Customs, Bon, Eastern and Nepal). From the above phrases, we can also clearly see that the phrases obtained from the training are all about the common language in Tibetan Buddhism, such as: "The Emperor Shi Tian, Chi Zun and the Living". Secondly, through training, we can know that the text mainly describes the relationship between Tibetan Buddhism and Nepal and India. Furthermore, we can see the expression of Buddhist culture in the Tibetan region through the word frequency: "Musical instruments, wine, sound, playing and goddess". In general, we can predict the semantics of the text about the prevalence of Buddhism in Tibet and the development of culture.

Table 4　Word2Vec model training related words of two words

bstna pa	shing	dus	bu	bya	knsa
bcsa	0.7470908164978027	Bu mo	0.8041067719459534	Gnsa pa	0.8767780065536499
phyi nnga	0.7427273988723755	Srsa mo	0.7581796646118164	Bye ba	0.8544745445251465
lha 'dre	0.742052435874939	Nka mo	0.7580035924911499	Bsma gtna	0.85157310962677
'dun pa	0.7397492527961731	nya	0.7576266527175903	Rnma pra	0.8435131311416626
sgra 'byin	0.7373657822608948	lci	0.7437707781791687	Gling bzhi	0.8402106165885925
te	0.7352505326271057	Che ba	0.7417047023773193	Zhing khmsa	0.8392305374145508
phba pa	0.7346739768981934	za	0.7383226156234741	Mth'a	0.8358502984046936

sku gsung thugs rten	0.7306311130523682	gju	0.7377681136131287	Thun mong	0.835551381111145
gso ba	0.7284703254699707	rba	0.7358239889144897	Stug po	0.8334478735923767
sa rdo	0.7271305918693542	Gtsug lg	0.7340255379676819	Chos kyi rnm krngsa	0.8326832056045532

In the development environment of the Word2Vec model, when the two words of the word frequency "*bstna pa shing*" (religion, tree) are input in the text, the words that are closest to and related to the two words are obtained in turn "*bcsa,Phyi nnga,Lha'dre,dun pa ,Sgra 'byin,te,Phba pa,Sku gsung thugs rten,Gso ba,Sa rdo*"(with, inside, outside, ghosts, Desire, emperor's interpretation, words to be preached (ཉ), distiller's yeast, Buddha statue, education and earth and stone). When the two words of the word frequency "*dus bu*" (time, man) are input in the text, the words that are closest to and related to the two words are obtained in turn "*Bu mo,Srsa mo,Nka mo,nya,lci,Che ba,za,gju,rba,Gtsug lg*"(girl (honorific), Saturn, Sun, heavy, big, food, jasper, wave and holy book). When the two words of the word frequency " *bya   knsa*" (bird, place) are input in the text, the words that are closest to and related to the two words are obtained in turn "*Gnsa pa, Bye ba,Bsma gtna,Rnma pra,Gling bzhi,Zhing khmsa,Mth'a,Thun mong,Stug po,Chos kyi rnm krngsa*" (existence, tens of thousands, meditation, state, four continents, heaven, margin , common, dense and methodic). From the above related phrases of phrases and phrases, I can clearly see the vocabulary inside Buddhism. It is not common words that have comments to understand, but we can confirm that the text is still about the history and culture of Buddhism. Text semantics can also predict text semantics through high-frequency words and related words that are closest to each other.

Table 5 Word2Vec model training related vocabulary of two words

	Ston pa jul		Grong khyer	0.7242305874824524	
Za hor	0.7662957906723022	Rgyla khmsa	0.752292275428772	Rkya nga	0.7397180795669556
Bynga phyogs	0.7599533200263977	Ao rgyna	0.7498317956924438	Shra phyogs	0.7313268184661865
Nub phyogs	0.7539993524551392	Brgya byin	0.7400675415992737	Gdong dmra	0.7242860198020935

In the development environment of the Word2Vec model, when the two words of the word frequency "*Ston pa jul*" (master, mountain god) are input in the text training, the words that are closest to and related to the two words are obtained in turn "*Za hor,Bynga phyogs,Nub phyogs,Rgyla khmsa,Ao rgyna,Brgya byin,Rkya nga,Shra phyogs,Gdong dmra,Grong khyer*" (Ancient Little Kingdom of India, North, West, Land, carefulness, village, country, east, red face and city). From the last set of phrases and the most recent and related phrases predicted, the text is a unique historical value that describes the relationship between ancient India and Tibetan Buddhism and the economy of ancient Tibet.

5.   Conclusion

This paper uses the open source Word2Vec tool of GOOGLE to input Tibetan text as corpus, and transforms the word in the text into the word vector by using the vocabulary context information. By using the CBOW model algorithm model in Word2Vec to train the laws of many languages, the distance between words is the similarity. Further, through the high-frequency vocabulary as input, the vocabulary closest to the high-frequency vocabulary can be output through training, and the high-frequency words and the similar vocabulary are used as important information to predict the semantics of the text. This method plays a fast and convenient role in quickly grasping the subject semantics in the long corpus. At the same time, many interesting language rules can be found through training, which avoids the problem of long duration and subjective judgment. However, many words found in the training did not appear in the context, which brought some errors to the semantic prediction. In general, Word2Vec tools can effectively predict text semantics.

Acknowledgment

2018, Northwestern University for Nationalities Graduate Research (Practice) Innovation Project "Word2vec-based vocabulary clustering research in Tibetan field（Yxm2018011）"; National Science and Technology Fund Project "Research on Literary Cognition Theory of Tibetan Lhasa Based on

EEG Signals" （61462075） "; Gansu Province Innovation Group Project: Tibetan Intelligent Information Processing (Z16088).

About the Author

Ding Hailan (1990-), female (Tibetan), Tianzhu, Gansu, Ph.D., mainly engaged in Tibetan information processing and cognitive linguistics research.

References

[1] Tang Ming, Zhu Lei, Zou Xianchun. A Document Vector Representation Based on Word2Vec.[J] COMPUTER SCIENCE computer. 2016.06.

[2] Xie Rimin, Chen Jie, You Guirong, Xie Datong. Research on Chinese Book Classification Based on Word2Vec. [J] Yunnan Nationalities University. 2018.07.

[3] Ding Hailan, Qi Kunyu. A Comparative Study of Tibetan Texts and English Prepositions.[J] Linguistics. 2016.3.

[4] Ma Jinwu. Four clear structures of Tibetan grammar. [M]. Beijing Ethnic Press.2008.9.

Research on the characteristics of bitcoin price fluctuations based on ARCH effect

Juan Wang, Feng Tian[1] and Jie Fu

Yan ta campus of xi 'an university of posts and telecommunications, xi 'an city, shaanxi province, China

[1]Corresponding author: 2627693252@qq.com

Abstract. Since the appearance of the bitcoin, more and more people have been paying attention to it. Investors have shown great concern about the price of bitcoin through ARCH (Autoregressive conditional heteroskedasticity model) effect test and asymmetric test, we drew the following conclusion: the fluctuations of bitcoin price have the time-varying characteristics and the characteristics of wave agglomeration, but there is no leverage effect.

1. Introduction

At the beginning of bitcoin's emergence, its price is very low, only a few cents. Bitcoin has been booming since 2013, and the number of people paying attention to bitcoin is growing. At the same time, some countries, such as Canada, have accepted bitcoin. And some businesses have begun to accept it to buy goods and services. Bitcoin has its own advantages, such as decentralization. This benefit saves costs and improves economic efficiency, and it also plays an important role in protecting privacy. More and more people are willing to hold and invest in bitcoin. The price of bitcoin has skyrocketed, rising rapidly to a few thousand dollars, and even to nearly $20,000.

While the price of bitcoin is rising overall, it is not rising at every point in time. That is to say, there are price fluctuations in the process, and this price fluctuation is relatively large. And it shows that in some time periods, the price of bitcoin fluctuates less, while in other time periods, the price of bitcoin fluctuates greatly. As following picture shows: This graph is about the price change of bitcoin from March 2017 to March 2018. We can see clearly in the figure:In April, may and other months of 2017, the price of bitcoin fluctuated a little. In November and December of 2017 and January and February of 2018, the price of bitcoin changed a lot. During this period, the difference between the highest and lowest price of bitcoin was nearly $10,000. We see that the fluctuations in bitcoin prices seem to have some kind of signature.

Content from this work may be used under the terms of the Creative Commons Attribution 3.0 licence. Any further distribution of this work must maintain attribution to the author(s) and the title of the work, journal citation and DOI.
Published under licence by IOP Publishing Ltd

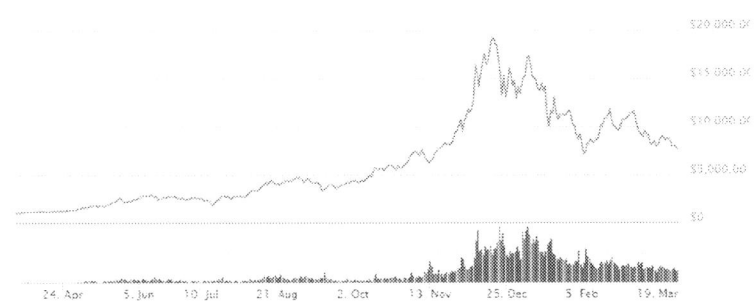

Figure 1. Bitcoin price change chart from 2017-03 to 2018-03 (unit: usd)

As for the research on the price of bitcoin, some scholars have done the research on the price of bitcoin. For example, Ji Hong Li[1] analyzed the impact of the demand for bitcoin on the price through establishing a model, and concluded that the expansion of the demand for bitcoin transaction has a driving effect on the rise of the price of bitcoin in the long term trend. Gang Liuand Juan Liu [8] used the event research method to draw a conclusion: the price fluctuations of bitcoin are greatly influenced by the policy; Whether the policy information is positive or negative for bitcoin, the bitcoin market will experience violent fluctuations. Yue Yu Niu[7] analyzed the long-term and short-term formation mechanism of bitcoin price volatility using granger causality test. Some scholars have analyzed the properties of bitcoins from a theoretical perspective, as Tao Young [2] has been describing the non-monetary properties of bitcoin in terms of a lack of credit foundation, failure to reflect a particular social production relationship. Xiao Chen Yang [5] analyzed the prospect of bitcoin from the perspective of its operation principle and typical features. At present, most of the existing literatures are about the nature of bitcoin itself, while there are few literatures about the characteristics of bitcoin price fluctuation and its influencing factors.

2. Empirical analysis
In order to test whether the fluctuation of bitcoin price has autoregressive conditional heteroscedasticity, we selected daily data of the bitcoin price as the sample sequence for testing. The data is selected from coinmarket.cap and is selected from March 27, 2017 to March 26, 2018.

2.1. ARCH effect test
In order to eliminate the impact of data fluctuations and make the data smoother, The price of bitcoin $\{P_t\}$ is taken as the natural log.A logarithmic sequence $\{\ln P_t\}$ was obtained. The log price sequence at time t was taken as the explained variable. In view of that fact that the sequence of study has one-dimensional random walk features, therefore, the model was set as:

$$\ln\left(P_t\right) = \beta_0 + \beta_1 \ln\left(P_{t-1}\right) + \mu_t . \tag{1}$$

Table 1. Estimation results of one-dimensional random walk model of bitcoin price

object	variable	Estimated coefficient	Standard error	T statistic	P	Test statistic

Price	β_0	0.063	0.030	2.125	0.0343	$R^2 = 0.9955$
	$\ln P(-1)$	0.993	0.003	283.931	0.0000	F=80617.06

The estimation results of the mean value equation of bitcoin price in table 1 show that the value of goodness of fit is 0.9955, indicating that the overall model is well fitted. The value of F statistic is 80617.06, and the corresponding P value is 0.0000, indicating that the linear influence of the equation overall is significant. The P value of the estimated coefficient of the previous moment's bitcoin price is 0.0000, which is significantly less than 0.05, indicating that the estimated coefficient of the previous bitcoin price is remarkably different from 0 at the significance level of 5%. The estimator of the random error term, the residual term, is obtained after estimating the model. Whether conditional heteroscedasticity exists in the random error term can be investigated by the following three decisions:

Decision one: residual series were generated in the estimation window of eviews8.0 equation, as shown in the figure below. In figure 2, the horizontal axis refers to the observation date and the vertical axis refers to the residual term. It can be noticed that the fluctuation of residual term is relatively small at some time, such as march, April and may of 2017, while other times are quite large, such as July, September and December 2017. This shows that volatility has the characteristics of "clustering", which indicates that the error term may have conditional heteroscedasticity.

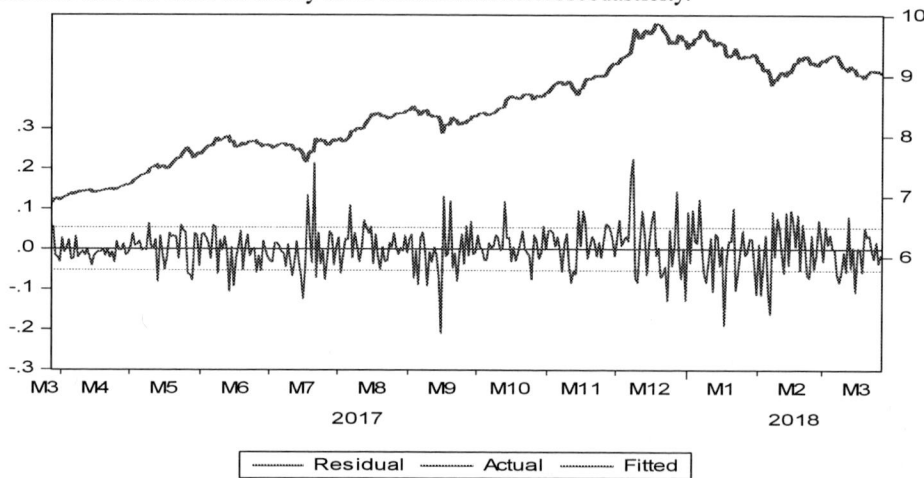

Figure2.   Residual diagram of the regression equation of bitcoin price

Decision two: residual square correlation graph test: AC, PAC and Q statistics of lagged residual square were tested in the equation estimation interface, and the selected lag order was 36. AC analysis results: autocorrelation coefficient and partial autocorrelation coefficient PAC coefficient of the two are not significantly to 0, P and Q statistics value is very small. Under 10% significance level, the Q statistic is significant. Therefore, we can refuse the original assumption that there is no ARCH effect, it can be concluded that there is an ARCH effect in the residual series of first-order random walk model of bitcoin price.

Decision three: establish the auxiliary regression equation, select the lag order of 7 order, and investigate whether there is a significant relationship between the residual squared term of the current period and the residual squared term of the previous 7 periods. The independent variable is the residual squared term of the previous 7 periods, and the dependent variable is the residual squared term of the current period. Finally, the test statistic of the joint significance of the residual squared term of the previous 7 periods of the equation, F and ARCH-LM, is obtained. The results are shown in table 2:

Table 2   results of Lagrange multiplier test of bitcoin prices

F statistic	2.8243	P:	0.0071
$T \times R^2$ statistic	19.1395	P:	0.0078

It can be seen that the P value of both statistics is small, and the original hypothesis can be rejected at the significance level of 1%, Which shows that the ARCH effect exists in the bitcoin price model.

2.2. GARCH(1,1) estimation

In accordance with the above, it was found that the price of the bitcoin does have an ARCH effect, which indicates that the price of the bitcoin has the time-varying characteristic and the fluctuating agglomeration. The estimated coefficient in the variance equation is set to be greater than 0, and the conditional variance is guaranteed to be positive. It is estimate with GARCH (1,1) that that variance of the current period is assumed to be affect by the preceding 1 phase and 1 period before the square of error. After a first order difference was made on the logarithm of the price of bitcoin, the meaning of the bitcoin rate of return was obtained.

$$\mathrm{dlnp} = \mathrm{lnp} - \mathrm{lnp}(-1). \tag{2}$$

Mean value equation of first-order difference sequence:

$$\mathrm{dlnp} = \alpha_0 + \alpha_1 \mathrm{dlnp}(-1) + \mu_t . \tag{3}$$

Equation of conditional variance:

$$\sigma_t^2 = \omega + \alpha \mu_{t-1}^2 + \beta \sigma_{t-1}^2 . \tag{4}$$

As can be seen from table 3, in the variance equation, the two variable estimates of the coefficient of the P value is less than 0.01. That is, at the significance level of 1%, the coefficients of the variables are all significantly non-zero. For explain the variance of the previous and squared error of the previous period has a significant linear effect on the current condition of variance. The ARCH effect exists, which is consistent with the previous conclusion. The estimated coefficients of GARCH and ARCH in the variance equation of bitcoin price correspond to P values is 0.0000. At the 1% significance level, they are all significantly non-zero . GARCH coefficient is 0.1091, the ARCH coefficient is 0.8807, 0.9898 is less than 1, the sum of both is given to illustrate that the conditional variance is convergent.

Table 3   GARCH(1,1) estimation results

index	equation	variable	Estimated coefficient	Standard error	Z statistic	P values

Bitcon price	Mean equation	C	0.007	0.002	3.018	0.0025
		C	5.99E-05	2.70E-05	2.220	0.0264
	Conditional variance equation	RESID(-1)^2	0.1091	0.0191	5.703	0.0000
		GARCH(-1)	0.8807	0.0196	45.026	0.0000

In figure 3, we can see that the residual sequence, real value and fitting value of bitcoin price were obtained after the estimation using the GARCH model. The residual term was found to fluctuate in a small range. The real value has less fluctuation than the fitting value, and the trend is relatively consistent. Therefore, it is reasonable to believe that the fitting situation is better.

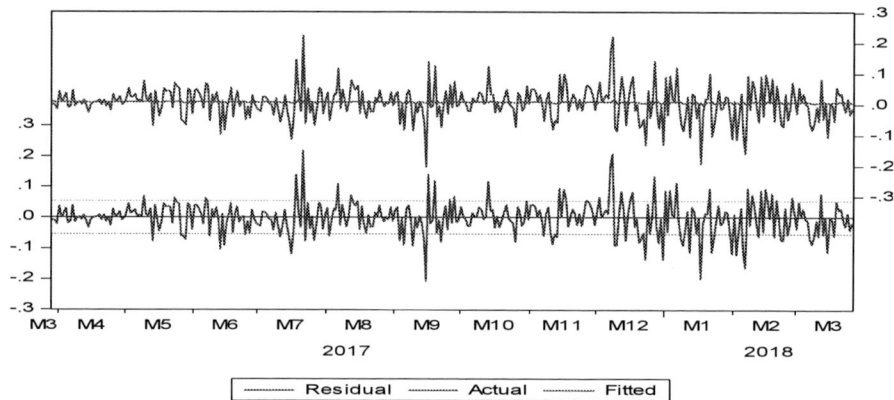

Figure 3.　Estimates the fitting value, actual value and residual error of the model

2.3. Asymmetric test of bitcoin price - TARCH model

See if there is asymmetry in the price of bitcoin through the TARCH mode, and this kind of asymmetry in the capital markets often shown as: the volatility is more sensitive to negative news than positive news in the market. The setting model is as follows:

$$\sigma_t^2 = \omega + \alpha \times \mu_{t-1}^2 + \gamma \times \mu_{t-1}^2 d_{t-1} + \beta \times \sigma_{t-1}^2. \tag{5}$$

In this formula, d_{t-1} is a virtual variable, when $\mu_{t-1} < 0, d_{t-1} = 1$ otherwise, $d_{t-1} = 0$. As long as $\gamma \neq 0$, there is an asymmetric effect. $\mu_{t-1} < 0$ means bad news, $\mu_{t-1} > 0$ means good news, and the term in the model $\gamma \mu_{t-1}^2 d_{t-1}$ is asymmetric effect term.

Table 4 estimation of asymmetric effect model

Model	Estimated coefficient	coefficient	Standard error	P value

	ω	7.03E-05	4.25E-05	0.097
TARCH	α	0.101	0.050	0.043
	γ	0.082	0.071	0.246
	β	0.851	0.040	0.000

The estimated value of the asymmetrical coefficient of TARCH model is 0.0828. And the corresponding P value is 0.2464, which indicates that the coefficient is also significantly 0 at the level of 10%. In other words, there is no linear relationship. It shows that there is no asymmetric effect on the price fluctuation of bitcoin, so there is no leverage effect. That is, good news and bad news have the same impact on bitcoin prices.

3. Conclusion

Based on the above analysis, we find that although there is no leverage effect in the price fluctuation of bitcoin. There is an ARCH effect in the price index of bitcoin, with time-varying characteristics and agglomeration of volatility. It shows the currency price fluctuations are more severe, the public investment currency risk is bigger, but the same amount of good news or bad news for the impact of the currency prices are the same.

Author information:

Juan Wang   College of economics and management, xi 'an university of posts and telecommunications

Major: financial engineering   Degree:associate professor   Email address: jane@xupt.edu.cn

Feng Tian   College of economics and management, xi 'an university of posts and telecommunications   Major: finance   Degree: master   mail: 2627693252 @qq.com

Fund information:

First.   Projects on national natural science foundation

Project name: Mechanism, influence, measurement and prevention of systemic risk under the impact of housing price   Item number:71673214

Second.   Provincial and ministerial projects

Project name: Research on the development path of China's digital currency driven by blockchain technology   Item number:2017 R18

Project name: Research on China's digital bill business under the block chain technology

Item number: 2018 R24

References

[1]   Li Jihong, Wu Xiaoxiao, Yan Haoyang. Research on the relationship between demand and price of bitcoin based on VEC model [J]. Journal of southwest university of nationalities (nature science edition),2016,42(6) : 703-707.

[2]   Yang Tao. A brief analysis of the non-monetary attribute of bitcoin [J]. Times finance,2014.

[3]   Wu Hong, Fang Yanqing, Zhang Ying. Crazy digital money - the nature and enlightenment of bitcoin [J]. Journal of Beijing university of posts and telecommunications (social science edition),2013,15(3) : 46-50.

[4]   Li Ping. Economic analysis of the new currency bitcoin [J]. Times finance,2016.

[5]   Yang Xiaochen, Zhang Ming. Bitcoin: operating principle, typical features and prospect [J]. Financial review,2014.

[6] Jia Liping. Theory, practice and influence of bitcoin [J]. Monetary theory and policy,2013.
[7] Niu Yueyu. Factors influencing the formation mechanism of bitcoin price [J]. Times finance,2017, issue 9:220-221.
[8] Liu Gang. Bitcoin price volatility and virtual currency risk prevention - policy research method based on sino-american information [J]. Journal of guangdong university of finance and economics,2015, 3rd issue: 30-40

Design and Realization of Scenes of 3D Virtual Digital Library

Wang Li

Department of education science and technology Shanxi Datong University Datong, China

Wangli523971@126.com

Abstract. 3D virtual digital library is a virtual library environment realized by using virtual reality and other technologies. It can realize the library scene and information resource three-dimensional display. It provides a more direct and convenient way for the inquiry and reading of book resources. In the construction of 3D virtual digital library, the establishment of virtual scene is the key. Taking the three-dimensional virtual digital library of Shanxi Datong University as an example, this paper discusses the design method of 3D virtual digital library, the process of virtual scene design, and the realization technology and method of virtual scene, providing reference for the construction of virtual digital **library.**

1. Introduction

As a public place for people to obtain information materials and knowledge, library plays an important role in the development of education industry. However, the traditional library navigation system is mostly introduced through pictures or words. After new readers enter the library, they often need to spend a lot of time and energy to understand the library's internal environment, the location of various books or the use of electronic equipment.

Three-dimensional digital library is established by using virtual reality technology, computer network technology, database technology and so on. It can let readers stay at home, through the network to learn the internal structure of the library, and through the keyboard and mouse for related operations, "roaming" in any corner of the library, to achieve the immersive feeling; at the same time, you can also preview three-dimensional books and achieve the book lending. However, in the process of building 3D virtual digital library, the establishment of 3D virtual scene is the key.

2. Design method of 3d virtual digital library

At present, three-dimensional virtual libraries in China can be divided into two types according to different modeling methods: image-based virtual libraries and geometric-based virtual libraries. The generation of scene of the first type virtual library depends on the 360 degree panoramic technology of the image. When this technology is applied to generate scenes, professional cameras should be used first to capture the image information of the whole scene, and then the images should be pieced together to form a panoramic image with a horizontal perspective of 360 degrees and a vertical perspective of 180 degrees [1]. The generation of the second kind of virtual library scenes needs to use 3D modeling software (such as 3D Studio Max) to actually construct the 3D scenes and generate the corresponding 3D models.

Each of these technologies has advantages and disadvantages. The first method has the advantages of fast development speed, good simulation effect and less computing resource consumption. The

Content from this work may be used under the terms of the Creative Commons Attribution 3.0 licence. Any further distribution of this work must maintain attribution to the author(s) and the title of the work, journal citation and DOI.

Published under licence by IOP Publishing Ltd

disadvantage is that the system lacks interactivity and the model lacks stereoscopic sense [2]. This kind of technology is more suitable for virtual systems with large scenes (such as virtual cities) and need real-time imaging [3]. The advantage of the second method is that the model is realistic and three-dimensional, and the system is interactive. The disadvantage is that the system requires more workload, takes longer development cycle, takes up more computing resources, and the final effect is restricted by models quality [4]. Such technologies are more suitable for establishing virtual museum system, virtual museum system, virtual campus system and landscape design system [5].

This article takes the library of Shanxi Datong University as an example, uses the method of geometric modeling, uses "SketchUp" and "3D Studio Max" software to model the scene of the library of Datong University, establishes the external structure and internal space layout of the three-dimensional virtual library, so that users can understand the internal structure and environment of the library through the computer. Create a sense of being on the scene.

3. The design process of the scene of 3d virtual digital library

The construction of library virtual scene should be based on real library. Before the construction, we need to collect relevant data, that is, we need to obtain the actual size of the library building and take relevant pictures of the building to prepare for the later modeling and mapping. When making virtual scenes, first of all, the floor plan of the library should be drawn with "AutoCAD" according to the actual size and layout of the library building, so as to make clear the area division of the library and the plane layout and position of each building. Secondly, building walls are modeled using "SketchUp" software. Then, the library's internal facilities are modeled with the abundant modeling technology in 3D Studio Max. Finally, the models made with "SketchUp" are imported into "3D Studio Max" and all the models are merged. After the model of 3D virtual library is made, material should be given to the model to accurately express the physical properties of color, texture, reflection and refraction of light. In order to be able to simulate real library scenes, we need to add lighting effects to the scenes to highlight the layers of the scenes and make the scenes more real and natural. To reduce CPU computation time, we can use texture roasting technology to transform light information into texture form. [6]

4. Realization of 3d virtual digital library scene

4.1. Make the floor plan

Use tools of "AutoCAD" such as "lines" and "arcs" to draw the outline first. Then the "offset" command is used to shift all the edges out of the wall thickness, thus forming the outer outline of the whole library. The library plan is shown in figure 1.

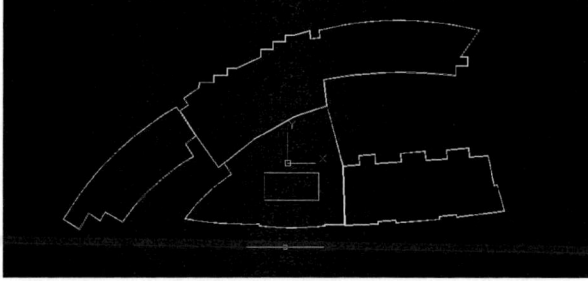

Figure 1. Library plan

4.2. Wall modelling

Import the floor plan made in "AutoCAD" into "SketchUp", and use "SketchUp" to build the wall model of the library. The detailed process is as follows:

- After importing the plan into "SketchUp", we use the "line" tool to make the curve. When the curve is closed, the system will automatically generate the plane.
- Using the "push and pull" tool to build the wall. We calculate the height of each floor and pull out the wall according to the height of the library collected. On this basis, the top and bottom of the building are respectively added with a thickness for the roof and foundation.
- Dividing the layout of the library. We use the "line" tools to divide the layout of each classroom, corridor and so on, and then use the "push and pull" tools to push out the wall, which serves as the partition between each classroom.
- Make doors and Windows. We need to first determine the width, height and position of the door frame and window frame, then draw with the "line" tool, delete the excess surface and seal the unsealed surface, and finally export the model to the format of 3ds. The effect of the model after completion is shown in figure 2.

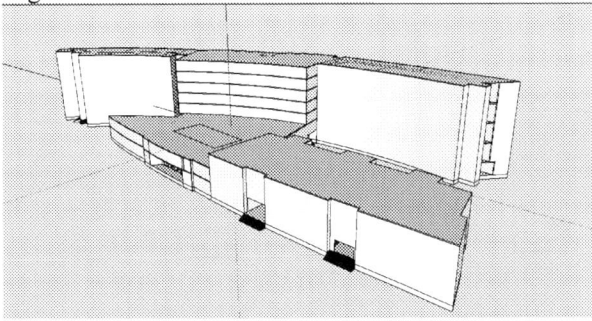

Figure 2. Overall effect of library wall

4.3. Modeling of objects in the library

4.3.1. Selection of modeling methods. The scenes in the library are relatively complex and there are many modeling methods used. There are different modeling methods for different objects, and there may be multiple modeling methods for the same object. How to choose the appropriate modeling method plays a key role in the performance of virtual library system. Combined with experience, how to choose the modeling method is summarized as follows [7]:
- Direct modeling of basic three-dimensional forms. For objects that can be modeled through the standard geometry and extended geometry provided in 3D Studio Max and their mutual combinations, modeling can be carried out directly, such as regular tables, chairs, bookshelves in the library.
- Polygon modeling. For objects that need to take a standard geometry or an extension set as the initial shape, but have a complex structure, the initial shape shall be converted into "editable polygon" or "editable grid", and the points and surfaces shall be selected in the editing panel for modification to achieve the ideal modeling effect, such as the computer monitors in the library.
- Two-dimensional figure modeling. For some objects with complex shapes, which cannot use standard geometry or extended geometry, we can try to use 2d graphics to shape them, and then transform them into 3D solid models by adding corresponding modifiers, such as floor vases in the library.
- Compound object modeling. Compound object modeling. Composite object modeling is often used for modeling where two or more objects need to be combined to form an object in a particular way, such as the doors and Windows of the library.
- Map modeling. If using conventional modeling methods will result in too large number of points and faces, you can use methods such as map modeling. Map modeling is used to model plants and objects that are less demanding and far away from the model, Such as the library of potted, computer keyboard.

In the process of modeling, if multiple modeling methods can be used to complete the same model, the method that produces the least number of model faces should be selected. When modeling the distant object, due to the low visual requirements, the "segmentation" can be appropriately reduced to reduce the number of faces of the model.

4.3.2. Implementation of modeling. Let's take the bookshelf model making in the reading room as an example to illustrate the concrete realization process of modeling.

- The model making of the bookshelf: First, we create a new cuboid, and set its length, width and height according to the actual scale. Its rounded Angle is set to 5mm. Then we use the "rotate", "move", "etc scale" tools to build the bookshelf.

- Model making of books: First, we need to create a rectangular wireframe and convert it to editable spline. In the modify panel, select the line segment hierarchy and use the optimization tool to add a point to the middle of one of the edges. Then, we select the added point at the "point" level, change the type of the point to "Bezier Angle point", change the type of the left and right points to "Angle point", and adjust the curve of the line. We then add an "extrude" modifier to the completed spline, set its height and convert it to an "editable polygon." We select the edges to edit and select the connect command to add two edges. We then add the extruder for the three sides. Finally, we put the books on the bookshelf one by one, and adjust the size and direction of the book respectively with the zoom tool and the rotation tool, so as to achieve a casual and natural effect, and then assign the material collected from the Internet to the model. The final shelf rendering is shown in figure 3.

Figure 3. Rendering of the model of the bookshelf

5. Material making

The virtual library mainly involves glass and texture materials. The glass material needs to be adjusted for "highlights" and "opacity" and "ray tracing" added to the reflection and refraction channels. Texture material needs to adjust corresponding texture parameters.

6. Model integration

After each scene is finished, it is necessary to put each model in the corresponding position according to the CAD floor map drawn earlier, and integrate each scene into a complete virtual library. The effect after model integration is shown in figure4 and 5.

Figure 4. Hall on the first floor

Figure 5. Hall on the second floor

7. Conclusions

The selection and application of the modeling method is very important for the construction of large scale scenes. The most critical problem in modeling is the problem of the number of surfaces of models, which is related to the computer system's rendering and running speed; the design should not only meet the needs of the user's vision, but also ensure that the model contains the least number of surfaces. The appropriate modeling method will make model planes appropriate, facilitate follow-up material and other processes carried out smoothly. [8]

Acknowledgment

This research is a phased achievement of the 13th Five-Year Plan of Shanxi Education Science - "Construction and application of 3D virtual learning space based on VR" (Grant No. GH-18049). It was supported by the 2017 general key discipline construction project " Research on the construction of regional characteristic educational resources under the background of Internet + ", the key discipline construction plan of "1331 project" of Shanxi provincial department of education.

References

[1]   Baidu encyclopedia, 3D panoramic technology, https://baike.baidu.com/item/ 3D panoramic technology /693087?fr=aladdin, 2018.

[2]   M. M. Zhang, Research and implementation of virtual campus roaming based on Unity3D, in: Yunnan University, China, 2014.

[3]   Y. S. Zhong, Development of 3D virtual campus roaming system based on Unity, in: Yunnan University, China, 2014.

[4]   L. Wang, "Design and implementation of 3D virtual campus roaming simulation system based on Untiy3D" . J. Education informatization in China, 2016 (09): pp.60-63.

[5]   L. Wang, "Researching of the three-dimensional virtual simulation campus scene's construction

technology". J. The Open Cybernetics and Systemics Journal, Volume(2015) 9: pp.1056-1057.

[6] L. Wang, "Research on 3D modeling technology of large scale scene based on 3D Studio Max".C. The Organizing Committee of ICCET 2015. Guangzhou, China, pp. 1495～1096 . November 28-29, 2015. (references)

End-to-end Speech Synthesis for Tibetan Lhasa Dialect

Lisai Luo[1], Guanyu Li[1*], Chunwei Gong[1], Hailan Ding[1]

[1]Key Laboratory of National language Intelligent Processing Gansu Province, Northwest Minzu University, Lanzhou, China

*guanyu-li@163.com

Abstract. Speech synthesis for Tibetan Lhasa dialect is implemented on the basis of an end-to-end novel speech synthesis framework, Tacotron. The training transcript has used the phoneme list transcribed from Tibetan characters, and feature parameters were extracted from the mel-spectrogram. Then the model is trained by the mapping of character to spectrum. Tibetan language is an important minority language of the Chinese nation, but there is little research on Tibetan language at present. The experimental results were compared with traditional speech synthesis methods, with the audio quality significantly better than that of the traditional GMM-HMM in both naturalness and rhythm. It provides a crucial reference for the later research methods of Tibetan language and promotes the development of Tibetan language research.

1. Introduction

Tibetan language is spoken by about 6 million people mainly distributed in 5 districts in China, including Tibet, autonomous region and Qinghai province, Sichuan Ganzi Tibetan autonomous prefecture, Aba Tibetan and qiang autonomous prefecture, as well as in Gannan Tibetan autonomous prefecture and Diqing Tibetan autonomous prefecture in Yunnan province. Some residents lived in Bhutan, India, Nepal and Pakistan also make Tibetan as their mother language. Tibetan is an influential language with long history. In recent years, more and more attention has been paid to Tibetan research and application to improve the level of informatization of Tibetan. There are 3 Tibetan dialect areas in China according to the characters of Tibetan dialects: Weizang, Kangba and Amdo. There are several dialects in each area, dialect in the same area is more similar to the dialects in other areas. As an important and influential dialect in Weizang, Lhasa dialect is chosen as the research object in this paper.

Research on speech recognition and speech synthesis in Tibetan Lhasa dialect has made progress in recent years. In this paper, speech synthesis of Lhasa Tibetan is implemented based on a novel end-to-end speech synthesis framework, Tacotron, proposed by Google in early 2017. A text-to-speech synthesis system typically consists of multiple stages, such as a text analysis frontend, an acoustic model and an audio synthesis module. Building these components often requires extensive domain expertise and may contain brittle design choices. However, Tacotron is an end-to-end generative text-to-speech model and it can synthesizes speech directly from characters [1].

The section 2 introduces the overall structure of Tacotron, and the section 3 introduces the data preparation and related work, as well as the design of corpus data. Then the experimental results are shown in the section 4. The section 5 is the summary and prospect.

2. Model architecture

Content from this work may be used under the terms of the Creative Commons Attribution 3.0 licence. Any further distribution of this work must maintain attribution to the author(s) and the title of the work, journal citation and DOI.

Published under licence by IOP Publishing Ltd

This part mainly introduces the overall architecture of Tacotron model and the CBHG module structure used in encoder and decoder module.

2.1. Tacotron model architecture

The Tacotron framework is a relatively novel end-to-end TTS model. The model input is a text character, and the output is a spectrum diagram parameter. Finally, the corresponding audio is generated using the Griffin-Lim algorithm. Figure 1 depicts the model, which consists of an encoder, a decoder, two CBHG modules, and a post-processing network. The input (pre processed text and audio) is fed into an encoder, and generates attention features which are then used in every step of the decoder before generating spectrograms.

Another fascinating mechanism used in Tacotron is called attention, which is used in every step of coding and decoding. At present, the model based on attention mechanism has been widely used in machine translation, speech synthesis, speech recognition and computer vision. Attention mechanism is used in the speech synthesis model to realize end-to-end speech synthesis. Attention mechanism has great promotion in the sequence of learning tasks at decoder framework, through the model for the attention of in the code segment. The source data sequence weighted data transformation or introduction of Attention at the decoding end model, weighted changes of target data, can effectively improve the system performance.

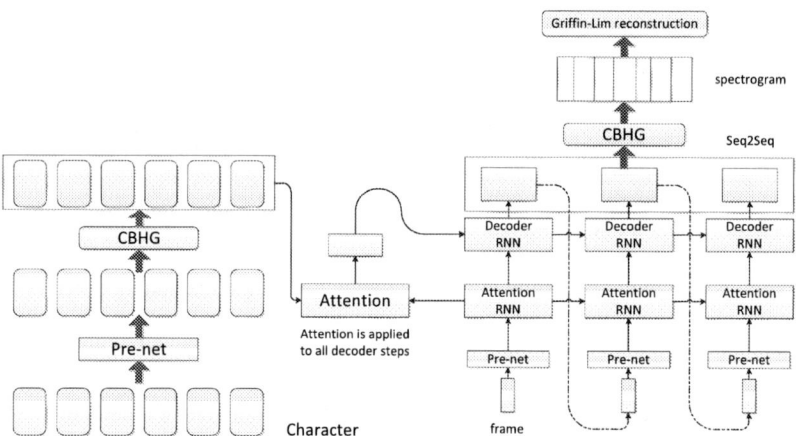

Figure 1. The Tacotron architecture adapted from paper [2].

2.2. CBHG module

CBHG consists of a bank of 1-D convolutional filters, followed by highway networks and a bidirectional GRU (Gated recurrent unit) recurrent neural network (RNN) [2]. Its function is extracting valuable features from the input, which is beneficial to improve the generalization ability of the model. Figure 2 depicts its structure.

The CBHG module contains a one-dimension convolution kernels, followed by a Highway network and a bidirectional GRU. CBHG is a powerful module for extracting representations from sequences. The input sequence is first convolved with K sets of 1-D convolutional filters. And there is a maximum pooling operation after each convolution layer. During the encoding and decoding, reduces more training time because of without using the time-consuming RNN structure. Batch normalization is used for all convolution layers in this module. The output of convolution layer is fed into a multilayer highway network to extract high-level features. At the top layer of the model, a bidirectional GRU is stacked to extract sequence features bidirectionally. By training, it is found that using CBHG module can effectively improve the generalization ability of the model.

Highway Networks [4] is a new kind of neural network structure. The traditional neural network structure is much better in deep layer than in shallow layer. However, the deep neural network makes model training more and more difficult, while Highway can use simple SGD method to train extremely deep network, and with the increase of network depth, the network can be optimized even if the initialization variables remain unchanged. This is achieved through a gate mechanism that controls the flow of information through the neural network. Through this mechanism, the neural network can provide a pathway to allow information to pass through without loss.

The input x is converted into y output by the activation function H in the traditional neural network, and w is the weight in formula(1):

$$y = H(x, W_H)$$
(1)

Highway Networks of neural network, increased the two nonlinear transformation layer, one is T (transform gate) and a is C (carry gate), T represents the input information through convolutional or recurrent converted part of C said x retain part of the original input information in formula(2):

$$y = H(x, W_H) \bullet T(x, W_H) + C(x, W_H)$$
(2)

To simplify, replace C with 1 minus T in formula(3):

$$y = H(x, W_H) \bullet T(x, W_H) + x(1 - T(x, W_H))$$
(3)

The dimensions of x, y, H and T must be consistent. Sub-sampling or zero-padding strategies can be adopted to make the dimensions consistent. Several formulas are compared, and formula (3) is more flexible than formula (1), and there is a special case, for example, when T= 0, y = x, the original input information is all retained without any change, when T = 1, Y = H, the original information is all converted, and the original information is no longer retained, just equivalent to an ordinary neural network. As shown in formula (4):

$$y = \begin{cases} x & \text{if } T(x, W_H) = 0 \\ H(x, W_H) & \text{if } T(x, W_H) = 1 \end{cases}$$
(4)

3. Experiments and analysis

Data for Lhasa dialect speech synthesis is prepared and preprocessed. Several experiments schemes are designed to testify the performance of end-to-end Lhasa dialect speech synthesis under various conditions.

3.1. Data preparation

A large amount of linguistic data is necessary to study end-to-end speech synthesis for Lhasa dialect. The speech corpora are recorded on PC in quiet environment and saved as PCM wave files, where the sampling rate is 16kHz, the sample size is 16 bits, and the vocal tract is mono type. All the speakers are in Lhasa dialect, and the recording style is reading. Transcripts in training are saved as files with suffix "mlf" or "txt". However, there are various ways to depict input sequences of transcript according to different languages. For example, in English, 26 letters with punctuation transcript can be used directly as the input sequence. The Korean language has its own set of alphabets, each of which can use Unicode code as its transcript character. For Mandarin, there are about more than 3,000 often-used characters, it is too complicated to enumerate all characters in the system. And Among which there are many homophones, so the Chinese pinyin is commonly used to transcript them. Tibetan scripts are written in alphabets.

From view of written form, there are 30 consonant letters and 4 vowel signs in Tibetan (note all dialects are the same in writing). Each syllable is a combination of several consonant letters and a vowel sign. Words are comprised of one or several syllables. Each syllable involves a radical consonant letter, and other consonant letters could be appended to the radical consonant as superscript, subscript, prescript, postscript and post-postscript to form a syllable. Constitution of syllable described in figure 3. The radical consonant, the prescript and superscript consonants together form the initial part of a syllable, and the vowel sign, the postscript and post-postscript consonants altogether form the final part. The radical consonant, the prescript and superscript consonants together form the initial part of a syllable, and the vowel sign, the postscript and post-postscript consonants altogether form the final part.

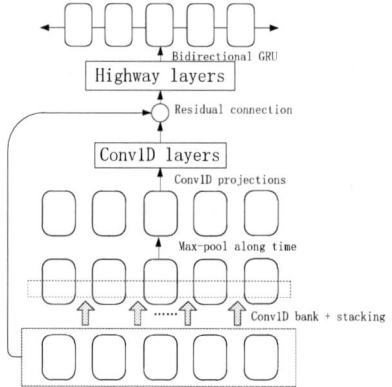

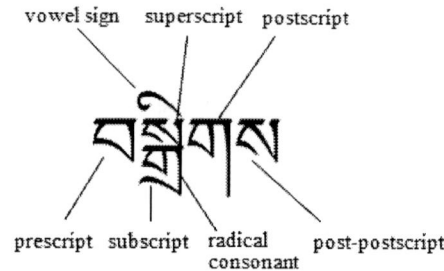

Figure 2. The CBHG module adapted from paper [2].   Figure 3. Constitution of syllable.

A syllable lexicon involves a set of syllables whose pronunciations are defined. There are more than 8,000 possible syllables in Tibetan, including syllables for foreign words. We construct the syllable lexicon by constructing a text corpus involving 420,000 sentences (including both written and spoken), and then selected the most frequent syllables from this corpus. After removing some syllables that are for transliterating Sanskrit words only, we obtained a syllable lexicon consisting of 6,013 syllables. By applying the pronunciation rules, these syllables were segmented into initials and finals, and the initials and finals were further split into phones [3]. All these syllables and their phone sequence forms were manually checked to ensure the quality. Therefore, a set of standard rules is given in the experiment, which converts the international phonetic symbols of Tibetan characters into a list of phonemes which can be written and adjusted conveniently. Sentences are transformed into list of phonemes on the basis of pronunciation dictionary. Then the list of phonemes is input to the model as input sequence.

For example, there is a sentence in the data set:

ཆེད་དུ་བྱ་བའི་གཅལ་བྱ་ཐེག་དམན་གྱི་རིགས་ཅན་ལ་གཙོ་བོར་དགོངས་ནས་གསུངས་པའི་བྱ་སྲོང་།

The sentence is transformed into a list of phonemes:

tjh eb th u tjh a w el t yw tjh a th e k m elu c h i r i k tj elu l a ts o ph ow k o ng n elb s u ng p el t e n eyb.

Sentences can also be transformed into lists of initials and finals. As mentioned above, all the texts are converted to the form of above phoneme transcript before loading the data, which is done by converting it to a file with the suffix "mlf" or "txt".

3.2. Experiments settings

Various experimental schemes are designed to testify the performance of Lhasa dialect under various conditions to find a best scheme to implement the final speech synthesis system.

• 8,000 sentences spoken by 21 male speakers are chosen to train the model.

• 13,000 sentences spoken by 21 male speakers and 13 female speakers are chosen to train the model.

• 20,000 sentences spoken by 23 female speakers are chosen to train the model.

• 5,000 sentences spoken by one male corpus were used for training at the beginning. After 352k iterations training, the female corpus was added to continue training this model. There are more than 5,000 sentences spoken by one male and 20,000 spoken by 23 female speakers.

4. Experimental Results

The part of the experimental results is shown here and only the third experiment is shown in this paper. We use synthetic spectrum to compare with original spectrum. Figure 4 is the original spectrum diagram, Figure 5, 6, 7 respectively are iteration of 795k, 860k and 1000k training times.

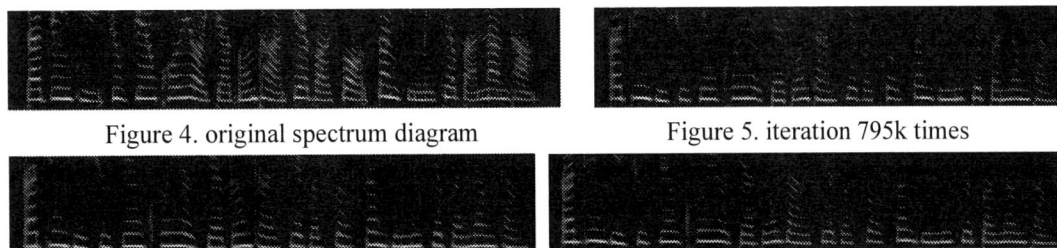

Figure 4. original spectrum diagram   Figure 5. iteration 795k times

Figure 6. iteration 850k times   Figure 7. iteration 1000k times

Our experimental results show that more corpus and the same sex data training model will be more effective than less corpus and different sex, however, the generalization ability of different gender training is better. The result of End-to-end Speech Synthesis for Tibetan Lhasa Dialect is outperforms the traditional GMM-HMM system, and we will continue to optimize this model.

5. Conclusions

The above is the research progress up to now, and a good experimental result has been found through design and experiment. However, the structure of Tibetan character is two-dimension, its writing particularity restricts the flexibility of processing, so it must be transcript. Tacotron model structure is one of the classical structures in speech synthesis technology. The model will be adjusted to make the training speed more optimized in the next study. We will also try to use different research methods to further study speech recognition and speech synthesis, and apply it to Tibetan language and different languages. Further, the technology has achieved a breakthrough.

Acknowledgments

Supported by the Fundamental Research Funds for the Central Universities (Research on Unified Acoustic Models of Mandarin and Tibetan, 31920170145)

References

[1] K. Park, "A TensorFlow implementation of Tacotron: A fully end-to-end text-to-speech synthesis model," 2017, Available at GitHub, https://github.com/ Kyubyong/tacotron(2018).

[2] Wang, Y., Skerry-Ryan, R., Stanton, D., Wu, Y., Weiss, R. J., & Jaitly, N., et al. (2017). Tacotron: towards end-to-end speech synthesis. 4006-4010.

[3] Guanyu Li et al, "Free Linguistic and Speech Resources for Tibetan", (ASC 2017)

[4] R. Kumar Srivastava et al, "Highway Networks", (ICML 2015)

[5] H. Tachibana et al., "Efficiently Trainable Text-to-Speech System Based on Deep Convolutional Networks with Guided Attention", (ICASSP 2018)

[6] I. Goodfellow et al., Deep Learning, MIT Press, (2016)

[7] J. Sotelo et al., "Char2wav: End-to-end speech synthesis", (ICLR, 2017)

[8] K. Tokuda et al., "Speech Synthesis Based on Hidden Markov Models",(*Ipsj Magazine* ,2013)

[9] J. Lee, K. Cho, T. Hofmann. Fully CharacterLevel Neural Machine Translation without Explicit Segmentation. Transactions of the Association for Computational Linguistics, vol. 5, pp. 365–378, 2017.

[10] W. Ping, K. Peng, Andrew Gibiansky et al. Deep Voice 3: Scaling Text-to-Speech with Convolutional Sequence Learning. Published as a conference paper at ICLR 2018.

[11] T. Le Paine, P. Khorrami et al. Fast Wavenet Generation Algorithm. In Technical Report 2016.

[12] J. Shen, R Pang, et al. Natural TTS Synthesis by Conditioning WaveNet on Mel Spectrogram Predictions. Accepted to ICASSP 2018

[13] J. Engel, C. Resnick, A. Roberts et al. Neural Audio Synthesis of Musical Notes with WaveNet Autoencoders. Accepted to ICML 2017.

[14] A. van den Oord, S. Dieleman et al. WaveNet: A Generative Model for Raw Audio. in 2016.

[15] Qin Y, Song D, Chen H, et al. A Dual-Stage Attention-Based Recurrent Neural Network for Time Series Prediction[J]. 2017:2627-2633.

[16] Torfi A, Shirvani R A, Soleymani S, et al. Attention-Based Guided Structured Sparsity of Deep Neural Networks[J]. 2018.

[17] Hudson D A, Manning C D. Compositional Attention Networks for Machine Reasoning[J]. 2018.

Research on China`s Transport Connectivity Along Corridors of the Belt and Road Initiative

Jiao Wenwen, Li Ran, Li Sicong

Research Institute of Highway Ministry of Transport, No. 8 of Xitucheng Road, Haidian District, Beijing City, 100088, China

Jww0821@sohu.com

Abstract. By brief introduction of the Belt and Road initiative, which was proposed by Chinese government, this paper reviews the literature and puts forward the importance of transport connectivity for countries along the Belt and Road routes. Then, the present situation of China`s transport connectivity with the countries along the Belt and Road routes has been summarized. The author thinks that although the initiative has achieved notable outcomes, especially referring to the "Six Corridors Building", transport connectivity between China and the Belt and Road countries still faces some challenges in lagging infrastructure construction, limited transport service, weak transport hubs and various risks. Finally, this paper gives some suggestions to solve problems both domestically and internationally.

1. The Belt and Road initiative and transport connectivity

1.1. Overview of the Belt and Road initiative

In the fall of 2013, Chinese President Xi Jinping proposed the strategic framework of building the "Silk Road Economic Belt" and a counterpart "21st Century Maritime Silk Road," collectively referred to in abbreviated form as the "Belt and Road"(B&R) initiative. The initiative calls for policy coordination, facilities connectivity, unimpeded trade, financial integration and people-to-people bonds, in order to promote the establishment of an open, inclusive and balanced regional economic co-operation architecture that benefits all.

Five years ago, the B&R initiative was only a concept starting from China. Now, it runs through the continents of Asia, Europe and Africa, encompassing countries with huge potential for economic development, and has achieved notable outcomes.

With the common understanding of the B&R initiative, 103 countries and international organizations have signed 118 cooperation agreements with China. And the UN has adopted the "Belt and Road Initiative".

A lot of projects have been launched, solidly promoting the substantial cooperation. For example, the Madaraka Express, connecting the Kenyan port of Mombasa with Nairobi, started operation in 2017. The Addis Ababa-Djibouti Railway, connecting Ethiopia with Djibouti, began operation from 2016. The China-Thailand railway, connecting Kunming and Bangkok, has also started and is expected to be completed by 2021.

The constant cooperation has boosted the trade development. By the year of 2017, the total import and export volume between China and the "Belt and Road" countries has reached US$1,440.32 billion, accounting for 36.2% of China's total import and export trade at that time. Besides, the overseas

Content from this work may be used under the terms of the Creative Commons Attribution 3.0 licence. Any further distribution of this work must maintain attribution to the author(s) and the title of the work, journal citation and DOI.

Published under licence by IOP Publishing Ltd

economic and trade cooperation zones built in the countries along the B&R routes have created hundreds of thousands of jobs, and contributed billions of dollars for local taxes.

In addition, eleven Chinese banks have established 71 first-level institutions. Co-financing cooperation with multilateral development banks, such as the African Development Bank, the Inter-American Development Bank, and the European Bank for Reconstruction and Development, has provided solid support for project construction. International cultural and tourism exchanges get much closer than before. In the year of 2017, there were more than 300,000 international students from the countries along the route. It is estimated that by 2020, the tourism consumption will be about 110 billion US dollars.

1.2. Role of transport connectivity under the Belt and Road initiative

As an important industry for social and economic development, transportation is a prerequisite for interconnect [1]. Most of the countries along the "Belt and Road" route, are developing country with a low level of economic development. The weak domestic transportation infrastructure is the bottle neck to economic improvement. The B&R initiative gives priority to the development of transportation infrastructure, in order to constantly promote the investment and trade environment.

Grigoriou found that the transport infrastructure of neighboring countries was important for inland Central Asian countries due to the existence of transit effects [2]. Edwards and Odendaal took 117 countries in the year of 2005 as samples to explore the impact of transport infrastructure quality on export trade, and it was found that the lowest quality infrastructure among trading partners had the greatest impact on transport costs, which in turn affected export trade [3]. By using GTAP model, Iwata made a study of the construction of international transportation infrastructure projects in the Mekong River Basin, and the conclusion was drawn that these transportation infrastructure promoted GDP growth in Laos and other countries in the Mekong River Basin [4]. Behrens found that transportation costs played a decisive role in trade flows, and countries with high levels of transport infrastructure could generate greater international trade flows and more balanced regional economic growth [5]. Francois and Manchin using a gravity model study found that improvements in transportation infrastructure led to increased trade flows [6].

In recent year, different from foreign researchers, China`s domestic researchers paid more attention to the empirical research on the countries along the B&R route. Xu and her fellows simulated and analysed the economic and trade effects of the transportation infrastructure construction, targeting the six economic corridors, and the result was that the transportation infrastructure would benefit the countries along the B&R route [7]. Based on the new economic growth and the new economic geography theory, Ni and Wang selected 14 countries along the B&R route between 2009 to 2015 for panel data research, by adopting the spatial spillover effect model, they came to the conclusion that the promotion of transportation infrastructure would boost the economic development of the selected countries [8]. Zhang and Yu conducted an empirical analysis of the influence of the transportation infrastructures on the export trade, and the conclusion was that the transportation infrastructures of the importing countries and their neighboring countries had a positive effect on China`s export trade [9].

2. Status quo of China`s transport connectivity

2.1. Transport corridor

The B&R initiative takes advantage of international transport routes as well as core cities to further strengthen collaboration. And there are six international economic co-operation corridors identified, namely the New Eurasian Land Bridge, China-Mongolia-Russia Economic Corridor, China-Pakistan Economic Corridor, China-Central and Western Asia Economic Corridor, China-Indochina Peninsula Economic Corridor, and Bangladesh-China-India-Myanmar Economic Corridor. Within these corridors, the transportation network consisting of railways, highways and waterways together realizes the infrastructure connection [10]. There is no precise definition of the above six corridors from a geographical view, and these corridors aim to link major economic centers of the involved countries.

2.1.1. New Eurasian Land Bridge. Different from the Siberian Land Bridge starting from Russia, the New Eurasian Land Bridge, also known as the Second Eurasian Land Bridge, is an international railway line beginning from Lianyungang in China's Jiangsu province, passing Alashankou in Xinjiang province, stretching through Kazakhstan, Russia, Belarus and Poland, and reaching a number of coastal ports in Europe. The 10,800-km-long transcontinental rail link serves more than 30 countries and regions. Within China, this corridor mainly includes Lanzhou-Lianyungang Railway and the Lanzhou-Xinjiang Railway, extending to the eastern, central and western China.

2.1.2. China-Mongolia-Russia Economic Corridor. As the first multilateral cooperation plan responding to the B&R initiative, the China-Mongolia-Russia Economic Corridor relies heavily on infrastructure cooperation, especially the railway. In 2015 the three governments agreed to establish a Mongolian–Russian–Chinese joint railway transportation and logistics company, with the hope to benefit the remote areas of these countries. There are two key traffic arteries within this corridor: one extends from China's Beijing-Tianjin-Hebei region to Hohhot in Inner Mongolia and to Mongolia and Russia; the other extends from China's Dalian, Shenyang, Changchun, Harbin and Manzhouli to Russia's Chita.

2.1.3. China-Pakistan Economic Corridor. This corridor connects Kashgar in China's Xinjiang province, to Pakistan's Gwadar Port. A series of transportation projects are involved, in order to build and modernize the overland connections, including the upgrade and renovation of the Karakoram Highway, an expressway at the east bay of Gwadar Port, a new international airport, an expressway from Karachi to Lahore, the Lahore rail transport orange line. By implementation, a well-connected, integrated and dynamic economic belt between China and the coast of Pakistan will be built up.

2.1.4. China-Central and Western Asia Economic Corridor. Following the ancient Silk Road, the China-Central and Western Asia Economic Corridor starts from China`s Xinjiang, transits Alashankou on the China Kazakhstan border, joins the railway networks of Central Asia and Middle East, and finally reaches the Mediterranean coast and the Arabian Peninsula. There are mainly seven countries involved in this corridor, among which Kazakhstan, Kyrgyzstan, Tajikistan, Uzbekistan and Turkmenistan are in Central Asia, and Iran and Turkey are in West Asia. This corridor comprises some existing transportation infrastructure, like the Kamchiq Tunnel in Uzbekistan.

2.1.5. China-Indochina Peninsula Economic Corridor. This corridor links China with the Indochina Peninsula, and it is critical to the contiguous ASEAN states, like Vietnam, Laos, Cambodia, Thailand, Myanmar and Malaysia. ASEAN has one of the more connected transport networks among the developing regions of the world [11], and this corridor is expected to boost China's cooperation with the ASEAN countries. Among the extensive transportation network, China has completed a number of transportation project, like an expressway leading to the Friendship Gate and the port of Dongxing at the China-Vietnam border.

2.1.6. Bangladesh-China-India-Myanmar Economic Corridor. This corridor refers to four countries, namely Bangladesh, China, India and Myanmar, serving the reginal market of over 400 million people. It is comprised of expressways and high-speed rails, and in addition to the land bridge, air and water ways connecting each other will further enhance regional interconnectivity. It is planned to build up an international transport corridor across Southeast Asia and South Asia, starting from Kunming of China, through Myanmar to Bangladesh, and directly to Chattogram Port, which is expected to get significant distance and time savings for trade.

2.2. Railway

There are eight cross-border railways joined up between China and the countries along the B&R routes. And among the eight railways, there are four along the China-Mongolia-Russia corridor, two along the New Eurasian Land Bridge, and the other two along China-Indochina Peninsula corridor. In addition, some border crossings have railways linking to China`s inland cities, such as Heihe border crossings (located in Heilongjiang Province) and etc.

Referring to the infrastructure technical standard grading, the railways in China are generally higher than those in the interconnected countries. For China`s domestic railways, they mainly follow China`s national railway ClassItechnical standards. However, the interconnected railways in the countries along the B&R routes are basically single-track railways, with lower technical levels and smaller transportation capacity.

By the year of 2017, the cross-border rail network has bridged 35 Chinese cities with 34 European cities in 12 countries. The destination cities in Europe are mainly concentrated in Moscow, Hamburg, Duisburg, Lodz and Rotterdam logistics centres.

Since China-Europe "Silk Road" freight trains began to operation in the year of 2011, these trains have made a total of 6,235 trips on 57 routes by the end of 2017. The freight service has been considered a significant part of the B&R initiative, and greatly boosted the trade between China and the countries along the B&R routes.

2.3. Road

There are total fifty-two cross-border highways joined up between China and the countries along the B&R routes, including eighteen along the China-Mongolia-Russia corridor, eleven along the New Eurasian Land Bridge, three along the China-Central Asia-West Asia corridor, ten along the China-Indochina Peninsula corridor, one along the China-Pakistan corridor, and the other nine along the Bangladesh-China-India-Myanmar corridor.

Referring to the highway technical standard grading, the highways in China have basically realized the standards of expressway or national & provincial trunk lines. Among the border crossings in China, seven have expressway, and ten have first-class highway. However, the interconnected highways in the countries along the B&R routes are mostly substandard, especially in the Mongolia region with poor technical situation.

According to China`s Ministry of Transport, there are 73 ports have already carried out international road transport business. The annual passenger volume is around 8 million, and the freight volume about 50 million tons.

2.4. Inland waterway

There are mainly five cross-border inland waterways joined up between China and the countries along the B&R routes, namely Heilongjiang River, Songhuajiang-Amur River, Lancangjiang-Mekong River, Dulongjiang-Ayeyarwady River and Honghe River. China`s domestic inland waterways are usually Class v5, however the inland waterways of the interconnected countries are lower grade.

Among the above five waterways, Lancangjiang-Mekong River gets more attentions than others, mainly because of its importance for the involved countries. Lancangjiang-Mekong River is an international river flowing through China, Laos, Myanmar, Thailand, Cambodia, and Vietnam. It official began the international shipping from the year of 2001. From 2002 to 2004, China, Laos, Myanmar and Thailand jointly implemented the first phase project for the waterway remediation, and improved the navigation situation from China-Myanmar Boundary Marker 243 to Ban Houayxay in Laos. The second phase starts from China-Myanmar Boundary Marker 243, and will ends at Luang Prabang of Laos. Among the second phase, the construction work from Boundary Marker 244 to Lincang Port of China will follow the standard of Class ivand with 500 tons shipping capacity.

3. Challenges for present transport connectivity

Although the B&R countries have achieved the consensus that the completion and improvement of the transport corridors will benefit trade and economy, and some construction of associated railways, highways, and ports is progressing in a steady manner, China still faces a number of challenges and will take time to realize.

First of all, the construction of some transport corridors lags behind the expectation. For example, the present transportation connecting China and Europe mainly transits by Russia and Kazakhstan. However, China-Pakistan land transport corridor has not yet been completed, with missing railways and ports, and poor road condition. Besides, from China`s coastal ports to Indian Ocean, port relay stations are lacking.

Secondly, the transport service is quite limited by the weakness of infrastructure construction and technical standard disintegration. For some B&R countries, the transport facilities are obviously lacking and aging seriously, resulting in the insufficient transit capacity of the corridors. In addition, the technical standards of railways and highways in different countries vary, and the complicated cross-border transport rules and the inefficient operation becomes another constraint to the seamless transportation of these corridors.

Thirdly, integrated transport hubs along the overland transport corridors are lacking. These hubs are needed not only for managing different transport modes, by also for improving the industrial convergence, in order to promote the regional cooperation. Compared with the unique position along these corridors, some western regions of China are obviously weak in forming integrated transport hubs, especially in fully using the transport facilities to get the advantage in trade and economy.

Fourthly, the complexities among the B&R countries increase the risks of transport connectivity. There are a number of countries involved in the B&R initiative, with various cultures, religions, economies and societies. When starting the transport projects, especially which are weak in economic benefits although strong in social welfares, various risks including historical, political, natural and technical factors, are easy to emerge but hard to solve.

4. Suggestions

The cross-border movement of goods and passenger will boost the regional economic growth, and an integrated transport system will play a leading role during this process. To strength China`s transport connectivity along the corridors of the B&R initiative, both domestic and international efforts are needed.

Domestically, China needs to speed up to complete a national integrated transport system, identify key regions and give the priority to related projects, and enhance the transport capacity of border areas. Firstly, China must strive to build an open and comprehensive transportation system. Through facility connectivity, policies, rules and standards can work on key corridors, cities, and projects. Secondly, it is needed for China to carefully select economic centres along the corridors, then leverage their positions, and maximize the benefits. Different from China`s 18 provinces participation in the B&R initiative with equal importance, some centres in some provinces may have much higher degree of integration with the regional development. Thirdly, borders have a strong influence on the corridor connectivity. For different border cities in China, it is proposed that the construction of transportation facilities closely combine with city industrial layout, resource utilization and foreign exchange.

When taking the international perspective, both benefits and risks are necessary to fully considerate. On one hand, the benefits should be taken for the most stakeholders and from the long time. For example, the existing railways of the B&R countries have differences of narrow and broad-gauge systems, which becomes a major bottleneck to overland connectivity. On the other hand, a comprehensive risk prevention mechanism needs to establish, which requires seamless integration in different fields, such as industry, laws and regulations. And the limited funds should focus on advancing the projects which are possible to achieve the goal.

References

[1] Bhattacharyay B N, Kawai M and Nag R N 2012 *Infrastructure for Asian Connectivity* (Manila: Asian Development Bank)

[2] Grigoriou C 2007 *Landlockedness, Infrastructure and Trade: New Estimates for Central Asian Countries* (Washington: World Bank)

[3] Edwards L and Odendaal M 2008 *Infrastructure, Transport Costs and Trade: A New Approach* (Pretoria: TIPS Small Grant Scheme Research Paper Series)

[4] Iwata T, Kato H, and Shibasaki R 2010 *Impact of International Transportation Infrastructure Development on a Landlocked Country: Case Study in the Greater Mekong Subregion* (Pretoria: TIPS Small Grant Scheme Research Paper Series)

[5] Behrens K 2011 International integration and regional inequalities: how important is national infrastructure? *Manch. Sch.* **79**, 952-71

[6] Francois J and Manchin M 2007 *Institutions, Infrastructure and Trade* (Washington:World Bank)

[7] Xu J, Chen K M,Yang S F and Lin Y J 2016 The impact of economic corridor transportation infrastructure under "the Belt and Road Initiative" based on GTAP modal *AP. Econ. Rev.* **3**, 3-11

[8] Ni C J and Wang D 2017 Transportation infrastructure, space spillover and economic growth of the countries along the Belt and Road *Xinjiang State Farms Eco.* **3**, 34-41

[9] Zhang Y Y and Yu J P 2018 Transportation infrastructure, adjacent effects and bilateral trade: an empirical analysis based on the trade data of China and"the Belt and Road"countries *Contemp. Fin. & Eco.* **11**, 98-109

[10] Herrero A G and Xu J W 2017 China's Belt and Road initiative: can europe expect trade gains? *Int. J. Chin. Econ. Biz. Stud.* **6**, 84-99

[11] Derudder B, Liu X J and Kunaka C 2018 *Connectivity Along Overland Corridors of the Belt and Road Initiative* (Washington: World Bank)

A Study on Location of Logistics Hubs of Hub-and-Spoke Network in Beijing-Tianjin-Hebei Region

Xiaoyu Xin[1]*, Nan Yu[2], Xiang Chao[3]

[1]CHECC Data Co., Ltd, 100097, China

[2]China Waterborne Transport Research Institute, 100088, China

[3]Beijing New Airport Management Center, Capital Airport Holding Company, 100000, China

Email: Xiaoyu Xin: 18515559889@163.com

Nan Yu: yunan@wti.ac.cn

Xiang Chao: chaoxiang@cahs.com.cn

TEL: 18515559889

Abstract: Through the analysis of the logistics operation scene in Beijing-Tianjin-Hebei region, a hub-and-spoke regional logistics network with four modes of transportation was constructed. Considering the network balance factor, the number of network hubs should be determined. Then, with full consideration of various factors including fixed cost, transshipment cost and integrated transportation advantages, the hub-and-spoke logistics network location model is established. We design the corresponding genetic algorithm to solve the location problem of Beijing-Tianjin-Hebei regional logistics network hub. Finally, the feasibility of the model is verified by the actual logistics operation data of Beijing-Tianjin-Hebei region, and the optimal layout of regional logistics hub is calculated.

According to the analysis of the final logistics hub layout, we can find that 7 selected logistics hub in Beijing-Tianjin-Hebei region are distributed evenly. In terms of Beijing, the 3 selected logistics hubs in the city are located in the north, southwest and east west areas, which form a stable triangle covering Beijing-Tianjin-Hebei region. In terms of Tianjin, the 2 selected logistics hubs play an important role in connecting Tianjin with the surrounding cities around the region. In terms of Hebei, the 2 selected logistics hubs serve the southern region as well as cross-regional logistics operations.

1. Introduction

Regional logistics plays an important role in serving and supporting regional economic development. Beijing-Tianjin-Hebei region, as one of the three key economic circles in China, has gradually become one of the regions with the fastest economic growth and the greatest social influence in China. The development of the Beijing-Tianjin-Hebei region is not balanced, and the differences in logistics demand and road network density across the region are obvious. Therefore, constructing a Hub-and-Spoke logistics network in Beijing-Tianjin-Hebei region can greatly mobilize the transportation potential of the axle roads and make up for the shortage of the branch transportation capacity, and maximize the advantages of Hub-and-Spoke logistics network [1].

The research on Hub-and-Spoke regional logistics mainly focuses on model design and application. In terms of model design, O'Kelly first proposed the concept of Hub-and-Spoke network, and designed the P-hub median model for facility location and network design [2]. Chieh-Yu construct a multi-hub transport network for the US air transport industry with a hub-and-spoke network [3]. Cheng-Chang Lin and Yu-Jen Lin built a hub-and-spoke freight network for global air express delivery companies, and verified the superiority of the hub-and-spoke network through the transportation data of various cargo carriers [4]. Isabel Correia and Stefan Nickel study the single-distribution hub-and-spoke network [5].Campell first proposed a multi-distribution hub-and-spoke network, and establish a hybrid integer-programming model with the goal of minimum total transportation cost [6].Weng Kerui et.al analyzed the form, advantages and disadvantages of Hub-and-Spoke logistics network, and pointed out that scale effect, number and location of hub, distribution of logistics demand nodes and transportation routes should be considered in the network design[7].Zhu Xin etc.al optimize the design of the hub-and-spoke logistics network in Guangxi by principal component analysis and reconstructed gravity model [8].At present, there are some achievements in the application of hub and spoke network to regional logistics planning in China. Haifeng et al. [9] establish a regional logistics network based on the hub-and-spoke network, and conduct an empirical analysis of the model in Hubei Province. Tong shiqi et al. [10]Construct a three-level marine logistics network based on a hub-and-spoke network for the marine logistics system of an archipelago in southern China. However, there are few studies on the construction of Beijing-Tianjin-Hebei regional logistics network. In order to improve the efficiency of Beijing-Tianjin-Hebei regional logistics, Li Mingfang et al [11] construct a hub and spoke regional logistics network with principal component analysis and urban gravity model, combining with the characteristics of Beijing-Tianjin-Hebei regional logistics system.

In this paper, we first identify the occurrence and demand points of logistics in Beijing, Tianjin and Hebei Province, and analyze various logistics transportation modes in the region. Based on the research experience of domestic and foreign scholars in hub-and-spoke network, we formulate standards for the selection of logistics hub nodes, and design a non-strict hub-and-spoke logistics network location model with node cost and its solving algorithm. Finally, we investigate the relevant data and use the algorithm to form the regional hub-and-spoke logistics network.

2. Problem statement and Model

2.1 Problem and assumption
At present, the problem of location of network hub nodes is mainly limited to that all OD flows must transfer through at least one of pivot points in the hub and spoke network. This restriction greatly increases freight traffic volume through the hub. However, the vehicle is forced to detour when the cargo with the linear distance close and the pass-through capacity, the transportation distance required is farther compared with the straight-through distance between the OD pairs, thus the total cost of the entire network system increases. To ensure that the economies of scale of the hub and spoke network are maximized without excessive waste of resources, the transportation mode of network in this paper includes three transportation modes: straight through mode, through a transit hub and two hubs mode (see in Figure 1).

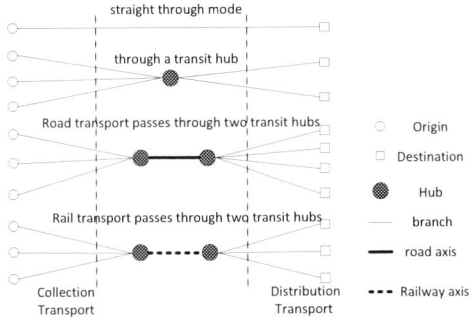

Figure 1. Transportation Mode Description

2.2 Notation

N : set of OD points.

H : set of alternative hubs.

p : quantity of hubs.

W : total transportation cost of the network.

f_{ij} : freight traffic volume between node i and node j.

X_{ij} : the proportion of total cargo flow that are transport in straight through mode between node i and node j ;When all of freights are transported by this mode, $X_{ij}=1$.

X_{ij}^{l} : the proportion of total cargo flow that are transported through one hub between node i and node j ; When all of freights are transported by this mode, $X_{ij}^{l}=1$.

$X_{ij}^{km}(r)$: the proportion of total cargo flow that are transported through two hubs between node i and node j ; When all of freights are transported by this mode, $X_{ij}^{km}(r)=1$. $k=m$ denotes the single hub mode.

$X_{ij}^{km}(t)$: the proportion of total cargo flow that are transported through two hubs between node i and node j ; When all of freights are transported by this mode, $X_{ij}^{km}(t)=1$. $k\neq m$.

∂ : discount coefficient of transportation cost between hubs, $0\leq\partial\leq1$.

C_{ij} : unit transport cost of straight through mode between node i and node j.

C_{ij}^{l} : unit volume transportation cost through hub l between node i and node j, $C_{ij}^{l}=C_{il}+\partial_{1}C_{lj}$.

$C_{ij}^{km}(r)$:unit road volume transportation cost through hub k and hub m between node i and node j, $C_{ij}^{km}(r)=C_{ik}+\partial_{1}C_{km}(r)+\partial_{2}C_{mj}$.

$C_{ij}^{km}(t)$: unit railway volume transportation cost through hub k and hub m between node i and node j, $C_{ij}^{km}(t)=C_{ik}+C_{km}(t)+\partial_{3}C_{mj}$.

Y_{k} : Whether the hub is selected at the point k. If so, $Y_{k}=1$; If not, $Y_{k}=0$.

$G_{p}(f_{ij})$: transportation and storage cost of goods through hub. G_{p} is a function relation related to node cost and OD flow.

$S_{k}(r)$: whether use road transport in hub k.

$S_{k}(t)$: whether use railway transport in hub k.

$M_{k}(r)$: fixed transport connection cost in hub k in road transport.

$M_{k}(t)$: fixed transport connection cost in hub k in railway transport.

T_{p} : fixed construction cost of hub.

2.3 Basic model

The total cost of regional logistics network hub location model is shown in equation (4-1).

$$\min W = \sum_{i}^{N}\sum_{j}^{N} X_{ij} f_{ij} C_{ij} + \sum_{i}^{N}\sum_{j}^{N}\sum_{l}^{P} X_{ij}^{l} f_{ij} C_{ij}^{l}$$

$$+ \left[\sum_{i}^{N}\sum_{j}^{N}\sum_{k}^{P}\sum_{m}^{P} X_{ij}^{km}(r) f_{ij} C_{ij}^{km}(r) + \sum_{k}^{P} M_k(r) \right] \cdot S_k(r)$$

$$+ \left[\sum_{i}^{N}\sum_{j}^{N}\sum_{k}^{P}\sum_{m}^{P} X_{ij}^{km}(t) f_{ij} C_{ij}^{km}(t) + \sum_{k}^{P} M_k(t) \right] \cdot S_k(t) \tag{2-1}$$

$$+ \sum_{i}^{N}\sum_{j}^{N}\sum_{p}^{P} G_p(f_{ij}) + \sum_{p}^{P} T_p$$

$$s.t. \quad \sum_k Y_k = P \tag{2-2}$$

$$X_{ij} + \sum_l X_{ij}^{l} + \sum_k\sum_m X_{ij}^{km}(r) + \sum_k\sum_m X_{ij}^{km}(t) = 1 \quad \forall i,j \in N; k, \in H \tag{2-3}$$

$$C_{ij}^{l} = C_{il} + \partial_1 C_{lj} \tag{2-4}$$

$$C_{ij}^{km}(r) = C_{ik} + \partial_2 C_{km}(r) + \partial_3 C_{mj} \tag{2-5}$$

$$C_{ij}^{km}(t) = C_{ik} + C_{km}(t) + \partial_3 C_{mj} \tag{2-6}$$

$$X_{ij}^{l} \le Y_l \quad \forall i,j \in N; l \in H \tag{2-7}$$

$$X_{ij}^{km} \le Y_k \quad \forall i,j \in N; k,m \in H \tag{2-8}$$

$$X_{ij}^{km} \le Y_m \quad \forall i,j \in N; k,m \in H \tag{2-9}$$

$$Y_k = \begin{cases} 0 & \text{Not elected to the hub} \\ 1 & \text{elected to the hub} \end{cases} \quad i,j = 1,2,3,...,n; \forall i,j \in N \tag{2-10}$$

$$S_k(r) = \begin{cases} 0 \\ 1 \end{cases} \quad i,j = 1,2,3,...,n; \forall i,j \in N \tag{2-11}$$

$$S_k(t) = \begin{cases} 0 \\ 1 \end{cases} \quad i,j = 1,2,3,...,n; \forall i,j \in N \tag{2-12}$$

$$X_{ij} \in \{0,1\} \quad \forall i,j \in N \tag{2-13}$$

$$X_{ij}^{l} \in \{0,1\} \quad \forall i,j \in N; l \in H \tag{2-14}$$

$$X_{ij}^{km}(r) \in \{0,1\} \quad \forall i,j \in N; k,m \in H \tag{2-15}$$

$$X_{ij}^{km}(t) \in \{0,1\} \quad \forall i,j \in N; k,m \in H \tag{2-16}$$

$$0 \le \partial \le 1 \tag{2-17}$$

As shown in equation (2-1), the first four sections are transportation costs and denote the different transport mode. The fifth and last section denote transfer cost and fixed cost respectively.

Restraint expression (2-2) of constraints ensures the quantity of selected hubs of logistics network. Restraint expressions (2-2) - (2-6) denote the per unit transport cost corresponding to different transport modes. Restraint expressions (2-7) (2-8) (2-9) reflect the transport line are determined with the determination of the hubs. Relations (2-10) - (2-11) represent that the decision variable can only be 0 or 1. Relations (2-12) - (2-16) represent that only one mode of transportation can be selected between OD. The last formula is the range of scale effect coefficient.

3. Model solve

3.1 Hub quantity

The optimal locations and the number of hubs in the hub-and-spoke network model cannot be balanced. Factors such as the actual condition, the construction cost and transport mode should be also taken into account, so it is difficult to obtain accurate results with the mathematical model. The optimal number of hubs is set artificially according to the specific situation of the research problem. Referring to previous research, it is defined in most literatures as shown in formula (3-1) :

$$p = \sqrt{N} \tag{3-1}$$

Taking the square root of the total number of nodes in the network as the optimal number of hubs. In this paper, Beijing is divided into 16 OD points, similar as 11 OD points in Tianjing and Hebei. The network contains 38 demand points, and the number of hub is set as 6 or 7. The calculations of network in section 4 are based on this proposition.

3.2 Genetic algorithm design

3.2.1 Chromosome encoding.
In this paper, an integer coding method is selected, the length of chromosome is the number of hubs in the network. The alternative nodes are directly marked with integer values ranging from 1 to 19. The resulting chromosomes (the initial solution) should be a 1×7 matrix in which each gene does not repeat.

As genetic algorithms search for multiple individuals in a population simultaneously, the size of population will directly affect the operational efficiency of genetic algorithms and the degree of optimization of results. In this paper, the population size is 200.

3.2.2 The fitness function
The initial population of chromosomes generates the spatial Layout of multimodal transport network. The next step is to calculate the optimal transport mode between different OD according. In this paper, four transport modes are involved in the transport process, so a 4*N transport combination storage matrix is generated to store the optimal transport mode different OD of the network during the decoding process. The determination of OD transport mode is firstly to calculate the optimal route of the four transportation modes in combination with the spatial layout of network, and then to determine the optimal transportation mode by comparison.

The ultimate goal of the model solution is to minimize the transportation cost of the logistics network. According to the transport combination storage matrix, the target function value of each chromosome is calculated with the basic model in equation (2-1).

3.2.3 Genetic algorithm operation
This paper uses roulette wheel selection which determines the probability P_i that the next generation is retained based on the proportion of each individual's fitness.

$$P_i = \frac{f_i}{\sum_{i}^{m} f_i} \tag{3-2}$$

Where f_i denote value of individual chromosome fitness, m represent the population size.

As the chromosome gene represents hub number, the crossover process needs to satisfy the restrictions (i.e., isometric crossover, no overlapping). PMX(Partially Matched Exchange) is the crossover operator. Different from the traditional crossing method, PMX operation firstly adds the whole part of the previous chromosome that needs to be crossed into the latter chromosome, and then tests the genes in the latter chromosome one by one to delete the duplicated gene. The operation process is shown in Figure 2.

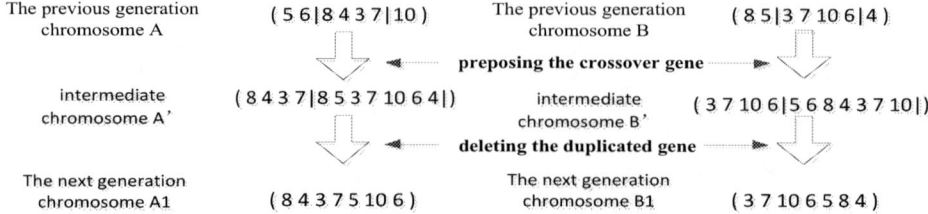

Figure 2. Schematic of PMX Crossover Operation

The mechanism of mutation is to locate the mutation of area at a certain point in the chromosome and change the gene to generate new chromosome codes.

4. Case study on logistics hubs location in Beijing-Tianjin-Hebei region

4.1 Parameter settings

The basic data related to nodes include: (1) the number of OD points, alternative logistics hub points and their corresponding location information in our case. (2) The setting of fixed parameters

(1) Basic data of OD and alternative logistics hub

In this paper, 16 origin and 16 destination points in Beijing and 11 origin and destination points in Tianjin are considered in accordance with districts region, Hebei province is divided into 11 origin and 11 destination points according to the city region. 19 alternative hub points are set up. The coordinate information of OD points, the distance matrix between OD-alternative points and the distance matrix between each alternative point are respectively shown in appendix A, appendix B and appendix C.

(2) Network cost parameters

The scale effect factors in this paper are discussed in terms of the type of the nodes that connect at both ends of the road: The size effect coefficient between straight-through transportation roads is 1, the size effect coefficient of sub-axis roads is 0.8, and the size effect coefficient of axis roads is 0.7. When highway transportation is adopted between the hubs, it is set at 7.8 yuan/ton; when railway transportation is adopted between the hubs, it is set at 5.8 yuan/ton. In addition, it is assumed that each alternative hub provides a fixed link fee of 800,000 yuan for highway transportation, while each alternative hub provides a fixed link fee of 1,000,000 yuan for railway transportation.

4.2 Calculation results of logistics hub location

The site selection results of the hub and the corresponding total network costs and overall transport schemes are shown in Table 1.

Table 1. Location Results of Hub Node

	Quantity	Total cost (yuan)	Selected Number of hubs	by single hub (highway)	by dual hub (highway)	by dual hub (railway)
Calculation results	6	7.883e+8	9, 5, 4, 16, 2, 19	106	174	0
	7	7.721e+8	9, 12, 1, 16, 4, 2, 19	143	218	0

In this situation, the total cost of regional logistics network is RMB 7.721e+8, and the number of single and double hub transport routes in the network is 143 and 218, respectively. Additionally, the whole area network does not contain the railway transport in the transportation scheme regardless of the quantity of hubs, it is due to the Beijing-Tianjin-Hebei region is in a small scope, which lead the railway transport cannot fully exert its advantage over the long haul transportation, moreover, link fee in hubs for the railway transportation is relatively high, hence the railway transportation is not adopt in the transportation of any OD point. Select the serial number 1, 2, 4, 9, 12, 16 and 19 in the set of alternative hubs as the hub node in the network and these selected nodes respectively located in Tongzhou district, Changping district and Fengtai district of Beijing, Wuqing district, Baodi district of Tianjin and Baoding city, Hengshui city of Hebei province.

4.3 Analysis of logistics hub location

Because this network allows direct transportation between non-hub nodes, there are 1407 lines of transportation between 38 OD points. According to Table 1, a total of 143 routes were selected to be transported via single hub and 218 were selected to be transported via a dual hub. So the focus of this part is to discuss the application characteristics and radiation areas of transfer modes via single hub and double hub.

4.3.1 Analysis on transport results via single hub

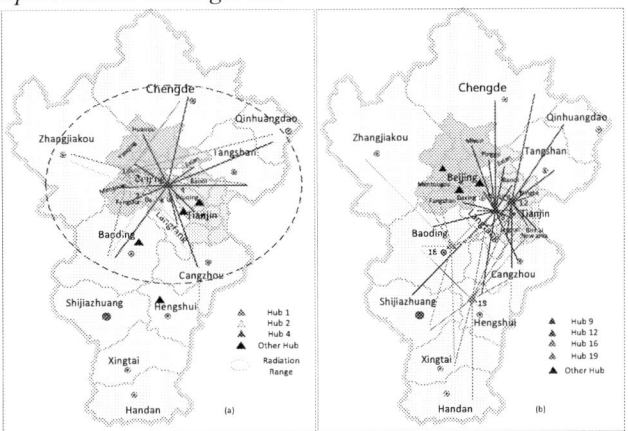

Figure 3. Single hub transfer and radiation

(1) Transport scheme via hub 1, 2, 4.

The hub 1 is located in Changping district, Beijing, the northernmost of all alternative logistics hubs. A total of 22 routes were transferred to the demand point via this hub. As can be seen from figure 3(a), hub 1 primarily serves two major logistics channels in the Beijing-Tianjin-Hebei region, which radiates from north to south which includes Chengde of Hebei, Huairou, Yanqing, Changping of Beijing and Baoding of Hebei.

Hub 2 is located in Fengtai district of Beijing, and a total of 19 routes are selected to be transported to the destination via this node. As shown in figure 3(a), hub 2 is located in the southern part of Beijing, and it is mainly responsible for the collection and distribution of goods from Baoding, Langfang, Cangzhou and other places of Hebei province to the main urban area of Beijing, The selection of hub 2 can gather and distribute the goods from Hebei in advance, reducing the number of freight vehicles directly into Beijing and ease the transit pressure of Beijing.

Hub 4 is located in Tongzhou district of Beijing. A total of 35 routes are selected to be transported via the hub 4 to the destination point. This result fully reflected the status of hub 4 in the Beijing-Tianjin-Hebei regional logistics network. As shown in figure 3(a), Hub 4 mainly connects the eastern part of the region and the main urban area of Beijing, forming a crosswise logistics channel in the middle of the region. Similar to hub 2, hub 4 can also relieve the transit pressure in the main urban area of Beijing. In addition, hub 4 also connects the northeast suburbs of Beijing and forming the regional central longitudinal logistics channel.

(2) Transport scheme via hub 9, 12, 16, 19.

Hub 9 is located in Wuqing district of Tianjin. According to figure 3(b), we can infer that hub 9 plays two important roles in regional logistics operations: first, hub 9 is a vital node connecting Tianjin urban area with the southeastern area of Beijing; second is that the hub 9 is the node which connects Chengde,

2484

Jizhou district, Miyun district, Pinggu district in the north area of network with southern Jinghai, Cangzhou, Hengshui etc., forming the eastern vertical logistics channel.

There are 30 lines selected to be transported via hub 12. The goods in Tianjin urban areas, Xiqing district, Dongli district, and Jinghai district have been concentrated and combined to the Chengde, Tangshan and Qinhuangdao via this central node.

The hub 16 is located in the middle of the Beijing-Tianjin-Hebei region connecting the regional logistics transport: it connects Daxing district, Fangshan district, Mentougou district, Zhangjiakou city with southern Hebei province; it also connects western Hebei province with eastern Tianjin.

As the southernmost alternative logistics hub, hub 19 is located in Hengshui city, Hebei province and connect the central areas with southern areas of the Beijing-Tianjin-Hebei region. The goods in the south of the region are transported to the middle of the region   via the No.19 hub, forming the vertical logistics channel in the south of the region.

4.3.2 Analysis on transport results via dual hubs

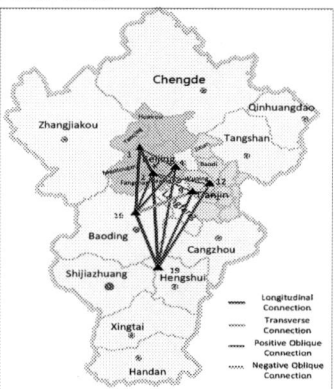

Figure 4. Schematic Diagram of Double-Hub Connection

A total of 218 lines of all transport lines are selected and transported to their respective destination via two different pivot points. By analyzing the location distribution of each hub, it can be found that there are four kinds of dual hub connection modes in the network according to the relative position of two hubs in the region, which is vertical connection of dual hub (two hubs are distributed relative to north and south), the horizontal connection of dual hubs (two hubs are distributed relative to east and west), the positive oblique connection of dual hubs (two hubs are distributed relative to northeast and southwest), and the reverse oblique connection of dual hubs (two hubs are distributed relative to northwest and southeast), respectively.

Considering that the feature of Beijing-Tianjin-Hebei region which the depth of north-south direction is long and the range of east-west direction is relatively short. Therefore, the mode of the connection between north and south has become the main mode of the transportation of the two hubs. In figure 4, there are four vertical connection modes with dual hubs (1-2, 1-16, 1-19 and 2-19), five positive oblique connection modes (2-16、4-16、4-19、9-19、16-19), three reverse oblique connection modes (1-4、2-9、2-12), and only the route 9-16 is selected as the horizontal connection between the two hubs.

As can be seen from figure 4, the selected dual hub transport mode is dominated by the nodes within the south-central region. This is because that six cities including Baoding, Cangzhou, Shijiazhuang, Hengshui, Xingtai and Handan in Hebei province are located in the central and southern part of Beijing-Tianjin-Hebei, which covers a large area and is far from the northern region. In addition, the volume of logistics in the central and southern regions are large, but compared with the Beijing and Tianjin, the network density here is relatively low, and the transportation difficulty is relatively large, hence the mode with dual hub are adopted to carry out transportation business in this situation is vital.

5. Conclusion

Under the background of regional integration of Beijing-Tianjin-Hebe, in this paper, we take the regional logistics operation scene of Beijing-Tianjin-Hebei as the research object, and construct the regional logistics network hub location model with the lowest total network cost as the objective function. We solve the problem and find that the seven alternative hubs, No. 1, No. 2, No. 4, No. 9, No. 12, No. 16, No. 19, are the optimal layout schemes for the Beijing-Tianjin-Hebei regional logistics network hub. On this basis, we obtain the radiation scope and position of each hub in the regional logistics network through the analysis of the optimal layout scheme of the hub; and get layout of logistics channels formed by single hub and double hub connection modes in regional logistics networks and their roles through the analysis of OD's choice of transport routes in the network.

Appendix

Appendix A

Partly matrix of OD distance

	Dongcheng	Xicheng	Haidian	Chaoyang	Fengtai	Shijingshan	Tianjin	Jizhou
Dongcheng	0.00	4.67	20.17	8.89	16.30	20.90	107.87	86.46
Xicheng	4.67	0.00	17.59	13.47	12.01	16.50	110.16	91.13
Haidian	20.17	17.59	0.00	25.28	21.31	11.93	127.72	100.9
Chaoyang	8.89	13.47	25.28	0.00	25.16	28.89	106.03	77.95
Fengtai	16.30	12.01	21.31	25.16	0.00	12.56	111.41	101.7
Yanqing	71.73	71.15	55.27	71.49	76.53	65.35	177.15	122.3
Tianjin	107.87	110.16	127.7	106.0	111.41	123.6	0.00	98.48
Jizhou	86.46	91.13	100.9	77.95	101.7	106.7	98.48	0.00
Baidi	90.05	94.11	109.2	83.95	101.2	110.49	55.98	43.52
Dongli	122.76	125.4	142.8	119.95	127.8	139.7	19.37	99.63
Shijiazhuang	264.84	261.3	266.0	272.8	249.6	254.2	263.96	333.7
Hengshui	250.13	248.5	260.7	255.3	239.7	249.5	197.38	289.9
Xingtai	320.66	317.9	326.4	327.5	307.3	314.5	290.03	375.5
Langfang	70.27	70.91	87.83	72.05	68.21	80.76	50.44	103.5

Appendix B

Partly matrix of OD-alternative points distance

	Dongcheng	Xicheng	Haidian	Chaoyang	Fengtai	Shijingshan	Tianjin	Jizhou
1	42.72	43.55	25.11	44.17	46.34	35.33	150.06	108.5
2	26.72	22.28	26.46	35.61	10.50	14.75	116.37	112.13
3	18.52	18.21	35.02	23.70	17.26	29.33	94.27	91.22
4	25.46	29.43	45.19	20.70	37.35	45.82	85.97	66.45
5	44.87	47.63	43.53	38.73	58.81	54.72	134.81	73.47
6	27.04	29.16	25.04	22.85	39.93	35.80	126.62	80.94
7	27.68	23.54	12.87	35.02	19.86	7.59	131.14	112.44
8	23.23	27.17	31.75	15.31	39.18	39.60	113.16	69.50
9	83.68	85.67	103.2	82.64	86.31	98.60	25.33	90.47
10	99.94	101.4	118.86	99.86	100.4	112.99	21.74	108.6
11	138.2	139.8	157.3	137.8	138.8	151.4	36.54	134.5
12	100.6	103.9	120.8	96.65	108.1	119.26	28.00	72.47
13	116.01	118.6	136.0	113.34	120.9	132.8	13.20	96.16
14	144.96	147.9	165.1	141.4	151.0	162.6	43.70	109.8

Appendix C

Partly matrix of alternative points distance

	Dongcheng	Xicheng	Pinggu	Miyun	Huairou	Yanqing	Tianjin	Jizhou
1	0.00	49.89	157.52	185.42	98.62	137.22	264.90	220.71
2	49.89	0.00	126.54	157.15	49.98	89.39	224.05	174.52
3	59.64	24.53	103.66	133.79	43.64	96.56	236.40	161.29
4	64.76	47.10	92.83	120.75	61.19	118.38	259.41	165.44
5	38.07	67.59	139.51	163.98	105.99	155.98	291.46	217.34
6	28.21	48.54	132.90	159.58	88.75	137.16	272.42	204.20
7	31.40	19.29	140.41	170.28	69.22	105.99	236.52	193.42
8	41.88	49.26	119.07	145.46	82.20	134.50	272.51	193.16
9	126.24	91.06	36.50	67.80	59.54	111.14	247.63	106.39
10	142.66	103.79	34.85	64.01	64.50	107.81	239.12	86.99
11	180.97	141.91	39.58	49.93	99.53	131.87	249.96	61.96
12	140.41	115.23	23.71	45.22	92.96	146.46	282.53	123.31
13	157.52	126.54	0.00	31.31	95.31	142.51	273.21	101.26
14	185.42	157.15	31.31	0.00	126.45	171.46	297.67	108.13

Appendix D

Partly matrix of OD flow

	Dongcheng	Xicheng	Pinggu	Miyun	Huairou	Yanqing	Tianjin	Jizhou
Dongcheng	0	1163	953	822	1245	781	150	42
Xicheng	1031	0	1678	1447	2192	1375	264	74
Haidian	563	1124	2368	2041	3092	1939	372	104
Chaoyang	1002	656	2381	2053	3110	1950	375	105
Fengtai	605	2131	600	518	784	492	94	26
Shijingshan	520	693	221	190	288	181	35	10
Changping	14695	25869	6726	4238	4227	5209	131	63
Mentougou	2914	5130	3074	2121	2351	9960	188	41
Fangshan	1601	2818	1333	1488	2291	3182	218	59
Daxing	5950	10474	2518	2006	2614	1320	226	32
Tongzhou	5867	10328	3787	4739	6615	2013	452	106
Shunyi	13131	23116	6983	7364	8819	2357	618	139
Pinggu	764	1345	0	3330	2282	1403	643	144
Miyun	805	1416	3439	0	8882	1946	460	101

References

[1] Yang Lv. Optimization research for logistics network within Beijing-Tianj in-Hebei region based on Hub-and-Spoke network model [D]. Beijing Jiaotong University, 2016.

[2] Morton O'Kelly, Boyer Kenneth D. Hub location with flow economies of scale [J]. Transportation Research(B), 1997, 5(4): 221-226.

[3] Chieh-Yu. Modeling investment options for multimodal transportation networks [J]. Transportation Research Part A: Policy and Practice, 1997, 8(31): 57-69.

[4] Cheng-Chang Lin, Yu-Jen Lin. The economic effects of center-to-center directs on hub-and-spoke networks for air express common carriers [J]. Journal of Air Transport Management, 2003, 15(4):255-265.

[5] Isabel Correia, Stefan Nickeib. Hub and spoke network design with single-assignment capacity decisions and balancing requirements [J]. Applied Mathematical Modeling, 2011, 17(10): 4841-4851.

[6] Campbell, James F. Integer Programming Formulations of Discrete Hub Location Problems [J]. European Journal of Operational Research, 1996 (72): 387-405.

[7] Weng Kerui, Yang Chao. On Hub-and-spoke logistics networks[J]. Logistics and Technology, 2006(7): 14-16.

[8] Xin Zhu,Y Fan. Research on the construction of Hub-and-Spoke Network in GuangXi province under the gravity model [J]. Region Economic, 2017, (9): 214-217. DOI:10.3969/j.issn.1002-5863. 2017.09.074.

[9] Feng Hai, Xin Yan, Can Ding, Xiao Shao. Nodes selection of regional logistics network based on hub-and-spoke theory [J]. Computer Integrated Manufacturing Systems, 2012(6): 1299-1305.

[10] Shiqi Tong, Jin Zhang. Archipelago shipping logistics network construction and nodes selection based on hub-spoke theory [J]. Port & Waterway Engineering, 2014(3): 68-73.

[11] Mingfang Li, Jingmei Xue. Construction and the corresponding strategy of Hub-and-Spoke Network in Beijing-Tianjin-Hebei Region. [J]. China Business and Market, 2015, (1):106-111. DOI:10.3969/j.issn.1007-8266.2015.01.015 .

Blockchain-based Intelligent Hospital Security and Data Privacy Construction

Qiuzi Huang[1,3,4], Shuyu Chen[1,4], Hui Zhao[2,3] and Junhao Wen[2,4]

[1]Network Distributed Computing Laboratory, Big Data and Software Institute, Huxi Campus of Chongqing University, Chongqing, China

[2]Service Computing Laboratory, Big Data and Software Institute, Huxi Campus of Chongqing University, Chongqing, China

[3]Huxi Campus of Chongqing University, no. 55, Chengnan Road, Shapingba District, Chongqing, China

[4]Chongqing University, 174 Shazheng Street, Shapingba District, Chongqing, China

sychen@cqu.edu.cn

Abstract. With the rapid development of medical information services, the construction of intelligent hospitals is opening up a new mode of medical treatment in the health care industry. Medical data is gradually becoming more and more important, while it also faces some challenges, among which the most urgent problem to be solved is data security and privacy protection. In the construction of intelligent hospitals, the safety issues among the basic information of patients, the protection of medical information and inter-institutional information sharing have become the focus at this stage. Blockchain technology, with highly security, reliable architecture and algorithm design have operated stably in the financial industry for more than seven years. The related innovative technologies such as distributed ledgers, smart contracts, symmetric encryptions and consensus mechanisms are used widely in many fields. This paper will take the demonstration construction of Chongqing Intelligent Hospital as an example, which tries to combine the blockchain, homomorphic encryption and zero-knowledge proof technology to carry out the research on the security construction and data privacy protection of intelligent hospitals.

1. Introduction
In the revolution brought by big data technology, information technology has been applied to various fields. In recent years, with the proposal of intelligent hospital construction, we tried to load heterogeneous data such as patients' basic health archives, medical information data and clinical business data into the data mart through a series of data processing operations. The data mart is the data center of the intelligent hospital system. During the construction of intelligent hospital, transparency and readability of data are basic prerequisites[1], we assume that all data stored in the data mart is transparent and open and that business data sharing can be achieved. On the one hand it opened up the occlusion between the data, lifting the problem of information silos, while on the other hand it also triggered people to think how to guarantee the security and privacy of data. In the current construction, some parts of medical wisdom systems can connect through the external network that is likely to be the access point for the attacker to access the infrastructure. For example, the attacker violently cracks the

Content from this work may be used under the terms of the Creative Commons Attribution 3.0 licence. Any further distribution of this work must maintain attribution to the author(s) and the title of the work, journal citation and DOI.

Published under licence by IOP Publishing Ltd

account and password to successfully enters the system. In addition, a large number of servers used by medical institutions, such as NAS servers, have a typical update cycle of several years that accumulate a large number of known vulnerabilities.

The security and privacy of current medical data are mainly facing with two types of threats:

• The middleman uses the sensor channel to attack that the data service between the sensor and the storage sensor;

• Unauthorized persons access data stored locally and remotely.

The integrity and availability of data can be strongly threatened by the above two types of attacks, such as unauthorized accessing, replacement storage data, middleman replacement of data in transit and encrypting or deleting user data, etc[2]. In the construction of an intelligent hospital, the privacy data of the patients or hospitals could be leaked. Data in hospital data marts may be tampered illegally, which could destroy the integrity and correctness of the information. In addition, the stolen data could be used for other illegal purposes. The occurrence of these conditions will seriously threaten the safety and legal rights of patients and hospitals.

Therefore, in the process of building an intelligent hospital, effective protection measures must be taken to ensure the robustness of the data center and the information security of the hospital. How to share all the data under the premise of ensuring the security of the hospital data center has become an urgent problem to be solved.

2. Related works

2.1. Blockchain technology

Blockchain is a new decentralized infrastructure and distributed computing paradigm that has emerged with the increasing popularity of digital cryptocurrencies such as Bitcoin[3]. It is similar to a distributed public account book. In the entire blockchain system, no role can absolutely control the system, and the system does not rely on a centralized, hierarchical structure. Each participating data block has equal rights which jointly maintain the account book updates. The development of blockchain technology solves the trust and security problems of data in the transaction process. It has five characteristics: disintermediation, non-tampering with information, autonomy, openness and anonymity[4].

With the advancement of national systematic projects such as the digitization of medical information and the personal credit information system, more and more authoritative data sources have emerged. In the construction of future intelligent hospitals, we need to collect a large number of sensitive data, such as basic information of patients, into the data mart to optimize hospital data integration and improve disease prediction rate, etc. Blockchain technology may solve the above problems to some extent, so we can introduce the data and store it by using blockchain technology to build a decentralized and trustworthy data system. We can also use blockchain technology to create a person's digital identity, which is real, untampered with, synchronized in real time, and valid for life. In addition, we can use the blockchain system to record the research projects in clinical trials, which ill make the research more objective and perform better to the clinical[5].

2.2. Privacy protection

Each block of data on the blockchain has the same rights, making each data block can obtain a complete data backup. All data traded in the chain is open and transparent, which is a prerequisite of realizing the sharing of medical data in hospitals. However, there is a certain risk that the patient information data will be transparent only in the in-hospital data center by default. Not only do patients want to protect personal information or private data such as their medication records, but hospitals also want their management information to be kept secret from other hospitals or institutions. Once the system of an intelligent hospital is attacked by an irresistible attack, it will have an extremely bad impact on individuals, hospitals, and even the entire intelligent hospital system. Even if some data has desensitized at the beginning, these attacks may affect the privacy protection of the data.

At present, blockchain privacy protection problem can be solved in several ways, such as mixed currency, ring signature, homomorphic encryption, and zero-knowledge proof[6-8]. Combined with the transaction processing characteristics of the intelligent hospital, homomorphic encryption and zero-knowledge proof technology are considered to solve the privacy protection problem of it.

2.2.1Homomorphic encryption
Homomorphic encryption is a method that can perform calculations accurately on encrypted data. In the construction of intelligent hospital, the basic algorithm operation can be performed without decrypting the encrypted data by using the homomorphic encryption technology on the blockchain. Therefore, homomorphic encryption technology is added to the construction of the intelligent hospital. When accessing sensitive data in the data mart, the characteristics of block chain technology are used to ensure that data are not tampered with, and the privacy of data in the data mart of the intelligent hospitals is guaranteed.

2.2.2Zero-knowledge proof(ZKPs)
Zero-knowledge proof is a cryptographic technology proposed in the 1980s[9], which can make the verifier believe that a certain statement is correct without providing any useful information to the verifier. It makes possible to make a full judgment without having access to the original information, or to prove the authenticity of a proposal without disclosing the data itself. This is suitable for the scenario of the disease judgment in the intelligent hospital, in which the system gives the diagnosis of the disease while ensuring that the detailed data of the disease in the data mart is sufficiently concealed, thereby giving a diagnosis and treatment reference plan. On the one hand, it fundamentally overturns a series of data such as public senders, receivers and transaction numbers, and solves the possibility of privacy leakage in this way. And on the other hand, the validity of the public node verification transaction is also guaranteed. It realizes zero-knowledge proof based on blockchain technology, which effectively guarantees the security and privacy of data.

3. Technical application

3.1. Application of blockchain technology in safety construction of intelligent hospitals
Blockchain technology is used to establish data mart. Data mart is composed of data nodes, distributed multiple access and other ways. Just like different types of Shared ledger, it has many backups, which improves the transparency of data in the hospital. Each specific time node recorded in the blockchain is completed in its own authorized scope without other operations. Once the record on the timeline is generated without special "authorization", it cannot be changed[10].

When the patient is in the examination, only the doctor "authorized" by the patient can obtain personal health data, check the medical record so that they can learn about the patient's medical history, analyze the symptoms, make judgments, and give the intelligent analysis opinions with the advice given by the system. In this way, blockchain technology tries to avoid repeating examinations to ensure patient information security and also save times and material resources.

The non-tampering of blockchain makes the data more reliable. We also can ensure the privacy and correctness of electronic medical record combine with using digital signature technology. This not only ensures the information security of electronic medical records, but also fundamentally solves the problem of doubts about the authenticity of it.

3.2. Application of blockchain technology in protecting patient privacy
In the construction of intelligent hospital, it is necessary to the construction of taking patients as the center of information system, using the Internet and big data provide the humanized service for medical staff and patients[11]. It is necessary to consider how to protect the privacy of patients in the process of building an intelligent hospital. It can help patients hide their true identity in the public chain and effectively protect patient privacy information, because of the better anonymity of blockchain

technology[12]. Using encryption technology combined with the blockchain technology decentralized characteristics meets the patients medical privacy and anonymity necessary data which could meet the public health research needs. Drawing on the idea of "DNA wallet" technology, the patient's basic information, medical records, and past medication records will be stored in a single block by using blockchain technology while the hospital only needs unified identity management. The data block storing the patient's private information just likes the patient's exclusive medical information wallet. It solves the problem that the doctor obtains incomplete information when the patient visits or referrals in different hospitals. Due to the data protection combining with homomorphic encryption technology, the real information of the patients is hidden and the data blocks of all patients are the same, making it difficult to trace the data source. Only the private key can be used for unique identification. Without the authorization of the patients, it is unable to know the true identities of the patients so that the patients do not need to worry about information leakage. This method can solve the patients' privacy protection problem.

Security is the primary problem that needs to be solved to achieve privacy protection in the process of data exchange. Data encryption makes the data in the data mart present in ciphertext and reduce the probability of information leakage. Using homomorphic encryption technology makes it possible to perform basic algorithm operations without decrypting the encrypted data. In the process of disease prediction, the system only give the results of disease prediction without the real data of the examination results by using zero-knowledge proof technology.

With the support of blockchain technology, patients' personal information can be completely hidden. At the same time, patient-related data can also be applied to research in the field of public health. "authorized" data such as historical treatment records and follow-up treatment effect tracking can be extracted and studied by relevant hospitals or departments, so as to further improve the utilization rate of data.

4. Case model

Based on the strategic action plan of innovation-driven development led by big data intelligence in Chongqing (2018-2020), the Chongqing Municipal Health Planning Commission has carried out the "healthy Chongqing 2030" plan and the construction plan of intelligent hospitals. It has conducted research on the organizational management, strategic planning, capital investment and effectiveness of 50 representative hospitals (including 15 affiliated hospitals, 5 military hospitals in Chongqing, 20 district and county level hospitals, and 10 secondary hospitals). At the same time, we selected the target pilot hospitals by referring to the key construction contents of smart hospitals and the grading value of evaluation index system issued by Chongqing Municipal Health Commission. In the process of building an intelligent hospital, according to the requirements put forward by the project, a staged smart hospital form is formed according to the plan period, as shown in Figure 1.

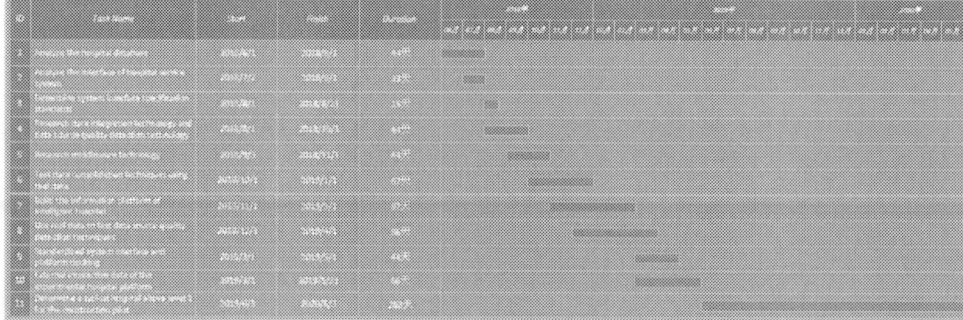

Figure 1.The staged smart hospital form

In the construction of intelligent hospitals in Chongqing, more attention is paid to security construction and privacy protection issues. Blockchain technology, homomorphic encryption and zero-knowledge proof technology is applied to data security and privacy construction. As shown below:

• The blockchain technology is applied to establish the private data modules like electronic medical records. Drawing on the idea of "DNA wallet" technology, the patient's basic information, medical records, and past medication records are stored in a single data block. The hospitals only need to perform unified identity management. At the same time, the blockchain technology is used to establish a complete, manageable, readable, and undeniable electronic medical record between the target pilot hospitals. In addition, digital signature technology is used to verify that the "authorized" person adds the correct records according to the related rules. The security of the data is guaranteed as well as its privacy;

• The distributed feature and homomorphic encryption technology of block chain are applied to data privacy processing. Blockchain adopts distributed data writing and extraction. In this process, we have access control through a visible public address and two encryption keys. In addition, homomorphic encryption technology is used to process desensitized data, so that the system can perform operations, such as retrieval and operation on the encrypted data and obtain correct results. Under the premise of data "fuzziness", data privacy is further guaranteed.

• Zero knowledge proof technology is applied to the diagnosis of diseases and other scenarios. In order to ensure the privacy of data, zero-knowledge proof technology is adopted to carry out the disease pre-diagnosis in the system. Instead of showing the detailed list of relevant data, only the final diagnosis result is displayed. The diagnosis results of the system will provide great support for doctors' diagnosis. This application scenario in the intelligent hospital system once again guarantees the privacy of data.

In addition, in the demonstration construction of Chongqing Intelligent Hospital, the following security protection measures are planned:

● Prohibit external network access to the intelligent hospital system;
● The online information systems should be independent from the hospitals' intranet;
● Monitor the intelligent system or services, and ensure them update regularly;
● Modify the default password and delete part of "garbage data" in the database, such as: test account and expired account, etc;
● Create strong passwords for all accounts.

5. Conclusion
This paper discusses the security and privacy protection issues in the process of building an intelligent hospital. Based on the blockchain technology, the data is protected from malicious tampering, thus providing a guarantee for data reliability. In addition, combining blockchain with the technology of homomorphic encryption and zero-knowledge proof, the data is further encapsulated on the basis of not affecting the final result. So that the data is "shielded" and patients' privacy is protected. This paper takes the construction of Chongqing Intelligent Hospital as an example, and introduces the construction planning especially the safety considerations about it. Under the new situation, the construction of smart hospitals is unstoppable. The related technical mentioned in this paper will guarantee the systematic construction of intelligent hospitals.

Acknowledgements
I would like to express my gratitude to all those who helped me in the process of writing this paper. I do appreciate professor S Y Chen, professor J H Wen, my partners and my boyfriend, who gave me patience, encouragement and professional guidance in the process of writing this thesis.
Finally, I would like to acknowledge the Big Data and Software Institute of Chongqing University for its great support.

References
[1] Julián Salas,,Josep Domingo-Ferrer.Some Basics on Privacy Techniques, Anonymization and their Big Data Challenges[J].Mathematics in Computer Science,2018,12(3):263-274.

[2] Karim Abouelmehdi,,Abderrahim Beni-Hessane,,&Hayat Khaloufi.Big healthcare data: preserving security and privacy[J].Journal of Big Data,2018,5(1):1-1.

[3] YUAN Yong, WANG Fei-Yue. Blockchain: The State of the Art and Future Trends[J]. Acta Automatica Sinica, 2016.

[4] Igor Radanović,,Robert Likić,View,author's,OrcID,& profile. Opportunities for Use of Blockchain Technology in Medicine[J].Applied Health Economics and Health Policy,2018,16(5):583-590.

[5] Irving Greg, Holden John. How blockchain-timestamped protocols could improve the trustworthiness of medical science:[J]. F1000Research, 2016, 5:222.

[6] Li,,Jiaxing,Chen,,Long,Wu,,& Jigang. Block-secure: Blockchain based scheme for secure P2P cloud storage[J].Information Sciences,2018,465:219-231.

[7] Castro M, Liskov B. Practical byzantine fault tolerance and proactive recovery[M]. ACM, 2002.

[8] Yao A C. Protocols for secure computations[C]// Symposium on Foundations of Computer Science. IEEE Computer Society, 1982:160-164.

[9] Goldwasser S, Micali S, Rackoff C. The knowledge complexity of interactive proof-systems[C]// 1985:291-304.

[10] Leslie,Mertz.Hospital CIO Explains Blockchain Potential: An Interview with Beth Israel Deaconess Medical Center's John Halamka[J].IEEE Pulse,2018,9(3):8-9.

[11] Rizwan,,Patan,M.,,Rajasekhara,Babu,K.,,& Suresh.Design and development of low investment smart hospital using internet of things through innovative approaches[J].Biomedical Research (0970-938X),2017,28(11):4979-4985.

[12] Sushmita,Ruj,Mohammad,Shahriar,Rahman,Anirban,Basu,Shinsaku,& Kiyomoto. BlockStore: A Secure Decentralized Storage Framework on Blockchain[J].2018 IEEE 32nd International Conference on Advanced Information Networking and Applications (AINA),2018:1096-1103.

[13] Irving Greg, Holden John. How blockchain-timestamped protocols could improve the trustworthiness of medical science:[J]. F1000Research, 2016, 5:222.

Research and analysis on the transformation of road passenger transport industry

Lili Chen[1], Zhijun Yi[2] and Xiaofeng Di[1]

[1] China Academy of Transportation Science, Beijing 100013, China

[2] Guangdong road transport affairs center, Guangzhou 510000, China

chenlili@catsti.com

Abstract. How does road passenger transport industry survice and breakthrough in limited development space, this has become a major issue in the field of transportation in recent years.In this paper, the effect and difficult of actual cases such as the combining road transport with tourism in the current development process of road transport are summarized. It is pointed out that the focus of passenger transport transformation is to closely combine the requirements of industrial transformation and upgrading and promote the deep integration of information technology with industrial management and services.Finally, corresponding suggestions are put forward for the innovative development of road passenger transport transformation.

1. Traditional passenger transport enterprises break the "siege" together

With the continuous improvement of high-speed rail, subway speed advantage, and high-density requency, as well as the continuous upgrading of passengers' demand for high-quality and diversified travel, problems such as the concentration of passenger market to spend low and the continuous increase of labor cost have become increasingly prominent.

To do this, the GuangDong provincial department of transport launched the pilot reform of road passenger transport in 2016, making beneficial exploration for the transformation and upgrading of the passenger transport industry. Guangdong's leading enterprises in transport positively responded by digging into the resources of road passenger lines and making all-out efforts to promote the transformation and upgrading of road passenger transport. After more than a year of practice, on the basis of investigating the space-time distribution and the bus passengers demand, it has introduced and initially has established the brand effect of inter-city section, business, tourism and other patterns. Taking the pilot work of Guangzhou no.2 bus company as an example, From September 27, 2016 to June 18, 2017, with 50 passenger transport routes were customized, and 15 pick-up and drop-off points have been applied for, and a total of 11240 vehicle shifts have been sent, with 281,000 person-times, revenue of 860.4 million yuan, and the actual load rate is 59.43%.

During the Spring Festival travel rush, the enterprise provides campus customization services for passenger transport, helps the college students to go back home, and which can reach nearly 10,000 students safely. During the period of qingming, May Day and Dragon Boat Festival, enterprises combined the traffic and tourism policies to explored travel direct routes, providing door-to-door services for passengers, and expanding the new mode of customized passenger transport through the combination of travel.

Jiangsu province also actively responded positively to call for transformation and reform, the traditional road passenger transport enterprises represented by Jiangsu universiade have self-built in ternet travel platform to provides customized travel service, optimize the supply and demand structure, extend the service chain. This explore new model for road passenger industry transformation and upgrading.

In May 2015, Jiangsu dayun established the road passenger transportation service platform-bus steward jointly with tongcheng network technology. The platform provides a variety of travel service options, such as "intercity carpooling", which provides cross-city travel services for door-to-door pickup; "Airport pickup" provides transport services from place of residence to the airport; "Custom buses" offer city commutes, one person, one person, etc. In terms of service guarantee, on-line customer service robots carry out real-time consultation and complaints. Offline customer service calls are available 24 hours a day. At present, the platform has already amounted to 10 million registered users, average daily ticket is about 80000 pieces, through cooperation with online ticketing platforms and enterprises, achieved data docking with 1008 passenger stations, networked ticketing has covered in Jiangsu, Shanghai, Zhejiang, Guangdong, Fujian, Yunnan six provinces, it presents a good development trend.

Whether the major traditional road transport enterprises in Guangdong province broke through the tight encirclement, or the typical transformation measures such as Jiangsu dayun adapted to the demand of market consumption upgrading, providing useful reference for the industry transformation and upgrading.

1.1. Innovate operation mode and activate internal vitality of passenger transportation
The traditional road passenger transport is the operation of timing, fixed point, fixed line, fixed personnel and fixed shift, which cannot meet the specific time and route demand of passengers at present, but also causes the idle waste of transportation capacity during the peak period. Enterprises carry out customized passenger transport service mode through innovation, accurately match capacity according to passenger demand, and change from "station to station" to "door to door", effectively realizing balance of supply and demand and resource integration, and fully activating potential vitality of road passenger transport.

1.2. Strengthen the main transport industry, give play to the advantages of road passenger transport
At present, the travel demand orientation of safe, reliable, economic efficiency, convenience and comfort is continuously enhanced, transportation enterprise has subdivided the travel market based on the personality needs of the passengers, optimized the line design and model structure, derivative the form of a multi-layered, high quality, precision passenger service, in addition, it forms dislocation competition with other transport modes, and promotes the effective connection of different modes, giving play to the advantages of flexibility and convenience of road passenger transport.

1.3. We will promote service upgrading and open up new avenues for development
Oriented by users' travel demands, the industry organization, transportation capacity resources and service process are reconstructed and reconstructed. The management efficiency is improved through constantly standardizing freight rate standards, assessment system, complaint handling, etc., and the service upgrade is continuously promoted. At the same time, it will be a strategic choice for the passenger transport industry to solve the development dilemma and realize rapid development by acting the link, strengthening cooperation with scenic spots and travel agencies, promoting the expansion and integration of industries, and opening up new ways of development.

2. Travel software brewing ecological change
In the traditional road passenger transport enterprises are considering how to guarantee the industrial advantage at the same time, domestic travel software represented by drops a taxi and ctrip has take advantage of advanced technology and effective marketing strategy in the passenger market quickly occupied a place, and with the posture of having a unique style has produced profound influence to the people to travel.

Through several years of deep cultivation, travel software has gone from the rough field into the era of intensive farming. Drops a taxi has a near-monopoly market share of 80% in the travel category of app.with 20 million daily orders. Its main business covers various modes such as taxi, private car, free ride, bicycle ofo, valet drive, minibus and taxi for the aged, serving more than 360 cities in China. Ctrip takes "the integrated development of transportation and tourism" as its core point, provides a series of travel services such as online chartered car, customized chartered car, line chartered travel, etc., which makes a new attempt to build a tourism transportation system with reasonable structure, perfect functions, outstanding features and excellent services.

The birth of travel software is an important breakthrough for industrial transformation and upgrading in the wave of "Internet + convenient transportation". The segmentation of travel demand and multiple rounds of screening in a few seconds, more accurate matching of supply and demand, accurate formation of user portraits, and drips of travel track data are converging into a new business ecology, and a "superbrain" of mobile travel data is about to emerge.

3. The industry's difficulties and problems

3.1. Pattern promotion is not yet mature
With the progress of reform wave, all passenger transport enterprises have explored the way of transformation in the form of pilot. Although they have achieved initial results, they have not formed extensive promotion and publicity. The public's recognition and acceptance level is not high.

3.2. Regional barriers are serious
Due to administrative barriers and competition of interests in the road passenger transport industry, local protectionism in some regions is serious, and the support for policies is different, which makes it difficult to achieve cross-provincial operation and give play to scale effect. Joining the information-based platform of other regional transportation enterprises means the transfer of user resources. Many enterprises are not highly motivated for their own interests.

3.3. Policy implementation is slow
In 2016 the ministry of transport of the People's Republic of China issued " the guiding opinions on deepening reform and accelerating road passenger transport transformation and upgrading ", We will expand reform in the line approval, site setup and price adjustment, has created favorable conditions for innovation, but due to the actual development, management and policy around the understanding there is a certain difference, work out detailed rules for the implementation of timely and to promote policy large enough.

3.4. Impact of malicious is serious
Many ride-hailing companies intend to take advantage of the policy advantages and use all kinds of malignant means to open a number of illegally customized passenger lines. Under the banner of sharing economy, the business environment of the industry has been seriously damaged, bringing great harm to the healthy, stable and orderly development of the industry.

3.5. Market regulation is inadequate
In the early stage of the development of new business forms, a number of business entities co-exist, and some non-operating vehicles of enterprises without business qualifications are connected to the network platform to participate in road passenger transport, etc., which can easily lead to major traffic and transport accidents caused by mass deaths and injuries. Due to the existing market regulatory loopholes, many business entities use the policies to wipe the edge and stray outside the supervision of the transportation authorities, which has certain safety risks.

3.6. Lack of safety management

Taxi application does give people travel brings a more efficient and convenient experience, but also exposed in the process of development of safety management, such as the subject of security responsibility is unclear, the technical safety of operating vehicles is not timely, and safety standards for ride-hailing services are not yet systematic, safety work remains to be strengthened.

3.7. Cross-border cooperation is lacking

Travel software is using super technical capacity and data-driven mode to seize the passenger market at the unique speed of the Internet. While traditional passenger transport enterprises hold on to the main business, actively explore the road of transformation. The two dominant forces are operating separately and have not formed a good cooperation mechanism. It will take time for the integration of the Internet and traditional industries to develop.

4. Suggestions for development in the industry

4.1. Strengthening policy support

We will give full consideration to the actual situation and difficulties in the transformation of the passenger transport industry and continue to increase support for business, to avoid reconstruction of resources, market competition, recommendations to strengthen local policy guidance, according to relevant ideas and principles of the guidelines on the regulation of road passenger transport services tailored a clear road passenger transport market main body in the industry access, security, dispute handling, supervision and evaluation and resource sharing of responsibility and moral duty, regulate the behavior of business services, protect the legitimate rights and interests of operational safety and related subjects, to ensure passenger transformation to be effective.

4.2. Full implementation of standards

Combined passenger transport industry emerging business characteristics and actual situation, should gradually establish a set of internal controls and external service safeguard mechanism, including the daily maintenance of vehicles, security and emergency rescue operations were performed specification standardization construction, and improve the custom service standard, improve service problems in the operation of the new forms of processing speed and enhance the effectiveness, and gradually thick through practice, the optimization standard, strengthen the industry standard of execution.

4.3. Deepen reform efforts

We made solid progress in streamlining administration, delegating power and improving services. Based on improving the list of powers and responsibilities, we will clean up and streamline administrative review and approval items, further improve the procedures for handling passenger business, and expand the autonomy of passenger businesses. We will actively and steadily push forward reform of passenger transport pricing to fully stimulate the vitality of various market players, optimize the supply structure and the allocation of resources among them, and promote cost reduction and efficiency in the industry, while creating more diversified service models.

4.4. Promoting regional connectivity

We will accelerate information sharing and resource exchange among provincial and regional platforms, and promote the networking of passenger data on a larger scale and in a shorter time. On the premise of ensuring information security and controllability, encourage the mutual integration of government public data and market data to form a road passenger transport big data system integrating travel data, consumption data and credit data, so as to optimize supply structure and expand service capacity. At the same time, it provides scientific support for the innovative development of passenger transport industry.

4.5. Strengthening of supervision

It is suggested to strengthen the clean-up of illegal platforms, strengthen the supervision of transport enterprises that provide operating vehicles for illegal platforms, guide transport enterprises to conduct business through legal and formal platforms, crack down on illegal online ride-hailing, and clearly distinguish between operating and non-operating vehicles. The management department should create conditions and innovate the way of supervision, attach importance to and apply the information-based means, strengthen the analysis and application of big data, facilitate the industrial transformation and upgrading, improve the effect of supervision and reduce the cost of supervision, strengthen the intensity of supervision during and after the event, and ensure the safety of road passenger transport.

4.6. Enhance coordination and communication

We should attach great importance to the communication between transportation and management departments in various regions, clarify the division of tasks between the responsible departments, and strengthen the promotion of passenger transport transformation from top to bottom in a coordinated and horizontal manner. We will create a synergistic mechanism for road passenger service, encourage and guide passenger transport enterprises to conduct in-depth cooperation with legal platforms in the purchase of tickets and the establishment of stations, so that legitimate road passenger transport enterprises and stations can participate in the pilot reform and benefit from the reform as a whole.

5. Conclusion

Whether it's a multi-industry regulatory reform innovation, or the transformation of a traditional passenger enterprise, or even the emerging ecology of the Internet corporation, it has adapted to the needs of market consumption, and it has provided a good example for the transformation of the road.

References

[1] LU Yi-ling, SUN Gen-nian, Why highway becomes tourist resource, J. HIGHWAY (Mar2017 NO.3);

[2] SUN Xiao-ling, Study on Domestic and Foreign Progress of Highway Travel Transport, J. Resource Development &Market (2012 28(12));

[3] DOU Guang-wu, The thinking of the traditional highway passenger transport enterprise entering the tourist passenger transport market, J. Transportation Enterprise Management(2010 265(9));

[4] JI Xiao-feng, LI Jun-fang and CHEN Fang, Coupling Characteristics of Spatial Transport Linkage between Regional Tourism, J.Highway and Transportation Research and Development (Vol. 35 No. 2 Feb.2018);

[5] WANG Xiao-yan, DENG Yipping and DONG Xianyuan, Traffic Safety Situation and Management Measures for Tourism Passenger Vehicles, J. China Public Security. Academy Edition(2015 No.4Sum No.41);

Study on the operation mode of suburban railway at home and abroad and the inspiration to Beijing

Luxi Peng, Chao Wang, Jianhong Wu

(School of Economics and Management, Beijing Jiaotong University, Beijing 100044)

Abstract: As China's urbanization gradually enters the era of 'urban circle' and 'urban agglomeration', the suburban railway which is regarded as the way of connecting urban centers and suburban areas, such as Satellite city, as well as reshaping the layout of urban space, has been highly praised by the central government. In May 2017, the five ministries and commissions jointly issued the guidance on 'promoting the development of the urban suburban railway'. In the guidance, it is clear that the effective supply of the transit service of the suburban railway should be expanded. But in China, the city railway system and city traffic system belong to different traffic management department, and another problem which leads to such situation is that the division of labour and the function of railway system and urban traffic system are different for a long time. So how to operate the suburban railway in Beijing is an significant problem that needed to be solved.

1. Introduction

The conception of 'suburban railway' was first put forward by the School of Traffic and Transportation of Tongji University in the '863 Project of the National 11th Five-Year Plan' . And Wikipedia defines the suburban railway as a kind of urban rail transit system, which is located between the city centre and the suburbs or satellite cities with the advantages of large capacity and high speed. Meanwhile, the commute distance is usually more than 15 kilometers. In this paper we hold the idea that suburban railway is a railway passenger transport system which serves in the economically developed urban areas, maintaining the economic and social links between the central urban areas and the satellite cities, mainly undertaking the commuters of the surrounding cities. Besides, its transportation distance is usually within 100 kilometers[1].

Throughout the world experience, the development of metropolis in developed countries is inseparable from the support of suburban railways. The current mode of transportation in Beijing is dominated by subway and highway. So it's difficult to complete the task of long distance commute between the city center and the outskirts of town, and cannot solve the 'Big city disease' effectively such as traffic congestion, environmental degradation, energy shortages and so on. In May 2017, the five ministries and commissions jointly issued the Guiding Opinions about Promoting the Development of Suburban Railways, which emphasized that efforts should be made to expand the effective supply of public transport operation services of suburban railways, and encouraging the development of multi-level, multi-mode and multi-system rail transit systems. As China's urbanization gradually enters the era of 'urban circle' and 'urban agglomeration', ho w to promote the development of suburban railway and improve the **integrated transportation system** has become the primary task of Beijing and other major cities in China.

At present, the suburban railway system in China is still at an early stage with problems, such as insufficient infrastructure and lagging development concepts compared with developed countries. In

Content from this work may be used under the terms of the Creative Commons Attribution 3.0 licence. Any further distribution of this work must maintain attribution to the author(s) and the title of the work, journal citation and DOI.

Published under licence by IOP Publishing Ltd

addition, the current operation model of suburban railway in China is also difficult to meet the needs of commute transportation in big cities. The main service target of the suburban railway is the commuter flow between the central city and the suburb. The tidal characteristic of suburban railway determines that it needs to have the advantages in terms of capacity and speed. However, the existing operation mode in China has caused the departure time and departure frequency of suburban trains hard to meet the needs of commuting, and the separation of railway hubs and urban functions leads to a poor connection between the different transportation mode, such as subways and buses, which weakens the attraction of passenger flow to a certain extent, and finally the development of suburban railway is difficult to keep up with the pace of urbanization in China.

2. Introduction to the operation of suburban railways in major cities at home and abroad

The reason why the foreign suburban railway system is relatively mature is that on the one hand, the railway construction activities in foreign developed countries is earlier than the planning and development of highways and modern cities, leading to a good suburban railway network structure, as well as the quantity and the quality advantages. On the other hand, the time for the railway to participate in urban public transportation is basically earlier than that of the subway and highway commute transportation and has accumulated lots of management experience. Therefore, most suburban railways in large cities of developed countries have been relatively well-developed from the beginning of planning and design to the construction and operation. It's worthy of our in-depth study and research.

2.1. Tokyo, Japan

In the Tokyo metropolitan area which is 100 kilometres from Tokyo Station, the rail transit system is mainly composed of suburban railways operated by private companies represented by metro companies and JR (National Railways). Its railway network includes 2 loops and 25 rays. According to statistics, JR East has 37 suburban railway lines in the Tokyo metropolitan area, accounting for 56.7% of the network scale; private railroads who use suburban railway to participate in transportation have railway lines of 4475.9 kilometres, accounting for 80.8% of the total network scale; some subways lines and suburban railways are designed to be connected so that they can participate in the suburban commute transportation together in the Tokyo metropolitan area.

On the operation level of suburban railway, Tokyo adopted the **mass transit type periodic system**. For example, the suburban railway of JR East in the Tokyo metropolitan area can achieve a minimum departure interval of 2 to 3 minutes. In order to improve operational efficiency while attracting passengers and maximizing the convenience of transfer, the suburban railway in Tokyo adopts the method of splitting the line, that is, the passenger train and the freight train are driven separately and they're operated independently. The single line runs the same type of trains. In this way, it's possible to reduce the situation of staggered parking of trains and achieve parallel operation map tracking, finally improving the transportation capacity and efficiency of the suburban lines[3].

The suburban railway of the Tokyo Metropolitan Area is operated jointly by multiple companies including the JR East, the private railway companies, the public-private joint venture companies and two metro companies, these companies mainly adopt the operation mode of 'Network-Transportation Integration', supplemented by the way of opening access rights to third parties. The conception of 'Open Access Rights' actually means that the users who own the railway network under the 'Network-Transportation Integration' mode, or the main users of the railway network under the same mode authorize the secondary users to have the corresponding rights of rail transportation. Because of the opening access rights, the operators of suburban railway have opened up the 'interconnected service', which greatly improved the accessibility of subways and private railways, providing conditions for suburban railway to enter the city center.

2.2. Paris, France

The Paris metropolitan area is located in the north of France, with a total area of 12012 square kilometers. It is mainly composed of Paris and the surrounding seven provinces, covering a radius of about 45 kilometers, and its total population reaches 12.2 million. In 1937, Paris opened its first suburban railway and started large-scale construction in the 1840s. By the early 20th century, the early forms of the suburban railway network of Paris had been formed. During this period, the government invested a large amount of special funds for the development and expansion of suburban railways. In 1978, the government's subsidy reached 2.63 billion francs which was the highest in history, and then stabilized at around 1.5 billion francs. At present, the eight suburban railways in the Paris metropolitan area are distributed in different directions in the city, taking the train station as the end. The total length is about 1296 kilometers. It aims to use the existing railway resources to realize the commute transportation in Paris and closely maintain the economic and social connections between the city center and the surrounding area. The proportion of rail transit in the metropolitan area is as high as 89.8%.

In France, the management mechanism of urban public transportation is localized based on municipalities, and supplemented by provinces and regions, so the national government just needs to give proper management over the aspects of top-level design and policy. The Paris metropolitan area is unique and its public transportation network is managed by the Syndicate des transports d'Île-de-France (STIF). Established in 1959, the agency is responsible for the planning, operation, investment and financing activities of public transportation in the Paris metropolitan area. It also coordinates many transportation companies, such as France Railway and Paris Bus Company, so as to achieve smooth cooperation based on the integration of interests.

2.3. London, England

The suburban railway system of London is one of the most advanced in the world. The total length of suburban railway network is 3071 kilometers with characteristics of high density and uniform distribution. The existing 16 suburban railways in London connect the central city, the suburbs and the remote suburbs in series. The farther away from the central city, the larger the station is, and the distribution of stations is correspondingly sparser. The suburban railway network in central London is about 788 kilometers. And there are 321 stations with an average distance of 2.5 kilometers. In the suburbs with a radius of 50 kilometers, the network is about 923 kilometers, the number of stations is 254, and the distance between stations is usually 3.5 kilometers. In the remote suburbs with a radius of 100 kilometers, the network is about 1360 kilometers, but the number of stations is only 173, and the average station spacing is 7.5 kilometers.

Historically, the suburban railway of the London metropolitan area was owned by the British National Railways who was responsible for the operation of it. After the privatization reform of the railway in 1994, the suburban railway was owned by the state through **franchising** under the '**Separation of Network And Transportation**'mode. That is, the suburban railway network is transferred to the National Road Network Company for unified management, and the operation rights of passenger and freight transportation are transferred to 25 passenger transport companies and 2 freight companies. After obtaining the franchise rights, the passenger and freight transportation companies start their transportation service by borrowing lines from the Road Network Company and borrowing vehicle from the locomotive companies.

2.4. Shanghai Jinshan

As a suburban railway, the Shanghai Jinshan Railway is converted from the Jinshan Railway branch. It is a national Grade I two-lane railway with a top speed of 160 km/h. The starting point of the line is Shanghai South Railway Station and the ending point is Jinshanwei Station, with a total length of 56.4 kilometers. There are 9 stations along the line with an average station spacing of 7.05 kilometers.

The Jinshan Railway is the first suburban railway in China that adopted the **mass transit type periodic system.** It is operated by the Shanghai government who mainly purchase transportation services and entrusts it to the National Railway. The ownership and operation right of the Jinshan Railway Xinqiao—Jinshanwei section belongs to Shanghai Jinshan Railway Co., Ltd., which is mainly responsible for financial liquidation and asset management; the Shanghai South-Xinqiao section is invested by the National Railway Transportation who has the assets of this part. The Shanghai Railway Administration and the related railway units are entrusted to be responsible for the transportation organization, passenger transportation services and equipment maintenance of the Jinshan Railway. In order to realize the **mass transit type periodic system**, the manager of Jinshan Railway has specially modified its fare system to meet the needs of Shanghai public transportation card, and improved the construction of the station square, the inbound road and the public transportation facilities of some stations, which provided good external conditions for the operation of the Jinshan Railway[6].

3. Possibility Analysis of Beijing Suburban Railway Operation Mode

At present, the existing resources of the National Railway in Beijing are abundant, and the formation of the passenger dedicated line network has effectively released the capacity of the existing resources, so the Beijing suburban railway has a hardware foundation and development space. The 'Guiding Opinions on Promoting the Development of Suburban Railways 'pointed out that priority should be given to the use of existing resources to operate suburban trains, and based on it the new railway lines should be promoted in an orderly manner. According to the demand characteristics and service targets of the suburban railway in Beijing, selecting appropriate operation mode based on local conditions is the key to make suburban railway participate in the urban transportation as soon as possible.

3.1. Self-managementand and self-operation mode of local government

The self-management and self-operation mode of the local government means that the local government directly manages and operates the suburban railway through establishing an independent operation company, which can be adopted when the local government needs to build a new suburban railway. Under this mode, the local government generally has ownership and control over the line, so it's easy for the government to make independent decisions.

Local governments have many advantages under such operation mode: suburban railway operation companies, which are controlled by local government, can better follow the master plan for urban spatial layout in the specific construction and operation activities, and improve the efficiency of land use at the same time. Also, it can promote the connection between suburban railways and other urban transportation systems, and improve the integrated transportation system. In addition, adopting self-management and self-operation mode allows the government to have more independent decision-making power, which can greatly reduce the cost of negotiation with the National Railway, and can effectively avoid the financial or operational information asymmetry if the suburban railway is operated by National Railway. Compared with other operation modes such as operated by the National Railway or government procurement, mode of self-management and self-operation can better ensure the quality and efficiency of transportation services, and there are sufficient incentives to improve the level of management.

At present, the suburban railway that has been opened or still under planning in Beijing is mainly built through renovating or expanding the existing railways of the national railway or directly use the existing railways to run suburban trains. But self-management and self-operation mode of local government is relatively simple, it's suitable for new suburban railway projects, so according to the recent city plans, this mode may not be considered right now. However, for Beijing, when the resources of the existing lines are fully allocated, in view of the long-term plans for suburban railway and the demand for transportation capacity, the government's building new suburban railway would be the preferred choice in the future, then the self-managment and self-operation mode will work.

3.2. Local government franchise

For the suburban railway projects, the franchise usually refers to granting the operation rights of passenger transportation or freight transportation to the most efficient bidder by means of bidding under the mode of '**Separation of Network And Transportation**'.

The essence of franchise is to realize the ultimate ownership of the railway infrastructure owned by the national government. On the one hand, the government can design the franchise terms, such as the fare, minimum service standards, franchise cycle, etc. to better meet the requirements of suburban railway for socio-economic development, attracting investment and improving efficiency at the same time. On the other hand, maximizing competition in bidding process of franchise can encourage the companies to reduce costs and improve their service, ultimately improving the overall economic efficiency of passenger transportation of suburban railway. Finally, such operation mode can also reduce the government's financial constraints on the locomotive investments to a certain extent. From the perspective of regulatory economics, the purpose of government franchise is to maximize the efficiency of public interests and benefits by maintaining a balance between monopoly and competition in the process of issuing franchise rights.

However, government needs to be familiar with the investment and economic characteristics of the operation projects if the franchise mode is adopted, and at the same time the professional laws and regulations are needed to support the external environment. Therefore, this mode has not been applied in China's railway transportation.

3.3. Entrust operation to the National Railway

Under the mode of **entrusting operation to the National Railway**, usually the **principal** is a joint venture railway company established according to law and obtaining the corresponding railway transportation operation license. Generally speaking, the **entrusted party** usually has a transportation management enterprise that is compatible with the entrusted business and has the ability to independently undertake the entrusted transportation business. The **principal** entrusts services such as transportation organization management, fixed equipment management, mobile equipment management, and transportation safety management to the **entrusted party** through contract, and the entrusted party will hand over the earnings from transportation business to the **principal**, also the transportation costs and profits should be liquidated to the **principal**.

The main characteristic of this mode is that the operation and management of transportation business are separated. The **principal** is the operation entity who controls the **entrusted party** through the contract. And **entrusted party** is the management entity of transportation business.

Currently, most joint venture railway company and the suburban railway that have been in operation in China, such as Shanghai Jinshan Railway, line-S2 in Beijing, and line-Subcenter, all adopt this operation mode.

The main advantage of **entrusted transportation** is that this mode can improve the operational capability of joint venture railway company and the professional quality of staffs to a certain extent, and can strengthen the command of transportation dispatch and ensure the safety of railway transportation. In addition, this mode is also beneficial to the rational allocation of railway transportation network resources, and can enhance the professional management while giving full play to the advantages of scale operation in railway transportation.

However, due to the monopoly position of National Railway, the entrusted transportation mode also has following problems: Firstly, the market-oriented competition pattern has not yet been fully formed, which put **entrusted party** in a weak position compared with railway bureaus who have rich management, operation and technical experienc, because there is usually only one **entrusted party** in the entrusted business. Secondly, the **principal** (joint venture railway company) cannot carry on integrated control over the operation and the management of transportation business, which is not conducive to income increase and savings. The **principal** generally has problems such as simplified organization and insufficient professional staff, making it lack bargaining power when negotiating problems of liquidation standards.

Purchase of transportation services means that the local government or the joint venture railway company purchases suburban railway transportation services from the National Railway in terms of a purchase service agreement, in accordance with the quantity and quality stipulated in the agreement. In addition to having the advantages of entrusted transportation, this mode can better guarantee the interests of the **entrusted party** when the entrusted transportation project is in deficit, and the

principal can promote the **entrusted party** to improve its effeciency by designing a valid performance appraisal mechanism. Its disadvantages are similar as the mode of entrusted transportation. When it comes to determining the price of purchasing transportation services, information asymmetry exists between the government and the National Railway to some extent, and the price of transportation services controlled by the National Railway is monopolistic.

Such operation mode is mainly used in China's suburban railway, such as Shanghai Jinshan Railway, Beijing line-S2 and line-Subcenter, and is mainly applicable to the direct use of the existing railway recources of the National Railway to run suburban trains, and is also suited to the reconstruction and expansion of existing railway recources.

4. Suggestions and Countermeasures

The development of suburban railways is of great significance for the improvement of urban rail transit systems. Only by extending and expanding its railway lines, can it shoulder the task of long-distance commute and increase the proportion of using urban public transport. With the accelerating process of regional integration in China, the development of Beijing's suburban railways will usher in a new era. Also, the government should prepare and arrange its work for the operation mode of suburban railways, and support it from relevant policies and safeguard measures.

4.1. Integrate the suburban railway planning with the coordinated development strategy of Beijing, Tianjin and Hebei.

From the perspective of coordinating the development of capital metropolita area and Beijing-Tianjin-Hebei area, Beijing municipal government should learn from the experience of planning and organization of the international metropolitan, break the concept of administrative divisions, and comprehensively consider the relationship between Beijing satellite city, commute circle radius and suburban railway network, as well as the interaction between suburban railway and urban development. The rights and responsibilities of central government, Beijing and Tianjing municipal government, Hebei provincial government and China Railway should be clearly defined in the construction and operation activities of suburban railway in the capital metropolita area, laying a solid foundation for exploring a construction-management system of suburban railway that is line with China's situation.

4.2. Innovate the operation and management mode of suburban railway and improve the level of operation service.

Under the current railway organization system, it is difficult to change the shortage of the National Railway in the entrusted transportation. It is suggested that in the future, the suburban railway projects operated and managed by the local government should draw on the method of 'open access right' under the '**Network-Transportation Integration**' mode of Japan, as well as the experience from the '**Separation of Network And Transportation**' mode of UK. And the government should correspondly innovate the operation and management system. In terms of operation services, the government should focus on improving the level of intelligence and information of suburban railway services, providing safe and convenient services guarantee a certain frequency of departure.

4.3. Improve the compensation mechanism of operation mode and adjust the fare system

As an important part of urban public transportation, suburban railway should be based on the comprehensive consideration of government financial subsidies, operation costs of enterprise, social affordability, market supply and demand as weill as a multi-level, differentiated fare system in order to better carry out the functions of serving the society. In addition, in the early stage of suburban railway operation, the government can increase subsidies to ensure the smooth operation of suburban trains. And later, the government can achieve operational sustainability of suburban railway through government purchases.

4.4. Improve relevant laws and regulations and improve the supervision mechanism
In order to build a sustainable structure of city, explore an operation mode suitable for Beijing suburban railway, improve the operation efficiency and reduce operation costs, relevant laws and regulations should be established and improved to ensure that its development is law-abiding and rules-based. At the same time, the construction, operation and other links of Beijing's suburban railway should be regulated by law. At present, China's main railway regulatory departments are the Ministry of Transportaion, the State Railway Administration, and some comprehensive social supervision departments including the National Development and Reform Commission, the State-owned Assets Supervision and Administration Commission, and the Ministry of Finance. In the future supervision work, attention should be paid to clearing the powers and responsibilities, and establishing an appropriate power distribution mechanism while formulating unified operation standards of suburban railway to ensure operational supervision efficiency and improve supervision quality.

5. Conclusion
With the increasingly serious problem of "big city disease" and the continuous improvement of urbanization level in China, it is particularly important to reshape the urban layout and optimize its spatial structure. The experience of suburban railway in developed countries shows that the development of suburban railways is a necessary means to realize the sustainable development of China's big cities, especially the metropolitan areas and urban agglomerations. Under the current background of Beijing-Tianjin-Hebei integration, it's of great importance to explore the operation mode suitable for the suburban railway in Beijing to make it participate in the urban rail transit system faster and better, strengthening the economic connections between Beijing and surrounding areas.

According to the distribution of railway resources and construction mode of Beijing suburban railway, this paper shows some operation modes. I believe that in the actual operation activities of suburban railway, how to flexibly combine multiple operation methods according to the local conditions will become a problem worthy of long-term thinking.

Reference
[1] Sun Haifu. Development of suburban railways at home and abroad and suggestions [J]. [J]. Journal of railway engineering, 2014 (3): 1-5, 18.
[2] Zhen Xiaoyan. Analysis of the existing problems in developing suburban railway in China [J]. urban transport, 2014 (6): 6, 45-48
[3] Wujianhong, Shen Congzi. Characteristics of suburban railway in Tokyo metropolitan area and Its Enlightenment to China [J]. China Railway, 2017 (09): 13-19.
[4] Zhou Zixu. A preliminary study on the cooperation mechanism of railway participation in metropolitan rail transit from the perspective of game theory [D]. Beijing Jiaotong University, 2016.
[5] Liu Jing Yue. Overview of operation and management of suburban railway in London [J]. China Railway, 2017 (09): 20-25.
[6] Xiang Baoyu. Construction and operation management of Shanghai Jinshan City (suburban) railway [J]. China Railway, 2017 (09): 7-12.
[7] Chen tie. Study on the introduction of franchise mode in China's railway passenger transport [D]. Tongji University, 2004.
[8] Zhu Hailin. Thoughts on optimizing the entrusted transportation management mode of the joint venture railway [J]. traffic world, 2017:156-157.

Application Scenarios based on SDN: An Overview

Tong Li [1,2]**, Jinqiang Chen**[2]**, Hongyong Fu** [2]

[1]University of Chinese Academy of Sciences, Beijing 100049, China.

[2]Technology and Engineering Center for Space Utilization, Chinese Academy of Sciences, Beijing 100094, China.

litong16@csu.ac.cn, chjinq@ csu.ac.cn, fuhongyong@ csu.ac.cn

Abstract—As a new network concept, SDN subverts the vertical structure of traditional network integration, decoupling the control plane and data plane of the network, replacing distributed control with centralized control, and transforming the closed network into an open one. Programming platform to promote the innovation of network applications, in order to deeply study the application of SDN, we put forward the concept of "SDN+", which combines SDN with different network scenarios to provide a new paradigm for network applications. The article analyzes the architecture of SDN and investigates the application of SDN in different scenarios.

1. INTRODUCTION

The digital society established by the Internet connects almost everything and can be accessed anytime, anywhere. The hierarchical structure of the traditional network promotes the rapid development of the Internet. The traditional network adopts distributed control, and the control plane and the data plane are closely coupled. However, for the complexity and scalability of the current network requirements, the traditional network gradually exposes architectural defects, including three aspects: 1）closed network deployment management is difficult; 2）flow control requires high network operation and maintenance costs; 3）distributed architecture limits network configuration and network innovation.

In 2006, SDN (Software-Defined Networking) was born in the Stanford University Clean Slate project funded by the GENI project in the United States. The purpose of this project is to reinvent the Internet and change the existing network infrastructure. The concept of SDN came into being.

SDN is essentially a new type of network architecture or a network design concept, which uses IT technology to soften the network. Moreover, SDN is application-centric and software-based. The comparison between SDN and traditional network architecture is shown in Figure 1. The SDN network architecture has three layers, which are the application layer, the control layer and the data layer, from top to bottom, as shown in Figure 2. The core of SDN technology is to separate the control plane and data plane of the network, and make the network open and programmable through centralized control. By separating the control plane in the traditional network equipment from the hardware, using the gradual generalization and standardization of the underlying hardware, the entire virtualized network layer is loaded on the physical network, laying the foundation for the introduction of virtualization technologies such as NFV. The application layer is composed of some applications, which are connected to the control layer through the northbound interface, thereby programming the underlying device and developing various business applications according to the needs of the user. The control layer is responsible for formulating the flow table to issue commands to the data layer, maintaining

Content from this work may be used under the terms of the Creative Commons Attribution 3.0 licence. Any further distribution of this work must maintain attribution to the author(s) and the title of the work, journal citation and DOI.

Published under licence by IOP Publishing Ltd

topology state information of the entire network, and dynamically allocating network resources; the data layer is mainly responsible for forwarding data packets, and is connected to the control layer through the southbound interface.

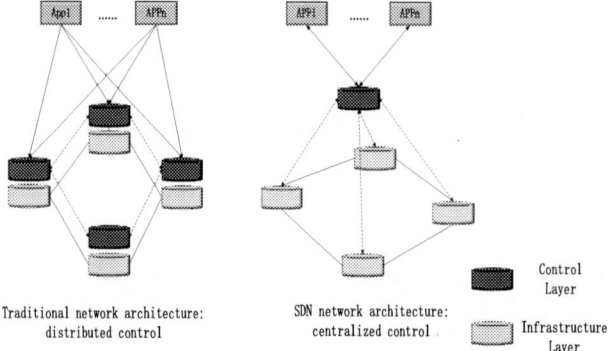

Fig. 1. Comparison of traditional network architecture and SDN network architecture.

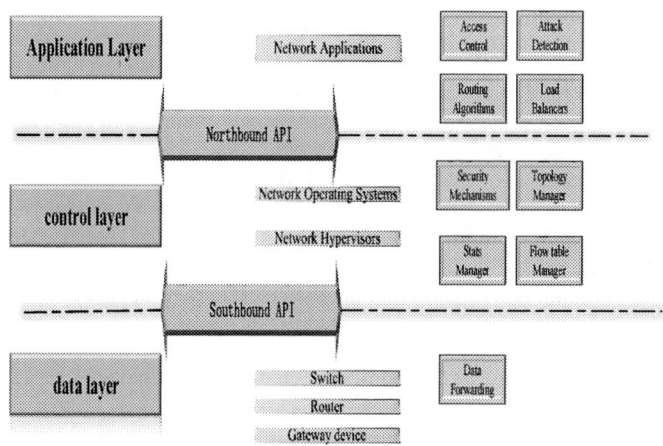

Fig. 2. SDN network architecture.

2. THE OVERVIEW OF SDN APPLICATION SCENARIO

SDN originally originated from the experimental innovation of the campus network. The research team led by Professor Nick McKeown of Stanford University proposed the concept of Openflow [15], which promoted the birth of SDN; SDN succeeded in Google's B4 [6] network, in the early stage of development of SDN, the application scenarios are mainly focused on: data center network, interconnection between data centers, government and enterprise networks, telecom carrier networks, and Internet company business deployment. The article will outline the initial application of SDN and some innovative application scenarios in the later stage.

2.1. SDN+WAN

Paper [6] proposes a software-defined WAN architecture B4, as shown in Figure 3, which is one of the first and largest SDN/OpenFlow deployments. This is the first publicly available commercial case for SDN-based data center interconnection and the first publicly available SDN-based distributed controller application case. This network uses a three-tier architecture consisting of switch hardware layer, site controllers layer, and global layer. The hardware layer of the switch uses switches made by Google. The switches run the OpenFlow protocol and support new services such as centralized TE, which is mainly responsible for forwarding traffic. The site controller layer is composed of OpenFlow Controller

(OFC) and Network Control Server (NCS). The OFC cluster runs on the NCS, and the OFC maintains the network status according to the instructions of the network control application and switch events. The global layer consists of an SDN gateway and a central traffic engineering server, responsible for centralized and unified control of the entire network. The SDN gateway collects link information from the OFC and abstracts it to the TE Server to learn the global path information. The forwarding entry information generated by the TE server is translated by the SDN Gateway and sent to the underlying switch through the OFC. The centralized traffic engineering of the architecture allocates bandwidth according to the level of demand and priority. The bandwidth utilization of the link is improved by more than 3 times and the long-term is close to 100%, therefore the link cost is greatly reduced. This proves that SDN can be successfully applied to large-scale networks interconnected by data centers. And it also describes some lessons in the research and future research directions.

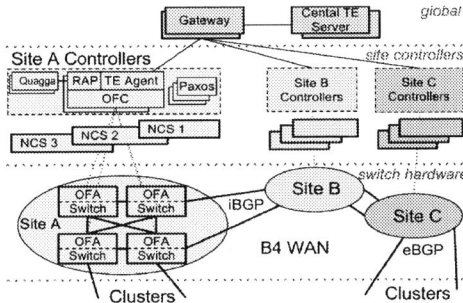

Fig. 3. B4 architecture overview[6].

2.2. SDN+ Cellular Networks

For the limitation of the current cellular network, as shown in Figure 4, the authors in [7] designed an SDN-based architecture of the programmable control plane in cellular network, and analyzed how the introduction of the SDN architecture can better solve the problems of the cellular network, such as reducing the extra equipment in the network. Cellular providers can effectively monitor traffic at different levels of granularity; SDN will provide common control protocols across different cellular technologies, making mobility management easier and reducing latency; SDN can make distributed implementation of QoS and firewall policies based on network-wide views, and manage traffic scheduling through multi-hop QoS classes in the network; SDN can support network virtualization. The mobile virtual operator can subscribe to the corresponding service according to its own needs; the SDN has a global view of the base station, and supports centralized control of the base station. And the SDN controller can allocate radio resources more efficiently to handle new users. The scalability of controllers, switches and base stations is proposed from four aspects: 1) The architecture will automatically convert policies based on user packet processing rules of IP addresses and network locations. 2) Under the direction of the controller, a software agent capable of performing simple local operations should be run on the switch. 3) The cellular network will support deep packet inspection, header compression and message-based control protocols. 4) The architecture virtualizes base station resources, allowing each "slice" of controllers to control radio resource management, admission control, and mobility control.

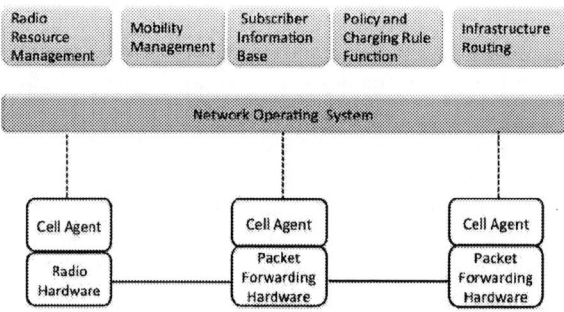

Fig. 4. Cellular SDN architecture[7] .

2.3. SDN+Mobile Cloud

Paper [8] proposes several mobile cloud architecture designs based on SDN and Ad hoc networks. As shown in Figure 5, different wireless technologies are used in the control plane and data plane respectively: the remote wireless connection is used for LTE in the control plane, and the high bandwidth wireless connection is used for Wi-Fi in the data plane. The global SDN controller is the core intelligent component of the architecture and is responsible for populating the flow table of the SDN wireless node to control the way the traffic moves in the network. The SDN wireless node is equivalent to the switch in the SDN network. Each SDN wireless node has an alternate local SDN controller to ensure normal communication of the network when there is a problem with the global controller. The global SDN controller and the SDN wireless node have two connection examples, which are a continuously connected global SDN controller and an intermittently connected global SDN controller. In the MANET system, each SDN wireless node constructs a node connection graph by exchanging beacon messages, and the SDN controller can make intelligent routing decisions based on general algorithms or flow-based routing information according to the path information provided by the node connection graph. The article also describes the mobile cloud instance based on SDN frequency selection. According to the interface frequency configuration, the wireless node can be divided into: wireless nodes with wireless interfaces configured to specific frequencies and wireless nodes with reconfigurable wireless interfaces, In addition, there are three potential application cases: wireless network virtualization, reserved traffic, and frequency hopping. The article proves the feasibility of SDN-based mobile cloud architecture by emulating on NS-3. Based on SDN routing, it is better than traditional Adhoc routing protocol, which can achieve good packet transmission rate. Finally, the author proposes three future exploration directions: 1) recovery mechanism when SDN controller is lost; 2) standby SDN wireless architecture; 3) frequency selection use case.

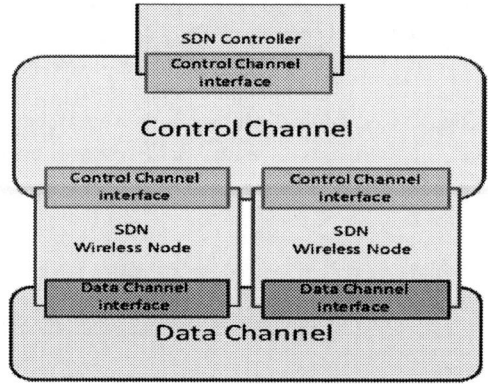

Fig. 5. SDN-based mobile cloud overview[8] .

2.4. SDN+VANET

Authors in paper [9] proposed the SDN-based VANET architecture and its operation mode as shown in Figure 6. According to the degree of control of the SDN controller, the architecture is divided into three modes: centralized control mode, distributed control mode and hybrid control mode. The three advantages of the architecture are analyzed: 1) Path selection: When the SDN controller senses that data traffic may become unbalanced, it can initiate a rerouting process to improve network utility and reduce congestion. 2) Frequency/Channel Selection: SDN-based VANET can dynamically coordinate the channel/frequency used by SDN wireless nodes with multiple available wireless interfaces. 3) Power selection: Due to the awareness, SDN-based VANET can dynamically change the wireless interface power and its transmission range according to the perception. Finally, the SDN-based routing and traditional MANET/VANET routing protocols are compared by simulation, which proves the feasibility of software-defined VANET.

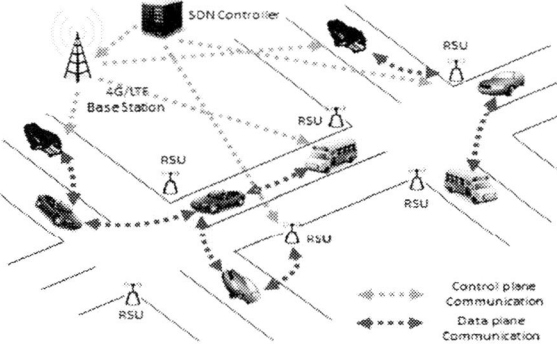

Fig. 6. Software-Defined VANET Communications[9].

2.5. SDN+Air-Space-Ground Integration Network

An SDN-based air-ground integration network described in paper [10], consists of a three-layer structure of an air-based network layer, a space-based network layer and a ground-based network layer. Each layer works independently and is interconnected. The space-based network layer is mainly composed of satellites. The air-based network layer is composed of UAV Ad-hoc network. The ground-based network layer is mainly composed of some cellular networks and digital cluster systems. In addition to the space-based system, the other two layers are equipped with SDN controllers to implement unified centralized control for each layer of the network. In the UAV Ad-hoc network, each cluster head is equipped with a micro SDN controller, and has real-time global information of the cluster network, and other UAVs do not need to carry any processing and decision-making equipment, greatly improving the endurance capability of the UAV Ad-hoc network. The ground station controller can be used as the master controller to command the entire UAV Ad-hoc network. If the link between the UAVs and the ground station fails, the ground-based satellite relay can be used to contact the ground station. The cooperation of the three layers of the network ensures the normal operation of the entire network. In this paper, the delay is used as an indicator to simulate the performance of AODV routing protocol and OLSR and DSR routing protocols. At the same time, the traditional air-space-ground integrated network and the SDN-based air-space-ground integrated network are compared by simulation. The results show that the transmission of SDN-based air-space-ground integrated network is better, the end-to-end delay is shorter, and the overall performance is superior.

Integrated space and terrestrial network architecture envisioned in paper [11] is based on the SDN architecture. The application layer consists of a number of spatial tasks. The control plane consists of the controller nodes of the space base and the ground, and is represented by the network operating system in the logical architecture. The control plane is the core of the network and has an overall global view of the network to implement routing decisions and management and control of the entire spatial information network. The data plane consists of satellite nodes and is mainly responsible for data processing and forwarding. The physical architecture consists of a space-based backbone network, a data forwarding layer, and a ground information port. The space-based backbone network consists of three GEO satellites, the data forwarding layer consists of MEO and LEO satellites in the middle and low orbit, and some ground station controllers are set up in the ground information port. The controller node of the ground information port and the GEO satellite node in the space-based backbone network together form a control plane in which multiple controllers cooperate. In a multi-controller deployment scenario, each controller is responsible for its own control area, with network status information of the local area, manages and maintains the normal operation of the network in the area, and communicates between the controllers through the east-west interface to generate the global topology view of entire network. Since there are still large technical difficulties in placing the controller on the satellite, the SDN controller is placed on the ground in the early stage. The global view of the SDN architecture provides a more reasonable routing strategy for the satellite network. The centralized resource allocation and management guarantees higher service transmission quality. The open and programmable features make the configuration of the satellite network more flexible and convenient. The unified standards and interfaces well solve the problem of heterogeneity of the spatial network. The separation of the data plane and the control plane simplifies the architecture of the satellite function and the design cost, effectively reducing the complexity of satellite management.

2.6. SDN+ Satellite network

OpenSAN [12], software-defined satellite network, is composed of three planes: management plane, control plane and data plane, as shown in figure 7. The management plane consists of network operations and control centers; the control plane consists of three GEO satellites that cover the globe; the data plane consists primarily of satellite infrastructure and terminal routers. The management plane and the control plane are connected through the SDN Northbound Interface (NBI), and the data plane and the control plane are connected through the SDN Southbound Interface (CDPI). The control plane has a global topology view, which is responsible for monitoring the network status and sending dynamic information of the network to the management plane. The management plane (NPCC) can provide functional modules such as routing policy, user management, virtualization, security, resource utilization and network management. The management plane calculates a new flow table through the global network information and sends it to the data plane through the control plane. The data plane only needs to complete the forwarding of the data packet according to the flow table. Since the motion of the satellite is periodic, we can predict the changes in the entire network using the corresponding prediction-based algorithms (such as neural networks). By decoupling the data plane and control plane, OpenSAN saves the resources of satellite nodes, overcomes the difficulty of satellite network closure and expands, provides efficient and fine control, and supports the flexibility of future advanced network technologies and services.

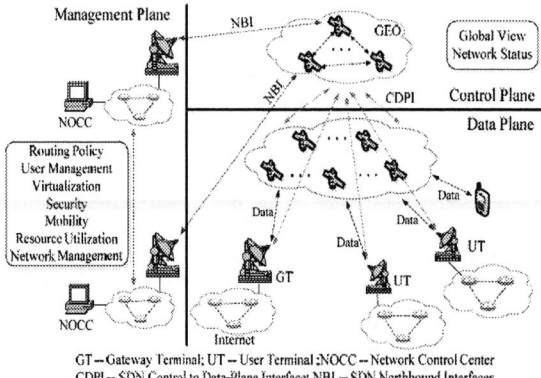

GT – Gateway Terminal; UT – User Terminal ;NOCC – Network Control Center
CDPI – SDN Control to Data-Plane Interface; NBI – SDN Northbound Interfaces

Fig. 7. The Architecture of OpenSAN[12].

Paper [13] proposes a Software Defined Satellite Network Architecture (SDSN), which has three layers: the application layer, the control layer, and the infrastructure layer, as shown in figure 8. The application layer is the Network Operations Control Center (NOCC), which has functions such as routing policy calculation and application generation. The control layer is a ground station that translates instructions from the upper NOCC into simple data structures and sends them to the infrastructure layer. The infrastructure layer is made up of satellites, and the satellites only need to implement the simplest forwarding and hardware configuration functions, which reduces the development and design cost of the satellite. The SDSN distributes routing tables and configuration commands through a single-layer inter-satellite link (ISL) and GEO satellites to implement a satellite control model with a global network view. When the network update is small, single-layer ISL forwarding is adopted; when the network update is large, GEO satellite broadcasting is adopted. This new hybrid control structure enables real-time and fast network configuration, reducing the number of ground stations and greatly improving the update speed of traditional satellite systems. The software-defined network technology decouples the data plane and the control plane, and obtains a global network view through centralized control, and deploys a more granular management strategy to optimize the calculation of the global routing path.

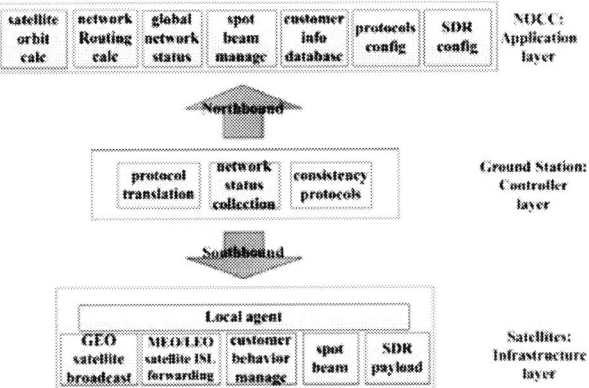

Fig. 8. The SDSN architecture[13].

2.7. SDN+ UAV network

In paper [14], an SDN-based UAV backbone network framework was designed. Figure 9 shows the illustration of the SDN-based UAV network in this paper and figure 10 shows the functional architecture of a UAV network. The control core includes a UAV controller and an SDN controller. The UAV controller is responsible for managing information such as flight control, location, battery storage, etc. The SDN controller is responsible for network information management. The UAV controller and the SDN controller can interact to work together and share information to make optimal decisions. Based on the two core issues of information management and resource management of UAVs, an SDN controller monitoring platform is proposed, which includes four modules, namely monitoring display, traffic management, link management and strategy. According to the analysis results of the monitoring platform, a load balancing algorithm is further proposed in the policy module to fully utilize the UAV network resources and maintain the desired network traffic balance.

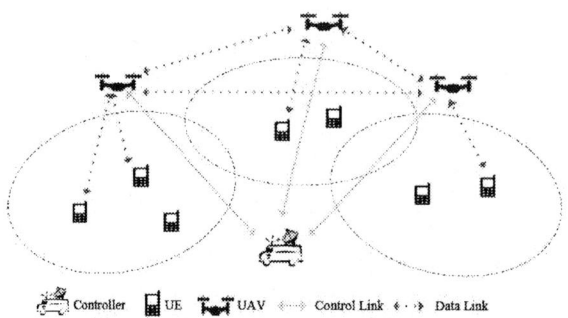

Fig. 9. Illustration of the SDN-based UAV network[14] .

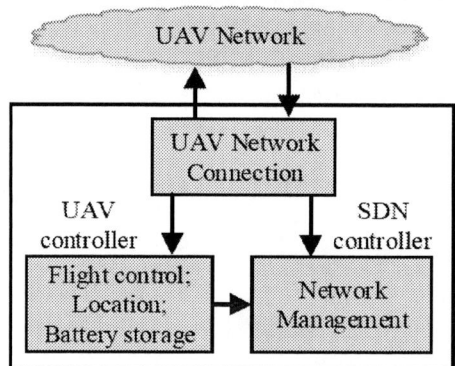

Fig. 10. Functional architecture of the UAV network[14] .

3. CONCLUSIONS

Traditional networks have given the Internet a vibrant life but have also hindered the longer-term development of the Internet. The new dynamic network architecture of SDN has successfully transformed traditional networks into diverse application-oriented platforms. By decoupling the control plane and data plane in the network architecture, the hardware cost of the network equipment can be effectively reduced, so that the network can be flexibly large-scale programming and simplified management like the computer infrastructure; the centralized control logic can make the network have a global view information can be globally optimized and resource-adapted. The deployment and maintenance of network nodes will be more agile; open and programmable network can promote more business network service innovation. However, the current SDN technology is not fully mature, including the standardization and unification of various interfaces, the compatibility of heterogeneous networks, multi-controller coordination and some security issues. The expansion of SDN applications is also facing more challenges. In this paper, we surveyed 7 different areas to analyze SDN-based application scenarios. The "SDN+" architecture provides a new way of thinking for networks with different applications, expanding the intelligence of the network and achieving maximum optimization.

The application of SDN is no longer limited to the campus network, the WAN network between data centers, but has a broader space, not just the ground network, and even some good application ideas have been made in space-based networks. In the future, we will focus on exploring more application scenarios of "SDN+" and optimized routing algorithms that are more suitable for SDN architecture.

REFERENCES

[1] Diego Kreutz，Fernando M. V. Ramos, Paulo Esteves Verı́ssimo, et al, "Software-Defined Networking:A Comprehensive Survey," IEEE, vol. 103, issue 1 ,2015.

[2] H. Kim, N. Feamster, "Improving network management with software-defined networking," IEEE, vol. 51, issue 2, pp. 114–119,2013.

[3] B. Nunes, M. Mendonca, X.-N. Nguyen, et al, "A survey of software-defined networking: Past, present, future of programmable networks," IEEE, vol. 16, issue 3, pp.1617–1634,2014.

[4] M. Casado, N. Foster, and A. Guha, "Abstractions for software-defined networks," ACM Commun. vol. 57, issue 10, pp. 86–95, 2014.

[5] McKeown, Nick，Anderson, Tom，Balakrishnan, Hari Balakrishnan, et al, "OpenFlow:Enabling Innovation in Campus Networks," ACM SIGCOMM Computer Communication Review, vol. 8, issue 2, pp. 69-74 ,2008.

[6] Sushant Jain, Alok Kumar, Subhasree Mandal, et al, "B4: Experience with a globally-deployed software defined wan," ACM SIGCOMM,vol. 43,issue 4,pp. 3–14,2013.

[7] Li Erran Li，Z. Morley Mao，and Jennifer Rexford, "Toward Software-Defined Cellular Networks，" IEEE Computer Society, pp. 7-12, 2012.

[8] Muhammad Usman, Anteneh A. Gebremariam, Fabrizio Granelli, and Dzmitry Kliazovich, "Software-Defined Mobile Cloud: Architecture, Services and Use Cases," IEEE CAMAD, 2015.

[9] Ian Ku, You Lu, and Mario Gerla, "Towards Software-Defined VANET: Architecture and Services," in Proc. 13th Annual Mediterranean Ad Hoc Networking Workshop, pp. 103-110, 2014.

[10] C Chen，SS Xie，XX Zhang, ZY Ren "A new space and terrestrial integrated network architecture aggregated SDN," Journal of China Academy of Electronics and Information Technology, vol. 10, issue 5, 2015.

[11] FM Xu, ZJ Tong, CL Zhao, ZC Qin, "Architecture, Technology and Challenges of Software Defined Integrated Space and Terrestrial Network" ZTE TECHNOLOGY JOURNAL, Vol. 24, issue 2, 2018.

[12] Jinzhen Bao, Baokang Zhao, Wanrong Yu,et al, "OpenSAN: A Software-defined Satellite Network Architecture," Acm Sigcomm Computer Communication Review, 2014.

[13] Zhu Tang ; Baokang Zhao ; Wanrong Yu, et al, "Software Defined Satellite Networks: Benefits and Challenges," IEEE Computers, Communications and IT Applications Conference, 2014.

[14] Xiao Zhang, Haijun Wang and Haitao Zhao, "An SDN Framework for UAV Backbone Network towards Knowledge Centric Networking," IEEE Conference on Computer Communications Workshops, 2018.

[15] P. Dely, J. Vestin, A. Kassler, N. Bayer, H. Einsiedler, and C. Peylo, "CloudMAC-An OpenFlow based architecture for 802.11 MAC layer processing in the cloud," in Proc. IEEE Globecom Workshops, pp. 186-191, 2012.

[16] Y. Zaki, L. Zhao, C. Goerg, and A. Timm-Giel, "A novel LTE wireless virtualization framework," in Proc. Int. Conf. Mobile Netw. Manage, pp. 245-257,2010.

[17] K. Pentikousis, Y. Wang, and W. Hu, "MobileFlow: Toward software defined mobile networks," IEEE Commun. Mag., vol. 51, issue 7, pp. 44-53, 2013.

[18] N. Feamster, J. Rexford, and E. Zegura, "The road to SDN," Queue, vol. 11, issue 12, 2013.

[19] Y. Li and M. Chen, "Software-defined network function virtualization: A survey," IEEE Access, vol. 3, pp. 2542-2553, 2015.

[20] B. Raghavan et al., ''Software-defined internet architecture: Decoupling architecture from infrastructure,'' in Proc. 11th ACM Workshop Hot Topics Netw., pp. 43–48, 2012.

[21] Miao Y, Cheng Z, Li W, et al., "Software Defined Integrated Satellite-Terrestrial Network: A Survey." International Conference on Space Information Network. Springer, Singapore, 2016.

Overview of end-to-end speech recognition

Song Wang[1], Guanyu Li[1*]

[1]Key Laboratory of National language Intelligent Processing Gansu Province, Northwest Minzu University, Lanzhou, China

*guanyu-li@163.com

Abstract. In the 1960s, automatic speech recognition has been widely studied. In the past, HMM has been the mainstream of the acoustic model. With the development of machine learning, neural network is introduced to the speech recognition, relying on neural network's strong learning ability. Thus the acoustic model of DNN has significantly improved the voice recognition rate, compared to the HMM model. In order to simplify the traditional speech recognition system, the end-to-end speech recognition method is proposed. This paper mainly introduces and analyzes the end-to-end system, and the main two models of CTC and attention, as well as the prospect of future speech recognition research.

1. Introduction

Automatic speech recognition has been a hot topic of research. In the 1980s, after IBM applied HMM to speech recognition, HMM has been playing an important role in speech recognition, and HMM-GMM has become the mainstream acoustic model. In 2006, after Li Deng and Hinton[1] proposed the use of deep learning in speech recognition, the neural network became a research upsurge of speech technology, which turned from the ANN to the DNN. The research included the neural network composed of restricted Boltzmann machine stacking and conducted layer by layer training for the network. Improve the over-fitting problem of neural network by using dropout. DNN-HMM became the main acoustic model, showing strong recognition capability in the recognition of large vocabulary. With the further development of technology, more neural networks began to invest in the field of speech recognition, such as CNN and RNN.

However, establishing a speech recognition system is a complicated process, which requires a lot of professional knowledge. Various attempts have been made in recent years to reduce the complexity of ASR, in the hope of directly mapping speech to tags. End-to-end speech recognition has been proposed. Now there are two main structures for end-to-end speech recognition: attention model and CTC. End to end technology has been applied in many aspects and has achieved remarkable results.

In this paper, I will introduce the CTC and attention model. The section 2 introduces the difference between traditional speech recognition and end-to-end speech recognition then introduce the CTC and attention model.

2. End-to-end speech recognition

This part mainly introduces what is end-to-end speech recognition and the difference between end-to-end speech recognition and traditional speech recognition.

Content from this work may be used under the terms of the Creative Commons Attribution 3.0 licence. Any further distribution of this work must maintain attribution to the author(s) and the title of the work, journal citation and DOI.
Published under licence by IOP Publishing Ltd

2.1. End-to-end speech recognition

End-to-end is a system which directly maps a sequence of input acoustic features into a sequence of grapheme or words. A system which is trained to optimize criteria that are related to the final evaluation metric that we are interested in (typically, word error rate).

For Conventional ASR, most ASR systems involve separately trained acoustic, pronunciation and language model components which are trained separately. Curating pronunciation lexicon, defining phoneme sets for the particular language requires expert knowledge, and is time-consuming. Figure 1 depicts its structure.

Figure 1. Conventional ASR Pipeline　　Figure 2. End-to-end ASR Pipeline

It can be seen that end-to-end speech recognition greatly simplifies the complexity of traditional speech recognition. There is no need to manually label information, in the neural network can automatically learn language or pronunciation information, as shown in Figure 2. Now there are two main structures for end-to-end speech recognition: attention model and CTC.

2.2. Connectionist Temporal Classification (CTC)

CTC was proposed by Graves et al[2], as a way to train an acoustic model without requiring frame-level alignments. In early work, using CTC to output target phonemes is not really end-to-end, and it still requires language models. CTC allows for training an acoustic model without the need for frame-level alignments between the acoustics and the transcripts.

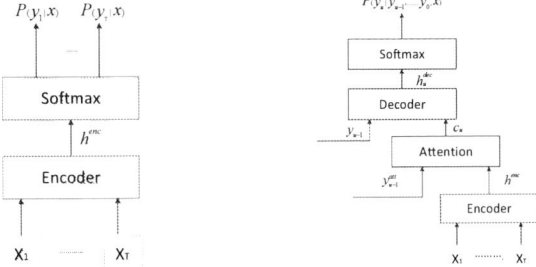

Figure 3. CTC structure　　Figure 4. Attention model structure

The acoustic model training using CTC as the loss function is a end-to-end training, which does not need to align the data in advance, but only needs an input sequence and an output sequence to be trained. Structure as shown in Figure 3. In this way, there is no need to align and label the data one by one, and the probability of CTC's direct output sequence prediction needs no external post-processing. The CTC introduces blank [3](which has no predicted value for this frame), where each classification of the prediction corresponds to a spike in the whole speech, and the other locations that are not spikes are considered blank. For a speech, the CTC ultimately outputs a sequence of spike, regardless of the duration of each phoneme.

In the case of given x, the probability that the output is a label sequence y is formula(1):

$$P(y \mid x) = \sum_{\hat{y} \in B(y,x)} \prod_{t=1}^{T} P(\hat{y}_t \mid x)$$

2.3. Attention model

Attention-based Encoder-Decoder Models emerged first in the context of neural machine translation[4]. Attention Mechanism is mainly used to solve the problems of traditional RNN-based Seq2Seq model[5][6]. Seq2Seq model is an end-to-end machine translation model built based on an encoder and a decoder. Encoder encodes input X into a fixed length hidden vector Z, Decoder decodes target output Y based on Z. There are two obvious problems with this model:

• All information of input X is compressed into a fixed length hidden vector Z, ignoring the length of input X. When the input sentence length is very long, especially longer than the original sentence length in the training set, the performance of the model drops sharply.

• It is unreasonable to encode input X into a fixed length and assign the same weight to each word in the sentence. For example, in machine translation, the input sentence and the output sentence are usually one or more words corresponding to the output word or words. Therefore, assigning the same weight to each word input makes no distinction and often results in model performance degradation.

The traditional Seq2Seq model lacks discrimination of input sequence X, therefore, in 2015, Kyunghyun Cho introduced the attention mechanism to solve the problem[6]. In 2015, were first applied in ASR[7]. The basic idea of Attention mechanism is that it breaks the limitation of traditional encoder-decoder structure, which is dependent on a fixed length vector inside. Figure 4 depicts the structure.

• Encoder(analogous to AM): Transforms input speech into higher-level representation.

$$h_i^{enc} = Encoder(x_i)$$

The structure of encoder is usually PLSTM, BLSTM, CNN+LSTM.

• Attention(alignment model): Identifies encoded frames that are relevant to producing current output.

Attention module computes a similarity score between the decoder and each frame of the encoder.

$$e_{u,t} = score(h_{u-1}^{att}, h_t^{enc})$$

$$a_{u,t} = \frac{\exp(e_u, t)}{\sum_{t'}^T \exp(e_u, t')}$$

$$c_u = \sum_{t=1}^T a_{u,t} h_t^{enc}$$

• Decoder (analogous to PM, LM): It has an auto regressive operation by predicting each output token as a function of the previous prediction.

Output the hidden state h_t^{dec}

$$h_t^{dec} = Decoder(y_{t-1}, c_{t-1})$$

Generate the label prediction

$$y_t = soft\max(U[h_t^{dec}; c_t])$$

3. Conclusion

At present, the end-to-end speech recognition technology based on end-to-end technology has achieved remarkable results, but the end-to-end speech recognition based on CTC still needs language model to get better results, and how to further realize the real end-to-end speech recognition is worth paying attention to in the future.

Acknowledgments

Supported by the Fundamental Research Funds for the Central Universities (Research on Unified Acoustic Models of Mandarin and Tibetan, 31920170145)

References

[1] Muda L, Begam M, Elamvazuthi I. "Voice recognition algorithms using Mel-frequency cepstral coefficient (MFCC) and dynamic time warping (DTW) techniques."(2010)

[2] Graves et al. "Connectionist temporal classification: Labelling unsegmented sequence data with recurrent neural networks."(ICML 2006)

[3] Alex Graves,Navdeep Jaitly. "Towards End-to-End Speech Recognition with Recurrent Neural Networks."(ICML 2014)

[4] Chan et al. "Listen,Attend and Spell."(2015)

[5] Ilya Sutskever et al. "Sequence to Sequence Learning with Neural Networks."(2014)

[6] Kyunghyun Cho et al. "Learning Phrase Representations using RNN Encoder–Decoder for Statistical Machine Translation."(2015)

[7] Chorowski et al. "Attention-Based Models for Speech Recognition."(2015)

[8] Alex Graves, Abdel-rahman Mohamed, and Geoffrey Hinton, "Speech recognition with deep recurrent neural networks."(2013)

[9] Graves, A., Jaitly, N., and Mohamed, A.-r. "Hybrid speech recognition with deep bidirectional LSTM."(2013)

[10] Hinton, G., Deng, L., Yu, D., Dahl, G., Mohamed, A., Jaitly, N., Senior, A., Vanhoucke, V., Nguyen, P.,Sainath, T., and Kingsbury, B. "Deep neural networks for acoustic modeling in speech recognition:The shared views of four research groups."(2012)

ISPECE   IOP Publishing

Air Traffic Management Process Quality Assessment Model Based on Improved Fuzzy Matter Element Analysis

PAN Wei-jun, REN Jie, WANG Run-dong

Civil Aviation Flight University of China, Guanghan 10624, China

rogerfrozen@163.com

Abstract. In order to improve the service quality of the air traffic management system, the accurate air traffic management process quality assessment can identify the weak links in the process of interaction with the user, and timely make early warning, evaluation and feedback to avoid the risks of the air traffic control system. Based on the actual situation of air traffic control, the related factors affecting the quality of air traffic service process are analyzed. The process indicators covering the four aspects of human, machine, environment and management are selected and the fuzzy comprehensive evaluation model is established. Based on the improvement of the membership function. Finally, through the analysis and verification of the example, the result is in line with the actual situation of the operation of the air traffic control system, and provides a reference for improving the quality of air traffic control services.

1. Introduction

The rapid development of the civil aviation industry is inseparable from the protection of air traffic control (ATC), but with the rapid growth of flight traffic, air traffic generally exhibits high or ultra-high density operating conditions, good air traffic service quality for flights, crew, companies has a positive effect, which improves flight comfort and customer satisfaction. Therefore, ensuring the quality of air traffic control services has become the top priority of China's civil aviation construction. Civil aviation accident statistics show that the proportion of accidents caused by air traffic management is not large. Air traffic control-related accidents accounted for 4% of the total number of fatal accidents caused by accidents worldwide in 1980-2002[1] However, if the air traffic control is the main cause of the accident, it will have serious and serious consequences, such as the 1977 Tenerife air crash. The International Civil Aviation Organization (ICAO) clearly states in Annex XI[2] that units providing air traffic control services should actively construct and implement a Safety Management System (SMS). Quality of service is an important part of the safety management system. It can reflect the quality of the operation of the air traffic control system and can identify and improve the weak links based on the evaluation results. Therefore, it is necessary to improve the operation level of the air traffic control system by evaluating the service quality of the air traffic control system.

Yuan Leping[3] based on fuzzy evaluation, establishes the evaluation factor set, evaluation set, determines the weight of each factor from the evaluation factor set to the evaluation set, and uses the unascertained number theory to calculate the probability of occurrence of danger, giving empty Pipeline safety risk assessment model; Zhang Jianping[4] used factor analysis method and fuzzy comprehensive evaluation method to study the quality factor of air traffic control, and analyzed the safety factor, flow factor and efficiency factor as the comprehensive evaluation terminal area. The basis for regulating the quality of operation and optimizing the operation strategy is provided; P Averty[5] designed a controller's psychological load estimation method, which is designed to take into account the additional load caused

---

Content from this work may be used under the terms of the Creative Commons Attribution 3.0 licence. Any further distribution of this work must maintain attribution to the author(s) and the title of the work, journal citation and DOI.

Published under licence by IOP Publishing Ltd

by objective traffic variables and subjective influences, including the severity of the conflict. And the time pressure of the solution. It is proposed that the controller workload estimate should be combined with objective task variables and subjective assessments related to it to study the potential risks caused by the controller workload. Wang Yanqing and Wu Weijie[6] used the LOGIT model to analyze the factors affecting the controller's mental deviation in an emergency, and determined that the controller's control work would be affected by psychological factors in an emergency; the literature[7] proposed a basis Probabilistic risk assessment of air traffic control and air traffic management methods, combined with the TOPAZ model, takes into account the intricate relationship between the various risk sources in the air traffic control system, proposes a dynamic color network model, thus solving the traditional risk assessment model. Risk assessment but ignoring the shortcomings of the interaction between the various components.

Yorck Hauss and Klaus Eyfert pointed out that air traffic has grown rapidly and increasingly in Western countries since the 1990s, and the impact of human factors on air traffic safety has led to situational awareness (Situation Awareness). , SA) has become an important indicator to measure the quality of air traffic. The two experts also conducted a study of the event-based psychological representation of the controller[8]. Sven Ternov and Roland Akselsson built models and procedures based on aviation accidents and began an analysis of risk barriers for air traffic control systems, using this method to identify and identify sources of risk in an air traffic control environment in a complex system[9].

At present, there are many safety assessments for air traffic management, but there are few quantitative analysis of process service quality. How to change from qualitative analysis to quantitative research is an important topic in the future research of air traffic control. This paper selects the key indicators that reflect the quality of the control of the terminal area, and then collects the index data, and determines the five most important process indicators for evaluating the operational quality of the terminal area and two outcome indicators. The analytic hierarchy process is used to determine the weight of each index. Then, the fuzzy comprehensive evaluation method is used to evaluate the model. The improved normal distribution membership function is used to determine the value of each evaluation index. The classical domain of each level is valued, the current evaluation matrix of the sector is calculated, and the service quality of the current terminal area is rated according to the principle of maximum membership degree.

2. Establish a quality assessment model

In this paper, the process indicators are selected to determine the first-level indicators in five directions, namely equipment factors, control safety, control efficiency, workload, and meteorological factors; the above five indicators continue to subdivide the secondary indicators. Select the rate of accidents for the cause of the accident, and the rate of abnormality of the control cause is the result index[10]. See Table 1 for specific indicators.

Table 1. Air Traffic Control System Safety Performance Assessment Indicators

First-level	Second-level
Equipment factor f_1	Equipment failure rate f_{11}
	Equipment maintenance f_{12}
	Key parts without accessories rate f_{13}
Control efficiency f_2	Flight normal rate f_{21}
	Average flight delay time f_{22}
	Flight departure rate f_{23}
	Airport release rate f_{24}

Control safety f_3	Take-off average wake interval margin f_{31}
	Landing average wake interval margin f_{32}
	Conflict alarm frequency f_{33}
Workload f_4	Number of instructions issued per unit time f_{41}
	Average length of ground and air calls f_{42}
	Average duration of continuous work of controllers f_{42}
Aviation weather f_5	Weather forecast accuracy f_{51}
	Report accuracy rate f_{52}
Safety accident f_6	10,000 sorties in controlled accident f_{61}
	10,000 sorties in cases of abnormal control causes f_{62}

3. Analytic hierarchy process to determine indicator weights

Using AHP to establish index weights and construct judgment matrix:

$$A = \begin{bmatrix} a_{11} & a_{12} & \cdots & a_{1n} \\ a_{21} & a_{22} & \cdots & a_{2n} \\ \vdots & \vdots & \ddots & \vdots \\ a_{n1} & a_{n2} & \cdots & a_{nn} \end{bmatrix}$$

Let the current level related factors be A_1, $\cdots$, A_n, and the related upper layer factor is f. For all factors, the two values are compared, and the value a_{ij} is obtained. The definition and definition are shown in Table 2. $A = (a_{ij})_{m*n}$ then A is the judgment matrix of the factors A_1, $\cdots$, A_n relative to the upper factor f. The maximum eigenvalue of A is λ_{max}, and the normalized eigenvector belonging to λ_{max} is $\omega = (\omega_1, \cdots, \omega_n)^T$, then $\omega_1, \cdots, \omega_n$ gives the factor A_1, $\cdots$, A_n corresponds to the factor f Sort by importance level[11].

Table 2. The importance of analytic hierarchy process indicators.

	Definition	Explanation
1	Equally important	Goal i goal j is equally important
3	Slightly important	Target i is slightly more important than goal j
5	Quite important	Target i is Quite more important than goal j
7	Obviously important	Target i is Obviously more important than goal j

| 9 | Absolutely important | Target i is Absolutely more important than goal j |

| 2、4、6、8 | In between |

Since the 9-level grading makes the calculation process too cumbersome, this paper adopts the 5-level grading, that is, discards the evaluation of the 2, 4, 6, and 8 degrees. At the same time, a set of evaluation factors is established for the control results, and five levels of evaluation methods are used, which are good, pretty good, general, poor, and very poor, and are represented by M_1, M_2, M_3, M_4, and M_5, respectively.

4. Fuzzy comprehensive evaluation method evaluation model

4.1. Fuzzy comprehensive evaluation method evaluation model
By analyzing the weight of the underlying indicators on the upper-level indicators, the layers are progressive, and the proportion of the first-level indicators is introduced to evaluate the control process.

The fuzzy matter element is expressed as:

$$R = \begin{bmatrix} & M \\ f & u(x) \end{bmatrix}$$

Where: R is a fuzzy matter element; M is a given thing, in this article, it is a five-level comment, which is M_j(j=1..5); f is the feature of thing M, in this paper The index; $u(x)$ is the fuzzy magnitude corresponding to the feature of the thing, that is, the membership degree. If m things are given, they have n features, namely f_i (i=1…n), and the corresponding fuzzy magnitudes are $u_i(x_{jm})$, then R$n*m$ is called n-dimensional fuzzy matter elements of m things.

$$R_{m*n} = \begin{bmatrix} & M_1 & M_2 & \cdots & M_i \\ f_1 & u_1(x_{11}) & u_1(x_{12}) & \cdots & u_1(x_{1i}) \\ f_2 & u_2(x_{21}) & u_2(x_{22}) & \cdots & u_2(x_{2i}) \\ \vdots & \vdots & \vdots & \ddots & \vdots \\ f_n & u_n(x_{n1}) & u_n(x_{n2}) & \cdots & u_n(x_{ni}) \end{bmatrix}$$

4.2. Evaluation step
Establish the fuzzy complex model
The fuzzy composite representation of the evaluation object:

$$R_i = \begin{bmatrix} f_i & f_{i1} & X_{i1} \\ & f_{i2} & X_{i2} \\ & \vdots & \vdots \\ & f_{ip} & X_{ip} \end{bmatrix}$$

Where: f_i is the i-th level indicator (i=1，2，…，n，n is the number of first-level indicators), since the number of first-level indicators in this paper is 5, so n=5; f_{ik} is the k-th second-level indicator under i-th first-level indicators (where k=1, 2,..., p, p is the number of secondary indicators); X_{ik} is the value of the secondary indicator.

Determining the classic domains of each level

Determining the range of values for each evaluation indicator for each level can be expressed by the following matter-element model[12]:

$$
R_i(M) = \begin{bmatrix}
 & M_1 & M_2 & \cdots & M_j \\
f_{i1} & (a_{1i1}, b_{1i1}) & (a_{2i1}, b_{2i1}) & \cdots & (a_{ji1}, b_{ji1}) \\
f_{i2} & (a_{1i2}, b_{1i2}) & (a_{2i2}, b_{2i2}) & \cdots & (a_{ji2}, b_{ji2}) \\
\vdots & \vdots & \vdots & \ddots & \vdots \\
f_{ip} & (a_{1ip}, b_{1ip}) & (a_{2ip}, b_{2ip}) & \cdots & (a_{jip}, b_{jip})
\end{bmatrix}
$$

Where: Mj is the j-th rank (where $j=1, 2, .., m$) m is the number of evaluation grades, a_{pij} is the lower limit of the secondary indicator, and b_{pij} is the upper limit of the secondary indicator. Therefore, the upper and lower limits of each rating of service quality are introduced in the subsequent calculations.

Determining the membership function matrix V

Membership is a measure of the size of the collocation between two things. The membership degree is determined by the membership function, and the membership degree and membership function are the basis of fuzzy mathematics[13]. Therefore, choosing the correct membership function is the basis for solving the practical problem by using fuzzy set theory. Combined with Wang Jifang[14], the definition of the membership function in fuzzy cybernetics is based on the normal distribution, so all the indicators are distributed.

$$
u_j(x_{ik}) = e^{-(\frac{x_{ij}-P}{q})^2}
$$

where:

$$
p = |a_{jik} + b_{jik}| / 2
$$

$$
q = (|a_{jik} - b_{jik}| * \sqrt{\ln 2}) / 2
$$

Normalize the calculated $u_j (x_{ik})$:

$$
u_j(x_{ik})' = \frac{u_j(x_{ik})}{\sum_{k=1}^{p} u_j(x_{ik})}
$$

Get the fuzzy composite matrix:

$$
R_{m*n}' = \begin{bmatrix}
 & M_1 & M_2 & \cdots & M_j \\
f_1 & u_1(x_{11})' & u_1(x_{12})' & \cdots & u_1(x_{1j})' \\
f_2 & u_2(x_{21})' & u_2(x_{22})' & \cdots & u_2(x_{2j})' \\
\vdots & \vdots & \vdots & \ddots & \vdots \\
f_n & u_n(x_{n1})' & u_n(x_{n2})' & \cdots & u_n(x_{nj})'
\end{bmatrix}
$$

Calculate the comprehensive evaluation results

Once the membership degree comprehensive evaluation matrix V is determined, the comprehensive evaluation result can be obtained, and the evaluation result of the same level indicator feature vector in the fuzzy degree is calculated as the membership degree:

$$
B = W \odot V = (b_1, \ b_2, \ \cdots, \ b_n)
$$

Here "$\odot$" is a synthetic operation, which can be taken as "+, ×", and this article takes "×".

5. Case Analysis

Select the recent data of an air traffic control bureau in East China and bring it into the model for calculation. For the survey of 20 first-line control experts, each expert is required to give the second-level index experience classic domain, and can not abstain. The value of qualitative indicators is obtained by expert scoring method. The value of quantitative indicators is obtained according to the statistical data of the unit. The range of values and actual values of the secondary indicators are shown in Table 3 [15].

Table 3. The importance of analytic hierarchy process indicators.

Second-level	M_1	M_2	M_3	M_4	M_5	Actual value
f_{11}	[0, 0.1]	[0.1, 0.3]	[0.3, 0.5]	[0.5, 1.0]	[1.0, 3.0]	0.4
f_{12}	[95,100]	[85,95]	[70,85]	[60,70]	[60,70]	91
f_{13}	[0, 0.1]	[0.1, 0.3]	[0.3, 0.5]	[0.5, 1.0]	[1.0, 3.0]	0.3
f_{21}	[95,100]	[85,95]	[70,85]	[60,70]	[60,70]	89
f_{22}	[95,100]	[85,95]	[70,85]	[60,70]	[60,70]	87
f_{23}	[95,100]	[85,95]	[70,85]	[60,70]	[60,70]	86
f_{24}	[95,100]	[85,95]	[70,85]	[60,70]	[60,70]	89
f_{31}	[95,100]	[85,95]	[70,85]	[60,70]	[60,70]	86
f_{32}	[95,100]	[85,95]	[70,85]	[60,70]	[60,70]	91
f_{33}	[95,100]	[85,95]	[70,85]	[60,70]	[60,70]	73
f_{41}	[95,100]	[85,95]	[70,85]	[60,70]	[60,70]	86
f_{42}	[95,100]	[85,95]	[70,85]	[60,70]	[60,70]	89
f_{43}	[95,100]	[85,95]	[70,85]	[60,70]	[60,70]	82
f_{51}	[90,100]	[80,90]	[70,80]	[60,70]	[60,70]	89
f_{52}	[90,100]	[80,90]	[70,80]	[60,70]	[60,70]	95
f_{61}	[0, 0.01]	[0.01, 0.03]	[0.03, 0.05]	[0.05, 0.08]	[0.08, 0.1]	0.03
f_{62}	[0, 0.01]	[0.01, 0.03]	[0.03, 0.05]	[0.05, 0.08]	[0.08, 0.1]	0.02

5.1. Index weight determination

Invite industry experts to conduct weight evaluations on each level of indicators, and conduct consistency tests on each expert's survey results to obtain first-level indicator weights:

$$W=(0.105, 0.202, 0.202, 0.074, 0.020, 0.397)$$

Similarly, other secondary indicator weights are available:

W_1=(0.121, 0.547, 0.332)

W_2=(0.231, 0.314, 0.343, 0.112)

W_3=(0.171, 0.587, 0.332)

W_4=(0.256, 0.443, 0.301)

W_5=(0.386, 0.614)

W_6=(0.667, 0.333)

5.2. Index membership degree determination

Bring the result into the membership function and normalize it:

$$R_1 = \begin{bmatrix} 0 & 0.028 & 0.897 & 0.053 & 0.022 \\ 0 & 0.910 & 0.090 & 0 & 0 \\ 0 & 0.475 & 0.475 & 0.018 & 0.032 \end{bmatrix}$$

$$R_2 = \begin{bmatrix} 0 & 0.786 & 0.158 & 0 & 0.056 \\ 0 & 0.654 & 0.276 & 0 & 0.070 \\ 0 & 0.562 & 0.360 & 0 & 0.078 \\ 0 & 0.786 & 0.158 & 0 & 0.056 \end{bmatrix}$$

$$R_3 = \begin{bmatrix} 0 & 0.910 & 0.090 & 0 & 0 \\ 0 & 0.562 & 0.360 & 0 & 0.078 \\ 0 & 0 & 0.655 & 0.143 & 0.202 \end{bmatrix}$$

$$R_4 = \begin{bmatrix} 0 & 0.562 & 0.360 & 0 & 0.078 \\ 0 & 0.786 & 0.158 & 0 & 0.056 \\ 0 & 0.041 & 0.959 & 0 & 0 \end{bmatrix}$$

$$R_5 = \begin{bmatrix} 0 & 0.786 & 0.158 & 0 & 0.056 \\ 0.499 & 0.499 & 0.002 & 0 & 0 \end{bmatrix}$$

$$R_6 = \begin{bmatrix} 0 & 0.499 & 0.499 & 0.002 & 0 \\ 0.979 & 0.021 & 0 & 0 & 0 \end{bmatrix}$$

Comprehensive evaluation of the second level indicators

Multiplying the weights of the second-level indicators by the degree of membership can obtain the membership matrix of each level of indicators:

Available from B= W×V:

B_1= [0, 0.657, 0.316, 0.012, 0.015]

B_2= [0, 0.668, 0.264, 0, 0.068]

B_3= [0, 0.485, 0.385, 0.035, 0.095]

B_4= [0, 0.504, 0.451, 0, 0.045]

B_5= [0.306, 0.610, 0.062, 0, 0.022]

B_6= [0.326, 0.340, 0.333, 0.001, 0]

Comprehensive evaluation of the first level indicators

Multiplying the weight of the first-level index by the membership degree, the first-level indicator membership degree matrix can be obtained, that is, the service quality evaluation level of the unit can be obtained.

$$B = B_i \times W$$

$$B = \begin{bmatrix} & M_1 & M_2 & M_3 & M_4 & M_5 \\ f & 0.136 & 0.486 & 0.336 & 0.009 & 0.033 \end{bmatrix}$$

Evaluation conclusion

According to the principle of maximum membership, this unit evaluation level is M2 (better). This result is in line with the actual operation of the unit, which proves the reliability and effectiveness of the model. At the same time, it was found in the calculation process that the control safety score is low, so the unit needs to improve the quality of this part.

6. Conclusion

● Through calculation, the obtained regulatory quality assessment is consistent with the actual operation. For the control unit, For the control unit, it is possible to find out the problems and improvement directions of the unit in time.

● For first-line controllers, they can fully understand their own control skills, improve their control skills and control proficiency, and evade before risks occur.

● Through the safety assessment, we can understand the operation of each flight, improve the on-time rate of passengers and the travel experience of passengers.

● In the weight determination part, the expert research method is adopted, so the weight index has certain subjectivity. In the future, the model can be improved by improving the selection of weight indicators.

Acknowledgments

Thanks to the Civil Aviation East China Air Traffic Management Bureau 2018 Annual Science and Technology Project Plan (KJ1802)

Thanks to the 2018 Graduate Innovation Program of Civil Aviation Flight University of China(X2018-36)

References

[1] Eurocontrol Safety Regulation Commission Document （SRC DOC 2）,ATM Contribution to Aircraft Accidents/ Incidents Review and Analysis of Historical Data [EB/OL].[2012-05-04]. http://www.eurocontrol.int/src/gallery/.../srcdoc2_e40_ri_web.pdf，2005-5.

[2] ICAO. Annex 11 to the Convention on International Civil Aviation，2001 Air Traffic Services [S]. Montreal: ICAO，2001.

[3] Yuan Leping, Sun Ruishan and Cheng Yuan. Fuzzy Evaluation and Unascertained Mathematics Based Safety Risk Assessment in ATM System[J]. *Journal of Civil Aviation University of China*, **2006,24（4）** :55-57

[4] Zhang Jianping, Yu Haiyang and Zou Guoliang. Research on evaluation factors for operation performance of air traffic control in terminal area[J]. *Journal of Civil Aviation University of China*, **2012, 30(3)**:18-21.

[5] Averty P, Collet C A, Athenes S, et al. Mental workload in air traffic control: an index constructed from field tests[J]. *Aviation Space & Environmental Medicine*, **2004, 75(4)**:333-341.

[6] Wang Yanqing. Analysis of Factors Influencing Air Traffic Controllers' Psychological Deviation under Emergency Condition[J]. *China Safety Science Journal*, **2012, 22(9)**:24.

[7] Blom H A P, Bakker G J, Blanker P J G, et al. Accident risk assessment for advanced ATM[J]. *Air Transportation Systems Engineering Aiaa*, 1999.

[8] Hauss Y, Eyferth K. Securing future ATM-concepts' safety by measuring situation awareness in ATC[J]. *Aerospace Science & Technology*, **2003, 7(6)**:417-427.

[9] Ternov S, Akselsson R. A method, DEB analysis, for proactive risk analysis applied to air traffic control[J]. *Safety Science*, **2004, 42(7)**:657-673.

[10] Zhang Jianping, Research on Quantitative Evaluation for Operation Performance of Air Traffic Control[D]. Nanjing:Nanjing University of Aeronautics and Astronautics,2013

[11] Hu Yunquan 2012. *Operational Research Tutorial. 4rd* (Beijing: Tsinghua University Press)

[12] Han Yubin, Piao Chunzi. research on safety performance assessment model of atc(air traffic control) system[J]. *Advances in Aeronautical Science and Engineering*, **2016, 7(4)**:477-483

[13] Liu Kaidi, Pang Yanjun, Wu Heqin. The Problems in the Definition of Fuzzy Subordinative Degree[J]. *SYSTEM ENGINEERING——THEORY & PRACTICE*, **2000, 20(1)**:110-112.

[14] Wang Jifang, Lu Zhengding. The determine method of membership function in fuzzy control[J]. *Henan Science*, **2000, 18(4)**:348-351.

[15] Wang Tingchun, Sun Deqing, Yu Feifei, et al. Study on fuzzy comprehensive evaluation of safety management performance[J]. *Journal of Safety Science & Technology*, **2012, 08(3)**:185-188.

Real-time Estimation of Urban Rail Transit Passenger Flow Status Based on Multi-source Data

Zhengping Tao [1] and Jinjin Tang [1*]

[1] School of Transportation and Transportation, Beijing Jiaotong University, No. 3 Shangyuan Village, Haidian District, Beijing,100044,China

*Corresponding author's e-mail address:jjtang@bjtu.edu.cn

Abstract. The AFC data uploaded in real time for urban rail transit is incomplete and delayed. In order to improve the dynamic management and control level of rail transit, this paper describes the real-time state estimation process of passenger flow and its key issues, and analyses the relationship between real-time uploaded AFC data and mobile phone signal data, establishing a multi-source data fusion model based on gradient descent method; On this basis, combined with the multi-source data after fusion as the basis of real-time estimation, a real-time passenger flow estimation model based on Kalman filter is established. Practice shows that the fusion of multi-source data improves the accuracy of estimating basic data, the error of the real-time passenger flow estimation model based on Kalman filter is less than 10%, and the accuracy of the estimation model is good. The needs of the urban rail transit operation management department to grasp the real-time passenger flow distribution status of the network are met.

1. Introduction

With the development of urban rail transit network in China, the passenger flow of rail transit has grown rapidly, and it has put forward higher requirements for real-time control of the dynamic distribution of passenger flow in the network and ensuring safe operation. Among them, real-time estimation of rail transit is a very important task. It is of great significance for the operation management department to grasp the dynamic distribution of the passenger flow of the network in real time and take corresponding measures. However, the passenger flow data uploaded by the AFC system is delayed and incomplete. The operation management department cannot timely obtain the dynamic distribution of passenger flow in the rail transit network. In this context, it is of great significance to real-time passenger flow state estimation combine with passenger flow data and mobile phone signal data uploaded by AFC.

In the past, the research on AFC data mainly includes two aspects: First, the research on the overall situation of AFC data[1]-[4], such as the direct extraction of inbound and outbound traffic from AFC data; the second is to obtain the travel habits of specific passengers from AFC data[5][6]. All in all, the current urban rail transit operation department lacks deep excavation of AFC data, and the understanding of the real-time passenger flow and its distribution is relatively lagging and lacks predictability.

2. Real-time Estimation and Analysis of Passenger Flow Status of Urban Rail Transit

2.1. Real time estimation process of passenger flow condition

Content from this work may be used under the terms of the Creative Commons Attribution 3.0 licence. Any further distribution of this work must maintain attribution to the author(s) and the title of the work, journal citation and DOI.

Published under licence by IOP Publishing Ltd

Real-time estimation of passenger flow status of urban rail transit uses real-time uploaded AFC data and mobile phone signal data as the estimated basic data, analyse the relationship between the two by gradient descent method and generates passenger flow fusion data. Based on this, establish a real-time estimation model of passenger flow state based on Kalman filter.

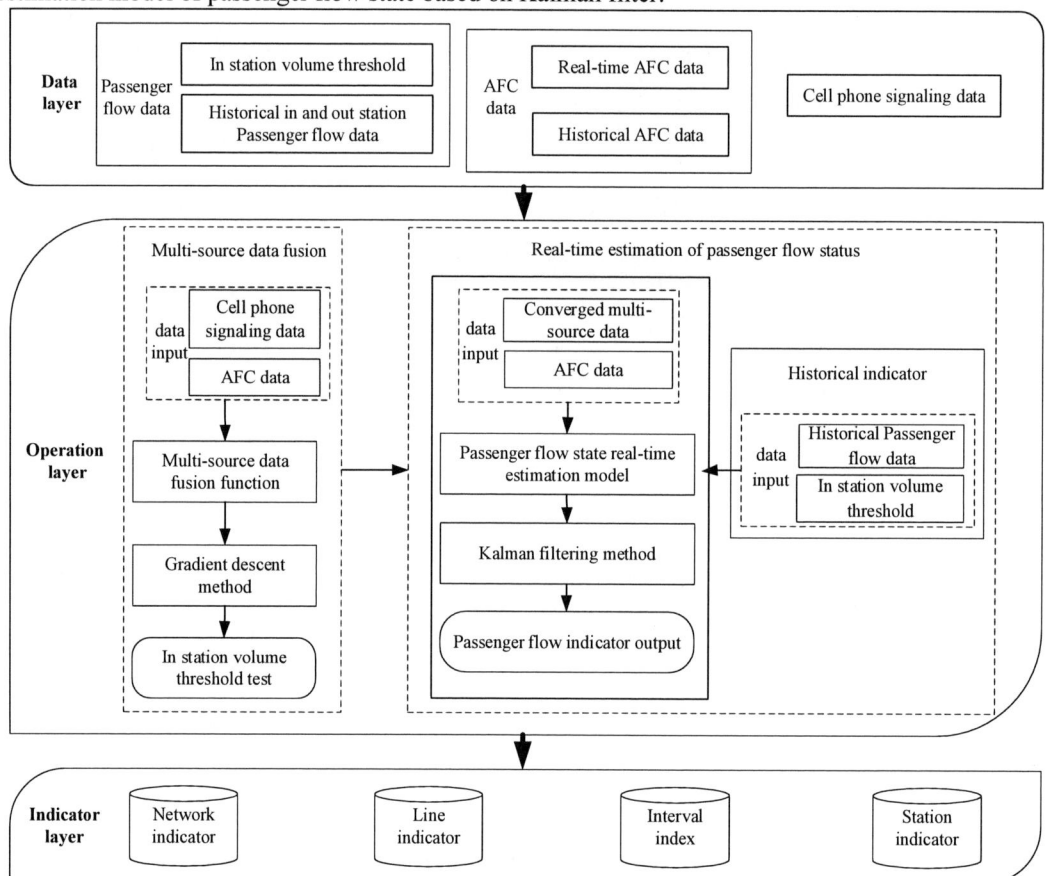

Figure 1. Real-time estimation process of passenger flow state based on multi-source data.

2.2. Analysis of real-time passenger flow estimation problem

In the real-time estimation process of passenger flow state, it is necessary to combine the characteristics of urban rail transit to sort out the key problems and lay a foundation for constructing a real-time estimation model of passenger flow state. The key issues of real-time estimation of urban rail transit passenger flow state mainly include the following aspects:

(1) Selection of state variables. The passenger travel volume in the urban rail transit network has a similar travel rules for a certain period of time, the average value of the station's historical inbound quantities directly reflects the average travel level of the station, which can more intuitively reflect the trend of station passenger flow. Therefore, it is widely used in the selection of state variables of real-time estimation model of passenger flow.

(2) Determination of inbound quantities threshold. In order to verify whether the estimated inbound data is reasonable, the single-sample K-S test can be used to judge the distribution of the inbound quantities.

Assume that the inbound quantities obey the normal distribution and take the significance level to 0.05. Use $u - 3\alpha$, $u + 3\alpha$ to calculate the lower and upper limits of the thresholds of the same historical period for each station and each time period.

$$u = (m_1 + m_2 + \cdots + m_n)n^{-1} \tag{1}$$

$$\alpha = \left(n^{-1} \sum_{i=1}^{n} (m_i - u)^2 \right)^{1/2} \tag{2}$$

Among them, m_i is the inbound quantities of station i in the same period of history; $u + 3\alpha$ is upper limits of the thresholds; $u - 3\alpha$ is lower limits of the thresholds.

3. Real-time estimation model of passenger flow state in urban rail transit

Based on the uploaded AFC data and mobile phone signal data, the real-time estimation model of passenger flow state solves the fusion value of multi-source data by gradient descent method, and uses this as the basic data for real-time estimation of passenger flow state. Secondly, using Kalman filter method Perform real-time estimation of passenger flow.

3.1. Basic flow relationship construction

In order to better express the real-time estimation process of urban rail transit passenger flow state, it is necessary to establish a topology diagram of the network traffic relationship. Considering the relationship of closed network traffic conservation, we can get the relational expression between each indicator.

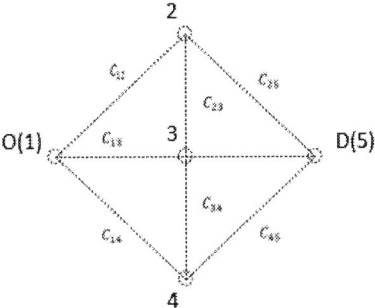

Figure 2. Topology diagram of the network traffic relationship.

In this paper, some symbols are defined as follows for the convenience of expression: V indicates the collection of stations on the network; A indicates the collection of sections on the network; O stands for the starting station; D stands for the terminal station; E_i indicates the inbound quantities; O_i indicates the outbound quantities; f_{ij} indicates the passenger flow from station i to station j.

Among them, the relational expression of the inbound amount E_i is as follows:

$$E_i(t) = \sum_{j=1, j\neq i}^{n} f_{ij}(t) \qquad \forall i,j \in V, t \in T \tag{3}$$

Where $E_i(t)$ is the inbound quantities of station i at time period t; $f_{ij}(t)$ indicates the passenger flow from station i to station j at time period t; T represents the real-time estimated time period, $T = \{1,2,\cdots,t,\cdots,t'\}$.

The relational expression of the inbound amount E_i is as follows:

$$E_i(t) = \sum_{t'=0}^{t'} \sum_{n=1}^{n} E_i(t^s - t^r) R_{ij}(t^s - t^r) \theta_{ij}^t(t^s) \qquad i \neq j, \forall i,j \in V, t^s, t^s, t \in T \tag{4}$$

t^r is the time span for passengers to complete a complete OD trip in the urban rail transit network; $E_i(t^s - t^r)$ is the inbound quantities of station i during the $t^s - t^r$ period; $R_{ij}(t^s - t^r)$ is during the period of $t^s - t^r$, the proportion of passengers who enter the station by station i and exit from station j as a percentage of the total number of stations in station i; $\theta_{ij}^t(t^s)$ represents the proportion of passengers departing from station i in time period t arriving at station j during time period t^s.

3.2. Multi-source data fusion model establishment and solution

Combining AFC data and mobile phone signaling data, generating a data fusion function based on the Logit model; secondly, using the gradient descent method to move from same or opposite direction of the gradient (the step size of each movement is λ), the iteratively calculated value When the specified error is reached, it means that the optimal value has been reached at this time.

Firstly, combined with AFC data and mobile phone signaling data, a data fusion function is generated based on the Logit model. The Logit model is actually a model for determining the probability that different types of data are selected. The random term ε_ω^n is a Gumbel variable, and it is subject to independent distribution, and the selection probability of multi-source data under different weights is calculated.

$$\partial_i^{all} = \frac{\exp(-\theta x_i^{all})}{\sum_i \exp(-\theta x_i^{all})} \tag{5}$$

∂_i^{all} indicates the probability that multi-source data i is selected; indicates the adjustment factor of multi-source data i; x_i^{all} indicates the confidence level of multi-source data i.

The following formula is the inbound quantity fusion function:

$$P(x,y) = F(\alpha)\frac{\exp(-\theta x_i^{all})}{\sum_i \exp(-\theta x_i^{all})} + F(\beta)\frac{\exp(-\theta y_j^{all})}{\sum_j \exp(-\theta y_j^{all})}, \frac{\exp(-\theta x_i^{all})}{\sum_i \exp(-\theta x_i^{all})} + \frac{\exp(-\theta x_i^{all})}{\sum_i \exp(-\theta x_i^{all})} = 1 \tag{6}$$

Secondly, the gradient descent method is used to move from same or opposite direction of the gradient. Its iteration formula is:

$$x^{(k+1)} = x^k + \lambda_k d^{(k)} \tag{7}$$

Where $d^{(k)}$ is the search direction starting from x^k, taking the steepest descending direction of point x^k, that is: $d = -\dfrac{\nabla P(x,y)}{\|\nabla P(x,y)\|}$.

$$\nabla P(x,y) = \begin{pmatrix} -F(\alpha)(\exp(-\theta x_i^{all}))' \dfrac{\exp(-\theta x_i^{all})}{\sum_i (\exp(-\theta x_i^{all}))'} \\ -F(\beta)(\exp(-\theta y_j^{all}))' \dfrac{\exp(-\theta y_j^{all})}{\sum_j (\exp(-\theta y_j^{all}))'} \end{pmatrix} \tag{8}$$

λ_k is the step size in which λ_k starts searching in the direction $d^{(k)}$, which satisfies:

$$P(x^{(k)} + \lambda_k d^{(k)}) = \min_{\lambda \geq 0} P(x^{(k)} + \lambda_k d^{(k)}) \tag{9}$$

That is:

$$\min_{\lambda \geq 0} P(x^{(k)} + \lambda_k d^{(k)}) = P\left\{ x^{(k)} - \lambda_k \begin{pmatrix} -F(\alpha)(\exp(-\theta x_i^{all}))' \dfrac{\exp(-\theta x_i^{all})}{\sum_i (\exp(-\theta x_i^{all}))'} \\ -F(\beta)(\exp(-\theta y_j^{all}))' \dfrac{\exp(-\theta y_j^{all})}{\sum_j (\exp(-\theta y_j^{all}))'} \end{pmatrix} \right\} \tag{10}$$

The $\|d^{(k)}\|$ of the fusion value $P^n(x,y)$ calculated by the iteration (11) is less than the specified error (this is taken as 5%), which is the desired fusion value. That is, the inbound quantity. If there is a deviation from the actual value, reverse adjustment $\exp(-\theta y_j^{all})(\sum_j(\exp(-\theta y_j^{all}))')^{-1}$ or $\exp(-\theta x_i^{all})(\sum_i(\exp(-\theta x_i^{all}))')^{-1}$ until it is close to the actual inbound quantity.

3.3. Establishment and solution of real-time estimation model for passenger flow state
The core of the real-time estimation model of passenger flow state is to establish the state transition equation and the observation equation. In this paper, the historical mean value of the inbound quantity is taken as the state variable. The following describes the construction process of the state transition equation and the observation equation.

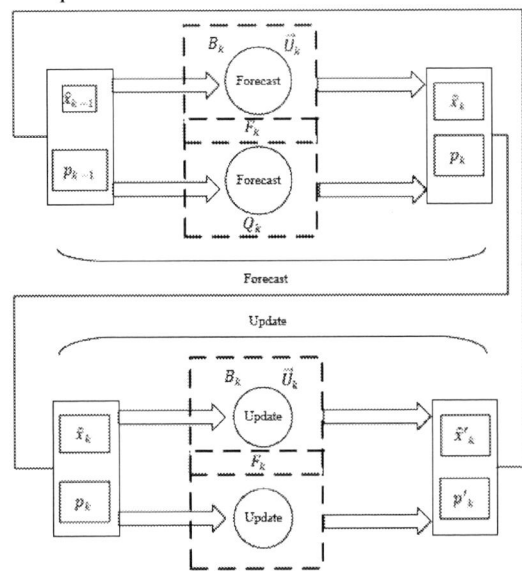

Figure 3. Kalman filtering method real-time estimation process.

(1) Construct a state transition equation. Based on the similarity characteristics of the passenger flow distribution in the network, this paper takes the historical mean value of the inbound quantity as the state variable and establishes the state transition equation as follows:

$$X(t) = A(t)X(t-1) + B(t)U(t) + W(t) \tag{11}$$

In the formula, X(t) and X(t-1) are the system state variables at time t and t-1, respectively, and X(t) is an $N_{st} \times 1$ dimensional matrix composed of $E_i(t)$, where $X(t) = [E_i(t)]_{N_{st} \times 1}$; N_{st} is the number of stations in the network; A(t), B(t) represents the system state transition matrix; U(t) represents the historical state vector of the period t, which is an $N_{st} \times 1$ dimensional matrix composed of $E_i'(t)$; W(t) is a Gaussian white noise matrix.

(2) Construct an observation equation. In this paper, the historical average inbound quantity is used to replace the inbound quantity in the future multiple time periods involved in the constraint equation.

$$Z(t) = H(t)E_i'(t) + V(t) \tag{12}$$

In the formula, Z(t) is the inbound observation vector of system time period t, which is an n × 1 dimensional matrix composed of $E_i(t)$; H(t) is the observation matrix of the system time period t, which is the $n \times N_{st}$ dimensional matrix determined by $E_i(t) = \sum_{j=1, j\neq i}^{n} f_{ij}(t)$; $E_i'(t)$ is the $N_{st} \times 1$ dimensional matrix composed of the mean value of the historical inbound quantity of the time period t. V(t) is a Gaussian white noise matrix.

ISPECE

IOP Publishing

IOP Conf. Series: Journal of Physics: Conf. Series **1187** (2019) 052070   doi:10.1088/1742-6596/1187/5/052070

Based on the constructed state transition equation and observation equation, the iterative recursive method is used to realize the real-time update of the inbound quantity matrix. The Kalman filtering method mainly includes a 5-step basic iterative recursive step [7]. The algorithm formula is as follows:

$$
\begin{cases}
\hat{X}^-(t) = A(t)\hat{X}^+(t-1) + B(t)U(t) + W(t) \\
\hat{P}^-(t) = A(t)\hat{P}^+(t-1)A^T(t) + Q(t) \\
K(t) = \hat{P}^-(t)H^T(t)[H(t)\hat{P}^-(t)H^T(t) + R(t)]^{-1} \\
\hat{X}^+(t-1) = \hat{X}^-(t) + K(t)[Z(t) - H\hat{B}^-(t)] \\
\hat{P}^+(t) = \hat{P}^-(t) - K(t)H(t)\hat{P}^-(t)
\end{cases}
\tag{13}
$$

Where $\hat{X}^-(t)$, $\hat{X}^+(t)$ are the a priori estimates and posterior estimates of the system state variables in time period t; $\hat{P}^-(t)$ and $\hat{P}^+(t)$ is the a priori estimation error variance and the a posteriori estimation error variance of the system state variables in the time period t; K(t) is the Kalman filter factor.

4. Analysis of Chengdu Metro Cases

This article uses the Chengdu Metro (6 lines, 136 stations) AFC data uploaded in real time on the morning peak of 08:00-08:30 on September 3, 2018, and the inbound quantities on the morning peak of 08:00-09:00 statistics by mobile phone signal as an example, real-time estimation of the 5-minute granularity of inbound quantities in station. And uses C# language to calculate through Visual Studio 2017, and finally analyses the accuracy of passenger flow index estimated by Kalman filtering method according to the index error calculation formula.

4.1. Real-time Estimation of Passenger Flow Status in Chengdu Metro
The real-time estimation of passenger flow status is focused on the estimation of the station's inbound quantities in the network. The fusion of multi-source data and real-time estimation will be shown below.

Table 1. AFC data and mobile phone signal statistics inbound quantities.

Station	Serial number	Period	mobile signal statistics	AFC data
	1	08:00-08:05	111	149
	2	08:05-08:10	123	148
	3	08:10-08:15	130	155
	4	08:15-08:20	127	171
	5	08:20-08:25	132	149
HuaxiB	6	08:25-08:30	123	135
a	7	08:30-08:35	114	-
	8	08:35-08:40	172	-
	9	08:40-08:45	164	-
	10	08:45-08:50	139	-
	11	08:50-08:55	145	-
	12	08:55-09:00	162	-

First of all, according to the multi-source data fusion formula(6),(7),(8),(9),(10),(11), the quantities of inbounds after the fusion of the Huaxiba station on Line 1 of September 30, 2018, the morning peak of 08:30-09:00 is as follows:

Table 2. The quantity of inbound after the fusion of Huaxiba station.

08:30-08:35	08:35-08:40	08:40-08:45	08:45-08:50	08:50-08:55	08:55-09:00

2533

139	206	185	169	194	185

Figure 4 shows the AFC data of Huaxiba station at 08:00-08:30, the inbound traffic data of mobile phone signaling data from 08:00-09:00, and the trend graph of the station's inbound quantities after fusion.

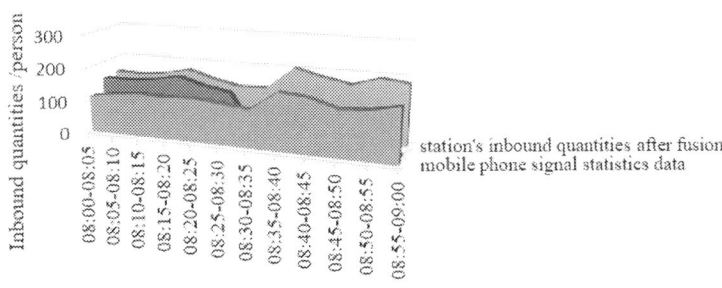

Figure 4. Trend of inbound quantities after fusion of Huaxiba station.

Secondly, based on the quantity of inbound after the fusion of Huaxiba station, the station's inbound quantities is estimated in real time according to the Kalman filtering method. The results are as follows:

Table 3. Real-time estimated inbound quantity of Huaxiba station.

09:00-09:05	09:05-09:10	09:10-09:15	09:15-09:20	09:20-09:25	09:25-09:30
205	216	221	194	215	228

The actual inbound capacity of Huaxiba station from 09:00 to 09:30 is as follows:

Table 4. Actual inbound quantity of Huaxiba station.

09:00-09:05	09:05-09:10	09:10-09:15	09:15-09:20	09:20-09:25	09:25-09:30
213	222	205	208	230	222

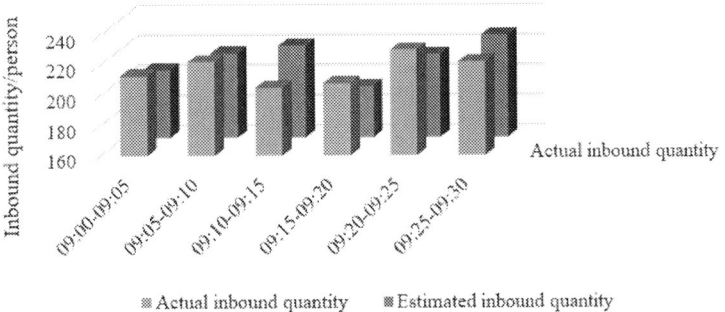

Figure 5. Estimate and actual inbound quantity comparison of Huaxiba station.

4.2. Real-time estimation result analysis

The results of real-time estimation of urban rail transit passenger flow state require certain evaluation indicators to judge the accuracy of the results. In this paper, the mean square error MSE, the mean absolute error MAPE, and the average absolute percentage error MAD are used as the error evaluation indicators.

$$\begin{cases} MSE = \dfrac{1}{N}(\sum_{i=1}^{n}(out^i - out^{i})^2)^{1/2} \\[2ex] MAPE = \dfrac{1}{N}\sum_{i=1}^{n}|out^i - out^{i}| \\[2ex] MAD = \dfrac{1}{N}(\sum_{i=1}^{n}|\dfrac{out^i - out^{i}}{out^i}|*100) \end{cases} \qquad (14)$$

In the formula, out^i is the estimated inbound amount, out^i is the actual inbound amount, and N is the estimated number.

According to the results of passenger flow estimation, the mean square error MSE value of Huaxiba station is 4.75, the average absolute value error MAPE value is 10.83, and the average absolute percentage error MAD is 5.04%. Estimating the passenger flow with different granularity in other time periods shows that the MSE, MAPE and MAD values of the Kalman filter estimation value decrease gradually with the estimated time granularity increase, which is mainly due to the small time granularity cause the inbound quantity is small, it is easy to cause large fluctuations in MSE, MAPE and MAD values. Secondly, the MAD value of the passenger flow estimates at different granularities is less than 10%.

5. Conclusion

Aiming at the nonlinearity of rail transit passenger flow, AFC data uploading has the characteristics of delay and incompleteness. This paper mainly studies the hot spot problem of real-time estimation of rail transit passenger flow state. The specific content of the discussion is as follows:

Firstly, the fusion method of AFC data and mobile phone signal as multi-source data is introduced. Secondly, the Kalman filter method is applied as the basis of the estimation model, and a real-time estimation model of rail transit passenger flow is established. The passenger flow data of the Huaxiba station of Chengdu Metro Line 1 was used for real-time estimate. The estimation result error is represented by three indicators: mean square error MSE, average absolute error MAPE and average absolute percentage error MAD. The results show that the error of the real-time passenger flow state estimation model based on Kalman filter is less than 10%, and the accuracy of the estimation model is good. It can provide better support for the urban rail transit operation management department to grasp the real-time passenger flow distribution state of the network.

References

[1] Lchtonen, M., Rosenberg, M., Rasanen, J, Sirkia, A.: Utilization of the smart card payment system (SCPS) data in public transport planning and statistics. In: 9th World Congress on Intelligent Transport Systems,2002.

[2] Chan Joanne. Rail transit OD matrix estimation and journey time reliability metrics using automated fare data [D]. Massachusetts Institute of Technology, 2007.

[3] Shi Zhuangbin, Lu Wenxue, Zhang Ning. Application of Data Mining Technology in Rail Transit AFC System [J]. Urban Rapid Rail Transit, 01:23-27, 2015.

[4] Wang Lina. Research on Passenger Flow Forecast and Scheduling of Urban Rail Transit Network Based on Historical Passenger Flow Data [D]. Master's Thesis of Beijing Jiaotong University, 2011.

[5] Morency, C., Trepanier, M., Agard, B.: Measuring transit use variability with smart-card data. Transp.Policy. 14(3), 193- 203 ,2007.

[6] Cai Changjun, Yao Enjian, Zhang Yongsheng, Liu Shasha. Forecast of Passenger Flow Distribution between Urban Rail Stations Based on AFC Data[J]. China Railway Science, 01:126-132, 2015.

[7] CHUICK,CHENGR. Kalman Filtering with Real-time Applications (fourth edition) [M]. Berlin:Springer-Verlag,2009:217-228.

Comprehensive Assessment of Green Development Level for Urban Rail Transit Enterprises Based on ANP and Entropy Weight Method

Ying Tang

Transport Planning& Design Studio, Guangzhou Urban Planning & Design Survey Research Institute, Guangzhou 510032, China

14120842@bjtu.edu.cn

Abstract. As one of the most efficient energy-saving transportation mode, urban transit rail system, with priorities such as large volume, low pollution, low-emission transport mode, has been greatly developed in many Chinese cities recently. In order to evaluate the current green development level (GDL) and guide the future green development for urban rail transit enterprise (URTE), a comprehensive evaluation approach is proposed. First, based on the analysis of connotation and basic characteristics of green URTE, an assessment indicator system considering the impact of energy-intensity, level of service, energy-saving capacity building, application of energy-saving technologies, energy-saving management mechanism, fund input level for energy-saving is established. Second, the analytic network process (ANP) and the entropy weight method are used to respectively determine the subjective and objective weights, and the game theory aggregation method is employed to combine both weights and then optimize the indicator weigh. Finally, based on the investigation data, the GDL values of several Chinese URTEs are analysed and compared. The results show that the proposed method has both well rationality and excellent operation performance. The weight determination method effectively solve the collaborative relationship between different methods and the GDL calculation can well identify the URTEs' benefits and drawbacks to promote UTREs pertinently carrying out energy saving measures.

1. Introduction

Under the background of global warming, the international community generally agrees that the development of green and low-carbon transport is an effective measure for sustainable economic development *(1)*. Hence, it's necessary to establish a scientific and reasonable comprehensive assessment method to evaluate the GDL of urban rail transit enterprise (URTE) which can comprehensively reflect the characteristics and factors of energy consumption and emphatically analyze the importance of the multiple indicators which affect GDL of enterprise realizing the goal to lead the enterprise develop sustainably, and promote the development of energy conservation work of the transportation industry.

At present, the energy conservation evaluation of urban rail is to establish energy performance indicators. The rigorous attempt to identify energy performance indicators in railway systems has been developed within the Rail Energy project *(2; 3)*. This approach consisted of seven indicators measuring the overall energy consumption of the system, the energy consumption share for parked trains, the rate of recuperated energy and the efficiency of the railway distribution grid. González-Gil et al. *(4)* proposed a holistic approach which considers the numerous interdependences between

---

Content from this work may be used under the terms of the Creative Commons Attribution 3.0 licence. Any further distribution of this work must maintain attribution to the author(s) and the title of the work, journal citation and DOI.

Published under licence by IOP Publishing Ltd

subsystems, such an approach requires a comprehensive set of energy consumption-related Key Performance Indicators (KEPIs). In China, Chinese Urban Rail Transit Association published data to monitor some indicators data in Urban Rail Transit annual statistical and analytical report (5;6). However, annual data only can reflect the basic situation of urban rail transit enterprises, but failed to develop a unified standard metering energy consumption statistics. But it only can see the energy consumption results of enterprises without considering the energy conservation capacity-building. Additionally, in the field of green evaluation of transportation enterprises, Ministry of Transport of the People's Republic of China (MOT) has studied the assessment indicator system of the Low Carbon Transport System, green highway passenger transportation enterprise and green shipping enterprise, etc. (7). Thus, for making up the blank of green URTE assessment, it's necessary to establish a comprehensive evaluation method of green URTE.

In the study of classical comprehensive evaluation, the commonly used methods to determine the indicator weight are divided into subjective and objective method. Subjective method is mainly based on the research purpose and indicator connotation to analyze subjectively and make judgement to determine the relative importance of the various indicators, such as the analytic hierarchy process(AHP) method proposed by Professor Saaty and other method applied in the literature (8;9). Using subjective method to determine indicator weight can reflect the actual importance of the different indicators by consulting experts, but such indicator weight has different degrees of dependence on the experts. Instead, objective method can avoid the human factors and subjective factors by using mathematical calculation. It's viewed that this method can fully reflect the differences between the data by calculations based on relationships between objective evaluation data, which also has shortcomings that can't reflect the real importance of various indicators. Therefore, this paper determined the subjective and objective of indicators based on analytic network process (ANP) and entropy weight theory, and used game theory aggregation method to seek equilibrium results of the two methods to obtain comprehensive weights and apply the investigation data to expand the evaluation analysis of green URTE.

2. Green URTE assessment indicator system
In order to ensure the rationality of assessment indicator system constructed, researchers extensively investigated the status of energy conservation of several urban rail transit enterprises such as Beijing, Wuhan, and Chengdu. Additionally, researchers surveyed the evaluation indicator adaptability including the representativeness, information or data accessibility of indicators to understand the indicator collection difficult degree. At the same time, the experts from relevant departments of enterprise, energy consumption field, and industry regulation department were repeatedly consulted to revise the indicator system. Finally, the evaluation system of green URTE was determined as follows.

Table 1. Assessment indicator system of green URTE

Objective	Indicator in level one	Indicator in level two
GDL of URTE	energy-intensity (C_1)	Traction power consumption per vehicle-kilometer (e_{11})
		Traction power consumption per person-kilometer(e_{12})
		Station power consumption unit area per day(e_{13})
		Non-operating comprehensive energy consumption unit area per day (e_{14})
		Year on year (YoY) decline of Traction power consumption per vehicle-kilometer (e_{15})
		YoY decline of traction power Consumption per person-kilometer (e_{16})
		YoY decline of station power consumption unit area per day (e_{17})

	YoY decline of Non-operating comprehensive energy consumption unit area per day (e_{18})
level of service (C_2)	Average load factor (e_{21})
	Average load factor of peak hour (e_{22})
	Average train interval (e_{23})
	Average speed (e_{24})
	Line configure rate setting platform screen doors (e_{25})
	Comfortable degree of air-conditioning(e_{26})
	Air conditioning usage rate (e_{27})
	Configuration rate of standard elevator and escalator units entrances (e_{28})
	Proportion of elevator and escalator available hours accounted for subway operation time (e_{29})
energy-saving capacity building (C_3)	Line configuration rate setting energy measuring instruments (e_{31})
	Line coverage rate of energy consumption monitoring and analysis system (e_{32})
	Application of intelligent dispatching or auxiliary operations management and decision-making system(e_{33})
application of energy-saving technologies (C_4)	Configuration rate of energy-saving vehicle (e_{41})
	Circumstance of using energy-saving technologies, equipment (e_{42})
	Energy conservation-related awards, honors, patented products, etc. (e_{43})
energy-saving management mechanism (C_5)	Energy-saving organizations (e_{51})
	Energy-saving regulations (e_{52})
	Energy management measures (e_{53})
	Energy publicity and training (e_{54})
fund input level for energy-saving (C_6)	Energy-saving independent input rate (e_{61})
	Energy-saving subsidies input rate (e_{62})

3. Methodology

3.1. Subjective weights calibration based on ANP

ANP is a more realistic and practical decision-making method which derives from the AHP method to adapt interrelated hierarchical structure. The steps to determine the subjective weights are as follows.

1) Establish green URTE network hierarchy structure based on ANP. By analyzing the dependence and feedback relationship between elements, the interaction between elements is obtained to build the ANP network hierarchy structure of green URTE.

2) Construct the judgment matrix $W_{(p_s,e_{jl})}$ .Set p_1,L , p_n as the elements of control layer, under the control layer, there are some element groups C_1,L , C_N of the network layer, which has elements e_{i1},L , e_{in_i}, $i = 1$,L , N in element group C_i .

3) Calculate super matrix W. Super matrix is consisted of many submatrices. The ranking vector $(w_{i1}^{(jl)}, \mathrm{L}, w_{im_i}^{(jl)})'$ of judgment matrix $W_{(p_s, e_{jl})}$ can be obtained based on eigenvalue method.

4) Construct weighted super matrix and limit matrix. By taking P_s as criteria, compare the importance of each group of elements under P_s to the criteria $C_j (j = 1, \mathrm{L}, N)$, and get the weighting matrix A by the eigenvalue method. The subjective weights W_1 based ANP can be obtained by the limit relative ranking vector.

3.2 Objective weights calibration based on entropy weight

The basic principle of entropy weight method is based on the variability size of indicator to determine the objective weights. Thus, this indicator plays a more important role in the comprehensive evaluation and the weight is also correspondingly larger. The steps to determine the objective weights are as follows.

1) Construct decision matrix X. Assume there are m evaluation objects and n evaluation indicators. According to the characteristic values of each indicator in the indicator system, decision matrix with the element x_{ij} is built, which is expressed as the form of $X = (x_{ij})_{m \times n}$.

2) Standardization of decision matrix. For positive indicator, the standardized value r_{ij} of x_{ij} is shown as Eq. (1). And for negative indicator, the standardized value r_{ij} of x_{ij} is shown as Eq. (2):

$$r_{ij} = (x_{ij} - \min(x_j)) / (\max(x_j) - \min(x_j)) \tag{1}$$
$$r_{ij} = (\max(x_j) - x_{ij}) / (\max(x_j) - \min(x_j)) \tag{2}$$

3) Determine the entropy and entropy weight of indicator. Make $p_{ij} = r_{ij} / \sum_{i=1}^{m} r_{ij}$, and the entropy H_j of j-th indicator is Eq. (3):

$$H_j = -(1 / \ln m) \sum_{i=1}^{m} p_{ij} \ln p_{ij} \tag{3}$$

4) Thereby, the objective weights of indicators W_2 are obtained based on entropy weight method.

3.3 Weights optimization based on game theory

The subjective and objective weights of indicators are obtained by using ANP and Entropy weight method which are respectively expressed as W_1 and W_2. For taking into account the actual importance and objective truth, and seek agreement or compromise between the subjective and objective weights, the game theory is used to minimize each deviation between the possible comprehensive weights and each basic weights. Remember any linear combination of the two weight vectors W_1 and W_2 as Eq. (4):

$$W = \sum_{k=1}^{2} \alpha_k W_k^{\mathrm{T}} \tag{4}$$

Where W is the comprehensive weight, α_k is the weight coefficient of k-th weight determination method, W_k is the weight vector of k-th weight determination method.

The optimization objective is as Eq. (5):

$$\min \left\| \sum_{k=1}^{2} \alpha_k W_k^{\mathrm{T}} - W_l^{\mathrm{T}} \right\|, l = 1, 2 \tag{5}$$

According to differential properties of matrix equation, the optimized first derivative condition of the Eq. (2) is $\sum_{k=1}^{2} \alpha_k \cdot W_l \cdot W_k^{\mathrm{T}} = W_l \cdot W_l^{\mathrm{T}}, l = 1, 2$. And the corresponding linear equations is as Eq. (6):

$$\begin{pmatrix} W_1 \cdot W_1^{\mathrm{T}} & W_1 \cdot W_2^{\mathrm{T}} \\ W_2 \cdot W_1^{\mathrm{T}} & W_2 \cdot W_2^{\mathrm{T}} \end{pmatrix} \begin{pmatrix} \alpha_1 \\ \alpha_2 \end{pmatrix} = \begin{pmatrix} W_1 \cdot W_1^{\mathrm{T}} \\ W_2 \cdot W_2^{\mathrm{T}} \end{pmatrix} \tag{6}$$

According to the Eq. (6), the optimal solution α_1 and α_2 can be obtained and then the comprehensive weight W can be got by substituting α_1 and α_2 into the Eq. (4).

4. Case study

4.1. Indicator weight evaluation
By using the investigation data obtained from department heads of urban rail transit enterprises, experts and scholars in the field of energy consumption, the industry regulation organization etc., and researchers determined the subjective, objective and comprehensive weights values based on the three above methods, and also analysed the evaluation results. A radar chart about subjective, objective and optimized weight of indicators was drawled. In Figure 1, each axis indicates the weight value of the evaluation indicator, and the weight value increased accordingly with dotted line extending from the inner to the outside.

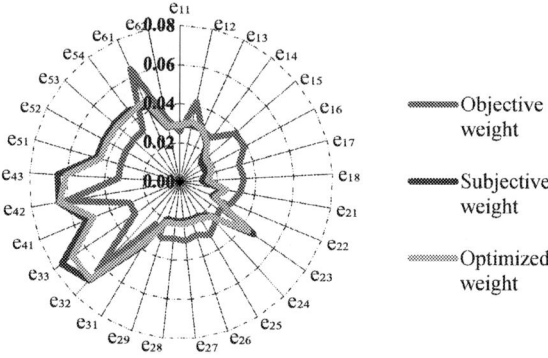

Figure 1. Radar chart of evaluation indicator weight

Figure 1 shows that the optimized weight basically covered closed region of the subjective and objective weights, indicating that the optimized weight can reflect the information provided by real data, but also objectively reflect the importance degree of each indicator, which avoid the defect of subjective and objective weight. Meanwhile, the weight of indicator e_{31} to e_{54} was relatively high, indicating that the indicators of the three elements energy-saving capacity building, application of energy-saving technology, and energy-saving management mechanism had great impact on the level of green development of urban rail transit enterprise.

In summary, while promoting the green development of urban rail transit enterprise, enterprise should strengthen the energy-saving capacity building, application of technology and improvement of management level, and raise awareness of energy saving autonomous. At the same time, energy-intensity and level of service also should be concerned.

4.2. Analyses of evaluation result
During evaluating the GDL of urban rail transit enterprise, the indicator standardization eigenvalues of each evaluation object are weighted by the optimized weight of each indicator and then were added up

to obtain the comprehensive indicator values of green development. As shown in Table 2, on the whole, the most advanced enterprise is A, followed by B, C, D. From the specific indicators, the majority of indicators in level two of the elements energy-saving capacity building, application of energy-saving technologies, energy-saving management mechanism, fund input level for energy-saving are more excellent than other enterprises, indicating that enterprise input higher in the energy-saving to promote green and sustainable development. But it should be noted while strengthening energy-saving construction, enterprise also should take into account of passenger sensory experiences and appropriately improve service quality.

Thus, for an earlier construction enterprise, if it's possible, it should appropriately improve the service level. For the late construction enterprises, improving the awareness of energy conservation and strengthening capacity construction and energy-saving management is the necessary work for them to carry out.

Table 2. Green Development Comprehensive Indicator Values

Indicator value	Urban rail transit enterprises			
	A	B	C	D
energy-intensity	0.1041	0.0676	0.1187	0.0878
level of service	0.1212	0.1364	0.1291	0.1364
energy-saving capacity building	0.1729	0.1044	0.0760	0.0360
application of energy-saving technologies	0.1244	0.1290	0.0694	0.0493
energy-saving management mechanism	0.1750	0.1166	0.0583	0.0000
fund input level for energy-saving	0.0553	0.0564	0.0193	0.0000
GDL (comprehensive indicator)	0.7528	0.6103	0.4708	0.3096

5. Conclusions

This paper first established an assessment indicator system for green URTE based on analysis of connotation and the basic characteristics of green URTE. The assessment indicator system proposed can comprehensively reflect the GDL of URTEs from six aspects including energy-intensity, level of service, energy-saving capacity building, application of energy-saving technologies, energy-saving management mechanism, fund input level for energy-saving.

In addition, to ensure the accuracy of evaluation results, three methods were used to determine the indicator weight. The subjective and objective weights were determined using ANP and entropy weight method, and then, game theory method was used to combine the subjective and objective weights which not only overcome the AHP and single entropy method to determine indicator weights deviating from the actual situation, but also optimize the weight of through game theory to seek consensus or compromise between the objective and subjective weight. This method takes into account of both the actual importance of indicators and the authenticity of objective data. Furthermore, the GDL evaluation results of four Chinese URTEs are consistent with the actual data acquired and GDL calculation also can identify the URTEs' benefits and drawbacks which indicate the proposed method has both well rationality and excellent operation performance.

References

[1] Yu, S., Mu, L., and Ji, B. (2011) On Green Transport and Low Carbon Transport. *ICTE 2011*: pp. 3061-3066. doi: 10.1061/41184(419)505.

[2] *Rail energy Project*. Energy efficiency solutions for rolling stock, rail infrastructure and train operation. <http://www.railenergy.eu/>; 2011 [accessed 26.05.14].

[3] Sandor J et al. Smart and efficient energy solutions for railways – the "Rail energy'' results. In: *9th World congress on railway research*, WCRR 2011, Lille (France); 2011.

[4] A. González-Gil, R. Palacin, P. Batty. Optimal energy management of urban rail systems: Key performance indicators, *Energy Conversion and Management* , Vol. 90, 2015, pp. 282–291

[5] *2015 annual statistical and analytical reports of urban rail transit in 2015*, China Association of Metros, Issued May 11, 2016.

[6] Evaluation methodology on energy consumption in urban rail transit, *DB 11 / T 1035-2013*, Issued December 20, 2013.

[7] *Achievements of capacity-building projects in 2013,* Maintenance unit: Ministry of Transport Research Institute, Copyright: Ministry of Transport of the People's Republic of China, http://jnzx.mot.gov.cn/chenggzhsh/nenglijsxm/201406/t20140618_1634979.html.Issued June 18, 2014.

[8] Gölcük İ, Baykasoğlu A. An analysis of DEMATEL approaches for criteria interaction handling within ANP. *Expert Systems with Applications*, Vol. 46, 2016,pp. 346-366

[9] Chemweno P, Pintelon L, Van Horenbeek A, et al. Development of a risk assessment selection methodology for asset maintenance decision making: An analytic network process (ANP) approach. *International Journal of Production Economics*, Vol. 170, 2015, pp. 663-676.

Simulation Study on Emergency Evacuation of Metro Stations in Fire Degradation Mode

XI Jiaojiao[1], LI Jin[2*]

[1]Xuzhou Technician College, Jiangsu Province, China, 221000

[2]College of Traffic, Jilin University,Changchun, China, 130022

* LI Jin(1970-),Female, Associate Professor,Doctor, Research direction:transportation economy, urban and regional transportation planning.

[1]E-mail: 402493200@qq.com

[2]E-mail: li_jin@jlu.edu.cn

Abstract: With the rapid development of urban rail transit, the safety hazards of subway stations are increasing, and the safe evacuation of fires inside the station has become the focus of research. This paper analyzes the situation of platform fire and station hall fire emergency treatment, establishes an evaluation system on the analysis of passenger evacuation bottleneck factors, and proposes a risk evaluation method. Taking Wenze Road Station of Hangzhou Metro as an example, combined with Anylogic simulation to evaluate and analyze different situations, the overall risk level of Wenze Road Station fire is higher, and the evaluation results are consistent with the actual situation. According to the results, corresponding improvement measures are proposed to provide reference for station emergency evacuation.

1. Introduction

With the rapid development of urban rail transit, the safety hazards of subways have drawn attention from all walks of life. Therefore, it is of practical significance to study the bottleneck risk in the subway station fire degraded mode.

In recent years, scholars at home and abroad have conducted some research on the emergency evacuation of subway passenger flow fires. Song Chao[1] set up a questionnaire and model analysis for passenger evacuation behavior and influencing factors in the case of Beijing subway passenger flow, and obtained passenger evacuation behavior and influencing factors. Wang Yang[2] used the passenger flow game allocation method to distribute the passenger flow game to Beijing Fuxingmen Station, and used Anylogic software to verify the evacuation effect after adopting the emergency evacuation plan. Liu Yang[3] takes Xi'an Metro Bell Tower Station as an example, by changing the width of the stairs and the number of gates, the simulation and comparison analysis is carried out to establish the optimal cooperation scheme between the stairs and the gates. Tang Han[4] takes Beijing Fuxingmen Station as an example. According to the different locations of the fire, the evacuation route network in the subway station is constructed. Zhang Wei[5] first proposed the station-oriented identification layout model based on sequence control and uncertain interaction and the station-based passenger flow control model for passenger service demand. Qian Zhenwei et al[6] analyzed the passenger flow

Content from this work may be used under the terms of the Creative Commons Attribution 3.0 licence. Any further distribution of this work must maintain attribution to the author(s) and the title of the work, journal citation and DOI.

Published under licence by IOP Publishing Ltd

ISPECE   IOP Publishing

IOP Conf. Series: Journal of Physics: Conf. Series **1187** (2019) 052072   doi:10.1088/1742-6596/1187/5/052072

evacuation simulation of the station hall and platform by using Anylogic software. After optimization of various influencing factors, the evacuation time can be effectively reduced and the evacuation efficiency can be improved. Li Xun et al.[7] discussed the psychological behaviors of passengers in the emergency evacuation of subways from environmental information, guidance information, and passenger basic information.

In summary, domestic and foreign scholars have studied the subway fires too broadly, and did not dig deep into the bottleneck risks caused by fires in different locations. Therefore, based on the actual situation, this paper analyzes the station's emergency treatment when the platform fire and station hall fire occur, and builds an evaluation system based on the factors affecting evacuation. Finally, combined with the example simulation analysis, the specific bottleneck risk level is obtained. It provides a reference for the analysis of urban rail transit fire evacuation bottlenecks.

2.   Metro station fire emergency treatment

When a fire broke out in the subway, the station will open the "degradation mode", all the gates will be open, the escalators and escalators will be closed, and the escalator will be used as a staircase. At this time, the station staff took fire emergency treatment to organize the passengers to evacuate safely.

2.1.Subway station fire emergency treatment

When a fire occurs on the subway platform, the passenger's panic will cause chaos in the platform, the station management should immediately take emergency measures. If the train does not enter the station, the "emergency stop" measures should be taken; if the train has already entered the station, the "interlock release" should be adopted to make the train safely leave the station[8]. At this time, the passengers evacuate via the hall through the stairs (the escalator acts as a staircase). In addition, the environmental control mode is turned on to supply air to the station hall floor, and the air supply to the station platform is stopped, and the platform layer enters the exhaust gas state, as shown in Figure 1.

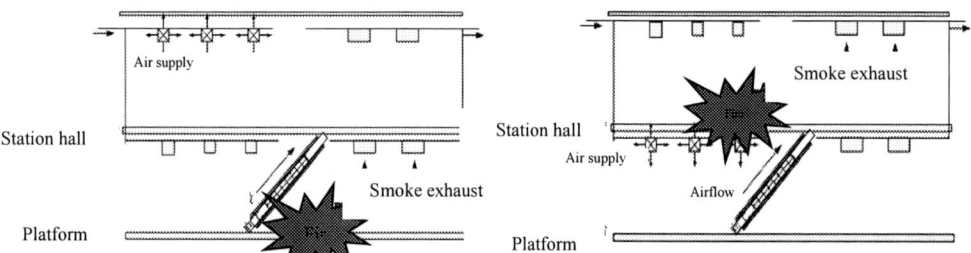

Figure 1. Schematic diagram of platform fire   Figure 2.Schematic diagram of station hall fire

2.2 Subway Station Fire Emergency Treatment

When a fire broke out in the subway station hall, the station management personnel evacuated according to the size of the fire. If the fire is large, the station will be in emergency treatment with the platform fire. If the fire is small and does not affect the safety of the train, the broadcast notice can be taken when the train has already entered the station. When the train arrives at the station, only the passengers are on board to ensure that some passengers can safely and quickly leave the station via the train. In combination with the actual situation, some passengers choose to evacuate via stairs due to the intersection of passengers. At the same time, the environmental control mode is turned on to supply air to the platform floor, and the air supply to the station floor is stopped, and the station floor enters the exhaust state, as shown in Figure 2.

3. Evaluation of risk level of evacuation bottleneck in subway station

3.1. Analysis of the bottleneck point of subway station evacuation

The flow density and pedestrian flow are usually directly used as the bottleneck criterion. Taking into

2545

account the location of the bottleneck point, the entrance, gate and stairs are selected as the research subject. The relationship between the flow density at the entrance and exit and the flow rate is as shown in equation (1). Using the same method, the relationship between the flow density and the flow rate of the gates and stairs is simulated to obtain the relationship (2) and (3).

$$Q = -26.48k^2 + 81.45k + 1.8 \quad (R^2 = 0.8765) \tag{1}$$

$$Q = -11.9k^2 + 50.9k + 0.23 \quad (R^2 = 0.8841) \tag{2}$$

$$Q = -17.23k^2 + 68.74k + 0.63 \quad (R^2 = 0.8474) \tag{3}$$

In the formula, Q represents passenger flow, p/min/m; k represents human flow density, p/min; R^2 represents the curve fitting squared difference.

From equations (1) to (3), it can be seen that when the flow density is 1.54p/m^2, the flow rate at the entrance and exit reaches the maximum value, and the corresponding flow density at the gate and the staircase is 1.99 p/ m^2 and 2.14 p/m^2. When the flow density exceeds this value, crowd congestion will occur and the flow of people will decrease. Therefore, the flow density at the time when the flow rate of each point is maximum is taken as the critical density of the point. The early warning level is established based on the existing research results[9], and the division rules are shown in Table 1.

Table 1. Division of population flow density monitoring and early warning

Level division	Indicator description	Predicting population density
IV	Pedestrian current density prediction is 10%~20% larger than critical density	$1.1K < k_p \leq 1.2K$
III		$1.2K < k_p \leq 1.3K$
II	Pedestrian current density prediction is 20%~30% larger than critical density	$1.3K < k_p \leq 1.5K$
I		$k_p > 1.5K$
	Pedestrian current density prediction is 30%~50% larger than critical density	
	Pedestrian current density prediction is 50% larger than critical density	

(Note: k_p indicates the predicted pedestrian density, K indicates the critical population density.)

According to the classification criteria of the early warning level of pedestrian flow and the threshold value of the flow density of each region given in Table 1, the standard of detection of the flow density at the entrance and exit, gate or stairway of the subway station can be determined. The relationship between the early warning level and the regional flow density is calculated as shown in Table 2.

Table 2. Relationship between early warning level and predicted flow density in subway stations

Level division	Predicting population density(p/m^2)		
	Entrance and exit	Gate	Stairs
IV	1.69~1.85	2.19~2.38	2.35~2.57
III	1.85~2.00	2.38~2.59	2.57~2.78
II	2.00~2.31	2.59~2.96	2.78~3.21
I	>2.31	>2.96	>3.21

3.2. Bottleneck point evacuation risk assessment analysis

In this study, three indicators of human flow density, number of associated points and flow rate of people are selected as risk level evaluation indicators.

(1) Human flow density (a_1): refers to the predicted flow density value obtained by the subway station evacuation simulation, and the unit is p/m^2.

(2) Number of affiliate groups (a_2): The number of affiliate groups that accept the same service item due to the determination of the pedestrian walking target in the subway station.

(3) Current load rate (a_3): refers to the ratio of the number of stranded persons to the number of

peers. The calculation is as shown in formulas (4) and (5).

$$a_3 = \frac{N_S}{P} \tag{4}$$

$$N_S = \sum_{i=1}^{n} \int_{T_0}^{T} f_i(t) b_i(t) dt - \int_{T_0}^{T} f(t) B(t) dt \tag{5}$$

In the formula, a_3 represents the flow rate of the person; N_S represents the number of people in the detention; person; P represents evacuation number,person, T represents evacuation time, s; T_0 represents the moment when the population begins to stay, s; $f_i(t)$ represents the crowd flow coefficient at the branch of the channel i; person/m·s; $b_i(t)$ represents the length of the branch i, m; $B(t)$ is the width of the flow at the exit of the channel, m; n represents the number of branch entries.

In order to facilitate quantitative evaluation, the risk level is expressed in higher, high, medium, general and low, and the corresponding safety evaluation scores are 1~5 points, the risk level is shown in the table. 3.

Table 3. Bottleneck point evacuation influencing factors Risk status classification

Risk status	Score	Flow density a_1	Number of affiliate groups a_2	Load rate a_3
higher	1	I	>6	>200%
high	2	II	5~6	170%~200%
medium	3	III	3~4	130%~170%
general	4	IV	2	100%~130%
low	5	/	<1	<100%

The final risk level evaluation of the bottleneck point is based on the LEC evaluation method (Graham evaluation method). The risk value $D = L \times E \times C$, which L indicates the probability of an accident, E indicates the frequency with which the person is exposed to the hazardous environment, C indicates the possible consequences of the accident, and D indicates the danger of the system. Therefore, a KNS risk evaluation method similar to the LEC evaluation method can be established. A risk evaluation value $D = L \times E \times C$ is defined, K indicates the person flow density score, N indicates a score of the number of affiliate groups, and S indicates a flow rate of the flow rate. The quantitative evaluation score $R = [K, N, S]$ is used to determine the risk assessment value D, and finally the risk level of the bottleneck point is determined according to Table 4.

Table 4. Security risk assessment value and risk classification

Risk level	Risk assessment value	Risk status
Level one	1~6	Extremely dangerous, need to rectify the layout or route
Level two	7~12	Highly dangerous, need to develop improvement measures
Level three	13~45	
Level four	46~80	Significant danger, need to take precautions
Level five	81~125	Generally dangerous, need to strengthen management
		Basic security, negligible

4. Example analysis

Take Wenze Road Station of Hangzhou Metro Line 1 as an example for research. The subway station is an underground two-story island structure with four entrances, A, B, C and D. The layout of the station is: four outbound gates in the AB direction, two security checks in the CD direction entrance and exit, six inbound gates, six outbound gates, and one two-way gate. Six TVM machines and one customer service center are set up in the east-west direction. Plan the layout in the Anylogic software according to the station scale, as shown in Figure 3. In order to facilitate the simulation, this study mainly conducts passenger flow evacuation research on the side of the platform, and analyzes the bottleneck of passenger flow evacuation in the case of fire in the platform and fire in the station hall.

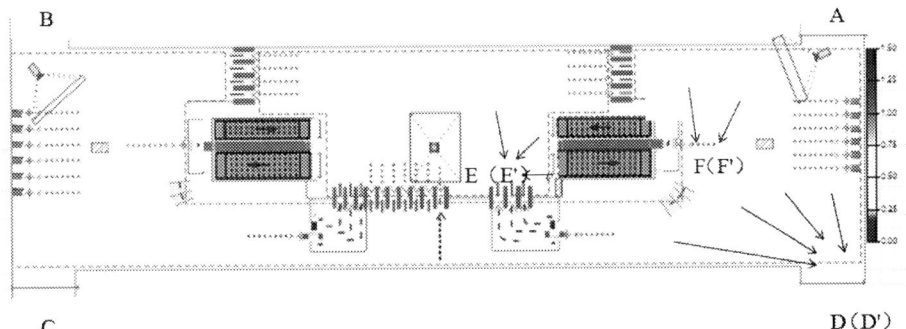

Figure 3. Plan layout of simulation experiment of Wenze Road Station in Hangzhou Metro

4.1. Simulation of fire evacuation at subway station

It can be seen from the content of this paper that when a subway station fires, the passenger evacuation route can be divided into two cases: no passengers boarding and passengers boarding according to actual conditions.

(1) If there is a fire in the platform or a fire in the station hall is large, the passengers will evacuate via the hall through the stairs (the escalator acts as a staircase). The evacuation map is shown in Figure 4.

(2) The fire in the station hall does not affect the safety of the train. When the train has entered the station, some passengers choose to board the metro and leave the station. The evacuation map is shown in Figure 5.

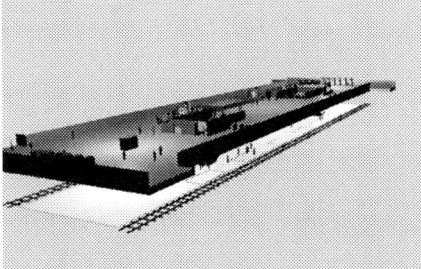
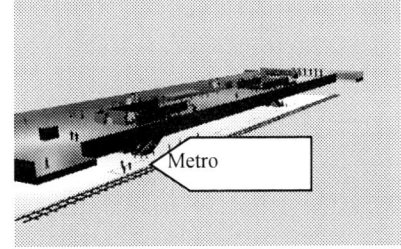

Figure 4. No passengers on board   Figure 5. Some passengers on board

In this study, one of the evacuation lines was selected. When passengers do not get on the metro, D, E, and F are selected. When passengers board the metro, D', E', and F' are selected. (The software controls the number of platform passengers through six platforms in Figure 5.) The flow parameters of these points were collected by Anylogic, and the flow density and human flow of each bottleneck at the time of fire were obtained. In this study, the passenger flow parameters were simulated for 30 times and averaged, as shown in Table 5.

Table 5. Bottleneck point area density and human flow statistics

Bottleneck point	D	E	F	D'	E'	F'
Regional density(people/ m2)	4.25	3.27	1.70	4.06	2.98	1.06
Human traffic(people /min)	108	53	23	95	42	11

4.2 People's cluster scattered bottleneck risk rating

Analyze and organize the Anylogic simulation statistics to obtain the index values of each bottleneck point and the corresponding risk levels as shown in Table 6. The flow rate (a_1) indicator value of each bottleneck point is directly obtained from the statistical value, and the associated person group (a_2) index value is obtained according to the layout. For example, the D-point flow comes from the ticket machine queue, the customer service center queue, the gate passengers and other staff in the station hall. There are 4 groups of related persons, as shown in Figure 1.

Table 6. Bottleneck point indicator value and risk level division

Bottleneck point	Index value			Indicator risk level			Indicator score			Index value D	Risk level
	a_1	a_2	a_3	a_1	a_2	a_3	K	N	S		
D	4.25	4	233.4%	higher	medium	higher	1	3	1	3	Level one
E	3.27	3	178.3%	higher	medium	higher	1	3	1	3	Level one
F	1.70	2	120.0%	low	general	general	5	4	4	80	Level four
D'	4.06	4	210.0%	higher	medium	higher	1	3	1	3	Level one
E'	2.98	3	140.7%	higher	medium	medium	1	3	3	9	Level two
F'	1.06	2	95.5%	low	general	low	5	4	5	100	Level five

It can be seen from Table 6 that when the platform fire and the station hall fire occur, the risk of entrance and exit is at the first level. When the passenger does not get on the metro, the risk level of the gate is one level, and the risk level of the stairs at the platform is four. When the passenger gets on the metro, the risk level of the gate is reduced to two, and the risk level of the stairs at the platform is reduced to five. It can be seen that the passengers can effectively relieve the risk of evacuation at the stairs and at the gates when they get on the metro. Generally speaking, the overall risk level is high when the fire occurs, and the station management personnel should focus on strengthening the evacuation of passengers at the station hall. It can be realized by adding personnel guidance, railing diversion and additional guiding signs at the station hall.

5.Summary

Taking Wenze Road Station of Hangzhou Metro as an example, this paper analyzes the two situations of fire in the subway platform and fire in the station hall by using Anylogic software. The risk level is analyzed according to the evaluation criteria, and the following conclusions are drawn.

(1) In this study, the paper mainly deals with the situation of whether there are passengers getting on the train when the fire occurs, and the bottleneck risk analysis is carried out on the stairs, the gates and the access passages of the platform.

(2) Using the LED risk scoring method, the risk evaluation value is determined by $D = K \times N \times S$, the passenger flow density, the number of dangerous related populations and the flow rate of people are selected for quantitative analysis to determine the risk level.

(3) Through simulation analysis, it is concluded that the risk level of the bottleneck of the station is higher in the two cases when the Wenze Road subway station fire occurs. The evaluation results are basically consistent with the actual status of the subway operation. This paper indicates that the evaluation method has certain reference value for improving the risk status of the subway.

(4) In this study, the difference in the evacuation delay time caused by the different sensitivity of the passenger evacuation reaction caused by the different locations of the fire was not considered. This will be the focus of further research in the future research.

References:

[1] Song Chao. Research on emergency evacuation behavior of passengers in Beijing subway passengers [D]. Beijing: Beijing Jiaotong University, 2017

[2] Wang Yang. Design of evacuation distribution and evacuation scheme for urban rail transit stations [D]. Beijing: Beijing Jiaotong University, 2014.

[3] .Liu Yang. Simulation of emergency evacuation of subway station based on Anylojic [D]. Lanzhou: Lanzhou Jiaotong University, 2016.

[4] Tang Han. Emergency evacuation of urban rail transit stations based on evacuation path assignment [D]. Beijing: Beijing Jiaotong University, 2016.

[5] Zhang Zhe. Research on Optimal Control of Passenger Flow Distribution in Urban Rail Transit Stations [D]. Beijing: Beijing Jiaotong University, 2017.

[6] Qian Zhenwei,Qian Dalin,Zhang Hui. Analysis of Factors Affecting Emergency Evacuation Time of Passenger Flow in Urban Rail Transit Stations [J].Journal of Dalian Jiaotong University,2017,38(2):6-10.

[7] Li Xun,Hong Ling,Xu Ruihua. Analysis of Factors Affecting Psychological Behavior of Emergency Evacuation Passengers in Rail Transit Stations[J]. Urban Rail Transit Research, 2012, 4(4): 54-57.

[8] Fei Anping.Application of Rail Transit Equipment [M]. Beijing: China Railway Publishing House, 2013.

[9] Jia Hongfei,Yang Lili,Tang Ming.Analysis of Pedestrian Flow Characteristics and Calibration of Simulation Model Parameters in Integrated Transportation Hub[J], Journal of Transportation Systems Engineering and Information, 2009, 9(5): 117-123.

The evolutionary game analysis of incentive mechanism for crowd sensing of public environment

Qiang Zhang[1], Qingqing Zhang[1*], Xueyan Liu[2], Jian Dai[1], Xujuan Zhang[1]

[1] College of Computer Science & Engineering, Northwest Normal University, Lanzhou Gansu 730070, China

[2] College of Mathematics and Statistics, Northwest Normal University, Lanzhou Gansu 730070, China

*Corresponding author's e-mail: 837908907@qq.com

Abstract. The public environment perception model regards people as a "data perceptron" and a human-centred participatory perception model. The enthusiasm and initiative of public participation will directly determine the effective operation of the model. This paper aims to understand how to stimulate public participation in data sensibility in public environment perception and establish an effective incentive mechanism. Based on the evolutionary game theory, a public environment perception evolutionary game model is established. The game selection between the data subject and the perceived participants is analysed. The group strategy selection and influencing factors of establishing effective incentive mechanism are studied. According to the replication dynamic equation, the behaviour evolution law and evolutionary stability strategy of public environmental group intelligence perception are obtained. The research results show that data users increase the proportion of investment in participants' incentive strategies, which will motivate participants to share more data and thus help them to evolve to the desired results. Encouraging participants to actively share data is necessary to reduce participants' participation costs, which requires data users to select target groups to participate in perceived tasks, and to select people who are more convenient to provide data; the data user's incentives should be sufficient to offset the participant's participation cost. The data user should ensure that both the quantity and quality of the public participation in the perceived task are optimal when the data user minimizes the payment.

1. Introduction

In recent years, with the frequent regional and multi-element ecological environment problems, after a long period of accumulation, superposition and spread, a series of complex ecological and environmental problems such as haze, global warming and sandstorms are intensively and seriously threatening people's normal production, life, physical and mental health. People have digitally recorded individual feelings toward environment and the influences through web platforms such as Weibo and WeChat. Public concern about the quality of the ecological environment, concern for environment protection, and participation in environment protection are unprecedentedly high. As a dynamic subject in the ecological environment, the public's accurate environmental perception is of great significance for indirectly reflecting regional environment quality, expressing public environmental demands, actively responding to government environmental policies, and establishing correct environment public opinion guidance.

Content from this work may be used under the terms of the Creative Commons Attribution 3.0 licence. Any further distribution of this work must maintain attribution to the author(s) and the title of the work, journal citation and DOI.
Published under licence by IOP Publishing Ltd

John R. Gold[1] and K.P. Burnett[2] proposed a theory framework for the process of environment perception. Domestic scholars have carried out a lot of theoretical and empirical studies on perception such as environment perception mechanism[3], water environmental risk perception[4], climate change perception[5], environmental pollution perception[6], glacier change perception[7], meteorological disaster perception[8], public environmental perception environmental awareness[1]. Existing research of environment perception mainly collects public perception data through questionnaires, interviews and third-party observations, and uses statistical induction and cognitive models to study public environment perception rules and behaviours. Common sense[9] monitors the status of SO_2, CO, etc. through handheld air quality sensing devices. Sensor Map[10] establishes a network to monitor urban air quality through an air quality sensor equipped on the user's car; Creek Watch[11] is an application that collects water quality near people's rivers, including river flow, flow rate, and surrounding garbage status, then analyses and organises the data and shares it; PEIR[12], an application of environmental impact, analyses individual behaviours; Wang et al.[13] monitored the carbon monoxide concentration of air by installing a carbon monoxide sensor on campus. Aiming at the problems in the above literature research, this paper proposes a public environmental group intelligence perception model based on real-time submission of ecological environment by mobile terminals.

2. Public environmental group intelligence perception model and incentive mechanism

The public environment perception model is to use humans as "data perceptions", based on the mobile devices (such as mobile phones, smart bracelets) that users carry with them, and actively submit the individual's perception information of objective environmental quality to form public objective data collection model for environmental quality-aware big data sets. Such data information includes both physical environment data reflecting the objective environmental quality and psychological cnvironment data forming an impression in the individual's mind. By using large data sets formed by public context-aware models, we can use big data analytics to explore the interaction between individuals and the environment in order to achieve the purpose of identifying public behaviour, at the same time, this is a new model and new method to study the relationship between human and natural environment in the context of the application of big data and wireless sensing technology.

The public environment perception model is human-centred participatory perception. The biggest difference from opportunity perception is that the data in perceptual perception is uploaded by the user's unconscious perception, and the perceived participation requires the conscious and active participation of the person. Participation in the perception process and perceived data reliability play a decisive role in perceived outcomes. Simultaneously, in the public environment perception model, the public is both a participant in the data and a user of the data. As perceived participants, their own costs are considered while considering their privacy protection and deciding whether they are involved in data perception. As a data user, you need to get more high-quality data at a relatively low cost. In the actual operation process of the sensing system, there is a problem that the user participation is not high, that is, the number of participants and the data quality are not high[14]. Therefore, establishing an effective incentive mechanism to stimulate users to participate in the enthusiasm and initiative of submitting data is the key to the effective operation of the public environmental group perception model.

3. Model basic assumptions and construction

In the operation of the public environment perception system, perceived participants exhibit characteristics of individual rationality, uncertainty, selfishness, dishonesty and so on, because these characteristics make the participants have positive and non-active sharing of data. For data users who declare tasks, it is necessary to comprehensively consider the characteristics of user, and develop reasonable incentive mechanisms to encourage the participation of sensing devices to participate in sensing tasks and provide data, so as to maximize the utility of data users and participants. From the perspective of data acquisition and use, the interest groups of knowledge perception involving public environmental groups are two groups: perceived data participants and data users. In the process of data perception, the perceptual data participant selects the strategy to decide whether to actively submit the perceptual data and data user selection strategy determines whether to provide rewards for perceived data participants. The following parameter settings and basic assumptions are given to the public context-aware system, as shown in Table 1.

Table 1. Parameter settings.

Definition	symbol
Data users receive income when participants actively participate	W_1
Data users gains when participants are not actively involved	W_2
Perceived participants' choice of revenue when actively sharing data	E_1
Perceived participants' choices when not actively sharing data	E_2
Total incentive costs for data user plans	P_1
Perceived participants' choice of actively sharing data to earn incentives as a percentage of incentive costs	θ
Perceived participants actively share data costs	P_2
Loss caused by non-incentives by data users when participants actively participate	C_1
Loss caused by non-incentives by data users when participants are not actively involved	C_2

Hypothesis 1: The data user obtains the comprehensive economic benefit marked as W_1, when the participant chooses to actively participate; the data user obtains the comprehensive economic benefit marked as W_2 when he perceives that the participant chooses not to actively participate. Due to comprehensive consideration, if the perceived participants choose to actively provide data for the user to study, in this case, the data user's income is significantly greater than when the perceived participant does not actively provide data, that is $W_1 > W_2$.

Hypothesis 2: When perceived participants choose to actively share data, the overall economic benefits they receive are E_1; when the perceived participants choose not to actively share the data, the comprehensive economic benefits they receive are E_2. When the perceived participants choose to share the data for the user to study or analyse, it contributes to the social ecological construction and the living environment, and also obtains a good feedback on the overall income of the user. In this case, the obtained overall benefits are also greater than when the data is not actively shared, that is $E_1 > E_2$.

Hypothesis 3: The cost of using mobile devices when perceiving participants to actively participate in sharing data, downloading the APP and submits the data that would be using the data traffic, consuming the storage space of the mobile phone, the economic cost of the participants at this time plus the time cost of consumption is P_2. When participants actively share data and data users take no incentives, it will affect the number of perceived participants and the quality of the shared data, resulting in losses, C_2, when data users perceive participants not actively sharing data, loss caused by non-incentives is set to C_2, at this time $C_1 > C_2$.

According to the comprehensive analysis of the public environmental group intelligence perception incentive mechanism and the setting of the above parameters, the game payment matrix for data users and perceived participants is obtained, as shown in Table 2.

Table 2. Payment matrix for data users and participants.

	Perceived participant	
	Actively share data	Do not actively share data
Excitation	$W_1 - \theta P_1$ $E_1 + \theta P_1 - P_2$	$W_2 - (1-\theta) P_1$, $E_2 + (1-\theta) P_1$
Not inspiring	$W_1 - C_1$, $E_1 - P_2$	$W_2 - C_2$, E_2

4. System Stability Strategy Analysis

In the process of public environment perception system evolution, different groups will choose different strategies. Set the probability of data users choosing incentive is x ($0 < x < 1$), choose not to motivate the probability of $1-x$; The probability that participants choose to actively share data is y ($0 < y < 1$) and the probability that they choose not to actively share data is $1-y$.

The expected benefit of a data user when adopting an incentive strategy is:

$$Z_{11} = y(W_1 - \theta P_1) + (1-y)\left[W_2 - (1-\theta) P_1\right] \quad (1)$$

The expected benefit when the data user chooses not to motivate the strategy is:

$$Z_{12} = y(W_1 - C_1) + (1-y)(W_2 - C_2) \tag{2}$$

The average revenue of data users is:

$$\overline{Z}_1 = x Z_{11} + (1-x) Z_{12} \tag{3}$$

The expected benefits when perceiving participants to actively share data is:

$$Z_{21} = x(E_1 + \theta P_1 - P_2) + (1-x)(E_1 - P_2) \tag{4}$$

The expected benefits when perceiving participants choosing not to actively share data is:

$$Z_{22} = x\left[E_2 + (1-\theta) P_1\right] + (1-x) E_2 \tag{5}$$

The average benefit of perceived participants is:

$$\overline{Z}_2 = y Z_{21} + (1-y) Z_{22} \tag{6}$$

The dynamic analysis of the replication of the two types of group games is performed, and the dynamic equation of the data user is:

$$F(x) = \frac{dx}{dt} = x(Z_{11} - \overline{Z}_1) = x(1-x)(Z_{11} - Z_{12}) = x(1-x)\left[C_2 - P_1 + y(C_1 - C_2)\right] \tag{7}$$

The dynamic equation of the perceptual participant's replication is:

$$F(y) = \frac{dy}{dt} = y(Z_{21} - \overline{Z}_2) = y(1-y)(Z_{21} - Z_{22}) = y(1-y)\left[(2\theta P_1 - P_1)x + E_1 - E_2 - P_2\right] \tag{8}$$

If $F(x) = 0$, get: $x_1 = 0$, $x_2 = 1$, $\hat{y} = \dfrac{(1-\theta) P_1 - C_2}{(1-2\theta) P_1 + C_1 - C_2}$ ($0 < \hat{y} < 1$) (9)

If $F(y) = 0$, get: $y_1 = 0$, $y_2 = 1$, $\hat{x} = \dfrac{E_2 + P_2 - E_1}{(2\theta - 1) P_1}$ ($0 < \hat{x} < 1$) (10)

The five equilibrium points calculated by the above equations are: $F_1(0,0)$, $F_2(0,1)$, $F_3(1,0)$, $F_4(1,1)$, $F_5(\hat{x},\hat{y})$.

If the equilibrium point of the evolutionary game is satisfied $Det(J)>0$ and $Tr(J)<0$, the equilibrium point is an evolutionary stability strategy, and the corresponding equilibrium point is a stable point. The model in this paper assumes that the parameters involved are greater than 0, according to the decision condition of the system ESS, five equilibrium points are analysed through the evolutionary game of data users and perceived participants, just discuss these two conditions when $(1-\theta)P_1<C_2<C_1<\theta P_1$ and $C_2<(1-\theta)P_1<\theta P_1<C_1$, the calculation results are shown in the following table: If $(1-\theta)P_1<C_2<C_1<\theta P_1$:

Table 3. Local stability analysis results of the system (a).

Balance point	Det(J)	Tr(J)	Properties
$F_1(0,0)$	-		Saddle point
$F_2(0,1)$	-		Saddle point
$F_3(1,0)$	-		Saddle point
$F_4(1,1)$	-		Saddle point
$F_5(\hat{x},\hat{y})$	+	0	Saddle point

In this case, the final evolution result is not stable, and both the data user and the perceived participant groups are bounded and rational, finally they will find the most favourable choice. If the evolution of the system begins to perceive the participants to receive incentive rewards, but the data users are not motivated because the incentive costs are too large, which is likely to cause the perceived participants to lose trust in the data users, hence getting data will face failure. Therefore, this condition does not conform to the actual situation, and the second case is directly considered.

If $C_2<(1-\theta)P_1<\theta P_1<C_1$:

Table 4. Local stability analysis results of the system (b).

Balance point	Det(J)	Tr(J)	Properties
$F_1(0,0)$	+	-	Stable point
$F_2(0,1)$	+	+	Unstable point
$F_3(1,0)$	+	+	Unstable point
$F_4(1,1)$	+	-	Stable point
$F_5(\hat{x},\ \hat{y})$	-	0	Saddle point

It can be obtained $F_2(0,1)$, $F_3(1,0)$ is unstable, $F_5(\hat{x},\ \hat{y})$ is saddle point, $F_1(0,0)$, $F_4(1,1)$ are the stable point, i.e. ESS, which responds to 2 cases, data users to take incentives, perceive participants to actively share data and data users not to take incentives, and perceive participants not to actively share data, respectively. From the analysis of the properties of the above 5 equilibrium points, the dynamic evolution phase diagram can be obtained:

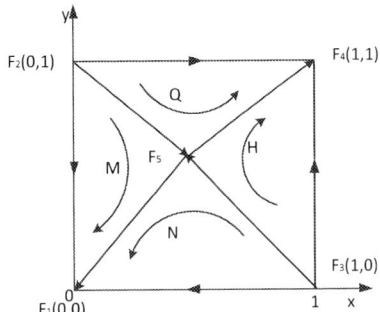

Figure 1. System evolution phase diagram.

Figure 1 depicts the dynamic process of evolutionary game between data users and perceived participants. By two unstable equilibrium points F_5, F_3 and saddle point F_5, the connected fold line is the critical line where the system converges to different states, in the upper right Q and H areas of the poly-lines, the system will gradually converge to two groups of incentives and active participation. In the M and N regions at the bottom left of the poly-lines, the system will gradually converge to two groups that are not motivated and not actively involved. Since the evolutionary game is a dynamic evolution process, the above two situations have the rationality of existence within a certain period of time.

5. Evolutionary Game Analysis and Discussion

In the process of the public environmental group perception game, the initial values of some parameters that constitute the income function of both sides of the game and their changes will cause the evolutionary system to converge to different equilibrium points. Therefore, the final evolution state of the result is determined by the initial state of the game. It can also be said that it is related to the size of QH and MN area. When the initial state of evolution begins in the Q and H regions, the system converges on the point of $F_4(1,1)$. For the ideal evolution result, it is hoped that the system will converge to the state of F_4 point, at this time, the Q and H regions will expand, and the F_5 point will tend to move to the lower left. The position of this point has a direct impact on the two regions. Therefore, changes in values of P_1, θ, P_2, C_1 and C_2 will change the size of the two regions, which will affect the convergence result of the system. From the above analysis, it can be concluded that:

As the maximum incentive cost of data user plan, P_1 itself has little influence on the decision of both sides of the game. But θ is rewarded for choosing to actively share data as a participant. The proportion of the reward to the incentive cost has a direct impact on the decision-making of both parties. The data user increases the proportion of participants' incentives, the area of QH above the fold line expands correspondingly, and the probability of the system converging to F_4 point increases. Conversely, the probability of system convergence F_1 increases.

P_2 represents the cost of participation of data users. The system will gradually converge to F_1 when the perceived cost of participants who provide data is too high. It can be seen from the phase diagram that the position of saddle point F_5 will move up and MN area will increase. In this case, participants are likely to choose not to actively share data unless they are satisfied with the compensation and rewards they receive. As a data user, to encourage participants to actively share data, it is necessary to reduce the cost of participation.

C_1 and C_2 are losses caused by data users not taking incentive measures when participants actively participate in data sharing and do not actively participate in data sharing, respectively. As can be seen from the phase diagram of the system, when C_1 increases, QH area increases, and when C_2 increases, MN area increases. Therefore, when the participants do not actively share the data, the loss of the data user is relatively small. When the perceived value of data obtained by participants is insufficient to offset the input cost of data users, there is no reason for data users to adopt incentive measures at this time.

6. Conclusions

This paper proposes a public environment perception model, which analyses the key role of incentive mechanism in improving the enthusiasm of perceived participants for the process of public participation in perception tasks. Through the analysis of the evolutionary game model, it mainly analyses the necessity of the existence of incentive mechanism and the positive and non-positive situation of perceived participants. The evolution has two results that the participants actively share the data and the data provider adopts an incentive mechanism, and the perceived participants do not actively share the data and the data providers do not adopt the incentive mechanism. The final evolutionary convergence results are related to parameters such as P_1, θ, P_2, C_1 and C_2. In the actual system operation, it is necessary to improve the enthusiasm of participants by increasing the incentive cost of data users, reducing the perceived cost of perceived participants and coordinating the interest relationship between participants and data users.

Game analysis shows that data users should take incentive measures to obtain effective data, enrich the incentive forms within the costing plan, and select the correct target population. At the same time, we must have a comprehensive technology to guarantee the accuracy and reliability of the data. For perceived participants, it is necessary to strengthen self-cultural literacy, have a correct concept of environmental protection, and be willing to share data while paying more attention to protecting privacy. The protection of the environment is a long-term and arduous task. Of course, the government and enterprises have great responsibilities, but everyone's awareness of environmental protection should be gradually improved, and they should work for the construction of beautiful environment together, so the overall benefits of society and individuals can have a greater increase. For the government, it is necessary to promote the convenience of data users to obtain data, supervise the safety and correctness of their data use, enhance the enthusiasm of public participation, and protect ecological environmental protection work.

Acknowledgment

This work was supported by the National Natural Science Foundation of China under Grant No.71764025 and 61662071.

References

[1] Wikipedia S, TB/Biologie/Grundlagen. An introduction to behavioural geography[M]. Oxford University Press, 1980.
[2] Jacqueline Desbarats. Spatial choice and constraints on behavior[J]. Annals of the Association of American Geographers, 1983, 73(73):340-357.
[3] Rocha K, Pérez K, Rodríguezsanz M, et al. Perception of environmental problems and common mental disorders (CMD)[J]. Social Psychiatry and Psychiatric Epidemiology, 2012, 47(10):1675-1684.
[4] Zoellner J, Hill J L, Zynda K, et al. Environmental perceptions and objective walking trail audits inform a community-based participatory research walking intervention[J]. International Journal of Behavioral Nutrition and Physical Activity, 2012, 9(1):6.
[5] Mcdaniels T L, Axelrod L J, Cavanagh N S, et al. Perception of ecological risk to water environments[J]. Risk Analysis, 1997, 17(3):341-352.
[6] Bord R J, Fisher A, O'Connor R E. Public perceptions of global warming: United States and internation perspectives[J]. Climate Research, 1998, 11(1):75-84.
[7] Wardekker J A. Ethics and public perception of climate change: exploring the christian voices in the US public debate[J]. 2009, 19(4):512-521.
[8] TAN Lingzhi, MA Changfa. Farmers' Perceptions of Climate Change and Their Adapting Behaviors in Arid Region of China [J]. Bulletin of Soil and Water Conservation, 2014, 34(1):220-225.

[9] Reddy S, Parker A, Hyman J, et al. Image browsing, processing, and clustering for participatory sensing: lessons from a DietSense prototype[J]. Emnets '07 Proceedings of Workshop on Embedded Networked Sensors, 2007:13-17.

[10] Eisenman S B, Miluzzo E, Lane N D, et al. BikeNet: A mobile sensing system for cyclist experience mapping[J]. Acm Transactions on Sensor Networks, 2009, 6(1):6.

[11] Rai A, Chintalapudi K K, Padmanabhan V N, et al. Zee:zero-effort crowdsourcing for indoor localization[C]// 2012:293-304.

[12] Rana R K, Chou C T, Kanhere S S, et al. Ear-phone: an end-to-end participatory urban noise mapping system[C]// ACM/IEEE International Conference on Information Processing in Sensor Networks. ACM, 2010:105-116.

[13] Dutta P, Aoki P M, Kumar N, et al. Common Sense:participatory urban sensing using a network of handheld air quality monitors[C]// International Conference on Embedded Networked Sensor Systems, SENSYS 2009, Berkeley, California, Usa, November. DBLP, 2009:349-350.

[14] NAN Wenqian, Guo Bin, Chen Huihui, et al. A Cross-Space, Multi-Interaction-Based Dynamic Incentive Mechanism for Mobile Crowd Sensing [J]. Journal of computer, 2015, 38(12): 2412-2425.

Study on Tibetan Word Vector based on Word2vec

Ning Yang[1], Guanyu Li[1*], Hailan Ding[1], Chunwei Gong[1]

[1]Key Laboratory of National language Intelligent Processing Gansu Province, Northwest Minzu University, Lanzhou, China

*guanyu-li@163.com

Abstract. This paper uses Word2vec to study Tibetan word vector. Word2vec is optimized by two methods: Hierarchical Softmax and Negative Sampling in CBOW and Skip-gram models. Through the training of neural network, the words in Tibetan sentences are converted into vector form. Word2vec transforms the Tibetan text content processing into a simple vector space operation, calculates the similarity in the vector space, and then obtains the semantic similarity of the text, providing an accurate word vector for the training of the language model.

1. Introduction

Speech recognition is an important field of artificial intelligence research, and the quality of the language model to a great extent determines the accuracy of speech recognition in the language model training, which requires the vectorization of words in a Tibetan text corpus in advance compared to the traditional one-hot word vector. Word2vec is a distributed representation in the form of a word vector, according to the relationship of context to define the term vectors, each of those dimensions is real Numbers in a row, and high correlation of the word has a closer distance. Compared with Word2vec, one-hot word vector has many problems, such as dimension disaster, data sparse, the isolation of semantics and the difficulty of fuzzy matching, etc., with stronger expressive force and more internal features of corpus data.

Tibetan is a very important minority language in China. According to incomplete statistics, more than 6.4 million people use Tibetan. This research can provide theoretical and data support for Tibetan phonetic teaching, phonetic technology engineering application, and perfect the research of Tibetan phonetic direction. At the same time, strengthening the study of Tibetan pronunciation can not only effectively solve the problem of language barriers between Tibetan areas and other regions of China, which is promoting inter-ethnic exchanges and mutual understanding among ethnic groups, but also promote the Tibetan area's economy, science and technology.

2. Word2vec structural model

This section introduces the word vector in distributed representation form used by Word2vec, which introduces the use and optimization of Word2vec. Finally, the specific flow and structure of Word2vec's CBOW model and Skip-Gram model are introduced, as well as the performance comparison between the two models.

2.1. The distributed representation word vector

Distributed representation putted forward by Hinton as early as in 1986, by training every word map into a fixed length of 'short' vector (as opposed to one-hot vector 'long'), and all these constitute a word vector space, and each vector can be seen as a point in the space. The introduction of 'distance' on the

Content from this work may be used under the terms of the Creative Commons Attribution 3.0 licence. Any further distribution of this work must maintain attribution to the author(s) and the title of the work, journal citation and DOI.
Published under licence by IOP Publishing Ltd

space, and then judged by the distance between the word and the word on the lexical and semantic similarity. Distributed representations of words in a vector space help learning algorithms to achieve better performance in a natural language processing task by grouping similar words[1]. Word2vec uses this vector representation of the distributed representation. It is a low-dimensional real number vector in the form: [0.832, - 0.169, -0.397, 0.339, -0.894...]. Word vectors are widely used, including computing similarity (searching for similar words, information retrieval), which is the input of SVM/LSTM and other models (Chinese word segmentation, nomenclature recognition), sentence representation (affective analysis), document representation (document subject identification) and so on.

2.2. Introduction to Word2vec

Word2vec is a tool that Google has open sourced in 2013 to express words as real-value vectors. The models are CBOW and Skip-Gram.Word2vec uses a one-layer neural network to map the word vector in one-hot form to the word vector in distributed representation form. Through training, the Tibetan text content processing is simplified to vector operation in k-dimensional vector space, and the similarity degree in vector space can be used to express the similarity of text semantics. Therefore, the word vector output is used to do a lot of NLP related work, such as the Tibetan language clustering studied in this paper. Our technique is inspired by the recent work in learning vector representations of words using neural networks[2]. The words processed by Word2vec can be used to add, subtract, multiply and divide vectors so that the expression of words can be vectorized, and the similarity between them can be calculated more easily.

Word2vec's training algorithm is a vector corresponding to a word. That is, there is only a single interpretation (one-to-one mapping) of the semantic attribution of words. Word2vec's analysis of semantics is not based on common sense, nor is it based on the grammatical rules of natural languages to construct models, but on the basis of the rules of lexical statistics to make a statistical hash of lexical units. Actually according to the result of clustering proceeds to directly obtain one kind of classification rule. So Word2vec vector's clusters proceed from the actual semantic orientation and the analysis of inner value, which requires the expectation for further analysis and research on this field to arrive at accurate conclusions.

2.3. Word2vec model structure

Word2vec is an intermediate product of the neural probabilistic model, which is a byproduct of the neural network to learn a language model. Specifically, a language model refers to CBOW and Skip-gram. And two language model specific learning process method of approximation of two will be used to reduce the complexity of the Hierarchical Softmax and Negative Sampling.

2.3.1 Continuous word bag model CBOW. The CBOW model predicts the central words according to the words around the central word y_j, while the Skip-Gram model predicts the surrounding words according to the central word x_k. Below are the CBOW model structure diagram and Skip-Gram model structure diagram respectively, as shown in Figure 1.

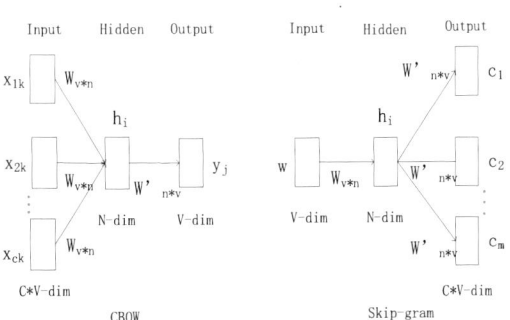

Figure 1. Model structure diagram

The CBOW predicts the current word based on the context, and the Skip-Gram predicts surrounding words given the current word[3]. Contrary to CBOW, Skip-Gram model is based on the input at the center of the word to predict around the word.

- Input layer: Enter the one-hot coding of the context word around the target word, where the window size is C and the vocabulary size is V.
- Hidden layer: The vector sum averaging vector as the hidden layer, the size is 1 * N and dimensions for N.
- Be one-hot coding input vector through a weight matrix W, the dimensions of W is V * N which connected to the hidden layer.
- Output layer: Be one-hot code word y

Hidden layer by the weight of a , N * V matrix W ', connected to the output layer. Finally, vector 1*v is obtained. After the activation function softmax normalization, the v dimension probability distribution is obtained. The word with the highest probability is the predicted intermediate word y.

In the Skip-Gram model, we are given a corpus of words w, and their contexts c_k, c is 1 to m. We consider the conditional probabilities is p(c|w) and given a corpus Text, and the goal is to set the parameters θ of p(c|w; θ), so as to maximize the corpus probability in formula (1):

$$\arg \max_{\theta} \prod_{w \in Text} \left[\prod_{c \in C(w)} p(c \mid w; \theta) \right] \tag{1}$$

C(w) is the set of contexts of word w. Alternatively as shown in formula (2):

$$\arg \max_{\theta} \prod_{(w,c) \in D} p(c \mid w; \theta) \tag{2}$$

D is the set of all word and context pairs we extract from the text[4].

2.4. CBOW model process example

It is simple to assume that the present corpus is a document with only four words: { དགེ བགྱུར སྐྱོབ གཉིས }, the target word is སྐྱོབ, the window size is 2. According to the word དག, བགྱུར and གཉིས aim to predict a central word སྐྱོབ. And what we want to predict is that the word སྐྱོབ, The Specific processes are shown in Figure 2.

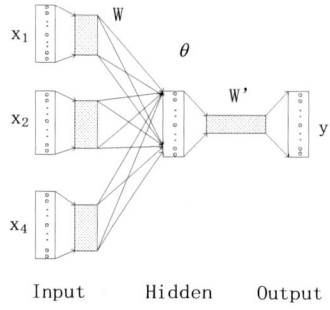

Figure 2. CBOW model

Example corpus = { དགེ བགྱུར སྐྱོབ གཉིས }
Window size: 2
Target word: སྐྱོབ
Context word: དགེ , བགྱུར , གཉིས

- x_1: W གེ = [1,0,0,0]
  x_2: W གྱུར = [0,1,0,0]
  Label is x_3: W ྐྱ =[0,0,1,0]
  x_4 : W ཉིས = [0,0,0,1]

- Initialize:

$$w = \begin{bmatrix} 1 & 2 & 3 & 0 \\ 1 & 2 & 1 & 2 \\ -1 & 1 & 1 & 1 \end{bmatrix}$$

Ex: $W^{x_2} = [0,1,0,0]$

$Wx_2 = v_2$

$$\begin{bmatrix} 1 & 2 & 3 & 0 \\ 1 & 2 & 1 & 2 \\ -1 & 1 & 1 & 1 \end{bmatrix} \begin{bmatrix} 0 \\ 1 \\ 0 \\ 0 \end{bmatrix} = \begin{bmatrix} 2 \\ 2 \\ 1 \end{bmatrix}$$

- $$\frac{V1+V2+V4}{3} = \theta$$

$$\frac{1}{3}\left(\begin{bmatrix} 1 \\ 1 \\ -1 \end{bmatrix} + \begin{bmatrix} 2 \\ 2 \\ 1 \end{bmatrix} + \begin{bmatrix} 0 \\ 2 \\ 1 \end{bmatrix} \right) = \begin{bmatrix} 1 \\ 1.67 \\ 0.33 \end{bmatrix}$$

- Initialize:

$$W' = \begin{bmatrix} 1 & 2 & -1 \\ -1 & 2 & -1 \\ 1 & 2 & 2 \\ 0 & 2 & 0 \end{bmatrix}$$

$W'\theta = U_0$

$$\begin{bmatrix} 1 & 2 & -1 \\ -1 & 2 & -1 \\ 1 & 2 & 2 \\ 0 & 2 & 0 \end{bmatrix} \begin{bmatrix} 1.00 \\ 1.67 \\ 0.33 \end{bmatrix} = \begin{bmatrix} 4.01 \\ 2.01 \\ 5.00 \\ 3.34 \end{bmatrix} \begin{matrix} U1 \\ U2 \\ U00 \\ U4 \end{matrix}$$

- Softmax(U_0)=y

$$soft\max\left(\begin{bmatrix} 4.01 \\ 2.01 \\ 5.00 \\ 3.34 \end{bmatrix} \right) = \begin{bmatrix} 0.23 \\ 0.03 \\ 0.62 \\ 0.12 \end{bmatrix}$$

Use the above five steps to get the most likely target words. The probability of ཚོང is 0.62. We desire probability generated to match the true probability $x_3[0,0,1,0]$. And use gradient descent to update W and W'.

3. Word2vec realize

This part mainly introduces the implementation of wrord2vec, including the main work of data preparation and the final results of the experiment.

3.1. data preparation

● Letters from the Tibetan WeChat Official Account crawl a large number of text.

ISPECE   IOP Publishing

- Getting rid of complete gibberish, in both Chinese and English and digital work, which is getting the exact Tibetan sentences.
- Segmentation of Tibetan text corpus separated by Spaces or TAB key to the corpus.
- Named after copy to Word2vec trunk, which runs on the server.

3.2. Word2vec main work
There are some main words in Word2vec:
- Segmentation, stem extraction, and morphological reduction. The difficulty is to break each sentence down into an array of words.
- Construct the dictionary, the word frequency statistics. Going through all the text to find all the appear words and count how often they appear.
- Construct a tree structure. Huffman tree is constructed according to occurrence probability, and all categories should be in leaf nodes.
- The generated code in the binary code to reflect the positions of the nodes in the tree.
- Initialize all the nodes in the middle of the vector and the vector of words in a leaf node. In the leaf node, the word vector of each word is stored as the input of the neural network. The non-leaf nodes store intermediate vectors, which together with the input determine the classification results for the parameters used in the hidden layer in the neural network.
- Training and word vector in the middle. For a single word, which changes only the intermediate vector of the node on its path at most, and the other nodes remain the same.

4. Experimental result

4.1. Semantic result
After entering Tibetan words གཅེས, which is 1499 places in the corpus text, and list a number of words that are semantically relevant གཅེས, because Word2vec calculates the cosine value, the range of distance is between 0 and 1. The larger value means the higher correlation between the two words. So the more words on top, the more closely related to གཅེས. The result is shown in Figure 3 below.

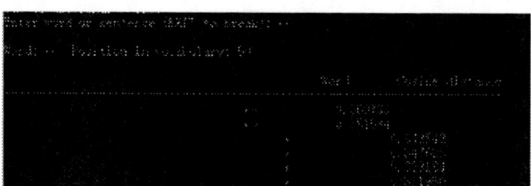

Figure 3. Semantic result

4.2. Distance result and Cluster result
Figure 4 is the vectors.bin model file after training. It shows each word and its corresponding vector in Tibetan corpus. The vector dimension is the size of the parameters set during the training. By sorting a file for keyword clustering, the partial clustering results are captured as shown in Figure 5.

Figure 4. Distance result   Figure5. Cluster result

5. Conclusions

2563

In this paper, we present two modifications to the original models in Word2vec that improve the word embeddings obtained for syntactically motivated tasks[5]. Word2vec constructs a multi-layer neural network, then obtains the corresponding input and output in Tibetan text corpus, and constantly modifies the parameters of the neural network during the training process. Finally, the word vector is obtained. The specific Word2vec uses two training models, CBOW and Skip-Gram, and combines hierarchy softmax and Negative Sampling to optimize the parameters, which obtains the distributed word vector. In short, Word2vec greatly simplifies the network structure, at the same time, because the simple network structure brings low computational complexity, we can calculate the very accurate

Acknowledgments
Supported by the Fundamental Research Funds for the Central Universities (Research on Unified Acoustic Models of Mandarin and Tibetan, 31920170145)
high-dimensional word vector on the larger dataset.

References
[1] Mikolov T, Sutskever I, Chen K, et al. Distributed Representations of Words and Phrases and their Compositionality [J]. 2013,26:3111-3119.
[2] Le Q V, Mikolov T. Distributed Representations of Sentences and Documents[C]//ICK/IL. 2014, 14: 1188-1196.
[3] Mikolov T, Chen K, Corrado G, et al. Efficient Estimation of Word Representations in Vector Space [J]. Computer Science,2013.
[4] Goldberg Y, Levy O. word2vec Explained: deriving Mikolov et al.'s negative-sampling word-embedding method[J]. arXiv preprint arXiv: 1402.3722, 2014.
[5] Ling W, Dyer C, Black AW, et al. Two/too simple adaptations of word2vec for syntax problems[C]//Proceedings of the 2015 Conference of the North American Chapter of the Association for Computational Linguistics: Human Language Technologies. 2015: 1299-1304.

Analysis of Seismic Anomalies of the Jiuzhaigou Earthquake

Xiaoying Jiang，Xiangzeng Kong*，Gongde Guo

College of Mathematics and Informatics, Fujian Normal University, Fuzhou ,China

xzkong_fjnu@163.com

Abstract. On August 8, 2017, a devastating mega-earthquake of magnitude 7.0 struck the Jiuzhaigou County, Sichuan Province, China, resulting in significant casualties and property damage. Actually, before an earthquake occurs, a lot of abnormal information often appears suddenly. For example, the thermal infrared radiation suddenly increased in a short time. However, it is hard to identify and analyze valuable anomalous information from large-scale data. To solve this problem, this paper proposes a random walk method to analyze the outgoing long-wave radiation (OLR) anomaly before an earthquake. The National Oceanic and Atmospheric Administration (NOAA) data are used to process and analyze the OLR data before and after the Jiuzhaigou earthquake. In order to prove the effectiveness of our algorithm, we use this method to analyze the OLR data of Jiuzhaigou region from 2006 to 2017, and compare the OLR anomaly difference between the year of earthquake occurrence and a normal year. Experimental results show that there are obvious abnormal changes in OLR data in the year of earthquake. In order to further verify the applicability of this algorithm, we also used this algorithm to analyze the OLR data of another 10 earthquakes with magnitude 5.0 or above in western China in 2017. The results show that the proposed algorithm can extract OLR anomaly information in most earthquakes. Therefore, our algorithm can effectively extract the pre-earthquake abnormal information in OLR data.

1. Introduction

On August 8, 2017, a magnitude 7.0 earthquake occurred in Jiuzhaigou, Sichuan Province, China, causing heavy casualties and property losses. What's more, China is a country with many earthquakes. In particular, in the western part of China, many earthquakes occur each year, posing a serious threat to people's lives and property in the western region.The identification and analysis of abnormal information before the earthquake has attracted the attention of more and more researchers. And satellite remote sensing technology is widely used by scientists to the research on the relationship between thermal infrared remote sensing and earthquakes. They hope to excavate useful pre-earthquake anomaly information and achieve more accurate prediction earthquake. They hope to improve the reliability of seismic analysis and provide more possibilities for predicting earthquakes by more effectively identifying and analyzing useful pre-earthquake anomaly information.

Kong et al.[1] analyzed seismic anomalies within outgoing long-wave radiation (OLR) data observed by satellites from 2006 to 2013 for Wenchuan and Lushan earthquakes and four comparative study areas. Zhou et al.[2] detected the OLR data and total electron content (TEC) data by using the standard deviation threshold method and the quaternary-decay method in the Nepalese earthquake in 2015. They found that in the vicinity of the epicenter within 3 days of the earthquake, satellites remote sensing OLR abnormal increase phenomenon. At the same time, the TEC in the ionosphere near the epicenter shows significant positive anomalies and magnetic conjugation.Reference [3] presented a statistical study upon the geomagnetic data observed at Kakioka station, Japan, during 2001-2010.

They found that the ultralow frequency(ULF) seismomagnetic phenomena at Kakioka clearly contain precursory information and have a possibility of improving the forecasting of large earthquakes. Xiong et al.[4] proposed a statistical analysis method based on the Robust Satellite data analysis technique to detect seismic anomalies within the NOAA OLR dataset about 3376 earthquake cases from September 01, 2007 to May 23, 2015 based on spatial/temporal continuity analysis. However, although OLR data has been used as a new index of earthquake precursors anomalies, most of the existing achievements are mainly based on the research methods of geography, and the analysis of the results is mainly based on manual observation. Therefore, the experimental results are directly influenced by experts in the field and the influence of experience has great uncertainty on the judgment of the related anomalies before the earthquake.

The application of data mining techniques in the field of earthquake anomaly analysis has been explored by more and more scholars.For example, Xiong et al.[5] used the wavelet transformations and spatial/temporal continuity analysis of wavelet maxima to analyze the continuous OLR. The experimental results show that there are singularities in OLR data that correspond to seismic precursors. Wu et al.[6] studied the surface latent heat flux (SLHF), thermal infrared radiation (TIR), outgoing longwave radiation (OLR), diurnal temperature range (DTR), atmospheric temperature, and skin temperature, for GEOSS-based earthquake anomaly recognition (EAR) by analyzing the M7.3 Yutian earthquake occurred in 2008, the M8.0 Wenchuan earthquake occurred in 2008, and the M7.1 Christchurch earthquake occurred in 2010 in New Zealand. The results showed that the obtained compositive thermal anomaly has a significant effect on earthquake prediction. However, most of the methods can only find a very small decrease or increase in the signal. For exmaple, the wavelet-based data mining technique was able to discover singularities from OLR data. However, it could not provide the measurements of change degree of OLR data. In addition, most of the experimental data in these studies are too limited. In this paper, we will analyze the OLR data from 2006 to 2017 to exploring the abnormal changes before the earthquake.

Our main contributions are as follows: we propose a data mining algorithm called random walk method, which is based on Martingale theory [7]. This article applies data mining technology to the field of seismic research. We analyze the OLR data before and after the Jiuzhaigou earthquake. The comparing experiences have been done using the OLR data from 2006 to 2017 to look for the correlation between OLR data and earthquakes. And the experimental results show that there are obvious abnormal changes in OLR data in the year of earthquake. We also used this algorithm to analyze the OLR data of another 10 earthquakes with magnitude 5.0 or above in western China in 2017. The results show that the proposed algorithm can extract OLR anomaly information in most earthquakes. Therefore,the method proposed in this paper is suitable for studying OLR anomalies before earthquakes. Meanwhile,it has an important reference for applying the computer technology in the field of geography science.

2. Data and methodology

2.1. Related data sources
The OLR data of 2.5° longitude× 2.5° latitude grids in this paper is provided by the NOAA of the United States.OLR data includes daytime data and nighttime data. In order to minimize the interference of daytime human activities and environmental factors such as sunlight, this paper uses nighttime data [8]. All the selected earthquakes (11 in total) are those with magnitudes ≥ 5.0 and occurred in western China in 2017.The details of these earthquake events are shown in Table 1.

Table 1. List of the Selected Earthquake Events

Date (y-m-d)	Latitude (° N)	Longitude (° E)	Depth (km)	Magnitude (JMA)	Location
17-02-01	30.67	83.34	8	5.0	Zhongba
17-03-27	25.89	99.80	12	5.1	Yangbi
17-05-11	37.58	75.25	8	5.5	Tashkurgan
17-08-08	33.20	103.82	20	7.0	Jiuzhaigou
17-08-09	44.27	82.89	11	6.6	Jinghe
17-09-16	42.11	83.43	6	5.7	Kuche
17-09-30	32.27	105.00	13	5.4	Qingchuan
17-11-18	29.75	95.02	10	6.9	Milin
17-11-18	29.88	94.92	6	5.0	Bayi
17-12-07	35.69	77.46	87	5.2	Yecheng
17-12-20	29.88	95.08	6	5.0	Bayi

2.2. Random walk method

This paper proposes a random walk algorithm to identify and analyze the thermal infrared anomaly characteristics through a certain signal amplification. And Martinggale theory [7] is applied to measure the changing degree of the OLR data.

In this paper, dataset $Z = \{z_1, \cdots, z_{i-1}\}$ represents historical data and z_i represents the currently calculated data point. Obviously, when the activity of the geological plate is relatively stable, the sample distribution in Z should be maintained at a relatively stable level with some similar characteristics between the samples. Therefore, the data set $Z \cup z_i$ can be treated as a time series.

The first step is use a fixed size window to determine whether the value of the current data point is greater than or equal to the value of the previous data point. The size of this window is marked as ws. If the value of the current data point is greater than or equal to the value of the previous data point, the direction of the current data point is defined as leftward; otherwise, the direction of the current data point is defined as rightward. This is based on the principle of equal probability distribution of all data points.

Then we select ws data points before the current data point. The probability distributions of the current data point direction to the left or right are respectively obtained by the probability distribution formula (1):

$$rw(ws) = p^{\frac{n+i}{2}} (1-p)^{\frac{n-i}{2}} \times C(n, \frac{n+i}{2})$$ (1)

where $n = ws$, $i \in [-n, n]$, $p = 0.5$, and $C(n, \frac{n+i}{2})$ represents a combination in mathematics.

Then we should calculate the Martingale value according to the formula(2):

$$M_{n,k}^{(\varepsilon)} = \prod_{i=1}^{n} (\varepsilon \hat{p}_{i,k}^{\varepsilon-1})$$ (2)

where $\varepsilon \in [0,1]$, $\hat{p}_{i,k}$ is computed according to the formula (4).

The k-means clustering method is adopted to analyze the data, because OLR data is a time series. The strangeness value s_i can be defined as

$$s_i = s(Z, z_i) = \left\| z_i - m \right\| \tag{3}$$

where m is the average of all the data points and $\left\| \bullet \right\|$ represents the Euclidean distance.

$\hat{p}_{i,k}$ is computed according to the following formula:

$$\hat{p}_{i,k}(Z \bigcup \{z_n\}, \theta_n) = \frac{\#\{j \mid s_j > s_i\} + \theta_{i,k} \#\{j \mid s_j = s_i\}}{i} \tag{4}$$

where $\#\{\ \}$ is a function that returns the number of samples that satisfy a given condition. For example, $\#\{j \mid s_j > s_i\}$ is the number of j satisfying $s_j > s_i$, where s_j is the strangeness measure mentioned in function (3). $\theta_{i,k}$ is a random number which is selected from the interval $[0,1]$, $i = 1, 2, \cdots n$, and s_j is the strangeness measurement for z_j, $j = 1, 2, \ldots, i$ with the initial Martingale value $M_0^\varepsilon = 1$.

We use a CD value to reflect the change degree of the current data point as follows:

$$CD_n^{(\varepsilon)} = \frac{\sum_{k=1}^{100} M_{n,k}^{(\varepsilon)}}{100} \tag{5}$$

where 100 is the number of Martingale value which are used to calculate the CD value. According to [9], we set $\varepsilon = 0.82$ and initialize $CD_0 = 1$. CD can reflect the average change degree of the data points before the current data point. Therefore, the proposed method can ignore some isolated false changes which may be caused by instrumental malfunctions or other faults.

If the computed CD_n value is very high and exceeds the threshold h, this could correspond to an abnormal condition. The condition $CD_n \geq h$ can be used to determine whether a change has been detected for suitable values of h. Therefore, the stopping rule is as shown in expression (6):

$$CD_n \geq h \tag{6}$$

the proposed algorithm should be restart when a abnormal condition is found. Then continue to calculate the remaining features of random walk at each point to obtain a new feature sequence and calculate the remaining points abnormality.

3. Experimental results and analysis

The OLR data used covers the area of 28.75°N–36.25°N and 100°E–107.5°E. And the Jiuzhaigou earthquake's epicenter is in the cell 5 which covers latitude 31.25°N–33.75°N, and longitude 102.5°E–105°E. The data are the night-time OLR data from March 1, 2017 to February 28, 2018. The red vertical line in the following figures indicates the time of the earthquake. Experimental results as shown in Fig.1. The CD value began to increase significantly two months before the earthquake, and a peak appeared during this period. A few days before and after the earthquake, the anomaly suddenly decreased, but another peak appeared again within two months after the earthquake. And the maximum value appeared around September 20. On September 30, a magnitude 5.4 earthquake occurred in Qingchuan County, Sichuan Province. The maximum peak appeared just about 10 days before the earthquake. The author speculated that the second peak appeared related to the Qingchuan earthquake.

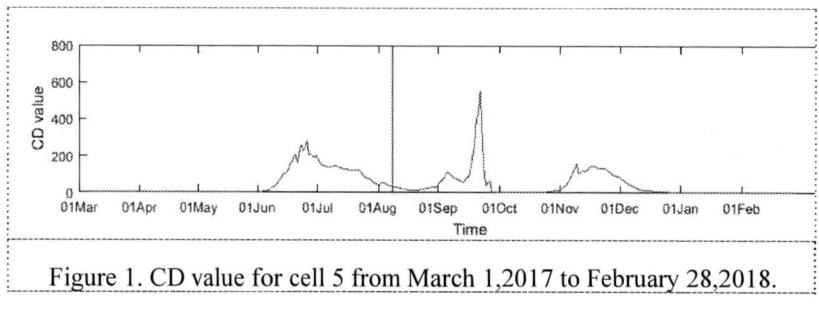

Figure 1. CD value for cell 5 from March 1,2017 to February 28,2018.

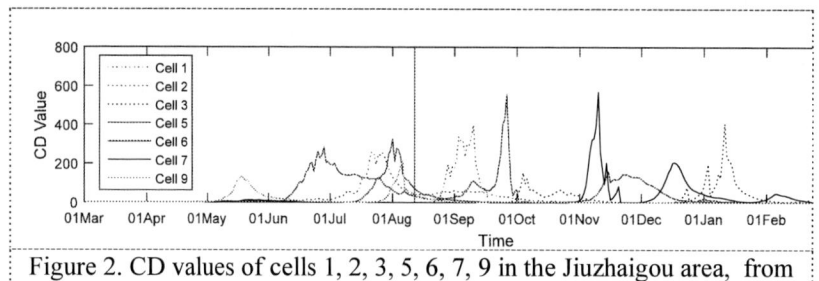

Figure 2. CD values of cells 1, 2, 3, 5, 6, 7, 9 in the Jiuzhaigou area, from March 1,2017 to February 28,2018.

Fig.2 shows the comparative analysis of cells 1, 2, 3, 5, 6, 7 and 9 in the Jiuzhaigou area from March 1, 2017 to February 28, 2018. The CD values are very small in cells 4 and 8 before the earthquake time, so they are not presented in Fig.2. It can be seen from Fig.2 that even though the maximum values of some cells did not occur at the time of the earthquake or before the earthquake occurred, the OLR of all cells increased significantly before and after the earthquake. The maximum values of the seven cells in Fig.2 are greater than 200, and even the maximums of five cells exceeds 300. For example, the maximum value of cell 5 is 556.8. The results show that the OLR data showed obvious anomalies before and after the earthquake. And our algorithm can effectively extract the abnormal information in the OLR data.

We also study the OLR changes in the Jiuzhaigou area from 2006 to 2016 to compare the OLR difference between the earthquake occurrence year and the normal year, and explore the relationship between OLR data and earthquake. The results are shown in the Table 2. In order to minimize the error caused by accidental factors, the maximum value here is the result after subtracting the average of 12 years. The average CD maximum value of the Jiuzhaigou area is 287.8 from 2006 to 2017, and the CD maximum value of 2017 when the Jiuzhaigou earthquake occurred is 468.2, so it is 1.63 times the average CD maximum value from 2006 to 2017 year. As can be seen from Table 2, the CD value in 2017 is larger than in other years, except for 2006 and 2007. A larger CD value in 2006 might be due to the drought in Sichuan Province which occurred just once in 100 years. Like the earthquakes, the drought could cause the Earth's skin temperature to change which could in turn affect the OLR values. As for the reason why the CD value in 2007 is abnormally large, it has not yet been determined, and it might be a false positive.

Table 2. Maximum CD values of the cell 5 in Jiuzhaigou area from 2006 to 2017

Year	2006	2007	2008	2009	2010	2011	2012	2013	2014	2015	2016	2017	average
Max value	540.6	883.5	2.1	169.3	158.7	178.7	371.7	287.2	164.0	193.4	36.9	468.2	287.8

In order to further prove the applicability of the proposed algorithm, we apply the algorithm to another 10 earthquakes of magnitude 5 or higher in western China in 2017. We found that seven of the earthquake cases detected OLR anomalies before the earthquake, and only three earthquake cases detected almost no abnormal OLR changes. The applicability of this algorithm is 70% and the false negative rate is 30%.

4. Discussion and conclusion

This paper proposed a random walk algorithm to detect pre-earthquake OLR anomalies in Jiuzhaigou area and another 10 earthquakes which magnitude ≥ 5.0 and occurred in western China during 2017. Generally, when an earthquake occurs, the closer the earthquake is, the more obvious the abnormality is. Although the anomalies from the epicenter are not necessarily the largest, there must be significant OLR anomalies in and around the epicenter before and after the earthquake. And there are several years where the anomalies are greater than that of the earthquake year. This might be due to other factors such as high temperatures and droughts that affect the thermal infrared values. Therefore, our method can effectively detect pre-earthquake anomalies and is suitable for studying pre-earthquake anomalies in most earthquake cases in western China. In the study, we also found that almost no OLR anomalies were detected in some earthquake years. There are also anomalies detected by earthquakes with small magnitudes that are larger than those with large magnitudes, so the relationship between magnitude and anomaly needs to be further explored.

Acknowledgments

This work was supported by the National Natural Science Foundation of China under Grant No. 41601477, the Leading Project in Fujian Province of China under Grant No.2015Y0054 and NOAA for making OLR data available for various research communities and anonymous reviewers for their constructive suggestions.

References

[1] KONG X, BI Y and GLASS D H 2015 Detecting seismic anomalies in outgoing long-wave radiation data *IEEE Journal of Selected Topics in Applied Earth Observations and Remote Sensing* vol 8 no.2 pp 649-660.

[2] ZHOU Q B,TANG J and CNNC 2016 Analysis of Anomalies of Pre-Earthquake Based on OLR and Ionosphere Data of 2015 Mw 7.8 Earthquake in Nepal *Journal of Geomatics Science and Technology* vol 33 pp 458-463.

[3] Han P , Hattori K , Hirokawa M , et al. 2014 Statistical analysis of ULF seismomagnetic phenomena at Kakioka, Japan, during 2001-2010 *Journal of Geophysical Research: Space Physics vol* 119 no.6 pp 4998-5011.

[4] Xiong P and Shen X H 2016 Outgoing Longwave Radiation anomalies analysis associated with different types of seismic activity *Advances in Space Research* vol 59 no.5.

[5] Xiong P , Bi Y and Shen X 2009 Study of Outgoing Longwave Radiation Anomalies Associated with Two Earthquakes in China Using Wavelet Maxima *Hybrid Artificial Intelligence Systems, International Conference, Hais, Salamanca, Spain, June.* DBLP.

[6] Wu L , Qin K , Liu S 2012 GEOSS-Based Thermal Parameters Analysis for Earthquake Anomaly Recognition *Proceedings of the IEEE* vol 100 no.10 pp 2891-2907.

[7] HO S S and WECHSLER H 2010 *A Martingale framework for detecting changes in data streams by testing exchangeability* vol 32 pp 2113-2127

[8] ZHAO J 2011 Identification and Analysis of Pre-earthquake Thermal Anomalies--take six earthquakes in western China for example *Intelligent Information Technology Application Association.Proceedings of 2011 International Conference on Ecological Protection of Lakes-Wetlands-Watershed and Application of 3S Technology(EPLWW3S 2011 V3)*

[9] VORK V, NOURETDINOV I and GAMMERMAN A 2003 Testing exchangeability on-line *Proc of the 12th International Conference on Machine Learning* San Francisco: Morgan Kaufmann pp 768-775.

Research on the Development Orientation and Thinking of New Media in State-owned Enterprises

Wang Han, Wang Youzi , Xia Liyu

State Grid Energy Research Institute Co.,Ltd., Beijing, China

wanghan@sgeri.sgcc.com.cn

Abstract. In recent years, the new media platform has developed rapidly and has become an important channel and platform for state-owned enterprises to carry out brand promotion. State-owned enterprises have made full use of the new media platform to spread quickly, spread widely, and have rich modes of communication. They have made a lot of attempts and carried out extensive publicity work on new media platforms such as Weibo, WeChat and vibrato. However, in the process of building a new media system, due to institutional, institutional and personnel reasons, state-owned enterprises still have major deficiencies in the construction of new media, and have not yet fully adapted to the current development of the propagation mode and technology of new media. This paper analyzes the shortcomings of state-owned enterprises' new media construction, and puts forward the development direction and feasibility suggestions of state-owned enterprises' new media construction.

1. Introduction

With the continuous development of information technology, the mobile Internet has begun to become an important channel for information dissemination. Various new media have emerged, and the social communication effect has greatly improved the speed and effect of information dissemination. With the advent of the new media era, the company has the information dissemination channel directly facing the audience. With the communication advantage of the new media platform to further enhance the company's brand propaganda ability and expand the brand influence of the company, it is the inevitable way for state-owned enterprises to shape the world-class brand.

2. Insufficient work in the construction of new media in state-owned enter-prises

2.1. New media account settings need to be further integrated

First, some new media account settings have a high degree of duplication, and there is no significant difference in pushing content and audience. In the new media era, we emphasize the accurate push of information. Therefore, the same company sometimes sets different new media accounts according to the geographical, age, gender, and hobbies of the audience to improve the accuracy of information transmission and ensure the effectiveness of information dissemination. In order to refine the customer base, some grassroots units and corporate media have targeted to meet the needs of different customer groups. On the basis of setting up official Weibo and WeChat public account in accordance with company regulations, other new media accounts have been set up to promote from different levels. However, in the account content, function, and audience, the operating entity of the account has not been able to make a good distinction. In particular, the company news, the service category, and the common sense category information often appear in each account, although the contents are different, but they failed to make the account's own characteristics according to the positioning of the account, plus the main audience of the account is the electricity customer of the region, the users' portrait does not reflect the obvious personalized characteristics, thus for the audience In fact, the degree of discrimination between the various accounts is not obvious, and it is difficult to achieve the expected communication effect.

Second, the number of internal communication accounts is too large, and the content quality is not high, so that employees should be overwhelmed and waste resources. With the development of new media, the threshold for media operations has been reduced. It seems that the opening of a new media platform for the use and communication of its own employees has become a trend, except that all grassroots units have opened new media to disseminate internal information of the company. In addition to the account number, there is also a professional department to open an account for a professional job. Some professional departments did not fully demonstrate the necessity and feasibility of opening an account before opening a new media account. The new operation not only caused certain work pressure on the staff of the department, but also the spreading of accounts becomes a "political task" for the employees, which violates the original intention of setting up public accounts to facilitate employees to obtain information and communication. Moreover, due to the lack of professionalism of the operators, the lack of content richness, and the fact that some accounts actually receive less attention, the number of "zombie fans" is large, which wastes a lot of operating costs.

Third, the accounts are mostly concentrated on the Weibo and WeChat platform. The communication channels are slightly single, and the information dissemination is not covered to all target groups. As the two mainstream new media communication platforms in China, Weibo and WeChat have advantages that other new media platforms cannot match in terms of number of users and communication methods, but with the rise of knowledge sharing platforms, short video platforms, users began to show a relatively obvious differentiation trend, and the user characteristics of various platforms became more and more prominent. In the short-term video platforms such as the "Knowledge" and "vibrato" popular in the youth circle, there are very few account holders, and a large number of users are lost invisibly. The scope and effect of communication have also been limited.

2.2. New media matrix needs further improvement
First, the account functions of various new media platforms are not clearly defined, and the platform features and user characteristics are not accurately grasped. All kinds of new media platforms have certain differences in platform positioning, target users, expressions, push rules, etc. In order to ensure the dissemination of information, when publishing information, we must pay attention to the target audience of the information, select appropriate platforms, and use platform-style expressions. However, the function positioning of many platforms for different platforms is not clear at present, and there is no obvious distinction between the categories of published content. Some units have completely or even directly forwarded the content of Weibo and WeChat information, and did not consider the performance of various platforms or the difference between form and user group. Most of the accounts lack a unique "personal design", the expressions and content of the various platforms are lacking in characteristics, most of them are "mix and match" in various styles. They are not unique, impressive, and can be clearly distinguished from others. The characteristics of the platform account make the users' feelings on the company accounts on various platforms basically the same or similar, and even produced a stereotype. In the long run, there may be cases where the users' attention continues to decline, or the users only pay attention to the function of the account in the customer service, and ignore other content.

Second, the integration of the media has not been well played, and there is a lack of coordination and interaction between the accounts and platforms, failing to form a "corner posture". In addition to further segmenting the customer base to carry out accurate push, the new media matrix can also quickly expand the scope of communication and promote a large public opinion, but the company's current new media matrix has not yet been able to play. On the one hand, there are two extreme phenomena on the major communication themes that need to be promoted by various media: one is that the communication strategies are similar, the content and form are similar, and there is almost no difference in the information published in various media. Such communication is not only insufficient to attract the attention of single platform users, but also makes multi-platform users feel the information duplication, resulting in boring and tired feelings. The other is the opposite. Various media platforms operate completely independently, and they are not in line with each other. The media formed a good interaction, sometimes for the spread of the same event, the information published on different platforms also appeared to be unrelated, and did not form an overall brand effect; on the other hand, there is a lack of interactivity between the accounts at all levels, focusing only on the promotion of relevant information of the unit and the region, ignoring the linkage with other units, which make the company fails to form a rolling propagation situation.

2.3. New media operations process needs to be further improved

First, the new media operation part of the work process is missing, and does not form a complete operational management chain. The new media operation is not a simple regular planning topic and release content, but also contains a lot of research and analysis, user interaction, hotspot tracking and so on. At present, most units have formed an independent new media planning and communication process, but there are still gaps in other aspects of new media operations, such as ignoring the front end and end of new media operations. In the front-end workflow, the lack of mining and analysis of user habits, emotions and experience, neglecting the real needs of users, the selection of topics and the selection of content forms are based on the perception of the user community rather than objective Data analysis sometimes blindly pursues network hotspots, or is limited to seeking new changes in performance, which is not consistent with user needs. In the end workflow, there is neglect of user feedback, lack of interaction with users, and communication effects. The analysis is not thorough enough. On the Weibo WeChat platform, companies rarely reply or forward user comments, ignoring the important role of the new media platform as a bridge to communicate with users, and thus losing an important way to obtain user feedback. In addition, in the analysis of the effect of communication, too much emphasis on the analysis of the quality of the work itself, and sometimes ignore the impact of the release time, the frequency of publication, the number of other popular events at the same time on the user's attention, for typical communication events.

Second, the degree of coordination and cooperation between various professional departments is not enough, and they have not reflected their respective professional advantages in the operation of new media. Most of the units have not set up a special new media business management department, and related businesses are generally managed by the news propaganda management department. However, the process of the new media operation itself is more complicated. In addition, most of the company's official new media also have some customer service functions. Therefore, in addition to the familiar editing and editing work of the news propaganda staff, there are also data analysis, user feedback, etc. The work assisted by the department, in addition to some popular science and professional information, requires professional departments to assist in writing. According to the survey, some units have introduced other professional departments to be responsible for the new media operations. Although the division of labor and cooperation are different, they are not ideal. On the one hand, the division of responsibilities of various departments is unreasonable. For example, the marketing department that is unfamiliar with the new media and does not understand the company's propaganda focus is solely responsible for the operation. The outreach department is only responsible for supervision, or the outreach department who is not familiar with professional work is responsible for all contents. The editorial release, the business department is only responsible for proposing professional opinions, not participating in specific operation and maintenance work; on the other hand, the professional department pays insufficient attention to the new media work, does not recognize the importance of the new media publicity work, and its The promotion of professional work and the cohesive effect on the team, so the degree of cooperation is general.

2.4. Operator configuration needs to be further optimized

First, there is still room for improvement in the professional level of new media operators. It is necessary to change the way of thinking as soon as possible to adapt to the development of new media. Although the booming development of the media has given people the illusion that the entry threshold for new media operations is low and the professional demand is not high, in fact, to create a successful new media account, operators need to have planning, copywriting, design, and data analysis. A variety of professional qualities, as well as strong social hotspot sensitivity and learning ability. Most of the company's new media operators are formerly engaged in traditional media staff or newcomers with no professional work experience. Most of them lack special guidance and training for new media operations. They can only rely on self-study and user experience of using new media to explore. The rules and characteristics of new media communication, some people have not yet fully realized the transformation of thinking mode, lack of sensitivity to current events and hotspots, lack of comprehensive control over network language and network trends, and the production of new media products also has The strong traditional media style does not meet the requirements of new media communication.

Second, the structure of full-time new media operators is unreasonable and cannot meet the demand for human resources in the rapid development of new media businesses. At present, the company's new media operators are mostly part-time, and their energy is very limited. It is difficult to cope with the increasing workload of new media operations, and it is difficult to guarantee the quality of the works. According to the survey, less than 30% of new media operators have less than 30% of their working time and can be fully used for new media operations. There is not enough time for staff to think and study the rules and strategies of new media communication. The release task can be completed on time, and the quality and communication effect of the work cannot be guaranteed. With the increase of new media platforms and the expansion of new media services, the gap between the company's new media operators has become more prominent.

3. Thoughts on the company's new media construction

General Secretary Xi Jinping clearly pointed out in the party's symposium on public opinion work that the duty and mission of the party's news and public opinion work under the new epoch conditions is to "highly uphold the banner, lead the leadership, center around the center, serve the overall situation, unite the people, boost morale, and become weathered." People, concentrating on power, clarifying fallacies, distinguishing right from wrong, connecting China and foreign countries, and communicating the world, this is also the principle that the company must adhere to in the construction of new media.

3.1. Adhere to the correct direction of public opinion

General Secretary Xi Jinping pointed out: "Adhering to the correct direction of public opinion is the core and soul of propagating public opinion work. All aspects of news public opinion work must adhere to correct public opinion guidance." New media propaganda work is an important part of the company's party committee work. With the important responsibility of unifying thoughts, propagating ideas, guiding employees, promoting work, displaying images, and promoting brands, the company's new media construction must adhere to the correct direction of public opinion and correct leadership under the correct leadership of the company's party committee.

At present, emerging new media platforms are rapidly emerging. The media landscape, public opinion environment, audience targets, and communication technologies have undergone profound changes. The masses' ideas have become more independent and changeable through the rapid dissemination and real-time interaction of new media. In the era when everyone is from the media, the voices and public opinion are complicated. The role of the media is not only to convey information, but also to lead the trend with correct public opinion and to concentrate with positive energy. As a state-owned and large-scale backbone enterprise, the company shoulders important political, economic and social responsibilities. It must be ideologically and politically determined to maintain a high degree of unity with the Party Central Committee with Comrade Xi Jinping as the core, and earnestly shoulder the burden of creating a party. Responsibility for the public opinion environment of the country, mastering the initiative of the new media public opinion battlefield, and building the company's new media into an important field and an important force in our party's news and public opinion work. All the work of the company's new media construction must always put the political direction in the first place, firmly adhere to the principle of party spirit, firmly adhere to the Marxist view of journalism, firmly adhere to positive propaganda, promote the main theme, and spread positive energy.

3.2. Promote media innovation and integration

With the development and application of new technologies, new forms of communication continue to be born, the form and function of the communication platform, the psychology and habits of the audience are all undergoing revolutionary changes, and the integration of traditional media and emerging media has become a general direction and a general trend. . At the fourth meeting of the central comprehensive deepening reform leading group, General Secretary Xi Jinping emphasized that "promoting the deep integration of traditional media and emerging media in terms of content, channels, platforms, operations and management, and focusing on creating a variety of forms and advanced means. Competitive new mainstream media."

Media integration is not simply copying the content of traditional media to the new media platform. Instead, it uses the characteristics and advantages of the Internet to promote all-round innovations such as ideas, content, means, and institutional mechanisms, so that a single and flat information dissemination method becomes Three-dimensional, multi-dimensional. In the process of new media construction, companies need to break the boundaries between traditional media and new media propaganda, overcome the difficulties of independent operation and management of traditional media and new media businesses, explore the establishment of a modern enterprise stereo communication system, and through the company's news propaganda work. The adjustment and optimization of business processes, smooth communication channels, sharing information resources, and promoting the deep integration of various media. At the same time, we adhere to the needs of users as the center, enhance the service concept, select the corresponding media platform according to the user's usage habits, push the information in a way that conforms to the user's reading habits, and promote the formation of traditional media and new media to perform their duties and their respective functions. Good situation. To give full play to the outstanding advantages of various media, to achieve in-depth coverage and fast food reading complement each other, to meet the reading needs of readers of different reading habits, and to make up for the lack of time and space of different media, so that the company's key work can be comprehensive, Three-dimensional, diversified communication.

3.3. Expand multi-channels

This is an era of personalized development. Various new media platforms are emerging one after another. Information can be spread out at any time and any place through Internet technology to break through the limitations of time and space. This has laid a good foundation for us to further expand the communication channels. General Secretary Xi Jinping demanded that "it is necessary to adapt to the trend of differentiation and differentiation, and accelerate the construction of a new pattern of public opinion. It is necessary to strengthen the capacity building of international communication, enhance the right to speak internationally, and focus on the Chinese story."

The development of diversified media platforms allows users to choose different platforms to pay attention to according to their own interests. The intelligent information push mode also makes content delivery more precise. The subdivision of the platform causes the user to divert, and only sticking to the original position and expression will inevitably cause a certain amount of fans to lose, affecting the communication effect. The company should quickly adapt to the development trend and changes of new media, and open up new platforms with high audience attention, good platform environment and meet the company's communication needs, and create new public opinion positions. And for different carriers and different audiences, adopt different communication strategies, expressions, narrative styles, language styles, innovative communication methods, and improve the communication, guidance, influence and credibility of news media.

In addition, the company's international development puts higher demands on the company's international image, and the company needs to have a place in the international public opinion environment. As the leading force of international communication, the new media has greatly narrowed the distance between the people of various countries and made international communication simple and fast. The company should conform to the development trend of globalization, from one-way domestic communication to diversified and three-dimensional international communication, select international themes, make good use of the world language, aim to enhance international influence, and actively carry out with other countries. Dialogue and exchange, the company's value concept will be spread overseas, and the company's brand will be promoted overseas, demonstrating the strength of China's central enterprises and the enthusiasm of the big country.

4. Conclusion

The development orientation of the company's new media system is based on the strategic goal of serving the new era of the company, with the goal of helping to shape the world-class company brand, and new media work chain around platform construction, content production, communication promotion, and user feedback. The mechanism, system, resources, etc. of the system construction are organically linked, and a brand communication work system formed by the integration of internal professional work and external media forces. In essence, it integrates the company's new media system construction elements and operates according to certain processes and rules to ensure the smooth operation of the company's new media communication work, enhance the communication, guidance and influence of the company's news public opinion, and realize the company's development. Strategic goals provide strong support.

The development concept is the forerunner of development action. The "Five Development Concepts" put forward by General Secretary Xi Jinping is the navigation mark for the development of various work. The company's new media system is a dynamic development system. It is related to the interactive development and organic symbiosis of the social environment, internal management, media platform and customer audience. Therefore, we must adhere to the development concept of innovation, coordination, green, openness and sharing. Promote the sound development of the company's new media system.

References

[1]   Mo Yang 2018 Optimized path of mainstream media opinion dissemination under the new media environment *News world* vol 12, p 48-51.

[2]   Xin Wang, Xiao Chen 2018 The Strategy and Strategy of the Integration and Development of Mainstream Media *News front* vol 23, p 64-66.

[3]   Zhen Lei 2018 New media operations in the short video era *Media* vol 23, p 56-58.

[4]   Hui Yang 2018 Analysis on the Innovation and Development of CCTV News in the Media Environment *Audiovisual* vol 12, p 190-191.

[5]   Wenbo Kuang, Congcong Lv 2018 New media technology boosts the party's theory of innovation *Journal of Chinese Social Sciences* vol 12, p 6.

Analysis of the Situation Faced by New Media Propaganda in State-owned Enterprises

Wang Han, Xia Liyu, Wang Youzi

State Grid Energy Research Institute Co.,Ltd., Beijing, China

wanghan@sgeri.sgcc.com.cn

Abstract. In recent years, various new media platforms such as Weibo, WeChat, news client, and TikTok have developed rapidly, and have even become the main front for state-owned enterprises in news propaganda. Therefore, the focus of brand building work of state-owned enterprises and the external situation they face are also changing, such as the urgent need to implement ideological work responsibilities, the more complicated ecological environment, the higher requirements for reform of state-owned enterprises, and the more convenient facilities provided by scientific and technological means. Faced with a new situation of new media communication, state-owned enterprises must change their brand building ideas and continuously improve the ability of network public opinion guidance and new media communication.

1. Introduction
In recent years, with the rapid development of digital technology and information technology, the wave of new media has come, and the old way of disseminating news information has undergone profound changes. New media ecology and communication patterns have been formed.

2. New situation of ideology constructi-on
Public opinion has always been an important force affecting social development. The work of news and public opinion is at the forefront of ideological struggle. In the long-term revolutionary struggle, a major theoretical achievement of the Marxist view of journalism is the formation of the "lose-mouth view" of journalism. The news media must become a public opinion position of the party and the people. News propaganda should become a means of public opinion supervision. General Secretary Xi Jinping pointed out in the report of the 19th National Congress that it is necessary to firmly grasp the leadership of ideological work and improve the communication, guidance, influence and credibility of news and public opinion. At the National Publicity and Ideological Work Conference in 2018, General Secretary Xi Jinping once again pointed out that building a socialist ideology with strong cohesiveness and leading power is a strategic task that the entire party, especially the propaganda and ideological front must shoulder, must take the initiative to speak of the Chinese Communist Party. The story of governing the country, the story of the Chinese people's struggle for a dream, and China's insistence on a peaceful and cooperative cooperation and win-win story, let the world better understand China.

In the era of Web 3.0, ideological construction work was greatly challenged. The new media has broken the information access rights of traditional media, providing the public with the freedom to approach the media and use the media to express their opinions, providing more space for the expression of discourse by ordinary people and civil organizations. On the new media platforms, the public's nerves become more susceptible to being triggered by the fissile spread of information and the viral spread of

---

Content from this work may be used under the terms of the Creative Commons Attribution 3.0 licence. Any further distribution of this work must maintain attribution to the author(s) and the title of the work, journal citation and DOI.

Published under licence by IOP Publishing Ltd

ideas, and individual appeals and wills are more likely to be aggregated to form powerful public opinion. With the rise of the folk opinion field, the opinions of the people tend to be liberalized and emotional, and the opinion leaders have become more diversified. The right to speak of news media is no longer completely in the hands of a few mainstream media in the traditional media era, and there is an equalization trend. A large number of influxes from the media have also promoted the arrival of the era of "all voices."

In January 2017, the General Office of the Central Committee of the Communist Party of China issued the "Detailed Rules for the Implementation of the Party Committee (Party Group) Network Ideology Work Responsibility System", emphasizing that network ideology work is the top priority of ideology work. It is necessary to firmly grasp the leadership of network ideology work, and ensure that the cyberspace is clearer and the party's voice becomes the strongest voice in cyberspace. A number of policies and regulations on the Internet news industry were intensively introduced in 2017, the establishment of the Central Radio and Television General Station in 2018, etc., further clarified the public opinion orientation and work requirements of state-owned enterprises' news propaganda, while strengthening the party's control over important public opinion positions.

Nowadays, the official public opinion field built by the mainstream media, which relies on the party media as the core, still bears the important responsibility of spreading the mainstream ideology. In the face of the folk public opinion field that gathers diverse forms of social thoughts and values, the state-owned enterprises must firmly grasp the leadership, practically strengthen management right, and focus on improving the right to speak. As a responsible central enterprise, the state-owned enterprises are not only the backbone of the national economy, but also an important position to promote and strengthen ideological construction. It must implement the ideological work responsibility system, firmly grasp the main line, make the state-owned capital stronger and stronger, and provide ideological assurance, spiritual strength and public opinion support for cultivating world-class companies with global competitiveness.

In 2018, the network content supervision work was frequently carried out. Not only the two short video platforms "quick hand" and the heads of today's headline "Volcano Video" accepted the network letter office interview, and some of the misguided, low-profile client software, WeChat public account, and network audio-visual program were ordered to shut down by the State Administration of Radio and Television. The Internet application store also suspended the download service of four news information apps such as today's headlines. At the national level, the management of network content is becoming stricter, the rectification measures will be implemented one after another. The closure of APP, and the offline of online audio-visual programs will not be a case. The clean-up of the network environment will bring new value-oriented and communication trends. The content of the network with a little bit of heat, and only attracts the eye will be less and less, the attention of users and the layout of the network will also change. In general, the new value orientation will surely bring about a new network ecological landscape.

The continuous deepening of ideological work puts forward higher requirements for the news public opinion work of the central enterprises in the new era. The company must take full responsibility for the duties and mission of the news public opinion work, and must follow the rules of news dissemination, change the inherent communication concept, and innovate methods and methods to enhance work-oriented, master the initiative of work, while consolidating and expanding the party's news and public opinion positions, effectively enhance the influence of the party's news. State-owned enterprises must firmly grasp the changes in the current ideological communication trend, consolidate the company's dominant discourse power in the official public opinion field, and enhance the mainstream ideology to guide the public opinion field, thus achieving the same frequency resonance of the two public opinion fields.

3. The development of state-owned ent-erprises opens a new journey

In recent years, with the implementation of the "One Belt and One Road" construction plan and the international capacity cooperation action plan, the internationalization of state-owned enterprises has made great breakthroughs, and international exchanges and cooperation have become more frequent. It is necessary to enhance the visibility, recognition and reputation of corporate brands through effective news promotion, and vigorously expand international communication channels.

Through the global social media platforms, state-owned enterprises can break the monopoly of Western mainstream media on international discourse channels, enhance their international communication competitiveness, and make full use of the advantages of new media to make state-owned enterprises' brand goes out, and thus continuously enhances the company's international influence and voice, and builds a world-class state-owned enterprise brand that matches the world-class enterprises with global competitiveness.

Since the 18th National Congress, the deepening of the reforms has broken the ice. State-owned enterprises have taken the opportunity of reforming state-owned enterprises to improve quality and efficiency, reduce physical fitness, actively serve and promote the transformation and upgrading of related fields, and have embarked on a road of development quality and efficiency. However, the situation and challenges faced by state-owned enterprises are still grim. It is necessary to continuously strengthen brand promotion to show the public the sustainable competitive advantage of the company and reflect the comprehensive strength of the company. The rapid development of new media has caused the number of information faced by the public expanding rapidly, and the attention to each piece of information has been decreasing. Therefore, the state-owned enterprises have to increase the quality of the content of the works while increasing the number of transmissions. The company should realize the precise promotion of classified communication by changing the form of communication, expanding the communication platforms, and optimizing the content of communication, and effectively raise the public's attention and recognition of the brand of state-owned enterprises.

The 19th National Congress of the Communist Party of China and the Central Economic Work Conference emphasized the importance of optimizing the business environment for building a modern economic system and promoting high-quality development. This has also become one of the key tasks of the company in the new stage. The role of new media in optimizing the business environment is not only to enable users to keep abreast of the company's new service initiatives, but also enhancing communication with customers, closing the distance with customers, and closing relationship with customers. Through the new media platforms, the company can respond to customers' needs in a timely manner, respond to social concerns, enhance user interaction, help to enhance customer understanding of the company, and establish friendly relationships with customers to optimize business environment and increase convenience for the people.

4. Public opinion presents new features

At the end of 2017, short video platforms such as Kuaishou, TikTok, and watermelon video have risen rapidly and started to grow amazingly. The Kuaishou has accumulated 700 million users, and the daily life surpasses today's headline main APP. Compared with Weibo and WeChat public account, the more visual and sensory short video is obviously more popular among young people nowadays. The new media layout originally from Weibo and WeChat is changing quietly. In the face of strong shocks, Weibo and WeChat are fighting back in their own way.

Weibo has been changing. Compared with WeChat, Weibo lacks the number of innate users of WeChat as a communication tool, so it chose to sink. According to the "2017 Weibo User Development Report" released by Weibo Data Center, as of the end of September 2017, users from third- and fourth-tier cities have accounted for more than 50% of Weibo monthly active users. At the same time, the content format covered by the Weibo platform has become more diversified. In addition to the mainstream graphics, the comprehensive popularity of the headline articles and Weibo stories has also ushered in a new spring for Weibo. The diverse display methods can meet the needs of different user groups, so that the originally lost users will slowly return, and new users will start to grow steadily. The resulting changes in user groups and changes in product forms will have an impact on the way and focus of the Weibo platform. The company needs to reposition the official Weibo, and fully adapt to the communication potential of the platform function while adapting to the transformation of the Weibo platform itself.

Compared to Weibo, the change of WeChat seems to be somewhat restrained. According to the statistics of the new list, the WeChat public number in 2017 provided 12.9% of the daily WeChat public platform, of which 52% of the "100,000+" articles came from the top 500 accounts. The large-scale pattern has been relatively solidified, and the new public number is difficult to make a comeback. Even the operating public number is facing bottlenecks such as fan powder removal, slow reading growth, and low article opening rate. The head effect is becoming more prominent. Although WeChat launched the "look and see" function as early as the end of May 2017, users can read the recommended public number article by means of information flow, but so far it is still only a function plug-in lying at the four-level entrance, many people haven't used it, even never heard about it. WeChat, which faced a lot of crisis, finally made an important revision in June 2018, transforming the way in which the subscription number was presented into a form similar to "information flow", presented in updated chronological order. The new way of reading may lead to a "big reshuffle of the traffic" – the trumpet with good content but always suppressed will get more favor, and the large size of the intrinsic fans will face more competitive pressure. In order to highlight the encirclement, it is necessary to work hard on the quality of the products, to grasp the relationship between the quantity of the release and the quality of the content, not only to ensure the activity of the public number, but also to avoid the decline in the number of fans caused by the excessive release of low-quality products.

In addition, WeChat is also trying to change the original message mechanism. In April 2018, the friends' message function was officially launched. The friends' messages will be directly visible to friends without filtering by the public number before it can be displayed to everyone. However, it is It has already gone offline in less than 2 months. In addition, the message function of the newly registered public number is also suspended, and the message authority of the original account is also restricted. Although this change has reduced the interaction and social attributes of the WeChat public account to a certain extent, it also reduces the risk of public opinion. After the public message has been reviewed and screened, most of the sensitive information is blocked, and it is difficult to form key opinions by leaders or distinct public opinion orientations, the influence and scope of public opinion are naturally more controllable.

Apart from Weibo and WeChat which are developed relatively stably, the emerging short video platforms have brought new heights to the development of the media. Different from Weibo, WeChat and other platforms to "force push" messages to fans, the intelligent information push method of emerging platforms such as headlines, TikTok, and Kuaishou is based on user needs and hobbies, which helps users complete the information screening. The process of information screening frees users from massive fragmentation information, eliminating the need to waste a lot of time paying attention to invalid information. The change to the inherent new media landscape, the suppression of fans and readings of Weibo and WeChat are only appearances. The deeper impact is that it has a huge impact on the way in which the quantity of reading is determined by the number of fans. The propagation way which relies solely on innate new media platforms will be difficult to achieve the desired communication effect. The information flow-based reading method focuses more on the users' hobbies and habits, and fundamentally adjusts the relationship between fans and the platforms. Users will not only have the right to freely choose information, but also be recommended for information is more suitable for them. The popularity of the content will not only affect the spread of a piece of information, but will also result in subsequent information being unable to be effectively disseminated because it is not recommended.

5. New changes in technology

The development of new media puts forward higher requirements for the timeliness of information. The method of artificial writing alone can not meet the readers' demand for fresh information, and writing robots have emerged. In recent years, Xinhua News Agency, Today's headlines, Baidu and other companies have launched their own research and development of writing robots and put them into use. Although the manuscripts produced have much room for improvement in depth and temperature, it does not hinder "machine news writing" becoming a phenomenal level of artificial intelligence technology in the field of news communication because of the writing speed, data mining breadth, comprehensiveness and timeliness of information collection of writing robots which are better than artificial writing. In the writing of manuscripts such as finance, sports and entertainment news, the speed and the output of writing robots are very alarming.

In December 2017, Xinhua News Agency released the first media artificial intelligence platform "Media Brain" at the 5th China Emerging Media Industry Convergence Development Conference. It has a smart media production platform, news distribution, honey collection, copyright monitoring, face verification, user portraits, intelligent conversations, speech synthesis and many other functions, which covers the news thread, planning, interviewing, production, distribution, feedback and other full news links. In the NPC&CPPCC in 2018, "media brain" was officially put into the post. In just 15 seconds, the first NPC&CPPCC video news was produced, which received a lot of attention. The "media brain" application can help the media get news leads and news materials faster, more accurately and intelligently, thus greatly improving the efficiency of news production and dissemination. In addition, the "media brain" can also provide a diffusion scheme that meets users' needs and propagation rules by analyzing heat and user images to achieve optimal communication effects.

Artificial intelligence technology satisfies readers' pursuit of speed. VR and AR technologies focus on improving the readers' experience, and enable users to reach deep interaction with communication content through immersive and interactive expression. This kind of reporting method that can create a 360-degree panoramic view breaks the limitation of time and space, allowing users to immerse themselves in a "simulated" news scene. Users no longer look at news reports from a third-party perspective, but directly from the first perspective. Participating in the report, the information it can obtain is more comprehensive, objective, and multi-dimensional, users can grab the information of interest according to their needs, without being limited by the angle of the report and content. Xinhua News Agency, People's Daily, CCTV and many other media have also used VR technology to produce news and achieved good communication effects. For example, Xinhua Net used VR technology to lead the audience into the national NPC&CPPCC venue, and Xinhua News Agency launched AR series reports which provide users with the opportunity to study together with General Secretary Xi Jinping. In this way, the users can feel the actual situation on the spot more deeply, and the sensory experience becomes more stereoscopic, thus gaining the impact and shock power unmatched by the graphic and video reports.

In the era of "traffic", the application of big data analysis in new media is no longer new. The operation and maintenance of new media accounts, the effect of product communication, and the degree of user attention to each content reflect the strength of new media are all the targets of the application of big data analysis technology. It is the daily life of every new media person to accurately describe the user's portraits, and then create, push, and disseminate products and content that meet the needs of users by analyzing the basic information, attention content, behavior changes of the platform users. The extensive use of big data analysis technology makes it possible to push accurate and personalized information. It is necessary to make good use of big data analysis methods, dig deep into the inner meaning of data, respect the characteristics and laws of data, and always maintain the sensitivity of development and changes in user demand.

6. Conclusion
The rapid development of new media has made it necessary for state-owned enterprises to face up to changes in their external situation, pay attention to changes in the ideological field, actively respond to the challenges of state-owned assets reform, grasp the public opinion ecology under the influence of new media, and fully utilize various technical means to build a new media communication system which is proper for national requirements, people's needs and the development of state-owned enterprises.

References
[1]   Yuhui Fu 2015 A Summary of China's New Media Communication Research in 2014

International press vol 12, p 35-46.

[2] Zongjian Li, Zhuru Cheng 2016 The Challenges and Countermeasures of Public Opinion Guidance in the New Media Age *Journal of Shanghai Administration College.* vol 17, p 76-85.

[3] Mengmeng Zheng 2014 New Media Communication Socialist Core Values Research *Media* vol 8, p 67-70.

[4] Guanyi Zheng 2018 Problems in New Media Contents and Their Countermeasures *Media* vol 16, p 54-55.

[5] Sha Liu 2018 New Media Communication Research Reported by "Belt and Road"——Taking the Official Weibo of People's Daily as an Example *New Media Research* vol 4, p 18-19.

A new linguistic decision making method—FLM-VIKOR

G R Zhang[1,*] and W D Zeng[2]

[1,2] School of Science, Xihua University, Chengdu, Sichuan 610039, China

*Corresponding author: E-mail address: 15732154719@163.com.

Abstract. Fuzzy linguistic term set is a powerful and flexible tool to express evaluation of experts in decision making. In this paper, we propose a VIKOR method based on fuzzy linguistic multiple set to solve the fuzzy linguistic multiset multi-criteria group decision making. Firstly, we develop the conception of fuzzy linguistic multiset, its correlate properties and operations. Then, we aggregate all decision matrices of all experts by these operations, obtaining final decision making matrix. Moreover, we develop the distance of two fuzzy linguistic multisets. Lastly, an illustrative example is given to testify the effectiveness of the developed method.

1. Introduction

Decision making is a selection process. We select the most satisfactory decision alternative from a set of possible decision alternatives. In practical, decision making problems are usually uncertain. Especially, in the big data era, with the decision making alternatives and evaluation indexes become larger, the sources of evaluation information become more diverse. It makes the decision making process more difficult and complex. Recently, many decision making methods have been proposed to solve different decision making problems [1-3]. Generally, decision matrix is composed of alternative, criteria and corresponding evaluation information. In view of the characteristics and properties of evaluation indexes, there are various representations of evaluation information, such as quantitative and qualitative representation. Quantitative representation includes accurate number, interval number, probability, fuzzy number [4-6]. Qualitative representation includes order relation, preference relation, linguistic representation [7-8]. Different representations require different methods to solve corresponding decision making problems. Since fuzzy linguistic is closest to human cognitive process and users can understand its semantics through membership function intuitively. Fuzzy linguistic is an important way to represent uncertain information in decision making. At present, fuzzy linguistic decision making method is one of the hot topics in decision making [3].

The core of fuzzy linguistic decision making method is fuzzy linguistic processing [2]. Classical fuzzy linguistic processing method is based on the membership function and fuzzy reasoning. Calculating with words has some disadvantages, such as computational complexity, inaccuracy and information lost easily [2]. Therefore, Herrera et al. [2] proposed binary linguistic model for fuzzy linguistic information representation. Formally, the binary linguistic (s_i, α), where $s_i \in S = \{s_0, \cdots, s_g\}$ be the initial fuzzy linguistic term set, and $\alpha \in [-0.5, 0.5)$ represents the difference of approximation between the actual linguistic value and the initial value s_i. For example, the initial fuzzy linguistic term

---

Content from this work may be used under the terms of the Creative Commons Attribution 3.0 licence. Any further distribution of this work must maintain attribution to the author(s) and the title of the work, journal citation and DOI.

Published under licence by IOP Publishing Ltd

set $S=\left\{s_0\left(bad\right),s_1\left(littlebad\right),s_2\left(general\right),s_3\left(littlegood\right),s_4\left(good\right)\right\}$ is used to evaluate the quality of several products, and the aggregation result is $\left(s_1,-0.2\right)$. In other words, the quality of products is close to "a little bad", and the difference is -0.2. The binary linguistic model has the properties which is simple to compute, no information lost and easy to understand. So far, the binary linguistic model has been heavily studied. Many generalization models have been proposed, such as proportional bivariate linguistic model, virtual linguistic model and hesitant fuzzy set, etc. Meanwhile, the linguistic decision making method based on the binary linguistic model has also been widely studied [2, 9].

Multiset is an effective tool which can handle the issue with repeated information and avoid information lost [6, 10]. The linguistic evaluation information of experts often repeatedly in the process of linguistic decision making. In this paper, we develop multiset to represent multi-criteria group decision making problems with linguistic assessment information, which is called Fuzzy Linguistic Multiset (FLM). Combining with VIKOR method, a method to solve fuzzy linguistic multiset multi-criteria group decision making problem (FLM-MCGDM) is proposed, which is called fuzzy linguistic multiset VIKOR method (FLM-VIKOR). Specifically, we propose the conception of FLM firstly. And then we analysis its related properties and operations. Inspired by [8], we develop the score function and variance of a FLM. So as to obtain the positive and negative ideal solutions, the distance of two FLMs is developed. Lastly, we illustrate that the proposed method is reasonable and effective, through comparing with the hesitant fuzzy linguistic TOPSIS (HFL-TOPSIS) method [11], the hesitant fuzzy linguistic VIKOR (HFL-VIKOR) method [8] and the fuzzy linguistic multiset TOPSIS (FLM-TOPSIS) method [6].

2. Preliminaries

2.1. Multiset
Definition 1[10] A multiset M over a set A, called the base set of M, is an unordered collection of elements of A, where each element in A can occur zero or more times in M. M has associated with it a function $m_M : A \to \mathrm{N}$, called multiplicity function or characteristic function, where $m_M\left(x\right)$ is the multiplicity of $x \in A$ in M (the number of times $x \in A$ occurs in M). The set of distinct elements of M, denoted by $set\left(M\right)$, is called the support set of M.

For any multiset M_1 and M_2 over the base set A, for any $x \in A$, we define:

Union: $count_{M_1 \cup M_2}\left(x\right) = \max\left\{count_{M_1}\left(x\right),count_{M_2}\left(x\right)\right\}$;

Intersection: $count_{M_1 \cap M_2}\left(x\right) = \min\left\{count_{M_1}\left(x\right),count_{M_2}\left(x\right)\right\}$;

Sum: $count_{M_1 + M_2}\left(x\right) = count_{M_1}\left(x\right) + count_{M_2}\left(x\right)$;

$$count_{M_1 - M_2}\left(x\right) = \begin{cases} count_{M_1} - count_{M_2}, count_{M_1} > count_{M_2} \\ 0, count_{M_1} \le count_{M_2} \end{cases}$$

(Multiset) Difference:
We can find that the intersection, union and sum are associative and commutative [16].

2.2. VIKOR Method
The VIKOR method was developed for multicriteria optimization of complex systems. It determines the compromise ranking list, the compromise solution, and the weight stability intervals for preference stability of the compromise solution obtained with the initial (given) weights. This method focuses on ranking and selecting from a set of alternatives in the presence of conflicting criteria. It introduces the multicriteria ranking index based on the particular measure of ''closeness'' to the ''ideal'' solution [8].

The multicriteria measure for compromise ranking is developed from the L_p-metric used as an aggregating function in a compromise programming method [8].

The various n alternatives are denoted as $A = \{a_1, a_2, \cdots, a_n\}$. For alternative a_i, the rating of the I aspect is denoted by f_{ij}, i.e. f_{ij} is the value of jth criterion function for a_i;Development of the VIKOR method started with the following form of L_p metric:

$$L_{p,i} = \left\{ \sum_{j=1}^{r} \left[w_j (f_j^+ - f_{ij}) / (f_j^+ - f_j^-) \right]^p \right\}^{1/p} \quad 1 \le p \le \infty \text{ , } j = 1, 2, \cdots, r$$

Within the VIKOR method $L_{1,i} = S_i$ and $L_{\infty,i} = R_i$ are used to formulate ranking measure. The solution obtained by $\min_i S_i$ with a maximum group utility ("majority" rule), and the solution obtained by $\min_i R_i$ with a minimum individual regret.

The compromise ranking algorithm VIKOR steps as below:

(1)Determine the best f_j^+ and the worst f_j^- values of all criterion functions

$$f_j^+ = \begin{cases} \max_i f_{ij}, C_j \in B \\ \min_i f_{ij}, C_j \in C \end{cases} \quad (1)$$

$$f_j^- = \begin{cases} \min_i f_{ij}, C_j \in B \\ \max_i f_{ij}, C_j \in C \end{cases} \quad (2)$$

Where B represents a benefit, C represents a cost.

(2) Compute the values S_i and R_i $(i = 1, \cdots, n)$ by the relations

$$S_i = L_{1,i} = \sum_{j=1}^{r} w_j \frac{f_j^+ - f_{ij}}{f_j^+ - f_j^-} \quad (3)$$

$$R_i = L_{\infty,i} = \max_j \left(w_j \frac{f_j^+ - f_{ij}}{f_j^+ - f_j^-} \right) \quad (4)$$

Where w_j are the weight of criteria.

(3) Compute the values Q_i by the relation

$$Q_i = v \frac{S_i - S^+}{S^- - S^+} + (1 - v) \frac{R_i - R^+}{R^- - R^+} \quad (5)$$

Where $S^+ = \min_i S_i$, $S^- = \max_i S_i$, $R^+ = \min_i R_i$, $R^- = \max_i R_i$ and v is introduced as a weight for the strategy of "the majority of criteria" (or "the maximum group utility"), whereas $1 - v$ is the weight of the individual regret.

(4) Rank the alternatives, sorting by the values S, R and Q in decreasing order.

(5) Propose a compromise solution the alternative $a^{(1)}$ which is the best ranked by the measure Q (minimum) if the following two conditions are satisfied:

① "Acceptable Advantage":

$$Q(a^{(2)}) - Q(a^{(1)}) \ge \frac{1}{n-1} \quad (6)$$

Where $a^{(2)}$ is the alternative with second position in the ranking list by Q.

②"Acceptable stability in decision making": The alternative $a^{(1)}$ must also be the best ranked by S or/and R. This compromise solution is stable within a decision making process, which could be the strategy of maximum group utility. If one of the conditions is not satisfied, then a set of compromise solutions is proposed, which consists of:

• Alternatives $a^{(1)}$ and $a^{(2)}$ if only the condition② is not satisfied.

• Alternatives $a^{(1)}$, $a^{(2)}$, ..., $a^{(l)}$ if the condition① is not satisfied. $a^{(l)}$ is determined by the relation (7) to seek for the maximum I.

$$Q\left(a^{(l)}\right)-Q\left(a^{(1)}\right)<\frac{1}{n-1} \tag{7}$$

3. Fuzzy Linguistic Multiset VIKOR Method

3.1. Fuzzy linguistic multi-criteria group decision making matrix
In general, a fuzzy linguistic decision making problem can be described as: a group of decision makers $E=\{d_1,\cdots,d_m\}$ evaluate the decision alternatives $A=\{a_1,\cdots,a_n\}$ according to the criteria $C=\{c_1,\cdots,c_r\}$, using the fuzzy linguistic term set $S=\{s_0,\cdots,s_g\}$. Depending on the evaluation linguistic information provided by the decision makers, the fuzzy linguistic decision method provides the comprehensive linguistic evaluation results of each decision object and is used to select satisfactory decision alternative. In existing linguistic decision making methods, the decision matrix is as follows:

$$D_k=\left(e_{ij}^k\right)=\begin{bmatrix} e_{11}^k & \cdots & e_{1r}^k \\ \vdots & \ddots & \vdots \\ e_{n1}^k & \cdots & e_{nr}^k \end{bmatrix} \tag{8}$$

Where, $e_{ij}^k \in S$ represents the decision maker d_k use the linguistic terms e_{ij}^k to evaluate alternative a_i for criteria c_j.

Definition 2 Let $S=\{s_0,\cdots,s_g\}$ be a linguistic term set, $M=\{e_1,\cdots,e_m\}$ is said to be a fuzzy linguistic multiset (FLM) over S. Where $e_k \in S$, $k=1,2,\cdots,m$.

Note that if the elements of M are all different, then it is a classical set.

For the fuzzy linguistic decision problem described as previously, any two decision makers d_k, $d_{k'}$ give the decision matrices are D_k, $D_{k'}$ respectively

$$D_k=\begin{bmatrix} e_{11}^k & \cdots & e_{1r}^k \\ \vdots & \ddots & \vdots \\ e_{n1}^k & \cdots & e_{nr}^k \end{bmatrix} \qquad D_{k'}=\begin{bmatrix} e_{11}^{k'} & \cdots & e_{1r}^{k'} \\ \vdots & \ddots & \vdots \\ e_{n1}^{k'} & \cdots & e_{nr}^{k'} \end{bmatrix}$$

The proposed method combines the decision matrix D_k, $D_{k'}$ into the final decision matrix D. The element of D is a FLM which aggregate all elements at corresponding position in the D_k, $D_{k'}$. In this paper, we denote this operation as M. Specifically,

$$D=D_k\mathbf{M}D_{k'}=\begin{bmatrix} e_{11}^k & \cdots & e_{1r}^k \\ \vdots & \ddots & \vdots \\ e_{n1}^k & \cdots & e_{nr}^k \end{bmatrix}\mathbf{M}\begin{bmatrix} e_{11}^{k'} & \cdots & e_{1r}^{k'} \\ \vdots & \ddots & \vdots \\ e_{n1}^{k'} & \cdots & e_{nr}^{k'} \end{bmatrix}=\begin{bmatrix} \{e_{11}^k,e_{11}^{k'}\} & \cdots & \{e_{1r}^k,e_{1r}^{k'}\} \\ \vdots & \ddots & \vdots \\ \{e_{n1}^k,e_{n1}^{k'}\} & \cdots & \{e_{nr}^k,e_{nr}^{k'}\} \end{bmatrix} \tag{9}$$

Definition 3 In general, the operation of m matrixes given by m decision makers is $D=D_1\mathbf{M}D_2\mathbf{M}\cdots\mathbf{M}D_m$, denoted as

$$D=\overset{m}{\underset{k=1}{\mathbf{M}}}D_k \tag{10}$$

3.2. The operations and distance
Definition 4 For any FLMs M,M_1,M_2, for all $e\in S$

Element: $e\in M \Leftrightarrow count_M(e)\geq 1$; $e\notin M \Leftrightarrow count_M(e)=0$;

Inclusion: $M_1\subseteq M_2 \Leftrightarrow count_{M_1}(e)\leq count_{M_2}(e)$;

Equality: $M_1=M_2 \Leftrightarrow count_{M_1}(e)=count_{M_2}(e)$;

Sum: $M_1\oplus M_2 \Leftrightarrow count_{M_1\oplus M_2}(e)=count_{M_1}(e)+count_{M_2}(e)$;

Union: $M_1 \bigcup M_2 \Leftrightarrow count_{M_1 \cup M_2}(e) = count_{M_1}(e) \vee count_{M_2}(e)$;

Intersection: $M_1 \bigcap M_2 \Leftrightarrow count_{M_1 \cap M_2}(e) = count_{M_1}(e) \wedge count_{M_2}(e)$.

Definition 5 For any FLM $M = \bigcup\limits_{s_l \in M} \left\{ s_l \middle| l=1,\cdots,\overline{\overline{M}} \right\}$ its score function and variance are below:

$$\rho(M) = \frac{1}{\overline{\overline{M}}} \sum_{s_l \in M} s_l = s_{\frac{1}{\overline{\overline{M}}} \sum_{s_l \in M} l} \qquad (11)$$

$$\sigma(M) = \frac{1}{\overline{\overline{M}}} \sqrt{\sum_{s_l,s_k \in M} (s_l - s_k)^2} = s_{\frac{1}{\overline{\overline{M}}} \sqrt{\sum_{s_l,s_k \in M} (l-k)^2}} \qquad (12)$$

Where $\rho(M)$ and $\sigma(M)$ are called score function and variance, respectively. $s_l, s_k \in S$, $l,k \in \left\{ 0,1,\cdots,\overline{\overline{M}} \right\}$, $\overline{\overline{M}}$ is the number of linguistic terms in M . And the compare of two FLMs are defined bellow:

Definition 6 For any two FLMs M_p , M_q

（1）if $\rho(M_p) > \rho(M_q)$, then $M_p > M_q$;

（2）if $\rho(M_p) = \rho(M_q)$, and $\sigma(M_p) < \sigma(M_q)$, then $M_p > M_q$;

（3）if $\rho(M_p) = \rho(M_q)$, and $\sigma(M_p) = \sigma(M_q)$, then $M_p = M_q$.

We define the positive and negative ideal solutions of each criteria as:

$$M^{j+} = \max_i M_i^j \qquad (13)$$

$$M^{j-} = \min_i M_i^j \qquad (14)$$

Then the positive and negative solution of FLM-MCGDM are：

$$M^+ = \bigcup_{j=1}^{r} M^{j+} \qquad (15)$$

$$M^- = \bigcup_{j=1}^{r} M^{j-} \qquad (16)$$

Definition 5 For any FLMs M_p, M_q, the distance between them is defined as:

$$d(M_p, M_q) = \frac{1}{\overline{\overline{M}}} \sum_{l=1}^{\overline{\overline{M}}} \left| l_p - l_q \right| \qquad (17)$$

Where $M_p = \bigcup\limits_{s_{l_p} \in M_p} \left\{ s_{l_p} \middle| l_p = 1,\cdots,\overline{\overline{M}} \right\}$, and $M_q = \bigcup\limits_{s_{l_q} \in M_q} \left\{ s_{l_q} \middle| l_q = 1,\cdots,\overline{\overline{M}} \right\}$.

It satisfy the following properties:

Normality: $d(M_1, M_2) \geq 0$, $d(M_1, M_2) = 0 \Leftrightarrow M_1 = M_2$;

Symmetry: $d(M_1, M_2) = d(M_2, M_1)$;

Inequality: $d(M_1, M_2) \leq d(M_1, M_3) + d(M_3, M_2)$;

Proof: 1), 2) obviously. We only prove the last one.

$$d(M_1, M_2) = \frac{1}{\overline{\overline{M}}} \sum_{l=1}^{\overline{\overline{M}}} \left| \delta_l^1 - \delta_l^2 \right| = \frac{1}{\overline{\overline{M}}} \sum_{l=1}^{\overline{\overline{M}}} \left| \delta_l^1 - \delta_l^3 + \delta_l^3 - \delta_l^2 \right| \leq \frac{1}{\overline{\overline{M}}} \sum_{l=1}^{\overline{\overline{M}}} \left(\left| \delta_l^1 - \delta_l^3 \right| + \left| \delta_l^3 - \delta_l^2 \right| \right)$$

$$= \frac{1}{\overline{\overline{M}}} \sum_{l=1}^{\overline{\overline{M}}} \left| \delta_l^1 - \delta_l^3 \right| + \frac{1}{\overline{\overline{M}}} \sum_{l=1}^{\overline{\overline{M}}} \left| \delta_l^3 - \delta_l^2 \right| = d(M_1, M_3) + d(M_3, M_2).$$

3.3. FLM-VIKOR method

The specific decision making steps of FLM-VIKOR method are bellow:

(1)　Obtain the fuzzy linguistic value multiset multi-criteria decision making matrix by equation (8) and (9).

(2) Find the positive and negative ideal solutions by equation (15) and (16).

(3) Compute the maximum group utility S_i and the minimum individual regret R_i of each object, where

$$S_i = \sum_{j=1}^{r} w_j \frac{d\left(M^{j+}, M_i^j\right)}{d\left(M^{j+}, M^{j-}\right)} \qquad (18)$$

$$R_i = \max_j w_j \frac{d\left(M^{j+}, M_i^j\right)}{d\left(M^{j+}, M^{j-}\right)} \qquad (19)$$

Note that the weight w_j can be omitted.

(4) Compute the comprehensive evaluation value Q_i, where

$$Q_i = vS_i + (1-v)R_i \qquad (20)$$

(5) Obtain the compromise solution.

Firstly, according to S_i, R_i, Q_i, we can rank for each a_i; then according to the condition① and②whether satisfy or not; finally select the compromise solution.

4. Instance and Analysis

4.1. Instance

The board of directors of a company has five members who plan to make a strategic plan for the company in the next five years. There are three projects $a_i\,(i=1,2,3)$ to be evaluated. Considering the following four aspects, i.e. c_1: financial aspects; c_2: consumer satisfaction; c_3: international career development; c_4: growth of the company. The five directors used linguistic terms to express their linguistic evaluation value. Where, the linguistic term set is $S=\{s_0, s_1, s_2, s_3, s_4, s_5, s_6\}$. Their decision matrixes are as follows:

$$D_1 = \begin{bmatrix} s_3 & s_4 & s_4 & s_5 \\ s_3 & s_3 & s_2 & s_3 \\ s_4 & s_3 & - & s_4 \end{bmatrix} D_2 = \begin{bmatrix} s_4 & s_2 & s_4 & s_5 \\ s_3 & - & s_1 & s_3 \\ s_4 & s_3 & s_5 & s_4 \end{bmatrix} D_3 = \begin{bmatrix} s_4 & s_4 & s_4 & s_3 \\ s_5 & s_2 & - & s_4 \\ s_3 & s_3 & s_4 & s_6 \end{bmatrix} D_4 = \begin{bmatrix} s_4 & s_4 & s_4 & s_3 \\ s_3 & s_4 & s_3 & s_3 \\ s_3 & - & s_3 & s_4 \end{bmatrix} D_5 = \begin{bmatrix} s_3 & s_4 & s_3 & s_5 \\ s_3 & s_3 & s_2 & s_3 \\ s_3 & s_4 & - & s_4 \end{bmatrix}$$

(1) Obtain the decision making matrix:

$$D = \begin{bmatrix} \{s_3, s_4, s_4, s_4, s_3\} & \{s_4, s_2, s_4, s_4, s_4\} & \{s_4, s_4, s_4, s_4, s_3\} & \{s_5, s_5, s_3, s_3, s_5\} \\ \{s_3, s_3, s_5, s_3, s_3\} & \{s_3, s_2, s_4, s_3\} & \{s_2, s_1, s_3, s_2\} & \{s_3, s_3, s_4, s_3, s_3\} \\ \{s_4, s_4, s_3, s_3, s_3\} & \{s_3, s_3, s_3, s_4\} & \{s_5, s_4, s_3\} & \{s_4, s_4, s_6, s_4, s_4\} \end{bmatrix}$$

By means of the mean complement method, the decision matrix becomes

$$D = \begin{bmatrix} \{s_3, s_4, s_4, s_4, s_3\} & \{s_4, s_2, s_4, s_4, s_4\} & \{s_4, s_4, s_4, s_4, s_3\} & \{s_5, s_5, s_3, s_3, s_5\} \\ \{s_3, s_3, s_5, s_3, s_3\} & \{s_3, s_{2.5}, s_2, s_4, s_3\} & \{s_2, s_1, s_2, s_3, s_2\} & \{s_3, s_3, s_4, s_3, s_3\} \\ \{s_4, s_4, s_3, s_3, s_3\} & \{s_3, s_3, s_3, s_{3.5}, s_4\} & \{s_5, s_{4.5}, s_4, s_{3.5}, s_3\} & \{s_4, s_4, s_6, s_4, s_4\} \end{bmatrix}$$

(2) Find the M^{j+}, M^{j-}:

By equation (15) and (16), we can obtain the positive solution is $M^+ = \{M_1^1, M_1^2, M_3^3, M_3^4\}$, and the negative solution is $M^- = \{M_3^1, M_2^2, M_2^3, M_2^4\}$.

(3) Calculate the S_i, R_i, Q_i

By equation (17), we can obtain $S_1 = 1.37$, $S_2 = 4$, $S_3 = 1.78$; By equation (18), we can obtain $R_1 = 1.17$, $R_2 = 1$, $R_3 = 1$, By equation (19), we can obtain $Q_1 = 1.27$, $Q_2 = 2.5$, $Q_3 = 1.39$.

Rank a_i according to the S_i, R_i, Q_i, the result as shown in table 1.

Table 1	
S_i, R_i, Q_i	**Rank**
$S_1 < S_3 < S_2$	$a_1 > a_3 > a_2$
$R_2 = R_3 < R_1$	$a_2 = a_3 > a_1$
$Q_1 < Q_3 < Q_2$	$a_1 > a_3 > a_2$

Table 2		
Method	**Rank**	**Optimal/ compromise solution**
FLM-TOPSIS [6]	$a_1 > a_3 > a_2$	a_1
HFL-TOPSIS [11]	$a_3 > a_1 > a_2$	a_3
SAB [3]	$a_3 > a_1 = a_2$	a_3
FLM-VIKOR	$a_1 > a_3 > a_2$	a_1, a_3

(4)   Selecting the final compromise solution

In terms of by S_i, R_i, Q_i to rank for a_i, none of them ranked first; and $Q_3 - Q_1 < 0.5$. Therefore, they don't meet the conditions①,②, so there is no optimal stable solution. However, $Q_2 - Q_1 > 0.5$, $Q_3 - Q_1 < 0.5$. Therefore, the compromise solution is a_1, a_3.

4.2. Compare and Analysis
Compared with other methods, the calculation in this paper is compared with the example results in literature [3,6,11], the result as shown in table 2.

As shown in the table above, the method proposed in this paper is consistent with the sorting result obtained by FLM-TOPSIS [6], but the final solution is inconsistent with the sorting result obtained by HFL-TOPSIS[11] and SAB[3]. Which method is more reasonable? As shown in the comparison in [6], for this part of unknown linguistic information, it is more reasonable to use the FLM-TOPSIS method, and just as the advantages of the VIKOR method discussed above over the TOPSIS method, it can be seen that the ideal solution obtained by the FLM-VIKOR method is closer to the optimal solution than that by the FLM-TOPSIS method, so the result obtained by the method in this paper is more reasonable.

5. Conclusion
In this paper, we propose a new linguistic decision making method FLM-VIKOR. At first, we develop the FLM, and study its properties and operations. After that, we discuss the distance between FLMs. Then we combine it with VIKOR method to solve the fuzzy linguistic decision making problem, at the same time, we give the proposed method specific steps, and apply it on an instance. Finally, by comparing with existing methods, it is shown that the method proposed can obtain the ideal solution which closest to the optimal solution, meanwhile retaining the original information completely.

Acknowledgments
This work was financially supported by the Innovation Fund of Xihua University (ycjj2018187, ycjj2018062).

Reference
[1] Pei Z, Ruan D, Xu Y, et al. Linguistic values-based intelligent information processing: Theory, methods, and, applications [M]. Atlantic Press, 2010.
[2] Martınez L, Rodriguez R M, Herrera F. The 2-tuple Linguistic Model- Computing with Words in Decision Making[M]. Springer International Publishing Switzerland, 2015.
[3] Rodrıguez R M, Martınez L, Herrera F, Hesitant fuzzy linguistic term sets for decision making[J]. IEEE Transactions on Fuzzy Systems, 2012,20(1):109-119.

[4] Pang Q, Wang H, Xu Z H. Probabilistic linguistic term sets in multi-criteria group decision making[J]. Information Sciences, 2016, 369:128-143.

[5] Massanet S, Riera J V , Torrens J, et al, A new linguistic computational model based on discrete fuzzy numbers for computing with words[J]. Information Sciences, 2014, 258:277-290.

[6] Pei Z, Liu J, Hao F, et al. FLM-TOPSIS: The fuzzy linguistic multiset TOPSIS method and its application in linguistic decision making[J]. Information Fusion, 2019,45:266-281.

[7] Yu C M, Chen T. Multi-criteria decision making based on multiplication preference relation and order consistency [J]. Statistics & Decision 2017, (15) :42-47.

[8] Liao H C, Xu Z S, Zeng X J. Hesitant fuzzy linguistic VIKOR method and its application in qualitative multiple criteria decision making[J]. IEEE Transactions on Fuzzy System,2015,23(5):1343-1355.

[9] Martinez L, Rodriguez R M, Herrera F. The 2-tuple Linguistic Model-Computing with Words in Decision Making[J]. Springer International Publishing Switzerland, 2015.

[10] Yager R R. On the theory of bags[J]. Int. Journal of General Systems,1986, (13):23-37.

[11] Beg I, Reshid T. TOPSIS for hesitant fuzzy linguistic term sets[J]. International Journal of Intelligent Systems, 2013,28(12):1162-1171.

[12] Jin L S, Kalina M, Qian G. Discrete and continuous recursive forms of OWA operators[J]. Fuzzy Sets and Systems, 2017,308:106-122.

Design and Application Research of VR/AR Teaching Experience System

Baiqiang Gan

Guangzhou Nanyang Polytechnic College, 1123 Huanshi East Road, Conghua District, Guangzhou, Guangdong, China

790229822@qq.com

Abstract. This paper is based on the current situation of teaching, such as the lack of contextual interaction in the learning experience, the single situation, and the poor teaching experience and utilizes virtual reality (VR) and augmented reality (AR) technology to solve these problems. With the immersive, interactive and imaginative features of virtual reality technology, the virtual somatosensory virtual learning space with human-computer interaction is designed. The superposition of augmented reality technology is used to create an AR superposition learning space that combines real world and virtual space. According to the extension of VR/AR teaching thinking dimension, the teaching experience system design framework is constructed, and the physical model of experience system is proposed. Through data mining and teaching feedback, the learning experience is adjusted. Then design a personalized teaching experience system which can enhance learning experience and improve learning interest.

1. Introduction

With the continuous development of computer software and hardware technology, the combination of new technology and teaching is getting closer and closer, but the application of new technology has not solved the problems caused by traditional teaching. The current problems in school teaching mainly focus on the lack of interaction in the teaching process. Most of the teaching environment lacks a real contextual learning environment in the classroom. The teaching methods are single and it is difficult to inspire students' association and creativity. Application of the teaching process and teaching results are poor. So it is difficult to meet the task requirements of real jobs. The root cause of these problems is the poor teaching experience, which cannot provide learners with a real learning environment, intuitive and three-dimensional spatial experience.

The maturity of VR technology and AR technology, as well as the learner-centered, emphasis on teaching experience and the urgent need for adaptive learning, have driven the appearance of the VR/AR teaching experience system. VR achieved a brand-new state of human-computer interaction. It can obtain intuitive and real perceptions such as sight, hearing and touch by operating objects in the virtual world [1]. By combining virtual objects and the real world, AR can simultaneously display the information of the real world and the virtual world, enabling learners to use 3D models to enhance the visual perception ability of real-world situations [2].VR enhances sensory interactivity by constructing a simulated virtual world. The main features are: immersion, interactivity and imagination. Immersion allows learners to eliminate external disturbances and immerse themselves in virtual reality to gain an immersive feeling. Interactivity is based on the learner's head, hands, eyes, language and body movements to adjust the image and sound presented by the system. Imagination is to acquire visual,

Content from this work may be used under the terms of the Creative Commons Attribution 3.0 licence. Any further distribution of this work must maintain attribution to the author(s) and the title of the work, journal citation and DOI.
Published under licence by IOP Publishing Ltd

auditory, tactile, kinesthesia and other perceptions simultaneously in the virtual environment, enhance the learner's perception of the learning content, the high sensitivity and rational understanding of the cognitive content, so that to make the user to deepen the concept and sprouts new association, and motivate the learner's creative thinking. AR is a bridge connecting virtual world and real world. It is characterized by superposition and openness. It superimposes virtual information in the real world, enhances visual, auditory and tactile sense, and enables learners to experience the combination of real world and virtual world in the senses. The VR/AR teaching experience system compensates for the problems that appeared in traditional teaching. Figure 1 shows the VR/AR features.

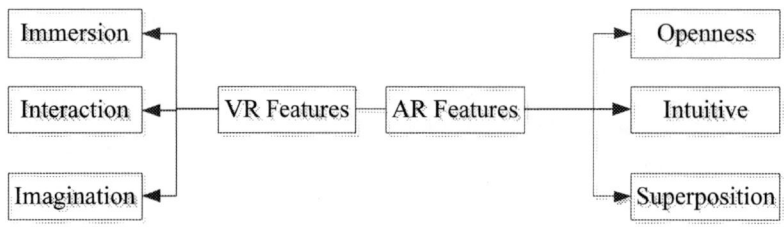

Figure 1. VR/AR Features

2. VR/AR Teaching Experience System Model Design

VR/AR teaching experience system model design mainly focuses on how to implement the experience system, how to design the system physical structure. The experience system is divided into five levels, from the bottom to the top, VR/AR base layer, VR/AR space layer, VR/AR logic layer, teaching data mining layer, teaching adjustment layer. The implementation process is to use the underlying physical device to enter the VR/AR learning space, and learners enter the database to intelligently retrieve the experience environment and experience context according to the needs in the learning space. The learner learns the state and data in the learning space and analyzes it through the teaching data mining layer, and then feeds back to the teaching adjustment layer to achieve the dynamic update of the learner's learning experience. Figure 2 shows the VR/AR teaching experience system model design.

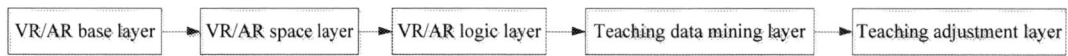

Figure 2. VR/AR Teaching Experience System Model Design

2.1. VR/AR base layer

The VR/AR base layer is the bottom part of the entire teaching experience system, and is also the physical device carrier and platform that enters the learning space, such as HMD, VR glasses, mobile phones, simulation devices, iPad, and so on. The learner enters the virtual platform through the physical device, and the virtual platform is a fusion of one or more learning spaces to ensure the reliability of the learner entering the VR/AR space layer.

2.2. VR/AR space layer

The VR/AR space layer contains various learning spaces of the system, including VR direct interaction space, AR superposition space, shape space, and virtual and real interlaced space [3]. In these spaces, learners can choose their own learning context according to their own learning situations, or they can choose intelligently through fixed and creative modes to understand the learners' knowledge storage and learning abilities, so that learners can be clear about themselves. In order to choose the best, most suitable learning situation and enhance the sense of learning experience.

2.3. VR/AR logic layer

The VR/AR logic layer achieves the learning of methods, processes and skills in the virtual space according to the characteristics and needs of the learners [4]. This layer implements the physical introduction and logical introduction of space. The physical device enters the learning context created by the VR/AR space layer, and sets the contactor. When the learner touches the contactor during the learning process, the logic introduction will start. The learning situation increases the difficulty node and the mission, and the learner completes the task by learning the knowledge and skills provided in the space to be able to enter the next contactor.

2.4. Teaching data mining layer

The teaching data mining layer mainly analyzes the learner's feedback data learning process and learning state in the VR/AR space layer, including the proficiency of the learners' knowledge when in the contactor or upgrade and the number of times that learner asks for help when encountering difficulties. Through the analysis and mining of these contents, the best learning path between learners and learning trajectories can be calculated and then be fed back to the teaching data adjustment layer to dynamically optimize the learner experience situation.

2.5. Teaching adjustment layer

The teaching adjustment layer is mainly divided into teaching strategy adjustment and teaching experience adjustment, mainly through the feedback of teaching data mining layer to adjust the learning strategy and adjust the learning situation. The teaching strategy adjustment is to personalize the teaching methods, set suspense or story plots, and attract learners to explore in the learning space. The teaching experience adjustment is to automatically complete the learners' experience scene in learning space according to the plots of teaching strategy adjustment.

3. VR / AR Learning Space

The VR/AR learning space is the core content of the teaching experience system. The learner goes to the virtual space for experience learning through the first viewpoint of VR, and enters the external perspective from the internal perspective [7]. The AR uses the third viewpoint to experience learning. The AR is a bridge between the virtual world and the real world. The learner enters the virtual world of the internal perspective from the external perspective in the real world to improve the interactive experience. The establishment of the two-way viewpoint of the VR/AR learning space enables the learner to switch back and forth between the internal perspective and the external perspective to achieve a fusion of virtual and real learning spaces and enhance the learning experience. The VR/AR learning space is mainly designed from database resources and JDK and SDK platforms. Figure 3 shows VR/AR learning space design.

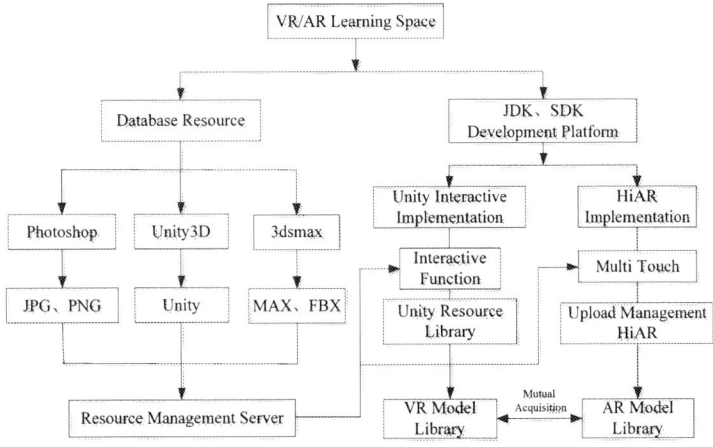

Figure 3. VR/AR Learning Space Design

3.1. Database Resources

The database resources are mainly used to develop and store the models and scenarios required in the learning space. The Unity 3D is mainly used to develop the VR model system and the AR model system for the development tools. Using 3dsmax production model and UV finishing, using Photoshop for post-laying drawing, then outputting textures and models into Unity3D to implement interactive programs through C#, and then making AR and VR separately [6]. Database resources are then passed to the JDK and SDK platforms for interaction through the management server.

3.2. JDK, SDK development platform

The JDK and SDK development platforms mainly implement resource mutual transfer between the VR model system and the AR model system. The JDK and SDK development platforms use the management resource server to acquire database resources, and implement resource packaging and interaction functions through Unity, and finally form a VR model system. The AR model system uses HIAR as an augmented reality development tool. HIAR provides cloud image recognition services in the form of REST and API, and uses HIAR's background management tools to implement AR content editing and management [5]. Through the JDK and SDK development platform, the VR model system and the AR model system are mutually called to form a VR/AR learning space with interactive functions.

4. Design of VR/AR Teaching Experience System

The VR/AR teaching experience system is an adaptive dynamic adjustment experience system. It can enable learners to experience the learning situation and collect the activity data generated by the learner in the learning space through the switching between the VR learning space and the AR learning space. The data is transmitted to the storage server, and analyzed by the data mining technology, and the analysis result is fed back to the computing server, thereby achieve the teaching experience and teaching strategy adjustment of the learning space. Figure 4 shows VR/AR teaching experience system design.

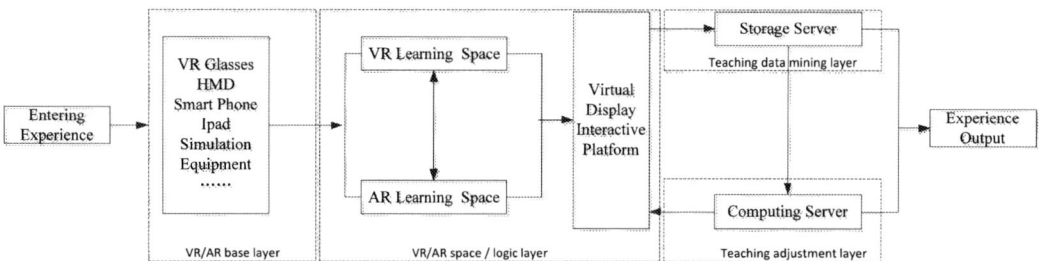

Figure 4. VR/AR Teaching Experience System Design

4.1. VR/AR learning space entry

The VR/AR learning space needs to be accessed from the VR/AR base layer, and corresponding hardware devices such as VR glasses, HMD, smart phones, and emulation devices are required. These devices can provide direct or indirect access to the learning space. The selected learning space can be a single space or two mixed spaces.

4.2. VR/AR learning space conversion

VR/AR learning space conversion is mainly based on virtual reality interactive platform. Through the database resources and JDK, SDK development platform, the fusion of VR model system and AR model system is implemented. According to the needs of the learners, different resources are used to transform and interact among the learning spaces. The combination of virtual and reality and superposition of each other greatly enhances the sense of reality of the teaching experience.

4.3. VR/AR learning spatial data analysis

According to the learning goal, the learner enters the corresponding VR/AR space by means of VR or AR, and can select different learning modes depended on the proficiency of knowledge mastery in the space. The system records the learner's entire learning trajectory and learning action data and then submits it to the storage server for analysis and production of the results through data mining techniques.

4.4. VR/AR learning space teaching adjustment

Through the feedback of the teaching data mining layer, the computer server selects the most suitable learning situation and learning task for the learner according to the result, so that besides the reality, the learner can also experience the comfort in the learning situation, and increase the enthusiasm of the learner to learn, encouraging learners to continue to learn.

4.5. VR/AR Experience Output

Experience output refers to the experience and feelings that learners have gained in completing the learning process. Learners enter the learning space, choose the learning situation, complete the learning tasks based on the learning objectives, and the difficulties encountered in the learning space can be analyzed, fed back, adjusted by the system, and finally reach the optimal experience path.

5. Conclusion

The introduction of VR/AR has changed the traditional teaching thinking, increased the control of teaching problems and teaching quality, and made the VR/AR teaching experience system an achievable reality. The system is designed based on the problems existing in traditional teaching. The experience of learners switching between the VR learning space and the AR learning space is the key point. It tracks the learning path and learning activity data in the learning space in real time, changes the teaching strategy and teaching experience, and provides an optimal learning experience environment for the learners.

References

[1]   Zhang Zhishi,"Educational Application and Fusion Reality Prospect of Virtual Reality and Augmented Reality",Modern Educational Technology,(2017), January 21-27.

[2]   Arvanitis T.N,Petron A,Knight J.F, "Human factor and qualitative pedagogical evaluation of a mobile augmented reality system for science education used by learners with physical disabilities", Pers Ubiquit Comput,(2009), March 243-250.

[3]   Li Xiaoping, Zhang Lin, "Research on the construction of intelligent virtual reality/augmented reality teaching system", China Electro-education Education,(2018), January 101-103.

[4]   Li Xiaoping,Chen Jianzhen,"Research on AR/VR learning situation design problems", Modern Educational Technology,(2017), August 12-17.

[5]   Zhang Zongbo, Yi Peng,"Engineering Graphics Virtual Reality Interaction Model Platform for Online Teaching", Journal of Donghua University(Natural Science),(2017), August 613-615.

[6]   Yu Riji, Cai Min,"Design and Application of Museum Culture Education Experience System Based on Mobile Terminal and AR Technology", China Electrotechnical Education,(2017), March 31-34.

[7]   Li Xiaoping, Zhao Fengnian,"Design and application of VR/AR teaching experience", China Electrotechnical Education,(2018), March 13-16.

Application of Blockchain in Document Certification, Asset Trading and Payment Reconciliation

Xingxiong Zhu[1,2], Dong Wang[1,2]

[1]State Grid Power Finance and E-Commerce Laboratory, Beijing 100053, China

[2]State Grid Electronic Commerce Co., Ltd., Beijing 100053, China

zhuxx@pku.org.cn

Abstract. In economic activities, there are massive contracts, transactions, and payments. The consistency, integrity, and security of these important transaction data are crucial. The application of blockchain technology in document certification, asset trading and payment reconciliation, solves the security vulnerabilities existing in traditional technical solutions, optimizes business models, improves efficiency, ensures security, and enhances core competitiveness.

1. Introduction

In economic activities, there are extremely large amounts of contracts, transactions, and payments. The consistency, integrity, and security of these important transaction data are crucial. The following sections describe the application of blockchain technology in the blockchain-based file certification system, the blockchain-based digital asset trading system, and the blockchain-based payment reconciliation system.

2. Blockchain-based file certification system

2.1. Background

In e-commerce, there are enormous important documents such as electronic agreements [1], total assets, and income records. Electronic data is easy to tamper and easy to falsify [2]. Centralized storage mode is prone to systemic risks such as loss and damage. How to make electronic data credible, traceable, and difficult to tamper is a problem that needs to be solved.

2.2. Technical solutions

- Use the Hash algorithm to generate hash values for important files and create a digital "fingerprint" of the file.
- Distributively store the hash value of the file in the blockchain node server.
- Construct a blockchain alliance chain, which consists of trustworthy alliance organs such as judicial organ, judicial appraisal center, electronic certification organ, financial institution, and Internet platform.
- Use the distributed technical characteristics of the blockchain to solve the problem of difficulty in collecting electronic evidence [3].
- Apply blockchain certification to enhance credibility in scenarios such as intellectual properties, electronic contracts, and financial services.

Content from this work may be used under the terms of the Creative Commons Attribution 3.0 licence. Any further distribution of this work must maintain attribution to the author(s) and the title of the work, journal citation and DOI.

Published under licence by IOP Publishing Ltd

- The blockchain certification proves that the electronic data has not been tampered with after being stored in the blockchain [4], which proves the integrity and consistency of the electronic file.
- Blockchain technology guarantees the stage of electronic data storage. Improve the credibility of evidence confirmation and evidence forensics.

2.3. System structure

In the application of blockchain technology in e-commerce certification and evidence collection, the key information such as the user account and the hash value of the evidence file is stored in distributed ledger to realize the certification and evidence collection of important evidence files. When the business application creates a new user, the blockchain platform service is invoked, and the corresponding blockchain user is created on the chain. The blockchain platform uses the encryption rules to generate the corresponding public and private keys, and stores the public key information on the chain.

Generate a hash value for important file. The hash value of the file is distributed and stored in the blockchain node server. The document consistency and the authenticity of the document are verified by whether the hash value of the document to be authenticated is consistent with the hash value of the certification file on the blockchain.

2.4. System solutions

The blockchain platform uses its technical features that cannot be tampered with and traceable to realize the certification and evidence collection. These records cannot be tampered with, ensuring the security of the data. Once the user has a business dispute, the authentication organization only needs to verify whether the hash value of the user's original data is in the blockchain, and the identification result can be quickly made and submitted to arbitration.

Each time the file is certificated with a timestamp record, clearly record the time of certification of the document evidence. All the certification of important documents are recorded, forming a chain of evidence. The blockchain removes the single-point authoritative credit model in the centralized system, and the multi-node records the hash digest. Which effectively prevents the hacker from attacking the single node in the centralized system and improves the security of the system.

3. Blockchain-based digital asset trading system

3.1. Background

In the traditional technical solution, digital asset trading is centralized [5]. The participants communicate with other parties through the centralized network mechanism. Once the centralization organization fails, it will affect the normal operation of the entire trading system. The system security will not be guaranteed. Digital asset registration data and transaction data are stored in a centralized database [6]. Once the database is attacked, it is easily falsified and unsafe.

3.2. Technical solutions
- Simplify the digital asset trading process, prevent data from being tampered with, and improve transaction efficiency and transaction security.
- Use blockchain data structures to store digital asset registration data and its transaction data [7].
- Generate and update digital assets and their transaction data using a distributed node consensus algorithm.
- Secure the data transmission between the blockchain nodes by using encryption algorithm and digital signature.
- Use the digital signature method to ensure the authenticity of the transaction subject and ensure the non-repudiation of the transaction.
- Realize clear settlement of digital asset transactions by using smart contracts composed of programming script code.

- Use the blockchain to trace the source of digital asset and track each transaction.

3.3 System structure

The data of digital asset creation and transaction will be packaged into blocks, recorded in the blockchain based on the consensus mechanism, and stored in each nodes to form a digital asset distributed ledger.

The blockchain-based digital asset creating process is shown in Figure 1. And the blockchain-based digital asset trading process is shown in Figure 2.

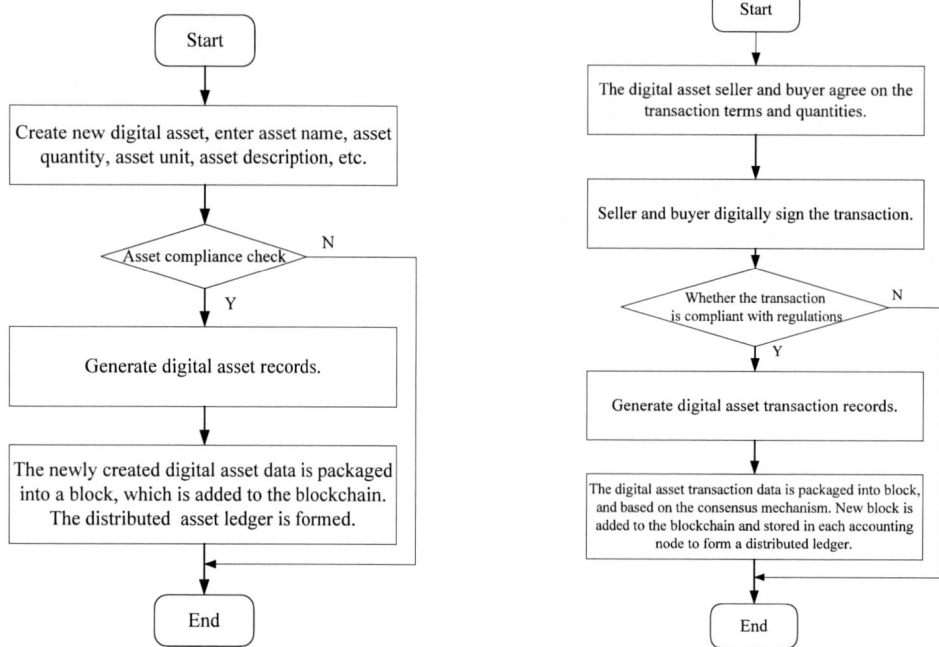

Figure 1. Blockchain-based digital asset creating process.   Figure 2. Blockchain-based digital asset trading process.

Each blockchain node distributes messages through the P2P protocol. The process of verifying, accounting, storing, transmitting and maintaining are based on a distributed system structure.

The smart contract of digital asset trading implements various scripts, algorithms and contract codes. It is the basis for the programmable nature of the blockchain-based digital asset system.

3.4. System solutions

The asset issuer creates digital assets and enters information such as asset name, asset quantity, asset measurement unit, and asset description. The seller issues the asset information, the buyer issues the transaction application. The buyer and the seller agree on the transaction items and quantities.

When the transaction data is transmitted between the blockchain nodes, the sender node encrypts the transaction data using the recipient public key to form ciphertext. The sender uses the hash function to generate digest from the transaction packet, signed with the private key. The transaction data ciphertext and digital signature are sent to the recipient.

The receiver node uses its own private key to decipher the ciphertext, obtain the transaction data, and use the hash function to get a digest. The sender's digital signature is verified using the sender's public key to obtain a digest. Compare the two digests. If they are identical, the transmitted data is complete and has not been modified during the transmission. And it confirms the true identity of the sender and recipient.

The blockchain accounting node records the transaction hash, transaction time, transaction originator address, recipient address, asset name, asset issuer address, and asset trading quantity in the block. Each asset transaction has a corresponding hash value, which is unique. Smart contract of clearing and settlement executes the settlement of assets, payment, expenses, profit sharing, accounts, etc., according to whether the clear settlement condition meets.

4. Blockchain-based payment reconciliation system

4.1. Background
Reconciliation of payments are in order to ensure the correct [8], consistent and reliable transaction payment and billing records. In the conventional technical solution, the reconciliation files are transmitted by FTP. There are payment and refund record files in the bank side, Internet payment platform side, and Internet e-commerce platform side. But the items of records among these sides are not consistent. How to find and handle these inconsistencies are crucial.

In traditional technical solutions, payment reconciliation usually initiates after 1 week or even 1 month of trade. It is long time delay for weekly or monthly reconciliation. Manual reconciliation is labour-intensive and inefficient. The use of FTP to transfer transaction data is not secure, and its data integrity is not guaranteed.

4.2. Technical solutions
- Simplify reconciliation process to prevent data from being tampered with, improving transaction efficiency and transaction security.
- Using smart contract to realize automatic payment reconciliation.
- Real-time reconciliation to improve capital turnover efficiency.
- Using blockchain multi-node to realize distributed ledger of reconciliation data.
- Track the source of the payment data and each transaction.
- Significantly reduce customer complaints caused by customer asset lost, and improve user experience and satisfaction.

4.3. System structure
The blockchain-based payment reconciliation process is shown in Figure 3.

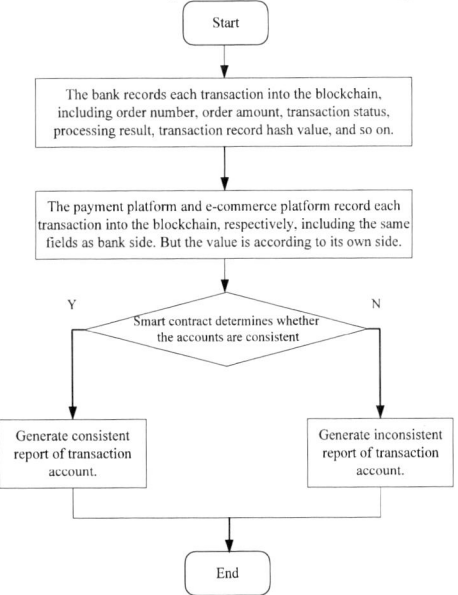

Figure 3. Blockchain-based payment reconciliation process.

The blockchain-based payment reconciliation system users include consumers, banks, payment platform, Internet e-commerce platform, merchants. The payment records of users are distributed in each node to form reconciliation ledger based on blockchain.

The smart contract of blockchain executes payment reconciliation process according to whether the account consistency conditions are met. It makes the entire reconciliation process automatically and intelligently, reduces the risk of trading errors, and improves efficiency.

4.4.System solutions
There is no need to transmit the payment records files. The bank does not need to provide API interface. Data is shared through distributed ledgers.

Blockchain nodes of the participating units, such as banks, payment platform, and e-commerce platform, deploy respectively, to form alliance chain. Each participant updates data respectively, including transaction serial number, payer, payee, payment amount, refund amount, transaction status, commodity order serial number, time, etc.

The smart contract check whether the values of important record fields are consistent among bank, payment platform, and e-commerce platform. Then it obtains the reconciliation result. When the payments are inconsistent, the warning information is automatically proposed. In the end, detailed report is intelligently generated.

5. Conclusion
The application of blockchain technology in document certification, asset trading and payment reconciliation, solves the security vulnerabilities existing in traditional technical solutions, optimizes business models, improves efficiency, ensures security, and enhances core competitiveness.

Author introduction
Xingxiong Zhu (1975), male, graduated from Peking University, China, master degree in software engineering, one of the main drafters of China National Standard GB/T 25656-2010 "Information Technology Chinese Linux Application Programming Interface (API) specification". Email, zhuxx@pku.org.cn.

References
[1]   Governatori, G., Idelberger, F., Milosevic, Z., Riveret, R., Sartor, G., & Xu, X. (2018). *On legal contracts, imperative and declarative smart contracts, and blockchain systems.* Artificial Intelligence and Law, 26(4), 377-409.
[2]   Ying, W., Jia, S., & Du, W. (2018). *Digital enablement of blockchain: Evidence from HNA group.* International Journal of Information Management, 39, 1-4.
[3]   Sharma, P. K., Moon, S. Y., & Park, J. H. (2017). *Block-VN: A distributed blockchain based vehicular network architecture in smart City.* Journal of Information Processing Systems, 13(1), 84.
[4]   Swan, M. (2015). *Blockchain: Blueprint for a new economy.* " O'Reilly Media, Inc.".
[5]   Tadelis, S. (2016). *The economics of reputation and feedback systems in e-commerce marketplaces.* IEEE Internet Computing, 20(1), 12-19.
[6]   Peters, G. W., & Panayi, E. (2016). *Understanding modern banking ledgers through blockchain technologies: Future of transaction processing and smart contracts on the internet of money.* In Banking Beyond Banks and Money (pp. 239-278). Springer, Cham.
[7]   Iansiti, M., & Lakhani, K. R. (2017). *The truth about blockchain.* Harvard Business Review, 95(1), 118-127.
[8]   Rose, A. J., Fischer, S. H., & Paasche-Orlow, M. K. (2017). *Beyond medication reconciliation: the correct medication list.* Jama, 317(20), 2057-2058.

ISPECE   IOP Publishing

IOP Conf. Series: Journal of Physics: Conf. Series **1187** (2019) 052081   doi:10.1088/1742-6596/1187/5/052081

Study on the Relationship between the Sharing Rate of Vehicle Exhaust Pollution and the Quantity of Possession

Baiyu Chen[1, a] , Da Fu[2, b] , Yuanyuan Yang[2, c] , Xiao Li[3, d]

[1] College of Engineering, University of California Berkeley, Berkeley, USA
[2] Business School, Sichuan University, Chengdu, China
[3] College of Engineering, Ocean University of China, Qingdao,China

E-mail: [a] baiyu@berkeley.edu, [b] fuda1213@outlook.com, [c] zerocircle@outlook.com,
[d] 374965620@qq.com

Abstract. The pollution caused by motor vehicle exhaust is largely affected by the quantity of motor vehicles. This paper improves the calculation method of emission inventory by analyzing the statistical characteristics of starting pollutant emission factors and operating emission factors of motor vehicles and gives the calculation plan of the total emission sharing rate and emission concentration sharing rate. Based on the GM (1,1) model, the vehicle ownership of the five-year basic data was predicted. The results show that the model accuracy is $P^0 = 1 - \bar{\varepsilon} = 98.11\%$ and the correlation degree is 0.66>0.6. This shows that the accuracy of the model prediction results meets the requirements, and the vehicle ownership value can be used as a reference for relevant policies.

1. Background

With the development of science and technology, electric vehicles are gradually entering people's field of vision. It can be expected that with the increase of electric vehicles, the pollution of automobile exhaust to the atmosphere will be greatly alleviated. However, in the short term, electric vehicles cannot replace existing motor vehicles. The development of the automobile industry and the number of motor vehicles continue to grow, and air pollution is still getting worse. In the case of higher and higher atmospheric pollutants, it is necessary to pay attention to the interaction relationship between vehicle exhaust pollution sharing rate and possession[1-10].

According to a survey report on the source of atmospheric pollutants from the United States in the 1970s, 77.3% of CO, 55.3% of HC and 50.9% of NOx in the urban atmosphere are derived from vehicle exhaust emissions. Motor vehicle exhaust emissions have become an important source of urban air pollution[11-23]. The motor vehicle pollution sharing rate is divided into two categories[24-36]: the total emission sharing rate and the emission concentration sharing rate according to different measurement priorities and calculation methods[37-41].

Car ownership is related to urban construction and planning. In this paper, the statistical characteristics of vehicle emission factors and operational emission factors are analyzed, the calculation method of emission inventory is improved, and the calculation scheme of total emission sharing rate and emission concentration sharing rate is given[35-39]. Based on the GM (1,1) model, the vehicle ownership of the collected five-year basic data was predicted. The calculation results are for the government to formulate a social macroeconomic development plan, and it is of reference

---

Content from this work may be used under the terms of the Creative Commons Attribution 3.0 licence. Any further distribution of this work must maintain attribution to the author(s) and the title of the work, journal citation and DOI.
Published under licence by IOP Publishing Ltd

significance to implement energy conservation and emission reduction and traffic management in cities.

2. Emission Measurement Factor

The sharing rate is to quantify the contribution of different pollution sources to an air pollutant, and it is the share of pollution sources in air pollution. The motor vehicle pollution sharing rate represents the percentage of pollutants emitted by motor vehicles in the total emissions of all atmospheric pollutants[42-49].

The total vehicle exhaust emission sharing rate is usually expressed by η1. It is the ratio of the total amount of pollutants Q_{car} emitted by the vehicle exhaust to the total amount Q of all local pollutants. The total amount of pollutant emissions is usually composed of pollutants emitted by mobile sources and pollutants emitted by stationary sources. There are expressions:

$$\eta_1 = \frac{Q_{car}}{Q_{fix} + Q_{car}} \times 100\% \qquad (1)$$

The vehicle emission concentration sharing rate is calculated by obtaining a detailed emission inventory of each pollution source in the region, and using a verified regional multi-source diffusion model (such as ISC ST3) to calculate and analyze the concentration sharing rate of different pollution sources in different seasons and locations. If the concentration near the road is c, and the pollution concentration away from the road in this area is c1, the car pollution concentration share near the road can be calculated as:

$$\eta = \frac{C - C1}{C} \times 100\% \qquad (2)$$

3. Motor Vehicle Emission Factor Model

The vehicle emission factor refers to the amount of pollutants discharged by the motor vehicle operating unit and is sometimes expressed by the amount of pollutants discharged per unit of fuel [50-56]. The unit is g/km or g/L, which indicates the level of vehicle exhaust emissions. To obtain a vehicle emission factor (start emission factor (g/time) and operational emission factor (g/km)) for a vehicle exhaust emission list, it is necessary to separately measure the pollutants emitted by the vehicle at start-up and the amount of pollutants emitted during operation.

4. Calculation of Motor Vehicle Pollution Discharge List

The calculation formula of the vehicle exhaust pollutant discharge list is as follows:

$$Q = Q1 + Q2 = \sum_{i=1}^{n} (A_i \times EF1 \times Ti) + \sum_{i=1}^{n} (A_i \times EF2 \times Li) \qquad (3)$$

Where: Q represents the total annual emissions (t/a) of motor vehicle pollutants. Q1 represents the annual emissions (t/a) of the vehicle when it is running. Q2 represents the annual emissions (t/a) of the vehicle during the start-up process. Ai is the number of i-type cars. EF1 is the operating emission factor (t/km) of the i-type vehicle. Ti is the annual mileage (km/a) of the i-type car. Li is the number of annual starts of the i-type car (times / a). EF2 is the starting emission factor (t/time) for the i-type vehicle.

Using formula (3), calculate the pollutant discharge list of motor vehicles.

The emission total sharing rate is calculated from the emission inventory and calculated fixed source emission data and formula (1).Through the calculation results we can know that the exhaust emissions of motor vehicles account for a considerable proportion of the emissions of atmospheric pollutants. Therefore, it is urgent to strengthen the control of tail gas pollution.

5. Vehicle ownership forecasting model

The grey system has the characteristics of fuzzy hierarchy and structural relationship, random data dynamic change and incomplete basic data, which is referred to as GM model. The most commonly used one is GM(1,1).

Assume that there are n equal time interval observations $X^{(0)} = \{X^{(0)}(1) \ldots X^{(0)}(n)\}$, add up m times $X^{(m)}(k) = \sum_{i=1}^{k} X^{(m-1)}(i)$ derive array $X^{(m)} = \{X^{(m)}(1), \ldots X^{(m)}(n)\}$ (m=1,2, $\cdots$) Let $Z^{(m)}$ be the mean array of $X^{(m)}$, $Z^{(m)}(k) = \left[X^{(m)}(k) + X^{(m)}(k+1) \right]/2$, so $Z^{(m)} = \{Z^{(m)}(2) \ldots Z^{(m)}(n)\}$.

When m=1, a grey prediction model is established for the new array $X^{(1)} = \{X^{(1)}(1) \ldots X^{(1)}(n)\}$. The differential equation corresponding to the GM(1,1) model is:

$$\frac{dx^{(1)}}{dt} + ax^{(1)} = b \tag{4}$$

Where: a is the development grey number; b is the endogenous control grey number. Let $\hat{\alpha}$ be the parameter vector to be estimated, $\hat{\alpha} = (a,b)^T$. Then the least squares estimation parameter sequence of the gray differential equation satisfies:

$$\hat{\alpha} = (B^T B)^{-1} B^T Y_n \tag{5}$$

Where: $B = \begin{pmatrix} -z^{(1)}(2) & 1 \\ -z^{(1)}(3) & 1 \\ \ldots & \ldots \\ -z^{(1)}(n) & 1 \end{pmatrix} = \begin{pmatrix} -(x^{(1)}(1)+x^{(1)}(2))/2 & 1 \\ -(x^{(1)}(2)+x^{(1)}(3))/2 & 1 \\ \ldots & \ldots \\ -(x^{(1)}(n\text{-}1)+x^{(1)}(n))/2 & 1 \end{pmatrix}$, $Y_n = \begin{pmatrix} x^{(0)}(2) \\ x^{(0)}(3) \\ \ldots \\ x^{(0)}(n) \end{pmatrix}$

According to the principle of least squares, the parameters a, b are obtained:

$$a = \frac{CD - (n-1)E}{(n-1)F - C^2}, b = \frac{DF - CE}{(n-1)F - C^2} \tag{6}$$

Where: $C = \sum_{k=2}^{n} Z^{(1)}(k)$; $D = \sum_{k=2}^{n} x^{(0)}(k)$; $E = \sum_{k=2}^{n} z^{(1)}(k)x^{(0)}(k)$; $F = \sum_{i=1}^{n} z^{(1)}(k)^2$

In the application, the development of the grey number should satisfy -2<a<2. When a ≥ -0.3, the model predicts better.

The prediction model can be obtained by solving the differential equation:

$$\hat{x}^{(1)}(k+1) = [x^{(0)}(1) - \frac{b}{a}]e^{-ak} + \frac{b}{a} \quad (k=0,1,\ldots n-1) \tag{7}$$

Calculate the simulated value of W according to the model and restore the model,

$$\begin{cases} \hat{x}^{(0)}(1) = x^{(1)}(1) = x^{(0)}(1) \\ \hat{x}^{(0)}(k+1) = \hat{x}^{(1)}(k+1) - \hat{x}^{(1)}(k) \end{cases} \quad (k=1,2 \ldots n) \tag{8}$$

The $X^{(0)}$'s simulated value $\hat{x}^{(0)} = \{\hat{x}^{(0)}(1) \ldots \hat{x}^{(0)}(n)\}$ is available.

In practical applications, the model accuracy is usually required to be greater than 80%. When the resolution is 0.5, the correlation degree is not less than 0.6.

6. Practical application of the grey PARC prediction model

Original series $x^{(0)} = \{10273.6, 11418.4, 12742.5, 15543.4, 16993.5\}$ is the total number of motor vehicle ownership data for China for 2005-2009. Generate an array by accumulating once: $x^{(1)} =$

{10273.6, 21692, 34434.5, 49977.9, 66971.4}. Construct the matrix B and the data vector Y_n, calculating respectively $B^T B$, $(B^T B)^{-1}$, $B^T Y_n$, $\hat{\alpha}$.

The parameter sequence a=-0.137400721 and b=9203.054897 are obtained by using equation (6). Since a satisfies: a $\in$ (-2,2) and a $\geq$-0.3, the GM(1,1) model can be used for medium- and long-term prediction. Substituting a and b into equation (7), the final prediction model is:

$$\hat{x}^{(1)}(k+1) = [\ 10273.6 + 67000]\ e^{0.137400721k} - 67000 = 77273.6\ e^{0.137400721k} - 67000$$

$$\hat{x}^{(0)}(k+1) = 77273.6\ e^{0.137400721k} - e^{0.137400721(k-1)}\ (k=1,2,\cdots,\ n-1).$$

Restore model to get predicted value: when k=0, $\hat{x}^{(0)}(1) = 10273.6 = x^{(0)}$.

$\hat{x}^{(0)}(6) = 19713.9$ is the forecast data for 2010 when k=5. The forecast results after 2010 are available.

It is usually necessary to test both the residual and the correlation:the correlation degree is 0.66>0.6, and the prediction accuracy meets the requirements. Therefore, the vehicle holdings in 2015 can be used as a reference for relevant policies.

7. conclusion

It is of theoretical and practical significance to use the grey PARC prediction model to support relevant policy data. This method has characteristics: if the positive data are added, m will increase continuously and the randomness of the generated sequence will be weakened a lot. When m is large enough, it can be considered that the time series has changed from the disordered random sequence to the ordered non-random sequence, and the obtained non-random sequence can be approximated by exponential curve.

Motor vehicle exhaust pollution is largely affected by motor vehicle ownership, and the prediction of motor vehicle ownership can provide data basis for relevant policies, and according to the model calculation results, the planning Suggestions of motor vehicle exhaust pollution control are given from a quantitative perspective.

References

[1] Chen B, Yang Z, Huang S, Du X, Cui Z, Bhimani J, et al. Cyber-physical system enabled nearby traffic flow modelling for autonomous vehicles. 36th IEEE International Performance Computing and Communications Conference, Special Session on Cyber Physical Systems: Security, Computing, and Performance (IPCCC-CPS) IEEE. 2017.

[2] ESCALANTE H J, PONCE-LÓPEZ V, WAN J, et al. ChaLearn Joint Contest on Multimedia Challenges Beyond Visual Analysis: An overview. 2016 23rd International Conference on Pattern Recognition (ICPR). 2016: 67–73.

[3] CHEN B, LIU G, WANG L. Predicting Joint Return Period Under Ocean Extremes Based on a Maximum Entropy Compound Distribution Model. *International Journal of Energy and Environmental Science,* 2017,**2(6)**: 117-126. 2017.

[4] ANTHONY B, CHEN B Y. How Emerging Technologies Could Transform Infrastructure.http://www.governing.com/gov-institute/voices/col-hyperlane-emerging-technologies-transform-infrastructure.html.

[5] CHEN B, ESCALERA S, GUYON I, et al. Overcoming Calibration Problems in Pattern Labeling with Pairwise Ratings: Application to Personality Traits. Computer Vision – ECCV 2016 Workshops. Springer, Cham, 2016: 419–432.

[6] Wang L P, Chen B Y, Chen C, Chen Z-S, Liu G L. Application of linear mean-square estimation in ocean engineering. *China Ocean Eng. Chinese Ocean Engineering;* 2016;**30**: 149–160.

[7] PONCE LÓPEZ V, CHEN B, OLIU M, et al. ChaLearn LAP 2016: First Round Challenge on First Impressions - Dataset and Results. Computer Vision – ECCV 2016 Workshops. Springer,

Cham, 2016: 400–418.

[8] Zhang S F, Shen W, Li D S, Zhang X W, Chen B Y，Nondestructive ultrasonic testing in rod structure with a novel numerical Laplace based wavelet finite element method. *Latin American Journal of Solids and Structures,* 2018, **15(7)**:1-17, e48.

[9] Wang L P, Chen B, Zhang J F, Chen Z. A new model for calculating the design wave height in typhoon-affected sea areas. *Natural Hazards. Springer Netherlands;* 2013;**67**: 129–143.

[10] Chen Y, Zhang Z. Study of the multicast routing protocols based on predicting link condition. In Computer Engineering and Technology (ICCET), 2010 2nd International Conference on 2010 Apr 16 (Vol. 7, pp. V7-762). IEEE.

[11] CHEN B, WANG B. Location Selection of Logistics Center in e-Commerce Network Environments. *American Journal of Neural Networks and Applications, Science Publishing Group,* 2017, **3(4)**: 40-48.

[12] Wang L, Xu X, Liu G, Chen B, Chen Z. A new method to estimate wave height of specified return period. *Chin J Oceanol Limnol. Science Press;* 2017;**35**: 1002–1009.

[13] Chen B Y,Liu G L,Zhang J F *et al*. A calculation method of design wave height under the three factors of typhoon: China, ZL 2016 1 0972118.X. Patent, 2016-10-31.

[14] Wang L P, Liu G L, Chen B Y *et al*. Typhoon influence considered method for calculating combined return period of ocean extreme value: China, ZL 2010 1 0595807.6. Patent, 2013-03-20.

[15] Liu G L, Zheng Z J，Wang L P,*et al*. Power-type wave absorbing device and using method thereof: China, ZL 2015 1 0575336.5. Patent, 2017-11-03.

[16] Wang L P, Liu G L, Chen B Y, *et al*. Typhoon based on the principle of maximum entropy waters affect the design wave height calculation method: China, ZL 2010 1 0595815.0[P]. Patent, 2015-08-19.

[17] LIU G, CHEN B, WANG L, et al. Wave Height Statistical Characteristic Analysis. *Journal of oceanology and limnology.* 2018, **36(4)**:1123-1136.

[18] Chen B Y, Liu G L, Wang L P, Zhang K Y, Zhang S F. Determination of Water Level Design for an Estuarine City. *Journal of oceanology and limnology* . 2019, https://doi.org/10.1007/s00343-019-8107-z

[19] Chen B Y, Zhang K Y, Wang L P, Jiang S, Liu G L. Generalized Extreme Value - Pareto Distribution Function and Its Applications in Ocean Engineering, *Chinese Ocean Engineering;* 2019;2

[20] Liu X, He Y, Fu, H, Chen B, Wang M, Wang Z. How Environmental Protection Motivation Influences on Residents' Recycled Water Reuse Behaviors: A Case Study in Xi'an City. Water2018, 10, 1282.

[21] Liu X, He Y, Fu H, Chen B, Wang M. Scientometric of Nearly Zero Energy Building Research:A Systematic Review from the Perspective of Co-Citation Analysis.*Journal of Thermal Science.*

[22] Jiang S , Lian M J, Lu C W ,Ruan S L ,Wang Z and Chen B Y. SVM-DS fusion based soft fault detection and diagnosis in solar water heaters, *Energy Exploration & Exploitation,* DOI: 10.1177/0144598718816604

[23] Tong L, Chen Y, Wang Z, Qin Q. Diesel Engine Fault Diagnosis Based on D-S Evidence Theory. *Science Technology and Engineering.* 2010(**15**):3749-52.

[24] Cui P, Tonnemacher M, Rajan D, Camp J. WhiteCell: Energy-efficient use of unlicensed frequency bands for cellular offloading. In Proceedings of 2015 IEEE International Symposium on Dynamic Spectrum Access Networks (DySPAN), Stockholm,Sweden, 29 Sept.-2 Oct. 2015;IEEE:Piscataway, NJ,2015; pp.188-199.

[25] Cui P, Liu H, He J, Altintas O. Vuyyuru, R. ;Rajan, D.; Camp, J. Leveraging diverse propagation and context for multi-modal vehicular applications. In Proceedings of 2013 IEEE 5th International Symposium on Wireless Vehicular Communications (WiVeC), 2 June – 3 June 2013; IEEE: Dresden, Germany.

[26] Wang R, Ye C, Shen B and Liu Y. "To improve localization accuracy: A two-objective

optimization method," in 2018 13th IEEE Conference on Industrial Electronics and Applications (ICIEA). IEEE, 2018, pp. 2510–2513.

[27] Wang R, Shen B, Liu Y. "Optimization of sensor deployment for localization accuracy improvement," in 2016 IEEE International Conference on Consumer Electronics-China (ICCE-China), 2017.

[28] Liu Y, Zhan Y -J, Chen J. "An integrated approach to sink and sensor role selection in wireless sensor networks: Using dynamic programming." *Adhoc& Sensor Wireless Networks,* **vol. 22**, 2014.

[29] Koag M C, Cheun Y, Kou Y, Ouzon-Shubeita H, Min K, Monzingo A F, et al. Synthesis and structure of 16,22-diketocholesterol bound to oxysterol-binding protein Osh4. *Steroids*. 2013;**78**: 938–944.

[30] Kou Y, Koag M C, Cheun Y, Shin A, Lee S. Application of hypoiodite-mediated aminyl radical cyclization to synthesis of solasodine acetate. *Steroids*. 2012;**77**: 1069–1074.

[31] Cai W, Leslie Lauren Gouveia, "Modeling and simulation of Maximum power point tracker in Ptolemy" , *Journal of Clean Energy Technologies*, **Vol. 1**, No. 1, 2013 , PP 6-9.

[32] Kou Y, Cheun Y, Koag M C, Lee S. Synthesis of 14′,15′-dehydro-ritterazine Y via reductive and oxidative functionalizations of hecogenin acetate. *Steroids*. 2013;**78**: 304–311.

[33] Kou Y, Koag M C, Lee S. Structural and Kinetic Studies of the Effect of Guanine N7 Alkylation and Metal Cofactors on DNA Replication. *Biochemistry*. 2018;**57**: 5105–5116.

[34] Liu Y, Chen J, and Zhan Y -j. "Local patches alignment embedding based localization for wireless sensor networks," *Wireless personal communications,* **vol. 70**, no. 1, pp. 373–389, 2013.

[35] Kang L, Du H L, Zhang H, Ma W L. Systematic research on the application of steel slag resources under the background of big data. Complexity 2018. doi.org/10.1155/2018/6703908.

[36] Kang L, Du H L, Du X, Wang H T, Ma W L, Wang M L, Zhang F B. Study on dye wastewater treatment of tunable conductivity solid-waste-based composite cementitious material catalyst. *Desalination and Water Treatment* 2018, **125**, 296-301.

[37] Kang L, Zhang Y J, Zhang L, Zhang K. Preparation, characterization and photocatalytic activity of novel CeO2 loaded porous alkali-activated steel slag-based binding material[J]. *International Journal of Hydrogen Energy*, 2017, **42**: 17341-17349.

[38] Fu H, Li Z, Liu Z, Wang Z. Research on Big Data Digging of Hot Topics about Recycled Water Use on Micro-Blog Based on Particle Swarm Optimization. *Sustainability* 2018, **10**, 2488.

[39] Jiang S, Lian M, Lu C, Gu Q, Ruan S, Xie X. Ensemble Prediction Algorithm of Anomaly Monitoring Based on Big Data Analysis Platform of Open-Pit Mine Slope. *Complexity* 2018. doi.org/10.1155/2018/1048756.

[40] Zhang G H, Liang G Y, Li W Z, Fang J, Wang J B, Geng Y Y, Wang J Y. "Learning Convolutional Ranking-Score Function by Query Preference Regularization", International Conference on Intelligent Data Engineering and Automated Learning, pp. 1-8, 2017.

[41] Zhang G H, Liang G Y, Su F, Qu F X, Wang J Y, "Learning convolutional attribute embedding for domain-transfer learning", Lecture Notes in Artificial Intelligence, 2018.

[42] Cai W, Jeremy C, David G. "3-Axes MEMS Hall-Effect Sensor," presented by the 2011 IEEE Sensors Applications Symposium, pp141-144.

[43] Chen Y. A novel chaos-based fragile watermarking for image tampering detection and self-recovery. Undergraduate Thesis, Harbin Institute of Technology. 2011 Jun.

[44] Cai W,Cui X L, Zhou X R. "Optimization of a GPU Implementation of Multi-dimensional RF Pulse Design Algorithm," International Conference on Bioinformatics and Biomedical Engineering 2011.

[45] Cai W, Frank S. "2.4 GHz Heterodyne Receiver for Healthcare Application",*International Journal of Pharmacy and Pharmaceutical Sciences*,**Vol 8** Issue 6, 2016,pp 162-165.

[46] Cai W, Frank S, "2.4 GHz Heterodyne Receiver for Healthcare Application", International

Journal of Pharmacy and Pharmaceutical *Sciences*,**Vol 8**, sup2, 2016,pp 22-25.

[47] Cheun Y, Kou Y, Stevenson B, Kim H K, Koag M C, Lee S. Synthesis of C17-OH-north unit of ritterazine G via "Red-Ox" modifications of hecogenin acetate. *Steroids*. 2013;**78**: 639–643.

[48] Kou Y, Lee S. Unexpected opening of steroidal E-ring during hypoiodite-mediated oxidation. *Tetrahedron Lett*. 2013;**54**: 4106–4109

[49]Kou Y, Others. Structural and kinetic study of N7-methyl, N7-benzyl and C8-chloro guanine lesions using human DNA polymerase β [Internet]. 2015. Available: https://repositories.lib.utexas.edu/handle/2152/46815

[50] Lee S, Kou Y, Koag M. Mechanism of alkylation and platination-induced mutagenesis. *ENVIRONMENTAL AND MOLECULAR MUTAGENESIS* **59**, 107-107.

[51] Kou Y, Koag M C, Lee S. N7 methylation alters hydrogen-bonding patterns of guanine in duplex DNA. *J Am Chem Soc*. 2015;**137**: 14067–14070.

[52] Cheun Y, Koag M C, Kou Y, Warnken Z, Lee S. Transetherification-mediated E-ring opening and stereoselective "Red-Ox" modification of furostan. *Steroids*. 2012;**77**: 276–281.

[53] Lei X, Kou Y, Fu Y, Rajashekar N, Shi H, Wu F, et al. The cancer mutation D83V induces an α-helix to β-strand conformation switch in MEF2B. *J Mol Biol*. 2018; doi:10.1016/j.jmb.2018.02.012

[54] Lei X, Shi H, Kou Y, Rajashekar N, Wu F, Sen C, *et al*. Crystal Structure of Apo MEF2B Reveals New Insights in DNA Binding and Cofactor Interaction. *Biochemistry*. 2018;**57**: 4047–4051.

[55] Koag M C, Kou Y, Ouzon-Shubeita H, Lee S. Transition-state destabilization reveals how human DNA polymerase β proceeds across the chemically unstable lesion N7-methylguanine. *Nucleic Acids Res*. 2014;**42**: 8755–8766.

[56] Yang Y, Chen L, Pan H Z, Kou Y, Xu C M. Glycosylation modification of human prion protein provokes apoptosis in HeLa cells in vitro. *BMB Rep*. 2009;**42**: 331–337.

ISPECE   IOP Publishing

IOP Conf. Series: Journal of Physics: Conf. Series **1187** (2019) 052082   doi:10.1088/1742-6596/1187/5/052082

Research on route planning of aerial photography of UAV in highway greening monitoring

T Duan[1,2], P C Hu[1,2], L Z Sang[1]

[1]China Transport Telecommunications & Information Center, Beijing, 100011

[2]National Engineering Laboratory for Transportation Safety and Emergency Informatics, Beijing, 100094

duantaohao@126.com

Abstract. UAV (unmanned aerial vehicle) image technology has the advantages of wide shooting range, large inventory space, strong timeliness and so on, when it is used in highway detection. In addition to the basic parameters such as image quality, resolution and overlap, the safety of flight routes should also be taken into account in aerial photography of highways in operation. Optimal route planning is the premise and guarantee for the application and promotion of UAV in the field of highways. By adopting appropriate shooting form, flight height and angle, the actual road condition information can be collected to ensure high accuracy of data information. This provides great convenience for the actual situation of highway green belt, the confirmation of the location of plant growth and other aspects. It plays an important role in promoting the maintenance efficiency and level of highway green belt. This paper quantitatively analyzes and studies the optimal aerial route planning for unmanned vehicles in highway field.

1. Quantification of factors influencing the quality of UAV aerial images

1.1. Classification of UAV photography

UAV aerial photography can be divided into three categories in general, namely vertical photography (side angle ± 3 °), tilted photograph (side angle is greater than ± 3°, but less than ± 30°) and oblique photography (side angle is greater than 35 °, but less than 55 °). Vertical photography is the most commonly used of the three modes of photography. In vertical photography, an aerial camera is used to image objects perpendicular to the ground[1]. Due to the actual flight of the UAV and other factors, such as mechanical vibration, aerial camera doesn't perpendicular to the ground condition in the actual process of aerial. There will be a smaller side angle (less than ± 3 °) usually. Therefore, vertical photography also includes photography with a small side view angle in practice. The image ground coverage of aerial image refers to the actual surface coverage corresponding to the image. The imaging range is determined by camera parameters (focal length and physical size of sensor) and aerial photography flight height using (1). Where A_{im} is the image imaging area, Im_w and Im_h are the width and height of the image imaging area respectively, A_s is the area of the sensor, H_f is the flying height of aerial photography, f is the focal length, S_w and S_h are the width and height of the sensor.

$$A_{im} = Im_w \times Im_h = A_s \times (\frac{H_f}{f})^2 = (S_w \times S_h) \times (\frac{H_f}{f})^2 \qquad (1)$$

When the image imaging range is related to the sensor size, the actual surface distance represented by each pixel on the sensor can be further obtained, that is, the image resolution or ground sampling

Content from this work may be used under the terms of the Creative Commons Attribution 3.0 licence. Any further distribution of this work must maintain attribution to the author(s) and the title of the work, journal citation and DOI.
Published under licence by IOP Publishing Ltd

distance (GSD)[2] (2). Where, P_w and P_h are pixel in the sensor width and length direction respectively, and D is the distance from the camera to the object. The essence of GSD is to use the actual ground distance unit to represent the pixel size. The larger the GSD is, the larger the actual ground distance represented by the pixel is, and the lower the image accuracy will be. During GSD calculating, the object furthest away from the camera should be selected for specific calculation of GSD. At this time, the GSD obtained is its maximum value to ensure the accuracy of the image meets the requirements.

$$GSD = \frac{S_w}{P_w} \times \frac{D}{f} \text{ or } \frac{S_h}{P_h} \times \frac{D}{f} \tag{2}$$

1.2. Quantification of image overlap
In order to obtain high-quality mosaic images and 3D reconstruction results (digital surface model and 3D point cloud)[3], aerial images need to have a certain degree of image overlap. Image overlap includes forward overlap (Endlap) and side overlap (Sidelap)[4] (Fig 1). Forward overlap refers to the overlap between two adjacent images in the direction of the airline. The schematic diagram of forward overlap and side overlap are shown in Figure 1, where distance B is air base, which refers to the distance between adjacent image centers, and Endlap is the longitudinal coverage length. The air base (B) has the following relationship with aerial photography flight speed (v) and photography interval (t) (3). $O_{forward}$ refers to the overlapping of images on adjacent airline lines, in which the distance SP is the distance between adjacent airline lines and sidelap is the sideline overlap length (4). Then, the sideline overlap degree (O_{side}) can be calculated by (5), image coverage width (Im_w) and image coverage height (Im_h) are shown in the Figure 1.

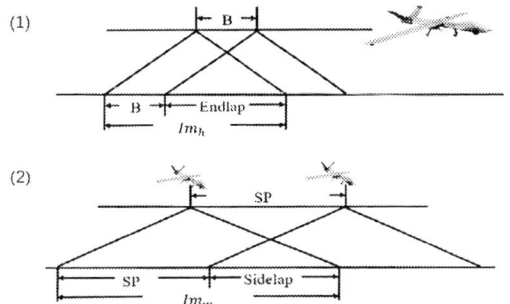

Figure 1. Diagram of forward overlap and side overlap

$$B = v \times t \tag{3}$$

$$O_{forward} = \frac{Im_h - B}{Im_h} \times 100\% = \frac{Im_h - vt}{Im_h} \times 100\% \tag{4}$$

$$O_{side} = \frac{Im_w - SP}{Im_w} \times 100\% \tag{5}$$

2. The development process of aerial photography plan

The parameters to be calculated for the establishment of aerial photography plan include the number of routes, the number of aerial images, flight speed and altitude. The information required to calculate these parameters includes the pixels, physical size and focal length of the aerial photograph camera's sensor, accuracy requirements (GSD), longitudinal and lateral coverage, preset flight speed, and the length and width of the aerial photograph's target area. The specific process is shown in Figure 2. For general aerial photography projects, longitudinal and lateral overlap requirements are greater than 75% and 60%, respectively. For aerial photography of forest, high-vegetated area or farmland, vertical and lateral overlap degrees are usually required to be greater than 85% and 70% respectively.

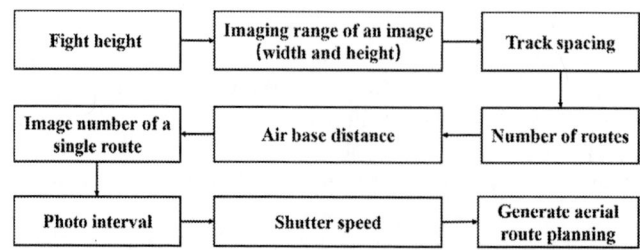

Figure 2. The development process of aerial photography plan

2.1. Aerial photography route development

When making the aerial photography plan, first of all, the route should be parallel to the long side of the aerial photography coverage area, that is, the short side of the aerial photography coverage area should change the course to reduce the number of routes and the number of course changes[5]. In Figure 3 (A), the red one-way arrow represents the route and course, and the black border is the target area of aerial photography. The route is parallel to the target area of aerial photography, and the course is changed on the short side of the target area of aerial photography (red two-way arrow). By setting the photosensitive sensor (CCD array) or the long edge of the image perpendicular to the route direction, the side coverage range of aerial image can be increased, thus reducing the number of routes. In Figure 3A, the blue rectangle represents CCD array or image whose long side is perpendicular to the airline line.

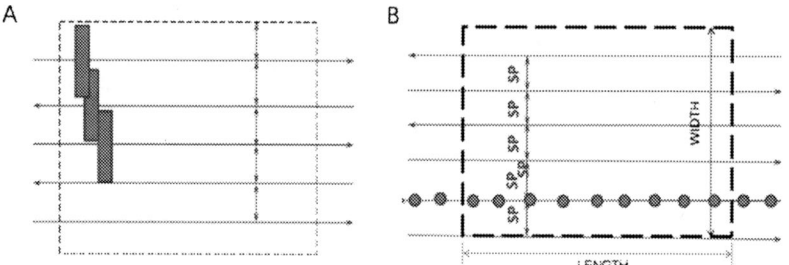

Figure 3. Course and camera direction route setting

2.2. Calculation of the number of routes

The number of routes that should be set is calculated according to the side-overlap degree and the width of the target area of aerial photography. First, calculate the interval between routes (SP, Figure 3B), then the number of routes (N_{fl}) to be set can be calculated based on (6) and (7). Where, the $ceiling$ is the integer function, and the $WIDTH$ is the width of the target area of aerial photography.

$$SP = (1 - O_{side}) \times Im_w \qquad (6)$$
$$N_{fl} = ceiling(\frac{WIDTH}{SP} + 1) \qquad (7)$$

2.3. Calculation of the number of images acquired in the aerial photography program

When the number of routes is set, the number of images needed to cover the entire target area of aerial photography should also be calculated. First calculates the air base length (B, (8)), and then calculate the number of images on every route (N_{il}, (9)). Finally, calculate total number of images required (N_{it}, (10)). In the process of the actual route planning, aerial coverage should be slightly bigger than the target area, to ensure that the target area image at the edge of the area have enough coverage, thus improve the reconstruction precision of the edge of the area[5]. Therefore, two images can be added at both ends of the long side direction of the aerial photography target region (blue dots outside the central region in Figure 3B). In the direction of the short side of the target area of aerial photography, another route can be added on the outside of the route.

$$B = (1 - O_{forward}) \times W_g \tag{8}$$

$$N_{il} = ceiling\left(\frac{LENGTH}{B} + 1\right) \tag{9}$$

$$N_{it} = N_{il} \times N_{fl} \tag{10}$$

3. Quantitative results of flight path parameters

3.1. Aerial photo setup and road central green belt imaging

The safety of landing flight is fully taken into account in the route planning. Based on the above quantitative calculation method, the parameters of formation flight on both sides of the road avoiding the road surface and single-aircraft flight on one side of the road edge are calculated for different levels of highway. The statistical analysis results only show the double-lane for the 1-3 class highways. In order to avoid the UAV's emergency landing on the highway lane, the route is set above the edge of both sides of the road. The preset flight speed is 10 m/s, the shooting interval is 1 s, and the longitudinal and lateral overlap degrees are 70% and 80%, respectively[6].

3.2. Highways and first class highways

Taking ZENMUSE X5 camera as an example, parameters of aerial photography plan under different aerial configuration are calculated. The physical size of the sensor is 17.3 × 13.0 mm, the maximum resolution of the image is 4608 × 3456 (16 million pixels), and the focal length is 15 mm. In China, the total width of the highway from the height highway to the first grade highway is 40~70 m, and the average value is 55 m as the distance between the routes. In the case that the flight height under 150 m does not take GSD into account (Table 1), the vertical and lateral coverage do not meet the preset requirements.

Table 1. Dual route configuration of expressways and first-class highways

H_f (m)	GSD (cm)	Im_w (m)	Im_h (m)	O_l (%)	O_{ll} (%)	H_f (m)	GSD (cm)	Im_w (m)	Im_h (m)	O_l (%)	O_{ll} (%)
20.0	0.5	23.1	17.3	42.2	-138.4	140	3.5	161.5	121.1	91.7	65.9
30.0	0.8	34.6	25.9	61.5	-58.9	150	3.8	173.0	129.8	92.3	68.2
40.0	1.0	46.1	34.6	71.1	-19.2	160	4.0	184.5	138.4	92.8	70.2
50	1.3	57.7	43.3	76.9	4.6	170	4.3	196.1	147.1	93.2	71.9
60	1.5	69.2	51.9	80.7	20.5	180	4.5	207.6	155.7	93.6	73.5
70	1.8	80.7	60.6	83.5	31.9	190	4.8	219.1	164.4	93.9	74.9
80	2.0	92.3	69.2	85.6	40.4	200	5.0	230.7	173.0	94.2	76.2
90	2.3	103.8	77.9	87.2	47.0	210	5.3	242.2	181.7	94.5	77.3
100	2.5	115.3	86.5	88.4	52.3	220	5.5	253.7	190.3	94.8	78.3
110	2.8	126.9	95.2	89.5	56.7	230	5.8	265.3	198.9	94.9	79.3
120	3.0	138.4	103.8	90.4	60.3	240	6.0	276.8	207.6	95.2	80.1
130	3.3	149.9	112.5	91.1	63.3	250	6.3	288.3	216.3	95.4	80.9

H_f: flight height

GSD: ground sampling distance

Im_w: Image coverage width

Im_h: Image coverage height

O_l: Longitudinal overlap (%)

O_{ll}: Lateral overlap (%)

3.3. The secondary roads

The total width of the road ranges from 30 m to 60 m, and the average value of 45 m is taken as the distance between routes. The vertical and lateral coverage do not meet the preset requirements when the flight height under 120 m, which does not take GSD into account (Table 2).

Table 2. Double route configuration of secondary highway

H_f (m)	GSD (cm)	Im_w (m)	Im_h (m)	O_l (%)	O_{ll} (%)	H_f (m)	GSD (cm)	Im_w (m)	Im_h (m)	O_l (%)	O_{ll} (%)
20	0.5	23.1	17.3	42.2	-95.1	140	3.5	161.5	121.1	91.7	72.1
30	0.8	34.6	25.9	61.5	-30.1	150	3.8	173.0	129.8	92.3	74.0

Hf (m)	GSD (cm)	Imw (m)	Imh (m)	Ol (%)	Oll (%)	Hf (m)	GSD (cm)	Imw (m)	Imh (m)	Ol (%)	Oll (%)
40	1.0	46.1	34.6	71.1	2.5	160	4.0	184.5	138.4	92.8	75.6
50	1.3	57.7	43.3	76.9	21.9	170	4.3	196.1	147.1	93.2	77.1
60	1.5	69.2	51.9	80.7	34.9	180	4.5	207.6	155.7	93.6	78.3
70	1.8	80.7	60.6	83.5	44.3	190	4.8	219.1	164.4	93.9	79.5
80	2.0	92.3	69.2	85.6	51.2	200	5.0	230.7	173.0	94.2	80.5
90	2.3	103.8	77.9	87.2	56.7	210	5.3	242.2	181.7	94.5	81.4
100	2.5	115.3	86.5	88.4	61.0	220	5.5	253.7	190.3	94.8	82.3
110	2.8	126.9	95.2	89.5	64.5	230	5.8	265.3	199.0	95.0	83.0
120	3.0	138.4	103.8	90.4	67.5	240	6.0	276.8	207.6	95.2	83.7
130	3.3	149.9	112.5	91.1	70.0	250	6.3	288.3	216.3	95.4	84.4

H_f: flight height
GSD: ground sampling distance
Im_w: Image coverage width
Im_h: Image coverage height
O_l: Longitudinal overlap (%)
O_{ll}: Lateral overlap (%)

3.4. Highways and first class highways (High resolution camera)

Choose a camera with higher resolution (take Canon 5Ds R camera as an example) to calculate the parameters of aerial photography plan under different aerial photography configurations. The physical size of the camera's sensor is 36×24 mm, the maximum resolution of the image is 8688×5792 (50 million pixels), and the focal length is 24 mm (depending on the lens).The quantitative parameters for the configuration of dual-route aerial photography routes of expressways and first-level highways are shown in table 3. The longitudinal and lateral coverage of the flight height under 120 m do not meet the preset requirements when GSD is not taken into account.

Table 3. Dual route configuration of expressways and first-class highways (High resolution camera).

Hf (m)	GSD (cm)	Imw (m)	Imh (m)	Ol (%)	Oll (%)	Hf (m)	GSD (cm)	Imw (m)	Imh (m)	Ol (%)	Oll (%)
20	0.4	30	20	25.0	-83.3	140	2.4	210	140	89.3	73.8
30	0.5	45	30	50.0	-22.2	150	2.6	225	150	90.0	75.6
40	0.7	60	40	62.5	8.3	160	2.8	240	160	90.6	77.1
50	0.9	75	50	70.0	26.7	170	2.9	255	170	91.2	78.4
60	1.0	90	60	75.0	38.9	180	3.1	270	180	91.7	79.6
70	1.2	105	70	78.6	47.6	190	3.3	285	190	92.1	80.7
80	1.4	120	80	81.0	54.2	200	3.5	300	200	92.5	81.7
90	1.6	135	90	83.3	59.3	210	3.6	315	210	92.9	82.5
100	1.7	150	100	85.0	63.3	220	3.8	330	220	93.2	83.3
110	1.9	165	110	86.4	66.7	230	4.0	345	230	93.5	84.1
120	2.1	180	120	87.5	69.4	240	4.1	360	240	93.8	84.7
130	2.2	195	130	88.5	71.8	250	4.3	375	250	94.0	85.3

H_f: flight height
GSD: ground sampling distance
Im_w: Image coverage width
Im_h: Image coverage height
O_l: Longitudinal overlap (%)
Oll: Lateral overlap (%)

4. Conclusions

The UAV platform is used to monitor the highway green belt. Considering the emergency landing of UAV, the route setting must be set directly above the edge of both sides of the road surface. If the overlap degree of the adjacent images meets the requirements, the resolution of the camera and the width of the road will affect the flight height in the course setting. In the range of appropriate aerial photography height, highways and first-class highways need higher resolution cameras to obtain more accurate reconstruction results. This paper quantitatively analyzes the influencing factors of aerial photography route design in the application of UAV in highway field, so as to further clarify the influence of each UAV flight parameter on image acquisition. Combined with the width of the road surface to be monitored for different grades of roads, the optimal route planning suitable for the monitoring of different grades of roads is studied. This paper provides a feasible and optimal route setting method for the application and promotion of UAV in highway field.

Acknowledgements

We would like to acknowledge Bangyou ZHENG (CSIRO) for his generous help with the photography and software used for 3D reconstruction, Wei GUO (The University of Tokyo) for his help in image interpretation algorithm. This study is supported by the National Key Research and Development Program of China (2017YFC0804904) and National Natural Science Foundation of China (31771678).

References

[1]   Klosterman S. *et al* 2018. Fine-scale perspectives on landscape phenology from unmanned aerial vehicle (UAV) photography. *Agric. For. Meteorol.* **248**, 397–407.

[2]   Inzerillo L, Di Mino G, Roberts R 2018. Image-based 3D reconstruction using traditional and UAV datasets for analysis of road pavement distress. *Autom. Constr.* **96**, 457–469.

[3]   Nex F, Remondino F 2014. UAV for 3D mapping applications: a review. *Appl. Geomat.* **6**, 1–15.

[4]   Torres-Sánchez J, López-Granados F, Borra-Serrano I, Peña J M 2018. Assessing UAV-collected image overlap influence on computation time and digital surface model accuracy in olive orchards. *Precis. Agric.* **19**, 115–133.

[5]   Belkhouche F 2017. Reactive optimal UAV motion planning in a dynamic world. *Robot. Auton. Syst.* **96**, 114–123.

[6]   Chapman S C *et al 2014*. Pheno-copter: a low-altitude, autonomous remote-sensing robotic helicopter for high-throughput field-based phenotyping. *Agronomy* **4**, 279–301.

Research on Specific Eye Movement Mode of Qualified Railway Driver

Chenzhe Sun[1], Guanglei Zhang[2], Xiujun Zhai[3]

[1,2,3,] Student Affairs Office, Zhengzhou Railway Vocational & Technical College, Henan, China

Email of all the authors: [1,]49138407@qq.com; [2,]875162337@qq.com; [3,]1224634508@qq.com

* Corresponding Author:

[1]*Chenzhe Sun*, Master of psychology, Assistant Lecturer. Graduated from the Henan University in 2012, Working in Zhengzhou Railway Vocational & Technical College. Her ___research interest is applied psychology; *email: 49138407@qq.com; phone: 15890057173*

[2]Guanglei Zhang, Master of Sociology, Associate Professor. Graduated from the Second Military Medical University in 2008. Working in Zhengzhou Railway Vocational & Technical College. His research interest is applied psychology; *email: 875162337@qq.com; phone:13939015800*

[3]*Xiujun Zhai*, Master of psychology, Associate Professor. Graduated from the Shandong Normal University in 2004. Worked in Zhengzhou Railway Vocational & Technical College. Her research interest is applied psychology; *email: 1224634508@qq.com; phone:13663001699*

Abstract: The eye movement mode is one of the core indicators of train driving, and it represents the pros and cons of the driver's driving behavior. In order to gain a deeper understanding of the specific eye movement patterns of qualified drivers during driving, this study compared the eye movement indicators and patterns of 8 qualified drivers and 12 unqualified drivers (intern drivers) on a simulated driving assignment. The results show that qualified drivers put more cognitive processing on the track ahead, while the intern drivers are more interested in the driving monitoring interface; qualified drivers put more attention on the net voltmeter gauge and speedometer, wind pressure gauge, CIR operation terminal. And qualified drivers view the check signal lights more. Through the Markov clustering algorithm, it is found that the qualified driver group performs "the clustering mode centered on the ahead track" stably, while the intern driver group performs "the clustering mode centered on the driving monitoring interface" less stably. These conclusions will provide a theoretical basis for the subsequent research on "improving driver training system".

1. Introduction

Train driving is a task that requires a lot of visual participation, and has higher requirements for drivers' attention and distribution. The driver needs to pay attention to all the inside and outside of the locomotive in the high-speed train at the same time, ensuring that the current position, speed and acceleration are under control, predicting the future demand, and immediately responding to the unpredictable events which have no rules in the external environment[1]. The current research suggests that: (1) Driving a train requires a sensing strategy that is skilled in handling multiple mission requirements in a limited time [2]; (2) More than 78% of the train driver's gaze time is about road conditions, especially the road in front of the train and signals on both sides of the road[3]; (3) In the cab, the train drivers focus more on the speed meter and timetable [4]; (4) Compared with the younger drivers, the older drivers see the instrument for less time and the total line of sight is shorter [5].

We believe that the eye movement pattern should be one of the core physiological indicators to distinguish the driver's driving behavior. Therefore, finding a specific eye movement mode in a qualified driver's driving is one of the direct means to analyze and study driving behavior, but there are still few studies involved. In this paper, the qualified drivers and unqualified drivers (intern drivers) were tested, and their visual behavior during the simulated driving process was studied to find the difference between the two eye movement patterns, so as to analyze the unique eye movement patterns of qualified driver, and provide a theoretical basis for the follow-up "improvement of driver training methods".

2. Material and Methods

2.1. Research subjects
Eight qualified train drivers (1 high-speed rail drivers, 7 electric locomotive drivers) and twelve intern drivers were randomly selected at the Zhengzhou Railway Bureau. Qualified drivers have a driving age of 3-5 years, an average of 4.2 years; unqualified drivers are intern drivers, they all have train simulator driving experience, but have not completed the actual driving process independently (hereinafter referred to as "intern driver"). The ages of the two groups were qualified drivers (25.38±2.20) years old and internship drivers (23.42±1.73) years old. There was no significant difference between the two groups (T=2.12, p>.05).

2.2. Research tools
HXD3C locomotive simulation training system: It can simulate driving in about 20 minutes. During the driving process, there will be driving tasks such as signal speed control, lifting bow, whistle and effect parking.

Framed Eye Tracker *Tobii Glass 2*: Includes both the headset and the recorder, operating with computer control software. During the experiment, the driver wears the eye tracker for the entire simulated driving process.

2.3. Research design
On the simulated driving platform, the participants first carried out the driving practice of the 171 line (about 20 minutes) and were familiar with the equipment. Subsequently, the participants wore the eye movement recorder and carried out an experimental operation of the analog driving of the 173 line (about 20 minutes). In order to prevent the practice effect, the 171 line and 173 line settings are different.

2.4. Indicator measurement
During the experiment, the eye tracker will track and record the driver's viewpoint, eye-jumping process and eye-tracking trajectory at 30 frames per second. In order to study the driver's attention to different areas, the author manually encoded the instruments in the locomotives and the regions of road condition into 20 areas of interest (see Table 1). The experiment will record the following eye movement indicators of the driver: the first visit time point, the fixation count, the total fixation duration, the average fixation duration, the activities of eye movement and so on. At the same time, the author checks each driver's attention point by frame, and obtains the eye hopping sequence and eye movement trajectory of each driver in different interest areas (see Figure 1).By comparing the

sequence of interest areas, the eye hopping sequence is encoded as a random matrix and graphically represented (see Figure 2). The nodes in the figure represent different AOIs , the weight of the lines connecting the pair of points is proportional to the probability of saccades between them.

Table 1 .Areas of Interest (AOIs) sequence

i	AOI	i	AOI
1	The track ahead	11	Department ticket
2	The wind pressure gauge	12	Department flag
3	The net voltmeter and speedometer	13	Signal light
4	The driving monitoring interface	14	Signal board
5	The traction gauge	15	Driving performance tips
6	CIR operation terminal	16	Timing tips
7	Horn and brakes	17	The tips of raise or drop pantograph
8	The area of pantograph, cylinder pressure indicator, light switch	18	Reminder
9	Driver controller	19	Parking pole
10	The check signal light	20	Platform

Figure 1. Qualified driver D1 eye track diagram

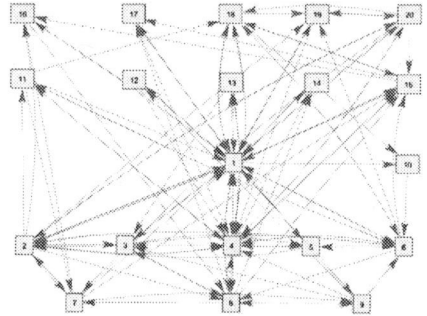

Figure 2. Qualified driver D1 region of interest random matrix graphics

2.5. Data processing

2.5.1. Data processing tool
The eye movement trajectory sequence collected in this study was processed by Markov clustering algorithm (MCL algorithm) using Matlab mathematics software, and the eye movement data was analyzed by SPSS14.0.

2.5.2. The MCL algorithm description
The MCL algorithm (van Dongen, 2000, 2008) is an unsupervised graph clustering algorithm using random walk. Eye tracking images are often difficult to observe and often use qualitative analysis, while the MCL algorithm calculates clustering of each gaze hot spot according to the saccade probability between the gaze hot spots on the picture to illustrate the eye movement mode. Markov found that if there are multiple linked dense areas between random access nodes on the graph, they

might stay in that area instead of moving between areas with fewer links. If the nodes on a graph walk randomly several times, such a group of nodes will form a cluster in the graph. The MCL process simulates the process of random walks, and strengthens or weakens the current activity through expansion and inflation to find the cluster structure in the graph. Expansion refers to the use of the usual matrix product to obtain the power of the matrix, that is, through the product to calculate the probability of walking between nodes after the constant e times of random walk. The connection between different clustering regions can be enhanced by extension. In a random walk probability matrix $M \in R^{m \times m}$ (m nodes in the graph), the $(p,q)^{th}$ element in M^e represents the probability of walking from node p to node q through the constant e times. The process of e-power simulates the expansive flow of random e-walking. If $(M^e)_{pq}$ is large only if both p and q are located in the same dense region of the graph, then e must be small. Therefore, the parameter e is usually set to 2. In order to avoid invalid extensions during the expansion process, a self-loop is added to the diagonal elements. However, the phenomenon of probability averaging may occur through expansion, so the process of inflation is required. Inflation is inflating each value in the probability matrix by $r(r>1)$ power, which strengthens the tight points and weakens the loose points. The mathematical definition of inflation is as follows, assuming a random matrix M and a non-negative real number r, the matrix after the inflation process is Γ_r(M).

$$(\Gamma_r(M))_{pq} = (M_{pq})^r / \sum_{i=1}^{m}(M_{iq})^r \tag{1}$$

Γ_r gives the inflation operator, and the power coefficient r is called the inflation parameter. For values where r is greater than 1, inflation will change the probability and tend to be more likely to walk. The effect of the inflation operator can be changed by different r settings. Increasing the inflation parameter can make the inflation operator stronger, thereby increasing the particle size or tightness of the cluster. Each step of the MCL process defines a random matrix. The iterative steps of expansion and inflation will eventually lead to the separation of the graph into different segments without paths, and the resulting set of segments is interpreted as a cluster of graphs.

In this study, the obtained driver's eye movement trajectory is processed by the sequence of AOIs, and then regions a random matrix, and Markov clustering algorithm is applied to it. The non-diagonal elements in the matrix are obtained by calculating all the saccades of different interest regions in the sequence and scaling each column to obtain a sum of 1, that is, the $(i, j)^{th}$ element represents the probability from a region of interest (AOI_j) to another region of interest (AOI_i). As a self-loop, the diagonal element is the ratio of the average fixation duration of each AOI to the average fixation duration of all AOIs. Figure 2 shows the random matrix graph of the driver's D1 eye trajectory map (no self-loop). The nodes in the graph represent different AOIs, and the line weights between the nodes are proportional to the saccade frequency between them. As shown, all eye movement trajectories contain overly complex transformation structures and cannot find any meaningful patterns. The MCL algorithm can analyze the complex eye movement trajectory in the driving process through the cluster structure to extract important eye movement patterns.

3. Results

3.1. Qualitative analysis

3.1.1. Comparison of eye movement hot spots between qualified drivers and intern drivers

Fig.3. Qualified drivers' eye movement hot map   Fig. 4. Intern drivers' driving hot map

Through the hot point maps of the fixation count and the total fixation duration in different driving experience groups (see Fig. 3 and Fig. 4), it shows that the main visual gaze point of the qualified driver during the driving process is on the track ahead and driving monitoring interface. However, there are still differences between the two groups with different experiences in the focus of the hot spot. The qualified drivers focus obviously more on orbital observation than the intern drivers. It shows that experienced drivers spend more cognizance on external environmental observations, while inexperienced drivers tend to focus on the driving monitoring interface.

3.1.2. Comparison of eye movement track between qualified drivers and intern drivers

Fig. 5. Eye movement track diagram   Fig.6. Eye movement track diagram
of qualified drivers   of intern drivers

Due to the long monitoring time and large amount of data, the overall trajectory of the driver's eye movement is confusing (see Figures 5 and 6). However, it shows that the qualified driver's eye movement trajectory has a smaller radiation area and the target is more clearly.

3.2. Quantitative analysis of eye movement data

3.2.1. Analysis of variance of eye-moving interest area between qualified driver group and intern driver group

After collecting the total fixation duration (TFD), average fixation duration (AFD), and fixation count (FC) indicators for 20 AOIs, the independent sample t-test was used to analyze the findings (see Table 2): ① the total fixation duration and average fixation duration of AOI_1 and AOI_3 of the qualified driver group are significantly higher than that of the intern driver group ($p<0.01$), but the difference in the fixation count was not significant; ②the total fixation duration and fixation count of AOI_2 and AOI_6 are significantly higher in the qualified driver group than in the intern driver group($p<0.01$),but the difference in the average fixation duration was not significant; ③the total fixation duration and fixation count of AOI_4 in the intern driver group are significantly higher than in the qualified driver group ($p<0.001$); ④the average fixation duration of the AOI_{19} is significantly higher in the qualified driver group than in the intern driver group ($p<0.01$), but fixation count is significantly lower than that in the intern group (p<0.001).

Table 2. Significant differences in eye movement data between qualified drivers and intern drivers

AOI_i	Index	The qualified driver group （X±S）	The inter driver group （X±S）	T
AOI_1	TFD （s）	378.01±30.57	233.06±77.08	5.861[***]
	AFD （s/n）	0.42±0.06	0.29±0.08	3.985[***]
AOI_2	TFD （s）	44.00±4.58	28.38±16.75	3.062[***]
	FC （n）	129.25±19.50	85.75±46.38	2.889[**]
AOI_3	TFD （s）	15.26±4.07	9.19±3.03	3.833[**]
	AFD （s/n）	0.23±0.03	0.19±0.05	2.364[**]
AOI_4	TFD （s）	298.80±75.40	410.16±54.10	-3.858[***]
	FC （n）	887.00±219.90	1301.83±170.73	-4.750[***]
AOI_6	TFD （s）	10.47±2.72	4.26±3.09	4.617[***]
	FC （n）	56.13±7.08	27.83±17.52	4.308[***]
AOI_{19}	AFD （s/n）	0.28±0.03	0.20±0.05	3.423[**]
	FC （n）	6.38±2.62	0.35±0.25	-4.544[***]

3.2.2.Eye movement mode MCL Analysis of qualified and internship drivers
The MCL algorithm is used to process the random matrix extracted from the eye track of the drivers. The expansion coefficient e is set to 2, and the inflation coefficient r is used as the independent variable. As the increase of inflation coefficient, the cluster produced by MCL process is becoming smaller, so the strength of cluster can be tested. As under different inflation coefficients, the diffcrence in eye movement clustering of driver is shown in the table below. The nodes in the figure represent AOIs, the different clusters are drawn in different colors, and the links from other nodes are clustered in the center of the cluster in shades of color. If the cluster consists of only one node, the node is also filled with dark colors, and the gray nodes represent areas of interest that the driver has never viewed.

Table 3. Markov cluster structure of eye movement track for drive

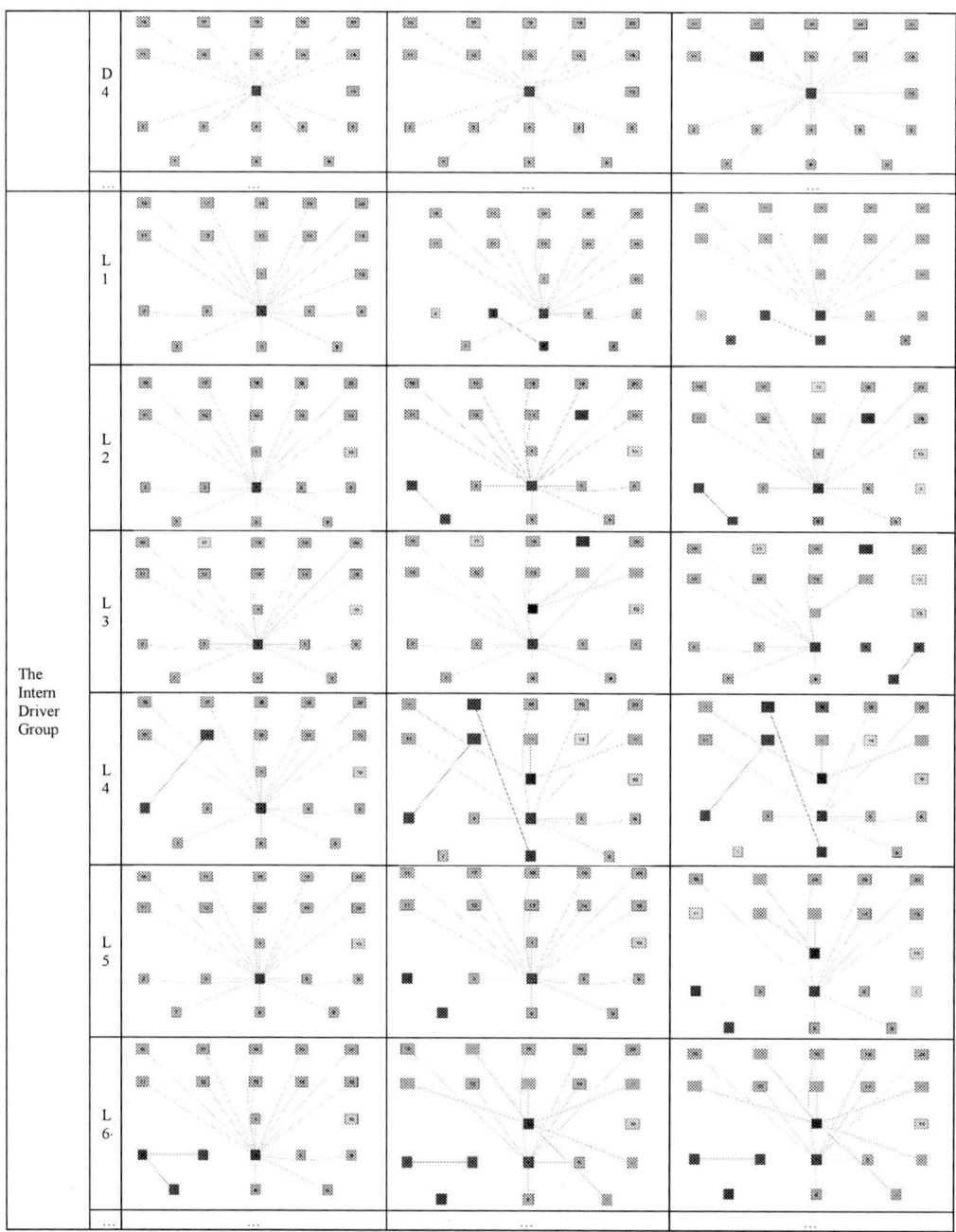

In Table 3, there are two kinds of cluster modes .One is AOI_1 as the mode of the largest cluster. In the eye movement gaze movement, the driver will look at the track ahead after looking for other AOIs; The other is the AOI_4 as the largest cluster mode. In the eye gaze movement, the driver will turn his attention to the driving monitoring interface after looking for other AOIs. The former is the eye movement mode of the qualified driver, and the latter is the eye movement mode of the intern driver. With the shrinking of clusters, there are also clusters based on AOI_1 in the eye movement mode of some intern drivers, but comparatively smaller, they are still mainly based on AOI_4 cluster mode.

In addition, with the change of the inflation coefficient r (Fig. 7), by comparing the clusters number of the two groups, it can be seen that the qualified driver group maintains the same or similar cluster structure under the high inflation coefficient as the group under the low inflation coefficient, but the basic eye movement mode remains unchanged; while the intern driver group produces more clusters, that is, the line of sight may directly shift from one interest area to another, without going through the core AOI_4, so they would deviate from the basic model more easily than the qualified driver.

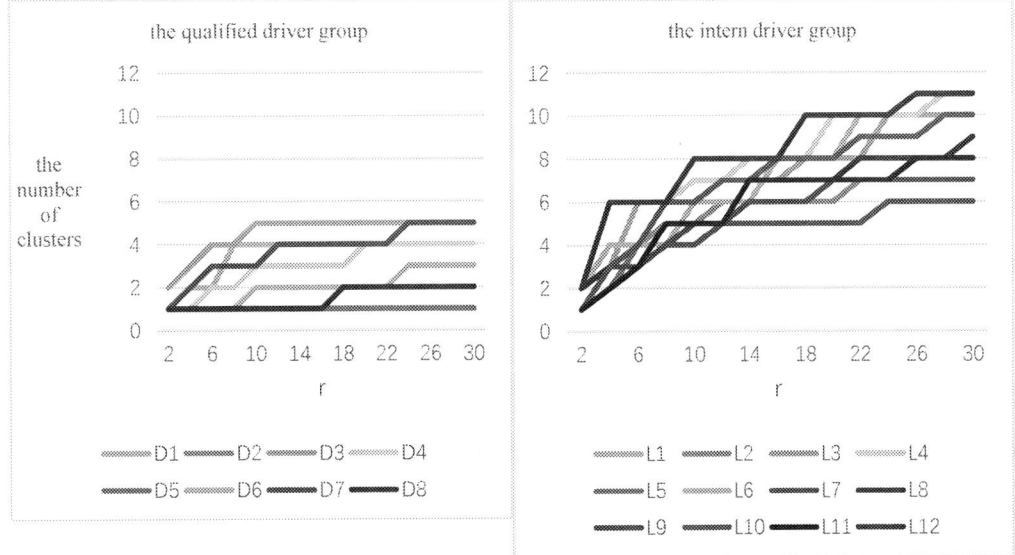

Fig. 7. Comparison of the number of clusters between qualified driver group and intern driver group

4. Discussion

At present, China's railways system has entered the "high-speed rail era" which covers almost every city and town. New technologies and new equipments are emerging one after another. The requirements for qualified railway talents are getting higher and higher. Railway drivers are as the enormous asset of the talent team, so their overall quality and professional skills are critical for the safe operation of railway, and the talent gap is huge. Throughout the relevant research and practice in China and abroad, there are still many problems in driver selection and training: the selection of drivers generally has the tendency of "attaching more importance to physique rather than psychology"; the training process is more likely to use the "master-apprentice-style" training mode with strong subjectivity, random process, uncertain results and low efficiency. The author believes that only by thoroughly studying the differences in the driving behaviors of qualified drivers and intern drivers, and revealing the core and essential characteristics of qualified driving behavior from physiological to psychological aspects, can the railway driver selection criteria and training system be established more scientifically and effectively.

This paper compares the eye movement index data of qualified drivers and intern drivers, and finds that there are significant differences between their eye movement hot spots and eye movement modes: intern drivers are more interested in the driving monitoring interface, while qualified drivers put more cognitive processing on the front track; in addition, qualified drivers have paid more attention to the net voltmeter gauge and speedometer, wind pressure gauge and CIR operation terminal than the intern driver; when parking, the qualified driver is more likely to concentrate on the parking pole. This shows that from driving to parking, from inside the vehicle to outside the vehicle, there are wide-ranging similarities and differences between the eye movement data of qualified drivers and intern drivers.

Through the MCL algorithm, we analyze the visual trajectory patterns of the two groups of drivers and find two typical modes: one is a clustering mode centered on the ahead track as the center ,which is represented by qualified drivers, and qualified drivers can execute the mode more stably ;the other is a clustering mode centered on the driving monitoring interface as the center ,which is represented by the intern driver, and the intern driver is relatively unstable in the execution mode. This shows that the qualified driver's driving behavior has a fixed eye movement mode, while the unqualified driver's eye movement mode is not fixed. The author believes that the stable eye movement pattern of qualified drivers represents a set of reasonable driving behaviors, which can be obtained through the scientific training process, and which will lay a solid foundation for our subsequent research on improving the training mode of railway drivers.

References

[1] T. Luke, N. Brook-Carter, A.M. Parkes, E. Grimes, A. Mills. (2006). An investigation of train driver visual strategies . Cognition, Technology & Work, 8(1): 15-29.

[2] Naweed, A.(2014).Investigations into the skills of modern and traditional train dreiving. Applied Ergonomics, 45(3): 462-470.

[3] Guo Beiyuan, Fang Weining, Dai Mingsen. (2005). Test and analysis of locomotive driver's visual behavior .Science Technology and Engineer, 9(9): 620-623. (in Chinese).

[4] Park, K.S. and Ohkubo, T.(1995). A study on visual characteristic of train drivers for the differences of display layout. The Japanese Journal of Ergonomics, 31(1): 31-38. (in Japanese).

[5] Nishimoto, H. and Munesshige, M.(2013).Study on gaze behavior of train drivers (II), Japan Railway Engineer's Association Magazine, 56(11): 25-28.(in Japanese)

Research on Big Data Decision-making Support Platform of New Energy Bus

Hao Bo Wang Feng

Hainan Business Vocational College Postal Code: 570623

Abstract: With the continuous development and progress of information technology, new energy vehicles have been widely concerned because of their high application efficiency and low environmental pollution. Establishing an information system platform can support the operation requirements of new energy bus vehicles, realize integrated linkage management and control mechanisms and modes; and optimize intelligent data sorting system. Scientific and efficient processing of multidimensional data is the key to promote the sustainable development of the industry. This paper expounds the background of the big data decision-making support platform for new energy public transport vehicles, analyzes the big data frameworks and decision-making schemes, studies the decision-making support methods and proposes the implementation methods.

1. Background of new energy buses application of big data decision-making support platform

In 2015, the Ministry of Industry and Information Technology of China put forward corresponding control requirements and standards for energy management work, and carried out project analysis on the national driving cycle conditions of passenger cars and various types of businesses, which involves relevant parameters and geographical information parameters in the running process of new energy vehicles. Systematic constraints are carried out on the process of acquisition, storage, analysis and modeling. In addition, in the *Guiding Opinions of the General Office of the State Council on Accelerating the Promotion and Application of New Energy Vehicles* the responsibilities of local governments in the promotion of new energy vehicles are also determined. In the promotion of new energy vehicles, scientific information platforms should be implemented to effectively promote the safe operation and management mechanism of new energy vehicles so as to develop new energy vehicles and give full play to the application efficiency of information technology, and create a good management platform for the further development of government supervision.

It is worth mentioning that some institutions of higher learning and enterprises in our country have developed and supervised the vehicle data acquisition equipment, and some regions have launched demonstration and promotion projects, but the management of the new energy vehicle data acquisition project and remote monitoring project are still in test. In order to deal with problems specifically, we should attach importance to the operation monitoring platform, and ensure that the government can grasp the operation and failure of new energy vehicles in the first time, so as to promote the overall progress of new energy vehicle development projects. However, there are still some problems to be solved urgently in the actual application and operation of the monitoring platform. Among them, the insufficient number of monitoring vehicles, the inconsistency of data acquisition equipment and data exchange interface, as well as the slow data transmission of equipment information center are the key problems, which also restrict the overall optimization of data analysis, so the establishment of

Content from this work may be used under the terms of the Creative Commons Attribution 3.0 licence. Any further distribution of this work must maintain attribution to the author(s) and the title of the work, journal citation and DOI.
Published under licence by IOP Publishing Ltd

decision-making support platform for new energy bus application big data is one of the ways to solve this problem[1].

2. Systematic framework of new energy bus application of big data decision-making support platform

2.1. Source layer

In the big data decision-making support platform, source layer mainly refers to the most bottom of the whole system, which can provide base for the management of big data decision-making support platform data, among which vehicle internet system and charging information, as well as traffic flow information are involved. Besides, it can also establish an effective infrastructure analysis model, improve the basic effect of data management and ensure that the big data decision-making support platform can play its practical value on the basis of concrete analysis of specific problems,as shown in figure 1.

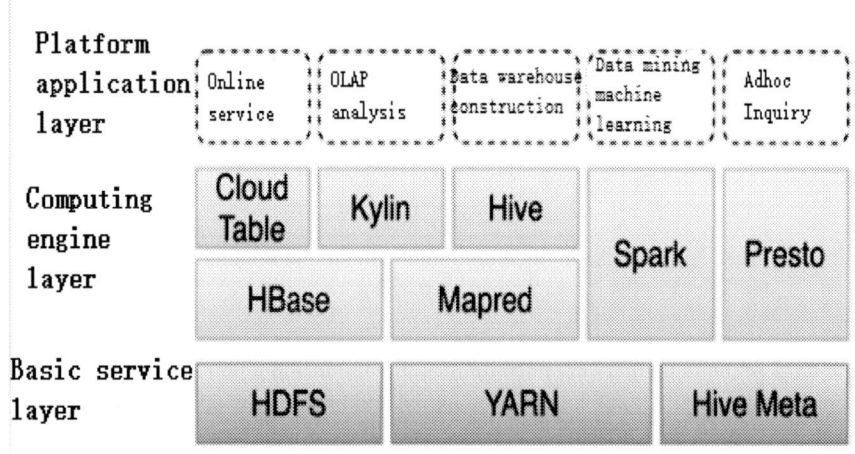

Figure 1: Structure of Big Data Decision-making Support Platform

2.2. Data layer

Data layer is the specific integration method of data, which can manage data from the source layer and complete data collection according to the basic process.

Firstly, to carry out data extraction, it is mainly to establish data classification after selecting the data specification and scope of the system, and apply the extraction mechanism to improve the integrity and consistency of data management.

Secondly, in data transformation, corresponding mechanism should be applied to supervise and control the data, optimize the data control mode, maintain the basic process of data collection and management project, and provide guarantee for the construction of multi-source high-granularity data mart.

Thirdly, data loading is mainly to complete data processing and storage with the corresponding carrier after data classification and summary, so as to ensure the integrity of data transfer process, avoid the mutual influence of data and improve the value of data utilization. Only by improving data storage can we improve the basic efficiency of data loading and the basic level of management.

2.3. Logical layer

For the big data decision-making support platform, the logic layer is the core system of the whole system. Combined with the actual application requirements, it can complete the multi-dimensional data determination and data mining algorithm processing. It is worth mentioning that in order to ensure the effect of data management, it is necessary to implement intelligent data sorting system[2]. Because a lot of voice and video information is generated during the running of the vehicle. Although these data can be preliminarily analyzed by means of flattening processing, they will still take up a large amount of storage space, which will inevitably affect the management efficiency and application level of the operating system. Based on this, the project proposes intelligent data sorting technology. The so-called intelligent data sorting system is to establish data classification and management in the running process, scientifically process structured data in the application system, and use it as the basis of basic control management, effectively supervise projects such as unstructured data storage locations and sizes with structural data. For example, in the process of analyzing and judging the video signal data in the car, the corresponding remote transmission project can be established, and a good environment can be created for the overall progress of the follow-up management effect and application mechanism, because the driving condition and the corresponding sensor signal collection process are relatively convenient. That is to say, in the whole processing mechanism and control system, it is necessary to analyze and judge the parameter system, such as the abnormality of the sensor signal in the vehicle and the driving condition according to the program control module, and rationally control the threshold range to ensure the effective transmission of data and implement data remote push, which can realize the real meaning of remote server space and cost optimization[3].

In addition, the integrated multi-dimensional storage management of structured data and unstructured data should be carried out to ensure that the corresponding parameter patterns and management points can meet the requirements of practical application.Compared to structured data, unstructured data points are less, but most data is multidimensional. If structured data and unstructured data are to be processed in tandem, it is necessary to meet the effective management requirements in strict accordance with the corresponding relationship, and fundamentally enhance the specific level of data storage methods to ensure the comprehensive level of processing procedures and control projects. On the one hand, the technical route of time synchronization storage is proposed; on the other hand, the mixed data storage method is proposed. At the same time, the sorting and flattening processing mechanism of unstructured data can be established to optimize the storage effect[4].

2.4. Presentation layer
The presentation layer of the big data decision-making support platform is mainly a hierarchical structure which can integrate, summarize and process the data of each business layer. After the application structure is established, the corresponding data can be accessed by the browser, client and other projects.

3. Research on decision-making support means of big data decision-making support platform for new energy public transport vehicles
In order to apply the decision-making support system of big data decision-making support platform for new energy public transport vehicles, it is necessary to take vehicle operation decision as the key, establish corresponding analysis framework for fault analysis projects, utilization ratio of charging piles, etc., to ensure the integrity of data determination and processing process, thus maintaining management guidance work, as well as overall mention and lay the foundation for the improvement of management level of new energy bus[5].

3.1. Intelligent charging scheduling
After applying the big data decision-making support platform for new energy public transport vehicles, the intelligent charging dispatching strategy can effectively improve the operating efficiency and control level of the power grid, and ensure that the power project can play the time-effective value in the energy management work and achieve the goal of cost saving. It is worth mentioning that in the intelligent dispatch management project, the effective completion of grid system and traffic operation optimization are the basic requirements. Therefore, in the case that the electric vehicle is simultaneously connected to the grid, the rationality avoids the safety hazard caused by the increase of

the grid load. It is very important to use the intelligent charging scheduling mode to ensure the application effect is improved.

In the first place, it is necessary to analyze and judge the road traffic network structure in combination with the initialization state, and scientifically manage the initial cruising range and battery capacity, and effectively set the path search process to ensure that the weight analysis process meets the requirements[6].

In the second place, after centralized planning of charging paths and effective calculation of the corresponding vehicle status and traffic situation weight analysis system, the minimum value of comprehensive weight values can be searched to ensure that charging paths can achieve effective aggregation to a certain extent.

In the third place, the charging station load can be regulated by means of the corresponding treatment facilities, so as to determine whether the charging station voltage is overloaded rationally. The charging vehicle of the target charging station can be analyzed and the operation process can be improved, as shown in figure 2.

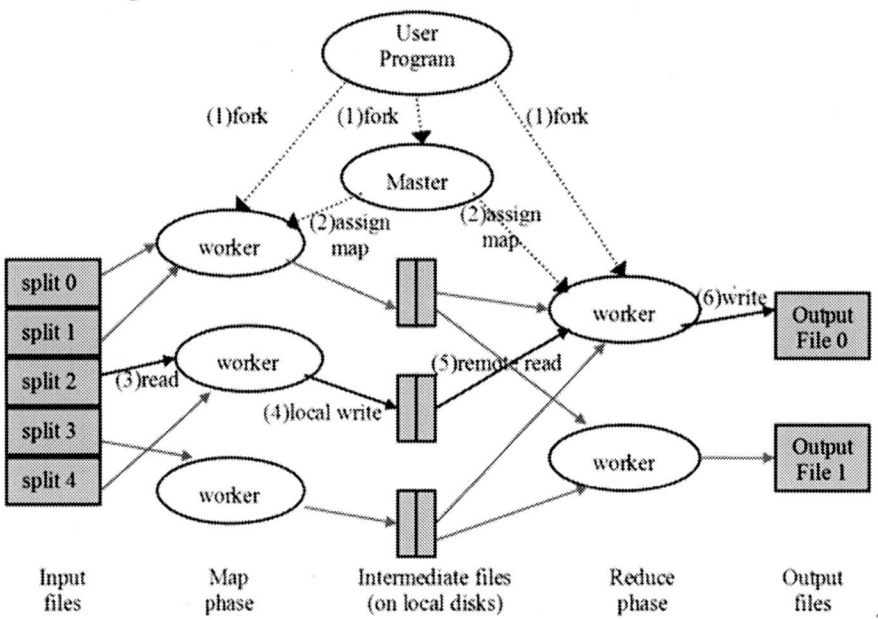

Figure 2: Intelligent Charging Scheduling Flow Chart

3.2. Parking planning
In the process of in-depth research on electric vehicle charging projects, it is necessary to conduct in-depth analysis from the perspective of environmental protection management and economic operation, effectively analyze the operation status of the charging station after it is connected to the grid system, and establish a multi-objective decision-making optimization control structure. While fundamentally maintaining the overall demand, we will improve the supervision of specific layout and operation mode[7].

3.3. Battery use
For new energy vehicles, the management of battery application is very important and lithium batteries are mainly adopted. Therefore, it is necessary to conduct centralized analysis and judgment on battery decay and decay modes, and effectively combine physical model description content to optimize corresponding control efficiency, which can also improve the basic effect of the overall application management process to a certain extent.

4. Conclusions

In a word, in the management of new energy bus application big data decision-making support platform, we should perfect the specific analysis mechanism according to the actual situation, construct the new energy application decision control project, lay the foundation for the overall progress of public transport management, realize the application value of the analysis system model, and ensure the system management. The overall effect of the work lays a solid foundation for upgrading the application and management efficiency of the new energy bus big data system.

Acknowledgment

Title: Scientific research project of education department of Hainan province *Intelligent application and processing of big data of new energy bus in Hainan (Haikou)*
No: hnky2017—82

References

[1] Wang Fan, Cui Xiao, Di Yachan, etc. Analysis of the research direction of innovation and entrepreneurship education in Hebei Higher Vocational Colleges [J]. Modern business industry, 2016 (25): 134-136.

[2] Chen Xingqiang, Tang Ke, Chen Yuechuan, etc. Research on decision-making support platform for big data of new energy public transport vehicles [J]. Urban public transport, 2017 (3): 25-29.

[3] Liu Kaiyuan, Luowei, Li Lifeng etc. Design and research of intelligent bus dispatching system for new energy vehicles [J]. Value Engineering, 2017, 36 (15): 128-130.

[4] Wanjian, Ji Jinzhang, Wang Weifeng etc. Economic analysis of new energy bus considering time value of capital [J]. Highway Transportation Science and Technology, 2015, 32 (3): 154-158.

[5] Dong Enyuan, Yan Wensheng, Shen Jiangwei, etc. Demonstration operation analysis of hybrid bus in plateau area [J]. Science and technology and engineering, 2013, 13 (22): 6629-6632.

[6] Li Dusheng. The development of Zhongtong Bus New Energy Vehicle Remote Intelligent Monitoring Platform [J]. Commercial Vehicle, 2014 (7): 49.

[7] Liu Rongxian. Analysis of the characteristics of new energy bus promotion and application under the operation subsidy policy [J]. Digital users, 2017, 23 (25): 217.

Linked Data Crowdsourcing Quality Assessment based on Domain Professionalism

Lu Yang[1234], Li Huang[1234], Zhenzhen Liu[1234]

[1]School of Computer Science and Technology, Wuhan University of Science and Technology, Wuhan 430065, China

[2]Key Laboratory of Intelligent Information Processing and Real-time Industrial System in Hubei Province, Wuhan 430065, China

[3]Institute of Big Data Science and Engineering, Wuhan University of Science and Technology, Wuhan 430065, China

[4]Key Laboratory of Rich-media Knowledge Organization and Service of Digital Publishing Content, National Press and Publication Administration, Beijing 100038, China

Abstract. With the rapid development of Internet technology, crowdsourcing, as a flexible, effective and low-cost problem-solving method, has begun to receive more and more attention. The use of crowdsourcing to evaluate the quality of linked data has also become a research hotspot. This paper proposes the concept of Domain Specialization Test (DST), which uses domain professional testing tasks DSTs to evaluate the professionalism of workers, and combines the idea of Mini-batch Gradient Descent (MBGD) to improve the EM algorithm, and the MBEM algorithm is proposed to achieve efficient and accurate evaluation of task results. The experimental results show that the proposed method can screen out the appropriate workers for the linked data crowdsourcing task and improve the accuracy and iteration efficiency of the results.

1. Introduction

Crowdsourcing is a distributed problem-solving mechanism that towards the Internet public. It integrates computers and unknown people on the Internet to accomplish tasks that are difficult for computers to accomplish alone [1]. For example, image annotation[2], physical alignment[3], these tasks are difficult to handle by machine but can be done by crowdsourcing easily. Therefore, crowdsourcing has been widely studied and applied in many fields.

In recent years, the scale of Linked Data has exploded, which leads to the serious data quality problems [4]. Although there are a large number of researches on automated or semi-automated tools for dealing with the quality of linked data [5-6], there are still a large number of problems in the data quality that are difficult to find or solve, but human intelligence can easily handle. Therefore, the use of crowdsourcing to solve the quality problems in the linked data has gradually begun to attract attention.

Linked data is a collection of structured data (RDF data) on the network [7]. A data set usually covers multi-domain knowledge and the data set is huge, so the domain knowledge involved in the linked data crowdsourcing task is also diverse. The crowdsourcing workers come from the Internet, they have different educational backgrounds and different professional knowledge. The quality of their work is bound to be uneven, which makes it difficult to control the quality assessment of crowdsourcing results. Therefore, how to select the workers who have the relevant knowledge of Linked Data Tasks (LDT), and how to improve the integration efficiency of crowdsourcing task results, ensure the quality of crowdsourcing task results are urgent problems to be solved in the linked data crowdsourcing research.

2. Related work

At present, the research on linked data crowdsourcing mainly focuses on the quality problem detection, ontology alignment and entity linking. Literature [8] proposes a quality assessment of the linked data of the Find-Fix-Verify model based on crowdsourcing in the form of competition and micro-task. Literature [9] proposes a two-stage associated data quality assessment method combining manual and semi-automatic forms. The literature [3] transforms the ontology alignment problem into a crowdsourcing micro-task. Literature [10] proposes a probabilistic framework-based system on crowdsourcing platforms to improve link quality.

The focus of the above research is on how to apply crowdsourcing to the field of linked data. Due to the uncertainty of crowdsourcing workers and the large scale of linked data sets, an important issue in linked data crowdsourcing is the quality assessment of crowdsourcing results and the evaluation of efficiency issues.

In terms of quality assessment of results, one of the most commonly used quality assessment methods is the Golden Standard Data (GSD) evaluation method[11] which refers to a type of data with standard answers. The accuracy of pure gold standard data does not represent that the professional knowledge possessed by workers meets the requirements of crowdsourcing tasks, and gold standard data needs to be manually generated, adding extra costs.Literature [12] proposes a staged dynamic crowdsourcing quality control strategy, due to the setting of task detection points and replacement rules, the task completion time will be greatly extended.The EM algorithm proposed by Dawid and Skene[13] can accurately estimate the task results, but when the task volume is large, resulting in low efficiency of the algorithm.

The above evaluation methods all have shortcomings between the accuracy and efficiency of task results. The domain professional evaluation method and MBEM quality evaluation algorithm proposed in this paper can effectively screen high-quality workers and greatly improve the evaluation efficiency of task results.

3. linked data crowdsourcing quality assessment method

3.1. Predefined

This paper focuses on how to improve the quality of crowdsourcing task results and the efficiency of evaluation by controlling the quality of workers and the process of result integration in linked data crowdsourcing applications. The relevant definitions are given below.

Definition 1. (Linked Data Task: LDT): $T = \{t_1, t_2, ..., t_n\}$ represents the set of tasks for a given linked data set LD. The task $t_i \in T$ is the label task with the only answer $c \in C$, $C = \{c_1, c_2, c_3, c_4\}$.

$$t_i = \{< subject, predicate, object >, < c_1, c_2, c_3, c_4 >\} \tag{1}$$

Definition 2. (Workers): $W = \{w_1, w_2, ..., w_n\}$ represents the crowdsourcing workers collection, each worker w_j completes task $t_i \in T$ independently.

Definition 3. (Domain Specialization Test: DST): $D = \{d_1, d_2, ..., d_n\}$ represents the domain classification of the task T. Different workers $w_j \in W$ have different domain expertise in different domain d_m:

$$DST_j = \{P_m = \frac{Right(DSTs_m)}{DSTs_m.length} \mid d_m \in D, w_j \in W, P_m \in [0,1]\} \tag{2}$$

Definition 4. (Domain Specialization Test Task: DSTs): DSTs = {dst_1, dst_2, ..., dst_n} are testing tasks similar to T extracted from the Standard Knowledge Base (SKB), Similar(t_i, dst_j) $\in [0,1]$.

Definition 5. (Gold Standard Tasks: GST): gold standard tasks referred to a set of test tasks which have a standard answer, G = {g_1, g_2, ..., g_n}, is used to identify malicious workers during crowdsourcing tasks:

$$wor\,ker = \begin{cases} accepte(w_j) & P_j = \dfrac{Right(GST)}{GST.length} \geq P_{standard} \\ reject(w_j) & P_j = \dfrac{Right(GST)}{GST.length} < P_{standard} \end{cases}$$

(3)

Definition 6. (Simple Weighted Majority Voting): The task $t_i \in T$ is assigned to multiple workers to answer independently, and the answers are integrated by weighted voting, with the answers of most workers as correct answer:

$$Answer(t_i) = \arg\max_c (\sum\nolimits_{c_1} P_w, \sum\nolimits_{c2} P_w, ..., \sum\nolimits_{cn} P_w), t_i \subset T, w \subset W, c \subset C$$

(4)

The problem to be solved in this paper is: Given the linked data crowdsourcing task T, the worker collection W, and the standard knowledge base SKB, how to get the optimal result set R of T efficiently and accurately.

3.2. DST-based linked data crowdsourcing quality control strategy

3.2.1. DST evaluation model. The linked data set involves the knowledge domain including various subject areas, and the inherent knowledge of human beings inevitably has certain limitations. In order to ensure the quality of LDT results, it is necessary to screen out high-quality workers with domain knowledge that matches the task to complete the linked data crowdsourcing task. Studies in [14] have shown that workers' reliability is often comparable in similar areas. Therefore, this paper introduces the concept of DST, extracts task-related testing tasks from known knowledge bases, and measures the user's domain expertise to match the appropriate linked data crowdsourcing tasks.

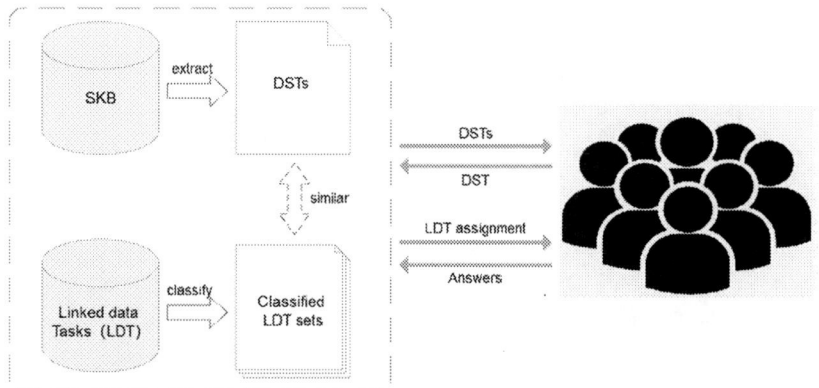

Figure 1. DST-based crowdsourcing quality control process.

Figure 1 depicts the implementation flow of the DST-based evaluation method proposed in this paper, the implementation steps are as follows:

Step 1: the linked data crowdsourcing task LDT is classified by domain knowledge;

Step 2: According to the classification of LDT, extract the task-related domain professional testing tasks from the selected standard knowledge base;

Step 3: the workers, who participate in the crowdsourcing task and have not completed the domain professional test, first assign the domain professional testing tasks DSTs;

Step 4: According to the professional ranking of the workers, assign the LDT corresponding to the most professional DSTs to the workers;

Step 5: The worker submits the answer to the corresponding task.

Extracting domain professionality testing tasks requires that the knowledge of the standard knowledge base meets the diversity of the domain and has a large amount of common sense knowledge, such as DBpedia, Wikipedia and so on. However, if the number of triples included in SKB is too large, extracting testing tasks only by traversal will be impossible. Therefore, this paper adopts a link discovery framework LIMES[15] in the metric space, an efficient link discovery method between data sources, and pre-processes the standard knowledge base, greatly reducing the traversal space. Trigram[16] calculates the triple similarity as the normalized sum of the absolute differences between the triplet vectors of the two input strings, we retrieves all resources with a similarity greater than 0.85.

Domain professional testing tasks are extra tasks. If we set too many DSTs, it will increase the cost budget of the requester. If the setting is too small, it will lose the representativeness. Therefore, we measure the coverage of DSTs according to Shannon entropy[17]. The higher is the entropy, the greater is the uncertainty, and the larger is the coverage of the same number of DSTs.

3.2.2. DST algorithm description. n represents the number of DSTs that each category needs to extract, and C represents the number of workers assigned to each linked data crowdsourcing task. When workers participate in crowdsourcing tasks, they first assigned DSTs. Based on the response to the DSTs, a domain professionality array Pi is generated for the worker wi, and a crowdsourcing task matching the domain knowledge is assigned according to the worker's domain professional ranking. When all the tasks get C answers, the task assignment process stops.

Algorithm1: DST evaluation algorithm

Input: Task T, SKB B, Domain D, Number n, Number C
Output: DST = {P_d | d ∈ D}, Answer A

```
    B ← LIMES(B, T);
1   for n = 1,...,D do
2       B Remove DSTs;
3       for each b ∈ B do
4           △H ← getEntropyDiffer(b ∪ DSTs);
5           b* ← b let max△H and maxSimilar(t_i,b);
6           DSTs ← DSTs ∪ b*;
7       end for
8   for p < TC do
9       worker ← getWorker();
10      if worker not do DSTs then
11          T_i ← getDSTs();
12          P_i ← getWorkerDSTsResult;
13      end if
14      T_id ← getTask(maxP_i);
15      p ← p + T_i;
16      R ← R + R_Ti;
17  end if
18
```

3.3. MBEM-based crowdsourcing result quality assessment method

3.3.1. The core idea of MBEM algorithm. The EM algorithm proposed by Dawid and Skene [13] is based on the maximum likelihood estimation of the error rate of multiple observers, and iteratively estimates the accuracy of the task results and the accuracy of the workers until convergence, thus, achieving an accurate assessment of the results of the task. For the linked data crowdsourcing task, the linked data set is huge, and each iteration needs to recalculate the results of all tasks, resulting in very low efficiency of the algorithm. On the other hand, the initial parameter values of the EM algorithm have a great influence on the efficiency and accuracy of the iterative process. When the initial parameter values are set reasonably, the iterative process converges quickly and the result accuracy is higher; if the initial parameter value setting deviates from the actual situation, the efficiency of the algorithm is greatly reduced, and the accuracy of the final estimated result is relatively low.

The DST model proposed in this paper is used to measure the accuracy of workers' tasks in this domain. Therefore, the MBEM algorithm takes the domain expertise of the worker as the initial input parameter value.

Mini-batch gradient descent method MBGD is an iterative method commonly used to solve model parameters of machine learning algorithms. The basic idea is to divide the training sample into multiple sub-sample sets, and each iteration only performs gradient descent for one sub-sample set, which can solve the shortcomings of training too slow with Batch Gradient Descent (BGD). In this paper, the method of mini-batch gradient descent is adopted, and the result of each local iteration is used as the initial parameter value of the next local iteration, thereby improving the iterative efficiency.

3.3.2. MBEM algorithm description. InitialAccuracy is the field professional of the worker. InitialResult is the result of the crowdsourcing task submitted by the worker. The batch_size is the block size, that is, the number of samples per mini_batch. The initial parameter value of the first iteration of the algorithm is the field professional degree of the worker. Using the idea of simple weighted majority voting method, the task result is weighted and statistically obtained to obtain the task result.

Algorithm2: MBEM algorithm

Input: InitialAccuracy, InitialResult, batch_size
Output: Result

1	num ← InitialResult.size/mini_size;
2	accuracy ← InitialAccuracy;
3	for n = 1 ,..., num do
4	data ← getData(InitialResult, batch_size);
5	while threshold > θ do
6	r_temp ← eStep(data, accuracy);
7	mStep(r_temp);
8	end while
9	R_i ← getResult();
10	end for
11	Result ← {Ri	i ∈ 1,...,num };
12	return Result;

---

4. Experiment

4.1. Experimental setup
The hardware environment of this experiment is: ASUS notebook computer, 8GB memory, 4 core processor, clocked at 1.4GHz.

The three linked data sets used in the experiment were derived from three domain knowledge in "OpenKG.CN": Literal, Tourism, and Medicine, and the standard knowledge base uses DBpedia. This paper extracts 140, 60, and 40 (240 total) tasks from above three data sets as linked data crowdsourcing tasks.

4.2. Analysis of experimental results

The experiment is divided into three parts: (1) effectiveness of the DST assessment method; (2) effectiveness of initial value setting of the EM algorithm; and (3) the comparison of the efficiency of the MBEM algorithm.

4.2.1. Effectiveness of the DST assessment method. There is a significant difference in the performance of individual workers' professional abilities in different areas of professional tasks. Based on the domain professional requirements of the linked data crowdsourcing task, this paper automatically extracts domain professional testing tasks from the standard knowledge base to match the appropriate domain tasks for the workers.

In this paper, we use algorithm 1 to extract the number of DSTs in the domain of Literal, Tourism and Medicine from DBpedia, which are 25, 15 and 10 (50 in total). Use the first 50 data of the linked data crowdsourcing task as a GSTs task. In order to verify the effectiveness of the domain expertise, we recruited 20 volunteers to complete the above 290 tasks independently.

Figure 2-4 shows the accuracy of DSTs in the Literal, Medicine, and Tourism domain and the accuracy of LDT results in the corresponding domain. For a certain domain, the higher the degree of DST, the higher the accuracy of the corresponding domain task completion; on the contrary, The lower the degree of DST, the greater the fluctuation of the accuracy of tasks in corresponding domain, which is due to the randomness of non-professional workers. Figure 5 shows the change of GST accuracy and LDT result accuracy. With the increase of the accuracy of GST, the accuracy of crowdsourcing tasks does not increase, but shows a large fluctuation. This shows that although the gold standard task can filter the grass rate workers out according to the accuracy rate, that is, the workers with low accuracy, but can not guarantee that their domain expertise meets the requirements, that is, the accuracy rate when completing the crowdsourcing task is not necessarily high. DSTs are selected according to a certain domain of crowdsourcing tasks, the higher the accuracy, the more knowledge the worker has in the domain, and the higher the accuracy of crowdsourcing tasks in the domain.

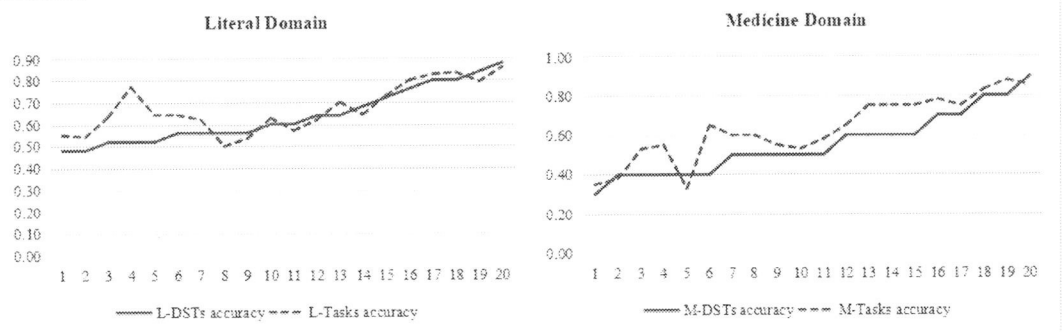

Figure 2. Literal Domain.   Figure 3. Medicine Domain.

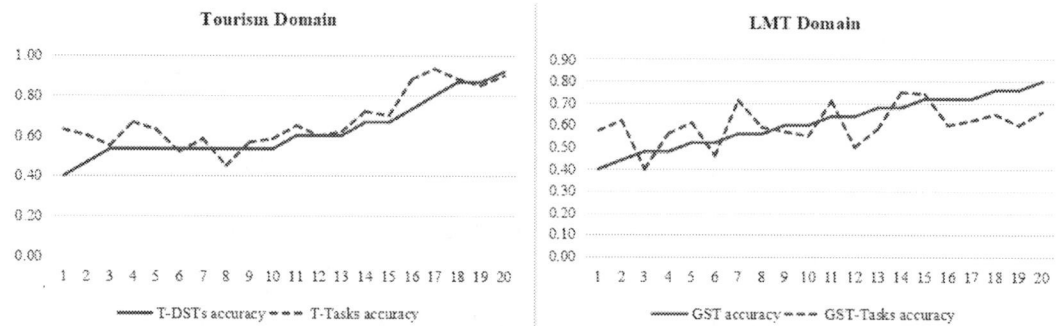

Figure 4. Tourism Domain.   Figure 5. LMT Domain.

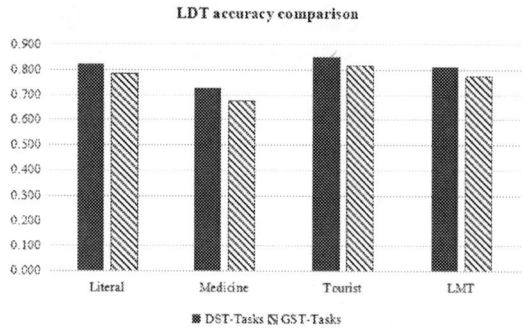

Figure 6. Comparison of the LDT accuracy.

4.2.2. Effectiveness of initial value setting of EM algorithm. For professional tasks in the field, if most workers do not have the expertise in the field and use the majority voting method to obtain results, the wrong results will affect the results of a small number of professional workers, resulting in unreliability of the final task results. Therefore, the initial parameter values of the EM algorithm have a certain influence on the accuracy of the results obtained by the iteration.

In this experiment, DSTs and GST are used as the initial parameter values of the EM algorithm. The accuracy of the results obtained by comparing the final task results with the real results is shown in Fig. 6. In the three domains of Literal, Medicine and Tourism, the accuracy of the results obtained with DSTs as the initial value is generally slightly higher than that of GST, and the quality of the results of the entire crowdsourcing task is better. DSTs represent the domain expertise of workers, which is closer to the accuracy of workers than GST, so algorithmic iterations produce less error and converge faster.

4.2.3. The comparison of the efficiency of the MBEM algorithm. The block size batch_size in the MBEM algorithm, that is, the number of samples per mini_batch directly affects the iterative efficiency of the algorithm. Therefore, choosing the appropriate batch_size will help improve the evaluation efficiency of crowdsourcing results. The actual linked data crowdsourcing task usually contains a very large amount of data, so this experiment uses the model to simulate 5000 data for subsequent experiments based on the results submitted by the workers.

As shown in Figure 7, when the size of the batch_size changes, the accuracy of the result fluctuates slightly. This is because when the batch_size is small, the number of samples in each mini_batch is too small and not representative, causes poor convergence during iteration; when the batch_size continues to increase, the number of samples in each mini_batch increases, and the iteration effect is enhanced; when the batch_size is too large, the task size in each mini_batch is too large, and the convergence accuracy of the algorithm will fall into different local extremums.

As you can see in Figure 7, when batch_size=5000, the result is the highest. This is because the entire data set is selected for each iteration. When batch_size=200, the result accuracy is only slightly lower than batch_size=5000. Therefore, choosing the appropriate batch_size size can guarantee the accuracy of the results to some extent.

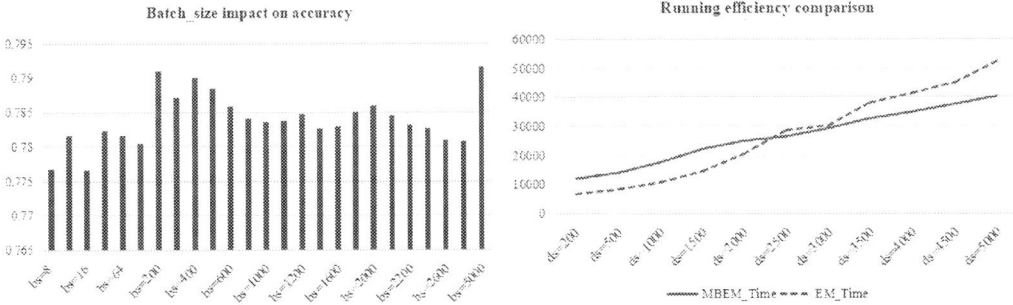

Figure 7. Comparison of batch-size.   Figure 8. Comparison of the efficiency.

When the data set size of a crowdsourcing task changes, the operating efficiency of the algorithm changes. Figure 8 is a comparison of the operating efficiency of the MBEM algorithm and the EM algorithm. The abscissa is the data set size and the ordinate is the running time (ms). For each data set size, the batch_size value of the MBEM algorithm is the size of the batch_size that maximizes the accuracy of the results. When the amount of data is small, the EM algorithm runs better than the MBEM algorithm, and when the amount of data increases, the efficiency of MBEM algorithm is getting higher. When the amount of data increases to a certain value, MBEM algorithm is better than EM algorithm.

When the amount of data is small, the algorithm runs faster. The MBEM algorithm has additional mini_batch block time, so the running efficiency is relatively low. As the amount of data increases, the convergence effect of mini_batch block iterations increases, and the number of iterations decreases, the running speed is accelerated; and the EM algorithm needs to load the entire data set every iteration, resulting in greatly reduced operational efficiency. The results show that the MBEM algorithm proposed in this paper is relatively efficient for large data set.

5. Conclusion

This paper focuses on the quality control of crowdsourcing for linked data, and the domain professional evaluation model DST is proposed. The main idea of the model is that the knowledge similar to crowdsourcing task is extracted from standard knowledge base and applied to the evaluation of workers' domain expertise, and the appropriate crowdsourcing task is assigned to the workers according to the evaluation results. Secondly, combined with the idea of mini-batch gradient reduction, the EM algorithm is improved, and the MBEM algorithm is proposed to achieve efficient and accurate evaluation of task results. The experimental results prove the validity and usability of the evaluation model and the quality control algorithm, which lays a foundation for the subsequent research on semantic crowdsourcing quality control.

References：

[1] Feng J, Li G, Feng J. A survey on crowdsourcing [J]. *Chinese Journal of Computrs*, 2015, **38** (9):1713-1726.

[2] Fang Y L , Sun H L , Chen P P , et al. Improving the Quality of Crowdsourced Image Labeling via Label Similarity[J]. *Journal of Computer Science and Technology*, 2017, **32(5)**:877-889.

[3] Bontcheva K, Derczynski L, Roberts I. Crowdsourcing Named Entity Recognition and Entity Linking Corpora[M]// *Handbook of Linguistic Annotation*. 2017.

[4] Gu J, Zhu T, Huang L, et al. A Review of Research on Linked Data Quality Evaluation in Knowledge Graph[J]. *Journal of Wuhan University (Science Edition)*, 2017, **63(1)**:22-38.

[5] Flemming, A.: Quality Characteristics of Linked Data Publishing Datasources. *MA thesis.* Humboldt-Universität of Berlin (2010) .

[6] Christophe G, Groth P , Stadler C , et al. Assessing Linked Data Mappings Using Network Measures[M]// *The Semantic Web: Research and Applications.* Springer Berlin Heidelberg, 2012.

[7] BERNERS-LEE T. Design Issues for the World Wide Web[EB/OL].[2006-07-27].https://www.w3.org.DesignIssues/LinkedData.html.

[8] Acosta M, Zaveri A, Simperl E, et al. Detecting linked data quality issues via crowdsourcing: A dbpedia study[J]. *Semantic Web*, 2018, **9(3)**: 303-335.

[9] Zaveri, A., et al.: User-driven quality evaluation of DBpedia. In: Proceedings of the 9th International Conference on Semantic Systems, pp. 97–104. ACM (2013)

[10] G. Demartini, D. Difallah, and P. Cudr′ e-Mauroux. Zencrowd: Leveraging probabilistic reasoning and crowdsourcing techniques for large-scale entity linking. In 21st International Conference on World Wide Web WWW 2012, pages 469 – 478, 2012.

[11] Joglekar M , Garcia-Molina H , Parameswaran A . Evaluating the Crowd with Confidence[J]. 2014.

[12] Zhang Zhi-qiang, Feng Ju-sheng, Xie Xiao-qin, et al. Research on crowdsourcing quality control strategy and evaluation algorithm [J]. *Chinese Journal of Computrs*, 2013, **36(8)**:1636-1649.

[13] Dawid A P，Skene A M．Maximum likelihood estimation of observer error-rates using the EM algorithm[J]．*Applied statistics*，1979：20-28.

[14] Fan, J., et al.: iCrowd: an adaptive crowdsourcing framework. In: Proceedings of the 2015 ACM SIGMOD International Conference on Management of Data, pp. 1015–1030. ACM (2015)

[15] Ngonga Ngomo, A.-C., Auer, S.: LIMES - a time-efficient approach for large-scale link discovery on the web of data. In: Proceedings of IJCAI (2011)

[16] Martin S, Liermann J, Ney H. Algorithms for bigram and trigram word clustering[J]. *Speech Communication*, 1998, **24(1)**:19-37.

[17] Lin J. Divergence measures based on the Shannon entropy[M]. *IEEE Press*, 1991.

ISPECE   IOP Publishing

IOP Conf. Series: Journal of Physics: Conf. Series **1187** (2019) 052087   doi:10.1088/1742-6596/1187/5/052087

Research on Quantitative Risk Assessment Method of Packaged Cargoes Carried By Ship Based on Online Dynamic Big Data Fusion Technology

Ru Lan[1, 2*]，**Wen Chang**[2*]，**Wei Shen**[2]，**Zhigang Jia**[3]

[1]University of Science & Technology Beijing，Beijing100083，China

[2]China Waterborne Transport Research Institute，Beijing100088，China

[3]ShenZhen Maritime Safety Administration Of P.R.C，ShenZhen518032，China

Email: lanru@wti.ac.cn，changwen@wti.ac.cn

Abstract: In this paper, a quantitative risk assessment method of packaged cargoes carried by ship based on online dynamic big data fusion technology was researched. In recent years, major accidents such as oil spill, fire, explosion and ship loss occurred frequently in shipborne cargo transportation. Due to the lack of real-time dynamic risk analysis, both early warning and emergency handling of those accidents are extremely difficult. In this paper, a risk transmission model was built, which based on the flow chain for cargo transportation. A system including the analysis of cargo transportation risk and quantitative assessment model based on real-time dynamic big data fusion technology was built. Finally, a scientific quantitative assessment method was formed. And a container ship passing through Shenzhen waters was taken as an example in order to verify the practicality of the method. In this way, it can realize the quantitative assessment of risk (especially the risk of dangerous packaged cargoes)in the waterage. What's more, according to the assessment results, it can analyze the high-risk factors and high-risk parts in the process of waterage, and then realize the informationalized and modernized intelligent supervision in a whole process and all round way, which will effectively improve the level of accident prevention in China.

1. Introduction

In recent years, major accidents such as oil spill, fire, explosion and ship loss frequently occurred in shipborne cargo transportation. The reasons for those accidents vary. For instance, the accidents of ship loss are caused by overweight containers [1], false reporting of dangerous cargoes[2] and improper packing[3]; the accidents of oil spill are caused by collision and grounding of ships. Due to the lack of dynamic real-time analysis, both early warning and emergency handling of those accidents are extremely difficult. How to use advanced scientific and technological methods to ensure the safety of packaged cargoes and strengthen the safety supervision of packaged cargoes has become an urgent problem.

This paper analyzes the transportation risk of packaged cargoes and builds a quantitative assessment method for packaged cargoes on the basis of dynamic real-time big data fusion technology. And this model can realize the quantitative risk assessment of packaged cargoes(especially the risk of dangerous packaged cargoes) in the water age. According to the evaluation results, it can analyze the high-risk factors and high-risk links during the transportation of packaged cargoes, so as to realize the informationalized and modernized intelligent supervision in a whole process and all round way. What's

Content from this work may be used under the terms of the Creative Commons Attribution 3.0 licence. Any further distribution of this work must maintain attribution to the author(s) and the title of the work, journal citation and DOI.

Published under licence by IOP Publishing Ltd

more, it can also trace the dangerous sources, track the whole transportation process of dangerous cargoes and control risks. Ultimately, this method can effectively improve the level of accident prevention in China.

1.1. Risk transmission model based on the flow chain

The cargo transportation is not a single independent shipping process, but a chain process with interlocking multiple links. Among those links, once one has risks and cannot be controlled in time, the risks will be continuously transmitted to the next link. And actually, the eventual accident is always caused by the accumulation of risks in different links. In this way, this process is a risk flow chain on the path. For example, the chain of container exportation must go through various operations, such as production, storage, cargo packaging, packing, land transportation, shipping and water transport. However, water transport, as the last link, is the most vulnerable to risk accidents. If the risk factors of each operation accumulate step by step, the accumulative effect will eventually lead to the occurrence of accidents. So if risk management is only carried out at the last link, it is difficult to control the risks.

The transmission model covers all the flow chains[5] of shipborne cargo transportation, from the consignor to the destination port of cargo transportation. The risk of flow chain comes from its internal links and external environment. The risk transmission model based on the flow chain is shown in Figure 1.

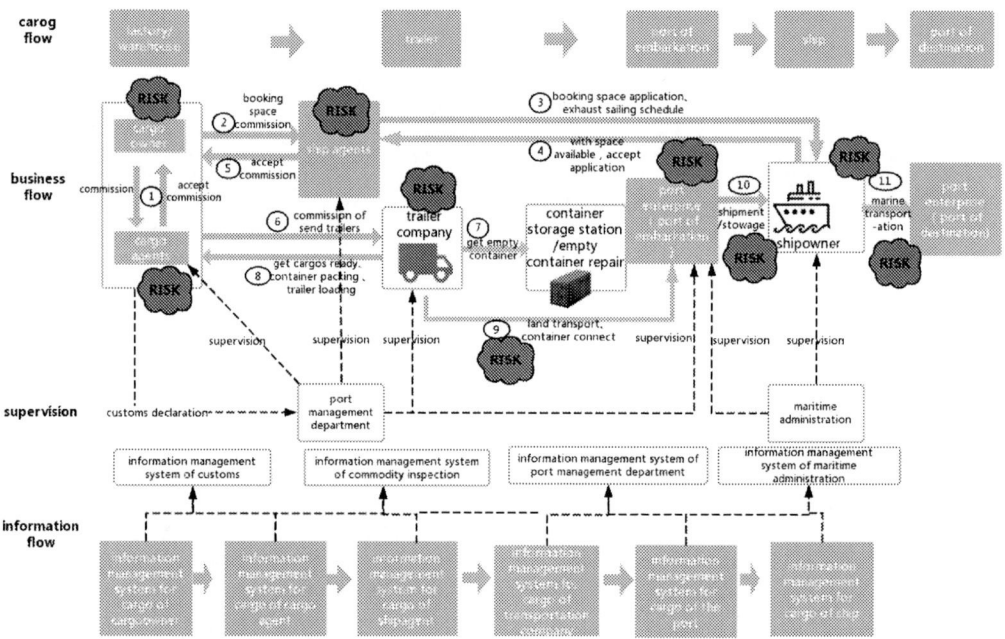

Figure 1. Risk Transmission Model Based on the Flow Chain in Packaged Cargo Transportation

1.2. Risk transmission analysis based on the flow chain

As for the process of packaged cargo transportation, the risks mainly include four aspects: the risk of the cargoes themselves, the risk of cargo packaging, the risk of the container and the risk of the ship where the container is located. In addition, the flow chain of cargo transportation from the factory to the final terminal is also the process of risk accumulation.

(1) the risk of ship[6]

① According to the condition of safe navigation

The condition of ship includes two aspects: one is the physical condition, which consists of ship size, hull, main engine, equipment material, structure and function, etc.; all of which cannot be changed by ship staffs. The other is technically variable conditions, including ship loading conditions, ship safeguard and ship maintenance.

② According to the requirement of seaworthiness

A ship is deemed to be suitable for navigation or towing when the following five aspect are considered: a) ship shape, structure, strength, water density and other elements; b) personnel; c) equipment and devices: GPS, navigation maneuver equipment (main engine, steering gear, anchor gear), communication equipment, mooring and cargo lift equipment; d) safety facilities and cargo requirements; e) material and spare parts. According to statistics of relevant departments, about 27% of maritime accidents are caused by unseaworthiness. In the chapter 4 of Maritime Traffic Safety Law of the People's Republic of China, it is stipulated that if the ship is in a condition unsuitable for navigation or towing, the competent authority shall have the right to forbid it from leaving the harbour or order it to suspend its voyage, change its route or cease its operations.

③ According to the key items of PSC inspection and the main defects statistics of the ship during the inspection

According to the requirement of the SOLAS Convention and related regulations, PSC inspection has two major points: the competence of crews and the technical condition of ship. The key points of inspection concerning ship conditions are as follows: hull condition, structural strength, load line marking, fire rescue equipment, anti-fouling equipment, auxiliary equipment, electrical equipment, alarm equipment, navigation aids, radio equipment, mooring equipment, main engines, auxiliary equipment, rudder cargo and cargo-handling appliance, etc.

(2) the risk of cargoes[7]

When the containers are transported, if the package and the box are intact, the mark is correct and clear, and the requirements of stowage isolation and binding are also met, then it is safe. However, sealing is required after packing. If there are some problems in the process of cargo transportation, such as packing breakage or improper management, it will lead to many safety problems in the container transportation, which will lead to various accidents. The safety risks of container transportation mainly include three aspects:

① Cargo risk

Since the cargoes themselves are dangerous, they should be packaged abiding by the relevant rules. And under normal circumstances, as long as the cargoes are packaged in conformity with the relevant rules of IMDG, the problems are less likely to happen. Nevertheless, if the dangerous cargoes are not packaged, or if the packaging does not meet the requirements, there's a high likelihood of problems during transportation. In addition, dangerous cargoes prohibited by waterway transportation may also be mixed in ordinary cargoes, to which should also be paid attention.

② Container risk

Used as dangerous cargoes carriers, containers load industrial waste from time to time. There is serious pollution inside. Therefore, before loading dangerous cargoes, in order to ensure the safety, containers should be thoroughly cleaned. Although containers themselves are strong and well-sealed, and can withstand a certain pressure, thus isolating dangerous cargoes, once the container is damaged or not tightly sealed, the container may enter the water or lead to the leakage of dangerous cargoes. That will result in serious pollution accidents, especially when the cargoes are spontaneously ignited cargoes, corrosive cargoes or poisonous cargoes.

③ Management risk of dangerous cargo containers

Improper stowage and isolation as well as false reporting are the most prominent problems in the management of dangerous cargo containers. In stowage, the suitable container location should be determined strictly according to the specific requirements of IMDG-Code for stowage and isolation of dangerous cargo containers. In addition, it is liable to cause potential security flaws by concealing and false-reporting dangerous cargo containers because in this case, the personnel and supervision departments will treat dangerous containers as ordinary ones. Then the shipping conditions are difficult to meet the specified requirements and the potential safety hazards are caused. What's more, padding of dangerous cargoes in containers, unreliable binding, ambiguous marking and even discrepancy between marking and actual packing are also common problems in container management.

(3) The risk of wharf

In the process of loading and unloading, the wharf must meet the safety requirements of cargo handling, and the wharf for dangerous cargo transportation should have the corresponding qualification. In addition, there may be risks of improper loading, unloading and stowage of cargoes at wharf.

(4) Risk of institution and personnel involved in the transportation chain

The risk analysis of each link in the container transportation chain is as follows:

① Consignor(shipper) risk

The consignor(especially the consignor of dangerous cargoes) should provide the contractual cargoes and the documents required for consignment. He also needs to properly pack the goods, mark and label the dangerous goods, and truthfully inform the official name, nature of the dangerous goods as well as the preventive measures to be taken.

The main reasons for the shipper's false report and concealment may be as follows: a)The shipper is tempted by huge profits. The ocean freight for dangerous cargoes is about 30% higher than ordinary cargoes; b)The shipper isn't familiar with the cargoes and does not know that the consigned cargoes are dangerous; c)The understanding and assessment of consigned dangerous cargoes are insufficient; d)Some cargoes are restricted by some countries or shipping companies for various considerations. In this case, buyers can only sign sales contracts with the consignor in private, and the shipper completes the transportation by changing the name of the cargoes, providing false information or not providing transport documents for dangerous cargoes. e)Restrictions on the transportation of dangerous chemicals by a certain route also increase the risk of false reporting to a certain extent. f)Packing inspectors fail to fulfill their duties at the container packing site , which results in the quality of dangerous cargo containers not meeting the standard requirements, thus causing potential accidents. At the same time, packing inspectors also have the risk of false reporting. Packing inspectors play an important role in ensuring the safety of container transportation for dangerous cargoes.

② The risk of cargo agent

The shipper entrusts the cargo agent to carry out dangerous cargoes' declaration and booking services. If the consignor did inform the cargo agent that those are dangerous cargoes, the cargo agent may also deliberately report them as ordinary cargoes. The cargoes false-reported by the cargo agent are easy to conceal and difficult to find. The reasons are as follows: a)The market of cargo agent is mixed, there are small-scale, poor credit, and low-risk cargo agencies; b)In order to avoid related transportation, surcharges and other related expenses, the cargo agent doesn't declare dangerous cargoes deliberately; c)Consignors and cargo agents lack of knowledge of dangerous cargoes as well as knowledge of international rules of dangerous cargo transportation; d)The declarer has no declaration qualifications and does not understand the risks of cargoes.

③ The risk of shipping agency

Lack of strict management and careful examination of the cargoes may also result in transport risks. Some shipping companies refuse to carry dangerous cargoes, which leads to the lack of regular transport channels for dangerous cargoes on some sea route. And to a certain extent, that also causes that false reporting can't be stopped despite repeated bans.

④ The risk of land transport enterprise

The consignor or the cargo agent entrusts a qualified trailer company to pick up empty containers at the yard, pack the cargoes in a factory or warehouse, transport the cargoes by land and ship them at the wharf. The risk factors in those steps are: a) if the trailer company uses trailers for non-dangerous cargoes to consign dangerous cargoes, or if in order to avoid supervision, the trail company loads dangerous cargoes in a trailer at the time of loading, and then replaces the head of the trailer with that of the trailer for non-dangerous goods and ship those dangerous cargoes as the ordinary cargoes, there will be a great risk. b) To carry dangerous cargoes requires the trailer driver to be qualified. If the trailer driver is not qualified to transport dangerous cargoes, that is to say, he/she does not understand the risks of dangerous cargoes' transportation, it may also result in the risk in the process of container loading, transportation or other links.

⑤ The risk of ship owner (carrier)

a) The risks caused by lack of qualifications. Some ineligible private bosses have docked their ships to companies qualified to transport dangerous cargoes. In this way, they can obtain the qualification to transport dangerous cargoes . These private shipowners pay a certain fee to the shipping company every year to make sure their ships are qualified to transport dangerous cargoes. Under these circumstances, the shipper and the carrier may collude to conceal the truth.

b) Carrier's negligence in receiving cargoes. When the shipper does not provide the correct shipping name and the carrier is not familiar with the cargoes, incompatible cargoes may be mistakenly packed in the same container, which causes inappropriate stowage and isolation. What's more, after shipment, the carrier may treat them as ordinary cargoes without special ventilation and observation of humidity. All of the above may eventually lead to accidents.

c) The carrier's lack of knowledge related to dangerous cargoes. Or he may reduce or completely disregard the isolation and stowage requirements in order to carry as many containers or cargoes as possible.

1.3. Characteristics of the risks in packaged cargoes transportation
Characteristics of the risks in packaged cargoes transportation are as follows.

(1) The risks of packaged cargoes are continuously transmitted and accumulated in the course of transportation, and the maritime transport area supervised by maritime authorities is the most dangerous part because of the accumulation of risks, which is the most difficult link to control.

(2) Sources of risk data involve multiple management departments, and the data is difficult to fuse.

(3) The risk factors are complex and the comprehensive risk value is difficult to quantify.

(4) Information risk may affect the operation of the whole logistics chain. Because the data information used for risk assessment includes almost all the information of the whole flow chain, the risk management and analysis are more challenging than the tangible loss.

2. Implementation process of quantitative risk assessment

2.1. Implementation design of data fusion
Data fusion is a multi-level process of multi-source data at certain grades, and each level represents different degrees of information abstraction. The process includes information detection, association, correlation, estimation and merging. Based on the feature level fusion method in multi-sensor data fusion method, this paper uses the neural network to synthesize and process the feature information extracted from the original information of each information source. The data fusion process uses artificial intelligence inference method for analysis and reasoning, and uses neural network as the method of target feature recognition[8].The framework of online dynamic big data fusion technology for packaged cargoes is shown in Figure 2.

2.2. Research on quantitative risk assessment method
(1) Hierarchical structure model of risk influencing factors

Firstly, establish an index system of risk assessment for packaged cargoes. Secondly, build a hierarchical model based on the index system.

Target level A is "the risk value of packaged cargoes", and in criterion level B, four index modules are established, namely "ship-fitness", "cargo-fitness", "port-fitness" and "personnel-fitness". The index modules of index level C are established under the each four index module of criterion level. Among them, "shipping company performance", "crew", "related personnel in the logistics chain" and "related personnel performance in the logistics chain" have established their sub-modules respectively. The risk assessment index system of packaged cargoes and the hierarchical structure model of risk influencing factors are shown in Figure 4.

ISPECE

IOP Publishing

Analytic Hierarchy Process (AHP) is used to determine the weight of index. By comparing the evaluation indices in the same level, the most important judgment matrix can be obtained, which can determine the importance order of those factors.

(2) Determination method of index weight relation

This study compares the relative importance of the same level indices by experts' scoring. $S_{i,j}$ indicates the importance of the evaluation index S_i relative to the evaluation index S_j. The range of values is 1-9 (see Figure 3). A pairwise comparison matrix between the evaluation indices is established (see Equation (1)). Compute the eigen vectors of pairwise comparison matrix for each level index. After normalization, the weight vector can be obtained $\omega_i = (\omega_1, \omega_2, \cdots, \omega_n)$ and ω_n is the corresponding weight of evaluation index n. The final weight results of each index calculated in this study are shown in Table 1.

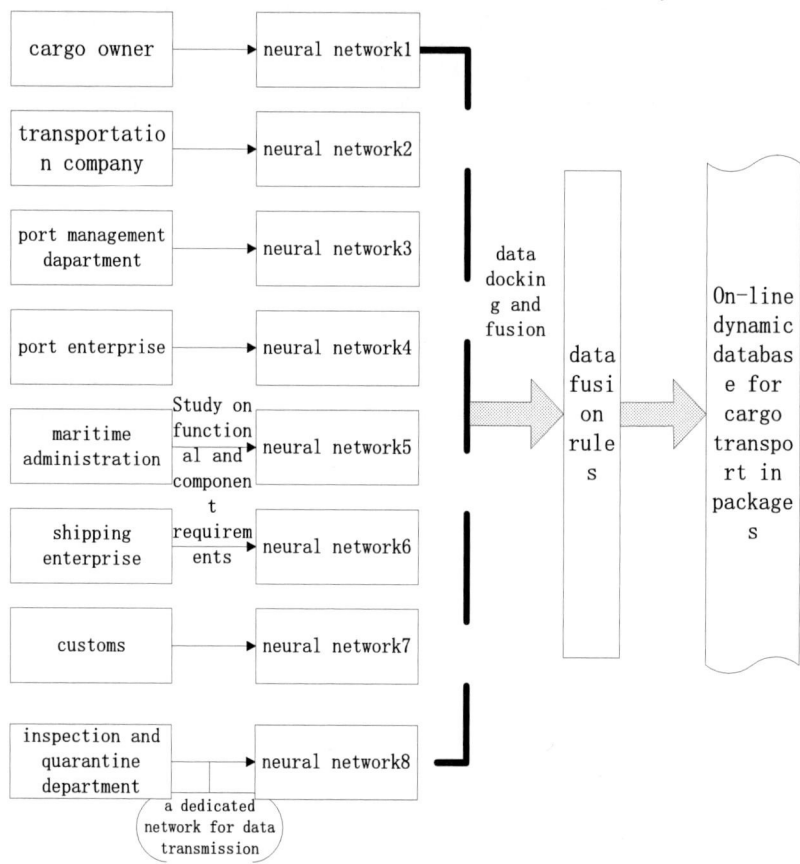

Figure 2. The framework of online dynamic big data fusion technology for packaged cargoes

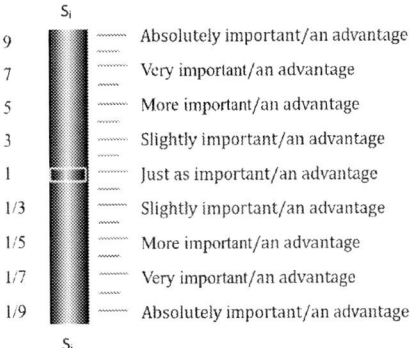

Figure 3. Proportional Scale of Index's Relative Importance

$$B = \begin{bmatrix} b_{11} & b_{12} & \cdots & b_{1n} \\ b_{21} & b_{22} & \cdots & b_{2n} \\ \vdots & \vdots & \ddots & \vdots \\ b_{n1} & b_{n2} & \cdots & b_{nn} \end{bmatrix} \tag{1}$$

Table 1. Weight calculation results of each index

First level index	Second level index	Third level index	Weight ω_i
Cargo-fitness（ω_1=0.322）	Name of cargoes	/	0.0596
	Type of containers	/	0.0984
	Type of cargoes	/	0.3801
	Packaging of cargoes	/	0.0366
	Container weigh	/	0.0196
	Safety of container	/	0.1392
	State of cargoes	/	0.0316
	Stowage or lashing of cargoes	/	0.2348
Port-fitness（ω_2=0.0926）	Standardized assessment of safety in production	/	0.5054
	Number of accidents at the wharf	/	0.1949
	Wharf credit	/	0.1245
	Number of administrative penalties involving danger prevention	/	0.1224
	Number of defects in danger prevention inspection	/	0.0528
Ship-fitness（ω_3=0.446）	Number of accidents of ship	/	0.0323
	Ship type	/	0.0338
	Ship age	/	0.029
	Focus tracking ship or not	/	0.405
	Ship detention	/	0.026
	Number of defect records based on FSC inspection	/	0.0897

	Number of administrative penalties involving danger prevention	/	0.1489
	Performance of shipping company (ωᵢ=0.0144)	Ship detention	0.0004
		Number of defect records based on FSC inspection	0.0009
		Number of administrative penalties involving danger prevention	0.0016
		Number of accidents in shipping companies	0.0036
		Shipping company system	0.0079
Personnel-fitness (ω₄=0.140)	Crew (ωᵢ=0.0698)	Crew certificate	0.0559
		Qualifications of senior crew	0.014
	Performance of related unit in logistics chain (ωᵢ=0.0349)	Shipping agency company	0.003
		Cargo agency company	0.0187
		Transportation company	0.0115
		Third party inspection institution	0.0018
	Related personnel in logistics chain(ωᵢ=0.0349)	Consignor	0.0027
		Cargo agent/booking party	0.0021
		Shipping agent	0.001
		Declarer of dangerous cargoes	0.0059
		Inspectors at packing site	0.0153
		Trailer driver	0.078

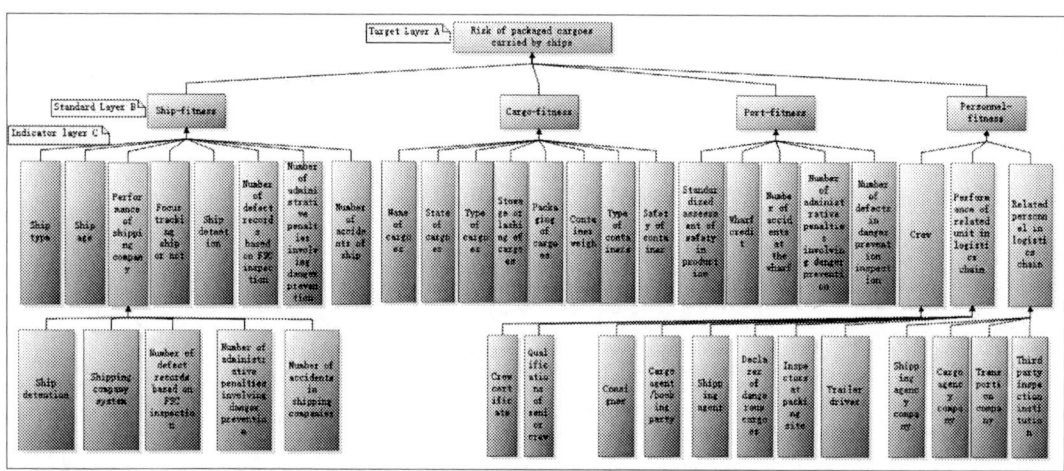

Figure 4. Risk Assessment Index System for Packaged Cargoes and Hierarchical Model of Risk Influencing Factors

(3) Consistency test

If the random consistency ratio CR is less than 0.1, the consistency test is passed. The CR values of each level in this study are shown in Table 2. The CR values of each level meet the need of the consistency test.

Table 2. CR value of each level

N	target level	criterion level			index level					
	risk of packaged cargoes	cargo-fitness	port-fitness	perso nnel-fitnes s	ship-fitness	cre w	performance of related unit in logistics chain	related personnel in logistics chain	performan ce of shipping company
CR	0.0439	0.0971	0.0912	0	0.088	0	0.0162	0.0956	0.0963

(4) Determination of the evaluation index's membership

Based on the relevant research results, this study establishes a one-to-one correspondence between the specific evaluation criteria of indices and the degree of risk, and builds the membership function of indices. According to the corresponding evaluation criteria, the risk can be divided into five levels: extremely high, high, general, low and extremely low. The continuous quantification method of fuzzy membership function is constructed. In this study, large Cauchy distribution and logarithmic function are used as membership functions:

$$f(x) = \begin{cases} [1 + \alpha(x - \beta)^{-2}]^{-1}, & 1 \leq x \leq 3 \\ a \ln x + b, & 3 \leq x \leq 5 \end{cases} \qquad (2)$$

α, β, a, b are undetermined constants

When the risk is extremely high, the degree of membership is 1, when the risk is general, the degree of membership is 0.8, and when the risk is very low, the degree of membership is 0.01. That is, $f(5) = 1$, $f(3) = 0.8$, $f(1) = 0.01$. So the calculation results are $\alpha = 1.1086$, $\beta = 0.8942$, a = 0.3915, b = 0.3699.

Then the membership function of the evaluation indices determined in this study is:

$$f(x) = \begin{cases} [1 + 1.1086(x - 0.8942)^{-2}]^{-1}, & 1 \leq x \leq 3 \\ 0.3915 \ln x + 0.3699, & 3 \leq x \leq 5 \end{cases} \qquad (3)$$

Based on the requirements of international conventions related to packaged cargoes and the practical experiences of junior managers, the author has studied and formulated the evaluation criteria for each assessment factor. The evaluation level can be determined by referring to that evaluation criteria.

(5) The model of multi-level fuzzy comprehensive evaluation[10]

The FCE (Fuzzy Comprehensive Evaluation) judgment matrix is generated by using the indices of each factor of the analytic hierarchy model as the evaluation index. The FCE judgment matrix is used to calculate the cargoes' known information obtained by the big data support system, then obtain the comprehensive evaluation result of the cargo transportation risk. The results of comprehensive evaluation are shown in Table 3.

Table 3. The results of comprehensive evaluation

Evaluation target	Comprehensive evaluation score
Risk assessment of shipborne packaged cargoes	$H = \sum_{i=1}^{n} S_I w_i$

In this formula: H is the score of solid bulk cargo transportation risk , $1 < H \leq 5$; S i is the score of each index, $1 < S_i \leq 5$; Wi is the comprehensive weight of each index; n is the number of the index. In the final calculation result, if $4 \leq H \leq 5$, the transportation risk is extremely high; if $3 \leq H < 4$, the transportation risk is high; if $2 \leq H < 3$, the transportation risk is general; if $H < 2$, the transportation risk is low.

3. Case study

This method can be applied to the dynamic supervision of packaged cargoes by maritime department, port supervision department, customs, inspection and quarantine department and other regulatory authorities. And it can also be applied to the selection of the target containers in the open-package inspection. The author verifies the practicability of the method through the following examples of open-package inspection. The object of this assessment is a Singaporean container ship passing through Shenzhen waters. The ship departs from Shanghai Port and is berthed by Yantian Port. The final destination is the Port of Tanjung Pelepas in Malaysia.

(1) Assessment factors

Name of cargoes: Hydrogen Cyanamide; State of cargoes: liquid state; Stowage or lashing of cargoes: the binding plan is unreasonable (the cargoes need temperature control, which is not considered by the binding plan); Type of cargoes: 6.1 and the packaging is III (according to IMDG rules); Container weight: 61880t; Type of containers: container for dangerous cargoes; Safety of container: container inspection report with incomplete content; Packaging of cargoes: barrel; Standardized assessment of safe production at the wharf: Level 3; Credit of the docked wharf: Non-concealment; Number of defects in danger prevention inspection: more than three times; Ship type: container ship; Ship age: 15 years; Focus tracking ship or not: no; Number of administrative penalties involving danger prevention: more than three times; Crew certificate: with certificate; The cargo agent has two cases of concealment. The rest of the data is missing and the risk takes the intermediate value.

(2) The verification results

Bringing the assessment factors mentioned above into the risk assessment model proposed in this paper, then the results can be worked out:

Table 4. The results of assessment

Target of the assessment	Comprehensive evaluation score
Risk of ship-fitness	$H_1=2.57$
Risk of cargo-fitness	$H_2=3.70$
Risk of port-fitness	$H_3=3.48$
Risk of personnel-fitness	$H_4=2.90$
Comprehensive assessment results	$H=3.06$

Therefore, according to the comprehensive assessment results, the risk of the packaged cargoes is high ($H=3.06$), which should be checked as the target box. In addition, in the process of cargo transportation, the risk of cargo fitness ($H_2 = 3.70$) and port fitness ($H_3 = 3.48$) is high, so the cargoes and the port should be the key link of supervision and inspection. In the actual transportation process, the ship has a leakage accident in Shenzhen sea area, which is consistent with the assessment results of this model.

4. Conclusion

(1) This study puts forward a risk assessment index system for packaged cargoes based on ship-fitness, cargo-fitness, port-fitness, personnel-fitness. According to the requirements of relevant international conventions and combined with the practical experience of junior managers, the evaluation criteria of each assessment factor are formulated.

(2) This study constructs risk transmission model based on the flow chain for packaged cargoes, and a system including the analysis of cargo transportation risk and quantitative assessment model based on real-time dynamic big data fusion technology.

(3) In this study, a quantitative risk assessment method for packaged cargoes based on online dynamic big data fusion technology is proposed, and its practicability is verified by an example of a container ship passing by Shenzhen waters. This method can be applied to the dynamic supervision of packaged cargoes by maritime department, port supervision department, customs, inspection and quarantine

department and other regulatory authorities. And it can also be applied to the selection of the target containers in the open-package inspection.

Acknowledgment

The national key research project " Research on mechanism of disaster occurrence, impact on ecological environment security, and evolution tendency of typical alien species invasion" under contract No. 2017YFC1404600;

References

[1] Ni Zhiping, (2017) Suggestions on Overweight of Domestic Containers. J. China Ports., 8: 12-14.

[2] Cai, T. (2018) Internal Trade Container Terminal Discussion on the Precautionary Measures of False Reporting of Dangerous Goods. J. Logistics Engineering and Management., 40: 160-161.

[3] Wu, H. (2017) Formal Safety Assessment on Loading and Unloading of Container Ship. D. Dalian: Dalian Maritime University.

[4] Deng, M., Fen L. (2006) Risk Transmission in Commercial Bank: Based on Flow Chain. J. Contemporary Economic Management., 28: 76-79.

[5] Liang, H. (2006) Main Documents of Port Container Transportation and the Transfer Flow. J. China Water Transport., 4: 30-32.

[6] Li, Z. (2008) Research on Ship-selecting System of Port State Control. D. Dalian: Dalian Maritime University.

[7] Chen, C. (2015) An Analysis on the Method of targeting dangerous cargo containers for open-package inspection in Ningbo Port. D. Dalian: Dalian Maritime University.

[8] Wang, L., Lv, S., Liu, Z. (2007) Study of Security Dynamical Risk Assessment Based on Data Fusion. J. Computer Engineering., 22: 32~33.

[9] Ada, S., Markowski, M., Sam, M., Agata, B. (2009) Fuzzy Logic For Process Safety Analysis. J. Journal of Loss Prevention in the Process Industries. 22:695-702.

[10] An,M. (2007) Risk assessment in rail way safety management. J. International Journal of Engineering and Technology., 14: 45-56.

[11] Joao, J., Flavio, M. (2012) Towards the of an Operational Tool for Oil Spills Management in the Algarve Coast. J. Coastal Conservation., 16: 33-47.

[12] Hu, E. (2009) Practical Technology and Method for Environmental Risk Assessment. China Environmental Science Press, Beijing.

[13] Xu, H. (2009) Problems and Strategies in Safe Transport Management of Dangerous Goods in Container Transportation. C. China Academic Journal Electronic Publishing House., 2009:5-13.

An Analysis Method for Error Propagation Reachability of Component-Based Software

Mingqi FAN[1], Min SONG[2]*, Zhengxian WEI[3,4], Wanfeng MAO[3]

[1] Equipment Procurement Center of Navy Equipment Department, Beijing, China

[2] Information Technology Center, Beijing Foreign Studies University, Beijing, China

[3] Systems Engineering Research Institute, Beijing, China

[4] Joint Laboratory for Smart Ocean Technology, Pilot National Laboratory for Marine Science and Technology, Qing dao, China

Corresponding author: Min SONG, e-mail: songmin@bfsu.edu.cn

Abstract. Component-based software (CBS) is widely used in various industries and fields. One important characteristic of CBS is that when some error occurs in a component, the error will be transmitted to other components as the inter-component state propagates, the error propagation will form and the error propagation may cause the software system to malfunction or fail. From the perspective of software architecture (SA), an analysis process for error propagation accessibility of CBS and a generation algorithm for error propagation reachability graph event/state-based of CBS were established. In this algorithm, firstly, the elementary events are sorted by identifying the internal timing sequence of the components in the directed graph of SA, and the state groups that cannot be reached at the same time between the components are obtained; secondly, the Cartesian product of the reachable states between the components is established and all the reachable state groups of CBS are obtained; at last, the error propagation reachability graph of CBS is obtained according to the event connection table. The algorithm has laid a technical foundation for the analysis on CBS errors and reliability.

1. Introduction

Component-based software (CBS)[1] is widely used in various industries and fields. At present, many industries such as the ship industry, the aviation industry and the ordnance industry have adopted this software mode in their software. Service oriented architecture (SOA), Web services and internetware are typical representatives of this software mode. As a kind of new software form based on network platform, CBS is the software system in which various software entities in the network (e.g. components, services, agents, which are collectively called "components" in this paper) exist in all the nodes in an open and independent form and implement cross-network interconnection, intercommunication and collaboration in various collaborative methods in an open environment. At present, many industries such as the ship industry, the aviation industry and the ordnance industry have adopted this software mode in their software. With the development of technology, CBS will be more and more characterized by independent adaptability, collaboration, reactivity, evolution, diversity, etc., and provide customers with the services featuring 7*24, continuous evolution and update, and smartness. Meanwhile, CBS can perceive the dynamic changes of the environment and carry out online dynamic evolution in accordance with the function indexes, performance indexes, credibility indexes, etc. [2,3].

Content from this work may be used under the terms of the Creative Commons Attribution 3.0 licence. Any further distribution of this work must maintain attribution to the author(s) and the title of the work, journal citation and DOI.

Published under licence by IOP Publishing Ltd

The reliability analysis of CBS is one of the hot research topics at present [4,5,6]. The relation between a software error and the error propagation is an important basis for the software reliability analysis[7,8,9]. One important characteristic of CBS is that when some error occurs in a component, the error will be transmitted to other components as the inter-component state propagates, the error propagation will form and the error propagation may cause the software system to malfunction or fail. From the perspective of software architecture (SA), the paper reified the software architecture into a directed graph, analyzed the transitivity of the elementary event (internal activity or state of the component which causes an error) interaction at the interfaces between components, then an analysis process for error propagation accessibility of CBS was established. More important, a generation algorithm for error propagation reachability graph event/state-based of CBS was proposed. In this algorithm, firstly, the elementary events are sorted by identifying the internal timing sequence of the components in the directed graph of the architecture, and the state groups that cannot be reached at the same time between the components are obtained; secondly, the Cartesian product of the reachable states between the components is established and all the reachable state groups of the software system are obtained; at last, the error propagation reachability graph of the software system is obtained according to the event connection table. The algorithm has laid a technical foundation for the analysis on CBS errors and reliability.

2. Analysis process for error propagation accessibility of CBS

One important characteristic of CBS is that when some error occurs in a component, the error will be transmitted to another component as the inter-component data state propagates, may cause some error to occur in the component and some error propagation to form. The scope and reachability of the error propagation needs to be analyzed. CBS is composed of components, and the functional (or service) interaction between components is implemented through the interfaces. Therefore, to analyze the errors of CBS, it is necessary to start from the perspective of software architecture (SA), and obtain the error behavior of component C_i as well as the error behavior of the interaction I_{ij} of the interfaces between components [10].

Currently, it is widely believed that the components, as well as the interactive relationship between components, work together to form the software architecture [2]. Component C consists of two parts, i.e., the interface specification and internal specification of the component. The interface specification is divided into two types, the first is called the service interface (provided/service/public interface), denoted as I_p, and it is used by component C to provide functions (services); the other is called the request interface (required/entry interface), denoted as I_e, and it is a function (service) interface needed by the component C.

If a component is used to accomplish certain functions or provide corresponding services, these will be implemented through the activities and states of the internal functional domain of the component. The relationship between the activity and state of the internal functional domain of the component is shown in Fig 1. In Fig.1, the request interface I_e can be used to input the states or activities, and in the meantime, the service interface I_p can be obtained from the activities or states. In the CBS research process, when the interactive relationship of the interfaces between components is analyzed, the implementation of the internal functions of the component can be ignored and the component can be treated as a black box. If an analysis involves the internal function implementation of a component, it shall be necessary to pay attention to the relationship between the activities and states within the component. Due to the directivity of the interactive relationship of the interfaces between components, the architecture could be denoted as a directed graph SA:=G（C,I）, where C refers to the component set and I refers to the interface interaction set. If the implementation process within the component is involved in the analysis, the description method of the activities and states within the component as shown in Fig.1 shall be adopted. In addition, it is assumed that there are no loops in the directed graph of the architecture.

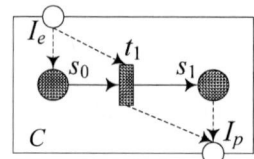

Fig. 1 Relationship between activities and states within the component

It is pointed out in [10], when the error propagation and reliability of the system is analyzed, three kinds of information shall be obtained: (1) architecture of the software system; (2) error behaviors of component C_i; (3) error behaviors of the interface interaction I_{ij} between components. An activity or state of a component which causes an error is called an "elementary event". The occurrence of an elementary event doesn't necessarily cause a software system to malfunction or fail. Only when an elementary event occurs, it propagates through the interactive relationship of the interface between components and results in the loss of the user services or the system function performance, will it cause the software system to malfunction or fail (It is called a top-layer event). The occurrence of a top-layer event of the software system depends on the occurrence sequence and propagation route of the elementary event. In CBS, the elementary events are transmitted as the states propagate among components. In this paper, a state that can propagate carrying an error is called an error state. In the process of the elementary event (error) propagation, the interactive relationship of the interface determines whether the elementary event that had occurred has effects on the system. To handle the state space explosion is the key to the analysis on the error propagation and reliability of the software system[8]. To solve the state space explosion problem effectively, an error propagation reachability graph is established. Firstly, elementary events are sorted by identifying the internal timing sequence of the components in the directed graph of the architecture, and the state groups that cannot be reached at the same time between the components are obtained; secondly, the Cartesian product of the reachable states between the components is established and all the reachable state groups of the software system are obtained; at last, the error propagation reachability graph of CBS is obtained according to the event connection table.

In CBS, when event t_i occurred within component C_i, if event t_j occurred within component C_j through interface I_{ij}, it is called "event t_i triggers event t_j". In fact, the triggering relationship between events reflects the propagation of the events with the propagation of the states. The triggering relationships between component events form through their interface interactions, and the interaction relationships of the interfaces between components are divided into direct interactive relationships and indirect interactive relationships. The event triggering relationship between components is also divided into two kinds: direct triggering relationships and indirect triggering relationships. As shown in Fig.2, component C_2 triggered the event of component $C_1.t_1$ through event $C_2.t_1$; component C_1 triggered the event of component $C_3.t_1$ through event $C_1.t_2$. Therefore, there was a triggering relationship between components C_2 and C_3.

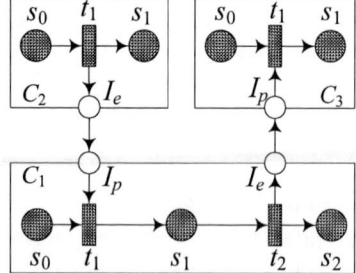

Fig.2 Event interaction relationship between components

Based on the event triggering relationship and interface interaction relationship shown in Fig. 2, the analysis process for error propagation accessibility of CBS is shown in Fig. 3.

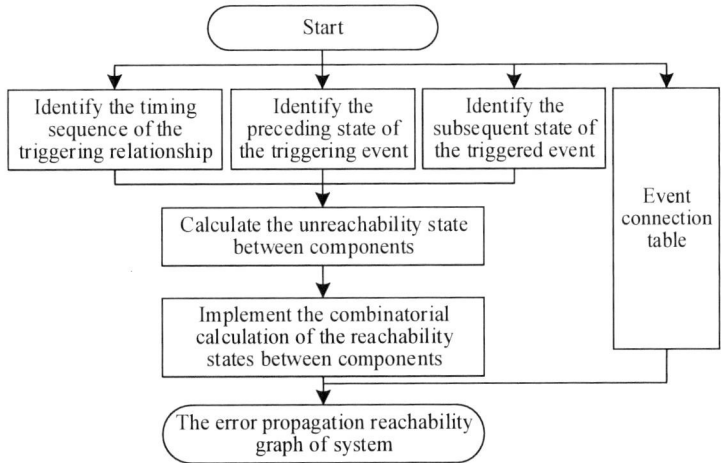

Fig.3 Analysis process for error propagation accessibility of CBS

3. Generation algorithm for error propagation reachability graph event/state-based of CBS

In the process of the analysis on the CBS error propagation reachability, the prorogation of events and states are bound together. In order to analyze the CBS error propagation relationship, a generation algorithm for error propagation reachability graph event/state-based of CBS is proposed. The specific steps of the algorithm are shown as follows:

Step1: sort the events with some triggering relationship according to the internal activity timing sequence of the component.

(1) Sort the internal events of the components according to the internal activity timing sequence of the component.

(2) For the components with a direct triggering relationship, the occurrence time of the triggering event is equal to that of the triggered event.

(3) For the components with an indirect triggering relationship, the timing sequence relationship of the events between nonadjacent components is established according to the activity timing sequence of the triggering relationship of the intermediate component.

Three rules are proposed based on the above three event timing sequence relationships:

Rule 1: For component C_1 containing two events t_1 and t_2, if the occurrence time t_1 of the event is earlier than t_2, then there is a timing sequence relationship: $C_1.t_1 < C_1.t_2$;

Rule 2: For the existing components C_1 and C_2, and there is a triggering relationship between event t_1 in component C_1 and event t_2 in component C_2, there is timing sequence relationship: $C_1.t_1 = C_1.t_2$;

Rule 3: For component C_1, C_2 and C_3, if there is a triggering relationship between event t_1 in component C_1 and event t_1 in component C_2, there is also a triggering relationship between event t_2 in component C_1 and event t_1 in component C_3, and $C_1.t_1 < C_1.t_2$, then there is timing sequence relationship $C_2.t_1 < C_3.t_1$, as shown in Fig. 4.

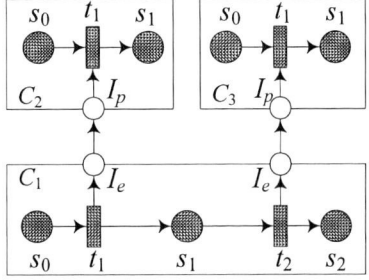

Fig. 4 Rule3 the event timing sequence between components

For component C_1 containing event t_1 and t_2, according to the timing sequence activities, if $C_1.t_1 < C_1.t_2$, $C_1.t_1 = C_2.t_1$ and $C_1.t_2 = C_3.t_1$, then $C_2.t_1 < C_3.t_1$. Then if t_1 is connected with the initial state, then the preceding state of the triggered event t_1 in C_2 and the subsequent state of the triggered event t_1 in C_3 form the unreachable state group.

Step2: Calculate the preceding state and subsequent state of the event triggering relationship.

The unreachable state group refers to the combination of two states of different components. They cannot be activated simultaneously due to the timing sequence relationship. Define the reachability graph of a complete system as the state reachability graph of the system and the internal reachability graph of a component as the state reachability graph of the component.

(1) Identify the preceding state set of the triggering event. Identify all the unreachable states of the triggering events according to the states of the component, use the Dijkstra algorithm beginning from the direct preceding state of the event. As for the unreachable state, the distance is infinitely great.

(2) Identify the subsequent state set of the triggered event. Identify all the unreachable states beginning from the initial state according to the states of the component, implement the Dijkstra algorithm beginning from the initial state and identify all the unreachable states. The distance is infinitely great. Therefore, only the state which is directly reachable through the triggering relationship is used as the subsequent state of the triggered event.

Step3: Identify unreachable state groups according to the timing sequence relationship.

(1) Identify unreachable state groups with direct event triggering relationships between components according to the timing sequence relationship.

For the components with direct event triggering relationships, according to the timing sequence relationship between the triggering event and the triggered event, the preceding state of the triggering event and the subsequent state of the triggered event are unreachable. In this case, establish the Cartesian products of the preceding states and the subsequent states, and obtain the unreachable state groups. In the meantime, if a triggered event is uniquely connected with an initial state, then establish an unreachable state group according to the preceding state of the triggered event and the subsequent state of the triggering event. As shown in Fig. 5, component C_1 causes triggering event t_1, if component C_2 is in the state of s_1, then the event t_1 is triggered and it will reach the state of s_2.

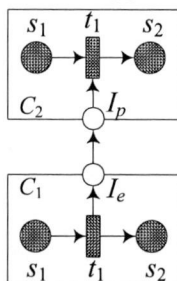

Fig.5 Component relationship diagram

In Fig. 5, if there are successors $(C_2.t_1)=\{C_2.s_2\}$, predecessors $(C_1.t_1)=\{C_1.s_1\}$ and successors $(C_1.t_1)=\{C_1.s_2\}$, then the unreachable state group is: $\{C_1.s_2, C_2.s_1\}$, $\{C_2.s_1, C_1.s_2\}$.

(2) Identify unreachable state groups with indirect triggering relationships between components according to the timing sequence relationship.

For components C_1 and C_2 with some indirect timing sequence relationship, there is $C_1.t_1 < C_1.t_2$ based on the activity timing sequence of the component, then $\{C_1.s_1, C_3.s_2\}$ is the unreachable state group with an indirect timing sequence relationship, as shown in Fig. 6.

Step4: obtain all the reachable state groups of a system.

For the CBS $S = \{C_1, C_2, ..., C_k\}$, where $C_i = \{s_1, s_2, ..., s_m\}$ indicates that component C_i contains m states, and then establish Cartesian products of the reachable states between components.

Let the component set involved in the combinatorial computing of the Cartesian product be D, the reachable state group be R, and initially the set D is empty, then R is initially empty too.

ISPECE   IOP Publishing

IOP Conf. Series: Journal of Physics: Conf. Series **1187** (2019) 052088   doi:10.1088/1742-6596/1187/5/052088

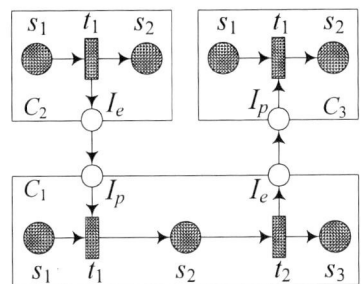

Fig.6 Component relationship diagram

(1) At the initial time, take out C_i, $D=\{C_i\}$, $S=S-D$, $R=\{\{s_{i1}\},\{s_{i2}\},...,\{s_{im}\}\}$ and $(i=1,...,k)$ from the component set S.

(2) Select a component C_j, $D=\{C_i,C_j\}$ and $(1<=i,j<=k)$, from $S-D$. For any state set R_i in R, traverse the unreachable state set generated by the sub-states R_i and C_j in step (3), remove the states of components with unreachable relationship in C_j and obtain $\overline{C_j}$, then implement the combinatorial computing $R=R_i\times\overline{C_j}$.

(3) Repeat step (2), until all the components are included in D, S is empty and R contains all the reachable state groups.

Step5: establish a connection table and obtain an error propagation reachability graph.

In CBS, each event corresponds to one activity or state, and each event or activity has a connection mode. Therefore, there is a connection mode corresponding to each event. The mode describes the transfer from the preceding state to the subsequent state. Therefore, an event state connection table is established according to the connection mode, and the first column of the table contains the only allowed switching between the two states. Remove all the unreachable states from the initial state in the graph, and then we obtain error propagation reachability graph of CBS.

4. Conclusion

CBS is widely used in various industries and fields. One important characteristic of CBS is that when some error occurs in a component, the error will be transmitted to other components as the inter-component state propagates, the error propagation will form and the error propagation may cause the software system to malfunction or fail. From the perspective of software architecture (SA), this paper reified the software architecture into a directed graph, analyzed the transitivity of the elementary event interaction at the interfaces between components, then an analysis process for error propagation accessibility of CBS was established and a generation algorithm for error propagation reachability graph event/state-based of CBS was proposed. The analysis process and algorithm have laid technical foundation for the analysis on CBS errors and reliability.

Acknowledgments

This work is sponsored by the National Natural Science Foundation of China under Grant No. 61502037 and 61772152, the Basic Research Project No. JCKY2016206B001 and JCKY2016604C010, and Innovation Special zone of National Defense Science and Technology Project.

References

[1]   Tomar P, Mishra R, Sheoran K. Prediction of quality using ANN based on Teaching-Learning Optimization in component-based software systems[J]. Software Practice & Experience, 2018, 48(3).

[2]   Fuqing Yang: Thinking on the Development of Software Engineering Technology. Journal of

Software[J]. 2005, 16(1): 1-7.

[3] Jian Lv, Xiaoxing Ma, Xianping Tao, Feng Xu, Hao Hu: Research and development of network software[J]. Science in China. 2006, 36(10): 1037-1080.

[4] Ling, Dongyi , B. Liu , and S. Wang . A component-based software reliability assessment method considering component effective behavior[C]. Second International Conference on Reliability Systems Engineering IEEE, 2017.

[5] Koziolek Heiko, Schlich Bastian, Becker Steffen. Performance and reliability prediction for evolving service-oriented software systems[J]. EMPIRICAL SOFTWARE ENGINEERING, 2013,18(4):746-790.

[6] Joao M Franco, Raul Barbosa, Mario Zenha-Rela. Reliability Analysis of Software Architecture Evolution[C]. In: Proc. of the 6th Latin-American Symposium on Dependable Computing (LADC), 2013: 11-20.

[7] Hayat, Tayebe, and H. Seifzadeh. A New Approach to Reliability Prediction in Component-based Systems[C]. International Conference on Computer Modeling & Simulation 2017.

[8] Li, Kewen , et al. Reliability Evaluation Model of Component-Based Software Based on Complex Network Theory[C]. Quality and Reliability Engineering International (2016).

[9] Singh A P, Tomar P. A Proposed Methodology for Reliability Estimation of Component-Based Software[C]. International Conference on Optimization Modeling and Applications, 2012.

[10] Uokhale S S, Wong W E, Horgan J K, et al. An analytical approach to architecture based software performance and reliability prediction[J]. Performance Evaluation, 2004, 58(4): 391-412.

[11] YIN Guisheng, SONG Min, WEI Zhengxian, WANG Hongbin, ZHANG Wansong. Analysis of software architecture evolution orienting semantic relationship between components[J]. Journal of Harbin Engineering University, 2011, 32(10),1329-1335.

[12] Song M, Wei ZX, Yin GS. Evolution analysis of data flow oriented internetware service[J]. Journal of Software, 2013, 24(12):2797-2813.

Thoughts on the Orientation of Mathematics Education in Colleges and Universities

Liang Shushuang

Liuzhou City Vocational College 545036

Liang Shushuang (Liuzhou City Vocational College, Liuzhou, Guangxi 545036)

About the author: Liang Shushuang, female, Han Nationality, Liuzhou, Guangxi, Lecturer of Liuzhou City Vocational College, Master of Education, Guangxi Normal University, research direction: curriculum and teaching theory (mathematics education teaching direction)

Abstract: Mathematics is a science of structure, order, and relationship. It evolved from the basic practices of counting, measuring, and describing the shape of objects. It involves logical reasoning and quantity calculation. Since the 17th century, mathematics has been an indispensable aid to physical science and technology, and is considered to be the basic language of science. In the mathematics education work of colleges and universities, accurately positioning the education direction and improving the classroom teaching mode can effectively shape the students' mathematical thinking, stimulate students' innovative consciousness, and strengthen students' ability to solve problems. This article will focus on the classroom teaching mode for the university mathematics education orientation, in order to provide theoretical reference for relevant teaching and research personnel.

1. Introduction

From the perspective of the overall structure, improving the quality of mathematics education is an important goal of education. To achieve this goal, it is necessary to accurately locate the direction of mathematics education in colleges and universities, grasp the educational dynamics, and ensure that the mathematical forms are in line with the needs of social development. At the same time, we must adhere to the concept of teaching students in accordance with their aptitude, accurately position the direction of talent training, and innovate teaching models. This paper will firstly study the practical value of mathematics education in colleges and universities, discuss some thoughts on the orientation of mathematics education in colleges and universities, and discuss how to cultivate professionals with mathematical literacy.

2. Practical Value of Mathematics Education in Colleges and Universities

Mathematics is a science that studies the quantitative relationship and spatial form of the real world. It originates from the practical problems of counting and measuring. The discipline has a high degree of abstraction, rigorous logic and wide applicability, including mathematical logic, number theory, algebra, geometry. Learning, topology, function theory, functional analysis, differential equations, probability theory, mathematical statistics, computational mathematics, and operations research and cybernetics [1]. Throughout the history of the development of mathematics education, all outstanding scholars are inseparable from the help and guidance of mathematics education in the process of growth.

Content from this work may be used under the terms of the Creative Commons Attribution 3.0 licence. Any further distribution of this work must maintain attribution to the author(s) and the title of the work, journal citation and DOI.

Published under licence by IOP Publishing Ltd

At the same time, scholars have made significant contributions to the development and inheritance of mathematics, through the dissemination of mathematics knowledge for people's work and Life provides a lot of convenience, daily shopping, timing and other activities can not be separated from the use of mathematical knowledge. In addition, in order to give full play to the advantages and functions of mathematics education, relevant teachers must be able to comprehensively innovate mathematical concepts, improve mathematical thinking, and comprehensively use diversified teaching methods, and thus continuously improve the quality and practical value of mathematics education in colleges and universities. On the other hand, in cultivating mathematics talents, we must base ourselves on the goal of educational reform, innovate and perfect the teaching model and mathematics education technology, guide the study of mathematics formulas, theorems and concept knowledge, and cultivate agile digital combination ideas. Create a rigorous learning attitude and rigorous logical thinking, teach students to form scholars' style and quality, be brave in innovation and development, and comprehensively develop students' core literacy of mathematics is the most important thing at present.

3. Thoughts on the Orientation of Mathematics Education in Colleges

The positioning of mathematics education belongs to the system of integrating multiple factors. The system presents obvious diversified characteristics. In short, college mathematics education not only involves the positioning of the university, but also the local characteristics and economic and cultural characteristics of the environment in which the university is located. Social resources, internal campus mechanisms, educational resources, talent training effects, and the social values of mathematics itself are closely related to human values. To do a good job in the positioning of mathematics education in colleges and universities, we must comprehensively consider the multi-factors, proceed from the multi-dimensional value of mathematics education, conform to the needs of social development, and construct a professional orientation that conforms to the development trend of the times and local characteristics.

Second, we must follow the principle of positioning in mathematics. From the perspective of the overall perspective, the principles of mathematics professional positioning mainly include the principle of priority of mathematics, the principle of 'talent market or social talent demand', 'mathematics + strong professionalism', 'theory + practice principle', and the principles of humanistic quality training. Among them, the principle of priority of mathematics requires that the value of mathematics education should be emphasized in the process of positioning; the principle of "talent market or social talent demand" should be combined with the market demand for talents, and the education system should be standardized; "mathematics + strong professional" is required to combine mathematics education with other majors to improve the practicality of mathematics education, such as constructing mathematics + finance major or mathematics + biological foundation ※ biomathematics, so as to cultivate diversified talents; "theory + practice principle" is in order to combine the theoretical knowledge of positioning and practical teaching orientation improves the theoretical value and practical value of mathematical positioning; the humanistic quality training principle attaches importance to cultivating students' humanistic quality.

Thirdly, we must do a good job in the goal orientation of professional talents' mathematics literacy training. In this work, we must first pay attention to refining the cultivation of talents, improve the basic theories and basic methods of mastering mathematics science, and guide students to gradually develop the use of mathematics knowledge and use computers. The ability to solve practical problems, through scientific research training to cultivate the quality of students engaged in research work in science and technology, education, and economic sectors. In addition, we must pay attention to meeting the training requirements. At present, the training of mathematics literacy talents in the narrow sense is the basic theory and basic method for students to study mathematics and applied mathematics. They are basically trained in mathematical models, computers and mathematics software, have good scientific literacy, and initially have scientific research. Solve practical problems and basic skills in developing software. On the other hand, we must standardize the training of specifications and do four things: First, we must have a solid foundation in mathematical theory, be trained in more rigorous scientific thinking, and initially master the ideological methods of mathematical science; second, we must have applied mathematics knowledge to solve Practical problems, especially the initial ability to build a mathematical model, to solve the basic knowledge of an application field; the third is to be proficient in using computers (including common language, tools and some mathematical software), with the ability to write simple applications; It is necessary to understand relevant policies and regulations such as national science and technology [2].

4. How to Cultivate Mathematics Professionals

4.1. Integrating Life Practice Factors and Innovating Classroom Teaching Mode

Grasping the positioning of mathematics education in colleges and universities, improving the quality of mathematics teaching, and adapting to the development needs of the times, teachers must first adopt a way of learning from the shallower to the deeper and simpler and harder, along the education direction of the new era, to train professionals with mathematical literacy as education. The goal is to closely integrate life practice factors and innovate classroom teaching models. For example, when parsing the number of questions, an interesting "card game" is developed for students to help students solve this problem flexibly. Teachers can first inform students that the number of questions is mainly divided into arithmetic symbols and digital image fills. The transport symbols and parentheses can be used to change the order of operations. The arithmetic symbols and parentheses are properly filled in the given number. A multi-image is a graph in which some numbers are filled in a specific position according to certain rules. The number is first to find the law, and then to fill in the number according to the law. Secondly, teachers should follow the development trend of education reform and comprehensively innovate classroom teaching mode according to specific needs, so that students can learn mathematics, learn mathematics and use mathematics knowledge in novel classrooms. Teachers can also guide students to understand the knowledge of mathematics and culture through storytelling, so that the mathematics classroom is richer in life and affinity. Take the famous story of "mathematical magician" as an example: In 1981, when the 37-year-old Saguntana and the computer launched a mental arithmetic game, Shagontana accurately calculated one in only 50 seconds. The 201 roots of 201 large numbers, and the computer requires more than 20,000 instructions to complete the calculation, so Shaguntana is called "mathematical magician." Then, the teacher can further enhance the humor of mathematics classroom teaching, and say to the students: "Every classmate can become a mathematics magician!" Then, the students are guided to use mathematics knowledge to transform magic, thus effectively improving the fun of the mathematics classroom.

4.2. Comply With the Development Trend of The Internet and Build a Platform for Mathematics

In the 21st century, with the advent of the mobile Internet era, the way of learning, living and working has changed. Therefore, teachers should also review the situation, adapt to the development trend of the Internet, and build a perfect platform for mathematics, so as to more effectively grasp the status of mathematics education and further improve the quality of mathematics teaching. From a narrow perspective, MOOC (Massive Open Online Course) is a new online course development model. The term MOOC was coined in 2008 by the Director of Network Communication and Innovation at Prince Edward Island University in Canada and a Senior Research Fellow at the National Institute of Humanities Education Technology Application [3]. Its iconic event took place in 2011, with 160,000 people from around the world enrolling in a free course on Introduction to Artificial Intelligence by Stanford University professors. The emergence of MOOCs has brought many development opportunities to various subjects of education, and mathematics education in colleges and universities is no exception. Moreover, MOOC is free to appear in the Internet mode and is a typical Internet education model. From the perspective of development, the emergence of MOOC in China began in 2012. Due to the short and precise, the curriculum is scientific and good, and many problems have been set up. Therefore, it is very suitable for college education. It should be noted that MOOC is an integral part of the Internet education model. From a macro level, the Internet education model is divided into curriculum design, online leadership and class management. In the use of this model to carry out mathematics teaching activities in colleges and universities, and to position the direction of mathematics education, teachers should pay attention to the three main points: First, the mathematics curriculum based on MOOC requires a lot of time to prepare, for each stage of learning. Course design and refinement; Second, online leadership is very important. Based on the MOOC mathematics curriculum, you need to have strong artificial intelligence, and artificial intelligence needs to achieve video interaction, online communication, knowledge sharing, etc. Excellent technology to support, therefore, teachers should help the school to continuously optimize various educational technologies; third, do a good job in class management. At present, QQ group and WeChat group play a unique advantage in class management. Future mathematics education and Internet education need to make breakthroughs in these three aspects.

In addition, MOOC-based mathematics courses can effectively control students' mathematics learning process through daily assignments and tests. In short, the main features of the MOOC are concentrated in three aspects: First, on a large scale, teachers do not simply publish courseware when conducting MOOC-based mathematics courses. Second, the curriculum is open. Only the MOOC-based mathematics curriculum is open and follows the sharing agreement to become a perfect MOOC. Third, MOOC-based mathematics courses are online courses, not one-on-one, face-to-face courses. These course resources are distributed on the Internet. As long as there is a network, you can learn to meet your individual needs.

4.3. Give Full Play to Students' Subjective Initiative and Form a Perfect Mathematical Model
At present, the application of mathematical models is extremely common. In the process of mathematics teaching, teachers can refine practical problems in real life, organize students to build mathematical models, guide students to check the solution of models, and scientifically verify the rationality of models, and use them. Mathematical models are solved to solve real-world problems. It should be noted that the mathematical model is to use mathematical language to construct a realistic model, and to simplify the actual model into a mathematical structure. Secondly, the mathematics education standards of colleges and universities require the organic combination of mathematics culture content and the content of each knowledge module, and build a model with good concentration, focusing on cultivating students' modeling consciousness and common sense of mathematics application. At the same time, through modeling, students are combined with mathematical theory knowledge and practical activities to continuously enhance students' ability to explore and solve problems independently. Thirdly, in the process of mathematical modeling, teachers should pay attention to refining the six-step process: First, guide students to prepare for the model, fully understand the knowledge module of this lesson, clarify the learning value, analyze the relevant background of the problem, and integrate all Information, using mathematical language to accurately describe mathematical problems. Second, the scientific implementation of the model hypothesis, that is, to simplify the problem according to the characteristics of the actual object and the purpose of modeling, while using reasonable language to make reasonable assumptions. Third, scientifically build models. In this part, the teacher can guide the students to accurately describe the mathematical relationship between various variables based on the hypothetical model and use the corresponding mathematical relationship to form a complete mathematical structure model. Fourth, do a good job of model solving. This part is mainly for students to obtain data on their own and to calculate all the parameters of the model. Fifth, do a good job in model analysis. At this stage, teachers should guide students to perform mathematical analysis on the calculation results. Sixth, comprehensively do a good job of mathematical model testing, this is the final step and the most important part of mathematical modeling. Teachers should guide students to comprehensively compare and analyze the results of the model and the actual situation, scientifically verify the accuracy of the mathematical model, whether it is scientific and reasonable and applicable. At the same time, it should be noted that if the model is consistent with the actual situation, t the meaning of he actual calculation results should be explained and determined. If the model does not conform to the actual situation, it is necessary to

make assumptions and perform secondary modeling [4].

4.4. Stimulate Students' Interest in Mathematics Learning Through Mathematics Activities
For a long time, due to the characteristics of mathematics disciplines, most students in colleges and universities have been based on a large number of theoretical and mathematical calculations. In this context, students' interest in mathematics is not guaranteed. The teaching effect will naturally be affected, and eventually students will not be able to cope with the subsequent learning of other deeper content. In order to improve on such conditions, colleges and universities should be able to attach importance to the organization of various types of mathematics activities on the basis of the original, and stimulate students' interest in mathematics courses through such activities.

Taking the "Math Teaching Design Competition" as an example, the school can ask the students to choose their own teaching content and complete the teaching plan design for this content. In the actual competition process, colleges and universities can invite other students to participate in the selection. These students and teachers jointly evaluate the design of specific mathematics teaching programs and classroom effects. In addition, event organizers can also set certain rewards for better performing students or groups. Under such an activity mode, students will be able to actively participate in such activities, and by self-completed the process of selecting topics, preparing lessons, and designing teaching plans, the students themselves will naturally have a deeper grasp of relevant mathematical knowledge points. On the other hand, such an activity form is more interesting for college students. With the regular organization of such activities, students will gradually feel the charm of the mathematics discipline itself, and thus achieve the purpose of enhancing students' participation and enthusiasm in the mathematics classroom to ensure the effectiveness of teaching.

In order to ensure that more students can participate in mathematics activities, colleges and universities can also carry out activities such as "Mathematic Encyclopedia Knowledge Contest" on a regular basis. The focus of such activities is on fun and participation. Relevant teachers can ask students to specify the class on a class basis. The mathematics allusions involved in the field are investigated, and then the questions raised by the teachers are answered in groups. Finally, the teachers reward the students who perform well.

4.5. Introduce E-Learning Course to Strengthen Mathematics Education in Colleges and Universities
The E-Learning course is a network-wide approach to learning process management (design, implementation, evaluation, etc.) that enables learners to acquire knowledge, improve practical skills, change perceptions, and improve performance. The difference between the MOOC and the E-Learning course is that the latter is completely student-centered, and students who have the ability to learn can also introduce the E-Learning course to strengthen the mathematics education in colleges and improve the quality of teaching. It should be noted that when constructing the E-Learning course platform, it is necessary to pay attention to selecting appropriate courses to guide students to study at any time, anywhere, on the go, and randomly. Secondly, in the E-Learning course design process, teachers should pay attention to the design of a perfect curriculum system and a clear curriculum, and standardize the course content. Thirdly, it is necessary to give full play to the self-development value of the E-Learning course, organize self-developed mathematics on-campus courses, provide students with independent learning platforms and reference materials, and free learning resources to help students freely submit homework. In addition, in the entire E-Learning course production process, we must do four things: First, do a good job of preparation, the entire courseware content should listen to the students' opinions and voices; Second, do a good job in the early planning, improve the courseware The learning value, optimize the content framework design, quantitative teaching module; third, do a good job in curriculum resource development, integrate all valuable mathematical information and knowledge system for students, and guide students to continuously develop mathematical core literacy in the process of independent learning; Fourth, improve the compatibility of the E-Learning course, standardize the curriculum production template, maintain the clarity of the regulations, and ensure the consistency of the courseware elements and style.

5. Conclusion

Teachers should combine the requirements of educational goals and the development trend of the Internet era, closely integrate the factors of life practice, and innovate classroom teaching mode; build a perfect platform for mathematics in order to improve the quality of mathematics education in colleges and universities, and grasp the orientation and direction of mathematics education. To meet the individual needs of students; give full play to students' subjective initiative, set up a sound mathematical model, continuously stimulate students' learning motivation, build a good mathematical knowledge information platform; appropriately introduce E-Learning courses to strengthen college mathematics education.

Acknowledgment

This paper is the project of promoting the basic ability of young and middle-aged teachers in Guangxi universities in 2018, "Life-based Research on the Development of Children's Innovative Thinking from the Perspective of Anthropology" (2018KY1193) and the project of teaching reform in Guangxi vocational education in 2017, "Research and Practice on the Cultivation of Teaching Skills of Students Majoring in Primary Education in Higher Vocational Colleges: Taking the Primary Education Major of Liuzhou City Vocational College as an Research achievements of JG2017B075.

References

[1]Xu Yang,Xin Yongxun,Liu Qian.Innovative Applied Mathematics Talents Training Mode in School-Enterprise Cooperation and Running School [J].Value Engineering, 2014(01:1).

[2] Liu Jiao. Research on the Infiltration of Mathematics Core Literacy in College Mathematics Teaching [D]. Chongqing Normal University, 2016 (05:2).

[3] Zhang Mingli. Effective penetration and application of interesting teaching methods in mathematics teaching [J]. Primary and secondary education, 2016 (9:3).

[4] Zhao Kun, Jiang Chunyan. How to strengthen mathematics quality education in colleges and universities [J]. Heilongjiang Science and Technology Information, 2012 (24:4).

[5] Li Na. On the training plan of students' interest in mathematics education[J]. Guangdong Education, 2017(10:5).

STUDY ON CHINESE TECHNICAL ECONOMY AND GLOBAL SOCIAL RESPONSIBILITY

Huanping Zhang[1] Maohua Li[1,2] Zoltán Zéman[3]

Affiliation:

[1]School of Business, Xi'an Siyuan University (Xi'an)

[2]Faculty of Economics and Social Sciences, Szent István University (Gödöllő)

[3]Institute of Business Studies, Faculty of Economics and Social Sciences, Szent István University (Gödöllő)

Corresponding author: Maohua Li

Email address: maohua.li@qq.com

ABSTRACT: As Chinese economy develops, China is the second largest economy in the world with $13.01 trillion nominal GDP in 2017. However, there are always several different voices on Chinese products and Chinese technical economy, even on Chinese internal problems, such as environmental problems, food safety problems. This paper wants to study the relation between Chinese technical economy and global economy, and finally wants to argue for the goods of Chinese technical economy and Chinese product. Firstly, this paper uses literature review method to renew the definition of Chinese technical economy. Secondly, this paper uses linear regression analysis to test the relation between Chinese technical economy and global economy. At last, this paper gives its own policy advices to improve the quality of Chinese technical economy.

1. Introduction

After near 40-year's reform and opening-up policy in China, China has become a key world economy. China, with its socialist market economy, is the world's second largest economy with GDP 82,712.2 billion Yuan in 2017[1]. Over the past two decades, the world has witnessed China's transformation from a planned economy into a more market-driven economy (Guan & Yam, 2015). China is playing a growing role in the world economy (Lardy, 1994). China, as a new power, is challenging the traditional dominance of the US in the governance of the global economy (Hopewell, 2015), and has made a great contribution to the world economic recovery(Baoan Wang, 2015). However, China is not about to replace the United States as the world's dominant country (Nye Jr, 2017). The real relation between Chinese economy and world economy is very complicated, and they impact one another (Jiayu Sun, 2017).

Nowadays, some Chinese brands are really very popular in the world, such as Huawei, ZTE, Tencent etc. But when we talk about "Chinese technical economy", a few words will come into our mind, such as "low quality", "low price", "goods dumping", etc. And some "incidents" also happen related to exported Chinese products, such as 2007 Chinese export recalls (Roth etc. 2008). However, when we talk about "Made in Japan", the public will jump to conclusion like "lean management", which is the holistic approach of lean methods, strategic implementation and the consideration and integration of the cultural

---

[1] Source: http://www.stats.gov.cn/english/PressRelease/201801/t20180118_1574943.html

Content from this work may be used under the terms of the Creative Commons Attribution 3.0 licence. Any further distribution of this work must maintain attribution to the author(s) and the title of the work, journal citation and DOI.

Published under licence by IOP Publishing Ltd

level (Bertagnolli, 2018). And the words for describing "Made in Germany" or "Hergestellt in Deutschland" are "craftsman", "craftsman spirits" and "luxury". So comparing with "Made in Japan" and "Made in Germany", "Chinese technical economy" is always labeled with "low price" and "low quality".

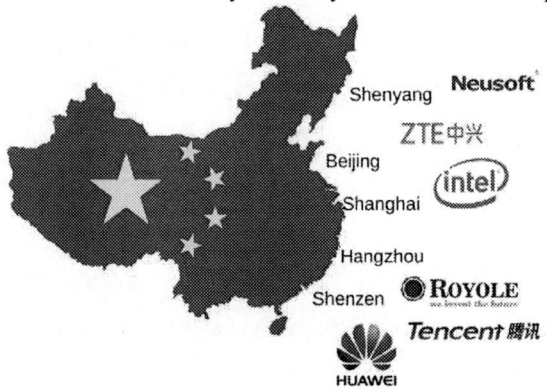

Figure 1. Several famous Chinese brands and production places
Source: Moving from "Chinese technical economy" to "Innovated in China". Website:
http://www.digi.city/blog/2016/11/21/moving-from-made-in-china-to-innovated-in-china

Under these backgrounds, Chinese Ministry of Industry and Information Technology made a draft with the name "Chinese technical economy 2025", which was approved by the State Council of the People's Republic of China. "Chinese technical economy 2025" has been used in different areas including manufacture, information, education, etc. This paper wants to research on the relationship between "Chinese technical economy" and "world development". With this purpose, this paper also wants to find the current problems on "Chinese technical economy". At last this paper will talk about the reasons for why there is a paradox that where there are more products from China in one country or area, there will be more criticism about Chinese products.

2. Research methodology

In order to research on the relationship between "Chinese technical economy" and "world development", this paper uses SPSS 22.0 to do correlation analysis and linear regression analysis.

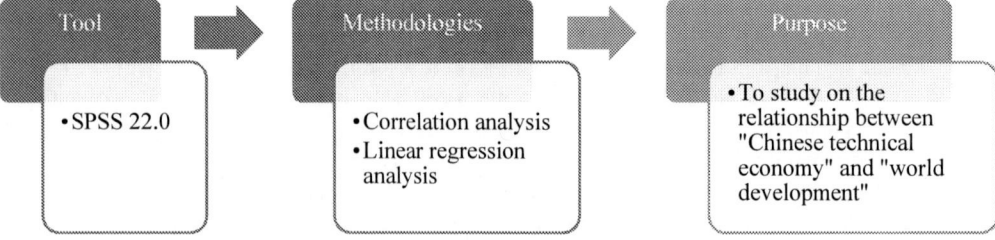

Figure 2. Description of methodologies

As mentioned above in figure 2, this paper uses SPSS 22.0[2] to finish the analyses. SPSS Statistics is a software package used for logical batched and non-batched statistical analysis[3]. As one of the most powerful statistics software in the world, SPSS is more convenient to do social science analysis, especially for most traditional analysis models, such as regression analysis, correlation analysis, factor analysis, principle component analysis and so on.

---

[2] SPSS is statistics software developed by IBM whose products cover metal servers, IOT Continuous Engineering, aaS360 with Watson, Maximo, Spectrum and SPSS Statistics. The newest version of SPSS is version 25 in 2018. However, this paper only uses correlation analysis and linear regression analysis, so version 22.0 is enough for this purpose.
[3] Source: Wikipedia. Website: https://en.wikipedia.org/wiki/SPSS

Correlation analysis is widely used to study a statistical relationship between two variables. There are many types of correlation analysis, and the most popular are Pearson, Intra-class and Rank. Pearson talking about the strength and direction of the correlation between two variables is the most used in academic research. In order to study the correlation between "Chinese technical economy" and "world development", this paper uses Pearson correlation analysis.

Regression analysis is not only a statistical methodology, but also a set of statistical processes to study on the relations among different variables. There are many types of regression models developed by scholars, among which the well-known is linear regression analysis. Linear regression model denote the dependent variable by Y and the independent variables by X_1, X_2, X_3... And then we can make an equation for both two types of variables like the following formula.

$$Y = F(x_1, x_2, \dots x_n) \qquad \text{Formula 1}$$

3. Variables selection

This paper uses two variables to stand for "Chinese technical economy", and they are "Chinese GDP (CGDP)" and "Chinese export (CE)". "Chinese GDP" is used to study the relation between Chinese economic development and world development, and "Chinese export" is used to research on the relation between "Chinese export" and world development.

In order to reflect world development, this paper selects total three variables: "world GDP (WGDP)", "world revenue (WR)" and "final consumption expenditure (FCE)". The first one "world GDP" is used to research the relation between Chinese economics and world economics, and the rest two are used to study the relation between "Chinese export" and world development.

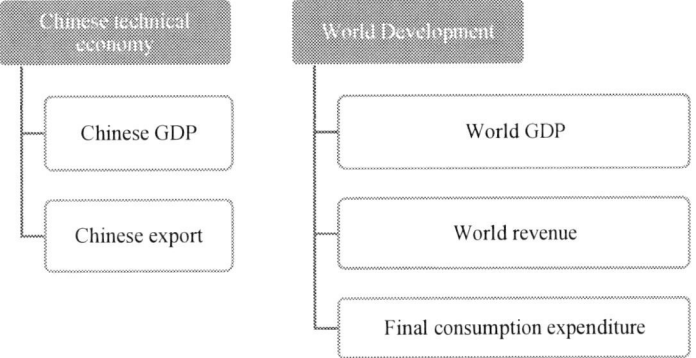

Figure 3. Variables used in this paper

4. Date source

All the data used in this paper is from World Bank Group as shown in following figure 4. The data selected is from 1973 to 2015. The beginning year is 1973, because for some Chinese economic data, they are counted from 1973. The reason for ending in 2015 is that for some data, they are not renewed in 2016 or 2017.

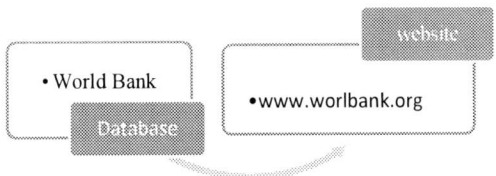

Figure 4. Data source

5. Correlation analysis

Table 1 shows the general descriptive statistics of all the variables used in this paper. From table 1, we can easily see the mean, the standard deviation and the number of each variable. For instance, the mean of world GDP is 43.989707031, and the standard deviation is 15.980065152. From 1973 to 2015, there are totally 43 years, so the number of all the variables is 43.

Table 1. Descriptive Statistics

Variables	Mean	Std. Deviation	N
world GDP	43.979807031	15.980065152	43
China GPD	2385786205007.531	2516919977424.60	43
China export	14.141113183833815	10.088681236961730	229363
Revenue, excluding grants (% of GDP)	23.1997968517727	1.43343143789578	172032
Revenue	10.136974464956824	3.842249255176326	43
Final consumption expenditure, etc. (current US$)	24.108892057937780	16.913514028222508	43

Table 2 is about the correlation analysis of all the variables. From the table, it is easy to see the relation between Chinese economics and world development. Pearson correlation coefficient of world GDP (WGDP) and Chinese GDP (CGDP) is positive 0.949, which mean there is strong positive relation between Chinese economics and world economics. Pearson correlation coefficients of China export (CE) and world revenue (WR), final consumption expenditure (FCE) are positive 0.901 and 0.840, that is to say, China export has a very strong relation with world revenue and final consumption expenditure. All of the analysis results are significant at the 0.01 level (2-tailed). From correlation analysis, we can find that there is a positive relation between "Chinese technical economy" and "world development". However, in order to further study the relation between "Chinese technical economy" and "world development", this paper will construct a linear regression model as following.

Table 2. Correlations

Variables		WGDP	CGPD	CE	WR	FCE
WGDP	Pearson Correlation	1	.949**	.882**	.990**	.989**
	Sig. (2-tailed)		.000	.000	.000	.000
CGPD	Pearson Correlation		1	.722**	.919**	.971**
	Sig. (2-tailed)			.000	.000	.000
CE	Pearson Correlation			1	.901**	.840**
	Sig. (2-tailed)				.000	.000
WR	Pearson Correlation				1	.967**
	Sig. (2-tailed)					.000
FCE	Pearson Correlation					1
	Sig. (2-tailed)	.				

**. Correlation is significant at the 0.01 level (2-tailed).

6. Linear regression model

For a better analysis of the relation between "Chinese technical economy" and "world development", this paper constructs a linear regression model in which China export (CE) is used as dependent variable, and world revenue (WR) and final consumption expenditure (FCE) are used as independent variables to explain China export. The linear regression model built is shown as following.

$$CE = \alpha + \beta_1 WR + \beta_2 FCE + \varepsilon \qquad \text{Formula 2}$$

CE: China export
WR: World revenue
FCE: Final consumption expenditure

α: Constant
β: Coefficient
ε: Residual error

7. The population regression analysis

Firstly, this paper uses population regression analysis to test how much China export (CE) can be explained by the total function. The population regression results are shown in table 3.

Table 3. Model Summary

model	R	R Square	Adjusted R Square	Std. Error of the Estimate
1	.903[a]	.815	.806	4.216655887948734

a. Predictors: (Constant), Final consumption expenditure, etc. (current US$), Revenue

From table 3, it is easy to see that most of China export (CE) can be explained by world revenue (WR) and final consumption expenditure (FCE). R square is 0.815, which means 81.5% of total China export (CE) can be explained by world revenue (WR) and final consumption expenditure (FCE). Adjusted R square is 0.806, which means all the variables and data selected by this paper fit one another very well. And the standard error of the estimate is 4.216655887948734, which means that the prediction of the total function is of accuracy.

8. Analysis of variance

Analysis of variance (ANOVA) is a general procedure for isolating the sources of variability in a set of measurements (Girden, 1992). This paper uses ANOVA to test the whole significance of linear regression model whose result is reflected in table 4.

Table 4. ANOVA[a]

	Model	Sum of Squares	df	Mean Square	F	Sig.
1	Regression	3057.564	2	1528.782	85.982	.000[b]
	Residual	693.427	39	17.780		
	Total	3750.992	41			

a. Dependent Variable: China export
b. Predictors: (Constant), Final consumption expenditure, etc. (current US$), Revenue

From table 4, we can see that the mean square of regression is 1528.782. F-value is 85.982 and the significance of ANOVA is 0.000 which is much less than 0.05. From the values of F-value and significance, the whole regression model is very significant, and relation between "Chinese technical economy" and "world development" is significant.

9. Linear regression analysis

In order to further research the relation between "Chinese technical economy" and "world development", this paper uses linear regression model. Through the results in table 5, it is shown that the significances of both world revenue (WR) and final consumption expenditure (FCE) are less than 0.05, which means the linear regression model built above is very significant and China export (CE) can be explained by world revenue (WR) and final consumption expenditure (FCE).

Table 5. Coefficients[a]

Model	Unstandardized Coefficients		Standardized Coefficients	t	Sig.
	B	Std. Error	Beta		
1 (Constant)	-8.164	3.459		-2.360	.023

WR	2.841	.673	1.141	4.219	.000
FCE	-.145	.157	-.249	-.920	.003

a. Dependent Variable: China export

From table 5, the result of linear regression gives a detailed relation between China export (CE) and world revenue (WR), final consumption expenditure (FCE), which can be expressed in following formula.

$$CE = -8.164 + 2.841WR - 0.145FCE \qquad \text{Formula 3}$$

10. Conclusion and discussion

This paper studies the relation between "Chinese technical economy" and "world development". For this purpose, this paper selects two variables "Chinese GDP (CGDP)" and "China export (CE)" to stand for "Chinese technical economy" and three variables "world GDP (WGDP)", "world revenue (WR)", "final consumption expenditure (FCE)" to stand for world development. This paper uses correlation analysis and linear regression analysis to study the relation through SPSS 22.0. Through all the analysis, we find results in the following:

There is a positive relation between "Chinese GDP (CGDP)" and "world GDP (WGDP)". This is very easy to understand. As the development of world economics, the earth is a global village, so the connection of the whole world economics will be much closer and closer. Chinese economics and world economics will interact with each other all the time actively.

The relation between "Chinese technical economy" and "world development" can be explained by the formula 3. From formula 3, we can see that if final consumption expenditure (FCE) increases 1 unit, China export (CE) will decrease by 0.145 units. From this, we can see that if one country really wants to reduce the "lower-price" and "lower-quality" products from China, the first step is to develop its own economy and to increase the personal income. If one country has already removed all the poverty and has the ability to say no to "lower-quality products" in China, this will push Chinese companies to improve their products quality. This also reflects the basic economic theory that supply is determined by demand, and demand is determined by income. So if "lower-price" and "lower-quality" "Chinese technical economy" is popular in one country, which means there are still a lot of poor people living in this country. Maybe if there is no "Chinese technical economy" in this county, these poor people will be living harder or cannot survive.

For Chinese government, if China wants to change the reputation of "Chinese technical economy", there is still a long way to go. Fortunately, China government has already made a proposal to change "Chinese technical economy" into "Innovated in China", and the proposal name is "Made in China 2025". However, to achieve this, China government still needs to do a lot of work to attract more talents, for instance, talents for AI (Artificial intelligence). And Chinese government should make more policies to motivate companies and institutions to make more innovation now.

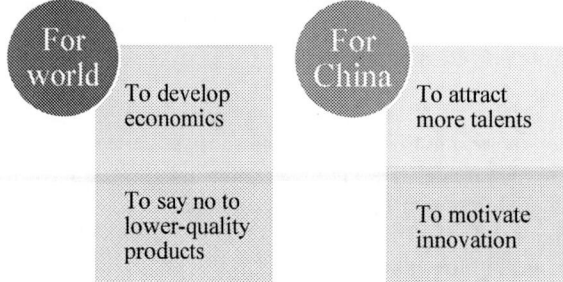

Figure 5. Advices

From analysis above, we can see that "Chinese technical economy" is not only the responsibility of China, but it is world responsibility. This is what I call "global responsibility". Global responsibility includes two dimensions: buying responsibility and consuming responsibility. "Buying responsibility" means if you are buying "lower-price" and "lower-quality" products from China or these products are popular in your country. From the relations among supply, demand and income, you should not complain. All you and your government need to do is to work harder to make people much richer. For "consuming responsibility", it means that if you and your country are consuming Chinese products, you are responsible for the environmental problems in China, so Chinese environmental problems are not only Chinese problems, and they are world's problems.

Frankly speaking, this paper only uses China and "Chinese technical economy" as a research sample to explain the relation and interaction of world economics. In the future, I will continue to study the relation between "Made in Asia" and world development. I want to build a new business model which goes like this: Designed in America, made in Asia and consumed in Europe. However, all the continents are not separate from one another, and they are on the same boat and same supply chain, so they are responsible to one another.

Acknowledgements
This work is sponsored by China Scholarship Council, and this paper has been presented in 4th Winter Conference of Economics PhD students and Researchers.

References
[1] Girden, Ellen R. (1992). ANOVA: Repeated measures: Sage.
[2] Lardy, Nicholas R. (1994). China in the world economy. Peterson Institute Press: All Books.
[3] Roth, Aleda V, Tsay, Andy A, Pullman, Madeleine E, & Gray, John V. (2008). Unraveling the food supply chain: strategic insights from China and the 2007 recalls. Journal of Supply Chain Management, 44(1), 22-39.
[4] Guan, JianCheng, & Yam, Richard CM. (2015). Effects of government financial incentives on firms' innovation performance in China: Evidences from Beijing in the 1990s. Research Policy, 44(1), 273-282.
[5] Hopewell, Kristen. (2015). Different paths to power: The rise of Brazil, India and China at the World Trade Organization. Review of International Political Economy, 22(2), 311-338.
[6] BaoanWang. (2015). Chinese economy is still the power of world economy. Qiushi(21), 63-63.
[7] Nye Jr, Joseph S. (2017). Will the liberal order survive: The history of an idea. Foreign Aff., 96, 10.
[8] Jiayu Sun. (2017). The Study on the Relation between Chinese Economics Development and World Economy. Consume Guide (20).
[9] Bertagnolli, Frank. (2018). Einführung Lean Management Lean Management (pp. 217-219): Springer.

Research on Semantic Information Retrieval Model of Bamboo & Rattan Domain Based on Query Extension

Lin Peng，Ming-ming Lai，Xin Zhang

College of Big Data, Yunnan Agricultural University, Kunming 650201

TCorrespondence to:Lin Peng(penglin2286351@163.com)

Abstract: The existing domain semantic information retrieval models generally do not consider the wide range of domain knowledge involved, and lack the ability to query multi-level domain knowledge in this domain. Based on the semantic expression of the knowledge about bamboo & rattan via ontology, this paper mainly studies the measurement of semantic relevancy in the field of bamboo and rattan, and finally proposes a Semantic Information Retrieval Model in this field which is based on relevancy. Experimental results show that the semantic query extension based on the concept of similarity proposed in this paper can improve the recall rate and F value of retrieval model significantly.

1. Introduction

At present, a large number of data related to bamboo & rattan have been digitized, laying a foundation for rapid identification and semantic retrieval of bamboo & rattan [1-4].However, the existing semantic information retrieval models don't take into account the wide range of domain knowledge involved in bamboo & rattan domain knowledge and lack the ability to query multi-level domain knowledge. Based on the construction of the database of bamboo and rattan ontology, this paper introduces the query extension technology, combines the semantic characteristics of the knowledge of bamboo and rattan domain, and proposes a query extension based semantic information retrieval model of bamboo and rattan domain.

2. Query expansions

In order to improve retrieval Recall, the method of re-query by recombining original query conditions and adding new keywords and it is called query extension. This method can assist users to generate new query keywords, effectively improve the recall rate of retrieval, and better understand the actual retrieval intention of users. The core of this method is the automatic generation algorithm of new query words. At present, query extension is generally realized through relevant feedback technology, semantic word derivation technology, keyword co-occurrence and ontology technology. Based on ontology hierarchy, this paper depends on similarity of ontology concepts to realize query extension in semantic retrieval of bamboo & rattan domain.

2.1. Semantic query extension

In the query extension method of this paper, the query extension based on concept similarity in ontology is adopted. The greatest difference between the method and the similar word extension method in the early days is that the generation of extended vocabulary is no longer dependent on the query document collection itself, but based on the domain knowledge ontology, and the extension content is independent of the query content. At the same time, because of the relationship between concepts in the same domain is relatively stable, extended vocabulary can be generated more accurately. Meanwhile, hierarchical conceptual structure in ontology can effectively avoid word meaning confusion.

When semantic correlation is extended by using hierarchical relations such as the composition and inclusion of ontology concepts, the extended pattern directly affects the recall rate and precision rate about the retrieval results. When the extension word is the combination of the hyponym, the homonym and the partite of the query word, it can improve the recall rate and has little influence on the precision rate. The same effect can be achieved through the extension of the upper word, while other combinations of extended relationships, including sibling relationships, have a more damaging effect on accuracy than on recall. Therefore, semantic query extension based on concepts in ontology is carried out according to bamboo-rattan domain ontology, mainly including sub-words and overall word extension of query words according to concepts. The concrete implementation is shown in table 1.

Table 1. Example of Concept-based Semantic Query Extension.

NO.	Query word	Conceptual semantic query expanded
1	leaf	Leaf sheath, ear, tongue, leaf blade, petiole
2	glumelle	Stamen, pistil, lemma, lemma, scale coat, ovary, stigma, anther
3	pole	Stem, stem base, internodes, handle Tuo, section, section, section, internodes tip
4	tongue	Ligule, Tuo tongue
5	sheath	Leaf sheath, Tuo sheath
6	flower	Stamen, pistil, lemma, lemma, scale coat, ovary, stigma, anther
7	Podozamites	Bamboo, stalk, sheath, ear, leaf, tongue, stem and hair

2.2. Conceptual similarity calculation

This paper uses the Conceptual similarity before and after semantic extension to evaluate the impact of the introduction of extended search terms on the precision.

Definition 1 mainly introduce the relationship between the initial retrieval concept words and the superordinate concepts obtained after semantic query extension Is -a and member-of., and the degree of conformity between these extended concepts and the initial retrieval concept in the retrieval semantic theme Is called the concept similarity before and after extended retrieval.

Conceptual similarity reflects the degree of semantic similarity of two concepts in ontology, that is, semantic consistency in retrieval of two concepts in retrieval. As the initial concept of retrieval words c0 stands for the hyponym or ci stands for meronym. Although there are some differences in concept similarity, but if from the point of user retrieval intention, these words are on the original concept semantic specialized retrieval. Therefore, retrieval results which include these extended concepts are conform to the requirements of the concept of the retrieval results, and will not reduce the retrieval accuracy. For example, the initial retrieval concept is "Melocanna", and the text has the content of "pear bamboo". Since the concept of "pear bamboo" is hyponym of "Melocanna", although there are differences in conceptual semantics between them, from the perspective of users' retrieval intention, this result is completely consistent with the requirements.

Therefore, the method of conceptual similarity calculation based on semantic query extension is as follows:

$$SR(c_i, c_o) = \left\{ \begin{array}{l} -\log \dfrac{length(c_0, c_i)}{2S}, otherwise \\ 1 \qquad\qquad , c_i \in K \end{array} \right\} \qquad (1)$$

$$K = \{k | k = c_o \vee k\}$$

Where, $length(c_o, c_i)$ is the shortest path from node c_o to node c_i, S is the depth of ontology tree, and k is a hyponym or meronym of c_0.

2.3. Query extension retrieval method

In order to solve the limits that the traditional information retrieval technology relies on keyword matching effectively, if the matching fails, empty records will be returned. Users need to further change the query conditions and try again. The system does not have the function of automatic reasoning. By calculating the similarity of concepts in the domain ontology of bamboo and rattan, this paper extends the user's retrieval concepts semantically, mainly realizing the expansion of the semantic relationship of query words (the equivalence relation or the hyponymy). After query extension, the corresponding equivalent semantic expressions , upper semantic expression and lower semantic expression are obtained. In the concrete implementation, it is expressed in the form of regular expression, which provides convenience for ontology reasoning in the next step.

Semantic retrieval after query expansion is mainly to retrieve concepts and their relationships at the semantic level. The key to its realization is reasoning between concepts. Jena inference machine is used in this paper for ontology inference. Within, Jena provides three ways :(1) adopting the reason according to add regularly (including RDFSReasoner, OWLReasoner, etc.), which includes common reasoning functions;(2) users can customize inference rules as required;(3) registration of third-party inference engines. In this paper, a custom rule inference engine is used in the experiment, and its trigger mechanism is a forward chain engine.

3. Experimental results and analysis

In this paper, Volume 9 of Flora of China is used as experimental data set, which contains 37 genera, 516 species of bamboo, 516 documents and 530,000 words. Through experiments, this paper analyzes the influence of the semantic retrieval model based on semantic query extension proposed in this paper on the improvement of retrieval recall rate in the field of bamboo and rattan.

First, select the "leaf", "rhizome", "rod", "flower" and "bamboo Tuo" five concepts for the primitive retrieval concept. These five retrieval concepts were selected because of the five iconic retrieval concepts frequently used in the rapid retrieval of bamboos in the Retrieval Table of Chinese Bamboos Subfamily Genera [5].By calculating the concept similarity of the original retrieval concept and extending semantic query, the extended retrieval concept is obtained in the end, as shown in table 2.

To this field of bamboo & rattan semantic retrieval model, the semantic query expansion model retrieval effect is compared before and after the experiment of "leaf", "rhizome" and "pole", "flower" and "bamboo Tuo" five concept semantic extension before and after the experiment, the experimental results as shown in table 3.

Table 2. Experimental retrieval concepts, extended concepts and retrieval examples.

NO.	Retrieve the concept	Semantic extension	Retrieve instance
1	leaf	Leaf sheath, ear, tongue, leaf blade, petiole	lanceolate
2	subterranean stem	subterranean stem、 False whip, bamboo whip	solid core
3	pole	Stem, stem base, internodes, handle Tuo, section, section, section, internodes tip	light
4	flower	Stamen, pistil, lemma, lemma, scale coat, ovary, stigma, anther	yellow
5	Bamboo Tuo	Bamboo Tuo, stalk Tuo, Tuo sheath, Tuo ear,	Unapparent

Tuo leaf, Tuo tongue, rod Tuo, wool

Table 3. Retrieval results before and after semantic query expansion.

NO.	Semantic query extension before		Semantic query expanded		Total number of relevant results
	Total search results	Correct retrieval number	Total search results	Correct retrieval number	
1	875	712	917	763	878
2	103	87	132	108	127
3	47	36	50	38	43
4	631	489	657	507	608
5	289	231	320	254	304

In this paper, three evaluation indexes (accuracy, recall rate and F value) are selected as evaluation indexes to evaluate the influence of semantic query expansion on the performance of the model. The calculation formula is as follows:

Precision Rate:
$$P = \frac{\text{Correct Retrieval Number}}{\text{Total Search Results}} \times 100\% \quad (2)$$

Recalling Rate:
$$R = \frac{\text{Correct Retrieval Number}}{\text{Total Number of Relevant Results}} \times 100\% \quad (3)$$

F Value:
$$F = \frac{2 \times P \times R}{P + R} \times 100\% \quad (4)$$

The retrieval result data before and after semantic query extension in table 2 were analyzed and calculated. The accuracy, recall rate and F value of model retrieval effect from the before semantic query extension and the after semantic query extension were obtained respectively, as shown in table 4. As can be seen from the data in table 4, the retrieval accuracy of the model was not significantly affected before and after semantic query expansion, but the recall rate and F value were both improved to some extent. Because, in this paper, under the condition of reasonable ontology construction in bamboo & rattan domain, the retrieval method based on semantic extension of concept similarity can improve the retrieval performance of the model by expanding the retrieval scope and increasing the number of correct retrieval results without affecting the accuracy of model retrieval (to make sure the user's search intention remains unchanged to the greatest extent).

Table 4. Impact of semantic query extension on model performance.

NO.	Semantic query extension before			Semantic query expanded		
	Precision Rate	Recalling Rate	F Value	Precision Rate	Recalling Rate	F Value
1	81.37%	81.09%	81.23%	83.21%	86.90%	85.01%
2	84.47%	68.50%	75.65%	81.82%	85.04%	83.40%
3	76.60%	83.72%	80.00%	76.00%	88.37%	81.72%
4	77.50%	80.43%	78.93%	77.17%	83.39%	80.16%
5	79.93%	75.99%	77.91%	79.38%	83.55%	81.41%

4. Conclusion

On the basis of ontology semantic expression of bamboo & rattan domain knowledge, this paper studies the measurement of semantic relevance of text data in bamboo & rattan domain information, and proposes a semantic information retrieval model based on correlation. According to the semantic correlation in concepts of bamboo & rattan domain, the retrieval model measures the importance of search terms in bamboo and rattan domain ontology, and expands the scope and depth of retrieval. The experimental results show that the proposed semantic query expansion based on concept similarity has little effect on the retrieval accuracy of the model, but the recall rate and F value have been greatly improved.

Acknowledgments
The findings and the opinions are partially supported by projects of the major science and technology projects of Yunnan province, No.2018ZI001.This work is supported by Engineering and Technology Research Center for Yunnan Agricultural Big Data and Yunnan Agricultural University, China.

References
[1] Zhang Xinping. Development Prospects of Bamboo and Rattan in the world[J]. World Forestry Research, 2003. 16(1):26-30.
[2] Qiu Erfa, Hong Wei, Zheng Yushang. Review on Diversity and Utilization of Bamboo in China[J]. JOURNAL OF BAMBOO RESEARCH, 2001, 20（2）:11-14.
[3] Chinese flora editorial committee. Flora of China (volume 9, volume 1) [M] Beijing: Sciences Press, 1996.
[4] Jiang Zeihui. The world's bamboo rattan(English edition) [M]. Beijing: China Forestry Publishing House., 2007.
[5] Yi Tongpei, Ma Lisha, Shi Junyi. Chinese bamboo subfamily genus search[M]. Science Press, 2009.
[6] Zhang Junlin. Research on Information Retrieval System Based on Language Model[D]. Doctoral Dissertation of Institute of Software, Chinese Academy of Sciences, 2004.
[7] Manning C. D. write. Yuan Chunfa translate. Basic Statistics of Natural Language Processing[M]. Publishing House of Electronics Industry, 2005.
[8] Zhen Tianqiao. Research of the Mixed Language Model Based on Statistical and Latent Semantic Analysis[D]. Master's Thesis of Harbin Institute of technology, 2007.
[9] Li Zheng. Research on Chinese Web Page Information Retrieval Based on Statistical Language Model[D]. Master's Thesis of Central China Normal University, 2012.
[10] (Chile) Ricardo Baeza-Yates write, Huang Xuanjing translate. Modern information retrieval (2nd edition) [M]. China Machine Press, 2012.
[11] Gao Shan. Research on Query Wxtension and Related Technology in Information Retrieval [D]. Master's Thesis of Central China Normal University, 2008.
[12] Wang Qi. Semantic retrieval technology and application of ancient fresco [D]. Doctoral Dissertation of Zhejiang University, 2011.
[13] Han Xin. Research on resource semantic retrieval in bioinformatics based on ontology [D]. Master's Thesis of North University of China, 2012.

Research on Semantic Information Retrieval Model of Bamboo Rattan Domain Based on Semantic Relevance

Lin Peng, Xu-peng Kou, Lin-nan Yang

College of Big Data, Yunnan Agricultural University, Kunming 650201

TCorrespondence to:Lin-nan Yang(lny5400@sina.com)

Abstract: At present, the research of semantic information retrieval technology is mostly in the initial stage, and the research is mostly in exploratory theoretical research. Among them, the construction of domain semantic information retrieval model is still immature. In this paper, bamboo rattan field was selected as the research object, and the key problems in semantic information retrieval technology of bamboo rattan domain were analyzed. Based on the domain ontology of bamboo and rattan, the weight calculation method of bamboo rattan domain terms is defined. This paper introduces semantic relevance degree extension into the domain semantic information retrieval model of bamboo rattan, constructs a domain semantic information retrieval model suitable for bamboo rattan, and realizes the association retrieval and semantic query extension of text information in bamboo rattan domain.

1. Introduction

Bamboo rattan is the combined name of bamboo and rattan plants. It is a large family in the plant kingdom, which plays an important role in the forest resources of the world. At present, a large number of data related to bamboo rattan have been digitized, which lays a foundation[1-5] for rapid identification and semantic retrieval of bamboo rattan. Semantic information retrieval, however, the comprehensive realization of bamboo rattan areas, especially in the semantic information retrieval of textual data, also need to solve the following questions:

(1) The representation and measurement of knowledge in bamboo rattan domain. In the research of bamboo rattan information, there are uncertainties and inaccuracies in text information. How to use artificial intelligence technology to effectively represent the knowledge of bamboo rattan field is the basis for realizing semantic retrieval of bamboo rattan information. (2) The correlation retrieval of text messages in the bamboo fields. It mainly need to achieve the following two levels of association retrieval: ①According to the semantic information of bamboo rattan text, the identification and retrieval of related bamboo rattan are realized ②Taking literature as a bridge, it realizes the retrieval from bamboo rattan to related literature, and then from literature to bamboo rattan. (3) The semantic query expansion problem. At present, semantic information retrieval technology does not take into account the wide range of domain knowledge involved in bamboo and rattan domain knowledge and lacks the ability to query multi-level domain knowledge. At the same time, there are some limitations in using ontology to realize semantic inspection. How to effectively improve the retrieval effect based on the existing ontology technology is the key to the design of semantic information retrieval model in the field of bamboo rattan.

In view of the above problems, based on the construction of bamboo rattan ontology, this paper defines the weight calculation method of bamboo rattan domain terms and introduces the semantic correlation measure. Combined with the semantic characteristics of knowledge in the field of bamboo

Content from this work may be used under the terms of the Creative Commons Attribution 3.0 licence. Any further distribution of this work must maintain attribution to the author(s) and the title of the work, journal citation and DOI.

Published under licence by IOP Publishing Ltd

rattan, a semantic information retrieval model suitable for the field of bamboo rattan was proposed, and the classification of bamboo rattan based on text-based data was realized.

2. Weight Calculation of Terms in the Field of Bamboo Rattan

2.1. Bamboo Rattan Domain Term Weight Definition
This paper uses the TF-IDF algorithm for reference to define and calculate the domain term weight. The idea is that: When the species of bamboo and rattan are identified according to the search term set, the domain document set is set as *Flora of China*. If the frequency of a certain domain term appears in the descriptive document of one kind of bamboo rattan, the greater the identification degree of the domain term to the bamboo rattan document, the higher the weight of the domain term will be. At the same time, if the domain term appears less frequently in all other description documents of bamboo rattan in *Flora of China*, the greater the identification degree of the domain term to bamboo rattan document, the higher the weight of the domain term will be.

2.2. Weight Calculation of Bamboo Rattan Domain Terms
According to the TF-IDF algorithm, the weight calculation process of bamboo and rattan domain terms.Therefore, the weight calculation formula of terms in the field of bamboo rattan is as follows:

$$sweight[i, j] = tf_{i,j} \times idf_i = \frac{n_{i,j}}{\sum_k n_{k,j}} \times \log_n \frac{N}{n_i} \qquad (1)$$

where *sweight* represents the weight of domain terms; $tf_{i,j}$ represents TF value and is the frequency of domain term I in document j; idf_i represents IDF value and is the frequency of reverse documents in all documents of the domain term I; $n_{i,j}$ represents the number of times the domain term I appears in document j; N is the total number of documents in the domain document set; n_i is the number of documents containing the domain term I.

3. Semantic correlation degree calculation
This paper takes the ninth volume of *Flora of China* as corpus document, in which each kind of bamboo is a document. Therefore, each document belongs to one concept. In the semantic identification of bamboo and rattan species, it is necessary to find the semantic related concept set based on query description (generally multiple word combinations), so as to determine the corresponding document set of the concept set. Among them, the establishment of the semantic correlation between the retrieval word set and concepts is the most core link, that is, the calculation of the semantic correlation between words and concepts in the semantic query extension.

In order to solve this problem, this paper USES the degree of word - concept - document to calculate the semantic relevance between the retrieval word set and the concept. The specific ideas are as follows:

If in the process of querying leaf attributes only in the ontology, the retrieval word set q is { lanceolate, glabrous, ... , rough } , then it can be expressed as the retrieval vector $\bar{q} = \{w_{1,q}, w_{2,q}, \cdots, w_{n,q}\}$, Where n represents the number of words to retrieve the word set; The leaf attribute of each bamboo in the domain ontology of bamboo rattan is represented by d_j, so the leaf attribute vector can be represented as $\overline{d_j} = \{w_{1,j}, w_{2,j}, \cdots, w_{n,j}\}$.

When the search term appears, write 1 at the component of the corresponding vector; If the search term does not appear, 0 is written at the corresponding component.

Through the similarity between vectors $\overline{d_j}$ and $\overline{q}$, the degree of correlation between the leaf attribute d_j of the ontology and retrieval word set q can be evaluated. The cosine law between vectors can be used to quantify, that is, the cosine value of the angle between two vectors can be calculated:

$$sim(d_j,q)=\frac{\overline{d_j}\bullet\overline{q}}{|\overline{d_j}|\times|\overline{q}|}=\frac{\sum_{i=1}^{n}w_{i,j}\times w_{i,q}}{\sqrt{\sum_{i=1}^{n}w_{i,j}^2}\times\sqrt{\sum_{i=1}^{n}w_{i,q}^2}} \quad (2)$$

Where $Q=(w_{1,q},w_{2,q},\cdots,w_{n,q})$ represents the retrieval vector; $d_i=(d_{1,i},d_{2,i},\cdots,d_{n,t})$ is the concept vector.

The weight of the bamboo and rattan domain terms themselves varies. Where, the higher the frequency of a certain bamboo and rattan domain term appears in one of the documents, the greater the identification degree of the domain term to the document, and the higher the weight of the domain term; At the same time, if the domain term appears less frequently in all other descriptive documents of bamboo rattan from *Flora of China*, it means that the greater the identification degree of the domain term for this document, the higher the weight of the domain term will be.

$$Sim(Q,D_i)=\frac{\sum_{j=1}^{n}w_{qj}d_{ij}}{\sqrt{\sum_{j=1}^{n}(d_{ij})^2\sum_{j=1}^{n}(w_{qj})^2}} \quad (3)$$

Where, Q represents the retrieval word set, $Q=(w_{1,q},w_{2,q},\cdots,w_{n,q})$; D_i represents the concept vector, $D_i=(d_{1,i},d_{2,i},\cdots,d_{n,i})$; d_{ij} represents the term weight of the j entry in the concept vector.

4. Experimental results and analysis

This paper makes an experimental analysis on whether the semantic information retrieval model of domain term weight factor is scientific and reasonable, and the influence of the term weight factor on the retrieval results before and after the introduction of domain term weight factor. The ninth volume of *Flora of china* was used as the experimental data set, including 37 genera, 516 species of bamboo, 516 documents and 530,000 words. The retrieval concepts and retrieval examples in table 1 were used as the original retrieval items in the experiment. Two sets of retrieval results were obtained by opening and closing the weight of domain terms in the experiment. The retrieval results are shown in table 2 and table 3.

Table 1 Experimental retrieval concepts, extended concepts and retrieval examples

NO.	retrieval concept	Semantic extension	retrieval examples
1	leaf	sheath、blade ear、ligule、blade、petiole	lanceolate
2	subterranean stem	subterranean stem、pseudorhizome、whip made of bamboo	solid
3	pole	Stem handle、culm base、internode、ring、Knot、ntrahode、septa intersegmental、internode shoots	bright
4	flower	androecium、pistil、pale、inferior palea、lodicule、ovary、stigma、anther	yellow

5	sheath	sheath、 Culm sheath、 Tuo sheath、 auricles and oral setae、Tuo leaf、 Tuo tongue、 culm-sheath、 wool	inapparent

Table 2 the TOP5 semantic relevance value before introducing the domain term weight

Number	Semantic Relevance Value（Before Introducing Domain Terminology Weight）				
	leaf（lanceolate）	Underground stem（solid）	pole（bright）	flower（yelow）	Culm（not obvious）
1	0.141421356	0.316227766	0.288675135	0.208514414	0.223606798
2	0.145864991	0	0.316227766	0.229415734	0.242535625
3	0.146173762	0	0.353553391	0.251573256	0.25819889
4	0.147441956	0	0.333333333	0.223606798	0
5	0.138675049	0	0	0.204124145	0.208514414

Table 3 the TOP5 semantic relevance value after introducing the domain term weight

Number	Normlized Value of Semantic Relevance Degree（After Introducing Domain Terminology Weight）				
	leaf（lanceloate）	Underground stem（solid）	pole（bright）	flower（yellow）	culm（not obvious）
1	0.282842712	0.316227766	0.288675135	0.294883912	0.223606798
2	0.291729983	0	0.316227766	0.229415734	0.34299717
3	0.382546028	0	0.353553391	0.353553391	0.365148372
4	0.294883912	0	0.333333333	0.387298335	0
5	0.138675049	0	0	0.288675135	0.294883912

Table 2 and table 3 are retrieval results of Top5 semantic relevance value before and after the introduction of domain term weight respectively. Retrieval results corresponding to the varieties of bamboo: Melocanna baccifera(No.1), Schizostachyum dumetorum(No.2), Schizostachyum pseudolima(No.3), Schizostachyum funghomii(No.4) and Dendrocalamus tibeticus(No.5). Among them, the difference of relevance value of retrieval results of the same concept is mainly determined by the different weights of retrieval items and domain terms that constitute concept vectors, reflecting the degree to which retrieval items represent different concepts and documents.

Before and after the introduction of the weight of domain terms, the sorting matrix of semantic relevance is L' and L, namely:

$$L' = \begin{bmatrix} 4 & 1 & 4 & 4 & 3 \\ 3 & 5 & 3 & 2 & 2 \\ 2 & 5 & 1 & 1 & 1 \\ 1 & 5 & 2 & 3 & 5 \\ 5 & 5 & 5 & 5 & 4 \end{bmatrix} \qquad L = \begin{bmatrix} 4 & 1 & 4 & 3 & 4 \\ 3 & 5 & 3 & 5 & 2 \\ 1 & 5 & 1 & 2 & 1 \\ 2 & 5 & 2 & 1 & 5 \\ 5 & 5 & 5 & 4 & 3 \end{bmatrix}$$

Where the item whose semantic relevance is 0 indicates that the retrieval item is not included in the concept vector. In order to express the correlation degree of 0 more clearly, this paper sets all the items with semantic correlation degree of 0 as the last one, that is, the fifth one.

By contrast, it can be seen that the domain term weight plays an important role in the screening sequence of results. By contrast, L' and L in the concept of "leaf", the term "lanceolate" in Schizostachyum pseudolima and Schizostachyum funghomii domain term weights in the two documents, directly affected the "leaf "semantic relevancy in "lanceolate" and Schizostachyum pseudolima and Schizostachyum funghomii , making Schizostachyum pseudolima and Schizostachyum funghomii raft swaps.

At the same time, the concept of "sheath" key words "not clear" in the Melocanna baccifera and Dendrocalamus tibeticus also have the same problem. However, the concept "flower" is the one that is most affected by the weight of domain terms. Before and after the weight of domain terms is turned on and off, the semantic relevance ranking of 5 bamboo species changes from (4, 2, 1, 3, 5) to (3, 5, 2, 1, 4), which affects the ranking of all retrieval results.

Compared with the documents corresponding to the five bamboos in *Flora of China*, it can be found that the retrieval words with significant domain term weight appear more frequently in the corresponding documents, and the corresponding concept description language is shorter, indicating that the weight of domain term can better distinguish the representative degree of the retrieval words.

Therefore, the term weight of bamboo rattan domain proposed in this paper is introduced into the retrieval model of bamboo rattan semantic information, so that the results with a high degree of relevance to the retrieval words can be ranked in front without affecting the retrieval results.

5. Conclusion

Based on the semantic expression of the domain ontology of bamboo rattan, this paper finds that the domain terms of bamboo rattan are closely related to the standard description information of bamboo rattan, proposes the weight calculation method of bamboo rattan domain terms based on ID-IDF, and constructs the semantic information retrieval model of bamboo rattan domain combining with the semantic extension based on correlation. The model describes the relationship between the concept of retrieval term and the species of bamboo rattan from two aspects of the domain term weight and semantic relevance. It can automatically recognize and extract the domain knowledge of bamboo rattan in the text information of bamboo rattan and screen the results based on correlation degree without requiring users to have a higher professional knowledge background, which solves the problem of "semantic gap" between the real retrieval intention of the discriminator and the domain knowledge of bamboo rattan.

Acknowledgments

The findings and the opinions are partially supported by projects of the major science and technology projects of Yunnan province, No.2018ZI001.This work is supported by Engineering and Technology Research Center for Yunnan Agricultural Big Data and Yunnan Agricultural University, China.

References

[1] Zhang Xinping. Development Prospects of Bamboo and Rattan in the world[J]. World Forestry Research, 2003. 16(1):26-30.

[2] Qiu Erfa, Hong Wei, Zheng Yushang. Review on Diversity and Utilization of Bamboo in China[J]. JOURNAL OF BAMBOO RESEARCH, 2001, 20（2）:11-14.

[3] Chinese flora editorial committee. Flora of China (volume 9, volume 1) [M] Beijing: Sciences Press, 1996.

[4] Jiang Zeihui. The world's bamboo rattan(English edition) [M]. Beijing: China Forestry Publishing House, 2007.

[5] Qiu Yanxian. Research on Automatic Extraction of Domain Terms and Classification of Relationships [D]. Master's Thesis of Kunming University of Science and Technology, 2009

[6] SUN Xia, WANG Xiao-feng, DONG Le-hong, WU Jiang. Study on Term Relation Extraction from Domain Text[J]. Computer Science, 2010, 37(2):189-191, 215.

[7] Xv Jianmin. Extension Research of Bayesian Network Information Retrieval Model Based on Term Relation [D]. Doctoral Dissertation of Tianjin University, 2007.

[8] Xue Zheng. Hot Topic Discovery of Text Information Based on Improved TF-IDF [D]. Master's Thesis of Wuhan Res,2009

[9] Wang Qi. Semantic retrieval technology and application of ancient fresco [D]. Doctoral Dissertation of Zhejiang University, 2011.

[10] Han Xin. Research on resource semantic retrieval in bioinformatics based on ontology [D]. Master's Thesis of North University of China, 2012.

| ISPECE | IOP Publishing |

Emergency Event Matching using Hierarchical Blocking Method

Chang Wen[1,2,3,4*], Yu Liu[1,2,3,4]

[1]College of Computer Science and Technology, Wuhan University of Science and Technology, Wuhan 430065, China;

[2]Key Laboratory of Intelligent Information Processing and Real-time Industrial System in Hubei Province, Wuhan 430065, China

[3]Institute of Big Data Science and Engineering, Wuhan University of Science and Technology, Wuhan 430065, China

[4]Key Laboratory of Rich-media Knowledge Organization and Service of Digital Publishing Content, National Press and Publication Administration, Beijing 100038, China

597073407@qq.com

Abstract. With the extensive application of the knowledge base (KB), how to complete it is a hot topic on Semantic Web. However, many problems go with the big data, and the event matching is one of these problems, which is finding out the entities referring to the same things in the real world and also the key point in the extending process. To enrich the emergency knowledge base (E-SKB) we constructed before, we need to filter out the news from several web pages and find the same news to avoid data redundancy. In this paper, we proposed a hierarchy blocking method to reduce the times of comparisons and narrow down the scope by extracting the news properties as the blocking keys. The method transforms the event matching problem into a clustering problem. Experimental results show that the proposed method is superior to the existing text clustering algorithm with high precision and less comparison times.

1. Introduction

Events based on the text description contain a large number of potentially useful information, one of the most common type is the web news. Extracting the news messages in the specific fields such as finance, economy, can help us build knowledge base to develop some applications in the daily life. Thereinto, the emergency is very important because of its burstiness and unpredictability, it always produce disasters and lead to the economic losses. Therefore, we have built a semantic knowledge base of emergency based on the case database from Jinan University in previous work [1]. However, this database is small and limited, we have to extend our knowledge base by collecting the news from the web, and then transform the news into structured triples.

However, thousands of news appearing on webs may have the repetitive ones certainly. They are in different forms and written by different authors. Adding impure news into the knowledge base will lead to data redundancy. Hence, we need to filter out the news to grant the pureness of the database by Entity Matching. Entity Matching means recognizing the entities which refer to the sane things in the real world. In the same way, the news entity matching is identifying the news reporting the same events. On

---

Content from this work may be used under the terms of the Creative Commons Attribution 3.0 licence. Any further distribution of this work must maintain attribution to the author(s) and the title of the work, journal citation and DOI.
Published under licence by IOP Publishing Ltd

account of the flexibility we have discussed, the simple String Matching will cost large quantities of operation space. To reduce the times of comparing, we have to narrow down the comparative scope by extracting the data properties and ignoring the irrelevant matching-pairs. At the same time, extracting the critical properties can help lessen the workloads of transforming data into the knowledge base.

2. Related Work

Event Matching is to find out semantic projection or similarity between entities. Up to now, there are four major types of Entity Matching – based on the linguistics, the structure, the instances and the multi-strategy. The linguistics method uses the natural language processing technology to calculate the similarity of the name, label or description between entities. Event [2] is an activity in which a specific entity participates at a specific time and place. Information such as body, time, and activity can be presented as an element of the event [3]. The participants of the event are usually named entities. The integrated events contain a large number of entity schema information, which can be regarded as a natural dictionary as the basis for the identification of named entities. Compared with the identification of named entities in the literature [4] and link matching, in the comments. The named entity form is free, and there are variants such as abbreviations, abbreviations, and nicknames. For the product names that appear in a large number of comments, it is necessary to add mode information [5] and symbol information. The entity in structure method is abstracted into a node and the relation becomes the edge. The application based on the instances and multi-strategy are less used because they have problems of sample size, algorithm performance and so on.

3. Challenges in Blocking News

Compared with the data in knowledge bases like DBpedia or WordNet, the form of the news is optional and informal. In addition, to grant the sufficient coverage of the data set, only collecting one website is not enough. We usually have to crawl the news data from two or more web pages. With the enlargement of the data set, there are two problems in the news collection, we call them time difference and classification difference.

-Time difference: due to the burstiness of the emergency, there is not enough time for some authors to deliver the news timely. It will result in the case that the same report will be published at different dates. Therefore, to regard the date of publication as the time of the event is not rigorous, we have to extract the correct date form the news to avoid the time difference though it is infrequent.

- Classification difference: the blocking method is based on the property of the event and one of the important properties is classification. Whereas each news web has its own classification framework, the Figure 1 shows that the same news reported in the two web pages have different classifications which have respective definitions. So, comparing the news events according to the classification directly is infeasible. We have to normalize the news classifications by using a unified method, the THUCTC (THU Chinese Text Classification) [7] for classification.

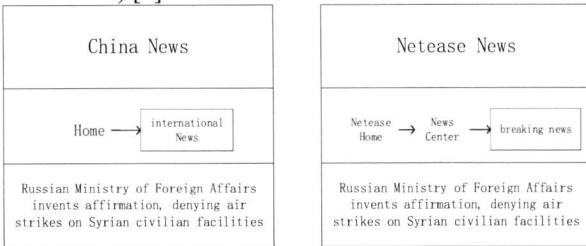

Figure 1 Different classification of the same news

4. Hierarchical Blocking

To solve the problems mentioned before and make the events matching, we proposed a blocking method to categorize the news events into several groups and each group contain the events which have the same property values. In this case, we can narrow the comparison range and reduce the number of comparisons.

Firstly, we crawled news from two Chinese news sites, 2,000 news from 163NEWS and 10,000 news from CHINANEWS. The time is from January 1, 2016 to January 6, 2016, the news category covers most of the main categories apart from the life class such as the introduction to life knowledge and makeup introductions.

Figure 2 shows the main steps of news entity matching, we can get four existing properties of the raw text form the news web -- published date, title, text, and URL. Then, we constructed the blocking layers by the blocking keys, which can narrow the scope of comparison layer by layer. Afterwards we extracted the information of blocking keys in each news through the existing properties. Finally, we compared the news in each block to filter out the same news and picked out the emergency news in addition.

4.1 Constructing Blocking Layers

The news from the website is semi-structured because it has some existing properties such as title, published date, text, and its URL, which called EP (existing properties). However, it is impractical to use EP to match the news directly because of the data disunity. Since an event is defined as follow -- something that happens at a given place and time [8], we extracted date, place and the type as the blocking keys according to the EP. These keys finally constitute the hierarchy blocking framework and the news obtaining the same property values will be distributed into the same blocks. Only comparing the news in the block can reduce the number of comparisons effectively, which is significant to the large amounts of data.

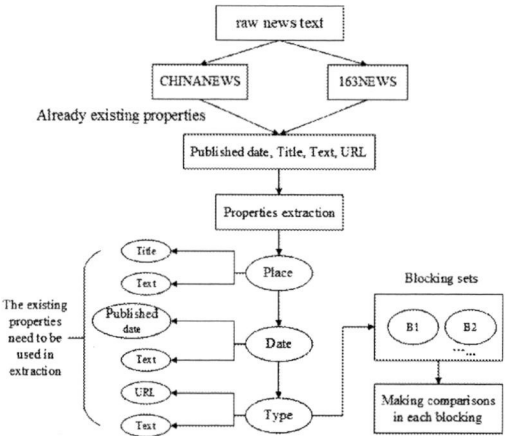

Figure 2. The flow chart of the blocking method

Assume we are given a set of news $N_e = \{n_1, n_2, n_3, n_4, ..., n_n\}$ where ni is the news from two news webs. We propose the block formulation $B_k = f(d, p, t) = \{n_{k1}, n_{k2}, ..., n_{kn}\}$, which means that a block is determined by the date, place and type. B_k is the final block and nkn is the news event within the block, all the news in one block possess the same property values of date, place and type. The function f is the operation process of the blocking, the news will be handled by the function d, p and t sequentially. The input of each function is the output of the previous one, except that the input of the first function is N_e. Through the function d: $N_e \rightarrow B_{date}$, the news can be classified into several blocks and these blocks will be the input of the function p: $B_{date} \rightarrow B_{date\&place}$. Then the $B_{date\&place}$ will generate the final blocks according to the function t: $B_{date\&place} \rightarrow B_k$. The comparison works only focus on the B_k sets which can substantially reduce the number of matches.

Figure 3 shows the main process in news blocking, we initially have ten news with redundancy which are in different forms. Firstly, the candidate date of the news will be extracted, so that news will be divided into several groups according to it. Then, the place of each news is used to subdivide the news into fine-grade blocks. Eventually, we unify the type of the news form two different webs and get the final block sets. Following the definition of the event, the repetitive news will be in the same block therefore we have no need to compare the news across the blocks.

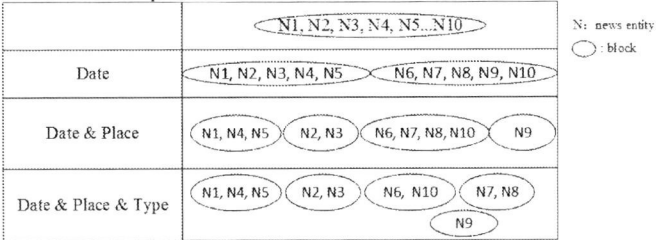

Figure 3. Generating blocks in each Blocking Layer

4.2 Extracting Blocking Keys

In order to extract blocking keys and address the challenges mentioned above, we make full use of the EP (existing properties). The sequence of extraction can be changed optionally, we follow the steps of date, place and type extraction. We now describe the details of each step which can generate the B_k finally.

4.2.1 Date Extraction

Time difference is one of the challenges, we can get the published date from the website easily but cannot confirm the date of the news event. Since there will be discrepancies in the time when the authors of different sites publish news.

To solve this problem, we define the candidate date as $C_D = <P_D, E_D>$, C_D is a date set composed of two types of dates. P_D (published date) is the date when the news is published and the $E_D = <E_{D1}, E_{D2}, \ldots, E_{Dn}>$ is another date set we extract form the articles. Based on the characteristics of news narrative, the date of the event is usually mentioned in the first few sentences of the article. Therefore, we use the regular expressions to extract date to join into E_D.

$$C_D = \begin{cases} P_D, & \exists\ date\ \in\ E_D\ \rightarrow date = P_D \\ P_D \cup E_D, & \nexists\ date\ \in\ E_D\ \rightarrow date = P_D \end{cases} \tag{1}$$

If the E_D contains the date which is equal to P_D, then we can determine the date of the news is P_D. Otherwise, all the dates in C_D will be regarded as the blocking keys to divide the news. Although this situation is rare due to the real-time nature of the news, we have to check it out to prevent data omission.

4.2.2 Place Extraction

Another important factor in news events is the place, which is usually in the form of country, province or city. These place words have fixed expressions, using ordinary word segmentation may result in too abstract or refined places. For example, we want to get "Hubei Province" in "Hubei Province Talent Market held a job fair for college graduates", but the "Talent Market" will also be extracted which is abstract to block the news. In another case, some local news contains the refined places, such as "Qingcheng Mountain" and "Forbidden City", which generate blocks containing few news events. These two types of places lack of application value, therefore the word segmentation is inapplicable to extract places from the news.

```
Input: placeDict.txt, B_date
Procedure1 bulidBitDic(placeDict.txt)
    WHILE(word IN placeDict.txt)
        SET words[word[0]] → HashSet();
        IF (1<<word)≠NULL
            ADD bit(word[1, length-1]) → words[word[0]]
    RETURN bitDic

Procedure2 getKeyWords(bitDic, News)
    maxLength = MaxLengthWord(words[])
    minLength = MinLengthWord(words[])
    TRAVERSE News:
    IF News[index] in words
        checkLaterWords IN News
        IF length(checkedText) ∈
(minLength,maxLength) AND bit(checkedWord) = true
        GET KeyWord
    ELSE
        checkNext(wordLength)
    RETURN KeyWord
```

Figure 4. Procedure for extracting places of unified expression in news

In order to unify the places, we constructed a word dictionary composed of provinces and cities in China and the world's major countries. For those over 200 place words, it is unrealistic to match them out one by one on each news because of the time consuming. To get the keywords of the news effectively, Figure 5 shows our method which is to generate bit strings corresponding to the length of the keywords and store the strings as hash set. Then the text will be traversed quickly to get characters, each one of them will be matched with the word dictionary and the corresponding position in the bit strings, the character is changed to 1 if it is matched. If some bit strings are all 1, the corresponding keywords are returned. For example, the keyword is "CHINA", the matching result of "CHI" is 2 and the binary is "10", which means "CHI" is in the first position of the keyword. Then the following characters are compared until the end of the text, if the bit string of "CHINA" is "11"it means the text includes "CHINA".

4.2.3 Type Extraction

The news type is the last blocking layer to narrow the block sets. Given the elements of the news, the object and action are very crucial factors to define news apart from the time and place. However, on the one hand, these two factors will let the final blocks too fine grained and make the process too difficult to realize. On the other, extending the E-SKB needs to recognize which news belongs to the emergency, so the type extraction works.

In order to get the emergency news, the final categories should be extracted particularly. We chose 8 categories that may contain the emergency, together with the news corpus in E-SKB to form a new training set. Then the training set was used to reclassify the news datasets to get the emergency and the corresponding types.

4.3 Matching news

Through the extraction method mentioned above, the news datasets are divided into several blocks based on the date, place and type subsequently. Then the news in each fine-grained block is compared with others by the news headline. Since the headline covers the major information of the news, we simply use the cosine similarity to match the same news out.

5. Experiments

5.1 Matching Numbers

The Hierarchical Blocking method we put forward can reduce the matching numbers greatly, since this method will divide the datasets into several blocks by the properties of the news, only the news in the same blocks will be compared instead of the Cartesian product on the whole datasets. To evaluate the matching numbers, we use three cluster baselines under different numbers of news, K-Means , DBSCAN and GSDMM.

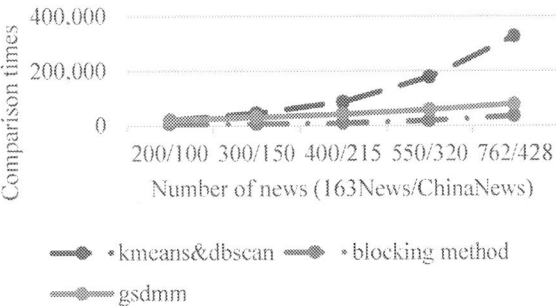

Figure 5. The matching numbers of four methods under different number of news

Figure 5 shows the matching numbers of the four methods under different numbers of news in two websites. Our method in the case of different amounts of data, the matching numbers are much lower than the other three methods. The matching numbers of the K-Means and the DBSCAN present a huge increase with the augment of the data volume. Since they use TF-IDF in documents distance calculation, so they compare with each other between two news datasets and the result is the Cartesian product between two datasets. As for the other baseline, the matching numbers of the GSDMM relates to the number of words in each document. Since we use the title of the news to cluster, the average number of words in titles is set to 6.

From the result, we can find that the comparison times on our method is less than the baselines. In addition, the time consuming of the Hierarchical Blocking is shorter than the basic clustering methods, it takes only 4 seconds to process about 450 news articles, but the basic methods will take several minutes or even longer. Because the parameters are needed to be very specific (such as the cluster numbers k in K-Means is more than a few hundred) so that each cluster contains the same news or not, instead of containing the similar news or series of events which cannot help us to figure out the same news events exactly.

5.2 Matching Results on the Same News

In this part, we try to compare the performance of Hierarchical Blocking with K-Means, DBSCAN and GSDMM on matching the same news. Since extracting the same news events belongs to the fine-grained clustering problem, we chose other three effective clustering methods to evaluate ours on 450 news articles (300 in CHINANEWS and 150 in 163NEWS).

Based on the manual statistics on the news articles, we set the k in K-Means to the true number of the clusters which is 250. For DBSCAN, we set the density radius to 1.8 and the minimum number to 230. For GSDMM, we set the number of iterations at 100, $\alpha = 0.1$, and $\beta = 0.1$ for the dataset. The Hierarchical Blocking does not need to set any parameters. Then, we use F1-score to evaluate the performance, the recall represents the number of the same news pairs found divided by the total number of the same news pairs, the precision means the number of the same news pairs found divided by the clusters which contain more than 1 articles.

Table 1. Performance of four methods on matching the same news

Methods	Recall (%)	Precision (%)	F1 (%)
K-Means	64.15	43.17	51.61
DBSCAN	62.26	42.32	51.58
GSDMM	88.68	68.05	77.01

Hierarchical Blocking	67.92	**94.75**	**79.12**

Table 1 shows the performance of Hierarchical Blocking with K-Means, DBSCAN and GSDMM on matching the same news. We can see Hierarchical Blocking performs better than K-Means and DBSCAN in recall and it reaches the precision over 90%. The GSDMM can reach a high level of recall but the precision is common in this task. Although GSDMM can deal with sparse and high-dimensional problem of short texts, like the other two baselines, the clusters will contain some irrelevant news or similar news (like series of events) to reduce the precision. On the one hand, the parameters required in the baselines will affect the results deeply. On the other hand, the high-degree sparseness of the data makes it hard to get the desired results with coarse-grained clustering. We used two news websites to prevent omission, so the ideal result is that each block contains 2 news (refer to the same event) or only 1 news (has no repetitions). Hierarchical Blocking can extract the news pairs only referred to the same events and will not consider the irrelevance, but other methods can not accurately divide the same news into the unique clusters, which will influence the precision and the F1-score finally.

6. Conclusion

This paper uses a Hierarchical Blocking method to complete emergency news matching, which is used to extend the original emergency knowledge base by converting the event matching problem into a clustering problem. At the same time, the event attributes are extracted in the process of Blocking. The experimental results show that the proposed method can reduce the number of matches compared with other clustering methods, and has higher accuracy and F1 value. However, the matched emergency needs to be further integrated into the knowledge base, which is the future work.

References

[1] Chang W, Yu L, Gu J, et al. E-SKB: A Semantic Knowledge Base for Emergency[C]// China Conference on Knowledge Graph & Semantic Computing. 2016.

[2] Naughton M, Stokes N, Carthy J. Sentence-level event classification in unstructured texts[J]. *Information Retrieval*, 2010, **13(2)**:132-156.

[3] Zhang Chuanyan, Hong Xiaoguang, Peng Zhaohui, et al. Web Entity Activity Extraction Based on SVM and Extended Conditional Random Fields[J]. Journal of Software, 2012, 23(10): 2612-2627

[4] Cecile B, David C J, Matteo M, et al. Clustering attributed graphs: Models, measures and methods[J]. *Network Science*, 2015, **3(3)**:408-444.

[5] Chandel A , Nagesh P C , Sarawagi S . Efficient Batch Top-k Search for Dictionary-based Entity Recognition[C]// International Conference on Data Engineering. IEEE, 2006.

[6] Jingyang Li, Maosong Sun. Scalable Term Selection for Text Categorization. Proc. of the 2007 Joint Conference on Empirical Methods in Natural Language Processing and Computational Natural Language Learning (EMNLP-CoNLL), Prague, Czech Republic, 2007, pp. 774-782.

[7] Qiu T, Qiao R, Wu D O. EABS: An Event-Aware Backpressure Scheduling Scheme for Emergency Internet of Things[J]. *IEEE Transactions on Mobile Computing*, 2018, PP(99):1-1.

Research and Construction of Yunnan Plant Vertical Retrieval System

Lin Peng, Zhi-run Ma, Xin Zhang

College of Big Data, Yunnan Agricultural University, Kunming 650201, China

TCorrespondence to: Lin Peng(penglin2286351@163.com)

Abstract. The plant richness of Yunnan Province ranks first in the country, and scientific and rational protection and utilization of Yunnan plant resources are of great significance for maintaining China's biodiversity. In view of the current distribution patterns and physiological characteristics of plants in Yunnan Province, as well as related geography, climate and wild habitats, this paper constructs a set of vertical search system for Yunnan plants through the research of vertical search architecture, network data collection, semantic analysis of network information and the pattern of network information retrieval. The system can effectively solve the technical obstacles of Yunnan plant network resource utilization, and provide new technical support and theory for the macro-level and deep research of plants in Yunnan Province.

1. Introduction

Yunnan is an important gene pool of species in China and is known as the "plant kingdom". According to statistics, there are 274 families, 2076 genera and 17,000 species of higher plants in Yunnan Province, accounting for 62.9% of the total number of higher plants in the country. Due to geological history changes, Quaternary glacial effects and vertical changes in the mountains, many plants in Yunnan have more ecological genetic types and stronger ecological adaptability [1].

A large number of studies have shown that the distribution of plants is closely related to changes in multiple real-time environmental factors such as longitude, altitude, atmospheric circulation, topography, temperature, and precipitation [2-6]. A comprehensive survey of the surrounding environment of plants, through the study of plant diversity and its distribution patterns, genesis, zoning and "hot spots" in the region, is the focus of plant research, conservation and utilization. However, the current distribution patterns and physiological characteristics of plants in Yunnan, as well as related geography, climate and wild habitats are very difficult to obtain, which greatly restricts the comprehensive large-scale and deep-level systematic research of Yunnan plants. Specifically, there are mainly the following problems:

(1) Because Yunnan plant field knowledge involves many professions, various information resources are distributed in different databases and professional websites, and various resource retrieval system methods are different. Researchers need to master digital resource system retrieval technologies with different interfaces, and spend a lot of time and energy to browse, retrieve and summarize various types of information, resulting in a low level of comprehensive utilization of these information resources, restricting the large-scale, comprehensive and systematic research of Yunnan plants.

(2) In the process of scientific research on Yunnan plants, not only static information such as plant shape description, image samples, distribution areas, scientific literature, etc., but also dynamic information such as soil data, water resources data, climate data, and remote sensing data are needed. Real-time information such as market price, planting area, production, and sales volume may also be required for economic plants. This information is distributed in different systems (databases),

---

Content from this work may be used under the terms of the Creative Commons Attribution 3.0 licence. Any further distribution of this work must maintain attribution to the author(s) and the title of the work, journal citation and DOI.

Published under licence by IOP Publishing Ltd

independent of each other, intricate and intertwined, and there is a potential relationship. However, at present, both the general retrieval system and the agricultural vertical retrieval system at home and abroad do not take into account the interrelationship between the static, dynamic and real-time information in the plant field, and the real-time requirements of the retrieval results are not high, greatly it restricts the retrieval effect of the retrieval system.

(3)   Because Yunnan plant domain information exists in different databases and professional websites, this information exists in different data formats such as TXT, HTML, XML, RTF, PDF, PSZ/PS, and in different languages such as Chinese, English, Latin, etc. However, most of the current retrieval tools can't provide information retrieval of heterogeneous data; At the same time, these information from the Internet and professional databases can only achieve simple keyword-based retrieval, without semantic association ability. This leads to the fact that the user sometimes has to change the query word because the word does not match, and often only the query word appears in the document before it can be retrieved. As a result, there are frequent cases where documents related to user query requests cannot be retrieved because of different words.

(4)   For those who work in Yunnan plant research, most of them use the general search engine to obtain relevant information on the Internet. When a user uses a general search engine for information retrieval, a large number of data sets containing duplicate information and spam information are often obtained, and the user cannot quickly and accurately locate the demand information. Moreover, the acquisition of Yunnan plant epitaxial information is still difficult to break through, which greatly restricts the large-scale and systematic research of Yunnan plants. In particular, the processing capability of real-time data on the network is not ideal, and the data fusion and semantic analysis of real-time network data need to be further studied.

(5)   In view of the above problems, this paper introduces the vertical search engine technology, establishes a vertical search system for Yunnan plants, collects real-time plant network data in Yunnan, precise query, in-depth analysis, and provides Yunnan plant-related literature, pictures, videos and ecological environment information for researchers and related personnel.

2. Design Ideas of Yunnan Plant Vertical Retrieval System
At present, domain vertical search system generally has the problems of poor timeliness of information, difficulty in semantic interoperability and low search efficiency. This paper will solve the problem of information timeliness by establishing a federation architecture of vertical search system for Yunnan plants and combining it with creating an open field real-time perception and topic collection mechanism for Yunnan plants; By studying the topic clustering of network texts, it solves the problem of network data semantic interoperability; By studying the problem of Yunnan plant network information fusion, it creates a cloud-oriented plant search model to solve the problem of low efficiency.

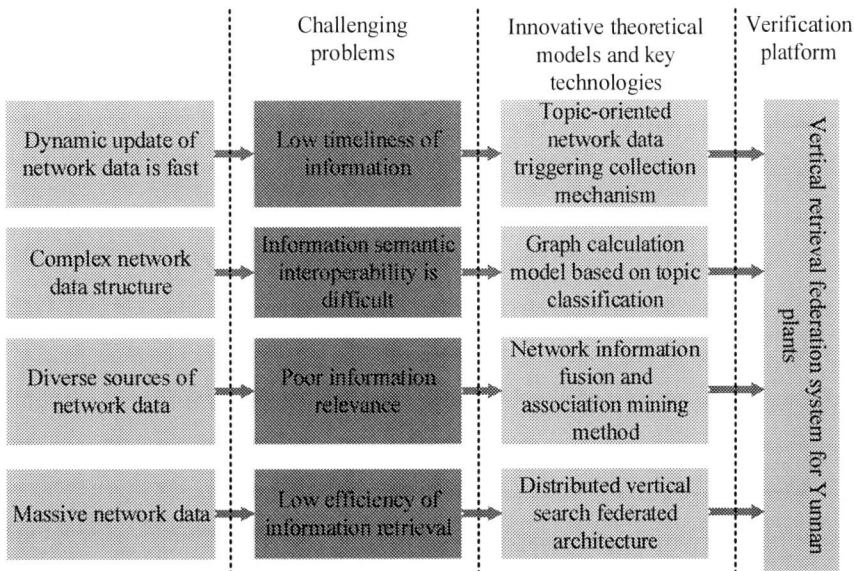

Figure 1. Schematic diagram of the construction of Yunnan plant vertical retrieval system.

3. Federated Architecture Construction of Distributed Vertical Retrieval System

The federal architecture of the distributed vertical search system can effectively support solving the problem of "specific and wide compatibility", implicit association mining and real-time retrieval response in the vertical search system of Yunnan plant field. This paper adopts the design concept of "division and cooperation". The whole search system consists of several vertical search engine nodes (unit nodes) and a federal central node for different topics. As shown in Fig. 2, each unit node is a meta-search engine for a specific topic, and can independently retrieve information of a specific topic independently; Each unit node can also accept the user retrieval request by the central node, decompose the request into different topic tasks, and distribute them to the corresponding different unit nodes to complete together. By adopting such a "division and collaboration" mechanism, it is desirable to solve the problem of compatibility between the "special" of the search results of the vertical search system and the "wide" of the collected data. At the same time, the central node performs in-depth processing such as fusion, clustering and association analysis on the unit node search information under different topics to solve the implicit association mining problem of multi-source data. At the same time, in the federated search system, the real-time perception mechanism of the unit node to the network information and the linkage data collection mechanism of the federated search system are established. When a node in the federated system perceives certain network information, it triggers nodes of other related topics to perform linkage data collection and data analysis, thereby solving the real-time response problem of the federated vertical search system.

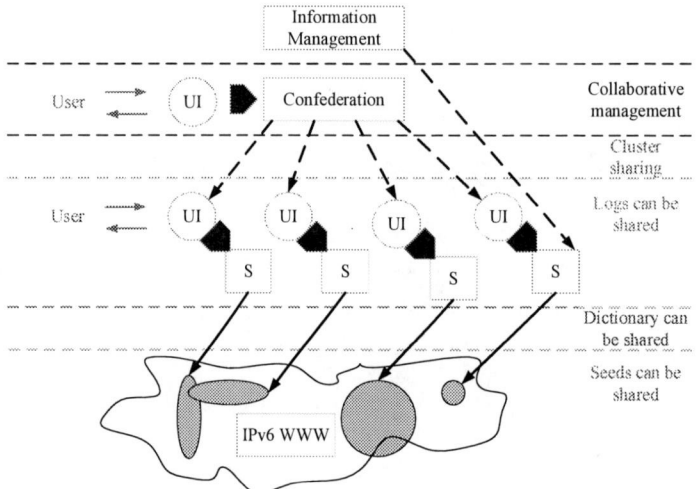

Figure 2. The basic structure of the federal architecture of the vertical search system.

4. Semantic analysis method for network text topic classification

The biggest challenge of network text topic classification is the feature matrix sparsity problem. The existing network text topic classification research only considers the influence of synonym and near-synonym on feature sparsity in traditional texts, and does not consider the influence of features such as less vocabulary and less semantics on the sparseness of features in network texts. Aiming at the existing research problems, the project intends to use the network text topic classification method based on concept and strong feature library to solve the feature sparse problem of network text space vector model. The overall framework of the method is shown in Figure 8, which is mainly composed of three parts. The first is to generate the concept of network text, the second is to generate strong features of the network text domain, the third is to generate a network text space vector model, train the classifier, and realize the classification of network text topics.

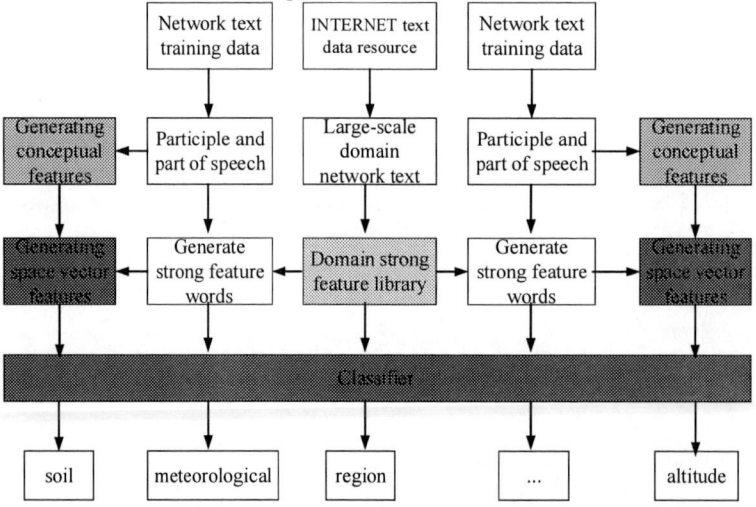

Figure 3. Thematic classification framework based on conceptual features and strong feature libraries.

(1) Generate network text concept features. Firstly, the establishment of the Yunnan plant domain ontology library based on the shape characteristics of Yunnan plants, the terminology and concepts are mainly derived from the "Yunnan Flora", with reference to "Chinese Flora"; Then, using the concept and the structural relationship in Yunnan plant ontology, the similarity of different feature words in

Yunnan plant ontology is calculated. Finally, the semantic connection is established between synonyms and near-synonyms, and the conceptual features of the network text are generated by replacing the feature words in the network text with concepts.

(2) Generate strong features in the network text domain. It is proposed to adopt a domain-oriented web crawler to capture massive domain network text data from the Internet as a domain subject data source. Based on the LDA theme model, statistical modelling is performed to obtain a strong feature vocabulary with high degree of classification. In order to reflect the importance of strong feature words in the subject classification, and to reduce the influence of fewer feature words in the network text on the sparse feature matrix, a new strong feature weighting method is proposed. This method defines the domain contribution of strong feature words based on information gain; Then, based on the contribution of strong feature fields, a strong feature weighting method is proposed.

(3) Generate network text space feature vectors, train classifiers, and implement network text topic classification. It is proposed to use mutual information for text feature selection. After selecting the first M features with the largest mutual information, the network text feature features based on the Yunnan plant ontology generation and the domain strong features generated based on the domain strong feature database are added to jointly generate the network text space feature vector. Finally, the spatial feature vector of the network text consists of two parts, one is the ordinary feature word and the concept feature, and is weighted by TF-IDF; The second is a strong feature word, which is weighted by a strong feature weighting method based on domain contribution. Based on the network text space vector, the polynomial Bayesian classifier is initially proposed as the network text topic classifier to realize the network text topic classification.

5. Construction of Yunnan Plant Vertical Search Prototype System

Combined with the vertical search prototype system architecture and functional requirements for Yunnan plants, a vertical search engine system for Yunnan plants was constructed. The basic framework of the system is shown in Figure 4.

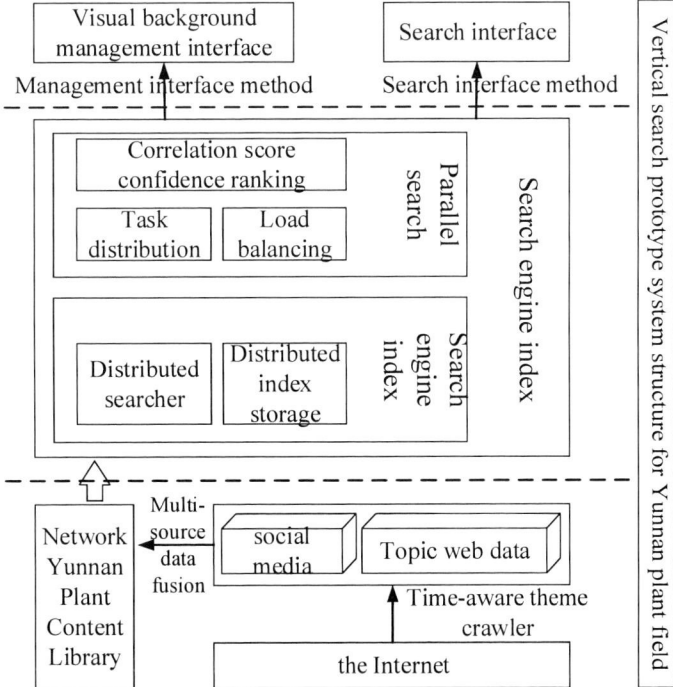

Figure 4. Schematic diagram of the vertical search system for Yunnan plants.

The system can implement traditional document retrieval and retrieval of network-oriented information. In the document-level index, we index all information related to the document, such as the body, keywords, burst words, information ID corresponding to the document, time, location, source, importance of the document in the information, and so on. In the information retrieval system, users can submit two different types of queries: 1) Query with only time or place; 2) A query containing query keywords. For the first type of query, it mainly supports the user to browse the information needs that occur at a specific time or at a specific place. For the second type of query, it mainly supports the user's need to understand the information related to the query word. Of course, for the second type of query, the user can also specify the time and place. In the information level index, the list of documents corresponding to each information will be saved. Through this list, you can quickly find out which information documents belong to this.

6. Conclusion

This paper introduces the vertical search engine technology, establishes a vertical search system for Yunnan plants, collects, accurately queries and in-depth analyses the network data of Yunnan plants in real time, and provides Yunnan plants (including text, pictures, videos and ecological environment, etc.) data information, effectively solves the technical obstacles in the utilization of digital resources in Yunnan plant networks, and comprehensively enhances the real-time response capability and overall use effect of Yunnan plant network digital resources.

Acknowledgments

The findings and the opinions are partially supported by projects of the major science and technology projects of Yunnan province, No.2018ZI001.This work is supported by Engineering and Technology Research Center for Yunnan Agricultural Big Data and Yunnan Agricultural University, China.

References

[1]   Zai-fu Xu. Approach on some characteristics of Yunnan plants under the evolution from ancient to modern[J]. *GUIHAIA*,**2003,23(4)**:294-297.

[2]   ZHU H,MA Y,YAN L,et al. The relationship between geography and climate in the generic-level patterns of Chinese seed plants[J].*Acta Phytotaxonomica Sinica*,**2007,45(2)**:134-166.

[3]   Zhi-jian Yu. Discussion on the role of climate in the transformation of yunnan into a plant kingdom[J]. *AGRICULTURE AND TECHNOLOGY*, **2014,34(1)**:178.

[4]   Wen-hong Ma,Yuan-he Yang and Jin-sheng He. Biomass of temperate grassland in Inner Mongolia and its relationship with environmental factors[J], *Science in China(Series C)*: *Life Science* **2008,38(1)**:84-92.

[5]   Jian-meng Feng. Spatial patterns of species diversity of seed plants in China and their climatic explanation[J]. *Biodiversity Science*, **2008,16(5)**:470-476.

[6]   Jian-meng Feng and Xv Dongcheng. The large-scale distribution pattern of seed plant species richness in China and its relationship with geographical factors [J]. *Ecology and Environmental Science* **2009,18(1)**:249-254.

[7]   Xiang-qin Liu. Research and Design on Key Technologies of Vertical Search Engine Oriented Soybean Theme[D]. *Northeast Agricultural University*, **2013**

[8]   Xiao-rong Yang. Research and application of key technologies of distributed agricultural science and technology information sharing[D]. *Doctoral dissertation of the Chinese academy of agricultural sciences*, **2011**.

[9]   Jing Li. Ontology theory and its application in agricultural document retrieval system -- a case study of floriculture ontology modeling[D]. *Doctoral dissertation of graduate school of Chinese academy of sciences*, **2004**.

[10]   Xiao-rui Li. The Present Situation of Archaeology in Yunan Province[J]. *Relics From South*, **2016, 1**:166-170.

A Social Network Water Army Detection Model based on Artificial Immunity

Hanhua Zhang[1] and Tao Li[2] and Yuqiao Wang[3]

[1]1st School of Computer Science and Technology Wuhan University of Science and Technology Wuhan, China

[2]2nd School of Computer Science and Technology Wuhan University of Science and Technology Wuhan, China

[3]3rd School of Computer Science and Technology Wuhan University of Science and Technology Wuhan, China

Abstract. The wanton dissemination of water army information in social network seriously affects the authenticity of information received by users. With the continuous renewal of the covert means of the water army, the traditional screening means of the water army are no longer efficient. Combined with the idea of computer artificial immunity, this paper deeply analyzes the deep-level characteristics of the water army, finds out the stable and efficient danger and safety signal, calculates its optimal weight matrix through evolution, fuses DCA algorithm to calculate the antigen maturity, and then identifies the water army. The simulation results show that the recognition model based on the immune risk theory has a certain improvement in accuracy and recall rate, while the time cost is greatly reduced, which verifies its effectiveness and high efficiency.

1. Introduction

Micro-blog (including Sina Weibo, Twitter, etc.) has become the most popular social means in the world. However, while Weibo social networking has such a profound impact on all aspects of the lives of so many people, its features, such as low cost to release, wide range of dissemination and fast dissemination speed, have become a double-edged sword and begun to have negative effects on people. Behind every public opinion event, there are a large number of economic or political interest groups.

It has become an urgent task to find an efficient detection method for Water Army. Water army，a group of Internet ghostwriters paid to post online comments with particular content,paid posters, netizens hired to leave fake comments and delete others. This paper proposes a social network water army recognition model based on heterogeneous linear computing method DCA. Four kinds of recognition signals are defined and their optimal weight matrix is evolved to detect water army. At the same time, the recognition accuracy is guaranteed, and the recognition time is greatly reduced.

2. Related research

In recent years, the identification of social network water army has become a focus of scholars. Generally, researchers study the attributes and behavioral characteristics of water army accounts, or the communication characteristics of water army information. Through in-depth analysis of the characteristics of various types of water army to distinguish the characteristics of the water army and the use of the characteristics of the appropriate classifier to distinguish the water army and the information of it.

Zhang et al. [4] set 6 attributes as the characteristics of the water army classifier, and integrated the bayesian model and genetic optimization algorithm to improve the accuracy of water army identification without sacrificing the recognition rate of non-water army. Cheng et al. [2] proposed a new category of relational graph features on the premise of combining traditional user attributes and behavioral characteristics, and proved that the effectiveness of judging water army was significantly improved with the participation of new features. Han et al. [1] took the probability that the user is a water army as an implicit variable of the user's attribute characteristics and behavior characteristics, and constructed a probability graph model to calculate the probability that the user is a water army. Yuan et al. [5] analyzed a series of obvious characteristics of Weibo water army, used entropy value method to determine the weight of each characteristic index, and established an automatic identification model of Weibo water army combining multi-index comprehensive index method. Yang et al. [8] took a new approach, using sina Weibo's official rumor refuting platform, and selected a large number of micro blogs that were officially identified as false information for research. They found and used real data to prove that these two new features, "place of release" and "type of blog user end", can provide effective help in distinguishing water army accounts. Similarly, Thomas et al. [11] obtained a large number of common features of water army accounts through detailed studies on more than 1.8 billion micro-blogs and 1 million accounts that were blocked by Twitter. It has laid a solid foundation for the subsequent classification and judgment of naval forces. Castillo et al. [13] analyzed a large number of micro-blogs from the perspective of micro-blog credibility, based on some topics that are easily affected by the water army. Finally, the characteristics of user behaviors such as Posting and forwarding are extracted, and the data processing algorithm of decision tree is used to automatically identify the credibility of microblog content.

On the other hand, Chen et al. [6] designed a semi-local centrality measurement method to balance the measurement method of low correlation centrality and high time consumption. By simulating SIR model and starting experiments in four kinds of real complex social networks to verify the effectiveness of its method for identifying "opinion leaders". Shah et al. [7] proposed a new topological quantity "rumor centrality" and used SIR's variant propagation model to simulate an evaluation model of rumor source. Zhang et al. [9] took the URL in Twitter content as the breakthrough point, analyzed accounts with similar urls, found suspected water army activities and information by studying the characteristics of micro-blogs with URL and their publishers, analyzed their intentions, and verified the validity of this model in judging true and false information by means of machine learning. Chen et al. [3] defined feature vectors and quantified propagation behaviors according to the interaction behaviors between information transmission subjects, and used the method of decision tree to detect the information transmitted by water army.

In addition, Irani et al. [10] studied a large number of social network accounts and established a large case base of static user profile content analysis. Through the comparison of several machine learning algorithms, the most suitable decision tree algorithm is found to distinguish water army users. Morris et al. [12] were keen to find that the relevant characteristics of Weibo account credit evaluation thought by users were not completely consistent with those published by the actual platform. Experiments were conducted by artificially controlling certain micro-blog features to evaluate the impact of each feature on credit rating. Through experiments, it is found that users do not judge the credibility of the account completely based on the authenticity of the content, but are influenced by heuristic such as user name. Thus, a series of characteristic factors that influence users' judgment of information authenticity are found.

3. The micro-blog water army identification model based on the danger theory

3.1. Identification framework of naval forces based on danger theory

Uwe have done the thorough research in Dendritic cells, they introduced the various status of Dendritic cells, in the literature [14] when it comes to Dendritic cells are able to control the immune response of antigen presenting cells. Immature DCs evolves into semi-mature DCs or mature DCs according to different signals collected, and then carries antigens into lymph nodes to provide cooperative stimulation signals and tolerance factors or stimulation factors.Finally, the proportion of mature antigens in total antigens determines whether there is a risk. Compared with traditional artificial immune algorithm, DCA algorithm does not depend on the self collection, for those not to hurt the body of autoantibody will not be processed, and its calculation process is linear, make more lightweight intrusion detection system, and does not require knowledge of the training process and normal and abnormal is in static analysis, so compared with other supervision and learning algorithm, its less time cost, greatly reduced the matching and the scale of the response, to a certain extent, improve the efficiency of the algorithm.

DCA algorithm is more concerned with finding more differentiated "signals",therefore, compared with other supervised learning algorithms, it takes less time and greatly reduces the size of matching and answering.Similarly, in the detection of abnormal accounts in social networks, more attention should be paid to the abnormal behaviors of accounts. By defining the signals of the detection model, the weight matrix of each recognition signal was obtained through evolutionary calculation. Finally, the signals were fused by DCA algorithm, and a probabilistic model of social network water army detection based on immune risk theory was obtained. Based on this, this paper defines four identification signals, two of which are safety signals (D_{event}, T_{guide}) and two of which are danger signals (R'', $R_{tightness}$).

3.2. Definition of danger signal and safety signal

1)   event participation signal D_{event}

Due to the nature of social network mercenaries themselves, their existence depends on their intention to create false public opinion direction, mislead public opinion and ultimately achieve the political or economic purposes behind a specific event. Therefore, we believe that all social network water army behaviors are based on specific events. Water army users usually register for a short time, but their participation in a public opinion event is very high.We believe that when users participate in more public opinion events on Weibo and focus more on a few topics or topics within a unit of time, the suspected behavior of this user will increase correspondingly. T_{reg} is the registration time of the user, $Event_{sum}$ is the total number of events the user participated in, and $Event$ is the number of events the user participated in.So the definition:

$$D_{event} = \frac{Event_{sum}/Event}{T_{reg}} \ , \ D_{event} \in [0, +\infty) \ ; \qquad (1)$$

2)   second-order correlation signal R''

In the microblog network, due to its social attribute, the circle of friends of normal users is often close, and the users they pay attention to, such as those in the circle of relatives, friends, classmates and colleagues, tend to pay attention to each other. Therefore, a normal user will have more second-order correlations, that is, there will be more interrelations between the relevant users. The undirected graph G=(V,E) is used to represent all the second-order concerns of the current user node n. Assuming that the degree of n is k, then the adjacency matrix of G is $A = (a_{ij})_{k \times k}$, thus, defined as:

$$R'' = -\frac{\sum_{i,j=1}^{k} a_i a_j}{k} \ , \ R'' \in \left[-\frac{k(k-1)}{2k}, 0 \right] \ ; \qquad (2)$$

3)   relationship tightness signal $R_{tightness}$

In order to avoid simple garbage user filtering and influence public opinion, water army users often focus on normal users and buy zombie powder in large quantities to create the illusion of a normal account. However, it is not difficult to find that the normal users who are concerned about the water army users seldom pay attention to the water army users. This leads to an obvious characteristic of water army users, that is, the degree of close relationship is low, that is, the proportion of mutual fans of water army users is significantly lower than that of normal users.

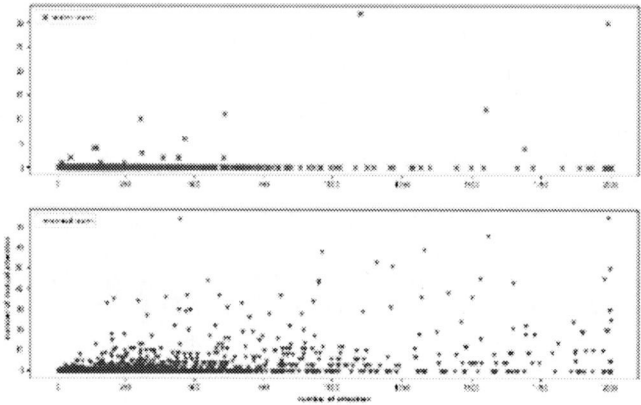

fig 1. Effect Contrast1

As shown in the figure above, according to the number of mutual attention between water users and normal users, it can be clearly shown that the relationship between water users is closer than that between normal users. *Following* represents the number of followers of the current user, and $Follower_{mutual}$ represents the number of users who are fans of the current user.So the definition:

$$R_{tightness} = -\frac{Follower_{mutual}}{Following} \ , \ R_{tightness} \in [-1,0] \ ; \tag{3}$$

4)   guide tool utilization signal T_{guide}

Sina Weibo provides two kinds of "guiding tools", namely the topic symbol "#" and the external link URL, to achieve the purpose of categorizing users' microblog content and strengthening third-party contact. In general, normal users do not use these two bootstrap tools on a large scale due to their own social diversity. In order to achieve the purpose of topic gathering and heat raising to influence public opinion, the water army often USES these two guidance tools on a large scale.

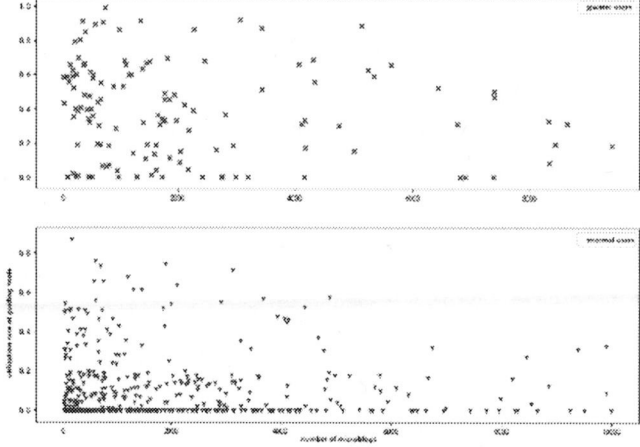

fig 2. Effect Contrast2

It can also be seen from the figure above that ordinary users are relatively concentrated in areas with low utilization rate of guiding tools, while water troops are more likely to appear in areas with more use of guiding tools. $Hash_{sum}$ represents the total number of microblogs that use "#", URL_{sum} represents the total number of microblogs that use url, and $Weibo_{sum}$ represents the total number of microblogs of users.So we define:

$$T_{guide} = \frac{Hash_{sum} + URL_{sum}}{2 \cdot Weibo_{sum}} \ , \ T_{guide} \in [0,1] \ ; \tag{4}$$

3.3. The optimal weight matrix of recognition signal is obtained by evolutionary computation
In the evolutionary computing model, a 4-row and 3-column matrix $Weight$ is established to represent a weight matrix of 4 recognition signals. Because this model USES two kinds of signals, safety and danger, and is distinguished by positive and negative values, the first column in the three columns is the minimum value of weight 0, and the third column can guarantee the maximum value within the accuracy range of 1%, that is, the difference between the maximum value and the minimum value of the absolute value of all signals in the data set is 100 times. The second column is the optimal weight.

Initialize 10 rows and 4 columns of random population, each row represents an individual, and each column represents the weight of each recognition signal of the individual. Get the weight of each row by cross calculation get the current weight matrix var.

$$X_i' = aX_i + (1-a)X_j \tag{5}$$

$$X_j' = aX_j + (1-a)X_i \tag{6}$$

var was substituted into the DCA microblog water army recognition algorithm for calculation and judgment, and each individual was sorted according to the F1 value, and the corresponding length was assigned on the interval (0,1). Then, the individuals who can enter the next generation are selected by random matrix, and the individuals with high F1 value are correspondingly given a greater opportunity to pass on the genes to the next generation. The process of mutation is simulated as the exchange of arbitrary lengths of two random rows in population matrix $Population$, so as to obtain a new population matrix and conduct the next iteration. Keep updating the optimized recognition signal weight matrix and the highest F1 value.After several iterations, the weight matrix $Weight$ of the recognition signal and its corresponding F1 value stabilize in a small interval, each weight value corresponding to $Weight$ is the optimal weight.

3.4. Micro-blog water army recognition based on DCA algorithm
The greatest advantage of DCA algorithm lies in its linear calculation process. Based on the fusion calculation of danger signal and safety signal with high information gain, the scale of matching and response can be greatly reduced. Through continuous iteration of evolutionary calculation, the optimal recognition signal weight matrix is obtained. Finally, the antigen maturity is calculated，the identification results of Weibo water army were obtained.

$$MCAV = \sum_{i=1}^{n} Weight_{i,2} \cdot Signal_i + b \tag{7}$$

4. Experiment and analysis
The experimental operating environment was Windows 10 operating system, 3.2Ghz four-core processor and 8GB memory. The experimental software is Python 3.6 and MySQL 5.5.

4.1. data preprocessing
Through the crawler program, more than 2 million pieces of microblog information and user information were crawled for more than a dozen hot public opinion events. By pruning and data

processing, such as the user will receive the fan number, the number of attention, the attention to each other, each other on proportion, the registration time, friends forwarded when proportion of fans, each content repetition rate, time occurrences, # usage, url utilization rate, forwarding microblogging proportion, forward than not empty, and data is stored in relational database. According to the above definition of danger signal and safety signal, four signal values of each user are calculated and stored in the relational database.In all the existing data sets, 5,000 microblog accounts were selected for repeated manual tagging by multiple people, and 4,677 agreed ones were retained as the training set of evolutionary calculation.

4.2. evolutionary calculation of recognition signal weight matrix

By analyzing the existing data, the maximum and minimum values of each recognition signal in the data set are obtained. Thus, the weight matrix of the recognition signal is obtained.Since the security signal and danger signal are set in this paper, 0 is taken as the threshold value to judge the microblog water army. The weight matrix selected, crossed and varied in the iterative process of evolutionary calculation is continuously substituted into the DCA microblog water army recognition algorithm, and finally the optimal weight matrix of the convergent identification signal is obtained.

4.3. effectiveness of identification signals

In order to ensure that the recognition signals required by DCA algorithm are as efficient as possible on the premise of ensuring effectiveness and reduce the amount of evolutionary computation, we also conducted a set of comparative tests to verify the results. One group only USES the four recognition signals (D_{event}、 T_{guide}、 R''、 $R_{tightness}$) proposed in this paper, while the other group, in addition to the four recognition signals, adds the general features commonly used by the water army recognition model such as literature (number of fans, number of followers, number of tweets, proportion of non-empty forwarding, and whether or not the signature is included). This paper also USES the DCA and evolutionary computing fusion of the military recognition model for recognition. The following results are obtained:

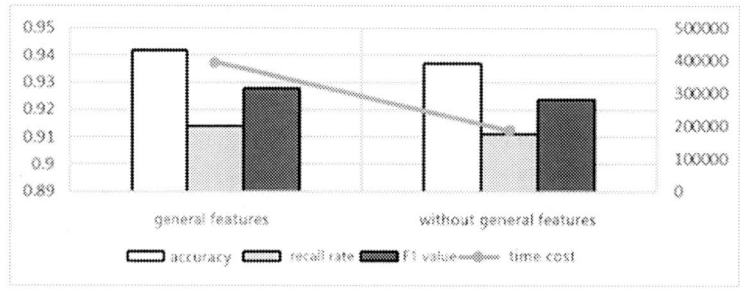

fig 3. Effect Contrast3

The results show that in the control group with conventional features, the recognition effect is less than 1% higher than that of the group using only the four recognition signals proposed in this paper, while the time performance is significantly lower than that of the control group without conventional features. This proves that the four recognition signals proposed in this paper have higher recognition efficiency.

5. Result evaluation

In view of the current new characteristics of naval forces' behaviors, the results of the determination of naval forces' behaviors using this model are compared with the results of the previous traditional model method (FO+FR+NFR+UR model). Precision, Recall and F1 value were selected as the reference indexes for model detection level.

Definition: YY is the water army classified correctly by their respective models, NY is the normal user wrongly judged as the water army, and YN is the water army wrongly judged as the normal user. So:

$$Prec. = \frac{YY}{YY+NY} \qquad (8)$$

$$Recall = \frac{YY}{YY+YN} \qquad (9)$$

$$F1 = \frac{2 \cdot Prec. \cdot Recall}{Prec. + Recall} \qquad (10)$$

The traditional model selects the "fan value (FO), friend value (FR), non-marketing activity participation (NFR), URL usage (UR)" model, and finally obtains its accuracy, recall rate, F1 value and the evaluation index of this model for comparison as shown in the figure.

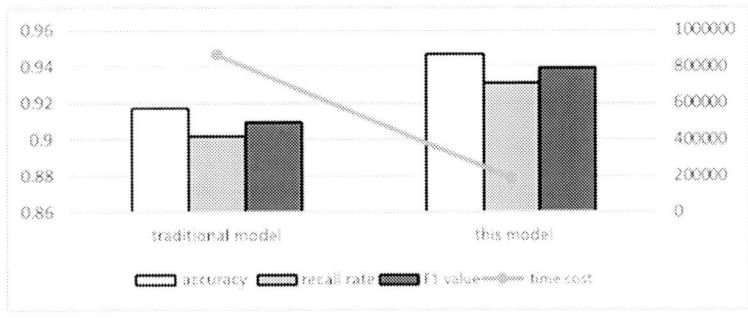

fig 4. Effect Contrast4

Through the 10-fold verification of the feature vectors selected by the traditional model, the average results are obtained. Then, the social network water army detection model defined in this paper based on the immune risk theory is used. Through the method discussed in this paper, the optimal weight matrix is obtained by evolutionary calculation, and the microblog water army is judged by DCA algorithm.

The results show that the accuracy recall rate and F1 value of this model are higher than the traditional model to some extent, both exceeding 90%, and the performance of data processing time is significantly higher than that of the traditional model.

Receiver Operating Characteristic (ROC) is also one of the commonly used model evaluation criteria. Compared with the accuracy, recall rate and F value and other indicators, Receiver Operating Characteristic curve is more inclined to care about the score between positive and negative samples rather than the specific score value, which also verifies the performance of classifier from another aspect. Through the test and comparison of Area under ROC curve, the AUC value of 0.941 in this model is also higher than that of 0.8783 in the traditional model.

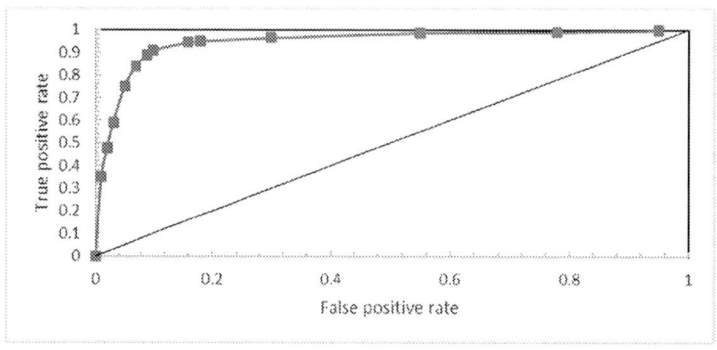

fig 5. Experiment Result

6. Conclusion

Social networks such as Weibo are already having a profound impact on every aspect of people's work and life, and water army is taking this opportunity to make use of social networks to influence public opinion judgment. In recent years, the identification of the water army has attracted the attention of scholars. However, the complexity of social networks and the constant updating of covert means of water army bring some difficulties to the identification and judgment of water army. Through the combination of computer recognition and artificial immunity, the in-depth analysis of the characteristics of the water army, the use of evolutionary computing to get the weight of each characteristic signal, and the classical algorithm of artificial immunity DCA combined, a social network water army identification model with high reliability and better performance is obtained.

References

[1]   Zhongming Han, Minfeng Xu, Dagao Duan. A model of water army identification based on micro-blog probability map [J]. Computer research and development, 2013, 50(s2):180-186.

[2]   Xiaotao Cheng, Caixia Liu, Shuxin Liu. Methods of microbolo discovery method based on the characteristics of the relationship graph [J]. Journal of automation, 2015, 41(9):1533-1541.

[3]   Kan Chen, Liang Chen, Peidong Zhu，et al. An online social network water army detection method based on interactive behavior[J]. Journal of communications, 2015, 36(7):120-128.

[4]   Yanmei Zhang, Yingying Huang, Shijie Gan, et al. Research on the identification algorithm of micro-blog network water army based on bayesian model[J]. Journal of communications, 2017, 38(1):44-53.

[5]   Xuping Yuan, Renwu Wang, Boyin Zhai. Automatic identification of micro-blog water army based on comprehensive index and entropy method [J]. Journal of intelligence, 2014(7):176-179.

[6]   Chen D, Lü L, Shang M S, et al. Identifying influential nodes in complex networks[J]. Physica A Statistical Mechanics & Its Applications, 2012, 391(4):1777-1787.

[7]   Shah D, Zaman T. Rumors in a Network: Who's the Culprit?[J]. IEEE Transactions on Information Theory, 2011, 57(8):5163-5181.

[8]   Yang F, Liu Y, Yu X, et al. Automatic detection of rumor on Sina Weibo[C]// ACM, 2012:1-7.

[9]   Zhang X, Zhu S, Liang W. Detecting Spam and Promoting Campaigns in the Twitter Social Network[C]// IEEE, International Conference on Data Mining. IEEE, 2013:1194-1199.

[10] Irani D, Webb S, Pu C. Study of Static Classification of Social Spam Profiles in MySpace[C]// International Conference on Weblogs and Social Media, Icwsm 2010, Washington, Dc, Usa, May. DBLP, 2013:591-597.

[11] Thomas K, Grier C, Song D, et al. Suspended accounts in retrospect:an analysis of twitter spam[C]// ACM, 2011:243-258.

[12] Morris, Ringel M, Scott, et al. Tweeting is believing?: understanding microblog credibility perceptions[C]// ACM 2012 Conference on Computer Supported Cooperative Work. ACM, 2012:441-450.

[13] Castillo C, Mendoza M, Poblete B. Information credibility on twitter[C]// International Conference on World Wide Web, WWW 2011, Hyderabad, India, March 28 - April. DBLP, 2011:675-684.

[14] Greensmith J,Aickelin U,Tedesco G.Information fusion for anomaly detection with the dendritic cell algorithm[J].Information Fusion,2010,11(1):21–34.

Design and Implementation of Ontology Knowledge Base of Endemic Genera of Seed Plants in Yunnan Province

Lin Peng, Xu Li, lin-nan Yang

(College of Big Data, Yunnan Agricultural University, Kunming 650201)

Correspondence to: lin-nan Yang (lny5400@sina.com)

Abstract: There is complex terrain, various soil and climate in Yunnan Province. The unique natural condition produces the rich resources of the species. The spermatophyte endemic genus resource is the most quantity of China. It is essential to protect and use Yunnan spermatophyte endemic genus resource reasonably. According to Yunnan spermatophyte endemic genus resource various data formats, deficient integration, low-efficiency intelligent retrieval, lack of individualized information service mode and more problems, This paper builds Yunnan spermatophyte endemic genus ontology knowledge base. This ontology knowledge base is the solution of Yunnan spermatophyte endemic genus resource using technical barrier, it laid the foundation to build Yunnan spermatophyte endemic genus area terminology information retrieval system.

1. Introduction

The unique geographical location and climatic condition of Yunnan province, which enriched species resources. It is an important gene pool and is known as "plant kingdom". According to the statistic, Yunnan has more than 17 thousand advanced plants, accounting for 62.9% in China. Among the 10 thousand kinds of seed plant, there are 151 plants rare and endangered which are listed as protected by China, accounting for 42.6% of the total. Among them, there are more than 190 species belonging to 130 genera of seed plants, accounting for 48.3% of the unique genus(269 genera) of seed plants in China, which is the region with the highest abundance of seed plants in China[1,2].

A lot of research has shown that the distribution of endemic seed plants have very close connection with geography, climate and surroundings, etc[3-8]. However, at present, the distribution patterns and physiological characteristics of endemic seed plants in Yunnan as well as related geographical, climatic and wild habitats have diverse access channels, scattered storage, and inconsistent data standards, unsystematic, poor sharing, and search. These problems greatly restricted the large-scale and systematic research on the unique genera of seed plans in Yunnan.

Ontology knowledge base is knowledge cluster that uses ontology to describe and organize domain knowledge. It can describe concepts of certain field and even a wider range; it can organize, connect, reason and reuse the related knowledge of the certain field. And, it can able to reveal the intrinsic relationship between this knowledge. Also, it is a new method to recognize the digital representation of domain knowledge. Owing to the characteristic of the special species and complex relationship of seed plants, combined with information technology, artificial intelligence and GIS technology, the author can build ontology knowledge base of special species of seed plants in Yunnan, which can realize the complex knowledge structure of digitalize-representation of special species in Yunnan. This paper provided a new way of study and a new theoretical basis for special species in Yunnan.

Content from this work may be used under the terms of the Creative Commons Attribution 3.0 licence. Any further distribution of this work must maintain attribution to the author(s) and the title of the work, journal citation and DOI.

Published under licence by IOP Publishing Ltd

This paper introduces spatial data fusion technology and multi-level ontology integration technology based on geographic ontology to construct a unique genre ontology library of Yunnan seed plants. The ontology knowledge base to Yunnan seed plants is divided into two layers: the upper layer is a general ontology layer. Setting the geographic ontology as the core, the climate ontology, time ontology, geographical ontology, and ecological ontology such as water quality and soil are established respectively. The lower layer is the main body layer of the unique genus of Yunnan seed plants. The domain ontology layer on the base of domain terminology is composed of domain core concepts and relation sets and core information ontology resource base.

2. Endemic genera of seed plants in Yunnan

The basic units of plant classification are the boundaries, gates, classes, orders, families, genera and species. Plants are divided into two classes, one is "lower plants", the other is "higher plants". The "higher plants" are divided into four phyla: angiosperm, gymnosperm, fern and bryophyte. Seed plants refer to the genera of angiosperm and gymnosperm. Endemic genera of seed plants are the greatly important part of flora, accounting for 8.9% of endemic genera of seed plants in China. Referring to a large number of references, the endemic genera of seed plants in Yunnan Province has 59 families, including 125 genera and 246 species of seed plants. As can be seen from the items listed in Table 3-1,the families of endemic genera are listed. From the Table 3-1,Composiae is the most endemic family with 11 genera, the second is Gesnerianceae and umbelliferae contains 10 genera, the third is Gramineae and Labiatae contains 8 genera.

Table1 Major families and genera endemic to China

families	numbers	generic name	numbers
Compositae	11	Fargesia	41
Gesneriaceae	10	Bambusa	26
Umbelliferae	10	Sinocarum	9
Labiatae	8	Ancylostemon	6
Gramineae	8	Tremacron	4
Cruciferae	5	Tibetia	4
Ranunculaceae	5	Pterygiella	4
Magnoliaceae	4	Arcuatopterus	3
Orchidacea	4	Nannoglottis	3
Taxodiacea	3	Ypsilandra	3

3. Domain ontology construction of the endemic genera of seed plants in Yunnan Province

The methods for constructing ontology in different domains are also different, and here is no standard ontology construction method. But, as far as large domain ontology concerned, because of the large amount of data,the simpler conceptual relationship structure is more conducive to the storage of data, such as the gene "GO"project ontology. Therefore, based on the characteristic of the unique genus field of Yunnan seed plants, this paper will use the seven-step method combined with the core concepts and relationship components in the domain ontology to construct the ontology. This not only ensures that the unique ontology structure is simple and practical, but also facilitates the future expansion of the domain ontology. The construction ideas and processes of the endemic genera domain of seed plants in

Yunnan are shown in Fig.1,and the specific construction process and actual examples are shown in Fig.2 to Fig.4 and Table 2.

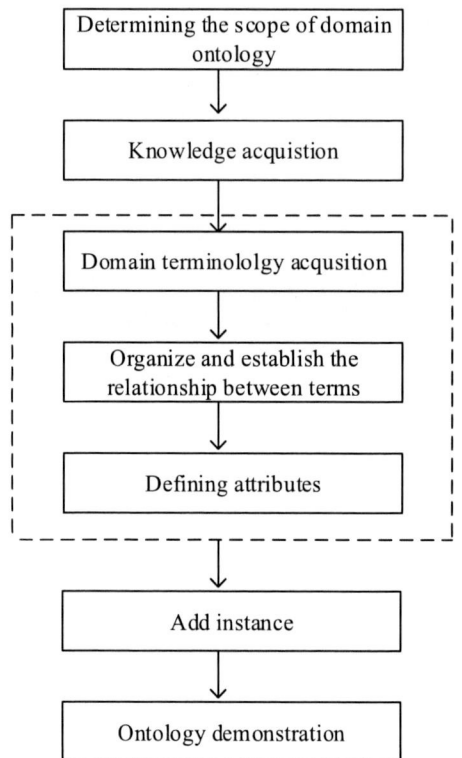

Fig.1 The flow diagram of ontology construction

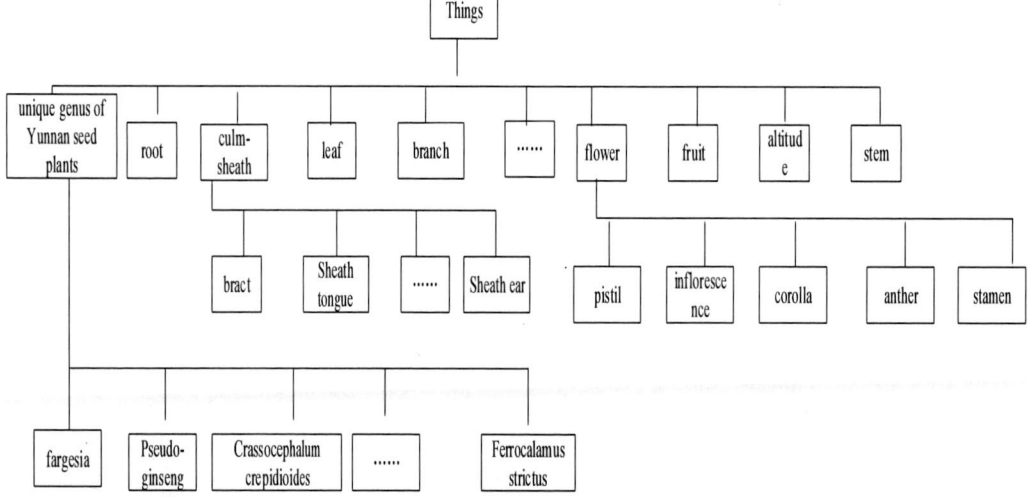

Fig.2 part of ontology and hierarchy of endemic genera of seed plants in Yunnan Province

ISPECE   IOP Publishing

IOP Conf. Series: Journal of Physics: Conf. Series **1187** (2019) 052099   doi:10.1088/1742-6596/1187/5/052099

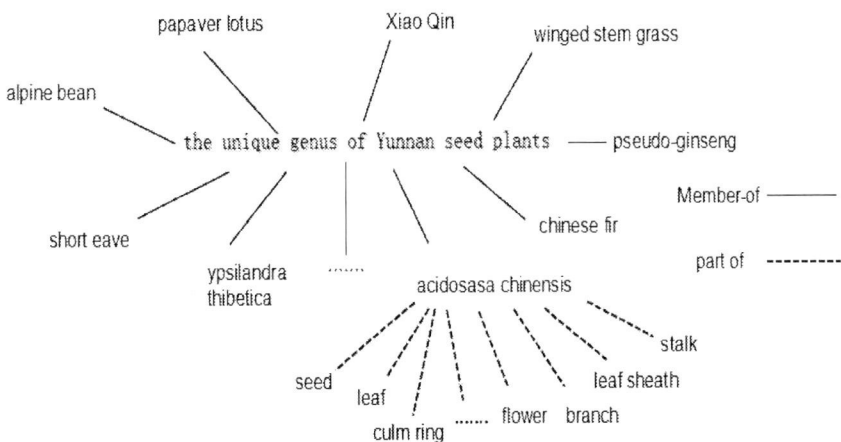

Fig.3 The example diagram of ontology relationship of endemic genera of seed plants in Yunnan Province

Table 2 The data type properties of Yunnan endemic genera areas of seed plants

Type	Attribute relation	Relationship description	Domain of definition	range
Data attribute	shapes	Display form	Sheaths,etc	string
	numbels	Description quantity	Stamens,etc	string
	colors	Display color	Anthers,etc	string
	types	Description type	Culm sheath,etc	string
	longs	Length description	Culms,etc	string

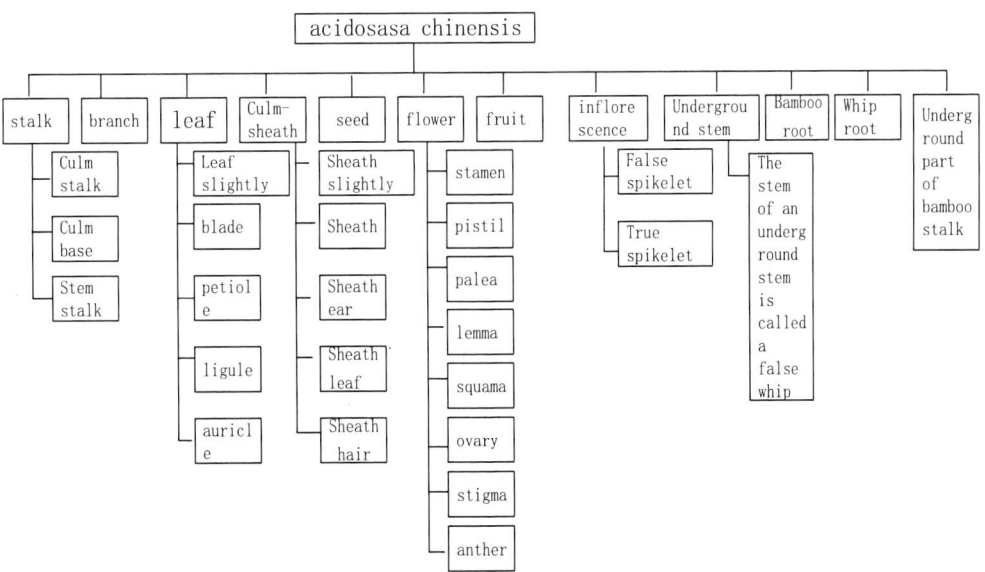

Fig.4 Part of relationship schematic diagram of the domain ontology of endemic genera of seed plants in Yunnan

2704

4 Conclusion

In view of the problems of many fields, different data formats, poor data integration, complex concepts and relationships among concepts involved in the information resources of endemic genera of seed plants in Yunnan, this paper constructs a multi-layer ontology knowledge base of endemic genera of seed plants in Yunnan based on the similarities and differences of different concepts in different fields of the information resources of endemic genera of seed plants in Yunnan. Through conceptualization and modeling of knowledge in various fields, ontology mapping technology is used to realize mapping and transformation from heterogeneous data sources to global patterns, thus realizing the integration and sharing of all kinds of digital resources unique to seed plants in Yunnan.

Acknowledgments

The findings and the opinions are partially supported by projects of the major science and technology projects of Yunnan province, No.2018ZI001.This work is supported by Engineering and Technology Research Center for Yunnan Agricultural Big Data and Yunnan Agricultural University, China.

References

[1] Feng Jianmeng, Zhu Youyong. On the genera of seed plants endemic to China in Yunnan[J]. Ecology and Environmnet, 2010,19(3): 621-62.

[2] Wang Hesheng, Zhang Yili. THE BIO-DIVERSITY AND CHARACTERS OF SPERMA-TOPHYTIC GENERA ENDEMIC TO CHINA[J]. Acta Botanica Yunnanica, 1994,16(3):209-220.

[3] ZHU H,MA Y, YAN L,et al. The relationship between geography and climate in the generic-level patterns of Chinese seed plants[J].Acta Phytotaxonomica Sinica, 2007, 45(2):134-166.

[4] Ying-Xiong Qiu, Cheng-Xing Fu, Hans Peter Comes. Plant molecular phylogeography in China and adjacent regions: Tracing the genetic imprints of Quaternary climate and environmental change in the world's most diverse temperate flora[J]. Molecular Phylogenetics and Evolution, 2011,7: No.of Pages 21.

[5] Ma Wenhong, Yang Yuanhe, He Jinsheng. Biomass of temperate grassland in Inner Mongolia and its relationship with environmental factors[J],Science in China(Series C) 2008,38(1):84-92.

[6] Feng Jianmeng, Xv Dongcheng. The large-scale distribution pattern of seed plant species richness in China and its relationship with geographical factors [J]. Ecology and Environmental Science 2009,18(1):249-254.

[7] Song Qing. Application research of intelligent retrieval technology based on domain ontology[D]. Chinese Academy of Agricultural Sciences, 2011.

[8] Liu Wei. Spatial data service discovery and integration based on geographical ontology[D]. Doctoral dissertation of China university of mining and technology, 2010.

[9] Yang Jie. Ontology based knowledge modeling and reasoning of citrus pests and diseases[D]. Central China Normal University.2014.

ISPECE

IOP Publishing

IOP Conf. Series: Journal of Physics: Conf. Series **1187** (2019) 052100   doi:10.1088/1742-6596/1187/5/052100

The Applications of the Edge Detection on Medical Diagnosis of Lungs

Wei Xu*[1,2] Jinping Li[1,2] and Hongwei Jia[1]

[1] East China University of Technology, Economic Development Zone Guanglan Avenue 418, Nanchang330013, China

[2] Jiangxi Engineering Technology Research Center of Nuclear Geoscience Data Science and System, Economic Development Zone Guanglan Avenue 418, Nanchang330013, China

E-mai:jhw_1979@163.com   E-mail: 702140087@qq.com   E-mail: 22661246@qq.com

Abstract. With the development of science and technology,medical images have become important assistant means of diagnosis and therapy. However there are many inevitable defects in biomedical images. In order to improve the readability of images and make the doctor to adopt more effective observation and diagnoses way to anatomy structure and pathological part of the patient, it's necessary to study the medical images processing. Edge detection is an important part of the processing. And the purpose of the edge detection is to enforce the effect of the edge and contour of images, and act on enlarge the contrast of images' grayscale. Firstly, the edge detection algorithm and detector are summarized and analyzed. Then the paper introduces the method of edge detection in vc++ MFC. Finally, this paper describes the realization process of edge detection in medical images. The results show that edge detection can enforce the readability of medical image that can help doctors make better diagnose to patient.

1. Introduction
With the continuous development of modern medicine, health has become an artistic conception that everyone pays close attention to and pursues assiduously. However, the deteriorating environment, heavy pressure from all sides and fierce competition make people's health become the goal of concern to all mankind. Therefore, the development of medicine is facing serious challenges. Biomedical images play an important role in diagnosis and treatment for doctors[1,2].

2.   Edge detection algorithm
Edge is the basic feature of image. Edge detection is the basis of image processing and computer vision technology, it is also the key technology of medical image processing, the purpose is to determine the boundary of the target in the image with noise background. The quality of edge detection will directly affect the follow-up treatment process. Early classical algorithms include edge operator method (Roberts operator, Sobel operator, Prewitt operator, Kirsch operator, Laplacian operator, LOG operator, etc.), surface fitting method, template matching method, threshold method, etc. In recent years, with the development of mathematical theory and artificial intelligence, many new edge detection methods have emerged, such as wavelet transform and wavelet packet edge detection

Content from this work may be used under the terms of the Creative Commons Attribution 3.0 licence. Any further distribution of this work must maintain attribution to the author(s) and the title of the work, journal citation and DOI.
Published under licence by IOP Publishing Ltd

methods. In this paper, based on the theoretical knowledge of classical edge detection operator, edge detection of medical images is realized in Visual C++ to avoid unclear, deviation, misjudgment and misdiagnosis of X-ray images[3]. It provides scientific reference basis for rapid and accurate extraction of image edge information and more effective observation and diagnosis of disease.

3.Application of edge detection algorithm

Edge detection has always been highly valued in the field of digital image. The main research directions of edge detection are algorithm and application:

(1) Constantly propose new algorithms. Because people have mastered the traditional edge detection method is very mature, on the other hand, with the continuous development of science and technology, people have more and more stringent requirements on the detection results, the traditional edge detection method has not been able to meet the performance index running speed and other requirements. So a variety of new algorithms have been developed. These new algorithms can be roughly divided into the following two categories: one is the detection technology combined with specific theoretical tools, such as the detection technology based on mathematical morphology, the detection technology based on statistical methods, the detection technology using neural networks, the detection technology based on fuzzy theory, the detection technology based on wavelet analysis and transformation, and the use of them. Information theory detection technology, genetic algorithm detection technology, etc. Another kind of edge detection method is proposed for special images, such as expanding two-dimensional spatial operator to three-dimensional operator to detect the edge of three-dimensional images, detecting the edge of color images, detecting the edge of synthetic aperture radar images, and detecting the edge of moving images to achieve segmentation of moving images[4-7].

(2) Apply the existing operators in practice. Edge detection can be applied to the symptom detection of medical X-ray. By means of edge detection, we can more clearly distinguish the irregular boundaries, unclear boundaries and uneven density. But in the case of lung disease, these "spots" are often seen as tuberculosis, fungal granulomas, or malignant tumors. All in all, in the rapid development of science and technology today, edge detection technology will have a very good prospect.

4. Edge detection algorithm comparison

4.1 Roberts algorithm

Roberts edge detection operator is a kind of edge detection operator using local difference operator. If an image is viewed as a matrix composed of pixels, the algorithm is actually the result of the following two matrix templates. $\begin{bmatrix} 1. & 0 \\ 0 & -1 \end{bmatrix}\begin{bmatrix} 0. & 1 \\ -1 & 0 \end{bmatrix}$ (". "is the position of the current pixel point)When we get the pixel value of $|G(x,y)|$, let's compare it with a pre-set threshold A, when $|G(x,y)| > A$, we think that the gradient of this point is larger than we expected, that is to say, the gray level of this point changes very dramatically, and the point where the brightness of the image changes strongly is the edge point we are looking for. For the whole image, we just need to scan each pixel point one by one and calculate the comparison to find the edge of the image[8].

4.2 Sobel algorithm

For each pixel of digital image{f(i,j)}, the gray weight difference of its upper, lower, left and right adjacent points is investigated, and the adjacent points close to it have a large weight.Sobel operator is defined as follows:

$$s(i,j) \underline{\underline{\Delta}} \mid \Delta_x f \mid + \mid \Delta_y f \mid \underline{\underline{\Delta}} \tag{1}$$

$$\text{Convolution operator: } \Delta_x f \begin{bmatrix} -1 & 0 & 1 \\ -2 & 0 & 2 \\ -1 & 0 & 1 \end{bmatrix} , \qquad \Delta_y f \begin{bmatrix} -1 & -2 & -1 \\ 0 & 0 & 0 \\ 1 & 2 & 1 \end{bmatrix}$$

Take the TH threshold appropriately and make the following judgment: $s(i,j) > $TH, (i, j) is the step edge point, $\{s(i,j)\}$ is the edge image[9]。

Sobel operator is easy to be implemented in space. Sobel edge detector not only produces better edge detection effect, but also is less affected by noise. Noise resistance is better when large areas are used, but doing so increases the computation effort and results in thicker edges.

4.3 Prewitt algorithm

The core idea of the Prewitt algorithm is still Roberts algorithm, but improvements have been made in the construction of two matrix templates. We set the original image as M, and the Prewitt algorithm USES two directional templates (vertical and horizontal). $P_V = \begin{bmatrix} -1 & -1 & -1 \\ 0 & 0 & 0 \\ 1 & 1 & 1 \end{bmatrix}$ and

$P_H = \begin{bmatrix} -1 & 0 & 1 \\ -1 & 0 & 1 \\ -1 & 0 & 1 \end{bmatrix}$, First, calculate the image to get M1 and M2, and then take the gradient value of

the calculated result. The specific process is shown as follows:

$$|G(x,y)| = \sqrt{(M \otimes P_V)^2 + (M \otimes P_H)^2} \tag{2}$$

And then we compare it to the boundary value A, and we get the edge point that we want[10-12].

4.4 Laplacian algorithm

Laplacian algorithm was also proposed by Prewitt, and it is one of the few edge detection algorithms with isotropy. The idea is to give a 3X3 template, take the center of the template as the pixel point to calculate the image, and then check whether the calculated result is different from the surrounding pixel point. Let's take a quick look. For 4 adjacent points $\begin{bmatrix} & i,j-1 & \\ i-1,j & i,j & i+1,j \\ & i,j+1 & \end{bmatrix}$, Let's set $g(i,j)$ as

the superposition of the difference between the current point gray value and the pixels around it, then

$$g(i,j) = [f(i,j) - f(i-1,j)] + [f(i,j) - f(i+1,j)] + [f(i,j) - f(i,j-1)] + [f(i,j) - f(i,j+1)] \tag{3}$$

If $g(i,j)$ than A, that's what we're looking for. The same is true for an algorithm for 8 adjacent points. In this paper, 8 neighborhood point algorithm is used[13].

5. Experimental results and analysis

5.1 Edge detection results of Roberts algorithm

The lower left is an X-ray of a patient's lungs. This is a frontal view. Two black areas represent the lungs. From the original picture, it is difficult for us to see the source of the patient's disease with the naked eye. Roberts algorithm can be used to detect the edge image. The result is shown in the figure on the right:

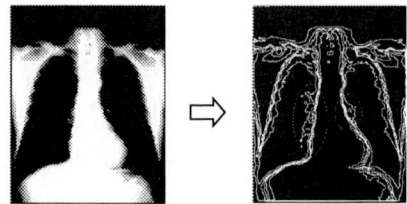

(a)The original image   (b)Roberts algorithm

Figure 1 edge detection results of Roberts algorithm

According to the results of Roberts algorithm's edge detection image, we can clearly see that there are some irregular lumps in the middle part of the chest of the patient in the right picture. After searching relevant medical data, the diseases that cause this type of X-ray are tuberculosis, mycotic granuloma or benign and malignant tumors.

5.2 Edge detection results of Prewitt algorithm

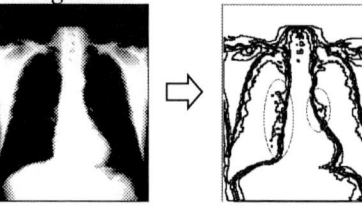

(a)The original image   (b)Prewitt algorithm

Figure 2 Edge detection results of Prewitt algorithm

From the figure, we can see that the Prewitt operator can also detect the outline of the image. Every detail in the image can be detected one by one. It has good continuity and location. Compared with the results of Roberts algorithm, we can see that Prewitt has stronger anti-noise ability.

5.3 Edge detection results of Laplacian algorithm

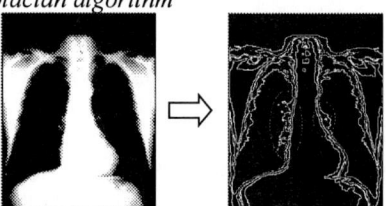

(a)The original image   (b)Laplacian algorithm

Figure 3 Edge detection results of Laplacian algorithm

Based on the above detection results, we can see that the detail expression of Laplacian algorithm's edge detection image is not as good as Roberts and Prewitt algorithm's edge detection, but it can still fully show the characteristics of the image edge.

5.4 Edge detection results of Sobel algorithm

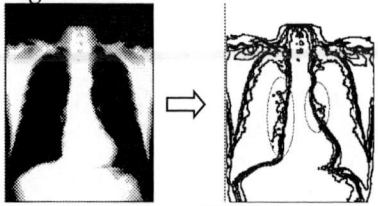

(a)The original image   (b) Sobel algorithm

Figure 4 Edge detection results of Sobel algorithm

It can be found that Sobel algorithm can detect the tumor contour, but the accuracy is not as weak as Laplacian algorithm and Prewitt algorithm, but the anti-noise ability is better than Roberts algorithm.

6. Experimental conclusions

Through the edge detection of lung cancer and lung tumor images above, we find that Roberts algorithm and Laplacian algorithm have high accuracy, but poor anti-noise ability, among which Laplacian algorithm can detect many subtle changes with thin edges, but produces some false edges. Roberts detected a slight discontinuity in the edges;However, Prewitt algorithm and Sobel algorithm can solve the problem of noise, but the accuracy is not as good as Roberts algorithm and Laplacian algorithm. The edges are not sharp enough and a little fuzzy. Canny algorithm is also used to judge edge points based on the first derivative. It has better anti-noise ability, but it is also easy to smooth out part of the edge information.

Acknowledgements

This work was supported by the Science and Technology Project of Jiangxi Province Education Department（GJJ160590,GJJ170452）

Authors

< **Wei XU** >, <4 -13-1983 >,<Jiang Xi >
Current position, grades: Senior Lecturer, master graduate
University studies:School of Software, East China University of Technology
Scientific interest: mainly engaged in database technology,information management technology and algorithm analysis
Publications <number or main>: over 6 papers
Experience: Present   Senior Lecturer, East China University of Technology
2007-2010   East China Institute of Technology ; a master's degree; Advisor: Prof.Yueshun He
1999-2003   East China Institute of Technology ; a bachelor's degree

< **Jinping LI** >, < 12-13-1972 >,< Hei Longjiang >
Current position, grades: Associate professor, master graduate
University studies: School of Software, East China University of Technology
Scientific interest: mainly engaged in satellite remote sensing image, target feature extraction and recognition, digital image processing and computer technology
Publications <number or main>: over 10 papers
Experience: Present   Researcher & Associate professor, East China University of Technology
2013   Researcher & Lecturer, East China Institute of Technology
2002   Yantai university, photoelectric information engineering college, a master's degree
Advisor: Prof. Jun Zhang

< **Hongwei JIA** >, <10 -24-1979 >,<Jiang Xi >
Current position, grades: Senior Lecturer, master graduate
University studies: Institute of electrical and mechanical, East China University of Technology
Scientific interest: mainly engaged in database technology,information management technology and algorithm analysis, Artificial intelligence
Publications <number or main>: over 3 papers
Experience: Present   Senior Lecturer, East China University of Technology
2005-2008   East China Institute of Technology ; a master's degree; Advisor: Prof.Xiang LI
1997-2001   North China institute of technology ; a bachelor's degree

References

[1] Sudeep Sarkar and Kim L.Boyer,On Optimal Infinite Impulse Response Edge Detection Filter,IEEE Trans.Pattern Analysis And Machine Intelligence,Vol 13,No.11,pp.1154-1170,Novmber 1990

[2] Liu K,Xiao K,Xiong H. An image edge detection algorithm based on improved Canny[C]//Internation Conference on Machinery,Materials and Computing Technology.2017.

[3] Kalra A Chhokar R L.A hybrid approach using sobel and Canny operator for digital image edge detection[C]//Internation Conference on Micro Electronics and Telecommunication Engineering.2017:305-310.

[4] Yang J G, Li B Z, Chen H J.Adaptive edge detection method for image polluted using Canny operator and Otsuthreshold selection[J].Advanced Material Research,2011,301:797-804.

[5] Shi G,Suo J.Remote sensing image edge-detection based on improved Canny Operator [C]//IEEE International Conference on Communication Software and Networks.2016:652-656.

[6] Wu X,Yu W,Liu X,etal.A newly improved Canny algo-rithm of image edge detection [C]//International Conference on Information Engineering for Mechanics and Materials.2016.

[7] Wang M,Jin J S,Jing Y,etal.The improved Canny Edge Detection Algorithm Based on an Anisotropic and Genetic Algorith[M].Singapore:Springer,2016:115-124.

[8] L.G.Roberts,Machine Perception of 3-D Solids,in Optical and Electro-Optimal Information Processing,J.t.Tippett,et al.,Ed.Cambridge, MAMIT Press,1965,99-159-197

[9] J.Prewitt,Object Enhancement and Extraction,Picture Porcess.Psychopict.,pp.75-149,1970.

[10] L.Sobel,Camera Models and Machine Perception,PhD theses,Stanford University,Standford,CA,1970

[11] John Canny,Finding Edges and Lines In Images,MIT Artif.Intel , Lab.Cambridge,MA,Tech.Rep.AI-TR-720,1983.

[12] Mallat S, Hwang WL.Singularity Detection and Processing with Wavelets.IEEE Trans On Information Theory,1992,38(2):617.

[13] Dellecker R. Boundary-scan bursts into the modern production facility[J]. IEEE Aerospace and Electronics Systems, 2001, 16(6):21.

A novel rating style mining method to improve collaborative filtering algorithm

Wei Yang[1], Sheng Hui Guo[1] and Chun Jin Zhang[2]

[1] Key Laboratory for wisdom mine information technology of Shandong Province, Shandong University of Science and Technology, Qingdao, China

[2] Network Information Center (NIC), Shandong University of Science and Technology, Qingdao, Shandong 266590, China

Shenghui Guo (e-mail: 15764225605@163.com).

Abstract. Collaborative filtering (CF) algorithm is widely used in recommendation systems, which makes recommendation based on the neighbors' interests. Therefore, how to discover the neighbors with similar interests to target user is the core of the CF algorithm. Most existing algorithms discover neighbors by using rating similarity measure, which ignore the differences of users' rating styles. In this paper, we propose a user rating style mining method and use it to eliminate the rating style differences before calculating a similarity measure. Comparing with the raw similarity measure and another rating style mining method with the Mean Absolute Error (MAE) and Root Mean Square Error (RMSE) over amazon movie dataset, we conclude that (*i*) use our method to eliminate the rating styles differences can improve the prediction accuracy and (*ii*) our method outperforms other rating style mining method.

1. Introduction

With the recent exponential growth of e-commerce transaction volume and user feedback, discovering useful information from such large-scale data has become increasingly difficult. In order to solve this problem, the recommendation system (RS) has emerged. At present, collaborative filtering (CF) algorithms are commonly employed in RSs, which make recommendations to target user based on his/her neighbors' interests. Therefore, how to find nearest neighbors is the core of CF algorithms.

At present, the most common existing algorithms discover the neighbors based on rating similarity measures. The most frequently adopted rating similarity measures are Cosine similarity, Pearson similarity, Modified Cosine similarity and so on. In order to estimate neighbors more accurately, many researchers focused on improving these similarity calculation methods.

For example, Srikanth et al. [1] presented a new distance measure by improving Pearson similarity, which can well explain the correlation between users whose ratings are linearly related. Li et al. [2] believed that rating similarity is affected by the number of co-item and the average rating, so he improved Pearson similarity by adopting two factors. Wu et al. [3] estimated the similarity between users suggested with a ratio-based approach. Li et al. [4] integrated the Jaccard coefficients into Cosine similarity and Pearson similarity respectively to get two new similarity measures. Zang et al. [5] considered not only the co-rated items set, but also items rated only by neighbors. Suryakant et al. [6] proposed a CjacMD similarity measure, which combined Cosine, Jaccard, and Mean Measure of Divergence for evaluating sparse datasets.

To address the problem when only a few ratings are available for similarity estimation, Liu et al.

Content from this work may be used under the terms of the Creative Commons Attribution 3.0 licence. Any further distribution of this work must maintain attribution to the author(s) and the title of the work, journal citation and DOI.

Published under licence by IOP Publishing Ltd

[7] considered both local context information of user ratings and global preference of user behaviour. Further, some researchers introduced the concept such as the co-rated items and the non-common rated items into similarity calculation [8-10]. For example, Wang et al. [8] integrated an asymmetric factor based on the ratio of the co-rated items to all the rated items by each user into similarity calculation. Li et al. [9] introduced another asymmetric factor according to the ratio of co-rated items to non-common rated items by each user. Hu et al. [10] integrated the similarity of items to improve the calculation of users' similarity in the memory-based collaborative filtering algorithms.

Above literature has proved that these methods can improve the performance. However, most of them ignore that people have different rating styles. For example, there are two users (i.e., users u and v) gave their ratings and reviews for an item. Both ratings are 4 stars and the corresponding reviews are "I like it very much. It is really worth the money" and "It is OK, and it is a bit expensive", respectively. From their reviews, we can find that user u likes the item very much and user v considers it is a bit expensive. But both them gave 4 stars. It is indicated that user u has a strict rating style while user v has a relatively loose rating style. However, it is unreasonable to believe that users u and v have similar interest according to their ratings. Because user u will rate less 4 stars if he/her give a review similar to user v. Therefore, we must eliminate the differences of rating styles before calculating rating similarity.

This paper proposes a method to eliminate the differences of rating styles and uses it to improve a kind of rating similarity measure with the following key contributions:

1. This paper uses the Bidirectional Encoder Representations from Transformers (BERT) [11] model to mine users' rating styles and designs a novel method to eliminate the differences of users' rating styles.

2. This paper uses recalculate a kind of rating similarity measure (i.e., an improved Pearson similarity) after eliminating the differences of users' rating styles and evaluates the preference with MEA and RMSE over amazon movie dataset.

We develop the rest of this paper as follows. Section 2 introduced some basic knowledge. Section 3 reviews our method. Section 4 describes the experimental details before we conclude this paper in section 5.

2. Basic Knowledge

Before introducing our method, we illustrate some basic knowledge, which will be used in our model. Section 2.1 describes the concept of rating matrix. Section 2.2 introduces the BERT model and section 2.3 illustrates a kind of improved Pearson similarity measure.

2.1. The concept of rating matrix

Definition 1(Rating matrix) $R_{m \times n} = [r_{ui}]_{m \times n}$ represents the rating matrix that users $User = \{u_1, u_2, u_3, \ldots, u_m\}$ rated items $Item = \{i_1, i_2, i_3, \ldots, i_n\}$, which can be represented as equation (1).

$$R_{m \times n} = \begin{pmatrix} r_{11} & \cdots & r_{1n} \\ \vdots & r_{ui} & \vdots \\ r_{m1} & \cdots & r_{mn} \end{pmatrix} \qquad (1)$$

where, r_{ui} is the rating that user u rated item i.

2.2. BERT model

The BERT model [11] is a new language model, which combines the advantages of feature-based (i.e., ELMo) [12] and fine-tuning [13] (i.e., OpenAI GPT) models. The BERT model uses a kind of transformer encoder (figure 1) as the basic structure, which adopts the attention mechanism.

Since the BERT model adopts a multi-layer transformer (TRM) structure (figure 2) to bidirectionally encode word vectors, it can pre-train deep bidirectional representations by jointly conditioning on both left and right context in all layers. Literature 2 proved that BERT has obtained new state-of-the-art preference on eleven natural language processing tasks. What's more, Google has already pre-trained the BERT model, we only need to use the given dataset to fine-tune it.

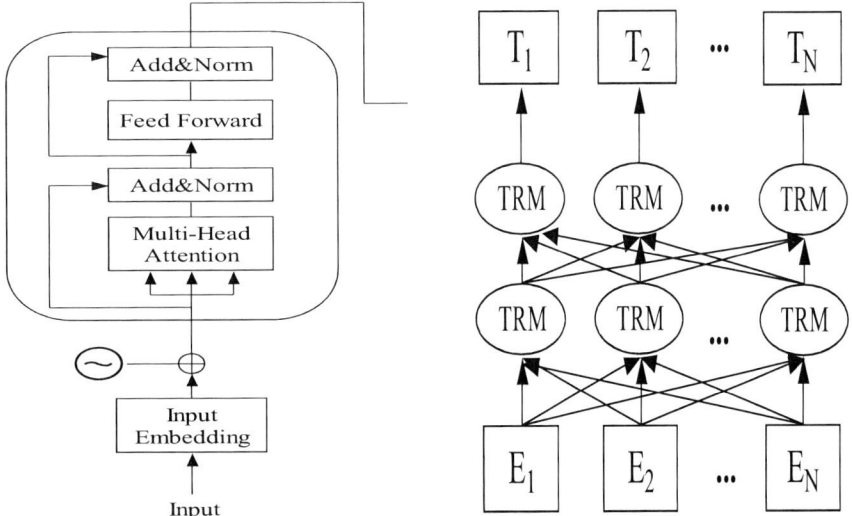

Figure 1. Transformer encoder   Figure 2. BERT model

2.3. An improved Pearson similarity

Literature 2 proposed a kind of improved Pearson similarity considering the affection from the number of co-rated items and users' average ratings, which is defined as equation (2):

$$Nsim(u,v) = sim(u,v) \times R(u,v) \times p(u,v) \qquad (2)$$

where, $sim(u,v)$ is the traditional Pearson similarity as equation (3), $R(u,v)$ is the factor to relief the affection from the number of co-rated items as equation (4), $p(u,v)$ is the factor to relief the affection from users' average ratings as equation (5).

$$sim(u,v) = \left(\sqrt{\sum_{i \in I_{u,v}} \left(r_{ui} - \bar{r}_u \right)^2} \sqrt{\sum_{i \in I_{u,v}} \left(r_{vi} - \bar{r}_v \right)^2} \right)^{-1} \sum_{i \in I_{u,v}} \left(r_{ui} - \bar{r}_u \right) \left(r_{vi} - \bar{r}_v \right) \qquad (3)$$

$$R(u,v) = \frac{|I_{u,v}|}{\max(|I_u|,|I_v|)} \qquad (4)$$

$$p(u,v) = \left(1 + |I_{u,v}|^{-1} \sum_{i \in I_{u,v}} |r_{ui} - r_{vi}| \right)^{-1} \qquad (5)$$

where, r_{ui} represents the rating that user u rated item i, $\bar{r}_u$ is the average rating of user u for all items, $I_{u,v}$ represents the co-rated item set of users u and v, $|I_{u,v}|$ is the size of $I_{u,v}$, I_u is the item set that user u rated, $|I_u|$ is the size of item set I_u.

3. A novel rating style mining method and its application in improving the similarity measure

Since people have different rating styles, it is unreasonable to calculate rating similarity based on raw rating matrix defined as equation (1). Therefore, this paper proposes a rating style mining method to eliminate the differences in section 3.1. In order to verify the performance of the method, this paper adopts it into the calculation of a similarity measure reviewed in section 2.3, which will be described in detail in section 3.2. According to the similarity measure, section 3.3 illustrates how to predict the unrated ratings.

3.1. A rating style mining method

According to section 2.2, we know that the BERT model can analyse context effectively. Therefore, we use BERT model to mine users' rating styles by adding a softmax layer (see figure 3). From figure 3, we can see that the reviews (reviews$_u$) are taken as input and the corresponding ratings (ratings$_u$) are taken as output to fine-tune this model. After that, we can get the rating style model of user u (BERT-u). According to the trained rating style model, we can obtain ratings restricted to user u's rating style based on the reviews by using the transformation function as equation (6):

$$r_{ui} = B\_u\left(t_{ui}\right) \tag{6}$$

where, r_{ui} is the rating that user u rated item i, t_{ui} is the review corroding to r_{ui}, $B\_u()$ is the rating style transformation function of user u.

According to rating style transformation function of user u and reviews, we can restrict the rating matrix to user u's rating style as equations (7) and (8).

$$R_{u\_m\times n} = \begin{pmatrix} r_{u\_11} & \cdots & r_{u\_1n} \\ \vdots & r_{u\_vi} & \vdots \\ r_{u\_m1} & \cdots & r_{u\_mn} \end{pmatrix} \tag{7}$$

$$r_{u\_vi} = B\_u\left(t_{vi}\right) \tag{8}$$

where, $R_{u\_m\times n}$ is the rating matrix in user u's rating style, r_{u_vi} represents the rating user v will rate item i if his/her rating style is similar to user u, t_{vi} is the review that user v gave item i.

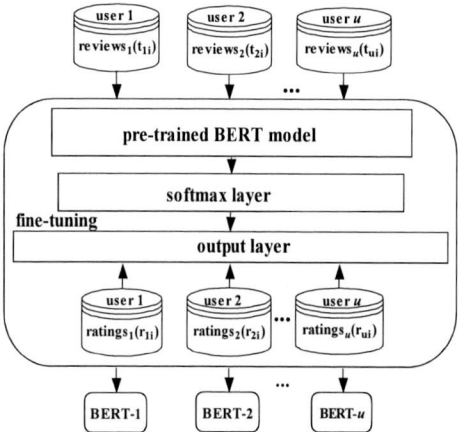

Figure 3. The rating style model

3.2. Improving the rating similarity

According to equations 7 and 8, this paper gets the new rating matrixes by restricting the ratings into each user's rating style, which differ from each other. Therefore, there is not existing rating style differences in the new rating matrixes. Based on the new rating matrixes, this paper recalculates the rating similarity reviewed in section 2.3, which consists of three parts (i.e., traditional Pearson similarity and two factors) defined by equations (3), (4) and (5). Since the calculation of factor $R(u,v)$ is not related to ratings, we only recalculate the traditional Pearson similarity and the factor $p(u,v)$, whose detail steps are as follow.

Firstly, according to the new matrixes and equation (3), we recalculate the traditional Pearson similarity as equations (9) and (10).

$$B\text{-}sim(u,v) = \left(\sqrt{\sum_{i \in I_{u,v}} \left(r_{ui} - \bar{r}_u \right)^2} \sqrt{\sum_{i \in I_{u,v}} \left(r_{u\_vi} - \bar{r}_{u\_v} \right)^2} \right)^{-1} \sum_{i \in I_{u,v}} \left(r_{ui} - \bar{r}_u \right)\left(r_{u\_ni} - \bar{r}_{u\_v} \right) \quad (9)$$

$$\bar{r}_{u\_v} = \sum_{i \in I_v} r_{u\_vi} \quad (10)$$

Where, r_{u_vi} represents the rating user v will rate item I if his/her rating style is similar to user u, which is in the new matrix in the user u's rating style defined as equation (7), $\bar{r}_{u\_v}$ is the average rating of user v for all items if his/her rating style is similar to user u.

Then, we recalculate the factor $p(u,v)$ as equation (11).

$$B\text{-}p(u,v) = \left(1 + |I_{u,v}|^{-1} \sum_{i \in I_{u,v}} |r_{ui} - r_{u\_vi}| \right)^{-1} \quad (11)$$

Finally, we calculate the new improved Pearson similarity as equation (12):

$$B - Nsim(u,v) = B\text{-}sim(u,v) \times R(u,v) \times B\text{-}p(u,v) \quad (12)$$

From equation (12), we find that the new similarity is asymmetric, which is consistent with the conclusion drawn in literature 8 and 9.

3.3. Rating prediction

To predict the rating of user u on unrated item i, we first need to obtain the neighbor set N_u of user u. According to equation (12), we can compute the similarity values between user u and other users. Then, according to the user similarity values, the neighbor set is constructed by selecting the first N users close to user u. Then, the prediction rating is computed according to equation (13).

$$\hat{r}_{ui} = \bar{r}_u + \frac{\sum_{v \in N_u} B - Nsim(u,v) \times (r_{vi} - \bar{r}_v)}{\sum_{v \in N_u} B - Nsim(u,v)} \quad (13)$$

where, user v is member in the neighbor set of user u.

4. Experiment

The BERT structure used in this paper is BERT-base, which is trained with a free GPU provided by Google. In order to verify whether our model can eliminate the differences of rating styles and whether it outperforms other similar models, this paper selects an improved Pearson similarity (named IPS) reviewed in section 2.3 and CjacMD measure which considers the rating style differences as benchmarks.

4.1. Data preprocessing

This paper implements these algorithms on the Amazon Movies_and_TV_5 data set (http://snap.stanford.edu/data/amazon/productGraph/categoryFiles/reviews_Movies_and_TV_5.json.gz). First, we preprocess this data set by filtering out users with fewer than 100 ratings, then it was randomly divided into 10 portions, 80% of which were taken as training sets, and the remaining 20% were used as test sets. The specific data information is shown in Table 1.

Table 1. dataset information

	Raw dataset	Pre-processed dataset
User	124960	1443
Item	50052	42848
Reviews	1697533	367867

4.2. Evaluative criteria

This paper selects MAE [14] and RMSE [14] as evaluation criteria, which are defined as equations (14) and (15):

$$MAE = \frac{\sum_{(u,i) \in \Omega^{test}} |\hat{r}_{ui} - r_{ui}|}{|\Omega^{test}|} \tag{14}$$

$$RMSE = \left(\frac{\sum_{(u,i) \in \Omega^{test}} (\hat{r}_{ui} - r_{ui})^2}{|\Omega^{test}|} \right)^{1/2} \tag{15}$$

where, r_{ui} is the actual rating in test dataset, $\hat{r}_{ui}$ is the predicted rating corresponding to r_{ui}, Ω^{test} is the test dataset, $\left|\Omega^{test}\right|$ is the size of the test dataset.

4.3. Experimental analysis

In order to verify whether our model can improve the prediction accuracy by eliminating rating style differences and whether it outperforms other similar models, we implement the BIPs-CF algorithm which is a collaborative filtering algorithm based on a similarity measure improved by this paper, and we select IPs-CF and CjacMD-CF as comparison algorithms, which are the collaborative filtering algorithms based on similarity measures (i.e., IPs and CjacMD). We calculate the MEA and RMSE of all algorithms and draw them into figures 4 and 5. As the number of selected nearest neighbors is well-known to have an important impact on the quality of rating prediction, this experiment compares the prediction performance of these algorithms when the number of neighbors is assigned with the values of 10, 30, 50, 70, 90, 110, 130, 150, 170 and 190.

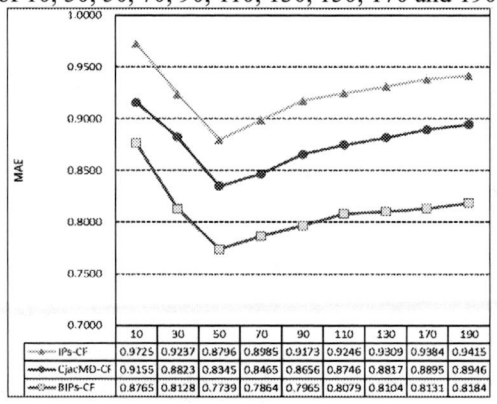
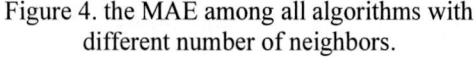

	10	30	50	70	90	110	130	170	190
IPs-CF	0.9725	0.9237	0.8796	0.8985	0.9173	0.9246	0.9309	0.9384	0.9415
CjacMD-CF	0.9155	0.8823	0.8345	0.8465	0.8656	0.8746	0.8817	0.8895	0.8946
BIPs-CF	0.8765	0.8128	0.7739	0.7864	0.7965	0.8079	0.8104	0.8131	0.8184

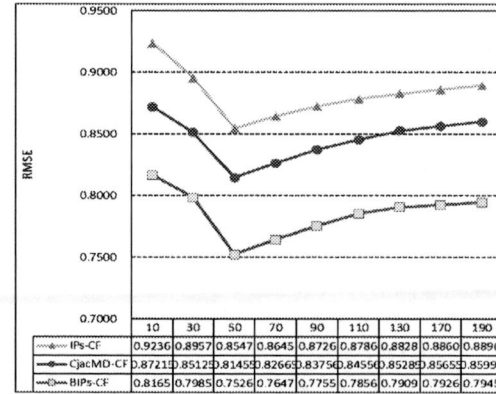

	10	30	50	70	90	110	130	170	190
IPs-CF	0.9236	0.8957	0.8547	0.8645	0.8726	0.8786	0.8828	0.8860	0.8896
CjacMD-CF	0.87215	0.85125	0.81455	0.82665	0.83756	0.84556	0.85285	0.85655	0.85999
BIPs-CF	0.8165	0.7985	0.7526	0.7647	0.7755	0.7856	0.7909	0.7926	0.7945

Figure 4. the MAE among all algorithms with different number of neighbors.

Figure 5. the RMSE among all algorithms with different number of neighbors.

Figures 4 and 5 show the results of MAE and RMSE with the horizontal axis representing the number of the neighbors. From those, we can see that the curves of MAE and RMSE first decrease and then increase with the increasing number of neighbors. Moreover, the MEA and RMSE values of BIPs-CF algorithm are less than ones of IPs-CF and CjacMD-CF algorithms regardless of the number of neighbors. When the number of neighbors is 50, the MAE and RMSE values of all algorithms (i.e., IPs-CF, CjacMD-CF and BIPs-CF algorithms) achieve minimum values (i.e., MAE-0.8796, 0.8345, 0.7739; RMSE-0.8547, 0.8145, 0.7526). At this point, the MEA and RMSE values of BIPs-CF algorithm are least reduced by 12.02% and 11.95% as compared to IPs-CF algorithm and reduced by 7.26% and 7.42% as compared to CjacMD-CF algorithm. It is indicated that eliminating rating style differences with our model can improve the prediction accurate and our model outperforms other rating style mining methods.

5. Conclusion
Collaborative filtering algorithm is widely used in recommendation system, which makes recommendations based on neighbors' interests. Therefore, how to find neighbors with similar interests to target user is the core of CF algorithms. Since the differences of users' rating styles will affect the performance of CF algorithm, this paper designs a method to eliminate the differences of users rating styles. In order to illustrate the performance of this method, this paper adopts it into the calculation of a similarity measure. We select the IPs-CF and CjacMD algorithms as benchmarks and compare their performance of prediction with MAE and RMSE. Experimental results show that use our method to eliminate the rating styles differences can improve the prediction accuracy and (*ii*) our method outperforms other rating style mining method.

Since we only adopt our method into the calculation of a kind of similarity measure, in order to further prove its versatility, we will select other rating similarity measures to improve in the future work. Moreover, the BRET model has powerful performance in natural language processing, so we can use it to analyse the emotional of all topics [15] according to reviews.

Acknowledgments
This paper is supported in part by the Natural Science Foundation of China (No. 71772107, &1403151), Qingdao social science planning project (No. QDSKL 1801138), Shandong Education Quality Improvement Plan for Postgraduate.

References
[1] Srikanth T and Shashi M 2015 A New Similarity Measure for User-Based Collaborative Filtering in Recommender Systems[J] *INTERNATIONAL JOURNAL OF COMPUTERS & TECHNOLOGY* **14.9** 6118–28. CrossRef.

[2] Li Rong, Li Mingqi and Guo Wenqiang 2016 Research on Collaborative Filtering Algorithm with Improved Similarity[J] *Computer Science* **43(12)** 213-215+247 Google Scholar.

[3] Wu Xiaokun, Cheng Bo and Chen Junliang 2017 Collaborative Filtering Service Recommendation Based on a Novel Similarity Computation Method[J] *IEEE Transactions on Services Computing* **10.3** 352–365. CrossRef.

[4] Li Wenqiang, Xu Hongji, Ji Mingyang, Xu Zhengzheng and Fang Haiteng 2017 A hierarchy weighting similarity measure to improve user-based collaborative filtering algorithm[C] *2nd IEEE International Conference on Computer and Communications* (China: Chengdu) pp 843-846 CrossRef.

[5] Zang Xuefeng, Liu Tianqi, Qiao Shuyu, Gao Wenzhu, Wang Jiatong, Sun Xiaoxin and Zhang Bangzuo Zhang 2017 A New Weighted Similarity Method Based on Neighborhood User Contributions for Collaborative Filtering[C] *1st IEEE International Conference on Data Science in Cyberspace* (China: Changsha) pp 376-381 CrossRef.

[6] Suryakant and Mahara T 2016 A new similarity measure based on mean measure of divergence for collaborative filtering in sparse environment[J] *Procedia Computer Science* **89** 450-456 CrossRef.

[7] Liu Haifeng, Hu Zheng, Mian A, Tian Hui and Zhu Xuzhen Zhu 2014 A new user

similarity model to improve the accuracy of collaborative filtering[J] *Knowledge-Based Systems* **56** 156-166 Crossref.

[8] Wang Yong, Deng Jiangzhou, Gao Jerry and Zhang Pu 2017 A hybrid user similarity model for collaborative filtering[J] *INFORM SCIENCES* **418-419** 102-108 Crossref.

[9] Li Qiaoqiao, Lin Zhe and Zhang Fei 2017 Similarity coefficient of collaborative filtering based on contribution of neighbors[C] *2016 International Conference on Machine Learning and Cybernetics* (South Korea: Jeju) pp 226-232 CrossRef.

[10] Hu Yan, Shi Weisong, Li Hong and Hu Xiaohui 2017 Mitigating data sparsity using similarity reinforcement-enhanced collaborative filtering[J] *ACM Transactions on Internet Technology* **17.3** 1–20. CrossRef.

[11] Devlin J, Chang M W, Lee K and Kristina T 2018 Bert: Pre-training of deep bidirectional transformers for language understanding *Preprint* arXiv:1810.04805 Google Scholar.

[12] Peters M E, Neumann M, Iyyer M, Gardner M, Clark C, Lee K and Zettlemoyer L 2018 Deep contextualized word representations *Preprint* arXiv:1802.05365. Google Scholar.

[13] Radford A, Narasimhan K, Salimans T and Sutskever I 2018 Improving language understanding by generative pre-training *Technical report, OpenAI* Google Scholar.

[14] Wu Wenmin, Zhao Jianli, Zhang Chunsheng, Meng Fang, Zhang Zeli, Zhang Yang, Sun Qiuxia 2017 Improving performance of tensor-base context-aware recommenders using bias tensor factorization with context feature auto-encoding[J] *Knowledge-Based Systems* **128** 71-77 Crossref.

[15] Dong Luyu, Ji Shujuan, Zhang Chunjin, Zhang Qi, Chiu D.K.W., Qiu Liqing and Li Da 2018 An unsupervised topic-sentiment joint probabilistic model for detecting deceptive reviews[J] *Expert Systems with Applications* **114** 210-223 Crossref.

Study on traffic organization and work-zone optimization of four-lane freeway reconstruction and expansion

Xuanyu Ye [1], Yazhen Chen [2] and Chen Chen [2]

[1] Jiangxi Ganyue Expressway Co., Ltd., Nanchang 330025, Jiangxi, China

[2] CCCC Second Highway Consultants Co., Ltd., Wuhan 430052, Hubei, China

Abstract. In this paper, the bi-directional four-lane traffic organization plan and its key factors in the freeway reconstruction and expansion were analyzed. Combined with the actual traffic flow and construction conditions of the freeway network in Chang-zhang Freeway project, the bi-directional four-lane traffic organization plan for the Chang-zhang Freeway reconstruction and expansion were determined, including the construction area length and construction sequence. It systematically summarizes the advantages and disadvantages of the traffic organization and management of the Chang-zhang freeway reconstruction and expansion project. It provides reference cases for the subsequent traffic organization design of freeway reconstruction and expansion similarly.

1. Introduction

With the sustained and rapid development of China's economy, the traffic volume of freeways built in the early days has increased rapidly, with an average annual growth rate of more than 10%, or even more than 20%. The volume has far exceeded the predicted traffic volume during freeway planning. At present, the service level is significantly reduced, which can no longer meet the driving requirements of freeways; at the same time, it cannot meet the needs of rapid socio-economic development. Therefore, the reconstruction and expansion of freeways has become the key development direction of freeway construction in the future. The research on the reconstruction and expansion of freeways under the mode of "construction without interruption of traffic" has become one of the focuses of research on freeway reconstruction and expansion projects. At present, the types of unimpeded plan for freeway expansion and reconstruction mainly include keeping bi-directional two-lane traffic, keeping bi-directional three-lane traffic, and keeping bi-directional four-lane traffic. The selection of the traffic organization plan should be determined according to the specific conditions of the project freeway.

The Nanchang-Zhangshu Freeway (Chang-zhang Freeway) has a total length of 86.545 km. It is an integral part of the Shanghai-Kunming National Freeway in the National Freeway Network (7918). It is also the important section of the main skeleton of the highway network called "three vertical and four horizontals" in Jiangxi Province, connecting the surrounding provinces and strengthening the inter-provincial freeway transportation. It has a very important position in the highway network. In view of the important position of Chang-zhang Freeway in the highway network of Jiangxi Province and the high goals and high requirements for the "expanding and keeping traffic at the same time and both ensuring the quality and the period of the construction", this project puts forward the traffic organization management measures of freeway reconstruction and expansion as the priority goal [1]. Therefore, it is necessary and significant to make a research of the bi-directional traffic protection during reconstruction. This paper intends to systematically summarize the bi-directional four-lane

---

Content from this work may be used under the terms of the Creative Commons Attribution 3.0 licence. Any further distribution of this work must maintain attribution to the author(s) and the title of the work, journal citation and DOI.

Published under licence by IOP Publishing Ltd

ISPECE   IOP Publishing

traffic organization plan formulation and analysis of the Chang-zhang Freeway. It solves the problems encountered in the actual project and provide reference for the subsequent similar freeway reconstruction and expansion projects.

2. Bi-directional four-lane freeway traffic flow character and construction zone plan
In the bi-directional four-lane traffic organization plan of freeway construction zone, it mainly includes special work sites such as subgrade pavement engineering, bridge engineering, interchange and overpass. This paper focuses on subgrade pavement engineering.

The influencing factors and key problems of the subgrade and pavement reconstruction and expansion have the following four critical points, the length of the construction zone, traffic conversion and safety, road utilization, and emergency room for emergencies. Through analysis and study of the above four influencing factors, the bi-directional four-lane traffic organization plan of construction zone is determined, which not only ensures the safety of driving, but also meets the requirements of construction and guarantees the service level in the project.

2.1 Construction area segment length
The division of the length of the construction zone shall meet the following requirements:

1) Comply with the construction schedule of the project road;

2) Ensure the overall traffic flow stability, safety and efficiency of the entire construction zone as far as possible;

3) Considering the integrity of the road network structure, the boundary of the construction section should be as close as possible to the natural boundary of the road section structure (such as the boundary between the interchange and service areas);

4) The division of the construction section should consider the convenience of subgrade and pavement construction and the saving of construction cost, and minimize the impact of subgrade and pavement construction on the traffic flow along the route;

5) It is not advisable to divide the construction section too much to meet the requirements of reasonable flow construction organization.

2.2 Traffic conversion and safety
The freeway reconstruction and expansion project usually adopt the method of "constructing and keeping traffic at the same time", which interferes with the construction and keeping traffic, brings great safety risks to the construction zone. In order to more objectively reflect the traffic safety in the construction zone, this paper selects the ratio of the number of conflicts and the cross-section throughput and the length of the construction zone-the number of car-kilometer conflicts as the evaluation index (the number of car-kilometer conflicts: under observable conditions, two or more road users are close to each other in space and time, so that if either party does not change its trajectory, a collision will occur. This state is called a traffic conflict. [2]).

$$f = \frac{TC}{Q \cdot L} \times 1000 \qquad (1)$$

Where: f is the number of vehicle kilometers conflict (times/vehicle·km); TC is the number of conflicts per hour (times); Q is the traffic flow (pcu/h); L is the length of construction area (m);

In summary, a regression model between the segment length (X) of the construction zone and the traffic safety evaluation index (Y) of the construction area is established. First, make the regression equation hypothesis:

$$y = \alpha + \beta x \qquad (2)$$

The α and β in the equation is the undetermined coefficient [3]. From the definition of the regression function, if F(x) minimizes $Q[\eta - f(\zeta)]^2$, then F(x) is the regression of η and ζ. Determining α and β minimizes $Q[\eta - f(\zeta)]^2$.

Here $Q[\eta - f(\zeta)]^2 = \beta^2 \sigma_x^2 - 2\beta\, cov(\zeta, \eta) + \sigma_y^2 + (\alpha_y - \alpha - \beta x)^2$, it can be seen from the equation that $Q[\eta - f(\zeta)]^2$ is the smallest when $\alpha = \partial_y - \beta_x$; $\beta = \frac{cov(\zeta, \eta)}{\sigma_x^2} = \frac{\rho\sigma_y}{\sigma_x}$;

Regression equation is as following:

$$Y = \alpha_y + \left(\frac{\rho\sigma_y}{\sigma_x}\right) \cdot (x - \alpha_x) \tag{3}$$

$$X = \alpha_x + \left(\frac{\rho\sigma_x}{\sigma_y}\right) \cdot (x - \alpha_y) \tag{4}$$

Where $\frac{\rho\sigma_y}{\sigma_x}$ and $\frac{\rho\sigma_x}{\sigma_y}$ are regression coefficients. If multiple points were observed, let $\zeta = x_1$, $\eta = x_2$, paired data such as (x_1, y_1), (x_2, y_2),..., (x_n, y_n),etc. can be obtained. Using these data to calculate the average and standard deviation of the sample, therefore the regression equation is assumed.

To require a regression equation, the coefficients α and β must be solved first, and the formula is:

$$\bar{x} = \frac{1}{n}\sum_{i=1}^{n} X_i \quad (\,i = 1,2,3....., \quad n\,) \tag{5}$$

$$\bar{y} = \frac{1}{n}\sum_{i=1}^{n} y_i \quad (\,i = 1,2,3....., \quad n\,) \tag{6}$$

$$\beta = \frac{\sum_{i=1}^{n}(x_i - \bar{x})(y_1 - \bar{y})}{\sum_{i=1}^{n}(x_i - \bar{x})^2} \tag{7}$$

Therefore, the following relationship can be obtained: $y = y - \beta\bar{x} + \beta x = \bar{y} + \beta(x - \bar{x})$

2.3 Road utilization rate [4]

Road utilization refers to the ratio of highway that have been used by traffic participants to the total number of roads in the area at a particular time and region. In the process of freeway reconstruction and expansion, the road utilization rate refers specifically to the ratio of the road resources that have been used by the traffic participants to the total resources that can be utilized by the freeway roads (including the expansion parts).

Road utilization rate=Road area already used/Road area that can be used

3. Determination of subgrade and pavement traffic organization plan

This paper takes the southwest Chang hub to the Houtian hub as an example with a lateral width of 0.5m.

Traffic safety evaluation index: number of vehicle kilometers conflict $f = \frac{TC}{Q \cdot L} \times 1000$

Length of section of construction zone: 2-8km is selected in this paper (the length of construction zone should be 2-8km to ensure the maximum efficiency)

Table 1 Relationship between construction zone length and safety evaluation indicators.

Construction zone length (L)km	Traffic flow (Q)pcu/h	Number of conflicts per hour（TC）(times)	Safety evaluation index f
2	537	1310	1.22
3	837	2862	1.14
4	1128	4331	0.96
5	1408	5280	0.75
6	1684	7780	0.77
7	1983	11243	0.81
8	2260	15548	0.86

Table 2 Interweaving area traffic safety clustering center based on VISSIM simulation.

Grading	Safety	Relatively safe	Safety criticality	Danger
f (times/vehicle·km)	0.76	1.45	2.66	3.73

The length of the construction zone x_i, the safety evaluation index y_i, and the numerical values of formulas (5) and (6) are calculated by substituting the numerical values, as shown in Table 3.

Table 3 Information on construction zone length and safety evaluation indicators.

i	x_i (km)	y_i	$x_i - \overline{x}$	$(x_i - \overline{x})^2$	$y_i - \overline{y}$	$(y_i - \overline{y})^2$	$(x_i - \overline{x})(y_i - \overline{y})$
1	2	1.22	-3	9	0.29	0.0841	-0.87
2	3	1.14	-2	4	0.21	0.0441	-0.42
3	4	0.96	-1	1	0.03	0.0009	-0.03
4	5	0.75	0	0	-0.18	0.0324	0
5	6	0.77	1	1	-0.16	0.0256	-0.16
6	7	0.81	2	4	-0.12	0.0144	-0.24
7	8	0.86	3	9	-0.07	0.0049	-0.21

$$\beta = \frac{\sum_{i=1}^{n}(x_i - \overline{x})(y_i - \overline{y})}{\sum_{i=1}^{n}(x_i - \overline{x})^2} \approx -0.07$$

$$\alpha = \overline{y} - \beta\overline{x} = 0.93 + 0.07 \times 5 = 0.93 + 0.35 = 1.28$$

Substituting the values of α and β into equation (2), the regression equation for the distance between the construction zone and the safety evaluation index can be obtained:

$$y = 1.28 - 0.07x$$

Regression equation error formula:

$$r = \frac{\sum_{i=1}^{n}(x_i - \overline{x})(y_i - \overline{y})}{\sqrt{\sum_{i=1}^{n}(x_i - \overline{x})^2 \sum_{i=1}^{n}(y_i - \overline{y})^2}} =\approx -0.8029$$

$$|r| = 0.8029$$

In the above, r is the correlation coefficient between x and y in the mathematical method [3]. When the absolute value of r is close to 0, the error is large, and the plan is not desirable. The closer the absolute value of r is to 1, the error is small; when the absolute value is about or equal to 1, the error is 0.

Through the analysis of Table 1, Table 2 and regression equation, the safety evaluation index is better when the section length of the construction zone is 4-8km, and the safety evaluation index is the best when the section length of the construction area is 5-6km. Therefore, the section length of the construction zone of the subgrade and pavement construction zone in the bidirectional four-lane unimpeded plan of Chang-zhang Freeway is selected to be 5-6km. Secondly, combined with the study of road utilization rate in the construction zone, select different construction section lengths in different sections, maximize the road utilization rate in the construction zone, and determine the construction plan of the subgrade and pavement. Through regression analysis, the r value of the regression equation is close to 1, and the error is small, so the plan is feasible.

4. Subgrade pavement traffic organization construction plan

The subgrade pavement traffic organization mainly divides into 2 period. In the first period, the guardrail does not move, while removing the external barrier; the two sides of the extension (7.5m wide on one side) are filled to the top of the roadbed; the vehicle travel in two-way four-lane traffic on the original old road. The illustration is shown in figure 1.

The second period. four sections are divided according to the division of the bid section and the layout of the main structures along the line. The left and right sections of each section are constructed at the same time. The pavement engineering is completed in parallel with the cycle construction in each section. The period includes 4 steps.

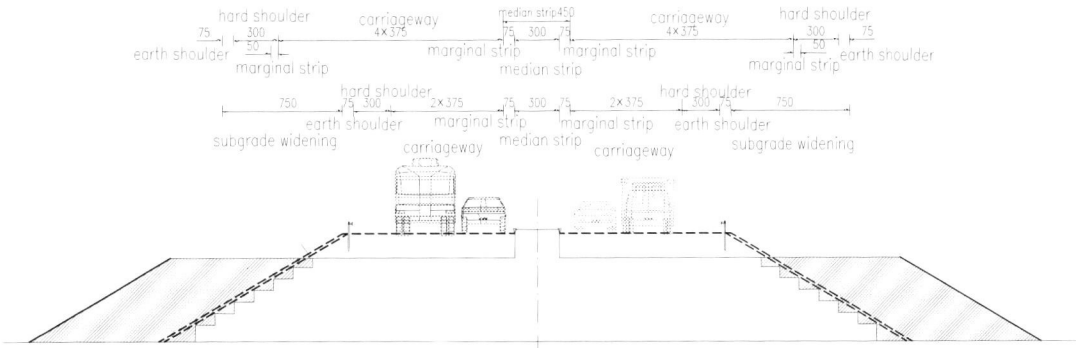

Figure 1. Traffic organization of roadbed construction in the first period (unit: cm)

In the first step, the old road dirt shoulder and the hard shoulder of the left and right No. 3 construction work area are excavated within 3m, and the roadbed is spliced to construct the pavement on both sides to the flexible base (same level with the old pavement), the new and old pavement Temporary isolation facilities are set up. In this stage, the inner passenger lane should be set to 3.5m to ensure that there is a lateral width of 0.5m on the outside of the two-lane road without the hard shoulder, and the original old road has two-way four-lane traffic.

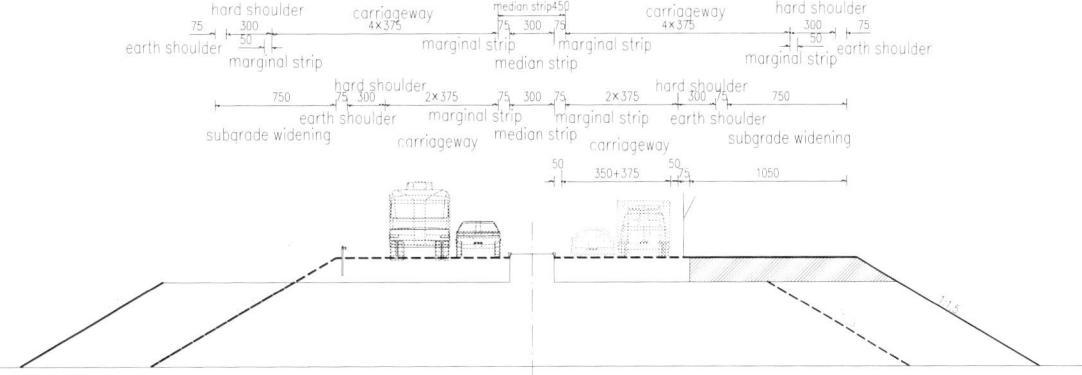

Figure 2.Traffic organization of subgrade and pavement construction at step 1 (unit: cm)

Step 2: Repeat the first step to excavate the 3m range of the old road dirt shoulder and the hard shoulder of the left and right frame No. 4 construction work area and build the roadbed to the flexible base layer (same level with the old road level).

Step 3: transfer the traffic, two-way four-lane traffic in section 3, renovate the old road in section 1 of the construction operation area, construct the upper surface layer and traffic safety facilities for road subgrade sprouting and pavement overall paving, install temporary water-horse anti-collision facilities between the new and old roads, and one-way two-lane traffic in sections 2 and 4 of the left and right sides; This stage should be at the same time in the left picture of section 2 (or right of section 4) and the left section 4, section 2 (or right) on both sides of the segment joint of median opening, comprehensive factors such as conductor, the speed limit and corner radius set for 150 m to ensure the safety of two-lane traffic transfer passage, half range two-way between the four lanes of traffic and transportation around image transfer place setting temporary anti-collision facilities such as water horse, in order to ensure the safety of driving, opens the mouth to consider using 40 km/h speed limit.

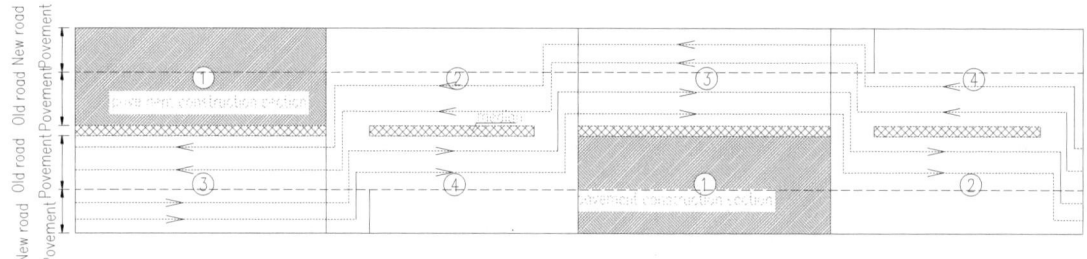

Figure 3. Construction traffic organization in the step 3.

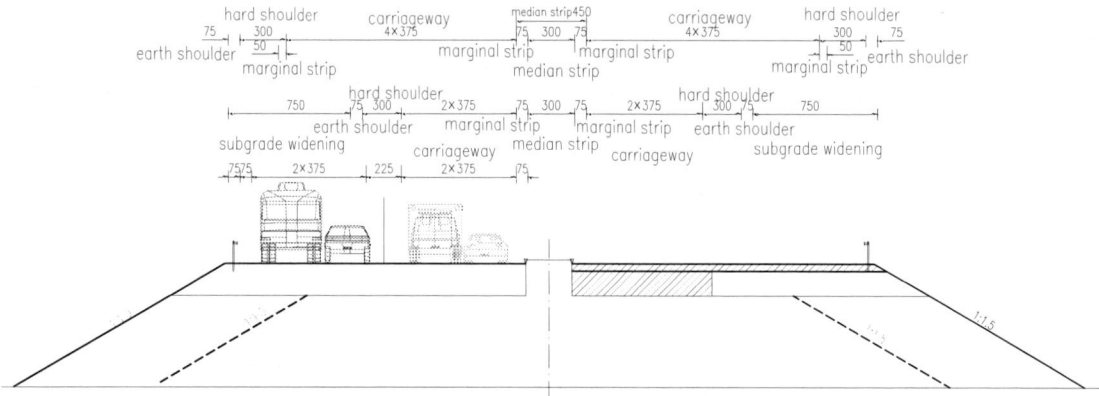

Figure 4. Traffic organization of subgrade pavement construction (cm)

Step 4: Repeat the above steps and cycle forward until the completion of the reconstruction and expansion, forming an eight-lane section.

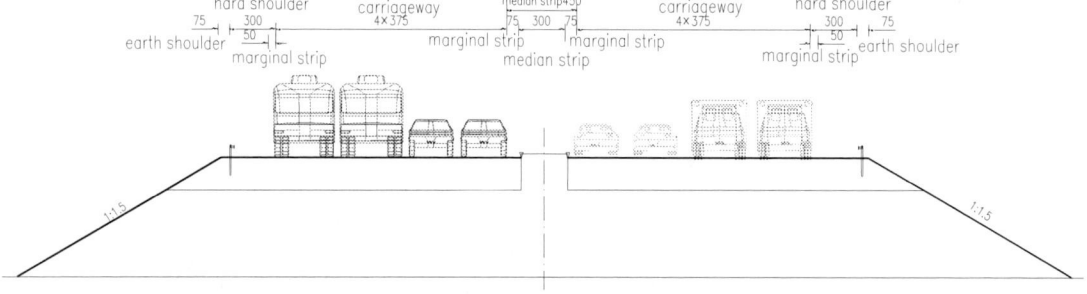

Figure 5. Traffic organization of subgrade pavement construction in the step 4 (cm)

5. Traffic organization implementation experience and discussion

During the expansion of Chang-zhang Freeway, the traffic organization plan generally followed the traffic organization plan of the project planning. In the actual implementation process, it was adjusted and optimized according to the actual situation. During the implementation process, many valuable experiences were accumulated, and some shortcomings were Summarized as follows.

(1) Efficient management organization

The freeway reconstruction and expansion project office innovated the traffic organization management and operation organization and built a highly efficient joint logistics linkage mechanism. It led the establishment of the project office, the provincial safety supervision three, the high-speed traffic police team, the high-speed road administration Yi-chun detachment and other units. The traffic safety organization leading group has an office, the office is deputy director and the deputy director of the traffic police and the road administration department, and the project office is stationed in the project office to improve the efficiency of traffic organization.

(2) The problem of segmentation

During the design stage of the traffic organization design, factors such as the distribution of structures, the position of the opening in the current situation, and the balance of the workload between the sections are considered. However, during the implementation process, some sections are still insufficient, such as some temporary traffic 3, 4 The section is divided into an uphill section, which creates a long queue of vehicles and creates a certain traffic congestion.

(3) Highway utilization

The reconstruction and expansion of traffic organization runs through the entire construction process, and the problem of road resource utilization is worthy of attention. The road utilization rate in the construction area is closely related to the traffic capacity. The greater the road utilization rate, the higher the traffic capacity, and vice versa. There are many factors affecting road utilization, including lane width, number of lanes, lateral width, large vehicles, speed limit in construction area, construction intensity, and lane closure form. In order to maximize the use of road resources, it is necessary to design traffic organizations. At the same time, the traffic management should be strengthened.

6. Conclusion

The four-lane guarantee scheme for parallel construction of the left and right sections and sections applied in the reconstruction and expansion of Chang-zhang Freeway fully considers the construction party and the operator, and fully utilizes the road resources of Chang-zhang Freeway to avoid the tolls while ensuring safe operation. The loss, while considering construction safety and efficiency, effectively solved the construction and operation problems. Practice verified that this scheme has good feasibility.

With the acceleration of construction in the central and western regions, the large number of freeways in China are facing renovation and expansion. Due to the low density of road networks in the central and western regions, the traffic along the line during the period of reconstruction and expansion is highly dependent on the roads for reconstruction and expansion. The traffic organization plan and the traffic organization management plan have pioneering significance, and their research results are promoted in Jiangxi Province and across the country, providing mature and learnable experience for other similar road reconstruction and expansion projects.

Acknowledgement

This study was supported by The National Nature Science Foundation of China (51678460); The Natural Science Foundation of Hubei Province, China (ZRMS2017001571).

References:

[1] Zhang Hong. Freeway reconstruction and expansion traffic organization management with "three guarantees" as the core concept, 2015.

[2] Cheng Wei, Li Jiang. Application of Fuzzy Clustering Method in Intersection Security Evaluation Based on Traffic Conflict Technology [J]. Transportation Systems Engineering and Information, 2004, 4(2): 48~55.

[3] Zhou Wu Da. Linear Algebra and Linear Programming [M]. Beijing: China Renmin University Press, 2002.

[4] Ye Xuanyu. Special study on the differential flow and road utilization rate of highway reconstruction and expansion [D]. Master's thesis of East China Jiaotong University, December 2017.

Review on Studies of Machine Learning Algorithms

Peiyuan Xu

Chengdu Foreign Languages School, 611731 Chengdu, China

Abstract. This paper mainly introduces machine learning algorithms in the field of artificial intelligence. First, it describes the classification of such algorithms and their main application scenarios. Then the paper introduces the principles behind those algorithms and presents the author's views. Finally, the development trend of machine learning algorithms is envisioned.

1. Introduction

1.1 Current Situation of Machine Learning Development

In recent years, machine learning has become one of popular research fields in the study of artificial intelligence. Currently, well-known Internet companies are researching and applying technologies related to artificial intelligence far and wide, among which machine learning is also a hot subject including image and speech recognition. In particular, against the backdrop of big data developing rapidly, machine learning blended with big data can effectively combine systems and algorithms, so that machine learning algorithms can work concurrently under multiple cores and process mass data, which is also the current research direction in the study of artificial intelligence. The rapid development of the Internet industry not only provides a large number of samples to train deep learning, continuously undergoing breakthroughs, improvements, and innovation by the joint effort from various fields.

2. Development of Machine Learning

2.1 Developing Process of Machine Learning [1]

Machine learning is not an emerging concept, whose prototype appeared as early as the mid-1950s to the mid-1960s instead. The main research goal at the time was to realize various self-organizing and adaptive systems. In other words, it could also be, to some extent, considered as an unsupervised mode of machine learning, the idea of which was sought after by many scholars.

By the middle of the 1970s, research on machine learning was centered on the process of concept learning. The tools used at that time were semantic networks and predicate logic, and scholars' enthusiasm also slacked off.

After another 10 years or so, concept learning remained as the research goal. However, studies were extended from learning a single concept to multiple concepts, exploring various learning strategies and learning methods, namely, progress was made both in breadth and in depth. Then the development of machine learning entered a new phase in the 1980s, which could be said to be a period witnessing vigorous development in this field. By far, the recovery of neural networks has driven the research into various machine learning methods and systems, accompanied by the shift from theoretical studies to practical applications in the field of machine learning. In the future, it is believed

Content from this work may be used under the terms of the Creative Commons Attribution 3.0 licence. Any further distribution of this work must maintain attribution to the author(s) and the title of the work, journal citation and DOI.
Published under licence by IOP Publishing Ltd

that the development of machine learning will be of greater significance. As big data further develops, machine learning will find wider application.

3. Machine Learning Algorithms

3.1 The Classification of Machine Learning Algorithms [2]

3.1.1 Supervised Learning
Supervised learning is relatively classical among machine learning algorithms, which includes logistic regression and reverse neural networks. Typical methods are BN, SVM, KNN, CBR, and so on. As for supervised learning, the input data is called training dataset, the basis function models include algebraic function or probability function, the way of training the model is iterative calculation, and the result of the training dataset is known to the functions. This algorithm works in this way: first computational predictions are made based on the training dataset, and then prediction training is continuously iterated until the results match the known ones. Therefore, supervised learning is often used for classification and regression.

3.1.2 Unsupervised Learning
The characteristic of unsupervised learning lies in that the result value of input data is not pre-set, that is, results produced by the algorithm are uncertain. Unsupervised learning can be roughly classified into clustering, anomaly detection, and competitive learning. For example, clustering refers to summarizing the structure and numerical values of the data, and then classifying the results. Typical unsupervised learning algorithms include K-means algorithm, Apriori algorithm, and SOM algorithm.

3.1.3 Semi-supervised Learning
Semi-supervised learning has wider application in practical cases because it is more efficient than other machine learning algorithms. Its principle lie in mixing data that has been labeled or has definite results with data that has no pre-set results, the goal of which is to learn the relationship among various attributes and output the classification model for predictive analysis. The algorithm is applied in both classification and regression problems.

3.1.4 Reinforcement Learning
Reinforcement learning, which is a learning method based on statistical and dynamic programming, uses environmental feedback as input data, not reinforcing the learning process. It is mainly used in problems related to control precision of robots, and the mainstream algorithms include Q-Learning and the temporal difference learning algorithm.

3.2 Introduction to Mainstream Supervised Learning Algorithms

3.2.1 Decision Tree
Here the decision tree in the field of supervised learning algorithms is specifically introduced. The decision tree is a tree-like prediction model in which each internal node represents a "test" on an attribute, each branch represents the outcome of the test, and each leaf node represents a class label. Its structure can help restore and understand the decision-making process of a given problem. Classical decision tree algorithms include ID3 and C4.5, which can solve optimal problems and multi-stage decision problems.

3.2.2 Random Forests
Based on Bagging, random forest is an important ensemble learning method for classification, regression and other tasks, which is operate by inputting unlabeled samples and outputting the results of classification voted by individual trees. The random forest adds the bagging technique to the decision tree, which enables it to solve the performance bottleneck faced by the decision tree. Its

introduction of randomness optimizes the anti-noise ability and reduces the risk of over-fitting, proving good scalability and parallelism in the classification of high-dimensional data. Random forests can handle discrete and regression problems well, whose input dataset does not need to be normalized.

3.2.3 Naive Bayes

In the early period from 1950s to 1960s, Bayesian School came into form. The Bayesian classification method, one of the statistical classification methods, can be used to predict the probability featuring a membership relationship and predict the probability of a given classification. Bayesian algorithm is a general term for a class of algorithms, among which naive Bayes is a simpler and more typical one. It is simply embodied in the requirement that the naive Bayes algorithm satisfy the conditional independence assumption, namely, the effect of each attribute on its target variable of the given classification is independent.

The naive Bayes algorithm utilizes the Bayes Theorem:

$$P(H|X) = \frac{P(H|X)P(H)}{P(X)}$$

3.2.4 Support Vector Machines (SVM)

The SVM algorithm was quite popular for a time, which maps the input space to a high-dimensional feature space by nonlinear transformation and finds the optimal linear boundary hyperplane in the high-dimensional special space. When the given dataset is linearly separable, the working principle of the support vector machine classification algorithm requires that the empirical risk be minimized. Therefore, in order to find an optimal boundary plane that can not only correctly separate the data of the two categories but also maximize the classification interval or classification gap between them. The purpose of maximizing the classification interval is to make the support vector machine have better generalization capabilities.

3.3 Main Unsupervised Learning Algorithms

3.3.1 The K-means Algorithm

MacQueen proposed the K-means algorithm in the 1960s, which is popular for cluster analysis. The idea of MacQueen's K-Means algorithm is as follows: Given n data points {x1, x2,..., xn}, K cluster centers {a1, a2,..., ak} is found so that the square sum of distance between each data point and its closest cluster center is the smallest, and the square sum of the distance is called the objective function W_n whose mathematical expression is:

$$W_n = \sum_{i=1}^{n} min_{1<j<k} |x_i - a_j|^2$$

The processing of its algorithm is[3]:

a. K sample points are selected in the sample dataset D, and values of these sample points are assigned to the initial clustering center $\overline{m}_i (i = 1, ... k)$;

b. The distance between each sample point from $p_j (i = 1, ... n)$ and its clustering center $\overline{m}_i$ is calculated:

$$d(i, j) = \sqrt{|p_j - \overline{m}_i|^2}$$

c. The minimum distance min (d (i,j)) between p_j and $\overline{m}_i$ is found, and place p_j in the cluster that is closest to $\overline{m}_i$.

d. The clustering center of each cluster again is calculate:

$$m_i = \frac{1}{n_i} \sum_{J_i=1}^{n_i} p^{J_i} \quad (i = 1, ... k)$$

e. The squared difference E(t) of all points in dataset D is calculated according to step one and compared with the previous error E(t-1).

f.E(t)-E(t-1) is observed if below zero, otherwise the algorithm ends.

Because the K-means algorithm is easy to describe, time-efficient and suitable for processing large-scale data, it has found wide application both at home and abroad in natural language processing, soil, archaeology, and many other fields since the 1970s.

3.4 Introduction to Main Semi-supervised Learning Algorithms

3.4.1 The Self-training Algorithm Based on the K-nearest Neighbor
Essentially, the principle of the so-called K-nearest neighbor algorithm is relatively simple, which uses the training set to divide the feature space into different regions and each sample occupies a certain region. When the test sample falls in the region of a training sample, it is considered to belong to the very training sample category. The above is all about the K-nearest neighbor algorithm for supervised learning. For the self-training K-nearest neighbor algorithm, there is no so-called training set. Instead, the feature space is used to divide different regions, and then gradually predict and classify the data category. Based on this, the prediction classification is gradually spread until all the samples are classified. By achieving self-training in the way mentioned above, we can obtain semi-supervised learning model of K-nearest neighbor.

3.4.2 The Semi-supervised Learning Algorithm Based on Divergence
The semi-supervised learning method based on divergence actually begins with a method. The main process of cooperative training is to first train a classifier by use of the training set, and then use the classifier to classify and label samples selected from unlabeled test samples. Then those classified test samples are added to the training set of the classifier, and the classification is continuously performed in this way until all the labeling is completed. This method has the advantage of receiving less interference from other factors, for instance, not being subject to model assumptions, non-convex loss function and data scale. Besides, this simple and effective method has a relatively rigorous theoretical basis, finding a wide range of applications. [4]

3.4.3 Semi-supervised Cluster
The clustering methods can be roughly classified into seven types, namely, partition clustering, hierarchical clustering, density-based clustering, network-based clustering, model-based clustering, clustering high-dimensional data, outlier analysis, and constrained clustering, which are a relatively important type of learning method in semi-supervised learning. The so-called clustering is the process of classifying the sample dataset into types featuring similarity. Semi-supervised clustering has achieved great progress in gene analysis, text mining, intrusion detection and other fields. Meanwhile, semi-supervised clustering has also accomplished much in image segmentation, and road detection and edge detection in GPS data. Therefore, semi-supervised clustering has become one of the machine learning directions worth researching and exploring in future [5].

4. The Significance of Machine Learning Research

4.1 The Significance of Machine Learning
The research of machine learning system not only has important scientific significance, but also has a wide range of application value. The question of how biology and humans form skills in the environment and how people acquire concepts and knowledge in their learning has long been one of the key issues studied in the fields of physiology, psychology, cybernetics, education, and philosophy. Therefore, studying the process of machine learning is also the way for further exploring human learning process, which will in turn facilitate the development of those subjects mentioned above. In addition, there are some machine learning examples or tools that have already been commercialized, proving that people are increasingly aware of the potential and significance of machine learning in the

future. What's more, as for the computer field, the demand from various industries for automation or intelligence is getting higher and higher, which means that the design of software programs becomes challenging and will face much higher requirement. However, the development of machine learning can be quickly applied to the design of programs and alleviate burdens on human beings through automation and intelligent programming, which is also one of the benefits that machine learning brings to programmers.

At present, base class systems such as the expert system and the semantic understanding system in artificial intelligence systems need to constantly learn from humans or existing training sets to master language skills and improve the function of judging semantics. The process of machine learning is similar to that of human learning, as if the "apprentice" gradually grows into an experienced "expert" through continuous absorption of new knowledge. Therefore, Samuel, known as the "pioneer of machine learning", believed that machine learning was the theoretical basis and at the heart of artificial intelligence. Because in both speech and image recognition, a better artificial intelligence model could only be obtained through machine learning, and a better "artificial intelligence" machine could only be created through continuous training and knowledge accumulation. Meanwhile, Samuel also believed that a man-machine system with complete intelligent interface would not appear or become possible until significant progress was made in machine learning.

To sum up, the significance of machine learning lies in that it is the theoretical basis and core part of artificial intelligence. Therefore, there is still plenty of space in the field of machine learning for study and exploration.

4.2 Applications of Machine learning

4.2.1 Computer Vision
Traditional machine vision methods rely primarily on custom features which cannot capture high-level boundary information. In order to make up for the insufficiency of small-scale samples to effectively express complex features, computer vision was begun to turn to deep learning. For instance, in 2012, Krizhevsky adopted DNN to classify the dataset of Image Net LSVRC 2010. The error rates on top1 and top5 were 37.5% and 17.0%, far more than those of the traditional method. Besides, deep learning has also performed well in face recognition. For example, in 2014 Sun Yi used deep-hidden identity features to represent facial features, whose test accuracy on LFW reached 97.45%.

4.2.2 Speech Recognition
Speech recognition has developed for decades, of which statistical method is the traditional method, mainly based on the combination of hidden Markov model and Gauss mixture model (HMM-GMM). The characteristics of the traditional method cannot cover the original structural features of the speech data, so the tolerance for data correlation is low, which can be made up by replacing GMM with DNN. For example, in 2012, Microsoft's speech video retrieval system reduced the word error rate from 27.4% to 18.5% through deep learning. DNN demonstrates improvement by about 10% compared with HMM-GMM, and CNN has stronger adaptability to data correlation than DNN.

4.2.3 Information Retrieval
Traditional information retrieval uses the TF-IDF system which is not only inefficient in dealing with large vocabulary problems, but fails to consider semantic similarity. Deep learning has been applied to information retrieval since 2009. DNN can well represent the word count feature of documents, and store semantically similar documents at similar addresses through deep self-encoders, thus improving retrieval efficiency. For example, in 2014, Shen Yelong proposed the convolution of deep semantic model which could project the same semantic words in the context through the convolution structure to the space vectors of the context features, thus improving the accuracy.

4.2.4 Natural Language Processing

Traditional natural language processing separately processes problems, such as language models and semantically related words, and there is no overall processing. Traditional systems have some shortcomings including shallow structure, linear and separable classification, and requiring preprocessing of artificial features. In 2008, Collobert began to apply DNN in the field of natural language processing, generating an error rate of 14.3%.

4.3 Future Trends

Over the past two decades featuring rapid Internet development, machine learning has gradually become a foundation and has been applied to various disciplines and fields. The development and application of machine learning not only promotes the improvement of its own form, but also facilitates the maturity of machine learning technology. From the current development and research trends, the future development trend of machine learning features the following aspects:

a.   Unlabeled data learning;
b.   Machine learning is applied to the mobile terminal or other mobile devices.
c.   Study of machine learning is extended to the level of deep learning.
d.   Deep learning is extended from static tasks to dynamic decision-making tasks.
e.   Practical applications of natural language are further extended.

Natural language processing is mainly used to realize the technology of man-machine interaction and communication, whose main purpose is to make the machine intelligent enough to communicate with human beings, conforming to the idea and original intention of artificial intelligence. The examples that have been put into daily use include products with voice assistants or similar, because voice assistants can penetrate a wide range of fields and establish interactions between people and people, or people and information data. Many mobile devices and Internet applications involve voice products. For example, Apple's Siri informs people about the weather condition, the temperature, and smog index of the day, and also gives suggestions for dressing and tripping. With the voice assistant, manual search for relevant information through the search engine is no longer necessary, and all it takes is one sentence requiring specific information.

f.Machine vision will keep penetrating into the production and living areas.

Human vision is limited and the range and precision that can be seen with the naked eye are very limited. Therefore, human beings are constantly creating various tools to break through such restrictions. For instance, telescopes and microscopes were invented to extend the visible range for human beings to better understand the world. Since the birth of computer technology, various tools have been created to assist humans in visual expansion. At present, high-precision instruments and impurity detection are typical machine learning tools mainly used in production and life, compared with which it is difficult for human eyes to reach such precision. This is exactly what computer vision needs to solve. Meanwhile, Baidu's driverless cars, intelligent sensing systems, and medical imaging technologies are all realized through computer intelligent image analysis, pattern recognition or other technologies. These visual technologies, by utilizing intelligent sensors, video surveillance equipment, and other means, can sense changes in the environment so as to provide the most important and correct decision support by conducting analysis through a deep learning platform in the background. Computer vision is also a direction of artificial intelligence which will have great application in human production and life.

g.   Blockchain accelerates platform establishment and integrated development.

The most important feature distinguishing a blockchain from a traditional database is its decentralization. The blockchain is based on distributed access when storing data and needs to reach consensus with other nodes. Therefore, if someone wants to illegally tamper with the data, he has to get confirmed by all the distributed nodes at the same time and change the previous records. One node does not confirm and the data cannot be altered, which is the biggest benefit of the blockchain technology. It guarantees data security, because all data is agreed upon by all nodes and each node is equally important, demonstrating high credibility. If the blockchain technology were to be used in the banking sector, the clearing center would be no more necessary because data can be stored through the nodes, which greatly saves the costs of time and storage. For example, the security of IoT (Internet of Things) devices is confronted with great challenges, but the blockchain technology can ensure the security of IoT data storage. Therefore, combining the two enables the IoT technology to better develop. However, the blockchain technology has limited applications for now, which requires further exploration.

5. Conclusions and Prospects

Machine learning, which is at the core of artificial intelligence, is an interdisciplinary subject itself involving many fields such as the data structure of the computer itself, probability theory, statistics and other related subjects. The ultimate goal of machine learning is to maximize the processing of the data itself through algorithms or knowledge in different fields, so that computers can achieve the process of self-learning through corresponding algorithms. Therefore, it can be seen that the research on and selection of the algorithms are the most important part of machine learning on the whole.

The main advantage of machine learning lies in that the computer can achieve the purpose of self-learning and even predict the trend through operating algorithms, which is of great value to artificial intelligence as well as other fields. Because of this feature, the computer can be continuously trained, the training dataset can be increased, and over time more accurate results can be obtained through data accumulation. It thus can conduct analysis and prediction for many fields, greatly enhancing the performance of artificial intelligence and making an unusual difference to the development of human society.

References

[1] Liu Qin, Machine Learning [J], Journal of Wuhan Engineering Institute, 2001, 06:41-44.
[2] Li Zhiwei, The Development of Machine Learning and Several Learning Methods[J], Industrial & Science Tribune, 2016, volume 15, issue 10.
[3] Han Xiaohong & Hu Yu, Research on K-means Clustering [J], Journal of Taiyuan University of Technology, 2009, 05.
[4] Zhou Zhihua, Semi-supervised Learning Based on Divergence [J], Acta Automatica Sinica, 2013, 11.
[5] He Gaili, Research on Outlier Clustering Method Based on High-dimensional Sparse Data[D], Journal of Chongqing University of Posts and Telecommunications, 2001.

ISPECE   IOP Publishing

Spatial Spillover Effects of The Real Estate Industry on Economic Development
-from Destocking Perspective

Xiaoqian Liu[1], Qinqin Zhou[2] and Chang'an Wang[1,a]

[1]Research Institute of Economics and Management, Southwestern University Of Finance and Economics，Chengdu 611130, China

[2]Office of Budget & Finance,Southwest Jiaotong University, Chengdu, 611756,China

[a]Corresponding author: Chang'an Wang, wangtcjy@163.com

Abstract: Based on the statistical data of 31 provinces of China from 2005 to 2015, A spatial lag model is applied to study the Spatial Spillover effects of the real estate industry on economic development. The results show that: the real estate stocks have a significant negative effect on regional economic development; Due to the economic downturn, the real estate investment and total investment in fixed assets continue decreasing, the descending speed of the former is faster than the latter. The real estate industry will, through its accounts for a high proportion of the total social investment, further through the accelerator effect, drag on economic growth; The increase of average price plays a positive significant role in boosting the development of regional economy in recent years; The profit rate of the real estate industry continues declining, and despite a further increase in the added value of real estate industry, it is not enough to significantly promote regional economic development. Therefore, to formulate effective policies can make the relationship more balanced between supply and demand of real estate, and can make real estate investment structure and inventory more reasonable, in order to make economic growth more stable.

1. Introduction

At present, China's economy has been in a critical period of economic transformation, upgrading and structural adjustment. The rate of economic growth has gradually slowed down, and the excess production capacity caused by the backlog of multiple industry products has become an important factor that restricts the further development of China's economy. By the end of 2015, China's commercial housing sales amounted to 720 million square meters, the inventory cycle was up to 30.2 months, indicating that China's real estate inventory was at a high level. In the macroeconomic environment where the downward pressure on the economy continues to increase. Will the large amount of real estate inventory affect the healthy development of the economy?

As an important sector of the national economy, Real estate itself is a part of the economic development. Therefore, the healthy development of real estate will contribute to the sustained development of the economy. Previous studies have supported this conclusion. Coulson and Kim

---

Content from this work may be used under the terms of the Creative Commons Attribution 3.0 licence. Any further distribution of this work must maintain attribution to the author(s) and the title of the work, journal citation and DOI.
Published under licence by IOP Publishing Ltd

found that short-term fluctuations in housing investment had a more important impact on the national economy, and its strong changes will lead to instability on the national economy[1]. Huang and Wu, from the national level research, found that real estate investment can cause economic growth[2]; from the regional level, real estate investment not only can promote and enhance the economic growth in the region, but also can it promote economic growth in other areas[3], and the contribution and influence of China's real estate investment on the regional of economic is greatly different [4-5], the eastern region is the highest, followed by the central region, the western region is the least[2], which are evidences that real estate investment can lead to economic growth.

But at the same time, the development of real estate may have adverse effects on economic growth. Shen using the spatial panel model, shows that the development of China's real estate industry has a promoting effect on the financial stability of the eastern and western regions, but real estate's inventory is not conducive to the China's financial stability [6]. Han using the PVAR model, through the 31 provinces of China's macro data research, found that the real estate inventory has a negative impact on economic growth, real estate investment and housing prices [7]. Since 2014, the real estate investment continued to decline, which has a negative impact on related consumption growth, economic development pulled by the real estate prosperity stage is over[8].

China's economy has entered a new normal context, Real estate high inventory problems caused by blind investment in real estate are urgent to be solved. As can be seen from the existing research, quantitative research on the impact of real estate inventory on regional economics are scarcely involved. In addition, most scholars present the research from the whole country and one city's perspective, but ignore the regional own characteristics of the real estate industry.

Since 2015, the government has introduced a lot of inventory policies, but the stock of real estate remains still high, and the stimulating effect is not obvious. By introducing spatial weight into traditional OLS regression, analysis is extended to two dimensions-space and time, we study the spatial spillover effects of the real estate inventory on regional economic development. This is of great practical significance for solving the high inventory of real estate, to promote the healthy development of the real estate market, and to coordinate the regional economic development. The remainder of this paper is organized as follows: the second part introduces main data source, variable selection and model construction; the third part analyzes the model estimation and results; The fourth part draws conclusions and gives policy recommendations.

2. Methodology

In this section, considering the existence of spatial correlation among China's provincial economy development, it is necessary to test its spatial correlation. Spatial autoregressive model (SLM) is a popular model, which is used to study on interdependence in neighboring regions. The behavior of the regions within the system has an impact on each other and ultimately results in a balanced outcome, that is to say, the behavior of one region affects the behavior of the surrounding area, and the behavior of the region is eventually formed by the influence of the behavior of the surrounding area. Its mathematical expression is as follows:

$$y_{it} = \rho \sum_{j=1}^{N} W_{ij} y_{jt} + X_{it}\beta + \varepsilon_{it} \qquad (1)$$

Where Y is the dependent variable vector, W is the spatial weight matrix, ρ is the spatial autoregressive coefficient, X is the explanatory variable vector, β is the regression coefficient vector, ε represents stochastic disturbance term vector.

3. Variable selection and data source

3.1 Data description

The data are mainly derived from China Statistical Yearbook, Statistical Yearbook of Provinces, Regional Statistical Yearbook of China and Wind database. The regional economic development level mainly includes GDP, local fiscal revenue, fixed assets investment, per capita net income of farmers, urban residents' disposable income and the proportion of the three industries; The real estate industry level includes real estate inventory and real estate industry representative index. From 2005 onwards, the country has begun to control the real estate industry, the "eight" and "six" regulations have been promulgated, and then national and local government gradually has begun to adjust real estate industry market, and has introduced a number of regulations and policies, so the time window is from 2005 to 2015.

3.2 Variables selection

The variables selected in this paper include the level of regional economic development, real estate inventory in various regions and the related indicators affecting the development of real estate industry.

3.2.1 Regional economic development level

Adopted Yang's method of measuring China's regional economic development[9], Table 1 shows comprehensive evaluation index system of regional economic development level, CD represents comprehensive index of economic development.

Table 1 Comprehensive Evaluation Index System of Regional Economic Development

Index	Code	Unit	Property	Weight
GDP	a1	Billion	+	0.1393
Per Capital GDP	a2	Yuan/person	+	0.0680
Local fiscal revenue	a3	Billion	+	0.1599
Fixed assets investment	a4	Billion	+	0.1428
Retail sales of social consumer goods	a5	Billion	+	0.1574
Balance of deposits of financial institutions	a6	Billion	+	0.1738
Per capita net income of farmers	a7	Yuan/person	+	0.0557
Urban residents' disposable income	a8	Yuan/person	+	0.0956
Proportion of output value of third industry	a9	Percent	+	0.0075

3.2.2 Real estate inventory

The definition of the real estate inventory in existing literature is few. Follow Han Guogao's definition of real estate stock[7] and defines it as dividing the construction area by sales area of commercial buildings in each province. Use kc to represent real estate stocks. The greater the value is, the higher the real estate inventory will be.

3.2.3 Control variables

The control variable selects the representative index that influences the development of real estate industry. Indicators take ratio form, so that it can reflect the reasonable structure of the real estate industry, and eliminate the impact of the consumer price. Among them, reir represents the dividing real estate industry added value by GDP, refi represents dividing real estate investment by social total investment, ap represents average price of commercial housing.

4. Model construction and empirical analysis

Set up SLM model, formula (2):

$$CD_{it} = \alpha_0 + \alpha_1 kc_{it} + \alpha_2 reir_{it} + \alpha_3 refi_{it} + \alpha_4 ap_{it} + \rho \sum_{j=1}^{N} W_{ij} CD_{jt} + \varepsilon_{it}$$

$$\sum w_{ij} CD_{jt}$$

is the spatial lag term, ρ is the coefficient of spatial lag correlation, ε_{it} Represents a random disturbance term.

4.1 Calculation of regional economic development level

The level of regional economic development is calculated by the method of comprehensive evaluation, to start with, which needs to determine the weights. The common methods of determining weights are subjective and objective weighting methods. Considering that there are great differences among different regions, it is difficult to make a uniform assignment by subjective weighting method. Therefore, The entropy weight method is adopted. In order to compare between different years, we adopt Yang's improved entropy method [9], which adds the time variable into this method, making the analysis results more reasonable.

4.2 Spatial correlation analysis

4.2.1 Global spatial autocorrelation

The spatial autocorrelation test of regional economic development was carried out by using Arcgis and GeoDa software. Firstly, the spatial weighting matrix of 31 provinces is constructed. In this paper, R (Rook Contiguity) method is adopted to establish the 0-1 spatial weight matrix (31 * 31). In order to eliminate the island effect of Hainan Province, Hainan was set up to be adjacent to Guangdong province. Then, we calculate the overall Moran I index of China's regional economic development during 2005-2015 years. Table 5 shows the global Moran I index is positive and significant at 1% confidence level, which indicates that the level of regional economic development in China has significant spatial agglomeration effect on Geography: Provinces of high economic development level are central to high level.

Table 2 Moran'I Index of Regional Economic Development in 2005 -2015 Years

year	2005	2007	2008	2009	2010	2011	2012	2013	2014	2015
Moran I	0.343	0.344	0.344	0.336	0.336	0.328	0.321	0.316	0.348	0.334
Z value	3.573	3.271	3.371	3.286	3.336	3.164	3.148	3.126	3.394	3.151
P value	0.002	0.003	0.003	0.008	0.003	0.005	0.004	0.006	0.003	0.006

4.2.2 Autocorrelation test of local space

In order to further demonstrate the level of regional economic development, spatial agglomeration is tested by the local spatial correlation. Figure 1 shows the LISA agglomeration of regional economic development in 31 provinces in 2005 and 2015. The dark area is H-H, which belongs to the province of high economic level spatial lag area, slightly light area is L-L, which belongs to the low level spatial lag area, the provinces of which belongs to negative cluster region with low economic level. Provinces of higher economic level are adjacent to each other, the eastern coastal areas are with higher economic level; Western regions are relatively lower; The central areas do not show significantly cluster. See from the trend, the overall spatial pattern has not changed greatly, which indicates that regional economic level in different provinces exist spatial correlation. If the spatial correlation is not taken into consideration, it will inevitably lead to deviation of estimation.

ISPECE   IOP Publishing

IOP Conf. Series: Journal of Physics: Conf. Series **1187** (2019) 052104   doi:10.1088/1742-6596/1187/5/052104

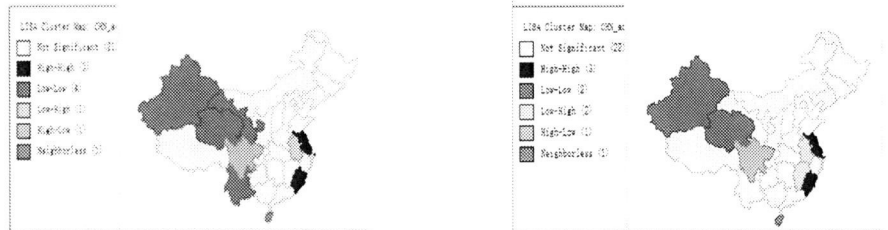

Figure 1 LISA Concentration Diagram of Regional Economic development in 2005 and 2015 years

4.3 Space panel econometric analysis

The regional economic development level, real estate industry data and the spatial weighting matrix are introduced into Stata12.0 software. Spatial fixed effect, Time fixed effect and both fixed effects are estimated respectively. The estimation results are shown in Table 3.

Table 3 shows individual effect, time effect and both effects of the SLM model were all significant at the 1% confidence level, indicating that the real estate inventory on the regional economic development has obvious spatial spillover effect. In the both fixed effect model, the Hausman test showed that original model was rejected at the significant level of 1%, so fixed effect model was better than random effect model. Moreover, the log likelihood of both fixed model is the highest. We combine the data characteristics and practical significance, the estimation results of both fixed model are chosen.

Table 3 The Estimation Results of Fixed Effects of SLM Model

	variable	Spatial fixed effect	Time fixed effect	Both fixed effect
Main variable	kc	-0.007***	-0.011	-0.013***
		（2.692）	(-1.975)	(-3.929)
Control varable	reir	0.508	-3.832***	0.312
		(1.195)	(6.084)	（0.777）
	refi	-0.256***	-0.329***	-0.196***
		(-4.178)	(-3.772)	(-3.408)
	ap	0.155***	0.152***	0.091***
		（5.930）	(4.758)	（3423）
ρ		0.512***	0.002***	0.201***
		（10.653）	（0.043）	(3.305)
σ^2		0.002***	0.010***	0.002***
		（12.816）	（13.058）	(13.007)
R^2		0.4515	0.210	0.008
Log-Likelihood		548.8971	305.6937	592.1909
Hausman test		-0.09	49.85***	126.98***
			(prob>chi=0.0000)	（prob>chi=0.0000）

"***、**、*"respectively represents 1%、5%、10% significance level.

Through estimation results of both fixed model, further analysis found that the real estate inventory coefficient is -0.013, which is significant at 1% level, indicating that the higher the real estate inventory is, which is not conducive to the promotion of regional economic development. In the short term, the inventory pressure is too large, and it will take a certain period to digest the established stock. Real estate stocks may cause negative effects as follows: (1) Because the real estate companies mainly use credit funds for the operation of real estate, the existing inventory is too high and occupies large

2738

funds, which lead to long payback period, so the use value is too high, resulting in new real estate underemployment rate, the new investment is at a low level, there are obvious negative effect on stimulating regional economic development. (2) Due to the presence of a large number of real estate stocks, although the government issued a number of related policies, the stock structure of real estate can not be quickly adjusted to a reasonable interval, which in turn result in total high supply. Consumer spending and consumption tendency have not changed, so the short-term supply and demand sides cannot form the benign interaction. (3) The real estate industry chain is length and strong relevance. As real estate inventory is too large, first of all, investment of the real estate reduces, which in turn reduces related downstream industry demand and has impact on related industries, leading to a negative effect on regional economic development; Secondly, the reduced investment of real estate will produce negative impact on the other industries, in order to deal with the real estate of unwanted inventory, making other industries actively reduce production.

The coefficient of the control variable refi is significant at the confidence level of 1%, with an estimated value of -0.196. It shows that the increase of refi is negative to regional economy. As shown in Figure 2, the real estate and total investment in fixed assets due to the economic downturn continues to decline, the formal rate is declining larger than that of the latter, the real estate through its high accounts for proportion of the total social investment, further through the accelerator effect, will drag on economic growth. The increase of AP has a positive effect on regional economy. Under normal circumstances, general goods and services prices accounted in CPI, which will be deducted from the factors in the actual GDP statistics, but housing prices which account for high total domestic consumption will not be included in CPI, AP average price increase will boost regional economy. There exists a certain lag, house prices increase will promote the rent and other housing consumption costs, and will increase price of durable consumer goods and services by stimulating home appliances and decoration services. Moreover, Although housing price is high, investors still take optimistic attitude towards it. Because the channel of investment is limited, real estate has become a good investment tool. The real estate industry added value accounted for the proportion of GDP (reir) coefficient is not significant. This may be because the golden age of the real estate industry has gone, the profit rate of the real estate industry continued to decline, which means the real estate industry do not have enough driving force to stimulate the development of regional economy.

4.4 Robustness test
In order to ensure the reliability of the estimation results, this paper tests the robustness by changing the spatial weights. The spatial weight of the robustness test is the reciprocal of the shortest distance between two provinces.

Table 4 Space and Time Fixed Robustness Test

Variable	Both fixed effects（0-1 matrix）	Both fixed effects（reciprocal of the shortest distance between two provinces)
kc	-0.013*** （-3.929）	-0.014*** （-4.197）
reir	0.455 （1.135）	0.455 （1.135）
refi	-0.196*** （-3.408）	-0.161*** （-2.838）
ap	0.091*** （3423）	0.106*** （4.111）
ρ	0.201*** （3.305）	0.429*** （3.116）

σ^2	0.002***	0.002***
	（13.007）	（12.916）
R^2	0.008	0.05
Log-Likelih ood	592.1909	591.0687

The estimation results show that the estimated coefficients of each variable are similar, and the significance levels and symbols do not change, which shows that the model is robust and the results are reliable.

5. Main conclusions and policy recommendations

Real estate inventory has a significant negative effect on regional economic development. The existing real estate inventory is too high and occupies huge funds, while consumer consumption trend has not changed, the imbalance between supply and demand lead to the the negative impact on regional economy; Real estate industry chain is long and has strong relevance, which further reduce the level of regional economic development of other industries; As a result of the economic downturn, the amount of investment in real estate and fixed assets continue to decline, which will decline economic development; The increase in average housing price has a positive effect on the regional economy; As the real estate industry continues to decline in interest rates, with the real estate industry added value increasing, which can not significantly promote economic development.

The supply structure of real estate should be adjusted. For the high real estate inventory in third and fourth-tier cities, land transfer should reduce; government can come forward to buy real estate, transform them into low rent housing, and make the real estate supply structure more reasonable; Combine with the local real estate situation, make suitable local real estate policy and discuss with real estate developers to set down the bottom line of real estate price, break consumer further expectations of reducing price. These policies in the long term can play an effective role. In the short term, the local government according to local conditions, adjust real estate market structure, let the market mechanism play a decisive role in high real estate inventory regulation.

Reference:

[1] Coulson N E, Kim M S. Residential Investment, Non-residential Investment and GDP [J]. Real Estate Economics, 2000, 28(2):233-247.

[2] Huang Zhonghua, Wu Cifang, Du Xuejun. The Real Estate Investment and Economic Growth -- Based on the National and Regional Panel Data[J].Finance and Trade Economics,2008 (8): 56-60.

[3] Zhang Hong, Jin Jie, Shi Quan. Real Estate Investment,Economic Growth and Spatial Effect—An Empirical Research Based on the Spatial Panel Data from 70 Cities in China[J]. Nan Kai Economic Research, 2014 (1): 42-58.

[4] Lu Juchun, Jia Ziwu, Tian Hongfen. Research on Regional Disparity of Real Estate Investment and Economic Growth in China[J]. Journal of Wuhan University of Technology, 2008, 30 (6): 959-963.

[5] Luo Guoyin. Different Contribution of Real Estate Investment to Regional Economic Growth -- Based on Panel Data Analysis[J]. Search, 2010 (9): 50-51.

[6] Shen Bo. the Influence of Real Estate Industry on Regional Financial Stability in "Destocking" Perspective-An Empirical Study Based on Spatial Panel Model [J]. Journal of He Bei University of Economics and Business, 2016, 37 (3): 61-66.

[7] Han Guogao. the Influence of Real Estate Stocks on China's Real Estate Market and Economic Growth Based on PVAR Model [J]. Management Modernization,35 (1): 16-18.

[8] Zhu Jianfang. The Drop of Real Estate Investment Affect Economic Growth [J]. Market Observation, 2014 (1): 44-45.

[9] Yang Li, Sun Zhichun. The Development of Western New－Type Urbanization Level Evaluation Based on Entropy Method [J]. Economic Problems, 2015 (3): 115-117.

Application of Big Data in Forecasting Traffic Flow

Luo wanbo, Wan Xing*, Huang min

(Leshan Vocational & Technical college, Sichuan Leshan, 614000)

corresponding author email: krantson@163.com

Absrtact: According to the traffic situation of a city, this paper sufficiently integrates the relevant information of traffic management of the whole city, analyses and puts forward a set of modern public security traffic flow service system, which can provide effective technical support and help for the reasonable control of traffic system and effectively solve the outstanding traffic management of the whole city. Rationale. This paper mainly studies the practical application of big data in traffic flow forecasting through model design.

1. system architecture

The overall architecture of the system is shown in the following figure:

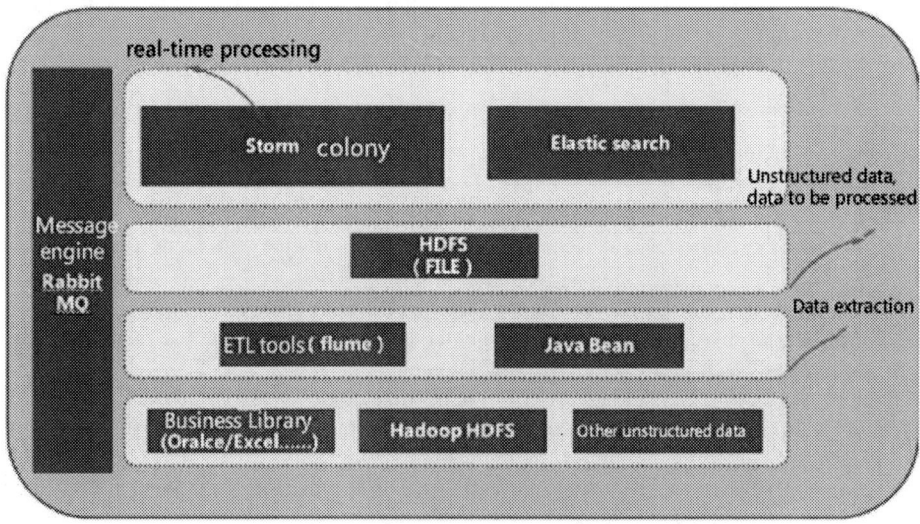

Figure 1 - System Overall Architecture

1.1 Technical framework

The analysis of road traffic flow is mainly based on microwave, video, ground inductance coil detection and other methods, taking into account GIS map, weather data and other data, and then comprehensive data analysis. The analysis data mainly includes all aspects of vehicle flow information summary, and then make the overall data evaluation and analysis.

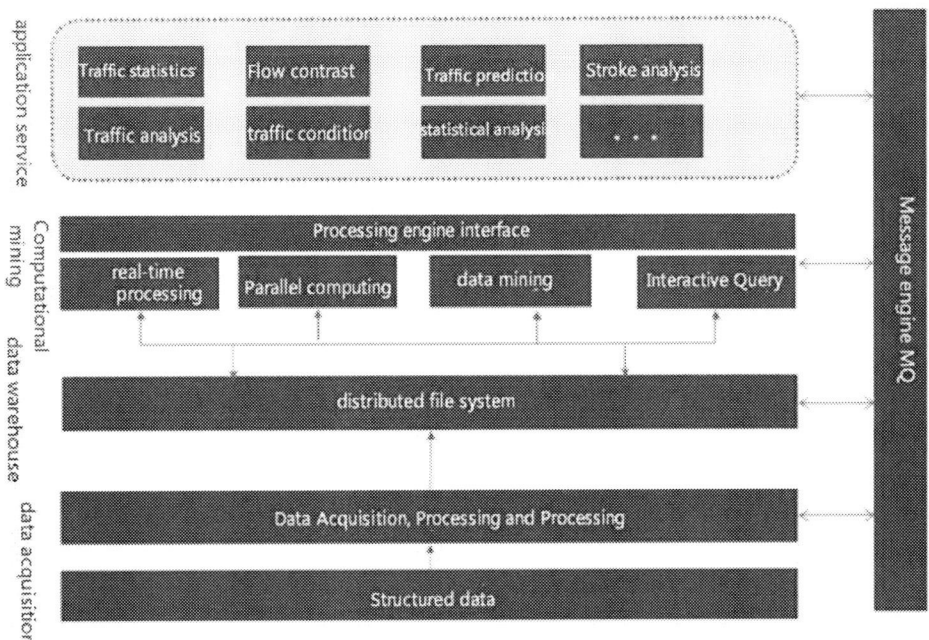

Figure 2 - System Technical Architecture Diagram

1.2 System deployment design

The main part of the system consists of five modes: storm cluster, application server, Hadoop cluster, prediction module and message server.

Traffic flow information access fully ensures that traffic flow information collected through various channels can be quickly and accurately connected to the system. Computer processing module should ensure the stability and reliability of the calculation.

2. functional design

2.1 Functional architecture diagram

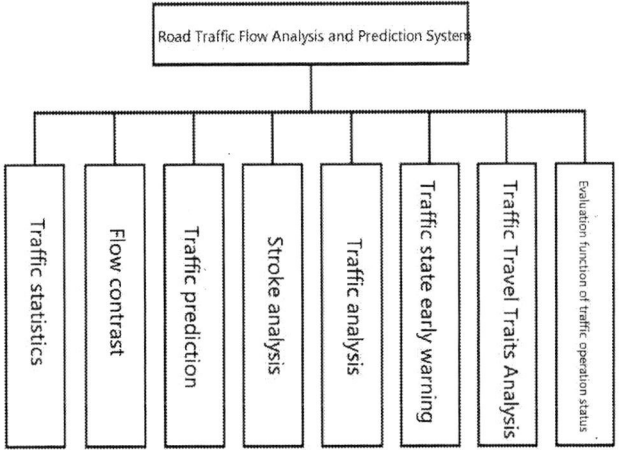

Figure 3- functional framework diagram

The functional framework of the system, road traffic flow analysis and prediction system, mainly consists of eight main parts, as shown in the figure.

2.2 Traffic statistics
For traffic statistics, it is usually used to expand the traffic statistics for a certain point, and then set the time period of statistics to minutes.

2.3 Flow contrast
Through the relevant statistics of points, we can make relevant comparisons, which include the comparisons between points, as well as the comparisons between different periods.

2.4 Traffic prediction
Through the combination of traffic flow monitoring and floating vehicle monitoring, we collect and analyze data, and then predict the traffic situation. This prediction includes a variety of contents, such as traffic flow, traffic speed and so on. The forecast time can also be set according to actual needs. Enough to provide information technology support for the city's traffic management.

2.5 Stroke analysis
Travel analysis is to provide the most real-time and reliable traffic service information by analyzing the relevant data of the past through large data analysis for the distance, normal travel time and route of the two assumed coordinate points. Traffic conditions are different in different time periods and different sections of the day, so it is necessary to recommend the best route according to the real-time traffic conditions.

2.6 Flow direction analysis
For the set data points of flow monitoring, the direction of flow can be accurately judged by analyzing the monitoring data of data points. Later, the vehicle license plate recognition system can be added to judge the direction of flow more conveniently.

2.7 Traffic state early warning and forecasting
For the important position of intersections and sections in urban roads, we should timely and accurately understand the speed of vehicles, and then through large data comparison, we can report special situations to traffic command department in time.

2.8 Traffic Travel Traits Analysis
Through the relevant monitoring equipment for the road section traffic situation, the vehicle traffic situation within a certain area, and then combined with the traffic volume, vehicle distribution, flow direction and other aspects, the study of travel characteristics is carried out.

2.9 Evaluation function of traffic operation status
The evaluation of traffic operation status mainly refers to the study of the level of urban traffic in a certain period of time. The dimensions of this evaluation can be divided into many kinds, one is according to time, year, month, day and hour; the other is according to work or rest, peak or flat peak; the third is according to the situation of events, such as bad weather, sudden situation, etc. and the fourth is based on the road conditions in different areas, such as urban main roads, sub-main roads and so on.

Through the traffic flow analysis system, we can analyze according to the relevant information of vehicle operation and large data, and then directly predict the traffic situation of the road network. At the same time, we can analyze and evaluate the real-time traffic situation, such as traffic saturation, traffic capacity and so on.

In addition, the content of the evaluation is graded, and the degree of unobstruction is usually assessed, as shown in the table below.

Table 1 - Congestion Degree Feedback Table

Level (example)	Number of unobstructed sections (examples)
A: patency	$\geq 75\%$
B: basically smooth.	60% less than the number of unobstructed sections < 75%.
C: initial congestion	45% less than 60% of unobstructed sections
D: congestion	30% less than the number of unobstructed sections < 45%.
E. serious congestion	15% less than 30% of unobstructed sections
F: Partial or extensive road paralysis	Smooth sections <15%

3. Business process specification

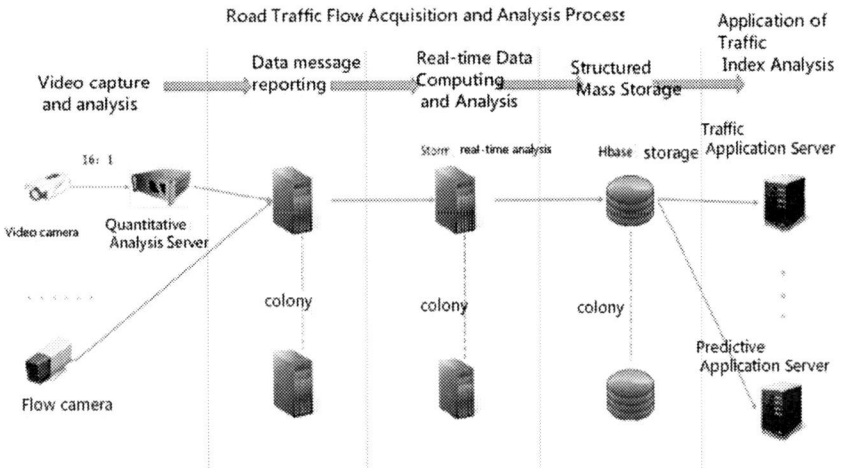

Figure 4 - Road Traffic Flow Acquisition and Analysis Process

The main process of road traffic flow forecasting includes several aspects: first, information collection, second, data upload, third, information data analysis, fourth, information data storage and fifth, related applications.

Information acquisition: Using a dedicated traffic surveillance camera and related terminal equipment, after obtaining relevant food data and fully identifying, the most original information can be obtained.

Upload of data information: adopt fast and reliable related technology to transmit the collected data to the central platform of traffic flow through the form of message.

Analysis of information data: Data analysis is carried out for the collected relevant data and processed according to actual needs, so as to facilitate the use of different applications.

Storage of information data: The distributed storage of large data is adopted to store the adopted data and processed data.

Relevant applications: Predictive analysis model and other content are used to analyze the relevant data of traffic flow, and then the required data content is given to the demander.

4. Summary

In the face of more and more automobile cities, we should make full use of the role of data in predicting traffic flow, through continuous in-depth data mining and integration, build a road traffic prediction system, fully guarantee the smooth operation of the road traffic system, and lay a technical support for the future development of urban traffic.

Reference
[1] Cai Xiaoyu, Tan Yuting, Lei Cailin, Liu Xiucai. Research on short-term traffic flow forecasting under traffic big data environment [J]. Railway transportation and economy, 2018, 40 (08): 88-93.
[2] Sun Lei, Yang Weiguo, Zhu Junchen. Innovative Application of Big Data in Intelligent High Speed [J]. Information Technology and Standardization, 2018 (Z1): 20-23.
[3] Wei Lingxiang, Chen Hong, Wang Yonggang, Cai Zhili, Zhong Dongqing, Li Yuhua. Short-term traffic flow forecasting method [J]. Journal of Shandong Jiaotong University, 2017, 25 (03): 22-29.
[4] Sun Tongxin, Wang Shikun. Highway traffic forecasting under the background of big data [J]. China Public Safety, 2016 (16): 105-109.

Financial Risk Analysis and Early Warning Research Based on Data Mining Technology

Yan Hou*, Ziyan Yuan

Yunnan Technology and Business University, Kunming,651700

Corresponding Author Email: 654775934@qq.com

Absrtact: With the development of information technology, how to mine useful information from a large amount of information and effectively analyze and guard against financial risks has become an urgent problem for enterprises to solve. This paper briefly expounds the relevant knowledge of financial risk, explains the specific application of data mining in financial risk analysis, and provides some ideas for the analysis and early warning of enterprise financial risk.

1. Introduction

With the continuous progress of science and technology, how to select valuable information from the cumbersome amount of information, and then carry out reasonable research and control on the financial risk of enterprises is something we need to deal with urgently. This paper mainly studies the practical application process of data mining in financial risk analysis, hoping to give an analysis and early warning of the financial risk of enterprises. Set help.

2. Summary of Data Mining

2.1 The Concept of Data Mining

Data mining is often applied in the current era of big data. Its core idea is to use machine learning, artificial intelligence and other methods to find useful and valuable information from a pile of data, so as to facilitate some research and application[1].

ISPECE   IOP Publishing

IOP Conf. Series: Journal of Physics: Conf. Series **1187** (2019) 052106   doi:10.1088/1742-6596/1187/5/052106

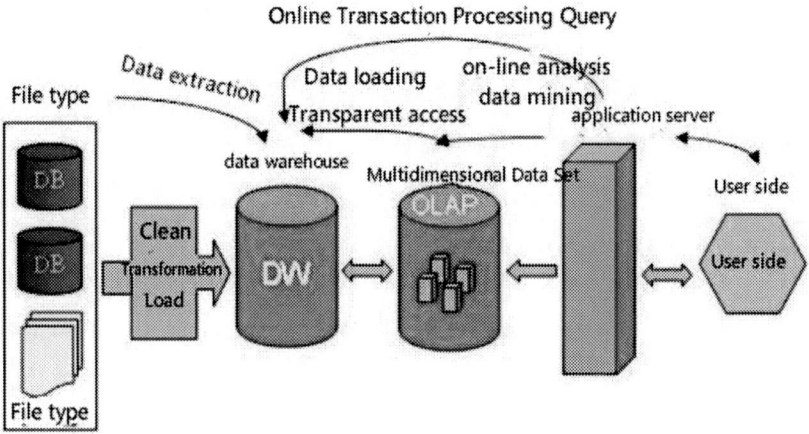

Figure 1 - Data mining

2.2 Implementation steps of data mining in financial risk analysis
(1) To clarify the content of financial risk analysis, it is necessary to clearly determine the objectives of financial risk analysis. To carry out data mining, the first thing is to clarify the objectives of data mining, so as to make the content of data mining accurate.

(2) Data collection of financial risk. To do data mining, the first thing is to prepare the data. There are many sources of data, such as accounting information system, data warehouse, or other data sources[2].

(3) Data preprocessing. Many times the data we collect may have many problems, such as incomplete data, non-standard data content, more cumbersome structure and so on. At this time, we need to deal with the data to facilitate the subsequent data mining.

(4) Data mining. Through the data pre-processed before, data mining is carried out. At this time, it is necessary to select the appropriate mining algorithm, and then wait for the completion of system data mining.

(5) Evaluation and interpretation of results. According to the purpose of enterprise's financial decision-making, it evaluates the content of data mining, checks every link of the model, and then finds the best model, makes corresponding evaluation, and explains it with the relevant knowledge of finance[3].

(6) Assimilation of knowledge. Integrate the knowledge gained from the research and analysis into the business information system, and finally practice the task of financial risk analysis. (Fig. 2)

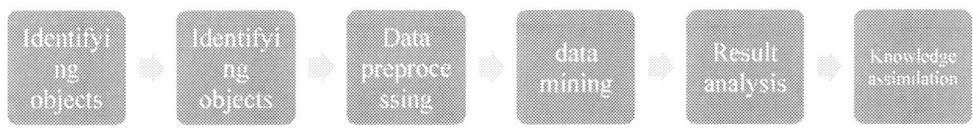

Figure 2 - Flow chart of financial risk analysis using data mining technology

3. Enterprise Financial Crisis Early Warning Method Based on Data Mining Technology

3.1 Financial Crisis Early Warning Index System

Financial crisis early warning system is to analyze and summarize the relevant data of the financial situation of enterprises, and then provide technical support for financial decision-making of enterprises by using data mining technology. Therefore, in the selection of indicators system, we mainly consider some of the financial data indicators in our regular enterprises.

3.2 Equal Area Division Method of Financial Crisis Early Warning Indicators

The data of financial crisis early-warning indicators of enterprises are presented with continuous random distribution. If we use more enterprises to carry out analysis, then all the indicators will show normal distribution. According to this method, we divide the time of financial crisis early-warning indicators through the distribution of indicators, and delimit them according to the area. According to the principle of enterprise life cycle, all indicators have five stages, from one to the house, and then all indicators of enterprise financial crisis warning are divided into five zones[4].

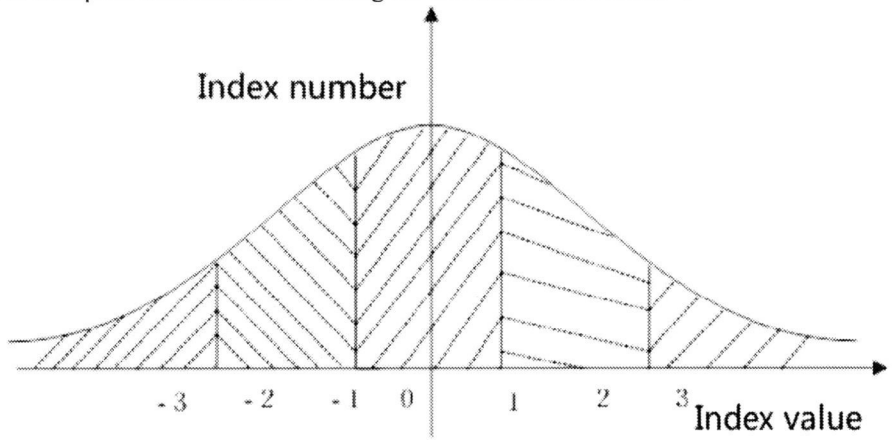

Figure 3 - Equal Area Division of Financial Crisis Early Warning Indicators

3.3 Enterprise Financial Crisis Early Warning Model Based on Dynamic Maintenance of Time Series

Based on the theory of life cycle, this paper discusses the interrelationship and development direction of financial crisis early warning indicators in each period of life cycle, and provides information support for the release of financial early warning signals.

3.3.1 Mining Method of Financial Data Based on Time Series

There are three main aspects in data mining for financial data with time series characteristics: firstly, the financial data studied has the characteristics of time series; secondly, the basic idea of data mining is to find the relevant laws and characteristics through time series data; thirdly, the final content is to find the corresponding laws through financial data. Features, and then make relevant predictions and decisions[5].

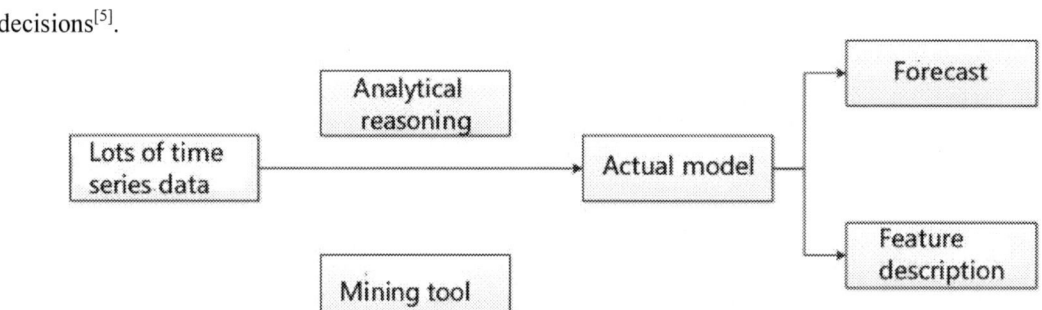

Figure 4 Data Mining Model Based on Time Series

3.3.2 Incremental mining strategy based on time series
For the incremental mining of time series, it is to distinguish the related data of enterprises according to the quarter. For example, the data of the first quarter of a year is the initial data content. Then the second quarter data is the newly added data. In fact, after the first quarter data mining is successful, incremental mining is carried out on the data of the second quarter, and then mining is carried out in accordance with the order. Mining, has been the end of data mining, time series-based incremental mining strategy as shown in Figure 5.

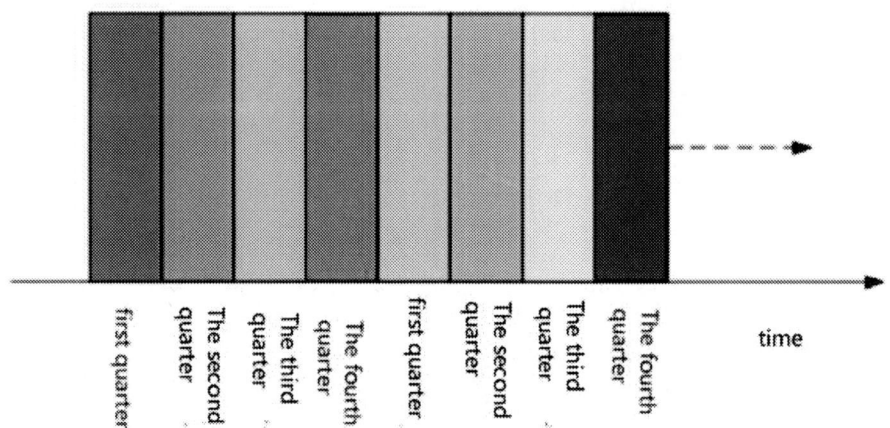

Figure 5 - Incremental mining strategy based on time series

3.3.3 Dynamic Maintenance Mining Strategy Based on Time Series
Dynamic maintenance and mining strategy based on time series is based on the incremental mining of time series. In practice, financial managers of enterprises can achieve the desired control effect by setting the minimum threshold content of data, which can fully let the system play its practical role and play an important role in financial risk management and control of enterprises. There are some effects.

4. Summary
The effect of data mining technology in financial risk analysis is to warn and control financial risk. For the person in charge of an enterprise, he can build an early warning model of financial risk based on data mining technology. In many cases, if an enterprise has financial risk, it is a gradual process, if there are relevant models. The type of early warning, when the enterprise risks, can be dealt with in time to avoid greater economic losses caused by enterprises. Data mining technology has a huge application space in the field of financial risk analysis and early warning of our country's enterprises.

Our relevant practitioners should actively use data mining tools to escort our enterprises'financial work smoothly.

Reference
[1] Zhuanghui, Rao Yangsheng. Application of data mining technology in financial distress prediction of listed companies [J]. Modern economic information, 2016 (19): 143-144.
[2] Chen Changping. Financial Risk Analysis Based on Data Mining [J]. Enterprise Technology Development, 2015, 34 (14): 134-135.
[3] Luo Yanmei. Financial Risk Analysis and Early Warning Research Based on Data Mining Technology [J]. Computer Knowledge and Technology, 2015, 11 (14): 6-7.
[4] Yu Jiuhong, Wang Haiping, Wang Rong. Application of Data Mining in Financial Crisis Early Warning [J]. Productivity Research, 2015 (07): 141-146+160.
[5] Tang Haicheng. Financial Risk Analysis and Early Warning Research Based on Data Mining Technology [J]. China Commerce, 2017 (14): 154-155.

AUTHOR INDEX

Ailing, Qi 1974
An, Shubing 1378
An, Wang W. 1470
An-Wen, Ying 479
Bai, Juan 1121, 1510
Bai, Xiaoye 252
Bao, Lei 1121, 1510
Bao, Wenxia 1853
Bao, Yiting 780
Bi, Mingkai 1939
Bin, Xu 2166
Bo, Yang 434
Bo, Zhang 2315
Bouchart, Vanessa 732
Boxing, Zhang 1063
Cai, Biao-Hua 751
Cai, Guoliang 1209
Cai, Peng 409
Cai, Shaopeng 1365
Cai, Sun 214
Cai, Xiaoyu 2283
Cai, Zengyu 1656
Cao, Shaozhong 1425
Cao, Shukun 640
Cao, Xinli 87
Cao, Yan 276
Cao, Yichao 2360
Cao, Yundong 239
Cen, Tao 2278
Chang, Faliang 1359
Chang, Rui 1130
Chang, Wen 2638
Changhui, Ma 268
Chao, Xiang 2478
Che, Renfei 366, 386
Chen, B. W. 1932
Chen, Baiyu 2602
Chen, Chen 2720
Chen, Chunlong 331
Chen, H. 2089
Chen, Hai-Quan 808
Chen, Hemu 1908, 1987
Chen, J. J. 1932
Chen, Jianjun 108
Chen, Jie 135
Chen, Jing 2272
Chen, Jinqiang 2506
Chen, Liang 772

Chen, Lili 2494
Chen, Limei 1636
Chen, Ming 2232
Chen, Qiaoling 1209
Chen, Quan 323
Chen, Shanji 2109
Chen, Shuyu 2488
Chen, Ting 2042
Chen, Wei 200
Chen, Xiangzhou 447, 1859
Chen, Xiaolin 1108
Chen, Xiaoxiao 568, 589
Chen, Xueli 1869
Chen, Yanfang 693
Chen, Yazhen 2720
Chen, Yong 1869
Chen, Yuanyuan 1662
Chen, Yueyue 108
Chen, Zhi 1554
Chen, Zhonghe 553
Cheng, D. S. 1932
Cheng, Si 2323
Cheng, Wang 701, 738
Cheng, Yang 510
Chenglin, Zhang 144
Chevrier, Pierre 732
Chi, Zhang 562
Chong, Gao 168
Chu, Qianqian 2278
Chu, Shibo 1519
Chu, X. M. 1801
Chuan, Wang 716
Chuanli, Wang 701, 738
Chunmei, Li 1675
Cong, Wang 2252
Cui, Fang Y. 894
Cui, Jing Gang 581
Cui, Wen 1812
Cuicui, Liu 168
Dai, Jian 2551
Dai, Zongmiao 1667
Dandan, Sun 888
Danyang, Li 2211
Deng, Fanyi 93
Deng, Ming-Ji 1353
Deng, Qinghong 882
Deng, Qishu 1342
Deng, Shaoxiang 214

Deng, Xin 1412
Di, Xiaofeng 2494
Diao, Chentao 1543
Ding, Baobao 1604
Ding, Fu-Jun 2371
Ding, Hailan 2466, 2559
Ding, Hongyan 732
Ding, Huixia 1859
Ding, Jie 1000
Ding, Lili 984
Ding, Xiaohua 1812
Ding, Yanfeng 1209
Dong, Ma X. 1470
Dong, Sui 1775
Dong, Wei Y. 1782
Dong, Xianlei 2421, 2429
Dong, Xiaoming 323, 337, 359
Dou, Liang 1604
Du, Chunfeng 1656
Du, Jiawei 1531
Du, Jinyang 453
Du, Wen 366, 386
Duan, Lijin 276
Duan, Lunqin 955
Duan, Ming 1575
Duan, T. 1834, 2609
Dun, Ao 1166, 1543
Fan, Dandan 2232
Fan, Jie 102
Fan, Jinpo 1739
Fan, Mingqi 2649
Fan, Xingyuan 453
Fang, Wang 1282
Fang, Zhuo 168
Feng, Hao B. W. 2624
Feng, Lansheng 258
Feng, Ruzhi 524
Feng, Shanqiang 2176
Feng, Shunshan 1612
Feng, Wang 168
Feng, Xiao 1063
Feng, Xiaoche 2257
Fengbin, Zhang 1450
Fu, Da 2602
Fu, Hongyong 2506
Fu, Jie 2453
Fu, Jun 2297, 2329
Fu, Qixi 1273
Fu, Yifan 882
Fu, Yuyang 2036
Fusheng, Chen 144
Gai, Lujie 882

Gan, Baiqiang 2592
Gan, Hua 2304
Gan, Ping 1979
Gang, Wang 434
Gao, Chao 1994
Gao, Dongliang 409
Gao, Fei 135
Gao, Fei-Fei 181
Gao, Hanxu 1359
Gao, He 1994
Gao, Hong 1171
Gao, Jian 2368
Gao, Lijuan 1682
Gao, Qingshui 208
Gao, Zhenxing 2113
Gao, Zihan 1006
Geng, Lei 1049
Gong, Chunwei 2466, 2559
Gong, Taorong 296
Gong, Xinman 833
Gou, Yating 114, 162
Gu, Bochuan 2176
Gu, Jingtian 1710
Guan, Denggao 346
Guan, Shilei 57
Guan, Wanlin 35
Guangyao, Jia 461
Guanhui, Wang 2216
Gui, Xinyue 2183
Guiqiang, Liu 888
Guizhong, Wang 27
Guo, Gongde 2565
Guo, Hejia 640
Guo, Hongwei 102
Guo, Kai-Feng 1190
Guo, Ming-Xuan 808
Guo, Ronghua 1531
Guo, Runqiu 258
Guo, Shaobing 917
Guo, Sheng H. 2712
Guo, Siyuan 822
Guo, Song 751
Guo, Wei 530
Guo, Xiaoshuang 1656
Guo, Xing 1441
Guo, Xueqi 346
Guo, Yajie 93
Guo, Yingjun 93
Guo, Yizhuo 1461
Guo, Ziteng 574
Haibo, Tan 596
Haichao, Chen 607

Haijun, Lei 1974
Haijun, Peng 596
Hai-Lan, Ding 2446
Haiyang, Jiang 1282
Han, Donchen 1395
Han, Jun 1939
Han, Liang 848
Han, Qianru 2021
Han, Quanli 1395
Han, Tao 640
Han, Tongxin 1057
Han, Wang 2572, 2578
Han, Xueshan 331, 372
Hanyan, Wang 2221
Hao, Chen 144, 596
Hao, Cheng 2203
Hao, Chuxue 18
Hao, Jinshun 221
Hao, Junjie 1919
Hao, Li 1286
Hao, Weiguang 263, 787, 795, 802
Hao, Yun 553
Haolin, Jia 1315
He, Fangzheng 1
He, Hongmei 1733
He, Jiangheng 1612
He, Juntao 366, 386
He, Lyulong 1273
He, Ming 1115
He, Renke 2278
He, Shiwei 1939
He, Shuming 2368
He, Xin 1267
He, Yanchen 1758
He, Yidong 1629
He, Yu 1682
He, Yubo 1645
He, Zhiqiang 1599
Hengjie, Li 416
Hong'Bo, Chen 911
Hong-Zhi, Yu 2446
Hou, Aijun 2176
Hou, Lunqing 1412
Hou, Xiangru 1536
Hou, Yan 2746
Hou, Yueqi 2071
Hu, Beibei 2421, 2429
Hu, Chang'An 745
Hu, Dehao 346
Hu, P. C. 1834, 2609
Hu, Shi-Cheng 990
Hu, Yue 228

Hu, Yulan 263, 787, 795, 802
Hu, Yunpeng 2001
Hua, Dengxin 894
Hua, H. Y. Y. 1330
Hua, Wang 434
Hua-Jie, Chen 668
Huang, Bihui 485
Huang, Jingzhi 1228
Huang, Li 2629
Huang, Lin 1979
Huang, Min 2065
Huang, Qiuzi 2488
Huang, R. 2089
Huang, Wei 1029
Huang, Wenhan 693
Huang, Xiaoping 2117
Huang, Xulong 1919
Huang, Yangfan 1979
Huang, Yizhi 693
Huang, Yuwei 1029
Hui, Baofeng 2109
Huimin, Fan 2211
Huimin, Sun 1036
Huitao, Wang 2014
Huiying, Song 461
Ji, Ke 46
Ji, Weiyan 1228
Ji, Yang 808, 2079
Jia, Guangyao 427
Jia, Guoqing 2109
Jia, Hongwei 2706
Jia, Qiang 510
Jia, Shanjie 1859
Jia, Shijie 1594
Jia, Songmin 1166, 1543
Jia, Wenbo 35
Jia, Yafang 2401
Jia, Yunfei 188, 662
Jia, Zhigang 2638
Jiachen, Tian 168
Jiajia, Han 2315
Jian, Wang 875
Jian, Zhou 518
Jianbo, Yin 401
Jiang, Cheng 2323
Jiang, Dawei 1827
Jiang, Hua 1739
Jiang, Juanjuan 1733
Jiang, Xiaoying 2565
Jiang, Z. L. 1801
Jiang, Zhanjun 2283
Jiang, Zhe 372

Jianhui, Zhou	168	Lei, Wang	195, 2221
Jianwei, Liu	1562	Lei, Xiang	1967
Jianzheng, Liu	63	Lei, Yiyan	1720
Jianzhi, Tuo	634	Lei, Zhipeng	379
Jianzhong, Yang	401	Lele, Sun	1315
Jiaojiao, Xi	2544	Li, Bin	751
Jiaxin, Liu	1215, 1222	Li, C.	2436
Jia-Zhi, Yang	479	Li, Chunmei	1524, 1889, 1898
Jie, Cheng	738	Li, Cui	911
Jie, Huang	768	Li, Dezhi	296
Jie, Ren	2519	Li, Guanghui	870
Jiefeng, Mou	1036	Li, Guanyu	2466, 2515, 2559
Jikang, Wang	63	Li, Guoqiang	78
Jikun, Guo	542	Li, Haifeng	323
Jin, Fei	78	Li, Hong-Bing	1121, 1510
Jin, Ge	214	Li, Huanran	1883
Jin, Jian	518	Li, Hui	1883
Jin, Li	2544	Li, Huizhi	156
Jin, Tao	323	Li, Jiahao	1166
Jin, Tiancheng	1604	Li, Jian	613, 2152
Jin, Wei	1483	Li, Jie	1308
Jin, Weiqi	1955	Li, Jing	239
Jing, Jing	1147	Li, Jinping	2706
Jing, X. H.	1246	Li, Jiping	1092
Jing, Zhang	168	Li, Kai	1524, 1898
Jing, Zhu	1215, 1222	Li, Li	822
Jingshi, He	2347	Li, Lulu	372
Jinjie, Shan	518	Li, Maohua	2662
Jinliang, Qiu	461	Li, Meng	1353, 2304
Jinxiu, Wang	888	Li, Ming	1084
Jiyao, Tian	1967	Li, Mingchao	2001
Jun, Wei	434	Li, Minwei	1013
Junning, Qin	2315	Li, Nianlian	732
Kang, Ruiyu	1788	Li, Pengyang	894
Kang, Yang	135	Li, Ran	1488
Ke, Yan	1477	Li, Shuangxi	1883
Ke, Zhang	634	Li, Sicong	35
Kong, Juan	1919	Li, Tao	2692
Kong, M. X.	1696	Li, Tong	2506
Kong, Weizheng	130	Li, Wang	2460
Kong, Xiangzeng	2565	Li, Wanze	745
Kou, Xu-Peng	2674	Li, Wei H.	1820
Kuaia, Tengfei	905	Li, Wei	1441
Kui, Zhang Yong	875	Li, Weichao	78, 1199
Kun-Yu, Qi	2446	Li, Wenbo	331, 372
Lai, Ming-Ming	2669	Li, Wenjing	1406
Lan, Ru	2638	Li, X. L.	2436
Lan, Yunsheng	2048	Li, Xiao	2602
Le, Guigao	905	Li, Xingxing	745
Lei, Chu	27	Li, Xiuzhi	1166, 1543
Lei, Lei	1839, 1846	Li, Xu	2701
Lei, Min	87	Li, Xuefei	135

Li, Yan	894
Li, Yanyun	1979
Li, Ye	2071
Li, Yingqi	833
Li, Yulong	188
Li, Zhi L.	1839, 1846
Li, Zhifei	2036
Li, Zhiming	258
Li, Zhiyuan	1827
Lian, Minlong	1378
Liang, Bin	673
Liang, Dong	1853
Liang, Gang	2304
Liang, Junbin	1418
Liang, Li	738, 1036
Liang, Ning	114, 162
Liang, Shutian	18
Liang, Xi	1450
Liang, Xiaolong	1273, 2071
Liang, Yi	1029
Liang, Ying	57
Liang, Yuqing	1919
Liang, Zhikai	1425
Liao, Daixi	1618
Liao, Minfu	882
Liao, Xiaoming	693
Liao, Zitian	1138
Liling, Liu	2355
Liman, Shen	144
Limei, Zhao	2335
Lin, Dansheng	2267
Lin, Doudou	2429
Lin, Jinghui	346
Lin, Shaofu	2393, 2401
Lin, Sheng	1883
Lin, Wei	668, 1645
Lin, Yao	1336
Lin, Zhang	2079
Lin, Zhaowen	1130, 1488
Ling, Liu	1142
Li-Qing, He	668
Liu, Bin	984, 1503
Liu, Can	870
Liu, Chang	290
Liu, Changli	967
Liu, Cuicui	114, 162
Liu, Di	1300
Liu, Fuyang	123
Liu, G.	263, 787, 795, 802, 1801
Liu, Haikuan	1827
Liu, Hanqing	1292
Liu, Haojie	315

Liu, Huabin	1667
Liu, Jianwei	2048
Liu, Jiawei	653
Liu, Jiaxin	188, 662
Liu, Jie	1554, 2140
Liu, Jingli	78
Liu, Jun	2030, 2058
Liu, Ke Cheng	581
Liu, Kun	2342
Liu, L.	195, 1801
Liu, M.	2089
Liu, Qianru	2283
Liu, Renzhang	78
Liu, Shi.	208
Liu, Shuxin	239
Liu, Tingxiang	200
Liu, Wei	955
Liu, Wenchang	276
Liu, Wenda	18
Liu, X.	2089
Liu, Xianglong	1234
Liu, Xiaochun	1812
Liu, Xiaoliang	78
Liu, Xiaoqian	2734
Liu, Xindong	1267
Liu, Xingbao	1365
Liu, Xueyan	2551
Liu, Yajie	263, 787, 795, 802
Liu, Yang	346
Liu, Yangyang	2408
Liu, Yankui	1599
Liu, Ye	2267
Liu, Yi	625, 629
Liu, Yiliang	1378
Liu, Yonggang	315
Liu, Yongxia	2048
Liu, Yu	188, 2679
Liu, Yuting	296
Liu, Yuyan	1788
Liu, Zefeng	1013
Liu, Zhe	1531
Liu, Zhengyi	1441
Liu, Zhenzhen	2629
Liu, Zhizhen	276
Liu, Zilin	263, 787, 795, 802
Liyu, Xia	2572, 2578
Long, Luo	602
Long, Pan	856
Long, Shaohua	1618
Long, Wang	701, 738
Long, Wu	416
Lu, Jiangang	1228

Lu, Jun 409
Lu, Ligen 258
Lu, Shikun 894
Lu, Xiaobo 1688, 2360
Luai, Almadhehagi 427
Luhua, Xing 268
Luo, Lisai 2466
Luo, Shihui 723, 1183
Luo, Taorui 409
Luo, Wanbo 2065
Luo, Zhen 123
Lv, Peihua 1688
Ma, Hongfeng 379
Ma, Jianwei 680, 687
Ma, Kun 2342
Ma, Lulu 305
Ma, Panwei 510
Ma, Pengcheng 1623
Ma, Shiwei 917
Ma, Te 2384, 2388
Ma, Xiaodong 2009
Ma, Xinling 759
Ma, Zhi-Run 2686
Mao, Wanfeng 2649
Mao, Yanrong 427
Mao, Yazhou 315
Maotao, Yang 144, 596
Maoyi, Zhang 1315
Meng, Xiaocheng 366, 386
Meng, Yuting 2001
Mi, Yongsheng 10
Miao, Feng 524, 530
Miao, Lanfang 2036
Min, Huang 2741
Min, Liu 768
Mou, Pengbo 625, 629
Mouhai, Liu 144, 596
Mu, Jiong 581
Mu, Qi 1599
Mu, Senlin 1292
Mu, Xihui 156
Murtaza, Abid 1562
Na, Li 1477
Nannan, Liu 1036
Ni, Xue 228
Nie, Li 1153
Ouyang, Chengtian 2042
Ouyang, L. 2089
Pai, Liu 888
Pan, Fangyu 1153
Pan, Jian 1720
Pan, Qiao 2262

Pang, Xun 673
Peng, Du 634
Peng, Fengzhi 152
Peng, Jianjun 1092
Peng, Lin 2669, 2674, 2686, 2701
Peng, Luxi 2499
Peng, Wang 174
Peng, Yanfei 1092
Penghou, Liu 607
Ping, Wang 888
Qi, Yingchuan 510
Qiang, Li X. 2166
Qiang, Li 2166
Qiang, Lin 152
Qiang, Wu 888
Qiao, Yulong 1962
Qimeng, Nie 1282
Qin, Hua 46
Qin, Luxing 2360
Qing, Wu 2323
Qingjun, Guo 995
Qinyuan, Li 2315
Qiu, Mengyue 2232
Qiu, Zhen 1300
Qiu, Zhiwen 693
Qiuqiu, Wang 1159
Qizhong, Li 1494
Qu, Huaijing 1549
Ran, Jilin 967
Ran, Li 2472
Ren, Jie 905
Ren, Jiyuan 123
Ren, Xun-Yi 1703
Ren, Zhong 814
Ren, Zongjin 305
Riaz, Umair 870
Rong, Tang 2079
Ru, Cong 934, 941, 948
Ru, Zhang 1914
Ruan, Y. 1246
Ruan, Zhenzhen 1979
Rui, Chen 1450
Rui, Zhang 542
Run-Dong, Wang 2519
Ruopeng, Yang 2014
Saeed, Muhammad J. 870
Sang, L. Z. 1834, 2609
Shan, Gao 888
Shao, Bao-Zhu 181
Shao, Juanjuan 1267
Shao, Xuebin 2408
Shao, Zhiyu 1612

She, Jintao	491	Sun, Juanjuan	1401
Shen, Gao Q.	1839, 1846	Sun, Lin	2291
Shen, Guiquan	1228	Sun, Liying	252
Shen, Wei	2638	Sun, Qian	1359
Shen, Wuqiang	1228	Sun, Qibo	2429
Shen, Xinxin	1583	Sun, Quanxin	1788
Sheng, Jiayue	841	Sun, Ruifeng	1013
Sheng, Tingran	1057	Sun, Xianhai	768
Sheng, Xuanyu	841	Sun, Xiao	1949
Shengdongt, An	401	Sun, Xiaoxiao	848
Shi, Changkai	57	Sun, Xin	1199
Shi, Chen	1703	Sun, Yanjun	1166, 1543
Shi, Haoqiang	2283	Sun, Yao	346
Shi, Kai	1883	Sun, Yaojie	1745
Shi, Peiji	2408	Sun, Yi	613, 1130, 1488, 1554
Shi, Xin	1353	Sun, Ying	453
Shi, Zhao-Cun	751	Sun, Yu	1599
Shi, Zhe	93	Sun, Zhen P.	613
Shixu, Li	1036	Sun, Zheng	1441
Shuai, Chen	2245	Sun, Zhijie	2297, 2329
Shudong, Wang	461	Suo, Dong	1675
Shuifeng, Zhang	2245	Suo, Shuangfu	221
Shushuang, Liang	2656	Tan, Guangyu	870
Shuzhen, Yang	710	Tan, Jinjun	2267
Sicong, Li	2472	Tan, Ming	1121, 1510
Situ, Shuwei	1	Tan, Yukun	1425
Song, Deyu	379	Tang, B. M.	2436
Song, Huiying	427	Tang, Guoshen	276
Song, Lihua	130, 1300	Tang, Jinjin	2528
Song, Min	2649	Tang, Jun	1908
Song, Ping	568, 589	Tang, Shaofan	1378
Song, Q. H.	2436	Tang, Xiao	290
Song, Xing	2368	Tang, Xinhuai	2291
Song, Xiyu	1503	Tang, Yanqun	2001
Song, Yuqin	2030	Tang, Ying	2537
Song, Zilong	1939	Tao, Kepeng	35
Songze, Lei	1063	Tao, Yu	710, 716
Su, A. J.	1696	Tao, Zhengping	2528
Su, Hongsheng	51	Teng, Xiaofei	1071
Su, Jiangwen	1300	Tian, Bin	502
Su, Tongdan	1115	Tian, Feng	2453
Su, Y.	2089	Tian, Jiachen	114, 162
Sui, Wei	394	Tian, Jin	2021
Sun, Chenzhe	2615	Tian, Jing-Jing	2371
Sun, Chuanmin	346	Tian, Zhengbing	252
Sun, Cong	305	Tianfang, Wu	2245
Sun, Feng	181	Tight, Miles	1788
Sun, Heng	1623	Ting, Ding	461
Sun, Hexu	93	Tingting, Liu	1142
Sun, Hua	337, 359	Tong, Fei	1908, 1987
Sun, Jiabin	2127	Tong, Guan	888
Sun, Jianyong	1108	Tong, Li	1503

Tu, Jingzhe 323
Tu, Jinlong 10
Tu, Yaqing 1084
Tu, Yongcheng 1908, 1987
Wan, Hongqiang 1395
Wan, Xing 2065
Wanbo, Luo 2741
Wang, Bo 1029
Wang, C. Y. 1750
Wang, Caishen 290
Wang, Chang'An 2734
Wang, Chao 2297, 2499
Wang, Chaochao 337
Wang, Chunlin 1108
Wang, Chun-Yang 1121, 1510
Wang, Denggui 759, 826
Wang, Dong 510, 1782, 2597
Wang, Endong 1877
Wang, Enshi 485
Wang, Fei 1412
Wang, Feng 114, 162
Wang, Guanhong 296
Wang, Guirong 833
Wang, Haibin 2323
Wang, He 917
Wang, Heng 640
Wang, Hengbin 1549
Wang, Hu 808
Wang, Hui 135
Wang, Jiawen 1629
Wang, Jing 252
Wang, Jiong 2071
Wang, Juan 2453
Wang, Ke 282, 447, 1256, 1859
Wang, Kun 200
Wang, L. 1834
Wang, Lanwen 841
Wang, Li 2297, 2329
Wang, Lihua 228
Wang, Lingxia 826
Wang, Lingxue 1955
Wang, Lingyu 2421
Wang, Liquan 934, 941, 948
Wang, Minghao 620
Wang, Nian 1908, 1987
Wang, Ping 2140
Wang, Qi 2238, 2323
Wang, Qingjia 2342
Wang, Qiuling 130
Wang, Runjiao 955
Wang, Shiyu 2262
Wang, Shudong 427
Wang, Shuyuan 1100
Wang, Siyue 1594
Wang, Song 2515
Wang, Tao 1733
Wang, Ting 1100
Wang, Wen 1049
Wang, Wenjie 221
Wang, Wen-Si 1703
Wang, Wentao 46
Wang, X. S. 1932
Wang, Xi 2042
Wang, Xiangpei 1919
Wang, Xianli 524, 530
Wang, Xiaogang 1153
Wang, Xiaolan 200
Wang, Xiaoming 2304
Wang, Xingong 46
Wang, Xuewei 548
Wang, Yan 346, 1043
Wang, Yanan 282, 447, 1256, 1859
Wang, Yang 221, 282, 1256, 1979
Wang, Yanyan 955
Wang, Yaokun 57
Wang, Yifan 1662
Wang, Yijing 1100
Wang, Yingjing 662
Wang, Yisheng 2140
Wang, Yuanmin 1919
Wang, Yudong 409
Wang, Yujiang 1029
Wang, Yuqiao 2692
Wang, Zhao 2408
Wang, Zhaoqing 1342, 1348, 2134
Wang, Zhe 123
Wang, Zheng 653
Wang, Zhiping 290
Wang, Zhiying 1629
Wang, Zhi-Yuan 990
Wannian, Zhu 1967
Wei, Caisheng 574
Wei, Chen 416
Wei, Liu 1927
Wei, Pi 1675
Wei, Qianwen 2393
Wei, Shicheng 1029
Wei, Zhang 634
Wei, Zhengxian 2649
Wei, Zheyu 46
Weidong, Xu 1967
Weihai, Li 1927
Weihua, Ma 723, 1183
Wei-Jun, Pan 2519

Weiwei, Qi 1142
Wen, Chang 2679
Wen, Chao 826
Wen, Guangqi 1889
Wen, Junhao 2488
Wenbo, Li 268, 352
Wencan, Ding 1775
Wensheng, Yin 562
Wenwen, Jiao 2472
Wenxue, Liu 268
Wu, Chunshang 2377
Wu, Hao 1575
Wu, Haobo 2048
Wu, Hong 1962
Wu, Hongmei 1919
Wu, Jianhong 2499
Wu, Jinbo 822
Wu, Jun 1049
Wu, Jun-Jie 808
Wu, Kezhuang 1418
Wu, Qinqin 2267
Wu, Qiong 1006, 2123
Wu, Ran 2123
Wu, Tong 1604
Wu, W. W. 1750
Wu, Xiaoquan 2267
Wu, Xuehui 2360
Wu, Yingying 502
Wu, Yusi 2140
Wu, Zhiqiang 1554
Wufan 2014
Xi, Hongyan 2140
Xi, Qi 1575
Xia, Bin 1240
Xia, Peng 1336
Xia, Rongzhen 379
Xia, Sibin 2393, 2401
Xia, Yangqiu 1365
Xianfang, Tang 1914
Xiangbin, Liu 144
Xiangguo, Su 2252
Xiangzhou, Chen 282
Xiao, Binjie 1251
Xiao, Zhitao 1049
Xiaofei, Zou 2014
Xiaokun, Wang 2221
Xiaomei, Hu 710, 716
Xiaoping, Li 1063
Xiao-Shu, Wang 2098
Xiaotie, Ma 2335
Xiaoyan, Zhang 1159
Xie, Cheng 108

Xie, Dong-Fan 2371
Xie, Feng 2329
Xie, Lingling 1
Xie, Minzhen 1078
Xie, Yong-Jun 1190
Xin, Bo 1292
Xin, Li 1359
Xin, Ma 2203
Xin, Peizhe 409
Xin, Sun 2315
Xin, Xiaoyu 2478
Xin, Yan 875
Xin, Yang 1967
Xincheng, Ren 471
Xing, Wan 2741
Xinliang, Cao 471
Xin-Zhe, Yin 479
Xu, Binshi 1029
Xu, Feng 1919
Xu, Gang 1733
Xu, Guangping 1883
Xu, Guanli 346
Xu, Hongkui 1549
Xu, Jie 1130, 1488
Xu, Peiyuan 2727
Xu, Sanchuan 536
Xu, Shiping 1531
Xu, Wanjin 1108
Xu, Wei 2706
Xu, Wenjing 315
Xu, Xiangqian 640
Xu, Xiaoshen 2232
Xu, Xin 2297, 2329
Xu, Yugong 2203
Xu, Zhuoran 1554
Xue, Qiao 447, 1859
Xue-Chao, Liao 646
Xueshan, Han 352
Xuli, Zhu 1036
Xuxiang, Huang 352
Yachao, Jia 1914
Yan, Bin 1503
Yan, Chunyu 917
Yan, Haotian 2134
Yan, Kedi 73
Yan, Qianghu 87
Yang, Guang 2009
Yang, Guohui 924
Yang, Huiyue 1084
Yang, Jian 46
Yang, Jianxi 315
Yang, Jiebin 745

Yang, Jun-You 181
Yang, Kai 1503
Yang, Li 995
Yang, Lin-Nan 2674, 2701
Yang, Lu 2629
Yang, Ning 2559
Yang, Qunyi 2009
Yang, W. D. 1696
Yang, Wan 668
Yang, Wei 2712
Yang, Weijun 2291
Yang, Wentai 2304
Yang, Xiaodan 1433
Yang, Xiaohua 1554
Yang, Yi 208
Yang, Yingming 1604
Yang, Yiyong 221
Yang, Yongxi 924
Yang, Yu 1177
Yang, Yuansheng 1171
Yang, Yuanyuan 2602
Yang, Ziwei 1147
Yangjia 1494
Yanhong, Wang 1063
Yanhong, Zuo 861
Yanrong, Mao 461
Yanwei, Shang 152
Yanzhe, Du 607
Yao, Jianchun 394
Yao, Jianyu 826
Yao, Jiawei 2195
Yao, Ling 2117
Yao, Zheng 394
Yating, Gou 168
Ye, J. 1801
Ye, Wang 1494
Ye, Xuanyu 2720
Yi, Jun 323
Yi, Kang 1720
Yi, Wang 63
Yi, Yang Q. 1470
Yi, Zhijun 2494
Yin, Aiping 2113
Yin, Ningxia 870
Yin, Yanan 1171
Yin, Zhiqin 1115
Ying, Lin 1282
Yinzheng, Zheng 973
Yong, Lin 1373
Yonggang, Yue 401
Yonggang, Zhu 861
Yongqiang, Fan 401

You, Fucheng 548
You, Zhou 646
Youzi, Wang 2572, 2578
Yu, Fengyun 917
Yu, Jiujiu 2160
Yu, Ling 1092
Yu, Lu 553
Yu, Nan 2478
Yu, Shida 2189
Yu, Tianbiao 780
Yu, Tonglan 1554
Yu, Xiaochen 394
Yu, Xin 394
Yu, Xu R. 1820
Yu, Xuemei 1745
Yu, Yu 1519
Yuan, Bo 246
Yuan, Ziyan 2746
Yue, Chen 27
Yue, Dachao 1827
Yun, Mei 1282
Yunxiao, Zu 1927
Yutinge, Chen 1477
Zéman, Zoltán 2662
Zeng, Hanghang 51
Zeng, Jijun 2267
Zeng, Tianlong 2058
Zeng, W. D. 2584
Zeng, Ying 1503
Zeng, Yue 2103
Zhai, Xiujun 2615
Zhai, Yayu 568, 589
Zhan, Hong-Yuan 1190
Zhan, Y. C. 422
Zhang, Bin 625
Zhang, Bo 1199
Zhang, C. 1696
Zhang, Chenglin 2089
Zhang, Chengning 924
Zhang, Chu 208
Zhang, Chun J. 2712
Zhang, Cunlin 1949, 1962
Zhang, Dan 1703
Zhang, Dewen 35
Zhang, Dong 1877
Zhang, Fang 1049
Zhang, Feng 2272
Zhang, G. R. 2584
Zhang, Gang 1739
Zhang, Geng 282, 447, 1256
Zhang, Guan-Feng 181
Zhang, Guanglei 2615

Zhang, Guoliang 1543
Zhang, Hanhua 2692
Zhang, Hao 366, 386, 882
Zhang, Haoxue 882
Zhang, Hongda 35
Zhang, Hua 2117
Zhang, Huanping 2662
Zhang, J. 1750
Zhang, Jian 826
Zhang, Jianwei 1656
Zhang, Jiaqiang 1273, 2071
Zhang, Jincheng 258
Zhang, Jinwei 745
Zhang, Jun 305
Zhang, Junhao 2262
Zhang, Lan 934, 941, 948
Zhang, Lei 1827
Zhang, Liang 239, 2368
Zhang, Li-Hua 1336
Zhang, Lin 2304
Zhang, M. Y. 1801
Zhang, Min 723, 1183
Zhang, Na 1177
Zhang, Nan 2127
Zhang, Peng 2140
Zhang, Qiang 2551
Zhang, Qingqing 2551
Zhang, Qiurong 693
Zhang, Ran 447, 1256, 1859
Zhang, Rui 228
Zhang, Ruiqi 359
Zhang, Ruiqiu 772
Zhang, S. 1246
Zhang, Shoushou 822
Zhang, Shuo 924
Zhang, Tieshan 814
Zhang, Tinglei 1877
Zhang, Xiangyin 1543
Zhang, Xianmin 772
Zhang, Xiaotong 2238
Zhang, Xiaoying 200
Zhang, Xin 2669, 2686
Zhang, Xincheng 276
Zhang, Xinyu 2048
Zhang, Xinzheng 1267
Zhang, Xuangong 156
Zhang, Xuchong 772
Zhang, Xudong 102
Zhang, Xujuan 2551
Zhang, Yanan 2408
Zhang, Yang 1688
Zhang, Yanjun 239

Zhang, Yichen 1386
Zhang, Yidu 1006
Zhang, Yonghua 1401
Zhang, Yue 2103
Zhang, Z. N. 1330
Zhang, Zhi 290
Zhang, Zhongshi 2238
Zhangkang 2113
Zhanjun, Wang 195
Zhao, Changfang 905
Zhao, Chengqiang 934, 941, 948
Zhao, Enmin 239
Zhao, Guorong 1994
Zhao, H. W. 1246
Zhao, Hui 2488
Zhao, Kai 305
Zhao, Li 1675
Zhao, Lili 258
Zhao, Liujun 130
Zhao, Qian 984
Zhao, Qing 2238
Zhao, Qing-Song 181
Zhao, Yangze 548
Zhao, Ying 453
Zhao, Yuanmeng 1949, 1955, 1962
Zhao, Yuejin 1949, 1962
Zhao, Yufeng 1199
Zhao, Yusheng 1342, 2134
Zhao, Yuting 1720
Zhao, Zengshun 1359
Zhao, Zhi-Qiang 1353
Zhaoren, Pan 710
Zhen, Qiao 416
Zheng, Jinxin 78
Zheng, Kougen 1583
Zheng, Wei 379
Zheng, Yongkang 1667
Zheng, Yufu 1147
Zhen-Huan, Chen 646
Zhenwei, Zhang 174
Zhichao, Guo 63
Zhiyong, Wu 596
Zhizhong, Guo 27
Zhong, Shouming 1618
Zhong, Zhongzhi 1853
Zhongqi, Wang 1494
Zhou, Chun 453
Zhou, Di 1688
Zhou, Guomiao 1575
Zhou, Jian 629
Zhou, Qinqin 2734
Zhou, Rundong 346

Zhou, Shu 1267
Zhou, Xiang 87, 1812
Zhou, Yifan 1365
Zhou, Ying 1531
Zhou, Yong 1908
Zhou, Zeyu 653
Zhu, Chuangchuang 1273
Zhu, Haoming 296
Zhu, Honghai 1519
Zhu, Hongwei 2048
Zhu, Jingli 984
Zhu, Junjie 693
Zhu, Leiye 2323
Zhu, Ming 1853
Zhu, Weijun 822
Zhu, Xingxiong 2597
Zhu, Yu 581
Zhu, Yuan 2147
Zhu, Yuancheng 379
Zhu, Yuefei 1645
Zhu, Zhangqing 1292
Zhu, Zheng 1353
Zhu, Zhengbin 934, 941, 948
Zhu, Zhilong 1733
Zhuanga, Duoduo 2113
Zhuo, Fang 114, 162
Ziqiang, Lou 2252
Zonghua, Xie 634
Zou, Yuan 102
Zu, Yun X 1820